2nd edition CPCR®

2nd edition CPCR®

Crop Protection Chemicals Reference

CHEMICAL AND PHARMACEUTICAL PRESS
a joint venture of

JOHN WILEY & SONS, New York · Chichester · Brisbane · Toronto · Singapore

and

CHEMICAL AND PHARMACEUTICAL PULBLISHING CORP.,
a subsidiary of OVP, Paris

made possible through the courtesy of agricultural chemi- rs and the information concerning each product has been rticipating manufacturers. Neither the inclusion nor exclu- duct in CPCR should be construed as a recommendation for se of such product by the Publisher.

care has been taken by the Publisher in compiling and check- ation and data contained in CPCR to insure that it is current at the time of publication, the Publisher disclaims any and all any errors, omissions, or inaccuracies in such information and her attributable to inadvertence or otherwise, and for any con- arising therefrom. The Publisher further disclaims any and all li- or inaccuracies in the information contained in CPCR due to revi- the product labels, from which the information contained in the is compiled, after January 1986. READERS OF CPCR ARE CAU- NED TO CONTACT MANUFACTURERS FOR THE MOST CURRENT ODUCT INFORMATION, INCLUDING SUPPLEMENTAL LABELING AND LN REGISTRATIONS. REFERENCE SHOULD ALWAYS BE MADE TO THE PRODUCT LABEL PRIOR TO THE USE OF ANY CROP PROTECTION CHEMICAL. IT IS A VIOLATION OF FEDERAL LAW TO USE ANY CROP PROTECTION CHEMICAL DESCRIBED IN CPCR IN A MANNER INCONSISTENT WITH ITS LABELING. It is further understood that THE PUBLISHER MAKES NO REPRESENTATIONS OR WARRANTIES OF ANY KIND, INCLUDING, BUT NOT LIMITED TO, THE WARRANTIES OF FITNESS FOR A PARTICULAR PURPOSE OR MERCHANTABILITY NOR ARE ANY SUCH REPRESENTATIONS OR WARRANTIES TO BE IMPLIED WITH RESPECT TO THE DATA FURNISHED, AND THE PUBLISHER TAKES NO RESPONSIBILITY WITH RESPECT TO THE USE OF THE INFORMATION AND DATA FURNISHED. THE PUBLISHER SHALL NOT BE LIABLE FOR ANY SPECIAL, CONSEQUENTIAL OR EXEMPLARY DAMAGES RESULTING, IN WHOLE OR IN PART, FROM CUSTOMER'S USE OF THE DATA. Further, no one should administer or recommend medical treatment in reliance upon the data contained in CPCR without first verifying such treatment with a knowledgeable medical professional.

Library of Congress Cataloging in Publication Data:

Library of Congress Catalog Number: 85-642883
ISBN 0-471-84824-7

Printed in the United States of America

10 9 8 7 6 5 4 3 2 1

Contents

2nd edition CPCR®

Abbott Laboratories

Chemical & Agricultural Products Division
14th & Sheridan Road
North Chicago, IL 60064
800-323-9597

DEVINE®

BIOLOGICAL HERBICIDE

ACTIVE INGREDIENT:

	W/W
Live Chlamydospores of *Phytophthora palmivora* MWV, 3.2×10^8 per pint. (6.7×10^5 live chlamydospores per ml)	0.8%
INERT INGREDIENTS	99.2%
TOTAL	100.0%

EPA Reg No. 275-39 EPA Est. No. 275-IL-1

CAUTION

Keep Out of Reach of Children

KEEP REFRIGERATED—DO NOT FREEZE

FOR USE IN FLORIDA ONLY
DO NOT USE IN CLAY, GULF, LIBERTY OR GADSDEN COUNTIES

DEVINE® is a mycoherbicide for control of *Morrenia odorata,* strangler or milkweed vine, in citrus groves. This fungus will initiate a root infection in milkweed vine plants that kills the vine in two to ten weeks following application, depending on the size and maturity of the vine. Populations of the vine generally continue to decrease over the period of a year after treatment. Complete control may not be achieved in the first year of treatment.
DEVINE will not harm citrus trees.

PRECAUTIONARY STATEMENTS

Environmental Hazard—Keep out of lakes, ponds and streams. Do not contaminate water by cleaning of equipment or disposal of wastes.

DIRECTIONS FOR USE

It is a violation of federal law to use this product in a manner inconsistent with its labeling.

STORAGE AND DISPOSAL

Do not contaminate water, food, or feed by storage or disposal.

STORAGE: KEEP REFRIGERATED. DO NOT FREEZE. Failure to maintain refrigeration causes rapid product deterioration and bottle leakage. Return unopened, still refrigerated, out-of-date product to your supplier immediately upon expiration for disposal.

PESTICIDE DISPOSAL: Wastes resulting from the use of this product may be disposed of on site or at an approved waste disposal facility.

CONTAINER DISPOSAL: Do not reuse empty containers. Triple rinse (or equivalent). Puncture and dispose of in a sanitary landfill, or incinerate, or, if allowed by state and local authorities, by burning. If burned, stay out of smoke.

MIXING DIRECTIONS

Triple flush the tank and lines of the herbicide boom sprayer system. Adjust nozzles for coverage of all the soil where the vine can grow. Use 50 mesh or coarser screens. Half-fill the tank with water, add the required amount of DEVINE® and agitate as the tank is filled.

APPLICATION DIRECTIONS

Use rate: 1 pint DEVINE per treated acre.

DEVINE may be applied in any type of citrus grove from May through September after the weed has germinated or is actively growing. Only one treatment per season is necessary.

Apply with a herbicide boom sprayer to achieve uniform coverage of the soil under the tree canopy. Use at least 50 gallons of spray water per treated acre. The surface of the soil *must* be wet at the time of application. Adequate moisture can be achieved with either 2 inches of water, rainfall or irrigation, prior to treatment or 1 inch water pretreatment followed by ½ inch on the third day after treatment.

APPLICATION RESTRICTIONS

DO NOT:
—mix with chlorinated water
—tank mix with a wetting agent, fertilizer, or any other pesticide
—apply in wind conditions favoring drift
—apply where the susceptible plants watermelon and periwinkle are grown
—apply within 100 ft. of the following plants: cucumber, squash, Begonia, Bougainvilla, boxwood, Hibiscus, oak areca palm, Pittosporum, snapdragon, Washingtonia and coconut palm, and hybrid Rhododendron.
—apply more than 1 pint per acre per season

NOTICE TO USER

Seller makes no warranty, express or implied, of merchantability, fitness or otherwise concerning the use of this product other than as indicated on the label. User assumes all risks of use, storage or handling not in strict accordance with accompanying directions.

DIPEL® Worm Killer

WETTABLE POWDER
BIOLOGICAL INSECTICIDE

ACTIVE INGREDIENT:

Bacillus thuringiensis, var. **kurstaki,** 16,000 International Units of Potency per mg. (7.26 billion International Units per pound)	3.2%
INERT INGREDIENTS	96.8%
TOTAL	100.0%

EPA Reg. No. 275-18 EPA Est. No. 33762-IA-1

CAUTION

Keep Out of Reach of Children

BENEFICIAL INSECTS

Honeybees foraging treated areas are not harmed by DIPEL use.

DIPEL does not interrupt the activities of beneficial and predacious arthropods in pest management programs.

ENVIRONMENTAL HAZARDS

Keep out of lakes, ponds or streams. Do not contaminate water by cleaning of equipment or disposal of wastes.

STORAGE AND DISPOSAL

Do not contaminate water, food or feed by storage or disposal.

STORAGE

Reclose containers of unused DIPEL. Store in a dry place.

PESTICIDE DISPOSAL

Wastes resulting from the use of this product may be disposed of on site or at an approved waste disposal facility.

CONTAINER DISPOSAL

Do not reuse empty container. Completely empty container into application equipment. Then dispose of empty container in a sanitary landfill or by incineration, or, if allowed by State and local authorities, by burning. If burned, stay out of smoke.

Days to harvest: There are no restrictions on applying DIPEL up to the time of harvest.

Sites: DIPEL may be used for any labeled pest in both field and greenhouse uses.

For Smaller Spray Volumes:

If Rate is	Use This Amount Per Gallon
¼ lb/acre or 100 gals	½ tsp.
½ lb/acre or 100 gals.	1 tsp.
1 lb/acre or 100 gals.	2 tsps.
2 lb/acre or 100 gals.	4 tsps.

DIRECTIONS FOR USE

It is a violation of Federal law to use this product in a manner inconsistent with its labeling.

DIPEL is a highly selective insecticide for use against listed caterpillars (larvae) of lepidopterous insects. Close scouting and early attention to infestation is highly recommended. Larvae must eat deposits of DIPEL to be affected. Always follow these directions:

- Treat when larvae are young (early instars) before the crop is extensively damaged.
- Larvae must be actively feeding on treated, exposed plant parts.
- Thorough spray coverage is needed to provide a uniform deposit of DIPEL at the site of larval feeding. For some crops directed drop nozzles by ground machine are required.
- Under heavy pest population pressure, use the higher label rates, shorten the spray interval, and/or raise gallonage to improve spray coverage.
- Repeat applications at an interval sufficient to maintain control, usually 3 to 14 days depending on plant growth rate, moth activity, rainfall after treating, and other factors. If attempting to control a pest with a single spray, make the treatment when egg hatch is essentially complete, but before extensive crop damage occurs.
- A spreader-sticker which has been approved for use on growing and harvested crops should be added for hard-to-wet crops such as cole crops, or to improve weather-fastness of the spray deposits.

After eating a lethal dose of DIPEL, larvae stop feeding within the hour, and will die within several days. Dying larvae move slowly, discolor, then shrivel, blacken and die.

DIPEL may be applied in conventional ground or aerial equipment with quantities of water sufficient to provide thorough coverage of infested plant parts. The amount of water needed per acre will depend on crop size, weather, spray equipment, and local experience. Unless otherwise indicated, use at least 2 gallons of water by air; except in the far west, 5 to 10 gallons is the usual minimum.

Add water to the spray or mixing tank at the level that provides maximum agitation. With the agitator running, slowly sprinkle in the DIPEL. Continue agitation. Then add other spray materials (if any). Add the balance of the water and agitate until mixed. Maintain the suspension while loading and spraying. Do not mix more DIPEL than can be used in a 12-hour period.

APPLICATION RATE

Crop	Pest	Pounds/Acre
Cole Crops and Vegetables Including Leafy, Root, Stalk, and Seed and Pod Types (Peas, Beans, Lentils, etc., dry or succulent), Potatoes, Cucurbits	Loopers Hornworms Imported Cabbageworm Diamondback Moth Green Cloverworm Webworm Armyworms*	½ to 1 ¼ to ½ ¼ to ½ ⅓ to ⅔ ¼ to ½ ½ to 1 1 to 2

* DIPEL may be used to suppress small armyworms (1st and 2nd instar) when populations are light and full coverage ground sprays are applied.

Tomatoes, Peppers, Eggplant	Loopers Hornworms Tomato Fruitworm* Variegated Cutworm Saltmarsh Caterpillars Armyworms**	½ to 1 ¼ to ½ 1 1 ½ to 1 1 to 2

* Apply weekly in a preventive program.
** DIPEL may be used to suppress small worms (1st and 2nd instar) when populations are light and full coverage ground sprays are applied.

Cucurbits: Melons, Squash, Cucumbers, etc.	Rindworm Complex* Melonworms	½ to 1 1 to 2

* Apply weekly in a preventative program

Tobacco	Tobacco Budworm Hornworms Loopers	½ to 1 ¼ to ½ ½ to 1

DIPEL may be tank mixed with sucker control products at recommended rates. Follow the application directions on the sucker control labels. Observe all label precautions for all products used.

Safflower, Sugar Beets, Mints	Loopers Saltmarsh Caterpillar	½ to 1 ½ to 1
Soybeans, Sunflowers, Peanuts	Loopers Green Cloverworm Velvetbean Caterpillar Podworm*	½ to 1 ¼ to ½ ⅓ to ⅔ ½ to 1

* DIPEL suppresses podworms under pest management conditions.

Alfalfa, Pastures, Hay, Small Grains and Forage Crops	Loopers Alfalfa Caterpillar European Skipper (Essex Skipper)	½ to 1 ⅛ to ¼ ¼ to ½
Turf	Sod Webworm	2 to 4
Rangeland	Range Caterpillar	⅒ to ¼

Use a minimum of ½ to 1 gallon water per acre.

Cotton	Tobacco Budworm* Cotton Bollworm* Loopers**	¼ to 1 ¼ to 1 ½ to 1

* Use to suppress light to moderate populations of newly hatched worms in pest management conditions. Use under close scouting when beneficial insects are active or building. Repeat treatments at 4 to 5-day intervals as long as necessary and results are acceptable. If 15% or more terminals have eggs, add ⅛ lb. AI chlordimeform[1] (Galecron® or Fundal®) per acre to ¼ lb. DIPEL. Increase chlordimeform to ¼ lb. AI if needed. Once beneficial insects are no longer a factor, DIPEL (at ¼ to ¾ lb./acre) may be tank mixed with methomyl (Lannate® or Nudrin® at .225 to .675 lbs. AI per acre to control light to moderate infestations of early instar larvae. Observe any rate, frequency, or boll load label restrictions for methomyl in your area. Methomyl will likely cause reddening of cotton foliage. Discontinue this tank mix if reddening becomes excessive. Before using any tank mix, read the product labels for all environmental and usage cautions. Chlordimeform is restricted to aerial application and to states where registered.

** Full plant coverage needed.

[1] The mixture of B.t. with chlordimeform for inhibiting the growth of lepidopterous larvae on plants is covered by Abbott U.S. patent 3,937,813.

Crop	Pest	Lbs./100 Gals.
Flowers and Ornamentals, (ground only)	Loopers Tobacco Budworm Omnivorous Looper Omnivorous Leafroller Diamondback Moth Armyworms* Ello Moth (Hornworm) Io Moth Oleander Moth	½ to 1 ½ to 1 ½ to 1 ½ to 1 ⅓ to ⅔ 1 to 2 ¼ to ½ ½ to 1 ½ to 1

* DIPEL may be used to suppress small armyworms (1st and 2nd instar) when populations are light and full coverage ground sprays are applied.

DIPEL FOR STORED AGRICULTURAL COMMODITIES

For the Control of Tobacco Moth on Flue-Cured Tobacco

Apply 0.4 ounce (approximately 5 teaspoonsful) of DIPEL in one quart of water per 100 pounds of tobacco as a fine mist spray. Avoid overwetting. Tobacco should have just enough moisture to be handled without shattering at the time of application.

Tobacco to be stored up to Twelve Months

Spray loose leaves as the tobacco is being bundled from the curing barn. For tobacco on sticks, treat both sides of leaves.

Stored Tobacco

For tobacco which is to be carried over, rebundle or restack sticks, fluff up tobacco and spray loose leaves.

For tobacco that has been stored over three weeks, apply at first signs of infestation, promptly open bundles, spray loose leaves, then bundle.

Treatment of Storage Barns

If tobacco has been treated, or is going to be treated, treatment of the floors and walls may be made to aid in control. Sweep out the area, especially cracks and corners, and all of the loose tobacco pieces in which the moth might breed. Make a spray mixture containing 1 oz. DIPEL per 2½ gallons of water. Apply this at a rate of ½ gallon per 1000 sq. ft. of surface area. Be sure to spray into cracks and between floorboards.

STORED GRAINS, SOYBEANS, SUNFLOWER SEED, PEANUTS, CROP SEED, POPCORN** AND BIRDSEED**

Pest	Rate
Indian Meal Moth, Almond Moth	¾ lb./100 bu (diluted and undiluted)*

* As a surface treatment, apply 1 lb. DIPEL in 10 gal. of water per 500 sq. ft. of grain surface area; mix into top 4 inches.

** For all states except California.

For the control and prevention of these pests, apply DIPEL in a constantly agitated water suspension to the top four inch surface layer of grain in the bin. Use a sprinkler can or sprayer to apply the dosage into the grain stream as the last (top) four inch layer is augered into the bin. Mix 1/10 lb. DIPEL per gallon of water. Apply 0.6 pint of this mixture per bushel as grain is augered into storage. Or, sprinkle the dosage onto the surface of the grain in the bin and mix thoroughly with a scoop or rake to the depth of four inches. More thorough coverage may be achieved by dividing the recommended dosage into three applications and mixing the grain between applications.

For the protection of bagged grain, apply the dosage to the entire grain mass and mix thoroughly prior to bagging.

Treatments can be applied to the stored grain at any time, but, for best results, make application immediately after harvest before moth activity occurs. In areas where late fall harvested grain is not subject to infestation because of low temperatures, application can be delayed until late winter or early spring before moth activity begins. Control for a full storage season should normally be expected; however, repeat application if infestation recurs.

This treatment controls the moth larvae. If an infestation is present when the grain is treated, moth emergency may continue for several days. If immediate control of severe infestations is desired, grain should be fumigated prior to application of this treatment. **DIPEL will not control weevils or other beetles.**

DIPEL is exempt from tolerance restrictions and the treated grain can be used at any time after treatment for any use.

DIPEL for Fruits, Nuts, Trees, Ornamentals

Crop	Pest		Pounds/ Acre
Grapes, Small Fruits, Cane and Bush Berries	Grapeleaf Skeletonizer (ground only)		⅔ to 1
	Grape Leaffolder		½ to ¾
	Achema Sphinx Moth (Hornworm)		⅓ to ¾
	Saltmarsh Caterpillar (ground only)		1
	Omnivorous Leafroller (ground only)		½ to 1
	Loopers		½ to 1
		Pounds/ 100 Gallons* (Ground Equipment)	**Pounds/ Acre (Aerial** Application)**
Stone Fruits, Nut Trees, Pomegranates	Redhumped Caterpillar	¼ to ½	⅓ to 1
	Tent Caterpillars	¼ to ½	½ to 1
	Omnivorous Leafroller	½ to 1	—
	Fall Webworm	¼ to ½	½
	Walnut Caterpillar	½ to 1	½ to 1
	Cankerworms	¼ to ½	½ to 1
Pome Fruits[1]	Cankerworms	¼ to ½	½ to 1
	Gypsy Moth	½ to 1	1
	Variegated Leafroller	½ to 1	—
	Redbanded Leafroller	½ to 1	—
	Tufted Apple Budmoth	½ to 1	—
	Tent Caterpillars	¼ to ½	½ to 1
Citrus[2]	Fruittree Leafroller	½ to 1	—
	Orangedog	¼ to ½	—
Tropical Fruits[3]	Hornworm	½ to 1	—
	Leafrollers	½ to 1	—
	Omnivorous Looper	½ to 1	—
	Loopers	½ to 1	—
Forest, Shade, Sugar Maple Trees and Ornamentals	Gypsy Moth	½ to 1	1
	Bagworm	½ to 1	½ to 1
	Redhumped Caterpillar	¼ to ½	¼ to 1
	Spring & Fall Cankerworm	¼ to ½	¼ to ½
	Fall Webworm	½	½
	Elm Spanworm	½ to 1	½ to 1
	Tent Caterpillars	¼ to ½	½
	California Oakworm	¼ to ½	¼ to ½
	Pine Butterfly	1	1
	Spruce Budworms[4]	1	1
	Saddle Prominent Caterpillar	½ to 1	—
	Douglas Fir Tussock Moth	1	1
	Western Tussock Moth	½ to 1	—
	Fruittree Leafroller	½ to 1	—
	Blackheaded Budworm	1	—
	Mimosa Webworm	½ to 1	—
	Jack Pine Budworm	½ to 1	¾ to 1
	Saddleback Caterpillar	½ to 1	—

* Rate for hydraulic sprayer. For mist blowers, mix the applicable amounts (lbs.) in 10 gallons of water.

** For aerial application, use in one to ten gallons of water depending on type and density of trees. For best results, spray systems which deliver droplet size of 200 microns should be used.

1 Use 400–450 gallons/acre to suppress listed pests in pest management programs.

2 Use 100 to 400 gallons/acre.

3 Use at least 200 gallons/acre.

4 Use against populations below 35 larvae per 18-inch branch tip. Use a minimum of ½ gallon/acre by air.

NOTICE TO USER

Seller makes no warranty, express or implied, of merchantability, fitness or otherwise concerning use of this product other than as indicated on the label. User assumes all risks of use, storage or handling not in strict accordance with accompanying directions.

® Fundal, Galecron, Lannate, Nudrin, are trademarks of companies other than Abbott Laboratories.

DIPEL® 2X Worm Killer

WETTABLE POWDER
BIOLOGICAL INSECTICIDE

ACTIVE INGREDIENT:
Bacillus thuringiensis, var. **kurstaki,**
32,000 International Units of Potency per mg.
(14.52 Billion International Units
per pound) 6.4%
INERT INGREDIENTS 93.6%
TOTAL 100.0%

Keep Out of Reach of Children
CAUTION

E.P.A. Reg. No. 275-37 E.P.A. Est. No. 33762-IA-1

BENEFICIAL INSECTS

Honeybees foraging treated areas are not harmed by DIPEL 2X use.

DIPEL 2X does not interrupt the activities of beneficial and predacious anthropods in pest management programs.

ENVIRONMENTAL HAZARDS

Keep out of lakes, ponds or streams. Do not contaminate water by cleaning of equipment or disposal of wastes.

STORAGE AND DISPOSAL

Do not contaminate water, food or feed by storage or disposal.

STORAGE

Reclose containers of unused DIPEL 2X. Store in a dry place.

PESTICIDE DISPOSAL

Wastes resulting from the use of this product may be disposed of on site or at an approved waste disposal facility.

CONTAINER DISPOSAL

Triple rinse (or equivalent). Then puncture and dispose of in a sanitary landfill, or by incineration, or, if allowed by state and local authorities, by burning. If burned, stay out of smoke.

DIRECTIONS FOR USE

It is a violation of Federal law to use this product in a manner inconsistent with its labeling.

Days to harvest: There are no restrictions on applying DIPEL 2X up to the time of harvest.

Sites: DIPEL 2X may be used for any labeled pest in both field and greenhouse uses.

For Smaller Spray Volumes:

If Rate is	Use This Amount Per Gallon
1/4 lb./acre or 100 gals.	1/2 tsp.
1/2 lb./acre or 100 gals.	1 tsp.
1 lb./acre or 100 gals.	2 tsps.
2 lb./acre or 100 gals.	4 tsps.

DIPEL 2X is a highly selective insecticide for use against listed caterpillars (larvae) of lepidopterous insects. Close scouting and early attention to infestations is highly recommended.

Larvae must eat deposits of DIPEL 2X to be affected. Always follow these directions:

- Treat when larvae are young (early instars) before the crop is extensively damaged.
- Larvae must be actively feeding on treated, exposed plant parts.
- Thorough spray coverage is needed to provide a uniform deposit of DIPEL 2X at the site of larval feeding. For some crops directed drop nozzles by ground machine are required.
- Under heavy pest population pressure, use the higher label rates, shorten the spray interval, and/or raise gallonage to improve spray coverage.
- Repeat applications at an interval sufficient to maintain control, usually 3 to 14 days depending on plant growth rate, moth activity, rainfall after treating, and other factors. If attempting to control a pest with a single spray, make the treatment when egg hatch is essentially complete, but before extensive crop damage occurs.
- A spreader-sticker which has been approved for use on growing and harvested crops should be added for hard-to-wet crops such as cole crops, or to improve weather-fastness of the spray deposits.

After eating a lethal dose of DIPEL 2X, larvae stop feeding within the hour, and will die within several days. Dying larvae move slowly, discolor, then shrivel, blacken and die.

DIPEL 2X may be applied in conventional ground or aerial equipment with quantities of water sufficient to provide thorough coverage of infested plant parts. The amount of water needed per acre will depend on crop size, weather, spray equipment, and local experience. Unless otherwise indicated, use at least 2 gallons of water by air; except in the far west, 5 to 10 gallons is the usual minimum. Add water to the spray or mixing tank at the level that provides maximum agitation. With the agitator running, slowly sprinkle in the DIPEL 2X. Continue agitation. Then add other spray materials (if any). Add the balance of the water and agitate until mixed. Maintain the suspension while loading and spraying. Do not mix more DIPEL 2X than can be used in a 12-hour period.

APPLICATION RATE

Crop	Pest	Pounds/Acre
Cole Crops and Vegetables including Leafy, Root, Stalk, and Seed and Pod Types (Peas, Beans, Lentils, etc., dry or succulent) Potatoes, Cucurbits	Loopers	1/4 to 1/2
	Hornworms	1/8 to 1/4
	Imported Cabbageworm	1/8 to 1/4
	Diamondback Moth	1/6 to 1/3
	Green Cloverworm	1/8 to 1/4
	Webworm	1/4 to 1/2
	Armyworms*	1/2 to 1

* DIPEL 2X may be used to suppress small armyworms (1st and 2nd instar) when populations are light and full coverage ground sprays are applied.

Crop	Pest	Pounds/Acre
Tomatoes, Peppers, Eggplant	Loopers	1/4 to 1/2
	Hornworms	1/8 to 1/4
	Tomato Fruitworm*	1/2
	Variegated Cutworm	1/2
	Saltmarsh Caterpillars	1/4 to 1/2
	Armyworms**	1/2 to 1

* Apply weekly in a preventive program.
** DIPEL 2X may be used to suppress small worms (1st and 2nd instar) when populations are light and full coverage ground sprays are applied.

Crop	Pest	Pounds/Acre
Cucurbits, Mellons, Squash, Cucumbers, etc.	Rindworm Complex*	1/4 to 1/2
	Melonworms	1/2 to 1

*Apply weekly in a preventive program.

Crop	Pest	Pounds/Acre
Tobacco	Tobacco Budworm	1/4 to 1/2
	Hornworms	1/8 to 1/4
	Loopers	1/4 to 1/2

DIPEL 2X may be tank mixed with sucker control products at recommended rates. Follow the application directions on the sucker control labels. Observe all label precautions for all products used.

Crop	Pest	Pounds/Acre
Safflower, Sugar beets, Mints	Loopers	1/4 to 1/2
	Saltmarsh Caterpillar	1/4 to 1/2
Soybeans, Sunflowers, Peanuts	Loopers	1/4 to 1/2
	Green Cloverworm	1/8 to 1/4
	Velvetbean Caterpillar	1/6 to 1/4
	Podworm*	1/4 to 1/2

* DIPEL 2X suppresses podworms under pest management conditions.

Crop	Pest	Pounds/Acre
Alfalfa, Pastures, Hay, Small Grains and Forage Crops	Loopers	1/4 to 1/2
	Alfalfa Catepillar	1/16 to 1/8
	European Skipper (Essex Skipper)	1/8 to 1/4
Rangeland Use a minimum of 1/2 to 1 gallon water per acre	Range Catepillar	1/20 to 1/8
Cotton	Tobacco Budworm*	1/8 to 1/2
	Cotton Bollworm*	1/8 to 1/2
	Loopers**	1/4 to 1/2

* Use to suppress light to moderate populations of newly hatched worms in pest management conditions. Use under close scouting when beneficial insects are active or building. Repeat treatments at 4 to 5-day intervals as long as necessary and results are acceptable. If 15% or more terminals have eggs, add 1/8 lb. AI Chlordimeform[1] (Galecron® or Fundal®) per acre to 1/8 lb. DIPEX 2X. Increase chlordimeform to 1/4 lb. AI if needed. Once beneficial insects are no longer a factor, DIPEL 2X (at 1/8 to 3/8 lb./acre) may be tank mixed with methomyl (Lannate® or Nudrin®) at .225 to .675 lbs. AI per acre to control light to moderate infestations of early instar larvae. Observe any rate, frequency, or boll load label restrictions for methomyl in your area. Methomyl with likely cause reddening of cotton foliage. Discontinue this tank mix if reddening becomes excessive.

Before using any tank mix, read the product labels for all environmental and usage cautions. Chlordimeform is restricted to aerial application and to states where registered.

** Full plant coverage needed.

1 The mixture of B.t. with chlordimeform for inhibiting the growth of lepidopterous larvae on plants is covered by Abbott U.S. patent 3,937,813.

Crop	Pest	Pounds/Acre
Turf	Sod Webworm	1 to 2

Crop	Pest	Lbs./100 Gals.
Flowers and Ornamentals (ground only)	Loopers	1/4 to 1/2
	Tobacco Budworm	1/4 to 1/2
	Omnivorous Looper	1/4 to 1/2
	Omnivorous Leafroller	1/4 to 1/2
	Diamondback Moth	1/6 to 1/3
	Armyworms*	1/2 to 1
	Ello Moth (Hornworm)	1/8 to 1/4
	Io Moth	1/4 to 1/2
	Oleander Moth	1/4 to 1/2

* DIPEL 2X may be used to suppress small armyworms (1st and 2nd instar) when populations are light and full coverage ground sprays are applied.

DIPEL 2X for Fruits, Nuts, Trees, Ornamentals

Crop	Pest	Pounds/Acre
Grapes, Small Fruits, Cane and Bush Berries	Grapeleaf Skeletonizer (ground only)	1/3 to 1/2
	Grape Leaffolder	1/4 to 3/8
	Achema Sphinx Moth (Hornworm)	1/6 to 3/8
	Saltmarsh Caterpillar (ground only)	1/2
	Omnivorous Leafroller (ground only)	1/4 to 1/2
	Loopers	1/4 to 1/2

Crop	Pest	Pounds/ 100 Gallons* (Ground Equipment)	Pounds/ Acre (Aerial** Application)
Stone Fruits, Nut Trees, Pomegranates	Redhumped Caterpillar	1/8 to 1/4	1/8 to 1/2
	Tent Caterpillars	1/8 to 1/4	1/4 to 1/2
	Omnivorous Leafroller	1/4 to 1/2	—
	Fall Webworm	1/8 to 1/4	1/4
	Walnut Caterpillars	1/4 to 1/2	1/4 to 1/2
	Cankerworms	1/8 to 1/4	1/4 to 1/2
Pome Fruits[1]	Cankerworms	1/8 to 1/4	1/4 to 1/2
	Gypsy Moth	1/4 to 1/2	1/2
	Variegated Leafroller	1/4 to 1/2	—
	Redbanded Leafroller	1/4 to 1/2	—
	Tufted Apple Budmoth	1/4 to 1/2	—
	Tent Caterpillars	1/8 to 1/4	1/4 to 1/2
Citrus[2]	Fruittree Leafroller	1/4 to 1/2	—
	Orangedog	1/8 to 1/4	—
Tropical Fruits[3]	Hornworm	1/4 to 1/2	—
	Leafrollers	1/4 to 1/2	—
	Omnivorous Looper	1/4 to 1/2	—
	Loopers	1/4 to 1/2	—
Forest, Shade, Sugar Maple Trees and Ornamentals	Gypsy Moth	1/4 to 1/2	1/2
	Bagworm	1/4 to 1/2	1/4 to 1/2
	Redhumped Caterpillar	1/8 to 1/4	1/8 to 1/2
	Spring and Fall Cankerworm	1/8 to 1/4	1/8 to 1/4
	Fall Webworm	1/4	1/4
	Elm Spanworm	1/4 to 1/2	1/4 to 1/2
	Tent Caterpillars	1/8 to 1/4	1/4
	California Oakworm	1/8 to 1/4	1/8 to 1/4
	Pine Butterfly	1/2	1/2
	Spruce Budworms[4]	1/2	1/2
	Saddle Prominent Caterpillar	1/4 to 1/2	—
	Douglas Fir Tussock Moth	1/2	1/2
	Western Tussock Moth	1/4 to 1/2	—
	Fruittree Leafroller	1/4 to 1/2	—
	Blackheaded Budworm	1/2	—
	Mimosa Webworm	1/4 to 1/2	—
	Jack Pine Budworm	1/4 to 1/2	3/8 to 1/2
	Saddleback Caterpillar	1/4 to 1/2	—

* Rate for hydraulic sprayer. For mist blowers, mix the applicable amount (lbs) in 10 gallons of water.

** For aerial application, use in one to ten gallons of water depending on type and density of trees. For best results, spray systems which deliver droplet size of 200 microns should be used.

1 Use 400–450 gallons/acre to suppress listed pests in pest management programs.

2 Use 100 to 400 gallons/acre.

3 Use at least 200 gallons/acre.

4 Use against populations below 35 larvae per 18-inch branch tip. Use a minimum of 1/2 gallon/acre by air.

DIPEL 2X FOR STORED AGRICULTURAL COMMODITIES

GRAINS, SOYBEANS SUNFLOWER SEED, CROP SEED, CONDIMENTAL SEEDS, SPICES, HERBS, BIRDSEED** and POPCORN**.

Pests	Rate
Indian Meal Moth, Almond Moth	3/8 lb/100 bu (undiluted and diluted)*

* As a surface treatment, apply 1/2 lb. DIPEL 2X in 10 gal. of water per 500 sq. ft. of grain surface area; mix into top 4 inches.**

** For all States except California.

For the control and prevention of these pests, apply DIPEL 2X in a constantly agitated water suspension to the top four surface layer of grain in the bin. Use a sprinkler can or sprayer to apply the dosage into the grain stream as the last (top) four inch layer is augered into the bin. Mix 1/20 lb. DIPEL 2X per gallon of water. Apply 0.6 pint in this mixture per bushel as grain is augered into storage. Or, sprinkle the dosage onto the surface of the grain in the bin and mix thoroughly with a scoop or rake to the depth of four inches. More thorough coverage may be achieved by dividing the recommended dosage into three applications and mixing the grain between applications.

For the protection of bagged grain, apply the dosage to the entire grain mass and mix thoroughly prior to bagging.

Treatments can be applied to the stored grain at any time, but, for best results, make application immediately after harvest before moth activity occurs. In areas where late fall harvested grain is not subject to infestation because of low temperatures, application can be delayed until late winter or early spring before moth activity begins. Control for a full storage season should normally be expected; however, repeat application if infestation recurs.

This treatment controls the moth larvae. If an infestation is present when the grain is treated, moth emergence may continue for several days. If immediate control of severe infestations is desired, grain should be fumigated prior to application of this treatment. **DIPEL 2X will not control weevils or other beetles.**

DIPEL 2X is exempt from tolerance restrictions and the treated grain can be used at any time after treatment for any use.

** For commodities coarser than shelled corn, increase depth of treatment according to the habit of the pest.

PEANUTS

Pest	Rate
Indian Meal Moth, Almond Moth	1/4 lb./ton*

* Apply this rate to the top four to eight feet of nuts when filling the warehouse.

To prevent and control these pests, spray an even coating of DIPEL 2X on the farmer stock peanuts while filling the warehouse. To make the spray solution, mix 3 3/4 lbs. DIPEL 2X per 5 gallons of water. Apply to 15 tons of commodity. Do not pre-mix more spray solution than will be used within 12 hours. Keep the spray suspension agitated during application, and use pressures and nozzles sufficient to handle this suspension.

Before filling the warehouse, clean thoroughly, then spray interior of the facility with a DIPEL 2X suspension at the rate of 1/2 lb. DIPEL 2X per 100 gallons water. Spray enough suspension to wet all cracks and crevices.

For bagged peanuts, treat the whole mass of commodity at the rate indicated above.

FLUE-CURED TOBACCO

Pest	Rate
Tobacco Moth	4/20 oz./100 lbs.*

* Apply 0.2 ounce (approximately 2 1/2 teaspoonsful) of DIPEL 2X in one quart of water per 100 pounds of tobacco as a fine mist spray. Avoid overwetting. Tobacco should have just enough moisture to be handled without shattering at the time of application.

Tobacco to be stored up to Twelve Months.

Spray loose leaves as the tobacco is being bundled from the curing barn. For tobacco on sticks, treat both sides of leaves.

Stored Tobacco.

For tobacco which is to be carried over, rebundle or restack sticks, fluff up tobacco and spray loose leaves.

For tobacco that has been stored over three weeks, apply at first signs of infestation, promptly open bundles, spray loose leaves, then bundle.

Treatment of Storage Barns.

If tobacco has been treated, or is going to be treated, treatment of the floors and walls may be made to aid in control. Sweep out the area, especially cracks and corners, and all of the loose tobacco pieces in which the moth might breed. Make a spray mixture containing 1/2 oz. DIPEL 2X per 2 1/2 gallons of water. Apply this at a rate of 1/2 gallon per 1000 sq. ft. of surface area. Be sure to spray into cracks and between floorboards.

NOTICE TO USER

Seller makes no warranty, express or implied, of merchantability, fitness or otherwise concerning the use of this product other than as indicated on the label. User assumes all risks of use, storage or handling not in strict accordance with accompanying directions.

® Fundal, Galecron, Lannate, Nudrin, are trademarks of companies other than Abbott Laboratories.

DIPEL® 4L Worm Killer

EMULSIFIABLE SUSPENSION BIOLOGICAL INSECTICIDE

ACTIVE INGREDIENT:
Bacillus thuringiensis, var. kurstaki,
8,800 International Units of Potency per mg.
(32 billion International Units per gallon) 1.76%
INERT INGREDIENTS 98.24%
TOTAL 100.0%

EPA Reg. No. 275-36 EPA Est. No. 33762-IA-1

CAUTION
Keep Out of Reach of Children

PRECAUTIONARY STATEMENTS

HAZARDS TO HUMANS (AND DOMESTIC ANIMALS)

Avoid contact with skin, eyes or clothing. In case of contact immediately flush eyes or skin with plenty of water. Get medical attention if irritation persists.

Beneficial Insects: Honeybees foraging treated areas are not harmed by DIPEL 4L use.

DIPEL 4L does not interrupt the activities of beneficial and predacious arthropods in pest management programs.

ENVIRONMENTAL HAZARDS

Keep out of lakes, ponds or streams. Do not contaminate water by cleaning of equipment or disposal of wastes.

STORAGE AND DISPOSAL

Do not contaminate water, food or feed by storage or disposal.

STORAGE

Keep containers tightly closed when not in use. At temperatures less than 0°F and greater than 100°F, DIPEL 4L should be stored under cover.

PESTICIDE DISPOSAL

Wastes resulting from the use of this product may be disposed of on site or at an approved waste disposal facility.

CONTAINER DISPOSAL

Triple rinse (or equivalent). Dispose of in a sanitary landfill, or by incineration if allowed by state and local authorities.

DIRECTIONS FOR USE

It is a violation of Federal law to use this product in a manner inconsistent with its labeling.

DIPEL 4L is a highly selective insecticide for use against listed caterpillars (larvae) of lepidopterous insects. Close scouting and early attention to infestations is highly recommended. Larvae must eat deposits of DIPEL 4L to be affected. Always follow these directions:

- Treat when larvae are young (early instars) before the crop is extensively damaged.
- Larvae must be actively feeding on treated, exposed plant parts.
- Thorough spray coverage is needed to provide a uniform deposit of DIPEL 4L at the site of larval feeding. For some crops directed drop nozzles by ground machine are required.
- Under heavy pest population pressure, use the higher label rates, shorten the spray interval, and/or raise gallonage to improve spray coverage.
- Repeat applications at an interval sufficient to maintain control, usually 3 to 14 days depending on plant growth rate, moth activity, rainfall after treating, and other factors. If attempting to control a pest with a single spray, make the treatment when egg hatch is essentially complete, but before extensive crop

damage occurs.

- A spreader-sticker which has been approved for use on growing and harvested crops should be added for hard-to-wet crops such as cole crops, or to improve weather-fastness of the spray deposits.

After eating a lethal dose of DIPEL 4L, larvae stop feeding within the hour and will die within several days. Dying larvae move slowly, discolor, then shrivel, blacken and die.

DIPEL 4L may be applied in conventional ground or aerial equipment with quantities of water sufficient to provide thorough coverage of infested plant parts. The amount of water needed per acre will depend on crop size, weather, spray equipment, and local experience. Unless otherwise indicated, use at least 2 gallons of water by air; except in the far west, 5 to 10 gallons is the usual minimum. Add water to the spray or mixing tank at the level that provides maximum agitation. With the agitator running, add the DIPEL 4L. Continue agitation. Then add other spray materials (if any). Add the balance of the water. Maintain the suspension while loading and spraying. Do not mix more DIPEL 4L than can be used in a 72-hour period. **CAUTION:** Rinse and flush spray equipment thoroughly following each use.

For Smaller Spray volumes:

If Rate is	Use This Amount Per Gallon
½ pt./acre or 100 gals.	½ tsp.
1 pt./acre or 100 gals.	1 tsp.
1 qt./acre or 100 gals.	2 tsps.
2 qts./acre or 100 gals.	4 tsps.

Days to Harvest: There are no restrictions on applying DIPEL 4L up to the time of harvest.

Sites: DIPEL 4L may be used for any labeled pest in both field and greenhouse uses.

CAUTION

DIPEL 4L should not be used in combination with Comite® Bravo® (except celery), Difolatan, Captan (except seed) or Dyrene®.

ALL STATES EXCEPT CALIFORNIA

The effects of combining DIPEL 4L with fungicides (other than benomyl, maneb, mancozeb or captan) and miticides (other than chlordimeform) are unknown.

FOR CALIFORNIA ONLY

The effects of combining DIPEL 4L with insecticides, fungicides and other spray materials are unknown. Tank mixes should be avoided except where specific recommendations are made in the label for use in California.

APPLICATION RATES

Crop	Pest	Pints/Acre
Cole Crops* [1]Celery Lettuce [1]Spinach	Loopers Imported Cabbageworm Diamondback Moth	1 to 2 ½ to 1 ⅔ to 1⅓

[1] For all states except California.
* Do no apply by air to plants after transplant or other stress before 6 weeks in the field. Use more than 25 gallons of water per acre by ground and 10 gallons of water per acre by air.

Soybeans [1]Peanuts [1]Peas	Loopers Green Cloverworm Velvetbean Caterpillar Podworm (Heliothis zea)* [1]Armyworms*	1 to 2 ½ to 1 ⅔ to 1⅓ 1 to 2 2 to 4

* DIPEL 4L suppresses early instar larvae of Podworms and Armyworms when populations are light to moderate.
[1] For all states except California.

[1]Tobacco	Tobacco Budworm Hornworms Loopers	1 ½ to 1 1

[1] For all states except California.

Crop	Pest	Pints/Acre
Tomatoes	Loopers Tomato Fruitworm* Variegated Cutworm Saltmarsh Caterpillar Hornworms Armyworms**	1 to 2 2 2 1 to 2 ½ to 1 2 to 4

* Apply weekly in a preventative program.
** DIPEL 4L may be used to suppress small Armyworms (1st and 2nd instar) when populations are light and full coverage ground sprays are applied.

DIPEL 4L[1] may be tank-mixed with benomyl, maneb or mancozeb for application on tomatoes. Observe all label precautions for all products used.

[1] For all states except California.

Alfalfa, Pastures, Hay, Small Grains and Forage Crops	Loopers Alfalfa Caterpillar European Skipper (Essex Skipper)	1 to 2 ¼ to ½ ½ to 1
Rangeland	Range Caterpillar	¼

Use in 1 to 2 qts. water per acre against 1st through 4th instar larvae.

[1]Avocados	Amorbia Moth Omnivorous Leafroller Omnivorous Looper Orange Tortix	Not more than 8

Use a minimum of 200 gallons water per acre by ground rig or 10 gallons by aircraft. Maximum of two applications per season.

[1] For all states except California

Cotton	Tobacco Budworm* Cotton Bollworm* Loopers**	½ to 2 ½ to 2 1 to 2

* Use to suppress light to moderate populations of newly hatched worms in pest management conditions. Use under close scouting when beneficial insects are active or building. Repeat treatments at 4 to 5-day intervals as long as necessary and results are acceptable. If 15% or more terminals have eggs, add ⅛ lb. AI chlordimeform[1] (Galecron® or Fundal® or ⅛ lbs. AI methomyl[2] (Lannate® or Nudrin®) per acre to ½ pt. DIPEL 4L. Increase chlordimeform to ¼ lb. AI if needed.

Once beneficial insects are no longer a factor, DIPEL 4L (at ½ to 1½ pts./acre) may be tank mixed with methomyl (Lannate® or Nudrin®) at .225 to .675 lbs. AI per acre to control light to moderate infestations of early instar larvae. Observe any rate, frequency, or boll load label restrictions for methomyl in your area. Methomyl will likely cause reddening of cotton foliage. Discontinue this tank mix if reddening becomes excessive.

Before using any tank mix, read the product labels for all environmental and usage cautions. Chlordimeform is restricted to aerial application and to states where registered.

DIPEL 4L[2] may be tank mixed with vegetable or cottonseed oil at the ratio of 1 pint 4L to 1 pint oil. DIPEL 4L may also be mixed at a ratio of 40 parts 4L to 60 parts water, and applied in total spray volume not to exceed 80 fl. oz. per acre.

** Full plant coverage needed.
1 The mixture of B.t. with chlordimeform for inhibiting the growth of lepidopterous larvae on plants is covered by Abbott U.S. patent 3,937,813.
2 All states except California.

CROP SEED[1]

Pest	Rate
Indian Meal Moth Almond Month	1½ pts/100 bu

Mix thoroughly prior to bagging.

DIPEL 4L is compatible with common seed treatments including captan, methoxychlor, carboxin (Vitavax) and malathion. Fumigation has not been found to decrease the effectiveness of DIPEL 4L.

[1] For all states except California.

Trees and Shrubs[2]

Crop	Pest	Pints/100 Gallons* (Ground Equipment)	Pints/Acre (Aerial** Application)
Forest, Shade Sugar Maple Trees and Shrubs[2]	Gypsy Moth	1[1] to 4	2[1] to 4
	Bagworm	1 to 2	1 to 2
	Redhumped Caterpillar	½ to 1	½ to 1
	Spring & Fall Cankerworm	½ to 1	½ to 1
	Fall Webworm	1	1
	Elm Spanworm	1 to 2	1 to 2
	Tent Caterpillars	½ to 1	1[1] to 2
	California Oakworm	½ to 1	½ to 1
	Pine Butterfly	2	2
	Spruce Budworms[3]	2[1] to 4	2[1] to 4
	Saddled Prominent Caterpillar	1 to 2	1 to 2[2]
	Douglas Fir Tussock Moth	2	2
	Western Tussock Moth	1 to 2	—
	Fruittree Leafroller	1 to 2	—
	Blackheaded Budworm	2	—
	Mimosa Webworm	1 to 2	—
	Jack Pine Budworm	1 to 2	1½ to 2
	Saddleback Caterpillar	1 to 2	—

* Rate for hydraulic sprayer in California. In other states, less water may be used if adequate coverage is provided. For mist blowers, mix the applicable amount (pints) in zero to 10 gallons of water; except in California, mix the applicable amount (pints) in 10 gallons of water.

** For aerial application, use in zero to 10 gallons (in California, against all pests except Spruce Budworms, use in 1 to 10 gallons) of water depending on type and density of trees. For best results, spray systems which deliver droplet size of 200 microns or less should be used.

[1] Use only the low rate in California.

[2] All states except in California.

[3] Use rates greater than 2 pints in Northern states for heavy populations. In California, tank mix ratio of water to DIPEL 4L must be no less than 50:50.

NOTICE TO USER

Seller makes no warranty, express or implied, of merchantability, fitness or otherwise concerning use of this product other than as indicated on the label. User assumes all risks of use, storage or handling not in strict accordance with accompanying directions.

DIPEL® 6AF Worm Killer

AQUEOUS FLOWABLE BIOLOGICAL INSECTICIDE

NOT REGISTERED IN CALIFORNIA

ACTIVE INGREDIENT:

Bacillus thuringiensis, var. kurstaki,
10,750 International Units of Potency per mg
(48 billion International Units per gallon) 2.15%
INERT INGREDIENTS 97.85%
TOTAL 100.0%

EPA Reg. No. 275–59 EPA Est. No. 33762-IA-1

CAUTION
Keep Out of Reach of Children

PRECAUTIONARY STATEMENTS

HAZARDS TO HUMANS (AND DOMESTIC ANIMALS)

Avoid contact with skin, eyes or clothing. In case of contact immediately flush eyes or skin with plenty of water. Get medical attention if irritation persists.

BENEFICIAL INSECTS

Honeybees foraging treated areas are not harmed by DIPEL 6AF use.

DIPEL 6AF does not interrupt the activities of beneficial and predacious arthropods in pest management programs.

STORAGE AND DISPOSAL

Do not contaminate water, food or feed by storage or disposal.

STORAGE

Keep containers tightly closed when not in use. At temperatures less than 0°F and greater than 100°F, DIPEL 6AF should be stored under cover.

PESTICIDE DISPOSAL

Wastes resulting from the use of this product may be disposed of on site or at an approved waste disposal facility.

CONTAINER DISPOSAL

Triple rinse (or equivalent). Puncture and dispose of in a sanitary landfill, or by other procedures approved by state and local authorities.

Days to harvest: There are no restrictions on applying DIPEL 6AF up to the time of harvest.

DIRECTIONS FOR USE

It is a violation of Federal law to use this product in a manner inconsistent with its labeling.

DIPEL 6AF is a highly selective insecticide for use against listed caterpillars (larvae) of lepidopterous insects. Close scouting and early attention to infestations is highly recommended. Larvae must eat deposits of DIPEL 6AF to be affected. Always follow these directions:

- Treat when larvae are young (early instars) before extensive damage has occurred.
- Larvae must be actively feeding on treated, exposed plant parts.
- Thorough spray coverage is needed to provide a uniform deposit of DIPEL 6AF at the site of larval feeding.
- Under heavy pest population pressure, use the higher label rates and/or raise gallonage to improve spray coverage.
- If attempting to control a pest with a single spray, make the treatment when egg hatch is essentially complete, but before extensive crop damage occurs.

- A spreader-sticker which has been approved for use on growing and harvested crops may be added to improve weather-fastness of the spray deposits.

After eating a lethal dose of DIPEL 6AF, larvae stop feeding within the hour and will die within several days. Dying larvae move slowly, discolor, then shrivel, blacken and die.

DIPEL 6AF may be applied in conventional ground or aerial equipment with quantities of water sufficient to provide thorough coverage of infested plant parts. The amount of water needed per acre will depend on weather, spray equipment, and local experience. Fill the mix tank or plane hopper with the desired quantity of water. Start the mechanical or hydraulic agitation to provide moderate circulation before adding DIPEL 6AF. Always add the sticker to the water prior to the addition of DIPEL 6AF. Add the desired volume of DIPEL 6AF to the tank or plane hopper and continue circulation. Include rinse water from containers. **Never add DIPEL 6AF to the mix tank before introducing the desired quantity of water.** Maintain the suspension while loading and spraying. Do not mix more DIPEL 6AF than can be used in a 144-hour period. **CAUTION:** Rinse and flush spray equipment thoroughly following each use.

DIPEL 6AF for Trees and Shrubs

Crop	Pest	Pints/ 100 Gallons* (Ground Equipment)	Pints/ Acre (Aerial** Application)
Forest, Shade, Sugar Maple Trees and Shrubs	Gypsy Moth	⅔ to 3⅓	1⅓ to 3⅓
	Browntail Moth	⅔ to 2⅔	1⅓ to 2⅔
	Bagworm	⅔ to 1⅓	⅔ to 1⅓
	Redhumped Caterpillar	⅓ to ⅔	⅓ to ⅔
	Spring & Fall Cankerworm	⅓ to ⅔	⅓ to ⅔
	Fall Webworm	⅔	⅔
	Elm Spanworm	⅔ to 1⅓	⅔ to 1⅓
	Tent Caterpillars	⅓ to ⅔	⅔ to 1⅓
	California Oakworm	⅓ to ⅔	⅓ to ⅔
	Pine Butterfly	1⅓	1⅓
	Spruce Budworms[1]	1⅓ to 3⅓	1⅓ to 3⅓
	Saddled Prominent Caterpillar	⅔ to 1⅓	⅔ to 1⅓
	Douglas Fir Tussock Moth	1⅓	1⅓
	Western Tussock Moth	⅔ to 1⅓	—
	Fruittree Leafroller	⅔ to 1⅓	—
	Blackheaded Budworm	1⅓	—
	Mimosa Webworm	⅔ to 1⅓	—
	Jack Pine Budworm	⅔ to 1⅓	1 to 1⅓
	Saddleback Caterpillar	⅔ to 1⅓	—

* Water dilution rate for hydraulic sprayer may be varied depending on coverage. For mist blowers, mix the applicable amount (pts.) in up to 10 gallons of water.

** For aerial application, use in up to 10 gallons of water depending on type and density of trees. For best results, spray systems which deliver droplet size of 200 microns or less should be used.

[1] Use rates greater than 1⅓ pints in Northern states for heavy populations. DIPEL 6AF may be sprayed undiluted for the control of spruce budworm.

NOTICE TO USER

Seller makes no warranty, express or implied, of merchantability, fitness or otherwise concerning use of this product other than as indicated on the label. User assumes all risks of use, storage or handling not in strict accordance with accompanying directions.

DIPEL® 6L Worm Killer

EMULSIFIBLE SUSPENSION BIOLOGICAL INSECTICIDE

ACTIVE INGREDIENT:
Bacillus thuringiensis, var. kurstaki,
13,200 International Units of Potency per mg
(48 billion International Units per gallon) 2.6%
INERT INGREDIENTS 97.4%
TOTAL 100.0%

EPA Reg. No. 275–48 EPA Est. No. 33762-IA-1

CAUTION
Keep Out of Reach of Children

PRECAUTIONARY STATEMENTS

HAZARDS TO HUMANS (AND DOMESTIC ANIMALS)

Avoid contact with skin, eyes or clothing. In case of contact immediately flush eyes or skin with plenty of water. Get medical attention if irritation persists.

BENEFICIAL INSECTS

Honeybees foraging treated areas are not harmed by DIPEL 6L use.

DIPEL 6L does not interrupt the activities of beneficial and predacious arthropods in pest management programs.

STORAGE AND DISPOSAL

Do not contaminate water, food or feed by storage or disposal.

STORAGE

Keep containers tightly closed when not in use. At temperatures less than 0°F and greater than 100°F, DIPEL 6L should be stored under cover.

PESTICIDE DISPOSAL

Wastes resulting from the use of this product may be disposed of on site or at an approved waste disposal facility.

CONTAINER DISPOSAL

Triple rinse (or equivalent). Puncture and dispose of in a sanitary landfill, or by other procedures approved by state and local authorities.

Days to harvest: There are no restrictions on applying DIPEL 6L up to the time of harvest.

DIRECTIONS FOR USE

It is a violation of Federal law to use this product in a manner inconsistent with its labeling.

DIPEL 6L is a highly selective insecticide for use against listed caterpillars (larvae) of lepidopterous insects. Close scouting and early attention to infestations is highly recommended. Larvae must eat deposits of DIPEL 6L to be affected. Always follow these directions:

- Treat when larvae are young (early instars) before extensive damage has occurred.
- Larvae must be actively feeding on treated, exposed plant parts.
- Thorough spray coverage is needed to provide a uniform deposit of DIPEL 6L at the site of larval feeding.
- Under heavy pest population pressure, use the higher label rates and/or raise gallonage to improve spray coverage.
- If attempting to control a pest with a single spray, make the treatment when egg hatch is essentially complete, but before extensive crop damage occurs.
- A spreader-sticker which has been approved for use on growing and harvested crops may be added to improve weather-fastness of the spray deposits.

After eating a lethal dose of DIPEL 6L, larvae stop feeding within the hour and will die within several days. Dying larvae move slowly, discolor, then shrivel, blacken and die.

DIPEL 6L may be applied in conventional ground or aerial equipment with quantities of water sufficient to provide thorough coverage of infested plant parts. The amount of water needed per acre will depend on weather, spray equipment, and local experience. Fill the mix tank or plane hopper with the desired quantity of water. Start the mechanical or hydraulic agitation to provide moderate circulation before adding DIPEL 6L. Always add the sticker to the water prior to the addition of DIPEL 6L. Add the desired volume of DIPEL 6L to the tank or plane hopper and continue circulation. Include rinse water from containers. **Never add DIPEL 6L to the mix tank before introducing the desired quantity of water.** Maintain the suspension while loading and spraying. Do not mix more DIPEL 6L than can be used in a 144-hour period. **CAUTION:** Rinse and flush spray equipment thoroughly following each use.

DIPEL 6L for Trees and Shrubs

Crop	Pest	Pints/ 100 Gallons* (Ground Equipment)	Pints/ Acre (Aerial** Application)
Forest, Shade, Sugar Maple Trees and Shrubs	Gypsy Moth	⅔ to 3⅓	1⅓ to 3⅓
	Browntail Moth	⅔ to 2⅔	1⅓ to 2⅔
	Bagworm	⅔ to 1⅓	⅔ to 1⅓
	Redhumped Caterpillar	⅓ to ⅔	⅓ to ⅔
	Spring & Fall Cankerworm	⅓ to ⅔	⅓ to ⅔
	Fall Webworm	⅔	⅔
	Elm Spanworm	⅔ to 1⅓	⅔ to 1⅓
	Tent Caterpillars	⅓ to ⅔	⅔ to 1⅓
	California Oakworm	⅓ to ⅔	⅓ to ⅔
	Pine Butterfly	1⅓	1⅓
	Spruce Budworms[1]	1⅓ to 3⅓	1⅓ to 3⅓
	Saddled Prominent Caterpillar	⅔ to 1⅓	⅔ to 1⅓
	Douglas Fir Tussock Moth	1⅓	1⅓
	Western Tussock Moth	⅔ to 1⅓	—
	Fruittree Leafroller	⅔ to 1⅓	—
	Blackheaded Budworm	1⅓	—
	Mimosa Webworm	⅔ to 1⅓	—
	Jack Pine Budworm	⅔ to 1⅓	1 to 1⅓
	Saddleback Caterpillar	⅔ to 1⅓	—

* Water dilution rate for hydraulic sprayer may be varied depending on coverage. For mist blowers, mix the applicable amount (pts.) in up to 10 gallons of water.

** For aerial application, use in up to 10 gallons of water depending on type and density of trees. For best results, spray systems which deliver droplet size of 200 microns or less should be used. DIPEL 6L should always be mixed with at least an equal amount of water for diluted applications.

[1] Use rates greater than 1⅓ pints in Northern states for heavy populations. DIPEL 6L may be sprayed undiluted for the control of spruce budworm.

NOTICE TO USER

Seller makes no warranty, express or implied, of merchantability, fitness or otherwise concerning use of this product other than as indicated on the label. User assumes all risks of use, storage or handling not in strict accordance with accompanying directions.

DIPEL® 8L Worm Killer

EMULSIFIABLE SUSPENSION BIOLOGICAL INSECTICIDE

ACTIVE INGREDIENT:
Bacillus thuringiensis, var. **kurstaki**,
17,600 International Units of Potency per mg
(64 Billion International Units per gallon) 3.5%
INERT INGREDIENTS 96.5%
TOTAL 100.0%

E.P.A. Reg. No. 275-51 E.P.A. Est. No. 33762-IA-1

CAUTION
Keep Out of Reach of Children

PRECAUTIONARY STATEMENTS

HAZARDS TO HUMANS (AND DOMESTIC ANIMALS)

Avoid contact with skin, eyes or clothing. In case of contact immediately flush eyes or skin with plenty of water. Get medical attention if irritation persists.

BENEFICIAL INSECTS

Honeybees foraging treated areas are not harmed by DIPEL 8L use.

DIPEL 8L does not interrupt the activities of beneficial and predacious arthropods in pest management programs.

STORAGE AND DISPOSAL

Do not contaminate water, food or feed by storage or disposal.

STORAGE

Keep containers tightly closed when not in use. At temperatures less than 0°F and greater than 100°F, Dipel 8L should be stored under cover. Roll or shake the drum before dispensing.

PESTICIDE DISPOSAL

Wastes resulting from the use of this product may be disposed of on site or at an approved waste disposal facility.

CONTAINER DISPOSAL

Triple rinse (or equivalent). Puncture and dispose of in a sanitary landfill, or by other procedures approved by state and local authorities.

Days to harvest: There are no restrictions on applying DIPEL 8L up to the time of harvest.

DIRECTIONS FOR USE

It is a violation of Federal law to use this product in a manner inconsistent with its labeling.

DIPEL 8L is a highly selective insecticide for use against listed caterpillars (larvae) of lepidopterous insects. Close scouting and early attention to infestations is highly recommended. Larvae must eat deposits of DIPEL 8L to be affected. Always follow these directions:

- Treat when larvae are young (early instars) before extensive damage has occurred.
- Larvae must be actively feeding on treated, exposed plant parts.
- Thorough spray coverage is needed to provide a uniform deposit of DIPEL 8L at the site of larval feeding.
- Under heavy pest population pressure, use the higher label rates and/or raise gallonage to improve spray coverage.
- If attempting to control a pest with a single spray, make the treatment when egg hatch is essentially complete, but before extensive crop damage occurs.
- A spreader-sticker which has been approved for use on growing and harvested crops may be added if rain is anticipated to improve weather-fastness of the spray deposits.

After eating a lethal dose of DIPEL 8L, larvae stop feeding within the hour and will die within several days. Dying larvae move slowly, discolor, then shrivel, blacken and die.

DIPEL 8L may be applied in conventional ground or aerial equipment with quantities of water sufficient to provide thorough coverage of infested plant parts. The amount of water needed per acre will depend on weather, spray equipment, and local experience. Fill the mix tank or plane hopper with the desired quantity of water. Start the mechanical or hydraulic agitation to provide moderate circulation before adding DIPEL 8L. Always add the sticker to the water prior to the addition of DIPEL 8L. Add the desired volume of DIPEL 8L to the tank or plane hopper and continue circulation. Include rinse water from containers. **Never add DIPEL 8L to the mix tank before introducing the desired quantity of water.** Maintain the suspension while loading and spraying. Do not mix more DIPEL 8L than can be used in a 144-hour period. **CAUTION:** Rinse and flush spray equipment thoroughly following each use.

DIPEL 8L for Trees and Shrubs

Crop	Pest	Pints/ 100 Gallons* (Ground Equipment)	Pints/ Acre (Aerial** Application)
Forest, Shade, Sugar Maple Trees and Shrubs	Gypsy Moth	½ to 2½	1 to 2½
	Browntail Moth	½ to 2½	1 to 2½
	Bagworm	½ to 1	½ to 1
	Redhumped Caterpillar	¼ to ½	¼ to ½
	Spring & Fall Cankerworm	¼ to ½	½
	Fall Webworm	½	½ to 1
	Elm Spanworm	½ to 1	½ to 1
	Tent Caterpillars	¼ to ½	¼ to ½
	California Oakworm	¼ to ½	1
	Pine Butterfly	1	1 to 2½
	Spruce Budworms[1]	1 to 2½	½ to 1
	Saddled Prominent Caterpillar	½ to 1	1
	Douglas Fir Tussock Moth	1	—
	Western Tussock Moth	½ to 1	—
	Fruittree Leafroller	½ to 1	—
	Blackheaded Budworm	1	—
	Mimosa Webworm	½ to 1	¾ to 1
	Jack Pine Budworm	½ to 1	—
	Saddleback Caterpillar	½ to 1	—

* Water dilution rate for hydraulic sprayer may be varied depending on coverage. For mist blowers, mix the applicable amount (pts.) in up to 10 gallons of water.

** For aerial application, use in up to 10 gallons of water depending on type and density of trees. For best results, spray systems which deliver droplet size of 200 microns or less should be used. DIPEL 8L should always be mixed with at least an equal amount of water for diluted applications.

[1] Use rates greater than 1 pint in Northern states for heavy populations. DIPEL 8L may be sprayed undiluted for the control of spruce budworm.

NOTICE TO USER

Seller makes no warranty, express or implied, of merchantability, fitness or otherwise concerning use of this product other than as indicated on the label. User assumes all risks of use, storage or handling not in strict accordance with accompanying directions.

DIPEL® 10G (Granular)

BIOLOGICAL INSECTICIDE

For control of European Corn Borer Larvae on Seed Corn, Field Corn, Popcorn, and Sweetcorn

ACTIVE INGREDIENT:

Bacillus thuringiensis, var. **kurstaki,**

1600 International Units of Potency per mg.* 0.32%

INERT INGREDIENTS 99.68%

TOTAL 100.00%

* This product contains 0.726 Billion IU/lb.

EPA Reg. No. 275–55 EPA Est. No. 34704-NB-1

CAUTION
Keep Out of Reach of Children

STORAGE AND DISPOSAL

Keep out of lakes, ponds or streams. Do not contaminate water by cleaning of equipment or disposal of wastes.

Do not contaminate water, food, or feed by storage or disposal.

STORAGE

Store in a cool, dry place.

PESTICIDE DISPOSAL

Wastes resulting from the use of this product may be disposed of on site or at an approved waste disposal facility.

CONTAINER DISPOSAL

Completely empty bag into application equipment, then dispose of bags in a sanitary landfill or by incineration, or, if allowed by state and local authorities, by burning. If burned, stay out of smoke.

GENERAL INFORMATION

DIPEL® 10G is a granular formulation of **Bacillus thuringiensis,** variety **kurstaki,** designed specifically as a selective insecticide for control of European Corn Borer infesting seed corn, field corn, popcorn and sweet corn. For practical determinations of active ingredient, DIPEL 10G can be expressed as a 10% formula of DIPEL WP.

The active ingredient in DIPEL 10G is a stomach poison and therefore must be eaten by larvae to be effective. Climatic extremes such as heavy rain or strong wind shortly after treatment may act to remove DIPEL 10G from the feeding zone making re-treatment necessary.

DIRECTIONS FOR USE

It is a violation of Federal law to use this product in a manner inconsistent with its labeling.

Application: (Ground or Aerial Equipment):

Crop	Pest	Rate
Seed Corn, Field Corn, Popcorn, and Sweet Corn	European corn Borer *(Ostrinia nubilalis),* First and Second Generation	10 lbs/acre

FIRST GENERATION

Apply DIPEL® 10G directly to corn whorls. Treat seed corn when 25 percent of plants show shot hole feeding in the whorls and field corn when 35 percent of whorls show feeding. Larvae should be small and present at this time. Satisfactory control cannot be expected if treatment is delayed to higher injury levels or after larvae have begun stalk boring.

SECOND GENERATION

Apply DIPEL 10G when a field count shows 100 egg masses per 100 plants and the first hatch is occurring. Under conditions of intensive population pressure, a second application may be necessary.

NOTICE

Local conditions may affect the use of DIPEL 10G. Consult State Agricultural Extension or Experiment Station specialists for specific recommendations related to local crop protection problems. Abbott makes no warranty, express or implied, including the warranties of commercially acceptable quality and/or fitness for any particular purpose concerning this material, other than those which are contained on this label.

PRO-GIBB® PLUS 2X

SOLUBLE POWDER

(Gibberellic Acid)

ACTIVE INGREDIENT:	W/W
Gibberellin A3	20%
INERT INGREDIENTS	80%
TOTAL	100%

Contains a total of 32 g. of Gibberellic Acid.

EPA Reg. No. 275-62 EPA Est. No. 48477-CA-02

CAUTION

Keep Out of Reach of Children

PRECAUTIONARY STATEMENTS

HAZARD TO HUMANS (AND DOMESTIC ANIMALS)

CAUTION

Powder causes eye irritation. In case of contact with eyes flush thoroughly with water.

Do not apply this product in such a manner as to directly or through drift expose workers or other persons. The area being treated must be vacated by unprotected persons.

REENTRY STATEMENT

Do not enter treated areas without protective clothing until sprays have dried.

Because certain states may require more restrictive reentry intervals for various crops treated with this product, consult your State Department of Agriculture for further informaton.

ENVIRONMENTAL HAZARDS

Keep out of lakes, ponds or streams. Do not contaminate water by cleaning of equipment or disposal of wastes.

DIRECTIONS FOR USE

It is a violation of Federal law to use this product in a manner inconsistent with its labeling.

IMPORTANT

Before application, read accompanying PRO-GIBB Spray Guide carefully and use only as directed.

STORAGE AND DISPOSAL

Do not contaminate water, food or feed by storage or disposal.

STORAGE

Keep containers tightly closed when not in use.

PESTICIDE DISPOSAL

Wastes resulting from the use of this product may be disposed of on site or at an approved waste disposal facility.

CONTAINER DISPOSAL

Do not reuse empty containers. Triple rinse (or equivalent). Then puncture and dispose of in a sanitary landfill, or by incineration, or, if allowed by state and local authorities, by burning. If burned, stay out of smoke.

PRO-GIBB® 4% LIQUID

CONCENTRATE

(Gibberellic Acid)

ACTIVE INGREDIENT:	W/W
Gibberellic Acid	4%
INERT INGREDIENTS	96%
TOTAL	100.0%

Contains a total of 121.6 g. of gibberellic acid/gallon.

Contains a total of 19.0 g. of gibberellic acid/20 fluid oz.

EPA Reg. No. 275-61 EPA Est. No. 35347-CA-1

CAUTION

Keep Out of Reach of Children

PRECAUTIONARY STATEMENTS

HAZARDS TO HUMANS (AND DOMESTIC ANIMALS)

CAUTION

PRO-GIBB 4% Liquid Concentrate may cause eye irritation and is harmful if swallowed or if vapors are breathed for prolonged periods. In case of contact with eyes, flush thoroughly with water.

Do not apply this product in such a manner as to directly or through drift expose workers or other persons. The area being treated must be vacated by unprotected persons.

REENTRY STATEMENT

Do not enter treated areas without protective clothing until sprays have dried. Because certain states may require more restrictive reentry intervals for various crops treated with this product, consult your State Department of Agriculture for further information.

ENVIRONMENTAL HAZARDS

Keep out of lakes, ponds or streams. Do not contaminate water by cleaning of equipment or disposal of wastes.

PHYSICAL OR CHEMICAL HAZARDS

FLAMMABLE! Keep away from heat and open flame.

DIRECTIONS FOR USE

It is a violation of Federal law to use this product in a manner inconsistent with its labeling.

IMPORTANT

Before application, read accompanying PRO-GIBB Spray Guide carefully and use only as directed.

STORAGE AND DISPOSAL

Do not contaminate water, food or feed by storage or disposal.

STORAGE

Keep containers tightly closed when not in use. Keep away from heat and open flame.

PESTICIDE DISPOSAL

Wastes resulting from the use of this product may be disposed of on site or at an approved waste disposal facility.

CONTAINER DISPOSAL

Do not reuse empty containers. Triple rinse (or equivalent). Puncture and dispose of in a sanitary landfill, or by incineration or, if allowed by state and local authorities, by burning. If burned, stay out of smoke.

NOTICE TO USER

Seller makes no warranty, express or implied, of merchantability, fitness or otherwise concerning the use of this product other than as indicated on the label. User assumes all risks of use, storage or handling not in strict accordance with accompanying directions.

PRO-GIBB®

SPRAYING GUIDE

(Gibberellic Acid)

CAUTION

Keep Out of Reach of Children

LIQUID FORMULATION

WARNING: FLAMMABLE! Keep away from heat and open flame. Keep container tightly closed when not in use.

CAUTION: Harmful if swallowed. Avoid breathing vapors. Avoid contact with eyes.

SOLUBLE POWDER FORMULATION

CAUTION: Powder causes eye irritation. If powder gets in eyes, flush thoroughly with water.

ENVIRONMENTAL HAZARDS

Avoid direct applications to any body of water. Do not contaminate water by disposal of waste or cleaning of equipment.

DIRECTIONS FOR USE

It is a violation of Federal law to use this product in a manner inconsistent with its labeling.

STORAGE AND DISPOSAL

See requirements for each product formulation.

REENTRY STATEMENT

Do not enter treated areas without protective clothing until sprays have dried. Because certain states may require more restrictive reentry intervals for various crops treated with this product, consult your State Department of Agriculture for further information.

FARMWORKER NOTIFICATION

Written or oral warnings must be given to workers who are expected to be in a treated area or in an area about to be treated with this product. Oral warnings must include the following information:

> Inform workers of areas or fields that must not be entered without appropriate protective clothing until sprays have dried. In case of accidental exposure, wash with plenty of water. If there is any irritation in eyes after washing, get medical attention.

When oral warnings are given, warnings shall be given in a language customarily understood by workers. Oral warnings must be given if there is reason to believe that written warnings cannot be understood by workers. Written warnings must include the following information:

> Area treated with PRO-GIBB on (date of application). Do not enter without appropriate protective clothing until sprays have dried. In case of accidental exposure, wash with plenty of water. If there is any irritation in eyes after washing, get medical attention.

WARNING: Gibberellic Acid is an extremely potent plant growth regulator. For best results, read all directions for use thoroughly. Consult your local experiment station specialist, distributor, or the Abbott agricultural specialist in your area for the spray schedule best suited to your conditions.

DIRECTIONS FOR USE

Discard any unused spray material at the end of each day. Prepare solution concentrations by mixing the required amount of product with water only in a clean, empty spray tank. For best results, applications should be made during cooler parts of the day.

Use only as directed. Good spray practices should be followed. The label should be read thoroughly and understood before making applications. Effectiveness requires that all parts of plant or crop must receive spray or desired result will not occur, so spray thoroughly. When a range of rates is indicated, use the concentration and spray volume recommended locally.

Gibberellic Acid is a naturally occurring compound, produced by Abbott Laboratories in a biological process.

Data concerning the compatibility of PRO-GIBB with other agricultural compounds is not available.

SPRAY GUIDELINES FOR GRAPES

For all grapes, application is recommended by ground sprayer. Use 100 to 500 gallons as a dilute spray according to foliage density, or 30 to 80 gallons as a concentrate spray, unless specified otherwise. Do not exceed maximum rates. It is important to wet all berries thoroughly.

Thompson Seedless

For cluster elongation ("Stretch"), looser cluster forms, and reducing cost of thinning, when used in conjunction with established girdling and thinning practices.

- **Guide:** Apply 3 to 8.5 grams*/A before bloom when flower clusters are 3 to 5 inches long.

* Refers to actual Gibberellic Acid. See Conversion Table to convert to amount of formulated Pro-Gibb needed.

For decreased berry set ("Thinning"), reducing hand-thinning costs, and hastened maturity.

- **Guide:** Apply 3 to 12 grams*/A during bloom. Higher amounts may cause an excess of shot berries or overthinning, except in high density plantings.

For larger berries ("Sizing") and larger clusters when used in conjunction with established girdling and thinning practices.

- **Guide:** Apply 32 to 64 grams*/A when average berry size is $^{13}/_{64}$" in diameter or as two applications of equal amounts with the first made 2 to 3 days at or just after shatter, followed during the next two weeks by the second application. Timing of the second spray will be dictated by experience in the vineyard to be sprayed and temperatures occurring during the interim between sprays. Potential effect will be reduced if the second spray occurs more than two weeks after the first application.

Thompson Seedless for Raisins

For decreasing berry set, with increased raisin quality, and hastened maturity.

- **Guide:** Apply 0.75 to 6 grams*/A during bloom.

Flame Seedless

For decreased berry set ("Thinning") and reducing hand-thinning costs.

- **Guide:** Apply 3 to 7.5 grams*/A during bloom.

Higher amounts may cause an excess of shot berries or overthinning.

For larger berries ("Sizing") and larger clusters when used in conjunction with established girdling and thinning practices.

- **Guide:** Apply 8 to 48 grams*/A as one application when berry diameter reaches 6 to 8 millimeters, or as two applications of equal amounts with the first made when berry diameter reaches 6 to 8 millimeters, followed during the next 5 to 10 days by the second application. Timing of the second spray will be dictated by experience in the vineyard to be sprayed and rate of berry growth during the interim between sprays.

Other Seedless Varieties such as Perlette, Seedless Tokay, Interlocken Series and Related Hybrids

For larger berries and larger clusters when used in conjunction with established girdling and thinning practices.

- **Guide:** Apply 8 to 48 grams*/A as one application at or just after shatter (usually 2 to 3 days later) or as two applications of equal amounts not to exceed a total of 48 grams*/A, with the first made at or just after shatter, followed during the next two weeks by the second application. Timing of the second spray with split application will be dictated by experience in the vineyard to be sprayed and temperatures occurring during the interim between sprays. Potential effect will be reduced if the second spray occurs more than two weeks after the first application.

Emperor

For reducing berry shrivel. This use can also increase berry size.

- **Guide:** Apply 20 grams*/A as one application in 200 to 250 gallons/A approximately two weeks after completion of shatter following bloom. This timing should correspond to a period when the predominant berry diameter ranges from 10 to 15 millimeters.

Black Corinth (Zante Currant)

For improving berry size.

- **Guide:** Apply spray containing 1 to 8 grams*/A 3 to 5 days after full bloom, but before shatter begins.

Concord

(Arkansas, Michigan, New York, Ohio and Pennsylvania)

For cluster elongation (Stretch), looser cluster forms, increased berry size, reduced number of green berries, increased soluble solids content, and increased yields, when used in conjunction with established girdling and thinning practices and a first-bloom application of daminozide (Alar®-85) to increase berry set.

- **Guide:** Apply 40–80 grams*/A in a postbloom spray at the berry shatter stage. Grape vines should have received a first-bloom application of daminozide (Alar®-85) at the recommended rate of 1 lb/A Alar®-85. See current Alar®-85 label for precautionary statements and other specific recommendations.

* Refers to actual Gibberellic Acid. See Conversion Table to convert to amount of formulated Pro-Gibb needed.

®Alar is the trade mark of a company other than Abbott Laboratories

Applications should not be made to vines considered to be in low vigor.

Apply in sufficient water to give uniform and complete coverage.

SPRAY GUIDELINES FOR CITRUS

NAVEL ORANGES

(California)

To delay aging of the rind and reduce rind disorders (e.g., rind staining, water spotting, sticky or tacky surface, puffy rind and rupture under pressure) and to produce a more orderly harvesting pattern.

EARLY SPRAY (October/November — before any color change).

Apply to groves where harvest is not anticipated before March 1. The delay in rind aging is greatest when the early spray is applied before a color change. This spray timing produces the firmest rind possible.

- **Guide:** Apply one spray in October or November before any color change. On large mature trees, apply 10 to 40 grams*/A in 400 to 500 gallons/A dilute, or 50 to 100 gallons/A concentrate.

CAUTION: Do not apply to groves that may be harvested before March 1 as a reduction in grade may result due to the delayed coloring. Do not apply in white wash sprays in which lime or other caustic material has produced a high pH in the spray tank.

LATE SPRAY (December/January — after marketable color is reached).

Apply to groves where harvest may be made before March 1 (or not known).

- **Guide:** Apply one spray in December or January just after marketable color has developed. On large mature trees, apply 10 to 40 grams*/A in 400 to 500 gallons/A dilute, or 50 to 100 gallons/A concentrate.

CAUTION: Sprays applied in late January/February may cause reduced production the following year. Do not apply within 10 days of harvest. Do not spray navel orange trees between February 15 and August 1.

VALENCIA ORANGES

(California)

To reduce rind creasing and to delay aging and softening of the rind.

- **Guide:** Apply a single spray in August or September to trees with a target crop of young fruit. On large mature trees, apply 40 to 80 grams*/A in approximately 500 gallons/A dilute or 100 gallons/A concentrate.

CAUTION: Some increased regreening, or slower color development, should be expected in the target crop. Some increased regreening of mature fruit, if present, may occur.

LEMONS

(California)

To decrease the amount of small tree ripe fruit and to produce a more desirable production pattern in relation to market demand.

- **Guide:** Apply in a single spray in November or December to control fruit maturity by delaying development of yellow colored fruit. Use 20 grams* in 500 gallons/A on large mature trees.

When applied two years in a row, an even larger difference in harvest pattern and maturity occurs.

CAUTION: Do not apply within one month of harvest. Do not apply in spring or summer.

TANGERINE HYBRIDS

(Florida)

To increase fruit set and yields on tangerine hybrids with pollination problems such as the Orlando, Robinson, and Minneola.

- **Guide:** Apply spray during full bloom. Be sure to wet the leaves sufficiently.

Fruits are generally seedless. Use 8 to 30 grams* in 400 to 500 gallons/A on large mature trees.

CAUTION: A slight increase in mature leaf drop occurs at concentrations above 25 ppm. Fruit sizes may be reduced and color development slightly retarded.

(California)

To delay disorders associated with rind aging of the Minneola tangelo; e.g., puffiness and softening, and to increase peel strength.

- **Guide:** Apply 20 to 40 grams*/A in 400–500 gallons (10 to 20 ppm) dilute spray two weeks prior to color break. For the San Joaquin Valley, apply in October; for San Diego County, apply in November.

CAUTION: Do not apply if early harvest is planned. Do not apply after coloring as pre-harvest rind staining may occur. Application during coloring may cause variation in rind color development.

GRAPEFRUIT

(Florida and Texas)

To delay disorders associated with rind aging; e.g., puffiness, softening, and orange coloration, to prevent pre-harvest drop of mature fruit, and to increase peel strength and reduce water loss during storage.

- **Guide:** Apply a single spray to fully colored fruit during the November through January period. Use 20 to 56 grams* in 500 to 700 gallons/A containing a suitable non-ionic surfactant at the manufacturer's recommended rate. It is advisable to spot pick heavy crops to aid early marketing and to avoid reduction of yields which generally follow late yield crops.

CAUTION: Applications made after January or when trees begin to break dormancy may adversely affect new crop. Do not use concentrate sprays. Results may vary season to season depending on environmental conditions.

GRAPEFRUIT, STAR RUBY VARIETY

(Texas)

To reduce early-season drop of small fruit of Star Ruby Variety thereby increasing yields.

- **Guide:** Apply a single spray during the bloom period.

* Refers to actual Gibberellic Acid. See Conversion Table to convert to amount of formulated Pro-Gibb needed.

Use PRO-GIBB 4% Liquid Concentrate. Use 25 fluid ounces (1¼ 20-ounce bottles) (25 grams*) in 250 gallons water final spray mixture per acre. A suitable surfactant may be used to enhance efficacy.

CAUTION: Do not tank-mix with other chemicals.

Do not apply concentrated solution.

Results may vary season to season depending on environmental conditions.

Maintain a well-balanced fertilization and watering program.

SPRAY GUIDELINES FOR FRUIT CROPS

BLUEBERRIES

For improving fruit set. For set problems due to insufficient natural honeybee pollination on varieties such as Coville, Jersey, Stanley, Earlieblue, Weymouth and others.

- **Guide:** Make a single foliage spray application at full bloom (when over 75 percent of all flowers are fully open). For Weymouth, application can be delayed up to two weeks after full bloom to affect sizing of shot berries.
  Use PRO-GIBB 4% Liquid Concentrate. Mix 80 fluid ounces in 100 gallons of water. Use of a spreader-sticker is recommended. Apply to the point of run-off, thoroughly wetting all parts of the plant. Total gallonage will depend on size and density of the plants.

CAUTION: Do not exceed 300 gallons/A. Although some varieties bloom closer to harvest than others — in no case should application be made closer than 40 days before harvest. Do not apply to plants in a low state of vigor.

SWEET CHERRIES

To delay harvesting, to produce a brighter colored, firmer fruit, and to increase size.

- **Guide:** Apply spray when the fruit is light green to straw colored. Apply spray to thoroughly wet the entire tree. Use 16 to 48 grams* in 400 to 600 gallons/A on large mature trees.

CAUTION: Do not apply within one week of harvest.

RED TART CHERRIES

(All states except California)

To maintain and extend high fruiting capacity of bearing tart cherry trees and reduce occurrence of "blind" nodes by stimulating lateral vegetative buds to develop a more productive balance of lateral shoots and spurs.

- **Guide:** Apply a foliar spray containing 4 to 8 fluid ounces of PRO-GIBB 4% in 100 gallons finished spray from 14 to 28 days after bloom (or up to 14 days after shuck split). Use full coverage sprays of 100 to 250 gal/A on medium to large bearing trees. Be sure entire trees receive good coverage. Use of a good horticultural wetting agent at the manufacturer's recommended rate will aid foliar wetting. PRO-GIBB must be applied annually to insure vegetative development and subsequent yield improvement year after year.

Note: PRO-GIBB works by affecting lateral bud differentiation which is apparent the year *after* application. Therefore, changes in shoot, spur, and flower production will not be evident until 2 or 3 years after program initiation. Once this period is satisfied, response will be yearly provided annual applications have been made.

CAUTION: Do not spray within one month of harvest. Adjust PRO-GIBB rate to complement vigor of trees. If trees are vigorous, use lowest recommended rates. Use higher rate for trees low in vigor and showing weak shoot and spur production. Excessive application rates on any tree will increase vegetative growth at the expense of fruit production the following year.

PRO-GIBB will *not* improve growth of trees under stress (nutritional, moisture, winter injury) or other factors inhibiting normal growth and development resulting from physical damage or unsound orchard practices. Best results from PRO-GIBB will be obtained when combined with good cultural practices.

YOUNG SOUR AND SWEET CHERRY TREES

(All states except California)

To reduce flowering and fruiting in young sour and sweet cherry trees to minimize the competitive effect of early fruiting on tree development.

- **Guide:** Apply two to four weeks after bloom with no more than two sprays. Use 5 to 20 grams* in 25 to 50 gallons/A (this assumes density of 100 trees per acre equivalent) or, about 1 quart of spray volume per tree.

Under conditions of low vigor, two applications are recommended. If two spray applications are made, allow at least a seven-day interval between sprays.

CAUTION: DO NOT SPRAY TREES IN THE FIRST YEAR. Treat in the second season for reduction of flowering in the third season, and again in the third season if reduction of flowering and fruiting is desired in the fourth season.

OLYMPUS STRAWBERRIES

(N.W. US Only; propagation stock)

To increase runner production of mother plants of the Olympus cultivar.

* Refers to actual Gibberellic Acid. See Conversion Table to convert to amount of formulated PRO-GIBB needed.

- **Guide:** Apply a single spray to mother plants 10 to 30 days after planting. At the time of spraying, plants should have 1 to 6 leaves. Apply 100 gallons/A to thoroughly wet new foliage to the point of run-off. Use 20 grams*/A.

CAUTION: Not for use on fruiting plants. Treatments may not be effective on plantings set out after mid-May.

FORCING RHUBARB

To increase yield of marketable forced rhubarb and to break dormancy on plants receiving insufficient chilling.

- **Guide:** Apply 2 fluid ounces (60 ml) of a solution containing 20 grams* in 10 gallons to each cleaned crown, when the rest period is not completely broken. When the rest period is broken by cold weather, apply 2 fluid ounces (60 ml) of a solution containing 10 grams* in 10 gallons.

CAUTION: Keep forcing house temperatures at 40° to 50°F for 24 hours after application. If house is warmer than 50°F, the crowns should be covered with plastic. Temperatures in the forcing house above 50°F will result in lower yields and poor stalk color.

SPRAY GUIDELINES FOR VEGETABLE CROPS

ARTICHOKES

(California)

To accelerate maturity of artichokes and to shift the harvest to an earlier date.

- **Guide:** Apply spray in the fall up to November 1. Be sure the entire plant (leaves, stems and buds) are covered to point of run-off. Use 3.5 to 5 grams* in 35 to 50 gallons/A.

CAUTION: Do not apply within seven days of harvest.

* Refers to actual Gibberellic Acid. See Conversion Table to convert to amount of formulated PRO-GIBB needed.

CELERY

To increase plant height and yield and overcome stress due to cold weather conditions, or saline soils and to obtain earlier maturity.

- **Guide:** Apply spray one to four weeks prior to harvest. Lower concentrations are applied at the three to four-week interval. Higher concentrations at the one to two-week interval. Use 2.5 to 10 grams* in 25 to 50 gallons/A.

CAUTION: Do not apply earlier than four weeks before harvest as Gibberellic Acid may induce bolting (seed stalk formation).

Applications made less than one week preharvest may result in residues.

Celery plants must be harvested when mature to ensure quality.

LETTUCE FOR SEED

To obtain uniform bolting and increase seed production.

- **Guide:** Apply the following spray schedule:

* Refers to actual Gibberellic Acid. See Conversion Table to convert to amount of formulated Pro-Gibb needed.

Growth Stage	ppm*	g.*/A	Gallon/Acre
4 leaf stage	10	0.4	10
8 leaf stage	10	1.6	40
12 leaf stage	10	4	100

CAUTION: Do not feed crop wastes to livestock.

SEED POTATOES

To stimulate uniform sprouting—for maximum production, more uniform development, fewer late maturing plants, and to break dormancy of newly harvested potatoes that have not had a full rest period.

- **Guide:** Dip freshly dug seed pieces in a solution containing 0.2 to 0.4 gram* in 100 gallons prior to planting.

CAUTION: If soil temperature is very high, avoid treating rested seed and use the minimum concentration for dormant seed.

SPINACH

(Arkansas, Oklahoma, and Texas)

To facilitate harvest, increase yield and improve quality of fall and over-winter spinach.

- **Guide:** Apply a single spray 10 to 14 days before harvest on fall or over-winter spinach, ideally when daytime temperatures are 40° to 70°F and during early morning hours when dew is present on crop.

Use PRO-GIBB 4% Liquid Concentrate.

In Arkansas and Oklahoma, mix 6 to 8 fluid ounces/A (6 to 8 grams*/A) in 25 to 50 gallons/A by ground sprayer or in a minimum of 10 gallons/A by air.

* Refers to actual Gibberellic Acid. See Conversion Table to convert to amount of formulated PRO-GIBB needed.

In Texas, mix 4 to 8 fluid ounces/A (4 to 8 grams*/A) in 25 to 50 gallons/A by ground sprayer or in a minimum of 5 gallons/A by air. Maximum benefit from PRO-GIBB is obtained when below normal temperatures predominate following application and growth would be otherwise allowed in untreated spinach. Applications of PRO-GIBB may be made as recommended above on successive crops following regrowth from preceding harvests.

CAUTION: Since Gibberellic Acid can promote bolting, do not apply to spinach after the mid-winter period or if temperatures may be expected to exceed 75°F within several days of application. Do not apply on spring-planted spinach.

SPRAY GUIDELINES FOR FLORICULTURE CROPS

POMPOM CHRYSANTHEMUMS

(Florida)

For elongating peduncles on pompom chrysanthemums.

- **Guide:** Apply a single spray 4 to 5 weeks after initiation of short day conditions.

Use PRO-GIBB 4% Liquid Concentrate. Use ½ to 1 fluid ounce (1/2 to 1 gram*) in 12 gallons for application to 1,000 sq. ft. of bed (20 to 40 fluid ounces equivalent to 20 to 40 grams* in 500 gallons/A).

Apply with overhead nozzles directing the spray to the flower buds.

CAUTION: Overuse or incorrect timing may cause long, spindly, and weak stems.

STATICE

(Florida)

To promote earlier flowering and to increase flower yield.

- **Guide:** Apply a single drench spray when plants are more than 10 inches in diameter (approximately 90 to 110 days after normal seeding time). Use 40 to 50 grams* in 25 gallons to provide 10 ml (5 mg*) solution per plant.

CAUTION: Do not exceed specified rates. Do not apply repeated sprays. Accelerated flowering is influenced by extended photoperiod, adequate nutrition, and reduced night temperature. Treatment with gibberellins lessens the requirement for the cold requirement and/or the long photoperiod.

SPRAY GUIDELINES FOR ADDITIONAL CROPS

BERMUDAGRASS GOLF TURF

(Florida)

To initiate or maintain growth and prevent color change during periods of cold stress and light frosts on golf course Bermudagrass (e.g., Tifdwarf, Tifgreen, etc.)

- **Guide:** Apply 10 grams* weekly or 25 grams* biweekly in 25 to 100 gallons/A.

* Refers to actual Gibberellic Acid. See Conversion Table to convert to amount of formulated Pro-Gibb needed.

Use PRO-GIBB 4% Liquid Concentrate. Mix ¼ to ⅔ fluid ounce (¼ to ⅔ gram*) in approximately 6 gallons appropriate for the spray equipment for application to 1,000 sq. ft. (10½ to 26½ fluid ounces/A equivalent to 10 to 25 grams*/A in 25 to 100 gallons/A).

CAUTION: Do not exceed specified rates.

Do not apply during extended warm periods where night temperatures exceed 65°F.

Maintain adequate moisture and proper fertilization programs recommended in local area.

Discontinue treatments if thinning is observed.

Do not apply the high rate more frequently than every two weeks. More frequent mowing may be necessary.

Do not use on dormant turf.

HOPS

For seeded and seedless Fuggle hops and similar varieties adapted to Oregon and the Northwest.

To increase yield and pickability.

- **Guide:** Apply spray when vine growth is five to eight feet in length. Use 4 to 6 grams* in 100 to 150 gallons/A.

CAUTION: Do not apply within three weeks of harvest.

* Refers to actual Gibberellic Acid. See Conversion Table to convert to amount of formulated PRO-GIBB needed.

SUGAR CANE

(Hawaii)

Use PRO-GIBB Plus 2X Soluble Powder

For increase in sucrose yield.

- **Guide:** Apply 28 to 84 grams* in 7 to 10 gallons/A of spray by airplane. Uniform coverage is essential for maximum response. Use 56 grams as a single treatment, or 28 grams two or three times in separate applications with 30 to 45-day intervals. Application may be made to cane during the first and/or second year of culture. Young cane should be at least three months old to avoid possible tiller reduction. Application should not be made less than 4 months prior to harvest.

Application should be made when growth rate is depressed by temperature. Cane grown below 1,500 feet elevation will benefit from applications made during November through March.

Cane should be treated when there is sufficient soil moisture from rain or irrigation to sustain a high growth rate for at least 30 days following each treatment. Lack of water will negate treatment effect.

CONVERSION TABLE

GRAMS OF ACTUAL GIBBERELLIC ACID PER ACRE TO AMOUNT OF PRO-GIBB® FORMULATION PER ACRE

Desired Actual Gibberellic Acid Concentration (Grams*) in Finished Spray (Per Acre)	PRO-GIBB Plus 2X Soluble Powder Contains (2.0 Grams*/10 Grams Formulated Product)	PRO-GIBB 4% Liquid Contains (1.0 Gram*/ Fluid Ounce of Formulated Product)
0.5	2.5 grams	0.5 oz.
1.0	5 grams	1 oz.
2.0	10 grams	2 oz.
4.0	20 grams (⅛ bottle)	4 oz.
5.0	25 grams	5 oz.
8.0	40 grams (¼ bottle)	8 oz.
10.0	50 grams	10 oz.
12.0	60 grams	12 oz.
16.0	80 grams (½ bottle)	16 oz.
20.0	100 grams	20 oz.
25.0	125 grams	25 oz.
32.0	160 grams (1 bottle)	32 oz.
40.0	200 grams	40 oz.
48.0	240 grams (1½ bottles)	48 oz.
50.0	250 grams	50 oz.

* Refers to actual Gibberellic Acid.

NOTICE TO USER

Seller makes no warranty, express or implied, of merchantability, fitness or otherwise concerning use of this product other than as indicated on the label. User assumes all risks of use, storage or handling not in strict accordance with accompanying directions.

* Refers to actual Gibberellic Acid. See Conversion Table to convert to amount of formulated PRO-GIBB needed.

PROMALIN®

PLANT GROWTH REGULATOR

ACTIVE INGREDIENTS	W/W
N-(phenylmethyl)-1H-purine-6-amine	1.8%
Gibberellins A_4A_7	1.8%
INERT INGREDIENTS:	96.4%
TOTAL	100.0%

EPA Reg. No. 275-32 EPA Est. No. 35347-CA-1

WARNING

Keep Out of Reach of Children

Causes eye irritation. Do not get in eyes. Harmful if swallowed. In case of contact, immediately flush eyes with plenty of water. Call a physician.

PROMALIN® is a plant growth regulator for use on apples*. When applied at specific spray timing and concentration, PROMALIN will provide the following effects:

1. PROMALIN improves the shape of Delicious* apples through elongation of fruit and development of more prominent calyx lobes. These desirable effects will be most evident in years when natural typiness is limited.
2. On Delicious and other varieties PROMALIN may increase weight of individual fruit and yield per tree. Some thinning may occur from the use of PROMALIN.
3. On red apple varieties, e.g. Delicious, Jonathan, Jonamac, MacIntosh, Paulared, Spartan, Empire, PROMALIN may be combined (in all States except California) with Alar®-85 to enhance expected desirable effects from Alar-85, including improved color, earlier coloring, increased fruit pressure and prevention of preharvest fruit drop.

**4. On most apple varieties, PROMALIN will increase lateral bud break, shoot growth, and improve branch angles on *non-bearing* apple trees, including nursery stock. This effect will provide a better tree framework for early cropping.

* Only Red Delicious apples in California.

** All States except California.

DIRECTIONS FOR USE

It is a violation of Federal law to use this product in a manner inconsistent with its labeling.

Keep out of lakes, ponds or streams. Do not contaminate water by cleaning of equipment or disposal of wastes.

STORAGE AND DISPOSAL

Do not contaminate water, food or feed by storage or disposal.

STORAGE

Store Below 75°F (23°C). Keep containers tightly closed when not in use.

PESTICIDE DISPOSAL

Wastes resulting from the use of this product may be disposed of on site or at an approved waste disposal facility.

CONTAINER DISPOSAL

Do not reuse empty containers. Triple rinse (or equivalent). Puncture and dispose of in a sanitary landfill, or by incineration or, if allowed by state and local authorities, by burning. If burned, stay out of smoke.

A. RATE AND TIMING OF APPLICATION—FRUIT DEVELOPMENT

1. Single Application for Improved Type.

Western U.S.: A single application of PROMALIN should be made from early king bloom (king blossoms opening and side blossoms in the balloon stage) to the early stages of petal fall of the king blossoms.

***PROMALIN-ELGETOL®** (DNOC, 19% AI) (for type and thinning on spur Delicious strains): Apply a tank mix of PROMALIN and Elgetol at the recommended Elgetol timing, i.e. 80% full king bloom on the north side of the tree. Apply 2 pints of PROMALIN plus 4 pints of Elgetol in 200 gallons of water per acre. Observe all precautionary statements on Elgetol label.

*All States except California.

Eastern U.S.: A single application of PROMALIN should be made from early king bloom to the early stages of petal fall of side blossoms.

For both eastern and western U.S. apply 1 to 2 pints of PROMALIN per acre in 50 to 200 gallons of spray mixture as a fine mist spray.

2. Split Applications for Improved Type and Size.

A split application may be utilized to improve typing and sizing responses. The first application should be made from early king bloom to the early stages of petal fall of the king bloom. The second application should be made 7 to 21 days later. Apply 0.5 to 1 pint of PROMALIN per acre in 50 to 200 gallons of spray mixture in each application. Do not, however, exceed thè maximum recommended rate of a total of 2 pints per acre for the combined sprays.

*3. Split Applications to Improve "Typiness" in Locations where the Bloom Period is Prolonged or Occurs over a Period of Several Days.

Make two PROMALIN applications three or more days apart. The first spray should be timed during the first flush of bloom. A second application should be made when the remainder of the tree canopy comes into bloom.

The time between PROMALIN applications may vary from orchard to orchard within a given growing area because of the influence of the local microclimate. In most instances a period from 3 to 7 days between sprays is adequate. The time interval between sprays should be lengthened should cool weather further prolong bloom. For the most efficient use of PROMALIN, bloom progression within individual orchards should be assessed at least every 48 hours.

Apply PROMALIN twice as two concentrate sprays each containing 0.5–1 pint of PROMALIN in 100 gpa. The spray machine should be calibrated to deliver 100 gpa and nozzled to direct the spray mist to those portions of the tree canopy currently in bloom. An appropriate wetting agent should be included so as to insure even thorough coverage of open blossoms.

*All States Except California.

*4. PROMALIN and ALAR®-85 (Plant Growth Regulants) for Commercial Applications to Apples.

PROMALIN and Alar-85 may be combined as a tank mix application during mid-summer (60–80 days pre-harvest) on apples. This tank mix may enhance the expected desirable effect from Alar-85 application, including improved color, earlier coloring, increased fruit pressure, and prevention of preharvest fruit drop.

Use 1 to 2 pints of PROMALIN and 2 to 4 pounds of Alar-85 per acre in 100–400 gallons of spray solution. Alar should be added to the tank mix prior to the addition of PROMALIN.

The PROMALIN and Alar-85 tank mix should be applied as a foliar spray in a manner that insures uniform and complete coverage. Do not exceed the per acre recommended rate. See current Alar-85 label for precautionary statements and other specific recommendations.

*All States Except California.

Mixing Instructions

Add the required amount of PROMALIN to a spray tank about half-filled with water. Agitate while bringing the total volume of water to the required level. New solutions should be mixed only in a clean, empty spray tank, and used within 24 hours. A nonionic wetting agent e.g. Tween®-20, Glyodin® or Triton X-100, may improve spray coverage and enhance absorption.

Application Considerations

1. Since Promalin is a plant growth regulator that must be absorbed by the plant to be effective, best performance can be expected with good absorption conditions and healthy, vigorous trees. Do not spray the recommended tank mixes on any low vigor trees or trees under stress as from drought, low fertility, winter injury, etc.
2. Apply PROMALIN spray mixture as a uniform fine mist spray.
3. Use a properly calibrated and adjusted sprayer that insures uniform and complete coverage of all foliage and blossoms. Spray volume should be adjusted to tree size and density.
4. Approximately 85% of the spray volume should be directed into the upper two-thirds of the trees.
5. To minimize excessive spray drift, apply when air is calm.
6. High relative humidity and slow drying conditions favor maximum absorption of PROMALIN. Application in morning or evening may be preferable. Conditions favoring rapid drying of the spray deposit should be avoided.
7. Rainfall within six hours after spraying may reduce activity of PROMALIN.
8. Do not apply PROMALIN when air temperatures are lower than 40°F or greater than 90°F.
9. not apply PROMALIN in more than 200 gallons of spray per acre.

B. RATE AND TIMING OF APPLICATION—

*Lateral Branching and Tree Development (Non-bearing trees)

A single foliar application of PROMALIN or a PROMALIN-latex spot application may be applied to non-bearing trees, including nursery stock, to increase lateral bud break, shoot growth, improve branch angles, and provide a better tree framework for early cropping.

*All States except California.

Foliar Applications

For *foliar* sprays, use one-quarter (0.25) to one (1) pint of PROMALIN per 5 gallons of spray solution to attain a concentration of 125 to 500 ppm of PROMALIN in the final spray solution. Preferably a non-ionic wetting agent, e.g. Tween-20, Triton X-100, Glyodin, etc. should be added to the tank-mix at a rate of 0.2–0.3% (1–2 oz./5 gal.) prior to the addition of PROMALIN. This practice will improve wetting and absorption through the waxy cuticular layers on the foliage and bark surfaces. In all cases the final spray should not be alkaline.

Uniformly apply the PROMALIN spray mixture with a pressurized hand sprayer, hand gun attachment to an airblast sprayer, or an airblast sprayer to thoroughly wet the bark and foliage surfaces. Approximately five to ten (5–10) gallons of spray mixture will treat an acre of non-bearing (1–4 year old) trees on a 12 by 12 spacing (310 trees).

Application of PROMALIN should be made during the period from one (1) to three (3) inches of new terminal shoot growth. This period generally coincides with a timing of one (1) to two (2) weeks after full bloom. Applications should be terminated when new terminal shoot growth ceases. Nursery stock should be treated after trees attain a terminal height at which the first lateral scaffold branches are desired (i.e., greater than 12 inches).

Latex Applications

For *latex* applications, use one-third (0.33) pint of PROMALIN per one (1) pint of interior flat latex paint to attain a concentration of 7500 ppm. Mix thoroughly. Preferably a non-ionic wetting agent, should be added to the tank-mix at the rate of 0.5 to 1.0% prior to the addition of the PROMALIN. This practice will improve the dispersion of PROMALIN in the latex paint and will also improve wetting and absorption through the waxy cuticular layers on the bark surfaces. The final application mixture should not be alkaline.

Uniformly apply the PROMALIN-latex mixture with a brush or sponge to thoroughly cover the bark surface. Apply only to one-year-old dormant branches.

Since this application is intended primarily for stimulating lateral growth from vegetative buds on one-year-old dormant wood, application of PROMALIN should be made in late April and early May depending on season and locality. A useful guide is to make the application just as the terminal buds on shoots or spurs are beginning to break. Applications after lateral buds have broken may cause some injury to the tender shoot tips and fail to promote continued shoot growth from that point.

Tank Mix Compatibility

1. PROMALIN may be applied in combination with plant nutrients.
2. A buffer should be added to alkaline spray water to reduce the pH below 8.

*3. Fungicides

Benlate®, Ferbam®, Polyram®, Thiram®, Glyodin®, Dithane® M-45, Orthocide®, wettable sulfur may be tank mixed with PROMALIN at the recommended timing associated with bloom applications as described under "Rate and Timing of Application" for PROMALIN. When using fungicides, refer to respective labels for directions and precautionary statements.

®Elgetol, Alar, Tween, Benomyl, Ferbam, Polyram, Thiram, *Dithane M-45, *Triton, Glyodin and Orthocide are trademarks of companies other than Abbott Laboratories.

*All States except California.

NOTICE TO USER

Seller makes no warranty, express or implied, of merchantability, fitness or otherwise concerning the use of this product other than as indicated on the label. User assumes all risks of use, storage or handling not in strict accordance with accompanying directions.

VECTOBAC®-AS

BIOLOGICAL MOSQUITO AND BLACK FLY LARVICIDE

AQUEOUS SUSPENSION

ACTIVE INGREDIENT:

Bacillus thuringiensis, Serotype, H-14, 600 International Toxic Units (ITU) per milligram (Equivalent to 2.19 billion ITU per gallon; 0.576 billion ITU per liter) 0.6%

INERT INGREDIENTS 99.4%

TOTAL 100.0%

EPA Reg. No. 275-52 EPA Est. No. 33762-IA-1

CAUTION

Keep Out of Reach of Children

VECTOBAC-AS is a highly selective insecticide for use against mosquito and black fly larvae.

PRECAUTIONARY STATEMENTS

HAZARD TO HUMANS AND DOMESTIC ANIMALS

As a precautionary measure in case of contact, flush eyes with plenty of water. In case of irritation, contact a physician.

DIRECTIONS FOR USE

It is a violation of Federal law to use this product in a manner inconsistent with its labeling.

STORAGE AND DISPOSAL

Do not contaminate water, food, or feed by storage or disposal.

STORAGE: Store in a cool place.

PESTICIDE DISPOSAL: Wastes resulting from use of this product may be disposed of on site or at an approved waste disposal facility.

CONTAINER DISPOSAL: Triple rinse (or equivalent). Then puncture and dispose of in a sanitary landfill, or by incineration, or, if allowed by state and local authorities, by burning. If burned, stay out of smoke. Do not reuse container.

APPLICATION DIRECTIONS

Mosquito Habitat	Suggested Range Rate*
(Such as the following examples):	
Irrigation ditches, roadside ditches, flood water, standing ponds, woodland pools, snow melt pools, pastures, catch basins, storm water retention areas, tidal water, salt marshes and rice fields.	0.5 pt-2 pts/acre
Polluted water (such as sewage lagoons, animal waste lagoons)	2 pts/acre

* Use higher rate in polluted water, and when late 3rd and early 4th instar larvae predominate, mosquito populations are high, water is heavily polluted, and/or algae are abundant.

Blackflies Habitat	Suggested Range Rate
Streams	1-50 mg/liter stream water** (=ppm)

** Use higher rate when stream contains high concentration of organic material, or dense aquatic vegetation.

Ground and Aerial Application

Vectobac®-AS may be applied in conventional ground or aerial application equipment with quantities of water sufficient to provide uniform coverage of the target area. The amount of water needed per acre will depend on weather, spray equipment, and mosquito habitat characteristics. Do not mix more Vectobac-AS than can be used in a 2-week period.

For most ground spraying, apply in 5-100 gallons per acre using hand-pump, airblast, mist blower, etc., spray equipment.

For aerial application, Vectobac-AS may be applied either undiluted or diluted with water. For undiluted applications, apply 0.5 to 2.0 pt/acre of Vectobac-AS through fixed wing or helicopter aircraft equipped with either conventional boom and nozzle systems or rotary atomizers.

For diluted application, fill the mix tank or plane hopper with the desired quantity of water. Start the mechanical or hydraulic agitation to provide moderate circulation before adding the Vectobac-AS. Maintain the agitation while loading or spraying. For all aerial applications, a droplet spectrum in the range of 50-100 microns (VMD) will produce best results.

Rinse and flush spray equipment thoroughly following each use.

For blackfly aerial applications, Vectobac-AS can be applied undiluted via fixed wing or helicopter aircraft equipped with either conventional boom and nozzle systems or open pipes. Rate of application will be determined by the stream discharge and the required amount of Vectobac-AS necessary to maintain a 1-50 ppm concentration in the stream water. Vectobac-AS can also be applied diluted with similar spray equipment. Do not mix more Vectobac-AS than can be used in a 2-week period.

SMALL QUANTITY DILUTION RATES FOR VECTOBAC-AS

Rates in Pints Per Acre	Gallons Spray Solution/Acre: 10 Gal/A	25 Gal/A	50 Gal/A
	(Ounces Needed Per Gallon of Spray)		
0.5	.8	.33	.16
1.0	1.6	.66	.32
2.0	3.2	1.30	.64

NOTICE TO USER

Seller makes no warranty, express or implied, of merchantability, fitness or otherwise concerning the use of this product other than as indicated on the label. User assumes all risks of use, storage or handling not in strict accordance with accompanying directions.

VECTOBAC®-G

BIOLOGICAL MOSQUITO LARVICIDE

GRANULES

ACTIVE INGREDIENT:

Bacillus thuringiensis, Serotype H-14, 200 International Toxic Units (ITU) per milligram (Equivalent to 0.091 billion ITU's per pound) 0.2%

INERT INGREDIENTS 99.8%

TOTAL 100.0%

EPA Registration No. 275-50 EPA Est. No. 2393-IL-3

Filled by weight, not volume.

CAUTION

Keep Out of Reach of Children

VECTOBAC®-G is a highly selective insecticide for use against mosquito larvae.

DIRECTIONS FOR USE

It is a violation of Federal law to use this product in a manner inconsistent with its labeling.

STORAGE AND DISPOSAL

Do not contaminate potable water, food, or feed by storage or disposal.

STORAGE: Store in a cool, dry place.

PESTICIDE DISPOSAL: Wastes resulting from use of this product may be disposed of on site or at an approved waste disposal facility.

CONTAINER DISPOSAL: Completely empty bag into application equipment. Then dispose of empty bag in a sanitary landfill or by incineration, or, if allowed by State and local authorities, by burning. If burned, stay out of smoke.

APPLICATION DIRECTIONS

MOSQUITOES

Habitat	Suggested Rate Range*
Irrigation ditches, roadside ditches, flood water, standing ponds, woodland pools, snow melt pools, pastures, catch basins, storm water retention areas, tidal water, salt marshes and rice fields.	2.5-10 lb/acre

* Use 10-20 lbs/acre when late 3rd and early 4th instar larvae predominate, mosquito populations are high, water is heavily polluted, (sewage lagoons, animal waste lagoons), and/or algae are a bundant.

Apply uniformly by aerial or ground conventional equipment.

A 7 to 14 day interval between applications should be employed. Where non-target aquatic invertebrate parasites and predators are present longer periods of suppression may result since these beneficials would be conserved to aid in mosquito population management.

NOTICE TO USER

Seller makes no warranty, express or implied of merchantability, fitness or otherwise concerning the use of this product other than as indicated on the label. User assumes all risks of use, storage or handling not in strict accordance with accompanying directions.

VECTOBAC®-SG

BIOLOGICAL MOSQUITO LARVICIDE™

BIOLOGICAL MOSQUITO LARVICIDE SAND GRANULES

ACTIVE INGREDIENT:
Bacillus thuringiensis,
Serotype H-14, 200 International Toxic Units (ITU) per milligram (Equivalent to 0.091 billion ITU's per pound) 0.2%
INERT INGREDIENTS: 99.8%
TOTAL 100.0%

EPA Reg. No. 275-60 EPA Est. No.
List No. 5008

CAUTION

Keep Out of Reach of Children

See side panels for precautions and additional information.

PRECAUTIONARY STATEMENTS

ENVIRONMENTAL HAZARDS
Do not contaminate water by cleaning of equipment or disposal of wastes.

DIRECTIONS FOR USE
It is violation of Federal law to use this product in a manner inconsistent with its labeling.

STORAGE AND DISPOSAL

Do not contaminate water, food or feed by storage or disposal.

STORAGE: Store in cool, dry place.

PESTICIDE DISPOSAL: Wastes resulting from the use of this product may be disposed of on site or at an approved waste disposal facility.

CONTAINER DISPOSAL: Completely empty bag into application equipment. Then dispose of empty bag in a sanitary landfill or by incineration, or, if allowed by State and local authorities, by burning. If burning, stay out of smoke.

Vectobac-SG is a highly selective insecticide for use against mosquito larvae.

APPLICATION DIRECTIONS

MOSQUITOES

Habitat	Suggested Rate Range*
Irrigation ditches, roadside ditches, flood water, standing ponds.	2-10 lb/acre
Catch basins, storm water retention areas, tidal water, salt marshes and rice fields.	2-10 lb/acre
Polluted water, sewage lagoons, etc.	10-20 lb/acre

* Use higher rate when late 3rd and 4th instar larvae predominate, mosquito populations are high, water is heavily polluted, and/or algae are abundant.

Apply uniformly by conventional aerial or ground equipment.

A 7 to 14 day interval between applications should be employed. Where non-target aquatic invertebrate parasites and predators are present, longer periods of suppression may result since these beneficials would be conserved to aid in mosquito population management.

NOTICE TO USER

Seller makes no warranty, express or implied of merchantability, fitness or otherwise concerning the use of this product other than as indicated on the label. User assumes all risks of use, storage or handling not in strict accordance with accompanying directions.

American Cyanamid Company

Agricultural Division
One Cyanamid Plaza
Wayne, NJ 07470
(201) 831-2000

ABATE® 1-SG

INSECTICIDE

For Control of Mosquito and Midge Larvae

ACTIVE INGREDIENT:

Temephos [0,0′-(thiodi-4, 1-phenylene) bis (0,0-dimethyl phosphorothioate))	1%
INERT INGREDIENTS:	99%
TOTAL	100%

EPA Reg. No. 241-152 EPA Est. No.
24678-01 D-40

CAUTION

Keep Out of Reach of Children

See Below For Other Cautions

AVISO

AL USUARIO: Si usted no lee inglés, no use éste producto hasta que la etiqueta le haya sido explicada ampliamente.

(ENGLISH TRANSLATION) CAUTION

(TO THE USER: If you cannot read English, do not use this product until the label has been fully explained to you.)

PRECAUTIONARY STATEMENTS

HAZARDS TO HUMANS AND DOMESTIC ANIMALS

Avoid breathing dust. Avoid contact with skin. Wash thoroughly with soap and water after handling. Do not contaminate food or feed products.

STATEMENT OF PRACTICAL TREATMENT

If in Eyes: Flush with plenty of water. Get medical attention if irritation persists.

If on Skin: Wash with plenty of soap and water after handling.

NOTE TO PHYSICIANS: This product may cause cholinesterase inhibition. Atropine is antidotal. Pralidozime chloride (2-PAM; PROTOPAM chloride) may be effective as an adjunct to atropine.

ENVIRONMENTAL HAZARDS

This product is toxic to birds and fish. Fish and other aquatic organisms in water treated with this product may be killed. You must consult your State Fish and Game Agency before applying this product to waters or wetlands. Do not contaminate water by cleaning of equipment or disposing of wastes.

STORAGE AND DISPOSAL

PROHIBITIONS

Do not contaminate water, food, or feed by storage or disposal.

PESTICIDE DISPOSAL

Wastes resulting from the use of this product may be disposed of on site or at an approved waste disposal facility.

CONTAINER DISPOSAL

Completely empty bag into application equipment. Then dispose of empty bag in a sanitary landfill or by incineration, or, if allowed by State and local authorities, by burning. If burned, stay out of smoke.

GENERAL

Consult Federal, State or local disposal authorities for approved alternative procedures.

DIRECTIONS FOR USE

It is a violation of Federal law to use this product in a manner inconsistent with its labeling.

For control of mosquito and midge larvae, apply ABATE 1-SG as specified below:

Areas of Treatment	Lbs. Per Acre
Standing water, shallow ponds, lakes, woodland pools, catch basins	5–10
Tidal waters, marshes, swamps and waters high in organic content	10–20

May be applied as a spot treatment to catch basins and similar areas where mosquitoes may breed.

Apply uniformly and repeat applications as necessary. Do not use on crops used for food, forage or pasture.

DISCLAIMER

The label instructions for the use of this product reflect the opinion of experts based on field use and tests. The directions are believed to be reliable and should be followed carefully. However, it is impossible to eliminate all risks inherently associated with use of this product. Crop injury, ineffectiveness or other unintended consequences may result because of such factors as weather conditions, presence of other materials, or the use or application of the product contrary to label instructions, all of which are beyond the control of American Cyanamid Company. All such risks shall be assumed by the user.

American Cyanamid Company warrants only that the material contained herein conforms to the chemical description on the label and is reasonably fit for the use therein described when used in accordance with the directions for use, subject to the risks referred to above.

Any damages arising from a breach of this warranty shall be limited to direct damages and shall not include consequential commercial damages such as loss of profits or values or any other special or indirect damages.

American Cyanamid Company makes no other express or implied warranty, including any other express or implied warranty of FITNESS or of MERCHANTABILITY.

In case of an emergency endangering life or property involving this product, call collect, day or night. Area Code 201–835-3100.

ABATE® 2-CG

INSECTICIDE

For Control of Mosquito and Midge Larvae

ACTIVE INGREDIENT:

Temephos (0,0′-(thiodi-4,1-phenylene) bis (0,0-dimethyl phosphorothioate))	2%
INERT INGREDIENTS:	98%
TOTAL	100%

EPA Reg. No. 241-151 EPA Est. No.
24680-20 D-40

CAUTION

Keep Out of Reach of Children

See Below For Other Cautions

AVISO

AL USUARIO: Si usted no lee inglés, no use éste producto hasta que la etiqueta le haya sido explicada ampliamente.

(ENGLISH TRANSLATION) CAUTION

(TO THE USER: If you cannot read English, do not use this product until the label has been fully explained to you.)

PRECAUTIONARY STATEMENTS

HAZARDS TO HUMANS AND DOMESTIC ANIMALS

Avoid breathing dust. Avoid contact with skin. Wash thoroughly with soap and water after handling. Do not contaminate food or feed products.

STATEMENT OF PRACTICAL TREATMENT

If in Eyes: Flush with plenty of water. Get medical attention if irritation persists.

If on Skin: Wash with plenty of soap and water after handling.

NOTE TO PHYSICIANS: This product may cause cholinesterase inhibition. Atropine is antidotal. Pralidozime chloride (2-PAM; PROTOPAM chloride) may be effective as an adjunct to atropine.

ENVIRONMENTAL HAZARDS

This product is toxic to birds and fish. Fish and other aquatic organisms in water treated with this product may be killed. You must consult your State Fish and Game Agency before applying this product to waters or wetlands. Do not contaminate water by cleaning of equipment or disposing of wastes.

STORAGE AND DISPOSAL

PROHIBITIONS

Do not contaminate water, food, or feed by storage or disposal.

PESTICIDE DISPOSAL

Wastes resulting from the use of this product may be disposed of on site or at an approved waste disposal facility.

CONTAINER DISPOSAL

Completely empty bag into application equipment. Then dispose of empty bag in a sanitary landfill or by incineration, or, if allowed by State and local authorities, by burning. If burned, stay out of smoke.

GENERAL

Consult Federal, State or local disposal authorities for approved alternative procedures.

DIRECTIONS FOR USE

It is a violation of Federal law to use this product in a manner inconsistent with its labeling.

Areas of Treatment	Insects (Larvae)	Lbs./Acre	Application
Standing water, shallow ponds, lakes, woodland pools, catch basins	Mosquito and Midge	2.5–5	Apply uniformly by air or ground equipment. Repeat applications as necessary
Tidal waters, marshes, swamps and waters high in organic content		10	
Highly polluted waters		25	
Marsh lands, margins of streams, intertidal zones of sandy beaches	Gnat Punkie Sandfly	5–10	

Do not use on crops used for food, forage or pasture.

DISCLAIMER

The label instructions for the use of this product reflect the opinion of experts based on field use and tests. The directions are believed to be reliable and should be followed carefully. However, it is impossible to eliminate all risks inherently associated with use of this product. Crop injury, ineffectiveness or other unintended consequences may result because of such factors as weather conditions, presence of other materials, or the use or application of the product contrary to label instructions, all of which are beyond the control of American Cyanamid Company. All such risks shall be assumed by the user.

American Cyanamid Company warrants only that the material contained herein conforms to the chemical description on the label and is reasonably fit for the use therein described when used in accordance with the directions for use, subject to the risks referred to above.

Any damages arising from a breach of this warranty shall be limited to direct damages and shall not include consequential commercial damages such as loss of profits or values or any other special or indirect damages.

American Cyanamid Company makes no other express or implied warranty, including any other express or implied warranty of FITNESS or of MERCHANTABILITY.

In case of an emergency endangering life or property involving this product, call collect, day or night, Area Code 201–835-3100.

ABATE® 5-CG

INSECTICIDE

For Control of Mosquito and Midge Larvae

ACTIVE INGREDIENT:

Temephos (0,0′-(thiodi-4,1-phenylene) bis (0,0-dimethyl phosphorothioate)) 5%

INERT INGREDIENTS: 95%

TOTAL 100%

EPA Reg. No. 241-150 EPA Est. No. 24656-01 D-40

CAUTION

Keep Out of Reach of Children

See Below For Other Cautions

AVISO

AL USUARIO: Si usted no lee inglés, no use éste producto hasta que la etiqueta le haya sido explicada ampliamente.

(ENGLISH TRANSLATION) CAUTION

(TO THE USER: If you cannot read English, do not use this product until the label has been fully explained to you.)

PRECAUTIONARY STATEMENTS

HAZARDS TO HUMANS AND DOMESTIC ANIMALS

Avoid breathing dust. Avoid contact with skin. Wash thoroughly with soap and water after handling. Do not contaminate food or feed products.

STATEMENT OF PRACTICAL TREATMENT

If in Eyes: Flush with plenty of water. Get medical attention if irritation persists.

If on Skin: Wash with plenty of soap and water after handling.

NOTE TO PHYSICIANS: This product may cause cholinesterase inhibition. Atropine is antidotal. Pralidozime chloride (2-PAM; PROTOPAM chloride) may be effective as an adjunct to atropine.

ENVIRONMENTAL HAZARDS

This product is toxic to birds and fish. Fish and other aquatic organisms in water treated with this product may be killed. You must consult your State Fish and Game Agency before applying this product to waters or wetlands. Do not contaminate water by cleaning of equipment or disposing of wastes.

STORAGE AND DISPOSAL

PROHIBITIONS

Do not contaminate water, food, or feed by storage or disposal.

PESTICIDE DISPOSAL
Wastes resulting from the use of this product may be disposed of on site or at an approved waste disposal facility.

CONTAINER DISPOSAL
Completely empty bag into application equipment. Then dispose of empty bag in a sanitary landfill or by incineration, or, if allowed by State and local authorities, by burning. If burned, stay out of smoke.

GENERAL
Consult Federal, State or local disposal authorities for approved alternative procedures.

DIRECTIONS FOR USE

It is a violation of Federal law to use this product in a manner inconsistent with its labeling.
For control of mosquito and midge larvae, apply ABATE 5-CG as specified below:

Areas of Treatment	Lbs. Per Acre
Standing water, shallow ponds, lakes, woodland pools	2
Tidal waters, marshes, swamps and waters high in organic content	4
Highly-polluted water	10

May be applied as a spot treatment to catch basins and similar areas where mosquitoes may breed.
Apply uniformly and repeat applications as necessary. Do not use on crops used for food, forage or pasture.

DISCLAIMER

The label instructions for the use of this product reflect the opinion of experts based on field use and tests. The directions are believed to be reliable and should be followed carefully. However, it is impossible to eliminate all risks inherently associated with use of this product. Crop injury, ineffectiveness or other unintended consequences may result because of such factors as weather conditions, presence of other materials, or the use or application of the product contrary to label instructions, all of which are beyond the control of American Cyanamid Company. All such risks shall be assumed by the user.

American Cyanamid Company warrants only that the material contained herein conforms to the chemical description on the label and is reasonably fit for the use herein described when used in accordance with the directions for use, subject to the risks referred to above.

Any damages arising from a breach of this warranty shall be limited to direct damages and shall not include consequential commercial damages such as loss of profits or values or any other special or indirect damages.

American Cyanamid Company makes no other express or implied warranty, including any other express or implied warranty of FITNESS or of MERCHANTABILITY.

In case of an emergency endangering life or property involving this product, call collect, day or night, Area Code 201–835-3100.

ABATE® 4-E

INSECTICIDE

For Control of Mosquito Larvae

ACTIVE INGREDIENT:
Temephos (O,O′-(thiodi-4,1-phenylene)bis (O,O-dimethyl phosphorothioate)) 44.6%
INERT INGREDIENTS: 55.4%
TOTAL 100.0%
(1 gallon contains 4 lbs. of Temephos)

EPA Reg. No. 241-132 EPA Est. No. 24619-09 D-40

WARNING

Keep Out of Reach of Children

See Below For Other Warnings

AVISO

AL USUARIO: Si usted no lee inglés, no use éste producto hasta que la etiqueta le haya sido explicada ampliamente.

(ENGLISH TRANSLATION) WARNING
(TO THE USER: If you cannot read English, do not use this product until the label has been fully explained to you.)

PRECAUTIONARY STATEMENTS

HAZARDS TO HUMANS AND DOMESTIC ANIMALS
Harmful if swallowed or absorbed through skin. Causes skin and eye irritation. Wash thoroughly with soap and water after handling. Avoid breathing spray mist. Do not contaminate food or feed products.

STATEMENT OF PRACTICAL TREATMENT

If Swallowed: Do not induce vomiting. Get medical attention.

If in Eyes: Flush with plenty of water. Get medical attention.

If on Skin: Wash with plenty of soap and water after handling. Get medical attention.

NOTE TO PHYSICIANS: This product may cause cholinesterase inhibition. Atropine is antidotal. Pralidozime chloride (2-PAM; PROTOPAM chloride) may be effective as an adjunct to atropine.

ENVIRONMENTAL HAZARDS
This product is toxic to birds and fish. Fish and other aquatic organisms in water treated with this product may be killed. You must consult your State Fish and Game Agency before applying this product to waters or wetlands. Do not contaminate water by cleaning of equipment or disposing of wastes. This product is toxic to bees and should not be applied when bees are actively visiting the treatment area.

STORAGE AND DISPOSAL

PROHIBITIONS
Do not contaminate water, food, or feed by storage or disposal.

PESTICIDE DISPOSAL
Wastes resulting from the use of this product may be disposed of on site or at an approved waste disposal facility.

CONTAINER DISPOSAL
Triple rinse (or equivalent). Then offer for recycling or reconditioning, or puncture and dispose of in a sanitary landfill, or by other approved State and local authorities.

GENERAL
Consult Federal, State or local disposal authorities for approved alternative procedures.

DIRECTIONS FOR USE

It is a violation of Federal law to use this product in a manner inconsistent with its labeling.

MOSQUITO LARVAE CONTROL

Apply ABATE 4-E as a uniform spray in sufficient water for good coverage at the rate of 0.5 to 1.5 fluid ounces per acre. Apply to standing water, moist areas, woodland pools, shallow ponds, around the edges of lakes, swamps, marshes, tidal waters, and catch basins. Use 0.5 fluid ounces per acre in clean open water and 1.0 to 1.5 fluid ounces per acre in waters high in organic matter content. The higher rate should be used in areas known to have organophosphate resistant mosquitoes. Repeat applications as necessary.

Do not use on crops used for food, forage or pasture.

DISCLAIMER

The label instructions for the use of this product reflect the opinion of experts based on field use and tests. The directions are believed to be reliable and should be followed carefully. However, it is impossible to eliminate all risks inherently associated with use of this product. Crop injury, ineffectiveness or other unintended consequences may result because of such factors as weather conditions, presence of other materials, or the use or application of the product contrary to label instructions, all of which are beyond the control of American Cyanamid Company. All such risks shall be assumed by the user.

American Cyanamid Company warrants only that the material contained herein conforms to the chemical description on the label and is reasonably fit for the use therein described when used in accordance with the directions for use, subject to the risks referred to above.

Any damages arising from a breach of this warranty shall be limited to direct damages and shall not include consequential commercial damages such as loss of profits or values or any other special or indirect damages.

American Cyanamid Company makes no other express or implied warranty, including any other express or implied warranty of FITNESS or of MERCHANTABILITY.

In case of an emergency endangering life or property involving this product, call collect, day or night, Area Code 201-835-3100.

AMDRO®

FIRE ANT INSECTICIDE

ACTIVE INGREDIENT:	BY WT.
Tetrahydro-5,5-dimethyl-2(1H)-pyrimidinone (3-[4-(trifluoromethyl)phenyl]-1-(2-[4-(trifluoromethyl) phenyl]-ethenyl)-2-propenylidene)hydrazone	0.88%
INERT INGREDIENTS:	99.12%
TOTAL	100.00%

EPA Reg. No. 241-260 EPA Est. No. 5905-AR-1

® Registered Trademark of American Cyanamid Company

CAUTION

Keep Out of Reach of Children

See Back Panel for Other Cautions

Product Code 24564-09 D-42

In case of an emergency endangering life or property involving this product, call collect, day or night, Area Code 201-835-3100.

READ ENTIRE LABEL. DO NOT USE THIS PRODUCT FOR ANY USES OTHER THAN THOSE SPECIFIED ON THIS LABEL. OBSERVE ALL PRECAUTIONARY STATEMENTS AND DIRECTIONS ON LABEL BEFORE USE. It is a violation of Federal Law to use this product in a manner inconsistant with its labeling.

DIRECTIONS

AMDRO is for use in lawns, turf, pasture, range grass and nonagricultural lands.

Treatment	Rate	Comments
Individual mound treatment: fire ants, harvester ants	5 level tablespoons	Distribute bait uniformly around the base of the mound. Do not disturb ant mounds. Do not apply directly to the top of the mound. Do not apply more than 1.5 lbs of bait per acre. Do not contaminate kitchen utensils by use or storage.
Broadcast treatment: fire ants, harvester ants	1/4–1/3 lb/10,000 square feet 1–1 1/2 lb/acre	Broadcast bait uniformly with ground equipment (spinning disk types).

NOTE: Do not use on vegetable or other food crops.

Apply when ants are active (typically when soil temperature is greater than 60°F).

AMDRO is a bait which is highly attractive to fire ants and harvester ants. It works by becoming part of the ant food chain. Worker ants will find the AMDRO bait, carry it back to the mound and pass it along to the queen and the other ants in the mound.

An effective ant insecticide must be slow acting so that it can be passed by the workers throughout the ant mound and eventually to the queen. AMDRO is a slow acting insecticide and is especially effective against the queen. Typically, in 1–4 weeks the queen and a number of ants are killed. Within 2–8 weeks, a visible reduction in mound activity is observed. Very large mounds may continue to be active for a period of time even though the queen is dead and no young are being produced. Retreatment may be desirable under these circumstances.

Further information regarding ants may be obtained from your Cooperative Agricultural Extension Service.

PRECAUTIONARY STATEMENTS

HAZARDS TO HUMANS

May be harmful if swallowed. Avoid any prolonged contact to skin or eyes. Wash thoroughly after handling.

FIRST AID: If swallowed, drink one or two glasses of water and induce vomiting by touching back of throat with finger. Do not induce vomiting or give anything by mouth to an unconscious person. Get medical attention.

HAZARDS TO ANIMALS

This product may be an attractant to pets and rodents. Store in a secure place. Keep pets away from treated areas for at least 24 hours after application.

ENVIRONMENTAL HAZARDS

This product is toxic to fish. Do not apply directly to lakes, ponds or streams.

STORAGE AND DISPOSAL

STORAGE

STORE IN A COOL, DRY SECURE PLACE AND KEEP CONTAINER TIGHTLY CLOSED. AMDRO is formulated in an oil bait that functions as an attractant to ants. Prolonged exposure to air may turn oil rancid and reduce the attractiveness of the bait. USE WITHIN 3 MONTHS AFTER OPENING.

DISPOSAL

Do not reuse empty container. Wrap container and put in trash collection.

DISCLAIMER

American Cyanamid Company warrants only that the material contained herein conforms to the chemical description on the label and is reasonably fit for the use therein described when used in accordance with the directions for use. American Cyanamid Company makes no other express or implied warranty, including any other express of implied warranty of FITNESS or of MERCHANTABILITY.

SUPPLEMENTAL LABELING

AMDRO® FIRE ANT INSECTICIDE
EPA Reg. No. 241-260

FOR CONTROL OF HARVESTER ANTS AND BIG-HEADED ANTS

FOR USE IN PASTURE AND RANGE GRASS, LAWNS, TURF AND NONAGRICULTURAL LANDS

In addition to the control of fire ants, AMDRO fire ant insecticide may be used in the control of harvester ants and big-headed ants.

AMDRO is formulated in an oil bait that functions as an attractant to ants. Prolonged exposure to air may turn oil rancid and reduce the attractiveness of the bait. Current data have shown that AMDRO can be used within 3 months after opening, when container is tightly closed after use and is stored in a cool, dry secure place.

Refer to the label on the container for specific directions regarding the rates, use, direction and application of AMDRO fire ant insecticide.

OBSERVE ALL PRECAUTIONARY STATEMENTS ON THE LABEL BEFORE USING

® Registered Trademark of American Cyanamid Company.

INFORMATION BULLETIN

AMDRO® FIRE ANT INSECTICIDE

AMDRO For Effective Control of Harvester Ants

Use in pasture and range grass, lawns, turf, and nonagricultural lands.

AMDRO® fire ant insecticide is formulated in a soybean oil bait which is highly attractive to harvester ants. Unlike ordinary insecticides that may or may not penetrate into the mound, AMDRO granules are carried directly into the mound by workers as food for all ants within the colony. As a result, the colony will die within a few days following treatment.

AMDRO has these technical benefits:

- Concentrated. One pound treats 17 mounds.
- Low toxicity to wildlife.
- Does not persist in the environment.
- Can be broadcast applied or used to treat individual mounds.

Application: For individual colony treatment, sprinkle 5 level tablespoons of AMDRO uniformly in the area devoid of vegetation that surrounds the entrance into the colony.

For broadcast application on small areas, use a hand-held spinning disk spreader. Apply at the rate of 1 to 1½ pounds per acre or ¼ to ⅓ pound per 10,000 square feet. For larger areas, AMDRO fire ant insecticide may be applied aerially or by using a Model M-3 Cyclone Spreader or equivalent.

Always read and follow label directions carefully.

AMINO TRIAZOLE

WEEDKILLER 90

ACTIVE INGREDIENT:

Amitrole*	90.0%
INERT INGREDIENTS:	10.0%
TOTAL	100.0%

*3-amino-1, 2, 4-triazole EPA Reg. No. 241-168-AA

CAUTION

Keep Out of Reach of Children

Harmful if swallowed. Avoid contact with skin, eyes or clothing.

PRECAUCIÓN

AL USUARIO: Si usted no lee inglés, no use este producto hasta que la etiqueta le haya sido explicada ampliamente.

(TO THE USER: If you cannot read English, do not use this product until the label has been fully explained to you.)

In case of an emergency endangering life or property involving this product, call collect, day or night, Area Code (201) 835-3100.

DO NOT USE THIS PRODUCT FOR ANY USES

OTHER THAN THOSE SPECIFIED ON THIS LABEL.

DO NOT SPRAY OR ALLOW SPRAY DRIFT TO CONTAMINATE EDIBLE CROPS OR WATER WHICH WILL BE USED FOR IRRIGATION, DRINKING OR OTHER DOMESTIC PURPOSES AS NO RESIDUE OF AMITROLE IS PERMISSIBLE ON CROPS. DO NOT SPRAY OR PERMIT DRIFT OF SPRAY ON LAWNS, SHRUBS, FLOWERS AND OTHER DESIRABLE VEGETATION SINCE THIS PRODUCT WILL CAUSE INJURY. DO NOT USE IN GREENHOUSES. WASH SPRAYER THOROUGHLY WITH CLEAN WATER IMMEDIATELY AFTER EACH USE TO AVOID CONTAMINATION OF OTHER SPRAYS. DO NOT CONTAMINATE FERTILIZERS, SEEDS, INSECTICIDES OR FUNGICIDES BY IMPROPER STORAGE. KEEP LIVESTOCK OFF TREATED AREAS. AVOID DRIFT OF SPRAY TO CROPLAND.

AMINO TRIAZOLE Weedkiller 90 is the most effective when applied directly to the foliage of actively growing plants. It is absorbed by the leaves and travels through the entire plant—down to the deepest roots. Leaves wilt within 10 days to 3 weeks, and the plants subsequently brown and die out.

AMINO TRIAZOLE Weedkiller 90 should be applied in sufficient water to thoroughly cover weed foliage. Retreatment may be necessary to control plants not thoroughly sprayed or plants arising from dormant seeds.

PROPER MIXING FOR OPTIMUM CONTROL

AMINO TRIAZOLE Weedkiller 90 is not a wettable powder; it is soluble in water. This unique feature allows a proper flow throughout the spray system without clogging nozzles or corroding spray equipment. When mixing AMINO TRIAZOLE Weedkiller 90 with simazine, first dissolve AMINO TRIAZOLE Weedkiller 90 and then add simazine wettable powder with constant agitation during mixing and spraying operations.

WEED CONTROL ON NONCROPLAND SUCH AS ALONG ROADSIDES, DRAINAGE DITCHBANKS, FENCE ROWS, RIGHTS-OF-WAY AND SIMILAR AREAS.

SPOT TREATMENT: For knapsack sprayer application of AMINO TRIAZOLE Weedkiller 90 for spot treatment of annual and perennial weeds specified in the table below, mix ½ to 1 ounce of AMINO TRIAZOLE Weedkiller 90 in 1 gallon of water (1 ounce equals 3 level tablespoonfuls). Use the higher rate for control of perennial weeds. One gallon of solution generally covers 1 square rod (272 square feet) thoroughly infested with weeds. In Intermountain states or other dryland areas, mix ½ to 1 ounce of this product in 2 gallons of water to cover 1 square rod. Apply when weeds are growing actively. Wet all foliage thoroughly. Repeat applications may be necessary to control deep-rooted perennials.

WEED CONTROL IN NONCROPLANDS such as along roadsides, fence rows, drainage ditchbanks, rights-of-way and similar areas.

WEEDS	AMOUNT PER ACRE*	WHEN TO APPLY
For control of ANNUAL GRASSES and BROADLEAF WEEDS such as: Foxtail, Rye Grass, Cheatgrass, Barnyard Grass, Annual Bluegrass, Pigweed, Puncture Vine, Kochia, Sunflower.	2 lbs. in 30–60 gals. water	Spray when weeds are 3–6 inches tall. KEEP LIVESTOCK OFF TREATED AREAS.
For control of PERENNIAL BROADLEAF WEEDS such as: White Top, Canada Thistle, Milkweed, Sow Thistle. For suppression of PERENNIAL GRASSES such as: Quack Grass, Ripgut Grass, Bermuda Grass, Reed Canary Grass, Wild Barley.	4 lbs. in 30–60 gals. water	When most weeds are 6–10 inches tall. Repeat if needed. KEEP LIVESTOCK OFF TREATED AREAS.
Cattails	9 to 11 lbs. in 300 to 400 gals. water for ground spray; 15 gals. water for air spray.	Spray when catkins are fully formed and up to 2 weeks before expected frost. Do not diturb plants after treatment. Spot treat any regrowth. Do not contaminate water used for irrigation or domestic purposes.

Use higher volumes on dense growth for good wetting.

WOODY PLANTS AND VINES

WEEDS	AMOUNT PER ACRE*	WHEN TO APPLY
Poison Ivy Poison Oak	2 lbs. in 100 gals. water	Spray when plants are in full leaf. Spray until plants are thoroughly wet to ground level.
Honeysuckle Kudzu Vine	5½ lbs. in 125 gals. water	Spray in June when leaves are fully developed. If regrowth appears, repeat application in August or following spring. Wet plants thoroughly.

* A strip 8 feet wide and 1 mile long equals approximately 1 acre.

HARDWOOD NURSERIES

WEEDS	AMOUNT PER ACRE	WHEN TO APPLY
Chrysanthemum	9 lbs. in 100 gals. water	Apply as a directed spray on young weed growth 4 to 6 inches high making sure to keep the spray off the foliage of hardwood nursery stock.

FOR GENERAL WEED CONTROL IN NURSERY STOCK*, ORNAMENTAL PLANTINGS* AND FOREST PLANTATIONS

AMINO TRIAZOLE Weedkiller 90	Simazine (80-W)	Gals. Water	Area
¼ lbs.	1 lb.	20	110 × 100
½ lbs.	2 lbs.	40	220 × 100
1 lb.	4 lbs.	80	1 acre

Apply mixture to small weeds in the spring or early summer. Weeds should be wet thoroughly. Cut or mow tall weeds to ground line, remove cuttings and spray regrowth 1 to 2 weeks later. Where Johnson grass, Bermuda grass or other weeds are not controlled by the first application, treat again with the same rate when regrowth is 6 to 12 inches tall (about 2 months).

* The following plant species may be treated. Use a directed spray. Balsam fir, Douglas fir, Fraser fir, Norway spruce, Red spruce, White spruce, Red pine, Scotch pine, Arbovitae (Thuja), Juniper, Yew (Taxus), Oleander, Pyracantha and Viburnum.

CAUTION: Lilac, privet, honeysuckle and barberry are susceptible to this mixture and may be injured if contacted by spray.

DISPOSAL OF BAGS: Destroy empty bags by burning. Stay away from smoke and fumes.

DISCLAIMER

The label instructions for the use of this product reflect the opinion of experts based on field use and tests. The directions are believed to be reliable and should be followed carefully. However, it is impossible to eliminate all risks inherently associated with use of this product. Crop injury, ineffectiveness or other unintended consequences may result because of such factors as weather conditions, presence of other materials, or the manner of use or application all of which are beyond the control of American Cyanamid Company. All such risks shall be assumed by the user.

- American Cyanamid Company warrants only that the material contained herein conforms to the chemical description on the label and is reasonably fit for the use therein described when used in accordance with the directions for use, subject to the risks referred to above.
- Any damages arising from breach of this warranty shall be limited to direct damages and shall not include consequential commercial damages such as loss of profits or values or any other special or indirect damages.
- American Cyanamid Company makes no other express or implied warranty, including any other express or implied warranty of FITNESS or of MERCHANTABILITY.

ARSENAL®

HERBICIDE

ACTIVE INGREDIENT:

2-[4,5-dihydro-4-methyl-4-(1-methylethyl)-5-oxo-1H-imidazol-2-yl]-3-pyridinecarboxylic acid with 2-propanamine (1:1) salt† 27.6%

INERT INGREDIENT: 72.4%

TOTAL 100.0%

† Equivalent to 22.6% 2-4,5-dihydro-4-methyl-4-(1-methylethyl)-5-oxo-1H-imidazol-2-yl-3-pyridinecarboxylic acid or 2 pounds acid per gallon.

EPA Reg. No. 241-273 EPA Est. No. 5905-AR-1

CAUTION

Keep Out of Reach of Children

See Side Panel for Other Warnings

PRECAUCION

PRECAUCION AL USUARIO: Si usted no lee inglés, no use este producto hasta que la etiqueta le haya sido explicada ampliamente.

In case of an emergency endangering life or property involving this product, call collect, day or night, Area Code 201–835-3100.

PRECAUTIONARY STATEMENTS

HAZARDS TO HUMANS

CAUTION

Avoid contact with skin, eyes or clothing. Avoiding breathing spray mist. Wash thoroughly with soap and water after handling. Remove contaminated clothing and wash before reuse.

FIRST AID

If on Skin: Wash with plenty of soap and water.
If in Eyes: Flush with plenty of water. Get medical attention if irritation persists.

PHYSICAL AND CHEMICAL HAZARDS

Spray solutions of ARSENAL should be mixed, stored and applied only in stainless steel, fiberglass, plastic and plastic-lined steel containers.

DO NOT mix, store or apply ARSENAL or spray solutions of ARSENAL in unlined steel (except stainless steel) containers or spray tanks.

ENVIRONMENTAL HAZARDS

DO NOT apply directly to any body of water. DO NOT contaminate water by cleaning of equipment or disposal of waste.

STORAGE AND DISPOSAL

PROHIBITIONS

DO NOT store below 10°F.

DO NOT contaminate water, food or feed by storage or disposal.

PESTICIDE DISPOSAL

Wastes resulting from the use of this product may be disposed of on site or at an approved waste disposal facility.

CONTAINER DISPOSAL

Triple rinse (or equivalent). Then offer for recycling or reconditioning, or puncture and dispose of in an approved sanitary landfill, or incineration, or, if allowed by state and local authorities, by burning. If burned, stay out of smoke.

IMPORTANT

DO NOT use on food or feed crops. DO NOT apply where runoff water may flow onto agricultural land as injury to crops may result. Keep from contact with fertilizers, insecticides, fungicides and seeds. DO NOT apply or drain or flush equipment on or near desirable trees or other plants, or on areas where their roots may extend, or in locations where the chemical may be washed or moved into contact with their roots. DO NOT use on lawns, walks, driveways, tennis courts, or similar areas. Prevent drift of spray to desirable plants. DO NOT use in California.

Thoroughly clean application equipment immediately after use. Flush tank, pump, hoses and boom with several changes of water after removing nozzle tips and screens (clean these parts separately).

GENERAL INFORMATION

ARSENAL herbicide is an aqueous solution to be mixed in water and applied as a spray for control of most annual and perennial grasses and broadleaf weeds on noncropland areas.

ARSENAL may be applied either preemergence of postemergence to the weeds; however, postemergence application is the method of choice in most situations, particularly for control of perennials. For maximum activity, weeds should be growing vigorously at the time of postemergence applications. The preemergence activity of ARSENAL will provide residual control of most weed species following a postemergence application.

ARSENAL is readily absorbed through foliage and roots and is translocated rapidly throughout the plant, with accumulation in the meristematic regions. Treated plants stop growing soon after spray application. Chlorosis appears first in the newest leaves, and necrosis spreads from this point. In perennials, the herbicide is translocated into and kills underground storage organs, thus preventing regrowth. Chlorosis and tissue necrosis may not be apparent in some plant species until two weeks after application. Complete kill of plants may not occur for several weeks.

DISCLAIMER

The label instructions for the use of this product reflect the opinion of experts based on field use and tests. The directions are believed to be reliable and should be followed carefully. However, it is impossible to eliminate all risks inherently associated with use of this product. Ineffectiveness or other unintended consequences may result because of such factors as weather conditions, presence of other materials, or the use or application of the product contrary to label instructions, all of which are beyond the control of American Cyanamid Company. All such risks shall be assumed by the user.

American Cyanamid Company warrants only that the material contained herein conforms to the chemical description on the label and is reasonably fit for the use therein described when used in accordance with the directions for use, subject to the risks referred to above.

Any damages arising from a breach of this warranty shall be limited to direct damages and shall not include consequential commercial damages such as loss of profits or values or any other special or indirect damages.

American Cyanamid Company makes no other express or implied warranty, including other express or implied warranty of FITNESS or of MERCHANTABILITY.

DIRECTIONS FOR USE

It is a violation of Federal Law to use this product in a manner inconsistent with its labeling.

ARSENAL herbicide should be used only in accordance with recommendations on the leaflet label attached to the container. Keep containers closed to avoid spills and contamination.

A postemergence application of ARSENAL is recommended for control of most annual and perennial grasses and broadleaf weeds on noncropland areas such as railroad, utility and pipeline rights-of-way, utility plant sites, petroleum tank farms, pumping installations, fence rows, storage areas, non-irrigation ditch banks and other similar areas.

MIXING INSTRUCTIONS

Mix the proper amount of ARSENAL in water in the spray tank with the agitator running. To minimize drift, a drift control agent may be added at the recommended label rate. A foam reducing agent may be added at the recommended label rate, if needed.

SPRAYING INSTRUCTIONS

Uniformly apply with properly calibrated ground equipment in 30 to 60 gallons of water per acre with a spray pressure of 25 to 50 psi. Spray volumes greater than 60 gallons per acre may be applied without reducing efficacy; however, additional nonionic surfactant such as Surfactant WK‡ or Ortho†‡ X-77 must then be added at the rate of 1 quart per 100 gallons of spray to provide optimum wetting and /or contact activity.

‡ Registered trademark of E.I. duPont de Nemours and Company.

†‡ Registered trademark of Chevron Chemical Company.

ARSENAL herbicide will provide postemergence control with residual control of the following weed species at the rates listed.

GRASSES

Apply 2-3 pints per acre†

Annual bluegrass *(Poa annua)*
Broadleaf signalgrass *(Brachiaria platyphylla)*
Canada bluegrass *(Poa compressa)*
Downy brome *(Bromus tectorum)*
Fescue *(Festuca spp.)*
Foxtail *(Setaria spp.)*
Lovegrass *(Eragrostis spp.)*
Orchardgrass *(Dactylis glomerata)*
Paragrass *(Brachiaria mutica)*
Quackgrass *(Agropyron repens)*
Sandbur *(Cenchrus spp.)*
Smooth brome *(Bromus inermis)*

Italian ryegrass *(Lolium multiflorum)*
Johnsongrass *(Sorghum halepense)*
Kentucky bluegrass *(Poa pratensis)*
Vaseygrass *(Paspalum urvillei)*
Wild oats *(Avena fatua)*
Witchgrass *(Panicum capillare)*

Apply 3-4 pints per acre†

Beardgrass *(Andropogon spp.)*
Cheat *(Bromus secalinus)*
Crabgrass *(Digitaria spp.)*
Fall panicum *(Panicum dichotomiflorum)*
Goosegrass *(Eleusine indica)*
Prairie threeawn *(Aristida oligantha)*
Torpedograss *(Panicum repens)*
Wild barley *(Hordeum spp.)*

Apply 4-6 pints per acre†

Bahiagrass *(Paspalum notatum)*
Bermudagrass *(Cynodon dactylon)*
Big bluestem *(Andropogon geradi)*
Cattail *(Typha spp.)*
Dallisgrass *(Paspalum dilatatum)*
Guineagrass *(Panicum maximum)*
Prairie cordgrass *(Spartina pectinata)*
Sand dropseed *(Sporobolus cryptandrus)*
Timothy *(Phleum pratense)*
Wirestem muhly *(Muhlenbergia frondosa)*

BROADLEAF WEEDS

Apply 2-3 pints per acre†

Burdock *(Arctium spp.)*
Camphorweed *(Heterotheca subaxillaris)*
Carpetweed *(Mullugo verticillata)*
Carolina geranium *(Geranium carolinianum)*
Clover *(Trifolium spp.)*
Common chickweed *(Stellaria media)*
Common ragweed *(Ambrosia artemisiifolia)*
Dandelion *(Taraxacum officinale)*
Dogfennel *(Eupatorium capillifolium)*
Filaree *(Erodium spp.)*
Fleabane *(Erigeron spp.)*
Hoary vervain *(Verbena stricta)*
Horseweed *(Conyza canadensis)*
Indian mustard *(Brassica juncea)*
Kochia *(Kochia scoparia)*
Lambsquarters *(Chenopodium album)*
Lespedeza *(Lespedeza spp.)*
Miners lettuce *(Montia perfoliata)*
Mullein *(Verbascum spp.)*
Nettleleaf goosefoot *(Chenopodium murale)*
Oxeye daisy *(Chrysanthemum leucanthemum)*
Pepperweed *(Lepidium spp.)*
Pigweed *(Amaranthus spp.)*
Plantain *(Plantago spp.)*
Smartweed *(Polygonum spp.)*
Sorrel *(Rumex spp.)*
Sunflower *(Helianthus spp.)*
Sweet clover *(Melilotus spp.)*
Tansymustard *(Descurainia pinnata)*
Western ragweed *(Ambrosia psilostachya)*
Wild carrot *(Daucus carota)*
Wild lettuce *(Lactuca spp.)*
Wild parsnip *(Pastinaca sativa)*
Wild turnip *(Brassica campestris)*
Wollyleaf bursage *(Ambrosia grayi)*
Yellow woodsorrel *(Oxalis stricta)*

Apply 3-4 pints per acre†

Bull thistle *(Circium vulgare)*
Cocklebur *(Xanthium strumarium)*
Dock *(Rumex spp.)*
Goldenrod *(Solidago spp.)*
Pokeweed *(Phytolacca americana)*
Purslane *(Portulaca spp.)*
Yellow starthistle *(Centaurea solstitialis)*

Apply 4-6 pints per acre†

Canada thistle *(Circium arvense)*
Giant ragweed *(Ambrosia trifida)*
Little mallow *(Malva parviflora)*
Milkweed *(Asclepias spp.)*
Primrose *(Oenothera kunthiana)*
Silverleaf nightshade *(Solanum elaeagnifolium)*
Sowthistle *(Sonchus spp.)*
Texas thistle *(Circium texanum)*

VINES AND BRAMBLES

Apply 2-3 pints per acre†

Field bindweed *(Convolvulus arvensis)*
Wild buckwheat *(Polygonum convolvulus)*

Apply 3-4 pints per acre†

Greenbriar *(Smilax spp.)*
Honeysuckle *(Lonicera spp.)*
Morningglory *(Ipomoea spp.)*
Poison ivy *(Rhus radicans)*
Redvine *(Brunnichia cirrhosa)*
Wild rose *(Rosa spp.)*

Apply 4-6 pints per acre†

Blackberry *(Rubus spp.)*
Dewberry *(Rubus spp.)*
Kudzu *(Pueraria lobata)*
Hedge bindweed *(Calystegia sepium)*
Trumpetcreeper *(Campsis radicans)*
Virginia creeper *(Parthenocissus quinquefolia)*
Wild grape *(Vitis spp.)*

BRUSH SPECIES

Apply 4-6 pints per acre†

Ash *(Fraxinus spp.)*
Cherry *(Prunus spp.)*
Dogwood *(Cornus spp.)*
Hawthorn *(Crataegus spp.)*
Hickory *(Carya spp.)*
Maple *(Acer spp.)*
Mulberry *(Morus spp.)*
Oak *(Quercus spp.)*
Poplar *(Populus spp.)*
Privet *(Ligustrum vulgare)*
Red Maple *(Acer rubrum)*
Sassafras *(Sassafras albidum)*
Sumac *(Rhus spp.)*
Sweetgum *(Liquidambar styraciflua)*
Willow *(Salix spp.)*

† The higher rates should be used where heavy or well established infestations occur.

AVENGE®

WILD OAT HERBICIDE

For Use in Barley & Wheat

ACTIVE INGREDIENT:
Difenzoquat (1,2-dimethyl-3,5-diphenyl-1H-pyrazolium methyl sulfate) 31.2%
INERT INGREDIENTS 68.8%
TOTAL 100.0%
1 gallon contains 2 lbs. of difenzoquat cation
EPA Reg.No.241-266 EPA Est. No.

DANGER

Keep Out of Reach of Children

(ENGLISH TRANSLATION)
DANGER!
(TO THE USER: If you cannot read English, do not use this product until the label has been fully explained to you.)

PELIGRO
(AL USUARIO: Si usted no lee inglés, no use este producto hasta que la etiqueta le haya sido explicada ampliamente.)

FIRST AID

In case of contact—immediately flush eyes with plenty of water. Call a physician.

If swallowed—drink one or two glasses of water and induce vomiting by touching the back of throat with finger. Do not induce vomiting or give anything by mouth to an unconscious person. Get medical attention.

See Side Panel for Additional Precautionary Statements

GENERAL INFORMATION

AVENGE is a selective herbicide for the postemergence control of wild oats in barley and wheat. See the section DIRECTIONS FOR USE for varietal restrictions.

If broadleaf weeds must be controlled in these crops, AVENGE may be tank-mixed with MCPA amine or ester, bromoxynil, MCPA plus bromoxynil, or 2,4-D amine or ester, or Glean. AVENGE alone or in combination can be applied postemergence by ground equipment or aircraft.

MAY BE FATAL IF SWALLOWED, INHALED OR ABSORBED THROUGH THE SKIN.

RAPIDLY ABSORBED THROUGH SKIN.

REPEATED INHALATION OR SKIN CONTACT MAY, WITHOUT SYMPTOMS, PROGRESSIVELY INCREASE SUSCEPTIBILITY TO POISONING.

See Back Panel for Antidote and Other Warnings

PELIGRO!
AL USUARIO: Si usted no lee inglés, no use este producto hasta que la etiqueta le haya sido explicada ampliamente.
(ENGLISH TRANSLATION)
DANGER!
(TO THE USER: If you cannot read English, do not use this product until the label has been fully explained to you.)

DANGER!

HAZARDS TO HUMANS AND DOMESTIC ANIMALS

Do Not Get In Eyes, On Skin, On Clothing

Wear freshly laundered, long-sleeved work clothing daily. While transferring from package to equipment, wear a clean cap and gloves (rubber or cotton). If cotton gloves are used, they must be laundered or discarded after each day's use. Rubber gloves should be washed with soap and water after each use. Do not wear the same gloves for other work. Destroy and replace gloves frequently.

In case of contact, immediately remove contaminated clothing and wash skin thoroughly with soap and water. Launder clothing and decontaminate shoes before reuse. Wash thoroughly with soap and water before eating or smoking. Bathe at the end of the work day and change clothing.

Do Not Breathe Dust
While emptying bags into equipment, pour downwind and allow as little free fall as possible. Do not pour at face level and do not allow dust to reach the breathing zone.

Do Not Contaminate Food or Feed Products
Once a bag has been opened, use it completely. Make sure the hoppers are emptied while still in the field. Refer to STORAGE AND DISPOSAL statement for further instructions.

KEEP ALL UNPROTECTED PERSONS OUT OF OPERATING AREAS.

KEEP OUT OF REACH OF DOMESTIC ANIMALS.

NOT FOR USE OR STORAGE IN OR AROUND THE HOME.

ENVIRONMENTAL HAZARDS
This product is toxic to fish, birds and other wildlife. Treated granules exposed on soil surface may be hazardous to birds and other wildlife. Keep out of any body of water. Do not apply where runoff is likely to occur. Do not contaminate water by cleaning of equipment or disposal of wastes.

In case of an emergency endangering life or property involving this product, call collect, day or night, Area Code 201-835-3100.

Antidote: Atropine is an antidote.

CALL A PHYSICIAN AT ONCE IN ALL CASES OF SUSPECTED POISONING

FIRST AID

If swallowed, drink one or two glasses of water and induce vomiting by touching back of throat with finger. Do not induce vomiting or give anything by mouth to an unconscious person. Get medical attention.

If inhaled, remove to fresh air. If not breathing, give artificial respiration, preferably mouth-to-mouth. If breathing is difficult, give oxygen.

In case of contact, immediately flush eyes or skin with plenty of water for at least 15 minutes and remove contaminated clothing and shoes. Wash clothing and decontaminate shoes before reuse.

NOTE TO PHYSICIANS: Warning symptoms include weakness, headache, tightness in chest, blurred vision, nonreactive pinpoint pupils, salivation, sweating, nausea, vomiting, diarrhea and abdominal cramps. Give atropine intramuscularly or intravenously, depending on severity of poisoning, 2 to 4 milligrams every 10 minutes until fully atropinized as shown by dilated pupils, dry flushed skin and tachycardia. Twenty to thirty milligrams, or more, may be required during the first 24 hours. Never give opiates or phenothiazine tranquilizers. Clear chest by postural drainage. Artificial respiration or oxygen administration may be necessary. Observe patient continuously for at least 48 hours. Allow no further exposure to any cholinesterase inhibitor until cholinesterase regeneration has taken place as determined by blood tests.

Pralidoxime chloride (2-PAM; PROTOPAM chloride) may be effective as an adjunct to atropine. Use according to label directions.

DISCLAIMER

The label instructions for the use of this product reflect the opinion of experts based on field use and tests. The directions are believed to be reliable and should be followed carefully. However, it is impossible to eliminate all risks inherently associated with use of this product. Crop injury, ineffectiveness or other unintended consequences may result because of such factors as weather conditions, presence of other materials, or the use or application of the product contrary to label instructions, all of which are beyond the control of American Cyanamid Company. All such risks shall be assumed by the user.

American Cyanamid Company warrants only that the material contained herein conforms to the chemical description on the label and is reasonably fit for the use therein described when used in accordance with the directions for use, subject to the risks referred to above.

Any damages arising from a breach of this warranty shall be limited to direct damages and shall not include consequential commercial damages such as loss of profits or values or any other special or indirect damages.

American Cyanamid Company makes no other express or implied warranty, including any other express or implied warranty of FITNESS or of MERCHANTABILITY.

STORAGE AND DISPOSAL

STORAGE
Store pesticide products in a secure locked area where children, unauthorized persons and animals cannot enter. Do not store in the same area with food or feed. Do not store opened bags.

PROHIBITIONS
Do not contaminate water, food or feed by storage or disposal. Open dumping is prohibited.

PESTICIDE DISPOSAL
Pesticide or spillage that cannot be used according to label instructions must be disposed of according to Federal, State or local procedures under the Resource Conservation and Recovery Act.

CONTAINER DISPOSAL
Completely empty bag by shaking and tapping sides and bottom to loosen any clinging material. Empty the residues into the application equipment. Dispose of bags in a sanitary landfill, or by other approved State and local procedures.

GENERAL
Consult Federal, State or local disposal authorities for approved alternative procedures such as limited open burning.

Row Width (inches)	Use Rates as Pounds per Acre* (based on various row spacings and application rates) Application Rates (Oz./1,000 Ft. of Row)						
	4 oz.	6 oz.	8 oz.	10 oz.	12 oz.	14 oz.	16 oz.
40"	3.3	4.9	6.5	8.2	9.8	11.4	13.1
38"	3.4	5.2	6.9	8.6	10.3	12.0	13.7
36"	3.6	5.4	7.3	9.1	10.9	12.7	14.5
34"	3.9	5.6	7.7	9.7	11.6	13.5	15.4

Row Width (inches)	Use Rates as Pounds per Acre* (based on various row spacings and application rates) Application Rates (Oz./1,000 Ft. of Row)						
32"	4.1	6.1	8.2	10.2	12.3	14.3	16.3
30"	4.4	6.5	8.7	10.9	13.1	15.2	17.4
28"	4.7	7.1	9.5	11.8	14.2	16.6	18.9
26"	5.0	7.6	10.2	12.7	15.3	17.9	20.4
24"	5.4	8.2	10.9	13.6	16.3	19.1	21.8
22"	5.9	8.9	11.9	14.9	17.9	21.0	24.0
20"	6.5	9.8	13.1	16.3	19.6	22.8	26.1

* See directions for specific use rates on each crop.

CALIBRATION INFORMATION

FIRST READ THE LABEL

It is important that applicator equipment be properly set to deliver the labeled rate. This chart will help determine the desired rate of application.

When label rate is 8 ounces per 1,000 feet of row, use these suggested starting gauge settings, regardless of row spacing:

Applicator	Planting Speed 3 mph	5 mph	7 mph
Gandy	18	24	28
New Noble (Square Hole)	8	12	16
Old Noble (Round Hole)	11	17	23
International Harvester	1/7.0	2/1.5	2/7.5
New International Harvester	1/9.0	2/6.0	3/2.0
New John Deere (Plastic Funnels)	13	19	25
Newest John Deere (Aluminum Funnels)	11	15	19
Old John Deere	1/28	2/6	2/15
Allis Chalmers	All Speeds-Gauge 7		

NOTE: These settings should be used as starting points only. Continually check the amount of COUNTER used against a known length of row or acreage and make further adjustments accordingly. Also, check calibration occasionally to make sure equipment wear, changing moisture conditions, etc., have not cause a change in flow rate.

DIRECTIONS FOR USE

It is a violation of Federal Law to use this product in a manner inconsistent with its labeling.

BEFORE USING, READ PRECAUTIONARY STATEMENTS.

COUNTER should be applied with a granular pesticide applicator properly calibrated to assure accurate placement and proper dosage. Cover granules that may be exposed on the ends of the treated rows and turns and loading areas by deep discing immediately after treating fields.

CROPS	PESTS CONTROLLED	RATES OF COUNTER	APPLICATION	REMARKS
FIELD CORN, POPCORN, AND SWEET CORN	Corn rootworms Wireworms White grubs	**Banded or In-Furrow** 8 oz. per 1,000 ft. of row for any row spacing (minimum 30-inch row spacing).	**Banded** Place granules in a 7-inch band directly behind the planter shoe in front of the press wheel.	
At Planting	Seedcorn maggots Seedcorn beetles Corn flea beetles Maize billbugs Southern corn billbugs Thrips Symphylans Nematodes: Lesion, Spiral, Stunt, Sting, Stubby root, Dagger, Lance Suppression of Cutworms Lesser cornstalk borer		**In-Furrow** Place granules directly in the seed furrow behind the planter shoe.	For use on conventional and conservation (i.e., minimum and no-till) tillage corn. In conservation tillage systems, where excessive crop debris can prevent proper distribution of banded granular products, in-furrow applications are recommended. Under dry soil conditions or heavy infestations, it may be necessary to apply an insecticide rescue treatment with another **registered insecticide** after corn emergence to control surviving cutworm larvae. If especially heavy infestations are expected, banded applications of up to 16 oz. per 1,000 ft. of row may be used.
Post-emergence Incorporated (Same as crops above)	Maize billbugs Southern corn billbugs	**Banded** 12–16 oz. per 1,000 ft. of row for any row spacing (minimum 30-inch row spacing).	Apply in a 7-inch band over the row of seedling corn plants and lightly incorporate into the soil when billbugs or damage are observed, usually in the 1–6 leaf stage of growth. Use cultivators or other suitable implement to lightly incorporate granules into soil.	Only one postemergence incorporated or one cultivation time treatment may be used in addition to treatment at planting time. Do not treat if more than 8 oz. per 1,000 ft. of row were applied at planting. Do not graze or cut for forage within 30 days of treatment. Consult your state experiment station, state extension service, or certified pest management specialist for proper rate and timing of application.

CROPS	PESTS CONTROLLED	RATES OF COUNTER	APPLICATION	REMARKS
At Cultivation (Same as crops above)	Corn rootworms	8 oz. per 1,000 ft. of row for any row spacing (minimum 20-inch row spacing).	Apply granules to base of plants or over the top of plants just ahead of cultivation shovels so as to cover granules with soil.	Same remarks apply as stated for Post-emergence Incorporated (above).
SUGAR BEETS At Planting	Sugar beet root maggots Wireworm	**Banded** 4–8 oz. per 1,000 ft. of row for any row spacing (minimum 30-inch row spacing).	**Banded** Apply in a 5 to 7-inch band over the row and lightly incorporate into the soil.	**At Planting** Do not place granules in direct contact with the seed as crop injury may occur. Power incorporation may be used. Do not incorporate deeper than 2 inches.
Post-emergence	Sugar beet root maggots			**Postemergence** Apply at the time of first fly emergence. Lightly incorporate into the soil.
At Planting	Suppression of cutworms	**Banded** 8 oz. per 1,000 ft. of row for any row spacing (Minimum 30 in. row spacing)		**Only one application per year may be used.**
GRAIN SORGHUM At Bedding	Greenbugs Corn leaf aphids	**Knifed-in** 8–16 oz. per 1,000 ft. of row for any row spacing (minimum 20-inch row spacing) or no more than 26 lbs. per acre.	**Knifed-in** Drill granules 1–4 inches directly below the seed OR 1–4 inches below the seed and up to 5 inches to the side of the seed.	
GRAIN SORGHUM At Planting	Greenbugs Corn leaf aphids	**Knifed-in or Banded** **Knifed-in** 8–16 oz. per 1,000 ft. of row for any row spacing (minimum 20-inch row spacing) or no more than 26 lbs. per acre. **Banded** 8–16 oz. per 1,000 ft. of row for any row spacing (minimum 20-inch row spacing).	**Knifed-in** Drill granules 1–4 inches directly below the seed OR 1–4 inches below the seed and up to 5 inches to the side of the seed. **Banded**[1] Place granules in a 5–7 inch band directly behind the planter shoe in front of the press wheel.	Do not place granules in direct contact with seed as crop injury may occur. [1]Do not use banded applications in New Mexico, West Texas and the Panhandle of Oklahoma. **Only one application per year may be used.**
At planting	Corn rootworms wireworms white grubs nematode	**Banded** 8–16 oz. per 1,000 ft. of row for any row spacing (minimum 20-inch row spacing).	**Banded**[1] Place granules in a 5–7 inch band directly behind the planter shoe in front of the press wheel.	**Only one application per year may be used.**

SUPPLEMENTAL LABELING

COUNTER® INSECTICIDE—NEMATICIDE

EPA Reg. No. 241–238
EPA SLN No. NE 830012

FOR DISTRIBUTION AND USE ONLY WITHIN THE STATE OF NEBRASKA

DIRECTIONS FOR USE IN FIELD CORN ONLY

It is a violation of Federal Law to use this product in a manner inconsistent with its labeling.

This product is highly toxic to honey bees and appropriate caution should be taken.

THIS LABEL MUST BE IN THE POSSESSION OF THE USER AT THE TIME OF PESTICIDE APPLICATION.

BEFORE USING, READ PRECAUTIONARY STATEMENT ON THE EPA REGISTERED LABEL.

CROP	PESTS CONTROLLED	RATES OF COUNTER® 15-G	APPLICATION	REMARKS
FIELD CORN Post-Emergence	European Corn Borer (1st or 2nd generation) Banks grass mite	6.7 lbs. per acre	Broadcast granules over the top of the plants by ground or air equipment prior to denting.	DO NOT enter field until 7 days post-treatment. DO NOT make more than one application after plant emergence. DO NOT graze or cut for forage within 30 days of treatment. Consult your state experiment station, state extension service, or pest management consultant for proper timing of application.

SUPPLEMENTAL LABELING

COUNTER® SYSTEMIC INSECTICIDE-NEMATICIDE

EPA Reg. No. 241-238 EPA SLN No. ID 85-0006

FOR DISTRIBUTION AND USE ONLY WITHIN THE STATE OF IDAHO

DIRECTIONS FOR USE IN SUGAR BEETS ONLY

It is a violation of Federal Law to use this product in a manner inconsistent with its labeling.

BEFORE USING, READ PRECAUTIONARY STATEMENTS ON THE EPA REGISTERED LABEL.

Crop	Pests Controlled	Rate of COUNTER	Application	Remarks
Sugar Beets At planting	Suppression of Sugar Beet cyst nematode	18 oz. per 1,000 ft. of row for any row spacing (minimum 20-inch row spacing) or no more than 29 lbs. per acre.	Knifed-in Drill granules 2 inches to the side of the seed and 2–4 inches below the seed.	Do not place granules in direct contact with the seed as crop injury may occur. Only one application per year may be used.

THIS LABEL SHOULD BE IN THE POSSESSION OF THE USER AT THE TIME OF PESTICIDE APPLICATION.

SUPPLEMENTAL LABELING

COUNTER® SYSTEMIC INSECTICIDE-NEMATICIDE

EPA Reg. No. 241-238 EPA SLN No. MT 85-0003

FOR DISTRIBUTION AND USE ONLY WITHIN THE STATE OF MONTANA

DIRECTIONS FOR USE IN SUGAR BEETS ONLY

It is a violation of Federal Law to use this product in a manner inconsistent with its labeling.

BEFORE USING, READ PRECAUTIONARY STATEMENTS ON THE EPA REGISTERED LABEL.

Crop	Pests Controlled	Rate of COUNTER	Application	Remarks
Sugar Beets At planting	Suppression of Sugar Beet cyst nematode	18 oz. per 1,000 ft. of row for any row spacing (minimum 20-inch row spacing) or no more than 29 lbs. per acre.	Knifed-in Drill granules 2 inches to the side of the seed and 2–4 inches below the seed.	Do not place granules in direct contact with the seed as crop injury may occur. Only one application per year may be used.

THIS LABEL SHOULD BE IN THE POSSESSION OF THE USER AT THE TIME OF PESTICIDE APPLICATION.

SUPPLEMENTAL LABELING

COUNTER® SYSTEMIC INSECTICIDE-NEMATICIDE

EPA Reg. No. 241-238 EPA SLN No. NE 85-0002

FOR DISTRIBUTION AND USE ONLY WITHIN THE STATE OF NEBRASKA

DIRECTIONS FOR USE IN SUGAR BEETS ONLY

It is a violation of Federal Law to use this product in a manner inconsistent with its labeling.

BEFORE USING, READ PRECAUTIONARY STATEMENTS ON THE EPA REGISTERED LABEL.

Crop	Pests Controlled	Rate of COUNTER	Application	Remarks
Sugar Beets At planting	Suppression of Sugar Beet cyst nematode	18 oz. per 1,000 ft. of row for any row spacing (minimum 20-inch row spacing) or no more than 29 lbs. per acre.	Knifed-in Drill granules 2 inches to the side of the seed and 2–4 inches below the seed.	Do not place granules in direct contact with the seed as crop injury may occur. Only one application per year may be used.

THIS LABEL SHOULD BE IN THE POSSESSION OF THE USER AT THE TIME OF PESTICIDE APPLICATION.

SUPPLEMENTAL LABELING

COUNTER® SYSTEMIC INSECTICIDE-NEMATICIDE

EPA Reg. No. 241-238 EPA SLN No. WY 85-0001

FOR DISTRIBUTION AND USE ONLY WITHIN THE STATE OF WYOMING

DIRECTIONS FOR USE IN SUGAR BEETS ONLY

It is a violation of Federal Law to use this product in a manner inconsistent with its labeling.

BEFORE USING, READ PRECAUTIONARY STATEMENTS ON THE EPA REGISTERED LABEL.

Crop	Pests Controlled	Rate of COUNTER	Application	Remarks
Sugar Beets At planting	Suppression of Sugar Beet cyst nematode	18 oz. per 1,000 ft. of row for any row spacing (minimum 20-inch row spacing) or no more than 29 lbs. per acre.	Knifed-in Drill granules 2 inches to the side of the seed and 2–4 inches below the seed.	Do not place granules in direct contact with the seed as crop injury may occur. Only one application per year may be used.

THIS LABEL SHOULD BE IN THE POSSESSION OF THE USER AT THE TIME OF PESTICIDE APPLICATION.

SUPPLEMENTAL LABELING

COUNTER® SYSTEMIC INSECTICIDE-NEMATICIDE

EPA Reg. No. 241-238 EPA SLN No. CO 85-0003

FOR DISTRIBUTION AND USE ONLY WITHIN THE STATE OF COLORADO

DIRECTIONS FOR USE IN SUGAR BEETS ONLY

It is a violation of Federal Law to use this product in a manner inconsistent with its labeling.

BEFORE USING, READ PRECAUTIONARY STATEMENTS ON THE EPA REGISTERED LABEL.

Crop	Pests Controlled	Rate of COUNTER	Application	Remarks
Sugar Beets At planting	Suppression of Sugar Beet cyst nematode	18 oz. per 1,000 ft. of row for any row spacing (minimum 20-inch row spacing) or no more than 29 lbs. per acre.	Knifed-in Drill granules 2 inches to the side of the seed and 2–4 inches below the seed.	Do not place granules in direct contact with the seed as crop injury may occur. Only one application per year may be used.

THIS LABEL SHOULD BE IN THE POSSESSION OF THE USER AT THE TIME OF PESTICIDE APPLICATION.

CYGON® 400

SYSTEMIC INSECTICIDE—MITICIDE

Do Not Store Below 45° F

ACTIVE INGREDIENT:

Dimethoate (0,0-dimethyl S-[N-(methylcarbamoyl)methyl] phosphorodithioate) 43.5%

INERT INGREDIENTS: 56.5%

TOTAL 100.0%

1 Gallon contains 4.0 pounds of Dimethoate

EPA Reg. No. 241-233-AA

WARNING

Keep Out of Reach of Children

See Side Panel For Antidote And Other Warnings

AVISO!

AL USUARIO: Si usted no lee inglés, no use este producto hasta

que la etiqueta le haya sido explicada ampliamente.
(ENGLISH TRANSLATION)
WARNING!
(TO THE USER: If you cannot read English, do not use this product until the label has been fully explained to you).

HARMFUL OR FATAL IF SWALLOWED
VAPOR HARMFUL
CONCENTRATED MATERIAL
CAUSES EYE IRRITATION
COMBUSTIBLE LIQUID AND VAPOR

HAZARDS TO HUMANS AND DOMESTIC ANIMALS

WARNING!

Avoid breathing vapor or spray mist.

Use only with adequate ventilation.

Keep container closed.

Avoid contact with skin and eyes.

Wash thoroughly after handling.

Do not contaminate food or feed products.

Do not use, pour, spill or store near heat or open flames.

Required Clothing and Equipment for Application:

All applicators, including homeowners, flaggers and all personnel involved with the mixing, loading, and transferring operations must wear the protective clothing and equipment enumerated below. Pilots are exempt from this requirement. The protective clothing nd equipment to be worn is as follows:

a. Impermeable gloves (for example, rubber or plastic covered reinforced gloves).
b. Boots or boot covers.
c. Long-sleeved shirt and long pants.
d. Wide-brimmed hat.
e. Respirators must be worn by flaggers and mixer/loaders.

FIRST AID: In case of contact with eyes, flush eyes with plenty of water. Get medical attention.

NOTE TO PHYSICIANS: This product may cause cholinesterase inhibition. Atropine is antidotal. Pralidoxime chloride (2-PAM; PROTOPAM chloride) may be effective as an adjunct to atropine. Use according to label directions.

DO NOT STORE BELOW 45°F

DISCLAIMER

The label instructions for the use of this product reflect the opinion of experts based on field use and tests. The directions are believed to be reliable and should be followed carefully. However, it is impossible to eliminate all risks inherently associated with use of this product. Crop injury, ineffectiveness or other unintended consequences may result because of such factors as weather conditions, presence of other materials, or the use or application of the product contrary to label instructions, all of which are beyond the control of American Cyanamid Company. All such risks shall be assumed by the user.

American Cyanamid Company warrants only that the material contained herein conforms to the chemical description on the label and is reasonably fit for the use therein described when used in accordance with the directions for use, subject to the risks referred to above.

Any damages arising from a breach of this warranty shall be limited to direct damages and shall not include consequential commercial damages such as loss of profits or values or any other special or indirect damages.

American Cyanamid Company makes no other express or implied warranty, including any other express or implied warranty of FITNESS or of MERCHANTABILITY.

STORAGE AND DISPOSAL

PROHIBITIONS
Do not contaminate water, food or feed by storage or disposal. Open dumping is prohibited.

PESTICIDE DISPOSAL
Pesticide, spray mixture or rinsate that cannot be used according to label instructions must be disposed of according to Federal, State or local procedures under the Resource Conservation and Recovery Act.

CONTAINER DISPOSAL
Drain drum completely. Add 5 gallons of water, 1 cup of detergent and 2 pounds of lye. Tighten bungs. Rotate drum to wet all surfaces and let stand for at least 15 minutes. Drain completely and rinse three times with water. Tighten bungs. Transport to a professional drum reconditioner having burning equipment for reconditioning, or dispose of in a sanitary landfill or by other approved State and local procedures.

GENERAL
Consult Federal, State or local disposal authorities for approved alternative procedures.

Toxic to birds and other wildlife. Keep out of lakes, streams or ponds. Do not apply where runoff is likely to occur. Do not apply when weather conditions favor drift from areas treated. Do not contaminate water by cleaning of equipment or disposal of wastes. Apply this product only as specified on this label.

This product is highly toxic to bees exposed to direct treatment or residues on blooming crops or weeds. Do not apply this product or allow it to drift to blooming crops or weeds if bees are visiting the treatment area. Protective information may be obtained from your Agricultural Extension Service.

DIRECTIONS FOR USE

AERIAL APPLICATION:
APPLY AT LEAST ONE GALLON OF FINISHED SPRAY PER ACRE. APPLY AT LEAST 5 GALLONS FINISHED SPRAY PER ACRE IN CALIFORNIA.
AUTOMATIC FLAGGING DEVICES SHOULD BE USED WHENEVER FEASIBLE.
IF HUMAN FLAGGERS ARE EMPLOYED THEY MUST WEAR THE PROTECTIVE CLOTHING AND RESPIRATOR SPECIFIED ON THIS LABEL.
IT IS A VIOLATION OF FEDERAL LAW TO USE THIS PRODUCT IN A MANNER INCONSISTENT WITH ITS LABELING.

BEFORE USING, READ WARNING STATEMENTS ON CONTAINER LABEL.
This product is intended for use in conventional hydraulic sprayers, ground applicators or airplane sprayers. When applying by ground equipment, use the recommended amount in sufficient water for thorough coverage; by aircraft, in 2 to 10 gallons of water unless otherwise specified. Do not apply when weather conditions favor drift of spray from areas treated. Repeat applications as necessary unless otherwise specified. Consult your state experiment station or state extension service for proper timing of applications.

See Table Following

CROP	PESTS CONTROLLED	RATE	INTERVAL (DAYS) BETWEEN LAST APPLICATION AND HARVEST
FRUIT: Apples	Apple maggot Codling moth*	1 pt./100 gals. water	28 Do not apply during the bloom period. Apply at petal-fall and every 10 to 14 days thereafter until control is achieved. Under heavy infestations, some sting injury may occur. *Midwest and eastern states only. Do not graze livestock in treated orchards.
Apples Pears	Aphids, leafhoppers, mites (except rust mites)	½ to 1 pt/100 gals water	28 Do not apply during the bloom period. Do not graze livestock in treated orchards.
Grapefruit Lemons Oranges Tangerines	Aphids	Ground Equipment: ½ to 1 pt./100 gals. water. Apply as an outside coverage spray. Aircraft Equipment: 1 to 2 qts./acre in 5 to 10 gals. water.	15
	Mites (except rust mites)	in Ground Equipment: ½ to 1 pt./100 gals water. Apply at a thorough distribution coverage spray.	15
Grapefuit Lemons Oranges Tangerines	Scales (Except black or snow)	Ground Equipment: 1 to 1½ pts./100 gals. water. Apply as a thorough coverage spray.	45
	Thrips	Ground Equipment: ½ to 1 pt./100 gals. water. Apply as a mist spray. Aircraft Equipment: 1 to 2 qts./acre in 5 to 10 gals. water.	15
	Whiteflies	Ground Equipment: 1 pt./100 gals. water. Apply as a thorough distribution coverage spray.	15
	Do not apply during bloom period. Do not use on citrus seedlings. Make no more than 2 applications to mature fruit. Do not enter treated groves within 4 days of last application. Do not graze livestock in treated orchards.		
NUTS: Pecans	Aphids, mites, leafhoppers	⅔ pt./acre	21 Do not graze livestock in treated groves.
VEGETABLE CROPS: Beans (Green, lima, snap, dry)	Aphids, grasshoppers, leafhoppers, leaf miners, lygus bugs, mites, bean leaf beetle, Mexican bean beetle	½ to 1 pt./acre	Beans may be harvested on day of application. Do not feed treated vines.
Broccoli Cauliflower	Aphids	½ to 1 pt./acre	7
Cabbage	Aphids	½ to 1 pt./acre	3
Head Lettuce	Aphids, leafhoppers, leaf miners	½ pt./acre	7
Celery (Florida)	Leaf miners	Ground Equipment: 1 pt./acre	7
Leaf Lettuce Spinach Collards Kale Turnip (greens and roots) Mustard Greens Swiss Chard Endive (Escarole)	Aphids, leafhoppers, leaf miners	½ pt./acre	14
Melons (except watermelons)	Aphids, leafhoppers, leaf miners, thrips	1 pt./acre	3
Watermelons	Aphids, leaf miners, leafhoppers	½ to 1 pt./acre	3
Peas	Aphids	⅓ pt./acre	Peas may be harvested on day of application. Do not feed or graze hay within 21 days after last application when a stationary viner is used. Do not feed or graze when a mobile viner is used.
Peppers	Aphids, leaf miners, maggots.	½ to ⅔ pt./acre	Peppers may be harvested on day of application.
Potatoes	Aphids, grasshoppers, leaf miners, leafhoppers	½ to 1 pt./acre	Potatoes may be harvested on day of application.
Tomatoes	Aphids, leaf miners, leafhoppers	½ to 1 pt./acre	7
FIELD CROPS: Alfalfa	Aphids, grasshoppers, leafhoppers, plant bugs including Lygus, reduction of alfalfa weevil larvae	½ to 1 pt./acre	Do not apply to alfalfa in the bloom period. Do not apply within 10 days of harvest or pasturing. Make only one application per cutting. Effective only on cutting to which applied.

CROP	PESTS CONTROLLED	RATE	INTERVAL (DAYS) BETWEEN LAST APPLICATION AND HARVEST
Field Corn	Banks grass mites (excluding Trans-Pecos area of Texas), aphids, beer beetle, corn rootworm adult, *two-spotted spider mite	⅔ to 1 pt./acre Aerial Application: Spray over the foliage when mites appear. Apply above rates in 1 or more gals. of water per acre.	14 Apply as necessary. Make no more than three applications per year. Do not feed or graze within 14 days of last application. Do not apply to corn when bees are in fields.
	Grasshoppers	1 pt./acre Ground Application: Apply above rate in 20 to 40 gals. of water per acre. Aerial Application: Apply above rate in 1 or more gals. of water per acre.	
Cotton (grown in California and Arizona)	Leafhoppers, fleahoppers, plant bugs including Lygus	½ to 1 pt./acre	14 Repeat applications should not be made at intervals closer than 14 days. Make only 2 applications per season at the higher rate. Do not feed treated forage or graze livestock on treated fields.
Cotton	Aphids, mites, thrips, fleahoppers	¼ to ½ pt./acre	14 Repeat applications should not be made at intervals closer than 14 days. Do not feed treated forage or graze livestock on treated fields.
	Plant bugs including Lygus	½ pt./acre	
Safflower (grown in California and Arizona)	Aphids, leafhoppers, thrips, plant bugs including Lygus	½ to 1 pt./acre	14 Repeat applications should not be made at intervals closer than 14 days. Make only 2 applications per season at the higher rate.
Sorghum (milo)	Aphids (Greenbugs)	½ to 1 pt./acre Ground Application: Apply above rates in 25 to 40 gals. of water per acre. Aerial Application: Apply above rates in 1 or more gals. of water per acre.	Do not feed or graze within 28 days of last application. Make no more than 3 applications as needed per season. Do not apply after heading.
	Banks grass mites (excluding Trans-Pecos area of Texas), spider mites	1 pt./acre Ground Application: Apply above rate in 25 to 40 gals. of water per acre. Aerial Application: Apply above rate in 1 or more gals. of water per acre.	
	Grasshoppers	1 pt./acre Ground Application: Apply above rate in 25 to 40 gals. of water per acre. Aerial Application: Apply above rate in 1 or more gals. of water per acre.	
	Sorghum midge	¼ to ½ pt./acre Aerial Application: Apply above rates in 1 or more gals. of water per acre.	
Soybeans	Mexican bean bettle, spider mites, bean leaf beetle, leafhoppers, three-cornered alfalfa hopper*	1 pt./acre Aerial Application: Apply recommended rate in a minimum of 2 gals. of water per acre.	21 Do not feed or graze within 5 days of last application.
	Grasshoppers	1 pt./acre Ground Application: Apply above rate in 25 to 40 gals. of water per acre. Aerial Application: Apply above rate in 1 or more gals. of ater per acre.	
Tobacco	Aphids	½ to ⅔ pt./acre	21
Wheat	Aphids (greenbugs)	½ to ¾ pt./acre	Do not apply within 14 days of grazing immature plant. Do not harvest grain within 60 days of last application. Do not make more than 2 applications per season.
	Brown wheat mite	⅓ to ½ pt./acre	
	Grasshoppers	¾ pt./acre	
SEED CROPS: Alfalfa	Aphids, leafhoppers, lygus bugs, grasshoppers, reduction of alfalfa weevil larvae	½ to 1 pt./acre	Do not apply to alfalfa in the bloom period. Do not feed or graze livestock in treated crops, hay, threshings or stubble within 10 days of application.

* Not registered in California.

ATTENTION:
DO NOT USE ON SEED ONIONS, SEED CARROTS, OR SEED BERMUDA GRASS.

FOREST INSECTS

Do not apply to Pine Trees used for Pine Nut or Pinyon Nut Production

TREE	PESTS CONTROLLED	RATE	DIRECTIONS
Pinyon pine	Pinyon needle scale	2 gals. in 100 gals. water	Apply spray to egg masses at the base of the trees and to all rough bark and crotches that can be reached from the ground. Make this bark application when crawlers start to emerge from the eggs. Use hydraulic or back-pack sprayer. Do not spray leaves or needles since phytotoxicity may result.
Douglas Fir	Fir cone midge	½ gal. in 100 gals. water	Make thorough coverage application when cones are closed and pendant. Use hydraulic or back-pack sprayer.
Pines	Loblolly pine sawfly	¼ to ½ pt./acre in 2 gals. water	Apply by aircraft when air is calm and the temperature is below 68°F. Apply when most larvae are in the second and third instars.

CITRUS TREES—NONBEARING AND NURSERY STOCK

Consult your state agricultural experimental station or state agricultural extension service for proper timing of applications.

PLANT	PESTS CONTROLLED	AMOUNT
CITRUS: (California, Arizona)	Aphids Thrips	Foliar Spray: 1 pint CYGON 400 per 100 gallons of water. Repeat applications as necessary. May be applied in the year grapefruit, lemon, orange and tangerine trees begin to bear fruit. Do not enter treated groves within 4 days of last application.
		Soil Drench (Trees 1 to 3 years old): 2 quarts CYGON 400 per acre applied in the furrow or basin around the base of tree. Apply when insect injury to new growth appears. Do not apply to trees that will bear fruit within one year.

The use of the within-contained material for systemically protecting living plants is covered by Patent No. 3,210,242. The selling price includes a royalty for the rights of any purchaser of said material to practice the methods claimed in said patent. A nontransferable license under said patent may be obtained from American Cyanamid Company on current royalty terms whatever the source of material used.

In case of an emergency endangering life or property involving this product, call collect, day or night, Area Code 201-835-3100.

ATTENTION
Do not use air pressure or apply heat with open flame to remove contents of this drum. After emptied, drum may retain solid, liquid, and/or vapor residues. Continue to observe all precautions, if any, on label. Do not cut, puncture, torch, or weld on or near the emptied drum. Do not use for other purposes.

SUPPLEMENTAL LABELING

CYGON 400 SYSTEMIC INSECTICIDE
EPA Reg. No. 241-233
EPA SLN No. SD-790008

FOR DISTRIBUTION AND USE ONLY WITHIN SOUTH DAKOTA

OBSERVE ALL PRECAUTIONARY STATEMENTS ON CONTAINER LABEL BEFORE USING

It is a violation of Federal law to use this product in a manner inconsistent with its labeling.

DIRECTIONS FOR CONTROL OF GRASSHOPPERS
Unless otherwise specified, when applying by ground equipment, use recommended rate in sufficient water for good coverage; by aircraft equipment, 2 to 10 gallons of water.

Site	Rate	Interval (Days) Between Last Application and Harvest
Alfalfa (grown for seed)	½ to 1 pt./acre	Do not apply to alfalfa in the bloom period. Do not apply within 10 days of harvest or pasturing. Make only one application per cutting. Effective only on cutting to which applied.
Alfalfa (grown for seed)	½ to 1 pt./acre	Do not apply to alfalfa in the bloom period. Do not feed or graze livestock in treated crops, hay, threshings or stubble within 10 days of application.
Dry Beans	½ to 1 pt./acre	Beans may be harvested on day of application. Do not feed treated vines.
Field Corn	*Aerial Application:* Apply 1 pt in 1 or more gallons of water per acre	Make no more than three applications per year. Do not feed or graze within 14 days of last application.
Sorghum (milo)	1 pt./acre *Ground Application:* Apply above rate in 25 to 40 gallons of water per acre. *Aerial Application:* Apply above rate in 1 or more gallons of water per acre.	Do not feed or graze within 28 days of last application. Make 3 applications as needed per season.
Soybeans	1 pt./acre *Aerial Application:* Apply recommended rate in a minimum of 3¾ gallons of water per acre.	Do not feed or graze within 5 days of last application. Do not harvest within 21 days of last application.
Wheat	¾ pt./acre	Do not apply within 14 days of grazing immature plant. Do not harvest grain within 60 days of last application.
Roadsides	1 pt./acre	Apply as necessary. Keep livestock off treated areas.

[1] This use is registered by EPA throughout the U.S.

SUPPLEMENTAL LABELING

CYGON 400 SYSTEMIC INSECTICIDE
EPA Reg. No. 241-233

THIS LABEL MUST BE IN POSSESSION OF THE USER AT TIME OF APPLICATION.

OBSERVE ALL PRECAUTIONARY STATEMENTS ON CONTAINER LABEL BEFORE USING

It is a violation of Federal law to use this product in a manner inconsistent with its labeling.

DIRECTIONS FOR CONTROL OF GRASSHOPPERS

Unless otherwise specified—When applying by ground equipment, use recommended rate in sufficient water for good coverage; by aircraft equipment, 2 to 10 gallons of water.

Site	Rate	Interval (Days) Between Last Application and Harvest
Beans (green, lima, snap, dry)	½ to 1 pt./acre	Beans may be harvested on day of application. Do not feed treated vines.
Field Corn	*Aerial Application:* Apply 1 pt. in 1 or more gallons of water per acre.	Make no more than 3 applications per year. Do not feed or graze within 14 days of last application.
Potatoes	½ to 1 pt./acre	Potatoes may be harvested on day of application.
Sorghum (milo)	1 pt./acre *Ground Application:* Apply above rate in 25 to 40 gallons of water per acre. *Aerial Application:* Apply above rate in 1 or more gallons of water per acre.	Do not feed or graze within 28 days of last application. Make 3 applications as needed per season.
Soybeans	1 pt./acre *Aerial Application:* Apply recommended rate in a minimum of 3¾ gallons of water per acre.	Do not feed or graze within 5 days of last application. Do not harvest within 21 days of last application.
Wheat	¾ pt./acre	Do not apply within 14 days of grazing immature plant. Do not harvest grain within 60 days of last application.
Alfalfa (grown for hay)	½ to 1 pt./acre	Do not apply to alfalfa in the bloom period. Do not apply within 10 days of harvest or pasturing. Make only one application per cutting. Effective only on cutting to which applied.
Alfalfa (grown for seed)	½ to 1 pt./acre	Do not apply to alfalfa in the bloom period. Do not feed or graze livestock in treated crops, hay, threshings or stubble within 10 days of application.

SUPPLEMENTAL LABELING

CYGON 400 SYSTEMIC INSECTICIDE

EPA Reg. No 241-233

EPA SLN No. CO 790025

FOR DISTRIBUTION AND USE ONLY WITHIN COLORADO

OBSERVE ALL PRECAUTIONARY STATEMENTS ON CONTAINER LABEL BEFORE USING

It is a violation of Federal law to use this product in a manner inconsistent with its labeling.

DIRECTIONS FOR CONTROL OF GRASSHOPPERS

Unless otherwise specified, when applying by ground equipment, use recommended rate in sufficient water for good coverage; by aircraft equipment, 2 to 10 gallons of water.

Site	Rate	Interval (Days) Between Last Application and Harvest
Alfalfa (grown for hay)	½ to 1 pt./acre	Do not apply to alfalfa in the bloom period. Do not apply within 10 days of harvest or pasturing. Make only one application per cutting. Effective only on cutting to which applied.
Alfalfa (grown for seed)[1]	½ to 1 pt./acre	Do not apply to alfalfa in the bloom period. Do not feed or graze livestock in treated crops, hay, threshings or stubble within 10 days of application.
Dry Beans[1]	½ to 1 pt./acre	Beans may be harvested on day of application. Do not feed treated vines.
Field Corn[1]	*Aerial Application:* Apply 1 pt. in 1 or more gallons of water per acre.	Make no more than three applications per year. Do not feed or graze within 14 days of last application.
Sorghum (milo)[1]	1 pt./acre *Ground Application:* Apply above rate in 25 to 40 gallons of water per acre. *Aerial Application:* Apply above rate in 1 or more gallons of water per acre.	Do not feed or graze within 28 days of 1st application. Make 3 applications as needed per season.
Soybeans[1]	1 pt./acre *Aerial Application:* Apply recommended rate in a minimum of 3¾ gallons of water per acre.	Do not feed or graze within 5 days of last application. Do not harvest within 21 days of last application.
Wheat[1]	¾ pt./acre	Do not apply within 14 days of grazing immature plant. Do not harvest grain within 60 days of last application.
Roadsides	1 pt./acre	Apply as necessary. Keep livestock off treated areas.

[1] This use is registered by EPA throughout the U.S.

SUPPLEMENTAL LABELING

CYGON 400 SYSTEMIC INSECTICIDE

EPA Reg. No. 241-233

EPA SLN KS 830012

FOR DISTRIBUTION AND USE ONLY WITHIN KANSAS

OBSERVE ALL PRECAUTIONARY STATEMENTS ON CONTAINER LABEL BEFORE USING

It is a violation of Federal law to use this product in a manner inconsistent with its labeling.

DIRECTIONS FOR CONTROL OF GRASSHOPPERS

Unless otherwise specified—When applying by ground equipment, use recommended rate in sufficient water for good coverage. Use a minimum of 10 gallons of water per acre and increase volume when vegetation is dense. When applying by aircraft equipment, use 2 to 3 gallons of water per acre. Use maximum rate of CYGON 400 for control of large nymphs and adults.

Site	Rate	Interval (Days) Between Last Application and Harvest
Alfalfa (grown for hay)[1]	¾ to 1 pt./acre	Do not apply to alfalfa in the bloom period. Do not apply within 10 days of harvest or pasturing. Make only one application per cutting. Effective only on cutting to which applied.

Site	Rate	Interval (Days) Between Last Application and Harvest
Alfalfa (grown for seed)[1]	3/4 to 1 pt./acre	Do not apply to alfalfa in the bloom period. Do not feed or graze livestock in treated crops, hay, threshings or stubble within 10 days of application.
Field Corn[1]	1 pt./acre	Make no more than three applications per year. Do not feed, graze or *harvest* within 14 days of last application.
Sorghum (milo)[1]	1 pt./acre	Do not feed or graze within 28 days of last application. Make 3 applications as needed per season.
Soybeans[1]	1 pt./acre	Do not feed or graze within 5 days of last application. Do not harvest within 21 days of last application.
Wheat[1]	3/4 pt./acre	Do not apply within 14 days of grazing immature plant. Do not harvest grain within 60 days of last application.
Roadsides	1 pt./acre	Apply as necessary. Keep livestock off treated areas.

[1] This use is registered by EPA throughout the U.S.

SUPPLEMENTAL LABELING

CYGON® 400 SYSTEMIC INSECTICIDE-MITICIDE

EPA Reg. No. 241-233
EPA SLN No. OR-840050

FOR DISTRIBUTION AND USE ONLY WITHIN OREGON

OBSERVE ALL PRECAUTIONARY STATEMENTS ON CONTAINER LABEL BEFORE USING

DIRECTIONS FOR USE

It is a violation of Federal law to use this product in a manner inconsistent with its labeling.

Crops	Pest Controlled	Rate	Remarks
Cherries	Western Cherry fruit fly	1 pt./100 gals. water not to exceed 400 gals. per acre.	Make a single application within 7 days of adult fruit fly emergence. Do not feed or graze livestock on cover crops in treated orchards. Do not apply within 21 days of harvest.

This label must be in the possession of the user at time of pesticide application.

24(c) Registrant:

SUPPLEMENTAL LABELING

CYGON® 400 SYSTEMIC INSECTICIDE-MITICIDE

EPA Reg. No. 241-233
EPA SLN No. WA-840068

FOR DISTRIBUTION AND USE ONLY WITHIN WASHINGTON

OBSERVE ALL PRECAUTIONARY STATEMENTS ON CONTAINER LABEL BEFORE USING

DIRECTIONS FOR USE

It is a violation of Federal law to use this product in a manner inconsistent with its labeling.

Crops	Pest Controlled	Rate	Remarks
Cherries	Western Cherry fruit fly	1 pt./100 gals. water not to exceed 400 gals. per acre.	For ground application only. Make a single application within 7 days of adult fruit fly emergence.
		4 pts. in a minimum of 50 gals. of water per acre.	For concentrate equipment only. **Do not** apply within 28 days of harvest.

* **Do not** feed or graze livestock on cover crops in treated orchards.

This label must be in the possession of the user at time of pesticide application.

24(c) Registrant:

CYPREX® 65–W

FRUIT FUNGICIDE

For control of certain fungus diseases of apples, cherries, peaches, pears, and pecans.

ACTIVE INGREDIENT:
Dodine (n-dodecylguanidine acetate) 65.0%
INERT INGREDIENTS 35.0%
TOTAL 100.0%

EPA Reg. No. 241-51-AA

DANGER

Keep Out of Reach of Children

See Side Panel For Additional Precautions

PELIGRO!

AL USUARIO: Si usted no lee inglés, no use este producto hasta que la etiqueta le haya sido explicada ampliamente.

(TO THE USER: If you cannot read English, do not use this product until the label has been fully explained to you.)

DANGER!

CAUSES EYE BURNS • HARMFUL IF SWALLOWED

Do not get in eyes or on skin • Wear gloves and goggles when handling • Avoid breathing dust • Keep container closed • Use with adequate ventilation • Wash thoroughly after handling • Do not contaminate food or feed products.

FIRST AID

In case of contact, immediately flush eyes with plenty of water for at least 15 minutes. Call a physician.

If swallowed, induce vomiting by sticking finger down throat or by giving strong soapy or salty water to drink. Repeat until vomit is clear.

Never give anything by mouth to an unconscious person.

This product is toxic to fish. Keep out of lakes, streams or ponds. Do not apply when weather conditions favor drift from areas treated. Do not contaminate water by cleaning equipment or disposal of wastes.

Observe all cautions and limitations on labeling of products used in mixtures. Apply this product only as specified on this label.

Destroy empty bags and cartons by burning. Stay away from smoke and fumes.

DILUTE AND CONCENTRATE SPRAYS

CYPREX 65–W may be used effectively in dilute or concentrate sprays when applied properly.

Fruit and foliage should not be excessively drenched with dilute sprays. Avoid heavy deposits when using concentrate sprays.

DISCLAIMER

The label instructions for the use of this product reflect the opinion of experts based on field use and tests. The directions are believed to be reliable and should be followed carefully. However, it is impossible to eliminate all risks inherently associated with use of this product. Crop injury, ineffectiveness or other unintended consequences may result because of such factors as weather conditions, presence of other materials, or the manner of use or application, all of which are beyond the control of American Cyanamid Company. All such risks shall be assumed by the user.

- American Cyanamid Company warrants only that the material contained herein conforms to the chemical description on the label and is reasonably fit for the use therein described when used in accordance with the directions for use, subject to the risks referred to above.
- Any damages arising from a breach of this warranty shall be limited to direct damages and shall not include consequential commercial damages such as loss of profits or values or any other special or indirect damages.
- American Cyanamid Company makes no other express or implied warranty, including any other express or implied warranty of FITNESS or of MERCHANTABILITY.

In case of an emergency endangering life or property involving this product, call collect, day or night, Area Code 201-835-3100.

INSTRUCTIONS FOR MIXING CYPREX 65-W

CYPREX 65-W, a fine wettable powder, should be added directly to the spray tank when half filled with water. Keep agitator running during filling and spraying. Do not wash wettable powder through filler screen into spray tank.

COMPATIBILITY

Experience indicates that CYPREX 65-W can be used effectively at recommended dosages with most of the commonly used fungicides, insecticides and acaricides such as:

Cygon* dimethoate	Imidan®
diazinon	Omite®
Karathane® dinocap	sulfurs
ferbam	zineb

—and wettable powder formulations of—

Guthion® azinphosmethyl	malathion
Sevin® carbaryl	Dimecron® phosphamidon
Mitox® chlorbenside	parathion
Kelthane® dicofol	ryania
Thiodan® endosulfan	

While CYPREX 65-W has been used satisfactorily with emulsifiable concentrate formulations and flowable (water-based) emulsions, it is desirable to check the compatibility of this type combination before materials are mixed in the spray tank. CYPREX 65-W is not compatible with Acaraben® chlorobenzilate, Bordeaux mixture, fixed copper, Fundal®, Galecron®, lime or lime sulfur. Observe all cautions and limitations on labeling of all products used in mixtures.

CYPREX 65-W and Oil

Tank-mix combinations of CYPREX 65-W and oils of 60–75 viscosity have been used effectively as a delayed dormant application on apples. Oils of the slow-breaking type are suggested for tank mixing with CYPREX 65-W. Liquid concentrate or flowable (water-based) emulsion formulations of the phosphate insecticides have been used satisfactorily in CYPREX 65-W oil systems. Wettable powder formulations of parathion, Guthion® azinphosmethyl, Trithion® carbophenothion and certain other insecticides have appeared suitable in limited trials. CYPREX 65-W oil mixtures are not compatible with wettable powder malathion, Sevin® carbaryl, zineb, ferbam and sulfurs.

Because uniform dispersibility and sprayability may be influenced by the particular oil or phosphate insecticide formulation used or by the local water supply, it is recommended that compatibility be determined before adding pesticides to the spray tank.

Procedure for tank-mixing CYPREX 65-W, insecticide and oil:

Maintain continuous paddle agitation from the time CYPREX 65-W is added to the spray tank until mixture is sprayed out:

1. CYPREX 65-W should be added to the tank when one-quarter to one-half full with water and allowed to mix thoroughly.
2. If insecticide is used, add as tank is filling.
3. Add self-emulsifying oil last when tank is nearly three-quarters full.

Consult your state agricultural experiment station or state extension service for information on self-emulsifying oils and pesticide formulations for tank mixing with CYPREX 65-W.

DIRECTIONS FOR USE

Do not use this product for any uses other than those specified on this label.

Consult your state experiment station or state extension service for proper timing of sprays. Do not graze cover crops in treated orchards.

Apple: Scab • Protection Schedule (Ground Application) • Use 1/4 to 1/2 lb. (1–2 lbs. per acre) CYPREX 65-W in 100 gallons of spray at 5 to 7-day intervals or as needed to maintain scab control from prebloom through the first cover applications. Use 1/4 lb. (1 lb. per acre) CYPREX 65-W in subsequent cover sprays as needed.

CYPREX 65-W may be used in a delayed dormant oil schedule. Use 1/4 to 1/2 lb. (1 to 2 lbs. per acre) CYPREX 65-W with 2 gallons of oils of 60–75 viscosity (8 gallons per acre) per 100 gallons of spray. For mixing instructions, refer to "CYPREX 65-W and Oil" section on label.

(Air Application) • Use 1 1/2 lbs. CYPREX 65-W per acre in 5 to 7 gallons of spray at 5 to 7-day intervals or as needed to maintain scab control from prebloom through the first cover sprays. Repeat application in subsequent cover sprays as needed.

"After-infection" Applications • After an infection period has occurred, use 3/4 lb. (3 lbs. per acre) CYPREX 65-W per 100 gallons of spray during or following a rain from prebloom through first cover applications and, in later cover sprays, as needed. While timing of the spray may vary considerably depending upon temperatures, it is recommended that application be made within 36 hours from the beginning of the rain or wet period which caused the infection.

Do not apply within 7 days of harvest.

Wettable sulfur can be used in combination with CYPREX 65-W for control of powdery mildew on apples. Consult your state extension service for proper rates and timing of sprays.

CYPREX 65-W TREATED APPLES SHOULD NOT BE USED IN THE MANUFACTURE OF APPLE POMACE FOR USE IN LIVESTOCK FEEDS. SUCH USE MAY RESULT IN ILLEGAL RESIDUES IN MILK OR MEAT.

Fruit color and finish on most apple varieties sprayed with CYPREX 65-W at recommended dosages have frequently been outstanding.

CYPREX 65-W may cause russetting of Golden Delicious and Grimes Golden apple varieties if applied during cool, slow drying weather conditions. Do not use on these varieties under these conditions.

Freezing or near-freezing temperatures, particularly when accompanied by slow-drying conditions may injure apple fruit or predispose the fruit to chemical injury. Do not exceed 1/2 lb. CYPREX 65-W per 100 gallons immediately before, during or immediately after occurrence of such conditions.

Cherry (Sour and Sweet): Leaf Spot • Use 1/4 to 1/2 lb. CYPREX 65-W in 100 gallons of spray at petal fall or when first leaves unfold. Repeat application thereafter at 7 to 10-day intervals. If conditions favor **leaf spot,** continue applications through harvest. **Post-Harvest Sprays:** Apply 1/4 to 1/2 lb. CYPREX 65-W per 100 gallons of spray immediately after harvest and as needed. **Blossom Brown Rot** (Western States Only): Use ½ lb. CYPREX 65-W per 100 gallons of spray at the time petals begin to open. Repeat application at full bloom and again at petal fall.

Wettable sulfur can be used in combination with CYPREX 65-W for control of powdery mildew on cherries. Consult your state extension service for proper rates and timing of sprays.

Applications of CYPREX 65-W to cherries when temperatures and humidity are excessively high may cause spotting to foliage or fruit.

Peach: Bacterial Spot • Under light to moderate disease conditions, use ½ lb. CYPREX 65-W and 1 lb. captan (50%) per 100 gallons of spray at shuck-split and repeat application at 7 to 10-day intervals through the cover sprays. Shorten intervals during wet periods as needed to maintain maximum protection against infection.

Do not use concentrate sprays higher than 3X.

Do not apply within 15 days of harvest.

Captan or the CYPREX-captan mixture is toxic to fish. Do not contaminate any body of water.

When it is necessary to use an insecticide with the combination, the wettable powder formulation of parathion is suggested. Limited experience indicates that wettable powder formulations of Guthion® azinphosmethyl, Kelthane® dicofol, Tedion® tetradifon, and Thiodan® endosulfan are compatible with the mixture. Do not combine CYPREX 65-W and captan with solvent formulations of insecticides or sulfur. Spray injury may result if the combination is applied with, immediately before, or closely after an oil or zinc sulfate-lime spray.

Peach: Leaf Curl (Western States Only) • Use 2 lbs. CYPREX 65-W per 100 gallons of spray just before buds swell in the spring. If **leaf curl** was moderate to severe the previous season, use 1 to 2 lbs. CYPREX 65-W per 100 gallons of spray 1 month before buds swell in the spring and again just prior to bud swelling. **Blossom Brown Rot** (Western States Only) • Use ½ lb. CYPREX 65-W per 100 gallons of spray at the time petals begin to open. Repeat application at full bloom and again at petal fall.

Do not apply after petal fall.

Pear: Scab (Western States Only) • Use ¾ lb. CYPREX 65-W per 100 gallons of spray at 5 to 7-day intervals or as needed to maintain scab control from prebloom through petal fall and again in first cover. Use ½ to ¾ lb. CYPREX 65-W in subsequent sprays as needed.

Do not apply within 7 days of harvest.

Pecan: Scab, liver spot, brown leaf spot, downy leaf spot, leaf blotch, downy mildew (Ground Application) • Use ½ to 1 lb. CYPREX 65-W per 100 gallons of spray. Begin applications when first leaves are showing and repeat at 10 to 14-day intervals through first cover. Repeat applications at 2 to 3-week intervals or as needed to maintain disease control. Do not apply more than 700 gallons per acre.

(Air Application) • Use 2 lbs. CYPREX 65-W per acre in a minimum of 20 gallons of spray. Begin applications when first leaves are showing and repeat at 10 to 14-day intervals through first cover. Repeat application at 2 to 3-week intervals or as needed to maintain disease control. Under severe scab conditions, use 4 lbs. CYPREX 65-W per acre in a minimum of 20 gallons of spray.

Do not apply after shucks have started to open.

Do not graze meat or dairy animals in treated groves.

CYPREX 65-W has been used effectively on most pecan varieties. Severe foliage injury has been experienced on Moore and Van Deman in the Southeast.

Strawberry: Leaf Scorch, Leaf Spot and Leaf Blight • Use 1½–2 lbs. CYPREX 65-W per acre in sufficient water (250–300 gallons) for thorough plant coverage. Make first application as soon as new growth starts and repeat sprays at weekly intervals.

Do not apply within 14 days of harvest.

Black Walnut: Anthracnose (Leaf Blight) • Use 1 lb. CYPREX 65-W per 100 gallons of spray when first leaves unfold. Repeat application at 10 to 14-day intervals. Do not graze meat or dairy animals in treated groves.

A license is granted to any owner hereof to use the within contents in practicing the invention of U.S. Patent No. 2,867,562 claiming methods of controlling fungus organisms which attack fruit trees, a prepaid royalty being included in the selling price of such contents to the orginal purchaser hereof. A nontransferable license under said patent may be obtained from American Cyanamid Company on current royalty terms whatever the source of the material used.

CYTHION®

INSECTICIDE

The Premium Grade Malathion®
EPA Reg. No. 241-208-AA

and

MALATHION ULV* CONCENTRATE

INSECTICIDE
EPA Reg. No. 241-110-AA

ACTIVE INGREDIENT:

Malathion**	91.0%
INERT INGREDIENTS	9.0%
TOTAL	100.0%

**0,0-dimethyl phosphorodithioate of diethyl mercaptosuccinate (1 gallon contains 9.33 pounds of malathion)

* Trademark of American Cyanamid Company

CAUTION

Keep Out of Reach of Children

HARMFUL BY SWALLOWING, INHALATION OR SKIN CONTACT

Avoid Breathing Spray Mist

Avoid Contact With Skin

Wash Thoroughly After Handling

Change Contaminated Clothing

Do Not Contaminate Food or Feed Products

CONTENTS OF THIS LEAFLET APPLY TO BOTH PRODUCTS

IMPORTANT! BEFORE USING! STOP! READ DIRECTIONS IN THIS LEAFLET!

PRECAUCIÓN
AL USUARIO: Si usted no lee inglés, no use este producto hasta que la etiqueta le haya sido explicada ampliamente.
ENGLISH TRANSLATION
CAUTION
(TO THE USER: If you cannot read English, do not use this product until the label has been fully explained to you.)

IN CASE OF AN EMERGENCY ENDANGERING LIFE OR PROPERTY INVOLVING THIS PRODUCT, CALL COLLECT, DAY OR NIGHT, AREA CODE 201-835-3100.

BEFORE USING, READ THE DIRECTIONS CONTAINED IN THIS LEAFLET FOR THE PROPER METHODS AND PROCEDURES WHICH MUST BE FOLLOWED TO ACHIEVE EFFECTIVE INSECT CONTROL AND AVOID PERMANENT DAMAGE TO AUTOMOBILE AND OTHER PAINT FINISHES.

This product is toxic to fish. Avoid direct applications to lakes, streams, ponds, tidal marshes and estuaries. Do not apply where runoff is likely to occur. Do not apply when weather conditions favor drift from areas treated. Do not contaminate water by cleaning of equipment, or disposal of wastes. Shrimp and crab may be killed at application rates recommended on this label. Do not apply where these are important resources. Apply this product only as specified on this label.

DISCLAIMER

The label instructions for the use of this product reflect the opinion of experts based on field use and tests. The directions are believed to be reliable and should be followed carefully. However, it is impossible to eliminate all risks inherently associated with the use of this product. Crop injury, automobile paint damage, ineffectiveness or other unintended consequences may result because of such factors as weather conditions, presence of other materials, or the use or application of the product contrary to label instructions, all of which are beyond the control of American Cyanamid Company. All such risks shall be assumed by the user.

American Cyanamid Company warrants only that the material contained herein conforms to the chemical description on the label and is reasonably fit for the use therein described when used in accordance with the directions for use, subject to the risks referred to above.

Any damages arising from a breach of this warranty shall be limited to direct damages and shall not include consequential commercial damages such as loss of profits or values or any other special or indirect damages.

American Cyanamid Company makes no other express or implied warranty, including any other express or implied warranty of FITNESS or of MERCHANTABILITY.

DIRECTIONS FOR USE

It is a violation of Federal Law to use this product in a manner inconsistent with its labeling.

Note to Formulators:

CYTHION or MALATHION ULV may be used for the preparation of malathion insecticides. Before using for this purpose, manufacturers should consult American Cyanamid Company for manufacturing and handling instructions.

The sale of this product does not include a license under any patent owned by the American Cyanamid Company.

STORAGE AND DISPOSAL

See container label for storage and disposal instructions.

AGRICULTURAL USES

Operating Instructions

CYTHION or MALATHION ULV is used undiluted in specially designed aircraft or ground equipment capable of applying ultra low volumes for control of the insects indicated. Aerial applications are most effective when made at a boom height of 5 feet and a swath width of 50 feet. Apply only when weather conditions are favorable. Wind and rising air currents may cause undesirable spray drift and reduce insect control.

Mist blowers and boom sprayers utilizing a controlled air flow to facilitate particle size and spray deposition may be used at a vehicle speed of 4 to 10 mph.

Mist blowers with a pump capable of producing up to 40 psi and blower speeds of 2600 rpm are satisfactory. Use flat fan nozzles, 8001 to 8002, place 30° into air blast or rotary atomizers into the air blast that produce an efficient spray particle with a mass median diameter of 40 to 100 microns. Swath widths should not exceed 30 feet. Apply only when weather conditions are favorable. Wind and rising air currents may cause undesirable spray drift and reduce insect control.

Boom sprayers with a filtered rotary air compressor, either PTO or gas engine driven or an air pump capable of producing at least 12 psi are satisfactory. Use air pressure on chemical tanks and an accurate metering valve to assure a calibrated flow of the pesticide. Air should be regulated with relief valve and gauge for proper air and liquid mixture. Pneumatic-type spray nozzles, as suggested by equipment manufacturer, should be used for spray particles with mass median diameter of 30 to 100 microns. Apply only when weather conditions are favorable. Wind and rising air currents may cause undesirable spray drift and reduce insect control.

Repeat applications should be made as necessary unless otherwise specified.

CROP	PESTS CONTROLLED	FLUID OUNCES PER ACRE	DAYS TO HARVEST OR GRAZING AND COMMENTS
Alfalfa	Alfalfa Caterpillar	8	0 day. Apply when larvae are small.
	Western Yellow Striped Armyworm	12	5 days. Apply when larvae are large or when foliage is dense.
	Alfalfa Weevil Larvae	16	5 days. Apply when day temperatures are expected to exceed 65° F. and when 50–75% of leaves show feeding damage.
	Beet Armyworm	8	0 day. Apply when larvae are small.
		16	5 days. Apply when larvae are large or when foliage is dense.
	Grasshoppers	8	0 day.

Do not apply to alfalfa in bloom. Do not apply to seed alfalfa.

Beans (lima, green, snap, Navy, red kidney, wax, dry, black-eye)	Mexican Bean Beetle Leafhoppers Green Cloverworm Japanese Beetle Lygus Bug	8	1 day.
Blueberries	Blueberry Maggot	10	0 day.
Cherries	Cherry Fruit Fly	12–16	1 day. Apply by aircraft only. Use higher rate when foliage is heavy or infestation is severe. Make first application as soon as flies appear.
Cereal Crops, (barley, corn, oats, wheat) and grasses	Cereal Leaf Beetle	4–8	Barley, oats, wheat: 7 days. Corn: 5 days. Grasses: 0 day.
Clover, Pasture and Range Grass, Grass, Grass Hay, Nonagricultural Land (wastelands, roadsides)	Blackgrass Bugs Grasshoppers	8–12	0 day. Do not apply to clover in bloom.
Corn	Adult Corn Rootworm	4	5 days.

CROP	PESTS CONTROLLED	FLUID OUNCES PER ACRE	DAYS TO HARVEST OR GRAZING AND COMMENTS	
Cotton	Early Season Insects Thrips Fleahoppers Leafhoppers	4–8	0 day	
	Boll Weevil	8–12	Early to midseason	0 day. May be tank mixed with once-refined cottonseed or vegetable oil sufficient to make a 1 qt. finished spray per acre
		16	Late season	
	Grasshoppers	8	0 day	
	Lygus Bug	8–12	0 day	
		16	Very heavy migrating populations. 0 day	
Flax	Grasshoppers	8	Apply once per season. Do not apply within 45 days of harvest. Do not graze or feed forage.	
Grain Crops (barley, corn, oats, rye, rice, grain sorghum and wheat)	Grasshoppers	8	7 days, except corn. Corn: 5 days	
Grain Sorghum	Sorghum Midge	8–12	7 days. Apply during the bloom stage.	
Peas (Northwest)	Pea Weevil	8	14 days.	
Rice—Grain Form (Louisiana, Texas)	Rice Stink Bug	8	7 days. Apply by aircraft only. Apply during early milk and dough stage of growing rice.	
Safflower	Grasshoppers Lygus Bug	8	3 days of harvesting seeds.	
Soybeans	Mexican Bean Beetle Grasshoppers Japanese Beetle Green Cloverworm	8	7 days.	
Sugar Beets	Grasshoppers Sugar Beet Root Maggot Adults	8	0 day. 7 days. If tops are to be used for food or feed.	
Nonagricultural Lands	Beet Leafhopper (on wild host plants)	8	0 day.	
Beef Cattle-Feed Lots and Holding Pens	Adult Flies and Mosquitoes	6–8	0 day.	

OTHER AGRICULTURAL USES

Alfalfa, Clover, Pasture and Range Grass, Grass and Grass Hay, Grain Crops, Beans, Rice, Tomatoes and Nonagricultural Lands (wasteland): Adult mosquitoes and flies-Apply CYTHION or MALATHION ULV at the rate of 2 to 4 fluid ounces for control of adult mosquitoes and at 6 to 8 fluid ounces per acre for control of adult flies and mosquitoes. Repeat applications as necessary. On alfalfa, clover, pasture and range grass, grass and grass hay, may be applied on day of harvest or grazing. Do not apply to alfalfa and clover in bloom. Do not use on seed alfalfa. On grain crops, make no application within 7 days of harvest or forage use; on corn, within 5 days of harvest or forage; on rice, within 7 days of harvest; on beans and tomatoes, within 1 day of harvest.

FOR CONTROL OF BOLL WEEVIL AND BOLLWORM/BUDWORMS
CYTHION + PYRETHROID TANK MIXES

PYRETHROIDS*	FLUID OUNCES/ACRE	COMMENTS
[1]PAY-OFF® insecticide 2.5 E.C.	1.3–4.0	Tank mix with 12–16 fl. oz. of CYTHION or Malathion ULV and apply as a 1 qt. finished spray per acre in once-refined cottonseed or vegetable oil. Consult the pyrethroid label for detailed directions.
[2]PYDRIN® insecticide 2.4 E.C.	5⅓–10⅔	
[3]POUNCE® insecticide 3.2 E.C.	2.0–8.0	For POUNCE tank mix with 12–16 fl. oz. of CYTHION or Malathion ULV. Add these materials to one quart of once-refined cottonseed or vegetable oil and apply this mixture with proper application equipment.

[1] Registered Trademark, American Cyanamid Company
[2] Registered Trademark, Shell Chemical Company
[3] Registered Trademark, FMC Corporation

IMPORTANT

Undiluted spray droplets of CYTHION or MALATHION ULV will permanently damage automobile paint. If accidental exposure does occur, the vehicle should be washed immediately.

Consult your state experiment station or state extension service for proper timing of sprays.

These products are highly toxic to bees exposed to direct treatment or residues on crops. Protective information may be obtained from your Cooperative Agricultural Extension Service.

All applicable directions, restrictions and precautions on the EPA registered labels are to be followed.

Operator should inspect and calibrate equipment to assure that the proper amounts of pesticide are being applied, and that uniform coverage is obtained.

This label must be in the possession of the user at time of pesticide application.

* Lower rates should be used only with close field supervision and low pest numbers.

FOREST INSECTS

Apply with aircraft equipped for ultra low volume application. Make application when air is calm and temperature is below 68° F. Do not allow spray to contact ferns, hickory and maples as injury may result. Do not spray on elms under extreme heat, drought and disease conditions.

TREE	PESTS CONTROLLED	FLUID OUNCES PER ACRE	COMMENTS
Douglas Fir True Fir Spruce	Spruce Budworm	13	Apply when highest percentage of larvae are in the fifth instar
Hemlock	Hemlock Looper	8	Apply when most larvae are in third and fourth instar
Pines	European Pine Sawfly	10	Apply when larvae are in the first or second instar or before they reach ½ inch in length.
	Saratoga Spittlebug		Apply when 95% of the population has become adult
Larch	Larch Casebearer	8	Apply in spring as soon as larvae break hibernation and begin feeding on new foliage.

Undiluted spray droplets of CYTHION® insecticide, 'The Premium Grade Malathion'® or MALATHION ULV* concentrate insecticide will permanently damage automobile paint unless these specific instructions for ground and aerial application are followed.

MOSQUITO CONTROL IN POPULATED AND RURAL AREAS

AERIAL APPLICATION

ADULT MOSQUITO CONTROL OVER CITIES, TOWNS, AND OTHER AREAS WHERE AUTOMOBILES, TRAILERS, TRUCKS AND PLEASURE BOATS ARE PRESENT: Apply 2.6 to 3.0 fluid ounces of CYTHION or MALATHION ULV per acre. Apply only when weather conditions are favorable. Wind and rising air currents may cause undesirable spray drift and reduce insect control.

IMPORTANT—Undiluted spray droplets of CYTHION or MALATHION ULV will permanently damage vehicle paint finishes unless the aircraft used for the ultra low volume application meets all of the specifications listed below.

FIXED WING AIRCRAFT

1. Aircraft is operated at 150 mph or more.
2. There are no leaks in the ultra low volume spray system.
3. Nozzles are placed on the boom at a 45° angle down and into the wind.
4. Diaphragm check valves are used on all nozzles to insure positive cut-off of the spray.
5. Dosage of CYTHION or MALATHION ULV does not exceed 3 fluid ounces per acre.
6. The spray system produces droplets of this product in the 50 to 60 mass median diameter (MMD) micron range, with no more than 10% of the droplets exceeding 100 microns, as determined by readings made from microscope slides coated with DRI-FILM* or TEFLON®.

* Trademark of General Electric Company

®Registered Trademark of E.I. duPont de Nemours & Co., Inc.

HELICOPTER

Equipment specifications

1. Rotary nozzle equivalent to Beecomist Spray Head Assembly Model No. 35C equipped with:
   a. a direct reading RPM tachometer or low RPM signal light readily visible to operator;
   b. a stainless steel porous metal sleeve, 20 micron pore size, dynamically balanced to the nozzle;
   c. a diaphragm check valve as near to the rotary nozzle as possible to insure positive cut off of the spray;
   d. nozzle on-off switch separate from main switch and pump switch.
2. Minimum no-load nozzle speed of 10,500 RPM.
3. A continuous nonpulsating metered flow must be maintained by a variable speed metering pump equipped with:
   a. a positive cut off valve between tank and pump;
   b. a flow gauge or tachometer visible to operator;
   c. a pump on-off switch separate from main switch and nozzle switch.
4. Maximum flow rate of 0.5 gallon per minute per nozzle.
5. Rotary nozzle must be mounted behind and below the boom with the sleeve directed toward the rear of the aircraft and parallel to the ground during flight. Nozzle must be positioned to minimize air turbulence and the collection of CYTHION or MALATHION ULV droplets on mounting brackets, feed lines, fittings, etc., or any part of the aircraft.

Operating Procedures

1. CYTHION or MALATHION ULV must be prefiltered through a 10 micron filter prior to transfer into helicopter tank. A 50 mesh stainless steel line strainer must be installed in the pump feed line.
2. Entire system, including tank, pump, nozzle and feed lines, to be used only for application of CYTHION or MALATHION ULV.
3. Entire system must be inspected daily to insure that there are no leaks.
4. Sleeve must be removed and cleaned immediately after each use by washing with hot water and blowing dry from outside in with clean air.
5. Rotating nozzle must be turned on and operating before turning on pump. For shut off, pump must be shut off and lines cleared prior to stopping nozzle rotation.
6. Dosage of CYTHION or MALATHION ULV does not exceed 3 fluid ounces per acre.
7. The spray system must produce droplets of CYTHION or MALATHION ULV with a mass median diameter (MMD) of less than 50 microns, with no more than 2.5% of the droplets exceeding 100 microns, as determined by readings made from microscope slides coated with DRI-FILM or TEFLON.

GROUND APPLICATION

Thermal Aerosols or Fogs

For control of adult mosquitoes with thermal aerosols or fogs, apply CYTHION or MALATHION ULV at the rate of 6–8 ozs. actual/gallon (3.9–5.2 gallons CYTHION or MALATHION ULV in 100 gallons finished solution*) by ground equipment delivering 40 gallons per hour at a vehicle speed of 5 miles per hour to treat a swath width of 300–400 feet.

* There is a great variation in the chemical composition of fuel oils which may be used as thermal fog solvents. These differences may cause sludge and/or affect the solubility of the CYTHION or MALATHION ULV. For more complete details on tests for sludge formation and solubility in thermal fog solutions, write American Cyanamid Company, One Cyanamid Plaza, Wayne, New Jersey 07470.

Nonthermal Aerosols

Over a 300-foot swath can be produced using the non-thermal ultra low volume aerosol method with CYTHION or MALATHION ULV.

Vehicle Speed Rate per Hour	Flow Rate of MALATHION Fluid Ounces per Minute	Maximum Flow Rate per Hour
5	1.0 to 2.1 fluid ounces	1 gallon
10	2.0 to 4.3 fluid ounces	2 gallons
15	3 to 6.45 fluid ounces	3 gallons
20	4 to 8.6 fluid ounces	4 gallons

For control of adult stable fly in populated and rural areas with non-thermal aerosols of CYTHION or MALATHION ULV using the ultra low volume method, use the following flow rates at the indicated vehicle speeds:

Vehicle Speed Miles per Hour	Flow Rate of MALATHION Fluid Ounces per Minute	Maximum Flow Rate per Hour
5	2.1 fluid ounces	1 gallon
10	4.3 fluid ounces	2 gallons

DROPLET SIZE

1. The Mass Median Diameter (MMD) of the droplets should not exceed 17 microns. The MMD is the drop diameter which divides the spray volume into two equal parts; i.e., 50% of the volume is in the drop sizes below the MMD and 50% is above the MMD.
2. Spray droplets should not exceed 32 microns in size. Three percent of the spray droplets (6 droplets out of 200) can exceed 32 microns providing the MMD does not exceed 17 microns and no droplets exceed a maximum of 48 microns. Larger droplets, when transported by natural air currents, impinge more readily on objects in their pathway and will permanently damage automobile-type paints.
3. More than one-half of the total spray mass must consist of droplets in the 6 to 18 micron range to achieve adequate dispersal of insecticide over a 300-foot swath.
4. A minimum of two-thirds, preferably four-fifths, of the total spray mass must consist of droplets not exceeding 24 microns in range.

OPERATING EQUIPMENT

Each Nonthermal Aerosol Generator used for dispersal of CYTHION or MALATHION ULV to control adult mosquitoes must have minimum capability of producing the droplet spectrum described under DROPLET SIZE. The initial determination of droplet size is made after the unit is installed in a vehicle and prior to its use in mosquito control operations. The unit should be rechecked as frequently as necessary to insure that proper droplet size is maintained for each operation. Determination of droplet size every two months is usually sufficient if the unit has been maintained in good operating condition. Equipment manufacturer's instructions setting forth cleaning and maintenance of the unit must be followed. The unit must be inspected before each operation to correct any leaks or obstructions in the spray system; to detect whether the nozzle, hoses, or other parts are worn and need replacement; to insure that the flow meter is properly calibrated; and to determine that the pressure recommended by the manufacturer is being maintained.

Flow Rate—must be regulated by accurate flow meter.

—not greater than 1 gallon per hour at 5 mph, 2 gallons per hour at 10 mph, 3 gallons per hour at 15 mph, or 4 gallons per hour at 20 mph.

Nozzle Direction—rear of the vehicle.

—upward at an angle of 45° or more.

Vehicle Speed—not greater than 20 miles per hour.

—shut off spray equipment when vehicle is stopped.

IMPORTANT—Spray droplets of undiluted CYTHION or MALATHION ULV will permanently damage automobile paint unless all the conditions described and recommended in this leaflet are met.

Directions for Determining the droplet size of CYTHION or MALATHION ULV nonthermal aerosols

Permanent records of each droplet size determination must be kept and made available to American Cyanamid upon request.

1. Preparation of Slides with DRI-FILM

CYTHION or MALATHION ULV droplet sizes are determined by depositing a sample of the aerosol on a coated glass slide and measuring the droplets under a high-power microscope. Ordinary 3″ × 1″ glass slides must be coated with silicone (General Electric SC-87 DRI-FILM) prior to sampling to prevent excessive spreading or coalescence of the droplets. The slides are dipped into a 10 percent solution of DRI-FILM in toluene, drained and dried at about 200° F. for 30 minutes, after which they are dipped in acetone, allowed to dry and stored in a tight slide box. Coating solution must be freshly prepared. Do not store coating solution because it will deteriorate. Slides are lightly polished with a soft tissue before using to remove any foreign particles.

2. Deposition of CYTHION or MALATHION ULV Droplets on Slides

Droplets should be collected under ideal operating conditions to insure representative sampling of droplets in the aerosol. A sample of the CYTHION or MALATHION ULV aerosol is deposited on a slide by waving the slide as rapidly as possible perpendicular through the aerosol cloud at a distance of 25 feet from the point of discharge. The slide velocity may be increased by attaching it to a 3 or 4 foot stick by means of a spring paper clip. At least two slides should be exposed to insure an adequate sample. Store slides in a tight slide box for transfer to a location where measurements can be made. Avoid excessive heat during transit and store in a cool place until measurements can be made.

Although label specifications require the aerosol nozzle to be angled upward at 45° or more during operation, it is more convenient to position the nozzle parallel to the ground for droplet sampling. If this is not possible it will be necessary to be positioned at a sufficient height to obtain a representative sample of the aerosol.

3. Determination of CYTHION or MALATHION ULV Droplet Sizes

A microscope with mechanical stage and an eyepiece micrometer are used to determine the size of the individual aerosol droplets. Prior to taking measurements, the divisions of the eyepiece micrometer must be calibrated into microns by means of a stage micrometer. In the example represented in Table 1, droplets were measured at 400x magnification. At that magnification each division of the eyepiece was calibrated to equal 3.5 microns.

At least 200 droplets should be measured. Usually this is easily accomplished on one slide. An accurate method is to measure all droplets that pass through the micrometer scale as the slide is moved from one edge to the other by using the mechanical stage. Measurements should not be taken along the margins of the slide. It is more convenient to measure in terms of the divisions of the eyepiece micrometer and then convert these divisions into microns.

The measurements converted into microns must then be corrected for the amount of spread that occurred on the slides. The CYTHION or MALATHION ULV spread factor for silicone-coated slides is 0.5. Therefore, in Table 1 each division of the eyepiece actually equals 1.75 microns (3.5 microns times the 0.5 spread factor).

The spread factor for TEFLON-coated slides is 0.69. The following procedure is given for silicone-coated slides, would be the same for TEFLON-coated slides once the value for each eyepiece division has been determined.

The measurements are tabulated and processed as in Table 1. The Maximum Diameter is calculated by converting the diameter of the largest droplet measured into microns. In Table 1, the largest droplet measured had a diameter of 19 eyepiece divisions. Therefore, the Maximum Diameter is 33.3 microns (19 × 1.75 = 33.3).

To determine the Mass Median Diameter (MMD), the accumulative percentages from the last column in Table 1 are plotted against the eyepiece divisions (D) on semilogarithmic paper as in Figure 1. Directly across from the 50 percent point on the line is the median droplet size in eyepiece divisions which must be converted to microns. In Figure 1, 9.2 eyepiece divisions times the conversion factor of 1.75 equals a Mass Median Diameter of 16.1 microns.

Table 1—Representative Count of CYTHION or MALATHION ULV Aerosol Droplets Impinged on Microscope Slides Coated with DRI-FILM.

Eyepiece Divisions (D)*	Number of Droplets (N)	DxN	% of Total DxN Σ (DxN)	Accumulative Percentages
1	5	5	0.31	0.31
2	10	20	1.22	1.53
3	9	27	1.65	3.18
4	12	48	2.93	6.11
5	15	75	4.58	10.69
6	12	72	4.40	15.09
7	25	175	10.70	25.79
8	14	112	6.85	32.64
9	28	252	15.40	48.04
10	19	190	11.61	59.65
11	14	154	9.41	65.06
12	10	120	7.33	76.39
13	6	78	4.77	81.16
14	4	56	3.42	84.58
15	11	165	10.09	94.67
16	2	32	1.96	96.63
18	2	36	2.20	98.83
19	1	19	1.16	99.99
Total	199	1636		

* Measurements were taken at 400x magnification. Each eyepiece division equals 1.75 microns (3.5 microns times the 0.5 spread factor).

Also for use in accordance with the recommendations and instructions issued by the United States Department of Agriculture for quarantine programs. To be used only by or under the direction of Federal/State personnel for quarantine treatments.

Figure 1.
Percentage of the total volume of aerosol samples below each stated droplet size (from Table 1). The Mass Median Diameter is determined from the 50 percent point on the line. The Mass Median Diameter (MMD) = 9.2 divisions times 1.75 = 16.1 microns.

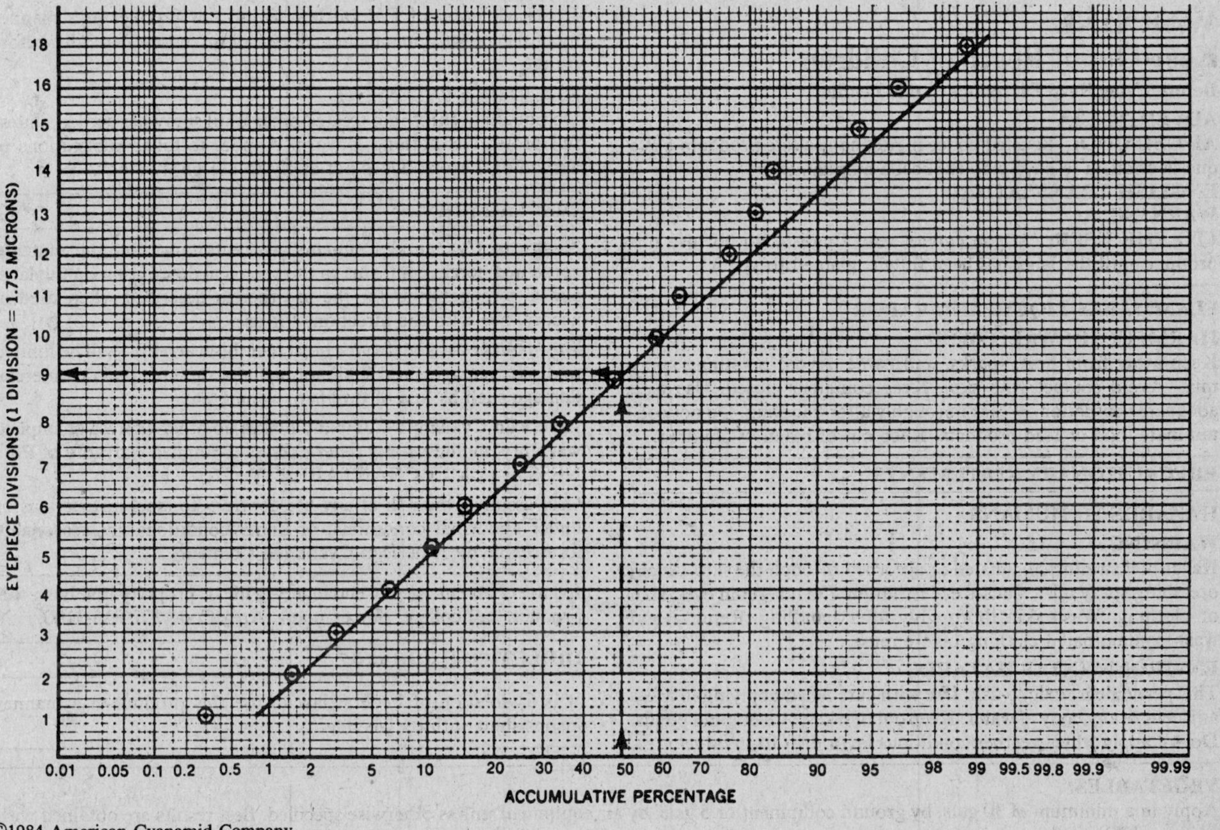

CYTHION®

57% EC

INSECTICIDE

'The Premium Grade Malathion'®

57% Emulsifiable Concentrate

ACTIVE INGREDIENT:

Malathion*	57.0%
Xylene	30.0%
INERT INGREDIENTS:	13.0%
TOTAL	100.0%

* 0,0-dimethyl phosphorodithioate of diethyl mercaptosuccinate (contains 5 lb Malathion per gallon)

EPA Reg. No. 241-47

WARNING

Keep Out of Reach of Children

Before Using Read Directions in this Leaflet.

ADVERTENCIA

AL USUARIO: Si usted no lee inglés, no use este producto hasta que la etiqueta le haya sido explicada ampliamente.

ENGLISH TRANSLATION

WARNING

(TO THE USER: If you cannot read English, do not use this product until the label has been fully explained to you.)

FLAMMABLE LIQUID AND VAPOR

HARMFUL IF SWALLOWED

Keep away from heat, sparks, and flame. Avoid breathing spray mist. Avoid contact with skin. Keep container closed. Use with adequate ventilation. Wash thoroughly after handling. Do not contaminate food or feed products. Keep out of reach of children.

PRECAUTIONARY STATEMENTS

HAZARDS TO HUMANS

WARNING

Harmful if swallowed, inhaled or absorbed through the skin. Avoid breathing spray mist. Causes eye irritation. Do not get in eyes, skin or clothing. Wash skin thoroughly after handling. Remove and wash contaminated clothing before reuse.

ENVIRONMENTAL HAZARDS

This pesticide is toxic to fish. Use with care when applying in areas adjacent to any body of water. Keep out of lakes, streams and ponds. Do not apply when weather conditions favor drift from target area.

This pesticide is highly toxic to bees exposed to direct treatment or to residues remaining on the treated area. Do not apply when bees are actively visiting the crop, cover crop, or weeds blooming in the treated area. Applications should be timed to provide the maximum possible interval between treatment and the next period of bee activity.

STATEMENT OF PRACTICAL TREATMENT

If swallowed, do not induce vomiting. Get medical attention! In case of contact, immediately flush eyes and skin with plenty of water for at least 15 minutes while removing contaminated clothing and shoes. Call a physician. Wash clothing before reuse.

NOTE TO PHYSICIAN: This product may cause cholinesterase inhibition. Atropine is antidotal. 2-PAM may be effective as an adjunct to atropine.

STORAGE AND DISPOSAL

Refer to container label for complete directions for pesticide storage and disposal.

DIRECTIONS FOR USE

It is a violation of Federal law to use this product in a manner inconsistent with its labeling. Read all directions carefully before using.

REENTRY STATEMENT

Do not allow entry into treated areas until spray has settled, unless full protective clothing is worn. Applies to foliar applications to agricultural crops. Does not apply to use on livestock.

DISCLAIMER

American Cyanamid Company warrants only that the material contained herein conforms to the chemical description on the label and is reasonably fit for the use therein described when used in accordance with the directions for use.

Any damages arising from a breach of this warranty shall be limited to direct damages, and shall not include consequential commercial damage such as loss of profits or values, etc.

American Cyanamid Company makes no other express or implied warranty, including any other express or implied warranty of FITNESS or of MERCHANTABILITY.

User assumes the risk of any use contrary to label instructions or under abnormal conditions, or under conditions not reasonably foreseeable by American Cyanamid Company.

In case of an emergency endangering life or property involving this product, call collect, day or night, Area Code 201-835-3100.

DIRECTIONS FOR USE

It is a violation of Federal law to use this product in a manner inconsistent with its labeling.

VEGETABLES:

Apply in a minimum of 30 gals. by ground equipment or 5 gals. by air equipment unless otherwise specified. Best results are obtained with uniform coverage.

CROP	PESTS CONTROLLED	RATE/A	INTERVAL (DAYS) BETWEEN LAST APPLICATION AND HARVEST
Asparagus	Asparagus aphid Asparagus beetle	2 pts.	Do not apply within 1 day of harvest.
	Thrips	1½–2 pts.	
Beans (Green Beans, Lima Beans, Navy Beans, Red Kidney Beans, Snap Beans, Wax Beans, Cowpeas and Blackeyed Peas)	Aphids Bean leaf beetles Blister beetles Cucumber beetles Potato leafhoppers Lygus bugs	2 pts.	Do not apply within 1 day of harvest.
	Japanese beetle Mexican bean beetle	1½–2 pts.	
	Red Spider mites	1½ pts.	
Dry Beans (California and Northwest only)	Lygus bugs	1½–2 pts. (ground) 1½ pts. (air) in 10 gallons of water	Do not apply within 1 day of harvest.

CROP	PESTS CONTROLLED	RATE/A	INTERVAL (DAYS) BETWEEN LAST APPLICATION AND HARVEST
Beets (Garden and Table)	Aphids	1½–2 pts.	Do not apply within 7 days of harvest if tops are to be used as food or feed.
Broccoli	Aphids Cabbage looper Imported cabbageworm	1–2 pts.	Do not apply within 3 days of harvest.
Brussels Sprouts	Aphids	1–2 pts.	Do not apply within 7 days of harvest.
Cabbage	Aphids Cabbage looper Imported cabbageworm	1–2 pts.	Do not apply within 7 days of harvest. For control of caterpillars on summer and fall plantings in the South, begin when true leaves appear. On other plantings, and for control of other insects, begin when insects appear.
	Diamondback moth Webworm	2 pts.	
Carrots	Aphids	1½–2 pts.	Do not apply within 7 days of harvest.
	Leafhoppers	2½ pts.	
Cauliflower	Diamondback moth	2–4 pts.	Do not apply within 7 days of harvest.
	Aphids	1–2 pts.	
Celery Anise	Aphids Spider mites	1½ pts.	To be applied to fresh leaves and stalks only. Do not use on crops grown for seed and oil. Do not apply within 7 days of harvest.
Collards	Aphids	1½–2 pts.	Do not apply within 7 days of harvest. For control of caterpillars on summer and fall plantings in the South, begin when true leaves appear. On other plantings, and for control of other insects, begin when insects appear.
	Harlequin cabbage bug	1 pt.	
Cucumbers	Aphids Pickleworms Spider mites	1½–2 pts.	Do not apply to cucumbers unless plants are dry. Do not apply within 1 day of harvest.
	Squash vine borer		
	Cucumber beetle Leaf miner	2 pts.	
Dandelion, Parsley, Parsnip, Swiss Chard, Watercress	Aphids	1½–2 pts.	Do not apply within 21 days of harvesting parsley or within 7 days of harvesting dan delion, parsnip, Swiss chard or watercress.
Eggplant	Aphids Spider mites	1 pt.	Do not apply within 3 days of harvest.
	Lacebugs	3 pts.	
Endive	Aphids Spider mites	1½–2 pts.	Do not apply within 7 days of harvest.
Garlic, Shallots	Aphids Thrips	1½–2 pts.	Do not apply within 3 days of harvest.
Hops	Aphids Spider mites	1 pt.	Do not apply within 10 days of harvest.
Kale	Aphids Cabbage looper Imported cabbageworm	1–2 pts.	Do not apply within 7 days of harvest. For control of caterpillars on summer and fall plantings in the South, begin when true leaves appear. On other plantings, and for control of other insects, begin when insects appear.
	Webworm Diamondback moth	2 pts.	
Kohlrabi	Aphids	1–2 pts.	Do not apply within 7 days of harvest.
	Cabbage looper Diamondback moth Imported cabbageworm Webworm	2 pts.	
Leeks	Aphids	1½–2 pts.	Do not apply within 3 days of harvest.
	Onion maggot flies	2½ pts.	
	Thrips	3 pts.	
Lentils	Cowpea aphid Pea aphid	1½ pts.	Do not apply within 3 days of harvest.
Lettuce	Aphids Leafhoppers Spider mites	2 pts.	Do not apply within 7 days of harvesting head lettuce or within 14 days of harvesting leaf lettuce.
	Cabbage looper	3 pts.	
Melons (Cantaloupe, Casaba, Crenshaw, Honeydew Melons, Honey Balls, Muskmelons, Persian Melons and hybrids of	Aphids Spider mites	1½–2 pts.	Do not apply to melons unless plants are dry. Do not apply within 1 day of harvest.
	Cucumber beetles	2–3 pts.	
	Leaf miners	2 pts.	
	Leafhoppers	1½–2 pts.	

CROP	PESTS CONTROLLED	RATE/A	INTERVAL (DAYS) BETWEEN LAST APPLICATION AND HARVEST
these, Watermelons and their hybrids)	Pickleworms	2 pts.	
	Squash vine bore	3 pts.	
Mushrooms	Mites Phorid and Sciarid flies	2½ pts. in 130 gallons of water; or, 2 tablespoons in 3 gallons of water per 1000 sq. ft. of bed	Make thorough applications as soon after picking as possible. Repeat applications as necessary, usually twice a week. Do not apply within 1 day of harvest.
Mustards	Aphids Cabbage looper Imported cabbageworm	2 pts.	For control of caterpillars on summer and fall plantings in the South, begin when true leaves appear. On other plantings, and for control of other insects, begin when insects appear. Do not apply within 7 days of harvest.
	Flea beetles	1½–2 pts.	
Okra	Aphids	1½ pts.	Do not apply within 1 day of harvest. Make no application after pods start to form.
	Japanese beetle	2 pts.	
Onions	Onion maggots	1½–2½ pts.	Do not apply within 3 days of harvest.
	Onion thrips	1½–2 pts.	
Peas	Aphids	1½–2 pts.	Do not apply within 7 days of harvest if vines are to be fed, or within 3 days of harvest if vines are not to be fed.
	Grasshoppers	2 pts.	
	Leafhoppers	1¼–4 pts.	
Peppermint, Spearmint	Aphids Flea beetles Leafhoppers Spider mites	1½ pts.	Do not apply within 7 days of harvest.
Peppers	Aphids	1–2½ pts.	Do not apply within 3 days of harvest.
	Pepper maggots	2½ pts.	
Potatoes	Aphids Grasshoppers Leafhoppers	2 pts.	May be applied on the day of harvest.
	False chinch bugs	1½ pts.	
	Mealy bugs	2–2½ pts.	
Pumpkins	Aphids Pickleworms Spider mites	2 pts.	Do not apply to pumpkins unless plants are dry. Do not apply within 3 days of harvest.
	Leafhoppers	1½–2 pts.	
	Cucumber beetles Squash vine borer	3 pts.	
Radish, Horseradish	Aphids	1½–2 pts.	Do not apply within 7 days of harvest.
Rice	Rice leaf miner	2½ pts.	Make first application shortly after the first rice blades appear on the surface of the water and repeat if necessary. Do not apply within 7 days of harvest.
	Rick stink bugs	1–1½ pts.	Apply by airplane in 2 gallons of water per acre during the early milk and dough stage of growing rice. Repeat applications as necessary. Do not apply within 7 days of harvest.
Rutabaga	Aphids	1½ pts.	Do not apply within 3 days of harvest.
Spinach	Aphids	2 pts.	Do not apply within 7 days of harvest.
Squash	Aphids Leaf miners Spider mites Pickleworm	2 pts.	Do not apply to squash unless plants are dry. Do not apply within 1 day of harvest.
	Cucumber beetles Squash vine borer	3 pts.	
Strawberries	Aphids Spider mites	1½ pts.	Do not apply within 3 days of harvest.
	Field crickets Lygus bugs Spittlebugs Thrips	1½ pts–3 pts.	
	Potato leafhopper Strawberry leafroller Strawberry root weevil Whiteflies	1½–2½ pts.	
Sweet Corn	Japanese beetle	2 pts.	Do not apply within 5 days of harvest. Injury may occur in the whorl or to the silks.
Sweet Potatoes	Leafhoppers	1½–2 pts.	Do not apply within 3 days of harvest.
	Morningglory leaf miner	2½–3 pts.	
Tomatoes	Aphids	1½ pts.	

CROP	PESTS CONTROLLED	RATE/A	INTERVAL (DAYS) BETWEEN LAST APPLICATION AND HARVEST
	Drosophila	2½ pts.	Do not apply within 1 day of harvest.
	Spiter mites	1½ pts.	
	Armyworms Fruitworms (California only)	2¾ qts.	Do not apply within 3 days of harvest.
Turnip	Aphids Cabbage looper Imported cabbageworm	1–2 pts.	For control of caterpillars on summer and fall plantings in the South, begin when true leaves appear. On other plantings, and for control of other insects, begin when insects appear. Do not apply within 7 days of harvest if tops are to be used for food or feed.

VEGETABLES GROWN IN COMMERCIAL GREENHOUSES:

Apply to give complete coverage.

CROP	PESTS CONTROLLED	RATE/100 GALS.	INTERVAL (DAYS) BETWEEN LAST APPLICATION AND HARVEST
Cucumbers	Aphids Armyworms Cabbage loopers Earwigs Garden fleahoppers Mealybugs Spider mites Thrips Whiteflies	1½–2 pts.	Spray plants thoroughly. Do not apply to cucumber unless plants are dry. Do not apply within 1 day of harvest.
Endive	Aphids Armyworms Cabbage loopers Serpentine leaf miners Thrips Whiteflies	1½–2 pts.	Spray plants thoroughly. Do not apply within 7 days of harvest.
Lettuce	Aphids Armyworms Cabbage loopers Serpentine leaf miners Thrips Whiteflies	1½–2 pts.	Spray plants thoroughly. Do not apply within 14 days of harvesting leaf lettuce or within 7 days of harvesting head lettuce.
Radish	Aphids Flea beetles	1½–2 pts.	Spray plants thoroughly. Do not apply within 7 days of harvest.
Tomato	Aphids Armyworms Cabbage loopers Cutworms Drosophilia Garden fleahoppers Mealybugs Serpentine leaf miners Spider mites Thrips Tomato russet mites Whiteflies	1½–2 pts.	Spray plants thoroughly. Do not apply within 1 day of harvest.
Watercress	Aphids Serpentine leaf miners Thrips	1½–2 pts.	Spray plants thoroughly. Do not apply within 7 days of harvest.

FRUIT AND NUT

CROP	PESTS CONTROLLED	RATE/ 100 GALS.	RATE/A	INTERVAL (DAYS) BETWEEN LAST APPLICATIONS AND HARVEST
Almonds	Aphids Spider mites	1–2 pts.	2–4 pts.	May be applied on the day of harvest.
	Peach twig borer	1–1½ pts.	2–3 pts.	Apply during the petal fall period or during the period appropriate for spraying larvae of the May brood. May be applied on the day of harvest.
	The rates for use on almonds are based on a standard of 200 gallons per acre dilute spray for mature trees. Do not apply more than 8 lbs. actual CYTHION per acre to almond trees.			
Apples	BudmothForbes scale Wooly apple aphid	1 pt.	4 pts.	
	Eastern tent caterpillar	1–1½ pts.	4–6 pts.	
	Green apple aphid Rosy apple aphid	1½ pts.	6 pts.	

CROP	PESTS CONTROLLED	RATE/ 100 GALS.	RATE/A	INTERVAL (DAYS) BETWEEN LAST APPLICATIONS AND HARVEST
	Codling moth Plum curculio Red-banded leafroller	2 pts.	8 pts.	Do not apply within 3 days of harvest.
	Mealybug Mites, such as Clover mite, European red mite, Twospotted spider mite, Willamette mite	1–2 pts.	4–8 pts.	
	The rates for use on apples are based on a standard of 400 gallons per acre dilute spray for mature trees pruned 20 to 22 feet high in rows 40 feet apart.			
	CYTHION Emulsifiable Concentrate may cause fruit injury to McIntosh and Cortland varieties if spray is applied within 4 weeks of harvest.			
Dormant or Delayed Dormant Sprays	Aphids Mites Red-banded leafrollers	1 pt. + 1 gal. superior oil	—	Make full coverage dormant or delayed dormant applications only.
Apricots	Aphids Codling moth European fruit lecanium Orange tortrix Soft brown scale Terrapin scale	1½–2 pts.	4½–6 pts.	Do not apply within 7 days of harvest.
	The rate for use on apricots is based on a standard of 300 gallons per acre dilute spray for mature trees.			
Avocado	Greenhouse thrips Latania scale Omnivorous looper Orange tortrix Soft brown scale	1½ pts.	7½ pts.	Do not apply within 7 days of harvest.
	The rate for use on avocado is based on a standard of 500 gallons per acre dilute spray for mature trees.			
Blackberry Boysenberry Dewberry Loganberry Raspberry	Aphids Rose scale chafers Japanese beetle Leafhoppers Mites Thrips	1½ pts.	3 pts.	Do not apply within 1 day of harvest.
	The rates for use on brambles are based on a standard of 200 gallons per acre dilute spray.			
Blueberries	Blueberry maggots	1 pt. + 1½ qts. Staley's Sauce Base No. 7	—	For use in the Northeast only. Apply in 100 gallons of water per acre. Do not apply within 8 hours of harvest.
	Cherry fruitworm	1 pt.	2 pts.	Do not apply within 1 day of harvest.
	Cranberry fruitworm	1 pt.	2 pts.	Make first application at egg hatch and repeat applications every 4 or 5 days until a total of four applications has been made. Do not apply within 1 day of harvest.
	Japanese beetle	¾ pt.	½ pts.	Make application when first berries turn blue and at 10 day intervals as long as infestation persists. Do not apply within 1 day of harvest.
	The rates for use on blueberries are based on a standard of 200 gallons per acre dilute spray.			
Cherry	Black cherry aphid Fruit-tree leafroller	1½ pts.	6 pts.	Do not apply within 3 days of harvest.
	Cherry fruit fly Japanese beetle	1 pt.	4 pts.	
	The rates for use on cherries are based on a standard of 400 gallons per acre dilute spray for mature trees. Do not apply more than 8 lbs. of actual CYTHION per acre to cherry trees.			
	Injury may occur on certain varieties of sweet cherries, particularly in the Northwest.			
Citrus Grapefruit Kumquat Lemon Lime Orange Tangelo	Black scale (single and off-brooded) California red scale Citricola scale Purple scale Soft brown scale Yellow scale	1–1½ pts.	—	Do not apply when trees are in bloom. Do not apply within 7 days of harvest.
	Florida purple scale Florida red scale	2 pts.	—	

CROP	PESTS CONTROLLED	RATE/ 100 GALS.	RATE/A	INTERVAL (DAYS) BETWEEN LAST APPLICATIONS AND HARVEST
	Citrus red mites Thrips	—	2½ pts.	Apply in 200 gallons of water per acre. Do not apply when trees are in bloom. Do not apply within 7 days of harvest.
	The rate for use on citrus is based on a standard of 500 gallons per acre dilute spray for mature trees.			
Currents Gooseberries	Japanese beetles	1½ pts.	3 pts.	Do not apply within 3 days of harvest.
	Rose chafers Mites	1 pt.	2 pts.	
	The rates for use on currants and gooseberries are based on a standard of 200 gallons per acre dilute spray.			
Figs	Dried fruit beetles Vinegar flies	—	2 qts. 1–2 gals. unsulfured molasses	Do not apply within 3 days of harvest.
Filberts	Apple mealybug Eye-spotted bud moth Filbert aphids Tingids	1 pt.	4 pts.	Do not apply within 3 days of harvest.
	Scales (crawler stages)	1 pt.	—	
	The rates for use on filberts are based on a standard of 400 gallons per acre dilute spray for mature trees. Do not apply more than 3 lbs. of actual CYTHION per acre to filbert trees.			
Grapes	Leafhoppers Spider mites	½ pts.	3 pts.	Do not apply within 3 days of harvest.
	European fruit lecanium	1½ pts.	—	Make full coverage applications when newly hatched nymphs are migrating over vines, usually shortly after bloom. Do not apply within 3 days of harvest.
	Mealybugs	—	1½ pts.	Apply in 50–100 gallons of water per acre. Do not apply within 3 days of harvest.
	The rates for use on grapes are based on a standard of 200 gallons per acre dilute spray.			
	Injury may occur on grapes of Almeria, Cardinal, Italia and Ribier varieties when sprays containing CYTHION are applied after clusters appear.			
Grape Vines (Overwintering on nursery stock only)	Grape phylloxera	3 pts.	—	Remove excess soil from the roots and dip in the CYTHION solution. Submerge the entire root system for 5 minutes. Keep the solution agitated at all times.
Macadamia Nuts	Green stink bugs	1½ pts.	—	Apply when insects start to feed on the nuts and repeat as necessary. May be applied during harvest.
	No more than 15 lbs. of actual CYTHION per acre should be applied to macadamia nut trees.			
Nectarines	Spider mites	1–2 pts.	3–6 pts.	Do not apply within 7 days of harvest.
	Plum curculio	2 pts.	6 pts.	
	The rates for use on nectarines are based on a standard of 300 gallons per acre dilute spray for mature trees.			
	CYTHION may cause fruit spotting on nectarines.			
Papaya	Aphids Mealybugs	1½–2 pts.	—	May be applied on the day of harvest.
Peaches	Black cherry aphid Black peach aphid European red mite Green peach aphid Rusty plum aphid Japanese beetle Spider mites	1 pt.	3 pts.	Do not apply within 7 days of harvest.
	Oriental fruit moth Plum curculio	2 pts.	3 qts.	
	Cottony peach scale European fruit lecanium Terrapin scale	2 pts.	—	Make full coverage applications at completion of hatching, about the last of June. Do not apply within 7 days of harvest.
	The rates for use on peaches are based on a standard of 300 gallons per acre dilute spray for mature trees. Do not apply more than 9 pounds of actual CYTHION per acre to peach trees.			

CROP	PESTS CONTROLLED	RATE/ 100 GALS.	RATE/A	INTERVAL (DAYS) BETWEEN LAST APPLICATIONS AND HARVEST
Pears	Aphids	1 pt.	3 pts.	Do not apply within 1 day of harvest.
	Mealybug Mites Pear psylla	1–2 pts.	3–6 pts.	
	Codling moth Fruit-tree leafroller Plum curculio Red-banded leafroller	2 pts.	6 pts.	
	The rates for use on pears are based on a standard of 300 gallons per acre dilute spray for mature trees.			
	Injury may occur under certain conditions in the Northeast on Bosc pears.			
Pecans	Aphids Spider mites Walnut aphid	1½–2 pts.	7½–10 pts.	May be applied on the day of harvest.
	European red mite	1¼ pts.	6¼ pts.	May be applied on the day of harvest.
	Pecan phylloxera	1¼ pts.	6¼ pts.	Apply when buds begin to develop and repeat as necessary. May be applied on the day of harvest.
	Pecan nut casebearer Walnut husk fly	1¼ pts.	6¼ pts.	Apply after first generation eggs begin to hatch. May be applied on the day of harvest.
	The rates for use on pecans are based on 500 gallons per acre dilute spray for mature trees 25-35 feet high.			
Pineapple	Mealybug	1 pt.	8 pts.	Do not apply within 7 days of harvest.
Plums and Prunes	Mealy plum aphid Plum curculio	1–1½ pts.	3 pts.	Do not apply within 3 days of harvest. Make full coverage applications to trunks and scaffold limbs 4 times at 21–day intervals beginning with emergence in June. In Georgia, two applications each 4 weeks apart for first (April and May) and second (August and September) broods. Do not apply within 3 days of harvest.
	The rates for use on plums and prunes are based on a standard 300 gallons per acre dilute spray for mature trees			
Quince	Codling moth Oriental fruit moth Plum curculio	2 pts.	6 pts.	Do not apply within 3 days of harvest.
	Spider mites	1–2 pts.	3–6 pts.	
	Forbes scale	1 pt.	—	Make full coverage applications when crawlers are present, about first or second cover and last half of July. Do not apply within 3 days of harvest.
	The rates for use on quince are based on a standard 300 gallons per acre dilute spray for mature trees			
Walnuts	Aphids European red mite Walnut aphid	½ pt.	1½–3 pts.	May be applied on the day of harvest
	Walnut husk fly	½ pt.	1½–3 pts	Apply with conventional sprayer. For bait sprays, combine Staley's Sauce Base No. 2 or No. 7 at the rate of 2 qts. per acre with CYTHION. May be applied on the day of harvest.
		—	3–4 pts.	Apply in 500 gallons of water per acre with air-carrier type sprayer. For bait sprays combine Staley's Sauce Base 2 or No. 7 at the rate of 2 qts. per acre with CYTHION. May be applied on the day of harvest.

FIELD CROPS AND PASTURE AND RANGE GRASSES

CROP	PESTS CONTROLLED	RATE/A	INTERVAL (DAYS) BETWEEN LAST APPLICATION AND HARVEST
Alfalfa	Alfalfa weevil larvae Aphids Grasshoppers Lygus bugs Potato leafhoppers Spider mites Spittle bugs Stink bugs	1½–2 pts.	Apply to alfalfa in bloom only in the evening or early morning when bees are not working in the field or are not hanging on outside of hives. May be applied on the day of harvest.
	Armyworms Clover leaf weevil	2 pts. 1½ pts.	
	Vetch bruchid	2–2½ pts.	

CROP	PESTS CONTROLLED	RATE/A	INTERVAL (DAYS) BETWEEN LAST APPLICATION AND HARVEST
Clover	Alfalfa weevil larvae Aphids Grasshoppers Lygus bugs Potato leafhoppers Spider mites Spittlebugs	1½–2 pts.	Do not apply to clover in bloom. May be applied on the day of harvest.
	Armyworms	2 pts.	
	Clover leaf weevil	1½ pts.	
Corn (Grain or Forage)	Aphids Corn earworm Corn rootworm adults Grasshoppers Sap beetle Thrips	1½ pts.	For control of corn earworm and sap beetles begin treatments when 10% of the ears show silk. Repeat applications at 3–5 day intervals until 4–5 applications have been made. Do not apply within 5 days of harvest or forage use.
	Armyworms	1½–2 pts.	
	Injury may occur in the whorl and silk stages with CYTHION.		
Cotton	Brown cotton Leafworm Cotton aphid Cotton leafworm Cotton leafperforator Desert spider mite Leafhoppers Lygus bugs Thrips Whiteflies	½–2 pts.	May be applied on the day of harvest. Consult local agricultural authorities for exact time of application.
	Boll weevil	2–4 pts.	
	Cotton fleahoppers	1-1½ pts.	
	Fall armyworms Garden webworms Grasshoppers	1½–3 pts.	May be applied on the day of harvest.
	Lygus bugs Thrips	1–4 pts.	
Grasses (Barnyardgrass, Canarygrass, Fescue, Orchardgrass, Red top, Timothy, Yellow foxtail)	Cereal leaf beetle	1–1½ pts.	May be applied on the day of harvest or grazing
Pasture and Range Grass, Grass and Grass Hay	Aphids Grasshoppers Leafhoppers	1½–2 pts. OR 1½ pts. in 1 gal. of diesel fuel oil per acre	May be applied on the day of harvest or grazing.
	Armyworms	2 pts. OR 1½ pts. in 1 ga. of diesel fuel oil per acre	
Peanuts	Potato leafhoppers	1½ pts.	May be harvested or grazed on the day of application.
Safflower	Aphid Grasshoppers Lygus bugs	1½–2 pts.	Do not apply within 3 days of harvesting seeds
Small Grains (Barley, Oats, Rye, Wheat)	Cereal leaf beetle		Do not apply within 7 days of harvest or forage use.
	English grain aphids Grasshoppers Greenbugs	1–1½ pts.	
	Winter grain mite	2 pts.	
Sorghum	Greenbugs	1½ pts.	Do not apply within 7 days of harvest or forage use.
Soybeans	Green cloverworms Mexican bean beetle	3 pts.	May be applied on the day of harvest or forage use.
Sugar Beets	Aphids Spider mites	1½–2 pts.	Do not apply within 7 days of harvest if tops are used as feed or forage.
	Grasshoppers	3 pts.	
Tobacco	Aphids	1½–2½ pts.	May be applied on the day of harvest. Use maximum dosage in plant beds.
Vetch	Omniverous leaf tier Pea aphid Vetch bruchid	1½–2 pts.	May be applied on the day of harvest or grazing.

STORED PRODUCTS

SITE	PESTS CONTROLLED	RATE	DIRECTIONS FOR USE
Stored in-shell Almonds	Indian meal moth Merchant grain beetle	4 fl. oz. in sufficient water to treat 10,000 lbs. of almonds	Use a suitable mechanical spray applicator that regulates the rate application to the flow of the almonds. Avoid spraying with a fine mist that drifts away. Shield the nozzle against wind and air currents.
Stored Peanuts	Confused flour beetle Flat grain beetle Granary weevil Indian meal moth Lesser grain borer Red flour beetle Rice weevil Rusty grain beetle Saw-toothed grain beetle	1 pt. in $2\frac{1}{2}$ gal. of water OR 8 pts. in 19 gal. of water.	Residual Warehouse Spray—Before Storing Peanuts: Apply as a coarse spray at the rate of 2 gal. per 1,000 sq. ft. of surface or to run-off. Clean warehouse thoroughly of trash and remains of old peanuts 1 to 2 weeks before new peanut crop is stored. Then, thoroughly spray the interior of the empty warehouse, especially cracks and protected places, treat outside walls to a height of 6 to 8 feet and the ground to a distance of 6 feet from the warehouse.
		$2\frac{1}{2}$ pts. in 5 gal. of water for each 15 tons of farmers stock peanuts.	Bulk Spray Treatment—Peanuts Going Into Storage: Use good spray equipment. Apply coarse spray uniformly. Preferably, use a suitable mechanical spray applicator that regulates the rate of application to the flow of peanuts. Adjust the operating pressure of spray pump and size of nozzle opening to correlate the amount of spray delivery with the rate of flow of peanuts being treated. Avoid spraying with a fine mist that drifts away, by using low nozzle pressure. Shield the nozzle against wind and air currents.
Nonmedicated cattle feed concentrate blocks	Cigarette beetles	8 pts. in 25 gal. water;	Before storing feed blocks thoroughly clean storage areas and remove and burn all debris and sweepings. Apply as a thorough application for a residual spray.
		4 fl. oz. in 1 qt. water	Treat paper on the side next to the feed concentrate at a rate of 100 mg/sq. ft., or 1 qt. of diluted CYTHION per 710 sq. ft.
Bagged citrus pulp Residual warehouse spray	Almond moth Angoumois grain beetle Cigarette beetle Confused flour beetle Flat grain beetle Indian meal moth Mediterranean flour moth Red flour beetle Saw-toothed grain beetle	1 pt. in sufficient water to make $2\frac{1}{2}$ gals of spray OR 8 pts. per 19 gal of water	Apply spray at the rate of 2 gallons per 1,000 sq. ft. of surface or to run-off. Before bagged citrus pulp is stored, thoroughly clean wharehouses by removing and burning all debris and sweepings. Thoroughly spray with sufficient pressure, interior of empty warehouse (including cracks and protected places), outside walls to a height of 6 to 8 feet and ground to a distance of about 6 ft. from warehouse.
	Do not use treated burlap bags other than for dried citrus pulp.		
Stored Grains (Barley, corn, grain sorghum, oats, rice, rye wheat) and field or garden seeds	Cereal leaf beetle Confused flour beetle Flat grain beetle Granary weevil Indian meal moth Lesser grain borer Maize weevil Red flour beetle Rice weevil Rusty grain beetle Saw-toothed grain beetle	8 pts. per 25 gal of water	Residual Spray Before Storing Grains: Before applying spray, clean elevators, box cars, etc. thoroughly. Remove and burn all sweepings and debris. For a residual wall, floor and machinery spray in grain elevators in treating truck beds, box cars and ships' holds, before loading grain make a thorough application.
		1 pt. in 2 to 5 gal of water per 1,000 bushels	Grans Going Into Storage: Apply as the grain is being loaded or turned into final storage.
		$\frac{1}{2}$ pt. in 1 to 2 gal of water per 1,000 sq ft of grain surface area.	As a Surface Treatment After Grains are Stored: Apply the spray evenly over the surface of the grain. Apply immediately after grain is loaded into storage and repeat if necessary.
Warehouse	Khapra Beetle	8 pts. per 20 gallons of water	Apply spray at the rate of 2 gallons per 1,000 sq ft of surface or to run-off. Thoroughly spray with sufficient pressure, interior of empty warehouse including cracks and protected places.

FLY AND MOSQUITO CONTROL

Fly Control: For use in and around buildings which house domestic animals, around yards, homes and meat and food-processing plants. Do not use in edible products areas of food processing plants, restaurants or other areas where food is commercially prepared or processed. Do not use in serving areas while food is exposed.

PESTS CONTROLLED	RATE	DIRECTIONS FOR USE
Adult flies	Straight Sprays: 5 Tablespoons + 1 gal water OR 1 cup + 2½ gal water OR 1 quart + 12 gal water.	Apply as a spray at the rate of 1 gallon per 1,000 sq. ft. on painted surfaces and 2 gals per 1,000 sq. ft. on unpainted surfaces where flies alight or congregate, such as walls, ceilings, stanchions, windows in dairy barns, fences, around garbage cans, etc.
Adult flies Fly maggots	Bait Sprays: 5 Tablespoons + 7 Tablespoons sugar or molasses (unsulfurized) or corn syrup + 1 gal water OR 1 cup + 1 cup sugar or molasses (unsulfurized) or corn syrup + 2½ gal water OR 1 quart + 2½ lbs sugar or 1 quart molasses (unsulfurized) or 1 quart corn syrup + 12 gal water.	Apply as a bait spray over the surface of manure or poultry droppings. In loafing sheds, spray the dry bedding within 18 inches of the walls and around upright braces. For effective control in and around dairy barns, fly-breeding sites such as manure and other waste material, should be eliminated. Do not apply to freshly whitewashed surfaces. Wait 14 days after whitewashing before applying

Repeat applications as necessary.
Avoid contamination of milk, milk equipment and water.
Avoid contamination of feed and food products, also drinking fountains and feed troughs.
Remove lactating dairy animals from buildings before treating. Also remove animals under one month of age before treating.
Avoid applying oil-based formulations to valuable ornamental plants as injury may occur.

Mosquito larvae	13 fl oz per acre	For use in standing water (intermittently flooded areas, Stagnant water, temporary rain pools). Mix in sufficient water or oil when applied by air or ground equipment.
Mosquito adults	1 part to 28 parts water, fuel oil or diesel oil.	Spray building foundations, shrubs, low trees and lawn areas.

LIVESTOCK PEST CONTROL

ANIMAL	PEST	RATE	REMARKS
Hogs	Lice	1 gal per 100 gals of water OR 6½ ozs per 5 gals of water.	Apply complete coverage spray to animals, pens and litter. One treatment may be sufficient. Repeat only if needed.
	Sarcoptic mange	1 gal per 100 gals of water OR 6½ ozs per 5 gals of water.	Apply complete coverage spray to all animals in herd, bedding and walls. Use extreme care to thoroughly cover all body surfaces of the animal, including inside of ears. One thorough spraying will usually control sarcoptic mange of swine, however, extensive cases will require second treatment about 10 days after first application.
			After spraying, swine should be kept out of sun and wind for a few hours.
			Residue tolerance: 4 ppm in meat, fat, and meat by-products.
			Avoid contamination of feed, food containers and watering troughs.
Sheep and Goats	Lice Ticks Keds	1 gal per 100 gals of water OR 6½ ozs per 5 gals of water.	Spray animals thoroughly. Repeat application after 2 or 3 weeks if needed. Do not apply to milk goats. Do not treat animals under one month of age.
			Residue tolerance: 4 ppm in meat, fat and meat by-products.
Horses, beef and non-milking cattle	Lice	1 gal per 100 gals of water OR 6½ ozs per 5 gals of water.	Apply complete coverage spray. One treatment may be sufficient. Repeat only if needed.
		BACK-RUBBING DEVICES: To reduce lice apply a mixute of 2% CYTHION (using 57% Emulsifiable Concentrate) in fuel oil. There may also be a reduction in hornflies. These devices should be made continuously accessible, one to each 35-45 head of cattle. Back-rubbing devices must be retreated every 2 to 3 weeks.	
	Ticks	1–2 gals per 100 gals of water OR 6½–13 ozs per 5 gals of water.	Apply complete coverage spray. Repeat at 2 week intervals if needed.
	Hornflies	1–1½ gals per 100 gals of water OR 6½–10 ozs per 5 gals of water.	Apply complete coverage spray. Repeat at 2 week intervals if needed.
			Do not apply to lactating dairy animals or nonlactating dairy animals within 2 weeks of freshening.
			Do not treat animals under one month of age.
			Residue tolerance: 4 ppm in meat, fat, and meat by-products.

ANIMAL	PEST	RATE	REMARKS
Poultry: Chickens, Ducks Geese, Turkeys	Northern fowl mite Poultry lice Chicken red mite	2 tbs per 1 gal water per 100-150 birds.	Repeat application in 4–8 weeks or when necessary. As a supplement to premise treatment for chicken red mite

ANIMAL	PEST	SITE	RATE	REMARKS
Poultry: Chickens, Ducks Geese, Turkeys	Northern fowl mite Chicken body lice Shaft lice	Tail dipping	$8\frac{1}{2}$ ozs per 15 gals water per 400 birds	Hold bird by wings and dip 3 to 4 inches of tail into solution. Treat vent and surrounding areas. Repeat in 7–10 days if necessary
	Chicken red mite Poultry lice	Roost paint	2–7 ozs per 1 gal water	Brush on at rate of 1 pt per 150 ft of roost.
	Northern fowl mite Chicken red mite Poultry lice Flies	Premise treatments	4 tbs per 1 gal water	Apply liberally to litter, walls, ceilings, roost nests and adjacent areas. Force spray into cracks and crevices
	Poultry ticks	Premise treatments	6–7 ozs per 1 gal water	Apply liberally to walls, ceilings and adjacent areas. Force spray into cracks and crevices.
	Chiggers	Range treatments	$1–1\frac{1}{2}$ pt per acre	Treat range thoroughly the day before placing poultry on range. Repeat every 2–3 weeks.
	Residue tolerance: 4 ppm in or on meat and meat by-products Eggs 0.1 ppm (from application to poultry)			
Domestic pets (dogs and cats)	Fleas Lice Ticks	—	1 oz/gal	Wet animal thoroughly. Repeat in 2 to 3 weeks if necessary.
	Fleas Ticks	Premise treatment	5 oz/gal	Apply per 1,000 sq ft of surface to pet quarters, yards and lawns. Remove manure or debris before treating. Repeat treatment in 3 to 4 weeks if necessary.

OUTDOOR ORNAMENTALS: FLOWERS, SHADE TREES, SHRUBS

Injury may occur on Ferns, Hickory, Viburnum, Lantana, Crassula and Canareti Juniper following the use of CYTHION Emulsifiable Concentrate, slight injury has also been reported on Boston, Pteris, and Maidenhair Ferns, Petunias, Small-Leaf Spirea, White Pine and Maples. Under extreme heat, drought and disease conditions the emulsifiable concentrates may cause slight damage to elms.

PEST	RATE	REMARKS
Aphids Spider mites	$1\frac{1}{2}$ pts. per 100 gal. OR 2 teaspoons per gal.	—
Bagworms Birch leaf miner Boxwood leaf miner	2 pts. per 100 gal. OR 2 teaspoons per gal.	—
European pine shoot moth Four-lined leaf bug Japanese beetle adult Potato leafhopper Rose leafhopper Tarnished plant bug Thrips	$1\frac{1}{2}$ pts. per 100 gal. OR 2 teaspoons per gal.	—
Lace bug	1 pt. per 100 gal. OR 2 teaspoons per gal.	—
Mealybugs Whiteflies	$1\frac{1}{2}$ pts. per 100 gal. OR 2 teaspoons per gallon	—
Oak kermes	2 pts. per 100 gal. OR 2 teaspoons per gal.	Apply when scale crawlers have settled on foliage.
Tent caterpillar	2 pts. per 100 gal. OR 2 teaspoons per gal.	—
Oyster shell scale	1 pt. per 100 gal. OR 2 teaspoons per gal.	Apply when scale crawlers have settled on foliage.
Euonymus scale Scurfy scale	$1\frac{1}{2}$ pt. per 100 gal. OR 2 teaspoons per gal.	—
Azalea scale Magnolia scale Pine leaf scale	2 pts. per 100 gal. OR 2 teaspoons per gal.	—
Fletcher scale	2 pts. per 100 gal. OR 2 teaspoons per gal.	Apply when scale crawlers have settled on foliage.
Florida red scale Juniper scale	2 pts. per 100 gal. OR 2 teaspoons per gal.	Apply when scale crawlers have settled on foliage.
Black scale crawlers Soft brown scale	$2\frac{1}{4}$ pts. per 100 gal. OR 2 teaspoons per gal.	—
Monterey pine scale	$2\frac{1}{2}$ pts. per 100 gal. OR 2 teaspoons per gal.	—
Pine needle scale	4 pts. per 100 gal. OR 4 teaspoons per gal.	—
Wax scale	4 pts. per 100 gal. OR 4 teaspoons per gal.	Apply in spring when crawlers are active. Repeat 1 or 2 full-coverage applications at 10-day intervals.

FOREST TREES

TREE	PEST	RATE	REMARKS
Deciduous Forest and Shade trees	Tent caterpillar	2 pts./100 gal. water	Make thorough coverage application before larvae reach third instar.
Pines	Pine needlesheath miner	1½ pts./acre	Apply by ground or aircraft equipment in 25 gals of water. Make application when air is calm and temperature is below 68°F. Apply when 75% of larvae have left overwintering mines.
Eastern pines	Pine tortoise scale	1½ pts./acre	Apply by aircraft in 2 gals. of water per acre when crawlers merge. Repeat application in 1 week under conditions of heavy infestations. Make application when air is calm and temperature is below 68°F.
Red pine	Red-headed pine sawfly	0.8 pt./acre	Apply by mist blower in 2 gallons of water per acre when egg hatch is complete.

AROUND THE HOME

SITE	PEST	RATE	REMARKS
In and around green-houses and gardens	Millipedes Sawbugs Springtails	1 tsp./gal.	Apply to 150 square feet of soil surface or where insects congregate. Repeat at 7 to 10 day intervals as needed.
In and around homes	Bedbugs	2 to 4 tablespoons per gallon of deodorized kerosene	Apply lightly to all mattress surfaces in sufficient quantity to "mist" the fabric and generously to beds and woodwork, with special care taken to wet all possible hiding places.
Lawns	Ant mounds	1½ pts./100 gal.	Spray ant hills thoroughly so that they are well soaked. For other small ants in flower beds, lawns, around trees, spray lightly in the infested areas. Repeat in 10 to 15 days if ants return.
	Ground pearls	3 to 4 qts./100 gal.	Make full coverage to soil surfact when ground pearl nymphs are in the pink, "crawler" or active stage and immediately wash into soil with additional water.

ON AND AROUND CULL FRUIT AND VEGETABLE DUMPS

PEST	RATE	REMARKS
Drosophila Dried fruit beetles	1½ gal per 100 gal water	Apply as a drench using 8 to 10 gal of spray per 100 sq. ft. For best results, dumps should not be over 18 inches deep. DO NOT FEED TREATED FRUIT AND VEGETABLES.

IN AND AROUND WINERIES AND PROCESSING PLANTS

PEST	RATE	REMARKS
Drosophila flies	3½ oz in 1 qt water	Paint all doors and window screens. AVOID CONTAMINATION OF WINE, FOOD, UTENSILS, EQUIPMENT AND WATER.

PLANTS PROCESSING DRY MILK

PEST	RATE	REMARKS
Black carpet beetles Trogoderma species	1 pt per 2½ gal	Clean premises thoroughly before applying and maintain good sanitation at all times. Use spray equipment and nozzles that will produce a coarse spray. Application must be made only by an experienced or trained person. Apply as a residual spray to all sections of the plant and warehouses where insects hide or crawl such as cracks, corners, edges of floors, lower parts of walls, floors under storage platforms and underneath and behind protected places. Avoid contamination of milk, dry milk, equipment, utensils, work surfaces, containers and liners. Repeat application as necessary.

CAUTION: CYTHION sprays may damage finished surfaces and fabrics. Avoid contamination of food, utensils, milk, milk equipment, and water. Do not use in milk processing rooms.

CRACK AND CREVICE TREATMENT IN FOOD HANDLING ESTABLISHMENTS IN FOOD AREAS: Includes areas for receiving, serving, storage, packing (canning, bottling, wrapping, boxing), preparing, edible waste storage and enclosed processing systems (mills, dairies, edible oils, syrups).

PEST	RATE	DIRECTIONS FOR USE
Ants Carpet beetles Cigarette beetles Clover mites Confused flour beetles Crickets	1 pt/2½ gal OR 1 part in 19 parts mixture of 4 parts kerosene-type solvent and 1 part aromatic hydro-carbon-type solvent.	Apply in small amounts directly into cracks and crevices using a pin stream of insecticide into points such as expansion joints between different elements of construction or between equipment bases and the floor, wall voids, motor housing, junction boxes or switch boxes, conduits or hollow equipment legs where the named insects hide. Care should be taken to

PEST	RATE	DIRECTIONS FOR USE
Drug store beetles Flat grain beetles Granary weevils Indian meal moths Red flour beetles Rice weevils Roaches Rusty grain beetles Saw-toothed grain beetles Silverfish Spiders		avoid depositing the product onto exposed surfaces or introducing the material into the air. Avoid contamination of food or food processing surfaces. Repeat application as necessary. Application must be made only by a commercial pest control operator.

IN NONFOOD AREAS: Includes garbage rooms, lavatories, floor drains (to sewers), entries and vestibules, offices, locker rooms, machine rooms, boiler rooms, garages, mop closets and storage.

PEST	RATE	DIRECTIONS FOR USE
Ants Carpet beetles Cigarette beetles Clover mites Confused flour beetles Crickets Drug store beetles Flat grain beetles Granary weevils Indian meal moths Red flour beetles Rice weevils Roaches Rusty grain beetles Saw-toothed grain beetles Silverfish Spiders	1 pt/2½ gal OR 1 part in 19 parts mixture of 4 parts kerosene-type solvent and 1 part aromatic hydro-carbon-type solvent.	Apply as a coarse spray to baseboard areas, around water pipes, surfaces behind and beneath sinks, lockers, tables, pallets and similar areas where insects hide or through which they may enter. Repeat application as necessary. Application must be made only by a commercial pest control operator.

CYTROL® Amitrole-T

LIQUID WEEDKILLER

ACTIVE INGREDIENT:

Amitrole*	21.1%
INERT INGREDIENTS	78.9%
TOTAL	100.0%

(1 gallon contains 2 pounds of Amitrole)

* 3-amino-1, 2, 4-triazole EPA Reg. No. 241-60-AA

CAUTION

Keep Out of Reach of Children

Harmful if swallowed. Avoid contact with skin, eyes or clothing.

PRECAUCIÓN

AL USUARIO: Si usted no lee inglés, no use este producto hasta que la etiqueta le haya sido explicada ampliamente.

(**TO THE USER:** If you cannot read English, do not use this product until the label has been fully explained to you.)

IMPORTANT

DO NOT USE THIS PRODUCT FOR ANY USES OTHER THAN THOSE SPECIFIED.

A method of killing weeds with the contained materials is covered by U.S. Patent No. 2,992,089. The price at which this material was originally sold included a prepaid royalty for the right of the ultimate consumer to use the same (alone without admixture or dilution except with water) in practicing with said patented method of killing weeds. A nontransferable license to practice said method with said material (alone without admixture or dilution except with water) from any source may be acquired from the patent owner on written request at the same royalty rate currently being charged.

Rinse spills on outside of bottle after use. When empty, wash thoroughly with water containing a household detergent. Recap and discard bottle. Never reuse.

DISCLAIMER

The label instructions for the use of this product reflect the opinion of experts based on field use and tests. The directions are believed to be reliable and should be followed carefully. However, it is impossible to eliminate all risks inherently associated with use of this product. Crop injury, ineffectiveness or other unintended consequences may result because of such factors as weather conditions, presence of other materials, or the use or application of the product contrary to label instructions, all of which are beyond the control of American Cyanamid Company. All such risks shall be assumed by the user.

- American Cyanamid Company warrants only that the material contained herein conforms to the chemical description on the label and is reasonably fit for the use therein described when used in accordance with the directions for use, subject to the risks referred to above.
- Any damages arising from a breach of this warranty shall be limited to direct damages and shall not include consequential commercial damages such as loss of profits or values or any other special or indirect damages.
- American Cyanamid Company makes no other express or implied warranty, including any other express or implied warranty of FITNESS or of MERCHANTABILITY.

DO NOT SPRAY OR ALLOW SPRAY DRIFT TO CONTAMINATE EDIBLE CROPS OR WATER WHICH WILL BE USED FOR IRRIGATION, DRINKING OR OTHER DOMESTIC PURPOSES AS NO RESIDUE OF CYTROL AMITROLE-T IS PERMISSIBLE ON CROPS. KEEP LIVESTOCK OFF TREATED AREAS. DO NOT SPRAY OR PERMIT DRIFT OF SPRAY

ON LAWNS, SHRUBS, FLOWERS AND OTHER DESIRABLE VEGETATION SINCE THE PRODUCT WILL CAUSE INJURY. DO NOT USE IN GREENHOUSES. WASH SPRAYER THOROUGHLY WITH CLEAN WATER IMMEDIATELY AFTER EACH USE TO AVOID CONTAMINATION OF OTHER SPRAYS. DO NOT CONTAMINATE FERTILIZERS, SEEDS, INSECTICIDES OR FUNGICIDES BY IMPROPER STORAGE.

CYTROL Amitrole-T is most effective when applied directly to the foliage of actively growing plants. It is absorbed by the leaves and travels through the entire plant—down to the deepest roots. Leaves wilt within 10 days to 3 weeks and the plants subsequently brown and die out.

CYTROL Amitrole-T should be applied in sufficient water to thoroughly cover weed foliage. Retreatment may be necessary to control plants not thoroughly sprayed or plants arising from dormant seeds.

WEED CONTROL ON NON-CROPLAND SUCH AS ALONG ROADSIDES, FENCE ROWS, DRAINAGE DITCH BANKS, RIGHTS-OF-WAY AND SIMILAR AREAS

DIRECTIONS FOR USE ON NON-CROPLAND

KEEP LIVESTOCK OFF TREATED AREAS, AVOID DRIFT OF SPRAY TO ADJACENT CROPLAND.

SPOT TREATMENT

Apply when growth is young and growing actively. Spray until weeds are thoroughly wet. Retreatment may be necessary to control plants not thoroughly sprayed or plants arising from dormant seeds.

For knapsack sprayer application, use a medium-size nozzle. A very coarse nozzle or watering can is not recommended for making spot treatments.

Mix ½ cup (4 fluid ounces) CYTROL **Amitrole-T** per gallon of water. One gallon of spray solution is usually sufficient to cover 272 square feet (1 square rod) thoroughly infested with weeds.

IN INTERMOUNTAIN STATES or other dryland areas, use ½ cup CYTROL **Amitrole-T** in 2 gallons of water to cover 1 square rod.

ACRE TREATMENTS
KEEP LIVESTOCK OFF TREATED AREAS

WEEDS	SPRAY MIXTURE GALS. CYTROL AMITROLE-T	GALS. WATER	WHEN TO APPLY
Quack grass (Couch, Witch)	2	30–60	Spray when vigorous young growth is 6 to 10 inches high. Repeat if needed.
Canada thistle Sow thistle	2	30–60	Spray when most thistles are up and in the bud to bloom stage. Do not mow treated plants. In the far West, treat during the bloom stage.
Horsetail rush	3	100	Spray when 6 to 8 inches tall and growing actively.
Bermuda grass	4	30–60	Treat from midsummer to early fall when shoots are growing vigorously. Spot treat any green regrowth with a solution of 1 gal. CYTROL Amitrole-T per 50 gals. of water.
Nutgrass (sedge)	2	40–60	Spray when shoots are 4 to 5 inches tall and growing actively. Retreat when dormant nuts germinate.
Leafy spurge (western states)	2	30–60	Spray in spring when weeds are in the bud to bloom stage. Do not mow treated plants. Spot treat any regrowth.
Horse nettle	2	100	In East and Mid-West, spray when horse nettle is in the bud to bloom stage. In the far West, treat from midsummer to early frost.
White top (Hoary cress)	2	30–60	Spray in spring when weeds are in the bud to bloom stage. Do not mow treated plants. Spot treat any regrowth.
Volunteer alfalfa	2	30–60	Spray when volunteer alfalfa is growing vigorously.
Dock	1	40–60	Apply any time when dock is actively growing, preferably before development of bloomstalk.
Milkweed	1	40–60	Spray in spring when weeds are in the bud to bloom stage. Do not mow treated plants. Spot treat any regrowth.

Use higher volumes for dense growth for good wetting.

IN INTERMOUNTAIN STATES: For Canada thistle—use 4 gals. CYTROL **Amitrole-T** in 100 gals. water per acre. All other weeds—use the rate of CYTROL **Amitrole-T** per acre as listed in the table above applied in a **minimum of 200 gallons of water per acre** or in sufficient water to thoroughly wet weed foliage.

WEEDS	SPRAY MIXTURE GALS. CYTROL AMITROLE-T	GALS. WATER	WHEN TO APPLY
Control Annual Grasses and Broadleaf Weeds such as: Foxtail, Ryegrass, Cheatgrass, Barnyard Grass, Annual Blu grass, Pigweed, Kochia, Sunflower	1	30–60	Spray when most weeds are 3 to 4 inches tall.
Control Perennial Broadleaf Weeds such as: Whitetop, Canada Thistle, Milkweed, Sow Thistle	2	30–60	Spray when most weeds are 6 to 10 inches tall.
Suppress Perennial Grasses such as: Reed Canarygrass, Ripgut Grass, Quackgrass, Wild Barley, Bermuda Grass	2	30–60	Spray when most weeds are 6 to 10 inches tall.

THESE SOLUTIONS SHOULD SPRAY 1 ACRE. A strip 8¼ feet wide and 1 mile long is about 1 acre. Repeat applications may be necessary in areas with long growing seasons. Use higher volumes for dense growth for good wetting.

WOODY PLANTS AND VINES

WEEDS	AMOUNTS TO USE GALS. CYTROL AMITROLE-T	GALS. WATER	WHEN TO APPLY
Poison Ivy Poison Oak	1	100 or (½ cup in 2 gals. water)	Treat from the time foliage is fully developed until plants begin to go dormant. For effective kill, all leaves, stems and suckers must be thoroughly wet to ground line.
Honeysuckle* Kudzu vine	2	125 or (1 cup in 2 gals. water)	
Salmon berry Western dewberry (blackberry) Big leaf maple	2	100 or (1 cup in 2 gals. water)	
Wild cherry Ash, Locust, Sumac	1½	100 or (¾ cup in 2 gals. water)	

* If regrowth appears, retreat in August or the following spring.

AQUATIC WEEDS IN MARSHES DRAINAGE DITCHES

	AMOUNTS PER ACRE GROUND SPRAYING	AERIAL SPRAYING	WHEN TO TREAT
Cattails North Central, Eastern States	3 gals. CYTROL Amitrole-T in 300 gals. water	3 gals. CYTROL Amitrole-T in 10 gals. water	Apply after catkins are fully formed but before frost. Do not disturb sprayed plants. Spot treat any regrowth.
Southwestern, Western States	4 gals. CYTROL Amitrole-T in 400 gals. water	5 gals. CYTROL Amitrole-T in 15 gals. water	
Phragmites North Central States	4 gals. CYTROL Amitrole-T in 100 gals. water	4 gals. CYTROL Amitrole-T in 5–10 gals. water	Apply after plants are 30 inches tall until the early fruiting stage. Use at least the volume of water suggested to assure complete wetting.
Atlantic Coast Region- flooded	10 gals. CYTROL Amitrole-T in 300 gals. water	10 gals. CYTROL Amitrole-T in 5 gals. water	
Nonflooded	7½ gals. CYTROL Amitrole-T in 300 gals. water	7½ gals. CYTROL Amitrole-T in 5 gals. water	
Water Hyacinth Southeastern States	½ to ¾ gal. CYTROL Amitrole-T in 200 gals. water	½ to ¾ gal. CYTROL Amitrole-T in 5 gals. water	Apply in April. Wet the weeds thoroughly, including small shoots under the main plant. Use the higher rate where growth is very dense.

DO NOT APPLY WHERE WATER WILL BE USED FOR IRRIGATING, DRINKING, FISHING OR OTHER DOMESTIC PURPOSES

HARDWOOD NURSERIES

Weeds	Amount Per Acre	When to Apply
Chrysanthemum	4 gals. in 100 gals. water	Apply as a directed spray on young weed growth 4 to 6 inches high making sure to keep the spray off the foliage of hardwood nursery stock.

FOR GENERAL WEED CONTROL IN NURSERY STOCK*, ORNAMENTAL PLANTINGS* AND FOREST PLANTATIONS*

CYTROL Amitrole-T	Simazine (80-W)	Gals. Water	Area
1 pt.	1 lb.	20	110 × 100
1 qt.	2 lbs.	40	220 × 100
½ gal.	4 lbs.	80	1 acre

Apply mixture to small weeds in the spring or early summer. Weeds should be wet thoroughly. Cut or mow tall weeds to ground line, remove cuttings and spray regrowth one or two weeks later. Where Johnson grass, Bermuda grass or other weeds are not controlled by the first application, treat again with the same rate when regrowth is 6 to 12 inches tall (about 2 months).

* The following plant species may be treated. Use a directed spray. Balsam fir, Douglas fir, Fraser fir, Norway spruce, Red spruce, White spruce, Red pine, Scotch pine, Arborvitae (Thuja), Juniper, Yew (Taxus), Oleander, Pyracantha and Viburnum.

CAUTION:
Lilac, privet, honeysuckle and barberry are susceptible to this mixture and may be injured if contacted by spray.

In the case of emergency endangering life or property involving this product, call collect, day or night. Area Code 201-835-3100.

MALATHION ULV* CONCENTRATE

INSECTICIDE

EPA Reg. No. 241-110-AA

and

CYTHION®

INSECTICIDE

The Premium Grade Malathion®

EPA Reg No. 241-208-AA

ACTIVE INGREDIENT:

Malathion**	91.0%
INERT INGREDIENTS:	9.0%
TOTAL	100.0%

**0,0-dimethyl phosphorodithioate of diethyl mercaptosuccinate

(1 gallon contains 9.33 pounds of malathion)

* Trademark of American Cyanamid Company

CAUTION

Keep Out of Reach of Children

HARMFUL BY SWALLOWING, INHALATION OR SKIN CONTACT

Avoid Breathing Spray Mist

Avoid Contact With Skin

Wash Thoroughly After Handling

Change Contaminated Clothing

Do Not Contaminate Food or Feed Products

CONTENTS OF THIS LEAFLET APPLY TO BOTH PRODUCTS.

IMPORTANT! BEFORE USING! STOP! READ DIRECTIONS IN THIS LEAFLET!

PRECAUCIÓN

AL USUARIO: Si usted no lee inglés, no use este producto hasta que la etiqueta le haya sido explicada ampliamente.

ENGLISH TRANSLATION

CAUTION

(**TO THE USER:** If you cannot read English, do not use this product until the label has been fully explained to you.)

IN CASE OF AN EMERGENCY ENDANGERING LIFE OR PROPERTY INVOLVING THIS PRODUCT, CALL COLLECT, DAY OR NIGHT, AREA CODE 201-835-3100.

BEFORE USING, READ THE DIRECTIONS CONTAINED IN THIS LEAFLET FOR THE PROPER METHODS AND PROCEDURES WHICH MUST BE FOLLOWED TO ACHIEVE EFFECTIVE INSECT CONTROL AND AVOID PERMANENT DAMAGE TO AUTOMOBILE AND OTHER PAINT FINISHES.

This product is toxic to fish. Avoid direct applications to lakes, streams, ponds, tidal marshes and estuaries. Do not apply where runoff is likely to occur. Do not apply when weather conditions favor drift from areas treated. Do not contaminate water by cleaning of equipment, or disposal of wastes. Shrimp and crab may be killed at application rates recommended on this label. Do not apply where these are important resources. Apply this product only as specified on this label.

DISCLAIMER

The label instructions for the use of this product reflect the opinion of experts based on field use and tests. The directions are believed to be reliable and should be followed carefully. However, it is impossible to eliminate all risks inherently associated with the use of this product. Crop injury, automobile paint damage, ineffectiveness or other unintended consequences may result because of such factors as weather conditions, presence of other materials, or the use or application of the product contrary to label instructions, all of which are beyond the control of American Cyanamid Company. All such risks shall be assumed by the user.

American Cyanamid Company warrants only that the material contained herein conforms to the chemical description on the label and is reasonably fit for the use therein described when used in accordance with the directions for use, subject to the risks referred to above.

Any damages arising from a breach of this warranty shall be limited to direct damages and shall not include consequential commercial damages such as loss of profits or values or any other special or indirect damages.

American Cyanamid Company makes no other express or implied warranty, including any other express or implied warranty of FITNESS or of MERCHANTABILITY.

DIRECTIONS FOR USE

It is a violation of Federal Law to use this product in a manner inconsistent with its labeling.

NOTE TO FORMULATORS:

CYTHION or MALATHION ULV may be used for the preparation of malathion insecticides. Before using for this purpose, manufacturers should consult American Cyanamid Company for manufacturing and handling instructions.

The sale of this product does not include a license under any patent owned by the American Cyanamid Company.

STORAGE AND DISPOSAL

See container label for storage and disposal instructions.

AGRICULTURAL USES

Operating Instructions

CYTHION or MALATHION ULV is used undiluted in specially designed aircraft or ground equipment capable of applying ultra low volumes for control of the insects indicated. Aerial applications are most effective when made at a boom height of 5 feet and a swath width of 50 feet. Apply only when weather conditions are favorable. Wind and rising air currents may cause undesirable spray drift and reduce insect control.

Mist blowers and boom sprayers utilizing a controlled air flow to facilitate particle size and spray deposition may be used at a vehicle speed of 4 to 10 mph.

Mist blowers with a pump capable of producing up to 40 psi and blower speeds of 2600 rpm are satisfactory. Use flat fan nozzles, 8001 to 8002, place 30° into air blast or rotary atomizers into the air blast that produce an efficient spray particle with a mass median diameter of 40 to 100 microns. Swath widths should not exceed 30 feet. Apply only when weather conditions are favorable. Wind and rising air currents may cause undesirable spray drift and reduce insect control.

Boom sprayers with a filtered rotary air compressor, either PTO or gas engine driven or an air pump capable of producing at least 12 psi are satisfactory. Use air pressure on chemical tanks and an accurate metering valve to assure a calibrated flow of the pesticide. Air should be regulated with relief valve and gauge for proper air and liquid mixture. Pneumatic-type spray nozzles, as suggested by equipment manufacturer, should be used for spray particles with mass median diameter of 30 to 100 microns. Apply only when weather conditions are favorable. Wind and rising air currents may cause undesirable spray drift and reduce insect control.

Repeat applications should be made as necessary unless otherwise specified.

CROP	PESTS CONTROLLED	FLUID OUNCES PER ACRE	DAYS TO HARVEST or GRAZING and COMMENTS
Alfalfa	Alfalfa Caterpillar Western Yellow Striped Armyworm	8	0 day. Apply when larvae are small.
		12	5 days. Apply when larvae are large or when foliage is dense.
	Alfalfa Weevil Larvae	16	5 days. Apply when day temperatures are expected to exceed 65° F. and when 50–75% of leaves show feeding damage.
	Beet Armyworm	8	0 day. Apply when larvae are small.
		16	5 days. Apply when larvae are large or when foliage is dense.
	Grasshoppers	8	0 day.
Do not apply to alfalfa in bloom. Do not apply to seed alfalfa.			
Beans (lima, green, snap, Navy, red kidney, wax, dry, blackeye)	Mexican Bean Beetle Leafhoppers Green Cloverworm Japanese Beetle Lygus Bug	8	1 day.
Blueberries	Blueberry Maggot	10	0 day.
Cherries	Cherry Fruit Fly	12–16	1 day. Apply by aircraft only. Use higher rate when foliage is heavy or infestation is severe. Make first application as soon as flies appear.
Cereal Crops, (barley, corn, oats, wheat) and grasses	Cereal Leaf Beetle	4–8	Barley, oats, wheat: 7 days. Corn: 5 days. Grasses: 0 day.
Clover, Pasture and Range Grass, Grass, Grass Hay, Nonagricultural Land (wastelands, roadsides)	Blackgrass Bugs Grasshoppers	8–12	0 day. Do not apply to clover in bloom.
Corn	Adult Corn Rootworm	4	5 days.
Cotton	Early Season Insects Thrips Fleahoppers Leafhoppers	4–8	0 day.
	Boll Weevil	8–12	Early to midseason
		16	Late season—0 day. May be tank mixed with once-refined cottonseed or vegetable oil sufficient to make a 1 qt. finished spray per acre.
	Grasshoppers	8	0 day.
	Lygus Bug	8–12	0 day.
		16	Very heavy migrating populations. 0 day.
Flax	Grasshoppers	8	Apply once per season. Do not apply within 45 days of harvest. Do not graze or feed forage.
Grain Crops (barley, corn, oats, rye, rice, grain sorghum and wheat)	Grasshoppers	8	7 days, except corn. Corn: 5 days.
Grain Sorghum	Sorghum Midge	8–12	7 days. Apply during the bloom stage.
Peas (Northwest)	Pea Weevil	8	14 days.
Rice—Grain Form (Louisiana, Texas)	Rice Stink Bug	8	7 days. Apply by aircraft only. Apply during early milk and dough stage of growing rice.
Safflower	Grasshoppers Lygus Bug	8	3 days of harvesting seeds.
Soybeans	Mexican Bean Beetle, Grasshoppers, Japanese Beetle, Green Cloverworm	8	7 days.
Sugar Beets	Grasshoppers Sugar Beet Root, Maggot Adults	8	0 day. 7 days. If tops are to be used for food or feed.
Nonagricultural Lands	Beet Leafhopper (on wild host plants)	8	0 day.
Beef Cattle—Feed Lots and Holding Pens	Adult Flies and Mosquitoes	6–8	0 day.

OTHER AGRICULTURAL USES

Alfalfa, Clover, Pasture and Range Grass, Grass and Grass Hay, Grain Crops, Beans, Rice, Tomatoes and Nonagricultural Lands (wasteland): Adult mosquitoes and flies-Apply CYTHION or MALATHION ULV at the rate of 2 to 4 fluid ounces for control of adult mosquitoes and at 6 to 8 fluid ounces per acre for control of adult flies and mosquitoes. Repeat applications as necessary. On alfalfa, clover, pasture and range grass, grass and grass hay, may be applied on day of harvest or grazing. Do not apply to alfalfa and clover in bloom. Do not use on seed alfalfa. On grain crops, make no application within 7 days of harvest or forage use; on corn, within 5 days of harvest or forage; on rice, within 7 days of harvest; on beans and tomatoes, within 1 day of harvest.

FOR CONTROL OF BOLL WEEVIL AND BOLLWORM/ BUDWORMS

CYTHION + Pyrethroid Tank Mixes

Pyrethroids*	Fluid Ounces/Acre	Comments
[1]PAY-OFF® insecticide 2.5 E.C.	1.3–4.0	Tank mix with 12–16 fl. oz. of CYTHION or Malathion ULV and apply as a 1 qt. finished spray per acre in
[2]PYDRIN® insecticide 2.4 E.C.	5⅓–10⅔	once-refined cottonseed or vegetable oil. Consult the pyrethroid label for detailed directions. For POUNCE tank mix
[3]POUNCE® insecticide 3.2 E.C.	2.0–8.0	with 12–16 fl. oz. of CYTHION or Malathion ULV. Add these materials to one quart of once-refined cottonseed or vegetable oil and apply this mixture with proper application equipment.

[1]Registered Trademark, American Cyanamid Company

[2]Registered Trademark, Shell Chemical Company

[3]Registered Trademark, FMC Corporation

IMPORTANT

Undiluted spray droplets of CYTHION or MALATHION ULV will permanently damage automobile paint. If accidental exposure does occur, the vehicle should be washed immediately.

Consult your state experiment station or state extension service for proper timing of sprays.

These products are highly toxic to bees exposed to direct treatment or residues on crops. Protective information may be obtained from your Cooperative Agricultural Extension Service.

All applicable directions, restrictions and precautions on the EPA registered labels are to be followed.

Operator should inspect and calibrate equipment to assure that the proper amounts of pesticide are being applied, and that uniform coverage is obtained.

This label must in the possession of the user at time of pesticide application.

* Lower rates should be used only with close field supervision and low pest numbers.

FOREST INSECTS

Apply with aircraft equipped for ultra low volume application. Make application when air is calm and temperature is below 68° F. Do not allow spray to contact ferns, hickory and maples as injury may result. Do not spray on elms under extreme heat, drought and disease conditions.

Tree	Pests Controlled	Fluid Ounces Per Acre	Comments
Douglas Fir True Fir Spruce	Spruce Budworm	13	Apply when highest percentage of larvae are in the fifth instar.
Hemlock	Hemlock Looper	8	Apply when most larvae are in third and fourth instar.
Pines	European Pine Sawfly	10	Apply when larvae are in the first or second instar or before they reach ½ inch in length.
	Saratoga Spittlebug		Apply when 95% of the population has become adult.
Larch	Larch Casebearer	8	Apply in spring as soon as larvae break hibernation and begin feeding on new foliage.

Undiluted spray droplets of CYTHION® insecticide, 'The Premium Grade Malathion'® or MALATHION ULV* concentrate insecticide will permanently damage automobile paint unless these specific instructions for ground and aerial application are followed.

MOSQUITO CONTROL IN POPULATED AND RURAL AREAS

AERIAL APPLICATION

ADULT MOSQUITO CONTROL OVER CITIES, TOWNS, AND OTHER AREAS WHERE AUTOMOBILES, TRAILERS, TRUCKS AND PLEASURE BOATS ARE PRESENT: Apply 2.6 to 3.0 fluid ounces of CYTHION or MALATHION ULV per acre. Apply only when weather conditions are favorable. Wind and rising air currents may cause undesirable spray drift and reduce insect control.

IMPORTANT—Undiluted spray droplets of CYTHION or MALATHION ULV will permanently damage vehicle paint finishes unless the aircraft used for the ultra low volume application meets all of the specifications listed below.

FIXED WING AIRCRAFT

1. Aircraft is operated at 150 mph or more.
2. There are no leaks in the ultra low volume spray system.
3. Nozzles are placed on the boom at a 45° angle down and into the wind.
4. Diaphragm check valves are used on all nozzles to insure positive cut-off of the spray.
5. Dosage of CYTHION or MALATHION ULV does not exceed 3 fluid ounces per acre.
6. The spray system produces droplets of this product in the 50 to 60 mass median diameter (MMD) micron range, with no more than 10% of the droplets exceeding 100 microns, as determined by readings made from microscope slides coated with DRI-FILM* or TEFLON®.

* Trademark of General Electric Company

® Registered Trademark of E.I. duPont de Nemours & Co., Inc.

HELICOPTER

Equipment specifications

1. Rotary nozzle equivalent to Beecomist Spray Head Assembly Model No. 35C equipped with:
   a. a direct reading RPM tachometer or low RPM signal light readily visible to operator;
   b. a stainless steel porous metal sleeve, 20 micron pore size, dynamically balanced to the nozzle;
   c. a diaphragm check valve as near to the rotary nozzle as possible to insure positive cut off of the spray;
   d. nozzle on-off switch separate from main switch and pump switch.
2. Minimum no-load nozzle speed of 10,500 RPM.
3. A continuous nonpulsating metered flow must be maintained by a variable speed metering pump equipped with:
   a. a positive cut off valve between tank and pump;
   b. a flow gauge or tachometer visible to operator;
   c. a pump on-off switch separate from main switch and nozzle switch.
4. Maximum flow rate of 0.5 gallon per minute per nozzle.
5. Rotary nozzle must be mounted behind and below the boom with the sleeve directed toward the rear of the aircraft and parallel to the ground during flight. Nozzle must be positioned to minimize air turbulence and the collection of CYTHION or MALATHION ULV droplets on mounting brackets, feed lines, fittings, etc., or any part of the aircraft.

Operating Procedures

1. CYTHION or MALATHION ULV must be prefiltered through

a 10 micron filter prior to transfer into helicopter tank. A 50 mesh stainless steel line strainer must be installed in the pump feed line.

2. Entire system, including tank, pump, nozzle and feed lines, to be used only for application of CYTHION or MALATHION ULV.
3. Entire system must be inspected daily to insure that there are no leaks.
4. Sleeve must be removed and cleaned immediately after each use by washing with hot water and blowing dry from outside in with clean air.
5. Rotating nozzle must be turned on and operating before turning on pump. For shut off, pump must be shut off and lines cleared prior to stopping nozzle rotation.
6. Dosage of CYTHION or MALATHION ULV does not exceed 3 fluid ounces per acre.
7. The spray system must produce droplets of CYTHION or MALATHION ULV with a mass median diameter (MMD) of less than 50 microns, with no more than 2.5% of the droplets exceeding 100 microns, as determined by readings made from microscope slides coated with DRI-FILM or TEFLON.

GROUND APPLICATION

Thermal Aerosols or Fogs

For control of adult mosquitoes with thermal aerosols or fogs, apply CYTHION or MALATHION ULV at the rate of 6–8 oz. actual/gallon (3.9–5.2 gallons CYTHION or MALATHION ULV in 100 gallons finished solution*) by ground equipment delivering 40 gallons per hour at a vehicle speed of 5 miles per hour to treat a swath width of 300–400 feet.

* There is a great variation in the chemical composition of fuel oils which may be used as thermal fog solvents. These differences may cause sludge and/or affect the solubility of the CYTHION or MALATHION ULV. For more complete details on tests for sludge formation and solubility in thermal fog solutions, write American Cyanamid Company, One Cyanamid Plaza, Wayne, New Jersey 07470.

Nonthermal Aerosols

Over a 300-foot swath can be produced using the non-thermal ultra low volume aerosol method with CYTHION or MALATHION ULV.

Vehicle Speed Rate per Hour	Flow Rate of MALATHION Fluid Ounces per Minute	Maximum Flow Rate per Hour
5	1.0 to 2.1 fluid ounces	1 gallon
10	2.0 to 4.3 fluid ounces	2 gallons
15	3 to 6.45 fluid ounces	3 gallons
20	4 to 8.6 fluid ounces	4 gallons

For control of adult stable fly in populated and rural areas with non-thermal aerosols of CYTHION or MALATHION ULV using the ultra low volume method, use the following flow rates at the indicated vehicle speeds:

Vehicle Speed Miles per Hour	Flow Rate of MALATHION Fluid Ounces per Minute	Maximum Flow Rate per Hour
5	2.1 fluid ounces	1 gallon
10	4.3 fluid ounces	2 gallons

DROPLET SIZE

1. The Mass Median Diameter (MMD) of the droplets should not exceed 17 microns. The MMD is the drop diameter which divides the spray volume into two equal parts; i.e., 50% of the volume is in the drop sizes below the MMD and 50% is above the MMD.
2. Spray droplets should not exceed 32 microns in size. Three percent of the spray droplets (6 droplets out of 200) can exceed 32 microns providing the MMD does not exceed 17 microns and no droplets exceed a maximum of 48 microns. Larger droplets, when transported by natural air currents, impinge more readily on objects in their pathway and will permanently damage automobile-type paints.
3. More than one-half of the total spray mass must consist of droplets in the 6 to 18 micron range to achieve adequate dispersal of insecticide over a 300-foot swath.
4. A minimum of two-thirds, preferably four-fifths of the total spray mass must consist of droplets not exceeding 24 microns in range.

OPERATING EQUIPMENT

Each Nonthermal Aerosol Generator used for dispersal of CYTHION or MALATHION ULV to control adult mosquitoes must have minimum capability of producing the droplet spectrum described under DROPLET SIZE. The initial determination of droplet size is made after the unit is installed in a vehicle and prior to its use in mosquito control operations. The unit should be rechecked as frequently as necessary to insure that proper droplet size is maintained for each operation. Determination of droplet size every two months is usually sufficient if the unit has been maintained in good operating condition. Equipment manufacturer's instructions setting forth cleaning and maintenance of the unit must be followed. The unit must be inspected before each operation to correct any leaks or obstructions in the spray system; to detect whether the nozzle, hoses, or other parts are worn and need replacement; to insure that the flow meter is properly calibrated; and to determine that the pressure recommended by the manufacturer is being maintained.

Flow Rate—must be regulated by accurate flow meter.
—not greater than 1 gallon per hour at 5 mph, 2 gallons per hour at 10 mph, 3 gallons per hour at 15 mph, or 4 gallons per hour at 20 mph.

Nozzle Direction—rear of the vehicle.
—upward at an angle of 45° or more.

Vehicle Speed—not greater than 20 miles per hour.
—shut off spray equipment when vehicle is stopped.

IMPORTANT—Spray droplets of undiluted CYTHION or MALATHION ULV will permanently damage automobile paint unless all the conditions described and recommended in this leaflet are met.

Directions for Determining the droplet size of CYTHION or MALATHION ULV nonthermal aerosols

Permanent records of each droplet size determination must be kept and made available to American Cyanamid upon request.

1. Preparation of Slides with DRI-FILM

CYTHION or MALATHION ULV droplet sizes are determined by depositing a sample of the aerosol on a coated glass slide and measuring the droplets under a high-power microscope. Ordinary 3″ × 1″ glass slides must be coated with silicone (General Electric SC-87 DRI-FILM) prior to sampling to prevent excessive spreading or coalescence of the droplets. The slides are dipped into a 10 percent solution of DRI-FILM in toluene, drained and dried at about 200° F. for 30 minutes, after which they are dipped in acetone, allowed to dry and stored in a tight slide box. Coating solution must be freshly prepared. Do not store coating solution because it will deteriorate. Slides are lightly polished with a soft tissue before using to remove any foreign particles.

2. Deposition of CYTHION or MALATHION ULV Droplets on Slides

Droplets should be collected under ideal operating conditions to insure representative sampling of droplets in the aerosol. A sample of the CYTHION or MALATHION ULV aerosol is deposited on a slide by waving the slide as rapidly as possible perpendicular through the aerosol cloud at a distance of 25 feet from the point of discharge. The slide velocity may be increased by attaching it to a 3 or 4 foot stick by means of a spring paper clip. At least two slides should be exposed to insure an adequate sample. Store slides in a tight slide box for transfer to a location where measurements can be made. Avoid excessive heat during transit and store in a cool place until measurements can be made.

Although label specifications require the aerosol nozzle to be angled upward at 45° or more during operation, it is more convenient to position the nozzle parallel to the ground for droplet sampling. If this is not possible it will be necessary to be positioned at a sufficient height to obtain a representative sample of the aerosol.

3. Determination of CYTHION or MALATHION ULV Droplet Sizes

A microscope with mechanical stage and an eyepiece micrometer are used to determine the size of the individual aerosol droplets. Prior to taking measurements, the divisions of the eyepiece micrometer must be calibrated into microns by means of a stage micrometer. In the example represented in Table 1, droplets were measured at 400x magnification. At that magnification each division of the eyepiece was calibrated to equal 3.5 microns.

At least 200 droplets should be measured. Usually this is easily accomplished on one slide. An accurate method is to measure all droplets that pass through the micrometer scale as the slide is moved from one edge to the other by using the mechanical stage. Measurements should not be taken along the margins of the slide. It is more convenient to measure in terms of the divisions of the eyepiece micrometer and then convert these divisions into microns.

The measurements converted into microns must then be corrected for the amount of spread that occurred on the slides. The CYTHION or MALATHION ULV spread factor for silicone-coated slides is 0.5. Therefore, in Table 1 each division of the eyepiece actually equals 1.75 microns (3.5 microns times the 0.5 spread factor).

The spread factor for TEFLON-coated slides is 0.69. The following procedure is given for silicone-coated slides, would be the same for TEFLON-coated slides once the value for each eyepiece division has been determined.

The measurements are tabulated and processed as in Table 1. The Maximum Diameter is calculated by converting the diameter of the largest droplet measured into microns. In Table 1, the largest droplet measured had a diameter of 19 eyepiece divisions. Therefore, the Maximum Diameter is 33.3 microns (19 × 1.75 = 33.3).

To determine the Mass Median Diameter (MMD), the accumulative percentages from the last column in Table 1 are plotted against the eyepiece divisions (D) on semilogarithmic paper as in Figure 1. Directly across from the 50 percent point on the line is the median droplet size in eyepiece divisions which must be converted to microns. In Figure 1, 9.2 eyepiece divisions times the conversion factor of 1.75 equals a Mass Median Diameter of 16.1 microns.

Figure 1.
Percentage of the total volume of aerosol samples below each stated droplet size (from Table 1). The Mass Median Diameter is determined from the 50 percent point on the line. The Mass Median Diameter (MMD) = 9.2 divisions times 1.75 = 16.1 microns.

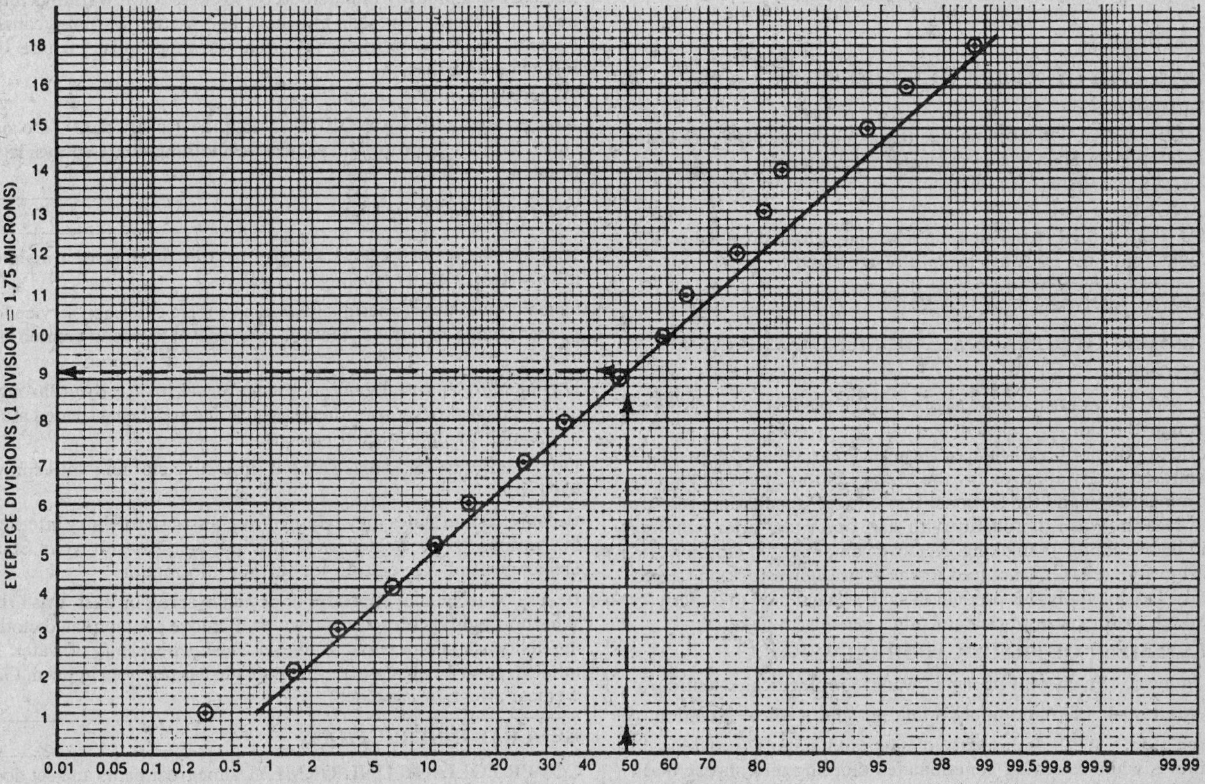

Table 1—Representative Count of CYTHION or MALATHION ULV Aerosol Droplets Impinged on Microscope Slides Coated with DRI-FILM.

Eyepiece Divisions (D)*	Number of Droplets (N)	D×N	% of Total D×N E (D×N)	Accumulative Percentages
1	5	5	0.31	0.31
2	10	20	1.22	1.53
3	9	27	1.65	3.18
4	12	48	2.93	6.11
5	15	75	4.58	10.69
6	12	72	4.40	15.09
7	25	175	10.70	25.79
8	14	112	6.85	32.64
9	28	252	15.40	48.04
10	19	190	11.61	59.65
11	14	154	9.41	65.06
12	10	120	7.33	76.39
13	6	78	4.77	81.16
14	4	56	3.42	84.58
15	11	165	10.09	94.67
16	2	32	1.96	96.63
18	2	36	2.20	98.83
19	1	19	1.16	99.99
Total	199	1636		

* Measurements were taken at 400× magnification. Each eyepiece division equals 1.75 microns (3.5 microns times the 0.5 spread factor).

Also for use in accordance with the recommendations and instructions issued by the United States Department of Agriculture for quarantine programs. To be used only by or under the direction of Federal/State personnel for quarantine treatments.

PAY-OFF®
INSECTICIDE

RESTRICTED USE PESTICIDE

For retail sale and use only by Certified Applicators or persons under their direct supervision and only for uses covered by the Certified Applicators certfication.

ACTIVE INGREDIENT:
Flucythrinate(±) Cyano (3-phenoxyphenyl)methyl (+)-4-(difluoro-methoxy)-alpha-(1-methylethyl)
benzeneacetate 30.0%
INERT INGREDIENTS: 70.0%

TOTAL 100.0%

EPA Reg. No. 241-259 EPA Est. No. 241-NJ-02

D-42

® Registered Trademark American Cyanamid Company

DANGER
Keep Out of Reach of Children

PELIGRO!

AL USUARIO: Si usted no lee inglés, no use este producto hasta que la etiqueta le haya sido explicada ampliamente.

ENGLISH TRANSLATION

DANGER

(TO THE USER: If you cannot read English, do not use this product until the label has been fully explained to you.)

PRECAUTIONARY STATEMENTS

HAZARDS TO HUMANS AND DOMESTIC ANIMALS

DANGER

Danger! Causes eye damage. May be fatal if swallowed or absorbed through the skin, or inhaled. Do not get in eyes, or on skin or clothing. Do not breathe vapor or mist. Wear a mask or pesticide respirator jointly approved by the Mining Enforcement and Safety Administration and the National Institute for Occupational Safety and Health. Wash thoroughly with soap and water after handling and before eating or smoking. Remove contaminated clothing and wash before reuse. Do not feed forage or graze livestock on treated fields. Do not apply this product in such a manner as to directly or through drift expose workers or other persons.

STATEMENT OF PRACTICAL TREATMENT

If in eyes: Flush with plenty of water. Get medical attention.

If swallowed: Do not induce vomiting. Get medical attention.

If on skin: Wash with plenty of soap and water for at least 15 minutes while removing contaminated clothing and shoes. Wash clothing before re-use. Occasionally, individuals will experience a flushing or burning sensation on the skin. The reaction usually subsides in several hours.

If inhaled: Remove victim to fresh air. If not breathing, give artificial respiration, preferably mouth-to-mouth. Get medical attention.

ENVIRONMENTAL HAZARDS

This pesticide is extremely toxic to fish and aquatic organisms. Do not apply it directly to water. Maintain a buffer zone (lay-off distance) of 100 feet from bodies of water. Do not contaminate water by the cleaning of equipment or disposal of wastes. Do not apply when conditions favor drift from the target area.

Notify State and/or Federal authorities and American Cyanamid Company immediately if you observe any adverse environmental effects due to the use of PAY-OFF insecticide.

This pesticide is highly toxic to bees exposed to direct treatment or residues on blooming crops or weeds. Do not apply it or allow it to drift to crops or weeds on which bees are actively foraging.

Written or oral warnings must be given to workers who are expected to be in a treated area or in an area about to be treated with this product. Inform workers of areas or fields that may not be entered without specific protective clothing, period of time field must be vacated and appropriate actions to take in case of accidental exposure. An example of such information is given under written warnings. When oral warnings are given, warnings shall be given in a language customarily understood by workers. Oral warnings must be given if there is a reason to believe that written warnings cannot be understood by workers. Written warnings must include the following information:

DANGER

Area treated with PAY-OFF® insecticide (insert date). Do not enter without appropriate protective clothing until sprays have dried.

DIRECTIONS FOR USE

It is a violation of Federal Law to use this product in a manner inconsistent with its labeling.

Consult your state experiment station or state extension service for proper timing of sprays. Do not graze cover crops in treated orchards.

Do not apply this product in such a manner as to directly or through a drift expose workers or other persons. The area being treated must be vacated by unprotected persons.

Do not enter treated areas without protective clothing until sprays have dried.

Because certain states may require more restrictive reentry intervals for various crops treated with this product, consult your State Department of Agriculture for further information.

In case of accidental exposure to pesticide spray or dust, wash the skin thoroughly with soap and water. Remove contaminated clothing and wash before reuse. If in eyes, flush with plenty of water. If inhaled, go to an area where the pesticide has not been applied. Get medical attention if needed.

GENERAL INSTRUCTIONS

COVER FOLIAGE UNIFORMLY! Time treatments and set dosages according to insect population or damage. Repeat treatments as necessary to maintain control. Consult your local extension service or professional consultant for treatment thresholds.

Ground Application: Use water for dilution and apply at least 5 gallons of finished spray per acre.

CROP	INSECT	APPLICATION RATES (FL. OZ. OF PAY-OFF® INSECTICIDE PER ACRE)		REMARKS
Cotton	Cotton Bollworm[1] Tobacco Budworm[1]	2.0–2.5 Moderate to Heavy Populations		
		1.3–2.0 Light to Moderate Populations. Use these rates in frequently scouted fields. Make treatments promptly after the decision to treat has been made.	2.5–4.0 Heavy to Extremely Heavy Populations. Use these rates and close intervals to bring populations under control.	DO NOT apply within 28 days of harvest DO NOT graze or feed cotton for forage

CROP	INSECT	APPLICATION RATES (FL. OZ. OF PAY-OFF® INSECTICIDE PER ACRE)	REMARKS
	Boll Weevil	2.0–4.0 These rates control weevils if applied at 3–4 day intervals, and suppress weevils if applied at the longer intervals typical of budworm/bollworm control programs.	
	Cabbage Looper Pink Bollworm Saltmarsh Caterpillar	1.3–3.0	
	Cotton Leafperforator Lygus Bugs	2.0–4.0 1.3–4.0	

[1] If once-refined vegetable oil is the diluent, ½ the label rate of PAY-OFF may be used but not less than 1.3 fl. oz. per acre.

Aerial Applications: If water is used for dilution, apply at least one gallon of finished spray per acre. Apply at least 3 gallons finished spray per acre in Arizona, and at least 5 gallons finished spray per acre in California.

If once-refined vegetable oil is used for dilution, apply at least one quart finished spray per acre. When once-refined vegetable oil is the diluent, applications may be made at one-half of the dose rates given in the following table, but not less than 1.3 fl. oz. of PAY-OFF insecticide per acre. Applications in once-refined vegetable oil are not registered for use in California.

TANK MIXES

For Control of Boll Weevil and Bollworm/Budworm in Cotton

PAY-OFF may be tank mixed with once-refined cottonseed or vegetable oil and CYTHION® insecticide 'The Premium Grade Malathion'® or Malathion ULV* concentrate insecticide for use on cotton in accordance with the more restrictive label limitations and precautions. No label dosage rates should be exceeded. This product cannot be mixed with any product containing a label prohibition against such mixing.

Use Rates

PAY-OFF	CYTHION or Malathion ULV
1.3 to 4 oz.	12–16 oz.

CROP	INSECT	APPLICATION RATES LB.AI/ACRE	FLUID OUNCES PER ACRE	REMARKS
Field Corn	Cutworms	0.04–0.08	2.0–4.0	Do not apply by air. Do not make more than 2 applications per season. Do not apply to dry cloddy soil when cutworms are underground and hard to control. Do not harvest corn grain, do not cut for fodder or silage, and do not allow grazing of treated fields within 60 days of the final application.

ROW CROPS

CROP	INSECT	APPLICATION RATES LB.AI/ACRE	FLUID OUNCES PER ACRE	REMARKS
Cabbage	Cabbage Looper Corn Earworm Diamondback Moth Imported Cabbageworm Saltmarsh Caterpillar	.025–.04	1.3–2.0	Repeat as needed to maintain control. Do not exceed 16 fl. oz. PAY-OFF per acre per season. Do not apply within 3 days of harvest. Do not feed treated plant parts to livestock.
Lettuce, Head	Diamondback Moth Cabbage Looper Corn Earworm Saltmarsh Caterpillar	0.025–0.04	1.3–2.0	Repeat as needed to maintain control. Do not exceed 10 fl. oz PAY-OFF per acre per season. Do not apply within 7 days of harvest. Do not feed treated plant parts to livestock.

FOR USE ON PEARS

CROP	INSECT	APPLICATION RATES LB.AI/ACRE	FLUID OUNCES PER ACRE	REMARKS
Pears	Pear Psylla[1]	1.25*	5.0	Apply during dormant to prebloom (white bud) stage only. Do not make more than 1 application per season. DO NOT GRAZE LIVESTOCK IN TREATED AREAS. Cover trees uniformly.

* Based on 400 gallons per acre spray solution.
[1] Supreme or superior type oil may be added to the spray solution at recommended rates.

FOR USE ON APPLES

CROP	INSECT	APPLICATION RATES LB.AI/ACRE	FLUID OUNCES PER ACRE	REMARKS
Apples	Rosy Apple Aphid White Apple Leafhopper Apple Maggot Fruit Tree Leafroller Green Apple Aphid Green Fruitworm Lesser Appleworm	0.1–0.18 0.12–0.18	8–9 6–9	Cover trees uniformly using 200–600 gals. of finished spray per acre. Repeat as needed to maintain control. Time applications for optimum insect control. [2] Do not exceed 45 fl. oz. of PAY-OFF per acre per season. Do not graze livestock in treated areas. Do not apply within 14 days of harvest.

CROP	INSECT	APPLICATION RATES LB.AI/ACRE	FLUID OUNCES PER ACRE	REMARKS
	Oblique-Banded Leafroller Periodical Cicada Plum Curculio Red-Banded Leafroller Spotted Tentiform Leafminer Tarnished Plant Bug Tufted Apple Budmoth Codling Moth	0.10–0.18	5–9	

USE INFORMATION

[1] Use high range of label rates for control of heavy infestations or when conditions exist that make insect control difficult. Lower range of label rates should be used only under good orchard management conditions for control of low to moderate insect infestations.

[2] PAY-OFF is not a systemic insecticide. Insect pest must come into contact with spray material to be controlled. Leafminer larvae in their mines and aphids or leafrollers protected by curled leaves will not be effectively controlled. For best control, time applications for maximum insect exposure.

PAY-OFF is compatible with other insecticides, acaricides, and fungicides.

PAY-OFF may be tank mixed with up to 10 gallons per acre superior oil for pre-bloom applications. Determine physical compatibility of tank mixtures BEFORE they are mixed in the spray tank. Observe all cautions and limitations on the labels of all product used in tank mixtures.

Conversion Chart

Rate fl. oz./acre	Approximate Rate pounds a.i./acre	Approximate acres treated by 1 quart	1 gallon
1.0	0.020	31.0	124.0
1.3	0.025	24.5	98.5
1.5	0.030	20.8	83.0
1.8	0.035	18.0	72.0
2.0	0.040	16.0	64.0
2.5	0.050	13.0	51.0
3.0	0.060	10.5	42.5
3.5	0.070	9.0	36.5
4.0	0.080	8.0	32.0
4.5	0.090	7.0	28.5
5.0	0.100	6.5	25.5
5.5	0.105	6.0	23.5
6.0	0.115	5.5	21.5
6.5	0.125	5.0	19.5
7.0	0.135	4.5	18.5
7.5	0.145	4.5	17.0
8.0	0.155	4.0	16.0
8.5	0.165	4.0	15.0
9.0	0.175	3.5	14.0
9.5	0.185	3.5	13.5
10.0	0.195	3.0	13.0

PHYSICAL OR CHEMICAL HAZARDS COMBUSTIBLE LIQUID AND VAPOR

Do not use or store near heat or open flame. Do not use in the vicinity of pilot lights. Keep container closed.

STORAGE AND DISPOSAL

STORAGE

DO NOT contaminate water, food, or feed by storage or disposal.

PESTICIDE DISPOSAL: Pesticide wastes are acutely hazardous. Improper disposal of excess pesticide, spray mixture, or rinsate is a violation of Federal law. If these wastes cannot be disposed of by use according to label instructions, contact your State Pesticide or Environmental Control Agency, or the Hazardous Waste representative at the nearest EPA Regional office for guidance.

CONTAINER DISPOSAL: If Metal Triple-rinse (or equivalent). Then offer for recycling or reconditioning, or puncture and dispose of in a sanitary landfill, or by other procedures approved by state and local authorities.

If Plastic: Triple-rinse (or equivalent). Then offer for recycling or reconditioning, or puncture and dispose of in a sanitary landfill, or incineration, or, if allowed by state and local authorities, by burning. If burned, stay out of smoke.

DISCLAIMER

American Cyanamid Company warrants only that the material contained herein conforms to the chemical description on the label and is reasonably fit for the use therein described when used in accordance with the directions for use.

Any damages arising from a breach of this warranty shall be limited to direct damages, and shall not include consequential commercial damage such as loss of profits or values, etc.

American Cyanamid Company makes no other express or implied warranty, including any other express or implied warranty of FITNESS or of MERCHANTABILITY.

User assumes the risk of any use contrary to label instructions or under abnormal conditions, or under conditions not reasonably foreseeable by American Cyanamid Company.

In case of an emergency endangering life or property involving this product, call collect, day or night, Area Code 201-835-3100.

Consult your local extenstion service or professional consultant for proper timing of PAY-OFF applications.

PROWL®

HERBICIDE

For use in Field Corn, Cotton, Soybeans, Sorghum, Rice, Potatoes, Tobacco, Peanuts and Sunflowers, Dry Beans, Non-Bearing Trees and Vines

For Control of Most Annual Grasses and Certain Broadleaf Weeds

ACTIVE INGREDIENT:

Pendimethalin (N-(1-ethylpropyl)-3,4-dimethyl-2, 6-dinitrobenzenamine)	42.3%
INERT INGREDIENTS	57.7%
TOTAL	100.0%

(1 gal. contains 4 lbs. of Pendimethalin)

EPA Reg.No. 241-243-ZA

WARNING

Keep Out of Reach of Children

PROWL SHOULD BE USED IN ACCORDANCE WITH THE DIRECTIONS IN THIS LEAFLET, OR IN SUPPLEMENTAL LABELS FOR REGISTERED USES NOT INCLUDED IN THIS LEAFLET AVAILABLE THROUGH LOCAL DEALERS. READ ALL DIRECTIONS CAREFULLY BEFORE USING.

In case of an emergency endangering life or property involving this product, call collect, day or night, Area Code 201-835-3100.

ADVERTENCIA!
AL USARIO: Si usted no lee inglés, no use este producto hasta que la etiqueta le haya sido explicada ampliamente.
(ENGLISH TRANSLATION)
WARNING!
(TO THE USER: If you cannot read English, do not use this product until the label has been fully explained to you).

Read all information on pages 3–5 before referring to specific crop use.

PRECAUTIONARY STATEMENTS

HAZARDS TO HUMANS

WARNING

HARMFUL IF SWALLOWED OR INHALED,

CAUSES EYE AND SKIN IRRITATION.

Avoid contact with eyes, skin, and clothing. Avoid breathing spray mist. Wash thoroughly after handling. Use with adequate ventilation. Keep container closed.

FIRST AID: If swallowed, DO NOT induce vomiting. Call a physician immediately. In case of contact, immediately flush eyes with plenty of water. Flush skin with water. Call a physician. This product contains monochlorobenzene.

ENVIRONMENTAL HAZARDS

This product is toxic to fish. Keep out of lakes, streams or ponds. DO NOT apply when weather conditions favor drift from target area. DO NOT contaminate water by cleaning of equipment or disposal of wastes.

PHYSICAL HAZARDS

FLAMMABLE LIQUID AND VAPOR

Flash Point 92°F

DO NOT use, pour, spill, or store near heat or open flame.

DISCLAIMER

The label instructions for the use of this product reflect the opinion of experts based on field use and tests. The directions are believed to be reliable and should be followed carefully. However, it is impossible to eliminate all risks inherently associated with use of this product. Crop injury, ineffectiveness or other unintended consequences may result because of such factors as weather conditions, presence of other materials, or the use or application of the product contrary to label instructions, all of which are beyond the control of American Cyanamid Company. All such risks shall be assumed by the user.

American Cyanamid Company warrants only that the material contained herein conforms to the chemical description on the label and is reasonably fit for the use therein described when used in accordance with the directions for use, subject to the risks referred to above.

Any damages arising from a breach of this warranty shall be limited to direct damages and shall not include consequential commercial damages such as loss of profits or values or any other special or indirect damages.

American Cyanamid Company makes no other express or implied warranty, including any other express or implied warranty of FITNESS or MERCHANTABILITY.

DIRECTIONS FOR USE

It is a violation of Federal Law to use this product in a manner inconsistent with its labeling.

Observe all cautions and limitations in this leaflet and the labels of products used in combination with PROWL. The use of PROWL not consistent with this label can result in injury to crops, animals, or persons. Keep containers closed to avoid spills and contamination.

STORAGE AND DISPOSAL

DO NOT STORE BELOW 40° F. Extended storage at temperatures below 40° F can result in the formation of crystals on the bottom of the container. If crystallization does occur, store the container on its side at room temperature (70° F) and rock occasionally until crystals redissolve. DO NOT store near food or feed products.

RINSE/DRAIN PROCEDURE: (1) Drain container into spray tank (after normal emptying) in a vertical position for 30 seconds. (2) Rinse carefully 3 times with the amount of rinse water listed below for each container size and drain into spray tank after each rinse. (3) Do not reuse container.

Container size	Rinse volume
1 gal.	1 qt for each rinse
5 gal.	1 gal. for each rinse
30 gal.	5 gal. for each rinse
55 gal.	5 gal. for each rinse

CONTAINER DISPOSAL

METAL PAILS: Preferred disposal: crush and recycle for scrap to a steel melting plant. If preferred disposal cannot be accomplished, container should be crushed and/or buried at an approved dump site according to local and state regulations.

PLASTIC CONTAINERS: Triple rinse (or equivalent). Offer for recycling or reconditioning, or dispose of in a sanitary landfill, or by incineration if allowed by state and local authorities.

GENERAL INFORMATION

Special Precautions

PROWL herbicide controls most annual grasses and certain broadleaf weeds as they germinate, but will not control established weeds. Destroy existing weeds before applying PROWL (except as recommended in specific postemergence combination treatments). Unusually cold, excessively wet, or hot and dry conditions that delay germination or extend germination over a long period of time can reduce weed control.

Applied according to label directions and under normal growing conditions, PROWL or PROWL tank-mix combinations will not cause crop injury. Over-application can result in crop stand loss, crop injury, or soil residues. Uneven application or improper soil incorporation can decrease weed control or cause crop injury. Soil incorporation deeper than recommended can reduce weed control.

Seedling diseases, cold weather, excessive moisture, deep planting, high soil pH, high soil salt concentration, or drought can weaken seedlings and plants and increase the possibility of crop damage from PROWL. Under these conditions, crop yields can be reduced.

PAGE 3

Cultural Practices Following Application

Should weeds develop, a shallow cultivation or rotary hoeing will generally result in better weed control. When cultivating for any reason, it should be shallow. A PROWL treatment may be followed by any registered herbicide to control weeds not listed on the PROWL label.

MIXING INSTRUCTIONS

Respective sections of this leaflet define those recommended PROWL or PROWL tank-mix treatments registered for application in water or liquid fertilizer. Prior to mixing PROWL or PROWL tank-mix treatments in the specific liquid fertilizer to be used, refer to liquid fertilizer compatibility determinations below.

1. Fill tank one-half to three-quarters full with clean water or liquid fertilizer and agitate.
2. PROWL Alone
   When using PROWL alone, add PROWL to the partially-filled tank while agitating and then fill the remainder of the tank with water or liquid fertilizer.
3. PROWL Tank Mixes
   Add the tank-mix ingredients in the order listed below prior to

adding PROWL (For tank mixtures with Paraquat CL see mixing instructions on page 16):

(a) Wettable powder formulations
Make a slurry of the wettable powder in water (1 part WP + 2 parts water). Add the slurry slowly into the partially-filled tank while agitating.

(b) Flowable formulations
Premix one part of the flowable and one part water and add this diluted mixture into the partially-filled tank while agitating.

(c) Water soluble concentration formulations
Add the water soluble concentration to the partially-filled tank while agitating.

(d) Dry flowable/water dispersible granule formulations
Add the granules to the partially-filled tank while agitating.

(e) Emulsifiable concentrate formulations
Add the emulsifiable concentrate to the partially-filled tank while agitating.
After complete mixing, add PROWL to the tank.

Fill the remainder of the tank with water or liquid fertilizer while agitating.

4. MAINTAIN CONTINUOUS AGITATION WHILE ADDING HERBICIDES AND UNTIL SPRAYING IS COMPLETED. If the spray mixture is allowed to settle for any period of time, thorough agitation is essential to resuspend the mixture before spraying is resumed. Continue agitation while spraying.

FORMULATION CONVERSION TABLES

The rate tables in this label for PROWL tank mixtures with atrazine, Bladex, Sencor/Lexone and Lorox list rates for one formulation of the respective herbicides. Use the following conversion tables to determine rates when using other formulations.

Atrazine Conversion Table

Rate of atrazine 80 WP	Rate of atrazine 4L	Rate of AAtrex Nine-O
1.25 lb.	2.0 pts.	1.1 lb.
1.50 lb.	2.4 pts.	1.3 lb.
2.00 lb.	3.2 pts.	1.8 lb.

Bladex Conversion Table

Rate of Bladex 80 WP	Rate of Bladex 4L
1.25 lb.	2.0 pts.
1.50 lb.	2.4 pts.
2.00 lb.	3.2 pts.
2.50 lb.	4.0 pts.
3.00 lb.	4.8 pts.

Sencor/Lexone Conversion Table

Rate of Sencor/ Lexone DF	Rate of Sencor/ Lexone 50 WP	Rate of Sencor 4F/ Lexone 4L
0.33 lb.	0.50 lb.	0.50 pt.
0.50 lb.	0.75 lb.	0.75 pt.
0.67 lb.	1.00 lb.	1.00 pt.

Lorox Conversion Table

Rate of Lorox L	Rate of Lorox 50W
1.0 pt.	1.0 lb.
1.5 pts.	1.5 lb.
2.0 pts.	2.0 lb.
2.5 pts.	2.5 lb.

LIQUID FERTILIZER COMPATIBILITY DETERMINATIONS

If liquid fertilizer and herbicide(s) mixture separates in the spray tank, clogged equipment and uneven application can result, which can cause poor weed control and crop injury. Always predetermine the compatibility of PROWL alone or with other herbicides in the specific liquid fertilizer to be used according to the following directions:

1. Add 1 pint of fertilizer to each of 2 one-quart jars.
2. Add ½ teaspoon of adjuvant* to one jar.
3. (a) When using PROWL alone, add to each jar the correct amount of PROWL as specified in the table below.
   (b) When using PROWL tank mixtures, first add the specified quantity of atrazine, Bladex or Cotoran (see table) and then add the correct amount of PROWL.
4. Close both jars and shake thoroughly for 10 seconds. Let them stand for 30 minutes and then observe the results. Look for signs of separation, an oily layer or globules, sludge, flakes or other precipitates.
5. Determine compatibility.
   (a) If the herbicide(s) and liquid fertilizer mixture without adjuvant does not separate, use this mixture in your spray tank.
   (b) If the mixture with adjuvant does not separate, but the one without adjuvant separates, use the adjuvant mixture in your spray tank. Add the adjuvant to the liquid fertilizer as directed on the manufacturer's label.
   (c) If either mixture separates, but mixes readily with shaking, the mixture can be used providing good agitation is maintained in the spray tank.
   (d) If separation of the mixture occurs and agitation and/or adjuvant does not correct this problem, DO NOT use PROWL and PROWL tank mixtures with that specific liquid fertilizer.

* Compex-Kalo Laboratories Incorporated, Kansas City, MO; Sponto 168-D-Witco-Retzloff Chemical Company, Houston, TX; Unite-Hopkins Agricultural Chemical Company, Madison, WI; or other comparable adjuvants.

Teaspoons of Specified Herbicide to be Added to 1 pint of Liquid Fertilizer Solution**

Gallons of Liquid Fertilizer to be Applied Per Acre	PROWL	Atrazine 80W	AAtrex Nine-O or Atrazine 4L	Bladex 80WP	Bladex 4L	Cotoran
20	2½	3	2	4½	3	3¾
30	1¾	2	1½	3	2	2½
40	1¼	1½	1	2¼	1½	1⅞

** Based on highest per acre rate used in this leaflet for each product

PREPLANT INCORPORATED APPLICATIONS WITH DRY BULK FERTILIZERS

PROWL may be impregnated on dry bulk fertilizers. When applied as directed, PROWL/dry bulk fertilizer mixtures provide weed control equal to that provided by the same rates of PROWL applied in water.

PAGE 4

Follow all PROWL label recommendations regarding rates per acre, registered crops, incorporation, special instructions and precautions. Apply PROWL/dry bulk fertilizer mixtures only for preplant incorporated uses with registered crops. PROWL impregnated on dry bulk fertilizer is NOT RECOMMENDED for use in corn. Apply PROWL/dry bulk fertilizer mixtures only with ground equipment.

All individual state regulations relating to dry bulk fertilizer blending, registration, labeling, and application are the responsibility of the individual and/or company selling the PROWL/dry bulk fertilizer mixture.

A minimum of 200 pounds and a maximum of 450 pounds of dry bulk fertilizer inpregnated with the recommended amount of PROWL must be applied per acre. DO NOT impregnate PROWL onto straight coated ammonium nitrate or straight limestone because these materials will not absorb the herbicide. Dry fertilizer blends containing mixtures of ammonium nitrate or limestone may be impregnated with PROWL.

Refer to appropriate crop section in PROWL leaflet label to determine the rate of PROWL per acre. Use the following table to determine the amount of PROWL to be impregnated on a ton of dry bulk fertilizer based on the rate of fertilizer which will be applied per acre.

RATE CHART FOR IMPREGNATION OF DRY BULK FERTILIZER WITH PROWL
(PINTS OF PROWL PER TON OF FERTILIZER)

FERTILIZER RATE LBS./ ACRE	PROWL RATE PER ACRE (RECOMMENDED RATE FOR CROP AND SOIL TEXTURE)				
	1 PINT	$1\frac{1}{2}$ PINTS	2 PINTS	$2\frac{1}{2}$ PINTS	3 PINTS
200	10 pts./ton	15 pts./ton	20 pts./ton	25 pts./ton	30 pts./ton
250	8 pts./ton	12 pts./ton	16 pts./ton	20 pts./ton	24 pts./ton
300	$6\frac{2}{3}$ pts./ton	10 pts./ton	$13\frac{1}{3}$ pts./ton	$16\frac{2}{3}$ pts./ton	20 pts./ton
350	$5\frac{3}{4}$ pts./ton	$8\frac{1}{2}$ pts./ton	$11\frac{1}{2}$ pts./ton	$14\frac{1}{4}$ pts./ton	17 pts./ton
400	5 pts./ton	$7\frac{1}{2}$ pts./ton	10 pts./ton	$12\frac{1}{2}$ pts./ton	15 pts./ton
450	$4\frac{1}{2}$ pts./ton	$6\frac{2}{3}$ pts./ton	9 pts./ton	11 pts./ton	$13\frac{1}{3}$ pts./ton

For those rates not listed in this table, calculate the pints of PROWL to be impregnated on a ton of dry bulk fertilizer using the following formula:

$$\frac{2000}{\text{pounds of dry fertilizer per acre}} \times \text{pints of PROWL per acre (recommended rate for crop and soil texture)} = \text{pints of PROWL per ton of fertilizer}$$

To impregnate PROWL on dry bulk fertilizer, use a closed rotary-drum mixer or other commonly used dry bulk fertilizer blender equipped with suitable spray equipment. Spray nozzles must be placed to provide uniform coverage of PROWL onto the fertilizer during mixing.

Apply the PROWL/dry bulk fertilizer mixture with an accurately calibrated dry fertilizer spreader. The PROWL/dry bulk fertilizer mixture must be spread uniformly on the soil surface. Uneven spreading can cause poor weed control and crop injury.

Refer to specific crop sections for directions regarding soil incorporation.

Spraying Instructions

GROUND APPLICATIONS

Use a properly calibrated low-pressure (20 to 40 psi) sprayer equipped with 8002 or larger size Tee-Jet or comparable nozzles to achieve uniform spray distribution and minimize drift. Keep the by-pass line on or near the bottom of the tank to minimize foaming. Nozzle screens must be no finer than 50 mesh. DO NOT apply PROWL during periods of gusty winds or when wind velocity is greater than 10 mph.

BROADCAST TREATMENT—Apply in water or liquid fertilizer as specified in the appropriate sections of this leaflet.

BAND TREATMENT—Apply the broadcast equivalent rate and volume per acre. To determine these:

$$\frac{\text{Band width in inches}}{\text{Row width in inches}} \times \text{Broadcast RATE per acre} = \text{Band RATE per acre}$$

$$\frac{\text{Band width in inches}}{\text{Row width in inches}} \times \text{Broadcast VOLUME per acre} = \text{Band VOLUME per acre}$$

AERIAL APPLICATIONS

Apply in 5 or more gallons of water per acre. To minimize drift, DO NOT apply during periods of gusty winds or winds in excess of 5 mph. It is recommended that a flagman or an automatic mechanical flagging unit on the aircraft be used to avoid overlapping and possible crop injury.

FOLLOW CROP RESTRICTIONS

1. PROWL-treated land can be planted to other crops the following year. See restrictions below for sugar beets, red beets, and spinach.
2. **Winter wheat, winter barley**
   Winter wheat and winter barley may be planted in the fall 120 days after a PROWL application in any registered crop. Winter wheat and winter barley may be planted in the fall 90 days after a PROWL CULTI-SPRAY+up,TM application in irrigated field corn or grain sorghum. The PROWL-treated crop must be grown to maturity and harvested before planting wheat or barley.

   In areas where irrigation is necessary to produce the PROWL-treated crop, DO NOT plant winter wheat or winter barley as follow crops if crop failure/destruction occurs and land is fallowed during the summer as crop injury may result.

   DO NOT plant winter wheat or winter barley as follow crops in PROWL-treated land if the follow crop is planted using no-tillage procedures as crop injury may result.

   DO NOT plant winter wheat or winter barley as follow crops in PROWL-treated land if higher PROWL rates were used for rhizome johnsongrass suppression in soybeans or cotton or for red rice control and itchgrass suppression in soybeans.

   DO NOT feed forage or graze livestock for 75 days after planting wheat or barley in PROWL-treated land.
3. **Sugar beets, red beets, spinach**
   To avoid crop injury, do not plant sugar beets, red beets or spinach for 12 months following a PROWL application. Land should be plowed to a depth of 12 inches prior to planting these crops.
4. When PROWL is used in tank-mix or sequential combinations, refer to labels of other herbicides for additional follow crop restrictions.

CORN—Field

PROWL can be applied in corn as follows:

Preemergence (surface treatment after planting)	Early Postemergence
PROWL + atrazine p. 6	PROWL + atrazine p. 6
PROWL + Bladex p. 6	PROWL + Bladex 80W p. 6
PROWL alone p. 7	
PROWL + Banvel p. 7	

DO NOT INCORPORATE

Refer to specific sections of the label for directions for each method of application and special instructions for field corn grown in minimum tillage systems on page 8.
Refer to page 17 for instructions on CULTI-SPRAY applications (layby treatment) in field corn.
Refer to page 5 for follow crop restrictions.

GENERAL INSTRUCTIONS

With the exception of minimum tillage systems, the seed bed should be firm and free of clods and trash. Use only where adequate tillage is practiced to provide good seed coverage. Plant corn AT LEAST $1\frac{1}{2}$ inches deep to ensure good seed coverage.

PROWL or PROWL tank mix combination treatments are most effective in controlling weeds when adequate rainfall or overhead irrigation is received within 7 days after application. If cultivation is necessary because of soil crusting, soil compaction, or weed germination before rain or irrigation, use shallow tillage (such as rotary hoe), and make certain corn seeds are below the tilled area. After early postemergence treatments, wait 7 to 10 days before cultivating.

PAGE 5

APPLICATIONS OF PROWL PLUS ATRAZINE MIXTURES OR PROWL PLUS BLADEX MIXTURES

PREEMERGENCE (surface treatment after planting)

Uniformly apply PROWL plus atrazine or PROWL plus Bladex tank mix treatments in 10 or more gallons of water or in 20 or more gallons of liquid fertilizer per acre by ground equipment or in 5 or more gallons of water by aircraft. Apply the tank mixtures preemergence, after planting, but before weeds or crop emerge.

(Refer to page 5 for spraying instructions. Refer to page 4 for liquid fertilizer compatibility determinations for ground applications. Refer to page 8 for instructions for use in minimum tillage systems.)

DO NOT INCORPORATE or serious corn injury can result.

EARLY POSTEMERGENCE

Uniformly apply PROWL plus atrazine or PROWL plus Bladex 80W tank mix treatments in 10 or more gallons of water by ground equipment or in 5 or more gallons of water by aircraft.

(Refer to page 5 for spraying instructions. Refer to page 8 for instructions for use in minimum tillage systems.)

DO NOT INCORPORATE or serious corn injury can result.

DO NOT apply in liquid fertilizer or following preemergence PROWL applications.

DO NOT apply PROWL alone postemergence.

DO NOT apply PROWL plus Bladex 80W early postemergence under extended cold (below 50°F), wet weather conditions, or when cold, wet conditions are predicted, or serious corn injury can result.

Tank mixtures of PROWL plus atrazine or PROWL plus Bladex 80W may be applied in water ONLY, after field corn emergence but no later than when corn is in the 2-leaf stage of growth and when weeds are no more than 1 inch tall except for Texas panicum which must be no larger than the 1–2 leaf stage. (DO NOT use Bladex 4L postemergence on field corn.)

WEED SPECIES CONTROLLED

The following grass and broadleaf weed species are controlled by treatments of PROWL plus atrazine or PROWL plus Bladex:

GRASSES CONTROLLED

- Barnyardgrass *(Echinochloa crus-galli)*
- Crabgrass *(Digitaria spp.)*
- Fall panicum *(Panicum dichotomiflorum)*
- Foxtail (giant) *(Setaria faberi)*
- Foxtail (green) *(Setaria viridis)*
- Foxtail (yellow) *(Setaria lutescens)*
- Goosegrass *(Eleusine indica)*
- Johnsongrass (from seed) *(Sorghum halepense)*
- Signalgrass *(Brachiaria platyphylla)*
- Texas panicum *(Panicum texanum)*
- Witchgrass *(Panicum capillare)*
- Woolly cupgrass *(Eriochloa villosa)*

BROADLEAF WEEDS CONTROLLED

- Black nightshade *(Solanum nigrum)*
- Carpetweed *(Mollugo verticillata)*
- Common ragweed *(Ambrosia artemisiifolia)*
- Florida pusley *(Richardia scabra)*
- Lambsquarters *(Chenopodium album)*
- Mustard *(Brassica spp.)*
- Pigweed *(Amaranthus spp.)*
- Purslane *(Portulaca oleracea)*
- Smartweed *(Polygonum pensylvanicum)*
- Velvetleaf (Buttonweed) *(Abutilon theophrasti)*
- Venice mallow *(Hibiscus trionum)*

Recommended tank mixtures of PROWL plus atrazine or PROWL plus Bladex will reduce competition from the following hard-to-control annual broadleaf weeds and grasses:

- Cocklebur *(Xanthium pensylvanicum)*
- Annual morningglory *(Ipomoea spp.)*
- Jimsonweed *(Datura stramonium)*
- Wild proso millet *(panicum miliaceum)*

Broadcast Rate Per Acre of PROWL Plus Atrazine in Field Corn

Soil Texture	Less than 1.5% Organic Matter[1] PROWL+ atrazine 80W*	1.5% to 3% Organic Matter PROWL+ atrazine 80W*	More than 3% Organic Matter PROWL+ atrazine 80W*
COARSE sands, loamy sands, sandy loams	1.5 to 2.0 pts. + 1.25 lbs.	2.0 pts. + 1.25 lbs.	3.0 pts. 1.25 lbs.
MEDIUM sandy clay loams**, sandy clays, loams, silts, silt loams	2.0 pts. + 1.25 to 1.5 lbs.	3.0 pts. + 1.25 to 1.5 lbs.	3.0 pts. + 1.5 to 2.0 lbs.
FINE silty clay loams** sandy clay loams** clay loams, silty clays, clays	2.0 pts. + 1.25 to 1.5 lbs.	3.0 pts. + 1.25 to 2.0 lbs.	3.0 pts. + 1.5 to 2.0 lbs.

[1] For use in states east of and including Minnesota, Iowa, Missouri, Arkansas and Louisiana.

* When using atrazine 4L or AAtrex Nine-O use equivalent rates. One quart of atrazine 4L equals 1.25 lbs. of atrazine 80W. One pound of AAtrex Nine-O equals 1.12 lbs. of atrazine 80W. See conversion chart on page 4.

** Sometimes considered transitional soils, and may be classified as medium or fine textured soils.

The high rate for each soil texture, where listed, should be used if heavy infestations of grass or broadleaf weeds are anticipated or if extensive crop residues were present prior to seedbed preparation. DO NOT use on peat or muck soils.

NOTE: If corn crop loss occurs due to weather conditions and corn is replanted, seeding depth must be below retilled area. Follow cropping restrictions on atrazine labels. Observe all precautions and limitations on atrazine labels.

Broadcast Rate Per Acre of PROWL Plus Bladex in Field Corn

Soil Texture	Less than 1.5% Organic Matter[1] PROWL + Bladex 80WP*	1.5% to 3% Organic Matter PROWL + Bladex 80WP*	More than 3% Organic Matter PROWL + Bladex 80WP*
COARSE[2] sands, loamy sands, sandy loams	1.5 to 2.0 pts. + 1.25 to 1.5 lbs.	2.0 pts. + 2.0 lbs.	3.0 pts. + 2.0 to 2.5 lbs.
MEDIUM sandy clay loams**, sandy clays, loams, silts, silt loams	2.0 pts. + 1.25 to 1.5 lbs.	3.0 pts. + 2.0 to 2.5 lbs.	3.0 pts. + 2.5 to 3.0 lbs.
FINE silty clay loams**, sandy clay loams**, clay loams, silty clays, clays	2.0 pts. + 2.0 to 2.5 lbs.	3.0 pts. + 2.5 to 3.0 lbs.	3.0 pts. + 2.5 to 3.0 lbs.

[1] For use in states east of and including Minnesota, Iowa, Missouri, Arkansas and Louisiana.

[2] DO NOT use this combination on sands or loamy sands with less than 1% organic matter.

* When using Bladex 4L use equivalent rates. One quart Bladex 4L equals 1.25 lbs. of Bladex 80WP. DO NOT use Bladex 4L early postemergence. See conversion chart on page 4.

** Sometimes considered transitional soils and may be classified as medium or fine textured soils.

The MAXIMUM RATE of Bladex 80WP for early postemergence treatments is 2.5 lbs. per acre.

The high rate for each soil texture, where listed, should be used if heavy infestations of grass or broadleaf weeds are anticipated or if extensive crop residues were present prior to seedbed preparation. DO NOT use on peat or muck soils.

NOTE: If corn crop loss occurs due to weather conditions and corn is replanted, seeding depth must be below retilled area. Observe all precautions and limitations on Bladex label.

PAGE 6

APPLICATIONS OF PROWL ALONE

PREEMERGENCE APPLICATION (Surface treatment after planting)

Uniformly apply the recommended PROWL preemergence treatment in 10 or more gallons of water or in 20 or more gallons of liquid fertilizer per acre by ground equipment or in 5 or more gallons of water by aircraft. (Refer to page 4 for liquid fertilizer compatibility determinations. Refer to page 5 for spraying instructions.)

Apply PROWL preemergence, after planting, but before weeds and crop emerge. DO NOT INCORPORATE or serious corn injury can result. DO NOT APPLY PROWL ALONE POSTEMERGENCE.

WEED SPECIES CONTROLLED

The following grass and broadleaf weed species are controlled by preemergence treatments of PROWL at the rates recommended for each soil texture listed below:

GRASSES CONTROLLED

Barnyardgrass *(Echinochloa crus-galli)*
Crabgrass *(Digitaria* spp.*)*
Fall panicum *(Panicum dichotomiflourm)*
Foxtail (giant) *(Setaria faberi)*
Foxtail (green) *(Setaria viridis)*
Foxtail (yellow) *(Setaria lutescens)*
Goosegrass *(Eleusine indica)*
Johnsongrass (form seed) *(Sorghum halepense)*
Signalgrass *(Brachiaria platyphylla)*
Texas panicum *(Panicum texanum)*
Witchgrass *(Panicum capillare)*
Woolly cupgrass *(Eriochloa villosa)*

BROADLEAF WEEDS CONTROLLED

Carpetweed *(Mollugo verticillata)*
Florida pusley *(Richardia scabra)*
Lambsquarters *(Chenopodium album)*
Pigweed *(Amaranthus* spp.*)*
Purslane *(Portulaca oleracea)*
Smartweed *(Polygonum pensylvanicum)*
Velvetleaf (Buttonwood) *(Abutilon theophrasti)*

Preemergence Broadcast Rate Per Acre of PROWL in Field Corn

Soil Texture	Less than 1.5% Organic Matter[1]	1.5% to 3% Organic Matter	More than 3% Organic Matter
COARSE sands, loamy sands, sandy loams	1.5 to 2.0 pts.	3.0 pts.	3.0 pts.
MEDIUM sandy clay loams*, sandy clays, loams, silts, silt loams	2.0 to 3.0 pts.	3.0 pts.	3.0 to 4.0 pts.
FINE silty clay loams*, sandy clay loams*, clay loams, silty clays, clays	2.0 to 3.0 pts.	3.0 to 4.0 pts.	4.0 pts.

[1] For use in states east of and including Minnesota, Iowa Missouri, Arkansas and Louisiana.

* Sometimes considered transitional soils and may be classified as either medium or fine textured soils.

The high rate for each soil texture above, where listed, should be used if heavy infestations of grass or broadleaf weeds are anticipated or if extensive crop residues were present prior to seedbed preparation.

DO NOT use on peat or muck soils.

NOTE: When PROWL alone is used, if corn crop loss occurs due to weather conditions, corn, cotton or soybeans can be replanted the same year without adverse effects. If corn is replanted, seeding depth must be below retilled area.

Applications of PROWL for Reduction of Wild Proso Millet

PROWL alone preemergence; PROWL plus atrazine preemergence tank mixture; PROWL plus Bladex preemergence tank mixture; PROWL plus atrazine early postemergence tank mixture; or PROWL plus Bladex 80W early postemergence tank mixture can be used in field corn to aid in the control of and reduce competition from:

Wild proso millet *(Panicum miliaceum)*

Wild proso millet is a prolific seed producer and light infestations can develop into heavy populations the following year. This weed is extremely competitive and can signficantly reduce yields. Wild proso millet germinates throughout the growing season making long control extremely difficult. Wild proso millet plants present at the end of the growing season produce seeds for reinfestation and can cause harvesting problems.

Effective management of wild proso millet can be achieved with a program involving herbicide applications and mechanical cultivation to eliminate weed escapes. PROWL early postemergence (no later than when corn is in the 2-leaf stage) treatments have provided the most consistent suppression of wild proso millet because: (1) early postemergence treatments provide a burndown of the initial germination flush of wild proso millet and (2) early postemergence treatments provide residual activity later into the growing season. Additional wild proso millet reduction can be obtained when PROWL treatments follow effective registered preplant incorporated herbicide treatments.

Refer to above for applications of PROWL alone preemergence. Refer to page 6 for applications of PROWL plus atrazine tank mixture or PROWL plus Bladex tank mixture (preemergence or early postemergence).

Preemergence Applications of PROWL Plus Banvel Mixtures in Field Corn

Apply recommended tank mixtures of PROWL plus Banvel preemergence (surface treatment) only, after planting, but before weeds or crop emerge. Apply in 10 or more gallons of water per acre by GROUND EQUIPMENT ONLY. DO NOT INCORPORATE or serious corn injury can result.

These tank mixtures are recommended for use ONLY in the states of Illinois, Indiana, Iowa, Kansas, Michigan, Minnesota, Missouri, Montana, Nebraska, New York, North Dakota, Ohio, Pennsylvania, South Dakota, and Wisconsin. (APPLY ONLY ON LEVEL PLANTED CORN.) Plant corn AT LEAST 1½ inches deep to ensure good seed coverage.

WEED SPECIES CONTROLLED

In addition to the weeds controlled by PROWL alone, the following weeds are controlled by PROWL plus Banvel at the rates recommended for each soil texture listed below.

Cocklebur *(Xanthium pensylvanicum)*

Venice mallow *(Hibiscus trionum)* Page 584 Table A

Preemergence Broadcast Rate Per Acre of PROWL Plus Banvel in Field Corn

Soil Texture[1]	1.5% to 3% Organic Matter PROWL + Banvel	More than 3% Organic Matter PROWL + Banvel
MEDIUM sandy clay loams*, sandy clays, loams, silts, silt loams	DO NOT USE	3.0 pts. + 0.75 pts.
FINE silty clay loams*, clay loams, silty clays, clays	3.0 pts. + 0.75 pts.	3.0 pts. + 0.75 pts.

[1] DO NOT use on sands, loamy sands, sandy loams, peat or muck soils. DO NOT use on medium soils containing less than 3% organic matter.

* Sometimes considered transitional soils and may be classified as either medium or fine textured soils.

AVOID drift of this mixture to soybeans or other sensitive crops.

DO NOT use on furrow irrigated corn.

DO NOT apply this tank mixture by aircraft.

When applying this tank mixture at planting, the spray nozzles must be far enough behind the planter to avoid contact of the spray mixture with the seeds to avoid crop injury.

NOTE: Follow cropping restrictions on Banvel label. Observe all precautions and limitations on Banvel label.

PAGE 7

FIELD CORN GROWN IN MINIMUM TILLAGE SYSTEMS

Broadcast Ground Applications of PROWL and

PROWL Tank Mixtures in Minimum Tillage Systems

PROWL and PROWL tank mixtures with atrazine or Bladex may be used in minimum tillage systems where adequate tillage is practiced to provide a seedbed which allows for good seed coverage. Plant corn AT LEAST 1½ inches deep to ensure good seed coverage. The use of no-till planters in minimum tillage corn under conditions which do not allow good soil coverage of the corn seed can result in reduced crop stand or injury if PROWL contacts the germinating corn seed. Check equipment to ensure good seed coverage.

Refer to the General Information Section regarding mixing instructions and liquid fertilizer compatibility determinations (page 4) and the Spraying Instructions Section regarding Ground Applications (page 5). Weed species listed in the application sections (pages 6–7) are controlled by PROWL or PROWL tank-mix combinations when applied at the proper treatment rates determined by soil textures. Refer to the appropriate table for the specific rates of application (pages 6–7).

PREEMERGENCE (surface treatment after planting)

In those minimum tillage systems where an adequate seedbed, without living vegetation, is prepared which allows for good seed coverage, apply PROWL, PROWL plus atrazine 80W*, or PROWL plus Bladex 80W** in 20 or more gallons of water or liquid fertilizer per acre with ground equipment only after planting but before the corn emerges. DO NOT INCORPORATE or serious corn injury can result.

EARLY POSTEMERGENCE

In those minimum tillage systems where an adequate seedbed is prepared which allows for good seed coverage, PROWL plus atrazine (80 W) and PROWL plus Bladex 80W tank mix combinations may be applied in 20 or more gallons of water per acre with ground equipment only after field corn emergence but no later than when corn is in the 2-leaf stage of growth and weeds are no more than 1 inch tall except for Texas panicum which must be no larger than the 1–2 leaf stage.

DO NOT incorporate or serious corn injury can result.

DO NOT apply early postemergence treatments in liquid fertilizer or following preemergence PROWL applications. DO NOT apply Bladex 4L postemergence to field corn. DO NOT apply PROWL alone postemergence.

DO NOT apply PROWL plus Bladex early postemergence under extended cold (below 50° F), wet weather conditions, or when cold, wet conditions are predicted, or serious corn injury can result.

* When using atrazine 4L or AAtrex Nine-O use equivalent rates. One quart of atrazine 4L equals 1.25 lbs. of atrazine 80W. One pound of AAtrex Nine-O equals 1.12 lbs. of atrazine 80W. See conversion chart on page 4.

** When using Bladex 4L use equivalent rates. One quart Bladex 4L equals 1.25 lbs. of Bladex 80W. See conversion chart on page 4.

COTTON

PROWL can be applied in cotton as follows:

Preplant Incorporated	Sequential	Preemergence
PROWL alone p. 9 PROWL + Caparol p. 9	PROWL preplant incorporated followed by Cotoran preemergence p. 9 PROWL preplant incorporated followed by Karmex preemergence p. 9	PROWL + Cotoran p. 10

Refer to specific sections of the label for directions for each method of application.

Refer to page 5 for follow crop restrictions.

Preplant Incorporated Applications

Uniformly apply the recommended PROWL treatment in 10 or more gallons of water or in 20 or more gallons of liquid fertilizer per acre by ground equipment. (Refer to page 4 for liquid fertilizer compatibility determinations.) Apply in 5 or more gallons of water per acre by aircraft. Refer to page 5 for spraying instruction. Refer to sections below for incorporation instructions. Refer to page 4 for PROWL/dry bulk fertilizer applications.

Thoroughly mix the previous crops residues into the soil to a depth of 4 to 6 inches by plowing or disking prior to application.

PROWL may be applied immediately before planting or up to 60 days prior to planting cotton in all states except Arizona, California, Louisiana, New Mexico, Mississippi, Oklahoma and Texas where PROWL may be applied in the fall, after October 15th (up to 140 days prior to planting cotton).

After application, rotary-hoeing, shallow cultivation, or hand-hoeing can be practiced without reducing weed control. When mechanically incorporated, PROWL is not dependent upon rainfall for effective weed control.

DO NOT APPLY PROWL AS A POSTEMERGENCE SPRAY IN COTTON.

Incorporation Instructions

1. Flat-planted cotton

PROWL is to be incorporated within 7 days of application prior to planting. When PROWL is applied to flat land that will not be bedded, mechanical incorporation should be into the top 1 or 2 inches of soil. Mechanical incorporation of PROWL can be achieved by the following:

(a) Disk harrow set to cut 3 to 4 inches deep and operated in two different directions at 4 to 6 mph.

(b) PTO-driven equipment (tillers, cultivators, hoes) set to cut 2 inches deep and operated one time at 4 mph or less.

(c) Rolling cultivator set to cut 2 to 3 inches deep and operated two times at 6 to 8 mph. Use only on coarse and medium textured soils.

(d) Field cultivator with 3 or 4 rows of sweeps, spaced at intervals of 7 inches or less and staggered so that no soil is left unturned (Do not use chisel points), set to cut 3 inches deep and operated two times at more than 5 mph with the second pass made at an angle to the first.

(e) LELY-ROTERRA[1] set and operated according to the manufacturer's directions to thoroughly incorporate PROWL to a depth of 1 to 2 inches.

(f) Do-All set and operated according to the manufacturer's directions to thoroughly incorporate PROWL to a depth of 1 to 2 inches.

[1] LELY-ROTERRA is a trademark of Lely.

2. Bedded Cotton—Application Over Beds

PROWL is to be incorporated within 7 days of application prior to planting. When PROWL is applied over beds, mechanical incorporation should be into the top 1 or 2 inches of soil. Mechanical incorporation can be achieved by the following:

PAGE 8

(a) Rolling cultivator set to cut 2 to 3 inches deep and operated two times at 6 to 8 mph. Use only on coarse and medium textured soils.

(b) PTO-driven equipment (tillers, cultivators, hoes) set to cut 2 inches deep and operated one time at 4 mph or less.

(c) Bed conditioner set to cut 2 to 3 inches deep and operated one time at 4 to 6 mph.

During planting, or if beds must be reshaped prior to planting, avoid tillage that will bring untreated soil to the surface or expose untreated soil on the seedbed or in the furrow.

3. Bedded Cotton—Application Prior To Bedding

PROWL is to be incorporated within 7 days of application prior to planting. When PROWL is applied to flat land prior to bedding, mechanical incorporation should be of sufficient depth so that bedding does not bring up untreated soil. Mechanical incorporation of PROWL can be achieved by the following:

(a) Disk harrow set to cut 4 to 6 inches deep and operated in two different directions at 4 to 6 mph.

(b) PTO-driven equipment (tillers, cultivators, hoes) set to cut 3 to 4 inches deep and operated one time at 4 mph or less.

During planting, or if beds must be reshaped prior to planting, avoid tillage that will bring untreated soil to the surface or expose untreated soil on the seedbed or in the furrow.

Preplant Incorporated Applications of PROWL Alone

Weed Species Controlled

The following grass and broadleaf weed species are controlled by preplant incorporated treatments of PROWL in cotton at the rates recommended for each soil texture listed below:

GRASSES CONTROLLED

Barnyardgrass *(Echinochloa crus-galli)*
Crabgrass *(Digitaria* spp.*)*
Crowfootgrass *(Dactyloctenium aegyptium)*
Fall panicum *(Panicum dichotomiflorum)*
Foxtail (giant) *(Setaria faberi)*
Foxtail (green) *(Setaria viridis)*
Foxtail (yellow) *(Setaria lutescens)*
Goosegrass *(Eleusine indica)*
Johnsongrass (from seed) *(Sorghum halepense)*
Signalgrass *(Brachiaria platyphylla)*
Texas panicum *(Panicum texanum)*

BROADLEAF WEEDS CONTROLLED

Carpetweed *(Mollugo verticillata)*
Florida pusley *(Richardia scabra)*
Kochia *(Kochia scoparia)*
Lambsquarters *(Chenopodium album)*
Pigweed *(Amaranthus* spp.*)*
Purslane *(Portulaca oleracea)*
Spurge, annuals *(Euphorbia* spp.*)*

PROWL will aid in the control of and reduce competition from:

Smartweed *(Polygonum pensylvanicum)*
Velvetleaf *(Abutilon theophrasti)*

Preplant Incorporated Broadcast Rate Per Acre of PROWL in Cotton

Soil Texture	PROWL
COARSE sands, loamy sands, sandy loams	1.0 to 1.5 pts.
MEDIUM sandy clay loams*, sandy clays, loams, silts, silt loams	1.5 to 2.0 pts.
FINE silty clay loams*, clay loams, silty clays, clays	1.5 to 3.0 pts.

* Sometimes considered transitional soils and may be classified as either medium or fine textured soils.

The high rate for each soil texture above should be used if heavy weed populations are anticipated, if PROWL is applied from 60 to 140 days before planting or if extensive crop residues were present prior to seedbed preparation. Use the 3 pint rate for heavy clay soils.

NOTE: DO NOT feed forage or graze livestock in treated cotton fields. If loss of cotton crop occurs due to adverse weather conditions, cotton or any crop registered for PROWL preplant incorporated use can be replanted the same year into PROWL treated soil without adverse effects. If replanting is necessary, DO NOT rework the soil deeper than the PROWL treated zone.

Preplant Incorporated Application of PROWL Followed by Preemergence Application of Cotoran or Karmex

PROWL preplant incorporated followed by an overlay preemergence application of Cotoran 80WP or Karmex may be applied for weed control in cotton.

Refer to the Cotoran or Karmex label for specific directions regarding rates, mixing, spraying and application. Use the Cotoran or Karmex alone preemergence rates. Follow all precautions and limitations on the Cotoran or Karmex label.

When applied as directed, in addition to those weeds controlled by PROWL alone, PROWL preplant incorporated followed by Cotoran or Karmex preemergence will control those additional broadleaf weeds listed on the Cotoran or Karmex label.

Preplant Incorporated Applications of PROWL Plus Caparol Mixtures

PROWL plus Caparol 80W tank mixture may be applied preplant incorporated for weed control in cotton grown in Arizona, California, New Mexico and the Upper and Lower El Paso Valley of Texas.

Uniformly apply the recommended tank mixtures of PROWL plus Caparol in 10 or more gallons of water per acre by ground equipment. PROWL plus Caparol may be applied prior to listing or over partially finished or finished beds. Incorporation should immediately follow application. Refer to page 8 for incorporation instructions. Maintain good agitation at all times until spraying is completed to keep the material in suspension. Mechanical agitation should be used in Arizona and California.

WEED SPECIES CONTROLLED

When applied as directed, tank mixtures of PROWL plus Caparol preplant incorporated will control the following grass and broadleaf weeds in addition to those controlled by PROWL alone:

GRASSES CONTROLLED

Junglerice *(Echinochloa colonum)*
Wild oat *(Avena fatua)*

BROADLEAF WEEDS CONTROLLED

Annual morningglory *(Ipomoea* spp.*)*
Black nightshade *(Solanum nigrum)*
Common Ragweed *(Ambrosia artemisiifolia)*
Groundcherry *(Physalis* spp.*)*
Malva *(Malva* spp.*)*
Mustard *(Brassica* spp.*)*
Prickly sida *(Sida spinosa)*
Smartweed *(Polygonum pensylvanicum)*

Recommended tank mixtures of PROWL plus Caparol preplant incorporated will control shallow-germinating seedlings of:

Cocklebur *(Xanthium pensylvanicum)*
Coffeeweed *(Daubentonia texana)*
Sandbur *(Cenchrus* spp.*)*

PAGE 9

Preplant Incorporated Broadcast[1] Rate Per Acre of PROWL Plus Caparol 80W in Cotton Grown in Arizona, California, New Mexico and the Upper and Lower El Paso Valley of Texas

	Soil Texture	PROWL + Caparol 80W
COARSE	Sands, loamy sands	DO NOT USE
	Sandy loams	1.0 to 1.5 pts. + 1.5 to 2.0 lbs.

MEDIUM	Loams	1.5 to 2.0 pts. + 2.0 lbs.
	Silt loams, silts, sandy clay loams*	1.5 to 2.0 pts. + 2.0 to 3.0 lbs.
FINE	Silty clay loams*, clay loams, clays	1.5 to 3.0 pts. + 2.0 to 3.0 lbs.

* Sometimes considered transitional soils and may be classified as either medium or fine textured soils.

The high rate for each soil texture above should be used if heavy weed populations are anticipated or if extensive crop residues were present prior to seedbed preparation. Use the 3 pint rate of PROWL for heavy clay soils.

[1]For calculation of band treatment rate, see page 5.

NOTE: DO NOT feed forage or graze livestock in treated cotton fields. If replanting is necessary, cotton may be planted in soil previously treated with PROWL plus Caparol. DO NOT rework the soil. Follow cropping restrictions on Caparol label.

PRECAUTIONS: DO NOT use this tank mixture in cut areas of newly leveled fields, in areas of excess salt or in areas where flooding over the bed is likely to occur as crop injury may result. Cotton should not be planted in tractor wheel depressions or crop injury may result. DO NOT use this tank mixture when cotton is irrigated up as crop injury may result. On mulch planted cotton, water back only after cotton seedlings are well established. Follow all precautions and limitations on Caparol label.

Preemergence Application PROWL Plus Cotoran Mixtures

PROWL plus Cotoran 80WP preemergence tank mixture may be applied immediately after planting for weed control in cotton grown in all states except Texas, Oklahoma, New Mexico, Arizona and California.

Uniformly apply recommended tank mixtures of PROWL plus Cotoran in 10 or more gallons of water or in 20 or more gallons of nitrogen solution per acre by ground equipment immediately after planting. Make sure cotton seeds are placed 1 inch or deeper below soil surface. Apply PROWL plus Cotoran preemergence tank mixture immediately after planting or crop injury may result. Refer to page 4 for liquid fertilizer compatibility determinations. Maintain good agitation at all times until spraying is completed. Refer to page 5 for spraying instructions.

PROWL plus Cotoran preemergence tank mixture is most effective in controlling weeds when adequate rainfall or overhead irrigation is received within 7 days after application. A shallow cultivation is recommended if soil crusting or soil compaction occurs. If weeds begin to germinate or adequate moisture is not received within 7 days after application, use shallow tillage (rotary hoe or light harrow) and make certain cotton seeds are below the tilled area.

Weed Species Controlled

When applied as directed, PROWL plus Cotoran preemergence tank mixture will control the following grass and broadleaf weed species:

GRASSES CONTROLLED

Barnyardgrass *(Echinochloa crus-galli)*
Crabgrass *(Digitaria* spp.*)*
Fall panicum *(Panicum dichotomiflorum)*,
Foxtail (giant) *(Setaria faberi)*
Foxtail (green) *(Setaria viridis)*
Foxtail (yellow) *(Setaria lutescens)*
Goosegrass *(Eleusine indica)*
Johnsongrass (from seed) *(Sorghum halepense)*
Signalgrass *(Brachiaria platphylla)*
Texas panicum *(Panicum texanum)*

BROADLEAF WEEDS CONTROLLED

Carpetweed *(Mollugo verticillata)*
Florida pusley *(Richardia scabra)*
Jimsonweed *(Datura stramonium)*
Kochia *(Kochia scoparia)*
Lambsquarters *(Chenopodium album)*
Pigweed *(Amaranthus* spp.*)*
Prickly sida *(Sida spinosa)*
Purslane *(Portulaca oleracea)*
Spurge, annuals *(Euphorbia* spp.*)*

Tank mixtures of PROWL plus Cotoran will reduce competition from the following hard-to-control annual broadleaf weeds:

Cocklebur *(Xanthium pensylvanicum)*
Annual Morningglory *(Ipomoea* spp.*)*

Broadcast Rates[1] Per Acre of PROWL plus Cotoran Preemergence Tank Mixture

Soil Texture	States	PROWL + Cotoran 80WP
COARSE Sands, loamy sands, sandy loams	Arkansas, Louisiana and Mississippi	1.0 to 1.5 pts. + 1.0 lb.
	Other states, except Texas, Oklahoma, New Mexico, Arizona, and California	1.0 to 1.5 pts. + 1.25 lbs.
MEDIUM sandy clay, loams*, sandy clays, loams, silts, silt, loams	All states, except Texas, Oklahoma, New Mexico, Arizona, and California	1.5 to 2.0 pts. + 1.25 to 2.0 lbs.
FINE silty clay loams*, clay loams, silty clays, clays	All states, except Texas, Oklahoma, New Mexico, Arizona, and California	1.5 to 3.0 pts. + 2.0 to 2.5 lbs.

[1] For calculation of band treatment rate, see page 5.
* Sometimes considered transitional soils and may be classified as either fine or medium textured soils.

The high rate for each soil texture above should be used if heavy weed populations are anticipated or if extensive crop residues were present prior to seedbed preparation. Use the 3 pint rate of PROWL for heavy clay soils.

NOTE: The use of Cotoran 80WP as a preemergence application, following the use of a systemic insecticide at planting, may result in injury to cotton. DO NOT plant crops other than cotton within 6 months of the last application of Cotoran or injury may result. DO NOT feed foliage from treated cotton plants or gin trash to livestock. Observe all precautions and limitations on the Cotoran label.

Preplant Incorporated Application of PROWL for Suppression of Rhizome Johnsongrass

PROWL applied preplant incorporated will provide suppression of rhizome johnsongrass (*Sorghum halepense*) in cotton at the rates recommended for soil textures below. This use is not recommended for Arizona, New Mexico and California.

PAGE 10

Before application, use a chisel plow or similar implement to bring johnsongrass rhizomes to the surface. Chop rhizomes into small pieces with a disk harrow set to cut 4 to 6 inches deep and operated in two different directions at 4 to 6 mph.

Uniformly apply PROWL in 10 or more gallons of water or in 20 or more gallons of liquid fertilizer per acre by ground equipment. (Refer to page 4 for liquid fertilizer compatibility determinations.) Apply in 5 or more gallons of water per acre by aircraft. PROWL may be applied immediately before planting or up to 60 days prior to planting cotton. Refer to page 5 for specific instructions for use with ground and aerial equipment.

PROWL is to be incorporated into the soil within 7 days after application prior to planting. (For maximum suppression of rhizome johnsongrass, incorporate PROWL as soon as possible after application.) Deep and thorough incorporation of PROWL is necessary for suppression of rhizome johnsongrass. Mechanical incorporation can be achieved by the following methods:

(a) Disk harrow set to cut 4 to 6 inches deep and operated in two different directions at 4 to 6 mph.

(b) PTO-driven equipment (tillers, cultivators, hoes) set to cut 3 to 4 inches deep and operated one time at 4 mph or less.

For johnsongrass escapes during the crop season, cultivation and/or application of registered postemergence herbicides are recommended. Follow the directions for use on the labels of the respective herbicides.

Preplant Incorporated Broadcast Rate Per Acre of PROWL in Cotton for Suppression of Rhizome Johnsongrass

Soil Texture—Up to 3% Organic Matter[1]	
COARSE sands, loamy sands, sandy loams	2.0 pts.
MEDIUM sandy clay loams*, sandy clays, loams, silt loams, silts	3.0 pts.
FINE silty clay loams*, clays, silty clays, clay loams	4.0 pts.

* Sometimes considered transitional soils and may be classified as either medium or fine textured soils.

[1] This use is not recommended for soils with more than 3% organic matter.

NOTE: DO NOT feed forage or graze livestock in PROWL-treated cotton fields.

If cotton crop loss occurs due to weather conditions, cotton or soybeans can be replanted the same year into PROWL-treated soil without adverse effects. If replanting is necessary, DO NOT rework the soil deeper than the PROWL-treated zone.

SOYBEANS

PROWL can be applied in soybeans as follows (DO NOT USE IN CALIFORNIA):

Preplant Incorporated	Sequential	Preemergence
PROWL alone p. 12 PROWL + Sencor/Lexone p. 12 PROWL + Amiben p. 13 PROWL + Amiben + Sencor/Lexone p. 13	PROWL preplant incorporated followed by preemergence applications of: Sencor/Lexone p. 13 Amiben p. 13 Lorox p. 14	PROWL + Sencor/Lexone p. 12 PROWL + Amiben p. 13 PROWL + Lorox p. 14 PROWL + Amiben + Sencor/Lexone p. 13

Refer to specific sections of the label for directions for each method of application.

Refer to page 16 for PROWL + Sencor/Lexone or Lorox + Paraquat CL treatments in minimum and no-till soybeans.

PROWL will provide control or suppression of the following special weeds:

Red rice, see page 15
Rhizome johnsongrass, see page 15
Shattercane, see page 15
Itchgrass, see page 15

Refer to page 5 for follow crop restrictions.

GENERAL APPLICATION INSTRUCTIONS

Uniformly apply the recommended PROWL or PROWL tank-mix combination treatment in 10 or more gallons of water or PROWL alone in 20 or more gallons of liquid fertilizer per acre by ground equipment. (Refer to page 4 for liquid fertilizer compatibility determinations). Apply in 5 or more gallons of water per acre by aircraft. Refer to page 5 for spraying instructions. Refer to page 4 for PROWL/dry bulk fertilizer applications.

Incorporation Instructions

1. Flat-planted soybeans

PROWL is to be incorporated within 7 days of application prior to planting. When PROWL is applied to flat land that will not be bedded, mechanical incorporation should be into the top 1 or 2 inches of soil. Mechanical incorporation of PROWL can be achieved by the following:

(a) Disk harrow set to cut 3 to 4 inches deep and operated in two different directions at 4 to 6 mph.

(b) PTO-driven equipment (tillers, cultivators, hoes) set to cut 2 inches deep and operated one time at 4 mph or less.

(c) Rolling cultivator set to cut 2 to 3 inches deep and operated two times at 6 to 8 mph. Use only on coarse and medium textured soils.

(d) Field cultivator with 3 or 4 rows of sweeps, spaced at intervals of 7 inches or less and staggered so that no soil is left unturned (Do not use chisel points), set to cut 3 inches deep and operated two times at more than 5 mph with the second pass made at an angle to the first.

(e) LELY-ROTERRA[1] set and operated according to the manufacturer's directions to thoroughly incorporate PROWL to a depth of 1 to 2 inches.

(f) Do-All set and operated according to the manufacturer's directions to thoroughly incorporate PROWL to a depth of 1 to 2 inches.

[1]LELY-ROTERRA is a trademark of Lely.

NOTE: PROWL can be incorporated in a single pass when the soil is of good tilth with moderate moisture and relatively free of clods and trash. Implements must be properly set up to thoroughly incorporate PROWL into the top 1 or 2 inches of soil. Recommended implements include: (1) C-shank or S-shank field cultivators equipped with flextine drag or rolling basket attachment and (2) the combination disk/field cultivator implement equipped with flextine drag or rolling basket attachment. Field cultivators must have 3 or 4 rows of sweeps, spaced at intervals of 7 inches or less and staggered so that no soil is left unturned. Implements must be operated at 6 to 8 miles per hour and set to cut 3 to 4 inches deep.

2. Bedded Soybeans—Application Over Beds

PROWL is to be incorporated within 7 days of application prior to planting. When PROWL is applied over beds, mechanical incorporation should be into the top 1 or 2 inches of soil. Mechanical incorporation can be achieved by the following:

(a) Rolling cultivator set to cut 2 to 3 inches deep and operated two times at 6 to 8 mph. Use only on coarse and medium textured soils.

(b) PTO-driven equipment (tillers, cultivators, hoes) set to cut 2 inches deep and operated one time at 4 mph or less.

(c) Bed conditioner set to cut 2 to 3 inches deep and operated one time at 4 to 6 mph.

During planting, or if beds must be reshaped prior to planting, avoid tillage that will bring untreated soil to the surface or expose untreated soil on the seedbed or in the furrow.

3. Bedded Soybeans—Application Prior To Bedding

PROWL is to be incorporated within 7 days of application prior to planting. When PROWL is applied to flat land prior to bedding, mechanical incorporation should be of sufficient depth so that bedding does not bring up untreated soil. Mechanical incorporation of PROWL can be achieved by the following:

(a) Disk harrow set to cut 4 to 6 inches deep and operated in two different directions at 4 to 6 mph.

(b) TO-driven equipment (tillers, cultivators, hoes) set to cut 3 to 4 inches deep and operated one time at 4 mph or less.

PAGE 11

During planting, or if beds must be reshaped prior to planting, avoid tillage that will bring untreated soil to the surface or expose untreated soil on the seedbed or in the furrow.

Preplant Incorporated Applications of PROWL alone

The following recommended treatments for preplant incorporated applications of PROWL alone are to be used when no tank mixtures or sequential combination treatments are planned. Thoroughly mix previous crop residues to a depth of 4 to 6 inches by plowing or disking prior to application. PROWL may be applied immediately prior to planting or up to 60 days prior to planting. Refer to incorporation instructions above.

Weed Species Controlled

The following grass and broadleaf weed species are controlled by preplant incorporated treatments of PROWL in soybeans at the rates recommended for each soil texture listed:

GRASSES CONTROLLED

Barnyardgrass *(Echinochloa crus-galli)*
Crabgrass *(Digitaria* spp.*)*
Crowfootgrass *(Dactyloctenium aegyptium)*
Fall panicum *(Panicum dichotomiflorum)*
Foxtail (giant) *(Setaria faberi)*
Foxtail (green) *(Setaria viridis)*
Foxtail (yellow) *(Setaria lutescens)*
Goosegrass *(Eleusine indica)*
Johnsongrass (from seed) *(Sorghum halepense)*
Signalgrass *(Brachiaria platyphylla)*
Texas panicum *(Panicum texanum)*
Witchgrass *(Panicum capillare)*

BROADLEAF WEEDS CONTROLLED

Carpetweed *(Mollugo verticillata)*
Florida pusley *(Richardia scabra)*
Kochia *(Kochia scoparia)*
Lambsquarters *(Chenopodium album)*
Pigweed *(Amaranthus* spp.*)*
Purslane *(Portulaca oleracea)*
Spurge, annuals *(Euphorbia* spp.*)*

PROWL will aid in the control of and reduce competition from:

Smartweed *(Polygonum pensylvanicum)*
Velvetleaf (Buttonweed) *(Abutilon theophrasti)*

Refer to page 15 for instructions for control or suppression of the following weeds:

Shattercane
Rhizome johnsongrass
Red Rice
Itchgrass

Preplant Incorporated Broadcast Rate Per Acre of PROWL Alone in Soybeans

	Southern States (including Eastern Coastal Plains)[1]	Northern States[1]	
Soil Texture	No Organic Matter Restriction	Up to 3% Organic Matter	More than 3% Organic Matter
COARSE sands, loamy sands, sandy loams	1.0 to 1.5 pts.	1.0 to 2.0 pts.	2.0 pts.
MEDIUM sandy clay loams*, sandy clays, loams, silts, silt loams	1.5 to 2.0 pts.	1.5 to 2.5 pts.	2.5 to 3.0 pts.
FINE silty clay loams*, clay loams, silty clays, clays	1.5 to 3.0 pts.	2.0 to 3.0 pts.	3.0 pts.

[1] See map on page 26 for specific states. DO NOT use in California.

* Sometimes considered transitional soils and may be classified as either medium or fine textured soils.

The high rate for each soil texture above, where listed, should be used if heavy weed populations are anticipated or if extensive crop residues were present prior to seedbed preparation. Use the 3 pint rate for heavy clay soils.

DO NOT USE on peat or muck soils.

NOTE: Livestock can graze or be fed soybean forage from PROWL-treated fields. If soybean crop loss occurs due to weather conditions, soybeans or any crop registered for PROWL preplant incorporated use can be replanted. If replanting is necessary, DO NOT rework the soil deeper than the PROWL-treated zone.

PROWL Plus Sencor/Lexone

The following PROWL plus Sencor/Lexone treatments may be used in soybeans. Uniformly apply PROWL or PROWL plus Sencor/ Lexone tank mixtures in 10 or more gallons of water per acre with ground equipment or in 5 or more gallons of water per acre with aircraft. See page 5 for spraying instructions. See page 11 for incorporation instructions. Use PROWL and Sencor/Lexone rates listed in table on page 13. See special precautions below.

1. PROWL Preplant Incorporated Followed by Sencor/Lexone Preemergence

Apply PROWL preplant incorporated. Follow with a preemergence treatment of Sencor/Lexone as directed on the Sencor or Lexone label.

2. Preplant Incorporated Tank Mixture of PROWL plus Sencor/ Lexone

PROWL plus Sencor/Lexone must be incorporated into the soil within 7 days after application and soybeans must be planted no later than 7 days after application. For best results, incorporation should be in the top 1 to 2 inches of soil.

3. Preemergence Tank Mixtures of PROWL plus Sencor/Lexone

Apply PROWL plus Sencor/Lexone at planting or up to 2 days after planting. Apply to a seedbed which is firm and free of clods and trash. DO NOT APPLY POSTEMERGENCE. These treatments are most effective in controlling weeds when adequate rainfall or overhead irrigation is received within 7 days after application.

If moisture is insufficient to activate the herbicides, a shallow cultivation (preferably with a rotary hoe) should be made after emergence of soybeans, but while weeds are small enough to be controlled by mechanical cultivation.

A shallow, thorough mixing of soil after herbicide application with a rotary hoe or similar implement will improve consistency of performance and reduce dependency on rainfall to activate herbicides.

Special Precautions for PROWL Plus Sencor or Lexone Combinations

Soybean seed should be planted at least 1½ inches below soil surface. Applications should not be made to soils having calcareous surface or a pH of 7.5 or higher. Soil incorporation deeper than recommended will reduce weed control and can result in crop injury.

The PROWL plus Sencor or Lexone treatment can be applied only once per cropping season. DO NOT USE on sensitive soybean varieties-Altona, Coker 102 and 156, Govan, Semmes, Tracy or Vansoy.

Follow cropping restrictions on Sencor and Lexone labels.

WEED SPECIES CONTROLLED

When applied as directed, tank mixtures of PROWL plus Sencor/ Lexone or PROWL preplant incorporated followed by Sencor/Lexone preemergence will control the following weed species in addition to those controlled by PROWL alone:

Common ragweed *(Ambrosia artemisiifolia)*
Jimsonweed *(Datura stramonium)*
Prickly sida *(Sida spinosa)*
Smartweed *(Polygonum pensylvanicum)*
Velvetleaf (Buttonweed) *(Abutilon theophrasti)*
Venice mallow *(Hibiscus trionum)*
Wild mustard *(Brassica kaber)* (Medium and fine soils only)

PROWL plus Sencor/Lexone will reduce competition from cocklebur *(Xanthium pensylvanicum)*. Use either a preplant incorporated application of PROWL followed by a preemergence application of Sencor/Lexone or a preemergence tank mix combination of PROWL plus Sencor/Lexone where severe cocklebur infestations are anticipated. See page 15 for PROWL plus Sencor/Lexone use for shattercane control.

Broadcast Rates Per Acre of Preplant Incorporated or Preemergence PROWL Plus Sencor/Lexone Tank Mixtures or PROWL Preplant Incorporated followed by Sencor/Lexone Preemergence

	Southern States (including Eastern Coastal Plains)[1]	Northern States[1]	
	PROWL + Sencor/ Lexone DF*	PROWL + Sencor/ Lexone DF*	PROWL + Sencor/ Lexone DF*
Soil Texture		Up to 3% Organic Matter	More than 3% Organic Matter
COARSE** loamy sands, sandy loams	1.0 to 1.5 pts. + 0.33 lbs.	1.0 to 1.5 pts. + 0.5 lbs.	1.5 pts. + 0.5 lbs.
MEDIUM sandy clay loams***, sandy clays, loams, silts, silt loams	1.5 to 2.0 pts. + 0.5 lbs.	1.5 to 2.0 pts. + 0.5 lbs.	1.5 to 2.0 pts. + 0.5 to 0.67 lbs.
FINE silty clay loams***, clay loams, silty clays, clays	1.5 to 2.0 pts. + 0.5 lbs.	1.5 to 2.0 pts. + 0.5 to 0.67 lbs.	2.0 to 2.5 pts. + 0.67 lbs.

[1] See map on page 26 for specific states. DO NOT USE IN CALIFORNIA.

* When using Sencor 4F, Sencor 50WP, Lexone 4L, or Lexone 50WP, use equivalent rates. One pint of the flowable formulation or 1 lb. of the wettable powder formulation equals 0.67 lb. of Sencor DF or Lexone DF. See conversion chart on page 4.

** DO NOT USE on sands. DO NOT USE on loamy sands or on sandy loams that contain less than 2% organic matter. DO NOT USE on peat or muck soils.

*** Sometimes considered transitional soils and may be classified as either medium or fine textured soils.

The high rates of PROWL and Sencor/Lexone for each soil texture, as stated above, should be used if heavy weed populations are anticipated or if extensive crop residues were present prior to seedbed preparation.

Note: In the event of a crop loss due to weather conditions, soybeans can be replanted without adverse effects the same year. If replanting is necessary, DO NOT rework the soil. Livestock can graze or be fed soybean forage from treated fields 40 days after application. Observe all precautions and limitations on Sencor and Lexone labels.

PROWL PLUS AMIBEN (NORTHERN STATES ONLY)

The following PROWL plus Amiben treatments may be used in soybeans in the Northern States. See the map on page 26 for specific states. Uniformly apply PROWL or PROWL plus Amiben tank mixtures in 10 or more gallons of water per acre with ground equipment or in 5 or more gallons of water per acre with aircraft. See page 5 for spraying instructions. See page 11 for incorporation instructions. Use PROWL and Amiben rates listed in table below.

1. PROWL Preplant Incorporated Followed by Amiben Preemergence.

Apply PROWL preplant incorporated. Follow with a preemergence treatment of Amiben as directed on the Amiben label.

2. Preplant Incorporated Tank Mixture of PROWL plus Amiben

PROWL plus Amiben must be incorporated into the soil within 7 days after application and soybeans must be planted no later than 7 days after application. For best results, incorporation should be in the top 1 to 2 inches of soil.

3. Preemergence Tank Mixtures of PROWL plus Amiben

Apply PROWL plus Amiben at planting or up to 2 days after planting. Apply to a seedbed which is firm and free of clods and trash. DO NOT APPLY POSTEMERGENCE. These treatments are most effective in controlling weeds when adequate rainfall or overhead irrigation is received within 7 days after application.

If moisture is insufficient to activate the herbicides, a shallow cultivation (preferably with a rotary hoe) should be made after emergence of soybeans, but while weeds are small enough to be controlled by mechanical cultivation.

A shallow, thorough mixing of soil after herbicide application with a rotary hoe or similar implement will improve consistency of performance and reduce dependency on rainfall to activate herbicides.

WEED SPECIES CONTROLLED

When applied as directed, tank mixtures of PROWL plus Amiben or PROWL preplant incorporated followed by Amiben preemergence will control the following weed species in addition to those controlled by PROWL alone:

Common ragweed *(Ambrosia artemisiifolia)*
Mustard *(Brassica* spp.*)*
Smartweed *(Polygonum pensylvanicum)*
Velvetleaf (Buttonweed) *(Abutilon theophrasti)*

Broadcast Rates Per Acre of Preplant Incorporated or Preemergence Prowl Plus Amiben Tank Mixtures or Prowl Preplant Incorporated Followed By Amiben Preemergence

Soil Texture	Up to 3% Organic Matter	More than 3% Organic Matter
COARSE sands, loamy sands, sandy loams	DO NOT USE	DO NOT USE
MEDIUM sandy clay loams*, sanday clays, loams, silts, silt loms	1.5 to 2.0 pts. + 1.0 gallon	1.5 to 2.0 pts. + 1.0 gallons
FINE silty clay loams*, clay loams, silty clays, clays	1.5 to 2.0 pts. + 1.0 gallon	2.0 to 2.5 pts. + 1.0 gallon

* Sometimes considered transitional soils and may be classified as either medium or fine textured soils.

DO NOT USE on peat or muck soils.

The high rate of PROWL for each soil texture above should be used if heavy weed populations are anticipated, if velvetleaf or smartweed infestations are anticipated, or if extensive crop residues were present prior to seedbed preparation.

Note: In the event of a crop loss due to weather conditions, soybeans can be replanted without adverse effects the same year. If replanting is necessary, DO NOT rework the soil. Livestock can graze or be fed soybean forage from treated fields. Observe all precautions and limitations on Amiben label.

PROWL Plus Amiben Plus Sencor/Lexone

The following PROWL plus Amiben plus Sencor/Lexone treatments may be used in soybeans in the Northern States. See page 26 for specific states. Uniformly apply PROWL plus Amiben plus Sencor/Lexone tank mixtures in 10 or more gallons of water per acre with ground equipment or in 5 or more gallons of water per acre with aircraft. When mixing add Sencor or Lexone first, then Amiben followed by PROWL. See page 4 for complete mixing instructions. See page 5 for spraying instructions. See page 11 for incorporation instructions. Use PROWL plus Amiben plus Sencor/Lexone rates listed in table on page 14. See special precautions on page 12 for PROWL plus Sencor/Lexone combinations.

1. Preplant Incorporated Tank Mixture of PROWL plus Amiben plus Sencor/Lexone

PROWL plus Amiben plus Sencor/Lexone must be incorporated into the soil within 7 days after application and soybeans must be planted no later than 7 days after application. For best results, incorporation should be in the top 1 to 2 inches of soil.

2. Preemergence Tank Mixtures of PROWL plus Amiben plus Sencor/Lexone

Apply PROWL plus Amiben plus Sencor/Lexone at planting or up to 2 days after planting. Apply to a seedbed which is firm and free of clods and trash. DO NOT APPLY POSTEMERGENCE. These treatments are most effective in controlling weeds when adequate rainfall or overhead irrigation is received within 7 days after application.

If moisture is insufficient to activate the herbicides, a shallow cultivation (preferably with a rotary hoe) should be made after emergence of soybeans, but while weeds are small enough to be controlled by mechanical cultivation.

A shallow, thorough mixing of soil after herbicide application with a rotary hoe or similar implement will improve consistency of performance and reduce dependency on rainfall to activate herbicides.

WEED SPECIES CONTROLLED

When applied as directed, tank mixtures of PROWL plus Amiben plus Sencor/Lexone will control the following weed species in addition to those controlled by PROWL alone:

Common ragweed *(Ambrosia artemisiifolia)*
Mustards *(Brassica* spp.*)*
Prickly sida *(Sida spinosa)*
Smartweed *(Polygonum pensylvanicum)*
Velvetleaf (Buttonweed) *(Abutilon theophrasti)*
Venice mallow *(Hibiscus trionum)*

PROWL plus Amiben plus Sencor/Lexone preemergence will reduce competition from cocklebur *(Xanthium pensylvanicum)*.

PROWL plus Amiben plus Sencor/Lexone preplant incorporated or preemergence will reduce competition from jimsonweed *(Datura stramonium)*.

Broadcast Rates Per Acre of Preplant Incorporated or Preemergence PROWL plus Amiben plus Sencor/Lexone tank mixtures

	PROWL + Amiben + Sencor/ Lexone DF*	PROWL + Amiben + Sencor/ Lexone DF*
Soil Texture	**Up to 3% Organic Matter**	**More than 3% Organic Matter**
COARSE** Loamy sands, sandy loams (greater than 2% organic matter only)	1.5 pts. + 6.0 pts. + 0.33lbs.	1.5 pts. + 6.0 pts. +0.33 lbs.
MEDIUM Sandy clay looms***, sandy clays, loams, silts, silt loams	1.5 to 2 pts. + 6.0 to 8.0 pts. +0.33 to 0.5 lbs.	1.5 to 2 pts. + 6.0 to 8.0 pts. + 0.33 to 0.5 lbs.
FINE Silty clay loams***, clay loams, silty clays, clays	1.5 to 2.0 pts. + 6.0 to 8.0 pts. + 0.33 to 0.5 lbs.	2.0 to 2.5 pts. + 8.0 pts. +0.5 lbs.

* When using Sencor 4F, Sencor 50WP, Lexone 4L, or Lexone 50 WP, use equivalent rates. One pint of the flowable formulation or 1 lb. of the wettable powder formulation equals 0.67 lb. of Sencor DF or Lexone DF. See conversion chart on page 4.

** DO NOT USE on sands. DO NOT USE on loamy sands or on sandy loans that contain less than 2% organic matter. DO NOT USE on peat or muck soils.

*** Sometimes considered transitional soils and may be classified as either medium or fine textured soils.

The high rates of PROWL, Amiben and Sencor/Lexone for each soil texture, as stated above, should be used if heavy weed populations are anticipated, if cocklebur, jimsonweed, smartweed or velvetleaf populations are expected or if extensive crop residues were present prior to seedbed preparation.

Note: In the event of a crop loss due to weather conditions, soybeans can be replanted without adverse effects the same year. If replanting is necessary, DO NOT rework the soil. DO NOT use foliage for livestock feed or forage. Observe all precautions and limitations on Amiben, Sencor and Lexone labels.

PROWL Plus Lorox Preemergence Tank Mixtures or PROWL Preplant Incorporated plus Lorox Preemergence

The following PROWL plus Lorox treatments may be used in soybeans in all states except California. Uniformly apply PROWL or PROWL plus Lorox tank mixtures in 10 or more gallons of water per acre with ground equipment or in 5 or more gallons of water per acre by aircraft. See page 5 for spraying instructions. See page 11 for incorporation instructions. Use PROWL and Lorox rates listed in table below.

1. PROWL Preplant Incorporated Followed by Lorox Preemergence.

Apply PROWL preplant incorporated. Follow with a preemergence treatment of Lorox as directed on the Lorox label.

2. Preemergence Tank Mixtures of PROWL plus Lorox.

Apply PROWL plus Lorox at planting or up to 2 days after planting. Apply to a seedbed which is firm and free of clods and trash. DO NOT APPLY POSTEMERGENCE. These treatments are most effective in controlling weeds when adequate rainfall or overhead irrigation is received within 7 days after application.

If moisture is insufficient to activate the herbicides, a shallow cultivation (preferably with a rotary hoe) should be made after emergence of soybeans, but while weeds are small enough to be controlled by mechanical cultivation.

A shallow, thorough mixing of soil after herbicide application with a rotary hoe or similar implement will improve consistency of performance and reduce dependency on rainfall to activate herbicides.

WEED SPECIES CONTROLLED

When applied as directed, tank mixtures of PROWL plus Lorox or PROWL preplant incorporated plus Lorox preemergence, will control the following weed species in addition to those controlled by PROWL alone:

Mustard *(Brassica* spp.*)*
Smartweed *(Polygonum pensylvanicum)*
Velvetleaf (Buttonweed) *(Abutilon theophrasti)*

Recommended tank mixes of PROWL plus Lorox, or PROWL preplant incorporated plus Lorox preemergence, will aid in the control of and reduce competition from:

Common ragweed *(Ambrosia artemisiifolia)*

Broadcast Rate Per Acre in Soybeans for PROWL Plus Lorox Preemergence Tank Mixtures or PROWL Preplant Incorporated Plus Lorox Preemergence

	PROWL + Lorox L*	PROWL + Lorox L*
Soil Texture	**½ to 3% Organic Matter**	**More than 3% Organic Matter**
COARSE sandy loams	1.0 to 1.5 pts. + 1.0 pts.	1.5 pts. + 1.0 to 1.5 pts.
MEDIUM Sandy clay looms**, sandy clays, loams, silts, silt loams	1.5 to 2.0 pts. + 1.0 to 1.5 pts.	1.5 to 2.0 pts. + 1.5 to 2.0 pts.
FINE Silty clay loams**, silty clays, clay loams, clays	1.5 to 2.0 pts. + 1.5 to 2.0 pts.	2.0 to 2.5 pts. + 2.0 to 2.5 pts.

* When using Lorox 50W use equivalent rates. One pint of the flowable formulation equals 1 lb. of the wettable powder formulation. See conversion charge on page 4.

** Sometimes considered transitional soils and may be classified as either medium or fine textured soils.

The high rates for each soil texture above, where listed, should be used if heavy weed populations are anticipated or if extensive crop residues were present prior to seedbed preparation.

DO NOT USE on peat or muck soils. DO NOT USE on sands or gravelly soils, or on soils containing less than ½% organic matter.

NOTE: Livestock may graze or be fed forage from PROWL plus Lorox treated soybean fields. If soybean crop loss occurs due to weather conditions, soybeans may be replanted. If replanting is necessary, thoroughly rework the soil before replanting. Observe all precautions and limitations on Lorox label.

SPECIAL WEEDS

1. Shattercane

Preplant incorporated treatments of PROWL alone; preplant incorporated PROWL tank-mix combinations with Sencor/Lexone; or preplant incorporated PROWL applications followed by registered sequential preemergence herbicides will control shattercane (wild cane) *(Sorghum bicolor)*.

Shattercane is difficult to control in many soybean growing areas because of its ability to germinate from a greater depth in the soil than most other weeds. Shattercane can emerge throughout the growing season depending on soil temperature, moisture, and seed dormancy.

Thoroughly mix previous crop residues into the soil and destroy any existing vegetation prior to herbicide application. Apply PROWL or PROWL combinations at the recommended rate listed in the chart below. See page 11 for incorporation instructions. Thoroughly and uniformly mechanically incorporate PROWL in the TOP 2 INCHES of soil. Acceptable results CANNOT be expected if PROWL is applied preemergence. Cultivation may be necessary to control escapes and late germinating shattercane during the crop season.

Preplant Incorporated Broadcast Rate Per Acre of PROWL for Control of Shattercane in Soybeans

PROWL alone or PROWL Sequential Treatments		
Soil Texture	Up to 3% Organic Matter	More than 3% Organic Matter
COARSE sands, loamy sands, sandy loams	2.0 pts.	2.0 pts.
MEDIUM Sandy clay loams*, sandy clays, loams, silts, silt loams	2.5 pts.	3.0 pts.
FINE Silty clay loams*, clay loams, silty clays, clays	3.0 pts.	3.0 pts.

* Sometimes considered transitional soils and may be classified as either medium or fine textured soils.

Preplant Incorporated Broadcast Rate Per Acre of PROWL plus Sencor/Lexone Tank-Mix for Control of Shattercane in Soybeans[1]

	PROWL + Sencor/ Lexone DF*	PROWL + Sencor/ Lexone DF*
Soil Texture	Up to 3% Organic Matter	More than 3% Organic Matter
COARSE Loamy sands, sandy loams	1.5 pts. + 0.5 lb.	1.5 pts. + 0.5 lb.
MEDIUM Sandy clay loams***, sandy clay, loams, silts, silt loams	2.0 pts + 0.5 lb.	2.0 pts. + 0.5 lb. to 0.67 lb.
FINE Silty clay loams***, clay loams, silty clays, clays	2.0 pts. + 0.5 lb. to 0.67 lb.	2.5 pts. + 0.67 lb.

1 Refer to page 12 for special precautions for PROWL plus Sencor/Lexone combinations.

* When using Sencor 4F, Sencor 50WP, Lexone 4L, or Lexone 50WP, use equivalent rates. One pint of the flowable formulation or 1 lb. of the wettable powder formulation equals 0.67 lb. of Sencor DF or Lexone DF. See conversion chart on page 4.

** DO NOT USE on sands. DO NOT USE on loamy sands, or on sandy loams that contain less than 2% organic matter. DO NOT USE on peat or muck soils.

*** Sometimes considered transitional soils and may be classified as either medium or fine textured soils.

2. Red Rice, Itchgrass

PROWL applied preplant incorporated will provide control of red rice *(Oryza sativa)* and suppression of itchgrass *(Rottboellia exaltata)* at the rates recommended for soil textures below.

Preplant Incorporated Broadcast Rate Per Acre of PROWL in Soybeans for Control of Red Rice and Suppression of Itchgrass

Soil Texture—Up to 3% Organic Matter[1]	
COARSE sands, loamy sands, sandy loams	2.0 to 3.0 pts.*
MEDIUM sandy clay loams**, sandy clays, loams silts, silt loams	3.0 pts.
FINE silty clay loams**, clay loams, silty clays, clays	4.0 pts.

1 This use is not recommended for soils with more than 3% organic matter.

* The higher rate should be used if heavy red rice or itchgrass populations are anticipated.

** Sometimes considered transitional soils and may be classified as either medium or fine textured soils.

NOTE: Livestock can graze or be fed from PROWL-treated soybean fields. If soybean crop loss occurs due to weather conditions, cotton or soybeans can be replanted the same year into PROWL-treated soil without adverse effects. If replanting is necessary, DO NOT rework the soil deeper than the PROWL treated zone.

3. Rhizome Johnsongrass

PROWL applied preplant incorporated will provide suppression of rhizome johnsongrass (*Sorghum halepense*) in soybeans at the rates recommended for soil textures below. This use is not recommended for Arizona, New Mexico and California.

Before application, use a chisel plow or similar implement to bring johnsongrass rhizomes to the surface. Chop rhizomes into small pieces with a disk harrow set to cut 4 to 6 inches deep and operated in two different directions at 4 to 6 mph.

Uniformly apply PROWL in 10 or more gallons of water or in 20 or more gallons of liquid fertilizer per acre by ground equipment. (Refer to page 4 for liquid fertilizer compatibility determinations.) Apply in 5 or more gallons of water per acre by aircraft. PROWL may be applied immediately before planting or up to 60 days prior to planting soybeans. Refer to page 5 for specific instructions for use with ground and aerial equipment.

PROWL is to be incorporated into the soil within 7 days after application prior to planting. (For maximum suppression of rhizome johnsongrass, incorporate PROWL as soon as possible after application.) Deep and thorough incorporation of PROWL is necessary for suppression of rhizome johnsongrass. Mechanical incorporation can be achieved by the following methods:

(a) Disk harrow set to cut 4 to 6 inches deep and operated in two different directions at 4 to 6 mph.

(b) PTO-driven equipment (tillers, cultivators, hoes) set to cut 3 to 4 inches deep and operated one time at 4 mph or less.

For johnsongrass escapes during the crop season, cultivation and/or application of registered postemergence herbicides are recommended. Follow the directions for use on the labels of the respective herbicides.

Preplant Incorporated Broadcast Rate Per Acre of PROWL in Soybeans for Suppression of Rhizome Johnsongrass

Soil Texture—Up to 3% Organic Matter[1]	
COARSE sands, loamy sands, sandy loams	2.0 pts.
MEDIUM sandy clay loams*, sandy clays, loams, silt loams, silts	3.0 pts.

FINE silty clay loams*, clays, silty clays, clay loams	4.0 pts.

* Sometimes considered transitional soils and may be classified as either medium or fine textured soils.

[1] This use is not recommended for soils with more than 3% organic matter. If soybean crops loss occurs due to weather conditions, cotton or soybeans can be replanted the same year into PROWL-treated soil without adverse effects. If replanting is necessary, DO NOT rework the soil deeper than the PROWL-treated zone.

NOTE: Livestock can graze or be fed from PROWL-treated soybean fields.

PROWL TANK MIXTURES IN MINIMUM AND NO-TILL SOYBEANS

PROWL may be applied as a preemergence tank mixture treatment in minimum or no-till soybeans as follows:

PROWL + Sencor/Lexone + Paraquat CL
PROWL + Lorox + Paraquat CL

Paraquat CL is used to kill existing vegetative growth. Paraquat CL should be deleted from the tank mixture if vegetation is absent at time of application.

APPLICATION INSTRUCTIONS

Uniformly apply the recommended PROWL tank-mix combination treatment in 20 to 60 gallons of water per acre with ground equipment. Use the higher gallonage for fields with heavy weed infestations or excessive crop residues. See page 5 for spraying instructions.

Apply the PROWL plus Sencor/Lexone plus Paraquat CL or PROWL plus Lorox plus Paraquat CL tank-mixture combination at planting or up to 2 days after planting. DO NOT APPLY POSTEMERGENCE to soybeans since soybeans exposed at time of application will be killed. These treatments are most effective in controlling weeds when adequate rainfall or overhead irrigation is received within 7 days after application. See special precautions on page 12 for PROWL plus Sencor/Lexone combinations.

If vegetation is absent at time of application, delete Paraquat CL from the tank mixture as it controls only emerged vegetation.

MIXING INSTRUCTIONS

Fill tank one-half to three-quarter full with clean water. MAINTAIN CONTINUOUS AGITATION WHILE ADDING HERBICIDES. Add Sencor, Lexone or Lorox formulation first, as follows:

When using Sencor/Lexone or Lorox Flowable Formulation: Premix one part of the flowable and one part water and add this diluted mixture into the partially filled tank while agitating.

When using Sencor/Lexone Dry Flowable Formulation: Add the granules to the partially filled tank while agitating, and allow the granules to disperse.

When using Sencor/Lexone or Lorox Wettable Powder Formulation: Make a slurry of the wettable powder in water (1 part WP + 2 parts water). Add the slurry slowly into the partially filled tank while agitating.

After Sencor, Lexone or Lorox herbicide has been added to the tank and mixing is complete, add PROWL to tank while agitating.

If Paraquat CL is included in the tank-mix treatment, add to tank after complete mixing of PROWL. As a last ingredient, add 8 oz of Ortho* X-77 spreader or similar non-ionic surfactant per 100 gal of total spray mixture.

Fill the remainder of the tank with water.

PROWL PLUS SENCOR/LEXONE PLUS PARAQUAT CL WEED SPECIES CONTROLLED

Controlled by Paraquat CL:

Paraquat CL will control most annual emerged weeds and suppress many emerged perennials. Refer to the Paraquat CL label for specific use recommendations.

PROWL plus Sencor/Lexone preemergence will control the following weed species in addition to those controlled by PROWL alone:

Common Ragweed *(Ambrosia artemisiifolia)*
Jimsonweed *(Datura stramonium)*
Prickly sida *(Sida spinosa)*
Smartweed *(Polygonum pensylvanicum)*
Velvetleaf (Buttonwood) *(Abutilon theophrasti)*
Venice Mallow *(Hibiscus trionum)*
Wild Mustard *(Brassica kaber)* (Medium and fine soils only)

PROWL Plus Sencor/Lexone will aid in the control of and reduce competition from cocklebur *(Xanthium pensylvanicum)*.

*Ortho is a trademark of Chevron Chemical Company.

BROADCAST RATES PER ACRE OF PROWL PLUS SENCOR/LEXONE PLUS PARAQUAT CL TANK MIXTURE[1]

	Southern States (including Eastern Coastal Plains)[2]	Northern States[2]	
	PROWL + Sencor/ Lexone DF[3] + Paraquat CL[4]	PROWL + Sencor[3]/ Lexone DF[3] + Paraquat CL[4]	PROWL + Sencor/ Lexone DF[3] + Paraquat CL[4]
Soil Texture		Up to 3% Organic Matter	More than 3% Organic Matter
COARSE* loamy sands, sandy loams	1.0 to 1.5 pts. + 0.33 lb. + 1 to 2 pts.	1.0 to 1.5 pts. + 0.5 lb. + 1 to 2 pt.	1.5 pts. + 0.5 lb. + 1 to 2 pts.
MEDIUM sandy clay loams**, sandy clays, loams, silts, silt loams	1.5 to 2.0 pts. + 0.5 lb. + 1 to 2 pts.	1.5 to 2.0 pts. + 0.5 lb. + 1 to 2 pts.	1.5 to 2.0 pts. + 0.5 to 0.67 lbs. + 1 to 2 pts.
FINE silty clay loams**, clay loams, silty clays, clays	1.5 to 2.0 pts. + 0.67 lb. + 1 to 2 pts.	1.5 to 2.0 pts. + 0.5 to 0.67 lbs. + 1 to 2 pts.	2.0 to 2.5 pts. + 0.67 lb. + 1 to 2 pts.

[1] Paraquat CL should be deleted from the tank-mixture if vegetation is absent at time of application.

[2] See map on page 26 for specific states. DO NOT USE IN CALIFORNIA.

[3] When using Sencor 4F, Sencor 50WP, Lexone 4L or Lexone 50WP, use equivalent rates. One pint of the flowable formulation or 1 lb. of the wettable powder formulation equals 0.67 lb. of Sencor DF or Lexone DF. See conversion chart on page 4.

[4] When Paraquat CL is included in the tank-mixture add 8 oz. of Ortho X-77 spreader or similar non-ionic surfactant per 100 gal. of spray mixture. See mixiing instructions above.

* DO NOT USE on sands. DO NOT USE on loamy sands or on sandy loams that contain less than 2% organic matter. DO NOT USE on peat or muck soils.

** Sometimes considered transitional soils and may be classified as either medium or fine textured soils.

The high rates of PROWL and Sencor/Lexone for each soil texture, as stated above, should be used if heavy weed populations are anticipated or if field is particularly trashy or cloddy. The high rate of Paraquat CL should be used if weeds are 4 to 6 inches tall. Weeds over 6 inches tall will not be controlled with this treatment.

NOTE: Livestock may graze or be fed forage from treated fields 40 days after application. Observe all precautions and limitations on Sencor, Lexone and Paraquat CL labels.

PROWL PLUS LOROX PLUS PARAQUAT CL

PROWL plus Lorox plus Paraquat CL treatment may be used in minimum and no-till soybeans in all states except California.

WEED SPECIES CONTROLLED

Controlled by Paraquat CL:

Paraquat CL will control most annual emerged weeds and suppress many emerged perennials. Refer to the Paraquat CL label for specific use recommendations.

PROWL plus Lorox preemergence will control the following weed species in addition to those controlled by PROWL alone:

- Mustards (*Brassica* spp.)
- Smartweed (*Polygonum pensylvanicum*)
- Velvetleaf (Buttonweed) (*Abutilon theophrasti*)

PROWL plus Lorox will aid in the control of and reduce competition from:

- Common ragweed (*Ambrosia artemisiifolia*).

PAGE 16

BROADCAST RATES PER ACRE OF PROWL PLUS LOROX PLUS PARAQUAT CL TANK MIXTURE[1]

Soil Texture	PROWL + Lorox L[2] + Paraquat CL[3] ½% to 3% Organic Matter	PROWL + Lorox L[2] + Paraquat CL[3] More than 3% Organic Matter
COARSE sandy loams	1.0 to 1.5 pts. + 1.0 pt. + 1 to 2 pts.	1.5 pts. + 1.0 to 1.5 pts. + 1 to 2 pts.
MEDIUM sandy clay loams,* sandy clays, loams, silts, silt loams	1.5 to 2.0 pts. + 1.0 to 1.5 pts. + 1 to 2 pts.	1.5 to 2.0 pts. + 1.5 to 2.0 pts. + 1 to 2 pts.
FINE silty clay loams,* clay loams, silty clays, clays	1.5 to 2.0 pts. + 1.5 to 2.0 pts. + 1 to 2 pts.	2.0 to 2.5 pts. 2.0 to 2.5 pts. + 1 to 2 pts.

[1] Paraquat CL should be deleted from the tank-mixture if vegetation is absent at time of application.

[2] When using Lorox 50W use equivalent rates. One pint of the flowable formulation equals 1 lb. of the wettable powder formulation. See conversion chart on page 4.

[3] When Paraquat CL is included in the tank-mixture add 8 oz. of Ortho X-77 spreader or similar non-ionic surfactant per 100 gal of spray mixture. See mixing instructions on page 16.

* Sometimes considered transitional soils and may be classified as either medium or fine textured soils.

The high rates of PROWL and Lorox for each soil texture, as stated above, should be used if heavy weed populations are anticipated or if field is particularly trashy or cloddy. The high rate of Paraquat CL should be used if weeds are 4 to 6 inches tall. Weeds over 6 inches tall will not be controlled with this treatment.

DO NOT USE on peat or muck soils. DO NOT USE on sands, loamy sands, or gravelly soils, or on soils containing less than ½% organic matter.

NOTE: Livestock may graze or be fed storage from PROWL plus Lorox treated fields. Observe all precautions and limitations on Lorox and Paraquat CL labels.

CULTI-SPRAY™ (POSTEMERGENCE INCORPORATED) FIELD CORN—GRAIN SORGHUM

PROWL CULTI-SPRAY (Postemergence incorporated) can be applied in field corn in all states except California as follows:

PROWL alone p. 17
PROWL + atrazine p. 18

PROWL CULTI-SPRAY (Postemergence incorporated) can be applied in grain sorghum in Colorado, Kansas, Nebraska, New Mexico, Oklahoma and Texas as follows:

PROWL alone p. 18
PROWL + atrazine p. 19

Winter wheat and winter barley may be planted in the fall 90 days after a PROWL CULTI-SPRAY application in irrigated field corn or grain sorghum or 120 days after a PROWL CULTI-SPRAY application in dry land field corn or grain sorghum. The PROWL-treated crop must be grown to maturity and harvested before planting wheat or barley. See page 5 for complete follow crop instructions.

General Information

PROWL or PROWL plus atrazine tank mixtures applied as a CULTI-SPRAY (postemergence incorporated) treatment will control most late season annual grasses and certain broadleaf weeds in field corn or grain sorghum. PROWL treatments can be applied from the 4-inch growth stage to as late as the last cultivation (layby) of field corn or grain sorghum. PROWL treatments will not control established weeds. EMERGED WEEDS MUST BE DESTROYED BY CULTIVATION PRIOR TO PROWL APPLICATION.

Additional Herbicide Applications

CULTI-SPRAY (postemergence incorporated) applications of PROWL or PROWL plus atrazine tank-mix can be applied in field corn or grain sorghum previously treated with herbicides registered for use in field corn or grain sorghum. Consult the labels of those herbicides for suggested treatments, rates to be used, and precautions or restrictions for use in field corn or grain sorghum and for follow crop restrictions.

CULTI-SPRAY (Postemergence Incorporated) APPLICATIONS

The field must be cultivated and all emerged weeds must be destroyed prior to application. Cultivation should be accomplished with a sweep-type or rolling cultivator operated at sufficient speed to throw at least one inch of soil over the bases of the field corn or grain sorghum plants. This will kill small weed seedlings growing in the field corn or grain sorghum row and will prevent direct contact of the zone of brace root formation by PROWL during application.

Uniformly apply the recommended PROWL or PROWL plus atrazine tank-mix treatment in 10 or more gallons of water per acre by ground equipment. Refer to page 5 for spraying instructions.

Uniform soil coverage must be obtained. Drop nozzles should be used if field corn or grain sorghum foilage will prevent uniform coverage of the soil surface within the rows.

PROWL treatments can be applied from the 4-inch growth stage to as late as the last cultivation (layby) of field corn or grain sorghum.

PROWL treatments must be thoroughly and uniformly incorporated into the soil (1) with a sweep-type or rolling cultivator set to provide thorough incorporation in the top 1 inch of soil OR (2) with adequate irrigation water or rainfall. For best results, PROWL treatments should be incorporated as soon as possible after application but incorporation must be completed within 7 days after application. If adequate moisture is not received within 7 days after application, PROWL must be incorporated with a sweep-type or rolling cultivator.

Under situations of low rainfall or soil moisture, when deep germinating weeds such as shattercane or field sandbur are anticipated, mechanical incorporation will provide best results.

If cultivation is needed after application and incorporation of PROWL, the depth of cut should be no deeper than the depth of cut used to incorporate.

DO NOT APPLY PROWL in field corn or grain sorghum more than once per crop season.

FIELD CORN

DO NOT APPLY PROWL preplant incorporated in field corn as serious crop injury can result.

Applications of PROWL Alone

The following grass and broadleaf weed species are controlled by CULTI-SPRAY (postemergence incorporated) treatments of PROWL in field corn at the rates recommended for each soil texture listed below:

GRASSES CONTROLLED

- Barnyardgrass (*Echinochloa crus-galli*)
- Crabgrass (*Digitaria* spp.)
- Fall panicum (*Panicum dichotomiflorum*)
- Field sandbur (*Cenchrus incertus*)
- Foxtail (giant) (*Setaria faberi*)
- Foxtail (green) (*Setaria viridis*)
- Foxtail (yellow) (*Setaria lutescens*)
- Goosegrass (*Eleusine indica*)
- Johnsongrass (from seed) (*Sorghum halepense*)
- Signalgrass (*Brachiaria platyphylla*)
- Shattercane (*Sorghum bicolor*)

Texas panicum *(Panicum texanum)*
Witchgrass *(Panicum capillare)*
Wild proso millet *(Panicum miliaceum)*

PAGE 17

BROADLEAF WEEDS CONTROLLED

Carpetweed *(Mollugo verticillata)*
Florida pusley *(Richardia scabra)*
Kochia *(Kochia scoparia)*
Lambsquarters *(Chenopodium album)*
Pigweed *(Amaranthus* spp.*)*
Purslane *(Portulaca oleracea)*
Spurge, annuals *(Euphorbia* spp.*)*

PROWL will aid in the control of and reduce competition from:
Smartweed *(Polygonum pensylvanicum)*
Velvetleaf (Buttonweed) *(Abutilon theophrasti)*

Broadcast Rate Per Acre for CULTI-SRAY (postemergence incorporated) Application of PROWL in Field Corn

Soil Texture	Southern[1] States	Northern[1] States
COARSE sands, loamy sands, sandy loams	1.0 to 1.5 pts.	1.5 to 2.0 pts.
MEDIUM sandy clay loams*, sandy clays, loams, silts, silt loams	1.5 to 2.0 pts.	2.0 to 3.0 pts.
FINE silty clay loams*, clay loams, silty clays, clays	1.5 to 3.0 pts.	2.0 to 3.0 pts.

[1] See map on page 26 for specific states.

* Sometimes considered transitional soils and may be classified as either medium or fine textured soils.

The high rate for each soil texture should be used if heavy weed populations are anticipated.
DO NOT use on peat or muck soils.

NOTE: Livestock can graze or be fed forage from PROWL treated field corn after 21 days following applications.

Applications of PROWL Plus Atrazine Mixtures

In addition to weed species controlled by PROWL alone, the following weed species are controlled by CULTI-SPRAY (postemergence incorporated) treatments of PROWL plus atrazine in field corn at the rates recommended for each soil texture listed below:
Velvetleaf (Buttonweed) *(Abutilon theophrasti)*
Venice mallow *(Hibiscus trionum)*
Smartweed *(Polygonum pensylvanicum)*

Broadcast Rate Per Acre for CULTI-SPRAY (postemergence incorporated) Application of Prowl Plus Atrazine in Field Corn

	PROWL + atrazine 80W*	
Soil Texture	Southern States[1]	Northern States[1]
COARSE sands, loamy sands, sandy loams	1.0 to 1.5 pts. + 1.25 lb.	1.5 to 2.0 pts. + 1.25 lb.
MEDIUM sandy clay loams**, sandy clays, loams, silts, silt loams	1.5 to 2.0 pts. + 1.25 to 1.5 lbs.	2.0 to 3.0 pts. + 1.25 to 1.5 lbs.
FINE silty clay loams**, clay loams, silty clays, clays	1.5 to 3.0 pts. + 1.25 to 1.5 lbs.	2.0 to 3.0 pts. + 1.25 to 1.5 lbs.

[1] See map on page 26 for specific states.

* When using atrazine 4L or AAtrex Nine-0 use equivalent rates. One quart of atrazine 4L equals 1.25 lbs. of atrazine 80W. One pound of AAtrex Nine-0 equals 1.2 lbs. of atrazine 80W. See conversion chart on page 4.

** Sometimes considered transitional soils and may be classified as either medium or fine textured soils.

The high rate for each soil texture above should be used if heavy weed populations are anticipated.
DO NOT use on peat or muck soils.

NOTE: Observe all precautions and limitations on atrazine labels. Follow cropping restrictions on atrazine labels. Livestock can graze or be fed forage from PROWL plus atrazine treated field corn after 21 days following applications.

GRAIN SORGHUM

DO NOT APPLY PROWL in grain sorghum preplant incorporated, preemergence or postemergence prior to 4 inch growth stage as serious crop injury can result.

DO NOT APPLY PROWL in grain sorghum which is planted in double row beds since this cultural practice does not permit adequate soil coverage on the bases of the grain sorghum plants with cultivation.

Applications of PROWL Alone

The following grass and broadleaf weed species are controlled by CULTI-SPRAY (postemergence incorporated) treatments of PROWL in grain sorghum at the rates recommended for each soil texture listed below:

GRASSES CONTROLLED

Barnyardgrass *(Echinochloa crus-galli)*
Crabgrass *(Digitaria* spp.*)*
Fall panicum *(Panicum dichotomiflorum)*
Field sandbur *(Cenchrus incertus)*
Foxtail (giant) *(Setaria faberi)*
Foxtail (green) *(Setaria viridis)*
Foxtail (yellow) *(Setaria lutescens)*
Goosegrass *(Eleusine indica)*
Johnsongrass (from seed) *(Sorghum halepense)*
Shattercane *(Sorghum bicolor)*
Signalgrass *(Brachiaria platyphylla)*
Texas panicum *(Panicum texanum)*
Witchgrass *(Panicum capillare)*
Wild proso millet *(Panicum millaceum)*

BROADLEAF WEEDS CONTROLLED

Carpetweed *(Mollugo verticillata)*
Florida pusley *(Richardia scabra)*
Kochia *(Kochia scoparia)*
Lambsquarters *(Chenopodium album)*
Pigweed *(Amaranthus* spp.*)*
Purslane *(Portulaca oleracea)*
Spurge, annuals *(Euphorbia* spp.*)*

PROWL will aid in the control of and reduce competition from:
Smartweed *(Polygonum pensylvanicum)*
Velvetleaf (Buttonweed) *(Abutilon theophrasti)*

Broadcast Rate Per Acre for CULTI-SPRAY (postemergence incorporated) Application of PROWL in Grain Sorghum

Soil Texture	Southern States[1]	Northern States[2]
COARSE sands, loam sands, sandy loams	1.0 to 1.5 pts.	1.5 to 2.0 pts.
MEDIUM sandy clay loams*, sandy clays, loams, silts, silt loams	1.5 to 2.0 pts.	2.0 to 3.0 pts.
FINE silty clay loams*, clay loams, silty clays, clays	1.5 to 3.0 pts.	2.0 to 3.0 pts.

[1] New Mexico, Oklahoma and Texas.
[2] Colorado, Kansas and Nebraska.
* Sometimes considered transitional soils and may be classified as either fine or medium textured soils.

The high rate for each soil texture above should be used if heavy weed populations are anticipated.

DO NOT use on peat or muck soils.

NOTE: Livestock can graze or be fed forage from PROWL-treated grain sorghum after 21 days following applications.

Applications of PROWL Plus Atrazine Mixtures

In addition to weed species controlled by PROWL alone, the following weed species are controlled by CULTI-SPRAY (postemergence incorporated) treatments of PROWL plus atrazine in grain sorghum at the rates recommended for each soil texture listed below:

Velvetleaf (Buttonweed) *(Abutilon theophrasti)*
Venice mallow *(Hibiscus trionum)*
Smartweed *(Polygonum pensylvanicum)*

Broadcast Rate Per Acre for CULTI-SPRAY (postemergence incorporated) Application of PROWL Plus Atrazine in Grain Sorghum

Soil Texture	PROWL + atrazine 80W*	
	Southern States[1]	Northern States[2]
COARSE sands, loamy sands, sandy loams	DO NOT USE	DO NOT USE
MEDIUM sandy clay loams**, sandy clays, loams, silts, silt loams	1.5 to 2.0 pts. + 1.25 lbs.	2.0 to 3.0 pts. + 1.25 lbs.
FINE silty clay loams**, clay loams, silty clays, clays	1.5 to 3.0 pts. + 1.25 lbs.	2.0 to 3.0 pts. + 1.25 lbs.

[1] New Mexico, Oklahoma and Texas.
[2] Colorado, Kansas and Nebraska.
* When using atrazine 4L or AAtrex Nine-O use equivalent rates. One quart of atrazine 4L equals 1.25 lbs. of atrazine 80W. One pound of AAtrex Nine-0 equals 1.12 lbs. of atrazine 80W. See conversion chart on page 4.
** Sometimes considered transitional soils, and may be classified as either medium or fine textured soils.

The high rate for each soil texture above should be used if heavy weed populations are anticipated.
DO NOT use on peat or muck soils.

NOTE: Observe all precautions and limitations on atrazine labels. Follow cropping restrictions on atrazine labels. Livestock can graze or be fed forage from PROWL plus atrazine treated grain sorghum fields after 21 days following applications.

RICE

PROWL plus Stam M-4 early postemergence tank mixture may be used in dry-seeded rice.

DO NOT use this treatment in water-seeded rice. DO NOT use in California.

Refer to page 5 for follow crop restrictions.

GENERAL INFORMATION

PROWL plus Stam M-4 is a postemergence treatment which combines the direct contact action of Stam M-4 and the residual activity of PROWL. Since the residual activity of PROWL provides preemergence control of certain annual grasses which can germinate after this tank-mix treatment is applied, flooding after application can be delayed.

For maximum weed control with this treatment, it is important to carefully follow the directions below for (1) adequate spray coverage of weeds and soil and (2) proper timing of application, when barn yardgrass (watergrass) is in the 1 to 3 leaf stage of growth with an occasional 4 leaf plant (make application when sprangletop is less than ½ inch in height.)

Spraying Instructions

AERIAL APPLICATIONS

For aerial application, apply the recommended rate in 10 to 12 gallons of water per acre to ensure adequate coverage. To minimize drift (see Stam M-4 label for sensitive crops), DO NOT apply during periods of gusty winds or winds in excess of 5 mph. Uniform spray distribution may not be achieved when wind velocity is greater than 5 mph. It is recommended that a flagman or an automatic mechanical flagging unit on the aircraft be used to avoid overlapping and possible crop injury.

GROUND APPLICATIONS

For ground equipment, apply the recommended rate in 15 to 25 gallons of water per acre to ensure adequate coverage. Use a properly calibrated low-pressure (20 to 40 psi) sprayer equipped with 8002 or larger size Tee-Jet or comparable nozzles to achieve uniform spray distribution and minimize drift. Keep the by-pass line on or near the bottom of the tank to minimize foaming. Nozzle screens must be no finer than 50 mesh. DO NOT apply PROWL during periods of gusty winds or when wind velocity is greater than 10 mph.

Early Postemergence Applications

The seedbed should be firm and free of clods and trash. The seedbed must be prepared to allow for good seed coverage. Previous crop residues should be thoroughly mixed into the soil to a depth of 4 to 6 inches by plowing or disking before planting rice.

Uniformly apply recommended PROWL plus Stam M-4 treatment by aircraft or ground equipment after rice emergence, according to spraying instructions above, when barnyardgrass is in the 1 to 3 leaf growth stage with an occasional 4 leaf plant (make application when sprangletop is less than ½ inch). THE GROWER SHOULD INSPECT FIELDS FREQUENTLY TO CHECK GROWTH OF BARNYARDGRASS AND/OR SPRANGLETOP TO DETERMINE PROPER APPLICATION TIMING. Timing of applications should be based on the growth stage of barnyardgrass and/or sprangletop and not on the growth stage of rice. If rice is too small to maintain a flood on the field, the treatment can still be applied since flooding can be delayed because of the residual activity of PROWL.

While the residual activity of PROWL allows flooding to be delayed, proper water management practices must be followed for normal rice growth. Flooding should not be delayed if weeds begin to develop after application.

Since soil and weeds must be completely exposed to spray coverage, no flood water should be on field at time of application. If necessary, fields may be flushed prior to treatment to produce vigorous rice and weed growth.

Since the residual activity of PROWL is activated by moisture, PROWL is most effective in controlling emerging weeds when adequate rainfall or irrigation (flush) is received within 7 days after application.

For maximum direct contact activity of Stam M-4, delay application if there is a chance of rain within 6 to 8 hours.

WEED SPECIES CONTROLLED

PROWL plus Stam M-4 tank mixture provides direct contact control of weeds listed on the Stam M-4 label. The following grass species are controlled by the residual activity of PROWL plus Stam M-4 tank mixture at the rates recommended for each soil texture listed below:

Barnyardgrass (watergrass) *(Echinochloa crus-galli* or *Echinochloa colonum)*
Crabgrass *(Digitaria* spp.*)*
Signalgrass *(Brachiaria platyphylla)*
Sprangletop *(Leptochloa* spp.*)*

PAGE 19

Broadcast Rate Per Acre of PROWL Plus Stam M-4

Soil Texture	PROWL + Stam M-4
COARSE sandy loams	1.5 pts. + 6.0 to 8.0 pts.
MEDIUM sandy clay loams*, sandy clays, loams, silts, silt loams	1.5 to 2.0 pts. + 6.0 to 8.0 pts.
FINE silty clay loams*, clay loams, silty clays, clays	1.5 to 2.0 pts. + 6.0 to 8.0 pts.

* Sometimes considered transitional soils and may be classified as either medium or fine textured soils.

The high rate of PROWL for medium and fine textured soils should be used if heavy weed populations are present. The high rate of Stam M-4 should be used in the following situations at the time of application: (1) in case of unseasonably cool weather, (2) if barnyardgrass has progressed to predominantly the 3 to 4 leaf stage, or (3) if emerged sprangletop (less and 1/2 inch) is present.

DO NOT apply this tank mixture within 14 days before or after insecticide applications because serious damage to rice may occur.

DO NOT apply this tank mixture in liquid fertilizer.

DO NOT make more than one application of PROWL per season.

DO NOT bale or use rice straw from PROWL-treated fields for feed or bedding.

Observe all precautions and limitations on the Stam M-4 label.

POTATOES

PROWL can be applied in potatoes as follows:

Preemergence	Preemergence Incorporated
PROWL alone p. 20	PROWL alone p. 20
PROWL + Sencor/Lexone p. 21	PROWL + Eptam p. 21

Refer to specific sections of the label for directions for method of application.

Refer to page 5 for follow crop restrictions.

Uniformly apply the recommended PROWL or PROWL tank-mix combination treatment in 10 or more gallons of water or PROWL alone in 20 or more gallons of liquid fertilizer per acre by ground equipment. (Refer to page 4 for liquid fertilizer compatibility determinations.) Apply in 5 or more gallons of water per acre by aircraft. Refer to page 5 for spraying instructions. Refer below for PROWL applications through sprinkler irrigation systems. DO NOT apply prior to planting crop or postemergence to crop and weeds.

The seed bed should be firm and free of clods and trash. Previous crop residues should be thoroughly mixed into the soil to a depth of 4 to 6 inches by plowing or disking before planting potatoes.

Applications of PROWL Alone

1. Preemergence

PROWL may be applied after planting but before potatoes and weeds emerge or after drag-off where this operation is practiced.

PROWL applied preemergence is most effective in controlling weeds when adequate rainfall or irrigation is received within 7 days after application.

2. Preemergence Incorporated

Apply PROWL and incorporate after planting but before potatoes and weeds emerge. Where drag-off is practiced, apply PROWL and incorporate before, at or after drag-off but before potatoes and weeds emerge. PROWL is to be incorporated within 7 days after application.

PROWL must be thoroughly and uniformly incorporated into the top 1 or 2 inches of soil. Care must be taken so that incorporation equipment does not damage seed pieces or elongating sprouts.

Mechanical incorporation is not required if adequate rainfall for good crop and weed emergence occurs or irrigation is received within 7 days after application.

Applications of PROWL Alone

Sprinkler Irrigation Systems

PROWL may be applied preemergence (after planting but before potatoes and weeds have emerged or after drag-off where this operation is practiced) through sprinkler irrigation systems, in all states except California. UNIFORM application requires:

1. The system must be free of leaks and clogged nozzles.
2. The system must be properly calibrated, with water only, to ensure the amount of PROWL applied per acre corresponds to the recommended PROWL rate (see below).
3. The system must apply a uniform amount of water over all areas. Uneven application may cause lack of weed control or crop injury.
4. Application must not be made during periods of gusty winds or when wind velocity is greater than 10 mph.
5. Agitation must be maintained in the nurse tank.

SPRINKLER IRRIGATION SYSTEMS MUST BE EQUIPPED WITH CHECK VALVES OR ANTISIPHON DEVICES AND AN AUTOMATIC SHUTOFF TO PREVENT BACKFLOW INTO THE WATER SUPPLY.

Maintain continuous agitation in nurse tanks during herbicide injection. Apply PROWL in 1/2 to 3/4 inches of water. When application is complete, flush the system with water.

WEED SPECIES CONTROLLED

The following grass and broadleaf weed species are controlled by treatments of PROWL at the rates recommended for each soil texture listed below:

GRASSES CONTROLLED

Barnyardgrass *(Echinochloa crus-galli)*
Crabgrass *(Digitaria* spp.*)*
Fall panicum *(Panicum dichotomiflorum)*
Foxtail (giant) *(Setaria faberi)*
Foxtail (green) *(Setaria viridis)*
Foxtail (yellow) *(Setaria lutescens)*
Goosegrass *(Eleusine indica)*

BROADLEAF WEEDS CONTROLLED

Kochia *(Kochia scoparia)*
Lambsquarters *(Chenopodium album)*
Pigweed *(Amaranthus* spp.*)*
Purslane *(Portulaca oleracea)*

PROWL will aid in the control of and reduce competition from:

Ladysthumb *(Polygonum persicaria)*
Smartweed *(Polygonum pensylvanicum)*
Velvetleaf (Buttonweed) *(Abutilon theophrasti)*

Broadcast Rate Per Acre of PROWL Alone in Potatoes

Soil Texture	Less than 3% Organic Matter	More than 3% Organic Matter
COARSE sands, loamy sands, sandy loams	1.5 pts.	1.5 pts.
MEDIUM sandy clay loams*, sandy clays, loams, silts, silt loams	2.0 pts.	3.0 pts.
FINE silty clay loams*, clay loams, silty clays, clays	3.0 pts.	3.0 pts.

* Sometimes considered transitional soils and may be classified as either medium or fine textured soils.

DO NOT use on peat or muck soils.

Preemergence Applications of PROWL Plus Sencor/Lexone

PROWL plus Sencor/Lexone preemergence tank-mixture may be applied after planting but before potatoes and weeds emerge or after drag-off where this operation is practiced. DO NOT incorporate. DO NOT apply postemergence to crop.

PROWL plus Sencor/Lexone preemergence tank-mixture is most effective in controlling weeds when adequate rainfall or irrigation is received within 7 days after application.

In addition to weed species controlled by PROWL alone, the following weed species are controlled by treatments of PROWL plus Sencor/Lexone at the rates recommended for each soil texture listed below:

Cocklebur *(Xanthium pensylvanicum)*
Common ragweed *(Ambrosia artemisiifolia)*
Smartweed *(Polygonum pensylvanicum)*
Velvetleaf (Buttonweed) *(Abutilon theophrasti)*

Tank mixtures of PROWL plus Sencor/Lexone will aid in the control of and reduce competition from:

Black nightshade *(Solanum nigrum)*
Hairy nightshade *(Solanum sarachoides)*

Broadcast Rate Per Acre of PROWL Plus Sencor/Lexone in Potatoes

Soil Texture	PROWL + Sencor/ Lexone DF* Less than 3% Organic Matter[1]	PROWL + Sencor/ Lexone DF* More than 3% Organic Matter
COARSE sands, loamy sands, sandy loams	1.0 to 1.5 pts. + 0.33 to 0.5 lb.	1.0 to 1.5 pts. + 0.33 to 0.5 lb.
MEDIUM sandy clay loams**, sandy clays, loams, silts, silt loams	1.5 to 2.0 pts. + 0.5 to 0.67 lb.	2.0 to 3.0 pts. + 0.5 to 0.67 lb.
FINE silty clay loams**, clay loams, silty clays, clays	2.0 to 2.0 pts. + 0.5 to 0.67 lb.	3.0 pts. + 0.5 to 0.67 lb.

* When using Sencor 4F, Sencor 50WP, Lexone 4L, or Lexone 50WP, use equivalent rates. One pint of the flowable formulation or 1 lb. of the wettable powder formulation equals 0.67 lb. of Sencor DF or Lexone DF. See conversion chart on page 4.

** Sometimes considered transitional soils and may be classified as either medium or fine textured soils.

[1] Observe organic matter restrictions on Sencor and Lexone labels.

The high rate for each soil texture above, where listed, should be used when heavy infestations of grass or broadleaf weeds are anticipated or if extensive crop residues were present prior to seedbed preparation.

DO NOT USE on peat or muck soils.

NOTE: Observe all precautions and limitations on Sencor and Lexone labels. Follow cropping restrictions on Sencor and Lexone labels.

Preemergence Incorporated Applications of PROWL Plus Eptam

PROWL plus Eptam tank-mixture may be applied and incorporated after planting but before potatoes and weeds emerge. In areas where potatoes are normally dragged-off, apply and incorporate PROWL plus Eptam following drag-off but before potatoes and weeds emerge. PROWL plus Eptam must be incorporated immediately after application.

PROWL plus Eptam must be thoroughly and uniformly incorporated into the top 1 or 2 inches of soil. Whenever possible, application and incorporation should be done in the same operation. Care must be taken so that incorporation equipment does not damage seed pieces or elongating sprouts.

In addition to weed species controlled by PROWL alone, the following weed species are controlled by treatments of PROWL plus Eptam at the rates recommended for each soil texture listed below:

Purple nutsedge *(Cyperus rotundus)*
Yellow nutsedge *(Cyperus esculentus)*
Black nightshade *(Solanum nigrum)*
Hairy nightshade *(Solanum sarachoides)*

Broadcast Rate Per Acre of PROWL Plus Eptam in Potatoes[1]

Soil Texture	PROWL + Eptam 7E Less than 3% Organic Matter	PROWL + Eptam 7E More than 3% Organic Matter
COARSE sands, loamy sands, sandy loams	1.0 to 1.5 pts. + 3.0 to 3.5 pts.	1.0 to 1.5 pts. + 3.0 to 3.5 pts.
MEDIUM sandy clay loams*, sandy clays, loams, silts, silt loams	1.5 to 2.0 pts. + 3.0 to 3.5 pts.	3.0 pts. + 3.0 to 3.5 pts.
FINE silty clay loams*, clay loams, silty clays, clays	3.0 pts. + 3.0 to 3.5 pts.	3.0 pts. + 3.0 to 3.5 pts.

* Sometimes considered transitional soils, and may be classified as either medium or fine textured soils.

[1] For nutsedge control, use 4.5 to 7.0 pints of Eptam 7E per acre.

The higher rate for each soil texture above, where listed, should be used when heavy infestations of grass or broadleaf weeds are anticipated or if extensive crop residues were present prior to seedbed preparation.

DO NOT use on peat or muck soils.

NOTE: The Superior variety potato is sensitive to Eptam and under stress conditions, early season stunting may occur. Observe all precautions and limitations on Eptam label.

TOBACCO

PROWL may be applied preplant incorporated for weed control in transplanted tobacco.

Uniformly apply the recommended PROWL treatment in 10 or more gallons of water per acre by ground equipment. Refer to page 5 for spraying instructions. Refer to page 4 for PROWL/dry bulk fertilizer applications.

Thoroughly mix the previous crop residues into the soil to a depth of 4 to 6 inches by plowing or disking prior to application.

PROWL may be applied immediately before transplanting or up to 60 days prior to transplanting tobacco. DO NOT APPLY PROWL AS A POST-TRANSPLANT SPRAY.

After application, shallow cultivation or hand-hoeing can be practiced without reducing weed control.

Refer to page 5 for follow crop restrictions.

Incorporation Instructions

1. Bedded Tobacco—Application Prior to Bedding

PAGE 21

PROWL is to be incorporated within 7 days of application prior to planting. When PROWL is applied to flat land prior to bedding, mechanical incorporation should be of sufficient depth so that bedding does not bring up untreated soil. Mechanical incorporation of PROWL can be achieved by the following:

(a) Disk harrow set to cut 4 to 6 inches deep and operated in two different directions at 4 to 6 mph.

(b) PTO-driven equipment (tillers, cultivators, hoes) set to cut 3 to 4 inches deep and operated one time at 4 mph or less.

During transplanting, or if beds must be reshaped prior to transplanting, avoid tillage that will bring untreated soil to the surface or expose untreated soil on the bed or in the furrow.

2. Flat-transplanted Tobacco

PROWL is to be incorporated within 7 days of application prior to planting. When PROWL is applied to flat land that will not be bedded, mechanical incorporation should be into the top 1 or 2 inches of soil. Mechanical incorporation of PROWL can be achieved by the following:

(a) Disk harrow set to cut 3 to 4 inches deep and operated in two different directions at 4 to 6 mph.

(b) PTO-driven equipment (tillers, cultivators, hoes) set to cut 2 inches deep and operated one time at 4 mph or less.

(c) Rolling cultivator set to cut 2 to 3 inches deep and operated two times at 6 to 8 mph. Use only on coarse and medium textured soils.

(d) Field cultivator with 3 or 4 rows of sweeps, spaced at intervals of 7 inches or less and staggered so that no soil is left unturned (Do not use chisel points), set to cut 3 inches deep and operated two times at more than 5 mph with the second pass made at an angle to the first.

(e) LELY-ROTERRA[1] set and operated according to the manufacturer's directions to thoroughly incorporate PROWL to a depth of 1 to 2 inches.

(f) Do-All set and operated according to the manufacturer's directions to thoroughly incorporate PROWL to a depth of 1 to 2 inches.

[1]LELY-ROTERRA is a trademark of Lely.

Weed Species Controlled Preplant Incorporated Applications of PROWL

The following grass and broadleaf weed species are controlled by preplant incorporated treatments of PROWL in tobacco at the rates recommended for each soil texture listed.

GRASSES CONTROLLED

Barnyardgrass *(Echinochloa crus-galli)*
Crabgrass *(Digitaria* spp.*)*
Crowfootgrass *(Dactyloctenium aegyptium)*
Fall panicum *(Panicum dichotomiflorum)*
Foxtail (giant) *(Setaria faberi)*
Foxtail (green) *(Setaria viridis)*
Foxtail (yellow) *(Setaria lutescens)*
Goosegrass *(Eleusine indica)*
Johnsongrass (from seed) *(Sorghum halepense)*
Signalgrass *(Brachiaria platyphylla)*
Texas panicum *(Panicum texanum)*
Witchgrass *(Panicum capillare)*

BROADLEAF WEEDS CONTROLLED

Carpetweed *(Mollugo verticillata)*
Florida pusley *(Richardia scabra)*
Lambsquarters *(Chenopodium album)*
Pigweed *(Amaranthus* spp.*)*
Purslane *(Portulaca oleracea)*
Spurge, annuals *(Euphorbia* spp.*)*

PROWL will aid in control of and reduce competiton from:
Smartweed *(Polygonum pensylvanicum)*
Velvetleaf (Buttonweed) *(Abutilon theophrasti)*

Preplant Incorporated Broadcast Rate Per Acre of PROWL in Transplanted Tobacco

Region	Soil Texture		PROWL
Maryland, Virginia, North Carolina, South Carolina, Georgia, Florida	COARSE	sands, loamy sands, sandy loams	1.5 to 2.0 pints
	MEDIUM	sandy clay loams, loams	1.5 to 2.0 pints
		silt loams, silts	2.0 to 2.5 pints
	FINE	clay loams, silty clay loams, clays	2.0 to 2.5 pints
Other tobacco-growing states	COARSE	sands, loamy sands, sandy loams	1.5 to 2.0 pints
	MEDIUM	sandy clay loams, loams, silt loams, silts	2.5 to 3.0 pints
	FINE	clay loams, silty clay loams, clays	2.5 to 3.0 pints

The high rate for each soil texture above should be used when heavy infestations of grass or broadleaf weeds are anticipated or if extensive crop residues were present prior to seedbed preparation.
DO NOT use on peat or muck soil.

NOTE: Applied according to directions and under normal growing conditions, PROWL will not harm transplanted tobacco. Under stress conditions for plant growth such as cold/wet or hot/dry weather, PROWL can produce a temporary retardation of tobacco development.

If loss of tobacco crop occurs due to adverse weather conditions, transplanted tobacco or any crop registered for PROWL preplant incorporated use can be replanted the same year into PROWL-treated soil without adverse effects. If replanting is necessary, DO NOT rework the soil deeper than the PROWL-treated zone.

PEANUTS

PROWL may be applied preplant incorporated in peanuts as follows:

PROWL alone page 22.
PROWL + Vernam page 23.

Refer to specific sections of the label for directions for method of application.

Preplant Incorporated Applications of PROWL Alone (see page 23)

Uniformly apply the recommended PROWL treatment in 10 or more gallons of water or in 20 or more gallons of liquid fertilizer per acre by ground equipment. (Refer to page 4 for liquid fertilizer compatibility determinations). Apply in 5 or more gallons of water per acre by aircraft. Refer to page 5 for spraying instructions.

Thoroughly mix the previous crop residues into the soil to a depth of 4 to 6 inches by plowing or disking prior to application. PROWL may be applied immediately before planting or up to 60 days prior to planting. Refer to sections below for incorporation instructions.

After application, shallow cultivation or hand-hoeing can be practiced without reducing weed control.

PAGE 22

Incorporation Instructions

1. Flat-planted peanuts

PROWL is to be incorporated within 7 days of application prior to planting. When PROWL is applied to flat land that will not be bedded, mechanical incorporation should be into the top 1 or 2 inches of soil. Mechanical incorporation of PROWL can be achieved by the following:

(a) Disk harrow set to cut 3 to 4 inches deep and operated in two different directions at 4 to 6 mph.

(b) PTO-driven equipment (tillers, cultivators, hoes) set to cut 2 inches deep and operated one time at 4 mph or less.

(c) Rolling cultivator set to cut 2 to 3 inches deep and operated two times at 6 to 8 mph. Use only on coarse and medium textured soils.

(d) Field cultivator with 3 or 4 rows of sweeps, spaced at intervals of 7 inches or less and staggered so that no soil is left unturned (Do not use chisel points), set to cut 3 inches deep and operated two times at more than 5 mph with the second pass made at an angle to the first.

(e) LELY-ROTERRA[1] set and operated according to the manufacturer's directions to thoroughly incorporate PROWL to a depth of 1 to 2 inches.

(f) Do-all set and operated according to the manufacturer's directions to thoroughly incorporate PROWL to a depth of 1 to 2 inches.

[1]LELY-ROTERRA is a trademark of Lely.

2. Bedded Peanuts—Application Over Beds

PROWL is to be incorporated within 7 days of application prior to planting. When PROWL is applied over beds, mechanical incorporation should be into the top 1 to 2 inches of soil. Mechanical incorporation can be achieved by the following:

(a) Rolling cultivator set to cut 2 to 3 inches deep and operated two times at 6 to 8 mph. Use only on coarse and medium textured soils.

(b) PTO-driven equipment (tillers, cultivators, hoes) set to cut 2 inches deep and operated one time at 4 mph or less.

(c) Bed conditioner set to cut 2 to 3 inches deep and operated one time at 4 to 6 mph.

During planting, or if beds must be reshaped prior to planting, avoid tillage that will bring untreated soil to the surface or expose untreated soil on the seedbed or in the furrow.

3. Bedded Peanuts—Application Prior to Bedding

PROWL is to be incorporated within 7 days of application prior to planting. When PROWL is applied to flat land prior to bedding, mechanical incorporation should be of sufficient depth so that bedding does not bring up untreated soil. Mechanical incorporation of PROWL can be achieved by the following:

(a) Disk harrow set to cut 4 to 6 inches deep and operated in two different directions at 4 to 6 mph.

(b) PTO-driven equipment (tillers, cultivators, hoes) set to cut 3 to 4 inches deep and operated one time at 4 mph or less.

During planting, or if beds must be reshaped prior to planting, avoid tillage that will bring untreated soil to the surface or expose untreated soil on the seedbed or in the furrow.

Weed Species Controlled

The following grass and broadleaf weed species are controlled by preplant incorporated treatments of PROWL in peanuts at the rates recommended for each soil texture listed:

GRASSES CONTROLLED

Barnyardgrass *(Echinochloa crus-galli)*
Crabgrass *(Digitaria* spp.)
Crowfootgrass *(Dactyloctenium aegyptium)*
Fall panicum *(Panicum dichotomiflorum)*
Foxtail (giant) *(Setaria faberi)*
Foxtail (green) *(Setaria viridis)*
Foxtail (yellow) *(Setaria lutescens)*
Goosegrass *(Eleusine indica)*
Johnsongrass (from seed) *(Sorghum halepense)*
Signalgrass *(Brachiaria platyphylla)*
Texas panicum *(Panicum texanum)*

BROADLEAF WEEDS CONTROLLED

Carpetweed *(Mollugo verticillata)*
Florida pusley *(Richardia scabra)*
Kochia *(Kochia scoparia)*
Lambsquarters *(Chenopodium album)*
Pigweed *(Amaranthus* spp.)
Purslane *(Portulaca oleracea)*
Spurge, annuals *(Euphorbia* spp.)

PROWL will aid in the control of and reduce competition from:

Smartweed *(Polygonum pensylvanicum)*
Velvetleaf (Buttonweed) *(Abutilon theophrasti)*

Preplant Incorporated Broadcast Rate Per Acre of PROWL in Peanuts

Region	PROWL
Texas, Oklahoma and New Mexico	1.0 to 1.5 pints
Other peanut-growing states	1.5 to 2.0 pints

The high rate for each region above should be used if heavy weed populations are anticipated or if extensive crop residues were present prior to seedbed preparation.

If loss of peanut crop occurs due to adverse weather conditions, peanuts or any crop registered for PROWL preplant incorporated use can be replanted the same year into PROWL-treated soil without adverse effects. If replanting is necessary, DO NOT rework the soil deeper than the PROWL treated zone.

Preplant Incorporated Applications of PROWL Plus Vernam

PROWL plus Vernam may be applied preplant incorporated for weed control in peanuts.

1. Application Instructions

Uniformly apply the recommended PROWL plus Vernam tank-mix combination treatment in 10 or more gallons of water per acre with ground equipment only. See page 5 for spraying instructions.

2. Incorporation Instructions

PROWL plus Vernam tank-mix combinations must be incorporated IMMEDIATELY (within minutes) after application to prevent loss of Vernam herbicide. Whenever possible, application and incorporation should be done in the same operation.

Use power-driven cultivation equipment set to cut to a depth of 2 to 3 inches or tandem disks set to cut to a depth of 4 to 6 inches, operated at 4 to 6 mph, followed by a spike-toothed harrow or some other leveling device which extends beyond the ends of the disk. For thorough mixing, disk in two different directions (cross disk). Prior to second disking, raise the disk to prevent cutting deeper than 4 to 6 inches.

3. Planting and Weed Control Instructions

Do not plant seed deeper than 2 inches. Seeding can be accomplished immediately after PROWL plus Vernam application or up to 10 days after application. If it is necessary to rework the soil after incorporation, but before planting, do not work the soil deeper than the depth of incorporation.

PAGE 23

For nutsedge control, existing stands must be turned under and chopped up thoroughly before the preplant incorporated application of PROWL plus Vernam. Cultivation is suggested in addition to PROWL plus Vernam treatment in fields with moderate or heavy nutsedge infestations. Do not cultivate deeper than the depth of incorporation.

Weed Species Controlled

When applied as directed, tank mixtures of PROWL plus Vernam will control the following weed species in addition to those controlled by PROWL alone:

Yellow nutsedge *(Cyperus esculentus)*
Purple nutsedge *(Cyperus rotundus)*

Preplant Incorporated Broadcast Rates Per Acre of PROWL Plus Vernam Tank Mixture

Region	Soil Texture	PROWL + Vernam 7E
Texas, Oklahoma and New Mexico	sands, loamy sands	1.0 to 1.5 pts. + 2⅓ pts.
	sandy loams and finer soils	1.0 to 1.5 pts. + 3 pts.
Other peanut-growing states	sands, loamy sands	1.5 to 2.0 pts. + 2⅓ pts.
	sandy loams and finer soils	1.5 to 2.0 pts. + 3 pts.

The high rate of PROWL should be used if heavy weed populations are anticipated or if extensive crop residues were present prior to seedbed preparation.

Follow all precautions and limitations on Vernam label.

SUNFLOWERS

PROWL may be applied preplant incorporated in sunflowers as follows:

PROWL alone page 24.
PROWL + Amiben page 25.

Refer to specific sections of the label for directions for method of application.
Refer to page 5 for follow crop restrictions.

Uniformly apply the recommended PROWL treatment in 10 or more gallons of water or in 20 or more gallons of liquid fertilizer per acre by ground equipment. (Refer to page 4 for liquid fertilizer compatibility determinations). Apply in 5 or more gallons of water per acre by aircraft. Refer to page 5 for spraying instructions.

Thoroughly mix the previous crop residues into the soil to a depth of 4 to 6 inches by plowing or disking prior to application. PROWL may be applied immediately before planting or up to 60 days prior to planting. Refer to sections below for incorporation instructions.

After application, rotary-hoeing, shallow cultivation, or hand-hoeing can be practiced without reducing weed control.

Incorporation Instructions

1. Flat-planted sunflowers

PROWL is to be incorporated within 7 days of application prior to planting. When PROWL is applied to flat land that will not be bedded, mechanical incorporation should be into the top 1 or 2 inches of soil. Mechanical incorporation of PROWL can be achieved by the following:

(a) Disk harrow set to cut 3 to 4 inches deep and operated in two different directions at 4 to 6 mph.

(b) PTO-driven equipment (tillers, cultivators, hoes) set to cut 2 inches deep and operated one time at 4 mph or less.

(c) Rolling cultivator set to cut 2 to 3 inches deep and operated two times at 6 to 8 mph. Use only on coarse and medium textured soils.

(d) Field cultivator with 3 or 4 rows of sweeps, spaced at intervals of 7 inches or less and staggered so that no soil is left unturned (Do not use chisel points), set to cut 3 inches deep and operated two times at more than 5 mph with the second pass made at an angle to the first.

(e) LELY-ROTERRA[1] set and operated according to the manufacturer's directions to thoroughly incorporate PROWL to a depth of 1 to 2 inches.

(f) Do-All set and operated according to the manufacturer's directions to thoroughly incorporate PROWL to a depth of 1 to 2 inches.

[1] LELY-ROTERRA is a trademark of Lely.

NOTE: PROWL can be incorporated in a single pass when the soil is of good tilth with moderate moisture and relatively free of clods and trash. Implements must be properly set up to thoroughly incorporate PROWL into the top 1 or 2 inches of soil. Recommended implements include (1) C-shank or S-shank field cultivators equipped with flextine drag or rolling basket attachment and (2) the combination disk/field cultivator implement equipped with flextine drag or rolling basket attachment. Field cultivators must have 3 or 4 rows of sweeps, spaced at intervals of 7 inches or less and staggered so that no soil is left unturned. Implements must be operated at 6 to 8 miles per hour and set to cut 3 or 4 inches deep.

2. Bedded Sunflowers—Application Over Beds

PROWL is to be incorporated within 7 days of application prior to planting. When PROWL is applied over beds, mechanical incorporation should be into the top 1 or 2 inches of soil. Mechanical incorporation can be achieved by the following:

(a) Rolling cultivator set to cut 2 to 3 inches deep and operated two times at 6 to 8 mph. Use only on coarse and medium textured soils.

(b) PTO-driven equipment (tillers, cultivators, hoes) set to cut 2 inches deep and operated one time at 4 mph or less.

(c) Bed conditioner set to cut 2 or 3 inches deep and operated one time at 4 to 6 mph.

During planting, or if beds must be reshaped prior to planting avoid tillage that will bring untreated soil to the surface or expose untreated soil on the seedbed or in the furrow.

3. Bedded Sunflowers—Application Prior to Bedding

PROWL is to be incorporated within 7 days of application prior to planting. When PROWL is applied to flat land prior to bedding, mechanical incorporation should be of sufficient depth so that bedding does not bring up untreated soil. Mechanical incorporation of PROWL can be achieved by the following:

(a) Disk harrow set to cut 4 to 6 inches deep and operated in two different directions at 4 to 6 mph.

(b) PTO-driven equipment (tillers, cultivators, hoes) set to cut 3 to 4 inches deep and operated one time at 4 mph or less.

During planting, or if beds must be reshaped prior to planting, avoid tillage that will bring untreated soil to the surface or expose untreated soil on the seedbed or in the furrow.

Preplant Incorporated Applications of PROWL Alone Weed Species Controlled

The following grass and broadleaf weed species are controlled by preplant incorporated treatments of PROWL in sunflowers at the rates recommended for each soil texture listed:

PAGE 24

GRASSES CONTROLLED

Barnyardgrass *(Echinochloa crus-galli)*
Crabgrass *(Digitaria* spp.*)*
Fall panicum *(Panicum dichotomiflorum)*
Foxtail (giant) *(Setaria faberi)*
Foxtail (green) *(Setaria viridis)*
Foxtail (yellow) *(Setaria lutescens)*
Goosegrass *(Eleusine indica)*
Johnsongrass (from seed) *(Sorghum halepense)*
Signalgrass *(Brachiaria platyphylla)*
Texas panicum *(Panicum texanum)*
Witchgrass *(Panicum capillare)*

BROADLEAF WEEDS CONTROLLED

Carpetweed *(Mollugo verticillata)*
Florida pusley *(Richardia scabra)*
Kochia *(Kochia scoparia)*
Lambsquarters *(Chenopodium album)*
Pigweed *(Amaranthus* spp.*)*
Purslane *(Portulaca oleracea)*
Spurge, annuals *(Euphorbia* spp.*)*

PROWL will aid in the control of and reduce competition from:

Smartweed *(Polygonum pensylvanicum)*
Velvetleaf (Buttonweed) *(Abutilon theophrasti)*

Preplant Incorporated Broadcast Rate Per Acre of PROWL In Sunflowers

	Southern States[1]	Northern States[1]	
Soil Texture	No Organic Matter Restrictions	Up to 3% Organic Matter	More than 3% Organic Matter
COARSE sands, loamy sands, sandy loams	1.0 to 1.5 pts.	1.0 to 2.0 pts.	2.0 pts.
MEDIUM sandy clay loams*, sandy clays, loams, silts, silt loams	1.5 to 2.0 pts.	1.5 to 2.5 pts.	2.5 to 3.0 pts.
FINE silty clay loams*, clay loams, clays loams, silty clays, clays	1.5 to 3.0 pts.	2.0 to 3.0 pts.	3.0 pts.

[1] See map on page 26 for specific states.

* Sometimes considered transitional soils and may be classified as either medium or fine textured soils.

The high rate for each soil texture above, where listed, should be used if heavy weed populations are anticipated or if extensive crop residues were present prior to seedbed preparation. Use the 3 pint rate for heavy clay soils.

DO NOT USE on peat or muck soils.

NOTE: DO NOT feed forage or graze livestock in treated sunflower fields. If sunflower crop loss occurs due to weather conditions, sunflowers or any crop registered for PROWL preplant incorporated use can be replanted. If replanting is necessary, DO NOT rework the soil deeper than the PROWL treated zone.

PROWL Plus Amiben

The following PROWL plus Amiben treatments may be used in sunflowers grown in Northern States. See map on page 26 for specific states. Uniformly apply PROWL or PROWL plus Amiben tank mixture in 10 or more gallons of water per acre with ground equipment or in 5 or more gallons of water per acre by aircraft. See page 5 for spraying instructions. See page 24 for incorporation instructions. Use PROWL and Amiben rates listed in Table below.

1. PROWL Preplant Incorporated Followed by Amiben Preemergence

Apply PROWL preplant incorporated. Follow with a preemergence treatment of Amiben as directed on the Amiben label.

2. Preplant Incorporated Tank Mixture of PROWL Plus Amiben

PROWL plus Amiben must be incorporated into the soil within 7 days after application and sunflowers must be planted no later than 7 days after application. For best results, incorporation should be in the top 1 to 2 inches of soil.

Weed Species Controlled

When applied as directed, PROWL plus Amiben preplant incorporated tank mixture or PROWL preplant incorporated followed by Amiben preemergence will control the following weeds in addition to those controlled by PROWL alone:

Common ragweed *(Ambrosia artemisiifolia)*
Mustard *(Brassica* spp.*)*
Smartweed *(Polygonum pensylvanicum)*
Velvetleaf (Buttonweed) *(Abutilon theophrasti)*

Broadcast Rate Per Acre in Sunflowers for PROWL Plus Amiben Preplant Incorporated Tank Mixture or PROWL Preplant Incorporated Plus Amiben Preemergence

Soil Texture	PROWL + Amiben Up to 3% Organic Matter	PROWL + Amiben More than 3% Organic Matter
COARSE loamy sands, sandy loams	DO NOT USE	
MEDIUM loams, silt loams, silts, sandy clays, sandy clay loams*	1.5 to 2.0 pts. + 1 gallon	1.5 to 2.0 pts. + 1 gallon
FINE silty clay loams*, clay loams, silty clays, clays	1.5 to 2.0 pts. + 1 gallon	2.0 to 2.5 pts. + 1 gallon

* Sometimes considered transitional soils and may be classified as either medium or fine textured soils.

The high rate of PROWL for each soil texture above should be used if heavy weed populations are anticipated, if velvetleaf and smartweed infestations are anticipated, or if extensive crop residues were present prior to seedbed preparation.

DO NOT USE on peat or muck soils.

NOTE: DO NOT feed forage, or graze livestock in PROWL plus Amiben treated sunflower fields. If sunflower loss occurs due to weather conditions, sunflowers can be replanted. If replanting is necessary, DO NOT rework the soil. Observe all precautions and limitations on Amiben label.

PAGE 25

REGIONAL MAP FOR RATE DETERMINATION

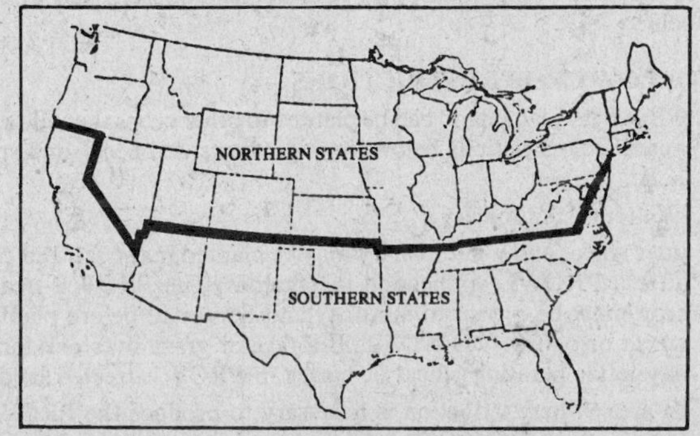

PAGE 26

SUPPLEMENTAL LABELING

PROWL HERBICIDE
EPA Reg. No. 241-243
EPA-SLN No. ID-830016
For Distribution and Use Only Within Idaho

FOR POSTEMERGENCE USE IN POTATOES

DIRECTIONS FOR USE

It is a violation of Federal law to use this product in a manner inconsistent with its labeling.

PROWL may be applied early postemergence in potatoes as follows:
PROWL alone
PROWL + Sencor[1]/Lexone[2]
Refer to specific sections of the label for directions for method of application.
Refer to page 5 for follow crop restrictions.

[1] Sencor is a trademark of Farbenfabriken Bayer GmbH, Leverkusen, West Germany.

[2] Lexone is a trademark of E.I.duPont de Nemours and Company.

GENERAL INFORMATION

Uniformly apply the recommended PROWL or PROWL tank-mix combination treatment in 10 or more gallons of water per acre with ground equipment or in 5 or more gallons of water per acre by aircraft. Refer to sections below for specific instructions for use with ground and aerial applications. Refer to page 2 for applications through sprinkler irrigation systems.

SPRAYING INSTRUCTIONS

GROUND APPLICATIONS

Use proper calibrated low-pressure (20 to 40 psi) sprayer equipped with 8002 or larger size Tee-Jet or comparable nozzles to achieve uniform spray distribution and minimize drift. Keep the by-pass line on or near the bottom of the tank to minimize foaming. Nozzle screens must be no finer than 50 mesh. DO NOT apply PROWL during periods of gusty winds or when wind velocity is greater than 10 mph.

PAGE 1

BROADCAST TREATMENT—Apply in water or liquid fertilizer as specified in this label.

BAND TREATMENT—Apply the broadcast equivalent rate and volume per acre. To determine these:

$$\frac{\text{Band width in inches}}{\text{Row width in inches}} \times \text{Broadcast RATE per acre} = \text{Band RATE per acre}$$

$$\frac{\text{Band width in inches}}{\text{Row width in inches}} \times \text{Broadcast VOLUME per acre} = \text{Band VOLUME per acre}$$

AERIAL APPLICATIONS

Apply in 5 or more gallons of water per acre. To minimize drift, DO NOT apply during periods of gusty winds or winds in excess of 5 mph. It is recommended that a flagman or an automatic mechanical flagging unit on the aircraft be used to avoid overlapping and possible crop injury.

SPRINKLER IRRIGATION APPLICATIONS

Uniformly apply the recommended PROWL or PROWL tank-mix combination treatment through sprinkler irrigation systems. UNIFORM application requires:

1. The system must be free of leaks and clogged nozzles.
2. The system must be properly calibrated, with water only, to ensure the amount applied per acre corresponds to the recommended treatment rate.
3. The system must apply a uniform amount of water over all areas. Uneven application may cause a lack of weed control or crop injury.
4. Application must not be made during periods of gusty winds or when wind velocity is greater than 10 mph.
5. Agitation must be maintained in the nurse tank.

SPRINKLER IRRIGATION SYSTEMS MUST BE EQUIPPED WITH CHECK VALVES OR ANTISIPHON DEVICES AND AN AUTOMATIC SHUTOFF TO PREVENT BACKFLOW INTO THE WATER SUPPLY.

Maintain continuous agitation in nurse tanks during herbicide injection. Apply recommended treatment in 1/2 to 3/4 inches of water. When application is complete, flush the system with water.

APPLICATIONS OF PROWL ALONE

PROWL may be applied from crop emergence to the 6-inch stage of growth. PROWL treatments will control most annual grasses and certain broadleaf weeds in potatoes. PROWL treatments will not control established weeds. EMERGED WEEDS MUST BE DE-

STROYED PRIOR TO APPLICATION. DO NOT make more than one application per season. PROWL treatments are most effective in controlling weeds when adequate rainfall or overhead irrigation is received within 7 days after application.

PAGE 2

DO NOT apply PROWL postemergence if potatoes are under stress from cold/wet or hot/dry conditions or crop injury may occur.

WEED SPECIES CONTROLLED

The following grass and broadleaf weed species are controlled by treatments of PROWL at the rates recommended for each soil texture listed below:

GRASSES CONTROLLED:

Barnyardgrass *(Echinochloa crus-galli)*
Crabgrass *(Digitaria spp.)*
Fall panicum *(Panicum dichotomiflorum)*
Foxtail (giant) *(Setaria faberi)*
Foxtail (green) *(Setaria viridis)*
Foxtail (yellow) *(Setaria lutescens)*
Goosegrass *(Eleusine indica)*
Witchgrass *(Panicum capillare)*

BROADLEAF WEEDS CONTROLLED:

Kochia *(Kochia scoparia)*
Lambsquarters *(Chenopodium album)*
Pigweed *(Amaranthus spp.)*
Purslane *(Portulaca oleracea)*

PROWL will aid in the control of and reduce competition from:

Ladysthumb *(Polygonum persicaria)*
Smartweed *(Polygonum pensylvanicum)*
Velvetleaf (Buttonweed) *(Abutilon theophrasti)*

Broadcast Rate Per Acre of Prowl Alone in Potatoes

Soil Texture	Less than 3% Organic Matter	More than 3% Organic Matter
COARSE sands, loamy sands, sandy loams	1.5 pts.	1.5 pts.
MEDIUM sandy clay loams*, sandy clays, loams, silts, silt loams	2.0 pts.	3.0 pts.
FINE silty clay loams*, clay loams, silty clays, clays	3.0 pts.	3.0 pts.

* Sometimes considered transitional soils and may be classified as either medium or fine textured soils.

DO NOT use on peat or muck soils.

APPLICATIONS OF PROWL PLUS SENCOR/LEXONE

PROWL Plus Sencor/Lexone postemergence tank-mixture may be applied from crop emergence to the 6-inch stage of growth. For optimum control of weeds in potatoes apply before weeds are 1-inch tall. DO NOT make more than one application of this tank mixture per season. DO NOT apply within 24 hours of application of other pesticides. DO NOT apply within three days after periods of cool, wet or cloudy weather or crop injury may occur. DO NOT apply if potatoes are under stress from cold/wet or hot/dry conditions or crop injury may occur. DO NOT use on early maturing smooth-skinned white or red-skinned varieties of potatoes. PROWL plus Sencor/ Lexone treatments are most effective in controlling weeds when adequate rainfall or overhead irrigation is received within 7 days of application.

PAGE 3

In addition to weed species controlled by PROWL alone, the following weed species are controlled by treatments of PROWL plus Sencor/Lexone at the rates recommended for each soil texture listed below:

Cocklebur *(Xanthium pensylvanicum)*
Common ragweed *(Ambrosia artemisiifolia)*
Russian thistle *(Salsola kali)*
Smartweed *(Polygonum pensylvanicum)*

Broadcast Rate Per Acre of PROWL Plus Sencor/Lexone in Potatoes

Soil Texture	PROWL + Sencor/ Lexone DF* Less than 3% Organic Matter[1]	PROWL + Sencor/ Lexone DF* More than 3% Organic Matter
COARSE sands, loamy sands, sandy loams	1.0 to 1.5 pts. + 0.33 to 0.5 lb.	1.0 to 1.5 pts. + 0.33 to 0.5 lb.
MEDIUM sandy clay loams**, sandy clays, loams, silts, silt loams	1.5 to 2.0 pts. 0.5 to 0.67 lb.	2.0 to 3.0 pts. + 0.5 to 0.67 lb.
FINE silty clay loams**, clay loams, silty clays, clays	2.0 to 3.0 pts. + 0.5 to 0.67 lb.	3.0 pts. + 0.5 to 0.67 lb.

* When using Sencor 4F, Sencor 50WP, Lexone 4L, or Lexone 50 WP, use equivalent rates. One pint of the flowable formulation or 1 lb. of the wettable powder formulation equals 0.67 lb. of Sencor DF or Lexone DF. Refer to leaflet label on top of PROWL container for conversion chart.

** Sometimes considered transitional soils and may be classified as either medium or fine textured soils.

[1] Observe organic matter restrictions on Sencor and Lexone labels.

The high rate for each soil texture above, where listed, should be used when heavy infestations of grass or broadleaf weeds are anticipated or if extensive crop residues were present prior to seedbed preparation.

DO NOT USE on peat or muck soils.

NOTE: Observe all precautions and limitations on Sencor and Lexone labels. Follow cropping restrictions on Sencor and Lexone labels.

PAGE 4

FOLLOW CROP RESTRICTIONS

1. PROWL-treated land can be planted to other crops the following year. See restrictions below for sugar beets, red beets, and spinach.
2. **Winter wheat, barley**

   inter wheat and winter barley may be planted in the fall 120 days after a PROWL application in potatoes. The PROWL-treated crop must be grown to maturity and harvested before planting wheat or barley. DO NOT feed forage or graze livestock for 75 days after planting wheat or barley in PROWL-treated land.

   In areas where irrigation is necessary to produce the PROWL-treated crop, DO NOT plant winter wheat or winter barley as follow crops if crop failure/destruction occurs and land is fallowed during the summer as crop injury may result.

   DO NOT plant winter wheat or winter barley as follow crops in PROWL-treated land if the follow crop is planted using no-tillage procedures as crop injury may result.
3. **Sugar beets, red beets, spinach**

   To avoid crop injury, DO NOT plant sugar beets, red beets or spinach for 12 months following a PROWL application. Land should be plowed to a depth of 12 inches prior to planting these crops.
4. When PROWL is used in tank-mix or sequential combinations, refer to labels of other herbicides for additional follow crop restrictions.

NOTE: OBSERVE ALL PRECAUTIONARY STATEMENTS IN THE PROWL CONTAINER LABEL BEFORE USING.

NOTE: This labeling must be in the possession of the user at the time of pesticide application.

24(c) Registrant:

PAGE 5

SUPPLEMENTAL LABELING

PROWL HERBICIDE
EPA Reg. No. 241-243
EPA-SLN No. OR-830023
For Distribution and Use Only Within Oregon

FOR POSTEMERGENCE USE IN POTATOES

DIRECTIONS FOR USE

It is a violation of Federal law to use this product in a manner inconsistent with its labeling.

PROWL may be applied early postemergence in potatoes as follows:

PROWL alone
PROWL + Sencor[1]/Lexone[2]

Refer to specific sections of the label for directions for method of application.
Refer to page 5 for follow crop restrictions.

[1] Sencor is a trademark of Farbenfabriken Bayer GmbH, Leverkusen, West Germany.
[2] Lexone is a trademark of E.I.duPont de Nemours and Company.

GENERAL INFORMATION

Uniformly apply the recommended PROWL or PROWL tank-mix combination treatment in 10 or more gallons of water per acre with ground equipment or in 5 or more gallons of water per acre by aircraft. Refer to sections below for specific instructions for use with ground and aerial applications. Refer to page 2 for applications through sprinkler irrigation systems.

SPRAYING INSTRUCTIONS

GROUND APPLICATIONS

Use a properly calibrated low-pressure (20 to 40 psi) sprayer equipped with 8002 or larger size Tee-Jet or comparable nozzles to achieve uniform spray distribution and minimize drift. Keep the by-pass line on or near the bottom of the tank to minimize foaming. Nozzle screens must be no finer than 50 mesh. DO NOT apply PROWL during periods of gusty winds or when wind velocity is greater than 10 mph.

BROADCAST TREATMENT—Apply in water or liquid fertilizer as specified in this label.

BAND TREATMENT—Apply the broadcast equivalent rate and volume per acre. To determine these:

$$\frac{\text{Band width in inches}}{\text{Row width in inches}} \times \text{Broadcast RATE per acre} = \text{Band RATE per acre}$$

$$\frac{\text{Band width in inches}}{\text{Row width in inches}} \times \text{Broadcast VOLUME per acre} = \text{Band VOLUME per acre}$$

AERIAL APPLICATIONS

Apply in 5 or more gallons of water per acre. To minimize drift, DO NOT apply during periods of gusty winds or winds in excess of 5 mph. It is recommended that a flagman or an automatic mechanical flagging unit on the aircraft to used to avoid overlapping and possible crop injury.

SPRINKLER IRRIGATION APPLICATIONS

Uniformly apply the recommended PROWL or PROWL tank-mix combination treatment through sprinkler irrigation systems. UNIFORM application requires:

1. The system must be free of leaks and clogged nozzles.
2. The system must be properly calibrated, with water only, to ensure the amount applied per acre corresponds to the recommended treatment rate.
3. The system must apply a uniform amount of water over all areas. Uneven application may cause lack of weed control or crop injury.
4. Application must not be made during periods of gusty winds or when wind velocity is greater than 10 mph.
5. Agitation must be maintained in the nurse tank

SPRINKLER AGITATION SYSTEMS MUST BE EQUIPPED WITH CHECK VALVES OR ANTISIPHON DEVICES AND AN AUTOMATIC SHUTOFF TO PREVENT BACKFLOW INTO THE WATER SUPPLY.

Maintain continuous agitation in nurse tanks during herbicide injection. Apply recommended treatment in 1/2 to 3/4 inches of water. When application is complete, flush the system with water.

APPLICATIONS OF PROWL ALONE

PROWL may be applied from crop emergence to the 6-inch stage of growth. PROWL treatments will control most annual grasses and certain broadleaf weeds in potatoes. PROWL treatments will not control established weeds. EMERGED WEEDS MUST BE DESTROYED PRIOR TO APPLICATION. DO NOT make more than one application per season. PROWL treatments are most effective in controlling weeds when adequate rainfall or overhead irrigation is received within 7 days after application.

DO NOT apply PROWL postemergence if potatoes are under stress from cold/wet or hot/dry conditions or crop injury may occur.

WEED SPECIES CONTROLLED

The following grass and boadleaf weed species are controlled by treatments of PROWL at the rates recommended for each soil texture listed below:

GRASSES CONTROLLED:

Barnyardgrass *(Echinochloa crus-galli)*
Crabgrass *(Digitaria* spp.*)*
Fall panicum *(Panicum dichotomiflorum)*
Foxtail (giant) *(Setaria faberi)*
Foxtail (green) *(Setaria viridis)*
Foxtail (yellow) *(Setaria lutescens)*
Goosegrass *(Eleusine indica)*
Witchgrass *(Panicum capillare)*

BROADLEAF WEEDS CONTROLLED:

Kochia *(Kochia scoparia)*
Lambsquarters *(Chenopodium album)*
Pigweed *(Amaranthus* spp.*)*
Purslane *(Portulaca oleracea)*

PROWL will aid in the control of and reduce competition from:

Ladysthumb *(Polygonum persicaria)*
Smartweed *(Polygonum pensylvanicum)*
Velvetleaf (Buttonweed) *(Abutilon theophrasti)*

Broadcast Rate Per Acre of PROWL Alone in Potatoes

Soil Texture	Less than 3% Organic Matter	More than 3% Organic Matter
COARSE sands, loamy sands, sandy loams	1.5 pts.	1.5 pts.
MEDIUM sandy clay loams*, sandy clays, loams, silts, silt loams	2.0 pts.	3.0 pts.
FINE silty clay loams*, clay loams, silty clays, clays	3.0 pts.	3.0 pts.

* Sometimes considered transitional soils and may be classified as either medium or fine textured soils.

DO NOT use on peat or muck soils.

APPLICATIONS OF PROWL PLUS SENCOR/LEXONE

PROWL Plus Sencor/Lexone postemergence tank-mixture may be applied from crop emergence to the 6-inch stage of growth. For optimum control of weeds in potatoes apply before weeds are 1-inch tall. DO NOT make more than one application of this tank mixture per season. DO NOT apply within 24 hours of application of other pesticides. DO NOT apply within three days after periods of cool, wet or cloudy weather or crop injury may occur. DO NOT apply if potatoes are under stress from cold/wet or hot/dry conditions or crop injury may occur. DO NOT use on early maturing smooth-skinned white or red-skinned varieties of potatoes. PROWL plus Sencor/ Lexone treatments are most effective in controlling weeds when adequate rainfall or overhead irrigation is received within 7 days of application.

PAGE 3

In addition to weed species controlled by PROWL alone, the following weed species are controlled by treatments of PROWL plus Sencor/Lexone at the rates recommended for each soil texture listed below:

Cocklebur *(Xanthium pensylvanicum)*
Common ragweed *(Ambrosia artemisiifolia)*
Russian thistle *(Salsola kali)*
Smartweed *(Polygonum pensylvanicum)*

Broadcast Rate Per Acre of PROWL Plus Sencor/Lexone in Potatoes

Soil Texture	PROWL + Sencor/ Lexone DF* Less than 3% Organic Matter[1]	PROWL + Sencor/ Lexone DF* More than 3% Organic Matter
COARSE		
sands, loamy sands sandy loams	1.0 to 1.5 pts. + 0.33 to 0.5 lb.	1.0 to 1.5 pts. + 0.33 to 0.5 lb.
MEDIUM		
sandy clay loams**, sandy clays, loans, silts, silt loams	1.5 to 2.0 pts. 0.5 to 0.67 lb.	2.0 to 3.0 pts. + 0.5 to 0.67 lb.
FINE		
silty clay loams**, clay loams, silty clays, clays	2.0 to 3.0 pts. + 0.5 to 0.67 lb.	3.0 pts. + 0.5 to 0.67 lb.

* When using Sencor 4F, Sencor 50WP, Lexone 4L, or Lexone 50 WP, use equivalent rates. One pint of the flowable formulation or 1 lb. of the wettable power formulation equals 0.67 lb. of Sencor DF or Lexone DF. Refer to leaflet lable on top of PROWL container for conversion chart.

** Sometimes considered transitional soils and may be classified as either medium or fine textured soils.

1 Observe organic matter restrictions on Sencor and Lexone labels.

The high rate for each soil texture above, where listed, should be used when heavy infestations of grass or broadleaf weeds are anticipated or if extensive crop residues were present prior to seedbed preparation.

DO NOT USE on peat or muck soils.

NOTE: Observe all precautions and limitations on Sencor and Lexone labels. Follow cropping restrictions on Sencor and Lexone labels.

PAGE 4

FOLLOW CROP RESTRICTIONS

1. PROWL-treated land can be planted to other crops the following year. See restrictions below for sugar beets, red beets, and spinach.
2. **Winter wheat, barley**

   Winter wheat and winter barley may be planted in the fall 120 days after a PROWL application in potatoes. The PROWL-treated crop must be grown to maturity and harvested before planting wheat or barley. DO NOT feed forage or graze livestock for 75 days after planting wheat or barley in PROWL-treated land.

   In areas where irrigation is necessary to produce the PROWL-treated crop, DO NOT plant winter wheat or winter barley as follow crops if crop failure/destruction occurs and land is fallowed during the summer as crop injury may result.

   DO NOT plant winter wheat or winter barley as follow crops in PROWL-treated land if the follow crop is planted using no-tillage procedures as crop injury may result.
3. **Sugar beets, red beets, spinach**

   To avoid crop injury, DO NOT plant sugar beets, red beets or spinach for 12 months following a PROWL application. Land should be plowed to a depth of 12 inches prior to planting these crops.
4. When PROWL is used in tank-mix or sequential combinations, refer to labels of other herbicides for additional follow crop restrictions.

NOTE: OBSERVE ALL PRECAUTIONARY STATEMENTS IN THE PROWL CONTAINER LABEL BEFORE USING.

NOTE: This labeling must be in the possession of the user at the time of pesticide application.

24(c) Registrant:

PAGE 5

SUPPLEMENTAL LABELING

PROWL HERBICIDE
EPA Reg. No. 241-243
EPA-SLN No. WA-830020
For Distribution and Use Only Within Washington

FOR POSTEMERGENCE USE IN POTATOES

DIRECTIONS FOR USE

It is a violation of Federal law to use this product in a manner inconsistent with its labeling.

PROWL may be applied early postemergence in potatoes as follows:
PROWL alone
PROWL + Sencor[1]/Lexone[2]
Refer to specific sections of the label for directions for method of application.
Refer to page 5 for follow crop restrictions.

1 Sencor is a trademark of Farbenfabriken Bayer GmbH, Leverkusen, West Germany

2 Lexone is a trademark of E.I.duPont de Nemours and Company.

GENERAL INSTRUCTIONS

Uniformly apply the recommended PROWL or PROWL tank-mix combination treatment in 10 or more gallons of water per acre with ground equipment or in 5 or more gallons of water per acre by aircraft. Refer to sections below for specific instructions for use with ground and aerial applications. Refer to page 2 for applications through sprinkler irrigation systems.

SPRAYING INSTRUCTIONS

GROUND APPLICATIONS

Use a properly calibrated low-pressure (20 to 40 psi) sprayer equipped with 8002 or larger size Tee-Jet or comparable nozzles to achieve uniform spray distribution and minimize drift. Keep the by-pass line on or near the bottom of the tank to minimize foaming. Nozzle screens must be no finer than 50 mesh. DO NOT apply PROWL during periods of gusty winds or when wind velocity is greater than 10 mph.

BROADCAST TREATMENT—Apply in water or liquid fertilizer as specified in this label.

BAND TREATMENT—Apply the broadcast equivalent rate and volume per acre. To determine these:

$$\frac{\text{Band width in inches}}{\text{Row width in inches}} \times \text{Broadcast RATE per acre} = \text{Band RATE per acre}$$

$$\frac{\text{Band width in inches}}{\text{Row width in inches}} \times \text{Broadcast VOLUME per acre} = \text{Band VOLUME per acre}$$

injury. Over-application can result in crop stand loss, crop injury, or soil residues. Uneven application or improper soil incorporation can decrease weed control or cause crop injury.

Seedling diseases, cold weather, excessive moisture, deep planting, low or high soil pH, high soil salt concentration, or drought can weaken seedlings and plants and increase the possibility of crop damage from PROWL. Under these conditions, crop yields can be reduced.

DIRECTIONS FOR USE

Uniformly apply PROWL in 10 or more gallons of water or in 20 or more gallons of liquid fertilizer per acre by ground equipment. Apply in 5 or more gallons of water per acre by aircraft. Refer to the Prowl leaflet label for specific directions regarding mixing, spraying, and liquid fertilizer compatibility determinations.

Apply PROWL preemergence tank mixtures at planting or up to 2 days after planting. Apply to a seedbed which is firm and free of clods and trash. DO NOT APPLY POSTEMERGENCE. Preemergence treatments are most effective in controlling weeds when adequate rainfall or overhead irrigation is received within 7 days after application.

If moisture is insufficient to activate the herbicides, a shallow cultivation (preferably with a rotary hoe) should be made after emergence of soybeans, but while weeds are small enough to be controlled by mechanical cultivation.

A shallow, thorough mixing of soil after herbicide application with a rotary hoe or similar implement will improve consistency of performance and reduce dependency on rainfall to activate herbicides.

PREEMERGENCE TANK MIXTURES OF PROWL PLUS SENCOR/LEXONE

Soybean seed should be planted at least 1½ inches below the soil surface.

DO NOT use tank mixtures containing Sencor or Lexone on soils having a calcareous surface or a pH of 7.5 or higher.

DO NOT use tank mixtures containing Sencor or Lexone on sensitive soybean varieties such as Altona, Coker 102, Coker 156, Govan, Semmes, Tracy, or Vansoy.

WEED SPECIES CONTROLLED

When applied as directed, tank mixtures of PROWL plus Sencor/Lexone applied preemergence will control the following weed species in addition to those listed in the PROWL leaflet label as being controlled by PROWL alone:

- Common ragweed *(Ambrosia artemisiifolia)*
- Jimsonweed *(Datura stramonium)*
- Prickly sida *(Sida spinosa)*
- Smartweed *(Polygonum pensylvanicum)*
- Velvetleaf (Buttonweed) *(Abutilon theophrasti)*
- Venice mallow *(Hibiscus trionum)*
- Wild mustard *(Brassica kaber)* (Medium and fine soils only)

PROWL plus Sencor/Lexone applied preemergence will reduce competition from cocklebur (*Xanthium strumarium*)

BROADCAST RATES PER ACRE FOR PROWL PLUS SENCOR/LEXONE PREEMERGENCE TANK MIXTURES

SOIL TEXTURE	PROWL + Sencor/Lexone DF* Up to 3% organic matter	PROWL + Sencor/Lexone DF* More than 3% organic matter
COARSE** loamy sands, sandy loams	1 to 1.5 pts. + 0.5 lbs.	1.5 pts. + 0.5 lbs.
MEDIUM sandy clay loams***, sandy clays, loams, silts, silt loams	1.5 to 2 pts. + 0.5 lbs.	1.5 to 2.0 pts. + 0.5 to 0.67 lbs.
FINE silty clay loams***, clay loams, silty clays, clays	1.5 to 2 pts. + 0.5 to 0.67 lbs.	2.0 to 2.5 pts. + 0.67 lbs.

* When using Sencor 4F, Sencor 50WP, Lexone 4L, or Lexone 50WP, use equivalent rates. One pint of the flowable formulation or 1 lb. of the wettable powder formulation equals 0.67 lb. of Sencor DF or Lexone DF.

** DO NOT USE on sands. DO NOT USE on loamy sands or on sandy loams that contain less than 2% organic matter, nor on other soils with less than ½% organic matter, as crop injury may occur.

*** Sometimes considered transitional soils and may be classified as either medium or fine-textured soils.

DO NOT USE on peat or muck soils.

The high rates of PROWL and Sencor/Lexone for each soil texture, as stated above, should be used if heavy weed populations are anticipated or if extensive crop residues were present prior to seedbed preparation.

NOTE: In the event of a crop loss due to weather conditions, soybeans can be replanted without adverse effects the same year. If replanting is necessary, DO NOT rework the soil. Livestock can graze or be fed soybean forage from treated fields 40 days after application. Observe all precautions and limitations on Sencor and Lexone labels.

PREEMERGENCE TANK MIXTURES OF PROWL PLUS AMIBEN

When applied as directed, tank mixtures of PROWL plus Amiben will control the following weed species in addition to those listed in the PROWL leaflet label as being controlled by PROWL alone:

- Common ragweed *(Ambrosia artemisiifolia)*
- Mustard *(Brassica* spp.*)*
- Smartweed *(Polygonum pensylvanicum)*
- Velvetleaf (Buttonweed) *(Abutilon theophrasti)*

BROADCAST RATES PER ACRE FOR PROWL PLUS AMIBEN PREEMERGENCE TANK MIXTURE

SOIL TEXTURE	PROWL + Amiben Up to 3% organic matter	PROWL + Amiben More than 3% organic matter
COARSE sands, loamy sands, sandy loams	DO NOT USE	DO NOT USE
MEDIUM sandy clay loams*, sandy clay, loams, silts, silt loams	1.5 to 2 pts. + 1.0 gallon	1.5 to 2.0 pts. + 1.0 gallon
FINE silty clay loams*, clay loams, silty clays, clays	1.5 to 2.0 pts. + 1.0 gallon	2.0 to 2.5 pts. + 1.0 gallon

* Sometimes considered transitional soils and may be classified as either medium or fine-textured soils.

DO NOT USE on peat or muck soils.

The high rates of PROWL for each soil texture above should be used if heavy weed populations are anticipated or if extensive crop residues were present prior to seedbed preparation.

NOTE: In the event of a crop loss due to weather conditions, soybeans can be replanted without adverse effects the same year. If replanting is necessary, DO NOT rework the soil. Livestock can graze or be fed soybean forage from treated fields. Observe all precautions and limitations on Amiben label.

PREEMERGENCE TANK MIXTURES OF PROWL PLUS LOROX

When applied as directed, tank mixtures of PROWL plus Lorox will control the following weed species in addition to those listed in the PROWL leaflet label as being controlled by PROWL alone:

- Mustard *(Brassica* spp.*)*
- Smartweed *(Polygonum pensylvanicum)*
- Velvetleaf (Buttonweed) *(Abutilon theophrasti)*

BROADCAST RATES PER ACRE FOR PROWL PLUS LOROX PREEMERGENCE TANK MIXTURES

SOIL TEXTURE	PROWL + Lorox L* ½ to 3% organic matter	PROWL + Lorox L* More than 3% organic matter
COARSE sandy loams	1.0 to 1.5 pts. + 1.0 pts.	1.5 pts. + 1.0 to 1.5 pts.
MEDIUM sandy clay loams**, sandy clays, loams, silts, silt loams	1.5 to 2 pts. + 1.0 to 1.5 pts.	1.5 to 2.0 pts. + 1.5 to 2.0 pts.
FINE silty clay loams**, clay loams, silty clays, clays	1.5 to 2.0 pts. + 1.5 to 2.0 pts.	2.0 to 2.5 pts. + 2.0 to 2.5 pts.

* When using Lorox 50W use equivalent rates. One pint of the flowable formulation equals 1 lb. of the wettable powder formulation.

** Sometimes considered transitional soils and may be classified as either medium or fine-textured soils.

DO NOT USE on peat or muck soils. **DO NOT USE** on sands or gravelly soils, or on soils containing less than ½% organic matter.

The high rates of PROWL plus Lorox for each soil texture above should be used if heavy weed populations are anticipated or if extensive crop residues were present prior to seedbed preparation.

NOTE: Livestock may graze or be fed forage from soybean fields treated with PROWL plus Lorox. If soybean crop loss occurs due to weather conditions, soybeans may be replanted. If replanting is necessary, thoroughly rework the soil before replanting. Observe all precautions and limitations on Lorox label.

PREEMERGENCE TANK MIXTURES OF PROWL PLUS AMIBEN PLUS SENCOR/LEXONE

When mixing, add Sencor or Lexone first, then Amiben followed by PROWL. Follow all precautions listed in the preceding sections for PROWL plus Amiben and PROWL plus Lexone/Sencor preemergence tank mixtures.

WEED SPECIES CONTROLLED

When applied as directed, tank mixtures of PROWL plus Amiben plus Sencor/Lexone applied preemergence will control the following weed species in addition to those listed in the PROWL leaflet label as being controlled by PROWL alone:

- Common ragweed *(Ambrosia artemisiifolia)*
- Mustards *(Brassica* spp.*)*
- Prickly sida *(Sida spinosa)*
- Smartweed *(Polygonum pensylvanicum)*
- Velvetleaf (Buttonweed) *(Abutilon theophrasti)*
- Venice mallow (*Hibiscus trionum)*

PROWL plus Amiben plus Sencor/Lexone applied preemergence will reduce competition from cocklebur (*Xanthium strumarium*) and jimsonweed (*Datura stramonium*).

DO NOT USE on peat or muck soils.

The high rates of PROWL, Amiben and Sencor/Lexone for each soil texture, as stated above, should be used if heavy weed populations are anticipated or if extensive crop residues were present prior to seedbed preparation.

NOTE: In the event of a crop loss due to weather conditions, soybeans can be replanted without adverse effects the same year. If replanting is necessary, DO NOT rework the soil.

DO NOT use foliage for livestock feed or forage from treated fields. Observe all precautions and limitations on Amiben, Sencor and Lexone labels.

BROADCAST RATES PER ACRE FOR PROWL PLUS AMIBEN PLUS SENCOR/LEXONE PREEMERGENCE TANK MIXTURES

SOIL TEXTURE	PROWL + Amiben + Sencor/Lexone DF* Up to 3% organic matter	PROWL + Amiben + Sencor/Lexone DF* More than 3% organic matter
COARSE** loamy sands, sandy loams (greater than 2% organic matter only)	1.5 pts. + 6.0 pts. + 0.33 lbs.	1.5 pts. + 6.0 pts. + 0.33 lbs.
MEDIUM sandy clay loams***, sandy clays, loams, silts, silt loams	1.5 to 2 pts. + 6.0 to 8.0 pts. + 0.33 to 0.5 lbs.	1.5 to 2.0 pts. + 6.0 to 8.0 pts. + 0.33 to 0.5 lbs.
FINE silty clay loams***, clay loams, silty clays, clays	1.5 to 2.0 pts. + 6.0 to 8.0 pts. + 0.33 to 0.5 lbs.	2.0 to 2.5 pts. + 8.0 pts. + 0.5 lbs.

* When using Sencor 4F, Sencor 50WP, Lexone 4L, or Lexone 50WP, use equivalent rates. One pint of the flowable formation or 1 lb. of the wettable powder formulations equals 0.67 lb. of Sencor DF or Lexone DF.

** DO NOT USE on sands. DO NOT USE on loamy sands or on sandy loams that contain less than 2% organic matter, nor on other soils with less than ½% organic matter, as crop injury may occur.

*** Sometimes considered transitional soils and may be classified as either medium or fine-textured soils.

PREEMERGENCE TANK MIXTURES OF PROWL PLUS LASSO

Apply within 5 days of the last preplant tillage. When applied as directed, PROWL plus Lasso will control the following weeds in addition to those controlled by PROWL alone:

- Black nightshade *(Solanum nigrum)*
- Hairy nightshade *(Solanum sarrachoides)*

BROADCAST RATES PER ACRE FOR PROWL PLUS LASSO PREEMERGENCE TANK MIXTURES

SOIL TEXTURE	PROWL + Lasso Up to 3% organic matter	PROWL + Lasso More than 3% organic matter
COARSE	1.0 to 1.5 pts. + 2.5 qts.	1.5 pts. + 2.5 qts.
MEDIUM	1.5 to 2.0 pts. + 3.0 to 4.0 qts.	1.5 to 2.0 pts. + 3.0 to 4.0 qts.
FINE	1.5 to 2.0 pts. + 3.0 to 4.0 qts.	2.0 to 2.5 pts. + 3.0 to 4.0 qts.

Observe all precautions and limitations on the Lasso label.

PREEMERGENCE TANK MIXTURES OF PROWL PLUS DUAL.

When applied as directed, PROWL plus Dual will control the following weeds in addition to those controlled by PROWL alone:

- Red rice *(Oryza sativa)*
- Yellow nutsedge *(Cyperus esculentus)*
- Black nightshade *(Solanum nigrum)*

BROADCAST RATES PER ACRE FOR PROWL PLUS DUAL PREEMERGENCE TANK MIXTURES

SOIL TEXTURE	PROWL + Dual 8E Up to 3% organic matter	PROWL + Dual 8E More than 3% organic matter
COARSE	1.0 to 1.5 pts. + 1.5 to 2.0 pts.	1.5 pts. + 2.0 pts.
MEDIUM	1.5 to 2.0 pts. + 2.0 to 2.5 pts.	1.5 to 2.0 pts. + 2.0 to 2.5 pts.
FINE	1.5 to 2.0 pts. + 2.0 to 2.5 pts.	2.0 to 2.5 pts. + 2.0 to 3.0 pts.

Observe all precautions and limitations on the Dual label.

SEE PAGE 26 FOR REGIONAL MAP FOR RATE DETERMINATION

® Registered Trademark of American Cyanamid Company

1 See map for specific states

2 Sencor is a trademark of Farbenfabriken Bayer GmbH, Leverkusen, West Germany

3 Lexone is a trademark of E.I. DuPont de Nemours and Company

4 Amiben is a trademark of Union Carbide Agricultural Products Company, Inc.

5 Lorox is a trademark of E.I. DuPont de Nemours and Company

6 Lasso is a trademark of Monsanto Agricultural Products Company

7 Dual is a trademark of Ciba-Geigy Corporation

SUPPLEMENTAL LABELING

PROWL® HERBICIDE
EPA Reg. No. 241-243
EPA SLN No. NE-850007

FOR DISTRIBUTION AND USE ONLY WITHIN NEBRASKA

OBSERVE ALL PRECAUTIONARY STATEMENTS IN THE PROWL® LEAFLET LABEL BEFORE USING.

PREPLANT INCORPORATED APPLICATION OF PROWL PLUS SENCOR[1]/LEXONE[2] FOLLOWED BY A PREEMERGENCE APPLICATION OF PROWL PLUS SENCOR/LEXONE

DIRECTIONS FOR USE

It is a violation of Federal law to use this product in a manner inconsistent with its labeling.

Uniformly apply the PROWL in 10 or more gallons of water or in 20 or more gallons of liquid fertilizer per acre by ground equipment. Apply in 5 or more gallons of water per acre by aircraft. Refer to the PROWL leaflet label for specific directions regarding mixing, spraying, and liquid fertilizer compatibility determinations.

PROWL plus Sencor/Lexone may be applied up to 7 days prior to planting soybeans; incorporation must be made before planting. Refer to PROWL leaflet label for complete incorporation instructions. Preemergence applications of PROWL plus Sencor/Lexone should be made after planting, before crop and weed emergence.

WEED SPECIES CONTROLLED

When applied as directed, tank mixtures of PROWL plus Sencor/Lexone applied preplant incorporated followed by a preemergence application will control the following weed species in addition to those listed in the PROWL leaflet label as being controlled by PROWL alone:

Common ragweed *(Ambrosia artemisiifolia)*
Jimsonweed *(Datura stramonium)*
Smartweed *(Polygonum pensylvanicum)*
Velvetleaf (Buttonweed) *(Abutilon theophrasti)*
Venice mallow *(Hibiscus trionum)*

SPECIAL PRECAUTIONS FOR PROWL PLUS SENCOR/LEXONE PREPLANT INCORPORATED FOLLOWED BY PROWL PLUS SENCOR/LEXONE PREEMERGENCE APPLICATIONS.

Soybean seed should be planted at least 1½ inches below the soil surface. Applications should not be made to soils having a calcareous surface area or a pH of 7.5 or higher. Soil incorporation deeper than recommended will reduce weed control and can result in crop injury.

Do not apply to sensitive soybean varieties—Altona, Coker 102 and 156, Govan, Semmes, Tracy or Vansoy.

Follow cropping restrictions on Sencor and Lexone labels.

BROADCAST RATES PER ACRE FOR PROWL PLUS SENCOR/LEXONE PREPLANT INCORPORATED (PPI) FOLLOWED BY PROWL PLUS SENCOR/LEXONE PREEMERGENCE (PE) TANK MIXTURE

SOIL TEXTURE	PROWL + SENCOR/ LEXONE DF Up to 3% Organic Matter	PROWL + SENCOR/ LEXONE DF More than 3% Organic Matter
COARSE	DO NOT USE	1.0 to 1.5 pts. + .25 lbs. (ppi)/ 0.5 pts. + .25 lbs. (pe)
MEDIUM	1.5 pts. + .25 lbs. (ppi)/ 1.0 pts. + .25 lbs. (pe)	1.5 to 2.0 pts. + .25 to .33 lbs. (ppi)/ 1.0 pts. + .25 to .33 lbs. (pe)
FINE	2.0 pts. + .25 to .33 lbs. (ppi)/ 1.0 pts. + .25 to .33 lbs. (pe)	2.0 pts. + .33 lbs. (ppi)/ 1.0 pts. + .33 lbs. (pe)

When using Sencor 4F, Sencor 50WP, Lexone 4L, or Lexone 50WP, use equivalent rates. One pint of the flowable formulation or 1 lb. of the wettable powder formulation equals 0.67 lb. of Sencor DF or Lexone DF.

DO NOT USE on sands. **DO NOT USE** on loamy sands or on sandy loams that contain less than 2% organic matter, nor on other soils with less than ½% organic matter, as crop injury may occur.

NOTE: In the event of a crop loss due to weather conditions, soybeans can be replanted without adverse effects the same year. If replanting is necessary, DO NOT rework the soil. Livestock can graze or be fed soybean forage from treated fields 40 days after application. Observe all precautions and limitations on Sencor and Lexone labels.

This label must be in the possession of the user at the time of pesticide application.

24(c)Registrant:

® Registered Trademark of American Cyanamid Company

1 Sencor is a trademark of Farbenfabriken Bayer GmbH, Leverkusen, West Germany

2 Lexone is a trademark of E.I. duPont de Nemours and Company

SUPPLEMENTAL LABELING

PROWL® HERBICIDE
EPA Reg. No. 241-243
EPA SLN No. ND-850003

FOR DISTRIBUTION AND USE ONLY WITHIN NORTH DAKOTA

OBSERVE ALL PRECAUTIONARY STATEMENTS IN THE PROWL® LEAFLET LABEL BEFORE USING.

PROWL OR PROWL PLUS AMIBEN[1] TANK MIXTURE FOR PREPLANT OR PREEMERGENCE APPLICATION IN NO-TILL SUNFLOWER PRODUCTION SYSTEMS

DIRECTIONS FOR USE

It is a violation of Federal law to use this product in a manner inconsistent with its labeling.

NOTE: Existing living vegetation must be controlled at or before the application of PROWL or PROWL plus Amiben. This may be accomplished by applying a contact herbicide registered for use in sunflowers, or by using PROWL or PROWL plus Amiben in combination with such a contact herbicide. Consult contact herbicide label for all directions, precautions and restrictions.

Uniformly apply the recommended PROWL tank mixture in 20 to 60 gallons of water per acre by ground equipment. Use the higher gallonage for fields with heavy weed infestations or excessive crop residues.

PROWL alone may be applied immediately after planting, or up to 30 days before planting. Sunflowers must be planted within 7 days after application of PROWL plus Amiben. DO NOT apply postemergence since sunflowers exposed at the time of application will be killed. Adequate rainfall within 7 days of application is essential for season-long weed control.

Refer to the PROWL leaflet label for specific directions regarding mixing, spraying, and liquid fertilizer compatibility determinations.

All applicable directions, restrictions, and precautions on the EPA-registered label(s) are to be followed.

WEED SPECIES CONTROLLED

When applied according to label directions, PROWL will control the following weeds in no-till sunflowers.

- Barnyardgrass *(Echinochloa crus-galli)*
- Crabgrass *(Digitaria* spp.*)*
- Carpetweed *(Mollugo Veticillata)*
- Green foxtail *(Setaria viridis)*
- Lambsquarters *(Chenopodium album)*
- Pigweed *(Amaranthus* spp.*)*
- Yellow foxtail *(Setaria glauca)*

A tank mix of PROWL plus Amiben will control the following weeds in no-till sunflowers in addition to those listed above:

- Kochia *(Kochia scoparia)*
- Mustard *(Brassica* spp.*)*
- Common ragweed *(Ambrosia artemisiifolia)*
- Pennsylvania smartweed *(Polygonum pensylvanicum)*

BROADCAST RATES PER ACRE FOR PROWL APPLIED PREPLANT OR PREEMERGENCE IN NO-TILL SUNFLOWERS

SOIL TEXTURE	Up to 3% Organic Matter	More than 3% Organic Matter
COARSE	2.5 to 3.0 pts.	3.0 pts.
MEDIUM	3.0 pts.	3.0 pts.
FINE	3.0 pts.	DO NOT USE

BROADCAST RATES PER ACRE FOR PROWL PLUS AMIBEN TANK MIXTURE APPLIED PREPLANT OR PREEMERGENCE IN NO-TILL SUNFLOWERS[1]

SOIL TEXTURE	Up to 3% Organic Matter	More than 3% Organic Matter
COARSE	DO NOT USE	DO NOT USE
MEDIUM	2.5 to 3.0 pts.† 4 to 6 qts.	3.0 pts. + 4 to 6 qts.
FINE	3.0 pts. + 4 to 6 qts.	3.0 pts. + 4 to 6 qts.

[1] Use the higher rates of Amiben if infestations of the following weeds are anticipated: common ragweed, Pennsylvania smartweed, wild mustard *(Brassica kaber)*.

DO NOT use on peat or muck soils.

NOTE: DO NOT feed forage or graze livestock on treated sunflower fields. In the event of a crop loss due to weather conditions, sunflowers may be replanted. If replanting is necessary, DO NOT work the soil. Observe all precautions and limitations on Amiben label.

This label must be in the possession of the user at the time of pesticide application.

24(c) Registrant:

® Registered Trademark of American Cyanamid Company

[1] Amiben is a registered trademark of Union Carbide Agricultural Products Company, Inc.

THIMET® 20-G

SOIL AND SYSTEMIC INSECTICIDE

RESTRICTED USE PESTICIDE

Due to acute oral and dermal toxicity and bird toxicity.

For retail sale to and use only by Certified Applicators or persons under their direct supervision.

ACTIVE INGREDIENT:

Phorate (O,O-diethyl S-[(ethylthio)methyl] phosphorodithioate)	20.0%
INERT INGREDIENTS	80.0%
TOTAL	100.0%

EPA Reg. No. 241-257-AA

®Registered Trademark of the American Cyanamid Company

DANGER–POISON

Keep Out of Reach of Children

See Side Panel For Antidote and Other Precautions

PELIGRO!

AL USARIO: Si used no lee inglés, no use este producto hasta que la etiqueta le haya sido explicada ampliamente.

(ENGLISH TRANSLATION)

DANGER!

(TO THE USER: If you cannot read English, do not use this product until the label has been fully explained to you.)

PRECAUTIONARY STATEMENTS

HAZARDS TO HUMANS (AND DOMESTIC ANIMALS)

DANGER!

Danger, fatal if swallowed, inhaled or absorbed through the skin, repeated inhalation or skin contact may, without symptoms, progressively increase susceptibility to poisoning.

Do not get in Eyes, On Skin, On Clothing

Wear freshly laundered, long-sleeved work clothing daily. While transferring from package to equipment, wear a clean cap and gloves (rubber or cotton). If cotton gloves are used, they must be laundered or discarded after each day's use. Rubber gloves should be washed with soap and water after each use. Do not wear the same gloves for other work. Destroy and replace gloves frequently.

In case of contact, immediately remove contaminated clothing and wash skin thoroughly with soap and water. Launder clothing before re-use. Wash thoroughly with soap and water before eating or smoking. Bathe at the end of the work day and change outer clothing.

DO NOT BREATHE DUST

Wear a face mask or other respiratory equipment while emptying bags of product into a hopper. While emptying bags into equipment, pour downwind and allow as little free fall as possible. Do not pour at face level and do not allow dust to reach the breathing zone.

Keep all unprotected persons out of operating areas.

KEEP OUT OF REACH OF DOMESTIC ANIMALS

ENVIRONMENTAL HAZARDS

This product is toxic to fish, shrimp, crab, birds and other wildlife. Birds and other wildlife in treated areas may be killed. Keep out of lakes, streams, ponds, tidal marshes and estuaries. Do not apply where runoff is likely to occur. Do not apply when weather conditions favor drift from areas treated. Do not contaminate water by cleaning of equipment or disposal of wastes. Shrimp and crab may be killed at application rate recommended on this label. Do not apply where these are important resources. Apply this product only as specified on this label.

This pesticide is toxic to bees exposed to direct application. Applications should be timed to coincide with periods of minimum bee activity, usually between late evening and early morning.

In case of emergency endangering life or property involving this product, call collect, day or night, Area Code 201-835-3100.

ANTIDOTE: Atropine is antidotal.

FIRST AID

If swallowed, drink one or two glasses of water and induce vomiting by touching back of throat with finger. Do not induce vomiting or give anything by mouth to an unconscious person. Get medical attention.

If inhaled, remove to fresh air. If not breathing, give artificial respiration, preferably mouth-to-mouth. If breathing is difficult, give oxygen.

In case of contact, immediately flush eyes or skin with plenty of water for at least 15 minutes while removing contaminated clothing and shoes. Wash clothing before reuse.

CALL A PHYSICIAN AT ONCE IN ALL CASES OF SUSPECTED POISONING

NOTE TO PHYSICIANS: Warning symptoms include weakness, headache, tightness of chest, blurred vision, nonreactive pinpoint pupils, salivation, sweating, nausea, vomiting, diarrhea and abdominal cramps. Give atropine intramuscularly or intravenously, depending on severity of poisoning, 2 to 4 milligrams (3 to 6 tablets, $\frac{1}{100}$ grain each) every 10 minutes until fully atropinized as shown by dilated pupils, dry flushed skin and tachycardia. Twenty to thirty milligrams, or more, may be required during the first 24 hours. Never give opiates or phenothiazine tranquilizers. Clear chest by postural drainage. Artificial respiration or oxygen administration may be necessary. Observe patient continuously for at least 48 hours. Allow no further exposure to any cholinesterase inhibitor until cholinesterase regeneration has taken place as determined by blood tests.

Pralidoxime chloride (2-PAM; PROTOPAM chloride) may be effective as an adjunct to atropine. Use according to label directions.

DISCLAIMER

The label instructions for the use of this product reflect the opinion of experts based on field use and tests. The directions are believed to be reliable and should be followed carefully. However, it is impossible to eliminate all risks inherently associated with the use of this product. Crop injury, ineffectiveness or other unintended consequences may result because of such factors as weather conditions, presence of other materials, or the use or application of the product contrary to label instructions, all of which are beyond the control of American Cyanamid Company. All such risks shall be assumed by the user.

American Cyanamid Company warrants only that the material contained herein conforms to the chemical description on the label and is reasonably fit for the use therein described when used in accordance with the directions for use, subject to the risks referred to above.

Any damages arising from a breach of this warranty shall be limited to direct damages and shall not include consequential commercial damages such as loss of profits or values or any other special or indirect damages.

American Cyanamid Company makes no other express or implied warranty, including any other express or implied warranty of FITNESS or of MERCHANTABILITY.

DIRECTIONS FOR USE

It is a violation of Federal Law to use this product in a manner inconsistent with its labeling.

Fields may be reentered on day of treatment with soil applications of THIMET 20-G. Do not enter fields within 7 days after plants have been treated with foliar applications. THIMET 20-G should be applied with a granular pesticide applicator properly calibrated to assure accurate placement and proper dosage.

Do not enter treated areas without protective clothing until treatments have been completed.

Because certain states may require more restrictive re-entry intervals for various crops treated with this product, consult your State Department of Agriculture for further information.

Written or oral warnings must be given to workers who are expected to be in a treated area or in an area about to be treated with this product. When oral warnings are given, warnings shall be given in a language customarily understood by workers. Oral warnings must be given if there is reason to believe that written warnings cannot be understood by workers. Written warnings must include the following information: DANGER. Area treated with THIMET 20-G on (date of application). Do not enter without protective clothing.

IMPORTANT! Read precautionary statements before using this product.

STORAGE AND DISPOSAL

PROHIBITIONS

Do not contaminate water, food or feed by storage or disposal. Open dumping is prohibited. Sweep up and bury spillage, whether it occurs indoors or in the field. Once a bag has been opened, use it completely or bury the remainder. Make sure that the hoppers of equipment are emptied while still in the field. Not for use or storage in or around the home.

PESTICIDE DISPOSAL

Pesticide or spillage that cannot be used according to label instructions must be disposed of according to federal, state or local procedures under the Resource Conservation and Recovery Act.

CONTAINER DISPOSAL

Completely empty bag by shaking and tapping sides and bottom to loosen any clinging material. Empty the residues into the application equipment. Dispose of bags in a sanitary landfill, or by other approved State and local procedures.

GENERAL

Consult federal, state or local disposal authorities for approved alternative procedures such as limited open burning.

TYPICAL APPLICATION RATES FOR VARIOUS ROW SPACINGS

Rate—Oz./1,000 Ft. of Row

		1.2 oz.	3 oz.	4.5 oz.	6 oz.	8 oz.	10 oz.	12 oz.	14 oz.	16 oz.	18 oz.
	40"			3.7	4.9	6.5	8.2	9.8	11.4	13.1	14.7
	38"			3.9	5.2	6.9	8.6	10.3	12.0	13.7	15.5
	36"			4.1	5.4	7.3	9.1	10.9	12.7	14.5	16.3
R	34"			4.3	5.8	7.7	9.6	11.5	13.5	15.4	17.3
O	32"			4.6	6.1	8.2	10.2	12.2	14.3	16.3	18.4
W	30"			4.9	6.5	8.7	10.9	13.1	15.2	17.4	19.6
	26"			5.7	7.5	10.0	12.6	15.1	17.6	20.1	22.6
W	24"			6.1	8.2	10.9	13.6	16.3	19.1	21.8	24.5
I	20"		4.9								
D	18"		5.4								
T	16"		6.1								
H	12"		8.2								
	8"	4.9	12.2								
	6"	6.5	16.3								

EQUIVALENT POUNDS PER ACRE

THIMET 20-G

CROP	PESTS CONTROLLED	RATE OF THIMET 20-G	APPLICATION	REMARKS
BEANS At Planting	Mexican bean beetles	4.5–9.4 ozs. per 1,000 ft. of row for any row spacing (minimum 30-inch spacing)	Drill granules to the side of the seed.	Do not place THIMET granules in direct contact with seed.
	Leafhoppers Aphids Lygus bugs Thrips Mites Seedcorn maggots	4.5–7.0 ozs. per 1,000 ft. of row for any row spacing (minimum 30-inch spacing)	Apply granules in a band over the row.	Do not feed the foliage of treated beans within 60 days of treatment.
FIELD CORN AND SWEET CORN (excluding popcorn) At Planting	Corn rootworms Wireworms White grubs Seedcorn maggots Seedcorn beetles Flea Beetles Mites	6 ozs. per 1,000 ft. of row for any row spacing (minimum 30-inch row spacing)	Place granules in a 7-inch band over the row directly behind the planter shoe in front of or behind the press wheel.	Do not place THIMET granules in direct contact with seed. One additional application can be applied at cultivation OR over corn later in the season.
FIELD CORN AND SWEET CORN (excluding popcorn) At Cultivation	Corn rootworms		Apply granules to base of plants or over the top of plants just ahead of cultivation shovels so as to cover granules with soil.	Do not make any applications of THIMET after cultivation treatment.
FIELD CORN (excluding sweet and popcorn) At Cultivation	Chinch bug nymphs (Colorado, Kansas, and Nebraska)		Apply granules to base of plants or over the top of plants just ahead of cultivation shovels so as to cover granules with soil. Apply as soon as chinch bug nymphs are observed at base of plants. Granules must be placed at the infested region of the plants and adequately covered with soil.	Later infestations of chinch bugs may not be controlled. Do not make any applications of THIMET after cultivation treatment.
FIELD CORN (excluding sweet and popcorn) Over the Plant	European corn borer (1st brood)	5.0 lbs. per acre	Apply granules into the whorl of the plant prior to tassle emergence.	Do not graze or cut for forage within 30 days of treatment. Do not make more than one application over the plant. Do not apply under prolonged drought conditions. Consult your state experiment or state extension service for proper timing of application.
	Corn leaf aphids Mites		Broadcast the granules evenly over the top of plants by air or ground equipment.	
Application at Cultivation	Corn rootworms		Apply granules at time of cultivation in a band at base of plants just ahead of cultivator shovels so as to cover granules with soil. Apply in June as soon as rootworm activity is observed.	Do not make any later applications after cultivation treatment.

CROP	PESTS CONTROLLED	RATE OF THIMET 20-G	APPLICATION	REMARKS
COTTON: At Planting Time	Early seaon thrips, aphids, mites & leaf miners	3.0 ozs. per 1,000 ft. of row for any row spacing (minimum) 36-inch spacing)	Distribute the granules evenly in the furrow at planting time. Soil temperature, moisture and tilth should be favorable for good germination and emergence.	Suggested for areas where a limited period of protection against early season pests is necessary for good cotton growth. Use 1.5 lbs. THIMET 20-G per acre for hill-dropped cotton.
	Mites, thrips, aphids, black cutworms, leafhoppers, white flies and leaf miners.	6.0–9.0 ozs. per 1,000 ft. of row for any row spacing (minimum 36-inch spacing)	Distribute the granules evenly in the furrow at planting time. Soil temperature, moisture and tilth should be favorable for good germination and emergence. Under adverse conditions plant stand may be affected.	Recommended for areas requiring an extended period of mite control as well as early season protection against other cotton pests. The higher rate is suggested for heavier soils.
Side-Dressing (Irrigated cotton only)	Mites from spring to early summer	9.0–12.0 ozs. per 1,000 ft. of row for any row spacing (minimum 36-inch spacing)	Incorporate granules into soil as a side-dress application deep enough to avoid disturbance by future cultivations.	Irrigate as soon as possible after treatment. Do not apply later than 60 days before harvest.
PEANUTS: At Planting Time	Thrips, leafhoppers	5.5 ozs. per 1,000 ft. of row for any row spacing (minimum 24-inch spacing)	Distribute the granules evenly in the furrow at planting time. For best results, plant peanuts when soil temperatures are favorable for establishing good peanut stands.	Do not graze or feed treated hay or forage.
At Pegging	Southern corn rootworms and Leafhoppers	11.0 ozs. per 1,000 ft. of row for any row spacing (minimum 24-inch spacing)	Distribute the granules as a band over the fruiting zone at pegging time. Work into the top few inches of soil immediately.	Do not graze or feed treated hay or forage.
POTATOES: At Planting Time	Aphids, leafhoppers, leaf miners, psyllids, wireworms, flea beetle larvae and reduction of flea beetle adults.	**Light or Sandy Soils:** 11.3 ozs. per 1,000 ft. of row for any row spacing (minimum 32-inch spacing) **Heavy or Clay Soils:** 17.3 ozs. per 1,000 ft. or 1 row for any row spacing (minimum 32–inch spacing)	Distribute the granules evenly in the furrow or granules may be banded on each side of the row at planting time.	Wait 90 days after treatment before harvesting potatoes. Do not use for Colorado potato beetle control in the Northeast United States.
	Colorado potato beetle (early season control)	**Heavy or Clay Soils:** 17.3 ozs. per 1,000 ft. of row for any row spacing (minimum 32-inch spacing)		
SORGHUM: At Planting Time	Aphids (Greenbugs)	6.0 ozs. per 1,000 ft. of row for any row spacing (minimum 30-inch spacing)	Place granules 2 inches to side and 1 inch below the seed at planting or as a 7-inch band over the row just prior to press wheel.	Do not place THIMET granules in direct contact with seed. A second application of granules may be applied topically later in the season.
Application at Cultivation	Chinch bug nymphs (Colorado, Kansas, Nebraska)		Apply granules at time of cultivation in a band at base of plants or over the row just ahead of cultivator so as to cover granules with soil. Apply as soon as chinch bug nymphs are observed at base of plants. Granules must be placed at the infested region of the plants and adequately covered with soil.	Later infestations of chinch bugs may not be controlled. Do not make any applications of THIMET after cultivation treatment. Do not feed foilage before grain harvest.
Application over the Plant	Aphids (Greenbugs) Banks grass mites	4.9 lbs. per acre	Broacast granules into the whorls of the plants by air or ground equipment.	Make only one application after plant emergence. Do not apply within 28 days of harvest.
SOYBEANS: At Planting	For the early season control of: Mexican bean beetles Leafhoppers Seedcorn maggots Thrips Aphids Mites	9.0 ozs. per 1,000 ft. of row for any row spacing (minimum 30-inch row spacing)	Drill granules to the side of the seed.	Do not place THIMET granules in direct contact with seed. Do not feed the foliage of treated soybeans. Crop injury may result when preplant incorporated of preemergence applications of metribuzine herbicides are used in conjunction with this product or other soil-applied organophosphate pesticides.
	Seedcorn maggots		Place granules in a 7-inch band over the row directly behind the planter shoe in front of the press wheel.	

CROP	PESTS CONTROLLED	RATE OF THIMET 20-G	APPLICATION	REMARKS
SUGAR BEETS:	Aphids Beet root maggots Leafhoppers Mites Leaf miners	4.5 ozs. per 1,000 ft. row for any row spacing (minimum 20- inch spacing)	Drill granules to the side of the seed OR place granules in a band over the row.	Do not place THIMET granules in direct contact with the seed. Do not apply within 30 days of harvest. Do not feed sugar beet tops or silage to dairy cattle. Consult your state experiment station or state extension service for proper timing of application in your area.
SUGAR BEETS Postemergence	Mites Aphids	4.9–7.5 lbs. per acre	Apply to foliage when plants are dry. Use the higher rate on dense foliage for good coverage. Foliar application by air may be used for aphid control when wet soil prohibits ground application.	
SUGAR CANE (Florida)	Wireworms	19.5 lbs. per acre	Apply in a 10–12 inch band directly on the seed piece and surrounding soil in the open furrow immediately before covering.	
WHEAT: Application at planting time	Grasshoppers Aphids (Western and North Central States only) Hessian fly	1.2 ozs. per 1,000 ft. of row for any row spacing (minimum 8-inch spacing)	Apply the granules in the seed furrow at planting time.	Do not feed or graze foliage within 45 days of treatment. Do not make any later applications after planting time treatment.
Application over the plant	Aphids (Western States only)	4.9 lbs. per acre	Broadcast the granules evenly over the field by air equipment when aphids first appear.	Do not apply over plants if application was made at planting time. Do not make more than one application over the plants. Do not apply within 70 days of harvest of grain. Do not feed or graze foliage within 28 days of treatment.

CALIBRATION INFORMATION

FIRST READ THE LABEL

It is important that applicator equipment be properly set to deliver the labeled rate. The table below contains suggested starting guage settings for calibration at planting speeds of 5 or 7 mph and application rates of 4.5, 6.0 or 9.0 oz./1,000 ft. of row.

NOTE: The settings in this table SHOULD ONLY BE USED AS STARTING POINTS. Continually check the amount of THIMET 20-G used against a known length of row and make further adjustments accordingly. Also check calibration occasionally to make sure equipment wear, changing moisture conditions, etc., have not caused a change in flow rate.

Suggested Granular Applicator Setting

	At 5 MPH			At 7 MPH		
Applicator	**4.5 oz.**	**6.0 oz.**	**9.0 oz.**	**4.5 oz.**	**6.0 oz.**	**9.0 oz.**
New Gandy	15	18	22	18	21	25
New Noble	6	8	11	8	11	14
Old Noble	8	11	16	12	15	22
International Harvester	1/5	1/7	2/1	1/7	2/0	2/8
New John Deere	9	13	20	13	19	26
Newest John Deere[2]	9	11	14	11	13	17
Old John Deere	1/23	1/26	2/3	1/26	1/32	2/14
Allis Chalmers	4	5	7	4	5	7

[1] Suggested starting point only.
[2] Aluminum funnel.

BASF Corporation

Agricultural Chemicals Division
100 Cherry Hill Road
Parsippany, NJ 07054
(201) 263-3400

BASAGRAN®

POSTEMERGENCE HERBICIDE

A soluble liquid formulation containing:

ACTIVE INGREDIENT:

Sodium salt of bentazon*	42.0%
INERT INGREDIENTS	58.0%
TOTAL	100.0%

* Equivalent to 4 pounds per gallon bentazon (3-(1-methylethyl)-1*H*-2,1,3-benzothiadiazin-4 (3*H*)-one 2,2-dioxide)

EPA Reg. No. 7969-45

CAUTION

Keep Out of Reach of Children

STATEMENT OF PRACTICAL TREATMENT

Avoid contact with eyes or skin. In case of contact immediately flush eyes or skin with plenty of water.

Get medical attention if irritation persists.

First Aid: If contacted, flush eyes immediately with water for at least 15 minutes. Call a physician.

DIRECTIONS FOR USE

—ALL CROPS

It is a violation of Federal law to use this product in a manner inconsistent with its labeling.

Read the precautionary statement, environmental hazards, storage and disposal statements, and Conditions of Sale and Warranty statement appearing on the container label.

GENERAL INFORMATION

BASAGRAN® herbicide is intended for selective postemergence control of certain broadleaf weeds and sedges. (See **Directions For Use** for specific crops and weeds.) BASAGRAN does not control grasses. BASAGRAN is effective mainly through contact action; therefore, weeds must be thoroughly covered with spray. Large crop-and-weed leaf canopies shelter smaller weeds and prevent adequate spray coverage. Labeled crops are tolerant to BASAGRAN; however, some leaf-speckling and leaf-bronzing may occur under certain conditions. (See **Restrictions and Limitations** for each crop.)

TIMING OF APPLICATIONS

Make postemergence applications of BASAGRAN early, when weeds are small and actively growing and before weeds reach the maximum size listed in the application rate tables for the individual crops.

Early application to weeds produces the most beneficial effect on weed control (exception, yellow nutsedge and Canada thistle), allows use of the lower rate (depending on weed species), and makes it easier to obtain thorough spray coverage. Delay in application which permits weeds to exceed the maximum size stated will result in inadequate control.

Do not cultivate within five days before or after application of BASAGRAN in the following northern and western states: AZ, CA, CO, CT, IA, ID, IL, IN, KS, KY, MA, ME, MI, MN, MO, MT, NB, ND, NH, NJ, NV, NY, OH, OR, PA, RI, SD, UT, VT, WA, WI, WV, WY.

WATER VOLUME AND SPRAY PRESSURE

Apply recommended rates of BASAGRAN as follows:

Ground equipment: Use a minimum of 20 gals. of water per broadcast acre and a minimum of 40 psi pressure (measured at the boom—not at the pump or in the line). When crop and weed foliage is dense use up to 50 gals. of water and up to 80 psi pressure. Use standard high pressure pesticide hollow cone or flat fan nozzles spaced 20 inches apart. Do not use flood, or whirl chamber, or controlled droplet applicator (CDA) nozzles.

Air equipment: Use a minimum of 5 gals. of water per acre (except 10 gals. for rice) and a maximum of 40 psi pressure. Use only diaphragm-type nozzles producing cone or fan spray patterns.

AERIAL APPLICATION—SPECIAL DIRECTIONS

To obtain uniform coverage and to avoid drift hazards, the following application equipment and practices should be used:

Nozzle height: Maximum of 10 feet above crop.

Nozzle orientation: Nozzles must be oriented so as to discharge straight back with the air stream (opposite the direction of travel of the aircraft) or at some angle between straight back and straight down. For optimal coverage when applying BASAGRAN by air in rice, orient all nozzles straight down.

Nozzles must not be located further out than three-fourths the distance from the center of the aircraft to the end of the wing or rotor.

Water volume and spray pressure: See **Air equipment.**

Do not apply BASAGRAN by aircraft when wind is blowing at a velocity above 10 mph (except above 5 mph in California). Coarse sprays (larger droplets) are less likely to drift.

Do not apply BASAGRAN by air within 200 feet of ornamentals or sensitive non-target crops such as cotton, sugar beets, sunflowers or okra.

Applicator must follow the most restrictive use cautions to avoid drift hazards, including those found in this labeling as well as applicable state and local regulations and ordinances.

SPECIAL INFORMATION FOR IRRIGATED AREAS

In irrigated areas, it may be necessary to irrigate prior to treatment with BASAGRAN to ensure that weeds are growing actively. Weeds growing under drought conditions usually are not satisfactorily controlled.

ADDITION OF OIL CONCENTRATE TO SPRAY TANK

A nonphytotoxic oil concentrate (commonly referred to as oil concentrate) should be added to the spray tank for certain weed problems as recommended in the directions for specific crops. The oil concentrate must contain either a petroleum or vegetable oil base and must meet the following criteria: 1) be nonphytotoxic, 2) contain only EPA-exempt ingredients, 3) provide good mixing quality in the jar test (see below), and 4) be successful in local experience.

The exact composition of suitable products will vary; however, vegetable and petroleum oil concentrates should contain emulsifiers which provide good mixing quality. For vegetable oil concentrates, it has been observed that highly refined vegetable oils are more satisfactory than unrefined vegetable oils. For additional information see "Jar Test for Estimating Suitability of Oil Concentrates" at the end of this section.

With the addition of oil concentrate to BASAGRAN on soybeans, beans and peanuts, a slight leaf burn may occur, but all new growth is normal and crop vigor is not reduced. The potential for leaf burn is increased when relative humidity and temperature are high. A few oil concentrates have exhibited excessive leaf burn. Refer to your supplier of BASAGRAN for information concerning successful local experience prior to purchasing any oil concentrate.

Do not add 28% nitrogen solution to Basagran when oil concentrate is included in the spray tank.

Do *not* add oil concentrate to BASAGRAN for use on peas.

Rate of Oil Concentrate:
Ground application—2 pints/acre (maximum)
Air application—1 pint/acre (maximum)
California—refer to additional information under the specific crop (rice, beans, corn/grain sorghum)

Mixing/Spraying: Fill tank of a thoroughly clean sprayer half to two-thirds full with clean water. Start agitation and add BASAGRAN; allow to mix thoroughly. Add oil concentrate and remaining volume of water. Maintain constant agitation during application.

Jar Test for Estimating Suitability of Oil Concentrates.

1. Water supply: Use only water from intended source and at the source temperature.

2. Amount of water in jar:
Ground Application—For 20 gal/A spray volume use 3¾ cups (800 ml) of water.
Air application—For 5 gal/A spray volume use ⅚ cup (200 ml) of water, or
For 10 gal/A spray volume use 1⅔ cups (400 ml) of water.
For other spray volumes, adjust proportionately to above.

3. Amount of herbicide(s) and oil concentrate to add: Add herbicide(s) and oil concentrate at the rate of 1 teaspoon (5 ml) for each pint of recommended label rate.

4. Add components in following sequence, gently mixing between component additions:
1) Dry products (dry flowables and wettable powders) when applicable.
2) BASAGRAN, and when applicable, other water miscible products (such as Blazer), liquid fertilizers and/or liquid flowables.
3) Oil concentrate.
4) Poast or other emulsifiable concentrates when applicable.

5. Cap jar, invert 10 cycles, let stand for 15 minutes, evaluate.

6. Evaluation: An ideal tank mix combination will be uniform; thus, the suitability of the oil concentrate is questionable if any of the following are observed:
Free oil at the surface—film or globules.
Flocculation—fine particles which may be suspended in the liquid or found as a precipitated layer at the bottom of the jar.
Clabbering—thickening texture (coagulated) resembling yogurt or a curd-like texture as with cottage cheese.

RESTRICTIONS AND LIMITATIONS

Do not apply BASAGRAN to crops listed on this labeling that have been subject to stress conditions such as hail damage, flooding, drought, injury from other herbicides or widely fluctuating temperatures, as crop injury may result.

Do not apply BASAGRAN if crops listed on this label show injury (leaf phytotoxicity and/or plant stunting) produced by any other prior herbicide applications, because this injury may be enhanced and/or prolonged.

Do not apply BASAGRAN during prolonged periods of drought or during unseasonably cold weather, as unsatisfactory weed control may result.

Rainfall or overhead irrigation soon after application (within 8 hours) may nullify the effectiveness of BASAGRAN.

Do not mix or apply BASAGRAN with any other pesticide or with fertilizer except as specifically recommended on this labeling.

Clean sprayer thoroughly prior to application of BASAGRAN, particularly if a herbicide was used which has the potential to injure the crop to be sprayed with BASAGRAN.

ENVIRONMENTAL HAZARDS

Do not apply directly to water.

Do not contaminate water by cleaning of equipment or disposal of wastes.

STORAGE AND DISPOSAL

Do not allow product to freeze.

Do not contaminate water, food, or feed by storage or disposal.

Wastes resulting from the use of this product may be disposed of on site or at an approved waste disposal facility.

Triple rinse container (or equivalent). Then offer for recycling or reconditioning, or puncture and dispose of in a sanitary landfill, or by incineration, or, if allowed by state and local authorities, by burning. If burned, stay out of smoke.

Do not reuse empty container.

FOR DIRECTIONS FOR USE—SPECIFIC CROPS—SEE FOLLOWING PAGES.

SOYBEANS—DIRECTIONS FOR USE

Applications of BASAGRAN should be made when weeds are small and actively growing and before weeds reach the maximum size listed in the **Application Rate Table for Soybeans.** Such applications generally correspond to the soybean growth stages of unifoliate to two expanded trifoliate leaves. Soybeans are tolerant to BASAGRAN at all stages of growth. Slight yellowing, bronzing, speckling, or burning of leaves may occur under certain conditions. Soybean plants generally outgrow this condition within 10 days.

RESTRICTIONS AND LIMITATIONS

Do not apply more than a total of 4 pints of BASAGRAN per acre in one season.

Application Rate Table for Soybeans

For Split Application to Selected Weeds, See Section Below

Weeds Controlled	Application Rates for Weed Growth Stages: 1½ Pts. Per Acre		2 Pts. Per Acre	
	Leaf Stage	Max. Height	Leaf Stage	Max. Height
Balloonvine	2–4	2″	4–6	3″
Beggarticks	up to 6	6″	6–8	8″
Bristly Starbur	Not Recommended		4–6	3″
Cocklebur	2–6*	6″	6–10	10″
Coffee Senna	Not Recommended		up to 1 pinnate **	2″
Common Lambsquarters†	Not Recommended		4–8 **	2″
Common Purslane	up to 4	1″	4–6	2″
Common Ragweed	Not Recommended		4–6 **	3″
Dayflower	up to 6	4″	6–10	8″
Devilsclaw	Not Recommended		up to 6 **	3″
Galinsoga	Not Recommended		cotyledon to 6 **	2″
Giant Ragweed‡	Not Recommended		up to 4	6″
Jimsonweed	up to 6	6″	6–10	10″
Ladysthumb	up to 6	6″	6–10	10″
Pennsylvania Smartweed	up to 6	6″	6–10	10″
Prickly Sida or Teaweed	up to 6	3″	6–8	4″
Redweed	4–6	6″	6–10	8″
Sesbania	Not Recommended		3–5 **	3″
Shepherdspurse†‡	up to 6	4″	6–10	8″
Spurred Anoda	up to 6	3″	6–8	4″
Tropic Croton	up to 2	2″	2–4	4″
Velvetleaf***	up to 4 **	2″	4–6 **	5″
Venice Mallow	up to 6	2″	6–10	4″
Wild Buckwheat	up to 4	3″	4–6	5″
Wild Mustard	up to 6	4″	6–10	8″

Wild Poinsettia	**		**	
	2–4	4″	4–8	6″
Wild Sunflower	up to 4	5″	4–6	8″

For additional weeds see Special Directions section following.

* Do not treat earlier than leaf stage shown and do not count cotyledon leaves.
** Add oil concentrate according to the DIRECTIONS FOR USE—All Crops.
*** Add 28% Nitrogen solution according to the Special Directions for Other Weed Problems (see page 5)° or add oil concentrate according to DIRECTIONS FOR USE—All Crops
† Control may be partial or inconsistent.
‡ If after the first application a second weed flush develops, re-treat according to this rate table.
†‡ Do not treat rosette before seed stalk appears.
° Not applicable in California.

Split Application in Soybeans
(One Pint Plus One Pint Per Acre)

Apply the first pint of Basagran to weeds indicated in the application table presented below, before weeds reach the maximum size or leaf stage indicated. Make a second application of 1 pint 10 to 14 days after the first application.

APPLICATION TABLE

Weeds Controlled	Leaf Stage	Maximum Height
Cocklebur	2-4*	4″
Jimsonweed	Up to 4	4″
Pennsylvania Smartweed	Up to 4	4″
Prickly Sida or Teaweed	Up to 4	2″
Velvetleaf**	Up to 3	2″
Venice Mallow	Up to 4	2″
Wild Mustard	Up to 4	2″
Wild Sunflower	Up to 2	3″

* Do not treat earlier than leaf stage shown and do not count cotyledon leaves.
** Except in California, add 28% nitrogen solution according to the Special Directions for other weed problems. Or add oil concentrate according to the DIRECTIONS FOR USE—All Crops.

SPECIAL DIRECTIONS FOR OTHER WEED PROBLEMS IN SOYBEANS

Annual Morningglories

South: (AL, AR, FL, GA, LA, MS, NC, OK, SC, TN, TX, VA). To control smallflower and cypressvine morningglories apply a single application of either 1½ pints of BASAGRAN per acre to plants not larger than 4 true leaves and 4 inches in height, OR 2 pints of BASAGRAN per acre to plants not larger than 6 true leaves and 6 inches in height.

To control palmleaf, pitted, tall (common), entireleaf, purple moonflower, and ivyleaf morningglories, apply 1½ pints of BASAGRAN per acre to plants not larger than 4 true leaves and 4 inches in height (14 to 18 days after morningglory emergence). Make a second application at the same rate 5 to 14 days later.

All states other than the South (see above): Apply 2 to 3 pints of BASAGRAN per acre to annual morningglories not larger than 4 true leaves. Control may be partial or inconsistent. Add oil concentrate to the spray solution of BASAGRAN/water, according to the DIRECTIONS FOR USE—All Crops.

Because morningglories grow very rapidly, it is important to watch the growth stage carefully and to be certain that BASAGRAN is applied to morningglories before they exceed the maximum size recommended on this label.

Canada Thistle

Apply 2 pints of BASAGRAN per acre when plants are from 8 inches tall to the bud stage. Make a second application at the same rate 7 to 10 days later.

Yellow Nutsedge

South: (AL ,AR ,FL, GA, LA, MS, NC, OK, SC, TN, TX, VA). Two applications are preferred for best results. Apply 1½ to 2 pints of BASAGRAN per acre when plants are 6 to 8 inches tall. If needed, make a second application at the same rate 7 to 10 days later.

All states other than the South (see above): Two applications are preferred for best results. Apply 1½ to 2 pints of BASAGRAN per acre when plants are 6 to 8 inches tall. If needed, make a second application at the same rate 7 to 10 days later. Add oil concentrate to the spray solution of BASAGRAN/water for each application according to the DIRECTIONS FOR USE—All Crops.

Field and Hedge Bindweed in KY, IL, IN, MI, OH *only.*
For suppression of field and hedge bindweed, apply 2 to 3 pints of BASAGRAN per acre when vines are a maximum of 10 inches long. Add oil concentrate to the spray solution of BASAGRAN/ water, according to the DIRECTIONS FOR USE—All Crops.

Late Cocklebur Rescue Treatment

This treatment is intended to provide only partial control of cocklebur in the event early postemergence treatments were not made. Very thorough spray coverage is essential. Apply a single application of 2 to 3 pints of BASAGRAN per acre to plants up to 24 inches tall or, for best results, apply 1½ pints of BASAGRAN per acre to plants up to 24 inches tall and repeat 10 to 14 days later.

Velvetleaf-Addition of 28% Urea Ammonium Nitrate (UAN) Solution*

In soybeans, 28% UAN solution (commonly referred to as 28% nitrogen solution) may be added to the spray tank in place of oil concentrate for improved control of velvetleaf. The 28% nitrogen solution is an agricultural grade fertilizer used by local dealers for agricultural applications. It may be added to the tank with Basagran plus Blazer when velvetleaf is the primary target weed. However, Basagran plus 28% nitrogen solution will not provide adequate control of common ragweed and common lambsquarters. If these weeds are present in addition to velvetleaf, then Basagran plus oil concentrate should be used.

With the addition of 28% nitrogen solution to Basagran on soybeans, a slight leaf burn may occur, but the new growth is normal and crop vigor is not reduced. Refer to your supplier of Basagran for information concerning successful local experience prior to using 28% nitrogen solution. Do not use brass or aluminum nozzles when spraying Basagran and 28% nitrogen solution.

Do not include Oil Concentrate with 28% nitrogen solution.

Do not add 28% nitrogen solution to Basagran for use on corn, sorghum, rice, peanuts, beans, peas, mint, or turf.

RATE OF 28% NITROGEN SOLUTION:
GROUND APPLICATION—1 gallon/acre
AIR APPLICATION-NOT RECOMMENDED

APPLICATIONS OF BASAGRAN®, POAST® AND BLAZER® 2L HERBICIDES FOR POSTEMERGENCE BROADLEAF AND GRASS WEED CONTROL IN SOYBEANS

GENERAL INFORMATION

For postemergence broadleaf and grass weed control in soybeans, tank mix or separate applications of BASAGRAN, Poast and/or Blazer 2L may be made. If broadleaf and grass weeds (see page 6) are present, all at the stage of growth recommended for treatment on the labels for BASAGRAN, BASAGRAN + Blazer 2L and Poast, then two options are available to control both broadleaf and grass weeds.

Option 1: Separate applications of BASAGRAN or BASAGRAN + Blazer 2L preceded or followed by Poast.

Applications of BASAGRAN or BASAGRAN tank mixed with Blazer 2L can be preceded or followed by Poast to obtain broad spectrum control of weeds listed on the respective product labels (refer to this label, page 6 and the labels for Poast, and BASAGRAN + Blazer 2L tank mix). Also refer to these product labels for timing, rate and other information for ground and aerial applications.

When BASAGRAN is applied alone or tank mixed with Blazer 2L, add a nonphytotoxic oil concentrate to the spray solution when recommended on the label for BASAGRAN or BASAGRAN + Blazer 2L.

When Poast is applied *always* add a nonphytotoxic oil concentrate to the spray solution at the rate of 2 pints per acre for ground and aerial applications.

For best results when making separate applications, a minimum period of time is recommended between applications, depending upon their order, according to the table below.

Order of Application		Minimum Time Between Applications
Product(s) Applied First	**Product(s) Applied Second**	
BASAGRAN	Poast	24 hours
BASAGRAN + Blazer 2L	Poast	7 days
Poast	BASADRAN or BASAGRAN + Blazer 2L	24 hours

Option 2: Tank Mix of Basagran with Poast and/or Blazer 2L. A tank mix of Basagran + Poast or Basagran + Poast + Blazer 2L may be made.

If all weeds to be controlled are not at the correct stage of growth for treatment at the same time, or if perennial grasses and certain other annual grasses listed below are a problem, then separate applications must be made. See **Option 1:** Separate applications.

Do not tank mix with Poast when grasses to be controlled include rhizome johnsongrass, quackgrass, bermudagrass, wirestem muhly, volunteer corn, shattercane, volunteer cereals, wild oats, red rice or itchgrass since unsatisfactory control will probably result. Separate applications should be made when these weeds are present.

Thorough coverage of actively growing weeds is essential. Large crop-and-weed leaf canopies shelter smaller weeds and can prevent adequate spray coverage. Soybeans are tolerant to the above tank mixes; however, under certain conditions soybeans may burn, crinkle and bronze. Soybeans at all stages of growth are tolerant to Poast.

TANK MIXES—Guide to weeds and grasses controlled.

BASAGRAN + Poast

BASAGRAN + Blazer 2L

BASAGRAN + Poast + Blazer 2L

Use the following chart as a guide to determine broadleaf weeds and grasses controlled by BASAGRAN alone and various tank mixes with BASAGRAN.

BASAGRAN controls the broad spectrum of weeds listed below:

BASAGRAN

Balloonvine	Ladysthumb
Beggarticks	Pennsylvania Smartweed
Bristly Starbur	Prickly Sida or Teawood
*Canada Thistle	Redweed
Cocklebur	Shepherdspurse
Coffee Senna	Smallflower Morningglory
Common Lambsquarters	Spurred Anoda
Common Purslane	Tropic Croton
Common Ragweed	Velvetleaf
Cypressvine Morningglory	Venice Mallow
Dayflower	Wild Buckwheat
Devilsclaw	Wild Mustard
Galinsoga	Wild Poinsettia
Giant Ragweed	Wild Sunflower
Jimsonweed	Yellow Nutsedge

* Requires two applications of BASAGRAN in accordance with this label.

Grass weeds contolled by tank mixing BASAGRAN with:

Poast	**Blazer 2L**
Wild Proso Millet	Common Ragweed
Barnyardgrass	Redroot Pigweed
Broadleaf Signalgrass	Smooth Pigweed
Fall Panicum	Black Nightshade
Giant Foxtail	Sesbania
Green Foxtail	Morningglories
Yellow Foxtail	Crotalaria
Seedling Johnsongrass	
Junglerice	
Red Sprangletop	
Texas Panicum	
Witchgrass	
Wooly Cupgrass	
Goosegrass	
Large Crabgrass	
Smooth Crabgrass	

TANK MIX OF BASAGRAN WITH POAST AND/OR BLAZER 2L

Include BASAGRAN, Poast and/or Blazer 2L in the tank mix when required according to the weed problem present. Follow rate recommendations indicated in the table below. All tank mixes with Poast must include 2 pints per acre of an oil concentrate. Do not substitute 28% nitrogen solution for oil concentrate when Poast is included in tank.

Application Rate Table

Tank Mix Combination	Product				
	Basagran	**Poast***	**Blazer 2L****	**Oil Concentrate**	**28% Nitrogen Solution**
Basagran + Poast	1½–2 pints/A according to weed species and size given on this label.	1½ pints/A***		2 pints/A	Not Recommended
Basagran + Blazer 2L	1½–2 pints/A according to weed species and size given on this label.		½ pint/A	2 pints/A	Not Recommended
			1 pint/A	† or	1 Gallon Acre††
Basagran + Poast + Blazer 2L	1½–2 pints/A according to weed species and size given on this label.	1½ pints/A***	½ or 1 pint/A	2 pints/A	Not Recommended

* The rate of Poast recommended in the tank mix is 50% greater than the rate of Poast used alone.

** See Time and Rate of Application for Blazer in Tank Mix for recommended rate, this page. Blazer 2L may also be included in the tank mix at rate up to 2 pints/acre; however, this will increase the severity and/or frequency with which soybean injury is observed.

*** For the control of wild proso millet only, include Poast in the tank mix at the rate of ¾ pints/A.

† **Addition of Oil Concentrate** (Recommended in all states except AL, AR, FL, GA, LA, MS, OK, TX): Add oil concentrate to the tank mix according to recommendations in the **Application Rate Table for BASAGRAN.** (Page 4)

†† Addition of 28% nitrogen solution as an alternate to oil concentrate. See Special Directions for Other Weed Problems; (Page 4). Do not use 28% nitrogen solution in California.

TIME OF APPLICATION

The timing of all applications of BASAGRAN and Poast should be in accordance with the weed growth stages indicated on each label and when weeds are actively growing. When Blazer 2L is included in the tank mix, the timing should be in accordance with the weed growth stages indicated in the table below and when weeds are actively growing. If weeds are not at the correct stage of growth for treatment at the same time or if perennial grasses and certain other annual grasses (see page 5) are a problem, then separate applications should be made. Delay in application which permits weeds to exceed the maximum size stated will result in inadequate control.

Time and Rate of Application for Blazer 2L in Tank Mix

Weeds Controlled	Leaf Stage	Max. Height	Blazer 2L Rate Per Acre	Oil Concentrate Rate Per Acre
Common Ragweed	Up to 10	6"	1 Pint	—
Black Nightshade	Up to 6	2"	1 Pint	—
Morningglories	Up to 4	4"	1 Pint	—
Crotalaria	Up to 6	6"	1 Pint	—
Sesbania	Up to 4 Pinnate	6"	1 Pint	—
Tall Waterhemp	Up to 6	3"	1 Pint	—
Redroot Pigweed	Up to 6	3"	1 Pint	—
Redroot Pigweed	Up to 6	3"	½ Pint	2 Pints
Smooth Pigweed	Up to 6	3"	1 Pint	—
Smooth Pigweed	Up to 6	3"	½ Pint	2 Pints

WATER VOLUME AND SPRAY PRESSURE

Ground equipment: For the tank mix of BASAGRAN + Poast and/or BASAGRAN + Poast + Blazer 2L, use 20 gallons of total spray solution per acre (broadcast basis) and a minimum of 40 psi pressure. Use standard high pressure hollow cone or flat fan nozzles spaced 20 inches apart. Do not use flood or whirl chamber nozzles.

Air equipment: Use a minimum of 5 gallons of total spray solution per acre; except when Blazer 2L is included in the tank mix, use a minimum of 10 gallons of total spray solution per acre.

MIXING

Fill the spray tank half full with water and add the recommended amount of product in the following order—BASAGRAN, Blazer 2L, Poast, oil concentrate—while the agitator is running. Then add the remaining quantity of water.

RESTRICTIONS AND LIMITATIONS (PARTIAL LIST) FOR TANK MIX OR SEPARATE APPLICATIONS

Do not apply if soybeans show injury (leaf phytotoxicity and/or plant stunting) produced by any other prior herbicide treatment because this injury may be enhanced and/or prolonged.

Do not make applications to grasses or broadleaf weeds under stress, such as stress due to lack of moisture, mechanical injury or unseasonably cold weather, since unsatisfactory control will result.

Do not apply to soybeans that have been subject to stress conditions such as hail damage, flooding, drought or unseasonably cold or widely fluctuating temperatures because injury may result.

Do not apply Blazer 2L within 50 days of harvest and do not apply Poast within 90 days of harvest.

Do not graze treated soybean fields and do not feed treated soybean forage, ensilage or hay to livestock (see labels for Blazer 2L and Poast).

TANK MIX WITH 2,4-DB FOR THE POSTEMERGENCE CONTROL OF ENTIRELEAF, TALL (COMMON), AND IVYLEAF (HAIRY LEAF) MORNINGGLORIES IN SOYBEANS

These directions are intended to provide the user of BASAGRAN with instructions for tank mixing with 2,4-DB (such as Butyrac® 200 herbicide or Butoxone® 200 herbicide) to control entireleaf, tall (common), and ivyleaf (hairy leaf) morningglories.

The tank mix is effective partly through contact action. Therefore, weeds must be thoroughly covered with spray.

Large crop-and-weed leaf canopies shelter smaller weeds and prevent adequate spray coverage.

TIME AND RATE OF APPLICATION

Apply the tank mix of BASAGRAN and 2,4-DB broadcast when weeds are actively growing and before entireleaf, tall (common), and ivyleaf (hairy leaf) morningglory vines are a maximum of 6 inches long in the South (AL, AR, FL, GA, LA, MS, NC, OK, SC, TN, TX, VA) and a maximum of 10 inches long in all other states.

Delay in application which permits weeds to exceed the maximum size stated will result in inadequate control.

To the label rate of BASAGRAN selected according to the other weed species present (see the **Application Rate Table for Soybeans**) add 2 fluid ounces per acre of Butyrac 200 or Butoxone 200 for annual morningglories. For other amine formulations of 2,4-DB add a quantity of product which contains 0.03 pounds of 2,4-DB acid equivalent per acre.

Water Volume and Spray Pressure (Refer to section entitled **Directions For Use—All Crops** for additional information.)

Ground equipment: For broadcast application, use a minimum of 20 gallons of total spray solution per acre and 40 psi pressure with flat fan or hollow cone nozzles spaced 20 inches apart.

Mixing: Fill the spray tank half full with water and add the recommended amount of BASAGRAN and 2,4-DB while the agitator is running, then finish filling. Then add the remaining quantity of water.

RESTRICTIONS AND LIMITATIONS FOR TANK MIX WITH 2,4-DB

Use only amine formulations of 2,4-DB.

Do not apply to or allow drift to any other adjacent crop.

Do not add oil or any other additives (including 28% nitrogen solution) to tank mix with 2,4-DB.

Do not apply tank mix if soybeans show injury (leaf phytotoxicity and/or plant stunting) produced by any other prior herbicide treatment because this injury may be enhanced and/or prolonged.

Do not apply tank mix during prolonged periods of drought or during unseasonably cold weather, as unsatisfactory weed control may result.

Rainfall soon after application (within 8 hours) may nullify the effectiveness of the tank mix.

Do not apply tank mix to soybeans that have been subject to stress conditions such as hail damage, flooding, drought, injury from other herbicides, or unseasonably cold or widely fluctuating temperatures because injury may result.

Do not apply more than 1 application of the tank mix per season.

Do not apply within 60 days of harvest (see label for 2,4-DB).

The use of this tank mix will cause soybean foliage injury (such as burning, bronzing or crinkling) and may reduce yields.

Do not use this tank mix on soybeans that show symptoms of disease such as phytophthora root rot (see label for 2,4-DB).

Observe all applicable directions, restrictions and precautions on labeling of both products used in this mixture.

CORN, GRAIN SORGHUM—DIRECTIONS FOR USE

Applications of BASAGRAN should be made when weeds are small and actively growing and before weeds reach the maximum size listed in **Application Rate Table for Corn, Grain Sorghum**. Such applications generally correspond to the crop growth stages of one to five leaves. Corn is tolerant to BASAGRAN at all stages of growth. Grain sorghum is tolerant to BASAGRAN at all stages of growth up to and including early boot stage. Very slight leaf-speckling of corn and grain sorghum may occur but plants generally outgrow this condition within 10 days. Corn types included are field, sweet, and popcorn; and corn grown for seed or silage.

RESTRICTIONS AND LIMITATIONS

Do not apply more than a total of 4 pints per acre in one season in corn, or 2 pints per acre in one season in grain sorghum.

Seed producers should consult the seed company regarding tolerance of seed production inbred lines to BASAGRAN.

Do not apply to grain sorghum that is heading or blooming.

California Only: Not recommended for control of yellow nutsedge in corn or grain sorghum.

Application Rate Table for Corn, Grain Sorghum

Weeds Controlled	Application Rates for Weed Growth Stages			
	$1\frac{1}{2}$ Pts. Per Acre		2 Pts. Per Acre	
	Leaf Stage	Max. Height	Leaf Stage	Max. Height
Beggarticks	up to 6	6″	6–8	8″
Bristly Starbur	up to 4	2″	4–6	3″
Cocklebur	2–6*	6″	6–10	10″
Common Lambsquarters†	Not Recommended		4–8 **	2″
Common Ragweed	Not Recommended		4–6 **	3″
Dayflower	up to 6	4″	6–10	8″
Galinsoga	Not Recommended		cotyledon to 6 **	2″
Giant Ragweed‡	Not Recommended		up to 4	6″
Jimsonweed	up to 6	6″	6–10	10″
Ladysthumb	up to 6	6″	6–10	10″
Pennsylvania Smartweed	up to 6	6″	6–10	10″
Prickly Sida or Teaweed	up to 6	3″	6–8	4″
Spurred Anoda	up to 6	3″	6–8	4″
Tropic Croton	up to 2	2″	2–4	4″
Velvetleaf	up to 4 **	2″	4–6 **	5″
Venice Mallow	up to 6	2″	6–10	4″
Wild Buckwheat	up to 4	3″	4–6	5″
Wild Mustard	up to 6	4″	6–10	8″
Wild Sunflower	up to 4	5″	4–6	8″

For additional weeds see Special Directions section following.

* Do not treat earlier than leaf stage shown and do not count cotyledon leaves.

** Add oil concentrate according to the DIRECTIONS FOR USE—All Crops. In California, add nonphytotoxic oil (containing 2-3% emulsifier) to the spray solution of BASAGRAN/water for each application by ground equipment at the rate of ½ gallon per acre in coastal valleys and 1 gallon per acre in central valleys. Add nonphytotoxic oil to the spray solution for each application by air equipment at the rate of 1% by volume (1 gallon per 100 gallons spray solution). The oil should have an unsulphonated residue rating of 90% or above.

† Control may be partial or inconsistent.

‡ If after the first application a second weed flush develops, re-treat according to this rate table (corn only).

SPECIAL DIRECTIONS FOR OTHER WEED PROBLEMS IN CORN

Annual Morningglories

South: (AL, AR, FL, GA, LA, MS, NC, OK, SC, TN, TX, VA). To control smallflower and cypressvine morningglories apply a single application of either 1½ pints of BASAGRAN per acre to plants not larger than 4 true leaves and 4 inches in height OR 2 pints of BASAGRAN per acre to plants not larger than 6 true leaves and 6 inches in height.

To control palmleaf, pitted, tall (common), entireleaf, ivyleaf and purple moonflower morningglories, apply 1½ pints of BASAGRAN per acre to plants not larger than 4 true leaves and 4 inches in height (14 to 18 days after morningglory emergence). Make a second application at the same rate 5 to 14 days later.

All states other than the South (see above): Apply 2 to 3 pints of BASAGRAN per acre to annual morningglories not larger than 4 true leaves. Control may be partial or inconsistent. Add oil concentrate to the spray solution of BASAGRAN/water, according to the DIRECTIONS FOR USE—All Crops.

Because Morningglories grow very rapidly, it is important to watch the growth stage carefully and to be certain that BASAGRAN is applied to morningglories before they exceed the maximum size recommended on this label.

Canada Thistle

Apply 2 pints of BASAGRAN per acre when plants are from 8 inches tall to the bud stage. Make a second application at the same rate 7 to 10 days later.

Yellow Nutsedge

South: (AL, AR, FL, GA, LA, MS, NC, OK, SC, TN, TX, VA). Two applications are preferred for best results. Apply 1½ to 2 pints of BASAGRAN per acre when plants are 6 to 8 inches tall. If needed, make a second application at the same rate 7 to 10 days later.

All states other than the South (see above): Two applications are preferred for best results. Apply 1½ to 2 pints of BASAGRAN per acre when plants are 6 to 8 inches tall. If needed, make a second application at the same rate 7 to 10 days later. Add oil concentrate to the spray solution of BASAGRAN/water for each application according to the DIRECTIONS FOR USE—All Crops.

Field and Hedge Bindweed in KY, IL, IN, MI, OH *only.*

For suppression of field and hedge bindweed, apply 2 to 3 pints of BASAGRAN per acre when vines are a maximum of 10 inches long. Add oil concentrate to the spray solution of BASAGRAN/water, according to the DIRECTIONS FOR USE—All Crops.

Late Cocklebur Rescue Treatment

This treatment is intended to provide only partial control of cocklebur in the event early postemergence treatments were not made. Very thorough spray coverage is essential. Apply a single application of 2 to 3 pints of BASAGRAN per acre to plants up to 24 inches tall or, for best results, apply 1½ pints of BASAGRAN per acre to plants up to 24 inches tall and repeat 10 to 14 days later.

SPECIAL DIRECTIONS FOR OTHER WEED PROBLEMS IN GRAIN SORGHUM

Annual Morningglories

Apply 2 pints of BASAGRAN per acre to annual morningglories not larger than 4 true leaves. Control may be partial or inconsistent. Add oil concentrate to the spray solution of BASAGRAN/water, according to the DIRECTIONS FOR USE—All crops.

Because morningglories grow very rapidly, it is important to watch the growth stage carefully and to be certain that BASAGRAN is applied to morningglories before they exceed the maximum size recommeded on this label.

Canada Thistle

Apply 2 pints of BASAGRAN per acre when plants are from 8 inches tall to the bud stage. Control may be partial or inconsistent.

Yellow Nutsedge

Apply 1½ to 2 pints of BASAGRAN per acre when plants are 6 to 8 inches tall. Add oil concentrate to the spray solution of BASAGRAN/water, for each application according to the DIRECTIONS FOR USE—All Crops. Control may be partial or inconsistent.

TANK MIX WITH ATRAZINE AND OIL CONCENTRATE FOR POSTEMERGENCE APPLICATIONS IN CORN AND GRAIN SORGHUM

The tank mix of BASAGRAN/atrazine/oil concentrate effectively controls a broad spectrum of broadleaf weeds included on the labeling of both products. For the control of annual morningglories, Canada thistle and yellow nutsedge, refer to the section entitled **Special Directions for Other Weed Problems in Corn or Grain Sorghum.**

Atrazine products compatible with BASAGRAN include AAtrex® 80W, AAtrex® 4L, and AAtrex® Nine-O herbicides. Refer to the respective atrazine labels for additional directions and limitations. *Always add oil concentrate* according to the **Directions For Use—All Crops.**

Mixing and spray equipment: Use intake, in-line, or nozzle screens no finer than 50 mesh. Fill tank of a thoroughly clean sprayer half to two-thirds full of clean water. Start agitation. Add atrazine and allow to wet and mix thoroughly. Maintain agitation and add BASAGRAN, allow to mix. Add oil concentrate and remaining quantity of water and mix thoroughly. Maintain constant agitation during application. Avoid allowing the mixture to stand overnight. Always clean sprayer thoroughly immediately after use by flushing the system with water and a strong detergent. Do not allow cleaning water to contaminate any streams or ponds.

Time and rate of application: Tank mix applications should be made when weeds are small and actively growing and before weeds reach the maximum size listed in the **Application Rate Table for Tank Mix of BASAGRAN + Atrazine for Corn and Grain Sorghum.** Such applications generally correspond to the crop growth stages of one to five leaves.

Corn is tolerant to the tank mix at all stages of growth. Grain sorghum is tolerant to the tank mix at all stages of growth up to and including early boot stage.

Very light leaf speckling may occur in corn and grain sorghum, but plants generally outgrow this condition within 10 days.

Refer to the Conversion Table below for application rates depending on formulation. A cultivation may be necessary if all weeds are not controlled or if regrowth of weeds occurs.

RESTRICTIONS AND LIMITATIONS FOR TANK MIX WITH ATRAZINE

Do not use tank mix when crop is under stress from prolonged cold, wet weather, poor fertility, or other factors or when crop is wet and succulent from recent rainfall as crop injury may occur.

Seed producers should consult the seed company regarding tolerance of seed production inbred lines to tank mix.

Do not apply to grain sorghum that is heading out or blooming.

Do not make more than one application of tank mix per season.

Do not graze treated area or feed treated forage to livestock for 21 days following application (see label for AAtrex).

Acreage Conversion Table

Tank Mix Rate Recommendation (lb ai/A)*	Amount of Formulated Product									
	BASAGRAN	Atrazine (AAtrex)								
	1 Acre	1 Acre			10 Acres			50 Acres		
	Pts.	80W lbs.	Nine O lbs.	4L or 4LC pts.	80W lbs.	Nine O lbs.	4L or 4LC pts.	80W lbs.	Nine O lbs.	4L or 4LC pts.
½ + ½	1	5/8	6/10	1	6¼	6	10	31½	30	50
¾ + ¾	1½	1	9/10	1½	10	9	15	50	45	75

* According to weed growth stage indicated in table below.

Application Rate Table for Tank Mix of BASAGRAN + Atrazine for Corn and Grain Sorghum

Weeds Controlled	Application Rates For Weed Growth Stages			
	½ + ½ lb. ai/A*		¾ + ¾ lb. ai/A*	
	Leaf Stage	Max. Height	Leaf Stage	Max. Height
Cocklebur	2–6**	4″	6–10	8″
Common Lambsquarters†	4–8	2″	8–12	4″
Common Ragweed	up to 4	2″	4–6	3″
Giant Ragweed	up to 4	4″	4–6	6″
Jimsonweed	up to 6	4″	6–10	8″
Ladysthumb	up to 6	4″	6–10	8″
Penna. Smartweed	up to 6	4″	6–10	8″
Redroot Pigweed	4–8	2″	8–12	4″
Smooth Pigweed	4–8	2″	8–12	4″
Velvetleaf	up to 4	2″	4–6	5″
Venice Mallow	up to 6	2″	6–10	4″
Wild Buckwheat	up to 4	3″	4–6	5″
Wild Mustard	up to 6	4″	6–10	8″
Wild Sunflower	up to 4	4″	4–6	8″

Other weeds: Other weeds listed on the label for BASAGRAN at the ¾ pound rate will also be controlled with the ¾ plus ¾ pound BASAGRAN/atrazine tank mix. Refer to page 8.

* Refer to Conversion Table for recommended rate of formulated product per acre. Always add oil concentrate according to the DIRECTIONS FOR USE—All Crops. In California, add nonphytotoxic oil (containing emulsifier) to the spray solution of BASAGRAN/ water for each application by ground equipment at the rate of ½ gallon per acre in coastal valleys and 1 gallon per acre in central valleys. Add nonphytotoxic oil to the spray solution for each application by air equipment at the rate of 1% by volume (1 gal. per 100 gallons spray solution). The oil should have an unsulphonated residue rating of 90% or above.

** Do not treat earlier than leaf stage shown and do not count cotyledon leaves.

† Control may be partial or inconsistent.

RICE—DIRECTIONS FOR USE

Make postemergence applications of Basagran early, before weeds exceed the maximum size listed in the **Application Rate Table for Rice.**

Early application produces the most beneficial effect on crop yields, allows use of the lower rate (depending on weed species), and makes it easier to obtain thorough spray coverage. Delay in application which permits weeds to exceed the maximum size for a given rate will result in inadequate control.

BASAGRAN has no adverse effect on rice when used according to directions and may be used on the first and second (ratoon) crops. If grasses are a problem, use propanil in tank mixture with BASAGRAN (see below).

For optimal coverage when applying BASAGRAN by air in rice, orient all nozzles straight down. For additional aerial application information, refer to **Directions For Use—All Crops.**

Alternate flooding culture: In Texas, Louisiana, Arkansas and Mississippi, weed growth stages generally correspond to rice that is tillering (stooling) and occur prior to the permanent flood. Application of BASAGRAN *must* be made when there is no water on the field and 24 hours or more prior to flooding. If BASAGRAN cannot be applied until after flooding, see directions under **Continuous flooding culture.**

Continuous flooding culture: In California and in other states using continuous flooding culture or when treating after permanent flood, treatment should be made only when weeds are above the surface of the water. Weeds submerged at the time of application are not adequately controlled.

For early treatment, water may be partly or completely drained to expose more weed growth to spray applications of BASAGRAN. Do not raise water level for at least 24 hours after application or unsatisfactory control may result. Do not use ground equipment for applications on flooded fields because splashing will wash BASAGRAN off weed leaf surfaces and ineffective control may result.

Addition of a nonphytotoxic oil or oil concentrate in California only: In California, add nonphytotoxic oil (containing emulsifier) to the spray solution of BASAGRAN/water for each application by ground or air at a range of 1 to 5% by volume (1–5 gallons/100 gallons spray solution). On days when the maximum temperature is predicted below 90°F, use the upper range of oil (i.e. 4 to 5%). On days when the maximum temperature is predicted to exceed 90°F, use the lower range of oil (i.e. 1 to 2%). The oil should have an unsulphonated residue rating of 90% or above.

Or, in California, add oil concentrate at the maximum rate of 2 pints per acre for ground and aerial application. Refer to section entitled **Addition of Oil Concentrate to Spray Tank** for additional information.

California only: Avoid applications of Basagran (a) during cold weather (day temperature below 75°F and night temperatures below 55°F for 2 to 5 days) as weed control may be nullified, or (b) when rapid temperature drops are forecast.

RESTRICTIONS AND LIMITATIONS

Rice straw may be fed to livestock.

Do not apply BASAGRAN to rice with ground equipment when field is flooded because splashing will wash BASAGRAN off weed leaf surfaces and ineffective control may result.

Do not apply more than 6 pints of BASAGRAN per acre in one season. (Maximum of 4 pints per acre in first crop and 2 pints per acre in second [ratoon] crop.)

In California, do not apply Basagran by air when temperature exceeds 90°F as reduced weed control may result.

TANK MIXTURE WITH PROPANIL

Use a tank mix of BASAGRAN/propanil by ground or air for the control of mixed populations of grasses and broadleaf weeds listed as susceptible on the two product labels. Prepare tank mixture by adding BASAGRAN to half the final volume of water with agitator running. Then add propanil and bring mix to final volume. Agitation must be continuous from time of mixing through spraying.

Apply BASAGRAN at the rate of 1½ pints per acre plus either 3 quarts (4 lbs./gal.) or 4 quarts (3 lbs./gal.) of propanil formulation* in the spray volume specified on this labeling.

Apply this tank mixture only to drained fields.

Do not use propanil on second crop (ratoon) rice.

When applying tank mix of BASAGRAN/propanil by air, orient all nozzles straight back in accordance with the propanil label.

Do not add nonphytotoxic oil or oil concentrate to the tank mix of BASAGRAN/propanil.

Do not apply the tank mix of BASAGRAN/propanil in prohibited areas of California.

Observe all cautions and limitations on the labels of both products.

* Propanil products compatible with BASAGRAN are STAM F-34 (3 lbs./gal.); STAM M-4 (4 lbs./gal.); Helena 4 (4 lbs./gal.) and Crystal 4E (4 lbs./gal.).

Application Rate Table for Rice—Drained Fields

Weeds Controlled (All States)	Application Rates for Weed Growth Stages			
	1½ Pts. Per Acre*		2 Pts. Per Acre*	
	Leaf Stage	Max. Height	Leaf Stage	Max. Height
Cocklebur	2–10	10″	10–15	15″
Dayflower	2–10	6″	10–15	10″
Ducksalad	Not Recommended		6–10 **	6″
Gooseweed	4–6	4″	6–10	8″
Redstem	up to 6	4″	6–10	8″
Redweed	4–6	6″	6–10	8″
Smartweed	2–10	6″	10–15	10″
Spikerush	2–6	6″	6–8	8″
Water Plantains Arrowhead Common Water Plantain	Not Recommended		up to 4	7″
Yellow Nutsedge	4–6	6″	6–8	10″
Weeds Controlled (California Only)				
River Bulrush	Not Recommended		6–8†	10″–30″
Roughseed Bulrush	Not Recommended		2–4‡	10″
Smallflower Umbrellaplant	Not Recommended		4–6‡	8″ 0

* If after the first application a second weed flush develops, re-treat according to this Rate Table.
** Control may be partial or inconsistent.
† Land preparation should be such that rhizomes are thoroughly cut up so there are no more than 2 tubers per rhizome section remaining. Apply BASAGRAN when 10–15% of the river bulrush plants are flowering.
‡ Apply BASAGRAN before weeds are flowering.

Application Rate Table for Rice—Flooded Fields

Weeds Controlled (All States)	Application Rates for Weed Growth Stages			
	1½ Pts. Per Acre*		2 Pts. Per Acre*	
	Maximum Height Above Soil	Minimum Height Range Above Water Level	Maximum Height Above Soil	Minimum Height Range Above Water Level
Cocklebur	10″	3″–6″	15″	6″–10″
Dayflower	6″	3″–5″	10″	5″–8″
Ducksalad	Not Recommended		Not Recommended	
Gooseweed	Not Recommended		Not Recommended	
Redstem	4″	2″–3″	8″	4″–6″
Redweed	Not Recommended		Not Recommended	
Smartweed	6″	3″–5″	10″	5″–8″
Spikerush	Not Recommended		Not Recommended	
Water Plantains Arrowhead Common Water Plantain	Not Recommended		7″	5″–6″
Yellow Nutsedge	6″	4″–5″	10″	6″–8″
Weeds Controlled (California Only)				
River Bulrush	Not Recommended		10″–30″†	10″–24″
Roughseed Bulrush	Not Recommended		10″‡	6″–8″
Smallflower Umbrellaplant	Not Recommended		8″‡	6″–8″

* If after the first application a second weed flush develops, re-treat according to this Rate Table.
† Land preparation should be such that rhizomes are thoroughly cut up so there are no more than 2 tubers per rhizome section remaining. Apply BASAGRAN when 10–15% of the river bulrush plants are flowering.
‡ Apply BASAGRAN before weeds are flowering.

PEANUTS—DIRECTIONS FOR USE

Applications of BASAGRAN should be made when weeds are small and actively growing and before weeds reach the maximum size listed in the **Application Rate Table for Peanuts.** Such applications generally correspond to the peanut growth stages of bunching to pegging.

Peanuts are tolerant to BASAGRAN at all stages of growth, but slight leaf-speckling may occur under certain conditions (see **Restrictions and Limitations**). Peanut plants generally outgrow this condition within 10 days.

RESTRICTIONS AND LIMITATIONS

Do not apply BASAGRAN if peanuts show injury (leaf phytotoxicity and/or plant stunting) produced by any prior herbicide applications (preplant incorporated, preemergence, cracking and/or postemergence), because this injury may be enhanced and/or prolonged. In the Southeast, in-furrow treatments of insecticides/nematicides may possibly predispose the peanuts to injury from BASAGRAN.

Do not apply more than a total of 4 pints of BASAGRAN per acre in one season.

Peanut hay and forage may be fed to livestock.

Application Rate Table for Peanuts

Weeds Controlled	Application Rates for Weed Growth Stages			
	1½ Pts. Per Acre		2 Pts. Per Acre	
	Leaf Stage	Max. Height	Leaf Stage	Max. Height
Balloonvine	2–4	2″	4–6	3″

Beggarticks	up to 6	6"	6–8	8"
Bristly Starbur	up to 4	2"	4–6	3"
Cocklebur	2–6*	6"	6–10	10"
Coffee Senna	Not Recommended		up to 1 pinnate **	2"
Common Ragweed	Not Recommended		4–6 **	3"
Dayflower	up to 6	4"	6–10	8"
Devilsclaw	Not Recommended		up to 6 **	3"
Giant Ragweed	Not Recommended		up to 4	6"
Jimsonweed	up to 6	6"	6–10	10"
Ladysthumb	up to 6	6"	6–10	10"
Pennsylvania Smartweed	up to 6	6"	6–10	10"
Prickly Sida or Teaweed	up to 6	3"	6–8	4"
Spurred Anoda	up to 6	3"	6–8	4"
Tropic Croton	up to 2	2"	2–4	4"
Velvetleaf	up to 4	2"	4–6	5"
Wild Sunflower	up to 4	5"	4–6	8"

For additional weeds see Special Directions section following.

* Do not treat earlier than leaf stage shown and do not count cotyledon leaves.

** Add oil concentrate according to the DIRECTIONS FOR USE—All Crops.

SPECIAL DIRECTIONS FOR OTHER WEED PROBLEMS IN PEANUTS

Annual Morningglories

To control smallflower and cypressvine morningglories apply a single application of either 1½ pints of BASAGRAN per acre to plants not larger than 4 true leaves and 4 inches in height, OR 2 pints of BASAGRAN per acre to plants not larger than 6 true leaves and 6 inches in height.

To control palmleaf, pitted, common, entireleaf, purple moonflower and ivyleaf morningglories, apply 1½ pints of BASAGRAN per acre to plants not larger than 4 true leaves and 4 inches in height (14 to 18 days after morningglory emergence). Make a second application at the same rate 5 to 14 days later.

Because morningglories grow very rapidly, it is important to watch the growth stage carefully and to be certain that BASAGRAN is applied to morningglories before they exceed the maximum size recommended on this label.

Yellow Nutsedge

Two applications are preferred for best results. Apply 1½ to 2 pints of BASAGRAN per acre when plants are 6 to 8 inches tall. In Texas and Oklahoma use 2 pints. If needed, make a second application at the same rate 7 to 10 days later. Add oil concentrate to the spray solution of BASAGRAN/water, according to the DIRECTIONS FOR USE—All Crops.

Late Cocklebur Rescue Treatment

This treatment is intended to provide only partial control of cocklebur in the event early postemergence treatments were not made. Very thorough spray coverage is essential. Apply a single application of 2 to 3 pints of BASAGRAN per acre to plants up to 24 inches tall or, for best results, apply 1½ pints of BASAGRAN per acre to plants up to 24 inches tall and repeat 10 to 14 days later.

TANK MIX WITH 2,4-DB FOR POSTEMERGENCE CONTROL OF MORNINGGLORIES IN PEANUTS

These directions are intended to provide the user of BASAGRAN with instructions for tank mixing with 2,4-DB, such as Butyrac® 200 herbicide or Butoxone® 200 herbicide, to control annual morningglories in addition to all the other weeds listed in **Application Rate Table for Peanuts.**

The tank mix is effective partly through contact action. Therefore, weeds must be thoroughly covered with spray.

Large crop-and-weed leaf canopies shelter smaller weeds and prevent adequate spray coverage. Peanuts are tolerant to the tank mix of BASAGRAN + 2,4-DB; however, under certain conditions peanuts may have a white, bleached appearance and the leaves may be slightly elongated.

Time and Rate of Application: Apply the BASAGRAN + 2,4-DB tank mix broadcast when weeds are actively growing and before annual morningglory vines are a maximum of 10 inches long.

Delay in application which permits weeds to exceed the maximum size stated will result in inadequate control.

To the label rate (1½–2 pints/acre) of BASAGRAN selected according to the other weed species present, add 8 fluid ounces per acre of Butyrac 200 or Butoxone 200 for annual morningglories.

To the label rate (1½-2 pints/acre) of Basagran selected according to the other weed species present, add 8 fluid ounces per acre of Butyrac 200 or Butoxone 200 for annual morningglories. For other formulations of 2,4-DB use a quantity of product which contains 0.125 pounds of 2,4-DB acid equivalent per acre.

WATER VOLUME AND SPRAY PRESSURE

Ground equipment: For broadcast application, use a minimum of 20 gallons of total spray mixture per acre and 40 psi pressure with flat fan or hollow cone nozzles spaced 20 inches apart.

Mixing: Fill the spray tank half full with water and add the recommended amount of BASAGRAN + 2,4-DB while the agitator is running. Then add the remaining quantity of water.

RESTRICTIONS AND LIMITATIONS FOR TANK MIX WITH 2,4-DB

Use only amine formulations of 2,4-DB.

Do not apply to or allow drift to any other adjacent crop.

Do not add oil or any other additives to tank mix.

Do not apply tank mix if peanuts show injury (leaf phytotoxity and/or plant stunting) produced by any other prior herbicide treatment or by disease because this injury may be enhanced and/or prolonged.

Do not apply tank mix during prolonged periods of drought or during unseasonably cold weather, as unsatisfactory weed control may result.

Rainfall soon after application (within 8 hours) may nullify the effectiveness of the tank mix.

Do not apply tank mix to peanuts that have been subject to stress conditions such as hail damage, flooding, drought, injury from other herbicides, or unseasonably cold or widely fluctuating temperatures because injury may result.

Do not apply more than 2 applications of the tank mix per season.

Do not apply within 30 days of harvest in Oklahoma, Texas and New Mexico or 45 days in the Virginia-Carolina area. (See label for 2,4-DB.)

Do not feed treated peanut vines and peanut hay to livestock. (See label for 2,4-DB.)

TANK MIX WITH BLAZER® 2L HERBICIDE FOR POSTEMERGENCE APPLICATION IN PEANUTS (EXCEPT TEXAS AND OKLAHOMA)

The tank mix of BASAGRAN + Blazer 2L will control the weeds listed below in **Tank Mix Time of Application Table** in addition to all the weeds listed in the **Application Rate Table for Peanuts.**

The tank mix is effective mainly through contact action. Therefore, weeds must be thoroughly covered with spray.

Large crop-and-weed leaf canopies shelter smaller weeds and prevent adequate spray coverage. Peanuts are tolerant to the tank mix of BASAGRAN + Blazer 2L; however, under certain conditions, peanuts may burn, crinkle and bronze.

Apply the BASAGRAN + Blazer 2L tank mix broadcast when weeds are actively growing and before weeds reach the maximum size listed below. Such applications generally correspond to the peanut growth stages of bunching to pegging.

Delay in application which permits weeds to exceed the maximum size stated will result in inadequate control.

Tank Mix Time of Application Table

Additional Weeds Controlled With Tank Mix

Weeds Controlled	Maximum Height	Number of Leaves
Common Ragweed	6″	up to 10
Redroot Pigweed	3″	up to 6
Smooth Pigweed	3″	up to 6
Black Nightshade	2″	up to 6
Sesbania	6″	up to 4 pinnate
Morningglories	4″	up to 4
Crotalaria	6″	up to 6

Choose the rate of BASAGRAN (1½ or 2 pints per acre) according to the size and species of the weeds to be controlled with BASAGRAN alone (See **Application Rate Table for Peanuts**). Then add Blazer 2L at the rate of 1 pint per acre, if needed, to control the additional weed species, up the maximum size, as shown in the **Tank Mix Time of Application Table** above. Blazer 2L may also be included in the tank mix at a rate of up to 2 pints per acre; however, this will increase the severity and/or frequency with which peanut injury is observed.

Tank Mix Rate of Application Table

Product	Rate
BASAGRAN	1½-2 pints/A according to weed species and size. (See **Application Rate Table for Peanuts.**)
Blazer 2L	1 pint/A
*Surfactant	If Coffee senna, common ragweed, devilsclaw or yellow nutsedge are present, add oil concentrate according to **Directions For Use in Peanuts.** If crotalaria or sesbania are present, add Triton® AG-98 at the rate of ½ pint per 100 gallons of spray solution. Do not include Triton AG-98 when oil concentrate is added to tank mix.

* The addition of oil concentrate when tank mixing with Blazer 2L will probably increase the severity and/or frequency with which peanut injury is observed.

WATER VOLUME AND SPRAY PRESSURE

Ground equipment: Use a minimum of 20 gallons of total spray mixture per acre (broadcast basis) and 40 psi pressure with standard high pressure hollow cone or flat fan nozzles spaced 20 inches apart.

Air equipment: Use a minimum of 10 gallons of water per acre.

Mixing: Fill the spray tank half full with water and add the recommended amount of BASAGRAN + Blazer 2L while the agitator is running. Then add the remaining quantity of water.

RESTRICTIONS AND LIMITATIONS (PARTIAL LIST) FOR TANK MIX WITH BLAZER®2L HERBICIDE

Do not apply tank mix if peanuts show injury (leaf phytoxicity and/or plant stunting) produced by any other prior herbicide treatment because this injury may be enhanced and/or prolonged.

Do not apply tank mix during prolonged periods of drought or during unseasonably cold weather, as unsatisfactory weed control may result.

Do not apply tank mix to peanuts that have been subjected to stress conditions such as hail damage, flooding, drought, or unseasonably cold or widely fluctuating temperatures because injury may result.

Do not add a surfactant or oil concentrate except where specifically recommended.

Observe all applicable directions, restrictions and precautions on labeling of both products used in this mixture.

BEANS (DRY OR SUCCULENT) —DIRECTIONS FOR USE

Apply BASAGRAN early postemergence when weeds are small and actively growing and before weeds reach the maximum size listed in the **Application Rate Table for Beans.** Such weed growth stages generally correspond to bean stages of greater than one expanded trifoliate leaf.

Beans are tolerant to BASAGRAN after the first trifoliate leaf has fully expanded. Snap bean injury can be very pronounced. Even at the tolerant stages yellowing, bronzing, speckling or burning of leaves may occur under certain conditions (see **Restrictions and Limitations**). This temporary injury is generally outgrown without delaying podset or maturity or reducing yield. The use of oil with BASAGRAN may increase injury and may reduce yields.

Tolerant bean types are navy, pinto, pinks, great Northern, kidney, red, whites, cranberry, black turtle soup, small limas, large limas, and snap beans.

Application Rate Table for Beans (Dry or Succulent)

Weeds Controlled	Application Rates for Weed Growth Stages			
	1½ Pts. Per Acre		2 Pts. Acre	
	Leaf Stage	Max. Height	Leaf Stage	Max. Height
Cocklebur (PNW)	2–6*	6″	6–10	10″
Common Lambsquarters†	Not Recommended		4–8 **	2″
Common Purslane	up to 4	1″	4–6	2″
Common Ragweed	Not Recommended		4–6 **	3″
Devilsclaw	Not Recommended		up to 6 **	3″
Galinsoga	Not Recommended		cotyledon to 6 **	2″
Giant Ragweed‡	Not Recommended		2–4	6″
Hairy Nightshade***	Not Recommended		2–6	4″
Jimsonweed	up to 6	6″	6–10	10″
Ladysthumb	up to 6	6″	6–10	10″
Pennsylvania Smartweed	up to 6	4″	6–10	10″
Prickly Sida or Teaweed	up to 6	3″	6–8	4″
Shepherdspurse‡†	up to 6	4″	6–10	8″
Velvetleaf	up to 4 **	2″	4–6 **	5″
Wild Mustard (PNW)	up to 6	4″	6–10	10″
Wild Sunflower	up to 4	5″	4–6	8″

For additional weeds see Special Directions section following.

* Do not treat earlier than leaf stage shown and do not count cotyledon leaves.

** Add oil concentrate according to the DIRECTION FOR USE—All Crops. In California, add nonphytotoxic oil (containing emulsifier) to the spray solution of BASAGRAN/water for each application by ground equipment at the rate of ½ gallon per acre in coastal valleys and 1 gallon per acre in central valleys. Add nonphytotoxic oil to the spray solution for each application by air equipment at the rate of 1% by volume (1 gal. per 100 gal. spray solution).
The oil should have an unsulphonated residue rating of 90% or above. The use of oil with BASAGRAN may increase injury and may reduce yields.

*** BASAGRAN does not adequately contol black nightshade.

† Control may be partial or inconsistent.

‡ If after the first application a second weed flush develops, re-treat according to this rate table.

‡† Do not treat rosette before seed stalk appears.

PNW—See special directions for Pacific Northwest.

WESTERN IRRIGATED AREAS

In the Western irrigated areas, it may be necessary to irrigate prior to treatment with BASAGRAN to ensure that weeds are growing actively. Weeds that are growing under moisture stress are not actively growing and are not satisfactorily controlled.

Avoid application of BASAGRAN during prolonged periods of cold weather (day temperature below 75°F and night temperature below 55°F for 2 to 5 days) because weed control may be nullified.

SPECIAL DIRECTIONS FOR OTHER WEED PROBLEMS IN BEANS

Yellow Nutsedge

Two applications are preferred for best results. Apply 1 ½ to 2 pints (except Pacific Northwest) of BASAGRAN per acre when plants are 6 to 8 inches tall. If needed, make a second application at the same rate 7 to 10 days later.

Add oil concentrate to the spray solution of BASAGRAN/water for each application according to the DIRECTIONS FOR USE — All Crops.

In California, apply 2 pints of BASAGRAN per acre when plants are 6 to 8 inches tall. Make a second application at the same rate 10 to 14 days later. Add nonphytotoxic oil (containing emulsifier) to the spray solution of BASAGRAN/water for each application by ground equipment at the rate of ½ gal. per acre in coastal valleys and 1 gal. per acre in central valleys. Add nonphytotoxic oil to the spray solution for each application by air equipment at the rate of 1% by volume (1 gal. per 100 gals. spray solution). The oil should have an unsulphonated residue rating of 90% or above. The use of oil with BASAGRAN may increase injury and may reduce yields.

Canada Thistle

Apply 2 pints of BASAGRAN per acre when plants are from 8 inches tall to the bud stage. Make a second application at the same rate 7 to 10 days later.

Field and Hedge Bindweed in KY, IL, IN, MI, OH *only.*

For suppression of field and hedge bindweed, apply 2 to 3 pints of BASAGRAN per acre when vines are maximum of 10 inches long. Add oil concentrate to the spray solution of BASAGRAN/water, according to the DIRECTIONS FOR USE—All Crops.

Pacific Northwest (ID, OR, WA)

For control of cocklebur, yellow nutsedge, and wild mustard, use only the 2 pint rate.

For cocklebur, treat when plants are in the 2–10 leaf stage and a maximum height of 10 inches.

For yellow nutsedge, follow the directions indicated above using only the 2 pint rate.

For wild mustard, treat when plants are up to the 10 leaf stage and a maximum height of 10 inches.

RESTRICTIONS AND LIMITATIONS

Do not apply BASAGRAN to bean fields until beans have *at least the first trifoliate leaf fully expanded because severe crop damage may occur.*

Do not apply BASAGRAN to blackeyes grown in California or to garbanzo beans at any stage of growth, as severe crop damage may occur.

Do not apply more than a total of 4 pints of BASAGRAN per acre in one season.

In succulent beans only, do not apply BASAGRAN within 30 days of harvest.

PEAS (DRY OR SUCCULENT)—DIRECTIONS FOR USE

Apply BASAGRAN early postemergence when weeds are small and actively growing and before weeds reach the maximum size listed in the **Application Rate Table for Peas.** Such weed growth stages generally correspond to pea stages of greater than 3 pairs of leaves (or 4 nodes).

Peas are tolerant to BASAGRAN after 3 pairs of leaves (or 4 nodes) are present. Pea injury can be very pronounced. Even at the tolerant stages yellowing, bronzing, speckling or burning of leaves may occur under certain conditions (see **Restrictions and Limitations**). This temporary injury is generally outgrown without delaying podset or maturity or reducing yield.

Tolerant pea types are garden peas and Southern peas.

WESTERN IRRIGATED AREAS

In the Western irrigated areas, it may be necessary to irrigate prior to treatment with BASAGRAN to ensure weeds are growing actively. Weeds that are growing under moisture stress are not actively growing and are not satisfactorily controlled.

Avoid application of BASAGRAN during prolonged periods of cold weather (day temperature below 75°F and night temperature below 55°F for 2 to 5 days) because weed control may be nullified.

RESTRICTIONS AND LIMITATIONS

Do not apply BASAGRAN to pea fields until *peas have at least 3 pairs of leaves (or 4 nodes) because severe crop damage may occur.*

In the Southeast, in-furrow treatments of insecticides/nematicides may possibly predispose the peas to injury from BASAGRAN.

Do not apply BASAGRAN to blackeyes grown in California or to garbanzo beans at any stage of growth, as severe crop damage may occur.

Do not apply more than a total of 4 pints of BASAGRAN per acre in one season.

Do not add oil to BASAGRAN for use on peas.

Application Rate Table for Peas (Dry or Succulent)

Weeds Controlled	Application Rates for Weed Growth Stages			
	1½ Pts. Per Acre		2 Pts. Per Acre	
	Leaf Stage	Max. Height	Leaf Stage	Max. Height
Cocklebur (PNW)	2–4*	6"	6–10	10"
Common Purslane	up to 4	1"	4–6	2"
Giant Ragweed†	Not Recommended		2–4	6"
Hairy Nightshade**	Not Recommended		2–6	4"
Jimsonweed	up to 6	6"	6–10	10"
Ladysthumb	up to 6	6"	6–10	10"
Pennsylvania Smartweed	up to 6	4"	6–10	10"
Prickly Sida or Teaweed	up to 6	3"	6–8	4"
Shepherdspurse‡	up to 6	4"	6–10	8"
Wild Mustard (PNW)	up to 6	4"	6–10	10"
Wild Sunflower	up to 4	5"	4–6	8"

For additional weeds see Special Directions section following.

* Do not treat earlier than leaf stage shown and do not count cotyledon leaves.

** BASAGRAN does not adequately control black nightshade.

† If after the first application a second weed flush develops, re-treat according to this rate table.

‡ Do not treat rosette before seed stalk appears.

PNW—See special directions for Pacific Northwest.

SPECIAL DIRECTIONS FOR OTHER WEED PROBLEMS IN PEAS

Canada Thistle

Apply 2 pints of BASAGRAN per acre when plants are from 8 inches tall to the bud stage. Make a second application at the same rate 7 to 10 days later.

Pacific Northwest (ID, OR, WA)

For control of cocklebur and wild mustard, use only the 2 pint rate.

For cocklebur, treat when plants are in the 2–10 leaf stage and a maximum height of 10 inches.

For wild mustard, treat when plants are up to 10 leaf stage and a maximum height of 10 inches.

ESTABLISHED PEPPERMINT AND SPEARMINT —DIRECTIONS FOR USE

Apply BASAGRAN early postemergence when weeds are small and actively growing and before weeds reach the maximum size listed in the **Application Rate Table for Peppermint and Spearmint.**

Peppermint and spearmint are tolerant to BASAGRAN; however, some leaf-burning may occur under certain conditions, such as when plants are growing very actively and have extensive new, succulent tissue. Mint plants generally outgrow this condition within 10 days.

IRRIGATED AREAS

In irrigated areas it may be necessary to irrigate prior to treatment with BASAGRAN to ensure that weeds are growing actively. Weeds growing under drought conditions or unseasonably cold weather usually are not satisfactorily controlled.

RESTRICTIONS AND LIMITATIONS

Do not apply more than a total of 8 pints of BASAGRAN per acre in one season.

Application Rate Table for Peppermint and Spearmint

Weeds Controlled	2 Pts. Per Acre			4 Pts. Per Acre		
	Leaf Stage		Max. Height	Leaf Stage		Max. Height
Common Lambsquarters*	4–8	**	2″	—		—
Common Ragweed	4–6	**	3″	—		—
Hairy Nightshade***	2–6		4″	6–10		6″
Kochia	NA	**	4″	NA	**	6″
Ladysthumb	6–10		10″	—		—
Pennsylvania Smartweed	6–10		10″	—		—
Wild Mustard	6–10		8″	—		—

For additional weeds see Special Directions section following.

* Control may be partial or inconsistent.

** Add oil concentrate according to the DIRECTIONS FOR USE—All Crops.

*** BASAGRAN does not adequately control black nightshade.

NA—Not applicable.

SPECIAL DIRECTIONS FOR OTHER WEED PROBLEMS IN PEPPERMINT AND SPEARMINT

Yellow Nutsedge

Apply 2 pints of BASAGRAN per acre when plants are 6 to 8 inches tall. Make a second application at the same rate 7 to 10 days later. Add oil concentrate to the spray solution of BASAGRAN/water for each application according to the DIRECTIONS FOR USE—All Crops.

Canada Thistle

Apply 4 pints of BASAGRAN per acre when plants are from 8 inches tall to the bud stage. Make a second application at the same rate 7 to 10 days later.

Western Goldenrod

Apply 4 pints of BASAGRAN per acre when plants are less than 8 inches tall. Make a second application at the same rate 7 to 10 days later.

Salsify

Apply 4 pints of BASAGRAN per acre when plants are 4 to 8 inches tall. Make a second application at the same rate 7 to 10 days later. Add oil concentrate to the spray solution of BASAGRAN/water for each application according to the DIRECTIONS FOR USE—All Crops.

Common Groundsel

Apply 2 to 3 pints of BASAGRAN per acre when plants are less than 3 inches tall. Add oil concentrate to the spray solution of BASAGRAN/water, according to the DIRECTIONS FOR USE—All Crops.

ESTABLISHED ORNAMENTAL TURF —DIRECTIONS FOR USE

BASAGRAN may be used on established bluegrass, fescue, bentgrass, bermudagrass, bahiagrass, centipedegrass, zoysiagrass, ryegrass, and St. Augustinegrass.

TIMING OF APPLICATIONS

Apply BASAGRAN® herbicide postemergently to yellow nutsedge when actively growing and under good soil moisture conditions. If desired control is not obtained with the first application, make additional applications at intervals of 10 to 14 days. Do not apply more than 6 pints per acre in one season.

In the northern United States, yellow nutsedge can emerge from May through July; whereas, in the southern United States, yellow nutsedge can emerge throughout the year. Therefore, initial applications should be planned when most plants have emerged. If new yellow nutsedge plants emerge later in the season, make additional applications of BASAGRAN in accordance with the label directions. In unmowed turf, make first application after emergence but before yellow nutsedge is 8 inches tall. Thorough spray coverage of yellow nutsedge is essential for maximum control.

For optimum control do not mow 3 to 5 days before or after application.

APPLICATION CONDITIONS

Water volume: Minimum of 1 gal./1000 square feet or 40 gallons per acre.

Spray pressure: Minimum of 40 psi.

Equipment: Use any spray equipment customarily used for spraying pesticides on turf. Clean equipment carefully before using BASAGRAN. Rinse with water after use.

SPRAYER CALIBRATION SUGGESTIONS FOR HOMEOWNERS

Hand sprayers:

1. Stake off a 400 sq. ft. area of turf for practice. This is an area 20′ (7 steps) × 20′.
2. Add a measured quantity (1½ gallons for example) of water to sprayer and uniformly spray the 400 sq. ft. area. Measure water remaining and thereby determine the amount applied per 400 sq. ft. (NOTE: A minimum of 3 pints/400 sq. ft. is recommended).
3. Prepare spray solution according to **Application Rate Table for Ornamental Turf.**

Example: Assume that in Step 2 the 400 sq. ft. area was uniformly covered with ½ gallon of water. Referring to the table, add BASAGRAN at the rate of 2 to 4 teaspoons per ½ gallon of water for each 400 sq. ft. of turf to be sprayed. (NOTE: Use of this mixture for spot spraying of individual nutsedge plants may result in an excessive dosage and possible turf injury.)

Hose-end applicators:

A procedure similar to the above may be followed for calibrating hose-end sprayers. Half-fill container with water to an even mark on the "Gallons" scale and note the gallonage level. Spray the 400 sq. ft. area, noting the new gallonage reading, and thereby determine the amount of water used to spray the area. Then proceed as in Step 3 above.

Application Rate Table for Ornamental Turf

Weed Controlled—Yellow Nutsedge Application Rate*			
Teaspoons per 200 sq. ft.	Teaspoons per 400 sq. ft.	Fluid Oz. Per 1000 sq. ft.	Pints per Acre
1 to 2	2 to 4	¾ to 1½	2 to 4

* If needed, make subsequent applications at 10–14 day intervals until yellow nutsedge is eliminated. Apply no more than 6 pints per acre in one season.

RESTRICTIONS AND LIMITATIONS

Do not apply BASAGRAN to any newly seeded or newly sprigged turf until seedlings or sprigs are well established, as injury may result.

Do not apply BASAGRAN to golf course greens or collars.

When treating turf with BASAGRAN, avoid over-the-top spraying of adjacent ornamental trees, shrubs, and flowers. Spraying near the base of established ornamental trees, shrubs, and flowers should not result in injury.

APPENDIX

The following are scientific names for the weeds listed in this section. For specific recommendations on control of these weeds, refer to the major crop and/or tank mix sections.

GRASSES

Common Name	Scientific Name
Barnyardgrass	*Echinochloa crus-galli*
Bermudagrass	*Cynodon dactylon*
Broadleaf Signalgrass	*Brachiaria platyphylla*
Crabgrass, Large	*Digitaria sanguinalis*
Smooth	*Digitaria ischaemum*
Cupgrass, Wooly	*Eriochloa villosa*
Foxtail, Giant	*Setaria faberi*
Green	*Setaria viridis*
Yellow	*Setaria lutescens*
Goosegrass	*Eleusine indica*
Itchgrass	*Rottboellia exaltata*
Johnsongrass	*Sorghum halepense*
Junglerice	*Echinochloa colonum*
Lovegrass	*Eragrostis cilianensis*
Panicum, Fall	*Panicum dichotomiflorum*
Texas	*Panicum texanum*
Quackgrass	*Agropyron repens*
Red Rice	*Oryza sativa*
Red Sprangletop	*Leptochloa filiformis*
Volunteer Barley	*Hordeum vulgare*
Corn	*Zea mays*
Oats	*Avena sativa*
Rye	*Secale cereale*
Wheat	*Triticum aestivum*
Wildcane/Shattercane	*Sorghum bicolor*
Wild Oats	*Avena fatua*
Wild Proso Millet	*Panicum miliaceum*
Wirestem Muhly	*Muhlenbergia frondosa*
Witchgrass	*Panicum capillare*

BROADLEAF WEEDS

Common Name	Scientific Name
Arrowhead	*Sagittaria* spp.
Balloonvine	*Cardiospermum halicacabum*
Beggarticks	*Bidens frondosa*
Bindweed, Field	*Convolvulus arvensis*
Hedge	*Convolvulus sepium*
Bristly Starbur	*Acanthospermum hispidum*
Butterprint (see Velvetleaf)	*Abutilon theophrasti*
Buttonweed (see Velvetleaf)	*Abutilon theophrasti*
Canada Thistle	*Cirsium arvense*
Cocklebur	*Xanthium pensylvanicum*
Coffee Senna	*Cassia occidentalis*
Common Lambsquarters	*Chenopodium album*
Common Purslane	*Portulaca oleracea*
Crotalaria	*Crotalaria spectabilis*
Dayflower	*Commelina* spp.
Devilsclaw	*Probiscidea louisianica*
Ducksalad	*Heteranthera limosa*
Galinsoga	*Galinsoga* spp.
Goldenrod, Western	*Solidago occidentalis*
Gooseweed	*Sphenoclea zeylanica*
Groundsel, Common	*Senacio vulgaris*
Jimsonweed	*Dutura stramonium*
Kochia	*Kochia scoparia*
Ladysthumb	*Polygonum persicaria*
Morningglory, Tall (Common)	*Ipomoea purpurea*
Cypressine	*Ipomoea quamoclit*
Entireleaf	*Ipomoea hederacea* var. *intergruscula*
Ivyleaf	*Ipomoea hederacea*
Palmleaf	*Ipomoea wrightii*
Pitted	*Ipomoea lacunosa*
Purple	*Ipomoea muricata*
Moonflower Smallflower	*Jacquemontia tamnifolia*
Nightshade, Black	*Solanum nigrum*
Hairy	*Solanum sarachoides*
Pennsylvania Smartweed	*Polygonum pensylvanicum*
Pigweed, Redroot	*Amaranthus retroflexus*
Smooth	*Amaranthus hybridis*
Prickly Sida or Teaweed	*Sida spinosa*
Ragweed, Common	*Ambrosia artemisiifolia*
Giant	*Ambrosia trifida*
Redstem	*Ammannia* spp.
Redweed	*Melochia corchorifolia*
Salsify	*Tragopogon* spp.
Sesbania	*Sesbania exaltata*
Shepherdspurse	*Capsella bursa-pastoris*
Spurred Anoda	*Anoda cristata*
Tropic Croton	*Croton glandulosus*
Velvetleaf	*Abutilon theophrasti*
Venice Mallow	*Hibiscus trionum*
Waterhemp, Tall	*Amaranthus tubercalatus*
Water Plantain, Common	*Alisma triviale*
Wild Buckwheat	*Polygonum convolvulus*
Wild Mustard	*Sinapsis arvensis*
Wild Poinsettia	*Euphorbia heterophylla*
Wild Sunflower	*Helianthus annuus*

SEDGES

Common Name	Scientific Name
Bulrush, River	*Scirpus fluviatilis*
Roughseed	*Scirpus mucronatus*
Spikerush	*Eleocharis macrostachya*
Umbrellaplant, Smallflower	*Cyperus difformis*
Yellow Nutsedge	*Cyperus esculentus*

CONDITIONS OF SALE AND WARRANTY

The Directions for Use of this product reflect the opinion of experts based on field use and tests. The directions are believed to be reliable and should be followed carefully. However, it is impossible to eliminate all risks inherently associated with use of this product. Crop injury, ineffectiveness or other unintended consequences may result because of such factors as weather conditions, presence of other materials, or use of the product in a manner inconsistent with its labeling, all of which are beyond the control of BASF WYANDOTTE CORPORATION ("BWC") or the Seller. All such risks shall be assumed by the Buyer.

BWC warrants that this product conforms to the chemical description on the label and is reasonably fit for the purposes referred to in the Directions for Use, subject to the inherent risks referred to above. BWC MAKES NO OTHER EXPRESS OR IMPLIED WARRANTY OF FITNESS OR MERCHANTABILITY OR ANY OTHER EXPRESS OR IMPLIED WARRANTY. IN NO CASE SHALL BWC OR THE SELLER BE LIABLE FOR CONSEQUENTIAL, SPECIAL OR INDIRECT DAMAGES RESULTING FROM THE USE OR HANDLING OF THIS PRODUCT. BWC and the Seller offer this product, and the Buyer and user accept it, subject to the foregoing Conditions of Sale and Warranty which may be varied only by agreement in writing signed by a duly authorized representative of BWC.

BASAGRAN and Poast are registered trademarks of BASF AG.

Butyrac is a registered trademark of Union Carbide Agricultural Products Co., Inc.

Butoxone is a registered trademark of Vertac Chemical Corporation.

AAtrex is a registered trademark of Ciba-Geigy.

Blazer, STAM and Triton are registered trademarks of the Rohm and Haas Company.

The purchase price of BASAGRAN® herbicide includes a royalty for a license to practice the method of U.S. Patent 3,708,277.

PIX®

PLANT REGULATOR

ACTIVE INGREDIENT*

N,N-dimethylpiperidinium chloride	4.2%
INERT INGREDIENTS	95.8%
TOTAL	100.0%

* Contains 0.35 pounds active ingredient per gallon.

EPA Reg. No. 7969-52

PIX is a registered trademark of BASF AG.

CAUTION

Keep Out of Reach of Children

Avoid contact with eyes. In case of contact immediately flush eyes with plenty of water.

Get medical attention if irritation persists.

DIRECTIONS FOR USE—COTTON

It is a violation of Federal law to use this product in a manner inconsistent with its labeling.

GENERAL INFORMATION

PIX is a foliar applied plant regulator which modifies the cotton plant by regulating its growth pattern and behavior. PIX may produce one or more of the following effects in cotton: darker leaf color, reduction of internodal length, more open canopy, better boll retention, earlier maturity. Some of these effects may favorably influence the yield potential of the cotton plant.

SPRAY VOLUME

When applying PIX with ground equipment, use a minimum of 20 gallons of water per acre. For aerial application, use a minimum of 5 gallons of water per acre. Regardless of method of application thorough coverage of the cotton foliage is required.

COMPATIBILITY

PIX has an aqueous base, and as such is compatible with most insecticides and miticides. If in doubt, test compatibility by adding a teaspoon of the insecticide or miticide product to a pint of ready-to-use spray solution of PIX.

RESTRICTIONS AND PRECAUTIONS

Insect or mite damage before, at or after an application of PIX can lead to yield decreases or other undesirable effects.

Do not apply PIX to cotton that is drought stressed, i.e. stressed due to lack of soil moisture. If the cotton is stressed due to lack of soil moisture at time of application or if severe drought conditions occur after application of PIX, results can be unsatisfactory and/or yield decreases can occur.

Do not use PIX on the short-staple variety Delta & Pineland 70 because rate and time of application have not been established.

Do not apply PIX at rates or times other than those specified on the label as yield decreases and/or delayed maturity may result.

Do not apply more than 1½ pints of PIX per acre per season.

Do not apply PIX within 30 days of harvest.

Do not graze or feed cotton forage to livestock within 30 days of application.

Do not apply PIX if rainfall is expected within 8 hours of application, as effectiveness of PIX may be reduced.

The pink color of PIX may disappear instantly in spray tanks if chlorinated water is used; however, effectiveness is not related to color of spray solution.

Do not tank mix with any products other than insecticides or miticides unless recommended on this label.

ENVIRONMENTAL HAZARDS

Keep out of lakes, ponds or streams.

Do not contaminate water by cleaning of equipment or disposal of wastes.

STORAGE AND DISPOSAL

Do not contaminate water, food, or feed by storage or disposal.

Wastes resulting from the use of this product may be disposed of on site or at an approved waste disposal facility.

Triple rinse container (or equivalent). Then offer for recycling or reconditioning, or puncture and dispose of in a sanitary landfill, or by incineration, or, if allowed by state and local authorities, by burning. If burned, stay out of smoke.

TIME AND RATE OF APPLICATION SHORT-STAPLE COTTON

The times and rates of application have been carefully researched and the Directions For Use should be observed as specified below.

Geographic Area	Time of Application	Rate Per Acre
CA, AZ, NM	**First Application:** Apply when cotton is in early bloom stage (5–6 white blooms per 25 row feet) and actively growing, but not when cotton is smaller than 18" or taller than 24". If no blooms are present and the cotton is 24" tall, apply PIX. See Restrictions and Precautions.	1 pint
	Second Application for Control of Excessive Vegetative Growth: If cotton field has a history of excessive growth, and/or conditions after the first application of PIX favor excessive growth, apply a second application 3 to 4 weeks after the first application. This may help to suppress any further excessive growth.	½ pint
Miss. Delta, Southeast	**First Application:** Optimum results will be achieved when plants are in early bloom stage (5–6 white blooms per 25 row feet) and an average of 24" tall. Treatments can also be made when cotton height averages a minimum of 20" and a maximum of 28" provided cotton is not more than 7 days beyond early bloom as defined above. See Restrictions and Precautions.	1 pint
	Second Application for Control of Excessive Vegetative Growth: If cotton field has a history of excessive growth, and/or conditions after the first application of PIX favor excessive growth, apply a second application 3 to 4 weeks after the first application.	½ pint
TX (Rio Grande Valley)	**First Application:** *Areas where excessive vegetative growth is a problem:* Optimum results will be achieved when plants are in early bloom stage (5–6 white blooms per 25 row feet) and an average of 24" tall. Treatments can also be made when cotton height averages a minimum of 20" and a maximum of 28" provided cotton is not more than 7 days beyond early bloom as defined above. See Restrictions and Precautions. *Areas where excessive vegetative growth is NOT a problem:* Apply when cotton is in early bloom stage (5–6 white blooms per 25 row feet) and actively growing. If no blooms are present and the cotton is 24" tall and actively growing, apply PIX. See Restrictions and Precautions.	1 pint
	Second Application for Control of Excessive Vegetative Growth: If cotton field has a history of excessive growth, and/or conditions after the first application of PIX favor excessive growth, apply a second application 3 to 4 weeks after the first application.	½ pint
OK, TX (Except Rio Grande Valley)	*Areas where excessive vegetative growth is NOT a problem:* **First Application:** Apply when cotton squares are between pinhead size and ¼ inch wide, and plants are actively growing. See Restrictions and Precautions.	½ pint
	Second Application: If conditions after the first PIX application favor excessive growth, apply a second application 3–5 weeks after the first application.	½ pint

	Areas where excessive vegetative growth IS a problem: **First Application:** Optimum results will be achieved when plants are in early bloom stage (5–6 white blooms per 25 row feet) and an average of 24" tall. Treatments can also be made when cotton height averages a minimum of 20" and a maximum of 28" provided cotton is not more than 7 days beyond early bloom as defined above. See Restrictions and Precautions.	1 pint
	Second Application for Control of Excessive Vegetative Growth: If cotton field has a history of excessive growth, and/or conditions after the first PIX application favor excessive growth, apply a second application 3 to 4 weeks after the first application.	½ pint
AZ, NM, TX	**First Application:** Apply when cotton is in early bloom stage (5–10 yellow blooms per 25 row feet) and actively growing, but not when cotton is smaller than 18" or taller than 24". Soil moisture levels must be at field capacity at time of application; therefore do not use on Pima cotton which is under moisture stress or where moisture stress is used to induce flower initiation.	1 pint
	Second Application for Control of Excessive Vegetative Growth: If cotton field has a history of excessive growth, and/or conditions after the first application of PIX favor excessive growth, apply a second application 3 to 4 weeks after the first application. This may help to suppress any further excessive growth.	½ pint

CONDITIONS OF SALE AND WARRANTY

The Directions for Use of this product reflect the opinion of experts based on field use and tests. The directions are believed to be reliable and should be followed carefully. However, it is impossible to eliminate all risks inherently associated with use of this product. Crop injury, ineffectiveness or other unintended consequences may result because of such factors as weather conditions, presence of other materials, or use of the product in a manner inconsistent with its labeling, all of which are beyond the control of BASF WYANDOTTE CORPORATION ("BWC") or the Seller. All such risks shall be assumed by the Buyer.

BWC warrants that this product conforms to the chemical description on the label and is reasonably fit for the purposes referred to in the Directions for Use subject to the inherent risks referred to above. BWC MAKES NO OTHER EXPRESS OR IMPLIED WARRANTY OF FITNESS OR MERCHANTABILITY OR ANY OTHER EXPRESS OR IMPLIED WARRANTY. IN NO CASE SHALL BWC OR THE SELLER BE LIABLE FOR CONSEQUENTIAL, SPECIAL OR INDIRECT DAMAGES RESULTING FROM THE USE OR HANDLING OF THIS PRODUCT. BWC and the Seller offer this product, and the Buyer and user accept it, subject to the foregoing Conditions of Sale and Warranty which may be varied only be agreement in writing signed by a duly authorized representative of BWC.

POAST®

POSTEMERGENCE GRASS HERBICIDE

ACTIVE INGREDIENT:	
2-[1-(ethoxyimino)butyl]-5-[2-(ethylthio)propyl]-3-hydroxy-2-cyclohexen-1-one*	18.0%
INERT INGREDIENTS	82.0%
TOTAL	100.0%

*Equivalent to 1.5 pounds per gallon.

EPA Reg. No. 7969-58

WARNING

Keep Out of Reach of Children

Causes eye irritation. Do not get in eyes. Avoid contact with skin. Harmful if swallowed.

If in eyes or on skin, immediately flush with plenty of water. Get medical attention if irritation persists.

ENVIRONMENTAL HAZARDS

Do not apply directly to lakes, ponds, or streams. Do not contaminate water by cleaning of equipment or disposal of wastes.

DIRECTIONS FOR USE—ALL CROPS

It is a violation of Federal law to use this product in a manner inconsistent with its labeling.

GENERAL INFORMATION

POAST is a selective broad spectrum postemergence herbicide for control of annual and perennial grass weeds. POAST does not control sedges or broadleaf weeds.

Since all grass crops such as sorghum, corn, small grains, and rice, as well as ornamental grasses such as turf, are extremely susceptible to POAST, avoid all direct or indirect contact with any desired grass plant.

Control symptoms: POAST rapidly enters the plant through the foliage and translocates throughout the plant. Control symptoms exhibited by the grass plant progress from a slowing or stopping of growth (generally within 2 days), to reddening of the foliage, and to leaf tip burn. Subsequently, burn-back of the foliage occurs. These symptoms will generally be observed within three weeks, depending on environmental conditions.

APPLICATION INFORMATION

Apply POAST® herbicide to actively growing grasses when they are at the proper growth stage as specified by the **Recommendations for Grass Control.**

Do not make applications to grasses under stress, such as stress due to lack of moisture, herbicide injury, mechanical injury or cold temperatures, since unsatisfactory control will probably result.

Ground equipment: Thorough spray coverage of grass foliage is essential. For broadcast application use standard high pressure pesticide hollow cone or flat fan nozzles. Do not use flood or whirl chamber nozzles. POAST application with control drop applicator (CDA) nozzles is not recommended due to erratic coverage which causes inconsistent weed control. Use a minimum volume of 5 gallons (10 in California) and a maximum volume of 20* gallons of spray solution per acre. When using standard high pressure hollow cone or flat fan nozzles adjust pressure to a minimum of 40 psi and a maximum of 60 psi at the nozzle.

Always adjust spray pressure, spray volume and height of spray boom to ensure penetration of plant canopy and thorough coverage of grasses to be controlled.

Do not use selective application equipment such as recirculating sprayers, wiper applicators, or shielded applicators.

*See Cotton, Region 2 for exception.

Air equipment:* Thorough spray coverage of grass foliage is essential. Use a minimum of 5 gallons of water per acre. Increase water volume to 10 gallons per acre if grass foliage and/or crop canopy is dense.

*Not applicable in California.

CULTIVATION INFORMATION

Do not cultivate 5 days prior to application of POAST or within 7 days following application.

A timely cultivation after 7 days may aid in providing season-long control (exception—quackgrass, see individual crop recommendations).

ADDITION OF OIL CONCENTRATE

A nonphytotoxic oil concentrate (commonly referred to as oil concentrate) should always be added to the spray tank. The oil concentrate must contain either a petroleum or vegetable oil base and must meet the following criteria: 1) be nonphytotoxic, 2) contain only EPA exempt ingredients, 3) provide good mixing quality in the jar test (see below), and 4) be successful in local experience.

The exact composition of suitable products will vary; however, vegetable and petroleum oil concentrates should contain emulsifiers which provide good mixing quality. For vegetable oil concentrates, it has been observed that highly refined vegetables oils are more satisfactory than unrefined vegetable oils. For additional information see **Jar Test for Estimating Suitability of Oil Concentrates** at the end of this section.

Rate of oil concentrate: Ground and air application—2 pints/acre.

MIXING/SPRAYING: Fill tank of a thoroughly clean sprayer half to two-thirds full with clean water. Start agitation and add oil concentrate; allow to mix thoroughly. Add POAST and remaining volume of water. Maintain constant agitation during application.

Jar Test for Estimating Suitability of Oil Concentrates

1. **Water supply:** Use only water from intended source and at the source temperature.
2. **Amount of water in jar:**

   For 20 gal/A spray volume use 3¾ cups (800 ml) of water.

   For 10 gal/A spray volume use 1⅔ cups (400 ml) of water.

   For 5 gal/A spray volume use ⅚ cup (200 ml) of water.

   For other spray volumes, adjust proportionately to above.
3. **Amount of herbicide(s) and oil concentrate to add:** Add herbicide(s) and oil concentrate at the rate of 1 teaspoon (5 ml) for each pint of recommended label rate.
4. **Add components in following sequence, gently mixing between component additions:**

   1) Water miscible products (such as Basagran, Blazer) when applicable.

   2) Oil concentrate.

   3) POAST
5. Cap jar, invert 10 cycles, let stand for 15 minutes, evaluate.
6. **Evaluation:** An ideal tank mix combination will be uniform; thus, the suitability of the oil concentrate is questionable if any of the following are observed;

Free oil at the surface—film or globules.

Flocculation—fine particles which may be suspended in the liquid or found as a precipitated layer at the bottom of the jar. Clabbering—thickening texture (coagulated) resembling yogurt or a curd-like texture as with cottage cheese.

SPOT OR SMALL AREA TREATMENT

When using knapsack sprayers or high volume spray equipment utilizing hand guns or other suitable nozzle arrangements, make 1% solution of POAST in water. A recommended oil concentrate must also be used at a concentration of 1%.

Apply to foliage of grasses on a spray-to-wet basis. Spray coverage should be uniform and complete. Do not spray to point of runoff.

Prepare the desired volume of spray solution by mixing the amount of POAST and the amount of oil concentrate in water according to the table below.

Desired Spray Solution Volume	Amount To Be Added To Obtain a 1% Spray Solution	
	POAST	**Oil Concentrate**
1 gallon	1¼ fl. oz.†	1¼ fl. oz.
25 gallons	1 qt.	1 qt.
50 gallons	2 qts.	2 qts.
100 gallons	4 qts.	4 qts.

† 2 tablespoons = 1 fl. oz.

In soybeans and cotton, spot or small area treatments should not exceed $^1/_{10}$ of an acre in size, and no more than 10% of any given acre should be treated.

In soybeans, do not make more than one spot or small area treatment to the same area within the same growing season. Also in soybeans, do not apply both broadcast and spot or small area treatments to the same area within the same growing season.

In cotton, do not make more than two spot or small area treatments to the same area within the same growing season.

RESTRICTIONS AND LIMITATIONS

Do not make applications to grasses under stress, such as stress due to lack of moisture, herbicide injury, mechanical injury or cold temperatures, since unsatisfactory control will probably result.

Do not apply if rainfall is expected within one hour following application as grass control will probably be unsatisfactory.

Do not mix or apply POAST with any other pesticide, additive, or fertilizer except as specifically recommended on this labeling.

ATTENTION! Clean Sprayer thoroughly Before and After Application of POAST

Clean sprayer thoroughly prior to application of POAST particularly if a herbicide was used which has the potential to injure the crop to be sprayed with POAST.

Failure to clean sprayer thoroughly after the application of POAST may result in injury to any grass crop subsequently sprayed, such as corn, sorghum, small grains, rice and turf. Fill the sprayer with clean water and add a commercial sprayer cleaner or a surfactant/adjuvant at the recommended rate on its label. Circulate through entire sprayer system. Spray approximately half the tank solution through the hoses, booms, and nozzles to clean these parts. Drain the tank and rinse the total system thoroughly several times with clean water.

STORAGE AND DISPOSAL

Do not contaminate water, food, or feed by storage or disposal.

Pesticide wastes are toxic. Improper disposal of excess pesticide, spray mixture, or rinsate is a violation of Federal law. If these wastes cannot be disposed of by use according to label instructions, contact your State Pesticide or Environmental Control Agency, or the Hazardous Waste representative at the nearest EPA Regional Office for guidance.

Triple rinse container (or equivalent). Then offer for recycling or reconditioning, or puncture and dispose of in a sanitary landfill, or by incineration, or, if allowed by state and local authorities, by burning. If burned, stay out of smoke.

SOYBEANS—RECOMMENDATIONS FOR GRASS CONTROL

- Apply to actively growing grasses at the sizes indicated below.
- Soybeans at all stages of growth are tolerant to POAST.
- Always follow recommendations given in **Application Information Section.**
- Always adjust spray pressure, spray volume and height of spray boom to ensure penetration of plant canopy and thorough coverage of grasses to be controlled.
- ALWAYS ADD 2 PINTS OF OIL CONCENTRATE PER ACRE.

ANNUAL GRASSES*

Group	Grass	Time of Application	POAST Rate Per Acre	Oil Concentrate Rate Per Acre Ground & Air
A	Wild Proso Millet	4–10″	½ pt.	2 pts.
B	Barnyardgrass Broadleaf Signalgrass Fall Panicum Foxtails: Giant, Green Yellow Johnsongrass, Seedling Junglerice Sprangletop Texas Panicum Witchgrass Wooly cupgrass	3–8″	1 pt.	2 pts.
	Goosegrass Large Crabgrass Smooth Crabgrass	3–6″		
	Wildcane/Shattercane If needed, retreat at the same rate and stage of growth.	6–18″		
	Volunteer Corn	6–20″		
C	Volunteer Cereals Barley Rye Oats Wheat POAST is not recommended for spring control of volunteer cereals that emerged the previous fall.	Before tillering, up to 6″ and prior to over-wintering	1½ pts.	2 pts.
	Wild Oats	2–4″		
D	Itchgrass Red Rice	Up to 4″	2 pts.	2 pts.

* For broad spectrum control of annual grasses in Groups A & B, use 1 pint of POAST per acre. When weed populations include grasses in Groups C and/or D, increase the rate of POAST as indicated. If a second flush of annual grasses emerges after the first application, make an additional application at the same rate and at the recommended stage of growth.

RESCUE TREATMENT FOR CONTROLLING SELECTED ANNUAL GRASSES

For best results, always apply POAST to grasses at the growth stage and rate specified in the above table. However, if POAST cannot be applied at the recommended time, larger annual grasses can be controlled with a later application by increasing the rate of POAST. For control of actively growing giant foxtail, green foxtail, yellow foxtail, or seedling johnsongrass up to a maximum height of 16″, apply POAST at a rate of 1½ pints per acre. For control of actively growing Texas panicum, fall panicum, barnyardgrass, or broadleaf signalgrass up to a maximum height of 12″, apply POAST at a rate of 1½ pints per acre. For control of actively growing large crabgrass, smooth crabgrass, or goosegrass up to a maximum height of 8″, apply POAST at a rate of 1½ pints per acre. For control of actively growing wild proso millet up to a maximum height of 24″, apply POAST at a rate of 1 pint per acre. Oil concentrate must always be added to the spray solution at the rate of 2 pints per acre.

PERENNIAL GRASSES

Grass	Application Time	POAST Rate Per Acre	Oil Concentrate Rate Per Acre Ground & Air
Bermudagrass • First Application	Before plant diameter exceeds 6″ or leaf height above ground exceeds 1″	1½ pts.	2 pts.
• Second Application If regrowth occurs or new plants emerge.	1–4″ length of new plants	1 pt.	2 pts.
Johnsongrass, Rhizome For best results, rhizomes should be thoroughly fragmented (less than 6″). • First Application When using 10–20 gallons of spray solution per acre.	15–25″ (15–20″ in no-till culture)	1½ pts.	2 pts.
When using 5–10 gallons of spray solution per acre. Maintain a ground speed of no more than 6 miles per hour.	15–25″ (15–20″ in no-till culture)	1⅓ pts.	2 pts.
• Second Application If regrowth occurs or new plants emerge.	6–12″	1 pt.	2 pts.
Quackgrass For best results, rhizomes should be thoroughly fragmented (less than 6″). • First Application	6–8″	2½ pts.	2 pts.
• Second Application If regrowth occurs or new plants emerge Depending upon environmental conditions and crop cultural system, season-long control may not always be obtained. However, competition of quackgrass with the crop will be reduced. Note: In conventional wide-row soybeans, a cultivation no sooner than 14 days after application but within 21 days of application will aid in control.	6–8″	1½ pts.	2 pts.
Wirestem Muhly If regrowth occurs, retreat at the same rate and stage of growth.	Up to 6″	1 pt.	2 pts.

RESTRICTIONS AND LIMITATIONS FOR SOYBEANS

Do not apply POAST to soybeans within 90 days of harvest.

Do not apply more than a total of 5 pints of POAST per acre to soybeans in one season.

Do not graze treated soybean fields and do not feed treated soybean forage, ensilage or hay to livestock.

Recommendations for Applying POAST® **and Basagran® herbicides with Blazer® 2L herbidide for Control of Broadleaf Weeds and Grasses in Soybeans.**

GENERAL INFORMATION

For postemergence broadleaf and grass weed control in soybeans, tank mix or separate applications of POAST, Basagran and/or Blazer 2L may be made. If broadleaf and grass weeds (see preceding charts) are present, all at the stage of growth recommended for treatment on the POAST, Basagran and Basagran + Blazer 2L labels, then two options are available to control both broadleaf and grass weeds.

Option A: Separate applications of POAST preceded or followed by Basagran or Basagran + Blazer 2L may be made.

Option B*: A tank mix of POAST + Basagran or POAST + Basagran + Blazer 2L may be made. If all weeds are not at the correct stage of growth for treatment at the same time or if perennial grasses and certain other annual grasses not listed on page 32 are a problem, then separate applications must be made.

Thorough coverage of actively growing weeds is essential. Large crop-and-weed leaf canopies shelter smaller weeds and can prevent adequate spray coverage. Soybeans are tolerant to the above tank mixes; however, under certain conditions soybean leaves may appear burned, crinkled or bronzed. Soybeans at all states of growth are tolerant to POAST.

*Not applicable in California.

Option A: Separate Applications

Applications of POAST can be preceded or followed by Basagran or Basagran tank mixed with Blazer 2L to obtain broad spectrum control of weeds listed on the respective product labels (refer to the label for POAST, Basagran, and Basagran + Blazer 2L tank mix). Also refer to these product labels for timing, rate and other information for ground and aerial applications.

Whether Basagran is applied alone or tank mixed with Blazer 2L, add a nonphytotoxic oil concentrate to the spray solution when recommended on the label for Basagran or Basagran + Blazer 2L.

When POAST is applied, *always* add a nonphytotoxic oil concentrate to the spray solution at the rate of 2 pints per acre for ground and aerial applications.

For best results when applying separately, a minimum period of time is required between applications, depending upon their order, according to the table below.

Order of Application		
Product(s) Applied First	**Product(s) Applied Second**	**Minimum Time Between Applications**
Basagran	POAST	24 Hours
Basagran + Blazer 2L	POAST	7 Days
POAST	Basagran or Basagran + Blazer 2L	24 Hours

Option B*: Tank Mixes

POAST + Basagran + Oil Concentrate

POAST + Basagran + Blazer 2L + Oil Concentrate

POAST® postemergence grass herbicide controls the broad spectrum of annual grasses listed below:**

Wild Proso Millet	Junglerice
Barnyardgrass	Red Sprangletop
Broadleaf Signalgrass	Texas Panicum
Fall Panicum	Witchgrass
Giant Foxtail	Wooly Cupgrass
Green Foxtail	Goosegrass
Yellow Foxtail	Large Crabgrass
Johnsongrass, Seedling	Smooth Crabgrass

Do not tank mix when grasses to be controlled include rhizome johnsongrass, quackgrass, bermudagrass, wirestem muhly, volunteer corn, shattercane, volunteer cereals, wild oats, red rice or itchgrass since unsatisfactory control will probably result. Option A should be followed then these weeds are present.

Basagran® postemergence broadleaf herbicide controls the broad spectrum of weeds listed below:**

Balloonvine	Ladysthumb
Beggersticks	Pennsylvania Smartweed
Bristly Starbar	Prickly Sida or Teaweed
Cocklebur	Redweed
Coffee Senna	Shepherdspurse
Common Lambsquarters	Smallflower Morningglory
Common Purslane	Spurred Anoda
Common Ragweed	Tropic Croton
Cypressvine Morningglory	Velvetleaf
	Venice Mallow
Canada Thistle***	Wild Buckwheat
Dayflower	Wild Mustard
Devilsclaw	Wild Poinsettia
Galinsoga	Wild Sunflower
Giant Ragweed	Yellow Nutsedge
Jimsonweed	

* Not applicable in California.
** For genus and species see pg. 40.
*** Requires two applications of Basagran in accordance with the label for Basagran.

POAST and Basagran may be tank mixed to obtain broad spectrum control of all the broadleaf and grass weeds listed above. Oil concentrate must always be added to the spray solution.

To control certain broadleaf weeds (see list below)** not controlled by Basagran alone, Blazer 2L may be included in the tank mix of POAST and Basagran. Oil concentrates must always be added to the spray solution.

Common ragweed	Black Nightshade
Redroot Pigweed	Sesbania morningglories
Smooth Pigweed	Crotalaria

Time of Application: The timing of all applications of POAST and Basagran should be in accordance with the weed growth stages indicated on each product label and when weeds are actively growing. When Blazer 2L is included in the tank mix, the timing should be in accordance with the weed growth stages indicated in the table below and when weeds are actively growing. If weeds are not at the correct stage of growth for treatment at the same time, or if perennial grasses and certain other annual grasses (see Option B) are a problem, then separate applications should be made (see Option A). Delay in application which permits weeds to exceed the maximum size stated will result in inadequate control.

Time of Application for Blazer 2L in Tank Mix

Weeds	Maximum Height	Number of Leaves
Common Ragweed	6″	Up to 10
Redroot Pigweed	3″	Up to 6
Smooth Pigweed	3″	Up to 6
Black Nightshade	2″	Up to 6
Sesbania	6″	Up to 4 pinnate
Morningglories	4″	Up to 4
Crotalaria	6″	Up to 6

Rate of Application: Include POAST, Basagran, or Blazer 2L in the tank mix when required according to the weed problem present. Follow rate recommendations indicated in the table below. Oil concentrate must *always* be added to the spray solution.

Rates of Application for Tank Mix

Tank Mix Combination	Product			
	Basagran	POAST*	Blazer 2L**	Oil Concentrate
POAST + Basagran	1½–2 pints per acre	1½ pints per acre***		2 pints per acre
POAST + Basagran + Blazer 2L	1½–2 pints per acre†	1½ pints per acre***	1 pint per acre	2 pints per acre

* The rate of POAST recommended in the tank mix is 50% greater than the rate of POAST used above.
** Blazer 2L may also be included in the tank mix at a rate up to 2 pints/acre; however, this will increase the severity and/or frequency with which soybean injury is observed.
*** For the control of wild proso millet only, include POAST in the tank mix at the rate of ¾ pints/A.
† According to weed species and size given on product label.

Addition of Oil Concentrate: *Always* add a nonphytotoxic oil concentrate to the tank mix of POAST + Basagran and POAST + Basagran + Blazer 2L, at the rate of 2 pints per acre for ground and aerial applications. See DIRECTIONS FOR USE—All Crops for additional information.

WATER VOLUME AND SPRAY PRESSURE

Ground equipment: For the tank mix of POAST + Basagran or POAST + Basagran + Blazer 2L, use 20 gallons of total spray solution per acre (broadcast basis) and a minimum of 40 psi pressure. Use standard high pressure hollow cone or flat fan nozzles spaced 20 inches apart. Do not use flood or whirl chamber nozzles.

Air equipment: Use minimum of 5 gallons of total spray solution per acre except when Blazer 2L is included in the tank mix, then use a minimum of 10 gallons of total spray solution per acre.

Mixing: Fill the spray tank half full with water and add the recommended amount of product in the following order—Basagran, Blazer 2L, oil concentrate, POAST—while the agitator is running. Then add the remaining quantity of water.

RESTRICTIONS AND LIMITATIONS (PARTIAL LIST) FOR TANK MIX OR SEPARATE APPLICATION

Do not apply if soybeans show injury (leaf phytotoxicity and/or plant stunting) produced by any other prior herbicide treatment because this injury may be enhanced and/or prolonged.

Do not make applications to grasses or broadleaf weeds under stress, such as stress due to lack of moisture, mechanical injury or unseasonably cold weather, since unsatisfactory control will result.

Do not apply to soybeans that have been subjected to stress conditions such as hail damage, flooding, drought or unseasonably cold or widely fluctuating temperatures because injury may result.

Do not apply Blazer 2L within 50 days of harvest and do not apply POAST within 90 days of harvest.

Do not graze treated soybean fields and do not feed treated soybean forage, ensilage or hay to livestock (see label for Blazer 2L and this label).

COTTON—REGIONAL USE MAP

All recommendations for cotton are based on growing region. Refer to the map below. Follow the recommendations for grass control for your region only.

This is a 150 pt. drop.

1. California, Arizona, and Western New Mexico
2. High and Rolling Plains of Texas, Oklahoma and Eastern New Mexico
3. Southern United States

COTTON—RECOMMENDATIONS FOR GRASS CONTROL

REGION 1: CALIFORNIA, ARIZONA, AND WESTERN NEW MEXICO

- Apply to actively growing grasses at the sizes indicated below.
- Cotton at all stages of growth is tolerant to POAST.
- Always follow recommendations given in **Application Information Section.**
- Always adjust spray pressure, spray volume and height of spray boom to ensure penetration of plant canopy and thorough coverage of grasses to be controlled.
- Drop nozzles may be needed to obtain adequate coverage of grasses to be controlled when cotton is taller than 24 inches.
- Do not apply to drought-stressed grass or grass which has gone through an extended dry period.
- For best results, apply POAST 2 to 4 days after an irrigation.
- ALWAYS ADD 2 PINTS OF OIL CONCENTRATE PER ACRE.

ANNUAL GRASSES* (Region 1)
(California, Arizona and Western New Mexico)

Group	Grass	Time of Application	POAST Rate Per Acre	Oil Concentrate Rate Per Acre Ground & Air
A	Barnyardgrass Broadleaf Signalgrass Fall Panicum Foxtails: Giant, Green, Yellow Johnsongrass, Seedling Junglerice Red Sprangletop Texas Panicum Witchgrass	3–8"	1½ pts.	2 pts.
	Goosegrass Large Crabgrass Smooth Crabgrass	Up to 4"		
	Wildcane/Shattercane If needed, retreat at the same rate and stage of growth	6–10"		
	Volunteer Corn	8–12"		
B	Volunteer Cereals Barley Rye Oats Wheat Volunteer cereals which emerge from late spring through early summer (May through July) may be partially or incompletely controlled due to unfavorable conditions at time of application.	Before tillering, up to 4"	2 pts.	2 pts.

* For broad spectrum control of annual grasses in Group A, use 1½ pints of POAST per acre.
When weed populations include additional grasses in Group B, increase the rate of POAST as indicated. If subsequent flushes of annual grasses emerge after the first application, make additional applications at the same rate and at the recommended stage of growth.

PERENNIAL GRASSES (Region 1)
(California, Arizona and Western New Mexico)

Grass	Time of Application	POAST Rate Per Acre	Oil Concentrate Rate Per Acre Ground & Air
Bermudagrass • First Application	Before Plant diameter exceeds 6" or leaf height above ground exceeds 1".	2½ pts.	2 pts.
• Second Application	21 days after first application	1½ pts.	2 pts.
• Third application When regrowth occurs or new plants emerge	1–4" length of regrowth or new plants	1½ pts.	2 pts.
Johnsongrass, Rhizome For best result, rhizomes should be thoroughly fragmented (less than 6"). • First Application	6–10"	2½ pts.	2 pts.
• Subsequent Applications When regrowth occurs or new plants emerge.	4–8"	1½ pts.	2 pts.

COTTON—RECOMMENDATIONS FOR GRASS CONTROL

REGION 2: HIGH AND ROLLING PLAINS OF TEXAS, OKLAHOMA AND EASTERN NEW MEXICO

- Apply to actively growing grasses at the sizes indicated below.
- Cotton at all stages of growth is tolerant to POAST.
- Always follow recommendations given in **Application Information Section.**
- Always adjust spray pressure, spray volume and height of spray boom to ensure penetration of plant canopy and thorough coverage of grasses to be controlled.
- Do not apply to drought-stressed grass or grass which has gone through an extended dry period.
- For best results, apply POAST 2 to 4 days after an irrigation.
- ALWAYS ADD 2 PINTS OF OIL CONCENTRATE PER ACRE.

ANNUAL GRASSES* (Region 2)
High and Rolling Plains of Texas, Oklahoma and Eastern New Mexico

Group	Grass	Time of Application	POAST Rate Per Acre	Oil Concentrate Rate Per Acre Ground & Air
A	Barnyardgrass Broadleaf Signalgrass Browntop Panicum Foxtails: Giant, Green Yellow Johnsongrass, Seedling Junglerice Red Sprangletop Texas Panicum Witchgrass	2–6"	1½ pts.	2 pts.
	Goosegrass Large Crabgrass Smooth Crabgrass	Up to 4"		
	Wildcane/Shattercane If needed, retreat at the same rate and stage of growth.	6–10"		
	Volunteer Corn	8–10"		
B	Volunteer Cereals Barley Rye Oats Wheat POAST is not recommended for spring control of volunteer cereals that emerged the previous fall.	Before tillering, up to 4" and prior to over-wintering	2 pts.	2 pts.

* For broad spectrum control of annual grasses in Group A, use 1½ pints of POAST per acre.
When weed populations include additional grasses in Group B, increase the rate of POAST as indicated. If subsequent flushes of annual grasses emerge after first application, make additional applications, at the same rate and at the recommended stage of growth.

PERENNIAL GRASSES (Region 2)
High and Rolling Plains of Texas, Oklahoma and Eastern New Mexico

Grass	Time of Application	POAST Rate Per Acre	Oil Concentrate Rate Per Acre Ground & Air
Bermudagrass • First Application	Before plant diameter exceeds 6" or leaf height above ground exceeds 1".	2 pts.	2 pts.
• Second Application	21 days after first application	1½ pts.	2 pts.
• Third application When regrowth occurs or new plants emerge	1–4" length of regrowth or new plants	1½ pts.	2 pts.
Johnsongrass, Rhizome For best result, rhizomes should be thoroughly fragmented (less than 6"). Adjust volume of spray solution to a maximum of 10 gallons and a minimum of 5 gallons per acre while maintaining a ground speed of no more than 6 miles per hour. • First Application	6–10"	1½ pts.	2 pts.
• Subsequent Applications When regrowth occurs or new plants emerge.	4–8"	1 pt.	2 pts.

COTTON—RECOMMENDATIONS FOR GRASS CONTROL

REGION 3: SOUTHERN UNITED STATES

- Apply to actively growing grasses at the sizes indicated below.
- Cotton at all stages of growth is tolerant to POAST.
- Always follow recommendations given in **Application Information Section.**
- Always adjust spray pressure, spray volume and height of spray boom to ensure penetration of plant canopy and thorough coverage of grasses to be controlled.
- ALWAYS ADD 2 PINTS OF OIL CONCENTRATE PER ACRE.

ANNUAL GRASSES* (Region 3)
Southern United States

Group	Grass	Time of Application	POAST Rate Per Acre	Oil Concentrate Rate Per Acre Ground & Air
A	Barnyardgrass Broadleaf Signalgrass Browntop Panicum Foxtails: Giant, Green, Yellow Johnsongrass, Seedling Junglerice Red Sprangletop Texas Panicum Witchgrass Wooly Cupgrass	3–8"	1 pt.	2 pts.
	Goosegrass Large Crabgrass Smooth Crabgrass	Up to 4"		
	Wildcane/Shattercane If needed, retreat at the same rate and stage of growth.	6–18"		
	Volunteer Corn	6–20"		
B	Volunteer Cereals Barley Rye Oats Wheat POAST is not recommended for spring control of volunteer cereals that emerge the previous fall.	Before tillering, up to 4" and prior to over-wintering	1½ pts.	2 pts.
	Wild Oats	2–4"		
C	Itchgrass Red Rice	Up to 4"	2 pts.	2 pts.

* For broad spectrum control of annual grasses in Group A, use 1 pint of POAST per acre.
When weed populations include additional grasses in Group B and/or C, increase the rate of POAST as indicated. If subsequent flushes of annual grasses emerge after first application, make additional applications at the same rate and at the recommended stage of growth.

PERENNIAL GRASSES (Region 3)
Southern United States

Grass	Time of Application	POAST Rate Per Acre	Oil Concentrate Rate Per Acre Ground & Air
Bermudagrass • First Application	Before plant diameter exceeds 6" or leaf height above ground exceeds 1".	1½ pts.	2 pts.
• Subsequent Applications When regrowth occurs or new plants emerge	1–4" length of regrowth or new plants	1 pt.	2 pts.

Johnsongrass, Rhizome For best result, rhizomes should be thoroughly fragmented (less than 6"). • First Application When using 10–20 gallons of spray solution per acre.	15–25"	1½ pts.	2 pts.
When using 5–10 gallons of spray solution per acre. Maintain a ground speed of no more than 6 miles per hour.	15–25"	1½ pts.	2 pts.
• Subsequent Applications If regrowth occurs or new plants emerge.	6–12"	1 pt.	2 pts.

RESCUE TREATMENT FOR CONTROLLING SELECTED ANNUAL GRASSES IN COTTON FOR SOUTHERN UNITED STATES (REGION 3)

For best results, always apply POAST to grasses at the growth stage and rate specified in the above table. However, if POAST cannot be applied at the recommended time, larger annual grasses can be controlled with a later application by increasing the rate of POAST. For control of actively growing giant foxtail, green foxtail, yellow foxtail, or seedling johnsongrass up to a maximum height of 16", apply POAST at a rate of 1½ pints per acre. For control of actively growing Texas panicum, fall panicum, barnyardgrass or broadleaf signalgrass up to a maximum height of 12", apply POAST at a rate of 1½ pints per acre. For control of actively growing large crabgrass, smooth crabgrass, or goosegrass up to maximum height of 8", apply POAST at a rate of 1½ pints per acre. Oil concentrate must always be added to the rate of 2 pints per acre.

RESTRICTIONS AND LIMITATIONS FOR COTTON

Do not apply POAST within 40 days of harvest.
Do not apply more than a total of 7½ pints of POAST per acre to cotton in one season.
Do not graze treated cotton fields and do not feed treated cotton forage to livestock.

SUGAR BEETS—RECOMMENDATIONS FOR GRASS CONTROL

CALIFORNIA, ARIZONA, AND NEW MEXICO

- Apply to actively growing grasses at the sizes indicated below.
- Sugar beets at all stages of growth are tolerant to POAST.
- Always follow recommendations given in the **Application Information Section.**
- Always adjust spray pressure, spray volume and height of spray boom to ensure penetration of plant canopy and thorough coverage of grasses to be controlled.
- In irrigated areas, apply POAST 2 to 4 days after an irrigation for best results.
- ALWAYS ADD 2 PINTS OF OIL CONCENTRATE PER ACRE.

ANNUAL GRASSES*

Group	Grass	Time of Application	POAST Rate Per Acre	Oil Concentrate Rate Per Acre Ground & Air
A	Barnyardgrass Broadleaf Signalgrass Fall Panicum Foxtails: Giant, Green, Yellow Johnsongrass, Seedling Junglerice Ryegrass, Annual Red Sprangletop Texas Panicum Witchgrass Southwestern Cupgrass	3–8"	1½ pts.	2 pts.
	Goosegrass Large Crabgrass Smooth Crabgrass	Up to 4"		
	Wildcane/Shattercane	6–18"		
B	Volunteer Cereals Barley Rye Oats Wheat Volunteer cereals which emerge from late spring through early summer (May through July) may be partially or incompletely controlled due to unfavorable environmental conditions at time of application	Before tillering, up to 4" and prior to over-wintering	2 pts.	2 pts.
D	Volunteer Corn	2–6"	2½ pts.	2 pts.

* For broad spectrum control of annual grasses in Group A, use 1½ pints of POAST per acre.
When weed populations include additional grasses in Groups B and/or C, increase the rate of POAST as indicated. If additional applications are needed, apply at the same rate and at the recommended stage of growth.

PERENNIAL GRASSES

Grass	Time of Application	POAST Rate Per Acre	Oil Concentrate Rate Per Acre Ground & Air
Bermudagrass • First Application	Before plant diameter exceeds 6" or leaf height above ground exceeds 1".	2½ pts.	2 pts.
• Second Application When regrowth occurs or new plants emerge.	1–4" length of regrowth or new plants.	1½ pts.	2 pts.
Johnsongrass, Rhizome • First Application	6–10"	2½ pts.	2 pts.
• Second Application When regrowth occurs. or new plants emerge.	4–8"	1½ pts.	2 pts.
Ryegrass, Perennial If regrowth occurs retreat at the same rate and stage of growth	3–8"	1½ pts.	2 pts.

SUGAR BEETS—RECOMMENDATIONS FOR GRASS CONTROL

STATES OTHER THAN CALIFORNIA, ARIZONA AND NEW MEXICO

- Apply to actively growing grasses at the sizes indicated below.
- Sugar beets at all stages of growth are tolerant to POAST.
- Always follow recommendations given in the **Application Information Section.**
- Always adjust spray pressure, spray volume and height of spray boom to ensure penetration of plant canopy and thorough coverage of grasses to be controlled.
- In irrigated areas, apply POAST 2 to 4 days after an irrigation for best results.
- ALWAYS ADD 2 PINTS OF OIL CONCENTRATE PER ACRE.

ANNUAL GRASSES*

Group	Grass	Time of Application	POAST Rate Per Acre	Oil Concentrate Rate Per Acre Ground & Air
A	Wild Proso Millet	4–10"	½ pt.	2 pts.
B	Barnyardgrass Broadleaf Signalgrass			

Group	Grass	Time of Application	POAST Rate Per Acre	Oil Concentrate Rate Per Acre Ground & Air
	Fall Panicum Foxtails: Giant, Green, Yellow Johnsongrass, Seedling Junglerice Ryegrass, Annual Red Sprangletop Texas Panicum Witchgrass Wooly Cupgrass	3–8″	1 pt.	2 pts.
	Goosegrass Large Crabgrass Smooth Crabgrass	Up to 4″		
	Wildcane/Shattercane	6–18″		
	Volunteer Corn	6–20″		
C	Volunteer Cereals Barley Rye Oats Wheat POAST is not recommended for spring control of volunteer cereals that emerged the previous fall.	Before tillering, up to 4″ and prior to overwintering	1½ pts.	2 pts.
	Wild Oats	2–4″		

* For broad spectrum control of annual grasses in Groups A & B, use 1 pint of POAST per acre.

When weed populations include additional grasses in Groups C, increase the rate of POAST as indicated. If additional applications are needed, apply at the same rate and at the recommended stage of growth.

PERENNIAL GRASSES

Grass	Time of Application	POAST Rate Per Acre	Oil Concentrate Rate Per Acre Ground & Air
Bermudagrass • First Application	Before plant diameter exceeds 6″ or leaf height above ground exceeds 1″	1½ pts.	2 pts.
• Subsequent Application When regrowth occurs or new plants emerge.	1–4″ length of regrowth or new plants	1 pt.	2 pts.
Johnsongrass, Rhizome For best results, rhizomes should be thoroughly fragmented (less than 6″). • First Application	15–25″	1½ pts.	2 pts.
• Subsequent Applications If regrowth occurs or new plants emerge.	6–12″	1 pt.	2 pts.
Quackgrass For best results, rhizomes should be thoroughly fragmented (less than 6″) • First Application	6–8″	2½ pts.	2 pts.
• Subsequent Applications When regrowth occurs or new plants emerge. Depending upon environmental conditions and crop competition, season-long control may not always be obtained. However, competition of quackgrass with the crop will be reduced. Note: A cultivation no sooner than 14 days after application but within 21 days of application will aid in control.	6–8″	1½ pts.	2 pts.
Wirestem Muhly If regrowth occurs, retreat at the same rate and stage of growth.	Up to 6″	1½ pts.	2 pts.
Ryegrass, Perennial If regrowth occurs, retreat at the same rate and stage of growth.	3–8″	1 pt.	2 pts.

RESTRICTIONS AND LIMITATIONS FOR SUGAR BEETS

Do not apply more than a total of 5 pints of POAST to sugar beets in one season.

Do not make applications to grasses under stress, such as stress due to lack of moisture, herbicide injury, mechanical injury or cold temperatures, since unsatisfactory control will probably result.

Do not apply if rainfall is expected within one hour following application as grass control will probably be unsatisfactory.

Do not mix or apply POAST with any other pesticide, additive, or fertilizer except as specifically recommended on this labeling.

NONBEARING FOOD CROPS; ORNAMENTAL, NURSERY, AND OTHER NON-FOOD CROPS DIRECTIONS FOR USE

Not intended for domestic use, except by professional applicators.

POAST should be applied when grasses are actively growing and before they reach the maximum size listed in **Recommendations for Grass Control.** Many seedling, newly transplanted and established non-grassy ornamentals, trees, shrubs, and ground covers are tolerant to POAST. Very slight leaf speckling has been observed on a few species with no reduction in vigor or growth.

The following plants are tolerant to POAST:

NONBEARING FOOD CROPS*

Do not apply to nonbearing food crops within 1 year of harvest.

Almonds	Limes
Apples	Macadamia
Apricots	Nectarines
Avocados	Olives
Blackberries	Oranges
Blueberries	Peaches
Cherries	Pears
Crabapples	Pecans
Cranberries	Pistachios
Dates	Plums
Figs	Pomegranates
Filberts	Prunes
Grapes	Raspberries
Grapefruit	Tangelos
Lemons	Tangerines
	Walnuts

* Approval for use of POAST on these plants is pending in California.

ORNAMENTAL CROPS

TREES:

Common Name	Scientific Name
Arborvitae (var: Teehny, Globe Pyramidalis	*Thuja occidentalis**
Ash, Green	*Fraxinus pennsylvanicum**
Dogwood, Flowering	*Cornus florida*
Dogwood, Red Osier	*Cornus sericea**
Fir, Douglas	*Pseudotsuga menziesii**
Fir, Frasier	*Abies fraseri*
Holly, Chinese (var: Bufordii, Rotunda)	*Ilex cornuta**
Holly, Japanese (var: Convexa, Compacta, Helleri)	*Ilex crenata*
Locust, Honey	*Gleditsia triacanthos**
Magnolia, Southern	*Magnolia grandiflora*

Maple, Red	*Acer rubrum*
Maple, Silver	*Acer saccharinum**
Oak, Water	*Quercus nigra**
Oak, Willow	*Quercus phellos*
Olive, Russian	*Elaeagnus anqustifolia**
Pine, Austrian	*Pinus nigra**
Pine Jack	*Pinus banksiana**
Pine Jap. Black	*Pinus thunbergii**
Pine Loblolly	*Pinus, taeda*
Pine, Mugho	*Pinus mugho*
Pine, Scotch	*Pinus sylvestris**
Pine, Virginia	*Pinus virginiana*
Pine, White	*Pinus strobus*
Poplar, Hybrid	*Populus alba*
Spruce, Black Halls (var: Densata)	*Picea glauca*
Spruce, Col. Blue	*Picea pungens**
Spruce, Norway	*Picea abies*
Spruce, White	*Picea glauca*
Sweet Gum	*Liquidambar styraciflua**
Sycamore	*Platanus occidentalis**

* Approval for use of POAST on these plants is pending in California.

SHRUBS:

Common Name	Scientific Name
Alpine Currant	*Ribes alpinum**
American Cranberry Bush	*Viburnum trilobum**
Boxwood	*Buxus sempervirens**
Camellia	*Camellia japonica**
Cotoneaster, Bearberry	*Cotoneaster dammerii*
Cotoneaster, Cranberry	*Cotoneaster apiculata**
Forsythia	*Forsythia viridissima Broxeniss*
Jojoba	*Simmondsia chinensis**
Juniper, Chinese (var: Maney, Hetzii)	*Juniperus chinensis**
Juniper, Creeping (var: wiltonii)	*Juniperus horizontalis*
Juniper, Pfitzer	*Juniperus* spp.
Juniper, Rocky Mountain (var: Blue Heaven, Welchii)	*Juniperus scropulorum**
Juniper, Shore (var: Compacta)	*Juniperus conferta**
Nandina	*Nandina domestica**
Nannyberry	*Viburnum lantago**
Ninebark	*Physocarpus opulifolius nanusa**
Rhododendron, Azalea (var: Hinocrimson, Hershey Red, Formosa Flame)	*Rhododendron* spp.
Sandcherry, Purpleleaf	*Prunus cistena**
Spindle tree	*Euonymus kiautschovica*
Spirea (var: Anthony Waterer)	*Spirea bumalda**
Yew	*Faxus cuspitata**

* Approval for use of POAST on these plants is pending in California.

ORNAMENTALS

Common Name	Scientific Name
Begonia	*Bogonia semperflorens**
Chrysanthemum	*Chrysanthemum indicum**
Coleus	*Coleus* spp.*
Gladiolus	*Gladiolus* spp.*
Impatiens	*Impatiens* spp.*
Iris	*Iris* spp.*
Marigold	*Tagetes* spp.*
Periwinkle	*Vinca minor*
Petunia	*Petunia* spp.*
Snapdragon	*Antirrhinum majus**
Zinnia	*Zinnia elegans**

GROUND COVERS

Common Name	Scientific Name
Ivy, English	*Hedera helix**
Pachysandra	*Pachysandra terminalis**

NOTICE TO USER

It is impossible to test every species and variety of ornamental or nursery plants. Plant tolerance to POAST has been found to be acceptable in the above specified genera, species, and varieties. Neither the manufacturer nor the seller has determined whether or not POAST can safely be used on plants not specified on this label. The professional user should determine if POAST can be used safely prior to use.

*Approval for use of POAST on these plants is pending in California.

RECOMMENDATIONS FOR GRASS CONTROL

- Apply to actively growing grasses.
- Follow water volume and spray pressure recommendations.
- Apply to grasses at the sizes indicated below.
- ALWAYS ADD 2 PINTS PER ACRE OF OIL CONCENTRATE.

ANNUAL GRASSES

Grass	POAST, Rate Per Acre*		Oil Concentrate Rate Per Acre
	Grass Up to 6″ Height	Grass Up To 12″ Height	
Barnyardgrass			
Broadleaf Signalgrass			
Fall Panicum			
Foxtails: Giant			
Green			
Yellow			
Goosegrass			
Johnsongrass, Seedling			
Junglerice			
Large Crabgrass	1½ pts.	2½ pts.	2 pts.
Lovegrass			
Orchardgrass, seedling			
Red Sprangletop			
Smooth Crabgrass			
Tall Fescue, seedling			
Texas Panicum			
Wildcane/Shattercane			
Wild Proso Millet			
Witchgrass			
Wooly Cupgrass			

PERENNIAL GRASSES

Grass Species	Maximum Size Range	POAST, Rate Per Acre*	Oil Concentrate Rate Per Acre
Bermudagrass	up to 6″ runners		
Johnsongrass, Rhizome	15-20″ height	2½ pts	2 pts.
Quackgrass	6-8″ height		
Wirestem Muhly	up to 6″ height	1½ pts.	

* Repeat applications as needed.

APPENDIX

The following are scientific names for the weeds listed in this section. For specific recommendations on control of these weeds, refer to the major crop and/or tank mix sections.

GRASSES:

Common Name		Scientific Name
Barnyardgrass		*Echinochloa crus-galli*
Bermudagrass		*Cynodon dactylon*
Broadleaf Signalgrass		*Brachiaria platyphylla*
Crabgrass,	Large	*Digitaria sanguinalis*
	Smooth	*Digitaria ischaemum*
Cupgrass,	Southwestern	*Erochloa gracilis*
	Wooly	*Erochloa villosa*
Foxtail,	Giant	*Setaria faberi*
	Green	*Setaria viridis*
	Yellow	*Setaria lutescens*

GRASSES:

Common Name		Scientific Name
Goosegrass		*Eleusine indica*
Itchgrass		*Rottboellia exaltata*
Johnsongrass		*Sorghum halepense*
Junglerice		*Echinochloa colonum*
Lovegrass		*Eragrostis cilianensis*
Orchardgrass		*Dactylis glomerata*
Panicum,	Browntop	*Panicum fasciculatum*
	Fall	*Panicum dichotomiflorum*
	Texas	*Panicum texanum*
Quackgrass		*Agropyron repens*
Red Rice		*Oryza sativa*
Red Sprangletop		*Leptochloa filiformis*
Ryegrass,	Annual	*Lolium multiflorum*
	Perennial	*Lolium perenne*
Tall Fescue		*Festuca arundinacea*
Volunteer	Barley	*Hordeum vulgare*
	Corn	*Zea mays*
	Oats	*Avena sativa*
	Rye	*Secale cereale*
	Wheat	*Triticum aestivum*
Wildcane/Shattercane		*Sorghum bicolor*
Wild Oats		*Avena fatua*
Wild Proso Millet		*Panicum miliaceum*
Wirestem Muhly		*Muhlenbergia frondosa*
Witchgrass		*Panicum capillare*

BROADLEAF WEEDS:

Common name		Scientific Name
Balloonvine		*Cardiospermum halicacabum*
Beggarticks		*Bidens frondosa*
Black Nightshade		*Solanum nigrum*
Bristly Starbur		*Acanthospermum hispidum*
Canada Thistle		*Cirsium arvense*
Cocklebur		*Xanthium pensylvanicum*
Coffee Senna		*Cassia occidentalis*
Common Lambsquarters		*Chenopodium album*
Common Purslane		*Portulaca oleracea*
Crotalaria		*Crotalaria spectabilis*
Dayflower		*Commelina* spp.
Devilsclaw		*Probisidea louisianica*
Galinsoga		*Galinsoga* spp.
Jimsonweed		*Datura stramonium*
Ladythumb		*Polygonum persicaria*
Morningglories		*Ipomoea* spp.
	Cypressvine	*Ipomoea quamoclit*
	Smallflower	*Jacquemontia tamnifolia*
Pennsylvania Smartweed		*Polygonum pensylvanicum*
Pigweed,	Redroot	*Amaranthus retroflexus*
	Smooth	*Amaranthus hybridis*
Prickly Sida or Teaweed		*Sida spinosa*
Ragweed,	Common	*Ambrosia artemisiifolia*
	Giant	*Ambrosia trifida*
Redweed		*Melochia corchorifolia*
Sesbania		*Sesbania exaltata*
Shepherdspurse		*Capsella bursa-pastoris*
Spurred Anoda		*Anoda cristata*
Tropic Croton		*Croton glandulosus*
Velvetleaf		*Abutilon theophrasti*
Venice Mallow		*Hibiscus trionum*
Wild Buckwheat		*Polygonum convolvulus*
Wild Mustard		*Brassica kaber*
Wild Poinsettia		*Euphorbia heterophylla*
Wild Sunflower		*Helianthus annuus*

SEDGES:

Common Name	Scientific Name
Yellow Nutsedge	*Cyperus esculentus*

CONDITIONS OF SALE AND WARRANTY

The Directions for Use of this product reflect the opinion of experts based on field use and tests. The directions are believed to be reliable and should be followed carefully. However, it is impossible to eliminate all risks inherently associated with use of this product. Crop injury, ineffectiveness or other unintended consequences may result because of such factors as weather conditions, presence of other materials, or use of the products in a manner inconsistent with its labeling, all of which are beyond the control of BASF WYANDOTTE CORPORATION ("BWC") or the Seller. All such risks shall be assumed by the Buyer.

BWC warrants that this product conforms to the chemical discription on the labels and is reasonably fit for the purposes referred to in the Directions for use subject to the inherent risks referred to above. BWC MAKES NO OTHER EXPRESS OR IMPLIED WARRANTY OF FITNESS OR MERCHANTABILITY OR ANY OTHER EXPRESS OR IMPLIED WARRANTY. IN NO CASE SHALL BWC OR THE SELLER BE LIABLE FOR CONSEQUENTIAL, SPECIAL OR INDIRECT DAMAGES RESULTING FROM THE USE OR HANDLING OF THIS PRODUCT. BWC and the Seller offer this product, and the Buyer and user accept it, subject to the foregoing Conditions of Sale and Warranty which may be varied only by agreement in writing signed by a duly authorized representative of BWC.

POAST and Basagran are registered trademarks of BASF AG.

Blazer is a registered trademark of Rohm and Haas Company.

PYRAMIN® FL

HERBICIDE

For Weed Control in Sugar Beets and Red Table Beets

A flowable containing:

ACTIVE INGREDIENTS*:		41.3%
Pyrazon (5-amino-4-chloro-2-phenyl-3(2H)-pyridazinone	35.5%	
Related aminochloro and dichlorophenyl pyridazinones	5.8%	
INERT INGREDIENTS		58.7%
TOTAL		100.0%

* Equivalent to 4.2 pounds per gallon of pyrazon and related aminochloro and dichlorophenyl pyridazinones. Pyrazon ANSI equals chloridazon BSI *et al.*

EPA Reg. No. 7969-50 U.S. Pat. No. 3,210,353

CAUTION

Keep Out of Reach of Children

Harmful if swallowed or absorbed through the skin. Avoid contact with skin, eyes or clothing. Wash thoroughly after handling. Remove and wash contaminated clothing before reuse.

STATEMENT OF PRACTICAL TREATMENT

In case of contact: for skin, wash with plenty of soap and water; for eyes, flush with plenty of water and get medical attention if irritation persists.

Shake well before using

DIRECTIONS FOR USE

It is a violation of Federal law to use this product in a manner inconsistent with its labeling.

Among the weeds controlled by PYRAMIN® FL herbicide are lambsquarters, pigweed, ragweed, shepherdspurse, fanweed, purslane, nightshade, mustard, henbit, smartweed, nettleleaf goosefoot and velvetleaf (Michigan and Ohio only).

PYRAMIN FL will not consistently provide acceptable control of annual grasses. Where grasses are a serious problem, follow directions on this label for control of grasses with specific combinations featuring PYRAMIN FL or use a separate treatment for grasses labeled for use in sugar beets or red table beets.

Do not use PYRAMIN FL as a broadcast treatment in the dry irrigated areas.

The correct method of application for PYRAMIN FL varies depending on area and cultural practices. Always follow the label recommendations which fit your area and cultural system. Check with local experts to ensure that you are applying PYRAMIN FL according to label directions.

See sections entitled **Precautions** and **Mixing Directions.**

For band treatments see section entitled **Band Treatment** in each crop section.

MIXING DIRECTIONS

Use 20 to 40 gallons of water per acre, broadcast basis, or proportionately less for banded applications.

Fill sprayer tank halfway with water. Start agitation with a mechanical or good by-pass agitator. Assure adequate hydraulic agitation. While agitating, add the required amount of PYRAMIN FL. Add water to fill tank to required volume.

Continue agitation during spraying. Mix only the amount of spray that will be used immediately.

PRECAUTIONS

Do not use this product on soils classified as sands or loamy sands because this may result in crop injury. Do not use preemergence on muck and peat.

Do not use on soils with organic matter higher than 5%, except in the Red River Valley (see special section) or less than 2½% in Michigan/Ohio for velvetleaf control (see special section), since this may result in unsatisfactory weed control or crop injury.

Do not use PYRAMIN FL as a broadcast treatment in the dry irrigated regions.

If the beet crop is lost due to climatic or soil conditions following application of PYRAMIN FL, do not plant other crops in the land treated with PYRAMIN FL during the same season. If fields are replanted to sugar beets, reseed into the treated band. Do not use PYRAMIN FL again as a preemergence treatment on the replanted beets because crop injury may result.

Do not use the same spray equipment for other purposes unless thoroughly cleaned. Coarse sprays are less likely to drift. Do not store near fertilizers, seeds, insecticides or fungicides. Do not contaminate water used for irrigation or domestic purposes.

Where this material is used in combination with TCA, do not use treated beef tops for feed or forage.

Observe all cautions and limitations on labeling of all products used in mixtures.

ENVIRONMENTAL HAZARDS

Keep out of lakes, ponds or streams. Do not contaminate water by cleaning of equipment or disposal of wastes.

STORAGE AND DISPOSAL

Do not contaminate water, food, or feed by storage or disposal.

Wastes resulting from the use of this product may be disposed of on site or at an approved waste disposal facility.

Triple rinse container (or equivalent). Then offer for recycling or reconditioning, or puncture and dispose of in a sanitary landfill, or by incineration, or, if allowed by state and local authorities, by burning. If burned, stay out of smoke.

SUGAR BEETS

Two applications of PYRAMIN FL can be used in the Central and Eastern states only. Do not apply more than a total of 7¼ quarts per acre per season as excessive chemical residue may result.

PREEMERGENCE APPLICATIONS

CENTRAL AND EASTERN STATES

(MI, MN, ND, OH)

GENERAL INFORMATION

Preemergence without Irrigation

Apply PYRAMIN FL at the rate of 3 to 3½ quarts per acre in enough water for good coverage. Use the lower rate on sandy loam soils. Spray immediately after beet seeds are planted and before beets and weeds emerge.

Where dry weather follows application and weeds emerge, use a shallow cultivation before weeds are 2 inches tall. Otherwise, do not disturb or cover the treated band.

MINNESOTA, NORTH DAKOTA

For Soils with Organic Matter of 5 to 7%

In the Red River Valley only, for the control of redroot and prostrate pigweeds, common lambsquarters and wild mustard in soils with organic matter content of 5 to 7%, apply PYRAMIN FL at the rate of 7¼ quarts per acre as a single preemergence application. Spray immediately after beet seeds are planted and before beets and weeds emerge. Do not follow with a postemergence treatment of PYRAMIN FL. Do not use the 7¼ quart rate on soils with organic matter higher than 7% or less than 5% since this may result in unsatisfactory weed control or crop injury, respectively.

Where dry weather follows application or where annual grasses are also a problem, refer to section entitled **Preemergence without Irrigation.**

MICHIGAN AND OHIO

For Velvetleaf Control

In Michigan and Ohio only, for the control of velvetleaf, apply PYRAMIN FL at the rate of 5¾ quarts per acre as a single preemergence application. Spray immediately after beet seeds are planted and before beets and weeds emerge. Do not use the 5¾ quart rate on soils with organic matter higher than 5% or less than 2½% since this may result in unsatisfactory weed control or crop injury, respectively.

Where dry weather follows application or where annual grasses are also a problem, refer to section entitled **Preemergence without Irrigation.**

PLAINS, MOUNTAIN AND WESTERN STATES

(CA, CO, ID, KS, NE, OR, TX, UT, WA, WY)

Preemergence with Sprinkler Irrigation

If sprinkler irrigation is used, apply PYRAMIN FL as a banded surface preemergence application. Do not incorporate PYRAMIN FL where sprinkler irrigation is to be used, as crop injury may result. Do not use broadcast treatments. Prepare the seedbed and pre-irrigate to the field capacity of the soil. Plant seed ¾ to one inch deep.

Immediately, or within 3 days after planting, apply PRYAMIN FL at the rate of 3 quarts per acre on low organic matter loam soils or 3½ quarts per acre on loam and clay soils. Follow directions for banded treatments (see section entitled **Band Treatment**) and adjust application rates accordingly. Follow with no more than ¾ inch of sprinkler irrigation per set before beets and weeds emerge, as greater amounts may cause beet injury. Repeat as needed to get good beet emergence. Do not exceed ¾ inch of sprinkler irrigation per set until the beets have 2 true leaves. (Do not count the cotyledonary leaves, which are the first 2 leaves to appear).

Preplant Incorporation with Furrow Irrigation

On fields to receive furrow irrigation, use the following planting procedure. Prepare seedbed or form beds for planting. On low organic matter, sandy loam soils, apply and incorporate PYRAMIN FL at the rate of 3 quarts per acre. On loams, silt loams and clays, use the 3½ quart per acre rate. Follow directions for banded treatments (see section entitled **Band Treatment**) and adjust application rates accordingly. Use a rotary tiller type of incorporator and incorporate not more than 2 inches deep. Plant beets and furrow irrigate. Since PYRAMIN FL must have moisture to control weeds effectively, irrigate until tops of beds are thoroughly wetted.

Repeat furrow irrigation as often as necessary to ensure good beet emergence and growth. Sprinkler irrigation can be substituted for furrow irrigation after the beets have developed 2 true leaves. (Do not count the cotyledonary leaves, which are the first 2 leaves to appear.)

To assure planting into treated bands, do the following things all in one tractor operation: (1) spray PYRAMIN FL, (2) incorporate, and (3) plant beets. Treat a band 1 or 2 inches wider than the tiller head used. Do not incorporate PYRAMIN FL with a disk because unsatisfactory weed control and sugar beet injury may result.

TANK MIXES

PYRAMIN FL + TCA

(Michigan, Minnesota, N. Dakota, Ohio)

Where annual grasses are also a problem, apply 4 to 8 lbs. per acre of TCA with the recommended rate of PYRAMIN FL. Follow label directions for both products.

PYRAMIN® FL + NORTRON® EC HERBICIDES FOR WINTER-GROWN SUGAR BEETS IN CALIFORNIA

In fields where wild oats and volunteer cereals are expected to be a problem, use a tank mixture of PYRAMIN FL plus Nortron at recommended rates listed in the following rate table. When mixing PYRAMIN FL in the spray tank with Nortron, add PYRAMIN FL first and agitate spray solution thoroughly; then add Nortron.

Tank Mix of PYRAMIN FL + Nortron EC
Application Rate Table

Soil Texture	10-Inch Band Width (1) Nortron EC Per Acre (2) 30″ Row	PYRAMIN FL Per Acre 30″ Row
Coarse Textured Soils: Sands, loamy sands, and sandy loams	NOT RECOMMENDED	
Medium Textured Soils: Silt loams, clay loams which contain less than 3% organic matter	$2^2/_3$ Pints	1 Quart
Fine Textured Soils: Clay loams which contain more than 3% organic matter and clays	$3^1/_2$ Pints	1 Quart

(1) For other band or row widths, adjust rates in proportion to the area actually treated. Do not apply this mixture broadcast.
(2) Nortron EC (1.5 lb. active ingredient per gallon).

Under sprinkler irrigation or where natural rainfall is adequate, apply this tank mix preemergence. See **Preemergence with Sprinkler Irrigation** above for directions and precautions regarding application of sprinkler irrigation. Where furrow irrigation is to be used, apply this tank mix preplant. See **Preplant Incorporation with Furrow Irrigation** above for directions and precautions regarding application of furrow irrigaton. Do not use the mixture under conditions where Nortron alone is not recommended. Before use, read label for Nortron for additional information and precautions.

PYRAMIN® FL + ANTOR® 4ES

(In California, Northwest, Intermountain States, Michigan and Ohio)

For control of a mixed population of weeds included on the labels for PYRAMIN® FL and Antor® 4ES herbicides, apply a tank mix according to the directions presented below by region. Observe all cautions and limitations on labeling of both products.

Mixing Directions: Add PYRAMIN FL to half the spray volume of water, agitate spray solution thoroughly, and then add Antor 4ES. Agitate while filling to final spray volume and during spraying.

CALIFORNIA

For preplant incorporation with furrow irrigation, apply the tank mix to the soil surface prior to, or at time of planting and incorporate to a depth of 1 to 2 inches using a rotary type tiller. Use the recommended rates in **Application Rate Table.** Irrigate until tops of beds are thoroughly wetted.

For preemergence application with sprinkler irrigation, apply the tank mix to the soil surface at time of planting or shortly after, but prior to weed germination, using the rates recommended in **Application Rate Table.** Irrigate prior to crop and weed germination. Follow directions and precautions on this label for use preemergence with sprinkler irrigation. Do not exceed $^3/_4$ inch of sprinkler irrigation per set until the sugar beets have two true leaves.

NORTHWEST, INTERMOUNTAIN STATES

For use preemergence under sprinkler irrigation only, apply the tank mix to the soil surface at the time of planting or shortly after, but prior to weed germination, using the rates recommended in **Application Rate Table.** Irrigate prior to crop and weed germination. Follow directions and precautions on this label for use preemergence with sprinkler irrigation. Do not exceed $^3/_4$ inch of sprinkler irrigation per set until the sugar beets have two true leaves.

MICHIGAN AND OHIO

For use preemergence without irrigation, apply the tank mix to the soil surface at the time of planting or shortly after, but prior to weed germination using rates recommended in **Application Rate Table.**

Tank Mix of PYRAMIN FL + Antor 4ES
Application Rate Table

Area	Soil Texture	Broadcast Rate (Quarts/Acre) (1) PYRAMIN FL	Antor 4ES
California, Northwest, Intermountain states, Michigan and Ohio	**Coarse:** Sandy loams only.	$2^1/_2$	2
	Medium: Loams, silt loams and clay loams containing less than 3% organic matter.	3	2 to 3 (2)
	Fine: Clay loams and clays containing 3–5% organic matter.	3	3

(1) See section of both labels on Band Application for calculation of proportional amounts.
(2) Use higher rate on clay loam soils.

POSTEMERGENCE APPLICATIONS

(ALL STATES)

GENERAL INFORMATION

Timing is very important. Observe weed growth daily. Treat after beets have 2 expanded true leaves (do not count the cotyledonary leaves, which are the first 2 leaves to appear) and before any weeds have more than 2 to 4 true leaves. Treatments on larger weeds will not be effective.

Good soil moisture *prior* to treatment is necessary. If soil is dry, irrigation is recommended prior to chemical application. For optimum performance, a postemergence application should be followed with moisture. Where furrow irrigation is possible, irrigate shortly after application to the point that top surface of beds has reached maximum water-holding capacity. If sprinkler irrigation is used, do not apply more than $^3/_4$ inch of water on the first set after the herbicide is applied.

Do not apply PYRAMIN FL if weeds are stressed by lack of moisture, excessive heat, high winds, frost or low temperature, as insufficient weed control may result. Wait until more favorable conditions prevail before applying.

In the Central and Eastern States, these combinations can follow preemergence use of PYRAMIN FL provided a total of $7^1/_4$ quarts PYRAMIN FL per acre per season is not exceeded.

The addition of a surfactant such as Citowett® Plus spreader sticker adjuvant or its equivalent often improves results of postemergence treatments. However, certain surfactants will seriously injure sugar beets. Consult your Experiment Station or Extension Service Weed Specialist for the suggested surfactant in your area.

TANK MIXES

PYRAMIN FL + NONPHYTOTOXIC EMULSIFIABLE OIL

Addition of a nonphytotoxic emulsifiable oil to PYRAMIN FL may improve broadleaf weed control and allow greater flexibility in timing of successful applications relative to size of weeds. Apply 3 to 3½ quarts PYRAMIN FL per acre according to soil type and organic matter content with 1 gallon of nonpytotoxic, emulsifiable oil per acre (broadcast basis). Use sufficient spray solution for good coverage, 40–100 gallons per acre treated. Use a nonphytotoxic oil containing 2% emulsifier and characterized as having an unsulphonated residue of over 95%.

Add PYRAMIN FL first to half the spray volume of water and mix thoroughly in the spray tank; then add the emulsifiable oil with the agitator running. Finally, fill the tank with water to required volume.

Do not use oil when maximum temperatures are expected to exceed 90°F at any time during the five-day period following the application as excessive crop injury may result.

For specific directions for using a nonphytotoxic oil with PYRAMIN FL in your locality and for recommended brands of oil to use, consult your local State Extension Weed Specialist.

Increased temporary sugar beet injury has been observed where PYRAMIN FL in combination treatments have followed preplant applications of Tillam® or RoNeet® herbicides.

PYRAMIN FL + DALAPON

Where grass and broadleaf weeds are present, use 3 to 3½ quarts of PYRAMIN FL per acre mixed with dalapon, according to state recommendations and directions on the dalapon product label.

Where broadleaf weeds are the principal problem, use PYRAMIN FL with 1 gallon per acre of nonphytotoxic, emulsifiable oil (for directions, see above). These mixtures may burn beet leaves or cause temporary stunting under some conditions. Do not apply when sugar beet foliage is wet.

PYRAMIN® FL + BETANAL® EC HERBICIDES

For broad spectrum weed control including pigweed, kochia, wild buckwheat, lambsquarters, smartweed, mustard, nightshade, ragweed and green and yellow foxtails, combine 3 quarts of PYRAMIN FL and 3 quarts of Betanal per acre. This combination provides residual as well as quick postemergence weed control. Follow the directions on the labels of both products. Add the Betanal to the half-full spray tank after the PYRAMIN FL is adequately suspended. Do not add additional surfactants to the mixture. Do not apply in more than 30 gallons of water per acre.

BAND TREATMENT

When spraying a band over the row, the amount of PYRAMIN FL applied per banded acre is reduced in proportion to the area actually treated. Example: to spray a 7-inch band over beets in rows 28 inches apart, the amount of PYRAMIN FL applied per banded acre is one-fourth the amount recommended for broadcast applications.

Three quarts/acre × ¼ of area banded = Three-quarters quart/banded acre.

Band and Row Width	Quarts PYRAMIN FL Per Banded Acre for Broadcast Rate of: 3 Quarts/Acre	3½ Quarts/Acre
7-inch band on 22 inch rows	1	1¼
7-inch band on 28 inch rows	¾	1
10-inch band on 22 inch rows	1¼	1¾
10-inch band on 28 inch rows	1	1¼

Using planter-mounted band sprayers ensures that all planted beets are treated, and that band width and rate of chemical are uniform.

RED TABLE BEETS

PREEMERGENCE APPLICATION

Apply 3 to 3½ quarts per acre of PYRAMIN FL in enough water for good coverage. Use lower rate on sandy loam soils. Spray after beet seeds are planted but before beets and weeds emerge.

If rain does not fall within 5 to 10 days after treatment, beets should be irrigated to activate PYRAMIN FL. If irrigation is not possible, use a shallow cultivation before weeds are 2 inches tall.

EARLY POSTEMERGENCE APPLICATION

Apply 3½ quarts per acre of PYRAMIN FL.

Timing is very important. Observe weed growth daily.

Treat after beets have 2 expanded true leaves and before any weeds have more than 2 to 4 true leaves. This is usually within 2 weeks after planting. Treatment on larger weeds will not be effective.

Note: PYRAMIN FL may be used both pre- and postemergence in the same season. Follow directions given on this label for each type of application.

BAND TREATMENT

When spraying a band over the row, the amount of PYRAMIN FL applied per acre is in proportion to the area actually treated.

Beet Rows 24 Inches apart and band width of:	Quarts PYRAMIN FL Per Banded Acre for Broadcast Rate of: 3 Quarts/Acre	3½ Quarts/Acre
6-inch band	¾	1
12-inch band	1½	1¾

CONDITIONS OF SALE AND WARRANTY

The Directions for Use of this product reflect the opinion of experts based on field use and tests. The directions are believed to be reliable and should be followed carefully. However, it is impossible to eliminate all risks inherently associated with use of this product. Crop injury, ineffectiveness or other unintended consequences may result because of such factors as weather conditions, presence of other materials, or use of the product in a manner inconsistent with its labeling, all of which are beyond the control of BASF WYANDOTTE CORPORATION ("BWC") or the Seller. All such risks shall be assumed by the Buyer.

BWC warrants that this product conforms to the chemical description on the label and is reasonably fit for the purposes referred to in the Directions for Use subject to the inherent risks referred to above. BWC MAKES NO OTHER EXPRESS OR IMPLIED WARRANTY OF FITNESS OR MERCHANTABILITY OR ANY OTHER EXPRESS OR IMPLIED WARRANTY. IN NO CASE SHALL BWC OR THE SELLER BE LIABLE FOR CONSEQUENTIAL, SPECIAL OR INDIRECT DAMAGES RESULTING FROM THE USE OR HANDLING OF THIS PRODUCT. BWC and the Seller offer this product, and the Buyer and user accept it, subject to the foregoing Conditions of Sale and Warranty which may be varied only by agreement in writing signed by a duly authorized representative of BWC.

Citowett and PYRAMIN are registered trademarks of BASF AG.

Antor, Betanal and Nortron are registered trademarks of NOR-AM Chemical Company.

Tillam and Ro-Neet are registered trademarks of Stauffer Chemical Company, U.S.A.

RONILAN®

FUNGICIDE 50W

A wettable powder containing:
ACTIVE INGREDIENT:
3-(3,5-dichlorophenyl)-5-ethenyl-5-methyl-2,4-oxazolidinedione 50%
INERT INGREDIENTS 50%
TOTAL 100%

EPA Reg. No. 7969-53 EPA Est. No. 279-NJ-1
RONILAN is a registered trademark of BASF AG.
Made in West Germany

CAUTION

Keep Out of Reach of Children

Causes Eye And Skin Irritation.

Do not get on skin, in eyes or on clothing.

STATEMENT OF PRACTICAL TREATMENT

If contacted flush eyes immediately with water for 15 minutes. Get medical attention. In case of contact with skin or clothing, remove contaminated clothing, wash skin thoroughly with soap and water. This product is a potential skin sensitizer. If irritation persists get medical attention.

See Additional Environmental Precautionary Statements.

Do not remove bags from carton except for immediate use.

DIRECTIONS FOR USE

It is a violation of Federal law to use this product in a manner inconsistent with its labeling.

Do not apply this product in such a manner as to directly or through drift expose workers or other persons. The area being treated must be vacated by unprotected persons.

Do not enter treated areas without protective clothing until sprays have dried.

Because certain states may require more restrictive reentry intervals for various crops treated with this product, consult your State Department of Agriculture for further information.

GENERAL INFORMATION

RONILAN is a contact fungicide for the control of botrytis fruit rot (gray mold) of strawberries and sclerotinia "drop" (watery soft rot) of head lettuce. If other diseases are a problem an additional fungicide will be needed.

STRAWBERRIES

TIME AND RATE OF APPLICATION FOR STATES OTHER THAN FLORIDA

Spray coverage of the developing fruit is essential. For full season control of botrytis disease the following spray program is recommended. The first application should be made not later than 10% primary bloom at rates indicated below (see table). The interval between subsequent applications will vary according to weather conditions and resultant disease pressure. A rate of $1^1/_2$ pounds product per acre is generally recommended. A 1 pound product per acre rate of RONILAN should be used only when low disease pressure can be predicted. A 2 pound product per acre rate should be used when the foliage is dense and/or disease pressure is high. If a heavy rainfall occurs any time during this spray program or if a wet period (light rain, fog or dew) lasting more than 24 hours occurs, immediate retreatment is necessary at a rate of $1^1/_2$–2 pounds product per acre as soon as conditions allow the spray to dry on the plants.

Application Rates for Strawberries for States Other Than Florida

Moisture Conditions	Spray* Interval (Days)	Rate (lbs. Product/A)	
		1st Year Plants Or Sparse Foliage	Dense Foliage
Frequent natural moisture (intermittent rain, fog, or dew) or when using sprinkler irrigation *(high disease pressure).*	7–9	1½	1½–2
Limited natural moisture or infrequent sprinkler irrigation *(low disease pressure).*	10–14	1	1½–2

* Use spray interval throughout the bearing cycle.

TIME AND RATE OF APPLICATION FOR FLORIDA

Spray coverage of the developing fruit is essential. For full season control of botrytis disease, applications should be made at seven day intervals throughout the production cycle. If a frost protection irrigation is planned, an application of RONILAN must be made immediately before and after the irrigation. If a heavy rainfall occurs any time during the spray program or if a wet period (light rain, fog or dew) lasting more than 24 hours occurs, immediate retreatment is necessary as soon as conditions allow the spray to dry on the plants.

Application Rate for Florida Strawberries

Moisture Conditions	Spray Interval (days)	Rate (lbs. Product/A)
Frequent natural moisture (intermittent rain, fog, or dew) or when using sprinkler irrigation *(high disease pressure).*	7	2

METHOD OF APPLICATION

Ground equipment: Application of RONILAN should be made in not less than 100 gallons of spray solution per acre to obtain thorough coverage of the developing fruit. An operating pressure of 60–150 psi is recommended to obtain adequate penetration of the spray through the canopy. Cone-type nozzles are recommended. Spray blooms with at least 3 nozzles per row (1 over row: 2 side drops) are recommended.

Air equipment: Application of RONILAN should be made in not less than 20 gallons of spray solution per acre. Thorough spray coverage of the developing fruit is essential.

RESTRICTIONS AND LIMITATIONS FOR STRAWBERRIES

Do not apply more than a total of 35 pounds of RONILAN per acre in one season.

RONILAN does not control rhizopus rot of strawberries in the field or in storage. There may be a competitive relationship between botrytis and rhizopus incidence such that control of botrytis may result in an increase of rhizopus rot in stored fruit. Rhizopus rot becomes a problem as a result of certain climatic conditions (such as prolonged warm, humid periods) and cultural practices (such as high nitrogen fertilization, which may lead to the production of softer fruit). If conditions are conducive for rhizopus development, including those described above, do not use RONILAN.

Lettuce (head types only) may be planted after strawberries.

Leafy vegetables may be planted 6 months after treatment that does not exceed 12 pounds active ingredient per acre (24 pounds product per acre).

Cucurbits may be planted 2 months after treatment that does not exceed 9 pounds active ingredient per acre (18 pounds product per acre).

Corn may be planted 2 months after treatment that does not exceed 9 pounds active ingredient per acre (18 pounds product per acre), provided only the corn grain is used for food and/or feed purposes.

Other grain crops may be planted 9 months after treatment that does not exceed 8 pounds active ingredient per acre (16 pounds product per acre).

Any rotational crops may be planted 20 days after treatment that does not exceed 3 pounds active ingredient per acre (6 pounds product per acre).

Do not use RONILAN as a plant dip as injury will occur.

LETTUCE (Head Types Only)

TIME AND RATE OF APPLICATION

For control of sclerotinia "drop" up to three applications may be made under certain conditions in one season as specified in the rate table below. Thorough spray coverage of plant parts to be protected is essential for effective disease control. *Do not disturb soil after application.*

Application Rates for Lettuce (Head Types Only)

Application	Timing	Rate (lbs. Product/A) *Low Disease Pressure	Rate (lbs. Product/A) *High Disease Pressure
First	Direct Seeded Lettuce—Application should be made immediately (within 2 days) after thinning. Transplanted Lettuce—Application should be made 7-10 days after transplanting	1	1½–2
Second	Application should be made 14 days after first spray if cool, wet conditions (which favor disease) occur for periods lasting 48 hours or more.		
Third	Direct Seeded Lettuce Only—Application should be made 14 days after second spray if conditions favoring disease occur.		

* Based on previous history of disease infestation in the field or adjacent fields.

METHOD OF APPLICATION

Application of RONILAN should be made in not less than 100 gallons of spray solution per acre (broadcast basis) to obtain thorough coverage of the lower leaves, plant stem and bed surface. An operating pressure of 50–100 psi is recommended. Use cone or flat fan nozzles. Flood type nozzles should not be used. To apply RONILAN to one-row beds spray booms should have at least 2 nozzles per row. For two-row beds use 3 nozzles or more per bed.

RESTRICTIONS AND LIMITATIONS FOR LETTUCE

Do not apply RONILAN within 28 days of harvest.

Do not use on leaf type lettuce.

Do not use RONILAN as a plant dip as injury will occur.

STORAGE AND DISPOSAL

Do not contaminate water, food, or feed by storage or disposal.

Wastes resulting from the use of this product may be disposed of on site or at an approved waste disposal facility.

Completely empty bag into application equipment. Then dispose of empty bag and box in a sanitary landfill or by incineration, or, if allowed by State and local authorities, by burning. If burned, stay out of smoke.

ENVIRONMENTAL HAZARDS

Do not apply directly to wetlands and other water bodies.

Do not contaminate water by cleaning of equipment or disposal of wastes.

In case of emergency:

In case of large scale spillage regarding this product, call:

CHEMTREC 800-424-9300

BASF Wyandotte 201-263-3400

In case of medical emergency regarding this product, call:

1. Your local doctor for immediate treatment.
2. Your local poison control center (hospital).
3. BASF Wyandotte 201-263-3400

CONDITIONS OF SALE AND WARRANTY

The Directions for Use of this product reflect the opinion of experts based on field use and tests. The directions are believed to be reliable and should be followed carefully. However, it is impossible to eliminate all risks inherently associated with use of this product. Crop injury, ineffectiveness or other unintended consequences may result because of such factors as weather conditions, presence of other materials, or use of the product in a manner inconsistent with its labeling, all of which are beyond the control of BASF WYANDOTTE CORPORATION ("BWC") or the Seller. All such risks shall be assumed by the Buyer.

BWC warrants that this product conforms to the chemical description on the label and is reasonably fit for the purposes referred to in the Directions for Use subject to the inherent risks referred to above. BWC MAKES NO OTHER EXPRESS OR IMPLIED WARRANTY OF FITNESS OR MERCHANTABILITY OR ANY OTHER EXPRESS OR IMPLIED WARRANTY. IN NO CASE SHALL BWC OR THE SELLER BE LIABLE FOR CONSEQUENTIAL, SPECIAL OR INDIRECT DAMAGES RESULTING FROM THE USE OR HANDLING OF THIS PRODUCT. BWC and the Seller offer this product, and the Buyer and user accept it, subject to the foregoing Conditions of Sale and Warranty which may be varied only by agreement in writing signed by a duly authorized representative of BWC.

Chevron Chemical Company

Ortho Agricultural Chemicals Division
575 Market Street
San Francisco, CA 94105
(415) 894-4300

BOLERO®8EC

HERBICIDE

ACTIVE INGREDIENT	BY WT.
*Thiobencarb	84%
INERT INGREDIENTS	16%
TOTAL	100%

*S-[(4-chlorophenyl)methyl] diethylcarbamothioate
Contains 8 pounds S-[(4-chlorophenyl)methyl] diethylcarbamothioate per gallon.

BOLERO®—Reg. TM of Kumiai Chemical Industry Co., Ltd. for Thiobencarb Herbicide.

Product 5281 Made in U.S.A.
Form 10129-F
EPA Reg. No. 239-2450 EPA Est. 239-MO-1

Keep Out of Reach of Children
CAUTION

PRECAUTIONARY STATEMENTS

HAZARDS TO HUMANS & DOMESTIC ANIMALS

CAUTION

Causes eye irritation. Harmful if swallowed. Avoid contact with eyes, skin and clothing. Avoid breathing spray mist.

STATEMENT OF PRACTICAL TREATMENT

In case of eye contact, wash eyes immediately with plenty of water. See a doctor if irritation persists. If swallowed, give a large amount of water to drink, make person vomit, and call a doctor.

Note to Physician: Emergency Information—call (415) 233-3737.

ENVIRONMENTAL HAZARDS

This product is toxic to shrimp. Do not apply directly to lakes, streams, ponds, tidal marshes and estuaries. Do not contaminate water by cleaning of equipment or disposal of wastes.

PHYSICAL OR CHEMICAL HAZARDS

Do not use or store near heat or open flame.

DIRECTIONS FOR USE

It is a violation of Federal law to use this product in a manner inconsistent with its labeling.

READ ENTIRE LABEL. USE STRICTLY IN ACCORDANCE WITH PRECAUTIONARY STATEMENTS AND DIRECTIONS, AND WITH APPLICABLE STATE AND FEDERAL REGULATIONS.

STORAGE AND DISPOSAL

PROHIBITIONS

Do not contaminate water, food or feed by storage, disposal or cleaning of equipment. Open dumping is prohibited.

STORAGE

Keep pesticide in original container.
Do not put concentrate or dilute into food or drink containers.
Store in cool, dry place.
Protect from excessive heat.
For help with any spill, leak, fire or exposure involving this material, call day or night (415)233-3737.

PESTICIDE DISPOSAL

Wastes resulting from the use of this product may be disposed of on site or at an approved waste disposal facility.

CONTAINER DISPOSAL

Triple rinse (or equivalent). Do not reuse container. Offer for recycling or reconditioning, or puncture and dispose of in a sanitary landfill, or dispose of by other procedures allowed by State and local authorities.

WEED CONTROL IN DRY-SEEDED RICE IN TEXAS, LOUISIANA, MISSISSIPPI AND ARKANSAS.

Application Directions—For Dry Seeded Rice in Texas, Louisiana, Mississippi and Arkansas.

Late Preemergence Application (Prior to rice and grass emergence)—For control of Watergrass (Echinochloa spp.) and Sprangletop (Leptochloa spp.). Apply BOLERO 8 EC 1 to 5 days prior to rice emergence; generally this would be 5 to 9 days after planting of the rice. Apply 4 pts. of BOLERO 8 EC per acre after soils have been sealed by rainfall or by flushing. Apply to well prepared seedbeds free of large clods. Soil should be wet at time of application. Do not apply BOLERO 8 EC as a preemergence treatment to cracked soil.

Early Postemergence Application—For control of emerged Watergrass. Apply 4 pts. of BOLERO 8 EC per acre when Watergrass has developed no further than the two-leaf stage. Rice may be fully emerged at time of application. Soil should be wet at time of application.

Do not apply BOLERO 8 EC to stressed rice. Do not apply BOLERO 8 EC to fields with exposed seed as exposed seed will be killed. Do not apply BOLERO 8 EC to second crop (stubble crop) rice.

Water Management—After application, flush the fields as often as necessary to prevent crusting and drying of the soil. Fields should be flooded as soon as the rice plants will tolerate permanent flooding.

Do not apply under conditions involving possible drift to food, forage or other plantings that might be damaged or the crops thereof rendered unfit for sale, use or consumption.

APPLICATION EQUIPMENT

Aircraft: Apply BOLERO 8 EC in no less than 10 gals. spray mix per acre.

Ground Sprayers: Apply in 10 to 20 gals. of total spray mix per acre.

Precautions: Do not apply more than 4 lbs. active ingredient per acre per year. Do not plant subsequent crops in treated fields within 6 months of last application. Do not use in fields where fall farming of crayfish will be practiced. Do not use adjacent to catfish ponds. In Texas and other areas where double cropping is the agricultural practice, do not apply to a second stubble rice crop. Water drained directly from treated fields must not be used to irrigate other crops. Application to stressed rice can result in stand reduction, chlorosis, growth inhibition, delayed maturity and/or leaf desiccation. Stress factors include but are not limited to the following: Daily temperatures below 65°F., or above 95°F., problem soils, (i.e., Zn deficiency, high salt content, high pH.), excessive moisture, (i.e., above field capacity while rice seed is germinating), drought conditions, poor field drainage or deep water after application.

CONDITIONS OF SALE

1. Chevron Chemical Company (Chevron) warrants that this material conforms to the chemical description on the label and is reasonably fit for use as directed hereon. Chevron neither makes, nor authorizes any agent or representative to make, any other warranty of FITNESS or of MERCHANTABILITY, guarantee or representation, express or implied, concerning this material.

2. Critical and unforeseeable factors beyond Chevron's control prevent it from eliminating all risks in connection with the use of chemicals. Such risks include, but are not limited to, damage to plants and crops to which the material is applied, lack of complete control, and damage caused by drift to other plants or crops. Such

risks occur even though the product is reasonably fit for the uses stated hereon and even though label directions are followed. Buyer and user acknowledge and assume all risks and liability (except those assumed by Chevron under 1 above) resulting from handling, storage and use of this material.

ORTHENE® 75 S Soluble Powder

INSECTICIDE

ACTIVE INGREDIENT	BY WT.
*Acephate (O,S-Dimethyl acetylphosphoramidothioate) ...	75%
INERT INGREDIENTS	25%
TOTAL	100%

*ORTHENE®, Acephate U.S. Pat. No. 3,716,600

EPA Reg. No. 239-2418-AA EPA Est. 239-CA-1

Form 8600-C2 Product 5332 Made in U.S.A.

CAUTION

Keep Out of Reach of Children

PRECAUTIONARY STATEMENTS

HAZARDS TO HUMANS & DOMESTIC ANIMALS

CAUTION

Harmful if swallowed. Causes eye irritation. Avoid contact with eyes, skin and clothing. Avoid breathing dust or spray mist. If swallowed, give a large amount of water to drink, make person vomit and call a doctor.

STATEMENT OF PRACTICAL TREATMENT

In case of eye contact, wash eyes with fresh water for 15 minutes. If irritation continues, see a doctor. Wash hands thoroughly after handling.

Note to Physicians: Emergency Information—call (415) 233-3737. Acephate is a cholinesterase inhibitor. If signs of cholinesterase inhibition appear, atropine is antidotal. 2-PAM may also be used in conjunction with atropine but should not be used alone.

ENVIRONMENTAL HAZARDS

This pesticide is toxic to birds.

Do not apply directly to water or wetlands. Cover or soil-incorporate spills.

Do not apply this product or allow it to drift to blooming crops or weeds if bees are visiting treatment area. This product is highly toxic to bees exposed to direct treatment or residues on blooming crops.

DIRECTIONS FOR USE

It is a violation of Federal law to use this product in a manner inconsistent with its labeling.

READ ENTIRE LABEL. USE STRICTLY IN ACCORDANCE WITH PRECAUTIONARY STATEMENTS AND DIRECTIONS, AND WITH APPLICABLE STATE AND FEDERAL REGULATIONS.

REENTRY STATEMENT

Keep all unprotected persons out of operating areas or vicinity where there may be drift.

Do not enter treated areas without protective clothing until sprays have dried.

Certain states may require more restrictive reentry intervals; consult your State Department of Agriculture for further information.

Written or oral warnings regarding use of protective clothing and accidental exposure must be given to workers who are expected to be in treated areas or in areas about to be treated.

STORAGE AND DISPOSAL

PROHIBITIONS

Do not contaminate water, food or feed by storage, disposal or cleaning of equipment.

Open dumping is prohibited.

STORAGE

Keep pesticide in original container.

Do not put concentrate or dilute into food or drink containers.

Store in a cool, dry place. Protect from excessive heat.

Do not contaminate food or foodstuffs.

Do not store or transport near feed or food.

For help with any spill, leak, fire or exposure involving this material, call day or night (415) 233-3737.

PESTICIDE DISPOSAL

Wastes resulting from the use of this product may be disposed of on site or at an approved waste disposal facility.

CONTAINER DISPOSAL

Completely empty liner by shaking and tapping sides and bottom to loosen clinging particles. Empty residue into application equipment. Do not reuse container. Dispose of liner in a sanitary landfill or by incineration if allowed by State and local authorities. If drum is contaminated and cannot be reused, dispose of in the same manner.

DIRECTIONS

Table of Equivalents

Amount of ORTHENE 75 S Soluble Powder	1⅓ lbs.	1 lb.	⅔ lb.	4 oz.	2 oz.
Amount Active Contained	1 lb.	¾ lb.	½ lb.	⅕ lb.	⅒ lb.

BEANS AND LIMA BEANS (Dry and Succulent forms): Fleahoppers, Grasshoppers—Apply ⅓ to ⅔ lb. per acre. **Aphids, Bean Leaf Beetle, Bean Leafroller, Cabbage Looper, Cutworms, Green Cloverworm, Leafhoppers, Mexican Bean Beetle, Plantbugs (Lygus), Soybean Looper, Thrips, Whitefly**—Apply ⅔ to 1⅓ lbs. per acre. **Armyworms, Corn Earworm, European Corn Borer**—Apply 1 to 1⅓ lbs. per acre. Use higher rates under severe insect infestation conditions. Apply in 20 to 100 gals. of water per acre with ground sprayers or in a minimum of 2 gals. by air. Apply at 7 to 10 day intervals or as needed. Begin applications when eggs or insects first appear. Do not apply to snapbeans or dry beans within 14 days of harvest. Lima beans harvested in succulent form may be treated up to the day of harvest. Do not feed treated vines.

BRUSSELS SPROUTS, CAULIFLOWER: Cabbage Looper, Diamondback Moth, Imported Cabbageworm—Apply 1⅓ lbs. per acre. **Green Peach Aphid**—Apply ⅔ to 1⅓ lbs. per acre. Apply in water at 25 to 150 gals. spray/acre by ground or a minimum of 5 gals. spray/acre by air. The high rate should be used when heavy infestations of Aphids are present. Begin applications when insects or eggs first appear. Repeat as needed. Do not apply more than 8 lbs. per acre per season to Brussels Sprouts. Do not apply within 14 days of harvest. Do not feed trimmings to livestock or allow animals to graze in treated areas.

CELERY: Green Peach Aphid—Apply ⅔ to 1⅓ lbs. per acre **Beet and Fall Armyworms and Cabbage Looper**—Apply 1⅓ lbs. per acre. Apply in 50 to 100 gals. water per acre with ground sprayers or in a minimum of 5 gals. per acre by air. Begin applications when eggs or insects appear and repeat at 3 to 10 day intervals as needed to maintain control. Do not apply within 21 days of harvest. Plants should be trimmed (tops removed) before shipping or use. Tops should not be used for food or feed.

COTTON: Armyworms and Cabbage Looper—Apply 1⅓ lbs. per acre. **Plantbugs (Lygus)**—Apply ⅓ to 1⅓ lbs. per acre. **Fleahopper**—Apply ⅓ lb. per acre. **Whitefly**—Apply ⅔ to 1⅓ lbs. per acre. **Thrips**—Apply 2 to 4 oz. per acre. Begin applications when eggs or insects appear and apply as needed to maintain control. **Bollworm and Tobacco Budworm (Florida, Alabama, Georgia, Arkansas, Louisiana, Mississippi and Texas Only)**—Apply ⅔ lb. per acre in early season under light infestations. Increase the rate to 1 to 1⅓ lbs. in mid and late season as infestations become moderate to severe.

(California and Arizona Only)—Apply 1⅓ lbs. per acre. Begin applications when bollworm and tobacco budworm eggs appear in the field and maintain a 3 to 7 day spray schedule. Apply in 10 to 25 gals. water with ground equipment or in 3 to 10 gals. with aerial equipment in Southern States and in 5 to 10 gals. water with aerial equipment in California and Arizona. Do not apply within 21 days of harvest. Do not feed gin trash to livestock or allow animals to graze on treated areas.

CRANBERRIES: Fireworms, Spanworms, Sparganothis—Apply 1⅓ lbs. per acre by air, ground or with sprinklers. Use a minimum of 2 gallons of water per acre by air or sufficient water to give thorough coverage with ground or sprinkler equipment. **NOTE:** Limit to 1 application per growing season. Do not apply from start of bloom until all berries are set. Do not apply within 90 days of harvest.

HEAD LETTUCE (Crisphead Type Only)—Green Peach Aphid — Apply ⅔ to 1⅓ lbs. per acre. **Cabbage Looper and Armyworms** —Apply 1⅓ lbs. per acre. Apply in 10 to 60 gals. of spray mix by ground as broadcast application or in 5 to 10 gals. of spray mix by air. Apply as necessary to maintain desired control. Do not apply more than 6⅔ lbs. per acre per season. Do not feed trimmings to livestock or allow animals to graze on treated areas. **For spring, summer and early fall crops in all areas and winter crops in Florida and Texas** do not apply within 21 days of harvest. **For late fall crops in Arizona and winter crops in Arizona and California** do not apply within 21 days of harvest except that for crops which germinate from mid-September through November in desert areas, do not apply after first head begins to form.

IMPORTED FIRE ANTS: Mound Treatment—Drench Method: Mix 1 oz. in 5 gallons of water. Apply 1 gallon of mix to each mound area by sprinkling the mound until it is wet and treat a four (4) foot diameter circle around the mound. **Dusting Method:** Evenly distribute 1 to 2 teaspoons over the mound. For best results apply the material in the early morning or late afternoon when the ants are most active. Applications made under prolonged hot and dry conditions may be ineffective due to the location of the ants deep within the nest. Grass in treated area may be injured. Treat a maximum of 13 mounds per acre. Do not treat mound more than once per season.

Do not apply when lactating dairy cattle are present. Do not pasture or feed treated hay to lactating dairy cattle within 21 days after application.

Remove meat animals from treated areas at least 1 day before slaughter if they were present at application or grazed areas within 21 days after application.

PASTURES, RANGELAND, WASTELAND: Grasshoppers— Apply ⅛ to ⅙ lb (0.094 to 0.125 lb. active ingredient) per acre in a minimum of ½ gal. spray solution by air or 10 to 20 gallons per acre with ground equipment. ORTHOTROL or NALCO-TROL® may be added to aid in controlling drift and reducing evaporation of aerial sprays. Use the higher rate of ORTHENE 75 S Soluble Powder in thicker vegetation or on larger insects. Do not apply when lactating dairy cattle are present. Do not pasture or feed treated hay to lactating dairy cattle within 21 days after application. Do not remove hay from treated area within 21 days after application. Remove meat animals from treated areas at least 1 day before slaughter if they were present at application or grazed treated areas within 21 days after application. Do not make more than 1 application per season.

NALCO-TROL®—Reg TM of Nalco Chemical Co.

PEANUTS—Grasshoppers—Apply ⅓ to ⅔ lb. per acre. **Thrips** — Apply ⅔ to 1 lb. per acre. **Corn Earworm, Fall Armyworm, Leafhoppers, Loopers, Velvetbean Caterpillar**—Apply 1 to 1⅓ lbs. per acre. Apply in 10 to 50 gals. water per acre with ground equipment or in 5 to 10 gals. water per acre by air. Begin applications when insects first appear and repeat as needed to maintain control. Do not apply within 14 days of digging. Do not feed treated forage or hay to livestock or allow animals to graze treated areas.

PEPPERMINT, SPEARMINT: Alfalfa Looper, Cutworms, Aphids—Apply 1⅓ lbs. per acre in 20 to 100 gals. of spray mix by ground or in 5 to 10 gals. of spray mix by air. Begin applications when eggs or insects first appear. Make one repeat application, if necessary, to maintain control. Do not apply more than twice per season. Do not apply within 14 days of harvest.

PEPPERS (Non-Bell Type—For Use in Midwestern and Eastern States Only: Aphids—Apply ⅔ lb. per acre and continue at 7 to 10 day intervals as needed throughout the growing season. Apply in 40 to 150 gals. of water per acre with ground equipment. Do not exceed 7 applications per season. Do not apply within 7 days of harvest.

SOYBEANS: Grasshoppers—Apply ⅓ to ⅔ lb. per acre. **Soybean Looper, Cabbage Looper, Green Cloverworm, Velvetbean Caterpillar, Bean Leaf Beetle, Mexican Bean Beetle**—Apply ⅔ to 1⅓ lbs. per acre. **Stinkbugs, Armyworms, Threecornered Alfalfa Hopper, Corn Earworm**—Apply 1 to 1⅓ lbs. per acre. Apply in 10 to 50 gals. of water with ground equipment or in 2 to 10 gals. with aerial equipment. Apply as needed but do not apply within 14 days of harvest. Do not graze or cut vines for hay or forage.

SWEET PEPPERS (Bell Type): Grasshoppers—Apply ⅓ to ⅔ lb. per acre. **Cabbage Looper, Green Peach Aphid, Tobacco Hornworm**—Apply ⅔ to 1⅓ lbs. per acre. **European Corn Borer**—Apply 1 to 1⅓ lbs. per acre. Apply in 25 to 150 gals. water with ground equipment or a minimum of 3 gals. per acre by air. Begin applications when eggs or insects appear and apply as needed until insects have been reduced below economic numbers. Do not apply within 7 days of harvest.

Do not apply under conditions involving possible drift to food, forage or other plantings that might be damaged or the crops thereof rendered unfit for sale, use or consumption.

NOTE: This product is sold by weight and package is full when packed but due to fluffy nature this material is inclined to settle.

CONDITIONS OF SALE

1. Chevron Chemical Company (Chevron) warrants that this material conforms to the chemical description on the label and is reasonably fit for use as directed hereon. Chevron neither makes, nor authorizes any agent or representative to make, any other warranty of FITNESS or of MERCHANTABILITY, guarantee or representation, express or implied, concerning this material.
2. Critical and unforeseeable factors beyond Chevron's control prevent it from eliminating all risks in connection with the use of chemicals. Such risks include, but are not limited to, damage to plants and crops to which the material is applied, lack of complete control, and damage caused by drift to other plants or crops. Such risks occur even though the product is reasonably fit for the uses stated hereon and even though label directions are followed. Buyer and user acknowledge and assume all risks and liability (except those assumed by Chevron under 1 above) resulting from handling, storage, and use of this material.

ORTHENE® Specialty Concentrate

ACTIVE INGREDIENT	BY WT.
Acephate (O,S-Dimethyl acetylphosphoramidothioate)	97%
INERT INGREDIENTS	3%
TOTAL	100%

ORTHENE®, Acephate U.S. Pat. No. 3,716,600

EPA Reg. No. 239-2486-AA EPA Est. 239-CA-1
Form 9839-K Product 5478
Made in U.S.A.

CAUTION
Keep Out of Reach of Children

PRECAUTIONARY STATEMENTS

HAZARDS TO HUMANS & DOMESTIC ANIMALS

CAUTION

Harmful if swallowed, inhaled or absorbed through the skin. Avoid contact with eyes, skin or clothing. Avoid breathing vapors. Remove and launder contaminated clothing before reuse. If swallowed, promptly drink a large quantity of water and induce vomiting. Get medical attention immediately. In case of eye contact, immediately flush eyes with fresh water for 15 minutes. Get medical attention if irritation persists. For skin contact, wash with soap and water.

Note to Physicians: Emergency Information—call (415) 233-3737.

Acephate is a cholinesterase inhibitor. If signs of cholinesterase inhibition appear, atropine is antidotal. 2-PAM may also be used in conjunction with atropine but should not be used alone.

ENVIRONMENTAL HAZARDS

This pesticide is toxic to birds. Do not apply directly to water or wetlands. Cover or soil-incorporate spills.

This product is highly toxic to bees exposed to direct treatment or residues on blooming crops or weeds. Do not apply this product or allow it to drift to blooming crops on weeds if bees are visiting treatment area.

DIRECTIONS FOR USE

It is a violation of Federal law to use this product in a manner inconsistent with its labeling.

READ ENTIRE LABEL. USE STRICTLY IN ACCORDANCE WITH PRECAUTIONARY STATEMENTS AND DIRECTIONS, AND WITH APPLICABLE STATE AND FEDERAL REGULATIONS.

STORAGE AND DISPOSAL

PROHIBITIONS

Do not contaminate water, food or feed by storage, disposal or cleaning of equipment.
Open dumping is prohibited.

STORAGE

Keep pesticide in original container.
Do not put concentrate or dilute into food or drink containers.
Store in cool, dry place.
Protect from excessive heat.
Do not contaminate food or foodstuffs.
Do not store or transport near feed or food.
For help with any spill, leak, fire or exposure involving this material, call day or night (415) 233-3737.

PESTICIDE DISPOSAL

Wastes resulting from the use of this product may be disposed of on site or at an approved waste disposal facility.

CONTAINER DISPOSAL

Completely empty liner by shaking and tapping sides and bottom to loosen clinging particles. Empty residue into application equipment. Do not reuse container. Dispose of liner in a sanitary landfill or by incineration if allowed by State and local authorities. If drum is contaminated and cannot be reused, dispose of in the same manner.

DIRECTIONS

PASTURES, RANGELAND AND WASTELAND: Grasshoppers, Mormon Crickets and Black Grass Bugs — Apply 0.094 to 0.125 lb. active ingredient per acre in 1 qt. water by air. ORTHO-TROL or NALCO-TROL® may be added to aid in controlling drift and reducing evaporation. Use the higher rate of ORTHENE in thicker vegetation or on larger insects. Do not apply when lactating dairy cattle are present. Do not pasture or feed treated hay to lactating dairy cattle within 21 days after application. Do not remove hay from treated area within 21 days after application. Remove meat animals from treated areas at least 1 day before slaughter if they were present at application or grazed treated areas within 21 days after application. Do not make more than 1 application per season.

NALCO-TROL®-Reg. TM of Nalco Chemical Co.

FOREST USES

To be used by or under supervision of personnel involved in state or federal control programs.

For Some Foliage Protection Against Gypsy Moth Larvae in Infested Forests—Apply with aircraft equipment using .52 lb. to .77 lb. (0.5 to 0.75 lb. active) in ½ gal. of water per acre. Apply after foliage expansion and insect hatch have begun. Use the low rate for suppression of light to moderate populations. Use the high rate for heavy populations and where extended residual control is desired. Prepare solution in mixing tank with good agitation prior to filling aircraft.

Aerial Applications for Suppression of Larval Populations of Douglas Fir Tussock Moth—Apply .77 lb. (0.75 lb. active) in 1 gal. of water per acre.

Aerial (Helicopter) Application for Suppression of Larval Populations of Western Spruce Budworm—Apply .52 lb. (0.5 lb. active) in 1 gal. water per acre.

Aerial Application for Suppression of Larval Populations of Spruce Budworm—Apply .52 lb. (0.5 lb. active) in ¼ to ½ gal. water per acre.

Do not apply under conditions involving possible drift to food, forage or other plantings that might be damaged or the crops thereof rendered unfit for sale, use or consumption.

CONDITIONS OF SALE

1. Chevron Chemical Company (Chevron) warrants that this material conforms to the chemical description on the label and is reasonably fit for use as directed hereon. Chevron neither makes, nor authorizes any agent or representative to make, any other warranty of FITNESS or of MERCHANTABILITY, guarantee or representation, express or implied, concerning this material.
2. Critical and unforeseeable factors beyond Chevron's control prevent it from eliminating all risks in connection with the use of chemicals. Such risks include, but are not limited to, damage to plants and crops to which the material is applied, lack of complete control, and damage caused by drift to other plants or crops. Such risks occur even though the product is reasonably fit for the uses stated hereon and even though label directions are followed. Buyer and user acknowledge and assume all risks and liability (except those assumed by Chevron under 1 above) resulting from handling, storage, and use of this material.

ORTHENE® Tobacco Insect Spray

INSECTICIDE

ACTIVE INGREDIENT	BY WT.
*Acephate (O,S-Dimethyl acetylphosphoramidothioate)	75%
INERT INGREDIENTS	25%
TOTAL	100%

*ORTHENE®, Acephate — U.S. Pat No. 3,716,600

EPA Reg. No. 239-2419 — EPA Est. 239-CA-1
Product 5248 — Made in U.S.A.
Farm 8730-Q

CAUTION

Keep Out of Reach of Children

PRECAUTIONARY STATEMENTS

HAZARDS TO HUMANS & DOMESTIC ANIMALS

CAUTION

Harmful if swallowed. Causes eye irritation. Avoid contact with eyes, skin and clothing. Avoiding breathing dust or spray mist. If swallowed, give a large amount of water to drink, make person vomit and call a doctor. In case of eye contact, wash eyes with fresh water for 15 minutes. If irritation continues, see a doctor. Wash hands thoroughly after handling.

Note to Physicians: Emergency Information—call (415) 233-3737. Acephate is a cholinesterase inhibitor. If signs of cholinesterase inhibition appear, atropine is antidotal. 2-PAM may also be used in conjunction with atropine but should not be used alone.

ENVIRONMENTAL HAZARDS

This pesticide is toxic to birds. Do not apply directly to water on wetlands. Cover or soil-incorporate spills.

This product is highly toxic to bees exposed to direct treatment or residues on blooming crops or weeds. Do not apply this product or allow it to drift to blooming crops or weeds if bees are visiting the treatment area.

DIRECTIONS FOR USE

It is a violation of Federal law to use this product in a manner inconsistent with its labeling.

READ ENTIRE LABEL. USE STRICTLY IN ACCORDANCE WITH PRECAUTIONARY STATEMENTS AND DIRECTIONS, AND WITH APPLICABLE STATE AND FEDERAL REGULATIONS.

REENTRY STATEMENT

Keep all unprotected persons out of operating areas or vicinity where there may be drift.

Do not enter treated areas without protective clothing until sprays have dried.

Certain states may require more restrictive reentry intervals; consult your State Department of Agriculture for further information.

Written or oral warnings regarding use of protective clothing and accidental exposure must be given to workers who are expected to be in treated areas or in areas about to be treated.

STORAGE AND DISPOSAL

PROHIBITIONS

Do not contaminate water, food, or feed by storage, disposal or cleaning of equipment. Open dumping is prohibited.

STORAGE

Keep pesticide in original container.

Do not put concentrate or dilute into food or drink containers.

Store in a cool, dry place.

Protect from excessive heat.

Do not store or transport near feed or food.

For help with any spill, leak, fire or exposure involving this material, call day or night (415) 233-3737.

PESTICIDE DISPOSAL

Wastes resulting from the use of this product may be disposed of on site or at an approved waste disposal facility.

CONTAINER DISPOSAL

Completely empty liner by shaking and tapping sides and bottom to loosen clinging particles. Empty residue into application equipment. Do not reuse container. Dispose of liner in a sanitary landfill or by incineration if allowed by State and local authorities. If drum is contaminated and cannot be reused, dispose of in the same manner.

DIRECTIONS

Transplant Water Application

Crop	Pest	Lbs. ORTHENE Tobacco Insect Spray Per Acre
Tobacco	Flea Beetle, Green Peach Aphid, Cutworm	1

Provides control of early season flea beetles, green peach aphids and cutworms for approximately 3–4 weeks after transplanting. For later season control of these insects, apply a foliar spray of ORTHENE Tobacco Insect Spray.

Apply in a minimum of 100 gallons of transplant water per acre.

DO NOT APPLY MORE THAN 1 LB. ORTHENE TOBACCO INSECT SPRAY PER ACRE AS A TRANSPLANT WATER APPLICATION AS SOME PHYTOTOXICITY MAY OCCUR.

Foliar Application

Crop	Pest	Lbs. ORTHENE Tobacco Insect Spray Per Acre
Tobacco (Flue-Cured, Air-Cured, Dark Fire-Cured)	Grasshoppers	1/3 to 2/3
	Green Peach Aphid, Flea Beetle, Hornworm	2/3
	Budworm, Cabbage Looper, Cutworm	1

Apply in 10 to 50 gals. water per acre with ground equipment or a minimum of 3 gals. per acre by air. Apply on a 7 day schedule or as needed. Do not apply within 3 days of harvest. Do not re-enter sprayed areas until the spray deposit has dried.

Do not apply under conditions involving possible drift to food, forage or other plantings that might be damaged or the crops thereof rendered unfit for sale, use or consumption.

NOTE: This product is sold by weight and package is full when packed, but due to fluffy nature this material is inclined to settle.

CONDITIONS OF SALE

1. Chevron Chemical Company (Chevron) warrants that this material conforms to the chemical description on the label and is reasonably fit for use as directed hereon. Chevron neither makes, nor authorizes any agent or representative to make, any other warranty of FITNESS or of MERCHANTABILITY, guarantee or representation, express or implied, concerning this material.
2. Critical and unforeseeable factors beyond Chevron's control prevent it from eliminating all risks in connection with the use of chemicals. Such risks include, but are not limited to, damage to plants and crops to which the material is applied, lack of complete control, and damage caused by drift to other plants or crops. Such risks occur even though the product is reasonably fit for the uses stated hereon and even though label directions are followed. Buyer and user acknowledge and assume all risks and liability (except those assumed by Chevron under 1 above) resulting from handling, storage, and use of this material.

PARAQUAT + PLUS®

RESTRICTED USE PESTICIDE

For retail sale to and use only by Certified Applicators or persons under their direct supervision and only for those uses covered by the Certified Applicator's certification.

A Weed and Grass Killer

Harvest Aid Chemical

ACTIVE INGREDIENT	BY WT.
Paraquat dichloride (1,1'-dimethyl-4,4'-bipyridinium dichloride)	29.1%
INERT INGREDIENTS	70.9%
TOTAL	100.0%

Contains 2 lbs. Paraquat cation per gallon.
EPA Reg. No. 239-2186-ZA

R Form CA-10220-K Product 1000-22

DANGER–POISON

POISON

DO NOT POUR FROM ORIGINAL CONTAINER EXCEPT FOR IMMEDIATE USE.

SUPPLEMENTAL PAMPHLET DIRECTIONS FOR USE.

TO PREVENT ACCIDENTS, NEVER PUT INTO FOOD, DRINK OR OTHER CONTAINERS.

DANGER:

ONE SWALLOW CAN KILL.

SYMPTOMS ARE PROLONGED AND PAINFUL.

ONSET OF SYMPTOMS MAY BE DELAYED FOR UP TO 3 DAYS AFTER SWALLOWING.

PELIGRO— PRECAUCION AL USUARIO:

Si usted no lee ingles, no use este producto hasta que la etiqueta le haya sido explicada ampliamente.

May Be Harmful or Fatal If Absorbed Through Skin or Inhaled. Harmful to Skin and Eyes.

DO NOT USE OR STORE IN OR AROUND THE HOME

KEEP OUT OF REACH OF CHILDREN

READ WORKER SAFETY RULES ON LAST PANEL

WASH AND DESTROY CONTAINER WHEN EMPTY. NEVER REUSE.

STATEMENT OF PRACTICAL TREATMENT

IF SWALLOWED

IMMEDIATELY induce vomiting by inserting finger in throat. Administer fluids and induce further vomiting. Get to a hospital or physician fast. PROMPT TREATMENT IS ESSENTIAL and must be initiated before signs and symptoms of injury appear.

Note to Physicians: Call the Chevron Poison Information Center (415) 233-3737 at any hour to obtain toxicology and medical management consultation, a supply of bentonite clay, and paraquat analysis. Symptoms following ingestion may be delayed up to 3 days. To be effective, treatment for paraquat poisoning must begin **IMMEDIATELY.** Treatment consists of binding paraquat in the gut with suspensions of bentonite clay or charcoal and removal of paraquat from the blood by prolonged charcoal hemoperfusion or continuous hemodialysis.

SKIN CONTACT

IMMEDIATELY wash with soap and water. Remove and wash all contaminated clothing with soap and hot water before reuse. Prolonged contact will cause severe irritation. Contact with irritated skin, or a cut, or repeated contact with intact skin may result in poisoning.

EYE CONTACT

IMMEDIATELY wash eyes with water for at least 15 minutes and get medical attention. Contact may cause eye injury.

INHALATION

IMMEDIATELY get away from spray mist. Exposure may cause irritation, nose bleeds and may lead to poisoning. Stop and check spray procedure. Follow Worker Safety Rules on Last panel.

DIRECTIONS FOR USE

It is a violation of Federal law and may endanger the applicator or third parties to use this product without complying with all label directions and warnings.

READ ENTIRE LABEL. USE STRICTLY IN ACCORDANCE WITH DANGER STATEMENTS AND DIRECTIONS, AND WITH APPLICABLE STATE AND FEDERAL REGULATIONS.

READ DANGER AND FIRST AID TREATMENT ON FRONT PANEL AND WORKER SAFETY RULES ON LAST PANEL.

NOTE: Do not use Around Home Gardens, Schools, Recreational Parks or Playgrounds.

Do not combine ORTHO PARAQUAT+PLUS with liquid fertilizer solutions except as specified below. DO NOT combine with suspension type liquid fertilizers as the activity of Paraquat will be reduced.

Use ORTHO X-77 Spreader (non-ionic). ORTHO X-77 Spreader is recommended, but a similar and suitable non-ionic surfactant may be substituted.

Do not use dirty or muddy water for Paraquat dilution as Paraquat will be inactivated.

DIRECTIONS
PREPLANT OR PREEMERGENCE USES

PREPLANT OR PREEMERGENCE USE—For Kill of Emerged Annual Broadleaf Weeds and Grasses and for Top Kill and Suppression of Perennials—Apply when the weeds and grasses are succulent and growth is from 1" to 6" high (larger plants are less affected by this treatment). Note: To prevent injury to germinating crop seedlings, do not apply preplant or preemergence to soils lacking clay minerals, i.e., peat, muck, pure sand, artifical planting media.

ALFALFA, CORN, LETTUCE, MELONS, PEPPERS, SORGHUM, SUGAR BEETS, TOMATOES, SAFFLOWER AND SUNFLOWER: Preplant or Preemergence Treatment—Apply 1 to 2 qts. per sprayed acre as a band treatment over the crop row, or as a broadcast treatment, prior to, during or after planting—but before emergence of the crop. **Aerial Application**—5 to 10 gals. water per acre. **Ground Application**—Use 20 to 60 gals. diluted spray. Add ORTHO X-77 Spreader (non-ionic) at 8 to 32 fl. oz. per 100 gals. of diluted spray. Seedbeds should be formed as far ahead of planting and treatment as possible to permit maximum weed and grass emergence. Seeding should be done with a minimum amount of soil disturbance. Weeds and grasses emerging after application will not be controlled. Crop plants emerged at time of application will be killed.

ASPARAGUS: Preplant or Preemergence—Apply 1 to 2 qts. per sprayed acre as a band treatment over the row or as a broadcast treatment prior to, during, or after planting, but before emergence of the crop. Add ORTHO X-77 Spreader (non-ionic) at 8 to 32 fl. oz. per 100 gals. spray. Apply 50 to 100 gals. spray per acre. Allow maximum weeds and grass emergence prior to treatment. Seeding or transplanting should be done with a minimum of soil disturbance. Weeds and grasses emerging after treatment will not be controlled.

ASPARAGUS (Plantings Established 2 Years or More): Preemergence—Apply 1 to 2 qts. per sprayed acre in 25 to 100 gals. of spray mix. Apply as a band treatment over the row or as a broadcast application prior to emergence of the crop and/or after the last harvest. Add ORTHO X-77 Spreader (non-ionic) at 8 to 32 fl. oz.

per 100 gals. of spray mix. For residual control add $2\frac{1}{2}$ to 5 lbs. PRINCEP® 80 W as a tank mix. Refer to PRINCEP® 80 W label for specific recommendations and cautions. Crop plants emerged at time of application will be severely injured. Do not apply within 6 days of harvest.

PRINCEP®—Reg. TM of CIBA-GEIGY.

BARLEY, WHEAT: Preplant and Preemergence Treatment — Apply 1 to 4 pts. per sprayed acre as a band treatment over the crop row, or as a broadcast treatment prior to, during or after planting—but before emergence of the crop. **Aerial Application**—5 to 10 gals. water per acre. **Ground Application**—Use 20 to 60 gals. diluted spray. Add ORTHO X-77 Spreader (non-ionic) at 8 to 32 fl. oz. per 100 gals. of diluted spray. Seedbeds should be formed as far ahead of planting and treatment as possible to permit maximum weed and grass emergence. Seeding should be done with a minimum amount of soil disturbance. Weeds and grasses emerging after application will not be controlled. Crop plants emerged at time of application may be killed. Do not graze livestock in treated area.

BEANS (LIMA, SNAP), BROCCOLI, CABBAGE, CARROTS, CAULIFLOWER, CHINESE CABBAGE, COLLARDS, CUCURBITS (WATERMELON, CANTALOUPE/MUSKMELON, PUMPKIN, SQUASH, CUCUMBER), EGGPLANT, PEAS, TURNIP: Preplant and Preemergence Treatment—Apply 1 to 2 qts. per sprayed acre as a band treatment over the crop row, or as a broadcast treatment before, during or after planting—but before emergence of the crop. Use sufficient water to give thorough coverage. Add ORTHO X-77 Spreader (non-ionic) at 8 to 32 fl oz. per 100 gals. diluted spray. Use the higher rate with heavier weed infestations. Seedbeds should be formed as far ahead of planting and treatment as possible to permit maximum weed and grass emergence. Seeding should be done with a minimum amount of soil disturbance. Weeds and grasses emerging after application will not be controlled. Crop plants emerged at time of application will be killed.

COTTON: Preplant Treatment—Ground Application—Apply 1 to 2 qts. per acre as a single broadcast application in 20 to 60 gals. water. Beds should be preformed to permit maximum weed and grass emergence prior to treatment. Seeding should be done with a minimum amount of soil disturbance. Weeds and grasses emerging after application will not be controlled. Add ORTHO X-77 Spreader (non-ionic) at 8 to 32 fl. oz. per 100 gals. spray.

ONIONS (Seeded) For Green or Dry Bulb Harvest: Preplant and Preemergence—Apply 1 to 2 qts. per sprayed acre as a broadcast treatment prior to, during or after planting, but before emergence of the crop. Use the high rate for heavier weed infestations and for control of Wild Oats. Add ORTHO X-77 Spreader (non-ionic) at 8 to 32 fl. oz. per 100 gals. spray. Apply in 20 to 100 gals. water per acre. Allow maximum weed and grass emergence prior to treatment. Seeding should be done with a minimum of soil disturbance. Weeds and grasses emerging after treatment will not be controlled. Crop plants emerged at time of application will be injured. Maximum of one application at the 2 qts. per acre dosage. Do not apply within 60 days of harvest. **In California**—Do not apply within 200 days of harvest.

POTATOES (Preplant or Preemergence Treatment)—Apply 1 to 2 pts. per acre as a broadcast application in 30 to 100 gals. water by ground application, or in 5 to 10 gals. by air. Add ORTHO X-77 Spreader (non-ionic) at 8 to 32 fl. oz. per 100 gals. of diluted spray. The application should be delayed to provide maximum weed and grass emergence but should be applied not later than ground cracking, before potatoes have emerged. Application made after crop emergence has reduced yields of certain sensitive varieties. Weeds and grasses emerging after application will not be controlled. Postemergence management should be performed as required to control subsequent weed and grass emergence.

FOR KILL OF EMERGED VOLUNTEER BARLEY (California Only)—Preplant in Cotton, Potato, Lettuce, Melon, Safflower, Sugar Beet and Tomato Fields—Apply $\frac{1}{2}$ to 1 pt. per acre broadcast as a preplant treatment to pre-formed seedbeds. **Aerial Application**—Use 5 to 10 gals. water per acre. Add ORTHO X-77 Spreader (non-ionic) at 8 to 32 fl. oz. per 100 gals. diluted spray.

SEEDBED PREPARATION FOR ESTABLISHING GRASSES FOR SEED PRODUCTION: To Control Grasses and Broadleaved Weeds—Prepare the seedbed and allow weed seeds to germinate. When broadleaved weeds and annual grasses are at the 3 to 5 leaf stage apply 1 qt. in 20 to 40 gals. water per acre (larger weeds will be less affected by this treatment). Add ORTHO X-77 Spreader (non-ionic) at 8 to 32 fl. oz. per 100 gals. diluted spray. This treatment can be repeated prior to or on day of seeding. Do not allow animals to graze on treated areas. Do not use seed from treated areas for animal feed. Do not use straw from treated areas for animal bedding or feed.

CONSERVATION TILLAGE USES

Paraquat and Paraquat tank mixes may be applied in liquid nitrogen and/or complete liquid fertilizer solutions (follow label recommendations for crop involved) when 20 to 60 gals. of dilute solution are applied per acre in combination with 1 to 2 pts. of ORTHO X-77 Spreader (non-ionic) per 100 gals.

Equipment must be adjusted and calibrated to obtain thorough coverage of undesirable vegetation. Inadequate coverage and/or the absence of ORTHO X-77 Spreader (non-ionic) will result in a severe reduction of Paraquat activity. Phosphate containing liquid fertilizer solutions will reduce the activity of Paraquat. When liquid fertilizer solutions containing phosphorus are used with Paraquat, and a more rapid kill is desired, use the high Paraquat label rate.

CORN ONLY: For Kill of Existing Vegetation and Residual Weed Control Where Corn Will Be Planted Directly Into a Cover Crop, Established Sod or in Previous Crop Residues.

Tank Mix with Atrazine:
Apply 1 to 2 pts. ORTHO PARAQUAT+PLUS plus $2\frac{1}{2}$ to $3\frac{3}{4}$ lbs. Atrazine 80 Wettable per acre, as a broadcast spray. Refer to Atrazine 80 Wettable label for directions, limitations and cautions. Apply 20 to 60 gals. diluted spray per acre. Add ORTHO X-77 Spreader (non-ionic) at 8 fl. oz. per 100 gals. diluted spray. Add Atrazine to spray tank first, mixing thoroughly in water. Then add Paraquat and surfactant. Constant agitation recommended at all times.

Tank Mix with AATREX® 80 W and PRINCEP® 80 W:
Apply 1 to 2 pts. ORTHO PARAQUAT+PLUS plus $1\frac{1}{4}$ to $2\frac{1}{2}$ lbs. PRINCEP 80 W plus $1\frac{1}{4}$ to $2\frac{1}{2}$ lbs. AATREX 80 W per sprayed acre prior to, during or after planting—but before crop emergence. Use the low rates of PRINCEP 80 W and AATREX 80 W on light soils, and the higher rates on heavy clay soils. Use 20 to 60 gals. of diluted spray per sprayed acre. Add ORTHO X-77 Spreader (non-ionic) at 8 fl. oz. per 100 gals. diluted spray. Add the AATREX 80 W to the spray tank while agitating, then add the PRINCEP 80 W, Paraquat and ORTHO X-77 Spreader (non-ionic). Continue agitation during application. Refer to the AATREX 80 W and PRINCEP 80 W labels for specific precautionary statements.

PRINCEP® and AATREX®—Reg. TMs of CIBA-GEIGY.

Tank Mix with BLADEX® 80 W:
Apply 1 to 2 pts. ORTHO PARAQUAT+PLUS plus $1\frac{1}{2}$ to 5 lbs. BLADEX 80 W, depending upon soil type, as a broadcast spray. Apply 20 to 60 gals. diluted spray per acre. Add ORTHO X-77 Spreader (non-ionic) at 1 qt. per 100 gals. diluted spray. Add BLADEX 80 W to spray tank first, mixing thoroughly in water. Then add Paraquat and surfactant. Constant agitation recommended at all times. Do not mix with liquid fertilizer. Refer to BLADEX 80 W label for directions, limitations and cautions.

BLADEX®—Reg. TM of Shell Chemical Co.

Tank Mix with LASSO® and Atrazine:
Apply 1 to 2 pts. ORTHO PARAQUAT+PLUS (0.25 to 0.5 lb. active) plus 4 to 6 pts. LASSO (2.0 to 3.0 lbs. active), plus 1.25 to 2.5 lbs. Atrazine 80 Wettable (1.0 to 2.0 lbs. active) per treated acre prior to, during, or after planting, but before crop emergence. Use 20 to 60 gals. of diluted spray per treated acre. Add ORTHO X-77 Spreader (non-ionic) at 8 oz. per 100 gals. of diluted spray. Add the

Atrazine to the spray tank while agitating, then add the LASSO, Paraquat and ORTHO X-77 Spreader (non-ionic). Continue agitation during application. Refer to Atrazine and LASSO labels for cautions and for specific rates for a given soil type.

LASSO®—Reg. TM of Monsanto Company.

Tank Mix with Dual® and AATREX®:

Apply 1 to 2 pts. ORTHO PARAQUAT + PLUS per acre plus DUAL plus AATREX as a broadcast ground spray before, during, or after planting but before crop emergence. Apply in 20 to 60 gals. of diluted spray per acre. Add ORTHO X-77 Spreader (non-ionic) at 8 fl. oz. per 100 gals. of diluted spray. Refer to DUAL and AATREX labels for rates of application, directions, limitations, cautions and for a list of weeds controlled by preemergence activity of DUAL and AATREX. Add AATREX to spray tank first, mixing thoroughly in carrier. Then add DUAL, ORTHO PARAQUAT + PLUS AND ORTHO X-77 Spreader (non-ionic). Constant agitation recommended at all times.

DUAL® and AATREX®—Reg. TMs of CIBA-GEIGY.

Tank Mix with DUAL® and BLADEX®:

Apply 2 pts. ORTHO PARAQUAT + PLUS per acre plus DUAL plus BLADEX as a broadcast ground spray before, during, or after planting but before crop emergence. Apply in 20 to 60 gals. of diluted spray per acre. Add ORTHO X-77 Spreader (non-ionic) at 1 to 2 qts. per 100 gals. of diluted spray. Refer to DUAL and BLADEX labels for rates of application, directions, limitations, cautions and for a list of weeds controlled by preemergence activity of DUAL and BLADEX. Add BLADEX to spray tank first, mixing thoroughly in carrier. Then add DUAL, ORTHO PARAQUAT + PLUS and ORTHO X-77 Spreader (non-ionic). Constant agitation recommended at all times.

BLADEX®—Reg. TM of Shell Chemical Co.

DUAL®—Reg. TM of CIBA-GEIGY.

GRAIN SORGHUM: For Desiccation and Residual Control of Annual Broadleaf Weeds and Grasses Where Sorghum Will Be Planted Directly Into Previous Crop Residues—Apply 1 to 2 pts. ORTHO PARAQUAT+PLUS plus 2 to 3 lbs. Atrazine 80 Wettable per acre as a broadcast ground spray before, during or after planting but before crop emergence. Refer to Atrazine 80 Wettable label for directions, limitations, cautions and for a list of weeds controlled by preemergence activity of Atrazine. Apply 20 to 60 gals. diluted spray per acre. Add ORTHO X-77 Spreader (non-ionic) at 16 to 32 fl. oz. per 100 gals. diluted spray. Water or nitrogen solutions may be used as a carrier. Add Atrazine to spray tank first, mixing thoroughly in carrier. Then add Paraquat and ORTHO X-77 Spreader (non-ionic). Constant agitation recommended at all times. This herbicide tank mixture will not control volunteer sorghum or shattercane which emerges after application.

Tank Mix with Atrazine and IGRAN® 80 W: (This tank mix should be used instead of Paraquat + Atrazine when better preemergence control of barnyardgrass, crabgrass and fall panicum is desired.)

Apply 1 to 2 pts. ORTHO PARAQUAT+PLUS plus 1 to 2 lbs. Atrazine 80 W or 1.6 to 3.2 pts. Atrazine 4L plus 2 to 2½ lbs. IGRAN 80 W per acre as a broadcast ground spray before, during or after planting but before crop emergence. Weed control is often superior from application made approximately one week prior to planting. Use low rates of Atrazine and IGRAN 80 W on medium soils, and the higher rates on heavy soils. Refer to Atrazine and IGRAN 80 W labels for directions, limitations, cautions and for a list of weeds controlled by preemergence activity of Atrazine and IGRAN 80 W. The addition of 2,4-D (ester) at 0.5 lb. active per sprayed acre may aid in the suppression of perennial broadleaf weeds and annual broadleaf weeds that have made excessive growth before application. Refer to the 2,4-D label for directions, limitations and cautions. Apply 20 to 40 gals. diluted spray per acre. Add ORTHO X-77 Spreader (non-ionic) at 16 to 32 fl. oz. per 100 gals. diluted spray. Use only water or nitrogen solutions as a carrier. Add IGRAN 80 W to spray tank first, mixing thoroughly in carrier. Then add Atrazine, 2,4-D (if used), Paraquat and ORTHO X-77 Spreader (non-ionic). Constant agitation recommended at all times.

IGRAN®—Reg. TM of CIBA-GEIGY.

SOYBEANS: Preplant or Preemergence Treatment—Apply 1 to 4 pts. of Paraquat per sprayed acre as a band treatment over the crop row, or as a broadcast treatment, prior to, during or after planting—but before emergence of the crop. Use the 1.0 pt. rate when spring annual broadleaf weeds and grasses are less than 4 inches high or when a directed spray or cultivation will be used within 3 weeks after planting. **Aerial Application**—5 to 10 gals. spray mix per acre. **Ground Application**—20 to 60 gals. spray mix per acre. Add ORTHO X-77 Spreader (non-ionic) at 8 fl. oz. per 100 gals. of diluted spray. Seeding should be done with a minimum amount of soil disturbance. Weeds and grasses emerging after application will not be controlled. Crop plants emerged at time of application may be killed and yields reduced.

SOYBEANS: Minimum Tillage Systems.

Tank Mix with LASSO® 4 EC and LOROX® 50 WP:

Use this tank mixture for control of most emerged annual weeds, suppression of many emerged perennial weeds and preemergence control of many annual weeds when soybeans will be planted directly into a cover crop, stale seed bed, or in previous crop residues such as wheat stubble. This tank mixture will not control regrowth from perennial weeds. Do not apply by air.

For weeds controlled preemergence, see the LASSO and LOROX labels. Refer to the LASSO and LOROX labels for precautionary statements, crop rotation restrictions, rates which are dependent upon soil textures, directions, limitations, and cautions.

Broadcast Treatment—Apply 1 to 2 pts. ORTHO PARAQUAT+PLUS in this tank mixture in 20 to 60 gals. water per acre immediately before, during or after planting but before crop emergence. Use the 2 pt. rate if any weeds are 4 to 6 inches tall. This mixture may not control weeds taller than 6 inches. As density of stubble, crop residue or weeds increases, spray gallonage should be increased within the recommended range for complete coverage. Add ORTHO X-77 Spreader (non-ionic) at 8 fl. oz. per 100 gals. of diluted spray. First, add the LOROX 50 WP to the spray tank while agitating, then add the LASSO, ORTHO PARAQUAT+PLUS and ORTHO X-77 Spreader (non-ionic). Continue agitation during application.

LASSO®—Reg. TM of Monsanto Company.

LOROX®—Reg. TM of E.I. du Pont de Nemours & Co., Inc.

SOYBEANS (Preemergence Treatment)—For Kill of Emerged Annual Broadleaf Weeds and Grasses and for Top Kill and Suppression of Emerged Perennials Where Soybeans Will Be Planted Directly into a Preformed Bed, Cover Crop, or in Previous Crop Residues.

Tank Mix with LOROX®:

Apply 1 pt. ORTHO PARAQUAT+PLUS in this tank mixture as a band treatment over the crop row or as a broadcast treatment after planting—but before emergence of the crop. Refer to the LOROX labels for precautionary statements, rates which are dependent upon soil textures, directions, limitations and cautions. **Aerial Application**—Apply 5 to 10 gals. spray mix per acre. **Ground Application**—Use 20 to 60 gals. diluted spray. Add ORTHO X-77 Spreader (non-ionic) at 8 fl. oz. per 100 gals. of diluted spray. Seeding should be done with a minimum amount of soil disturbance. Crop plants emerged at time of application will be killed. For band applications, compute amount of Paraquat and LOROX needed per acre by dividing the band width by the inches between crop rows and then multiplying the answer by the recommended broadcast rates. Continue agitation during application. This herbicide tank mixture WILL NOT CONTROL Rhizome Johnsongrass.

LOROX®—Reg. TM of E.I. du Pont de Nemours & Co., Inc.

Tank Mix with LASSO® and SENCOR®:

For Contact Action on Emerged Weeds and Preemergence Control of Smartweed, Pigweed and Cocklebur, Foxtail and Crabgrass. (This tank mix should be used instead of Paraquat + LOROX or Paraquat + SENCOR when better preemergence control of crabgrass and foxtail is desired.)

Apply 1 to 2 pts. ORTHO PARAQUAT+PLUS in this tank mixture per acre as a preemergence broadcast ground spray. Refer to LASSO and SENCOR labels for rates which are dependent upon soil textures, directions, limitations and cautions. Apply in 20 to 60 gals. spray mix per acre. Use the 2 pt. rate of Paraquat if weeds are 4 to 6 inches tall. This mixture will not control weeds taller than six inches. Add ORTHO X-77 Spreader (non-ionic) at 8 fl. oz. per 100 gals. of diluted spray. Add SENCOR to the spray tank first and mix thoroughly. Then add LASSO followed by ORTHO PARAQUAT+PLUS and ORTHO X-77 Spreader (non-ionic). Constant agitation is recommended at all times. Do not use treated vines for feed or forage.

LASSO®—Reg. TM of Monsanto Company.

SENCOR®—Reg. TM of the Parent Company of Farbenfabriken Bayer GmbH, Leverkusen.

Tank Mix with PROWL® and SENCOR® or LEXONE® and/or LOROX® (Except CA):

Use this tank mixture for control of most emerged annual weeds, suppression of many emerged perennial weeds and preemergence control of many annual weeds when soybeans will be planted directly into a cover crop, stale seed bed, or in previous crop residues such as wheat stubble. This tank mixture will not control regrowth from perennial weeds.

Apply 1 to 4 pts. ORTHO PARAQUAT + PLUS per acre plus PROWL plus SENCOR or LEXONE and/or LOROX as a broadcast ground spray before, during, or no more than 2 days after planting (before crop emergence). The higher Paraquat rates will be required under dry conditions or when weeds are 3-6 inches tall. Apply in 20 to 60 gals. of diluted spray per acre. Add ORTHO X-77 Spreader (non-ionic) at 2 to 4 pts. per 100 gals. of diluted spray. Refer to PROWL, SENCOR, LEXONE and LOROX labels for rates of application, directions, limitations, cautions and for a list of weeds controlled preemergence by each herbicide. Fill spray tank one-half to three-quarters full with water. Add SENCOR or LEXONE or LOROX to spray tank first, mixing thoroughly in water. Refer to respective labels for mixing directions. Then add PROWL, ORTHO PARAQUAT + PLUS and ORTHO X-77 Spreader (non-ionic). Constant agitation recommended at all times.

PROWL®—Reg. TM of American Cyanamid Company.

SENCOR®—Reg. TM of the Parent Company of Farbenfabriken Bayer, GmbH, Leverkusen.

LEXONE® and LOROX®—Reg. TMs of E.I. du Pont de Nemours & Co., Inc.

Tank Mix with SENCOR®:

For Contact Action on Emerged Weeds and Preemergence Control of Smartweed, Pigweed and Cocklebur. (This tank mix should be used instead of Paraquat + LOROX when infestations of emerged smartweed, pigweed and cocklebur are present at application and when better preemergence control of these species is desired.)

Apply 1 to 2 pts. ORTHO PARAQUAT+PLUS in this tank mixture per acre as a broadcast ground spray before, during, or after planting but before emergence of the crop. Use the 2 pt. rate of ORTHO PARAQUAT+PLUS if weeds are 4 to 6 inches tall. This mixture will not control weeds taller than six inches. Apply in 20 to 60 gals. spray mix per acre. Add ORTHO X-77 Spreader (non-ionic) at 1 qt. per 100 gals. of diluted spray. Do not use treated vines for feed or forage.

Refer to the SENCOR labels for cautions, limitations and recommended rates for specific soil textures. In using minimum tillage equipment, it is important to observe the proper planting depth caution stated on the SENCOR labels so application is not made directly onto the soybean seed.

SENCOR®—Reg. TM of the Parent Company of Farbenfabriken Bayer GmbH, Leverkusen.

Tank Mix with SURFLAN® and SENCOR® or with SURFLAN® and LEXONE®: For Contact Action on Emerged Weeds and Preemergence Control of Smartweed, Pigweed, Lambsquarters, Velvetleaf, Common Ragweed, Foxtail and Fall Panicum. (This tank mix should be used instead of Parquat + LASSO + SENCOR, Paraquat + SENCOR or Paraquat + LOROX when longer residual control of foxtail and fall panicum is desired.)

Apply 1 qt. ORTHO PARAQUAT+PLUS in this tank mixture per acre at planting or within 2 days after planting. Refer to SURFLAN and SENCOR or LEXONE labels for rates which are dependent upon soil textures, directions, limitations and cautions. Apply in 20–40 gals. spray mix per acre. Add ORTHO X-77 Spreader (non-ionic) at 8 fl. oz. per 100 gals. of diluted spray. Add SENCOR or LEXONE to the spray tank first and mix thoroughly. Then add SURFLAN followed by ORTHO PARAQUAT+PLUS and ORTHO X-77 Spreader (non-ionic). Constant agitation is recommended at all times. Do not use treated vines for feed or forage.

SURFLAN®—Reg. TM of Elanco Products Company.

SENCOR®—Reg. TM of the Parent Company of Farbenfabriken Bayer GmbH, Leverkusen.

LEXONE®—Reg. TM of E. I. du Pont de Nemours & Co., Inc.

Tank Mix with DUAL® and SENCOR® or LEXONE® or LOROX®:

Apply 1 to 2 pts. ORTHO PARAQUAT + PLUS per acre plus DUAL plus SENCOR or LEXONE or LOROX as a broadcast ground spray before, during, or after planting but before crop emergence. Apply in 20 to 60 gals. of diluted spray per acre. Add ORTHO X-77 Spreader (non-ionic) at 8 fl. oz. per 100 gals. of diluted spray. Refer to DUAL, SENCOR, LEXONE and LOROX labels for rates of application, directions, limitations, cautions, and for a list of weeds controlled by preemergence activity of DUAL, SENCOR, LEXONE and LOROX. Add SENCOR or LEXONE or LOROX to spray tank first, mixing thoroughly in carrier. Then add DUAL, ORTHO PARAQUAT + PLUS and ORTHO X-77 Spreader (non-ionic). Constant agitation recommended at all times.

DUAL®—Reg. TM of CIBA-GEIGY.

SENCOR®—Reg. TM of the Parent Company of Farbenfabriken Bayer, GmbH, Leverkusen.

LEXONE® and LOROX®—Reg. TMs of E.I. du Pont de Nemours & Co., Inc.

WHEAT: CHEMICAL FALLOW

For Desiccation of Most Emerged Annual Broadleaf Weeds and Grasses and Suppression of Many Emerged Perennial Weeds During the Fallow Period(s) of the Wheat-Fallow, Wheat-Annual Crop-Fallow, Wheat-Annual Crop and Continuous Wheat Cropping Systems—Apply 1 to 2 pts. ORTHO PARAQUAT + PLUS per acre as a broadcast spray. **Aerial Application**—Apply 5 to 10 gals. spray mix per acre. **Ground Application**—Use 20 to 60 gals. spray mix per acre. As density of stubble, crop residue, or weeds increases, spray gallonage should be increased within the recommended range for complete coverage. Add ORTHO X-77 Spreader (non-ionic) at 1 to 4 pts. per 100 gals. of diluted spray. Use only clean water or nitrogen solutions as a carrier. Weeds and grasses emerging after application and regrowth from perennial weeds will not be controlled. Weeds taller than 6 inches at application may not be controlled. The addition of 2,4-D or BANVEL® may aid in the suppression of perennial broadleaf weeds and annual broadleaf weeds that have made excessive growth before application. Refer to 2,4-D and BANVEL labels for directions, limitations and cautions. For extended weed control during the fallow period, tank mix with a registered residual herbicide or herbicide combination. Refer to residual herbicide label(s) for directions, limitations, cautions, and for a list of weeds controlled by preemergence activity of the residual herbicide(s). Add wettable powder herbicide(s) to spray tank first, mixing thoroughly in carrier. Then add liquid herbicide(s), ORTHO PARAQUAT + PLUS, and ORTHO X-77 Spreader (non-ionic). Constant agitation recommended at all times.

BANVEL®—Reg. TM of Velsicol Chemical Corp.

PASTURE AND RANGE RESEEDING

PASTURE AND RANGE RESEEDING: Suppression of Existing Sod and Undesirable Emerged Broadleaf Weeds and Grasses to Permit Pasture and Range Reseeding—Apply recommended rates in water to make 18 to 75 gals. diluted mix per sprayed acre. Add 8 to 32 fl. oz. ORTHO X-77 Spreader (non-ionic) per 100 gals. spray. Apply broadcast or in bands over the drill row prior to, or at time of seeding desired grasses and/or forage legumes. Apply as a band treatment when recovery of the sprayed sod is desired, such as in the case with desirable fescue grass. Apply as a broadcast treatment when existing grass species is undesirable and should be totally suppressed, such as in the case with run-down bluegrass sod.

West of Cascade and Sierra Nevada Mountains—1 to 2 pts. per sprayed acre prior to, or at time of seeding grasses such as Hardding-grass and Palestine Orchardgrass. Apply in October through December after first fall rains and after weeds have emerged and sod has started new growth. For best seeding results, apply on moderate to heavily grazed areas. Do not use in areas with heavy sod and weed growth. Do not pasture treated areas until growth of newly planted seedlings is 3 to 6 inches high.

East of Rocky Mountains—1 to 2 pts. per sprayed acre during the growing season. Use the 2 pts. rate to suppress vigorous and coarse sod species such as Bromegrass. See specific directions for use in the South and Southeast for suppression of Bermudagrass or Bahiagrass sods. Apply prior to, or at time of seeding grasses or forage legumes such as alfalfa, clover and birdsfoot trefoil. Apply only to grazed or mowed pastures which are not more than 2 to 3 inches in height at time of treatment. Do not graze treated areas until the newly planted seedlings have reached their recommended grazing height. Under good seedling emergence and growing conditions, allow a minimum of one month between treatment and first grazing. Seedling grasses, and forage legumes should be at least 6″ tall; Sudangrass must be 18″ tall and Sorghum-sudan 24″ tall at first grazing. Under late fall or winter conditions, it may take 3 to 5 months for seeded grasses and / or legumes to reach recommended grazing height.

South and Southeast—Bermudagrass or Bahiagrass Sods—Late Summer or Early Fall Application for Sod Suppression when Seeding Winter Annuals: Apply 1 pt. per acre in 20 to 30 gals. water in late summer or early fall to sod that does not exceed 3 inches in height. Apply prior to, or at time of seeding winter annuals. Add 1 qt. ORTHO X-77 Spreader (non-ionic) per 100 gals. spray. Do not pasture in treated areas until 60 days after treatment or until winter annual seedlings are 9 inches tall.

BERMUDAGRASS PASTURES: For the Suppression of Emerged Annual Broadleaf and Grassy Weeds including Little Barley, Bromes, Buttercup and Carolina Geranium in Dormant Bermudagrass Pastures—Apply 1 pt. per acre in 20 to 30 gals. water with ground equipment or 3 to 5 gals. water per acre with fixed wing aircraft during March. For control of Little Barley, apply before the mid boot stage. Add 1 qt. ORTHO X-77 Spreader (non-ionic) per 100 gals. spray mix. Do not pasture or mow for hay until 40 days after treatment.

ALFALFA, CLOVER

ALFALFA, CLOVER—Washington, Oregon, Colorado, Idaho, Montana, Utah, Nevada, Wyoming and California (counties of Del Norte, Siskiyou, Modoc, Shasta, Lassen, Plumas, Sierra and Nevada): Apply for Desiccation of Ryegrass, Bluegrass, Cheatgrass, Dogfennel, Chickweed and Tansy Mustard in Dormant Alfalfa and Clover, Except Annual Clover —Apply 2 to 3 pts. per acre (use high rate for cheatgrass) in 20 to 100 gals. spray mix by ground as a broadcast application. Apply by air in 5 to 10 gals. spray mix per acre. Apply to established stands after the last fall cutting when the crop is dormant but before spring growth starts. Do not apply if regrowth following the last fall cutting is more than 2″ tall. Alfalfa and clover foliage present at the time of application will be burned. Apply when the broadleaved weeds and grasses are succulent and growth is from 1″ to 6″ tall. Add ORTHO X-77 Spreader (non-ionic) at 8 to 32 fl. oz./100 gals. spray mix. Do not pasture animals in treated fields before first cutting. Do not graze, cut or harvest within 60 days of application for Clover or 42 days of application for alfalfa. Do not apply more than once per season.

ALFALFA: For use in states east of the Rockies and north of the southern borders of Colorado, Kansas, Missouri, Kentucky and Virginia—For control of Bluegrass, Chickweed, Henbit and Downy Brome and suppression of perennial grasses including Orchard Grass, Timothy and Smooth Brome. Apply 2 to 3 pts. per acre in 20 to 60 gals. water by ground equipment as a broadcast application. Apply by air in 5 to 10 gals. spray mix per acre. Use 3 pt. rate for kill or suppression of harder to control weeds and grasses, such as the perennial species. Do not apply if fall regrowth following the last fall cutting is more than 6″ tall. Apply to well-established stands (at least one year old) after the last fall cutting when the crop is dormant or before spring growth reaches one inch. Alfalfa foliage present at the time of application will be burned which may reduce the yield of the first cutting. Weeds and grasses should be succulent and growing at the time of application. Weeds that germinate after application will not be controlled. Add ORTHO X-77 Spreader (non-ionic) at 8 to 32 fl. oz./100 gals. spray mix. Do not graze, cut or harvest within 60 days of application for clover or 42 days of application for alfalfa. Do not apply more than once per season.

Tank Mix with PRINCEP® 80 W Simazine Weed Killer:
ALFALFA only—For control of Filaree, Jim Hill Mustard and Shepherdspurse, and for improved control of Bluegrass, Cheatgrass, Chickweed and Tansy Mustard. Apply 1 to 3 pts. ORTHO PARAQUAT+PLUS plus 1 to 2 lbs. PRINCEP 80 W in 20 to 100 gals. spray mix per acre with ground equipment or 5 to 10 gals. spray mix per acre by air. Add ORTHO X-77 Spreader (non-ionic) at 8 to 32 fl. oz. per 100 gals. spray mix. Use the 1 pt. rate of Paraquat when weeds and grasses are less than 4″ tall. Refer to PRINCEP 80 W label for cautions, limitations and recommended rates for specific soil types. Do not pasture animals in treated fields before first cutting. Do not graze, cut or harvest within 90 days of application. Do not apply more than once per season.

PRINCEP®—Reg. TM of CIBA-GEIGY.

Tank Mix with VELPAR L®:

Alfalfa only—For control of Chickweed, Downy Brome and Tansy-mustard (Idaho, Oregon, Utah and Washington only). Apply 1 to 2 pts. of ORTHO PARAQUAT + PLUS plus 1 to 2 qts. of VELPAR L per acre in 10 gals. spray mix by air or 20 gals. spray mix by ground equipment. Add ORTHO X-77 Spreader (non-ionic) at 16 to 32 fl. oz. per 100 gals. spray mix. Use the lower rate of VELPAR L on loamy sand or sandy loam soils. Use the 1 pt. rate of Paraquat when weeds and grasses are less than 4 inches tall. Refer to the VELPAR label for cautions, limitations and recommended rates for specific soil types. Apply once to established stands of alfalfa during the dormant season for alfalfa. Do not apply to regrowth that is more than 2 inches tall. Do not apply following cuttings during the growing season. Do not apply to alfalfa during the first season after seeding. Stress which may be caused in part by low fertility, disease, insects, winter kill, over cutting, drought or frost may increase the chances of crop injury. Temporary chlorosis may occur on alfalfa regrowth. Do not graze, cut or harvest within 60 days of application.

NOTE: Do not use on gravelly or rocky soils, exposed sub soils, hardpan, sand, poorly drained or alkali soils. Crop injury, including mortality, may result in fields having restricted root growth due to non-uniform soil profiles such as gravel bases, hardpans and clay lenses.

VELPAR L®—Reg. TM of E.I. du Pont de Nemours & Co., Inc.

ALFALFA only—Suppression and Control of Broadleaf Weeds and Grasses in New Alfalfa Seedlings Grown for Hay (California Only).

For Control of:	Use Rate Directions—Pints/Acre For Suppression	For Control
Spikeweed (4 in. tall or less)	½ pt.	1 to 2 pts.
Volunteer Small Grain (8 in. tall or less)	½ to 1 pt.	2 pts.
Fiddleneck (6 in. tall or less)	½ to 1 pt.	2 pts.
Shepherdspurse	1 to 2 pts.	—
Annual Bluegrass	—	1 to 2 pts.
Chickweed	—	1 to 2 pts.
Red Maids (6 in. tall or less)	—	1 to 2 pts.

Apply in 25 to 50 gals. water per acre by ground or in 5 to 10 gals. by air as a broadcast application during late winter or early spring. Add ORTHO X-77 Spreader (non-ionic) at 8 to 32 fl. oz./100 gals.

spray mix. Do not use the ½ pt. rate unless the alfalfa has at least 3 trifoliate leaves; the 1 pt. rate unless the alfalfa has 6 trifoliate leaves; or rates over 1 pt. unless there are 9 trifoliate leaves. Alfalfa foliage present at time of application will be burned. Do not pasture animals in treated fields before first cutting. Do not graze, cut or harvest within 70 days of application. Do not apply more than once during the first growing season. **CAUTION:** SEEDLING ALFALFA STANDS WILL BE REDUCED AND REPLANTING MAY BE NECESSARY. NOT RECOMMENDED IN SEEDLING ALFALFA GROWN FOR SEED.

HARVEST-AID USES

COTTON HARVEST AID

TO AID IN OPENING OF MATURE BOLLS AND TO DESICCATE GREEN LEAVES: ½ pt. in combination with 1 pt. of phosphate or 1 gal. of chlorate defoliant per acre. Apply when 80% or more of the bolls are open and the remaining bolls to be harvested are mature. Development of immature bolls will be inhibited.

TO AID IN DEFOLIATION AND OPENING OF MATURE BOLLS OF WESTERN COTTON (ARIZONA, CALIFORNIA) First Application—⅓ to ½ pt. Paraquat plus 1 to 2 pts. phosphate defoliant or 1 to 2 gals. chlorate defoliant per acre. Use higher dosages on rank cotton. (Caution—Do not use more than ½ pt. Paraquat in initial application, as excessive desiccation may occur.) Apply when 60% or more of the bolls are open and remaining bolls to be harvested are mature. Development of immature bolls will be inhibited. Do not use more than 4.0 lbs. actual sodium chlorate per acre or more than 1.5 lbs. actual phosphate type defoliants per acre.

Second Application—To aid in Opening of Mature Bolls at Second or Later Harvest Aid Treatment and to Desiccate Green Leaves—½ to 1 pt. Paraquat per acre (in desert cotton areas or on rank vigorous cotton, use the 1 pt. rate only) or ½ to 1 pt. Paraquat plus 1½ to 2 pts. phosphate defoliant or 1½ to 2 gals. chlorate defoliant. Apply when 75% or more of the bolls are open and the remaining bolls to be harvested are mature. Development of immature bolls will be inhibited.

TO DESICCATE STRIPPER COTTON: 1 to 2 pts. per acre. When foliage is dense, use 2 applications of 1 pt. if necessary. Apply when 85% or more of the bolls are open and the remaining bolls to be harvested are mature. Development of immature bolls will be inhibited.

AERIAL APPLICATION—Apply 3 to 10 gals. spray mix per acre. Use ORTHO X-77 Spreader (non-ionic) at 8 to 32 fl. oz. per 100 gals. spray mix. Do not apply during periods of thermal inversion to avoid drift.

GROUND APPLICATION—Apply 10 to 30 gals. spray mix per acre. Arrange nozzles to provide thorough coverage of the foliage. Use ORTHO X-77 Spreader (non-ionic) at 8 to 32 fl. oz. per 100 gals. spray mix.

May apply as split application. Do not pasture lactating dairy animals. Do not pasture livestock in treated fields within 15 days after treatment. Remove livestock from treated area 30 days before slaughter. Do not feed gin trash to livestock. Do not apply to cotton within 3 days before harvest. Repeat application if necessary. Do not make more than 2 applications or exceed a total of 2 pts. per acre. When combined with phosphate defoliants, observe livestock cautions listed on the phosphate label. When combined with chlorate defoliants, do not pasture livestock on treated areas or feed treated foliage or gin trash. Do not apply within 7 days when used with phosphate or chlorate defoliant.

GUAR—For Use as a Preharvest Crop Desiccant—Apply 1 qt. in 20 to 30 gals. of water per acre. Add ORTHO X-77 Spreader (non-ionic) at 1 to 2 qts. per 100 gals. water. Apply after pods are fully matured. Do not harvest until at least 4 days after application. Do not graze treated areas or feed treated forage to livestock.

POTATO VINES: PREHARVEST VINE KILLING—1 to 2 pts. per acre. Use the 2 pt. rate where quick vine kill is desired or 2 applications of 1 pt. when vine growth is dense. Apply in 50 to 100 gals. water per acre with thorough coverage. Add 8 to 32 fl. oz. of ORTHO X-77 Spreader (non-ionic) per 100 gals. water. Begin application when leaves are beginning to turn yellow. Immature potato foliage and drought stressed potato foliage are tolerant to Paraquat and desiccation will not be complete. Do not make more than 2 applications with a minimum of 5 days between applications. Do not apply Paraquat to potatoes within 3 days before harvest. Do not pasture livestock in treated potato fields. To avoid injury to subsequent crops, do not use on muck or peat soils. Do not use Paraquat for desiccating potato vines when the potatoes are to be stored or used for seed as this use may result in tuber decomposition and failure of seed pieces to germinate and grow normally.

SOYBEAN HARVEST AID—Apply ½ to 1 pt. per acre on Broadleaf Weeds and Grasses. Use high rate on Cocklebur. Add 1 qt. ORTHO X-77 Spreader (non-ionic) per 100 gals. spray. **Aerial Application**—2 to 5 gals. spray per acre. **Ground Application**—20 to 40 gals. spray per acre. **Indeterminate Varieties:** Apply when at least 65% of the seed pods have reached a mature brown color or when seed moisture is 30% or less. **Determinate Varieties:** Apply when soybean plants are mature, i.e., beans are fully developed, at least ½ of leaves have dropped, and leaves left on plants are turning yellow. Immature soybeans will be injured. Mature cockleburs, especially drought-stressed plants, are tolerant to Paraquat and desiccation will not be complete. Do not pasture livestock within 15 days of treatment. Remove livestock from treated fields at least 30 days before slaughter.

SUNFLOWER (OILSEED VARIETIES ONLY)—For Desiccation of Sunflower Plants and Broadleaf Weeds and Grasses to Faciliate Harvesting. Apply 1 to 2 pts. ORTHO PARAQUAT+PLUS per acre in 5 gals. spray mix by air or in 20 to 40 gals. by ground application equipment. Use the higher rate when crop stands or weed infestations are heavy. Apply when the sunflower seeds reach physiological maturity (when seed moisture is 35 percent or lower). For many varieties, this corresponds to the time when the back of the heads are yellow and bracts are turning brown. Harvest 7 to 21 days after application. Add ORTHO X-77 Spreader (non-ionic) at 1 qt. per 100 gals. water. Do not graze treated areas or feed treated forage to livestock.

SUGARCANE—Preharvest Crop Desiccation (Florida and Texas Only)—Apply ½ to 1 pt. per acre. Apply to mature plant or stubble cane using 4 to 10 gals. water per acre. Use higher rate under cool, cloudy weather conditions. ORTHO X-77 Spreader (non-ionic) should be added to all sprays at the rate of 1 pt. per 100 gals. water. Apply with aircraft sprayer 3 to 14 days before burning and harvest. For rapid results, apply when weather is clear and warm. Apply as a coarse spray at low pressure when air is calm to prevent spray drift.

DIRECTED SPRAYS

ORCHARDS, VINEYARDS, WINDBREAK, SHADE AND ORNAMENTAL TREES: For kill of annual broadleaf weeds and grasses and for top kill and suppression of perennials and green suckers under Filbert trees, in the interspaces and around the base of trees and vines—Acerola, Almonds, Apples, Apricots, Avocados, Bananas, Cherries, Citrus, Coffee, Figs, Filberts, Grapes, Kiwifruit, Macadamia Nuts, Nectarines, Olives, Papayas, Peaches, Pears, Pecans, Pistachios, Plums, Prunes, Walnuts and other Trees such as Arborvitae, Ash, Elm, Fir, Oak and Pine—Apply as a directed spray at 1 to 2 qts. per sprayed acre in 30 to 150 gals. water (use 30 to 50 gals. water for pecans) to thoroughly cover the weeds and grasses. Add ORTHO X-77 Spreader (non-ionic) at 8 to 32 fl. oz. per 100 gals. Apply when the weeds and grasses are succulent and the new growth is from 1 to 6 inches high. For mature woody weeds, green suckers, late germinating weeds and grasses, and for perennials—retreatment or spot treatment may be necessary. **CAUTION:** Do not allow spray to contact green stems, fruit or foliage as injury may result. Do not spray under windy conditions and use a shield for young trees or vines. Do not allow animals to graze on treated areas. Do not apply when nuts to be harvested are on the ground. For Kiwifruit, do not treat more than 3 times per year and do not apply within 14 days of harvest. For Pistachios, do not harvest within 7 days of treatment and do not exceed two applications after shells split.

ALMONDS, GRAPES AND WALNUTS (California Only); ORANGES (California and Florida Only); SOUR CHERRIES (Michigan Only); APPLES, PEACHES, PEARS: Tank Mix With PRINCEP® 80 W Simazine Weed Killer—For Top Kill and Residual Control of Annual Broadleaf Weeds and Grasses and for Suppression of Perennial Weeds—For broadcast application apply 1 to 2 qts. ORTHO PARAQUAT+PLUS per acre and PRINCEP 80 W at the following rates per acre:

Crop	Rate PRINCEP 80 W Per Sprayed Acre	Precautions
Almonds and Peaches (California Only)	1¼–2½ lbs. in 2–4 ft. band on each side of tree row. 1¼ lbs. provides residual control of burclover, common chickweed, wild mustard and shepherdspurse.	Do not treat trees established less than 3 years. Do not treat Mission variety of almonds. Do not apply to almond trees propagated on plum root stalks. Do not replant almonds or peaches in treated soil. Do not apply on gravelly sand or loamy sand soil or on soil with less than 1% organic matter. Do not treat areas where water will accumulate.
Apples Pears Sour Cherries (Michigan Only) Walnuts (California Only)	2½–5 lbs.	Do not apply to sandy soil. Leveling and furrowing operations after application will lessen effectiveness of weed control.
Peaches (Arkansas, Louisiana, Missouri, Oklahoma, Texas and states East of the Mississippi River)	2–5 lbs.	Do not apply to sandy or gravelly soil.
Grapes (California Only)	2½–6 lbs.	Do not use in vineyards established less than 3 years.
Oranges (California Only) Except Imperial, Coachella and Palo Verde Valleys	2½–5 lbs.	Do not use in nurseries. To avoid possible injury, do not apply to trees under stress from freeze damange for one year following the freeze.
(Florida Only)—Bedded	4 lbs.	

Use the low rate of PRINCEP 80 W on coarse textured soils and low organic matter soils. Use the high rate on fine textured soils and high organic matter soils. Apply as a tank mix in 50 to 200 gals. of water per acre to the orchard floor avoiding contact with fruit, foliage or stems. Add ORTHO X-77 Spreader (non-ionic) at 8 to 32 fl. oz. per 100 gals. diluted spray. Apply when weeds and grasses are succulent and the new growth is from 1 to 6 inches tall.

For band applications or spot applications around trees, reduce the broadcast rate and the amount of water in proportion to the area actually sprayed.

Make only one application per year. Apply in orchards where trees have been established one year or more. Do not spray under windy conditions and use a shield for young trees. Do not apply to almonds and walnuts when nuts are on the ground during the harvest period. Do not allow animals to graze treated areas.

Refer to the PRINCEP 80 W Simazine label for general information, caution and warning statements.

PRINCEP®—Reg. TM of CIBA-GEIGY.

HOPS (Washington, Oregon and Idaho Only): For Kill of Annual Bluegrass, Ryegrass, Barnyardgrass, Pigweed and Chickweed—Apply as a directed spray in the interspaces and around the base of hop plantings using 1 qt. in 20 to 100 gals. water per acre to thoroughly cover weeds and grasses. Add ORTHO X-77 Spreader (non-ionic) at 8 to 32 fl. oz. per 100 gals. spray mix. Do not apply around hop vines less than 10 ft. tall. Apply when weeds and grasses are succulent and growth is from 1 to 6 inches tall. Retreatment or spot treatment may be necessary. **For Suckering and Stripping (Cascade, Yakima Cluster and Bullion Varieties Only):** Spray only basal 2 ft. of vines using 1 qt. in 20 to 100 gals. water per acre. Add ORTHO X-77 Spreader (non-ionic) at 8 to 32 fl. oz. per 100 gals. spray mix. Do not apply to hop vines less than 10 ft. tall. Repeat as necessary. **CAUTION:** Do not apply more than 3 times per season. Do not apply within 14 days of harvest. Do not allow spray to contact green stems, foliage, flowers or cones, as injury may result. Do not allow animals to graze in treated hopyards. Hop vine refuse and silage may be fed to livestock.

GUAVA—Apply as directed spray at 2 qts. per sprayed acre in 50 to 200 gals. to thoroughly cover the weeds and grasses. Add ORTHO X-77 Spreader (non-ionic) at 8 to 32 fl. oz. per 100 gals. Apply when the weeds and grasses are succulent and the new growth is from 1 to 6 inches high. For mature woody weeds, late germinating weeds and grasses and for perennials—retreatment or spot treatment may be necessary. **CAUTION:** Do not allow spray to contact green stems, fruit or foliage as injury may result. Do not spray under windy conditions and use a shield for young trees. Do not allow animals to graze on treated areas.

PASSION FRUIT—Apply as a directed spray at 2 qts. per sprayed acre in 50 to 200 gals. to thoroughly cover the weeds and grasses. Add ORTHO X-77 Spreader (non-ionic) at 8 to 32 fl. oz. per 100 gals. Apply when the weeds and grasses are succulent and the new growth is from 1 to 6 inches high. For mature woody weeds, late germinating weeds and grasses and for perennials—retreatment or spot treatment may be necessary. **CAUTION:** Do not allow spray to contact green stems, fruit or foliage as injury may result. Do not spray under windy conditions and use a shield for young vines. If bark is still green, wrap vine prior to application to prevent injury. If application is to be made during harvesting season, apply Paraquat only after picking fruits off the ground. Do not allow animals to graze on treated areas.

PINEAPPLES: For Kill of Annual Broadleaf Weeds and Grasses and Top Kill and Suppression of Perennials—Apply as a directed spray using 1 to 2 qts. in 100 to 200 gals. water per sprayed acre to thoroughly cover the weeds and grasses. Add ORTHO X-77 Spreader (non-ionic) at 8 to 32 fl. oz. per 100 gals. Apply when weeds and grasses are succulent and new growth is from 1 to 6 inches high. Retreatment may be necessary on more mature weeds. Do not apply within 20 days of harvest.

SMALL FRUITS (Blackberries, Boysenberries, Raspberries, Blueberries)—For Kill of Annual Broadleaf Weeds and Grasses and for Top Kill and Suppression of Perennial Weeds in the Interspaces and Around the Base of Bushes—Apply as a coarse directed spray at 1 to 2 qts. per sprayed acre in 50 to 200 gals. water to thoroughly wet the weeds and grasses. Add ORTHO X-77 Spreader (non-ionic) at 8 to 32 fl. oz. per 100 gals. water. Apply before emergence of new canes or shoots as injury to those canes or shoots can result. Apply as a coarse spray to avoid drift injury from fine spray mist.

STRAWBERRIES: Postemergence Directed Spray—For Control of Emerged Annual Broadleaf Weeds and Grasses and for Top Kill and Suppression of Emerged Perennial Weeds Between the Rows After Crop Plant Emergence at Establishment—Apply 1 qt. ORTHO PARAQUAT + PLUS per sprayed acre in 20 to 60 gals. spray mix. Apply with conventional ground equipment directing spray between the rows and using shields to prevent spray contact with crop plants. Add ORTHO X-77 Spreader (non-ionic) at 8 fl. oz. per 100 gals. spray mix. For best results, apply when weeds and grasses are succulent and weed growth is 1 to 6 inches high. Weeds and grasses emerging after application will not be controlled. Do not allow spray to contact strawberry plants as injury or excessive residues may result. Do not apply more than 3 times per crop season. Do not apply within 21 days of harvest. Do not graze livestock in treated areas.

GRAIN SORGHUM: Postemergence Directed Spray—For Kill of Emerged Crabgrass and Pigweed 3 Inches or Less in Height—1 to 2 pts. ORTHO PARAQUAT+PLUS in 20 to 40 gals. water per sprayed acre. Use higher gallonage of water for larger weeds. For band applications, 1 pt. ORTHO PARAQUAT+PLUS will cover 2 land or crop acres when a 20 inch band is treated in 40 inch rows. Treat when grain sorghum is naturally standing 12 inches tall or more. Add ORTHO X-77 Spreader (non-ionic) at 1 qt. per 100 gals. of spray. Do not exceed 30 psi; high pressure contributes to spray drift.

For best results, apply ORTHO PARAQUAT+PLUS at 1 pt. per sprayed acre to grass and weeds less than 2 inches tall. Grass and weeds 2 to 3 inches tall should be sprayed with 2 pts. ORTHO PARAQUAT+PLUS per sprayed acre. Use precision application equipment with spray nozzles mounted on skid shoes, oiling shoes, fenders or cultivators with gauge wheels and adjusted so that only the lower 3 inches or less of the sorghum stalk is sprayed.

For best coverage on single rows, use 4 flat fan nozzles per row, 2 nozzles on each side of the row. The 2 forward nozzles should point forward and downward and rear nozzles should point to the rear and downward. For double row sorghum use one skid between the two rows with one nozzle spraying backward and downward.

Some sheath and lower leaf spotting and desiccation will occur. Basal spotting does not penetrate beneath outer sheath. Degree of injury is related to precision of application and wind velocity. Do not spray under windy conditions.

FIELD CORN: Postemergence Directed Spray using ORTHO PARAQUAT + PLUS Alone or ORTHO PARAQUAT+PLUS + Atrazine Tank Mix—For Control of Emerged Annual Broadleaf Weeds and Annual Grasses less than 4 inches tall—Apply 1 pt. ORTHO PARAQUAT+PLUS per acre in 20 to 40 gals. of spray mix. Apply when corn plants are at least 10 inches tall. Corn plants shorter than 10 inches may be injured and not recover. Corn height measured from soil surface to top of whorl. Apply with directed spray application equipment. Arrange nozzles to spray no higher than the lower three inches of corn stalks to provide for maximum contact of weeds with minimum contact of corn plants to prevent excessive sheath and lower leaf spotting and desiccation. Add 1 qt. ORTHO X-77 Spreader (non-ionic) per 100 gals. of spray. Do not mix with liquid fertilizer. **For Improved Control of Weeds, Especially Broadleaves, Tank Mix Paraquat and AATREX®:** Add 5/8 to 1¼ lbs. AATREX 80W or 1 to 2 pts. AATREX 4L in combination with 1 pt. ORTHO PARAQUAT+PLUS per acre. Add AATREX 80W or AATREX 4L to spray tank first, mixing thoroughly in water. Then add PARAQUAT and surfactant. Constant agitation recommended at all times. Refer to AATREX labels for limitation and cautions.

Do not graze treated areas or feed treated forage to livestock.

AATREX®—Reg. TM of CIBA-GEIGY.

CORN: For use in the USDA Witchweed eradication program to kill grassy weeds. Apply 1 qt. ORTHO PARAQUAT+PLUS in 10 to 20 gals. spray per acre. Add ORTHO X-77 Spreader (non-ionic) at 8 fl. oz. per 100 gals. spray mix. Apply as a spray directed onto the grassy weeds at the base of the corn stalks in the hills or row. Initiate sprays in early July and repeat in early August if regrowth of grassy weeds occurs. Do not graze livestock in treated area.

SOYBEANS: Postemergence Directed Spray—For control of seedling Johnsongrass, Crabgrass, Goosegrass, Brachiaria, Echinochloa spp. (Texas Millet) and pigweed less than 2 inches tall apply ORTHO PARAQUAT+PLUS at 4 fl. oz. (.063 lb. active) per acre. Use proportionately less for band treatment . . .for example on 40 inch rows and 16 inch band apply 1 gal. ORTHO PARAQUAT+PLUS to 80 acres.

For 2 to 4 inch grass and 2 to 3 inch pigweed apply ORTHO PARAQUAT+PLUS at 8 fl. oz. (.125 lb. active) per acre . . .for example on 40 inch rows and 16 inch band apply 1 gal. ORTHO PARAQUAT+PLUS to 40 acres. Soybeans should be 8 inches tall at first application. If needed, make second application 7 to 14 days later. Do not treat more than twice.

Apply 20 gals. spray mix per treated acre. Use proportionately less for band treatments . . .for example 8 gals. spray mix per 16 inch band on 40 inch rows.

For best coverage, use 4 flat fan nozzles per row, 2 nozzles on each side of row. The two forward nozzles should point forward and downward and rear nozzles should point to the rear and downward. Do not exceed 30 psi; high pressure contributes to drift. Adjust nozzles to spray no higher than the lower three inches of the soybean plant. Do not treat if soybeans are below 8 inches tall; soybeans treated topically will be killed. Add ORTHO X-77 Spreader (non-ionic) at 1 to 2 qts. per 100 gals. of spray.

Do not graze treated areas or feed treated forage to livestock.

SUGARCANE—Directed Spray: For Kill of Emerged Annual Broadleaf Weeds and Grasses and for Top Kill of Perennials—Apply as a shielded or directed spray when weeds are 2 to 6 inches high. Avoid contact with cane foliage to prevent serious leaf burn and reduced yields. Make a second and final application, if necessary, when new weed growth is 2 to 6 inches high. Add ORTHO X-77 Spreader (non-ionic) at 8 fl. oz. per 100 gals. spray mix.

Florida Only—Apply 1 qt. in 50 to 100 gals. of water per acre. For optimum results, apply early in the season (March-April) when weeds are small. Do not apply after June 1st as cane growth may be stunted and yields reduced.

Hawaii Only—Apply 1 qt. in 20 to 200 gals. of water per acre. Do not apply after cane rows have closed in.

Louisiana Only—Apply ½ to 1 qt. in 20 to 200 gals. of water per acre. For tiller control, apply when tillers are less than 18" high. Use the higher rate of Paraquat with heavier weed infestations or tiller growth. Do not apply within 30 days of harvest.

TOMATOES AND PEPPERS (Middle Atlantic, Southeast, South Central Regions and Southwest to the Western Boundary of Texas): Postemergence Directed Spray—For Kill of Emerged Annual Broadleaf Weeds and Grasses and for Top Kill and Suppression of Emerged Perennial Weeds between Plastic Mulch Covered Rows after Crop Plant Emergence or Establishment—Apply as directed spray using 1 qt. per sprayed acre in 20 to 100 gals. spray mix. Apply with conventional ground equipment directing spray between the rows and using shields to prevent spray contact with crop plants. Add ORTHO X-77 Spreader (non-ionic) at 8 fl. oz. per 100 gals. spray mix. Apply when weeds and grasses are succulent and weed growth is 1 to 6 inches high. Weeds and grasses emerging after application will not be controlled. Do not allow spray to contact tomato or pepper plants as injury or excessive residues may result. Do not apply more than 3 applications per crop season. Do not allow animals to graze on treated areas.

DORMANT USE

PEPPERMINT, SPEARMINT: For Suppression of Weeds Such as Italian Ryegrass, Prickly Lettuce, Groundsel, Chickweed, Downy Brome and Bluegrasses During the Dormant Season—Apply 1½ to 3 pts. per acre as a broadcast spray in 25 to 65 gals. spray mix. Use the higher rate with heavier weed infestations. Add ORTHO X-77 Spreader (non-ionic) at 8 fl. oz. per 100 gals. spray mix. Apply when peppermint or spearmint is dormant and before spring growth begins. Do not apply to weeds which are greater than 6" tall. Do not apply more than 3 pts. per acre per dormant season as injury or excessive residues may result. **For Improved Contact Activity and Residual Control of Italian Ryegrass, Prickly Lettuce and Groundsel**—Add 1 to 2 lbs. of SINBAR® (Terbacil) Weed Killer as a tank mix. Use the higher tank mix rate with heavier weed infestations. Apply tank mix treatment no more than once per season. Use a minimum of 25 gals. spray mix per acre. Refer to the SINBAR label for specific recommendations and cautions.

SINBAR®—Reg. TM of E.I. du Pont de Nemours & Co., Inc.

RHUBARB: For Kill of Emerged Annual Broadleaved Weeds and Grasses and for Top Kill and Suppression of Emerged Perennial Weeds in Dormant Rhubarb—Apply 1 to 2 qts. per sprayed acre by ground application equipment. Use 50 to 150 gals. spray mix per acre. Add ORTHO X-77 Spreader (non-ionic) at 8 to 32 fl. oz. per

100 gals. spray mix. Use the higher rate with heavier weed infestations. Apply during the dormant season before buds in crown begin to grow. Do not make more than two applications per season.

NONCROP USES

NONCROP AREAS (Public Airports, Electric Transformer Stations and Substations, Pipeline Pumping Stations, Around Commercial Buildings, Storage Yards and Other Installations, Fence Lines, and Similar Noncrop Areas)—For the Control of Annual Broadleaved Weeds and Grasses (such as Burclover, Chickweed, Filaree, Groundsel, Nettle, Pigweed, Plantain, Puncturevine, Purslane, Red Clover, Shepherdspurse, Thistle, Wild Mustard, Wild Radish, Wild Oats, Blue Grass, Cheat Grass, Crabgrass) and Top Kill and Suppression of Perennial Weeds (such as Bermudagrass, Johnsongrass, and Morningglory)—1 qt. per acre. Apply for full coverage and thorough weed contact (50 to 100 gals. diluted spray per acre). Best results are obtained when applications are made to young, succulent weeds and grasses. Mature, woody weeds are less susceptible. Repeat as needed. Add ORTHO X-77 Spreader (nonionic) at 8 oz. per 100 gals. Paraquat is compatible with many residual herbicides. Avoid spray contact with foliage or fruit of food crops and ornamentals.

RESIN SOAKING

RESIN SOAKING IN PINES (Loblolly, Shortleaf, Longleaf, Slash, Virginia, Pond, Pitch and Spruce Pines):

Tree Selection—Select trees to be treated from stands on sites not subject to periods of extreme drought stress as the desiccating effect of Paraquat to pines is accentuated during such periods, causing a reduction in the amount of oleoresin deposited in the xylem. Select trees to be treated from vigorous, non-stagnated stands, either natural or planted. In stagnated stands or commercial timber stands, plan treating with Paraquat not sooner than three years after a commercial thinning.

Application Directions—Apply ORTHO PARAQUAT+PLUS diluted in water (0.2% to 5% cation wt./wt. basis, depending on the type of treatment method) to a suitable wound in the tree trunk to bring the treatment into contact with the xylem (sapwood). One of the following methods may be used:

Bark Streaks or Cuts: This type of wound is made using a standard or rotary bark hack or a chainsaw chipping tool employed in naval stores work to remove a single one-inch wide streak of bark about 1 to 2 feet from ground level. The total length should not exceed 1/3 of the tree circumference. Multiple streaks or cuts can result in serious girdling of the trunk and premature death of the tree. A coarse spray (about 2 to 6 ml.) of Paraquat solution (1% to 5% cation, wt./wt. basis) should be applied to runoff to the exposed xylem, using a plastic acid bottle or a low-pressure sprayer. Wear rubber gloves when using the acid bottle. The amount of spray required per cut depends on tree circumference and the length of cut or streak (1/3 of circumference). For a 9-inch diameter tree, 3 ml. of spray will cover the 1 inch wide streak. Using 3 ml. of a 2% or 4% Paraquat solution will result in application of 60 or 120 mg. Paraquat per streak.

Tree Injection: Pressure injections of 1% to 4% Paraquat solution (cation, wt./wt. basis) should be applied beneath the bark and just beneath the cambium layer. Injections should be made 1 to 2 inches apart around 1/3 of the tree circumference or no closer than 3 inches apart all around the tree. From 0.2 ml. to 0.4 ml. of Paraquat solution is applied per injection. Using 2% to 4% Paraquat solution, this results in 36 to 144 mg. Paraquat applied per 9-inch diameter tree, for holes spaced 1 inch apart around 1/3 of the tree or 3 inches apart all around the tree.

Bore Holes: 3/8 to 5/8 inch diameter holes 2 to 6 inches deep, depending on tree diameter, are bored sloping slightly downward and tangentially inward rather than directly towards the center of the tree. Solutions of 0.3% to 5% Paraquat can be used. Lower concentrations (0.3%) are usually applied at higher volumes of 15 to 35 ml. per tree, resulting in application of 45 to approximately 100 mg. Paraquat per tree. Using 2% to 4% solutions and applying 5 ml. per single hole per tree results in application of 100 or 200 mg. Paraquat per tree. Bore hole applications in the range of 45 to 100 mg. Paraquat per tree have resulted in effective oleoresin induction.

Time of Treatment: Resin soaking can occur from treatments made any time of the year; however, cool season treatments under non-drought conditions usually result in less severe pine beetle infestations and longer tree life.

Interval Between Treatment and Tree Harvest: The interval between application of Paraquat and tree harvest should be a minimum of six months and preferably from 12 to 24 months. Intervals of over six months may not be possible under conditions of drought or serious pine beetle attacks, which may make early harvest necessary. The Paraquat treatment may encourage beetle attack, or may cause premature death of the tree. Desiccation of the xylem tissue, rather than the desired resin soaking, may occur, and is more likely at higher dosage rates.

Effect on Stem Growth:
Paraquat treatment can result in reduced stem growth during the interval between treatment and tree harvest.

Dilution Table for ORTHO PARAQUAT+PLUS
(2 lbs. cation per gallon)

Concentration of Cation Desired (Wt./Wt. Basis)	To 1 Gal. of ORTHO PARAQUAT+PLUS, add the following No. Gals. of Water:
0.2%	118.8
0.5%	46.8
1.0%	22.9
2.0%	10.9
3.0%	6.9
4.0%	4.9
5.0%	3.7

WORKER SAFETY RULES/

REENTRY STATEMENTS

USE STRICTLY IN ACCORDANCE WITH DANGER STATEMENTS AND DIRECTIONS, AND WITH APPLICABLE STATE AND FEDERAL REGULATIONS.
DO NOT get on skin, eyes or clothing.
DO NOT inhale spray mist.
WASH SPLASHES from skin and eyes immediately.
REMOVE and wash contaminated clothing.
WASH before eating, smoking and drinking.
WEAR full face shield, rubber gloves and apron and waterproof footwear when handling or mixing concentrate.
WEAR waterproof footwear and clothing when spraying or when contacting vegetation wet with spray.
DO NOT enter treated areas without protective clothing until sprays have dried.
AVOID working in spray mist. If there is risk of exposure wear goggles and approved face mask capable of filtering spray droplets.
KEEP all unprotected persons out of operating areas or vicinity where there may be danger of drift.
Certain states may require more restrictive reentry intervals; consult your State Department of Agriculture for further information.
Written or oral warnings regarding use of protective clothing and accidental exposure must be given to workers who are expected to be in treated areas or in areas about to be treated.

IMPORTANT: The effect of swallowing Paraquat is far more severe than that which usually results from skin contact or inhalation of spray mist. No opportunity for mist or product to cause serious injury or death should arise when used in strict compliance with these rules but the opportunity may arise in case of gross violations so strictly follow all these rules **as if your life depends on it.**

ENVIRONMENTAL SAFETY CAUTIONS

WILDLIFE
This product is toxic to wildlife. Birds and other wildlife in treated areas may be killed. Do not apply directly to water.

DRIFT

ORTHO PARAQUAT+PLUS is a contact herbicide that kills all green plant tissue. Do not apply under conditions involving possible drift to food, forage or other plantings that might be damaged or the crops thereof rendered unfit for sale, use or consumption. Do not apply when weather conditions favor drift from areas treated. For aerial applicators use, do not apply during periods of thermal inversion to avoid drift.

CHEMICAL HAZARD

This product reacts with aluminum to produce hydrogen gas which may form a highly combustible gas mixture. Such gas could flash or explode, causing serious personal injury, if ignited by open flame, spark, welder's torch, lighted cigarette or other ignition source. Do not mix or store in containers, spray tanks, nurse tanks or such systems made of aluminum or having aluminum fittings. After dilution with water to spray concentrations, spray systems made of, or containing aluminum parts, may be used but should not be used for extended (overnight) storage.

EQUIPMENT/CONTAINERS

Flush all spray equipment with water after use each day. ORTHO PARAQUAT+PLUS is corrosive to aluminum. Aluminum spray equipment and aluminum aircraft structures that are exposed to spray solution or spray drift should be flushed thoroughly with water immediately after use.

STORAGE AND DISPOSAL

PROHIBITIONS

Do not contaminate water, food or feed by storage, disposal or cleaning of equipment. Open dumping is prohibited.

STORAGE

Store at temperatures above 32°F. Do not contaminate feed, foodstuffs or drinking water. Do not store next to feed or food, or transport in or on vehicles containing foodstuffs or feeds.

For help with any spill, leak, fire or exposure involving this material, call day or night (415) 233-3737.

PESTICIDE DISPOSAL

This product is acutely hazardous. Improper disposal of excesss pesticide, spray mixture, or rinsate is a violation of Federal Law. If these wastes cannot be disposed of by use according to label instructions, contact your State Pesticide or Environmental Control Agency, or the Hazardous Waste representative at the nearest EPA Regional Office for guidance.

CONTAINER DISPOSAL

See Container Label For Container Disposal Directions.

CONDITIONS OF SALE

1. Chevron Chemical Company (Chevron) warrants that this material conforms to the chemical description on the label and is reasonably fit for use as directed hereon. Chevron neither makes, nor authorizes any agent or representative to make, any other warranty of FITNESS or of MERCHANTABILITY, guarantee or representation, express or implied, concerning this material.
2. Critical and unforeseeable factors beyond Chevron's control prevent it from eliminating all risks in connection with the use of chemicals. Such risks include, but are not limited to, damage to plants and crops to which the material is applied, lack of complete control, and damage caused by drift to other plants or crops. Such risks occur even though the product is reasonably fit for the uses stated hereon and even though label directions are followed. Buyer and user acknowledge and assume all risks and liability (except those assumed by Chevron under 1 above) resulting from handling, storage and use of this material.

CIBA-GEIGY Corporation

Agricultural Division
P.O. Box 18300
Greensboro, NC 27419
(919) 292-7100

AATREX® 4L

HERBICIDE

For season-long weed control in corn and sorghum

For weed control in certain other crops; in noncrop areas; and industrial sites

ACTIVE INGREDIENTS:

Atrazine: 2-chloro-4-ethylamino-6-isopropylamino-*s*-triazine	40.8%
Related compounds	2.2%
INERT INGREDIENTS	57.0%
TOTAL	100.0%

EPA Reg. No. 100-497 EPA Est. 100-LA-1

AAtrex® trademark of CIBA—GEIGY for atrazine

AAtrex 4L contains 4 lbs. active ingredients per gallon.

Shake well before using.

Use entire contents at one time.

Keep Out of Reach of Children.

Caution

See additional precautionary statements at end of label booklet.

See directions for use inside booklet.

DIRECTIONS FOR USE AND CONDITIONS OF SALE AND WARRANTY

IMPORTANT: Read the entire **Directions for Use** and the **Conditions of Sale and Warranty** before using this product.

CONDITIONS OF SALE AND WARRANTY

The **Directions for Use** of this product reflect the opinion of experts based on field use and tests. The directions are believed to be reliable and should be followed carefully. However, it is impossible to eliminate all risks inherently associated with use of this product. Crop injury, ineffectiveness, or other unintended consequences may result because of such factors as weather conditions, presence of other materials, or the manner of use or application all of which are beyond the control of CIBA—GEIGY or the Seller. All such risks shall be assumed by the Buyer.

CIBA—GEIGY warrants that this product conforms to the chemical description on the label and is reasonably fit for the purposes referred to in the Directions for Use subject to the inherent risks referred to above. **CIBA—GEIGY makes no other express or implied warranty of Fitness or Merchantability or any other express or implied warranty. In no case shall CIBA—GEIGY or the Seller be liable for consequential, special, or indirect damages resulting from the use or handling of this product.** CIBA—GEIGY and the Seller offer this product, and the Buyer and user accept it, subject to the foregoing Conditions of Sale and Warranty, which may be varied only by agreement in writing signed by a duly authorized representative of CIBA—GEIGY.

DIRECTIONS FOR USE

It is a violation of federal law to use this product in a manner inconsistent with its labeling.

FAILURE TO FOLLOW THE DIRECTIONS FOR USE AND PRECAUTIONS ON THIS LABEL MAY RESULT IN POOR WEED CONTROL, CROP INJURY, OR ILLEGAL RESIDUES.

Do not apply this product in such a manner as to directly or through drift expose workers or other persons, except those knowingly involved in the application. The area being treated must be vacated by unprotected persons.

Do not enter treated areas without protective clothing until sprays have dried.

Because certain states may require more restrictive reentry intervals for various crops treated with this product, consult your State Department of Agriculture for further information.

Written or oral warnings must be given to workers who are expected to be in a treated area or in an area about to be treated with this product. Oral warnings must be given which inform workers of areas or fields that may not be entered without specific protective clothing until sprays have dried, and appropriate actions to take in case of accidental exposure as described under **Precautionary Statements** on this label. When oral warnings are given, warnings shall be given in a language customarily understood by workers. Oral warnings must be given if there is reason to believe that written warnings cannot be understood by workers. Written warnings must include the following information: "CAUTION. Area treated with AAtrex 4L on *(date of application)*. Do not enter without appropriate protective clothing until sprays have dried. In case of accidental exposure, flush eyes or skin with plenty of water. Call a physician if irritation persists. Remove and wash contaminated clothing before reuse."

GENERAL INFORMATION

This herbicide controls many annual broadleaf and grass weeds in corn, sorghum, sugarcane, pineapple, and certain other crops specified on this label. It is also effective in noncrop areas and industrial sites for control of most annual and many perennial broadleaf and grass weeds. This product may be applied before or after weeds emerge.

Following many years of continuous use of this product and chemically related products, biotypes of some of the weeds listed on this label have been reported which cannot be effectively controlled by this and related herbicides. Where this is known or suspected and weeds controlled by this product are expected to be present along with resistant biotypes, we recommend the use of this product in combinations or in sequence with other registered herbicides which are not triazines. If only resistant biotypes are expected to be present, use a registered nontriazine herbicide. Consult with your state Agricultural Extension Service for specific recommendations.

Since this product acts mainly through root absorption, its effectiveness depends on moisture to move it into the root zone. If weeds develop, a shallow cultivation or rotary hoeing will generally result in better weed control.

This product is nonflammable.

Avoid using near adjacent desirable plants or in greenhouses, or injury may occur.

To avoid spray drift, do not apply under windy conditions. Avoid spray overlap, as crop injury may result.

Where the use directions give a range of rates, use the lower rate on coarse-textured soil and soil low in organic matter; use the higher rate on fine-textured soil and soil high in organic matter.

Note: CIBA—GEIGY does not recommend applications in combination with other herbicides or oils, except as specifically described on the label or in literature published by CIBA—GEIGY.

APPLICATION PROCEDURES

Ground application: Use conventional ground sprayers equipped with nozzles that provide accurate and uniform application. Be certain that nozzles are uniformly spaced and are the same size. Calibrate sprayer before use and recalibrate at the start of each season and when changing carriers. Unless otherwise specified, use a minimum of 10 gal of spray mixture/A for all preplant incorporated, preplant surface, preemergence, and postemergence applications (with or without oil or surfactant) with ground equipment.

Use a pump with capacity to (1) maintain 35-40 psi at nozzles, (2) provide sufficient agitation in tank to keep mixture in suspension, and (3) to provide a minimum of 20% bypass at all times. Use centrifugal pumps which provide propeller shear action for dispersing and mixing this product. The pump should provide a minimum of 10 gal/minute/100 gal tank size circulated through a correctly positioned sparger tube or jets.

Use screens to protect the pump and to prevent nozzles from clogging. Screens placed on the suction side of the pump should be *16-mesh or coarser.* Do not place a screen in the recirculation line. Use 50-mesh or coarser screens between the pump and boom, and where required, at the nozzles. Check nozzle manufacturer's recommendations.

For band applications, calculate amount to be applied per acre as follows:

$$\frac{\text{band width in inches}}{\text{row width in inches}} \times \text{broadcast rate per acre} = \text{amount needed per acre of field}$$

Aerial application: Use aerial application only where broadcast applications are specified. Apply in a minimum of 1 qt of water for each quart of this product applied per acre. For postemergence treatments on corn and sorghum, apply recommended rate in a minimum of 2 gal of water/A. Avoid applications under conditions where uniform coverage cannot be obtained or where excessive spray drift may occur.

Avoid application to humans or animals. Flagmen and loaders should avoid inhalation of spray mist and prolonged contact with skin, and should wash thoroughly before eating and at the end of each day's operation.

Application in water or liquid fertilizer: Nitrogen solution or complete liquid fertilizer may replace all or part of the water as a carrier for preemergence, preplant incorporated, or preplant surface ground application on corn and sorghum. Check the compatibility of this product with liquid fertilizer and/or nitrogen solution as shown below before use. Do not apply in nitrogen solution or complete liquid fertilizer after corn or sorghum emerges, except as noted under **Lay-by treatment** for corn, or crop injury may occur.

Compatibility Test: Since liquid fertilizers can vary, even within the same analysis, always check compatibility with herbicide(s) each time before use. Be especially careful when using complete suspension or fluid fertilizers as serious compatibility problems are more likely to occur. Commercial application equipment may improve compatibility in some instances. The following test assumes a spray volume of 25 gal per acre. For other spray volumes, make appropriate changes in the ingredients. Check compatibility using this procedure:

1. Add 1 pint of fertilizer to each of 2 one-quart jars with tight lids.
2. To **one** of the jars add 1/4 tsp or 1.2 milliliters of a compatibility agent approved for this use, such as Compex® or Unite® (1/4 tsp is equivalent to 2 pt per 100 gal spray). Shake or stir gently to mix.
3. To **both** jars add the appropriate amount of herbicide(s). If more than one herbicide is used, add them separately with dry herbicides first, flowables next, and emulsifiable concentrates last. After each addition, shake or stir gently to thoroughly mix. The appropriate amount of herbicides for this test follows:

Dry herbicides:	For each pound to be applied per acre, add 1.5 level teaspoons to each jar.
Liquid herbicides:	For each pint to be applied per acre, add 0.5 teaspoon or 2.5 milliliters to each jar.

4. After adding all ingredients, put lids on and tighten, and invert each jar ten times to mix. Let the mixtures stand 15 minutes and then look for separation, large flakes, precipitates, gels, heavy oily film on the jar, or other signs of incompatibility. Determine if the compatibility agent is needed in the spray mixture by comparing the two jars. If either mixture separates, but can be remixed readily, the mixture can be sprayed as long as good agitation is used. If the mixtures are imcompatible, test the following methods of improving compatibility: (A) slurry the dry herbicides(s) in water before addition, or (B) add 1/2 of the compatibility agent to the fertilizer and the other 1/2 to the emulsifiable concentrate or flowable herbicide before addition to the mixture. If incompatibility is still observed, do not use the mixture.

Application in water plus emulsifiable oil or oil concentrate: Adding emulsifiable oil (petroleum-derived, petroleum-derived oil concentrate, or single or mixed crop-derived oil concentrate) to postemergence water-based sprays in corn and sorghum may improve weed control. However, under certain conditions, the use of either type of oil may seriously injure the crop. To minimize this possibility, observe the following directions:

Use one of the following properly emulsified:

1. A suitable oil concentrate containing at least 1% but not more than 20% suitable emulsifier or surfactant blend.
2. Petroleum-derived oil containing at least 1% suitable emulsifier.

Note: In the event of a compatibility problem when mixing oil with this product and water, a compatibility agent such as Compex or Unite should be used. Any of the above oils contaminated with water or other materials can cause compatibility problems and/or crop injury.

Mixing procedures—all uses: (1) Be sure sprayer is clean and not contaminated with any other materials, or crop injury or sprayer clogging may result. (2) Fill tank 1/4 full with clean water, nitrogen solution, or complete liquid fertilizer. (3) Start agitation. (4) Be certain that the agitation system is working properly and creates a rippling or rolling action on the liquid surface. (5) Pour product directly from container into tank. (6) Continue filling tank until 90% full. Increase agitation if necessary to maintain surface action. (7) Add emulsifiable oil, oil concentrate, or tank mix herbicide(s) after this product is thoroughly suspended. (8) Finish filling tank. (9) Empty tank as completely as possible before refilling to prevent buildup of oil or emulsifiable concentrate residue. Maintain agitation to avoid separation of materials. (10) If an oil or emulsifiable concentrate film starts to build up in tank, drain it and clean with strong detergent solution or solvent. (11) Clean sprayer thoroughly immediately after use by flushing system with water containing a detergent.

Center pivot sprinkler application

Preemergence or postemergence: Apply this product alone, or in tank mixtures with other herbicides on this label which are registered for center pivot application, with irrigation water. Apply either after planting before crop and weeds emerge, or after crop emergence, but before lay-by (20-30 inches) and before weeds exceed 1.5 inches in height. Apply at rates recommended on this label. Prepare mixture with a minimum of 1 part water to 1 part product. Injecting a larger volume of a more dilute slurry per hour will assure more accurate calibration of metering equipment. Maintain sufficient agitation to keep herbicide in suspension. Meter slurry into irrigation water during entire period. Apply in 1/2 -1 inch of water. Use the lower water volume on coarser textured soils, the higher volume on finer textured soils. More than 1 inch of water may reduce weed control by moving herbicide below the effective zone in the soil. Inject dilute slurry into system through a positive displacement pump.

Refer to supplemental literature published by CIBA—GEIGY for more information on calibrating.

Precautions: *(1) Apply only through irrigation systems containing anti-siphon and check valves to prevent contamination of well during shutdown and overflow of solution tank. (2) Inject ahead of any right angle turn in the main line to insure adequate mixing. (3) Chemical injection pumps and water pumps must have interlocking controls to insure simultaneous shutoff. (4) Application when drift may occur from windy conditions, when system joints and connections are leaking, or when nozzles are not providing uniform distribution may cause crop injury. (5) Where sprinkler distribution patterns do not overlap sufficiently, unacceptable weed control may result. Where sprinkler distribution patterns overlap excessively, crop injury may result.*

AATREX 4L APPLIED ALONE—CORN OR GRAIN SORGHUM

Preplant Surface-Applied, Preplant Incorporated, or Preemergence (or Postemergence at 4 pt/A with Oil)

Broadleaf or Grass Weeds Controlled

barnyardgrass (watergrass)	cocklebur	nightshade
giant foxtail	groundcherry	pigweed
green foxtail	jimsonweed	purslane
large (hairy) crabgrass	kochia	ragweed
wild oats	lambsquarters	velvetleaf (buttonweed)
witchgrass *(Panicum capillare)*	annual morningglory	
yellow foxtail	mustards	

Postemergence with Emulsifiable Oil or Oil Concentrate in Water (2 + 2.4 pt/A)

Broadleaf Weeds Controlled

annual morningglory	lambsquarters	ragweed
cocklebur	mustards	smartweed
jimsonweed	pigweed	wild buckwheat
		velvetleaf

CORN

Preplant Surface-Applied (Broadleaf and grass control): Use on medium- and fine-textured soil with minimum-tillage or no-tillage systems only in CO, IL, IN, IA, KS, KY, MN, MO, MT, NE, ND, SD, WI, WY. Apply ⅔ the recommended rate of AAtrex shown in Table 1 for the soil texture as a split treatment 30-45 days prior to planting and the remainder at planting. Applications made less than 30 days prior to planting may be as either a split or single treatment. On *coarse-textured* soils, do not apply more than two weeks prior to planting.

If weeds are present at the time of treatment, apply in tank mix combination with a contact herbicide (for example, paraquat or Roundup). Observe directions for use, precautions, and restrictions on the label of the contact herbicide.

Note: To the extent possible, do not move treated soil out of the row or move untreated soil to the surface during planting, or weed control will be diminished.

Preplant Incorporated (Broadleaf and grass control): Broadcast in spring after plowing at rate in Table 1. Apply to the soil and incorporate before, during, or after final seedbed preparation. Avoid deep incorporation. For best results, apply within two weeks prior to planting.

Preemergence (Broadleaf and grass control): Apply during or shortly after planting before weed emergence, at rate in Table 1.

Postemergence (Broadleaf and grass control): Apply before weeds exceed 1.5 inches in height, at rate in Table 1.

Table 1: Broadleaf and Grass Weed Control on Corn*

Soil texture	Broadcast rate/A
COARSE Sand, loamy sand, sandy loam	4 pt
MEDIUM Loam, silt loam, silt, clay loam, sandy clay loam, silty clay loam, sandy clay, or silty clay with low organic matter	4.75 pt
FINE Loam, silt loam, silt, clay loam, sandy clay loam, silty clay loam, sandy clay, or silty clay with medium to high organic matter, and clay (including the dark prairie soils of the Corn Belt)	6 pt
Peat, muck, and high organic clay (apply postemergence only)	6 pt

**Broadleaf control (eastern CO, western KS, western NE, NM, OK Pan Handle, west TX, and eastern WY):* On sand, loamy sand, sandy loam, mild to strongly alkaline soil, and all recently leveled soil, apply no more than 2.4 pt/A, either preplant surface, preplant incorporated, or preemergence. On other soils in these areas, apply rate in Table 1 for broadleaf and grass control.

Lay-by treatment (Broadleaf and grass control): Broadcast 2-4 pt/A before weeds are 1.5 inches tall and before corn is 20-30 inches tall. When using nitrogen solutions, direct the spray to lower 3-4 inches of cornstalks to avoid corn foliage injury.

Postemergence with emulsifiable oil or oil concentrate in water: Add the following volume of one of the type oils indicated for aerial or ground application unless the oil label specifies otherwise:

Type Oil	Ground Application	Aerial Application
Oil concentrate (Crop or Petroleum-derived)	1 qt/A	½-1 qt/A
Petroleum-derived oil	1 gal/A	2 qt/A

Note: Crop-derived or petroleum-derived oil concentrates should contain at least 1% but not more than 20% suitable emulsifier or surfactant blend. Petroleum-derived oils should contain at least 1% suitable emulsifier.

Broadleaf and grass control: For postemergence control of those weeds listed under **Preplant Incorporated and Preemergence,** broadcast 4 pt/A plus emulsifiable oil or oil concentrate after weed emergence, but before weeds reach 1.5 inches in height.

Broadleaf control: For postemergence control of those weeds listed under **Postemergence with emulsifiable oil or oil concentrate in water,** broadcast 2.4 pt/A plus emulsifiable oil or oil concentrate before pigweed and lambsquarters reach 6 inches in height and before all other weeds reach 4 inches in height. A cultivation may be necessary if all weeds are not controlled or if weeds regrow.

Precautions for applications with emulsifiable oil or oil concentrate in water: *(1) Inbred lines or any breeding stock may be severely injured by applications with emulsifiable oil or oil concentrate. (2) Adding other insecticides, herbicides, liquid fertilizers, or other materials is not recommended, because they may cause compatibility problems or crop injury. (3) Store and handle emulsifiable oil and oil concentrate carefully. Oil contaminated with even a small amount of water may not emulsify properly when added to the tank. To avoid crop injury. (4) Do not apply when crop is under stress from prolonged cold, wet weather, poor fertility, or other factors, or when crop is wet and succulent from recent rainfall. (5) Do not make more than one application per season except as recommended for control of yellow nutsedge, quackgrass, and Canada thistle.*

Problem weeds

Yellow nutsedge and Canada thistle: This product will control yellow nutsedge *(Cyperus esculentus)* and Canada thistle *(Cirsium arvense)* when applied according to directions. For best results, apply each year until yellow nutsedge or Canada thistle is eliminated or reaches a level of infestation where neither weed is an economic problem. If yellow nutsedge or Canada thistle regrow following last application, cultivate once.

Four alternative methods of use for control of yellow nutsedge and Canada thistle are listed below in order of preference. If other weed species, such as annual grasses, are also expected, use alternative 2 or 4.

1. Broadcast 4 pt/A plus 1 gal of emulsifiable oil or 1 qt of oil concentrate. Apply after crop and yellow nutsedge or Canada thistle emerge, but before yellow nutsedge reaches a height of 3 inches or Canada thistle reaches a height of 6 inches. Repeat application before lay-by (20-30 inches), 10-20 days after the first application.
2. Broadcast 4 pt/A preplant incorporated for control of yellow nutsedge, or broadcast 4 pt/A during or shortly after planting, but prior to crop or weed emergence for control of Canada thistle. Follow with an application of 4 pt/A plus 1 gal of emulsifiable oil or 1 qt of oil concentrate after corn and weeds emerge, but before yellow nutsedge or Canada thistle reaches a height of 6 inches.
3. Broadcast 8 pt/A plus 1 gal of emulsifiable oil or 1 qt of oil concentrate after crop emerges, but prior to lay-by (20-30 inches). Apply after yellow nutsedge and Canada thistle emerge,

but before yellow nutsedge is 3 inches tall or Canada thistle is 6 inches tall.

4. Broadcast 8 pt/A before, during, or shortly after planting, but before crop or weed emergence (yellow nutsedge control only).

Note: Do not use emulsifiable oil or oil concentrate when corn is wet or under stress, especially when using 8 pt/A of this product, or crop injury may occur. See "Precautions for applications with emulsifiable oil or oil concentrate in water" for additional directions.

Quackgrass control on land going into corn production
Split application: Broadcast 4 pt/A in fall or spring. Plow 1-3 weeks later. Apply a second application at 4 pt/A in spring before, during, or after planting, but before weeds are 1.5 inches high. This split application will control quackgrass and most annual broadleaf and grass weeds.

Single application: Broadcast 6-8 pt/A in fall or spring. Plow 1-3 weeks later.

TANK MIXTURES FOR CORN

This product may be tank mixed with these herbicides for control of certain broadleaf and grass weeds in corn:

Dual® (metolachlor)	Lasso® or Lasso EC (alachlor)
Dual + paraquat	Lasso or Lassc EC + Roundup
Dual + Roundup® (glyphosate)	Lasso or Lasso EC + paraquat
Dual + Princep®	Paraquat
Dual + Princep + paraquat	Princep
Dual + Princep + Roundup	Princep + paraquat
Bexton® or Ramrod® (propachlor)	Princep + Roundup
	Roundup
	Sutan + ®

Use tank mix directions appearing on the labels of the above herbicides when tank mixing with this product. Observe all precautions and limitations on labeling of products used in a particular tank mix.

Note: When the labels of the above herbicides refer to atrazine 80W, use equivalent rates of AAtrex 4L. One lb of 80W equals 1.6 pt of 4L.

Princep 80W, Princep 4L (4LC), or Princep Caliber® 90
In addition to the weeds listed under **AAtrex 4L Applied Alone—Corn and Grain Sorghum—Preplant Surface-Applied, Preplant Incorporated, or Preemergence** this combination also controls crabgrass, fall panicum, and carpetweed.

Broadcast tank mix before planting, at planting, or after planting, but before crop and weeds emerge, at rates in Table 2. Use the 1:1 ratio for control of most weeds. Use the 1:2 ratio for expected heavy infestations of crabgrass and fall panicum. Cultivate shallowly if weeds develop.

Preplant Surface-Applied: Use on medium- and fine-textured soils with minimum-tillage or no-tillage systems only in CO, IL, IN, IA, KS, KY, MN, MO, MT, NE, ND, SD, WI, and WY. Apply ⅔ the recommended rate of AAtrex and Princep shown in Table 2 for the soil texture as a split treatment 30-45 days prior to planting and the remainder at planting. Applications made less than 30 days prior to planting may be as either a split or single treatment. On *coarse-textured* soils, do not apply more than two weeks prior to planting. Refer to the AAtrex Alone—Preplant Surface-Applied section of the corn label for additional details.

If weeds are present at time of treatment, apply in a tank mix combination with a contact herbicide (for example, paraquat or Roundup). Observe directions for use, precautions, and restrictions on the label of the contact herbicide.

Note: To the extent possible, do not move treated soil out of the row or move untreated soil to the surface during planting, or weed control will be diminished.

Preplant Incorporated: Apply to the soil and incorporate in the spring before, during, or after final seedbed preparation. Avoid deep incorporation. For best results, apply within two weeks prior to planting.

Preemergence: Apply during or shortly after planting, but before crop and weeds emerge.

Refer to Corn sections of this label and to Princep 80W, Princep Caliber 90, or Princep 4L (4LC) labels for further directions, limitations, and precautions.

Table 2: Tank Mixtures with Princep on Corn

	Broadcast rate/A			
	1:1 Ratio*		1:2 Ratio**	
Soil texture	This product	Princep 80W[1]	This product	Princep 80W[1]
Sand, loamy sand, sandy loam	2 pt	1.25 lb	1.32 pt	1.67 lb
Loam, silt loam, silt, clay loam, sandy clay loam, silty clay loam, sandy clay, or silty clay with low organic matter	2.4 pt	1.5 lb	1.6 pt	2 lb
Loam, silt loam, silt, clay loam, sandy clay loam, silty clay loam, sandy clay, or silty clay with medium to high organic matter, and clay (including dark prairie soils of the Corn Belt)	3 pt	1.8 lb	1.92 pt	2.4 lb

* For control of most weeds.

** For control of expected heavy infestations of crabgrass and fall panicum.

† When using Princep Caliber 90 or Princep 4L (4LC), use equivalent rates. One lb of Princep 80W equals 0.9 lb of Princep Caliber 90 or 1.6 pt of Princep 4L (4LC).

Princep 80W, Princep 4L (4LC), or Princep Caliber 90 plus Roundup: Use as tank mixture for preemergence and postemergence control of certain broadleaf and grass weeds where corn will be planted directly into a cover crop, established sod, or in previous crop residues. Refer to Roundup label for all directions, weeds controlled, precautions, and limitations.

Princep 80W, Princep 4L (4LC), or Princep Caliber 90 plus Paraquat: Use as tank mixture with Princep and paraquat to kill existing vegetation and for residual weed control where corn will be planted directly into a cover crop, established sod, or in previous crop residues. Add this product and Princep to water in spray tank, agitating until thoroughly mixed. Then add paraquat and a nonionic surfactant, such as X-77®. Continue agitation during application. Broadcast 2-4 pt of this product plus 1.25-2.5 lb Princep 80W [or 2-4 pt Princep 4L (4LC), or 1.1-2.2 lb Princep Caliber 90] plus 1-2 pt paraquat in 20-60 gal of water per sprayed acre. Apply before, during, or after planting, but before corn emerges. Add 0.5 pt of a nonionic surfactant, such as X-77, per 100 gal of spray mixture. Use the 2 pt rate of paraquat if existing vegetation is 4-6 inches tall. This mixture will not control weeds taller than 6 inches.

Refer to further limitations and precautions on labels for this product, Princep, and paraquat.

Precautions for all applications to corn: *(1) To avoid crop injury and illegal residues, do not apply more than 8 pt/A of this product per year. (2) Following harvest, plow (moldboard or disk-plow) and thoroughly till soil in fall or spring to minimize possible injury to spring-seeded rotational crops, regardless of rate used.*

Note for all applications to corn: Do not graze or feed forage from treated areas for 21 days following application, or illegal residues may result.

Rotational crops: (1) Do not rotate to any crop except corn or sorghum until the following year, or injury may occur. (2) If applied after June 10, do not rotate with crops other than corn or sorghum the next year, or crop injury may occur. (3) If the broadcast rate, or equivalent band application rate, exceeds 6 pt/A of AAtrex 4L, or 6 pt/A of AAtrex 4L plus Princep 4L (or equivalent of another Princep formulation), a crop of untreated corn or sorghum should precede the next rotational crop. (4) In the High Plains and Intermountain areas of the West where rainfall is sparse and erratic or where irrigation is required, use only when corn or sorghum is to follow corn or sorghum or when a crop of untreated corn or sorghum is to precede other rotational crops. (5) In eastern parts of the Dakotas, KS, western MN, and NE, do not rotate to soybeans if the rate applied to corn or sorghum was more than 4 pt/A or equivalent band application rate, or soybean injury may occur. (6) Injury may occur to soybeans planted the year following application on soils having a calcareous surface layer. (7) Do not plant sugar beets, tobacco, vegetables (including dry beans), spring-seeded small grains, or small-seeded legumes and grasses the year following application, or injury may occur.

SORGHUM AND SORGHUM-SUDAN HYBRIDS (GRAIN AND FORAGE TYPES)

Preplant Surface-Applied (Broadleaf and grass control): Use on medium- and fine-textured soil with minimum-tillage or no-tillage systems only in CO, IL, IN, IA, KS, KY, MN, MO, MT, NE, ND, SD, WI, WY. Apply ⅔ the recommended rate of AAtrex shown in Table 3 for the soil texture as a split treatment 30-45 days prior to planting and the remainder at planting. Applications made less than 30 days prior to planting may be as either a split or single treatment. Under dry conditions, irrigation after application is recommended to move AAtrex into the soil.

If weeds are present at time of treatment, apply in a tank mix combination with a contact herbicide (for example, paraquat or Roundup). Observe directions for use, precautions, and restrictions on the label of the contact herbicide.

Note: To the extent possible, do not move treated soil out of the row or move untreated soil to the surface during planting, or weed control will be diminished.

Preplant Incorporated (Broadleaf and grass control): Broadcast in spring after plowing at rate shown in Table 3. Apply before, during, or after final seedbed preparation. If soil is tilled or worked after application, avoid deep incorporation. For best results, apply within two weeks prior to planting.

Preemergence (Broadleaf and grass control): Apply during or shortly after planting, but prior to weed or crop emergence at rate shown in Table 3.

Table 3: Preplant Surface, Preplant Incorporated, and Preemergence Broadleaf and Grass Weed Control in Sorghum*

Soil texture	Organic	Broadcast rate/A
COARSE Sand, loamy sand sandy loam	any level	DO NOT USE (except for preemergence use on bedded sorghum in AZ and CA)
MEDIUM AND FINE	less than 1%	
Loam, silt loam, silt, clay loam, sandy clay loam, silty clay loam,	1-1.5%	3.2-4 pt
sandy clay, silty clay, clay	more than 1.5%	4-4.75 pt

* Do not apply preplant surface or preplant incorporated in AL, AR, FL, GA, LA, MS, NC, NM, OK, SC, TN, or TX. Do not apply preemergence in NM, OK, or TX, except in northeast OK and the TX Gulf Coast.

In case of planting failure, sorghum or corn may be replanted. Do not make a second broadcast application, or injury may occur. If originally applied in a band and sorghum or corn is replanted in untreated row middles, this product may be applied in a band to the second planting.

Preemergence broadleaf weed control in furrow irrigated bedded sorghum (AZ and CA only): For preemergence control of many broadleaf weeds, broadcast 1.6-2.4 pt/A after bed preparation, during or after planting, but before sorghum and weeds emerge and before the first furrow irrigation. Follow with several regular irrigations, making sure to thoroughly wet all soil.

Precautions: *For preemergence application to furrow irrigated bedded sorghum in AZ and CA: To avoid possible sorghum injury, do not use on sand or loamy sand soil or on sorghum planted in furrows. Applications to sorghum growing on alkali soils or where cuts, fills, or erosion have exposed calcareous or alkali subsoils may cause crop injury. In case of crop failure, do not replant sorghum for 8 months following application. Corn may be planted immediately.*

Postemergence (Broadleaf and grass control): Apply at rate shown in Table 4 before weeds exceed 1.5 inches in height. Apply up to close-in.

Table 4: Postemergence Broadleaf and Grass Weed Control in Sorghum

Soil texture	Minimum sorghum height at treatment	Broadcast rate/A
Sand or loamy sand	DO NOT USE	
Sandy loam	See directions for postemergence broadleaf weed control	
Silt loam to sandy clay loam	Completely emerged	4-4.75 pt
Olton and Pullman clay soil	6 inches	4-4.75 pt
Silty clay loam and finer fextured soil	Completely emerged	6 pt

Postemergence broadleaf weed control with emulsifiable oil or oil concentrate in water: Broadcast 2.4 pt/A for control of many broadleaf weeds. Apply before pigweed and lambsquarters reach 6 inches in height and before all other weeds reach 4 inches in height. In CO, western KS, NM, OK, TX, and desert regions of AZ and CA, apply when sorghum is 6-10 inches in height, but before it reaches boot stage. In all other areas, apply after sorghum reaches the 3-leaf stage. Add 1 gal of emulsifiable oil/A for ground application and 0.5 gal/A for aerial application, or add 1 qt/A of oil concentrate for ground application. A cultivation may be necessary if all weeds are not controlled or if weeds regrow.

For the list of weeds controlled, see **AAtrex 4L Applied Alone—Corn or Grain Sorghum—Postemergence with Emulsifiable Oil or Oil Concentrate in Water.**

Precautions for applications with emulsifiable oil or oil concentrate in water: *See "Precautions for applications with emulsifiable oil or oil concentrate in water" in Corn section.*

Postemergence broadleaf weed control with surfactant (CO, western KS, NM, OK, TX, and desert regions of AZ and CA only): Broadcast 2.4 pt/A plus 0.75-1.5 pt of surfactant after sorghum reaches 6 inches in height, but before weeds exceed 1.5 inches in height. Apply only on sandy loam and finer textured soil.

Precautions for all applications to sorghum: *(1) Heavy rain immediately following application tends to cause excessive concentrations of herbicide in seed furrow, resulting in possible crop injury. Do not apply to furrow-planted sorghum until furrows are leveled (plowed in). Level deep planter marks or seed furrows before application. (2) Application to sorghum growing under stress caused by minor element deficiency or to sorghum growing on highly calcareous soil may result in crop injury. (3) Following harvest, plow (moldboard or disk-plow) and thoroughly till soil in fall or spring to minimize possible injury to spring-seeded rotational crops, regardless of rate used. (4) Injury may occur if both this herbicide, preplant surface, preplant incorporated, or preemergence, and an at-planting systemic insecticide applied in-furrow are used.*

Note: Do not graze or feed forage from treated areas for 21 days following application, or illegal residues may result.

Rotational crops: See "Rotational crops" in Corn section.

TANK MIXTURES FOR GRAIN SORGHUM

Igran® 80W: Use as tank mixture for control of certain broadleaf and grass weeds in grain sorghum. Refer to Igran 80W label for all directions, weeds controlled, precautions, and limitations.

Dual 8E: Use as tank mixture with Dual 8E for control of those weeds listed on the Dual label, as well as on this label. Use this tank mixture only on sorghum seed treated with Concep®. Refer to the Dual 8E label for all directions, precautions, and limitations.

WINTER WEED CONTROL IN TEXAS

For postemergence control of winter weeds only, such as henbit, seedling dock, and annual thistle on fall bedded land in the Gulf Coast and Blacklands of Texas. Apply 0.8-1 qt/A postemergence to the weeds in November or December to land that will be planted to corn, grain sorghum, or forage sorghum the following spring. For best results, add a suitable surfactant, such as X-77, at the rate of 0.5% of the spray volume, an emulsifiable oil at the rate of 1.0% of the spray volume, or an oil concentrate at the rate of 1 qt per acre.

Normal weed control programs may be used in the following corn, grain sorghum or forage sorghum crop.

Note: Do not plant any crops except corn, grain sorghum, or forage sorghum the spring following this treatment, or illegal residues may result.

AATREX 4L ALONE—CHEMICAL FALLOW

Wheat-Sorghum-Fallow: To control annual broadleaf and grass weeds following wheat harvest and in the following sorghum crop when grown under minimum tillage, broadcast 6 pt/A to wheat stubble immediately following wheat harvest. If weeds are present at application, remove them with a sweep plow or other suitable implement after application, or use an approved contact herbicide before or after the application of AAtrex. Plant sorghum into wheat stubble the following spring with minimum soil disturbance. Use a surface planter or a planter leaving a shallow furrow. If weeds are present at planting, remove them with a sweep plow or other suitable implement before planting.

For the list of weeds controlled, see **AAtrex 4L Applied Alone—Corn or Grain Sorghum—Preplant Surface-Applied, Preplant Incorporated,** or **Preemergence.**

Precautions: *(1) Use only on silt loam or finer textured soil, or crop injury may result. (2) Wheat-sorghum-fallow cropping sequence must be followed. (3) Do not apply following sorghum harvest.*

Note: To avoid illegal residues, do not graze or feed forage from treated area to livestock. To avoid illegal residues and crop injury, do not plant any crop other than those on this label within 18 months following treatment.

Wheat-Corn-Fallow (KS, NE): This product controls cheatgrass (downy brome, chess), kochia, mustards, pigweed, Russian thistle, wild lettuce, wild sunflower, and volunteer wheat during period after wheat harvest. Weed control may extend into following corn crop grown under minimum tillage.

Follow directions for use, notes, and precautions in the "Wheat-Sorghum-Fallow" section above, substituting corn for references to sorghum.

Wheat-Fallow-Wheat (CO, KS, MT, NE, ND, SD, and WY): For preemergence control of cheatgrass (downy brome, chess), common lambsquarters, field pennycress, kochia, mustard, Russian thistle, wild lettuce, and suppression of volunteer wheat during fallow period of a wheat-fallow-wheat rotation, broadcast 1-2 pt/A on all soils except those listed under "Precautions." For control of pigweed and wild sunflower, use the higher rate. Apply to stubble ground. Treat only once during same fallow period.

TANK MIXTURES FOR CHEMICAL FALLOW

Wheat-Sorghum-Fallow or Wheat-Corn-Fallow (KS, NE)

Igran 80W: Use as tank mixture for control of certain broadleaf and grass weeds in wheat-sorghum-fallow or wheat-corn-fallow rotations. Refer to Igran 80W label for directions, weeds controlled, precautions, and limitations.

Paraquat: If weeds are present at application, a tank mix with paraquat may be used. Broadcast 6 pt of AAtrex 4L plus 1-2 pt of paraquat in 20-60 gal of water/A by ground equipment. Add 0.5-1 pt of a nonionic surfactant, such as X-77, per 100 gal of spray mixture. Add AAtrex to spray tank first and thoroughly mix with water. Then add paraquat, followed by surfactant. Use the 2 pt rate of paraquat if weeds are 4-6 inches tall. This mixture will not control weeds taller than 6 inches. Apply to stubble ground. Treat only once during same fallow period. Refer to paraquat label for further directions, precautions, and limitations.

Wheat-Fallow-Wheat (CO, KS, MT, NE, ND, SD, and WY)

Igran 80W: Use as tank mixture for control of certain broadleaf and grass weeds in wheat-fallow-wheat rotations. Refer to Igran 80W label for directions, weeds controlled, precautions, and limitations.

Paraquat: If weeds are present at application, a tank mix with paraquat may be used. Broadcast 1-2 pt of AAtrex 4L plus 1-2 pt of paraquat in 20-60 gal of water/A by ground equipment. Add 0.5-1 pt of a nonionic surfactant, such as X-77, per 100 gal of spray mixture. Add AAtrex to spray tank first and thoroughly mix with water. Then add paraquat, followed by surfactant. Use the 2 pt rate of paraquat if weeds are 4-6 inches tall. This mixture will not control weeds taller than 6 inches. Apply to stubble ground. Treat only once during same fallow period. Refer to paraquat label for further directions, precautions, and limitations.

If weeds are present at application and this product is used alone, use either an approved contact herbicide before or after treatment, or tillage after treatment.

Use tillage to control weeds which escape during fallow period. Till before planting. For this product applied alone or in tank mixture with paraquat, plant at least 2 inches deep and 12 months or more after application.

Precautions: *To avoid crop injury, (1) Do not use on sand soil. (2) Do not treat eroded hillsides, caliche and rocky outcroppings, or exposed calcareous subsoil. (3) Do not treat soils of the Rosebud and Canyon Series in western NE and adjoining counties in CO and WY. (4) Do not treat soils with calcareous surface layers. (5) Avoid spray overlap.*

Note: Do not graze treated areas within 6 months after application, or illegal residues may result.

Aerial application: In order to assure that spray will be controllable within the target area when used according to label directions, make applications at a maximum height of 10 ft. using low drift nozzles at a maximum pressure of 40 psi, and restrict application to periods when wind speed does not exceed 10 mph. To assure that spray will not adversely affect adjacent sensitive nontarget plants, apply AAtrex 4L alone by aircraft at a minimum upwind distance of 400 ft from sensitive plants, or apply AAtrex 4L + Igran 80W at a minimum upwind distance of 500 ft from sensitive plants. When applying AAtrex 4L + Igran 80W by aircraft, use a minimum of 5 gal of water per acre.

PROSO MILLET

For preplant incorporated or preemergence control of foxtails, kochia, common lambsquarters, redroot pigweed, Russian thistle, slimleaf lambsquarters and spotted spurge, apply 1-2 pt/A on soils with less than 2% organic matter, or apply 2-4 pt/A on soils with 2% or higher organic matter.

Precaution: *To avoid crop injury and illegal residues, make only one application per year.*

Rotational crops: See "Rotational crops" in Corn section.

RANGELAND

To control certain weeds, including cheatgrass (downy brome, chess), common (annual) broomweed, little barley, medusahead, sagewort, and tumble mustard. Use lower rate where cheatgrass is major problem weed; use higher rate for other weeds.

Poisonous plants: Tansymustard, pigweeds, lambsquarters, tumble mustard, cocklebur, and fiddleneck are controlled at 2 pt/A. Other mustards (*Brassica* spp.), groundcherry, (annual), broomweed, and threeawn or poverty grass (*Aristida* spp.) are partially controlled at 2 pt/A and controlled at 4 pt/A. Sneezeweed and dogbane are partially controlled at 4 pt/A.

Aerial application: In order to assure that spray will be controllable within the target area when used according to label directions, make applications at a maximum height of 10 ft above vegetation, using low drift nozzles at a maximum pressure of 40 psi, and restrict application to periods when wind speed does not exceed 10 mph. To assure that spray will not adversely affect adjacent sensitive nontarget plants, apply AAtrex 4L by aircraft at a minimum upwind distance of 400 ft from sensitive plants.

Establishment of perennial range grasses in West (CA, ID, NV, OR, UT, WA): Broadcast 1.6-2 pt/A in fall after rains begin but before ground freezes. Perennial range grass may be seeded when rain begins the next fall following treatment. Plant in deep furrows such as those made by a deep furrow rangeland drill. Treated areas may be grazed 7 months after seeding. However, to provide sufficient time for establishment of desirable range grasses, treated areas should not be grazed until the fall following seeding.

Renovation of existing stands of perennial range grasses
Central Great Plains (CO, KS, MT, NE, ND, SD, WY): Broadcast 1-2 pt/A in fall before ground freezes.

Southwest (AR, KS, LA, MS, NM, OK, TX): Apply 2 pt/A AAtrex 4L in early spring, before regrowth, to established rangeland or pasture species such as Asiatic (Old World) bluestems, little and big bluestems, buffalograss, switchgrass, and Indiangrass.

West (CA, ID, NV, OR, UT, WA): Do not graze area to be treated for 6 months prior to application. Broadcast 1-1.5 pt/A by ground equipment in fall before ground freezes.

Bluegrass and intermediate wheatgrass may be injured by this product at recommended rates.

Notes: To avoid illegal residues, (1) Do not apply more than once per year. (2) Apply only on neutral to acid soils. (3) Do not cut or feed range grass hay. (4) Do not graze treated areas within 3 months following spring application, or within 7 months following fall application. (5) Do not dump or spill product or dispose of container within reach of livestock.

Pure stands of Switchgrass and Big Bluestem (IA, IL, IN, KS, MN, MO, ND, NE, OH, SD, WI): For control of those weeds listed under **AAtrex 4L Applied Alone—Corn or Grain Sorghum—Preplant Surface-Applied, Preplant Incorporated, or Preemergence.**

Use only on loam, silt loam, silty clay loam, clay loam, and silty clay soils with at least 1 percent organic matter.

Establishment: Broadcast 4 pt/A after planting but before weeds and crop emerge. Prepare a good, firm seedbed. Plant ½ inch deep with a grassland drill (preferred method) or a conventional drill. If a conventional drill is used on prepared seedbeds, remove all tension from the disk openers. For best results, cultipack or roll after planting. Clip weeds that escape in July or August. Avoid clipping switchgrass or big bluestem seedlings.

Renovation of existing stands: Broadcast 4 pt/A in early spring, before regrowth of the crop, and before weed emergence.

Notes: To avoid illegal residues, (1) Do not cut for hay. (2) Do not graze treated areas within 4 months following application at seeding or 3 months following application to established switchgrass or big bluestem. (3) Treat only once per year.

Rotational crops: See "Rotational crops" in Corn section. Do not plant small-seeded legumes the year following application, or injury may occur.

ROADSIDES

To control certain annual weeds in established perennial grasses along roadsides in CO, KS, MT, NE, ND, SD, and WY, including cheatgrass (downy brome, chess), common (annual) broomweed, little barley, medusahead, sagewort, and tumble mustard, broadcast 2 pt/A in a minimum of 10 gallons of water by ground equipment in the fall before ground freezes, or after thawing in the spring, but before the established grasses green-up and before weeds emerge. Examples of desirable established grasses include big bluestem, bluegrama, bromegrass, buffalo grass, crested wheatgrass, Indiangrass, little bluestem, side-oats grama, switchgrass, and Western wheatgrass. Apply only once per year. Temporary discoloration or other form of injury to the desirable perennial grasses may occur following application.

Note: To avoid illegal residues. (1) Do not cut or feed roadside grass hay. (2) Do not allow livestock to graze treated areas.

PERENNIAL RYEGRASS GROWN FOR SEED
(Pacific Northwest only)

For control of annual ryegrass and rattail fescue, broadcast 2.4 pt/A after first fall rain.

Precautions: *To avoid crop injury. (1) Apply only to perennial ryegrass from which seed crop has been harvested. (2) Do not apply to ryegrass more than 2 years in succession.*

Notes: To avoid illegal residues. (1) Do not graze treated land within 10 days after application. (2) Do not plant treated land to other crops for 16 months.

ESTABLISHED ORCHARDGRASS GROWN FOR SEED
(Pacific Northwest only)

AAtrex aids in control of annual ryegrass, rattail fescue, and volunteer orchardgrass. Broadcast 4 pt/A of AAtrex 4L in at least 15 gal of water to field which has been burned after first fall rain.

Precautions: *Some yellowing or crop stunting can occur. To avoid crop injury. (1) Apply only to orchardgrass from which seed crop has been harvested. (2) Do not apply to orchardgrass more than two years in succession.*

Notes: To avoid illegal residues, (1) Do not graze treated land within 6 months of application. (2) Do not plant treated land to other crops for 16 months.

SUGARCANE

For control of many broadleaf and grass weeds, including amaranths, crabgrass, fireweed, Flora's paintbrush, foxtails, junglerice and wiregrass, broadcast 4-8 pt/A of AAtrex 4L at time of planting or ratooning, but before sugarcane emerges. Broadcast aerially in a minimum of 5 gal of spray per acre, or broadcast or band by ground equipment in a minimum of 20 gal per acre, unless indicated otherwise. One additional application may be made over the sugarcane as it emerges, and two additional applications may be made interline after emergence as directed sprays. Repeat treatments, where needed, may be applied broadcast, band, or interline as suggested with the final application being prior to close-in. Do not exceed the rate of herbicide suggested for any one crop of sugarcane.

Note: Where high rates of AAtrex are used alone, apply in a minimum of 1 qt of water for each 1 qt of AAtrex 4L applied per acre.

Aerial application: In order to assure that spray will be controllable within the target area when used according to label directions, make applications at a maximum height of 10 ft, using low drift nozzles at a maximum pressure of 40 psi, and restrict application to periods when wind speed does not exceed 10 mph. To assure that spray will not adversely affect adjacent sensitive nontarget plants, apply AAtrex 4L alone by aircraft at a minimum upwind distance of 400 ft from sensitive plants.

Florida
For control of emerged pellitory weed: Apply 0.8-1.2 pt/A in at least 40 gal of water as a directed spray by ground equipment prior to close-in. Add 4 qt of surfactant for each 100 gal of spray. Thoroughly cover weed foliage.

For control of alexandergrass, large crabgrass, pellitory (artillery) weed, and spiny amaranth, use one of the following methods at planting or ratooning:

1. Apply 8 pt/A preemergence. Follow with one or two applications, as needed, postemergence to sugarcane and weeds, at 4 pt/A. Treat before weeds exceed 1.5 inches in height.
2. Apply 1-3 times, as needed, at 4 pt/A postemergence to sugarcane and weeds. Treat before weeds exceed 1.5 inches in height.

Louisiana
For control of annual weeds during the summer fallow period, apply 4 pt A to weedfree beds immediately after bed formation. Follow normal weed control after planting.

Precautions: *To avoid crop injury, (1) Do not apply more than 10 qt A to any one crop of sugarcane. (2) If making 4 pt/A application during summer fallow period, do not exceed 8 qt/A during the remainder of the growing season, or illegal residues may result.*

Texas
Use AAtrex 4L for control of barnyardgrass, pigweed, purslane and sunflower, in plant or ratoon sugarcane.

Apply 8 pt/A of AAtrex 4L preemergence. Follow with one or two applications, as needed, at 6 pt/A postemergence to sugarcane and weeds.

For best results when weeds are emerged, add a nonionic surfactant at a concentration of 2 qt/100 gal to the spray and apply before weeds exceed 1.5 inches in height.

Precautions: *(1) Injury to sugarcane may occur when under moisture stress, when soil is of low adsorptive capacity, or when land is first cropped to sugarcane. (2) Do not apply after close-in. (3) Do not apply more than 20 pt/A to any one crop of sugarcane, or crop injury may result.*

TURF GRASSES FOR SOD (Florida only)

St. Augustinegrass, Centipedegrass, and Zoysia Grass
Broadcast 4-8 pt/A according to soil texture to control those weeds listed under

AAtrex 4L Applied Alone—Corn or Grain Sorghum—Preplant Surface-Applied, or Preplant Incorporated, or Preemergence.

Muck or peat	8 pt	Old beds: Within 2 days after lifting sod.
		New beds: 3-4 days after sprigging or plugging.
Sandy soil	4 pt	Old beds: Within 2 days after lifting sod.
		New beds: 7-10 days after sprigging or plugging.

If weeds regrow, apply an additional 4 pt/A on muck or peat, or 2 pt/A on sandy soil.

Precautions: *To avoid crop injury, (1) Do not apply within 30 days prior to cutting or lifting. (2) Do not apply in combination with surfactants or other spray additives. (3) Use only on turfgrass reasonably free of infestations of insects, nematodes, and diseases. (4) On newly sprigged turfgrass, temporary slowing of growth may follow application.*

Turfgrass for Fairways, Lawns, Sod Production* and Similar Areas

* In states other than Florida. For use on turfgrass for sod in Florida, see "Turfgrasses for Sod (Florida only)" section above.

Bermudagrass, Centipedegrass, St. Augustinegrass, and Zoysia Grass
Apply AAtrex 4L after October 1 before emergence of winter annual weeds for control of annual bluegrass, burclover, carpet burweed, chickweed, corn speedwell, henbit, hop clover, and spurweed. AAtrex 4L will control annual bluegrass even if it is emerged at time of treatment. For control of summer annual weeds listed in the preemergence portion of the AAtrex 4L Applied Alone—Corn or Grain Sorghum section of this label, also apply AAtrex 4L in late winter before the weeds emerge. Apply in a minimum of 15 gal of water per acre or 1 gal per 1,000 sq ft.

Where annual bluegrass is the major weed, use 1 qt/A (22 ml or 0.75 fl oz per 1,000 sq ft). Use 2 qt/A (44 ml or 1.5 fl oz per 1,000 sq ft) for control of the other weeds named above. Do not exceed 1 qt/A per treatment on newly sprigged turfgrass or on hybrid bermudagrass such as Tiflawn, Tifway, and Ormond.

For continued summer annual weed control, apply another 1 qt/A at least 30 days after the previous application, but not after April 15. However, do not make more than two applications of this product per year.

Precautions: *On newly sprigged turfgrass and hybrid bermudagrass, temporary slowing of growth and yellowing may occur following application. To avoid turf injury, (1) Use only on turfgrass reasonably free of infestations of insects, nematodes, and diseases. (2) Do not use on golf greens. (3) Do not use north of NC (except may be used in Virginia Coastal Plains) or west of the high rainfall areas of eastern OK and eastern TX. (4) Do not use on muck or alkaline soils. (5) Do not apply over the rooting area of trees or ornaments not listed on this label. (6) Do not overseed with desirable turfgrass within 4 months before or 6 months after treatment.*

Note: Do not graze or feed turf clippings to animals, or illegal residues may result.

MACADAMIA NUTS

For preemergence control of many broadleaf and grass weeds, including crabgrass, foxtail, wiregrass, Flora's paintbrush, spanishneedles, and fireweed, broadcast 4-8 pt/A before harvest and before weeds emerge. Repeat as necessary. Do not spray when nuts are on ground during harvest period. Do not apply by air.

PINEAPPLE

For control of purslane, spanishneedles, annual grasses, annual bindweed, ageratum, amaranth, rattlepod, Flora's paintbrush, fireweed, spurge, indigo, and papalo, apply up to 12.8 pt/A as a blanket spray immediately after planting or following harvest. Make additional blanket or interspace applications of up to 3.2 pt/A at 1-2 month intervals as needed, prior to differentiation. Apply in 20-40 gal of water/A to assure thorough coverage.

Precautions: *(1) Do not apply more than 60 pt (7.5 gal)/A per cycle, as crop injury or illegal residues may result. (2) Repeated monthly applications to plant foliage may slow plant growth and delay fruiting.*

Note: Do not apply within 45 days of fruit harvest, or forage harvest if forage is to be fed to livestock, or illegal residues may result.

GUAVA

Use only on established plantings which are at least 18 months old. Apply as a directed spray at 4-8 pt/A of AAtrex 4L in 20-50 gal of spray mix preemergence or early postemergence to weeds. When applying postemergence, the use of a surfactant and greater spray volume (80-100 gal of spray mix per acre) may enhance weed control. This product controls many annual broadleaf and grass weeds, including fireweed, purslane, scarlet pimpernel, spanishneedles, and sowthistle.

Notes: To avoid illegal residues, (1) Do not allow spray to contact foliage or fruit. (2) Do not apply more frequently than at 4 month intervals. (3) Do not apply more than 16 pt of AAtrex 4L per year.

CONIFERS

For control of annual broadleaf and grass weeds prior to transplanting, after transplanting or in established conifers [including Douglas fir, grand fir, noble fir, white fir, Austrian pine, bishop pine, Jeffrey pine, knobcone pine, loblolly pine, lodgepole pine (shore pine), monterey pine, ponderosa pine, Scotch pine, slash pine, blue spruce, and Sitka spruce]: Broadcast 4-8 pt in a minimum of 5 gal of water per acre by air or 10 gal by ground before weeds are 1.5 inches tall. Apply to established trees between fall and early spring while trees are dormant. For new transplants, apply during or soon after transplanting. For applications prior to transplanting, allow sufficient precipitation to activate AAtrex before transplanting. In areas where spring and summer rainfall is inadequate to activate AAtrex, apply during fall prior to spring transplanting.

For the list of weeds controlled, see **AAtrex 4L Applied Alone—Corn or Grain Sorghum—Preplant Surface-Applied, Preplant Incorporated, or Preemergence.**

Quackgrass control: Broadcast 8 pt in a minimum of 5 gal of water per acre by air or 10 gal by ground between fall and early spring while trees are dormant and before quackgrass is more than 1.5 inches tall.

Precautions: *Temporary injury to trees may occur following the use of AAtrex on coarse-textured soil. To avoid crop injury, (1) Do not apply to seedbeds. (2) Apply only once per year.*

Note: Do not graze treated areas, or illegal residues may result.

Aerial application: In order to assure that spray will be controllable within the target area when used according to label directions, make applications at a maximum height of 10 ft above vegetation, using low drift nozzles at a maximum pressure of 40 psi, and restrict application to periods when wind speed does not exceed 10 mph. To assure that spray will not adversely affect adjacent sensitive nontarget plants, apply AAtrex 4L by aircraft at a minimum upwind distance of 400 ft from sensitive plants.

NONSELECTIVE WEED CONTROL ON NONCROP LAND

This product provides long-term weed control on industrial sites, highway medians and shoulders, railroad rights-of-way, lumberyards, petroleum tank farms, and in noncrop areas on farms, such as around buildings, equipment and fuel storage areas, along fences, and lanes. Do not use near adjacent desirable plants or in greenhouses, or injury may occur.

Apply before or soon after weeds begin growth. Make postemergence applications when weeds are young and actively growing. Use the higher rates on fine clay and muck soils. Use sufficient water to assure thorough coverage. Use at least 1 gal of water for each qt of product; use more if practical.

Note: Do not allow livestock to graze treated areas, or illegal residues may result.

Annual broadleaf and grass weeds (including barnyardgrass, cheatgrass, crabgrass, lambsquarters, foxtail, ragweed, puncturevine, and turkey mullein): Broadcast 4.8-10 qt/A.

Hard-to-kill annual and many perennial broadleaf and grass weeds (including bluegrass, burdock, Canada thistle, dogfennel, orchardgrass, plantain, quackgrass, purpletop, redtop, and smooth brome): Broadcast 10-20 qt/A.

Hard-to-kill biennial and perennial weeds (including bullthistle and sowthistle): Broadcast 20-40 qt/A.

For longer residual control in regions of high rainfall and a long growing season, broadcast 20-40 qt/A.

For small areas, 0.28 pt per 1,000 sq ft is equivalent to 10 qt/A.

STORAGE AND DISPOSAL

PESTICIDE DISPOSAL

Do not contaminate water, food, or feed by storage or disposal, or cleaning of equipment. Open dumping is prohibited. Wastes resulting from the use of this product are acutely toxic. Improper disposal of unused pesticide, spray mixture, or rinsate is a violation of federal law. Pesticide, spray mixture, or rinsate that cannot be used according to label instructions must be disposed of according to federal, state or local procedures. For guidance in proper disposal methods, contact your State Pesticide or Environmental Control Agency, or the Hazardous Waste representative at the nearest EPA Regional Office.

CONTAINER DISPOSAL

Triple rinse (or equivalent) and offer for recycling or reconditioning, or puncture and dispose of in a sanitary landfill, or by incineration, or by open burning, if allowed by state and local authorities. If burned, keep out of smoke.

For minor spills, leaks, etc., follow all precautions indicated on this label and clean up immediately. Take special care to avoid contamination of equipment and facilities during cleanup procedures and disposal of wastes. In the event of a major spill, fire or other emergency, call (919) 292-7100 day or night.

PRECAUTIONARY STATEMENTS

HAZARDS TO HUMANS AND DOMESTIC ANIMALS

CAUTION

Harmful if swallowed, inhaled, or absorbed through skin. Do not breathe vapors or spray mist. Avoid contact with eyes, skin or clothing.

First Aid: If swallowed, contact your local poison control center, hospital, or physician. If the patient is unconscious, maintain breathing and heartbeat (cardiopulmonary resuscitation). If the patient is conscious and alert, induce vomiting (syrup of ipecac or stimulate the back of the throat with a finger). NEVER GIVE ANYTHING BY MOUTH TO AN UNCONSCIOUS PERSON!

In case of skin contact, remove contaminated clothing and wash with soap and water. In case of contact with eyes, flush with *plenty* of water. Get medical attention if irritation persists.

Note to Physician: There is no specific antidote for atrazine. If this product is ingested, induce emesis or lavage stomach. The use of an aqueous slurry of activated charcoal may be considered.

ENVIRONMENTAL HAZARDS

Do not apply directly to any body of water. Do not apply where runoff is likely to occur. Do not contaminate water by cleaning of equipment or disposal of wastes. Do not apply when weather conditions favor drift from treated areas.

AAtrex® trademark of CIBA—GEIGY for atrazine

Bexton® trademark of Dow Chemical Co. for propachlor

Caliber® trademark of CIBA—GEIGY

Compex® trademark of KALO Agricultural Chemicals, Inc.

Concep® trademark of CIBA—GEIGY

Dual® trademark of CIBA—GEIGY for metolachlor U.S. Patent No. 3,937,730

Igran® trademark of CIBA—GEIGY for terbutryn

Lasso® trademark of Monsanto Company for alachlor

Princep® trademark of CIBA—GEIGY for simazine

Ramrod® trademark of Monsanto Company for propachlor

Roundup® trademark of Monsanto Company for glyphosate herbicide

Sutan +® trademark of Stauffer Chemical Co. for butylate

Unite® trademark of Hopkins Agricultural Chemical Co.

X-77® trademark of KALO Agricultural Chemicals, Inc.

CGA 7L38T 035

AATREX®80W

HERBICIDE

For season-long weed control in corn and sorghum

For weed control in certain other crops; in noncrop areas ; and industrial sites

AAtrex 80W is a wettable powder

ACTIVE INGREDIENTS:	
Atrazine:	
2-chloro-4-ethylamino-6-isopropylamino-*s*-triazine	76%
Related compounds	4%
INERT INGREDIENTS:	20%
TOTAL:	100%

EPA Reg. No. 100-439 EPA Est. 100-AL-1

AAtrex® trademark of CIBA—GEIGY for atrazine

Keep Out of Reach of Children.

Caution

See additional precautionary statements at end of label booklet.

See directions for use inside booklet.

DIRECTIONS FOR USE AND CONDITIONS OF SALE AND WARRANTY

IMPORTANT: Read the entire **Directions for Use** and the **Conditions of Sale and Warranty** before using this product.

CONDITIONS OF SALE AND WARRANTY

The **Directions for Use** of this product reflect the opinion of experts based on field use and tests. The directions are believed to be reliable and should be followed carefully. However, it is impossible to eliminate all risks inherently associated with use of this product. Crop injury, ineffectiveness, or other unintended consequences may result because of such factors as weather conditions, presence of other materials, or the manner of use or application all of which are beyond the control of CIBA—GEIGY or the Seller. All such risks shall be assumed by the Buyer.

CIBA—GEIGY warrants that this product conforms to the chemical description on the label and is reasonably fit for the purposes referred to in the **Directions for Use** subject to the inherent risks referred to above. **CIBA—GEIGY makes no other express or implied warranty of Fitness or Merchantability or any other express or implied warranty. In no case shall CIBA—GEIGY or the Seller be liable for consequential, special, or indirect damages resulting from the use or handling of this product.** CIBA—GEIGY and the Seller offer this product, and the Buyer and user accept it, subject to the foregoing **Conditions of Sale and Warranty,** which may be varied only by agreement in writing signed by a duly authorized representative of CIBA—GEIGY.

DIRECTIONS FOR USE

It is a violation of federal law to use this product in a manner inconsistent with its labeling.

FAILURE TO FOLLOW THE DIRECTIONS FOR USE AND PRECAUTIONS ON THIS LABEL MAY RESULT IN POOR WEED CONTROL, CROP INJURY, OR ILLEGAL RESIDUES.

Do not apply this product in such a manner as to directly or through drift expose workers or other persons, except those knowingly involved in the application. The area being treated must be vacated by unprotected persons.

Do not enter treated areas without protective clothing until sprays have dried.

Because certain states may require more restrictive reentry intervals for various crops treated with this product, consult your State Department of Agriculture for further information.

Written or oral warnings must be given to workers who are expected to be in a treated area or in an area about to be treated with this product. Oral warnings must be given which inform workers of areas or fields that may not be entered without specific protective clothing until sprays have dried, and appropriate actions to take in case of accidental exposure as described under **Precautionary Statements** on this label. When oral warnings are given, warnings shall be given in a language customarily understood by workers. Oral warnings must be given if there is reason to believe that written warnings cannot be understood by workers. Written warnings must include the following information: "CAUTION. Area treated with AAtrex 80W on *(date of application)*. Do not enter without appropriate protective clothing until sprays have dried. In case of accidental exposure, flush eyes or skin with plenty of water. Call a physician if irritation persists. Remove and wash contaminated clothing before reuse."

GENERAL INFORMATION

This herbicide controls many annual broadleaf and grass weeds in corn, sorghum, sugarcane, pineapple, and certain other crops specified on this label. It is also effective in noncrop areas and industrial sites for control of most annual and many perennial broadleaf and grass weeds. This product may be applied before or after weeds emerge.

Following many years of continuous use of this product and chemically related products, biotypes of some of the weeds listed on this label have been reported which cannot be effectively controlled by this and related herbicides. Where this is known or suspected and weeds controlled by this product are expected to be present along with resistant biotypes, we recommend the use of this product in combinations or in sequence with other registered herbicides which are not triazines. If only resistant biotypes are expected to be present, use a registered non-triazine herbicide. Consult with your state Agricultural Extension Service for specific recommendations.

Since this product acts mainly through root absorption, its effectiveness depends on moisture to move it into the root zone. If weeds develop, a shallow cultivation or rotary hoeing will generally result in better weed control.

This product is noncorrosive to equipment and metal surfaces, nonflammable, and has low electrical conductivity.

Avoid using near adjacent desirable plants or in greenhouses, or injury may occur.

To avoid spray drift, do not apply under windy conditions. Avoid spray overlap, as crop injury may result.

Where the use directions give a range of rates, use the lower rate on coarse-textured soil and soil low in organic matter; use the higher rate on fine-textured soil and soil high in organic matter.

Note: CIBA—GEIGY does not recommend applications in combination with other herbicides or oils, except as specifically described on the label or in literature published by CIBA—GEIGY.

APPLICATION PROCEDURES

Ground application: Use conventional ground sprayers equipped with nozzles that provide accurate and uniform application. Be certain that nozzles are uniformly spaced and are the same size. Calibrate sprayer before use and recalibrate at the start of each season and when changing carriers. Unless otherwise specified, use a minimum of 10 gal of spray mixture/A for all preplant incorporated, preplant surface, preemergence, and postemergence applications (with or without oil or surfactant) with ground equipment.

Use a pump with capacity to (1) maintain 35-40 psi at nozzles, (2) provide sufficient agitation in tank to keep mixture in suspension, and (3) to provide a minimum of 20% bypass at all times. Use centrifugal pumps which provide propeller shear action for dispersing and mixing this product. The pump should provide a minimum of 10 gal/minute/100 gal tank size circulated through a correctly positioned sparger tube or jets.

Use screens to protect the pump and to prevent nozzles from clogging. Screens placed on the suction side of the pump should be *16-mesh or coarser.* Do not place a screen in the recirculation line. Use 50-mesh or coarser screens between the pump and boom, and where required, at the nozzles. Check nozzle manufacturer's recommendations.

For band applications, calculate amount to be applied per acre as follows:

$$\frac{\text{band width in inches}}{\text{row width in inches}} \times \text{broadcast rate per acre} = \text{amount needed per acre of field}$$

Aerial application: Use aerial application only where broadcast applications are specified. Apply in a minimum of 1 gal of water for each 1 lb of this product applied per acre. For postemergence treatments on corn and sorghum, apply recommended rate in a minimum of 2 gal of water/A. Avoid applications under conditions where uniform coverage cannot be obtained or where excessive spray drift may occur.

Avoid application to humans or animals. Flagmen and loaders should avoid inhalation of spray mist and prolonged contact with skin, and should wash thoroughly before eating and at the end of each day's operation.

Application in water or liquid fertilizer: Nitrogen solution or complete liquid fertilizer may replace all or part of the water as a carrier for preemergence, preplant incorporated, or preplant surface ground application on corn and sorghum. Check the compatibility of this product with liquid fertilizer and/or nitrogen solution as shown below before use. Do not apply in nitrogen solution or complete liquid fertilizer after corn or sorghum emerges, except as noted under **Lay-by treatment** for corn, or crop injury may occur.

Compatibility Test: Since liquid fertilizers can vary, even within the same analysis, always **check compatibility with herbicide(s) each time before use.** Be especially careful when using **complete** suspension or fluid fertilizers as serious compatibility problems are more likely to occur. Commercial application equipment may improve compatibility in some instances. The following test assumes a spray volume of 25 gal. per acre. For other spray volumes, make appropriate changes in the ingredients. Check compatibility using this procedure:

1. Add 1 pint of fertilizer to each of 2 one-quart jars with tight lids.
2. To **one** of the jars add $1/4$ tsp or 1.2 milliliters of a compatibility agent approved for this use, such as Compex® or Unite® ($1/4$ tsp is equivalent to 2 pt. per 100 gals. spray). Shake or stir gently to mix.
3. To **both** jars add the appropriate amount of herbicide(s). If more

than one herbicide is used, add them separately with dry herbicides first, flowables next, and emulsifiable concentrates last. After each addition, shake or stir gently to thoroughly mix. The appropriate amount of herbicides for this test follows:

Dry herbicides: For each pound to be applied per acre, add 1.5 level teaspoons to each jar.

Liquid herbicides: For each pint to be applied per acre, add 0.5 teaspoon or 2.5 milliliters to each jar.

4. After adding all ingredients, put lids on and tighten, and invert each jar ten times to mix. Let the mixtures stand 15 minutes and then look for separation, large flakes, precipitates, gels, heavy oily film on the jar, or other signs of incompatibility. Determine if the compatibility agent is needed in the spray mixture by comparing the two jars. If either mixture separates, but can be remixed readily, the mixture can be sprayed as long as good agitation is used. If the mixtures are incompatible, test the following methods of improving compatibility: (A) slurry the dry herbicide(s) in water before addition, or (B) add ½ of the compatibility agent to the fertilizer and the other ½ to the emulsifiable concentrate or flowable herbicide before addition to the mixture. If incompatibility is still observed, do not use the mixture.

Application in water plus emulsifiable oil or oil concentrate: Adding emulsifiable oil (petroleum-derived, petroleum-derived oil concentrate, or single or mixed crop-derived oil concentrate) to postemergence water-based sprays in corn and sorghum may improve weed control. However, under certain conditions, the use of either type of oil may seriously injure the crop. To minimize this possibility, observe the following directions:

Use one of the following properly emulsified:

1. A suitable oil concentrate containing at least 1% but not more than 20% suitable emulsifier or surfactant blend.
2. Petroleum-derived oil containing at least 1% suitable emulsifier.

Note: In the event of a compatibility problem when mixing oil with this product and water, a compatibility agent such as Compex or Unite should be used. Any of the above oils contaminated with water or other materials can cause compatibility problems and/or crop injury.

Mixing procedures—all uses: (1) Be sure sprayer is clean and not contaminated with any other materials, or crop injury or sprayer clogging may result. (2) Fill tank ¼ full with clean water, nitrogen solution, or complete liquid fertilizer. (3) Start agitation. (4) Be certain that the agitation system is working properly and creates a rippling or rolling action on the liquid surface. (5) Pour product directly from bag into tank. (6) Continue filling tank until 90% full. Increase agitation if necessary to maintain surface action. (7) Add emulsifiable oil, oil concentrate, or tank mix herbicide(s) after this product is thoroughly suspended. (8) Finish filling tank. (9) Empty tank as completely as possible before refilling to prevent buildup of oil or emulsifiable concentrate residue. Maintain agitation to avoid separation of materials. (10) If an oil or emulsifiable concentrate film starts to build up in tank, drain it and clean with strong detergent solution or solvent. (11) Clean sprayer thoroughly immediately after use by flushing system with water containing a detergent.

Center pivot sprinkler application:

Preemergence or postemergence: Apply this product alone, or in tank mixtures with other herbicides on this label which are registered for center pivot application, with irrigation water. Apply either after planting before crop and weeds emerge, or after crop emergence, but before lay-by (20-30 inches) and before weeds exceed 1.5 inches in height. Apply at rates recommended on this label. Prepare mixture with a minimum of 1 part water to 1 part product. Injecting a larger volume of a more dilute slurry per hour will assure more accurate calibration of metering equipment. Maintain sufficient agitation to keep herbicide in suspension. Meter slurry into irrigation water during entire period. Apply in ½-1 inch of water. Use the lower water volume on coarser textured soils, the higher volume on finer textured soils. More than 1 inch of water may reduce weed control by moving herbicide below the effective zone in the soil. Inject dilute slurry into system through a positive displacement pump.

Refer to supplemental literature published by CIBA—GEIGY for more information on calibrating.

Precautions: *(1) Apply only through irrigation systems containing anti-siphon and check valves to prevent contamination of well during shutdown and overflow of solution tank. (2) Inject ahead of any right angle turn in the main line to insure adequate mixing. (3) Chemical injection pumps and water pumps must have interlocking controls to insure simultaneous shutoff. (4) Application when drift may occur from windy conditions, when system joints and connections are leaking, or when nozzles are not providing uniform distribution may cause crop injury. (5) Where sprinkler distribution patterns do not overlap sufficiently, unacceptable weed control may result. Where sprinkler distribution patterns overlap excessively, crop injury may result.*

AATREX 80W APPLIED ALONE—CORN OR GRAIN SORGHUM

Preplant Surface-Applied, Preplant Incorporated, or Preemergence (or Postemergence at 2.5 lb/A with oil)

Broadleaf and Grass Weeds Controlled

barnyardgrass (watergrass)	witchgrass (*Panicum capillare*)	cocklebur	mustards
giant foxtail	yellow foxtail	groundcherry	nightshade
green foxtail		jimsonweed	pigweed
large (hairy) crabgrass		kochia	purslane
wild oats		lambsquarters	ragweed
		annual morningglory	velvetleaf (buttonweed)

Postemergence with Emulsifiable Oil or Oil Concentrate in Water (at 1.5 lb/A)

Broadleaf Weeds Controlled

annual morningglory	lambsquarters	ragweed
cocklebur	mustards	smartweed
jimsonweed	pigweed	wild buckwheat
		velvetleaf

CORN

Preplant Surface-Applied (Broadleaf and grass control): Use on medium- and fine-textured soil with minimum-tillage or no-tillage systems only in CO, IL, IN, IA, KS, KY, MN, MO, MT, NE, ND, SD, WI, WY. Apply ⅔ the recommended rate of AAtrex shown in Table 1 for the soil texture as a split treatment 30-45 days prior to planting and the remainder at planting. Applications made less than 30 days prior to planting may be as either a split or single treatment. On *coarse-textured* soils, do not apply more than two weeks prior to planting.

If weeds are present at the time of treatment, apply in tank mix combination with a contact herbicide (for example, paraquat or Roundup). Observe directions for use, precautions, and restrictions on the label of the contact herbicide.

Note: To the extent possible, do not move treated soil out of the row or move untreated soil to the surface during planting, or weed control will be diminished.

Preplant Incorporated (Broadleaf and grass control): Broadcast in spring after plowing at rate in Table 1. Apply to the soil and incorporate before, during, or after final seedbed preparation. Avoid deep incorporation. For best results, apply within two weeks prior to planting.

Preemergence (Broadleaf and grass control): Apply during or shortly after planting before weed emergence, at rate in Table 1.

Postemergence (Broadleaf and grass control): Apply before weeds exceed 1.5 inches in height, at rate in Table 1.

Table 1: Broadleaf and Grass Weed Control on Corn*

Soil texture	Broadcast rate/A
COARSE Sand, loamy sand, sandy loam	2.5 lb

MEDIUM Loam, silt loam, silt, clay loam, sandy clay loam, silty clay loam, sandy clay, or silty clay with low organic matter	3 lb
FINE Loam, silt loam, silt, clay loam, sandy clay loam, silty clay loam, sandy clay, or silty clay with medium to high organic matter, and clay (including the dark prairie soils of the Corn Belt)	3.75 lb
Peat, muck, and high organic clay (apply postemergence only)	3.75 lb

* *Broadleaf control (eastern CO, western KS, western NE, NM, OK Pan Handle, west TX, and eastern WY):* On sand, loamy sand, sandy loam, mild to strongly alkaline soil, and all recently leveled soil, apply no more than 1.5 lb/A, either preplant surface, preplant incorporated, or preemergence. On other soils in these areas, apply rate in Table 1 for broadleaf and grass control.

Lay-by treatment (Broadleaf and grass control): Broadcast 1.25-2.5 lb/A before weeds are 1.5 inches tall and before corn is 20-30 inches tall. When using nitrogen solutions, direct the spray to lower 3-4 inches of cornstalks to avoid corn foliage injury.

Postemergence with emulsifiable oil or oil concentrate in water: Add the following volume of one of the type oils indicated for aerial or ground application unless the oil label specifies otherwise:

Type Oil	Ground Application	Aerial Application
Oil concentrate (Crop or Petroleum-derived)	1 qt/A	½-1 qt/A
Petroleum-derived oil	1 gal/A	2 qt/A

Note: Crop-derived or petroleum-derived oil concentrates should contain at least 1% but not more than 20% suitable emulsifier or surfactant blend. Petroleum-derived oils should contain at least 1% suitable emulsifier.

Broadleaf and grass control: For postemergence control of those weeds listed under **Preplant Incorporated and Preemergence,** broadcast 2.5 lb/A plus emulsifiable oil or oil concentrate after weed emergence, but before weeds reach 1.5 inches in height.

Broadleaf control: For postemergence control of those weeds listed under **Postemergence with emulsifiable oil or oil concentrate in water,** broadcast 1.5 lb/A plus emulsifiable oil or oil concentrate before pigweed and lambsquarters reach 6 inches in height and before all other weeds reach 4 inches in height. A cultivation may be necessary if all weeds are not controlled or if weeds regrow.

Precautions for applications with emulsifiable oil or oil concentrate in water: *(1) Inbred lines or any breeding stock may be severely injured by applications with emulsifiable oil or oil concentrate. (2) Adding other insecticides, herbicides, liquid fertilizers, or other materials is not recommended, because they may cause compatibility problems or crop injury. (3) Store and handle emulsifiable oil and oil concentrate carefully. Oil contaminated with even a small amount of water may not emulsify properly when added to the tank. To avoid crop injury, (4) Do not apply when crop is under stress from prolonged cold, wet weather, poor fertility, or other factors, or when crop is wet and succulent from recent rainfall. (5) Do not make more than one application per season except as recommended for control of yellow nutsedge, quackgrass, and Canada thistle.*

Problem weeds

Yellow nutsedge and Canada thistle: This product will control yellow nutsedge *(Cyperus esculentus)* and Canada thistle *(Cirsium arvense)* when applied according to directions. For best results, apply each year until yellow nutsedge or Canada thistle is eliminated or reaches a level of infestation where neither weed is an economic problem. If yellow nutsedge or Canada thistle regrow following last application, cultivate once.

Four alternative methods of use for control of yellow nutsedge and Canada thistle are listed below in order of preference. If other weed species, such as annual grasses, are also expected, use alternative 2 or 4.

1. Broadcast 2.5 lb/A plus 1 gal of emulsifiable oil or 1 qt of oil concentrate. Apply after crop and yellow nutsedge or Canada thistle emerge, but before yellow nutsedge reaches a height of 3 inches or Canada thistle reaches a height of 6 inches. Repeat application before lay-by (20-30 inches), 10-20 days after the first application.
2. Broadcast 2.5 lb/A preplant incorporated for control of yellow nutsedge, or broadcast 2.5 lb/A during or shortly after planting, but prior to crop or weed emergence for control of Canada thistle. Follow with an application of 2.5 lb/A plus 1 gal of emulsifiable oil or 1 qt of oil concentrate after corn and weeds emerge, but before yellow nutsedge or Canada thistle reaches a height of 6 inches.
3. Broadcast 5 lb/A plus 1 gal of emulsifiable oil or 1 qt of oil concentrate after crop emerges, but prior to lay-by (20-30 inches). Apply after yellow nutsedge and Canada thistle emerge, but before yellow nutsedge is 3 inches tall or Canada thistle is 6 inches tall.
4. Broadcast 5 lb/A before, during, or shortly after planting, but before crop or weed emergence (yellow nutsedge control only).

Note: Do not use emulsifiable oil or oil concentrate when corn is wet or under stress, especially when using 5 lb/A of this product, as crop injury may occur. See "Precautions for applications with emulsifiable oil or oil concentrate in water" for additional directions.

Quackgrass control on land going into corn production

Split application: Broadcast 2.5 lb/A in fall or spring. Plow 1-3 weeks later. Apply a second application at 2.5 lb/A in spring before, during, or after planting, but before weeds are 1.5 inches high. This split application will control quackgrass and most annual broadleaf and grass weeds.

Single application: Broadcast 3.75-5 lb/A in fall or spring. Plow 1-3 weeks later.

TANK MIXTURES FOR CORN

This product may be tank mixed with these herbicides for control of certain broadleaf and grass weeds in corn:

Dual® (metolachlor)	Lasso or Lasso EC + Roundup
Dual + paraquat	Lasso or Lasso EC + paraquat
Dual + Roundup® (glyphosate)	Paraquat
Dual + Princep®	Princep
Dual + Princep + paraquat	Princep + paraquat
Dual + Princep + Roundup	Princep + Roundup
Bexton® or Ramrod® (propachlor)	Roundup
Lasso® or Lasso EC (alachlor)	Sutan + ®

Use tank mix directions appearing on the labels of the above herbicides when tank mixing with this product. Observe all precautions and limitations on labeling of products used in a particular tank mix.

Princep 80W, Princep 4L (4LC), or Princep Caliber® 90

In addition to the weeds listed under **AAtrex 80W Applied Alone—Corn and Grain Sorghum—Preplant Surface-Applied, Preplant Incorporated, or Preemergence,** this combination also controls crabgrass, fall panicum, and carpetweed.

Broadcast tank mix before planting, at planting, or after planting, but before crop and weeds emerge, at rates in Table 2. Use the 1:1 ratio for control of most weeds. Use the 1:2 ratio for expected heavy infestations of crabgrass and fall panicum. Cultivate shallowly if weeds develop.

Preplant Surface-Applied: Use on medium- and fine-textured soils with minimum-tillage or no-tillage systems only in CO, IL, IN, IA, KS, KY, MN, MO, MT, NE, ND, SD, WI, and WY. Apply ⅔ the recommended rate of AAtrex and Princep shown in Table 2 for the soil texture as a split treatment 30-45 days prior to planting and the remainder at planting. Applications made less than 30 days prior to planting may be as either a split or single treatment. On *coarse-textured* soils, do not apply more than two weeks prior to planting. Refer to the AAtrex Alone—Preplant Surface-Applied section of the corn label for additional details.

If weeds are present at time of treatment, apply in a tank mix combination with a contact herbicide (for example, paraquat or

Roundup). Observe directions for use, precautions, and restrictions on the label of the contact herbicide.

Note: To the extent possible, do not move treated soil out of the row or move untreated soil to the surface during planting, or weed control will be diminished.

Preplant Incorporated: Apply to the soil and incorporate in the spring before, during, or after final seedbed preparation. Avoid deep incorporation. For best results, apply within two weeks prior to planting.

Preemergence: Apply during or shortly after planting, but before crop and weeds emerge.

Refer to Corn sections of this label and to Princep 80W, Princep Caliber 90, or Princep 4L (4LC) labels for further directions, limitations, and precautions.

Table 2: Tank Mixtures with Princep on Corn

	Broadcast rate/A			
	Ratio*		1:2 Ratio**	
Soil texture	This product	Princep 80W[1]	This product	Princep 80W[1]
Sand, loamy sand, sandy loam	1.25 lb	1.25 lb	0.83 lb	1.67 lb
Loam, silt loam, silt, clay loam, sandy clay loam, silty clay loam, sandy clay, or silty clay with low organic matter	1.5 lb	1.5 lb	1 lb	2 lb
Loam, silt loam, silt, clay loam, sandy clay loam, silty clay loam, sandy clay, or silty clay with medium to high organic matter, and clay (including dark prairie soils of the Corn Belt)	1.8 lb	1.8 lb	1.2 lb	2.4 lb

* For control of most weeds.

** For control of expected heavy infestations of crabgrass and fall panicum.

[1] When using Princep Caliber 90 or Princep 4L (4LC), use equivalent rates. One lb of Princep 80W equals 0.9 lb of Princep Caliber 90 or 1.6 pt of Princep 4L (4LC).

Princep 80W, Princep 4L (4LC), or Princep Caliber 90 plus Roundup: Use as tank mixture for preemergence and postemergence control of certain broadleaf and grass weeds where corn will be planted directly into a cover crop, established sod, or in previous crop residues. Refer to Roundup label for all directions, weeds controlled, precautions, and limitations.

Princep 80W, Princep 4L (4LC), or Princep Caliber 90 plus Paraquat: Use as tank mixture with Princep and paraquat to kill existing vegetation and for residual weed control where corn will be planted directly into a cover crop, established sod, or in previous crop residues. Add this product and Princep to water in spray tank, agitating until thoroughly mixed. Then add paraquat and a nonionic surfactant, such as X-77®. Continue agitation during application. Broadcast 1.25-2.5 lb of this product plus 1.25-2.5 lb Princep 80W [or 2-4 pt Princep 4L (4LC), or 1.1-2.2 lb Princep Caliber 90] plus 1-2 pt paraquat in 20-60 gal of water per sprayed acre. Apply before, during, or after planting, but before corn emerges. Add 0.5 pt of a nonionic surfactant, such as X-77, per 100 gal of spray mixture. Use the 2 pt rate of paraquat if existing vegetation is 4-6 inches tall. This mixture will not control weeds taller than 6 inches.

Refer to further limitations and precautions on labels for this product, Princep, and paraquat.

Precautions for all applications to corn: *(1) To avoid crop injury and illegal residues, do not apply more than 5 lb/A of this product per year. (2) Following harvest, plow (moldboard or disk-plow) and thoroughly till soil in fall or spring to minimize possible injury to spring-seeded rotational crops, regardless of rate used.*

Note for all applications to corn: Do not graze or feed forage from treated areas for 21 days following application, or illegal residues may result.

Rotational crops: (1) Do not rotate to any crop except corn or sorghum until the following year, or injury may occur. (2) If applied after June 10, do not rotate with crops other than corn or sorghum the next year, or crop injury may occur. (3) If the broadcast rate, or equivalent band application rate, exceeds 3.75 lb/A of AAtrex 80W, or 3.75 lb/A of AAtrex 80W plus Princep 80W (or equivalent of another Princep formulation), a crop of untreated corn or sorghum should precede the next rotational crop. (4) In the High Plains and Intermountain areas of the West where rainfall is sparse and erratic or where irrigation is required, use only when corn or sorghum is to follow corn or sorghum or when a crop of untreated corn or sorghum is to precede other rotational crops. (5) In eastern parts of the Dakotas, KS, western MN, and NE, do not rotate to soybeans if the rate applied to corn or sorghum was more than 2.5 lb/A or equivalent band application rate, or soybean injury may occur. (6) Injury may occur to soybeans planted the year following application on soils having a calcareous surface layer. (7) Do not plant sugar beets, tobacco, vegetables (including dry beans), spring-seeded small grains, or small-seeded legumes and grasses the year following application, or injury may occur.

SORGHUM AND SORGHUM-SUDAN HYBRIDS (Grain and Forage Types)

Preplant Surface-Applied (Broadleaf and grass control): Use on medium- and fine-textured soil with minimum-tillage or no-tillage systems only in CO, IL, IN, IA, KS, KY, MN, MO, MT, NE, ND, SD, WI, WY. Apply ⅔ the recommended rate of AAtrex shown in Table 3 for the soil texture as a split treatment 30-45 days prior to planting and the remainder at planting. Applications made less than 30 days prior to planting may be as either a split or single treatment. Under dry conditions, irrigation after application is recommended to move AAtrex into the soil.

If weeds are present at time of treatment, apply in a tank mix combination with a contact herbicide (for example, paraquat or Roundup). Observe directions for use, precautions, and restrictions on the label of the contact herbicide.

Note: To the extent possible, do not move treated soil out of the row or move untreated soil to the surface during planting, or weed control will be diminished.

Preplant Incorporated (Broadleaf and grass control): Broadcast in spring after plowing at rate shown in Table 3. Apply before, during, or after final seedbed preparation. If soil is tilled or worked after application, avoid deep incorporation. For best results, apply within two weeks prior to planting.

Preemergence (Broadleaf and grass control): Apply during or shortly after planting, but prior to weed or crop emergence at rate shown in Table 3.

Table 3: Preplant Surface, Preplant Incorporated, and Preemergence Broadleaf and Grass Weed Control in Sorghum*

Soil texture	Organic matter	Broadcast rate/A
COARSE Sand, loamy sand, Sandy loam	any level	DO NOT USE (except for preemergence use on bedded sorghum in AZ and CA)
MEDIUM AND FINE Loam, silt loam, silt, clay loam, sandy clay loam, silty clay loam, sandy clay, silty clay, clay	less than 1%	
	1-1.5%	2-2.5 lb
	more than 1.5%	2.5-3 lb

* Do not apply preplant surface or preplant incorporated in AL, AR, FL, GA, LA, MS, NC, NM, OK, SC, TN, or TX. Do not apply preemergence in NM, OK, or TX, except in northeast OK and the TX Gulf Coast.

In case of planting failure, sorghum or corn may be replanted. Do not make a second broadcast application, or injury may occur. If originally applied in a band and sorghum or corn is replanted in untreated row middles, this product may be applied in a band to the second planting.

Preemergence broadleaf weed control in furrow irrigated bedded sorghum (AZ and CA only): For preemergence control of many broadleaf weeds, broadcast 1-1.5 lb/A after bed preparation, during or after planting, but before sorghum and weeds emerge and before the first furrow irrigation. Follow with several regular irrigations, making sure to thoroughly wet all soil.

Precautions for preemergence application to furrow irrigated bedded sorghum in AZ and CA: *To avoid possible sorghum injury, do not use on sand or loamy sand soil or on sorghum planted in furrows. Applications to sorghum growing on alkali soils or where cuts, fills, or erosion have exposed calcareous or alkali subsoils may cause crop injury. In case of crop failure, do not replant sorghum for 8 months following application. Corn may be planted immediately.*

Postemergence (Broadleaf and grass control): Apply at rate shown in Table 4 before weeds exceed 1.5 inches in height. Apply up to close-in.

Table 4: Postemergence Broadleaf and Grass Weed Control in Sorghum

Soil texture	Minimum sorghum height at treatment	Broadcast rate/A
Sand or loamy sand	DO NOT USE	
Sandy loam	See directions for postemergence broadleaf weed control.	
Silt loam to sandy clay loam	Completely emerged	2.5-3 lb
Olton and Pullman clay soil	6 inches	2.5-3 lb
Silty clay loam and finer textured soil	Completely emerged	3.75 lb

Postemergence broadleaf weed control with emulsifiable oil or oil concentrate in water: Broadcast 1.5 lb/A for control of many broadleaf weeds. Apply before pigweed and lambsquarters reach 6 inches in height and before all other weeds reach 4 inches in height. In CO, western KS, NM, OK, TX, and desert regions of AZ and CA, apply when sorghum is 6-10 inches in height, but before it reaches boot stage. In all other areas, apply after sorghum reaches the 3-leaf stage. Add 1 gal of emulsifiable oil/A for ground application and 0.5 gal/A for aerial application, or add 1 qt/A of oil concentrate for ground application. A cultivation may be necessary if all weeds are not controlled or if weeds regrow.

For the list of weeds controlled, see **AAtrex 80W Applied Alone - Corn or Grain Sorghum — Postemergence with Emulsifiable Oil or Oil Concentrate in Water.**

Precautions for applications with emulsifiable oil or oil concentrate in water: *See "Precautions for applications with emulsifiable oil or oil concentrate in water" in Corn section.*

Postemergence broadleaf weed control with surfactant (CO, western KS, NM, OK, TX, and desert regions of AZ and CA only) Broadcast 1.5 lb/A plus 0.75-1.5 pt of surfactant after sorghum reaches 6 inches in height, but before weeds exceed 1.5 inches in height. Apply only on sandy loam and finer textured soil.

Precautions for all applications to sorghum: *(1) Heavy rain immediately following application tends to cause excessive concentrations of herbicide in seed furrow, resulting in possible crop injury. Do not apply to furrow-planted sorghum until furrows are leveled (plowed in). Level deep planter marks or seed furrows before application. (2) Application to sorghum growing under stress caused by minor element deficiency or to sorghum growing on highly calcareous soil may result in crop injury. (3) Following harvest, plow (moldboard or disk-plow) and thoroughly till soil in fall or spring to minimize possible injury to spring-seeded rotational crops, regardless of rate used. (4) Injury may occur if both this herbicide, preplant surface, preplant incorporated, or preemergence, and an at-planting systemic insecticide applied in-furrow are used.*

Note : Do not graze or feed forage from treated areas for 21 days following application, or illegal residues may result.

Rotational crops : See "Rotational crops" in Corn section.

TANK MIXTURES FOR GRAIN SORGHUM

Igran® 80W: Use as tank mixture for control of certain broadleaf and grass weeds in grain sorghum. Refer to Igran 80W label for all directions, weeds controlled, precautions, and limitations.

Dual 8E: Use as tank mixture with Dual 8E for control of those weeds listed on the Dual label, as well as on this label. Use this tank mixture only on sorghum seed treated with Concep®. Refer to the Dual 8E label for all directions, precautions, and limitations.

WINTER WEED CONTROL IN TEXAS

For postemergence control of winter weeds only, such as henbit, seedling dock, and annual thistle on fall bedded land in the Gulf Coast and Blacklands of Texas. Apply 1-1.25 lb./A postemergence to the weeds in November or December to land that will be planted to corn, grain sorghum, or forage sorghum the following spring. For best results, add a suitable surfactant, such as X-77, at the rate of 0.5% of the spray volume, an emulsifiable oil at the rate of 1.0% of the spray volume, or an oil concentrate at the rate of 1 qt. per acre.

Normal weed control programs may be used in the following corn, grain sorghum or forage sorghum crop.

Note: Do not plant any crops except corn, grain sorghum, or forage sorghum the spring following this treatment, or illegal residues may result.

AATREX ALONE -CHEMICAL FALLOW

Wheat-Sorghum-Fallow : To control annual broadleaf and grass weeds following wheat harvest and in the following sorghum crop when grown under minimum tillage, broadcast 3.75 lb/A to wheat stubble immediately following wheat harvest. If weeds are present at application, remove them with a sweep plow or other suitable implement after application, or use an approved contact herbicide before or after the application of AAtrex. Plant sorghum into wheat stubble the following spring with minimum soil disturbance. Use a surface planter or a planter leaving a shallow furrow. If weeds are present at planting, remove them with a sweep plow or other suitable implement before planting.

For the list of weeds controlled, see **AAtrex 80W Applied Alone — Corn or Grain Sorghum — Preplant Surface-Applied, Preplant Incorporated, or Preemergence.**

Precautions : *(1) Use only on silt loam or finer textured soil, or crop injury may result. (2) Wheat-sorghum-fallow cropping sequence must be followed. (3) Do not apply following sorghum harvest.*

Note: To avoid illegal residues, do not graze or feed forage from treated area to livestock. To avoid illegal residues and crop injury, do not plant any crop other than those on this label within 18 months following treatment.

Wheat-Corn-Fallow (KS, NE): This product controls cheatgrass (downy brome, chess), kochia, mustards, pigweed, Russian thistle, wild lettuce, wild sunflower, and volunteer wheat during period after wheat harvest. Weed control may extend into following corn crop grown under minimum tillage.

Follow directions for use, notes, and precautions in the "Wheat-Sorghum-Fallow" section above, substituting corn for references to sorghum.

Wheat-Fallow-Wheat (CO, KS, MT, NE, ND, SD, and WY): For preemergence control of cheatgrass (downy brome, chess), common lambsquarters, field pennycress, kochia, mustard, Russian thistle, wild lettuce, and suppression of volunteer wheat during fallow period of a wheat-fallow-wheat rotation, broadcast 0.6-1.25 lb/A on all soils except those listed under "Precautions." For control of pigweed and wild sunflower, use the higher rate. Apply to stubble ground. Treat only once during same fallow period.

TANK MIXTURES FOR CHEMICAL FALLOW

Wheat-Sorghum-Fallow or Wheat-Corn-Fallow (KS, NE) Igran 80W: Use as tank mixture for control of certain broadleaf and grass weeds in wheat-sorghum-fallow or wheat-corn-fallow rotations. Refer to Igran 80W label for directions, weeds controlled, precautions, and limitations.

Paraquat: If weeds are present at application, a tank mix with paraquat may be used. Broadcast 3.75 lb of AAtrex 80W plus 1-2 pt of paraquat in 20-60 gal of water/A by ground equipment. Add 0.5-1 pt of a nonionic surfactant, such as X-77, per 100 gal of spray mixture. Add AAtrex to spray tank first and thoroughly mix with water. Then add paraquat, followed by surfactant. Use the 2 pt rate of paraquat if weeds are 4-6 inches tall. This mixture will not control weeds taller than 6 inches. Apply to stubble ground. Treat only once during same fallow period. Refer to paraquat label for further directions, precautions, and limitations.

Wheat-Fallow-Wheat (CO, KS, MT, NE, ND, SD, and WY) Igran 80W: Use as tank mixture for control of certain broadleaf and grass weeds in wheat-fallow-wheat rotations. Refer to Igran 80W label for directions, weeds controlled, precautions, and limitations.

Paraquat: If weeds are present at application, a tank mix with paraquat may be used. Broadcast 0.6-1.25 lb of AAtrex 80W plus 1-2 pt of paraquat in 20-60 gal of water/A by ground equipment. Add 0.5-1 pt of a nonionic surfactant, such as X-77, per 100 gal of spray mixture. Add AAtrex to spray tank first and thoroughly mix with water. Then add paraquat, followed by surfactant. Use the 2 pt rate of paraquat if weeds are 4-6 inches tall. This mixture will not control weeds taller than 6 inches. Apply to stubble ground. Treat only once during same fallow period. Refer to paraquat label for further directions, precautions, and limitations.

If weeds are present at application and this product is used alone, use either an approved contact herbicide before or after treatment, or tillage after treatment.

Use tillage to control weeds which escape during fallow period. Till before planting. For this product applied alone or in tank mixture with paraquat, plant at least 2 inches deep and 12 months or more after application.

Precautions: *To avoid crop injury, (1) Do not use on sand soil. (2) Do not treat eroded hillsides, caliche and rocky outcroppings, or exposed calcareous subsoil. (3) Do not treat soils of the Rosebud and Canyon Series in western NE and adjoining counties in CO and WY. (4) Do not treat soils with calcareous surface layers. (5) Avoid spray overlap.*

Note : Do not graze treated areas within 6 months after application, or illegal residues may result.

Aerial application : In order to assure that spray will be controllable within the target area when used according to label directions, make applications at a maximum height of 10 ft, using low drift nozzles at a maximum pressure of 40 psi, and restrict application to periods when wind speed does not exceed 10 mph. To assure that spray will not adversely affect adjacent sensitive nontarget plants, apply AAtrex 80W alone by aircraft at a minimum upwind distance of 400 ft from sensitive plants, or apply AAtrex 80W + Igran 80W at a minimum upwind distance of 500 ft from sensitive plants. When applying AAtrex 80W + Igran 80W by aircraft, use a minimum of 5 gal of water per acre.

PROSO MILLET

For preplant incorporated or preemergence control of foxtails, kochia, common lambsquarters, redroot pigweed, Russian thistle, slimleaf lambsquarters and spotted spurge, apply 0.6-1.25 lb/A on soils with less than 2% organic matter, or apply 1.25-2.5 lb/A on soils with 2% or higher organic matter.

Precaution: *To avoid crop injury and illegal residues, make only one application per year.*

Rotational crops: See "Rotational crops" in Corn section.

RANGELAND

To control certain weeds, including cheatgrass (downy brome, chess), common (annual) broomweed, little barley, medusahead, sagewort, and tumble mustard. Use lower rate where cheatgrass is major problem weed; use higher rate for other weeds.

Poisonous plants: Tansymustard, pigweeds, lambsquarters, tumble mustard, cocklebur, and fiddleneck are controlled at 1.25 lb/A. Other mustards (*Brassica* spp.), groundcherry (annual), broomweed, and threeawn or poverty grass (*Aristida* spp.) are partially controlled at 1.25 lb/A and controlled at 2.5 lb/A. Sneezeweed and dogbane are partially controlled at 2.5 lb/A.

Aerial application: In order to assure that spray will be controllable within the target area when used according to label directions, make applications at a maximum height of 10 ft above vegetation, using low drift nozzles at a maximum pressure of 40 psi, and restrict application to periods when wind speed does not exceed 10 mph. To assure that spray will not adversely affect adjacent sensitive nontarget plants, apply AAtrex 80W by aircraft at a minimum upwind distance of 400 ft from sensitive plants.

Establishment of perennial range grasses in West (CA, ID, NV, OR, UT, WA): Broadcast 1-1.25 lb/A in fall after rains begin but before ground freezes. Perennial range grass may be seeded when rain begins the next fall following treatment. Plant in deep furrows such as those made by a deep furrow rangeland drill. Treated areas may be grazed 7 months after seeding. However, to provide sufficient time for establishment of desirable range grasses, treated areas should not be grazed until the fall following seeding.

Renovation of existing stands of perennial range grasses
Central Great Plains (CO, KS, MT, NE, ND, SD, WY): Broadcast 0.6-1.25 lb/A in fall before ground freezes.

Southwest (AR, KS, LA, MS, NM, OK, TX): Apply 1.25 lb/A AAtrex 80W in early spring, before regrowth, to established rangeland or pasture species such as Asiatic (Old World) bluestems, little and big bluestems, buffalograss, switchgrass, and Indiangrass.

West (CA, ID, NV, OR, UT, WA): Do not graze area to be treated for 6 months prior to application. Broadcast 0.6-0.9 lb/A by ground equipment in fall before ground freezes.

Bluegrass and intermediate wheatgrass may be injured by this product at recommended rates.

Notes : To avoid illegal residues, (1) Do not apply more than once per year. (2) Apply only on neutral to acid soils. (3) Do not cut or feed range grass hay. (4) Do not graze treated areas within 3 months following spring application, or within 7 months following fall application. (5) Do not dump or spill product or dispose of container within reach of livestock.

Pure stands of Switchgrass and Big Bluestem (IA, IL, IN, KS, MN, MO, ND, NE, OH, SD, WI): For control of those weeds listed under **AAtrex 80W Applied Alone—Corn or Grain Sorghum—Preplant Surface-Applied, Preplant Incorporated, or Preemergence.**

Use only on loam, silt loam, silty clay loam, clay loam, and silty clay soils with at least 1 percent organic matter.

Establishment: Broadcast 2.5 lb/A after planting but before weeds and crop emerge. Prepare a good, firm seedbed. Plant ½ inch deep with a grassland drill (preferred method) or a conventional drill. If a conventional drill is used on prepared seedbeds, remove all tension from the disk openers. For best results, cultipack or roll after planting. Clip weeds that escape in July or August. Avoid clipping switchgrass or big bluestem seedlings.

Renovation of existing stands: Broadcast 2.5 lb/A in early spring before regrowth of the crop, and before weed emergence.

Notes : To avoid illegal residues, (1) Do not cut for hay. (2) Do not graze treated areas within 4 months following application at seeding or 3 months following application to established switchgrass or big bluestem. (3) Treat only once per year.

Rotational crops: See "Rotational crops" in Corn section. Do not plant small-seeded legumes the year following application, or injury may occur.

ROADSIDES

To control certain annual weeds in established perennial grasses along roadsides in CO, KS, MT, NE, ND, SD, and WY, including cheatgrass (downy brome, chess), common (annual) broomweed, little barley, medusahead, sagewort, and tumble mustard, broadcast 1.25 lb/A in a minimum of 10 gallons of water by ground equipment in the fall before ground freezes, or after thawing in the spring, but before the established grasses green-up and before weeds emerge. Examples of desirable established grasses include big bluestem, bluegrama, bromegrass, buffalograss, crested wheatgrass, Indiangrass, little bluestem, side-oats grama, switchgrass, and Western wheatgrass. Apply only once a year. Temporary discoloration or other form of injury to the desirable perennial grasses may occur following application.

Notes : To avoid illegal residues, (1) Do not cut or feed roadside grass hay. (2) Do not allow livestock to graze treated areas.

PERENNIAL RYEGRASS GROWN FOR SEED (Pacific Northwest only)

For control of annual ryegrass and rattail fescue, broadcast 1.5 lb/A after first fall rain.

Precautions: *To avoid crop injury, (1) Apply only to perennial ryegrass from which seed crop has been harvested. (2) Do not apply to ryegrass more than 2 years in succession.*

Notes : To avoid illegal residues, (1) Do not graze treated land within 10 days after application. (2) Do not plant treated land to other crops for 16 months.

ESTABLISHED ORCHARDGRASS GROWN FOR SEED (Pacific Northwest Only)

AAtrex aids in control of annual ryegrass, rattail fescue, and volunteer orchardgrass. Broadcast 2.5 lb/A of AAtrex 80W in at least 15 gal of water to field which has been burned after first fall rain.

Precautions: *Some yellowing or crop stunting can occur. To avoid crop injury, (1) Apply only to orchardgrass from which seed crop has been harvested. (2) Do not apply to orchardgrass more than two years in succession.*

Notes : To avoid illegal residues, (1) Do not graze treated land within 6 months of application. (2) Do not plant treated land to other crops for 16 months.

SUGARCANE

For control of many broadleaf and grass weeds, including amaranths, crabgrass, fireweed, Flora's paintbrush, foxtails, junglerice and wiregrass, broadcast 2.5-5 lb of AAtrex 80W at time of planting or ratooning, but before sugarcane emerges. Broadcast aerially in a minimum of 5 gal of spray per acre, or broadcast or band by ground equipment in a minimum of 20 gal per acre, unless indicated otherwise. One additional application may be made over the sugarcane as it emerges, and two additional applications may be made interline after emergence as directed sprays. Repeat treatments, where needed, may be applied broadcast, band, or interline as suggested with the final application being prior to close-in. Do not exceed the rate of herbicide suggested for any one crop of sugarcane.

Note : Where high rates of AAtrex are used, apply in a minimum of 1 gal of water for each 1 lb of product applied per acre.

Aerial application : In order to assure that spray will be controllable within the target area when used according to label directions, make applications at a maximum height of 10 ft, using low drift nozzles at a maximum pressure of 40 psi, and restrict application to periods when wind speed does not exceed 10 mph. To assure that spray will not adversely affect adjacent sensitive nontarget plants, apply AAtrex 80W alone by aircraft at a minimum upwind distance of 400 ft from sensitive plants.

Florida

For control of emerged pellitory weed: Apply 0.5-0.75 lb/A in at least 40 gal of water as a directed spray by ground equipment prior to close-in. Add 4 qt of surfactant for each 100 gal of spray. Thoroughly cover weed foliage.

For control of alexandergrass, large crabgrass, pellitory (artillery) weed, and spiny amaranth, use one of the following methods at planting or ratooning:

1. Apply 5 lb/A preemergence. Follow with one or two applications, as needed, postemergence to sugarcane and weeds, at 2.5 lb/A. Treat before weeds exceed 1.5 inches in height.
2. Apply 1-3 times, as needed, at 2.5 lb/A postemergence to sugarcane and weeds. Treat before weeds exceed 1.5 inches in height.

Louisiana

For control of annual weeds during the summer fallow period, apply 2.5 lb/A to weed-free beds immediately after bed formation. Follow normal weed control program after planting.

Precautions : *To avoid crop injury, (1) Do not apply more than 12.5 lb/A to any one crop of sugarcane. (2) If making 2.5 lb/A application during the summer fallow period, do not exceed 10 lb/A during the remainder of the growing season, or illegal residues may result.*

Texas

Use AAtrex 80W for control of barnyardgrass, pigweed, purslane and sunflower, in plant or ratoon sugarcane.

Apply 5 lb/A of AAtrex 80W preemergence. Follow with one or two applications, as needed, at 3.75 lb/A postemergence to sugarcane and weeds.

For best results when weeds are emerged, add a nonionic surfactant at a concentration of 2 qt/100 gal to the spray and apply before weeds exceed 1.5 inches in height.

Precautions : *(1) Injury to sugarcane may occur when under moisture stress, when soil is of low adsorptive capacity, or when land is first cropped to sugarcane. (2) Do not apply after close-in. (3) Do not apply more than 12.5 lb/A to any one crop of sugarcane, or crop injury may result.*

TURF GRASSES FOR SOD (Florida only)

St. Augustinegrass, Centipedegrass, and Zoysia Grass

Broadcast 2.5-5 lb/A according to soil texture to control those weeds listed under **AAtrex 80W Applied Alone—Corn or Grain Sorghum—Preplant Surface-Applied, Preplant Incorporated, or Preemergence.**

Muck or peat	5 lb	Old beds: Within 2 days after lifting sod
		New beds: 3-4 days after sprigging or plugging
Sandy soil	2.5 lb	Old beds: Within 2 days after lifting sod
		New beds: 7-10 days after sprigging or plugging

If weeds regrow, apply an additional 2.5 lb/A on muck or peat, or 1.25 lb/A on sandy soil.

Precautions : *To avoid crop injury, (1) Do not apply within 30 days prior to cutting or lifting. (2) Do not apply in combination with surfactants or other spray additives. (3) Use only on turfgrass reasonably free of infestations of insects, nematodes, and diseases. (4) On newly sprigged turfgrass, temporary slowing of growth may follow application.*

TURFGRASS FOR FAIRWAYS, LAWNS, SOD PRODUCTION* AND SIMILAR AREAS

* In states other than Florida. For use on turfgrass for sod in Florida, see "Turfgrasses for Sod (Florida only)" section above.

Bermudagrass, Centipedegrass, St. Augustinegrass, and Zoysia Grass

Apply AAtrex 80W after October 1 before emergence of winter annual weeds for control of annual bluegrass, burclover, carpet burweed, chickweed, corn speedwell, henbit, hop clover, and spurweed. AAtrex 80W will control annual bluegrass even if it is emerged at time of treatment. For control of summer annual weeds listed in the preemergence portion of the AAtrex 80W Applied Alone—Corn or Grain Sorghum section of this label, also apply AAtrex 80W in late winter before the weeds emerge. Apply in a minimum of 15 gal of water per acre or 1 gal per 1,000 sq ft.

Where annual bluegrass is the major weed, use 1.25 lb of AAtrex 80W per acre (0.5 oz per 1,000 sq ft). Use 2.5 lb/A (1 oz per 1,000 sq ft) for control of the other weeds named above. Do not exceed 1.25 lb/A per treatment on newly sprigged turfgrass or on hybrid bermudagrass such as Tiflawn, Tifway, and Ormond.

For continued summer annual weed control, apply another 1.25 lb/A at least 30 days after the previous application, but not after April 15. However, do not make more than two applications on this product per year.

Precautions: *On newly sprigged turfgrass and hybrid bermudagrass, temporary slowing of growth and yellowing may occur following application. To avoid turf injury, (1) Use only on turfgrass reasonably free of infestations of insects, nematodes, and diseases. (2) Do not use on golf greens. (3) Do not use north of NC (except may be used in Virginia Coastal Plains) or west of the high rainfall areas of eastern OK and eastern TX. (4) Do not use on muck or alkaline soils. (5) Do not apply over the rooting area of trees or ornamentals not listed on this label. (6) Do not overseed with desirable turfgrass within 4 months before or 6 months after treatment.*

Note: Do not graze or feed turf clippings to animals, or illegal residues may result.

MACADAMIA NUTS

For preemergence control of many broadleaf and grass weeds, including crabgrass, foxtail, wiregrass, Flora's paintbrush, spanishneedles, and fireweed, broadcast 2.5-5 lb/A before harvest and before weeds emerge. Repeat as necessary. Do not spray when nuts are on ground during harvest period. Do not apply by air.

PINEAPPLE

For control of purslane, spanishneedles, annual grasses, annual bindweed, ageratum, amaranth, rattlepod, Flora's paintbrush, fireweed, spurge, indigo, and papalo, apply up to 8 lb/A as a blanket spray immediately after planting or following harvest. Make additional blanket or interspace applications of up to 2 lb/A at 1-2 month intervals as needed, prior to differentiation. Apply in 20-40 gal of water/A to assure thorough coverage.

Precautions: *(1) Do not apply more than 37.5 lb/A per cycle as crop injury or illegal residues may result. (2) Repeated monthly applications to plant foliage may slow plant growth and delay fruiting.*

Note: Do not apply within 45 days of fruit harvest, or forage harvest if forage is to be fed to livestock, or illegal residues may result.

GUAVA

Use only on established plantings which are at least 18 months old. Apply as a directed spray at 2.5-5 lb/A of AAtrex 80W in 20-50 gal of spray mix preemergence or early postemergence to weeds. When applying postemergence, the use of a surfactant and greater spray volume (80-100 gal of spray mix per acre) may enhance weed control. This product controls many annual broadleaf and grass weeds, including fireweed, purslane, scarlet pimpernel, spanishneedles, and sowthistle.

Notes: To avoid illegal residues, (1) Do not allow spray to contact foliage or fruit. (2) Do not apply more frequently than at 4 month intervals. (3) Do not apply more than 10 lb of AAtrex 80W per year.

CONIFERS

For control of annual broadleaf and grass weeds prior to transplanting, after transplanting or in established conifers [including Douglas fir, grand fir, noble fir, white fir, Austrian pine, bishop pine, Jeffrey pine, knobcone pine, loblolly pine, lodgepole pine (shore pine), monterey pine, ponderosa pine, Scotch pine, slash pine, blue spruce, and Sitka spruce]: Broadcast 2.5-5 lb in a minimum of 5 gal of water per acre by air or 10 gal by ground before weeds are 1.5 inches tall. Apply to established trees between fall and early spring while trees are dormant. For new transplants, apply during or soon after transplanting. For applications prior to transplanting, allow sufficient precipitation to activate AAtrex before transplanting. In areas where spring and summer rainfall is inadequate to activate AAtrex, apply during fall prior to spring transplanting.

For the list of weeds controlled, see **AAtrex 80W Applied Alone—Corn or Grain Sorghum—Preplant Surface-Applied, Preplant Incorporated, or Preemergence.**

Quackgrass control: Broadcast 5 lb in a minimum of 5 gal of water per acre by air or 10 gal by ground between fall and early spring while trees are dormant and before quackgrass is more than 1.5 inches tall.

Precautions: *Temporary injury to trees may occur following the use of AAtrex on coarse-textured soil. To avoid crop injury, (1) Do not apply to seedbeds. (2) Apply only once per year.*

Note: Do not graze treated areas, or illegal residues may result.

Aerial application: In order to assure that spray will be controllable within the target area when used according to label directions, make applications at a maximum height of 10 ft above vegetation, using low drift nozzles at a maximum pressure of 40 psi, and restrict application to periods when wind speed does not exceed 10 mph. To assure that spray will not adversely affect adjacent sensitive nontarget plants, apply AAtrex 80W by aircraft at a minimum upwind distance of 400 ft from sensitive plants.

NONSELECTIVE WEED CONTROL ON NONCROP LAND

This product provides long-term weed control on industrial sites, highway medians and shoulders, railroad rights-of-way, lumberyards, petroleum tank farms, and in noncrop areas on farms, such as around buildings, equipment and fuel storage areas, along fences, and lanes. Do not use near adjacent desirable plants or in greenhouses, or injury may occur.

Apply before or soon after weeds begin growth. Make postemergence applications when weeds are young and actively growing. Use the higher rates on fine clay and muck soils. Use sufficient water to assure thorough coverage. Use at least 1 gal of water for each pound of product; use more if practical.

Note: Do not allow livestock to graze treated areas, or illegal residues may result.

Annual broadleaf and grass weeds (including barnyardgrass, cheatgrass, crabgrass, lambsquarters, foxtail, ragweed, puncturevine, and turkey mullein): Broadcast 6-12.5 lb/A.

Hard-to-kill annual and many perennial broadleaf and grass weeds (including bluegrass, burdock, Canada thistle, dogfennel, orchardgrass, plantain, quackgrass, purpletop, redtop, and smooth brome): Broadcast 12.5-25 lb/A.

Hard-to-kill biennial and perennial weeds (including bullthistle and sowthistle): Broadcast 25-50 lb/A.

For longer residual control in regions of high rainfall and a long growing season, broadcast 25-50 lb/A.

For small areas, 4.5 oz per 1,000 sq ft is equivalent to 12.5 lb/A.

STORAGE AND DISPOSAL

PESTICIDE STORAGE AND DISPOSAL

Store in a dry place. Do not contaminate water, food, or feed by storage, disposal, or cleaning of equipment. Open dumping is prohibited. Improper disposal of unused pesticide, spray mixture, or rinsate is a violation of federal law. Pesticide, spray mixture, or rinsate that cannot be used according to label instructions must be disposed of according to federal, state, or local procedures. For guidance in proper disposal methods, contact your State Pesticide or Environmental Control Agency, or the Hazardous Waste representative at the nearest EPA Regional Office.

CONTAINER DISPOSAL

Completely empty bag into application equipment. Dispose of empty bag in a sanitary landfill or by incineration, or by open burning, if allowed by state and local authorities. If burned, keep out of smoke.

For minor spills, leaks, etc., follow all precautions indicated on this label and clean up immediately. Take special care to avoid contamination of equipment and facilities during cleanup procedures and disposal of wastes. In the event of a major spill, fire or other emergency, call (919) 292-7100 day or night.

PRECAUTIONARY STATEMENTS

HAZARDS TO HUMANS AND DOMESTIC ANIMALS

CAUTION

Harmful if swallowed, inhaled, or absorbed through skin. Do not breathe dust or spray mist. Avoid contact with eyes, skin or clothing.

First Aid: If swallowed, contact your local poison control center, hospital, or physician. If the patient is unconscious, maintain breathing and heartbeat (cardiopulmonary resuscitation). If the patient is conscious and alert, induce vomiting (syrup of ipecac or stimulate the back of the throat with a finger). NEVER GIVE ANYTHING BY MOUTH TO AN UNCONSCIOUS PERSON!

In case of skin contact, remove contaminated clothing and wash with soap and water. In case of contact with eyes, flush with *plenty* of water. Get medical attention if irritation persists.

Note to Physician: There is no specific antidote for atrazine. If this product is ingested, induce emesis or lavage stomach. The use of an aqueous slurry of activated charcoal may be considered.

ENVIRONMENTAL HAZARDS

Do not apply directly to any body of water. Do not apply where runoff is likely to occur. Do not contaminate water by cleaning of equipment or disposal of wastes. Do not apply when weather conditions favor drift from treated areas.

AAtrex® trademark of CIBA—GEIGY for atrazine

Bexton® trademark of Dow Chemical Co. for propachlor

Caliber® trademark of CIBA—GEIGY

Compex® trademark of KALO Agricultural Chemicals, Inc.

Concep® trademark of CIBA—GEIGY

Dual® trademark of CIBA—GEIGY for metolachlor U.S. Patent No. 3,937,730

Igran® trademark of CIBA—GEIGY for terbutryn

Lasso® trademark of Monsanto Company for alachlor

Princep® trademark of CIBA—GEIGY for simazine

Ramrod® trademark of Monsanto Company for propachlor

Roundup® trademark of Monsanto Company for glyphosate herbicide

Sutan + ® trademark of Stauffer Chemical Co. for butylate

Unite® trademark of Hopkins Agricultural Chemical Co.

X-77® trademark of KALO Agricultural Chemicals, Inc.

CGA 7L91D 064

AATREX®NINE-O®

HERBICIDE

For season-long weed control in corn and sorghum

For weed control in certain other crops; in noncrop areas; and industrial sites

AAtrex Nine-O is a water dispersible granule

ACTIVE INGREDIENTS:	
Atrazine: 2-chloro-4-ethylamino-6-isopropyl-amino-*s*-triazine	85.5%
Related compounds	4.5%
INERT INGREDIENTS:	10.0%
TOTAL:	100.0%

EPA Reg. No. 100-585 EPA Est. 100-LA-1

AAtrex® trademark of CIBA-GEIGY for atrazine

Keep Out of Reach of Children.

Caution

See additional precautionary statements at end of label booklet.

See directions for use inside booklet.

DIRECTIONS FOR USE AND CONDITIONS OF SALE AND WARRANTY

IMPORTANT: Read the entire **Directions for Use** and the **Conditions of Sale and Warranty** before using this product.

CONDITIONS OF SALE AND WARRANTY

The **Directions for Use** of this product reflect the opinion of experts based on field use and tests. The directions are believed to be reliable and should be followed carefully. However, it is impossible to eliminate all risks inherently associated with use of this product. Crop injury, ineffectiveness, or other unintended consequences may result because of such factors as weather conditions, presence of other materials, or the manner of use or application all of which are beyond the control of CIBA-GEIGY or the Seller. All such risks shall be assumed by the Buyer.

CIBA-GEIGY warrants that this product conforms to the chemical description on the label and is reasonably fit for the purposes referred to in the **Directions for Use** subject to the inherent risks referred to above. **CIBA-GEIGY makes no other express or implied warranty of Fitness or Merchantability or any other express or implied warranty. In no case shall CIBA-GEIGY or the Seller be liable for consequential, special, or indirect damages resulting from the use or handling of this product.** CIBA-GEIGY and the Seller offer this product, and the Buyer and user accept it, subject to the foregoing **Conditions of Sale and Warranty,** which may be varied only by agreement in writing signed by a duly authorized representative of CIBA-GEIGY.

DIRECTIONS FOR USE

It is a violation of federal law to use this product in a manner inconsistent with its labeling.

FAILURE TO FOLLOW THE DIRECTIONS FOR USE AND PRECAUTIONS ON THIS LABEL MAY RESULT IN POOR WEED CONTROL, CROP INJURY, OR ILLEGAL RESIDUES.

Do not apply this product in such a manner as to directly or through drift expose workers or other persons, except those knowingly involved in the application. The area being treated must be vacated by unprotected persons.

Do not enter treated areas without protective clothing until sprays have dried.

Because certain states may require more restrictive reentry intervals for various crops treated with this product, consult your State Department of Agriculture for further information.

Written or oral warnings must be given to workers who are expected to be in a treated area or in an area about to be treated with this product. Oral warnings must be given which inform workers of areas or fields that may not be entered without specific protective clothing until sprays have dried, and appropriate actions to take in case of accidental exposure as described under **Precautionary Statements** on this label. When oral warnings are given, warnings shall be given in a language customarily understood by workers. Oral warnings must be given if there is reason to believe that written warnings cannot be understood by workers. Written warnings must include the following information: "CAUTION. Area treated with AAtrex Nine-O on (date of application). Do not enter without appropriate protective clothing until sprays have dried. In case of accidental exposure, flush eyes or skin with plenty of water. Call a physician if irritation persists. Remove and wash contaminated clothing before reuse."

GENERAL INFORMATION

This herbicide controls many annual broadleaf and grass weeds in corn, sorghum, sugarcane, pineapple, and certain other crops specified on this label. It is also effective in noncrop areas and industrial sites for control of most annual and many perennial broadleaf and grass weeds. This product may be applied before or after weeds emerge.

Following many years of continuous use of this product and chemically related products, biotypes of some of the weeds listed on this label have been reported which cannot be effectively controlled by this and related herbicides. Where this is known or suspected and weeds controlled by this product are expected to be present along with resistant biotypes, we recommend the use of this product in combinations or in sequence with other registered herbicides which are not triazines. If only resistant biotypes are expected to be present, use a registered non-triazine herbicide. Consult with your state Agricultural Extension Service for specific recommendations.

Since this product acts mainly through root absorption, its effectiveness depends on moisture to move it into the root zone. If weeds develop, a shallow cultivation or rotary hoeing will generally result in better weed control.

This product is noncorrosive to equipment and metal surfaces, nonflammable, and has low electrical conductivity.

Avoid using near adjacent desirable plants or in greenhouses, or injury may occur.

To avoid spray drift, do not apply under windy conditions. Avoid spray overlap, as crop injury may result.

Where the use directions give a range of rates, use the lower rate on coarse-textured soil and soil low in organic matter; use the higher rate on fine-textured soil and soil high in organic matter.

Note: CIBA-GEIGY does not recommend applications in combination with other herbicides or oils, except as specifically described on the label or in literature published by CIBA-GEIGY.

APPLICATION PROCEDURES

Ground application: Use conventional ground sprayers equipped with nozzles that provide accurate and uniform application. Be certain that nozzles are uniformly spaced and are the same size. Calibrate sprayer before use and recalibrate at the start of each season and when changing carriers. Unless otherwise specified, use a minimum of 10 gal of spray mixture/A for all preplant incorporated, preplant surface, preemergence, and postemergence applications (with or without oil or surfactant) with ground equipment.

Use a pump with capacity to (1) maintain 35-40 psi at nozzles, (2) provide sufficient agitation in tank to keep mixture in suspension, and (3) to provide a minimum of 20% bypass at all times. Use centrifugal pumps which provide propeller shear action for dispersing and mixing this product. The pump should provide a minimum of 10 gal/minute/100 gal tank size circulated through a correctly positioned sparger tube or jets.

Use screens to protect the pump and to prevent nozzles from clogging. Screens placed on the suction side of the pump should be 16-mesh or coarser. Do not place a screen in the recirculation line. Use 50-mesh or coarser screens between the pump and boom, and where required, at the nozzles. Check nozzle manufacturer's recommendations.

For band applications, calculate amount to be applied per acre as follows:

$$\frac{\text{band width in inches}}{\text{row width in inches}} \times \text{broadcast rate per acre} = \text{amount needed per acre of field}$$

Aerial application: Use aerial application only where broadcast applications are specified. Apply in a minimum of 1 gal of water for each 1 lb of AAtrex Nine-O applied per acre. For postemergence treatments on corn and sorghum, apply recommended rate in a minimum of 2 gal of water/A. Avoid applications under conditions where uniform coverage cannot be obtained or where excessive spray drift may occur.

Avoid application to humans or animals. Flagmen and loaders should avoid inhalation of spray mist and prolonged contact with skin, and should wash thoroughly before eating and at the end of each day's operation.

Application in water or liquid fertilizer: Nitrogen solution or complete liquid fertilizer may replace all or part of the water as a carrier for preemergence, preplant incorporated, or preplant surface ground application on corn and sorghum. Check the compatibility of this product with liquid fertilizer and/or nitrogen solution as shown below before use. Do not apply in nitrogen solution or complete liquid fertilizer after corn or sorghum emerges, except as noted under **Lay-by treatment** for corn, or crop injury may occur.

Compatibility Test: Since liquid fertilizers can vary, even within the same analysis, always **check compatibility with herbicide(s) each time before use.** Be especially careful when using **complete** suspension or fluid fertilizers as serious compatibility problems are more likely to occur. Commercial application equipment may improve compatibility in some instances. The following test assumes a spray volume of 25 gal per acre. For other spray volumes, make appropriate changes in the ingredients. Check compatibility using this procedure:

1. Add 1 pint of fertilizer to each of 2 one-quart jars with tight lids.
2. To **one** of the jars add 1/4 tsp or 1.2 milliliters of a compatibility agent approved for this use, such as Compex® or Unite® (1/4 tsp is equivalent to 2 pt per 100 gal spray). Shake or stir gently to mix.
3. To **both** jars add the appropriate amount of herbicide(s). If more than one herbicide is used, add them separately with dry herbicides first, flowables next, and emulsifiable concentrates last. After each addition, shake or stir gently to thoroughly mix. The appropriate amount of herbicides for this test follows:

Dry herbicides:	For each pound to be applied per acre, add 1.5 level teaspoons to each jar.
Liquid herbicides:	For each pint to be applied per acre, add 0.5 teaspoon or 2.5 milliliters to each jar.

4. After adding all ingredients, put lids on and tighten, and invert each jar ten times to mix. Let the mixtures stand 15 minutes and then look for separation, large flakes, precipitates, gels, heavy oily film on the jar, or other signs of incompatibility. Determine if the compatibility agent is needed in the spray mixture by comparing the two jars. If either mixture separates, but can be remixed readily, the mixture can be sprayed as long as good agitation is used. If the mixtures are incompatible, test the following methods of improving compatibility: (A) slurry the dry herbicides(s) in water before addition, or (B) add 1/2 of the compatibility agent to the fertilizer and the other 1/2 to the emulsifiable concentrate or flowable herbicide before addition to the mixture. If incompatibility is still observed, do not use the mixture.

Application in water plus emulsifiable oil or oil concentrate: Adding emulsifiable oil (petroleum-derived, petroleum-derived oil concentrate, or single or mixed crop-derived oil concentrate) to postemergence water-based sprays in corn and sorghum may improve weed control. However, under certain conditions, the use of either type of oil may seriously injure the crop. To minimize this possibility, observe the following directions:

Use one of the following properly emulsified:

1. A suitable oil concentrate containing at least 1% but not more than 20% suitable emulsifier or surfactant blend.
2. Petroleum-derived oil containing at least 1% suitable emulsifier.

Note: In the event of a compatibility problem when mixing oil with AAtrex and water, a compatibility agent such as Compex or Unite should be used. Any of the above oils contaminated with water or other materials can cause compatibility problems and/or crop injury.

Mixing procedures—all uses: (1) Be sure sprayer is clean and not contaminated with any other materials, or crop injury or sprayer clogging may result. (2) Fill tank 1/4 full with clean water, nitrogen solution, or complete liquid fertilizer. (3) Start agitation. (4) Be certain that the agitation system is working properly and creates a rippling or rolling action on the liquid surface. (5) Pour product directly from bag into tank. (6) Continue filling tank until 90% full. Increase agitation if necessary to maintain surface action. (7) Add emulsifiable oil, oil concentrate, or tank mix herbicide(s) after this product is thoroughly suspended. (8) Finish filling tank. (9) Empty tank as completely as possible before refilling to prevent buildup of oil or emulsifiable concentrate residue. Maintain agitation to avoid separation of materials. (10) If an oil or emulsifiable concentrate film starts to build up in tank, drain it and clean with strong detergent solution or solvent. (11) Clean sprayer thoroughly immediately after use by flushing system with water containing a detergent.

Center pivot sprinkler application

Preemergence or postemergence: Apply AAtrex Nine-O alone, or in tank mixtures with other herbicides on this label which are registered for center pivot application, with irrigation water. Apply either after planting before crop and weeds emerge, or after crop emergence, but before lay-by (20-30 inches) and before weeds exceed 1.5 inches in height. Apply at rates recommended on this label. Prepare mixture with a minimum of 1 part water to 1 part product. Injecting a larger volume of a more dilute slurry per hour will assure more accurate calibration of metering equipment. Maintain sufficient agitation to keep herbicide in suspension. Meter slurry into irrigation water during entire period. Apply in 1/2 -1 inch of water. Use the lower water volume on coarser textured soils, the higher volume on finer textured soils. More than 1 inch of water may reduce weed control by moving herbicide below the effective zone in the soil. Inject dilute slurry into system through a positive displacement pump.

Refer to supplemental literature published by CIBA-GEIGY for more information on calibrating.

Precautions: *(1) Apply only through irrigation systems containing anti-siphon and check valves to prevent contamination of well during shutdown and overflow of solution tank. (2) Inject ahead of any right angle turn in the main line to insure adequate mixing. (3) Chemical injection pumps and water pumps must have interlocking controls to insure simultaneous shutoff. (4) Application when drift may occur from windy conditions, when system joints and connections are leaking, or when nozzles are not providing uniform distribution may cause crop injury. (5) Where sprinkler distribution patterns do not overlap sufficiently, unacceptable weed control may result. Where sprinkler distribution patterns overlap excessively, crop injury may result.*

AATREX NINE-O APPLIED ALONE—CORN OR GRAIN SORGHUM

Preplant Surface-Applied, Preplant Incorporated, or Preemergence (or Postemergence at 2.2 lb/A with Oil)

Broadleaf and Grass Weeds Controlled

barnyardgrass (watergrass)	cocklebur	mustards
giant foxtail	groundcherry	nightshade
green foxtail	jimsonweed	pigweed
large (hairy) crabgrass	kochia	purslane
wild oats	lambsquarters	ragweed
witchgrass (*Panicum capillare*)	annual morningglory	sicklepod
yellow foxtail	mustards	velvetleaf (buttonweed)

Postemergence with Emulsifiable Oil or Oil Concentrate in Water (at 1.3 lb/A)

Broadleaf Weeds Controlled

annual morningglory	lambsquarters	ragweed
cocklebur	mustards	smartweed
jimsonweed	pigweed	wild buckwheat
		velvetleaf

CORN

Preplant Surface-Applied (Broadleaf and grass control): Use on medium- and fine-textured soils with minimum-tillage or no-tillage systems only in CO, IL, IN, IA, KS, KY, MN, MO, MT, NE, ND, SD, WI, WY. Apply 2/3 the recommended rate of AAtrex shown in Table 1 for the soil texture as a split treatment 30-45 days prior to planting and the remainder at planting. Applications made less than 30 days prior to planting may be as either a split or single treatment. On *coarse-textured* soils, do not apply more than two weeks prior to planting.

If weeds are present at the time of treatment, apply in tank mix combination with a contact herbicide (for example, paraquat or Roundup). Observe directions for use, precautions, and restrictions on the label of the contact herbicide.

Note: To the extent possible, do not move treated soil out of the row or move untreated soil to the surface during planting, or weed control will be diminished.

Preplant Incorporated (Broadleaf and grass control): Broadcast in spring after plowing at rate in Table 1. Apply to the soil and incorporate before, during, or after final seedbed preparation. Avoid deep incorporation. For best results, apply within two weeks prior to planting.

Preemergence (Broadleaf and grass control): Apply during or shortly after planting before weed emergence, at rate in Table 1.

Postemergence (Broadleaf and grass control): Apply before weeds exceed 1.5 inches in height, at rate in Table 1.

Table 1: Broadleaf and Grass Weed Control on Corn*

Soil texture	Broadcast rate/A
COARSE Sand, loamy sand, sandy loam	2.2 lb
MEDIUM Loam, silt loan, silt, clay loam, sandy clay loam, silty clay loam, sandy clay, or silty clay with low organic matter	2.6 lb
FINE Loam, silt loam, silt, clay loam, sandy clay loam, silty clay loam, sandy clay, or silty clay with medium to high organic matter, and clay (including the dark prairie soils of the Corn Belt)	3.3 lb
Peat, muck, and high organic clay (apply postemergence only)	3.3 lb

* *Broadleaf control (eastern CO, western KS, western NE, NM, OK Pan Handle, west TX, and eastern WY):* On sand, loamy sand, sandy loam, mild to strongly alkaline soil, and all recently leveled soil, apply no more than 1.3 lb/A, either preplant surface, preplant incorporated, or preemergence. On other soils in these areas, apply rate in Table 1 for broadleaf and grass control.

Lay-by treatment (Broadleaf and grass control): Broadcast 1.1-2.2 lb/A before weeds are 1.5 inches tall and before corn is 20-30 inches tall. When using nitrogen solutions, direct the spray to lower 3-4 inches of cornstalks to avoid corn foliage injury.

Postemergence with emulsifiable oil or oil concentrate in water: Add the following volume of one of the type oils indicated for aerial or ground application unless the oil label specifies otherwise:

Type Oil	Ground Application	Aerial Application
Oil concentrate (Crop or Petroleum-derived)	1 qt/A	½-1 qt/A
Petroleum-derived oil	1 gal/A	2 qt/A

Note: Crop-derived or petroleum-derived oil concentrates should contain at least 1% but not more than 20% suitable emulsifier or surfactant blend. Petroleum-derived oils should contain at least 1% suitable emulsifier.

Broadleaf and grass control: For postemergence control of those weeds listed under **Preplant incorporated and Preemergence,** broadcast 2.2 lb/A plus emulsifiable oil or oil concentrate after weed emergence, but before weeds reach 1.5 inches in height.

Broadleaf control: For postemergence control of those weeds listed under **Postemergence with emulsifiable oil or oil concentrate in water** broadcast 1.3 lb/A plus emulsifiable oil or oil concentrate before pigweed and lambsquarters reach 6 inches in height and before all other weeds reach 4 inches in height. A cultivation may be necessary if all weeds are not controlled or if weeds regrow.

Precautions for applications with emulsifiable oil or oil concentrate in water: *(1) Inbred lines or any breeding stock may be severely injured by applications with emulsifiable oil or oil concentrate. (2) Adding other insecticides, herbicides, liquid fertilizers, or other materials is not recommended, because they may cause compatibility problems or crop injury. (3) Store and handle emulsifiable oil and oil concentrate carefully. Oil contaminated with even a small amount of water may not emulsify properly when added to the tank. To avoid crop injury, (4) Do not apply when crop is under stress from prolonged cold, wet weather, poor fertility, or other factors, or when crop is wet and succulent from recent rainfall. (5) Do not make more than one application per season except as recommended for control of yellow nutsedge, quackgrass, and Canada thistle.*

Problem weeds

Yellow nutsedge and Canada thistle: This product will control yellow nutsedge *(Cyperus esculentus)* and Canada thistle *(Cirsium arvense)* when applied according to directions. For best results, apply each year until yellow nutsedge or Canada thistle is eliminated or reaches a level of infestation where neither weed is an economic problem. If yellow nutsedge or Canada thistle regrow following last application, cultivate once.

Four alternative methods of use for control of yellow nutsedge and Canada thistle are listed below in order of preference. If other weed species, such as annual grasses, are also expected, use alternative 2 or 4.

1. Broadcast 2.2 lb/A plus 1 gal of emulsifiable oil or 1 qt of oil concentrate. Apply after crop and yellow nutsedge or Canada thistle emerge, but before yellow nutsedge reaches a height of 3 inches or Canada thistle reaches a height of 6 inches. Repeat application before lay-by (20-30 inches), 10-20 days after the first application.
2. Broadcast 2.2 lb/A preplant incorporated for control of yellow nutsedge, or broadcast 2.2 lb/A during or shortly after planting, but prior to crop or weed emergence for control of Canada thistle. Follow with an application of 2.2 lb/A plus 1 gal of emulsifiable oil or 1 qt of oil concentrate after corn and weeds emerge, but before yellow nutsedge or Canada thistle reaches a height of 6 inches.
3. Broadcast 4.4 lb/A plus 1 gal of emulsifiable oil or 1 qt of oil concentrate after crop emerges, but prior to lay-by (20-30 inches). Apply after yellow nutsedge and Canada thistle emerge, but before yellow nutsedge is 3 inches tall or Canada thistle is 6 inches tall.
4. Broadcast 4.4 lb/A before, during, or shortly after planting, but before crop or weed emergence (yellow nutsedge control only).

Note: Do not use emulsifiable oil or oil concentrate when corn is wet or under stress, especially when using 4.4 lb/A of this product, as crop injury may occur. See "Precautions for applications with emulsifiable oil or oil concentrate in water" for additional directions.

Quackgrass control on land going into corn production:

Split application: Broadcast 2.2 lb/A in fall or spring. Plow 1-3 weeks later. Apply a second application at 2.2 lb/A in spring before, during, or after planting, but before weeds are 1.5 inches high. This split application will control quackgrass and most annual broadleaf and grass weeds.

Single application: Broadcast 3.3-4.4 lb/A in fall or spring. Plow 1-3 weeks later.

TANK MIXTURES FOR CORN

This product may be tank mixed with these herbicides for control of certain broadleaf and grass weeds in corn:

Dual® (metolachlor)	Lasso® or Lasso EC (alachlor)
Dual + paraquat	Lasso or Lasso EC + Roundup
Dual + Roundup® (glyphosate)	Lasso or Lasso EC + paraquat
Dual + Princep®	Paraquat
Dual + Princep + paraquat	Princep
Dual + Princep + Roundup	Princep + paraquat
Bexton® or Ramrod® (propachlor)	Princep + Roundup
	Roundup
	Sutan+®

Use tank mix directions appearing on the labels of the above herbicides when tank mixing with this product. Observe all precautions and limitations on labeling of products used in a particular tank mix.

Note: When the labels of the above herbicides refer to atrazine 80W, use equivalent rate of AAtrex Nine-O. One lb of 80W equals 0.9 lb Nine-O.

PRINCEP 80W, PRINCEP 4L (4LC), OR PRINCEP CALIBER® 90

In addition to the weeds listed under **AAtrex Nine-O Applied Alone — Corn and Grain Sorghum — Preplant Surface-Applied, Preplant Incorporated, or Preemergence** this combination also controls crabgrass, fall panicum, and carpetweed.

Broadcast tank mix before planting, at planting, or after planting, but before crop and weeds emerge, at rates in Table 2. Use the 1:1 ratio for control of most weeds. Use the 1:2 ratio for expected heavy infestations of crabgrass and fall panicum. Cultivate shallowly if weeds develop.

Preplant Surface-Applied: Use on medium- and fine-textured soils with minimum-tillage or no-tillage systems only in CO, IL, IN, IA, KS, KY, MN, MO, MT, NE, ND, SD, WI, and WY. Apply ⅔ the recommended rate of AAtrex and Princep shown in Table 2 for the soil textures as a split treatment 30-45 days prior to planting and the remainder at planting. Applications made less than 30 days prior to planting may be as either a split or single treatment. On *coarse-textured* soils, do not apply more than two weeks prior to planting. Refer to the AAtrex Alone — Preplant Surface-Applied section of the corn label for additional details.

If weeds are present at time of treatment, apply in a tank mix combination with a contact herbicide (for example, paraquat or Roundup). Observe directions for use, precautions, and restrictions on the label of the contact herbicide.

Note: To the extent possible, do not move treated soil out of the row or move untreated soil to the surface during planting, or weed control will be diminished.

Preplant Incorporated: Apply to the soil and incorporate in the spring before, during, or after final seedbed preparation. Avoid deep incorporation. For best results, apply within two weeks prior to planting.

Preemergence: Apply during or shortly after planting, but before crop and weeds emerge.

Refer to Corn sections of this label and to Princep 80W, Princep Caliber 90, or Princep 4L (4LC) labels for further directions, limitations, and precautions.

Table 2: Tank Mixtures with Princep on Corn

Soil texture	Broadcast rate/A 1:1 Ratio* This product	1:1 Ratio* Princep 80W[1]	1:2 Ratio** This product	1:2 Ratio** Princep 80W[1]
Sand, loamy sand, sandy loam	1.1 lb	1.25 lb	0.73 lb	1.67 lb
Loam, silt loam, silt, clay loam, sandy clay loam, silty clay loam, sandy clay, or silty clay with low organic matter	1.3 lb	1.5 lb	0.88 lb	2 lb
Loam, silt loam, silt, clay loam, sandy clay loam, silty clay loam, sandy clay, or silty clay with medium to high organic matter, and clay (including dark prairie soils of the Corn Belt)	1.6 lb	1.8 lb	1.07 lb	2.4 lb

* For control of most weeds.
** For control of expected heavy infestations of crabgrass and fall panicum.
[1] When using Princep Caliber 90 or Princep 4L (4LC), use equivalent rates. One lb of Princep 80W equals 0.9 lb of Princep Caliber 90 or 1.6 pt of Princep 4L (4LC).

Princep 80W, Princep 4L (4LC) or Princep Caliber 90 plus Roundup: Use as tank mixture for preemergence and postemergence control of certain broadleaf and grass weeds where corn will be planted directly into a cover crop, established sod, or in previous crop residues. Refer to Roundup label for all directions, weeds controlled, precautions, and limitations.

Princep 80W, Princep 4L (4LC), or Princep Caliber 90 plus Paraquat: Use as tank mixture with Princep and paraquat to kill existing vegetation and for residual weed control where corn will be planted directly into a cover crop, established sod, or in previous crop residues. Add this product and Princep to water in spray tank, agitating until thoroughly mixed. Then add paraquat and a nonionic surfactant, such as X-77®. Continue agitation during application. Broadcast 1.1-2.2 lb of this product plus 1.25-2.5 lb Princep 80W [or 2-4 pt Princep 4L(4LC), or 1.1-2.2 lb Princep Caliber 90] plus 1-2 pt paraquat in 20-60 gal of water per sprayed acre. Apply before, during, or after planting, but before corn emerges. Add 0.5 pt of a nonionic surfactant, such as X-77, per 100 gal of spray mixture. Use the 2 pt rate of paraquat if existing vegetation is 4-6 inches tall. This mixture will not control weeds taller than 6 inches.

Refer to further limitations and precautions on labels for this product, Princep, and paraquat.

Precautions for all applications to corn: *(1) To avoid crop injury and illegal residues, do not apply more than 4.4 lb/A of this product per year. (2) Following harvest, plow (moldboard or disk-plow) and thoroughly till soil in fall or spring to minimize possible injury to spring-seeded rotational crops, regardless of rate used.*

Note for all applications to corn: Do not graze or feed forage from treated areas for 21 days following application, or illegal residues may result.

Rotational crops: (1) Do not rotate to any crop except corn or sorghum until the following year, or injury may occur. (2) If applied after June 10, do not rotate with crops other than corn or sorghum the next year, or crop injury may occur. (3) If the broadcast rate, or equivalent band application rate, exceeds 3.3 lb/A of AAtrex Nine-O, or 3.3 lb/A of AAtrex Nine-O plus Princep Caliber 90 (or equivalent of another Princep formulation), a crop of untreated corn or sorghum should precede the next rotational crop. (4) In the High Plains and Intermountain areas of the West where rainfall is sparse and erratic or where irrigation is required, use only when corn or sorghum is to follow corn or sorghum or when a crop of untreated corn or sorghum is to precede other rotational crops. (5) In eastern parts of the Dakotas, KS, western MN, and NE, do not rotate to soybeans if the rate applied to corn or sorghum was more than 2.2 lb/A or equivalent band application rate, or soybean injury may occur. (6) Injury may occur to soybeans planted the year following application on soils having a calcareous surface layer. (7) Do not plant sugar beets, tobacco, vegetables (including dry beans), spring-seeded small grains, or small-seeded legumes and grasses the year following application, or injury may occur.

SORGHUM AND SORGHUM-SUDAN HYBRIDS
(Grain and Forage Types)

Preplant Surface-Applied (Broadleaf and grass control): Use on medium- and fine-textured soils with minimum-tillage or no-tillage systems only in CO, IL, IN, IA, KS, KY, MN, MO, MT, NE, ND, SD, WI, WY. Apply ⅔ the recommended rate of AAtrex shown in Table 3 for the soil texture as a split treatment 30-45 days prior to planting and the remainder at planting. Applications made less than 30 days prior to planting may be as either a split or single treatment. Under dry conditions, irrigation after application is recommended to move AAtrex into the soil.

If weeds are present at time of treatment, apply in a tank mix combination with a contact herbicide (for example, paraquat or Roundup). Observe directions for use, precautions, and restrictions on the label of the contact herbicide.

Note: To the extent possible, do not move treated soil out of the row or move untreated soil to the surface during planting, or weed control will be diminished.

Preplant Incorporated (Broadleaf and grass control): Broadcast in spring after plowing at rate shown in Table 3. Apply before, during, or after final seedbed preparation. If soil is tilled or worked after application, avoid deep incorporation. For best results, apply within two weeks prior to planting.

Preemergence (Broadleaf and grass control): Apply during or shortly after planting, but prior to weed or crop emergence at rate shown in Table 3.

Table 3: Preplant Surface, Preplant Incorporated, and Preemergence Broadleaf and Grass Weed Control in Sorghum*

Soil texture	Organic matter	Broadcast rate/A
COARSE Sand, loamy sand, sandy loam	any level	DO NOT USE (except for preemergence use on bedded sorghum in AZ and CA)
MEDIUM AND FINE Loam, silt loam, silt, clay loam, sandy clay loam, silty clay loam, sandy clay, silty clay, clay	less than 1%	DO NOT USE (except for preemergence use on bedded sorghum in AZ and CA)
	1-1.5%	1.7-2.2 lb
	more than 1.5%	2.2-2.6 lb

* Do not apply preplant surface or preplant incorporated in AL, AR, FL, GA, LA, MS, NC, NM, OK, SC, TN, or TX. Do not apply preemergence in NM, OK, or TX, except in northeast OK and the TX Gulf Coast.

In case of planting failure, sorghum or corn may be replanted. Do not make a second broadcast application, or injury may occur. If originally applied in a band and sorghum or corn is replanted in untreated row middles, this product may be applied in a band to the second planting.

Preemergence broadleaf weed control in furrow irrigated bedded sorghum (AZ and CA only): For preemergence control of many broadleaf weeds, broadcast 0.9-1.3 lb/A after bed preparation, during or after planting, but before sorghum and weeds emerge and before the first furrow irrigation. Follow with several regular irrigations, making sure to thoroughly wet all soil.

Precautions for preemergence application to furrow irrigated bedded sorghum in AZ and CA: To avoid possible sorghum injury, do not use on sand or loamy sand soil or on sorghum planted in furrows. Applications to sorghum growing on alkali soils or where cuts, fills, or erosion have exposed calcareous or alkali subsoils may cause crop injury. In case of crop failure, do not replant sorghum for 8 months following application. Corn may be planted immediately.

Postemergence (Broadleaf and grass control): Apply at rate shown in Table 4 before weeds exceed 1.5 inches in height. Apply up to close-in.

Table 4: Postemergence Broadleaf and Grass Weed Control in Sorghum

Soil texture	Minimum sorghum height at treatment	Broadcast rate/A
Sand or loamy sand	DO NOT USE	
Sandy loam	See directions for postemergence broadleaf weed control	
Silt loam to sandy clay loam	Completely emerged	2.2-2.6 lb
Olton and Pullman clay soil	6 inches	2.2-2.6 lb
Silty clay loam and finer textured soil	Completely emerged	3.3 lb

Postemergence broadleaf weed control with emulsifiable oil or oil concentrate in water: Broadcast 1.3 lb/A for control of many broadleaf weeds. Apply before pigweed and lambsquarters reach 6 inches in height and before all other weeds reach 4 inches in height. In CO, western KS, NM, OK, TX, and desert regions of AZ and CA, apply when sorghum is 6-10 inches in height, but before it reaches boot stage. In all other areas, apply after sorghum reaches the 3-leaf stage. Add 1 gal of emulsifiable oil/A for ground application and 0.5 gal/A for aerial application, or add 1 qt of oil concentrate for ground application. A cultivation may be necessary if all weeds are not controlled or if weeds regrow.

For the list of weeds controlled, see **AAtrex Nine-O Applied Alone—Corn or Grain Sorghum—Postemergence with Emulsifiable Oil or Oil Concentrate in Water.**

Precautions for applications with emulsifiable oil or oil concentrate in water: **See** *"Precautions for applications with emulsifiable oil or oil concentrate in water" in Corn section.*

Postemergence broadleaf weed control with surfactant (CO, western KS, NM, OK, TX, and desert regions of AZ and CA only): Broadcast 1.3 lb/A plus 0.75-1.5 pt of surfactant after sorghum reaches 6 inches in height, but before weeds exceed 1.5 inches in height. Apply only on sandy loam and finer textured soil.

Precautions for all applications to sorghum: *(1) Heavy rain immediately following application tends to cause excessive concentrations of herbicide in seed furrow, resulting in possible crop injury. Do not apply to furrow-planted sorghum until furrows are leveled (plowed in). Level deep planter marks or seed furrows before application. (2) Application to sorghum growing under stress caused by minor element deficiency or to sorghum growing on highly calcareous soil may result in crop injury. (3) Following harvest, plow (moldboard or disk-plow) and thoroughly till soil in fall or spring to minimize possible injury to spring-seeded rotational crops, regardless of rate used. (4) Injury may occur if both this herbicide, preplant surface, preplant incorporated, or preemergence, and an at-planting systemic insecticide applied in-furrow are used.*

Note: Do not graze or feed forage from treated areas for 21 days following application, or illegal residues may result.

Rotational crops: See "Rotational crops" at end of Corn section.

TANK MIXTURES FOR GRAIN SORGHUM

Igran® 80W: Use as tank mixture for control of certain broadleaf and grass weeds in grain sorghum. Refer to Igran 80W label for all directions, weeds controlled, precautions, and limitations.

Dual 8E: Use as tank mixture with Dual 8E for control of those weeds listed on the Dual label, as well as on this label. Use this tank mixture only on sorghum seed treated with Concep®. Refer to the Dual 8E label for all directions, precautions, and limitations.

WINTER WEED CONTROL IN TEXAS

For postemergence control of winter weeds only, such as henbit, seedling dock, and annual thistle on fall bedded land in the Gulf Coast and Blacklands of Texas. Apply 0.9-1.1 lb/A postemergence to the weeds in November or December to land that will be planted to corn, grain sorghum, or forage sorghum the following spring. For best results, add a suitable surfactant, such as X-77, at the rate of 0.5% of the spray volume, an emulsifiable oil at the rate of 1.0% of the spray volume, or an oil concentrate at the rate of 1 qt per acre.

Normal weed control programs may be used in the following corn, grain sorghum or forage sorghum crop.

Note: Do not plant any crops except corn, grain sorghum, or forage sorghum the spring following this treatment, or illegal residues may result.

AATREX ALONE — CHEMICAL FALLOW

Wheat-Sorghum-Fallow: To control annual broadleaf and grass weeds following wheat harvest and in the following sorghum crop when grown under minimum tillage, broadcast 3.3 lb/A to wheat stubble immediately following wheat harvest. If weeds are present at application, remove them with a sweep plow or other suitable implement after application, or use an approved contact herbicide before or after the application of AAtrex. Plant sorghum into wheat stubble the following spring with minimum soil disturbance. Use a surface planter or a planter leaving a shallow furrow. If weeds are present at planting, remove them with a sweep plow or other suitable implement before planting.

For the list of weeds controlled, see **AAtrex Nine-O Applied Alone — Corn or Grain Sorghum — Preplant Surface-Applied, Preplant Incorporated, or Preemergence.**

Precautions: *(1) Use only on silt loam or finer textured soil, or crop injury may result. (2) Wheat-sorghum-fallow cropping sequence must be followed. (3) Do not apply following sorghum harvest.*

Notes: To avoid illegal residues, do not graze or feed forage from treated area to livestock. To avoid illegal residues and crop injury, do not plant any crop other than those on this label within 18 months following treatment.

Wheat-Corn-Fallow (KS, NE): This product controls cheatgrass (downy brome, chess), kochia, mustards, pigweed, Russian thistle, wild lettuce, wild sunflower, and volunteer wheat during period after wheat harvest. Weed control may extend into following corn crop grown under minimum tillage.

Follow directions for use, notes, and precautions in the "Wheat-Sorghum-Fallow" section above, substituting corn for references to sorghum.

Wheat-Fallow-Wheat (CO, KS, MT, NE, ND, SD, and WY): For preemergence control of cheatgrass (downy brome, chess), common lambsquarters, field pennycress, kochia, mustard, Russian thistle, wild lettuce, and suppression of volunteer wheat during fallow period of a wheat-fallow-wheat rotation, broadcast 0.5-1.1 lb/A on all soils except those listed under "Precautions." For control of pigweed and wild sunflower, use the higher rate. Apply to stubble ground. Treat only once during same fallow period.

TANK MIXTURES FOR CHEMICAL FALLOW

Wheat-Sorghum-Fallow or Wheat-Corn-Fallow (KS, NE) Igran 80W: Use as tank mixture for control of certain broadleaf and grass weeds in wheat-sorghum-fallow or wheat-corn-fallow rotations. Refer to Igran 80W label for directions, weeds controlled, precautions, and limitations.

Paraquat: If weeds are present at application, a tank mix with paraquat may be used. Broadcast 3.3 lb of AAtrex Nine-O plus 1-2 pt of paraquat in 20-60 gal of water/A by ground equipment. Add 0.5-1 pt of a nonionic surfactant, such as X-77, per 100 gal of spray mixture. Add AAtrex to spray tank first and thoroughly mix with water. Then add paraquat, followed by surfactant. Use the 2 pt rate of paraquat if weeds are 4-6 inches tall. This mixture will not control weeds taller than 6 inches. Apply to stubble ground. Treat only once during same fallow period. Refer to paraquat label for further directions, precautions, and limitations.

Wheat-Fallow-Wheat (CO, KS, MT, NE, ND, SD, and WY) Igran 80W: Use as tank mixture for control of certain broadleaf and grass weeds in wheat-fallow-wheat rotations. Refer to Igran 80W label for directions, weeds controlled, precautions, and limitations.

Paraquat: If weeds are present at application, a tank mix with paraquat may be used. Broadcast 0.5-1.1 lb of AAtrex Nine-O plus 1-2 pt of paraquat in 20-60 gal of water/A by ground equipment. Add 0.5-1 pt of a nonionic surfactant, such as X-77, per 100 gal of spray mixture. Add AAtrex to spray tank first and thoroughly mix with water. Then add paraquat, followed by surfactant. Use the 2 pt rate of paraquat if weeds are 4-6 inches tall. This mixture will not control weeds taller than 6 inches. Apply to stubble ground. Treat only once during same fallow period. Refer to paraquat label for further directions, precautions, and limitations.

If weeds are present at application and this product is used alone, use either an approved contact herbicide before or after treatment, or tillage after treatment.

Use tillage to control weeds which escape during fallow period. Till before planting. For this product applied alone or in tank mixture with paraquat, plant at least 2 inches deep and 12 months or more after application.

Precautions: *(1) To avoid crop injury, (1) Do not use on sand soil. (2) Do not treat eroded hillsides, caliche and rocky out-croppings, or exposed calcareous subsoil. (3) Do not treat soils of the Rosebud and Canyon Series in western NE and adjoining counties in CO and WY. (4) Do not treat soils with calcareous surface layers. (5) Avoid spray overlap.*

Note: Do not graze treated areas within 6 months after application, or illegal residues may result.

Aerial application: In order to assure that spray will be controllable within the target area when used according to label directions, make applications at a maximum height of 10 ft, using low drift nozzles at a maximum pressure of 40 psi, and restrict application to periods when wind speed does not exceed 10 mph. To assure that spray will not adversely affect adjacent sensitive nontarget plants, apply AAtrex Nine-O alone by aircraft at a minimum upwind distance of 400 ft from sensitive plants, or apply AAtrex Nine-O + Igran 80W at a minimum upwind distance of 500 ft from sensitive plants. When applying AAtrex Nine-O + Igran 80W by aircraft, use a minimum of 5 gal of water per acre.

PROSO MILLET

For preplant incorporated or preemergence control of foxtails, kochia, common lambsquarters, redroot pigweed, Russian thistle, slimleaf lambsquarters and spotted spurge, apply 0.6-1.1 lb/A on soils with less than 2% organic matter, or apply 1.1-2.2 lb/A on soils with 2% or higher organic matter.

Precaution: To avoid crop injury and illegal residues, make only one application per year.

Rotational crops: See "Rotational crops" in Corn section.

RANGELAND

To control certain weeds, including cheatgrass (downy brome, chess), common (annual) broomweed, little barley, medusahead, sagewort, and tumble mustard: Use lower rate where cheatgrass is major problem weed; use higher rate for other weeds.

Poisonous plants: Tansymustard, pigweeds, lambsquarters, tumble mustard, cocklebur, and fiddleneck are controlled at 1.1 lb/A. Other mustards (*Brassica* spp.), groundcherry (annual), broomweed, and threeawn or poverty grass (*Aristida* spp.) are partially controlled at 1.1 lb/A and controlled at 2.2 lb/A. Sneezeweed and dogbane are partially controlled at 2.2 lb/A.

Aerial application: In order to assure that spray will be controllable within the target area when used according to label directions, make applications at a maximum height of 10 ft above vegetation, using low drift nozzles at a maximum pressure of 40 psi, and restrict application to periods when wind speed does not exceed 10 mph. To assure that spray will not adversely affect adjacent sensitive nontarget plants, apply AAtrex Nine-O by aircraft at a minimum upwind distance of 400 ft from sensitive plants.

Establishment of perennial range grasses in West (CA, ID, NV, OR, UT, WA): Broadcast 0.9-1.1 lb/A in fall after rains begin but before ground freezes. Perennial range grass may be seeded when rain begins the next fall following treatment. Plant in deep furrows such as those made by a deep furrow rangeland drill. Treated areas may be grazed 7 months after seeding. However, to provide sufficient time for establishment of desirable range grasses, treated areas should not be grazed until the fall following seeding.

Renovation of existing stands of perennial range grasses

Central Great Plains (CO, KS, MT, NE, ND, SD, WY): Broadcast 0.6-1.1 lb/A in fall before ground freezes.

Southwest (AR, KS, LA, MS, NM, OK, TX): Apply 1.1 lb/A AAtrex Nine-O in early spring, before regrowth, to established rangeland or pasture species such as Asiatic (Old World) bluestems, little and big bluestems, buffalograss, switchgrass, and Indiangrass.

West (CA, ID, NV, OR, UT, WA): Do not graze area to be treated for 6 months prior to application. Broadcast 0.6-0.8 lb/A in fall before ground freezes.

Bluegrass and intermediate wheatgrass may be injured by this product at recommended rates.

Notes: To avoid illegal residues, (1) Do not apply more than once per year. (2) Apply only on neutral to acid soils. (3) Do not cut or feed range grass hay. (4) Do not graze treated areas within 3 months following spring application, or within 7 months following fall application. (5) Do not dump or spill product or dispose of container within reach of livestock.

Pure stands of Switchgrass and Big Bluestem (IA, IL, IN, KS, MN, MO, ND, NE, OH, SD, WI): For control of those weeds listed under **AAtrex Nine-O Applied Alone — Corn or Grain Sorghum — Preplant Surface-Applied, Preplant Incorporated, or Preemergence.**[1]

Use only on loam, silt loam, silty clay loam, clay loam, and silty clay soils with at least 1 percent organic matter.

Establishment: Broadcast 2.2 lb/A after planting but before weeds and crop emerge. Prepare a good, firm seedbed. Plant ½ inch deep with a grassland drill (preferred method) or a conventional drill. If a conventional drill is used on prepared seedbeds, remove all tension from the disk openers. For best results, cultipack or roll after planting. Clip weeds that escape in July or August. Avoid clipping switchgrass or big bluestem seedlings.

Renovation of existing stands: Broadcast 2.2 lb/A in early spring before regrowth of the crop, and before weed emergence.

Notes: To avoid illegal residues, (1) Do not cut for hay. (2) Do not graze treated areas within 4 months following application at seeding or 3 months following application to established switchgrass or big bluestem. (3) Treat only once per year.

Rotational crops: See "Rotational crops" in Corn section. Do not plant small-seeded legumes the year following application, or injury may occur.

ROADSIDES

To control certain annual weeds in established perennial grasses along roadsides in CO, KS, MT, NE, ND, SD, and WY, including cheatgrass (downy brome, chess), common (annual) broomweed, little barley, medusahead, sagewort, and tumble mustard, broadcast 1.1 lb/A in a minimum of 10 gallons of water by ground equipment in the fall before ground freezes, or after thawing in the spring, but before the established grasses green-up and before weeds emerge. Examples of desirable established grasses include big bluestem, bluegrama, bromegrass, buffalo grass, crested wheatgrass, Indiangrass, little bluestem, side-oats grama, switchgrass, and Western wheatgrass. Apply only once per year. Temporary discoloration or other form of injury to the desirable perennial grasses may occur following application.

Notes: To avoid illegal residues, (1) Do not cut or feed roadside grass hay. (2) Do not allow livestock to graze treated areas.

PERENNIAL RYEGRASS GROWN FOR SEED

(Pacific Northwest only)

For control of annual ryegrass and rattail fescue, broadcast 1.33 lb/A after first fall rain.

Precautions: *To avoid crop injury, (1) Apply only to perennial ryegrass from which seed crop has been harvested. (2) Do not apply to ryegrass more than 2 years in succession.*

Notes : To avoid illegal residues, (1) Do not graze treated land within 10 days after application. (2) Do not plant treated land to other crops for 16 months.

ESTABLISHED ORCHARDGRASS GROWN FOR SEED

(Pacific Northwest Only)

AAtrex aids in control of annual ryegrass, rattail fescue, and volunteer orchardgrass. Broadcast 2.2 lb of AAtrex Nine-O in at least 15 gal of water to field which has been burned after first fall rain.

Precautions: *Some yellowing or crop stunting can occur. To avoid crop injury, (1) Apply only to orchardgrass from which seed crop has been harvested. (2) Do not apply to orchardgrass more than two years in succession.*

Notes : To avoid illegal residues, (1) Do not graze treated land within 6 months of application. (2) Do not plant treated land to other crops for 16 months.

SUGARCANE

For control of many broadleaf and grass weeds, including amaranths, crabgrass, fireweed, Flora's paintbrush, foxtails, junglerice and wiregrass, broadcast 2.2-4.4 lb/A of AAtrex Nine-O at time of planting or ratooning, but before sugarcane emerges. Broadcast aerially in a minimum of 5 gal of spray per acre, or broadcast or band by ground equipment in a minimum of 20 gal per acre, unless indicated otherwise. One additional application may be made over the sugarcane as it emerges, and two additional applications may be made interline after emergence as directed sprays. Repeat treatments, where needed, may be applied broadcast, band, or interline as suggested with the final application being prior to close-in. Do not exceed the rate of herbicide suggested for any one crop of sugarcane.

Note: Where high rates of AAtrex are used, apply in a minimum of 1 gal of water for each 1 lb of product applied per acre.

Aerial application: In order to assure that spray will be controllable within the target area when used according to label directions, make applications at a maximum height of 10 ft, using low drift nozzles at a maximum pressure of 40 psi, and restrict application to periods when wind speed does not exceed 10 mph. To assure that spray will not adversely affect adjacent sensitive nontarget plants, apply AAtrex Nine-O alone by aircraft at a minimum upwind distance of 400 ft from sensitive plants.

Florida

For control of emerged pellitory weed: Apply 0.4-0.6 lb/A in at least 40 gal of water as a directed spray by ground equipment prior to close-in. Add 4 qt of surfactant for each 100 gal of spray. Thoroughly cover weed foliage.

For control of alexandergrass, large crabgrass, pellitory (artillery) weed, and spiny amaranth, use one of the following methods at planting or ratooning:

1. Apply 4.4 lb/A preemergence. Follow with one or two applications, as needed, postemergence to sugarcane and weeds, at 2.2 lb/A. Treat before weeds exceed 1.5 inches in height.
2. Apply 1-3 times, as needed, at 2.2 lb/A postemergence to sugarcane and weeds. Treat before weeds exceed 1.5 inches in height.

Louisiana

For control of annual weeds during the summer fallow period, apply 2.2 lb/A to weed-free beds immediately after bed formation. Follow normal weed control program after planting.

Precautions: *To avoid crop injury, (1) Do not apply more than 11 lb/A to any one crop of sugarcane. (2) If making a 2.2 lb/A application during the summer fallow period, do not exceed 8.8 lb/A during the remainder of the growing season, or illegal residues may result.*

Texas

Use AAtrex Nine-O for control of barnyardgrass, pigweed, pursiane and sunflower, in plant or ratoon sugarcane.

Apply 4.4 lb/A of AAtrex Nine-O preemergence. Follow with one or two applications, as needed, at 3.3 lb/A postemergence to sugarcane and weeds.

For best results when weeds are emerged, add a nonionic surfactant at a concentration of 2 qt/100 gal to the spray and apply before weeds exceed 1.5 inches in height.

Precautions: *(1) Injury to sugarcane may occur when under moisture stress, when soil is of low adsorptive capacity, or when land is first cropped to sugarcane. (2) Do not apply after close-in. (3) Do not apply more than 11 lb/A to any one crop of sugarcane, or crop injury may result.*

TURF GRASSES FOR SOD (Florida only)

St. Augustinegrass, Centipedegrass, and Zoysia Grass

Broadcast 2.2-4.4 lb/A according to soil texture to control those weeds listed under **AAtrex Nine-O Applied Alone -Corn or Grain Sorghum — Preplant Surface-Applied, Preplant Incorporated, or Preemergence.**

Muck or peat	4.4 lb	Old beds: Within 2 days after lifting sod
		New beds: 3-4 days after sprigging or plugging
Sandy soil	2.2 lb	Old beds: Within 2 days after lifting sod
		New beds: 7-10 days after sprigging or plugging

If weeds regrow, apply an additional 2.2 lb/A on muck or peat, or 1.1 lb/A on sandy soil.

Precautions: *To avoid crop injury, (1) Do not apply within 30 days prior to cutting or lifting. (2) Do not apply in combination with surfactants or other spray additives. (3) Use only on turfgrass reasonably free of infestations of insects, nematodes, and diseases. (4) On newly sprigged turfgrass, temporary slowing of growth may follow application.*

TURFGRASS FOR FAIRWAYS, LAWNS, SOD PRODUCTION* AND SIMILAR AREAS

*In states other than Florida. For use on turfgrass for sod in Florida, see "Turfgrasses for Sod (Florida only)" section above.

Bermudagrass, Centipedegrass, St. Augustinegrass, and Zoysia Grass

Apply AAtrex Nine-O after October 1 before emergence of winter annual weeds for control of annual bluegrass, burclover, carpet burweed, chickweed, corn speedwell, henbit, hop clover, and spurweed. AAtrex Nine-O will control annual bluegrass even if it is emerged at time of treatment. For control of summer annual weeds listed in the preemergence portion of the AAtrex Nine-O Applied Alone—Corn or Grain Sorghum section of this label, also apply AAtrex Nine-O in late winter before the weeds emerge. Apply in a minimum of 15 gal of water per acre or 1 gal per 1,000 sq ft.

Where annual bluegrass is the major weed, use 1.1 lb of AAtrex Nine-O per acre (0.4 fl oz per 1,000 sq ft). Use 2.2 lb/A (0.8 fl oz per 1,000 sq ft) for control of the other weeds named above. Do not exceed 1.1 lb/A per treatment on newly sprigged turfgrass or on hybrid bermudagrass such as Tiflawn, Tifway, and Ormond.

For continued summer annual weed control, apply another 1.1 lb/A at least 30 days after the previous application, but not after April 15. However, do not make more than two applications of this product per year.

Precautions: *On newly sprigged turfgrass and hybrid bermudagrass, temporary slowing of growth and yellowing may occur following application. To avoid turf injury, (1) Use only on turfgrass reasonably free of infestations of insects, nematodes, and diseases. (2) Do not use on golf greens. (3) Do not use north of NC (except may be used in Virginia Coastal Plains) or west of the high rainfall areas of eastern OK and eastern TX. (4) Do not use on muck or alkaline soils. (5) Do* *not apply over the rooting area of trees or ornamentals not listed on this label. (6) Do not overseed with desirable turfgrass within 4 months before or 6 months after treatment.*

Note: Do not graze or feed turf clippings to animals, or illegal residues may result.

MACADAMIA NUTS

For preemergence control of many broadleaf and grass weeds, including crabgrass, foxtail, wiregrass, Flora's paintbrush, spanishneedles, and fireweed, broadcast 2.2-4.4 lb/A before harvest and before weeds emerge. Repeat as necessary. Do not spray when nuts are on ground during harvest period. Do not apply by air.

PINEAPPLE

For control of purslane, spanishneedles, annual grasses, annual bindweed, ageratum, amaranth, rattlepod, Flora's paintbrush, fireweed, spurge, indigo, and papalo, apply up to 7.1 lb/A as a blanket spray immediately after planting or following harvest. Make additional blanket or interspace applications of up to 1.7 lb/A at 1-2 month intervals as needed, prior to differentiation. Apply in 20-40 gal of water/A to assure thorough coverage.

Precautions: *(1) Do not apply more than 33.3 lb/A per cycle, as crop injury or illegal residues may result. (2) Repeated monthly applications to plant foliage may slow plant growth and delay fruiting.*

Note : Do not apply within 45 days of fruit harvest, or forage harvest if forage is to be fed to livestock, or illegal residues may result.

GUAVA

Use only on established plantings which are at least 18 months old. Apply as a directed spray at 2.2-4.4 lb/A of AAtrex Nine-O in 20-50 gal of spray mix preemergence or early postemergence to weeds. When applying postemergence, the use of a surfactant and greater spray volume (80-100 gal of spray mix per acre) may enhance weed control. This product controls many annual broadleaf and grass weeds, including fireweed, purslane, scarlet pimpernel, spanishneedles, and sowthistle.

Notes: To avoid illegal residues, (1) Do not allow spray to contact foliage or fruit. (2) Do not apply more frequently than at 4 month intervals. (3) Do not apply more than 8.8 lb of AAtrex Nine-O per year.

CONIFERS

For control of annual broadleaf and grass weeds prior to transplanting, after transplanting or in established conifers [including Douglas fir, grand fir, noble fir, white fir, Austrian pine, bishop pine, Jeffrey pine, knobcone pine, loblolly pine, lodgepole pine (shore pine), monterey pine, ponderosa pine, Scotch pine, slash pine, blue spruce, and Sitka spruce]: Broadcast 2.2-4.4 lb in a minimum of 5 gal of water per acre by air or 10 gal by ground before weeds are 1.5 inches tall. Apply to established trees between fall and early spring while trees are dormant. For new transplants, apply during or soon after transplanting. For applications prior to transplanting, allow sufficient precipitation to activate AAtrex before transplanting. In areas where spring and summer rainfall is inadequate to activate AAtrex, apply during fall prior to spring transplanting.

For the list of weeds controlled, see **AAtrex Nine-O Applied Alone — Corn or Grain Sorghum — Preplant Surface-Applied, Preplant Incorporated, or Preemergence.**

Quackgrass control: Broadcast 4.4 lb in a minimum of 5 gal of water per acre by air or 10 gal by ground between fall and early spring while trees are dormant and before quackgrass is more than 1.5 inches tall.

Precautions: *Temporary injury to trees may occur following the use of AAtrex on coarse-textured soil. To avoid crop injury, (1) Do not apply to seedbeds. (2) Apply only once per year.*

Note: Do not graze treated areas, or illegal residues may result.

Aerial application: In order to assure that spray will be controllable within the target area when used according to label directions, make applications at a maximum height of 10 ft above vegetation, using low drift nozzles at a maximum pressure of 40 psi, and restrict application to periods when wind speed does not exceed 10 mph. To assure that spray will not adversely affect adjacent sensitive nontarget plants, apply AAtrex Nine-O by aircraft at a minimum upwind distance of 400 ft from sensitive plants.

NONSELECTIVE WEED CONTROL ON NONCROP LAND

This product provides long-term weed control on industrial sites, highway medians and shoulders, railroad rights-of-way, lumberyards, petroleum tank farms, and in noncrop areas on farms, such as around buildings, equipment and fuel storage areas, along fences, and lanes. Do not use near adjacent desirable plants or in greenhouses, or injury may occur.

Apply before or soon after weeds begin growth. Make postemergence applications when weeds are young and actively growing. Use the higher rates on fine clay and muck soils. Use sufficient water to assure thorough coverage. Use at least 1 gal of water for each pound of product; use more if practical.

Note : Do not allow livestock to graze treated areas, or illegal residues may result.

Annual broadleaf and grass weeds (including barnyardgrass, cheatgrass, crabgrass, lambsquarters, foxtail, ragweed, puncturevine, and turkey mullein): Broadcast 5.3-11.1 lb/A.

Hard-to-kill annual and many perennial broadleaf and grass weeds including bluegrass, burdock, Canada thistle, dogfennel, orchardgrass, plantain, quackgrass, purpletop, redtop, and smooth brome): Broadcast 11.1-22.2 lb/A.

Hard-to-kill biennial and perennial weeds (including bullthistle and sowthistle): Broadcast 22.2-44.4 lb/A.

For longer residual control in regions of high rainfall and a long growing season, broadcast 22.2-44.4 lb/A.

For small areas, 4 oz per 1,000 sq ft is equivalent to 11.1 lb/A.

Tank Mixture with Princep 80W or Princep Caliber 90

Use only for long-term control of broadleaf and grass vegetation in areas listed in respective labels. Do not use on cropland, or near desirable plants, or in greenhouses, or injury may occur.

This tank mixture has similar activity and moisture requirements as described under above General Information section, except that Princep provides preemergence control only.

For best results, apply to the soil shortly before weed growth begins or apply postemergence to young, actively growing plants not over 6 inches tall.

Use sufficient water for thorough soil and plant coverage. Use at least 1 gal of water for each lb of tank mixture herbicide. More water may be needed for heavy weed growth. Maintain agitation during mixing and application.

Broadcast tank mixtures at the following rates to control those weeds listed. Use the lower rates for light weed infestations and the higher rates for heavier infestations.

Use 2.9-5.6 lb/A of this product plus 3.25-6.25 lb/A of Princep 80W (or 2.9-5.6 lb/A of Princep Caliber 90) to control weeds listed in the General Information and the several crop sections of the Princep 80W and Princep Caliber 90 labels.

The above rates will also control California burclover, *Orthocarpus purpurescens,* ripgut brome, smooth catsear, summer lupine, and turkey mullein.

Use 5.6 lb/A of this product plus 6.25 lb/A of Princep 80W (or 5.6 lb/A of Princep Caliber 90) to control burdock, daisy fleabane, goldenrod, horseweed, little mallow, milkweed, orchardgrass, quackgrass, Russian thistle, shortpod mustard, wild barley, and wood sorrel.

Use 11.1 lb/A of this product plus 12.5 lb/A of Princep 80W (or 11.1 lb/A of Princep Caliber 90) to control eveningprimrose, shepherds-needle, and tall fescue.

For small areas, ½ cup (4 oz) of this product per 1,000 sq ft is equivalent to 11.1 lb/A; ⅔ cup (4.5 oz) of Princep 80W or ½ cup (4 oz) of Princep Caliber 90 per 1,000 sq ft is equivalent to 12.5 lb/A of Princep 80W or 11.1 lb/A of Princep Caliber 90, respectively.

Tank Mixture with Diuron 80W

Use only for long-term control of broadleaf and grass vegetation in areas listed in respective labels. Do not use on cropland, or near desirable plants, or in greenhouses, or injury may occur.

This tank mixture has activity and moisture requirements similar to that described in above General Information section and in the diuron 80W label. For best results, apply to soil shortly before weed growth begins or apply to young, actively growing plants not over 6 inches tall. For increased contact activity, add a surfactant, such as Surfactant WK, at 2 qt per 100 gal of spray volume. Add surfactant as last ingredient to nearly full spray tank.

Use sufficient water for thorough ground and plant coverage. Use at least 1 gal of water for each lb of tank mixture. More water may be needed for heavy weed growth. Maintain agitation during mixing and application.

Broadcast tank mixtures at the following rates to control those weeds listed. Use the lower rates for light weed infestations and the higher rates for heavier infestations.

Use 4.4-5.6 lb/A of this product plus 5-6.25 lb/A of diuron 80W, for example, Karmex® 80W, to control weeds listed in crop and non-crop Directions for Use on the diuron 80W label.

Use 5.6 lb/A of this product plus 6.25 lb/A of diuron 80W to control broomsedge, buckhorn plantain, Carolina geranium, chicory, downy brome, eveningprimrose, fleabane, goldenrod, horseweed, lespedeza, little mallow, orchardgrass, Russian thistle, shortpod mustard, tall fescue, wild barley, and wood sorrel.

Use 11.1 lb/A of this product plus 12.5 lb/A of diuron 80W to control alfalfa, black nightshade, leafy spurge, and shepherds-needle.

To determine compatibility of tank mixtures, thoroughly mix products in a small container at the ratio of 1 qt of water to 1.75 oz of this product plus 2 oz of diuron 80W. Let stand for 5 minutes. If the mixture remains mixed, or can be remixed readily, it is compatible and can be sprayed.

For small areas, ½ cup (4 oz) of this product or 1.25 cups of diuron 80W, such as Karmex 80W, per 1,000 sq ft is equivalent to 11.1 lb/A of this product or 12.5 lb/A of diuron 80W.

STORAGE AND DISPOSAL

PESTICIDE STORAGE AND DISPOSAL

Store in a dry place. Do not contaminate water, food, or feed by storage, disposal, or cleaning of equipment. Open dumping is prohibited. Wastes resulting from the use of this product are acutely toxic. Improper disposal of unused pesticide, spray mixture, or rinsate is a violation of federal law. Pesticide, spray mixture, or rinsate that cannot be used according to label instructions must be disposed of according to federal, state or local procedures. For guidance in proper disposal methods, contact your State Pesticide or Environmental Control Agency, or the Hazardous Waste representative at the nearest EPA Regional Office.

CONTAINER DISPOSAL

Completely empty bag into application equipment. Dispose of empty bag in a sanitary landfill or by incineration, or by open burning, if allowed by state and local authorities. If burned, keep out of smoke.

For minor spills, leaks, etc., follow all precautions indicated on this label and clean up immediately. Take special care to avoid contamination of equipment and facilities during cleanup procedures and disposal of wastes. In the event of a major spill, fire or other emergency, call (919) 292-7100 day or night.

PRECAUTIONARY STATEMENTS

HAZARDS TO HUMANS AND DOMESTIC ANIMALS

CAUTION

Harmful if swallowed, inhaled, or absorbed through skin. Do not breathe dust or spray mist. Avoid contact with eyes, skin or clothing.

First Aid: If swallowed, contact your local poison control center, hospital, or physician. If the patient is unconscious, maintain breathing and heartbeat (cardiopulmonary resuscitation). If the patient is conscious and alert, induce vomiting (syrup of ipecac or stimulate the back of the throat with a finger). NEVER GIVE ANYTHING BY MOUTH TO AN UNCONSCIOUS PERSON!

In case of skin contact, remove contaminated clothing and wash with soap and water. In case of contact with eyes, flush with *plenty* of water. Get medical attention if irritation persists.

Note to Physician: There is no specific antidote for atrazine. If this product is ingested, induce emesis or lavage stomach. The use of an aqueous slurry of activated charcoal may be considered.

ENVIRONMENTAL HAZARDS

Do not apply directly to any body of water. Do not apply where runoff is likely to occur. Do not contaminate water by cleaning of equipment or disposal of wastes. Do not apply when weather conditions favor drift from treated areas.

AAtrex® trademark of CIBA-GEIGY for atrazine

Bexton® trademark of Dow Chemical Co. for propachlor

Caliber® trademark of CIBA-GEIGY

Compex® trademark of KALO Agricultural Chemicals, Inc.

Concep® trademark of CIBA-GEIGY

Dual® trademark of CIBA-GEIGY for metolachlor U.S. Patent No. 3,937,730

Igran® trademark of CIBA-GEIGY for terbutryn

Karmex® trademark of E. I. duPont de Nemours & Co. for diuron

Lasso® trademark of Monsanto Company for alachlor

Nine-O® trademark of CIBA-GEIGY

Princep® trademark of CIBA-GEIGY for simazine

Ramrod® trademark of Monsanto Company for propachlor

Roundup® trademark of Monsanto Company for glyphosate herbicide

Sutan+® trademark of Stauffer Chemical Co. for butylate

Unite® trademark of Hopkins Agricultural Chemical Co.

X-77® trademark of KALO Agricultural Chemicals, Inc.

CGA 7L101C 064

APRON® 2E

FUNGICIDE

A seed treatment chemical for control of systemic downy mildew, Pythium seed rot, Pythium damping-off, and early season Phytophthora diseases of certain crops

For use only by commercial seed treaters

ACTIVE INGREDIENT:
Metalaxyl: *N*-(2, 6-dimethylphenyl)-*N*-(methoxyacetyl) alanine methyl ester 25.1%
INERT INGREDIENTS: 74.9%
TOTAL: 100.0%

Apron 2E contains 2 lbs. active ingredient per gallon.

EPA Reg. No. 100-626

Keep Out of Reach of Children

WARNING

See additional precautionary statements at back of this booklet.

DIRECTIONS FOR USE AND CONDITIONS OF SALE AND WARRANTY

IMPORTANT: Read the entire **Directions for Use** and the **Conditions of Sale and Warranty** before using this product.

CONDITIONS OF SALE AND WARRANTY

The **Directions for Use** of this product reflect the opinion of experts based on field use and tests. The directions are believed to be reliable and should be followed carefully. However, it is impossible to eliminate all risks inherently associated with use of this product. Crop injury, ineffectiveness, or other unintended consequences may result because of such factors as weather conditions, presence of other materials, or the manner of use or application all of which are beyond the control of CIBA-GEIGY or the Seller. All such risks shall be assumed by the Buyer.

CIBA-GEIGY warrants that this product conforms to the chemical description on the label and is reasonably fit for the purposes referred to in the **Directions for Use** subject to the inherent risks referred to above. **CIBA-GEIGY makes no other express or implied warranty of Fitness or Merchantability or any other express or implied warranty. In no case shall CIBA-GEIGY or the Seller be liable for consequential, special, or indirect damages resulting from the use or handling of this product.** CIBA-GEIGY and the Seller offer this product, and the Buyer and user accept it, subject to the foregoing **Conditions of Sale and Warranty,** which may be varied only by agreement in writing signed by a duly authorized representative of CIBA-GEIGY.

GENERAL INFORMATION

Apron is a systemic fungicide for use as a seed dressing.

Note: Apron is a systemic fungicide having a specific mode of action. Apron could be subject to development of resistant strains of fungi. Development of resistance cannot be predicted. Therefore, CIBA-GEIGY cannot assume liability for crop damage resulting from resistant strains of fungi. Consult with your State Agricultural Experiment Station or Extension Service Specialist for guidance and ways to control any possible Apron resistant strains of fungi which may occur.

Apron 2E may be applied as a water-based slurry with other registered seed treatment insecticides and fungicides through standard slurry- or mist-type commercial seed treatment equipment.

DIRECTIONS FOR USE

It is a violation of federal law to use this product in a manner inconsistent with its labeling.

FAILURE TO FOLLOW THE DIRECTIONS FOR USE AND PRECAUTIONS ON THIS LABEL MAY RESULT IN CROP INJURY OR POOR DISEASE CONTROL.

Grain Sorghum	For Pythium damping-off control: Apply Apron 2E as a seed treatment at the rate of 1-2 fluid oz. per 100 lbs. of seed.
Soybeans	For Pythium damping-off and early season Phytophthora control: Apply Apron 2E as a seed treatment at the rate of 1-2 fluid oz. per 100 lbs. of seed.
Sunflowers	For control of systemic downy mildew: Apply Apron 2E as a seed treatment at the rate of 4 fluid oz. per 100 lbs. of seed.

Note: Federal law requires that bags containing treated seeds shall be labeled with the following information: "This seed has been treated with metalaxyl fungicide. Do not use for feed, food or oil purposes. Store away from feeds and foodstuffs." Use with an EPA-approved dye that imparts an unnatural color to the seed.

STORAGE AND DISPOSAL

PESTICIDE DISPOSAL

Do not contaminate water, food, or feed by storage or disposal. Open dumping is prohibited. Wastes resulting from the use of this product are acutely toxic. Improper disposal of unused pesticide, spray mixture, or rinsate that cannot be used according to label instructions must be disposed of according to federal, state, or local procedures. For guidance in proper disposal methods, contact your State Pesticide or Environmental Control Agency, or the Hazardous Waste representative at the nearest EPA Regional Office.

CONTAINER DISPOSAL

Do not reuse empty container. Triple rinse (or equivalent), puncture and dispose of in a sanitary landfill, or by incineration, or by open burning, if allowed by state and local authorities. If burned, keep out of smoke.

Store at temperatures above 40°F. Crystals may form at lower storage temperatures. If this occurs, place the product in a warm room (68°F or above) and roll or shake the container at frequent intervals until all crystals are dissolved.

PRECAUTIONARY STATEMENTS

HAZARDS TO HUMANS AND DOMESTIC ANIMALS

WARNING

Causes eye injury. Do not get in eyes. Avoid contact with skin or clothing. Harmful if swallowed, inhaled, or absorbed through the skin. Avoid breathing vapors or spray mist.

Practical Treatment: In case of contact with eyes, immediately flush with plenty of water. Get medical attention. In case of contact with skin, wash thoroughly with soap and water. Get medical attention if irritation occurs. If swallowed, immediately contact a physician, hospital or local poison control center. Do not induce vomiting. If inhaled, move to fresh air. If not breathing, give artificial respiration, (preferably mouth to mouth) and get medical attention. **Notes to Physician:** If Apron 2E is ingested, lavage stomach. A slurry of activated charcoal in water can be left in the stomach. Give a saline laxative and supportive therapy.

ENVIRONMENTAL HAZARDS

Do not apply directly to water. Apply only as specified on this label. Do not apply when weather conditions favor drift from treated areas. Do not contaminate water by cleaning of equipment or disposal of wastes.

PHYSICAL AND CHEMICAL HAZARDS

Do not use or store near heat or open flame.

Apron® trademark of CIBA-GEIGY for metalaxyl
U.S. Patent No. 4,151,299

CGA 66L3F 115

APRON®25W

FUNGICIDE

A seed treatment chemical for control of systemic downy mildew, seed rot and damping-off diseases of certain crops

For use only by commercial seed treaters

ACTIVE INGREDIENT:
Metalaxyl:
N-(2,6-dimethylphenyl)-*N*-(methoxyacetyl) alanine methyl ester 25.0%
INERT INGREDIENTS: 75.0%
TOTAL: 100.0%

EPA Reg. No. 100-639 Control No. 52180

Keep Out of Reach of Children.

Caution

See additional precautionary statements on back of this bag.
See directions for use on back of this bag.

DIRECTIONS FOR USE AND CONDITIONS OF SALE AND WARRANTY

IMPORTANT: Read the entire **Directions for Use** and the **Conditions of Sale and Warranty** before using this product.

CONDITIONS OF SALE AND WARRANTY

The **Directions for Use** of this product reflect the opinion of experts based on field use and tests. The directions are believed to be reliable and should be followed carefully. However, it is impossible to eliminate all risks inherently associated with use of this product. Crop injury, ineffectiveness, or other unintended consequences may result because of such factors as weather conditions, presence of other materials, or the manner of use or application all of which are beyond the control of CIBA-GEIGY or the Seller. All such risks shall be assumed by the Buyer.

CIBA-GEIGY warrants that this product conforms to the chemical description on the label and is reasonably fit for the purposes referred to in the **Directions for Use** subject to the inherent risks referred to above. **CIBA-GEIGY makes no other express or implied warranty of Fitness or Merchantability or any other express or implied warranty. In no case shall CIBA-GEIGY or the Seller be liable for consequential, special, or indirect damages resulting from the use or handling of this product.** CIBA-GEIGY and the Seller offer this product, and the Buyer and user accept it, subject to the foregoing **Conditions of Sale and Warranty,** which may be varied only by agreement in writing signed by a duly authorized representative of CIBA-GEIGY.

GENERAL INFORMATION

Apron is a systemic fungicide seed dressing specifically for control of systemic downy mildews and diseases caused by soilborne *Pythium* and *Phytophthora* spp. For control of other soilborne diseases, such as *Rhizoctonia* spp., Apron 25W should be applied in combination with other registered seed dressing fungicides.

Note: Apron is a systemic fungicide having a specific mode of action. Apron could be subject to development of resistant strains of fungi. Development of resistance cannot be predicted. Therefore, CIBA-GEIGY cannot assume liability for crop damage resulting from resistant strains of fungi. Consult with your State Agricultural Experiment Station or Extension Service Specialist for guidance and ways to control any possible Apron resistant strains of fungi which may occur.

Apron 25W may be applied as a water-based slurry with other registered seed treatment insecticides and fungicides through standard slurry- or mist-type commercial seed treatment equipment.

DIRECTIONS FOR USE

It is a violation of federal law to use this product in a manner inconsistent with its labeling.

FAILURE TO FOLLOW DIRECTIONS FOR USE AND PRECAUTIONS ON THIS LABEL MAY RESULT IN CROP INJURY OR POOR DISEASE CONTROL.

Cotton	For Pythium seed rot and damping-off: Apply Apron 25W as a seed treatment to machine-delinted or acid-delinted cottonseed at the rate of 1-2 ounces per 100 lbs. of seed.
Forage Grasses	Grasses grown for hay grazing or silage, corn (fodder or silage), and small grain grown for hay grazing or silage. For Pythium damping-off control: Apply Apron 25W as a seed treatment at the rate of 1-2 ounces per 100 lbs. of seed.
Forage Legumes	Alfalfa, clover, lespedeza, beans (for forage), soybeans (for forage), soybean hay, peas (for forage), pea vine hay, cowpeas (for forage), cowpea hay, trefoil, vetch and velvet beans (for forage). For Pythium damping-off and early season Phytophthora control: Apply Apron 25W as a seed treatment at the rate of 2 ounces per 100 lbs. of seed.
Garden Beets (including sugar beets)	For control of Pythium damping-off: Apply Apron 25W as a seed treatment at the rate of 2 ounces per 100 lbs. of seed.
Grain Crops	Wheat, barley, rye, oats, millet, rice, corn (field corn, sweet corn, and popcorn), and buckwheat. For Pythium damping-off control: Apply Apron 25W as a seed treatment at the rate of 1-2 ounces per 100 lbs. of seed.
Ryegrass Bentgrass and Fescues for Turf	For Pythium damping-off control on golf courses and home lawns: Apply Apron 25W as a seed treatment at the rate of 1-2 ounces per 100 lbs. of seed.
Seed and Pod Vegetables	Black-eyed peas, cowpeas, dill, edible soybeans, field beans, field peas, garden peas, green beans, kidney beans, lima beans, navy beans, okra, peas, pole beans, snap beans, string beans, wax beans, and lentils. For Pythium damping-off and early season Phytophthora control: Apply Apron 25W as a seed treatment at the rate of 2 ounces per 100 lbs. of seed. For control of Pythium damping-off and systemic downy mildew of peas: Apply Apron 25W as a seed treatment at the rate of 5 ounces per 100 lbs. of seed.
Sorghum	For control of Pythium damping-off and systemic downy mildew: Apply Apron 25W as a seed treatment at the rate of 4 ounces per 100 lbs. of seed.
Soybeans	For Pythium damping-off and early season Phytophthora control: Apply Apron 25W as a seed treatment at the rate of 2 ounces per 100 lbs. of seed.
Sunflowers	For control of systemic downy mildew: Apply Apron 25W as a seed treatment at the rate of 4 ounces per 100 lbs. of seed.

Note: Federal law requires that bags containing treated seeds shall be labeled with the following information: "This seed has been treated with metalaxyl fungicide. Do not use for feed, food or oil purposes. Store away from feeds and foodstuffs." Use with an EPA-approved dye that imparts an unnatural color to the seed.

STORAGE AND DISPOSAL

PESTICIDE STORAGE AND DISPOSAL

Store in a dry place. Do not contaminate water, food or feed by storage, disposal, or cleaning of equipment. Open dumping is prohibited. Wastes resulting from the use of this product are acutely toxic. Improper disposal of unused pesticide, spray mixture, or rinsate is a violation of federal law. Pesticide, spray mixture, or rinsate that cannot be used according to label instructions must be disposed of according to federal, state or local procedures. For guidance in proper disposal methods, contact your State Pesticide or Environmental Control Agency, or the Hazardous Waste representative at the nearest EPA Regional Office.

CONTAINER DISPOSAL

Completely empty bag into application equipment. Dispose of empty bag in a sanitary landfill or by incineration, or by open burning, if allowed by state and local authorities. If burned, keep out of smoke.

For minor spills, leaks, etc., follow all precautions indicated on this label and clean up immediately. Take special care to avoid contamination of equipment and facilities during cleanup procedures and disposal of wastes. In the event of a major spill, fire or other emergency, call (919) 292-7100 day or night.

PRECAUTIONARY STATEMENTS

HAZARDS TO HUMANS AND DOMESTIC ANIMALS

CAUTION

Harmful if swallowed, inhaled, or absorbed through the skin. Causes eye irritation. Avoid contact with eyes, skin, or clothing. Avoid breathing dust or spray mist. Wash thoroughly with soap and water after handling. Remove and wash contaminated clothing before reuse. Do not store near food or feed.

Practical Treatment: In case of contact, flush eyes or skin with plenty of water. Get medical attention if irritation persists. If swallowed, give large quantities of water to drink and induce vomiting. Never give anything by mouth to an unconscious person. Call a physician immediately. If inhaled, move to fresh air; if not breathing, give artificial respiration and get medical attention.

ENVIRONMENTAL HAZARDS

Do not apply directly to water. Apply only as specified on this label. Do not apply when weather conditions favor drift from treated areas. Do not contaminate water by cleaning of equipment or disposal of wastes.

PHYSICAL AND CHEMICAL HAZARDS

Do not use or store near heat or open flame.

Apron® trademark of CIBA-GEIGY for metalaxyl

CGA 66L7C 115

AQUAZINE®

ALGICIDE

For control of algae and certain weeds in ponds

ACTIVE INGREDIENT:

Simazine: 2-chloro-4,6-bis (ethylamino)-*s*-triazine	80%
INERT INGREDIENTS:	20%
TOTAL:	100%

EPA Reg. No. 100-570

Aquazine® trademark of CIBA-GEIGY for simazine

Keep Out of Reach of Children.

Caution

See additional precautionary statements on panel at left.

See directions for use on panel at right.

1. **Broad spectrum algae control.**
2. Usually season-long control.
3. May be used in ponds containing fish.
4. No special equipment needed when applied to water.

DIRECTIONS FOR USE AND CONDITIONS OF SALE AND WARRANTY

IMPORTANT: Read the entire **Directions for Use** and the **Conditions of Sale and Warranty** before using this product.

CONDITIONS OF SALE AND WARRANTY

The **Directions for Use** of this product reflect the opinion of experts based on field use and tests. The directions are believed to be reliable and should be followed carefully. However, it is impossible to eliminate all risks inherently associated with use of this product. Ineffectiveness or other unintended consequences may result because of such factors as weather conditions, presence of other materials, or the manner of use or application all of which are beyond the control of CIBA-GEIGY or the Seller. All such risks shall be assumed by the Buyer.

CIBA-GEIGY warrants that this product conforms to the chemical description on the label and is reasonably fit for the purposes referred to in the **Directions for Use** subject to the inherent risks referred to above. **CIBA-GEIGY makes no other express or implied warranty of Fitness or Merchantability or any other express or implied warranty. In no case shall CIBA-GEIGY or the Seller be liable for consequential, special, or indirect damages resulting from the use or handling of this product.** CIBA-GEIGY and the Seller offer this product, and the Buyer and user accept it, subject to the foregoing **Conditions of Sale and Warranty,** which may be varied only by agreement in writing signed by a duly authorized representative of CIBA-GEIGY.

DIRECTIONS FOR USE

It is a violation of federal law to use this product in a manner inconsistent with its labeling.

GENERAL INFORMATION

Aquazine may be used in ponds containing water or it may be applied to the bottoms of drained ponds for control of several aquatic weeds and algae. Control of weeds is usually season-long. Algae may be controlled for one to three months or may require retreatment later in the season. In ponds containing fish, applications should not be made to water during prolonged periods of hot weather because oxygen levels would normally be low at these times and treatment could induce additional oxygen stress to fish, resulting occasionally in fish kills. If fish stress is observed, aerate the water by any method to reduce the chances of a fish kill. Use only in ponds which will have little or no outflow after treatment. Do not use Aquazine as a spot treatment.

Apply fertilizer 7-10 days after an Aquazine application in ponds that have a regular fertilization program. A few days delay in bloom of planktonic algae and a reduction in the extent of bloom may occur when Aquazine is used in ponds that receive fertilization.

APPLICATION TO WATER

When to Apply: Treat ponds after seasonal flow has ceased and early in the weed and algae growth period. Treat when 5-10% of the pond surface is covered with algae mats or scum and/or while submerged aquatic weeds are actively growing but before they reach the surface of the water. Usually, this will be between April 1 and May 15 in southern areas and between May 1 and June 15 in northern areas.

High water temperatures cause more rapid decay of dead weeds and algae. This leads to reduced oxygen content which can cause fish stress and/or death. Apply Aquazine before water temperature exceeds 75°F (22°C).

Do not treat ponds with heavy infestations of weeds and/or algae, such as occurs in mid and late summer, since decomposition of heavy vegetative growth greatly reduces oxygen content of water which can cause fish stress and/or death.

How to Apply: After determining the amount of Aquazine required to treat the entire pond, mix convenient portions such as 5 or 10 lbs. with enough water in a pail or bucket to form a thin paste or slurry. Toss the paste or slurry into the pond from the shoreline. Repeat this operation at several locations around the pond until the required amount has been applied. It is not necessary to use spray equipment or to evenly distribute Aquazine over the surface since it will dissolve in water and evenly distribute itself throughout the entire pond. Do not spill paste or slurry on pond banks as injury to vegetation will occur.

Rates of Application: The appropriate rate of Aquazine depends on the types of algae or weeds present and the extent of infestation. Rates are given in lbs. of Aquazine per acre foot of water. Determine the number of acre feet of water in a pond by multiplying the surface acres by the average depth in feet.

Algae Control: Most unicellular algae (algae which impart a green, yellow, red, or brown turbid color to the water) and many floating mat (pond moss) algae may be controlled with 1.7 to 3.4 lbs. of Aquazine per acre foot of water. If the algae infestation is light, use 1.7 lbs. If moderate, use 2.5 to 3.4 lbs. of Aquazine per acre foot.

Cladophora, Chara, Hydrodictyon (water net algae), Pithophora, and most other algae which form on the pond bottom (benthic) or which produce underwater plumes that break loose and float on the surface, require 4.25 lbs. of Aquazine per acre foot of water. Two applications at 4.25 lbs. per acre foot of water are usually required for control of Pithophora. If a second treatment is needed, apply at the first sign of algae buildup.

Algae control should occur within three to fourteen days after application, depending upon type of algae. Unicellular algae are usually controlled sooner than benthic filamentous algae.

Due to uncontrollable factors such as infestation with later appearing algae, some ponds may require re-treatment in mid or later summer. If re-treatment becomes necessary, apply at the first signs that algae (mats or moss) is building up. Do not treat in the fall or just before freezing is expected, since water oxygen levels may be lowered during ice cover.

Weed Control: Submerged weeds, including pond weeds (*Potamogeton* spp.), naiad (*Najas* spp.), watermilfoil (*Myriophyllum* spp.), and coontail (*Ceratophyllum* spp.) may be controlled with 3.4-6.8 lbs. of Aquazine per acre foot of water. Use 3-4 lbs. for light infestations and 5.1 to 6.8 lbs. per acre foot of water for moderate infestations. For control of fanwort (*Cabomba* spp.) use 8.5 lbs. of Aquazine per acre foot.

Control of submerged weeds, except coontail, occurs in 4-6 weeks. Coontail control may require 10 weeks.

Floating weeds, including duckweed (*Lemna* spp.) and watermeal (*Wolffia* spp.) may be controlled with 3.4-6.8 lbs. of Aquazine per acre foot of water. Within this range use lower rates for light infestations and higher rates for moderate infestations.

Where watermeal occurs alone or with other weeds, control is most effective with a split application. Apply one-half the recommended rate initially, followed in 3-4 weeks with a repeat application if control is not satisfactory.

Control of duckweed occurs in 1-5 weeks, and watermeal in 5-9 weeks.

APPLICATION TO BOTTOMS OF FISH HATCHERY, FISH REARING AND OTHER PONDS WITH DRAINING CAPABILITIES

Treat bottoms of drained ponds having a minimum average depth of two feet before regrowth of existing submerged weeds begins, or before new weeds emerge.

Allow at least 72 hours for Aquazine to become adsorbed to the soil before refilling the pond. Sprinkler irrigation or gentle rainfall after applying Aquazine will help move Aquazine into the surface soil, enhancing its effectiveness. Heavy rainfall during this period may erode treated soil, resulting in poor weed control.

Mixing Instructions: Make a slurry by adding Aquazine to a small amount of water in a separate container. Then pour the slurry into the spray tank partially filled with clean water, and add the rest of the water. Apply Aquazine with constant agitation in 20-40 gals. of water per acre. Do not spill the slurry on pond banks as injury to vegetation will occur.

Spray Equipment: Use conventional spray equipment with fan-type nozzles. Screens in nozzles as well as those in suction and in-line strainers, should be no finer than 50-mesh.

Use a pump with capacity to maintain 35-40 psi at the nozzles. If hydraulic agitation is used, the pump should also provide sufficient agitation in the tank to keep the mixture in suspension.

Wash sprayer thoroughly with clean water immediately after use.

Avoid application under conditions where uniform coverage cannot be obtained or where excessive spray drift may occur.

Algae Control: Apply 10-12.5 lbs. of Aquazine per acre as a broadcast spray for control of the algae indicated under the section, Application to Water. Use 10 lbs. per acre for expected light infestations and 12.5 lbs. per acre for expected heavy infestations.

Weed Control: Apply 10-15 lbs. of Aquazine per acre for control of American lotus, duckweeds, fanwort, naiad, pond weeds and watermilfoil.

Within the rate range, use the lower rate of Aquazine for soils with large amounts of sand or where there have been light infestations of weeds, or where algae dominates. Use the higher rate for soils high in silt and clay and where there have been heavy infestations of weeds.

Note: Do not apply more than 10 lbs of Aquazine per acre where striped bass fry or fingerlings will be cultured immediately following the application of Aquazine. Do not apply more than once each year to bottoms of drained ponds. Where heavy infestations of weeds and/or algae are treated, do not restock ponds with fish until adequate oxygen levels are restored.

USE OF FISH AND WATER

Fish taken from treated ponds may be used for food.

Treated ponds may be used for swimming 4 hours after application.

Water from treated ponds may not be used for irrigation or spraying of agricultural crops, lawns, or ornamental plantings, or for watering cattle, goats, hogs, horses, poultry, or sheep or for human consumption until 12 months following treatment.

Precaution: Do not treat ponds which have bordering trees with roots visibly extended into the water, since injury to these trees may occur.

STORAGE AND DISPOSAL

Store in a dry place. Do not contaminate water, food, or feed by storage or disposal. Improper disposal of unused pesticide, pesticide mixture, or rinse water is a violation of federal law. If these wastes cannot be used according to label instructions, contact your State Pesticide or Environmental Control Agency, or the Hazardous Waste representative at the nearest EPA Regional Office for guidance in proper disposal methods.

Completely empty bag into application equipment. Dispose of empty bag in a sanitary landfill or by incineration, or if allowed by state and local authorities, by burning. Stay out of smoke from burning bags.

For minor spills, leaks, etc., follow all precautions indicated on this label and clean up immediately. Take special care to avoid contamination of equipment and facilities during cleanup procedures and disposal of wastes. In the event of a major spill, fire or other emergency, call (919) 292-7100 day or night.

PRECAUTIONARY STATEMENTS

HAZARDS TO HUMANS AND DOMESTIC ANIMALS

CAUTION

Causes eye irritation. Harmful if inhaled or absorbed through the skin. Avoid breathing dust or spray mist. Avoid contact with eyes, skin, or clothing. Wash thoroughly with soap and water after handling. Remove and wash contaminated clothing before reuse.

First Aid: If in eyes, flush with plenty of water. Get medical attention if irritation persists. If on skin, wash with soap and water. If inhaled, remove victim to fresh air. If not breathing, give artificial respiration, preferably mouth to mouth; get medical attention.

ENVIRONMENTAL HAZARDS

Do not contaminate domestic, livestock, or irrigation water supplies, or lakes or streams. Do not contaminate water by cleaning of equipment or disposal of wastes.

Aquazine® trademark of CIBA-GEIGY for simazine

CGA 47L1G 034

ATRATOL®8P

PELLETED HERBICIDE

For control of weeds on noncrop land

ACTIVE INGREDIENTS:

Atrazine: 2-chloro-4-ethylamino-6-isopropylamino-*s*-triazine	7.6%
Atrazine-related compounds	0.4%
Sodium chlorate ($NaClO_3$)	40.0%
Sodium metaborate* ($Na_2B_2O_4$-$5H_2O$)	47.0%
INERT INGREDIENTS:	5.0%
TOTAL:	100.0%

EPA Reg. No. 100-475 Control No. 47235

*Equivalent to 4.58% boron expressed as elemental boron

Keep Out of Reach of Children.

Caution

See additional caution statements on back of bag.

See directions for use on back of bag.

DIRECTIONS FOR USE AND CONDITIONS OF SALE AND WARRANTY

IMPORTANT: Read the entire **Directions for Use** and the **Conditions of Sale and Warranty** before using this product.

CONDITIONS OF SALE AND WARRANTY

The **Directions for Use** of this product reflect the opinion of experts based on field use and tests. The directions are believed to be reliable and should be followed carefully. However, it is impossible to eliminate all risks inherently associated with use of this product. Crop injury, ineffectiveness, or other unintended consequences may result because of such factors as weather conditions, presence of other materials, or the manner of use or application all of which are beyond the control of CIBA-GEIGY or the Seller. All such risks shall be assumed by the Buyer.

CIBA-GEIGY warrants that this product conforms to the chemical description on the label and is reasonably fit for the purposes referred to in the **Directions for Use** subject to the inherent risks referred to above. **CIBA-GEIGY makes no other express or implied warranty of Fitness or Merchantability or any other express or implied warranty. In no case shall CIBA-GEIGY or the Seller be liable for consequential, special, or indirect damages resulting from the use or handling of this product.** CIBA-GEIGY and the Seller offer this product, and the Buyer and user accept it, subject to the foregoing **Conditions of Sale and Warranty,** which may be varied only by agreement in writing signed by a duly authorized representative of CIBA-GEIGY.

DIRECTIONS FOR USE

It is a violation of federal law to use this product in a manner inconsistent with its labeling.

Atratol 8P is a nonselective herbicide that can be applied before or after plant growth begins. Since Atratol enters plants through their roots, its effectiveness is dependent on rainfall to move the chemical into the root zone. Very dry soil conditions and lack of rainfall may result in poor weed control.

Important: Use only in areas where complete control of all vegetation is desired, such as industrial sites, rights-of-way, lumberyards, petroleum tank farms, around farm buildings, along fence lines, and similar areas. When applied to the soil, this product usually inhibits all plant growth for a year or more. It should not be used on land to be cropped, or where adjacent desirable trees, shrubs, or plants might be injured.

To control annual broadleaf weeds and grasses: Apply $\frac{1}{4}$-$\frac{1}{2}$ lb. per 100 sq. ft.

To control most perennial broadleaf weeds and grasses: Apply $\frac{1}{2}$-1 lb. per 100 sq. ft. Where deep-rooted perennials are present, use the 1 lb. rate.

The higher rates listed above should be used in areas having a long growing season or high rainfall. The higher rates will also provide longer residual control.

STORAGE AND DISPOSAL

Store in a dry place. Do not contaminate water, food, or feed by storage or disposal. Pesticide wastes are acutely hazardous. Improper disposal of unused pesticide is a violation of federal law. If these wastes cannot be used according to label instructions, contact your State Pesticide or Environmental Control Agency, or the Hazardous Waste representative at the nearest EPA Regional Office for guidance in proper disposal methods.

Completely empty bag into application equipment. Dispose of empty bag in a sanitary landfill or by incineration, or if allowed by state and local authorities, by burning. Stay out of smoke from burning bags.

For minor spills, leaks, etc., follow all precautions indicated on this label and clean up immediately. Take special care to avoid contamination of equipment and facilities during cleanup procedures and disposal of wastes. In the event of a major spill, fire or other emergency, call (919) 292-7100 day or night.

CAUTION

Keep out of reach of children.

Harmful if swallowed. Avoid contamination of food and feed. Avoid inhalation of dust. Avoid contact with eyes, skin, and clothing. Wash clothing after application. Wash out equipment after application. Do not use equipment to apply insecticides, fungicides, or fertilizers. Do not contaminate water intended for domestic or irrigation use.

Atratol® trademark of CIBA-GEIGY

CGA 20L1M 035

ATRATOL®90

HERBICIDE

Bare ground weed control with high suspensibility and easy wetting.

For nonselective weed control on industrial sites and noncrop areas

Do not use on agricultural lands

Atratol 90 is a water dispersible granule

ACTIVE INGREDIENTS:	
Atrazine: 2-chloro-4-ethylamino-6-isopropyl-amino-s-triazine	85.5%
Related compounds	4.5%
INERT INGREDIENTS:	10.0%
TOTAL	100.0%

EPA Reg. No. 100-622 EPA Est. 100-LA-1

Control No. 53160

Atratol® of CIBA-GEIGY

Keep Out of Reach of Children.

Caution

See additional precautionary statements on panel below.

See directions for use on panel below.

DIRECTIONS FOR USE AND CONDITIONS OF SALE AND WARRANTY

IMPORTANT: Read the entire **Directions for Use** and the **Conditions of Sale and Warranty** before using this product.

CONDITIONS OF SALE AND WARRANTY

The **Directions for Use** of this product reflect the opinion of experts based on field use and tests. The directions are believed to be reliable and should be followed carefully. However, it is impossible to eliminate all risks inherently associated with use of this product. Crop injury, ineffectiveness, or other unintended consequences may result because of such factors as weather conditions, presence of other materials, or the manner of use or application all of which are beyond the control of CIBA-GEIGY or the Seller. All such risks shall be assumed by the Buyer.

CIBA-GEIGY warrants that this product conforms to the chemical description on the label and is reasonably fit for the purposes referred to in the **Directions for Use** subject to the inherent risks referred to above. **CIBA-GEIGY makes no other express or implied warranty of Fitness or Merchantability or any other express or implied warranty. In no case shall CIBA-GEIGY or the Seller be liable for consequential, special, or indirect damages resulting from the use or handling of this product.** CIBA-GEIGY and the Seller offer this product, and the Buyer and user accept it, subject to the foregoing **Conditions of Sale and Warranty,** which may be varied only by agreement in writing signed by a duly authorized representative of CIBA-GEIGY.

DIRECTIONS FOR USE

It is a violation of federal law to use this product in a manner inconsistent with its labeling.

FAILURE TO FOLLOW THE DIRECTIONS FOR USE AND PRECAUTIONS ON THIS LABEL MAY RESULT IN POOR WEED CONTROL AND/OR INJURY TO DESIRABLE PLANTS.

GENERAL INFORMATION

Atratol 90 is a nonselective herbicide for use on noncrop areas. Atratol 90 can be applied before or after plant growth begins. Although Atratol 90 has some foliage activity, most of its activity is through root absorption. Therefore, moisture is required to move it into the root zone. Very dry soil conditions and lack of sufficient rainfall may result in poor weed control. Postemergence applications should be made when weeds are young and growing actively.

Following many years of continuous use of this product and chemically related products, biotypes of some of the weeds listed on this label have been reported which cannot be effectively controlled by this and related herbicides. Where this is known or suspected and weeds controlled by this product are expected to be present along with resistant biotypes, we recommend the use of this product in registered combinations or in sequence with other registered herbicides which are not triazines. If only resistant biotypes are expected to be present, use a registered non-triazine herbicide. Consult with your state Agricultural Extension Service for specific recommendations.

Use only in areas where complete control of broadleaf and grass vegetation is desired, such as (1) railroad, highway, and utility rights-of-way, (2) petroleum tank farms, (3) industrial sites, (4) lumberyards, (5) around warehouses and chemical plants, and (6) along industrial fence lines. When applied to the soil, this product usually inhibits plant growth for one year or more. It should not be used on land to be cropped or near adjacent desirable trees, shrubs or plants, or in greenhouses, as injury may occur. Do not allow livestock to graze treated areas.

Atratol 90 is noncorrosive to equipment.

MIXING PROCEDURES

(1) Be sure sprayer is clean and not contaminated with any material, as sprayer clogging may result. (2) Fill tank 1/4 full with clean water, and start agitation. (3) Be certain that the agitation system is working properly and creates a rippling or rolling action on the liquid surface. (4) Pour product directly from bag into tank. (5) Continue filling tank until 90% full. Increase agitation if necessary to maintain surface action. (6) Add tank mix herbicides. (7) Complete filling tank, maintaining sufficient agitation at all times to ensure surface action. This applies to both spray and nurse tanks. (8) Empty tank as completely as possible before refilling to prevent buildup of emulsifiable concentrate residue. Always maintain agitation to avoid separation. (9) If an emulsifiable concentrate film starts to build up in tank, drain and clean with strong detergent solution or solvent. (10) Immediately after use, clean sprayer thoroughly by flushing system with water containing a detergent.

APPLICATION EQUIPMENT

Use conventional ground sprayers equipped with nozzles that provide accurate and uniform application. Be certain that nozzles are uniformly spaced and the same size. Calibrate sprayer before use and recalibrate at the start of each season.

Use a pump with capacity to (1) maintain 35-40 psi at nozzles, (2) provide sufficient agitation in tank to keep mixture in suspension, and (3) to provide a minimum of 20% bypass at all times. Use centrifugal pumps which provide propeller shear action for dispersing and mixing this product. The pump should provide a minimum of 10 gals./minute/100 gals. tank size circulated through a correctly positioned sparger tube or jets.

Use screens to protect the pump and to prevent nozzles from clogging. Screens placed on suction side of the pump should be *16-mesh or coarser.* Do not place a screen in the recirculation line. Use 50-mesh or coarser screens between the pump and boom, and where required, at the nozzles. Check nozzle manufacturer's recommendations.

APPLICATION RATES — ATRATOL 90 APPLIED ALONE

Use sufficient water to assure thorough ground and plant coverage. Use at *least* 1 gallon of water for each pound of Atratol 90; more water may be necessary where weed growth is heavy.

To control annual broadleaf and grass weeds (such as barnyardgrass, cheatgrass, crabgrass, foxtail, lambsquarters, puncturevine, and ragweed) and biennials (such as turkey mullein): Broadcast 5.3-11.1 lbs. of Atratol 90 per acre.

To control most annual and many perennial broadleaf and grass weeds (such as bluegrass, burdock, Canada thistle, dogfennel, orchardgrass, plantain, purpletop, quackgrass, redtop, and smooth brome): Broadcast 11.1-22.2 lbs. of Atratol 90 per acre.

To control hard-to-kill biennial and perennial weeds (such as bullthistle and sowthistle): Broadcast 22.2-44.4 lbs. of Atratol 90 per acre.

For longer residual control in regions of high rainfall and a long growing season, apply 22.2-44.4 lbs of Atratol 90 per acre.

For small areas, 4 oz. per 1,000 sq. ft. is equivalent to 11.1 lbs. per acre.

TANK MIXTURES

Atratol 90 plus Princep® 80W or Princep Caliber® 90

Use only for long-term control of broadleaf and grass vegetation in areas listed in respective labels. Do not use on cropland, or near desirable trees, shrubs or other plants, or in greenhouses, as injury may occur.

This tank mixture has similar activity and moisture requirements as described in above General Information section, except that Princep provides preemergence control only.

For best results, apply to soil shortly before weed growth begins or apply postemergence to young, actively growing plants not over 6 inches tall.

Use sufficient water for thorough soil and plant coverage. Use at least 1 gal. of water for each lb. of Atratol 90 plus Princep 80W or Caliber 90 combination. More water may be needed for heavy weed growth. Maintain agitation during mixing and application.

Broadcast tank mixtures at the following rates to control weeds listed. Use the lower rates for light weed infestations and the higher rates for heavier infestations.

Use 2.9-5.6 lbs. of Atratol 90 plus 3.25-6.25 lbs. of Princep 80W or 2.9-5.6 lbs. of Princep Caliber 90 per acre to control weeds listed in the General Information and the several crop sections of the Princep 80W and Princep Caliber 90 labels.

The above rates will also control California burclover, *Orthocarpus purpurescens,* ripgut brome, smooth catsear, summer lupine, and turkey mullein.

Use 5.6 lbs. of Atratol 90 plus 6.25 lbs. of Princep 80W or 5.6 lbs. of Princep Caliber 90 per acre to control burdock, daisy fleabane, goldenrod, horseweed, little mallow, milkweed, orchardgrass, quackgrass, Russian thistle, shortpod mustard, wild barley, and wood sorrel.

Use 11.1 lbs. of Atratol 90 plus 12.5 lbs. of Princep 80W or 11.1 lbs. of Princep Caliber 90 per acre to control eveningprimrose, shepherds-needle, and tall fescue.

For small areas, 1/2 cup (4 oz.) of Atratol 90 or 2/3 cups (4.5 oz.) of Princep 80W or 1/2 cup (4 oz.) of Princep Caliber 90, per 1,000 sq. ft. is equivalent to 11.1 lbs. of Atratol 90 or 12.5 lbs. of Princep 80W or 11.1 lbs. of Princep Caliber 90 per acre, respectively.

ATRATOL 90 PLUS DIURON 80W

Use only for long-term control of broadleaf and grass vegetation in areas listed in respective labels. Do not use on cropland, or near desirable trees, shrubs or other plants, or in greenhouses, as injury may occur.

This tank mixture has activity and moisture requirements similar to that described in above **General Information** section and in the Diuron 80W label. For best results, apply to soil shortly before weed growth begins or apply to young, actively growing plants not over 6 inches tall. For increased contact activity, add a surfactant, such as Surfactant WK, at 2 qts. per 100 gals. of spray volume. Add surfactant as last ingredient to nearly full spray tank.

Use sufficient water for thorough ground and plant coverage. Use at least 1 gal. of water for each lb. of tank mixture. More water may be needed for heavy weed growth. Maintain agitation during mixing and application.

Broadcast tank mixtures at the following rates to control weeds listed. Use the lower rates for light weed infestations and the higher rates for heavier infestations.

Use 4.4-5.6 lbs. of Atratol 90 plus 5-6.25 lbs. of Diuron 80W, for example, Karmex® 80W, per acre to control weeds listed in crop and noncrop Directions for Use in the Diuron 80W label.

Use 5.6 lbs. of Atratol 90 plus 6.25 lbs. of Diuron 80W per acre to control broomsedge, buckhorn plantain, Carolina geranium, chicory, downy brome, eveningprimrose, fleabane, goldenrod, horseweed, lespedeza, little mallow, orchardgrass, Russian thistle, shortpod mustard, tall fescue, wild barley, and wood sorrel.

Use 11.1 lbs. of Atratol 90 plus 12.5 lbs. of Diuron 80W per acre to control alfalfa, black nightshade, leafy spurge, and shepherds-needle.

To determine compatibility of tank mixtures, thoroughly mix products in a small container at the ratio of 1 qt. of water to 1.75 oz. of Atratol 90 plus 2 oz. of Diuron 80W. Let stand for 5 minutes. If the mixture remains mixed, or can be remixed readily, it is compatible and can be sprayed.

For small areas, ½ cup (4 oz.) of Atratol 90 or 1.25 cups of Diuron 80W, such as Karmex 80W, per 1,000 sq. ft. is equivalent to 11.1 lbs. of Atratol 90 or 12.5 lbs. of Diuron 80W per acre.

ATRATOL 90 PLUS ONTRACK® 8E

Use only for control of annual grass weeds, or annual broadleaf and grass weeds on railroad rights-of-way.

Broadcast the tank mixture of Atratol 90 plus Ontrack 8E at rates of 5.3-11.1 lbs. plus 4 pts., respectively, in 50-100 gals. of water per acre to control barnyardgrass, crabgrass, fall panicum, giant foxtail, kochia, little barley, pigweed, ragweed, ripgut brome, Russian thistle, ryegrass, sprangletop, volunteer wheat, wild oats, and witchgrass. Use the lower rates in the rate range for light weed infestations and the higher rates for heavier infestations. Use no more than 4 pts. Ontrack 8E per acre in tank mixtures with Atratol.

For best results, apply immediately prior to weed emergence. The combination of Atratol 90 plus Ontrack 8E may also be applied after weeds emerge, but before they exceed 6 inches in height.

Precautions: Do not use near desirable trees, shrubs, plants or in greenhouses, or injury may occur.

ATRATOL 90 PLUS OUST®

Use this tank mixture to control the weeds listed below with either preemergence, early postemergence, or postemergence applications. Use the higher rates within the ranges given under conditions of heavy weed growth, on soil with over 2½ % organic matter, on areas with heavy decaying plant residues, or on high moisture areas such as along road edges or railroad shoulders. To improve wetting and/or contact activity on emerged weeds, add a nonionic surfactant such as duPont Surfactant WK or X-77® at 1 qt. per 100 gals. of spray. Do not apply by aircraft. Provide constant agitation during mixing and application to keep the mixture in suspension. Refer to the labels of both herbicides for specific noncrop sites and further directions, precautions, and limitations.

Preemergence to Early Postemergence: To control the weeds listed below apply 4.4-8.9 lbs. of Atratol 90 plus 3-5 oz. of Oust in a minimum of 25 gals. of water per acre.

Broadleaf Weeds (Up to 3-4" in Height or Diameter)

bouncingbet	pigweed
buckhorn plantain	prickly lettuce
burclover	puncturevine
Carolina geranium	purple starthistle
common chickweed	ragweed
common dandelion	Russian thistle
common speedwell	sowthistle
common yarrow	St. Johnswort
crimson clover	sunflower
dogfennel	sweet clover
filaree	tansymustard
hoary cress	tansy ragwort
horsetail	tumble mustard
kochia	turkey mullein
lambsquarters	vetch
little mallow	wild carrot
mustard	yellow rocket
ox-eye daisy	

Grasses (Up to 6-12" in Height)

Alta fescue	Italian ryegrass
annual bluegrass	johnsongrass (short-term control)
annual ryegrass	little barley
bahia grass	red brome
barnyardgrass	red fescue
cheatgrass	reed canarygrass
crabgrass	ripgut brome
downy brome	ryegrass
fall panicum	smooth brome
fescue	sprangletop
foxtails	witchgrass
Indiangrass	

Postemergence: To control actively growing weeds listed below apply 8.9-17.8 lbs. of Atratol 90 plus 6-12 oz. of Oust in a minimum of 25 gals. of water per acre.

Actively Growing Weeds

bedstraw	musk thistle
burdock	plantain
Canada thistle	poison ivy
curly dock	spanishneedles
dewberry	Virginia pepperweed
fiddleneck	wild blackberry
fleabane	johnsongrass (season-long control)
goldenrod	orchardgrass
hemlock	quackgrass
honeysuckle	redtop
Jerusalem artichoke	yellow nutsedge
kudzu	
mayweed	

Atratol 90 plus Arsenal®

Use this tank mixture to provide postemergence and residual control of the weeds listed in the various categories below. This tank mixture provides quicker postemergence weed control over that of Arsenal applied alone. Use the higher rates within the rate ranges given where heavy or well established infestations of weeds occur. Apply uniformly to the noncrop land area to be treated in 30-60 gals. of water per acre with a spray pressure of 25 to 30 psi. Refer to the Arsenal label for the addition of surfactant when spray volumes over 60 gallons per acre are applied. Also refer to the Atratol 90 and Arsenal lable for specific noncrop sites and further directions, precautions and limitations.

Annual Weeds

Apply 4.4-11.1 lbs. Atratol 90 plus 2-3 pts. of Arsenal per acre.

barnyardgrass (*Echinochloa crusgalli*)	lespedeza (*Lespedeza* spp.)
broadleaf signalgrass (*Brachiaria platyphylla*)	marestail (*Erigeron canadensis*)
carpetweed (*Mollugo verticillata*)	nightshade (*Solanum* spp.)
cheat (*Bromus* spp.)	pigweed (*Amaranthus* spp.)
common ragweed (*Ambrosia artemisifolia*)	plaintain (*Plantago* spp.)
curly dock (*Rumex crispus*)	puncturevine (*Tribulus terrestris*)
	purslane (*Portulaca* spp.)
	ragweed (*Ambrosia* spp.)

downy brome (*Bromus tectorum*)
fleabane (*Erigeron* spp.)
foxtails (*Setaria* spp.)
goldenrod (*Solidago* spp.)
hoary vervain (*Verbena stricta*)
jimsonweed (*Datura stramonium*)
kochia (*Kochia scoparia*)
lambsquarters (*Chenopodium alba*)
sicklepod (*Cassia obtusifolia*)
smartweed (*Polygonum* spp.)
sorrel (*Rumex* spp.)
sowthistle (*Sonchrus* spp.)
sunflower (*Helianthus annuus*)
velvetleaf (*Abutilon theophrasti*)
wild buckwheat (*Polygonum convolvulus*)
wild mustard (*Brassica kaber*)
wild oats (*Avena fatua*)
yellow woodsorrel (*Oxalis stricta*)

Apply 4.4-11.1 lbs. Atratol 90 plus 3-4 pts. of Arsenal per acre.

cocklebur (*Xanthium pennsylvanicum*)
crabgrass (*Digitaria* spp.)
crotolaria (*Crotolaria* spp.)
goosegrass (*Eleusine indica*)
morningglory (*Ipomoea* spp.)

Biennial/Perennial Weeds

Apply 5.3-11.1 lbs. Atratol 90 plus 2-3 pts. of Arsenal per acre.

dandelion (*Taraxacum officinale*)
field bindweed (*Convolvulus arvensis*)
guineagrass (*Panicum maximum*)
honeylocust (*Gleditsia triacanthos*)
johnsongrass (*Sorghum halepense*)
multiflora rose (*Rosa multiflora*)
ox-eye daisy (*Chrysanthemum leucathemum*)
paragrass (*Brachiaria mutica*)
quackgrass (*Agropyron repens*)
sandspur (*Cenchrus* spp.)
tall fescue (*Festuca arundinacea*)
vaseygrass (*Paspalum urvillei*)
wild carrot (*Daucus carota*)

Apply 5.3-11.1 lbs. Atratol 90 plus 3-4 pts of Arsenal per acre.

dewberry (*Rubus* spp.)
greenbriar (*Smilax* spp.)
honeysuckle (*Lonicera* spp.)
opossum grape (*Cissus sicyoides*)
poison ivy (*Rhus radicans*)
redvine (*Brunnichia cirrhosa*)
trumpetcreeper (*Campsis radicans*)
Virginia creeper (*Parthenocissus quinquefolia*)

Apply 5.3-11.1 lbs. Atratol 90 plus 4-6 pts. of Arsenal per acre.

bermudagrass (*Cynodon dactylon*)
blackberry (*Rubus* spp.)
Canada thistle (*Cirsium arvense*)
mulberry (*Morus* spp.)
sumac (*Rhus* spp.)

STORAGE AND DISPOSAL

Store in a dry place. Do not contaminate water, food, or feed by storage or disposal. Pesticide wastes may be toxic. Improper disposal of unused pesticide, spray mixture, or rinse water is a violation of federal law. If these wastes cannot be used according to label instructions, contact your State Pesticide or Environmental Control Agency, or the Hazardous Waste representative at the nearest EPA Regional Office for guidance in proper disposal methods.

Completely empty bag into application equipment. Dispose of empty bag in a sanitary landfill or by incineration, or if allowed by state and local authorities, by burning. Stay out of smoke from burning bags.

For minor spills, leaks, etc., follow all precautions indicated on this label and clean up immediately. Take special care to avoid contamination of equipment and facilities during cleanup procedures and disposal of wastes. In the event of a major spill, fire or other emergency, call (919) 292-7100 day or night.

PRECAUTIONARY STATEMENTS

HAZARDS TO HUMANS AND DOMESTIC ANIMALS

CAUTION

Harmful if swallowed, inhaled, or absorbed through the skin. Avoid contact with eyes, prolonged contact with skin, and inhalation of dust. Wash thoroughly after handling and before eating or smoking. Remove and wash contaminated clothing before reuse.

First Aid: In case of contact with skin, wash with plenty of soap and water; with eyes, flush with water for 15 minutes and get medical attention if irritation persists. In case of inhalation exposure, move from contaminated area.

ENVIRONMENTAL HAZARDS

Do not contaminate domestic or irrigation water supplies, or lakes, streams, or ponds. Do not contaminate water by cleaning of equipment or disposal of wastes. Do not apply when weather conditions favor drift from treated areas.

Arsenal® trademark of American Cyanamid Company
Atratol® trademark of CIBA-GEIGY
Caliber® trademark of CIBA-GEIGY
Ontrack® trademark of CIBA-GEIGY
Oust® trademark of E. I. duPont de Nemours and Co., Inc.
Princep® trademark of CIBA-GEIGY for simazine
Karmex® trademark of E. I. duPont de Nemours and Co., Inc. for diuron
X-77® trademark of KALO Agricultural Chemicals, Inc.

CGA 20L6C 065

BICEP®
HERBICIDE

For weed control in corn and grain or forage sorghum

ACTIVE INGREDIENTS:
Atrazine: 2-chloro-4-ethylamino-6-isopropylamino-s-triazine 20.8%
Atrazine related compounds 1.1%
Metolachlor: 2-chloro-*N*-(2-ethyl-6-methylphenyl)-*N*-(2-methoxy-1-methylethyl) acetamide 27.5%
INERT INGREDIENTS: 50.6%
TOTAL: 100.0%

EPA Reg. No. 100-590 EPA Est. 100-LA-1

Bicep® trademark of CIBA–GEIGY
U.S. Patent No. 3,937,730 (metolachlor)

Bicep contains 4.5 lbs. active ingredients per gallon.

Keep Out of Reach of Children.
Caution

See additional precautionary statements at end of label booklet.
See directions for use inside booklet

DIRECTIONS FOR USE AND CONDITIONS OF SALE AND WARRANTY

IMPORTANT: Read the entire **Directions for Use** and the **Conditions of Sale and Warranty** before using this product.

CONDITIONS OF SALE AND WARRANTY

The **Directions for Use** of this product reflect the opinion of experts based on field use and tests. The directions are believed to be reliable and should be followed carefully. However, it is impossible to eliminate all risks inherently associated with use of this product. Crop injury, ineffectiveness, or other unintended consequences may result because of such factors as weather conditions, presence of other materials, or the manner of use or application all of which are beyond the control of CIBA–GEIGY or the Seller. All such risks shall be assumed by the Buyer.

CIBA–GEIGY warrants that this product conforms to the chemical description on the label and is reasonably fit for the purposes referred to in the **Directions for Use** subject to the inherent risks referred to above. **CIBA–GEIGY makes no other express or implied warranty of Fitness or Merchantability or any other express or implied warranty. In no case shall CIBA–GEIGY or the Seller be liable for consequential, special, or indirect damages resulting from the use or handling of this product.** CIBA–GEIGY and the Seller offer this product, and the Buyer and user accept it, subject to the foregoing **Conditions of Sale and Warranty,** which may be varied only by agreement in writing signed by a duly authorized representative of CIBA–GEIGY.

DIRECTIONS FOR USE

It is a violation of federal law to use this product in a manner inconsistent with its labeling.

FAILURE TO FOLLOW ALL PRECAUTIONS ON THIS LABEL MAY RESULT IN POOR WEED CONTROL, CROP INJURY, OR ILLEGAL RESIDUES.

Do not apply this product in such a manner as to directly or through drift expose workers or other persons, except those knowingly involved in the application. The area being treated must be vacated by unprotected persons.

REENTRY STATEMENT

Do not enter treated areas without protective clothing until sprays have dried.

Because certain states may require more restrictive re-entry intervals for various crops treated with this product, consult your State Department of Agriculture for further information.

Written or oral warnings must be given to workers who are expected to be in a treated area or in an area about to be treated with this product. Oral warnings must be given which inform workers of areas or fields that may not be entered without specific protective clothing until sprays have dried, and appropriate actions to take in case of accidental exposure as described under **Precautionary Statements** on this label. When oral warnings are given, warnings shall be given in a language customarily understood by workers. Oral warnings must be given if there is reason to believe that written warnings cannot be understood by workers. Written warnings must include the following information: "CAUTION. Area treated with Bicep on *(date of application)*. Do not enter without appropriate protective clothing until sprays have dried. In case of accidental exposure, flush eyes or skin with plenty of water. Call a physician if irritation persists. Remove and wash contaminated clothing before reuse."

Bicep is a selective herbicide recommended as a preplant surface-applied, preplant incorporated, preemergence, or early postemergence treatment for control of most annual grasses and broadleaf weeds in corn (field corn, silage corn, sweet corn, or popcorn) and as a preplant surface-applied, preplant incorporated, or preemergence treatment for control of most annual grasses and broadleaf weeds in grain or forage sorghum provided the sorghum seed has been properly treated by the seed company with Concep®. This product may be tank mixed with Bladex® (4L, 80W, 90WDG), Dual® 8E, AAtrex® 4L (4LC, 80W, Nine-O®), or Princep® 4L (80W or Caliber® 90) for weed control in conventional tillage corn. This product may also be tank mixed with either paraquat or Roundup® alone or in combination with AAtrex, Bladex, Dual or Princep, in minimum-tillage or no-tillage corn, or tank mixed with either paraquat or Roundup, in minimum-tillage or no tillage sorghum.

Precautions: *1) If sorghum seed is not properly pretreated with Concep, Bicep will severely injure the crop. 2) Injury may occur to corn or sorghum following the use of Bicep under abnormally high soil moisture conditions during early development of the crop.*

Bicep alone or in tank mixture with AAtrex, Bladex, Dual, or Princep may be applied preplant incorporated, preplant surface, or preemergence on corn in water or fluid fertilizer. Bicep may be applied in tank mix combination with paraquat or Roundup with or without the above herbicides preplant surface or preemergence to corn. Apply the early post-emergence treatment on corn in water only. Bicep alone may also be applied on sorghum preplant incorporated, preplant surface, or preemergence in water or in fluid fertilizer.

Bicep may be applied in water by aircraft. Applications in fluid fertilizer should be only by ground equipment.

To avoid spray drift, do not apply under windy conditions. Avoid spray overlap, as crop injury may result.

This product may also be applied in irrigation water through center pivot irrigation systems.

Dry weather following preemergence application of Bicep or a tank mixture may reduce effectiveness. Cultivate if weeds develop in conventional tillage corn or sorghum.

Observe all precautions and limitations on the label of each product used in tank mixtures.

Thoroughly clean sprayer prior to use. Do not use a sprayer contaminated with any other materials, or crop damage or sprayer clogging may result.

Note: (Bicep Alone and in Tank Mixtures): Use only on corn and grain or forage sorghum.

Mixing Instructions

Shake well before using. Bicep is a liquid that may be mixed with water or fluid fertilizer and applied as a spray. Bicep may also be sprayed onto dry bulk granular fertilizer and applied with the granular fertilizer.

DRY BULK GRANULAR FERTILIZERS

Many dry bulk granular fertilizers may be impregnated or coated with Bicep and used to control weeds in corn or Concep-treated sorghum.

All individual state regulations relating to dry bulk granular fertilizer blending, registration, labeling, and application are the responsibility of the individual and/or company selling the herbicide/fertilizer mixture.

Prepare the herbicide/fertilizer mixture by using any closed drum, belt, ribbon, or other commonly used dry bulk fertilizer blender. Nozzles used to spray Bicep onto the fertilizer must be placed to provide uniform spray coverage.

If the herbicide/fertilizer mixture is too wet, use a highly absorptive powder, such as Microcel E (Johns-Manville Products Corporation), diatomaceous earth or finely powdered clay, to obtain a dry free-flowing mixture. Add the absorptive powder separately and uniformly to the herbicide/fertilizer mixture and blend to form a suitable free-flowing mixture. Generally, less than 2% by weight of absorptive powder will be needed.

Calculate the amount of Bicep to be used by the following:

$$\frac{2{,}000}{\text{lbs. of fertilizer per acre}} \times \text{qts. of Bicep per acre} = \text{qts. of Bicep per ton of fertilizer}$$

Precautions: *To avoid potential for explosion, 1) Do not impregnate Bicep on ammonium nitrate, potassium nitrate, or sodium nitrate either alone or in blends with other fertilizers. 2) Do not combine Bicep with a single superphosphate (0-20-0) or treble superphosphate (0-46-0). 3) Do not use Bicep on straight limestone, since absorption will not be achieved. Fertilizer blends containing limestone can be impregnated.*

Application

Apply 200-450 pounds of the herbicide/fertilizer mixture per acre. For best results, apply the mixture uniformly to the soil with properly calibrated equipment immediately after blending. Uniform application of the herbicide/fertilizer mixture is essential in order to prevent possible crop injury or injury to subsequent rotational crops. Non-uniform application may also result in unsatisfactory weed control. In areas where conventional tillage is practiced, a shallow incorporation of the mixture into the soil is recommended to obtain satisfactory weed control. On fine or medium-textured soils in areas where soil incorporation is not planned, i.e., reduced tillage situations or in some conventional till situations, make applications approximately 30 days before planting to allow moisture to move the herbicide/fertilizer mixture into the soil. On coarse-textured soils, make applications approximately 14 days prior to planting.

Precautions: *1) To help avoid rotational crop injury, make applications as early as possible since Bicep impregnated onto dry bulk granular fertilizers can be expected to last longer in the soil than when Bicep is applied as a spray in water or fluid fertilizer. 2) Do not use on crops where bedding occurs, or crop injury may occur.*

APPLICATION IN WATER OR FLUID FERTILIZERS

Bicep Alone: Fill the spray tank one-half to three-fourths full with water or fluid fertilizer, add the proper amount of Bicep, then add the rest of the water or fluid fertilizer. Provide sufficient agitation during mixing and application to maintain a uniform suspension.

Tank Mixtures: Fill the spray tank one-half to three-fourths full with water or fluid fertilizer, add the proper amount of Bicep, then add AAtrex or Princep, Dual, paraquat or Roundup, and finally add the rest of the water or fluid fertilizer. When Bladex is used in the tank mixture, add it before Bicep unless otherwise specified. (See Bladex section under Bicep Combinations-Corn for further mixing instructions.) Provide sufficient agitation during mixing and application to maintain a uniform suspension.

Compatibility Test: Check the compatibility of Bicep and tank mixtures in fluid fertilizer by mixing proportionate quantities in a small container, as described below, before mixing in the spray tank. Nitrogen solutions or complete fluid fertilizers may replace all or part of the water in the spray. Since liquid fertilizers can vary, even within the same analysis, **always check compatibility each time before reuse.** Be especially careful when using **complete** suspension or fluid fertilizers as serious compatibility problems are more apt to occur. Commercial application equipment may improve compatibility in some instances. The following test assumes a spray volume of 25 gals. per acre. For other spray volumes, make appropriate changes in the ingredients. Check compatibility using this procedure:

1. Add 1 pint of fertilizer to each of 2 one-quart jars with tight lids.
2. To **one** of the jars add $\frac{1}{4}$ tsp. or 1.2 milliliters of a compatibility agent approved for this use, such as Compex® or Unite® ($\frac{1}{4}$ tsp. is equivalent to 2 pts. per 100 gals. spray). Shake or stir gently to mix.
3. To **both** jars add the appropriate amount of herbicide(s). If more than one herbicide is used, add them separately with dry herbicides first, flowables next, and emulsifiable concentrates last. After each addition, shake or stir gently to thoroughly mix. The appropriate amount of herbicides for this test follows:

Dry herbicides:	For each pound to be applied per acre add 1.4 teaspoons to each jar.
Liquid herbicides:	For each pint to be applied per acre add 0.5 teaspoon or 2.5 milliliters to each jar.

4. After adding all ingredients, put lids on and tighten, and invert each jar ten times to mix. Let the mixtures stand 15 minutes and then look for separation, large flakes, precipitates, gels, heavy oily film on the jar, or other signs of incompatibility. Determine if the compatibility agent is needed in the spray mixture by comparing the two jars. If either mixture separates, but can be remixed readily, the mixture can be sprayed as long as good agitation is used. If the mixtures are incompatible, test the following methods of improving compatibility: (A) slurry the dry herbicide(s) in water before addition, or (B) add $\frac{1}{2}$ of the compatibility agent to the fertilizer and the other $\frac{1}{2}$ to the emulsifiable concentrate or flowable herbicide before addition to the mixture. If still incompatible, do not use the mixture.

APPLICATION PROCEDURES

Ground Application: Use sprayers that provide accurate and uniform application. Screens in nozzles and in suction and in-line strainers should be no finer than 50-mesh. Use a pump with capacity to (1) maintain 35-40 psi at the nozzles, and (2) provide sufficient agitation in tank to keep mixture in suspension. Unless otherwise specified, use a minimum of 10 gals. of spray mixture per acre. Rinse sprayer thoroughly with clean water immediately after use.

For band applications, calculate amount to be applied per acre as follows:

$$\frac{\text{band width in inches}}{\text{row width in inches}} \times \text{broadcast rate per acre} = \text{amount needed per acre of field}$$

Aerial Application (for Bicep alone): Use aerial application only where broadcast applications are specified. Apply a minimum of 1 gal. of water for each 1 gal. of this product applied per acre, but for rates below 1 gal./acre, use in sufficient water to equal 2 gals./acre of total spray. Avoid applications under conditions where uniform coverage cannot be obtained or where excessive spray drift may occur. In order to assure that spray will be controllable within the target area when used according to label directions, make applications at a maximum height of 10 ft., using low drift nozzles at a maximum pressure of 40 psi, and restrict application to periods when wind speed does not exceed 10 mph. To assure that spray will not adversely affect adjacent sensitive nontarget plants, apply Bicep by aircraft at a minimum upwind distance of 400 ft. from sensitive plants.

Avoid application to humans or animals. Flagmen and loaders should avoid inhalation of spray mist and prolonged contact with skin, and should wash thoroughly before eating and at the end of each day's operation.

Center Pivot Irrigation Application: Bicep alone or in mixtures with other herbicides which are registered for center pivot application may be applied in irrigation water preemergence (after planting but before weeds or crop emerge) at rates recommended on this label. Use only center pivot systems that apply water uniformly. Prepare a mixture with a minimum of 1 part of water to 1 part herbicide(s), and inject this mixture into the center pivot system using a positive displacement pump. Injecting a larger volume of a more dilute mixture per hour will usually provide more accurate calibration of metering equipment. Maintain sufficient agitation to keep the herbicide in suspension. Apply in $\frac{1}{2}$-1 inch of water. Use the lower water volume ($\frac{1}{2}$ inch) on coarser textured soils and the higher volume (1 inch) on finer textured soils. More than 1 inch of water at application may reduce weed control by moving the herbicide below the effective zone in the soil. Refer to supplemental literature published by CIBA–GEIGY for more information on calibration.

Precautions for center pivot applications: *1) Apply only through irrigation systems containing anti-siphon and check valves to prevent contamination of well during shutdown and overflow of solution tank. 2) Inject ahead of any right angle turn in the main line to insure adequate mixing. 3) Chemical injection pumps and water pumps must have interlocking controls to insure simultaneous shutoff. 4) Application when drift may occur from windy conditions, or when system joints and connections are leaking, or when nozzles are not providing uniform distribution may cause crop injury. 5) Where sprinkler distribution patterns do not overlap sufficiently, unacceptable weed control may result. Where sprinkler distribution patterns overlap excessively, crop injury may result.*

Within rate ranges in all tables on this label, use the lower rate on soil relatively coarse-textured or low in organic matter; use the higher rate on soil relatively fine-textured or high in organic matter.

BICEP APPLIED ALONE — CORN, GRAIN SORGHUM, OR FORAGE SORGHUM

PREPLANT SURFACE-APPLIED, PREPLANT INCORPORATED, OR PREEMERGENCE

Weeds Controlled		*Weeds Partially Controlled**
barnyardgrass (watergrass)	black nightshade	sandbur
browntop panicum	carpetweed	seedling johnsongrass
crabgrass	cocklebur	shattercane
	common purslane	volunteer sorghum
	Florida pusley	

crowfootgrass	galinsoga
fall panicum	hairy nightshade
giant foxtail	jimsonweed
goosegrass	lambsquarters
green foxtail	morningglory
red rice	pigweed
signalgrass (Brachiaria)	ragweed
	smartweed
southwestern cupgrass	velvetleaf
witchgrass	
yellow foxtail	
yellow nutsedge	

* Control of these weeds can be erratic due partially to variable weather conditions. Control may be improved by following these suggested procedures:

1. **Thoroughly till moist soil** to destroy germinating and emerged weeds. If Bicep is to be applied preplant incorporated, this tillage may be used to incorporate Bicep if uniform 2 inch incorporation is achieved as recommended under **Application Procedures.**
2. Plant crop into moist soil **immediately after tillage.** If Bicep is to be used preemergence, apply at planting or immediately after planting.
3. If available, **sprinkler irrigate** within 2 days after application. Apply ½-1 inch of water. Use lower water volume (½ inch) on coarse-textured soils. Also, refer to the section on **Center Pivot Irrigation Application** for this method of applying Bicep.
4. If irrigation is not possible and rain does not occur within 2 days after planting and application, weed control may be decreased. Under these conditions, a uniform, shallow cultivation is recommended as soon as weeds emerge.

APPLICATION TIMING

Preplant Surface-Applied (Corn): Use on medium- and fine-textured soils with minimum-tillage or no-tillage systems in CO, IN, IL, IA, KS, KY, MN, MO, MT, NE, ND, SD, WI, and WY. Apply ⅔ the recommended rate of Bicep (3.2-4 qts./A on *medium* soils and 4-5 qts./A on *fine* soils) as a split treatment 30-45 days before planting and the remainder at planting. Applications made less than 30 days prior to planting may be as either a split or single treatment. Use the lower rate for light expected weed infestations and the higher rate for heavy expected weed infestations. On *coarse-textured* soils, apply 3.2 qts./A not more than 2 weeks prior to planting. The above procedure also may be followed if AAtrex or Dual 8E or Princep is used in tank mixtures with Bicep. If weeds are present at the time of treatment, apply in a tank mixture combination with a contact herbicide (for example, paraquat or Roundup). Observe directions for use, precautions, and restrictions on the label of the contact herbicide.

Note: To the extent possible, do not move treated soil out of the row or move untreated soil to the surface during planting, or weed control will be diminished.

Preplant Surface-Applied (Sorghum-Seed Treated with Concep): For minimum tillage and no-tillage systems only, Bicep may be applied up to 45 days before planting grain sorghum in IL, IA, eastern KS, MO, NE, and SD. Use only split applications for treatments made 30 to 45 days before planting with ⅔ of the recommended rate applied initially and the remaining ⅓ at planting. On medium soils with greater than 1.5% organic matter, and on fine soils with less than 1.5% organic matter, apply 3.2-3.6 qts./A of Bicep; on fine soils with greater than 1.5% organic matter, apply 3.6-4 qts./A of Bicep. Applications made less than 30 days prior to planting may be as either a split or a single application.

If weeds are present at the time of treatment, apply in a tank mixture combination with a contact herbicide (for example, paraquat or Roundup). Observe directions for use, precautions, and restrictions on the label of the contact herbicide. Under dry conditions, irrigation after application is recommended to move Bicep into the soil.

Note: To the extent possible, do not move treated soil out of the row or move untreated soil to the surface during planting, or weed control will be diminished. Do not use on coarse soils. Do not use on medium soils with less than 1.5% organic matter.

Preplant Incorporated or Preemergence (Corn or Sorghum-Seed Treated with Concep): Apply Bicep preplant incorporated or preemergence using the appropriate rates from Table 1 for corn, or from Table 2 for sorghum.

Preplant Incorporated: Apply to the soil and incorporate into the top 2 inches of the soil within 14 days prior to planting using a finishing disk harrow, rolling cultivator, or similar implement capable of providing uniform 2-inch incorporation. Use preplant incorporated if furrow irrigation is used or when a period of dry weather after application is expected. If crop is to be planted on beds, apply and incorporate after bed formation.

Preemergence: Apply to the soil surface at planting (behind the planter), or after planting but before weeds or crop emerge.

Table 1: Bicep Preplant Surface-Applied, Preplant Incorporated, or Preemergence—Corn (Field Corn, Silage Corn, Sweet Corn, Popcorn)

	Broadcast rate per acre	
Soil texture	**Less than 3% organic matter**	**3% organic matter or greater**
COARSE: Sand, loamy sand, sandy loam	2 qts.	2.4 qts.
MEDIUM: Loam, silt loam, silt	2.4 qts.	3.2 qts.
FINE: Silty clay loam, sandy clay loam, silty clay, sandy clay, clay loam, clay	3.2 qts.	3.2-4 qts.*
Muck or peat soils	DO NOT USE	

* For cocklebur, yellow nutsedge, and velvetleaf control on fine-textured soils above 3% organic matter: Apply 4 qts. of Bicep per acre.

Table 2: Bicep—Grain or Forage Sorghum † (Seed treated with Concep)

Soil texture	Organic matter	Broadcast rate per acre
COARSE: Sand, loamy sand, sandy loam	any level	DO NOT USE
MEDIUM AND FINE: Loam, silt loam, silt, sandy clay loam,	less than 1%	DO NOT USE
clay loam, silty clay loam,	1-1.5%	2.4 qts.
sandy clay, silty clay, clay	more than 1.5%	2.8-3.2 qts.

† Do not use in NM, OK, or TX except in northeast OK and Texas Gulf Coast and Blacklands areas. Do not apply preplant incorporated in AZ or the Imperial Valley of CA.

Precautions: *To avoid possible crop injury, 1) Do not apply Bicep on highly alkaline soils or on eroded areas where calcareous subsoils are exposed. 2) Do not apply Bicep when sorghum is planted in deep furrows because heavy rains following application can cause excessive concentrations of herbicide in the furrow. 3) Do not apply Bicep preplant surface, preplant incorporated, or preemergence, and an at-planting systemic insecticide applied in-furrow. 4) Do not apply to sorghum grown under dry mulch tillage. 5) In addition, sorghum growing under stress caused by minor element deficiency may be injured by Bicep.*

Rotational Crops: 1) If treated crop is lost due to poor germination, hail, flood, insects, etc., corn may be replanted immediately or sorghum may be replanted immediately provided the seed has been properly treated with Concep. Do not make a second broadcast application. If the original application was banded and the second crop is planted in the untreated row middles, a second banded treatment may be applied. 2) Corn, sorghum, soybeans, cotton, or peanuts may be planted the spring following treatment. Do not graze or feed forage or fodder from cotton to livestock. Injury may occur to soybeans planted in north-central and northwest Iowa, south-central and southwest Minnesota, northeast Nebraska, and southeast South Dakota the year following application on soils having a calcareous surface layer. 3) If applied after June 10, do not rotate with crops other than corn or sorghum the next year, or crop injury may occur. 4) In the High Plains and Intermountain areas of the West where rainfall is sparse and erratic or where irrigation is required, use only when corn or sorghum is to be planted the following year, or a crop of untreated corn or sorghum is to precede other rotational crops. 5) Small grains may be planted 15 months following treatment. 6) All other crops may be planted 18 months after application.

POSTEMERGENCE — CORN

Weeds Controlled		*Weeds Partially Controlled*
barnyardgrass (watergrass)	jimsonweed	cocklebur
crabgrass	kochia	morningglory
crowfootgrass	lambsquarters	yellow nutsedge
fall panicum	mustard	
giant foxtail	pigweed	
green foxtail	prickly sida	
yellow foxtail	purslane	
	ragweed	
	smartweed	
	velvetleaf	

Application: Apply early postemergence using the appropriate rate from Table 3. Apply this treatment before grass and broadleaf weeds pass the 2-leaf stage and before corn exceeds 5 inches in height. Application to weeds larger than the 2-leaf stage will generally result in unsatisfactory control. Occasional corn leaf burn may result, but this should not affect later growth or yield. Do not apply postemergence in fluid fertilizer or severe crop injury may occur.

Table 3: Postemergence—Corn

Soil texture	Broadcast rate per acre
COARSE: Sand, loamy sand, sandy loam	2.4 qts.
MEDIUM: Loam, silt loam, silt	3.2 qts.
FINE: Sandy clay loam, silty clay loam, clay loam, sandy clay, silty clay, clay	3.2-4 qts.*

* For better control of cocklebur, morningglory, velvetleaf, and yellow nutsedge on fine-textured soils above 3% organic matter. apply 4 qts. of Bicep per acre.

Rotational Crops: Follow the preceding crop rotation instructions for Bicep Preplant Surface-Applied Preplant Incorporated, or Preemergence.

BICEP COMBINATIONS — CORN

Tank Mixture with AAtrex, Princep, Dual 8E, or Bladex® — Conventional Tillage

AAtrex (4L, 4LC, 80W, or Nine-O): Add up to 1 qt. of AAtrex 4L or 4LC (1.25 lbs. 80W or 1.1 lbs. Nine-O) per acre to the rate of Bicep recommended in Table 1 in the southeastern U.S. where high rainfall can shorten the duration of control of broadleaf weeds, and in all areas where heavy infestations of cocklebur, morningglory, velvetleaf, or other broadleaf weeds claimed are expected.

Dual 8E: Add up to ½ pt. of Dual 8E per acre to the rate of Bicep recommended in Table 1 when heavy infestations of yellow nutsedge, sandbur, or seedling johnsongrass are expected.

Princep (4L, 80W, or Caliber 90): Add up to 1 qt. of Princep 4L (1.25 lbs. 80W or 1.1 lbs. Caliber 90) per acre to the rate of Bicep recommended in Table 1 in the northeastern U.S. where heavy infestations of crabgrass or fall panicum are expected.

Bladex (4L, 80W, or 90WDG) [Field Corn and Silage Corn Only]: Add up to 2 qts. of Bladex 4L (2.5 lbs. 80W or 2.2 lbs. 90WDG) per acre. When Bladex is added, the rate of Bicep suggested in Table 1 should be reduced by not more than 25% for a given soil texture in the Clarion-Nicollet-Webster soil association in northern Iowa and southern Minnesota, or in other areas where soybean rotational concerns exist.

Note: Check the compatibility of Bicep tank mixtures with Bladex before mixing in spray tank by using the procedure described under **Application in Water or Fluid Fertilizers.** Always use Unite (or an equivalent compatibility agent) at 2 pts./100 gals. when using Bladex 80W or Bladex 90WDG in tank mixtures with Bicep. Compatibility testing is especially critical when using Bladex 80W or Bladex 90WDG combinations. Always add Bladex to the spray tank before Bicep, regardless of which Bladex formulation is used.

TANK MIXTURES OF BICEP ALONE OR BICEP PLUS AATREX, BLADEX, DUAL 8E, OR PRINCEP, WITH PARAQUAT OR ROUNDUP FOR MINIMUM-TILLAGE OR NO-TILLAGE SYSTEMS

In minimum-tillage or no-tillage systems where corn is planted directly into a cover crop, stale seedbed, established sod, or previous crop residues, the contact herbicides paraquat or Roundup may be tank mixed with Bicep alone or with Bicep plus AAtrex, Bladex, Dual 8E, or Princep. When used as directed, the paraquat portion of the tank mixture controls most emerged annual weeds and suppresses many perennial weeds. Roundup combinations will control emerged annual and perennial weeds when applied as directed on the Roundup label. The Bicep portion of the tank mixture provides preemergence control of the weeds listed on this label in the Bicep alone section for corn. The addition of AAtrex, Bladex, Dual 8E, or Princep offers the advantage indicated for each under "Conventional Tillage" above.

Application: Apply before, during, or after planting but before corn emerges, at the appropriate rate in Table 4. Up to 1 qt. of AAtrex 4L/4LC (1.25 lbs 80W or 1.1 lbs. Nine-O), or 2 qts. of Bladex 4L (2.5 lbs. 80W or 2.2 lbs. 90WDG), or ½ pt. of Dual 8E, or 1 qt. of Princep 4L (1.25 lbs. 80W or 1.1 lbs. Caliber 90) per acre may be added to the rate of Bicep recommended in Table 4. Add paraquat or Roundup at the following broadcast rates:

Paraquat: 1-2 pts.* per acre plus 8 oz. of X-77® Spreader per 100 gals. of spray mixture. Use the lower rate for control of annual weeds less than 4 inches tall and the higher rate for weeds 4-6 inches tall. This treatment will not consistently control weeds taller than 6 inches.

Roundup: 1.5 qts. per acre for control of existing annual weeds, or 2-4 qts. per acre for existing perennial weeds. See the Roundup label for weeds controlled and recommended rates for specific weeds.

* Based on a product containing 2 lbs. ai/gal.

Apply in 20-60 gals. of water per acre with conventional spray equipment.

TANK MIXTURES OF BICEP ALONE OR BICEP PLUS AATREX OR BLADEX, WITH 2,4-D OR 2,4-D PLUS BANVEL FOR MINIMUM-TILLAGE OR NO-TILLAGE SYSTEMS

In minimum-tillage or no-tillage systems where corn is planted directly into a cover crop, stale seedbed, established sod, of previous crop residues, Bicep may be applied in combination with AAtrex or Bladex. When used as directed, the Bicep portion of the tank mixture provides preemergence control of the weeds listed on this label in the Bicep Alone section for corn. The addition of AAtrex or Bladex offers the advantage indicated for each under "Conventional Tillage" above.

Application: Apply Bicep before, during, or after planting but before corn emerges, at the appropriate rate in Table 4. Up to 1 qt. of AAtrex 4L/4LC (1.25 lbs. 80W or 1.1 lbs. Nine-O) or 2 qts. of Bladex 4L (2.5 lbs. 80W or 2.2 lbs. 90WDG) per acre may be added to the rate of Bicep recommended in Table 4.

Where heavy crop residues exist, add 0.75-1.5 pts./A of an appropriately labeled 4 lb. ai/gal. 2,4-D amine (or equivalent amount of another 2,4-D amine such as Weedar 64, Weedar 64A, DMA-4 Herbicide, or Formula 40 at 0.8-1.6 pts./A) to the spray tank last and apply in a minimum of 25 gals. of carrier/A.

As carriers, nitrogen solutions and complete liquid fertilizers, applied before corn emergence, enhance burndown of existing weeds, and therefore are recommended instead of water. Add X-77 surfactant at 1-2 qts./100 gals. of diluted spray, or another surfactant cleared for use on growing crops at its recommended rate. Apply before weeds exceed 3 inches in height. If alfalfa is present, add Banvel to the spray mixture at 0.33-0.5 pt./A and apply before alfalfa exceeds 6 inches in height.

For fields with existing sod grasses (e.g., bromegrass, orchardgrass, rye or timothy), when existing weeds exceed 3 inches in height or when very dry conditions exist, add paraquat at the rate of 2 pts./A in place of or in addition to 2,4-D as indicated above. Do not apply paraquat in suspension-type liquid fertilizer. Observe all directions for use, precautions and limitations on the respective product labels when applying these products in tank mix combination.

Note: When applying Bicep and Bladex in tank mix combination, follow directions under "Conventional Tillage" above to ensure compatibility of these products in the tank mixture.

Tank Mixture with Linuron for Control of Lambsquarters and Pigweed

For prolonged control of lambsquarters and pigweed in DE, MD, NJ, NY, PA, VA and WV. Bicep may be applied preemergence in combination with linuron. Apply Bicep according to the rates in Table 4 and linuron according to the rate below.

Soil texture	Broadcast rate per acre
Sandy loam (1-3% organic matter)	.67 lb. linuron 50W*
Sandy loam (3-6% organic matter)	1.0 lb. linuron 50W*
Medium-and fine-textured soils (1-6% organic matter)	1.0 lb. linuron 50W*

* When using linuron L (4 lbs. a.i. per gallon), use equivalent rates. One pt. of linuron L equals one lb. of linuron wettable powder.

Follow instructions and precautions on the Bicep and linuron lables when tank mixing these products.

Table 4: Bicep for Minimum-Tillage or No-Tillage Corn

Soil texture	Broadcast rate per acre
COARSE: Sand, loamy sand, sandy loam	2.4 qts.
MEDIUM: Loam, silt loam, silt	3.2 qts.
FINE: Sandy clay loam, silty clay loam, clay loam, sandy clay, silty clay, clay	3.2-4 qts.*
Muck or peat soils	DO NOT USE

* For cocklebur, yellow nutsedge, and velvetleaf control on fine-textured soils above 3% organic matter, apply 4 qts. of Bicep per acre.

Rotational Crops: Follow the crop rotation instructions in the Bicep Alone section for corn.

BICEP COMBINATIONS — GRAIN SORGHUM (SEED TREATED WITH CONCEP)

Tank Mixtures of Bicep with Paraquat or Roundup for Minimum-Tillage or No-Tillage Systems

In minimum-tillage or no-tillage systems where grain sorghum is planted directly into a cover crop, stale seedbed, established sod, or previous crop residues, the contact herbicides paraquat or Roundup may be tank mixed with Bicep. When used as directed, the paraquat portion of the tank mixture controls most emerged annual weeds and suppresses many perennial weeds. Roundup combinations will control emerged annual and perennial weeds when applied as directed on the Roundup label. The Bicep portion of the tank mixture provides preemergence control of the weeds listed on this label in the Bicep Applied Alone section.

Refer to the label of each product used in combination and observe the planting details, restrictions, and all other precautions and limitations.

Application: Apply before, during, or after planting but before grain sorghum emerges, at the appropriate rate in Table 5. Add paraquat or Roundup at the following broadcast rates:

Paraquat: 1-2 pts.* per acre plus 8 oz. of X-77 Spreader per 100 gals. of spray mixture. Use the lower rate for control of annual weeds less than 4 inches tall and the higher rate for weeds 4-6 inches tall. This treatment will not consistently control weeds taller than 6 inches.

Roundup: 1.5 qts. per acre for control of existing annual weeds, or 2-4 qts. per acre for existing perennial weeds. See the Roundup label for weeds controlled and recommended rates for specific weeds.

* Based on a product containing 2 lbs. ai/gal.

Apply in a minimum of 20 gals. of water per acre with conventional spray equipment.

Table 5: Bicep for Minimum-Tillage or No-Tillage Grain Sorghum† (Seed treated with Concep)

Soil texture	Organic matter	Broadcast rate per acre
COARSE: Sand, loamy sand, sandy loam	any level	DO NOT USE
MEDIUM AND FINE: Loam, silt loam, silt, sandy clay loam, clay loam, silty clay loam, sandy clay, silty clay, clay	less than 1%	DO NOT USE
	1-1.5%	2.4 qts.
	more than 1.5%	2.8-3.2 qts.

† Do not use in NM, OK, or TX except in northeast OK and Texas Gulf Coast and Blacklands areas. Do not apply preplant incorporated in AZ or the Imperial Valley of CA.

Rotational Crops: Follow the crop rotation instructions in the Bicep Alone Section.

STORAGE AND DISPOSAL

Pesticide Disposal

Do not contaminate water, food, or feed by storage or disposal, or cleaning of equipment. Open dumping is prohibited. Improper disposal of unused pesticide, spray mixture, or rinsate is a violation of federal law. Pesticide, spray mixture, or rinsate that cannot be used according to label instructions must be disposed of according to federal, state, or local procedures. For guidance in proper disposal methods, contact your State Pesticide or Environmental Control Agency, or the Hazardous Waste Representative at the nearest EPA Regional Office.

Container Disposal

Do not reuse empty container. Triple rinse (or equivalent) and offer for recycling or reconditioning, or puncture and dispose of in a sanitary landfill, or by incineration, or by open burning, if allowed by state and local authorities. If burned, keep out of smoke.

For minor spills, leaks, etc., follow all precautions indicated on this label and clean up immediately. Take special care to avoid contamination of equipment and facilities during cleanup procedures and disposal of wastes. In the event of a major spill, fire or other emergency, call (919) 292-7100 day or night.

PRECAUTIONARY STATEMENTS

HAZARDS TO HUMANS AND DOMESTIC ANIMALS

CAUTION

Harmful if swallowed, inhaled or absorbed through the skin. Avoid breathing vapors or spray mists. Causes eye and skin irritation. May be a skin sensitizer. Avoid contact with eyes, skin or clothing. Do not contaminate food or feed.

Practical Treatment: If swallowed, contact your local poison control center, hospital, or physician. If the patient is unconscious, maintain breathing and heartbeat (cardiopulmonary resuscitation). If the patient is conscious and alert, induce vomiting (syrup of ipecac or stimulate the back of the throat with a finger). NEVER GIVE ANYTHING BY MOUTH TO AN UNCONSCIOUS PERSON.

In case of contact, immediately flush eyes or skin with plenty of water. Remove and wash contaminated clothing before reuse. Get medical attention if irritation persists.

Note to Physician: There is no specific antidote. If swallowed, induce emesis or lavage stomach. The use of an aqueous slurry of activated charcoal should be considered.

Environmental Hazards

Do not apply directly to any body of water. Do not apply where runoff is likely to occur. Do not contaminate water by cleaning of equipment or disposal of wastes. Do not apply when weather conditions favor drift from areas treated.

AAtrex® trademark of CIBA-GEIGY for atrazine

Banvel® trademark of Velsicol Chemical Corporation

Bicep® trademark of CIBA-GEIGY

U.S. Patent No. 3,937,730 (metolachlor)

Bladex® trademark of Shell Chemical Company for cyanazine

Brominal® trademark of Union Carbide Corporation

Buctril® trademark of Rhone-Poulenc Chemical Company

Caliber® trademark of CIBA-GEIGY

Compex® trademark of KALO Agricultural Chemicals, Inc.

Concep® trademark of CIBA-GEIGY

U.S. Patent No. 4,070,389

Dual® trademark of CIBA-GEIGY for metolachlor

U.S. Patent No. 3,937,730

Nine-O® trademark of CIBA-GEIGY

Princep® trademark of CIBA-GEIGY for simazine

Roundup® trademark of Monsanto Company for glyphosate herbicide

Unite® trademark of Hopkins Agricultural Chemical Company

X-77® trademark of KALO Agricultural Chemicals, Inc.

CGA 56L1P 055

CAPAROL®80W

HERBICIDE

For selective weed control in cotton, celery, and pigeon peas

Caparol 80W is a wettable powder

ACTIVE INGREDIENT:

Prometryn: 2, 4-bis (isopropylamino)-6-(methylthio)-s-triazine	80%
INERT INGREDIENTS:	20%
TOTAL:	100%

EPA Reg. No. 100-471 EPA Est. 100-AL-1

Caparol® trademark of CIBA-GEIGY for prometryn

Keep Out of Reach of Children.

Caution

See additional precautionary statements on left panel.

See directions for use on panel at right.

DIRECTIONS FOR USE AND CONDITIONS OF SALE AND WARRANTY

IMPORTANT: Read the entire **Directions for Use** and the **Conditions of Sale and Warranty** before using this product.

CONDITIONS OF SALE AND WARRANTY

The **Directions for Use** of this product reflect the opinion of experts based on field use and tests. The directions are believed to be reliable and should be followed carefully. However, it is impossible to eliminate all risks inherently associated with use of this product. Crop injury, ineffectiveness, or other unintended consequences may result because of such factors as weather conditions, presence of other materials, or the manner of use or application all of which are beyond the control of CIBA-GEIGY or the Seller. All such risks shall be assumed by the Buyer.

CIBA-GEIGY warrants that this product conforms to the chemical description on the label and is reasonably fit for the purposes referred to in the **Directions for Use** subject to the inherent risks referred to above. **CIBA-GEIGY makes no other express or implied warranty of Fitness or Merchantability or any other express or implied warranty. In no case shall CIBA-GEIGY or the seller be liable for consequential, special, or indirect damages resulting from the use or handling of this product.** CIBA-GEIGY and the Seller offer this product, and the Buyer and user accept it, subject to the foregoing, **Conditions of Sale and Warranty,** which may be varied only by agreement in writing signed by a duly authorized representative of CIBA-GEIGY.

DIRECTIONS FOR USE

It is a violation of federal law to use this product in a manner inconsistent with its labeling.

FAILURE TO FOLLOW PRECAUTIONS ON THIS LABEL MAY RESULT IN CROP INJURY, POOR PERFORMANCE, AND/OR ILLEGAL RESIDUES.

Do not apply this product in such a manner as to directly or through drift expose workers or other persons except those knowingly involved in the application. The area being treated must be vacated by unprotected persons.

REENTRY STATEMENT

Do not enter treated areas without protective clothing until sprays have dried.

Because certain states may require more restrictive reentry intervals for various crops treated with this product, consult your State Department of Agriculture for further information.

Written or oral warnings must be given to workers who are expected to be in a treated area or in an area about to be treated with this product. Oral warnings must inform workers of areas or fields that may not be entered without specific protective clothing until sprays have dried, and appropriate actions to take in case of accidental exposure, as described under **Precautionary Statements** on this label. When oral warnings are given, warnings shall be given in a language customarily understood by workers. Oral warnings must be given if there is reason to believe that written warnings cannot be understood by workers. Written warnings must include the following information: CAUTION Area treated with Caparol 80W on (date of application). Do not enter without appropriate protective clothing until sprays have dried. In case of accidental exposure, flush eyes or skin with plenty of water. Call physician if irritation persists. Remove and wash contaminated clothing before reuse.

GENERAL INFORMATION

Caparol 80W is a selective herbicide that may be applied either before or after weeds emerge for control of most annual broadleaf weeds and grasses, including groundcherry, lambsquarters, annual morningglory, malva, mustard, black nightshade, pigweed (carelessweed), purslane, Florida pusley, ragweed, smartweed, teaweed (prickly sida), barnyardgrass (watergrass), crabgrass, foxtail, goosegrass, junglerice, *Panicum* spp., signalgrass (and other *Brachiaria* spp.), and wild oats. Caparol also controls shallow-germinating seedlings of cocklebur, coffeeweed, and sandbur. It does not control johnsongrass, bermudagrass, other established perennials, or sprangletop at selective rates.

When applied before weeds emerge. Caparol enters weeds through their roots. Thus, its effectiveness depends on moisture to move it into the soil. Under very dry soil conditions after application, a shallow cultivation or rotary hoeing will generally result in better weed control.

When applied to emerged weeds, Caparol provides foliar knockdown and or residual control of later germinating weeds, depending on the rate applied.

Following many years of continuous use of this product and chemically related products, biotypes of some of the weeds listed on this label have been reported which cannot be effectively controlled by this and related herbicides. Where this is known or suspected, we recommend the use of this product in combination with other registered herbicides which are not triazines. Consult with your State Agricultural Extension Service for specific recommendations.

To avoid spray drift, do not apply under windy conditions. Avoid spray overlap, as crop injury may result.

EQUIPMENT AND MIXING

Thoroughly clean sprayer prior to use. Do not use a sprayer contaminated with 2, 4-D or other materials, as crop damage or sprayer clogging may result.

Use conventional spray equipment with hydraulic or mechanical agitation except in California and Arizona where only mechanical agitators are recommended. Screens in nozzles and in suction and in-line strainers should be no finer than 50-mesh. Use a pump with capacity to maintain 35-40 psi at the nozzles. If hydraulic agitation is used, the pump should also provide sufficient agitation in the tank to keep the mixture in suspension.

Nozzles: For preplant incorporated or preemergence application, use flat fan nozzle tips. For postemergence band application, use off-center nozzle tips. For postemergence broadcast application, use flat fan or off-center nozzle tips. Use flood nozzle tips only in Arizona and California for lay-by treatment in cotton at least 18 inches tall.

Mixing: Mix the proper amount of Caparol 80W with water in a clean pail to form a slurry. Pour the slurry into the spray tank partially filled with water, and then add the rest of the water. Provide agitation during mixing and application to maintain a uniform suspension.

Cleaning: Wash sprayer thoroughly with clean water immediately after use. Do not use the same sprayer on sensitive crops, as even small residues of Caparol in the tank may cause injury to these crops.

SEEDBED PREPARATION

To insure proper placement of Caparol 80W, seedbeds must be well prepared and as free as possible from trash and clods. A firm seedbed is best for obtaining effective weed control. Uniformity in height and width of seedbed is essential for proper postemergence applications of Caparol 80W. Beds should be low and flat. Take care to avoid planter marks. Wide plant packer wheels or rollers are recommended. Wheel furrows should be uniform in depth. Mount the sprayer so that it follows the same rows as the planter.

BAND TREATMENT

For band treatment, calculate the amount of Caparol 80W needed by the formula:

$$\frac{\text{band width in inches}}{\text{row width in inches}} \times \text{broadcast rate per acre} = \text{amount needed per acre of field}$$

COTTON

Caparol 80W may be applied preplant incorporated or preemergence and/or postemergence as recommended in the following tables. The postemergence applications may follow preplant incorporated or preemergence treatments of Caparol 80W.

Apply Caparol 80W in a minimum of 20 gals. of water per acre as a broadcast or band application at the rates indicated in the following tables.

Do not use on glandless cotton varieties as crop injury will occur.

Note: Do not feed treated forage to livestock, or graze treated areas.

Preplant Incorporation (Arizona, California, and New Mexico)

Apply Caparol 80W at the appropriate rate in Table 1 as a broadcast or band treatment. If broadcast, treat the flat soil surface prior to listing. If banded, apply over partially finished or finished beds. Incorporate up to 4 inches deep immediately after application with PTO-driven equipment, double disk rolling cultivator, or bed conditioner.

Table 1: Preplant Incorporation

Region	Soil texture	Broadcast rate per acre
Arizona, California, and New Mexico	sand, loamy sand	DO NOT USE
	sandy loam (AZ & CA only)	1.5-2 lbs.
	sandy loam, loams (NM only)	2 lbs.
	silt loam, clay	3 lbs.

Precautions: *Do not use Caparol 80W in cut areas of newly leveled fields, in areas of excess salt, or in areas where flooding over the beds is likely to occur as crop injury may result. Do not plant cotton in tractor wheel depressions as crop injury may result. On mulch planted cotton, water back only after cotton seedlings are well established. In New Mexico, apply either preplant incorporated or preemergence (not both)—see Preemergence section.*

PREEMERGENCE

Apply at planting or shortly after planting at the appropriate rate in Table 2. Caparol 80W may be used on cotton planted flat, on beds, or in furrows. To avoid concentration of Caparol 80W in the seed furrow, do not make broadcast applications to cotton planted in furrows deeper than 2 inches. Band applications may be made to cotton planted in furrows deeper than 2 inches, but band width should not exceed the width of the bottom of the furrow. If banded, do not cover treated bands with soil while cultivating untreated row middles. Do not use on sand or loamy sand, on shallow soils with caliche subsoils, or in areas with caliche outcroppings.

Cotton may be replanted in soil previously treated with Caparol 80W. Do not apply a second preemergence application of Caparol as injury may occur.

Table 2. Preemergence

Region	Soil texture	Broadcast rate per acre
Mid-South & Southeast other than Mississippi River Delta in Miss.	sandy loam	2-2.5 lbs.
	silt and clay loam	3 lbs.
	Sharkey clay (Arkansas only)	3.5 lbs.
Miss. River Delta in Miss.	sandy loam	2.5-3 lbs.
	silt and clay loam	3.5 lbs.
	Sharkey clay	DO NOT USE
Blacklands of Texas & Okla., Texas Gulf Coast, & Texas Coastal Bend	loam	1.5 lbs.
	clay	3 lbs.
Rio Grande Valley of Texas*	loam	2 lbs.
	clay	3 lbs.
High Plains, Rolling Plains and Edwards Plateau of Texas, Southwest Texas and New Mexico**	sand, loamy sand	DO NOT USE
	sandy loam	1 lb.
	loam, sandy clay loam	1.5 lbs.
	other clay soils	2 lbs.
Arizona and California	DO NOT USE	

* *Rio Grande Valley of Texas*—Furrow irrigation cotton—If adequate rain does not fall soon after application, a shallow cultivation will insure good weed control.

** *New Mexico*—Apply either preplant incorporated or preemergence (not both)—See Preplant Incorporation section.

Winter Weed Control in Texas

For control of winter weeds only, such as henbit (purpletop) and seedling dock on fall bedded cotton land in the Texas Gulf Coast and Blacklands of Texas, apply 0.75-1 lb. of Caparol 80W per acre in the fall or winter to land that will be planted to cotton the following spring. For best results, apply before weeds emerge. Caparol 80W will give effective control of emerged henbit if applied before it reaches 4-5 inches tall. For postemergence henbit control, add a suitable surfactant, such as X-77®, at 0.5% of spray volume or an emulsifiable oil at 1.0% of spray volume.

Postemergence Directed

Be especially careful in applying Caparol 80W postemergence to prevent contact of the spray with cotton leaves, or injury may occur. Use precision application equipment so the spray is accurately directed to the base of the cotton plants. Apply during calm periods to prevent drift. Use leaf lifters or shields if leaf contact cannot be avoided merely by directing the spray. Apply only when all plants have exceeded the minimum recommended height. Apply to level, well prepared surfaces such as relatively clod-free beds made with bed-shapers.

To avoid crop injury, do not apply to furrow-planted cotton until furrows are leveled (plowed in). Do not treat cotton under stress from drought, cultivator damage, or fertilizer application.

When applying to emerged weeds, add 2 qts. of surfactant per 100 gals. of spray mixture. Use a surfactant that is compatible with Caparol 80W when applied in cotton and is approved by EPA for use on food and feed crops. Examples include X-77, Tronic, and Triton.

Chemical Hoe (Emerged Weeds only): Apply Caparol 80W at the appropriate rate in Table 3, two or three times if necessary. In cotton 3-6 inches tall, be extremely careful to avoid spray contact with cotton leaves by applying Caparol 80W with a precision applicator equipped with fenders or shields such as Bell Row Shield, Dickey Fenders, or W&A Fenders. In cotton less than 10 inches tall, apply only if cotton is bed or flat-planted.

Table 3: Chemical Hoe

Height of cotton and area of use	Height of weeds	Broadcast rate per acre
3 to 6 inches (Arkansas, Louisiana, Mississippi, Missouri Tennessee, and Texas)	less than 1 inch	0.6 lb.
6 or more inches (all regions)	less than 2 inches	0.6-0.8 lb.

Lay-by (Emerged Weeds and Germinating Weeds): Apply Caparol 80W at the appropriate rate in Table 4, once per season when cotton is at least 12 inches tall (18 inches where flood nozzles are used in Arizona and California). Apply before weeds are 2 inches tall.

Table 4: Lay-by

Region	Soil texture	Broadcast rate per acre
Mid-South and Southeast	sandy	1.5 lbs.
	loam	1.75 lbs.
	clay	2 lbs.
Blacklands of Texas and Oklahoma	loam	1 lb.
	clay	2 lbs.
High Plains of Texas and New Mexico	sandy	1 lb.
	loam and clay	1.5 lbs.
Southwest Texas	loam	1.5 lbs.
	clay	2 lbs.
Rio Grande Valley of Texas	DO NOT USE	
Arizona and California (Do not use in the Coachella Valley)	sand and loamy sand	DO NOT USE
	sandy loam	1.5-2 lbs.
	loam	2 lbs.

Rotational Crops

The following vegetable and cover crops may be planted in the fall when Caparol 80W was applied on cotton by no more than one of these methods that year; preplant incorporated, preemergence, or only one chemical hoe treatment.

Vegetables

cabbage, okra, onions, peas, red beets, sweet corn

Cover Crops (must be plowed down and not used for food or feed) oats, sorghum, winter barley, winter rye, winter wheat.

Spring-seeded crops in California and Arizona and spring-seeded vegetables in the Rio Grande Valley of Texas should not be planted until after April 1.

CAPAROL 80W COMBINATIONS FOR COTTON

Prowl® (Arizona, California, New Mexico, and the upper and lower El Paso Valley of Texas)

This preplant incorporated tank mixture controls all weeds listed on this label and on the Prowl cotton label. Apply prior to listing or over partially finished or finished beds and incorporate immediately. Refer to the Prowl label for specific mixing, spraying, and incorporation methods. Continuous agitation in the spray tank is required to keep the material in suspension.

Apply the tank mixture with ground equipment in at least 10 gals. of water per acre at the appropriate rates from Table 5.

Table 5. Tank Mixture with Prowl

Soil texture	Broadcast rate per acre	
	Prowl	Caparol 80W
sand, loamy sand	DO NOT USE	
sandy loam	1-1.5 pts.	1.5-2 lbs.
loam	1.5-2 pts.	2 lbs.
silt loam, silt, sandy clay loam	1.5-2 pts.	2-3 lbs.
clay loam, silty clay loam, clay	1.5-3 pts.	2-3 lbs.

Use the high rate for each soil texture above if heavy weed populations are anticipated. Use the 3 pt. rate of Prowl for heavy clay soils.

Precautions: *Do not use in cut areas of newly leveled fields, in areas of excess salt, or in areas where flooding over the bed is likely to occur as crop injury may result. Do not plant cotton in tractor wheel depressions or crop injury may result. Do not use this tank mixture when cotton is irrigated up as crop injury may result. On mulch planted cotton, water back only after cotton seedlings are well established.*

Note: Do not feed treated forage to livestock, or graze treated areas.

Rotational Crops: If treated crop is lost, cotton may be replanted. Do not rework the soil. Refer to the Prowl label and the Cotton section of this label for rotational crop restrictions.

Treflan® EC Tank Mixture (Arizona, California, New Mexico and the upper and lower El Paso Valley of Texas)

This combination controls weeds listed on this label and on the Treflan label. This combination also controls shallow-germinating seedlings of cocklebur and coffeeweed.

Follow procedures on the Treflan label for soil preparation and incorporation. Apply the tank mix combination to the flat soil before disking.

Mix Caparol 80W in a clean pail with water to form a slurry, pour into spray tank one-half to three-fourths full of water, allow it to disperse with agitation, add Treflan EC, and then add the rest of the water. Continuous agitation in the spray tank is required to keep the material in suspension. Apply the tank mixture in at least 20 gals. of water per acre at the appropriate rates from Table 6.

Table 6. Tank Mixture with Treflan EC

Soil texture	Broadcast rate per acre	
	Treflan EC	Caparol 80W
sand, loamy sand	DO NOT USE	
sandy loam	1 pt.	1½-2 lbs.*
medium soils	1½ pts.	2½ lbs.
fine soils	2 pts.	2½ lbs.
muck or peat	DO NOT USE	

* Use less than 2 lbs. per acre only in Arizona and California.

Precautions: *Do not use in cut areas of newly leveled fields, in areas of excess salt, or in areas where flooding over the bed is likely to occur as crop injury may result. Do not plant cotton in tractor wheel depressions or crop injury may result. On mulch planted cotton, water back only after cotton seedlings are well established.*

Note: Do not feed treated forage to livestock, or graze treated areas.

Rotational Crops: Cabbage, celery, okra, onions, and peas may be planted in the fall after a spring application of Treflan + Caparol. Winter barley, rye and wheat can be planted in the fall if they are plowed down and not used for food or feed. Refer to the Treflan EC label for other directions and precautions.

Treflan EC Split Application (Arizona and California)

Apply a preplant incorporated application of Treflan EC as directed on that label, except use the appropriate rate from Table 6. Do not apply Treflan EC before January 1. Follow at planting or just before planting with a preplant incorporated treatment of Caparol 80W as directed in the Cotton section of this label, except use the appropriate rate from Table 6.

DSMA or MSMA

For faster knockdown of the weeds controlled by Caparol 80W alone, apply 0.6-0.8 lb. of Caparol 80W plus 3 lbs. active ingredient of DSMA or 2 lbs. active ingredient of MSMA per acre, following the same directions, precautions, and limitations as given on this label for Caparol 80W applied alone postemergence directed (chemical hoe). If needed, make a second application 1-3 weeks after the first application. Do not apply after first bloom.

Several formulations of DSMA and MSMA are available under various trade names from several manufacturers. Observe the directions, limitations, and precuations on the label of the product used.

CELERY

Seedbeds (Florida)

Broadcast 0.75-1 lb. in a minimum of 20 gals of water per acre after celery has 2-5 true leaves. Application may be made over the celery. Apply only after seedbed covers have been removed from seedbeds for at least one week. Apply only once per year to seedbeds.

Direct-seeded Celery (California only)

Apply Caparol 80W at rates given below in a minimum of 20 gals of water per acre. Within the rate ranges given, use the lower rates on coarse-textured soils and soils low in organic matter; use the higher rates on fine-textured soils and soils high in organic matter.

Preemergence: Broadcast 1.5-2 lbs. per acre at planting or shortly after planting before celery emerges.

Postemergence: Broadcast 1-1.25 lbs. per acre after celery has 2-5 true leaves. Application may be made over the celery. Apply before weeds are 2 inches tall.

To Avoid Injury to Direct-seeded Celery: *(1) Make either one preemergence or one postemergence application (not both) per celery crop. (2) Do not use on sand or loamy sand. (3) Do not apply if celery is under water stress. (4) Do not apply postemergence treatments of Caparol 80W with other pesticides. Apply only after foliar applications of other pesticides are dry. (5) Do not apply within two weeks after an application of a herbicidal oil, such as "carrot" oil.*

Transplants

Apply one application at the appropriate rate from Table 7 in a minimum of 20 gals. of water per acre during the 2-6 week period after transplanting. Within the rate ranges given, use the lower rate on relatively coarse-textured soils and soils low in organic matter; use the higher rate on relatively fine-textured soils and soils high in organic matter. Application may be made over the celery. Apply before weeds are 2 inches tall.

Table 7: Transplanted Celery

State	Soil	Broadcast rate per acre
Florida	sandy or muck	1-2 lbs.
California	coarse-textured	1.25-2 lbs.
	fine-textured	2-2.5 lbs.
Hawaii	coarse-textured	2-3 lbs.
	fine-textured	3-4 lbs.
Michigan	fine-textured or muck	1.25-2.5 lbs.
Wisconsin	fine-textured	2-2.5 lbs.

Rotational Crops

The following crops may be seeded 5 months after applying no more than 2.5 lbs. Caparol 80W per acre on celery; cabbage, celery, corn, okra, onions, peas, and red beets.

PIGEON PEAS (PUERTO RICO ONLY)

For preemergence control of annual weeds, such as horse purslane, jungle rice, wild spider flower, jimsonweed, spurge, pigweed and Florida pusley, apply 2.5 lbs. of Caparol 80W per acre on loam soils, or 3.75 per acre on clay soils. Apply at planting or immediately after planting before the crop or weeds emerge.

Precautions: *To avoid crop injury, 1) Make only one application per year. 2) Do not use on sandy or loamy sand soils.*

Note: Do not graze or feed forage or hay to livestock, or illegal residues may result.

Rotational Crops

The following crops may be seeded 5 months after appplying no more than 2.5 lbs. Caparol 80W per acre on pigeon peas: cabbage, celery, corn, onions, peas, and red beets.

STORAGE AND DISPOSAL

Do not contaminate water, food, or feed by storage or disposal. Improper disposal of unused pesticide spray mixture, or rinse water is a violation of federal law. If these wastes cannot be used according to label instructions, contact your State Pesticide or Environmental Control Agency, or the Hazardous Waste representative at the nearest EPA Regional Office for guidance in proper disposal methods.

Completely empty bag into application equipment. Dispose of empty bag in a sanitary landfill or by incineration, or if allowed by state and local authorities, by burning. Stay out of smoke from burning bags.

For minor spills, leaks, etc., follow all precautions indicated on this label and clean up immediately. Take special care to avoid contamination of equipment and facilities during cleanup procedures and disposal of wastes. In the event of a major spill, fire or other emergency, call (919)292-7100 day or night.

PRECAUTIONARY STATEMENTS

HAZARDS TO HUMANS AND DOMESTIC ANIMALS

CAUTION

Harmful if swallowed, inhaled, or absorbed through skin. Causes eye irritation. Avoid breathing vapors, dust, or spray mist. Avoid contact with eyes, skin, or clothing. Wash thoroughly with soap and water after handling. Remove and wash contaminated clothing before reuse.

First Aid: If swallowed, contact your local poison control center, hospital, or physician. If the patient is unconscious, maintain breathing and heartbeat (cardiopulmonary resuscitation). If the patient is conscious and alert, induce vomiting (syrup of ipecac or stimulate the back of the throat with a finger). Never give anything by mouth to an unconscious person!

If in eyes, flush with plenty of water. If on skin, wash with plenty of soap and water. Get medical attention if irritation persists. If inhaled, remove victim to fresh air.

Note to the Physician: If ingested, there is no specific antidote. Induce emesis or lavage stomach. Give a saline laxative and supportive therapy.

ENVIRONMENTAL HAZARDS

Do not contaminate water by cleaning of equipment or disposal of wastes. Do not apply where runoff is likely to occur. Do not apply when weather conditions favor drift from areas treated.

Caparol® trademark of CIBA-GEIGY for prometryn

Prowl® trademark of American Cyanamid for pendimethalin

Treflan® trademark of Elanco Products for trifluralin

X-77® trademark of KALO Agricultural Chemicals, Inc.

CGA 17L1T 065

CAPAROL®4L

HERBICIDE

For selective weed control in cotton, celery, and pigeon peas

ACTIVE INGREDIENT:

Prometryn:

2, 4-bis (isopropylamino)-6-(methylthio)-s-triazine....... 44.4%

INERT INGREDIENTS: 55.6%

TOTAL: 100.0%

EPA Reg. No. 100-620 EPA Est. 100-LA-1

Caparol® trademark of CIBA-GEIGY for prometryn

Caparol 4L contains 4 lbs. active ingredient per gallon

Shake well before using.

Keep Out of Reach of Children.

Caution

See additional precautionary statements at end of label booklet.

See directions for use inside booklet.

DIRECTIONS FOR USE AND CONDITIONS OF SALE AND WARRANTY

IMPORTANT: Read the entire **Directions for Use** and the **Conditions of Sale and Warranty** before using this product.

CONDITIONS OF SALE AND WARRANTY

The **Directions for Use** of this product reflect the opinion of experts based on field use and tests. The directions are believed to be reliable and should be followed carefully. However, it is impossible to eliminate all risks inherently associated with use of this product. Crop injury, ineffectiveness, or other unintended consequences may result because of such factors as weather conditions, presence of other materials, or the manner of use or application all of which are beyond the control of CIBA-GEIGY or the Seller. All such risks shall be assumed by the Buyer.

CIBA-GEIGY warrants that this product conforms to the chemical description on the label and is reasonably fit for the purposes referred to in the **Directions for Use** subject to the inherent risks referred to above. **CIBA-GEIGY makes no other express or implied warranty of Fitness or Merchantability or any other express or implied warranty. In no case shall CIBA-GEIGY or the Seller be liable for consequential, special, or indirect damages resulting from the use or handling of this product.** CIBA-GEIGY and the Seller offer this product, and the Buyer and user accept it, subject to the foregoing **Conditions of Sale and Warranty,** which may be varied only by agreement in writing signed by a duly authorized representative of CIBA-GEIGY.

DIRECTIONS FOR USE

It is a violation of federal law to use this product in a manner inconsistent with its labeling.

FAILURE TO FOLLOW THE DIRECTIONS FOR USE AND PRECAUTIONS ON THIS LABEL MAY RESULT IN CROP INJURY, POOR PERFORMANCE, AND/OR ILLEGAL RESIDUES.

Do not apply this product in such a manner as to directly or through drift expose workers or other persons except those knowingly involved in the application. The area being treated must be vacated by unprotected persons.

REENTRY STATEMENT

Do not enter treated areas without protective clothing until sprays have dried.

Because certain states may require more restrictive reentry intervals for various crops treated with this product, consult your State Department of Agriculture for further information.

Written or oral warnings must be given to workers who are expected to be in a treated area or in an area about to be treated with this product. Oral warnings must inform workers of areas or fields that may not be entered without specific protective clothing until sprays have dried, and appropriate actions to take in case of accidental exposure, as described under **Precautionary Statements** on this label. When oral warnings are given, warnings shall be given in a language customarily understood by workers. Oral warnings must be given if there is reason to believe that written warnings cannot be understood by workers. Written warnings must include the following information: CAUTION. Area treated with Caparol 4L on (date of application). Do not enter without appropriate protective clothing until sprays have dried. In case of accidental exposure, flush eyes or skin with plenty of water. Call physician if irritation persists. Remove and wash contaminated clothing before reuse.

GENERAL INFORMATION

Caparol 4L is a selective herbicide that may be applied either before or after weeds emerge for control of most annual broadleaf weeds and grasses, including groundcherry, lambsquarters, annual morningglory, malva, mustard, black nightshade, pigweed (carelessweed), purslane, Florida pusley, ragweed, smartweed, teaweed (prickly sida), barnyardgrass (watergrass), crabgrass, foxtail, goosegrass, junglerice, *Panicum* spp., signalgrass (and other *Brachiaria* spp.), and wild oats. Caparol also controls shallow-germinating seedlings of cocklebur, coffeeweed, and sandbur. It does not control johnsongrass, bermudagrass, other established perennials, or sprangletop at selective rates.

When applied before weeds emerge, Caparol enters weeds through their roots. Thus, its effectiveness depends on moisture to move it into the soil. Under very dry soil conditions after application, a shallow cultivation or rotary hoeing will generally result in better weed control.

When applied to emerged weeds, Caparol provides foliar knockdown and/or residual control of later germinating weeds, depending on the rate applied.

Following many years of continuous use of this product and chemically related products, biotypes of some of the weeds listed on this label have been reported which cannot be effectively controlled by this and related herbicides. Where this is known or suspected and weeds controlled by this product are expected to be present along with resistant biotypes, we recommend the use of this product in registered combinations or in sequence with other registered herbicides which are not triazines. If only resistant biotypes are expected to be present, use a registered non-triazine herbicide. Consult with your State Agricultural Extension Service for specific recommendations.

To avoid spray drift, do not apply under windy conditions. Avoid spray overlap, as crop injury may result.

EQUIPMENT AND MIXING

Thoroughly clean sprayer prior to use. Do not use a sprayer contaminated with 2,4-D or other materials, as crop damage or sprayer clogging may result.

Use conventional spray equipment with hydraulic or mechanical agitation except in California and Arizona where only mechanical agitators are recommended. Screens in nozzles and in suction and in-line strainers should be no finer than 50-mesh. Use a pump with capacity to maintain 35-40 psi at the nozzles. If hydraulic agitation is used, the pump should also provide sufficient agitation in the tank to keep the mixture in suspension.

Nozzles: For preplant incorporated or preemergence application, use flat fan nozzle tips. For postemergence band application, use off-center nozzle tips. For postemergence broadcast application, use flat fan or off-center nozzle tips. Use flood nozzle tips only in Arizona and California for lay-by treatment in cotton at least 18 inches tall.

Mixing: 1) Fill spray tank $\frac{1}{2}$ to $\frac{3}{4}$ full with clean water. 2) Start agitation. 3) Pour the product directly from the container into the spray tank partially filled with water, and then add the rest of the water. 4) Provide agitation during mixing and application to maintain a uniform suspension.

Cleaning: Wash sprayer thoroughly with clean water immediately after use. Do not use the same sprayer on sensitive crops, as even small residues of Caparol in the tank may cause injury to these crops.

Seedbed preparation

To insure proper placement of Caparol 4L, seedbeds must be well prepared and as free as possible from trash and clods. A firm seedbed is best for obtaining effective weed control. Uniformity in height and width of seedbed is essential for proper postemergence applications of Caparol 4L. Beds should be low and flat. Take care to avoid planter marks. Wide planter packing wheels or rollers are recommended. Wheel furrows should be uniform in depth. Mount the sprayer so that it follows the same rows as the planter.

BAND TREATMENT

For band treatment, calculate the amount of Caparol 4L needed by the formula:

$$\frac{\text{band width in inches}}{\text{row width in inches}} \times \text{broadcast rate per acre} = \text{amount needed per acre of field}$$

COTTON

Caparol 4L may be applied preplant incorporated or preemergence and/or postemergence as recommended in the following tables. The postemergence applications may follow preplant incorporated or preemergence treatments of Caparol 4L.

Apply Caparol 4L in a minimum of 20 gals. of water per acre as broadcast or band application at the rates indicated in the following tables.

Do not use on glandless cotton varieties or crop injury will occur.

Note: Do not feed treated forage to livestock, or graze treated areas, or illegal residues may result.

Preplant Incorporation (Arizona, California, and New Mexico)

Apply Caparol 4L at the appropriate rate in Table 1 as a broadcast or band treatment. If broadcast, treat the flat soil surface prior to listing. If banded, apply over partially finished or finished beds. Incorporate up to 4 inches deep immediately after application with PTO-driven equipment, double disk, rolling cultivator, or bed conditioner.

Table 1: Preplant Incorporation

Region	Soil texture	Broadcast rate per acre
Arizona, California, and New Mexico	sand, loamy sand	DO NOT USE
	sandy loam (AZ & CA only)	2.4-3.2 pts.
	sandy loam, loams (NM only)	3.2 pts.
	silt loam, clay	4.8 pts.

Precautions: *To avoid crop injury, 1) Do not use Caparol 4L in cut areas of newly leveled fields, in areas of excess salt, or in areas where flooding over the beds is likely to occur. 2) Do not plant cotton in tractor wheel depressions. 3) On mulch planted cotton, water back only after cotton seedlings are well established. 4) In New Mexico, apply either preplant incorporated or preemergence (not both)—see Preemergence section.*

Preemergence

Apply at planting or shortly after planting at the appropriate rate in Table 2. Caparol 4L may be used on cotton planted flat, on beds, or in furrows. To avoid concentration of Caparol 4L in the seed furrow, do not make broadcast applications to cotton planted in furrows deeper than 2 inches. Band applications may be made to cotton planted in furrows deeper than 2 inches, but band width should not exceed the width of the bottom of the furrow. If banded, do not cover treated bands with soil while cultivating untreated row middles. To avoid crop injury, do not use on sand or loamy sand, on shallow soils with caliche subsoils, or in areas with caliche outcroppings.

Cotton may be replanted in soil previously treated with Caparol 4L. Do not apply a second preemergence application of Caparol or injury may occur.

Table 2: Preemergence

Region	Soil texture	Broadcast rate per acre
Mid-South and Southeast other than Mississippi River Delta in Mississippi	sandy loam	3.2-4 pts.
	silt and clay loam	4.8 pts
	Sharkey clay (Arkansas only)	5.6 pts.
Mississippi River Delta in Mississippi	sandy loam	4-4.8 pts.
	silt and clay loam	5.6 pts.
	Sharkey clay	DO NOT USE
Blacklands of Texas and Oklahoma, Texas Gulf Coast, and Texas Coastal Bend	loam	2.4 pts.
	clay	4.8 pts.
Rio Grande Valley of Texas*	loam	3.2 pts.
	clay	4.8 pts.
High Plains, Rolling Plains and Edwards Plateau of Texas, Southwest Texas and New Mexico**	sand, loamy sand	DO NOT USE
	sandy loam	1.6 pts.
	loam, sandy clay loam	2.4 pts.
	other clay soils	3.2 pts.
Arizona and California	DO NOT USE	

* Rio Grande Valley of Texas—Furrow irrigation cotton—if adequate rain does not fall soon after application, a shallow cultivation will insure good weed control.

** New Mexico—Apply either preplant incorporated or preemergence (not both)—See Preplant Incorporation section.

Winter Weed Control in Texas

For control of winter weeds **only,** such as henbit (purpletop) and seedling dock on fall bedded cotton land in the Texas Gulf Coast and Blacklands of Texas, apply 1.2-1.6 pts. of Caparol 4L per acre in the fall or winter to land that will be planted to cotton the following spring. For best results, apply before weeds emerge. Caparol 4L will give effective control of emerged henbit if applied before it reaches 4-6 inches tall. For post-emergence henbit control, add a suitable surfactant, such as X-77®, at 0.5% of spray volume or an emulsifiable oil at 1.0% of spray volume to the spray tank.

Winter Weed Control in California

For control of winter weeds on fall bedded cotton land, apply Caparol 4L after bedding either preemergence or postemergence to weeds less than 2 inches tall. Winter weeds controlled include:

chickweed	London rocket	redmaids
fiddleneck	mustards	shepherdspurse
filarees	pineappleweed	sowthistle, annual

On sandy loam soil, apply 3.2 pts. per acre; on medium or fine soil, apply 4 pts. per acre. To avoid crop injury, do not use on sand or loamy sand. For postemergence weed control, add a suitable surfactant, such as X-77, at 0.5% of spray volume or an emulsifiable oil at 1.0% of spray volume. Rainfall or sprinkler irrigation is necessary to activate the preemergence activity of Caparol 4L.

After pre-irrigation and before planting in the spring, knock off the top $\frac{1}{3}$ to $\frac{1}{2}$ of the seedbed. Then make a preplant application of Caparol 4L over the surface of the seedbed using a power-tiller, rolling cultivator, or similar implement that will provide uniform incorporation. Refer to Table 1, for preplant incorporation rates of Caparol 4L in California. To avoid crop injury, do not cultivate treated soil back toward the cotton until after cotton emergence and just before the first irrigation.

Precautions: *To avoid crop injury, do not use Caparol 4L for winter weed control in areas of excess salt or calcareous soil.*

Note: To avoid illegal residues, do not use more than 10.3 pts. of Caparol 4L on sandy loam soil or 11.9 pts. Caparol 4L on medium or fine soil per acre per year including winter weed control, preplant incorporatión, chemical hoe, and lay-by applications.

Postemergence Directed

Be especially careful in applying Caparol 4L postemergence to prevent contact of the spray with cotton leaves, or injury may occur. Use precision application equipment so the spray is accurately directed to the base of the cotton plants and still thoroughly covers the soil and weeds beneath the cotton plants. Apply during calm periods to prevent drift. Use leaf lifters or shields if leaf contact cannot be avoided merely by directing the spray. Apply only when all plants have exceeded the minimum recommended height. Apply to level, well prepared surfaces such as relatively clod-free beds made with bedshapers.

To avoid crop injury, do not apply to furrow-planted cotton until furrows are leveled (plowed in). Do not treat cotton under stress from drought, cultivator damage, or fertilizer application.

When applying to emerged weeds, add 2 qts. of surfactant per 100 gals. of spray mixture. Use a surfactant that is compatible with Caparol 4L when applied in cotton and is approved by EPA for use on food and feed crops. Examples include X-77, Tronic, and Triton.

Chemical Hoe (Emerged Weeds only): Apply Caparol 4L at the appropriate rate in Table 3, two or three times if necessary. In cotton 3-6 inches tall, be extremely careful to avoid spray contact with cotton leaves by applying Caparol 4L with a precision applicator equipped with fenders or shields such as Bell Row Shield, Dickey Fenders, or W&A Fenders. In cotton less than 10 inches tall, apply only if cotton is bed or flat-planted.

Table 3: Chemical Hoe

Height of cotton and area of use	Height of weeds	Broadcast rate per acre
3 to 6 inches (Arkansas, Louisiana, Mississippi, Missouri, Tennessee, and Texas)	less than 1 inch	1 pt.
6 or more inches (all regions)	less than 2 inches	1-1.3 pts.

Lay-by (Emerged Weeds and Germinating Weeds): Apply Caparol 4L at the appropriate rate in Table 4, once per season when cotton is at least 12 inches tall (18 inches where flood nozzles are used in Arizona and California). Apply before weeds are 2 inches tall.

Table 4: Lay-by

Region	Soil texture	Broadcast rate per acre
Mid-South and Southeast	sandy	2.4 pts.
	loam	2.8 pts.
	clay	3.2 pts.
Blacklands of Texas and Oklahoma	loam	1.6 pts
	clay	3.2 pts.
High Plains of Texas and New Mexico	sandy	1.6 pts.
	loam and clay	2.4 pts.
Southwest Texas	loam	2.4 pts.
	clay	3.2 pts.
Rio Grande Valley of Texas	DO NOT USE	
Arizona and California (Do not use in the Coachella Valley)	sand and loamy sand	DO NOT USE
	sandy loam	2.4-3.2 pts.
	loam	3.2 pts.

Rotational Crops

The following vegetable and cover crops may be planted in the fall when Caparol 4L was applied on cotton by no more than one of these methods that year; preplant incorporated, preemergence, or only one chemical hoe treatment.

Vegetables

Cabbage, okra, onions, peas, red beets, sweet corn

Cover Crops (must be plowed down and not used for food or feed)
Oats, sorghum, winter barley, winter rye, winter wheat

Spring-seeded crops in California and Arizona and spring-seeded vegetables in the Rio Grande Valley of Texas should not be planted until after April 1.

CAPAROL 4L COMBINATIONS FOR COTTON

Prowl® (Arizona, California, New Mexico, and the upper and lower El Paso Valley of Texas)

This preplant incorporated tank mixture controls all weeds listed on this label and on the Prowl cotton label. Apply prior to listing or over partially finished or finished beds and incorporate immediately. Refer to the Prowl label for specific mixing, spraying, and incorporation methods. Continuous agitation in the spray tank is required to keep the material in suspension.

Apply the tank mixture with ground equipment in at least 10 gals. of water per acre at the appropriate rates from Table 5.

Table 5: Tank Mixture with Prowl

Soil Texture	Broadcast rate per acre	
	Prowl	Caparol 4L
sand, loamy sand	DO NOT USE	
sandy loam	1-1.5 pts.	2.4-3.2 pts.
loam	1.5-2 pts	3.2 pts.
silt loam, silt, sandy clay loam	1.5-2 pts.	3.2-4.8 pts.
clay loam, silty clay loam, clay	1.5-3 pts.	3.2-4.8 pts.

Use the high rate for each soil texture above if heavy weed populations are anticipated. Use the 3 pt. rate of Prowl for heavy clay soils.

Precautions: *To avoid crop injury, 1) Do not use in cut areas of newly leveled fields, in areas of excess salt, or in areas where flooding over the bed is likely to occur. 2) Do not plant cotton in tractor wheel depressions. 3) Do not use this tank mixture when cotton is irrigated up. 4) On mulch planted cotton, water back only after cotton seedlings are well established.*

Note: Do not feed treated forage to livestock, or graze treated areas, or illegal residues may result.

Rotational Crops: If treated crop is lost, cotton may be replanted. Do not rework the soil. Refer to the Prowl label and the Cotton section of this label for rotational crop restrictions.

Treflan® EC Tank Mixture (Arizona, California, New Mexico and the upper and lower El Paso Valley of Texas)

This combination controls weeds listed on this label and on the Treflan label. This combination also controls shallow-germinating seedlings of cocklebur and coffeeweed.

Follow procedures on the Treflan label for soil preparation and incorporation. Apply the tank mix combination to the flat soil before disking.

Pour Caparol 4L directly into spray tank one-half to three-fourths full of water, allow it to disperse with agitation, add Treflan EC, and then add the rest of the water. Continuous agitation in the spray tank is required to keep the material in suspension. Apply the tank mixture in at least 20 gals. of water per acre at the appropriate rates from Table 6.

Table 6: Tank Mixture with Treflan EC

Soil texture	Broadcast rate per acre	
	Treflan EC	Caparol 4L
sand, loamy sand	DO NOT USE	
sandy loam	1 pt.	2.4-3.2 pts.*
medium soils	1½ pts.	4 pts.
fine soils	2 pts.	4 pts.
muck or peat	DO NOT USE	

* Use less than 3.2 pts. per acre only in Arizona and California.

Precautions: *To avoid crop injury, 1) Do not use in cut areas of newly leveled fields; in areas of excess salt, or in areas where flooding over the bed is likely to occur. 2) Do not plant cotton in tractor wheel depressions. 3) On mulch planted cotton, water back only after cotton seedlings are well established.*

Note: Do not feed treated forage to livestock, or graze treated areas, or illegal residues may result.

Rotational Crops: Cabbage, celery, okra, onions, and peas may be planted in the fall after a spring application of Treflan + Caparol. Winter barley, rye, and wheat can be planted in the fall if they are plowed down and not used for food or feed. Refer to the Treflan EC label for other directions and precautions.

Treflan EC Split Application (Arizona and California)
Apply a preplant incorporated application of Treflan EC as directed on that label, except use the appropriate rate from Table 6. Do not apply Treflan EC before January 1. Follow at planting or just before planting with a preplant incorporated treatment of Caparol 4L as directed in the Cotton section of this label, except use the appropriate rate from Table 6.

DSMA or MSMA
For faster knockdown of the weeds controlled by Caparol 4L alone, apply 1-1.3 pts. of Caparol 4L plus 3 lbs. active ingredient of DSMA or 2 lbs. active ingredient of MSMA per acre, following the same directions, precautions, and limitations as given on this label for Caparol 4L applied alone postemergence directed (chemical hoe). If needed, make a second application 1-3 weeks after the first application. Do not apply after first bloom.

Several formulations of DSMA and MSMA are available under various trade names from several manufacturers. Observe the directions, limitations, and precautions on the label of the product used.

CELERY

Seedbeds (Florida)
Broadcast 1.2-1.6 pts. in a minimum of 20 gals. of water per acre after celery has 2-5 true leaves. Application may be made over the celery. Apply only after seedbed covers have been removed from seedbeds for at least one week. Apply only once per year to seedbeds.

Direct-seeded Celery (California only)
Apply Caparol 4L at rates given below in a minimum of 20 gals. of water per acre. Within the rate ranges given, use the lower rates on coarse-textured soils and soils low in organic matter; use the higher rates on fine-textured soils and soils high in organic matter.

Preemergence: Broadcast 2.4-3.2 pts. per acre at planting or shortly after planting before celery emerges.

Postemergence: Broadcast 1.6-2 pts. per acre after celery has 2-5 true leaves. Application may be made over the celery. Apply before weeds are 2 inches tall.

Precautions: *To avoid injury to direct-seeded celery: 1) Make either one preemergence or one postemergence application (not both) per celery crop. 2) Do not use on sand or loamy sand. 3) Do not apply if celery is under water stress. 4) Do not apply postemergence treatments of Caparol 4L with other pesticides. Apply only after foliar applications of other pesticides are dry. 5) Do not apply within two weeks after an application of a herbicidal oil, such as "carrot" oil.*

Transplants
Apply one application at the appropriate rate from Table 7 in a minimum of 20 gals. of water per acre during the 2-6 week period after transplanting. Within the rate ranges given, use the lower rate on relatively coarse-textured soils and soils low in organic matter; use the higher rate on relatively fine-textured soils and soils high in organic matter. Application may be made over the celery. Apply before weeds are 2 inches tall.

Table 7: Transplanted Celery

State	Soil	Broadcast rate per acre
Florida	sandy or muck	1.6-3.2 pts.
California	coarse-textured	2-3.2 pts.
	fine-textured	3.2-4 pts.
Hawaii	coarse-textured	3.2-4.8 pts.
	fine-textured	4.8-6.4 pts.
Michigan, Ohio	fine-textured or muck	2-4 pts.
Wisconsin	fine-textured	3.2-4 pts.

Rotational Crops
The following crops may be seeded 5 months after applying no more than 4 pts. Caparol 4L per acre on celery: cabbage, celery, corn, onions, peas, and red beets.

PIGEON PEAS (PUERTO RICO ONLY)

For preemergence control of annual weeds, such as horse purslane, jungle rice, wild spider flower, jimsonweed, spurge, pigweed and Florida pusley, apply 4 pts. of Caparol 4L per acre on loam soils, or 6 pts. per acre on clay soils. Apply at planting or immediately after planting before the crop or weeds emerge.

Precautions: *To avoid crop injury 1) Make only one application per year. 2) Do not use on sandy or loamy sand soils.*

Note: Do not graze or feed forage or hay to livestock, or illegal residues may result.

Rotational Crops
The following crops may be seeded 5 months after applying no more than 4 pts. Caparol 4L per acre on pigeon peas: cabbage, celery, corn, onions, peas, and red beets.

STORAGE AND DISPOSAL

Do not contaminate water, food, or feed by storage or disposal. Pesticide wastes may be toxic. Improper disposal of unused pesticide, spray mixture, or rinse water is a violation of federal law. If these wastes cannot be used according to label instructions, contact your State Pesticide or Environmental Control Agency, or the Hazardous Waste representative at the nearest EPA Regional Office for guidance in proper disposal methods.

Triple rinse container (or equivalent). Then offer for recycling or reconditioning, or puncture and dispose of in a sanitary landfill or by other procedures approved by state and local authorities.

For minor spills, leaks, etc., follow all precautions indicated on this label and clean up immediately. Take special care to avoid contamination of equipment and facilities during cleanup procedures and disposal of wastes. In the event of a major spill, fire or other emergency, call (919) 292-7100 day or night.

PRECAUTIONARY STATEMENTS

HAZARDS TO HUMANS AND DOMESTIC ANIMALS

CAUTION

Harmful if absorbed through skin or inhaled. Causes moderate eye irritation. Avoid contact with eyes, skin, or clothing. Avoid repeated skin contact, as this product may cause skin sensitization reactions in certain individuals. Avoid breathing vapor or spray mist. Wash thoroughly with soap and water after handling. Remove contaminated clothing and wash before reuse.

First Aid: If in eyes, flush with plenty of water. If on skin, wash with plenty of soap and water. Get medical attention if eye or skin irritation persists.

ENVIRONMENTAL HAZARDS

Do not contaminate any body of water by direct application, cleaning of equipment, or disposal of wastes. Do not apply where runoff is likely to occur. Do not apply when weather conditions favor drift from areas treated.

Caparol® trademark of CIBA-GEIGY for prometryn

Prowl® trademark of American Cyanamid for pendimethalin

Treflan® trademark of Elanco Products for trifluralin

X-77® trademark of KALO Agricultural Chemicals, Inc.

CGA 17L7C 065

COTORAN®80W

HERBICIDE

For preemergence and postemergence weed control in cotton and sugarcane

Cotoran 80W is a wettable powder

ACTIVE INGREDIENT:
Fluometuron:
1,1-dimethyl-3-(α,α,α-trifluoro-*m*-tolyl) urea 80%
INERT INGREDIENTS: 20%

TOTAL 100%

EPA Reg. No. 100-609

Cotoran® trademark of CIBA-GEIGY for fluometuron

Keep Out of Reach of Children.

Warning

See additional precautionary statements on side of bag.

See directions for use on panel at right.

DIRECTIONS FOR USE AND CONDITIONS OF SALE AND WARRANTY

IMPORTANT: Read the entire **Directions for Use** and the **Conditions of Sale and Warranty** before using this product.

CONDITIONS OF SALE AND WARRANTY

The **Directions for Use** of this product reflect the opinion of experts based on field use and tests. The directions are believed to be reliable and should be followed carefully. However, it is impossible to eliminate all risks inherently associated with use of this product. Crop injury, ineffectiveness, or other unintended consequences may result because of such factors as weather conditions, presence of other materials, or the manner of use or application all of which are beyond the control of CIBA-GEIGY or the Seller. All such risks shall be assumed by the Buyer.

CIBA-GEIGY warrants that this product conforms to the chemical description on the label and is reasonably fit for the purposes referred to in the **Directions for Use** subject to the inherent risks referred to above. **CIBA-GEIGY makes no other express or implied warranty of Fitness or Merchantability or any other express or implied warranty. In no case shall CIBA-GEIGY or the Seller be liable for consequential, special, or indirect damages resulting from the use or handling of this product.** CIBA-GEIGY and the Seller offer this product, and the Buyer and user accept it, subject to the foregoing **Conditions of Sale and Warranty,** which may be varied only by agreement in writing signed by a duly authorized representative of CIBA-GEIGY.

GENERAL INFORMATION

Cotoran 80W is effective for both annual grass and annual broadleaf weed control in cotton and sugarcane.

WEED CONTROL

Cotoran 80W applied according to directions will selectively control annual weeds in cotton and sugarcane, such as:

Annual Grass Weeds

Barnyardgrass	Fall panicum
Brachiaria	Foxtail
Crabgrass	Goosegrass
Crowfootgrass	Ryegrass

Annual Broadleaf Weeds

Buttonweed	Pigweed
Cocklebur	Prickly sida (Ironweed or teaweed)
Florida pusley	Purslane
Goathead	Ragweed
Groundcherry (Wright)	Sesbania
Jimsonweed	Sicklepod
Lambsquarters	Smartweed
Morningglory	Tumbleweed

Cotoran 80W at rates up to 2.5 lbs. per acre in cotton and up to 5 lbs. per acre in sugarcane is not effective for the control of established perennial weeds, such as johnsongrass from rhizomes.

APPLICATION PROCEDURES

MIXING INSTRUCTIONS

Cotoran 80W is highly concentrated and should be thoroughly mixed as a paste or thick slurry in water in a separate container before adding to the spray tank. The spray tank should contain ⅓-½ the total amount of water (cotton or sugarcane) or nitrogen solution (cotton only) to be used before adding the slurry.

To determine compatibility, add one tablespoon of Cotoran 80W to one pint of nitrogen solution (equals 2.5 lbs. in 25 gals; if using other application rates, adjust compatibility test amounts proportionally). Stir or shake thoroughly. Let stand 5 minutes. If it remains mixed or can be remixed readily, the mixture is compatible and can be used. If Cotoran 80W settles out when mixed with nitrogen solution, a suspensibility agent, such as Compex® at 1-2 pts. per 100 gals. fertilizer, may improve compatibility. Add ⅛-¼ teaspoon of Compex to the compatibility test to determine its effectiveness.

Use bypass or mechanical agitation in the spray tank. Avoid leaving spray mixture in tank without constant agitation. If bypass agitation is used, it should terminate at the bottom of the tank to minimize foaming. Wash sprayer thoroughly with clean water immediately after use.

EQUIPMENT

Thoroughly clean sprayer prior to use. Do not use a sprayer contaminated with 2,4-D or other materials as crop damage or sprayer clogging may result.

For uniform distribution use fan-type nozzles. Use 50-mesh or coarser screens in nozzles and strainers. Use a suitable pump with capacity to deliver (1) the necessary volume to the nozzles at 35-40 psi, and (2) an additional volume to maintain the mixture in suspension in the spray tank. Apply in 20-40 gals. of spray mixture per acre.

BAND TREATMENT

For band treatment, calculate the amount of Cotoran 80W needed by the formula:

$$\frac{\text{band width in inches}}{\text{row width in inches}} \times \text{broadcast rate per acre} = \text{amount needed per acre of field}$$

DIRECTIONS FOR USE

COTTON

Cotoran 80W may be applied as a preemergence treatment to the surface of the soil, or lightly incorporated immediately after planting, or as a directed, semi-directed, or over-the-top postemergence application. For preemergence treatment only, Cotoran may be tank-mixed with most liquid nitrogen fertilizers. Cotoran may be combined with other herbicides in cotton as directed in the section, "Combinations."

Cotoran 80W is most effective when used as follows:

1st Application: Preemergence broadcast. See Preemergence Treatment below.

2nd Application: Postemergence, directed, semi-directed, or over-the-top when cotton is 3-6 inches high. See Postemergence Treatments below.

3rd Application: If needed, postemergence, directed, or semi-directed, at lay-by (last normal cultivation), using no more than 1.25 lbs. per acre.

PREEMERGENCE TREATMENT

Apply Cotoran 80W at a rate of 1.25-2.5 lbs. per acre. On light silt and sandy soils, low in organic matter, use the lower rate.

Cotoran may be applied to the surface of the soil, or lightly incorporated into the soil, after planting cotton.

PREPLANT INCORPORATED USE (Arizona, California, and New Mexico)

Cotoran 80W may be applied preplant as a broadcast spray at 2 lbs. per acre after the beds for preplant irrigation have been formed. Apply before the beds are dragged off (using a bed shaper set to remove 1-2 inches from beds) in preparation for planting. Furrow irrigation should precede the application and planting should follow as soon as possible. Plant into untreated soil after drag-off and return treated soil to the beds when furrows are re-formed. Do not allow furrow irrigation water to pond over the beds.

POSTEMERGENCE TREATMENTS

Cotoran 80W may be applied postemergence to cotton but preemergence to weeds, or postemergence to both cotton and weeds. Cotoran 80W should be applied at rates of 1.25-2.5 lbs. per acre. Apply as a directed, semi-directed, or over-the-top spray. Cotoran may be applied to cotton plants at any stage of growth from 3 inches high to lay-by. Use the higher rate when applying postemergence to weeds; weeds should be 2 inches high or less. Use of a surfactant that is approved by FDA for use on food and feed crops and is compatible with Cotoran 80W when applied to cotton may enhance the postemergence herbicidal activity of Cotoran. The surfactant rate should not exceed 0.5% of the carrier volume (1 pt. per 25 gals. of water).

Precautions: *Rainfall that is sufficient to germinate seeds or for growing crops is adequate to activate the herbicidal activity of Cotoran; where dry weather conditions prevail, the herbicidal activity of Cotoran may be delayed or reduced.*

Do not make more than three applications of Cotoran to the same crops or field in any one year. Do not plant crops other than cotton within 6 months of the last application of Cotoran or injury may result.

The use of Cotoran 80W as a preemergence or early postemergence application, following the use of a systemic insecticide at planting, may result in injury to cotton.

West Texas only: Do not use Cotoran on sandy, loamy sand, or fine sandy loam soils. Do not use Cotoran on cotton planted in furrows.

Arkansas, Louisiana, and Mississippi only: Use 1 lb. of Cotoran 80W per acre on sandy loam soils low in organic matter.

Arizona, California, and New Mexico only: Cotton, corn, and grain sorghum can be planted the next spring. Do not plant treated areas to other crops until one year after last application. Do not use on sandy or coarse-textured soil of less than 1% organic matter.

Note: Do not apply within 60 days of harvest. Do not feed foliage from treated cotton plants or gin trash to livestock.

COMBINATIONS

Tolban® 4E: Use in a tank mixture or in an overlay combination with Cotoran 80W in cotton for control of all weeds listed on both labels. Refer to the Tolban 4E label for all directions, precautions, and limitations.

Treflan® EC: Use in a tank mixture with Cotoran 80W (all areas except Arizona and California) or before an overlay with Cotoran 80W in cotton for control of all weeds listed on both labels. For the **tank mixture,** refer to the Treflan EC label for directions, precautions, and limitations. For the **overlay,** apply Cotoran 80W as a preemergence surface treatment as recommended on this label at 1.25-2.5 lbs. per acre after a preplant soil-incorporated treatment of Treflan EC. Apply Treflan EC as recommended on its label at 1-2.5 pts. per acre.

Prowl®: Apply Cotoran 80W as a preemergence surface treatment as recommended on this label at 1.25-2.5 lbs. per acre after a preplant soil-incorporated treatment of Prowl for control of all weeds listed on both labels. Apply Prowl as recommended on its label at 1-3 pts. per acre.

DSMA or MSMA: Cotoran may be tank-mixed with DSMA or MSMA and applied as a directed spray to cotton from three inches high to first bloom. Use the broadcast rate of 1.25-2.5 lbs. of Cotoran per acre and the necessary amount of DSMA or MSMA plus 0.5% (1 pt. per 25 gals.) of a surfactant that is approved by FDA for use on food and feed crops and is compatible with Cotoran 80W + DSMA or Cotoran 80W + MSMA when applied to cotton. When using this mixture, do not allow spray to contact cotton foliage. Do not graze treated areas or feed treated forage or gin trash to livestock. Adhere strictly to the dosage and limitations on the labels of the products.

SUGARCANE

Cotoran 80W may be applied as a preemergence application to sugarcane at any time after planting, and a second application may be made postemergence to sugarcane 12-18 inches high.

Cotoran 80W is most effective in sugarcane when used as follows:

PLANT CANE

First Application: Preemergence broadcast, at a rate of 2.5-5 lbs. per acre to the surface of the soil after planting sugarcane, or any time before cane and weeds emerge. On muck soils, the maximum rate of Cotoran 80W may not give adequate weed control.

Second Application: Postemergence, directed, semi-directed, or over-the-top of the cane at a rate of 2.5-5 lbs. per acre when sugarcane is 12-18 inches high. This application should be made preemergence to weeds, or to weeds less than 2 inches in height.

RATOON CANE

First Application: Preemergence broadcast, at a rate of 2.5-5 lbs. per acre after harvest and any time before new cane shoots and weeds emerge.

Second Application: Postemergence, directed, semi-directed, or over-the-top of the cane at a rate of 2.5-5 lbs. per acre when cane is 12-18 inches high. This application should be made preemergence to weeds, or to weeds less than 2 inches in height.

Precaution: *Do not apply more than 10 lbs. of Cotoran 80W to the soil during any one year. Do not plant crops other than sugarcane or cotton within one year after the last application or injury may result.*

Note: Do not apply within 180 days of harvest. Do not graze treated fields with livestock.

STORAGE AND DISPOSAL

Store in a dry place. Do not contaminate water, food, or feed by storage or disposal. Improper disposal of unused spray mixture, or rinse water is a violation of federal law. If these wastes cannot be used according to label instructions, contact your State Pesticide or Environmental Control Agency, or the Hazardous Waste representative at the nearest EPA Regional Office for guidance in proper disposal methods.

Completely empty bag into application equipment. Dispose of empty bag in a sanitary landfill or by incineration, or if allowed by state and local authorities, by burning. Stay out of smoke from burning bags.

For minor spills, leaks, etc., follow all precautions indicated on this label and clean up immediately. Take special care to avoid contamination of equipment and facilities during cleanup procedures and disposal of wastes. In the event of a major spill, fire or other emergency, call (919) 292-7100 day or night.

PRECAUTIONARY STATEMENTS

HAZARDS TO HUMANS AND DOMESTIC ANIMALS

WARNING

Causes eye irritation. Do not get in eyes. Harmful if swallowed, inhaled, or absorbed through the skin. Wear protective clothing, as this product may cause skin rash after repeated exposure. Avoid contact with skin or clothing. Avoid breathing dust or spray mist.

First Aid: In case of contact with the eyes, flush with water for at least 15 minutes and get medical attention if irritation persists. In case of contact with skin, wash immediately with plenty of soap and water. If swallowed, drink promptly a large quantity of milk, egg whites, or gelatin solution; if these are not available, drink large quantities of water. Induce vomiting. Avoid alcohol. Call a physician immediately. In case of inhalation exposure, move from the contaminated area.

Note to Physician: There is no specific antidote for Cotoran 80W. Induce emesis and lavage stomach. Give a saline laxative and supportive therapy.

ENVIRONMENTAL HAZARDS

Do not contaminate water used for irrigation or domestic purposes. Do not contaminate water by cleaning of equipment or disposal of wastes.

Cotoran® trademark of CIBA-GEIGY for fluometuron

Compex® trademark of KALO Agricultural Chemicals, Inc.

Prowl® trademark of American Cyanamid for pendimethalin

Tolban® trademark of CIBA-GEIGY for profluralin

U.S. Patent Nos. 3,546,295; 3,672,864

Treflan® trademark of Elanco Products for trifluralin

CGA 33L11B 084

COTORAN® 4L

HERBICIDE

For preemergence and postemergence weed control in cotton

Cotoran 4L contains 4 lbs. active ingredient per gallon.

Use entire contents at one time.

ACTIVE INGREDIENT:

Fluometuron: 1,1-dimethyl-3-(a,a,a-trifluro-m-tolyl) urea	41.7%
INERT INGREDIENTS:	58.3%
TOTAL:	100.0%

EPA Reg. No. 100-642

Keep Out of Reach of Children.

Caution

See additional precautionary statements at end of label booklet.

See directions for use inside booklet.

DIRECTIONS FOR USE AND CONDITIONS OF SALE AND WARRANTY

IMPORTANT: Read the entire **Directions for Use** and the **Conditions of Sale and Warranty** before using this product.

CONDITIONS OF SALE AND WARRANTY

The **Directions for Use** of this product reflect the opinion of experts based on field use and tests. The directions are believed to be reliable and should be followed carefully. However, it is impossible to eliminate all risks inherently associated with use of this product. Crop injury, ineffectiveness, or other unintended consequences may result because of such factors as weather conditions, presence of other materials, or the manner of use or application all of which are beyond the control of CIBA-GEIGY or the Seller. All such risks shall be assumed by the Buyer.

CIBA-GEIGY warrants that this product conforms to the chemical description on the label and is reasonably fit for the purposes referred to in the **Directions for Use** subject to the inherent risks referred to above. **CIBA-GEIGY makes no other express or implied warranty of Fitness or Merchantability or any other express or implied warranty. In no case shall CIBA-GEIGY or the Seller be liable for consequential, special, or indirect damages resulting from the use or handling of this product.** CIBA-GEIGY and the Seller offer this product, and the Buyer and user accept it, subject to the foregoing **Conditions of Sale and Warranty,** which may be varied only by agreement in writing signed by a duly authorized representative of CIBA-GEIGY.

DIRECTIONS FOR USE

It is a violation of federal law to use this product in a manner inconsistent with its labeling.

FAILURE TO FOLLOW THE DIRECTIONS FOR USE AND PRECAUTIONS ON THIS LABEL MAY RESULT IN CROP INJURY, POOR WEED CONTROL, AND/OR ILLEGAL RESIDUES.

GENERAL INFORMATION

Cotoran 4L is effective for both annual grass and annual broadleaf weed control in cotton.

Weed Control

Cotoran 4L applied according to directions will selectively control annual weeds in cotton, such as:

Annual Grass Weeds

Barnyardgrass	Fall panicum	Ryegrass
Crabgrass	Foxtail	Signalgrass *(Brachiaria)*
Crowfootgrass	Goosegrass	

Annual Broadleaf Weeds

Buttonweed	Lambsquarters	Ragweed
Cocklebur	Morningglory	Sesbania
Florida pusley	Pigweed	Sicklepod
Groundcherry (Wright)	Prickly sida (teaweed)	Smartweed
Jimsonweed	Purslane	Tumbleweed

Cotoran 4L at rates up to 4 pts. per acre in cotton is not effective for the control of established perennial weeds, such as johnsongrass from rhizomes.

APPLICATION PROCEDURES

MIXING INSTRUCTIONS

(1) Fill spray tank ½ to ¾ full with clean water or nitrogen solution. (2) Start agitation. (3) Pour Cotoran 4L directly from the container into the partially filled spray tank. (4) If Cotoran 4L is to be tank mixed with another herbicide, add that product after the Cotoran 4L is thoroughly mixed in the partially filled spray tank. (5) Add the rest of the water or nitrogen solution. (6) Provide agitations during mixing and application to maintain a uniform suspension.

To determine compatability, add two teaspoons of Cotoran 4L to one pint of nitrogen solution (equals 4 pts. in 25 gals.; if using other application rates, adjust compatibility test amounts proportionally). Stir or shake thoroughly. Let stand 5 minutes. If it remains mixed or can be remixed readily, the mixture is compatible and can be used. If Cotoran 4L settles out when mixed with a nitrogen solution, a suspensibility agent, such as Unite®, Compex®, or equivalent at 1-3 pts. per 100 gals. fertilizer, may improve compatibility. Add ⅛-⅜ teaspoon of Unite to the compatibility test to determine its effectiveness.

EQUIPMENT

Thoroughly clean sprayer prior to use. Do not use a sprayer contaminated with 2, 4-D or other materials as crop damage or sprayer clogging may result.

Use properly calibrated conventional spray equipment that will apply spray uniformly and has hydraulic or mechanical agitation, except in California or Arizona where only mechanical agitators are recommended. Screens in nozzles, in suction, and in-line strainers should be no finer than 50-mesh. Use a pump with capacity to maintain 35-40 psi at the nozzles. If hydraulic agitation is used, the pump should also provide sufficient agitation in the tank to keep the mixture in suspension. Avoid leaving spray mixture in tank without constant agitation. If bypass agitation is used, it should terminate at the bottom of the tank to minimize foaming. Wash sprayer thoroughly with clean water immediately after use.

Apply in a minimum of 10 gals. per acre for preplant incorporated or preemergence applications, or apply in a minimum of 20 gals per acre for postemergence application.

Band Treatment

For band treatment, calculate the amount of Cotoran 4L needed by the formula:

$$\frac{\text{band width in inches}}{\text{row width in inches}} \times \text{broadcast rate per acre} = \text{amount needed per acre of field}$$

COTTON

Cotoran 4L may be applied as a preemergence treatment to the surface of the soil, or lightly incorporated immediately after planting, or as a directed, semidirected, or over-the-top postemergence application. For preemergence treatment only, Cotoran 4L may be tank-mixed with most liquid nitrogen fertilizers. Cotoran 4L may be combined with other herbicides in cotton as directed in the section, "Combinations."

Cotoran 4L is most effective when used as follows:

1st Application: Preemergence broadcast. See Preemergence Treatment below.

2nd Application: Postemergence, directed, semidirected, or over-the-top when cotton is 3-6 inches high. See Postemergence Treatments below.

3rd Application: If needed, postemergence, directed, or semidirected, at lay-by (last normal cultivation), using no more than 2 pts. per acre.

Preemergence Treatment

Apply Cotoran 4L at a rate of 2-4 pts. per acre. On light silt and sandy soils, low in organic matter, use the lower rate.

Cotoran 4L may be applied to the surface of the soil, or lightly incorporated into the soil, after planting cotton.

Preplant Incorporated Use (Arizona, California, and New Mexico)

Cotoran 4L may be applied preplant as a broadcast spray at 3.2 pts. per acre after the beds for preplant irrigation have been formed. Apply before the beds are dragged off (using a bed shaper set to remove 1-2 inches from beds) in preparation for planting. Furrow irrigation should precede the application and planting should follow as soon as possible. Plant into untreated soil after drag-off and return treated soil to the beds when furrows are reformed. Do not allow furrow irrigation water to pond over the beds.

Postemergence Treatments

Cotoran 4L may be applied postemergence to cotton but preemergence to weeds, or postemergence to both cotton and weeds. Cotoran 4L should be applied at rates ot 2-4 pts. per acre. Apply as a directed, semidirected, or over-the-top spray. Cotoran 4L may be applied to cotton plants at any stage of growth from 3 inches high to lay-by. Use the higher rate when applying postemergence to weeds; weeds should be 2 inches high or less. Use of a surfactant that is approved by FDA for use on food and feed crops and is compatibile with Cotoran 4L when applied to cotton may enhance the postemergence herbicidal activity of Cotoran 4L. The surfactant rate should not exceed 0.5% of the carrier volume (1 pt. per 25 gals. of water).

Precaution: *Rainfall that is sufficient to germinate seeds or for growing crops is adequate to activate the herbicidal activity of Cotoran 4L; where dry weather conditions prevail, the herbicidal activity of Cotoran 4L may be delayed or reduced. Cultivate if weeds develop.*

Do not make more than three applications of Cotoran 4L to the same crops or field in any one year. Do not plant crops other than cotton or sugarcane within 6 months of the last application of Cotoran 4L or injury may result.

The use of Cotoran 4L as a preemergence or early postemergence application, following the use of a systematic insecticide at planting, may result in injury to cotton.

West Texas only: Do not use Cotoran 4L on sandy, loamy sand, or fine sandy loam soils. Do not use Cotoran 4L on cotton planted in furrows.

Arkansas, Louisiana, and Mississippi only: Use 1.6 pts. of Cotoran 4L per acre on sandy loam soils low in organic matter.

Arizona, California, and New Mexico only: Cotton, corn and grain sorghum can be planted the next spring. Do not plant treated areas to other crops until one year after last application. Do not use on sandy or coarse-textured soil of less than 1% organic matter.

Note: Do not apply within 60 days of harvest. Do not feed foliage from treated cotton plants or gin trash to livestock.

COMBINATIONS

Dual® 8E: Use in a tank mixture with Cotoran 4L preemergence in cotton for control of all weeds as listed on both lables. Apply to the soil surface at planting or after planting, but before weeds or crop emerge, using the appropriate rates from the table below.

Cotoran 4L + Dual 8E—Cotton (AR, LA, MS, Eastern OK, TN, Bootheel of MO, and Gulf Coast, Rio Grande Valley and Eastern TX)

Soil texture	Broadcast rates per acre	
	Cotoran 4L	Dual 8E
Sand, loamy sand	DO NOT USE	
Sandy loam*	2 pts.	1.25-1.5 pts.
Loam, silt, silt loam	2-3.2 pts.	1.5-2 pts.
Silty clay loam, sandy clay loam, silty clay, sandy clay, clay loam, clay	3.2-4 pts.	2 pts.

* Use on sandy loam soils only in eastern OK and Gulf Coast, Rio Grande Valley, and eastern TX.

Precautions: *(1) Do not apply Cotoran 4L + Dual 8E on sand or loamy sand soils or in areas where water is likely to "pond" over the bed, or crop injury may occur. (2) Do not apply broadcast to furrow-planted cotton. (3) The use of Cotoran 4L following the use of a systemic insecticide at planting may result in crop injury. (4) Do not use on Taloka silt loam, or crop injury may occur.*

Refer to the Dual 8E label for further instructions, precautions, and limitations.

Note: Do not feed treated forage or gin trash to livestock, or graze treated areas.

Rotational Crops: Refer to the crop rotation instructions for Cotoran 4L alone on this label and for Dual 8E alone on the Dual 8E label.

Treflan® EC: Use in a tank mixture with Cotoran 4L (all areas except Arizona and California) or before an overlay with Cotoran 4L in cotton for control of all weeds listed on both labels.

For the **tank mixture,** follow soil preparation and incorporation procedures for Treflan EC. Apply Cotoran 4L + Treflan EC tank mix at rates specified below in a minimum of 15 gals. of water per acre.

Cotoran 4L + Treflan EC Tank Mixture—Cotton (All areas except Arizona and California)

Soil texture	Broadcast rates per acre	
	Cotoran 4L	Treflan EC
Sand, loamy sand, sandy loam	2 pts.	1 pt.
Loam, silty clay loam*, silt loam, silt, sandy clay loam*	3.2 pts.	1.5 pts.
Clay, clay loam, silty clay loam*, silty clay, sandy clay, sandy clay loam*	4 pts.	2 pts.

* Silty clay loam and sandy clay loam are transitional soils. If these soils are predominantly sand or silt, they are usually classified as medium-textured soils. If predominantly clay, they are classified as fine-textured soils.

For the **overlay**, apply Cotoran 4L as a preemergence soil surface treatment as recommended on this label at 2-4 pts. per acre after a preplant soil incorporated treatment of Treflan EC. Apply Treflan EC as recommended on its label at 1-2.5 pts. per acre.

Precautions: Do not use the tank mix in Arizona and California. Do not apply the tank mix of Cotoran 4L + Treflan EC in liquid fertilizer. Refer to the Treflan EC label for directions, precautions, and limitations for use of that product, as well as the directions, precautions, and limitations on this Cotoran 4L label.

Prowl®: Apply Cotoran 4L as a preemergence surface treatment as recommended on this label at 2-4 pts. per acre after a preplant soil-incorporated treatment of Prowl for control of all weeds listed on both labels. Apply Prowl as recommended on its label at 1-3 pts. per acre. Refer to the Prowl label for directions, precautions, and limitations for use of that product.

DSMA or MSMA: Cotoran 4L may be tank-mixed with DSMA or MSMA and applied as a directed spray to cotton from three inches high to first bloom. Use the broadcast rate of 2-4 pts. of Cotoran 4L per acre and the necessary amount of DSMA or MSMA plus 0.5% (1 pt. per 25 gals.) of a surfactant that is approved by FDA for use on food and feed crops and is compatible with Cotoran 4L + DSMA or Cotoran 4L + MSMA when applied to cotton. When using this mixture, do not allow spray to contact cotton foliage. Do not graze treated areas or feed treated forage or gin trash to livestock. Adhere strictly to the dosage, precautions and limitations on the labels of the products.

STORAGE AND DISPOSAL

Do not contaminate water, food, or feed by storage or disposal. Pesticide wastes may be toxic. Improper disposal of unused pesticide, spray mixture, or rinse water is a violation of federal law. If these wastes cannot be used according to label instructions, contact your State Pesticide or Environmental Control Agency, or the Hazardous Waste representative at the nearest EPA Regional Office for guidance in proper disposal methods.

Triple rinse container (or equivalent). Then offer for recycling or reconditioning, or puncture and dispose of in a sanitary landfill or by incineration, or if allowed by state and local authorities, by burning. Stay out of smoke from burning jugs.

For minor spills, leaks, etc., follow all precautions indicated on this label and clean up immediately. Take special care to avoid contamination of equipment and facilities during cleanup procedures and disposal of wastes. In the event of a major spill, fire or other emergency, call (919) 292-7100 day or night.

PRECAUTIONARY STATEMENTS

HAZARDS TO HUMANS AND DOMESTIC ANIMALS

CAUTION

Harmful if swallowed, inhaled, or absorbed through the skin. Avoid contact with skin, eyes, or clothing. Avoid breathing dust or spray mist. Wash thoroughly with soap and water after handling. Remove and wash contaminated clothing before reuse.

First Aid: In case of contact with eyes, immediately flush with plenty of water. Get medical attention if irritation occurs. In case of contact with skin, wash thoroughly with soap and water. Get medical attention if irritation occurs. If swallowed, drink water and induce vomiting by placing fingers at back of throat. Immediately contact a physician, hospital, or local poison control center. In case of inhalation exposure, move from the contaminated area.

Notes to Physician: If Cotoran 4L is ingested, lavage stomach or induce vomiting. A slurry of activated charcoal in water can be left in the stomach. Give a saline laxative and supportive therapy.

ENVIRONMENTAL HAZARDS

Do not apply directly to water or wetlands. Do not contaminate water by cleaning of equipment or disposal of wastes.

Cotoran® trademark of CIBA-GEIGY for fluometuron

Compex® trademark of KALO Agricultural Chemicals, Inc.

Dual® trademark of CIBA-GEIGY for metolachlor

Prowl® trademark of American Cyanamid for pendimethalin

Treflan® trademark of Elanco Products for trifluralin

Unite® trademark of Hopkins Agricultural Chemical Co.

CGA 33L14A 055

CURACRON®6E

INSECTICIDE-MITICIDE

RESTRICTED USE PESTICIDE

For Retail Sale to and Use Only by Certified Applicators or Persons Under Their Direct Supervision. And Only for Those Uses Covered by the Certified Applicators Certification

For insect and mite control in cotton

ACTIVE INGREDIENT:

Profenofos: 0-(4-bromo-2-chlorophenyl)0-ethyl *S*-propyl phosphorothioate	59.4%
INERT INGREDIENTS:	40.6%
TOTAL:	100.0%

EPA Reg. No. 100-599 EPA Est. 100-AL-1

Curacron® trademark of CIBA-GEIGY for profenofos U.S. Patent No. 3,992,533

Curacron 6E contains 6 lbs. active ingredient per gallon

Keep Out of Reach of Children.

WARNING

See additional precautionary statements at end of label booklet.

See directions for use inside booklet.

DIRECTIONS FOR USE AND CONDITIONS OF SALE AND WARRANTY

IMPORTANT: Read the entire **Directions for Use** and the **Conditions of Sale and Warranty** before using this product.

CONDITIONS OF SALE AND WARRANTY

The **Directions for Use** of this product reflect the opinion of experts based on field use and tests. The directions are believed to be reliable and should be followed carefully. However, it is impossible to eliminate all risks inherently associated with use of this product. Crop injury, ineffectiveness, or other unintended consequences may result because of such factors as weather conditions, presence of other materials, or the manner of use or application all of which are beyond the control of CIBA-GEIGY or the Seller. All such risks shall be assumed by the Buyer.

CIBA-GEIGY warrants that this product conforms to the chemical description on the label and is reasonably fit for the purposes referred to in the **Directions for Use** subject to the inherent risks referred to above. **CIBA-GEIGY makes no other express or implied warranty of Fitness or Merchantability or any other express or implied warranty. In no case shall CIBA-GEIGY or the Seller be liable for consequential, special, or indirect damages resulting from the use or handling of this product.** CIBA-GEIGY and the Seller offer this product, and the Buyer and user accept it, subject to the foregoing **Conditions of Sale and Warranty,** which may be varied only by agreement in writing signed by a duly authorized representative of CIBA-GEIGY.

GENERAL INFORMATION

Curacron 6E is an emulsifiable insecticide-miticide which is effective for control of cotton bollworm, tobacco budworm, certain other insects, and mites on cotton.

Best results can be expected from fields under Pest Management/Scouting programs. NOTE: When using Curacron 6E as a larvicide, best results will be obtained when application is timed to control first to third instar larvae.

Mixing Instructions

Prepare no more spray mixture than is needed for the immediate operation. Thoroughly clean spray equipment before using this product. Agitate thoroughly before and during application. Good bypass agitation is sufficient to keep Curacron 6E in emulsion.

Curacron 6E Alone: Add 2/3 of the required amount of water to the spray tank with the agitator running, then add the desired amount of Curacron 6E. Continue agitation while adding the remainder of the water.

Curacron 6E + Tank Mixtures: Add 2/3 of the required amount of water to the spray tank. With the agitator running if possible, add the desired amount of the other products recommended on this label. Then add the desired amount of Curacron 6E. Continue agitation while adding the remainder of water.

Note: Do not let spray mixture stand overnight in the spray tank. Rinse spray tank at the end of the day.

DIRECTIONS FOR USE

It is a violation of federal law to use this product in a manner inconsistent with its labeling.

FAILURE TO FOLLOW ALL DIRECTIONS AND PRECAUTIONS ON THIS LABEL MAY RESULT IN CROP INJURY, POOR CONTROL, AND/OR ILLEGAL RESIDUES.

COTTON

Apply Curacron 6E at the specified rates below in a minimum of 5 gallons of water per acre with ground equipment or in a minimum of 1 gallon of water per acre by air.

For better coverage of late season "rank" cotton increase gallonage. To avoid spray drift, do not apply under windy conditions. Avoid spray overlap because crop injury may result.

If the need to control boll weevil becomes apparent while using Curacron, Guthion® or methyl parathion may be tank mixed with Curacron. Refer to the Guthion or methyl parathion labels for directions, limitations, and precautions.

Aphids, plant bugs/lygus bugs*. Apply 2/3 pt. of Curacron 6E per acre at 5-7 day intervals as needed starting when economic thresholds are reached. Economic thresholds will vary by locality. Consult your pest control advisor or cotton consultant to determine economic thresholds in your area.

* In Arizona and California, Curacron 6E offers suppression of lygus bugs.

Cotton bollworm, tobacco budworm: Apply Curacron 6E **at 5-7 day intervals as needed beginning when infestations first appear.** Use 2/3 to 1 pt. per acre for light to moderate infestations and 1-1 1/3 pts. per acre for moderate to heavy infestations.

Eggs may be controlled by tank mixing Curacron 6E with Galecron® SP or 4E. Refer to the Galecron SP or 4E label section for directions, limitations, or precautions.

Cotton leafperforator, beet armyworm: Apply Curacron 6E **at 5-7 day intervals as needed** starting when larvae first appear. Use 2/3-1 pt. per acre for light to moderate infestations and 1-1 1/3 pts. per acre for moderate to heavy infestations.

Mites: Apply 2/3-1 pt. of Curacron 6E per acre **at 5-7 day intervals as needed** starting when mites first appear.

Whitefly suppression: To aid in the reduction of whitefly populations, apply 1-1 1/3 pts. of Curacron 6E **at 5-7 day intervals as needed** starting when whiteflies first appear.

Curacron 6E + Tank Mixtures (All cotton growing states): To broaden the spectrum of insect control and make a more useful spray application, if multiple pests occur on cotton, it may be advantageous to tank mix Curacron 6E with other EPA-registered insecticides.

Curacron 6E has been found to be compatible in tank mixtures with the products listed below. When using Curacron 6E in tank mixtures, observe all limitations, precautions, and rate recommendations which appear on the labels of these products: Ambush®, Ammo®, Azodrin®, Galecron, Guthion, Methomyl, Methyl parathion, Pay-Off®, Pounce® and Pydrin®.

Note: To avoid illegal residues: 1) Do not apply more than 4 qts. of Curacron 6E per acre per season. 2) Do not apply Curacron 6E within 14 days of harvest. 3) Do not allow livestock to graze treated cotton. 4) Do not feed gin trash or foliage from treated cotton plants to livestock.

Refer to Guthion and methyl parathion labels for further directions, limitations, and precautions.

Fields treated with Curacron may be rotated to other crops.

STORAGE AND DISPOSAL

Do not contaminate water, food, or feed by storage or disposal.

Pesticide disposal: Pesticide wastes are **Toxic**. Improper disposal of excess pesticide, spray mixture, or rinsate is a violation of Federal law. If these wastes cannot be disposed of by use according to label instructions, contact your State Pesticide or Environmental Control Agency or the Hazardous Waste Representative at the nearest EPA Regional Office for guidance.

Container disposal: Triple rinse (or equivalent). Then puncture container and dispose of in a sanitary landfill, or incinerate. The container may be disposed of by open burning if allowed by state and local authorities. Stay out of smoke from burning container.

Emergency Response
In the event of a major spill, fire, or other emergency, call (919) 292-7100.

PRECAUTIONARY STATEMENTS

HAZARDS TO HUMANS AND DOMESTIC ANIMALS WARNING

May be fatal if absorbed through skin. Harmful if swallowed or inhaled. Causes moderate eye irritation. Do not get in eyes, on skin, or on clothing. Avoid breathing vapors or spray mist. Wear protective clothing and rubber gloves when handling. Wash thoroughly with soap and water before eating or smoking. Remove contaminated clothing and wash before reuse.

STATEMENT OF PRACTICAL TREATMENT

IF IN EYES immediately flush with plenty of water. Get medical attention.

IF ON SKIN wash with plenty of soap and water. Remove contaminated clothing. Get medical attention.

IF SWALLOWED call a physician or poison control center. Give water or milk if conscious. May present aspiration hazard, DO NOT induce vomiting. Get medical attention immediately.

IF INHALED remove to fresh air.

Note to Physician: If symptoms of cholinergic poisoning occur, atropine plus 2-PAM may be helpful. If swallowed, lavage stomach, taking care to avoid aspiration of stomach contents into the lungs.

ENVIRONMENTAL HAZARDS

Due to the hazard to endangered fish, the application of this product is prohibited in Reeves County, Texas and within one mile of the Dexter National Fish Hatchery in New Mexico. This pesticide is toxic to fish and wildlife. Use with care when applying to areas frequented by wildlife or adjacent to any body of water. Do not apply directly to water. Do not apply with aircraft within 300 feet upwind of impounded water. Do not apply when weather conditions favor drift from treated areas. Do not apply with aircraft when wind speed is greater than 10 mph. Apply this pesticide only as specified on this label. Do not contaminate water by cleaning of equipment or disposal of wastes.

This product is highly toxic to bees exposed to direct treatment or residues on crops or weeds. Do not apply this product or allow it to drift to crops or weeds on which bees are foraging. Additional information may be obtained from your Cooperative Agricultural Extension Service.

Guthion® trademark of the Parent Company of Farbenfabriken Bayer GmbH, Leverkusen for azinphos-methyl

Pay Off® trademark of American Cyanamid for flucythrinate

Pounce® trademark of FMC Corporation for permethrin

Pydrin® trademark of Shell Chemical Company for fenvalerate

CGA 55L6C

065

D·Z·N®
diazinon 14G
GRANULAR INSECTICIDE

For control of certain soil insects including wireworms, cutworms and other crop pests

D•z•n diazinon 14G is a granular insecticide

ACTIVE INGREDIENT:
Diazinon:
O.O-diethyl O-(2-isopropyl-6-methyl-4-pyrimidinyl) phosphorothioate 14.3%
INERT INGREDIENTS: 85.7%
TOTAL: 100.0%

EPA Reg. No. 100-469 Control No. 06850

D•Z•N® trademark of CIBA-GEIGY for diazinon

Keep Out of Reach of Children.
Caution

See additional caution statements on back of bag.

See directions for use on back of bag.

DIRECTIONS FOR USE AND CONDITIONS OF SALE AND WARRANTY

IMPORTANT: Read the entire **Directions for Use** and the **Conditions of Sale and Warranty** before using this product.

CONDITIONS OF SALE AND WARRANTY

The **Directions for Use** of this product reflect the opinion of experts based on field use and tests. The directions are believed to be reliable and should be followed carefully. However, it is impossible to eliminate all risks inherently associated with use of this product. Crop injury, ineffectiveness, or other unintended consequences may result because of such factors as weather conditions, presence of other materials, or the manner of use or application all of which are beyond the control of CIBA-GEIGY or the Seller. All such risks shall be assumed by the Buyer.

CIBA-GEIGY warrants that this product conforms to the chemical description on the label and is reasonably fit for the purposes referred to in the **Directions for Use,** subject to the inherent risks referred to above. **CIBA-GEIGY makes no other express or implied warranty of Fitness or Merchantability or any other express or implied warranty. In no case shall CIBA-GEIGY or the Seller be liable for consequential, special, or indirect damages resulting from the use or handling of this product.** CIBA-GEIGY and the Seller offer this product and the Buyer and user accept it, subject to the foregoing **Conditions of Sale and Warranty** which may be varied only by agreement in writing signed by a duly authorized representative of CIBA-GEIGY.

DIRECTIONS FOR USE

It is a violation of federal law to use this product in a manner inconsistent with its labels.

REENTRY STATEMENT FOR AGRICULTURAL USES

No user shall permit the application of this product in such a manner as to directly or indirectly expose workers or other persons except those knowingly involved in the application. The area being treated must be vacated by unprotected persons.

Do not enter treated areas without protective clothing until dusts have settled.

Because certain states may require more restrictive reentry intervals for various crops treated with this product, consult your State Department of Agriculture for further information.

Written or oral warnings must be given to workers who are expected to be in a treated area or in an area about to be treated with this product. Oral warnings must be given which inform workers of areas or fields that may not be entered without specific protective clothing until dusts have settled and appropriate actions to take as described under **Caution** statements at the end of this label. When oral warnings are given, warnings shall be given in a language customarily understood by workers. Oral warnings must be given if there is reason to believe that written warnings cannot be understood by workers. Written warnings must include the following information: "CAUTION. Area treated with Diazinon 14G on (date of application). Do not enter without appropriate protective clothing until dusts have settled. In case of accidental exposure, flush eyes or skin with plenty of water. Call a physician if irritation persists. Remove and wash contaminated clothing before reuse."

Applications to the soil should be worked into the top few inches immediately after application with a rotary hoe, cultivator, disk, harrow, or other suitable means for greatest effectiveness.

FAILURE TO FOLLOW ALL DIRECTIONS FOR USE AND PRECAUTIONS ON THIS LABEL MAY RESULT IN CROP INJURY, POOR INSECT CONTROL, AND/OR ILLEGAL RESIDUES.

FIELD AND FORAGE INSECTS

BEANS (Pinto)

Lesser Cornstalk Borers: Apply 7-14 lbs. per acre in a 6-inch band centered over the row when beans are $\frac{1}{2}$-1 inch high.

CORN

Corn Rootworm Larvae: (Postemergence Basal Treatment)

Apply $4\frac{1}{2}$-$8\frac{1}{2}$ oz. per 1,000 linear feet of row in a band over the row at first sign of rootworms, but not later than June 24 in the Middle Corn Belt or July 1 in the Northern Corn Belt. Cover treated band with 2-3 inches of soil.

Apply amount per acre recommended according to row spacing chart for basal treatment.

Lesser Cornstalk Borers: Apply $8\frac{1}{2}$-17 oz. per 1,000 linear feet of row in a 10-inch band centered over the row. Make application at planting time or when crop begins to emerge. Cover the treated band with 1-2 inches of soil.

Apply amount per acre recommended according to row spacing chart for basal treatment.

Suggested Rates According to Corn Row Spacing

Row spacing	Linear feet of corn row per acre	Amount of D•z•n Diazinon 14G per acre: Postemergence basal treatment
48"	10,890	3-6 lbs.
42"	12,455	$3\frac{1}{4}$-$6\frac{1}{2}$ lbs.
40"	13,068	$3\frac{1}{2}$-7 lbs.
38"	13,939	$3\frac{3}{4}$-$7\frac{1}{2}$ lbs.
36"	14,520	4-8 lbs.
30"	17,424	$4\frac{3}{4}$-$9\frac{1}{2}$ lbs.
24"	21,780	$5\frac{3}{4}$-$11\frac{1}{2}$ lbs.
20"	26,136	7-14 lbs.
15"	34,848	$9\frac{1}{4}$-$18\frac{1}{2}$ lbs.

Note: For row spacings not listed, calculate the appropriate dosage rates based on the 40-in. row figures.

European Corn Borers: Apply 7-14 lbs. per acre over the corn so granules fall into whorls, when 75% of plants show first-generation borers feeding. Make a second application 7 days later if needed. To control second-generation corn borers, apply 7-14 lbs. per acre, as above, when field count shows 100 egg masses per 100 plants.

D•z•n Diazinon 14G may be applied by aircraft for the above use. This material should be uniformly applied to the area to be treated at the recommended rates given above.

Fall Armyworms: Apply 7-14 lbs. per acre over the corn so granules fall into whorls when armyworms first appear. Repeat application if necessary.

Southwestern Corn Borers: Apply 7-14 lbs. per acre over the corn so granules fall into whorls when second-generation borers first appear. Repeat application one week later.

Note: Do not feed treated fodder to dairy or beef cattle or sheep for 10 days following application. Corn may be picked immediately. Temporary spotting of foliage sometimes occurs following application.

Cutworms: Broadcast 14-28 lbs. per acre. For surface cutworms apply just prior to planting or when larvae first appear and work into soil 1-2 inches. Do not apply after lay-by. For subterranean cutworms apply just prior to planting and work into soil 3-6 inches.

Seed Corn Maggots: Broadcast 14-28 lbs. per acre just prior to planting. Work into soil 2 inches.

Garden Symphylans (Centipedes): Broadcast 70 lbs. per acre just prior to planting and work into soil 4-8 inches. This preplant treatment will also control other soil insects such as cutworms (surface and subterranean) and wireworms.

Wireworms: Broadcast 21-28 lbs. per acre before planting. Work into soil 4-6 inches.

PEANUTS

Southern Corn Rootworms: Broadcast 28 lbs. per acre or the equivalent in a band over the fruiting zone, just prior to pegging. Work into top few inches of soil.

Cutworms: Apply 14 lbs. per acre in a 12-inch band over the row when cutworms first appear. Up to three additional applications at three-week intervals may be necessary. Do not apply within 21 days of peanut harvest.

Lesser Cornstalk Borers: Apply 14-21 lbs. per acre in a band over the fruiting zone just prior to pegging. Cover the band with 1-2 inches of soil.

Note: Livestock (1) may be fed peanut hulls (2) may be fed peanut forage 7 days following treatment and (3) may be fed peanut hay 21 days following treatment.

SORGHUM

Cutworms: Broadcast 14-28 lbs. per acre just prior to planting. Work into soil 2-3 inches to control surface cutworms; 3-6 inches to control subterranean cutworms.

Lesser Cornstalk Borers: Apply 7-14 lbs. per acre in a 10-inch band over the row at planting time or as crop emerges. Cover the band with 1-2 inches of soil.

Note: Do not apply directly to the seed furrow as reduction of sorghum seed germination may occur.

SOYBEANS

Cutworms: Broadcast 14-28 lbs. per acre just prior to planting. Work into soil 2-3 inches to control surface cutworms; 3-6 inches to control subterranean cutworms.

Lesser Cornstalk Borers: Apply 7-14 lbs. per acre in a 6-inch band over the row just prior to planting or when soybeans are 1-2 inches high. Cover the band with 1-2 inches of soil.

SUGAR BEETS

Sugar Beet Root Maggots: Apply 7-14 lbs. per acre in a 5-7 inch band (based on row width of 22 in.) or 4.7-9.4 oz. per 1,000 linear feet of row in a 5-7 inch band either incorporated 2-3 inches at planting or applied above the seed after the seed is covered. Apply ahead of the press wheel. If a heavy maggot infestation is anticipated or a wet spring occurs, a second application may be made, early postemergence, when the adult maggots (flies) first start to appear. If a second application is necessary, apply 7-10 $\frac{1}{2}$ lbs. per acre in a 5-7 inch band (based on row width of 22 in.) or 4.7-7.1 oz. per 1,000 linear feet of row in a 5-7 inch band directly over the row.

Note: D•z•n diazinon 14G should not come in direct contact with the seed or injury will occur.

Wireworms: Broadcast 21-28 lbs. per acre just prior to planting. Work into soil 4-8 inches.

SUGARCANE

Sugarcane Borers: Broadcast 10$\frac{1}{2}$ -17$\frac{1}{2}$ lbs. per acre after joints have begun to form and when at least 5% of plants show borers feeding in or under the leaf sheath, but which have not bored into the stalks. Additional applications may be necessary to assure continuing control.

Note: Do not make more than four applications per growing season. Allow 28 days between last application and harvest. At that time bagasse and forage may be fed to livestock.

Wireworms: Apply 42 lbs. per acre in a 12-inch band over the open furrow, directly to seed pieces.

TOBACCO

Cutworms: Broadcast 14-21 lbs. per acre before transplanting. Work into soil 2-4 inches.

Mole Crickets: Apply 7 lbs. per acre when insects are present. Make application to plant beds or fields 1-2 days prior to sowing of seed or transplanting and work into soil 2-1 inches.

Seed Corn Maggots: Broadcast 10$\frac{1}{2}$ lbs. per acre 3-4 weeks before transplanting. Work into soil 2 inches.

Wireworms: Broadcast 14-21 lbs. per acre 3-4 weeks before transplanting. Work into soil 6-9 inches.

FORAGE CROPS (Alfalfa, Clover, Cowpeas, and Lespedeza)

Cutworms: Broadcast 14-28 lbs. per acre just prior to planting. Work into soil 2-3 inches to control surface cutworms, 3-6 inches to control subterranean cutworms.

VEGETABLE CROP INSECTS

BROCCOLI, BRUSSELS SPROUTS, CABBAGE, AND CAULIFLOWER

Root Maggots: Broadcast 14-21 lbs. per acre before planting or transplanting. Work into soil 3-4 inches.

ONIONS

Onion Maggots: Apply 14-28 lbs. per acre as a broadcast soil treatment just prior to planting. Thoroughly mix the chemical into the top 3-4 inches of soil by disking, harrowing, or other suitable means.

RADISHES

Root Maggots: Apply 10$\frac{1}{2}$-14 lbs. per acre (2$\frac{1}{2}$-2$\frac{4}{5}$ oz. per 1,000 linear feet of row) in the seed furrow at planting time.

SWEET CORN

See Corn under Field and Forage Insects.

TOMATOES

Vinegar Flies (*Drosophila* spp.): Broadcast 7 lbs. per acre when flies first appear. Repeat in 7-10 day intervals or as required.

OTHER VEGETABLE INSECTS

CUTWORMS (Surface and Subterranean)

Broccoli	Endive (escarole)	Pole Beans
Brussels Sprouts	Kale	Snap Beans
Cabbage	Lettuce	Spinach
Cantaloupes	Lima Beans	Summer Squash
Cauliflower	Muskmelons	Swiss Chard
Celery	Parsley	Tomatoes
Collards	Peas	Watermelons
Cucumbers	Peppers	Winter Squash

Broadcast 14-28 lbs. per acre just prior to planting. Work into soil 2-3 inches to control surface cutworms, 3-6 inches to control subterranean cutworms.

Potatoes: Broadcast 14-28 lbs. per acre before planting. Work into soil 4-6 inches.

GARDEN SYMPHYLANS (Centipedes)

Cabbage	Pole Beans	Tomatoes
Carrots	Radishes	Turnips
Lettuce	Red Beets	
Peas	Snap Beans	

Broadcast 70 lbs. per acre just prior to planting. Work into soil 4-8 inches. This preplant treatment will also control other soil insects such as cutworms (surface and subterranean), mole crickets, root maggots, and wireworms.

MOLE CRICKETS

Broccoli	Endive (escarole)	Radishes
Brussels Sprouts	Kale	Tomatoes

Aphid eggs (including apple, black cherry, mealy plum); Mealybugs: Apply as a dormant spray $1\frac{1}{4}$ lbs. D•z•n diazinon 50W plus 3 gals. dormant oil or $1\text{-}1\frac{1}{2}$ gals. superior type oil per 100 gals. of water.

Eyespotted bud moth, Fruittree leafroller: Mix 1 lb. per 100 gals. of water. Apply in cover sprays as necessary, beginning with petal fall.

Leafhoppers: Apply 1 lb. per 100 gals. of water in 3-5 applications beginning at petal fall and continuing through the growing season, or apply as infestations develop.

Cherry fruit fly: Mix $\frac{1}{2}$-1 lb. per 100 gals. of water. Make 3 or 4 applications, beginning when adult flies begin to emerge and repeating at 10-day intervals.

Black cherry aphid: Mix 1 lb. per 100 gals. of water. Apply as a cover spray when aphids appear and repeat after 10 days.

San Jose scale crawler: Mix 1 lb. per 100 gals. of water. Apply during the pre-pink stage or when crawlers first appear.

Cherry rust mite: Mix $\frac{1}{2}$-1 lb. per 100 gals. of water. Apply as a supplemental spray after harvest.

Do not apply within 10 days of harvest.

CRANBERRIES

Blackheaded fireworm, *Rhopobota Naevana:* Apply 4 lbs. in 400 gals. of water per acre.

Cranberry fruitworm: Apply 6 lbs. in 400 gals. of water per acre.

Do not apply within 7 days of harvest.

FIGS

Vinegar fly *(Drosophila spp.)*. Dried fruit beetle: Apply 1 lb. per 100 gals. of water. Do not apply within 5 days of harvest.

FILBERTS

Filbert leafroller, Aphids: Apply 2-4 lbs. in 250-400 gals. of water per acre, as a thorough cover spray.

CITRUS FRUITS (GRAPEFRUIT, LEMONS, LIMES, ORANGES, TANGELOS, TANGERINES, CITRUS CITRON, KUMQUATS, AND HYBRIDS OF THESE)

Citrus aphid, Cottonycushion scale crawlers, and Soft scale crawlers: Apply $\frac{1}{2}$-1 lb. per 100 gals. of water (maximum 1000 gals. per acre) as a thorough cover spray when aphids and crawler stages of scale insects first appear.

Fruittree leafroller: Mix $\frac{1}{2}$-1 lb. per 100 gals. of spray (maximum 1000 gals. per acre) when an average of 50 percent of eggs in egg masses have hatched in the spring.

Citrus thrip: Apply 1 lb. in 100 gals. of water (maximum 1000 gals. per acre) when thrips first appear.

Citrus snow scale: Apply 1-2 lbs. per 100 gals. of water (maximum 1000 gals. per acre) as a postbloom spray and again as a summer spray. For heavy snow scale infestations, use the higher dosage rate. Thorough coverage of tree trunk, limbs, and twigs is essential.

Do not apply during bloom. Do not apply to citrus within 21 days of harvest.

GRAPES

Pacific spider mite, Leafhoppers, Grape berry moth, Grape leaffolder: Mix 1 lb. per 100 gals. of water and apply as a thorough cover spray, or apply 2 lbs. per acre in sufficient water for thorough coverage. Do not apply within 18 days of harvest.

Drosophila spp.: Mix 6 oz. per 100 gals. of water and apply as a thorough cover spray, or apply a maximum of 18 oz. per acre in sufficient water for thorough coverage when insects first appear. Make additional applications at weekly intervals as necessary. Do not apply within 7 days of harvest. If a spreader-sticker is used, do not apply within 18 days of harvest.

OLIVES

Olive scale: Mix $\frac{3}{4}$-1 lb. of D•z•n diazinon 50W plus $1\frac{1}{2}$ gals. of light medium summer horticultural oil per 100 gals. of water. Apply as a cover spray in June or July.

Do not apply during bloom to avoid injury to pollinating insects or within 75 days of harvest.

PEACHES AND NECTARINES

Scales (including San Jose, parlatoria, black, brown, apricot); Mite eggs (including European red mite, brown mite); Twig borers; Aphid eggs (including apple, black cherry, mealy plum); Mealybugs: Apply as a dormant spray 1 lb. D•z•n diazinon 50W plus 2-3 gals. dormant oil or $1\text{-}1\frac{1}{2}$ gals. superior type oil per 100 gals. of water.

Peach twig borer, Parlatoria scale: Mix 1 lb. D•z•n diazinon 50W plus 1 gal. of light medium summer horticultural oil per 100 gals. of water. Apply at petal fall.

Aphids, Brown mite (Clover mite), Twospotted spider mite, Olive scale crawlers, San Jose scale crawlers: Mix 1 lb. per 100 gals. of water. Apply when aphids, mites, and scale crawlers first appear.

Olive scale crawlers: Mix $\frac{1}{2}$ lb. of D•z•n diazinon 50W plus $1\frac{1}{2}$ gals. light medium horticultural oil per 100 gals. of water. Apply when scale crawlers are present.

Oriental fruit moth: Mix 1 lb. per 100 gals. of water and apply at 100% petal fall. Make 2 additional applications at 10-15 day intervals. Time the last application to peak adult Oriental fruit moth flight, if possible.

Peach twig borer: Mix 1 lb. per 100 gals. of water and apply to peaches in May and June.

Leafhoppers (Peaches only): Apply 1 lb. per 100 gals. of water in 3-5 applications beginning at petal fall and continuing through the growing season, or apply as infestations develop.

White peach scale (Peaches only): As a postharvest spray apply $1\frac{1}{2}$-2 lbs. per 100 gals. of water timed to coincide with peak crawler and immature scale activity. Under conditions of heavy infestations, 2 applications may be necessary.

Do not apply to peaches within 20 days, or to nectarines within 10 days, of harvest.

PEARS

Scales (including San Jose, parlatoria, black, brown, apricot); Mite eggs (including European red mite, brown mite, pearleaf blister mite); Twig borers; Aphid eggs (including apple, black cherry, mealy plum); Mealybugs: Apply as a dormant spray 1 lb. of D•z•n diazinon 50W plus 2 gals. dormant oil or $1\text{-}1\frac{1}{2}$ gals. superior type oil per 100 gals. of water.

Pear psylla: Mix $\frac{1}{2}$-1 lb. per 100 gals. of water. Apply as a cover spray at pre-pink stage. Make 1-2 additional applications during the first and second cover sprays if necessary.

Tentiform leafminer, Mealybug: Mix 1 lb. per 100 gals. of water. Apply in 3 cover sprays, beginning in April.

Refer also to the Apples section.

Do not apply within 14 days of harvest.

PECANS

Aphids, Mites, Pecan nut casebearer, Walnut caterpillar: Apply 2-6 lbs. per acre as a foliage spray. Do not apply after husks crack.

PINEAPPLE

Pineapple scale: Apply 10 lbs. D•z•n diazinon 50W plus 2 gals. of light medium horticultural oil in 500 gals. of water per acre to the vegetative slips on the plant following harvest of the pineapple.

Pineapple mealybug: Apply 4 lbs. per acre. May be applied on day of harvest.

Pineapple forage and refuse may be fed to livestock 7 days following last application.

PLUMS AND PRUNES

Scales (including San Jose, parlatoria, black, brown, apricot); Mite eggs (including European red mite, brown mite, pearleaf blister mite); Twig borers; Aphid eggs (including apple, black cherry, mealy plum); Mealybugs: Apply as a dormant spray 1 lb. of D•z•n diazinon 50W plus 2-3 gals. dormant oil or $1\text{-}1\frac{1}{2}$ gals. superior type oil per 100 gals. of water. For late April or May applications, mix 1 lb. of D•z•n diazinon 50W plus 1 gal. of light medium summer horticultural oil per 100 gals. of water.

Brown mite (Clover mite), European red mite: Apply 1 lb. in 100 gals. of water.

Leafcurl plum aphid, Mealy plum aphid, Thistle aphid: Apply ½-1 lb. per 100 gals. of water.

Do not apply within 10 days of harvest.

STRAWBERRIES

Aphids, Twospotted spider mite: Apply 1 lb. in 100 gals. of water per acre.

Cyclamen mite: Apply 2 lbs. in 100 gals. of water per acre, directing spray to plant crowns. Plants should be sufficiently agitated to assure thorough coverage of the foliage. Make additional applications, if necessary, after harvest.

Mole cricket: Broadcast 2 lbs. per acre when insects are present. Apply 1-2 days before transplanting and mix into the top 1-2 inches of soil immediately.

Strawberry leafroller: Mix ¾-1 lb. per 100 gals. of water. Apply 200 gals. spray per acre when blossoms show color.

Do not apply within 5 days of harvest.

WALNUTS

Aphids, Mites: Apply 2 lbs. per acre as a foliage spray.

Codling moth, Scale crawlers, Walnut caterpillar: Apply 6 lbs. per acre.

Do not apply after the husks open.

VEGETABLE FOLIAR INSECTS

D•z•n diazinon 50W should always be applied as a spray in sufficient water to assure thorough coverage of the foliage.

BEANS (Pole Beans, Snap Beans, Lima Beans)

Black bean aphid, Mexican bean beetle, Leafminers: Apply 1-1½ lbs. per acre.

Cucumber beetle: Apply ½-¾ lb. in 100 gals. of water per acre.

Do not apply within 7 days of bean harvest. Bean forage or bean hay may be fed to dairy and beef cattle and sheep. Wait 4 days following treatment if forage is to be cut for hay.

LIMA BEANS

Dipterous leafminer, Leafhoppers, Mites: Apply 1 lb. per acre. Do not apply within 7 days of bean harvest. Bean forage or bean hay may be fed to dairy and beef cattle and sheep. Wait 4 days following treatment if forage is to be cut for hay.

BROCCOLI, CABBAGE, CAULIFLOWER

Aphids, Diamondback moth, Imported cabbageworm, Dipterous leafminer: Apply ½-1 lb. per acre. Applications at 7-10 day intervals will usually be required. Do not apply to broccoli or cauliflower within 5 days, or to cabbage within 7 days, of harvest.

BRUSSELS SPROUTS

Aphids, Diamondback moth, Imported cabbageworm: Apply 1 lb. per acre. Applications at 7-10 day intervals will usually be required. Do not apply to brussels sprouts within 7 days of harvest.

CARROTS

Aphids, Flea beetles: Apply 1 lb. per acre. Do not apply within 10 days of harvest.

CELERY

Aphids, Flea beetles: Apply 1 lb. in 100 gals. of water per acre.

Dipterous leafminer: Apply ½-1 lb. per acre in sufficient water to achieve thorough coverage when infestations first appear. Repeat as necessary.

Do not apply within 10 days of harvest. Plants should be market trimmed (tops removed) before being shipped or used. Tops should not be used for feed or food.

COLLARDS, KALE, SWISS CHARD, TURNIP TOPS

Aphids, Imported cabbageworm, Diamondback moth, Harlequin cabbage bug, Corn earworm: Apply ½-1 lb. per acre. Do not apply to collards, kale, or turnip tops within 10 days, or to Swiss chard within 12 days, of harvest.

CUCUMBERS

Aphids, Serpentine leafminer, Thrips: Apply 1 lb. per acre. Do not apply within 7 days of harvest.

LETTUCE AND ENDIVE

Aphids, Dipterous leafminer: Apply ½-1 lb. per acre. Do not apply within 10 days of harvest.

MELONS AND SQUASH (Cantaloupes, Casabas, Crenshaws, Honeydew Melons, Muskmelons, Persian Melons, and hybrids of these; Watermelons and their hybrids; Summer and Winter Squash)

Aphids, Leafhoppers, Mites: Apply 1-1½ lbs. per acre.

Serpentine leafminer, Thrips: Apply ½-1 lb. per acre.

Do not apply to melons or winter squash within 3 days, or to summer squash within 7 days, of harvest.

MUSHROOM HOUSES

Mushroom flies (Phorids and Sciarids): Mix 4-8 lbs. in 50 gals. of water (0.5% to 1.0% actual diazinon) and spray on outside and inside walls and ventilators of mushroom houses, after compost in beds has been pasteurized by heating. Do not spray compost in beds or after spawning.

To treat wooden sides of beds, posts, doors, and door frames after pasteurizing, mix ½ lb. in 6 gals. of water and apply with a brush.

Do not treat compost.

ONIONS

Onion thrips: Apply 1 lb. per acre. Do not apply within 10 days of harvest.

PEAS

Aphids, Leafminers: Apply ¾-1 lb. per acre. Peas may be picked immediately. Pea vines and pea vine hay may be fed to dairy and beef cattle and sheep. Wait 4 days following treatment if forage is to be cut for hay.

PEPPERS

Aphids, Serpentine leafminer: Apply ½ lb. per acre. Do not apply within 5 days of harvest.

POTATOES

Aphids, Flea beetles, Colorado potato beetle, Dipterous leafminer: Apply ½-¾ lb. per acre.

Banded cucumber beetle, Leafhoppers, Southern armyworm: Apply ¾-1 lb. per acre.

Do not apply within 35 days of harvest.

RADISHES, PARSNIPS, TURNIPS

Aphids, Flea beetles, Dipterous leafminer: Apply ½-1 lb. per acre. Do not apply within 10 days of harvest.

SPINACH AND BEETS

Aphids, Dipterous leafminer: Apply ½-1 lb. per acre. Do not apply to spinach within 10 days, or to beets within 14 days of harvest.

SQUASH (Summer and Winter)

See Melons and Squash section.

SWEET CORN

Corn earworm: Apply 2-2½ lbs. per acre when silks first appear. Two or three repeat applications at 2-3 day intervals will usually be necessary. Corn may be picked immediately following last application. Corn forage may be fed to dairy and beef cattle and sheep.

Refer also to Corn under Field and Forage Insects.

TOMATOES

Aphids, Dipterous leafminer: Apply ½ lb. per acre. To control leafminers weekly applications are usually required.

Banded cucumber beetle, Fall armyworm, Southern armyworm: Apply ¾-1 lb. per acre.

Vinegar fly (*Drosophila* spp.): Apply ½-1½ lbs. per acre. Weekly applications are usually required.

Do not apply within 1 day of harvest.

WATERCRESS (Hawaii only)

Cyclamen mite: Apply 1 lb. in 100 gals. of water per acre. Direct spray to crowns and growing tips of plants. If infestation is heavy, make a second application no later than 10 days after the first application. Repeat applications as necessary. Do not apply within 5 days of harvest.

VEGETABLE SOIL INSECTS

D•z•n diazinon 50W should always be applied as a spray in sufficient water to assure thorough coverage of the soil.

CARROT RUST FLY

Carrots: Apply 4 lbs. per acre as a drench to the seed furrow at planting time.

CUTWORMS (Surface and Subterranean)

Broccoli	Kale	Snap Beans
Brussels Sprouts	Lettuce	Spinach
Cabbage	Lima Beans	Summer Squash
Cantaloupes	Muskmelons	Sweet Corn
Cauliflower	Parsley	Swiss Chard
Celery	Peas	Tomatoes
Collards	Peppers	Watermelons
Cucumbers	Potatoes	Winter Squash
Endive (escarole)		

Broadcast 4-8 lbs. per acre just prior to planting. Work into the soil immediately, 2-3 inches for surface cutworms, 3-6 inches for subterranean cutworms.

GARDEN SYMPHYLANS (Centipedes)

Cabbage	Peas	Snap Beans
Carrots	Pole Beans	Tomatoes
Corn	Radishes	Turnips
Lettuce	Red Beets	

Broadcast 20 lbs. per acre just prior to planting. Work into soil 4-8 inches immediately. This preplant treatment will also control other soil insects: cutworms (surface and subterranean), mole crickets, root maggots, and wireworms.

MOLE CRICKETS

Broccoli	Endive (escarole)	Peppers
Brussels Sprouts	Kale	Radishes
Cabbage	Lettuce	Tomatoes
Cauliflower	Mustard	Turnips
Collards		

Broadcast 2 lbs. per acre when insects are present. Make application to plant beds or fields 1-2 days prior to sowing of seed or transplanting and work into soil 1-2 inches immediately.

ONION MAGGOTS

Onions: Furrow Application: Apply 2 lbs. per acre in sufficient water to drench the seed furrow at planting time.

Broadcast Application: Broadcast 4-8 lbs. per acre just prior to planting. Work into soil 3-4 inches immediately.

ROOT MAGGOTS

Broccoli, Brussels sprouts, Cabbage, and Cauliflower: (1) Broadcast Application: Broadcast 4-6 lbs. per acre and immediately work into the soil 3-4 inches before seeding, either in the seedbed or in the field or before transplanting. (2) Transplant Water (Drench) Application: Mix ¼-½ lb. in 50 gals. of transplant water, when 200-300 gals. of water are used per acre. May be applied at the rate of ½-1 cupful per plant, either by hand or by tractor-mounted sprayer equipped with drop nozzles to direct spray to the base of the plant. Transplant water treatments may result in stand reduction due to plant stress at time of transplanting.

Radishes: Mix 2 lbs. in 50 gals of water. Apply 80-100 gals. per acre. (1-1⅓ gals. per 1000 linear feet of row) in the seed furrow at planting time.

WIREWORMS

Broccoli	Endive (escarole)	Snap Beans
Brussels Sprouts	Kale	Spinach
Cabbage	Lettuce	Summer Squash
Cantaloupes	Lima Beans	Sweet Corn
Carrots	Muskmelons	Sweet Potatoes
Cauliflower	Onions	Swiss Chard
Celery	Parsley	Tomatoes
Collards	Peas	Watermelons
Cucumbers	Peppers	Winter Squash

Broadcast 6-8 lbs. per acre prior to planting, and work into soil 4-8 inches immediately.

Potatoes: Broadcast 6-12 lbs. per acre prior to planting. Immediately disk or harrow into soil 4-8 inches.

SOUTHERN POTATO WIREWORM

Potatoes (Southeastern U.S.): For potatoes on mineral soils, broadcast 4-8 lbs. per acre prior to planting, and immediately work into soil 3-6 inches depending on location of wireworms.

FIELD AND FORAGE INSECTS

D•z•n diazinon 50W should always be applied as a spray in sufficient water to assure thorough coverage of the foliage or soil, depending on the type of application.

ALFALFA, CLOVER AND TREFOIL

Spotted alfalfa aphid, Yellow clover aphid, Pea aphid, Leafhoppers, Plant bug nymphs, Mites: Apply ¾-1 lb. per acre.

Grasshoppers: Apply 1 lb. per acre.

Cutworms (Surface and Subterranean): Broadcast 4-8 lbs. per acre just prior to planting. Work into the soil immediately, 2-3 inches for surface cutworms, 3-6 inches for subterranean cutworms.

Dairy and beef cattle and sheep (1) may be grazed immediately following application, (2) may be fed green alfalfa or clover immediately following cutting, and (3) may be fed hay provided alfalfa or clover is not cut for hay for 7 days following treatment.

Alfalfa weevil larvae: For control in the spring, apply 2-3 lbs. per acre when feeding damage to growing tips becomes evident. Repeat application, if necessary. If damage is already extensive, cut the hay and make application to the stubble. Do not graze dairy and beef animals or sheep for 2 days. Do not cut for hay for 10 days following treatment.

Do not spray livestock directly except as indicated below under Livestock Insects, with either air or ground equipment.

Do not treat during bloom to avoid injury to pollinating insects.

CORN

Corn rootworm larvae —
(Postemergence Basal Treatment)

Apply 1¼-2½ oz. per 1000 linear feet of corn row in sufficient water to wet base of corn plant and a narrow band of soil next to the plant. Refer to row spacing chart for amounts recommended per acre. Cover treated band with 2-3 inches of soil using a disk hiller or cultivator immediately after spraying. Application should be made at the first sign of corn rootworm larvae. Application should not be made later than June 24 in the Middle Corn Belt or July in the Northern Corn Belt.

SUGGESTED RATES ACCORDING TO ROW SPACING FOR CORN ROOTWORM LARVAE CONTROL

Spacing	Linear Feet of Corn Row/Acre	Amount of D•z•n diazinon 50W Per Acre — Postemergence Basal Treatment
48"	10,890	14-28 oz.
42"	12,455	15½-31 oz.
40"	13,068	16-32 oz. (1-2 lbs.)
38"	13,939	17½-35 oz.
36"	14,250	18-36 oz.
30"	17,424	22-44 oz. (1½-2¾ lbs.)
24"	21,780	27½-55 oz.
20"	26,136	32-64 oz. (2-4 lbs.)
15"	34,848	44-88 oz.

Note: For row spacings not listed, calculate the appropriate dosage rates based on the 40-inch row figures.

Seedcorn maggot: Broadcast 4-8 lbs. per acre just prior to planting. Work into 2 inches of soil immediately.

Cutworms (Surface and Subterranean): Broadcast 4-8 lbs. per acre. For surface cutworms apply just prior to planting or when larvae first appear and work into 1-2 inches of soil immediately. Do not apply later than lay-by. For subterranean cutworms, apply just prior to planting and work into 3-6 inches of soil immediately.

Wireworms: Apply 6-8 lbs. per acre prior to planting and work into 4-8 inches of soil immediately.

Corn leaf aphid: Apply 1-2 lbs. per acre in sufficient water to provide complete coverage of the upper portions of plants. Repeat application as necessary.

Grasshoppers: Apply 1 lb. per acre

Sap beetle: Apply 2-2½ lbs. per acre. Add 1-2 gals. of soluble mineral oil per acre to the spray mixture. Consult local agricultural authorities for proper timing of sprays.

Corn may be picked immediately following last application. Corn forage may be fed to dairy and beef cattle and sheep.

COTTON

Spider mites, Aphids, Leafhoppers, Lygus bug, Cotton leafworm, Cotton leafperforator: Apply 1-2 lbs. in 5-25 gals. of water per acre. Repeat as infestation warrants.

Do not apply within 14 days of harvest. Do not graze treated cotton plants, or feed gin wastes to livestock.

COWPEAS

Cutworms (Surface and Subterranean): Broadcast 4-8 lbs. per acre, just prior to planting. Work into 2-3 inches of soil for surface cutworms, 3-6 inches for subterranean cutworms.

HOPS

Aphids, Mites: Apply 2 lbs. per acre. Do not apply within 14 days of harvest.

LESPEDEZA

Cutworms (Surface and Subterranean): Broadcast 4-8 lbs. per acre just prior to planting. Work into 2-3 inches of soil immediately for surface cutworms, 3-6 inches for subterranean cutworms.

PEANUTS

Cutworms: Apply 4 lbs. per acre in a 12-inch band over the row. Use sufficient water to achieve thorough coverage. Begin application when small larvae appear. Two or three additional applications at 3-week intervals may be necessary. Do not apply within 21 days of peanut harvest.

Note: Livestock (1) may be fed peanut hulls, (2) may be fed peanut forage 7 days following treatment, and (3) may be fed peanut hay 21 days following treatment.

SORGHUM

Sorghum midge: Apply ½-1 lb. per acre when 90% of the heads have emerged from the boot. Repeat application within 4 days.

Crop may be fed to livestock immediately following treatment. Do not apply within 7 days of harvesting for grain.

Cutworms (Surface and Subterranean): Broadcast 4-8 lbs. per acre just prior to planting. Work into 2-3 inches of soil immediately for surface cutworms, 3-6 inches for subterranean cutworms.

SOYBEANS

Cutworms (Surface and Subterranean): Broadcast 4-8 lbs. per acre just prior to planting. Work into 2-3 inches of soil immediately for surface cutworms, 3-6 inches for subterranean cutworms.

SUGAR BEETS

Beet leafhopper, Dipterous leafminer, Aphids: Apply ¾-1 lb. per acre.

Grasshoppers: Apply 1 lb. per acre.

Wireworms: Broadcast 6-8 lbs. per acre just prior to planting, and work into 4-8 inches of soil immediately. Tops may be fed to beef and dairy animals.

SUGARCANE

Aphids: Apply ½-1 lb. per acre.

Note: Allow 1 day between last application and harvest. Bagasse and forage may be fed to livestock.

Wireworms: Apply 8 lbs. per acre on a 12-inch band centered over the furrow in sufficient water to obtain thorough coverage. Make application directly to seed pieces in open furrow at covering time.

TOBACCO

Aphids, Flea beetles: Apply ½-¾ lb. per acre. Do not apply within 3 days of harvest.

Cutworms: Broadcast 4-6 lbs. per acre before transplanting and work into 2-4 inches of soil immediately.

Green June beetle larvae: Mix ¼ lb. in 50 gals. of water and apply as a drench (100 gals. per 100 sq. yds.) to uprooted areas of plant beds with sprinkling can.

Mole cricket: Broadcast 2 lbs. per acre, when insects are present. Make application to beds or fields 1-2 days prior to sowing seed or transplanting, and work into 1-2 inches of soil immediately.

Seedcorn maggot: Broadcast Application: Broadcast 3 lbs. per acre, 3-4 weeks before transplanting. Work into 2 inches of soil immediately.

Transplant Water Application: Mix 1½-3 oz. in 50 gals. of transplant water.

Wireworms: Broadcast Application: Broadcast 4-6 lbs. per acre, 3-4 weeks before transplanting. Work into 6-9 inches of soil immediately.

Transplant Water Application: Mix 1½-3 oz. in 50 gals. of transplant water.

RANGE, PASTURE, AND GRASSLAND INSECTS

BERMUDAGRASS

Thrips, Whiteflies, Grasshoppers: Apply 1 lb. per acre.

RANGELAND, PASTURE, GRASS FORAGE, DITCH BANKS, ROADSIDES, WASTELAND, NONCROP AREAS, BARRIER STRIPS

Grasshoppers: Apply ¾-1 lb. per acre in sufficient water or oil to obtain thorough coverage when insects first appear, preferably in the nymphal stage.

Dairy and beef cattle and sheep (1) may be grazed immediately following application, (2) may be fed green forage immediately following cutting, and (3) may be fed hay, provided grass is not cut for hay for 21 days if the spray is a water solution, or for 30 days if the spray is an oil solution. Do not spray livestock directly with either air or ground equipment except as indicated below under Livestock Insects. Do not repeat application for 30 days.

LIVESTOCK INSECTS

SHEEP "TICKS" (KEDS), LICE

High gallonage, high pressure spray: Mix ½ lb. in 100 gals. of water, and apply at the rate of 1 gal. per animal.

Low gallonage, low pressure spray: Mix 1 lb. in 100 gals. of water. Apply at the rate of 1 qt. per animal.

Sprinkler Can: Mix 1 oz. in 6 gals. of water. Crowd sheep into a small pen and sprinkle 1 qt. of this mixture over the head, neck, top, and side of each sheep. Stir or agitate mixture frequently.

Note: Do not slaughter sheep within 14 days after treatment. Repeat applications as necessary.

ORNAMENTAL INSECTS

To control certain insects on ornamentals, such as Arborvitae, Azalea, Birch, Boxwood, Camelia, Carnation, Chrysanthemum, Douglas Fir, Elm, Gladioli, Hawthorn, Holly, Juniper, Lilac, Locust, Maple, Oak, Pine, Plum, Poplar, Rhododendron, Rose, Spruce, and Willow, apply the recommended rates indicated below.

Insects	Amount D•z•n diazinon 50W to Use: per 3 gals. water	per 100 gals. water
Aphids, Bagworms, Carnation bud mites, Carnation shoot mites, Clover mites, Cyclamen mites, Dipterous leafminers, European pine shoot moths, European red mites, Flea beetles, Holly bud moths, Leafhoppers, Obscure root weevils, Omnivorous leaftiers, Privet mites, Scale crawlers: (Cottonycushion, Lecanium, Pine needle, San Jose, Soft scale), Thrips, Twospotted mites, Webworms, Whiteflies	½ oz.	1 lb.
Apple-and-thorn skeletonizers, Cotoneaster webworms, Fall webworms, Hemlock chermes, Oak loopers, Obliquebanded leafrollers, Pear slugs, Tent caterpillars	1½ oz.	3 lbs.

Do not use on Ferns, Poinsettia, Hibiscus, Papaya, Pilea, and Gardenia, since injury to the plants may occur.

Mimosa webworms: Mix ½ oz. in 3 gals. of water (or 1 lb. in 100 gals. of water) and apply as a thorough foliar spray when adults first appear. Make a second and third application at 4-5 week intervals.

LAWN INSECTS

To control insects listed below, apply recommended rates. For insects such as Ants, Brown dog ticks, Clover mites, Crickets, Earwigs, Fleas, Millipedes, Sowbugs, and Springtails (Collembola), which are commonly found near the house and may enter the house, spray a 5-ft. band of soil around the house as well as the house foundation wall to a height of 2-3 ft. Repeat application if necessary.

Insects	Rate/1000 sq. ft.: D•z•n diazinon 50W	Water	Remarks
Lawn chinch bugs	2-3 oz. or 3-6 oz.	25 gals.*	Use higher rate for longer residual control and in lawns of dense growth, such as St. Augustine lawns.
Ants, Armyworms, Clover mites, Springtails (Collembola), Crickets, Cutworms, Digger wasps, Earwigs, Frit Flies, Lawn billbugs, Sod webworms (Lawn moths), Sowbugs	4 oz.	3 gals	Spot spray ant hills and wasp ground nest openings. For lawn billbugs, irrigate the lawn thoroughly immediately after treatment. For frit flies, mow grass and water well before treatment, but delay application until grass is dry. Do not water again until necessary.
Brown dog ticks, Bermudagrass mites. Chiggers, Fleas, Leafhoppers	1¼ oz.	3 gals.	Do not apply to animals. For brown dog ticks, spray grass and under shrubbery, particularly near house.
Millipedes	8 oz.	3 gals.	
Rhodesgrass scales	oz.	25 gals.*	Apply when crawlers first emerge.

* Application can be made in 3 gals. of water per 1000 sq. ft. Thoroughly water treated area immediately following application.

Boxelder bugs: Spray bugs outside the home with 1½-2½ oz. in 1 gal. of water. At the same time, spray thoroughly around foundation, windows, doors, tree trunks, and into cracks and other places where the insects hide.

DUST FORMULATIONS BASED ON D•Z•N DIAZINON 50W

D•z•n diazinon 50W is recommended for use in formulating dilute dust products. Formulators should contact the Technical Sales Department, Agricultural Division, for information regarding proper diluents and suggested formulae.

SEED TREATMENT—PLANTER-BOX TREATMENT

CORN

Seedcorn beetle: Mix 3 oz. per bushel of seed as a dry mixture in the planter box as a seed treatment prior to planting.

Corn, Peas, Soybeans, and Beans (Lima, Snap, White, and Kidney)

Seedcorn maggot: Mix ½ oz. per bushel of seed as a dry mixture in the planter box as a seed treatment prior to planting.

Planter-box Mixing Instructions: Stir seed slightly using a mixing stick. Additional mixing will result from the operation of the planter. DO NOT MIX WITH BARE HANDS.

Note: If seed has not been treated with a fungicide, use Captan or Thiram in combination with D•z•n diazinon 50W at rates indicated on fungicide package. Failure to use a fungicide with D•z•n diazinon 50W may result in injury to the seed. If the seed treatment increases the friction between seeds sufficiently to reduce the rate of delivery through the planter, ½ oz. of graphite per bushel of seed should be added as a drill lubricant. Adjust the planter to give the required rate of seed delivery with the treated seed.

Caution: D•z•n diazinon 50W treated grain cannot be used for feed for livestock, or for the preparation of products for human consumption.

FLY CONTROL

Fly control in barns and animal sleeping quarters (except dairy barns, milk rooms, and poultry houses).

Residual Sprays: Remove animals from buildings prior to treatment and keep animals out for at least four hours. Mix 2 lbs. of D•z•n diazinon 50W in 25 gals. of water and apply as an overall spray to ceilings and walls of livestock sheds, calf barns, hog barns, loafing sheds, and other farm buildings except poultry houses, dairy barns, and milk rooms. Areas where flies congregate, such as garbage dumps and corrals, should be sprayed thoroughly. Longer residual control may be obtained by increasing the rate to 4 lbs. in 25 gals. of water. Applications should be made to point of run-off. One gal. of the above spray should cover from 350-750 sq. ft. of area, depending on the surface treated. Repeat applications as fly populations warrant.

Bait Sprays: Mix the appropriate amount of D•z•n diazinon 50W with 1 lb. of sugar (or 2 cups of syrup or molasses) for the desired application.

For knapsack or similar type sprayer, mix ½ lb. with the sugar in 2½ gals. of water and spot treat areas frequented by flies including cracks, crevices, doorways, around windows, or other areas where flies congregate. Repeat applications as fly populations warrant.

For sprinkling can application, mix 2 oz. with the sugar in 5 gals. of water and sprinkle lightly over the floor in areas frequented by flies. Applications every day or two will generally be required.

Do not contaminate milk, feed, drinking water, or apply to animals or poultry. Do not use in edible product areas of food processing plants or other areas where food is commercially prepared or processed. Do not use in dwellings.

STORAGE AND DISPOSAL

Do not contaminate water, food, or feed by storage or disposal. Open dumping is prohibited.

PESTICIDE

Pesticide wastes are toxic. Improper disposal of excess pesticide, spray mixture, or rinsate is a violation of Federal law. If pesticide, spray mixture or rinsates cannot be disposed of by use according to label instructions, contact your State Pesticide or Environmental Control Agency, or the Hazardous Waste representative at the nearest EPA Regional Office for guidance.

CONTAINER
Completely empty bag into application equipment. Then dispose of empty bag in a sanitary landfill or by incineration or, if allowed by state and local authorities, by burning. If burned, stay out of smoke.

PRECAUTIONARY STATEMENTS

HAZARDS TO HUMANS AND DOMESTIC ANIMALS
WARNING
May be fatal if swallowed. May be absorbed through skin. Do not breathe dust or spray mist. Avoid contact with skin, eyes, and clothing. Wash thoroughly after handling and before eating or smoking. Avoid contamination of food and feed. Keep out of reach of domestic animals. Do not use on humans or household pets. Do not permit children or pets to go onto sprayed grass until spray has completely dried.

STATEMENT OF PRACTICAL TREATMENT

If swallowed:	Call a physician or Poison Control Center immediately. Drink one or two glasses of water and induce vomiting by touching back of throat with finger. Repeat until vomit fluid is clear. Do not induce vomiting or give anything by mouth to an unconscious person.
If inhaled:	Remove victim to fresh air. Apply artificial respiration if indicated.
If on skin:	Remove contaminated clothing and wash affected areas thoroughly with soap and water.
If in eyes:	Flush eyes with plenty of water. Call a physician if irritation develops or persists.

Note to Physician: Gastric lavage is indicated. Atropine sulfate and 2-PAM are specific antidotes.

ENVIRONMENTAL HAZARDS
This product is toxic to fish, birds, and other wildlife. Birds feeding on treated areas may be killed. Keep out of lakes, streams, or ponds. Do not apply where runoff is likely to occur. Do not contaminate water by cleaning of equipment or disposal of wastes. Apply this product only as specified on this label.

This pesticide is highly toxic to bees exposed to direct treatment or to residues remaining on the treated areas. Do not apply when bees are actively visiting the crop, cover crop, or weeds blooming in the treatment area. Applications should be timed to provide the maximum possible interval between treatment and the next period of bee activity.

CGA 1L66H 015

D·Z·N®
diazinon AG500
INSECTICIDE

For Agricultural or Commercial Use Only

For control of certain insects on fruits, nuts, vegetables, field crops, and ornamentals

ACTIVE INGREDIENTS:	
Diazinon	
O.O-diethyl O-(2-isopropyl-6-methyl-4-pyrimidinyl) phosphorothioate	48.0%
Xylene range aromatic solvent	38.4%
INERT INGREDIENTS:	13.6%
TOTAL:	100.0%

EPA Reg. No. 100-461 EPA Est. 100-AL-1

D•z•n diazinon AG500 contains 4 lbs. diazinon per gallon.

Keep Out of Reach of Children
WARNING/AVISO

PRECAUCIÓN AL USUARIO: Si usted no lee inglés, no use este producto hasta que la etiqueta haya sido explicado ampliamente.

See additional precautionary statements inside booklet.

See directions for use inside booklet.

DIRECTIONS FOR USE AND CONDITIONS OF SALE AND WARRANTY

IMPORTANT: Read the entire **Directions for Use** and the **Conditions of Sale and Warranty** before using this product.

CONDITIONS OF SALE AND WARRANTY

The **Directions for Use** of this product reflect the opinion of experts based on field use and tests. The directions are believed to be reliable and should be followed carefully. However, it is impossible to eliminate all risks inherently associated with use of this product. Crop injury, ineffectiveness or other unintended consequences may result because of such factors as weather conditions, presence of other materials, or the manner of use or application all of which are beyond the control of CIBA-GEIGY or the Seller. All such risks shall be assumed by the Buyer.

CIBA-GEIGY warrants that this product conforms to the chemical description on the label and is reasonably fit for the purposes referred to in the **Directions for Use,** subject to the inherent risks referred to above. **CIBA-GEIGY makes no other express or implied warranty of Fitness or Merchantability or any other express or implied warranty. In no case shall CIBA-GEIGY or the Seller be liable for consequential, special, or indirect damages resulting from the use or handling of this product.** CIBA-GEIGY and the Seller offer this product, and the Buyer and user accept it, subject to the foregoing **Conditions of Sale and Warranty,** which may be varied only by agreement in writing signed by a duly authorized representative of CIBA-GEIGY.

GENERAL INFORMATION

D•z•n diazinon AG500 is an emulsifiable solution which, when diluted with water according to directions, is intended for the control of insect pests of agricultural crops. Do not use in barns or dwellings. Do not use on livestock.

In the directions for use which follow, best control is obtained when application is made at the first sign of insects. Application should be repeated as necessary to maintain control.

DIRECTIONS FOR USE

It is a violation of federal law to use this product in a manner inconsistent with its labeling.

Do not apply this product in such a manner as to directly or through drift expose workers or other persons except those knowingly involved in the application. The area being treated must be vacated by unprotected persons.

FAILURE TO FOLLOW ALL PRECAUTIONS ON THIS LABEL MAY RESULT IN POOR INSECT CONTROL, CROP INJURY, OR ILLEGAL RESIDUES.

REENTRY STATEMENT

Do not enter treated areas without protective clothing until sprays have dried.

Because certain states may require more restrictive reentry intervals for various crops treated with this product, consult your State Department of Agriculture for further information. (See **Note** below.)

Written or oral warnings must be given to workers who are expected to be in a treated area or in an area about to be treated with this product. Oral warnings must inform workers of areas or fields that may not be entered without specific protective clothing until sprays have dried, and appropriate actions to take in case of accidental exposure, as described under Precautionary Statements on this label. When oral warnings are given, warnings shall be given in a language customarily understood by workers. Oral warnings must be given if there is reason to believe that written warnings cannot be understood by workers. Written warnings must include the following information: WARNING. Area treated with D•z•n diazinon AG500 on (date of application). Do not enter without appropriate protective clothing until sprays have dried. In case of accidental exposure, flush eyes or skin with plenty of water. Call physician if irritation persists. Remove and wash contaminated clothing before reuse.

Note: The California Department of Agriculture has set a 5-day reentry for diazinon on citrus, grapes, peaches, and nectarines. Five days should elapse between the time of application and the time a worker may enter the field to engage in any activity requiring substantial contact with treated foliage. When a mixture of two or more organic phosphate pesticides is applied, the interval shall be prolonged by adding to the longest applicable interval an additional 50% of that interval.

FRUIT AND NUT CROPS

Note: Unless otherwise stated, the following rates should be applied as a dilute full cover spray.

ALMONDS

Mites, Twig Borers: Apply 1-1½ pts. per 100 gals. of water.

Parlatoria Scales, San Jose Scales: Mix 1-1½ pts. D•z•n diazinon AG500 plus 2-3 gals. dormant oil per 100 gals. of water and apply as a dormant spray.

Almond hulls may be used for feed for livestock.

APRICOTS

Aphids, Brown Mites (Clover Mites), Twospotted Spider Mites, Olive Scale Crawlers, San Jose Scale Crawlers: Mix 1 pt. per 100 gals. of water. Apply when aphids, mites, and scale crawlers first appear.

Olive Scale Crawlers: Mix ½ pt. of D•z•n diazinon AG500 plus 1½ gals. of light medium horticultural oil per 100 gals. of water. Apply when scale crawlers are present.

Apricot Mealybugs: Mix 1 pt. per 100 gals. of water. Apply as a cover spray from petal fall to June.

Do not apply within 10 days of harvest.

BLUEBERRIES

Cranberry Fruitworms, Cherry Fruitworms, Blueberry Maggots, Aphids, Thrips, Twospotted Mites: Mix 1 pt. per 100 gals. of water. Apply up to 200 gals. of spray per acre. Do not apply within 7 days of harvest.

CANEBERRIES

Blackberries, Boysenberries, Dewberries, Loganberries, Raspberries

Leafhoppers, Thrips, Aphids, Twospotted Mites, Raspberry Sawflies: Apply 2 pts. per acre in a minimum of 100 gals. of water.

Raspberry Fruitworms: Apply at the above rates when blossom buds separate and repeat application immediately before blossoms open.

Dryberry Mites: Apply 1-2 qts. per acre in a minimum of 100 gals. of water. Apply as a dormant spray in early spring, when canes are still on the ground and buds are closed. Make a second application when the buds are well developed, but still closed.

Do not apply within 7 days of harvest.

Raspberry Crown Borers (Raspberry Root Borers): Apply 2 qts. per acre in a minimum of 100 gals. of water as a drench to the crown area and to the lower canes. Apply either in the spring before berries begin to form, or in the fall after harvest.

CHERRIES

Eyespotted Bud Moths, Fruittree Leafrollers: Mix 1 pt. per 100 gals. of water. Apply in cover sprays as necessary beginning with petal fall.

Cherry Fruit Flies: Mix ½-1 pt. per 100 gals. of water. Make 3 or 4 applications, beginning when adult cherry fruit flies begin to emerge and repeat at 10-day intervals.

Leafhoppers: Apply 1 pt. per 100 gals. of water in 3-5 applications beginning at petal fall and continuing through the growing season, or apply as infestations develop.

Black Cherry Aphids: Mix 1 pt. per 100 gals. of water. Apply as a cover spray when aphids appear and repeat after 10 days.

San Jose Scale Crawlers: Mix 1 pt. per 100 gals. of water. Apply during the pre-pink stage or when crawlers first appear.

Do not apply within 10 days of harvest.

Cherry Rust Mites: Mix ½-1 pt. per 100 gals. of water. Apply as a supplemental spray after harvest.

CITRUS FRUITS

Grapefruit, Lemons, Limes, Oranges, Tangelos, Tangerines, Citrus Citron, Kumquats, and hybrids of these

Citrus Aphids, Fruittree Leafrollers, Cottonycushion Scale Crawlers, Soft Scale Crawlers: Apply ½-1 pt. per 100 gals. of water (maximum 1,000 gals. per acre) as a thorough cover spray when aphids and crawler stages of scale insects first appear.

Fruittree Leafrollers: Use ½-1 pt. per 100 gals. of spray (maximum 1,000 gals. per acre) when an average of 50 percent of eggs in egg masses have hatched in the spring.

Citrus Thrips: Apply 1 pt. in 100 gals. of water (maximum 1,000 gals. per acre) when thrips first appear.

Citrus Snow Scales: Apply 1–2 pts. per 100 gals. of water (maximum 1,000 gals. per acre) as a postbloom spray and again as a summer spray. For heavy snow scale infestations, use the higher dosage rate. Thorough coverage of tree trunks, limbs, and twigs is essential.

Do not apply during bloom. Do not apply to citrus within 21 days of harvest. Shrimp and crab may be killed at application rates recommended on this label. Do not apply where these are important resources.

CRANBERRIES

Blackheaded Fireworms *(Rhopobota naevana):* Apply 2 qts. in 400 gals. of water per acre.

Cranberry Fruitworms: Apply 3 qts. in 400 gals. of water per acre.

Do not apply within 7 days of harvest.

FIGS

Vinegar Flies (*Drosophila* spp.), Dried Fruit Beetles: Apply 1 pt. per 100 gals. of water. Do not apply within 5 days of harvest.

FILBERTS

Filbert Leafrollers, Aphids: Apply 1–2 qts. in 250 400 gals. of water per acre as a thorough cover spray.

GRAPES

Pacific Spider Mites, Leafhoppers, Grape Berry Moths, Grape Leaffolders: Apply 1–2 pts. per acre in a minimum of 100 gals. of water as a thorough cover spray. Do not apply within 18 days of harvest.

Drosophila spp: Apply 12-18 fl. oz. per acre in a minimum of 100 gals. of water as a thorough cover spray, or apply when insects first appear. Make additional applications at weekly intervals as necessary.

Do not apply within 7 days of harvest.

OLIVES

Olive Scales: Mix $3/4$-1 pt. of D•z•n diazinon AG500 plus $1\frac{1}{2}$ gals. of light medium summer horticultural oil per 100 gals. of water. Apply as a cover spray in June or July.

Do not apply during bloom to avoid injury to pollinating insects or within 75 days of harvest.

PEACHES, NECTARINES

Aphids, Brown Mites (Clover Mites), Twospotted Spider Mites, Olive Scale Crawlers, San Jose Scale Crawlers: Mix 1 pt. per 100 gals. of water. Apply when aphids, mites, and scale crawlers first appear.

Olive Scale Crawlers: Mix $1/2$ pt. of D•z•n diazinon AG500 plus $1\frac{1}{2}$ gals. of light medium horticultural oil per 100 gals. of water. Apply when scale crawlers are present.

Oriental Fruit Moths: Mix 1 pt. per 100 gals. of water and apply at 100% petal fall. Make 2 additional applications at 10–15 day intervals. Time the last application to peak adult oriental fruit moth flight, if possible.

Peach Twig Borers: Mix 1 pt. per 100 gals. of water and apply to peaches in May and June.

Leafhoppers (Peaches only): Apply 1 pt. per 100 gals. of water in 3-5 applications beginning at petal fall and continuing through the growing season, or apply as infestations develop.

White Peach Scale (Peaches only): As a post-harvest spray apply $1\frac{1}{2}$-2 pts. per 100 gals. of water timed to coincide with peak crawler and immature scale activity. Under conditions of heavy infestations, two applications may be necessary.

Do not apply to peaches within 20 days, or to nectarines within 10 days, of harvest.

PECANS

Aphids, Mites, Pecan Nut Casebearers, Walnut Caterpillars: Apply 1–3 qts. per acre as a foliage spray in a minimum of 100 gals. of water. Do not apply after husks crack.

PINEAPPLE

Scales: Apply 5 qts. of D•z•n diazinon AG500 plus 2 gals. of light medium horticultural oil in 500 gals. of water per acre to the vegetative slips on the plants following harvest of pineapple.

Mealybugs: Apply 2 qts. per acre in a minimum of 100 gals. of water. May be applied on day of harvest.

Pineapple forage and refuse may be fed to livestock 7 days following last application.

PLUMS, PRUNES

Brown Mites (Clover Mites), European Red Mites: Apply 1 pt. per 100 gals. of water.

Leafcurl Plum Aphids, Mealy Plum Aphids, Thistle Aphids: Apply $1/2$-1 pt. per 100 gals. of water.

Do not apply within 10 days of harvest.

STRAWBERRIES

Aphids, Twospotted Spider Mites: Apply 1 pt. per 100 gals. of water per acre.

Cyclamen Mites: Apply 1 qt. per 100 gals. of water per acre, directing spray to plant crowns. Plants should be sufficiently agitated to assure thorough coverage of the foliage. Make additional applications, if necessary, after harvest.

Mole Crickets: Broadcast 1 qt. per acre when insects are present. Apply 1–2 days before transplanting and immediately mix into the top 1–2 inches of soil.

Strawberry Leafrollers: Mix $3/4$-1 pt. per 100 gals. of water. Apply 200 gals. of spray per acre when blossoms show color.

Do not apply within 5 days of harvest.

WALNUTS

Aphids, Mites: Apply 1 qt. per acre as a foliage spray in a minimum of 100 gals. of water.

Codling Moths, Scale Crawlers, Walnut Caterpillars: Apply 3 qts. per acre as a foliage spray in a minimum of 100 gals. of water.

Do not apply after the husks open.

VEGETABLE FOLIAR INSECTS

Unless otherwise stated, foliar applications of D•z•n diazinon AG500 should always be applied to vegetable crops as a spray in a minimum of 2 gals. of water per acre for aerial applications and a minimum of 10 gals. of water per acre for ground applications.

BEANS

Pole Beans, Snap Beans, Lima Beans

Black Bean Aphids, Mexican Bean Beetles, Leafminers: Apply 1-$1\frac{1}{2}$ pts. per acre.

Cucumber Beetles: Apply $3/4$ pt. in 100 gals. of water per acre.

Do not apply within 7 days of bean harvest. Bean forage or bean hay may be fed to dairy and beef cattle and sheep. Wait 4 days following treatment if forage is to be cut for hay.

LIMA BEANS

Dipterous Leafminers, Leafhoppers, Mites: Apply 1 pt. per acre. Do not apply within 7 days of harvest. Bean forage or bean hay may be fed to dairy and beef cattle and sheep. Wait 4 days following treatment if forage is to be cut for hay.

BROCCOLI, CABBAGE, CAULIFLOWER

Aphids, Diamondback Moths, Imported Cabbageworms: Apply $1/2$-1 pt. per acre. Applications at 7–10 day intervals will usually be required. Do not apply to broccoli or cauliflower within 5 days, or to cabbage within 7 days, of harvest.

BRUSSELS SPROUTS

Aphids, Diamondback Moths, Imported Cabbageworms: Apply 1 pt. per acre. Generally, applications at 7–10 day intervals will be required. Do not apply to brussels sprouts within 7 days of harvest.

CARROTS

Aphids: Apply 1 pt. per acre. Do not apply within 10 days of harvest.

CELERY

Aphids, Flea Beetles: Apply 1 pt. in 100 gals. of water per acre. Do not apply within 10 days of harvest. Plants should be market trimmed (tops removed) before being shipped or used. Tops should not be used for feed or food.

COLLARDS, KALE, SWISS CHARD, TURNIP TOPS

Aphids, Imported Cabbageworms, Diamondback Moths, Harlequin Cabbage Bugs: Apply $1/2$-1 pt. per acre. Do not apply to collards, kale, or turnip tops within 10 days, or to Swiss chard within 12 days, of harvest.

CUCUMBERS

Aphids, Serpentine Leafminers, Thrips: Apply 1 pt. per acre. Do not apply within 7 days of harvest.

LETTUCE, ENDIVE

Aphids, Dipterous Leafminers: Apply $1/2$-1 pt. per acre. Do not apply within 10 days of harvest.

MELONS, SQUASH

Cantaloupes, Casabas, Crenshaws, Honeydew Melons, Muskmelons, Persian Melons, and hybrids of these: Watermelons and their hybrids; Summer and Winter Squash

Aphids, Leafhoppers, Mites: Apply 1-$1\frac{1}{2}$ pts. per acre.

Serpentine Leafminers, Thrips: Apply $1/2$-1 pt. per acre.

Do not apply to melons and winter squash within 3 days or to summer squash within 7 days, of harvest.

MUSHROOM HOUSES

Mushroom Flies (Phorids and Sciarids): Mix $1/2$-1 gal. in 50 gals. of water (0.5% to 1.0% actual diazinon) and spray on outside and inside walls and ventilators of mushroom houses, after compost in beds has been pasteurized by heating. Do not spray compost in beds or after spawning.

To treat wooden sides of beds, posts, doors, and door frames after pasteurizing, mix ½ pt. in 6 gals. of water and apply with a brush.

Do not treat compost.

ONIONS
Onion Thrips: Apply 1 pt. per acre. Do not apply within 10 days of harvest.

PEAS
Aphids, Leafminers: Apply ¾-1 pt. per acre. Peas may be picked immediately. Pea vines and pea vine hay may be fed to dairy and beef cattle and sheep. Wait 4 days following treatment if forage is to be cut for hay.

PEPPERS
Aphids, Serpentine Leafminers: Apply ½ pt. per acre. Do not apply within 5 days of harvest.

POTATOES
Aphids, Flea Beetles, Colorado Potato Beetles, Dipterous Leafminers: Apply at the rate of ½-¾ pt. per acre.

Banded Cucumber Beetles, Leafhoppers, Southern Armyworms: Apply ¾-1 pt. per acre.

Do not apply within 35 days of harvest.

RADISHES, PARSNIPS, TURNIPS
Aphids, Flea Beetles, Dipterous Leafminers: Apply ½-1 pt. per acre. Do not apply within 10 days of harvest.

SPINACH, BEETS
Aphids, Dipterous Leafminers: Apply ½-1 pt. per acre. Do not apply to spinach within 10 days, or to beets within 14 days, of harvest.

SQUASH (Summer and Winter)
See Melons and Squash section.

SWEET CORN
Corn Earworms: Apply 1-1¼ qts. per acre when silks first appear. Two or three repeat applications at 2-3 day intervals will usually be necessary. Corn may be picked immediately following last application. Corn forage may be fed to dairy and beef cattle and sheep.

Refer also to Corn under Field and Forage Foliar Insects and Field and Forage Soil Insects.

TOMATOES
Aphids, Dipterous Leafminers: Apply ½ pt. per acre. To control leafminers, weekly applications are usually required.

Banded Cucumber Beetles, Fall Armyworms, Southern Armyworms: Apply ¾-1 pt. per acre.

Vinegar Flies (*Drosophila* spp.): Apply ½-1½ pts. per acre. Weekly applications are usually required.

Do not apply within 1 day of harvest.

WATERCRESS (Hawaii only)
Cyclamen Mites: Mix 1 pt. per 100 gals. of water and apply to one acre as a thorough cover spray. Spray should be directed into the crowns and growing tips of plants. If infestation is heavy, make a second application no later than 10 days after the first application. Repeat applications as necessary. Do not apply within 5 days of harvest.

VEGETABLE SOIL INSECTS

Unless otherwise stated, soil applications of D•z•n diazinon AG500 should always be applied to vegetable crops as a spray in a minimum of 5 gals. of water per acre for ground applications.

CUTWORMS (Surface and Subterranean)

Broccoli	Kale	Snap Beans
Brussels Sprouts	Lettuce	Spinach
Cabbage	Lima Beans	Summer Squash
Cantaloupes	Muskmelons	Sweet Corn
Cauliflower	Parsley	Swiss Chard
Celery	Peas	Tomatoes
Collards	Peppers	Watermelons
Cucumbers	Potatoes	Winter Squash
Endive (Escarole)		

Broadcast 2–4 qts. per acre just prior to planting. Work into the soil immediately, 2–3 inches for surface cutworms, 3–6 inches for subterranean cutworms.

GARDEN SYMPHYLANS (Centipedes)

Cabbage	Peas	Snap Beans
Carrots	Pole Beans	Tomatoes
Corn	Radishes	Turnips
Lettuce	Red Beets	

Broadcast 10 qts. per acre just prior to planting. Immediately work into soil 4-8 inches. This preplant treatment will also control other soil insects: cutworms (surface and subterranean), mole crickets, root maggots, and wireworms.

MOLE CRICKETS

Broccoli	Endive (Escarole)	Radishes
Brussels Sprouts	Kale	Tomatoes
Cabbage	Lettuce	Turnips
Cauliflower	Mustard	
Collards	Peppers	

Broadcast 1 qt. per acre when insects are present. Make application to plant beds or field 1–2 days prior to sowing of seed or transplanting and immediately work into soil 1–2 inches.

ONION MAGGOTS

Furrow Application: Apply 1 qt. per acre in sufficient water to drench the seed furrow at planting time.

Broadcast Application: Broadcast 2–4 qts. per acre just prior to planting. Immediately work into soil 3–4 inches.

ROOT MAGGOTS
Broccoli, Brussels Sprouts, Cabbage, Cauliflower: Broadcast Application: Broadcast 2–3 qts. per acre and immediately work into the soil 3–4 inches, before seeding, either in the seedbed or in the field, or before transplanting.

Transplant Water (Drench) Application: Mix ¼-½ pt. in 50 gals. of transplant water when 200–300 gals. of water are used per acre. May be applied at the rate of ½-1 cupful per plant, either by hand or by tractor-mounted sprayer equipped with drop nozzles to direct spray to the base of the plant. Transplant water treatments may result in stand reduction due to plant stress at time of transplanting.

Radishes: Mix 1 qt. in 50 gals. of water. Apply 80–100 gals. per acre (1-1⅓ gals. per 1,000 linear feet of row) in the seed furrow at planting time.

WIREWORMS

Broccoli	Endive (Escarole)	Potatoes
Brussels Sprouts	Kale	Snap Beans
Cabbage	Lettuce	Spinach
Cantaloupes	Lima Beans	Summer Squash
Carrots	Muskmelons	Sweet Corn
Cauliflower	Onions	Sweet Potatoes
Celery	Parsley	Swiss Chard
Collards	Peas	Tomatoes
Cucumbers	Peppers	Watermelons
		Winter Squash

Broadcast 3–4 qts. per acre prior to planting, and immediately work into soil 4–8 inches.

Potatoes: Broadcast 3–6 qts. per acre prior to planting. Immediately disk or harrow into soil 4–8 inches.

SOUTHERN POTATO WIREWORMS (SOUTHEASTERN U.S.)

Potatoes

For use to control southern potato wireworms in potatoes grown on mineral soils in the Southeast, apply 2-4 qts. per acre as a soil treatment prior to planting, and immediately mix thoroughly by disking, or other suitable means into the top 3-6 inches of soil, depending on location of wireworms.

FIELD AND FORAGE FOLIAR INSECTS

Unless otherwise stated, foliar applications of D•z•n diazinon AG500 should always be applied to field and forage crops as a spray in a minimum of 2 gals. of water per acre for aerial applications and a minimum of 10 gals. of water per acre for ground applications.

ALFALFA, CLOVER

Spotted Alfalfa Aphids, Yellow Clover Aphids, Pea Aphids, Leafhoppers, Plant Bug Nymphs, Mites: Apply $^{3}/_{4}$-1 pt. per acre.

Grasshoppers: Apply 1 pt. per acre.

Following the above applications, dairy and beef cattle and sheep (1) may be grazed immediately following application, (2) may be fed green alfalfa or clover immediately following cutting, and (3) may be fed hay provided alfalfa or clover is not cut for hay for 7 days following treatment.

Alfalfa Weevil Larvae: For control in the spring, apply 2-3 pts. per acre when feeding damage to growing tips becomes evident. Repeat application, if necessary. If damage is already extensive, cut the hay and make application to the stubble. Do not graze dairy and beef animals or sheep for 2 days. Do not cut for hay for 10 days following treatment.

Do not spray livestock directly with either air or ground equipment. Do not treat during bloom to avoid injury to pollinating insects.

CLOVER

Striped Ground Crickets: Broadcast 1-2 pts. per acre following seeding in the fall but prior to clover emergence.

CORN

Use a minimum of 1 gal. of water for aerial applications and 5 gals. of water for ground applications.

Corn Rootworm Adults: Apply $^{1}/_{2}$-1 pt. per acre.

Corn Leaf Aphids: Apply 1-2 pts. per acre. Complete coverage of the upper portions of the plants is essential. Repeat applications as necessary.

Mites: Apply 1 pt. per acre when infestations first appear. Repeat applications as necessary.

Flea Beetles: Apply 1 pt. per acre when insects first appear. Repeat applications as necessary.

Grasshoppers: Apply 1 pt. per acre.

Sap Beetles: Apply 2-2$^{1}/_{2}$ pts. per acre. Add 1-2 gals. of soluble mineral oil per acre to the spray mixture. Consult State Agricultural Experiment Stations or State Extension Service Specialists for proper timing of sprays.

Corn may be picked immediately following last application. Corn forage may be fed to dairy and beef cattle and sheep.

COTTON

Spider Mites, Aphids, Leafhoppers, Lygus Bugs, Cotton Leafworms, Cotton Leafperforators: Apply 1-2 pts. in 5-25 gals. of water per acre. Repeat as infestation warrants.

Do not apply within 14 days of harvest. Do not graze treated cotton plants, or feed gin waste to livestock.

GUAR (TEXAS AND OKLAHOMA ONLY)

Midge: Apply $^{1}/_{2}$-1 pt. per acre in a minimum of 10 gals. of water for ground application or 2 gals. of water for aerial application when 15% of the blooms are infested. Should the infestation warrant, repeat applications should be made in 5-6 days.

Do not make more than 3 applications to the same crop.

Do not apply within 12 weeks of harvest.

Do not graze or feed treated forage to livestock within 12 weeks of harvest.

HOPS

Aphids, Mites: Apply 1 qt. per acre. Do not apply within 14 days of harvest.

SORGHUM

Use a minimum of 1 gal. of water for aerial applications and 5 gals. of water for ground applications.

Sorghum Midge: Apply $^{1}/_{2}$-1 pt. per acre when 90% of heads have emerged from the boot. Repeat application within 4 days.

Mites: Apply 1 pt. per acre when infestation first appears and repeat applications as necessary.

Greenbug: Apply at the rates indicated below when infestation first appears and repeat as necessary.

Texas and Oklahoma—Apply $^{1}/_{2}$-1 pt. per acre. All Other Areas—Apply 1 pt. per acre.

Crop may be fed to livestock immediately following treatment. Do not apply within 7 days of harvesting for grain.

SUGAR BEETS

Beet Leafhoppers, Dipterous Leafminers, Aphids: Apply $^{3}/_{4}$-1 pt. per acre.

Grasshoppers: Apply 1 pt. per acre.

Tops may be fed to beef and dairy animals.

SUGARCANE

Aphids: Apply $^{1}/_{2}$-1 pt. per acre. Allow 1 day between last application and harvest. Bagasse and forage may be fed to livestock.

TOBACCO

Aphids, Flea Beetles: Apply $^{1}/_{2}$-$^{3}/_{4}$ pt. per acre. Do not apply within 3 days of harvest.

Grasshoppers: Apply 1 pt. per acre. Do not apply within 14 days of harvest.

FIELD AND FORAGE SOIL INSECTS

Unless otherwise stated, soil applications of D•z•n diazinon AG500 should always be applied to field and forage crops as a spray in a minimum of 5 gals. of water per acre for ground applications.

ALFALFA, CLOVER

Cutworms (Surface and Subterranean): Broadcast 2-4 qts. per acre just prior to planting. Work into the soil immediately, 2-3 inches for surface cutworms, 3-6 inches for subterranean cutworms.

CORN

Corn Rootworm Larvae: Postemergence Basal Treatment: Apply 1$^{1}/_{4}$-2$^{1}/_{2}$ oz. per 1,000 linear feet of corn row in sufficient water to thoroughly wet base of corn plant and a narrow band of soil next to the plant. Refer to the row spacing chart for amounts recommended per acre. Cover treated band with 2-3 inches of soil using a disk hiller or cultivator immediately after spraying. Application should be made at the first sign of corn rootworm larvae, but not later than June 24 in the Middle Corn Belt, or July 1 in the Northern Corn Belt.

Suggested Rates According to Row Spacing for Corn Rootworm Larvae Control

Row Spacing	Linear Feet of Corn Row/Acre	Postemergence Basal Treatment (Fl. oz.)
48"	10,890	14-28
42"	12,455	15.5-31
40"	13,068	16-32 ($^{1}/_{2}$-1 qt.)
38"	13,939	17.5-35
36"	14,250	18-36
30"	17,424	22-44
24"	21,780	27.5-55
20"	26,136	32-64 (1-2 qts.)
15"	34,848	44-88

Note: For row spacings not listed, calculate the appropriate dosage based on the 40-inch row figure.

Seedcorn Maggots: Broadcast 2-4 qts. per acre just prior to planting. Work immediately into soil 2 inches.

Cutworms (Surface and Subterranean): Broadcast 2-4 qts. per acre. For surface cutworms apply just prior to planting or when larvae first appear, and immediately work into soil 1-2 inches. Do not apply later than lay-by.

For subterranean cutworms apply just prior to planting and immediately work into soil 3-6 inches.

Wireworms: Apply 3-4 qts. per acre prior to planting and immediately work into the soil 4-8 inches.

Corn may be picked immediately following last application. Corn forage may be fed to dairy and beef cattle and sheep.

COWPEAS

Cutworms (Surface and Subterranean): Broadcast 2-4 qts. per acre, just prior to planting. Work into the soil immediately, 2-3 inches for surface cutworms, 3-6 inches for subterranean cutworms.

LESPEDEZA

Cutworms (Surface and Subterranean): Broadcast 2-4 qts. per acre just prior to planting. Work into the soil immediately, 2-3 inches for surface cutworms, 3-6 inches for subterranean cutworms.

PEANUTS

Cutworms: Apply 2 qts. per acre in a 12-inch band over the row. Thorough coverage of the foliage is essential. Begin application when small larvae appear. Two or three additional applications at three-week intervals may be necessary. Do not apply within 21 days of peanut harvest.

Livestock (1) may be fed peanut hulls, (2) may be fed peanut forage 7 days following treatment, and (3) may be fed peanut hay 21 days following treatment.

SORGHUM

Cutworms (Surface and Subterranean): Broadcast 2-4 qts. per acre just prior to planting. Work into the soil immediately, 2-3 inches for surface cutworms, 3-6 inches for subterranean cutworms.

SOYBEANS

Cutworms (Surface and Subterranean): Broadcast 2-4 qts. per acre just prior to planting. Work into the soil immediately, 2-3 inches for surface cutworms, 3-6 inches for subterranean cutworms.

SUGAR BEETS

Wireworms: Broadcast 3-4 qts. per acre just prior to planting, and immediately work into soil 4-8 inches.

Tops may be fed to beef and dairy animals.

SUGARCANE

Wireworms: Apply 1 gal. per acre in a 12-inch band centered over the furrow. Make application directly to seed pieces in open furrow at covering time.

TOBACCO

Cutworms: Broadcast 2-3 qts. per acre before transplanting and immediately work into the soil 2-4 inches.

Green June Beetle Larvae: Mix 1/4 pt. in 50 gals. of water and apply as a drench (100 gals. per 100 sq. yds.) to uprooted areas of plant beds with sprinkling can.

Mole Crickets: Broadcast 1 qt. per acre, when insects are present. Make application to beds or fields 1-2 days prior to sowing of seed or transplanting, and immediately work into soil 1-2 inches.

Seedcorn Maggots: Broadcast 3 pts. per acre, 3-4 weeks before transplanting. Immediately work into soil 2 inches.

Wireworms: Broadcast 2-3 qts. per acre, 3-4 weeks before transplanting. Immediately work into soil 6-9 inches.

RANGE, PASTURE, AND GRASSLAND INSECTS

Unless otherwise stated, D•z•n diazinon AG500 should always be applied as a spray in a minimum of 1 gal. of water per acre for aerial applications and a minimum of 5 gals. of water per acre for ground applications.

BERMUDAGRASS AND FORAGE GRASSES (OTHER THAN RANGE AND PASTURE GRASSES)

Thrips, Whiteflies, Grasshoppers: Apply 1 pt. per acre.

Dairy and beef cattle and sheep (1) may be grazed immediately following application, (2) may be fed green forage immediately following cutting. Forage grasses may be cut for hay 7 days following treatment. Do not repeat application for 30 days.

RANGELAND, PASTURE, DITCH BANKS, ROADSIDES, WASTELAND, NONCROP AREAS, BARRIER STRIPS

Grasshoppers: Apply 3/4-1 pt. per acre in water or oil (minimum of 1 gal. of oil per acre) when insects first appear, preferably in the nymphal stage. Thorough coverage of the foliage is essential.

Dairy and beef cattle and sheep (1) may be grazed immediately following application, (2) may be fed green forage immediately following cutting, and (3) may be fed hay provided grass is not cut for hay for 21 days if the spray is a water solution, or for 30 days if the spray is an oil solution. Do not spray livestock directly with either air or ground equipment. Do not repeat application for 30 days.

ORNAMENTAL INSECT CONTROL

To control certain insects on Arborvitae, Azalea, Birch, Boxwood, Camellia, Carnation, Chrysanthemum, Douglas Fir, Elm, Gladioli, Hawthorn, Holly, Juniper, Lilac, Locust, Maple, Oak, Pine, Plum, Poplar, Rhododendron, Rose, Spruce, and Willow, apply the recommended rates indicated below:

Insects	Rate per 3 gals. water	Rate per 100 gals. water
Aphids, Bagworms, Carnation Bud Mites, Carnation Shoot Mites, Clover Mites, Cyclamen Mites, Dipterous Leafminers, European Pine Shoot Moths, European Red Mites, Flea Beetles, Holly Bud Moths, Leafhoppers, Obscure Root Weevils, Omnivorous Leaftiers, Privet Mites, Scale Crawlers (Cottonycushion, Lecanium, Pine Needle, San Jose, Soft Scale), Thrips, Twospotted Mites, Webworms, Whiteflies	1/2 fl. oz.*	1 pt.**
Apple-and-Thorn Skeletonizers, Cotoneaster Webworms, Fall Webworms, Hemlock Chermes, Oak Loopers, Obliquebanded Leafrollers, Pear Slugs, Tent Caterpillars	1 1/2 fl. oz.*	3 pts.**

* 1 fl. oz. = 2 tablespoons ** 16 fl. oz. = 1 pint

Do not use on Ferns, Poinsettia, Hibiscus, Papaya, Pilea, and Gardenia, since injury to the plants may occur.

Mimosa Webworms: Mix 1/2 fl. oz. in 3 gals. of water (or 1 pt. in 100 gals. of water) and apply as a thorough foliar spray when adults first appear. Make a second and third application at 4–5 week intervals.

LAWN INSECT CONTROL

To control insects listed below, apply recommended rates. For control of Ants, Brown Dog Ticks, Clover Mites, Crickets, Earwigs, Fleas, Millipedes, Sowbugs, and Springtails (Collembola), which are commonly found near the house and may enter the house, spray a 5-ft. band of soil around the house, as well as the house foundation wall to a height of 2-3 ft. Repeat application if necessary.

Insects	Rate/1,000 sq. ft.		Remarks
	D•z•n diazinon AG500	Water	
Lawn Chinch Bugs	2-3 fl. oz. or 3-6 fl. oz.	25 gals.	Use higher rate for longer residual control and in lawns of dense growth, such as St. Augustine lawns.
Ants, Armyworms, Clover Mites, Springtails (Collembola), Crickets, Cutworms, Digger Wasps, Earwigs, Frit Flies, Lawn Billbugs, Sod Webworms (Lawn Moths), Sowbugs	4 fl. oz.	3 gals.	Spot spray ant hills and wasp ground nest openings. For lawn billbugs, irrigate the lawn thoroughly immediately after treatment. For frit flies, mow grass and water well before treatment but delay application until grass is dry. Do not water again until necessary.

Brown Dog Ticks, Bermudagrass Mites, Chiggers, Fleas, Leafhoppers	1¼ fl. oz.	3 gals.	Do not apply to animals. For brown dog ticks, spray grass and under shrubbery, particularly near house.
Millipedes	8 fl. oz.	3 gals.	
Rhodesgrass Scales	5 fl. oz.	25 gals.	Apply when crawlers first emerge

* Or make application in 3 gals. of water per 1,000 sq. ft. and thoroughly water immediately following application.

Boxelder bugs: Spray bugs outside the home with 1¼-2½ fl. oz. (2½-5 tablespoons) in 1 gal. water. At the same time, thoroughly spray around foundation, windows, doors, tree trunks, and into cracks and other places where the insects hide.

Billbugs, White Grubs (such as Japanese Beetle Larvae, European Chafer, Southern Chafer): Apply 4 fl. oz. per 1,000 sq. ft. of lawn. Water grass thoroughly after application. For grub control make application any time between late July and early October. For billbug control apply when activity is first observed or when chewed or brown grass is present, indicating damage from this insect.

Hyperodes Weevil (New York only): Apply 3 fl. oz. per 1,000 sq. ft. of grass. Water grass thoroughly after application. Treat problem areas in mid-April and again in mid-May.

FOREST INSECTS

PINE (SEED ORCHARDS, PLANTATIONS, FORESTS)

Arkansas Sawfly (Loblolly Sawfly): Apply ½ pt. per acre over the tree canopy. Application should be made at or near the completion of egg hatch. Aerial applications should be made with a minimum of 2 gals. of spray per acre, in swaths not more than 90 ft. wide. Repeat application as needed.

STORAGE AND DISPOSAL

Do not contaminate water, food, or feed by storage or disposal. Open dumping is prohibited.

PESTICIDE

Pesticide wastes are toxic. Improper disposal of excess pesticide, spray mixture, or rinsate is a violation of Federal law. If pesticide, spray mixture or rinsate cannot be disposed of by use according to label instructions, contact your State Pesticide or Environmental Control Agency, or the Hazardous Waste representative at the nearest EPA Regional Office for guidance.

CONTAINER

Triple rinse (or equivalent). The offer for recycling or reconditioning, or puncture and dispose of in a sanitary landfill, or by incineration, or, if allowed by state and local authorities, by burning. If burned, stay out of smoke.

PRECAUTIONARY STATEMENTS

HAZARDS TO HUMANS AND DOMESTIC ANIMALS

WARNING

May be fatal if swallowed. May be absorbed through skin. Do not breathe spray mist. Do not get in eyes, on skin, or on clothing. Wash thoroughly after handling. Avoid contamination of food. Do not use on humans or permit children to go onto sprayed grass until spray has completely dried.

Keep out of reach of domestic animals. Do not contaminate feed or use on household pets or livestock. Do not permit pets to go onto sprayed grass until spray has completely dried.

STATEMENT OF PRACTICAL TREATMENT

If swallowed, call a physician immediately. DO NOT INDUCE VOMITING UNLESS NO OTHER TREATMENT IS AVAILABLE. Vomiting may cause aspiration pneumonia. If it is necessary to induce vomiting, give victim one or two glasses of water and insert finger in back of throat. Repeat until vomit fluid is clear. Do not induce vomiting or give anything by mouth to an unconscious person.

If inhaled, remove victim to fresh air and apply artificial respiration if indicated.

If on skin, wash promptly with soap and water. Rinse thoroughly.

If in eyes, rinse eyes with plenty of water and call a physician immediately.

Note to Physician: Gastric lavage is indicated. Atropine sulfate and 2-PAM are specific antidotes.

ENVIRONMENTAL HAZARDS

This product is toxic to fish, birds, and other wildlife. Birds feeding on treated areas may be killed. Keep out of lakes, streams, ponds, tidal marshes, and estuaries (except for effluence from treated watercress beds). Do not apply to water that will be used for recreational purposes and human and livestock consumption. Shrimp and crab may be killed at application rates recommended on this label. Do not apply where fish, shrimp, crab, and other aquatic life are important resources. This product must not be used in areas where impact on threatened endangered species is likely. Contact your State Fish and Game Agency for information pertaining to areas having threatened endangered species. Apply this product only as specified on this label.

This pesticide is highly toxic to bees exposed to direct treatment or to residues remaining on the treated area. Do not apply when bees are actively visiting the crop, cover crop, or weeds blooming in the treatment area. Applications should be timed to provide the maximum possible interval between treatment and the next period of bee activity.

PHYSICAL OR CHEMICAL HAZARDS

Do not use, pour, spill, or store near heat or open flame.

D•z•n® trademark of CIBA-GEIGY for diazinon

CGA 1L128 035

DUAL®8E

HERBICIDE

For weed control in corn, cotton, peanuts, pod crops, potatoes, safflowers, grain or forage sorghum, soybeans, and woody ornamentals.

ACTIVE INGREDIENT

Metolachlor:

2 chloro-*N*-(2-ethyl-6-methylphenyl)-*N*-(2-methoxy-1-methylethyl) acetamide 86.4%

INERT INGREDIENTS: 13.6%

TOTAL 100.0%

EPA Reg. No. 100-597 CGA49L12P 075

Dual® trademark of CIBA-GEIGY for metolachlor

U.S. Patent No. 3,937,730

Dual 8E contains 8 lbs. active ingredient per gallon

Keep Out of Reach of Children

WARNING

See additional precautionary statements inside booklet.

AVISO

PRECAUCIÓN AL USUARIO:

Si usted no lee inglés, no use este producto hasta que la etiqueta haya sido explicado ampliamente.

See directions for use inside booklet.

DIRECTIONS FOR USE AND CONDITIONS OF SALE AND WARRANTY

IMPORTANT: Read the entire **Directions for Use** and the **Conditions of Sale and Warranty** before using this product.

CONDITIONS OF SALE AND WARRANTY

The **Directions for Use** of this product reflect the opinion of experts based on field use and tests. The directions are believed to be reliable and should be followed carefully. However, it is impossible to eliminate all risks inherently associated with use of this product. Crop injury, ineffectiveness, or other unintended consequences may result because of such factors as weather conditions, presence of other materials, or the manner of use or application all of which are beyond the control of CIBA-GEIGY or the Seller. All such risks shall be assumed by the Buyer.

CIBA-GEIGY warrants that this product conforms to the chemical description on the label and is reasonably fit for the purposes referred to in the **Directions for Use** subject to the inherent risks referred to above. **CIBA-GEIGY makes no other express or implied warranty of Fitness or Merchantability or any other express or implied warranty. In no case shall CIBA-GEIGY or the Seller be liable for consequential, special, or indirect damages resulting from the use or handling of this product.** CIBA-GEIGY and the Seller offer this product, and the Buyer and user accept it, subject to the foregoing **Conditions of Sale and Warranty,** which may be varied only by agreement in writing signed by a duly authorized representative of CIBA-GEIGY.

DIRECTIONS FOR USE

It is a violation of federal law to use this product in a manner inconsistent with its labeling.

FAILURE TO FOLLOW ALL PRECAUTIONS ON THIS LABEL MAY RESULT IN POOR WEED CONTROL, CROP INJURY, OR ILLEGAL RESIDUES.

Do not apply this product in such a manner as to directly or through drift expose workers or other persons except those knowingly involved in the application. The area being treated must be vacated by unprotected persons.

REENTRY STATEMENT

Do not enter treated areas without protective clothing until sprays have dried.

Because certain states may require more restrictive reentry intervals for various crops treated with this product, consult your State Department of Agriculture for further information.

Written or oral warnings must be given to workers who are expected to be in a treated area or in an area about to be treated with this product. Oral warnings must inform workers of areas or fields that may not be entered without specific protective clothing until sprays have dried, and appropriate actions to take in case of accidental exposure, as described under **Precautionary Statements** on this label. When oral warnings are given, warnings shall be given in a language customarily understood by workers. Oral warnings must be given if there is reason to believe that written warnings cannot be understood by workers. Written warnings must include the following information: WARNING. Area treated with Dual 8E on (date of application). Do not enter without appropriate protective clothing until sprays have dried. In case of accidental exposure, flush eyes or skin with plenty of water. Call physician if irritation persists. Remove and wash contaminated clothing before reuse.

To avoid spray drift, do not apply under windy conditions. Avoid spray overlap, as crop injury may result.

A. GENERAL INFORMATION

Observe all precautions and limitations on the labels of each product used in tank mixtures.

Dual 8E is a selective herbicide recommended as a preplant surface-applied, preplant incorporated, or preemergence treatment in water or fluid fertilizer for control of most annual grasses and certain broadleaf weeds in corn, cotton, peanuts, pod crops, potatoes, safflowers, grain or forage sorghum, soybeans, and woody ornamentals.

Where directions specify a Dual 8E tank mixture with AAtrex formulations, other brands of atrazine may be used. Follow the rates, recommendations, and limitations on the respective product label, if other brands of atrazine are used.

If Dual 8E is incorporated, any supplemental tillage before planting must not exceed the specified depth of incorporation.

Dry weather following preemergence application of Dual 8E or a tank mixture may reduce effectiveness. Cultivate if weeds develop.

Where reference is made to weeds partially controlled, partial control can either mean erratic control from good to poor, *or* consistent control at a level below that generally considered acceptable.

Precaution: *Injury may occur following the use of Dual 8E under abnormally high soil moisture conditions during early development of the crop.*

1. SOIL TEXTURES AND HERBICIDE RATES

Where rates are based on coarse- medium- or fine-textured soils, it is understood that soil textural classes are generally categorized as follows:

Coarse	Medium	Fine	
Sand	Loam	Silty clay loam	Sandy clay
Loamy sand	Silt	Sandy clay loam	Clay loam
Sandy loam	Silt loam	Silty clay	Clay

Within rate ranges in the rate tables and elsewhere on this label, use the lower rate on soils relatively coarse-textured or low in organic matter; use the higher rate on soils relatively fine-textured or high in organic matter.

Note: Dual 8E may be applied preemergence alone, or in combination with tank mix partners specified on this label, following preplant incorporated herbicides when used according to their label recommendations, provided that such use is not prohibited on the respective labels.

Thoroughly clean sprayer prior to use. Do not use a sprayer contaminated with any other materials, or crop damage or sprayer clogging may result.

2. MIXING INSTRUCTIONS

Dual 8E Alone: Mix Dual 8E with water or fluid fertilizer and apply as a spray. Fill the spray tank one-half to three-fourths full with water or fluid fertilizer, add the rest of the water or fluid fertilizer. Provide sufficient agitation during mixing and application to maintain a uniform emulsion.

Tank Mixtures: Fill the spray tank one-fourth full with water, and start agitation; add AAtrex®, Amiben®, Balan®, Banvel®, Bladex®, Caparol 4L®, Dyanap®, Eptam®, Lexone®, Lorox®, Milogard®, Premerge 3®, Princep®, AAtrex + Princep®, Sencor® or Treflan®, and allow it to become dispersed, then add Dual 8E, then add paraquat or Roundup if these products are being used, and finally the rest of the water. If Igran or Bladex is to be used in the tank mixture, first add a compatibility agent, such as Unite® or Compex®, at 2-3 pts./100 gals. of spray mixture. Then add Igran or Bladex and allow it to become dispersed. Then add Dual 8E, and finally add the rest of the water. Agitate during mixing and application to maintain a uniform suspension. For tank mixtures with AAtrex, Amiben, Banvel, Bladex, Caparol 4L, Eptam, Igran, Lexone, Lorox (but not Lorox L), Milogard, Princep, AAtrex + Princep, Sencor or Treflan, fluid fertilizers may replace all or part of the water as carrier, except in the AAtrex postemergence and the Banvel postemergence tank mixes. Do not apply tank mixtures of Dual 8E + Dyanap, Dual 8E + Lorox L, or Dual 8E + Premerge 3 in fluid fertilizers. For tank mixtures with AAtrex, see additional mixing instructions on the AAtrex label. Check compatibility of mixture with fluid fertilizer, as described below, before mixing in spray tank.

Tank mixtures with Bladex should not be allowed to stand without agitation. If there is any evidence of incompatibility with Bladex in tank mixtures, add Compex, Unite, or an equivalent compatibility agent, to the mixture as with Igran or Bladex above. Check the compatibility of tank mixtures containing Bladex closely if fluid fertilizers are used as carriers. Empty the spray tank during the spraying operation.

Compatibility Test: Since liquid fertilizers can vary, even within the same analysis, always **check compatibility with herbicide(s) each time before use.** Be especially careful when using complete suspension or fluid fertilizers as serious compatibility problems are more likely to occur. Commercial application equipment may improve compatibility in some instances. The following test assumes a spray volume of 25 gals. per acre. For other spray volumes, make appropriate changes in the ingredients. Check compatibility using this procedure:

1. Add 1 pint of fertilizer to each of 2 one-quart jars with tight lids.
2. To **one** of the jars add 1/4 tsp. or 1.2 milliliters of a compatibility agent approved for this use, such as Compex or Unite (1/4 tsp. is equivalent to 2 pts. per 100 gals. spray). Shake or stir gently to mix.
3. To **both** jars add the appropriate amount of herbicide(s). If more than one herbicide is used, add them separately with dry herbicides first, flowables next, and emulsifiable concentrates last. After each addition, shake or stir gently to thoroughly mix. The appropriate amount of herbicides for this test follows:

   Dry herbicides: For each pound to be applied per acre, add 1.5 level teaspoons to each jar.

   Liquid herbicides: For each pint to be applied per acre, add 0.5 teaspoon or 2.5 milliliters to each jar.

   Note: For Dual 8E tank mixtures with AAtrex + Princep, use 1/3 - 1/2 the amount of AAtrex specified above and the remainder as Princep, depending on whether the 1:2 or 1:1 ratio of AAtrex to Princep is to be applied.
4. After adding all ingredients, put lids on and tighten, and invert each jar ten times to mix. Let the mixtures stand 15 minutes and then look for separation, large flakes, precipitates, gels, heavy oily film on the jar, or other signs of incompatibility. Determine if the compatibility agent is needed in the spray mixture by comparing the two jars. If either mixture separates, but can be remixed readily, the mixture can be sprayed as long as good agitation is used. If the mixtures are incompatible, test the following methods of improving compatibility: (A) slurry the dry herbicide(s) in water before addition, or (B) add 1/2 of the compatibility agent to the fertilizer and the other 1/2 to the emulsifiable concentrate or flowable herbicide before addition to the mixture. If incompatibility is still observed, do not use the mixture.

3. APPLICATION PROCEDURES

Application Timing

Dual alone or in some tank mixtures with other labeled herbicides may be applied for weed control in certain crops at various times. Refer to the given crop section of the label to determine if application timings listed below are recommended.

1. **Preplant Surface-Applied:** For minimum-tillage or no-tillage systems only..Dual 8E alone and some Dual 8E tank mixtures may be applied up to 45 days before planting certain crops. Use only split applications for treatments made 30-45 days before planting, with 2/3 the recommended broadcast rate for the crop and soil texture applied initially and the remaining 1/3 at planting. Treatments less than 30 days before planting may be made either as a split or a single application. Refer to individual crop to determine if early preplant surface application is recommended. If weeds are present at the time of treatment, apply in a tank mixture combination with a contact herbicide (for example, paraquat or Roundup). Observe directions for use, precautions, and restrictions on the label of the contact herbicide. To the extent possible, do not move treated soil out of the row or move untreated soil to the surface during planting, or weed control will be diminished.
2. **Preplant Incorporated:** Apply Dual 8E to the soil and incorporate into the top 2 inches of soil within 14 days before planting using a finishing disk, harrow, rolling cultivator, or similar implement capable of providing uniform 2 inch incorporation. Use a preplant incorporated application if furrow irrigation is used or when a period of dry weather after application is expected. If crop will be planted on beds, apply and incorporate Dual 8E after bed formation unless specified otherwise.
3. **Preemergence:** Apply Dual 8E during planting (behind the planter) or after planting but before weeds or crops emerge.

Special Application Procedures—California Only (Corn, Safflower, Pod Crops) *Preplant Incorporated:* Broadcast Dual 8E to the soil and thoroughly incorporate with a disk or similar implement set to till 4-6" deep. For more thorough incorporation, till the soil in two different directions (cross-till). Crops may be planted on flat surface or on beds. Caution should be used when forming the beds that only soil from the Dual 8E treated zone is used (i.e., untreated soil should not be brought to soil surface). If the application is made to preformed beds, incorporate Dual 8E with a tillage implement set to till 2-4" deep. Care should be taken during tilling to keep the tilled (Dual 8E treated) soil on the beds. *Preemergence:* Apply Dual 8E after planting. Water with sprinkler or flood irrigation within 7-10 days.

Ground Application: Apply Dual 8E alone or in tank mixtures by ground equipment in a minimum of 10 gals. of spray mixture per acre unless otherwise specified.

Use sprayers that provide accurate and uniform application. For Dual 8E tank mixtures with wettable powder or dry flowable formulations, screens and strainers should be no finer than 50-mesh. Rinse sprayer thoroughly with clean water immediately after use.

Calculate the amount of herbicide needed for band treatment by the formula:

$$\frac{\text{band width in inches}}{\text{row width in inches}} \times \begin{matrix}\text{broadcast rate}\\ \text{per acre}\end{matrix} = \begin{matrix}\text{amount needed}\\ \text{per acre of field}\end{matrix}$$

Aerial Application: Apply Dual 8E in water alone or in tank mixtures with AAtrex, Lorox, Lexone, or Sencor in a minimum total volume of 2 gals. per acre by aircraft. Avoid application under conditions where uniform coverage cannot be obtained or where excessive spray drift may occur. In order to assure that spray will be controllable within the target area when used according to label directions, make applications at a maximum height of 10 ft., using low drift nozzles at a maximum pressure of 40 psi, and restrict application to periods when wind speed does not exceed 10 mph. To assure that spray will not adversely affect adjacent sensitive nontarget plants, apply Dual 8E alone or Dual 8E + AAtrex by aircraft at a minimum upwind distance of 400 ft. from sensitive plants, or apply Dual 8E + Lexone, Sencor, or Lorox at a minimum upwind distance of 300 ft. from sensitive plants.

Avoid application to humans or animals. Flagmen and loaders should avoid inhalation of spray mist and prolonged contact with skin and should wash thoroughly before eating and at the end of each day's operation.

Center Pivot Irrigation Application: Dual 8E alone or tank mixtures with AAtrex, or other herbicides on this label which are registered for center pivot application, may be applied in irrigation water preemergence (after planting but before weeds or crop emerge) at rates recommended on this label. Use only center pivot systems that apply water uniformly. Prepare a mixture with a minimum of 1 part of water to 1 part herbicide(s), and inject this mixture into the center pivot system using a positive displacement pump. Injecting a larger volume of a more dilute mixture per hour will usually provide more accurate calibration of metering equipment. Maintain sufficient agitation to keep the herbicide(s) in suspension. Apply in 1/2 -1 inch of water. Use the lower water volume (1/2 inch) on coarser textured soils and the higher volume (1 inch) on finer textured soils. More than 1 inch of water at application may reduce weed control by moving the herbicide below the effective zone in the soil. Refer to supplemental literature published by CIBA-GEIGY for more information on calibration.

Precautions for center pivot applications: *1) Apply only through irrigation systems containing anti-syphon and check valves to prevent contamination of well during shutdown and overflow of solution tank. 2) Inject ahead of any right angle turn in the main line to insure adequate mixing. 3) Chemical injection pumps and water pumps must have interlocking controls to insure simultaneous shutoff. 4) Application when drift may occur, such as from windy conditions, or when system joints and connections are leaking, or when nozzles are not providing uniform distribution, may cause crop injury. 5) Where sprinkler distribution patterns do not overlap sufficiently, unac-*

ceptable weed control may result. Where sprinkler distribution patterns overlap excessively, crop injury may result.

4. DRY BULK GRANULAR FERTILIZERS

Many dry bulk granular fertilizers may be impregnated or coated with Dual 8E alone or Dual 8E tank mixtures registered for preplant incorporated application to control weeds in crops on the Dual 8E label.

Labels, supplementary literature, and precautions regarding rates per acre, approved crops, soil texture, application, incorporation, crop rotation and other directions must be followed for Dual 8E and other herbicides used in tank mixtures with Dual 8E.

All individual state regulations relating to dry bulk granular fertilizer blending, registration, labeling, and application are the responsibility of the individual and/or company selling the herbicide/fertilizer mixture.

Prepare the granular herbicide/fertilizer mixtures by using any closed drum, belt, ribbon, or other commonly used dry bulk fertilizer blender. Nozzles used to spray Dual 8E and Dual 8E mixtures onto the fertilizer must be placed to provide uniform spray coverage.

If the herbicide/fertilizer mixture is too wet, use a highly absorptive powder, such as Microcel E (Johns-Manville Products Corporation), diatomaceous earth or finely powdered clay, to obtain a dry free-flowing mixture. Add the absorptive powder separately and uniformly to the herbicide/fertilizer mixture and blend to form a suitable free-flowing mixture. Generally, less than 2% by weight of absorptive powder will be needed.

Calculate amounts of Dual 8E, AAtrex, AAtrex + Princep, Bladex, Igran, Milogard, Princep, or Sencor (Lexone) by the following formula:

$$\frac{2{,}000}{\text{lbs. of fertilizer per acre}} \times \text{pts./acre of liquid or flowable product} = \text{pts. of liquid or flowable product per ton of fertilizer}$$

$$\frac{2{,}000}{\text{lbs. of fertilizer per acre}} \times \text{lbs/acre of dry product} = \text{lbs. of dry product per ton of fertilizer}$$

Precautions: *To avoid potential for explosion, (1) Do not impregnate Dual 8E or Dual 8E mixtures on ammonium nitrate, potassium nitrate or sodium nitrate either alone or in blends with other fertilizers. (2) Do not combine mixtures of Dual 8E plus any other herbicide with single superphosphate (0-20-0) or treble superphosphate (0-46-0). (3) Do not use Dual 8E or Dual 8E mixtures on straight limestone, since absorption will not be achieved. Fertilizer blends containing limestone can be impregnated.*

Application

Apply 200-450 pounds of the herbicide/fertilizer mixture per acre. For best results, apply the mixture uniformly to the soil with properly calibrated equipment immediately after blending. Uniform application of the herbicide/fertilizer mixture is essential to prevent possible crop injury. Nonuniform application may also result in unsatisfactory weed control. In areas where conventional tillage is practiced, a shallow incorporation of the mixture into the soil may improve weed control. On fine- or medium-textured soils in areas where soil incorporation is not planned, i.e., reduced tillage situations or in some conventional till situations, make applications approximately 30 days before planting to allow moisture to move the herbicide/fertilizer mixture into the soil. On coarse-textured soils, make applications approximately 14 days prior to planting.

Precaution: *To avoid crop injury, do not use the herbicide/fertilizer mixture on crops where bedding occurs.*

B. DUAL 8E APPLIED ALONE

1. *Weeds Controlled*

barnyardgrass (watergrass)	green foxtail	yellow foxtail
crabgrass	prairie cupgrass	yellow nutsedge
crowfootgrass	red rice	
fall panicum	signalgrass (*Brachiaria*)	black nightshade
foxtail millet	southwestern cupgrass	carpetweed
giant foxtail	witchgrass	Florida pusley
goosegrass		galinsoga
		pigweed

Weeds Partially Controlled:* common purslane, Florida beggarweed,** hairy nightshade, sandbur, seedling johnsongrass, shattercane, Texas panicum***, volunteer sorghum.

* See General Information section. Control of these weeds can be erratic due partially to variable weather conditions. Control may be improved by following these suggested procedures:

1. **Thoroughly till moist soil** to destroy germinating and emerged weeds. If Dual 8E is to be applied preplant incorporated, this tillage may be used to incorporate Dual 8E if uniform 2 inch incorporation is achieved as recommended under **Application Procedures.**
2. Plant crop into moist soil **immediately after tillage.** If Dual 8E is to be used preemergence, apply at planting or immediately after planting.
3. If available, **sprinkler irrigate** within 2 days after application. Apply ½ -1 inch of water. Use lower water volume (½ inch) on coarse-textured soils and higher volume (1 inch) on finer textured soils. Also, refer to the section on **Center Pivot Irrigation Application** for this method of applying Dual 8E.
4. If irrigation is not possible and rain does not occur within 2 days after planting and application, weed control may be decreased. Under these conditions, a uniform, shallow cultivation is recommended as soon as weeds emerge.

** For partial control of this weed, use a minimum of 2 pts./A and apply preemergence.

*** For partial control of this weed, use a minimum of 2 pts./A and apply through a center pivot irrigation system.

2. Rotational Crops: 1) If crop treated with Dual 8E alone is lost, corn, cotton, soybeans, peanuts, potatoes, safflowers, grain or forage sorghum treated with Concep, or the listed pod crops may be replanted immediately. Do not make a second broadcast application of Dual 8E. If the original application was banded and the second crop is planted in the untreated row middles, a second banded treatment may be applied. 2) Barley, oats, rye, or wheat may be planted 4½ months following treatment. Corn, cotton, soybeans, peanuts, potatoes, safflowers, sorghum, pod crops, root crops, barley, buckwheat, milo, oats, rice, rye, or wheat may be planted in the spring following treatment. Do not graze or feed forage or fodder from cotton or pod crops to livestock. All other rotational crops may be planted 18 months after application.

Note: For **Rotational Crops** restrictions for Dual 8E used in tank mixtures, refer to the statements/restrictions above for Dual 8E and to the respective product labels of any mixing partner(s) for additional statements/restrictions.

C. CORN—DUAL 8E ALONE

Apply Dual 8E either preplant surface, preplant incorporated, or preemergence using the appropriate rate specified below.

Preplant Surface-Applied: Refer to instructions for use of Dual 8E alone under **Application Procedures.** Use on medium- and fine-textured soils with minimum-tillage or no-tillage systems in CO, IL, IN, IA, KS, KY, MN, MO, MT, NE, ND, SD, WI, WY. Apply ⅔ the recommended rate of Dual 8E (2.5 pts./A on *medium soils* and 3 pts./A on *fine soils*) as a split treatment 30-45 days before planting and the remainder at planting. Applications made less than 30 days prior to planting may be as either a split or single treatment. Apply 2 pts./A on *coarse soils* not more than 2 weeks prior to planting.

On medium- and fine-textured soils with minimum- or no-tillage systems in DE, MD, MI, NY, OH, PA, VA, and WV, preplant surface applications may be applied following the directions for use above. If the amount of rainfall results in unsatisfactory length of weed control following the earlier treatment, a post-emergence application of an appropriately labeled broadleaf and/or grass weed herbicide may be used, i.e., AAtrex, 2, 4-D, Banvel®, bromoxynil (Brominal® or Buctril®). If the postemergence treatment includes the herbicide used preplant surface applied, do not exceed the labeled rate for corn on a given soil texture. Observe all directions for use, precautions and limitations on the label of the postemergent herbicide.

Preplant Incorporated or Preemergence: Follow instructions for use of Dual 8E alone under **Application Procedures.** On *coarse soils,* apply 1.5-2 pts./A of Dual 8E if organic matter content is less than 3%, or 2 pts./A if organic matter content is 3% or greater. On *medium soils,* apply 2-2.5 pts./A of Dual 8E. On *fine soils,* apply 2-2.5 pts./A of Dual 8E if organic matter content is less than 3%, or 2.5-3 pts./A if organic matter content is 3% or greater.

Notes: (1) Do not apply more than the labeled application rate for a given soil texture per year either as a single or split treatment or illegal residues may result. (2) In corn, Dual 8E may be used up to 4 pts./A as either a preplant surface, preplant incorporated, or preemergence treatment on soils having an organic matter content between 6% and 20%.

D. CORN—DUAL 8E COMBINATIONS

Dual 8E alone or any tank mixture for corn (except Dual + AAtrex postemergence and Dual + Banvel postemergence) may be applied in water or fluid fertilizer. Use only water in the Dual + AAtrex or the Dual + Banvel postemergence tank mixes.

1. TANK MIXTURE WITH AATREX OR PRINCEP, OR AATREX PLUS PRINCEP—PREPLANT SURFACE, PREPLANT INCORPORATED, OR PREEMERGENCE

In addition to the weeds controlled by Dual 8E alone, Dual 8E + AAtrex or Princep, or Dual 8E + AAtrex + Princep, applied preplant surface, preplant incorporated, or preemergence, also controls the following weeds: browntop panicum, cocklebur, common purslane, hairy nightshade, lambsquarters, morningglory, ragweed, smartweed, and velvetleaf.

Apply Dual 8E + AAtrex or Princep, or Dual 8E + AAtrex + Princep either preplant surface, preplant incorporated, or preemergence.

Preplant Surface-Applied: Follow instructions for use of Dual 8E alone under **Application Procedures** and under application instructions for Dual 8E alone on corn. Apply Dual 8E + AAtrex or Princep, or Dual 8E + AAtrex + Princep on *medium soils* (2.5 pts./A Dual 8E) 3.2-4 pts./A of AAtrex 4L or Princep 4L, or AAtrex 4L + Princep 4L combined) and on *fine soils* (2.5-3 pts./A Dual 8E + 4-5 pts./A of AAtrex 4L or Princep 4L, or AAtrex 4L + Princep 4L combined in minimum-tillage and no-tillage systems in CO, IL, IN, IA, KS, KY, MN, MO, MT, NE, ND, SD, WI, WY. Apply the tank mixtures as a split or single treatment in those states and as indicated in the Dual 8E alone Preplant Surface-Applied section of the label for corn. On *coarse soils,* apply 2 pts./A of Dual 8E and 3.2 pts./A of AAtrex 4L or Princep 4L, or AAtrex 4L + Princep 4L combined.

Preplant Incorporated or Preemergence: Follow instructions for use of Dual 8E alone under **Application Procedures.** Apply Dual 8E + AAtrex or Princep, or Dual 8E + AAtrex + Princep using the appropriate rates from Table 1.

Note: Do not apply more than the labeled rate for a given soil texture per year either as a split or single treatment, or illegal residues may result.

Table 1: Dual 8E + AAtrex or Princep, or Dual 8E + AAtrex + Princep, Preplant Incorporated or Preemergence — Corn

Soil texture	Broadcast rates per acre: Less than 3% organic matter: Dual 8E + AAtrex Nine-O®* or Princep Caliber® 90*	OR Dual 8E + AAtrex Nine-O** + Princep Caliber 90**	3% organic matter or greater: Dual 8E + AAtrex Nine-O* or Princep Caliber 90*	OR Dual 8E + AAtrex Nine-O** + Princep Caliber 90**
COARSE	1.25-1.5 pts. + 1.1-2.2 lbs.	1.25-1.5 pts. + .6-1.1 lbs. + .6-1.1 lbs.	1.5 pts. + 1.3-2.2 lbs.	1.5 pts. + .7-1.1 lbs. + .7-1.1 lbs.
MEDIUM	1.5-2 pts. + 1.3-2.2 lbs.	1.5-2 pts. + .7-1.1 lbs. + .7-1.1 lbs.	2 pts. + 1.8-2.2 lbs.	2 pts. + .9-1.1 lbs. + .9-1.1 lbs.
FINE	2 pts. + 1.8-2.2 lbs.	2 pts. + .9-1.1 lbs. + .9-1.1 lbs.	2-2.5 pts. + 1.8-2.2 lbs.***	2-2.5 pts. + .9-1.1 lbs.*** + .9-1.1 lbs.***
muck or peat (soils with more than 20% organic matter)	DO NOT USE			

* When using AAtrex 4L (or Princep 4L), AAtrex 4LC, or AAtrex 80W (or Princep 80W), use equivalent rates. One lb. of Nine-O or Caliber 90 equals 1.8 pts. of 4L (4LC) or 1.1 lbs. 80W. Use Princep in preference to AAtrex when heavy infestations of crabgrass or fall panicum are expected. On soils having between 6% and 20% organic matter, Dual 8E may be used up to 3.5 pts./A in tank mix combination with 3.3 lbs./A of AAtrex Nine-0, or equivalent rates of AAtrex 80W or AAtrex 4L (4LC).

** When using the tank mixture of Dual 8E + AAtrex Nine-O + Princep Caliber 90, use equal rates of each as shown when heavy broadleaf weed infestations are expected. When heavy infestations of crabgrass or fall panicum are expected, use a 1.2 ratio of AAtrex + Princep instead of the 1:1 ratio given in Table 1 above. (*Example:* Total AAtrex Nine-O + Princep Caliber 90 = 1.2 lbs./A, use 0.4 lb. of AAtrex plus 0.8 lb. of Princep, respectively.) Refer to the asterisk (*) above for AAtrex 4L, 4LC and 80W, and Princep 4L and 80W conversions.

*** For cocklebur, yellow nutsedge, and velvetleaf control on fine-textured soils above 3% organic matter, apply 2.25 lbs./A of AAtrex Nine-0, or equivalent rates of AAtrex 4L, 4LC or 80W, or the same total amount of AAtrex + Princep with 2-2.5 pts./A of Dual 8E.

2. TANK MIXTURE WITH AATREX POSTEMERGENCE

Weeds Controlled

barnyardgrass (watergrass)	yellow foxtail	prickly sida
crabgrass	jimsonweed	purslane
crowfootgrass	kochia	ragweed
fall panicum	lambsquarters	smartweed
giant foxtail	mustard	velvetleaf
green foxtail	pigweed	

Weeds Partially Controlled: cocklebur, morningglory, yellow nutsedge.

Apply 1.5 pts./A of Dual 8E + 1.3 lbs./A of AAtrex Nine-O* on *coarse soils,* 2 pts./A of Dual 8E + 1.8 lbs./A of AAtrex Nine-O on *medium soils,* or 2-2.5 pts./A of Dual 8E + 1.8-2.2 lbs./A** of AAtrex Nine-O on *fine soils.* Apply this tank mixture before grass and broadleaf weeds pass the 2-leaf stage and before corn exceeds 5 inches in height. Application to weeds larger than the 2-leaf stage will generally result in unsatisfactory control. Occasionally some corn leaf burn may result, but this should not affect later growth or yield. Do not apply this post-emergence tank mixture in fluid fertilizer, or severe crop injury may occur.

* When using AAtrex 4L, AAtrex 4LC, or AAtrex 80W, use equivalent rates. One lb. of Nine-O equals 1.8 pts. of 4L (4LC) or 1.1 lbs. 80W.

** For better control of cocklebur, morningglory, velvetleaf, and yellow nutsedge on fine-textured soils above 3% organic matter, apply 2.2 lbs./A of AAtrex Nine-O, or equivalent rate of AAtrex 80W or AAtrex 4L (4LC), with 2-2.5 pts./A of Dual 8E.

3. TANK MIXTURE WITH BLADEX

(Field Corn and Silage Corn Only)

In addition to the weeds controlled by Dual 8E alone, Dual 8E + Bladex, applied preplant incorporated or preemergence, also controls the following broadleaf weeds: jimsonweed, lambsquarters, mustard, ragweed, smartweed, cocklebur*, morningglory*, and velvetleaf*.

* Partially controlled.

Preplant Incorporated or Preemergence: Apply Dual 8E + Bladex preplant incorporated or preemergence using the appropriate rate from Table 2. Follow instructions for use of Dual 8E alone under **Application Procedures.**

Table 2: Dual 8E + Baldex — Field Corn and Silage Corn Only — Broadcast Rates Per Acre

Soil texture	Percent organic matter in soil: Less than 1% (Pts. Dual 8E + Qts. Bladex 4L*)	1-2.5% (Pts. Dual 8E + Qts. Bladex 4L*)	2.5-4% (Pts. Dual 8E + Qts. Bladex 4L*)	Over 4% (Pts. Dual 8E + Qts. Bladex 4L*)
COARSE Sand, loamy sand	DO NOT USE	1.25-1.5 + .8-1.4	1.5-1.75 + 1.4-1.6	2 + 1.75
COARSE Sandy loam	1.25 + .8	1.25-1.5 + 1.2-1.4	1.5-1.75 + 1.4-1.75	2 + 2
MEDIUM	1.5 + 1.2	1.5-1.75 + 1.4-1.75	1.75-2 + 1.75-2	2.25 + 2.25
FINE	1.5 + 1.4	1.75-2 + 1.6-2	2-2.25 + 2-2.25	2.5 + 2.5
muck or peat soils (soils with more than 20% organic matter)	DO NOT USE			

* When using Bladex 80W, use equivalent rates. One qt. of 4L equals 1.25 lbs. of 80W.

4. TANK MIXTURE WITH BANVEL

Preemergence: Use this tank mixture only on field corn which is flat-planted (no furrows) in IL, IN, IA, MN, NE, OH, SD, and WI.

In addition to the weeds controlled by Dual 8E alone, Dual 8E + Banvel, applied preemergence, also controls lambsquarters, ragweed, smartweed, cocklebur*, jimsonweed*, morningglory*, and velvetleaf*.

* Partially controlled.

Apply Dual 8E + Banvel preemergence. Broadcast 1 pt./A of Banvel with 2 pts./A of Dual 8E on *medium soils,* or with 2-2.5 pts./A of Dual 8E on *fine soils.* Do not apply on *coarse soils* or on soils with less than 2.5% organic matter. Apply this tank mixture to the soil surface at planting or after planting but before corn emerges. Avoid drift to soybeans or other desirable plants. Do not apply with aircraft. Plant corn at least 1.5 inches deep and apply behind planting equipment avoiding incorporation by the planter wheel or other seed covering device. Do not incorporate before corn emergence. If it is necessary to rotary hoe to break the soil crust, do not disturb the soil more than one-half inch deep.

Postemergence for Control of Pigweed (Mid-Atlantic states including DE, MD, PA, VA, and WV): Apply 1.5-2.25 pts. of Dual 8E plus 0.5-1 pt./A of Banvel or 1-2 pt./A of Banvel II by ground equipment when pigweed plants are less than 3 inches tall and before corn exceeds 5 inches in height in a minimum of 20 gallons of spray per acre.

Use the lower rate on coarse-textured and low organic matter soils. Use the higher rate on fine-textured and high organic matter soils.

Precaution: Avoid drift to sensitive nontarget plants during application, or injury may occur.

5. Tank Mixture with AAtrex and Linuron for Control of Lambsquarters and Pigweed

For prolonged control of lambsquarters and pigweed in DE, MD, NJ, NY, PA, VA and WV, Dual 8E may be applied preemergence in tank mix combination with AAtrex + linuron. Apply Dual 8E and AAtrex according to the rates in Table 1 and linuron according to the rates below.

Soil Texture	Broadcast Rate/A
Sandy loam (1-3% organic matter)	.67 lb. linuron 50W*
Sandy loam (3-6% organic matter)	1.0 lb. linuron 50W*
Medium- and fine-textured soils (1-6% organic matter)	1.0 lb. linuron 50W*

* When using linuron L (4 lbs. ai per gallon), use equivalent rates. One pt. of linuron L equals one lb. of linuron wettable powder.

Observe all directions for use, precautions and limitations on the Dual, AAtrex and linuron labels when applying these products in tank mix combinations.

6. Tank Mixture with AAtrex, Bladex, or Princep, AAtrex plus Bladex or AAtrex plus Princep, with Paraquat or Roundup for Minimum-Tillage or No-Tillage Systems

In minimum-tillage or no-tillage systems where corn is planted directly into a cover crop, stale seedbed, established sod, or previous crop residues, the contact herbicides paraquat or Roundup may be added to a tank mix of Dual 8E + AAtrex, Bladex or Princep, Dual 8E + AAtrex + Bladex or Dual 8E + AAtrex + Princep. When used as directed, the paraquat portion of the tank mixture controls most emerged annual weeds and suppresses many perennial weeds. Roundup combinations will control emerged annual and perennial weeds when applied as directed on the Roundup label. The Dual 8E + AAtrex or Princep, or Dual 8E + AAtrex + Princep portion of the tank mixture provides preemergence control of the weeds listed on this label in the tank mixture section for Dual 8E + AAtrex or Princep, or Dual 8E + AAtrex + Princep Preplant Surface, Preplant Incorporated, or Preemergence. The Dual 8E + Bladex or Dual 8E + AAtrex + Bladex tank mixture provides preemergence or postemergence control of weeds listed on this label in the tank mixture sections with AAtrex and Bladex.

Application: Apply before, during, or after planting, but before the corn emerges, at the rates specified below. Add paraquat or Roundup at the following broadcast rates:

Paraquat: 1-2 pts./A* plus 8 oz. of X-77® Spreader per 100 gals. of spray mixture. Use the lower rate for control of annual weeds less than 4 inches tall and the higher rate for weeds 4-6 inches tall. This treatment will not consistently control weeds taller than 6 inches.

Note: Do not apply combinations containing paraquat in suspension type liquid fertilizers as the activity of paraquat will be reduced.

* Based on products containing 2 lbs. active ingredient per gal.

Roundup: 1.5 qts./A for existing annual weeds, or 2-4 qts. per acre for existing perennial weeds. See the Roundup label for weeds controlled and recommended rates for specific weeds.

Apply in 20-60 gals. of water or fluid fertilizer per acre with ground equipment.

On *coarse soils,* apply 1.5 pts./A of Dual 8E with 1.3 lbs. AAtrex Nine-O* or Princep Caliber 90*, or with 0.7 lb. Nine-O** plus 0.7 lb. Caliber 90**. On *medium soils,* apply 2 pts./A of Dual 8E with 1.8 lbs. AAtrex Nine-O or Princep Caliber 90, or with 0.9 lb. Nine-O plus 0.9 lb. Caliber 90. On *fine soils,**** apply 2-2.5 pts./A of Dual 8E with 1.8-2.2 lbs. AAtrex Nine-O or Princep Caliber 90, or with 0.9-1.1 lbs. of Nine-O plus 0.9-1.1 lbs. Caliber 90.

* When using AAtrex 4L (or Princep 4L), AAtrex 4LC, or AAtrex 80W (or Princep 80W), use equivalent rates. One lb. of Nine-O or Caliber 90 equals 1.8 pts. of 4L (4LC) or 1.1 lbs. of 80W. Use Princep in preference to AAtrex when heavy infestations of crabgrass or fall panicum are expected.

** When using the tank mixture of Dual 8E + AAtrex Nine-O + Princep Caliber 90, use equal rates of AAtrex and Princep as shown when heavy broadleaf weed infestations are expected. When heavy infestations of crabgrass or fall panicum are expected, use a 1:2 ratio of AAtrex + Princep instead of the 1:1 ratio given above. (*Example:* Total AAtrex Nine-O + Princep Caliber 90 = 1.8 lbs./A, use 0.6 lb. of AAtrex plus 1.2 lbs. of Princep, respectively.) Refer to the asterisk (*) above for AAtrex 4L, 4LC and 80W, and Princep 4L and Princep 80W conversions.

*** For cocklebur, yellow nutsedge, and velvetleaf control on fine-textured soils above 3% organic matter, apply 2.25 lbs./A of AAtrex Nine-O, or equivalent rates of AAtrex 80W or AAtrex 4L (4LC), or the same total amount of AAtrex + Princep, with 2-2½ pts./A of Dual 8E.

7. Tank Mixture with AAtrex or AAtrex plus 2,4-D or AAtrex plus 2,4-D plus Banvel for Minimum-Tillage or No-Tillage Systems

In minimum-tillage or no-tillage systems where corn is planted directly into a cover crop, stale seedbed, established sod, or previous crop residues, Dual 8E applied in combination with AAtrex will kill most emerged small annual weeds. Apply Dual 8E + AAtrex before, during, or after planting, but before corn emerges according to the rates in Table 1.

Where heavy crop residues exist, add 0.8-1.6 pts./A of an appropriately labeled 3.8 lb. ai/gal. 2,4-D amine (such as Weedar 64, Weedar 64A, DMA-4 Herbicide, or Formula 40) to the spray tank last and apply in a minimum of 25 gals. of carrier/A.

As carriers, nitrogen solutions and complete liquid fertilizers, applied before corn emergence, enhance burndown of existing weeds, and therefore are recommended instead of water. Add X-77 surfactant at 1.0-2.0 qts./100 gals. of diluted spray, or another appropriate surfactant at its recommended rate. Apply before weeds exceed 3 inches in height. If alfalfa is present, add Banvel to the spray mixture at 0.33-0.5 pt./A and apply before alfalfa exceeds 6 inches in height.

For fields with existing sod grasses (e.g., bromegrass, orchardgrass, rye or timothy), when existing weeds exceed 3 inches in height or when very dry conditions exist, add paraquat at the rate of 2 pts./A in place of or in addition to 2,4-D as indicated above. Do not apply paraquat in suspension-type liquid fertilizer. Observe all directions for use, precautions and limitations on the respective product labels when applying these products in tank mix combination.

8. Tank Mixture with Bladex or Bladex plus AAtrex plus 2,4-D or Bladex plus AAtrex plus 2,4-D plus Banvel for Minimum-Tillage or No-Tillage Systems

In minimum-tillage or no-tillage systems where corn is planted directly into a cover crop, stale seedbed, established sod, or previous crop residues, Dual 8E applied in combination with Bladex or Bladex + AAtrex will provide control of emerged small annual weeds, and extended control of those species on the Dual and Bladex labels.

Apply Dual 8E in tank mix combination with Bladex according to the **"Tank Mixtures with Bladex"** and **"Tank Mixtures with AAtrex. . ."** sections of this label or the preemergence section of the Bladex and AAtrex labels. Where heavy corn residues exist at the time of treatment, follow directions for use in the **"Bladex Plus Dual"** tank mixture in the **Minimum- or No-Till Weed Control** section of the Bladex label. Using a minimum of 25 gals./A of carrier, add 0.8-1.6 pts./A of an appropriately labeled 3.8 ai/gal. 2,4-D amine (such as Weedar 64, Weedar 64A, DMA-4 Herbicide, or Formula 40) to the spray tank last.

As carriers, nitrogen solutions and complete liquid fertilizers, applied before corn emergence, enhance burndown of existing weeds, and therefore are recommended instead of water. Add X-77 surfactant at 1.0 to 2.0 qts./100 gals. of diluted spray, or another appropriate surfactant at its recommended rate. Apply before weeds exceed 3 inches in height. If alfalfa is present, add Banvel to the spray mixture at 0.33-0.5 pt./A and apply before alfalfa exceeds 6 inches in height.

For fields with existing sod grasses (e.g., bromegrass, orchardgrass, rye or timothy), when existing weeds exceed 3 inches in height or when very dry conditions exist, add paraquat at the rate of 2.0 pts./A in place of or in addition to 2,4-D as indicated above. Do not apply paraquat in suspension-type liquid fertilizer. Observe all directions for use, precautions and limitations on the respective product labels when applying these products in tank mix combination.

E. COTTON—DUAL 8E ALONE—AR, LA, MS, NM, OK, TN, TX, AND BOOTHEEL OF MO

Application: Apply Dual 8E either preplant incorporated or preemergence at the rate of 1.5 pts./A on sandy loams*, 1.5-2 pts./A on *medium soils,* or 2 pts./A on *fine soils. Do not use on sands and loamy sands.*

* Use on sandy loam soils only in NM, OK, and TX.

Preplant Incorporated: Apply to the soil and incorporate into the top inch of soil immediately before planting, at planting, or after planting but before crop or weeds emerge. Use a rolling cultivator or similar implement to uniformly incorporate not more than one inch deep. Use a preplant incorporated application if furrow irrigation is used or when a period of dry weather after application is expected. Where furrow irrigation is used, wet the top of the bed for best results. If the crop is to be planted on beds, apply and incorporate after bed formation. Cotton should be planted below the zone of incorporation; i.e., at least 1 inch on fine soils and 1.5 inches on coarse and medium soils. If incorporated prior to planting, use a planter that will result in a minimum of soil disturbance.

Preemergence: Apply to the soil surface at planting, or after planting but before weeds or crop emerge.

Note: For best control of yellow nutsedge and suppression of seedling johnsongrass, apply Dual preplant incorporated at the maximum rate for the soil texture, whether applied alone or mixed with Caparol 4L.

Precautions: *1) Do not apply Dual on sand or loamy sand soils or in areas where water is likely to "pond" over the bed, or crop injury may occur. 2) Do not apply broadcast to furrow-planted cotton, or injury may occur. 3) Do not apply on Taloka silt loam, or injury may occur.*

F. COTTON—DUAL 8E COMBINATIONS

1. TANK MIXTURE WITH CAPAROL 4L

Dual 8E tank mixtures with Caparol 4L may be applied in water or fluid fertilizer. When fluid fertilizer is used as a carrier for Dual, either alone or in combination with Caparol 4L, mix only the amount that will be sprayed in one operation. These mixtures should not be allowed to stand without agitation.

In addition to those weeds controlled by Dual 8E alone, Dual 8E + Caparol 4L, applied preplant incorporated or preemergence, also controls the following weeds: junglerice, wild oats, annual morningglory, groundcherry, hairy nightshade, lambsquarters, malva, mustard, prickly sida (teaweed), purslane, ragweed, and shallow-germinating seedlings of cocklebur and coffeeweed.

Preplant Incorporated or Preemergence: Apply Dual 8E + Caparol 4L either preplant incorporated or preemergence using the appropriate rate from Table 3.

Table 3. Dual 8E + Caparol 4L — Cotton (NM, OK, TX)

Use areas	Soil texture	Broadcast rates per acre	
		Dual 8E	Caparol 4L
ALL	Sand, loamy sand	DO NOT USE	
Oklahoma, and Blacklands and Gulf Coast of Texas	loams	1.25-2 pts.	2.4 pts.
	clays	2 pts.	4.8 pts.
Rio Grande Valley of Texas	loams	1.25-2 pts.	3.2 pts.
	clays	2 pts.	4.8 pts.
New Mexico and High Plains, Rolling Plains, Edwards Plateau of Texas and Southwest Texas	sandy loam	1.25-1.5 pts.	1.6 pts.
	loams	1.25-2 pts.	2.4 pts.
	sandy clay loams	2 pts.	2.4 pts.
	other clay soils	2 pts.	3.2 pts.

Precautions: *(1) To avoid concentration in the seed furrow, do not make broadcast applications of Dual 8E + Caparol 4L to cotton planted in furrows more than two inches deep. Band applications may be made to cotton planted in furrows deeper than two inches, but band width should not exceed the width of the bottom of the furrow. To avoid crop injury, (2) Do not apply on sand or loamy sand soils, or in areas where water is likely to "pond" over the bed; (3) Do not apply in cut areas of newly leveled fields, or in areas of excess salt; (4) Do not apply to glandless cotton varieties; and (5) Do not apply on Taloka silt loam.*

Note: *Do not feed treated forage to livestock, or graze treated areas, or illegal residues may result.*

Refer to the Caparol 4L label for further instructions and limitations.

G. PEANUTS—DUAL 8E ALONE

Apply Dual 8E either preplant incorporated, postplant incorporated or preemergence using the appropriate rate specified below. *Preplant Incorporated or Preemergence:* Follow instructions for use of Dual 8E alone under **Application Procedures.** *Postplant Incorporated:* Apply and shallowly incorporate Dual 8E into the soil after planting, but before peanut germination. Incorporation depth and incorporating implements must be kept above the seed or seed will be damaged.

Apply Dual 8E alone preplant incorporated, postplant incorporated or preemergence at a broadcast rate of 1.5-2 pts./A in the Southeast* and 1.25-2 pts./A in OK, TX and NM.

* In the Southeast, use 2-3 pts./A and apply preemergence for partial control of Florida beggarweed.

Note: Dual 8E alone may be applied as directed after any of the following preplant incorporated herbicides when used according to their label recommendations: Balan at 3-4 qts./A; Treflan E.C. at 1 pt./A; or Vernam® at 2.33-3 pts./A.

H. PEANUTS—DUAL 8E COMBINATIONS

1. TANK MIXTURE WITH BALAN L.C.

Dual 8E + Balan tank mixture applied preplant incorporated controls those weeds listed under **Dual 8E Applied Alone** and those weeds as listed on the Balan label.

Apply 1.5-3 pts./A of Dual 8E + 3-4 qts./A of Balan in a minimum of 10 gals. of spray volume per acre for ground application or in a minimum of 5 gals. of spray volume per acre for aerial application. Follow the recommended procedures for Balan on the Balan label for soil preparation and incorporation of this tank mix. Apply and incorporate Dual 8E + Balan up to 14 days prior to planting.

Note: Follow all restrictions and precautions on the Balan label.

2. Tank Mixture and Sequential Application with Dyanap (or equivalent formulation of sodium N-1-naphthylphthalamate and sodium salt of dinoseb)

In addition to the weeds controlled by Dual 8E alone, Dual + Dyanap, when applied as directed, also controls the following broadleaf weeds: Florida beggarweed (beggarlice), chickweed, cocklebur, coffeeweed*, groundcherry, jimsonweed, lambsquarters, morningglory, mustards, purslane, ragweed, shepherdspurse, sicklepod*, smartweed, velvetleaf, and wild sunflower.

* Partially controlled.

Tank Mixtures: Apply Dual 8E + Dyanap at a broadcast rate/A of 1.5-3 pts.* of Dual 8E + 6 qts. of Dyanap in the Southeast and at 1.5-2 pts. of Dual + 6 qts. of Dyanap in OK, TX, and NM. Apply just as the peanuts begin to crack the ground or "cracking time."

Notes: (1) Do not feed peanut forage, hay, or hulls to livestock if peanuts are treated with the Dual 8E + Dyanap tank mixture or illegal residues may result. (2) Only water may be used as a carrier for Dual 8E + Dyanap tank mixtures.

* In the Southeast, use 2-3 pts./A for better weed control in areas of heavy weed infestations and better control of Florida beggarweed.

Sequential Treatments: Apply Dual 8E preplant incorporated or preemergence according to the directions for Dual 8E applied alone, and follow with a Dyanap preemergence to cracking time treatment using 6 qts. per acre as specified for use alone on that label.

If ½-1 inch of rainfall does not occur within 7 days after application, shallow incorporation (½-1 inch) is recommended. Cultivate if weeds develop. Refer to the Dyanap label for planting details and information on all other factors affecting its use.

Note: The tank mixture and sequential treatments with Dyanap may be applied as directed after any of the following preplant incorporated herbicides when used according to their label recommendations: Balan at 3-4 qts./A; Treflan E.C. at 1 pt./A; or Vernam at 2.33-3 pts./A.

I. POD CROPS—DUAL 8E ALONE

Pod crops including: garbanzo, great northern beans, kidney beans, lima beans, mung beans, navy beans, peas (English*; southern peas, such as blackeye, pinkeye, crowder, etc.), pinto beans, and snap beans (green, wax, string).

Apply Dual 8E either preplant incorporated or preemergence using the appropriate rate specified below. *Preplant Incorporated or Preemergence:* Follow instructions for use of Dual 8E alone under **Application Procedures.** On *coarse soils* with less than 3% organic matter, apply 1.5-2 pts./A of Dual 8E or 2 pts./A if organic matter is 3% or greater. On *medium soils,* apply 2-2.5 pts./A of Dual 8E. On *fine soils,* apply 2-2.5 pts./A of Dual 8E if organic matter content is less than 3%, or 2.5-3 pts./A if organic matter content is 3% or greater.

* On English peas, use only preemergence applications. Do not use on English peas in northeastern U.S., or injury may occur.

J. POD CROPS—DUAL 8E COMBINATIONS

1. TANK MIXTURE AND SEQUENTIAL APPLICATIONS WITH EPTAM—BEANS (GREEN OR DRY)

This mixture controls all weeds controlled by Dual 8E alone and by Eptam alone. Refer to the **Dual 8E Applied Alone** section of this label for weeds controlled by Dual 8E alone and to the Eptam label for weeds controlled by Eptam.

Preplant Incorporated: Follow instructions for use of Dual 8E alone under **Application Procedures.** *Sequential:* Apply Eptam alone preplant incorporated, as specified on that label. Follow with a preemergence application of Dual, at rates specified for Dual 8E alone, during planting (behind the planter) or after planting before the weeds or crop emerge.

Refer to the **General Information** section of this label and to the Eptam label for weather, cultural practices, and all other precautions and limitations that affect performance of these products.

Apply 3.5-4.5 pts./A of Eptam 7E* with Dual 8E as specified. On *coarse soils,* apply 1.25 pts./A of Dual 8E if organic matter content is less than 3%, or 1.5 pts./A if organic matter content is 3% or greater. On *medium soils,* apply 1.5 pts./A of Dual 8E if organic matter content is less than 3%, or 2 pts./A if organic matter content is 3% or greater. On *fine soils,* apply 2 pts./A of Dual 8E per acre if organic matter is less than 3%, or 2-2.5 pts./A if organic matter is 3% or greater.

* Refer to the Eptam label for rate limits depending on geographical area, and for species and varietal restrictions.

Precaution: *Do not exceed 3.5 pts./A Eptam 7E on small white beans or green beans grown on coarse-textured soils.*

2. TANK MIXTURE AND SEQUENTIAL APPLICATIONS WITH PREMERGE 3—BEANS (FIELD, LIMA AND SNAP)

This mixture controls all weeds controlled by Dual 8E alone and by Premerge 3 alone. Refer to the **Dual 8E Applied Alone** section of this label for weeds controlled by Dual 8E alone, and to the Premerge 3 label for weeds controlled by Premerge 3.

Note: Only water may be used as a carrier for Dual 8E + Premerge 3 tank mixtures.

Preemergence: Using the appropriate rates as specified below, apply the tank mixture during planting (behind the planter) or after planting, but before weeds or crop emerge; or application may be delayed until just before or during early emergence when beans are in, but not beyond the "crook" stage. *Sequential:* Apply Dual 8E preplant incorporated as specified for Dual 8E alone. Follow with a preemergence or at emergence application of Premerge 3, or Dual may be applied preemergence as specified for Dual 8E alone, followed by an at emergence application of Premerge 3 according to rates and instructions specified on that label.

Refer to the **General Information** section of this label and to the Premerge 3 label for weather, cultural practices, and all other precautions and limitations that affect performance of these products.

Apply 1-1.5 gals./A of Premerge 3* with Dual 8E as specified. On *coarse soils*, apply 1.25 pts./A of Dual 8E if organic matter content is less than 3%, or 1.5 pts./A if organic matter content is 3% or greater. On *medium soils*, apply 1.5 pts./A of Dual 8E if organic matter content is less than 3%, or 2 pts./A if organic matter content is 3% or greater. On *fine soils*, apply 2 pts./A of Dual 8E if organic matter content is less than 3%, or 2-2.5 pts./A if organic matter content is 3% or greater.

* In the Premerge 3 rate range, use the high rate for preemergence application and the low rate for the at emergence application.

Precaution: Do not use on light sandy soils having little or no organic matter.

Note: Do not graze or feed forage or fodder from pod crops treated with Dual, Dual + Eptam, or Dual + Premerge 3 to livestock.

3. TANK MIXTURE WITH TREFLAN—BEANS (DRY—KIDNEY, NAVY, PINTO, ETC., LIMA AND SNAP)

Dual 8E + Treflan tank mix applied preplant incorporated controls those weeds listed under **Dual 8E Applied Alone** and those weeds listed for Treflan alone on the Treflan E.C. label. Dual + Treflan E.C. may be applied by ground or aerial equipment and incorporated up to 14 days prior to planting. Follow the recommended procedures on this label and on the Treflan E.C. label using equipment that provides uniform 2 inch incorporation.

Apply Dual + Treflan tank mix using the appropriate Dual 8E rate specified for Dual 8E alone, and the Treflan rate from the **Dry Beans** and the **Lima and Snap Beans** sections of the Treflan E.C. label. Choose the product rate for the specific soil texture/organic matter classification and weed species expected.

Note: Follow all restrictions and precautions on the Treflan E.C. label and on the Dual alone portion of this label.

4. TANK MIXTURE WITH AMIBEN—BEANS (WHITE, NAVY, KIDNEY, PINTO, AND LIMA)

Dual 8E + Amiben tank mix applied either preplant incorporated or preemergence controls those weeds listed under **Dual 8E Applied Alone** and those weeds listed for Amiben alone on the Amiben label.

Preplant Incorporated or Preemergence: Apply Dual 8E and Amiben either preplant incorporated or preemergence using the Dual rate as specified for Dual 8E alone, and the Amiben rate as specified on that label according to the soil texture/organic matter classification and weed species expected. Follow instructions for use of Dual 8E alone under **Application Procedures.**

Refer to the **General Information** section of this label and to the Amiben label for weather and cultural practices that affect performance of these products.

K. POTATOES—DUAL 8E ALONE

Apply Dual 8E either incorporated, preemergence, or after hilling/lay-by according to directions specified below for control of weeds listed under the **General Information** section. Within a rate range, use the lower rate on soils relatively coarse-textured or low in organic matter; use the higher rate on soils relatively fine-textured or high in organic matter. Effectiveness will be reduced if later cultural practices expose untreated soil.

Incorporated: Apply Dual 8E at 1.5-3 pts./A to the soil and incorporate into the top 3 inches before planting using a finishing disk, harrow, rolling cultivator, or similar implement. Planting and later cultural practices should not bring untreated soil to the surface. Postplant incorporated application may be made any time after planting to drag-off, but before potato emergence. Use an implement that evenly distributes Dual in the top 2 inches of soil. Do not damage potato seed pieces or sprouts with incorporation equipment.

Preemergence: Apply Dual 8E at 1.5-3 pts./A either after planting as a preemergence, delayed preemergence, after drag-off or hilling treatment, but before weeds emerge. Up to 4 pts./A of Dual 8E alone may be used where soil organic matter is between 6% and 20%.

After Hilling/Lay-by: Apply 2.5 pts./A Dual 8E after hilling/at lay-by to control Dual-sensitive species for remainder of the growing season. This application will not control emerged weeds. It may be applied over a previous Dual 8E application, but do not apply more than 5.5 pts./A of Dual 8E in a single crop season.

Precaution: *(1) Do not use on muck or peat soils. If cool, wet soil conditions occur after application, Dual may delay maturity and/or reduce yield of Superior and other early maturing potato varieties. To avoid crop injury, (2) Do not use on sweet potatoes or yams; (3) Do not apply both as a preemergence and an incorporated treatment; and (4) Do not use in Kern County, California.*

Note: Potatoes treated with Dual 8E should not be harvested within 60 days after the at planting to drag-off application, or within 40 days after a lay-by application, or illegal residues may result.

L. POTATOES—DUAL 8E COMBINATIONS

1. TANK MIXTURE WITH SENCOR OR LEXONE

In addition to those weeds controlled by Dual 8E alone, Dual 8E + Sencor or Lexone, applied preplant incorporated, postplant incorporated or preemergence, also controls the following broadleaf weeds: cocklebur*, hairy nightshade*, hemp sesbania, jimsonweed*, lambsquarters, prickly sida, ragweed, smartweed, velvetleaf, Venice mallow, and wild mustard.

* Partially controlled.

Dual 8E at 1.5-3 pts./A plus the labeled Sancor/Lexone use rate may be used preemergence through after last hilling. Apply 1.5-2 pts./A of Dual 8E on *coarse soils* and 2-3 pts./A on other soil textures. Within this rate range, use the lower rate on soils relatively coarse-textured or low in organic matter; use the higher rate on soils relatively fine-textured or high in organic matter. Effectiveness will be reduced if later cultural practices expose untreated soil. Dual 8E will not control emerged weeds.

Refer to the Sencor or Lexone labels for precautionary statements, restrictions, application information and weeds controlled.

Precautions: *(1) Postemergence applications to potatoes should be made only as a directed or semi-directed spray to avoid chlorosis, minor necrosis or leaf distortion. To avoid crop injury, (2) Do not use Dual + Sencor or Lexone on potatoes in Kern County, California, and (3) Do not apply to sweet potatoes or yams. (4) Do not use this tank mixture on muck or peat soils.*

Notes: (1) Potatoes treated with Dual 8E in tank mixture with Sencor or Lexone cannot be harvested within 60 days after application or illegal residues may result. (2) Potatoes may not be harvested within 40 days after a lay-by application of Dual 8E, or illegal residues may result.

2. DUAL 8E + LOROX TANK MIXTURE (EAST OF ROCKY MOUNTAINS)

Dual 8E may be applied in a tank mix combination with any of the registered Lorox formulations as a preemergence broadcast application to potatoes. Apply to the soil surface after planting and before emergence of the crop or after final drag-off according to the rates specified in Table 4.

Table 4: Dual 8E + Lorox — Potatoes (East of Rocky Mountains)

	Broadcast rate per acre			
	1 to less than 3% organic matter		3 to 5% organic matter	
Soil texture	Dual (Pts.)	Lorox* (Lbs.)	Dual (Pts.)	Lorox* (Lbs.)
COARSE Sandy loam	1.5	1-1.5	2	1.5-2
MEDIUM Loam, silt loam, silt	2	1.5-2	2.5-3	2-2.5

* When using Lorox L, use equivalent rates. One pt. of Lorox L equals one lb. of Lorox wettable powder.

Precautions: *To avoid crop injury, (1) Do not use on sands or loamy sands, and (2) Do not incorporate or spray over the top of emerged potatoes, (3) Only water should be used as a carrier for Dual 8E + Lorox L.*

Refer to the **General Information** section of this label and to the Lorox label for precautionary statements, restrictions, application information and weeds controlled.

3. TANK MIXTURE WITH PREMERGE 3

Dual 8E may be applied in a tank mix combination with Premerge 3 as a preemergence broadcast application 1 to 3 days before potatoes emerge. Apply 1.5-3 pts./A of Dual 8E + 4-6 qts./A of Premerge 3.

Precautions: *(1) To avoid crop injury, do not use on light sandy soils with less than 1.0% organic matter. (2) Only water should be used as a carrier for Dual 8E + Premerge 3.*

Refer to the **General Information** section of this label and to the Premerge 3 label for precautionary statements, restrictions, application information and weeds controlled.

M. SAFFLOWERS—DUAL 8E ALONE

Preplant Incorporated or Preemergence: Follow instructions for use of Dual 8E alone under **Application Procedures.**

On *coarse soils,* apply 1.5-2 pts./A of Dual 8E if organic matter content is less than 3%, or 2 pts./A if organic matter is 3% or greater. On *medium soils,* apply 2-2.5 pts./A of Dual 8E. On *fine soils,* apply 2-2.5 pts./A of Dual 8E if organic matter content is less than 3%, or 2.5-3 pts./A if organic matter content is 3% or greater.

N. GRAIN OR FORAGE SORGHUM (SEED TREATED WITH CONCEP)—DUAL 8E ALONE

Apply Dual 8E either preplant surface, preplant incorporated, or preemergence using the appropriate rate specified below.

Preplant Surface-Applied: Refer to instructions for use of Dual 8E under **Application Procedures.** For minimum tillage or no-tillage systems only. Dual 8E may be applied up to 45 days before planting in CO, IL, IA, KS, MO, NE, and SD. Use only split applications for treatments made 30-45 days prior to planting, with ⅔ of the broadcast rate applied initially and the remaining ⅓ at planting. Apply 2.25 pts./A of Dual 8E on medium soils or 2.5 pts./A on fine soils. Treatments less than 30 days prior to planting may be made either as a split or single application. Apply 2 pts./A of Dual 8E on coarse soils not more than 2 weeks prior to planting.

Under dry conditions, irrigation after application is recommended to move Dual 8E into the soil.

Preplant Incorporated or Preemergence: Refer to instructions for use of Dual 8E under **Application Procedures.** Broadcast 1.5-2 pts./A of Dual 8E on coarse soils, 2-2.25 pts./A on medium soils, or 2-2.5 pts./A on fine soils.

Precautions: *1) If sorghum seed is not properly treated with Concep, Dual 8E will severely injure the crop. 2) Under high soil moisture conditions prior to sorghum emergence, injury may ocur following the use of Dual 8E. The crop will normally outgrow this effect. 3) Do not use Dual 8E on sorghum grown under dry mulch tillage or injury may occur. 4) Do not make more than one application per year, or illegal residues may result.*

O. GRAIN OR FORAGE SORGHUM (SEED TREATED WITH CONCEP)—DUAL 8E COMBINATIONS

Dual 8E tank mixtures with AAtrex, Milogard, or Igran may be applied in water or fluid fertilizer. Apply Dual 8E alone or in the tank mixtures only when the sorghum seed has been properly treated by the seed company with Concep.

Precautions: *1) Applications of Dual 8E + AAtrex, Dual 8E + Igran, or Dual 8E + Milogard on highly alkaline soils or on eroded areas where calcareous subsoils are exposed may cause sorghum injury. 2) If sorghum seed is not properly treated with Concep, Dual 8E + AAtrex, Dual 8E + Igran, or Dual 8E + Milogard may severely injure the crop. 3) Under high soil moisture conditions prior to sorghum emergence, injury may occur following the use of Dual 8E + AAtrex, Dual 8E + Igran, or Dual 8E + Milogard. The crop will normally outgrow this effect. 4) Do not use Dual 8E + AAtrex, Dual 8E + Igran, or Dual 8E + Milogard on sorghum grown under dry mulch tillage or injury may occur. 5) Do not make more than one application per year, or illegal residues may result.*

1. TANK MIXTURE WITH AATREX

In addition to the weeds controlled by Dual 8E alone, Dual 8E + AAtrex also controls the following broadleaf weeds when applied, either preplant surface, preplant incorporated, or preemergence: cocklebur, common purslane, hairy nightshade, lambsquarters, morningglory, ragweed, smartweed, and velvetleaf.

Preplant Surface-Applied: Refer to instructions for use of Dual 8E under **Application Procedures.** For minimum-tillage or no-tillage systems only, Dual 8E + AAtrex may be applied up to 45 days prior to planting in IL, IA, eastern KS, MO, NE and SD. Use only split applications for treatments made 30-45 days prior to planting, with ⅔ of the broadcast rate applied initially and the remaining ⅓ at planting. Apply 2.25 pts./A of Dual 8E + 1.7-2 lbs./A of AAtrex Nine-O* on *medium soils* with 1.5% organic matter or greater. Apply 2.25 pts./A of Dual 8E + 1.7-2 lbs./A of AAtrex Nine-O* on *fine soils* with less than 1.5% organic matter, or apply 2.5 pts./A of Dual 8E + 2-2.2 lbs./A AAtrex Nine-O* on *fine soils* with 1.5% organic matter or greater. Treatments less than 30 days prior to planting may be made either as a split or single application. Under dry conditions, irrigation after application is recommended to move Dual 8E + AAtrex into the soil.

Precautions: *To avoid crop injury, (1) Do not use on coarse soils, and (2) Do not use on medium soils with less than 1.5% organic matter.*

Preplant Incorporated or Preemergence: Follow instructions for use of Dual 8E under **Application Procedures.** On *medium soils* with 1.5% organic matter or greater, apply 1.5 pts./A of Dual 8E + 1.3 lbs./A of AAtrex Nine-O*. On *fine soils* with less than 1.5% organic matter, apply 1.5 pts./A of Dual 8E plus 1.3 lbs./A of AAtrex Nine-O; on *fine soils* with 1.5% organic matter or greater, apply 1.75-2 pts./A of Dual 8E plus 1.6-1.8 lbs./A of AAtrex Nine-O.

* When using AAtrex 80W, AAtrex 4L, or AAtrex 4LC, use equivalent rates. One lb. of AAtrex Nine-O equals 1.1 lbs. of AAtrex 80W or 1.8 pts. of AAtrex 4L (4LC).

Precautions: *To avoid crop injury, (1) Do not use on coarse soils; (2) Do not use on medium soils with less than 1.5% organic matter; (3) Do not use in NM, OK, or TX except in northeast OK and the Texas Gulf Coast and Blacklands areas, and (4) Do not apply preplant incorporated in AZ or the Imperial Valley or CA.*

2. TANK MIXTURE WITH IGRAN 80W (GRAIN SORGHUM ONLY)

In addition to the weeds controlled by Dual 8E alone, Dual 8E + Igran 80W also controls common purslane, hairy nightshade, kochia, lambsquarters, morningglory, puncturevine, ragweed, and velvetleaf. Dual 8E + Igran 80W may not give adequate control of foxtail millet and Florida pusley.

Preplant Incorporated or Preemergence: Refer to instructions for use of Dual 8E alone under **Application Procedures.** On *coarse soils**, apply 1.25 pts./A of Dual 8E + 1.5 lbs./A of Igran 80W. On *medium soils,* apply 1.5-1.75 pts./A of Dual 8E + 2-2.2 lbs./A of Igran 80W. On *fine soils,* apply 1.75-2 pts./A of Dual 8E + 2.2-2.5 lbs./A of Igran 80W.

* On sandy loams, apply 1.25-1.5 pts./A of Dual 8E + 1.5-2 lbs./A of Igran 80W.

3. TANK MIXTURE WITH MILOGARD

In addition to the weeds controlled by Dual 8E alone, Dual 8E + Milogard applied either preplant surface, preplant incorporated, or preemergence also controls browntop panicum, common purslane, hairy nightshade, lambsquarters, and morningglory.

Preplant Surface-Applied: Refer to instructions for use of Dual 8E under **Application Procedures.** For minimum-tillage or no-tillage systems only. Dual 8E + Milogard 80W* may be applied up to 45 days prior to planting in central and western KS and eastern CO. Use only split applications for treatments made 30-45 days prior to planting, with ⅔ of the broadcast rate applied initially and the remaining ⅓ at planting. On sandy loams, apply 2 pts./A of Dual 8E + 0.75-1 lb./A of Milogard 80W*. On medium soils, apply 2.25 pts./A of Dual 8E + 1-1.2 lbs./A of Milogard 80W*. On fine soils, apply 2.5 pts./A of Dual 8E + 1.2-1.3 lbs./A of Milogard 80W*. Treatments less than 30 days prior to planting may be made either as a split or single application. Under dry conditions, irrigation after application is recommended to move Dual 8E + Milogard 80W* into the soil.

Preplant Incorporated or Preemergence: Refer to instructions for use of Dual 8E under **Application Procedures.** On *sandy loams,* apply 1.25-1.5 pts./A of Dual 8E + 0.75-1 lb./A of Milogard 80W*. On *medium soils,* apply 1.5-1.9 pts./A of Dual 8E + 1-1.2 lbs./A of Milogard 80W. On *fine soils,* apply 1.9-2.1 pts./A of Dual 8E + 1.2-1.3 lbs./A of Milogard 80W.

* When using Milogard 4L or Milogard Maxx® 90, use equivalent rates. One lb. of Milogard 80W equals 0.9 lb. of Milogard Maxx 90 or 1.6 pts. of Milogard 4L.

Do not apply on sands and loamy sands.

Precautions: *To avoid crop injury, (1) Do not apply on sands and loamy sands and (2) Do not incorporate on sandy loam soils in AR, GA, OK, NM, TX, and the Gulf Coast states.*

4. TANK MIXTURE OF DUAL 8E OR DUAL 8E + AATREX, WITH PARAQUAT OR ROUNDUP FOR MINIMUM-TILLAGE OR NO-TILLAGE SYSTEMS

In minimum-tillage or no-tillage systems where sorghum (seed treated with Concep) is planted directly into a cover crop, stale seedbed, established sod or previous crop residues, the contact herbicides paraquat or Roundup may be tank mixed with Dual 8E or Dual 8E + AAtrex. When used as directed, the paraquat portion of the tank mixture controls most emerged annual weeds and suppresses many perennial weeds. Roundup combinations will control emerged annual and perennial weeds when applied as directed on the Roundup label. The Dual 8E or Dual 8E + AAtrex portion of the tank mixture provides preemergence control of the weeds listed on this label under the respective sections.

Refer to the label of each product used in combination and observe the planting details, restrictions, and all other precautions and limitations.

Application: Apply before, during, or after planting but before sorghum emerges, at the appropriate rates listed under **Grain or Forage Sorghum—Dual 8E Alone or Dual 8E Combinations—Dual 8E + AAtrex,** respectively. Add paraquat or Roundup at the following broadcast rates:

Paraquat: 1-2 pts.* per acre plus 8 oz. of X-77 Spreader per 100 gals. of spray mixture. Use the lower rate for control of annual weeds less than 4 inches tall and the higher rate for weeds 4-6 inches tall. This treatment will not consistently control weeds taller than 6 inches.

* Based on products containing 2 lbs. active ingredient per gal.

Roundup: 1.5 pts. per acre for control of existing annual weeds, or 2-4 qts. per acre for existing perennial weeds. See the Roundup label for weeds controlled and recommended rates for specific weeds.

Apply in a minimum of 20 gals. of water per acre with conventional spray equipment.

P. SOYBEANS—DUAL 8E ALONE

Apply Dual 8E either preplant surface-applied, preplant incorporated, or preemergence using the appropriate rate specified below. *Preplant Surface Applied, Preplant Incorporated or Preemergence:* Follow instructions for use of Dual 8E alone under **Application Procedures.**

Preplant Surface-Applied: Use on *medium* and *fine soils* with minimum-tillage or no-tillage systems in CO, IL, IN, IA, KS, KY, MN, MO, MT, NE, ND, SD, WI, WY. Apply ⅔ the recommended rate of Dual 8E (2.5 pts./A on *medium soils* and 3 pts./A on *fine soils*) as a split treatment 30-45 days prior to planting and the remainder at planting. Applications made less than 30 days before planting may be as either a split or single treatment. Apply 2 pts./A on *coarse soils* not more than 2 weeks prior to planting.

Preplant Incorporated or Preemergence: On *coarse soils,* apply 1.5-2 pts./A of Dual 8E if organic matter content is less than 3%, or 2 pts./A if organic matter content is 3% or greater. On *medium soils,* apply 2-2.5 pts./A of Dual 8E. On *fine soils,* apply 2-2.5 pts./A of Dual 8E if organic matter content is less than 3%, or 2.5-3 pts./A if organic matter content is 3% or greater.

Q. SOYBEANS—DUAL 8E COMBINATIONS

Water or fluid fertilizer may be used as carrier for Dual 8E in combination with Amiben, Sencor, Lexone, or Lorox 50W. Only water may be used as carrier for Dual 8E + Dyanap, Dual 8E + Lorox L, or Dual 8E + Premerge 3.

1. TANK MIXTURE WITH SENCOR OR LEXONE

In addition to those weeds controlled by Dual 8E alone, Dual + Sencor or Lexone, when applied as directed, also controls the following broadleaf weeds: cocklebur*, hairy nightshade, hemp sesbania, jimsonweed*, lambsquarters, prickly sida, ragweed, smartweed, velvetleaf, Venice mallow, and wild mustard.

* Partially controlled.

Apply Dual 8E and Sencor or Lexone preplant incorporated or preemergence using the appropriate rates from Table 5. *Preplant Incorporated or Preemergence:* Follow instructions for use of Dual 8E alone under **Application Procedures.** *Sequential:* Apply Dual 8E alone preplant incorporated, as specified in Table 5 for this tank mixture. Follow with a preemergence application of Sencor or Lexone during planting (behind the planter) or after planting before weeds or soybeans emerge.

Refer to the Sencor or Lexone labels for planting details and soybean variety restrictions.

Table 5: Dual 8E + Sencor or Lexone — Soybeans

	Broadcast rates per acre			
	0.5% to less than 3% organic matter		3% organic matter or greater	
Soil texture**	**Dual 8E +**	**Sencor 50WP* or Lexone**	**Dual 8E +**	**Sencor 50WP* or Lexone**
COARSE Loamy sand (over 2% organic matter), sandy loam	1.25-1.5 pts. +	.5 lb	1.5 pts. +	.75 lb.
MEDIUM	1.5-2 pts. +	.75 lb.	2 pts. +	1 lb.***
FINE	2 pts. +	1 lb.	2-2.5 pts. +	1 lb.
Mississippi Delta only: Silty clay, clay	2 pts. +	1.5 lbs.	2-2.5 pts. +	1.5 lbs.
Muck or Peat soils (soils with more than 20% organic matter)	DO NOT USE			

* When using Sencor 4 or Lexone 4L, substitute pints for pounds (1 pint equals 1 pound). For Sencor DF or Lexone DF, multiply pounds 50WP by 0.67.

** On all sand and on loamy sand with less than 2% organic matter, do not use this tank mixture preemergence, or the sequential treatment. Do not use the tank mixture preplant incorporated on any sand, loamy sand, or sandy loam, or crop injury may occur.

*** Use 0.75 lb./A if applied preplant incorporated.

Precautions: *1) Do not use the tank mix or sequential application on soil with less than 0.5% organic matter or on alkaline soil with a pH over 7.4, or crop injury may occur. 2) If heavy rain occurs soon after application, crop injury may result, especially in poorly drained areas where water stands for several days.*

2. TANK MIXTURE WITH LOROX

In addition to those weeds controlled by Dual 8E alone, Dual 8E + Lorox, applied preemergence, also controls the following broadleaf weeds: cocklebur*, jimsonweed*, lambsquarters, morningglory*, prickly sida, ragweed, smartweed, velvetleaf*, Venice mallow, and wild mustard.

* Partially controlled.

Preemergence: Apply during planting (behind planter) or after planting but before weeds or soybeans emerge. Refer to the Lorox label for planting details. Apply the appropriate rates from Table 6.

Table 6: Dual 8E + Lorox — Soybeans

Soil texture*	Broadcast rates per acre					
	0.5% to less than 3% organic matter			3% organic matter or greater		
	Dual 8E	+	Lorox***	Dual 8E	+	Lorox***
COARSE**	1.25 pts.	+	1 lb.	1.5 pts.	+	1-1.5 lbs.
MEDIUM	1.5 pts.	+	1-1.5 lbs.	2 pts.	+	1.5-2 lbs.
FINE	2 pts.	+	2 lbs.	2-2.5 pts.	+	2.5-3 lbs.
Muck or Peat soils (soils with more than 20% organic matter)	DO NOT USE					

* Do not use on sand, gravelly soils, or exposed subsoils.

** Do not use on loamy sand, except in the northeastern U.S. on loamy sand with over 1% organic matter.

*** When using Lorox L, use equivalent rates. One pt. of Lorox L equals one lb. of Lorox wettable powder.

Precaution: *1) Do not use on soil with less than 0.5% organic matter or crop injury may occur. 2) Only water should be used as a carrier for Dual 8E + Lorox L.*

3. TANK MIXTURE WITH DYANAP

In addition to those weeds controlled by Dual 8E alone, Dual 8E + Dyanap, when applied as directed, also controls the following broadleaf weeds: cocklebur*, jimsonweed, lambsquarters, morningglory*, prickly sida, purslane, ragweed, smartweed, and wild mustard.

* Cocklebur and morningglory controlled with crook stage application only.

Apply Dual 8E + Dyanap in 20-40 gals. of spray mixture per acre during the interval preemergence during planting (behind the planter) through crook stage (from soil cracking to before the soybean true leaves open) according to the rates specified below. Do not apply this tank mixture in fluid fertilizer.

On *coarse soils,* apply 1.25 pts./A of Dual 8E plus 4.5 qts./A of Dyanap if organic matter content is less than 3%, or apply 1.5 pts./A of Dual 8E + 4.5 qts./A of Dyanap if organic matter content is 3% or greater. On *medium soils,* apply 1.5 pts./A of Dual 8E + 6 qts./A of Dyanap if organic matter content is less than 3%, or apply 2 pts./A of Dual 8E + 6 qts./A of Dyanap if organic matter content is 3% or greater. On *fine soils,* apply 2 pts./A of Dual 8E + 6 qts./A of Dyanap if organic matter content is less than 3%, or apply 2-2.5 pts./A of Dual 8E + 6 qts./A of Dyanap if organic matter content is 3% or greater.

If ½-1 inch of rainfall does not occur within 7 days after application, shallow incorporation (½ -1 inch) is recommended. Cultivate if weeds develop. Refer to the Dyanap label for planting and other product use information.

4. TANK MIXTURE WITH AMIBEN

This tank mixture controls all weeds controlled by Dual 8E alone and by Amiben alone. Refer to the **Dual 8E Applied Alone** section of this label for weeds controlled by Dual 8E alone and to the Amiben label for weeds controlled by Amiben.

Preplant Incorporated or Preemergence: Apply 1 gal./A of Amiben plus either 1.5 pts./A of Dual 8E on *coarse soils,* 1.5-2 pts./A of Dual 8E on *medium soils,* or 2-2.5 pts./A of Dual 8E on *fine soils.* Use 1.5 gals. of Amiben/A on *fine-textured high organic matter soils* or where a dense stand of difficult to control broadleaf weeds, such as black nightshade, velvetleaf, or wild mustard is expected to occur. *Preplant Incorporated:* Apply the tank mixture to the soil and incorporate into the top 2 inches of soil just before planting. Follow instructions for use, other than the preplant interval, of Dual 8E alone under **Application Procedures.** *Preemergence:* Apply the tank mixture during planting (behind the planter) or after planting before weeds or soybeans emerge. *Sequential:* Apply Dual 8E alone preplant incorporated, as specified above for this tank mixture. Follow with a preemergence application of Amiben during planting (behind the planter) or after planting before weeds or soybeans emerge.

Refer to the **General Information** section of this label and to the Amiben label for weather and cultural practices that affect performance of these products.

5. TANK MIXTURE AND SEQUENTIAL APPLICATION WITH PREMERGE 3

This mixture controls those weeds controlled by Dual 8E alone and by Premerge 3 alone. Refer to the **Dual 8E Applied Alone** section of this label for weeds controlled by Dual 8E alone and to the Premerge 3 label for weeds controlled by Premerge 3.

Preemergence: Using the appropriate rates specified below, apply the tank mixture during planting (behind the planter) or after planting, but before weeds or crop emerge; or application may be delayed to ground cracking. *Sequential:* Apply Dual 8E preplant incorporated as specified for Dual 8E alone. Follow with a preemergence or ground crack application of Premerge 3; or Dual 8E may be applied preemergence as specified for Dual 8E alone followed by a ground crack application of Premerge 3. *At Emergence* (when soybeans are emerging and are still in the cotyledon stage before leaves open to expose the terminal bud): To control small emerged weeds, use the Dual rate specified for Dual 8E alone plus 2-3 qts./A of Premerge 3.

Precaution: *Do not apply this tank mix in fluid fertilizer.*

On *coarse soils,* apply 1.5 pts./A of Dual 8E + 4 qts./A of Premerge 3. On *medium soils,* apply 1.5 pts./A of Dual 8E + 6 qts./A of Premerge 3 if organic matter content is less than 3%, or apply 2 pts./A of Dual 8E + 6 qts./A of Premerge 3 if organic matter content is 3% or greater. On *fine soils,* apply 2 pts./A of Dual 8E + 6 qts./A of Premerge 3 if organic matter content is less than 3%, or apply 2-2.5 pts./A of Dual 8E + 6 qts./A of Premerge 3 if organic matter content is 3% or greater.

Refer to the Premerge 3 label for weather, temperature, cultural practices, and all other precautions and limitations that affect performance and crop tolerance of this product.

Precaution: *Do not use on light sandy soils having little or no organic matter, or injury may occur.*

6. TANK MIXTURE WITH AMIBEN AND SENCOR OR LEXONE

This tank mixture effectively controls those weeds as listed on the current Dual 8E, Amiben, and Lexone WP/4L/DF or Sencor WP/4/DF labels.

Apply Dual + Amiben + Lexone or Sencor preplant incorporated or preemergence using the appropriate rates from Table 7. *Preplant Incorporated or Preemergence:* Follow instructions for use of Dual 8E alone under **Application Procedures.**

Table 7: Dual 8E + Amiben + Sencor or Lexone — Soybeans

Soil texture	Broadcast rates per acre*				
	Dual 8E	+	Amiben	+	Sencor 50WP** or Lexone
COARSE Loamy sand (over 2% organic matter), sandy loam	1.5-2 pts.	+	3-4 qts.	+	.5 lb.
MEDIUM	2-2.5 pts.	+	4-5 qts.	+	.5-.75 lb.***
FINE	2.5-3 pts.	+	5-6 qts.	+	.75-1 lb.***
Muck or Peat soils (soils with more than 20% organic matter)	DO NOT USE				

* Refer to the Amiben + Dual + Sencor/Lexone section of the Amiben label for determining product use rate within rate ranges.

** When using Sencor 4 or Lexone 4L, substitute pints for pounds (1 pint equals 1 pound). For Sencor DF or Lexone DF, multiply pounds 50WP by 0.67.

*** On Clarion-Webster or similar alkaline (calcareous) soils with a pH of 7.5 or above, apply the WP at 0.5 lb./A on medium soils and 0.5-0.75 lb./A on fine soils.

Special Precautions: Refer to the Dual + Sencor/Lexone section of this label, the Amiben + Dual + Sencor/Lexone section of the Amiben label and the Sencor/Lexone labels for planting details, variety limitations, and all cautions and restrictions concerning use of these products on soybeans.

7. TANK MIXTURE WITH TREFLAN

Dual 8E + Treflan tank mix applied preplant incorporated controls those weeds listed under **Dual 8E Applied Alone** and those weeds listed for Treflan alone on the Treflan E.C. label. Dual + Treflan may be applied by ground or aerial equipment and incorporated up to 14 days prior to planting. Follow the recommended procedures on the Treflan E.C. and Dual labels using equipment that provides uniform 2 inch incorporation.

Apply Dual + Treflan tank mix using the appropriate rate from the Dual alone section of this label and the Treflan alone section of the Treflan E.C. label for the specific soil texture/organic matter classification and weed species expected.

Note: Follow all restrictions and precautions on the Treflan E.C. label and on the Dual alone portion of this label.

8. TANK MIXTURE WITH LEXONE, SENCOR, OR LOROX PLUS PARAQUAT OR ROUNDUP FOR MINIMUM-TILLAGE OR NO-TILLAGE SYSTEMS

In minimum-tillage or no-tillage systems where soybeans are planted directly into a cover crop, stale seedbed, established sod, or previous crop residues, the contact herbicides paraquat or Roundup may be added to a tank mix of either Dual 8E + Sencor or Lexone, or Dual 8E + Lorox. When used as directed, the paraquat portion of the tank mixture controls most emerged weeds and suppresses many perennial weeds. Roundup combinations will control emerged annual and perennial weeds when applied as directed on the Roundup label. The Dual 8E + Sencor/Lexone or Lorox portion of the tank mixture provides preemergence control of the weeds listed on this label in the tank mixture section for Dual 8E + Sencor/Lexone, or Dual 8E + Lorox, respectively.

Refer to the label of each product used in combination and observe the planting details, soybean variety restrictions and all other precautions and limitations.

Refer below for rates of Dual 8E + Sencor/Lexone or Dual 8E + Lorox, respectively.

Refer to the section for tank mixtures with AAtrex or Princep, or AAtrex plus Princep with paraquat or Roundup on corn for rates of paraquat or Roundup, for volume of water, and other instructions and limitations.

DUAL 8E + LEXONE/SENCOR + PARAQUAT OR ROUNDUP:

On *loamy sand* with over 2% organic matter, apply 1.5 pts./A of Dual 8E + 0.5-0.75 lb./A of Sencor 50WP* or Lexone. On *medium soils,* apply 2 pts./A of Dual 8E + 0.75-1 lb./A of Sencor 50WP or Lexone. On *fine soils,* apply 2-2.5 pts./A of Dual 8E + 1 lb./A of Sencor 50WP or Lexone.

* When using Sencor 4 or Lexone 4L, substitute pints for pounds (1 pint equals 1 pound). For Sencor DF or Lexone DF, multiply pounds 50W by 0.67.

Precautions: *To avoid crop injury, (1) Do not use this tank mixture on soil with less than 0.5% organic matter, on alkaline soil with a pH over 7.4, or on all sand and on loamy sand with less than 2% organic matter. (2) If heavy rain occurs soon after application, crop injury may result, especially in poorly drained areas where water stands for several days or where the seeding slit has not been properly closed.*

DUAL 8E + LOROX + PARAQUAT OR ROUNDUP:

On *coarse soils**, apply 1.5 pts./A of Dual 8E + 1-1.5 lbs./A of Lorox**. On *medium soils,* apply 2 pts./A of Dual 8E + 1-2 lbs./A of Lorox. On *fine soils,* apply 2-2.5 pts./A of Dual 8E + 2-3 lbs./A of Lorox.

* Do not use on loamy sand, except in the northeastern U.S. on loamy sand with over 1% organic matter, or injury may occur. Do not use on sand, gravelly soils, or exposed subsoils, or injury may occur.

** When using Lorox L, use equivalent rates. One pt. of Lorox L equals one lb. of Lorox wettable powder.

Precautions: *1) Do not use on soil with less than 0.5% organic matter or crop injury may occur. 2) Only water should be used as a carrier for Dual 8E + Lorox L + paraquat or Roundup.*

R. DUAL 8E ALONE AND COMBINATIONS—FIELD AND LINER GROWN WOODY ORNAMENTALS

To control the weeds indicated below for approximately 60 days, use one of the following:

1. **Dual 8E:** Barnyardgrass, crowfootgrass, fall panicum, giant foxtail, goosegrass, green foxtail, large crabgrass, redroot pigweed, smooth crabgrass, yellow foxtail, yellow nutsedge, and other weeds listed for Dual 8E alone on this label.
2. **Dual 8E + Princep:** Galinsoga, lambsquarters, purslane, ragweed, smartweed, yellow nutsedge, Venice mallow and other weeds listed for Dual 8E alone on this label.

Apply Dual 8E alone or in combination with Princep at the rates shown below in a minimum of 15 gals. of water per acre. Direct the spray towards the base of established ornamentals or ornamentals transplanted a minimum of 10 days. Apply before weeds emerge or after removing existing weeds. Use the higher rate of Dual alone on fine-textured soil and where heavy infestations of grass weeds are expected. Use the lower rate of Dual on coarse-textured soil and where light infestations of grass weeds are expected. Use the higher rate of Dual + Princep on fine-textured soils and where broadleaf weeds are expected to form a significant part of the infestation. Use the lower rate of Dual + Princep on coarse-textured soil and where light infestations of broadleaf weeds are expected.

Apply Dual 8E at 2-3 pts./A or Dual 8E at 2-3 pts./A plus Princep 4L at 1.6-2 pts./A posttransplant to the field or liner grown woody ornamentals listed below. For band applications, use a proportional amount of herbicide. When using Princep 80W or Princep Caliber 90, use equivalent rates; 2 pts. of 4L equals 1.25 lbs. of 80W or 1.1 lbs. of Caliber 90.

Apply Dual 8E alone or Dual 8E + Princep as recommended above to the following field and liner grown ornamentals:

Dual 8E Alone

Arborvitae (*Thuja* spp.)*
Aucuba (*Aucuba* spp.)*
Azalea (*Rhododendron* sp.), Hershey Red, Herbert, Purple Splendor
Barberry (*Berberis* spp.), Mentor, Warty, Wintergreen
Box Tree (*Buxus* spp.)
Cotoneaster (*Cotoneaster* spp.)*
Crabapple (*Malus* spp.)*
Dogwood (*Cornus* sp.), Red Stem
Euonymus (*Euonymus* spp.), including Purple Winter Creeper
Firethorn (*Pyracantha* spp.)
Forsythia (*Forsythia* spp.), including Spring Glory and Golden Bells
Holly (*Ilex* spp.), American, Boxleaf Japanese*, Japanese*, Green Lustre Japanese
Honeysuckle (*Lonicera* spp.)
Hybrid Tea Rose (*Rosa* spp.)
Ivy (*Hedera* sp.), English
Japanese Andromeda (*Pieris* spp.)*
Japanese Red Maple (*Acer* spp.)
Juniper (*Juniperus* spp.)*, Andorra, Creeping, Shore, Bar Harbor, Blue Rug
Leucothoe (*Leucothoe* spp.)*
Lilac (*Syringa* spp.), James McFarland
Liriope (*Liriope* spp.)*, Silver Sunproof
Maple Whips (*Acer* spp.)*
Miniature Rose (*Rosa* spp.)
Mountain Laurel (*Kalmia* spp.)
Osmanthus (*Osmanthus* spp.)*
Pachysandra (*Pachysandra* spp.)
Privet (*Ligustrum* spp.)
Red Oak (*Quercus* spp.)*
Rhododendron (*Rhododendron* spp.)*
Rosebay (*Rhododendron* spp.)
Spirea (*Spirea* spp.)
Viburnum (*Virburnum* spp.)
Weigela (*Weigela* spp.)
Yew (*Taxus* spp.), Dwarf Japanese*, Hicks

* Use Dual 8E + Princep on these field and liner grown ornamentals only.

Note: Length of weed control may be reduced when continuous wet soil conditions follow herbicide application.

Precautions: *Do not use on seedbeds or on unrooted cuttings, or injury may occur. Do not use in greenhouses or other enclosed areas, or injury may occur.*

S. STORAGE AND DISPOSAL

PESTICIDE DISPOSAL
Do not contaminate water, food or feed by storage or disposal. Open dumping is prohibited. Wastes resulting from the use of this product are acutely toxic. Improper disposal of unused pesticide, spray mixture, or rinsate is a violation of federal law. Pesticide, spray mixture, or rinsate that cannot be used according to label instructions must be disposed of according to federal, state, or local procedures. For guidance in proper disposal methods, contact your State Pesticide or Environmental Control Agency, or the Hazardous Waste representative at the nearest EPA Regional Office.

CONTAINER DISPOSAL
Do not reuse empty container. Triple rinse (or equivalent), or puncture and dispose of in a sanitary landfill, or by incineration, or by open burning, if allowed by state and local authorities. Keep out of smoke from burning containers.

This product may be stored at temperatures down to 30 degrees below 0°F.

For minor spills, leaks, etc., follow all precautions indicated on this label and clean up immediately. Take special care to avoid contamination of equipment and facilities during cleanup procedures and disposal of wastes. In the event of a major spill, fire or other emergency, call (919) 292-7100 day or night.

T. PRECAUTIONARY STATEMENTS

HAZARDS TO HUMANS AND DOMESTIC ANIMALS

WARNING
The active ingredient, metolachlor, may cause skin sensitization reactions in certain individuals. Wear protective clothing (coveralls and gloves) while handling or using this product. Causes skin and eye irritation. Do not get in eyes, on skin, or on clothing. Harmful if inhaled. Avoid breathing spray mist.

Harmful if swallowed or absorbed through the skin. Wash thoroughly after handling. Avoid contamination of food.

First Aid: In case of contact, immediately flush eyes and/or skin with plenty of water. Call a physician. Remove and wash contaminated clothing before reuse.

If inhalation occurs, the victim should be moved to fresh air, and medical attention should be sought.

If swallowed, contact your local Poison Control Center, hospital, or physician immediately. If patient is unconscious, maintain breathing and heartbeat (CPR: cardiopulmonary resuscitation). If patient is conscious, induce vomiting (syrup of ipecac; if not available, stimulate back of throat with finger). Never give anything by mouth to an unconscious person!

Note to Physician: If swallowed, there is no specific antidote. Induce emesis and lavage stomach. Treat symptomatically! The use of an aqueous slurry of activated charcoal (such as Norit A) and a saline cathartic should be considered.

ENVIRONMENTAL HAZARDS
Do not apply directly to any body of water. Do not apply where runoff is likely to occur. Do not contaminate water by cleaning of equipment or disposal of wastes. Do not apply when weather conditions favor drift from areas treated.

AAtrex®, Caliber®, Caparol®, Concep®, Dual®, Igran®, Maxx®, Milogard®, Nine-O® and Princep® are trademarks of CIBA-GEIGY Corporation

Amiben® trademark of Union Carbide Agricultural Products, Inc. (formerly Amchem Products, Inc.)

Balan® trademark of Elanco Products for benefin

Banvel® trademark of Velsicol Chemical Corp. for dicamba

Bladex® trademark of Shell Chemical Company for cyanazine

Brominal® trademark of Union Carbide Corporation for bromoxynil

Buctril® trademark of Rhone-Poulenc Chemical Company for bromoxynil

Compex® trademark of KALO Agricultural Chemicals, Inc.

Dyanap® trademark of Uniroyal Chemical for Alanap®-Dinitro

Eptam® trademark of Stauffer Chemical Company for EPTC

Lexone® trademark of E.I. duPont de Nemours and Co. for metribuzin

Lorox® trademark of E.I. duPont de Nemours and Co. for linuron

Premerge® trademark of the Dow Chemical Company for dinoseb

Roundup® trademark of Monsanto Company for glyphosate herbicide

Sencor® trademark of the Parent Company of Farbenfabriken Bayer GmbH, Leverkusen, for metribuzin

Treflan® trademark of Elanco Products for trifluralin

Unite® trademark of Hopkins Agricultural Chemical Company for compatibility agent HA-914

Vernam® trademark of Stauffer Chemical Company for vernolate

X-77® trademark of KALO Agricultural Chemicals, Inc.

DUAL®25G

GRANULAR HERBICIDE

For weed control in corn, peanuts, pod crops, potatoes, soybeans, and ornamentals

Dual 25G is a granular herbicide

ACTIVE INGREDIENT: METOLACHLOR: 2-chloro-*N*-(2-ethyl-6-methylphenyl)-*N*-(2-methoxy-1-methylethyl) acetamide	25.0%
INERT INGREDIENTS:	75.0%
TOTAL:	100.0%

EPA Reg. No. 100-638 Control No. 52600

Dual® trademark of CIBA-GEIGY for metolachlor

U.S. Patent No. 3,937,730

Keep Out of Reach of Children.

Caution

See additional precautionary statements on back of bag.

See directions for use on back of bag.

DIRECTIONS FOR USE AND CONDITIONS OF SALE AND WARRANTY

IMPORTANT: Read the entire **Directions for Use** and the **Conditions of Sale and Warranty** before using this product.

CONDITIONS OF SALE AND WARRANTY

The **Directions for Use** of this product reflect the opinion of experts based on field use and tests. The directions are believed to be reliable and should be followed carefully. However, it is impossible to eliminate all risks inherently associated with use of this product. Crop injury, ineffectiveness, or other unintended consequences may result because of such factors as weather conditions, presence of other materials, or the manner of use or application all of which are beyond the control of CIBA-GEIGY or the Seller. All such risks shall be assumed by the Buyer.

CIBA-GEIGY warrants that this product conforms to the chemical description on the label and is reasonably fit for the purposes referred to in the **Directions for Use** subject to the inherent risks referred to above. **CIBA-GEIGY makes no other express or implied warranty of Fitness or Merchantability or any other express or implied warranty. In no case shall CIBA-GEIGY or the Seller be liable for consequential, special, or indirect damages resulting from the use or handling of this product.** CIBA-GEIGY and the Seller offer this product, and the Buyer and user accept it, subject to the foregoing **Conditions of Sale and Warranty,** which may be varied only by agreement in writing signed by a duly authorized representative of CIBA-GEIGY.

DIRECTIONS FOR USE

It is a violation of federal law to use this product in a manner inconsistent with its labeling.

No user shall permit the application of this product in such a manner as to directly or indirectly expose workers or other persons except those knowingly involved in the application. The area being treated must be vacated by unprotected persons.

REENTRY STATEMENT

Do not enter treated areas without protective clothing until dusts have settled.

Because certain states may require more restrictive reentry intervals for various crops treated with this product, consult your State Department of Agriculture for further information.

Written or oral warnings must be given to workers who are expected to be in a treated area or in an area about to be treated with this product. Oral warnings must be given which inform workers of areas or fields that may not be entered without specific protective clothing until dusts have settled, and appropriate actions to take as described under **Precautionary Statements** at the end of this label. When oral warnings are given, warnings shall be given in a language customarily understood by workers. Oral warnings must be given if there is reason to believe that written warnings cannot be understood by workers. Written warnings must include the following information: "CAUTION. Area treated with Dual 25G on (date of application). Do not enter without appropriate protective clothing until dusts have settled. In case of accidental exposure, flush eyes or skin with plenty of water. Call a physician if irritation persists. Remove and wash contaminated clothing before reuse."

FAILURE TO FOLLOW THE DIRECTIONS FOR USE AND PRECAUTIONS ON THIS LABEL MAY RESULT IN POOR WEED CONTROL, CROP INJURY, AND/OR ILLEGAL RESIDUES.

GENERAL INFORMATION

Dual 25G will control many annual grasses and certain broadleaf weeds in corn, peanuts, pod crops, potatoes, soybeans, and ornamentals. It should be applied prior to the emergence of weeds. It may be preplant surface-applied (corn and soybeans only), or incorporated into the surface 1-2 inches of soil prior to planting or left on the soil surface during or after planting. Dry weather following application of Dual 25G may reduce its effectiveness. Shallow cultivate if weeds develop or if tillage is necessary to improve the physical condition of the soil.

Precaution: *Injury may occur following the use of Dual 25G under abnormally high moisture conditions during early development of the crop.*

Weeds Controlled

barnyardgrass (watergrass)	green foxtail	yellow foxtail
crabgrass	prairie cupgrass	yellow nutsedge
crowfootgrass	red rice	
fall panicum	signalgrass *(Brachiaria)*	black nightshade
foxtail millet	southwestern cupgrass	carpetweed
giant foxtail	witchgrass	Florida pusley
goosegrass		galinsoga
		pigweed

Weeds Partially Controlled*: common purslane, hairy nightshade, sandbur, seedling johnsongrass, shattercane and volunteer sorghum.

* Control of these weeds can be erratic due partially to variable weather conditions. Control may be improved by following these suggested procedures:

1. **Thoroughly till moist soil** to destroy germinating and emerged weeds. If Dual 25G is to be applied preplant incorporated, this tillage may be used to incorporate Dual 25G if uniform 2 inch incorporation is achieved as recommended under **Application Procedures.**
2. Plant crop into moist soil **immediately after tillage.** If Dual 25G is to be used preemergence, apply at planting or immediately after planting.
3. If available, **sprinkler irrigate** within 2 days after application. Apply ½ -1 inch of water. Use lower water volume (½ inch) on coarse-textured soils and higher volume (1 inch) on finer textured soils.
4. If irrigation is not possible and rain does not occur within 2 days after planting and application, weed control may be decreased. Under these conditions, a uniform, shallow cultivation is recommended as soon as weeds emerge.

APPLICATION PROCEDURES

Dual 25G may be applied either broadcast or banded.

Broadcast Treatment—Dual 25G may be applied with a broadcast type granular applicator or with a cyclone type spreader. Use applicators and techniques that will apply the granules uniformly to the soil.

Band Treatment—Dual 25G may be applied with any standard granular applicator capable of uniformly distributing a predetermined amount of granules to a desired area. Band treatment made at seeding should only be applied behind the press wheel.

Note: Caution should be taken with seeding equipment so that any track (soil depression) made by the planter press wheel is minimal. This will reduce concentration of granules in the press wheel track by heavy rain which can result in crop injury. Soil tillage should not follow the use of band applications unless it is adapted so as not to displace the band and hence reduce weed control in the band.

Applicator settings are given for band applications in Chart 1. These settings were determined under laboratory conditions to ±5% accuracy. Note that the settings may differ depending on speed of application, as well as rate per acre. **It is important that the user calibrate each individual granular applicator prior to its use. Since the actual setting for each individual granular applicator may vary, the settings shown should serve only as guidelines for calibration.**

Calculate the amount of Dual 25G needed for band treatment as follows:

$$\frac{\text{Band width in inches}}{\text{Row width in inches}} \times \text{Broadcast rate per acre} = \text{Band rate per acre}$$

Example: If 16 lbs. per acre is recommended for broadcast application and you wish to treat a band 14 inches wide in crop rows spaced 30 inches apart:

$$\frac{14}{30} \times 16 = \text{7.5 lbs. of granules needed for each acre that is to be treated.}$$

Do not apply Dual 25G during periods of high winds. Wind may disrupt placement of the granules and reduce weed control. Locate granular diffusers close to the ground to reduce granular displacement. Wind shield attachments are advisable.

Chart 1: Applicator Settings For 14-Inch Band Width*

Applicator Brand	Speed MPH	Equivalent Broadcast Rate Per Acre: 6 lbs.	8 lbs.	10 lbs.	12 lbs.	16 lbs.
Gandy Two-Opening (For 14" band) Model No. S-902	4	8.5	10	11	12	14
	5	9.5	11	12	13	15.5
	6	10.5	12	13	15	17
International Harvester 1980 & Before Allis Chalmers 1979 & After	4	.80	1.0	1.1	1.2	1.35
	5	.92	1.1	1.15	1.38	1.6
	6	1.0	1.13	1.25	1.45	1.9
International Harvester 1981 & After	4	1.65	1.75	1.85	1.9	2.1
	5	1.75	1.85	1.95	2.0	2.25
	6	1.85	1.95	2.0	2.1	2.45
John Deere (Metal Box)	4	1/19	1/21	1/23	1/25	1/29
	5	1/21	1/23	1/25	1/28	2/1
	6	1/22	1/24	1/27	1/30	2/5

John Deere 1978 & Before Meter Scale 0..10..20..etc.	4	7	8	10	11	14
	5	8	9	11	13	17
	6	9	10	12	14	18
John Deere 1979 & Later Meter Scale 0..5..15..etc.	4	5	6	7	9	12
	5	6	7	9	10	13
	6	7	8	10	11	15
Noble 1972 & Before	4	4	5	6	8	10
	5	5	6	8	9	12
	6	6	7	9	11	13
Noble 1973 & Later	4	3	4	5	6	7
	5	4	5	6	7	9
	6	5	6	7	8	10

* Settings are for application of the equivalent broadcast rates per acre indicated in a 14-inch band at speed (miles per hour) shown. If other than a 14-inch band is to be used, these settings do not apply and additional calibration will be necessary.

Calibration of Band Applicators

To calibrate a granular applicator for a standard 14-inch band, a sample has to be collected while the applicator is either running or simulating a run over a distance of 500 ft.

Procedure:

1. Fill applicator box with Dual 25G.
2. Refer to Chart 1 and adjust the setting on your applicator to correspond to the desired rate per acre.
3. Attach a plastic bag or similar container to the applicator spout after detaching the delivery tube.
4. Make the 500 ft. run at planting speed.
   Speed/Time required to cover 500 feet:
   4 mph = 1 min. 25 sec.
   5 mph = 1 min. 8 sec.
   6 mph = 57 sec.
5. Weigh granules collected (or measure volume), and if necessary, adjust and make successive runs until the setting delivers the correct number of ounces per 500 ft. of row, as shown in Chart 2.
6. Repeat this procedure for each granular applicator box.

Chart 2: The Equivalent Broadcast Rate Per Acre in Volume and Weight for 500 Feet on the Basis of a 14-Inch Band

Broadcast Rate Per Acre	Volume Collected In 14-inch Band 500 Linear Feet	Weight Collected In 14-Inch Band 500 Linear Feet
6 lbs.	1.6 fl. oz.	1.3 oz.
8 lbs.	2.2 fl. oz.	1.7 oz.
10 lbs.	2.8 fl. oz.	2.2 oz.
12 lbs.	3.2 fl. oz.	2.6 oz.
16 lbs.	4.5 fl. oz.	3.5 oz.

For band widths other than 14 inches, the calibration distance will differ from 500 ft.

For other band widths, calculate the relative calibration distance in which to collect the corresponding weight or volume of Dual 25G in Chart 2 as follows:

$$\frac{14 \text{ (Inches)}}{\text{Desired band width (Inches)}} \times 500 = \text{Calibration distance in feet}$$

Example: A band width of 10 inches is to be used, instead of 14 inches, to apply Dual 25G at an equivalent rate of 10 lbs./broadcast acre.

A. $\frac{14 \text{ Inches}}{10 \text{ Inches}} \times 500 \text{ Ft.} = 700 \text{ Ft.}$

B. Collect 2.8 fl. oz. by volume or 2.15 oz. by weight for the newly calculated calibration distance of 700 ft. for a 10-inch band.

CORN

For control of annual grasses and broadleaf weeds listed under the General Information section, apply Dual 25G either preplant surface, preplant incorporated or preemergence to corn before weeds emerge.

CONVENTIONAL TILLAGE

Preplant Incorporated: Broadcast Dual 25G to the soil and incorporate in the top 1-2 inches of soil within 14 days before planting, using appropriate surface incorporating equipment.

Preemergence: Band or broadcast Dual 25G to the appropriate rate from Table 1 during or after planting and before weeds emerge. For band treatment, apply granules behind the press wheel.

Table 1: Dual 25G—Corn**

Soil texture	Broadcast rate per acre: Less than 3% organic matter	Broadcast rate per acre: 3-6%* organic matter
COARSE: Sand, loamy sand, sandy loam	6-8 lbs.	8 lbs.
MEDIUM: Loam, silt loam, silt	8-10 lbs.	8-10 lbs.
FINE: Silty clay loam, sandy clay loam, silty clay, sandy clay, clay loam, clay	8-10 lbs.	10-12 lbs.
Muck or peat soils	DO NOT USE	

* Dual 25G may be used up to 16 lbs. per acre, broadcast basis, on soils having an organic matter content greater than 6%.

** Dual 25G may be used sequentially in place of Dual 8E with Bladex® or Banvel® applied at the tank mix rate specified on the respective product labels.

Reduced-Tillage and No Tillage

Preplant Surface-Applied: Broadcast Dual 25G up to 45 days before planting corn on medium- and fine-textured soils in CO, IL, IN, IA, KS, KY, MN, MO, MT, NE, ND, SD, WI, and WY. Apply ⅔ the recommended broadcast rate of Dual 25G from Table 1 as a split treatment 30-45 days before planting and the remainder immediately before planting, or at planting. Applications less than 30 days before planting may be made either as a split or as a single treatment. Apply 8 lbs./A on coarse soils not more than 2 weeks before planting. If weeds are present at the time of treatment apply a labeled contact herbicide (for example, paraquat or Roundup®). Observe the directions for use, precautions, and restrictions on the label of the contact herbicide. To the extent possible, do not move treated soil out of the row or move untreated soil to the surface during planting, or weed control will be diminished. To avoid reduced weed control, do not apply to frozen soil as rainfall may change the pattern of the Dual granules. Injury may occur from movement of Dual 25G into the planter slit if the slit is not properly closed during planting.

Preemergence: Broadcast Dual 25G at the appropriate rate from Table 1 during or after planting, before weeds emerge, or after weeds present have been killed with a labeled contact. Injury may occur from movement of Dual 25G into the planter slit if the slit is not properly closed during planting.

PEANUTS

Dual 25G Single Applications — All Peanut Growing Areas

Apply Dual 25G either preplant incorporated, postplant incorporated or preemergence, using the appropriate rate specified below. **Preplant Incorporated:** Broadcast Dual 25G to the soil surface and incorporate into the top 1-2 inches of soil before planting using a finishing disk, harrow, rolling cultivator, or similar implement capable of providing uniform 1-2 inch incorporation. Use a preplant incorporated application if furrow irrigation is to be used, or when a period of dry weather after application is expected. If crop will be planted on beds, apply and incorporate Dual 25G after bed formation. **Postplant Incorporated:** Broadcast or band Dual 25G to the soil surface and shallowly incorporate into the soil after planting, but before peanut germination. Incorporation depth and incorporating implements must be kept above the seed or seed will be damaged. **Preemergence:** Broadcast or band Dual 25G to the soil surface during planting (behind the planter) or after planting, but before weeds or crop emerge.

Apply Dual 25G at a broadcast rate of 6-12 lbs. per acre. Within the rate range, use the lower rate on soils relatively coarse-textured or low in organic matter; use the higher rate on soils relatively fine-textured or high in organic matter.

Note: Dual 25G may be applied as directed after any of the following preplant incorporated herbicides when used according to their label recommendations: Balan® at 3-4 qts. per acre; Treflan® at 1 pt. per acre; or Vernam® at 2.3-3 pts. per acre. Dual 25G may be applied as directed prior to a sequential treatment of Dyanap® applied preemergence to cracking time using 6 qts. Dyanap per acre as specified for use on that label.

POD CROPS

Pod crops, including garbanzo, great northern beans, kidney beans, lima beans, mung beans, navy beans, okra, peas (English*, southern peas, such as blackeye, pinkeye, crowder, etc.), pinto beans and snap beans (green, wax, string).

For control of annual grasses and broadleaf weeds listed under the General Information section, apply Dual 25G either preplant incorporated or preemergence to the crop before weeds emerge.

Preplant: Broadcast Dual 25G to the soil and incorporated within the surface 1-2 inches of soil using the appropriate rate from Table 2 within 14 days before planting using appropriate surface incorporating equipment.

Preemergence: Band or broadcast Dual 25G to the soil surface during or after planting and before weeds emerge using the appropriate rate from Table 2. For band treatment, apply granules behind the press wheel.

Table 2: Dual 25G—Pod Crops*

Soil texture	Broadcast rate per acre: Less than 3% organic matter	Broadcast rate per acre: 3% organic matter or greater
COARSE: Sand, loamy sand, sandy loam	6-8 lbs.	8 lbs.
MEDIUM: Loam, silt loam, silt	8-10 lbs.	8-10 lbs.
FINE: Silty clay loam, sandy clay loam, silty clay, sandy clay, clay loam, clay	8-10 lbs.	10-12 lbs.
Muck or peat soils	DO NOT USE	

* On English peas, use only preemergence applications. Do not use on English peas in northern U.S., crop injury may occur.

SOYBEANS

For control of annual grasses and broadleaf weeds in the General Information section, apply Dual 25G either preplant surface, preplant incorporated, or preemergence to soybeans before weeds emerge.

Conventional Tillage

Preplant Incorporated: Broadcast Dual 25G to the soil surface and incorporate in the top 2 inches of soil within 14 days before planting, using a finishing disk, harrow, or other appropriate tillage equipment at the appropriate rate from Table 3.

Preemergence: Band or broadcast Dual 25G to the soil surface during or after planting and before weeds emerge, using the appropriate rate from Table 3. For band treatment, apply granules behind the press wheel.

Table 3: Dual 25G—Soybeans*

Soil texture	Broadcast rate per acre: Less than 3% organic matter	Broadcast rate per acre: 3-6% organic matter
COARSE: Sand, loamy sand, sandy loam	6-8 lbs.	8 lbs.
MEDIUM: Loam, silt loam, silt	8-10 lbs.	8-10 lbs.
FINE: Silty clay loam, sandy clay loam, silty clay, sandy clay, clay loam, clay	8-10 lbs.	10-12 lbs.
Muck or peat soils	DO NOT USE	

* Dual 25G may be used sequentially in place of Dual 8E with Sencor®/Lexone®, Lorax®, Amiben®, Dyanap, or Premerge® 3 applied at the tank mix rate specified on the respective product labels.

Reduced-Tillage and No Tillage

Preplant Surface-Applied: Broadcast Dual 25G up to 45 days before planting soybeans on medium- and fine-textured soils in CO, IL, IN, IA, KS, KY, MN, MO, MT, NE, ND, SD, WI, and WY. Apply ⅔ the recommended broadcast rate of Dual 25G as a split treatment 30-45 days before planting and the remainder immediately before planting, or at planting. Applications less than 30 days before planting may be made either as a split or as a single treatment. Apply 8 lbs./A on coarse soils not more than 2 weeks before planting. If weeds are present at the time of treatment, apply a labeled contact herbicide (for example, paraquat or Roundup®). Observe the directions for use, precautions, and restrictions on the label of the contact herbicide. To the extent possible, do not move treated soil out of the row or move untreated soil to the surface during planting, or weed control will be diminished. To avoid reduced weed control, do not apply to frozen soil as rainfall may change the pattern of the Dual granules. Injury may occur from movement of Dual 25G into the planter slit if the slit is not properly closed during planting.

Preemergence: Broadcast Dual 25G at the appropriate rate from Table 3 during or after planting, before weeds emerge, or after weeds present have been killed with a labeled contact. Injury may occur from movement of Dual 25G into the planter slit if the slit is not properly closed during planting.

POTATOES

Apply Dual 25G either incorporated, preemergence, or after hilling/lay-by according to the directions specified below for control of weeds listed under the General Information section. Within a rate range, use the lower rate on soils relatively coarse-textured or low in organic matter; use the higher rate on soils relatively fine-textured or high in organic matter. Effectiveness will be reduced if later cultural practices expose untreated soil.

Incorporated: Apply Dual 25G at 6-12 lbs. per acre to the soil and incorporate into the surface of the soil 1-2 inches using appropriate tillage equipment. Planting and later cultural practices should not bring untreated soil to the surface. Postplant application may be made any time after planting to drag-off, but before potato emergence. Incorporate using an implement that distributes Dual within the top 1-2 inches of soil. Do not damage potato seed pieces or sprouts with incorporation equipment.

Preemergence: Apply Dual 25G at 6-12 lbs. per acre either after planting as a preemergence, delayed preemergence, after drag-off or hilling treatment, but before weeds emerge. Up to 16 lbs. per acre of Dual 25G alone may be used where soil organic matter is greater than 6%.

After Hilling/Lay-by: Apply 10 lbs. per acre of Dual 25G after hilling/at lay-by to control Dual sensitive weeds for remainder of the growing season. This application will not control emerged weeds. It may be applied over a previous Dual 25G application, but do not apply more than 22 lbs. of Dual 25G per acre in a single crop season.

Note: Do not use on muck or peat soils. If cool, wet soil conditions occur after application, Dual may delay maturity and/or reduce yield of Superior and other early maturing potato varieties. Do not use on sweet potatoes or yams. Do not apply both preemergence and incorporated treatments.

To avoid illegal residues, potatoes treated with Dual should not be harvested within 60 days after a planting to drag-off application, or within 40 days after a lay-by application. Do not use in Kern County, California.

ORNAMENTALS (CONTAINER, FIELD AND LINER GROWN)

Apply Dual 25G using a cyclone or similar type applicator at rates indicated for control of annual grasses and yellow nutsedge.

Broadcast Dual 25G at the rate of 8-16 lbs. per acre on established ornamentals, or to ornamentals at least 5 days after transplanting, but before grass weeds or yellow nutsedge emerge, or after existing weeds or nutsedge plants have been removed. A second application may be needed to provide grass and yellow nutsedge control for an extended period. Do not make more than 2 applications per year.

Use the higher rate of Dual 25G on fine-textured soils, on high-organic matter soils, in potting mixes, and where yellow nutsedge and heavy infestations of grass weeds are expected. Use the lower rate on coarse-textured soils with low organic matter content, and where light infestations of grass weeds are expected.

If band applications are to be used, apply an amount proportional to the fraction of an acre treated.

To control annual broadleaf weeds, a broadcast application of Princep® 4G at 20-25 pounds per acre may be made immediately before or after applying Dual 25G. See the Princep 4G label for application instructions.

Precautions: *Do not use Dual 25G alone or in conjunction with Princep 4G on seedbeds or on unrooted cuttings or prior to transplanting. Do not apply Dual 25G to ornamentals whose foliage is wet for the granules may adhere to the foliage and cause injury. Should granules adhere to the foliage, irrigate soon after application to remove the granules and reduce potential damage.*

DO NOT USE IN GREENHOUSES OR OTHER ENCLOSED STRUCTURES.

CONTAINER GROWN ORNAMENTALS

Use in aucuba (*Aucuba* spp.), holly (*Ilex* spp.), juniper (*Juniperus* spp.) and liriope (*Liriope* spp.) var. Silver Sunproof.

Princep 4G may be used in conjunction with Dual 25G in aucuba, holly, juniper, and liriope.

FIELD AND LINER GROWN ORNAMENTALS

Use Dual 25G in the following:

Arborvitae (*Thuja* spp.)*
Aucuba (*Aucuba* spp.)*
Azalea (*Rhododendron* sp.), Hershey Red, Herbert, Purple Splendor
Barberry (*Berberis* spp.), Mentor, Warty, Wintergreen
Box Tree (*Buxus* spp.)
Cotoneaster (*Cotoneaster* spp.)*
Crabapple (*Malus* spp.)*
Dogwood (*Cornus* sp.), Red Stem
Euonymus (*Euonymus* spp.), including Purpleleaf Winter Creeper
Firethorn (*Pyracantha* spp.)
Forsythia (*Forsythia* spp.), including Spring Glory and Golden Bells
Holly (*Ilex* spp.), American, Boxleaf Japanese*, Japanese*, Green Lustre Japanese
Honeysuckle (*Lonicera* spp.)
Hybrid Tea Rose (*Rosa* spp.)
Ivy (*Hedera* sp.), English
Japanese Andromeda (*Pieris* spp.)*
Japanese Red Maple (*Acer* spp.)
Juniper (*Juniperus* spp.)*, Andorra, Creeping, Shore, Bar Harbor, Blue Rug
Leucothoe (*Leucothoe* spp.)*
Lilac (*Syringa* spp.), James McFarland
Liriope (*Liriope* spp.)*, Silver Sunproof
Maple whips (*Acer* spp.)*
Miniature Rose (*Rosa* spp.)
Mountain Laurel (*Kalmia* spp.)
Osmanthus (*Osmanthus* spp.)*
Pachysandra (*Pachysandra* spp.)
Privet (*Ligustrum* spp.)
Red Oak (*Quercus* spp.)*
Rhododendron (*Rhododendron* spp.)*
Rosebay (*Rhododendron* spp.)
Spirea (*Spiraea* spp.)
Viburnum (*Viburnum* spp.)
Weigela (*Weigela* spp.)*
Yew (*Taxus* spp.), Dwarf Japanese*, Hicks

* Dual 25G may be applied in conjunction with Princep 4G on these field and liner grown ornamentals.

ROTATIONAL CROPS

1) If crop treated with a preplant or preemergence application of Dual 25G is lost, corn, cotton, soybeans, safflower, peanuts, potatoes, grain sorghum*, or pod crops** may be replanted immediately. Do not make a second broadcast application of Dual 25G. If the original application was banded and the second crop is planted in the untreated row middles, a second banded treatment may be applied. 2) Barley, oats, rye, or wheat may be planted $4\frac{1}{2}$ months following a preplant or preemergence treatment. Corn, cotton, soybeans, peanuts, potatoes, sorghum, pod crops, root crops, barley, buckwheat, milo, oats, rice, rye, or wheat may be planted in the spring following treatment. Do not graze or feed forage or fodder from cotton or pod crops to livestock. All other rotational crops may be planted 18 months after application. 3) Corn, cotton, soybeans, peanuts, potatoes, safflower, sorghum or pod crops may be planted in the spring following a lay-by treatment applied the previous season. All other rotational crops may be planted 18 months after a lay-by application.

* Provided the seed has been properly treated with Concep® seed protectant.

STORAGE AND DISPOSAL

PESTICIDE STORAGE AND DISPOSAL

Store in a dry place. Do not contaminate water, food, or feed by storage, disposal, or cleaning of equipment. Open dumping is prohibited. Improper disposal of unused pesticide, spray mixture, or rinsate is a violation of federal law. If these wastes cannot be used according to label instructions, contact your State Pesticide or Environmental Control Agency, or the Hazardous Waste representative at the nearest EPA Regional Office for guidance in proper disposal methods.

CONTAINER DISPOSAL

Completely empty bag into application equipment. Dispose of empty bag in a sanitary landfill or by incineration, or if allowed by state and local authorities, by burning. Stay out of smoke from burning bags.

For minor spills, follow all precautions indicated on this label and clean up immediately. Take special care to avoid contamination of equipment and facilities during cleanup procedures and disposal of wastes. In the event of a major spill, fire or other emergency, call (919) 292-7100 day or night.

PRECAUTIONARY STATEMENTS

HAZARDS TO HUMANS AND DOMESTIC ANIMALS

CAUTION

The active ingredient, metolachlor, may cause skin sensitization reactions in certain individuals. Wear protective clothing (coveralls and gloves) while handling or using this product. Harmful if swallowed, inhaled or absorbed through the skin. Avoid breathing dust. Causes eye irritation. Avoid contact with eyes, skin or clothing. Do not contaminate food or feed.

Practical Treatment: In case of contact with eyes or skin, wash immediately with plenty of water. Get medical attention if irritation persists. If inhaled, move to clean air. If swallowed, give one or two glasses of water and induce vomiting by placing fingers in back of throat, and contact a local poison control center, hospital or physician immediately. Never give anything by mouth to an unconscious person.

ENVIRONMENTAL HAZARDS

Do not apply directly to any body of water. Do not apply where runoff is likely to occur. Do not contaminate water by cleaning of equipment or disposal of wastes.

Dual® trademark of CIBA-GEIGY for metolachlor
U.S. Patent No. 3,937,730

Amiben® trademark of Union Carbide Agricultural Products, Inc.

AAtrex® trademark of CIBA-GEIGY for atrazine

Balan® trademark of Elanco Products for benefin

Banvel® trademark of Velsicol Chemical Corp. for dicamba

Bladex® trademark of Shell Chemical Company for cyanazine

Brominal® trademark of Union Carbide Corporation for bromoxynil

Buctril® trademark of Rhone-Poulenc Chemical Company for bromoxynil

Concep® trademark of CIBA-GEIGY
U.S. Patent No. 4,070,389

Dyanap® trademark of Uniroyal Chemical for Alanap®-Dinitro

Lexone® trademark of E. I. duPont de Nemours and Co. for metribuzin

Lorox® trademark of E. I. duPont de Nemours and Co. for linuron

Premerge® trademark of the Dow Chemical Company for dinoseb

Princep® trademark of CIBA-GEIGY for simazine

Roundup® trademark of Monsanto Company for glyphosate herbicide

Sencor® trademark of the Parent Company of Farbenfabriken Bayer GmbH, Leverkusen, for metribuzin

Treflan® trademark of Elanco Products for trifluralin

Vernam® trademark of Stauffer Chemical Company for vernolate

CGA 49L34A 045

EVIK®80W

HERBICIDE

For weed control in bananas, corn, grapefruit, oranges, pineapple, sugarcane, and noncrop areas

For use as a potato vine desiccant

Evik 80W is a wettable powder

ACTIVE INGREDIENTS:	
Ametryn:	
2-ethylamino-4-isopropylamino-6-methylthio-*s*-triazine	76%
Related compounds	4%
INERT INGREDIENTS:	20%
TOTAL:	100%

EPA Reg. No. 100-473

Evik® trademark of CIBA-GEIGY for Ametryn

Keep Out of Reach of Children.

Caution

See additional precautionary statements on left panel.

See directions for use on panel at right.

DIRECTIONS FOR USE AND CONDITIONS OF SALE AND WARRANTY

IMPORTANT: Read the entire **Directions for Use** and the **Conditions of Sale and Warranty** before using this product.

CONDITIONS OF SALE AND WARRANTY

The **Directions for Use** of this product reflect the opinion of experts based on field use and tests. The directions are believed to be reliable and should be followed carefully. However, it is impossible to eliminate all risks inherently associated with use of this product. Crop injury, ineffectiveness, or other unintended consequences may result because of such factors as weather conditions, presence of other materials, or the manner of use or application all of which are beyond the control of CIBA-GEIGY or the Seller. All such risks shall be assumed by the Buyer.

CIBA-GEIGY warrants that this product conforms to the chemical description on the label and is reasonably fit for the purposes referred to in the **Directions for Use** Subject to the inherent risks referred to above. **CIBA-GEIGY makes no other express or implied warranty of Fitness or Merchantability or any other express or implied warranty. In no case shall CIBA-GEIGY or the Seller be liable for consequential, special, or indirect damages resulting from the use or handling of this product.** CIBA-GEIGY and the Seller offer this product, and the Buyer and user accept it, subject to the foregoing **Conditions of Sale and Warranty,** which may be varied only by agreement in writing signed by a duly authorized representative of CIBA-GEIGY.

DIRECTIONS FOR USE

It is a violation of federal law to use this product in a manner inconsistent with its labeling.

FAILURE TO FOLLOW ALL PRECAUTIONS ON THIS LABEL MAY RESULT IN CROP INJURY, POOR WEED CONTROL, AND/OR ILLEGAL RESIDUES.

Do not apply this product in such a manner as to directly or through drift expose workers or other persons except those knowingly involved in the application. The area being treated must be vacated by unprotected persons.

REENTRY STATEMENT

Do not enter treated areas without protective clothing until sprays have dried.

Because certain states may require more restrictive reentry intervals for various crops treated with this product, consult your State Department of Agriculture for further information.

Written or oral warnings must be given to workers who are expected to be in a treated area or in an area about to be treated with this product. Oral warnings must inform workers of areas or fields that may not be entered without specific protective clothing until sprays have dried, and appropriate actions to take in case of accidental exposure, as described under **Precautionary Statements** on this label. When oral warnings are given, warnings shall be given in a language customarily understood by workers. Oral warnings must be given if there is reason to believe that written warnings cannot be understood by workers. Written warnings must include the following information: CAUTION. Area treated with Evik 80W on (date of application). Do not enter without appropriate protective clothing until sprays have dried. In case of accidental exposure, flush eyes or skin with plenty of water. Call physician if irritation persists. Remove and wash contaminated clothing before reuse.

GENERAL INFORMATION

Evik 80W controls most annual broadleaf and grass weeds (see list under each crop). When applied before weed emergence, Evik 80W kills weeds as they germinate by entering roots. On existing weeds, it is effective through leaf contact.

Following many years of continuous use of this product and chemically related products, biotypes of some of the weeds listed on this label have been reported which cannot be effectively controlled by this and related herbicides. Where this is known or suspected, we recommend the use of this product in combination with other registered herbicides which are not triazines. Consult with your state Agricultural Extension Service for specific recommendations.

Avoid using Evik 80W where adjacent desirable plants may be injured.

Evik 80W is noncorrosive to equipment and metal surfaces, nonflammable, and has low electrical conductivity.

APPLICATION PROCEDURES

Ground application: Use conventional ground sprayers equipped with nozzles that provide accurate and uniform application. Be certain that nozzles are uniformly spaced and the same size. Calibrate sprayer before use and recalibrate at the start of each season and when changing carriers.

Use a pump with capacity to (1) maintain 35-40 psi at nozzles, (2) provide sufficient agitation in tank to keep mixture in suspension, and (3) to provide a minimum of 20% bypass at all times. Use centrifugal pumps which provide propeller shear action for dispersing and mixing this product. The pump should provide a minimum of 10 gals./minute/100 gal. tank size circulated through a correctly positioned sparger tube or jets.

Use screens to protect the pump and to prevent nozzles from clogging. Screens placed on the suction side of the pump should be *16 mesh or coarser.* Do not place a screen in the recirculation line. Use 50-mesh or coarser screens between the pump and boom, and where required, at the nozzles. Check nozzle manufacturer's recommendations.

For band applications, calculate amount to be applied per acre as follows:

$$\frac{\text{band width in inches}}{\text{row width in inches}} \times \text{broadcast rate per acre} = \text{amount needed per acre of field}$$

Aerial application: Use aerial application only where specified in the use directions. Avoid applications under conditions where uniform coverage cannot be obtained or where excessive spray drift may occur.

Avoid application to humans or animals. Flagmen and loaders should avoid inhalation of spray mist and prolonged contact with skin, and should wash thoroughly before eating and at the end of each day's operation.

Mixing procedures — all uses: (1) Be sure sprayer is clean and not contaminated with any other materials, or crop injury or sprayer clogging may result. (2) Fill tank ¼ full with clean water or nitrogen solution (may be used in corn). (3) Start agitation. (4) Be certain that the agitation system is working properly and creates a rippling or rolling action on the liquid surface. (5) Make a slurry by adding Evik 80W to a small amount of water in a separate container and pour slurry into tank. (6) Continue filling tank until 90% full. Increase agitation if necessary to maintain surface action. (7) Add tank mix herbicide(s) after this product is thoroughly suspended. (8) Finish filling tank. Maintain agitation to avoid separation of materials. (9) Empty tank as completely as possible before refilling to prevent buildup of emulsifiable concentrate residue from possible tank mix herbicides. (10) If an emulsifiable concentrate film starts to build up in tank, drain it and clean with strong detergent solution or solvent. (11) Clean sprayer thoroughly immediately after use by flushing system with water containing a detergent.

Compatibility test: Nitrogen solutions may replace all or part of the water in the spray in corn. Since nitrogen solutions can vary, even within the same analysis, check compatibility each time before use. Commercial application equipment may improve compatibility in some instances. The following test assumes a spray volume of 25 gals. per acre. For other spray volumes, make appropriate changes in the ingredients. Check compatibility using this procedure.

1. Add 1 pint of fertilizer to each of 2 one-quart jars with tight lids.
2. To **one** of the jars add ¼ tsp. or 1.2 milliliters of a compatibility agent approved for this use (¼ tsp. is equivalent to 2 pts. per 100 gals. spray). Shake or stir gently to mix. Examples of compatibility agents include Compex® and Unite®.
3. To both jars add 1.4 teaspoons of Evik 80W for each pound per acre to be applied.
4. After adding all ingredients, put lids on and tighten, and invert each jar ten times to mix. Let the mixtures stand 15 minutes and then look for separation, large flakes, precipitates, gels, heavy oily film on the jar, or other signs of incompatibility. Determine if the compatibility agent is needed in the spray mixture by comparing the two jars. If either mixture separates, but can be remixed readily, the mixture can be sprayed as long as good agitation is used. If the mixtures are incompatible, try slurrying the Evik 80W in water before addition. If still incompatible, do not use the nitrogen solution and Evik 80W in the same spray tank.

BANANAS AND PLANTAINS

In Hawaii, Evik 80W controls broadleaf and grass weeds including rattlebox (*Crotalaria* spp.), dallisgrass, goosegrass *(Eleusine indica),* Japanese tea, kukaipuaa and other crabgrass species (*Digitaria* spp.), pualele (sowthistle), common purslane, *Richardia* spp., spanishneedles, wild pea bean, *Amaranthus* spp., Flora's paintbrush, foxtail, junglerice, fireweed, and *Panicum* spp.

In Puerto Rico, Evik 80W controls broadleaf and grass weeds including *Commelina diffusa, Paspalum* spp., *Axonopus compressus, Panicum* spp., *Digitaria* spp., and *Setaria* spp.

Apply Evik 80W in Hawaii or Puerto Rico as a directed basal spray immediately after setting banana or plantain plants, or any time thereafter. Apply before weeds emerge or to emerged, growing weeds. Use the lower rates for applications made before weed emergence and for young, easy-to-kill weeds. Use the higher rates for more residual control and for controlling hard-to-kill weeds. Repeat as needed at 3-4 month intervals.

	Rates of application per acre	
Area	Per application (minimum of 30 gals. water/A)	Maximum per year
Hawaii	5-10 lbs.	30 lbs.
Puerto Rico	4-6 lbs.	16 lbs.

In Central America, apply 2-4 lbs. of Evik 80W in a minimum of 20 gals. of water/A covering all vegetation thoroughly. Use the lower rate for application to young weeds, and the higher rate for application to heavier growth of older, hard-to-kill weeds. Repeat application 30 days later, covering all vegetation. Apply the third and subsequent applications as needed, covering only living vegetation. Use a maximum of 16 lbs. of Evik 80W/A per year.

CORN (FIELD CORN, SWEET CORN, POPCORN)

Evik 80W controls annual broadleaf and grass weeds including Texas panicum (Texas millet), fall panicum, signalgrass (*Brachiaria* spp.), goosegrass, crabgrass, barnyardgrass, giant foxtail, yellow and green foxtails, nutsedge, shattercane, wild proso millet, cocklebur, lambsquarters, Florida pusley, morningglory, pigweed, wild mustard, ragweed, velvetleaf, and smartweed. Weeds taller than specified in the rate table will not be controlled.

Apply Evik 80W as a postemergence directed spray to weeds after the smallest corn is at least 12 inches tall (measured to the highest leaf surface on free standing plants).

Precaution: *Do not spray over top of corn or injury will occur. Do not apply within 3 weeks of tasseling.*

Apply in a minimum of 20 gals. of water/A to assure uniform coverage (nonpressure nitrogen solution may be substituted for all or part of the water). The entire weed must receive spray to be killed. Add a surfactant such as X-77®, duPont WK, or Tronic at the rate of 2 qts./100 gals. of spray mixture (0.5% of spray volume).

It is recommended that gauge wheels and/or leaf lifter equipment be used to prevent leaf contact with the spray. Drop nozzles may be used, but extreme care must be taken to keep the spray or drift from contacting the leaves and especially the whorl of the corn plant. Be sure the entire spray pattern is directed downward. Apply at a spray pressure of 30 psi or less to prevent "bounce back"—spray bouncing off soil or weeds and settling on corn leaves.

Use rates of application according to the following geographical areas:

Area 1

Arizona	Michigan	Ohio
California	Minnesota	Oregon
Colorado	Missouri	Pennsylvania
Connecticut	Montana	Rhode Island
Idaho	Nebraska	South Dakota
Illinois	Nevada	Utah
Indiana	New Hampshire	Vermont
Iowa	New Jersey	Washington
Kansas	New Mexico	Wisconsin
Maine	New York	Wyoming
Massachusetts	North Dakota	

Area 2

Alabama	Louisiana	Tennessee
Arkansas	Maryland	Texas
Delaware	Mississippi	Virginia
Florida	North Carolina	West Virginia
Georgia	Oklahoma	
Kentucky	South Carolina	

Where a rate range is indicated, use the higher rate for relatively dense weed infestations.

Rates of Application for Corn

Weed height	Weeds controlled*	Broadcast rate per acre: Area 2	Broadcast rate per acre: Area 1
Up to 2 inches	*Brachiara* broadleafs	75 lb.	
	Texas panicum fall panicum barnyardgrass goosegrass	1.25 lbs.	2 lbs.
	crabgrass	2-2.5 lbs.	
	foxtail	2-2.5 lbs.	
	shattercane nutsedge	Do not apply	2.5 lbs.
Up to 4 inches	wild proso millet		
2-4 inches	*Brachiaria* broadleafs	1.25 lbs.	
	Texas panicum fall panicum barnyardgrass goosegrass	2-2.5 lbs.	
	crabgrass	Do not apply	2.5 lbs
	foxtail	2-2.5 lbs.	
	shattercane nutsedge	Do not apply	
4-6 inches	*Brachiaria* broadleafs	2-2.5 lbs.	2.5 lbs.
	fall panicum nutsedge	Do not apply	

* A mechanical cultivation may be required if weeds regrow.

Suggestions For Crop Rotation: Small grains, such as wheat, oats, and rye, may be planted in the fall following the recommended application. Do not plant any rotational crop other than small grains until the following year.

Note: Allow 30 days after the application before harvesting, grazing, or feeding forage to livestock.

GRAPEFRUIT AND ORANGES

Florida: Apply Evik 80W twice per year (see following table) in a minimum of 40 gals./A as a directed spray to thoroughly cover existing weeds. Weeds controlled include common bermudagrass *(Cynodon dactylon)*, spanishneedles *(Bidens bipinnata)*, and Florida pusley *(Richardia scabra)*. Mow 7-10 days before application to obtain most effective control of dense stands of common bermudagrass. Evik is most effective when good moisture conditions exist.

Make the first application in the spring when weeds have resumed active growth, but before they exceed 4 inches in height. Make the second application 90-100 days later to obtain season-long control. For heavy bermudagrass infestations, use the higher rates in the first application.

Florida application	Broadcast rate per acre: Light weed infestation	Broadcast rate per acre: Heavy weed infestation
First (spring)	5-6 lbs.	7-8 lbs.
Second (summer)	5-6 lbs.	7 lbs.

Texas: Apply Evik 80W in a minimum of 40 gals./A as a directed spray to thoroughly cover existing weeds. Use 3 lbs./A for control of heavy weed infestations and for top-kill of johnsongrass and marinevine (ivy treebine). Weeds controlled include barnyardgrass, buffelgrass, sandbur, Texas panicum (Coloradograss), silver leaf nightshade, pigweed, common purslane, and sunflower. Use 2 lbs./A to control light weed infestations. For both rates, add a nonionic surfactant, such as duPont WK, X-77, or LOC, at the rate of 2 qts./100 gals. of spray mixture (0.5% of spray volume).

Make the first application in the spring when weeds have resumed active growth, but before they exceed 4 inches in height. Make additional applications as needed up to 15 lbs./A per year. Evik is most effective when weeds are actively growing and 1-4 inches in height.

Precautions (Florida and Texas): *1) Do not apply where trees have been established in the grove less than two years. 2) Apply no more than 15 lbs. of Evik 80W/A per year. 3) Do not allow spray to contact stems, fruit, or foliage on trees. 4) Do not apply where uniform coverage cannot be obtained or where excessive spray drift may occur.*

Notes (Florida and Texas): 1) Do not apply within 30 days of harvest. 2) Do not graze treated areas.

PINEAPPLE

For control of broadleaf and grass weeds including rattlebox (*Crotalaria* spp.), dallisgrass, goosegrass *(Eleusine indica)*, Japanese tea, kukaipuaa and other crabgrass species (*Digitaria* spp.), pualele (sowthistle), common purslane, *Richardia* spp., spanishneedles, wild pea bean, *Amaranthus* spp., Flora's paintbrush, foxtail, junglerice, fireweed, and *Panicum* spp., apply up to 10 lbs. of Evik 80W/A as a blanket spray immediately after planting, or after plant crop harvest is completed and before weeds emerge. Additional blanket or interspace applications may be made at 1-2 month intervals prior to differentiation as needed, but should not exceed 2½ lbs./A per application. Apply in a minimum of 30 gals. of water/A.

Note: 1) Do not apply more than 37 ½ lbs. Evik 80W/A per crop cycle. 2) Do not make the last application within 160 days of harvest.

SUGARCANE

To control weeds specified in the various states, apply Evik 80W alone or in tank mix combinations. Broadcast aerially in a minimum volume of 5 gals. of spray/A, or broadcast or band by ground in a minimum of 20 gals./A, unless indicated otherwise. Repeat treatments, where needed, may be applied broadcast, band, or interline as recommended, with final application prior to close-in.

Aerial Application: Use aerial application only where broadcast applications are specified. Unless specified otherwise, apply a minimum of 1 gal. of water for each 1 lb. of product applied/A, but not less than 5 gals. total volume/A. Avoid applications under conditions where uniform coverage cannot be obtained or where excessive spray drift may occur. In order to assure that spray will be controllable within the target area when used according to label directions, make applications at a maximum height of 10 ft., using low drift nozzles at a maximum pressure of 40 psi, and restrict application to periods when wind speed does not exceed 10 mph. To assure that spray will not adversely affect adjacent sensitive nontarget plants, apply Evik 80W alone by aircraft at a minimum upwind distance of 800 ft. from sensitive plants, or apply Evik 80W + Karmex® at a minimum upwind distance of 500 ft. from sensitive plants.

Florida: Apply 0.5-1.5 lbs. of Evik 80W in a minimum of 20 gals. of water/A directed to the base of plant or ratoon sugarcane to emerged weeds. Avoid wetting sugarcane foliage or injury may occur. Use nozzle tips which will minimize atomization or spray drift. Use the higher rate for high grass populations. Apply up to 2 repeat applications, if needed, at 30-day intervals before close-in. To control alexandergrass *(Brachiaria plantaginea),* apply at the 3-4 leaf stage or before 3 inches tall. For mixed weed infestations, use 1.5 lbs. of Evik 80W plus 0.5 lb. acid equivalent of 2,4-D amine/A and/ or 2 qts. of surfactant, such as X-114 or ACL 209, for each 100 gals. of spray to improve weed control. Observe all precautions and limitations on labeling of all products used in mixtures.

Hawaii: Use one of the following methods in plant or ratoon sugarcane for control of ageratum, rattlebox (*Crotalaria* supp.), dallisgrass, fireweed, goosegrass, *(Eleusine indica),* guineagrass, Japanese tea, kukaipuaa, and other crabgrass species (*Digitaria* spp.), morningglory, pualele (sowthistle), common purslane, *Richardia* spp., spanishneedles, wild pea bean, *Amaranthus* spp., Flora's paintbrush, foxtail, junglerice, and swollen fingergrass.

1. Apply 5-10 lbs./A before weeds or sugarcane emerge. A second and third application at 2.5-5 lbs./A may be made, as needed, to emerged weeds prior to close-in.
2. Apply 2.5-5 lbs. of Evik 80W plus 2.5-5 lbs. of Karmex 80W before sugarcane and weeds emerge.

   A second application at 2.5-5 lbs. of Evik 80W plus 2.5-5 lbs. of Karmex 80W may be made, as needed, postemergence to sugarcane and weeds. A third application at 2.5-3 lbs. of Evik 80W plus 2.5-3 lbs. of Karmex 80W may be applied prior to close-in.

   For best results when Karmex 80W is used on emerged weeds, add a nonionic surfactant to the spray at the rate of 1-2 qts. per 100 gals. and apply as a directed spray.

Note: Use the minimum preemergence rates on nonirrigated sugarcane (high rainfall areas), on land first cropped to sugarcane, and for light weed infestations.

Precautions: *(1) Sugarcane growing in areas of exposed subsoil, in rocky areas, or in soils of low adsorptive capacity may show temporary chlorosis following treatment. (2) Injury to sugarcane may occur when under moisture stress. (3) Certain sugarcane varieties may show a temporary chlorosis or stunting as a result of over-the-top application.*

Louisiana: Use directions below for control of these weeds:

Weed height	Weeds controlled
Up to 3 inches	*itchgrass (raoulgrass)
Up to 4 inches	barnyardgrass, crabgrass, fall panicum, foxtail, goosegrass, Texas panicum
Up to 5 inches	annual sowthistle, common chickweed, henbit, paleseed plantain, swinecress
Up to 6 inches	*Brachiaria* spp., browntop panicum, cocklebur, Florida pusley, common lambsquarters, morningglory, pigweed, ragweed, smartweed, velvetleaf, wild mustard

* Controls emerged itchgrass. May not control itchgrass germinating after treatment.

Broadcast or band by ground equipment over the top of plant or ratoon sugarcane 2.5 lbs. of Evik 80W plus 0.5 lb. acid equivalent of 2,4-D amine plus 1 qt. of crop oil concentrate, such as Agri-Dex, Amoco, or Unico (or 1 pt. of nonionic surfactant, such as duPont WK, X-77, or LOC) in a minimum of 20 gals. of water/A.

Follow with one or two repeat over-the-top or directed applications, as needed, using 3 lbs. of Evik 80W plus 0.5 lb. of 2,4-D amine plus 1 qt. of crop oil concentrate (or 1 pt. of nonionic surfactant) in a minimum of 20 gals. of water/A. To avoid injury, do not apply over the top of sugarcane after April 10 or after sugarcane exceeds 20 inches in height. If needed, follow with one or two additional applications directed to the base of sugarcane (using same rates as in second or third application) before close-in. Observe all precautions and limitations on labeling of all products used in mixtures.

Precaution: *Temporary yellowing of sugarcane leaves may follow over-the-top applications.*

Puerto Rico: Use one of the following methods for control of *Amaranthus* spp., crabgrass, dallisgrass, foxtail, goosegrass *(Eleusine indica),* itchgrass (raoulgrass), junglerice, milkweed, morningglory, *Panicum* spp., pigweed, purpletop, purslane, *Richardia* spp., spanishneedles, sandbur, and sowthistle in plant and/or ratoon sugarcane as specified:

1. **Plant or Ratoon Sugarcane:** Apply 5-10 lbs. of Evik 80W/A before sugarcane or weeds emerge. Apply one or two additional applications, as needed, at 2.5-5 lbs./A before close-in, directed to the base of sugarcane.
2. **Plant Sugarcane:** Apply 5 lbs./A before sugarcane emerges and before weeds exceed 5 inches in height. A second application at 4 lbs./A may be made, as needed, over-the-top of sugarcane after the sugarcane and weeds emerge but prior to close-in.
3. **Ratoon Sugarcane:** Apply 4 lbs./A interline or over-the-top of sugarcane and weeds prior to close-in.

In plant or ratoon sugarcane, add 1 pt. of a nonionic surfactant to the spray volume to be applied/A where weeds have emerged at application. Treat before weeds exceed 5 inches in height.

Precautions: *(1) Sugarcane growing in areas of exposed subsoil, in rocky areas or during time of water stress may show a temporary chlorosis. (2) Do not apply Evik 80W in combination with other herbicides when making over-the-top applications. (3) Temporary chlorosis or stunting may occur after over-the-top applications.*

Texas: Use directions below for control of these weeds:

Weed height	Weeds controlled
Up to 2 inches	fall panicum, Texas panicum
Up to 4 inches	barnyardgrass, *Brachiaria* spp., cocklebur, Florida pusley, common lambsquarters, morningglory, ragweed, smartweed, velvetleaf, wild mustard
Up to 6 inches	pigweed, sunflower

Broadcast 1.5-2.5 lbs./A preemergence or postemergence to sugarcane or weeds. Add a nonionic surfactant at the rate of 2 qts./100 gals. of spray mixture.

Follow with one or two repeat applications, as needed. Make the final application before close-in. Use the higher rates of Evik 80W for heavier weed infestations.

Precaution: *Temporary yellowing of sugarcane leaves may follow over-the-top applications.*

POTATO VINE DESICCANT

To kill potato vines and weed growth, apply Evik 80W 10-14 days before harvest at the rates in the following table. Apply in a minimum of 100 gals. of water/A. Thorough coverage must be obtained in order to achieve a satisfactory kill. Complete vine kill is normally obtained in 10-14 days in the Southeast and in 14-17 days elsewhere. Vine kill may be delayed by unseasonably cold weather.

Note: Do not graze livestock on treated areas.

Area	Rate Evik 80W per acre
Florida	1.25-2.5 lbs.
Southeast (NC, TN, and south except FL)	1.25-2.5 lbs. or 1 lb. *plus* 2 gals. nonphytotoxic oil
Northeast (VA, KY, and north), Midwest, North Central, CO, and WY	2.5-3 lbs.

Noncrop Land

Evik 80W provides weed control on industrial sites, railroad rights-of-way, around highway guard rails, lumberyards, petroleum tank farms, utility substations and airport runways. Use only where control of all vegetation is desired. Do not use near adjacent desirable trees, shrubs or plants, or in greenhouses, as injury may occur.

Evik 80W controls barnyardgrass, crabgrass, fall panicum, foxtails (giant, green, yellow), goosegrass, nutsedge, shattercane, signalgrass (*Brachiaria* spp.), Texas panicum (Texas millet), cocklebur, Florida pusley, lambsquarters, morningglory, wild mustard, pigweed, ragweed, smartweed, and velvetleaf.

Apply to emerged actively growing weeds at the rates recommended in the rate chart "Rates of Application for Corn" on this label. Apply in at least 20 gals. of water/A and add a surfactant such as X-77, duPont WK, or Tronic at the rate of 2 qts./100 gals. of spray mixture (0.5% of spray volume). Cover weeds thoroughly with spray for good control.

Do not plant desirable plants in treated areas for one year.

STORAGE AND DISPOSAL

Store in a cool, dry place. Do not contaminate water, food, or feed by storage or disposal. Improper disposal of unused pesticide, spray mixture, or rinse water is a violation of federal law. If these wastes cannot be used according to label instructions, contact your State Pesticide or Environmental Control Agency, or the Hazardous Waste representative at the nearest EPA Regional Office for guidance in proper disposal methods.

Completely empty bag into application equipment. Dispose of empty bag and box in a sanitary landfill or by incineration, or if allowed by state and local authorities, by burning. Stay out of smoke from burning containers.

For minor spills, leaks, etc., follow all precautions indicated on this label and clean up immediately. Take special care to avoid contamination of equipment and facilities during cleanup procedures and disposal of wastes. In the event of a major spill, fire or other emergency, call (919) 292-7100 day or night.

PRECAUTIONARY STATEMENTS

HAZARDS TO HUMANS AND DOMESTIC ANIMALS

CAUTION

Harmful if swallowed, inhaled, or absorbed through the skin. Causes eye irritation. Avoid contact with skin, eyes, or clothing. Avoid breathing vapors, dust, or spray mist. Remove contaminated clothing and wash before reuse. Wash thoroughly with soap and water after handling.

First Aid: If swallowed, contact your local poison control center, hospital, or physician. If the patient is unconscious, maintain breathing and heart beat (cardiopulmonary resuscitation). If the patient is conscious and alert, induce vomiting (syrup of ipecac or stimulate the back of the throat with a finger). Never give anything by mouth to an unconscious person!

If inhaled, move to clean air. If in eyes, flush with plenty of water. If on skin, wash with plenty of soap and water. If irritation persists, seek medical attention.

Note to Physician: If ingested, there is no specific antidote. Induce emesis or lavage stomach. Give saline laxative and supportive therapy.

ENVIRONMENTAL HAZARDS

Do not apply directly to any body of water. Do not contaminate water by cleaning of equipment or disposal of wastes.

Evik® trademark of CIBA-GEIGY for ametryn

Karmex® trademark of E.I. duPont de Nemours and Co. for diuron

X-77® trademark of KALO Agricultural Chemicals, Inc.

CGA 15L5N 024

GALECRON®4E

INSECTICIDE-OVICIDE

RESTRICTED USE PESTICIDE

For retail sale to and use only by Certified Applicators or persons under their direct supervision and only for those uses covered by the Certified Applicator's certification

For control of certain insects on cotton

NOTE: This product has been shown to cause cancer in laboratory animals. The user must read and follow all precautionary statements and instructions for use.

ACTIVE INGREDIENTS:	
Chlordimeform:	
N'-(4-chloro-*o*-tolyl)-*N,N*-dimethyl-formamidine	48.5%
Xylene range aromatic solvent	43.4%
INERT INGREDIENTS:	8.1%
TOTAL:	100.0%

EPA Reg. No. 100-551 EPA Est. 100-AL-1
Control No. 52100

Galecron® trademark of CIBA–GEIGY for chlordimeform
U.S. Patent Nos. 3,378,437; 3,502,720; 3,487,156; 3,629,460; 3,462,537

Galecron 4E contains 4 lbs. chlordimeform per gallon

Keep Out of Reach of Children.
WARNING

See additional precautionary statements left of container seam

See directions for use on panel at right.

DIRECTIONS FOR USE AND CONDITIONS OF SALE AND WARRANTY

IMPORTANT: Read the entire **Directions for Use** and the **Conditions of Sale and Warranty** before using this product.

CONDITIONS OF SALE AND WARRANTY

The **Directions for Use** of this product reflect the opinion of experts based on field use and tests. The directions are believed to be reliable and should be followed carefully. However, it is impossible to eliminate all risks inherently associated with use of this product. Crop injury, ineffectiveness, or other unintended consequences may result because of such factors as weather conditions, presence of other materials, or the manner of use or application all of which are beyond the control of CIBA–GEIGY or the Seller. All such risks shall be assumed by the Buyer.

CIBA–GEIGY warrants that this product conforms to the chemical description on the label and is reasonably fit for the purposes referred to in the **Directions for Use** subject to the inherent risks referred to above. **CIBA–GEIGY makes no other express or implied warranty of Fitness or Merchantability or any other express or implied warranty. In no case shall CIBA–GEIGY or the Seller be liable for consequential, special, or indirect damages resulting from the use or handling of this product.** CIBA–GEIGY and the Seller offer this product, and the Buyer and user accept it, subject to the foregoing **Conditions of Sale and Warranty,** which may be varied only by agreement in writing signed by a duly authorized representative of CIBA–GEIGY.

GENERAL INFORMATION

Galecron 4E is an insecticide and ovicide.

Galecron 4E acts primarily as an ovicide for the control of bollworm and tobacco budworm on cotton. Best results can be expected from fields under Pest Management/Scouting programs.

Galecron 4E used alone will have little or no effect on boll weevil (cotton) or on predators, beneficial insects, and bees.

Mixing Instructions
Prepare no more spray mixture than is needed for the immediate operation. Thoroughly clean spray equipment before using this product. Agitate thoroughly before and during application. Good bypass agitation is sufficient to keep Galecron 4E in emulsion.

Galecron 4E Alone: Add $^2/_3$ of the required amount of water to the spray tank with the agitator running, then add the desired amount of Galecron 4E. Continue agitation while adding the remainder of the water.

Galecron 4E + Tank Mixtures: Add $^2/_3$ of the required amount of water to the spray tank. With the agitator running if possible, add the desired amount of the other products recommended on this label. Then add the desired amount of Galecron 4E. Continue agitation while adding the remainder of water.

Note: Do not let spray mixture stand overnight in the spray tank. Rinse spray tank at the end of the day.

To avoid spray drift, do not apply under windy conditions. Avoid spray overlap, as crop injury may result.

DIRECTIONS FOR USE

It is a violation of federal law to use this product in a manner inconsistent with its labeling.

FAILURE TO FOLLOW THE DIRECTIONS FOR USE AND PRECAUTIONS ON THIS LABEL MAY RESULT IN CROP INJURY, POOR INSECT CONTROL, AND/OR ILLEGAL RESIDUES.

COTTON

Apply the specified rate below in a minimum of 1 gal. of water with ground or aerial application equipment using low drift nozzles. Apply on a 3-5 day schedule when moth flights begin and/or when eggs appear. Continue applications as long as egg laying occurs. Apply the lower rate when insect populations are low or egg deposition is just starting. If larvae reach economic levels, use a tank mix as described below.

Note: Do not apply just before a rain or poor control may result.

Galecron 4E Alone (All cotton growing states except CA and AZ)

Bollworm: $^1/_4$-$^1/_2$ pt./A

Tobacco budworm: $^1/_2$ pt./A

Galecron 4E + Tank Mixtures (All cotton growing states)
Galecron 4E acts primarily as an ovicide for control of bollworm and/or tobacco budworm on cotton. Best results can be expected in fields under a regular scouting program. In some cases it may be advantageous to apply Galecron 4E in a tank mix with other insecticides; for example, when egg hatch occurs because of delayed application or other causes, or when it is necessary to begin application in a field where bollworm/tobacco budworm larvae or other insect pests listed on the other product labels have reached economic levels. Use the above rates of Galecron 4E per acre except in CA and AZ where a minimum of $^1/_2$ pt. is recommended. Mix Galecron 4E with other insecticides recommended on this label according to the above mixing directions. Apply tank mixes immediately after mixing.

Galecron 4E has been found to be effective and compatible in tank mixture with the listed products. When using Galecron 4E in tank mixture, observe all limitations, precautions, and rate recommendations which appear on the labels for these products. Under conditions of low to moderate bollworm/budworm infestation levels and in closely scouted and supervised fields, the pyrethroid products may be tank mixed at the lowest labeled rate.

Ambush®
Ammo®
Azodrin®
Bacillus thuringiensis
Bolstar®
Bidrin®
Curacron®
Cymbush®
Dimethoate
Guthion®
Imidan®
Lorsban®
Methomyl
Methyl parathion
Methyl parathion + toxaphene
Orthene®
Pay-Off®
Dimilin®
Elcar®
EPN + Methyl parathion
Pounce®
Pydrin®
Supracide®

Notes: To avoid illegal residues, (1) Do not apply within 21 days of picking. (2) Do not feed foliage from treated cotton plants or gin trash to livestock. (3) Do not allow livestock to graze treated areas.

Yield Enhancement (All Cotton Growing Areas)
Applications of Galecron at 5-7 day intervals have resulted in cotton yields above that expected from insect control. For a yield enhancement effect, apply Galecron 4E at $^1/_4$-$^1/_2$ pt./A 4-6 times beginning at first pin head square formation (5-7 true leaf stage).

Note: Yield increase may not occur if plants are under stress during treatment. Applications may be combined with other insecticides as needed.

STORAGE AND DISPOSAL

Do not contaminate water, food, or feed by storage or disposal. Open dumping is prohibited.

STORAGE
Do not store at temperatures below 32°F, since crystals may form. If allowed to freeze, place in a warm room (50-80°F) and roll and shake the container frequently for several days until crystals are dissolved.

PESTICIDE DISPOSAL
Pesticide wastes may be acutely toxic. Improper disposal of unused pesticide, spray mixture, or rinse water is a violation of federal law. If these wastes cannot be used according to label instructions, contact your State Pesticide or Environmental Control Agency, or the Hazardous Waste representative at the nearest EPA Regional Office for guidance in proper disposal methods.

CONTAINER DISPOSAL
Do not reuse this container. Triple rinse (or equivalent) and offer for recycling or reconditioning, or puncture and dispose of in a sanitary landfill or by other procedures approved by state and local authorities.

SPECIAL PRECAUTIONS TO PREVENT EXPOSURE

Wear minimum protective clothing when working in the loading area. Wear special protective clothing as indicated below to ensure against possible exposure during all operations associated with handling, mixing, and application.

Minimum Protective Clothing and Equipment: Heavy-duty, long-sleeve, one-piece work suit; cloth cap; underwear including T-type undershirt; heavy-duty fabric, rubber or neoprene-type gloves; socks; heavy-duty work boots (waterproofed), and goggles or face shield.

- Wear clean clothes each day and launder before reusing. Place contaminated clothing in a specially designated container until laundering.
- Keep a complete clean change of clothes at the work site for use if needed.
- Replace boots and gloves if contaminated.

SPECIAL PROTECTIVE CLOTHING AND EQUIPMENT

- Wear rubber boots (slip-on, knee-length, waterproof) and heavy-duty rubber or neoprene gloves when handling Galecron 4E during wet conditions, for limited cleaning or adjustment of equipment, in the case of a minor spill, or when otherwise appropriate for added protection.
- Wear a complete waterproof rubberized or plastic rain suit with head covering when cleaning or handling application equipment or the operation area, in the case of a spill or accident, or when otherwise appropriate for maximum protection.
- Wear an approved pesticide respirator, such as Willson 1200 facepiece equipped with Willson R-21 cartridges and R-15 filters, during application of Galecron 4E, during cleanup operations, in the event of spills or accidents, or when otherwise exposed to vapors or spray mist. Clean mask after each use and

change cartridges and filters daily not to exceed 8 hours of use, or if breathing becomes difficult.

- Clean all protective clothing and equipment before reuse.

PERSONAL PRECAUTIONS

- If Galecron 4E concentrate or spray gets on clothing, remove contaminated clothing immediately and wash affected parts of body with soap and water. If there is any question as to the extent of contamination, bathe entire body thoroughly. Change to clean clothing.
- Wash hands with soap and water before eating, drinking, or smoking.
- At the end of the workday, bathe entire body with soap and plenty of water.

SPECIAL EQUIPMENT

- Galecron 4E must be transported, transferred, and mixed using closed system equipment. A closed system is defined as appropriate connections, meters, pumps, and plumbing designed to eliminate human contact. Do not use open mixing vats or tanks, or open pouring.
- All plumbing connections must be sufficiently tight to prevent leakage.
- Use hoses constructed of chemically resistant, heavy-duty, reinforced materials. Avoid excessive pressures to minimize the possibility of rupture. If any sign of weakness appears, flush the hose thoroughly with clean water and replace immediately.
- Transport and storage vessels, and transfer equipment must be equipped with dry-break valves.
- Any equipment which has been contaminated with Galecron 4E must be thoroughly cleaned and flushed with clean water prior to maintenance or use for other purposes. Wear appropriate protective clothing when performing equipment maintenance operations.
- Spray equipment must be equipped with low drift nozzles.

ACCIDENTS

- For minor spills, leaks, etc., follow all precautions indicated above and clean up immediately. Take special care to avoid contamination of equipment and facilities during cleanup procedures and disposal of wastes.
- In the event of a major spill or accident, contact CIBA–GEIGY for assistance at (919) 292-7100 day or night.

APPLICATION AND REENTRY

- Do not apply this product in any manner that would directly or through drift expose unprotected persons. Unprotected persons must vacate the area being treated.
- Do not apply near an occupied dwelling.
- Do not enter treated areas for at least one day (24 hrs.) after application.

PRECAUTIONARY STATEMENTS

HAZARDS TO HUMANS AND DOMESTIC ANIMALS

WARNING

May be harmful or fatal if swallowed or absorbed through skin. Causes substantial but temporary eye irritation. Do not get in eyes, on skin or on clothing. Wear goggles or face shield when handling and protective clothing as specified. Avoid breathing of vapors or spray mist. Wash thoroughly with soap and water after handling and before eating, drinking, or smoking. Remove contaminated clothing immediately and wash before reuse.

STATEMENT OF PRACTICAL TREATMENT

If swallowed, call a physician or Poison Control Center immediately. DO NOT INDUCE VOMITING UNLESS NO OTHER TREATMENT IS AVAILABLE. Vomiting may cause aspiration pneumonia. If it is necessary to induce vomiting, do so under medical supervision. Do not induce vomiting or give anything by mouth to an unconscious person.

If in eyes, immediately flush with plenty of water. Get medical attention.

If on skin, immediately remove contaminated clothing, wiping away excess material from skin. Wash skin with plenty of soap and water. Get medical attention.

If inhaled, move to fresh air.

Note to Physician: There is no specific antidote for Galecron ingestion. Do not induce emesis. Lavage stomach, give a saline laxative and supportive therapy. Sympathomimetic drugs should be avoided as these may have potentiated effects (e.g. hypertensive crises).

ENVIRONMENTAL HAZARDS

This product is toxic to fish and wildlife. Do not contaminate any body of water by direct application, cleaning of equipment, or disposal of wastes. Do not apply where runoff is likely to occur. Do not apply when weather conditions favor drift from areas treated.

Physical or Chemical Hazards

Do not use, pour, spill or store near heat or open flame.

Ambush® trademark of ICI Americas, Inc.
Ammo® trademark of ICI Americas, Inc.
Azodrin® trademark of Shell Chemical Company
Bolstar® trademark of Mobay Chemical Corporation
Bidrin® trademark of Shell Chemical Company
Curacron® trademark of CIBA-GEIGY for profenofos
U.S. Patent No. 3,992,533
Cymbush® trademark of ICI Corporation
Dimilin® trademark of Thompson-Hayward Chemical Company
Elcar® trademark of Sandoz, Inc.
Galecron® trademark of CIBA-GEIGY for chlordimeform
U.S. Patent Nos. 3,378,437; 3,502,720; 3,487,156; 3,629,460; 3,462,537
Guthion® trademark of Mobay Chemical Corporation
Imidan® trademark of Stauffer Chemical Company
Lorsban® trademark of Dow Chemical Company
Orthene® trademark of Chevron Chemical Company
Pay-Off® trademark of American Cyanamid
Pounce® trademark of FMC Corporation
Pydrin® trademark of Shell Chemical Company
Supracide® trademark of CIBA-GEIGY

CGA 35L31D 045

GALECRON®SP
INSECTICIDE-OVICIDE

RESTRICTED USE PESTICIDE

For retail sale to and use only by certified applicators or persons under their direct supervision, and only for those uses covered by the certified applicator's certification.

For control of certain insects on cotton

Galecron SP is a soluble powder

NOTE: This product has been shown to cause cancer in laboratory animals. The user must read and follow all precautionary statements and instructions for use.

This outer protective carton contains Galecron SP in an inner water-soluble bag. Entire inner bag and contents dissolve in water. After opening outer carton, immediately dump entire unopened inner bag into the partially filled sprayer tank or closed-system mix tank. Do not handle soluble bag or expose it to moisture, since this may cause rupturing.

ACTIVE INGREDIENT:
Chlordimeform hydrochloride:
N'-(4-chloro-*o*-tolyl)-*N,N*-dimethyl-
formamidine mono-hydrochloride 95%
INERT INGREDIENTS: 5%

TOTAL 100%

EPA Reg. No. 100-554

Galecron® trademark of CIBA-GEIGY for chlordimeform
U.S. Patent Nos. 3,378,437; 3,502,720; 3,487,156; 3,629,460; 3,462,537

Keep Out of Reach of Children.
WARNING

PRECAUTIONARY STATEMENTS

Hazards to Humans and Domestic Animals
May be harmful or fatal if swallowed or absorbed through skin. Do not get in eyes, on skin or on clothing. Wear goggles or face shield when handling and other protective clothing as specified. Do not contaminate food or feed. Avoid breathing dust or spray mist. Wash thoroughly with soap and water after handling and before eating, drinking, or smoking. Remove contaminated clothing immediately and wash before reuse.

STATEMENT OF PRACTICAL TREATMENT

If swallowed, contact a physician or Poison Control Center immediately. Promptly drink at least 1 to 2 glasses of water and induce vomiting by touching back of throat with finger. Do not induce vomiting or give anything by mouth to an unconscious person.

If on skin, immediately remove contaminated clothing, wiping away excess material from skin. Wash skin with plenty of soap and water. Get medical attention.

If in eyes, immediately flush with plenty of water. Get medical attention.

If inhaled, move to fresh air.

Note to Physician: There is no specific antidote for Galecron ingestion. Induce emesis or lavage stomach. Give a saline laxative and supportive therapy. Sympathomimetic drugs should be avoided as these may have potential effects (e.g., hypertensive crisis).

ENVIRONMENTAL HAZARDS

This product is toxic to fish and wildlife. Do not contaminate any body of water by direct application, cleaning of equipment, or disposal of wastes and containers. Do not apply where runoff is likely to occur. Do not apply when weather conditions favor drift from areas treated.

See directions for use on panel at right.

DIRECTIONS FOR USE AND CONDITIONS OF SALE AND WARRANTY

IMPORTANT: Read the entire **Directions for Use** and the **Conditions of Sale and Warranty** before using this product.

CONDITIONS OF SALE AND WARRANTY

The **Directions for Use** of this product reflect the opinion of experts based on field use and tests. The directions are believed to be reliable and should be followed carefully. However, it is impossible to eliminate all risks inherently associated with use of this product. Crop injury, ineffectiveness, or other unintended consequences may result because of such factors as weather conditions, presence of other materials, or the manner of use or application all of which are beyond the control of CIBA-GEIGY or the Seller. All such risks shall be assumed by the Buyer.

CIBA-GEIGY warrants that this product conforms to the chemical description on the label and is reasonably fit for the purposes referred to in the **Directions for Use** subject to the inherent risks referred to above. **CIBA-GEIGY makes no other express or implied warranty of Fitness or Merchantability or any other express or implied warranty. In no case shall CIBA-GEIGY or the Seller be liable for consequential, special, or indirect damages resulting from the use or handling of this product.** CIBA-GEIGY and the Seller offer this product, and the Buyer and user accept it, subject to the foregoing **Conditions of Sale and Warranty,** which may be varied only by agreement in writing signed by a duly authorized representative of CIBA-GEIGY.

GENERAL INFORMATION

Galecron SP is an insecticide and ovicide.

Galecron SP acts primarily as an ovicide for the control of bollworm and tobacco budworm on cotton. Best results can be expected from fields under Pest Management/Scouting programs.

Galecron SP where used alone will have little or no effect on boll weevil (cotton) or on predators, beneficial insects, and bees.

Mixing Instructions
Prepare no more spray mixture than is needed for the immediate operation. Thoroughly clean spray equipment before using this product. Agitate thoroughly before and during application. Good bypass agitation is sufficient to keep Galecron SP thoroughly mixed. Flush spray tank thoroughly with water daily after use and dispose of pesticide rinsate by application to an already treated area.

Galecron SP Alone: Add $^2/_3$ of the required amount of water to the spray or mixing tank. With the agitator running, drop in the required number of **unopened** $2^1/_2$ lb. soluble bags of Galecron SP in the tank. Continue agitation while adding the remainder of water. If mixing is done in a separate mixing tank, the Galecron SP dilution must be transferred to the sprayer tank by a closed system.

Galecron SP + Tank Mixtures: Add $^2/_3$ of the required amount of water to the spray or mixing tank. With the agitator running if possible, drop in the required number of **unopened** $2\ ^1/_2$ lb. soluble bags of Galecron SP in the tank. Then add the desired amount of the other products recommended on the Galecron SP label. Continue agitation while adding the remainder of water. If mixing is done in a separate mixing tank, the Galecron SP tank mixtures must be transferred to the sprayer tank by a closed system.

Note: Do not let spray mixture stand overnight in the spray tank. Rinse spray tank at the end of the day.

It is not necessary to use a spreader-sticker with Galecron SP. However, if one is used, it is suggested that the tolerance of the crop to such a mixture be determined by first testing it on a small area.

To avoid spray drift, do not apply under windy conditions. Avoid spray overlap, as crop injury may result.

DIRECTIONS FOR USE

It is a violation of federal law to use this product in a manner inconsistent with its labeling.

FAILURE TO FOLLOW THE DIRECTIONS FOR USE AND PRECAUTIONS ON THIS LABEL MAY RESULT IN CROP INJURY, POOR INSECT CONTROL AND/OR ILLEGAL RESIDUES.

COTTON

Apply the specified rate below in a minimum of 1 gal. of water with ground or aerial application equipment only, using low drift nozzles. Apply on a 3-5 day schedule when moth flights begin and/or when eggs appear. Continue applications as long as egg laying occurs. Apply the lower rate when insect populations are low or egg deposition is just starting. If larvae reach economic levels, use a tank mix as described below.

Note: Do not apply just before rain or poor control may result.

Galecron SP Alone (All cotton growing states except CA and AZ)

Pest	Rate of Application
Bollworm	One bag treats 20 acres (at 1/8 lb./acre) or 10 acres (at 1/4 lb./acre)
Tobacco budworm	One bag treats 10 acres (at 1/4 lb./acre)

Galecron SP + Tank Mixtures (All cotton growing states)

Galecron SP acts primarily as an ovicide for control of bollworm and/or tobacco budworm on cotton. Best results can be expected in fields under a regular scouting program. In some cases it may be advantageous to apply Galecron SP in a tank mix with other insecticides; for example, when egg hatch occurs because of delayed application or other causes, or when it is necessary to begin application in a field where bollworm/tobacco budworm larvae or other pests listed on the other product labels have reached economic levels. Use the above rates of Galecron SP per acre except in CA and AZ where a minimum of 1/4 lb. is recommended. Mix Galecron SP with other insecticides recommended on this label for control of other insect pests according to the above mixing directions. Apply tank mixes immediately after mixing.

Galecron SP has been found to be effective and compatible in tank mixture with the listed products. When using Galecron SP in tank mixture, observe all limitations, precautions, and rate recommendations which appear on the labels for these products. Under conditions of low to moderate bollworm/budworm infestation levels and in closely scouted and supervised fields, the pyrethroid products may be tank mixed at the lowest labeled rate.

Ambush®	Guthion®
Ammo®	Imidan®
Azodrin®	Lorsban®
Bacillus thuringiensis	Methomyl
	Methyl parathion
Bolstar®	Methyl parathion + toxaphene
Bidrin®	
Curacron®	Orthene®
Cymbush®	Pay-Off®
Dimethoate	Pounce®
Dimilin®	Pydrin®
Elcar®	Supracide®
EPN + Methyl parathion	

Notes: To avoid illegal residues, (1) Do not apply within 21 days of picking. (2) Do not feed foliage from treated cotton plants or gin trash to livestock. (3) Do not allow livestock to graze treated areas.

Yield Enhancement

(All Cotton Growing Areas)

Applications of Galecron at 5-7 day intervals have resulted in cotton yields above that expected from insect control. For a yield enhancement effect, apply Galecron SP at 1/8-1/4 lb./A 4-6 times beginning at first pin head square formation (5-7 true leaf stage).

Note: Yield increase may not occur if plants are under stress during treatment. Applications may be combined with other insecticides as needed.

SPECIAL PRECAUTIONS TO PREVENT EXPOSURE

Wear minimum protective clothing when working in the loading area. Wear special protective clothing as indicated below to ensure against possible exposure during all operations associated with handling, mixing, and application.

Minimum Protective Clothing and Equipment: Heavy-duty, long-sleeve, one-piece work suit; cloth cap; underwear including T-type undershirt; heavy-duty fabric, rubber or neoprene-type gloves; socks; heavy-duty work boots (waterproofed), and goggles or face shield.

- Wear clean clothes each day and launder before reusing. Place contaminated clothing in a specially designated container until laundering.
- Keep a complete clean change of clothes at the work site for use if needed.
- Replace boots and gloves if contaminated.

Special Protective Clothing and Equipment

- Wear rubber boots (slip-on, knee-length, waterproof) and heavy-duty rubber or neoprene gloves when handling Galecron SP during wet conditions, for limited cleaning or adjustment of equipment, in the case of a minor spill, or when otherwise appropriate for added protection.
- Wear a complete waterproof rubberized or plastic rain suit with head covering when cleaning or handling application equipment or the operation area, in the case of a spill or accident, or when otherwise appropriate for maximum protection.
- Wear an approved pesticide respirator, such as Willson 1200 facepiece equipped with Willson R-21 cartridges and R-15 filters, during application of Galecron SP, during cleanup operations, in the event of spills or accidents, or when otherwise exposed to vapors or spray mist. Clean mask after each use and change cartridges and filters daily not to exceed 8 hours of use, or if breathing becomes difficult.
- Clean all protective clothing and equipment before reuse.

Personal Precautions

- If Galecron SP powder, mixture or spray gets on clothing, remove contaminated clothing immediately and wash affected parts of body with soap and water. If there is any question as to the extent of contamination, bathe entire body thoroughly. Change to clean clothing.
- Wash hands with soap and water before eating, drinking, or smoking.
- At the end of the workday, bathe entire body with soap and plenty of water.

Special Equipment

- If Galecron SP is mixed in tanks separate from the applicator (e.g. sprayer tank), transfer from the mixing tank to the sprayer tank must be done by a closed system. A closed system is defined as appropriate connections, meters, pumps, and plumbing designed to eliminate human contact. Do not use open mixing vats. After introduction of Galecron SP into the mixing tank, close lids and open only if prescribed protective clothing is in use.
- All plumbing connections must be sufficiently tight to prevent leakage.
- Use hoses constructed of chemically resistant, heavy-duty, reinforced materials. Avoid excessive pressures to minimize the possibility of rupture. If any sign of weakness appears, flush the hose thoroughly with clean water and replace immediately.
- Transport and storage vessels, and transfer equipment must be equipped with dry-break valves.
- Any equipment which has been contaminated with Galecron SP must be thoroughly cleaned and flushed with clean water prior to maintenance or use for other purposes. Wear appropriate protective clothing when performing equipment maintenance operations.
- Spray equipment must be equipped with low drift nozzles.

Accidents

- For minor spills, leaks, etc., follow all precautions indicated above and clean up immediately. Take special care to avoid contamination of equipment and facilities during cleanup procedures and disposal of wastes.
- In the event of a major spill or accident, contact CIBA-GEIGY for assistance at (919) 292-7100 day or night.

Application and Reentry

- Do not apply this product in any manner that would directly or through drift expose unprotected persons. Unprotected persons must vacate the area being treated.
- Do not apply near an occupied dwelling.
- Do not enter treated areas for at least one day (24 hrs.) after application.

STORAGE AND DISPOSAL

Do not contaminate water, food, or feed by storage or disposal. Open dumping is prohibited.

Storage
Do not store this product under wet conditions. Handle carton carefully when stored at temperatures of less than 50°F to avoid breakage of inner soluble bag.

Pesticide Disposal
Dispose of pesticide rinsate by application to an already treated area. Pesticide, spray mixture, or rinsate that cannot be used according to label instructions must be disposed of according to federal, state, or local procedures. Contact your State Pesticide or Environmental Control Agency or the Hazardous Waste representative at the nearest EPA Regional Office for guidance. Because pesticide wastes may be toxic, improper disposal is a violation of Federal Law.

Container Disposal
Do not reuse outer carton. Dispose of empty outer carton in an incinerator or in a sanitary landfill approved for pesticide containers or bury in a safe place. Consult federal, state, or local disposal authorities for approved alternative procedures such as limited open burning.

Ambush® trademark of ICI Americas, Inc.
Ammo® trademark of ICI Americas, Inc.
Azodrin® trademark of Shell Chemical Co.
Bolstar® trademark of Mobay Chemical Corp.
Bidrin® trademark of Shell Chemical Co.
Curacron® trademark of CIBA-GEIGY for profenofos
U.S. Patent No. 3,992,533
Cymbush® trademark ICI Corporation
Dimilin® trademark of Thompson-Hayward Chemical Company
Elcar® trademark of Sandoz, Inc.
Galecron® trademark of CIBA-GEIGY for chlordimeform
U.S. Patent Nos. 3,378,437; 3,502,720; 3,487,156; 3,629,460; 3,462,537
Guthion® trademark of Mobay Chemical Corp.
Imidan® trademark of Stauffer Chemical Co.
Lorsban® trademark of Dow Chemical Co.
Orthene® trademark of Chevron Chemical Co.
Pay-Off® trademark of American Cyanamid
Pounce® trademark of FMC Corporation
Pydrin® trademark of Shell Chemical Co.
Supracide® trademark of CIBA-GEIGY

CGA 35L32F 045

IGRAN®80W
HERBICIDE

For weed control in winter wheat, winter barley, grain sorghum, and noncrop land areas

Igran 80W is a wettable powder

ACTIVE INGREDIENTS:
Terbutryn:
2-*tert*-butylamino-4-ethylamino-6-methylthio-*s*-triazine 76%
Related compounds 4%
INERT INGREDIENTS: 20%
TOTAL: 100%

EPA Reg. No. 100-496 EPA Est. 100-AL-1

Igran® trademark of CIBA-GEIGY for terbutryn U.S. Patent No. 3,634,062

Keep Out of Reach of Children.
WARNING

See additional precautionary statements on left panel.

See directions for use on panel at right.

DIRECTIONS FOR USE AND CONDITIONS OF SALE AND WARRANTY

IMPORTANT: Read the entire **Directions for Use** and the **Conditions of Sale and Warranty** before using this product.

CONDITIONS OF SALE AND WARRANTY

The **Directions for Use** of this product reflect the opinion of experts based on field use and tests. The directions are believed to be reliable and should be followed carefully. However, it is impossible to eliminate all risks inherently associated with use of this product. Crop injury, ineffectiveness, or other unintended consequences may result because of such factors as weather conditions, presence of other materials, or the manner of use or application all of which are beyond the control of CIBA-GEIGY or the Seller. All such risks shall be assumed by the Buyer.

CIBA-GEIGY warrants that this product conforms to the chemical description on the label and is reasonably fit for the purposes referred to in the **Directions for Use** subject to the inherent risks referred to above. **CIBA-GEIGY makes no other express or implied warranty of Fitness or Merchantability or any other express or implied warranty. In no case shall CIBA-GEIGY or the Seller be liable for consequential, special, or indirect damages resulting from the use or handling of this product.** CIBA-GEIGY and the Seller offer this product, and the Buyer and user accept it, subject to the foregoing **Conditions of Sale and Warranty,** which may be varied only by agreement in writing signed by a duly authorized representative of CIBA-GEIGY.

DIRECTIONS FOR USE

It is a violation of federal law to use this product in a manner inconsistent with its labeling.

GENERAL INFORMATION

Igran 80W is a selective herbicide for control of many annual broadleaf weeds and grasses. The weeds controlled are as follows:

Wheat and Barley

bedstraw
blue lips
bur buttercup
catchweed
chickweed (common and umbellate)
China lettuce
coastal fiddleneck (tarweed)
corn gromwell
Jacob's ladder
mayweed (dogfennel)
miners lettuce
Pacific meadow foxtail
pineappleweed
prickly lettuce
prostrate knotweed
redstem filaree

corn spurry	shepherdspurse
cow cockle	speedwell (snow weed)
falseflax *(Camelina)*	spring beauty
field pennycress	Jim Hill mustard*
henbit	blue mustard*

Sorghum

annual groundcherry	pigweed
carpetweed	puncturevine
crabgrass	ragweed
foxtail	Russian thistle
junglerice	smartweed
kochia	velvetleaf
lambsquarters	barnyardgrass (watergrass)
morningglory	

Noncrop Areas

barnyardgrass	kochia
blue mustard	nightshade
carpetweed	pepperweed
chickweed	pigweed
China lettuce	purslane
cocklebur	tarweed
dogfennel	teaweed
fiddleneck	tumble mustard
field pennycress	tumbleweed
green foxtail*	white sweet clover
gromwell	wild buckwheat*
henbit	wild mustard

* Some mustards (Jim Hill and blue mustard) can be effectively controlled if treated while they are very small. However, late germinating mustard seedlings may not be controlled. In fallowed noncrop areas, Igran gives partial control of green foxtail and wild buckwheat.

Following many years of continuous use of this product and chemically related products, biotypes of some of the weeds listed on this label have been reported which cannot be effectively controlled by this and related herbicides. Where this is known or suspected, and weeds controlled by this product are expected to be present along with resistant biotypes, we recommend the use of this product in registered combinations or in sequence with other registered herbicides which are not triazines. If only resistant biotypes are expected to be present, use a registered nontriazine herbicide. Consult with your State Agricultural Extension Service for specific recommendations.

When applied before weeds emerge, Igran 80W enters weeds through their roots after germination. Preemergence effectiveness depends, therefore, on rainfall or irrigation to move Igran 80W into the soil. Very dry soil conditions may necessitate a shallow cultivation. When applied to emerged weeds, Igran 80W provides foliar knockdown of existing weeds and residual control of later germinating weeds.

In each case where a range of rates is given, the lower rate should be used on relatively coarse-textured soils or soils low in organic matter; the higher rate should be used on relatively fine-textured soils or soils high in organic matter.

For Igran 80W alone on winter wheat or sorghum, nonpressurized liquid fertilizer may replace all or part of the water as carrier for ground application.

FAILURE TO FOLLOW THE PRECAUTIONS AND DIRECTIONS ON THIS LABEL MAY RESULT IN CROP INJURY, POOR PERFORMANCE, AND/OR ILLEGAL RESIDUES.

MIXING INSTRUCTIONS

Igran 80W is a wettable powder formulation to be mixed with water or nonpressurized liquid fertilizer solutions and applied as a spray. For sprayers or nurse tanks which have continuous agitation (mechanical or hydraulic), Igran 80W does not have to be premixed as a slurry. Fill the tank ½ full of water. Start agitation and add the proper amount of Igran 80W to the tank. Allow the Igran 80W time to wet. Continue to fill the tank with water. Continue agitation for several minutes prior to spraying, as well as during the spraying operation.

For sprayers which have agitation dependent upon the forward motion of the spray rig, the Igran 80W should be premixed as a slurry. Make a slurry by adding Igran 80W to a small amount of water in a separate container. Then pour the slurry into the spray tank partially filled with clean water, and add the rest of the water. The forward motion of the spray rig should not stop until the tank is empty as agitation is required during the spraying operation to keep the material in suspension.

It is best to empty the spray tank completely before shutting off the pump and agitator. However, if the spray mixture stands in the tank before use, agitate for several minutes to obtain a uniform suspension before starting to spray. Wash sprayer thoroughly with clean water immediately after use.

Ground Application: For uniform distribution, use conventional spray equipment with fan-type nozzles. Screens in nozzles and in suction and in-line strainers should be no finer than 50-mesh. Use hydraulic or mechanical agitation during mixing and application to maintain a uniform suspension. Use a pump with capacity to maintain 35-40 psi at the nozzles. If hydraulic agitation is used, the pump should also provide sufficient agitation in the tank to keep the mixture in suspension. Unless otherwise specified, use a minimum of 20 gals. of water or nonpressurized liquid fertilizer per acre for ground application.

For band applications in sorghum, calculate the amount to be applied per acre as follows:

$$\frac{\text{Band width in inches}}{\text{Row width in inches}} \times \text{Rate/A for broadcast treatment} = \text{Amount needed for band treatment}$$

Aerial Application: Use aerial application only where broadcast applications are specified. Apply Igran 80W alone or in combinations in a minimum of 5 gals. of water per acre. Avoid applications under conditions where uniform coverage cannot be obtained or where excessive spray drift may occur. In order to assure that spray will be controllable within the target area when used according to label directions, make applications at a maximum height of 10 ft., using low drift nozzles at a maximum pressure of 40 psi, and restrict application to periods when wind speed does not exceed 10 mph. To assure that spray will not adversely affect adjacent sensitive nontarget plants, apply Igran 80W alone, or in combination with Milogard®, by aircraft at a minimum upwind distance of 300 ft. from sensitive plants, or apply Igran 80W + AAtrex® and Igran 80W + AAtrex + paraquat at a minimum upwind distance of 500 ft. from sensitive plants. For additional precautions, see Precautionary Statements section at the end of the label.

WINTER WHEAT AND BARLEY

Preemergence (Winter Wheat Only — Washington, West of the Cascades Only)

To avoid crop injury apply to winter wheat after planting and before emergence.

Igran 80W Alone: Apply at rates listed in Table 1.

Postemergence (Winter Wheat — Washington, Oregon, Idaho and Utah; Winter Barley — Washington, Oregon, and Idaho)

Apply only after the crop is fully tillered and as soon as the crop has well developed secondary (crown) roots (minimum two inches in length) but prior to jointing. If the crop does not reach this stage of development in the fall, Igran 80W alone or tank mixed with Sencor® or Lexone® may be applied in early spring. Maximum weed control effectiveness is achieved from early applications; however, treatment prior to the development of the secondary root system may result in increased winter injury in the event of severe weather conditions.

Igran 80W Alone: For effective weed control, apply at rates in Table 1 before weed rosettes have a 3 in. diameter or weeds exceed 4 in. in height.

Igran 80W + Sencor (50WP, 4, DF) or Lexone (50W, 4L, DF) Tank mix Igran 80W + Sencor or Lexone at rates in Table 2 to control many of the weeds listed on the Igran and Sencor or Lexone labels. Refer to each label for the list of weeds controlled, precautions, and variety and geographical restrictions. This tank mix will only give partial control of downy brome. For optimum control, apply prior to the three-tiller stage of downy brome. *Note:* To avoid crop injury, do not use on irrigated wheat or barley.

Igran 80W + 2,4-D (amines, salts, and esters), MCPA amines, salts, and esters, bromoxynil (Brominal or Buctril), Banvel, or Glean Tank mix Igran 80W with 2,4-D (amines, salts, and esters), MCPA (amines, salts and esters), bromoxynil (Brominal or Buctril), Banvel, or Glean at rates in Table 3. Observe all directions and precautions on the labels of the tank mix components. Tank mixtures of Igran 80W with emulsifiable concentrates (EC) may cause temporary yellowing of wheat or barley when applied under conditions of freezing nights and warm days. When tank mixing with EC formulations, Igran 80W should be dispersed in the spray tank before adding the EC. Compatibility agents added to the tank after Igran 80W and before the EC will improve compatibility.

Table 1

Winter Wheat — Postemergence: Washington, Oregon, Idaho, Utah
— Preemergence*: Washington, West of Cascades Only
Winter Barley — Postemergence Only: Washington, Oregon, Idaho

Lbs./A of Igran 80W

Soil Texture	Wheat	Barley
Sand, loamy sand	DO NOT USE	DO NOT USE
Sandy loam, silts, clay loam	1.50-2.75 lbs.	1.50-2.25 lbs.

* Preemergence applications of Igran 80W may be mixed with either water or nonpressurized fertilizer solutions as a carrier on wheat. Nonpressurized liquid fertilizer solutions should not be used as a carrier for Igran 80W on barley or injury may occur.

Table 2

Tank Mix—Postemergence
Winter Wheat (Washington, Oregon, Idaho, Utah)
Winter Barley (Washington, Oregon, Idaho)

Soil Texture	Lbs./A Igran 80W	Lbs./A Sencor DF or Lexone DF*
Sand, loamy sand	DO NOT USE	DO NOT USE
Sandy loam, silts, clay loam	0.75-1.00	0.33

* 0.33 lb./A Sencor or Lexone DF is equivalent to 0.5 pt./A Sencor 4 or Lexone 4L, or 0.5 lb./A Sencor 50WP or Lexone 50W.

Table 3

Tank Mixes—Postemergence
Winter Wheat (Washington, Oregon, Idaho, Utah)
Winter Barley (Washington, Oregon, Idaho)

Soil Texture	Lbs. Active Ingredient/A					
	Igran*	2,4-D and MCPA amines and salts	2,4-D and MCPA esters	bromoxynil	Banvel	Glean
Sand, loamy sand	DO NOT USE	DO NOT USE	DO NOT USE	DO NOT USE	DO NOT USE	DO NOT USE
Sandy loam, silts, clay loam	0.8 lb.	0.5 lb.	0.25 lb.	0.25 lb.	0.125 lb.	0.25 oz.

* 0.8 lb. active ingredient/A is equivalent to 1.0 lb. Igran 80W/A.

Precautions: *Wheat and Barley (Postemergence) — To avoid crop injury: 1) Do not apply before there is a well developed secondary root system (crown roots) on the wheat or barley plants or crop damage may occur. 2) Do not make more than one application per crop cycle. 3) Do not make postemergence applications after temperatures exceed 70°F or crop damage may occur. 4) To avoid crop injury and illegal residues, fields on which Igran 80W has been applied should not be planted to any crop other than wheat or barley for 9 months following application. 5) Wheat and barley growing on thinly covered or exposed subsoil areas (clay knobs), are more susceptible to yellowing from applications of Igran 80W. This yellowing is usually temporary.*

Note: Do not feed green forage or graze treated areas to livestock for six months following application.

GRAIN SORGHUM

Igran 80W may be applied alone or in tank mix combination with either AAtrex or Milogard prior to planting with shallow incorporation, at planting time, or immediately following planting, but before weeds and sorghum emerge. Igran 80W may also be applied in tank mix combination with AAtrex and paraquat where grain sorghum will be planted directly into cover crop, established sod, or in previous crop residues. Apply during or after planting, but before sorghum emerges. Do not apply to emerged sorghum. When applying preplant with shallow incorporation, apply within 2 weeks of planting and incorporate into the top 1-2 inches of soil using a rolling cultivator, rotary hoe, spike-tooth harrow, or similar implement. For sorghum planted on beds, apply and incorporate after bed formation.

In case of planting failures, sorghum may be replanted into soil previously treated with Igran 80W or Igran combinations. Do not apply a second broadcast application, or injury may occur. If Igran 80W or Igran combinations are applied in a band, and sorghum is replanted in the untreated row middles, Igran 80W or Igran combinations may be applied in a band to the second planting.

If irrigation is necessary, irrigate during the first 36 hours after planting or wait until the sorghum is at least two inches tall, or injury may occur.

Igran 80W Alone: Apply Igran 80W at the rate indicated in Table 4 or 5.

Table 4

Grain Sorghum

Annual Broadleaf and Grass Weed Control

Region	Soil Texture	Broadcast Rate Per Acre
All sorghum growing regions of the U.S.; except states shown below	sand, loamy sand	DO NOT USE
	sandy loam to loam	2.5 lbs.
	silt loam to clay	3 lbs.
Eastern Colorado, South Dakota, Illinois, Iowa, Kansas, Minnesota, Missouri, and Nebraska	sand	DO NOT USE
	loamy sand	2-2.5 lbs.
	sandy loam to loam	2.5 lbs.
	silt loam to clay	3 lbs.

Table 5

Grain Sorghum		
Annual Broadleaf Weed Control Only		
Region	**Soil Texture**	**Broadcast Rate Per Acre**
Kansas, Oklahoma, New Mexico, Texas, and the desert regions of Arizona and California	sand	1 lb.
	loamy sand	1.5 lbs.
	all other soil textures	2 lbs.

Igran 80W + Milogard or Igran 80W + AAtrex: A tank mixture of Igran 80W + Milogard or Igran 80W + AAtrex can be used in the regions indicated below, for more consistent weed control than when Igran 80W is used alone. Apply the tank mixture at the rate indicated in Table 5. Do not use either combination on sand or loamy sand soils.

Table 6

Grain Sorghum				
Annual Broadleaf and Grass Weed Control				
		Broadcast Rate Per Acre		
Region	Soil Texture	Igran 80W	Milogard Maxx® 90*	AAtrex Nine-O®*
Illinois, Iowa, Eastern Kansas, Minnesota, Missouri, and Eastern and Central Nebraska, South Dakota	sandy loam	2 lbs.	0.45 lb.	DO NOT USE
	loam and finer	2-2.5 lbs.	DO NOT USE	0.9 lb.
Central Kansas, Western Nebraska, Oklahoma, and Texas	sandy loam	2 lbs.	0.45 lb.	DO NOT USE
	loam and finer	2-2.5 lbs.	0.9 lb.	DO NOT USE
Eastern Colorado, and Western Kansas	sandy loam and finer	2 lbs.	0.45 lb.	DO NOT USE

*When using Milogard 4L, Milogard 80W, AAtrex 4L, AAtrex 4LC or AAtrex 80W, use equivalent rates. 0.9 lb. of Nine-O or Maxx 90 equals 1 lb. of 80W. 0.9 lb. of Nine-O or Maxx 90 equals 1.6 pts. of 4L or 4LC.

Igran 80W + AAtrex + Paraquat: Apply the tank mixture at the rates indicated in Table 7. Refer to paraquat label concerning directions, limitations, and cautions for use of that product. The addition of 2,4-D (ester) at 0.5 lb. active ingredient per sprayed acre may aid in the suppression of perennial broadleaf weeds and large annual broadleaf weeds present before application. Refer to 2,4-D labels registered for use on grain sorghum for directions, limitations, and cautions. Apply in 20-40 gals. diluted spray per acre. Add a nonionic surfactant such as X-77® at 1 to 2 pts. per 100 gals. diluted spray. Use only water or nitrogen solutions as a carrier. Add Igran to spray tank first, mixing thoroughly in carrier. Next, add AAtrex, 2,4-D (if used), paraquat and X-77. Maintain agitation of carrier at all times.

Table 7

Grain Sorghum			
Broadleaf and Grass Control			
	Broadcast Rate Per Acre		
Soil Texture	**Igran 80W**	**AAtrex Nine-O***	**Paraquat**
COARSE: Sand, loamy sand	DO NOT USE	DO NOT USE	DO NOT USE
MEDIUM: Loam, silt loam, silt	2 lbs.	0.9 lb.	1-2 pts.
FINE: Silty clay loam, sandy clay loam, clay loam, sandy clay, silty clay, clay	2.5 lbs.	1.8 lbs.	1-2 pts.

* When using AAtrex 4L, AAtrex 4LC or AAtrex 80W, use equivalent rates. 0.9 lb. of Nine-O equals 1.6 pts. of 4L or 4LC. 0.9 lb. of Nine-O equals 1 lb. of 80W.

Winter Weed Control in Texas: For postemergence control of winter weeds *only,* such as henbit (purple top), seedling dock, and annual thistle on fall bedded land in the Texas Gulf Coast and Blacklands of Texas, apply 0.75 to 1 lb. of Igran 80W per acre, postemergence to the weeds in the fall or winter to land that will be planted to sorghum the following spring. For best results, add a suitable surfactant, such as X-77, at the rate of 0.5% of the spray volume, an emulsifiable oil at the rate of 1.0% of the spray volume, or an oil concentrate at the rate of 1 qt. per acre.

Normal programs of weed control may be used in the following sorghum crop.

Note: To avoid crop injury or illegal residues, do not plant any crop except grain sorghum the spring following this treatment.

Precautions: *Igran 80W alone and Igran 80W combinations — To avoid crop injury, (1) apply only on sorghum planted after the minimum soil temperature has reached 60°F or more for at least three consecutive days, (2) do not wait more than two days after planting to apply, (3) do not apply over emerged sorghum, (4) do not apply to millets, sudan-sorghum hybrids, or sorghum breeding stocks, (5) do not apply to sorghum growing on alkali soils, caliche outcroppings, or where cuts, fills, or erosion have exposed calcareous subsoil, (6) and to avoid concentration of herbicide in the seed furrow, do not make a broadcast application to sorghum planted in furrows deeper than two inches. Width of the band should not exceed the width of the bottom of the furrow, (7) do not apply a second application to the same crop. (8) Rainfall or overhead irrigation at the time of sorghum emergence and until sorghum is 2 inches tall may cause crop injury.*

Note (Igran 80W alone and Igran 80W combinations): Do not graze or feed forage from treated areas.

Suggestions for Crop Rotation

Following Igran 80W Alone:

1. Sorghum stubble should be plowed (disk or moldboard) prior to planting the next crop.
2. Rotational crops may be planted as follows:

   4 months: winter wheat

   7 months or longer: all other crops following normal agricultural practices and planting dates.

Following Igran 80W + AAtrex and Igran 80W + AAtrex + paraquat: To avoid crop injury: (1) land treated with Igran 80W + AAtrex should not be planted to any crop except corn or sorghum until the following year. (2) do not rotate with crops other than corn or sorghum the next year if Igran 80W was applied after June 10. (3) do not plant sugar beets, tobacco, vegetables (including dry beans), spring-seeded legumes, and grasses the following year. (4) Injury may occur to soybeans planted in north central and northwest Iowa, south central and southwest Minnesota, southeast South Dakota, and northeast Nebraska the year following an application on soils having a calcareous surface layer.

Following Igran 80W + Milogard:

Sorghum, cotton, or corn may be planted 12 months after application. All other crops may be planted 18 months after application.

When applying Igran combinations, see the General Information section, caution statements, and precautions listed on the AAtrex and Milogard labels for further information.

CHEMICAL FALLOW

WHEAT-SORGHUM-FALLOW

AAtrex 4L (4LC), AAtrex 80W, or AAtrex Nine-O may be used to control annual weeds after wheat harvest and in the following sorghum crop when grown under minimum tillage. Refer to the AAtrex label for directions, precautions, and limitations.

Tank Mixtures for Wheat-Sorghum-Fallow

If weeds are present at time of application (in late July or August following wheat harvest), Igran 80W may be tank mixed with AAtrex to control existing annual weeds. Broadcast 2-2.5 lbs. of Igran 80W plus 6 pts. of AAtrex 4L (4LC) (or 3.75 lbs. of AAtrex 80W, or 3.3 lbs. of AAtrex Nine-O) per acre. Add 1-2 pts. of a nonionic surfactant (such as X-77) per 100 gals. of spray mixture. Add Igran 80W to the spray tank first and thoroughly mix with the water. Then add the AAtrex, followed by the surfactant. Use the 2.5 lb. rate of Igran 80W if weeds are 4-6 inches tall. This mixture will not control weeds taller than 6 inches. *Plant sorghum into the wheat stubble the following spring with minimum disturbance of the soil.* Use a surface planter or a planter leaving a shallow furrow. If weeds are present at the time of sorghum planting, remove them with a sweep plow or other suitable implement before planting. Refer to the AAtrex label for further directions, precautions, and limitations.

Note: To avoid illegal residues, do not graze or feed forage from treated areas.

WHEAT-CORN-FALLOW (KS AND NE)

AAtrex 4L (4LC), AAtrex 80W, or AAtrex Nine-O may be used to control annual weeds after wheat harvest and in the following corn crop when grown under minimum tillage. Refer to the AAtrex label for directions, precautions, and limitations.

Tank Mixtures for Wheat-Corn-Fallow

(KS and NE)

If weeds are present at time of application (in late July or August following wheat harvest), Igran 80W may be tank mixed with AAtrex and applied to control existing annual weeds. Follow directions for use in the "Wheat-Sorghum-Fallow" section above, substituting corn for references to sorghum.

WHEAT-FALLOW-WHEAT (CO, KS, MT, NE, ND, SD, AND WY)

AAtrex 4L (4LC), AAtrex 80W, or AAtrex Nine-O may be used for preemergence control of annual weeds during the fallow period of a wheat-fallow-wheat rotation. If AAtrex is applied alone in late summer after wheat harvest, Igran 80W may be applied the following spring at 2-2.5 lbs./A to control escaping weeds. Add 1-2 pts. of a nonionic surfactant (such as X-77) per 100 gals. of spray mixture. Use the 2.5 lb. rate if weeds are 4-6 inches tall. Igran 80W will not control weeds taller than 6 inches. Treat with Igran 80W only once during the same fallow period, and *plant wheat no sooner than 4 months after application of the Igran 80W alone and no sooner than 12 months after the AAtrex application.* Refer to the AAtrex label for directions, precautions, and limitations.

Tank Mixtures for Wheat-Fallow-Wheat (CO, KS, MT, NE, ND, SD, and WY)

If weeds are present at time of application (normally in late July or August following wheat harvest), Igran 80W may be tank mixed with AAtrex to control existing annual weeds. Broadcast 2-2.5 lbs. of Igran 80W plus 1-2 pts. of AAtrex 4L (4LC) (or 0.6-1.25 lbs. of AAtrex 80W, or 0.5-1.1 lbs. of AAtrex Nine-O) per acre. Add 1-2 pts. of a nonionic surfactant (such as X-77) per 100 gals. of spray mixture. Add Igran 80W to the spray tank first and thoroughly mix with the water. Then add the AAtrex, followed by the surfactant. Use the 2.5 lb. rate of Igran 80W if weeds are 4-6 inches tall. This mixture will not control weeds taller than 6 inches. Treat with this tank mixture only once during the same fallow period. Use tillage to control weeds which escape during the fallow period. *Plant wheat no sooner than 12 months after application.* Refer to the AAtrex label for further directions, precautions, and limitations.

Note: To avoid illegal residues, do not feed forage or graze treated areas to livestock for six months following application.

WHEAT-FALLOW-WHEAT, SORGHUM-FALLOW-WHEAT (GREAT PLAINS, INCLUDING OK, N. TX)

Apply Igran 80W preemergence or postemergence at 2-2.5 lbs./A in the spring at the start of the fallow period (as soon as possible following harvest in a winter wheat-fallow-wheat rotation). Use the higher rate when weeds are 4-6 inches tall. Igran 80W will not control weeds taller than 6 inches. Postemergence weed control is improved by adding a nonionic surfactant, such as X-77 at 1-2 pts. per 100 gals. of spray mixture, an emulsifiable oil at 2-4 qts./A or an oil concentrate at 1-2 pts./A. Treat only once during the fallow period, and do not plant wheat sooner than 4 months after application (2 months in Kansas, Oklahoma, and northern Texas). Shallow cultivation or paraquat at 1-2 pts./A + X-77 surfactant at 1-2 pts. per 100 gals. of spray solution may be used to control weeds which escape or reinfest during the fallow season.

Note: To avoid illegal residues, do not feed green forage or graze treated areas for 6 months following application.

Tank Mixtures for Wheat-Fallow-Wheat (Great Plains, including OK, N. TX)

For control of established weeds, tank mix Igran 80W with one of the following herbicides, or stubble plow to control large established weeds. The following tank mixes will not control weeds taller than 6 inches. Use the higher rate of each product when weeds are 4-6 inches tall. Treat only once during the fallow period and do not plant wheat sooner than 4 months after application (2 months in Kansas, Oklahoma, and northern Texas). Shallow cultivation may be used to control weeds which escape during the fallow season. A nonionic surfactant, such as X-77 at 1-2 pts. per 100 gals. of spray mixture, should be added to each tank mix.

Note: To avoid illegal residues, do not feed green forage or graze treated areas for 6 months following application.

Paraquat: Apply 1-2 pts./A of paraquat in tank mix with 2-2.5 lbs./A of Igran 80W if weeds are present at application. Refer to the paraquat label for further directions, precautions, and wheat rotation restrictions.

2,4-D [various labeled amine and low volatile ester formulations]: Apply 0.5-1 lb. active ingredient per acre of 2,4-D in tank mix with 2-2.5 lbs./A of Igran 80W. Refer to the 2,4-D labels for further directions, precautions, and wheat rotation restrictions.

Bladex® 80W: Apply 2 lbs./A of Bladex 80W in tank mix with 2-2.5 lbs./A of Igran 80W. Refer to the Bladex label for further directions, precautions, and wheat rotation restrictions.

Sencor [50WP, 4, DF] (CO, KS, MT, NE, WY only): Apply 0.5-0.67 lb./A of Sencor DF, 0.75-1 pt./A of Sencor 4, or 0.75-1 lb./A of Sencor 50WP in tank mix with 2-2.5 lbs./A of Igran 80W. To avoid crop injury or illegal residues, do not plant wheat sooner than 4 months after application. Refer to the Sencor labels for further directions, precautions, and wheat rotation restrictions.

NONCROP LAND

Igran 80W provides weed control on railroad rights-of-way and on other noncrop land sites.

Do not use this product for nonselective weed control on land to be cropped, near adjacent desirable trees, shrubs, or plants, or in greenhouses, or injury may occur.

Apply Igran 80W to emerged actively growing weeds not more than 6 inches in height at 2.5-3 lbs./A. Apply in at least 20 gals. of water per acre. A surfactant such as X-77, duPont WK, or Tronic may be added to the spray mixture at the rate of 2 qts./100 gals. of spray (0.5% of spray volume). Cover weeds thoroughly with spray for good control. For extended weed control, Igran 80W may be applied in tank mix combination with Atratol® formulations 90 and 80W, Princep® formulations (4L, 80W, and Caliber® 90). Refer to the Atratol and Princep labels for use instructions for these products.

Note: Agitate Igran 80W and Igran 80W tank mixtures during the tank filling and spraying process. Empty the spray tank during the spraying operation.

STORAGE AND DISPOSAL

Store in a cool, dry place. Do not contaminate water, food, or feed by storage or disposal.

Pesticide
Pesticide wastes are toxic. Improper disposal of excess pesticide, spray mixture, or rinsate is a violation of Federal Law. If these wastes cannot be disposed of by use according to label instructions, contact your State Pesticide or Environmental Control Agency, or the Hazardous Waste Representative at the nearest EPA Regional Office for guidance.

Container
Completely empty bag into application equipment. Then dispose of empty bag in a sanitary landfill or by incineration, or if allowed by state and local authorities, by burning. If burned, stay out of smoke.

For minor spills, etc., follow all precautions indicated on this label and clean up immediately. Take special care to avoid contamination of equipment and facilities during clean-up procedures and disposal of wastes. In the event of a major spill, fire, or other emergency, call (919) 292-7100 day or night.

PRECAUTIONARY STATEMENTS

HAZARDS TO HUMANS AND DOMESTIC ANIMALS

WARNING
Causes eye irritation. Do not get in eyes. Harmful if swallowed. Avoid contact with eyes, contact with skin, inhalation of dust, contamination of food and feed, and application to animals or humans. When mixing or loading the product, wear gloves and long-sleeved coveralls. Flagmen and loaders should avoid inhalation of spray mist and contact with skin, and should wash thoroughly before eating and at the end of each day's operation.

First Aid: In case of contact with eyes, flush with plenty of water and get medical attention if irritation persists. In case of contact with skin, wash with soap and water. If accidentally ingested, induce vomiting. In case of inhalation, move from contaminated area.

Note to Physician: There is no specific antidote. Induce emesis or lavage stomach. Give a saline laxative and supportive therapy.

Environmental Hazards
Do not contaminate domestic or irrigation water supplies, or lakes, streams, or ponds. Do not apply directly to water or wetlands.

Atratol® trademark of CIBA-GEIGY

AAtrex® trademark of CIBA-GEIGY for atrazine

Banvel® trademark of Velsicol Chemical Corporation for dicamba

Bladex® trademark of Shell Chemical Company for cyanazine

Brominal® trademark of Union Carbide Corporation for bromoxynil

Buctril® trademark of Rhone-Poulenc Chemical Company for bromoxynil

Caliber® trademark of CIBA-GEIGY

Glean® trademark of E. I. duPont de Nemours & Co., Inc. for chlorsulfron

Igran® trademark of CIBA-GEIGY for terbutryn U.S. Patent No. 3,634,062

Lexone® trademark of E. I. duPont de Nemours & Co., Inc. for metribuzin

Maxx® trademark of CIBA-GEIGY

Milogard® trademark of CIBA-GEIGY for propazine

Nine-O® trademark of CIBA-GEIGY

Princep® trademark of CIBA-GEIGY for simazine

Sencor® trademark of the Parent Company of Farbenfabriken Bayer GmbH, Leverkusen, for metribuzin

X-77® trademark of KALO Agricultural Chemicals, Inc.

CGA 26L1Q 045

MILOGARD®4L

HERBICIDE

For weed control in sorghum (milo and sweet sorghum)

ACTIVE INGREDIENT:
Propazine:
2-chloro-4,6-bis (isopropylamino)-*s*-triazine 43%
INERT INGREDIENTS: 57%
TOTAL: 100%

EPA Reg. No. 100-584 EPA Est. 100-LA-1

Milogard® trademark of CIBA-GEIGY for propazine

Milogard 4L contains 4 lbs. active ingredient per gallon

Keep Out of Reach of Children.

Caution

See additional caution statements at end of label booklet.

See directions for use inside booklet.

DIRECTIONS FOR USE AND CONDITIONS OF SALE AND WARRANTY

IMPORTANT: Read the entire **Directions for Use** and the **Conditions of Sale and Warranty** before using this product.

CONDITIONS OF SALE AND WARRANTY

The **Directions for Use** of this product reflect the opinion of experts based on field use and tests. The directions are believed to be reliable and should be followed carefully. However, it is impossible to eliminate all risks inherently associated with use of this product. Crop injury, ineffectiveness, or other unintended consequences may result because of such factors as weather conditions, presence of other materials, or the manner of use or application all of which are beyond the control of CIBA-GEIGY or the Seller. All such risks shall be assumed by the Buyer.

CIBA-GEIGY warrants that this product conforms to the chemical description on the label and is reasonably fit for the purposes referred to in the **Directions for Use** subject to the inherent risks referred to above. **CIBA-GEIGY makes no other express or implied warranty of Fitness or Merchantability or any other express or implied warranty. In no case shall CIBA-GEIGY or the Seller be liable for consequential, special, or indirect damages resulting from the use or handling of this product.** CIBA-GEIGY and the Seller offer this product, and the Buyer and user accept it, subject to the foregoing **Conditions of Sale and Warranty,** which may be varied only by agreement in writing signed by a duly authorized representative of CIBA-GEIGY.

GENERAL INFORMATION

Milogard 4L is a selective herbicide that may be applied before planting or after planting before sorghum or weeds emerge. Milogard 4L controls annual morningglory, carpetweed, lambsquarters, pigweed, ragweed, foxtail, smartweed, and velvetleaf.

Following many years of continuous use of this product and chemically related products, biotypes of some of the weeds listed on this label have been reported which cannot be effectively controlled by this and related herbicides. Where this is known or suspected, we recommend the use of this product in combination with other registered herbicides which are not triazines. Consult with your State Agricultural Extension Service for specific recommendations.

Since Milogard enters the weeds mainly through their roots, incorporation or moisture is needed to move it into the weed root zone. Should weeds develop, a shallow cultivation or rotary hoeing will generally result in better weed control. Do not use for weed control in greenhouses.

MIXING INSTRUCTIONS

Mix this product with water, nitrogen solutions, or complete liquid fertilizers and apply as a spray. Pour this product into the tank during or after filling. Hydraulic (jet) or mechanical agitation is recommended during mixing and application to keep the material in suspension. All return lines to the tank must discharge below liquid level and agitation should not be so violent as to cause air bubbles to form in the liquid. Wash sprayer thoroughly after use.

To determine compatibility, pour the product into a small container of liquid fertilizer or nitrogen solution in approximate proportion to be mixed in spray tank. Stir or shake thoroughly. Let stand 5 minutes. If it remains mixed, or can be remixed readily, the mixture is compatible and can be used.

GROUND APPLICATION

For the most uniform distribution of broadcast applications use flat fan-type nozzles. For band applications, use flat fan even spray nozzles. Screens and strainers should be no finer than 50-mesh. Use a pump with capacity to (1) provide sufficient hydraulic agitation during mixing and application to keep the material in suspension and (2) maintain 30-40 psi operating pressure. Use a minimum of 10 gals. of water per acre. For band treatments, use a proportional volume.

For band applications, calculate the amount to be applied per acre as follows:

$$\frac{\text{band width in inches}}{\text{row width in inches}} \times \text{broadcast rate per acre} = \text{amount needed per acre of field}$$

AERIAL APPLICATION

Apply in a minimum of 5 gals. of water per acre. Do not apply where uniform coverage cannot be obtained or where excessive spray drift may occur. Avoid application directly to humans or animals. Flagmen or loaders should avoid inhalation of spray mist or contact with skin and should wash thoroughly before eating and at the end of each day's operation.

DIRECTIONS FOR USE

For preplant and preemergence applications, nitrogen solutions or complete liquid fertilizers may replace all or part of the water as a carrier with this product. See mixing instructions, under General Information.

Preplant (Kansas, Nebraska, eastern Colorado, and South Dakota)

Apply in the spring after plowing. If sorghum is planted on a flat soil surface, apply during or after final seedbed preparation. If sorghum is planted on beds, apply only after bed formation. Best weed control results are obtained when this product is applied within 2 weeks before planting. Shallow incorporation, not more than 2 inches deep, following application will generally result in better weed control, particularly under dry or minimum moisture conditions.

Broadleaf and grass weed control

Soil texture	Broadcast* rate per acre
Sand, loamy sand	DO NOT USE
Sandy loam, loam	4 pts.
Silt loam, clay loam	4.8 pts.
Heavy clay or high organic matter soil	DO NOT USE

* See the General Information section to calculate amount per acre for band treatment.

Preplant (Oklahoma, New Mexico, Texas, the Gulf Coast states, Arkansas, and Georgia)

Apply this product in the spring after plowing. If sorghum is planted on a flat soil surface, apply during or after seedbed preparation. If sorghum is planted on beds, apply only after bed formation. Best weed control results are obtained when this product is applied within 4 weeks before planting. Shallow incorporation, not more than 2 inches deep, following application will generally result in better weed control, particularly under dry or minimum moisture conditions. Do not incorporate preplant applications on sandy loam soils in the states listed in this section.

Soil texture	Broadleaf weed control	Broadleaf and grass weed control
	Broadcast* rate per acre	Broadcast* rate per acre
Sand, loamy sand	DO NOT USE	DO NOT USE
Sandy loam, loam	2.4 pts.	4 pts.**
Silt loam, clay loam	2.4 pts.	4.8 pts.
Heavy clay or high organic matter soil in the Gulf Coast states, Arkansas, and Georgia	2.4 pts.	6.4 pts.

* See the General Information section to calculate amount per acre for band treatment.

** To avoid crop injury, use only 2.4 pts. per acre on sandy loam in Oklahoma, New Mexico, and Texas (except Texas Gulf Coast). Controls broadleaf weeds only.

PREEMERGENCE

Apply the amount shown in the following table at planting time or immediately after planting before weeds and sorghum emerge.

Soil texture	Broadleaf weed control (Oklahoma, New Mexico, Texas and desert regions of Arizona and California only)	Broadleaf and grass weed control (all areas except as noted)
	Broadcast* rate per acre	Broadcast* rate per acre
Sand, loamy sand	DO NOT USE	DO NOT USE
Sandy loam, loam	2.4 pts.	4 pts.**
Silt loam, clay loam	2.4 pts.	4.8 pts.
Heavy clay or high organic matter soil in the Gulf Coast states, Arkansas, and Georgia	2.4 pts.	6.4 pts.

* See the General Information section to calculate amount per acre for band treatment.

** To avoid crop injury, use only 2.4 pts. per acre on sandy loam in Oklahoma, New Mexico, and Texas (except Texas Gulf Coast). Controls broadleaf weeds only.

Precaution (both preplant and preemergence): *Crop damage may result from applications on highly alkaline soil, or on eroded areas.*

Rotational Crops: *Do not rotate to any crop other than sorghum except: 1) In Arkansas, Louisiana, the Texas Gulf Coast, Texas Blacklands, and the Southeast, cotton, soybeans, or corn may be planted 12 months after treatment. Do not plant other crops for 18 months after treatment. 2) In Oklahoma, New Mexico, and West Texas, cotton or corn may be planted 12 months after a broadcast application of 2.4 pts. or after a 10-20 inch band application which was proportional to a broadcast rate of 4 pts. or less. If a higher rate per acre was used in these areas, only corn may be planted 12 months after treatment. Do not plant other crops for 18 months after treatment. 3) In all other sorghum growing regions, corn may be planted 12 months after treatment. Do not plant other crops for 18 months after treatment. 4) If replanting is necessary, sorghum may be replanted in soil treated with this product. Do not make a second application as injury may occur.*

STORAGE AND DISPOSAL

Do not contaminate water, food, or feed by storage or disposal.

PESTICIDE DISPOSAL

Pesticide wastes are toxic. Improper disposal of excess pesticide, spray mixture, or rinsate is a violation of Federal law. If these wastes cannot be disposed of by use according to label instructions, contact your State Pesticide or Environmental Control Agency or the Hazardous Waste Representative at the nearest EPA Regional Office for guidance.

CONTAINER DISPOSAL

Triple rinse container (or equivalent). Then offer container for recycling or reconditioning, or puncture and dispose of in a sanitary landfill, or by incineration, or if allowed by state and local authorities by burning. If burned, stay out of smoke.

For minor spills, leaks, etc., follow all precautions indicated on this label and clean up immediately. Take special care to avoid contamination of equipment and facilities during cleanup procedures and disposal of wastes. In the event of a major spill, fire, or other emergency, call (919) 292-7100 day or night.

CAUTION

Keep Out of Reach of Children.

Harmful if swallowed. Prolonged or repeated inhalation of mist or contact with skin may be harmful.

Do not contaminate domestic or irrigation water supplies, lakes, streams, or ponds.

Milogard® trademark of CIBA-GEIGY for propazine

CGA 10L10D 014

MILOGARD® MAXX 90

HERBICIDE

For weed control in sorghum (milo and sweet sorghum)

Milogard Maxx 90 is a water dispersible granule

ACTIVE INGREDIENT:

Propazine:

2-chloro-4, 6-bis (isopropylamino)-*s*-triazine 90%

INERT INGREDIENTS: 10%

TOTAL: 100%

EPA Reg. No. 100-606 EPA Est. 100-LA-1

Milogard® trademark of CIBA-GEIGY for propazine

Keep Out of Reach of Children.

CAUTION

See additional precautionary statements on back of bag.

See directions for use on panel below.

DIRECTIONS FOR USE AND CONDITIONS OF SALE AND WARRANTY

IMPORTANT: Read the entire **Directions for Use** and the **Conditions of Sale and Warranty** before using this product.

CONDITIONS OF SALE AND WARRANTY

The **Directions for Use** of this product reflect the opinion of experts based on field use and tests. The directions are believed to be reliable and should be followed carefully. However, it is impossible to eliminate all risks inherently associated with use of this product. Crop injury, ineffectiveness, or other unintended consequences may result because of such factors as weather conditions, presence of other materials, or the manner of use or application all of which are beyond the control of CIBA-GEIGY or the Seller. All such risks shall be assumed by the Buyer.

CIBA-GEIGY warrants that this product conforms to the chemical description on the label and is reasonably fit for the purposes referred to in the **Directions for Use** subject to the inherent risks referred to above. **CIBA-GEIGY makes no other express or implied warranty of Fitness or Merchantability or any other express or implied warranty. In no case shall CIBA-GEIGY or the Seller be liable for consequential, special, or indirect damages resulting from the use or handling of this product.** CIBA-GEIGY and the Seller offer this product, and the Buyer and user accept it, subject to the foregoing **Conditions of Sale and Warranty,** which may be varied only by agreement in writing signed by a duly authorized representative of CIBA-GEIGY.

GENERAL INFORMATION

FAILURE TO FOLLOW ALL PRECAUTIONS ON THIS LABEL MAY RESULT IN POOR WEED CONTROL, CROP INJURY, OR ILLEGAL RESIDUES.

This product is a selective herbicide that can be applied before planting or after planting before sorghum or weeds emerge for the control of most annual broadleaf weeds and grasses, including annual morningglory, carpetweed, lambsquarters, pigweed, ragweed, foxtail, smartweed, and velvetleaf.

Following many years of continuous use of this product and chemically related products, biotypes of some of the weeds listed on this label have been reported which cannot be effectively controlled by this and related herbicides. Where this is known or suspected, we recommend the use of this product in combination with other registered herbicides which are not triazines. Consult with your State Agricultural Extension Service for specific recommendations.

Since this product enters the weeds mainly through their roots, incorporation, rainfall, or irrigation is needed to move it into the weed root zone. Should weeds develop, a shallow cultivation or rotary hoeing will generally result in better weed control.

MIXING PROCEDURES

This product is a water dispersible granule formulation to be mixed with water or fluid fertilizer and applied as a spray.

Unless otherwise specified, use a minimum of 10 gals. of spray mixture/A for preplant and preemergence applications with ground equipment. For band treatments, a proportional volume should be used. For band applications, calculate the amount to be applied per acre as follows:

$$\frac{\text{Band width in inches}}{\text{Row width in inches}} \times \text{Rate/A broadcast treatment} = \text{Amount needed for band treatment}$$

Water: (1) Be sure sprayer is clean and not contaminated with any material, as crop injury or sprayer clogging may result. (2) Fill tank 1/4 full with clean water, and start agitation. (3) Be certain that the agitation system is working properly and creates a rippling or rolling action on the liquid surface. (4) Pour product directly from bag into tank. (5) Continue filling tank until 90% full. Increase agitation if necessary to maintain surface action. (6) Complete filling tank, maintaining sufficient agitation at all times to ensure surface action. This applies to both spray and nurse tanks. (7) Immediately after use, clean sprayer thoroughly by flushing system with water containing a detergent.

Liquid fertilizer: (1) Nitrogen solution or complete liquid fertilizer may replace all or part of the water as a carrier for preemergence or preplant application on sorghum. Do not apply fertilizer mixtures after sorghum emerges as crop injury may occur. Before mixing, always check for compatibility. To do this, put proportionate quantities of this product and the liquid fertilizer being used into a quart jar. If 2.6 lbs. is being applied in 25 gals. of liquid fertilizer per acre, mix 1/4 oz. of this product in 1 pt. of liquid fertilizer. After shaking vigorously for 30 seconds, let stand for 5 minutes. If the fertilizer/herbicide combination remains mixed or can be remixed readily, the mixture is compatible and can be sprayed. If it is not compatible, make a water slurry with the herbicide before adding fertilizer and use the mixture only if this test shows it is compatible. Always maintain agitation. (2) If the product is compatible with liquid fertilizer, follow the mixing procedures for water-steps 1-7.

APPLICATION EQUIPMENT

Ground equipment: Use conventional ground sprayers equipped with nozzles that provide accurate and uniform application. Be certain that nozzles are uniformly spaced and the same size. Calibrate sprayer before use and recalibrate at the start of each season and when changing carriers.

Use a pump with capacity to (1) maintain 35-40 psi at nozzles, (2) provide sufficient agitation in tank to keep mixture in suspension, and (3) provide a minimum of 20% bypass at all times. Use centrifugal pumps which provide propeller shear action for dispersing and mixing this product. The pump should provide a minimum of 10 gals./minute/100 gals. tank size circulated through a correctly positioned sparger tube or jets.

Use screens to protect the pump and to prevent nozzles from clogging. Screens placed on suction side of pump should be *16-mesh or coarser.* Do not place a screen in the recirculation line. Use 50-mesh or coarser screens between the pump and boom, and where required, at the nozzles. Check nozzle manufacturer's recommendations.

AERIAL APPLICATION PROCEDURES

Use only where broadcast applications are specified. Apply a minimum of 1 gal. of water for each 1-1.5 lbs. applied per acre, but never use less than 2 gals. of water per acre. Avoid applications under conditions where uniform coverage cannot be obtained or where excessive spray drift may occur.

Instructions for Safe Handling

Do not apply directly to humans or animals. Although it is unnecessary for flagmen or loaders to wear special protective clothing or equipment, they should avoid inhalation of dust or spray mist and prolonged contact with skin. Flagmen and loaders should wash thoroughly before eating and at the end of each day's operation.

DIRECTIONS FOR USE

It is a violation of federal law to use this product in a manner inconsistent with its labeling.

PREPLANT APPLICATIONS (KS, NE, EASTERN CO, SD)

For use in Kansas, Nebraska, eastern Colorado, and South Dakota, apply this product in the spring after plowing. To sorghum planted on a flat soil surface, application can be made during or after final seedbed preparation. To sorghum planted on beds, application should only be made after bed formation. Best weed control results are obtained when this product is applied within 2 weeks prior to planting. Shallow incorporation, not more than 2 inches deep following application, will generally result in better weed control particularly under dry or minimum moisture conditions.

For Broadleaf and Grass Weed Control

Soil texture	Broadcast rate of this product per acre
Sand or loamy sand	DO NOT USE
Sandy loam and loam	2.2 lbs.
Silt loam, clay loam	2.7 lbs.
Heavy clay and high organic matter soils	DO NOT USE

PREPLANT APPLICATIONS (OK, NM, TX, THE GULF COAST STATES, AR, GA)

For use in Oklahoma, New Mexico, Texas, the Gulf Coast states, Arkansas, and Georgia, apply this product in the spring after plowing. To sorghum planted on flat soil surface, application can be made during or after seedbed preparation. To sorghum planted on beds, application should be made only after bed formation. Best weed control results are obtained when this product is applied within 4 weeks prior to planting. Shallow incorporation, not more than 2 inches deep following application, will generally result in better weed control particularly under dry or minimum moisture conditions. Do not incorporate preplant applications on sandy loam soils.

For Broadleaf Weed Control (Oklahoma, New Mexico, and Texas, except Texas Gulf Coast)

Soil texture	Broadcast rate of this product per acre
Sand or loamy sand	DO NOT USE
All other soil textures	1.33 lbs.

For Broadleaf and Grass Weed Control

Soil texture	Broadcast rate of this product per acre
Sand or loamy sand	DO NOT USE
Sandy loam*, loam	2.2 lbs.
Silt loam, clay loam	2.7 lbs.
Heavy clay and high organic matter soils of Arkansas, Georgia, and the Gulf Coast states including Texas Gulf Coast	3.6 lbs.

* Except in the states of Oklahoma, New Mexico, and Texas (except Texas Gulf Coast) where a rate of 1.33 lbs. of this product broadcast or equivalent per acre is to be applied to sandy loam soil. See the information above on Broadleaf Weed Control for specific recommendations for these areas.

PREEMERGENCE APPLICATIONS

Apply this product at planting time or immediately following planting before weeds and sorghum emerge.

For Broadleaf Weed Control (Oklahoma, New Mexico, Texas, and desert regions of Arizona and California only)

Soil texture	Broadcast rate of this product per acre
Sand or loamy sand	DO NOT USE
All other soil textures	1.33 lbs.

For Broadleaf and Grass Weed Control (All areas except as noted below)

Soil texture	Broadcast rate of this product per acre
Sand or loamy sand	DO NOT USE
Sandy loam†, loam	2.2 lbs.
Silt loam, clay loam	2.7 lbs.
Heavy clay and high organic matter soils of the Gulf Coast states, Arkansas, and Georgia	3.6 lbs.

† In the states of Oklahoma, New Mexico, and Texas (except Texas Gulf Coast), use a rate of 1.33 lbs. of this product on sandy loam soil. See the information above on Broadleaf Weed Control for specific recommendations for these areas.

See the General Information section to calculate amount per acre for band treatment.

Precautions: *1) Crop damage may result from applications made on highly alkaline soils, or on eroded spots in the field. 2) Do not incorporate on sandy loam soils in Oklahoma, New Mexico, Texas, the Gulf Coast states, Arkansas, and Georgia.*

CROP ROTATION SUGGESTIONS

Fields on which this product has been applied should not be planted to any crop other than sorghum except as noted below:

1. In Arkansas, Louisiana, the Texas Gulf Coast, Texas Blacklands, and the Southeast, fields treated with this product may be planted to cotton, soybeans, or corn 12 months after treatment. Other crops may be planted 18 months following treatment.
2. In Oklahoma, New Mexico, and west Texas, fields treated with this product may be planted to cotton or corn 12 months or more after a broadcast application of 1.33 lbs. or less, or where a 10-20 inch band application was made and the rate applied was proportional to a broadcast rate of 2.2 lbs. or less. If a higher rate per acre was used as a band treatment, or if broadcast applications were made in this area, corn may be planted 12 months following

treatment. Other crops may be planted 18 months following treatment.

3. In all other sorghum growing regions, corn may be planted in rotation 12 months after treatment. Other crops may be planted 18 months following treatment.

If replanting is necessary, sorghum may be replanted in soil previously treated with this product. Do not make a second application of this product or injury may occur.

STORAGE AND DISPOSAL

Store this product in a dry place. Do not contaminate water, food, or feed by storage or disposal.

PESTICIDE

Pesticide wastes are toxic. Improper disposal of excess pesticide, spray mixture, or rinsate is a violation of Federal Law. If these wastes cannot be disposed of by use according to label instructions, contact your State Pesticide or Environmental Control Agency, or the Hazardous Waste Representative at the nearest EPA Regional Office for guidance.

CONTAINER

Completely empty bag into application equipment. Then dispose of empty bag in a sanitary landfill or by incineration, or, if allowed by state and local authorities, by burning. If burned, stay out of smoke.

PRECAUTIONARY STATEMENTS

HAZARDS TO HUMANS AND DOMESTIC ANIMALS

CAUTION

Harmful if swallowed, inhaled, or absorbed through the skin. Avoid contact with skin, eyes, or clothing. Avoid breathing dust or spray mist.

First Aid: In case of contact with skin, wash immediately with plenty of soap and water. In case of contact with the eyes, flush with water for at least 15 minutes and get medical attention if irritation persists. If swallowed, drink promptly a large quantity of milk, egg whites, or gelatin solution, or if these are not available, drink large quantities of water. Induce vomiting. Avoid alcohol. Call a physician immediately. In case of inhalation exposure, move from the contaminated area. If unconscious, give artificial respiration (preferably mouth-to-mouth).

Note to Physician: There is no specific antidote. Induce emesis or lavage stomach. Give a saline laxative and supportive therapy.

ENVIRONMENTAL HAZARDS

Do not contaminate domestic or irrigation water supplies, or lakes, streams, or ponds.

Maxx® trademark of CIBA-GEIGY

Milogard® trademark of CIBA-GEIGY for propazine

CGA 10L20C 034

PRAMITOL®25E

HERBICIDE

For weed control on industrial sites, noncrop areas around-the-farm, and for use in and under asphalt.

Controls johnsongrass, bindweed, and other hard-to-kill weeds

Bare ground weed control

around . . . buildings, storage areas, fences, pumps, machinery, fuel tanks, drainage ditch banks, recreational areas, roadways, guard rails, airports, military installations, highway medians, pipelines, railroads, lumberyards, industrial sites, rights-of-way . . . or similar areas

ACTIVE INGREDIENT:

Prometon: 2,4-bis (isopropylamino)-6-methoxy-*s*-triazine . . . 25%

INERT INGREDIENTS: 75%

TOTAL: 100%

EPA Reg. No. 100-443 EPA Est. 100-AL-1

Control No. 06550

Pramitol® trademark of CIBA-GEIGY for prometon

Pramitol 25E contains 2 lbs. prometon per gallon.

Keep Out of Reach of Children.

DANGER

See additional precautionary statements left of container seam.

See directions for use on panel at right.

DIRECTIONS FOR USE AND CONDITIONS OF SALE AND WARRANTY

IMPORTANT: Read the entire **Directions for Use** and the **Conditions of Sale and Warranty** before using this product.

CONDITIONS OF SALE AND WARRANTY

The **Directions for Use** of this product reflect the opinion of experts based on field use and tests. The directions are believed to be reliable and should be followed carefully. However, it is impossible to eliminate all risks inherently associated with use of this product. Crop injury, ineffectiveness, or other unintended consequences may result because of such factors as weather conditions, presence of other materials or the manner of use or application all of which are beyond the control of CIBA-GEIGY or the Seller. All such risks shall be assumed by the Buyer.

CIBA-GEIGY warrants that this product conforms to the chemical description on the label and is reasonably fit for the purposes referred to in the **Directions for Use** subject to the inherent risks referred to above. **CIBA-GEIGY makes no other express or implied warranty of Fitness or Merchantability or any other express or implied warranty. In no case shall CIBA-GEIGY or the Seller be liable for consequential, special, or indirect damages resulting from the use or handling of this product.** CIBA-GEIGY and the Seller offer this product, and the Buyer and user accept it, subject to the foregoing **Conditions of Sale and Warranty,** which may be varied only by agreement in writing signed by a duly authorized representative of CIBA-GEIGY.

DIRECTIONS FOR USE

It is a violation of federal law to use this product in a manner inconsistent with its labeling.

GENERAL INFORMATION

Pramitol 25E is a nonselective herbicide that can be applied in water or oil before or after plant growth begins. Although Pramitol 25E has considerable activity through foliar contact, much of its activity is through roots; therefore, its effectiveness is dependent on rainfall to move it into the root zone. Very dry soil conditions and lack of sufficient rainfall may result in poor weed control.

Use only in areas where complete control of all vegetation is desired, such as industrial sites, rights-of-way, lumberyards, petroleum tank farms, around farm buildings, along fence lines, etc. When applied to the soil, this product usually inhibits plant growth for a year or more. It should not be used on land to be cropped, or near adjacent desirable trees, shrubs or plants, or injury may occur.

Do not use for weed control in greenhouses.

Following many years of continuous use of this product and chemically related products, biotypes of some of the weeds listed on this label have been reported which cannot be effectively controlled by this and related herbicides. Where this is known or suspected, we recommend the use of this product in combination with other registered herbicides which are not triazines. Consult with your State Agricultural Extension Service for specific recommendations.

It is generally noncorrosive to equipment, but may cause swelling or more rapid deterioration of hoses and fittings containing natural rubber.

INDUSTRIAL SITES AND NONCROP AREAS

Pramitol 25E can be applied prior to or up to 3 months after weed emergence. Application rates vary from 5-30 gals. of Pramitol 25E per acre, depending on climatic conditions, soil type, the weeds present, and the stage of growth of the weeds. In the following rate recommendations, the higher rates are intended for use on heavier soils where weed growth is heavy, and where rainfall is expected to be relatively high. The higher rates should also be used where longer residual control is desired in regions with a long growing season.

For best results, apply Pramitol 25E prior to weed emergence or when weeds are young and actively growing.

For control of annual and susceptible perennial weeds (such as downy bromegrass, oatgrass, goosegrass, quackgrass, puncturevine, goldenrod, and plantain): Broadcast 5-7½ gals. Pramitol 25E in 50-100 gals. of water per acre. For small areas, apply 1-1½ pts. Pramitol 25E per 1,000 sq. ft. in sufficient water to give thorough and uniform coverage.

For control of hard-to-kill perennial weeds and grasses (such as johnsongrass, bindweed, and wild carrot): Broadcast 20-30 gals. Pramitol 25E in 50-100 gals. of water per acre. For small areas, use 4-5½ pts. Pramitol 25E per 1,000 sq. ft. in sufficient water to give thorough and uniform coverage.

For faster top-kill of existing vegetation, apply Pramitol 25E in diesel oil, fuel oil, or weed oil. Use oils at rates of 100-200 gals. per acre. Tall dense vegetation such as johnsongrass will generally require 150 or more gals. of oil per acre for thorough coverage.

ASPHALT PAVEMENT

Pramitol 25E may be used to extend the useful life of asphalt pavement by preventing weeds from emerging through it. Pramitol 25E may be applied to the ground before laying asphalt, or it may be mixed with cutback asphalts.

Pramitol 25E should be used only where the area to be treated has been prepared according to good construction practices. If rhizomes, tubers, or other vegetative parts are present in the site, they should be removed by scalping with a grader blade. Recompaction of the site following disking or plowing will allow good penetration of Pramitol 25E, and will help to ensure proper asphalt curing and good weed control.

Pramitol 25E will not control woody vegetation such as small trees, brush, or woody vines. Roots of such species should be grubbed and removed from the site prior to application.

Under Asphalt: Pramitol 25E may be applied to the ground before laying rapid, medium, or slow curing asphalt coatings such as those used on parking lots, highway shoulders and median strips, roadways, and other industrial sites. Pramitol 25E should be applied just prior to the laying of the asphalt coating to prevent possible lateral movement of the herbicide by rainfall or other mechanical means. Sprayers should be equipped to provide continuous agitation of the spray mixture during application to ensure application of a uniform spray emulsion.

Apply Pramitol 25E uniformly at the rate of 20-25 gals. per acre (0.53-0.66 fl. oz. per sq. yd.) in a minimum of 100 gals. of water. Use the lower rate to control annual and susceptible perennial weeds such as downy bromegrass, oatgrass, goosegrass, quackgrass, puncturevine, goldenrod, and plantain. Use the higher rate to control hard-to-kill perennial weeds and grasses such as johnsongrass, bindweed, and wild carrot.

Precaution: Do not use Pramitol 25E under asphalt coatings less than 2 inches thick.

Mixed with Cutback Asphalt: Pramitol 25E may be mixed with standard and special cutback asphalts (such as RC, MC, and SC) for weed control in areas being stabilized, such as highway shoulders, highway median strips, under highway fences and guard rails, along airport runways and lights, soil and small stone aggregates on shoulders, and cracked asphalt surfaces being resurfaced.

Pramitol 25E should be simultaneously mixed with the cutback asphalt as it is being pumped into the applicator or transport truck to assure a homogeneous mix. Temperatures in the applicator or transport truck should not exceed 170°F.

Pramitol 25E should be used at the rate of 0.53-0.80 fl. oz. per sq. yd. (equivalent to 20-30 gals./A of Pramitol 25E). Use the lower rate on ordinary weed and grass complexes such as dallisgrass, King Ranch bluestem, knotroot bristlegrass, and the higher rate where johnsongrass, bermudagrass, or other deep-rooted perennials are present.

In arid areas where deep-rooted perennials are a problem, best results can be obtained by making the application from midsummer through late fall or prior to the rainy season.

Precautions: *1) Do not apply cutback asphalt mixtures containing Pramitol 25E to soil saturated with water. The condition of the soil should be favorable for penetration or proper asphalt curing may be affected. 2) Do not apply Pramitol 25E with asphalt emulsions such as SS.*

STORAGE AND DISPOSAL

Do not contaminate water, food, or feed by storage or disposal. Pesticide wastes may be acutely hazardous. Improper disposal of unused pesticide, spray mixture, or rinse water is a violation of federal law. If these wastes cannot be used according to label instructions, contact your State Pesticide or Environmental Control Agency, or the Hazardous Waste representative at the nearest EPA Regional Office for guidance in proper disposal methods.

Triple rinse container (or equivalent). Then offer for recycling or reconditioning, or puncture and dispose of in a sanitary landfill or by other procedures approved by state and local authorities.

For minor spills, leaks, etc., follow all precautions indicated on this label and clean up immediately. Take special care to avoid contamination of equipment and facilities during cleanup procedures and disposal of wastes. In the event of a major spill, fire or other emergency, call (919) 292-7100 day or night.

PRECAUTIONARY STATEMENTS

HAZARDS TO HUMANS AND DOMESTIC ANIMALS

DANGER

Corrosive—causes eye damage. Wear goggles or face shield and rubber gloves when handling. Causes skin irritation. Harmful if swallowed, inhaled or absorbed through the skin. Avoid breathing spray mists. Do not get in eyes, on skin, or on clothing. Avoid contamination of food.

First Aid: In case of contact with eyes, immediately flush with plenty of water for 15 minutes and get medical attention; with skin, wash immediately with plenty of soap and water. If swallowed, drink *promptly* a large quantity of milk, egg whites, or gelatin solution; or if these are not available, drink large quantities of water. Avoid alcohol. Call a physician immediately. In case of inhalation exposure, remove from contaminated areas. Wash thoroughly after handling and before eating and smoking. Remove and wash contaminated clothing before reuse.

Note to Physician: Probable mucosal damage may contraindicate the use of gastric lavage. Treat for circulatory shock, respiratory depression, and convulsions if needed.

ENVIRONMENTAL HAZARDS

Do not contaminate water used for irrigation or domestic purposes.

Do not contaminate water by cleaning of equipment, or disposal of wastes. Apply this product only as specified on this label.

PHYSICAL OR CHEMICAL HAZARDS

Do not use, pour, spill, or store near heat or open flame. Store at temperatures above 32°F.

Pramitol® trademark of CIBA-GEIGY for prometon

PRAMITOL®5PS

PELLETED HERBICIDE

For control of weeds in noncrop areas

Bare ground weed control around . . . buildings, storage areas, fences, pumps, machinery, fuel tanks, drainage ditch banks, recreational areas, roadways, guard rails, airports, military installations, highway medians, pipelines, railroads, lumberyards, industrial sites, rights-of-way, or similar areas.

ACTIVE INGREDIENTS:

Prometon:	
2,4-bis (isopropylamino)-6-methoxy-s-triazine	5.00%
Simazine:	
2-chloro-4, 6-bis (ethylamino)-*s*-triazine	0.75%
Sodium chlorate ($NaClO_3$)	40.00%
Sodium metaborate*	
($Na_2B_2O_4$-$5H_2O$)	50.00%
INERT INGREDIENTS:	4.25%
TOTAL:	100.00%

Equivalent to 4.87% boron expressed as elemental boron

EPA Reg. No. 100-479 EPA Est. 33560-TN-1

Pramitol® trademark of CIBA-GEIGY for prometon

Keep Out of Reach of Children.

Danger

See additional precautionary statements on back of bag.

See directions for use on back of bag.

DIRECTIONS FOR USE AND CONDITIONS OF SALE AND WARRANTY

IMPORTANT: Read the entire **Directions for Use** and the **Conditions of Sale and Warranty** before using this product.

CONDITIONS OF SALE AND WARRANTY

The **Directions for Use** of this product reflect the opinion of experts based on field use and tests. The directions are believed to be reliable and should be followed carefully. However, it is impossible to eliminate all risks inherently associated with use of this product. Crop injury, ineffectiveness, or other unintended consequences may result because of such factors as weather conditions, presence of other materials, or the manner of use or application all of which are beyond the control of CIBA-GEIGY or the Seller. All such risks shall be assumed by the Buyer.

CIBA-GEIGY warrants that this product conforms to the chemical description on the label and is reasonably fit for the purposes referred to in the **Directions for Use** subject to the inherent risks referred to above. **CIBA-GEIGY makes no other express or implied warranty of Fitness or Merchantability or any other express or implied warranty. In no case shall CIBA-GEIGY or the Seller be liable for consequential, special, or indirect damages resulting from the use or handling of this product.** CIBA-GEIGY and the Seller offer this product, and the Buyer and user accept it, subject to the foregoing **Conditions of Sale and Warranty,** which may be varied only by agreement in writing signed by a duly authorized representative of CIBA-GEIGY.

DIRECTIONS FOR USE

It is a violation of federal law to use this product in a manner inconsistent with its labeling.

GENERAL INFORMATION

Pramitol 5PS is a nonselective herbicide that can be applied before or after plant growth begins. Since Pramitol enters plants through their roots, its effectiveness is dependent on rainfall to move the chemical into the root zone. Very dry soil conditions and lack of rainfall may result in poor weed control.

Important: Use only in areas where complete control of all vegetation is desired, such as industrial sites, rights-of-way, lumberyards, petroleum tank farms, around farm buildings, along fence lines, and similar areas. When applied to the soil, this product usually inhibits all plant growth for a year or more. It should not be used on land to be cropped, or near adjacent desirable trees, shrubs, plants, or in greenhouses or injury may occur.

To control annual broadleaf weeds and grasses: Apply ½-1 lb. per 100 sq. ft.

To control most perennial broadleaf weeds and grasses: Apply 1-2 lbs. per 100 sq. ft. Where deeprooted perennials are present, use the 2 lb. rate.

The higher rates listed above should be used in areas having a long growing season or high rainfall. The higher rates will also provide longer residual control.

Wash out equipment after application. Do not use equipment to apply insecticides, fungicides, or fertilizers.

STORAGE AND DISPOSAL

Store in a dry place. Do not contaminate water, food, or feed by storage or disposal. Pesticide wastes may be acutely hazardous. Improper disposal of unused pesticide is a violation of federal law. If these wastes cannot be used according to label instructions, contact your State Pesticide or Environmental Control Agency, or the Hazardous Waste representative at the nearest EPA Regional Office for guidance in proper disposal methods.

Completely empty bag into application equipment. Dispose of empty bag in a sanitary landfill or by incineration, or if allowed by state and local authorities, by burning. Stay out of smoke from burning bags.

For minor spills, leaks, etc., follow all precautions indicated on this label and clean up immediately. Take special care to avoid contamination of equipment and facilities during cleanup procedures and disposal of wastes. In the event of a major spill, fire or other emergency, call (919) 292-7100 day or night.

PRECAUTIONARY STATEMENTS

HAZARDS TO HUMANS AND DOMESTIC ANIMALS

DANGER

Corrosive -causes eye and skin damage. Do not get in eyes, on skin or clothing. Wash clothing after handling. Wear goggles or face shield and rubber gloves when handling. Harmful if swallowed, inhaled or absorbed through the skin. Do not breathe dusts. Do not contaminate food or feed.

First Aid: In case of contact with skin, wash immediately with plenty of soap and water. In case of contact with the eyes, flush with water for at least 15 minutes and get medical attention. If swallowed drink *promptly* a large quantity of milk, egg whites, gelatin solution or if these are not available, drink large quantities of water. Avoid alcohol. Call a physician immediately. In case of inhalation exposure, move from the contaminated area.

Note to Physician: Probable mucosal damage may contraindicate the use of gastric lavage. Treat for circulatory shock, respiratory depression, and convulsion if needed.

ENVIRONMENTAL HAZARDS

Do not apply directly to water or wetlands. Do not contaminate water by cleaning of equipment or disposal of wastes.

Simazine is a chemical which can travel (seep or leach) through soil and can contaminate ground water which may be used as drinking water. Simazine has been found in ground water as a result of agricultural use. Users are advised not to apply simazine where the water table (ground water) is close to the surface and where the soils are very permeable, i.e., well-drained soils such as loamy sands. Your local agricultural agencies can provide further information on the type of soil in your area and the location of ground water.

Pramitol® trademark of CIBA-GEIGY for prometon

CGA 27L3L 035

PRIME + ®

A PLANT GROWTH REGULATOR

For Control of Tobacco Suckers

ACTIVE INGREDIENT:

2-chloro-*N*-[2,6-dinitro-4-(trifluoromethyl) phenyl]-*N*-ethyl-6-fluoro-benzene-methanamine 15%

INERT INGREDIENTS: 85%

TOTAL: 100%

EPA Reg. No. 100-640 EPA Est. 100-LA-1

Prime +® trademark of CIBA-GEIGY

Prime + contains 1.2 lbs. active ingredient per gallon.

Keep Out of Reach of Children.

DANGER/PELIGRO

PRECAUCIÓN AL USUARIO:

Si usted no lee inglés, no use este producto hasta que la etiqueta haya sido explicado ampliamente.

See additional precautionary statements inside booklet.

See directions for use inside booklet.

DIRECTIONS FOR USE AND CONDITIONS OF SALE AND WARRANTY

IMPORTANT: Read the entire **Directions for Use** and the **Conditions of Sale and Warranty** before using this product.

CONDITIONS OF SALE AND WARRANTY

The **Directions for Use** of this product reflect the opinion of experts based on field use and tests. The directions are believed to be reliable and should be followed carefully. However, it is impossible to eliminate all risks inherently associated with use of this product. Crop injury, ineffectiveness, or other unintended consequences may result because of such factors as weather conditions, presence of other materials, or the manner of use or application all of which are beyond the control of CIBA-GEIGY or the Seller. All such risks shall be assumed by the Buyer.

CIBA-GEIGY warrants that this product conforms to the chemical description on the label and is reasonably fit for the purposes referred to in the **Directions for Use** subject to the inherent risks referred to above. **CIBA-GEIGY makes no other express or implied warranty of Fitness or Merchantability or any other express or implied warranty. In no case shall CIBA-GEIGY or the Seller be liable for consequential, special, or indirect damages resulting from the use or handling of this product.** CIBA-GEIGY and the Seller offer this product, and the Buyer and user accept it, subject to the foregoing **Conditions of Sale and Warranty,** which may be varied only by agreement in writing signed by a duly authorized representative of CIBA-GEIGY.

DIRECTIONS FOR USE

It is a violation of federal law to use this product in a manner inconsistent with its labeling.

FAILURE TO FOLLOW THE DIRECTIONS FOR USE AND PRECAUTIONS ON THIS LABEL MAY RESULT IN POOR SUCKER CONTROL AND/OR CROP INJURY.

Do not apply this product in such a manner as to directly or through drift expose workers or other persons except those knowingly involved in the application. The area being treated must be vacated by unprotected persons.

REENTRY STATEMENT

Do not enter treated areas without protective clothing until sprays have dried.

Because certain states may require more restrictive reentry intervals for various crops treated with this product, consult your State Department of Agriculture for further information.

Written or oral warnings must be given to workers who are expected to be in a treated area or in an area about to be treated with this product. Oral warnings must inform workers of areas or fields that may not be entered without specific protective clothing until sprays have dried, and appropriate actions to take in case of accidental exposure, as described under **Precautionary Statements** on this label. When oral warnings are given, warnings shall be given in a language customarily understood by workers. Oral warnings must be given if there is reason to believe that written warnings cannot be understood by workers. Written warnings must include the following information: "DANGER. Area treated with Prime + on (date of application). Do not enter without appropriate protective clothing until sprays have dried. In case of accidental eye exposure, flush eyes with plenty of water. Get medical attention. In case of accidental skin contact, wash skin with soap and water. Call physician if irritation persists. Remove and wash contaminated clothing before reuse."

GENERAL INFORMATION

Prime + is a plant growth regulator for the control of sucker growth on flue-cured, burley, dark, Maryland, and cigar tobacco.

One application of Prime + will provide full season control of excessive sucker development in tobacco. Read this label carefully and observe all directions and precautions.

MIXING INSTRUCTIONS

Mix 1 gal. of Prime + in 49 gals. of water or 2.5 fl. oz. Prime + per one gallon of water. Provide sufficient agitation during mixing and application to maintain a uniform mixture.

Note: Do not mix Prime + with insecticides or fungicides.

APPLICATION INSTRUCTIONS

Time of Application

Apply Prime + when most tobacco plants are in the elongated button to early flower stage, either before or after topping. If Prime + is applied after topping, apply within 24 hours of topping. During topping, remove all suckers one inch or longer from all plants. (See notes listed below.)

Precision Hand Application

Mix 1 gal. of Prime + in 49 gals. of water and apply as a coarse spray or drench to the top of the tobacco stalk using a handheld drop line, or mix 2.5 fl. oz. of Prime + per gallon of water when using the jug method or backpack sprayer. Apply the spray mixture at a rate sufficient to run down the stalk contacting each leaf axil and just reach the ground line. Spray volume per plant will generally be ⅓-⅔ fl. oz. depending on plant height. (See notes listed below.)

Application with Powered Spray Equipment (3 nozzles per row)

Mix 1 gal. of Prime + in 49 gals. of water and apply at a rate of 30 to 50 gals. spray volume per acre. Use the high spray volume in tall types of tobacco, such as flue-cured, burley, Maryland, and cigar. Use the low spray volume only in short types of tobacco, such as dark. Spray volume should be sufficient to run down the stalk contacting each leaf axil and just reach the ground line. (See notes listed below.)

Apply Prime + as a coarse spray directed downward to the top of the tobacco stalk. This is best accomplished with three full cone nozzles per row; one nozzle directed downward over the center of the row and one 7-9 inches to each side of the center, directed at the top of the stalk. Maintain nozzles at the height of 12-16 inches above the top of the stalk during application. Maintain boom pressure at 20-25 psi, not exceeding 30 psi.

Notes:

1. Timing is important; applications should be made at timing described above so that number and size of suckers will be small. Applications made before the elongated button stage of growth may result in chemical topping or distortion of leaves that were too immature at time of application. Spraying too late may result in poor sucker control.
2. Prime + will not control large suckers. All suckers one inch or longer on all plants must be removed before application of Prime +. Suckers that escape treatment should be removed by hand about 10-14 days after application.
3. Where a contact (fatty alcohol) is commonly used to control early suckers and provide a more uniform crop, the normal sequence would be: (a) apply contact material according to its label, and (b) about 5 days after application of contact material, apply Prime +. Timing of Prime + application should be during the elongated button to early flower stage.
4. Where a contact (fatty alcohol) is not used and a wide range of flowering exists, early flowering plants should be topped and suckers removed by hand. Then, when most plants are in the elongated button to early flower stage, remove all suckers one inch or longer, and apply Prime +. If Prime + is applied using precision hand application, early flowering plants can be topped and treated selectively; later flowering plants can be topped and treated as they reach the proper treatment stage; and a contact may not be needed.
5. For proper sucker control, Prime + must contact the small suckers on the axil where the leaf and stalk connect. Prime + will not control suckers on plants that are leaning at time of application. Applications with powered spray equipment during windy periods are not recommended. High wind will disrupt the spray pattern of nozzles on boom sprayers and turn leaves, causing improper contact and flow of the spray solution, resulting in poor sucker control. Application to wet plants, or drought stressed plants may result in poor sucker control.
6. Carefully follow the use rates and spray volume recommended, and spray field uniformly. Volume should be sufficient to run down the stalk, contact each leaf axil, and just reach the ground line.
7. Rain occurring within 2 hours after spraying may reduce effectiveness.

Precautions: *(1) To avoid illegal residues and possible injury to crops planted in the fall or following spring, do not apply excessive spray volume to the point of runoff onto the soil. Prime + residues in the soil may interact with dinitroaniline herbicides such as Balan®, Basalin®, Paarlan®, Prowl®, or Treflan® to increase early season crop stunting. Thorough disking and deep tillage in the fall are recommended where excessive amounts of Prime + may have reached the soil. (2) When tobacco is in a state of very rapid, succulent growth due to high nitrogen and rainfall, Prime + may cause injury. Under these conditions, use only the precision hand application method and avoid contacting leaves with the spray solution.*

STORAGE AND DISPOSAL

PESTICIDE DISPOSAL

Do not contaminate water, food, or feed by storage or disposal. Open dumping is prohibited. Wastes resulting from the use of this product are acutely hazardous. Improper disposal of unused pesticide, spray mixture, or rinsate that cannot be used according to label instructions must be disposed of according to federal, state, or local procedures. For guidance in proper disposal methods, contact your State Pesticide or Environmental Control Agency, or the Hazardous Waste representative at the nearest EPA Regional Office.

CONTAINER DISPOSAL

Do not reuse empty container. Triple rinse (or equivalent) and offer for recycling or reconditioning, or puncture and dispose of in a sanitary landfill, or by incineration or by open burning, if allowed by state and local authorities. If burned, keep out of smoke.

For minor spills, leaks, etc., follow all precautions indicated on this label and clean up immediately. Take special care to avoid contamination of equipment and facilities during cleanup procedures and disposal of wastes. In the event of a major spill, fire, or other emergency, call (919) 292-7100 day or night.

Store at temperatures above 32°F.

PRECAUTIONARY STATEMENTS

HAZARDS TO HUMANS AND DOMESTIC ANIMALS

DANGER

Corrosive -causes eye damage. Causes skin irritation or skin sensitization on repeated contact. Do not get in eyes, on skin, or on clothing. Wear goggles or face shield and protective clothing when handling. Harmful if swallowed, inhaled, or absorbed through the skin. Avoid breathing spray mist.

First Aid: In case of contact with the eyes, immediately flush with plenty of water and get medical attention. In case of contact with skin, wash immediately with soap and water. If swallowed, drink plenty of water. Do not induce vomiting. Avoid alcohol. Call a physician immediately. In case of inhalation exposure, move from the contaminated area. **Note to Physician:** There is no specific antidote for Prime +. Lavage stomach. Give a saline laxative or slurry of activated carbon in water and supportive therapy.

ENVIRONMENTAL HAZARDS

This pesticide is toxic to fish. Do not apply directly to lakes, streams, or ponds. Do not apply when weather conditions favor drift from treated areas. Do not contaminate water by cleaning of equipment or disposal of wastes.

Balan® trademark Elanco Products for benefin

Basalin® trademark of BASF/Wyandotte Corp. for fluchloralin

Paarlan® trademark of Elanco Products for isopropalin

Prime +® trademark of CIBA-GEIGY

Prowl® trademark of American Cyanamid for pendimethalin

Treflan® trademark of Elanco Products for trifluralin

CGA 73L9C 095

PRINCEP®4G

GRANULAR HERBICIDE

For weed control in certain crops, ornamental plantings, on industrial sites, and around-the-farm

Princep 4G is a granular herbicide

ACTIVE INGREDIENT:	
Simazine:	
2-chloro-4,6-bis (ethylamino)-*s*-triazine	4%
INERT INGREDIENTS:	96%
TOTAL:	100%

EPA Reg. No. 100-435 CGA 6L13V 095

Princep® trademark of CIBA-GEIGY for simazine

Keep Out of Reach of Children.

Caution

See additional precautionary statements at end of label booklet.

See directions for use inside booklet.

DIRECTIONS FOR USE AND CONDITIONS OF SALE AND WARRANTY

IMPORTANT: Read the entire **Directions for Use** and the **Conditions of Sale and Warranty** before using this product.

CONDITIONS OF SALE AND WARRANTY

The **Directions for Use** of this product reflect the opinion of experts based on field use and tests. The directions are believed to be reliable and should be followed carefully. However, it is impossible to eliminate all risks inherently associated with use of this product. Crop injury, ineffectiveness, or other unintended consequences may result because of such factors as weather conditions, presence of other materials, or the manner of use or application all of which are beyond the control of CIBA-GEIGY or the Seller. All such risks shall be assumed by the Buyer.

CIBA-GEIGY warrants that this product conforms to the chemical description on the label and is reasonably fit for the purposes referred to in the **Directions for Use** subject to the inherent risks referred to above. **CIBA-GEIGY makes no other express or implied warranty of Fitness or Merchantability or any other express or implied warranty. In no case shall CIBA-GEIGY or the Seller be liable for consequential, special, or indirect damages resulting from the use or handling of this product.** CIBA-GEIGY and the Seller offer this product, and the Buyer and user accept it, subject to the foregoing **Conditions of Sale and Warranty,** which may be varied only by agreement in writing signed by a duly authorized representative of CIBA-GEIGY.

DIRECTIONS FOR USE

It is a violation of federal law to use this product in a manner inconsistent with its labeling.

FAILURE TO FOLLOW THE DIRECTIONS FOR USE AND PRECAUTIONS ON THIS LABEL MAY RESULT IN CROP INJURY, POOR WEED CONTROL, AND/OR ILLEGAL RESIDUES.

Do not apply this product in such a manner as to directly expose workers or other persons except those knowingly involved in the application. The area being treated must be vacated by unprotected persons.

Written or oral warnings must be given to workers who are expected to be in an area about to be treated with this product. Oral warnings must inform workers of areas or fields that may not be entered without specific protective clothing during application and appropriate actions to take in case of accidental exposure, as described under **Precautionary Statements** on this label. When oral warnings are given, warnings shall be given in a language customarily understood by workers. Oral warnings must be given if there is reason to believe that written warnings cannot be understood by workers. Written warnings must include the following information: CAUTION. Area will be treated with Princep 4G. Do not enter without appropriate protective clothing during application. In case of accidental exposure, flush eyes or skin with plenty of water. Call physician if irritation persists. Remove and wash contaminated clothing before reuse.

GENERAL INFORMATION

Apply Princep 4G herbicide before weeds emerge or following removal of weed growth. Princep controls a wide variety of annual broadleaf and grass weeds when used at selective rates in agricultural crops and ornamental plantings. When used at higher, nonselective rates in noncrop areas, it also controls many perennial broadleaf and grass weeds.

Since Princep enters weeds mainly through their roots, moisture is needed to move it into the root zone. Very dry soil conditions and lack of rainfall following application may necessitate shallow cultivation.

Princep 4G is noncorrosive to equipment, nonflammable, and has low electrical conductivity.

Annual Weeds Controlled

alyssum
annual bluegrass
annual morningglory
annual ryegrass
common lambsquarters
nightshade
pepperweed
barnyardgrass (watergrass)
burclover
carelessweed
carpetweed
common chickweed
crabgrass (*Digitaria* spp.)
downy brome (cheatgrass)
fall panicum
fiddleneck
filaree
fireweed
fivehook bassia
Flora's paintbrush
Florida pusley
foxtails
goosegrass
groundsel
henbit
junglerice
knawel (German moss)
pigweed
pineappleweed
prickly lettuce
common purslane
quackgrass
ragweed
rattail fescue
redmaids
Russian thistle
shepherdspurse
signalgrass (*Brachiaria* spp.)
silver hairgrass
smartweed
spanishneedles
speedwell
tansymustard
wild mustard
wild oats
witchgrass
yellow flower pepperweed

Following many years of continuous use of this product and chemically related products, biotypes of some of the weeds listed on this label have been reported which cannot be effectively controlled by this and related herbicides. Where this is known or suspected and weeds controlled by this product are expected to be present along with resistant biotypes, we recommend the use of this product in sequential combination with other registered herbicides which are not triazines. If only resistant biotypes are expected to be present, use a registered non-triazine herbicide. Consult with your State Agricultural Extension Service for specific recommendations.

SELECTIVE WEED CONTROL

Where a range of dosage rates is given, use the low rate on coarser textured soil and soil lower in organic matter; use the high rate on finer textured soil and soil higher in organic matter. For band applications, reduce the rate of Princep 4G in proportion to the area treated. For example: when applying a 2-foot band to a crop planted in rows 6 feet apart, apply one-third the recommended rate of Princep 4G per acre of field.

Distribute Princep 4G uniformly over the soil surface by means of a cyclone-type spreader or other suitable application equipment. For treatment of small areas, application by hand may be adequate.

After treatment with Princep 4G, do not plant any crop other than those specified on this label until the following year or injury may occur. Do not rotate with oats, sugar beets, or vegetables.

FRUIT CROPS

BLUEBERRIES

Apply 50-100 lbs. per acre in the spring before bud break and weed emergence, **or** apply a split application of 50 lbs. per acre in the spring before bud break and weed emergence, and 50 lbs. per acre in the fall. On plantings less than 6 months old, use ½ the above rate.

To control quackgrass, apply 100 lbs. per acre in the fall **or** split the application, applying 50 lbs. per acre in the fall and 50 lbs. per acre in the early spring, when quackgrass is actively growing and before bud break.

Note: Do not apply when fruit is present, or illegal residues may result.

CANEBERRIES; i.e., blackberries, boysenberries, loganberries, raspberries

Apply 50-100 lbs. per acre in the spring before bud break and weed emergence. On plantings less than 6 months old, use ½ the above rate.

To control quackgrass, apply 100 lbs. per acre in the fall **or** split the application, applying 50 lbs. per acre in the fall and 50 lbs. per acre in the early spring when quackgrass is actively growing.

Note: Do not apply when fruit is present, or illegal residues may result.

CRANBERRIES

Massachusetts: Apply up to 100 lbs. per acre either in the spring before cranberry growth or in the fall after harvest.

Other areas: Apply 50 lbs. per acre in the spring before new cranberry growth.

GRAPEFRUIT AND ORANGES

Florida: Apply 160-240 lbs. per acre in the spring (March or April) prior to weed emergence. Repeat annually.

Texas: Apply 100-125 lbs. per acre in the spring prior to weed emergence. Repeat annually.

Precautions: *To avoid crop injury, 1) Do not use in nurseries. 2) Do not apply to bedded citrus. 3) Use only where trees have been established in the grove for one year or more. 4) Do not apply to trees under stress from freeze damage for one year after the freeze. 5) Avoid contact with fruit or foliage.*

NURSERIES, CHRISTMAS TREE PLANTINGS, SHELTERBELTS

NURSERIES (SEE LIST BELOW)

Apply 50-75 lbs. per acre in fall or spring.

Precaution: *To avoid plant injury, do not apply for at least one year after transplanting.*

CHRISTMAS TREE PLANTINGS AND SHELTERBELTS (SEE LIST BELOW)

Remove weed growth before application. Apply 50-100 lbs. per acre after transplanting. Use the same rate for annual maintenance applications.

For quackgrass control, apply 100 lbs. per acre in the fall or apply a split application of 50 lbs. per acre in the fall plus 50 lbs. per acre in early spring, after quackgrass begins growth.

Precautions: *To avoid tree injury, 1) Do not use Princep 4G on seedbeds or cutting beds. 2) In CA, OR, and WA, do not apply to Christmas trees or shelterbelts sooner than one year after transplanting. In other areas, do not apply to Christmas tree or shelterbelt transplants less than three years of age. 3) Do not use until soil is firmly settled around roots. 4) Do not apply more than once a year, except as directed for quackgrass control.*

Apply Princep 4G to these species of trees and shrubs, as recommended above:

Conifers

arborvitae	Scotch pine	Fraser fir
Austrian pine	white pine	white fir
*Bishop pine	blue spruce	hemlock
*knobcone pine	Norway spruce	juniper
lodgepole pine (shore pine)	red spruce	red cedar
*Monterey pine	white spruce	white cedar
Mugho pine	balsam fir	yew (*Taxus* spp.)
red pine (Norway pine)	Douglas fir	

Deciduous Trees and Woody Ornamentals

American elm	Oregon grape (*Mahonia* spp.)
Siberian elm	*Pieris* spp. (max. 80 lbs./A)
barberry	red oak
boxelder	Russian olive
bush honeysuckle	*oleander
caragana	*palm (50 lbs./A)
cotoneaster	*bottle brush
dogwood	*carob
eucalyptus	
holly (*Ilex* spp.) (max. 80 lbs./A)	
honey locust	

*For California only

TREE SEEDLINGS IN NORTH CENTRAL U.S.

Broadcast 50-100 lbs. per acre before or after transplanting tree species listed below, but before weeds emerge or following removal of weed growth. Do not use on seed nurseries or injury may occur. Apply only once per year. For band or spot treatments use a proportional amount of herbicide. Use on the following species of trees:

American beech	red oak
balsam fir	red pine (Norway pine)
black locust	red spruce
black walnut	Russian olive
caragana	Scotch pine
dogwood	sweetgum
eastern red cedar	white ash
honey locust	white cedar
maple	white oak
Norway spruce	white pine
pecan	white spruce

TURFGRASS FOR FAIRWAYS, LAWNS, AND SIMILAR AREAS

BERMUDAGRASS, CENTIPEDEGRASS, ST. AUGUSTINEGRASS, AND ZOYSIA GRASS

Apply Princep 4G after October 1 before emergence of winter annual weeds for control of annual bluegrass, burclover, carpet burweed, chickweed, corn speedwell, henbit, hop clover, and spurweed. Princep 4G will control annual bluegrass even if it is emerged at time of treatment. For control of summer annual weeds listed in the General Information section of this label, also apply Princep 4G in late winter before the weeds emerge.

Where annual bluegrass is the major weed, use 25 lbs. of Princep 4G per acre (9 oz. per 1,000 sq. ft.). Use 50 lbs. per acre (18 oz. per 1,000 sq. ft.) for control of the other weeds named above. Do not exceed 25 lbs. per acre per treatment on newly sprigged turfgrass or on hybrid bermudagrass such as Tiflawn, Tifway, and Ormond.

For continued summer annual weed control, apply another 25 lbs. per acre at least 30 days after the previous application, but not after April 15. However, do not make more than two applications of this product per year.

Precautions: *On newly sprigged turfgrass and hybrid bermudagrass, temporary slowing of growth and yellowing may occur following application. To avoid turf injury, 1) Use only on turfgrass reasonably free of infestations of insects, nematodes, and disease. 2) Do not use on golf greens. 3) Do not use north of North Carolina or west of the high rainfall areas of eastern Oklahoma and eastern Texas. 4) Do not use on muck or alkaline soils. 5) Do not apply over the rooting area of trees or ornamentals not listed on this label. 6) Do not overseed with desirable turfgrass within 4 months before or 6 months after treatment.*

NONSELECTIVE WEED CONTROL ON NONCROP LAND

Apply Princep 4G at rates recommended below. Use the higher rates on heavy clay and muck soils. Best results are obtained when Princep 4G is applied before weeds emerge.

Do not contaminate domestic or irrigation water supplies, or lakes, streams, or ponds.

Princep 4G provides long-term weed control on industrial sites, highway medians and shoulders, railroad rights-of-way, lumberyards, petroleum tank farms, and in noncrop areas on farms, such as around buildings, equipment and fuel storage areas, along fences, roadsides, and lanes.

Precaution: *Do not use this product for nonselective weed control on land to be cropped, near adjacent desirable trees, shrubs or plants, or in greenhouses, or injury may occur.*

To control most annual broadleaf and grass weeds (such as barnyardgrass, cheatgrass, crabgrass, lambsquarters, foxtail, ragweed, puncturevine, and turkey mullein): Broadcast 250 lbs. per acre.

To control most annual and many perennial broadleaf and grass weeds (including quackgrass, bluegrass, redtop, burdock, Canada thistle, orchardgrass, purple top, smooth brome, dogfennel, and plantain): Broadcast 250-500 lbs. per acre.

To control hard-to-kill perennial weeds (such as bullthistle and sowthistle): Broadcast 500-1,000 lbs. per acre.

For longer residual control in regions of high rainfall and a long growing season: Broadcast 500-1,000 lbs. per acre.

For small areas, 5¾ lbs. per 1,000 sq. ft. is equivalent to 250 lbs. per acre.

STORAGE AND DISPOSAL

Store in a dry place. Do not contaminate water, food, or feed by storage or disposal. Improper disposal of unused pesticide is a violation of federal law. If wastes cannot be used according to label instructions, contact your State Pesticide or Environmental Control Agency, or the Hazardous Waste representative at the nearest EPA Regional Office for guidance in proper disposal methods.

Completely empty bag into application equipment. Dispose of empty bag and box in a sanitary landfill or by incineration, or if allowed by state and local authorities, by burning. Stay out of smoke from burning containers.

For minor spills, follow all precautions indicated on this label and clean up immediately. Take special care to avoid contamination of equipment and facilities during cleanup procedures and disposal of wastes. In the event of a major spill, fire or other emergency, call (919) 292-7100 day or night.

PRECAUTIONARY STATEMENTS

HAZARDS TO HUMANS AND DOMESTIC ANIMALS

CAUTION

Causes eye irritation. Harmful if swallowed, inhaled, or absorbed through the skin. Avoid breathing dust. Avoid contact with skin, eyes, or clothing. Wash thoroughly with soap and water after handling. Remove and wash contaminated clothing before reuse.

First Aid: If in eyes, flush with plenty of water. Get medical attention if irritation persists. If on skin, wash with plenty of soap and water. If inhaled, remove victim to fresh air. If not breathing, give artificial respiration, preferably mouth-to-mouth; get medical attention.

ENVIRONMENTAL HAZARDS

Do not apply directly to water or wetlands. Do not contaminate water by cleaning of equipment or disposal of wastes. Simazine is a chemical which can travel (seep or leach) through soil and can contaminate ground water which may be used as drinking water. Simazine has been found in ground water as a result of agricultural use. Users are advised not to apply simazine where the water table (ground water) is close to the surface and where the soils are very permeable, i.e., well-drained soils such as loamy sands. Your local agricultural agencies can provide further information on the type of soil in your area and the location of ground water.

Princep® trademark of CIBA-GEIGY for simazine

CGA 6L13V 095

PRINCEP®4L

HERBICIDE

For weed control in certain crops, ornamental plantings, on industrial sites, and around-the-farm

ACTIVE INGREDIENT:

Simazine:

2-chloro-4,6-bis (ethylamino)-*s*-triazine 41.9%

INERT INGREDIENTS: 58.1%

TOTAL: 100.0%

EPA Reg. No. 100-526 EPA Est. 100-LA-1

Princep® trademark of CIBA-GEIGY for simazine

Princep 4L contains 4 lbs. active ingredient per gallon.

Use entire contents at one time.

Keep Out of Reach of Children.

Caution

See additional precautionary statements at end of label booklet.

See directions for use inside booklet.

DIRECTIONS FOR USE AND CONDITIONS OF SALE AND WARRANTY

IMPORTANT: Read the entire **Directions for Use** and the **Conditions of Sale and Warranty** before using this product.

CONDITIONS OF SALE AND WARRANTY

The **Directions for Use** of this product reflect the opinion of experts based on field use and tests. The directions are believed to be reliable and should be followed carefully. However, it is impossible to eliminate all risks inherently associated with use of this product. Crop injury, ineffectiveness, or other unintended consequences may result because of such factors as weather conditions, presence of other materials, or the manner of use or application all of which are beyond the control of CIBA-GEIGY or the Seller. All such risks shall be assumed by the Buyer.

CIBA-GEIGY warrants that this product conforms to the chemical description on the label and is reasonably fit for the purposes referred to in the **Directions for Use** subject to the inherent risks referred to above. **CIBA-GEIGY makes no other express or implied warranty of Fitness or Merchantability or any other express or implied warranty. In no case shall CIBA-GEIGY or the Seller be liable for consequential, special, or indirect damages resulting from the use or handling of this product.** CIBA-GEIGY and the Seller offer this product, and the Buyer and user accept it, subject to the foregoing **Conditions of Sale and Warranty,** which may be varied only by agreement in writing signed by a duly authorized representative of CIBA-GEIGY.

DIRECTIONS FOR USE

It is a violation of federal law to use this product in a manner inconsistent with its labeling.

FAILURE TO FOLLOW THE DIRECTIONS FOR USE AND PRECAUTIONS ON THIS LABEL MAY RESULT IN CROP INJURY, POOR WEED CONTROL, AND/OR ILLEGAL RESIDUES.

Do not apply this product in such a manner as to directly or through drift expose workers or other persons except those knowingly involved in the application. The area being treated must be vacated by unprotected persons.

REENTRY STATEMENT

Do not enter treated areas without protective clothing until sprays have dried.

Because certain states may require more restrictive reentry intervals for various crops treated with this product, consult your State Department of Agriculture for further information.

Written or oral warnings must be given to workers who are expected to be in a treated area or in an area about to be treated with this product. Oral warnings must inform workers of areas or fields that may not be entered without specific protective clothing until sprays have dried, and appropriate actions to take in case of accidental exposure, as described under **Precautionary Statements** on this label. When oral warnings are given, warnings shall be given in a language customarily understood by workers. Oral warnings must be given if there is reason to believe that written warnings cannot be understood by workers. Written warnings must include the following information: CAUTION. Area treated with Princep 4L on *(date of application)*. Do not enter without appropriate protective clothing until sprays have dried. In case of accidental exposure, flush eyes or skin with plenty of water. Call physician if irritation persists. Remove and wash contaminated clothing before reuse.

GENERAL INFORMATION

Apply Princep 4L Herbicide before weeds emerge or after removal of weed growth. Princep controls a wide variety of annual broadleaf and grass weeds when used at selective rates in agricultural crops and ornamental plantings. When used at higher, nonselective rates in noncrop areas, it also controls many perennial broadleaf and grass weeds.

Where a range of application rates is given, use the low rate on coarser textured soil and soil lower in organic matter; use the high rate on finer textured soil and soil higher in organic matter.

To avoid spray drift, do not apply under windy conditions. Avoid spray overlap, as crop injury may result.

Since Princep enters weeds mainly through their roots, moisture is needed to move it into the root zone. Very dry soil conditions following application may necessitate shallow cultivation or rotary hoeing.

Princep 4L is nonflammable.

Annual Weeds Controlled

alyssum	fireweed	common purslane
annual bluegrass	fivehook bassia	ragweed
annual morningglory	Flora's paintbrush	rattail fescue
annual ryegrass	Florida pusley	redmaids
barnyardgrass (watergrass)	foxtails	Russian thistle
burclover	goosegrass	shepherdspurse
carelessweed	groundsel	signalgrass (*Brachiaria* spp.)
carpetweed	henbit	silver hairgrass
common chickweed	junglerice	smartweed
crabgrass (*Digitaria* spp.)	knawel (German moss)	spanishneedles
downy brome (cheatgrass)	common lambsquarters	speedwell
fall panicum	nightshade	tansymustard
fiddleneck	pepperweed	wild mustard
filaree	pigweed	wild oats
	pineappleweed	witchgrass
	prickly lettuce	yellow flower pepperweed

Following many years of continuous use of this product and chemically related products, biotypes of some of the weeds listed on this label have been reported which cannot be effectively controlled by this and related herbicides. Where this is known or suspected and weeds controlled by this product are expected to be present along with resistant biotypes, we recommend the use of this product in registered combinations or in sequence with other registered herbicides which are not triazines. If only resistant biotypes are expected to be present, use a registered non-triazine herbicide. Consult with your State Agricultural Extension Service for specific recommendations.

APPLICATION PROCEDURES

Ground application: Use conventional ground sprayers equipped with nozzles that provide accurate and uniform application. Be certain that nozzles are uniformly spaced and the same size. Calibrate sprayer before use and recalibrate at the start of each season and when changing carriers. Unless otherwise specified, use a minimum of 20 gals. of spray mixture per acre.

Use a pump with capacity to (1) maintain 35-40 psi at nozzles, (2) provide sufficient agitation in tank to keep mixture in suspension, and (3) to provide a minimum of 20% bypass at all times. Use centrifugal pumps which provide propeller shear action for dispersing and mixing this product. The pump should provide a minimum of 10 gals./minute/100 gal. tank size circulated through a correctly positioned sparger tube or jets.

Use screens to protect the pump and to prevent nozzles from clogging. Screens placed on the suction side of the pump should be *16-mesh or coarser.* Do not place a screen in the recirculation line. Use 50-mesh or coarser screens between the pump and boom, and where required, at the nozzles. Check nozzle manufacturer's recommendations.

For band applications, calculate amount to be applied per acre as follows:

$$\frac{\text{band width in inches}}{\text{row width in inches}} \times \text{broadcast rate per acre} = \text{amount needed per acre of field}$$

Aerial application: Use aerial application only where specified in the use directions. Apply in a minimum of 1 gal. of water for each quart of Princep 4L applied per acre. Avoid applications under conditions where uniform coverage cannot be obtained or where excessive spray drift may occur.

Avoid application to humans or animals. Flagmen and loaders should avoid inhalation of spray mist and prolonged contact with skin, and should wash thoroughly before eating and at the end of each day's operation.

Mixing procedures—all uses: (1) Be sure sprayer is clean and not contaminated with any other materials, or crop injury or sprayer clogging may result. (2) Fill tank ¼ full with clean water, nitrogen solution, or complete fluid fertilizer. (3) Start agitation. (4) Be certain that the agitation system is working properly and creates a rippling or rolling action on the liquid surface. (5) If any dry herbicides are being tank mixed, add them to tank and allow them to become thoroughly suspended. (6) Then pour Princep 4L directly from jug into tank. (7) Continue filling tank until 90% full. Increase agitation if necessary to maintain surface action. (8) Add other liquid tank mix herbicide(s) after this product is thoroughly suspended. (9) Finish filling tank. Maintain agitation to avoid separation of materials. (10) Empty tank as completely as possible before refilling to prevent buildup of emulsifiable concentrate residue from possible tank mix herbicides. (11) If an emulsifiable concentrate film starts to build up in tank, drain it and clean with strong detergent solution or solvent. (12) Clean sprayer thoroughly immediately after use by flushing system with water containing a detergent.

Compatibility test: To determine the tank mix compatibility of Princep 4L with liquid fertilizer, crop oil, spreaders, or other recommended pesticides, use this test method. Nitrogen solutions or complete fluid fertilizers may replace all or part of the water in the spray in corn. Since liquid fertilizers can vary, even within the same analysis, check compatibility each time before use. Be especially careful when using **complete** suspension or fluid fertilizers as serious compatibility problems are more apt to occur. Commercial application equipment may improve compatibility in some instances. Check compatibility using this procedure:

1. Add 1 pint of fertilizer to each of 2 one-quart glass jars with tight lids.
2. To **one** of the jars add ¼ tsp. of a compatibility agent approved for this use (¼ tsp. is equivalent to 2 pts. per 100 gals. spray). Cap and shake until mixed. Examples of compatibility agents include Compex® and Unite®.
3. To **both** jars add the appropriate amount of herbicide(s). If more than one herbicide is used, add them separately with dry herbicides first, flowables next, and emulsifiable concentrates last. After each addition, cap and shake until thoroughly mixed. The

appropriate amount of herbicides for this test follows (assuming a spray volume of 25 gals. per acre):

Dry herbicides: **For each pound per acre** add 1.5 teaspoons to each jar. Fluff up wettable powder products before measuring.

Liquid herbicides: **For each pint per acre,** add 0.5 teaspoon to each jar.

For a spray volume other than 25 gals. per acre, change the teaspoons added to each jar as follows:

$$\frac{\text{25 gals. x no. teaspoons given above}}{\text{desired gals. spray volume per acre}} = \text{teaspoons to add to each jar}$$

4. After adding all ingredients, put lids on and tighten. Shake jars vigorously one minute. Let the mixtures stand 15 minutes and then look for separation, large flakes, precipitates, grease, gels, medium to heavy oily film on the jar, or other signs of incompatibility. Determine if the compatibility agent is needed in the spray mixture by comparing the two jars. If either mixture separates, but can be remixed readily, the mixture can probably be sprayed as long as good agitation is used. If the mixtures are incompatible, test the following methods of improving compatibility: (A) slurry the dry herbicide(s) in water before addition, or (B) add ½ of the compatibility agent to the fertilizer and the other ½ to the emulsifiable concentrate or flowable herbicide before addition to the mixture. If still incompatible, do not use the materials mixed in the same spray tank.

ROTATIONAL CROPS AND PERENNIAL CROP REPLANTING

To avoid crop injury, observe the following precautions: 1) If rotating treated land the year following application, plant only corn, unless otherwise stated in this label. 2) If replanting perennial crops or if rotating land to crops other than corn, do not apply Princep in the year preceding planting these crops.

FRUIT AND NUT CROPS

Apply the spray to the orchard or vineyard floor avoiding contact with fruit, foliage, or stems. Recommended rates are based on broadcast treatment. For band applications or spot applications around trees in fruit or nut plantings, reduce the broadcast rate of Princep 4L and water per acre in proportion to the area actually sprayed.

Precautions (All fruit and nut crops): *To avoid crop injury, 1) Apply only to orchards or groves where trees have been established 12 months or more unless specified differently. 2) Make only one application per year, except as noted otherwise. 3) Do not use on gravelly, sand, or loamy sand soil. 4) Immediately following application, limit overhead sprinkler irrigation to* ½ inch.

ALMONDS AND PEACHES (CA only)

Apply 1-2 qts. of Princep 4L per treated acre in a 2-4 ft. band on each side of the tree row. Apply before weeds emerge in late fall or early winter. Weeds controlled by 1 qt. of Princep 4L include burclover, common chickweed, wild mustard, and shepherdspurse. Apply only once per year.

Precautions: *To avoid crop injury, 1) Do not treat trees established in the grove less than 3 years. 2) Do not treat the Mission (Texas) variety of almonds. 3) Do not apply to almond trees propagated on plum rootstocks. 4) Do not replant almonds or peaches in treated soil for 12 months after treatment. 5) Do not apply on soil with less than 1% organic matter. 6) Do not treat areas where water will accumulate.*

APPLES, PEARS, SOUR CHERRIES

Apply 2-4 qts. per acre.

AVOCADOS

CA and FL only: Apply 2-4 qts. per acre after final preparation of grove.

BLUEBERRIES AND CANEBERRIES (blackberries, boysenberries, loganberries, raspberries)

Apply 2-4 qts. per acre in the spring or apply a split application of 2 qts. per acre in the spring plus 2 qts. per acre in the fall. Apply in a minimum of 40 gals. of water per acre.

On plantings less than 6 months old, use ½ the above rate.

To control quackgrass, apply 4 qts. per acre in the fall or split the application applying 2 qts. per acre in the fall plus 2 qts. per acre in the spring, when quackgrass is actively growing.

Note: Do not apply when fruit is present, or illegal residues may result.

CRANBERRIES

MA: Apply up to 4 qts. per acre either before spring growth begins or in the fall after harvest.

Other areas: Apply 2 qts. per acre before spring growth begins.

FILBERTS

Apply 2-4 qts. per acre in the fall or apply a split application of 2 qts. per acre in the fall plus 2 qts. per acre in the spring.

Precaution: *If trees are planted on a hillside, excessive soil erosion may result from the elimination of weeds.*

Note: Do not apply when nuts are on the ground during the harvest period, or illegal residues may result.

GRAPES

Apply 2-4.8 qts. per acre any time between harvest and early spring.

Precaution: *Do not use in vineyards established less than three years, or crop injury may occur.*

GRAPEFRUIT, LEMONS, ORANGES

AZ (Lemons and Oranges only): Apply a split application of 1.6 qts. per acre in the spring plus 1.6 qts. per acre in the fall.

CA: Apply 2-4 qts. per acre in a single application; or apply 2 qts. per acre in the fall and 2 qts. per acre in the spring.

Precaution: *Do not use in the Imperial, Coachella, or Palo Verde Valleys, or crop injury may occur.*

FL (Grapefruit and Oranges only): Apply only once per year. For control of weeds listed under General Information, apply 6.4-9.6 qts. per acre in the spring; to control milkweed vine, use 8-9.6 qts. per acre either preemergence or before vines are 12 inches tall.

TX (Grapefruit and Oranges only): Apply 4-4.8 qts. per acre.

Precautions (All areas): *To avoid crop injury, 1) Do not use in nurseries. 2) Do not apply to bedded grapefruit, lemons, or oranges. 3) Do not apply to trees under stress from freeze damage for one year after the freeze.*

MACADAMIA NUTS

Apply 2-4 qts. in 50 gals. of water per acre before harvest and just prior to weed emergence. Repeat application as necessary.

Note: Do not apply when nuts are on the ground during the harvest period, as illegal residues may result.

OLIVES

Apply 2-4 qts. per acre following grove preparation in the fall. Repeat annually in midwinter.

PEACHES, PLUMS, SWEET CHERRIES

Apply 1.6-4 qts. per acre. Apply in late fall to early spring prior to weed emergence.

Precautions: *To avoid crop injury, 1) Peaches: use only in AR, LA, MO, OK, TX, and states east of the Miss. River. For California, see specific directions in the section "Almonds and Peaches (CA only)." 2) Plums and sweet cherries: use only in MO and states east of the Miss. River except TN.*

PECANS

Apply 2-4 qts. per acre before weeds emerge in the spring.

Precautions: *To avoid crop injury, 1) Do not use west of the Pecos River in TX or in NM, AZ, or CA. 2) Do not make applications to transplanted trees that have been established less than two years in the grove.*

Note: Do not apply when nuts are on the ground, or illegal residues may result. Do not allow animals to graze treated areas.

STRAWBERRIES

Oregon and Washington: For control of chickweed, groundsel, mustard, and shepherdspurse, apply broadcast 1 qt. per acre. In fields where overhead irrigation is used to activate this product, apply after harvest at time of bed renovation. In fields where overhead irrigation is not available, apply during early October through November.

Precautions: *To avoid crop injury, 1) Make only one application per growing season. 2) Do not apply within 4 months after transplanting.*

WALNUTS

Apply 2-4 qts. per acre. Leveling and furrowing operations after application will lessen effectiveness of weed control.

Note: Do not apply when nuts are on the ground, or illegal residues may result.

PRINCEP 4L PLUS EVIK® 80W FOR GRAPEFRUIT AND ORANGES

Use in grapefruit and oranges in FL and TX for control of weeds listed in the following directions and under General Information. Apply the rates given below in a minimum of 40 gals. of water per acre thoroughly covering existing vegetation, but avoiding contact with citrus fruit, foliage, or stems. Apply when weeds have resumed active growth in the spring, but before they exceed 4 inches in height (except as noted for vines). Observe all precautions and limitations on the labeling of Princep 4L and Evik 80W. When mixing, add Evik 80W slurry to water in spray tank, agitate thoroughly, then add Princep 4L and agitate thoroughly again.

FL (Ridge Area): Apply 4 qts. of Princep 4L plus 5 lbs. of Evik 80W per acre; controls milkweed vine when applied either preemergence or on vines up to 24 inches tall.

FL (Bedded Areas): Apply 3.2 qts. of Princep 4L plus 6 lbs. of Evik 80W per acre; controls balsamapple when applied either preemergence or on vines up to 12 inches tall.

TX: Apply 3.2-4 qts. of Princep 4L plus 2-3 lbs. of Evik 80W per acre plus 2 qts. nonionic surfactant, such as X-77®, per 100 gals. of spray mixture (add surfactant to spray tank last); controls buffelgrass, marinevine (ivy treebine), Texas panicum, sandbur, and sunflower, and gives top-kill of johnsongrass. Use 4 qts. of Princep 4L per acre for longer residual control; use 3 lbs. of Evik 80W per acre for control of heavy weed growth and for top-kill of johnsongrass; use the higher rate of both products for control of marinevine.

Precautions (All areas): *To avoid injury, 1) Do not use in nurseries, or around trees that have been established in the grove less than two years, or where trees are under stress from freeze damage for one year after the freeze. 2) Apply only once per year.*

Note (All areas): Do not apply within 30 days before harvest, or illegal residues may result. Do not graze treated areas.

PRINCEP 4L PLUS BROMACIL 80W FOR GRAPEFRUIT AND ORANGES (Florida Only)

Use in grapefruit and oranges in FL for control of balsamapple, black nightshade, carpetweed, crabgrass, cudweed, dayweed, Florida pusley, horseweed, pepperweed, pigweed, poorjoe, ragweed, rattlebox, spanishneedles and sandbur, and for partial control of bermudagrass, bahiagrass, pangolagrass, paragrass, and torpedograss. Apply 3.2-6.4 qts. of Princep 4L plus 3-4 lbs. of bromacil 80W per acre beneath trees in a minimum of 40 gals. of water per acre before or soon after weed growth begins. When mixing, add bromacil 80W slurry to water in spray tank, agitate thoroughly, then add Princep 4L and agitate thoroughly again. Use the lower rates for light weed infestations or all applications in bedded citrus areas. Use the higher rates for heavy weed infestations only in ridge grown citrus areas. Temporary yellowing of citrus leaves may occur following treatment.

Precautions: *To avoid crop injury, 1) Do not use in nurseries, or around trees that have been established in the grove less than four years, or where trees are under stress from freeze damage for one year after the freeze. 2) Do not use on soil with less than 1% organic matter or on poorly drained soil. 3) Do not treat trees planted in irrigation furrows. 4) Do not treat diseased trees such as those with foot rot. 5) Do not use in groves interplanted with other trees or desirable plants, nor in home grapefruit or orange plantings, or in areas where roots of other valuable plants or trees may extend. 6) Treated areas may be planted to citrus trees one year after application. 7) Do not rotate to other crops within two years after application.*

Note: Apply only once per year and avoid contact with foliage and fruit with spray or mist, or illegal residues may result.

PRINCEP 4L PLUS PARAQUAT

This tank mix is effective in the following fruit and nut crops for kill of existing vegetation and for residual control of the annual broadleaf and grass weeds claimed for Princep 4L applied alone. This combination is also effective for top-kill and suppression of perennial weeds.

Use this tank mix on these crops:

almonds (CA only)	macadamia nuts
apples	olives
avocados (CA and FL[1] only)	oranges[2]
cherries (sour and sweet[4])	peaches[3]
filberts	pears
grapefruit (CA, FL[2] and TX only)	pecans
grapes	plums[4]
lemons (AZ and CA only)	walnuts

Apply the Princep 4L rate given under the appropriate crop on this label plus 1-2 qts. paraquat* in 50-200 gals. (30-50 gals. for pecans) of water per acre to the orchard floor avoiding contact with fruit, foliage, or stems. Add a nonionic surfactant, such as X-77, at 0.5 pt. per 100 gals. of spray. Apply when weeds are succulent and new growth is 1-6 inches tall. For mature woody weeds or difficult to control perennial weeds, re-treat or spot treat with paraquat if regrowth occurs.

Add Princep 4L to the spray tank first (refer to Mixing Procedures section of this label), then add paraquat, and add the surfactant last. Provide constant agitation during mixing and application to keep the mixture in suspension.

Precautions: *To avoid crop injury, 1) Apply the tank mix only once per year. 2) Use a shield for young trees or vines.*

Refer to the Princep 4L and paraquat labels for further directions, specific weeds controlled, and precautions and limitations on each crop.

[1]In avocados in Florida, this tank mix also controls balsamapple, rattail amaranth, and at the higher rate of each herbicide, it suppresses coral vine.

[2]In grapefruit and oranges in Florida, apply 3.2 qts. Princep 4L per acre in bedded areas and 6.4-9.6 qts. Princep 4L per acre in ridge areas.

[3]Limited to CA, AR, LA, MO, OK, TX, and states east of the Miss. River. As appropriate, refer to Princep 4L sections "Almonds and Peaches (CA only)" or "Peaches, Plums, Sweet Cherries" for rate of Princep 4L and other information.

[4]Limited to MO and states east of the Miss. River except TN.

*Based on a product containing 2 lbs. paraquat cation per gal.

TANK MIXTURE WITH ROUNDUP®

This tank mixture is effective in grape vineyards and in the following bearing or nonbearing tree crops for control of existing vegetation and for residual control of the annual broadleaf and grass weeds claimed for this product applied alone. This combination is also effective for partial control of perennial weeds contacted by the spray mixture during application. Use this tank mixture on these crops:

almonds** (CA only)	macadamia nuts**
apples*	oranges* (AZ, CA, FL[1] and TX)
avocados* (CA and FL only)	peaches*[2]
cherries (sour and sweet[3])	pears*
filberts**	pecans**
grapefruits* (CA, FL[1] and TX)	plums*[3]
grapes*	walnuts**
lemons* (AZ and CA)	

Use the appropriate rate given elsewhere on this label for Princep applied alone to the crop being treated. Add to the spray tank 1-5 qts. of Roundup per acre depending on weeds present and their growth stage. Also add an agriculturally approved nonionic surfactant at 0.5% by volume of spray solution. Apply the mixture in 10-40 gals. of water per acre as a postemergence spray to the weeds at the appropriate weed growth stage given on the Roundup label.

Add Princep to the spray tank first, then add Roundup. Provide constant agitation during mixing and application to keep the mixture in suspension. Refer to Application Procedures section of this label for further directions.

Precautions: *To avoid crop injury, take extreme care to avoid contact of herbicide solution, spray, drift, or mist with foliage or green bark of trunk, branches, suckers, fruit, or other parts of trees or vines.*

Refer to the labels of both herbicides for further directions, specific weeds controlled, precautions and limitations on each crop.

*Allow a minimum of 14 days between last application and harvest.

**Allow a minimum of 21 days between last application and harvest of these crops.

[1]In grapefruit and oranges in Florida, apply 3.2 qts. of this product per acre in bedded areas and 6.4-9.6 qts. per acre in ridge areas.

[2]Limited to CA, AR, LA, MO, OK, TX, and states east of the Mississippi River. For CA, see specific directions in the section "Almonds and Peaches (CA only)."

[3]Plums and sweet cherries: limited to MO and states east of the Mississippi River except TN.

PRINCEP 4L PLUS SURFLAN®

Use this tank mix in the following crops for preemergence control of all weeds claimed on both labels:

almonds	cherries	lemons	pecans
apples	filberts	oranges	plums
avocados	grapes	peaches	English walnuts
caneberries	grapefruit	pears	

Apply the Princep 4L rate given under the appropriate crop on this label plus 2.67-5.33 lbs. of Surflan 75W or 2-4 qts. of Surflan A.S. in 20-40 gals. of water per acre. Refer to the Surflan 75W or Surflan A.S. label for complete tank mix directions, and use 0.8 qt. of Princep 4L for each pound of Princep 80W. Observe all precautions and limitations on the Princep and Surflan labels.

FIELD CROPS

CORN

Nitrogen solution or complete liquid fertilizer may replace all or part of the water in the spray. Determine the physical compatibility of Princep 4L with fertilizer before use. Refer to the General Information section of this label for a compatibility test procedure. Do not apply after corn has emerged, as there is danger of liquid fertilizer causing crop injury. Use a minimum of 10 gals. of spray mixture per acre.

Preemergence: Apply before weeds and corn emerge. Use the appropriate rate in the following table.

Preplant: Broadcast in the spring after plowing at the appropriate rate in the following table. Apply before, during, or after final seedbed preparation. If soil is tilled or worked after application, avoid deep incorporation of Princep 4L. Best results will be obtained when Princep is applied within 2 weeks before planting. Under dry weather conditions, preplant applications may give better weed control than preemergence. If weeds develop, particularly under relatively dry conditions, a shallow cultivation will generally result in better weed control.

Soil texture	Broadcast* rate per acre
Sand, silt, and loam that is low in organic matter	4 pts.
Soil containing moderate amounts of clay and organic matter	4.8 pts.
Loam that is high in organic matter and clay (including dark prairie soils of Corn Belt)	6 pts.
Peat, muck and high-organic clay	8 pts.

* For calculation of band treatment rate, see the General Information section.

Quackgrass Control: Broadcast 6-8 pts. per acre in the fall. Plow 2-3 weeks later, or if erosion is a problem, delay plowing until spring.

Precautions: *1) Do not apply more than 8 pts. of Princep 4L per acre of corn per year. 2) Do not plant any crop except corn until the following year or injury may occur. 3) Following harvest, plow and thoroughly till the soil in fall or spring to minimize possible injury to spring-seeded rotational crops, regardless of the rate used. 4) If Princep 4L is used at a rate higher than 6 pts. per acre or equivalent band rate, a crop of untreated corn should precede the next rotational crop. 5) Do not apply Princep 4L preplant incorporated for weed control in corn in the High Plains and Intermountain areas of the West (including central and western KS, western NE, western OK, and the Panhandle of TX) where rainfall is sparse and erratic or where irrigation is required. 6) In the High Plains and Intermountain areas of the West where rainfall is sparse and erratic or where irrigation is required, use Princep 4L only when corn is to follow corn or when a crop of untreated corn is to precede other rotational crops. 7) In western MN and eastern parts of the Dakotas, NE and KS, do not rotate to soybeans if the rate applied was more than 4 pts. per acre or equivalent band rate or injury may occur. 8) Injury may occur to soybeans planted in north-central and northwest IA, south-central and southwest MN, northeast NE, southeast SD, and other areas the year following application on soils having a calcareous surface layer. 9) Do not plant sugar beets, tobacco, vegetables (including dry beans), spring-seeded small grains, or small-seeded legumes and grasses the year following Princep 4L application or injury may occur.*

Note: Do not graze treated areas, or illegal residues may result.

TANK MIXTURES ON CORN

AAtrex®: Use Princep 4L in a preplant or preemergence tank mixture with AAtrex Nine-O®, AAtrex 80W, or AAtrex 4L (4LC) for control of many annual weeds, including carpetweed, crabgrass, fall panicum, foxtail, lambsquarters, morningglory, pigweed, ragweed, and velvetleaf.

Apply at the rates given below. Use the 1:1 ratio for most weeds and the 2:1 ratio for expected heavy infestations of crabgrass and fall panicum.

	Broadcast rate per acre			
	1:1 Ratio		2:1 Ratio	
Soil texture	Princep 4L	AAtrex 80W*	Princep 4L	AAtrex 80W*
Sand, loamy sand, sandy loam	1 qt.	1.25 lbs.	1.33 qts.	0.83 lb.
Loam, silt loam, silt, clay loam, sandy clay loam, silty clay loam, sandy clay, or silty clay with low organic matter	1.2 qts.	1.5 lbs.	1.6 qts.	1 lb.
Loam, silt loam, silt, clay loam, sandy clay loam, silty clay loam, sandy clay, or silty clay with medium to high organic matter, and clay (including dark prairie soils of the Corn Belt)	1.44 qts.	1.8 lbs.	1.92 qts.	1.2 lbs.

*When using AAtrex 4L (4LC) or AAtrex Nine-O, use equivalent rates. One lb. of AAtrex 80W equals 0.8 qt. of AAtrex 4L (4LC) or 0.9 lb. of AAtrex Nine-O.

Refer to the AAtrex label for complete directions. Also refer to the AAtrex label and the Corn section of this label for precautions and limitations.

Eradicane®: Use in a preplant incorporated tank mixture for control of all weeds claimed on both the Princep and Eradicane labels including partial control (suppression) of shattercane (wild cane). Fluid fertilizer may replace all or part of the water in the spray. Check the physical compatibility of mixture with fertilizer before use. Refer to the General Information section of this label for a compatibility test procedure. Use a minimum of 20 gals. of spray volume per acre. Refer to the Eradicane label for incorporation directions. Use the higher rate of Princep on fine-textured soil and where heavy broadleaf infestations are expected. Use the higher rate of Eradicane for heavy nutsedge and bermudagrass infestations. For partial control (suppression) of shattercane, broadcast and incorporate **immediately** before planting 2-4 pts. of Princep 4L plus 7.33 pts. of Eradicane 6.7E (or 8 pts. Eradicane Extra) per acre. For control of the other weeds claimed on both labels, broadcast and incorporate 2-6 pts. of Princep 4L plus 4.75-7.33 pts. of Eradicane 6.7E (or 5.33 pts. of Eradicane Extra) per acre. Observe all precautions and limitations on the Princep 4L and Eradicane labels.

Paraquat: Use in a tank mixture where corn will be planted directly in a cover crop, established sod, or previous crop residues. This combination controls existing vegetation and provides residual control of the annual broadleaf and grass weeds listed under General Information.

Add Princep 4L to the spray tank, mix thoroughly with water, and then add paraquat and a nonionic surfactant. Provide constant agitation during mixing and application to keep the mixture in suspension.

Apply 4-6 pts. of Princep 4L plus 1-2 pts. of paraquat* in 20-60 gals. of water per acre as a broadcast spray either before or after planting but before corn emerges. Add a nonionic surfactant at the rate of 0.5 pt. per 100 gals. of spray volume.

For further information, see general information, caution and warning statements, precautions, and notes on the Princep 4L and the paraquat labels.

*Based on a product containing 2 lbs. paraquat cation per gal.

Sutan +®: Use in a preplant incorporated tank mixture for control of all weeds claimed on both the Princep and Sutan + labels including partial control (suppression) of shattercane (wild cane). Fluid fertilizer may replace all or part of the water in the spray. Check the physical compatibility of mixture with fertilizer before use. Refer to the General Information section of this label for a compatibility test procedure. Use a minimum of 20 gals. of spray volume per acre. Refer to the Sutan + label for incorporation directions. Use the higher rate of Princep on fine-textured soil and where heavy broadleaf infestations are expected. For partial control (suppression) of shattercane, broadcast and incorporate **immediately** before planting 2-4 pts. of Princep 4L plus 7.33 pts. of Sutan + 6.7E per acre. For control of the other weeds claimed on both labels, broadcast and incorporate 2-6 pts. of Princep 4L plus 4.75 pts. of Sutan + 6.7E per acre. Observe all precautions and limitations on the Princep 4L and Sutan + labels.

SUGARCANE

Apply 2-4 qts. per acre at planting or after harvest if ratoon crop is to be grown, but before weeds or sugarcane emerge. Princep 4L may also be applied interline-directed or over the sugarcane to provide continuing control of germinating weeds.

Precautions: *To avoid crop injury or illegal residues, 1) Do not apply more than one 2-4 qt. application over the plants or more than two interline-directed applications. 2) Do not apply after close-in. 3) Total application should not exceed 10 qts. per acre crop cycle.*

Note: Do not feed or graze treated forage to livestock.

GRASSES GROWN FOR SEED (Pacific Northwest Only)

Perennial ryegrass, Highland or Astoria bentgrasses, orchardgrass, tall fescue, and fine fescues such as Illahee, Rainier, Pennlawn, Chewings, and related species. For the control of broadleaf weeds and annual grasses including annual ryegrass, rattail fescue, silver hairgrass, and downy brome (cheat), apply 2 qts. Princep 4L in a minimum of 15 gals. of water per acre as soon as fall rains start. Early application is important.

Precaution: *To avoid crop injury, apply only to grasses from which at least one seed crop has been cut.*

Note: To avoid illegal residues, allow 30 days between last application and grazing of dairy and beef cattle and sheep, and allow 60 days between the last application and cutting for hay.

NURSERIES, CHRISTMAS TREE PLANTINGS, SHELTERBELTS

NURSERIES (see list below)

Apply 2-3 qts. in at least 25 gals. of water per acre in fall or spring.

Precaution: *To avoid plant injury, do not apply for at least one year after transplanting.*

CHRISTMAS TREE PLANTINGS AND SHELTERBELTS (see list below)

Remove weed growth before application. Apply 2-4 qts. in at least 25 gals. of water per acre after transplanting. Use the same rate for annual maintenance applications.

For quackgrass control apply 4 qts. per acre in the fall or apply a split application of 2 qts. per acre in the fall plus 2 qts. per acre in early spring, after quackgrass begins growth.

Precautions: *To avoid tree injury, 1) Do not use Princep 4L on seedbeds or cutting beds. 2) In CA, OR, and WA, do not apply to Christmas trees or shelterbelts sooner than one year after transplanting. In other areas, do not apply to Christmas trees or shelterbelt transplants less than three years of age. 3) Do not use until soil is firmly settled around roots. 4) Do not apply more than once a year, except as directed for quackgrass control.*

Apply Princep 4L to these species of trees and shrubs, as recommended above:

Conifers		Deciduous Trees and Woody Ornamentals	
arborvitae	Norway spruce	American elm	honey locust
Austrian pine	red spruce	Siberian elm	Oregon grape (*Mahonia* spp.)
*Bishop pine	white spruce	barberry	*Pieris* spp. (max. 3.2 qts./A)
*knobcone pine	balsam fir	boxelder	red oak
lodgepole pine (shore pine)	Douglas fir	bush honeysuckle	Russian olive
*Monterey pine	Fraser fir	caragana	*oleander
Mugho pine	white fir	cotoneaster	*palm (2qts./A)
red pine (Norway pine)	hemlock	dogwood	*bottle brush
Scotch pine	juniper	eucalyptus	*carob
white pine	red cedar	holly (*Ilex* spp.) (max. 3.2 qts./A)	
blue spruce	white cedar		
	yew (*Taxus* spp.)		

* For California only

Surflan Tank Mix: On Christmas tree plantings, use this tank mix for preemergence control of weeds listed on the Surflan 75W (or Surflan A.S.) and Princep 4L labels. Use on field grown conifer species listed on the labels for each herbicide plus grand fir, alpine fir, Engelmann spruce, black spruce, Colorado blue spruce, Coulter pine, giant redwood, and Veitchi fir. Broadcast the mixture as a directed spray to the soil surface or as an overtop spray using 2-4 qts. of Princep 4L and 2.67-5.33 lbs. of Surflan 75W (or 2-4 qts. of Surflan A.S.). Apply in sufficient water per acre to uniformly treat the area. Follow overtop sprays with sprinkler irrigation to move the herbicide from leaf surfaces to the soil. Remove weed growth before application. Mix weed residues, prunings or trash into the soil, or remove them before treatment. Soil should be in good tilth and free of clods at time of application. Shallow cultivation (1-2 inches) after treatment will not reduce weed control. Observe all precautions and limitations on the Princep 4L and Surflan labels.

Dual® 8E Tank Mix: On field and liner grown woody ornamentals, use this tank mix for preemergence control of yellow nutsedge, barnyardgrass, crowfootgrass, fall panicum, giant foxtail, green foxtail, yellow foxtail, goosegrass, large crabgrass, smooth crabgrass, redroot pigweed, galinsoga, Venice mallow, lambsquarters, purslane, ragweed, smartweed and other weeds listed on the Dual 8E alone label. Control should last about 60 days. Apply 1.6-2 pts. of Princep 4L plus 2-3 pts. of Dual 8E in a minimum of 15 gals. of water per acre. For band applications use a proportional amount of herbicide. Direct the spray toward the base of established ornamentals or ornamentals transplanted at least 10 days. Apply before weeds emerge or after removing existing weeds. Use the higher rates on fine-textured soil and where broadleaf weeds are expected to form a significant part of the infestation. Use the lower rates on coarse-textured soil and where light infestations of broadleaf weeds are expected. Apply post-transplant to the following field and liner grown ornamentals:

Arborvitae (*Thuja* spp.)	Leucothoe (*Leucothoe* spp.)
Cotoneaster (*Cotoneaster* spp.)	Osmanthus (*Osmanthus* spp.)
Crabapple (*Malus* spp.)	Weigela (*Weigela* spp.)

Holly (*Ilex* spp.), Boxleaf Japanese, Japanese
Juniper (*Juniperus* spp.), Andorra, Creeping, Shore
Yew (*Taxus* spp.), Dwarf Japanese

Note: Length of weed control may be reduced when continuous wet soil conditions follow herbicide application.

Precautions: *To avoid plant injury, 1) Do not use on seedbeds or on unrooted cuttings. 2) Do not use in greenhouses or other enclosed areas.*

TREE PLANTATIONS FOR TIMBER

Apply before weeds emerge or following removal of weed growth. Apply before or after seedling transplanting. If applying after transplanting, application may be made over the top of tree seedlings before new growth appears. If new growth is present, direct the spray to avoid tree foliage contact.

The following rates are based on broadcast treatment in 20-100 gals. of water per acre. For band or spot applications, reduce the broadcast rate in proportion to the area actually sprayed.

Precautions: *To avoid tree injury, 1) Do not use on seed nurseries. 2) Apply only once per year.*

BLACK WALNUTS

Apply 2-4 qts. per acre.

SLASH PINE AND WHITE PINE

Use for establishment in the South only. Apply 4-6 qts. per acre.

LOBLOLLY PINE

Use for establishment in the South only. Apply either 4-6 qts. of Princep 4L per acre or a tank mix of 3.2-4.8 qts. of Princep 4L plus 1-2 lbs. of AAtrex 80W (or 0.9-1.75 lbs. AAtrex Nine-O or 0.8-1.6 qts. AAtrex 4L or 4LC) per acre. The tank mix provides broad spectrum annual weed control (see list in General Information section of this label) and better weed control in limited rainfall areas than Princep 4L alone.

TREE SEEDLINGS IN NORTH CENTRAL U.S.

Apply 2-4 qts. per acre on the following species of trees:

American beech	maple	Scotch pine
balsam fir	Norway spruce	*sweet gum
black locust	pecan	*white ash
*black walnut	*red oak	white cedar
caragana	red pine (Norway pine)	*white oak
dogwood		white pine
eastern red cedar	red spruce	white spruce
honey locust	Russian olive	

* Instead of Princep 4L alone, on these species a tank mix of 2 qts. of Princep 4L + 2.5 lbs. of AAtrex 80W (or 2.2 lbs. AAtrex Nine-O or 2 qts. AAtrex 4L or 4LC) per acre may be applied in a minimum of 25 gals. of water.

Precautions: *To avoid tree injury, 1) Apply only once per year. 2) On sandy soil, do not use more than 2 qts. Princep 4L per acre and do not use the tank mix. 3) Do not apply the tank mix directly to seedlings after the start of bud break.*

TURFGRASSES FOR SOD (Florida only)

ST. AUGUSTINEGRASS, CENTIPEDEGRASS, AND ZOYSIA GRASS

Apply 2-4 qts. per acre, according to soil texture as indicated below.

Muck or peat	4 qts.	Old beds	within 2 days after lifting of sod
		New beds	3-4 days after sprigging or plugging
Sandy soil	2 qts.	Old beds	within 2 days after lifting of sod
		New beds	7-10 days after sprigging or plugging

If weeds regrow, apply an additional 2 qts. on muck or peat, or 1 qt. on sandy soil.

Precautions: *On newly sprigged turfgrass, temporary slowing of growth or yellowing may occur following application. To avoid turf injury, 1) Do not apply within 30 days before cutting or lifting. 2) Use only on turfgrass reasonably free of infestations of insects, nematodes, and diseases.*

TURFGRASS FOR FAIRWAYS, LAWNS, AND SIMILAR AREAS

BERMUDAGRASS, CENTIPEDEGRASS, ST. AUGUSTINEGRASS, AND ZOYSIA GRASS

Apply Princep 4L after October 1 before emergence of winter annual weeds for control of annual bluegrass, burclover, carpet burweed, chickweed, corn speedwell, henbit, hop clover, and spurweed. Princep 4L will control annual bluegrass even if it is emerged at time of treatment. For control of summer annual weeds listed in the General Information section of this label, also apply Princep 4L in late winter before the weeds emerge. Apply in a minimum of 15 gals. of water per acre or 1 gal. per 1,000 sq. ft.

Where annual bluegrass is the major weed, use 1 qt. of Princep 4L per acre (22 ml. or 0.75 fl. oz. per 1,000 sq. ft.) Use 2 qts. per acre (44 ml. or 1.5 fl. oz. per 1,000 sq. ft.) for control of the other weeds named above. Do not exceed 1 qt. per acre per treatment on newly sprigged turfgrass or on hybrid bermudagrass such as Tiflawn, Tifway, and Ormond.

For continued summer annual weed control, apply another 1 qt. per acre at least 30 days after the previous application, but not after April 15. However, do not make more than two applications of this product per year.

Precautions: *On newly sprigged turfgrass and hybrid bermudagrass, temporary slowing of growth and yellowing may occur following application. To avoid turf injury, 1) Use only on turfgrass reasonably free of infestations of insects, nematodes, and diseases. 2) Do not use on golf greens. 3) Do not use north of NC or west of the high rainfall areas of eastern OK and eastern TX. 4) Do not use on muck or alkaline soils. 5) Do not apply over the rooting area of trees or ornamentals not listed on this label. 6) Do not overseed with desirable turfgrass within 4 months before or 6 months after treatment.*

VEGETABLE CROPS

ASPARAGUS (Established)

Apply 2-4 qts. per acre to beds established one year or more. Apply once per year either after disking in the spring before weeds emerge, or following harvest after removal of existing weed growth. Spring application must be made at least 3 days before first cutting.

Precautions (WA and OR only): *To avoid crop injury, 1) Do not apply on gravelly, sand, or loamy sand soil. 2) Do not allow irrigation water to stand or to form pools in the field.*

ARTICHOKES

Apply 2-4 qts. per acre as a directed spray after the last fall tillage.

Precautions: *To avoid crop injury, 1) Do not apply on gravelly, sand, or loamy sand soil. 2) Apply only once per year.*

NONSELECTIVE WEED CONTROL ON NONCROP LAND

Princep 4L provides long-term weed control on industrial sites, highway medians and shoulders, railroad rights-of-way, lumberyards, petroleum tank farms, and in noncrop areas on farms, such as around buildings, equipment and fuel storage areas, along fences, roadsides, and lanes.

Best results are obtained when Princep 4L is applied before weeds emerge. If weed growth has begun and removal is not feasible, include a contact or translocated herbicide in the spray. Aerial applications may be made where it is feasible. Use at least 1 gal. of water for each 1 qt. of Princep 4L for both ground and aerial applications. Use more water if practical.

Precautions: *1) Do not use this product for nonselective weed control on land to be cropped, near adjacent desirable trees, shrubs, or plants, or in greenhouses, or injury may occur. 2) Do not use in areas accessible to livestock or allow livestock to graze treated weed foliage.*

Annual broadleaf and grass weeds (including barnyardgrass, cheat, crabgrass, lambsquarters, foxtail, ragweed, puncturevine, and turkey mullein): Apply 4.8-10 qts. per acre.

Annual and perennial broadleaf and grass weeds (including quackgrass, bluegrass, redtop, burdock, Canada thistle, orchardgrass, purpletop, smooth brome, dogfennel, and plantain): Apply 2.5-5 gals. per acre.

Hard-to-kill perennial weeds (including bullthistle and sowthistle): Apply 5-10 gals. per acre.

For longer residual control in regions of high rainfall and a long growing season, apply 5-10 gals. per acre.

TANK MIXTURE WITH ROUNDUP

This tank mixture will provide control of emerged annual weeds and residual control of weeds listed on the Princep label. The combination also will partially control emerged perennial weeds listed on the Roundup label. Add to the spray tank the appropriate rate of Princep noted in the above section for Princep alone. Then add 1-5 qts. of Roundup per acre depending on weeds present and their growth stage. Also add an agriculturally approved nonionic surfactant at 0.5% by volume of spray solution. Apply the mixture in 10-40 gals. of water per acre as a postemergence spray at the appropriate weed growth stage given on the Roundup label. Provide constant agitation during mixing and application to keep the mixture in suspension. Refer to the Application Procedures section of this label for further directions. Refer to the labels of both herbicides for specific noncrop sites, rates, weeds controlled, and further directions, precautions, and limitations.

TANK MIXTURE WITH OUST®

Use this tank mixture to control the weeds listed below with either preemergence, early postemergence, or postemergence applications. Use the higher rates within the ranges given under conditions of heavy weed growth, on soil with over 2½% organic matter, on areas with heavy decaying plant residues, or on high moisture areas such as along road edges or railroad shoulders. To improve wetting and/or contact activity on emerged weeds, add a nonionic surfactant such as duPont Surfactant WK or X-77 at 1 qt. per 100 gals. of spray. Do not apply by aircraft. Provide constant agitation during mixing and application to keep the mixture in suspension. Refer to the labels of both herbicides for specific noncrop sites and further directions, precautions, and limitations.

Preemergence to Early Postemergence: To control the weeds listed below apply 1-2 gals. of Princep 4L plus 3-5 ounces of Oust in a minimum of 25 gals. of water per acre.

Broadleaf Weeds (Up to 3-4″ in Height or Diameter)

bouncingbet	horsetail	sowthistle
buckhorn plantain	kochia	St. Johnswort
burclover	lambsquarters	sunflower
Carolina geranium	little mallow	sweet clover
common chickweed	mustard	tansymustard
common dandelion	ox-eye daisy	tansy ragwort
common speedwell	pigweed	tumble mustard
common yarrow	prickly lettuce	turkey mullein
crimson clover	puncturevine	vetch
dogfennel	purple starthistle	wild carrot
filaree	ragweed	yellow rocket
hoary cress	Russian thistle	

Grasses (Up to 6-12″ In Height)

Alta fescue	fall panicum	red brome
annual bluegrass	fescue	red fescue
annual ryegrass	foxtails	reed canarygrass
bahia grass	Indiangrass	ripgut brome
barnyardgrass	Italian ryegrass	ryegrass
cheatgrass	johnsongrass (short-term control)	smooth brome
crabgrass		sprangletop
downy brome	little barley	witchgrass

Postemergence: To control actively growing weeds listed below apply 2-4 gals. of Princep 4L plus 6-12 ounces of Oust in a minimum of 25 gals. of water per acre.

Actively Growing Weeds

bedstraw	Jerusalem artichoke	johnsongrass (season-long control)
Canada thistle	kudzu	
curly dock	mayweed	
dewberry	musk thistle	yellow nutsedge
fiddleneck	poison ivy	
fleabane	spanishneedles	
goldenrod	turkey mullein	
hemlock	Virginia pepperweed	
honeysuckle	wild blackberry	

STORAGE AND DISPOSAL

Do not contaminate water, food, or feed by storage or disposal. Pesticide wastes may be toxic. Improper disposal of unused pesticide, spray mixture, or rinse water is a violation of federal law. If these wastes cannot be used according to label instructions, contact your State Pesticide or Environmental Control Agency, or the Hazardous Waste representative at the nearest EPA Regional Office for guidance in proper disposal methods.

Triple rinse container (or equivalent). Then offer for recycling or reconditioning, or puncture and dispose of in a sanitary landfill or by other procedures approved by state and local authorities.

For minor spills, leaks, etc., follow all precautions indicated on this label and clean up immediately. Take special care to avoid contamination of equipment and facilities during cleanup procedures and disposal of wastes. In the event of a major spill, fire or other emergency, call (919) 292-7100 day or night.

PRECAUTIONARY STATEMENTS

HAZARDS TO HUMANS AND DOMESTIC ANIMALS

CAUTION

Causes eye irritation. Harmful if inhaled or absorbed through the skin. Avoid breathing vapors or spray mist. Avoid contact with eyes, skin, or clothing. Wash thoroughly with soap and water after handling. Remove and wash contaminated clothing before reuse.

First Aid: If in eyes, flush with plenty of water. Get medical attention if irritation persists. If on skin, wash with plenty of soap and water. If inhaled, remove victim to fresh air. If not breathing, give artificial respiration, preferably mouth-to-mouth; get medical attention.

ENVIRONMENTAL HAZARDS

Do not apply directly to water or wetlands. Do not contaminate water by cleaning of equipment or disposal of wastes. Simazine is a chemical which can travel (seep or leach) through soil and can contaminate ground water which may be used as drinking water. Simazine has been found in ground water as a result of agricultural use. Users are advised not to apply simazine where the water table (ground water) is close to the surface and where the soils are very permeable, i.e., well-drained soils such as loamy sands. Your local agricultural agencies can provide further information on the type of solid in your area and the location of ground water.

AAtrex® trademark of CIBA-GEIGY for atrazine

Compex® trademark of KALO Agricultural Chemicals, Inc.

Dual® trademark of CIBA-GEIGY for metolachlor

Eradicane® trademark of Stauffer Chemical Co.

Evik® trademark of CIBA-GEIGY for ametryn

Nine-O® trademark of CIBA-GEIGY

Oust® trademark of E. I. duPont de Nemours Co., Inc.

Princep® trademark of CIBA-GEIGY for simazine

Roundup® trademark of Monsanto Company for glyphosate herbicide

Surflan® trademark of Elanco Products Co. for oryzalin

Sutan +® trademark of Stauffer Chemical Co.

Unite® trademark of Hopkins Agricultural Chemical Co.

X-77® trademark of KALO Agricultural Chemicals, Inc.

CGA 6L32G 095

PRINCEP® 80W

HERBICIDE

For weed control in certain crops, ornamental plantings, on industrial sites, and around-the-farm

Princep 80W is a wettable powder

ACTIVE INGREDIENT: SIMAZINE

2-chloro-4,6-bis (ethylamino)*s*-triazine 80%
INERT INGREDIENTS: 20%

TOTAL: 100%

EPA Reg. No. 100-437 EPA Est. 100-AL-1

Princep® trademark of CIBA-GEIGY for simazine

Keep Out of Reach of Children.

Caution

See additional precautionary statements at end of booklet.

See directions for use inside booklet.

DIRECTIONS FOR USE AND CONDITIONS OF SALE AND WARRANTY

IMPORTANT: Read the entire **Directions for Use** and the **Conditions of Sale and Warranty** before using this product.

CONDITIONS OF SALE AND WARRANTY

The **Directions for Use** of this product reflect the opinion of experts based on field use and tests. The directions are believed to be reliable and should be followed carefully. However, it is impossible to eliminate all risks inherently associated with use of this product. Crop injury, ineffectiveness, or other unintended consequences may result because of such factors as weather conditions, presence of other materials, or the manner of use or application all of which are beyond the control of CIBA-GEIGY or the Seller. All such risks shall be assumed by the Buyer.

CIBA-GEIGY warrants that this product conforms to the chemical description on the label and is reasonably fit for the purposes referred to in the **Directions for Use** subject to the inherent risks referred to above. **CIBA-GEIGY makes no other express or implied warranty of Fitness or Merchantability or any other express or implied warranty. In no case shall CIBA-GEIGY or the Seller be liable for consequential, special, or indirect damages resulting from the use or handling of this product.** CIBA-GEIGY and the Seller offer this product, and the Buyer and user accept it, subject to the foregoing **Conditions of Sale and Warranty,** which may be varied only by agreement in writing signed by a duly authorized representative of CIBA-GEIGY.

DIRECTIONS FOR USE

It is a violation of federal law to use this product in a manner inconsistent with its labeling.

FAILURE TO FOLLOW THE DIRECTIONS FOR USE AND PRECAUTIONS ON THIS LABEL MAY RESULT IN CROP INJURY, POOR WEED CONTROL, AND/OR ILLEGAL RESIDUES.

Do not apply this product in such a manner as to directly or through drift expose workers or other persons except those knowingly involved in the application. The area being treated must be vacated by unprotected persons.

REENTRY STATEMENT

Do not enter treated areas without protective clothing until sprays have dried.

Because certain states may require more restrictive reentry intervals for various crops treated with this product, consult your State Department of Agriculture for further information.

Written or oral warnings must be given to workers who are expected to be in a treated area or in an area about to be treated with this product. Oral warnings must inform workers of areas or fields that may not be entered without specific protective clothing until sprays have dried, and appropriate actions to take in case of accidental exposure, as described under **Precautionary Statements** on this label. When oral warnings are given, warnings shall be given in a language customarily understood by workers. Oral warnings must be given if there is reason to believe that written warnings cannot be understood by workers. Written warnings must include the following information: CAUTION. Area treated with Princep 80W on *(date of application)*. Do not enter without appropriate protective clothing until sprays have dried. In case of accidental exposure, flush eyes or skin with plenty of water. Call physician if irritation persists. Remove and wash contaminated clothing before reuse.

GENERAL INFORMATION

Apply Princep 80W Herbicide before weeds emerge or after removal of weed growth. Princep controls a wide variety of annual broadleaf and grass weeds when used at selective rates in agricultural crops and ornamental plantings. When used at higher, nonselective rates in noncrop areas, it also controls many perennial broadleaf and grass weeds.

Where a range of application rates is given, use the low rate on coarser textured soil and soil lower in organic matter; use the high rate on finer textured soil and soil higher in organic matter.

To avoid spray drift, do not apply under windy conditions. Avoid spray overlap, as crop injury may result.

Since Princep enters weeds mainly through their roots, moisture is needed to move it into the root zone. Very dry soil conditions and lack of rainfall following application may necessitate shallow cultivation.

Princep 80W is noncorrosive to equipment, nonflammable, and has low electrical conductivity.

Annual Weeds Controlled

alyssum	fivehook bassia	ragweed
annual bluegrass	Flora's paintbrush	rattail fescue
annual morningglory	Florida pusley	redmaids
annual ryegrass	foxtails	Russian thistle
barnyardgrass (watergrass)	goosegrass	shepherdspurse
burclover	groundsel	signalgrass (*Brachiaria* spp.)
carelessweed	henbit	silver hairgrass
carpetweed	junglerice	smartweed
common chickweed	knawel (German moss)	spanishneedles
crabgrass (*Digitaria* spp.)	common lambsquarters	speedwell
downy brome (cheatgrass)	nightshade	tansymustard
fall panicum	pepperweed	wild mustard
fiddleneck	pigweed	wild oats
filaree	pineappleweed	witchgrass
fireweed	prickly lettuce	yellow flower pepperweed
	common purslane	

Following many years of continuous use of this product and chemically related products, biotypes of some of the weeds listed on this label have been reported which cannot be effectively controlled by this and related herbicides. Where this is known or suspected and weeds controlled by this product are expected to be present along with resistant biotypes, we recommend the use of this product in registered combinations or in sequence with other registered herbicides which are not triazines. If only resistant biotypes are expected to be present, use a registered nontriazine herbicide. Consult with your State Agricultural Extension Service for specific recommendations.

APPLICATION PROCEDURES

Ground application: Use conventional ground sprayers equipped with nozzles that provide accurate and uniform application. Be certain that nozzles are uniformly spaced and the same size. Calibrate sprayer before use and recalibrate at the start of each season and when changing carriers. Unless otherwise specified, use a minimum of 20 gals. of spray mixture per acre.

Use a pump with capacity to (1) maintain 35-40 psi at nozzles, (2) provide sufficient agitation in tank to keep mixture in suspension, and (3) to provide a minimum of 20% bypass at all times. Use centrifugal pumps which provide propeller shear action for dispersing and mixing this product. The pump should provide a minimum of 10 gals./minute/100 gals. tank size circulated through a correctly positioned sparger tube or jets.

Use screens to protect the pump and to prevent nozzles from clogging. Screens placed on suction side of pump should be *16-mesh or coarser.* Do not place a screen in the recirculation line. Use 50-mesh or coarser screens between the pump and boom, and where required, at the nozzles. Check nozzle manufacturer's recommendations.

For band applications, calculate amount to be applied per acre as follows:

$$\frac{\text{band width in inches}}{\text{row width in inches}} \times \text{broadcast rate per acre} = \text{amount needed per acre of field}$$

Aerial application: Use aerial application only where specified in the use directions. Apply in a minimum of 1 gal. of water for each 1-1.5 lbs. of wettable powder applied per acre. Avoid applications under conditions where uniform coverage cannot be obtained or where excessive spray drift may occur.

Avoid application to humans or animals. Flagmen and loaders should avoid inhalation of spray mist and prolonged contact with skin, and should wash thoroughly before eating and at the end of each day's operation.

Mixing procedures—all uses: (1) Be sure sprayer is clean and not contaminated with any other materials, or crop injury or sprayer clogging may result. (2) Fill tank ¼ full with clean water, nitrogen solution, or complete fluid fertilizer. (3) Start agitation. (4) Be certain that the agitation system is working properly and creates a rippling or rolling action on the liquid surface. (5) Pour product directly from bag into tank. (6) Continue filling tank until 90% full. Increase agitation if necessary to maintain surface action. (7) Add tank mix herbicide(s) after this product is thoroughly suspended. (8) Finish filling tank. Maintain agitation to avoid separation of materials. (9) Empty tank as completely as possible before refilling to prevent buildup of emulsifiable concentrate residue from possible tank mix herbicides. (10) If an emulsifiable concentrate film starts to build up in tank, drain it and clean with strong detergent solution or solvent. (11) Clean sprayer thoroughly immediately after use by flushing system with water containing a detergent.

Compatibility test: To determine the tank mix compatibility of Princep 80W with liquid fertilizer, crop oil, spreaders, or recommended pesticides, use this test method. Nitrogen solutions or complete fluid fertilizers may replace all or part of the water in the spray in alfalfa and corn. Since liquid fertilizers can vary, even within the same analysis, check compatibility each time before use. Be especially careful when using complete suspension or fluid fertilizers as serious compatibility problems are more apt to occur. Commercial application equipment may improve compatibility in some instances. Check compatibility using this procedure:

1. Add 1 pint of fertilizer to each of 2 one-quart glass jars with tight lids.
2. To **one** of the jars add ¼ tsp. of a compatibility agent approved for this use (¼ tsp. is equivalent to 2 pts. per 100 gals. spray). Cap and shake until mixed. Examples of compatibility agents include Compex® and Unite®.
3. To **both** jars add the appropriate amount of herbicide(s). If more than one herbicide is used, add them separately with dry herbicides first, flowables next, and emulsifiable concentrates last. After each addition, cap and shake until thoroughly mixed. The appropriate amount of herbicides for this test follows (assuming a spray volume of 25 gals. per acre):

Dry herbicides:	**For each pound per acre** add 1.5 teaspoons to each jar. Fluff up wettable powder products before measuring.
Liquid herbicides:	**For each pint per acre** add 0.5 teaspoon to each jar.

   For a spray volume other than 25 gals. per acre, change the teaspoons added to each jar as follows:

   $$\frac{\text{25 gal.} \times \text{no. teaspoons given above}}{\text{desired gals. spray volume per acre}} = \text{teaspoons to add to each jar}$$

4. After adding all ingredients, put lids on and tighten. Shake jars vigorously one minute. Let the mixtures stand 15 minutes and then look for separation, large flakes, precipitates, grease, gels, medium to heavy oily film on the jar, or other signs of incompatibility. Determine if the compatibility agent is needed in the spray mixture by comparing the two jars. If either mixture separates, but can be remixed readily, the mixture probably can be sprayed as long as good agitation is used. If the mixtures are incompatible, test the following methods of improving compatibility: (A) slurry the dry herbicide(s) in water before addition, or (B) add ½ of the compatibility agent to the fertilizer and the other ½ to the emulsifiable concentrate or flowable herbicide before addition to the mixture. If still incompatible, do not use the materials mixed in the same spray tank.

ROTATIONAL CROPS AND PERENNIAL CROP REPLANTING

To avoid crop injury, observe the following precautions: (1) If rotating treated land the year following application, plant only corn, unless otherwise stated in this label. (2) If replanting perennial crops or if rotating land to crops other than corn, do not apply Princep in the year preceding planting these crops.

FRUIT AND NUT CROPS

Apply the spray to the orchard or vineyard floor avoiding contact with fruit, foliage, or stems. Recommended rates are based on broadcast treatment. For band applications or spot applications around fruit or nut plantings, reduce the broadcast rate of Princep 80W and water per acre in proportion to the area actually sprayed.

Precautions (All fruit and nut crops): *To avoid crop injury, 1) Apply only to orchards or groves where trees have been established 12 months or more unless specified differently. 2) Make only one application per year, except as noted otherwise. 3) Do not use on gravelly, sand, or loamy sand soil. 4) Immediately following application, limit overhead sprinkler irrigation to ½ inch.*

ALMONDS AND PEACHES (CA only)

Apply 1.25-2.5 lbs. of Princep 80W per treated acre in a 2-4 ft. band on each side of the tree row. Apply before weeds emerge in late fall or early winter. Weeds controlled by 1.25 lbs. of Princep 80W include burclover, common chickweed, wild mustard, and shepherdspurse. Apply only once per year.

Precautions: *To avoid crop injury, 1) Do not treat trees established in the grove less than 3 years. 2) Do not treat the Mission (Texas) variety of almonds. 3) Do not apply to almond trees propagated on plum rootstocks. 4) Do not replant almonds or peaches in treated soil for 12 months after treatment. 5) Do not apply on soil with less than 1% organic matter. 6) Do not treat areas where water will accumulate.*

APPLES, PEARS, SOUR CHERRIES

Apply 2.5-5 lbs. per acre.

AVOCADOS

CA and FL only: Apply 2.5-5 lbs. per acre after final preparation of grove.

BLUEBERRIES AND CANEBERRIES (blackberries, boysenberries, loganberries, raspberries)

Apply 2.5-5 lbs. per acre in the spring or apply a split application of 2.5 lbs. per acre in the spring plus 2.5 lbs. per acre in the fall. Apply in a minimum of 40 gals. of water per acre.

On plantings less than 6 months old, use ½ the above rate.

To control quackgrass, apply 5 lbs. per acre in the fall or split the application applying 2.5 lbs. per acre in the fall plus 2.5 lbs. per acre in the spring, when quackgrass is actively growing.

Note: Do not apply when fruit is present, or illegal residues may result.

CRANBERRIES

MA: Apply up to 5 lbs. per acre either before spring growth begins or in the fall after harvest.

Other areas: Apply 2.5 lbs. per acre before spring growth begins.

FILBERTS

Apply 2.5-5 lbs. per acre in the fall or apply a split application of 2.5 lbs. per acre in the fall plus 2.5 lbs. per acre in the spring.

Precaution: *If trees are planted on a hillside, excessive soil erosion may result from the elimination of weeds.*

Note: Do not apply when nuts are on the ground during the harvest period, or illegal residues may result.

GRAPES

Apply 2.5-6 lbs. per acre any time between harvest and early spring.

Precaution: *Do not use in vineyards established less than three years, or crop injury may occur.*

GRAPEFRUIT, LEMONS, ORANGES

AZ (Lemons and Oranges only): Apply a split application of 2 lbs. per acre in the spring plus 2 lbs. per acre in the fall.

CA: In grapefruit, lemons, and oranges, apply 2.5-5 lbs. per acre in a single application; or apply 2.5 lbs. per acre in the fall and 2.5 lbs. per acre in the spring.

Precaution: *Do not use in the Imperial, Coachella, or Palo Verde valleys, or crop injury may occur.*

FL (Grapefruit and Oranges only): Apply only once per year. For control of weeds listed under General Information, apply 8-12 lbs. per acre in the spring; to also control milkweed vine, use 10-12 lbs. per acre either preemergence or before vines are 12 inches tall.

TX (Grapefruit and Oranges only): Apply 5-6 lbs. per acre.

Precautions (All areas): *To avoid crop injury, 1) Do not use in nurseries. 2) Do not apply to bedded grapefruit, lemons, or oranges. 3) Do not apply to trees under stress from freeze damage for one year after the freeze.*

MACADAMIA NUTS

Apply 2.5-5 lbs. in 50 gals. of water per acre before harvest and just prior to weed emergence. Repeat application as necessary.

Note: Do not apply when nuts are on the ground during the harvest period, as illegal residues may result.

OLIVES

Apply 2.5-5 lbs. per acre following grove preparation in the fall. Repeat annually in midwinter.

PEACHES, PLUMS, SWEET CHERRIES

Apply 2-5 lbs. per acre. Apply in late fall to early spring prior to weed emergence.

Precautions: *To avoid crop injury, 1) Peaches: use only in AR, LA, MO, OK, TX, and states east of the Miss. River. For CA, see specific directions in the section "Almonds and Peaches (CA only)." 2) Plums and sweet cherries: use only in MO and states east of the Miss. River except TN.*

PECANS

Apply 2.5-5 lbs. per acre before weeds emerge in the spring.

Precautions: *To avoid crop injury, 1) Do not use west of the Pecos River in TX or in NM, AZ, or CA. 2) Do not make applications to transplanted trees that have been established less than two years in the grove.*

Note: Do not apply when nuts are on the ground, or illegal residues may result. Do not allow animals to graze treated areas.

STRAWBERRIES

Oregon and Washington: For control of chickweed, groundsel, mustard, and shepherdspurse, apply broadcast 1.25 lbs. per acre. In fields where overhead irrigation is used to activate this product, apply after harvest at time of bed renovation. In fields where overhead irrigation is not available, apply during early October through November.

Precautions: *To avoid crop injury, 1) Make only one application per growing season. 2) Do not apply within 4 months after transplanting.*

WALNUTS

Apply 2.5-5 lbs. per acre. Leveling and furrowing operations after application will lessen effectiveness of weed control.

Note: Do not apply when nuts are on the ground, or illegal residues may result.

PRINCEP 80W PLUS EVIK® 80W FOR GRAPEFRUIT AND ORANGES

Use in grapefruit and oranges in FL and TX for control of weeds listed in the following directions and under General Information. Apply the rates given below in a minimum of 40 gals. of water per acre thoroughly covering existing vegetation, but avoiding contact with citrus fruit, foliage, or stems. Apply when weeds have resumed active growth in the spring, but before they exceed 4 inches in height (except as noted for vines). Observe all precautions and limitations on the labeling of this product and Evik 80W.

FL (Ridge Area): Apply 5 lbs. of Princep 80W plus 5 lbs. of Evik 80W per acre; controls milkweed vine when applied either preemergence or on vines up to 24 inches tall.

FL (Bedded Areas): Apply 4 lbs. of Princep 80W plus 6 lbs. of Evik 80W per acre; controls balsamapple when applied either preemergence or on vines up to 12 inches tall.

TX: Apply 4-5 lbs. of Princep 80W plus 2-3 lbs. of Evik 80W per acre plus 2 qts. nonionic surfactant, such as X-77®, per 100 gals. of spray mixture (add surfactant to spray tank last); controls buffelgrass, marinevine (ivy treebine), Texas panicum, sandbur, and sunflower, and gives top-kill of johnsongrass. Use 5 lbs of Princep 80W per acre for longer residual control; use 3 lbs. of Evik 80W per acre for control of heavy weed growth and for top-kill of johnsongrass; use the higher rate of both products for control of marinevine.

Precautions (All areas): *To avoid injury, 1) Do not use in nurseries, or around trees that have been established in the grove less than two years, or where trees are under stress from freeze damage for one year after the freeze. 2) Apply only once per year.*

Note (All areas): Do not apply within 30 days before harvest, or illegal residues may result. Do not graze treated areas.

PRINCEP 80W PLUS BROMACIL 80W FOR GRAPEFRUIT AND ORANGES (Florida Only)

Use in grapefruit and oranges in FL for control of balsamapple, black nightshade, carpetweed, crabgrass, cudweed, dayweed, Florida pusley, horseweed, pepperweed, pigweed, poorjoe, ragweed, rattlebox, spanishneedles and sandbur, and for partial control of bermudagrass, bahiagrass, pangolagrass, paragrass, and torpedograss. Apply 4-8 lbs. of Princep 80W plus 3-4 lbs. of bromacil 80W per acre beneath trees in a minimum of 40 gals. of water per acre before or soon after weed growth begins. Use the lower rates for light weed infestations or all applications in bedded citrus areas. Use the higher rates for heavy weed infestations only in ridge grown citrus areas. Temporary yellowing of citrus leaves may occur following treatment.

Precautions: *To avoid crop injury, 1) Do not use in nurseries, or around trees that have been established in the grove less than four years, or where trees are under stress from freeze damage for one year after the freeze. 2) Do not use on soil with less than 1% organic matter or on poorly drained soil. 3) Do not treat trees planted in irrigation furrows. 4) Do not treat diseased trees such as those with foot rot. 5) Do not use in groves interplanted with other trees or desirable plants, nor in home grapefruit or orange plantings, or in areas where roots of other valuable plants or trees may extend. 6) Treated areas may be planted to citrus trees one year after application. 7) Do not rotate to other crops within two years after application.*

Note: Apply only once per year and avoid contact with foliage and fruit with spray or mist, or illegal residues may result.

PRINCEP 80W PLUS PARAQUAT

This tank mix is effective in the following fruit and nut crops for kill of existing vegetation and for residual control of the annual broadleaf and grass weeds claimed for Princep 80W applied alone. This combination is also effective for top-kill and suppression of perennial weeds.

Use this tank mix on these crops:

almonds (CA only)	grapefruit	olives
apples	(CA, FL[2] and TX only	oranges[2]

avocados (CA and FL[1] only)	grapes	peaches[3]
cherries (sour and sweet[4])	lemons (AZ and CA only)	pears
filberts	macadamia nuts	pecans
		plums[4]
		walnuts

Apply the Princep 80W rate given under the appropriate crop on this label plus 1-2 qts. paraquat* in 50-200 gals. (30-50 gals. for pecans) of water per acre to the orchard floor avoiding contact with fruit, foliage, or stems. Add a nonionic surfactant, such as X-77, at 0.5 pt. per 100 gals. of spray. Apply when weeds are succulent and new growth is 1-6 inches tall. For mature woody weeds or difficult to control perennial weeds, re-treat or spot treat with paraquat if regrowth occurs.

Add Princep 80W to the spray tank first (refer to Mixing Instructions section of this label), then add paraquat, and add the surfactant last. Provide constant agitation during mixing and application to keep the mixture in suspension.

Precautions: *To avoid crop injury, 1) Apply the tank mix only once per year. 2) Use a shield for young trees or vines.*

Refer to the Princep 80W and paraquat labels for further directions, specific weeds controlled, and precautions and limitations on each crop.

1 In avocados in Florida, this tank mix also controls balsamapple vine, rattail amaranth, and at the higher rate of each herbicide, it suppresses coral vine.

2 In grapefruit and oranges in Florida, apply 4 lbs. Princep 80W per acre in bedded areas and 8-12 lbs. Princep 80W per acre in ridge areas.

3 Limited to CA, AR, LA, MO, OK, TX, and states east of the Miss. River. As appropriate, refer to Princep 80W sections "Almonds and Peaches (CA only)" or "Peaches, Plums, Sweet Cherries" for rate of Princep 80W and other information.

4 Limited to MO and states east of the Miss. River except TN.

* Based on a product containing 2 lbs. paraquat cation per gal.

TANK MIXTURE WITH ROUNDUP®

This tank mixture is effective in grape vineyards and in the following bearing or nonbearing tree crops for control of existing vegetation and for residual control of the annual broadleaf and grass weeds claimed for this product applied alone. This combination is also effective for partial control of perennial weeds contacted by the spray mixture during application. Use this tank mixture on these crops:

almonds** (CA only)	grapefruits* (CA, FL[1] and TX)	peaches*[2]
apples*	grapes*	pears*
avocados* (CA and FL only)	lemons* (AZ and CA)	pecans**
cherries (sour and sweet[3])	macadamia nuts**	plums*[3]
filberts**	oranges* (AZ, CA, FL[1] and TX)	walnuts**

Use the appropriate rate given elsewhere on this label for Princep applied alone to the crop being treated. Add to the spray tank 1-5 qts. of Roundup per acre depending on weeds present and their growth stage. Also add an agriculturally approved nonionic surfactant at 0.5% by volume of spray solution. Apply the mixture in 10-40 gals. of water per acre as a postemergence spray to the weeds at the appropriate weed growth stage given on the Roundup label.

Add Princep to the spray tank first, then add Roundup. Provide constant agitation during mixing and application to keep the mixture in suspension. Refer to Application Procedures section of this label for further directions.

Precaution: *To avoid crop injury, take extreme care to avoid contact of herbicide solution, spray, drift, or mist with foliage or green bark of trunk, branches, suckers, fruit, or other parts of trees or vines.*

Refer to the labels of both herbicides for further directions, specific weeds controlled, precautions and limitations on each crop.

* Allow a minimum of 14 days between last application and harvest.

** Allow a minimum of 21 days between last application and harvest of these crops.

1 In grapefruit and oranges in Florida, apply 4 lbs. of this product per acre in bedded areas and 8-12 lbs. per acre in ridge areas.

2 Limited to CA, AR, LA, MO, OK, TX, and states east of the Mississippi River. For CA, see specific directions in the section "Almonds and Peaches (CA only)."

3 Plums and sweet cherries: limited to MO and states east of the Mississippi River except TN.

PRINCEP 80W PLUS SURFLAN®

Use this tank mix in the following crops for preemergence control of all weeds claimed on both labels:

almonds	cherries	lemons	pecans
apples	filberts	oranges	plums
avocados	grapes	peaches	English walnuts
caneberries	grapefruit	pears	

Apply the Princep 80W rate given under the appropriate crop on this label plus 2.67-5.33 lbs. of Surflan 75W or 2-4 qts. of Surflan A.S. in 20-40 gals. of water per acre. Refer to the Surflan 75W or Surflan A.S. label for complete tank mix directions. Observe all precautions and limitations on the Princep and Surflan labels.

FIELD AND FORAGE CROPS

ALFALFA

Pure alfalfa at least 12 months old, pure alfalfa less than 12 months old, and alfalfa-orchardgrass mixtures at least 12 months old

For ground application, apply Princep 80W in a minimum of 15 gals. of water per acre. Fluid fertilizer may replace all or part of the water used as a carrier, except fluid fertilizers should not be used when alfalfa is actively growing, as injury may occur. Check the physical compatibility of Princep 80W with fertilizer before use. Refer to the General Information section of this label for a compatibility test procedure.

For aerial application, apply Princep 80W in a minimum of 2.5 gals. of water per acre.

Pure Alfalfa At Least 12 Months Old (All Areas Listed)

Apply once per year in the fall after the last cutting but prior to permanently frozen ground conditions. Alfalfa may be actively growing, semidormant, or dormant when the application is made.

Alfalfa must be growing for at least 12 months before treatment.

Alfalfa regions	Weeds controlled	Soil texture	Rate per acre
WA, OR, NV, UT, ID, MT, WY, CO, and CA counties of Del Norte, Siskiyou, Modoc, Shasta, Lassen, Plumas, Sierra, and Nevada	Downy brome (cheatgrass), tansymustard, prickly lettuce, pepperweed, seedling shepherdspurse, yellow flower pepperweed	Sand and loamy sand	DO NOT USE
		Sandy loam	1 lb.
		Loam soil	1-1.5 lbs.
		Clay soil	1.5-2 lbs.
ND, SD, MN, WI, NE, KS, IA, MI, and NY	Henbit, wild mustard, common chickweed, knawel (German moss), alyssum, downy brome (cheatgrass), wild oats, pepperweed, tansy-mustard, pigweed, speedwell, and seedlings of yellow rocket, shepherdspurse, white cockel, fanweed (penny-cress), dandelion, Russian thistle, kochia	Sand and loamy sand	DO NOT USE
		Sandy loam	1 lb.
		Loam soil	1.5 lbs.
		Clay soil	2 lbs.

IL, IN, OH, KY, TN, NC*, VA, PA, NJ, DE, MD, WV, VT, MA, CT, ME, NH, RI, and MO *Use only on loam and clay soils	Pigweed, common chickweed, mouseear chickweed, speedwell, henbit, knawel (German moss), pepperweed, downy brome, alyssum, common lambsquarters, wild oats, and seedlings of shepherdspurse, fanweed (pennycress), yellow rocket (wintercress), Russian thistle, dandelion kochia, white cockle	Sand and loamy sand	DO NOT USE
		Sandy loam	1 lb.
		Loam soil	1.25 lbs.
		Clay soil	1.5 lbs.

Pure Alfalfa Less Than 12 months Old (Northeastern U.S. Only)
Pure alfalfa which has been seeded in the spring before June 1 and has become well established may be treated in the fall after the last cutting but before permanently frozen ground conditions. Alfalfa may be actively growing, semidormant, or dormant when Princep 80W is applied. Apply 1 lb. of Princep 80W per acre to control henbit, wild mustard, common lambsquarters, common chickweed, German moss, alyssum, downy brome (cheatgrass), wild oats, tansymustard, pigweed, speedwell, and pepperweed.

Established Alfalfa-Orchardgrass Mixtures (Northeastern U.S. Only)
Apply only to alfalfa-orchardgrass mixtures which have been established at least 12 months. Make a single application in the fall after the alfalfa-orchardgrass mixture has gone dormant. Applications should be uniformly broadcast at the rate of 1 lb. of Princep 80W per acre to control henbit, wild mustard, common lambsquarters, common chickweed, German moss, alyssum, downy brome (cheatgrass), wild oats, tansymustard, pigweed, speedwell, and pepperweed.

Precautions (For Each Use): *1) Do not use on gravelly, sand, or loamy sand soil, or on soil low in organic matter, or crop injury may occur. 2) Applications made to alfalfa growing on highly calcareous or highly alkaline soil may result in crop injury. 3) Use only in areas where frost occurs and alfalfa goes into a semidormant or dormant state, or crop injury may occur. 4) Do not apply on snow, or poor weed control may result. 5) Princep 80W will not control dodder and established perennial or biennial weeds such as dandelion, white cockle, quackgrass, cinquefoil, wild carrot, and dock. 6) Temporary crop chlorosis may appear on areas where overlapping occurs. The sprayer should be shut off when turning around at the end or corners of the field. 7) Do not rotate to any crops except corn for the growing season following application as injury may occur.*

Note: Allow 30 days between application and grazing of dairy and beef cattle and sheep, and allow 60 days between application and cutting for hay, or illegal residues may result.

Princep 80W Plus Paraquat for Alfalfa (WA, OR, NV, UT, ID, MT, WY, CO, and the California counties of Del Norte, Sisklyou, Modoc, Shasta, Lassen, Plumas, Sierra and Nevada)
Controls filaree, Jim Hill mustard, and shepherdspurse, and gives improved control of bluegrass, cheatgrass, common chickweed, and tansymustard. Apply 1-2 lbs. of Princep 80W plus 1-3 pts. of paraquat* in 20-100 gals. of water per acre with ground equipment or in 5-10 gals. of water per acre aerially. Refer to the Princep 80W rate chart for rates of this product on specific soil textures. Use 1 pt. of paraquat when weeds and grasses are less than 4 inches tall. Apply when weeds are succulent and growth is 1-6 inches tall. Apply to established alfalfa after the last fall cutting when the crop is dormant and before spring growth starts. Observe all precautions and limitations on the labeling of Princep 80W and paraquat.

* Based on a product containing 2 lbs. paraquat cation per gal.

Precautions: *To avoid crop injury, 1) Do not apply if regrowth following the last fall cutting is more than 2 inches tall. Alfalfa foliage present at application will be burned. 2) Apply once per growing season.*

Note: Do not pasture animals in treated fields before first cutting, or illegal residues may result. Do not graze, cut, or harvest within 90 days after application.

FORAGE BERMUDAGRASSES (Such as Coastal and Midland)
For preemergence control of most annual broadleaf and grassy weeds including barnyardgrass, crabgrass, *Brachiaria* spp., goosegrass, foxtail, carelessweed, carpetweed, annual morningglory, pigweed, common purslane, Russian thistle, common chickweed, henbit, and burclover, apply Princep 80W shortly after sprigging. Apply it to established bermudagrass from fall to spring depending on the weed problem, but before weed emergence. Use the appropriate rate in the following table in a minimum of 15 gals. of water per acre for ground application and in a minimum of 1 gal. of water for each 1-1.5 lbs. of wettable powder per acre for aerial application.

Soil texture	Rate per acre
Fine sandy soil low in organic matter	1 lb.
Sandy loam soil	1.5-2.5 lbs.
Loam and clay soil low in organic matter	2.5-3 lbs.
Clay soil	3-4 lbs.

Precautions: *To avoid crop injury, 1) Do not use Princep 80W on bermudagrass in NM and the High Plains of TX. 2) Do not apply within 7 days after cutting.*

Note: To avoid illegal residues, 1) Do not graze livestock for 30 days after 1-3 lbs. per acre application or for 60 days after 4 lbs. per acre application. 2) Do not cut for hay for 60 days after 1-3 lbs. per acre application or for 90 days after 4 lbs. per acre application.

CORN
Nitrogen solutions or complete fluid fertilizers may replace all or part of the water as a carrier in the spray. Determine the physical compatibility of Princep 80W with fertilizer before use. Refer to the General Information section of this label for a compatibility test. Do not apply after corn has emerged as there is danger of liquid fertilizers causing crop injury.

Preemergence: Apply before weeds and corn emerge. Use the appropriate rate in the table below.

Preplant: Apply in the spring after plowing at the appropriate rate in the table below. Apply before, during, or after final seedbed preparation. If soil is tilled or worked after application, avoid deep incorporation of Princep 80W. Best results will be obtained when Princep 80W is applied within 2 weeks before planting. Under dry weather conditions, preplant applications may give better weed control. If weeds develop, particularly under relatively dry conditions, a shallow cultivation will generally result in better weed control.

Soil texture	Broadcast* rate per acre
Coarse-textured soil: Sand, silt, and loam that is low in organic matter	2.5 lbs.
Medium-textured soil: Soil containing a moderate amount of clay and organic matter	3 lbs.
Fine-textured soil: Loam that is high in organic matter and clay (including dark prairie soils of Corn Belt)	3.75 lbs.
Organic soil: Peat, muck, and high-organic clay	5 lbs.

* For calculation of band treatment rate, see the General Information section.

To control quackgrass: Apply 3.75-5 lbs. per acre in the fall. Plow two to three weeks later, or if erosion is a problem, delay plowing until spring.

Precautions: *1) Do not apply more than 5 lbs. of Princep 80W per acre to corn per year. 2) Do not rotate to any crop except corn until the following year or injury may occur. 3) After harvest, plow and thoroughly till the soil in fall or spring to minimize possible injury to spring-seeded rotational crops, regardless of the rate used. 4) If more* *than 3.75 lbs. of Princep 80W is used per acre (or equivalent rate in a band), a crop of untreated corn should precede the next rotational crop. 5) Do not apply Princep 80W preplant incorporated in corn in the High Plains and Intermountain areas of the West (including central and western KS, western NE, western OK, and the Panhandle of TX) where rainfall is sparse and erratic or where irrigation is required. 6) In the High Plains and Intermountain areas of the West where rainfall is sparse and erratic or where irrigation is required, use Princep 80W to control weeds in corn only when corn is to follow corn or when a crop of untreated corn is to precede another rotational crop. 7) In western MN and eastern parts of the Dakotas, NE, and KS, do not plant soybeans following corn treated with Princep if more than 2.5 lbs. of Princep 80W per acre (or equivalent rate in a band) was applied or injury may occur. 8) Injury may occur to soybeans planted in north-central and northwest IA, south-central and southwest MN, northeast NE, southeast SD, and other areas the year following application on soils having a calcareous surface layer. 9) Do not plant sugar beets, tobacco, vegetables (including dry beans), spring-seeded small grains, or small-seeded legumes and grasses the year after Princep 80W application, or injury may occur.*

Note: Do not graze treated areas, or illegal residues may result.

TANK MIXTURES ON CORN

AAtrex®: Use Princep 80W in a preplant or preemergence tank mixture with AAtrex Nine-O®, AAtrex 80W, or AAtrex 4L (4LC) for control of many annual weeds, including carpetweed, crabgrass, fall panicum, foxtail, lambsquarters, morningglory, pigweed, ragweed, and velvetleaf.

Apply at the rates given below. Use the 1:1 ratio for most weeds and the 2:1 ratio for expected heavy infestations of crabgrass and fall panicum.

	Broadcast rate per acre			
	1:1 Ratio		2:1 Ratio	
Soil texture	Princep 80W	AAtrex 80W*	Princep 80W	AAtrex 80W*
Sand, loamy sand, sandy loam	1.25 lbs.	1.25 lbs.	1.67 lbs.	0.83 lb.
Loam, silt loam, silt, clay loam sandy clay loam, silty clay loam, sandy clay, or silty clay with low organic matter	1.5 lbs.	1.5 lbs.	2 lbs.	1 lb.
Loam, silt loam, silt, clay loam, sandy clay loam, silty clay loam, sandy clay, or silty clay with medium to high organic matter, and clay (including dark prairie soils of the Corn Belt)	1.8 lbs.	1.8 lbs.	2.4 lbs.	1.2 lbs.

* When using AAtrex 4L (4LC) or AAtrex Nine-O, use equivalent rates. One lb. of AAtrex 80W equals 0.8 qt. of AAtrex 4L (4LC) or 0.9 lb. of AAtrex Nine-O.

Refer to the AAtrex label for complete directions. Also refer to the AAtrex label and the Corn section of this label for precautions and limitations.

Eradicane®: Use in a preplant incorporated tank mixture for control of all weeds claimed on both the Princep and Eradicane labels including partial control (suppression) of shattercane (wild cane). Fluid fertilizer may replace all or part of the water in the spray. Check the physical compatibility of mixture with fertilizer before use. Refer to the General Information section of this label for a compatibility test procedure. Use a minimum of 20 gals. of spray volume per acre. Refer to the Eradicane label for incorporation directions. Use the higher rate of Princep on fine-textured soil and where heavy broadleaf infestations are expected. Use the higher rate of Eradicane for heavy nutsedge and bermudagrass infestations. For partial control (suppression) of shattercane, broadcast and incorporate **immediately** before planting 1.25-2.5 lbs. of Princep 80W plus 7.33 pts. of Eradicane 6.7E (or 8 pts. Eradicane Extra) per acre. For control of the other weeds claimed on both labels, broadcast and incorporate 1.25-3.75 lbs. of Princep 80W plus 4.75-7.33 pts. of Eradicane 6.7E (or 5.33-8 pts. of Eradicane Extra) per acre. Observe all precautions and limitations on the Princep 80W and Eradicane labels.

Paraquat: Use in a tank mixture where corn will be planted directly in a cover crop, established sod, or previous crop residues. This combination controls existing vegetation and provides residual control of the annual broadleaf and grass weeds listed under General Information.

Add Princep 80W to the spray tank, mix thoroughly with water, and then add paraquat and a nonionic surfactant. Provide constant agitation during mixing and application to keep the mixture in suspension.

Apply 2.5-3.75 lbs. of Princep 80W plus 1-2 pts. of paraquat* in 20-60 gals. of water per acre as a broadcast spray either before or after planting but before corn emerges. Add a nonionic surfactant at the rate of 0.5 pt. per 100 gals. of spray volume.

For further information, see general information, caution and warning statements, precautions, and notes on the Princep 80W and the paraquat labels.

* Based on a product containing 2 lbs. paraquat cation per gal.

Sutan+®: Use in a preplant incorporated tank mixture for control of all weeds claimed on both the Princep and Sutan+ labels including partial control (suppression) of shattercane (wild cane). Fluid fertilizer may replace all or part of the water in the spray. Check the physical compatibility of mixture with fertilizer before use. Refer to the General Information section of this label for a compatibility test procedure. Use a minimum of 20 gals. of spray volume per acre. Refer to the Sutan+ label for incorporation directions. Use the higher rate of Princep on fine-textured soil and where heavy broadleaf infestations are expected. For partial control (suppression) of shattercane, broadcast and incorporate **immediately** before planting 1.25-2.5 lbs. of Princep 80W plus 7.33 pts. of Sutan+ 6.7E per acre. For control of the other weeds claimed on both labels, broadcast and incorporate 1.25-3.75 lbs. of Princep 80W plus 4.75 pts. of Sutan+ 6.7E per acre. Observe all precautions and limitations on the Princep 80W and Sutan+ labels.

SUGARCANE

Apply 2.5-5 lbs. per acre at planting or after harvest if ratoon crop is to be grown, but before weeds or sugarcane emerge. Princep 80W may also be applied interline-directed or over the sugarcane to provide continuing control of germinating weeds.

Precautions: *To avoid crop injury or illegal residues, 1) Do not apply more than one 2.5-5 lbs. application over the plants or more than two interline-directed applications. 2) Do not apply after close-in. 3) Total application should not exceed 12.5 lbs. per acre crop cycle.*

Note: Do not feed or graze treated forage to livestock.

GRASSES GROWN FOR SEED (Pacific Northwest Only)

Perennial ryegrass, Highland or Astoria bentgrasses, orchardgrass, tall fescue, and fine fescues such as Illahee, Rainier, Pennlawn, Chewings, and related species. For the control of broadleaf weeds and annual grasses including annual ryegrass, rattail fescue, silver hairgrass, and downy brome (cheat), apply 2.5 lbs. Princep 80W in a minimum of 15 gals. of water per acre as soon as fall rains start. Early application is important.

Precaution: *To avoid crop injury, apply only to grasses from which at least one seed crop has been cut.*

Note: To avoid illegal residues, allow 30 days between last application and grazing of dairy and beef cattle and sheep, and allow 60 days between the last application and cutting for hay.

NURSERIES, CHRISTMAS TREE PLANTINGS, SHELTERBELTS

NURSERIES (see list below)
Apply 2.5-3.75 lbs. in at least 25 gals. of water per acre in fall or spring.

Precaution: *To avoid plant injury, do not apply for at least one year after transplanting.*

CHRISTMAS TREE PLANTINGS AND SHELTERBELTS (see list below)
Remove weed growth before application. Apply 2.5-5 lbs. in at least 25 gals. of water per acre after transplanting. Use the same rate for annual maintenance applications.

For quackgrass control apply 5 lbs. per acre in the fall or apply a split application of 2.5 lbs. per acre in the fall plus 2.5 lbs. per acre in early spring, after quackgrass begins growth.

Precautions: *To avoid tree injury, 1) Do not use Princep 80W on seedbeds or cutting beds. 2) In CA, OR, and WA do not apply to Christmas trees or shelterbelts sooner than one year after transplanting. In other areas, do not apply to Christmas trees or shelterbelt transplants less than three years of age. 3) Do not use until soil is firmly settled around roots. 4) Do not apply more than once a year, except as directed for quackgrass control.*

Apply Princep 80W to these species of trees and shrubs, as recommended above:

Conifers

arborvitae	Scotch pine	hemlock
Austrian pine	white pine	juniper
*Bishop Pine	blue spruce	red cedar
*knobcone pine	Norway spruce	white cedar
lodgepole pine (shore pine)	red spruce	yew (*Taxus* spp.)
	white spruce	
*Monterey pine	balsam fir	
Mugho pine	Douglas fir	*For CA only
red pine (Norway pine)	Fraser fir	
	white fir	

Deciduous Trees and Woody Ornamentals

American elm	eucalyptus	Russian olive
Siberian elm	holly (*Ilex* spp.) (max 4 lbs./A)	*oleander
barberry		*palm (2.5 lbs./A)
boxelder	honey locust	
bush-honeysuckle	Oregon grape (*Mahonia* spp.)	*bottle brush
		*carob
caragana	Pieris spp. (max. 4 lbs./A)	
cotoneaster		*For CA only
dogwood	red oak	

Surflan Tank Mix: On Christmas tree plantings, use this tank mix for preemergence control of weeds listed on this label and the Surflan 75W (or Surflan A.S.) label. Use on field grown conifer species listed on the labels for each herbicide plus grand fir, alpine fir, Engelmann spruce, black spruce, Colorado blue spruce, Coulter pine, giant redwood, and Veitchi fir. Broadcast the mixture as a directed spray to the soil surface or as an overtop spray using 2.5-5 lbs. of Princep 80W and 2.67-5.33 lbs. of Surflan 75W (or 2-4 qts. of Surflan A.S.). Apply in sufficient water per acre to uniformly treat the area. Follow overtop sprays with sprinkler irrigation to move the herbicide from leaf surfaces to the soil. Remove weed growth before application. Mix weed residues, prunings or trash into the soil, or remove them before treatment. Soil should be in good tilth and free of clods at time of application. Shallow cultivation (1-2 inches) after treatment will not reduce weed control. Observe all precautions and limitations on the Princep 80W and Surflan labels.

Dual® 8E Tank Mix: On field and liner grown woody ornamentals, use this tank mix for preemergence control of yellow nutsedge, barnyardgrass, crowfootgrass, fall panicum, giant foxtail, green foxtail, yellow foxtail, goosegrass, large crabgrass, smooth crabgrass, redroot pigweed, galinsoga, Venice mallow, lambsquarters, purslane, ragweed, smartweed and other weeds listed on the Dual 8E alone label. Control should last about 60 days. Apply 1-1.25 lbs. of Princep 80W plus 2-3 pts. of Dual 8E in a minimum of 15 gals. of water per acre. For band applications use a proportional amount of herbicide. Direct the spray toward the base of established ornamentals or ornamentals transplanted at least 10 days. Apply before weeds emerge or after removing existing weeds. Use the higher rates on fine-textured soil and where broadleaf weeds are expected to form a significant part of the infestation. Use the lower rates on coarse-textured soil and where light infestations of broadleaf weeds are expected. Apply post-transplant to the following field and liner grown ornamentals:

Arborvitae (*Thuja* spp.)	Leucothoe (*Leucothoe* spp.)
Cotoneaster (*Cotoneaster* spp.)	Osmanthus (*Osmanthus* spp.)
Crabapple (*Malus* spp.)	
Holly (*Ilex* spp.), Boxleaf Japanese, Japanese	Weigela (*Weigela* spp.)
Juniper (*Juniperus* spp.), Andorra, Creeping, Shore	Yew (*Taxus* spp.), Dwarf Japanese

Note: Length of weed control may be reduced when continuous wet soil conditions follow herbicide application.

Precautions: *To avoid plant injury, 1) Do not use on seedbeds or on unrooted cuttings. 2) Do not use in greenhouses or other enclosed areas.*

TREE PLANTATIONS FOR TIMBER

Apply before weeds emerge or following removal of weed growth. Apply before or after seedling transplanting. If applying after transplanting, application may be made over the top of tree seedlings before new growth appears. If new growth is present, direct the spray to avoid tree foliage contact.

The following rates are based on broadcast treatment in 20-100 gals. of water per acre. For band or spot applications, reduce the broadcast rate in proportion to the area actually sprayed.

Precautions: *To avoid tree injury, 1) Do not use on seed nurseries. 2) Apply only once per year.*

BLACK WALNUTS
Apply 2.5-5 lbs. per acre.

SLASH PINE AND WHITE PINE
Use for establishment in the South only. Apply 5-7.5 lbs. per acre.

LOBLOLLY PINE
Use for establishment in the South only. Apply either 5-7.5 lbs. of Princep 80W per acre or a tank mix of 4-6 lbs. of Princep 80W plus 1-2 lbs. of AAtrex 80W (or 0.9-1.75 lbs. AAtrex Nine-O or 0.8-1.6 qts. AAtrex 4L or 4LC) per acre. The tank mix provides broad spectrum annual weed control (see list in General Information section of this label) and better weed control in limited rainfall areas than Princep 80W alone.

TREE SEEDLINGS IN NORTH CENTRAL U.S.
Apply 2.5-5 lbs. per acre on the following species of trees:

American beech	maple	Scotch pine
balsam fir	Norway spruce	*sweet gum
black locust	pecan	*white ash
*black walnut	*red oak	white cedar
caragana	red pine (Norway pine)	*white oak
dogwood		white pine
eastern red cedar	red spruce	white spruce
honey locust	Russian olive	

* Instead of Princep 80W alone, on these species a tank mix of 2.5 lbs. of Princep 80W + 2.5 lbs. of AAtrex 80W (or 2.2 lbs. AAtrex Nine-O or 2 qts. AAtrex 4L or 4LC) per acre may be applied in a minimum of 25 gals. of water.

Precautions: *To avoid tree injury, 1) Apply only once per year. 2) On sandy soil, do not use more than 2.5 lbs. of Princep 80W per acre and do not use the tank mix. 3) Do not apply the tank mix directly to seedlings after the start of bud break.*

TURF GRASSES FOR SOD (Florida only)

ST. AUGUSTINEGRASS, CENTIPEDEGRASS, AND ZOYSLA GRASS

Apply 2.5-5 lbs. per acre, according to soil texture as indicated below.

Muck or peat	5 lbs.	Old beds	within 2 days after lifting of sod
		New beds	3-4 days after sprigging or plugging
Sandy soil	2.5 lbs.	Old beds	within 2 days after lifting of sod
		New beds	7-10 days after sprigging or plugging

If weeds regrow, apply an additional 2.5 lbs. on muck or peat, or 1.25 lbs. on sandy soil.

Precautions: *On newly sprigged turfgrass, temporary slowing of growth or yellowing may occur following application. To avoid turf injury, 1) Do not apply within 30 days before cutting or lifting. 2) Use only on turfgrass reasonably free of infestations of insects, nematodes, and diseases.*

TURFGRASS FOR FAIRWAYS, LAWNS, AND SIMILAR AREAS

BERMUDAGRASS, CENTIPEDEGRASS, ST. AUGUSTINEGRASS, AND ZOYSIA GRASS

Apply Princep 80W after October 1 before emergence of winter annual weeds for control of annual bluegrass, burclover, carpet burweed, chickweed, corn speedwell, henbit, hop clover, and spurweed. Princep 80W will control annual bluegrass even if it is emerged at time of treatment. For control of summer annual weeds listed in the General Information section of this label, also apply Princep 80W in late winter before the weeds emerge. Apply in a minimum of 15 gals. of water per acre or 1 gal. per 1,000 sq. ft.

Where annual bluegrass is the major weed, use 1.25 lbs. of Princep 80W per acre (0.5 oz. per 1,000 sq. ft.). Use 2.5 lbs. per acre (1 oz. per 1,000 sq. ft.) for control of the other weeds named above. Do not exceed 1.25 lbs. per acre per treatment on newly sprigged turfgrass or on hybrid bermudagrass such as Tiflawn, Tifway, and Ormond.

For continued summer annual weed control, apply another 1.25 lbs. per acre at least 30 days after the previous application, but not after April 15. However, do not make more than two applications of this product per year.

Precautions: *On newly sprigged turfgrass and hybrid bermudagrass, temporary slowing of growth and yellowing may occur following application. To avoid turf injury, 1) Use only on turfgrass reasonably free of infestations of insects, nematodes, and diseases. 2) Do not use on golf greens. 3) Do not use north of NC or west of the high rainfall areas of eastern OK and eastern TX. 4) Do not use on muck or alkaline soils. 5) Do not apply over the rooting area of trees or ornamentals not listed on this label. 6) Do not overseed with desirable turfgrass within 4 months before or 6 months after treatment.*

VEGETABLE CROPS

ASPARAGUS (Established)

Apply 2.5-5 lbs. per acre to beds established one year or more. Apply once per year either after disking in the spring before weeds emerge, or following harvest after removal of existing weed growth. Spring application must be made at least 3 days before first cutting.

Precautions: (WA and OR only): *To avoid crop injury, 1) Do not apply on gravelly, sand, or loamy sand soil. 2) Do not allow irrigation water to stand or to form pools in the field.*

ARTICHOKES

Apply 2.5-5 lbs. per acre as a directed spray after the last fall tillage.

Precautions: *To avoid crop injury, 1) Do not apply on gravelly, sand, or loamy sand soil. 2) Apply only once per year.*

NONSELECTIVE WEED CONTROL ON NONCROP LAND

Best results are obtained when Princep 80W is applied before weeds emerge. If weed growth has begun and removal is not feasible, include a contact or translocated herbicide in the spray.

Princep 80W provides long-term weed control on industrial sites, highway medians and shoulders, railroad rights-of-way, lumberyards, petroleum tank farms, and in noncrop areas on farms, such as around buildings, equipment and fuel storage areas, along fences, roadsides, and lanes.

Aerial applications may be made where it is feasible. Use at least 1 gal. of water for each 1 lb. of Princep 80W for both ground and aerial applications. Use more water if practical.

Precautions: *1) Do not use this product for nonselective weed control on land to be cropped, near adjacent desirable trees, shrubs, or plants, or in greenhouses, or injury may occur. 2) Do not use in areas accessible to livestock or allow livestock to graze treated weed foliage.*

To control annual broadleaf and grass weeds (including barnyardgrass, cheat, crabgrass, common lambsquarters, foxtail, ragweed, puncturevine, and turkey mullein) apply 6-12.5 lbs. per acre.

To control most annual and many perennial broadleaf and grass weeds (including quackgrass, bluegrass, redtop, burdock, Canada thistle, orchardgrass, purpletop, smooth brome, dogfennel, and plantain) apply 12.5-25 lbs. per acre.

To control hard-to-kill perennial weeds (including bullthistle and sowthistle) apply 25-50 lbs. per acre.

For longer residual control in regions of high rainfall and a long growing season, apply 25-50 lbs. per acre.

TANK MIXTURE WITH ROUNDUP

This tank mixture will provide control of emerged annual weeds and residual control of weeds listed on the Princep label. The combination also will partially control emerged perennial weeds listed on the Roundup label. Add to the spray tank the appropriate rate of Princep noted in the above section for Princep alone. Then add 1-5 qts. of Roundup per acre depending on weeds present and their growth stage. Also add an agriculturally approved nonionic surfactant at 0.5% by volume of spray solution. Apply the mixture in 10-40 gals. of water per acre as a postemergence spray at the appropriate weed growth stage given on the Roundup label. Provide constant agitation during mixing and application to keep the mixture in suspension. Refer to the Application Procedures section of this label for further directions. Refer to the labels of both herbicides for specific noncrop sites, rates, weeds controlled, and further directions, precautions, and limitations.

TANK MIXTURE WITH OUST®

Use this tank mixture to control the weeds listed below with either preemergence, early postemergence, or postemergence applications. Use the higher rates within the ranges given under conditions of heavy weed growth, on soil with over $2\frac{1}{2}$% organic matter, on areas with heavy decaying plant residues, or on high moisture areas such as along road edges or railroad shoulders. To improve wetting and/or contact activity on emerged weeds, add a nonionic surfactant such as duPont Surfactant WK or X-77 at 1 qt. per 100 gals. of spray. Do not apply by aircraft. Provide constant agitation during mixing and application to keep the mixture in suspension. Refer to the labels of both herbicides for specific noncrop sites and further directions, precautions, and limitations.

Preemergence to Early Postemergence: To control the weeds listed below apply 5-10 lbs. of Princep 80W plus 3-5 ounces of Oust in a minimum of 25 gals. of water per acre.

Broadleaf Weeds (Up to 3-4" in Height or Diameter)

bouncingbet	horsetail	sowthistle
buckhorn plantain	kochia	St. Johnswort
burclover	lambsquarters	sunflower
Carolina geranium	little mallow	sweet clover
common chickweed	mustard	tansymustard
common dandelion	ox-eye daisy	tansy ragwort
common speedwell	pigweed	tumble mustard
common yarrow	prickly lettuce	turkey mullein

crimson clover	puncturevine	vetch
dogfennel	purple starthistle	wild carrot
filaree	ragweed	yellow rocket
hoary cress	Russian thistle	

Grasses (Up to 6-12" In Height)

Alta fescue	fescue	red brome
annual bluegrass	foxtails	red fescue
annual ryegrass	Indiangrass	reed canarygrass
bahia grass	Italian ryegrass	ripgut brome
barnyardgrass	johnsongrass (short-term control)	ryegrass
cheatgrass		smooth brome
crabgrass		sprangletop
downy brome	little barley	witchgrass
fall panicum		

Postemergence: To control actively growing weeds listed below apply 10-20 lbs. of Princep 80W plus 6-12 ounces of Oust in a minimum of 25 gals. of water per acre.

Actively Growing Weeds

bedstraw	Jerusalem artichoke	johnsongrass (season-long control)
Canada thistle	kudzu	
curly dock	mayweed	
dewberry	musk thistle	yellow nutsedge
fiddleneck	poison ivy	
fleabane	spanishneedles	
goldenrod	turkey mullein	
hemlock	Virginia pepperweed	
honeysuckle	wild blackberry	

STORAGE AND DISPOSAL

Store in a dry place. Do not contaminate water, food, or feed by storage or disposal. Improper disposal of unused spray mixture, or rinse water is a violation of federal law. If these wastes cannot be used according to label instructions, contact your State Pesticide or Environmental Control Agency, or the Hazardous Waste representative at the nearest EPA Regional Office for guidance in proper disposal methods.

Completely empty bag into application equipment. Dispose of empty bag in a sanitary landfill or by incineration, or if allowed by state and local authorities, by burning. Stay out of smoke from burning bags.

For minor spills, leaks, etc., follow all precautions indicated on this label and clean up immediately. Take special care to avoid contamination of equipment and facilities during cleanup procedures and disposal of wastes. In the event of a major spill, fire or other emergency, call (919) 292-7100 day or night.

PRECAUTIONARY STATEMENTS

HAZARDS TO HUMANS AND DOMESTIC ANIMALS

CAUTION

Causes eye irritation. Harmful if inhaled or absorbed through the skin. Avoid breathing dust or spray mist. Avoid contact with eyes, skin, or clothing. Wash thoroughly with soap and water after handling. Remove and wash contaminated clothing before reuse.

First Aid: If in eyes, flush with plenty of water. Get medical attention if irritation persists. If on skin, wash with soap and water. If inhaled, remove victim to fresh air. If not breathing, give artificial respiration, preferably mouth-to-mouth; get medical attention.

ENVIRONMENTAL HAZARDS

Do not apply directly to water or wetlands. Do not contaminate water by cleaning of equipment or disposal of wastes. Simazine is a chemical which can travel (seep or leach) through soil and can contaminate ground water which may be used as drinking water. Simazine has been found in ground water as a result of agricultural use. Users are advised not to apply simazine where the water table (ground water) is close to the surface and where the soils are very permeable, i.e., well-drained soils such as loamy sands. Your local agricultural agencies can provide further information on the type of soil in your area and the location of ground water.

AAtrex® trademark of CIBA-GEIGY for atrazine
Compex® trademark of KALO Agricultural Chemicals, Inc.
Dual® trademark of CIBA-GEIGY for metolachlor
Eradicane® trademark of Stauffer Chemical Co.
Evik® trademark of CIBA-GEIGY for ametryn
Nine-O® trademark of CIBA-GEIGY
Oust® trademark of E. I. duPont de Nemours & Co., Inc.
Princep® trademark of CIBA-GEIGY for simazine
Roundup® trademark of Monsanto Company
for glyphosate herbicide
Surflan® trademark of Elanco Products Co. for oryzalin
Sutan+® trademark of Stauffer Chemical Co.
Unite® trademark of Hopkins Agricultural Chemical Co.
X-77® trademark of KALO Agricultural Chemicals, Inc.

CGA 6L46D 095

PRINCEP® CALIBER® 90

HERBICIDE

For weed control in certain crops, ornamental plantings, on industrial sites, and around-the-farm

Princep Caliber 90 is a water dispersible granule

ACTIVE INGREDIENT:
Simazine:
2-chloro-4,6-bis (ethylamino)-*s*-triazine 90%
INERT INGREDIENTS: 10%
TOTAL: 100%

EPA Reg. No. 100-603 EPA Est. 100-LA-1

Princep® trademark of CIBA-GEIGY for simazine

Keep Out of Reach of Children.

Caution

See additional precautionary statements at end of booklet.

See directions for use inside booklet.

DIRECTIONS FOR USE AND CONDITIONS OF SALE AND WARRANTY

IMPORTANT: Read the entire **Directions for Use** and the **Conditions of Sale and Warranty** before using this product.

CONDITIONS OF SALE AND WARRANTY

The **Directions for Use** of this product reflect the opinion of experts based on field use and tests. The directions are believed to be reliable and should be followed carefully. However, it is impossible to eliminate all risks inherently associated with use of this product. Crop injury, ineffectiveness, or other unintended consequences may result because of such factors as weather conditions, presence of other materials, or the manner of use or application all of which are beyond the control of CIBA-GEIGY or the Seller. All such risks shall be assumed by the Buyer.

CIBA-GEIGY warrants that this product conforms to the chemical description on the label and is reasonably fit for the purposes referred to in the **Directions for Use** subject to the inherent risks referred to above. **CIBA-GEIGY makes no other express or implied warranty of Fitness or Merchantability or any other express or implied warranty. In no case shall CIBA-GEIGY or the Seller be liable for consequential, special, or indirect damages resulting from the use or handling of this product.** CIBA-GEIGY and the Seller offer this product, and the Buyer and user accept it, subject to the foregoing **Conditions of Sale and Warranty,** which may be varied only by agreement in writing signed by a duly authorized representative of CIBA-GEIGY.

DIRECTIONS FOR USE

It is a violation of federal law to use this product in a manner inconsistent with its labeling.

FAILURE TO FOLLOW THE DIRECTIONS FOR USE AND PRECAUTIONS ON THIS LABEL MAY RESULT IN CROP INJURY, POOR WEED CONTROL, AND/OR ILLEGAL RESIDUES.

Do not apply this product in such a manner as to directly or through drift expose workers or other persons except those knowingly involved in the application. The area being treated must be vacated by unprotected persons.

REENTRY STATEMENT

Do not enter treated areas without protective clothing until sprays have dried.

Because certain states may require more restrictive reentry intervals for various crops treated with this product, consult your State Department of Agriculture for further information.

Written or oral warnings must be given to workers who are expected to be in a treated area or in an area about to be treated with this product. Oral warnings must inform workers of areas or fields that may not be entered without specific protective clothing until sprays have dried, and appropriate actions to take in case of accidental exposure, as described under **Precautionary Statements** on this label. When oral warnings are given, warnings shall be given in a language customarily understood by workers. Oral warnings must be given if there is reason to believe that written warnings cannot be understood by workers. Written warnings must include the following information: CAUTION. Area treated with Princep Caliber 90 on *(date of application)*. Do not enter without appropriate protective clothing until sprays have dried. In case of accidental exposure, flush eyes or skin with plenty of water. Call physician if irritation persists. Remove and wash contaminated clothing before reuse.

GENERAL INFORMATION

Apply this herbicide before weeds emerge or after removal of weed growth. Princep controls a wide variety of annual broadleaf and grass weeds when used at selective rates in agricultural crops and ornamental plantings. When used at higher, nonselective rates in noncrop areas, it also controls many perennial broadleaf and grass weeds.

Where a range of application rates is given, use the low rate on coarser textured soil and soil lower in organic matter; use the high rate on finer textured soil and soil higher in organic matter.

To avoid spray drift, do not apply under windy conditions. Avoid spray overlap, as crop injury may result.

Since this product enters weeds mainly through their roots, moisture is needed to move it into the root zone. Very dry soil conditions and lack of rainfall following application may necessitate shallow cultivation.

This product is noncorrosive to equipment, nonflammable, and has low electrical conductivity.

Annual Weeds Controlled

alyssum
annual blue grass
annual morningglory
annual ryegrass
barnyardgrass (watergrass)
burclover
carelessweed
carpetweed
common chickweed
crabgrass (*Digitaria* spp.)
downy brome (cheatgrass)
fall panicum
fiddleneck
filaree
fireweed
fivehook bassia
Flora's paintbrush
Florida pusley
foxtails
goosegrass
groundsel
henbit
junglerice
knawel (German moss)
common lambsquarters
nightshade
pepperweed
pigweed
pineappleweed
prickly lettuce
common purslane
ragweed
rattail fescue
redmaids
Russian thistle
shepherdspurse
signalgrass (*Brachiaria* spp.)
silver hairgrass
smartweed
spanishneedles
speedwell
tansymustard
wild mustard
wild oats
witchgrass
yellow flower pepperweed

Following many years of continuous use of this product and chemically related products, biotypes of some of the weeds listed on this label have been reported which cannot be effectively controlled by this and related herbicides. Where this is known or suspected and weeds controlled by this product are expected to be present along with resistant biotypes, we recommend the use of this product in registered combinations or in sequence with other registered herbicides which are not triazines. If only resistant biotypes are expected to be present, use a registered non-triazine herbicide. Consult with your State Agricultural Extension Service for specific recommendations.

APPLICATION PROCEDURES

Ground application: Use conventional ground sprayers equipped with nozzles that provide accurate and uniform application. Be certain that nozzles are uniformly spaced and the same size. Calibrate sprayer before use and recalibrate at the start of each season and when changing carriers. Unless otherwise specified, use a minimum of 20 gals. of spray mixture per acre.

Use a pump with capacity to (1) maintain 35-40 psi at nozzles, (2) provide sufficient agitation in tank to keep mixture in suspension, and (3) to provide a minimum of 20% bypass at all times. Use centrifugal pumps which provide propeller shear action for dispersing and mixing this product. The pump should provide a minimum of 10 gals./minute/100 gals. tank size circulated through a correctly positioned sparger tube or jets.

Use screens to protect the pump and to prevent nozzles from clogging. Screens placed on suction side of pump should be *16-mesh or coarser.* Do not place a screen in the recirculation line. Use 50-mesh or coarser screens between the pump and boom, and where required, at the nozzles. Check nozzle manufacturer's recommendations.

For band applications, calculate amount to be applied per acre as follows:

$$\frac{\text{band width in inches}}{\text{row width in inches}} \times \text{broadcast rate per acre} = \text{amount needed per acre of field}$$

Aerial application: Use aerial application only where specified in the use directions. Apply in a minimum of 1 gal. of water for each 1-1.5 lbs. of herbicide applied per acre. Avoid application under conditions where uniform coverage cannot be obtained or where excessive spray drift may occur.

Avoid application to animals or humans. Flagmen and loaders should avoid inhalation of spray mist and prolonged contact with skin, and should wash thoroughly before eating and at the end of each day's operation.

Mixing procedures—all uses: (1) Be sure sprayer is clean and not contaminated with any other materials, or crop injury or sprayer clogging may result. (2) Fill tank ¼ full with clean water, nitrogen solution or complete fluid fertilizer. (3) Start agitation. (4) Be certain that the agitation system is working properly and creates a rippling or rolling action on the liquid surface. (5) Pour product directly from bag into tank. (6) Continue filling tank until 90% full. Increase agitation if necessary to maintain surface action. (7) Add tank mix herbicide(s) after this product is thoroughly suspended. (8) Finish filling tank. Maintain agitation to avoid separation of materials. (9) Empty tank as completely as possible before refilling to prevent buildup of emulsifiable concentrate residue from possible tank mix herbicides. (10) If an emulsifiable concentrate film starts to build up in tank, drain it and clean with strong detergent solution or solvent. (11) Clean sprayer thoroughly immediately after use by flushing system with water containing a detergent.

Compatibility test: To determine the tank mix compatibility of Princep Caliber 90 with liquid fertilizer, crop oil, spreaders, or recommended pesticides, use this test method. Nitrogen solutions or complete fluid fertilizers may replace all or part of the water in the spray in alfalfa and corn. Since liquid fertilizers can vary, even within the same analysis, check compatibility each time before use. Be especially careful when using complete suspension or fluid fertilizers as serious compatibility problems are more apt to occur. Commercial application equipment may improve compatibility in some instances. Check compatibility using this procedure:

1. Add 1 pint of fertilizer to each of 2 one-quart glass jars with tight lids.
2. To **one** of the jars add ¼ tsp. of a compatibility agent approved for this use (¼ tsp. is equivalent to 2 pts. per 100 gals. spray). Cap and shake until mixed. Examples of compatibility agents include Compex® and Unite®.
3. To **both** jars add the appropriate amount of herbicide(s). If more than one herbicide is used, add them separately with dry herbicides first, flowables next, and emulsifiable concentrates last. After each addition, cap and shake until thoroughly mixed. The appropriate amount of herbicides for this test follows (assuming a spray volume of 25 gals. per acre):

   Dry herbicides: **For each pound per acre** add 1.5 teaspoons to each jar. Fluff up wettable powder products before measuring.

   Liquid herbicides: **For each pint per acre** add 0.5 teaspoon to each jar.

   For a spray volume other than 25 gals. per acre, change the teaspoons added to each jar as follows:

   $$\frac{\text{25 gals.} \times \text{no. teaspoons given above}}{\text{desired gals. spray volume per acre}} = \text{teaspoons to add to each jar}$$
4. After adding all ingredients, put lids on and tighten. Shake jars vigorously one minute. Let the mixtures stand 15 minutes and then look for separation, large flakes, precipitates, grease, gels, medium to heavy oily film on the jar, or other signs of incompatibility. Determine if the compatibility agent is needed in the spray mixture by comparing the two jars. If either mixture separates, but can be remixed readily, the mixture probably can be sprayed as long as good agitation is used. If the mixtures are incompatible, test the following methods of improving compatibility: (A) slurry the dry herbicide(s) in water before addition, or (B) add ½ of the compatibility agent to the fertilizer and the other ½ to the emulsifiable concentrate or flowable herbicide before addition to the mixture. If still incompatible, do not use the materials mixed in the same spray tank.

ROTATIONAL CROPS AND PERENNIAL CROP REPLANTING

To avoid crop injury, observe the following precautions: (1) If rotating treated land the year following application, plant only corn, unless otherwise stated in this label. (2) If replanting perennial crops or if rotating land to crops other than corn, do not apply this product in the year preceding planting these crops.

FRUIT AND NUT CROPS

Apply the spray to the orchard or vineyard floor avoiding contact with fruit, foliage, or stems. Recommended rates are based on broadcast treatment. For band applications or spot applications around fruit or nut trees, reduce the broadcast rate of this product and water per acre in proportion to the area actually sprayed.

Precautions (All fruit and nut crops): *To avoid crop injury, 1) Apply only to orchards or groves where trees have been established 12 months or more unless specified differently. 2) Make only one application per year, except as noted otherwise. 3) Do not use on gravelly, sand, or loamy sand soil. 4) Immediately following application, limit overhead sprinkler irrigation to ½ inch.*

ALMONDS AND PEACHES (CA only)

Apply 1.1-2.2 lbs. per treated acre in a 2-4 ft. band on each side of the tree row. Apply before weeds emerge in late fall or early winter. Weeds controlled by 1.1 lbs. include burclover, common chickweed, wild mustard, and shepherdspurse. Apply only once per year.

Precautions: *To avoid crop injury, 1) Do not treat trees established in the grove less than 3 years. 2) Do not treat the Mission (Texas) variety of almonds. 3) Do not apply to almond trees propagated on plum rootstocks. 4) Do not replant almonds or peaches in treated soil for 12 months after treatment. 5) Do not apply on soil with less than 1% organic matter. 6) Do not treat areas where water will accumulate.*

APPLES, PEARS, SOUR CHERRIES

Apply 2.2-4.4 lbs. per acre.

AVOCADOS

CA and FL only: Apply 2.2-4.4 lbs. per acre after final preparation of grove.

BLUEBERRIES AND CANEBERRIES (blackberries, boysenberries, loganberries, raspberries)

Apply 2.2-4.4 lbs. per acre in the spring or apply a split application of 2.2 lbs. per acre in the spring plus 2.2 lbs. per acre in the fall. Apply in a minimum of 40 gals. of water per acre.

On plantings less than 6 months old, use ½ the above rate.

To control quackgrass, apply 4.4 lbs. per acre in the fall or split the application applying 2.2 lbs. per acre in the fall plus 2.2 lbs. per acre in the spring, when quackgrass is actively growing.

Note: Do not apply when fruit is present, or illegal residues may result.

CRANBERRIES

MA: Apply up to 4.4 lbs. per acre either before spring growth begins or in the fall after harvest.

Other areas: Apply 2.2 lbs. per acre before spring growth begins.

FILBERTS

Apply 2.2-4.4 lbs. per acre in the fall or apply a split application of 2.2 lbs. per acre in the fall plus 2.2 lbs. per acre in the spring.

Precaution: *If trees are planted on a hillside, excessive soil erosion may result from the elimination of weeds.*

Note: Do not apply when nuts are on the ground during the harvest period, or illegal residues may result.

GRAPES

Apply 2.2-5.3 lbs. per acre any time between harvest and early spring.

Precaution: *Do not use in vineyards established less than three years, or crop injury may occur.*

GRAPEFRUIT, LEMONS, ORANGES

AZ (Lemons and Oranges only): Apply a split application of 1.75 lbs. per acre in the spring plus 1.75 lbs. per acre in the fall.

CA: In grapefruit, lemons, and oranges, apply 2.2-4.4 lbs. per acre in a single application; or apply 2.2 lbs. per acre in the fall and 2.2 lbs. per acre in the spring.

Precaution: *Do not use in the Imperial, Coachella, or Palo Verde valleys, or crop injury may occur.*

FL (Grapefruit and Oranges only): Apply only once per year. For control of weeds listed under General Information, apply 7.1-10.6 lbs. per acre in the spring; to also control milkweed vine, use 8.9-10.6 lbs. per acre either preemergence or before vines are 12 inches tall.

TX (Grapefruit and Oranges only): Apply 4.4-5.3 lbs. per acre.

Precautions (All areas): *To avoid crop injury, 1) Do not use in nurseries. 2) Do not apply to bedded grapefruit, lemons, or oranges. 3) Do not apply to trees under stress from freeze damage for one year after the freeze.*

MACADAMIA NUTS

Apply 2.2-4.4 lbs. in 50 gals. of water per acre before harvest and just prior to weed emergence. Repeat application as necessary.

Note: Do not apply when nuts are on the ground during the harvest period, as illegal residues may result.

OLIVES

Apply 2.2-4.4 lbs. per acre following grove preparation in the fall. Repeat annually in midwinter.

PEACHES, PLUMS, SWEET CHERRIES

Apply 1.75-4.4 lbs. per acre. Apply in late fall to early spring prior to weed emergence.

Precautions: *To avoid crop injury, 1) Peaches : use only in AR, LA, MO, OK, TX, and states east of the Miss. River. For CA, see specific* *directions in the section "Almonds and Peaches (CA only)." 2) Plums and sweet cherries : use only in MO and states east of the Miss. River except TN.*

PECANS

Apply 2.2-4.4 lbs. per acre before weeds emerge in the spring.

Precautions: *To avoid crop injury, 1) Do not use west of the Pecos River in TX or in NM, AZ, or CA. 2) Do not make applications to transplanted trees that have been established less than two years in the grove.*

Note: Do not apply when nuts are on the ground, or illegal residues may result. Do not allow animals to graze treated areas.

STRAWBERRIES

Oregon and Washington: For control of chickweed, groundsel, mustard, and shepherdspurse, apply broadcast 1.1 lbs. per acre. In fields where overhead irrigation is used to activate this product, apply after harvest at time of bed renovation. In fields where overhead irrigation is not available, apply during early October through November.

Precautions: *To avoid crop injury, 1) Make only one application per growing season. 2) Do not apply within 4 months after transplanting.*

WALNUTS

Apply 2.2-4.4 lbs. per acre. Leveling and furrowing operations after application will lessen effectiveness of weed control.

Note: Do not apply when nuts are on the ground, or illegal residues may result.

TANK MIXTURE WITH EVIK® 80W FOR GRAPEFRUIT AND ORANGES

Use in grapefruit and oranges in FL and TX for control of weeds listed in the following directions and under General Information. Apply the rates given below in a minimum of 40 gals. of water per acre thoroughly covering existing vegetation, but avoiding contact with citrus fruit, foliage, or stems. Apply when weeds have resumed active growth in the spring, but before they exceed 4 inches in height (except as noted for vines). Observe all precautions and limitations on the labeling of this product and Evik 80W.

FL (Ridge Area): Apply 4.4 lbs. of this product plus 5 lbs. of Evik 80W per acre; controls milkweed vine when applied either preemergence or on vines up to 24 inches tall.

FL (Bedded Areas): Apply 3.5 lbs. of this product plus 6 lbs. of Evik 80W per acre; controls balsamapple when applied either preemergence or on vines up to 12 inches tall.

TX: Apply 3.5-4.4 lbs. of this product plus 2-3 lbs. of Evik 80W per acre plus 2 qts. nonionic surfactant, such as X-77®, per 100 gals. of spray mixture (add surfactant to spray tank last); controls buffelgrass, marinevine (ivy treebine), Texas panicum, sandbur, and sunflower, and gives top-kill of johnsongrass. Use 4.4 lbs. of this product per acre for longer residual control; use 3 lbs. of Evik 80W per acre for control of heavy weed growth and for top-kill of johnsongrass; use the higher rate of both products for control of marinevine.

Precautions (All areas): *To avoid injury, 1) Do not use in nurseries, or around trees that have been established in the grove less than two years, or where trees are under stress from freeze damage for one year after the freeze. 2) Apply only once per year.*

Note (All areas): Do not apply within 30 days before harvest, or illegal residues may result. Do not graze treated areas.

TANK MIXTURE WITH BROMACIL 80W FOR GRAPEFRUIT AND ORANGES (Florida Only)

Use in grapefruit and oranges in FL for control of balsamapple, black nightshade, carpetweed, crabgrass, cudweed, dayweed, Florida pusley, horseweed, pepperweed, pigweed, poorjoe, ragweed, rattlebox, spanishneedles and sandbur, and for partial control of bermudagrass, bahiagrass, pangolagrass, paragrass, and torpedograss. Apply 3.5-7.1 lbs. of this product plus 3-4 lbs. of bromacil 80W per acre beneath trees in a minimum of 40 gals. of water per acre before or soon after weed growth begins. Use the lower rates for light weed infestations or all applications in bedded citrus areas. Use the higher rates for heavy weed infestations only in ridge grown citrus areas.

Temporary yellowing of citrus leaves may occur following treatment.

Precautions: *To avoid crop injury, 1) Do not use in nurseries, or around trees that have been established in the grove less than four years, or where trees are under stress from freeze damage for one year after the freeze. 2) Do not use on soil with less than 1% organic matter or on poorly drained soil. 3) Do not treat trees planted in irrigation furrows. 4) Do not treat diseased trees such as those with foot rot. 5) Do not use in groves interplanted with other trees or desirable plants, nor in home grapefruit or orange plantings, or in areas where roots of other valuable plants or trees may extend. 6) Treated areas may be planted to citrus trees one year after application. 7) Do not rotate to other crops within two years after application.*

Note: Apply only once per year and avoid contact with foliage and fruit with spray or mist, or illegal residues may result.

TANK MIXTURE WITH PARAQUAT

This tank mix is effective in the following fruit and nut crops for kill of existing vegetation and for residual control of the annual broadleaf and grass weeds claimed for this product applied alone. This combination is also effective for top-kill and suppression of perennial weeds.

Use this tank mix on these crops:

almonds (CA only)	grapefruit	oranges[2]
apples	(CA, FL[2] & TX only)	peaches[3]
avocados (CA & FL[1] only)	grapes	pears
cherries (sour and sweet[4])	lemons (AZ & CA only)	pecans
filberts	macadamia nuts	plums[4]
	olives	walnuts

Apply the rate given under the appropriate crop on this label plus 1-2 qts. paraquat* in 50-200 gals. (30-50 gals. for pecans) of water per acre to the orchard floor avoiding contact with fruit, foliage, or stems. Add a nonionic surfactant, such as X-77, at 0.5 pt. per 100 gals. of spray. Apply when weeds are succulent and new growth is 1-6 inches tall. For mature woody weeds or difficult to control perennial weeds, re-treat or spot treat with paraquat if regrowth occurs.

Add this product to the spray tank first (refer to Mixing procedures section of this label), then add paraquat, and add the surfactant last. Provide constant agitation during mixing and application to keep the mixture in suspension.

Precautions: *To avoid crop injury, 1) Apply the tank mix only once per year. 2) Use a shield for young trees or vines.*

Refer to the labels of both products for further directions, specific weeds controlled, and precautions and limitations on each crop.

1 In avocados in Florida, this tank mix also controls balsamapple vine, rattail amaranth, and at the higher rate of each herbicide, it suppresses coral vine.

2 In grapefruit and oranges in Florida, apply 3.5 lbs. of this product per acre in bedded areas and 7.1-10.6 lbs. per acre in ridge areas.

3 Limited to CA, AR, LA, MO, OK, TX, and states east of the Miss. River. As appropriate, refer to the sections "Almonds and Peaches (CA only)" or "Peaches, Plums, Sweet Cherries" for rate of this product and other information.

4 Limited to MO and states east of the Miss. River except TN.

* Based on a product containing 2 lbs. paraquat cation per gal.

TANK MIXTURE WITH ROUNDUP®

This tank mixture is effective in grape vineyards and in the following bearing or nonbearing tree crops for control of existing vegetation and for residual control of the annual broadleaf and grass weeds claimed for this product applied alone. This combination is also effective for partial control of perennial weeds contacted by the spray mixture during application. Use this tank mixture on these crops:

almonds** (CA only)	grapefruits* (CA, FL[1] and TX)	peaches*[2]
apples*	grapes*	pears*
avocados* (CA and FL only)	lemons* (AZ and CA)	pecans**
cherries (sour and sweet[3])	macadamia nuts**	plums*[3]
filberts**	oranges* (AZ, CA, FL[1] and TX)	walnuts**

Use the appropriate rate given elsewhere on this label for Princep applied alone to the crop being treated. Add to the spray tank 1-5 qts. of Roundup per acre depending on weeds present and their growth stage. Also add an agriculturally approved nonionic surfactant at 0.5% by volume of spray solution. Apply the mixture in 10-40 gals. of water per acre as a postemergence spray to the weeds at the appropriate weed growth stage given on the Roundup label.

Add Princep to the spray tank first, then add Roundup. Provide constant agitation during mixing and application to keep the mixture in suspension. Refer to Application Procedures section of this label for further directions.

Precaution: *To avoid crop injury, take extreme care to avoid contact of herbicide solution, spray, drift, or mist with foliage or green bark of trunk, branches, suckers, fruit, or other parts of trees or vines.*

Refer to the labels of both herbicides for further directions, specific weeds controlled, precautions and limitations on each crop.

* Allow a minimum of 14 days between last application and harvest.

** Allow a minimum of 21 days between last application and harvest of these crops.

1 In grapefruit and oranges in Florida, apply 3.5 lbs. of this product per acre in bedded areas and 7.1-10.6 lbs. per acre in ridge areas.

2 Limited to CA, AR, LA, MO, OK, TX, and states east of the Mississippi River. For CA, see specific directions in the section "Almonds and Peaches (CA only)."

3 Plums and sweet cherries: limited to MO and states east of the Mississippi River except TN.

TANK MIXTURE WITH SURFLAN®

Use this tank mix in the following crops for preemergence control of all weeds claimed on both labels:

almonds	cherries	lemons	pecans
apples	filberts	oranges	plums
avocados	grapes	peaches	English walnuts
caneberries	grapefruit	pears	

Apply the Princep Caliber 90 rate given under the appropriate crop on this label plus 2.67-5.33 lbs. of Surflan 75W or 2-4 qts. of Surflan A.S. in 20-40 gals. of water per acre. Refer to the Surflan 75W or Surflan A.S. label for complete tank mix directions. Observe all precautions and limitations on the Princep and Surflan labels.

FIELD AND FORAGE CROPS

ALFALFA

Pure alfalfa at least 12 months old, pure alfalfa less than 12 months old, and alfalfa-orchardgrass mixtures at least 12 months old

For ground application, apply this product in a minimum of 15 gals. of water per acre. Fluid fertilizer may replace all or part of the water used as a carrier, except fluid fertilizers should not be used when alfalfa is actively growing, as injury may occur. Check the physical compatibility of this product with fertilizer before use. Refer to the General Information section of this label for a compatibility test procedure.

For aerial application, apply this product in a minimum of 2.5 gals. of water per acre.

PURE ALFALFA AT LEAST 12 MONTHS OLD (All Areas Listed)

Apply once per year in the fall after the last cutting but prior to permanently frozen ground conditions. Alfalfa may be actively growing, semidormant, or dormant when the application is made.

Alfalfa must be growing for at least 12 months before treatment.

Alfalfa regions	Weeds controlled	Soil texture	Rate per acre
WA, OR, NV, UT, ID, MT, WY, CO, and the California counties of Del Norte, Siskiyou, Modoc Shasta, Lassen, Plumas, Sierra, and Nevada	Downy brome (cheatgrass), tansymustard, prickly lettuce, pepperweed, seedling shepherdspurse, yellow flower pepperweed	Sand and loamy sand	DO NOT USE
		Sandy loam	0.9 lb.
		Loam soil	0.9-1.3 lbs.
		Clay soil	1.3-1.75 lbs.
ND, SD, MN, WI, NE, KS, IA, MI, and NY	Henbit, wild mustard, common chickweed, knawel (German moss), alyssum, downy brome (cheatgrass), wild oats, pepperweed, tansymustard, pigweed, speedwell, and seedlings of yellow rocket, shepherdspurse, white cockle, fanweed (pennycress), dandelion, Russian thistle, kochia	Sand and loamy sand	DO NOT USE
		Sandy loam	0.9 lb.
		Loam soil	1.3 lbs.
		Clay soil	1.75 lbs.
IL, IN, OH, KY, TN, NC*, VA, PA, NJ, DE, ND, WV, VT, MA, CT, ME, NH, RI, and MO * Use only on loam and clay soils	Pigweed, common chickweed, mouseear chickweed, speedwell, henbit, knawel (German moss), pepperweed, downy brome, alyssum, common lambsquarters, wild oats, and seedlings of shepherdspurse, fanweed (pennycress), yellow rocket (wintercress), Russian thistle, dandelion, kochia, white cockle	Sand and loamy sand	DO NOT USE
		Sandy loam	0.9 lb.
		Loam soil	1.1 lbs.
		Clay soil	1.3 lbs.

PURE ALFALFA LESS THAN 12 MONTHS OLD (Northeastern U.S. Only)

Pure alfalfa which has been seeded in the spring before June 1 and has become well established may be treated in the fall after the last cutting but before permanently frozen ground conditions. Alfalfa may be actively growing, semidormant, or dormant when this product is applied. Apply 0.9 lb. per acre to control henbit, wild mustard, common lambsquarters, common chickweed, German moss, alyssum, downy brome (cheatgrass), wild oats, tansymustard, pigweed, speedwell, and pepperweed.

ESTABLISHED ALFALFA-ORCHARDGRASS MIXTURES (Northeastern U.S. Only)

Apply only to alfalfa-orchardgrass mixtures which have been established at least 12 months. Make a single application in the fall after the alfalfa-orchardgrass mixture has gone dormant.

Applications should be uniformly broadcast at the rate of 0.9 lb. per acre to control henbit, wild mustard, common lambsquarters, common chickweed, German moss, alyssum, downy brome (cheatgrass), wild oats, tansymustard, pigweed, speedwell, and pepperweed.

Precautions (For Each Use): *1) Do not use on gravelly, sand, or loamy sand soil, or on soil low in organic matter, or crop injury may occur. 2) Applications made to alfalfa growing on highly calcareous or highly alkaline soil may result in crop injury. 3) Use only in areas where frost occurs and alfalfa goes into a semidormant or dormant state, or crop injury may occur. 4) Do not apply on snow, or poor weed control may result. 5) This product will not control dodder and established perennial or biennial weeds such as dandelion, white cockle, quackgrass, cinquefoil, wild carrot, and dock. 6) Temporary crop chlorosis may appear on areas where overlapping occurs. The sprayer should be shut off when turning around at the end or corners* *of the field. 7) Do not rotate to any crops except corn for the growing season following application as injury may occur.*

Note: Allow 30 days between application and grazing of dairy and beef cattle and sheep, and allow 60 days between application and cutting for hay or illegal residues may result.

TANK MIXTURE WITH PARAQUAT FOR ALFALFA (WA, OR, NV, UT, ID, MT, WY, CO, and the California counties of Del Norte, Siskiyou, Modoc, Shasta, Lassen, Plumas, Sierra and Nevada)

Controls filaree, Jim Hill mustard, and shepherdspurse, and gives improved control of bluegrass, cheatgrass, common chickweed, and tansymustard. Apply 0.9-1.75 lbs. of this product plus 1-3 pts. of paraquat* in 20-100 gals. of water per acre with ground equipment or in 5-10 gals. of water per acre aerially. Refer to the alfalfa regions rate chart for rates of this product on specific soil textures. Use 1 pt. of paraquat when weeds and grasses are less than 4 inches tall. Apply when weeds are succulent and growth is 1-6 inches tall. Apply to established alfalfa after the last fall cutting when the crop is dormant and before spring growth starts. Observe all precautions and limitations on the labeling of this product and paraquat.

* Based on a product containing 2 lbs. paraquat cation per gal.

Precautions: *To avoid crop injury, 1) Do not apply if regrowth following the last fall cutting is more than 2 inches tall. Alfalfa foliage present at application will be burned. 2) Apply once per growing season.*

Note: Do not pasture animals in treated fields before first cutting, or illegal residues may result. Do not graze, cut, or harvest within 90 days after application.

FORAGE BERMUDAGRASSES (Such as Coastal and Midland)

For preemergence control of most annual broadleaf and grassy weeds including barnyardgrass, crabgrass, *Brachiaria* spp., goosegrass, foxtail, carelessweed, carpetweed, annual morningglory, pigweed, common purslane, Russian thistle, common chickweed, henbit, and burclover, apply this product shortly after sprigging. Apply it to established bermudagrass from fall to spring depending on the weed problem, but before weed emergence. Use the appropriate rate in the following table in a minimum of 15 gals. of water per acre for ground application and in a minimum of 1 gal. of water for each 0.9-1.3 lbs. of this product per acre for aerial application.

Soil Texture	Rate per acre
Fine sandy soil low in organic matter	0.9 lb.
Sandy loam soil	1.3-2.2 lbs.
Loam and clay soil low in organic matter	2.2-2.6 lbs.
Clay soil	2.6-3.5 lbs.

Precautions: *To avoid crop injury, 1) Do not use on bermudagrass in NM and the High Plains of TX. 2) Do not apply within 7 days after cutting.*

Note: To avoid illegal residues, 1) Do not graze livestock for 30 days after 0.9-2.6 lbs. per acre application or for 60 days after 3.5 lbs. per acre application. 2) Do not cut for hay for 60 days after 0.9-2.6 lbs. per acre application or for 90 days after 3.5 lbs. per acre application.

CORN

Nitrogen solutions or complete fluid fertilizers may replace all or part of the water as a carrier in the spray. Determine the physical compatibility of this product with fertilizer before use. Refer to the General Information section of this label for a compatibility test. Do not apply after corn has emerged as there is danger of liquid fertilizers causing crop injury.

Preemergence: Apply before weeds and corn emerge. Use the appropriate rate in the table below.

Preplant: Apply in the spring after plowing at the appropriate rate in the table below. Apply before, during, or after final seedbed preparation. If soil is tilled or worked after application, avoid deep incorporation. Best results will be obtained when this product is applied within 2 weeks before planting. Under dry weather conditions, preplant applications may give better weed control. If weeds develop, particularly under relatively dry conditions, a shallow cultivation will generally result in better weed control.

Soil Texture	Broadcast* rate per acre
Coarse-texture soil: Sand, silt, and loam that is low in organic matter	2.2 lbs.
Medium-textured soil: Soil containing a moderate amount of clay and organic matter	2.6 lbs.
Fine-textured soil: Loam that is high in organic matter and clay (including dark prairie soils of Corn Belt)	3.3 lbs.
Organic soil: Peat, muck, and high-organic clay	4.4 lbs.

* For calculation of band treatment rate, see the General Information section.

To control quackgrass: Apply 3.3-4.4 lbs. per acre in the fall. Plow two to three weeks later, or if erosion is a problem, delay plowing until spring.

Precautions: *1) Do not apply more than 4.4 lbs. per acre to corn per year. 2) Do not rotate to any crop except corn until the following year or injury may occur. 3) After harvest, plow and thoroughly till the soil in fall or spring to minimize possible injury to spring-seeded rotational crops, regardless of the rate used. 4) If more than 3.3 lbs. is used per acre (or equivalent rate in a band), a crop of untreated corn should precede the next rotational crop. 5) Do not apply preplant incorporated in corn in the High Plains and Intermountain areas of the West (including central and western KS, western NE, western OK, and the Panhandle of TX) where rainfall is sparse and erratic or where irrigation is required. 6) In the High Plains and Intermountain areas of the West where rainfall is sparse and erratic or where irrigation is required, use this product to control weeds in corn only when corn is to follow corn or when a crop of untreated corn is to precede another rotational crop. 7) In western MN and eastern parts of the Dakotas, NE, and KS, do not plant soybeans following corn treated with this product if more than 2.2 lbs. per acre (or equivalent rate in a band) was applied or injury may occur. 8) Injury may occur to soybeans planted in north-central and northwest IA, south-central and southwest MN, northeast NE, southeast SD, and other areas the year following application on soils having a calcareous surface layer. 9) Do not plant sugar beets, tobacco, vegetables (including dry beans), spring-seeded small grains, or small-seeded legumes and grasses the year after an application, or injury may occur.*

Note: Do not graze treated areas, or illegal residues may result.

TANK MIXTURES ON CORN

AAtrex®: Use this product in a preplant or preemergence tank mixture with AAtrex Nine-O®, AAtrex 80W, or AAtrex 4L (4LC) for control of many annual weeds, including carpetweed, crabgrass, fall panicum, foxtail, lambsquarters, morningglory, pigweed, ragweed, and velvetleaf.

Apply at the rates given below. Use the 1:1 ratio for most weeds and the 2:1 ratio for expected heavy infestations of crabgrass and fall panicum.

	Broadcast rate per acre			
	1:1 Ratio		2:1 Ratio	
Soil texture	Princep Caliber 90	AAtrex 80W*	Princep Caliber 90	AAtrex 80W*
Sand loamy sand, sandy loam	1.1 lbs.	1.25 lbs.	1.5 lbs.	0.83 lb.
Loam, silt loam, silt, clay loam, sandy clay loam, silty clay loam, sandy clay, or silty clay with low organic matter	1.33 lbs.	1.5 lbs.	1.8 lbs.	1 lb.

Loam, silt loam, silt, clay loam, sandy clay loam, silty clay loam, sandy clay, or silty clay with medium to high organic matter, and clay (including dark prairie soils of the Corn Belt)	1.6 lbs.	1.8 lbs.	2.1 lbs.	1.2 lbs.

* When using AAtrex 4L (4LC) or AAtrex Nine-O, use equivalent rates. One lb. of AAtrex 80W equals 0.8 qt. of AAtrex 4L (4LC) or 0.9 lb. of AAtrex Nine-O.

Refer to the AAtrex label for complete directions, and use 0.9 lb. of this product for each pound of Princep 80W. Also refer to the AAtrex label and the Corn section of this label for precautions and limitations.

Eradicane®: Use in a preplant incorporated tank mixture for control of all weeds claimed on both the Princep and Eradicane labels including partial control (suppression) of shattercane (wild cane). Fluid fertilizer may replace all or part of the water in the spray. Check the physical compatibility of mixture with fertilizer before use. Refer to the General Information section of this label for a compatibility test procedure. Use a minimum of 20 gals. of spray volume per acre. Refer to the Eradicane label for incorporation directions. Use the higher rate of Princep on fine-textured soil and where heavy broadleaf infestations are expected. Use the higher rate of Eradicane for heavy nutsedge and bermudagrass infestations. For partial control (suppression) of shattercane, broadcast and incorporate **immediately** before planting 1.1-2.2 lbs. of this product plus 7.33 pts. of Eradicane 6.7E (or 8 pts. Eradicane Extra) per acre. For control of the other weeds claimed on both labels, broadcast and incorporate 1.1-3.3 lbs. of this product plus 4.75-7.33 pts. of Eradicane 6.7E (or 5.33-8 pts. of Eradicane Extra) per acre. Observe all precautions and limitations on the Princep and Eradicane labels.

Paraquat: Use in a tank mixture where corn will be planted directly in a cover crop, established sod, or previous crop residues. This combination controls existing vegetation and provides residual control of the annual broadleaf and grass weeds listed under General Information.

Add this product to the spray tank, mix thoroughly with water, and then add paraquat and a nonionic surfactant. Provide constant agitation during mixing and application to keep the mixture in suspension.

Apply 2.2-3.3 lbs. of this product plus 1-2 pts. of paraquat* in 20-60 gals. of water per acre as a broadcast spray either before or after planting but before corn emerges. Add a nonionic surfactant at the rate of 0.5 pt. per 100 gals. of spray volume.

For further information, see general information, caution and warning statements, precautions, and notes on the Princep and the paraquat labels.

* Based on a product containing 2 lbs. paraquat cation per gal.

Sutan+®: Use in a preplant incorporated tank mixture for control of all weeds claimed on both the Princep and Sutan+ labels including partial control (suppression) of shattercane (wild cane). Fluid fertilizer may replace all or part of the water in the spray. Check the physical compatibility of mixture with fertilizer before use. Refer to the General Information section of this label for a compatibility test procedure. Use a minimum of 20 gals. of spray volume per acre. Refer to the Sutan+ label for incorporation directions. Use the higher rate of Princep on fine-textured soil and where heavy broadleaf infestations are expected. For partial control (suppression) of shattercane, broadcast and incorporate **immediately** before planting 1.1-2.2 lbs. of this product plus 7.33 pts. of Sutan+ 6.7E per acre. For control of the other weeds claimed on both labels, broadcast and incorporate 1.1-3.3 lbs. of this product plus 4.75 pts. of Sutan+ 6.7E per acre. Observe all precautions and limitations on the Princep and Sutan+ labels.

SUGARCANE

Apply 2.2-4.4 lbs. per acre at planting or after harvest if ratoon crop is to be grown, but before weeds or sugarcane emerge. This product may also be applied interline-directed or over the sugarcane to provide continuing control of germinating weeds.

Precautions: *To avoid crop injury or illegal residues, 1) Do not apply more than one 2.2-4.4 lbs. application over the plants or more than two interline-directed applications. 2) Do not apply after close-in. 3) Total application should not exceed 11.1 lbs. per acre crop cycle.*

Note: Do not feed or graze treated forage to livestock.

GRASSES GROWN FOR SEED (Pacific Northwest Only)

Perennial ryegrass, Highland or Astoria bentgrasses, orchardgrass, tall fescue, and fine fescues such as Illahee, Rainier, Pennlawn, Chewings, and related species. For the control of broadleaf weeds and annual grasses including annual ryegrass, rattail fescue, silver hairgrass, and downy brome (cheat), apply 2.2 lbs. in a minimum of 15 gals. of water per acre as soon as fall rains start. Early application is important.

Precaution: *To avoid crop injury, apply only to grasses from which at least one seed crop has been cut.*

Note: To avoid illegal residues, allow 30 days between last application and grazing of dairy and beef cattle and sheep, and allow 60 days between the last application and cutting for hay.

NURSERIES, CHRISTMAS TREE PLANTINGS, SHELTERBELTS

NURSERIES (see list below)

Apply 2.2-3.4 lbs. in at least 25 gals. of water per acre in fall or spring.

Precaution: *To avoid plant injury, do not apply for at least one year after transplanting.*

CHRISTMAS TREE PLANTINGS AND SHELTERBELTS (See List Below)

Remove weed growth before application. Apply 2.2-4.4 lbs. in at least 25 gals. of water per acre after transplanting. Use the same rate for annual maintenance applications.

For quackgrass control apply 4.4 lbs. per acre in the fall or apply a split application of 2.2 lbs. per acre in the fall plus 2.2 lbs. per acre in early spring, after quackgrass begins growth.

Precautions: *To avoid tree injury, 1) Do not use on seedbeds or cutting beds. 2) In CA, OR, and WA do not apply to Christmas trees or shelterbelts sooner than one year after transplanting. In other areas, do not apply to Christmas trees or shelterbelt transplants less than three years of age. 3) Do not use until soil is firmly settled around roots. 4) Do not apply more than once a year, except as directed for quackgrass control.*

Apply to these species of trees and shrubs, as recommended above:

Conifers

arbovitae	Mugho pine	red spruce	juniper
Austrian pine	red pine (Norway pine)	white spruce	red cedar
*Bishop pine	Scotch pine	balsam fir	white cedar
*knobcone pine	white pine	Douglas fir	yew (*Taxus* spp.)
lodgepole pine (shore pine)	blue spruce	Fraser fir	
*Monterey pine	Norway spruce	white fir	
		hemlock	*For CA only

Deciduous Trees and Woody Ornamentals

American elm	eucalyptus	red oak
Siberian elm	holly (*Ilex* spp.) (max. 3.5 lbs./A)	*Russian olive
barberry		*oleander
boxelder	honey locust	*palm (2.2 lbs./A)
bush honeysuckle	Oregon grape (*Mahonia* spp.)	*bottle brush
caragana		*carob
cotoneaster	*Pieris* spp. (max. 3.5 lbs./A)	
dogwood		*For CA only

Surflan Tank Mix: On Christmas tree plantings, use this tank mix for preemergence control of weeds listed on this label and the Surflan 75W (or Surflan A.S.) label. Use on field grown conifer species listed on the labels for each herbicide plus grand fir, alpine fir,

Engelmann spruce, black spruce, Colorado blue spruce, Coulter pine, giant redwood, and Veitchi fir. Broadcast the mixture as a directed spray to the soil surface or as an overtop spray using 2.2-4.4 lbs. of Princep Caliber 90 and 2.67-5.33 lbs. of Surflan 75W (or 2-4 qts. of Surflan A.S.). Apply in sufficient water per acre to uniformly treat the area. Follow overtop sprays with sprinkler irrigation to move the herbicide from leaf surfaces to the soil. Remove weed growth before application. Mix weed residues, prunings or trash into the soil, or remove them before treatment. Soil should be in good tilth and free of clods at time of application. Shallow cultivation (1-2 inches) after treatment will not reduce weed control. Observe all precautions and limitations on the Princep and Surflan labels.

Dual® 8E Tank Mix: On field and liner grown woody ornamentals, use this tank mix for preemergence control of yellow nutsedge, barnyardgrass, crowfootgrass, fall panicum, giant foxtail, green foxtail, yellow foxtail, goosegrass, large crabgrass, smooth crabgrass, redroot pigweed, galinsoga, Venice mallow, lambsquarters, purslane, ragweed, smartweed and other weeds listed on the Dual 8E alone label. Control should last about 60 days. Apply 0.9-1.1 lbs. of Princep Caliber 90 plus 2-3 pts. of Dual 8E in a minimum of 15 gals. of water per acre. For band applications use a proportional amount of herbicide. Direct the spray toward the base of established ornamentals or ornamentals transplanted at least 10 days. Apply before weeds emerge or after removing existing weeds. Use the higher rates on fine-textured soil and where broadleaf weeds are expected to form a significant part of the infestation. Use the lower rates on coarse-textured soil and where light infestations of broadleaf weeds are expected. Apply post-transplant to the following field and liner grown ornamentals:

Arborvitae (*Thuja* spp.)	Leucothoe (*Leucothoe* spp.)
Cotoneaster (*Cotoneaster* spp.)	Osmanthus (*Osmanthus* spp.)
Crabapple (*Malus* spp.)	Weigela (*Weigela* spp.)
Holly (*Ilex* spp.), Boxleaf Japanese, Japanese	Yew (*Taxus* spp.), Dwarf Japanese
Juniper (*Juniperus* spp.), Andorra, Creeping, Shore	

Note: Length of weed control may be reduced when continuous wet soil conditions follow herbicide application.

Precautions: *To avoid plant injury, 1) Do not use on seedbeds or on unrooted cuttings. 2) Do not use in greenhouses or other enclosed areas.*

TREE PLANTATIONS FOR TIMBER

Apply before weeds emerge or following removal of weed growth. Apply before or after seedling transplanting. If applying after transplanting, application may be made over the top of tree seedlings before new growth appears. If new growth is present, direct the spray to avoid tree foliage contact.

The following rates are based on broadcast treatment in 20-100 gals. of water per acre. For band or spot applications, reduce the broadcast rate in proportion to the area actually sprayed.

Precautions: *To avoid tree injury, 1) Do not use on seed nurseries. 2) Apply only once per year.*

BLACK WALNUTS

Apply 2.2-4.4 lbs. per acre.

SLASH PINE AND WHITE PINE

Use for establishment in the South only. Apply 4.4-6.6 lbs. per acre.

LOBLOLLY PINE

Use for establishment in the South only. Apply either 4.4-6.6 lbs. of this product per acre or a tank mix of 3.5-5.3 lbs. of this product plus 1-2 lbs. of AAtrex 80W (or 0.9-1.75 lbs. AAtrex Nine-O or 0.8-1.6 qts. AAtrex 4L or 4LC) per acre. The tank mix provides broad spectrum annual weed control (see list in General Information section of this label) and better weed control in limited rainfall areas than this product alone.

TREE SEEDLINGS IN NORTH CENTRAL U.S.

Apply 2.2-4.4 lbs. per acre on the following species of trees:

American beech	maple	*sweet gum
balsam fir	Norway spruce	*white ash
black locust	pecan	white cedar
*black walnut	*red oak	*white oak
caragana	red pine (Norway pine)	white pine
dogwood	red spruce	white spruce
eastern red cedar	Russian olive	
honey locust	Scotch pine	

* Instead of this product alone, on these species a tank mix of 2.2 lbs. of this product + 2.5 lbs. of AAtrex 80W (or 2.2 lbs. AAtrex Nine-O or 2 qts. AAtrex 4L or 4LC) per acre may be applied in a minimum of 25 gals. of water.

Precautions: *To avoid tree injury, 1) Apply only once per year. 2) On sandy soil, do not use more than 2.2 lbs. of this product per acre and do not use the tank mix. 3) Do not apply the tank mix directly to seedlings after the start of bud break.*

TURF GRASSES FOR SOD (Florida Only)

ST. AUGUSTINEGRASS, CENTIPEDEGRASS, AND ZOYSIA GRASS

Apply 2.2-4.4 lbs. per acre, according to soil texture as indicated below.

Muck or peat	4.4 lbs.	Old beds	within 2 days after lifting of sod
		New beds	3-4 days after sprigging or plugging
Sandy soil	2.2 lbs.	Old beds	within 2 days after lifting of sod
		New beds	7-10 days after sprigging or plugging

If weeds regrow, apply an additional 2.2 lbs. on muck or peat, or 1.1 lbs. on sandy soil.

Precautions: *On newly sprigged turfgrass, temporary slowing of growth or yellowing may occur following application. To avoid turf injury, 1) Do not apply within 30 days before cutting or lifting. 2) Use only on turfgrass reasonably free of infestations of insects, nematodes, and diseases.*

TURFGRASS FOR FAIRWAYS, LAWNS, AND SIMILAR AREAS

BERMUDAGRASS, CENTIPEDEGRASS, ST. AUGUSTINEGRASS, AND ZOYSIA GRASS

Apply this product after October 1 before emergence of winter annual weeds for control of annual bluegrass, burclover, carpet burweed, chickweed, corn speedwell, henbit, hop clover, and spurweed. This product will control annual bluegrass even if it is emerged at time of treatment. For control of summer annual weeds listed in the General Information section of this label, also apply this product in late winter before the weeds emerge. Apply in a minimum of 15 gals. of water per acre or 1 gal. per 1,000 sq. ft.

Where annual bluegrass is the major weed, use 1.1 lbs. of this product per acre (0.4 oz. per 1,000 sq. ft.). Use 2.2 lbs. per acre (0.8 oz. per 1,000 sq. ft.) for control of the other weeds named above. Do not exceed 1.1 lbs. per acre per treatment on newly sprigged turfgrass or on hybrid bermudagrass such as Tiflawn, Tifway, and Ormond.

For continued summer annual weed control, apply another 1.1 lbs. per acre at least 30 days after the previous application, but not after April 15. However, do not make more than two applications of this product per year.

Precautions: *On newly sprigged turfgrass and hybrid bermudagrass, temporary slowing of growth and yellowing may occur following application. To avoid turf injury, 1) Use only on turfgrass reasonably free of infestations of insects, nematodes, and diseases. 2) Do not use on golf greens. 3) Do not use north of NC or west of the high rainfall areas of eastern OK and eastern TX. 4) Do not use on muck or alkaline soils. 5) Do not apply over the rooting area of trees or ornamentals not listed on this label. 6) Do not overseed with desirable turfgrass within 4 months before or 6 months after treatment.*

VEGETABLE CROPS

ASPARAGUS (Established)
Apply 2.2-4.4 lbs. per acre to beds established one year or more. Apply once per year either after disking in the spring before weeds emerge, or following harvest after removal of existing weed growth. Spring application must be made at least 3 days before first cutting.

Precautions: (WA and OR only): *To avoid crop injury, 1) Do not apply on gravelly, sand, or loamy sand soil. 2) Do not allow irrigation water to stand or to form pools in the field.*

ARTICHOKES
Apply 2.2-4.4 lbs. per acre as a directed spray after the last fall tillage.

Precautions: *To avoid crop injury, 1) Do not apply on gravelly, sand, or loamy sand soil. 2) Apply only once per year.*

NONSELECTIVE WEED CONTROL ON NONCROP LAND

Best results are obtained when this product is applied before weeds emerge. If weed growth has begun and removal is not feasible, include a contact or translocated herbicide in the spray.

This product provides long-term weed control on industrial sites, highway medians and shoulders, railroad rights-of-way, lumberyards, petroleum tank farms, and in noncrop areas on farms, such as around buildings, equipment and fuel storage areas, along fences, roadsides, and lanes.

Aerial applications may be made where it is feasible. Use at least 1 gal. of water for each 0.9 lb. of this product for both ground and aerial applications. Use more water if practical.

Precautions: *1) Do not use this product for nonselective weed control on land to be cropped, near adjacent desirable trees, shrubs, or plants, or in greenhouses, or injury may occur. 2) Do not use in areas accessible to livestock or allow livestock to graze treated weed foliage.*

To control annual broadleaf and grass weeds (including barnyardgrass, cheat, crabgrass, common lambsquarters, foxtail, ragweed, puncturevine, and turkey mullein) apply 5.3-11.1 lbs. per acre.

To control most annual and many perennial broadleaf and grass weeds (including quackgrass, bluegrass, redtop, burdock, Canada thistle, orchardgrass, purpletop, smooth brome, dogfennel, and plantain) apply 11.1-22.2 lbs. per acre.

To control hard-to-kill perennial weeds (including bullthistle and sowthistle) apply 22.2-44.4 lbs. per acre.

For longer residual control in regions of high rainfall and a long growing season, apply 22.2-44.4 lbs. per acre.

TANK MIXTURE WITH ROUNDUP
This tank mixture will provide control of emerged annual weeds and residual control of weeds listed on the Princep label. The combination also will partially control emerged perennial weeds listed on the Roundup label. Add to the spray tank the appropriate rate of Princep noted in the above section for Princep alone. Then add 1-5 qts. of Roundup per acre depending on weeds present and their growth stage. Also add an agriculturally approved nonionic surfactant at 0.5% by volume of spray solution. Apply the mixture in 10-40 gals. of water per acre as a postemergence spray at the appropriate weed growth stage given on the Roundup label. Provide constant agitation during mixing and application to keep the mixture in suspension. Refer to the Application Procedures section of this label for further directions. Refer to the labels of both herbicides for specific noncrop sites, rates, weeds controlled, and further directions, precautions, and limitations.

TANK MIXTURE WITH OUST®
Use this tank mixture to control the weeds listed below with either preemergence, early postemergence, or postemergence applications. Use the higher rates within the ranges given under conditions of heavy weed growth, on soil with over 2½% organic matter, on areas with heavy decaying plant residues, or on high moisture areas such as along road edges or railroad shoulders. To improve wetting and/or contact activity on emerged weeds, add a nonionic surfactant such as duPont Surfactant WK or X-77 at 1 qt. per 100 gals. of spray. Do not apply by aircraft. Provide constant agitation during mixing and application to keep the mixture in suspension. Refer to the labels of both herbicides for specific noncrop sites and further directions, precautions, and limitations.

Preemergence to Early Postemergence To control the weeds listed below apply 4.4-8.9 lbs. of Princep Caliber 90 plus 3-5 ounces of Oust in a minimum of 25 gals. of water per acre.

Broadleaf Weeds (Up to 3-4" in Height or Diameter)

bouncingbet	horsetail	sowthistle
buckhorn plantain	kochia	St. Johnswort
burclover	lambsquarters	sunflower
Carolina geranium	little mallow	sweet clover
common chickweed	mustard	tansymustard
common dandelion	ox-eye daisy	tansy ragwort
common speedwell	pigweed	tumble mustard
common yarrow	prickly lettuce	turkey mullein
crimson clover	puncturevine	vetch
dogfennel	purple starthistle	wild carrot
filaree	ragweed	yellow rocket
hoary cress	Russian thistle	

Grasses (Up to 6-12" in Height)

Alta fescue	fall panicum	red brome
annual bluegrass	fescue	red fescue
annual ryegrass	foxtails	reed canarygrass
bahia grass	Indiangrass	ripgut brome
barnyardgrass	Italian ryegrass	ryegrass
cheatgrass	johnsongrass	smooth brome
crabgrass	(short-term control)	sprangletop
downy brome	little barley	witchgrass

Postemergence To control actively growing weeds listed below apply 8.9-17.8 lbs. of Princep Caliber 90 plus 6-12 ounces of Oust in a minimum of 25 gals. of water per acre.

Actively Growing Weeds

bedstraw	Jerusalem artichoke	johnsongrass
Canada thistle	kudzu	(season-long
curly dock	mayweed	control)
dewberry	musk thistle	yellow nutsedge
fiddleneck	poison ivy	
fleabane	spanishneedles	
goldenrod	turkey mullein	
hemlock	Virginia pepperweed	
honeysuckle	wild blackberry	

STORAGE AND DISPOSAL

Store in a dry place. Do not contaminate water, food, or feed by storage or disposal. Pesticide wastes may be toxic. Improper disposal of unused pesticide, spray mixture, or rinse water is a violation of federal law. If these wastes cannot be used according to label instructions, contact your State Pesticide or Environmental Control Agency, or the Hazardous Waste representative at the nearest EPA Regional Office for guidance in proper disposal methods.

Completely empty bag into application equipment. Dispose of empty bag in a sanitary landfill or by incineration, or if allowed by state and local authorities, by burning. Stay out of smoke from burning bags.

For minor spills, leaks, etc., follow all precautions indicated on this label and clean up immediately. Take special care to avoid contamination of equipment and facilities during cleanup procedures and disposal of wastes. In the event of a major spill, fire or other emergency, call (919) 292-7100 day or night.

PRECAUTIONARY STATEMENTS

HAZARDS TO HUMANS AND DOMESTIC ANIMALS

CAUTION
Harmful if swallowed, inhaled or absorbed through the skin. Avoid breathing dust or spray mist. Avoid contact with eyes, skin, or clothing.

First Aid In case of contact, immediately flush eyes or skin with plenty of water. Get medical attention if irritation persists.

ENVIRONMENTAL HAZARDS

Do not apply directly to water or wetlands. Do not contaminate water by cleaning of equipment or disposal of wastes. Simazine is a chemical which can travel (seep or leach) through soil and can contaminate ground water which may be used as drinking water. Simazine has been found in ground water as a result of agricultural use. Users are advised not to apply simazine where the water table (ground water) is close to the surface and where the soils are very permeable, i.e., well-drained soils such as loamy sands. Your local agricultural agencies can provide further information on the type of soil in your area and the location of ground water.

AAtrex® trademark of CIBA-GEIGY for atrazine

Caliber® trademark of CIBA-GEIGY

Compex® trademark of KALO Agricultural Chemicals, Inc.

Dual® trademark of CIBA-GEIGY for metolachlor

Eradicane® trademark of Stauffer Chemical Co.

Evik® trademark of CIBA-GEIGY for ametryn

Nine-O® trademark of CIBA-GEIGY

Oust® trademark of E. I. duPont de Nemours & Co., Inc.

Princep® trademark of CIBA-GEIGY for simazine

Roundup® trademark of Monsanto Company for glyphosate herbicide

Surflan® trademark of Elanco Products Co. for oryzalin

Sutan+® trademark of Stauffer Chemical Co.

Unite® trademark of Hopkins Agricultural Chemical Co.

X-77® trademark of KALO Agricultural Chemicals, Inc.

CGA 6L48A 095

RIDOMIL® 2E

FUNGICIDE

For the control of certain diseases in various crops caused by the Oomycete class of fungi

ACTIVE INGREDIENT:

Metalaxyl: *N*-(2,6-dimethylphenyl)-*N*-(methoxyacetyl) alanine methyl ester 25.1%

INERT INGREDIENTS: 74.9%

TOTAL: 100.0%

EPA Reg. No. 100-607 EPA Est. 100-AL-1

Ridomil® trademark of CIBA-GEIGY for metataxyl

U.S. Patent No. 4,151,299

Ridomil 2E contains 2 lbs. active ingredient per gallon

Keep Out of Reach of Children.

WARNING/AVISO

PRECAUCIÓN AL USUARIO:

Si usted no lee inglés, no use este producto hasta que la etiqueta haya sido explicado ampliamente.

See additional precautionary statements inside booklet.

See directions for use inside booklet.

DIRECTIONS FOR USE AND CONDITIONS OF SALE AND WARRANTY

IMPORTANT: Read the entire **Directions for Use** and the **Conditions of Sale and Warranty** before using this product.

CONDITIONS OF SALE AND WARRANTY

The **Directions for Use** of this product reflect the opinion of experts based on field use and tests. The directions are believed to be reliable and should be followed carefully. However, it is impossible to eliminate all risks inherently associated with use of this product. Crop injury, ineffectiveness, or other unintended consequences may result because of such factors as weather conditions, presence of other materials, or the manner of use or application all of which are beyond the control of CIBA-GEIGY or the Seller. All such risks shall be assumed by the Buyer.

CIBA-GEIGY warrants that this product conforms to the chemical description on the label and is reasonably fit for the purposes referred to in the **Directions for Use** subject to the inherent risks referred to above. **CIBA-GEIGY makes no other express or implied warranty of Fitness or Merchantability or any other express or implied warranty. In no case shall CIBA-GEIGY or the Seller be liable for consequential, special, or indirect damages resulting from the use or handling of this product.** CIBA-GEIGY and the Seller offer this product, and the Buyer and user accept it, subject to the foregoing **Conditions of Sale and Warranty,** which may be varied only by agreement in writing signed by a duly authorized representative of CIBA-GEIGY.

DIRECTIONS FOR USE

It is a violation of federal law to use this product in a manner inconsistent with its labeling.

FAILURE TO FOLLOW DIRECTIONS FOR USE AND PRECAUTIONS ON THIS LABEL MAY RESULT IN CROP INJURY OR POOR DISEASE CONTROL.

Do not apply this product in such a manner as to directly, or through drift, expose workers or other persons, except those knowingly involved in the application. The area being treated must be vacated by unprotected persons.

REENTRY STATEMENTS

Do not enter treated areas without protective clothing until sprays have dried.

Because certain states may require more restrictive reentry intervals for various crops treated with this product, consult your State Department of Agriculture for further information.

Written or oral warnings must be given to workers who are expected to be in the treated area or in an area about to be treated with this product. Oral warnings must be given which inform workers of areas or fields that may not be entered without specific protective clothing until sprays have dried and appropriate actions to take in case of accidental exposure, as described under **Precautionary Statements** at the back of this label. When oral warnings are given, warnings shall be given in a language customarily understood by workers. Oral warnings must be given if there is reason to believe that written warnings cannot be understood by workers. Written warnings must include the following information: "WARNING. Area treated with Ridomil 2E on (*date of application*). Do not enter without appropriate protective clothing until sprays have dried. Flush eyes or skin with plenty of water. Call a physician if irritation persists. Remove contaminated clothing and wash before reuse."

GENERAL INFORMATION

Ridomil is a systemic fungicide for use on selected crops to control certain diseases caused by members of the Oomycete class of fungi. Other fungicides must be used to control diseases incited by other classes of fungi.

THIS PRODUCT IS NOT TO BE USED IN FOLIAR APPLICATIONS OR IN SOLUTIONS USED TO DIP PLANTS.

Note: Ridomil is a systemic fungicide having a specific mode of action and could be subject to development of resistant strains of fungi. Development of resistance cannot be predicted. Therefore, CIBA-GEIGY cannot assume liability for crop damage resulting from resistant strains of fungi. If treatment is not effective following the use of Ridomil as recommended, a resistant strain of fungi may be present. If the treatment is ineffective due to the presence of a Ridomil resistant strain of fungi, neither Ridomil nor any other fungicide with similar action will effectively control that disease. Consideration should then be given to the prompt use of other types of suitable fungicides. Consult with your State Agricultural Experiment Station or Extension Service Specialist for guidance on your particular crop and disease control situation.

Do not make foliar applications to field grown tobacco, or other crops since this practice may encourage more rapid development of resistance.

To avoid spray drift, do not apply under windy conditions. Avoid spray overlap, as crop injury may result.

MIXING INSTRUCTIONS
Add one-fourth to one-half of the required amount of water to the spray tank, add the proper amount of Ridomil 2E, then add the rest of the water. When tank mixing other products with Ridomil 2E, follow the proper sequence of adding products to the spray tank. Wettable powders or water dispersible granules should be added to the water in the tank first, followed by flowable products, with emulsifiable concentrates, such as Ridomil 2E, added last. Provide sufficient agitation during mixing and application to maintain a uniform emulsion.

Ridomil 2E is usually compatible with Balan®, Bravo®, Dasanit® + DiSyston®, Dithane® M-22, Dithane M-45, Dasanit + Nemacur®, Dasanit, D•z•n®, Difolatan®, Enide®, Furadan®, Manzate®, Manzate 200, Mocap®, Mocap Plus 4-2EC, Paarlan®, and Tillam®.

To assure the compatibility of Ridomil 2E with these and other products, pour the products into a small container of water in the correct proportions. After thorough mixing, let stand for five minutes. If the combination remains mixed, or can be remixed readily, the mixture is compatible.

Calculate the amount of Ridomil 2E needed for band treatment by the formula:

$$\frac{\text{band spacing in inches}}{\text{row spacing in inches}} \times \text{broadcast rate per acre} = \text{amount needed per acre of field}$$

APPLES

Use of Ridomil 2E will aid in the control of crown rot caused by *Phytophthora cactorum* when used in conjunction with good cultural practices and rootstocks that are most tolerant to the disease. Applications should be made before symptoms appear especially in areas of the orchard favorable for disease development. Ridomil 2E will not revitalize trees showing moderate to severe crown rot symptoms.

Mix 1 qt. of Ridomil 2E with 100 gals. of water. Apply the amount of diluted mixture indicated in the table below around the trunk of each tree. Make one application at the time of planting or in the spring before growth starts. Make another application in the fall after harvest.

Trunk Diameter	Qts. of Diluted Mixture/Tree
< 1 inch	1 qt.
1-3 inches	2 qts.
3-5 inches	3 qts.
> 5 inches	4 qts.

Note: Do not dip roots of trees in solutions containing Ridomil 2E.

AVOCADOS

Root Rot—*Phytophthora cinnamomi*

Begin applications at the start of the growing season or at transplanting. Two additional applications should be made at three-month intervals. Applications are not needed during the winter months of November through February. Ridomil 2E may be applied as a soil surface spray under sprinkler irrigation systems, as a directed spray under drip emitters or injected into the irrigation water.

Sprinkler Irrigation: Apply as a soil surface spray under the canopy of the tree in sufficient water to obtain uniform coverage. See the table for the amount of Ridomil 2E to use based on the diameter of the tree canopy. Start applications at the beginning of the growing season or at transplanting and continue at three-month intervals.

Drip Irrigation. Apply the recommended amount of Ridomil 2E (see table) to the soil directly under the drip emitter at each tree. If there is more than one emitter per tree, distribute the total amount of Ridomil 2E needed among the emitters.

Injection into Irrigation Water (Sprinkler or Drip Irrigation only): Inject Ridomil 2E into the irrigation water at a rate of 2-4 fl. oz./1,000 gals. ($3\frac{3}{4}$-$7\frac{1}{2}$ ppm active ingredient) at each irrigation. If Ridomil is not applied at each irrigation, use the table below to determine how much Ridomil 2E should be injected into the irrigation water. If Ridomil is to be used more frequently than every three months, adjust the rates so that no more than the specified amount is applied during each three-month period.

Precautions: *(1) Apply only through irrigation systems containing anti-syphon and check valves to prevent contamination of well during shutdown and overflow of solution tank. (2) Inject ahead of any right angle turn in the main line to insure adequate mixing. (3) Chemical injection pumps and water pumps must have interlocking controls to insure simultaneous shutoff. (4) Application when drift may occur, such as from windy conditions, or when system joints and connections are leaking, or when nozzles are not providing uniform distribution, may cause crop injury.*

Diameter of Tree Canopy (Ft.)	Amount of Ridomil 2E Per Ten Trees/3 Months
2	½-1 fl. oz.
5	3-6 fl. oz.
10	13-26 fl. oz.
15	29-58 fl. oz.
20	51-103 fl. oz.

Notes: (1) For best results, use Ridomil 2E as soon as soil tests indicate the presence of Phytophthora. (2) For new plantings, the use of Phytophthora resistant root stocks with Ridomil is recommended. Mature trees in moderate to advanced stages of decline cannot be cured with Ridomil. (3) Do not apply more than 24 gals./A of Ridomil 2E/year or make an application within 28 days of harvest, or illegal residues may result.

CITRUS

Use Ridomil 2E on citrus for control of citrus foot rot, root rot, and trunk canker caused by *Phytophthora* spp. Apply to the soil as a drench or as a spray in a banded application.

Citrus in Nurseries
Make the first application of Ridomil 2E at the time of planting. Make repeat applications at three-month intervals during the period when trees are actively growing.

Soil Drench: Apply 4-6 fl. oz./100 gals. of water as a drench over the row at the rate of 100-250 gals./1,000 feet of row. The width of the drench treatment should be wide enough to cover the root systems of the plants.

Soil Surface Spray: Apply 2 gals./A of soil treated in a broadcast or banded surface spray to seedbeds, liners, or bedded stock in sufficient water to obtain uniform coverage. If applications are banded, the treated area should be wide enough to cover the root systems of the plants. Follow with a ½ inch irrigation.

Citrus Resets or New Plantings
Make the first application of Ridomil 2E to citrus resets or new plantings at the time of transplanting. Make up to three repeat applications, at three-month intervals to coincide with root growth flushes during the growing season.

Water Ring Drench: Mix 4-6 fl. oz./100 gals. of water. Apply 5 gals. of the mix around the base of each tree within the watering ring.

Soil Surface Spray: Apply 1-2 gals./A (3-6 fl. oz./1,000 sq. ft.) in sufficient water to obtain uniform coverage of the soil surface. Apply spray to the soil surface beneath the tree canopy.

Citrus in Groves

Soil Application (Control of Phytophthora Foot and Root Rot): Apply 1-2 gals./A of soil treated (3-6 fl. oz./1,000 sq. ft.) in a banded surface spray under the canopy of the trees. Two additional applications may be made per year. Use the table as a guide for treating individual trees.

Diameter of Tree Canopy (Ft.)	Fl. Oz. Ridomil 2E per Ten Trees
5	½- 1
10	2½- 5
15	5-10
20	10-20

Trunk Spray (Gummosis caused by *Phytophthora* spp.): Add one gallon of Ridomil 2E to 15 gallons of water and spray the surface of the trunks using enough spray to thoroughly wet the lesions. Up to three applications per year may be made.

Note: Trunk sprays for use only in Texas, California, and Arizona.

COLE CROPS (BROCCOLI, CABBAGE, AND CAULIFLOWER)

Ridomil 2E applied as a soil application at seeding will control damping-off caused by *Pythium* spp. and basal stem rot caused by *Phytophthora* spp. Applications may be banded over the rows, broadcast or incorporated.

Broadcast and Banded Applications: Apply 4 pts./treated acre at planting in 20-50 gals. of water. For banded applications, 2 pts./13,000 linear ft. of row in a 7-inch band is recommended. For Pythium control, use ½-1 pt./13,000 linear ft. of row.

Incorporated application: Apply 4-8 pts./A as a broadcast soil application in 20-50 gals. of water and incorporate in the top 2 inches of soil. If plantings are made on beds, broadcast the Ridomil and incorporate before forming beds. For control of Pythium damping-off only use 1-2 pts./A.

CONIFERS IN NURSERIES

Ridomil provides control of Phytophthora root rot of conifers.

Seedbeds and Plug-Plantings
Apply 2½ pts./A as a broadcast spray over the beds in at least 50 gals. of water at the time of seeding or transplanting. Additional applications should be made at approximately six-month intervals.

2-0 Transplants*
Apply 5 pts./A as a broadcast spray over the beds in at least 50 gals. of water at the time of transplanting. Additional applications should be made at approximately six-month intervals.

*Newly transplanted two-year old seedlings.

COTTON

Seed Rots and Seedling Diseases of Cotton *(Pythium* spp.)

Apply ¼-½ pt./13,000 linear feet of row as an in-furrow spray in 5-15 gals. of water at planting. Mount the spray nozzle so the spray is directed into the furrow over the seed just before the seeds are covered.

For broader spectrum disease control, Ridomil 2E can be applied with 2-4 qts. Terraclor® 2E/13,000 linear feet of row as an in-furrow spray.

Note: When Ridomil is applied with Terraclor 2E, observe all precautions and restrictions that appear on the Terraclor 2E label.

CUCUMBERS, MELONS, AND SQUASH

Ridomil 2E applied at seeding will provide control of damping-off and cottony leak caused by *Pythium* spp. Applications may be made in a band over the row at the time of seeding or broadcast and lightly incorporated. If planting on beds, Ridomil 2E may be broadcast and incorporated before bedding.

Broadcast and Banded Application: Apply 4-8 pts./treated acre at planting in 20-50 gals. of water. For banded applications, a 7-inch band is recommended. To calculate the amount of Ridomil 2E needed per acre for banded applications, use the formula in the **General Information** section of this label.

Incorporated application: Apply 4-8 pts./A as a broadcast soil application in 20-50 gals. water and incorporate into the top 2 inches of soil. If plantings are made on beds, Ridomil 2E may be broadcast before bedding provided it is not incorporated more than 4 inches deep.

DECIDUOUS FRUITS AND NUTS IN NURSERIES AND FIELD PLANTINGS (NONBEARING)

Phytophthora Root Rot
Make the first application of Ridomil 2E at the time of planting. Additional applications should be made at three-month intervals throughout the growing season. For established plantings, the first application should be made before the plants start growth in the spring.

Apply 2-4 gals./treated acre (6-12 fl. oz./1,000 sq. ft.) in sufficient water to obtain thorough coverage of the soil under the canopy of the trees. Sufficient surface area should be treated in nurseries to cover the root zone of the plants. Use the high rate in areas known to be infested with Phytophthora.

Notes: (1) Do not apply to plantings that will bear harvestable fruit within 12 months of application, or illegal residues may result. (2) When applying Ridomil 2E in apples, see **Apples** section of this label for specific recommendations.

HOPS

Downy Mildew *(Pseudoperonospora humuli):* Apply 1 qt./A in a minimum of 20 gals. of water to the soil surface over the crowns after pruning, but before training. Early application before shoots are six inches long is preferable.

Notes: (1) Do not apply after training or make more than one application per year and (2) Do not feed hop refuse to livestock, or illegal residues may result.

LEAFY VEGETABLES (HEAD LETTUCE AND SPINACH)

Ridomil 2E applied as a soil application at seeding will control damping-off caused by *Pythium* spp. Applications may be banded over the rows, broadcast or incorporated.

Broadcast and Banded Applications: Apply 4-8 pts./treated acre at planting in 20-50 gals. of water. For banded applications, a 7-inch band is recommended. To calculate the amount of Ridomil 2E needed per acre, use the formula in the **General Information** section of this label.

Incorporated application: Apply 4-8 pts./A as a broadcast soil application in 20-50 gals. of water and incorporate in the top 2 inches of soil. If plantings are made on beds, broadcast the Ridomil and incorporate before forming beds.

LEGUME VEGETABLES (SUCCULENT OR DRIED)

Includes field beans, French beans, kidney beans, lima beans, mung beans, navy beans, pinto beans, runner beans, snap beans, wax beans, broad beans (fava beans), chickpeas (garbanzo beans), lentils, garden peas, field peas, sugar peas, southern peas (blackeyed peas, crowder peas, cowpeas, catjang), and edible soybeans.

Pythium Damping-Off and Root Rot: Ridomil 2E applied at seeding will control damping-off and root rot caused by *Pythium* spp. Applications can be made in the seed furrow or banded at planting.

Application: Apply 1 pt./13,000 linear feet of row in-furrow or in a 7-inch band at the time of planting.

ONIONS—DRY BULB, GREEN, AND ONIONS GROWN FOR SEED

Ridomil 2E applied at seeding will control damping-off caused by *Pythium* spp. Applications may be made in a band over the rows, broadcast or incorporated.

Broadcast or Banded application: Apply 4-8 pts./A as a broadcast application at planting in 20-50 gals. of water. For banded applications, a 7-inch band is recommended. To calculate the amount of Ridomil 2E needed per acre for banded applications, use the formula in the **General Information** section of this label.

Incorporated application: Apply 4-8 pts./A as a broadcast soil application in 20-50 gals. of water and incorporate in the top 2 inches of soil. If plantings are made on beds, Ridomil 2E may be broadcast before planting provided it is not incorporated more than 4 inches deep.

PEANUTS

Ridomil 2E is a soil-applied systemic fungicide for use in control of seedling and pod diseases of peanuts incited by *Pythium* spp.

Seedling Diseases: Apply 1 pt./A in-furrow in a 7-inch band at the time of planting.

Pod Rot: Apply 2 qts./A at early pod set through the irrigation water. Apply early in set to insure adequate flushing into soil.

Note: Where *Rhizoctonia* and *Fusarium* are a problem, fungicides that control these pathogens should be used with Ridomil.

PINEAPPLES

Ridomil 2E, applied as a "seed piece" dip, provides effective control of heart rot disease of pineapple caused by *Phytophthora* spp.

Apply as a preplant crown dip at the rate of 2 qts./100 gals. of water. Use 75 to 100 gals. of dip solution per planted acre, depending on crown size, plant density and dipping techniques.

Note: If there is a crop failure within one year of planting treated crowns, do not harvest plant material for animal feed.

RASPBERRIES

Phytophthora Root Rot *(Phytophthora* spp.): Apply 1 pt./1,000 ft. of row to the soil surface in a three foot band over the row (5.5 pts./A on broadcast basis). One application may be made in the fall after harvest, and a second application in the spring.

Note: Do not apply less than 45 days prior to harvest, or illegal residues may result.

SOYBEANS

Ridomil 2E is a soil-applied systemic fungicide for use in the control of Phytophthora root and stem rot and Pythium damping-off. Ridomil 2E may be applied broadcast, banded or in the seed furrow before the seeds are covered. The seed furrow applications will provide more consistent results if rain is not expected before the seeds germinate.

For best results against Phytophthora root and stem rot, use Ridomil 2E with soybean varieties that have some tolerance to the races of Phytophthora present in the field. The high rate of Ridomil 2E should be used in areas with a history of heavy Phytophthora damage. Under heavy late season Phytophthora pressure, Ridomil 2E may not provide complete control.

Broadcast and Banded Applications: Apply $2\frac{3}{4}$-$5\frac{1}{2}$ pts./treated acre at planting. For banded applications, a 7-inch band is recommended. Use the formula in the **General Information** section of this label to calculate the amount of Ridomil 2E needed per acre.

In-furrow Applications: Apply $\frac{3}{5}$-$1\frac{1}{5}$fl. oz./1,000 linear feet of row in the seed furrow at planting. Use the table below to determine the amount of Ridomil 2E needed per acre based on row spacing.

Suggested Rates Per Acre According to Row Spacing

Row Spacing	Linear Feet of Soybean Row Per Acre	Amount of Ridomil 2E Needed Per Acre
38 in.	13,756	$8\frac{1}{3}$-$16\frac{2}{3}$ fl. oz.
36 in.	14,520	$8\frac{2}{3}$-$17\frac{1}{2}$ fl. oz.
30 in.	17,424	$10\frac{1}{2}$-21 fl. oz.
24 in.	21,780	13-26 fl. oz.
20 in.	26,136	16-32 fl. oz.

Note: Ridomil 2E is specific for Pythium and Phytophthora and will not control other diseases that may attack soybeans.

TOBACCO

Ridomil 2E is a soil-applied systemic fungicide for use in the field before transplanting for control of black shank *(Phytophthora parasitica,* var. *Nicotianae)* and blue mold *(Peronospora tabacina)* on all types of tobacco. Ridomil 2E may also be applied in tobacco plant beds for control of blue mold and damping-off caused by *Pythium* spp. Ridomil 2E will not control Anthracnose; continue using a preventative foliar fungicide control program.

For best results against black shank, use Ridomil 2E on tobacco varieties with some resistance to black shank. In fields where there is a history of severe black shank, use the highest rate and plant a variety that is resistant to the race of Phytophthora present. (Note: Burley L8 hybrids are only resistant to Phytophthora Race 0.) Ridomil 2E is not recommended for use in high black shank areas on very susceptible flue-cured varieties such as Hicks, Virginia Gold, and White Gold.

Notes: (1) The preferred method of blue mold control is to use Ridomil 2E in the plant bed and follow with an application in the field prior to transplanting. (2) Failure to adequately control nematodes in fields treated with Ridomil 2E may result in poor control of black shank.

TOBACCO PLANT BED

For control of blue mold and damping-off caused by *Pythium* spp., apply Ridomil 2E only as a preplant broadcast soil application at the rate of 1 qt./A in 50 gals. of water (1 fl. oz. or 2 tablespoons/150 sq. yds. of bed in 2 gals. of water). Apply **before or at the time of seeding** to the surface of the plant bed and lightly incorporate or follow with one-half inch sprinkler irrigation water. In Pennsylvania, use 2 qts./A (2 fl. oz. or 4 tablespoons/150 sq. yds. of bed in 2 gals. of water).

Note: These treatments are only adequate to control blue mold in the plant bed and will not control blue mold in the field following transplanting. Follow the directions below for field planted tobacco.

FIELD PLANTED TOBACCO

Blue Mold: Use Ridomil 2E as a broadcast soil application and incorporate in the top 2-4 inches of soil and form beds. For flue-cured tobacco, use 1-2 qts./A, depending on disease pressure and length of control desired. Under low disease pressure or for early season control, use 1 qt./A; for burley and other tobacco types, use 2 qts./A of Ridomil 2E.

Black Shank: Apply a broadcast application of Ridomil 2E using the appropriate rate in the table below. Apply with a conventional ground sprayer in a minimum of 15 gals. of water per acre. Incorporate Ridomil 2E in the top 2-4 inches of soil and form beds.

Type of Tobacco*	Disease Level in Field	Rate of Ridomil 2E Per Acre
Flue-Cured	Low to Moderate (Less than 6% Disease)	2 qts.
	High (More than 6% Disease)	4 qts.**
Burley and Other***	Low to Moderate (Less than 6% Disease)	4 qts.
	High (More than 6% Disease)	6 qts.

*See **General Information** for recommendations on the use of resistant varieties.

**Florida and Georgia—Use 6 qts./A of Ridomil 2E in fields with very high black shank levels (greater than 60% disease).

***Pennsylvania—Do not use Ridomil 2E for black shank control.

For prolonged control of blue mold or black shank in field planted tobacco, make a supplemental application of 1 qt./A as a broadcast soil application at the last cultivation. Do not make this lay-by application if more than 2 qts./A of Ridomil 2E were applied at transplanting.

Note: Do not use Ridomil 2E in transplant water or in foliar application on field-grown tobacco, or injury may occur.

TOMATOES

Soil applications of Ridomil 2E at seeding will provide control of damping-off caused by *Pythium* spp. and soil applications applied 4-12 weeks before harvest under the vines will reduce fruit and root rot caused by *Pythium* spp. and *Phytophthora* spp.

Seedling Damping-Off *(Pythium* spp.): Apply 4-8 pts./A as a broadcast surface spray immediately before or after planting in 20-50 gals. of water. Lightly incorporate with mechanical equipment if application is made prior to planting or with sprinkler irrigation if applied after planting. For banded applications, use the formula shown in the **General Information** section to calculate the amount of Ridomil 2E needed per acre.

Fruit and Root Rot *(Pythium* spp. and *Phytophthora* spp.): Apply 4 pts./A as a soil surface application under the vines 4-12 weeks before harvest in sufficient water to obtain thorough coverage. Follow as soon as possible with an irrigation. For banded applications, use the formula shown in the **General Information** section to determine the amount of Ridomil 2E needed per acre.

Note: Do not apply more than 12 pts./A/season, or illegal residues may result.

VEGETABLE BEDDING PLANTS

Pythium Damping-Off of Broccoli, Cabbage, Cauliflower, Cucumbers, Lettuce, Melons, Spinach and Squash
Apply 2-4 pts./A as a preplant broadcast spray in 50 gals. of water (1-2 fl. oz. or 2-4 tablespoons/150 sq. yds. of bed in 2 gals. water) before or at the time of seeding to the surface of the beds and lightly incorporate or follow with one-half inch sprinkler irrigation water.

ROTATIONAL CROPS

If replanting is necessary following a soil application, tobacco or cotton may be replanted immediately. Tomatoes, broccoli, cabbage, cauliflower, cucumbers, melons, squash, onions, lettuce, or spinach may be replanted immediately provided that no more than 4 qts. per acre of Ridomil were applied to the soil. Do not make a second application of Ridomil. Wheat, or any crop on this label may be planted during the fall following application of Ridomil. Other small grain cover crops may also be planted during the fall following application provided they are plowed down and not used for food or feed. Corn, soybeans, root crops, or any crop on this label may be planted the year following treatment. Other crops may be planted 18 months following application.

STORAGE AND DISPOSAL

Store at temperatures above 40°F. Crystals may form at lower storage temperatures. If this occurs, place the product in a warm room (68°F or above) and roll or shake the container at frequent intervals until all crystals are dissolved.

PESTICIDE DISPOSAL

Do not contaminate water, food or feed by storage or disposal. Open dumping is prohibited. Wastes resulting from the use of this product are acutely toxic. Improper disposal of unused pesticide, spray mixture, or rinsate is a violation of federal law. Pesticide, spray mixture, or rinsate that cannot be used according to label instructions must be disposed of according to federal, state or local procedures. For guidance in proper disposal methods, contact your State Pesticide or Environmental Control Agency, or the Hazardous Waste representative at the nearest EPA Regional Office.

CONTAINER DISPOSAL

Do not reuse empty container. Triple rinse (or equivalent) and offer for recycling or reconditioning, or puncture and dispose of in a sanitary landfill, or by incineration, or by open burning, if allowed by state and local authorities. If burned, keep out of smoke.

PRECAUTIONARY STATEMENTS

HAZARDS TO HUMANS AND DOMESTIC ANIMALS

WARNING

Causes eye injury. Do not get in eyes. Avoid contact with skin or clothing. Harmful if swallowed, inhaled or absorbed through the skin. Avoid breathing vapors or spray mist.

Practical Treatment: In case of contact with eyes, immediately flush with plenty of water. Get medical attention. In case of contact with skin, wash thoroughly with soap and water. Get medical attention if irritation occurs. If swallowed, immediately contact a physician, hospital or local Poison Control Center. Do not induce vomiting. If inhaled, move to fresh air. If not breathing, give artificial respiration (preferably mouth to mouth) and get medical attention.

Notes to Physician: If Ridomil 2E is ingested, lavage stomach. A slurry of activated charcoal in water can be left in the stomach. Give a saline laxative and supportive therapy.

ENVIRONMENTAL HAZARDS

Do not apply directly to water. Apply only as specified on this label. Do not apply when weather conditions favor drift from treated areas. Do not contaminate water by cleaning of equipment or disposal of wastes.

PHYSICAL OR CHEMICAL HAZARDS

Do not use or store near heat or open flame.

Ridomil® trademark of CIBA-GEIGY for metalaxyl U.S. Patent No. 4,151,299

D•z•n® trademark of CIBA-GEIGY for diazinon

Balan® trademark of Elanco Products for benefin

Bravo® trademark of SDS Biotech for chlorothalonil

Dasanit® trademark of the Parent Company of Farbenfabriken Bayer GmbH, Leverkusen, for fensulfothion

Difolatan® trademark of Chevron Chemical Company for captafol

Di-Syston® trademark of the Parent Company of Farbenfabriken Bayer GmbH, Leverkusen, for disulfoton

Dithane® trademark of Rohm and Haas Company for maneb/mancozeb

Enide® trademark of Tuco Division of the Upjohn Company for diphenamid

Furadan® trademark of FMC Corporation for carbofuran

Manzate® trademark of E.I. duPont de Nemours and Company for maneb/mancozeb

Mocap® trademark of Rhone Poulenc, Inc. for ethoprop

Nemacur® trademark of the Parent Company of Farbenfabriken Bayer GmbH, Leverkusen, for fenamiphos

Paarlan® trademark of Elanco Products for isopropalin

Terraclor® trademark of UniRoyal, Inc. for PCNB

Tillam® trademark of Stauffer Chemical Co. for pebulate

CGA 65L14C 105

RIDOMIL® MZ58

FUNGICIDE

For control of certain diseases of potatoes, tomatoes, onions, cucumbers, melons, and squash

Ridomil MZ58 is a wettable powder

ACTIVE INGREDIENTS:	
Metalaxyl:	
N-(2,6-dimethylphenyl)-*N*-(methoxyacetyl)	
alanine methyl ester	10.0%
Zinc ion and manganese	
ethylene bisdithiocarbamate,*	48.0%
a coordination product of manganese++	9.6%
zinc++	1.2%
ethylene bisdithiocarbamate ion	37.2%
INERT INGREDIENTS:	42.0%
TOTAL:	100.0%

*Same as the active ingredient found in Dithane® M-45 and Manzate® 200.

EPA Reg. No. 100-629

Ridomil® trademark of CIBA-GEIGY for metalaxyl

U.S. Patent No. 4,151,299

Keep Out of Reach of Children.

WARNING/AVISO

PRECAUCIÓN AL USUARIO:
Si usted no lee inglés, no use este producto hasta que la etiqueta haya sido explicado ampliamente.

See additional precautionary statements on side of bag.

See directions for use on panel at right.

DIRECTIONS FOR USE AND CONDITIONS OF SALE AND WARRANTY.

IMPORTANT: Read the entire **Directions for Use** and the **Conditions of Sale and Warranty** before using this product.

CONDITIONS OF SALE AND WARRANTY

The **Directions for Use** of this product reflect the opinion of experts based on field use and tests. The directions are believed to be reliable and should be followed carefully. However, it is impossible to eliminate all risks inherently associated with use of this product. Crop injury, ineffectiveness, or other unintended consequences may result because of such factors as weather conditions, presence of other materials, or the manner of use or application all of which are beyond the control of CIBA-GEIGY or the Seller. All such risks shall be assumed by the Buyer.

CIBA-GEIGY warrants that this product conforms to the chemical description on the label and is reasonably fit for the purposes referred to in the **Directions for Use** subject to the inherent risks referred to above.

CIBA-GEIGY makes no other express or implied warranty of Fitness or Merchantability or any other express or implied warranty. In no case shall CIBA-GEIGY or the Seller be liable for consequential, special, or indirect damages resulting from the use or handling of this product. CIBA-GEIGY and the Seller offer this product, and the Buyer and user accept it, subject to the foregoing **Conditions of Sale and Warranty,** which may be varied only by agreement in writing signed by a duly authorized representative of CIBA-GEIGY.

DIRECTIONS FOR USE

It is a violation of federal law to use this product in a manner inconsistent with its labeling.

Do not apply this product in such a manner as to directly or through drift expose workers or other persons, except those knowingly involved in the application. The area being treated must be vacated by unprotected persons.

FAILURE TO FOLLOW ALL PRECAUTIONS ON THIS LABEL MAY RESULT IN POOR DISEASE CONTROL OR CROP INJURY, OR ILLEGAL RESIDUES.

Do not enter treated areas without protective clothing until sprays have dried.

Because certain states may require more restrictive reentry intervals for various crops treated with this product, consult your State Department of Agriculture for further information.

Written or oral warnings must be given to workers who are expected to be in a treated area or in an area about to be treated with this product. Oral warnings must inform workers of areas or fields that may not be entered without specific protective clothing until sprays have dried, and appropriate actions to take in case of accidental exposure, as described under **Precautionary Statements** on this label. When oral warnings are given, warnings shall be given in a language customarily understood by workers. Oral warnings must be given if there is reason to believe that written warnings cannot be understood by workers. Written warnings must include the following information: "WARNING. Area treated with Ridomil MZ58 on (date of application). Do not enter without appropriate protective clothing until sprays have dried. In case of accidental eye exposure, flush eyes with plenty of water. Get medical attention. In case of contact with skin, wash thoroughly with soap and water. Call physician if irritation occurs. Remove and wash contaminated clothing before reuse."

GENERAL INFORMATION

Ridomil MZ58 contains 10% of the active ingredient metalaxyl, and 48% of the active ingredient mancozeb. Metalaxyl is a systemic fungicide for use on selected crops to control diseases caused by members of the Phycomycete family of fungi. Mancozeb is effective against a wide range of fungal pathogens.

Note: Metalaxyl is a systemic fungicide having a specific mode of action and could be subject to development of resistant strains of fungi. Development of resistance cannot be predicted. Therefore, CIBA-GEIGY cannot assume liability for crop damage resulting from resistant strains of fungi. If treatment is not effective following the use of Ridomil MZ58 as recommended, a resistant strain of fungi may be present. If the treatment is ineffective due to the presence of a metalaxyl resistant strain of fungi, neither Ridomil MZ58 nor any other fungicide with similar action will effectively control that disease. Consideration should then be given to the prompt use of other types of suitable fungicides. Consult with your State Agricultural Experiment Station or Extension Service Specialist for guidance in your particular crop and disease control situation.

MIXING INSTRUCTIONS

Add one-half to three-fourths of the required amount of water to the spray tank, add the proper amount of Ridomil MZ58, then add the rest of the water. Provide sufficient agitation during mixing and application to maintain a uniform emulsion.

Ridomil MZ58 is usually compatible with Bravo®, Dithane M-22, Dithane M-45, D•z•n®, Difolatan®, Manzate, and Manzate 200.

To assure the compatibility of Ridomil MZ58 with these and other products, pour the products into a small container of water in the correct proportions. After thorough mixing, let stand for five minutes. If the combination remains mixed, or can be remixed readily, the mixture is compatible.

IMPORTANT

Do not allow Ridomil MZ58 to become wet or overheated during storage. This may lead to chemical changes which will impair the fungicidal effectiveness of Ridomil MZ58 and may also generate flammable vapors. Keep bag closed when not in use.

CUCUMBERS, MELONS, AND SQUASH

Ridomil MZ58 provides effective control of downy mildew caused by *Pseudoperonospora cubensis* when used as a foliar spray in a preventative disease control program.

Apply 1½-2 lbs. Ridomil MZ58 per acre in sufficient water to obtain thorough coverage. Start when the plants are in the two-leaf stage and continue at 14-day intervals throughout the season. Under heavy disease pressure, use the higher rate of Ridomil MZ58.

Note: Do not apply more than 20 lbs. Ridomil MZ58 per acre per year or make the last application within 5 days of harvest.

ONIONS (DRY BULB AND SEED ONIONS)

Ridomil MZ58 provides effective control of downy mildew caused by *Peronospora destructor* when used as a foliar spray in a preventative disease control program.

Apply 1½-2 lbs. Ridomil MZ58 per acre in sufficient water to obtain thorough coverage. Start applications when conditions are favorable for disease development and continue at 14-day intervals throughout the season. Use a suitable spreader-sticker at rates recommended on the product label. Under heavy disease pressure, use the higher rate of Ridomil MZ58.

Note: Do not apply more than 12.5 lbs. Ridomil MZ58 per acre per season or make the last application within 7 days of harvest.

POTATOES

Ridomil MZ58, when used as a foliar application in a preventative disease control program, will provide effective control of late blight and tuber rot caused by *Phytophthora infestans.*

Apply 1½-2 lbs. Ridomil MZ58 per acre starting when plants are six inches high or when conditions are favorable for late blight development and continue at 14-day intervals throughout the season. Under heavy disease pressure, use the higher rate of Ridomil MZ58. Use sufficient water to obtain thorough coverage for ground applications and a minimum of 3 gals. of water per acre for aerial applications.

For supplemental control of early blight, add 0.8 lb. of Manzate 200 80W, Dithane M-45 80W, Manzate 80W or Dithane M-22 80W to the Ridomil MZ58 spray or make an additional application of any registered early blight fungicide between each Ridomil MZ58 spray.

Where early blight disease pressure is heavy, use early blight fungicides according to product use directions, and use Ridomil MZ58 only when late blight threatens. Do not apply more than 0.8 lb. per acre per application of Manzate 200 80W, Dithane M-45 80W, Manzate 80W, or Dithane M-22 80W in combination with Ridomil MZ58.

Note: To avoid illegal residues do not apply Ridomil MZ58 within 7 days of harvest. Do not apply more than 22.5 lbs. of Ridomil MZ58 per year. Do not tank mix with amobam or Polyram.®.

TOMATOES

Ridomil MZ58, when used as a foliar application in a preventative disease control program, will provide effective control of late blight caused by *Phytophthora infestans.*

Apply 1½-2 lbs. Ridomil MZ58 per acre as a foliar spray in sufficient water to provide thorough coverage. Start applications when plants are six inches high or when conditions are favorable for disease development, and continue at 14-day intervals throughout the season. Under heavy disease pressure, use the higher rate of Ridomil MZ58.

Note: When conditions are favorable for development of diseases other than late blight, any registered fungicides which control those diseases may be applied according to their label directions. When making applications of Manzate 200, Dithane M-45, Manzate or Dithane M-22, 1½ lbs./A of Ridomil MZ58 may be added to the spray mix every 14 days for control of late blight. Do not apply more than 2 lbs. per acre per application of Dithane M-45 80W, Manzate 200 80W, Dithane M-22 80W, or Manzate 80W in combination with Ridomil MZ58. Observe all precautions and restrictions that appear on the labels of other products applied with Ridomil MZ58. Do not apply Ridomil MZ58 within 5 days of harvest. Do not tank mix with amobam, Polyram, nabam, thiram, zineb, or ziram.

ROTATIONAL CROPS

Any crop on this label may be planted during the fall following application of Ridomil MZ58. Small grain cover crops may also be planted during the fall following application provided they are plowed down and not used for food or feed. Corn, root crops, or any crop on this label may be planted the year following treatment. Other crops may be planted 18 months following application.

STORAGE AND DISPOSAL

PESTICIDE AND STORAGE DISPOSAL

Store in a dry place. Do not contaminate water, food, or feed by storage, disposal, or cleaning of equipment. Open dumping is prohibited. Wastes resulting from the use of this product are acutely toxic. Improper disposal of unused pesticide, spray mixture, or rinsate is a violation of federal law. Pesticide, spray mixture, or rinsate that cannot be used according to label instructions must be disposed of according to federal, state, or local procedures. For guidance in proper disposal methods, contact your State Pesticide or Environmental Control Agency, or the Hazardous Waste representative at the nearest EPA Regional Office.

CONTAINER DISPOSAL

Completely empty bag into application equipment. Dispose of empty bag in a sanitary landfill or by incineration, or by open burning, if allowed by state and local authorities. If burned, keep out of smoke.

For minor spills, leaks, etc., follow all precautions indicated on this label and clean up immediately. Take special care to avoid contamination of equipment and facilities during cleanup procedures and disposal of wastes. In the event of a major spill, fire or other emergency, call (919) 292-7100 day or night.

PRECAUTIONARY STATEMENTS

HAZARDS TO HUMANS AND DOMESTIC ANIMALS

WARNING

Causes eye irritation. Do not get in eyes.

Harmful if swallowed, inhaled or absorbed through the skin. Avoid breathing dust or spray mist. Avoid contact with skin or clothing. Protective clothing (long pants, long sleeve shirt, gloves, hat, and boots) must be worn during mixing and loading. This product may cause allergic skin reaction in some individuals.

First Aid: In case of contact, immediately flush eyes or skin with plenty of water. Get medical attention if irritation persists. If swallowed, drink large quantities of water and induce vomiting by placing finger in back of throat, and contact local poison control center. Avoid alcohol. Never give anything by mouth to an unconscious person. In case of inhalation exposure, move from the contaminated area.

Note to Physician: There is no specific antidote for Ridomil MZ58. Lavage stomach. Give a saline laxative and supportive therapy.

ENVIRONMENTAL HAZARDS

This product is toxic to fish. Do not apply directly to water. Apply only as specified on this label. Do not apply when weather conditions favor drift from treated areas. Do not contaminate water by cleaning of equipment or disposal of wastes.

Ridomil® trademark of CIBA-GEIGY for metalaxyl U.S. Patent No. 4,151,299

Bravo® trademark of SDS Biotech Corp. for chlorothalonil

Difolatan® trademark of Chevron Chemical Company for captafol

Dithane® trademark of Rohm and Haas Company for maneb/mancozeb

D•z•n® trademark of CIBA-GEIGY for diazinon

Manzate® trademark of E. I. duPont de Nemours and Company for maneb/mancozeb

Polyram® trademark of BASF Corporation for metiram

CGA 65L12C 075

RIDOMIL®5G

FUNGICIDE

For control of Phytophthora and Pythium diseases of avocados, citrus, cotton, peanuts, raspberries, and soybeans

Ridomil 5G is a granular fungicide

ACTIVE INGREDIENT:

Metalaxyl: *N*-(2,6-dimethylphenyl)-*N*-(methoxyacetyl) alanine methyl ester 5.0%

INERT INGREDIENTS: 95.0%

TOTAL: 100.0%

EPA Reg. No. 100-628

EPA Est. 100-AL-1

Control No. 51120

Ridomil® trademark of CIBA-GEIGY

Keep Out of Reach of Children.

Caution

See additional precautionary statements on back of bag.

See directions for use on back of bag.

DIRECTIONS FOR USE AND CONDITIONS OF SALE AND WARRANTY

IMPORTANT: Read the entire **Directions for Use** and the **Conditions of Sale and Warranty** before using this product.

CONDITIONS OF SALE AND WARRANTY

The **Directions for Use** of this product reflect the opinion of experts based on field use and tests. The directions are believed to be reliable and should be followed carefully. However, it is impossible to eliminate all risks inherently associated with use of this product. Crop injury, ineffectiveness, or other unintended consequences may result because of such factors as weather conditions, presence of other materials, or the manner of use or application all of which are beyond the control of CIBA-GEIGY or the Seller. All such risks shall be assumed by the Buyer.

CIBA-GEIGY warrants that this product conforms to the chemical description on the label and is reasonably fit for the purposes referred to in the **Directions for Use** subject to the inherent risks referred to above. **CIBA-GEIGY makes no other express or implied warranty of Fitness or Merchantability or any other express or implied warranty. In no case shall CIBA-GEIGY or the Seller be liable for consequential, special, or indirect damages resulting from the use or handling of this product.** CIBA-GEIGY and the Seller offer this product, and the Buyer and user accept it, subject to the foregoing **Conditions of Sale and Warranty,** which may be varied only by agreement in writing signed by a duly authorized representative of CIBA-GEIGY.

DIRECTIONS FOR USE

It is a violation of federal law to use this product in a manner inconsistent with its labeling.

FAILURE TO FOLLOW ALL PRECAUTIONS ON THIS LABEL MAY RESULT IN POOR DISEASE CONTROL OR CROP INJURY.

Do not apply this product in such a manner as to directly or indirectly expose workers or other persons except those knowingly involved in the application. The area being treated must be vacated by unprotected persons.

REENTRY STATEMENT

Do not enter treated areas without protective clothing until dust has settled.

Because certain states may require restrictive reentry intervals for various crops treated with this product, consult your State Department of Agriculture for further information.

Written or oral warnings must be given to workers who are expected to be in a treated area or in an area about to be treated with this product. Oral warnings must inform workers of areas or fields that may not be entered without specific protective clothing until dust has settled, and appropriate actions to take in case of accidental exposure, as described under **Precautionary Statements** on this label. When oral warnings are given, warnings shall be given in a language customarily understood by workers. Oral warnings must be given if there is reason to believe that written warnings cannot be understood by workers. Written warnings must include the following information: "CAUTION. Area treated with Ridomil 5G on (date of application). Do not enter without appropriate protective clothing until dust has settled. In case of accidental exposure, flush eyes or skin with plenty of water. Call physician if irritation persists. Remove and wash contaminated clothing before reuse."

GENERAL INFORMATION

Ridomil 5G is a systemic fungicide which provides control of certain diseases caused by members of the Oomycete class of fungi on selected crops.

Note: Ridomil is a systemic fungicide having a specific mode of action. Ridomil could be subject to development of resistant strains of fungi. Development of resistance cannot be predicted. Therefore, CIBA-GEIGY cannot assume liability for crop damage resulting from resistant strains of fungi. Consult with your State Agricultural Experiment Station or Extension Service Specialist for guidance or ways to control any possible Ridomil resistant strains of fungi which may occur.

Calculate the amount of Ridomil 5G needed for band treatment by the formula:

$$\frac{\text{band width in inches}}{\text{row spacing in inches}} \times \text{broadcast rate} = \text{amount needed per acre of field}$$

AVOCADOS

Ridomil 5G is a systemic fungicide which provides control of avocado root rot caused by *Phytophthora cinnamomi*. For best results, Ridomil should be used before or as soon as early symptoms of disease appear. For new plantings, the use of Phytophthora resistant root stocks with Ridomil 5G is recommended. Mature trees in advanced stages of decline cannot be cured with Ridomil. Applications are not needed during the winter months of November through February.

Sprinkler Irrigation: Apply ¾-1½ oz./sq. yd. evenly to the soil surface under the canopy of the tree. The table below shows the amount of Ridomil 5G to use based on the diameter of the tree canopy. Start applications at the beginning of the growing season or at transplanting and make two additional applications at three-month intervals.

Drip Irrigation: Place the recommended amount of Ridomil 5G (see table below) on the soil surface directly around the drip emitter at each tree. If there is more than one emitter per tree, distribute the total amount of Ridomil 5G among the emitters. Start applications at the beginning of the growing season or at transplanting and make two additional applications at three-month intervals.

Notes: (1) Do not take more than three applications per year and do not make the last application within 28 days of harvest or illegal residues may result. (2) Do not apply more than 960 lbs. Ridomil 5G per acre per year, or illegal residues may result.

Diameter of Tree Canopy in Feet	Amount of Ridomil 5G Per Tree
5	1½-3 oz.
10	6-12 oz.
15	14-28 oz.
20	25-50 oz.

CITRUS

Use Ridomil 5G on citrus to control citrus foot rot, root rot, and trunk canker caused by *Phytophthora* spp. Apply to the soil as a broadcast application under the canopy of the trees. If rain is not expected within three days of application, water in with ½ inch of irrigation. For banded applications, use the formula in the **General Information** section for calculating the amount of Ridomil 5G needed per acre of trees.

Citrus in Nurseries

Make the first application of Ridomil 5G at the time of planting and make two additional applications at three-month intervals during the period when trees are actively growing. Apply 80 lbs./A as a broadcast or banded application to the seedbeds, liners or bedded stock. For banded applications, use a band wide enough to cover the root system of the plants. Follow the application with ½ inch of irrigation if rainfall is not expected within three days.

Citrus Resets or New Plantings

Make the first application of Ridomil 5G to citrus resets or new plantings at the time of transplanting. Make up to three repeat applications at three-month intervals to coincide with root growth flushes during the growing season.

Water Ring Treatment: Uniformly spread 1-1½ oz. within the water ring, then drench with five gallons water per tree.

Soil Surface Treatment: Apply 40-80 lbs./A (15-30 oz./1,000 sq. ft.) under the canopy of the trees. If natural rainfall is not expected within three days, lightly irrigate with ½-1 inch of water.

Established Plantings

Apply 40-80 lbs./A of treated soil (15-30 oz./1,000 sq. ft.) as a banded surface treatment under the canopy of the trees. Two additional applications may be made per year. Use the table as a guide for treating individual trees.

Diameter of Tree Canopy in Feet	Amount of Ridomil 5G Per Tree
5	¼-½ oz.
10	1¼-2½ oz.
15	2½-5 oz.
20	5-10 oz.

COTTON

Ridomil 5G provides control of seed rot and seedling diseases of cotton caused by *Pythium* spp. It does not control seedling diseases caused by other fungi.

Seed Rots and Seedling Disease of Cotton Caused by *Pythium* spp. Apply 1¼-2½ lbs./13,000 linear feet of row as an in-furrow application. Mount the tubes so the granules are directed into the furrow over the seed just before the seeds are covered.

For broader spectrum disease control, Ridomil 5G can be applied with 10-20 lbs. Terraclor® 10G per 13,000 linear feet of row as an in-furrow application.

Note: When Ridomil 5G is applied with Terraclor 10G, observe all precautions and restrictions that appear on the Terraclor 10G label.

PEANUTS

Ridomil 5G is a systemic fungicide for use in control of seedling and pod diseases of peanuts caused by *Pythium* spp.

Seedling Diseases: Apply 5 lbs./A as an in-furrow or banded application at the time of planting.

Pod Rot: Apply 10 lbs./12,400 linear feet in an 8 to 12 inch band over the rows at early pegging. For best results, an irrigation should follow the application.

Note: Do not apply Ridomil 5G to wet foliage as foliar phytotoxicity and/or poor control may result. Ridomil will not control diseases caused by *Rhizoctonia* or *Fusarium*. Where *Rhizoctonia* and *Fusarium* are a problem, fungicides that control these pathogens should be used with Ridomil.

RASPBERRIES

Ridomil 5G is a systemic fungicide for use in the control of Phytophthora root rot.

Apply 5 lbs./1,000 ft. of row to the soil surface in a three foot band over the row (27.5 lbs./A on a broadcast basis). One application may be made in the fall after harvest, and a second application in the spring.

Note: Do not apply within 45 days of harvest, or illegal residues may result.

SOYBEANS

Ridomil 5G is a systemic fungicide for use in the control of Phytophthora root and stem rot and Pythium damping-off. Ridomil 5G may be applied in a 7-inch band over the row at planting or directly in the seed furrow before the seeds are covered. The seed furrow application will provide more consistent results if rain is not expected before the seeds germinate.

For best results against Phytophthora root and stem rot, use Ridomil 5G with soybean varieties that have some tolerance to the races of Phytophthora present in the field. The high rate of Ridomil 5G should be used in areas with a history of heavy Phytophthora damage. Under heavy late season Phytophthora pressure, Ridomil 5G may not provide complete control.

Pythium and Phytophthora: Apply 3-6 oz./1,000 linear feet of row in the seed furrow or as a 7-inch band over the row at planting. Apply amount recommended per acre according to the row spacing chart. Use the high rate in areas with a history of Phytophthora problems.

Suggested Rates Per Acre According to Row Spacing

Row Spacing	Linear Feet of Soybean Row Per Acre	Amount of Ridomil 5G Needed Per Acre
38 in.	13,756	2⅔-5⅓ lbs.
36 in.	14,520	2¾-5½ lbs.
30 in.	17,424	3⅓-6⅔ lbs.
24 in.	21,780	4⅙-8⅓ lbs.
20 in.	26,136	5-10 lbs.

Note: Ridomil 5G is specific for Pythium and Phytophthora and will not control other diseases that may attack soybeans.

STORAGE AND DISPOSAL

Store in a dry place. Do not contaminate water, food, or feed by storage, disposal, or cleaning of equipment. Open dumping is prohibited.

PESTICIDE DISPOSAL
Wastes resulting from the use of this product are acutely toxic. Improper disposal of unused pesticide, spray mixture, or rinsate is a violation of federal law. Pesticide, spray mixture, or rinsate that cannot be used according to label instructions must be disposed of according to federal, state, or local procedures. For guidance in proper disposal methods, contact your State Pesticide or Environmental Control Agency, or the Hazardous Waste representative at the nearest EPA Regional Office.

CONTAINER DISPOSAL
Completely empty bag into application equipment. Dispose of empty bag in a sanitary landfill or by incineration, or by open burning, if allowed by state and local authorities. If burned, keep out of smoke.

For minor spills, leaks, etc., follow all precautions indicated on this label and clean up immediately. Take special care to avoid contamination of equipment and facilities during cleanup procedures and disposal of wastes. In the event of a major spill, fire or other emergency, call (919) 292-7100 day or night.

PRECAUTIONARY STATEMENTS

HAZARDS TO HUMANS AND DOMESTIC ANIMALS

CAUTION
Harmful if inhaled or absorbed through the skin. Causes moderate eye irritation. Avoid breathing dust. Avoid contact with skin, eyes or clothing. Wash thoroughly with soap and water after handling. Remove contaminated clothing and wash before reuse.

First Aid: In case of contact with eyes, flush with plenty of water. If on skin, wash with plenty of soap and water. Get medical attention if irritation persists. If swallowed, drink large quantities of water and induce vomiting by placing finger in back of throat; call a physician or contact your local Poison Control Center. Never give anything by mouth to an unconscious person. Avoid alcohol. In case of inhalation exposure, remove victim to fresh air.

ENVIRONMENTAL HAZARDS
Do not apply to water bodies or wetland. Do not contaminate water by cleaning of equipment or disposal of waste.

Ridomil® trademark of CIBA-GEIGY

Terraclor® trademark of UniRoyal, Inc. for PCNB

CGA 65L7C 115

SANCAP® 80W

HERBICIDE

For preemergence weed control in cotton

Sancap 80W is a wettable powder

ACTIVE INGREDIENT:
Dipropetryn:
2-(ethylthio)-4, 6-bis (isopropylamino)-*s*-triazine 80%
INERT INGREDIENTS: 20%
TOTAL: 100%

EPA Reg. No. 100-522

Sancap® trademark of CIBA-GEIGY for dipropetryn

Keep Out of Reach of Children.
Caution

See additional caution statements on back of bag.

See directions for use on panel at right.

DIRECTIONS FOR USE AND CONDITIONS OF SALE AND WARRANTY

IMPORTANT: Read the entire **Directions for Use** and the **Conditions of Sale and Warranty** before using this product.

CONDITIONS OF SALE AND WARRANTY

The **Directions for Use** of this product reflect the opinion of experts based on field use and tests. The directions are believed to be reliable and should be followed carefully. However, it is impossible to eliminate all risks inherently associated with use of this product. Crop injury, ineffectiveness, or other unintended consequences may result because of such factors as weather conditions, presence of other materials, or the manner of use or application all of which are beyond the control of CIBA-GEIGY or the Seller. All such risks shall be assumed by the Buyer.

CIBA-GEIGY warrants that this product conforms to the chemical description on the label and is reasonably fit for the purposes referred to in the **Directions for Use** subject to the inherent risks referred to above. **CIBA-GEIGY makes no other express or implied warranty of Fitness or Merchantability or any other express or implied warranty. In no case shall CIBA-GEIGY or the Seller be liable for consequential, special, or indirect damages resulting from the use or handling of this product.** CIBA-GEIGY and the Seller offer this product, and the Buyer and user accept it, subject to the foregoing **Conditions of Sale and Warranty,** which may be varied only by agreement in writing signed by a duly authorized representative of CIBA-GEIGY.

DIRECTIONS FOR USE

It is a violation of federal law to use this product in a manner inconsistent with its labeling.

GENERAL INFORMATION

Sancap 80W is a selective herbicide for preemergence weed control in cotton grown on the sandy soils of Arizona, New Mexico, Oklahoma, and Texas. Sancap 80W will control pigweed (carelessweed) and Russian thistle.

Since Sancap acts mainly through root absorption, its effectiveness depends on rainfall or irrigation to move it into the root zone. Should weeds develop, a shallow cultivation or rotary hoeing will generally result in improved weed control.

Following many years of continuous use of this product and chemically related products, biotypes of some of the weeds listed on this label have been reported which cannot be effectively controlled by this and related herbicides. Where this is known or suspected, we recommend the use of this product in combination with other registered herbicides which are not triazines. Consult with your State Agricultural Extension Service for specific recommendations.

APPLICATION INSTRUCTIONS

Premix Sancap 80W with a small quantity of water to make a slurry before adding to the spray tank. Pour this slurry into the spray tank during or after filling.

Ground Application: For broadcast ground applications, use a minimum of 20 gals. of water per acre. For band treatment, use a proportional volume of water per acre. Use conventional spray equipment with fan-type nozzles. Screens in nozzles and strainers should be no finer than 50-mesh. Use hydraulic or mechanical agitation during mixing and application to maintain a uniform suspension. Use a suitable pump with capacity to maintain 35-40 psi at the nozzles. If hydraulic agitation is used, the pump should also provide sufficient agitation in the tank to keep the mixture in suspension.

The amount of Sancap 80W needed for band treatment may be calculated by the formula:

$$\frac{\text{band width in inches}}{\text{row width in inches}} \times \text{broadcast rate per acre} = \text{amount needed per acre of field}$$

Aerial Application: For aerial application, use a minimum of 2 gals. of water per acre. Avoid application under conditions where uniform coverage cannot be obtained or where excessive spray drift may occur. In order to assure that spray will be controllable within the target area when used according to label directions, make applications at a maximum height of 10 ft., using low drift nozzles at a maximum pressure of 40 psi, and restrict application to periods when wind speed does not exceed 10 mph. To assure that spray will not adversely affect adjacent sensitive nontarget plants, apply Sancap by aircraft at a minimum upwind distance of 300 ft. from sensitive plants.

Avoid application to humans or animals. Flagmen and loaders should avoid inhalation of spray mist and prolonged contact with skin, and should wash thoroughly before eating and at the end of each day's operation.

APPLICATION RATE

Apply Sancap 80W at the rates shown in the table. Apply to flat-planted, bedded, or furrow-planted cotton at planting or within two days following planting.

Either broadcast or band applications can be made to flat-planted or bedded cotton. Only band applications, not exceeding the width of the bottom of the furrow, should be made to furrow-planted cotton.

If replanting is necessary, do not make a second broadcast application or a second band application in the same furrow, or crop injury may occur. A second band application may be made to furrow-planted cotton if the new furrows are formed down the untreated row middles.

Annual Weed Control

Region	Soil texture	Broadcast* rate per acre
Arizona	Sand	1.5 lbs
New Mexico, Oklahoma, and Texas	Loamy sand	2 lbs.
	Fine sandy loam	2.5 lbs

* For calculation of band treatment rate, see General Information section.

Precautions: *1) Cotton growing on alkali soils, caliche outcroppings, or where cuts, fills, or erosions have exposed calcareous subsoil may be injured by an application of Sancap. 2) Do not apply over emerged cotton or crop injury will occur. 3) To avoid concentration of Sancap in seed furrow, do not make broadcast applications to cotton planted in furrows. Width of band should not exceed the width of the bottom of the furrow.*

Note: Do not feed treated forage to livestock or graze treated areas.

ROTATIONAL CROPS

Rotational crops may be planted six months following a Sancap 80W application to cotton. Cotton stalks should be disked or plowed under prior to planting a rotational crop.

STORAGE AND DISPOSAL

Store in a dry place. Do not contaminate water, food, or feed by storage or disposal. Improper disposal of unused spray mixture, or rinse water is a violation of federal law. If these wastes cannot be used according to label instructions, contact your State Pesticide or Environmental Control Agency, or the Hazardous Waste representative at the nearest EPA Regional Office for guidance in proper disposal methods.

Completely empty bag into application equipment. Dispose of empty bag in a sanitary landfill or by incineration, or if allowed by state and local authorities, by burning. Stay out of smoke from burning bags.

For minor spills, leaks, etc., follow all precautions indicated on this label and clean up immediately. Take special care to avoid contamination of equipment and facilities during cleanup procedures and disposal of wastes. In the event of a major spill, fire or other emergency, call (919) 292-7100 day or night.

CAUTION

Keep out of reach of children.

Harmful if swallowed. Avoid contact with eyes and skin, inhalation of dust, and contamination of food and feed.

Do not contaminate domestic or irrigation water supplies, or lakes, streams, or ponds.

Sancap® trademark of CIBA-GEIGY for dipropetryn

CGA 41L1F 084

SUPRACIDE® 2E

INSECTICIDE-MITICIDE

RESTRICTED USE PESTICIDE

For retail sale to and use only by certified applicators or persons under their direct supervision and only for those uses covered by the certified applicators certification.

For control of certain insects of alfalfa; artichokes; certain citrus, fruits and nuts; cotton; nursery stock; olives; safflowers; sunflowers; sorghum; and tobacco

ACTIVE INGREDIENTS:	
Methidathion: 0,0-dimethyl phosphorodithioate, S-ester with 4-(mercaptomethyl)-2- methoxy-Δ^2-1,3,4-thiadiazolin-5-one	24.4%
Xylene range aromatic solvent	65.1%
INERT INGREDIENTS:	10.5%
TOTAL:	100.0%

EPA Reg. No. 100-501 EPA Est. 100-AL-1
Control No. 52330

Supracide® trademark of CIBA-GEIGY for methidathion

Supracide 2E contains 2 lbs. active ingredient per gallon.

Keep Out of Reach of Children.
DANGER/PELIGRO

See additional precautionary statements left of container seam.

PRECAUCIÓN AL USUARIO:
Si usted no lee inglés, no use este producto hasta que la etiqueta haya sido explicado ampliamente.

See directions for use on panel at right.

DIRECTIONS FOR USE AND CONDITIONS OF SALE AND WARRANTY

IMPORTANT: Read the entire **Directions for Use** and the **Conditions of Sale and Warranty** before using this product.

CONDITIONS OF SALE AND WARRANTY

The **Directions for Use** of this product reflect the opinion of experts based on field use and tests. The directions are believed to be reliable and should be followed carefully. However, it is impossible to eliminate all risks inherently associated with use of this product. Crop injury, ineffectiveness, or other unintended consequences may result because of such factors as weather conditions, presence of other materials, or the manner of use or application all of which are beyond the control of CIBA-GEIGY or the Seller. All such risks shall be assumed by the Buyer.

CIBA-GEIGY warrants that this product conforms to the chemical description on the label and is reasonably fit for the purposes referred to in the **Directions for Use** subject to the inherent risks referred to above. **CIBA-GEIGY makes no other express or implied warranty of Fitness or Merchantability or any other express or implied warranty. In no case shall CIBA-GEIGY or the Seller be liable for consequential, special, or indirect damages resulting from the use or handling of this product.** CIBA-GEIGY and the Seller offer this product, and the Buyer and user accept it, subject to the foregoing **Conditions of Sale and Warranty,** which may be varied only by agreement in writing signed by a duly authorized representative of CIBA-GEIGY.

DIRECTIONS FOR USE

It is a violation of federal law to use this product in a manner inconsistent with its labeling.

FAILURE TO FOLLOW THE DIRECTIONS FOR USE AND PRECAUTIONS ON THIS LABEL MAY RESULT IN POOR INSECT AND MITE CONTROL, CROP INJURY, AND/OR ILLEGAL RESIDUES.

GENERAL INFORMATION

Supracide 2E is an emulsifiable concentrate which when applied according to the Directions for Use will control certain insect pests and mites on agricultural crops.

Wear protective clothing as described in the USDA/EPA Guide for Commercial Applicators during the preparation of the spray mixture and during application.

Do not apply this product in such a manner as to directly or through drift expose workers or other persons except those knowingly involved in the application. Do not treat while unprotected humans or domestic animals are present in the treatment area.

REENTRY STATEMENT

Thirty days must elapse between the time of application in grapefruit, lemon, and orange groves and the time a worker enters the grove without appropriate protective clothing to engage in any activity requiring substantial contact with treated foliage.

On other crops, two days must elapse between the time of application and the time a worker enters the treated field without wearing protective clothing as described above, for applicators to engage in any activity requiring substantial contact with treated foliage. Between 3 and 7 days following application, personnel reentering fields where there is an opportunity for significant contact with treated foliage should wear clothing which has long sleeves and long pants. Wash thoroughly before eating and at the end of each day's operation. Change and wash clothes daily.

Because certain states may require more restrictive reentry intervals for various crops treated with this product, consult your State Department of Agriculture for further information.

Written or oral warnings must be given to workers who are expected to be in a treated area or in an area about to be treated with this product. Oral warnings must inform workers of areas or fields that may not be entered without specific protective clothing until the intervals stated above have elapsed, and appropriate actions to take in case of accidental exposure, as described under **Precautionary Statements** on this label. When oral warnings are given, warnings shall be given in a language customarily understood by workers. Oral warnings must be given if there is reason to believe that written warnings cannot be understood by workers. Written warnings must include the following information: "DANGER. Area treated with Supracide 2E on (date of application). Do not enter without appropriate protective clothing until (date based on intervals stated above). In case of accidental exposure, flush eyes or skin with plenty of water. Call physician if irritation persists. Remove and wash contaminated clothing before reuse."

MIXING PROCEDURES

(1) Be sure sprayer is clean; if it is contaminated with any material, crop injury or sprayer clogging may result. (2) Fill tank 1/4–1/2 full with clean water, and start agitation. (3) Be certain that the agitation system is working properly and creates a rippling or rolling action on the liquid surface. (4) Add Supracide to the tank. Then add tank mix ingredients. (5) Complete filling tank, maintaining sufficient agitation at all times to ensure surface action. This applies to both spray and nurse tanks. (6) Do not leave tank mixtures in the tank without adequate agitation. Always maintain agitation to avoid separation and buildup of residues. If a residue starts to build up in tank, drain and clean with strong detergent and ammonia solution and rinse thoroughly with clean water. (7) Immediately after use, clean sprayer thoroughly by flushing system with water containing a detergent and rinse with clean water.

ALFALFA (Pure Stands or Stands Containing Clover or Grass)
Apply at the rates listed below in a minimum of 10 gals. of water per acre with ground equipment or 2 gals. of water per acre by air. Foliage applications for alfalfa weevil and Egyptian alfalfa weevil control should be made to standing alfalfa when 20-30% of the growing tips show feeding damage. Stubble applications should be made immediately following a cutting. For the other insects listed on alfalfa, apply when insects first appear. Make no more than one stubble application and no more than one foliage application per alfalfa cutting. Use the low rate for light infestations and the high rate for moderate to heavy infestations.

TABLE 1: Alfalfa—Rates Per Acre

Insects Controlled	North-eastern States	Southern States	California	Pacific Inter-mountain States Excluding CA
Alfalfa weevil	2-4 pts.	3-4 pts.	2 pts.	2-4 pts.
Aphids	2-4 pts.	3-4 pts.	2 pts.*	2-4 pts.
Egyptian alfalfa weevil	—	—	2 pts.*	2-4 pts.
Grass scale *(Eriococcus insignis)*	—	—	—	4 pts. (Washington only)
Lygus	—	—	2 pts.	2-4 pts.
Leafhoppers	2-4 pts.	2-4 pts.	2 pts.	2-4 pts.
Spittlebugs	2-4 pts.	—	—	—

* Egyptian alfalfa weevil and pea aphid: (California only) Use 1-2 pts. where integrated control is practiced. Apply as above. When Egyptian alfalfa weevil larval counts exceed 20 per sweep, use the 2 pts. rate only.

Note: To avoid injury to pollinating insects, do not apply during bloom. To avoid illegal residues, do not apply within 14 days of cutting, harvesting, grazing, or feeding to livestock.

ALMONDS
Dilute and Concentrate Sprays

Peach twig borer, San Jose scale, brown almond mite: Apply during the dormant or delayed dormant period, or as a cover spray. Mix a dilute spray at the rate of 2 pts. per 100 gals. of water and apply 300-600 gals. per acre. If applying as a concentrate spray*, apply 6-12 pts. per acre in a minimum of 50 gals. of water per acre.

When concentrate applications are made, proper equipment must be used to obtain uniform coverage.

Aerial Sprays

Peach twig borer, San Jose scale: Apply during the dormant or delayed dormant period and use 6-12 pts. per acre as an aerial spray* in a minimum of 20 gals. of spray per acre. Spray swath should be no wider than the distance between the tree rows or the length of the spray boom.

EPA-registered oils may be added to the Supracide spray mixture in the dormant/delayed dormant period. Do not use more than 1.5 gals. of oil per 100 gals. spray as a dilute spray or 6 gals. of oil per acre as a concentrate spray or 5 gals. of oil per acre as an aerial spray.

* The rate of Supracide for concentrate and aerial sprays is based on a dilute spray with conventional ground equipment. A dilute spray is considered to be 300-600 gals. per acre. This amounts to a minimum of 6 pts. and a maximum of 12 pts. of Supracide per acre.

Note: To avoid illegal residues, (1) Do not make more than one application during the dormant or delayed dormant period nor more than one cover spray each season. (2) Do not apply within 80 days of harvest. (3) Do not graze treated areas or feed cover crops grown in treated areas to livestock.

Precaution: *Apply the dormant/delayed dormant application before any blossoms open, or injury may occur. Do not combine with oils in the cover spray, or injury or excessive residues may occur. The cover spray of Supracide may be slightly phytotoxic on some almond varieties.*

ARTICHOKES
Apply in a minimum of 50 gals. of water with ground equipment or a minimum of 10 gals. of water by air. Begin applications when insects appear in newly planted fields or when shoots emerge following complete cutback of the plants. Continue applications at 14 day intervals until buds begin to appear. Do not plant crops except those listed on this label within one year after last application.

Artichoke plume moth, chrysanthemum leafminer: Apply 4 pts. per acre.

Note: To avoid illegal residues, (1) Do not apply after buds begin to form. (2) Avoid application under conditions which favor drift to adjacent artichoke fields where buds are present. (3) Do not exceed a total of 8 applications per crop cycle.

COTTON
Apply in a minimum of 10 gals. of water with ground equipment or in a minimum of 3 gals. of water by air unless otherwise specified.

Spider mites, boll weevil, bollworm: Apply 4 pts. per acre. Repeat applications at 5-7 day intervals if necessary.

Lygus bug: (Arizona, California only) Apply 2-4 pts. per acre. Use the low rate for light to moderate infestations and the high rate for moderate to heavy infestations. Repeat applications if necessary.

Pink bollworm: (Arizona, California only) Apply 2 pts. per acre. Repeat applications at 5-7 day intervals if necessary.

Bandedwing whitefly: (mid-South and southeast United States only) Apply 1-2 pts. per acre. Use the low rate for light to moderate infestations and the high rate for moderate to heavy infestations. Repeat applications if necessary.

Note: To avoid illegal residues, (1) Do not apply more than 16 pts. per acre during any one growing season. (2) Do not apply within 14 days of harvest. (3) Do not apply more than the 2 pt. per acre rate to open bolls. (4) Do not graze treated cotton plants or feed gin trash to livestock.

DECIDUOUS FRUIT
Dilute and Concentrate Sprays

See Table 2 for specific crops and insects. Apply during the dormant or delayed dormant period. Use a rate of 1-2 pts. per 100 gals. of water as a dilute spray (300-600 gals. per acre) or a rate of 3-12 pts. per acre as a concentrate spray* (a minimum of 50 gals. of water per acre). The lower rates may be used for light infestations and the higher rates for moderate to heavy infestations. For rates less than 2 pts. of Supracide per 100 gals., tank mix with an EPA-registered oil.

When concentrate applications are made, proper equipment must be used to obtain uniform coverage.

TABLE 2: Deciduous Fruits

Crop	Insects
Apples	San Jose scale, rosy apple aphid
Cherries, pears	San Jose scale
Nectarines, peaches, plums, prunes	Peach twig borer, San Jose scale, Green Peach aphid[1]
Apricots	Peach twig borer
Pears†	Pear psylla[2]

† California only

[1] For green peach aphid, use 8 pts. Supracide + 4 gals. oil per acre or 12 pts. Supracide per acre alone.

[2] For pear psylla, use 8-12 pts. Supracide + 4-6 gals. oil in a minimum of 100 gals. total spray per acre.

Aerial Sprays

See Table 3 for specific crops and insects. Apply during the dormant or delayed dormant period and use 3-12 pts. per acre as an aerial spray* in a minimum of 20 gals. total spray per acre. Spray swath should be no wider than the distance between the tree rows or the length of the spray boom. The lower rates may be used for light infestations and the higher rates for moderate to heavy infestations.

For rates less than 2 pts. of Supracide per 100 gals., tank mix with an EPA-registered oil.

TABLE 3: Deciduous Fruits

Crop	Insects
Apples	San Jose scale, rosy apple aphid
Peaches, plums, prunes, nectarines, apricots	Peach twig borer, San Jose scale
Cherries, pears	San Jose scale

EPA-registered oils may be added to the Supracide spray mixture in the dormant/delayed dormant period. Do not use more than 1.5 gals. of oil per 100 gals. spray as a dilute spray or 6 gals. of oil per acre as a concentrate spray or 5 gals. of oil per acre as an aerial spray.

* The rate of Supracide for concentrate and aerial sprays is based on a dilute spray with conventional ground equipment. A dilute spray is considered to be 300-600 gals. per acre. This amounts to a minimum of 3 pts. and a maximum of 12 pts. of Supracide per acre.

Note: To avoid illegal residues, (1) Apply before any blossoms open or injury may occur. (2) Do not graze treated areas or feed cover crops grown in treated areas to livestock.

Precaution: Do not use Supracide on Red Beaut plums.

GRAPEFRUIT AND ORANGES

Supracide Alone

Apply at the recommended rates as a postbloom and/or summer spray. Use the low rate for light infestations and the high rate for moderate to heavy infestations. Thorough coverage is essential for good control.

Make no more than 2 applications per growing season and allow at least 45 days between applications. Severe infestations may require 2 treatments.

California red scale: (Arizona and California only) Apply 1-2 pts. per 100 gals. of water.

Black scale: (California only) Apply 1-2 pts. per 100 gals. of water.

Chaff, Glovers, purple, snow, yellow scale, and mealybug: (Florida only) Apply 1/2-2 pts. per 100 gals. of water.

Black scale, whiteflies, and citrus blackfly: (Florida only) Apply 1-2 pts. per 100 gals. of water.

Brown soft, California red, chaff scale, and mealybug: (Texas only) Apply 1/2-2 pts. per 100 gals. of water.

The minimum amount of Supracide applied per acre is dependent upon the number of trees per acre and upon tree height. See example below.

Example: Mature trees of normal foliage require a minimum of 5 gals. of spray plus one additional gallon for each foot of tree height for adequate coverage. Eighty trees per acre 20 feet tall require 80 × 25 = 2,000 gals. of dilute spray per acre.

Note: To avoid illegal residues, (1) Do not apply more than 40 pts. per acre. (2) Do not graze treated areas. (3) Make no more than 2 applications per growing season nor apply within 14 days of harvest.

Supracide Plus Oil

In Florida and Texas, EPA-registered citrus spray oils at rates of 2-4 qts. per 100 gals. of water may be tank mixed with Supracide following the rates and timing listed above for Supracide alone.

In California, EPA-registered citrus spray oils at rates listed below may be tank mixed with Supracide. See the above directions for rates and timing. A rate of 1/2 pt. of Supracide per 100 gals. of water may be used if tank mixed with oil.

Type of oil	Amount per 100 gals. of water
Light-medium grade	1.6-1.8 gals.
Medium grade	1.6-1.8 gals.
Narrow-range 415 grade	1.2-1.4 gals.

For further details, consult the current *University of California Citrus Treatment Guide.*

Note: To avoid illegal residues, (1) Do not apply more than 40 pts. of Supracide per acre in any one season. (2) Do not graze treated areas. (3) Make no more than 2 applications with oil per growing season nor apply within 60 days of harvest.

LEMONS

Supracide Alone

Apply at the recommended rates as a postbloom and/or summer spray. Use the low rate for light infestations and the high rate for moderate to heavy infestations. Thorough coverage is essential for good control.

Make no more than 2 applications per growing season and allow at least 45 days between applications. Severe infestations may require 2 treatments.

California red scale: (Arizona and California only) Apply 1-2 pts. per 100 gals. of water.

Black scale: (California only) Apply 1-2 pts. per 100 gals. of water.

Whiteflies, citrus blackfly, mealybug, scale (includes black, chaff, Glovers, purple, snow, and yellow): (Florida only) Apply 1-2 pts. per 100 gals. of water.

The minimum amount of Supracide applied per acre is dependent upon the number of trees per acre and on the tree height. See example in Grapefruit and Oranges section.

Note: To avoid illegal residues, (1) Do not apply more than 40 pts. per acre per season. (2) Do not graze treated areas. (3) Do not apply within 14 days of harvest.

Supracide Plus Oil

See above directions for application and pests controlled. Apply a maximum of 1 pt. of Supracide per 100 gals. of water.

EPA-registered citrus spray oils at rates of ½-1 gal. per 100 gals. of water may be tank mixed with Supracide in Florida. Use the following rates of oil in California:

Type of oil	Amount per 100 gals. of water
Light-medium grade	1.6-1.8 gals.
Medium grade	1.6-1.8 gals.
Narrow-range 415 grade	1.2-1.4 gals.

For further details, consult the current *University of California Citrus Treatment Guide.*

Note: To avoid illegal residues, (1) Do not apply more than one application of Supracide per growing season, or exceed 10 pts. per acre per season, if tank mixed with oil (2) Do not apply tank mix with oil within 60 days of harvest. (3) Do not graze treated areas.

MANGOES

Apply during the postharvest to bloom stage with ground equipment in 100 gals. of spray mixture per acre. Spray foliage to thoroughly wet upper and lower leaf surfaces. Repeat as needed at 3 week intervals.

False oleander scale, *Pseudaulacaspis cockerelli* (cooley): Apply ½-1 pt. and 0.5% oil per acre.

Note: To avoid illegal residues, (1) Do not exceed 5 applications per season. (2) Do not graze or feed cover crops from treated groves.

NURSERY STOCK (For Professional Use Only)

Apply at the rate listed below in a thorough cover spray to runoff. Thorough coverage and timing of application are essential for effective control.

Apply once per season after scale crawlers have hatched. Apply either in early spring, April-May (when lilac is in full bloom), for spring-generation crawlers, or in summer, June-July, for second generation summer crawlers. The summer spray will also control pine tortoise scale if present (when golden chain tree is blooming).

Pine needle scale: (Scotch, Mugho, and Red Pines, northeast U.S. and Ohio) Apply 2 pts. per 100 gals. of water.

OLIVES

Apply once during the postharvest or prebloom season at the rate listed below as a full cover spray in a minimum of 400 gals. of water per acre. Supreme or superior type oil may be added to the spray mixture at the rate of 1-1.5 gals. per 100 gals. of water.

Black scale: Apply 2 pts./100 gals.

Note: To avoid illegal residues, (1) Do not apply more than 12 pts. per acre per year. (2) Do not apply after buds break. (3) Do not graze treated areas or feed cover crops grown in treated areas to livestock.

PECANS (Southeastern U.S. Only)

Apply the appropriate rate listed below as a foliar or full cover spray in sufficient water for thorough coverage.

Yellow pecan aphid: Apply 2 pts. per 100 gals. and repeat as necessary at 14-21 day intervals.

Hickory shuckworm and pecan serpentine leafminer: Apply 2-4 pts. per 100 gals. when insects first appear or leafminer damage first appears and repeat as necessary at 14-21 day intervals. Use the 4 pt. per 100 gals. rate when infestations are heavy.

Precaution: *Do not tank mix Supracide with Cyprex® as phytotoxicity may result.*

Note: To avoid illegal residues, (1) Do not apply within 60 days of harvest. (2) Do not graze treated areas.

SAFFLOWERS

Apply in a minimum of 20 gals. of water by ground equipment or in a minimum of 5 gals. of water by air.

Lygus bugs and thrips: Apply 2 pts. per acre. Begin applications either at bud stage, 12-14 days before onset of bloom, or as infestations develop. Additional applications can be made on a 7-14 day schedule or as further infestations develop.

Note: To avoid illegal residues, (1) Do not apply within 28 days of harvest. (2) Do not make more than three applications per year. (3) Do not graze treated areas.

SUNFLOWERS

Apply 2 pts. per acre. Make ground application in a minimum of 10 gals. of water per acre; aerial application in a minimum of 5 gals. of water per acre. Thorough coverage of the heads is necessary. One to three applications may be made on a 7-day schedule. The number of applications will be determined by the severity of the infestations.

Stem weevil*: Begin applications in early season at vegetative stage 2.3 (third alternate leaf developed) to vegetative stage 3.1 (inflorescence head visible but tightly surrounded by young leaves).

Sunflower moth, seed weevil**, banded sunflower moth, and sunflower maggot complex: Begin applications at the onset of bloom as the infestations develop.

Sunflower seed midge: Begin applications during early bud development.

* Cylindrocopturus adspersus and Apion occidentalis.
** Smicronyx fulvus and Smicronyx sordidus.

Note: To avoid illegal residues, (1) Do not apply within 50 days of harvest. (2) Do not graze treated areas or feed treated forage to livestock.

GRAIN SORGHUM GROWN FOR FORAGE AND GRAIN (Colorado, Kansas, Oklahoma, and Texas Only)

Apply in a minimum of 10 gals. of water per acre with ground equipment or 2 gals. of water per acre by air. Up to three applications may be made on a 10-14 day schedule. The number of applications will be determined by the severity of the infestation.

Banks grass mite, sorghum greenbug: Apply 2 pts. per acre.

Precaution: *Application may result in phytotoxicity to certain sorghum varieties. This phytotoxicity is a temporary spotting, reddening, or chlorosis of the leaves.*

Note: To avoid illegal residues, do not apply within 30 days of harvesting or feeding to livestock.

TOBACCO

Apply at the rates listed below in a minimum of 25 gals. of water per acre when insects appear. Apply 1-3 repeat applications on a 7-14 day interval as needed.

Flea beetles, hornworms: Apply 2-3 pts. per acre.

Budworms: Apply 4 pts. per acre.

Precaution: *When applying Supracide with hand sprayers, avoid concentrating the spray in the bud area since phytotoxicity may occur. Do not apply Supracide to tobacco plants which are under stress as injury may occur.*

Note: Do not apply in mixture with other pesticides or within 3 days of tobacco harvest.

WALNUTS

Dilute and Concentrate Sprays

Walnut scale, frosted scale, Italian pear scale: Apply during the dormant or delayed dormant period or as a cover spray when crawlers appear.

Codling moth, walnut aphid: Apply as a cover spray according to moth activity or when aphids appear.

Use a rate of 2 pts. per 100 gals. of water as a dilute spray (400 gals. per acre), or a rate of 8 pts. per acre as a concentrate spray* (a minimum of 100 gals. of water per acre).

When concentrate applications are made, proper equipment must be used to obtain uniform coverage. If scale infestations are severe, dilute applications should be used.

Aerial Sprays

Walnut scale, frosted scale: Apply during the dormant or delayed dormant period and use 8 pts. per acre as an aerial spray* in a minimum of 20 gals. of spray per acre. Spray swath should be no wider than the distance between the tree rows or the length of the spray boom.

EPA-registered oils may be added to the Supracide spray mixture in the dormant/delayed dormant period. Do not use more than 1 gal. of oil per acre in concentrate and aerial sprays or 4 gals. of oil per acre in dilute sprays.

* The rate of Supracide for concentrate and aerial sprays is based on a dilute spray with conventional ground equipment. A dilute spray is considered to be 400 gals. per acre. This amounts to 8 pts. of Supracide per acre.

Note: To avoid illegal residues, (1) Make no more than one application during the dormant or delayed dormant period nor more than two cover sprays each season. (2) Do not apply within 7 days of harvest. (3) Do not graze treated areas or feed cover crops grown in treated areas to livestock.

Precaution: *Do not add oil to Supracide sprays to walnuts in southern or central California or to trees that may be subject to stress due to drought, drying winds, disease, or severe insect infestations.*

STORAGE AND DISPOSAL

Do not contaminate water, food, or feed by storage or disposal.

STORAGE

Store at temperatures above 40°F. Crystals may form at lower storage temperatures. If this occurs, place the product in a warm room (68°F or above) and roll or shake the container at frequent intervals until all crystals are dissolved.

Pouring the product through a 50-mesh or finer screen will indicate adequacy of or need to repeat above procedure. Do not store near food, feed, or fertilizer. Do not use or store in or around the home.

PESTICIDE AND CONTAINER DISPOSAL

Pesticide wastes are acutely hazardous. Improper disposal of unused pesticide, spray mixture, or rinse water is a violation of federal law. If these wastes cannot be used according to label instructions, contact your State Pesticide or Environmental Control Agency, or the Hazardous Waste representative at the nearest EPA Regional Office for guidance in proper disposal methods.

Triple rinse container (or equivalent) and offer for recycling or reconditioning, or puncture and dispose of in a sanitary landfill, or by other procedures approved by state and local authorities.

For minor spills, leaks, etc., follow all precautions indicated on this label and clean up immediately. Take special care to avoid contamination of equipment and facilities during cleanup procedures and disposal of wastes. In the event of a major spill, fire or other emergency, call (919) 292-7100 day or night.

PRECAUTIONARY STATEMENTS

HAZARDS TO HUMANS AND DOMESTIC ANIMALS

DANGER

Corrosive. Causes irreversible eye damage. May cause skin irritation. Do not get in eyes, on skin, or on clothing. Wear goggles or face shield and rubber gloves when mixing. May be harmful or fatal if swallowed or absorbed through skin. Do not breathe vapor or spray mist. Prolonged or repeated inhalation may be harmful. Wear a mask or pesticide respirator jointly approved by the Mining Enforcement Safety Administration (MESA) or the National Institute for Occupational Safety and Health (NIOSH) during periods of prolonged or repeated use. Wash thoroughly with soap and water after handling and before eating, drinking, or smoking. Remove contaminated clothing and wash before reuse or if not washable, discard. Keep out of reach of children and domestic animals.

STATEMENT OF PRACTICAL TREATMENT

If in eyes: Flush with plenty of water. Get medical attention immediately.

If swallowed: Call a physician or poison control center immediately. DO NOT INDUCE VOMITING UNLESS NO OTHER TREATMENT IS AVAILABLE. Vomiting may cause aspiration pneumonia. If it is necessary to induce vomiting, do so under medical supervision. Give victim one or two glasses of water and insert finger in back of throat. Repeat until vomit fluid is clear. Do not induce vomiting or give anything by mouth to an unconscious person.

If on skin: Remove contaminated clothing and wash skin with plenty of soap and water. Get medical attention.

If inhaled: Remove victim to fresh air. If not breathing, give artificial respiration, preferably mouth-to-mouth unless exposure was due to swallowing; then use cardiopulmonary resuscitation (CPR). Get medical attention.

Note to physician: Lavage stomach with care to prevent aspiration into lung. After lavage, instill and leave 30 gm. activated charcoal in 3-4 oz. water. Atropine sulfate and 2-PAM are specific antidotes. Morphine, theophylline, aminophylline, and large amounts of intravenous fluids are contraindicated.

ENVIRONMENTAL HAZARDS

This product is toxic to fish, birds, and other wildlife. Do not apply directly to lakes, streams, ponds, or wetlands. Do not apply where runoff is likely to occur, as runoff from treated areas may be hazardous to aquatic organisms in neighboring areas. Do not apply when weather conditions favor drift from treated areas. Do not contaminate any body of water by cleaning of equipment or disposal of wastes. Apply this product only as specified on this label. This pesticide is highly toxic to bees exposed to direct treatment or to residues remaining on the treated area. Do not apply when bees are actively visiting the crop, cover crop, or weeds blooming in the treatment area. Applications should be timed to provide the maximum possible interval between treatment and the next period of bee activity.

PHYSICAL AND CHEMICAL HAZARDS

Do not use, pour, spill, or store near heat or open flame.

Supracide® trademark of CIBA-GEIGY for methidathion

Cyprex® trademark of American Cyanamid for dodine

CGA 30L1T 015

TILT®

FUNGICIDE

For control of certain diseases on grasses grown for seed

ACTIVE INGREDIENT:
1-[2-(2,4-dichlorophenyl)-4-propyl-1,3-dioxolan-2-yl] methyl-1*H*-1,2,4-triazole 41.8%
INERT INGREDIENTS: 58.2%
TOTAL: 100.0%

EPA Reg. No. 100-617 EPA Est. 100-AL-1
U.S. Patent No. 4,079,062
CGA 72L1D 084

Tilt® trademark of CIBA-GEIGY

Tilt contains 3.6 lbs. active ingredient per gallon.

Keep Out of Reach of Children.

WARNING

See additional precautionary statements at end of booklet.

See directions for use inside booklet.

DIRECTIONS FOR USE AND CONDITIONS OF SALE AND WARRANTY

IMPORTANT: Read the entire **Directions for Use** and the **Conditions of Sale and Warranty** before using this product.

CONDITIONS OF SALE AND WARRANTY

The **Directions for Use** of this product reflect the opinion of experts based on field use and tests. The directions are believed to be reliable and should be followed carefully. However, it is impossible to eliminate all risks inherently associated with use of this product. Crop injury, ineffectiveness, or other unintended consequences may result because of such factors as weather conditions, presence of other materials, or the manner of use or application all of which are beyond the control of CIBA-GEIGY or the Seller. All such risks shall be assumed by the Buyer.

CIBA-GEIGY warrants that this product conforms to the chemical description on the label and is reasonably fit for the purposes referred to in the **Directions for Use** subject to the inherent risks referred to above. **CIBA-GEIGY makes no other express or implied warranty of Fitness or Merchantability or any other express or implied warranty. In no case shall CIBA-GEIGY or the Seller be liable for consequential, special, or indirect damages resulting from the use or handling of this product.** CIBA-GEIGY and the Seller offer this product, and the Buyer and user accept it, subject to the foregoing **Conditions of Sale and Warranty,** which may be varied only by agreement in writing signed by a duly authorized representative of CIBA-GEIGY.

DIRECTIONS FOR USE

It is a violation of federal law to use this product in a manner inconsistent with its labeling.

GENERAL INFORMATION

Tilt is a broad spectrum fungicide for the control of rusts (*Puccinia* spp.) and powdery mildew and Selenophoma stem eyespot in grasses grown for seed in Oregon, Washington, Idaho, and Minnesota, including perennial ryegrass, red fescue, chewings fescue, bluegrass, and orchardgrass.

FAILURE TO FOLLOW DIRECTIONS AND PRECAUTIONS ON THIS LABEL MAY RESULT IN CROP INJURY, POOR DISEASE CONTROL, OR ILLEGAL RESIDUES.

SPRAY EQUIPMENT

Thorough coverage is necessary to provide good disease control.

Equip sprayers with nozzles that provide accurate and uniform application. Be certain that nozzles are the same size and uniformly spaced across the boom. Calibrate sprayer before use and recalibrate at the start of each season.

Use a centrifugal pump with capacity to (1) maintain 35-40 psi at nozzles, and (2) provide sufficient agitation in tank to keep mixture in suspension—this requires a minimum of 10 gals./minute/100 gals. tank mixture recirculated back into the tank. Use a jet agitator or liquid sparge tube for agitation. Do not air sparge.

Use screens to protect the pump and to prevent nozzles from clogging. Screens placed on suction side of pump should be 16-mesh or coarser. Do not place a screen in the recirculation line. Use 50-mesh or coarser screens between the pump and boom, and where required, at the nozzles. Check nozzle manufacturer's recommendations.

For more information on spray equipment and calibration, consult sprayer manufacturers and state recommendations.

MIXING PROCEDURES

(1) Be sure sprayer is clean; if it is contaminated with any material, crop injury or sprayer clogging may result. (2) Fill tank 1/4-1/2 full with clean water, and start agitation. (3) Be certain that the agitation system is working properly and creates a rippling or rolling action on the liquid surface. (4) Add product to the tank. (5) Complete filling tank, maintaining sufficient agitation at all times to ensure surface action. This applies to both the spray and nurse tanks. (6) Do not leave tank mixtures in the tank without adequate agitation. Always maintain agitation to avoid separation and build-up of residues. If a residue starts to build up in tank, drain and clean with strong detergent and ammonia solution, and rinse thoroughly with clean water. (7) Immediately after use, clean sprayer thoroughly by flushing system with water containing a detergent and rinse with clean water.

AERIAL APPLICATION PROCEDURES

Avoid applications under conditions where uniform coverage cannot be obtained or where excessive spray drift may occur. Do not apply directly to humans or animals. Flagmen and loaders should wash thoroughly before eating and at the end of each day's operation.

APPLICATION (Oregon, Washington, Idaho, and Minnesota only)

Mix 1/4 to 1/2 pt. of Tilt (maximum 1/4 pt. on bluegrass) in a minimum of 20 gals. of water per acre for ground equipment, or in a minimum of 10 gals. of water per acre for aerial application. Apply when powdery mildew and Selenophoma infections or rust pustules are noticeable and increasing in number in late spring or early summer. Repeat at 14 to 21 day intervals. Under severe rust pressure, use the higher rate of 1/2 pt. per acre (except on bluegrass), or make more frequent applications to maintain rust control until the seed is mature. Make the last application about two weeks before seed matures. For bluegrass, it is important to begin applications in early season.

Note: Do not graze livestock in treated areas. Do not use any part of crop or crop wastes for feed or bedding.

STORAGE AND DISPOSAL

Store at temperatures above 0°F. Do not contaminate water, food or feed by storage or disposal. Pesticide wastes may be toxic. Improper disposal of unused pesticide, spray mixture, or rinse water is a violation of federal law. If these wastes cannot be used according to label instructions, contact your State Pesticide or Environmental Control Agency, or the Hazardous Waste representative at the nearest EPA Regional Office for guidance in proper disposal methods.

Triple rinse container (or equivalent). Then offer for recycling or reconditioning, or puncture and dispose of in a sanitary landfill, by incineration, or, if allowed by state and local authorities, by burning. Stay out of smoke from burning containers.

For minor spills, leaks, etc., follow all precautions indicated on this label and clean up immediately. Take special care to avoid contamination of equipment and facilities during cleanup procedures and disposal of wastes. In the event of a major spill, fire, or other emergency, call (919) 292-7100 day or night.

PRECAUTIONARY STATEMENTS

HAZARDS TO HUMANS AND DOMESTIC ANIMALS

WARNING

This product may cause skin sensitization reactions in certain individuals. Wear rubber gloves and long sleeves when mixing or handling the concentrated product. Remove contaminated clothing and wash before reuse.

Causes eye irritation. Do not get in eyes, on skin, or on clothing. Harmful if swallowed, inhaled, or absorbed through skin. Wash thoroughly with soap and water after handling. Avoid breathing spray mist.

First Aid: If in eyes, flush with plenty of water and get medical attention if irritation persists. If on skin, wash with plenty of soap and water. Get medical attention if irritation persists. If swallowed, drink promptly a large quantity of milk, egg whites, or gelatin solution; if these are not available, drink large quantities of water. Never give anything by mouth to an unconscious person. Do not induce vomiting. Call a physician or Poison Control Center immediately. In case of inhalation exposure, move to fresh air.

Note to Physician: There is no specific antidote for this product. Induce emesis or lavage stomach, taking care to avoid aspiration of stomach contents into lungs. Give a saline laxative and supportive therapy.

ENVIRONMENTAL HAZARDS

This product is toxic to fish. Do not apply directly to water. Do not contaminate water by cleaning of equipment or disposal of wastes.

PHYSICAL OR CHEMICAL HAZARDS

Do not use, pour, spill, or store near heat or open flame.

Tilt® trademark of CIBA-GEIGY

U.S. Patent No. 4,079,062

CGA 72L1D 084

TRIGARD® 75W

INSECTICIDE

For leafminer control in celery and lettuce

For aerial application only

ACTIVE INGREDIENT:

Cyromazine: *N*-cyclopropyl-1,3,5-triazine-2,4,6-triamine 75%

INERT INGREDIENTS: 25%

TOTAL: 100%

Trigard 75W is a wettable powder

EPA Reg. No. 100-654

EPA Est. 100-AL-1

Keep Out of Reach of Children.

CAUTION

See additional precautionary statements on back of bag.

This outer protective bag contains Trigard 75W in an inner water-soluble bag. Entire inner bag and contents dissolve in water. After opening outer bag, immediately dump *entire* unopened inner bag into the partially filled sprayer or mix tank. Do not handle the soluble bag or expose it to moisture, since this may cause rupturing.

See directions for use on back of bag.

DIRECTIONS FOR USE AND CONDITIONS OF SALE AND WARRANTY

IMPORTANT: Read the entire **Directions for Use** and the **Conditions of Sale and Warranty** before using this product.

CONDITIONS OF SALE AND WARRANTY

The **Directions for Use** of this product reflect the opinion of experts based on field use and tests. The directions are believed to be reliable and should be followed carefully. However, it is impossible to eliminate all risks inherently associated with use of this product. Crop injury, ineffectiveness, or other unintended consequences may result because of such factors as weather conditions, presence of other materials, or the manner of use or application all of which are beyond the control of CIBA-GEIGY or the Seller. All such risks shall be assumed by the Buyer.

CIBA-GEIGY warrants that this product conforms to the chemical description on the label and is reasonably fit for the purposes referred to in the **Directions for Use** subject to the inherent risks referred to above. **CIBA-GEIGY makes no other express or implied warranty of Fitness or Merchantability or any other express or implied warranty. In no case shall CIBA-GEIGY or the Seller be liable for consequential, special, or indirect damages resulting from the use or handling of this product.** CIBA-GEIGY and the Seller offer this product, and the Buyer and user accept it, subject to the foregoing **Conditions of Sale and Warranty,** which may be varied only by agreement in writing signed by a duly authorized representative of CIBA-GEIGY.

DIRECTIONS FOR USE

It is a violation of federal law to use this product in a manner inconsistent with its labeling.

FAILURE TO FOLLOW THE DIRECTIONS FOR USE AND PRECAUTIONS ON THIS LABEL MAY RESULT IN CROP INJURY AND/OR POOR INSECT CONTROL.

Trigard is a 75% wettable powder which, when mixed with water according to the directions for use given below, will control certain insect species on designated plants.

Do not apply this product in such a manner as to directly or through drift expose workers or other persons, except those knowingly involved in the application. The area being treated must be vacated by unprotected persons.

REENTRY STATEMENT

Do not enter treated areas without protective clothing until seven days after last application.

Because certain states may require more restrictive reentry intervals for various crops treated with this product, consult your State Department of Agriculture for further information.

Written or oral warnings must be given to workers who are expected to be in a treated area or in an area about to be treated with this product. Oral warnings must be given which inform workers of areas or fields that may not be entered without specific protective clothing until seven days after last application, and appropriate actions to take in case of accidental exposure as described under **Precautionary Statements** on this label. When oral warnings are given, warnings shall be given in a language customarily understood by workers. Oral warnings must be given if there is reason to believe that written warnings cannot be understood by workers. Written warnings must include the following information: "CAUTION. Area treated with Trigard 75W on (date of application). Do not enter without appropriate protective clothing until seven days after last application. In case of accidental exposure, flush eyes or skin with plenty of water. Call a physician if irritation persists. Remove and wash contaminated clothing before reuse.

MIXING INSTRUCTIONS

Prepare no more spray mixture than is needed for the immediate operation. Thoroughly clean spray equipment before using this product. Agitate thoroughly before and during application. Good bypass agitation is sufficient to keep Trigard 75W thoroughly mixed. Flush spray tank thoroughly with water daily after use and dispose of pesticide rinsate by application to an already treated area.

Trigard 75W Alone. Add 2/3 of the required amount of water to the spray or mixing tank. With the agitator running, drop the required number of **unopened** soluble bags of Trigard 75W into the tank. (One 2-lb. water-soluble bag of Trigard 75W will treat 12 acres of celery or lettuce.) Continue agitation while adding the remainder of the water.

Note: Do not let spray mixture stand overnight in the spray tank. Rinse spray tank at the end of the day.

To avoid spray drift, do not apply under windy conditions. Avoid spray overlap, as crop injury may result.

CELERY AND LETTUCE (HEAD ONLY)

For leafminer control, apply 1/6 lb. Trigard 75W per acre as a foliar spray in a minimum of 5 gals. of water by air as leafminers first appear. Repeat applications at 7-day intervals or as necessary to maintain control.

Notes: 1) Do not make more than 12 applications to one crop of celery or 8 applications to one crop of lettuce, and 2) Do not make the last application within 7 days of harvest, or illegal residues may result.

Rotational Crops

Do not rotate to any crop other than sweet corn or radishes. Do not plant within 3 months after last application.

STORAGE AND DISPOSAL

Store in a cool, dry place. Do not contaminate water, food, or feed by storage or disposal. Do not use in animal feeds.

STORAGE. Do not store this product under wet conditions. Handle outer bag carefully when stored at temperatures less than 50°F to avoid breakage of inner soluble bag.

PESTICIDE DISPOSAL. Improper disposal of unused pesticide, spray mixture, or rinsate is a violation of federal law. Pesticide, spray mixture, or rinsate that cannot be used according to label instructions, must be disposed of according to federal, state or local procedures. For guidance in proper disposal methods, contact your State Pesticide or Environmental Control Agency, or the Hazardous Waste representative at the nearest EPA Regional Office.

CONTAINER DISPOSAL. Do not reuse outer bag. Dispose of empty outer bag in a sanitary landfill or by incineration or, if allowed by state and local authorities, by burning. If burned, stay out of smoke.

For minor spills, leaks, etc., follow all precautions indicated on this label and clean up immediately. Take special care to avoid contamination of equipment and facilities during cleanup procedures and disposal of wastes. In the event of a major spill, fire, or other emergency, call (919) 292-7100 day or night.

PRECAUTIONARY STATEMENTS

HAZARDS TO HUMANS AND DOMESTIC ANIMALS

CAUTION

Causes moderate eye irritation. Harmful if swallowed, absorbed through the skin or inhaled. Avoid contact with eyes, skin or clothing. Wear long-sleeved shirt and pants when handling or applying this product. Avoid breathing dust. Wash thoroughly with soap and water after handling. Remove contaminated clothing and wash before reuse.

STATEMENT OF PRACTICAL TREATMENT

If swallowed: Call a physician or Poison Control Center. Drink 1 or 2 glasses of water and induce vomiting by touching back of throat with finger. Do not induce vomiting or give anything by mouth to an unconscious person.

If in eyes: Flush with plenty of water. If irritation develops or persists, get medical attention.

If on skin: Wash with plenty of soap and water. If irritation develops or persists, get medical attention.

If inhaled: Move to fresh air.

Environmental Hazards

Do not apply directly to water or wetlands. Do not contaminate water by cleaning of equipment or disposal of wastes. Do not apply when weather conditions favor drift from target area.

CGA 69L30 105

The Dow Chemical Company

Agricultural Products Department
P.O. Box 1706
Midland, MI 48640
(517) 636-1000

CROSSBOW* Weed and Brush Killer

LOW VOLATILE BRUSH AND WEED HERBICIDE

For the Control of Most Kinds of Unwanted Trees and Brush, as Well as Annual and Perennial Broadleaf Weeds on Fence rows, Roadsides, Other Non-crop Areas and Industrial Sites

ACTIVE INGREDIENT(S):	
2,4–Dichlorophenoxyacetic Acid, Butoxyethyl Ester	34.4%
Triclopyr (3,5,6–trichloro–2–pyridinyloxyacetic acid), Butoxyethyl Ester	16.5%
INERT INGREDIENTS	49.1%
TOTAL	100%

Contains Petroleum Distillates.
2,4–D Acid Equivalent 23.7%—2 pounds per gallon
Triclopyr Acid Equivalent 11.9% —1 pound per gallon

EPA Reg. No. 464-589 Est. 464-MI-1

*Trademark of The Dow Chemical Company.

CAUTION
Keep Out of Reach of Children
AVISO

PRECAUCION AL USUARIO: Si usted no lee inglés, no use este producto hasta que la etiqueta le haya sido explicada ampliamente.

PRECAUTIONARY STATEMENTS

HAZARDS TO HUMANS AND DOMESTIC ANIMALS

• HARMFUL IF SWALLOWED, INHALED, OR ABSORBED THROUGH SKIN •

CAUSES EYE IRRITATION

Avoid Contact With Skin, Eyes, Or Clothing • Avoid Breathing Vapor • Wash Thoroughly With Soap and Water After Handling and Before Eating or Smoking.

STATEMENT OF PRACTICAL TREATMENT

In case of contact, flush skin or eyes with plenty of water. Get medical attention if irritation persists. Remove and wash contaminated clothing before reuse. If swallowed, do not induce vomiting. CALL A PHYSICIAN OR POISON CONTROL CENTER.

PHYSICAL OR CHEMICAL HAZARDS

COMBUSTIBLE
Do Not Use or Store Near Heat or Open Flame
Do Not Cut or Weld Container

ENVIRONMENTAL HAZARDS

This pesticide is toxic to fish. Do not apply directly to bodies of water. Do not contaminate irrigation ditches or water used for irrigation or domestic purposes.

NOTICE

Read the entire label. Use only according to label directions.

Before buying or using this product, read "WARRANTY LIMITATIONS AND DISCLAIMER" elwewhere on this label. If terms are not acceptable, return **unopened** package at once to seller for full refund of purchase price paid. Otherwise, use by the buyer or any other user constitutes acceptance of the terms under "WARRANTY LIMITATIONS AND DISCLAIMER."

In case of an emergency endangering life or property involving this product, call collect **517–636-4400**

AGRICULTURAL CHEMICAL

Do Not Ship or Store with Food, Feeds, Drugs or Clothing

FOR TREE, BRUSH AND WEED CONTROL

Crossbow Weed and Brush Killer is recommended for control of most species of unwanted woody plants, as well as annual and perennial broadleaf weeds, growing on fence rows, roadsides, other non-crop areas, and industrial sites.

Woody Species Controlled: Alder, ash, aspen, birch, blackberry, blackgum, cherry, elderberry, hawthorn, hazel, maples, Multiflora rose, oaks, pine, salmonberry, sumac, sweetgum, tamarack, willow, and many others.

Weeds Controlled: Broadleaf weeds such as burdock, Canada thistle, cocklebur, dandelion, docks, goldenrod, kochia, lambsquarters, marshelder, milkweed, oxales, plaintains, Queen Anne's lace, ragweeds, tall ironweed, sunflower, vetch, and many others.

DIRECTIONS FOR USE

It is a violation of Federal law to use this product in a manner inconsistent with its labeling.

Foliage sprays should be applied during warm weather when brush and weeds are actively growing. Application under drought conditions may provide less than desirable results. Use low spray pressures to minimize spray drift. **Apply Crossbow Weed and Brush Killer in a manner to avoid contacting nearby susceptible crops or other desirable plants and to avoid contaminating water intended for irrigation or domestic use. Read and follow all Use Precautions given on this label.**

PREPARING THE SPRAY

Crossbow Weed and Brush Killer in water forms an emulsion (not a solution), and separation may occur unless the spray mixture is agitated continuously.

Water Spray: Fill the spray tank about half full with clean water. Then add the **Crossbow Weed and Brush Killer** and complete filling the tank with agitation running. Mix thoroughly and continue moderate agitation while spraying.

APPLICATION DIRECTIONS

WOODY PLANT CONTROL

For **high volume applications through handguns,** apply as a foliar wetting spray containing 1 to $1\frac{1}{2}$ or 1 to 1.5 gallons of this product in sufficient water to make 100 gallons of total spray. The lower rate may be used for the more sensitive species such as aspen, blackberry and willow. Use the higher rate for other woody plants. Spray to give thorough coverage, wetting all leaves, stems, and root collars.

For broadcast sprays with ground equipment, apply 1.5 to 4 gallons of this product in enough water to deliver 15 to 30 gallons total spray per acre. Use the lower chemical rates for more sensitive species such as blackberry and willow and the higher chemical rates and spray volumes for more tolerant species such as ash, oaks, and maples.

GENERAL WEED CONTROL

Broadcast Treatment: Use 1 to 2 quarts of Crossbow Weed and Brush Killer per acre in the amount of water needed for uniform coverage to control annual and perennial broadleaf weeds. Use the 2 quart rate for deep rooted perennial broadleaf weeds.

Spot Treatment: To control broadleaf weeds in small areas with a hand sprayer, use 1/4 pint of Crossbow Weed and Brush Killer in 3 gallons of water and spray to thoroughly wet all foliage.

STORAGE AND DISPOSAL

Do not contaminate water, food or feed by storage or disposal.

STORAGE

Store above 10°F or agitate before use

DISPOSAL

Pesticide Disposal—Pesticide wastes are toxic. Improper disposal of excess pesticide, spray mixture, or rinsate is a violation of Federal Law. If these wastes cannot be disposed of by use according to label instructions, contact your State Pesticide or Environmental Control Agency, or the Hazardous Waste representative at the nearest EPA Regional Office for guidance.

Container Disposal—Triple rinse (or equivalent). Then offer for recycling or reconditioning, or puncture and dispose of in a sanitary landfill, or by other approved state and local procedures. Plastic containers, after triple rinsing, may be incinerated if allowed by state and local authorities. If burned, stay out of smoke.

General—Consult federal, state, or local disposal authorities for approved alternative procedures.

Be sure that use of this product conforms to all applicable regulations.

USE PRECAUTIONS

Apply this product only as specified on this label.

Do not apply Crossbow Weed and Brush Killer directly to, or otherwise permit it to come into direct contact with cotton, grapes, tobacco, vegetable crops, flowers, fruit or ornamental trees, or other desirable broadleaf plants and do not permit spray mists containing it to drift onto them.

Avoid Injurious Spray Drift. Applications should be made only when hazards from spray drift are at a minimum. Very small quantities of spray, which may not be visible may seriously injure susceptible plants. Do not spray when wind is blowing toward susceptible crops or ornamental plants near enough to be injured. Spray drift can be reduced by adding a spray thickening agent such as Nalco-Trol† or its equivalent to the spray mixture. If a spray thickening agent is used, follow all use recommendations and precautions on the product label.

†Nalco-Trol Trademark of Nalco Chemical Company.

With ground broadcast equipment, drift can be reduced by keeping the spray boom as low as possible; by applying no less than 20 gallons of spray per acre; by using no more than 20 pounds spraying pressure with flat fan or flooding flat fan nozzle tips; and by spraying only when wind velocity is less than 5 miles per hour. In handgun application, use gun pressure no higher than 50 psi and apply in such manner as to confine the spray to the target area.

At high temperatures, vapors from this product may injure susceptible crops growing nearby. Excessive amounts of this herbicide in the soil may temporarily inhibit seed germination and plant growth.

Do not graze treated areas or feed treated forage.

WARRANTY LIMITATIONS AND DISCLAIMER

The Dow Chemical Company warrants that this product conforms to the chemical description on the label and is reasonably fit for the purposes stated on the label when used in strict accordance with the directions therein under normal conditions of use. THIS IS THE ONLY WARRANTY MADE ON THIS PRODUCT. NO OTHER EXPRESS AND NO IMPLIED WARRANTY OF MERCHANTABILITY OR FITNESS FOR A PARTICULAR PURPOSE IS MADE OUTSIDE OF THIS LABEL. Therefore, neither this warranty nor any other warranty of merchantability or fitness for a particular purpose, express or implied, extends to the use of this product contrary to label instructions (including conditions noted on the label, such as unfavorable temperatures, soil conditions, etc.), under abnormal conditions (such as excessive rainfall, drought, tornadoes, hurricanes, etc.) or under conditions not reasonably foreseeable to or beyond the control of seller.

When buyer or user suffers losses or damages resulting from the use or handling of this product (including claims based on contract, negligence, strict liability, or other legal theories), buyer or user must promptly notify in writing The Dow Chemical Company of any claims to be eligible to receive either remedy given below. The EXCLUSIVE REMEDY OF THE BUYER OR USER and the LIMIT OF LIABILITY of The Dow Chemical Company or any other seller will be one of the following, at the election of The Dow Chemical Company:

(1) **Refund of purchase price** paid by buyer or user for product brought **or**

(2) **Replacement** of amount of product used.

The seller **will not be liable for consequential or incidental damages or losses.**

The terms of this Warranty Limitations and Disclaimer cannot be varied by any written or verbal statements or agreements. Any employee or sales agent of the seller is not authorized to vary or exceed the terms of this Warranty Limitations And Disclaimer in any manner.

GRAZON* P + D
HERBICIDE

RESTRICTED USE PESTICIDE

For retail sale to and use only by Certified Applicators or persons under their direct supervision and only for those uses covered by the Certified Applicator's certification.

For the Control of Broadleaf Annual and Perennial Weeds, and Certain Woody Species On Rangeland and Permanent Grass Pastures

For Distribution and Use Only Within New Mexico, Oklahoma, and Texas

ACTIVE INGREDIENT(S):	
Picloram (4-amino-3,5,6-trichloropicolinic acid) as the triisopropanol-amine salt	10.2%
2,4-Dichlorophenoxyacetic Acid as the triisopropanol-amine salt	39.6%
INERT INGREDIENTS:	50.2%
TOTAL	100.0%
ACID EQUIVALENTS:	
Picloram (4-amino-3,5,6-trichloropicolinic acid)	5.7%–0.54 lb/gal
2,4-Dichlorophenoxyacetic acid	21.2%–2 lb/gal

EPA Est. 464-MI-1 U.S. Patent No. 3,285,925
New Mexico EPA SLN No. NM-820003
Oklahoma EPA SLN No. OK-820005
Texas EPA SLN No. TX-820007

*Trademark of The Dow Chemical Company

CAUTION
Keep Out of Reach of Children
AVISO

PRECAUCION AL USUARIO: Si usted no lee inglés, no use este producto hasta que la etiqueta le haya sido explicada ampliamente.

PRECAUTIONARY STATEMENTS

HAZARDS TO HUMANS AND DOMESTIC ANIMALS
HARMFUL IF SWALLOWED
CAUSES EYE INJURY
MAY CAUSE SKIN IRRITATION

Avoid Contact with Eyes, Skin or Clothing • Wash Thoroughly After Handling • Keep Container Closed • When Handling Concentrate Wear Suitable Eye Protection

STATEMENTS OF PRACTICAL TREATMENT

In case of eye contact, immediately flush eyes with plenty of water for at least 15 minutes and get medical attention promptly.

If swallowed, induce vomiting immediately by giving two glasses of water and sticking finger down throat. Call a physician. Do not induce vomiting or give anything by mouth to an unconscious person.

In case of skin contact, flush skin with plenty of water. Remove contaminated clothing and wash before reuse.

PHYSICAL OR CHEMICAL HAZARDS

Do Not Cut or Weld Container

ENVIRONMENTAL HAZARDS

Do not apply directly to water. Do not apply where runoff is likely to occur. Do not contaminate water by cleaning of equipment or disposal of wastes. Do not contaminate irrigation ditches or water used for irrigation or domestic purposes.

NOTICE

Read the entire label. Use only according to label directions.

Before buying or using this product, read "WARRANTY LIMITATIONS AND DISCLAIMER" elsewhere on this label. If terms are not acceptable, return **unopened** package at once to seller for full refund of purchase price paid. Otherwise, use by the buyer or any other user constitutes acceptance of the terms under the Limit of Warranty and Liability.

IN CASE OF AN EMERGENCY endangering life or property involving this product, call collect **517–636-4400**

AGRICULTURAL CHEMICAL
Do Not Ship or Store with Food, Feeds, Drugs or Clothing

GENERAL INFORMATION

GRAZON P + D Herbicide is a water soluble liquid product containing picloram and 2,4-D. Very small amounts can kill or damage many broadleaved plants. To prevent damage to crops and other desirable plants, **carefully follow all directions and precautions.**

Use GRAZON P + D Herbicide in rangeland and permanent grass pastures to selectively control annual, biennial, and perennial herbaceous weeds and certain woody species, including bitterweed, broom snakeweed, goldenrod, horsenettle, locoweed, Macartney and Multiflora rose, prickly lettuce, ragweeds (bur, common, and western), thistles and many other species susceptible to 2,4-D.

DIRECTIONS FOR USE

Do not use for manufacturing or formulating.
It is a violation of Federal law to use this product in a manner inconsistent with its labeling.

Annual Broadleaf Weeds such as annual broomweed, common ragweed and bitterweed: For **early season** (March–April), use 1 to 2 pints GRAZON P + D Herbicide in enough water or oil-water emulsion to give a total spray volume of 10 to 20 gallons per acre with ground equipment or 1 to 4 gallons per acre when applied by aircraft. The lower rate is effective only in very early season when weeds are no more than 2 inches tall and conditions are favorable for good plant growth. For **mid to late season** (May–June) when weeds range from 3 inches to early flowering, use 2 to 4 pints GRAZON P + D Herbicide in enough water or oil-water emulsion to give a total spray volume of 10 to 20 gallons per acre with ground equipment or 1 to 4 gallons per acre when using aerial application.

Perennial Broadleaf Weeds such as broom snakeweed (turpentine weed): Treat when weeds are actively growing, using 2 to 4 pints GRAZON P + D Herbicide in enough water or oil-water emulsion to give a total spray volume of 10 to 20 gallons per acre with ground equipment or 1 to 4 gallons per acre when applied by aircraft. The higher rate should be used when weeds are larger than 4 inches tall. Applications made during drouth stress or other conditions preventing active growth may not provide acceptable control.

For **common and Drummond's goldenweed** *(Isocoma spp.)*, use 2 to 4 pints GRAZON P + D Herbicide in enough water plus 0.25 to 0.5% (vol./vol.) agricultural surfactant or in enough oil-water emulsion to give a total spray volume of 15 to 20 gallons per acre with ground equipment or 4 to 5 gallons per acre when applied by aircraft. One to 2 quarts of surfactant per 100 gallons of total spray mix provides the surfactant concentrations of 0.25 to 0.5%. Apply in the spring (April–June) when good growing conditions produce substantial canopy development. Good spray wetting of well-developed foliage is essential.

Poisonous Broadleaf Plants such as woolly loco garbancillo (Wooton loco), and groundsel *(Senecio spp.)*: Use 3 to 4 pints GRAZON P + D Herbicide in enough water, or **preferably** in enough oil-water emulsion to give total spray volume of 10 to 20 gallons per acre with ground equipment or 1 to 4 gallons per acre when applied by aircraft. Excellent control results from fall and winter application in Southwest Texas (Trans-Pecos). Since locoweeds are difficult to wet, add an agricultural surfactant such as ORTHO† X-77 at 0.25 to 0.5% (vol./vol.), which is 1 to 2 quarts surfactant per 100 gallons of total spray mix.

† Trademark of Kalo Laboratories, Inc.

Macartney and Multiflora Rose: Individual plant treatment for control of large undisturbed clumps or smaller regrowth, treat with a hand-gun using a spray mix containing 1 to 2 gallons GRAZON P + D Herbicide plus 2 quarts of commercial agricultural surfactant (such as ORTHO X-77) per 100 gallons spray solution. Coverage should be thorough to wet all leaves and stems using sufficient pressure to penetrate to the center of large clumps. The use of NALCO-TROL† drift control additive is recommended to reduce drift during application. **Broadcast ground application,** where such equipment can operate, should provide a complete wetting spray coverage of Macartney or Multiflora rose foliage. Apply 1 gallon GRAZON P + D Herbicide in enough water to give 20 to 25 gallons total spray volume per acre. Add a commercial agricultural surfactant at 0.5% (vol./vol.), which is 2 quarts of surfactant per 100 gallons total spray mix. **Aerial Application** of previously mowed or sprayed Macartney or Multiflora rose may be done using 1 gallon GRAZON P + D Herbicide plus 0.5% agricultural surfactant in enough water or oil-water emulsion to give 5 gallons total volume per acre. Swath width should be such as to provide adequate overlap between swaths to facilitate good coverage. Increased deposition and drift control can be achieved with the addition of a drift control additive such as NALCO-TROL at recommended rates. **Note:** Macartney or Multiflora rose may be treated in the fall or spring if conditions are favorable for good plant growth. Comply with all local regulations relative to application of hormone-type herbicides. Consult the University Extension Service for more information.

† Trademark of the Nalco Chemical Company

APPLICATION DIRECTIONS

Apply GRAZON P + D Herbicide at recommended rates in water alone or as an oil-water emulsion in a total volume of 1 to 4 gallons per acre by air or 10 to 25 gallons per acre by ground. To provide more complete wetting of the foliage, agriculture surfactants such as ORTHO X-77 may be used at recommended rates. The use of a drift control additive such as NALCO-TROL is strongly recommended for drift reduction and improved deposition.

MIXING INSTRUCTIONS

Ground Application—For Use With Water Alone: Start with about half the required amount of water in the spray tank. With agitation operating, add the required amount of GRAZON P + D Herbicide. If a surfactant is needed, it should be added as the remainder of the required water is added to complete the spray mix. When using a drift control additive such as NALCO-TROL, following the manufacturer's directions carefully. Complete dispersion of the additive such as obtained by passage through a pump with moderate to high shearing action is essential for proper performance of the drift control product.

For Use With Water and Oil: Add oil to the total spray mix at the ratio of 1 part oil to 5 parts water (15% to 20% of the total mix) using agricultural spray emulsifiers and mixing procedures given below for aerial application.

Aerial application—For Use With Water Alone: Follow procedures given for ground application above.

Aerial Application—For Use With Water and Oil (Batch Mixing): With continuous, vigorous agitation:

1. Add half the amount of water to be used to the spray tank.
2. Add the amount of GRAZON P + D Herbicide required for the total volume of spray being mixed.
3. Premix the required amount of oil with an emulsifier such as SPONTO† 712 or TRITON†† X-100 brands using 1 ounce of emulsifier per gallon of oil. Add the oil-emulsifier premix to the spray tank.
4. Finally, add the remaining amount of water required to bring the spray batch to the desired total volume.
5. Maintain agitation in the spray tank during application.

† Trademark of The Witco Chemical Corporation.

†† Trademark of Rohm & Haas Co.

Aerial Application—For Use With Water and Oil (Meter Mixing): Water-oil sprays containing GRAZON P + D Herbicide may also be prepared using the spray tank "loading" pump to mix the ingredients while filling the tank. This procedure involves simultaneous metering of the required amounts of water-herbicide and oil-emulsifier mixtures into the intake line of the pump during operation. Since emulsification is accomplished in the pump, it is important that a pump with moderate to high shearing action be used. For proper results always introduce the water-herbicide premix into the intake line before introducing the oil emulsifier, or a thick, invert emulsion may result. Maintain agitation in the spray tank during application.

Note: Consult the following table for amounts of product, oil and water to mix 100 gallons of total spray to be applied by air at the rate of 1 to 4 gallons total volume per acre.

TOTAL SPRAY VOL./ACRE (GALLONS)	GRAZON P + D RATE/ACRE (PINTS)	GALLONS			†EMULSIFIER (OZ.)	‡N.T. (OZ.)	TOTAL SPRAY MIX (GALLONS)
		GRAZON P + D	OIL	WATER			
	1	12.5	14.6	72.9	14.6	4.4	100
	2	25.0	12.5	62.5	12.5	3.8	100
1	3	37.5	10.4	52.1	10.4	3.1	100
	4	50.0	8.3	41.7	8.3	2.5	100
	1	6.2	15.6	78.1	15.6	4.7	100
	2	12.5	14.6	72.9	14.6	4.4	100
2	3	18.8	13.5	67.7	13.5	4.1	100
	4	25.0	12.5	62.5	12.5	3.8	100
	1	4.2	16.0	79.8	16.0	4.0	100
	2	8.3	15.3	76.4	15.3	4.6	100
3	3	12.5	14.6	72.9	14.6	4.4	100
	4	16.7	13.9	69.4	13.9	4.2	100
	1	3.1	16.2	80.8	16.2	4.8	100
	2	6.3	15.6	78.1	15.6	4.7	100
4	3	9.4	15.1	75.5	15.1	4.5	100
	4	12.5	14.6	72.9	14.6	4.4	100

† An emulsifier such as SPONTO 712 or TRITON X-100 should always be added at the rate of 1 ounce per gallon of oil in the total spray mix. The oil-water ratio is 1:5 for all spray mixes.

†† Amounts of NALCO-TROL (N.T.) are calculated on the basis of 6 oz. of N.T./100 gals. of water.

USE PRECAUTIONS

Use this product only as specified on this label. Observe any special use and application restrictions and limitations, including method of application and permissible areas of use as promulgated by state or local authorities.

Do Not Contaminate Nontarget Land Areas, Cropland, Water, or Irrigation Ditches. Do not contaminate water to be used for irrigation or domestic purposes. Do not apply where surface water from treated areas can run off to adjacent cropland, either planted or to be planted, or into streams, irrigation ditches, irrigation ponds or wells. Do not clean containers nor application equipment on or near these areas. Do not apply to frozen or saturated ground.

Do not apply on or in the vicinity of susceptible crops or desirable plants including alfalfa, beans, cotton, melons, potatoes, sugar beets, sunflower, tobacco, tomatoes and other vegetable crops, flowers, fruit plants, ornamentals or shade trees.

Avoid Spray Drift. Application should be made only when there is no hazard from spray drift since very small quantities of the spray, which may not be visible, may injure susceptible crops during both growing and dormant periods. Use coarse sprays to minimize drift. A drift control additive such as NALCO-TROL may be used with this product to aid in reducing spray drift. If used follow all use recommendations and precautions on the product label. Application with mist-type equipment is not recommended.

Ground Equipment: With ground equipment, spray drift can be lessened by keeping the spray boom as low as possible; by applying 20 gallons or more of spray per acre; by using no more than 25 pounds spraying pressure with large droplet producing nozzle tips; by spraying when wind velocity is low; and, by stopping all spraying when wind exceeds 10 miles per hours. Do not apply with hollow cone-type insecticide or other nozzles that produce a fine-droplet spray.

Aerial Application: With aircraft, drift can be lessened by applying a coarse spray; by using no more than 25 pounds spray pressure at nozzles; by using nozzles which do not produce fine spray droplets; by using a spray boom no longer than 3/4 the wing span of the aircraft; and, by spraying only when wind velocity is less than 10 mph.

Determine air movement and direction before foliar application: Use a smoke generator or other means at or near the application site for the detection of air movement, air stability or temperature inversions. Use appropirate drift control measures or avoid application when smoke is moving toward nearby desirable susceptible plants or sensitive areas. Avoid application when smoke forms layers or otherwise indicates inversion conditions which would prevent spray droplets from dispersing or falling to the intended spray site.

Do not rotate treated rangeland or pasture to other crop uses, as long as residues in the soil may affect susceptible crops.

Do not spray pastures if the forage legume component is desired. GRAZON P + D Herbicide may injure or kill legumes. Also new legume seedings may not be successful if made within 2 years following application of this herbicide.

Do not move treated soil to other areas. Do not use it to grow plants, unless an adequate sensitive bioassay shows that no detectable picloram is present in the soil.

Do not transfer livestock from treated grass areas onto broadleaf crop areas without first allowing 7 days of grazing on untreated grass pasture. Otherwise, urine may contain enough picloram to cause injury to sensitive broadleaf plants.

Be sure that use of this product conforms to all applicable regulations.

Be Sure To Follow All Use Precautions Given on This Label And Remember These Key Points:

1. Use only the recommended amounts of GRAZON P + D Herbicide at the recommended times. 2. Picloram remains active in the soil for extended periods. Treated soil should not be moved out of treated areas. 3. The potassium salt of picloram is water soluble. It can move with water in surface runoff, or in irrigation and drainage ditches. 4. Spray drift can cause injury to certain crops, ornamentals, trees, and numerous other desirable plants.

Do not mix with dry fertilizer.

STORAGE AND DISPOSAL

Do not contaminate water, food or feed by storage or disposal.

PESTICIDE DISPOSAL

Pesticide wastes are toxic. Improper disposal of excess pesticide, spray mixture, or rinsate is a violation of Federal Law. If these wastes cannot be disposed of by use according to label instructions, contact your State Pesticide or Environmental Control Agency, or the Hazardous Waste representative at the nearest EPA Regional Office for guidance.

CONTAINER DISPOSAL

Do not reuse containers. Puncture and dispose of them in a sanitary landfill after triple rinsing or equivalent, or by other approved state and local procedures. Plastic containers may also be disposed of by incineration or, if allowed by state and local authorities, by burning. If burned, stay out of smoke.

WARRANTY LIMITATIONS AND DISCLAIMER

The Dow Chemical Company warrants that this product conforms to the chemical description on the label and is reasonably fit for the purposes stated on the label when used in strict accordance with the directions therein under normal conditions of use. THIS IS THE ONLY WARRANTY MADE ON THIS PRODUCT. NO OTHER EXPRESS AND NO IMPLIED WARRANTY OF MERCHANTABILITY OR FITNESS FOR A PARTICULAR PURPOSE IS MADE OUTSIDE OF THIS LABEL. Therefore, neither this warranty nor any other warranty of merchantability or fitness for a particular purpose, express or implied, extends to the use of this product contrary to label instructions (including conditions noted on the label, such as unfavorable temperatures, soil conditions, etc.), under abnormal conditions (such as excessive rainfall, drought, tornadoes, hurricanes, etc.) or under conditions not reasonably foreseeable to or beyond the control of seller.

When buyer or user suffers losses or damages resulting from the use or handling of this product (including claims based on contract, negligence, strict liability, or other legal theories), buyer or user must promptly notify in writing The Dow Chemical Company of any claims to be eligible to receive either remedy given below. The EXCLUSIVE REMEDY OF THE BUYER OR USER and the LIMIT OF LIABILITY of The Dow Chemical Company or any other seller will be one of the following, at the election of The Dow Chemical Company:

(1) **Refund of purchase price** paid by buyer or user for product bought, **or**

(2) **Replacement** of amount of product used.

The seller **will not be liable for consequential or incidental damages or losses.**

The terms of this Limit of Warranty and Liability cannot be varied by any written or verbal Statements or agreements. Any employee or sales agent of the seller is not authorized to vary or exceed the terms of this Warranty Limitations And Disclaimer in any manner.

GRAZON* PC
HERBICIDE

RESTRICTED USE PESTICIDE

For retail sale to and use only by Certified Applicators or persons under their direct supervision and only for those uses covered by the Certified Applicator's certification.

For Distribution and Use Only Within New Mexico, Oklahoma, and Texas

For the Control of Broadleaf Annual and Perennial Weeds, Pricklypear, Mesquite and Various Other Woody Species On Rangeland and Permanent Grass Pastures

ACTIVE INGREDIENT	
Picloram (4-amino-3,5,6-trichloropicolinic acid) as the potassium salt	24.4%
INERT INGREDIENTS:	75.6%
TOTAL	100.0%

Acid Equivalent: Picloram (4-amino-3,5,6-trichloropicolinic acid): 21.1%—2 lbs./gallon.

EPA Est. 464-MI-1
New Mexico EPA SLN No. NM-820002
Oklahoma EPA SLN No. OK-820004
Texas EPA SLN No. TX-820006

Keep Out of Reach of Children

CAUTION
AVISO

PRECAUCION AL USUARIO: Si usted no lee inglés, no use este producto hasta que la etiqueta le haya sido explicada ampliamente.

PRECAUTIONARY STATEMENTS

HAZARDS TO HUMANS AND DOMESTIC ANIMALS

MAY CAUSE IRRITATION

Avoid Contact with Skin and Eyes

Avoid Breathing Spray Mist

Keep Container Closed

PHYSICAL OR CHEMICAL HAZARDS

Do Not Cut or Weld Container

ENVIRONMENTAL HAZARDS

Do not apply directly to water. Do not apply where runoff is likely to occur. Do not contaminate water by cleaning of equipment or disposal of wastes. Do not contaminate irrigation ditches or water used for irrigation or domestic purposes.

NOTICE

Read the entire label. Use only according to label directions.

Before buying or using this product, read "WARRANTY LIMITATIONS AND DISCLAIMER" elsewhere on this label. If terms are not acceptable return **unopened** package at once to seller for full refund of purchase price paid. Otherwise, use by the buyer or any other user constitutes acceptance of the terms under the Limit of

Warranty and Liability.

In case of an emergency endangering life or property involving this product, call collect **517–636-4400**

AGRICULTURAL CHEMICAL

Do Not Ship or Store with Food, Feeds, Drugs or Clothing

GENERAL INFORMATION

GRAZON PC Herbicide is a water soluble liquid product containing picloram. Very small amounts can kill or injure many broadleaved plants. To prevent damage to crops and other desirable plants, **follow all directions and precautions.**

Use GRAZON PC Herbicide in rangeland and permanent grass pastures to selectively control annual, biennial, and perennial herbaceous weeds such as bitterweed, bursage (bur ragweed, woolyleaf povertyweed), broomweed (annual), horsenettle, locoweed, pricklypear, ragweed (common, lanceleaf, western), snakeweed (broom snakeweed), tasajillo, and thistles. For equal or improved control of such weeds along with many species affected by 2,4-D or 2,4,5-T, use GRAZON PC Herbicide in tank mixtures with 2,4-D low volatile esters such as ESTERON* 99* Concentrate, or with 2,4-D amines such as FORMULA 40* or with 2,4,5-T low volatile esters such as ESTERON* 245 herbicides. Tank mixtures provide improved control of herbaceous species such as garbancillo (Wooton loco), goldenrod (*Solidago* spp.), goldenweed (*Isocoma* spp.), groundsel (*Senecio* spp.), woolly loco, and woody species such as blackbrush, catclaw acacia, granjeno (spiny hackberry), huisache, Macartney and multiflora rose, mesquite, oaks (blackjack, post and shinnery), persimmon (eastern), twisted acacia, and winged elm.

Applications made according to label directions usually give excellent control of most annual and perennial herbaceous broadleaf weed species. Certain more resistant herbaceous and woody species may require repeat treatment in subsequent years.

Note: May forbs are susceptible to GRAZON PC Herbicide, 2,4-D and 2,4,5-T. Do not spray pastures containing desirable forbs, especially legumes, unless injury to such plants can be tolerated. However, the growth of established grasses is improved after spraying, especially when rainfall is adequate and grazing is deferred.

FOR DISTRIBUTION AND USE ONLY WITHIN NEW MEXICO AND TEXAS

NEW MEXICO .... EPA SLN No. NM-820002

TEXAS .... EPA SLN No. TX-820006

For the Control of Susceptible Broadleaf Weeds and Woody Plants on Rangeland and Permanent Grass Pastures

DIRECTIONS FOR USE

Do not use for manufacturing or formulating.

It is a violation of Federal law to use this product in a manner inconsistent with its labeling.

Read Complete use directions and precautions on this label and container label before using.

Mix the required amount of GRAZON PC Herbicide in sufficient water for uniform coverage and apply as a coarse, low-pressure spray (under 25 psi) with ground or aerial application equipment. The use of a drift control additive such as NALCO-TROL† is strongly recommended, especially when water alone is used as the carrier.

† Trademark of the Nalco Chemical Company

For the control of most susceptible broadleaf weeds such as annual and perennial broomweed (broom snakeweed), bitterweed, bursage (bur ragweed), certain poisonous plants such as locoweed and senecio, and cactus species such as pricklypear and tasajillo, apply GRAZON PC Herbicide broadcast at the rate of 1 to 2 pints per acre or as a spot treatment at the rate of $\frac{1}{4}$ to $\frac{1}{2}$ teaspoon per 100 sq. ft. Individual plant treatments may be made using a 0.5 to 1% solution (2 to 4 qt GRAZON PC Herbicide per 100 gal water) applied as a thoroughly wetting spray.

For best results, treat when the weeds are growing actively in the spring before full bloom or in late summer or fall. Treatments during full bloom or seed stage of some weeds may not give good control. Applications for pricklypear control can be made at any time of the year. Herbicidal activity on pricklypear is very slow and may continue for two years or longer.

Consult your range or weed extension specialist for additional information.

TANK MIXTURES

GRAZON PC Herbicide plus 2,4,5-T Low Volatile Ester

Use GRAZON PC Herbicide in tank mixtures with 2,4,5-T low volatile esters such as ESTERON 245 herbicide, for control of mesquite and associated woody and herbaceous broadleaf plant species infesting rangeland and permanent grass pastures. Woody species controlled include:

blackbrush	oaks
catclaw acacia	pricklypear cactus
granjeno (spiny hackberry)	skunkbush
huisache	tasajillo
mesquite	winged elm

Recommended rates usually provide moderate to good topkill but only partial root-kill of the following species:

lotebush	whitebrush
twisted acacia	yaupon
shinnery oak	

DOSAGE RECOMMENDATIONS

GRAZON PC Herbicide in tank mixtures with 2,4,5-T low volatile ester herbicides is recommended for use at the rate of 1 to 2 pints GRAZON PC Herbicide per acre plus 0.5 to 1 pint per acre ESTERON 245 or equivalent 2,4,5-T low volatile herbicide. Use the higher rate in areas with a high percentage of mixed brush species. Use the lower rate where mesquite is the principal species to be controlled. However, where mesquite control together with a more complete control of pricklypear is desired, the higher rate (2 pints/acre) of GRAZON PC Herbicide may be tank mixed with the lower rate (0.5 pint/acre) of 2,4,5-T low volatile herbicide. The ratio between the two herbicides may be varied within the recommended rates.

MIXING INSTRUCTIONS

GRAZON PC Herbicide in tank mixtures with 2,4,5-T low volatile ester herbicides may be used in water alone or in oil-water mixes. For oil-water mixes, use diesel fuel, fuel oil or kerosene. To ensure proper mixing and spraying, maintain continuous, vigorous agitation during the entire mixing and spraying period. Regulate agitation and the addition of water to minimize foaming.

For Use With Water Alone (ground or aerial): Start with about half the required amount of water in the spray tank. With agitation, add the required amount of GRAZON PC Herbicide. After it is mixed in solution, add the required amount of 2,4,5-T low volatile ester with continued agitation. Add the remainder of the required amount of water to complete the spray tank mix. The use of a drift control additive such as NALCO-TROL is strongly recommended.

For Use With Water and Oil (Batch Mixing): With continuous vigorous agitation:

1. Add half the amount of water to be used to the spray tank.
2. Add the amount of GRAZON PC Herbicide required for the total volume of spray being mixed.
3. Add the amount of 2,4,5-T low volatile ester required for the total volume of spray being mixed.
4. Premix the required amount of oil with a good agricultural spray emulsifier such as SPONTO† 712 or TRITON†† X-100 brands using 1 ounce of emulsifier per gallon of oil. Add the oil-emulsifier premix to the spray tank. NOTE: the use of agricultural surfactants in place of emulsifiers may not provide satisfactory emulsions.

† Trademark of The Witco Chemical Corporation

‡ Trademark of Rohm and Haas Co.

5. Finally, add the remaining amount of water required to bring the spray batch to the desired total volume.
6. Maintain agitation in the spray tank during application.

For Use With Water and Oil (Meter Mixing): Water-oil sprays containing GRAZON PC Herbicide in tank mixtures with 2,4,5-T low volatile ester herbicides may also be prepared using the spray tank "loading" pump to mix the ingredients while filling the tank. This procedure involves simultaneous metering of the required amounts of water and GRAZON PC Herbicide with an oil-emulsifier premix and 2,4,5-T low volatile ester herbicide mixtures into the intake line of the pump during operation. Since emulsification is accomplished in the pump, it is important that a pump with moderate to high shearing action be used. For proper results, always introduce the water and GRAZON PC Herbicide into the intake line before introducing the oil emulsifier premix and 2,4,5-T low volatile ester herbicide, or a thick, invert emulsion may result. Maintain vigorous agitation in the spray tank during application.

APPLICATION DIRECTIONS

For best results, especially on mesquite, apply in the spring during the period 40 to 90 days after the first green growth appears and when soil moisture is adequate for good growth. Fall application is not effective on mesquite but can be effective for control of blackbrush, granjeno, and small huisache when soil moisture is adequate for good plant growth.

Aerial Application: Use 1 to 5 pints of oil and enough water to give 1 to 4 gallons of total spray per acre. The ratio of oil to water in the total spray mixture should be approximately 1:5 (1 part oil to 5 parts water). For mixing up 100 gallons of spray mix as an oil-water emulsion, consult table below:

TOTAL SPRAY VOL./A (GAL)	RATE IN PINTS/ACRE		GALLONS				OUNCES		GAL. TOTAL SPRAY MIX
	GRAZON PC HERBICIDE	ESTERON 245†	GRAZON PC HERBICIDE	ESTERON 245†	OIL	WATER	EMUL-SIFIER	N.T. ‡	
1	1	0.5	12.5	6.2	13.5	67.7	13.5	4.1	100
1	2	1.0	25.0	12.5	10.4	52.1	10.4	3.1	100
2	1	0.5	6.2	3.1	15.1	75.5	15.1	4.5	100
2	2	1.0	12.5	6.2	13.5	67.7	13.5	4.1	100
3	1	0.5	4.2	2.1	15.6	78.1	15.6	4.7	100
3	2	1.0	8.3	4.2	14.6	72.9	14.6	4.4	100
4	1	0.5	3.1	1.6	15.9	79.4	15.9	4.8	100
4	2	1.0	6.2	3.1	15.1	75.5	15.1	4.5	100

† ESTERON 245 or equivalent 2,4,5-T low volatile ester herbicide.

‡ Amounts of NALCO-TROL (N.T.) are calculated on the basis of 6 oz. of N.T. per 100 gal. of water.

Ground Application—Broadcast Spraying: To control low-growing species and regrowth, apply 1 to 2 pints GRAZON PC Herbicide plus ½ to 1 pint ESTERON 245 or equivalent 2,4,5-T ester herbicide in 10 to 25 gallons of water per acre. Add 1 to 2 quarts agricultural surfactant per 100 gallons of mixed spray to provide wetting of weed and brush foliage.

Ground Application—Spot Spraying: To control small patches of mesquite, small individual redberry juniper, pricklypear, and other susceptible species, apply a spray containing 2 quarts GRAZON PC Herbicide plus 1 quart ESTERON 245 or equivalent 2,4,5-T ester herbicide plus 1–2 quarts agricultural surfactant per 100 gallons of water. Spray to thoroughly wet foliage, stems, and trunks.

Ground Application—Special Equipment: To control annual and perennial weed species and low regrowth brush using special low-volume, minimum-drift equipment, such as the hooded Forage Chemical Mower, apply 1 to 2 pints GRAZON PC Herbicide alone or in combination with ½ to 1 pint ESTERON 245 or equivalent 2,4,5-T ester herbicide in total volumes ranging from 1 gallon to 5 gallons per acre in water alone or as an oil-water emulsion at a 1:5 and 1:4 oil-to-water ratio for 1-gallon and 5-gallon-per-acre solutions, respectively.

Note: Repeat applications may be made in successive years if more complete control is desired. Consult your Extension Weed Specialist for local recommendations.

GRAZON PC Herbicide Plus 2,4-D

For selective control of a wide range of broadleaf weeds on rangeland and permanent grass pastures, use GRAZON PC Herbicide in tank mixtures with 2,4-D low volatile esters such as ESTERON 99 Concentrate, or with 2,4-D amines such as FORMULA 40 herbicide. These tank mixtures are effective on many herbaceous and certain woody species including annual and perennial broomweed (broom snakeweed), bitterweed, goldenrod, horsenettle, locoweed, Macartney rose, prickly lettuce, ragweeds (bur, common, lanceleaf, and western), thistles and many other species susceptible to 2,4-D.

DOSAGE RECOMMENDATIONS

Annual Broadleaf Weeds such as annual broomweed, common ragweed and bitterweed: For **early season** (March–April), use 0.25 to 0.5 pint GRAZON PC Herbicide plus 0.5 to 1 pint ESTERON 99 Concentrate or FORMULA 40 herbicide in enough water or oil-water emulsion to give a total spray volume of 10 to 20 gallons per acre with ground equipment or 1 to 4 gallons per acre when using aerial application. Lower tank mix rates are effective only in very early season when weeds are no more than 2 inches tall and conditions are favorable for good plant growth. For **mid to late season** (May–June) when weeds range from 3 inches tall to early flowering, use 0.5 to 1 pint GRAZON PC Herbicide plus 1 to 2 pints ESTERON 99 Concentrate or FORMULA 40 herbicide in enough water or oil-water emulsion to give a total spray volume of 10 to 20 gallons per acre with ground equipment or 1 to 4 gallons per acre when applied by aircraft.

Perennial Broadleaf Weeds such as broom snakeweed (turpentine weed): Treat when weeds are actively growing, using 0.5 to 1 pint GRAZON PC Herbicide plus 1 to 2 pints ESTERON 99 Concentrate or FORMULA 40 herbicide in enough water or oil-water emulsion to give a total spray volume of 10 to 20 gallons per acre with ground equipment or 1 to 4 gallons per acre when applied by aircraft. The higher rate whould be used when weeds are larger than 4 inches tall. Applications made during drought stress or other conditions preventing active growth may not provide acceptable control.

For **common and Drummond's goldenweed** (*Isocoma* spp.) in the Texas Gulf Coast and Rio Grande Plains, use the same herbicide rates in water plus 0.25 to 0.5% (vol./vol.) agricultural surfactant or in enough oil-water emulsion to give a total spray volume of 15 to 20 gallons per acre with ground equipment or 4 to 5 gallons per acre when applied by aircraft. One to 2 quarts of surfactant per 100 gallons of total spray mix provides the surfactant concentrations of 0.25 to 0.5%. Apply in the spring (April–June) when good growing conditions produce substantial canopy development. Good spray wetting of well-developed foliage is essential.

Poisonous Broadleaf Plants such as woolly loco, garbancillo (Wooton loco), and groundsel (*Senecio* spp.): Use 0.75 to 1 pint GRAZON PC Herbicide plus 1.5 to 2 pints ESTERON 99 Concentrate or FORMULA 40 herbicides in enough water, or **preferably** in enough oil-water emulsion to give a total spray volume of 10 to 20 gallons per acre with ground equipment or 1 to 4 gallons per acre when applied by aircraft. Excellent control results from fall and winter application in Southwest Texas (Trans-Pecos). Since locoweeds are difficult to wet, add an agricultural surfactant such as ORTHO† X-77 at 0.25 to 0.5% (vol./vol.), which is 1 to 2 quarts surfactant per 100 gallons of total spray mix.

† Trademark of Kalo Laboratories, Inc.

Macartney and Multiflora Rose: Individual plant treatment for control of large undisturbed clumps or small regrowth: treat with a hand-gun using a spray mix containing 1 to 2 quarts GRAZON PC Herbicide plus 2 to 4 quarts FORMULA 40 (containing 2,4-D amine) plus 2 quarts of commercial agricultural surfactant (such as ORTHO X-77) per 100 gallons spray solution. Coverage should be thorough to wet all leaves and stems using sufficient pressure to penetrate to the center of large clumps. The use of a drift control additive such as NALCO-TROL is recommended to reduce drift during application. **Broadcast ground applications,** where such equipment can operate, should provide a complete wetting spray coverage of Macartney or multiflora rose foliage. Apply 1 quart GRAZON PC Herbicide plus 2 quarts ESTERON 99 Concentrate in enough water to give 20 to 25 gallons total spray volume per acre. Add a commercial agricultural surfactant at 0.5% (vol./vol.), which is 2 quarts per 100 gallons total spray mix. **Aerial application** of previously mowed or sprayed Macartney or multiflora rose may be done using 1 quart GRAZON PC Herbicide plus 2 quarts ESTERON 99 Concentrate plus 0.5% agricultural surfactant in enough water or oil-water emulsion to give 4 gallons total volume per acre. Swath width should be such as to provide adequate overlap between swaths to facilitate good coverage. Increased deposition and drift control can be achieved with the addition of a drift control additive such as NALCO-TROL at recommended rates. **Note:** Macartney or multiflora rose may be treated in the fall or spring if conditions are favorable for good plant growth. Comply with all local regulations relative to application of hormone-type herbicides. Consult the University Extension Service for more information.

APPLICATION DIRECTIONS

Apply GRAZON PC Herbicide tank mixed with ESTERON 99 Concentrate or FORMULA 40 or equivalent herbicide at recommended rates in water alone or as an oil-water emulsion in a total volume of 1 to 4 gallons per acre by air or 10 to 25 gallons per acre by ground. To provide more complete wetting of the foliage, agriculture surfactants such as ORTHO X-77 may be used at recommended rates. The use of a drift control additive such as NALCO-TROL is strongly recommended for drift reduction and improved deposition.

MIXING INSTRUCTIONS

Ground Application—For Use With Water Alone: Start with about half the required amount of water in the spray tank. With agitation operating, add the required amount of GRAZON PC Herbicide. Then add the required amount of ESTERON 99 Concentrate or FORMULA 40 or equivalent 2,4-D herbicide. If a surfactant is needed it should be added as the remainder of the required water is added to complete the spray mix. When using a drift control additive such as NALCO-TROL or equivalent, follow the manufacturer's directions carefully. Complete dispersion of the additive such as obtained by passage through a pump with moderate to high shearing action, is essential for proper performance of the product.

Ground Application—For Use With Water and Oil: Add oil to the total spray mix at the ratio of 1 part oil to 5 parts water (15% to 20% of the total mix) using agricultural spray emulsifiers and mixing procedures given below for aerial application.

Aerial Application—For Use With Water Alone: Follow procedures given for ground application above.

Aerial Application—For Use With Water and Oil (Batch Mixing): With continuous, vigorous, agitation:

1. Add half the amount of water to be used to the spray tank.
2. Add the amount of GRAZON PC Herbicide required for the total volume of spray being mixed.
3. Add the amount of ESTERON 99 Concentrate or FORMULA 40 or equivalent herbicides required for the total spray volume being mixed.
4. Premix the required amount of oil with an emulsifier such as SPONTO 712 or TRITON X-100 brands using 1 ounce of emulsifier per gallon of oil. Add the oil-emulsifier premix to the spray tank.
5. Finally, add the remaining amount of water required to bring the spray batch to the desired total volume.
6. Maintain agitation in the spray tank during application.

Aerial Application—for Use With Water and Oil (Meter Mixing): Water-oil sprays containing tank mixes of GRAZON PC Herbicide plus ESTERON 99 Concentrate or FORMULA 40 herbicide may also be prepared using the spray tank "loading" pump to mix the ingredients while filling the tank. This procedure involves simultaneous metering of the required amounts of water and GRAZON PC Herbicide with an oil-emulsifier and 2,4-D into the intake line of the pump during operation. Since emulsification is accomplished in the pump, it is important that a pump with moderate to high shearing action be used. For proper results, always introduce the water and GRAZON PC Herbicide into the intake line before introducing the oil-emulsifier premix, or a thick, invert emulsion may result. Maintain vigorous agitation in the spray tank during application.

Note: Consult the following table for amounts of product, oil and water to mix 100 gallons of total spray as an oil-water emulsion to be applied by air at the rate of 1 to 4 gpa total volume per acre.

TOTAL SPRAY VOL./ ACRE (GAL.)	PINTS		GALLONS						
	GRAZON PC	ESTERON 99 CONCENTRATE OR FORMULA 40	GRAZON PC	ESTERON 99 CONCENTRATE OR FORMULA 40	OIL	WATER	†EMULSIFIER (OZ.)	‡N.T. (OZ.)	TOTAL SPRAY (GALLONS)
	0.25	0.5	3.1	6.2	15.1	75.5	15.1	4.5	100
1	0.5	1	6.2	12.5	13.5	67.7	13.5	4.1	100
	1	2	12.5	25.0	10.4	52.1	10.4	3.1	100
	2	4	25.0	50.0	4.2	20.8	4.2	1.2	100
	0.25	0.5	1.6	3.1	15.9	79.4	15.9	4.8	100
2	0.5	1	3.1	6.2	15.1	75.5	15.1	4.5	100
	1	2	6.2	12.5	13.5	67.7	13.5	4.1	100
	2	4	12.5	25.0	10.4	52.1	10.4	3.1	100
	0.25	0.5	1.0	2.1	16.1	80.7	16.1	4.8	100
3	0.5	1	2.1	4.2	15.6	78.1	15.6	4.7	100
	1	2	4.2	8.3	14.6	72.9	14.6	4.4	100
	2	4	8.3	16.7	12.5	62.5	12.5	3.8	100
	0.25	0.5	0.8	1.6	16.2	81.4	16.2	4.9	100
4	0.5	1	1.6	3.1	15.9	79.4	15.9	4.8	100
	1	2	3.1	6.2	15.1	75.5	15.1	4.5	100
	2	4	6.2	12.5	13.5	67.7	13.5	4.1	100

† An emulsifier such as SPONTO 712 or TRITON X-100 should always be added at the rate of 1 ounce per gallon of oil in the total spray mix.

‡ Amounts of Nalco-Trol (N.T.) are calculated on the basis of 6 oz. of N.T./100 gals. of water.

READ AND FOLLOW ALL USE PRECAUTIONS AND STORAGE AND DISPOSAL INSTRUCTIONS

FOR DISTRIBUTION AND USE ONLY WITHIN OKLAHOMA

OKLAHOMA .... EPA SLN No. OK-820004

For the Control of Susceptible Broadleaf Weeds and Woody Plants on Rangeland and Permanent Grass Pastures

DIRECTIONS FOR USE

Do not use for manufacturing or formulating.

It is a violation of Federal law to use this product in a manner inconsistent with its labeling.

Read complete use directions and precautions on this label and container label before using.

Mix the required amount of GRAZON PC Herbicide in sufficient water for uniform coverage and apply as a coarse, low-pressure spray (under 25 psi) with ground or aerial application equipment. The use of a drift control additive such as NALCO-TROL† is strongly recommended.

† Trademark of the Nalco Chemical Company

For the control of most susceptible broadleaf weeds such as annual and perennial broomweed (broom snakeweed), ragweed (bur, common, and lanceleaf), bursage (bur ragweed), locoweed, and pricklypear. Apply GRAZON PC Herbicide broadcast at the rate of 1 to 2 pints per acre or as a spot treatment at the rate of 1/4 to 1/2 teaspoon per 100 sq. ft. Individual plant treatments may be made using a 0.5 to 1% solution (2 to 4 qt. GRAZON PC Herbicide per 100 gal water) applied as a thoroughly wetting spray.

For best results, treat when the weeds are growing actively in the spring before full bloom or in late summer or fall if soil moisture is adequate. Treatments during full bloom or seed stage of some weeds may not give good control.

Consult your range or weed extension specialist for additional information.

TANK MIXTURES

GRAZON PC Herbicide plus 2,4,5-T Low Volatile Ester

Use GRAZON PC Herbicide in tank mixtures with 2,4,5-T low volatile esters such as ESTERON 245 herbicide, for control of post oak, blackjack oak, winged elm, pricklypear, and associated woody and herbaceous broadleaf species infesting rangeland and permanent grass pastures.

DOSAGE RECOMMENDATIONS

Post Oak—Blackjack Oak—Winged Elm—Associated Species: For control of post oak, blackjack oak, winged elm, and certain associated species, apply a tank mix of 2 to 4 pints GRAZON PC Herbicide with 1 to 2 quarts of ESTERON 245 or an equivalent low-volatile 2,4,5-T ester formulation per acre as an oil-water emulsion spray. Use the lower rates only when oak-winged elm regrowth is small. Select sufficient application spray volume to insure uniform coverage, whether applied by air or ground equipment. Ground spray volumes are usually higher than those in the table. For best results, apply immediately prior to or at full oak leaf expansion, usually May 20 to June 20. Soil moisture must be adequate for good plant growth or results may be unsatisfactory.

Note: Do not graze areas treated with more than 2 pints GRAZON PC Herbicide per acre for 60 days from date of application. For areas treated with GRAZON PC Herbicide at 2 pints per acre or less, read Use Precautions for grazing restrictions.

Annual and Perennial Weed Control/Pricklypear: Apply 1 to 2 pints GRAZON PC Herbicide alone or in combination with 0.5 to 1 pint ESTERON 245 (or equivalent 2,4,5-T) per acre in water plus a drift control additive such as NALCO-TROL, or as an oil-water emulsion spray. Select sufficient application spray volume to insure uniform coverage, whether applied by air or ground equipment. Application for pricklypear control may be made throughout the year; however, where weed control is desired, applications should be made only when soil moisture is good and weed species are actively growing and succulent.

APPLICATION DIRECTIONS

Apply GRAZON PC Herbicide tank mixed with ESTERON 245 or equivalent 2,4,5-T herbicide at recommended rates in water alone or as an oil-water emulsion in a total volume of 1 to 5 gallons per acre by air or 10 to 25 gallons per acre by ground. To provide more complete wetting of the foliage, agriculture surfactants such as ORTHO X-77† may be used at recommended rates. The use of a drift control additive such as NALCO-TROL is strongly recommended for drift reduction and improved deposition.

† Trademark of Kalo Laboratories, Inc.

Aerial Application: Use 1 pint to 6 pints of oil and enough water to give 1 to 5 gallons of total spray per acre. The ratio of oil to water in the total spray mixture should be approximately 1:5 (1 part oil to 5 parts water). Refer to table for correct amounts of chemical, oil and water to use with the different total spray volumes per acre.

Total Spray Volume/A (Gallons)	Amount to Use Per Acre (Pints)			
	GRAZON PC	ESTERON 245	Oil[1]	Water[2]
1	1	0.5	1.0	Sufficient to bring to desired spray volume shown under "Total Spray Volume per Acre"
	2	1.0	0.8	
2	1	0.5	2.4	
	2	1.0	2.2	
	2	4.0	1.7	
	4	4.0	1.3	
3	1	0.5	3.8	
	2	1.0	3.5	
	2	4.0	3.0	
	4	4.0	2.7	
4	1	0.5	5.1	
	2	1.0	4.8	
	2	4.0	4.3	
	4	4.0	4.0	
5	1	0.5	6.4	
	2	1.0	6.2	
	2	4.0	5.7	
	4	4.0	5.3	

[1] Add emulsifier such as SPONTO 712 or TRITON X-100 at the rate of 1 oz. per gal. oil in total spray mix.

[2] Add drift control additive such as NALCO-TROL at the rate of 6 oz. per 100 gal. of water.

Note: Repeat applications may be made in successive years if more complete control is desired; however, do not apply more than a total or 4 pints of GRAZON PC Herbicide per acre during any 5-year period. Consult your Extension Weed Specialist for local recommendations. Be sure that use of this product conforms to all applicable State and Federal regulations regarding use of herbicides.

Ground Application—Broadcast Spraying: To control low-growing species and regrowth, apply recommended rates of GRAZON PC Herbicide alone or in combination with ESTERON 245 or equivalent 2,4,5-T ester herbicide in 10 to 25 gallons of water per acre. Add 1 to 2 quarts agricultural surfactant per 100 gallons of mixed spray to provide wetting of weed and brush foliage. **For Use With Water and Oil:** Add oil to the total spray mix at the ratio of 1 part oil to 5 parts water (15% to 20% of the total mix) using agricultural spray emulsifiers and mixing procedures given below for aerial application.

Ground Application—Spot Spraying: To control small patches of mesquite, oak sprouts, small individual juniper, pricklypear and other susceptible species, apply a spray containing 2 quarts GRAZON PC Herbicide plus 1 quart ESTERON 245 or equivalent 2,4,5-T ester herbicide plus 1–2 quarts agricultural surfactant per 100 gallons of water. Spray to thoroughly wet foliage, stems, and trunks.

Ground Application—Special Equipment: To control annual and perennial weed species and low regrowth brush using special low-volume, minimum-drift equipment, such as the hooded Forage Chemical Mower, apply 1 to 2 pints GRAZON PC Herbicide alone or in combination with ½ to 1 pint ESTERON 245 or equivalent 2,4,5-T ester herbicide in total volume ranging from 1 gallon to 5 gallons per acre in water alone or as an oil-water emulsion at a 1:5 and 1:4 oil-to-water ratio for 1-gallon and 5-gallon-per-acre solutions, respectively.

MIXING INSTRUCTIONS

GRAZON PC Herbicide in tank mixtures with 2,4,5-T low volatile ester herbicides may be used in water alone or in oil-water mixes. For oil-water mixes, use diesel fuel, fuel oil or kerosene. To ensure proper mixing and spraying maintain continuous, vigorous agitation during the entire mixing and spraying period. Regulate agitation and the addition of water to minimize foaming.

For Use With Water Alone (ground or aerial): Start with about half the required amount of water in the spray tank. With agitation, add the required amount of GRAZON PC Herbicide. After it is mixed in solution, add the required amount of 2,4,5-T low volatile ester with continued agitation. Add the remainder of the required amount of water to complete the spray tank mix. The use of a drift control additive such as NALCO-TROL is strongly recommended.

For Use With Water and Oil (Batch Mixing): With continuous vigorous agitation:

1. Add half the amount of water to be used to the spray tank.
2. Add the amount of GRAZON PC Herbicide required for the total volume of spray being mixed.
3. Add the amount of 2,4,5-T low volatile ester required for the total volume of spray being mixed.
4. Premix the required amount of oil with a good agricultural spray emulsifier such as SPONTO† 712 or TRITON‡ X-100 brands using 1 ounce of emulsifier per gallon of oil. Add the oil-emulsifier premix to the spray tank. Note: The use of agricultural surfactants in place of emulsifiers may not provide satisfactory emulsions.

† Trademark of The Witco Chemical Corporation

‡ Trademark of Rohm and Haas Co.

5. Finally, add the remaining amount of water required to bring the spray batch to the total desired volume.
6. Maintain agitation in the spray tank during application.

For Use With Water and Oil (Meter Mixing): Water-oil sprays containing GRAZON PC Herbicide in tank mixtures with 2,4,5-T low volatile ester herbicides may also be prepared using the spray tank "loading" pump to mix the ingredients while filling the tank. This procedure involves simultaneous metering of the required amounts of water and GRAZON PC Herbicides with an oil-emulsifier premix and 2,4,5-T low volatile ester herbicide mixtures into the intake line of the pump during operation. Since emulsification is accomplished in the pump, it is important that a pump with moderate to high shearing action be used. For proper results, always introduce the water and GRAZON PC Herbicide into the intake line before introducing the oil-emulsifier premix and 2,4,5-T low volatile ester herbicide, or a thick, invert emulsion may result. Maintain vigorous agitation in the spray tank during applications.

GRAZON PC Herbicide Plus 2,4,-D

For selective control of a wide range of broadleaf weeds on rangeland and permanent grass pastures, use GRAZON PC Herbicide in tank mixtures with 2,4-D low volatile esters such as ESTERON 99 Concentrate, or with 2,4-D amines such as FORMULA 40 herbicides. These tank mixtures are effective on many herbaceous species including bitterweed, broom snakeweed, goldenrod, horsenettle, locoweed, prickly lettuce, ragweeds (bur, common, and western), thistles and many other species susceptible to 2,4-D, as well as some woody species.

DOSAGE RECOMMENDATIONS

Annual Broadleaf Weeds such as annual broomweed, common ragweed and bitterweed: for **early season** (March–April), use 0.25 to 0.5 pint GRAZON PC Herbicide plus 0.5 to 1 pint ESTERON 99 Concentrate or FORMULA 40 herbicide in enough water or oil-water emulsion to give a total spray volume of 10 to 20 gallons per acre with ground equipment or 1 to 4 gallons per acre when using aerial application. Lower tank mix rates are effective only in very early season when weeds are no more than 2 inches tall and conditions are favorable for good plant growth. For **mid to late season** (May–June), use 0.5 to 1 pint GRAZON PC Herbicide plus 1 to 2 pints ESTERON 99 Concentrate or FORMULA 40 herbicide in enough water or oil-water emulsion to give a total spray volume of 10 to 20 gallons per acre with ground equipment or 1 to 4 gallons per acre when applied by aircraft.

Perennial Broadleaf Weeds such as broom snakeweed (turpentine weed): Treat when weeds are actively growing using 0.5 to 1 pint GRAZON PC Herbicide plus 1 to 2 pints ESTERON 99 Concentrate or FORMULA 40 herbicide in enough water or oil-water emulsion to give a total spray volume of 10 to 20 gallons per acre with ground equipment or 1 to 4 gallons per acre when applied by aircraft. The higher rate should be used when weeds are larger than 4 inches tall. Applications made during drought stress or other conditions preventing active growth may not provide acceptable control.

Macartney and Multiflora Rose: Individual plant treatment for control of large undisturbed clumps or smaller regrowth, treat with a hand-gun using a spray mix containing 1 to 2 quarts GRAZON PC Herbicide plus 2 to 4 quarts FORMULA 40 (containing 2,4-D amine) plus 2 quarts of commercial agricultural surfactant (such as ORTHO X-77) per 100 gallons spray solution. Coverage should be thorough to wet all leaves and stems using sufficient pressure to penetrate to the center of large clumps. The use of a drift control additive such as NALCO-TROL is recommended to reduce drift during application. **Broadcast ground applications,** where such equipment can operate, should provide a complete wetting spray coverage of Macartney or multiflora rose foliage. Apply 1 quart GRAZON PC Herbicide plus 2 quarts of ESTERON 99 Concentrate in enough water to give 20 to 25 gallons total spray volume per acre. Add a commercial agricultural surfactant at 0.5% (vol./vol.) which is 2 quarts per 100 gallons total spray mix. **Aerial application** of previously mowed or sprayed Macartney or multiflora rose may be done using 1 quart GRAZON PC Herbicide plus 2 quarts ESTERON 99 Concentate plus 0.5% agricultural surfactant in enough water or oil-water emulsion to give 4 gallons total volume per acre. Swath width should be such as to provide adequate overlap between swaths to facilitate good coverage. Increased deposition and drift control can be achieved with the addition of a drift control additive such as NALCO-TROL at recommended rates. **Note:** Macartney or multiflora rose may be treated in the fall or spring if conditions are favorable for good plant growth. Comply with all local regulations relative to application of hormone-type herbicides. Consult the University Extension Service for more information.

Poisonous Broadleaf Plants such as woolly loco, garbancillo (Wooten loco), and groundsel (*Senecio* spp.): Use 0.75 to 1 pint GRAZON PC Herbicide plus 1.5 to 2 pints ESTERON 99 Concentrate or FORMULA 40 herbicide in enough water, or **preferably** in enough oil-water emulsion to give a total spray volume of 10 to 20 gallons per acre with ground equipment or 1 to 4 gallons per acre when applied by aircraft. Since locoweeds are difficult to wet, add an agricultural surfactant per 100 gallons total spray mix.

APPLICATION DIRECTIONS

Apply GRAZON PC Herbicide tank mixed with ESTERON 99 Concentrate or FORMULA 40 or equivalent herbicide at recommended rates in water alone or as an oil-water emulsion in a total volume of 1 to 4 gallons per acre by air or 10 to 25 gallons per acre by ground. To provide more complete wetting of the foliage, agricultural surfactants such as ORTHO X-77 may be used at recommended rates. The use of a drift control additive such as NALCO-TROL is strongly recommended for drift reduction and improved deposition.

MIXING INSTRUCTIONS

Ground Application—For Use With Water Alone: Start with about half the required amount of water in the spray tank. With agitation operating, add the required amount of GRAZON PC Herbicide. Then add the required amount of ESTERON 99 Concentrate or FORMULA 40 or equivalent 2,4-D herbicide. If a surfactant is needed it should be added as the remainder of the required water is added to complete the spray mix. When using a drift control additive such as NALCO-TROL or equivalent, follow the manufacturer's directions carefully. Complete dispersion of the additive such as obtained by passage through a pump with moderate to high shearing action, is essential for proper performance of the product.

Ground Application—For Use With Water and Oil: Add oil to the total spray mix at the ratio of 1 part oil to 5 parts water (15% to 20% of the total mix) using agricultural spray emulsifiers and mixing procedures given below for aerial application.

Aerial Application—For Use With Water Alone: Follow procedures given for ground application above.

Aerial Application—For Use With Water and Oil (Batch Mixing): With continuous, vigorous agitation:

1. Add half the amount of water to be used to the spray tank.
2. Add the amount of GRAZON PC Herbicide required for the total volume of spray being mixed.
3. Add the amount of ESTERON 99 Concentrate or FORMULA 40 or equivalent herbicides required for the total spray volume being mixed.
4. Premix the required amount of oil with an emulsifier such as SPONTO 712 or TRITON X-100 brands using 1 ounce of emulsifier per gallon of oil. Add the oil-emulsifier premix to the spray tank.
5. Finally, add the remaining amount of water required to bring the spray batch to the desired total volume.
6. Maintain agitation in the spray tank during application.

Aerial Application—For Use With Water and Oil (Meter Mixing): Water-oil sprays containing tank mixes of GRAZON PC Herbicide plus ESTERON 99 Concentrate or FORMULA 40 herbicide may also be prepared using the spray tank "loading" pump to mix the ingredients while filling the tank. This procedure involves simultaneous metering of the required amounts of water and GRAZON PC Herbicide with an oil-emulsifier and 2,4-D into the intake line of the pump during operation. Since emulsification is accomplished in the pump, it is important that a pump with moderate to high shearing action be used. For proper results, always introduce the water and GRAZON PC Herbicide into the intake line before introducing the oil-emulsifier premix, or a thick, invert emulsion may result. Maintain vigorous agitation in the spray tank during application.

Note: Consult the following table for amounts of product, oil and water to mix 100 gallons of total spray as an oil-water emulsion to be applied by air at the rate of 1 to 4 gpa total volume per acre.

TOTAL SPRAY VOL./ ACRE (GAL.)	PINTS		GALLONS						
	GRAZON PC	ESTERON 99 CONCENTRATE OR FORMULA 40	GRAZON PC	ESTERON 99 CONCENTRATE OR FORMULA 40	OIL	WATER	†EMULSIFIER (OZ.)	‡N.T. (OZ.)	TOTAL SPRAY (GALLONS)
1	0.25	0.5	3.1	6.2	15.1	75.5	15.1	4.5	100
	0.5	1	6.2	12.5	13.5	67.7	13.5	4.1	100
	1	2	12.5	25.0	10.4	52.1	10.4	3.1	100
	2	4	25.0	50.0	4.2	20.8	4.2	1.2	100
2	0.25	0.5	1.6	3.1	15.9	79.4	15.9	4.8	100
	0.5	1	3.1	6.2	15.1	75.5	15.1	4.5	100
	1	2	6.2	12.5	13.5	67.7	13.5	4.1	100
	2	4	12.5	25.0	10.4	52.1	10.4	3.1	100
3	0.25	0.5	1.0	2.1	16.1	80.7	16.1	4.8	100
	0.5	1	2.1	4.2	15.6	78.1	15.6	4.7	100
	1	2	4.2	8.3	14.6	72.9	14.6	4.4	100
	2	4	8.3	16.7	12.5	62.5	12.5	3.0	100
4	0.25	0.5	0.8	1.6	16.2	81.4	16.2	4.9	100
	0.5	1	1.6	3.1	15.9	79.4	15.9	4.8	100
	1	2	3.1	6.2	15.1	75.5	15.1	4.5	100
	2	4	6.2	12.5	13.5	67.7	13.5	4.1	100

† An emulsifier such as SPONTO 712 or TRITON X-100 should always be added at the rate of 1 ounce per gallon of oil in the total spray mix.

‡ Amounts of Nalco-Trol (N.T.) are calculated on the basis of 6 oz. of N.T./100 gals. of water.

READ AND FOLLOW ALL USE PRECAUTIONS AND STORAGE AND DISPOSAL INSTRUCTIONS GIVEN ON PAGES 8-9

FOR DISTRIBUTION AND USE ONLY WITHIN OKLAHOMA

OKLAHOMA ....EPA SLN No. OK-820004

For the Control of Susceptible Broadleaf Weeds and Woody Plants on Rangeland and Permanent Grass Pastures

DIRECTIONS FOR USE

Do not use for manufacturing or formulating.

It is a violation of Federal law to use this product in a manner inconsistent with its labeling.

Read complete use directions and precautions on this label and container label before using.

Mix the required amount of GRAZON PC Herbicide in sufficient water for uniform coverage and apply as a coarse, low-pressure spray (under 25 psi) with ground or aerial application equipment. The use of a drift control additive such as NALCO-TROL† is strongly recommended.

† Trademark of The Nalco Chemical Company

For the control of most susceptible broadleaf weeds such as annual and perennial broomweed (broom snakeweed), ragweed (bur, common, and lanceleaf), bursage (bur ragweed), locoweed, and pricklypear. Apply GRAZON PC Herbicide broadcast at the rate of 1 to 2 pints per acre or as a spot treatment at the rate of $\frac{1}{4}$ to $\frac{1}{2}$ teaspoon per 100 sq. ft. Individual plant treatments may be made using a 0.5 to 1% solution (2 to 4 qt. GRAZON PC Herbicide per 100 gal water) applied as a thoroughly wetting spray.

For best results, treat when the weeds are growing actively in the spring before full bloom or in late summer or fall if soil moisture is adequate. Treatments during full bloom or seed stage of some weeds may not give good control.

Consult your range or weed extension specialist for additional information.

TANK MIXTURES

GRAZON PC Herbicide plus 2,4,5-T Low Volatile Ester

Use GRAZON PC Herbicide in tank mixtures with 2,4,5-T low volatile esters such as ESTERON 245 herbicide, for control of post oak, blackjack oak, winged elm, pricklypear, and associated woody and herbaceous broadleaf species infesting rangeland and permanent grass pastures.

DOSAGE RECOMMENDATIONS

Post Oak-Blackjack Oak-Winged Elm-Associated Species: For control of post oak, blackjack oak, winged elm, and certain associated species, apply a tank mix of 2 to 4 pints GRAZON PC Herbicide with 1 to 2 quarts of ESTERON 245 or an equivalent low-volatile 2,4,5-T ester formulation per acre as an oil-water emulsion spray. Use the lower rates only when oak-winged elm regrowth is small. Select sufficient application spray volume to ensure uniform coverage, whether applied by air or ground equipment. Ground spray volumes are usually higher than those in the table. For best results, apply immediately prior to or at full oak leaf expansion, usually May 20 to June 20. Soil moisture must be adequate for good plant growth or results may be unsatisfactory.

Note: Do not graze areas treated with more than 2 pints GRAZON PC Herbicide per acre for 60 days from date of application. For areas treated with GRAZON PC Herbicide at 2 pints per acre or less, read Use Precautions for grazing restrictions.

Annual and Perennial Weed Control/Pricklypear: Apply 1 to 2 pints GRAZON PC Herbicide alone or in combination with 0.5 to 1 pint ESTERON 245 (or equivalent 2,4,5-T) per acre in water plus a drift control additive such as NALCO-TROL, or as an oil-water emulsion spray. Select sufficient application spray volume to ensure uniform coverage, whether applied by air or ground equipment. Application for pricklypear control may be made throughout the year; however, where weed control is desired, applications should be made only when soil moisture is good and weed species are actively growing and succulent.

APPLICATION DIRECTIONS

Apply GRAZON PC Herbicide tank mixed with ESTERON 245 or equivalent 2,4,5-T herbicide at recommended rates in water alone or as an oil-water emulsion in a total volume of 1 to 5 gallons per acre by air or 10 to 25 gallons per acre by ground. To provide more complete wetting of the foliage, agriculture surfactants such as ORTHO X-77† may be used at recommended rates. The use of a drift control additive such as NALCO-TROL is strongly recommended for drift reduction and improved deposition.

† Trademark of Kalo Laboratories, Inc.

Aerial Application: Use 1 pint to 6 pints of oil and enough water to give 1 to 5 gallons of total spray per acre. The ratio of oil to water in the total spray mixture should be approximately 1:5 (1 part oil to 5 parts water). Refer to table for correct amounts of chemical, oil and water to use with the different total spray volumes per acre.

Total Spray Volume/A (gallons)	Amount to Use Per Acre (Pints)			
	GRAZON PC	ESTERON 245	Oil[1]	Water[2]
1	1	0.5	1.0	Sufficient to bring to desired spray volume shown under "Total Spray Volume per Acre"
	2	1.0	0.8	
2	1	0.5	2.4	
	2	1.0	2.2	
	2	4.0	1.7	
	4	4.0	1.3	
3	1	0.5	3.8	
	2	1.0	3.5	
	2	4.0	3.0	
	4	4.0	2.7	
4	1	0.5	5.1	
	2	1.0	4.8	
	2	4.0	4.3	
	4	4.0	4.0	
5	1	0.5	6.4	
	2	1.0	6.2	
	2	4.0	5.7	
	4	4.0	5.3	

[1] Add emulsifier such as SPONTO 712 or TRITON X-100 at the rate of 1 oz. per gal. oil in total spray mix.

[2] Add drift control additive such as NALCO-TROL at the rate of 6 oz. per 100 gal. of water.

Note: Repeat applications may be made in successive years if more complete control is desired; however, do not apply more than a total of 4 pints of GRAZON PC Herbicide per acre during any 5-year period. Consult your Extension Weed Specialist for local recommendations. Be sure that use of this product conforms to all applicable State and Federal regulations regarding use of herbicides.

Ground Application—Broadcast Spraying: To control low-growing species and regrowth, apply recommended rates of GRAZON PC Herbicide alone or in combination with ESTERON 245 or equivalent 2,4,5-T ester herbicide in 10 to 25 gallons of water per acre. Add 1 to 2 quarts agricultural surfactant per 100 gallons of mixed spray to provide wetting of weed and brush foliage. **For Use With Water and Oil:** Add oil to the total spray mix at the ratio of 1 part oil to 5 parts water (15% to 20% of the total mix) using agricultural spray emulsifiers and mixing procedures given below for aerial application.

Ground Application—Spot Spraying: To control small patches of mesquite, oak sprouts, small individual juniper, pricklypear and other susceptible species, apply a spray containing 2 quarts GRAZON PC Herbicide plus 1 quart ESTERON 245 or equivalent 2,4,5-T ester herbicide plus 1-2 quarts agricultural surfactant per 100 gallons of water. Spray to thoroughly wet foliage, stems, and trunks.

Ground Application—Special Equipment: To control annual and perennial weed species and low regrowth brush using special low-volume, minimum-drift equipment, such as the hooded Forage Chemical Mower, apply 1 to 2 pints GRAZON PC Herbicide alone or in combination with ½ to 1 pint ESTERON 245 or equivalent 2,4,5-T ester herbicide in total volume ranging from 1 gallon to 5 gallons per acre in water alone or as an oil-water emulsion at a 1:5 and 1:4 oil-to-water ratio for 1-gallon and 5-gallon-per-acre solutions, respectively.

MIXING INSTRUCTIONS

GRAZON PC Herbicide in tank mixtures with 2,4,5-T low volatile ester herbicides may be used in water alone or in oil-water mixes. For oil-water mixes, use diesel fuel, fuel oil or kerosene. To ensure proper mixing and spraying maintain continuous, vigorous agitation during the entire mixing and spraying period. Regulate agitation and the addition of water to minimize foaming.

For Use With Water Alone (Ground or Aerial): Start with about half the required amount of water in the spray tank. With agitation, add the required amount of GRAZON PC Herbicide. After it is mixed in solution, add the required amount of 2,4,5-T low volatile ester with continued agitation. Add the remainder of the required amount of water to complete the spray tank mix. The use of a drift control additive such as NALCO-TROL is strongly recommended.

For Use With Water and Oil (Batch Mixing): With continuous vigorous agitation:

1. Add half the amount of water to be used to the spray tank.
2. Add the amount of GRAZON PC Herbicide required for the total volume of spray being mixed.
3. Add the amount of 2,4,5-T low volatile ester required for the total volume of spray being mixed.
4. Premix the required amount of oil with a good agricultural spray emulsifier such as SPONTO† 712 or TRITON†† X-100 brands using 1 ounce of emulsifier per gallon of oil. Add the oil-emulsifier premix to the spray tank. Note: The use of agricultural surfactants in place of emulsifiers may not provide satisfactory emulsions.

† Trademark of The Witco Chemical Corporation

†† Trademark of Rohm & Haas Co.

5. Finally, add the remaining amount of water required to bring the spray batch to the total desired volume.
6. Maintain agitation in the spray tank during application.

For Use With Water and Oil (Meter Mixing): Water-oil sprays containing GRAZON PC Herbicide in tank mixtures with 2,4,5-T low volatile ester herbicides may also be prepared using the spray tank "loading" pump to mix the ingredients while filling the tank. This procedure involves simultaneous metering of the required amounts of water and GRAZON PC Herbicides with an oil-emulsifier premix and 2,4,5-T low volatile ester herbicide mixtures into the intake line of the pump during operation. Since emulsification is accomplished in the pump, it is important that a pump with moderate to high shearing action be used. For proper results, always introduce the water and GRAZON PC Herbicide into the intake line before introducing the oil-emulsifier premix and 2,4,5-T low volatile ester herbicide, or a thick, invert emulsion may result. Maintain vigorous agitation in the spray tank during applications.

GRAZON PC HERBICIDE PLUS 2,4-D

For selective control of a wide range of broadleaf weeds on rangeland and permanent grass pastures, use GRAZON PC Herbicide in tank mixtures with 2,4-D low volatile esters such as ESTERON 99 Concentrate, or with 2,4-D amines such as FORMULA 40 herbicides. These tank mixtures are effective on many herbaceous species including bitterweed, broom snakeweed, goldenrod, horsenettle, locoweed, prickly lettuce, ragweeds (bur, common, and western), thistles and many other species susceptible to 2,4-D, as well as some woody species.

DOSAGE RECOMMENDATIONS

Annual Broadleaf Weeds such as annual broomweed, common ragweed and bitterweed: For early season (March-April), use 0.25 to 0.5 pint GRAZON PC Herbicide plus 0.5 to 1 pint ESTERON 99 Concentrate or FORMULA 40 herbicide in enough water or oil-water emulsion to give a total spray volume of 10 to 20 gallons per acre with ground equipment or 1 to 4 gallons per acre when using aerial application. Lower tank mix rates are effective only in very early season when weeds are no more than 2 inches tall and conditions are favorable for good plant growth. For **mid to late season** (May-June), use 0.5 to 1 pint GRAZON PC Herbicide plus 1 to 2 pints ESTERON 99 Concentrate or FORMULA 40 herbicide in enough water or oil-water emulsion to give a total spray volume of 10 to 20 gallons per acre with ground equipment or 1 to 4 gallons per acre when applied by aircraft.

Perennial Broadleaf Weeds such as broom snakeweed (turpentine weed): Treat when weeds are actively growing using 0.5 to 1 pint GRAZON PC Herbicide plus 1 to 2 pints ESTERON 99 Concentrate or FORMULA 40 herbicide in enough water or oil-water emulsion to give a total spray volume of 10 to 20 gallons per acre with ground equipment or 1 to 4 gallons per acre when applied by aircraft. The higher rate should be used when weeds are larger than 4 inches tall. Applications made during drought stress or other conditions preventing active growth may not provide acceptable control.

Macartney and Multiflora Rose: Individual plant treatment for control of large undisturbed clumps or smaller regrowth, treat with a hand-gun using a spray mix containing 1 to 2 quarts GRAZON PC Herbicide plus 2 to 4 quarts FORMULA 40 (containing 2,4-D amine) plus 2 quarts of commercial agricultural surfactant (such as ORTHO X-77) per 100 gallons spray solution. Coverage should be thorough to wet all leaves and stems using sufficient pressure to penetrate to the center of large clumps. The use of a drift control additive such as NALCO-TROL is recommended to reduce drift during application. **Broadcast ground applications,** where such equipment can operate, should provide a complete wetting spray coverage of Macartney or multiflora rose foliage. Apply 1 quart GRAZON PC Herbicide plus 2 quarts of ESTERON 99 Concentrate in enough water to give 20 to 25 gallons total spray volume per acre. Add a commercial agricultural surfactant at 0.5% (vol./vol.) which is 2 quarts per 100 gallons total spray mix.

Aerial application of previously mowed or sprayed Macartney or multiflora rose may be done using 1 quart GRAZON PC Herbicide plus 2 quarts ESTERON 99 Concentrate plus 0.5% agricultural surfactant in enough water or oil-water emulsion to give 4 gallons total volume per acre. Swath width should be such as to provide adequate overlap between swaths to facilitate good coverage. Increased deposition and drift control can be achieved with the addition of a drift control additive such as NALCO-TROL at recommended rates. **Note:** Macartney or multiflora rose may be treated in the fall or spring if conditions are favorable for good plant growth. Comply with all local regulations relative to application of hormone-type herbicides. Consult the University Extension Service for more information.

Poisonous Broadleaf Plants such as woolly loco, garbancillo (Wooton loco), and groundsel (*Senecio* spp.): Use 0.75 to 1 pint GRAZON PC Herbicide plus 1.5 to 2 pints ESTERON 99 Concentrate or FORMULA 40 herbicide in enough water, or **preferably** in enough oil-water emulsion to give a total spray volume of 10 to 20 gallons per acre with ground equipment or 1 to 4 gallons per acre when applied by aircraft. Since locoweeds are difficult to wet, add an agricultural surfactant per 100 gallons total spray mix.

APPLICATION DIRECTIONS

Apply GRAZON PC Herbicide tank mixed with ESTERON 99 Concentrate or FORMULA 40 or equivalent herbicide at recommended rates in water alone or as an oil-water emulsion in a total volume of 1 to 4 gallons per acre by air or 10 to 25 gallons per acre by ground. To provide more complete wetting of the foliage, an agricultural surfactant such as ORTHO X-77 may be used at recommended rates. The use of a drift control additive such as NALCO-TROL is strongly recommended for drift reduction and improved deposition.

MIXING INSTRUCTIONS

Ground Application—For Use With Water Alone: Start with about half the required amount of water in the spray tank. With agitation operating, add the required amount of GRAZON PC Herbicide. Then add the required amount of ESTERON 99 Concentrate or FORMULA 40 or equivalent 2,4-D herbicide. If a surfactant is needed, it should be added as the remainder of the required water is added to complete the spray mix. When using a drift control additive such as NALCO-TROL or equivalent, follow the manufacturer's directions carefully. Complete dispersion of the additive such as obtained by passage through a pump with moderate to high shearing action is essential for proper performance of the product.

Ground Application—For Use With Water and Oil: Add oil to the total spray mix at the ratio of 1 part to 5 parts water (15% to 20% of the total mix) using agricultural spray emulsifiers and mixing procedures given below for aerial application.

Aerial Application—For Use With Water Alone: Follow procedures given for ground application above.

Aerial Application—For Use With Water and Oil (Batch Mixing): With continuous, vigorous agitation:

1. Add half the amount of water to be used to the spray tank.
2. Add the amount of GRAZON PC Herbicide required for the total volume of spray being mixed.
3. Add the amount of ESTERON 99 Concentrate or FORMULA 40 or equivalent herbicides required for the total spray volume being mixed.
4. Premix the required amount of oil with an emulsifier such as SPONTO 712 or TRITON X-100 brands using 1 ounce of emulsifier per gallon of oil. Add the oil-emulsifier premix to the spray tank.
5. Finally, add the remaining amount of water required to bring the spray batch to the desired total volume.
6. Maintain agitation in the spray tank during application.

Aerial Application—For Use With Water and Oil (Meter Mixing): Water-oil sprays containing tank mixes of GRAZON PC Herbicide plus ESTERON 99 Concentrate or FORMULA 40 herbicide may also be prepared using the spray tank "loading" pump to mix the ingredients while filling the tank. This procedure involves simultaneous metering of the required amounts of water-GRAZON PC Herbicide and oil 2,4-D emulsifier mixtures into the intake line of the pump during operation. Since emulsification is accomplished in the pump, it is important that a pump with moderate to high shearing action be used. For proper results, always introduce the water and GRAZON PC Herbicide premix into the intake line before introducing the oil and 2,4-D emulsifier, or a thick, invert emulsion may result. Maintain agitation in the spray tank during application.

Note: Consult the following table for amounts of product, oil and water to mix 100 gallons of total spray to be applied by air at the rate of 1 to 4 gpa total volume per acre.

TOTAL SPRAY VOL./ ACRE (GAL.)	PINTS		GALLONS				†EMULSIFIER (OZ.)	††N.T. (OZ.)	TOTAL SPRAY (GALLONS)
	GRAZON PC	ESTERON 99 CONCENTRATE OR FORMULA 40	GRAZON PC	ESTERON 99 CONCENTRATE OR FORMULA 40	OIL	WATER			
1	0.25	0.5	3.1	6.2	15.1	75.5	15.1	4.5	100
	0.5	1	6.2	12.5	13.5	67.7	13.5	4.1	100
	1	2	12.5	25.0	10.4	52.1	10.4	3.1	100
	2	4	25.0	50.0	4.2	20.8	4.2	1.2	100
2	0.25	0.5	1.6	3.1	15.9	79.4	15.9	4.8	100
	0.5	1	3.1	6.2	15.1	75.5	15.1	4.5	100
	1	2	6.2	12.5	13.5	67.7	13.5	4.1	100
	2	4	12.5	25.0	10.4	52.1	10.4	3.1	100
3	0.25	0.5	1.0	2.1	16.1	80.7	16.1	4.8	100
	0.5	1	2.1	4.2	15.6	78.1	15.6	4.7	100
	1	2	4.2	8.3	14.6	72.9	14.6	4.4	100
	2	4	8.3	16.7	12.5	62.5	12.5	3.0	100
4	0.25	0.5	0.8	1.6	16.2	81.4	16.2	4.9	100
	0.5	1	1.6	3.1	15.9	79.4	15.9	4.8	100
	1	2	3.1	6.2	15.1	75.5	15.1	4.5	100
	2	4	6.2	12.5	13.5	67.7	13.5	4.1	100

† An emulsifier such as SPONTO 712 or TRITON X-100 should always be added at the rate of 1 ounce per gallon of oil in the total spray mix.

†† Amounts of NALCO-TROL (N.T.) are calculated on the basis of 6 oz. of N.T./100 gals. of water.

READ AND FOLLOW ALL USE PRECAUTIONS AND STORAGE AND DISPOSAL INSTRUCTIONS GIVEN BELOW

USE PRECAUTIONS

Use this product only as specified on this label. Observe any special use and application restrictions and limitations, including method of application and permissible areas of use as promulgated by state or local authorities. Observe labels of other products used in tank mixtures and follow precautions and recommendations thoroughly.

Do Not Contaminate Nontarget Land Areas, Cropland, Water, or Irrigation Ditches. Do not contaminate water to be used for irrigation or domestic purposes. Do not apply where surface water from treated areas can run off to adjacent cropland, either planted or to be planted, or into streams, irrigation ditches, irrigation ponds, or wells. Do not clean containers nor application equipment on or near these areas. Do not apply to frozen or saturated ground.

Do not apply on or in the vicinity of susceptible crops or desirable plants including alfalfa, beans, cotton, melons, potatoes, sugar beets, sunflower, tobacco, tomatoes and other vegetable crops, flowers, fruit plants, ornamentals or shade trees.

Avoid Spray Drift. Applications should be made only when there is no hazard from spray drift since very small quantities of the spray, which may not be visible, may injure susceptible crops during both growing and dormant periods. Use coarse sprays to minimize drifts. A drift control additive, such as NALCO-TROL, may be used with this product to aid in reducing spray drift. If used, follow all use recommendations and precautions on the product label. Appliation with mist-type equipment is not recommended.

Ground Equipment: With ground equipment, spray drift can be lessened by keeping the spray boom as low as possible; by applying 20 gallons or more of spray per acre; by using no more than 25 pounds spraying pressure with large droplet producing nozzle tips; by spraying when wind velocity is low; and, by stopping all spraying when wind exceeds 10 miles per hour. Do not apply with hollow cone-type insecticide or other nozzles that produce a fine-droplet spray.

Aerial Application: With arcraft, drift can be lessened by applying a coarse spray; by using no more than 25 pounds spray pressure at nozzles; by using nozzles which do not produce fine spray droplets; by using a spray boom no longer than 3/4 the wing span of the aircraft; and, by spraying only when wind velocity is less than 10 mph.

Determine air movement and directions before foliar application: Use a smoke generator or other means at or near the application site for the detection of air movement, air stability or temperature inversions. Use appropriate drift control measures or avoid application when smoke is moving toward nearby desirable susceptible plants or sensitive areas. Avoid application when smoke forms layers or otherwise indicates inversion conditions which would prevent spray droplets from dispersing or falling to the intended spray site.

Do not rotate treated rangeland or pasture to other crop uses, as long as residues in the soil may affect susceptible crops.

Do not spray pastures if the forage legume component is desired. GRAZON PC Herbicide may injure or kill legumes. Also new legume seedings may not be successful if made within 2 years following application of this herbicide.

Do not move treated soil to other areas. Do not use it to grow plants, unless an adequate sensitive bioassay shows that no detectable picloram is present in the soil.

Do not transfer livestock from treated grass areas onto broadleaf crop areas without first allowing 7 days of grazing on untreated grass pasture. Otherwise, urine may contain enough picloram to cause injury to sensitive broadleaf plants.

Be sure that use of this product conforms to all applicable regulations.

Be Sure To Follow All Use Precautions Given on This Label And Remember These Key Points:

1. Use only the recommended amounts of GRAZON PC Herbicide at the recommended times. 2. Picloram remains active in the soil for extended periods. Treated soil should not be moved out of treated areas. 3. The potassium salt of picloram is water soluble. It can move with water in surface runoff, or in irrigation and drainage ditches. 4. Spray drift can cause injury to certain crops, ornamentals, trees, and numerous other desirable plants.

STORAGE AND DISPOSAL

Do not contaminate water, food or feed by storage or disposal.

PESTICIDE DISPOSAL

Pesticide, spray mixture or rinsate that cannot be used according to label instructions must be disposed of according to Federal, State, or local procedures under the Resource Conservation and Recovery Act.

CONTAINER DISPOSAL

Do not reuse containers. Puncture and dispose of them in a sanitary landfill after triple rinsing or equivalent, or by other approved state and local procedures. Plastic containers may also be disposed of by incinerator or, if allowed by state and local authorities, by burning. If burned, stay out of smoke.

WARRANTY LIMITATIONS AND DISCLAIMER

The Dow Chemical Company warrants that this product conforms to the chemical description on the label and is reasonably fit for the purposes stated on the label when used in strict accordance with the directions therein under normal conditions of use. THIS IS THE ONLY WARRANTY MADE ON THIS PRODUCT. NO OTHER EXPRESS AND NO IMPLIED WARRANTY OF MERCHANTABILITY OR FITNESS FOR A PARTICULAR PURPOSE IS MADE OUTSIDE OF THIS LABEL. Therefore, neither this warranty nor any other warranty of merchantability or fitness for a particular purpose, express or implied, extends to the use of this product contrary to label instructions (including conditions noted on the label, such as unfavorable temperatures, soil conditions, etc.), under abnormal conditions (such as excessive rainfall, drought, tornadoes, hurricanes, etc.) or under conditions not reasonably foreseeable to or beyond the control of seller.

When buyer or user suffers losses or damages resulting from the use or handling of this product (including claims based on contract, negligence, strict liability, or other legal theories), buyer or user must promptly notify in writing The Dow Chemical Company of any claims to be eligible to receive either remedy given below. The EXCLUSIVE REMEDY OF THE BUYER OR USER and the LIMIT OF LIABILITY of The Dow Chemical Company or any other seller will be one of the following, at the election of The Dow Chemical Company:

(1) **Refund of purchase price** paid by buyer or user for product bought, **or**

(2) **Replacement** of amount of product used.

The seller **will not be liable for consequential or incidental damages or losses.**

The terms of this Limit of Warranty and Liability cannot be varied by any written or verbal statements or agreements. Any employee or sales agent of the seller is not authorized to vary or exceed the terms of this Warranty Limitations And Disclaimer in any manner.

* Trademark of THE DOW CHEMICAL COMPANY

39173-L5 A1184

GRAZON* 10K

PELLETS HERBICIDE

RESTRICTED USE PESTICIDE

For retail sale to and use only by Certified Applicators or persons under their direct supervision and only for those uses covered by the Certified Applicator's certification.

For the Control of Unwanted Woody Plants and Various Broadleaf Weeds on Permanent Grass Pastures and Rangeland

ACTIVE INGREDIENT:
Picloram
(4-amino-3,5,6-trichloropicolinic acid)
potassium salt 11.6%
INERT INGREDIENTS 88.4%
TOTAL 100.0%
Picloram acid equivalent 10.0%

EPA Reg. No. 464-320 EPA Est. 464-MI-1

CAUTION

Keep Out of Reach of Children

AVISO

PRECAUCION AL USUARIO: Si usted no lee inglés, no use este producto hasta que la etiqueta le haya sido explicada ampliamente.

PRECAUTIONARY STATEMENTS

ENVIRONMENTAL HAZARDS

Keep out of lakes, ponds, and streams. Do not contaminate water by cleaning of equipment or disposal of wastes.

In case of an emergency endangering life or property involving this product, call collect **517–636-4400**

AGRICULTURAL CHEMICAL

Do Not Ship or Store with Food, Feeds, Drugs or Clothing

NOTICE

Read the entire label. Use only according to label directions.

Before buying or using this product, read "WARRANTY LIMITATIONS AND DISCLAIMER" elsewhere on this label. If terms are not acceptable, return **unopened** package at once to seller for full refund of purchase price paid. Otherwise, use by the buyer or any other user constitutes acceptance of the terms under the Limit of Warranty and Liability.

GRAZON 10K Pellets, a highly effective herbicide for control of many undesirable broadleaf weeds and woody plants, is designed to be applied to the soil for the control of undesirable woody plants. It is intended for use on permanent grass pastures, rangeland, and fencerows. GRAZON 10K Pellets may be applied as spot or broadcast treatments. The application method and equipment used should assure uniform distribution of the recommended dosage over the treated area. Rapid response should not be expected. Maximum effects of treatment with GRAZON 10K Pellets do not become apparent until after the chemical has been carried by moisture into the root zones of the woody plants. GRAZON 10K Pellets are noncorrosive to application equipment, non-volatile, and non-flammable.

USE DIRECTIONS

Do not use for manufacturing or formulating.

It is a violation of Federal law to use this product in a manner inconsistent with its labeling.

READ AND FOLLOW ALL USE PRECAUTIONS ON THIS LABEL.

FOR PERMANENT GRASS PASTURE AND RANGELAND, use only in Oklahoma, Texas and New Mexico.

GRAZON 10K Pellets may be applied at anytime soil is not frozen. However, best results are obtained from application in the spring or fall or during periods of vigorous growth when subsequent rainfall can be expected. Treatment during hot, dry periods is not advised. Distribute GRAZON 10K Pellets uniformly by spot or broadcast treatment to the soil over the roots of woody plants to be controlled.

Use GRAZON 10K Pellets to control undesirable broadleaf weeds and woody plants applied at rates indicated in the table of Application Rates. The maximum use rate is not to exceed 40 pounds per acre (1 pound product per 1000 square feet or 1½ ounces per 100 square feet). Treatments will not be effective until rain occurs after application. Some spot retreatment the following year or later may be needed.

For best timing of applications, consult your Experiment Station or Extension Range Management Specialists.

GRAZON 10K Pellets, especially at the higher rates, may injure or suppress certain desirable grasses such as smooth bromegrass, western wheatgrass, and switchgrass, therefore, do not apply where possible injury to such grasses cannot be tolerated. However, injury to established grass usually is followed by increased growth and development due to release from brush competition. Grass seedlings may be suppressed or killed in treated areas for up to 2 years after application. Reseeding during this period is not recommended.

Application Rates

	Broadcast treatment		Spot treatment
	Pounds product per acre		Ounces of product per 1000 sq. ft.
Species	Aerial	Ground	
Broom snakeweed (Perennial broomweed, Turpentine weed)	5–10	10[1]	3.75
Pricklypear, Tasajillo, and Cholla cactus	5–10	5–10[1]	3.75
Whitebrush reduction	10–15	10–15	5.5
Sand shinnery oak suppression	10–20	10–20	7.5

Species	Broadcast treatment: Pounds product per acre, Aerial	Broadcast treatment: Pounds product per acre, Ground	Spot treatment: Ounces of product per 1000 sq. ft.
Junipers (cedars)	20	20	7.5
South Texas mixed brush Huisache and shallow-rooted woody plants, e.g., Blackbrush, Granjeno, Guajillo, Large pricklypear (Mesquite not controlled)	25–30	25–30	11
Macartney rose	30–40[2]	30–40[2]	16
Multiflora rose	40	40	16
Ceanothus species	40	40	16
Shrub live oak, Gambel oak	40	40	16
Locust	40	40	16
Sumac	40	40	16

[1] Difficult to distribute less than 10 lb. uniformly by ground equipment.
[2] Use higher rate on heavy clay soils.

Broom snakeweed (Perennial broomweed, Turpentine weed): Apply GRAZON 10K Pellets Herbicide during or prior to periods of rainfall or when moisture in the form of snow is sufficient to move the chemical into the soil profile. Do not apply to frozen soils. Apply liquid formulations in early spring (following winter snow or rainfall) or in fall following normal rainfall.

Pricklypear, Tasajillo, and Cholla Cactus: Use the higher rate of application for the control of large picklypear species (nopal) in South Texas. Control of catcus species is slow and may require two years or longer for maximum response. Uniform product distribution is essential. The higher application rate may be required to obtain adequate coverage, especially when applying with ground equipment.

Juniper (Alligator, Redberry, Blueberry and other Junipers); Eastern Red Cedar and Piñion Pine: For individual trees, use 2 tsp. product per 5 square feet of area within the dripline. Apply out to dripline on level ground and on uphill side of plants growing on slopes. Results with Blueberry juniper over 3 feet tall are variable. Utah, Western and One-seed juniper and Piñion pine can be controlled when under 12 feet tall.

South Texas Mixed Brush: GRAZON 10K Pellets is recommended where mesquite is not the major problem, but mixed brush species and prickly pear control is desired. Whitebrush can be reduced with product rates of 10 to 15 pounds per acre. Improved control can be obtained at 25 to 30 pounds product per acre. GRAZON 10K Pellets is not effective for mesquite control.

Macartney Rose and Multiflora Rose: Uniform product distribution over root zone is required. For Macartney Rose, 30 pounds product per acre is sufficient except for heavy clay soils where the 40-pound rate should be used. Consult your Agricultural Extension Service for local recommendations.

RESTRICTIONS FOR PASTURF AND RANGELAND USE

Limit coverage to no greater than 25% of an applicator's acreage, found in any particular watershed.

Do not use where a sandy porous surface and substrate overlie ground water 10 feet or less below the surface.

Where watersheds have significant slope and where rapid runoff can occur, **use spot treatment only.**

Kill or injury may occur to desirable forbs, trees or shrubs such as blackberry, cherry, dogwood, locust, maple, oak, poplar, sycamore, and sumac, from root uptake. If such effects cannot be tolerated, do not apply on or near such desirable plants.

Do not apply to cropland used for production of desirable crops other than forage grasses. Do not rotate treated rangeland or pastures to other crops. Do not treat pastures or rangeland if the forage legume component is substantial and desired.

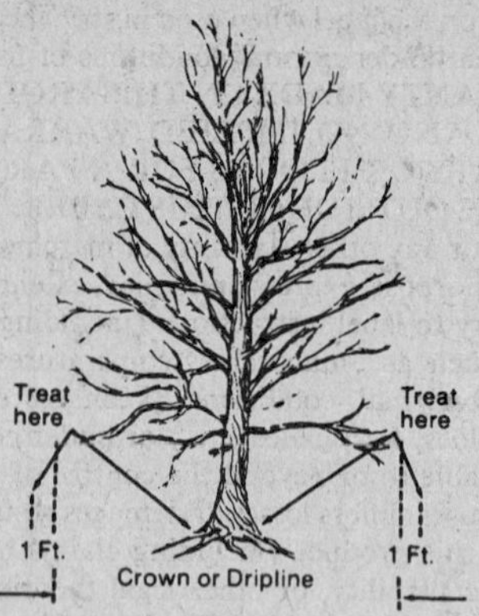

USE PRECAUTIONS

Apply this product only as specified on this label.

The active ingredient in GRAZON 10K Pellets is water soluble and should not be applied where surface water from treated areas is likely run off to crop lands either planted or to be planted.

Avoid use near desirable plants: This herbicide is water soluble, highly active and can remain in the soil for more than one growing season. Very small amounts can injure broadleaf plants, especially **tobacco and potatoes.** Therefore, do not apply on or near tobacco, potatoes, beans, cotton, alfalfa, peas, peanuts, blueberries, melons or other fruit plants, ornamentals, shade trees, soybeans, or other vegetable crops. Injurious amounts of picloram can be absorbed by roots of trees invading treated areas.

Avoid movement of treated soil: Picloram may remain in treated soil for an extended period. Do not move treated soil to other areas and do not use such soil to grow plants unless adequate sensitive bioassay or chemical tests show that no detectable picloram residues are present in the soils.

Avoid transfer of livestock from a treated area to a broadleaf crop area without first allowing 7 days of grazing on untreated pasture. Otherwise, urine may contain enough picloram to cause injury to sensitive broadleaf plants. Do not use manure from animals grazing treated areas to fertilize soil or fields used to grow susceptible broadleaf crops.

Avoid water contamination: Do not allow GRAZON 10K Pellets to contaminate water used for drinking, irrigation, or other domestic purposes. Do not apply on or in irrigation ditches. Do not clean containers or application equipment on or near these areas.

Do Not Mix or Blend with Fertilizers

STORAGE AND DISPOSAL

Do not contaminate water, food or feed by storage or disposal.

Avoid improper storage and equipment use: Do not store near fertilizers, seeds, insecticides, fungicides or other pesticides. Containers and equipment used for GRAZON 10K Pellets should not be used for other agricultural chemicals since small residues of GRAZON 10K Pellets herbicide can damage desirable plants.

Avoid improper disposal:

CONTENTS DISPOSAL

Wastes resulting from the use of this product may be disposed of on site or at an approved waste disposal facility. Rinse equipment and dispose of waste by burying in non-cropland away from water supplies.

CONTAINER DISPOSAL

Do not reuse containers: Puncture and dispose of in a sanitary landfill, or incineration, or, if allowed by state and local authorities, by burning. If burned, stay out of smoke.

Be sure that use of this product conforms to all applicable regulations.

WARRANTY LIMITATIONS AND DISCLAIMER

The Dow Chemical Company warrants that this product conforms to the chemical description on the label and is reasonably fit for the purposes stated on the label when used in strict accordance with the directions therein under normal conditions of use. THIS IS THE ONLY WARRANTY MADE ON THIS PRODUCT. NO OTHER EXPRESS AND NO IMPLIED WARRANTY OF MERCHANTABILITY OR FITNESS FOR A PARTICULAR PURPOSE IS MADE OUTSIDE OF THIS LABEL. Therefore, neither this warranty nor any other warranty of merchantability or fitness for a particular purpose, express or implied, extends to the use of this product contrary to label instructions (including conditions noted on the label, such as unfavorable temperatures, soil conditions, etc.), under abnormal conditions (such as excessive rainfall, drought, tornadoes, hurricanes, etc.) or under conditions not reasonably foreseeable to or beyond the control of seller.

When buyer or user suffers losses or damages resulting from the use or handling of this product (including claims based on contract, negligence, strict liability, or other legal theories), buyer or user must promptly notify in writing The Dow Chemical Company of any claims to be eligible to receive either remedy given below. THE EXCLUSIVE REMEDY OF THE BUYER OR USER and the LIMIT OF LIABILITY of The Dow Chemical Company or any other seller will be one of the following, at the election of The Dow Chemical Company:

(1) **Refund of purchase price** paid by buyer or user for product bought, **or**

(2) **Replacement** of amount of product used.

The seller **will not be liable for consequential or incidental damages or losses.**

The terms of this Limit of Warranty and Liability cannot be varied by any written or verbal statements or agreements. Any employee or sales agent of the seller is not authorized to vary or exceed the terms of this Warranty Limitations And Disclaimer in any manner.

* Trademark of THE DOW CHEMICAL COMPANY

87256-L2 A1284

LORSBAN* 4E

INSECTICIDE

For Control of Various Insects Infesting Certain Field, Fruit, Nut and Vegetable Crops

ACTIVE INGREDIENT(S):	
Chlorpyrifos [O,O-diethyl O-(3,5,6-trichloro-2-pyridyl) phosphorothioate]	40.7%
Aromatic Petroleum Derivative Solvent	22.8%
INERT INGREDIENTS	36.5%
TOTAL	100.0%

Contains 4 pounds of chlorpyrifos per gallon

EPA Reg. No. 464-448 EPA Est. 464-MI-1

* Trademark of The Dow Chemical Company

WARNING

Keep Out of Reach of Children

AVISO

PRECAUCION AL USUARIO: Si usted no lee inglés, no use este producto hasta que la etiqueta le haya sido explicada ampliamente.

PRECAUTIONARY STATEMENTS

HAZARDS TO HUMANS AND DOMESTIC ANIMALS

MAY BE FATAL IF SWALLOWED • HARMFUL IF ABSORBED THROUGH SKIN • CAUSES SUBSTANTIAL BUT TEMPORARY EYE INJURY • CAUSES SKIN IRRITATION

Avoid Breathing Vapor Or Spray Mist • Do Not Get In Eyes, On Skin, Or On Clothing • Wear Protective Clothing, Rubber Gloves, And Goggles • Wash Thoroughly With Soap And Water After Handling And Before Eating Or Smoking • Immediately Remove Contaminated Clothing And Wash Before Reuse • Destroy Contaminated Shoes • Do Not Apply This Product In Such A Manner As To Directly Or Through Drift Expose Workers Or Other Persons • The Area Being Treated Must Be Vacated By Unprotected Persons • Keep Away From Food, Feedstuffs And Water Supplies

STATEMENTS OF PRACTICAL TREATMENT

If Swallowed: Do not induce vomiting. Contains aromatic petroleum solvent. Call a physician or poison control center immediately.

If In Eyes: Flush with plenty of water for at least 15 minutes. Get medical attention.

If On Skin: Wash with plenty of soap and water. Get medical attention if irritation persists.

If Inhaled: Remove to fresh air if effects occur. Get medical attention.

NOTE TO PHYSICIAN:

Chlorpyrifos is a cholinesterase inhibitor. Treat symptomatically. Atropine only by injection is an antidote. If exposed, plasma and red blood cell cholinesterase tests may indicate significance of exposure (baseline data are useful). Atropine, only by injection, is the preferable antidote. Oximes, such as 2-PAM/protopam, may be therapeutic if used early; however, use only in conjunction with atropine. In case of severe acute poisoning, use antidote immediately after establishing an open airway and respiration.

PHYSICAL OR CHEMICAL HAZARDS

COMBUSTIBLE • Do Not Use or Store Near Heat or Open Flame. • Do Not Cut or Weld Container.

ENVIRONMENTAL HAZARDS

This pesticide is toxic to birds and wildlife, and extremely toxic to fish and aquatic organisms. Do not apply directly to water. Drift and runoff from treated areas may be hazardous to aquatic organisms in adjacent aquatic sites. Cover or incorporate spills. Do not contaminate water by cleaning of equipment or disposal of waste.

This product is highly toxic to bees exposed to direct treatment or residues on blooming crops or weeds. Avoid use when bees are actively foraging. Protective information may be obtained from your Cooperative Agricultural Extension Service.

REENTRY STATEMENT

Do not allow reentry into treated areas within 24 hours unless protective clothing is worn. Protective clothing consists of at least a hat or other suitable head covering, a long-sleeved shirt and long-legged trousers or a coverall type garment (all of closely woven fabric covering the body, including the arms and legs), shoes and socks. Because certain states may require more restrictive reentry intervals for various crops treated with this product, consult your State Department of Agriculture for further information.

Written or oral warnings must be given to workers who are expected to be in a treated area or in an area about to be treated with this product. These oral warnings shall inform workers of areas or fields that may not be entered without protective clothing and actions to take in case of accidental exposure as shown by the Statement of Practical Treatment section on this label. When oral warnings are given, warnings shall be given in a language customarily understood by workers. Written warnings must include the following information: WARNING. Area treated with LORSBAN 4E insecticide on (insert date of application). Do not enter without protective clothing for 24 hours. In case of accidental exposure (insert actions to take as shown by the Statement of Practical Treatment section on this label).

Do not use on turf or for structural pest control. Do not formulate this product into other end use products.

NOTICE:

Read the entire label. Use only according to label directions.

Before buying or using this product, read "WARRANTY LIMITATIONS AND DISCLAIMER" elsewhere on this label. If terms are not acceptable, return **unopened** package at once to seller for full refund of purchase price paid. Otherwise, use by the buyer or any other user constitutes acceptance of the terms under the Limit of Warranty and Liability.

In case of an emergency endangering life or property involving this product, call collect **517-636-4400**.

AGRICULTURAL CHEMICAL

Do Not Ship or Store with Food, Feeds, Drugs, or Clothing

Do not use on turf for structural pest control. Do not formulate this product into other end use products.

DIRECTIONS FOR USE

It is a violation of Federal law to use this product in a manner inconsistent with its labeling.

GENERAL INFORMATION

LORSBAN 4E insecticide forms an emulsion when diluted with water and is suitable for use in all conventional hand- and power-operated ground spray equipment. Aerial spray equipment also may be used where specified.

MIXING INSTRUCTIONS

To prepare the spray, add a portion of the required amount of water to the spray tank and with agitation add the LORSBAN 4E. Complete filling the tank with the balance of water needed. Maintain sufficient agitation during both mixing and application to ensure uniformity of the spray mixture.

LORSBAN 4E insecticide may also be used in tank mixtures with certain herbicides and/or with non-pressure fertilizer solutions as recommended under specific crop use directions. Prepare tank mixtures in the same manner as recommended above for use of LORSBAN 4E alone. When tank mixtures of LORSBAN 4E and herbicides are involved, add wettable powders first, flowables second, and emulsifiable concentrates last. Where a fertilizer solution is involved, it is strongly recommended that a fertilizer pesticide compatibility agent such as Unite(1) or Compex(2) be used. Maintain constant agitation during both mixing and application to ensure uniformity of the spray mixture.

NOTE: Text compatibility of the intended tank mixture before adding LORSBAN 4E to the spray or mix tank. Add proportionate amounts of each ingredient to a pint or quart jar, cap, shake, and let set 15 minutes. Formation of precipitates that do not readily redisperse indicates an incompatible mixture that should not be used.

(1) Trademark of The Hopkins Agricultural Chemical Company.

(2) Trademark of Kalo Laboratories, Inc.

OVERHEAD SPRINKLER IRRIGATION

The following use directions are to be followed when LORSBAN 4E is applied through overhead sprinkler irrigation systems. Thoroughly clean the injection system and tank of any fertilizer or chemical residues, and dispose of the residues according to state and federal laws. Flush the injector with soap and water. Calibrate the injector system. Prepare a mixture of non-emulsifiable oil, LORSBAN 4E and water, if used, in a steel mix tank after determining the amount of insecticide that will be injected per hour at the desired rate and time of operation. Pump the required LORSBAN 4E insecticide into the steel tank, start mechanical or hydraulic agitation, and add the non-emulsifiable oil and then water, if used. Continually agitate the LORSBAN 4E plus oil mixture. Set the sprinkler system to deliver the desired inches of water per acre. Start the water pump and sprinkler, and let the system achieve the desired pressure and speed before starting the injector. The LORSBAN 4E plus oil mixture must be injected continuously and uniformly into the irrigation water line as the sprinkler is moving. This procedure is necessary to deliver the desired rate per acre in a uniform manner. When application is finished, allow the entire irrigation and injector system to be thoroughly flushed clean before stopping the system.

The following use precautions will result in a safe and successful application of LORSBAN 4E plus oil mixtures:

1. In water dilutions of less than 1 part LORSBAN 4E to 14 parts water, use only in steel mix tanks.
2. **USERS OF LORSBAN 4E INSECTICIDE THROUGH OVERHEAD SPRINKLER IRRIGATION SYSTEMS MUST COMPLY WITH ALL FIRE REGULATIONS THAT APPLY TO CLASS 1C FLAMMABLE LIQUIDS. MINIMUM DISTANCES AND PROPER ELECTRICAL CONSTRUCTION IS ESSENTIAL TO SAFELY HANDLE LORSBAN 4E. ELECTRICAL INSTALLATIONS MUST MEET CLASS 1 GROUP D DIV. 1 OR 2 REQUIREMENTS AS SPECIFIED IN THE NATIONAL ELECTRICAL CODE 70 (NEC 70). THIS TYPE OF WIRING IS GENERALLY REFERRED TO AS EXPLOSION PROOF. CONSULT WITH A LICENSED ELECTRICIAN TO CORRECTLY WIRE THE INJECTION SYSTEM AS DEFINED BY NEC 70. WHEN LORSBAN 4E INSECTICIDE IS DILUTED WITH AT LEAST 2 PARTS WATER THE CHEMICAL INJECTOR MOTOR AND WIRING NEED NOT BE EXPLOSION PROOF. NON-EXPLOSION PROOF ELECTRICAL EQUIPMENT MUST EITHER BE OUTSIDE THE HAZARD ZONE OR DISCONNECTED FROM THE POWER SOURCE WHILE MAKING THE DILUTION.**
3. The irrigation pumping plant and the chemical injection pump must be interlocked so that if the irrigation pumping plant stops, the chemical injection pump also will stop. This procedure will prevent the possibility of filling the entire irrigation pipeline with the chemical mixture from the supply tank.
4. **Properly functioning** check and vacuum relief valves (antisiphon devices) are required in the irrigation pipeline to keep water and/or a mixture of water and chemical from draining or siphoning back into the irrigation well and polluting the ground water. Both of these valves are located between the irrigation pump discharge and the place of chemical injection into the irrigation pipeline.
5. A **properly functioning** check valve in the chemical injection line is required to stop the flow of water from the irrigation system into the chemical supply tank.
6. A positive displacement pump is required to provide uniform injection into the water line. The injector pump must provide a greater pressure than that of the irrigation system at the point of injection. The pump must contain Viton seals and be rated for use in hazardous areas. All electrical service inside the hazard zone must meet National Fire Protection Association 30 (NFPA 30) and NEC 70.
7. To insure uniform mixing of the insecticide into the water line, inject the mixture into the fertilizer injection port or just ahead of an elbow or tee in the irrigation line so that the turbulence created at those points will assist in mixing. It is suggested that the injection point be higher than the insecticide tank to prevent siphoning.
8. The steel tank holding the insecticide mixture should be large enough to allow the system to complete a revolution with one filling. It should be free of rust, fertilizer, sediment, and foreign material, and equipped with an in-line strainer situated between the tank and the injector pump.
9. In order to calibrate the irrigation system and injector to apply the LORSBAN 4E plus oil mixture, determine the following: 1) calculate the number of acres covered by the system, 2) the number of minutes for the system to make one revolution, 3) the number of inches of water per acre delivered by the system, 4) the number of minutes required for the injector pump to inject one gallon of insecticide mixture, and 5) the total gallons of insecticide mixture needed to cover the desired acreage. Divide the total gallons of insecticide mixture needed to cover the irrigated area by the number of minutes to make one circle. This value equals the gallons per minute output that the injector must deliver. It is suggested that the system be calibrated at least twice before operation, and the system should be monitored during operation.

10. Do not apply a pesticide solution when wind speed or direction results in unacceptable drift from the application area.
11. Do not allow irrigation runoff to collect or runoff and pose a hazard to livestock, wells, or adjoining crops.
12. Allow foliage to dry before reentering the field.
13. Do not apply through sprinkler systems without a reasonably uniform watering pattern such as certain water drive units.

USE AND DOSAGE RECOMMENDATIONS

ALFALFA: Use LORSBAN 4E to control the following pests at the dosages indicated by application as a broadcast, foliar spray:

Corn rootworm adults (spotted cucumber beetle),
Aphids .... ½ pint per acre;
Grasshoppers .... ½ to 1 pint per acre;
Alfalfa blotch leafminer, alfalfa looper, alfalfa weevil larvae and adults, armyworms, cutworms, Egyptian alfalfa weevil larvae and adults, plant bugs, leafhoppers, spittlebugs .... 1 to 2 pints per acre.

Note: Use higher rates to control spotted alfalfa aphid in California and Nevada. Stubble spray may be applied to control leafhopper in the Northeast.

Mix the required dosage with enough water to ensure thorough coverage of crop foliage and apply using aerial (fixed-wing or helicopter) or power-operated ground spray equipment. For aerial application use 2 to 5 gallons of water per acre. Control may be reduced at low spray volumes under high temperature and wind conditions. Treat when field counts or crop injury indicates that damaging pest populations are developing or present; however, do not apply more than once per crop cutting. Some reduction in insect control may be evident under excessively cool conditions.

LORSBAN 4E may also be applied through **overhead sprinkler irrigation systems** as a postemergence broadcast application to control the above listed foliar pests. For best results, use the recommended rate of LORSBAN 4E per acre. Maintain vigorous tank agitation to assure uniformity of the application throughout the injection period.

LORSBAN 4E insecticide should not be tank-mixed with pesticides, surfactants, or fertilizer formulations unless prior use has shown the combination non-injurious under your current conditions of use. Some phytotoxic symptoms may be observed on young, tender rapidly growing alfalfa when treated with LORSBAN 4E. Alfalfa will outgrow the symptoms and no yield loss should be expected.

This product is highly toxic to bees exposed to direct treatment on alfalfa. Do not apply if nearby bees are clustered outside of hives and bees are foraging. Protective information may be obtained from your Agricultural Extension Service.

Restrictions: Do not cut or graze treated alfalfa within 7 days after application of ½ pint of LORSBAN 4E per acre, within 14 days after application of 1 pint per acre, or within 21 days after application of rates above 1 pint per acre. Do not make more than 4 applications per year or apply more than once per crop cutting.

ASPARAGUS: Use LORSBAN 4E to control cutworms, asparagus aphids, and asparagus beetles by application at the rate of 2 pints per acre. Mix the specified dosage in sufficient water to ensure thorough coverage of treated plants and apply as a broadcast, foliar spray. For cutworms, it is preferable to apply LORSBAN 4E when the soil is moist and worms are active on or near the soil surface. Applications may be made during the fern stage for control of asparagus beetles and asparagus aphids when field counts or crop injury indicates that damaging pest populations are developing or present.

Restrictions: Do not make more than one preharvest application per season or apply within one day of harvest. Do not make more than two postharvest applications during the fern stage. Based on available residue data, the use of LORSBAN 4E on asparagus is limited to the Midwest and Pacific Northwest.

CHERRIES: Use LORSBAN 4E for the control of lesser peachtree borer by application as a trunk spray. Mix 1½ to 3 quarts of LORSBAN 4E with 100 gallons of water and apply as a coarse, low-pressure spray to give uniform coverage of tree trunks and lower limbs. Make a second application two weeks after the first one and a third application after harvest. Consult your State Agricultural Experiment Station or Extension Service Specialist for proper time to treat in your area.

Restrictions: Make only three applications per year. Do not apply within 6 days before harvest. Do not allow meat or dairy animals to graze in treated orchards. Based on available residue data, use of chlorpyrifos on cherries is restricted to Michigan, New York, Ohio, Pennsylvania, and Wisconsin.

CITRUS FRUITS: Use LORSBAN 4E insecticide at the rates indicated to control the following pests. Use the lower rates for light infestations and increase the dosage for heavier infestations.

CROP	PEST	DOSAGE LORSBAN 4E FOR SPRAY COVERAGE: DILUTE (PINTS/100 GAL. WATER)	DOSAGE LORSBAN 4E FOR SPRAY COVERAGE: DILUTE OR CONCENTRATE (PINTS/ACRE)	GALLONS OF SPRAY PER ACRE FOR DILUTE APPLICATIONS[3]	REMARKS
Grapefruit Lemons Oranges and Other Citrus Fruits	Aphids, Katydids Lepidopterous larvae Cutworms Fruit tree leafroller Orange tortrix Western tussock moth	Outside Coverage Spray (OC) ½ to 1	2 to 7	200 to 700	Use directions apply in all states including California.
	Citrus thrips[1], Mealy bugs[1] Scale insects[2] Black scale Brown soft scale California red scale Snow scale, Chaff scale	½ to 1	2 to 7	200 to 700	Use directions apply in all states except California (See footnotes 1 & 2)
	Citrus rust mite (suppression)		4 to 7	100 to 700	Use directions apply in Florida, only.

[1] In California apply as Outside Coverage (OC) using 6 to 12 pints of LORSBAN 4E per acre in 100 to 300 gallons of finished spray per acre.
[2] In California apply as Thorough Coverage (TC) using 8 to 12 pints of LORSBAN 4E per acre. For dilute sprays apply 1200 to 2400 gallons of spray per acre but do not apply more than 12 pints of LORSBAN 4E per acre or use less than ½ pint of LORSBAN 4E per 100 gallons of finished spray per acre.
[3] Gallonage per acre based on average size of mature trees. Adjust to tree size per acre.

For dilute applications, use the specified dosage in 100 gallons of water for the spray type indicated for pest to be controlled. For **concentrate sprays,** use the same amount of LORSBAN 4E per acre that would be needed for application as a dilute spray, using appropriate application/coverage procedures. A petroleum spray oil recommended for use on citrus may be added to dilute spray mixtures only at a rate of up to 1.8 gallons per 100 gallons of water to improve control of aphids, mealybugs, scale insects, and thrips. Treat when insects become a problem or in accordance with the local spray schedule recommended by your State Extension Service Specialist. Do not apply when temperature exceeds 95°F. LORSBAN 4E is highly toxic to bees exposed to direct treatment and should not be applied when bees are actively visiting the area. Apply late in evening or early morning to groves where honey bees are expected to visit.

Precautions: Observe local use directions for tank mix combinations especially in regard to applications of LORSBAN 4E plus spray oil. Consult with a County Farm Advisor, County Agent, Extension Service personnel, or Agricultural Commissioner for such information regarding a given locality. LORSBAN 4E should not be used in combination with spray oil when temperatures are expected to exceed 95°F the day of application or for several consecutive days thereafter. LORSBAN 4E should not be used in combination with PLICTRAN* 50W miticide except in dilute applications. For concentrate sprays in Florida, do not use more than 1 pint of LORSBAN 4E per 15 gallons of finished spray. Do not apply when trees are stressed by drought or high temperature.

Restrictions: Do not apply more than 2 applications per fruit year or more than 15 pints of LORSBAN 4E insecticide per acre per fruit year. Do not make second application within 30 days of the first application. Do not treat within 21 days before harvest for applications up to 7 pints of LORSBAN 4E per acre nor within 35 days for application of rates above 7 pints per acre. Do not pick fruit or do other work involving contact with trees within 2 days after treatment. Do not allow livestock to graze in treated areas.

CITRUS ORCHARD FLOORS: Use LORSBAN 4E to control red imported fire ants and other ant species by applying the specified dose in 25 or more gallons of water with ground application equipment that will uniformly apply the spray to the orchard floor. To control foraging ants and suppress mounds, apply LORSBAN 4E to the orchard floor at the rate of 3/4 to 1 quart per acre. Retreat as needed. For best insect control, uniform coverage of the orchard floor is necessary. Do not apply where weed growth or other obstructions would impede uniform coverage of the orchard floor. Do not apply in tank mixtures with Roundup[1] or Paraquat herbicides. Foliar applications of LORSBAN 4E may be made in addition to the orchard floor treatments.

[1] Trademark of Monsanto Company

Restrictions: Do not apply more than 10 quarts of LORSBAN 4E per acre per season. Do not apply last treatment within 28 days before harvest. Do not allow livestock to graze in treated areas.

CRANBERRIES: Use LORSBAN 4E by application as a broadcast, foliar spray to control brown spanworm, cranberry fruitworm, cranberry weevil, cutworms, fireworms, and Sparganothis fruitworms at the rate of 3 pints per acre. Mix the specified dosage in enough water to ensure thorough coverage and apply no less than 5 gallons of spray per acre when using aerial equipment or no less than 100 gallons of spray per acre when using ground equipment. For weevil control, apply once at flower bud development (late May, early June) and, if weevils are present, once after 100% bloom (early to mid July). For other insects, treat when field counts indicate damaging insect populations are developing or present. Apply only after the winter flood has been removed. To avoid pesticide contamination of flood waters, make no applications while bogs are flooded.

Restrictions: Do not make more than two applications per year or apply within 60 days before harvest.

FIELD CORN, POPCORN, SWEET CORN: For use to control cutworms, armyworms, corn earworm, corn rootworm adults, chinch bugs, grasshoppers, wireworms, flea beetle larvae and adults, aphids, billbugs, grubs, western bean cutworm, corn borers, symphylans, common stalk borer, and lesser cornstalk borer.

Preplant Incorporation Treatment: Use LORSBAN 4E at the following rates by application in sufficient water to the soil surface and incorporate into the soil:

Cutworms, symphylans .... 2 to 4 pints per acre
Wireworms, billbugs, flea beetle larvae,
grubs .... 4 pints per acre.

Use recommended rate in not less than 10 gallons of water per acre and apply as a broadcast spray to the soil surface using suitable power-operated ground spray equipment. On the same day of treatment, incorporate the insecticide into the top 2 to 4 inches of soil using a disc, field cultivator or equivalent equipment.

LORSBAN 4E insecticide may also be applied in tank mixtures with non-pressure fertilizer solutions and/or with Bladex[1], Eradicane[2], Sutan[3], Lasso[4], Dual[5], and atrazine herbicides. See the Mixing Instructions section of this label for further information. Read and carefully follow all applicable directions, restrictions, and precautions on labeling for the other products used in combination with LORSBAN 4E.

[1] Trademark of Shell Chemical Company

[2, 3] Trademark of Stauffer Chemical Company

[4] Trademark of Monsanto Company

[5] Trademark of Ciba-Geigy Corporation

Preplant, At-Plant or Preemergence Treatment in Conservation Tillage. Use LORSBAN 4E at the following rates by application in sufficient water to surface trash and exposed soil:

Cutworms, armyworms .... 1½ to 2 pints per acre.

Use recommended rate in not less than 20 gallons of water per acre and apply as a broadcast spray using suitable power-operated ground spray equipment.

LORSBAN 4E may also be applied in tank mixtures with non-pressure fertilizer solutions and/or with paraquat and Roundup[1]. See the Mixing Instructions section of this label for further information. Read and carefully follow all applicable directions, restrictions, and precautions on labeling for the other products used in combination with LORSBAN 4E.

[1] Trademark of Monsanto Company

Cultivation Time Treatment: Use LORSBAN 4E at the rate of 2 pints per acre to control corn rootworm larvae. Apply LORSBAN 4E as a water emulsion on both sides of the row at the base of the plants just ahead of the cultivator shovels. Cover the insecticide with soil around the brace roots. The best time to apply a basal treatment of a soil insecticide with cultivation is near the beginning of egg hatch. A cultivation application of LORSBAN 4E may be made in addition to an at planting application of LORSBAN 15G granular insecticide.

Postemergence Treatment: Use LORSBAN 4E at the following rate by application in sufficient water to ensure thorough coverage of treated plants:

Grasshoppers .... ½ to 1 pint per acre;
Armyworms, chinch bugs, aphids, corn rootworm
adults, webworms .... 1 to 2 pints per acre;
European and southwestern corn borer, corn
earworm .... 1½ to 2 pints per acre;
Cutworms, billbugs, lesser cornstalk borer,
flea beetle adults, western bean cutworm,
common stalk borer .... 2 to 3 pints per acre.

Note: The recommended dosage will control silk clipping by corn rootworm adults.

Treat when field counts indicate that pests are or may become a problem. For best billbug, chinch bug, and flea beetle control, apply with sufficient water to ensure a minimum spray volume of 20 to 40 gallons per acre and 40 psi using **ground spray equipment.** On corn less than 6 inches tall, apply the insecticide spray in a 9 to 12 inch wide band over the row. On corn greater than 6 inches tall, apply the insecticide spray using drop nozzles directed to the base of the plant. When chinch bugs continue to immigrate to corn over a prolonged period or under extreme pressure, a second application of LORSBAN 4E may be needed.

For cutworm, webworm, western bean cutworm, armyworm, aphid, European and southwestern corn borer, grasshopper, lesser cornstalk borer, corn rootworm adult, corn earworm, and common stalk borer control, apply as a broadcast spray using either aerial (fixed-wing or helicopter) or power-operated ground spray equipment. For aerial application use 2 to 5 gallons of spray per acre. Control may be reduced at low spray volumes under high temperature and wind conditions. For cutworms, it is preferable to apply LORSBAN 4E when soil is moist and worms are active on or near the soil surface. If ground is dry, cloddy or crusty at time of treatment, worms may be protected from the spray and effectiveness will be reduced. If such conditions exist, shallow incorporation using a rotary hoe or other suitable equipment immediately before or soon after treatment may improve control. Consult your Agricultural Experiment Station or Extension Service specialist for additional information concerning control practices in your area. For first-generation European corn borer control, treat when 25 to 50 percent of the corn plants show pinhole feeding or leaf-feeding scars. Ground applications should be directed into the corn whorls. For second-generation European corn borer control, treat when field counts of egg masses indicate pests are or may become a problem. For southwestern corn borer control, treat when field counts of egg masses indicate pests are or may become a problem. A second application may be applied 10 to 14 days later, if needed due to reinfestation. For common stalk borer control, treat approximately 11 days after application of Roundup[1] herbicide or after complete burndown with paraquat herbicide (3 to 5 days). Do not use LORSBAN 4E in combination with the burndown herbicide for control of common stalk borer.

[1] Trademark of Monsanto Company

LORSBAN 4E insecticide may also be applied through **overhead sprinkler irrigation systems** as a postemergence broadcast application to control the above listed foliar insects. For best results, use the recommended rate of LORSBAN 4E in a tank mix with 2 pints per acre of nonemulsifiable oil. Maintain vigorous tank agitation to assure uniformity of the LORSBAN 4E plus oil mixture throughout the injection period. See the Overhead Sprinkler Irrigation section of this label for further information.

Restrictions: Do not apply within 35 days before harvest of grain. Do not apply more than a total of 15 pints of LORSBAN 4E per acre per season. Do not allow livestock to graze in treated areas nor harvest treated corn silage as feed for meat or dairy animals within 14 days after last treatment. Do not feed treated corn fodder to meat or dairy animals within 35 days after last treatment.

SWEET CORN Grown Only in Florida and Georgia: Use LORSBAN 4E to control infestations of beet armyworm, fall armyworm and corn earworm by application as a broadcast, foliar spray at the rate of 1 to 2 pints per acre. Mix the specified dosage in enough water to ensure thorough coverage and apply using suitable aerial or ground spray equipment. For aerial application, use at least 2 gallons of spray per acre. Treat when field counts indicate damaging pest populations are developing or present. Re-treat as necessary to maintain control but do not apply more than twenty-two 1-pint or eleven 2-pint treatments per season.

LORSBAN 4E insecticide may also be applied through **overhead sprinkler irrigation systems** as a postemergence broadcast application to control the above listed foliar insects. For best results, use the recommended rate of LORSBAN 4E in a tank mix with 2 pints per acre of nonemulsifiable oil. Maintain vigorous tank agitation to assure uniformity of the LORSBAN 4E plus oil mixture throughout the injection period. See the Overhead Sprinkler Irrigation section of this label for further information.

Restrictions: Do not apply more than 22 pints of LORSBAN 4E per acre per season. Do not harvest corn ears, allow livestock to graze in treated areas, nor feed treated silage, fodder, or grain to meat or dairy animals within 21 days after treatment. Do not use in conjunction with post-plant broadcast, foliar applications of LORSBAN 15G granular insecticide.

COTTON: Use LORSBAN 4E for control of the following pests in all states except Arizona and California at the dosages indicated:

Cotton fleahopper, plant bus (*Lygus, Mirids*) .... 3/8 to 1 pint per acre;

Cotton aphid, fall armyworm, grasshoppers, spider mites, thrips, yellow-striped armyworm .... 1/2 to 1 pint per acre;

Beet armyworm, bollworms (*Heliothis spp.*), boll weevil, cutworms, pink bollworm, salt, marsh caterpillar .... 1 1/2 to 2 pints per acre.

NOTE: The recommended dosage rate of 3/8 pint per acre will not achieve the high degree of control of the higher label rate, but will minimize the damage done by plant bugs and cotton fleahopper and allow the beneficial insects to survive, build up; and be available to aid in the control of bollworms infesting cotton.

Use LORSBAN 4E for control of the following pests in Arizona and California at the dosages indicated:

Armyworms, cotton aphid, cotton fleahopper, *Lygus*, salt marsh caterpillar, thrips .... 1 to 2 pints per acre;

Bollworms, (*Heliothis spp.*), boll weevil, cutworms, pink bollworm .... 2 pints per acre.

NOTE: The 2 pint rate will aid in the suppression of cotton leafperforator and spider mites.

Mix the required dosage with sufficient water to ensure thorough coverage of plants and apply using aerial or power-operated ground spray equipment. For aerial application, use at least 1 gallon of spray per acre. Treat when field counts indicate damaging insect populations are developing or present. Re-treat as necessary to maintain control.

LORSBAN 4E insecticide may also be applied through **overhead sprinkler irrigation systems** as a postemergence broadcast application to control the above listed foliar pests. For best results, use the recommended rate of LORSBAN 4E per acre. Maintain vigorous tank agitation to assure uniformity of the application throughout the injection period. See the Overhead Sprinkler Irrigation section of this label for further information.

For effective control of spider mites when large numbers of eggs are present, apply a second spray 3 to 5 days after initial treatment to control newly hatched nymphs.

For best results on bollworms, it is suggested that fields be scouted twice per week and treatments made when worms are 1/4 inch or less in length. The following table illustrates the size of worms in relation to age and stage of development (instar) as a guide to timing of treatments for best control.

From the table it can be seen that a scouting schedule of only once per week will not be satisfactory since the worms may be too big to control effectively by the seventh or eighth day.

Timing for the best worm control

	Age (days)	Size	Instar
Get the worms at this stage	Hatch	1/16"	Hatch
	3	3/32"	I
	5	9/32"	II
	6	7/16"	III
	8	11/16"	IV

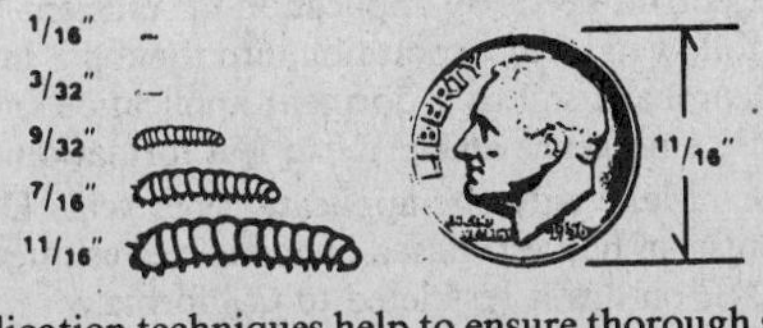

Proper application techniques help to ensure thorough spray coverage and correct dosage and are thus important in obtaining good control of pests. Consider these suggestions when applying LORSBAN 4E insecticide on cotton.

Aerial Application: Shorten boom length to avoid spray entering the vortices at the wing tips. Swath width should be reduced when wind direction is the same as direction of spraying.

The proper nozzle arrangement and swath width to avoid skips and vortices effect can be checked out by flying over a paper tape (adding machine paper) using water with or without soluble dye. (The dye gives a permanent record.)

Flying at a height of 5–15 feet above the target results in the best coverage.

Nozzle orientation of the boom is important. More breakup occurs when nozzles are pointed straight down versus the straight back position. Desired droplet size (100–200 microns) can be obtained by angling the nozzles somewhere in this range.

Marking of swath by flagging or permanent markers is essential.

Ground Application: Orient the boom and nozzles so that uniform coverage is obtained. The swath width should not be wider than the boom; drift spray is wasted spray so do not depend on it. Use flat fan or disc-core hollow cone nozzles with maximum spacing of 20 inches and a spray pressure of 40–60 psi with a droplet size of 100–200 microns.

Restrictions: Do not apply within 14 days before harvest. Do not allow livestock to graze in treated areas. Do not feed gin trash or treated forage to livestock.

LORSBAN 4E insecticide may also be used in tank mixtures in all states to control armyworms, bollworms *(Heliothis spp.)*, boll weevil, cotton fleahopper, plant bugs *(Lygus, Mirids)*, and spider mites as indicated in the following table.

Tank Mixture	Rate/Acre
LORSBAN 4E plus Ambush[1]2E	1 to 2 pints plus 1/5 to 2/5 pint
LORSBAN 4E plus Pounce[2]3.2E	1 to 2 pints plus 1/8 to 1/4 pint
LORSBAN 4E plus Pydrin[3]2.4E	1 to 2 pints plus 1/6 to 1/3 pint
LORSBAN 4E plus Bolstar[4]6E	1/2 to 1 pint plus 2/3 to 1 1/3 pints
LORSBAN 4E plus methyl parathion[5] Plus EPN[5]	1 pint plus 0.6 lb a.i. plus 0.6 lb a.i.

[1] Trademark of ICI Americas Inc.
[2] Trademark of FMC Corporation.
[3] Trademark of Shell Oil Company.
[4] Trademark of Farbenfabriken Bayer Gmbh Leverkusen.
[5] Use only emulsifiable formulations containing these active ingredients (a.i.).

Mix the tank mixture components in sufficient water to ensure thorough coverage of plants (at least 1 gallon of spray per acre) and apply using aerial spray equipment. Treat when field counts indicate damaging insect populations are developing or present. Re-treat as necessary to maintain control.

Restrictions: Make no applications of these tank mixtures closer to harvest than the longest pre-harvest interval shown for any of the products in the tank mixture. Do not allow livestock to graze in treated areas. Do not feed gin trash or treated forage to livestock. Workers entering the field within 48 hours after application should wear protective clothing.

FIGS: Use LORSBAN 4E at the rate of 2 quarts per acre for control of driedfruit beetle by application in sufficient water to the soil surface followed by incorporation into the top 3 inches of soil. Apply to fig orchard soil as a dormant application in late winter prior to beetle emergence and prior to leaf formation.

Restrictions: Make only one application per year. Do not apply within 7 months of harvest. Based on available residue data, use of LORSBAN 4E on figs is restricted to California.

GRAIN SORGHUM: Use LORSBAN 4E Insecticide for control of the following pests at the dosages indicated:

Sorghum midge .... 1/2 pint per acre
(Apply when 30 to 50 percent of the seed heads are in bloom, repeat at 3-day intervals if necessary.);

Greenbugs, grasshoppers, yellow sugar cane aphid, and other aphids .... 1/2 to 1 pint per acre;

Chinch bugs, lesser cornstalk borer .... 1 to 2 pints per acre
(Apply as a directed spray toward the base of the plant using power-operated ground spray equipment with sufficient water to ensure coverage of an 8 to 12 inch band centered on the row. On plants less than 6 inches high, apply an 8 to 12 inch band over the row);

Armyworm, corn earworm, cutworms .... 1 to 2 pints per acre;

European and southwestern corn borer .... 1 1/2 to 2 pints per acre;

Webworms .... 2 pints per acre.

Mix the specified dosage in enough water to ensure thorough coverage and apply using suitable aerial or ground spray equipment. To minimize chemical injury, do not apply LORSBAN 4E to drought stressed grain sorghum within 3 days following irrigation or rain except where the product is applied in irrigation water.

LORSBAN 4E insecticide may also be applied through **overhead sprinkler irrigation systems** as a postemergence broadcast application to control the above listed foliar pests. For best results, use the recommended rate of LORSBAN 4E per acre. Maintain vigorous tank agitation to assure uniformly of the application throughout the injection period. See the Overhead Sprinkler Irrigation section of this label for further information.

Restrictions: The treated crop is not to be used for forage, fodder, hay, or silage within 30 days after application of 1 pint of LORSBAN 4E per acre or within 60 days after application of rates above 1 pint per acre. Do not treat sweet varieties of sorghum. Do not apply more than 3 pints of LORSBAN 4E per acre per season.

GRAPES: Use LORSBAN 4E insecticide for control of grape root borer by application just before the pest emerges from the soil. Mix 4 1/2 pints of LORSBAN 4E with 100 gallons of water and apply 2 quarts of the diluted spray mixture to the soil surface on a 15-square foot area around the base of each vine. Do not allow spray to contact fruit or foliage.

Restrictions: Do not make more than one application per season or apply within 35 days before harvest. Based upon available residue data, the use of LORSBAN 4E in grapes is restricted to states East of the Rocky Mountains.

MINT: Use LORSBAN 4E by application as a broadcast, foliar spray to control cutworms at the rate of 2 to 4 pints per acre and mint root borer at the rate of 4 pints per acre. Mix the specified dosage in water to give no less than 10 gallons of spray per acre and apply using ground spray equipment. For cutworm control, treat during May and June when field counts indicate damaging insect populations are developing or present. When larvae are less than 3/4 inch in length, use the 2 pint rate. When larvae are 3/4 inch or more in length, use the higher rate. **Make only one application during the growing season. Do not apply within 90 days before harvest.** For **mint root borer** control, apply post-harvest when field counts indicate damaging insect populations are developing or present. Follow treatment with approximately 1 acre inch of sprinkler irrigation immediately after application to incorporate the insecticide into the soil. **Make only one post-harvest application per season.**

NECTARINES, PEACHES: Use LORSBAN 4E insecticide for the control of peach tree borers by application as a trunk spray before newly hatched borers enter the trees. Mix 3 quarts of LORSBAN 4E with 100 gallons of water and apply as a coarse, low-pressure spray to give uniform coverage of tree trunks. Thoroughly wet all bark areas from ground level to scaffold limbs. Do not allow spray to contact fruit. Consult your State Agricultural Experiment Station or Extension Service Specialist for proper time to treat in your area.

Restrictions: Make only one application per season. Do not apply within 14 days before harvest. Do not allow meat or dairy animals to graze in treated orchards.

ONIONS (dry bulb): Use LORSBAN 4E insecticide to control onion maggot by application as an in-furrow drench at seeding. Apply LORSBAN 4E at the rate of 1.1 fluid ounce per 1000 linear feet of row at an 18-inch row spacing. Use a minimum of 40 gallons of total drench per acre. Incorporate to a depth of 1 to 2 inches.

Restrictions: For use on direct seeded onions only. Do not make more than one application per year.

PEANUTS: Use LORSBAN 4E for control of the following pests at the dosage indicated:

Lesser cornstalk borer .... 2 to 4 pints per acre
(as a directed spray to the base of the plants in an 8 to 10 inch wide band);

Armyworms, corn earworm, cutworms, leafhoppers, mites, rednecked peanutworm, thrips .... 1 to 2 pints per acre;
Velvetbean caterpillar, green cloverworm .... ½ to 1 pint per acre;
(as a broadcast, foliar spray using ground or aerial equipment when field counts indicate damaging insect populations are developing or present).

Mix the required dosage in enough water to ensure thorough coverage.

LORSBAN 4E insecticide may also be applied through **overhead sprinkler irrigation systems** as a postemergence broadcast application to control the above listed foliar pests. For best results, use the recommended rate of LORSBAN 4E per acre. Maintain vigorous tank agitation to assure uniformity of the application throughout the injection period. See the Overhead Sprinkler Irrigation section of this label for further information.

LORSBAN 4E should not be tank mixed with pesticides, surfactants, or fertilizer formulations unless prior use has shown the combination noninjurious under your current conditions of use. To minimize chemical injury, do not apply LORSBAN 4E to drought stressed peanuts within 3 days following irrigation or rain except where the product is applied in irrigation water.

Restrictions: Do not make more than one application per season as a directed spray. Directed and foliar sprays should not exceed 8 pints of LORSBAN 4E per acre or 4 pounds of chlorpyrifos (active ingredient) per acre per season. Do not harvest within 21 days after treatment. Do not feed treated peanut forage or hay to meat or dairy animals.

SOYBEANS: For use to control armyworms, bean leaf beetle, corn earworm, cutworms, European corn borer, grasshoppers, green cloverworm, lesser cornstalk borer, Mexican bean beetle, saltmarsh caterpillar and other woollybears, southern green stink bug, spider mites, and velvetbean caterpillar.

Soil Treatment: Use LORSBAN 4E at the rate of 1 to 2 pints per acre to control cutworms and lesser cornstalk borer. Mix the specified dosage in a minimum of 10 gallons of spray per acre and apply to the soil surface using suitable ground spray equipment. Equivalent rates of insecticide spray required per 100 feet of row for various row spacing are given in the accompanying table. For at-plant treatments apply the insecticide over the row in a 4 to 6 inch band in front of the planter shoe or press wheel or after the press wheel followed by a drag chain for light incorporation. Do not apply as an in-furrow treatment. For postemergence rescue treatments, apply as a directed spray in a 9 to 12 inch band at the base of the plant. To plants under 6 inches high apply over the top in a 6 to 12 inch band. Treat when field counts or conditions indicate that pests are or may become a problem.

Volume of Spray Per Acre	Fluid Ounces of Spray Required Per 100 Feet of Row for Various Row Spacing			
	36"	32"	28"	24"
10 Gallons	8.8	7.9	6.9	5.9
15 Gallons	13.2	11.8	10.3	8.8
20 Gallons	17.6	15.7	13.7	11.8

Foliar Treatment: Use LORSBAN 4E at the following rate by applicaton in sufficient water to ensure thorough coverage of treated plants:

European corn borer, southern green stink bug .... 2 pints per acre
Bean leaf beetle, cutworms, corn earworm, saltmarsh caterpillar and other woollybears .... 1 to 2 pints per acre;
Mexican bean beetle, armyworms .... 1 to 1½ pints per acre;
Velvetbean caterpillar, grasshoppers, green cloverworm, spider mites .... ½ to 1 pint per acre.

Apply as a broadcast spray using either aerial or ground equipment when field counts indicate damaging insect populations are developing or present; retreat as necessary to maintain control. For effective control of spider mites when large numbers of eggs are present, apply a second spray 3 to 5 days after initial treatment to control newly hatched nymphs. **On determinate soybeans do not apply more than one application after pod set.**

LORSBAN 4E insecticide may also be applied through **overhead sprinkler irrigation systems** as a postemergence broadcast application to control the above listed foliar pests. For best results, use the recommended rate of LORSBAN 4E per acre. Maintain vigorous tank agitation to assure uniformity of the application throughout the injecton period. See the Overhead Sprinkler Irrigation section of this label for further information.

Restrictions: Do not apply more than 6 pints of LORSBAN 4E per acre of 3 pounds of chlorpyrifos (active ingredient) per acre per season. Do not apply last treatment within 28 days before harvest nor apply last two treatments closer than 14 days apart. Do not allow livestock to graze in treated areas nor otherwise feed treated soybean forage to meat or dairy animals within 14 days after applicaton. Do not feed straw from treated soybeans to meat or dairy animals within 28 days after application.

STRAWBERRIES: Use LORSBAN 4E insecticide by application as a broadcast foliar spray to control strawberry bud weevil at the rate of 1 quart per acre. Apply in a minimum of 40 gallons of spray per acre when buds first appear and 10 to 14 days later. Do not apply after berries start to form or when berries are present. LORSBAN 4E should not be tank mixed with pesticides, surfactants, or fertilizer formulations unless prior use has shown the combination noninjurious under your current conditions of use. Phytotoxicity may occur when LORSBAN 4E is applied to strawberries experiencing high temperature and drought stress.

Restrictions: For pre-bloom use only. Do not make more than two applications per season or apply within 21 days before harvest.

SUNFLOWERS: For use to control cutworms, sunflower beetle larvae and adults, stem weevil, sunflower moth, banded sunflower moth, woollybears, seed weevil, and grasshoppers.

Preplant Incorporation Treatment: Use LORSBAN 4E insecticide at the following rates by application in sufficient water to the soil surface and incorporate into the soil:

Cutworms .... 2 to 4 pints per acre.

Use recommended rate in not less than 10 gallons of water per acre and apply as a broadcast spray to the soil surface using suitable power-operated ground spray equipment. On the same day of treatment, incorporate the insecticide into the top 2 to 4 inches of soil using a disc, field cultivator or equivalent equipment.

Postemergence Treatment: Use LORSBAN 4E for control of the following pests at the dosage indicated by application in sufficient water to ensure thorough coverage of treated plants:

Cutworms .... 2 to 3 pints per acre:
Sunflower beetle larvae and adults, stem weevil, sunflower moth, banded sunflower moth, woollybears, and seed weevil .... 1 to 1½ pints per acre:
Grasshoppers .... 1 pint per acre.

Apply as a broadcast spray using ether aerial (fixed-wing or helicopter) or power-operated ground spray equipment when fields counts indicate that pests are or may become a problem. For cutworm control, a second treatment may be made 7 to 10 days later, if needed. For stem weevil control, optimal treatment time is within 5 to 7 days after adult weevils begin to appear. For sunflower moth control, make first application during early 1 to 5 percent bloom stage. A second treatment may be made 7 days later, if needed. For seed weevil control, treat when field counts indicate there are 10 to 12 adults per plant for oil crops and 1 to 3 adults per plant on confectionary crops. Additional treatments should be made at successive 7 to 10 day intervals if field counts indicate need to re-treat. For sunflower beetle larvae or adult control, treat when field counts indicate there are 10 larvae or 1 to 2 adults per seedling. Additional treatments may be made at successive 7 to 10 day intervals if field counts indicate need to re-treat.

Restrictions: Do not apply more than 9 pints of LORSBAN 4E per acre per season. Do not apply within 42 days before harvest. Do not allow livestock to graze in treated areas.

SUGAR BEETS: Use LORSBAN 4E by application as a broadcast, foliar spray to control beet armyworms at a rate of 1½ to 2 pints per acre and cutworms at a rate of 2 pints per acre. Treat when field counts indicate that damaging insect populations are or may be a problem. Mix the specified dosage with 2 to 5 gallons of water per acre and apply using suitable aerial spray equipment or with 10 to 30 gallons of water per acre when using power-operated ground spray equipment. Re-treat as necessary to maintain control.

Restrictions: Do not apply within 30 days before harvest of beet roots and tops. Do not apply more than a total of 8 pints of LORSBAN 4E per acre per season. Do not allow livestock to graze in treated areas nor harvest treated beet tops as feed for meat or dairy animals within 30 days after last treatment.

SWEET POTATOES: Use LORSBAN 4E to control *Conoderus* wireworms, *Systena* flea beetles and the sweet potato flea beetle by preplant application at the rate of 4 pints per acre as a broadcast (overall) spray to the soil surface followed by incorporation. Mix the specified dosage with enough water to obtain uniform coverage of the treated area and apply as a coarse spray using suitable ground spray equipment. After application, incorporate the insecticide into the soil to a depth of 4 to 6 inches using a rotary hoe, disc cultivator, or other suitable incorporation equipment. Plant the crop in the usual manner as soon as possible after treatment.

Restrictions: Do not make more than one application per season.

TOBACCO: Use LORBSAN 4E for preplant treatment to control larvae of cutworms, flea beetles, mole crickets, root maggots, and wireworms. Apply 2 to 3 quarts of LORSBAN 4E per acre in not less than 10 gallons of water as a broadcast (overall) spray to the soil surface one week before transplanting. Immediately following application, incorporate the insecticide into the soil to a depth of 2 to 4 inches using suitable equipment.

To control infestations of nematodes, apply LORSBAN 4E in a tank mix with Nemacur[1] 3 at the rate of 2 quarts of LORSBAN 4E plus 4 quarts of Nemacur 3 per acre. Apply the specified dosage in not less than 10 gallons of water as a broadcast (overall) spray to the soil surface one week before bedding and transplanting. Immediately following application, incorporate into the soil to a depth of at least 4 inches using suitable equipment. Where the nematode species *Meloidogyne arenaria* or *M. javanica* are present increase the amount of Nemacur 3 in the tank mix to 6.67 quarts per acre. Read and carefully follow all applicable directions, restrictions, and precautions on labeling for Nemacur 3 used in combination with LORSBAN 4E.

Restrictions: Do not make more than one application per season.

[1] Trademark of Farbenfabriken Bayer GmbH, Leverkusen.

TREE FRUITS: Use LORSBAN 4E as a dormant or delayed dormant spray at the rates indicated to control the following insects on the crops listed.

Crop	Insect	Dosage LORSBAN 4E
Apples	Rosy apple aphid San Jose scale Lygus Pandemis leafroller Climbing cutworms	
Pears	San Jose scale Climbing cutworms Pear psylla adults	½ to 1 pint
Plums Prunes	San Jose scale Mealy plum aphid Climbing cutworms Peach twig borer	
Almonds Peaches Nectarines	San Jose scale Peach twig borer Climbing cutworms	

Tank mix the specified dosage with 1 to 2 gallons of a petroleum spray oil recommended for dormant use in 100 gallons of water and spray the entire tree by application to runoff using suitable ground spray equipment. For low volume (concentrate) sprays (40 to 100 gallons of spray mixture per acre) use the same amounts of LORSBAN 4E and spray oil per acre required for application as a dilute spray and apply in a manner that will ensure thorough coverage of the trees. Use the higher dosage of LORSBAN 4E for severe infestations. Use oil as recommended by your State Agricultural Experiment Station or Extension Service Specialist.

Precautions: Because cold or dry conditions may cause LORSBAN 4E plus oil sprays to infuse trees resulting in bud damage or drop, do not apply until winter rains or irrigation has replenished soil moisture such that bark and twigs are not desiccated.

Restrictions: Make only one application during the dormant season. Do not allow meat or dairy animals to graze in treated orchards. Do not apply on almonds in the following counties in California: Butte, Colusa, Glenn, Solano, Sutter, Tehama, Yolo, and Yuba.

TREE NUTS: Use LORSBAN 4E at the dosages indicated by application as a foliar spray to control pests listed in the following table. Mix the required dosage in sufficient water to ensure thorough and complete coverage of the foliage and crop and apply as a concentrate or dilute spray using conventional, power-operated spray equipment. For dilute sprays applied to tree nut crops, mix the required dosage in sufficient water to allow for spray to runoff. For concentrate sprays, apply an equivalent amount of LORSBAN 4E per acre. Treat when pests appear or in accordance with local conditions. Consult your State Agricultural Experiment Station or Extension Service specialist for specific use information in your area.

Crop	Insects Controlled	Dosage LORSBAN 4E	Restrictions
Almonds	Navel orangeworm Peach twig borer San Jose scale	4 pints per acre	Make no more than 3 applications per season on almonds and filberts and no more than 2 applications per season on walnuts. Do not apply within 14 days of harvest. Do not allow livestock to graze in treated orchards.
Filberts	Eyespotted bud moth Filbert aphid Filbert leafroller Filbertworm Obliquebanded leafroller Omniverous leaftier Winter moth	3 to 4 pints per acre	
Walnuts	Codling moth Walnut scale	4 pints per acre	
Pecans	Black pecan aphid Fall webworm Hickory shuckworm Pecan leaf scorch mite Pecan nut casebearer Pecan weevil Phylloxera spp. Spittlebug	2 pints per 100 gallon	Make no more than 5 applications per season. Do not apply within 28 days of harvest. Do not allow livestock to graze in treated orchards.

VEGETABLES: Use LORSBAN 4E to control root maggots on the following crops at the dosages indicated:

Cauliflower .... 1.6 to 2.4 fl oz per 1000 linear feet of row;

Broccoli, Brussels Sprouts, Cabbage, Chinese Cabbage, Collards, Kale, Kohlrabi, Turnips .... 1.6 to 2.75 fl oz per 1000 linear feet of row.

For **direct seeded crops** apply the specified dosage in a water-based spray as a 4-inch wide band over the row at planting time, behind the planter shoe and in front of the press wheel to achieve shallow incorporation. Apply in a minimum of 40 gallons of total spray per acre.

For **transplanted crops,** apply LORSBAN 4E as a water-based spray directed to the base of the plants immediately after setting. Use a minimum of 40 gallons of total spray per acre. Do not add any additional adjuvants, surfactants, or spreader stickers. **Do not apply as a foliage application.**

Restrictions: Do not apply more than 2 pints of LORSBAN 4E to cauliflower planted in 40-inch rows. Use proportional amounts to other row spacings not to exceed 4 pints of LORSBAN 4E per acre. Do not apply more than 2.6 pints of LORSBAN 4E per acre to broccoli, Brussels sprouts, cabbage, Chinese cabbage, collards, kale, kohlrabi, and turnips planted in 40-inch rows or more than $4\frac{1}{2}$ pints of LORSBAN 4E per acre to these crops in 20-inch rows (or two rows per bed). Use proportional amounts for other row spacings not to exceed $4\frac{1}{2}$ pints of LORSBAN 4E per acre. Do not make more than one application per season or apply within 30 days before harvest.

Radishes .... 1.0 fl oz per 1000 linear feet of row.

Apply the specified dosage as a water-based drench in the seed furrows with the seed at planting time. Use a minimum of 40 gallons of total drench per acre.

Restrictions: Do not apply more than $5\frac{1}{2}$ pints of LORSBAN 4E per acre or make more than one application per season.

Rutabagas .... 1.6 to 3.3 fl oz per 1000 linear feet of row.

Apply the specified dosage in a water-based spray as a 4-inch wide band over the row at planting time, behind the planter shoe and in front of the press wheel to achieve shallow incorporation. Use a minimum of 40 gallons of total spray per acre.

Restrictions: Do not apply more than $4\frac{1}{2}$ pints of LORSBAN 4E per acre or make more than one application per season. Do not use rutabaga tops for food or feed purposes.

To avoid phytotoxicity in vegetable crops, do not mix with other pesticide products or treat plants that are under extreme heat and drought stress.

STORAGE AND DISPOSAL

Do not contaminate water, food or feed by storage or disposal. Open dumping is prohibited.

PESTICIDE DISPOSAL

Pesticide wastes are toxic. Improper disposal of excess pesticide, spray mixture, or rinsate is a violation of Federal law. If these wastes cannot be disposed of by use according to label instructions, contact your State Pesticide or Environmental Control Agency, or the Hazardous Waste representative at the nearest EPA Regional Office for guidance.

CONTAINER DISPOSAL

Triple rinse (or equivalent). Then offer for recycling or reconditioning, or puncture and dispose of in a sanitary landfill, or by other procedures approved by state and local authorities. For reusable portable containers, do not rinse or put any other material into container. When returning reusable portable container, seal all openings and return to owner.

OR

CONTAINER DISPOSAL

Triple rinse (or equivalent). Then puncture and dispose of in a sanitary landfill, or by incineration, or, if allowed by state and local authorities, by burning. If burned, stay out of smoke.

ATTENTION

Do Not Cut or Weld Container

WARRANTY LIMITATIONS AND DISCLAIMER

The Dow Chemical Company warrants that this product conforms to the chemical description on the label and is reasonably fit for the purposes stated on the label when used in strict accordance with the directions therein under normal conditions of use. THIS IS THE ONLY WARRANTY MADE ON THIS PRODUCT. NO OTHER EXPRESS AND NO IMPLIED WARRANTY OF MERCHANTABILITY OR FITNESS FOR A PARTICULAR PURPOSE IS MADE OUTSIDE OF THIS LABEL. Therefore, neither this warranty nor any other warranty of merchantability or fitness for a particular purpose, express or implied, extends to the use of this product contrary to label instructions (including conditions noted on the label, such as unfavorable temperatures, soil conditions, etc.), under abnormal conditions (such as excessive rainfall, drought, tornadoes, hurricanes, etc.) or under conditions not reasonably foreseeable to or beyond the control of seller.

When buyer or user suffers losses or damages resulting from the use or handling of this product (including claims based on contract, negligence, strict liability, or other legal theories), buyer or user must promptly notify in writing The Dow Chemical Company of any claims to be eligible to receive either remedy given below. The EXCLUSIVE REMEDY OF THE BUYER OR USER and the LIMIT OF LIABILITY of The Dow Chemical Company or any other seller will be one of the following, at the election of The Dow Chemical Company:

(1) Refund of purchase price paid by buyer or user for product bought, **or**

(2) Replacement of amount of product used.

The seller **will not be liable for consequential or incidental damages or losses.**

The terms of this Limit of Warranty and Liability cannot be varied by any written or verbal statements or agreements. Any employee or sales agent of the seller is not authorized to vary or exceed the terms of this Warranty Limitations and Disclaimer in any manner.

SPECIAL LOCAL NEED REGISTRATIONS

FIELD AND SWEET CORN: For use only within **Florida** (sweet corn only) (EPA SLN No. FL-780046) and **Georgia** (EPA SLN No. GA-790019) to control beet armyworms, fall armyworm, and corn earworm larvae and **Colorado** (sweet corn only) (EPA SLN No. CO-810017), **Maryland** (sweet corn only) (EPA SLN No. MD-840004, **New Jersey** (sweet corn only) (EPA SLN No. NJ-840008), **Pennsylvania** (sweet corn only) (EPA SLN No. PA-840008) and **Delaware** (sweet corn only) (EPA SLN No. DE-840003) to control armyworms and corn earworm larvae, use 1 pint of LORSBAN 4E in sufficient water to ensure thorough coverage of plants and apply using either aerial (fixed-wing or helicopter) or power-operated ground spray equipment. For aerial application use at least 3 gallons of water per acre. Treat when field counts indicate that worms are or may become a problem. Repeat the application as needed, but do not exceed 13 applications per crop season.

Restrictions: Do not treat within 7 days before harvest. Do not allow livestock to graze in treated areas. Do not feed treated corn silage, forage, or fodder to meat or dairy animals.

For use only within **Florida** (field and sweet corn grown only for planting seed) (EPA SLN No. FL-770025) to control beet armyworm, fall armyworm, and corn earworm, use 1 to 2 pints of LORSBAN 4E per acre in sufficient water to ensure thorough coverage of plants and apply using either aerial (fixed-wing or helicopter) or power-operated ground spray equipment. For aerial application use at least 3 gallons of water per acre. Treat when field counts indicate that worms are or may become a problem. Repeat the application as needed.

Restrictions: Apply only to corn to be harvested for seed. Do not allow livestock to graze in treated areas. Do not feed treated corn grain, silage, or forage to meat or dairy animals.

COTTON: For use only within **Mississippi** (EPA SLN No. MS-840004) to control larvae of wireworms and cutworms, apply 2 pints of LORSBAN 4E per acre for wireworm control and 1 to 2 pints of LORSBAN 4E per acre for cutworm control in not less than 10 gallons of water in a 16 to 18 inch band over the row prior to or at planting. Immediately following application incorporate the insecticide into the soil to a depth of 3 to 4 inches using suitable equipment.

Restriction: Do not make more than one application per season.

For use only within **Mississippi** (EPA SLN No. MS-830025) to control and suppress spider mites, use $\frac{1}{2}$ to 1 pint of LORSBAN 4E per acre in tank mixtures with pyrethroid insecticides at the dosage rate shown on their product label and sufficient cottonseed oil or soybean oil to give a finished spray volume of 1 quart per acre. Apply a minimum of 1 quart of spray per acre with aircraft calibrated to give adequate coverage of plants. Treat when field counts indicate damaging insect populations are developing or present.

Restrictions: Make no applications of these tank mixtures closer to harvest than the longest pre-harvest interval shown for any of the products in the tank mixture. Do not allow livestock to graze in treated areas. Do not feed gin trash or treated forage to livestock.

GRAPES: For use only within **Michigan** (EPA SLN No. MI-840002) to control climbing cutworms, use 1 quart of LORSBAN 4E per acre in a minimum of 25 gallons of water and apply as a prebloom, dilute, or concentrate spray. LORSBAN 4E should not be tank mixed with pesticides, surfactants, or fertilizer formulations unless prior use has shown the combination noninjurious under your conditions of use. Consult your State Agricultural Station or Extension Service specialist for proper time to treat in your area.

Restrictions: Do not make more than one application per season. Do not use in conjunction with a soil surface spray application of LORSBAN 4E for grape root borer control.

NON-BEARING GRAPES: For use only within **Idaho** (EPA SLN No. ID-820018), **Oregon** (EPA SLN No. OR-820001), and **Washington** (EPA SLN No. WA-810036) to control cutworms, apply 1 quart of LORSBAN 4E per acre as a broadcast spray using either aerial or power-operated ground spray equipment. For aerial application use at least three gallons of spray per acre. If sprinkler irrigation is available, apply $\frac{1}{4}$ to $\frac{1}{2}$ inch of water after application to help incorporate the material. Treat when cutworms first become active and when field counts indicate damaging insect populations are developing or present. Re-treat as necessary to maintain control. Consult your State Agricultural Experiment Station or Extension Service specialist for additional information concerning cutworm control practices in your area.

Restrictions: Do not harvest grapes in the same growing season as application.

PERENNIAL GRASS SEED CROPS: For use only within **Idaho** (EPA SLN No. ID-830005) and **Oregon** (EPA SLN No. OR-830007) to control billbugs, cutworms, and aphids and **Washington** (EPA SLN No. WA-830009) to control cutworms and aphids, use 1 quart of LORSBAN 4E per acre in sufficient water to ensure good coverage and apply as a broadcast spray using power-operated ground application equipment. For improved results, the thatch should be moist at the time of application and the insecticide washed into the thatch immediately after application. If sprinkler irrigation is available, apply $\frac{1}{4}$ to $\frac{1}{2}$ inch of water after application to help incorporate the material. Otherwise, apply to early morning dew-covered fields or during or immediately following a light rain. For billbug control, apply in early April when adult billbugs are active. For cutworm control, apply in either spring or fall when larvae or larval damage first appear. Fall applications should follow sufficient irrigation and/or rainfall. Heavy carbon residue from field burning will prevent penetration of LORSBAN 4E into the plant crown where the insects feed. Re-treat as necessary to maintain control. To control aphids, apply 1 to 2 pts of LORSBAN 4E per acre as a broadcast spray using power-operated ground application equipment. Treat as colonies begin forming before heading. Consult your Cooperative Agricultural Extension Service for additional information concerning billbug, cutworm, and aphid control practices in your area. This product is highly toxic to bees exposed to direct treatment or residues on treated crops or weeds. Protective information may be obtained from your Cooperative Agricultural Extension Service.

Restrictions: Apply only to perennial grass seed crops. Do not allow livestock to graze in treated areas. Do not feed treated grass cuttings or seed screenings to livestock nor use hay for livestock bedding.

STRAWBERRIES: For use only within **Idaho** (EPA SLN No. ID-830028), **Oregon** (EPA SLN No. OR-830017), and **Washington** (EPA SLN No. WA-830033) to control garden symphylans, use 2 quarts of LORSBAN 4E per acre with enough water to ensure good coverage of the soil surface and apply pre-plant as a broadcast spray to the soil surface using suitable power-operated ground application equipment. On the same day of treatment, incorporate the insecticide into the top 2 to 4 inches of soil using a disc, field cultivator or equivalent equipment. LORSBAN 4E should not be tank mixed with pesticides, surfactants, or fertilizer formulations unless prior use has shown the combination noninjurious under local conditions of use. Consult your Cooperative Agricultural Extension Service for additional information concerning garden symphylan control practices in your area.

Restrictions: Apply only as a soil pre-plant incorporation application in the spring for protection of strawberries the following year.

For use only within **Oregon** (EPA SLN No. OR-840021) to control strawberry crown moth, apply 1 quart of LORSBAN 4E per acre as a directed spray to crowns of strawberry plants immediately after harvest and after plants are topped. Apply in a minimum of 100 gallons of spray per acre. **Do not** sprinkle irrigate for one week following application. A second application may be required two to two and one half weeks later. Phytotoxicity may occur when LORSBAN 4E is applied to plants experiencing high temperatures and water stress. LORSBAN 4E should not be tank mixed with pesticides, surfactants, or fertilizer formulations unless prior use has shown the combination noninjurious under local conditions. Consult your Cooperative Agricultural Extension Service for additional information concerning strawberry crown moth control practices in your area.

Restrictions: Use post-harvest only. Do not make more than two applications per year.

SUGAR BEETS GROWN FOR SEED: For use only within **Idaho** (EPA SLN No. ID-830021) and **Oregon** (EPA SLN No. OR-830018) to control garden symphylans, use 2 quarts of LORSBAN 4E per acre with enough water to ensure good coverage of the soil surface and apply pre-plant as a broadcast spray to the soil surface using suitable power-operated ground application equipment. LORSBAN 4E should not be tank mixed with pesticides, surfactants, or fertilizer formulations unless prior use has shown the combination noninjurious under local conditions of use. Consult your Cooperative Agricultural Extension Service for additional information concerning garden symphylan control practices in your area.

Restrictions: Apply as a soil pre-plant incorporation application in the fall for protection of sugar beets the following year.

TREE FRUITS: For use only within **California** (EPA SLN No. CA-790238) to control the brown almond mite and European red mite infesting almonds, apples, nectarines, peaches, pears, plums, and prunes, use $\frac{1}{2}$ to 1 pint of LORSBAN 4E as a dormant or delayed dormant spray. Tank mix the specified dosage with 1 to 2 gallons of petroleum spray oil recommended for dormant use in 100 gallons of water and spray the entire tree by application to runoff using suitable ground spray equipment.

For low volume (concentrate) sprays (40 to 100 gallons of spray mixture per acre) use the same amounts of LORSBAN 4E and spray oil per acre required for application as a dilute spray and apply in a manner that will ensure thorough coverage of the trees. Use the higher dosage of LORSBAN 4E for severe infestations. Use oil as recommended by your State Agricultural Experiment Station or Extension Service specialist.

For aerial application sprays, use 4 pints LORSBAN 4E plus an appropriate amount of spray oil in 20 gallons spray volume per acre. Use a spraying system that will ensure thorough coverage of the trees. Use oil as recommended by your State Agricultural Experiment Station or Extension Service specialist.

Restrictions: Make only one application during the dormant season. Do not allow meat or dairy animals to graze in treated orchards. Avoid contact with treated trees until spray has dried.

LORSBAN* 15G

GRANULAR INSECTICIDE

For Control of Various Insects Infesting Certain Field and Vegetable Crops

ACTIVE INGREDIENT

Chlorpyrifos 0,0–diethyl 0–(3,5,6–trichloro-2–pyridyl)phosphorothioate	15.0%
INERT INGREDIENTS	85.0%
TOTAL	100.0%

EPA Reg. No. 464-523 EPA Est. 464-MI-1
US Patent No. 4,380,537

*Trademark of The Dow Chemical Company

CAUTION
Keep Out of Reach of Children
AVISO

PRECAUCION AL USUARIO: Si usted no lee inglés, no use este producto hasta que la etiqueta le haya sido explicada ampliamente.

PRECAUTIONARY STATEMENTS

HAZARDS TO HUMANS AND DOMESTIC ANIMALS

HARMFUL IF SWALLOWED • CAUSES EYE IRRITATION

Avoid Contact With Skin, Eyes, Or Clothing • Wash Thoroughly With Soap and Water After Handling • Remove Contaminated Clothing And Wash Before Reuse • Keep Away From Food, Feedstuffs and Water Supplies

STATEMENTS OF PRACTICAL TREATMENT

If Swallowed: Call a physician or Poison Control Center. Drink 1 or 2 glasses of water and induce vomiting by touching back of throat with finger. Do not induce vomiting or give anything by mouth to an unconscious person.

If In Eyes: Flush with plenty of water for 5 minutes. Get medical attention if irritation persists.

If On Skin: Wash with plenty of soap and water. Get medical attention if irritation persists.

NOTE TO PHYSICIANS: Chlorpyrifos is a cholinesterase inhibitor. Treat symptomatically. If exposed, plasma and red blood cell cholinesterase tests may indicate significance of exposure (baseline data are useful). Atropine, only by injection, is the preferable antidote. Oximes, such as 2-PAM/protopam, may be therapeutic if used early; however, use only in conjunction with atropine. In case of severe acute poisoning, use antidote immediately after establishing an open airway and respiration.

ENVIRONMENTAL HAZARDS

This pesticide is toxic to birds and wildlife and extremely toxic to fish and aquatic organisms. Do not apply directly to water. Drift and runoff from treated areas may be hazardous to aquatic organisms in adjacent aquatic sites. Cover or incorporate spills. Do not contaminate water by cleaning of equipment or disposal of waste.

NOTICE

Read the entire label. Use only according to label directions.

Before buying or using this product, read "WARRANTY LIMITATIONS AND DISCLAIMER" elsewhere on this label. If terms are not acceptable, return **unopened** package at once to seller for full refund of purchase price paid. Otherwise, use by the buyer or any other user constitutes acceptance of the terms under the Limit of Warranty and Liability.

In case of an emergency endangering life or property involving this product, call collect **517-636-4400.**

AGRICULTURAL CHEMICAL

Do Not Ship or Store with Foods, Feeds, Drugs or Clothing

STORAGE AND DISPOSAL

Do not contaminate water, food or feed by storage or disposal. Open dumping is prohibited.

PESTICIDE DISPOSAL

Wastes resulting from the use of this product may be disposed of on site or at an approved waste disposal facility.

CONTAINER DISPOSAL

Completely empty bag into application equipment. Then dispose of empty bag in a sanitary landfill or by incineration or, if allowed by state and local authorities, by burning. If burned, stay out of smoke.

LORSBAN 15G Insecticide—Rate Per 1000 Ft of Row

Amount of LORSBAN 15G Per 1,000 Feet of Row	Pounds of LORSBAN 15G Required Per Acre from Various Row Spacing							
	40″	38″	36″	34″	32″	30″	22″	18″
3.7 ounces	3.0	3.2	3.4	3.6	3.8	4.0	5.5	6.7
4 ounces	3.3	3.4	3.6	3.8	4.1	4.4	5.9	7.3
6 ounces	4.9	5.2	5.4	5.8	6.1	6.5	8.9	10.9
6.5 ounces	5.3	5.6	5.9	6.2	6.6	7.1	9.7	11.8
7.5 ounces	6.1	6.4	6.8	7.2	7.7	8.2	11.1	13.6
8 ounces	6.5	6.9	7.3	7.7	8.2	8.7	11.9	17.5
9 ounces	7.4	7.7	8.2	8.6	9.2	9.8	13.4	16.3
12 ounces	9.8	10.3	10.9	11.5	12.3	13.1	17.8	21.8
15 ounces	12.3	12.9	13.9	14.4	15.3	16.3	22.3	27.2
16 ounces	13.1	13.8	14.5	15.4	16.3	17.4	23.8	29.0

GENERAL INSTRUCTIONS FOR EQUIPMENT CALIBRATION

Caution: The following chart lists suggested initial gauge settings for application of LORSBAN 15G Insecticide with one hopper opening per row. Be sure to check the actual application rate under your operating conditions.

1. Fill hopper.
2. Attach a plastic bag to tube opening.
3. Set your planter to the initial settings shown on chart.
4. Measure off 1,000 row feet and drive your planter the pre-measured distance at your desired speed.
5. Each bag should contain 6 to 8 ounces (wt.) of granules depending on your desired rate.
6. If the result is over or under the desired rate—adjust the settings and repeat the calibration.

LORSBAN 15G Insecticide—Calibration

	Speed, mph									
	4		5		6		7		8	
	Application Rate, oz. Per 1000 row ft.									
	8	16	8	16	8	16	8	16	8	16
Planter Type	**Gauge Setting**									
Gandy[1]	21.4	30.2	23.7	32.4	26.0	36.0	27.7	39.0	30.2	41.0
John Deere[1] 7000 Maxi-Emerge (Odd)	14	22	16	24	18	26	19	28	21	30
John Deere[1] 7000 Maxi-Emerge (Even)	17	30	20	33	24	35	26	36	28	38
John Deere[2] 71 Flexi-Planter	$\frac{1}{30}$	$\frac{2}{17}$	$\frac{2}{5}$	$\frac{2}{22}$	$\frac{2}{9}$	$\frac{2}{27}$	$\frac{2}{13}$	$\frac{2}{31}$	$\frac{2}{16}$	$\frac{3}{1}$
Allis Chalmers[3] 70 Series	8	13	8	13	8	13	8	13	8	13
Allis Chalmers[3] 78 & 79 Series	$\frac{1}{9.0}$	$\frac{3}{3.0}$	$\frac{2}{3.3}$	$\frac{3}{9.5}$	$\frac{2}{6.0}$	$\frac{4}{3.0}$	$\frac{2}{9.0}$	$\frac{5}{4.0}$	$\frac{3}{2.5}$	$\frac{6}{0}$

LORSBAN 15G Insecticide—Calibration

	Speed, mph									
	4		5		6		7		8	
	Application Rate, oz. Per 1000 row ft.									
	8	16	8	16	8	16	8	16	8	16
Planter Type	Gauge Setting									
Noble[1] (New)	11	19	14	22	16	25	17	28	19	31
Noble[1] (Old)	14		17		21		24		27	
International Harvester[4]	$\frac{1}{9.0}$	$\frac{3}{3.0}$	$\frac{2}{3.3}$	$\frac{3}{9.5}$	$\frac{2}{6.0}$	$\frac{4}{3.0}$	$\frac{2}{9.0}$	$\frac{5}{4.0}$	$\frac{3}{2.5}$	$\frac{6}{0}$
Buffalo All-Flex[5] (Fleischer Mfg.)	$4\frac{7}{8}$	10	$4\frac{7}{8}$	10	$4\frac{7}{8}$	10	—	—	—	—

[1] Gauge setting

[2] Gauge setting with range 1 & 2—number shown is $\frac{\text{range}}{\text{notch}}$

[3] Gauge setting is constant regardless of speed

[4] Gauge setting shown with stem gates & dial settings—number shown is $\frac{\text{gate}}{\text{dial}}$

[5] Number of turns open on the adjustment nut

WARRANTY LIMITATIONS AND DISCLAIMER

The Dow Chemical Company warrants that this product conforms to the chemical description on the label and is reasonably fit for the purposes stated on the label when used in strict accordance with the directions therein under normal conditions of use. THIS IS THE ONLY WARRANTY MADE ON THIS PRODUCT. NO OTHER EXPRESS AND NO IMPLIED WARRANTY OF MERCHANTABILITY OR FITNESS FOR A PARTICULAR PURPOSE IS MADE OUTSIDE OF THIS LABEL. Therefore, neither this warranty nor any other warranty of merchantability or fitness for a particular purpose, express or implied, extends to the use of this product contrary to label instructions (including conditions noted on the label, such as unfavorable temperatures, soil conditions, etc.), under abnormal conditions (such as excessive rainfall, drought, tornadoes, hurricanes, etc.) or under conditions not reasonably foreseeable to or beyond the control of seller.

When buyer or user suffers losses or damages resulting from the use or handling of this product (including claims based on contract, negligence, strict liability, or other legal theories), buyer or user must promptly notify in writing The Dow Chemical Company of any claims to be eligible to receive either remedy given below. The EXCLUSIVE REMEDY OF THE BUYER OR USER and the LIMIT OF LIABILITY of The Dow Chemical Company or any other seller will be one of the following, at the election of The Dow Chemical Company:

(1) **Refund of purchase price** paid by buyer or user for product bought, **or**

(2) **Replacement** of amount of product used.

The seller **will not be liable for consequential or incidental damages or losses.**

The terms of this Limit of Warranty and Liability cannot be varied by any written or verbal statements or agreements. Any employee or sales agent of the seller is not authorized to vary or exceed the terms of this Warranty Limitations and Disclaimer in any manner.

DIRECTIONS FOR USE

It is a violation of Federal law to use this product in a manner inconsistent with its labeling.

Crop	Pest	Amount of LORSBAN 15G per 1,000 feet of row — Planting Time Treatment: T-Band	Band	In-Furrow	Postplant Treatment	Amount of LORSBAN 15G per acre for Preplant Broadcast Treatment	Use Directions	Restrictions
Corn: Field, Popcorn, Sweet Corn	Northern, Southern and Western Corn Rootworm Larvae	8 oz	8 oz	—	8 oz	—	**At Plant Band Treatment:** In conventional, minimum and no-till corn, place the granules in a 6 to 7 inch wide band over the row behind the planter shoe, **in front of the press wheel**, and incorporate into the top one inch of soil using tines or chains or other suitable equipment. For cutworm control under very dry soil conditions, best results will be obtained through proper incorporation. In soils over 10% organic matter, use the highest rate given. **At Plant T-Band Treatment:** In conventional, minimum and no-till corn, for control of rootworm, cutworm and billbug, in addition to lesser cornstalk borer, symphylans, wireworm, seed corn beetle, seed corn maggot and grubs, apply as a T-band over an open seed furrow. **At Plant In-Furrow Treatment:** Apply LORSBAN 15G at planting time as an in-furrow treatment in conventional, minimum and no-till corn. Direct the granules into the planter shoe with the seed or place the applicator tube directly behind the planter shoe so that the granules drop into the seed furrow or place the granular band applicator behind the planter shoe so that the granules fall on the soil surface and into the open seed furrow and are covered with soil. Equivalent rates of LORSBAN 15G per acre for various row spacings are given in the accompanying table. **Postplant Treatment:** Apply LORSBAN 15G insecticide at cultivation time to control corn rootworm larvae. Place the granules over the row and cultivate into the soil or sidedress both sides of the row and cultivate into the soil. For best rootworm con-	For soil insect control, do not exceed the equivalent of 16 ounces of LORSBAN 15G per 1,000 feet of row or 13.5 pounds of LORSBAN 15G per acre per crop season. For foliar insect control, do not exceed the equivalent of 16 ounces of LORSBAN 15G per 1,000 feet of row or 13 pounds of LORSBAN 15G per acre per crop season. Do not apply within 35 days before harvest of grain. Do not allow livestock to graze in treated areas nor harvest treated corn silage as feed for meat or dairy animals within 14 days after last treatment. Do not feed treated corn fodder to meat or dairy animals within 35 days after last treatment.
	Cutworms	8 oz	8 oz	8 to 16 oz	—	6.75 to 13.5 lb		
	Wireworms[1]	16 oz	8 to 16 oz	8 to 16 oz	—	13.5 lb		
	Billbugs	8 to 16 oz	8 to 16 oz	—	—	13.5 lb		
	Seed Corn Maggots	8 oz	—	8 oz	—	—		
	Grubs	16 oz	—	8 to 16 oz	—	13.5 lb		
	Lesser Cornstalk Borer	8 to 16 oz	8 to 16 oz	—	—	—		
	Seed Corn Beetle	8 to 16 oz	—	8 to 16 oz	—	—		
	Flea Beetle Larvae	—	—	—	—	13.5 lb		
	Symphylans	8 to 12 oz	8 to 12 oz	—	—	—		
	European & Southwestern Corn Borer[2]	—	—	—	6 to 8 oz	—		

[1] 8 oz rate—suppression only.

[2] Use 5.0 to 6.5 lb. of LORSBAN 15G per acre applied broadcast by air.

Crop	Pest	Amount of LORSBAN 15G per 1,000 feet of row — Planting Time Treatment — T-Band	Band	In-Furrow	Postplant Treatment	Amount of LORSBAN 15G per acre for Preplant Broadcast Treatment	Use Directions	Restrictions
							trol, place the granules at the base of the plants on both sides of the row just ahead of cultivation shovels and cover with soil. Consult your State Agricultural Experiment Station or extension service specialists for proper time to treat in your area. Equivalent rates of LORSBAN 15G per acre for various row spacings are given in the accompanying table. Consult your State Agricultural Experiment Station or extension service specialist for further information concerning soil insect control in your area. Equivalent rates of LORSBAN 15G per acre for various row spacings are given in the accompanying table. **Postplant Broadcast Treatment:** Use LORSBAN 15G for control of European and southwestern corn borers by uniformly broadcasting the granules over the corn plants by aerial application or applying the granules into the corn whorls by ground application. Consult your State Agricultural Experiment Station or extension service specialist for local threshold information. Scouting for insect damage is strongly encouraged. **Preplant Broadcast Treatment:** Use equipment that will evenly distribute the granules over the treated area. Incorporate the granules into the soil using a S tine, C tine, disc cultivator, or other suitable incorporation equipment set to cut to a depth of 4 to 6 inches. Plant the crop in the usual manner as soon as possible after treatment.	
Citrus Orchard Floors	Fire Ants, and other ant species	—	—	—	6.7 lb./A	—	To control foraging ants and suppress mounds, apply granules with ground application equipment using a suitable granular applicator, such as a cyclone fertilizer spreader, that will uniformly broadcast the granules over the orchard floor. Retreat as needed. Do not apply where weed growth or other obstructions would impede uniform coverage of the orchard floor.	Do not apply more than 67 pounds of LORSBAN 15G per acre per season or apply last treatment within 28 days before harvest. Do not allow livestock to graze in treated areas.
Grain Sorghum	Lesser Cornstalk Borer	4 to 12 oz	4 to 12 oz	—	—	—	Apply in a 6 to 8 inch band over the row and lightly incorporate by using the press wheel or drag chain. Equivalent rates of LORSBAN 15G per acre for various row spacings are given in the accompanying table. Use the lowest rate for lesser cornstalk borer control when protection is desired for 2 to 3 weeks and higher rates for longer residual activity. It is absolutely necessary to incorporate the granules especially at lower rates	Do not make more than one application per season.
	Corn Rootworm and Cutworms	8 oz	8 oz	—	—	—		
Onions (Dry Bulb)	Onion Maggot	—	—	3.7 oz	—	—	Apply the granules at the rate of 3.7 ounces per 1,000 feet of row at 18-inch row spacing by placing in the furrow at seeding.	Do not make more than one application per year.
Peanuts	Cutworms, Lesser Cornstalk Borer, Southern Corn Rootworm Larvae, Southern Blight (White Mold)	—	—	—	15 oz	—	Preventative Treatment: Use LORSBAN 15G for control of cutworms, lesser cornstalk borer, and southern corn rootworm larvae and to inhibit the growth and development of white mold (southern blight) disease caused by the fungus *Sclerotium rolfsii*. Apply the insecticide just prior to or during pegging in a 10 to 18 inch wide band over the fruiting zone. Irrigation or rain following application will enhance activity. Under conditions of heavy white mold pressure, a	Do not make more than one application per season. Do not harvest within 21 days after treatment. Do not feed peanut forage or hay to meat or dairy animals.

Crop	Pest	Amount of LORSBAN 15G per 1,000 feet of row: Planting Time Treatment T-Band	Band	In-Furrow	Postplant Treatment	Amount of LORSBAN 15G per acre for Preplant Broadcast Treatment	Use Directions	Restrictions
							suitable fungicide may also be required and should be applied separately.	
	Lesser Cornstalk Borer	—	—	—	7.5 to 15 oz	—	Rescue Treatment: Use LORSBAN 15G for the control of lesser cornstalk borer when the Insect first appears usually just prior to or at pegging. Apply LORSBAN 15G in a 10 to 18 inch band over the fruiting zone. Equivalent rates of LORSBAN 15G per acre for various row spacings are given in the accompanying table.	
Soybeans	Lesser Cornstalk Borer, Cutworms	8 oz	8 oz	—	8 oz	—	Use LORSBAN 15G insecticide to control larvae of the lesser cornstalk borer and cutworms by application at planting time or postemergence as a band (row) treatment at the rate of 8 ounces per 1000 linear feet of row. In the Southeast apply 4 to 8 ounces per 100 linear feet of row as an at-plant treatment. Equivalent rates of LORSBAN 15G per acre for various row spacing are given in the accompanying table. When applied at planting time incorporate the granules into the top 1 inch of soil by placing in a 4 to 10 inch band over the row behind the planter shoe and ahead of the press wheel. A drag chain can also be used for incorporation. For postemergence treatment when insects first appear incorporate the granules in a 4 to 10 inch band to a depth of ½ to 1 inch using a suitable cultivator. Apply LORSBAN 15G with equipment that will provide uniform distribution of the granules. Do not apply as an in-furrow treatment.	Do not make more than one application per season.
Sugarbeets	Sugarbeet Root Maggot	6.5 to 9.0 oz	6.5 to 9.0 oz	4.5 to 9.0 oz	6.5 to 9.0 oz	—	Use LORSBAN 15G insecticide to control larvae of the sugarbeet root maggot and cutworms as a band treatment applied over the row at planting time or postemergence (up to 2 to 4 true leaf stage of the plant). Apply the granules by placing in a 5-inch wide band treatment applied over the drill row (at planting) or plants (postemergence). At plant treatments are recommended for cutworm control. For best control, when used at planting, the granules should be applied behind the planter shoe, over the drill row, and in **front of the press wheel** (T-band). Incorporate the granules into the top 1 to 2 inches of soil using suitable equipment or devices. For postemergent treatment, incorporate the banded granules to a depth of ½ to 1 inch using a suitable cultivator. Apply LORSBAN 15G with equipment that will provide uniform distribution of the granules.	Do not make more than one application per year.
	Cutworms	6.5 to 9.0 oz	6.5 to 9.0 oz	—	—	—		
Sunflowers	Cutworms	8 oz	—	—	—	—	Place the granules in a 7 inch wide band over the row behind the planter shoe in front of the press wheel and incorporate into the top one inch of soil using suitable equipment.	Do not make more than one application per season.
Sweet Potatoes	Wireworms (*Conoderus*), Flea Beetles (*Systena*), and Sweetpotato Beetle	—	—	—	—	13.5 lb	Use equipment that will evenly distribute the granules over the treated area. After application, incorporate the granules into the soil to a depth of 4 to 6 inches using a rotary hoe, disc cultivator, or other suitable incorporation equipment. Plant the crop in the usual manner as soon as possible after treatment.	Do not make more than one application per season.
Tobacco	Cutworms, Flea Beetle Larvae, Root Maggots, and Wireworms	—	—	—	—	13.5 to 20 lb	Apply the granules one week before transplanting, using equipment that will evenly distribute the granules over a treated area. Immediately following application, incorporate the granules into the soil to a depth of 2 to 4 inches using suitable equipment.	Do not make more than one application per season.

Crop	Pest	Amount of LORSBAN 15G per 1,000 feet of row: Planting Time Treatment T-Band	Planting Time Treatment Band	Planting Time Treatment In-Furrow	Postplant Treatment	Amount of LORSBAN 15G per acre for Preplant Broadcast Treatment	Use Directions	Restrictions
Vegetables Broccoli, Brussels Sprouts, Cabbage, Cauliflower, Chinese Cabbage, Collards, Kale, Kohlrabi, Rutabagas, Turnips	Root Maggots	—	4.6 to 9.2 oz	—	—	—	Place LORSBAN 15G across the seed row in a 4-inch wide band behind the planter shoe and in front of the press wheel to achieve shallow incorporation. This application requires a spreader or splitter on the end of the drop tube from the granule applicator.	Do not apply more than 7½ pounds of LORSBAN 15G per acre to crops planted in 40 inch rows or more than 15 pounds of LORSBAN 15G per acre to crops planted in 20-inch rows (or two rows per bed). Use proportional amounts for other row spacings not to exceed 15 pounds of LORSBAN 15G per acre. Do not make more than one application per season. Do not use rutabaga tops for food or feed purposes.
Radishes	Root Maggots	—	—	3.3 oz	—	—	Apply the granules in the seed furrow with the seed at planting time.	Do not apply more than 18.3 pounds of LORSBAN 15G per acre or make more than one application per season.

LORSBAN* 50W

WETTABLE POWDER INSECTICIDE

For Control of Various Insects on Certain Fruit and Nut Crops

ACTIVE INGREDIENT

Chlorpyrifos[O,O–diethyl O–(3,5,6–trichloro 2 pyridyl)phosphorothioate] 50%

INERT INGREDIENTS 50%

TOTAL 100%

EPA Reg. No. 464-552 EPA Est. No. 477-TX-1

* Trademark of The Dow Chemical Company

Keep Out of Reach of Children

WARNING—AVISO

PRECAUCION AL USUARIO:

Si usted no lee inglés, no use este producto hasta que la etiqueta le haya sido explicada ampliamente.

PRECAUTIONARY STATEMENTS

HAZARDS TO HUMANS AND DOMESTIC ANIMALS

HARMFUL IF ABSORBED THROUGH SKIN

MAY BE FATAL IF SWALLOWED

CAUSES EYE IRRITATION

Avoid Contact with Eyes, Skin, or Clothing • Avoid Breathing Dust or Spray Mist • Wash Thoroughly with Soap and Water after Handling and Before Eating or Smoking • When Loading the Spray Tank or Handling the Powder, Wear a Pesticide Respirator Jointly Approved by the Mining Enforcement and Safety Administration (formerly the U.S. Bureau of Mines) and by the National Institute for Occupational Safety and Health Under the Provisions of 30 CFR Part II • Wear Protective Clothing such as Long Pants, Long Sleeve Shirt, Gloves, Hat, and Boots when Loading the Spray Tank or Handling the Powder and During Application of the Spray Mixture • Work to Windward to Stay Out of Spray Drift or Mist • Remove Contaminated Clothing and Wash Before Reuse • Do Not Apply This Product in Such a Manner as to Directly or Through Drift Expose Workers or Other Persons • The Area Being Treated Must Be Vacated by Unprotected Persons • Keep Away from Food, Feedstuffs and Water Supplies

STATEMENT OF PRACTICAL TREATMENT: If Swallowed: Call a physician or Poison Control Center. Drink 1 or 2 glasses of water and induce vomiting by touching back of throat with finger. Do not induce vomiting or give anything by mouth to an unconscious person. **If On Skin:** Wash with plenty of soap and water. Get medical attention if irritation persists. **If In Eyes:** Flush with plenty of water for 5 minutes. Get medical attention if irritation persists. **If Inhaled:** Remove to fresh air if effects occur. Get medical attention.

NOTE TO PHYSICIANS: Chlorpyrifos is a cholinesterase inhibitor. Treat symptomatically. If exposed, plasma and red blood cell cholinesterase tests may indicate significance of exposure (baseline data are useful). Atropine, only by injection, is the preferable antidote. Oximes, such as 2-PAM/protopam, may be therapeutic if used early; however, use only in conjunction with atropine. In case of severe acute poisoning, use antidote immediately after establishing an open airway and respiration.

ENVIRONMENTAL HAZARDS

This pesticide is toxic to birds and wildlife, and extremely toxic to fish and aquatic organisms. Do not apply directly to water. Drift and runoff from treated areas may be hazardous to aquatic organisms in adjacent aquatic sites. Cover or incorporate spills. Do not contaminate water by cleaning of equipment or disposal of waste. This product is highly toxic to bees exposed to direct treatment or residues on blooming crops or weeds. Avoid use when bees are actively foraging. Protective information may be obtained from your Cooperative Agricultural Extension Service.

NOTICE

Read and understand the entire label before using. Use only according to label directions.

Before buying or using this product, read "WARRANTY LIMITATIONS AND DISCLAIMER" elsewhere on this label. If terms are not acceptable, return **unopened** package at once to seller for full refund of purchase price paid. Otherwise, use by the buyer or any other user constitutes acceptance of the terms under WARRANTY LIMITATIONS AND DISCLAIMER.

In case of an emergency endangering life or property involving this product, call collect **517–636-4400.**

AGRICULTURAL CHEMICAL

Do Not Ship or Store with Food, Feeds, Drugs or Clothing

DIRECTIONS FOR USE

It is a violation of federal law to use this product in a manner inconsistent with its labeling.

GENERAL INFORMATION

LORSBAN 50W insecticide is a wettable powder formulation designed for use as a cover spray to control various pests. The pests controlled are listed in the accompanying table. LORSBAN 50W is compatible with insecticides, miticides, and fungicides commonly recommended in orchard spray schedules except for alkaline materials such as Bordeaux mixture and lime. Test compatibility of the intended tank mixture before adding LORSBAN 50W to the spray or mix tank. Add proportionate amounts of each ingredient to a pint or quart jar, cap, shake, and let set 15 minutes. Formation of precipitates that do not readily redisperse indicates an incompatible mixture that should not be used.

MIXING INSTRUCTIONS

LORSBAN 50W mixes readily with water to form a suspension. Mix the required amount of LORSBAN 50W with a small quantity of water and add this premix while filling the spray tank. Alternatively, fill the tank with the required amount of water and then add the premix. Maintain sufficient agitation during both mixing and application to ensure uniformity of the spray mixture. LORSBAN 50W contains a wetting agent. Do not use additional wetting agents, spreaders or stickers.

USE AND DOSAGE RECOMMENDATIONS

Use LORSBAN 50W at the dosages indicated by application as a foliar spray to control pests listed in the following table. Mix the required dosage in sufficient water to ensure thorough and complete coverage of the foliage and crop and apply as a concentrate or dilute spray using conventional power-operated ground spray equipment. For dilute sprays applied to tree fruits and nuts, mix the required dosage in sufficient water to allow for spray to run off. For concentrate sprays, apply an equivalent amount of LORSBAN 50W per acre. Treat when pests appear or in accordance with local conditions. Consult your State Agricultural Experiment Station or Extension Service specialist for specific use information in your area.

DOSAGE AND USE RECOMMENDATIONS FOR LORSBAN 50W WETTABLE POWDER INSECTIDES

CROP	INSECTS CONTROLLED	DOSAGE LORSBAN 50W PER ACRE	RESTRICTIONS
Apples	Apple aphid, Apple maggot, Codling moth, European apple sawfly, Eye-spotted bud moth, Fruittree leafroller, Green fruitworm, Lesser appleworm, Obliquebanded leafroller, Orange tortrix, Pandemis leafroller, Oriental fruit moth, Plum curculio, Redbanded leafroller, Rosy apple aphid, San Jose scale, Tarnished plant bug, Tufted apple budmoth, Variegated leafroller, Woolly apple aphid, Dogwood borer (trunk spray)	3 lb per acre	Make no more than 8 applications per season. Do not apply last treatment within 28 days before harvest or apply last two treatments closer than 21 days apart. Do not allow livestock to graze in treated orchards. In California do not do work involving contact with trees within 2 days after treatment.
Almonds	Navel orangeworm, Peach twig borer, San Jose scale	4 lb per acre	Make no more than 3 applications per season on almonds and no more than 2 applications per season on walnuts. Do not apply within 14 days of harvest. Do not allow livestock to graze in treated orchards.
Walnuts	Codling moth, Walnut scale	4 lb per acre	
Filberts	Eyespotted bud moth, Filbert aphid, Filbert leafroller, Filbertworm, Obliquebanded leafroller, Omniverous leaftier, Winter moth	3 to 4 lb per acre	
Pecans	Black pecan aphid, Hickory shuckworm, Pecan leaf scorch mite, Pecan nut casebearer, Pecan weevil, Phylloxera spp.	2 lb per 100 gallon	Make no more than 5 applications per season. Do not apply within 28 days of harvest. Do not allow livestock to graze in treated orchards.

REENTRY STATEMENT

Do not allow reentry into treated areas within 24 hours unless protective clothing is worn. Protective clothing consists of at least a hat or other suitable head covering, a long-sleeved shirt and long-legged trousers or a coverall type garment (all of closely woven fabric covering the body, including the arms and legs), shoes and socks. Because certain states may require more restrictive reentry intervals for various crops treated with this product, consult your State Department of Agriculture for further information.

Written or oral warnings must be given to workers who are expected to be in a treated area or in an area about to be treated with this product. These oral warnings shall inform workers of areas or fields that may not be entered without protective clothing and actions to take in case of accidental exposure as shown by the Statement of Practical Treatment section on this label. When oral warnings are given, warnings shall be given in a language customarily understood by workers. Written warnings must include the following information: WARNING. Area treated with LORSBAN 50W wettable powder insecticide on (insert date of application). Do not enter without protective clothing for 24 hours. In case of accidental exposure (insert actions to take as shown by the Statement of Practical Treatment section on this label).

STORAGE AND DISPOSAL

Do not contaminate water, food or feed by storage or disposal. Open dumping is prohibited.

PESTICIDE DISPOSAL

Wastes resulting from the use of this product may be disposed of on site or at an approved waste disposal facility.

CONTAINER DISPOSAL

Completely empty bag into application equipment. Then dispose of empty bag in a sanitary landfill or by incineration, or, if allowed by state and local authorities, by burning. If burned, stay out of smoke.

WARRANTY LIMITATIONS AND DISCLAIMER

The Dow Chemical Company warrants that this product conforms to the chemical description on the label and is reasonably fit for the purposes stated on the label when used in strict accordance with the directions therein under normal conditions of use. THIS IS THE ONLY WARRANTY MADE ON THIS PRODUCT. NO OTHER EXPRESS AND NO IMPLIED WARRANTY OF MERCHANTABILITY OR FITNESS FOR A PARTICULAR PURPOSE IS MADE OUTSIDE OF THIS LABEL. Therefore, neither this warranty nor any other warranty of merchantability or fitness for a particular purpose, express or implied, extends to the use of this product contrary to label instructions (including conditions noted on the label, such as unfavorable temperatures, soil conditions, etc.), under abnormal conditions (such as excessive rainfall, drought, tornadoes, hurricanes, etc.) or under conditions not reasonably foreseeable to or beyond the control of seller.

When buyer or user suffers losses or damages resulting from the use or handling of this product (including claims based on contract, negligence, strict liability, or other legal theories), buyer or user must promptly notify in writing The Dow Chemical Company of any claims to be eligible to receive either remedy given below. The EXCLUSIVE REMEDY OF THE BUYER OR USER and the LIMIT OF LIABILITY of The Dow Chemical Company or any other seller will be one of the following, at the election of The Dow Chemical Company:

(1) Refund of purchase price paid by buyer or user for product bought, or

(2) Replacement of amount of product used.

The seller **will not be liable for consequential or incidental damages or losses.**

The terms of this Warranty Limitations And Disclaimer cannot be varied by any written or verbal statements or agreements. Any employee or sales agent of the seller is not authorized to vary or exceed the terms of this Warranty Limitations And Disclaimer in any manner.

45386-L1 D1185

*Trademark of THE DOW CHEMICAL COMPANY

PLICTRAN* 50W

MITICIDE

Wettable Powder Formulation

For Control of Plant-Feeding Mites on Almonds, Apples, Citrus, Nectarines, Peaches, Pears, Plums (prunes), Strawberries, Walnuts, Ornamentals and Hops

ACTIVE INGREDIENT:

Cyhexatin (tricyclohexylhydroxystannane)	50.0%
INERT INGREDIENTS	50.0%
TOTAL	100%

EPA Reg. No. 464–393

EPA Est. 464–MI–1+up ^{MM}; 42761–MS–1+up ^{PV}

Superscript used corresponds to letter in Lot No.

* Trademark of The Dow Chemical Company

WARNING

Keep Out of Reach of Children

AVISO

PRECAUCION AL USUARIO: Si usted no lee inglés, no use este producto hasta que la etiqueta le haya sido explicada ampliamente.

PRECAUTIONARY STATEMENTS

HAZARDS TO HUMANS AND DOMESTIC ANIMALS

MAY BE FATAL IF SWALLOWED • CAUSES SUBSTANTIAL BUT TEMPORARY EYE INJURY

Do Not Take Internally • Do Not Get In Eyes, On Skin, Or On Clothing • Do Not Breathe Dust Or Spray Mist • Wear Goggles • Wear A Mask Or Pesticide Respirator Jointly Approved By The Mining Enforcement And Safety Administration And The National Institute For Occupational Safety And Health. • Wash Thoroughly With Soap And Water After Handling And Before Eating Or Smoking • Remove Contaminated Clothing And Wash Before Reuse • Do Not Apply This Product In Such A Manner As To Directly Or Through Drift Expose Workers Or Other Persons • The Area Being Treated Must Be Vacated By Unprotected Persons

STATEMENTS OF PRACTICAL TREATMENT

If Swallowed: Call a Physician or Poison Control Center. Drink 1 or 2 glasses of water and induce vomiting by touching back of throat with finger. Do not induce vomiting or give anything by mouth to an unconscious person.

If In Eyes: Flush with plenty of water for 15 minutes. Get medical attention.

If On Skin: Wash with plenty of soap and water. Get medical attention if irritation persists.

If Inhaled: Remove to fresh air if effects occur. Get medical attention.

ENVIRONMENTAL HAZARDS

This product is toxic to fish, shrimp, crab, and other aquatic organisms. Keep out of lakes, streams, ponds, tidal marshes, and estuaries. Do not apply where runoff is likely to occur. Do not apply when weather conditions favor drift from areas treated. Do not contaminate water by cleaning of equipment or disposal of wastes. Use this product only as specified on this label.

NOTICE: Read the entire label. Use only according to label directions.

Before buying or using this product, read "WARRANTY LIMITATIONS AND DISCLAIMER" on the side panel. If terms are not acceptable, return **unopened** package at once to seller for full refund of purchase price paid.

Otherwise, use by the buyer or any other user constitutes acceptance of the terms under the Limit of Warranty and Liability.

IN CASE OF AN EMERGENCY endangering life or property involving this product, call collect **517-636-4400.**

AGRICULTURAL CHEMICAL

Do Not Ship or Store with Food, Feeds, Drugs, or Clothing

DIRECTIONS FOR USE

It is a violation of Federal law to use this product in a manner inconsistent with its labeling.

GENERAL INFORMATION: PLICTRAN 50W miticide is recommended for control of motile forms of plant-feeding mites, including strains both susceptible and resistant to other miticides. Many species of mites are controlled as indicated for each crop in the Table. PLICTRAN 50W is not highly injurious to certain predatory mites and insects, and to honeybees. PLICTRAN 50W is compatible with insecticides and fungicides commonly recommended in orchard spray schedules except for oil on certain crops. (See Application Notes in the Table).

When to Apply: Apply when mites appear. Repeat as necessary to prevent build-up of mites to damaging numbers. If all lifestages of mites are present at the time of treatment, another application should be considered in 10 to 14 days. Observe the limitations indicated in the application notes for specific crops.

Amount and How to Use: PLICTRAN 50W miticide mixes readily with water to form a suspension that can be applied using conventional spray equipment. PLICTRAN 50W contains a wetting agent. Do not use additional wetting agents, spreaders or stickers. Either dilute or concentrate sprays can be used. Follow the dosage recommendations in the Table. **Thorough and complete coverage of foliage and fruit is essential for good mite control.**

USE PRECAUTIONS

Do not graze or feed livestock on cover crops growing in treated areas.

REENTRY STATEMENT

Do not enter treated areas without protective clothing until sprays have dried (or, if appropriate, dusts have settled). Because certain states may require more restrictive reentry intervals for various crops treated with this product, consult your State Department of Agriculture for further information.

Written or oral warnings must be given to workers who are expected to be in a treated area or in an area about to be treated with this product. These oral warnings shall inform workers of areas or fields that may not be entered without protective clothing until sprays have dried (or, if appropriate, dusts have settled) and actions to take in case of accidental exposure as shown by the Statements of Practical Treatment section on this label. When oral warnings are given, warnings shall be given in a language customarily understood by workers. Oral warnings must be given if there is reason to believe that written warnings cannot be understood by workers. Written warnings must include the following information: WARNING: Area treated with PLICTRAN 50W miticide on (insert date of application). Do not enter without protective clothing until sprays have dried (or, if appropriate, dusts have settled). In case of accidental exposure (insert actions to take as shown by the Statements of Practical Treatment section on this label).

STORAGE AND DISPOSAL

Do not contaminate water, food, or feed by storage or disposal. Open dumping is prohibited.

PESTICIDE DISPOSAL

Wastes resulting from the use of this product may be disposed of on site or at an approved waste disposal facility.

CONTAINER DISPOSAL

Completely empty bag into application equipment. Then dispose of empty bag in a sanitary landfill or by incineration, or, if allowed by state and local authorities, by burning. If burned, stay out of smoke.

WARRANTY LIMITATIONS AND DISCLAIMER

The Dow Chemical Company warrants that this product conforms to the chemical description on the label and is reasonably fit for the purposes stated on the label when used in strict accordance with the directions therein under normal conditions of use. THIS IS THE ONLY WARRANTY MADE ON THIS PRODUCT. NO OTHER EXPRESS AND NO IMPLIED WARRANTY OF MERCHANTABILITY OR FITNESS FOR A PARTICULAR PURPOSE IS MADE OUTSIDE OF THIS LABEL. Therefore, neither this warranty nor any other warranty of merchantability or fitness for a particular purpose, express or implied, extends to the use of this product contrary to label instructions (including conditions noted on the label, such as unfavorable temperatures, soil conditions, etc.), under abnormal conditions (such as excessive rainfall, drought, tornadoes, hurricanes, etc.) or under conditions not reasonably foreseeable to or beyond the control of seller.

When buyer or user suffers losses or damages resulting from the use or handling of this product (including claims based on contract, negligence, strict liability, or other legal theories), buyer or user must promptly notify in writing The Dow Chemical Company of any claims to be eligible to receive either remedy given below. The EXCLUSIVE REMEDY OF THE BUYER OR USER and the LIMIT OF LIABILITY of The Dow Chemical Company or any other seller will be one of the following, at the election of The Dow Chemical Company:

(1) **Refund of purchase price** paid by buyer or user for product bought, **or**

(2) **Replacement** of amount of product used.

The seller **will not be liable for consequential or incidental damages or losses.**

The terms of this Limit of Warranty and Liability cannot be varied by any written or verbal statements or agreements. Any employee or sales agent of the seller is not authorized to vary or exceed the terms of this Warranty Limitations and Disclaimer in any manner.

DOSAGE RECOMMENDATIONS PER APPLICATION

CROP	MITES CONTROLLED	FOR CONCENTRATE SPRAYS PLICTRAN 50W PER ACRE GROUND	FOR CONCENTRATE SPRAYS PLICTRAN 50W PER ACRE AIR	FOR DILUTE SPRAYS PLICTRAN 50W PER 100 GALLONS WATER	FOR DILUTE SPRAYS SPRAY MIXTURE PER ACRE	APPLICATION NOTES
Apples	European red; Two-spotted spider; Schoene Spider; Pacific spider; Yellow spider; Apple rust; McDaniel	1–3 lb	1–2 lb	4–6 oz	300–800 gallons	Frequently, the first application of PLICTRAN 50W miticide will be needed at or soon after petalfall. Do not apply a total of more than 12 lb of PLICTRAN 50W per acre per season by ground application. For aerial application, use a minimum of 5 gallons of water per acre and appropriate application equipment and operating conditions to provide thorough coverage of foliage. Limit aerial applications to 3 per season, with a total of no more than 6 lb PLICTRAN 50W per acre per season. Do not apply within 14 days of harvest.
Pears	European red; Two-spotted spider; Schoene spider; Pacific spider; Yellow spider; Apple rust; Pear rust	1–2.5 lb	1–2lb	4–6 oz	300–800 gallons	Frequently, the first application of PLICTRAN 50W miticide will be needed at or soon after petalfall. Do not apply a total of more than 10 lb of PLICTRAN 50W per acre per season by ground application. Do not use PLICTRAN 50W and oil sprays as a tank mix or separately within 28 days of each other because of possible injury to the fruit. For aerial application, use a minimum of 5 gallons of water per acre and appropriate application equipment and operating conditions to provide thorough coverage of foliage. Limit aerial applications to 3 per season, with a total of no more than 6 lb PLICTRAN 50W per acre per season. Do not apply within 14 days of harvest.
Citrus includes: grapefruit; lemon; lime; orange; tangerine	Citrus red Texas citrus Citrus rust Pacific spider	2–4 lb	2–4 lb	4–6 oz	800–1600 gallons	Do not tank mix PLICTRAN 50W miticide with oil in citrus sprays. Do not apply oil within 4 weeks before or after application of PLICTRAN 50W *in Florida* or within 8 weeks in other states. Do not apply to lemon trees when yellow fruit is present. To prevent a "pimpling" type injury on foliage that has not "hardened off," avoid application of PLICTRAN 50W to "feather" and new growth flushes. *In California and Arizona,* DO NOT APPLY PLICTRAN 50W to bearing citrus trees during the time period between the appearance of first blooms and until fruit has "hardened off," normally about mid-August, because of possible injury to the fruit. For aerial application, use minimum of 10 gallons of water per acre and appropriate application equipment and operating conditions to provide thorough coverage of foliage.

CROP	MITES CONTROLLED	FOR CONCENTRATE SPRAYS PLICTRAN 50W PER ACRE GROUND	AIR	FOR DILUTE SPRAYS PLICTRAN 50W PER 100 GALLONS WATER	SPRAY MIXTURE PER ACRE	APPLICATION NOTES
Peaches Plums (prunes) Nectarines	European red; Two-spotted spider; Peach silver; McDaniel; Pacific	1–2.5 lb		4–6 oz	300–800 gallons	Apply no more than 6 lb PLICTRAN 50W per acre per season. Do not apply summer spray oils during the same growing season, otherwise some foliar injury may be experienced.
Almonds Walnuts	Pacific; European red; Two-spotted spider	1–2.5 lb		4–6 oz	300–800 gallons	Apply no more than 6 lb PLICTRAN 50W per acre per season. Do not apply summer spray oils during the same growing season, otherwise some foliar injury may be experienced.
Strawberries	Two-spotted spider			12–16 oz	150–250 gallons	Use the higher rates in plantings with dense foliage or to prolong the time between applications. Insure thorough coverage of foliage. In *California and Florida,* it will sometimes be desirable to treat at 1 to 2 week intervals during the periods of intense mite pressure. Apply a minimum of 24 oz PLICTRAN 50W per acre per application. Do not harvest treated berries on the same day that PLICTRAN 50W is sprayed.
Hops	Two-spotted spider	2.2–3.5 lb		16–28 oz	150–225 gallons	Use 2.2 lb in 150 gallons of water per acre for first application and 3.5 lb in 225 gallons of water per acre for second and third applications. Do not apply more than 9 lb PLICTRAN 50W per acre per season. Do not apply within 21 days of harvest. Do not feed hop cones or refuse to livestock.
Ornamentals	Two-spotted spider			4–6 oz	Spray to wet	Includes greenhouse grown carnations, chrysanthemums, poinsettias. Apply PLICTRAN 50W to the foliage only of chrysanthemums (prebloom) and poinsettias (prebract) to avoid possible injury to blossoms and bracts.

TELONE* II

SOIL FUMIGANT

A Clean, Clear, Non-Clogging Liquid for Preplant Treatment of Crop Lands to Control Plant Parasitic Nematodes and Certain Other Soil Pests

ACTIVE INGREDIENT:

1,3-Dichloropropene	94%
INERT INGREDIENTS:	6%
TOTAL	100%

1 gallon of TELONE 11 weighs 10.1 lb.

E.P.A. Registration No. 464-511

E.P.A. Est. No. 464-TX-1^{TB}; 454-MI-1$^{(MM)}$; 464-CA-1^{WP}; 05770-CO-1^{V};
33780-SC-1^{J}; 33776-WA-1^{W}; 14775-FL-1^{B}.

Superscript used corresponds to letter in Lot No.

*Trademark of The Dow Chemical Company

DANGER
Keep Out of Reach of Children
PELIGRO

PRECAUCION AL USUARIO: Si usted no lee inglés, no use este producto hasta que la etiqueta le haya sido explicada ampliamente.

PRECAUTIONARY STATEMENTS

HAZARDS TO HUMANS AND DOMESTIC ANIMALS

HAZARDOUS LIQUID AND VAPOR • MAY BE FATAL IF INHALED, ABSORBED THROUGH SKIN, OR SWALLOWED • CAUSES BURNS OF SKIN OR EYES • MAY PRODUCE KIDNEY AND LIVER DAMAGE UPON PROLONGED EXPOSURE

Do Not Breathe Vapor• Do Not Get in Eyes, on Skin, on Clothing • Do Not Take Internally • Use Only With Adequate Ventilation • Wear Eye and Skin Protection Necessary to Prevent Contact When Handling TELONE II • Wash Thoroughly After Handling and Before Eating or Smoking • If protective gear, such as boots or gloves, becomes contaminated, immediately wash with soap and water. Never wear protective gear having the odor of 1,3-dichloropropene. Aerate and wash all protective gear thoroughly after use until all odor is gone. Render unusable and dispose of contaminated leather goods, including shoes. • Do not apply this product in such a manner as to directly or through drift expose workers or other personnel. The area being treated must be vacated by unprotected persons. • SEE FURTHER INSTRUCTIONS UNDER HANDLING PRECAUTIONS SECTION.

STATEMENT OF PRACTICAL TREATMENT

If inhaled, remove to fresh air. If not breathing, give artificial respiration, preferably mouth-to-mouth. If breathing is difficult, give oxygen. Call a physician.

If On Skin, immediately flush skin with plenty of water for at least 15 minutes while removing contaminated clothing and shoes. Call a physician. If water is not immediately available, remove excess chemical from skin with sorbent material such as towel or dry soil, then proceed at once to location where water is available and thoroughly wash contaminated skin with plenty of water. Call a physician.

If In Eyes, immediately flush eyes with plenty of water for at least 15 minutes. Call a physician.

If Swallowed, immediately induce vomiting by giving two glasses of water and sticking finger down throat. Call a physician. Never give anything by mouth to an unconscious person.

PHYSICAL OR CHEMICAL HAZARDS

COMBUSTIBLE

Do Not Use, Pour, Spill, or Store Near Heat or Open Flame
Do Not Cut or Weld Container

ENVIRONMENTAL HAZARDS

This product is toxic to fish and wildlife. • Do not apply directly to water. • Do not contaminate water by cleaning of equipment or disposal of wastes.

NOTICE: Read the entire label. Use only according to label directions.

Before buying or using this product, read "WARRANTY LIMITATIONS AND DISCLAIMER" elsewhere on this label. If terms are not acceptable, return **unopened** package at once to seller for full refund of purchase price paid. Otherwise, use by the buyer or any other user constitutes acceptance of the terms under the Limit of Warranty and Liability.

In case of an emergency endangering life or property involving this product, call collect **517–636–4400**

AGRICULTURAL CHEMICAL
Do Not Ship or Store with Food, Feeds, Drugs, or Clothing

DIRECTIONS FOR USE

It is a violation of Federal law to use this product in a manner inconsistent with its labeling.

HANDLING PRECAUTIONS

DO NOT INHALE VAPORS: NIOSH- or MSHA-approved respiratory protection should be worn when TELONE II liquid soil fumigant is exposed to the atmosphere or when conducting operations which vent to the atmosphere. A NIOSH- or MSHA-approved half-face respirator with chemical worker's goggles or full-face respirators shall be used during small spills, repairs, calibrations, transfers, sampling and when working in poorly ventilated areas. When in use cannisters or cartridges shall be replaced daily or sooner if specified by manufacturer or at first sign of odor breakthrough, whichever comes first. NIOSH-approved cartridges, such as a Welsh 7400-IL, will be adequate for short-term situations such as listed above.

Where very high concentrations of vapors might be expected (such as large spills in poorly ventilated areas) a self-contained or air-supplied respirator should be used.

DO NOT GET ON SKIN: When handling or working with TELONE II wear clean body covering including gloves and heavy footwear. Immediately remove contaminated coverings. Aerate and wash all protective clothing and gear thoroughly after use.

There are no protective clothing materials that are *completely* impervious. Rubber, vinyl protective gear, thin layers of polyethylene (minimum 1 mil), give short-term protection and shall be immediately discarded upon contamination. Heavy (3+ mil) polyethylene, rubber, neoprene provide longer term protection. Leather gives no protection. If gear becomes contaminated, immediately wash with soap and water. Never wear protective gear having odor of TELONE II. Wash and aerate all protective gear thoroughly after use until all odor is gone.

DO NOT GET IN EYES: Wear eye protection such as chemical worker's goggles or full-face respirator when handling TELONE II.

DO NOT SWALLOW: Do not use the mouth to siphon TELONE II from containers or to blow out clogged lines, nozzles, etc.

STORAGE, SHIPMENT AND DISPOSAL

STORAGE
Store in tightly-closed original container in a cool place away from dwellings. Do not allow contamination of seeds, plants, fertilizers, or other pesticide chemicals. Do not contaminate food, feedstuffs, drugs, or domestic water supplies. In outside storage, store drums on sides to avoid accumulation of rain water in top or bottom recessed areas.

SHIPMENT
Do not ship or store with food, feeds, drugs, or clothing.

DISPOSAL
Triple rinse containers and dispose of rinsate in field just treated. After aeration offer containers to qualified reconditioner or destroy by puncturing or crushing and dispose of in an approved sanitary landfill or bury in the user's cropland or dispose of by other procedures approved by state and local authorities. Do not pollute surface or underground water supplies.

GENERAL INFORMATION

Use TELONE II soil fumigant only as a **preplanting soil treatment** to control plant parasitic nematodes [root knot, meadow (lesion), citrus, cyst formers (golden, sugar beet, soybean), burrowing, ring, spiral, sting, pin, stubby root, stylet, dagger]; also to control wireworms and garden centipedes (symphylans). Fumigate land to be planted to the crops listed below by applying TELONE II under the conditions, and at the rates recommended under DIRECTIONS FOR USE, DOSAGE AND USE RECOMMENDATIONS and USE PRECAUTIONS. Read the entire label before using TELONE II.

Vegetable Crops:

asparagus	califlower	horseradish	parsnips	shallots
beans	celery	kale	peas	spinach
beets	collards	kohlrabi	peppers	squash (summer)
blackeyed peas	corn	leeks	pimentoes	squash (winter)
broccoli	cowpeas	lettuce	potatoes	sweet potatoes
brussels sprouts	cucumbers	melons	pumpkins	swiss chard
cabbage	egg plant	mustard greens	radishes	tomatoes
cantaloupe	endive	okra	rutabaga	turnips
carrots	garlic	onions	salsify	watermelons

Field Crops:

alfalfa	flax	oats	sorghum
barley	grasses	pasture grass	soybeans
birdsfoot trefoil	hops	peanuts	sugar beets
buckwheat	lespedeza	popcorn	sugarcane
clover	millet	rice	tobacco
corn	milo	rye	vetch
cotton	mint	safflower	wheat

Citrus Fruit Tree Planting Sites:
grapefruit kumquats lemons limes oranges tangerines tangelos

Deciduous Fruit and Nut-Tree Planting Sites:

almonds	dates	olives	plums
apples	figs	peaches	pomegranates
apricots	filberts	pears	prunes
cashew nuts	hazelnuts	pecans	quince
cherries	hickory nuts	persimmons	walnuts
chestnuts	nectarines	pineapple	

Brush and Vine Planting Sites:

blackberries	currants	huckleberries	youngberries
blueberries	dewberries	loganberries	
boysenberries	gooseberries	raspberries	
cranberries	grapes	strawberries	

Nursery Crops including floral plants, ornamentals, shrubs, and bushes; forest, shade, fruit and nut trees and vine and bramble fruits of all types.

APPLICATION DIRECTIONS

WHEN TO TREAT: Apply TELONE II either in the spring or fall, whenever soil type and conditions permit. For best results with annual crops, treat the soil each year. In northern states, late summer, or early fall treatment (before October 15) is best for land to be planted to early spring crops, especially transplanted crops such as celery, tomatoes, and nursery and orchard stock. Early fall treatment permits planting a fall cover crop. **Note:** Treat muck soils only in the early fall and plant as late as possible in the spring; treat fine textured (clay) soils only when they are near or at the wilting point. Do not use TELONE II to treat any type of soil when it is cold and/or wet.

SOIL PREPARATIONS: TELONE II gives best results when conditions permit rapid diffusion of the fumigant through the soil and the soil surface can be sealed to prevent excessive fumigant loss during the exposure period. The soil should be in good seed bed condition, free of clods and undecomposed plant material, moisture at about one-half of field capacity, and temperature between 40° and 80°F at the depth of injection. If undecomposed plant debris is present, it should be plowed down and allowed to decompose before applying TELONE II. Tillage deeper than 12 inches is necessary for good fumigant penetration in soils where a hard or "plow" pan occurs at plow depth. Where deep tillage is used, the tillage equipment may often be modified to simultaneously apply the fumigant and thus avoid going over the field twice.

APPLICATION: TELONE II may be applied either as an **overall (broadcast)** or **row treatment,** using suitable application equipment that will ensure placement of the fumigant at least 6 to 8 inches below the final soil surface. **For overall application** use either plow-sole or chisel equipment with the chisels spaced 12 inches apart. When the fumigant is injected at a depth of 12 inches or more (deep tillage), the chisel spacing may be up to twice the application depth but should not exceed 30 inches. Application may be made in the same direction or at an angle to the direction of the planting row, whichever is most convenient. **For row application,** use chisel equipment with one chisel per row or two chisels spaced 12 inches apart to treat only the soil where the crop is to be planted. When one chisel per row is used, adjust the fumigant flow rate to distribute about 1⅓ times more fumigant per chisel than is recommended for overall application. When two chisels are used per row, apply at the same flow rate per chisel as for overall. In both cases, the amount of fumigant required per acre will decrease as the distance between rows is increased and vice versa. At time of planting, avoid placing the seed row directly over the furrow left by the applicator chisel. When a single chisel is used per row, place the seed 3 to 4 inches to one side of the chisel furrow; when two chisels are used, plant in the center of the area between the chisel furrows.

SEALING: Immediately after application, compact the soil surface to prevent excessive fumigant loss. After chisel application, use a roller, cultipacker, or similar sealing device. After plow-sole application, disk the land, then compact it by floating or rolling. Sealing after row application can be accomplished by the tractor wheel, by listing, or by bedding so that the fumigant will be 12 to 14 inches below the top of the bed. When fumigating listed rows, seal in the fumigant with ring rollers, press sealers, or by relisting.

EXPOSURE PERIOD: After application and sealing, leave the soil undisturbed for 7 to 14 days. A longer exposure period will be required if the soil becomes excessively cold or wet during the exposure period.

AERATION AND PREPARATION OF SOIL BEFORE PLANTING: At the end of the exposure period allow the soil to aerate completely before planting the crop. Aeration is usually complete when the odor of TELONE II is no longer evident. Under optimum soil and weather conditions, allow one week of aeration time for each 10 gallons of TELONE II applied per acre. When TELONE II is used for treating deep-rooted tree and shrub planting sites, a 3 to 6 months aeration period should be allowed. To hasten aeration, **especially if heavy rains or low temperatures occur during the exposure period,** work the soil to the depth of the treatment zone. After row treatment use a knife-like chisel in the bed without turning the soil, thus reducing possible recontamination of the treated soil. To hasten aeration after overall treatment, plow or deep cultivate to the depth of the treated zone. This is especially desirable in northern states after fall fumigation of muck soils.

REENTRY

Do not enter treated areas without heavy synthetic rubber foot gear until after 3 days post treatment. Because certain states may require more restrictive reentry intervals for various crop areas treated with this product, consult your State Department of Agriculture for further information.

DOSAGE AND USE RECOMMENDATIONS TO CONTROL NEMATODES, SYMPHYLANS†, AND WIREWORMS†

CROPS (CONSULT LIST OF INDIVIDUAL CROPS UNDER GENERAL INFORMATION)	TYPE OF TREATMENT	SOIL TYPE	DOSAGE: GALLONS PER ACRE[1]	DOSAGE: FL. OZ./1000 FT. ROW PER CHISEL
Shallow Rooted Plants: Field Crops, Floral Crops, Grasses and Turf, Small Fruits, Vegetables, Ornamentals	Row (42″)[6]	Mineral	4.5 to 6	46 to 62
		Muck or peat	9 to 12	93 to 123
	Overall (Broadcast)	Mineral	9 to 15[2]	26 to 44
		Muck or Peat	24[3] to 36	71 to 106
Strawberries	Overall	Mineral	24 to 36	71 to 106
Sugar Beets: Root Knot Nematode	Row (42″)	Mineral	6	62
	Overall	Mineral	12 to 15	35 to 44
Sugar Beet Nematode	Row (42″)	Mineral	9	93
	Overall	Mineral	12 to 18	35 to 53
Pineapple[4]	Row	Mineral	24 to 36	—
Citrus—Florida[5]	Overall	Mineral	36	106

Nursery and Field:[7] Citrus Fruit Trees, Deciduous Fruit Trees[8], Forest Trees, Grapes, Nut Trees, Ornamentals (deep rooted)	Overall Gallons Per Acre to Penetrate Various Depths			
Mineral Soils	3 ft.	4 ft.	5 ft.	6 ft.
Sand	15	21	27	33
Sandy Loom	27	30	36	48
Silt Loam	42	51	63	75
Clay Loam	54	69	84	102

1 Use the higher rates in heavier soil.
2 For cyst-forming nematodes increase dosage to 18 gallons per acre (53 fl. oz./1000 ft. row per chisel).
3 For muck soils containing less than 30% organic matter use 18 gallons per acre.
4 For Hawaiian pineapple, application may be made at time of or just before planting.
5 For burrowing nematode in citrus inject on 18-inch centers, 12 inches deep. Keep free of plants susceptible to burrowing nematodes for 2 years before replanting to citrus.
6 Row treatment is not recommended for potatoes in irrigated areas of western and northwestern states.
7 Tree planting sites prepared by backhoeing may be treated by pouring 27 fl. oz. of TELONE II into the hole during backfilling. For best results prepare and treat sites in the fall and plant in the spring.
8 Use TELONE II as an aid in the control of bacterial canker and decline of peach trees by application as a preplant, overall treatment of light (sandy) soils at the rate of 24 to 36 gallons per acre (71 to 106 fl. oz./1000 ft. row per chisel) preferably in the fall when the soil is warm (50-85°F at 6-inch depth) and moist. Inject the fumigant at a depth of 10 to 12 inches with chisels mounted on 12 inch centers. Follow directions for soil preparation, sealing, exposure, and aeration as specified elsewhere on this label.

† **NOTE:** To control symphylans (garden centipedes) use only overall at 18 or more gallons per acre, and apply during late summer or early fall when the soil is warm. To control wireworms use dosages recommended for nematodes in overall or broadcast treatments.

Written or oral warnings must be given to workers who are expected to be in a treated area or in an area about to be treated with this product. These oral warnings shall inform workers of areas of fields that may not be entered without heavy synthetic rubber foot gear until after 3 days post treatment. In case of accidental exposure follow directions as shown by the Statement of Practical Treatment section on this label. When oral warnings are given, warnings shall be given in a language customarily understood by workers. Written warnings must include the following information:

DANGER. Area treated with TELONE II soil fumigant on (insert date of application). Do not enter without heavy synthetic rubber foot gear until after 3 days post treatment. In case of accidental exposure take actions as shown by the Statement of Practical Treatment section on this label.

White Potatoes in Northwestern States: Use TELONE II as a spring or preferably a fall treatment to control quackgrass and for suppression of the damaging effects of Verticillium wilt in fields to be planted to white potatoes. Apply as an overall treatment according to the following directions:

Time of Treatment	Gallons per Acre	Fl. Oz./1000 Ft. Row per Chisel
Spring	17 to 25	50 to 73
Fall	25 to 34	73 to 100

Mint in Northwestern States: Use 59 gallons of TELONE II per acre (173 fl. oz./1000 ft. row per chisel) as an overall treatment in the spring, or preferably in the fall, to aid in the reduction of the damaging effects of Verticillium wilt in disease infested land to be used for mint production. After treatment allow at least 7 to 8 weeks or until the odor of the fumigant has left the soil before planting. Consult State Agriculture Experiment Station or State Extension Service Specialists for the use of other practices such as flaming the stubble, weed control, and cultural practices when using TELONE II as an aid to reducing damage caused by Verticillium wilt.

USE PRECAUTIONS

Important — Note Carefully: Fumigation may temporarily raise the level of ammonia nitrogen and soluble salts in the soil. This is most likely to occur when heavy rates of fertilizer and fumigant are applied to soils that are either cold, wet, acid, or high in organic matter. To avoid injury to plant roots, fertilize as indicated by soil tests made after fumigation. To avoid ammonia injury or nitrate starvation, or both, to crops on high organic soils do not use fertilizers containing ammonium salts and use only fertilizers containing nitrates, until after the crop is well established and the soil temperature is above 65°F.

Certain crops including cotton, sugarcane, and pineapple are tolerant to ammonia and the above rule does not apply to them. When using high rates of TELONE II as required by certain state nursery regulations, liming of highly acid soils before fumigation may stimulate nitrification and reduce the possibility of ammonia toxicity. Certain nursery crops such as citrus seedlings. *Cornus sp., Crataegus sp.,* spruce, and vegetable crops such as cauliflower have shown evidence of phosphorus deficiency following fumigation. To avoid this possible effect, it is suggested that additional phosphate fertilizer be used on soils where experience indicates a deficiency may occur.

Attention: To avoid reinfestation of treated soil do not use irrigation water, transplants, tools, seed pieces, or crop remains that could carry soilborne pests from infested land. Clean equipment carefully before using.

Since TELONE II soil fumigant is corrosive under certain conditions, flush all applicators with fuel oil or kerosene immediately after use. DO NOT USE WATER. **Do not use containers, pumps, or other transfer equipment made of aluminum, magnesium or their alloys, as under certain conditions TELONE II may be severely corrosive to such metals.**

Use this product only as specified on this label.

WARRANTY LIMITATIONS AND DISCLAIMER

The Dow Chemical Company warrants that this product conforms to the chemical description on the label and is reasonably fit for the purposes stated on the label when used in strict accordance with the directions therein under normal conditions of use. THIS IS THE ONLY WARRANTY MADE ON THIS PRODUCT. NO OTHER EXPRESS AND NO IMPLIED WARRANTY OF MERCHANTABILITY OR FITNESS FOR A PARTICULAR PURPOSE IS MADE OUTSIDE OF THIS LABEL. Therefore, neither this warranty nor any other warranty of merchantability or fitness for a particular purpose, express or implied, extends to the use of this product contrary to label instructions (including conditions noted on the label, such as unfavorable temperatures, soil conditions, etc.), under abnormal conditions (such as excessive rainfall, drought, tornadoes, hurricanes, etc.) or under conditions not reasonably foreseeable to or beyond the control of seller.

When buyer or user suffers losses or damages resulting from the use or handling of this product (including claims based on contract, negligence, strict liability, or other legal theories), buyer or user must promptly notify in writing The Dow Chemical Company of any claims to be eligible to receive either remedy given below. The EXCLUSIVE REMEDY OF THE BUYER OR USER and the LIMIT OF LIABILITY of The Dow Chemical Company or any other seller will be one of the following, at the election of The Dow Chemical Company:

(1) **Refund of purchase price** paid by buyer or user for product bought **or**

(2) **Replacement** of amount of product used.

The seller **will not be liable for consequential or incidental damages or losses.**

The terms of this Limit of Warranty and Liability cannot be varied by any written or verbal statements or agreements. Any employee or sales agent of the seller is not authorized to vary or exceed the terms of this Warranty Limitations And Disclaimer in any manner.

TELONE* C-17

SOIL FUNGICIDE AND NEMATICIDE

RESTRICTED USE PESTICIDE

For retail sale to and use only by Certified Applicators or persons under their direct supervision and only for those uses covered by the Certified Applicator's certification.

A Multi-Purpose, Preplant Soil Fumigant for Use to Control Nematodes, Symphylans, Wireworms and Certain Soil Borne Diseases in Crop Lands

ACTIVE INGREDIENTS:

1,3-Dichloropropene 74.0%
Chloropicrin 16.5%
INERT INGREDIENTS: 9.5%
TOTAL 100.0%

One gallon of TELONE C-17 weighs about 10.5 lb.

Contains 7.82 pounds of 1,3-dichloropropene and 1.79 pounds of chloropicrin per gallon.

E.P.A. Reg.No. 464-379

E.P.A. Est. 464-CA-1$^{(WP)}$; 464-AR-1$^{(AM)}$; 464-MI-1$^{(MM)}$.

Superscript used corresponds to letter in Lot No.

* Trademark of The Dow Chemical Company

DANGER–POISON

Keep Out of Reach of Children
PELIGRO

PRECAUCION AL USUARIO: Si usted no lee inglés, no use este producto hasta que la etiqueta le haya sido explicada ampliamente.

PRECAUTIONARY STATEMENTS

HAZARDS TO HUMANS AND DOMESTIC ANIMALS

HAZARDOUS LIQUID AND VAPOR • MAY BE FATAL IF INHALED, ABSORBED THROUGH SKIN, OR SWALLOWED • CAUSES BURNS OF SKIN OR EYES • MAY PRODUCE KIDNEY AND LIVER DAMAGE UPON PROLONGED EXPOSURE • VAPOR EXTREMELY IRRITATING

Do Not Breathe Vapor • Do Not Get in Eyes, on Skin, on Clothing • Do Not Take Internally • Use Only With Adequate Ventilation • Wear Eye and Skin Protection Necessary to Prevent Contact When Handling TELONE C-17 • Wash Thoroughly After Handling and Before Eating or Smoking • If protective gear, such as boots or gloves, becomes contaminated, immediately wash with soap and water. Never wear protective gear having the odor of 1,3-dichloropropene. • Aerate and wash all protective gear thoroughly after use until all odor is gone. • Render unusable and dispose of contaminated leather goods, including shoes. • Do not apply this product in such a manner as to directly or through drift expose workers or other personnel. The area being treated must be vacated by unprotected persons. • SEE FURTHER INSTRUCTIONS UNDER HANDLING PRECAUTIONS SECTION

STATEMENT OF PRACTICAL TREATMENT

IF INHALED, remove to fresh air. If not breathing, give artificial respiration, preferably mouth-to-mouth. If breathing is difficult, give oxygen. Call a physician.

IF ON SKIN, *immediately* flush skin with plenty of water for at least 15 minutes while removing contaminated clothing and shoes. Call a physician. If water not immediately available, remove excess chemical from skin with sorbent material such as towel or dry soil, then proceed at once to location where water is available and thoroughly wash contaminated skin with plenty of water. Call a physician.

IF IN EYES, *immediately* flush eyes with plenty of water for at least 15 minutes. Call a physician.

IF SWALLOWED, *immediately* induce vomiting by giving two glasses of water and sticking finger down throat. Call a physician. Never give anything by mouth to an unconscious person.

PHYSICAL OR CHEMICAL HAZARDS

COMBUSTIBLE

Do Not Use, Pour, Spill, ot Store Near Heat or Open Flame • Do Not Cut or Weld Container

ENVIRONMENTAL HAZARDS

This product is toxic to fish and wildlife. Do not apply directly to water. Do not contaminate water by cleaning of equipment or disposal of wastes.

In case of an emergency endangering life or property involving this product, call collect **517–636-4400.**

AGRICULTURAL CHEMICAL

Do Not Ship or Store With Food, Feeds, Drugs, or Clothing

DIRECTIONS FOR USE

It is a violation of Federal law to use this product in a manner inconsistent with its labeling.

HANDLING PRECAUTIONS

DO NOT INHALE VAPORS: NIOSH- or MSHA-approved respiratory protection should be worn when TELONE C-17 liquid soil fumigant is exposed to the atmosphere or when conducting operations which vent to the atmosphere. A NIOSH- or MSHA-approved full-face respirator shall be used during small spills, repairs, calibrations, transfers, sampling and when working in poorly ventilated areas. When in use cannisters or cartridges shall be replaced after each half hour of use or sooner if specified by manufacturer or at first sign of odor breakthrough, whichever comes first. NIOSH-approved cartridges, such as a Welsh 7400-IL, will be adequate for short-term situations such as listed above. Where very high concentrations of vapors might be expected (such as large spills in poorly ventilated areas) a self-contained or air-supplied respirator should be used.

DO NOT GET ON SKIN: When handling or working with TELONE C-17, wear clean body covering including gloves and heavy footwear. Immediately remove contaminated coverings. Aerate and wash all protective clothing and gear thoroughly after use.

There are no protective clothing materials that are *completely* impervious. Rubber, vinyl protective gear, thin layers of polyethylene (minimum 1 mil), give short-term protection and shall be immediately discarded upon contamination. Heavy (3+ mil) polyethylene, rubber, neoprene provide longer term protection. Leather gives no protection. If gear becomes contaminated, immediately wash with soap and water. Never wear protective gear having odor of TELONE C-17. Wash and aerate all protective gear thoroughly after use until all odor is gone.

DO NOT GET IN EYES: Wear eye protection such as full-face respirator when handling TELONE C-17.

DO NOT SWALLOW: Do not use the mouth to siphon TELONE C-17 from containers or to blow out clogged lines, nozzles, etc.

STORAGE, SHIPMENT AND DISPOSAL

STORAGE

Store in tightly-closed original container in a cool place away from dwellings. Do not allow contamination of seeds, plants, fertilizers, or other pesticide chemicals. Do not contaminate food, feedstuffs, drugs, or domestic water supplies. In outside storage, store drums on sides to avoid accumulation of rain water in top or bottom recessed areas.

SHIPMENT

Do not ship or store with food, feeds, drugs, or clothing.

DISPOSAL

Triple rinse containers and dispose of rinsate in field just treated. After aeration offer containers to qualified reconditioner or destroy by puncturing or crushing and dispose of in an approved sanitary landfill or bury in the user's cropland or dispose of by other procedures approved by state and local authorities. Do not pollute surface or underground water supplies.

GENERAL INFORMATION

Apply TELONE C-17 soil fungicide and nematicide as a preplanting soil treatment to control or to aid in reducing the damaging effects of certain *soil borne diseases* [soil rot (soil pox) of sweet potatoes; Granville (bacterial) wilt, black root rot, black shank diseases of tobacco; Verticillium wilt of white potatoes and mint, pink root of onions, pod rot of peanuts]; *plant parasitic nematodes* [root-knot, meadow (lesion), citrus, cyst formers (golden, sugar beet, soybean), burrowing, ring, spiral, sting, pin, stubby root, stylet, dagger and certain others]; *symphylans* (garden centipedes) and *wireworms.* Treat land to be planted to the crops listed below by applying TELONE C-17 under the conditions and at the rates recommended under DIRECTIONS FOR USE. Read this entire label before using TELONE C-17.

VEGETABLE CROPS: Asparagus, Beans, Beets, Blackeyed Peas, Broccoli, Brussels Sprouts, Cabbage, Cantaloupe, Carrots, Cauliflower, Celery, Collards, Corn, Cowpeas, Cucumbers, Egg Plant, Endive, Garlic, Horseradish, Kale, Kohlrabi, Leeks, Lettuce, Melons, Mustard Greens, Okra, Onions, Parsnips, Peas, Peppers, Pimentoes, Potatoes, Pumpkins, Radishes, Rutabaga, Salsify, Shallots, Spinach, Squash (Summer), Squash (Winter), Sweet Potatoes, Swiss Chard, Tomatoes, Turnips, Watermelons.

FIELD CROPS: Alfalfa, Barley, Birdsfoot Trefoil, Buckwheat, Clover, Corn, Cotton, Flax, Grasses, Hops, Lespedeza, Millet, Milo, Mint, Oats, Peanuts, Popcorn, Rice, Rye, Safflower, Sorghum, Soybeans, Sugar Beets, Sugar Cane, Tobacco, Vetch, Wheat.

CITRUS FRUIT TREE PLANTING SITES: Grapefruit, Kumquats, Lemons, Limes, Oranges, Tangerines, Tangelos.

DECIDUOUS FRUIT AND NUT-TREE PLANTING SITES: Almonds, Apples, Apricots, Cashew Nuts, Cherries, Chestnuts,

Dates, Figs, Filberts, Hazelnuts, Hickory Nuts, Nectarines, Olives, Peaches, Pears, Pecans, Persimmons, Pineapple, Plums, Pomegranates, Prunes, Quince, Walnuts.

BRUSH AND VINE PLANTING SITES: Blackberries, Blueberries, Boysenberries, Cranberries, Currants, Dewberries, Gooseberries, Grapes, Huckleberries, Loganberries, Raspberries, Strawberries, Youngberries.

Nursery Crops including floral plants, ornamentals, shrubs and bushes, forest shade, fruit and nut trees and vine and bramble fruits of all types.

APPLICATION DIRECTIONS

When to Treat: Apply TELONE C-17 whenever soil conditions are suitable but preferably in the late summer or fall or in the spring up to 2 to 3 weeks before planting the crop. Land to be planted to early spring crops, especially crops to be set out, such as celery, tomatoes, nursery and orchard stock, should be treated only in the late summer or early fall. For best results on annual crops use the treatment once each year.

Soil Preparation: At time of application the soil should be in good seed bed condition, free of clods and undecomposed plant material and with soil moisture at about one-half of field capacity. Soil temperature should be 40° to 80°F at the depth of injection. If undecomposed plant debris is present, it should be plowed down and allowed to decompose before applying TELONE C-17. Deep tillage, 12 to 18 inches, often improves results. **NOTE: Do not use TELONE C-17 in extremely heavy clay or muck soils.**

Application: TELONE C-17 may be applied as a **row** or as an **overall (broadcast) treatment** using a chisel-type applicator to inject the fumigant below the soil surface. **For row treatment** spring application is preferable. Fall treatment should be made only when land can be fumigated and bedded for early spring planting. Use a single chisel per row to treat the soil directly beneath the area where the crop will be planted. Inject the material at a depth of 10 to 12 inches below the soil surface. Immediately after application seal the soil by ridging or bedding the fumigated row area with enough soil to bring the soil surface 14 to 16 inches above the point of injection. **For overall (broadcast) treatment** apply whenever soil conditions are suitable. Space applicator chisels 12 inches apart and inject the fumigant 10 to 12 inches below the soil surface. Immediately after application seal the soil by compacting the surface with a roller, cultipacker or similar soil sealing device. When overall treatment is made in late summer or early fall, the field may be planted to a fall cover crop to be plowed under before planting the spring crop.

Rates to Use: Consult the following Dosage and Use Recommendations tables for dosage and application details to control nematodes, symphylans, wireworms and certain soil borne diseases in crops listed.

Exposure Period: After application and sealing, leave the soil undisturbed for 7 to 14 days. Cold, wet soil retards diffusion of TELONE C-17, requiring a longer exposure period.

Aeration of Soil Before Planting: At the end of the exposure period allow the soil to aerate completely before planting the crop. Aeration is usually complete when the odor of TELONE C-17 is no longer evident. Under optimum soil and weather conditions, allow one week of aeration time for each 10 gallons of TELONE C-17 applied per acre. When TELONE C-17 is used for treating deep-rooted tree and shrub planting sites, a 3 to 6 months aeration period should be allowed. **To hasten aeration, especially if heavy rains or low temperatures occur during the exposure period, work the soil to the depth of the treatment zone.** After row treatment use a knife-like chisel in the bed without turning the soil, thus reducing possible recontamination of the treated soil. To hasten aeration after overall treatment, plow or deep cultivate to the depth of the treatment zone.

DOSAGE AND USE RECOMMENDATIONS TO CONTROL NEMATODES, SYMPHYLANS, AND WIREWORMS†

CROPS (CONSULT LIST OF INDIVIDUAL CROPS UNDER GENERAL INFORMATION)	TYPE OF TREATMENT	SOIL TYPE	DOSAGE: GALLONS PER ACRE[1]	DOSAGE: FL. OZ./1000 FT. ROW PER CHISEL
Shallow Rooted Plants: Field Crops, Floral Crops, Grasses and Turf, Small Fruits, Vegetables, Ornamentals	Row (42″)[5]	Mineral	5.1 to 6.9	52.5 to 71
		Muck or Peat	10.3 to 13.7	106 to 141
	Overall (Broadcast)	Mineral	10.3 to 17.1[2]	30.3 to 50.2
		Muck or Peat	27.4[3] to 41.1	80.5 to 120.8
Strawberries	Overall	Mineral	27.4 to 41.1	80.5 to 120.8
Sugar Beets Root Knot Nematode	Row (42″)	Mineral	6	61.7
	Overall		12.8 to 17.1	37.6 to 50.2
Sugar Beet Nematode	Row (42″)		10.3	106
	Overall		12.8 to 21.4	37.6 to 62.9
Pineapple	Row	Mineral	25.7 to 51.4	—
Citrus—Florida[4]	Overall	Mineral	34.2	100.5

Nursery and Field: Citrus Fruit Trees, Deciduous Fruit Trees, Forest Trees, Grapes, Nut Trees, Ornamentals (deep rooted)	Overall: gallons Per Acre to Penetrate Various Depths			
	Mineral Soils	3 ft.	4 ft.	5 ft.	6 ft.
	Sand	17.1	23.9	30.8	37.6
	Sandy Loam	30.8	34.2	41.1	54.7
	Silt Loam	47.9	58.1	71.8	85.5
	Clay Loam	61.6	78.7	95.8	116.3

[1] Use the higher rates in heavier soil.
[2] For cyst-forming nematodes increase dosage to 21.4 gallons per acre (62.9 fl. oz./1000 ft row per chisel)
[3] For muck soils containing less than 30% organic matter use 21.4 gallons per acre.
[4] For burrowing nematode in citrus inject on 18-inch centers, 12 inches deep. Keep free of plants susceptible to burrowing nematodes for 2 years before replanting to citrus.
[5] Row treatment is not recommended for potatoes in irrigated areas of western and northwestern states.
† **NOTE:** To control symphylans (garden centipedes) use only overall at 21.4 or more gallons per acre, and apply during late summer or early fall when the soil is warm. To control wireworms use dosages recommended for nematodes in overall or broadcast treatments.

Dosage and Use Recommendations to Control Soil Borne Disease†

Crop	Type of Treatment	Dosage: Gallons Per Acre	Dosage: Fl. Oz./1000 ft. Row Per Chisel
Sweet Potatoes	Row (42")	8.5 to 10.5	88[1] to 108[1]
	Overall	24 to 27.5	70 to 81
Tobacco	Row (42")	8.5 to 10.5	88[1] to 108[1]
	Overall	24 to 27.5	70 to 81
Peanuts	Row (42")	2.5 to 5	26[1] to 52[1]
White Potatoes	Overall only	24 to 27.5	70 to 81
Mint	Overall only	24 to 27.5	70 to 81
Onions	Overall only	24 to 27.5	70 to 81

[1] Use the same flow rate per chisel for other single chisel row spacings. As the distance between rows increases, the gallons required per acre will decrease and vice versa.

† **NOTE:** Consult **General Information** to determine diseases controlled in crops listed. Nematodes also will be controlled.

REENTRY

Do not enter treated areas without heavy synthetic rubber foot gear until after 3 days post treatment. Because certain states may require more restrictive reentry intervals for various crop areas treated with this product, consult your State Department of Agriculture for further information.

Written or oral warnings must be given to workers who are expected to be in a treated area or in an area about to be treated with this product. These oral warnings shall inform workers of areas or fields that may not be entered without heavy synthetic rubber foot gear until after 3 days post treatment. In case of accidental exposure follow directions as shown by the Statement of Practical Treatment section on this label. When oral warnings are given, warnings shall be given in a language customarily understood by workers. Written warnings must include the following information:

DANGER

Area treated with TELONE C-17 soil fungicide and nematicide on (insert date of application). Do not enter without heavy synthetic rubber foot gear until after 3 days post treatment. In case of accidental exposure take actions as shown by the Statement of Practical Treatment section on this label.

Attention: It is very important to avoid reinfestation of treated soil. Use only planting materials known to be free from disease and parasitic nematodes. It is good practice to cut sweet potato plants at or above the soil line and to use these "cut plants" for making field plantings. Do not use tools or crop residues that could carry soil borne pests from infested land to treated land.

Consult State Agricultural Experiment Station or Extension Service authorities for information on other practices such as post-harvest destruction of crop residues, weed control and cultural practices, and use of nematode and disease resistant crop varieties that also may aid in reducing crop losses caused by soil borne pests.

USE PRECAUTIONS

Important—Note Carefully: The treatment may temporarily raise the level of ammonia nitrogen and soluble salts in the soil. This is most likely to occur when heavy rates of fertilizer and TELONE C-17 are applied to soils that are cold, wet, acid, or high in organic matter. To avoid injury to plant roots, fertilize as indicated by soil tests made after treatment. To avoid ammonia injury or nitrate starvation, or both, to crops on high organic soils do not use fertilizers containing ammonium salts and use only fertilizers containing nitrates, until after the crop is well established and the soil temperature is above 65°F. In mineral soils do not apply more than ⅔ of the nitrogen requirements from fertilizers containing ammonium salts until the crop is well established and soil temperature is above 65°F. Liming highly acid soils before treatment stimulates nitrification and reduces the possibility of ammonia and/or nitrite toxicity. Certain crops including cotton, sugarcane, and pineapple are tolerant to ammonia and the above rule does not apply to them. Citrus seedlings, *Cornus sp., Crataegus sp.,* spruce, vegetable crops such as cauliflower and certain other crops have shown evidence of phosphorus deficiency following fumigation. To avoid this possible effect, it is suggested that additional phosphate fertilizer be used on soils which tend to be deficient in this nutrient.

Under certain conditions TELONE C-17 soil fungicide and nematicide may be corrosive to application equipment; flush equipment with fuel oil or kerosene immediately after use. **DO NOT USE WATER. Do not use containers, pumps, or other transfer equipment made of aluminum, magnesium or their alloys, as under certain conditions TELONE C-17 may be severely corrosive to such metals.**

WARRANTY LIMITATIONS AND DISCLAIMER

The Dow Chemical Company warrants that this product conforms to the chemical description on the label and is reasonably fit for the purposes stated on the label when used in strict accordance with the directions therein under normal conditions of use. THIS IS THE ONLY WARRANTY MADE ON THIS PRODUCT. NO OTHER EXPRESS AND NO IMPLIED WARRANTY OF MERCHANTABILITY OR FITNESS FOR A PARTICULAR PURPOSE IS MADE OUTSIDE OF THIS LABEL. Therefore, neither this warranty nor any other warranty of merchantability or fitness for a particular purpose, express or implied, extends to the use of this product contrary to label instructions (including conditions noted on the label, such as unfavorable temperatures, soil conditions, etc.), under abnormal conditions (such as excessive rainfall, drought, tornadoes, hurricanes, etc.) or under conditions not reasonably foreseeable to or beyond the control of seller.

When buyer or user suffers losses or damages resulting from the use or handling of this product (including claims based on contract, negligence, strict liability, or other legal theories), buyer or user must promptly notify in writing The Dow Chemical Company of any claims to be eligible to receive either remedy given below. THE EXCLUSIVE REMEDY OF THE BUYER OR USER and the LIMIT OF LIABILITY of The Dow Chemical Company or any other seller will be one of the following, at the election of The Dow Chemical Company:

(1) **Refund of purchase price** paid by buyer or user for product bought, **or**

(2) **Replacement** of amount of product used.

The seller **will not be liable for consequential or incidental damages or losses.**

The terms of this Limit of Warranty and Liability cannot be varied by any written or verbal statements or agreements. Any employee or sales agent of the seller is not authorized to vary or exceed the terms of this Warranty Limitations And Disclaimer in any manner.

TORDON* 2K

PELLETS HERBICIDE

RESTRICTED USE PESTICIDE

For retail sale to and use only by Certified Applicators or persons under their direct supervision and only for those uses covered by the Certified Applicator's certification.

ACTIVE INGREDIENTS:	
Picloram (4–amino–3,5,6-trichloropicolinic acid), as the potassium salt	2.3%
INERT INGREDIENTS:	97.7%
TOTAL	100.0%

Picloram acid equivalent—2.0%

EPA Reg. No. 464-333 EPA Est. 464-MI-1

Keep Out of Reach of Children

CAUTION

AVISO

PRECAUCION AL USARIO: Si usted no lee ingles, no use este producto hasta que la etiqueta le haya sido explicada ampliamente.

PRECAUTIONARY STATEMENTS

DUST CAUSES IRRITATION

MAY BE HARMFUL IF SWALLOWED

Avoid Skin and Eye Contact

Wash After Handling

ENVIRONMENTAL HAZARDS

Keep out of lakes, ponds, and streams. Do not contaminate water by cleaning of equipment or disposal of wastes.

NOTICE

Read the entire label. Use only according to label directions.

Before buying or using this product, read "WARRANTY LIMITATIONS AND DISCLAIMER" on back panel. If terms are not acceptable, return **unopened** package at once to seller for full refund of purchase price paid. Otherwise, use by the buyer or any other user constitutes acceptance of the terms under the Limit of Warranty and Liability.

IN CASE OF AN EMERGENCY endangering life or property involving this product, call collect **517–636-4400**

AGRICULTURAL CHEMICAL

Do Not Ship or Store with Food, Feeds, Drugs or Clothing

FEDERAL (EPA) LABEL

TORDON 2K Pellets herbicide applied to the soil over plant roots is highly effective for the control of broadleaved perennial and annual weeds and undesirable woody plants on utility, highway and other right-of-ways, fencerows, headlands around farm and industrial buildings and storage sites.

USE DIRECTIONS

Do not use for manufacturing or formulating

It is a violation of Federal law to use this product in a manner inconsistent with its labeling.

For Control of Broadleaved Perennial and Annual Weeds: Apply TORDON 2K Pellets uniformly anytime during the normal growing season where sufficient moisture is available to carry the herbicide into the soil. In areas where little or no summer rainfall occurs, application should be made in late summer or early fall. Maximum effects of the treatment do not become apparent until the chemical has been carried by moisture into the soil.

APPLICATION RATES

Weeds Controlled†	TORDON 2K Pellets—Amount to apply	Remarks
Docks Larkspur Pigweed Povertyweed Sowthistle (perennial) Sunflower Tansy Thistle (plumeless) Toadflax (dalmation)	50 to 100 lb. per acre 19 to 37 oz. per 1000 sq. ft. 5 to 10 oz. per square rod	Use lower rates in low rainfall areas in the northern states such as Idaho, Montana, North Dakota, Oregon, South Dakota, Wyoming, and Washington. Higher rates should be used where rainfall is greater or in southern states such as Arizona, Arkansas, Kansas, Missouri, New Mexico, Oklahoma and Texas.
Bindweed (field) Bursage (bur ragweed woolyleaf povertyweed) Knapweed (Russian) Milkweed Spurge (leafy) Thistle (Canada)	100 to 150 lb. per acre 37 to 56 oz. per 1000 sq. ft. 10 to 16 oz. per square rod	

† These are typical examples of weeds controlled.

TORDON 2K Pellets herbicide is effective against a wide range of weeds. Local conditions may affect the use of herbicides. Consult your State Agricultural Experiment Station or Extension Service weed specialists for local recommendations. Be sure that the use of this product conforms to all applicable regulations.

For Control of Woody Plants such as maple, locust, aspen, conifers, other woody trees, shrubs, wild rose, brambles, wild grapes and other vines, apply TORDON 2K Pellets uniformly to the soil over the root zone. Apply anytime during the normal growing season where sufficient moisture is available to carry the herbicide into the soil. In areas where little or no summer rainfall occurs applications should be made at "bud break" in late winter or early spring. Use at the rate of 300 to 400 pounds per acre (equivalent to approximately 7½ to 10 lb per 1000 square feet, 2 to 2½ lb per square rod, or ¾ to 1 lb per 100 sq ft). Maximum effects of the treatment do not become apparent until the chemical has been carried by moisture into the soil in the root zone of the plants.

USE PRECAUTIONS

Avoid Improper Application: This herbicide is highly active against most broadleaved plants. Small quantities may cause damage to plants whether applied during the growing or dormant season. Do not apply or otherwise permit TORDON 2K Pellets to contact desirable plants such as vegetables, flowers, grapes, fruit trees, ornamentals, cotton, beans, soybeans and other valuable broadleaved plants, nor the soil containing roots of such plants growing there on or nearby or where such plants are to be grown.

Avoid Water Contamination: To avoid crop or other plant injury, do not treat inner banks or bottom of irrigation and drainage ditches. Do not contaminate water to be used for drinking or other domestic purposes.

Avoid Movement of Treated Soil: Avoid the movement of treated soil into untreated areas.

Other Precautions: Do not store near food, feedstuffs, fertilizer, seeds, insecticides, fungicides or other pesticides. To avoid injury to desirable plants, containers and equipment used for TORDON 2K Pellets should not be re-used to contain or apply other materials.

Do Not Mix or Blend with Fertilizers.

STORAGE AND DISPOSAL

Do not contaminate water, food or feed by storage or disposal.

PESTICIDE DISPOSAL

Wastes resulting from the use of this product may be disposed of on site or at an approved waste disposal facility.

CONTAINER DISPOSAL

Refer to instructions on container for proper disposal information.

SPECIAL LOCAL NEEDS

IDAHO (A) ....EPA SLN No. ID-790025

NEVADA ....EPA SLN No. NV-790008

N. DAKOTA ....EPA SLN No. ND-800009

OREGON (A) ....EPA SLN No. OR-790057

S. DAKOTA.......................EPA SLN. No. SD-800003

UTAH ....EPA SLN No. UT-790016

For the Control of Susceptible Broadleaf Weeds and Woody Plants on Rangeland, Forests, and Permanent Grass Pastures.

DIRECTIONS FOR USE

It is a violation of Federal law to use this product in a manner inconsistent with its labeling.

TORDON 2K Pellets herbicide is designed for application to soil for control of susceptible herbaceous and woody plants by absorption through root pick-up. Rainfall is needed after application to leach the picloram to the roots. Application can be made by hand or broadcast equipment. Generally uniform distribution over the root-zone of the plants in the intended site is desirable, however, certain plants may be treated by concentrating the dose near the stem of the target plants. Best results are usually obtained when rain follows shortly after application and shortly before or during active growth. Do not apply TORDON 2K Pellets to frozen or saturated soil.

SUGGESTED USE RATES

	lb/A	oz/100 sq. ft
Yellow starthistle, Scotch thistle, musk thistle, spotted or diffuse knapweeds, lupines, locoweeds.	25–50	1–2
Rush skeletonweed, Russian knapweed, Canada thistle, larkspurs, rabbitbrushes, burrowweed, snakeweed, fringed snakewort, milkweeds, artichoke thistle, tansy ragwort, common tansy, pricklypear and cholla cacti.	50–100	2–4
Leafy spurge, Utah, Western and one-seed juniper.	100–150	4–6
Wild Peach, rose, manzanita.	150–200	6–8

Utah, Western and one-seed junipers, pinyon, and several susceptible woody plants can be controlled by placing the required rates of TORDON 2K Pellets herbicide around the stem of the plants; use 2 oz. TORDON 2K Pellets herbicide for each 3-4 feet of tree height on juniper or pinyon trees not over 12 feet tall.

TORDON 2K Pellets herbicide at rates over about 75 lb per acre may suppress certain grasses, such as wheatgrass. Usually later grass growth will be improved by release from competition. Grass seedlings may be suppressed or killed up to 2 years after application at higher rates. Broadleaf forage plants, especially legumes, in treated areas may be injured or killed and may not grow for 1 or 2 years.

RESTRICTIONS FOR PASTURE AND RANGELAND USE

For use rates above 150 lbs per acre, do not graze treated areas or feed treated forage for 16 weeks after application.

Limit coverage to no greater than 25% of an applicators acreage, found in any particular watershed.

Do not use where a sandy porous surface and substrate overlie ground water closer than 10 feet below the surface.

Where watersheds have significant slope and where rapid runoff can occur, use spot treatment only. Do not apply within ½ mile of where stream or pond water which drains from the treated watershed may be drawn to irrigate susceptible broadleaf crops, especially beans and potatoes. Do not clean containers or application equipment on or near these areas.

Kill or injury may occur to desirable forbs, trees or shrubs, such as blackberry, cherry, locust, poplar, mountain mahogany, bitterbrush and sumac, from root uptake. If such effects cannot be tolerated, do not apply on or near such desirable plants.

Do not apply to cropland used for production of desirable crops other than forage species. Do not rotate treated rangeland or pastures to other crops until residues of picloram have reached a nonphytotoxic level. Forage legumes on the treated areas may be injured and may not grow for two years or more after treatment.

Read and follow all other precautions on this label.

USE PRECAUTIONS

Apply this product only as specified on this label. The active ingredient in TORDON 2K Pellets herbicide is water soluble and should not be applied where surface water from treated areas can run off to cropland either planted or to be planted.

Avoid use near desirable plants. This herbicide is water soluble, highly active and can remain in the soil for more than one growing season. Very small amounts can injure broadleaf plants such as potatoes, peas, beans, sugarbeets or alfalfa; therefore, do not apply on or near these or other susceptible plants, ornamentals, shade trees or vegetable crops. Do not plant these crops or plants in soil that may have injurious amounts of this herbicide.

Avoid movement of treated soil. Picloram may remain in treated soil for an extended period. Do not move treated soil to other areas and do not use such soil to grow plants until residues have reached a non-phytotoxic level.

Avoid transfer of livestock from a treated area to a broadleaf crop area without first allowing 7 days of grazing on untreated pasture for the first 12 months after application. Otherwise, urine may contain enough picloram to cause injury to sensitive broadleaf plants. Do not use manure from animals grazing treated areas to fertilize soil or fields used to grow susceptible broadleaf crops.

Avoid water contamination. Do not allow TORDON 2K Pellets herbicide to contaminate water used for drinking, irrigation or other domestic purposes. Do not apply on inner banks or bottoms of irrigation ditches. Do not clean containers or application equipment on or near these areas.

Avoid improper storage and equipment use. Do not store near fertilizers, seeds, insecticides, fungicides or other pesticides. Containers and equipment used for TORDON 2K Pellets herbicide should not be used for other agricultural chemicals since small residues of TORDON 2K Pellets herbicide can damage desirable plants.

Avoid improper disposal. Rinse equipment and dispose of waste by burying in non-cropland away from water supplies. Do not reuse containers. Bury them with waste or dispose in a sanitary landfill or follow official container disposal regulations.

Be sure that use of this product conforms to all applicable state and federal regulations.

IDAHO(B) .... EPA SLN No. ID-830020

OREGON(B) .... EPA SLN No. OR-830034

WASHINGTON(A) .... EPA SLN No. WA-830027

For Control of Susceptible Broadleaf Weeds in Non-Irrigated Fallow Grainland East of the Cascade Mountains.

DIRECTIONS FOR USE

It is a violation of Federal law to use this product in a manner inconsistent with its labeling.

Read complete use directions and precautions on this label and container label before using.

SPOT TREATMENT FOR BROADLEAF WEED CONTROL IN FALLOW GRAINLAND

To control patches of perennial broadleaf weeds, such as bindweed, Canada thistle, or rush skeletonweed, in fallow grainland where susceptible plants will not be planted, use 75 to 100 lbs. of TORDON 2K Pellets after grain harvest and before rain ceases in spring. Patches of susceptible annual broadleaf weeds, such as Russian thistle, can be controlled with 20 to 25 lbs. of TORDON 2K Pellets per acre. Rain is needed to leach the picloram into the soil. The treated area should extend at least 10 feet beyond the visible weeds. Treated areas can be relocated more easily if marked with flags or stakes. **A permanent record** of treated sites with rates and dates of treatment should be made by the grower to aid in planning future crop rotations.

Apply with calibrated hand distribution equipment or by hand uniformly over patch and about 10 feet beyond edge. Four oz. of TORDON 2K Pellets per 100 sq. feet equals 100 lbs. per acre. No more than 10% of field should be treated in any one year.

DO NOT APPLY WITH AIRCRAFT.

NOTE: Small grains planted the next fall or spring may be adversely affected to varying degrees depending on rate, rainfall and temperature. A sensitive bioassay for a prospective crop may be used to determine that residues of picloram have disappeared. Contact your county agent or Dow representative for additional information.

Livestock may graze treated fields.

USE PRECAUTIONS

Apply this product only as specified on the label. The active ingredient in TORDON 2K Pellets herbicide is water soluble and should not be applied where surface water from treated areas can run off to croplands either planted or to be planted.

Avoid use near desirable plants. This herbicide is water soluble, highly active and can remain in the soil for more than one growing season. Very small amounts can injure broadleaf plants such as potatoes, peas, beans, sugarbeets or alfalfa; therefore, do not apply on or near these or other susceptible plants, ornamentals, shade trees or vegetable crops. Do not plant these crops or plants in soil that may have injurious amounts of this herbicide.

Avoid movement of treated soil. Picloram may remain in treated soil for an extended period. Do not move treated soil to other areas and do not use such soil to grow plants until residues have reached a non-phytotoxic level.

Avoid transfer of livestock from a treated area to sensitive broadleaf crop areas for 12 months after application or until picloram has disappeared from the soil, without first allowing 7 days on untreated pasture. Otherwise, urine may contain enough picloram to cause injury to sensitive broadleaf plants. Do not use manure from animals grazing treated areas to fertilize soil or fields used to grow susceptible broadleaf crops.

Do Not Contaminate Water Intended for Irrigation or Domestic Purposes. Do not treat or allow pellets to fall onto innerbanks or bottom of irrigation ditches, either dry or containing water, or other channels that may carry water that may be used for irrigation purposes.

STORAGE AND DISPOSAL

Avoid improper storage and equipment use. Do not contaminate fertilizers, seeds, insecticides, fungicides or other pesticides. Containers and equipment used for TORDON 2K Pellets herbicide should not be used for other agricultural chemicals since small residues of TORDON 2K Pellets herbicide can damage desirable plants.

Avoid improper disposal. Rinse equipment and dispose of waste by burying in non-cropland away from water supplies. Do not reuse containers. Bury them with waste or dispose of in a sanitary landfill or follow official container disposal regulations.

Be sure that use of this product conforms to all applicable regulations.

Do not use where a sandy porous surface and substrate overlie groundwater 10 feet or less below the surface.

MINNESOTA .... EPA SLN No. MN-820003

For the Control of Susceptible Broadleaf Weeds and Woody Plants on Rangeland, Forests, and Permanent Grass Pastures.

DIRECTIONS FOR USE

It is a violation of Federal law to use this product in a manner inconsistent with its labeling.

TORDON 2K Pellets herbicide is designed for application to soil for control of susceptible herbaceous and woody plants by absorption through root pick-up. Rainfall is needed after application to leach the picloram to the roots. Application can be made by hand or broadcast equipment. Generally uniform distribution over the root-zone of the plants in the intended site is desirable; however, certain plants may be treated by concentrating the dose near the stem of the target plants. Best results are usually obtained when rain falls shortly after application and shortly before or during active growth. Do not apply TORDON 2K Pellets to frozen or saturated soil.

SUGGESTED USE RATES

	lb/A	oz/100 sq. ft.
Yellow starthistle, Scotch thistle, musk thistle, spotted or diffuse knapweeds, lupines, locoweeds bracken fern, golden rod, absynth wormwood.	25–50	1–2
Rush skeletonweed, Russian knapweed, Canada thistle, larkspurs, rabbitbrushes, burrowweed, snakeweed, fringed snakewort, milkweeds, artichoke thistle, tansy ragwort, common tansy, pricklypear and cholla cacti	50–100	2–4
Leafy spurge, Utah, Western and one-seed juniper.	100–150	4–6
Wild Peach, rose, manzanita	150–200	6–8

Utah, Western and one-seed junipers, pinyon, and several susceptible woody plants can be controlled by placing the required rates of TORDON 2K Pellets herbicide around the stem of the plants; use 2 oz. TORDON 2K Pellets herbicide for each 3-4 feet of tree height on juniper or pinyon trees not over 12 feet tall.

TORDON 2K Pellets herbicide at rates over about 75 lb per acre may suppress certain grasses, such as wheatgrass. Usually later grass growth will be improved by release from competition. Grass seedlings may be suppressed or killed up to 2 years after application at higher rates. Broadleaf forage plants, especially legumes, in treated areas may be injured or killed and may not grow for 1 or 2 years.

RESTRICTIONS FOR PASTURE AND RANGELAND USE

For use at rates above 150 lbs per acre, do not graze treated areas or feed treated forage for 16 weeks after application.

Limit coverage to no greater than 25% of an applicators acreage, found in any particular watershed.

Do not use where a sandy porous surface and substrate overlie ground water closer than 10 feet below the surface.

Where watersheds have significant slope and where rapid runoff can occur, **use spot treatment only**. Do not apply within ½ mile of where stream or pond water which drains from the treated watershed may be drawn to irrigate susceptible broadleaf crops, especially beans and potatoes. Do not clean containers or application equipment on or near these areas.

Kill or injury may occur to desirable forbs, trees or shrubs, such as blackberry, cherry, locust, poplar, mountain mahogany, bitterbrush and sumac, from root uptake. If such effects cannot be tolerated, do not apply on or near such desirable plants.

Do not apply to cropland used for production of desirable crops other than forage species. Do not rotate treated rangeland or pastures to other crops until residues of picloram have reached a non-phytotoxic level. Forage legumes on the treated areas may be injured and may not grow for two years or more after treatment.

Read and follow all other use precautions on this label.

USE PRECAUTIONS

Apply this product only as specified on this label. The active ingredient in TORDON 2K Pellets herbicide is water soluble and should not be applied where surface water from the treated areas can run off to croplands either planted or to be planted.

Avoid use near desirable plants. This herbicide is water soluble, highly active and can remain in the soil for more than one growing season. Very small amounts can injure broadleaf plants such as potatoes, peas, beans, sugarbeets or alfalfa; therefore, do not apply on or near these or other susceptible plants, ornamentals, shade trees or vegetable crops. Do not plant these crops or plants in soil that may have injurious amounts of this herbicide.

Avoid movement of treated soil. Picloram may remain in treated soil for an extended period. Do not move treated soil to other areas and do not use such soil to grow plants until residues have reached a non-phytotoxic level.

Avoid transfer of livestock from a treated area to a broadleaf crop without first allowing 7 days of grazing on untreated pasture for the first 12 months of application. Otherwise, urine may contain enough picloram to cause injury to sensitive broadleaf plants. Do not use manure from animals grazing treated areas to fertilize soil or fields used to grow susceptible broadleaf crops.

Avoid water contamination. Do not allow TORDON 2K Pellets herbicide to contaminate water used for drinking, irrigation or other domestic purposes. Do not apply on inner banks or bottoms of irrigation ditches. Do not clean containers or application equipment on or near these areas.

Avoid improper storage and equipment use. Do not store near fertilizers, seeds, insecticides, fungicides or other pesticides. Containers and equipment used for TORDON 2K Pellets herbicide should not be used for other agricultural chemicals since small residues of TORDON 2K Pellets herbicide can damage desirable plants.

Avoid improper disposal. Rinse equipment and dispose of waste by burying in non-cropland away from water supplies. Do not reuse containers. Bury them with waste or dispose of in a sanitary landfill or follow official container disposal regulations.

Be sure that use of this product conforms to all applicable state and federal regulations.

MONTANA .... EPA SLN No. MT-800011

For the Control of Susceptible Broadleaf Weeds and Woody Plants on Rangeland, Forests, and Permanent Grass Pastures.

DIRECTIONS FOR USE

It is a violation of Federal law to use this product in a manner inconsistent with its labeling.

TORDON 2K Pellets Herbicide is designed for application to soil for control of susceptible herbaceous and woody plants by absorption through root pick-up.

Rainfall is needed after application to leach the picloram to the roots. Application can be made by hand or broadcast equipment. Generally uniform distribution over the root-zone of the plants in the intended site is desirable; however, certain species may be controlled by concentrating the dose near the stem. Best results are usually obtained when rain follows shortly after application and shortly before or during active growth. Do not apply TORDON 2K Pellets to frozen or saturated soil.

SUGGESTED USE RATES

	lb/A	oz/100 sq. ft.
Yellow starthistle, Scotch thistle, musk thistle, spotted and diffuse knapweeds, lupines, locoweeds.	25–50	1–2
Rush skeletonwood, Russian knapweed, Canada thistle, Tall larkspurs, rabbitbrush, burrowweed, snakeweed, fringed sagebrush, common milkweed, artichoke thistle, tansy ragwort, common tansy, pricklypear and cholla cacti.	50–100	2–4
Leafy spurge.	100–150	4–6

TORDON 2K Pellets Herbicide at rates over about 75 lb per acre may suppress certain grasses, such as wheatgrass, bromegrass, buffalograss and bluegramma. Usually, later grass growth will be improved by release from competition. Grass seedlings may be suppressed or killed up to 2 years after application at higher rates. Broadleaf forage plants, especially legumes, in treated areas may be injured or killed and may not grow for 1 to 2 years.

RESTRICTIONS FOR PASTURE AND RANGELAND USE

Limit coverage to no greater than 25% of an applicators acreage found in any particular watershed.

Do not use where a sandy porous surface and substrate overlie ground water closer than 10 feet below the surface.

Where watersheds have significant slope and where rapid runoff can occur, use spot treatment only. Do not apply within ½ mile of where stream or pond water which drains from the treated watershed may be drawn to irrigate susceptible broadleaf crops, especially beans and potatoes. Do not clean containers or application equipment on or near these areas.

Kill or injury may occur to desirable forbs, trees or shrubs, such as blackberry, cherry, locust, poplar, mountain mahogany, bitterbrush and sumac, from root uptake. If such effects cannot be tolerated, do not apply on or near such desirable plants.

Do not apply to cropland used for production of desirable crops other than forage species. Do not rotate treated rangeland or pastures to other crops until residues of picloram have reached a non-phytotoxic level. Forage legumes on the treated areas may be injured and may not grow for two years or more after treatment.

Read and follow all other use precautions on this label.

USE PRECAUTIONS

Apply this product only as specified on this label. The active ingredient in TORDON 2K Pellets herbicide is water soluble and should not be applied where surface water from treated areas can run off to croplands either planted or to be planted.

Avoid use near desirable plants. This herbicide is water soluble, highly active and can remain in the soil for more than one growing season. Very small amounts can injure broadleaf plants such as potatoes, peas, beans, sugarbeets or alfalfa; therefore, do not apply on or near these or other susceptible plants, ornamentals, shade trees or vegetable crops. Do not plant these crops or plants in soil that may have injurious amounts of this herbicide.

Avoid movement of treated soil. Picloram may remain in treated soil for an extended period. Do not move treated soil to other areas and do not use such soil to grow plants until residues have reached a non-phytotoxic level.

Avoid transfer of livestock from a treated area to a broadleaf crop area without first allowing 7 days of grazing on untreated pasture for the first 12 months after application. Otherwise, urine may contain enough picloram to cause injury to sensitive broadleaf plants. Do not use manure from animals grazing treated areas to fertilize soil or fields used to grow susceptible broadleaf crops.

Avoid water contamination. Do not allow TORDON 2K Pellets herbicide to contaminate water used for drinking, irrigation or other domestic purposes. Do not apply on inner banks or bottoms of irrigation ditches. Do not clean containers or application equipment on or near these areas.

Avoid improper storage and equipment use. Do not store near fertilizers, seeds, insecticides, fungicides or other pesticides. Containers and equipment used for TORDON 2K Pellets herbicide should not be used for other agricultural chemicals since small residues of TORDON 2K Pellets herbicide can damage desirable plants.

Avoid improper disposal. Rinse equipment and dispose of waste by burying in non-cropland away from water supplies. Do not reuse containers. Bury them with waste or dispose in a sanitary landfill or follow official container disposal regulations.

Be sure that use of this product conforms to all applicable state and federal regulations.

WASHINGTON(B) .... EPA SLN No. WA-810094

For the Control of Susceptible Broadleaf Weeds and Woody Plants on Rangeland, Forests, and Permanent Grass Pastures

DIRECTIONS FOR USE

It is a violation of Federal law to use this product in a manner inconsistent with its labeling.

TORDON 2K Pellets herbicide is designed for application to soil for control of susceptible herbaceous and woody plants by absorption through root pick-up. Rainfall is needed after application to leach the picloram to the roots. Application can be made by hand or broadcast equipment. Generally uniform distribution over the root-zone of the plants in the intended site is desirable; however, certain plants may be treated by concentrating the dose near the stem of the target plants. Best results are usually obtained when rain follows shortly after application and shortly before or during active growth. Do not apply TORDON 2K Pellets to frozen or saturated soil.

SUGGESTED USE RATES

	lb/A	oz/100 sq. ft.
Yellow starthistle, Scotch thistle, musk thistle, spotted or diffuse knapweeds, lupines, locoweeds.	25–50	1–2
Rush skeletonweed, Russian knapweed, Canada thistle, larkspurs, rabbitbrushes, burrowweed, snakeweed, fringed snakewort, milkweeds, artichoke thistle, tansy ragwort, common tansy and pricklypear.	50–100	2–4
Leafy spurge, and Western juniper.	100–150	4–6
Wild Peach, rose, manzanita.	150–200	6–8

Utah, Western and one-seed junipers, pinyon, and several susceptible woody plants can be controlled by placing the required rates of TORDON 2K Pellets herbicide around the stem of the plants; use 2 oz. TORDON 2K Pellets herbicide for each 3–4 feet of tree height on juniper or pinyon trees not over 12 feet tall.

TORDON 2K Pellets herbicide at rates over about 75 lb per acre may suppress certain grasses, such as wheat grass. Usually later grass growth will be improved by release from competition. Grass seedlings may be suppressed or killed up to 2 years after application at higher rates. Broadleaf forage plants, especially legumes, in treated areas may be injured or killed and may not grow for 1 or 2 years.

RESTRICTIONS FOR PASTURE AND RANGELAND USE

For use rates above 150 lbs per acre, do not graze treated areas or feed treated forage for 16 weeks after application.

Limit coverage to no greater than 25% of an applicators acreage, found in any particular watershed.

Do not use where a sandy porous surface and substrate overlie ground water closer than 10 feet below the surface.

Where watersheds have significant slope and where rapid runoff can occur, **use spot treatment only.** Do not apply within ½ mile of where stream or pond water which drains from the treated watershed may be drawn to irrigate susceptible broadleaf crops, especially beans and potatoes. Do not clean containers or application equipment on or near these areas.

Kill or injury may occur to desirable forbs, trees, or shrubs, such as blackberry, cherry, locust, poplar, mountain mahogany, bitterbrush and sumac, from root uptake. If such effects cannot be tolerated, do not apply on or near such desirable plants.

Do not apply to cropland other than pasture, rangeland, and forest. Do not rotate treated rangeland or pastures to other crops until residues of picloram have reached a non-phytotoxic level. Forage legumes on the treated areas may be injured and may not grow for two years or more after treatment.

Read and follow all other use precautions on this label.

USE PRECAUTIONS

Apply this product only as specified on this label. The active ingredient in TORDON 2K Pellets herbicide is water soluble and should not be applied where surface water from treated areas can run off to croplands either planted or to be planted.

Avoid use near desirable plants. This herbicide is water soluble, highly active and can remain in the soil for more than one growing season. Very small amounts can injure broadleaf plants such as potatoes, peas, beans, sugarbeets or alfalfa; therefore, do not apply on or near these or other susceptible plants, ornamentals, shade trees or vegetable crops. Do not plant these crops or plants in soil that may have injurious amounts of this herbicide.

Avoid movement of treated soil. Picloram may remain in treated soil for an extended period. Do not move treated soil to other areas and do not use such soil to grow plants until residues have reached a non-phytotoxic level.

Avoid transfer of livestock from a treated area to a broadleaf crop area without first allowing 7 days of grazing on untreated pastures for the first 12 months after application. Otherwise, urine may contain enough picloram to cause injury to sensitive broadleaf plants. Do not use manure from animals grazing treated areas to fertilize soil or fields used to grow susceptible broadleaf crops.

Avoid water contamination. Do not allow TORDON 2K Pellets herbicide to contaminate water used for drinking, irrigation or other domestic purposes. Do not apply on inner banks or bottoms of irrigation ditches. Do not clean containers or application equipment on or near these areas.

Avoid improper storage and equipment use. Do not store near fertilizers, seeds, insecticides, fungicides or other pesticides. Containers and equipment used for TORDON 2K Pellets herbicide should not be used for other agricultural chemicals since small residues of TORDON 2K Pellets herbicide can damage desirable plants.

Avoid improper disposal. Rinse equipment and dispose of waste by burying in non-cropland away from water supplies. Do not reuse containers. Bury them with waste or dispose in a sanitary landfill or follow official container disposal regulations.

Be sure that use of this product conforms to all applicable state and federal regulations.

WYOMING .... EPA SLN No. WY-800009

For the control of Susceptible Broadleaf Weeds and Woody Plants on Rangeland, Forests, and Permanent Grass Pastures.

DIRECTIONS FOR USE

It is a violation of Federal law to use this product in a manner inconsistent with its labeling.

TORDON 2K Pellets herbicide is designed for application to soil for control of susceptible herbaceous and woody plants by absorption through root pick-up. Rainfall is needed after application to leach the picloram to the roots. Application can be made by hand, broadcast, or aerial equipment. Generally uniform distribution over the rootzone of the plants in the intended site is desirable; however, certain species may be controlled by concentrating the dose near the stem. Best results are usually obtained when rain follows shortly after application and shortly before or during active growth. Do not apply TORDON 2K Pellets to frozen or saturated soil.

SUGGESTED USE RATES

	lb/A	oz/100 sq. ft.
Yellow starthistle, Scotch thistle, musk thistle, plumeless thistle, spotted knapweed, diffuse knapweed, lupine, locoweed.	25–50	1–2
Rush skeletonweed, tansy ragwort, Canada thistle, field bindweed, Russian knapweed, tansy, skeletonleaf bursage.	50–100	2–4
Leafy spurge, Common Milkweed, Dalmation toadflax, yellow toadflax.	100	4
Utah juniper, western juniper, oneseed juniper.	100–150	4–6

Utah, western and one-seed junipers, pinyon, and several other susceptible woody plants can be controlled by placing the required rates of TORDON 2K Pellets herbicide around the stem of the plants; use 2 oz TORDON 2K Pellets herbicide for each 3 to 4 feet of tree height on juniper or pinyon trees not over 12 feet tall.

TORDON 2K Pellets herbicide at rates over about 75 lb per acre may suppress certain grasses, such as wheatgrass, bromegrass, buffalograss and bluegramma. Usually, later grass growth will be improved by release from competition. Grass seedlings may be suppressed or killed up to 2 years after application at higher rates. Broadleaf forage plants, especially legumes, in treated areas may be injured or killed and may not grow for 1 to 2 years.

RESTRICTIONS FOR PASTURE AND RANGELAND USE

Limit coverage to no greater than 25% of an applicators acreage, found in any particular watershed.

Do not use where a sandy porous surface and substrate overlie ground water closer than 10 feet below the surface.

Where watersheds have significant slope and where rapid runoff can occur, **use spot treatment only**. Do not apply within 1/2 mile of where stream or pond water which drains to irrigate susceptible broadleaf crops, especially beans and potatoes. Do not clean containers or application equipment on or near these areas.

Kill or injury may occur to desirable forbs, trees and shrubs, such as blackberry, cherry, locust, poplar, mountain mahogany, bitterbrush and sumac, from root uptake. If such effects cannot be tolerated, do not apply on or near these or other desirable plants.

Do not apply to cropland used for production of desirable crops other than forage species. Do not rotate treated rangeland or pastures to other crops until residues of picloram have reached a non-phytotoxic level. Forage legumes on the treated areas may be injured and may not grow for two years or more after treatment.

Read and follow all other use precautions on this label.

USE PRECAUTIONS

Apply this product only as specified on this label. The active ingredient in TORDON 2K Pellets herbicide is water soluble and should not be applied where surface water from treated areas can run off to croplands either planted or to be planted.

Avoid use near desirable plants. This herbicide is water soluble, highly active and can remain in the soil for more than one growing season. Very small amounts can injure broadleaf plants such as potatoes, peas, beans, sugarbeets or alfalfa; therefore, do not apply on or near these or other susceptible plants, ornamentals, shade trees or vegetable crops. Do not plant these crops or plants in soil that may have injurious amounts of this herbicide.

Avoid movement of treated soil. Picloram may remain in treated soil for an extended period. Do not move treated soil to other areas and do not use such soil to grow plants until residues have reached a non-phytotoxic level.

Avoid transfer of livestock from a treated area to a broadleaf crop area without first allowing 7 days of grazing on untreated pasture for the first 12 months after application. Otherwise, urine may contain enough picloram to cause injury to sensitive broadleaf plants. Do not use manure from animals grazing treated areas to fertilize soil or fields used to grow susceptible broadleaf crops.

Avoid water contamination. Do not allow TORDON 2K Pellets herbicide to contaminate water used for drinking, irrigation or other domestic purposes. Do not apply on inner banks or bottoms of irrigation ditches. Do not clean containers or application equipment on or near these areas.

Avoid improper storage and equipment use. Do not store near fertilizers, seeds, insecticides, fungicides or other pesticides. Containers and equipment used for TORDON 2K Pellets herbicide should not be used for other agricultural chemicals since small residues of TORDON 2K Pellets herbicide can damage desirable plants.

Avoid improper disposal. Rinse equipment and dispose of waste by burying in non-cropland away from water supplies. Do not reuse containers. Bury them with waste or dispose in a sanitary landfill or follow official container disposal regulations.

Be sure that use of this product conforms to all applicable state and federal regulations.

WARRANTY LIMITATIONS AND DISCLAIMER

The Dow Chemical Company warrants that this product conforms to the chemical description on the label and is reasonably fit for the purposes stated on the label when used in strict accordance with the directions therein under normal conditions of use. THIS IS THE ONLY WARRANTY MADE ON THIS PRODUCT. NO OTHER EXPRESS AND NO IMPLIED WARRANTY OF MERCHANTABILITY OR FITNESS FOR A PARTICULAR PURPOSE IS MADE OUTSIDE OF THIS LABEL. Therefore, neither this warranty nor any other warranty of merchantability or fitness for a particular purpose, express or implied, extends to the use of this product contrary to label instructions (including conditions noted on the label, such as unfavorable temperatures, soil conditions, etc.), under abnormal conditions (such as excessive rainfall, drought, tornadoes, hurricanes, etc.) or under conditions not reasonably foreseeable to or beyond the control of seller.

When buyer or user suffers losses or damages resulting from the use or handling of this product (including claims based on contract, negligence, strict liability, or other legal theories), buyer or user must promptly notify in writing The Dow Chemical Company of any claims to be eligible to receive either remedy given below. The EXCLUSIVE REMEDY OF THE BUYER OR USER and the LIMIT OF LIABILITY of The Dow Chemical Company or any other seller will be one of the following, at the election of The Dow Chemical Company:

(1) **Refund of purchase price** paid by buyer or user for product bought, or

(2) **Replacement** of amount of product used.

The seller **will not be liable for consequential or incidental damages or losses.**

The terms of this Limit of Warranty and Liability cannot be varied by any written or verbal statements or agreements. Any employee or sales agent of the seller is not authorized to vary or exceed the terms of this Warranty Limitations And Disclaimer in any manner.

* Trademark of THE DOW CHEMICAL COMPANY

87242-L5 784

TORDON* 10K
PELLETS HERBICIDE

RESTRICTED USE PESTICIDE

For retail sale to and use only by Certified Applicators or persons under their direct supervision and only for those uses covered by the Certified Applicator's certification.

For the Control of Unwanted Woody Plants and Various Broadleaf Weeds on Permanent Grass Pastures and Rangeland

ACTIVE INGREDIENT	
Picloram (4–amino-3,5,6–trichloropicolinic acid) potassium salt	11.6%
INERT INGREDIENTS	88.4%
TOTAL	100.0%
Picloram acid equivalent	10%

EPA Reg. No. 464–320 EPA Est. 464-MI-1

Keep Out of Reach of Children
CAUTION
AVISO

PRECAUCION AL USUARIO: Si usted no lee inglés, no use este producto hasta que la etiqueta le haya sido explicada ampliamente.

PRECAUTIONARY STATEMENTS

Environmental Hazards

Keep out of lakes, ponds, and streams. Do not contaminate water by cleaning of equipment or disposal of wastes.

NOTICE

Read the entire lable. Use only according to label directions.

Before buying or using this product, read "WARRANTY LIMITATIONS AND DISCLAIMER" elsewhere on this label. If terms are not acceptable, return **unopened** package at once to seller for full refund of purchase price paid. Otherwise, use by the buyer or any other user constitutes acceptance of the terms under the Limit of Warranty and Liability.

IN CASE OF AN EMERGENCY endangering life or property involving this product, call collect **517–636-4400.**

AGRICULTURAL CHEMICAL

Do Not Ship or Store with Food, Feeds, Drugs or Clothing

READ THE ENTIRE LABEL BEFORE USING THIS PRODUCT.

FEDERAL (EPA) LABEL

For the Control of Unwanted Woody Plants and Various Broadleaf Weeds on Permanent Grass Pastures and Rangeland

TORDON 10K Pellets, a highly effective herbicide for control of many undesirable broadleaf weeds and woody plants, is designed to be applied to the soil for the control of undesirable woody plants. It is intended for use on permanent grass pastures, rangeland, fencerows, and in forests. TORDON 10K Pellets may be applied as spot or broadcast treatments. The application method and equipment used should assure uniform distribution of the recommended dosage over the treated area. Rapid response should not be expected. Maximum effects of treatment with TORDON 10K Pellets do not become apparent until after the chemical has been carried by moisture into the root zones of the woody plants. TORDON 10K Pellets are non-corrosive to application equipment, non-volatile, and non-flammable.

USE DIRECTIONS

Do Not Use for Manufacturing or Formulating.
It is a violation of Federal law to use this product in a manner inconsistent with its labeling.

READ AND FOLLOW ALL USE PRECAUTIONS ON THIS LABEL

FOR PERMANENT GRASS PASTURE AND RANGELAND, USE only in **West Virginia, Virginia, North Carolina, South Carolina, Georgia, Kentucky, Tennessee, Alabama, Mississippi, Missouri, Arkansas, Louisiana, Kansas, Oklahoma, Texas, Colorado, and New Mexico.**

TORDON 10K Pellets may be applied at anytime soil is not frozen. However, best results are obtained from application in the spring or fall or during periods of vigorous growth when subsequent rainfall can be expected. Treatment during hot, dry periods is not advised. Distribute TORDON 10K Pellets uniformly by spot or broadcast treatment to the soil over the roots of woody plants to be controlled.

Use TORDON 10K pellets to control undesirable broadleaf weeds and woody plants applied at rates indicated in the table of Application Rates. The maximum use rate is not to exceed 40 pounds per acre (1 pound product per 1000 square feet or $1\frac{1}{2}$ ounces per 100 square feet). Treatments will not be effective until rain occurs after application. Some spot retreatment the following year or later may be needed. For best timing of applications, consult your Experiment Station or Extension Range Management Specialists.

TORDON 10K Pellets, especially at the higher rates, may injure or suppress certain desirable grasses such as smooth bromegrass, western wheatgrass, and switchgrass; therefore, do not apply where possible injury to such grasses cannot be tolerated. However, injury to established grass usually is followed by increased growth and development due to release from brush competition. Grass seedlings may be suppressed or killed in treated areas for up to 2 years after application. Reseeding during this period is not recommended.

APPLICATION RATES

Species	Broadcast Treatment Pounds product per acre Aerial	Ground	Spot Treatment Ounces of product per 1,000 sq. ft.
Broom Snakeweed (Perennial broomweed, Turpentine weed)	5–10	10[1]	3.75
Pricklypear, Tasajillo, and Cholla cactus	5–10	5–10[1]	3.75
Whitebrush reduction	10–15	10–15	5.5
Sand shinnery oak suppression	10–20	10–20	7.5
Junipers (cedars)	20	20	7.5
South Texas mixed brush, Huisache and shallow-rooted woody plants, e.g. Blackbrush Granjeno, Guajillo, Large pricklypear (Mesquite not controlled).	25–30	25–30	11
Macartney rose	30–40[2]	30–40[2]	16
Multiflora rose	40	40	16
Ceanothus species	40	40	16
Shrub live oak, Gambel oak	40	40	16
Locust	40	40	16
Kudzu	40	40	16
Sumac	40	40	16

[1] Difficult to distribute less than 10 lb. uniformly by ground equipment.
[2] Use higher rate on heavy clay soils.

Broom snakeweed (Perennial broomweed, Turpentine weed): Apply TORDON 10K Pellets Herbicide during or prior to periods of rainfall or when moisture in the form of snow is sufficient to move the chemical into the soil profile. Do not apply to frozen soils. Apply liquid formulations in early spring (following winter snow or rainfall) or in fall following normal rainfall.

Pricklypear, Tasajillo, and Cholla Cactus: Use the higher rate of application for the control of large pricklypear species (nopal) in South Texas. Control of cactus species is slow and may require two years or longer for maximum response. Uniform product distribution is essential. The higher application rate may be required to obtain adequate coverage, especially when applying with ground equipment.

Juniper (Alligator, Redberry, Blueberry and other Junipers); Eastern Red Cedar and Piñion Pine: For individual trees, use 2 tsp. product per 5 square feet of area within the dripline. Apply out to dripline on level ground and on uphill side of plants growing on slopes. Results with Blueberry juniper over 3 feet tall are variable. Utah, Western and one-seed juniper and Piñion pine can be controlled when under 12 feet tall.

South Texas Mixed Brush: TORDON 10K Pellets is recommended where mesquite is not the major problem, but mixed brush species and prickly pear control is desired. Whitebrush can be reduced with product rates of 10 to 15 pounds per acre. Improved control can be obtained at 25 to 30 pounds product per acre. TORDON 10K Pellets is not effective for mesquite control.

Macartney Rose and Multiflora Rose: Uniform product distribution over root zone is required. For Macartney Rose, 30 pounds product per acre is sufficient except for heavy clay soils where the 40 pound rate should be used. Consult your Agricultural Extension Service for local recommendations.

RESTRICTIONS FOR PASTURE AND RANGELAND USE

Limit coverage to no greater than 25% of an applicator's acreage, found in any particular watershed.

Do not use where a sandy porous surface and substrate overlie ground water 10 feet or less below the surface.

Where watersheds have significant slope and where rapid runoff can occur **use spot treatment only.**

Kill or injury may occur to desirable forbs, trees or shrubs such as blackberry, cherry, dogwood, locust, maple, oak, poplar, sycamore, and sumac, from root uptake. If such effects cannot be tolerated, do not apply on or near such desirable plants.

Do not apply to cropland used for production of desirable crops other than forage grasses. Do not rotate treated rangeland or pastures to other crops. Do not treat pastures or rangeland if the forage legume component is substantial and desired.

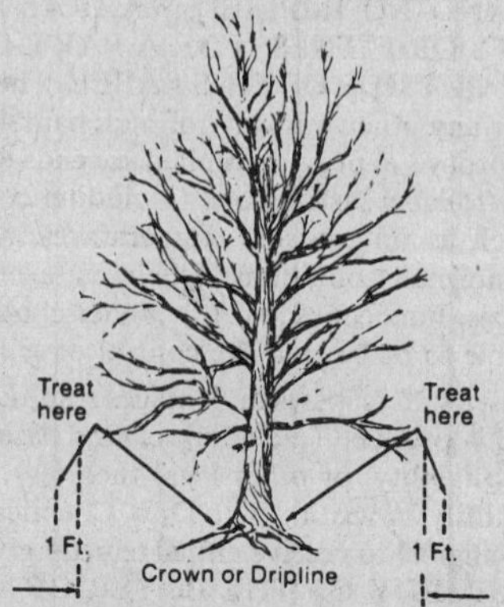

USE PRECAUTIONS

Apply this product only as specified on this label.
The active ingredient in TORDON 10K Pellets is water soluble and should not be applied where surface water from treated areas is likely to run off to crop lands either planted or to be planted.

Avoid use near desirable plants: This herbicide is water soluble, highly active and can remain in the soil for more than one growing season. Very small amounts can injure broadleaf plants, especially **tobacco and potatoes.** Therefore, do not apply on or near tobacco, potatoes, beans, cotton, alfalfa, peas, peanuts, melons, blueberries or other fruit plants, ornamentals, shade trees, soybeans, or other vegetable crops. Injurious amounts of picloram can be absorbed by roots of trees invading treated areas.

Avoid movement of treated soil: Picloram may remain in treated soil for an extended period. Do not move treated soil to other areas and do not use such soil to grow plants unless adequate sensitive bioassay or chemical tests show that no detectable picloram residues are present in the soil.

Avoid transfer of livestock from a treated area to a broadleaf crop area without first allowing 7 days of grazing on untreated pasture. Otherwise, urine may contain enough picloram to cause injury to sensitive broadleaf plants. Do not use manure from animals grazing treated areas to fertilize soil or fields used to grow susceptible broadleaf crops.

Avoid water contamination: Do not allow TORDON 10K Pellets to contaminate water used for drinking, irrigation, or other domestic purposes. Do not apply on or in irrigation ditches. Do not clean containers or application equipment on or near these areas.

Do Not Blend or Mix With Fertilizers.

STORAGE AND DISPOSAL

Do Not Contaminate water, food or feed by storage or disposal.

Avoid improper storage and equipment use: Do not store near fertilizers, seeds, insecticides, fungicides or other pesticides. Containers and equipment used for TORDON 10K Pellets should not be used for other agricultural chemicals since small residues of TORDON 10K Pellets herbicide can damage desirable plants.

Avoid improper disposal: Rinse equipment and dispose of waste by burying in non-cropland away from water supplies. Do not reuse containers. Completely empty box into application equipment. Then dispose of box in a sanitary landfill, or by burning if allowed by State and local authorities. If burned, stay out of smoke.

Be sure that use of this product conforms to all applicable regulations.

SPECIAL LOCAL NEEDS

ARIZONA .... EPA SLN No. AZ-780013

For the Control of Susceptible Woody Plants and Broadleaf Weeds on Rangeland and Permanent Grass Pastures as Itemized Below

TORDON 10K Pellets, a highly effective brush killer, is designed to be applied to the soil for the control of undesirable woody plants. It is intended for use on permanent grass pastures and rangeland, and in forests. TORDON 10K Pellets may be applied as spot or broadcast treatments. The application method and equipment used should assure uniform distribution of the recommended dosage over the treated area. Rapid response should not be expected. Maximum effects of treatment with TORDON 10K Pellets do not become apparent until after the chemical has been carried by moisture into the root zones of the woody plants. TORDON 10K Pellets are noncorrosive to application equipment, non-volatile, and non-flammable.

DIRECTIONS FOR USE

It is a violation of Federal law to use this product in a manner inconsistent with its labeling.

TORDON 10K Pellets herbicide may be applied at any time soil is not frozen. However, best results are obtained from application in the spring before growth begins or during periods of vigorous growth when subsequent rainfall can be expected.

Distribute TORDON 10K Pellets uniformly by spot or broadcast treatment to the soil over the roots of woody plants to be controlled. READ AND FOLLOW ALL USE PRECAUTIONS ON THIS LABEL.

FOR USE ON RANGELANDS AND PERMANENT GRASS PASTURES by applicators that have received instruction by federal, state or county extension personnel in cooperation with The Dow Chemical Company in the proper use of this product.

Use TORDON 10K Pellets herbicide to control woody plants at the rate of 10 to 40 lbs. per acre ($^1/_4$ to 1 lb. per 1,000 sq. ft. or $^3/_8$ to $1^1/_2$ oz. per 100 sq. ft.) and spread uniformly over the plant root areas except as noted below. Use 10 to 30 lbs. per acre to control or suppress certain species, such as prickly pear and cholla cactus, catclaw acacia, wait-a-minute bush and susceptible broadleaf weeds, such as snakeweed. Use 30 to 40 lbs. per acre to suppress or control junipers, and chaparral species such as skunk bush, sumac, mountain mahogany, locust, Ceanothus species, shrub live oak and Gambel oak patches less than 5 ft. high. Alligator, Utah and oneseed junipers can be controlled by applying 1 level tablespoon ($^1/_2$ oz.) of TORDON 10K Pellets per 3 to 4 ft. of height of trees less than 12 ft. tall. Response of taller trees is variable. Make applications only to noxious plants, clumps, or dense infestations. Consult your local Experimental Station or Extension Range Management Specialist for additional information.

TORDON 10K Pellets, especially at the higher rates, may injure or suppress certain desirable grasses, such as wheatgrass; therefore, do not apply where possible injury to such grasses cannot be tolerated. However, injury to established grass usually is followed by increased growth and development due to release from brush competition. Grass seedlings may be suppressed or killed in treated areas for up to 2 years after application. Reseeding during this period is not recommended. Broadleaf range plants including forage legumes in treated areas may be injured or killed and may not grow for 2 to 3 years.

Restrictions for Pasture and Rangeland Use

Limit coverage to no greater than 25% of an applicator's acreage, found in any particular watershed.

Do not use where a sandy porous surface and substrate overlie ground water for 10 feet or less below the surface.

Where watersheds have significant slope and where rapid runoff can occur, **use spot treatment only.** Do not apply within $^1/_2$ mile of where stream or pond water which drains from the treated watershed may be drawn to irrigate susceptible broadleaf crops, especially **beans and potatoes.** Do not clean containers or application equipment on or near these areas.

Kill or injury may occur to desirable forbs, trees or shrubs such as blackberry, cherry, locust, oak, poplar, mountain mahogany, bitterbrush and sumac, from root uptake. If such effects cannot be tolerated, do not apply on or near such desirable plants.

Do not apply to cropland used for production of desirable crops other than forage species. Do not rotate treated rangeland or pastures to other crops until residues of picloram have reached a non-phytotoxic level. Forage legumes on the treated areas may be injured and may not grow for two years or more after treatment.

USE PRECAUTIONS

Apply this product only as specified on this label. The active ingredient in TORDON 10K Pellets herbicide is water soluble and should not be applied where surface water from treated areas can run off to crop lands either planted or to be planted.

Avoid use near desirable plants: This herbicide is water soluble, highly active and can remain in the soil for more than one growing season. Very small amounts can injure broadleaf plants, such as potatoes, sugarbeets, or alfalfa. Therefore, do not apply on or near potatoes, beans, or other fruit plants, ornamentals, shade trees, or vegetable crops.

Avoid movement of treated soil: Picloram may remain in treated soil for an extended period. Do not move treated soil to other areas and do not use such soil to grow plants until residues have reached a non-phytotoxic level.

Avoid transfer of livestock from a treated area to a broadleaf crop area without first allowing 7 days of grazing on untreated pasture for the first 12 months after application. Otherwise, urine may contain enough picloram to cause injury to sensitive broadleaf plants. Do not use manure from animals grazing treated areas to fertilize soil or fields used to grow susceptible broadleaf crops.

Avoid water contamination: Do not allow TORDON 10K Pellets to contaminate water used for drinking, irrigation, or other domestic purposes. Do not apply on inner banks or bottoms of irrigation ditches. Do not clean containers or application equipment on or near these areas.

Avoid improper storage and equipment use: Do not store near fertilizers, seeds, insecticides, fungicides or other pesticides. Containers and equipment used for TORDON 10K Pellets should not be used for other agricultural chemicals since small residues of TORDON 10K Pellets herbicide can damage desirable plants.

Avoid improper disposal: Rinse equipment and dispose of waste by burying in non-cropland away from water supplies. Do not reuse containers. Bury them with waste or dispose in a sanitary landfill. Follow official local container disposal regulations.

Be sure that use of this product conforms to all applicable regulations.

CONNECTICUT .... EPA SLN No. CT-770006

For the Control of Unwanted Woody Plants on Permanent Grass Pastures

TORDON 10K Pellets, a highly effective brush killer, is designed to be applied to the soil for the control of undesirable woody plants. TORDON 10K Pellets may be applied as spot or broadcast treatments. The application method and equipment used should assure uniform distribution of the recommended dosage over the treated area. Rapid response should not be expected. Maximum effects of treatment with TORDON 10K Pellets do not become apparent until after the chemical has been carried by moisture into the root zones of the woody plants. TORDON 10K Pellets are non-corrosive to application equipment, non-volatile, and non-flammable.

DIRECTIONS FOR USE

It is a violation of Federal law to use this product in a manner inconsistent with its labeling.

TORDON 10K Pellets herbicide may be applied at any time soil is not frozen. However, best results are obtained from application in the spring before growth begins or during periods of vigorous growth when subsequent rainfall can be expected.

Distribute TORDON 10K Pellets uniformly by spot or broadcast treatment to the soil over the roots of woody plants to be controlled. **READ AND FOLLOW ALL USE PRECAUTIONS ON THIS LABEL.**

FOR PERMANENT GRASS PASTURE USE

For use only in **Connecticut** where training of applicators by state personnel or other state-supervised programs are conducted in cooperation with The Dow Chemical Company. For use only by (1) a certified commercial ag-plant applicator or (2) by private applicators on land previously inspected by USDA Agricultural Stabilization and Conservation Service (ASCS) where an ASCS program approval is required.

Use TORDON 10K Pellets herbicide to control undesirable woody species at the rate of 40 pounds per acre (1 pound product per 1000 square feet or 1½ ounces per 100 square feet) and spread uniformly over the plant root areas. Use 40 pounds per acre to suppress or control multiflora rose, hawthorn, juniper, and sumac. Applications should be made only to individual noxious plants, to clumps, and to larger dense infestations where grasses already are suppressed. Treatments will not be effective until rain occurs after application. Some spot retreatment the following year or later may be needed. Do not retreat areas by broadcast application for at least three years. For best timing of applications, consult your Experiment Station or Extension Specialists.

TORDON 10K Pellets herbicide may injure or suppress certain desirable forage grasses such as smooth bromegrass; therefore, do not apply where possible injury to such grasses cannot be tolerated. However, injury to established grass usually is followed by increased growth and development due to release from brush competition. Grass seedlings may be suppressed or killed in treated areas for up to 2 years after application. Reseeding during this period is not recommended.

Restrictions for Pasture Use

Limit coverage to no greater than 25% of an applicator's acreage, found in any particular watershed.

Do not use where a sandy porous surface and substrate overlie ground water 10 feet or less below the surface.

Where watersheds have significant slope and where rapid runoff can occur, **use spot treatment only.** Do not apply within ½ mile of where stream or pond water which drains from the treated watershed may be drawn to irrigate susceptible broadleaf crops, especially **tobacco, tomatoes and potatoes,** and where the water is used in greenhouses. Do not clean containers or application equipment on or near these areas.

Kill or injury may occur to desirable forbs, trees or shrubs such as blackberry, cherry, dogwood, locust, maple, oak, poplar, sycamore, and sumac, from root uptake. If such effects cannot be tolerated, do not apply on or near such desirable plants.

Do not apply to cropland used for production of desirable crops other than forage grasses. Do not rotate treated pastures to other crops. Do not treat pastures if the forage legume component is substantial and desired.

Read and follow all other use precautions on this label.

USE PRECAUTIONS

Apply this product only as specified on this label. The active ingredient in TORDON 10K Pellets herbicide is water soluble and should not be applied where surface water from treated areas can run off to croplands either planted or to be planted.

Avoid use near desirable plants: This herbicide is water soluble, highly active and can remain in the soil for more than one growing season. Very small amounts can injure broadleaf plants, especially **tobacco, tomatoes and potatoes.** Therefore, do not apply on or near tobacco, tomatoes, potatoes, beans, blueberries or other fruit plants, ornamentals, shade trees, soybeans, or other vegetable crops.

Avoid movement of treated soil: Picloram, the active ingredient in TORDON 10K Pellets, may remain in treated soil for an extended period. Do not move treated soil to other areas and do not use such soil to grow plants.

Avoid transfer of livestock from a treated area to a broadleaf crop area without first allowing 7 days of grazing on untreated pasture. Otherwise, urine may contain enough picloram to cause injury to sensitive broadleaf plants. Do not use manure from animals grazing treated areas to fertilize soil or fields used to grow susceptible broadleaf crops.

Avoid water contamination: Do not allow TORDON 10K Pellets herbicide to contaminate water used for drinking, irrigation, or other domestic purposes. Do not apply on inner banks or bottoms of irrigation ditches. Do not clean containers or application equipment on or near these areas.

Avoid improper storage and equipment use: Do not store near fertilizers, seeds, insecticides, fungicides or other pesticides. Containers and equipment used for TORDON 10K Pellets should not be used for other agricultural chemicals since small residues of TORDON 10K Pellets herbicide can damage desirable plants.

Avoid improper disposal: Rinse equipment and dispose of waste by burying in non-cropland away from water supplies. Do not reuse containers. Bury them with waste or dispose in a sanitary landfill. Follow official local container disposal regulations.

Be sure that use of this product conforms to all applicable state and federal regulations.

ILLINOIS .... EPA SLN No. IL-840008
NEW YORK .... EPA SLN No. NY-790004

For the Control of Unwanted Woody Plants on Permanent Grass Pastures and Rangeland

TORDON 10K Pellets herbicide, a highly effective brush killer, is designed to be applied to the soil for the control of undesirable woody plants. TORDON 10K Pellets may be applied as spot or broadcast treatments. The application method and equipment used should assure uniform distribution of the recommended dosage over the treated area. Rapid response should not be expected. Maximum effects of treatment with TORDON 10K Pellets do not become apparent until after the chemical has been carried by moisture into the root zones of the woody plants. TORDON 10K Pellets are non-corrosive to application equipment, non-volatile, and non-flammable.

DIRECTIONS FOR USE

It is a violation of Federal law to use this product in a manner inconsistent with its labeling.

TORDON 10K Pellets herbicide may be applied at any time soil is not frozen. However, best results are obtained from application in the spring before growth begins or during periods of vigorous growth when subsequent rainfall can be expected.

Distribute TORDON 10K Pellets uniformly by spot or broadcast treatment to the soil over the roots of woody plants to be controlled. **READ AND FOLLOW ALL USE PRECAUTIONS ON THIS LABEL.**

FOR PERMANENT GRASS PASTURE AND RANGELAND USE

Use TORDON 10K Pellets to control undesirable woody species at the rate of 40 pounds per acre (1 pound product per 1000 square feet or 1½ ounces per 100 square feet) and spread uniformly over the plant root areas. Use 40 pounds per acre to suppress or control multiflora rose, hawthorn, juniper, and sumac. Applications should be made only to individual noxious plants, to clumps, and to larger dense infestations where grasses already are suppressed. Treatments will not be effective until rain occurs after application. Some spot retreatment the following year or later may be needed. Do not retreat areas by broadcast application for at least three years. For best timing of applications, consult your Experiment Station or Extension Specialists.

TORDON 10K Pellets herbicide may injure or suppress certain desirable forage grasses such as smooth bromegrass; therefore, do not apply where possible injury to such grasses cannot be tolerated. However, injury to established grass usually is followed by increased growth and development due to release from brush competition. Grass seedlings may be suppressed or killed in treated areas for up to 2 years after application. Reseeding during this period is not recommended.

Restrictions for Pasture and Rangeland Use

Limit coverage to no greater than 25% of an applicator's acreage, found in any particular watershed.

Do not use where a sandy porous surface and substrate overlie ground water 10 feet or less below the surface.

Where watersheds have significant slope and where rapid runoff can occur, **use spot treatment only.** Do not apply within ½ mile of where stream or pond water which drains from the treated watershed may be drawn to irrigate susceptible broadleaf crops, especially **tobacco, tomatoes and potatoes,** and where the water is used in greenhouses. Do not clean containers or application equipment on or near these areas.

Kill or injury may occur to desirable forbs, trees or shrubs such as blackberry, cherry, dogwood, locust, maple, oak, poplar, sycamore, and sumac, from root uptake. If such effects cannot be tolerated, do not apply on or near such desirable plants.

Do not apply to cropland used for production of desirable crops other than forage grasses. Do not rotate treated pastures or rangeland to other crops. Do not treat pastures or rangeland if the forage legume component is substantial and desired.

Read and follow all other use precautions on this label.

USE PRECAUTIONS

Apply this product only as specified on this label. The active ingredient in TORDON 10K Pellets herbicide is water soluble and should not be applied where surface water from treated areas can run off to crop lands either planted or to be planted.

Avoid use near desirable plants: This herbicide is water soluble, highly active and can remain in the soil for more than one growing season. Very small amounts can injure broadleaf plants, especially **tobacco, tomatoes, and potatoes.** Therefore, do not apply on or near tobacco, tomatoes, potatoes, beans, blueberries or other fruit plants, ornamentals, shade trees, soybeans or other vegetable crops.

Avoid movement of treated soil: Picloram, the active ingredient in TORDON 10K Pellets, may remain in treated soil for an extended period. Do not move treated soil to other areas and do not use such soil to grow plants.

Avoid transfer of livestock from a treated area to a broadleaf crop area without first allowing 7 days of grazing on untreated pasture. Otherwise, urine may contain enough picloram to cause injury to sensitive broadleaf plants. Do not use manure from animals grazing treated areas to fertilize soil or fields used to grow susceptible broadleaf crops.

Avoid water contamination: Do not allow TORDON 10K Pellets to contaminate water used for drinking, irrigation, or other domestic purposes. Do not apply on inner banks or bottoms of irrigation ditches. Do not clean containers or application equipment on or near these areas.

Avoid improper storage and equipment use: Do not store near fertilizers, seeds, insecticides, fungicides, or other pesticides. Containers and equipment used for TORDON 10K Pellets should not be used for other agricultural chemicals since small residues of TORDON 10K Pellets herbicide can damage desirable plants.

Avoid improper disposal: Rinse equipment and dispose of waste by burying in non-cropland away from water supplies. Do not reuse containers. Bury them with waste or dispose in a sanitary landfill. Follow official local container disposal regulations.

Be sure that use of this product conforms to all applicable state and federal regulations.

INDIANA .... EPA SLN No. IN-780013

OHIO .... EPA SLN No. OH-770001

For the Control of Unwanted Woody Plants on Permanent Grass Pastures

TORDON 10K Pellets, a highly effective brush killer, is designed to be applied to the soil for the control of undesirable woody plants. TORDON 10K Pellets may be applied as spot or broadcast treatments. The application methods and equipment used should assure uniform distribution of the recommended dosage over the treated area. Rapid response should not be expected. Maximum effects of treatment with TORDON 10K Pellets do not become apparent until after the chemical has been carried by moisture into the root zones of the woody plants. TORDON 10K Pellets are non-corrosive to application equipment, non-volatile, and non-flammable.

DIRECTIONS FOR USE

It is a violation of Federal law to use this product in a manner inconsistent with its labeling.

TORDON 10K Pellets herbicide may be applied at any time soil is not frozen. However, best results are obtained from application in the spring before growth begins or during periods of vigorous growth when subsequent rainfall can be expected.

Distribute TORDON 10K Pellets uniformly by spot or broadcast treatment to the soil over the roots of woody plants to be controlled. **READ AND FOLLOW ALL USE PRECAUTIONS ON THIS LABEL.**

FOR PERMANENT GRASS PASTURE AND RANGELAND USE

For use only in **Ohio** and **Indiana** where training of applicators by state personnel or other state supervised programs are conducted in cooperation with The Dow Chemical Company.

Use TORDON 10K Pellets herbicide to control undesirable woody species at the rate of 40 pounds per acre (1 pound product per 1000 square feet or 1½ ounces per 100 square feet) and spread uniformly over the plant root areas. Use 40 pounds per acre to suppress or control multiflora rose, hawthorn, juniper and sumac. Applications should be made only to individual noxious plants, to clumps, and to larger dense infestations where grasses already are suppressed. Treatments will not be effective until rain occurs after application. Some spot retreatment the following year or later may be needed. Do not retreat areas by broadcast application for at least three years. For best timing of applications, consult your Experiment Station or Extension Specialists.

TORDON 10K Pellets herbicide may injure or suppress certain desirable forage grasses such as smooth bromegrass; therefore, do not apply where possible injury to such grasses cannot be tolerated. However, injury to established grass usually is followed by increased growth and development due to release from brush competition. Grass seedlings may be suppressed or killed in treated areas for up to 2 years after application. Reseeding during this period is not recommended.

Restrictions for Pasture Use

Limit coverage to no greater than 25% of an applicator's acreage, found in any particular watershed.

Do not use where a sandy porous surface and substrate overlie ground water 10 feet or less below the surface.

Where watersheds have significant slope and where rapid runoff can occur, **use spot treatment only.** Do not apply within ½ mile of where stream or pond water which drains from the treated watershed may be drawn to irrigate susceptible broadleaf crops, especially **tobacco, tomatoes and potatoes,** and where the water is used in greenhouses. Do not clean containers or application equipment on or near these areas.

Kill or injury may occur to desirable forbs, trees or shrubs such as blackberry, cherry, dogwood, locust, maple, oak, poplar, sycamore, and sumac, from root uptake. If such effects cannot be tolerated, do not apply on or near such desirable plants.

Do not apply to cropland used for production of desirable crops other than forage grasses. Do not rotate treated pastures to other crops. Do not treat pastures if the forage legume component is substantial and desired.

Read and follow all other use precautions on this label.

USE PRECAUTIONS

Apply this product only as specified on this label. The active ingredient in TORDON 10K Pellets herbicide is water soluble and should not be applied where surface water from treated areas can run off to croplands either planted or to be planted.

Avoid use near desirable plants: This herbicide is water soluble, highly active and can remain in the soil for more than one growing season. Very small amounts can injure broadleaf plants, especially **tobacco, tomatoes and potatoes.** Therefore, do not apply on or near tobacco, tomatoes, potatoes, beans, blueberries or other fruit plants, ornamentals, shade trees, soybeans, or other vegetable crops.

Avoid movement of treated soil: Picloram, the active ingredient in TORDON 10K Pellets, may remain in treated soil for an extended period. Do not move treated soil to other areas and do not use such soil to grow plants.

Avoid transfer of livestock from a treated area to a broadleaf crop area without first allowing 7 days of grazing on untreated pasture. Otherwise, urine may contain enough picloram to cause injury to sensitive broadleaf plants. Do not use manure from animals grazing treated areas to fertilize soil or fields used to grow susceptible broadleaf crops.

Avoid water contamination: Do not allow TORDON 10K Pellets herbicide to contaminate water used for drinking, irrigation, or other domestic purposes. Do not apply on inner banks or bottoms of irrigation ditches. Do not clean containers or application equipment on or near these areas.

Avoid improper storage and equipment use: Do not store near fertilizers, seeds, insecticides, fungicides or other pesticides. Containers and equipment used for TORDON 10K Pellets should not be used for other agricultural chemicals since small residues of TORDON 10K Pellets herbicide can damage desirable plants.

Avoid improper disposal: Rinse equipment and dispose of waste by burying in non-cropland away from water supplies. Do not reuse containers. Bury them with waste or dispose in a sanitary landfill. Follow official local container disposal regulations.

Be sure that use of this product conforms to all applicable state and federal regulations.

IOWA .... EPA SLN No. 1A-770002

For the Control of Unwanted Woody Plants on Permanent Grass Pastures and Rangeland

TORDON 10K Pellets, a highly effective brush killer, is designed to be applied to the soil for the control of undesirable woody plants. TORDON 10K Pellets may be applied as spot or broadcast treatments. The application method and equipment should assure uniform distribution of the recommended dosage over the treated area. Rapid response should not be expected. Maximum effects of treatment with TORDON 10K Pellets do not become apparent until after the chemical has been carried by moisture into the root zones of the woody plants. TORDON 10K Pellets are non-corrosive to application equipment, non-volatile and non-flammable.

USE DIRECTIONS

It is a violation of Federal law to use this product in a manner inconsistent with its labeling.

TORDON 10K Pellets may be applied at any time soil is not frozen. However, best results are obtained from application in the spring before growth begins or during periods of vigorous growth when subsequent rainfall can be expected. Distribute TORDON 10K Pellets uniformly by spot or broadcast treatment to the soil over the roots of woody plants to be controlled. **READ AND FOLLOW ALL USE PRECAUTIONS ON THIS LABEL.**

FOR PERMANENT GRASS PASTURE AND RANGELAND USE

For use only in **Iowa** where training of applicators by state personnel or other state supervised programs are conducted in cooperation with The Dow Chemical Company.

Use TORDON 10K Pellets to control undesirable woody species at the rate of 40 pounds per acre (1 pound product per 1000 square feet or 1½ ounces per 100 square feet) and spread uniformly over the plant root areas. Use 40 pounds per acre to suppress or control multiflora rose, hawthorn, juniper, and sumac. Applications should be made only to individual noxious plants, to clumps, and to larger dense infestations where grasses already are suppressed. Treatments will not be effective until rain occurs after application. Some spot retreatment the following year or later may be needed. Do not retreat areas by broadcast application for at least three years. For best timing of applications, consult your Experiment Station or Extension Specialists.

TORDON 10K Pellets may injure or suppress certain desirable forage grasses such as smooth bromegrass; therefore, do not apply where possible injury to such grasses cannot be tolerated. However, injury to established grass usually is followed by increased growth and development due to release from brush competition. Grass seedlings may be suppressed or killed in treated areas for up to 2 years after application. Reseeding during this period is not recommended.

Restrictions for Pasture and Rangeland Use

Limit coverage to no greater than 25% of an applicator's acreage, found in any particular watershed.

Do not use where a sandy porous surface and substrate overlie ground water 10 feet or less below the surface.

Where watersheds have significant slope and where rapid runoff can occur, **use spot treatment only.** Do not apply within ½ mile of where stream or pond water which drains from the treated watershed may be drawn to irrigate susceptible broadleaf crops, especially **tobacco, tomatoes and potatoes,** and where the water is used in greenhouses. Do not clean containers or application equipment on or near these areas.

Kill or injury may occur to desirable forbs, trees or shrubs such as blackberry, cherry, dogwood, locust, maple, oak, poplar, sycamore, and sumac, from root uptake. If such effects cannot be tolerated, do not apply on or near such desirable plants.

Do not apply to cropland used for production of desirable crops other than forage grasses. Do not rotate treated rangeland or pastures to other crops. Do not treat pastures or rangeland if the forage legume component is substantial and desired.

Read and follow all other use precautions on this label.

USE PRECAUTIONS

Apply this product only as specified on this label. The active ingredient in TORDON 10K Pellets is water soluble and should not be applied where surface water from treated areas can run off to crop lands either planted or to be planted.

Avoid use near desirable plants: This herbicide is water soluble, highly active and can remain in the soil for more than one growing season. Very small amounts can injure broadleaf plants, especially **tobacco, tomatoes, and potatoes.** Therefore, do not apply on or near tobacco, tomatoes, potatoes, beans, blueberries or other fruit plants, ornamentals, shade trees, soybeans, or other vegetable crops.

Avoid movement of treated soil: Picloram, the active ingredient in TORDON 10K Pellets, may remain in treated soil for an extended period. Do not move treated soil to other areas and do not use such soil to grow plants.

Avoid transfer of livestock from a treated area to a broadleaf crop area without first allowing 7 days of grazing on untreated pasture. Otherwise, urine may contain enough picloram to cause injury to sensitive broadleaf plants. Do not use manure from animals grazing treated areas to fertilize soil or fields used to grow susceptible broadleaf crops.

Avoid water contamination: Do not allow TORDON 10K Pellets to contaminate water used for drinking, irrigation, or other domestic purposes. Do not apply on inner banks or bottoms of irrigation ditches. Do not clean containers or application equipment on or near these areas.

Avoid improper storage and equipment use: Do not store near fertilizers, seeds, insecticides, fungicides or other pesticides. Containers and equipment used for TORDON 10K Pellets should not be used for other agricultural chemicals since small residues of TORDON 10K Pellets herbicide can damage desirable plants.

Avoid improper disposal: Rinse equipment and dispose of waste by burying in non-cropland away from water supplies. Do not reuse containers. Bury them with waste or dispose in a sanitary landfill. Follow official container disposal regulations.

Be sure that use of this product conforms to all applicable state and federal regulations.

MARYLAND .... EPA SLN No. MD-760001
PENNSYLVANIA .... EPA SLN No. PA-760003

For the Control of Unwanted Woody Plants on Permanent Grass Pastures

TORDON 10K Pellets, a highly effective brush killer, is designed to be applied to the soil for the control of undesirable woody plants. TORDON 10K Pellets may be applied as spot or broadcast treatments. The application method and equipment used should assure uniform distribution of the recommended dosage over the treated area. Rapid response should not be expected. Maximum effects of treatment with TORDON 10K Pellets do not become apparent until after the chemical has been carried by moisture into the root zones of the woody plants. TORDON 10K Pellets are non-corrosive to application equipment, non-volatile, and non-flammable.

USE DIRECTIONS

It is a violation of Federal law to use this product in a manner inconsistent with its labeling.

TORDON 10K Pellets may be applied at any time soil is not frozen. However, best results are obtained from application in the spring before growth begins or during periods of vigorous growth when subsequent rainfall can be expected. Distribute TORDON 10K Pellets uniformly by spot or broadcast treatment to the soil over the roots of woody plants to be controlled. **READ AND FOLLOW ALL USE PRECAUTIONS ON THIS LABEL**

FOR PERMANENT GRASS PASTURE AND RANGELAND USE

For use only in **Maryland** and **Pennsylvania** where training of applicators by state personnel or other state supervised programs are conducted in cooperation with The Dow Chemical Company.

Use TORDON 10K Pellets to control undesirable woody species at the rate of 40 pounds per acre (1 pound product per 1000 square feet or 1½ ounces per 100 square feet) and spread uniformly over the plant root areas. Use 40 pounds per acre to suppress or control multiflora rose, hawthorn, juniper, and sumac. Applications should be made only to individual noxious plants, to clumps, and to larger dense infestations where grasses already are suppressed. Treatments will not be effective until rain occurs after application. Some spot retreatment the following year or later may be needed. Do not retreat areas by broadcast application for at least three years. For best timing of applications, consult your Experiment Station or Extension Range Management Specialists.

TORDON 10K Pellets, may injure or suppress certain desirable grasses such as smooth bromegrass, western wheatgrass, and switchgrass; therefore, do not apply where possible injury to such grasses cannot be tolerated. However, injury to established grass usually is followed by increased growth and development due to release from brush competition. Grass seedlings may be suppressed or killed in treated areas for up to 2 years after application. Reseeding during this period is not recommended.

Restrictions for Pasture and Rangeland Use

Limit coverage to no greater than 25% of an applicator's acreage, found in any particular watershed.

Do not use where a sandy porous surface and substrate overlie ground water 10 feet or less below the surface.

Where watersheds have significant slope and where rapid runoff can occur, **use spot treatment only.** Do not apply within ½ mile of where stream or pond water which drains from the treated watershed may be drawn to irrigate susceptible broadleaf crops, especially **tobacco and potatoes.** Do not clean containers or application equipment on or near these areas.

Kill or injury may occur to desirable forbs, trees or shrubs such as blackberry, cherry, dogwood, locust, maple, oak, poplar, sycamore, and sumac, from root uptake. If such effects cannot be tolerated, do not apply on or near such desirable plants.

Do not apply to cropland used for production of desirable crops other than forage grasses. Do not rotate treated rangeland or pastures to other crops. Do not treat pastures or rangeland if the forage legume component is substantial and desired.

Read and follow all other use precautions on this label.

USE PRECAUTIONS

Apply this product only as specified on this label. The active ingredient in TORDON 10K Pellets is water soluble and should not be applied where surface water from treated areas can run off to crop lands either planted or to be planted.

Avoid use near desirable plants: This herbicide is water soluble, highly active and can remain in the soil for more than one growing season. Very small amounts can injure broadleaf plants, especially **tobacco and potatoes.** Therefore, do not apply on or near tobacco, potatoes, beans, blueberries or other fruit plants, ornamentals, shade trees, soybeans or other vegetable crops.

Avoid movement of treated soil: Picloram may remain in treated soil for an extended period. Do not move treated soil to other areas and do not use such soil to grow plants.

Avoid transfer of livestock from a treated area to a broadleaf crop area without first allowing 7 days of grazing on untreated pasture. Otherwise, urine may contain enough picloram to cause injury to sensitive broadleaf plants. Do not use manure from animals grazing treated areas to fertilize soil or fields used to grow susceptible broadleaf crops.

Avoid water contamination: Do not allow TORDON 10K Pellets to contaminate water used for drinking, irrigation, or other domestic purposes. Do not apply on inner banks or bottoms of irrigation ditches. Do not clean containers or application equipment on or near these areas.

Avoid improper storage and equipment use: Do not store near fertilizers, seeds, insecticides, fungicides, or other pesticides. Containers and equipment used for TORDON 10K Pellets should not be used for other agricultural chemicals since small residues of TORDON 10K Pellets herbicide can damage desirable plants.

Avoid improper disposal: Rinse equipment and dispose of waste by burying in non-cropland away from water supplies. Do not reuse containers. Bury them with waste or dispose in a sanitary landfill. Follow official local container disposal regulations.

Be sure that use of this product conforms to all applicable regulations.

NEW JERSEY .... EPA SLN No. NJ-840016

For the Control of Unwanted Woody Plants on Permanent Grass Pastures

TORDON 10K Pellets, a highly effective brush killer, is designed to be applied to the soil for the control of undesirable woody plants. TORDON 10K Pellets may be applied as spot or broadcast treatments. The application method and equipment used should assure uniform distribution of the recommended dosage over the treated area. Rapid response should not be expected. Maximum effects of treatment with TORDON 10K Pellets do not become apparent until after the chemical has been carried by moisture into the root zones of the woody plants. TORDON 10K Pellets are non-corrosive to application equipment, non-volatile, and non-flammable.

DIRECTIONS FOR USE

It is a violation of Federal law to use this product in a manner inconsistent with its labeling.

TORDON 10K Pellets herbicide may be applied at any time soil is not frozen. However, best results are obtained from application in the spring before growth begins or during period of vigorous growth when subsequent rainfall can be expected.

Distribute TORDON 10K Pellets uniformly by spot or broadcast treatment to the soil over the roots of woody plants to be controlled. **READ AND FOLLOW ALL USE PRECAUTIONS ON THIS LABEL.**

FOR PERMANENT GRASS PASTURE USE

For use only in **New Jersey** where training of applicators by state personnel or other state-supervised programs are conducted in cooperation with The Dow Chemical Company. All application sites must be inspected by personnel of the Cooperative Extension Service prior to use.

Use TORDON 10K Pellets herbicide to control undesirable woody species at the rate of 40 pounds per acre (1 pound product per 1000 square feet or $1\frac{1}{2}$ ounces per 100 square feet) and spread uniformly over the plant root areas. Use 40 pounds per acre to suppress or control mulitflora rose, hawthorn, juniper, and sumac. Applications should be made only to individual noxious plants, to clumps, and to larger dense infestations where grasses already are suppressed. Treatments will not be effective until rain occurs after application. Some spot retreatment the following year or later may be needed. Do not retreat areas by broadcast application for at least three years.

For best timing of applications, consult your Experiment Station or Extension Specialists.

TORDON 10K Pellets herbicide may injure or suppress certain desirable forage grasses such as smooth bromegrass; therefore, do not apply where possible injury to such grasses cannot be tolerated. However, injury to established grass usually is followed by increased growth and development due to release from brush competition. Grass seedlings may be suppressed or killed in treated areas for up to 2 years after application. Reseeding during this period is not recommended.

Restrictions for Pasture Use

Limit coverage to no greater than 25% of an applicator's acreage, found in any particular watershed.

Do not use where a sandy porous surface and substrate overlie ground water 10 feet or less below the surface.

Where watersheds have significant slope and where rapid runoff can occur, **use spot treatment only.** Do not apply within ½ mile of where stream or pond water which drains from the treated watershed may be drawn to irrigate susceptible broadleaf crops, especially **tobacco, tomatoes and potatoes,** and where the water is used in greenhouses. Do not clean containers or application equipment on or near these areas.

Kill or injury may occur to desirable forbs, trees or shrubs such as blackberry, cherry, dogwood, locust, maple, oak, poplar, sycamore, and sumac, from root uptake. If such effects cannot be tolerated do not apply on or near such desirable plants.

Do not apply to cropland used for production of desirable crops other than forage grasses. Do not rotate treated pastures to other crops. Do not treat pastures if the forage legume component is substantial and desired.

Read and follow all other use precautions on this label.

USE PRECAUTIONS

Apply this product only as specified on this label. The active ingredient in TORDON 10K Pellets herbicide is water soluble and should not be applied where surface water from treated areas can run off to croplands either planted or to be planted.

Avoid use near desirable plants: This herbicide is water soluble, highly active and can remain in the soil for more than one growing season. Very small amounts can injure broadleaf plants, especially **tobacco, tomatoes and potatoes.** Therefore, do not apply on or near tobacco, tomatoes, potatoes, beans, blueberries or other fruit plants, ornamentals, shade trees, soybeans or other vegetable crops.

Avoid movement of treated soil: Picloram, the active ingredient in TORBON 10K Pellets, may remain in treated soil for an extended period. Do not move treated soil to other areas and do not use such soil to grow plants.

Avoid transfer of livestock from a treated area to a broadleaf crop area without first allowing 7 days of grazing on untreated pasture. Otherwise, urine may contain enough picloram to cause injury to sensitive broadleaf plants. Do not use manure from animals grazing treated areas to fertilize soil or fields used to grow susceptible broadleaf crops.

Avoid water contamination: Do not allow TORDON 10K Pellets to contaminate water used for drinking, irrigation, or other domestic purposes. Do not apply on inner banks or bottoms of irrigation ditches. Do not clean containers or application equipment on or near these areas.

Avoid improper storage and equipment use: Do not store near fertilizers, seeds, insecticides, fungicides, or other pesticides. Containers and equipment used for TORDON 10K Pellets should not be used for other agricultural chemicals since small residues of TORDON 10K Pellets herbicide can damage desirable plants.

Avoid improper disposal: Rinse equipment and dispose of waste by burying in non-cropland away from water supplies. Do not reuse containers. Bury them with waste or dispose in a sanitary landfill. Follow official local container disposal regulations.

Be sure that use of this product conforms to all applicable state and federal regulations.

TEXAS .... SLN No. TX-760002

For the Control of Unwanted Woody Plants on Permanent Grass Pastures and Rangeland

TORDON 10K Pellets, a highly effective brush killer, is designed to be applied to the soil for the control of undesirable woody plants. TORDON 10K Pellets may be applied as spot or broadcast treatments. The application method and equipment used should assure uniform distribution of the recommended dosage over the treated area. Rapid response should not be expected. Maximum effects of treatment with TORDON 10K Pellets do not become apparent until after the chemical has been carried by moisture into the root zones of the woody plants. TORDON 10K Pellets are non-corrosive to application equipment, non-volatile, and non-flammable.

USE DIRECTIONS

It is a violation of Federal law to use this product in a manner inconsistent with its labeling.

TORDON 10K Pellets may be applied at any time soil is not frozen. However, best results are obtained from application in the spring before growth begins or during periods of vigorous growth when subsequent rainfall can be expected. Distribute TORDON 10K Pellets uniformly by spot or broadcast treatment to the soil over the roots of woody plants to be controlled. **READ AND FOLLOW ALL USE PRECAUTIONS ON THIS LABEL.**

FOR PERMANENT GRASS PASTURE AND RANGELAND USE

For use only in **Texas** where training of applicators by state personnel or other state supervised programs are conducted in cooperation with The Dow Chemical Company.

Use TORDON 10K Pellets by air or ground applications to control undesirable woody species at the rate of 10 to 40 pounds per acre (¼ to 1 pound product per 1000 square feet or ⅓ ounce to 1½ ounces per 100 square feet) and spread uniformly over the plant root areas. Use 10 to 30 pounds per acre to suppress or control Macartney rose. Use 20 to 30 pounds per acre to suppress or control huisache and pricklypear cactus. Use 30 pounds per acre to suppress or control whitebrush. Use 40 pounds per acre to suppress or control multiflora rose, hawthorn, juniper, kudzu and sumac. Applications should be made only to individual noxious plants, to clumps, and to larger dense infestations where grasses already are suppressed. Treatments will not be effective until rain occurs after application. Some spot retreatment the following year or later may be needed. Do not retreat areas by broadcast application for at least three years. For best timing of applications, and for lower use rates consult your Texas Experiment Station or Extension Range Management Specialists.

TORDON 10K Pellets, may injure or suppress certain desirable grasses such as smooth bromegrass, western wheatgrass, and switchgrass; therefore, do not apply where possible injury to such grasses cannot be tolerated. However, injury to established grass usually is followed by increased growth and development due to release from brush competition. Grass seedlings may be suppressed or killed in treated areas for up to 2 years after application. Reseeding during this period is not recommended.

Restrictions for Pasture and Rangeland Use

Limit coverage to no greater than 25% of an applicator's acreage, found in any particular watershed.

Do not use where a sandy porous surface and substrate overlie ground water 10 feet or less below the surface.

Where watersheds have significant slope and where rapid runoff can occur, **use spot treatment only.** Do not apply within ½ mile of where stream or pond water which drains from the treated watershed may be drawn to irrigate susceptible broadleaf crops, especially **tobacco and potatoes.** Do not clean containers or application equipment on or near these areas.

Kill or injury may occur to desirable forbs, trees or shrubs such as blackberry, cherry, dogwood, locust, maple, oak, poplar, sycamore, and sumac, from root uptake. If such effects cannot be tolerated, do not apply on or near such desirable plants.

Do not apply to cropland used for production of desirable crops other than forage grasses. Do not rotate treated rangeland or pastures to other crops. Do not treat pastures or rangeland if the forage legume component is substantial and desired.

Read and follow all other use precautions on this label.

USE PRECAUTIONS

Apply this product only as specified on this label. The active ingredient in TORDON 10K Pellets is water soluble and should not be applied where surface water from treated areas can run off to crop lands either planted or to be planted.

Avoid use near desirable plants: This herbicide is water soluble, highly active and can remain in the soil for more than one growing season. Very small amounts can injure broadleaf plants, especially **tobacco and potatoes.** Therefore, do not apply on or near tobacco, potatoes, beans, blueberries or other fruit plants, ornamentals, shade trees, soybeans, or other vegetable crops.

Avoid movement of treated soil: Picloram may remain in treated soil for an extended period. Do not move treated soil to other areas and do not use such soil to grow plants.

Avoid transfer of livestock from a treated area to a broadleaf crop area without first allowing 7 days of grazing on untreated pasture. Otherwise, urine may contain enough picloram to cause injury to sensitive broadleaf plants. Do not use manure from animals grazing treated areas to fertilize soil or fields used to grow susceptible broadleaf crops.

Avoid water contamination: Do not allow TORDON 10K Pellets to contaminate water used for drinking, irrigation, or other domestic purposes. Do not apply on inner banks or bottoms of irrigation ditches. Do not clean containers or application equipment on or near these areas.

Avoid improper storage and equipment use: Do not store near fertilizers, seeds, insecticides, fungicides or other pesticides. Containers and equipment used for TORDON 10K Pellets should not be used for other agricultural chemicals since small residues of TORDON 10K Pellets herbicide can damage desirable plants.

Avoid improper disposal: Rinse equipment and dispose of waste by burying in non-cropland away from water supplies. Do not reuse containers. Bury them with waste or dispose in a sanitary landfill. Follow official local container disposal regulations.

Be sure that use of this product conforms to all applicable regulations.

SUPPLEMENTAL USE LABELS

CALIFORNIA .... EPA No. 464-07103-CA

For Use on Rangeland

A permit issued by the appropriate County Agricultural Commissioner is required before using this product on grazing lands and all applications of this product shall be made under the direct supervision of employees of a County Agricultural Commissioner or Control and Eradication Unit, Division of Plant Industry, California Department of Food and Agriculture.

GENERAL INFORMATION

FOR CONTROL OF: Undesirable woody plants, such as juniper, brambles and conifers, and for some broadleaf weeds, such as artichoke thistle.

CONSULT YOUR COUNTY AGRICULTURAL COMMISSIONER OR EXTENSION WEED SPECIALIST FOR ADDITIONAL INFORMATION ON THIS PROGRAM

USE DIRECTIONS

Apply from early spring to early summer before growth begins or during vigorous vegetative growth when subsequent rainfall can be expected for best results. Fall application when rain is expected within a few weeks also gives good results. Distribute TORDON 10K Pellets herbicide uniformly by spot or broadcast treatment to the soil over the roots of plants to be controlled.

USE RATES

Use TORDON 10K Pellets herbicide to control undesirable woody and herbaceous species at rates of 20 to 60 pounds per acre (½ to 1½ pounds of product per 1000 square feet) and distribute evenly over the entire area where vegetation is to be controlled. The lower rates may be used for perennial broadleaf weeds and for easy-to-control brush species, while 30–40 pounds per acre may be required to control moderately susceptible species, such as juniper, and 40–60 pounds for brambles, conifers and other more difficult species.

Since applications, especially at the higher rates, may suppress certain range grasses, do not apply unless possible injury can be tolerated. Any initial injury to established grass usually is followed by increased growth due to release from brush and weed competition. In most cases, applications should be made to individual noxious plants, to clumps, and to larger dense infestations where grasses already are suppressed.

USE PRECAUTIONS

Apply this product only as specified on this label.

Avoid use in certain areas: Do not apply to cropland used for production of desirable crops other than forage grasses. Do not rotate treated rangeland or pasture to other crops. Do not treat pastures or rangeland if the forage legume component is substantial and desired. Do not cut grass for hay in treated areas.

Avoid use near desirable plants: This herbicide is water soluble, highly active and can remain in the soil for more than one growing season. Very small amounts can injure broadleaf plants. Therefore, do not apply on or near beans, blueberries, or other fruit plants, ornamentals, potatoes, shade trees, soybeans, tobacco, or vegetable crops. **Do not apply where surface water from treated areas can run off to cropland planted or to be planted to broadleaf crops susceptible to picloram.** Do not apply to areas containing roots of desirable trees or shrubs growing nearby.

Avoid movement of treated soil: Picloram may remain in treated soil for an extended period. Do not move treated soil to other areas and do not use such soil to grow plants.

Avoid transfer of livestock from a treated area to a broadleaf crop area without first allowing 7 days of grazing on untreated pasture. Otherwise, urine may contain enough picloram to cause injury to sensitive broadleaf plants.

Avoid water contamination: Do not allow TORDON 10K Pellets to contaminate water used for irrigation, drinking, or other domestic purposes. Do not apply on inner banks or bottoms of irrigation ditches. Do not apply on areas from which surface run off water may go into streams, ponds, or wells supplying irrigation water. Do not clean containers or application equipment on or near these areas.

Avoid improper storage and equipment use: Do not store near fertilizers, seeds, insecticides, fungicides or other pesticides. Containers and equipment used for TORDON 10K Pellets should not be used for other agricultural chemicals since small residues of TORDON 10K Pellets herbicide can damage desirable plants.

Avoid improper disposal: Rinse equipment and containers and dispose of waste by burying in non-cropland away from water supplies. Do not reuse containers. Bury them with waste or dispose in a sanitary landfill. Where required, follow official local container disposal regulations.

BE SURE THAT USE OF THIS PRODUCT CONFORMS TO ALL APPLICABLE REGULATIONS

IDAHO .... EPA No. 464-07102-ID

For the Control of Juniper on Rangeland

The Idaho State Department of Agriculture has accepted the following use of TORDON 10K Pellets for control of juniper on rangelands as specified in the directions below and by the regulations of the State of Idaho Department of Agriculture.

CONSULT YOUR COUNTY WEED SPECIALIST OR EXTENSION WEED SPECIALIST FOR ADDITIONAL INFORMATION ON THIS PROGRAM.

USE DIRECTIONS

Read all Use Precautions before using this product.

Apply TORDON 10K Pellets to the soil under juniper trees using the rates indicated below.

	Amount per Tree
Trees less than 5 feet high	1 tablespoonful
Trees 5 to 10 feet high	2 tablespoonfuls
Trees 10 to 15 feet high	3 tablespoonfuls
Trees taller than 15 feet high ...	4 tablespoonfuls

On smaller trees TORDON 10K Pellets may be applied directly over the top to ensure distribution on all sides of the plants. On larger trees the pellets could be used at higher rates and uniform distribution around the tree should be insured. Applications made in early summer when rain is expected within a few weeks or in late summer or fall before freeze up give the best results. Do not apply on frozen soil.

USE PRECAUTIONS

Do not apply to cropland used for production of desirable crops other than forage grasses. Do not rotate treated rangeland or pasture to other crop uses.

Do not treat pastures if the forage legume component is substantial and desired or if pastures are to be harvested for hay.

Do not move treated soil. Do not go over treated areas with land levelers, cultivation or harvesting equipment nor move the soil to untreated areas by any other means. Do not use treated soil to grow plants.

Do not transfer livestock from a treated grass area onto a broadleaf crop area without first allowing 7 days of grazing on untreated grass pasture. Otherwise, urine may contain enough picloram to cause injury to sensitive broadleaf plants.

Do not apply or otherwise permit TORDON 10K Pellets to contact desirable plants nor the soil containing roots of desirable plants such as tobacco, potatoes, soybeans, beans, vegetable crops, shade trees, ornamentals, flowers, fruit plants, or nurseries.

Do not allow TORDON 10K Pellets to contaminate water used for irrigation, drinking, or other domestic purposes. Do not apply on inner banks or bottoms of irrigation ditches.

TORDON 10K Pellets herbicide is highly active and water soluble.

Do not apply where surface water from treated areas can run off to croplands, either planted or to be planted or to streams, ponds, or wells. Do not clean containers or application equipment near these areas.

Do not store near fertilizers, seeds, insecticides, fungicides, other pesticides, or nursery stock.

Do not reuse containers for TORDON 10K Pellets for other purposes. Bury them with waste in non-cropland areas away from water supplies or dispose in a sanitary landfill.

Observe any special use and application restrictions and limitations, including method of application and permissible areas of use as promulgated by state or local authorities.

EASTERN OREGON .... EPA No. 464-07101-OR

For the Control of Juniper on Rangeland

The Oregon State Department of Agriculture approves the following use of TORDON 10K Pellets for control of juniper on rangelands east of the crest of the Cascade Mountains as specified in the directions below and by the regulations of the State of Oregon Department of Agriculture.

CONSULT YOUR COUNTY WEED SPECIALIST OR EXTENSION WEED SPECIALIST FOR ADDITIONAL INFORMATION ON THIS PROGRAM.

USE DIRECTIONS

Read all Use Precautions before using this product.

Apply TORDON 10K Pellets to the soil under juniper trees using the rates indicated below:

	Amount per Tree
Trees less than 5 feet high	1 tablespoonful
Trees 5 to 10 feet high	2 tablespoonfuls
Trees 10 to 15 feet high	3 tablespoonfuls
Trees taller than 15 feet high	4 tablespoonfuls

On smaller trees TORDON 10K Pellets may be applied directly over the top to ensure distribution on all sides of the plants. On larger trees the pellets could be used at higher rates and uniform distribution around the tree should be insured. Applications made in early summer when rain is expected within a few weeks or in late summer or fall before freeze up give the best results. Do not apply on frozen soil.

USE PRECAUTIONS

Use only east of the crest of the Cascade Mountains in Oregon!

Do not treat pastures if the forage legume component is substantial and desired or if pastures are to be harvested for hay.

Do not move treated soil. Do not go over treated areas with land levelers, cultivation or harvesting equipment nor move the soil to untreated areas by any other means. Do not use treated soil to grow plants.

Do not transfer livestock from a treated grass area onto a broadleaf crop area without first allowing 7 days of grazing on untreated grass pasture. Otherwise, urine may contain enough picloram to cause injury to sensitive broadleaf plants.

Do not apply or otherwise permit TORDON 10K Pellets to contact desirable plants nor the soil containing roots of desirable plants such as tobacco, potatoes, soybeans, beans, vegetable crops, shade trees, ornamentals, flowers, fruit plants, or nurseries.

Do not allow TORDON 10K Pellets to contaminate water used for irrigation, drinking, or other domestic purposes. Do not apply on inner banks or bottoms of irrigation drainage ditches.

TORDON 10K Pellets herbicide is highly active and water soluble.

Do not apply where surface water from treated areas can run off to croplands, either planted or to be planted or to streams, ponds, or wells. Do not clean containers or application equipment near these areas.

Do not store near fertilizers, seeds, insecticides, fungicides, other pesticides, or nursery stock.

Do not reuse containers for TORDON 10K Pellets for other purposes. Bury them with waste in noncropland areas away from water supplies or dispose in a sanitary landfill.

Observe any special use and application restrictions and limitations, including method of application and permissible areas of use as promulgated by state or local authorities.

WARRANTY LIMITATIONS AND DISCLAIMER

The Dow Chemical Company warrants that this product conforms to the chemical description on the label and is reasonably fit for the purposes stated on the label when used in strict accordance with the directions therein under normal conditions of use. THIS IS THE ONLY WARRANTY MADE ON THIS PRODUCT. NO OTHER EXPRESS AND NO IMPLIED WARRANTY OF MERCHANTABILITY OR FITNESS FOR A PARTICULAR PURPOSE IS MADE OUTSIDE OF THIS LABEL. Therefore, neither this warranty nor any other warranty of merchantability or fitness for a particular purpose, express or implied, extends to the use of this product contrary to label instructions (including conditions noted on the label, such as unfavorable temperatures, soil conditions, etc.), under abnormal conditions (such as excessive rainfall, drought, tornadoes, hurricanes, etc.) or under conditions not reasonably foreseeable to or beyond the control of seller.

When buyer or user suffers losses or damages resulting from the use or handling of this product (including claims based on contract, negligence, strict liability, or other legal theories), buyer or user must promptly notify in writing The Dow Chemical Company of any claims to be eligible to receive either remedy given below. THE EXCLUSIVE REMEDY OF THE BUYER OR USER and the LIMIT OF LIABILITY of The Dow Chemical Company or any other seller will be one of the following, at the election of The Dow Chemical Company:

(1) **Refund of purchase price** paid by buyer or user for product bought, or

(2) **Replacement** of amount of product used.

The seller **will not be liable for consequential or incidental damages or losses.**

The term of this Limit of Warranty and Liability cannot be varied by any written or verbal statements or agreements. Any employee or sales agent of the seller is not authorized to vary or exceed the terms of this Warranty Limitations And Disclaimer in any manner.

* Trademark of THE DOW CHEMICAL COMPANY

87094-L4 A185

TORDON* 22K
WEED KILLER

RESTRICTED USE PESTICIDE

For retail sale to and use only by Certified Applicators or persons under their direct supervision and only for those uses covered by the Certified Applicator's certification.

ACTIVE INGREDIENT	
Picloram (4–amino–3,5,6–trichloropicolinic acid) as the potassium salt	24.4%
INERT INGREDIENTS	75.6%
TOTAL	100.0%

Acid Equivalent: Picloram (4–amino–3,5,6–trichloropicolinic acid) 21.1%—2 lbs./gallon.

EPA Reg. No. 464-323 EPA Est. 464-MI-1

CAUTION
Keep Out of Reach of Children
AVISO

PRECAUCION AL USUARIO: Si usted no lee ingles, no use este producto hasta que le etiqueta le haya sido explicada ampliamente

PRECAUTIONARY STATEMENTS

HAZARDS TO HUMANS AND DOMESTIC ANIMALS

MAY CAUSE IRRITATION

Avoid Contact With Skin and Eyes • Avoid Breathing Spray Mist • Keep Container Closed

STATEMENTS OF PRACTICAL TREATMENT

In case of contact, immediately flush eyes or skin with plenty of water. Get medical attention if irritation persists.

If swallowed, induce vomiting immediately by giving two glasses of water and sticking finger down throat. Call a physician. Do not induce vomiting or give anything by mouth to an unconscious person.

PHYSICAL OR CHEMICAL HAZARDS

Do Not Cut or Weld Container.

ENVIRONMENTAL HAZARDS

Do not apply directly to water. Do not apply where runoff is likely to occur. Do not contaminate water by cleaning of equipment or disposal of wastes. Do not contaminate irrigation ditches or water used for irrigation or domestic purposes.

NOTICE

Read the entire label. Use only according to label directions.

Before buying or using this product, read "WARRANTY LIMITATIONS AND DISCLAIMER" elsewhere on this label. If terms are not acceptable, return **unopened** package at once to seller for full refund of purchase price paid. Otherwise, use by the buyer or any other user constitutes acceptance of the terms under the Limit of Warranty and Liability.

In case of an emergency endangering life or property involving this product, call collect **517–636-4400.**

AGRICULTURAL CHEMICAL

Do Not Ship or Store with Food, Feeds, Drugs or Clothing

READ THE FEDERAL LABEL AND APPROPRIATE STATE LABELS BEFORE USING THIS PRODUCT

GENERAL INFORMATION

Use TORDON 22K Weed Killer on non-cropland areas such as fence rows, around farm buildings, equipment pathways and roadsides, to control annual and deep rooted perennial weeds such as Absynth wormwood • Bracken fern • Buffalo bur • Bur ragweed • Burroweed • Cactus species • Camel thorn • Catclaw acacia • Chapparral species • Dock • Firebrush • Field bindweed (perennial morningglory) • Fringed sagebrush • Gambel oak • Goldenrod • Gorse • Guava • Hau • Horsenettles, Carolina, white • Java plum • Junipers • Knapweeds, diffuse, Russian, spotted squarrose • Lantana • Larkspurs • Leafy spurge • Live oak • Locoweeds • Lupines • Melastoma • Mesquite • Ox-eye daisy • Milkweed • Pamakani • Poison Oak • Pinyon • Pricklypear cactus • Rabbitbrush • Rush skeletonweed • Scotch broom • Snakeweeds • Sowthistle • Starthistles, Iberian, purple, yellow • Tansy ragwort • Toadflaxes • Thistles, artichoke, Beaumont, Canada, distaff, golden, italian, musk, scotch, wavyleaf.

Picloram is highly potent, persistent and water soluble. Tiny amounts can kill or injure many broadleaved plants. To prevent damage to desirable crops and plants **follow all directions and precautions.**

DIRECTIONS FOR USE

Do not use for manufacturing or formulating

It is a violation of Federal law to use this product in a manner inconsistent with its labeling.

Mix the required amount of TORDON 22K Weed Killer in water and apply as a coarse low pressure spray using ground equipment or aircraft. For best results treat when the weeds are growing actively in the spring before full bloom or late summer into fall. Treatments during full bloom or seed stage of some weeds may not give good control.

For General Use on Perennial Weeds on Non-Cropland, use 1 to 1½ gallons of TORDON 22K Weed Killer per acre in 50 to 100 gallons of water and spray to wet weed foliage and soil. **NOTE:** Local conditions may affect the use of herbicides. State agricultural experiment stations or extension service weed specialists in many states issue recommendations to fit local conditions. Be sure that use of this product conforms to all applicable regulations.

For Use As A Spot Treatment on Perennial Weeds. Mix at the rate of 1 gallon of TORDON 22K per 100 gallons of water. Apply at the rate of 100 gallons of spray mixture per acre. This will provide a rate of 2 pounds of picloram per acre. For small amounts use 2½ fluid ounces TORDON 22K per 2 gallons of water. For round patches apply as indicated in the table.

Feet across Round Patch to be treated (weed area plus 10 foot border)	Gallons of spray mixture to apply
25	1.0
50	4.5
75	10.0
100	18.0
235 or (1 acre)	100.0

Tank Mixture for Spot or Broadcast Treatment of susceptible Annual and Perennial Broadleaf Weeds: TORDON 22K may also be tank mixed with 2,4-D products such as ESTERON® 99® Concentrate, FORMULA 40®, DMA® 4, or ESTERON 6E herbicides for use on areas having mixed species including those which respond well to 2,4-D. Read and follow all directions and use precautions on other product labels.

Be Sure You Follow All Use Precautions Given on This Label and Remember These Key Points

1. Use only the recommended amounts.
2. Picloram is persistent. It will carry over in the soil.
3. TORDON 22K is water soluble. It can move with water in irrigation or drainage ditches.
4. Spray drift can damage crops.

USE PRECAUTIONS

Use this product only as specified on this label. Observe any special use and application restrictions and limitations, including method of application and permissable areas of use as promulgated by state or local authorities.

Do not apply or Otherwise Permit TORDON 22K or Sprays Containing TORDON 22K to Contact Crops or Other Desirable Broadleaf Plants including but not limited to alfalfa, beans, grapes, melons, peas, potatoes, safflower, soybeans, sugarbeets, sunflower, tomatoes, and other vegetable crops, flowers, fruit plants, ornamentals, shade trees nor the soil containing roots of nearby valuable plants.

Avoid Injurious Spray Drift: Applications should be made only when there is no hazard from spray drift since very small quantities of the spray, which may not be visible, may severely injure susceptible crops during both growing and dormant periods. To minimize spray drift use low spray pressure, under 30 psi; spray when wind velocity is less than 10 mph; and apply as a coarse spray. To aid in further reducing spray drift, a drift control and deposition aid such as NALCO-TROL[1] may be used with this product if such a drift control aid is used, follow all use recommendations and precautions on the product label.

[1] NALCO-TROL-Trademark of NALCO Chemical Company

Ground Equipment: With ground equipment spray drift can be lessened by keeping the spray boom as low as possible; by applying 20 gallons or more of spray per acre; by using no more than 30 pounds spraying pressure with large droplet producing nozzle tips; by spraying when wind velocity is low; and by stopping all spraying when wind exceeds 10 miles per hour. Do not apply with hollow cone-type insecticide or other nozzles that produce fine-droplet spray.

Aerial Application: With aircraft, drift can be lessened by using straight stream nozzles directed straight back, and by using a spray boom no longer than ¾ the wing span of the aircraft.

Determine Air Movement and Direction Before Foliar Application: Do not spray when wind is blowing toward susceptible crops or ornamental plants near enough to be injured. It is suggested that a continuous smoke column at or near the spray site or a smoke generator on the spray equipment be used to detect air movements, lapse conditions, or temperature inversions (stable air). If the smoke layers or indicates a potential of hazardous spray drift, do not spray.

Do Not Contaminate Water Intended for Irrigation or Domestic Purposes. To avoid injury to crops or other desirable plants do not treat or allow spray drift to fall onto innerbanks or bottom of irrigation ditches or other channels that carry water that may be used for irrigation purposes.

Do Not Move Treated Soil To Other Areas. Do not use it to grow plants, unless adequately sensitive bioassay or chemical tests show that no detectable picloram is present in the soil.

Do Not Mix with Dry Fertilizer

STORAGE AND DISPOSAL

PROHIBITIONS

Do not contaminate water, food, or feed by storage or disposal. Open dumping is prohibited.

PESTICIDE DISPOSAL

Pesticide, spray, mixture, or rinsate that cannot be used or chemically reprocessed should be disposed of in a landfill approved for pesticides or buried in a safe place away from water supplies.

CONTAINER DISPOSAL

Do not re-use containers for TORDON 22K Herbicide for any purpose. Dispose by punching holes in them and burying with waste or by taking to an approved landfill. Where indicated, follow official local container disposal regulations. Plastic containers may also be disposed of by incineration or, if allowed by State and local authorities, by burning. If burned, stay out of smoke.

Rinse application equipment after use, at least three times with water, and dispose of rinse water in a non-cropland area away from water supplies.

GENERAL

Consult federal, state or local disposal authorities for approved alternative procedures.

SPECIAL LOCAL NEEDS

ARIZONA .... **EPA SLN NO. AZ-800001**

For Spot Treatment of Broom Snakeweed, Thistles, Locoweeds, Lupines, Larkspurs, Horsenettle, Burroweed, Cactus Species, Mesquite, Catclaw, Acacia, Chapparral Species, Juniper, Pinyon, Rabbitbrush and Other Woody and Herbaceous Plants on Rangeland and Permanent Grass Pastures.

DIRECTIONS FOR USE

It is a violation of Federal law to use this product in a manner inconsistent with its labeling.

Read complete use directions and precautions on this label and container label before using.

Consult your range or weed extension specialist for additional information on this program.

Mix the required amount of TORDON 22K Weed Killer in water and apply as a coarse, low pressure spray using ground equipment or aircraft. For best results, treat when the weeds are growing actively in the spring before full bloom or late in summer or fall. Treatments during full bloom or seed stage of some weeds may not give good weed control.

SPOT TREATMENT: Suggested rates to control several broadleaf weeds and brush are shown in the table below. Apply only once during any one growing season. Retreat in subsequent years as needed using similar rates. TORDON 22K Weed Killer may be tank mixed with ESTERON® 99® Concentrate, FORMULA 40®, DMA 4 or ESTERON 6E herbicides (phenoxy) or other herbicides cleared for use in rangelands to control mixed species. In tank mix combinations, use ¼ to 1 qt. of TORDON 22K Weed Killer with 1 to 2 qt. of ESTERON 99 Concentrate, FORMULA 40 or DMA 4, or with ⅔ to 1⅓ qt. ESTERON 6E per acre. **Read and follow all directions and use precautions on other product labels.**

Weed or Brush Species	Rate of TORDON 22K Weed Killer per Treated Acre	Rate of TORDON 22K Weed Killer per 100 sq. ft.
Annual thistles, rabbitbrushes[1], lupines, locoweeds, broom snakeweed, mesquite	½–1 qt.	1.1–2.2 ml. or ¼–½ tsp.
Prickly pear and cholla cactus, burroweed, plains larkspur, catclaw acacia	1–2 qt.	2.2–4.5 ml. or ½–1 tsp.
White horsenettle, Russian knapweed, pinyon, juniper, chaparral species[2]	2–3 qt.	4.5–6.8 ml. or 1–1½ tsp.
Tall larkspur, Gambel oak[2], live oak brush[2]	3–4 qt.	6.8–9.0 ml. or 1½–2 tsp.

[1] Add 1½ to 2 qt. ESTERON 99 Concentrate per acre for sagebrush control also.

[2] Add 2-3 qt. ESTERON 99 Concentrate herbicide.

GROUND APPLICATION: When applying TORDON 22K Weed Killer with ground equipment, use coarse, low pressure spray (under 30 psi) and apply uniformly to provide good coverage of the weeds. Apply only when weather conditions are favorable for keeping spray on the target area. Do not allow spray drift to contact off-target susceptible plants, or areas to be planted to susceptible crops. Do not apply when wind velocity exceeds 10 mph. Where desirable susceptible plants, such as potatoes, beans and other vegetable crops, ornamentals or legumes are growing or may be planted within ½ mile, apply TORDON 22K Weed Killer only if air movement is continously from a definite direction and away from these plants.

AERIAL APPLICATION: When applying by aircraft, pilot must comply with all applicable state and local regulations. When applying TORDON 22K Weed Killer by air, use coarse, low pressure spray (under 30 psi) and apply uniformly to provide good coverage of weeds. The distance between outer-most nozzles should not exceed ¾ of the wing or rotor length. Apply only when weather conditions are favorable for keeping spray on target area.

Do not allow spray drift to contact off-target susceptible plants. Do not apply when wind velocity exceeds 10 mph, or as required by state regulations. Do not spray when air temperature exceeds 85°F. Where desirable susceptible plants, such as potatoes, beans, and other vegetable crops, ornamentals or legumes are growing or may be planted within 1 mile for helicopters under 55 mph or 4 miles for fixed wing aircraft, apply TORDON 22K Weed Killer only if air movement is continuously from a definite direction and away from these plants.

For treatment of small patches of broadleaf weeds, use ½ to 4 quarts of TORDON 22K Weed Killer in 100 gallons of water and spray weed foliage uniformly using 50 to 100 gallons of spray per treated acre. Mixtures with other herbicides referred to above can be made to control more species. NOTE: For treating small areas, 1 qt. TORDON 22K Weed Killer in 100 gals. of water per acre is equivalent to 2 teaspoonsful per gallon of water applied to a 500 square foot area. Basal applications of undiluted TORDON 22K Weed Killer can be made using 2 ml. (½ teaspoon) per 3 ft. of tree height of juniper or pinyon, treat shortly before a rain to leach the herbicide into the soil.

GRAZING RESTRICTIONS: Do not graze animals on treated areas within 30 days after application.

USE PRECAUTIONS

Use this product only as specified on this label. Observe any special use and application restrictions and limitations, including method of application and permissible areas of use as promulgated by state or local authorities.

Do Not Contaminate Nontarget Land Areas, Cropland, Water or Irrigation Ditches. Do not apply directly to standing or running water. Do no apply where surface water from treated areas can run off to adjacent cropland, either planted or to be planted, or into streams, irrigation ditches, irrigation ponds, or wells. Do not clean containers nor application equipment on or near these areas. Do not apply to frozen or saturated ground.

Do not apply on or in the vicinity of susceptible crops or desirable plants including alfalfa, beans, cotton, melons, potatoes, sugar beets, sunflower, tomatoes and other vegetable crops, flowers, fruit plants, ornamentals or shade trees.

Avoid Spray Drift. Applications should be made only when there is no hazard from spray drift since very small quantities of the spray, which may not be visible, may severely injure susceptible crops during both growing and dormant periods. Use coarse sprays to minimize drift since, under adverse weather conditions, fine spray droplets may drift a mile or more. The spray thickening agent, NALCO-TROL[1] may be used with this product to aid in reducing spray drift. If used, follow all use recommendations and precautions on the product label.

[1] NALCO-TROL—Trademark of NALCO Chemical Company

Ground Equipment: With ground equipment, spray drift can be lessened by keeping the spray boom as low as possible; by applying 20 gallons or more of spray per acre; by using no more than 30 pounds spraying pressure with large droplet producing nozzle tips; by spraying when wind velocity is low, and, by stopping all spraying when wind exceeds 10 miles per hour. Do not apply with hollow cone-type insecticide or other nozzles that produce a fine-droplet spray.

Aerial Application: With aircraft, drift can be lessened by applying a coarse spray; by using no more than 30 pounds spray pressure at nozzles; by using straight stream nozzles directed straight back; by using a spray boom no longer than ¾ the wing span of the aircraft; and, by spraying only when wind velocity is 10 mph or less.

Do Not Apply By Aircraft When An Air Temperature Inversion Exists: Such a condition is characterized by little or no wind and with air temperature lower near the ground than at higher levels. The use of a continuous smoke column at or near site of application or use of a smoke generating device on the aircraft is suggested to indicate direction and velocity of air movement, and to indicate a temperature inversion by layering of the smoke.

Do not rotate treated rangeland or pasture to other crop uses, as long as residues which persist in the soil may affect susceptible crops.

Do not spray pastures if the forage legume component is desired. TORDON 22K Weed Killer may injure or kill legumes. Also new legume seedlings may not be successful if made within 2 years following application of this herbicide.

Do not move treated soil to other areas. Do not use it to grow plants, unless adequate sensitive bioassay or chemical tests show that no detectable picloram is present in the soil.

Do not transfer livestock from treated grass areas onto broadleaf crop areas without first allowing 7 days of grazing on untreated grass pasture. Otherwise, urine may contain enough picloram to cause injury to sensitive broadleaf plants.

Do not re-use containers for TORDON 22K Weed Killer for any purpose. Dispose by punching holes in them and burying with waste or by taking to an approved landfill. Where indicated, follow official local container disposal regulations.

Rinse application equipment after use, preferably at least three times with water, and dispose of rinse water in a non-cropland area away from water supplies.

Be sure that use of this product conforms to all applicable regulations.

IDAHO .... EPA SLN NO. ID-780009
NEVADA .... EPA SLN NO. NV-790007
OREGON .... EPA SLN NO. OR-780012
UTAH .... EPA SLN NO. UT-820011
WASHINGTON .... EPA SLN NO. WA-820068

For control of Yellow Starthistle, Scotch, Musk, and Canada Thistles, Spotted, Diffuse and Russian Knapweeds, Rush Skeletonweed, Larkspur, Leafy Spurge, Rabbitbrush spp. Tansy Ragwort, Field Bindweed, Poison Oak, Gorse, Dalmation Toadflax, Buffalo Bur, Henbane and other Susceptible Poisonous Plants, Broadleaf Weeds and Woody Plants on Rangeland and Permanent Grass Pastures.

DIRECTIONS FOR USE

It is a violation of Federal law to use this product in a manner inconsistent with its labeling.

Read complete use directions and precautions on this label and container label before using.

Mix the required amount of TORDON 22K Weed Killer in water and apply as a coarse, low pressure spray using ground equipment or aircraft.

For best results, treat when the weeds are growing actively in the spring before full bloom or in late summer or fall. Treatments during full bloom or seed stage of some weeds may not give good control.

BROADCAST TREATMENT: TORDON 22K Weed Killer can be applied as a broadcast treatment at rates of 1 quart or less per acre; higher rates can be used on patches but the total areas treated should be regulated so no more than 25 gals. of TORDON 22K Weed Killer is used on any 100 acre block in a single season. Suggested rates to control several broadleaf weeds are shown in the table below. Weeds requiring higher rates may be suppressed with rates of 1 quart per acre. Apply as a single broadcast spray during any one growing season. Retreat in subsequent years as needed using similar rates. TORDON 22K Weed Killer may be tank mixed with ESTERON® 99® Concentrate, FORMULA 40®, DMA® 4 or ESTERON 6E herbicides (2,4-D products) for use on areas having mixed species including those which respond well to 2,4-D, such as big sagebrush. In tank mix combinations, use ¼ to 1 quart of TORDON 22K with 1 to 2 quarts ESTERON 99 Concentrate, FORMULA 40 or DMA 4 or with ⅔ to 1⅓ quarts ESTERON 6E per acre.

Weed Species	Rates of TORDON 22K per Treated Acre
Yellow Starthistle†, Scotch Thistle, Musk Thistle, Ox-eye Daisy	¼ to ½ quart
Diffuse Knapweed, Spotted Knapweed, Henbane, Buffalo Bur, Rabbitbrush spp, Tansy Ragwort	½ to 1 quart
Canada Thistle, Rush Skeletonweed, Russian Knapweed, Dalmation Toadflax, Gorse	2-3 quarts††
Leafy Spurge†, Larkspur†, Field Bindweed, Poison Oak†	4 quarts††

† Denotes poisonous plants.
†† Lower rates may be used for short term suppression.

GROUND APPLICATION: When applying TORDON 22K Weed Killer with ground equipment, use coarse, low pressure spray (under 30 psi) and apply uniformly to provide good coverage of the weeds. Apply only when weather conditions are favorable for keeping spray on the target area. Do not allow spray drift to contact off-target susceptible plants, or areas to be planted to susceptible crops. Do not apply when wind velocity exceeds 10 mph. Where desirable susceptible plants such as potatoes, beans, peas, and other vegetable crops, ornamentals or legumes **are growing or may be planted within ½ mile,** apply TORDON 22K **only if air movement is continuously from a definite direction and away from these plants.**

AIR APPLICATION: When applying by aircraft, pilot must comply with all applicable state and local regulations. When applying TORDON 22K Weed Killer by air, use coarse, low pressure spray (under 25 psi) and apply uniformly to provide good coverage of the weeds. Apply only when weather conditions are favorable for keeping spray on target area. Do not allow spray drift to contact off-target susceptible plants. Do not apply when wind velocity exceeds 10 mph, (or as required by state regulations.) The distance between outermost nozzles should not exceed ¾ of the wing or rotor length. Do not spray when air temperature exceeds 85°F. Where desirable susceptible plants such as potatoes, beans, peas and other vegetable crops, ornamentals, or legumes **are growing or may be planted within 1 mile,** apply TORDON 22K Weed Killer by helicopter (only if air movement is continuously from a definite direction and away from these plants.) Fixed wing aircraft may be used only in areas east of the crest of the Cascade Mountains under the above conditions provided that fields or homesites where susceptible plants may be growing or be planted are over 3 miles **downwind** at the time of application.

SPOT TREATMENT: For spot treatment of small patches of broadleaf weeds, use ½ to 4 quarts of TORDON 22K in 100 gallons of water and spray weed foliage uniformly using 50 to 100 gallons of spray per treated acre.

NOTE: For treating small areas, 1 quart TORDON 22K Weed Killer in 100 gallons of water per acre is equivalent to 2 teaspoonsful per gallon of water applied to a 500 square foot area.

GRAZING RESTRICTIONS: (Where rates greater than 1 quart per acre are applied.) Do not graze dairy animals on treated areas within 2 weeks after application. Other animals should be withdrawn from treated areas at least 3 days before slaughter. Observe grazing restrictions on other product labels when using tank mixtures.

USE PRECAUTIONS

Use this product only as specified on this label. Observe any special use and application restrictions and limitations, including method of application and permissible areas of use as promulgated by state or local authorities.

Do Not Contaminate Nontarget Land Areas, Cropland, Water, or Irrigation Ditches. Do not apply directly to standing or running water. Do not apply where surface water from treated areas is likely to run off to adjacent cropland, either planted or to be planted, or into streams, irrigation ditches, irrigation ponds, or wells. Do not clean containers nor application equipment on or near these areas. Do not apply on inner banks or bottom of irrigation ditches. Do not apply to frozen ground.

Do not apply on or in the vicinity of susceptible crops or desirable plants including alfalfa, beans, clovers, grapes, melons, peas, potatoes, safflower, soybeans, sugar beets, sunflower, tomatoes and other vegetable crops, flowers, fruit plants, ornamentals or shade trees.

Avoid Spray Drift: Applications should be made only when there is no hazard from spray drift since very small quantities of the spray, which may not be visible, may severely injure susceptible crops during both growing and dormant periods. Use coarse sprays to minimize drift since, under adverse weather conditions, fine spray droplets may drift a mile or more. The spray thickening agent, NALCO-TROL[1] may be used with this product to aid in reducing spray drift. If used follow all use recommendations and precautions on the product label.

[1] NALCO-TROL—Trademark of NALCO Chemical Company

Ground Equipment: With ground equipment, spray drift can be lessened by keeping the spray boom as low as possible; by applying 20 gallons or more of spray per acre; by using no more than 30 pounds spraying pressure with the large droplet-producing nozzle tips; by spraying when wind velocity is low; and by stopping all spraying when wind exceeds 10 miles per hour. Do not apply with hollow cone-type insecticide or other nozzles that produce a fine-droplet spray.

Aerial Application: Drift can be lessened by applying a coarse spray; by using no more than 25 pounds spray pressure at nozzles; by using nozzles which do not produce hazardous amounts of fine spray droplets; by using a spray boom no longer than ¾ of the rotor or wing length, and by spraying only when wind velocity is less than 10 mph or as required by state regulations.

Determine Air Movement and Direction Before Foliar Application: Use a smoke generator or other means at or near the applicaton site or on aircraft for detection of air movement, air stability, or temperature inversions. Such a condition exists when there is little or no wind and air temperature is lower near the ground than at higher levels. Use appropriate drift control measures or avoid application when smoke is moving toward nearby desirable susceptible plants or sensitive areas.

Do not rotate treated rangeland or pasture to other crop uses.

Do not spray pastures if the forage legume component is desired, TORDON 22K Weed Killer may injure or kill legumes. Also, new legume seedlings may not be successful if made within 2 years following application of this herbicide.

Do not move treated soil to other areas: Do not use it to grow plants unless adequate sensitive bioassay or chemical tests show that no detectable picloram is present in the soil.

Do not transfer livestock from treated grass areas onto broadleaf crop areas without first allowing 7 days of grazing on untreated grass pasture. Otherwise, urine may contain enough picloram to cause injury to sensitive broadleaf plants.

Do not re-use containers for TORDON 22K Weed Killer for any purpose. Dispose by punching holes in them and burying with waste or by taking to an approved landfill. Where indicated, follow official local container disposal regulations.

Rinse application equipment after use preferably at least three times with water, and dispose of rinse water in a non-cropland area away from water supplies.

Be sure that use of this product conforms to all applicable regulations.

IDAHO .... EPA SLN NO. ID-820030
OREGON .... EPA SLN NO. OR-820061
WASHINGTON .... EPA SLN NO. WA-820061

For Control of Susceptible Perennial Broadleaf Weeds in Non-Irrigated Fallow Cropland East of the Cascade Mountains

DIRECTIONS FOR USE

It is a violation of Federal law to use this product in a manner inconsistent with its labeling.

Read complete use directions and precautions on this label and container label before using.

PERENNIAL BROADLEAF WEED CONTROL IN FALLOW GRAINLAND

To control perennial broadleaf weeds, such as bindweed and Canada thistle, in fallow grainland, follow a treatment schedule and apply TORDON 22K Weed Killer at 1 to 1-½ quarts per acre after grain harvest. The addition of 2,4-D at 1 to 2 lbs. a.e. per acre may significantly improve weed control. The treated area should extend at least 10 feet beyond the visible weeds. During the next season, retreat the same fallow area with 1 pint of TORDON 22K. Adding 2,4-D to the tank mix may improve weed control. Treated areas can be relocated more readily if marked with flags or painted stakes. A **permanent record** of rates of TORDON 22K Weed Killer and dates of treatment should be made by the grower to aid in planning future crop rotations.

When starting treatment in the fallow season before planting, apply 1 pint of TORDON 22K per acre when weeds are growing well either early in the season or treating regrowth after removing early weeds by cultivation. Retreat the same fallow area with 1 to 1-½ quarts of TORDON 22K Weed Killer per acre after harvest of grain the next summer.

Treatments of TORDON 22K at rates over 1.5 quarts per acre can be made on fallow, non-irrigated cropland if the treated areas comprise less than 10% of the immediate field in any one year. TORDON 22K should not be applied to cropland at rates exceeding 1 gallon per acre.

To prepare the spray, mix the required amount of TORDON 22K Weed Killer and 2,4-D or other approved herbicides or additives in water and apply as a coarse, low pressure spray in sufficient volume for adequate coverage.

For Ground Application: Use enough spray volume to provide adequate coverage without producing hazardous amounts of drift. Care must be used to prevent off-target injury to plants especially with very low volume rates. Where desirable susceptible plants, such as potatoes, legumes, other vegetable crops or ornamentals, are growing or may be planted in the vicinity, apply TORDON 22K only if air movement is from a definite direction away from these plants at time of application.

DO NOT APPLY WITH AIRCRAFT

NOTE: Small grains planted after fallow land treatment with TORDON 22K may be adversely affected to varying degrees, depending on rate of application, rainfall and temperature. Sensitive crops such as peas, lentils, potatoes or beans, must not be planted until crop injurious residues of picloram have disappeared.

Livestock may graze treated fields with the following restrictions when applying more than 1 quart TORDON 22K per acre: (1) Meat animals grazing for up to two weeks after treatment should be removed from treated areas 3 days prior to slaughter. (2) Do not graze dairy animals on treated areas within 2 weeks after treatment.

USE PRECAUTIONS

Use this product only as specified on this label and product label. Observe any special use and application restrictions and limitations, including method of application and permissible areas of use as promulgated by state and local authorities.

Do Not Contaminate Water Intended for Irrigation or Domestic Purposes. Do not treat or allow spray drift to fall onto innerbanks or bottom of irrigation ditches, either dry or containing water or other channels that carry water that may be used for irrigation purposes.

Do not apply on or in the vicinity of susceptible crops or desirable plants including alfalfa, beans, grapes, melons, peas, potatoes, safflower, sugar beets, sunflower, tomatoes and other vegetable crops, flowers, fruit plants, ornamentals or shade trees.

Avoid Spray Drift: Applicatons should be made only when there is no hazard from spray drift since very small quantities of the spray, which may not be visible, may severely injure susceptible crops during both growing and dormant periods. Use coarse sprays to minimize drift since, under adverse weather conditions such as temperature inversions, fine spray droplets may drift a mile or more. Spray thickening agents, such as NALCO-TROL[1], may be used with this product to aid in reducing spray drift. If used follow all use recommendations and precautions on the product label.

[1] NALCO-TROL-Trademark of NALCO Chemical Company

Ground Equipment: With ground equipment, spray drift can be lessened by keeping the spray boom as low as possible; by applying 20 gallons or more of spray per acre; by using no more than 30 pounds spraying pressure with large droplet producing nozzle tips; by spraying when wind velocity is low; and by stopping all spraying when wind exceeds 10 miles per hour. Do not apply with hollow cone-type insecticide or other nozzles that produce fine-droplet spray.

Do not move treated soil to other areas. Do not use it to grow plants, unless adequately sensitive bioassays show that no detectable picloram is present in the soil.

Do not transfer livestock from treated grazing areas onto broadleaf crop areas during first full season after treatment without allowing 7 days of grazing on an untreated area. Otherwise, urine may contain enough picloram to cause injury to sensitive broadleaf plants.

Do not re-use containers for TORDON 22K Weed Killer for any purpose. Dispose by punching holes in them and burying with waste or by taking to an approved landfill. Where indicated, follow official local container disposal regulations.

Rinse application equipment after use, preferably at least three times with water, and dispose of rinse water in a non-cropland area away from water supplies.

Be sure that use of this product conforms to all applicable regulations.

Do not use where a sandy porous surface and substrate overlie groundwater 10 feed or less below the surface.

READ AND FOLLOW MIXING AND USE INSTRUCTIONS AND PRECAUTIONS ON PRODUCT CONTAINER LABEL

IOWA .... EPA SLN NO. IA-790007
KANSAS .... EPA SLN NO. KS-800018

For the Control of Musk Thistle in Rangeland and Permanent Grass Pastures.

GENERAL INFORMATION

Use TORDON 22K Weed Killer to control musk thistle growing in rangeland or permanent grass pastures. Picloram, the active ingredient in TORDON 22K Weed Killer, is highly active, persistent, and water soluble. To prevent damage to desirable crops and plants follow all directions on this label.

DIRECTIONS FOR USE

It is a violation of Federal law to use this product in a manner inconsistent with its labeling.

Mix the required amount of TORDON 22K Weed Killer in the specified amount of water and apply broadcast with conventional ground equipment. Use enough spray volume to provide uniform coverage of the weeds, usually 10 to 20 gallons per acre. Do not apply by air or with a mist blower.

Apply when musk thistle plants are in the rosette stage of growth or before flower stock elongates over 2 inches above leaves in the spring or in the fall prior to soil freeze-up. Treat no more than once in any 12-month period and no more than twice in any three-year period.

Use 6 to 8 fl oz ($^3/_8$ to $^1/_2$ pt) of TORDON 22K Weed Killer in 10 to 20 gallons of water per acre and apply as a coarse, low-pressure (20 to 30 psi) spray. Apply only when weather conditions are favorable for keeping spray on target area. Do not apply when wind velocity exceeds 10 mph. Where desirable susceptible plants (see **Use Precautions**) are growing within $^1/_2$ mile of the application site, apply TORDON 22K only if air movement is continuous from a definite direction away from these plants.

USE PRECAUTIONS

Use this product only as specified on this label. Observe any special use and application restrictions and limitations including method of application and permissible areas of use, promulgated by state or local authorities.

Handle and Use Correctly. Small amounts of TORDON 22K Weed Killer can injure sensitive broadleaf plants. To avoid problems, observe the following precautions:

1. **Do Not Contaminate Nontarget Land Areas, Cropland, Water, or Irrigation Ditches.** Do not apply directly to standing or running water. Do not apply where surface water from treated areas can run off to adjacent croplands, either planted or to be planted, or into drainage channels, streams, irrigation ditches, ponds, or wells. Do not clean containers or application equipment on or near these areas. Do not apply on inner banks or bottoms of irrigation ditches.
2. **Do Not Apply to or in the Vicinity of Sensitive Crops or Desirable Plants** including alfalfa, beans, cotton, grapes, melons, peanuts, peas, potatoes, safflower, sugar beets, soybeans, sunflower, tobacco, tomatoes, other field or vegetable crops, flowers, fruit plants, ornamentals, or shade trees. Do not apply on areas which may be planted at a later time to such crops or plants.
3. **Do Not Apply to Cropland Used for Production of Desirable Crops Other Than Forage Grasses.** Do not rotate treated rangeland or pasture to other crop uses.
4. **Do Not Spray Pastures if the Forage Legume Component is Substantial and Desired.** TORDON 22K Weed Killer may injure or kill forage legumes. Also, new legume seedling establishment may not be successful if seeded within 2 years following application of this herbicide.
5. **Do Not Move Treated Soil to Other Areas.** Do not use it to

grow plants, unless adequately sensitive bioassay or chemical tests show that no detectable picloram is present in the soil.

6. **Do Not Transfer Livestock** from treated grass area onto a broadleaf crop area without first allowing 7 days of grazing on untreated grass pasture. Otherwise, urine may contain enough picloram to cause injury to sensitive broadleaf plants.
7. **Do Not Store Near Food,** feedstuff, fertilizer, seed, insecticides, fungicides, or other pesticides.
8. **Do Not Re-use Containers for TORDON 22K Weed Killer for Any Purpose.** Dispose by punching holes in them and burying with waste or by taking to an approved sanitary landfill. Rinse application and handling equipment thoroughly after use, preferably a minimum of three times with clean water, and dispose of rinse water in a non-cropland area away from water supplies. Where indicated, follow official local container disposal regulations.
9. **Do Not Allow Spray Drift.** Do not permit any spray or spray drift to contact desirable plants off the area treated or the soil containing roots of such plants.

To avoid spray drift, use low spray pressure, under 30 psi and apply as a coarse spray. Do not use pressures nor spray nozzles that produce fine droplets which would present a drift hazard. Spray TORDON 22K Weed Killer only when wind velocity is less than 10 mph and **when there is no hazard from spray drift off the target area. Do not spray when the wind is blowing toward susceptible crops or ornamental plants that are one mile from the target area.**

10. **Do Not Use Where a Sandy Porous Surface and Substrate Overlie Ground Water Ten Feet or Less Below the Surface.**
11. **Be Sure That Use of This Product Conforms to All Applicable Regulations.**

KANSAS........................EPA SLN NO. KS-820024

For the Control of Musk Thistle in Rangeland and Permanent Grass Pastures.

GENERAL INFORMATION

Use TORDON 22K Weed Killer to control musk thistle growing in rangeland or permanent grass pastures. Picloram, the active ingredient in TORDON 22K Weed Killer, is highly active, persistent, and water soluble. To prevent damage to desirable crops and plants follow all directions on this label.

DIRECTIONS FOR USE

It is a violation of Federal law to use this product in a manner inconsistent with its labeling.

Mix the required amount of TORDON 22K Weed Killer in the specified amount of water and apply broadcast by air or ground equipment. Apply as a coarse, low-pressure spray using enough spray volume to provide uniform coverage of weeds. Do not apply with a mist blower.

Apply in the fall when musk thistle plants are in the rosette stage of growth. **Treat prior to soil freeze up,** October 1 through December 1. For spring application, apply by ground only. Treat when musk thistle plants are in the rosette stage of growth and before the flower stalk elongates over two inches above the rosette leaves. Spraying after this time is likely to result in unsatisfactory control. Treat no more than once in any 12-month period, no more than twice in any three-year period.

Use six to eight ounces (3/8-1/2 pint) of TORDON 22K Weed Killer usually in 1 to 4 gallons of water per acre by air or in 10 to 20 gallons of water per acre with ground application. The higher rate is preferred if flower stalks are nearing the two-inch height.

Apply only when weather conditions are favorable for keeping spray on the target area. Do not treat by aerial application any musk thistle infestation in the understory of areas designated as forest (land supporting timber stands 120 feet wide), wooded strips or wooded pasture where the tree component is desirable. These areas should be treated by spot application only. Where wooded areas or other desirable susceptible plants are growing within 1/2 mile of the application site, apply TORDON 22K only if air movement is continuous in a direction away from these plants. (See Use Precautions.)

USE PRECAUTIONS

Use this product only as specified on this label. Observe any special use and application restrictions and limitations, including method of application and permissible areas of use as promulgated by state or local authorities.

Do Not Contaminate Water Intended for Irrigation or Domestic Purposes. Do not treat or allow spray drift to fall onto innerbanks or bottom of irrigaton ditches, either dry or containing water or other channels that carry water that may be used for irrigation purposes.

Do not apply on or in the vicinity of susceptible crops or desirable plants including alfalfa, beans, grapes, melons, peas, potatoes, safflower, soybeans, sugar beets, sunflower, tomatoes and other vegetable crops, flowers, fruit plants, ornamentals, forest and shade trees or the soil containing roots of nearby valuable lands.

AVOID SPRAY DRIFT: Applications should be made only when there is no hazard from spray drift since very small quantities of the spray, which may not be visible, may severely injure susceptible crops during both growing and dormant periods. To avoid spray drift use low spray pressure, apply as a coarse spray, and spray when wind velocity is low. A spray thickening agent such as NALCO-TROL[1] may be used with this product to aid in reducing spray drift. If a thickening agent is used, follow all use recommendations and precautions on the product label.

[1] NALCO-TROL-Trademark of NALCO Chemical Company

Ground Application: With ground equipment, spray drift can be lessened by keeping the spray boom as low as possible; by applying 20 gallons or more of spray per acre; by using no more than 30 pounds spraying pressure with large droplet producing nozzle tips; by spraying when wind velocity is low; and by stopping all spraying when wind exceeds 10 miles per hour. Do not apply with hollow cone-type insecticide or other nozzles that produce a fine-droplet spray.

Aerial Application: Drift can be lessened by applying a coarse spray; by using no more than 25 pounds spray pressure at nozzles; by using nozzles which do not produce hazardous amounts of fine spray droplets; by using a spray boom no longer than 3/4 of the rotor or wing length; and by spraying only when wind velocity is less than 10 mph, or as required by state regulations.

Determine Air Movement and Direction Before Foliar Application: Use a smoke generator or other means at or near the application site or on aircraft for detection of air movement, air stability, or temperature inversions. Such a condition exists when there is little or no wind and air temperature is lower near the ground than at higher levels. Use appropriate drift control measures or avoid application when smoke is moving toward nearby desirable susceptible plants or sensitive areas.

Do not rotate treated rangeland or pasture to other crop uses.

Do not spray pastures if the forage legume component is substantial and desired. TORDON 22K Weed Killer may injure or kill legumes. Also, new legume seedings may not be successful if made within two years following application of this herbicide.

Do not move treated soil to other areas. Do not use it to grow plants, unless adequately sensitive bioassay or chemical tests show that no detectable picloram is present in the soil.

Do not transfer livestock from treated grass areas onto broadleaf crop areas without first allowing 7 days of grazing on an untreated grass pasture. Otherwise, urine may contain enough picloram to cause injury to sensitive broadleaf plants.

Do not re-use containers for TORDON 22K Weed Killer for any purpose. Dispose by punching holes in them and burying with waste or by taking to an approved landfill. Where indicated, follow official local container disposal regulations.

Rinse application equipment after use, preferably at least three times with water, and dispose of rinse water in a non-cropland area away from water supplies.

Be sure that use of this product conforms to all applicable regulations.

Do not use where a sandy porous surface and substrate overlie ground water ten feet or less below the surface.

READ AND FOLLOW MIXING AND USE INSTRUCTIONS AND PRECAUTIONS ON PRODUCT CONTAINER LABEL

MINNESOTA .... EPA SLN NO. MN-820002

For the Control of Broadleaf Weeds in Rangelands, Permanent Grass Pastures, Spring Barley and Oats, and Spring and Winter Wheat.

DIRECTIONS FOR USE

It is a violation of Federal law to use this product in a manner inconsistent with its labeling.

Read complete use directions and precautions on this label and container label before using.

RANGELAND AND PASTURE

Use TORDON 22K Weed Killer to control broadleaf annual and perennial weeds such as Canada and other thistles, field bindweed, leafy spurge, Russian knapweed, spotted knapweed, bracken fern, golden rod, and absynth wormwood, yellow toadflax on rangeland and permanent grass pastures. Treat when weeds are growing well using low pressure sprays. Retreatment at the same rate may be necessary the following year.

Spot Treatment: (ground application only) Use TORDON 22K at rates of 2 to 4 quarts in 5 to 100 gallons of water per acre and apply as a spray to the foliage. Use the higher rates to control leafy spurge, larkspur, toadflax and knapweed. Use the lower rates for bindweed and thistles. For a 1,000 square foot infestation, apply about 2¼ fluid ounces of TORDON 22K in 1 gallon of water (equivalent to 3 quarts per acre of TORDON 22K).

NOTE: When spot treating, do not cut grass for feed within 2 weeks after treatment. Meat animals grazing for up to two weeks after treatment should be removed from treated areas 3 days prior to slaughter. Do not graze dairy animals on treated areas within 2 weeks after treatment.

Broadcast Treatment: (ground or aerial application) To suppress the growth of many perennial broadleaf weeds, apply as a broadcast spray using 1 quart of TORDON 22K in 1 to 4 gallons of water per acre by air or in 5 to 100 gallons of water per acre by ground equipment and apply as a broadcast spray during the growing season when weeds are growing well. (Many seedling annual weeds can be controlled using 1 pint per acre.)

Tank Mix for Spot or Broadcast Treatments: TORDON 22K may also be tank mixed with 2,4-D products such as ESTERON® 99® Concentrate, FORMULA 40®, DMA 4 or ESTERON 6E herbicides for use on areas having mixed species including those which respond well to 2,4-D, such as buckbrush. In tank mix combinations, use 1 pint to 1 quart of TORDON 22K with 1 to 2 quarts Esteron 99 Concentrate, FORMULA 40 or DMA 4, or with ⅔ to 1⅓ quarts of ESTERON‡ 6E per acre, in spray volumes specified above. **Read and follow all directions and use precautions on other product labels.**

SPRING BARLEY AND OATS, SPRING AND WINTER WHEAT NOT UNDERSEEDED WITH A LEGUME

For the control of wild buckwheat and other annual broadleaf weeds normally controlled with 2,4-D or MCPA such as wild mustard, Russian thistle, pennycress, lambsquarters and pigweed, in spring wheat and barley and winter wheat, use TORDON 22K as a tank mix with a 2,4-D or MCPA formulation such as DMA 4, FORMULA 40, ESTERON 99 Concentrate, ESTERON‡ 6E, or MCP Amine Herbicides. For use on spring oats, tank mix only with MCP Amine Herbicide. **Read and follow all directions and use precautions on other product labels.**

Spring Wheat, Barley and Oats: Apply during the 3 through 5 leaf stage of growth. Application of TORDON 22K occasionally causes slight head malformations and straw shortening but normally this does not affect yield.

Durum Wheat: Do not treat durum since at least some varieties appear to be more sensitive than other wheat.

Winter Wheat: Apply after resumption of active growth in the spring and before the early boot stage.

For aerial or ground treatment, use enough total spray volume to provide adequate spray coverage. Apply 1 to 4 gpa by air and 5 to 20 gpa by ground. Spray pressure should not exceed 30 psi. Use coarse spray to minimize spray drift.

To prepare the spray, mix only with water. Add about half the desired amount of water in the spray tank. Then with agitation, add the recommended amount of TORDON 22K and 2,4-D or MCPA as outlined in the table. Finally, with continued agitation, add the rest of the water.

The dosages recommended equate to ¼ oz. picloram + 4 oz. 2,4-D or MCPA ae/acre when weeds are small or ⅜ + 6 oz/acre when weeds are more advanced or when dry soil conditions exist.

Use Rates for Barley, Oats, and Wheat

Weed Growth Stage	Amounts of Each Product Per Acre† TORDON 22K *plus*	MCP Amine *or*	DMA 4, FORMULA 40 ESTERON 99 Concentrate or ESTERON 6E
weeds 1–3 inches tall	1fl. oz.	½ pt.	⅓ pt.
weeds 3–6 inches tall or under dry conditions	1½ fl. oz.	¾ pt.	½ pt.

† When measuring small amounts of TORDON 22K weed killer, special care should be taken not to exceed suggested rates.

NOTE: Use only on land that will be planted the following year to grass or grain crops such as small grains, corn, sorghum, or flax. Do not apply more than ⅜ ounce picloram per acre during any 12-month period.

USE PRECAUTIONS

Use this product only as specified on this label. Observe any special use and application restrictions and limitations, including method of application and permissible areas of use as promulgated by state or local authorities.

Do Not Contaminate Nontarget Land Areas, Cropland, Water, or Irrigation Ditches. Do not apply directly to standing or running water. Do not apply where surface water from treated areas can run off to adjacent cropland, either planted or to be planted, or into streams, irrigation ditches, irrigation ponds, or wells. Do not clean containers nor application equipment on or near these areas. Do not apply on inner banks or bottom of irrigation ditches.

Do not apply on or in the vicinity of susceptible crops or desirable plants including alfalfa, beans, grapes, melons, peas, potatoes, safflower, soybeans, sugar beets, sunflower, tomatoes and other vegetable crops, flowers, fruit plants, ornamentals or shade trees.

Avoid Spray Drift: Applications should be made only when there is no hazard from spray drift since very small quantities of the spray, which may not be visible, may severely injure susceptible crops during both growing and dormant periods. Use coarse sprays to minimize drift since, under adverse weather conditions, fine spray droplets may drift a mile or more. The spray thickening agent, NALCO-TROL[1], may be used with this product to aid in reducing spray drift. If used follow all use recommendations and precautions on the product label.

[1] NALCO-TROL—Trademark of NALCO Chemical Company

Ground Equipment: With ground equipment, spray drift can be lessened by keeping the spray boom as low as possible; by applying 20 gallons or more of spray per acre; by using no more than 30 pounds spraying pressures with large droplet producing nozzle tips; by spraying when wind velocity is low; and by stopping all spraying when wind exceeds 10 miles per hour. Do not apply with hollow cone-type insecticide or other nozzles that produce a fine-droplet spray.

Aerial Application: With aircraft, drift can be lessened by applying a coarse spray; by using no more than 30 pounds spray pressure at the nozzles; by using straight stream nozzles directed straight back; by using a spray boom no longer than 3/4 the wing span of the aircraft; and by spraying only when wind velocity is less than 10 mph.

Do Not Apply By Aircraft When An Air Temperature Inversion Exists: Such a condition is characterized by little or no wind and with air temperature lower near the ground than at higher levels. The use of a continuous smoke column at or near the site of application or use of a smoke generating device on the aircraft is suggested to indicate direction and velocity of air movement, and to indicate a temperature inversion by layering of the smoke.

Do not rotate treated rangeland or pasture to other crop uses.

Do not spray pastures or grain if the forage legume component is desired. TORDON 22K weed killer may injure or kill legumes. Also new legume seedlings may not be successful if made within 2 years following application of this herbicide.

Do not move treated soil to other areas. Do not use it to grow plants, unless adequate sensitive bioassay or chemical tests show that no detectable picloram is present in the soil.

Do not transfer livestock from treated grass areas onto broadleaf crop areas without first allowing 7 days of grazing on untreated grass pastures. Otherwise, urine may contain enough picloram to cause injury to sensitive broadleaf plants.

Do not re-use containers for TORDON 22K Weed Killer for any purpose. Dispose by punching holes in them and burying with waste or by taking to an approved landfill. Where indicated, follow official local container disposal regulations.

Rinse application equipment after use, preferably at least three times with water, and dispose of rinse water in a non-cropland area away from water supplies.

Be sure that use of this product conforms to all applicable regulations.

Do not use where a sandy porous surface and substrate overlie ground water 10 feet or less below the surface.

MISSOURI .... EPA SLN NO. MO-790004

For the Control of Musk Thistle, Goldenrod, Ironweed, and other Broadleaf Weeds in Rangeland and Permanent Grass Pastures.

GENERAL INFORMATION

Use TORDON 22K Weed Killer to control musk thistle growing in rangeland or permanent grass pastures. Picloram, the active ingredient in TORDON 22K Weed Killer, is highly active, water soluble, and has moderate soil residual. To prevent damage to desirable crops and plants follow all directions on this label.

DIRECTIONS FOR USE

It is a violation of Federal law to use this product in a manner inconsistent with its labeling.

Broadcast Treatment: To control broadleaf weeds, treat when weeds are growing well and apply broadcast with conventional ground equipment. Do not apply by air or with mist blower.

For control of **musk thistle:** Apply when musk thistle plants are in the rosette stage of growth in the fall from October 1 to soil freeze-up or in the spring before flower stock elongates or "bolts" above the rosette leaves. Use 6 to 8 fluid ounces (3/8 to 1/2 pint) of TORDON 22K in 10 to 20 gallons of water. Generally, most consistent control is received from fall applications.

For control of **goldenrod:** Apply when plants are growing well at the rate of 6 to 8 fluid ounces (3/8 to 1/2 pint) of TORDON 22K in 10 to 20 gallons of water. Best results are obtained with applications made prior to early bud stage.

For control of **ironweed:** Apply when plants are growing well at the rate of 1 pint of TORDON 22K in 10 to 20 gallons of water. For best results, apply prior to early bud stage.

Wick Applications: Mix 1 part of TORDON 22K with 2 parts of water to prepare a 33% solution. Apply when weeds are actively growing and are above most desirable plants. For musk thistle, apply before seed head develops into purple flower for best results. Later applications may not stop early season seed production. For ironweed and goldenrod, best results are obtained with applications made prior to early bud stage.

Wick applicator should be drained and cleaned after each use. Ropes should be changed when flow is reduced from wear, extended use, poor cleaning or intermittent use.

USE PRECAUTIONS

Use this product only as specified on this label. Observe any special use and application restrictions and limitations, including method of application and permissible areas of use, as promulgated by state or local authorities.

Do not apply on or in the vicinity of susceptible crops or desirable plants including alfalfa, beans, grapes, melons, peas, potatoes, safflower, soybeans, sugar beets, sunflower, tomatoes and other vegetable crops, flowers, fruit plants, ornamentals, shade trees, or the soil containing roots of nearby valuable plants.

Avoid Spray Drift: Applications should be made only when there is no hazard from spray drift since very small quantities of the spray, which may not be visible, may severely injure susceptible crops during both growing and dormant periods. To avoid spray drift use low spray pressure, under 30 psi; spray when wind velocity is less than 10 mph; and apply as a coarse spray. Spray drift can be lessened by keeping the spray boom as low as possible; by applying 20 gallons or more of spray per acre; by using large droplet producing nozzle tips; and by stopping all spraying when wind exceeds 10 miles per hour. Do not apply with hollow cone-type insecticide or other nozzles that produce fine-droplet spray. Do not apply to cropland used for production of desirable crops other than forage grasses. Do not rotate treated rangeland or pasture to other crop uses.

Do not treat pasture or rangeland if injury to the forage legume cannot be tolerated. TORDON 22K Weed Killer may injure or kill forage legumes when exposed to sprays or residues from wicked plants. Also, new legume seedling establishment may not be successful if seeded within 2 years following application of herbicide.

Do Not Contaminate Water intended for irrigation, drinking or other domestic purposes. To avoid crop or other plant injury, do not treat or allow spray drift to fall onto inner banks or bottom of irrigation and drainage ditches.

Do not move treated soil to other areas or use it to grow plants unless adequately sensitive bioassay shows that no detectable picloram is present in the soil.

Do not transfer livestock from treated grass area onto a broadleaf crop area without first allowing 7 days of grazing on an untreated grass pasture. Otherwise, urine may contain enough picloram to cause injury to sensitive broadleaf plants.

Do not use where a sandy porous surface and substrate overlie ground water 10 feet or less below the surface.

STORAGE AND DISPOSAL

Do not contaminate water, food or feed by storage or disposal.

Pesticide Disposal

Pesticide, spray mixture or rinsate that cannot be used according to label instructions must be disposed of according to Federal, state or local procedures under the Resource Conservation and Recovery Act.

Container Disposal

Do not reuse containers. Dispose of them in a sanitary landfill or by other approved state and local authorities.

READ AND FOLLOW MIXING AND USE INSTRUCTIONS AND PRECAUTIONS ON PRODUCT CONTAINER LABEL

MONTANA .... EPA SLN NO. MT-780004

For the Control of Broadleaf Weeds on Non-Cropland.

DIRECTIONS FOR USE

It is a violation of Federal law to use this product in a manner inconsistent with its labeling.

Read complete use directions and precautions on this label and container label before using.

Use TORDON 22K Weed Killer by ground application to control broadleaf annual and perennial weeds such as Canadian and other thistles, field bindweed, leafy spurge, Russian knapweed, spotted knapweed, tall larkspur, yellow toadflax, locoweed, snakeweed and lupines on non-cropland areas such as on roadside or other rights-of-way, and along fence rows, around farm buildings and outer banks of ditches. Treat when weeds are growing well using low pressure sprays. Retreatment at the same rate may be necessary the following year.

Spot Treatment: Use TORDON 22K at rates of 2 to 4 quarts in 20 to 100 gallons of water per acre and apply as a spray to the foliage. Use the higher rates to control leafy spurge, larkspur, toadflax and knapweed. Use the lower rates for bindweed and thistles. For a 1,000 square foot infestation, use about 2¼ fluid ounces of TORDON 22K per 1 gallon of water (equivalent to 3 quarts per acre of TORDON 22K).

Broadcast Treatment: To suppress the growth of many perennial broadleaf weeds, apply as a broadcast spray using 1 quart of TORDON 22K in 20 to 100 gallons of water per acre and apply as a broadcast spray during the growing season when weeds are growing well. (Many seedling annual weeds can be controlled using 1 pint per acre.)

Tank Mixture for Spot or Broadcast Treatments: TORDON 22K may also be tank mixed with 2,4-D products such as ESTERON® 99® Concentrate, FORMULA 40®, DMA 4 or ESTERON 6E herbicides for use on areas having mixed species including those which respond well to 2,4-D. In tank mix combinations, use 1 pint to 1 quart of TORDON 22K with 1 to 2 quarts ESTERON 99 Concentrate, FORMULA 40 or DMA 4, or with ⅔ to 1⅓ quarts of ESTERON 6E per acre, in 20 to 100 gallons of water per acre. **Read and follow all directions and use precautions on other product labels.**

USE PRECAUTIONS

Use this product only as specified on this label. Observe any special use and application restrictions and limitations, including method of application and permissible areas of use as promulgated by state or local authorities.

Do Not Contaminate Nontarget Land Areas, Cropland, Water, or Irrigation Ditches. Do not apply directly to standing or running water. Do not apply where surface water from treated areas can run off to adjacent cropland, either planted or to be planted, or into streams, irrigation ditches, irrigation ponds, or wells. Do not clean containers nor application equipment on or near these areas. Do not apply on inner banks or bottom of irrigation ditches.

Do not apply on or in the vicinity of susceptible crops or desirable plants including alfalfa, beans, grapes, melons, peas, potatoes, safflower, soybeans, sugar beets, sunflower, tomatoes and other vegetable crops, flowers, fruit plants, ornamentals or shade trees.

Avoid Spray Drift: Applications should be made only when there is no hazard from spray drift since very small quantities of the spray, which may not be visible, may severely injure susceptible crops during both growing and dormant periods. Use coarse sprays to minimize drift since, under adverse weather conditions, fine spray droplets may drift a mile or more.

Spray drift can be lessened by keeping the spray boom as low as possible; by applying 20 gallons or more of spray per acre; by using no more than 30 pounds spraying pressure with large droplet producing nozzle tips; by spraying when wind velocity is low; and by stopping all spraying when wind exceeds 10 miles per hour. Do not apply with hollow cone-type insecticide or other nozzles that produce a fine droplet spray.

The spray thickening agent, NALCO-TROL[1] may be used with this product to aid in reducing spray drift. If used follow all use recommendations and precautions on the product label.

[1] NALCO-TROL—Trademark of NALCO Chemical Company.

Do Not Apply When An Air Temperature Inversion Exists: Such a condition is characteristized by little or no wind and with air temperature lower near the ground than at higher levels. The use of a continuous smoke column at or near site of application is suggested to indicate direction and velocity of air movement, and to indicate a temperature inversion by layering of the smoke.

Do not move treated soil to other areas. Do not use it to grow plants, unless adequate sensitive bioassay or chemical tests show that no detectable picloram is present in the soil.

Do not re-use containers for TORDON 22K Weed Killer for any purpose. Dispose by punching holes in them and burying with waste or by taking to an approved landfill. Where indicated, follow official local container disposal regulations.

Rinse application equipment after use, preferably at least three times with water, and dispose of rinse water in a non-cropland area away from water supplies.

Be sure that use of this product conforms to all applicable regulations.

MONTANA .... EPA SLN NO. MT-780003

For the Control of Broadleaf Weeds in Rangelands, Permanent Grass Pastures, Spring Barley and Oats, and Spring and Winter Wheat.

DIRECTIONS FOR USE

It is a violation of Federal law to use this product in a manner inconsistent with its labeling.

Read complete use directions and precautions on this label and container label before using.

RANGELAND AND PASTURE

Use TORDON 22K Weed Killer to control broadleaf annual and perennial weeds such as Canadian and other thistles, field bindweed, leafy spurge, Russian knapweed, spotted knapweed, tall larkspur, yellow toadflax, locoweed, snakeweed and lupines on rangeland and permanent grass pastures. Treat when weeds are growing well using low pressure sprays. Retreatment at the same rate may be necessary the following year.

Spot Treatment: (ground application only) Use TORDON 22K at rates of 2 to 4 quarts in 20 to 100 gallons of water per acre and apply as a spray to the foliage. Use the higher rates to control leafy spurge, larkspur, toadflax and knapweed. Use the lower rates for bindweed and thistles. For a 1,000 square foot infestation, apply about 2¼ fluid ounces of TORDON 22K in 1 gallon of water (equivalent to 3 quarts per acre of TORDON 22K).

NOTE: When spot treating, do not cut grass for feed within 2 weeks after treatment. Meat animals grazing for up to 2 weeks after treatment should be removed from treated areas 3 days prior to slaughter. Do not graze dairy animals on treated areas within 2 weeks after treatment.

Broadcast Treatment: (ground or aerial application) to suppress the growth of many perennial broadleaf weeds, apply as a broadcast spray using 1 quart of TORDON 22K in 1 to 4 gallons of water per acre by air or in 20 to 100 gallons of water per acre by ground equipment and apply as a broadcast spray during the growing season when weeds are growing well. (Many seedling annual weeds can be controlled using 1 pint per acre.)

Tank Mixture for Spot or Broadcast Treatments: TORDON 22K may also be tank mixed with 2,4-D products such as ESTERON® 99® Concentrate, FORMULA 40®, DMA 4 or ESTERON 6E herbicides for use on areas having mixed species including those which respond well to 2,4-D, such as big sagebrush. In tank mix combinations, use 1 pint to 1 quart of TORDON 22K with 1 to 2 quarts ESTERON 99 Concentrate, FORMULA 40 or DMA 4, or with ⅔ to 1⅓ quarts of ESTERON‡ 6E per acre, in spray volumes specified above. **Read and follow all directions and use precautions on other product labels.**

SPRING BARLEY AND OATS, SPRING AND WINTER WHEAT NOT UNDERSEEDED WITH A LEGUME

For the control of wild buckwheat and other annual broadleaf weeds normally controlled with 2,4-D or MCPA such as wild mustard, Russian thistle, pennycress, lambsquarters and pigweed, in spring wheat and barley and winter wheat, use TORDON 22K as a tank mix with a 2,4-D or MCPA formulation such as DMA 4, FORMULA 40, ESTERON 99 Concentrate, ESTERON‡ 6E, or MCP Amine Herbicides. For use on spring oats, tank mix only with MCP Amine Herbicide. **Read and follow all directions and use precautions on other product labels.**

Spring Wheat, Barley and Oats: Apply during the 3 through 5 leaf stage of growth. Application of TORDON 22K occasionally causes slight head malformations and straw shortening but normally this does not affect yield.

Durum Wheat: Do not treat durum since at least some varieties appear to be more sensitive than other wheat.

Winter Wheat: Apply after resumption of active growth in the spring and before the early boot stage.

For aerial or ground treatment, use enough total spray volume to provide adequate spray coverage. Apply 1 to 4 gpa by air and 5 to 20 gpa by ground. Spray pressure should not exceed 30 psi. Use a coarse spray to minimize spray drift.

To prepare the spray, mix only with water. Add about half the desired amount of water in the spray tank. Then with agitation, add the recommended amount of TORDON 22K and 2,4-D or MCPA as outlined in the table. Finally, with continued agitation, add the rest of the water.

The dosages recommended equate to ¼ oz. picloram + 4 oz. 2,4-D or MCPA ae/acre when weeds are small or ⅜ + 6 oz./acre when weeds are more advanced or when dry soil conditions exist.

Use Rates for Barley, Oats, and Wheat

Weed Growth Stage	Amounts of Each Product Per Acre†		
	TORDON 22K *plus*	DMA 4, FORMULA 40 ESTERON 90 Concentrate or MCP Amine *or*	ESTERON 6E
weeds 1–3 inches tall	1 fl. oz.	½ pt.	⅓ pt.
weeds 3–6 inches tall or under dry conditions	1½ fl. oz.	¾ pt.	½ pt.

† When measuring small amounts of TORDON 22K weed killer, special care should be taken not to exceed suggested rates.

NOTE: Use only on land that will be planted the following year to grass or grain crops such as small grains, corn, sorghum, or flax. Do not apply more than ⅜ ounce picloram per acre during any 12-month period.

USE PRECAUTIONS

Use this product only as specified on this label. Observe any special use and application restrictions and limitations, including method of application and permissible areas of use as promulgated by state or local authorities.

Do Not Contaminate Nontarget Land Areas, Cropland, Water, or Irrigation Ditches. Do not apply directly to standing or running water. Do not apply where surface water from treated areas can run off to adjacent cropland, either planted or to be planted, or into streams, irrigation ditches, irrigation ponds, or wells. Do not clean containers nor application equipment on or near these areas. Do not apply on inner banks or bottom of irrigation ditches.

Do not apply on or in the vicinity of susceptible crops or desirable plants including alfalfa, beans, grapes, melons, peas, potatoes, safflower, soybeans, sugar beets, sunflower, tomatoes and other vegetable crops, flowers, fruit plants, ornamentals or shade trees.

Avoid Spray Drift: Applications should be made only when there is no hazard from spray drift since very small quantities of the spray, which may not be visible, may severely injure susceptible crops during both growing and dormant periods. Use coarse sprays to minimize drift since, under adverse weather conditions, fine spray droplets may drift a mile or more. The spray thickening agent, NALCO-TROL[1] may be used with this product to aid in reducing spray drift. If used follow all use recommendations and precautions on the product label.

[1] NALCO-TROL—Trademark of NALCO Chemical Company

Ground Equipment: With ground equipment, spray drift can be lessened by keeping the spray boom as low as possible; by applying 20 gallons or more of spray per acre; by using no more than 30 pounds spraying pressure with large droplet producing nozzle tips; by spraying when wind velocity is low; and by stopping all spraying when wind exceeds 10 miles per hour. Do not apply with hollow cone-type insecticide or other nozzles that produce a fine-droplet spray.

Aerial Application: With aircraft, drift can be lessened by applying a coarse spray; by using no more than 30 pounds spray pressure at nozzles; by using straight stream nozzles directed straight back, by using a spray boom no longer than ¾ the wing span of the aircraft; and by spraying only when wind velocity is less than 10 mph.

Do Not Apply By Aircraft When An Air Temperature Inversion Exists: Such a condition is characterized by little or no wind and with air temperature lower near the ground than at higher levels. The use of a continuous smoke column at or near site of application or use of a smoke generating device on the aircraft is suggested to indicate direction and velocity of air movement, and to indicate a temperature inversion by layering of the smoke.

Do not rotate treated rangeland or pasture to other crop uses.

Do not spray pastures or grain if the forage legume component is desired. TORDON 22K Weed Killer may injure or kill legumes. Also new legume seedlings may not be successful if made within 2 years following application of this herbicide.

Do not move treated soil to other areas: Do not use it to grow plants, unless adequate sensitive bioassay or chemical tests show that no detectable picloram is present in the soil.

Do not transfer livestock from treated grass areas onto broadleaf crop areas without first allowing 7 days of grazing on untreated grass pasture. Otherwise, urine may contain enough picloram to cause injury to sensitive broadleaf plants.

Do not re-use containers for TORDON 22K Weed Killer for any purpose. Dispose by punching holes in them and burying with waste or by taking to an approved landfill. Where indicated, follow official local container disposal regulations.

Rinse application equipment after use, preferably at least three times with water, and dispose of rinse water in a non-cropland area away from water supplies.

Be sure that use of this product conforms to all applicable regulations.

Do not use where a sandy porous surface and substrate overlie ground water 10 feet or less below the surface.

NEBRASKA .... EPA SLN NO. NE-790023

For the Control of Leafy Spurge, Musk Thistle and other Broadleaf Weeds in Rangeland and Permanent Grass Pastures.

DIRECTIONS FOR USE

It is a violation of Federal law to use this product in a manner inconsistent with its labeling.

Read complete use directions and precautions on this label and the container label before using.

GENERAL INFORMATION

Use TORDON 22K Weed Killer to control leafy spurge, musk thistle and other broadleaf weeds growing in rangeland or permanent grass pastures. Picloram, the active ingredient in TORDON 22K Weed Killer, is highly active, persistent, and water soluble. To prevent damage to desirable crops and plants follow all directions and precautions on this label.

Treat when weeds are growing well using low pressure sprays. Treat no more than once in any 12-month period and no more than twice in any 3-year period. Do not apply with a mist blower.

For applications at rates greater than 8 fl. oz. per acre, limit coverage to no more than 10% or 10 acres, whichever is greater, of an operator's acreage found in any particular watershed.

For control of broadleaf annual and perennial weeds such as leafy spurge, Canadian and other thistles, field bindweed, Russian knapweed, spotted knapweed, tall larkspur, locoweed, snakeweed and lupines: (Spot Treatment—Ground Application Only) Use TORDON 22K at rates of 2 to 4 quarts in 5 to 100 gallons of water per acre and apply as a spray to the foliage. Use the higher rates to control leafy spurge, larkspur, toadflax and knapweed. Use the lower rates for bindweed and thistles. For a 1,000 square foot infestation, apply about 2¼ fluid ounces of TORDON 22K in 1 gallon of water (equivalent to 3 quarts per acre of TORDON 22K).

NOTE: When spot treating, do not cut grass for feed within 2 weeks after treatment. Meat animals grazing for up to 2 weeks after treatment should be removed from treated areas 3 days prior to slaughter. Do not graze dairy animals on treated areas within 2 weeks after treatment.

(Broadcast Treatment—Ground or Aerial Application) To suppress the growth of many perennial broadleaf weeds, apply as a broadcast spray using 1 quart of TORDON 22K in 1 to 4 gallons of water per acre by air or in 5 to 100 gallons of water per acre by ground equipment and apply as a broadcast spray during the growing season when weeds are growing well. (Many seedling annual weeds can be controlled using 1 pint per acre.)

(Tank Mixture for Spot or Broadcast Treatments) TORDON 22K Weed Killer may also be tank mixed with 2,4-D products such as ESTERON® 99® Concentrate, FORMULA 40®, DMA 4 or ESTERON 6E Herbicides for use on areas having mixed species including those which respond well to 2,4-D, such as big sagebrush. In tank mix combinations, use 1 pint to 1 quart of TORDON 22K with 1 to 2 quarts ESTERON 99 Concentrate, FORMULA 40 or DMA 4, or with ⅔ to 1⅓ quarts of ESTERON 6E per acre, in spray volumes specified above. **Read and follow all directions and use precautions on other product labels.**

For control of musk thistle: Use 6 to 8 fluid ounces (⅜ to ½ pint) of TORDON 22K in 1 to 4 gallons of water per acre by air or in 10 to 20 gallons of water per acre by ground equipment. Apply in the spring before May 15, or when musk thistle plants are in the rosette stage of growth or before flower stock elongates over 2 inches above leaves, whichever comes first. Apply in the fall after October 1, but before soil freeze-up or December 1, whichever is earlier.

USE PRECAUTIONS

Use this product only as specified on this label. Observe any special use and application restrictions and limitations, including method of application and permissible areas of use as promulgated by state or local authorities.

Do Not Contaminate Nontarget Land Areas, Cropland, Water, or Irrigation Ditches. Do not apply directly to standing or running water. Do not apply where surface water from treated areas can run off to adjacent cropland, either planted or to be planted, or into streams, irrigation ditches, irrigation ponds, or wells. Do not clean containers nor application equipment on or near these areas. Do not apply on inner banks or bottom of irrigation ditches.

Do not apply on or in the vicinity of susceptible crops or desirable plants including alfalfa, beans, grapes, melons, peas, potatoes, safflower, soybeans, sugar beets, sunflower, tomatoes and other vegetable crops, flowers, fruit plants, ornamentals or shade trees.

Avoid Spray Drift: Applications should be made only when there is no hazard from spray drift since very small quantities of the spray, which may not be visible, may severely injure susceptible crops during both growing and dormant periods. Use coarse sprays to minimize drift since, under adverse weather conditions, fine spray droplets may drift a mile or more. The spray thickening agent, NALCO-TROL[1] may be used with this product to aid in reducing spray drift. If used, follow all use recommendations and precautions on the product label.

[1] NALCO-TROL—Trademark of NALCO Chemical Company

Ground Equipment: With ground equipment, spray drift can be lessened by keeping the spray boom as low as possible; by applying 20 gallons or more of spray per acre; by using no more than 30 pounds spraying pressure with large droplet producing nozzle tips; by spraying when wind velocity is low; and by stopping all spraying when wind exceeds 10 miles per hour. To not apply with hollow cone-type insecticide or other nozzles that produce a fine-droplet spray.

Aerial Application: With aircraft, drift can be lessened by applying a coarse spray; by using no more than 30 pounds spray pressure at nozzles; by using straight stream nozzles directed straight back, by using a spray boom no longer than ¾ the wing span of the aircraft, and by spraying only when wind velocity is less than 10 mph.

Do Not Apply By Aircraft When An Air Temperature Inversion Exists: Such a condition is characterized by little or no wind and with air temperature lower near the ground than at higher levels. The use of a continuous smoke column at or near the site of application or use of a smoke generating device on the aircraft is suggested to indicate direction and velocity of air movement, and to indicate a temperature inversion by layering of the smoke.

Do not rotate treated rangeland or pasture to other crop uses.

Do not spray pastures if the forage legume component is desired. TORDON 22K Weed Killer may injure or kill legumes. Also, new legume seedlings may not be successful if made within 2 years following application of this herbicide.

Do not move treated soil to other areas: Do not use it to grow plants, unless adequate sensitive bioassay or chemical tests show that no detectable picloram is present in the soil.

Do not transfer livestock from treated grass areas onto broadleaf crop areas without first allowing 7 days of grazing on untreated grass pasture. Otherwise, urine may contain enough picloram to cause injury to sensitive broadleaf plants.

Do not re-use containers for TORDON 22K Weed Killer for any purpose. Dispose by punching holes in them and burying with waste or by taking to an approved landfill. Where indicated, follow official local container disposal regulations.

Rinse application equipment after use, preferably at least three times with water, and dispose of rinse water in a non-cropland area away from water supplies.

Be sure that use of this product conforms to all applicable regulations.

Do not use on flood, gravity, sprinkler, or sub-irrigated land.

Do not use where a sandy porous surface and substrate overlie ground water which may be 15 feet or less below the surface at any time.

NORTH DAKOTA .... EPA SLN NO. ND-780002

For the Control of Broadleaf Weeds in Rangelands, Permanent Grass Pastures, Spring Barley and Oats, and Spring and Winter Wheat and Flax.

DIRECTIONS FOR USE

It is a violation of Federal law to use this product in a manner inconsistent with its labeling.

Read complete use directions and precautions on this label and container label before using.

SPRING BARLEY AND OATS, SPRING AND WINTER WHEAT NOT UNDERSEEDED WITH A LEGUME

For the control of wild buckwheat, volunteer sunflower and other annual broadleaf weeds normally controlled with 2,4-D or MCPA, such as wild mustard, Russian thistle, pennycress, lambsquarters and pigweed, in spring wheat and barley and winter wheat, use TORDON 22K Weed Killer as a tank mix with a 2,4-D or MCPA® formulation such as FORMULA 40®, DMA® 4, ESTERON® 99 Concentrate, ESTERON 6E, or MCP Ester, or MCP Amine Herbicides. For use on spring oats, tank mix only with MCP Ester, or MCP Amine Herbicide. **Read and follow all directions and use precautions on other product labels.**

Spring Wheat, Barley and Oats: Apply during the 3 through 5 leaf stage of growth. Application of TORDON 22K occasionally causes slight head malformations and straw shortening, but normally this does not affect yield.

Durum Wheat: Do not treat durum since at least some varieties appear to be more sensitive than other wheat.

Winter Wheat: Apply after resumption of active growth in the spring and before the early boot stage.

For aerial or ground treatment, use enough total spray volume to provide adequate spray coverage. Apply 1 to 4 gpa by air or 5 to 20 gpa by ground.

To prepare the spray, mix only with water. Add about half the desired amount of water in the spray tank. Then with agitation, add the recommended amount of TORDON 22K and 2,4-D or MCPA as outlined in the table. Finally, with continued agitation, add the rest of the water.

USE RATES FOR WHEAT, BARLEY, AND OATS*

WEEDS	WEED GROWTH STAGE**	AMOUNTS OF EACH PRODUCT PER ACRE: TORDON 22K	*PLUS* FORMULA 40, DMA 4, ESTERON 99 CONCENTRATE, MCP ESTER, OR MCP AMINE	*OR* ESTERON 6E
More susceptible species such as lambsquarters, pennycress, and wild mustard.	1–3 inches 3–6 inches	1 fl. oz. 1½ fl. oz.	½ pt. ¾ pt.	⅓ pt. ½ pt.
Less susceptible species such as volunteer sunflower, wild buckwheat, Russian thistle, pigweed, and top growth supression of Canada thistle.	1–6 inches	1½ fl. oz.	¾ pt.	½ pt.

* For oats, tank mix only with MCP amine or MCP Ester Herbicides.
** For best results, treat when weeds have 2 to 4 leaves and are actively growing.

NOTE: Use only on land that will be planted the following year to grass or grain crops such as small grains, corn, sorghum, or flax. Do not spray a small grain crop more than once with TORDON 22 K Weed Killer.

Flax

For control or suppression of wild buckwheat, volunteer sunflower, pigweed and other susceptible broadleaf weeds, treat flax when 2 to 6 inches tall, before buds begin to form. Apply a tank mixture that delivers 1 fluid ounce TORDON 22K plus ½ pint MCP Amine herbicide per acre. For best results and crop safety, apply when flax is 2 to 3 inches tall and the weeds are young and growing well. Treat wild buckwheat before it begins to form a vine. For aerial or ground treatment, use enough water dilution to provide adquate spray coverage, usually 1 to 4 gpa by air or 5 to 20 gpa by ground equipment. If seedling foxtail is a problem, DOWPON* M Grass Killer at 1 pound product per acre may be added to the tank mix, but risk of crop injury is increased. Observe the higher spray volume requirements for DOWPON M. **Read and follow all directions and use precautions on labels of all products contained in the tank mixture.**

NOTE: The use of some preplant herbicides such as Eptam or Treflan, or the presence of crop stress from drought, disease or other elements, may increase risk of crop injury and poor weed control from postemergence herbicide treatment. Also, flax varieties may vary in tolerance to herbicides. Do not treat flax unless the possibility of crop injury can be tolerated. Consult local Extension or University Extension Specialists for further information.

Rangeland and Permanent Grass Pastures

Use TORDON 22K Weed Killer to control broadleaf annual and perennial weeds such as Canada Thistle and other thistles, field bindweed, leafy spurge, Russian knapweed, spotted and diffuse knapweed, plains or Geyer larkspur, Dalmation and yellow toadflax, locoweed, and lupines on rangeland and permanent grass pastures. Treat when weeds are growing well using low pressure sprays. Retreatment at the same rate may be necessary the following year.

SPOT TREATMENT: Use TORDON 22K at rates shown in the table for broadcast treatment and apply in water with sufficient dilution to obtain adequate spray coverage. For perennial weeds, spray the weed infested area plus a 10 foot border around the infestation. Consult the product container label for guidance to prepare spray mixes to spot treat small patches.

BROADCAST TREATMENT: Apply as a broadcast spray on hard to control weeds at rates shown in the following table. Apply the selected rate of TORDON 22K in 5 gallons of total spray volume per acre (water diluent only) by air, or in 20 to 100 gallons of water per acre by ground equipment. For rates exceeding 1 quart of TORDON 22K per acre, the total area treated in a single season should not exceed 25% of a landowner's acreage found in any particular watershed.

Weed Species	Rates of Tordon 22K per Sprayed Acre**
Diffuse Knapweed Spotted Knapweed Many Seedling annual weeds	1 pint
Plains or Geyer Larkspur* Locoweed* Lupines*	1 quart
Russian Knapweed	2 quarts
Canada Thistle	2 quarts—Fall rosette & prebolt 3 quarts—Any other time
Dalmation Toadflax Yellow Toadflax Field Bindweed	3 quarts
Leafy Spurge*	1 quart—Repeat annually for 3 to 4 years. 2 quarts—Treat 2 consecutive years. 4 quarts—Follow as needed with spot touchup treatment.

* Denotes poisonous plants

** NOTE: When applying more than 1 quart of TORDON 22K per acre, do not cut grass for feed within 2 weeks after treatment. Meat animals grazing for up to two weeks after treatment should be removed from treated areas 3 days prior to slaughter. Do not graze dairy animals on treated areas within 2 weeks after treatment.

USE PRECAUTIONS

Use this product only as specified on this label. Observe any special use and application restrictions and limitations, including method of application and permissible areas of use as promulgated by state or local authorities.

Do Not Contaminate Water Intended for Irrigation or Domestic Purposes. To avoid injury to crops or other desirable plants, do not treat or allow spray drift to fall onto inner banks or bottom of irrigation ditches or other channels that carry water that may be used for irrigation purposes.

Do not apply on or in the vicinity of susceptible crops or desirable plants including alfalfa, beans, grapes, melons, peas, potatoes, safflower, soybeans, sugar beets, sunflower, tomatoes, and other vegetable crops, flowers, fruit plants, ornamentals or shade trees.

Avoid Spray Drift: Applications should be made only when there is no hazard from spray drift since very small quantities of the spray, which may not be visible, may severely injure susceptible crops during both growing and dormant periods. Use coarse sprays to minimize drift since, under adverse weather conditions, fine spray droplets may drift a mile or more. The spray thickening agent, NALCO-TROL[1] may be used with this product to aid in reducing spray drift. If used, follow all use recommendations and precautions on the product label.

[1] NALCO-TROL—Trademark of NALCO Chemical Company

Ground Equipment: With ground equipment, spray drift can be lessened by keeping the spray boom as low as possible; by applying 20 gallons or more of spray per acre; by using no more than 30 pounds spraying pressure with large droplet producing nozzle tips; by spraying when wind velocity is low; and by stopping all spraying when wind exceeds 10 miles per hour. Do not apply with hollow cone-type insecticide or other nozzles that produce a fine-droplet spray.

Aerial Application: Drift can be lessened by applying a coarse spray; by using no more than 25 pounds spray pressure at nozzles; by using nozzles which do not produce hazardous amounts of fine spray droplets; by using a spray boom no longer than 3/4 of the rotor or wing length; and by spraying only when wind velocity is less than 10 mph, or as required by state regulations.

Determine Air Movement and Direction Before Foliar Application: Use a smoke generator or other means at or near the application site or on aircraft for detection of air movement, air stability, or temperature inversions. Such a condition exists when there is little or no wind and air temperature is lower near the ground than at higher levels. Use appropriate drift control measures or avoid application when smoke is moving toward nearby desirable susceptible plants or sensitive areas.

Do not rotate treated rangeland or pasture to other crop uses.

Do not spray pastures if the forage legume component is desired. TORDON 22K Weed Killer may injure or kill legumes. Also new legume seedlings may not be successful if made within 2 years following application of this herbicide.

Do not move treated soil to other areas. Do not use it to grow plants, unless adequate sensitive bioassay or chemical tests show that no detectable picloram is present in the soil.

Do not transfer livestock from treated grass areas onto sensitive broadleaf crop areas for 12 months after application or until picloram has disappeared from the soil, without first allowing 7 days of grazing on untreated grass pasture. Otherwise, urine may contain enough picloram to cause injury to sensitive broadleaf plants.

Do not re-use containers for TORDAN 22K Weed Killer for any purpose. Dispose by punching holes in them and burying with waste or by taking to an approved landfill. Where indicated, follow official local container disposal regulations.

Rinse application equipment after use, preferably at least three times with water, and dispose of rinse water in a non-cropland area away from water supplies.

Be sure that use of this product conforms to all applicable regulations.

Do not use where a sandy porous surface and substrate overlie ground water 10 feet or less below the surface.

This registration has been issued by the North Dakota State Laboratories Department pursuant to the provisions of section 24(c) of the Federal Insecticide, Fungicide, and Rodenticide Act, as amended.

OKLAHOMA .... EPA SLN No. OK-840005

For Control of Annual and Perennial Broadleaf Weeds Such As Field Bindweed in Grainland Between Crops of Small Grains.

GENERAL INFORMATION

Use TORDON 22K Weed Killer tank mixed with 2,[illegible]icides by ground application to control or reduce infestations of broadleaf annual and perennial weeds such as field bindweed on grainland during the period following crop harvest and prior to planting winter wheat, barley, or oats. Picloram, the active ingredient in TORDON 22K Weed Killer, is highly active, water soluble and has moderate soil residual. To prevent damage to desirable crops and plants follow all directions on this label.

DIRECTIONS FOR USE

It is a violation of Federal law to use this product in a manner inconsistent with its labeling.

Read complete use directions and precautions on this label and container label before using.

For reduction of perennial weeds such as field bindweed, use ground application equipment to broadcast treat infested areas using 0.5 to 1 pint TORDON 22K Weed Killer tank mixed with 1 to 2 pints of 3.8 lb/gal. 2,4-D amine or low-volatile ester products or with 1/3 to 2/3 quarts of 5.8 lb/gallon comparable 2,4-D formulations, in 10 or more gallons of water per acre.

To maintain control of perennial weed infestations, treat annually between crops of small grains. For best results with continuous grain production treat field bindweed infested areas in June or July following small grain harvest, when bindweed is actively growing, applying the lower rate of TORDON 22K Weed Killer and either rate of 2,4-D **(higher rate of TORDON 22K Weed Killer may be used but greater crop injury to fall planted grain may result compared to that which may occur with the lower rate)**. For best results with small grain-fallow production treat field bindweed infested areas in June or July following grain harvest, applying the higher rates of TORDON 22K Weed Killer and 2,4-D; then around mid-May during the fallow season treat the field bindweed infested area with the lower rate of TORDON 22K Weed Killer and either rate of 2,4-D **(higher rate of TORDON 22K Weed Killer may be used but greater crop injury to fall planted grain may result compared to what may occur from the lower rate)**.

Do not treat with TORDON 22K Weed Killer more than once each calendar year. Additional treatment with 2,4-D alone may be made anytime weeds are present with adequate foliage and in good growing condition.

To control volunteer grain or grass weeds that may be present, use cultivation or appropriate grass control herbicides alone or in combination with broadleaf weed control treatments. Consult the University Experiment Station or Extension Service for recommendations to fit local conditions.

Wheat or other small grain crops may be injured when planted at varying periods following application of TORDON 22K Weed Killer. Do not treat with TORDON 22K if the risk of injury to the small grain crop cannot be tolerated.

For long residual control of perennial weeds such as field bindweed, spot treat scattered infestations following use directions on the container label. When using these higher rate treatments of TORDON 22K do not treat more than 10% of the area of an infested field in any calendar year. Wheat or other small grain crops may be planted in such treated fields, but crop injury in the spot treated areas may occur for one or more seasons after treatment.

USE PRECAUTIONS

Use this product only as specified on this label. Observe any special use and application restrictions and limitations, including method of application and permissible areas of use as promulgated by state and local authorities.

Handle and Use Correctly. Small amounts of TORDON 22K Weed Killer can injure sensitive broadleaf plants. To avoid problems, observe the following precautions:

1. **Do Not Contaminate Water Intended for Irrigation or Domestic Purposes.** Do not treat or allow spray drift to fall onto innerbanks or bottom of irrigation ditches, either dry or containing water.

2. **Do Not Apply to or in the Vicinity of Sensitive Crops or Desirable Plants** including alfalfa, beans, cotton, grapes, melons, peanuts, peas, potatoes, sunflower, sugarbeets, soybeans, tobacco, tomatoes, other field or vegetable crops, flowers, fruit plants, ornamentals, or shade trees.

3. **Use a bioassay to determine that picloram residues are no longer detectable in the treated soil.** Contact State Extension Specialists for further information on bioassay testing.

4. **Do Not Move Treated Soil to Other Areas** where susceptible desirable plants may be exposed and damaged while phytotoxic residues are present.

5. **Avoid Spray Drift.** Applications should be made only when there is no hazard from spray drift since very small quantities of the spray, which may not be visible, may severely injure susceptible crops during both growing and dormant periods. To avoid spray drift apply only when wind velocity is less than 10 mph and is not blowing toward susceptible crops, keep spray boom as low as possible; apply 20 gallons or more of spray per acre; use no more than 30 psi spraying pressure with large droplet producing tips.

Do Not Apply With Hollow Cone-Type Insecticide or Other Nozzles that Produce Fine-Droplet Spray. A spray thickening agent such as NALCO-TROL[1] may be used with this product to aid in reducing spray drift. If a drift reducing agent is used, follow use directions and precautions on the product label.

[1]NALCOTROL—Trademark of NALCO Chemical Company

6. **Do Not Use Where a Sandy Porous Surface and Substrate Overlie Ground Water Ten Feet or Less Below the Surface.**

7. **Be Sure That Use of This Product Conforms to All Applicable Regulations.**

8. **Grazing Restrictions: None.**

9. **Follow Storage and Disposal Requirements on Container Label.**

SOUTH DAKOTA .... EPA SLN NO. SD-780001

For the Control of Broadleaf Weeds in Rangelands, Permanent Grass Pastures, Spring Barley and Oats, and Spring and Winter Wheat.

DIRECTIONS FOR USE

It is a violation of Federal law to use this product in a manner inconsistent with its labeling.

Read complete use directions and precautions on this label and container label before using.

RANGELAND AND PASTURE

Use TORDON 22K Weed Killer to control broadleaf annual and perennial weeds such as Canadian and other thistles, field bindweed, leafy spurge, Russian knapweed, spotted knapweed, tall larkspur, yellow toadflax, locoweed, snakeweed and lupines on rangeland and permanent grass pastures. Treat when weeds are growing well using low pressure sprays. Retreatment at the same rate may be necessary the following year.

Spot Treatment: (ground application only) Use TORDON 22K at rates of 2 to 4 quarts in 20 to 100 gallons of water per acre and apply as a spray to the foliage. Use the higher rates to control leafy spurge, larkspur, toadflax and knapweed. Use the lower rates for bindweed and thistles. For a 1,000 square foot infestation, apply about 2¼ fluid ounces of TORDON 22K in 1 gallon of water (equivalent to 3 quarts per acre of TORDON 22K).

NOTE: When spot treating, do not cut grass for feed within 2 weeks after treatment. Meat animals grazing for up to 2 weeks after treatment should be removed from treated areas 3 days prior to slaughter. Do not graze dairy animals on treated areas within 2 weeks after treatment.

Broadcast Treatment: (ground or aerial application) To suppress the growth of many perennial broadleaf weeds, apply as a broadcast spray using 1 quart of TORDON 22K in 1 to 4 gallons of water per acre by air or in 20 to 100 gallons of water per acre by ground equipment and apply as a broadcast spray during the growing season when weeds are growing well. (Many seedling annual weeds can be controlled using 1 pint per acre).

Tank Mixture for Spot or Broadcast Treatments: TORDON 22K may also be tank mixed with 2,4-D products such as ESTERON® 99® Concentrate, FORMULA 40®, DMA® 4 or ESTERON 6E herbicides for use on areas having mixed species including those which respond well to 2,4-D, such as big sagebrush. In tank mix combinations, use 1 pint to 1 quart of TORDON 22K with 1 to 2 quarts ESTERON 99 Concentrate. FORMULA 40 or DMA 4, or with ⅔ to 1⅓ quarts of ESTERON 6E per acre, in spray volumes specified above. **Read and follow all directions and use precautions on other product labels.**

For Control of Musk Thistle: Use 6 to 8 fluid ounces (⅜ to ½ pint) of TORDON 22K in 1 to 4 gallons of water per acre by air or in 10 to 20 gallons of water per acre by ground equipment, and apply in the spring or fall when musk thistle plants are in the rosette stage of growth.

SPRING BARLEY AND OATS, SPRING AND WINTER WHEAT NOT UNDERSEEDED WITH A LEGUME

For the control of wild buckwheat and other annual broadleaf weeds normally controlled with 2,4-D or MCPA such as wild mustard, Russian thistle, pennycress, lambsquarters and pigweed, in spring wheat and barley and winter wheat use TORDON 22K as a tank mix with a 2,4-D or MCPA formulation such as DMA 4, FORMULA 40, ESTERON 99 Concentrate, ESTERON 6E, or MCP Amine Herbicides. For use on spring oats, tank mix only with MCP Amine Herbicide. **Read and follow all directions and use precautions on other product labels.**

Spring Wheat, Barley and Oats: Apply during the 3 through 5 leaf stage of growth. Application of TORDON 22K occasionally causes slight head malformations and straw shortening but normally this does not affect yield.

Duram Wheat: Do not treat duram since at least some varieties appear to be more sensitive than other wheat.

Winter Wheat: Apply after resumption of active growth in the spring and before the early boot stage.

For aerial or ground treatment, use enough total spray volume to provide adequate spray coverage. Apply 1 to 4 gpa by air and 5 to 20 gpa by ground. Spray pressure should not exceed 30 psi. Use a coarse spray to minimize spray drift.

To prepare the spray, mix only with water. Add about half the desired amount of water in the spray tank. Then with agitation, add the recommended amount of TORDON 22K and 2,4-D or MCPA as outlined in the table. Finally, with continued agitation, add the rest of the water.

The dosages recommended equate to ¼ oz. picloram + 4 oz. 2,4-D or MCPA ae/acre when weeds are small or ⅜ + 6 oz/acre when weeds are more advanced or when dry soil conditions exist.

Use Rates for Barley, Oats, and Wheat

Weed Growth Stage	Amounts of Each Product Per Acre†		
	TORDON 22K *plus*	DMA 4, FORMULA 40 ESTERON 99 Concentrate or MCP Amine *or*	ESTERON 6E
weeds 1–3 inches tall	1 fl. oz.	½ pt.	⅓ pt.

Weed Growth Stage	Use Rates for Barley, Oats, and Wheat — Amounts of Each Product Per Acre† TORDON 22K *plus*	DMA 4, FORMULA 40 ESTERON 99 Concentrate or MCP Amine *or*	ESTERON 6E
weeds 3–6 inches tall or under dry conditions	1½ fl. oz.	¾ pt.	½ pt.

† When measuring small amounts of TORDON 22K weed killer, special care should be taken not to exceed suggested rates.

NOTE: Use only on land that will be planted the following year to grass or grain crops such as small grains, corn, sorghum, or flax. Do not apply more than ⅜ ounce picloram per acre during any 12-month period.

USE PRECAUTIONS

Use this product only as specified on this label. Observe any special use and application restrictions and limitations, including method of application and permissible areas of use as promulgated by state or local authorities.

Do Not Contaminate Nontarget Land Areas, Cropland, Water, or Irrigation Ditches. Do not apply directly to standing or running water. Do not apply where surface water from treated areas can run off to adjacent cropland, either planted or to be planted, or into streams, irrigation ditches, irrigation ponds, or wells. Do not clean containers nor application equipment on or near these areas. Do not apply on inner banks or bottom of irrigation ditches.

Do not apply on or in the vicinity of susceptible crops or desirable plants including alfalfa, beans, grapes, melons, peas, potatoes, safflower, soybeans, sugar beets, sunflower, tomatoes and other vegetable crops, flowers, fruit plants, ornamentals or shade trees.

Avoid Spray Drift: Applications should be made only when there is no hazard from spray drift since very small quantities of the spray, which may not be visible, may severely injure susceptible crops during both growing and dormant periods. Use coarse sprays to minimize drift since, under adverse weather conditions, fine spray droplets may drift a mile or more. The spray thickening agent, NALCO-TROL[1] may be used with this product to and in reducing spray drift. If used follow all use recommendations and precautions on the product label.

[1] NALCO-TROL—Trademark of NALCO Chemical Company.

Ground Equipment: With ground equipment, spray drift can be lessened by keeping the spray boom as low as possible; by applying 20 gallons or more of spray per acre; by using no more than 30 pounds spraying pressure with large droplet producing nozzle tips; by spraying when wind velocity is low; and by stopping all spraying when wind exceeds 10 miles per hour. Do not apply with hollow cone-type insecticide or other nozzles that produce a fine-droplet spray.

Aerial Application: With aircraft, drift can be lessened by applying a coarse spray; by using no more than 30 pounds spray pressure at nozzles; by using straight stream nozzles directed straight back, by using a spray boom no longer than ¾ the wing span of the aircraft; and by spraying only when wind velocity is less than 10 mph.

Do Not Apply By Aircraft When An Air Temperature Inversion Exists: Such a condition is characterized by little or no wind and with air temperature lower near the ground than at higher levels. The use of a continuous smoke column at or near site of application or use of a smoke generating device on the aircraft is suggested to indicate direction and velocity of air movement, and to indicate a temperature inversion by layering of the smoke.

Do not rotate treated rangeland or pasture to other crop uses.

Do not spray pastures or grain if the forage legume component is desired. TORDON 22K Weed Killer may injure or kill legumes. Also new legume seedlings may not be successful if made within 2 years following application of this herbicide.

Do not move treated soil to other areas. Do not use it to grow plants, unless adequate sensitive bioassay or chemical tests show that no detectable picloram is present in the soil.

Do not transfer livestock from treated grass areas onto broadleaf crop areas without first allowing 7 days of grazing on untreated grass pasture. Otherwise, urine may contain enough picloram to cause injury to sensitive broadleaf plants.

Do not re-use containers for TORDON 22K Weed Killer for any purpose. Dispose by punching holes in them and burying with waste or by taking to an approved landfill. Where indicated, follow official local container disposal regulations.

Rinse application equipment after use, preferably at least three times with water, and dispose of rinse water in a non-cropland area away from water supplies.

Be sure that use of this product conforms to all applicable regulations.

Do not use where a sandy porous surface and substrate overlie ground water 10 feet or less below the surface.

TEXAS .... EPA SLN NO. TX-830003

For Control of Annual and Perennial Broadleaf Weeds Such as Field Bindweed in Grainland Between Crops of Small Grains.

DIRECTIONS FOR USE

It is a violation of Federal law to use this product in a manner inconsistent with its labeling.

Read complete use directions and precautions on this label and container label before using.

Use TORDON 22K Weed Killer by ground application to control or reduce infestations of broadleaf annual and perennial weeds such as field bindweed on grainland during the period following crop harvest and prior to planting winter wheat, barley, or oats.

For **reduction of perennial** weeds such as field bindweed, broadcast treat infested areas using 0.5 to 1 pint TORDON 22K tank mixed with 1 to 2 pints ESTERON® 99® Concentrate, FORMULA 40®, DMA® 4 herbicides, or with ⅓ to ⅔ quart ESTRON 6E herbicide in 10 or more gallons of water per acre. To maintain or further reduce perennial weed infestations, treat annually between crops. For best results, treat in June or July following small grain harvest, applying the higher rates of TORDON 22K and 2,4-D. Initial treatment alternatively may be made soon after harvesting a fall maturing crop such as grain sorghum or corn. During the next calendar year, retreat the same area with the lower rate of TORDON 22K, preferably in May or June during the fallow period prior to fall planting of the next small grain crop.

Do not treat with TORDON 22K Weed Killer more than once each calendar year. Additional treatment with 2,4-D alone may be made anytime weeds are present with adequate foliage and in good growing condition.

To control volunteer grain or grass weeds that may be present, use cultivation or appropriate grass control herbicides alone or in combination with broadleaf weed control treatments. Consult the University Experiment Station or Extension Service for recommendations to fit local conditions.

Wheat or other small grain crops may be injured when planted at varying periods following application of TORDON 22K Weed Killer. Do not treat with TORDON 22K if the risk of injury to the small grain crop cannot be tolerated.

For long residual control of perennial weeds such as field bindweed, spot treat scattered infestations following use directions on the container label. When using these higher rate treatments of TORDON 22K, do not treat more than 10% of the area of an infested field in one calendar year. Wheat or other small grain crops may be planted in such treated fields, but crop injury in the spot treated areas may occur for one or more seasons after treatment.

USE PRECAUTIONS

Use this product only as specified on this label. Observe any special use and application restrictions and limitations, including method of application and permissible areas of use as promulgated by state and local authorities.

Do not apply on or in the vicinity of susceptible crops or desirable plants including alfalfa, beans, grapes, melons, peas, potatoes, safflower, soybeans, sugar beets, sunflower, tomatoes and other vegetable crops, flowers, fruit plants, ornamentals, shade trees, or the soil containing roots of nearby valuable plants.

AVOID SPRAY DRIFT: Applications should be made only when there is no hazard from spray drift since very small quantities of the spray, which may not be visible, may severely injure susceptible crops during both growing and dormant periods. To avoid spray drift use low spray pressure, under 30 psi; spray when wind velocity is less than 10 mph; and apply as a coarse spray. The spray thickening agent NALCO-TROL[1] may be used with this product to aid in reducing spray drift. If this thickening agent is used, follow all use recommendations and precautions on the product label.

[1] NALCO-TROL—Trademark of NALCO Chemical Company

Ground Equipment: With ground equipment, spray drift can be lessened by keeping the spray boom as low as possible; by applying 20 gallons or more of spray per acre; by using no more than 30 pounds spraying pressure with large droplet producing nozzle tips; by spraying when wind velocity is low; and by stopping all spraying when wind exceeds 10 miles per hour. Do not apply with hollow cone-type insecticide or other nozzles that produce fine-droplet spray.

Do Not Contaminate Water Intended for Irrigation or Domestic Purposes. Do not treat or allow spray drift to fall onto innerbanks or bottom of irrigation ditches, either dry or containing water or other channels that carry water that may be used for irrigation purposes.

Do not move treated soil to other areas where susceptible desirable plants may be exposed and damaged while phytotoxic residues are present.

UTAH .... EPA SLN NO. UT-840004

For the Control of Broadleaf Weeds in Rangeland, Forest and Permanent Grass Pastures, Spring Barley and Oats, and Spring and Winter Wheat.

DIRECTIONS FOR USE

It is a violation of Federal law to use this product in a manner inconsistent with its labeling.

Read complete use directions and precautions on this label and container label before using.

SPRING BARLEY AND OATS, SPRING AND WINTER WHEAT NOT UNDERSEEDED WITH A LEGUME

For the control of wild buckwheat, volunteer sunflower and other annual broadleaf weeds such as wild mustard, Russian thistle, pennycress, lambsquarters, pigweed and sowthistle in spring wheat and barley and winter wheat, use TORDON 22K Weed Killer as a tank mix with amine or ester formulations of 2,4-D or MCPA. For use on spring oats, tank mix only with MCP Amine Herbicide. READ AND FOLLOW ALL DIRECTIONS AND USE PRECAUTIONS AND USE PRECAUTIONS ON OTHER PRODUCT LABELS.

SPRING WHEAT, BARLEY AND OATS: Apply during the 3 through 5-leaf stage of growth. Application of TORDON 22K occasionally causes slight head malformations and straw shortening but normally this does not affect yield.

DURUM WHEAT: Do not treat durum since at least some varieties appear to be more sensitive than other wheat.

WINTER WHEAT: Apply after resumption of active growth in the spring and before the early boot stage.

For aerial or ground treatment, use enough total spray volume to provide adquate spray coverage. Apply 2 to 4 gpa by air and 5 to 10 gpa by ground. Spray pressure should not exceed 40 psi. Use coarse spray to minimize spray drift.

To prepare the spray, mix only with water. Add about half the desired amount of water in the spray tank. Then with agitation, add the recommended amount of TORDON 22K and 2,4-D or MCPA as outlined in the table. Finally, with continued agitation, add the rest of the water. Be sure to maintain agitation during application of tank mixes with 2,4-D or MCPA ester products.

Do not apply on or near susceptible crops or desirable plants including alfalfa, beans, grapes, melons, peas, potatoes, safflower, soybeans, sugar beets, sunflower, tomatoes and other vegetable crops, flowers, fruit plants, ornamentals or shade trees.

Avoid Spray Drift: Applications should be made only when there is no hazard from spray drift since very small quantities of spray, which may not be visible, may severely injure susceptible crops such as those listed above, during both growing and dormant periods. Use coarse sprays to minimize drift since, under adverse weather conditions, fine spray droplets may drift a mile or more. A spray thickening agent, such as NALCO-TROL[1] may be used with this product to aid in reducing spray drift. If used follow all use recommendations and precautions on the product label.

[1] NALCO-TROL—Trademark of NALCO Chemical Company

Ground Equipment: With ground equipment, spray drift can be lessened by keeping the spray boom to a height recommended for the nozzle tips in use and by spraying only when wind velocity is less than 5 miles per hour. Do not apply with nozzles that produce a fine-droplet spray.

Aerial Application: Drift can be lessened by applying a coarse spray, by using no more than 40 pounds spray pressure at nozzles; by using nozzles which do not produce hazardous amounts of fine spray droplets; by using a spray boom no longer than 3/4 of the rotor or wing length; and by spraying only when wind velocity is less than 5.0 mph.

Determine Air Movement and Direction Before Foliar Application: Use a smoke generator or other means at or near the application site or on aircraft for detection of air movement, air stability, or temperature inversions. Such a condition exists when there is little or no wind and air temperature is lower near the ground than at higher levels. Use appropriate drift control measures or stop application when smoke is moving toward nearby susceptible plants or sensitive areas.

DO NOT ROTATE TREATED RANGELAND OR PASTURE to susceptible crops until phyotoxic residues have disappeared from the soil.

DO NOT SPRAY PASTURES OR GRAIN IF THE FORAGE LEGUME COMPONENT IS DESIRED. TORDON 22K weed killer may injure or kill legumes. Also new legume seedlings may not be successful if made within 2 years following application of this herbicide.

Avoid movement of treated soil. Picloram may remain in treated soil for an extended period. Do not move treated soil to other areas and do not use such soil to grow susceptible plants until residues have reached a non-phytotoxic level.

LIVESTOCK TRANSFER RESTRICTIONS. If livestock are moved from a treated area within the first 12 months after application onto a sensitive broadleaved area, allow 7 days of grazing on untreated pasture. Otherwise, urine may contain enough picloram to cause injury to sensitive broadleaf plants.

Rinse application equipment and containers thoroughly, preferably at least three times with water, and dispose of rinse water in a noncropland area away from water supplies.

DO NOT RE-USE CONTAINERS FOR TORDON 22K WEED KILLER for any purpose. Containers should be disposed of by punching holes in them and burying (in a non-crop area away from water) or by taking to an approved landfill. Where indicated, follow official container disposal regulations.

BE SURE THAT USE OF THIS PRODUCT CONFORMS TO ALL APPLICABLE REGULATIONS.

WYOMING .... EPA SLN NO. WY-84001

For the Control of Broadleaf Weeds in rangeland, forest and permanent grass pastures, Spring Barley and Oats, and Spring and Winter Wheat.

DIRECTIONS FOR USE

It is a violation of Federal law to use this product in a manner inconsistent with its labeling.

Read complete use directions and precautions on this label and container label before using.

SPRING BARLEY AND OATS, SPRING AND WINTER WHEAT NOT UNDERSEEDED WITH A LEGUME

For the control of wild buckwheat, volunteer sunflower and other annual broadleaf weeds such as wild mustard, Russian thistle, pennycress, lambsquarter and pigweed, in spring wheat and barley and winter wheat, use TORDON 22K weed killer as a tank mix with amine or ester formulations of 2,4-D or MCPA. For use on spring oats, tank mix only with MCP Amine Herbicide. **Read and follow all directions and use precautions on other product labels.**

SPRING WHEAT, BARLEY AND OATS: Apply during the 3 through 5 leaf stage of growth. Application of TORDON 22K occasionally causes slight head malformations and straw shortening but normally this does not affect yield.

DURUM WHEAT: Do not treat durum since at least some varieties appear to be more sensitive than other wheat.

WINTER WHEAT: Apply after resumption of active growth in the spring and before the early boot stage.

For aerial or ground treatment, use enough total spray volume to provide adequate spray coverage. Apply 2 to 4 gpa by air and 5 to 10 gpa by ground. Spray pressure should not exceed 40 psi. Use coarse spray to minimize spray drift.

To prepare the spray, mix only with water. Add about half the desired amount of water in the spray tank. Then with agitation, add the recommended amount of TORDON 22K and 2,4-D or MCPA as outlined in the table. Finally, with continued agitation, add the rest of the water. Be sure to maintain agitation during application of tank mixes with 2,4-D ester products.

Use Rates For Wheat, Barley and Oats[1]

Weeds	Weed Growth Stage[2]	Amounts of Each Product Per Acre: TORDON 22K	plus Products Containing 3.8 lb 2,4-D or MCPA a.e. per gal.	or 5.6 lb 2,4-D a.e. per gal.
More susceptible species such as lambsquarters	1–3 inches	1 fl. oz.	½ pt.	⅓ pt.
	3–6 inches	1½ fl. oz.	¾ pt.	½ pt.
Less susceptible species such as volunteer sunflower, wild buckwheat, Russian thistle, annual and perennial sowthistle, pigweed, and top growth suppression of Canada thistle	1–6 inches	1½ fl. oz.	¾ pt.	½ pt.

1. For oats, tank mix only with MCP Amine or MCP Ester Herbicides.
2. For best results, treat when weeds have 2 to 4 leaves and are actively growing.

NOTE: Use only on land that will be planted the following year to grass or grain crops such as small grains, corn, sorghum, or flax.

Do not spray a small grain crop more than once with TORDON 22K Weed Killer, during a single season.

RANGELAND, FOREST AND PERMANENT GRASS PASTURES

Use TORDON 22K Weed Killer to control broadleaf annual and perennial weeds such as (but not limited to) Canada thistle and other thistles, field bindweed, leafy spurge, Russian knapweed, spotted and diffuse knapweed, plains or Geyer larkspur, Dalmation and yellow toadflax, fringed sagebrush, snakeweed, rabbit brush, pricklypear cactus, skeletonleaf bursage, locoweed, and lupines on rangeland and permanent grass pastures. Treat when weeds are actively growing. Retreatment may be necessary the following year.

SPOT TREATMENT: Use TORDON 22K rates shown in the following table and apply in sufficient water to obtain adequate spray coverage. For perennial weeds, spray the weed infested area plus a 10-foot border around the infestation. Consult the product container label for guidance to prepare spray mixes to spot treat small patches.

BROADCAST TREATMENT: Apply as a broadcast spray on hard to control weeds at rates shown in the following table. Apply the selected rate of TORDON 22K in 2-4 gallons of total spray volume per acre (water diluent only) by air, or in 5 to 30 gallons of water per acre by ground equipment. For rates exceeding 1 quart of TORDON 22K per acre, the total area treated in a single season should not exceed 25% of a landowner's acreage found in any particular watershed.

TANK MIXTURES FOR SPOT OR BROADCAST TREATMENTS: TORDON 22K may also be tank mixed with 2,4-D products for use on areas having mixed species including those which respond well to 2,4-D. In tank mix combinations, use 1 pint to 4 quarts of TORDON 22K with 1 to 2 quarts of low volatile ester or amine products containing 3.8 lb. 2,4-D acid equivalent per gallon or with ⅔ to 1⅓ quarts of comparable products containing 5.6 lb. 2,4-D acid equivalent per gallon in spray volumes specified below. When tank mixing with 2,4-D ester products, agitation should be maintained throughout preparation and application. READ AND FOLLOW ALL DIRECTIONS AND USE PRECAUTIONS ON OTHER PRODUCT LABELS.

NOTE: TORDON 22K at rates over 1 quart may suppress certain grasses, such as bromegrass, bluegramma and buffalograss. Usually later grass growth will be improved by release from competition.

GRAZING RESTRICTIONS: When applying more than 1 quart of TORDON 22K per acre, do not cut grass for feed within two weeks after treatment. Meat animals grazing for up to two weeks after treatment should be removed from treated areas three days prior to slaughter. Do not graze lactating dairy animals on treated areas within two weeks after treatment.

Weed Species[1]	Rates of TORDON 22K Per Sprayed Area[2]
Diffuse Knapweed; Spotted Knapweed; Rabbitbrush; Fringed Sagebrush	1 pint
Musk, Plumeless & Scotch Thistles; Tall or Geyer Larkspur[3,4]; Locoweed[3,4]; Lupines[4] Snakeweed; Pricklypear Cactus	1 to 2 pints
Russian Knapweed; Skeletonleaf Bursage	1 to 2 quarts
Canada Thistle; Dalmation Toadflax; Yellow Toadflax; Field Bindweed	2 to 4 quarts
Leafy Spurge[4]	1 quart Repeat annually for 3-4 years 2 quarts Treat 2 consecutive years 4 quarts Single application

1. For additional species or more specific rates consult the current WY Weed Control Guide.
2. NOTE: Follow all above treatments with spot touch-up treatment as needed using TORDON 22K or TORDON 2K Pellets herbicide.
3. Use higher rates.
4. Denotes poisonous plants.

USE PRECAUTIONS

Use this product only as specified on this label. Observe any special use and application restrictions and limitations, including method of application and permissible areas of use as promulgated by state or local authorities.

Avoid Water Contamination: Do not allow TORDON 22K Weed Killer to contaminate water used for drinking, irrigation or other domestic purposes. Do not apply on inner banks or bottoms of irrigation ditches. Do not clean containers or application equipment on or near these areas.

Do not apply on or near susceptible crops or desirable plants including alfalfa, beans, grapes, melons, peas, potatoes, safflower, soybeans, sugarbeets, sunflower, tomatoes and other vegetable crops, flowers, fruit plants, ornamentals or shade trees.

Avoid Spray Drift: Applications should be made only when there is no hazard from spray drift since very small quantities of the spray, which may not be visible, may severely injure susceptible crops such as those listed above, during both growing and dormant periods. Use coarse sprays to minimize drift since, under adverse weather conditions, fine spray droplets may drift a mile or more. A spray thickening agent, such as NALCO-TROL[1] may be used with this product to aid in reducing spray drift. If used follow all use recommendations and precautions on the product label.

[1]NALCO-TROL—Trademark of NALCO Chemical Company

Ground Equipment: With ground equipment, spray drift can be lessened by keeping the spray boom to a height recommended for nozzle tips in use and by spraying only when wind velocity is less than 10 miles per hour. Do not apply with nozzles that produce a fine-droplet spray.

Aerial Application: Drift can be lessened by applying a coarse spray; by using no more than 40 pounds spray pressures at nozzles; by using nozzles which do not produce hazardous amounts of fine spray droplets; by using a spray boom no longer than $^3/_4$ of the rotor or wing length; and by spraying only when wind velocity is less than 7.5 mph.

Determine Air Movement and Direction Before Foliar Application: Use a smoke generator or other means at or near the application site or on aircraft for detection of air movement, air stability, or temperature inversions. Such a condition exists when there is little or no wind and air temperature is lower near the ground than at higher levels. Use appropriate drift control measures or stop application when smoke is moving toward nearby susceptible plants or sensitive areas.

DO NOT ROTATE TREATED RANGELAND OR PASTURE to other sensitive crop uses.

Do not spray pastures or grain if the forage legume component is desired: TORDON 22K Weed Killer may injure or kill legumes. Also, new legume seeding may not be successful if made within 2 years following application of this herbicide.

Avoid Movement of Treated Soil. Picloram may remain in treated soil for an extended period. Do not move treated soil to other areas and do not use such soil to grow susceptible plants until residues have reaches a non-phytotoxic level.

LIVESTOCK TRANSFER RESTRICTIONS. If livestock are moved from a treated area within the first 12 months after application onto a sensitive broadleaf area, allow 7 days of grazing on untreated pasture. Otherwise, urine may contain enough picloram to cause injury to sensitive broadleaf plants.

Rinse application equipment and containers thoroughly, preferably at least three times with water, and dispose of rinse water in a noncropland area away from water supplies.

Do not re-use containers for TORDON 22K Weed Killer for any purpose. Containers should be disposed of by punching holes in them and burying (in a non-cropland area away from water) or by taking to an approved landfill. Where indicated, follow official container disposal regulations.

BE SURE THAT USE OF THIS PRODUCT CONFORMS TO ALL APPLICABLE REGULATIONS.

SUPPLEMENTAL USE LABELS

CALIFORNIA ...... EPA REGISTRATION NO. 464-323-ZA

For Use on Rangeland and Permanent Grass Pasture.

A permit issued by the appropriate County Agricultural Commissioner is required before using this product on grazing lands and all applications of this product shall be made under the direct supervision of employees of a County Agricultural Commissioner or Control and Eradication Unit, Division of Plant Industry, California State Department of Food and Agriculture.

GENERAL INFORMATION

FOR CONTROL OF: Artichoke, Scotch, Golden, Musk, Canada, Beaumont, Wavyleaf, Yellowstar, Purple Star, Iberian, Distaff and Italian thistles; Spotted, Russian, Diffuse and Squarrose knapweeds; Skeletonweed, Camelthorn, Dalmation Toadflax, White and Carolina horsenettle; Biddy Biddy, Leafy Spurge, Field Bindweed, Scotch Broom and Gorse.

CONSULT YOUR COUNTY AGRICULTURAL COMMISSIONER OR EXTENSION WEED SPECIALIST FOR ADDITIONAL INFORMATION ON THIS PROGRAM.

USE DIRECTIONS

TO PREVENT DAMAGE TO BROADLEAF CROPS AND OTHER DESIRABLE PLANTS, FOLLOW CAREFULLY ALL DIRECTIONS, MIXING AND USE INSTRUCTIONS AND PRECAUTIONS ON THE PRODUCT LABEL.

GRAZING RESTRICTIONS: DO NOT graze dairy animals on treated areas within 4 weeks after spot treating with rates higher than 1 quart TORDON 22K per acre. Animals grazing treated areas within 4 weeks after spot treating with rates higher than 1 quart TORDON 22K per acre should be withdrawn from treated areas at least 3 days before slaughtering. During the first growing season after application of TORDON 22K, livestock grazing treated areas should be held at least 7 days on an untreated pasture area before moving to crop areas since urine may contain enough picloram to cause crop injury. Where 2,4-D is used, do not graze dairy animals on treated areas within 7 days after application.

MIXING INSTRUCTIONS: Mix the required amounts of TORDON 22K Weed Killer in the specified amounts of water and apply as a coarse, low pressure spray (less than 30 psi) to weed and brush foliage. If applied by helicopter, use spray pressure under 10 psi. See helicopter application below. **DO NOT APPLY WITH FIXED WING AIRCRAFT.**

For best results, treat when the weeds are growing actively in the spring before full bloom or in the late summer or fall. Treatments during full bloom or seed stage of some weeds may not give good control. DO NOT APPLY to frozen ground nor to wet soil with free water on the surface.

GROUND APPLICATION: When applying TORDON 22K Weed Killer with ground equipment, use coarse, low pressure sprays (20–30 psi) and apply a minimum of 20 gallons of spray per acre. Apply only when weather conditions are favorable for keeping spray on the target area. Do not apply when wind velocity exceeds 7 mph. Do not allow spray drift to contact off-target susceptible plants. Do not apply to areas which have susceptible broadleaf crops, such as potatoes, vegetable crops or legumes, or other desirable broadleaf plants growing within $^1/_2$ mile downwind or within $^1/_8$ mile upwind.

HELICOPTER APPLICATION: All applications by helicopter must be approved by the California Department of Food and Agriculture in addition to the permit to apply TORDON 22K Weed Killer on range and pastureland from the local County Agricultural Commissioner. Applications by helicopter must conform with all applicable state regulations. Additionally, the pilot must have participated in a training session for this program. Aerial applications may be made only when there is little wind and **no hazard from spray drift.** Do not apply by helicopter to areas which have susceptible crops or other valuable plants within one mile downwind or within $^1/_4$ mile upwind.

Equipment must be designed and operated in a manner to minimize spray drift. Spray volume should be 8 gallons or more per acre. Speed of helicopter while spraying must be under 45 mph. Length of spray boom must not exceed $^2/_3$ the span of the rotor. Application should be made using only nozzles equipped with round orifices, 0.059 inch or larger, without swirling devices, directed straight back from line of flight or down as much as 90°. Spraying pressures should be less than 10 psi at the orifice. The sprayer should be equipped with check valves at each nozzle and a suck back or other device to prevent drip after shut-off of boom. Application should be made on a contour or slightly rising flight pattern at levels between

20 and 40 feet above the ground. Treatments should be made only when winds at flight level are from a definite direction and speeds are less than 5 mph. No application should be made above a temperature inversion, nor when temperature five feet (5′) above the ground exceeds 80° F.

USE RATES

BROADCAST TREATMENTS: To control yellowstar thistle, distaff and Italian thistles, use ¼ to ½ quart TORDON 22K Weed Killer per acre. To control artichoke thistle, purple star thistle and Iberian star thistle, use ½ to 1 quart TORDON 22K Weed Killer per acre. To suppress the growth of other broadleaf perennial weeds, use ½ to 1 quart per acre. Apply as a single broadcast spray during any one growing season. Retreat in subsequent years using similar rates. TORDON 22K Weed Killer may be tank mixed with ESTERON® 99 Concentrate, FORMULA 40®, DMA® 4, or ESTERON® 6E Herbicides (2,4-D products) for use on areas having mixed species including those which respond well to 2,4-D. In tank mix combinations, use ¼ to 1 quart TORDON 22K Weed Killer with 1 to 2 quarts ESTERON 99 Concentrate, FORMULA 40, or DMA 4 or with ⅔ to 1⅓ quarts ESTERON 6E per acre. NOTE: Where 2,4-D is used, do not graze treated areas by dairy animals within 7 days after application.

SPOT TREATMENTS: For spot treatments of scattered patches of broadleaf weeds, use ¼ to 4 quarts of TORDON 22K Weed Killer in 100 gallons of water and spray weed foliage uniformly using 100 gallons of spray per treated acre. Refer to the table below for dosage recommendations for different species.

Weeds to be Controlled	Rate of TORDON 22K Per Treated Acre
Yellowstar Thistle, Distaff Thistle, Italian Thistle	¼ to ½ quart
Artichoke Thistle, Purple Star Thistle, Iberian Thistle	½ to 1 quart
Scotch Thistle, Golden Thistle, Musk Thistle, Spotted Knapweed, Diffuse Knapweed, Squarrose Knapweed	1 to 2 quarts
Russian Knapweed, Canada Thistle, Field Bindweed, Rush Skeltonweed, Dalmation Toadflax, Scotch Broom, Gorse	2 to 3 quarts
Leafy Spurge, Camelthorn, White Horsenettle, Carolina Horsenettle, Beaumont Thistle, Wavyleaf Thistle	4 quarts

In any one growing season, when using the rate of 2 quarts per acre for spot treating, spray no more than 10 acres of any contiguous 20 acre block, or no more than one half the total area involved; at 3 quarts per acre, spray no more than 7 acres of a 20 acre block or one third of the total area; at 4 quarts per acre, spray no more than 5 acres of a 20 acre block, or one fourth of the total area. On any size area, follow these percentages of the total area treated in accordance with the rate used. These rates will provide a maximum of ½ pound of picloram per acre on range and pastureland.

For treating small areas, mix the amount of TORDON 22K Weed Killer indicated below for the desired dosage in 1 gallon of water and spray weed foliage thoroughly, covering about 500 square feet.

Rate of TORDON 22K Per Acre	Amount of TORDON 22K Per Gallon of Spray
4 quarts	3 tablespoons (1.5 fl. oz.)
2 quarts	1½ tablespoons
1 quart	2 teaspoons
½ quart	1 teaspoon

Some injury to certain range grasses, particularly smooth bromegrass and western wheatgrass, may occur at the higher use rates. If such injury is unacceptable, do not apply.

NOTE: Do not graze dairy animals on treated areas within 4 weeks after spot treating with rates higher than 1 quart per acre. Animals grazing treated areas within 4 weeks after spot treating with rates higher than 1 quart per acre should be withdrawn from treated areas at least 3 days before slaughtering.

USE PRECAUTIONS

USE THIS PRODUCT ONLY AS SPECIFIED ON THIS LABEL. DO NOT APPLY WHEN WEATHER CONDITIONS FAVOR DRIFT FROM AREAS TREATED.

Do not apply close to desirable plants. TORDON 22K Weed Killer is water soluble and highly active. Do not apply on or near desirable plants nor on ground likely to contain their roots. Such plants include shade trees, ornamentals, flowers, fruit plants, field crops and vegetable crops.

Do not apply on cropland, nor close to ponds, reservoirs, or streams, nor on sites where surface water from heavy rains can be expected to run off to these areas.

Do not move treated soil. Picloram may remain in treated soil for an extended period of time. Do not move treated soil to other areas and do not use it to grow plants.

Do not apply to land used for production of any crop other than forage grasses. Legumes may be killed by sprays containing TORDON 22K Weed Killer and may not grow in treated areas for two years or longer.

Do not allow spray drift. Small amounts of picloram may damage plants if applied during either the growing or dormant period. To avoid spray drift, do not use high pressure sprays. Coarse sprays are less likely to drift. Do not permit spray drift to contact desirable plants nor the soil containing roots of such plants. Spray TORDON 22K Weed Killer only when there is little wind (less than 7 mph) and **no hazard from spray drift.** Do not spray when the wind is blowing towards susceptible crops or ornamental plants.

DO NOT TRANSFER LIVESTOCK from a treated grass area onto a broadleaf crop area without first allowing 7 days of grazing on an untreated pasture. Otherwise, urine may contain sufficient amounts of picloram to cause injury to broadleaf crop plants.

Do not allow spray or spray drift to contaminate water intended for irrigation, drinking, or other domestic uses. Do not apply on inner banks or bottoms of irrigation ditches. Dike around and do not irrigate through treated areas. Do not apply on areas from which surface runoff water may go into cropland, other than areas used for production of forage grasses, nor into streams or ponds supplying irrigation or drinking water. Do not clean containers or application equipment on or near these areas.

OTHER PRECAUTIONS: Do not store near food, feedstuff, fertilizer, seeds, insecticides, fungicides, or other pesticides.

To avoid injury to desirable plants, containers for TORDON 22K Weed Killer should not be reused since small residues of picloram can damage desirable plants. Rinse equipment and containers thoroughly with water and dispose of rinse water by burying in noncroplands away from water supplies. Dispose of containers by punching holes in them and burying with waste. Where required, follow official local container disposal regulations.

BE SURE THAT USE OF THIS PRODUCT CONFORMS TO ALL APPLICABLE REGULATIONS.

COLORADO .... EPA NO. 464-07098-CO

To Control Broadleaf Weeds in Rangelands, Permanent Grass Pastures.

USE DIRECTIONS

Do not apply where spray drift may be a problem due to proximity of susceptible crops or other desirable plants. Do not apply aerially when wind exceeds 5 mph or when picloram-susceptible crops or other plants are located within one mile downwind from the application site. Such crops and plants include field beans, soybeans, sunflowers, peas, potatoes, safflower, sugar beets, alfalfa, vegetables, fruit plants, flowers, and ornamentals.

FOR RANGELAND AND PASTURES: Use TORDON 22K Weed Killer to control broadleaf annual and perennial weeds including Canadian and other thistles, field bindweed, leafy spurge, Russian knapweed, tall larkspur, yellow toadflax, locoweed, snakeweed, and lupines on rangeland and permanent grass pastures. Treat when weeds are growing well using low pressure sprays.

Spot Treatment: Use TORDON 22K at rates of 2 to 4 quarts in 20 to 100 gallons of water per acre and apply as a spray to the foliage. Use the higher rates to control leafy spurge, larkspur, toadflax, and knapweed. Use the lower rates for bindweed and thistles. For a 1,000 square foot infestation, use about 2¼ fluid ounces of TORDON 22K per 1 gallon of water (equivalent to 3 quarts per acre of TORDON 22K).

NOTE: During a season, do not use more than 10 gallons of TORDON 22K for any 100 acre area and do not treat more than 20 acres of any 100 acre area. Do not graze dairy animals on treated areas within four weeks after spot treating of rates greater than one quart per acre.

Broadcast Treatment: To suppress the growth of many perennial broadleaf weeds, use 1 quart of TORDON 22K weed killer in 20 to 100 gallons of water per acre and apply as a broadcast spray during the growing season when weeds are growing well. (Many seedling annual weeds can be controlled using 1 pint per acre.) Retreatment at the same rate may be necessary the following year.

USE PRECAUTIONS

Use this product only as specified on this label. Observe any special use and application restrictions and limitations, including method of application and permissible areas of use as promulgated by state or local authorities.

Do Not Contaminate Nontarget Land Areas, Cropland, Water, or Irrigation Ditches. Do not apply directly to standing or running water. Do not apply where surface water from treated areas can run off to adjacent cropland, either planted or to be planted, or into drainage channels, streams, irrigation ditches, ponds, or wells. Do not clean containers nor application equipment on or near these areas. Do not apply on inner banks or bottoms of irrigation ditches.

Do not apply on or in the vicinity of sensitive crops or desirable plants including alfalfa, beans, cotton, grapes, melons, peas, potatoes, saffflower, sugar beets, sunflower, tobacco, tomatoes, other field and vegetable crops, flowers, fruit plants, ornamentals or shade trees. Do not apply on areas which may be planted at a later time to such crops or plants.

Do not allow spray drift. Do not permit any spray or spray drift to contact desirable plants off the area treated or the soil containing roots of such plants. To avoid spray drift, use low spray pressure, under 30 psi for ground equipment and under 10 psi for aircraft, and apply as a coarse spray. Do not use pressures nor spray nozzles that produce fine droplets which could present a drift hazard. Spray TORDON 22K Weed Killer only when wind velocity is less than 10 mph and **when there is no hazard from spray drift off the target area. Do not spray when the wind is blowing toward susceptible crops or ornamental plants that are closer than one mile from the target area.** Do not spray above an air inversion. Consult local authorities on possible inversion situation.

Do not rotate treated rangeland or pasture to other crop uses.

Do not spray pastures if the forage legume component is desired. TORDON 22K Weed Killer may injure or kill legumes. Also, new legume seedlings may not be successful if made within 2 years following application of this herbicide.

Do not move treated soil to other areas. Do not use it to grow plants unless adequately sensitive bioassay or chemical tests show that no detectable picloram is present in the soil.

Do not transfer livestock from treated grass areas onto broadleaf crop areas without first allowing 7 days of grazing on untreated grass pasture. Otherwise, urine may contain enough picloram to cause injury to sensitive broadleaf plants.

Do not store near food, feedstuff, fertilizer, seed, insecticides, fungicides, or other pesticides.

Do not re-use containers for TORDON 22K Weed Killer for any purpose. Dispose by punching holes in them and burying with waste or by taking to an approved landfill. Rinse application equipment after use, preferably at least three times with water, and dispose of rinse water in a non-cropland area away from water supplies. Where indicated follow official local container disposal regulations.

Be sure that use of this product conforms to all applicable regulations.

HAWAII .... EPA NO. 464-07097-HI

For Use in the State of Hawaii to Control Certain Troublesome Woody Plants in Rangelands and Permanent Grass Pastures.

RESTRICTED TO PERMIT HOLDERS

USE DIRECTIONS

Use TORDON 22K Weed Killer to control troublesome woody plants such as lantana, melastoma, guava, Java plum, pamakani, cat's claw, Humuula gorse, firebush and hau as well as many other woody plants, broadleaf weeds and vines infesting rangelands and permanent grass pastures in the state of Hawaii. Treat anytime during the year when the plants are actively growing.

Spot Treatment: Mix TORDON 22K at the rate of 2 quarts per 100 gallons of water. Add 2 quarts of surfactant to the spray mixture. Apply as a low pressure (10 to 30 psi) spray to thoroughly wet the leaves, stems and trunks of the brush. For smaller amounts of spray mixtures, mix 2½ fluid ounces of TORDON 22K and 2½ fluid ounces of surfactant into 4 gallons of water and apply as above.

Re-treat areas in subsequent years as needed.

BROADCAST TREATMENT:

Ground Application: Use TORDON 22K at the rate of 2 quarts per 40 to 100 gallons of water per acre. Add 2 quarts of surfactant per 100 gallons of spray mixture. Apply at low pressure (10 to 30 psi) and keep the spray boom as low as possible consistent with good coverage.

Aerial Application: Use TORDON 22K at the rate of 2 quarts per 7 to 12 gallons of water per acre. Add 2 quarts of surfactant per 100 gallons of spray mixture. Conform to the aerial application regulations of the state of Hawaii.

Re-treat areas in subsequent years as needed.

Do not graze dairy animals on treated areas within two weeks after application. Do not slaughter meat animals grazing on treated areas until two weeks have elapsed after application.

Cut Surface Treatments: For the control of firebush, hau, Java plum and guava with trunks of 3 inches in diameter or larger. Use TORDON 22K Weed Killer diluted 1 to 4 in water as directed below.

Tree Injector: Application should be made by injecting 1 milliliter of the diluted solution through the bark at intervals of 3 inches between the edges of the injector wound. The injections should completely surround the tree at any convenient height.

Frill or Girdle: Make a single girdle through the bark completely around the tree at a convenient height. Wet the cut surface with the diluted solution.

USE PRECAUTIONS

Do not allow spray drift. TORDON herbicides are highly active. Very small amounts may cause damage to plants if applied during either growing or dormant periods. Do not apply or otherwise permit TORDON 22K or sprays containing it to contact desirable plants such as pineapple, papaya, macadamia, coffee, vegetables, flowers, grapes, fruit trees, ornamentals, tomatoes, potatoes, beans of all types and other valuable broadleaf plants, nor the soil containing roots of nearby valuable plants. Apply TORDON 22K only when there is little or no wind or no hazard from spray drift.

Do not contaminate water. To avoid injury to crops or other desirable plants, do not contaminate irrigation ditches or water intended for irrigation or domestic purposes. Do not treat or allow spray drift to fall onto inner banks or bottom of irrigation ditches, either dry or containing water, or other channels that carry water that may be used for irrigation purposes.

Do not transfer livestock directly from treated areas onto broadleaf crop areas without allowing 7 days on untreated grass pastures. Urine may contain enough picloram to cause crop injury.

Do not use manure from animals grazing treated areas on land used for growing broadleaf crops, ornamentals, orchard or other valuable plants. Manure may contain enough picloram to cause crop injury.

Do not mix with other pesticides except other herbicides in accordance with University of Hawaii suggestions.

Other precautions. Do not store near food, feedstuff, fertilizer, seeds, insecticides, fungicides or other pesticides.

Empty containers. Rinse with water, pour in spray tank. Repeat twice. Dispose of containers by punching holes in them and burying in non-cropland away from water supplies or in approved sanitary landfills.

Cleaning spray equipment. Flush spray tank with water and drain completely. Follow by thorough rinsing of tank, hoses and nozzles with soap and water. Dispose of all wastes by burying in non-cropland away from water supplies.

BE SURE THAT ALL USE OF TORDON 22K WEED KILLER CONFORMS TO LOCAL REGULATIONS.

NEBRASKA .... EPA NO. 464-07093-NE

To Control Broadleaf Weeds on Land Planted to Spring Barley and Spring and Winter Wheat.

USE PRECAUTIONS

Use this product only as specified on this label. Observe any special use and application restrictions and limitations, including method of application and permissible areas of use as promulgated by state or local authorities.

Note: Do not apply on or in the vicinity of sensitive crops or desirable plants including *alfalfa, beans, cotton, grapes, melons, peas, potatoes, safflower, sugar beets, sunflower, tobacco, tomatoes, other field and vegetables crops, flowers, fruit plants, ornamentals or shade trees.* Do not apply on areas which may be planted at a later time to such crops or plants.

Do Not Contaminate Nontarget Land Areas, Cropland, Water, or Irrigation Ditches. Do not apply directly to standing or running water. Do not apply where surface water from treated areas can run off to adjacent cropland, either planted or to be planted, or into drainage channels, streams, irrigation ditches, ponds, or wells. Do not clean containers nor application equipment on or near these areas. Do not apply on inner banks or bottoms of irrigation ditches.

Do not allow spray drift. Do not permit any spray or spray drift to contact desirable plants off the area treated or the soil containing roots of such plants. To avoid spray drift, use low spray pressure, under 30 psi for ground equipment and under 10 psi for aircraft, and apply as a coarse spray. Do not use pressures nor spray nozzles that produce fine droplets which could present a drift hazard. Spray TORDON 22K Weed Killer only when wind velocity is less than 10 mph and **when there is no hazard from spray drift off the target area. Do not spray when the wind is blowing toward susceptible crops or ornamental plants that are closer than one mile from the target area.** Do not spray above an air inversion. Consult local authorities on possible inversion situation.

Do not rotate treated grain fields to other crop uses.

Do not spray grain if the forage legume component is desired. TORDON 22K Weed Killer may injure or kill legumes. Also, new legume seedlings may not be successful if made within 2 years following application of this herbicide.

Do not move treated soil to other areas. Do not use it to grow plants, unless adequately sensitive bioassay or chemical tests show that no detectable picloram is present in the soil.

Do not store near food, feedstuff, fertilizer, seed, insecticides, fungicides, or other pesticides.

Do not re-use containers for TORDON 22K weed killer for any purpose. Dispose by punching holes in them and burying with waste or by taking to an approved landfill. Rinse application equipment after use, preferably at least three times with water, and dispose of rinse water in a non-cropland area away from water supplies. Where indicated, follow official local container disposal regulations.

Be sure that use of this product conforms to all applicable regulations.

Do not use where a sandy porous surface and substrate overlie ground water ten feet or less below the surface.

USE DIRECTIONS

Do not apply where spray drift may be a problem due to proximity of susceptible crops or other desirable plants. Do not apply aerially when wind exceeds 5 mph or when picloram-susceptible crops or other plants are located within one mile downwind from the application site. Such crops and plants include field beans, soybeans. sunflowers, peas, potatoes, safflower, sugar beets, alfalfa, vegetables, fruit plants, flowers, and ornamentals.

FOR SPRING BARLEY, SPRING AND WINTER WHEAT NOT UNDERSEEDED WITH A LEGUME: For the control of wild buckwheat as well as broadleaf weeds normally controlled with 2,4-D such as wild mustard, kochia. Russian thistle, fanweed, lambsquarter, and pigweed, use TORDON 22K Weed Killer as a tank mix only with a 2,4-D amine formulation containing 4 lb. acid equivalent per gallon, such as DMA® 4 or FORMULA 40® herbicides.

For aerial or ground treatment, use enough total spray volume to provide adequate spray coverage. Apply 1 to 4 gpa by air and 5 to 20 gpa by ground. Spray pressure should not exceed 20 psi. Coarse sprays are less likely to drift.

NOTE: Do not apply aerially when wind exceeds 5 mph or when picloram-susceptible crops, including field beans, soybeans, peas, potatoes, sugar beets, safflower or alfalfa, are located within one mile downwind from the application site.

To prepare the spray, mix only with water. Add about half the desired amount of water in the spray tank. Then with agitation, add the recommended amount of TORDON 22K and 2,4-D as outlined in the table. Finally with continued agitation, add the rest of the water.

Apply to spring barley and wheat during the period between the 4-leaf and early boot stages of growth. Apply to winter wheat after resumption of active growth in the spring and before the early boot stage.

Do not apply more than 3/8 ounce picloram per acre during any 12 month period.

Growth Stage	Amount of Products Per Acre†	
	TORDON 22K +	4 lbs./gal. 2,4-D Amine
4 to 6 leaf (weeds small)	1 fl. oz.	0.5 pt.
6 leaf to early boot (weeds more advanced or under dry conditions)	1.5 fl. oz.	0.75 pt.

† When measuring small amounts of TORDON 22K weed killer, special care should be taken not to exceed suggested rates.

The above table equates to 1/4 oz. picloram + 4 oz 2,4-D ae/acre for grains in the 4 to 6 leaf stage and 3/8 + 6 oz/acre for grains beyond the 6 leaf stage or when dry soil conditions exist.

NOTE: Since a specific growing season can be unusually dry and cool, to assure no carry-over problem, use only on land which will be fallowed or replanted to grass or grain crops the following year.

WARRANTY LIMITATIONS AND DISCLAIMER

The Dow Chemical Company warrants that this product conforms to the chemical description on the label and is reasonably fit for the purposes stated on the label when used in strict accordance with the directions therein under normal conditions of use. THIS IS THE ONLY WARRANTY MADE ON THIS PRODUCT. NO OTH-

ER EXPRESS AND NO IMPLIED WARRANTY OF MERCHANTABILITY OR FITNESS FOR A PARTICULAR PURPOSE IS MADE OUTSIDE OF THIS LABEL. Therefore, neither this warranty nor any other warranty of merchantability or fitness for a particular purpose, express or implied, extends to the use of this product contrary to label instructions (including conditions noted on the label, such as unfavorable temperatures, soil conditions, etc.), under abnormal conditions (such as excessive rainfall, drought, tornadoes, hurricanes, etc.) or under conditions not reasonably foreseeable to or beyond the control of seller.

When buyer or user suffers losses or damages resulting from the use or handling of this product (including claims based on contract, negligence, strict liability, or other legal theories), buyer or user must promptly notify in writing The Dow Chemical Company of any claims to be eligible to receive either remedy given below. The EXCLUSIVE REMEDY OF THE BUYER OR USER and the LIMIT OF LIABILITY of The Dow Chemical Company or any other seller will be one of the following, at the election of The Dow Chemical Company:

(1) **Refund of purchase price** paid by buyer or user for product bought, **or**

(2) **Replacement** of amount of product used.

The seller **will not be liable for consequential or incidental damages or losses.**

The terms of this Limit of Warranty and Liability cannot be varied by any written or verbal statements or agreements. Any employee or sales agent of the seller is not authorized to vary or exceed the terms of the Warranty Limitations and Disclaimer in any manner.

* Trademark of DOW CHEMICAL COMPANY

TORDON* RTU

HERBICIDE

For Controlling Unwanted Trees Via Cut Surface Treatments

ACTIVE INGREDIENTS:

Picloram (4–amino–3,5,6–trichloropicolinic acid) as the triisopropanol-amine salt	5.4%
2,4–Dichlorophenoxyacetic Acid as the triisopropanol-amine salt	20.9%
INERT INGREDIENTS	73.7%
TOTAL	100.0%

Acid Equivalents: Picloram—3.0%
2,4–Dichlorophenoxyacetic acid—11.2%

EPA Reg. No. 464-510 EPA Est. 464-MI-1

WARNING
Keep Out of Reach of Children
AVISO

PRECAUCION AL USUARIO: Si usted no lee inglés, no use este producto hasta que la etiqueta le haya sido explicada ampliamente.

PRECAUTIONARY STATEMENTS

HAZARDS TO HUMANS AND DOMESTIC ANIMALS

HARMFUL OR FATAL IF SWALLOWED • CAUSES EYE INJURY • MAY CAUSE SKIN IRRITATION

Avoid Contact with Eyes, Skin and Clothing • Wash Thoroughly After Handling • Keep Container Closed • Do Not Store Near Food or Feed • Do Not Drink Solution • Do Not Store in Open or Unlabeled Containers

STATEMENTS OF PRACTICAL TREATMENT

Contains ethylene glycol.

If swallowed, induce vomiting immediately by giving two glasses of water and sticking finger down throat. Get medical attention. Never give anything by mouth to an unconscious person.

In case of eye contact, promptly flush with plenty of water; and get medical attention. Remove contaminated clothing and wash before reuse.

PHYSICAL OR CHEMICAL HAZARDS

COMBUSTIBLE

Keep Away From Heat And Open Flame.

NOTICE

Read the entire label. Use only according to label directions. **Before** buying or using this product, read "WARRANTY LIMITATIONS AND DISCLAIMER" elsewhere on this label. If terms are not acceptable, return **unopened** package at once to seller for full refund of purchase price paid. Otherwise, use by the buyer or any other user constitutes acceptance of the terms under the Limit of Warranty and Liability.

IN CASE OF AN EMERGENCY endangering life or property involving this product, call collect **517–636-4400.**

AGRICULTURAL CHEMICAL

Do Not Ship or Store with Food, Feeds, Drugs or Clothing.

GENERAL INFORMATION

TORDON RTU Herbicide is a **ready-to-use product** that is effective in cut surface applications for killing unwanted trees and preventing undesirable sprouting of cut trees in forest and other non-crop areas such as fence rows and right-of-ways. Among the many species that can be controlled are:

ailanthus, alder, aspen, birch, cedar, cherry, dogwood, elm, firs, green ash, gum, hawthorn, hickory, hornbeam, maples, oaks, pecan, persimmon, serviceberry, sourwood, sweetbay.

USE DIRECTIONS

Use undiluted as indicated below.

TREE INJECTION TREATMENT: Inject 1 milliliter of undiluted TORDON RTU Herbicide solution through the bark completely around the tree trunk at intervals of 2 to 3 inches between edges of the injector wounds. Make injections near ground level when using the tree injector or 2 to 4 feet above the ground when using a HYPOHATCHET Injector or similar device. Treatments can be made in any season. Maples should not be treated during the spring sap flow. With some difficult to control species such as dogwood, hickory, sugar maple, bigleaf maple, tanoak, and some firs, application to a continuous cut rather than to spaced cuts may be more uniformly effective.

FRILL OR GIRDLE TREATMENT: Make a single hack girdle or "frill" of overlapping axe cuts through the bark completely around the tree as close to the ground as feasible. Spray or paint the injured surface with undiluted TORDON RTU Herbicide, using enough volume to wet treated areas.

STUMP TREATMENT: Spray or paint the cut surfaces of freshly cut stumps and stubs with undiluted TORDON RTU Herbicide. The cambium area next to the bark is the most vital area to wet.

USE PRECAUTIONS

Use this product only as specified on this label. TORDON RTU Herbicide is highly active as a herbicide, is water soluble, can move with surface runoff water, and can remain phytotoxic for a year or more if it gets into the soil. Untreated trees occasionally are affected by root uptake of herbicide leaches into the soil or excreted by the roots of nearby treated trees. If such injury cannot be tolerated, do not use TORDON RTU close to desirable trees.

Do not contaminate cropland, water, or irrigation ditches.

Do not apply on or in the vicinity of sensitive crops of desirable broadleaf plants. Small amounts can cause injury to plants.

Do not allow careless application or spray drift. Do not permit any spray or spray drift to contact desirable plants.

Use only gravity flow or very low spray pressure.

Do not remove product from original container except to apply as indicated under USE DIRECTIONS.

Do not store near food, feedstuff, fertilizer, seeds, insecticides, fungicides, or other pesticides.

STORAGE AND DISPOSAL

Do not contaminate water, food or feed by storage or disposal.

PESTICIDE DISPOSAL

Pesticide wastes are toxic. Improper use of excess pesticide, spray mixture, or rinsate is a violation of Federal law. If these wastes cannot be disposed of by use according to label instructions, contact your State Pesticide or Environmental Control Agency, or the Hazardous Waste representative at the nearest EPA Regional Office for guidance.

CONTAINER DISPOSAL

Do not reuse containers of TORDON RTU herbicide for any purpose. Puncture and dispose of in a sanitary landfill or by incineration, or, if allowed by State and local authorities, by burning. If burned, stay out of smoke. Rinse application and handling equipment thoroughly after use at least three times with clean water. Dispose of rinse water by burying in a non-cropland area away from water supplies. Where indicated, follow official local container disposal regulations.

Be sure that use of this product conforms to all application regulations.

WARRANTY LIMITATIONS AND DISCLAIMER

The Dow Chemical Company warrants that this product conforms to the chemical description on the label and is reasonably fit for the purposes stated on the label when used in strict accordance with the directions therein under normal conditions of use. THIS IS THE ONLY WARRANTY MADE ON THIS PRODUCT. NO OTHER EXPRESS AND NO IMPLIED WARRANTY OF MERCHANTABILITY OR FITNESS FOR A PARTICULAR PURPOSE IS MADE OUTSIDE OF THIS LABEL. Therefore, neither this warranty nor any other warranty of merchantability or fitness for a particular purpose, express or implied, extends to the use of this product contrary to label instructions (including conditions noted on the label, such as unfavorable temperatures, soil conditions, etc.), under abnormal conditions (such as excessive rainfall, drought, tornadoes, hurricanes, etc.) or under conditions not reasonably foreseeable to or beyond the control of seller.

When buyer or user suffers losses or damages resulting from the use or handling of this product (including claims based on contract, negligence, strict liability, or other legal theories), buyer or user must promptly notify in writing The Dow Chemical Company of any claims to be eligible to receive either remedy given below. The EXCLUSIVE REMEDY OF THE BUYER OR USER and the LIMIT OF LIABILITY of The Dow Chemical Company or any other seller will be one of the following, at the election of The Dow Chemical Company:

(1) **Refund of purchase price** paid by buyer or user for product bought, **or**

(2) **Replacement** of amount of product used.

The seller **will not be liable for consequential or incidental damages or losses.**

The terms of this Limit of Warranty and Liability cannot be varied by any written or verbal statements or agreements. Any employee or sales agent of the seller is not authorized to vary or exceed the terms of this Warranty Limitations and Disclaimer in any manner.

E.I. DuPont de Nemours Co., Inc.

Agricultural Chemical Division
Wilmington, DE 19898
(800) 441-7500

AMMATE® X-NI

WEED & BRUSH KILLER

ACTIVE INGREDIENT:

Ammonium Sulfamate	95%
INERT INGREDIENTS	5%
TOTAL	100%

EPA Reg. No. 352-311

Keep Out of Reach of Children

PRECAUTIONARY STATEMENTS

HAZARDS TO HUMANS

CAUTION!

MAY IRRITATE EYES, NOSE, THROAT, AND SKIN.

Avoid breathing dust or spray mist. Avoid contact with skin, eyes and clothing. In case of contact with eyes, immediately flush with plenty of water. Get medical attention if irritation persists. Wash thoroughly after handling. Remove and wash contaminated clothing before reuse. Do not apply this product in such a manner as to directly or through drift expose workers or other persons. The area treated must be vacated by unprotected persons.

ENVIRONMENTAL HAZARDS

Do not apply directly to water. Do not contaminate water by cleaning of equipment or disposal of wastes.

GENERAL INFORMATION

Du Pont AMMATE X-NI Weed and Brush Killer is highly effective for killing undesirable woody plants and for use as a contact spray for control of weeds and grasses. It is recommended for use on land adjacent to and surrounding domestic water supply reservoirs, supply streams, lakes and ponds, as well as on right-of-ways, roadsides, drainage ditches, rangelands, pastures, fence rows, industrial sites, tennis courts, driveways and similar areas. It is also recommended for selective control of poison ivy in apple and pear orchards.

Temporary nonproductivity of soil may be caused by heavy applications of AMMATE X-NI; this condition usually disappears in the "over-winter" period. Some species of woody plants and perennial weeds are difficult to control, and retreatment may be necessary if regrowth occurs.

AMMATE X-NI is water-soluble, nonvolatile and nonflammable.

IMPORTANT

AMMATE X-NI is non-selective; do not apply (except as recommended for orchard use), or drain, or flush equipment on desirable plants or vegetation, as injury or loss may result. Do not drain or flush equipment near domestic waters.

Wash sprayer thoroughly after use to remove all AMMATE X-NI and to reduce corrosion. Equipment should be coated with an asphaltic base paint; or thoroughly wash exterior of equipment at end of each day and apply protective coating of oil each week or as often as practical. When equipment is down for weekend, wash exterior and coat with oil; for end-of-season storage, wash inside and outside and coat with oil.

NOTICE OF WARRANTY

Du Pont warrants that this product conforms to the chemical description on the label thereof and is reasonably fit for purposes stated on such label only when used in accordance with the directions under normal use conditions. It is impossible to eliminate all risks inherently associated with the use of this product. Crop injury, ineffectiveness or other unintended consequences may result because of such factors as weather conditions, presence of other materials or the manner of use or application, all of which are beyond the control of Du Pont. In no case shall Du Pont be liable for consequential, special or indirect damages resulting from the use or handling of this product. All such risks shall be assumed by the Buyer. DUPONT MAKES NO WARRANTIES OF MERCHANTABILITY OR FITNESS FOR A PARTICULAR PURPOSE NOR ANY OTHER EXPRESS OR IMPLIED WARRANTY EXCEPT AS STATED ABOVE.

DIRECTIONS FOR USE

It is a violation of federal law to use this product in a manner inconsistent with its labeling.

AMMATE X-NI Weed & Brush Killer should be used only in accordance with recommendations on this label, or in separate published Du Pont recommendations available through local dealers.

Du Pont will not be responsible for losses or damages resulting from use of this product in any manner not specifically recommended by Du Pont. User assumes all risks associated with such nonrecommended use.

Do not enter treated areas without protective clothing until sprays have dried or dusts have settled. Because certain states may require more restrictive reentry intervals for various crops treated with this product, consult your State Department of Agriculture for further information.

Written or oral warnings must be given to workers who are expected to be in a treated area or in an area about to be treated with this product. When oral warnings are given, warnings shall be given in a language customarily understood by workers. Oral warnings must be given if there is reason to believe that written warnings cannot be understood by workers. Warnings must include the following information: "CAUTION. Area treated with AMMATE X-NI on (date of application). Do not enter without appropriate protective clothing until sprays have dried or dusts have settled. In case of contact, flush skin or eyes with plenty of water; for eyes, get medical attention."

BRUSH CONTROL

AMMATE X-NI is highly effective for killing most woody plants, including hardwood and coniferous species such as alder, ash, birch, cedar, elm, gum, hickory, maple, oak, pine and willow. Application may be made as a foliage spray; by frill, notch or cup method; or as stump treatment.

Foliage Spray Application—Prepare spray solution of AMMATE X-NI in water; use 60 lbs. per 100 gals. for hydraulic equipment, and 100 to 400 lbs. per 100 gals. for air blast equipment. To improve wetting of foliage, add 1 qt. Du Pont Surfactant WK per 100 gals. of spray.

Apply any time after brush has reached the full-leaf stage until foliage begins to discolor, preferably during periods of high temperatures and high humidity. Apply as a full-coverage spray to foliage, stems, limbs, and base of brush; thorough coverage is essential for best results. On tall, dense brush, it is often necessary to spray from both sides to obtain adequate coverage. Spraying away from domestic waters and away from crops planted close to right-of-ways will aid in preventing spray or drift into these areas; coarse sprays are less likely to drift.

Stump Treatment—On New Construction; Following Cutting Crews: Treat stumps as soon after cutting as possible. Sprinkle crystals of AMMATE X-NI liberally on freshly cut surface, or spray stump thoroughly with solution of 7 to 10 lbs. AMMATE X-NI in 2 gals. water; be sure to wet outer growth ring. A water-soluble wood dye may be added to spray solution to mark treated stumps. Stumps under 2 inches are best treated by crystal method.

Treatment of Undesirable Hardwood Trees—To kill hardwood species, such as blackjack oak, sweetgum, poplar, pecan, maple, ash, red oak, post oak and hickory, treat at any time of year as follows:

Frill Method—At convenient chopping height, make a continuous cut completely around tree with downward axe strokes, cutting well into sapwood. Saturate frilled area with solution of 7 to 10 lbs. AMMATE X-NI in 2 gals. water. Water-soluble wood dye may be added to the solution to identify treated trees.

Notch or Cup Method—Make a notch or cup by two downward axe cuts, one above the other, prying out chip. Notches should be at base of tree as near ground as possible and on main roots, if any show. Cut two notches on trees 3 to 6 inches in diameter; space notches every 4 to 6 inches around circumference of larger trees. Apply ½ oz. AMMATE X-NI in each notch. For trees under 3 inches, cut close to ground and use stump treatment.

WEED AND GRASS CONTROL

AMMATE X-NI, in water solution, is an effective contact spray for control of: **Woody Perennials** such as poison ivy, poison oak, and poison sumac; **Perennials** such as leafy spurge, bitter dock, goldenrod, perennial ragweed, milkweed, and blueweed; and **Annuals** such as crabgrass, broomweed, chickweed, cocklebur, jimsonweed, lambsquarters, larkspur, prickly lettuce, ragweed and shepherdspurse.

Use AMMATE X-NI at the rate of 100 lbs. per 100 gals. of water; add 1 qt. Du Pont Surfactant WK per 100 gals. of spray to improve wetting of foliage. Use hydraulic equipment; thoroughly wet foliage and stems of undesired vegetation. The degree of control and duration of effect will vary with weed species, rainfall, temperature and other conditions.

POISON IVY CONTROL

SELECTIVE USE IN APPLE AND PEAR ORCHARDS

Use 60 lbs. AMMATE X-NI plus 1 qt. Du Pont Surfactant WK per 100 gals. water. Apply as a coarse spray to poison ivy under and between trees when ivy plants are in full leaf; wet ivy foliage thoroughly. Spray volume depends upon density and size of poison ivy plants; usually 1 gal. will cover an area of 200 to 250 sq. ft. For poison ivy growing on tree trunks, spray only if bark of tree is well developed and shows no green; avoid excessive wetting of bark as injury may result. Do not allow spray or drift to contact tree foliage or fruit as injury will result.

STORAGE AND DISPOSAL

STORAGE

Store product in original container only, away from other pesticides, fertilizer, food or feed.

DISPOSAL

Do not contaminate water, food or feed by storage or disposal. Wastes resulting from the use of this product may be disposed of on site or at an approved waste disposal facility.

Completely empty bag into application equipment. Then dispose of empty bag in a sanitary landfill or by incineration, or, if allowed by State and local authorities, by burning. If burned, stay out of smoke.

BENLATE®

FUNGICIDE

ACTIVE INGREDIENT:
Benomyl
[Methyl 1-(butylcarbamoyl)-2-benzimidazolecarbamate] .. 50%
INERT INGREDIENTS 50%
TOTAL 100%

EPA Reg. No. 352-354 U.S. Pats. 3,541,213 & 3,631,176

Keep Out of Reach of Children

PRECAUTIONARY STATEMENTS

HAZARDS TO HUMANS

CAUTION

MAY IRRITATE EYES, NOSE, THROAT, AND SKIN.

Avoid breathing dust or spray mist. Avoid contact with skin, eyes, and clothing. Wash thoroughly after using. Do not apply this product in such a manner as to directly, or through, drift expose workers or other persons. The area being treated must be vacated by unprotected persons.

FIRST AID

In case of contact, flush skin or eyes with plenty of water; for eyes, get medical attention.

AERIAL MIXER-LOADERS

Wear a cloth or disposable paper dust mask during handling and mixing, as this product may be harmful if inhaled.

ENVIRONMENTAL HAZARDS

This product is toxic to fish. Do not apply directly to water or wetlands. Do not apply where runoff is likely to occur. Do not apply when weather conditions favor drift from areas treated.

PHYSICAL OR CHEMICAL HAZARDS

Keep away from fire or sparks.

NOTICE OF WARRANTY

DuPont warrants that this product conforms to the chemical description on the label thereof and is reasonably fit for purposes stated on such label only when used in accordance with directions under normal use conditions. It is impossible to eliminate all risks inherently associated with use of this product. Crop injury, ineffectiveness or other unintended consequences may result because of such factors as weather conditions, presence of other materials or the manner of use or application, all of which are beyond the control of DuPont. In no case shall DuPont be liable for consequential, special or indirect damages resulting from the use or handling of this product. All such risks shall be assumed by the buyer. DUPONT MAKES NO WARRANTIES OF MERCHANTABILITY OR FITNESS FOR A PARTICULAR PURPOSE NOR ANY OTHER EXPRESS OR IMPLIED WARRANTY EXCEPT AS STATED ABOVE.

DIRECTIONS FOR USE

It is a violation of federal law to use this product in a manner inconsistent with its labeling.

DuPont BENLATE should be used only in accordance with recommendations on this label, or in separate published DuPont recommendations available through local dealers.

DuPont will not be responsible for losses or damages resulting from use of this product in any manner not specifically recommended by DuPont. User assumes all risk associated with such nonrecommended use.

Do not enter treated areas without protective clothing until sprays have dried. Because certain states may require more restrictive reentry intervals for various crops treated with this product, consult your State Department of Agriculture for further information.

Written or oral warnings must be given to workers who are expected to be in a treated area or in an area about to be treated with this product. When oral warnings are given, warnings shall be given in a language customarily understood by workers. Oral warnings must be given if there is reason to believe that written warnings cannot be understood by workers. Warnings must include the following information: "CAUTION. Area treated with BENLATE on (date of application). Do not enter without appropriate protective clothing until sprays have dried. In case of contact, flush skin or eyes with plenty of water; for eyes, get medical attention."

BENLATE is a systemic fungicide recommended for the control of many important plant diseases. A resistant strain of the fungus may be present if treatment is not effective following use of BENLATE as recommended. If treatment is ineffective due to the presence of a benomyl resistant strain, then neither BENLATE, nor any other benzimidazole or thiophanate type fungicide will effectively control that disease; consideration should be given to prompt use of other types of suitable fungicides.

The repeated exclusive use of BENLATE may lead to buildup of resistant strains of fungi and loss of disease control. A spray program using other fungicides may delay resistant strain buildup. Consult your state extension specialist or official state recommendations for guidance on your particular crop and disease control situation.

Note: Do not tank mix or alternate BENLATE with benzimidazole or thiophanate products such as Mertect[1] or Topsin[2].

[1] Registered trademark of Merck & Co.

[2] Registered trademark of Nippon Soda Co., Japan

Apply as a spray with ground equipment (except as otherwise directed), using sufficient water to obtain thorough coverage of the plants. Under severe disease conditions use the higher rate and shorter interval specified for each crop; also, for tree crops, use the higher rate for large mature trees. For aerial application (listed crops only) use the following gals. per acre: Rice and Soybeans, 3 to 10; Cabbage (seed crop), Celery, Cucurbits, Peanuts and Sugar Beets, 5 to 10; Almonds, Avocados, Beans, Pecans, Stone Fruits and Strawberries, 10 to 20; Grapes, 15 to 20; Roses, Flowers, Ornamental and Shade Trees, 20 gals. per acre minimum.

For use in small gardens and orchards (less than 1 acre), application rates may be converted to lbs. per 100 gals. by dividing the lbs. per acre rate in half, and applying the resulting spray mixture at the rate of 4.5 gals. per 1000 sq. ft. (1 lb. BENLATE per 100 gals. equals 1 tablespoonful per gal.).

Add required amount of BENLATE to necessary volume of water in spray tank agitated by hydraulic or mechanical means; continuous agitation is required to keep the material in suspension. Do not tank mix BENLATE with lime or alkaline pesticides such as Bordeaux mixture or lime sulfur.

Where use of spray oil is recommended (apples, peanuts, pecans, stone fruits), use a nonphytotoxic superior-type (60 to 70 second viscosity) spray oil; add as last ingredient to spray tank. Before applying other pesticides in conjunction with spray oil, or immediately before or after oil application, consult product labels. Observe all cautions and limitations on labeling of all products used in mixtures.

ALMONDS: Brown Rot Blossom Blight—Apply 1 to 1½ lbs. per acre at pink bud. Under severe disease conditions and on highly susceptible varieties, make a second application during half-to-full bloom.

APPLES: For applications through cover sprays, use BENLATE as a tank mixture as detailed below. Apply 200 to 500 gals. of spray per acre with hydraulic ground equipment or equivalent amount of products per acre with concentrate sprayers. Do not graze livestock in treated orchards.

BENLATE + Manzate 200 Fungicide: Scab, Powdery Mildew, Sooty Blotch, Flyspeck, Cedar Apple Rust, Quince Rust, Bitter Rot, Black Rot, Brown Rot—Use 2 to 3 ozs. BENLATE plus 12 ozs. Manzate 200 per 100 gals. of water; 1 qt. spray oil may be added per 100 gals. Apply at ½" green tip and repeat at 7 to 14-day intervals (or as needed) through the cover sprays.

Use the 3 oz. rate of BENLATE and add spray oil to the spray mixture for varieties more susceptible to powdery mildew, and for scab if an application is missed during an infection period (apply as soon as possible after infection period in order to deactivate scab and to prevent further infection). Do not apply within 30 days of harvest.

BENLATE + Captan: Scab, Powdery Mildew, Sooty Blotch, Flyspeck, Bitter Rot, Black Rot—Use 2 to 3 ozs. BENLATE plus 12 to 16 ozs. Captan 50WP Fungicide (or 7½ to 10 ozs. Captan 80WP) per 100 gals. of water. Apply at ½" green tip and repeat at 7 to 14-day intervals (or as needed) through the cover sprays. Use the 3 oz. rate of BENLATE for varieties more susceptible to powdery mildew. If an application is missed during an infection period, apply the higher rates as soon as possible after the infection period in order to deactivate scab and to prevent further infection.

Note: Spray injury may result if Captan is used with, immediately before or closely following an oil spray.

Postharvest Fruit Rots (*Botrytis* spp., *Penicillium* spp., *Gloeosporium* spp.)—Make a single application of 6 ozs. BENLATE per 100 gals., anytime from 3 weeks before harvest up to day of harvest. For additional protection of fruit to be held in storage, thoroughly wet harvested fruit by dipping or spraying at 8 ozs. per 100 gals.

AVOCADOS (Florida): Scab, Cercospora Spot, Anthracnose—Apply 1 to 2 lbs. per acre; begin when buds swell and repeat at 3 to 4-week intervals. Do not apply within 30 days of harvest.

BEANS: White Mold *(Sclerotinia),* Gray Mold *(Botrytis)*—Use on beans grown as fresh vegetables, for processing or for the dry bean market. Apply 1½ to 2 lbs. per acre at 25% to 50% bloom; repeat at peak bloom. For narrow-row (20–40") irrigated dry beans in Montana, Nebraska, Colorado and Wyoming, apply at initial bloom and repeat 7 to 10 days later; only partial control of white mold may result.

Do not apply within 14 days of harvest (28 days for lima beans); do not use where crop is grown only for forage purposes.

BLUEBERRIES: Apply 1 lb. per acre. Do not make more than 4 applications before harvest; do not apply within 21 days of harvest. **Mummy Berry, Botrytis Blossom Blight**—Apply at green tip and repeat at 7 to 10-day intervals through petal fall. **Anthracnose Leafspot**—Apply when disease first appears and make one additional application 14 days later. After harvest, make up to 4 applications to the bushes at 14-day intervals as needed.

CABBAGE (Seed Crop, Pacific Northwest): White Blight (Sclerotinia Stalk Rot)—Apply 2 lbs. per acre by aircraft in 5 to 10 gals. of water; add a spreader-sticker to aid in wetting plants. Make first application at first petal fall; make two additional applications at 14-day intervals if conditions favor development of disease. **Note:** Do not graze treated areas; do not use seed or plant parts for food or feed purposes.

CANEBERRIES—RASPBERRIES, BLACKBERRIES, BOYSENBERRIES, LOGANBERRIES, DEWBERRIES: Botrytis, Powdery Mildew, Penicillium Rots—Apply ¾ lb. per acre at early bloom (5 to 10%) and at full bloom; make up to 3 additional applications at 14-day intervals as needed. Do not apply within 3 days of harvest.

CELERY: Early Blight *(Cercospora),* Late Blight *(Septoria)*—Apply ¼ to ½ lb. per acre; begin when disease first appears and repeat at 7 to 10-day intervals. Do not apply within 7 days of harvest.

CITRUS: Scab—Apply 1½ to 3 lbs. per acre. Under conditions of severe disease pressure, apply at pinhead stage (just prior to first flush) and repeat at ⅔ petal fall; otherwise, make a single application at ⅔ petal fall.

Greasy Spot—Make a single application of 1½ to 3 lbs. per acre during the period mid-June to mid-July.

Fruit Decay (Green Mold, Blue Mold, Stem-end Rot)—Preharvest Spray—Make a single application of 1 to 2 lbs. per acre anytime from 3 weeks prior to harvest up to day of harvest. **Postharvest**—Apply as a dip, flood or spray using 1 to 2 lbs. per 100 gals.; do not immerse fruit for more than 5 min. When citrus wax is used, BENLATE may be incorporated into the wax spray. Use the higher rate on more susceptible fruits and when excessive inoculum levels are present. For control of sporulation (*Penicillium* spp.), apply as a spray in citrus wax using 4 lbs. BENLATE per 100 gals.

Note: Do not graze livestock in treated groves.

CUCURBITS—CUCUMBERS, MELONS, PUMPKINS, SUMMER AND WINTER SQUASH: Target Spot (Cucumbers), Gummy Stem Blight, Powdery Mildew, Anthracnose—Apply $1/4$ to $1/2$ lb. per acre; for aerial application, use $1/2$ lb. per acre. Begin applications when plants begin to run or when disease first appears, and repeat at 7 to 14-day intervals as needed. For target spot, use 7-day intervals as needed.

GRAPES: Botrytis Bunch Rot—Apply 1 to $1\frac{1}{2}$ lbs. per acre at first bloom (no later than 5% bloom) and repeat 14 days later if severe disease conditions persist. Make an additional application 3 to 4 weeks before harvest or when sugar begins to build; repeat 14 days later if conditions favorable for disease persist. BENLATE does not control bunch rots caused by other organisms such as *Rhizopus* spp., *Alternaria* spp., and *Diplodia* spp.; these rots occur most frequently in high temperature areas such as the San Joaquin and Sacramento Valleys of California. **Powdery Mildew, Black Rot, Bitter Rot — East of Rockies**—Apply $3/4$ to $1\frac{1}{2}$ lbs. per acre when foliage first develops and repeat at 14 to 21-day intervals (or as needed) until berries are full size.

Note: Do not apply within 7 days of harvest.

MACADAMIA NUTS (Hawaii): Botrytis Blossom Blight—Apply $1\frac{3}{4}$ lbs. per acre; a surfactant may be added to the spray to improve wetting of foliage. Begin applications 1 to 2 weeks prior to bloom, and repeat at 7 to 14-day intervals through the bloom period.

MANGOES: Anthracnose—Apply 1 to 2 lbs. per acre. Begin applications at first appearance of panicles (approx. 2" long), and repeat at weekly intervals until all fruits are set. Continue at 3 to 4-week intervals. Do not apply within 14 days of harvest.

MUSHROOMS: Verticillium Spot (Dry Bubble)—Use 1 lb. per 100 gals. and apply to bed surface at the rate of $12\frac{1}{2}$ gals. per 1000 sq. ft. Apply immediately after casing and repeat at pinning; alternatively, if disease has occurred, apply to beds after picking and repeat 10 days later. Do not apply within 2 days of harvest.

PEANUTS: BENLATE + Manzate 200 Fungicide: Cercospora Leafspot, Rust, Ascochyta Web Blotch—Apply $1/4$ lb. BENLATE plus $1\frac{1}{2}$ lbs. Manzate 200 per acre; spray oil may be added at the rate of 1 pt. to 1 qt. per acre. Do not use BENLATE alone. Begin applications 35 to 40 days after planting or when disease first appears. Repeat at following intervals: for Cercospora leafspot, 10 to 14 days; for rust, 7 to 10 days; for Ascochyta web blotch, 7 to 14 days. Do not apply within 14 days of harvest; do not graze or feed treated vines, hay or hulls to livestock.

PEARS: Scab, Powdery Mildew, Sooty Blotch, Flyspeck—Use 4 to 6 ozs. per 100 gals. of water; apply 200 to 500 gals. of spray per acre with hydraulic ground equipment or equivalent amounts of BENLATE per acre with concentrate sprayers. Apply at $1/2$" green tip and repeat at 7 to 14-day intervals (or as needed) through the cover sprays. If an application is missed during an infection period, use 6 ozs. per 100 gals. and apply as soon as possible after the infection period in order to deactivate scab and to prevent further infection. Do not graze livestock in treated orchards.

Postharvest Fruit Rots (*Botrytis* spp., *Penicillium* spp., *Gloeosporium* spp.)—Make a single application of 6 ozs. per 100 gals. anytime from 3 weeks before harvest up to day of harvest. For additional protection of fruit to be held in storage, thoroughly wet harvested fruit by dipping or spraying at 8 ozs. per 100 gals.

Overwintering Scab—Apply 8 ozs. per 100 gals. after harvest but before leaf drop. Thorough wetting of foliage is necessary.

PECANS: Pecan Scab, Brown Leafspot, Downy Spot, Powdery Mildew, Liverspot, Zonate Leafspot, Fungal Leaf Scorch—Apply $1/2$ to 1 lb. per acre; use the higher rate on trees over 30 ft. tall. For aerial application (Ark., La., Miss., Okla., Tex. only), use 1 lb. per acre. Spray oil may be added at the rate of 1 to 2 gals. per acre. Apply at prepollination when young leaves are unfolding, when small nuts are forming and thereafter at 3 to 4-week intervals. Do not apply after shucks split.

PINEAPPLE: Thielaviopsis Rot (Fresh Fruit)—Use 2 to 4 lbs. per 100 gals. of water. Immediately after harvest, immerse or spray fruit to give thorough wetting and allow to drain; do not immerse for more than 5 min. **Pineapple Butt Rot** *(Thielaviopsis paradoxa)* — Use $1\frac{1}{4}$ lbs. per 100 gals. of water as a preplant dip treatment. Immerse seedpieces to give thorough wetting; remove and allow to drain.

RICE (Except California): Rice Blast, Stem Rot—Apply 1 to 2 lbs. per acre at booting and repeat at heading. Do not apply within 21 days of harvest. Do not apply to stubble rice. Do not apply to fields where crayfish or catfish farming is practiced, nor drain water from treated areas into areas where such farming is practiced. Water drained from treated areas must not be used to irrigate other crops.

SOYBEANS: Diaporthe Pod-and-Stem Blight, Anthracnose, Septoria Brown Spot, Cercospora Frogeye Leafspot, Purple Seed Stain—Apply $1/2$ to 1 lb. per acre. For **determinate** varieties (generally grown in the South), apply at early pod set when majority of pods are $1/8$ to $1/2$" in length; for **indeterminate** varieties (generally grown in the North), apply when pods near the top of the plant are $1/2$ to 1" in length. Make one additional application 14 to 21 days later. Do not apply within 35 days of harvest; do not graze or feed treated soybean vines or hay to livestock.

STONE FRUITS—APRICOTS, CHERRIES, NECTARINES, PEACHES, PLUMS, PRUNES: Treatment is most effective if applied just before rainfall; for aerial application, fly over every row or center.

EAST OF ROCKY MOUNTAINS—Use $3/4$ to $1\frac{1}{2}$ lbs. per acre on trees up to 12 ft. tall; over 12 ft., use $1\frac{1}{2}$ to 2 lbs. per acre.

Brown Rot Blossom Blight—Apply at early bloom stages (apricots—red bud; peaches, nectarines—pink bud; cherries—early popcorn; plums and prunes—green tip); for this application only, BENLATE may be used in combination with spray oil. Make a second application at 75% to 100% bloom. If blossoming is prolonged or conditions favorable for disease persist, apply at petal fall.

Fruit Brown Rot—After blossom blight sprays, make two preharvest applications beginning 3 weeks before harvest up to day of harvest.

Peach Scab, Powdery Mildew—Use same schedule as for Brown Rot Blossom Blight plus applications at shuck split, shuck fall and 14 days later.

Cherry Leaf Spot—Use same schedule as for Brown Rot Blossom Blight and continue at 10 to 14-day intervals through harvest. Make an additional application 2 to 3 weeks after harvest.

WEST OF ROCKY MOUNTAINS—Use $1\frac{1}{2}$ to 2 lbs. per acre.

Brown Rot Blossom Blight—Apply at early bloom stages (apricots—red bud; peaches, nectarines—pink bud; cherries—early popcorn; plums and prunes—green tip); for this application only, BENLATE may be used in combination with spray oil. If blossoming is prolonged or conditions favorable for disease persist, make a second application 14 days later.

Fruit Brown Rot—After blossom blight sprays, make a preharvest application (before rain) anytime from 3 weeks before harvest to day of harvest. Make a second application if conditions favorable for disease persist or harvest is prolonged. Preharvest applications are most effective when applied with ground equipment, using sufficient volume to provide thorough and uniform coverage of fruit.

Powdery Mildew—Use same schedule as for Brown Rot Blossom Blight plus applications at shuck split, shuck fall and 14 days later.

Cherry Leaf Spot—Use same schedule as for Brown Rot Blossom Blight and continue at 10 to 14-day intervals through harvest. Make an additional application 2 to 3 weeks after harvest.

POSTHARVEST FRUIT ROTS (U.S.)—Dip or spray fruit thoroughly as soon as possible after harvest; use $1/2$ lb. per 100 gals. When wax is used, BENLATE may be incorporated into the wax spray.

NOTE: BENLATE does not control peach leaf curl, shot hole *(Coryneum blight)*, bacterial blast, or fruit rots caused by *Rhizopus* spp. and *Alternaria* spp. Do not graze livestock in treated orchards.

STRAWBERRIES: Gray Mold *(Botrytis)*, Powdery Mildew, Leaf Scorch, Leaf Blight, Leaf Spot—Apply 1 lb. per acre at 10% bloom and at full bloom; continue at 10 to 14-day intervals, using ½ lb. per acre. **Anthracnose**—Apply 1 lb. per acre when plants are established (plant bed or field) and repeat at 7-day intervals.

Transplants: Botrytis Crown Rot, Leaf Spot—Use ½ lb. per 100 gals. of water. Immerse plants to give thorough wetting; remove and allow to drain.

SUGAR BEETS: Cercospora Leafspot—Apply ⅜ to ½ lb. per acre. Begin application when disease first appears and repeat at 14 to 21-day intervals as needed. Do not apply within 21 days of harvest.

SUGARCANE (HAWAII): Pineapple Disease *(Ceratocystis paradoxa)* —Apply to cut seedpieces either as a cold dip or hot dip. Cold Dip—Use ½ lb. per 100 gals. of water (1:1600). Immerse seedpieces to give thorough wetting; remove and allow to drain.

Hot Dip—Use ¼ lb. per 100 gals. of water (1:3200). Maintain temperature of the dip at 50°C. Soak seedpieces for 20 to 30 minutes; remove and allow to drain.

NOTE: Do not use treated seedpieces for food or feed purposes.

TOMATOES—Field and Greenhouses: Gray Mold *(Botrytis)*, Leaf Mold *(Cladosporium)*, White Mold *(Sclerotinia)*, Cercospora Leaf Spot, Phoma Leaf Spot—For field tomatoes, apply ½ to 1 lb. per acre; for greenhouse use ½ to 1 lb. per 100 gals. of water. Begin applications when disease first appears and repeat at 7 to 14-day intervals as needed.

ROSES, FLOWERS, ORNAMENTALS, SHADE TREES—Field and Greenhouse: Foliar Spray—Begin applications when disease first appears and repeat at 10 to 14-day intervals throughout the growing season; shorten interval during humid, rainy weather. Use at the following rates: **½ lb. per 100 gals.** (1 tablespoonful per 2 gals.)—For Powdery Mildew, Botrytis Gray Mold. **1 lb. per 100 gals.**—For Anthracnose (for shade trees and woody ornamentals, begin at bud break and make 2 or 3 additional applications at 10 to 14-day intervals); Black Spot of roses; Cercospora, Entomosporium, Ramularia, and Septoria Leaf Spots; Ascochyta and Phomopsis, blights; Didymellina Leaf Spot of iris; Corynespora Leaf Spot of Ligustrum; Ovulinia Blight of azalea and rhododendron (begin as flowers open); Scab of pyracantha and flower crab. Addition of a surfactant to the spray mixture improves distribution of the spray on hard-to-wet plants such as roses. For aerial application, use ½ to 1 lb. per acre.

Drench Treatment—Botrytis, Fusarium, Rhizoctonia and Sclerotinia stem, crown and root rots on herbaceous annuals, perennials and bedding plants; Cylindrocladium and Thielaviopsis rots on woody ornamentals such as azaleas, rhododendrons, conifers and poinsettias—Use 1 lb. per 100 gals.; apply as a drench or heavy spray (1 to 2 pts. per sq. ft.) after transplanting into propagation beds or containers. Repeat at 2 to 4-week intervals during periods favorable for disease. BENLATE does not control *Pythium* spp. or *Phytophthora* spp.

Preplant Dip Treatment—For disease listed under Drench Treatment, use 1 lb. per 100 gals. of water; immerse plants or cuttings for 10 to 15 min.; remove and allow to drain.

BULBS (Easter Lily, Tulip, Gladiolus, Daffodil, Iris): Fusarium and Penicillium Rots—Use 1⅔ lbs. per 100 gals. of water (2 tablespoonfuls per gal.). Soak cleaned bulbs for 15 to 30 minutes in warm dip (80 to 85°F.), preferably within 48 hours after digging. Dry bulbs after treatment. If bulbs are for forcing, treat after bulbs have been heat-cured.

NOTICE TO BUYER: Purchase of this material does not confer any rights under patents of countries outside of the United States.

STORAGE AND DISPOSAL

STORAGE

Never allow BENLATE to become wet during storage. This may lead to certain chemical changes which will reduce the effectiveness of BENLATE as a fungicide. Keep container tightly closed when not in use. Store product in original container only, away from other pesticides, fertilizers, food or feed.

DISPOSAL

Do not contaminate water, food or feed by storage or disposal. Wastes resulting from the use of this product may be disposed of on site or at an approved waste disposal facility. Completely empty bag into application equipment. Then dispose of empty bag in a sanitary landfill, or by incineration, or, if allowed by state and local authorities, by burning. If burned, stay out of smoke.

SUPPLEMENTAL LABELING

BENLATE® FUNGICIDE

EPA Reg. No. 352–354

FOR CONTROL OF CROWN ROT AND SURFACE MOLDS OF BANANAS

DuPont BENLATE Fungicide is recommended for post harvest use on bananas for the control of crown rot and surface molds caused by fungi such as Gloeosporium musarum, Colletotrichum musae and certain species of Penicillium and Fusarium.

HOW TO USE

Use ½ to 1 lb. BENLATE per 100 gals. water (equivalent to 300 to 600 ppm, active ingredient basis). Apply as a post harvest dip or spray to thoroughly wet crowns and fruit.

BENLATE must be kept in suspension at all times by continuous agitation; agitate by mechanical or hydraulic means.

IMPORTANT

BEFORE USING BENLATE, READ AND CAREFULLY OBSERVE THE CAUTIONARY STATEMENTS AND ALL OTHER INFORMATION APPEARING ON THE PRODUCT LABEL.

This bulletin contains new or supplemental instructions for use of this product which may not appear on package label. Follow the instructions carefully.

SUPPLEMENTAL LABELING

BENLATE® FUNGICIDE

EPA Reg. No. 352–354

FOR CONTROL OF SIGATOKA DISEASE IN BANANAS

DuPont BENLATE Fungicide, in combination with nonphytotoxic spray oil, is recommended for the control of Sigatoka disease in bananas.

HOW TO USE

Apply BENLATE at the rate of 2 to 4 ozs. per acre as a mixture in oil-water emulsion, using 2 qts. of nonphytotoxic spray oil per acre. Apply at 2 to 3-week intervals throughout the complete crop cycle; use the shorter interval when conditions favor disease.

Use a minimum of 2.5 gals. of spray mixture per acre for aerial application, and at least 5 gals. per acre with mist blowers.

Prepare spray mixture by adding proper amount of oil to spray tank half-filled with water; add emulsifier at rate of 1% of oil volume and agitate thoroughly. Add proper amount of BENLATE and balance of water. Agitate continuously from start of mixing until end of spraying operation.

IMPORTANT

BEFORE USING BENLATE, READ AND CAREFULLY OBSERVE THE CAUTIONARY STATEMENTS AND ALL OTHER INFORMATION APPEARING ON THE PRODUCT LABEL.

This bulletin contains new or supplemental directions for use of this product which may not appear on package label. Follow the instructions carefully.

SUPPLEMENTAL LABELING

BENLATE® FUNGICIDE
EPA Reg. No. 352–354

FOR CONTROL OF EARLY BLIGHT AND LATE BLIGHT OF CELERY

DuPont BENLATE Fungicide is recommended for use on celery for control of early blight *(Cercospora)* and late blight *(Septoria)*.

HOW TO USE
Apply 1/4 to 1/2 lb. per acre, using sufficient water to obtain thorough coverage (5 to 10 gals. per acre for aerial application). Begin applications when disease first appears and repeat at 7 to 10-day intervals. Under severe disease conditions, use the higher rate and shorter interval. Do not apply within 7 days of harvest.

Note: Do not use alkaline pesticides, such as basic copper sulfate, Bordeaux mixture or lime sulfur, as a tank mixture with BENLATE.

IMPORTANT
BEFORE USING BENLATE, READ AND CAREFULLY OBSERVE THE CAUTIONARY STATEMENTS AND ALL OTHER INFORMATION APPEARING ON THE PRODUCT LABEL.

This bulletin contains new or supplemental instructions for use of this product which may not appear on package label. Follow the instructions carefully.

SUPPLEMENTAL LABELING

BENLATE® FUNGICIDE
EPA Reg. No. 352-354

FOR CONTROL OF WHITE MOLD ON SNAP BEANS AND LIMA BEANS IN MINNESOTA AND WISCONSIN

DuPont BENLATE Fungicide is recommended for use on snap beans and lima beans in Minnesota and Wisconsin to control white mold *(Sclerotinia)*.

HOW TO USE
Apply 1 1/2 to 2 lbs. BENLATE per acre at 25 percent to 50 percent bloom, and repeat at peak bloom. Use sufficient water to obtain thorough coverage of plants. For aerial application, apply BENLATE in 4 to 10 gallons of water per acre.

Note: For snap beans, do not apply within 14 days of harvest. For lima beans, do not apply within 28 days of harvest.

Do not use where crop is grown only for forage purposes.

IMPORTANT
BEFORE USING BENLATE, READ AND CAREFULLY OBSERVE THE CAUTIONARY STATEMENTS AND ALL OTHER INFORMATION APPEARING ON THE PRODUCT LABEL.

This bulletin contains new or supplemental instructions for use of this product which may not appear on package label. Follow the instructions carefully.

SUPPLEMENTAL LABELING

BENLATE® FUNGICIDE
EPA Reg. No. 352–354

FOR STEM AND POD DISEASES OF SOYBEANS GROWN FOR SEED IN THE MIDWEST

DIRECTIONS FOR USE
DuPont BENLATE Fungicide is recommended for use in control of Diaporthe Pod and Stem Blight on Soybeans grown for seed in the Midwest.

HOW TO USE
Make a single application of 1 to 2 lbs. BENLATE per acre. Apply when beans are developing pods at one of the four uppermost nodes with a completely unrolled leaf.

Note: Consult BENLATE product labeling for recommendations for soybeans raised for feed. Do not graze or feed vines or hay to livestock. Do not apply within 35 days of harvest.

IMPORTANT
BEFORE USING BENLATE, READ AND CAREFULLY OBSERVE THE CAUTIONARY STATEMENTS AND ALL OTHER INFORMATION APPEARING ON THE PRODUCT LABEL.

This bulletin contains new or supplemental instructions for use of this product which may not appear on package label. Follow the instructions carefully.

SUPPLEMENTAL LABELING

BENLATE® FUNGICIDE
EPA Reg. No. 352–354

FOR CONTROL OF WHITE MOLD ON CARROTS

DuPont BENLATE Fungicide is recommended for use on carrots to control white mold *(Sclerotinia)*.

HOW TO USE
Apply 1/4 to 1 lb. BENLATE per acre using sufficient water to obtain thorough coverage (3 to 10 gallons per acre for aerial application). Begin applications when disease first appears and repeat at 7 to 10-day intervals. Under severe disease conditions, use the higher rate and shorter interval. Do not apply within 4 days of harvest.

Note: Do not tank mix BENLATE with lime or alkaline pesticides such as Bordeaux mixture or lime sulfur.

IMPORTANT
BEFORE USING BENLATE, READ AND CAREFULLY OBSERVE THE CAUTIONARY STATEMENTS AND ALL OTHER INFORMATION APPEARING ON THE PRODUCT LABEL.

This bulletin contains new or supplemental instructions for use of this product which may not appear on package label. Follow the instructions carefully.

SUPPLEMENTAL LABELING

BENLATE® FUNGICIDE
EPA Reg. No. 352–354

FOR CONTROL OF PENCILLIUM CLOVE ROT ON GARLIC IN CALIFORNIA

DIRECTIONS FOR USE
DuPont BENLATE Fungicide is recommended for use on garlic cloves used as seed pieces in California to control Pencillium clove rot.

HOW TO USE
Use 1 lb. of BENLATE per 100 gals. of water. Add the BENLATE to the required volume of water in a dip tank agitated by hydraulic or mechanical means. Continuous agitation is needed to keep the material in solution. Completely immerse the garlic cloves in the BENLATE suspension for a minimum period of 5 minutes. 150 lbs. of garlic cloves can be treated per 100 gals. of seed treatment solution.

After treatment, remove the cloves from the solution and allow to drain over the tank. Dry the cloves after treatment and prior to planting.

IMPORTANT
BEFORE USING BENLATE, READ AND CAREFULLY OBSERVE THE CAUTIONARY STATEMENTS AND ALL OTHER INFORMATION APPEARING ON THE PRODUCT LABEL.

This label bulletin contains new or supplemental instructions for use of this product which may not appear on package label. Follow the instructions carefully.

5/3184

SPECIAL LOCAL NEED REGISTRATIONS [24(c)] FOR BENLATE® FUNGICIDE*

CROP	STATE	EPA SLN. NO.	PRODUCTS	REMARKS
Almonds	California	CA-770002	BENLATE + Maneb or Captan	Tank mixtures
Apples	Vermont	VT-770004	BENLATE + MANZATE 200	Aerial application
	Vermont	VT-770005	BENLATE + Captan	
Apricots	California	CA-770003	BENLATE	
	New York	NY-810015	BENLATE + Captan	
	Washington	WA-790083	BENLATE	
Asparagus	North Carolina	NC-830005	BENLATE	Seed treatment
	Oregon	OR-780055	BENLATE	Pre-plant crown dips
	Washington	WA-770040	BENLATE	Seed treatment
	Washington	WA-770041	BENLATE	
Avocados	Puerto Rico	PR-820008	BENLATE	
Beans, Snap	Idaho	ID-810039	BENLATE	White mold and sprinkler irrigation
	Oregon	OR-780021	BENLATE	Planter Box treatment
	Oregon	OR-790030	BENLATE	In-furrow treatment
	Oregon	OR-810049	BENLATE	White mold and sprinkler irrigation
	Washington	WA-810035	BENLATE	Application through irrigation systems
Brassica Crops	Arizona	AZ-790004	BENLATE	Added diseases controlled
	California	CA-820079	BENLATE	
	New York	NY-820012	BENLATE	
	Oregon	OR-820046	BENLATE	Seed borne
	Washington	WA-790016	BENLATE	Back leg
	Washington	WA-820037	BENLATE	Seed crop only
Celery	California	CA-800098	BENLATE	Increase rate
Citrus	Florida	FL-770050	BENLATE + oil	Tank mixtures
Conifers	Delaware	DE-810002	BENLATE	Pine
	Idaho	ID-800059	BENLATE	Nurseries, greenhouses & seed orchards
	Maryland	MD-810012	BENLATE	Pine
	New Jersey	NJ-800007	BENLATE	Austrian pine
	New York	NY-810005	BENLATE	Pine
	Oregon	OR-770041	BENLATE	Nurseries, greenhouses and seed orchards
	Oregon	OR-800080	BENLATE	Douglas fir
	Oregon	OR-800082	BENLATE + Manzate 200	Douglas fir
	Pennsylvania	PA-800001	BENLATE	Red & Austrian pine
	Pennsylvania	PA-810012	BENLATE	Douglas fir
	Washington	WA-770034	BENLATE	Nurseries, greenhouses and seed orchards
	Washington	WA-800055	BENLATE	Douglas fir
	Washington	WA-800057	BENLATE + Manzate 200	Douglas fir
Ginger	Hawaii	HI-790001	BENLATE	Root seed piece treatment
Grapes	California	CA-770533	BENLATE	Tank mixtures
	California	CA-800106	BENLATE	Eutypa dieback
	Florida	FL-780051	BENLATE	Added diseases
	Michigan	MI-820005	BENLATE	
	New York	NY-810025	BENLATE	
	Washington	WA-820002	BENLATE	
Nectarines	California	CA-770003	BENLATE	
	Florida	FL-800026	BENLATE + sulfur	
	Georgia	GA-770004	BENLATE + sulfur	
	Georgia	GA-770005	BENLATE + Captan	
	Michigan	MI-770002	BENLATE + Captan	
	New Jersey	NJ-790014	BENLATE + Captan	Tank mixtures
	New Jersey	NJ-790015	BENLATE + Manzate D	
	New Jersey	NJ-790016	BENLATE + sulfur	
	New York	NY-810015	BENLATE + Captan	
	Pennsylvania	PA-780008	BENLATE + Captan	
	South Carolina	SC-770002	BENLATE + Captan	
	Virginia	VA-800031	BENLATE + Captan	
	Virginia	VA-800030	BENLATE + sulfur	
	Washington	WA-790083	BENLATE	
	West Virginia	WV-800006	BENLATE + Captan	
	West Virginia	WV-800007	BENLATE + Manzate D	
	West Virginia	WV-800008	BENLATE + sulfur	
Onions	California	CA-760211	BENLATE	Seed crop only
Peaches	California	CA-770003	BENLATE	
	Colorado	CO-780017	BENLATE	
	Florida	FL-800026	BENLATE + sulfur	
	Florida	FL-800027	BENLATE + Manzate D	
	Georgia	GA-770004	BENLATE + sulfur	
	Georgia	GA-770005	BENLATE + Captan	
	Illinois	IL-810010	BENLATE	

SPECIAL LOCAL NEED REGISTRATIONS [24(c)] FOR BENLATE® FUNGICIDE*

CROP	STATE	EPA SLN. NO.	PRODUCTS	REMARKS
	Michigan	MI-770002	BENLATE + Captan	
	Mississippi	MS-770004	BENLATE + Captan	
	New Jersey	NJ-790014	BENLATE + Captan	Tank mixtures only
	New Jersey	NJ-790015	BENLATE + Manzate D	
	New Jersey	NJ-790016	BENLATE + sulfur	
	New Jersey	NJ-810008	BENLATE	
	New York	NY-810015	BENLATE + Captan	
	New York	NY-810028	BENLATE	
	Pennsylvania	PA-780008	BENLATE + Captan	
	South Carolina	SC-770001	BENLATE + Manzate D	
	South Carolina	SC-770002	BENLATE + Captan	
	Virginia	VA-800030	BENLATE + sulfur	
	Virginia	VA-800031	BENLATE + Captan	
	Washington	WA-790083	BENLATE	
	West Virginia	WV-800006	BENLATE + Captan	
	West Virginia	WV-800007	BENLATE + Manzate D	
	West Virginia	WV-800008	BENLATE + sulfur	
Peanuts	New Mexico	NM-810019	BENLATE	
	North Carolina	NC-810031	BENLATE + sulfur	
	Oklahoma	OK-800014	BENLATE + Manzate 200	Application through sprinkler irrigation
	Texas	TX-770007	BENLATE + Manzate 200	
	Virginia	VA-770006	BENLATE + sulfur	Tank mixtures
	Virginia	VA-790008	BENLATE + Manzate 200	Provides for an increase in rate of BENLATE in tank mixture
Pecans	Georgia	GA-800025	BENLATE + Du-Ter	
	Georgia	GA-800026	BENLATE + Super Tin 4L	
Plums	California	CA-770003	BENLATE	
	Florida	FL-800026	BENLATE + sulfur	
	Georgia	GA-770004	BENLATE + sulfur	
	Georgia	GA-770005	BENLATE + Captan	
	Michigan	MI-780001	BENLATE + Captan 50WP	Tank mixtures
	Michigan	MI-770002	BENLATE + Captan	
	New York	NY-810015	BENLATE + Captan	
	Pennsylvania	PA-780008	BENLATE + Captan	
	South Carolina	SC-770002	BENLATE + Captan	
	Virginia	VA-800031	BENLATE + Captan	
	Washington	WA-790083	BENLATE	
	West Virginia	WV-800006	BENLATE + Captan	
Prunes	California	CA-770003	BENLATE	
	Michigan	MI-770002	BENLATE + Captan	
	Michigan	MI-780001	BENLATE + Captan	Tank mixtures
	New York	NY-810015	BENLATE + Captan	
	Washington	WA-790083	BENLATE	
Rice	Arkansas	AR-810010	BENLATE	
	Louisiana	LA-800020	BENLATE	Changes in rate and timing and additional diseases controlled
	Louisiana	LA-830001	BENLATE	
	Texas	TX-770004	BENLATE	
Seed Crops	California	CA-820120	BENLATE	
Soybeans	Illinois	IL-820016	BENLATE	
	Iowa	IA-820002	BENLATE	For seed beans
	Kentucky	KY-800021	BENLATE	For seed beans
	Louisiana	LA-820017	BENLATE	Aerial blight
	Missouri	MO-820022	BENLATE	
	New York	NY-780011	BENLATE + ARASAN 50-RED	Dry seed treatment
	Ohio	OH-780005	BENLATE	For seed beans
Strawberries	California	CA-770539	BENLATE	Tank mixtures
	Delaware	DE-820011	BENLATE	
	Michigan	MI-780001	BENLATE + Captan	Tank mixtures
	Michigan	MI-820005	BENLATE + Captan	Tank mixtures
	Missouri	MO-810017	BENLATE + Captan	
	New Jersey	NJ-800006	BENLATE	Verticellium wilt
Stone Fruits	California	CA-770003	BENLATE	
	Michigan	MI-770002	BENLATE + Captan	
	Pennsylvania	PA-780008	BENLATE + Captan	
	Virginia	VA-800031	BENLATE + Captan	
	Washington	WA-790083	BENLATE	
	West Virginia	WV-800006	BENLATE + Captan	
Sugarcane	Puerto Rico	PR-780005	BENLATE	Seed piece treatment
Wheat	Oregon	OR-840031	BENLATE + Manzate 200 FLOWABLE	Tank mixture

* Copies of label available upon request to nearest DuPont Agricultural Chemicals Office. For distribution and use only within the state specified.

BENLATE® PNW

FUNGICIDE

ACTIVE INGREDIENT:

Benomyl [Methyl 1-(butylcarbamoyl)-2-benzimidazolecarbamate] 50%
INERT INGREDIENTS 50%

TOTAL 100%

U.S. Pats. 3,541,213 & 3,631,176
EPA Reg. No. 352-434

Keep Out of Reach of Children

PRECAUTIONARY STATEMENTS

HAZARDS TO HUMANS

CAUTION!

MAY IRRITATE EYES, NOSE, THROAT, AND SKIN.

Avoid breathing dust or spray mist. Avoid contact with skin, eyes, and clothing.

Wash thoroughly after using.

This product may cause a temporary allergic skin reaction in a few susceptible persons. This condition should be treated as an allergic dermatitis. There is no evidence of after effects or permanent injury. If sensitivity occurs, protective clothing should be worn during periods of exposure such as during application or when contacting treated foliage.

First Aid: In case of contact, flush skin or eyes with plenty of water; for eyes, get medical attention.

Aerial Mixer-Loaders: Wear a cloth or disposable paper dust mask during handling and mixing as this product may be harmful if inhaled.

ENVIRONMENTAL HAZARDS

This product is toxic to fish. Do not apply directly to water or wetlands. Do not apply where runoff is likely to occur.

Do not apply when weather conditions favor drift from areas treated.

PHYSICAL OR CHEMICAL HAZARDS

Keep away from fire or sparks.

NOTICE OF WARRANTY

Du Pont warrants that this product conforms to the chemical description on the label thereof and is reasonably fit for purposes stated on such label only when used in accordance with the directions under normal use conditions. It is impossible to eliminate all risks inherently associated with the use of this product. Crop injury, ineffectiveness or other unintended consequences may result because of such factors as weather conditions, presence of other materials, or the manner of use or application, all of which are beyond the control of Du Pont. In no case shall Du Pont be liable for consequential, special or indirect damages resulting from the use or handling of this product. All such risks shall be assumed by the Buyer. DU PONT MAKES NO WARRANTIES OF MERCHANTABILITY OR FITNESS FOR A PARTICULAR PURPOSE NOR ANY OTHER EXPRESS OR IMPLIED WARRANTY EXCEPT AS STATED ABOVE.

STORAGE AND DISPOSAL

STORAGE Never allow BENLATE PNW to become wet during storage. This may lead to certain chemical changes which will reduce the effectiveness of BENLATE PNW as a fungicide. Keep container tightly closed when not in use. Store product in original container only, away from other pesticides, fertilizer, food or feed.

DISPOSAL Do not contaminate water, food, or feed by storage or disposal. Wastes resulting from the use of this product may be disposed of on site or at an approved waste disposal facility. Completely empty bag into application equipment. Then dispose of empty bag in a sanitary landfill or by incineration, or, if allowed by state and local authorities, by burning. If burned, stay out of smoke.

NOTICE

Do not enter treated areas without protective clothing until sprays have dried. Because certain states may require more restrictive reentry intervals for various crops treated with this product, consult your State Department of Agriculture for further information.

Written or oral warnings must be given to workers who are expected to be in a treated area or in an area about to be treated with this product. When oral warnings are given, warnings shall be given in a language customarily understood by workers. Oral warnings must be given if there is reason to believe that written warnings cannot be understood by workers. Warnings must include the following information: "CAUTION. Area treated with BENLATE PNW on (date of application). Do not enter without appropriate protective clothing until sprays have dried. In case of contact, flush skin or eyes with plenty of water; for eyes, get medical attention."

DIRECTIONS FOR USE

It is a violation of Federal law to use this product in a manner inconsistent with its labeling.

Du Pont BENLATE PNW should be used only in accordance with recommendations on this label or in separate published Du Pont recommendations available through local dealers.

Du Pont will not be responsible for losses or damages resulting from use of this product in any manner not specifically recommended by Du Pont. User assumes all risk associated with such nonrecommended use.

BENLATE PNW is a systemic fungicide recommended for the control of powdery mildew of cucumbers. If treatment is not effective following use of BENLATE PNW as recommended, a resistant strain of the fungus may be present. If treatment is ineffective due to the presence of a benomyl-resistant strain, then neither BENLATE PNW nor any other benzimidazole or thiophanate type fungicide will effectively control this disease; consideration should be given to prompt use of other types of suitable fungicides.

The repeated exclusive use of BENLATE PNW may lead to buildup of resistant strains of fungi and loss of disease control. A spray program using other fungicides may delay resistant strain buildup. Consult your state extention specialist or official state recommendations for guidance.

NOTE: Do not tank mix or alternate BENLATE PNW with benzimidazole or thiophanate products such as Mertect®[1] or Topsin®[2].

Add required amount of BENLATE PNW to necessary volume of water in spray tank agitated by hydraulic or mechanical means; continuous agitation is required to keep the material in suspension. Do not tank mix BENLATE PNW with lime or alkaline pesticides such as Bordeaux mixture or lime sulfur.

Observe all cautions and limitations on labeling of all products used in mixtures.

CUCUMBERS: Powdery mildew—Apply 1/4 to 1/2 lb. BENLATE PNW per acre. Begin applications when plants begin to run or when disease first appears and repeat at 7- to 14-day intervals as needed.

NOTICE TO BUYER—Purchase of this material does not confer any rights under patents of countries outside of the United States.

[1] Registered trademark of Merck & Co.

[2] Registered trademark of Nippon Soda Co.

3-84A 3/9/84 BO-9-118A

SUPPLEMENTAL LABELING

BENLATE® PNW

EPA REG. NO. 352-434

This page contains new or supplemental instructions for use of this product which may not appear on the package label. Follow the instructions carefully.

WHEAT (Washington, Oregon, Idaho and Utah): BENLATE PNW for control of pseudocercosporella foot rot of wheat.

DIRECTIONS FOR USE

It is a violation of federal law to use this product in a manner inconsistent with its labeling.

Du Pont BENLATE PNW Fungicide is recommended for use as a foliar spray for the control of Pseudocercosporella Foot Rot (Strawbreaker Foot Rot), in the states of Washington, Utah, Oregon and Idaho.

HOW TO USE

Apply 1 to 2 lbs. of BENLATE PNW per acre, using sufficient water to provide thorough and uniform coverage. For aerial application, use 5 to 10 gals. of water per acre. For fall-seeded winter wheat, make a single application when wheat is tillered but before elongation has occurred. For best results, apply in the spring. Use the higher dosage rate under conditions of severe disease pressure.

Add the required amount of BENLATE PNW to the necessary volume of water in the spray tank agitated by hydraulic or mechanical means; continuous agitation is required to keep the material in suspension. Do not tank BENLATE PNW with lime or alkaline pesticides such as Bordeaux mixture or lime sulfur.

NOTE: Do not apply within 21 days of harvest. Do not allow livestock to forage in treated fields.

Use with Fertilizer and Herbicide Tank Mixtures

BENLATE PNW may be applied in mixtures with a 50% solution of Urea Ammonium Nitrate (UAN) containing 30 to 32% nitrogen (Uran[1], Solution 32, etc.) and with certain herbicides, including Bronate[2] (bromoxynil), Chiptox[2] (MCPA amine), Dow Formula 40[3] amine, Glean®[4], Lexone® DF[4], Igran[5] and others. The mixing ability of these combinations will be improved by adding tank mixture compatibility agents such as Compex®[6]. Follow the directions on the container.

For compatibility of other products in mixtures of BENLATE PNW and UAN fertilizer, check by the following procedure:

1. Put 1 pint of UAN fertilizer solution (at concentration to be used) in a quart jar.
2. Add 2 level teaspoons of BENLATE PNW to 1 pint of UAN fertilizer solution.
3. Close jar and shake well.
4. Add the other products separately to the BENLATE PNW to 1 pint of UAN fertilizer solution mixture. The quantity of product to add depends on the recommended application rate of that product. A general guide is: 1 level teaspoon of powder in a pint equals about 2 to 3 lbs. per 100 gals.; 1 teaspoon of liquid in a pint equals about 1 gallon per 100 gallons.
5. Close jar and shake well after adding each ingredient to the mixture.
6. Watch mixture for several seconds; check again in 30 minutes. Some slight settling of wettable powder may occur after 30 minutes. This is normal and should resuspend after shaking.
7. Do not use mixture if it has excessive foam, gel, lumps or shows separation which cannot be readily suspended.

NOTE: When checking the compatibility of water-based spray mixtures, use the above procedure but substitute water for UAN solution.

Provided the above procedure shows the mixture to be compatible, prepare the tank mixture as follows: Add the UAN fertilizer solution to the spray tank first, filling the sprayer at least half full. With the agitator running, add the required amount of BENLATE PNW and allow sufficient time to mix thoroughly. Then add the required amount of herbicide(s) and mix thoroughly. When mixing more than one herbicide, mixing order is important. Always mix wettable powder and dry flowable formulations first. Liquid formulations should be added last (flowable and emulsifiable concentrates). Mixtures with some herbicides may be improved by making a 1:2 slurry (product:water) prior to adding to the spray tank.

Continuous agitation is required to keep the materials in suspension. Prolonged periods without agitation may result in settling and accumulation of material on the bottom or sides of the tank. For optimum mixing and spraying, tanks should be sprayed completely empty at least once daily and preferably before each new tankful is mixed.

[1] Registered trademark of Allied Chemical Company
[2] Registered trademark of Rhone-Poulenc
[3] Registered trademark of Dow Chemical Company
[4] Registered trademark of E. I. du Pont de Nemours & Co., Inc.
[5] Registered trademark of Ciba-Geigy Corporation
[6] Registered trademark of Kalo Laboratories

091184

IMPORTANT

BEFORE USING BENLATE PNW OR ANY OF THE PRODUCTS USED IN MIXTURES, READ AND CAREFULLY OBSERVE THE CAUTIONARY STATEMENTS AND ALL OTHER INFORMATION APPEARING ON THE RESPECTIVE PRODUCT LABELS.

GLEAN®

HERBICIDE
DRY FLOWABLE

ACTIVE INGREDIENT:
Chlorsulfuron
2-Chloro-N-[(4-methoxy-6-methyl-1,3,5-triazin-2-yl)aminocarbonyl]benzenesulfonamide 75%
INERT INGREDIENTS 25%
TOTAL 100%

EPA Reg. No. 352-404 U.S. Pat. 4,127,405

Keep Out of Reach of Children

PRECAUTIONARY STATEMENTS

HAZARDS TO HUMANS

CAUTION!

MAY IRRITATE EYES, NOSE, THROAT, AND SKIN.

Avoid breathing dust or spray mist. Avoid contact with skin, eyes, and clothing. In case of contact with eyes, immediately flush with plenty of water. Get medical attention if irritation persists. Wash thoroughly after handling. Remove and wash contaminated clothing before reuse.

ENVIRONMENTAL HAZARDS

Do not apply directly to any body of water. Do not contaminate water by cleaning of equipment or disposal of wastes.

IMPORTANT INFORMATION—READ BEFORE USING

GLEAN is recommended for use only on land dedicated to the long term production of wheat, barley, or spring oats. GLEAN should not be used in any area where crop rotation is frequently practiced except as indicated for the states of TX, OK, KS, LA and AR. (See Crop Rotation Guidelines.)

In areas having soil pH greater than 7.0, a short growing season, prolonged periods of low soil temperature, and low annual rainfall, GLEAN can remain in the soil for 3 to 4 years or more and cause severe injury to crops other than wheat, barley, or spring oats.

Before using GLEAN, carefully consider your crop rotation plans and options. For rotational flexibility, do not treat all your wheat, barley, spring oats, or fallow acres. See Recropping Intervals and Crop Rotation Guidelines for details.

Injury to or loss of desirable trees or vegetation may result from failure to observe the following: Do not apply, drain or flush equipment on or near desirable trees or other plants, or on areas where their roots may extend, or in locations where the chemical may be washed or moved into contact with their roots. Do not use on lawns, walks, driveways, tennis courts, or similar areas. Prevent drift of spray to desirable plants. Do not contaminate any body of water, including irrigation water that may be used on other crops. Carefully observe sprayer clean-up instructions, as spray tank residue may damage crops other than wheat, spring oats, or barley.

GENERAL INFORMATION

GLEAN Herbicide is a dry flowable granule to be mixed in water and applied as a uniform broadcast spray for selective weed control in wheat (including durum), spring oats, and barley. GLEAN is also recommended for fallow weed control (to reduce tillage) preceding wheat. It is noncorrosive, nonflammable, nonvolatile, does not freeze and has no grazing restriction.

Prior to using GLEAN, careful consideration should be given to crop rotation plans. Crops other than wheat, spring oats, and barley can be extremely sensitive to low concentrations of GLEAN in the soil.

In the states of CA, WA, OR, ID, and UT, GLEAN is recommended for use on land having a soil pH of 7.5 or lower and dedicated to the long term production of wheat, barley, or spring oats. The soil residual activity of GLEAN can injure crops other than wheat, barley, or spring oats for 2 to 3 years or more. GLEAN should not be used on soils higher than pH 7.5 as extended soil residual activity could adversely affect crop rotation options beyond normal intervals. (See Minimum Recropping Intervals and Crop Rotation Guidelines.)

In the states of MT, ND, SD, WY, and MN, GLEAN is recommended for use on land having a soil pH of 7.5 or less and dedicated to the long term production of wheat, barley, or spring oats. The soil residual activity of GLEAN can injure crops other than wheat, barley, or spring oats for 3 to 4 years or more. GLEAN should not be used on soils higher than pH 7.5 as extended soil residual activity could adversely affect crop rotation options beyond normal intervals. (See Minimum Recropping Intervals and Crop Rotation Guidelines.)

In the states of TX, CO, NM, LA, AR, KS, OK, and NE, GLEAN is recommended for use on land having a soil pH of 7.9 or lower. Unless otherwise specified, in the Crop Rotation Guidelines section of this label, in the low rainfall, dryland cropping areas of TX, NM, CO, KS, OK, and NE, GLEAN should only be used on land dedicated to the long term production of wheat, barley, or spring oats. The soil residual activity of GLEAN can injure crops other than wheat, barley, or spring oats for 2 to 4 years or more. GLEAN should not be used on soils higher than pH 7.9 as extended soil residual activity could adversely affect crop rotation options beyond normal intervals. (See Minimum Recropping Intervals and Crop Rotation Guidelines.)

Rainfall, soil temperature, and soil pH are important factors affecting GLEAN breakdown in soil. GLEAN breakdown is more rapid under conditions of low soil pH, high soil temperature, and moist soil. The breakdown process is slow under conditions of high soil pH, low soil temperature and dry soil.

IMPORTANT: UNLESS OTHERWISE SPECIFIED IN THE "CROP ROTATION GUIDELINES" SECTION OF THIS LABEL, land previously treated with GLEAN cannot be rotated to crops other than wheat, spring oats, or barley until a FIELD BIOASSAY confirms that residues of GLEAN are not present. A FIELD BIOASSAY involves growing a test strip of the crop(s) intended for production the following year in fields previously treated with GLEAN. Crop response will indicate whether or not to rotate to the crop(s) used in the test strip. See separate Du Pont bulletin 'GLEAN Field Bioassay". Failure to follow these instructions could result in injury to subsequent crops.

IMPORTANCE OF APPLICATION TIMING RELATIVE TO GROWING CONDITIONS AND RAINFALL

Growing Conditions—Postemergence applications to weeds are most effective when GLEAN is applied to young, actively growing weeds which are less than 2" tall or across. Warm, moist growing conditions promote active weed growth and enhance the activity of GLEAN by allowing maximum foliar uptake and contact activity. If cold, dry conditions exist, delay postemergence treatment until weather conditions promote active weed growth. Avoid postemergence applications to weeds which are not actively growing due to adverse weather conditions. Weeds hardened off by cold weather or drought stress may not be controlled.

Rainfall After Treatment—Rainfall after treatment will affect GLEAN performance when applied postemergence or preemergence to weeds.

Postemergence treatments control or suppress weeds through both foliar and root uptake. To maximize GLEAN activity on existing weeds, sufficient rainfall is needed soon after treatment to move GLEAN into the weed root zone before weeds develop an established root system and grow beyond the seedling stage.

For best preemergence results, it is important to apply GLEAN when you can expect at least $\frac{1}{2}$ to 1" (clay soils may require more) of rain or sprinkler irrigation to move GLEAN 2 to 3" deep into the soil profile before weed seeds germinate or develop an established root system. Weeds that germinate after treatment and develop an established root system before rainfall moves GLEAN into the soil profile may not be controlled.

When weed emergence is uneven, control of weeds that germinate after treatment will be dependent on the timing and amount of rainfall following application. Sufficient rainfall is needed to move GLEAN 2 to 3" deep into the weed root zone before weeds that germinate after treatment can develop an established root system. When favorable growing conditions exist, rainfall may be needed within a few days after treatment.

GLEAN rapidly inhibits growth of susceptible weeds. However, typical symptoms (discoloration) of dying weeds may not be noticeable for 1 to 3 weeks after application depending on growing conditions and weed susceptibility.

Degree of control, and duration of effect depend on: a) rate use, b) weed spectrum, c) weed size, d) growing conditions at and following time of treatment, e) soil pH, f) precipitation, g) soil organic matter, and h) length of growing season.

NOTICE OF WARRANTY

DuPont warrants that this product conforms to the chemical description on the label thereof and is reasonably fit for purposes stated on such label only when used in accordance with directions under normal use conditions. It is impossible to eliminate all risks inherently associated with the use of this product. Crop injury, ineffectiveness, or other unintended consequences may result because of such factors as weather conditions, presence of other materials, or the manner of use or application, all of which are beyond the control of DuPont. In no case shall DuPont be liable for consequential, special or indirect damages resulting from the use or handling of this product. All such risks shall be assumed by the buyer. DUPONT MAKES NO WARRANTIES OF MERCHANTABILITY OR FITNESS FOR A PARTICULAR PURPOSE NOR ANY OTHER EXPRESS OR IMPLIED WARRANTY EXCEPT AS STATED ABOVE.

DIRECTIONS FOR USE

It is a violation of federal law to use this product in a manner inconsistent with its labeling.

GLEAN Herbicide should be used only in accordance with recommendations on this label or in separate published DuPont recommendations available through local dealers.

DuPont will not be responsible for losses or damages resulting from the use of this product in any manner not specifically recommended by DuPont. User assumes all risks associated with such nonrecommended use.

MAXIMUM USE RATES AND SOIL PH LIMITATIONS

In the states of CA, WA, OR, ID, UT, MT, ND, SD, MN and WY, the maximum use rate is $\frac{1}{3}$ oz/A on soils having a pH of 6.6 to 7.5. The maximum use rate on soils with a pH of 6.5 or less is $\frac{1}{2}$ oz/A. Do not use GLEAN on soils with a pH greater than 7.5. Do not apply more than $\frac{1}{2}$ oz/A per crop or fallow period on soils pH 6.5 or lower. Do not apply more than $\frac{1}{3}$ oz/A per crop or fallow period on soils pH 6.6 to 7.5. See Minimum Recropping Intervals and Crop Rotation Guidelines section below.

In the states of TX, CO, NM, LA, AR, KS, OK and NE, the maximum use rate is $\frac{1}{2}$ oz/A on soils having a pH of 7.9 or less. Do not use on soils with a pH greater than 7.9. Do not apply more than $\frac{1}{2}$ oz/A per crop or fallow period. See Minimum Recropping Intervals and Crop Rotation Guidelines section below.

Note: Prior to use of GLEAN, take representative soil samples at 0–4" depth and determine soil pH by laboratory analysis using a 1:1, soil:water suspension. Consult local extension publications for recommended soil sampling procedures.

WEED CONTROL/USE RATE TABLE

For maximum weed control or suppression, always use the highest recommended rate for your area, soil pH, and weed problem. Do not use less than $\frac{1}{6}$ oz/A.

For best results with postemergence application, apply GLEAN to small (less than 2" tall or across), actively growing weeds. Add a surfactant of at least 80% active ingredient at the rate of 1-2 quarts/100 gallons of spray solution. The higher rate of surfactant is particularly useful with spray volumes of 5 GPA or less and when using low rates of GLEAN.

Sufficient rainfall after preemergence or postemergence treatment is necessary to move GLEAN 2–3" into the weed root zone before weed seeds germinate and develop an established root system or existing weeds grow beyond the seedling stage. In most areas, fall treatments provide the best opportunity for rainfall activation and most consistent residual weed control. Late spring applications may not receive enough rainfall after treatment resulting in poor weed control.

Weeds Controlled or Suppressed*		Rate Ounces Per Acre	Acres Treated Per 18-Ounce Container
Blue mustard Conical catchfly Field pennycress Hempnettle Henbit Jim Hill mustard Lambsquarters Mayweed Miners Lettuce Pepperweed	Pineappleweed Prostrate pigweed Redroot pigweed Shepherdspurse Smooth pigweed Tansy mustard Treacle mustard Wild mustard	$\frac{1}{6}$ thru $\frac{1}{3}$ Ounce	108 thru 54 Acres
Above list plus: Annual ryegrass* Bedstraw* Bur beakchervil Buttercup Canada thistle* Common chickweed Common groundsel* Corn spurry Cow cockle False chamomile False flax Fiddleneck (tarweed) Flixweed* Giant foxtail* Green foxtail* (pigeongrass) Gromwell* Kochia* Ladysthumb	Mouseear chickweed Pennsylvania smartweed* Prickly lettuce* Prostrate knotweed* Purslane (common) Russian thistle* Speedwell* Sunflower White cockle Wild buckwheat* Wild carrot Wild garlic** Wild onion** Wild radish Wild turnip Yellow starthistle Yellow foxtail*	$\frac{1}{3}$ thru $\frac{1}{2}$ Ounce	54 thru 36 Acres

* Weeds marked with an asterisk are suppressed. Weed suppression is a visual reduction in weed competition (reduced population and/or vigor) as compared to an untreated area. Degree of suppression will vary with rate used, size of weeds, and environmental conditions following treatment. See Specific Weed Problem section concerning Canada thistle, Russian thistle, wild buckwheat, kochia, annual ryegrass and foxtail.
** Aerial bulblet control.

SPECIFIC WEED PROBLEMS

Canada thistle: Apply GLEAN plus surfactant after the majority of thistles have emerged and while they are small (rosette stage to 4–6" tall) but actively growing. A single application will effectively inhibit the ability of Canada thistle to compete with the crop. For maximum long-term effect, yearly treatment may be required.

Annual ryegrass: (northeast Texas/southeast Oklahoma/ northwest Louisiana/southwest Arkansas): For best results, apply GLEAN at $\frac{1}{2}$ oz/A preemergence to ryegrass. $\frac{1}{2}$ to 1" of rainfall is needed to move GLEAN into the weed root zone prior to ryegrass emergence. Remove grazing cattle during wet (muddy) field conditions to avoid disturbing the herbicide barrier.

Foxtail (green, yellow and giant): In ND, SD, MT and MN fall applications at $\frac{1}{3}$ oz/A prior to planting spring wheat will control these foxtail species. Applications made in the spring give suppression. For best results in the spring, apply preemergence to the foxtail at $\frac{1}{3}$ to $\frac{1}{2}$ oz/A. Postemergence applications at $\frac{1}{3}$ to $\frac{1}{2}$ oz/A should be made with a surfactant no later than the 2 leaf stage. $\frac{1}{2}$ to 1" of rainfall is needed to move GLEAN into the weed root zone before the foxtail is beyond the 2 to 3 leaf stage.

Wild buckwheat: In ND, SD, MT, and MN, fall applications at $\frac{1}{3}$ oz/A prior to planting spring wheat will control wild buckwheat. Applications made in the spring give suppression. For best results in the spring, apply GLEAN with surfactant just after seedlings have emerged and are actively growing. Late spring applications may not receive enough rainfall after treatment, resulting in poor residual weed suppression.

In OK, TX and KS, best results are from fall or early spring applications made preemergence to wild buckwheat. Post emergence application should be made with surfactant just after seedlings have emerged and are actively growing.

In all other states, fall applications should provide best results. If treatment is delayed until spring, apply GLEAN with surfactant just after seedlings have emerged and are actively growing.

Rainfall immediately after a postemergence treatment may wash GLEAN off weed foliage and result in poor postemergence weed control.

Russian thistle, kochia, and sunflower: Winter wheat/fallow—For best results in NM, OK (panhandle) and TX (panhandle), apply preemergence in early spring (early as possible) to allow for timely and adequate rainfall to move GLEAN into the weed root zone before weed seeds germinate or develop an established root system.

In all other states, fall applications provide best results. If treatment is delayed until spring, apply preemergence or early postemergence. Postemergence applications should be made with surfactant, just after seedlings have emerged and are actively growing. Thorough coverage is important. Spring applications may not provide adequate residual control due to insufficient and/or untimely rainfall activation. Rainfall immediately after application may wash GLEAN off weed foliage resulting in poor postemergence weed control.

Spring wheat—(including Durum)—Fall application of GLEAN at $\frac{1}{3}$ oz/A prior to planting spring wheat will control these weeds in ND, SD, MT and MN. Applications made early in the spring at $\frac{1}{3}$ to $\frac{1}{2}$ oz/A will provide suppression. For best results in the spring, apply post emergence with surfactant just after seedlings have emerged and are actively growing. Thorough coverage is important. Rainfall immediately after application may wash GLEAN off the weed foliage resulting in poor postemergence weed control. Preemergence applications require rainfall soon after treatment to move GLEAN into the weed root zone before weed seeds germinate or develop an established root system.

Flixweed, Tansy Mustard: East of the Rockies, apply $\frac{1}{6}$ to $\frac{1}{3}$ oz/A for control of these weeds. West of the Rockies, $\frac{1}{3}$ to $\frac{1}{2}$ oz/A will provide suppression. For best results with post emergence applications, apply GLEAN with a surfactant when weeds are small (less than 2" tall or across) and actively growing. If weeds are not actively growing due to adverse weather conditions (cold, dry weather), delay application until conditions allow active growth to resume. For best results with a fall or preemergence application, GLEAN should be applied at highest recommended rate to provide residual activity.

Wild Garlic/Wild Onion: GLEAN will provide aerial bulblet control only.

Wild Radish: Postemergence application will provide best results.

Prostrate Knotweed: For best results apply in the fall.

TANK MIXTURES FOR SPECIFIC WEED COMPLEXES

Du Pont Karmex® Herbicide plus GLEAN: Where gromwell, wild buckwheat, annual ryegrass and annual bluegrass are the main weed problems in the Pacific Northwest, apply Karmex at 1 to 1½ pounds per acre with GLEAN at ⅙ to ⅓ ounce per acre to increase the spectrum of weeds controlled. Apply preemergence, or postemergence to actively growing weeds less than 2" tall or across. ½ to 1" of rainfall is needed within 1 to 2 weeks after application. Follow all restrictions on the Karmex label. (Do not use Karmex plus GLEAN in California.)

Du Pont Lexone® DF Herbicide plus GLEAN: Where cheatgrass, downy brome, gromwell, speedwell and jagged chickweed are the main problems in the Pacific Northwest, an application of LexoneDF at ⅓ to ½ pound per acre with GLEAN at ⅙ to ⅓ ounce per acre is recommended for best results. In TX, OK, and KS, LexoneDF at ⅓ to ½ pound per acre is recommended for downy brome and cheatgrass control and may be mixed with GLEAN at ⅙ to ⅓ ounce per acre to increase the spectrum of weeds controlled. Apply after wheat or barley is well tillered and has 2" secondary root system established throughout the field. ½ to 1" of rainfall is needed within 1 to 2 weeks of application. LexoneDF is recommended for use on winter wheat/barley only in the states of WA, OR, ID, UT, KS, OK, and TX. Follow all restrictions on the LexoneDF label. (Do not use LexoneDF plus GLEAN in California.)

MINIMUM RECROPPING INTERVAL AND ROTATION GUIDELINES FOR THE STATES OF CA, WA, OR, ID, UT, ND, SD, MN, WY

IMPORTANT

Recropping to Wheat, Spring Oats, and Barley
Recropping plans are determined by soil pH, rate of GLEAN applied and a minimum recropping interval. The minimum recropping interval is from time of last application to the anticipated date of planting.

Soil pH*	Use Rate oz./acre	Minimum Recropping Interval (Months) Wheat	Spring Oats/ Barley
6.5 or lower	⅙ thru ⅓	0	10
6.5 or lower	½	4	10
6.6 to 7.5	⅙ thru ⅓	0	16
7.5 or lower**	½**	4	16
above 7.5	Do not use	Not Applicable	

* Soil pH is to be determined by laboratory analysis using the 1:1, soil to water suspension method on representative soil samples taken at 0–4" depth. Consult local extension publications for recommended soil sampling procedures.

** Applicable to Oklahoma/Texas annual ryegrass areas only.

Rotation to Crops Other Than Wheat, Spring Oats, and Barley
The season before planting a crop other than wheat, spring oats or barley, a successful field bioassay must be completed. A successful field bioassay means growing to maturity a test strip of the crop(s) intended for production the following year, (See separate Du Pont bulletin, "GLEAN—Field Bioassay"). The field bioassay will detect small quantities of GLEAN which can remain in the soil and injure rotational crops. At soil pH of 6.5 or lower a 1 to 2 year interval or more may be required before completion of a successful field bioassay and subsequent planting of the next crop. At soil pH of 6.6 to 7.5 a 36- to 48-month or longer interval may be required.

For crop rotation flexibility, do not use GLEAN on all your wheat, spring oats, barley or fallow acreage.

MINIMUM RECROPPING INTERVALS AND CROP ROTATION GUIDELINES FOR THE STATES OF TX, CO, NM, LA, AR, KS, OK AND NE

IMPORTANT

Recropping to Wheat, Spring Oats and Barley
Recropping plans are determined by soil pH, rate of GLEAN applied and a minimum recropping interval. The minimum recropping interval is from time of last application to the anticipated date of planting.

Soil pH*	Use Rate oz/acre	Minimum Recropping Interval (Months) Wheat	Spring Oats	Barley
7.9 or lower	⅙ to ⅓	0	10	10
7.9 or lower	½	4	10	16
above 7.9	Do not use	—Not Applicable—		

* Soil pH is to be determined by laboratory analysis using the 1:1, soil to water suspension method on representative soil samples taken at 0–4" depth. Consult local extension publications for recommended soil sampling procedures.

Rotation to Crops Other Than Wheat, Spring Oats, and Barley
In dedicated wheat/fallow/wheat or continuous wheat production areas of Nebraska, Kansas, Oklahoma, Texas, Northeast New Mexico and Colorado, a successful field bioassay must be completed prior to rotating GLEAN treated fields to crops other than wheat, spring oats or barley, UNLESS OTHERWISE SPECIFIED under CROP ROTATION INTERVALS FOR SPECIFIC AREAS IN THE SOUTH CENTRAL STATES.

A successful field bioassay means growing to maturity a test strip of the crop(s) intended for production the following year. (See separate Du Pont bulletin, "GLEAN—Field Bioassay".) The field bioassay will detect small quantities of GLEAN which can remain in the soil and injure rotational crops. At soil pH of 6.5 or lower a 1 to 2 year interval or more may be required before completion of a successful field bioassay and subsequent planting of the next crop. At soil pH of 6.6 to 7.9 a 3 to 4 year or longer interval may be required.

For crop rotation flexibility, do not use GLEAN on all your wheat, spring oats, barley or fallow acreage.

CROP ROTATION INTERVALS FOR SPECIFIC AREAS IN THE SOUTH CENTRAL STATES

Texas:
In the indicated counties of eastern Texas on nonirrigated land, grain sorghum, soybeans, mungbeans, and cotton may be planted 14 months after a GLEAN application of ⅙ to ½ oz/A if accumulated rainfall is 25" or more during the crop rotation interval.

Archer	Kaufman
Bell	Lamar
Bosque	Limestone
Bowie	McLennan
Camp	Milam
Cass	Montague
Çlay	Morris
Collin	Navarro
Cooke	Palo Pinto
Coryell	Parker
Dallas	Rains
Delta	Red River
Denton	Robertson
Ellis	Rockwall
Falls	Somervell
Fannin	Tarrant
Franklin	Titus
Grayson	Upshur

Hill	Van Zandt
Hood	Wichita
Hopkins	Williamson
Hunt	Wise
Jack	Wood
Johnson	Young

In the indicated counties of central Texas on non-irrigated land, grain sorgum and cotton may be planted 14 months after a GLEAN application of $\frac{1}{6}$ to $\frac{1}{3}$ oz/A if accumulated rainfall is 25" or more during the crop rotation interval.* If use rate exceeds $\frac{1}{3}$ oz/A, a successful field bioassay must be completed before rotating to a crop other than wheat, barley or spring oats.

Baylor	Knox
Callahan	Shackelford
Eastland	Stephens
Foard	Throckmorton
Hardeman	Wilbarger
Haskell	

In counties other than those listed above, a field bioassay must be completed before rotating to any crop other than wheat, barley or spring oats.

Oklahoma:

In central and eastern Oklahoma (generally east of Highway 183) on non-irrigated land, grain sorghum, mungbeans, and cotton may be planted 14 months after a GLEAN application of $\frac{1}{6}$ to $\frac{1}{2}$ oz/A if accumulated rainfall is 25" or more during the crop rotation interval.*

In western Oklahoma (generally west of Highway 183 and east of the Panhandle) on non-irrigated land, grain sorghum and cotton may be planted 14 months after a GLEAN application of $\frac{1}{6}$ to $\frac{1}{3}$ oz/A if accumulated rainfall is 25" or more during the crop rotation interval.* If use rate exceeds $\frac{1}{3}$ oz/A, a successful field bioassay must be completed before rotating to a crop other than wheat, barley or spring oats.

In southeast Oklahoma on non-irrigated land, grain sorghum, soybeans, and cotton may be planted 14 months following a GLEAN application of $\frac{1}{6}$ to $\frac{1}{2}$ oz/A if accumulated rainfall is 25" or more during the crop rotation interval.*

In the Oklahoma panhandle, a field bioassay must be completed before rotating to any crop other than wheat, barley or spring oats.

Kansas:

In central Kansas (generally east of Highway 183 and west of Flinthills) on non-irrigated land, grain sorghum may be planted 14 months after a GLEAN application of $\frac{1}{6}$ to $\frac{1}{2}$ oz/A if accumulated rainfall is 25" or more during the crop rotation interval.*

In western Kansas (generally west of Highway 183) on non-irrigated land, grain sorghum may be planted 34 months after a GLEAN application of $\frac{1}{6}$–$\frac{1}{3}$ oz/A. If use rate exceeds $\frac{1}{3}$ oz/A, a successful field bioassay must be completed before rotating to a crop other than wheat, barley or spring oats.

Arkansas, Louisiana:

In southwest Arkansas and northwest Louisiana on non-irrigated land, grain sorghum, soybeans, and cotton may be replanted 14 months after a GLEAN application of $\frac{1}{6}$ to $\frac{1}{2}$ oz/A if accumulated rainfall is 25" or more during the crop rotation interval.*

* Should accumulated rainfall not be sufficient to meet the indicated amounts, do not rotate to the indicated crops until the following growing season.

** Do not rotate to crops other than those specified until a successful field bioassay has been conducted. Do not use on soils with a pH greater than 7.9.

TIMING OF APPLICATION FOR WHEAT, BARLEY, SPRING OATS, AND REDUCED TILLAGE FALLOW

• POSTEMERGENCE TO WINTER AND SPRING WHEAT (including durum*), SPRING OATS, AND BARLEY

Apply GLEAN ($\frac{1}{6}$–$\frac{1}{2}$ oz./A) in the fall or spring any time after wheat, spring oats, durum (varieties other than VIC) or barley is in the 2–3 leaf stage, but before boot stage. For maximum performance, apply to actively growing weeds before they are 2" tall or 2" in diameter prior to canopy closing in. Under cool, wet conditions apply to barley after tillering has begun.

***Note:** Apply to VIC durum after early tillering but before boot stage.

• PREEMERGENCE TO WHEAT (including durum)

Apply GLEAN ($\frac{1}{6}$–$\frac{1}{3}$ oz./A) after planting, but before crop emergence. For the best results on annual ryegrass, in northeast Texas and southeast Oklahoma, northwest Louisiana and southwest Arkansas GLEAN may be applied up to $\frac{1}{2}$ oz./A preemergence to wheat. Rainfall or sprinkler irrigation ($\frac{1}{2}$ to 1") within 2 weeks following treatment is necessary to activate GLEAN. Wheat must be planted at least 1" deep. For best results apply GLEAN uniformly to a smooth seedbed.

Do not make a preemergence treatment when cold and/or dry weather can delay seedling emergence and reduce seedling vigor. If these conditions exist, delay treatment until weather conditions allow active wheat growth and wheat is showing good vigor.

When environmental conditions cause delayed seedling emergence and/or poor seedling vigor, delay post treatment irrigation until after the wheat is actively growing and is showing good vigor.

Note: Do not apply preemergence to barley or spring oats as crop injury may result.

SPLIT-TREATMENT TO WHEAT

Apply GLEAN postplant preemergence plus postemergence, or early postemergence plus late postemergence to the crop. Allow at least 30 days between treatments. Do not apply more than the maximum use rate, per crop year, as indicated in the MAXIMUM USE RATES AND SOIL PH LIMITATIONS section of label. Do not make more than 2 treatments per crop. Apply before bootstage. Base recropping interval on date of last application and total amount of GLEAN used.

• FALL APPLICATION PRIOR TO PLANTING SPRING WHEAT (including durum)

In Montana, the Dakotas and Minnesota apply GLEAN ($\frac{1}{3}$ oz/A) in the fall to undisturbed stubble where straw is spread evenly, or after cultivation to a uniform soil surface. Shallow tillage, not more than 4" deep, may be done after application. In the spring use shallow tillage to prepare a seedbed. Do not moldboard plow. Fall application is not effective for Canada thistle emerging the following spring. See Canada thistle under SPECIFIC WEED PROBLEMS. Do not plant spring barley or spring oats after a fall application of GLEAN.

REDUCED TILLAGE FALLOW (preceding wheat)

Several options are available for use of GLEAN to reduce tillage in fallow. Use $\frac{1}{3}$ to $\frac{1}{2}$ oz./acre/fallow period. (See MAXIMUM USE RATE AND SOIL PH LIMITATIONS and MINIMUM RECROPPING INTERVAL AND CROP ROTATION GUIDELINES.) Application should be made before broadleaf weeds are 2" tall or across. If weed control is unsatisfactory because weeds were too large at application, or if weeds grow due to insufficient rainfall activation of GLEAN, a shallow cultivation is recommended. Rainfall following treatment to wet the soil 2-3" deep (usually $\frac{1}{2}$ to 1") is necessary to move GLEAN into the weed root zone before weed seeds germinate or existing weeds grow beyond the seedling stage.

1) Spring (in crop prior to fallow)—Apply GLEAN in the spring before wheat, spring oats, or barley are in the boot stage. This treatment is effective for post-harvest broadleaf weed control but may not provide weed control into the following spring.

2) Fall (post-harvest)—Apply GLEAN preemergence or early postemergence to the first flush of germinating weeds after harvest.

3) Spring (during fallow)—For best results apply preemergence early in the spring to ensure adequate rainfall activation prior to weed seed germination. Post emergence treatments should be applied just after most seedlings have emerged and are actively growing. When weed emergence is uneven, control of weeds that germinate after treatment will be dependent on the timing and amount of rainfall following application. Sufficient rainfall is needed to move

GLEAN 2-3" deep into the weed root zone before weeds that germinate after treatment can develop an established root system.

• **GLEAN PLUS ROUNDUP**[1]

GLEAN (1/3–1/2 oz./acre) plus Roundup (12–16 fl. oz./acre) applied as a tank mix is recommended for the control of emerged populations of broadleaf and grassy weeds. For best results, apply this mixture to young actively growing broadleaf weeds less than 2" tall or across while grassy weeds are 6" tall or less. GLEAN plus Roundup should be applied in 5–10 gallons spray volume per acre (GPA) with ground equipment using flat fan nozzles, or 3–5 GPA by air. If broadleaf and grassy weed stages are not appropriate for a tank mix application, GLEAN and Roundup should be applied separately as recommended for each product.

Follow all use instructions, warnings and precautions, and surfactant recommendations on the Roundup label.

SPRAY PREPARATION/TANK MIXTURES

Mix the proper amount of GLEAN into the necessary volume of water in the spray tank with the agitator running. Agitation is required for uniform mixing and application. Use GLEAN spray preparations immediately or product degradation may occur. Thoroughly reagitate before using.

SURFACTANT: Use a surfactant of at least 80% active ingredient in postemergence applications to weeds to improve wetting and/or contact activity of GLEAN. Add surfactant at the rate of 1 to 2 quarts/100 gallons of spray as the last ingredient. The higher rate of surfactant is particularly useful with spray volumes of 5 GPA or less and when using low rate of GLEAN.

LIQUID FERTILIZER: To apply GLEAN with liquid fertilizer, slurry the GLEAN in water; then thoroughly mix the slurry into the liquid fertilizer. Do not add a surfactant. Run a tank mix compatibility test before mixing GLEAN in fertilizer solutions.

TANK MIXTURES: Use a suitable registered companion herbicide if weeds larger than 2" tall or across are present or if weeds and grasses other than those listed for GLEAN are present or anticipated (follow manufacturer's label). GLEAN must be in suspension before adding the companion herbicide. Follow the surfactant recommendation on the companion herbicide.

EQUIPMENT—SPRAY VOLUMES

Apply using properly calibrated air or ground equipment. Select a spray volume and delivery system that will insure thorough coverage and a uniform spray pattern. For ground application, flat fan nozzles are recommended (minimum 3 GPA). When using flood jet or "Raindrop" nozzles, use higher spray volume (minimum 20 GPA) to ensure thorough coverage. Use at least 1 gallon spray volume per acre by air. Use 50-mesh screens or larger.

Use higher spray volumes to obtain better coverage, when the crop canopy is dense.

Continuous agitation is required to keep GLEAN in suspension. Avoid overlapping, and shut off spray booms while starting, turning, slowing or stopping, or injury to the crop may result.

Note: Do not allow spray to drift onto adjacent crops, or onto agricultural land scheduled to be planted to crops other than wheat, spring oats, or barley as injury to the crop may occur. Extreme care must be taken to prevent drift to desirable plants or non-target agricultural land.

SPRAYER CLEANUP

To avoid subsequent injury to crops other than wheat, spring oats, or barley, immediately after spraying thoroughly remove all traces of GLEAN from mixing and spray equipment as follows:

1) Drain tank; then flush tank, boom and hoses with clean water for a minimum of 10 minutes.
2) Fill the tank with clean water then add 1/2 gallon chlorine bleach (containing 5 1/4% sodium hypochlorite) per 100 gallons of water. Flush through boom and hoses, then allow to sit for 15 minutes with agitation; then drain.
3) Repeat Step 2.
4) Nozzles and screens should be removed and cleaned separately. To remove traces of chlorine bleach, rinse the tank thoroughly with clean water and flush through hoses and boom.

CAUTION: Do not use chlorine bleach with ammonia. All traces of liquid fertilizer containing ammonia, ammonium nitrate or ammonium sulphate must be rinsed from the mixing and application equipment with water before adding chlorine bleach solution. Failure to do so will release a gas with a musty chlorine odor which can cause eye, nose, throat and lung irritation. Do not clean equipment in an enclosed area.

PRECAUTIONS

Because varieties of wheat, spring oats, and barley differ in their tolerance to herbicides, limit first use of GLEAN to a small area prior to adoption as a field practice.

Do not apply GLEAN to wheat, spring oats, or barley that are stressed by severe weather conditions, drought, low fertility, water saturated soil, disease, or insect damage as crop injury may result. Severe winter stress, drought, disease, or insect damage following application may also result in crop injury.

In the states of CA, WA, OR, ID, UT, MT, ND, SD, MN and WY, do not use on soils with pH greater than 7.5. Do not exceed 1/2 oz/A per crop or fallow period on soils pH 6.5 or lower; or 1/3 oz/A per crop or fallow period on soil pH 6.5 to 7.5.

In the states of TX, CO, NM, AR, LA, KS, OK and NE, do not exceed 1/2 oz/A per crop or fallow period on soils pH 7.9 or lower. Do not use on soils with pH greater than 7.9.

Do not apply to wheat, spring oats, or barley undersown with legumes and grasses as injury to the forages will result.

Do not apply to frozen ground where surface runoff may occur.

Do not apply to irrigated land where tail water will be used to irrigate crop land.

Do not use on fields that have variable soil conditions where large areas are gravelly or sandy, have eroded knolls, calcium deposits or widely variable pH readings or organic matter content. Use of GLEAN on fields with these conditions may result in crop injury or adversely affect crop rotation.

Under certain conditions such as heavy rainfall and prolonged cool weather soon after treatment, temporary discoloration and/or crop injury may occur. GLEAN and methyl or ethyl parathion tank mixtures may also cause temporary discoloration.

To prevent weather related crop injury, avoid making applications during winter months when weather conditions are unpredictable and can be severe.

Weed control or suppression may be unsatisfactory on soils containing 5% or more organic matter.

Fall application on sandy soil may not provide adequate control or suppression of spring germinating weeds.

NOTE TO BUYER: Purchase of this material does not confer any rights under patents of countries outside of the United States.

STORAGE AND DISPOSAL

STORAGE

Store product in original container only, away from other pesticides, fertilizers, food, or feed.

DISPOSAL

Do not contaminate water, food, or feed by storage or disposal. Wastes resulting from the use of this product may be disposed of on site or at an approved waste disposal facility.

Triple rinse (or equivalent) the container. Then offer for recycling or reconditioning, or puncture and dispose of in a sanitary landfill, or incineration, or, if allowed by state and local authorities, by burning. If burned, stay out of smoke.

NOTE TO BUYER: Purchase of this material does not confer any rights under patents of countries outside of the United States.

[1] Registered trademark of Monsanto Co.

[2] Registered trademark of Chevron Chemical Co.

HYVAR® X

HERBICIDE

ACTIVE INGREDIENT:

Bromacil (5-bromo-3-sec-butyl-6-methyluracil)	80%
INERT INGREDIENTS	20%
TOTAL	100%

EPA Reg. No. 352-287 U.S. Pat. 3,352,862

Keep Out of Reach of Children

PRECAUTIONARY STATEMENTS

HAZARDS TO HUMANS

CAUTION!

MAY IRRITATE EYES, NOSE, THROAT, AND SKIN.

Avoid breathing dust. Avoid contact with skin, eyes or clothing. Wash thoroughly with soap and water after handling. Do not apply this product in such a manner as to directly or through drift expose workers or other persons. The area being treated must be vacated by unprotected persons.

STATEMENT OF PRACTICAL TREATMENT

If in eyes: Flush with plenty of water. Get medical attention if irritation persists.

If on skin: Wash with plenty of soap and water. Get medical attention if irritation persists.

ENVIRONMENTAL HAZARDS

Do not apply directly to water or wetlands.

Do not contaminate water by cleaning of equipment or disposal of wastes.

IMPORTANT

Injury to, or loss of, desirable trees or other plants may result from failure to observe the following: Do not apply (except as recommended for crop use), or drain or flush equipment on or near desirable trees or other plants, or on areas where their roots may extend, or in locations where the chemical may be washed or moved into contact with their roots. Do not use on lawns, walks, driveways, tennis courts or similar areas. Prevent drift of dry powder or spray to desirable plants. Do not contaminate domestic waters. Keep from contact with fertilizers, insecticides, fungicides and seeds. Thoroughly clean all traces of HYVAR X from application equipment immediately after use. Flush tank, pump, hoses and boom with several changes of water after removing nozzle tips and screens (clean these parts separately).

GENERAL INFORMATION

DuPont HYVAR X Herbicide is a wettable powder to be mixed in water and applied as a spray for control of weeds and brush. It is non-corrosive to equipment, nonflammable, and nonvolatile.

HYVAR X is an effective general herbicide for the control of many annual weeds at low rates and perennial weeds and brush at higher rates and is particularly useful for control of perennial grasses. It may be used on noncropland for nonselective weed and brush control and for selective weed control in certain crops.

Effects are slow to appear and may not become apparent until the chemical has been carried into the root zone of the weeds by moisture. The degree of control and duration of effect will vary with the amount of herbicide applied, soil texture, rainfall, and other conditions.

NOTICE OF WARRANTY

DuPont warrants that this product conforms to the chemical description on the label thereof and is reasonably fit for purposes stated on such label only when used in accordance with directions under normal use conditions. It is impossible to eliminate all risks inherently associated with use of this product. Crop injury, ineffectiveness or other unintended consequences may result because of such factors as weather conditions, presence of other materials, or the manner of use or application, all of which are beyond the control of DuPont. In no case shall DuPont be liable for consequential, special or indirect damages resulting from the use or handling of this product. All such risks shall be assumed by the buyer. DUPONT MAKES NO WARRANTIES OF MERCHANTABILITY OR FITNESS FOR A PARTICULAR PURPOSE NOR ANY OTHER EXPRESS OR IMPLIED WARRANTY EXCEPT AS STATED ABOVE.

DIRECTIONS FOR USE

It is a violation of federal law to use this product in a manner inconsistent with its labeling.

DuPont HYVAR X Herbicide should be used only in accordance with recommendations on this label, or in separate published DuPont recommendations available through local dealers.

DuPont will not be responsible for losses or damages resulting from use of this product in any manner not specifically recommended by DuPont. User assumes all risks associated with such nonrecommended use.

Do not enter treated areas without protective clothing until sprays have dried. Because certain states may require more restrictive reentry intervals for various crops treated with this product, consult your State Department of Agriculture for further information.

Written or oral warnings must be given to workers who are expected to be in a treated area or in an area about to be treated with this product. When oral warnings are given, warnings shall be given in a language customarily understood by workers. Oral warnings must be given if there is reason to believe that written warnings cannot be understood by workers. Warnings must include the following information: "CAUTION. Area treated with HYVAR X on (date of application). Do not enter without appropriate protective clothing until sprays have dried. In case of contact, flush eyes with plenty of water; wash skin with plenty of soap and water. Get medical attention if irritation persists."

Apply HYVAR X as a spray just before or during the period of active growth of plants to be controlled and when rainfall can be expected for soil activation. If dense growth is present, results will be improved if vegetation is removed before treatment. Do not apply when ground is frozen.

Before spraying, calibrate equipment to determine quantity of water necessary to uniformly cover measured area to be treated. Weigh the proper amount of HYVAR X and mix into necessary volume of water.

For crop use, apply with a fixed-boom power sprayer properly calibrated to a constant speed and rate of delivery. Use sufficient water (min. 40 gals. per acre) to provide thorough and uniform coverage of the ground. Spray booms must be shut off while starting, turning, slowing or stopping, or injury to the crop or successive crops may result.

For non-crop use, application also may be made with a handgun sprayer, using at least 200 gals. spray per acre to insure uniform coverage. For small areas, a hand sprayer or sprinkling can may be used.

Nozzle screens should be 50 mesh or larger. Continuous agitation in the spray tank is required to keep the material in suspension. Agitate by mechanical or hydraulic means in the spray tank. If bypass or return line is used, it should terminate at bottom of tank to minimize foaming. Do not use air agitation.

NON-CROP USE

WEED CONTROL

To control most weeds for an extended period of time on noncropland areas such as railroad, highway and pipeline right-of-ways, petroleum tank farms, lumberyards, storage areas and industrial plant sites:

Apply 3 to 6 lbs. per acre to control annuals such as cheatgrass, crabgrass, downy brome, foxtail, lambsquarters, puncturevine, ragweed, ryegrass, turkey mullein and wild oats. When applied just prior to or after emergence of annuals, rates as low as 2 lbs. per acre control many annual weeds and grasses in low rainfall areas, and give short-term control in higher rainfall areas.

Apply 7 to 12 lbs. per acre to control perennials such as aster, bahiagrass, bluegrass, broomsedge, dandelion, dog fennel, goldenrod, plantain, purpletop, natalgrass (redtop), quackgrass, smooth brome and wild carrot. In areas with low or seasonal rainfall, rates as low as 5 lbs. per acre control many perennial weeds and grasses.

Apply 15 to 30 lbs. per acre to control bermudagrass, bouncingbet, bracken fern, dallisgrass, dogbane, horsetail, johnsongrass, nutsedge, saltgrass and vaseygrass. Where limited rainfall (usually less than 4 inches) occurs during the active growth period, such as in some areas of the West, HYVAR X usually will not provide satisfactory control of hard-to-kill, deep-rooted perennial weeds such as johnsongrass.

Note—Use the higher levels of the dosage ranges on adsorptive soils (those high in organic matter or carbon).

Retreatment—Apply 2 to 6 lbs. per acre when annual weeds and grasses reappear on sites where weed growth has been controlled.

For Small Areas—¼ cupful of HYVAR X per 250 sq. ft. is approximately 15 lbs. per acre.

BRUSH CONTROL

To control undesirable woody plants on non-cropland areas such as railroad right-of-ways, storage areas, industrial plant sites and drainage ditches:

Apply in spring or summer as a broadcast or basal (spot) treatment; for use on drainage ditches, apply as a basal (spot) treatment only.

Note: For effective brush control and prevention of damage to desirable vegetation: do not apply to brush standing in water; do not use water from treated ditches for irrigation; do not use in irrigation ditches nor on right-of-ways or other sites where marketable timber or other desirable trees or shrubs are immediately adjacent to the treated area.

Broadcast Treatment—Apply 7 to 12 lbs. per acre to control oak, pine, sweet gum and willow; apply 15 to 30 lbs. per acre to control brush such as American elm, cottonwood, hackberry, sumac and winged elm. Use the higher levels of the dosage ranges on adsorptive soils (those high in organic matter or carbon).

Basal (Spot) Treatment—Mix 2½ lbs. in 5 gals. of water and apply at the rate of 1 to 2 fl. ozs. per stem 2 to 4" in basal diameter; wet base of stem to run-off. Treatment controls woody plants such as cottonwood, hackberry, maple, oak, poplar, red bud, sweet gum, wild cherry, willow and winged elm.

SELECTIVE USE IN CROPS

All dosages of HYVAR X are expressed as broadcast rates. For band treatment, use proportionately less; for example, use ⅓ of the broadcast rate when band treating ⅓ of the area.

Moisture is necessary to activate the chemical; best results are obtained if moisture is supplied by rainfall or irrigation within two weeks after application.

CITRUS

Apply as a band or broadcast treatment beneath and/or between trees. Avoid contact of foliage and fruit with spray or mist. Temporary yellowing of citrus leaves may occur following treatment. Do not use on citrus in Kern County, California.

Because injury to citrus trees may result: do not use on soils low in organic matter (less than 1%) nor on poorly drained soils; do not apply more than 8 lbs. per acre per year; do not treat trees planted in irrigation furrows; do not treat diseased trees such as those with foot rot. Do not use in citrus orchards interplanted with other trees or desirable plants, nor in home citrus plantings or in areas where roots of other valuable plants or trees may extend as plant injury may result. Do not graze cattle in treated areas. Treated areas may be planted to citrus trees one year after last application; do not replant to other crops within 2 years after last application as plant injury may result.

Trees Established for Four Years or More

Annual Weeds—Including barnyardgrass (watergrass), crabgrass, crowfootgrass, Florida pusley, henbit, lambsquarters, mustard, natalgrass (red top), puncturevine, purslane, sandspur, sedge (annual), sprangletop, Texas panicum (Colorado grass), and turkey mullein, apply 2 to 4 lbs. per acre. Apply anytime of the year, preferably shortly before or after weed growth begins when adequate moisture is available.

Perennial Weeds—Best results are obtained if application is made shortly before or shortly after weed growth begins; if dense growth is present, remove tops and spray the ground. Effects on perennial weeds are slow to appear, usually progressing over a period of several months.

Make a single application per year during the period from winter to early summer; use at the following rates:

Soil Texture	Lbs. HYVAR X Per Acre
Sand, loamy sand	4 to 5
Sandy loam	5 to 6
Silt loam, clay loam	6 to 8

Alternatively, make two applications of 3 to 4 lbs. HYVAR X per acre per year. In Florida, Texas, and Louisiana, apply in spring and summer; in California and Arizona, apply in fall and spring.

Note: Partial control usually occurs with a single treatment; repeat applications are required to control perennials. HYVAR X controls the following:

Bermudagrass . . . . All areas U.S.
Torpedograss, paragrass,
pangolagrass, Bahiagrass . . . . Florida
Johnsongrass . . . . Texas
Nutsedge . . . . Texas, California

Control of perennials may be improved by cultivation prior to treatment; otherwise, avoid working the soil as long as weed control continues or else effectiveness of the treatment may be reduced.

Florida Only—Trees Established Less Than Four Years.

Trees Established Less Than One Year—Apply 2 to 3 lbs. per acre as needed to maintain weed control. Do not apply at intervals less than sixty days; do not apply more than 6 lbs. per acre within any 6-month period or more than 8 lbs. per acre during the first year.

Trees Established One to Four Years—Apply 2 to 4 lbs. HYVAR X per acre as needed for weed control. Use the lower rate for 1- to 2-year old trees and the higher rate for trees more than 2 years old. A second application may be made if needed, but do not apply more than 4 lbs. HYVAR X per acre per year as injury to the trees may result.

Torpedograss Control—Barrier Strip Treatment

For control of torpedograss adjacent to citrus groves to prevent spread of the weed into groves, apply 30 lbs. per acre. Treat a border strip 10 to 20 ft. wide adjacent to the citrus grove, but not closer than 10 ft. to the dripline of citrus trees. Examine treated area every four months after application and spot-treat invading or surviving torpedograss at 30 lbs. per acre (11 ozs. per 1000 sq. ft.). For best results apply in late winter or early spring after the torpedograss has broken dormancy and is actively growing. For later season application where growth is rank, mow or disk the area prior to treatment.

PINEAPPLE

Do not replant treated areas to any crop other than pineapple within 2 years after last application as injury to subsequent crops may result.

Hawaii and Florida—For control of seedling weeds such as amaranthus, balsamapple, chloris, crabgrass, Flora's paintbrush, foxtail, goosegrass (wiregrass), and Hialoa, apply 2 to 6 lbs. per acre broadcast before the planting material begins to grow. Use the lower rates in low rainfall areas (5 to 10 inches annually) and on clean-culture fields; use the higher rates in high rainfall areas (above 10 inches annually) and for trashmulch fields. Follow with 2 to 4 lbs. per acre as directed interline spray as needed prior to differentiation, or apply 2 lbs. per acre broadcast as needed after plants are 8 months old but before differentiation. Do not apply more than 16 lbs. total per acre per plant crop.

For ratoon crop, apply 1 to 4 lbs. per acre broadcast after harvesting plant crop but before differentiation. Do not apply more than 4 lbs. total per acre per ratoon crop.

Puerto Rico—For control of seedling weeds such as crabgrass, goosegrass, jungle rice, pigweed, and purslane, apply 2 to 4 lbs. per acre broadcast immediately after planting and before planting material begins to grow.

NOTICE TO BUYER—Purchase of this material does not confer any rights under patents of countries outside of the United States.

SUPPLEMENTAL LABELING

HYVAR X BROMACIL HERBICIDE

EPA Reg. No. 352-287

AS A TANK MIXTURE WITH OTHER HERBICIDES FOR NON-CROP WEED CONTROL

DuPont HYVAR X Bromacil Herbicide with DuPont Surfactant WK is recommended for use as a tank mixture with amitrole, amitrole T, Banval®* D, 2,4-D amine, dalapon, Dowpon®** C, MSMA, paraquat, and Phytar®*** 560 (Na salt of cacodylic acid) for control of undesirable vegetation in non-crop areas such as railroad, highway, pipeline and utility right-of-ways, petroleum tank farms, lumberyards, storage areas and industrial plant sites.

DIRECTIONS

For non-cropland use, application may be made with a hand gun or boom sprayer using sufficient water (40 to 200 gals. of spray per acre) to insure uniform coverage. For smaller areas, a hand sprayer may be used. Nozzle screens should be 50 mesh or larger. Agitation may be either mechanical or hydraulic. If a bypass or return line is used, it should terminate at the bottom of the tank to minimize foaming. Do not use air agitation.

Spray Preparation—Fill spray tank half full of water. Start agitation and continue during entire mixing process. Add required amount of HYVAR X and allow to mix. If additional product is wettable powder, add to tank and allow to mix thoroughly. If additional product is a liquid, add slowly while filling remainder of tank with water. Dilute proper amount of Surfactant WK with 10 parts of water and add to tank last and allow contents to mix thoroughly. Continuous agitation in the spray tank is required to keep the materials in suspension.

Rates for Weed Control—HYVAR® X—Use HYVAR X at the following rates with Surfactant WK (1 to 2 qts. per 100 gals. spray) for non-crop weed control:

Apply 4 to 6 lbs. per acre to control annual weeds and grasses such as bromegrass, cheatgrass, crabgrass, foxtail, lambsquarters, puncturevine, ragweed, ryegrass, turkey mullein and wild oats.

Apply 7 to 12 lbs. per acre to control perennial weeds and grasses such as aster, bahiagrass, bluegrass, broomsedge, dandelion, dog fennel, goldenrod, plantain, purpletop, quackgrass, redtop, smooth brome, and wild carrot.

Apply 15 to 30 lbs. per acre to control hard-to-kill perennial weeds and grasses such as Bermudagrass, bouncingbet, bracken fern, Dallisgrass, dogbane, horsetail, Johnsongrass, nutsedge, saltgrass, and vaseygrass.

Rates for Products to be Tank Mixed with HYVAR® X and Surfactant WK—Use the lower rate of HYVAR X (See Note) recommended to provide control of weeds to be sprayed. Prepare a tank mixture with one of the following herbicides selected as appropriate to improve control of the weed population.

Amitrole or Amitrole T—Add 1 to 3 gals. per acre to HYVAR X to improve control of Bermudagrass, bluegrass, Canada thistle, horsetail rush, nutsedge, and quackgrass.

Banvel D *(dicamba)*—Add 1 to 4 pts. per acre to HYVAR X to improve control of bindweed, dewberry, smartweed, thistle, and many other broadleaved weeds.

2,4-D Amine—Add 2 to 4 lbs. active per acre to HYVAR X to improve control of blackberry, burdock, peppervine, trumpetvine, Virginia creeper and other broadleaved weeds.

Dalapon (Dowpon or Dowpon C)—Add 15 to 30 lbs. per acre to HYVAR X to improve control of bahiagrass, Bermudagrass, broomsedge, bunchgrass, Johnsongrass and quackgrass.

MSMA—Add 3½ to 7 lbs. active per acre to HYVAR X to aid in control of bahiagrass, broomsedge, Johnsongrass, paspalum and nutsedge.

Paraquat—Add 1 to 2 qts. per acre to HYVAR X for non-selective burndown of annual weeds and top kill of perennial vines and grasses.

Phytar 560 (Na salt of cacodylic acid)—Add 1½ to 4 gals. per acre to HYVAR X for non-selective burndown of annual weeds and top kill of perennial vines and grasses.

MSMA + 2,4-D Amine—Add recommended rates to HYVAR X for a 3-way tank mixture to improve control of weeds listed for all three products.

Note: Use the higher rates of HYVAR X on heavier textured soils or on soils high in organic matter or carbon. Use the higher rates of other herbicides mixed with HYVAR X on heavy, dense vegetation.

IMPORTANT

BEFORE USING **HYVAR X**, SURFACTANT WK, AMITROLE, BANVEL* D, 2,4-D AMINE, DALAPON, MSMA, PARAQUAT OR PHYTAR 560, READ AND CAREFULLY OBSERVE THE CAUTIONARY STATEMENTS AND ALL OTHER INFORMATION APPEARING ON THE PRODUCT LABELS.

This bulletin contains new or supplemental instructions for use of these products which may not appear on package labels. Follow the instructions carefully.

* Registered trademark of Velsicol Chemical Corp.
** Registered trademark of Dow Chemical Company.
*** Registered trademark of The Ansul Company.

SUPPLEMENTAL LABELING

HYVAR® X HERBICIDE

EPA Reg. No. 352-287

FOR WEED CONTROL UNDER ASPHALT, AND POND LINERS

DIRECTIONS FOR USE

It is a violation of federal law to use this product in a manner inconsistent with its labeling.

DuPont HYVAR X Herbicide is recommended for weed control under asphalt, concrete, and pond liners to extend the useful life of the surfacing material by preventing weeds from emerging through it.

HOW TO USE

Remove vegetation and debris from the subgrade. Uniformly apply HYVAR X to the area to be surfaced, including the shoulder areas, at the following rates:

- 6–10 pounds per acre for control of annuals such as cheatgrass, crabgrass, downy brome, foxtail, lambsquarters, puncturevine, ragweed, ryegrass, turkey mullein, and wild oats;
- 10–15 pounds per acre for control of perennials such as bahiagrass, broomsedge, dandelion, dogfennel, goldenrod, plantain, purpletop, quackgrass, redtop, smooth brome, and wild carrot;
- 25–30 pounds per acre for control of bermudagrass, bouncingbet, bracken fern, dallisgrass, dogbane, horsetail, johnsongrass, nutsedge, saltgrass, and vaseygrass.

For Herbicide Activation

- **On moist subsoils,** apply HYVAR X after final grading and immediately before laying asphalt or liner surface. Apply uniformly using at least 100 gallons of water per acre.
- **If moisture is not present,** incorporation of HYVAR X is need-

ed for activation. HYVAR X can be incorporated into the soil to a depth of 4–6 inches using a rototiller or disk. Rainfall or irrigation of 2 inches will also provide uniform incorporation. Do not allow treated soil to wash or move into untreated area.

Notes: HYVAR X is intended for use under asphalt, pond liners, and other pavement **only** in industrial sites or where the pavement has a suitable barrier along the perimeter that prevents encroachment of roots of desirable plants.

The product is not recommended for use under pavement in residential properties such as driveways or parking lots, nor is it recommended for use in recreational areas such as under jogging or bike paths, tennis courts, or golf cart paths, or where landscape plantings could be anticipated.

Injury or death of desirable plants may result if this product is applied where roots are present or where they may extend into treated area. Roots of trees and shrubs may extend a considerable distance beyond the branch extremities or so-called "dripline."

Where hard-to-control perennials such as peppervine, trumpet-creeper, field bindweed, and Smilax spp. (greenbrier) are in the treated area, another suitable herbicide should be applied.

Do not use in California.

IMPORTANT

BEFORE USING HYVAR X, READ AND CAREFULLY OBSERVE THE CAUTIONARY STATEMENTS AND ALL OTHER INFORMATION APPEARING ON THE PRODUCT LABEL.

This bulletin contains new or supplemental instructions for use of this product which may not appear on the package label. Follow the instructions carefully.

This labeling must be in possession of the user at the time of pesticide application.

SPECIAL LOCAL NEED REGISTRATION [24(c)] FOR HYVAR® X HERBICIDE*

Crop	State	EPA SLN No.	Remarks
Pineapple	Hawaii	HI-770027	Expands time of application

* Copies of label available upon request to nearest DuPont Agricultural Chemicals Office. For distribution and use only within the state specified.

STORAGE AND DISPOSAL

STORAGE

Store product in original container only, away from other pesticides, fertilizer, food, or feed.

DISPOSAL

Do not contaminate water, food, or feed by storage or disposal. Wastes resulting from the use of this product may be disposed of on site or at an approved waste disposal facility.

Completely empty bag into application equipment. Then dispose of empty bag in a sanitary landfill or by incineration, or, if allowed by state and local authorities, by burning. If burned, stay out of smoke.

HYVAR® X-L

HERBICIDE
WATER SOLUBLE LIQUID

1 Gallon Contains 2 Pounds Bromacil

ACTIVE INGREDIENT:
Lithium salt of bromacil
(5-bromo-3-sec-butyl-6-methyluracil) 21.9%
INERT INGREDIENTS 78.1%

TOTAL 100.0%

EPA Reg. No. 352-346 U.S. Pats. 3,352,862 & 3,909,236

Keep Out of Reach of Children

PRECAUTIONARY STATEMENTS

HAZARDS TO HUMANS

WARNING!

HARMFUL OR FATAL IF SWALLOWED. COMBUSTIBLE. CAUSES EYE IRRITATION. MAY IRRITATE NOSE, THROAT AND SKIN.

Do not get in eyes. Avoid contact with skin and clothing. Avoid breathing spray mist. Keep away from heat and open flame. Wash thoroughly after handling.

STATEMENT OF PRACTICAL TREATMENT

If in eyes: Flush with plenty of water. Get medical attention if irritation persists.

If on skin: Wash with plenty of soap and water. Get medical attention if irritation persists.

ENVIRONMENTAL HAZARDS

Do not contaminate water by cleaning of equipment or disposal of wastes.

IMPORTANT

Injury to or loss of desirable trees or other plants may result from failure to observe the following: Do not apply, or drain or flush equipment on or near desirable trees or other plants or on areas where their roots may extend, or in locations where the chemical may be washed or moved into contact with their roots. Do not use on lawns, walks, driveways, tennis courts or similar areas. Prevent drift of spray to desirable plants. Do not contaminate domestic waters. Do not use water from treated ditches for irrigation. Keep from contact with fertilizers, insecticides, fungicides and seeds. Thoroughly clean all traces of HYVAR X-L from application equipment immediately after use. Flush tank, pump, hoses and boom with several changes of water after removing nozzle tips and screens (clean these parts separately).

NOTICE OF WARRANTY

DuPont warrants that this product conforms to the chemical description on the label thereof and is reasonably fit for purposes stated on such label only when used in accordance with directions under normal use conditions. It is impossible to eliminate all risks inherently associated with use of this product. Ineffectiveness or other unintended consequences may result because of such factors as weather conditions, presence of other materials or the manner of use or application, all of which are beyond the control of DuPont. In no case shall DuPont be liable for consequential, special or indirect damages resulting from the use or handling of this product. All such risks shall be assumed by the buyer. DUPONT MAKES NO WARRANTIES OF MERCHANTABILITY OR FITNESS FOR A PARTICULAR PURPOSE NOR ANY OTHER EXPRESS OR IMPLIED WARRANTY EXCEPT AS STATED ABOVE.

GENERAL INFORMATION

DuPont HYVAR X-L Herbicide is an effective general herbicide for use on noncropland areas for control of brush and many annual and perennial weeds, and is particularly useful for control of perennial grasses. It is a water soluble liquid to be mixed in water and applied as a spray; HYVAR X-L may also be applied undiluted for basal (soil) treatment of brush. Effects are slow to appear and may not become apparent until the chemical has been carried into the root zone of the weeds by moisture. The degree of control and duration of effect will vary with the amount of chemical applied, soil type, rainfall and other conditions.

HYVAR X-L, after mixing in water, is nonvolatile, nonflammable, and noncorrosive to metals, except aluminum. Do not use with aluminum spray nozzles.

DIRECTIONS FOR USE

It is a violation of federal law to use this product in a manner inconsistent with its labeling.

HYVAR X-L Herbicide should be used only in accordance with recommendations on this label, or in separate published DuPont recommendations available through local dealers. Do not use on any crop in Kern County, California.

DuPont will not be responsible for losses or damages resulting from use of this product in any manner not specifically recommended by DuPont. User assumes all risks associated with such nonrecommended use.

Apply HYVAR X-L before or during the period of active growth of plants to be controlled when rainfall can be expected for soil activation. If dense growth is present, results will be improved if vegetation is removed before treatment. Do not apply when ground is frozen.

Before spraying, calibrate equipment to determine quantity of water necessary to uniformly cover measured area to be treated. Measure the proper amount of HYVAR X-L and mix into the necessary volume of water. After HYVAR X-L has been thoroughly mixed in the spray tank, agitation of the spray solution is not required.

Use a fixed-boom power sprayer properly calibrated to a constant speed and rate of delivery. For application with a hand-gun sprayer, use at least 200 gals. spray per acre. For treating small areas, a tank-type hand sprayer or sprinkling can may be used. For basal (soil) treatment of brush, HYVAR X-L may be applied undiluted with an exact delivery hand-gun applicator.

WEED CONTROL

To control most weeds for an extended period of time on noncropland areas such as railroad, highway and pipeline right-of-ways, petroleum tank farms, lumberyards, storage areas and industrial plant sites:

Apply 1½ to 3 gals. HYVAR X-L per acre to control ANNUAL WEEDS and GRASSES such as foxtail, ryegrass, wild oats, crabgrass, cheatgrass, bromegrass, ragweed, lambsquarters, puncturevine, orchardgrass and turkey mullein. When applied just before or soon after emergence of annuals, rates as low as ¾ gal. per acre control many annual weeds and grasses in low rainfall areas and give short-term control in higher rainfall areas.

Apply 3 to 6 gals. HYVAR X-L per acre to control PERENNIAL WEEDS and GRASSES such as smooth brome, bahiagrass, bluegrass, redtop, purpletop, quackgrass, broomsedge, aster, dandelion, dog fennel, goldenrod, plantain and wild carrot. In areas with low or seasonal rainfall, rates as low as 2¼ gals. per acre control many perennial weeds and grasses.

Apply 6 to 12 gals. HYVAR X-L per acre to control JOHNSONGRASS; use at the same rate for OTHER HARD-TO-KILL PERENNIAL WEEDS and GRASSES such as bermudagrass, dallisgrass, nutsedge, vaseygrass, saltgrass, bouncingbet, dogbane, bracken fern and horsetail. Where limited rainfall (usually less than 4 inches) occurs during the active growth period, such as some areas of the West, HYVAR X-L will not provide satisfactory control of hard-to-kill, deep-rooted perennial weeds such as johnsongrass.

Use the higher levels of the dosage ranges on adsorptive soils (usually those high in organic matter or carbon).

Retreatment—Apply ¾ to 3 gals. HYVAR X-L per acre when annual weeds and grasses reappear on sites where weed growth has been controlled.

For Small Areas—Three fl. ozs. of HYVAR X-L per 250 sq. ft. is approximately 4 gals. per acre.

BRUSH CONTROL

To control undesirable woody plants on noncropland areas such as railroad right-of-ways, storage areas, industrial plant sites and drainage ditches:

Apply as a broadcast or basal (soil) treatment; for use on drainage ditches, apply as a basal (soil) treatment only.

Note: For effective brush control and prevention of damage to desirable vegetation: do not apply to brush standing in water; do not use water from treated ditches for irrigation; do not use in irrigation ditches nor on right-of-ways or other sites where marketable timber or other desirable trees or shrubs are immediately adjacent to the treated areas.

Broadcast Treatment—Apply 2¼ to 5 gals. per acre to control oak, willow, sweet gum and pine; apply 6 to 12 gals. per acre to control brush such as American elm, winged elm, hackberry, sumac and cottonwood. Use the higher levels of the dosage ranges on adsorptive soils (those high in organic matter or carbon).

Basal (Soil) Treatment—Controls woody plants such as cottonwood, hackberry, maple, oak, poplar, red bud, sweet gum, wild cherry, willow, winged elm, American elm and sumac.

Diluted—Mix 1 gal. HYVAR X-L in 5 gals. of water and apply at the rate of 1 to 2 fl. ozs. per stem 2 to 4" in basal diameter; wet base of stem to run-off.

Undiluted—Apply HYVAR X-L undiluted with an exact delivery hand-gun applicator; this equipment delivers a thin stream of a predetermined volume when triggered. Apply at the rate of 5 to 10 milliliters per stem 2 to 4" in basal diameter; direct the treatment to a spot at the base of the brush (root collar area). When treating large stems and more than one delivery of solution is needed per stem, apply on the opposite side of the stem. **Note:** HYVAR X-L is a combustible product. Do not smoke while applying the undiluted product, and keep away from heat and open flame.

NOTICE TO BUYER: Purchase of this material does not confer any rights under patents of countries outside of the United States.

SUPPLEMENTAL LABELING

HYVAR® X-L HERBICIDE

EPA Reg. No. 352-346

FOR WEED CONTROL UNDER ASPHALT, CONCRETE AND POND LINERS

DIRECTIONS FOR USE

It is a violation of federal law to use this product in a manner inconsistent with its labeling.

DuPont HYVAR X-L Herbicide is recommended for weed control under asphalt, concrete and pond liners to extend the useful life of the surfacing material, by preventing weeds from emerging through it.

HOW TO USE

Remove vegetation and debris from the subgrade. Uniformly apply HYVAR X-L to the area to be surfaced, including the shoulder areas, at the following rates:

- 2.5 to 4 gallons per acre for control of annuals such as cheat, crabgrass, downy brome, foxtail, lambsquarters, puncturevine, ragweed, ryegrass, turkey mullein and wild oats;
- 4 to 6 gallons per acre for control of perennials such as bahiagrass, broomsedge, dandelion, dogfennel, goldenrod, plantain, purpletop, quackgrass, redtop, smooth brome and wild carrot;
- 10 to 12 gallons per acre for control of bermudagrass, bouncingbet, bracken fern, dallisgrass, dogbane, horsetail, johnsongrass, nutsedge, saltgrass and vaseygrass.

For Herbicide Activation:

- **On moist subsoils,** apply HYVAR X-L after final grading and immediately before laying asphalt or liner surface. Apply uniformly using at least 100 gallons of water per acre.
- **If moisture is not present,** incorporation of HYVAR X-L is needed for activation. HYVAR X-L can be incorporated into the soil to a depth of 4–6 inches using a rototiller or disk. Rainfall or irrigation of 2 inches will also provide uniform incorporation. Do not allow treated soil to wash or move into untreated area.

Notes: HYVAR X-L is intended for use under asphalt, pond liners and other pavement **only** in industrial sites or where the pavement has a suitable barrier along the perimeter that prevents encroachment of roots of desirable plants.

The product is not recommended for use under pavement in residential properties such as driveways or parking lots, recreational areas such as under jogging or bike paths, tennis courts, or golf cart paths or where landscape plantings could be anticipated.

Injury or death of desirable plants may result if this product is applied where roots are present or where they may extend into treated area. Roots of trees and shrubs may extend a considerable distance beyond the branch extremities or so-called "dripline."

Where hard-to-control perennials such as peppervine, trumpet-creeper, field bindweed and **Smilax** spp. (greenbrier) are in the treated area, another suitable herbicide should be applied.

Do not use in California.

IMPORTANT

BEFORE USING HYVAR X-L, READ AND CAREFULLY OBSERVE THE CAUTIONARY STATEMENTS AND ALL OTHER INFORMATION APPEARING ON THE PRODUCT LABEL.

This bulletin contains new or supplemental instructions for use of this product which may not appear on the package label. Follow the instructions carefully.

This labeling must be in possession of the user at the time of pesticide application.

STORAGE AND DISPOSAL

STORAGE

Store product in original container only, away from other pesticides, fertilizer, food or feed. Keep container closed when not in use.

DISPOSAL

Do not contaminate water, food or feed by storage or disposal. Wastes resulting from the use of this product may be disposed of on site or at an approved waste disposal facility. Triple rinse (or equivalent). Then offer for recycling or reconditioning, or puncture and dispose of in a sanitary landfill, or by incineration, or, if allowed by state and local authorities, by burning. If burned, stay out of smoke.

HYVAR® L

CITRUS HERBICIDE
WATER SOLUBLE LIQUID

For Use in the State of Florida Only

1 Gallon Contains 3.2 Pounds Bromacil

ACTIVE INGREDIENT:	
Lithium salt of bromacil (5-bromo-3-sec-butyl-6-methyluracil)	33.8%
INERT INGREDIENTS	66.2%
TOTAL	100.0%

EPA Reg. No. 352-425 U.S. Pats. 3,352,862 & 3,909,236

Keep Out of Reach of Children

PRECAUTIONARY STATEMENTS

HAZARDS TO HUMANS

WARNING

HARMFUL OR FATAL IF SWALLOWED. CAUSES EYE IRRITATION. MAY IRRITATE NOSE, THROAT, AND SKIN. Do not get in eyes. Avoid contact with skin and clothing. Avoid breathing vapors. Wash thoroughly with soap and water after handling and before eating or smoking. Do not apply this product in such a manner as to directly or through drift expose workers or other persons. The area being treated must be vacated by unprotected persons.

STATEMENT OF PRACTICAL TREATMENT

If in eyes: Flush with plenty of water. Get medical attention if irritation persists.

If on skin: Wash with plenty of soap and water. Get medical attention if irritation persists.

ENVIRONMENTAL HAZARDS

Do not apply directly to water. Do not contaminate water by cleaning of equipment or disposal of wastes.

PHYSICAL AND CHEMICAL HAZARDS

Combustible. Keep container closed when not in use.

IMPORTANT

Injury to or loss of desirable trees or other plants may result from failure to observe the following: Do not apply, drain, or flush equipment on or near desirable trees or other plants or on areas where their roots may extend, or in locations where the chemical may be washed or moved into contact with their roots. Do not use on lawns, walks, driveways, tennis courts, or similar areas. Prevent drift of spray to desirable plants. Do not contaminate domestic waters. Thoroughly clean all traces of HYVAR L Citrus Herbicide from application equipment immediately after use. Flush tank, pump, hoses, and boom with several changes of water after removing nozzle tips and screens (clean these parts separately).

GENERAL INFORMATION

Du Pont HYVAR L Citrus Herbicide is an effective general herbicide for use on citrus for control of many annual and perennial weeds, and is particularly useful for control of perennial grasses. It is a water soluble liquid to be mixed in water and applied as a spray. Effects are slow to appear and may not become apparent until the chemical has been carried into the root zone of the weeds by moisture. The degree of control and duration of effect will vary with the amount of chemical applied, soil type, rainfall, and other conditions.

HYVAR L, after mixing in water, is non-volatile, non-flammable, and non-corrosive to metals except aluminum. Do not use with aluminum spray nozzles.

NOTICE OF WARRANTY

Du Pont warrants that this product conforms to the chemical description on the label thereof and is reasonably fit for purposes stated on such label only when used in accordance with directions under normal use conditions. It is impossible to eliminate all risks inherently associated with use of this product. Ineffectiveness or other unintended consequences may result because of such factors as weather conditions, presence of other materials, or the manner of use or application, all of which are beyond the control of Du Pont. In no case shall the seller be liable for consequential, special or indirect damages resulting from the use or handling of this product. All such risks shall be assumed by the buyer. DU PONT MAKES NO WARANTIES OF MERCHANTABILITY OR FITNESS FOR A PARTICULAR PURPOSE NOR ANY OTHER EXPRESS OR IMPLIED WARRANTY EXCEPT AS STATED ABOVE.

DIRECTIONS FOR USE

It is a violation of federal law to use this product in a manner inconsistent with its labeling.

HYVAR L should be used only in accordance with recommendations on this label, or in separate published Du Pont recommendations available through local dealers.

Du Pont will not be responsible for losses or damages resulting from use of this product in any manner not specifically recommended by Du Pont. User assumes all risks associated with such nonrecommended use.

Do not enter treated areas without protective clothing until sprays have dried. Because certain states may require more restrictive reentry intervals for various crops treated with this product, consult your State Department of Agriculture for further information.

Written or oral warnings must be given to workers who are expected to be in a treated area or in an area about to be treated with this product. When oral warnings are given, warnings shall be given in a language customarily understood by workers. Oral warnings must be given if there is reason to believe that written warnings cannot be understood by workers. Warnings must include the following information: "WARNING. Area treated with HYVAR L on (date of application). Do not enter without appropriate protective clothing until sprays have dried. In case of contact, flush eyes with plenty of water; wash skin with plenty of soap and water. Get medical attention if irritation persists."

Apply HYVAR L before or during the period of active growth of plants to be controlled when rainfall can be expected for soil activation. If dense growth is present, results will be improved if vegetation is removed before treatment.

Before spraying, calibrate equipment to determine quantity of water necessary to uniformly cover measured area to be treated. Measure the proper amount of HYVAR L and mix into the necessary volume of water. After HYVAR L has been thoroughly mixed in the spray tank, agitation of the spray solution is not required. However, if HYVAR L is tank mixed with wettable powders, adequate, continuous agitation is required.

Use a fixed-boom power sprayer properly calibrated to a constant speed and rate of delivery.

NOTE: Do not tank mix HYVAR L with acidic herbicides such as Roundup[1] or with liquid fertilizers.

[1] Registered trademark of Monsanto Company.

WEED CONTROL

Annual weeds controlled include barnyardgrass (watergrass), crabgrass, crowfootgrass, Florida pusley, henbit, lambsquarters, mustard, natalgrass (red top), puncturevine, purslane, sandspur, sedge (annual), sprangletop, Texas panicum (Colorado grass), and turkey mullein. For annual weed control apply any time of the year, preferably shortly before or after weed growth begins when adequate moiture is available.

Perennial weeds controlled include bermudagrass, torpedograss, paragrass, pangolagrass, Bahiagrass, Johnsongrass, and nutsedge. Best results are obtained if application is made shortly before or shortly after weed growth begins; if dense growth is present, remove tops and spray the ground. Effects on perennial weeds are slow to appear, usually progressing over a period of several months.

Trees Less Than One Year Old

In newly developed groves or in groves that have been heavily reset, apply 2 to 3 quarts HYVAR L per acre as needed to maintain weed control. Do not apply at less than 60-day intervals; do not apply more than 6-quarts per acre within any 6-month period, or more than 8 quarts per acre during the first year.

Trees Established One to Three Years

Apply 2 to 3 quarts per acre as needed to maintain weed control. Do not apply more than 6 quarts per acre within any 6-month period, or more than 8 quarts per acre per year.

Trees Established for Four Years or More

For annual weed control apply 2 to 4 quarts per acre.

For perennial weed control make a single application per year during the period from winter to early summer; use at the following rates:

Soil Texture	Qts. HYVAR L Per Acre
Sand, loamy sand	4 to 5
Sandy loam	5 to 6
Silt loam, clay loam	6 to 8

Alternatively, make two applications of 3 to 4 qts. HYVAR L per acre per year in spring and summer.

NOTE: Partial control usually occurs with a single treatment; repeat applications are required to control perennials.

Apply as a band or broadcast treatment beneath and/or between trees. Avoid contact of foliage and fruit with spray or mist. Temporary yellowing of citrus leaves may occur following treatment.

Because injury to citrus trees may result: Do not use on soils low in organic matter (less than 1%) nor on poorly drained soils; do not apply more than 8 qts. per acre per year; do not treat trees planted in irrigation furrows; do not use in citrus orchards interplanted with other trees or desirable plants, nor in home citrus plantings or in areas where roots of other valuable plants or trees may extend as plant injury may result. Do not graze cattle in treated areas. Treated areas may be planted to citrus trees one year after last application; do not replant to other crops within 2 years after last application as plant injury may result.

Control of perennials may be improved by cultivation prior to treatment; otherwise, avoid working the soil as long as weed control continues or else effectiveness of the treatment may be reduced.

All dosages of HYVAR L are expressed as broadcast rates.

Moisture is necessary to activate the chemical; best results are obtained if moisture is supplied by rainfall or irrigation within two weeks after application.

Torpedograss Control—Barrier Strip Treatment

For control of torpedograss adjacent to citrus groves to prevent spread of the weed into groves, apply 30 qts. per acre. Treat a border strip 10 to 20 ft. wide adjacent to the citrus grove, but not closer than 10 ft. to the dripline of citrus trees. Examine treated area every four months after application and spot-treat invading or surviving torpedograss at 30 qts. per acre. For best results apply in late winter or early spring after the torpedograss has broken dormancy and is actively growing. For later season application where growth is rank, mow or disk the area prior to treatment.

STORAGE AND DISPOSAL

STORAGE

This product should not be stored at temperatures below 32°F. Store product in original container only, away from other pesticides, fertilizer, food, or feed. Keep away from heat and open flame.

DISPOSAL

Do not contaminate water, food, or feed by storage or disposal. Wastes resulting from the use of this product may be disposed of on site or at an approved waste disposal facility. Triple rinse (or equivalent) the container. Then offer for recycling or reconditioning, or puncture and dispose of in a sanitary landfill, or by incineration, or, if allowed by state and local authorities, by burning. If burned, stay out of smoke.

NOTICE TO BUYER: Purchase of this material does not confer any rights under patents of countries outside of the United States.

KARMEX®

HERBICIDE
WETTABLE POWDER

ACTIVE INGREDIENT:

Diuron [3-(3,4-dichlorophenyl)-1,1-dimethylurea]	80%
INERT INGREDIENTS	20%
TOTAL	100%

EPA Reg. No. 352-247

Keep Out of Reach of Children

PRECAUTIONARY STATEMENTS

HAZARDS TO HUMANS

CAUTION!

MAY IRRITATE EYES, NOSE, THROAT, AND SKIN.

Avoid breathing dust or spray mist. Avoid contact with skin, eyes, and clothing. Wash thoroughly with soap and water after handling. Do not apply this product in such a manner as to directly or through drift expose workers or other persons. The area being treated must be vacated by unprotected persons.

STATEMENT OF PRACTICAL TREATMENT

If on skin: Wash with plenty of soap and water; get medical attention if irritation persists.

If in eyes: Flush with plenty of water; get medical attention if irritation persists.

ENVIRONMENTAL HAZARDS

Do not apply directly to water. Do not contaminate water by cleaning of equipment or disposal of wastes. Cover or incorporate spills.

IMPORTANT

Injury to, or loss of, desirable trees or other plants may result from failure to observe the following:

Do not apply (except as recommended for crop use), or drain or flush equipment, on or near desirable trees or other plants, or on areas where their roots may extend, or in locations where the chemical may be washed or moved into contact with their roots. Do not use on home plantings of trees, shrubs or herbaceous plants, nor on lawns, walks, driveways, tennis courts or similar areas. Prevent drift of dry powder or spray to desirable plants. Do not contaminate any body of water. Keep from contact with fertilizers, insecticides, fungicides, and seeds.

Thoroughly clean all traces of KARMEX from application equipment immediately after use. Flush tank, pump, hoses and boom with several changes of water after removing nozzle tips and screens (clean these parts separately).

GENERAL INFORMATION

DuPont KARMEX Herbicide is a wettable powder to be mixed in water and applied as a spray for selective control of weeds in certain crops, and for nonselective weed control on noncropland areas. It is noncorrosive to equipment, nonflammable and nonvolatile.

KARMEX may be applied to soil prior to emergence of weeds to control susceptible weed seedlings for an extended period of time. The degree of control and duration of effect will vary with the amount of chemical applied, soil texture, rainfall and other conditions. Soils high in clay or organic matter require higher dosages than soil low in clay or organic matter, to obtain equivalent herbicide performance. Moisture is required to activate the chemical; best results occur if rainfall (or sprinkler irrigation) occurs within 2 weeks of application.

KARMEX applied preemergence, before emergence of crop and weeds, is an effective procedure because susceptible weeds are controlled in an early, vulnerable seedling state before they compete with the crop. With favorable moisture conditons, KARMEX continues to control weeds for some time as the treatment in significant numbers, secondary weed control procedures should be implemented; these include cultivation and postemergence herbicide application.

KARMEX may also be used to control emerged weeds. Results vary with rate applied and environmental conditions; best results are obtained on succulent weeds growing under conditions of high humidity and temperatures of 70°F or higher. Addition of a surfactant such as DuPont Surfactant WK to the spray (where recommended) increases contact effects of KARMEX. KARMEX may be used as a directed postemergence application, where spray nozzles are adjusted so the weeds are sprayed but the crop is not, on the following crops: artichoke, corn (field), cotton, sorghum, gooseberries, citrus, grapes, macadamia nuts, olives, papayas, peaches, pears, pecans, walnuts and certain tree plantings.

Under specified conditions (see Directions for Use), KARMEX without surfactant may be applied over the top of alfalfa (established, dormant or semidormant), asparagus (established, dormant), sugarcane, wheat, pineapple and plumosus fern (established, mowed).

Weed species vary in susceptibility to KARMEX and they may be more difficult to control when under stress. Combinations of KARMEX with other herbicides (as registered) increase the number of weed species controlled; consult labels of the companion product for this and other information. Because the effect of KARMEX varies with soils, uniformity of application and environmental conditions, it is suggested that growers limit their first use to small areas. Observe all cautions and limitations on labeling of all products used in mixtures.

NOTICE OF WARRANTY

DuPont warrants that this product conforms to the chemical description on the label thereof and is reasonably fit for purposes stated on such label only when used in accordance with the directions under normal use conditions. It is impossible to eliminate all risks inherently associated with the use of this product. Crop injury, ineffectiveness or other unintended consequences may result because of such factors as weather conditions, presence of other materials, or the manner of use or application, all of which are beyond the control of DuPont. In no case shall DuPont be liable for consequential, special or indirect damages resulting from the use or handling of this product. All such risks shall be assumed by the Buyer. DUPONT MAKES NO WARRANTIES OF MERCHANTABILITY OR FITNESS FOR A PARTICULAR PURPOSE NOR ANY OTHER EXPRESS OR IMPLIED WARRANTY EXCEPT AS STATED ABOVE.

DIRECTIONS FOR USE

It is a violation of federal law to use this product in a manner inconsistent with its labeling.

DuPont KARMEX Herbicide should be used only in accordance with recommendations on this label, or in separate published DuPont recommendations available through local dealers.

DuPont will not be responsible for losses or damages resulting from use of this product in any manner not specifically recommended by DuPont. User assumes all risk associated with such non-recommended use.

Do not use in Kern County, California, except for noncropland and citrus weed control.

Do not enter treated areas without protective clothing until sprays have dried. Because certain states may require more restrictive reentry intervals for various crops treated with this product, consult your State Department of Agriculture for further information.

Written or oral warnings must be given to workers who are expected to be in a treated area or in an area about to be treated with this product. When oral warnings are given, warnings shall be given in a language customarily understood by workers. Oral warnings must be given if there is reason to believe that written warnings cannot be understood by workers. Warnings must include the following information: "CAUTION. Area treated with KARMEX on (date of application). Do not enter without appropriate protective clothing until sprays have dried. In case of contact, flush eyes with plenty of water; wash skin with plenty of soap and water. Get medical attention if irritation to skin or eyes persists."

SELECTIVE USE IN CROPS

PREEMERGENCE USE (Germinating Weeds): KARMEX, at recommended rates, controls annual weeds such as:

Broadleaves

3/4 to 1 Lb./Acre	1½ to 2 Lbs./Acre	2 to 6 Lbs./Acre
Lambsquarters	Annual groundcherry	Ageratum
Pigweed	Annual morningglory	Annual smartweed
Purslane	Chickweed	Annual sowthistle
Ragweed	Corn spurry	Corn speedwell
	Dogfennel	Dayflower
	Fiddleneck (amsinckia)	Flora's paintbrush
		Hawksbeard
	Gromwell	Horseweed
	Knawel	Kochia
	Pennycress	Marigold
	Shepherdspurse	Mexican clover
	Tansy-mustard	Pineappleweed
	Wild buckwheat	Pokeweed

Broadleaves

3/4 to 1 Lb./Acre	1½ to 2 Lbs./Acre	2 to 6 Lbs./Acre
	Wild lettuce Wild mustard	Rabbit tobacco Spanishneedles Velvetleaf (buttonweed) Wild radish

Grasses

3/4 to 1 Lb./Acre	1½ to 2 Lbs./Acre	2 to 6 Lbs./Acre
Barnyardgrass (watergrass) Crabgrass	Annual bluegrass Annual sweet vernalgrass Foxtail Rattail fescue Red sprangletop Velvetgrass	Annual lovegrass Annual ryegrass Kyllinga Orchardgrass Peppergrass Ricegrass Sandbur Seedling johnsongrass

Partial control of the following weeds usually occurs at rates stated:

Broadleaves

1Lb./Acre	4 Lbs./Acre
Annual morningglory Cocklebur Prickly sida (teaweed) Sesbania Sicklepod	Horsenettle

Grasses

4 Lbs./Acre	8 to 10 Lbs./Acre
Quackgrass	Guineagrass Maidencane Pangolagrass

POSTEMERGENCE USE (Emerged Seedling Weeds): KARMEX, at recommended rates, controls annual weeds such an annual morningglory, barnyardgrass (watergrass), crabgrass, crowfoot, goosegrass, pigweed and purslane. Addition of Surfactant WK to the spray (where recommended) increases contact effects of KARMEX. Best results are obtained on succulent weeds growing under conditions of high humidity and temperatures of 70°F or higher.

EQUIPMENT—SPRAY VOLUMES AND PRESSURES: Use a boom power sprayer properly calibrated to a constant speed and rate of delivery. Openings in screens should be equal to or larger than 50 mesh. Continuous agitation in the spray tank is required to keep the material in suspension. Agitate by mechanical or hydraulic means; if bypass or return line is used, it should terminate at bottom of tank to minimize foaming. Avoid overlapping, and shut off spray booms while starting, turning, slowing or stopping, or injury to the crop may result.

For preemergence application, use 25 to 40 gals. per acre and spray pressure of 30 to 40 psi. For postemergence application, use sufficient volume (min. 25 gals. per acre) for thorough coverage of weed foliage; use spray pressure of 20 to 25 psi to keep spray drift to a minimum.

Aerial: For alfalfa, asparagus, barley (winter), cotton (preplant or preemergence only), grass seed crops, pineapple, sugarcane and wheat (winter), application may be made by aircraft (5 to 10 gals. per acre); avoid overlapping of spray swath and application under conditions where excessive drift may occur. Where land is bedded, make application parallel to rows.

SPRAY PREPARATION: Mix proper amount of KARMEX into necessary volume of water; where use of DuPont Surfactant WK is recommended, dilute with 10 parts of water and add as last ingredient to nearly full tank.

USE RATES: All dosages of KARMEX are expressed as broadcast rates; for band treatment, use proportionately less. For example, use ⅓ of the broadcast rate when treating a 14" band where row spacing is 42". Where a range of dosages is given, use the lower rate on coarse textured soils (low in clay or organic matter) and the higher rate on fine textured soils (high in clay or organic matter); for postemergence application, use the lower rate on smaller weeds and the higher rate on larger weeds.

SOIL LIMITATIONS: Crop injury may result from failure to observe the following:

Unless otherwise directed, do not use on sand, loamy sand, gravelly soils or exposed subsoils; nor on pecans where organic matter is less than ½%; nor on alfalfa, apples, artichoke, barley (winter), bermudagrass pasture, citrus, cotton, grapes, oats, olives, papayas, peaches, pears, plumosus fern, sorghum, sugarcane, walnuts and wheat (winter) where organic matter is less than 1%; nor on blueberries, birdsfoot trefoil, caneberries, gooseberries, macadamia nuts and peppermint where organic matter is less than 2%. Preemergence weed control will be reduced on high organic matter soils (greater than 5%, such as peat or muck).

REPLANTING: Unless otherwise directed, do not replant treated areas to any crop within 2 years after last application, as injury to subsequent crops may result.

FIELD CROPS (See Soil Limitations)

A good seedbed must be prepared before preemergence use of KARMEX, as crop injury may result if application is made to ground which is cloddy or compacted resulting in improperly planted seed. Plant seed to depth specified. Unless otherwise directed, surface of the soil should not be cultivated or disturbed after application of KARMEX and before emergence of the crop as weed control may be reduced and crop injury may result. However, if moisture is insufficient to activate the herbicide, a shallow cultivation (rotary hoe preferred) should be made after emergence of crops while weeds are small enough to be controlled by mechanical means.

ALFALFA: Treat only stands established for 1 year or more. Do not apply to seedling alfalfa nor to alfalfa grass mixtures; do not apply to alfalfa under stress from disease, insect damage, shallow root penetration (such as on shallow hard pans), alkali spots; nor to flooded fields, as crop injury may result. Do not spray on snow covered or frozen ground.

Idaho, Oregon, Washington: Use 1½ to 3 lbs. per acre; for control of volunteer alfalfa, use 4 lbs. per acre. Apply in fall after alfalfa becomes dormant, but no later than mid-December.

California (Dormant and Semi-Dormant Varieties): Use 1½ to 3 lbs. per acre; for control of volunteer alfalfa, use 4 lbs. per acre. Apply in fall or winter, after alfalfa becomes dormant or semi-dormant, but before growth begins in the spring. Crop injury may result if application is made to actively growing alfalfa. For best results, apply before weeds have emerged or become established (2" in height or diameter). Control of established weeds is improved by applying KARMEX with a suitable contact herbicide registered for such use. Sufficient rainfall for soil activation of KARMEX is unlikely in California after February 1. Treated areas may be replanted to any crop after one year from last application, if rate does not exceed 2 lbs. per acre.

Arizona, Nevada: Use 1½ to 3 lbs. per acre; apply in fall after alfalfa becomes dormant but no later than January.

Eastern Colorado, Kansas: For control of tansymustard, apply 1 lb. per acre shortly after emergence of mustard in the fall or winter; use 2 lbs. per acre if weeds are 2" to 4" in height. Alternatively, if other annual weeds are present, apply 2 to 3 lbs. per acre in February or March.

Other Areas Where Alfalfa Becomes Winter Dormant: Use 1½ to 3 lbs. per acre (1½ to 2 lbs. per acre East of Appalachian Mountains). Apply in March or early April, but before spring growth begins.

ARTICHOKE—California: Apply 2 to 4 lbs. per acre in late fall or early winter after the last cultivation. Apply before weeds germinate or to emerging seedlings. Direct spray to cover the area between the rows and at the base of artichoke plants, keeping contact with crop plants at a minimun.

ASPARAGUS: Apply as a band or broadcast treatment. Do not apply to young plants during the first growing season (except as noted below), nor to newly seeded asparagus, nor on plants with exposed roots, as severe crop injury may result. Preemergence weed control will be reduced on high organic matter soils (greater than 5%).

Established Plantings: On light sandy soils, and other soils low in clay or organic matter, apply 1 to 2 lbs. per acre. On soils high in clay or organic matter, use 2 to 4 lbs. per acre. Two applications may be used; the first application should be made before weeds become established, but no earlier than 4 weeks before spear emergence and no later than the early cutting period (if weeds are controlled into the cutting period by cultural practices, application may be delayed until immediately after the last cultivation); a second application may be made immediately following completion of harvest, provided rainfall is expected. When two applications are used in one season, do not exceed 3 lbs. per acre per application. In Washington (irrigated crop), apply a single treatment of 4 lbs. per acre. If treatment is delayed until late winter or early spring, incorporation of the chemical in the top 1" to 2" of soil may substitute for lack of rain to activate the herbicide.

Newly Planted Crowns—California (San Joaquin Delta): Make a single application of 2 to 4 lbs. per acre on soils high in clay or organic matter; use the lower rate on clay loams and the higher rate on peat soils. Do not use on soils containing less than 2% organic matter. Soil must be settled by rainfall or irrigation prior to treatment. Do not treat crowns planted to a depth of less than 2".

BARLEY, WINTER (Drill-Planted)—Western Oregon and Western Washington: Make a single application of 1½ to 2 lbs. per acre as soon as possible after planting but before emergence of barley. Do not replant treated areas to any crop within 1 year after last application, as injury to subsequent crops may result.

BERMUDAGRASS PASTURES (Newly Sprigged): Apply 1 to 3 lbs. after planting and before emergence of bermudagrass or weeds. Alternatively, for control of emerged annual weeds up to 4" in height, apply ½ to 1 lb. per acre; add 1 pt. Surfactant WK per 25 gals. of spray. If bermudagrass has emerged at time of treatment, temporary burn of exposed plant parts may occur.

Plant sprigs (stolons) 2" deep in a well-prepared seedbed; do not treat areas where sprigs are planted less than 2" deep, as crop injury may result. Do not graze or feed foliage from treated areas to livestock within 70 days after application.

BIRDSFOOT TREFOIL (Lotus)—Western Oregon: Treat only stands established for at least 1 year; do not apply to seedling trefoil, as injury may result. Make a single application of 2 lbs. per acre when trefoil is dormant (October 15 to December 15). Do not replant treated areas to any crop within 1 year after last application, as injury to subsequent crops may result.

CORN (FIELD)

Postemergence—Make a single application of ¾ lb. per acre in combination with non-pressure nitrogen solution. If nitrogen solution is not used, apply 1 lb. per acre; add 1 pt. Surfactant WK per 25 gals. of spray. Apply as a directed spray when corn is at least 20" high and weeds are no taller than 3". DO NOT APPLY OVER TOP OF CORN. Do not replant to any crop within 1 year, except that cotton, corn and grain sorghum may be planted the spring following treatment.

Preemergence—Arkansas, Louisiana, Mississippi and Tennessee: Make a single application of ⅔ to 1 lb. per acre as a broadcast or band treatment, after planting but before corn emerges. Plant corn at least 1½" deep. Do not replant treated areas to crops other than corn or cotton within 4 months following band treatment and 6 months following broadcast treatment, as crop injury may result.

COTTON: During a single crop season, do not exceed the following amounts of KARMEX per acre, as injury to subsequent crops may result: 1 lb. on loamy sand; 1½ lbs. on sandy loam; 2 lbs. on clay loam; 2¾ lbs. on clay. Injury may occur if KARMEX is used in conjunction with soil-applied organic phosphate pesticides. Do not allow livestock to graze treated cotton.

Preplant—Arizona and California: Use KARMEX alone, or apply as a separate operation following preplant broadcast treatment with Treflan* (incorporated according to directions on Treflan label). Apply KARMEX as a broadcast spray after beds are formed, preirrigated, and final seed beds prepared. Prior to planting, drag-off the tops of the beds and plant in moist soil not treated with KARMEX. Treated soil is returned to the bed after planting, when irrigation furrows are reformed after cotton has emerged. If more than two furrowing-out operations are made prior to lay-by, or deep furrows are made early, weed control may be reduced in furrow bottoms. Use at the following rates:

*Registered trademark of Elanco Products Co.

KARMEX Alone: 1 to 2½ lbs. per acre.

KARMEX Following Treflan:

Soil Texture	Product Per Acre Preplant	
	Treflan	KARMEX
Sandy loam, loam, silt loam, silt	1 pt.	⅔ to 1 lb.
Sandy clay loam, clay loam, silty clay loam, sandy clay, clay	1½ pts.	1 to 1¼ lbs.

Note: Seedling disease may weaken plants and increase the possibility of injury from the use of Treflan followed by KARMEX. These treatments should be used only in conjunction with a standard fungicide seed treatment, plus a good supplemental soil fungicide program such as captan-PCNB mixture.

Preemergence—U.S. except Arizona, California, and areas west of Interstate 35 or 35W in Texas and Oklahoma: Use KARMEX alone, or apply as a separate operation following preplant treatment with Treflan. Apply KARMEX after planting, but before cotton emerges. Do not treat cotton in deep furrows, as crop injury may result; use only where cotton is planted on flat or raised seedbeds. Shallow incorporation (no deeper than ¼") with a rotary hoe or similar equipment following planting usually improves results, especially during dry weather. A wide press wheel should be used on the planter to provide a level seed bed for subsequent early season postemergence treatments. If moisture is insufficient to activate KARMEX, or if soil becomes crusted before crop emerges, a shallow rotary hoeing (no deeper than ¼") should be made before weeds become established.

KARMEX Alone: Make a single application as a broadcast or band spray, using the following broadcast rates; for band treatment, use proportionately less.

Soil Texture**	Lbs. KARMEX Per Acre
Loamy sand	⅔
Sandy loam, loam, silt loam, silt	1
Sandy clay loam, clay loam, silty clay loam, sandy clay	1¼
Silty clay, clay	2

** Do not use on soils with less than 1% organic matter, as crop injury may result.

KARMEX Following Treflan Preplant: Apply Treflan prior to planting as a broadcast or band treatment; incorporate according to directions on Treflan label. As a separate operation, apply KARMEX as a band treatment (14" to 20" wide) after planting but before cotton emerges. Use at the following broadcast rates; for band treatment, use proportionately less. See "Note" under Preplant above.

Soil Texture**	Product Per Acre	
	Preplant Treflan	Preemergence KARMEX
Loamy sand	½ pt.	⅔ lb.
Sandy loam, loam, silt loam, silt	1 pt.	1 lb.
Sandy clay loam, clay loam, silty clay loam, sandy clay, silty clay, clay	1½ pts.	2 lb.

** Do not use on soils with less than 1% organic matter, as crop injury may result.

Postemergence, U.S.: Apply only as a directed spray to cover weed foliage; adjust nozzles to minimize contact of cotton leaves with spray or drift or crop injury may result. DO NOT SPRAY OVER TOP OF COTTON.

Early Season—Apply when cotton is at least 6″ tall, and when weeds are actively growing and do not exceed 2″ in height. Apply as a band treatment at following rate of 1 pt. Surfactant WK for each 25 gals. of spray. Two applications may be made if needed.

Weed Problem (Up to 2″ Tall)	Lbs. KARMEX Per Acre (Broadcast Basis)
Annual grasses	½
Pigweed	¼

For control of seedling perennial grasses such as johnsongrass and partial control of nutsedge, or when weed growth is under drought stress or as high as 4″, add 2 to 3½ lbs. disodium methylarsonate (DSMA; 63% anhydrous or equivalent) to above spray mixture. If DSMA is used, do not apply after first bloom.

Late Season (Lay-By)—Apply 1 to 1½ lbs. per acre (1 to 2 lbs. in Arizona and California) when cotton is at least 12″ tall (at least 20″ tall for Pima S-2). For control of germinating weed seedlings, apply to soil beneath cotton plants and between rows immediately after last cultivation. In irrigated cotton, best weed control is obtained if the field is irrigated within 3 to 4 days after application; thoroughly wet the surface of the ground over the row to carry the herbicide into the root zone of germinating weeds. Alternatively, for control of emerged annual weeds (up to 4″ in height) at lay-by time, make a single application in combination with Surfactant WK (1 pt. per 25 gals. spray), or use ½ to ¾ lb. KARMEX (plus surfactant) per acre and repeat later if needed.

Replanting: If initial seeding fails to produce a stand, cotton may be replanted in soil treated preplant or preemergence with KARMEX, alone or following Treflan. Wherever possible, avoid disturbing original bed. If necessary to rework soil before replanting, use shallow cultivation such as discing; do not relist nor move soil into the original drill area. Plant seed at least 1″ deep. Do not retreat field with a second preplant or preemergence application during the same crop year, as injury to the crop may result.

Subsequent Crops:

KARMEX—Type of Application	Crops That May Follow Treated Cotton
Band preemergence or postemergence	Any crop 4 months after last application.
Band preemergence plus postemergence —or— Broadcast preemergence (and preplant) —or— Broadcast preemergence plus band postemergence	Cotton, soybeans, corn or grain sorghums (not sorgos or forage sorghums nor grass sorghums) the next spring. Do not replant treated areas to any other crop within one year after last application, as injury to subsequent crops may result.
Broadcast postemergence (lay-by)	Cotton, corn, grain sorghums (not sorgos or forage sorghums nor grass sorghums) the next spring. Do not replant treated areas to any other crop within one year after last application, as injury to subsequent crops may result.

For subsequent crops in fields where Treflan is used, follow instructions on Treflan label.

GRASS SEED CROPS (Perennial): Except as noted, apply only to established plantings at least 1 year old.

Colorado, Kansas, New Mexico and Oklahoma: On sand bluestem, side oats grama and switchgrass, apply 2 to 3 lbs. per acre during the dormant period shortly before weed seedlings emerge. Do not apply after crop begins growth in the spring, as crop injury may result. In fields where ash residues have accumulated from burning straw, use 3 lbs. per acre; spread unburned chaff or straw with a harrow or chopper before application.

Western Oregon: On alta fescue, Astoria bentgrass, Highland bentgrass, Kentucky bluegrass (Merion bluegrass) and orchardgrass, apply 2 to 4 lbs. per acre between October 1 and November 15. In fields where ash residues have accumulated from burning straw, use 3 to 4 lbs. per acre; spread unburned chaff or straw with a harrow or chopper before application. If perennial velvetgrass *(Holcus lanatus)* is a problem, use 4 lbs. per acre. For best results, apply as soon as possible after fall rains start. Established weeds (beyond 2 to 4 leaf stage) should be removed prior to treatment.

Well established vigorous stands of spring-planted alta fescue, Kentucky bluegrass and orchardgrass may be treated the following fall, provided the crop is planted before April 1 and treatment is not applied before October 15; use 2 lbs. per acre.

Oregon: For use in newly planted bentgrass, Chewings fescue, Kentucky bluegrass, perennial ryegrass, orchardgrass and tall fescue. During planting operation, spray Aqua Nu Char† or Gro-Safe†† or other suitable brand of activated charcoal as a 1″ band on soil surface at rate of 300 lbs. per acre (broadcast basis; equivalent to 15 lbs. per acre of crop where row spacing is 20″). Mount nozzles to apply directly over seed rows to prevent crop injury. Follow with KARMEX as a single broadcast spray at rate of 2½ to 3 lbs. per acre; apply as soon as possible after planting, but before crops or weeds emerge and before rains or sprinkler irrigation. Fall or spring plantings may be treated; best results usually occur with early fall plantings. Treatment will not control downy brome or wild oats.

†Reg. trademark of Westvaco Corp.

††Reg. trademark of ICI United States Inc.

OATS (Drill-Planted): Do not replant treated areas to any crop within one year after last application, as injury to subsequent crops may result.

Spring Oats—Idaho, Eastern Oregon, Eastern Washington: Use in areas where average annual rainfall exceed 16″. Make a single application of 1 to 1½ lbs. per acre after planting, either before or after oats emerge, but within 6 weeks of planting. Best results are usually obtained when application is made 3 to 4 weeks after planting. Apply before weeds are 3″ to 4″ tall.

Winter Oats and Mixtures with Peas or Vetch—Western Oregon and Western Washington: Make a single application of 1½ to 2 lbs. per acre as soon as possible after planting, but before emergence of the crop.

PEPPERMINT—Pacific Northwest: Apply 3 lbs. per acre just after the last cultivation in the spring, prior to emergence of peppermint. Do not apply to newly planted (less than 1 year) nor to emerged peppermint as injury may result.

RED CLOVER—Western Oregon: Make a single application of two pounds per acre on established red clover stands (at least 9 months). Apply KARMEX when red clover is dormant (October 15 to December 15). Do not apply to seedling red clover, and do not replant treated area to any crop within one year after last application.

Treatment will control annual weeds such as bluegrass, chickweed, hawksbeard, rattail fescue, rye grass, and velvet grass.

SORGHUM (GRAIN)—Southwestern States: Apply ¼ to ½ lb. per acre; add 1 pt. Surfactant WK per 25 gals. of spray. Apply as directed postemergence broadcast or band spray after sorghum is 15″ tall to control weeds 2″ to 4″ in height. DO NOT SPRAY OVER TOP OF SORGHUM. Use the lower rate on broadleaved weeds up to 2″ tall; use the higher rate on grasses up to 2″ and broadleaved weeds up to 4″ tall. When the lower rate is used, a second application may be made if needed provided the amount applied in one crop year does not exceed ½ lb. per acre. Treatment of weeds under drought stress is usually ineffective.

Do not replant treated areas to crops other than cotton or corn within 4 months following band treatment and 6 months following broadcast treatment as crop injury may result.

SUGARCANE: To prevent possible crop injury on new cane varieties, tolerance to KARMEX should be determined prior to adoption as field practice. Do not treat sugarcane growing on thinly covered subsoils or rocky areas as crop injury may result. Temporary chlorosis of the crop may result from application over emerged cane; to minimize chlorosis, use directed postemergence sprays.

Florida Preemergence—For high organic soils, apply 2 to 4 lbs. per acre as a broadcast or band spray prior to weed emergence after planting or after harvesting plant crop (for ratoon crop). **Postemergence**—Make 1 or 2 applications of 2 lbs. per acre as needed by directed spray inter-row. Alternatively, for panicum control, make up to 3 applications of ½ to 1 lb. per acre as a directed spray after cane has emerged, but before panicum exceeds 2" in height; add 1 qt. Surfactant WK per 100 gals. of spray. Adjust nozzles to spray beneath cane plants and between rows to cover weed foliage and to minimize contact of cane leaves with spray or drift. Do not apply more than 6 lbs. total per acre between planting (or ratooning) and harvest.

Hawaii and Puerto Rico: Apply 4 to 8 lbs. per acre as a broadcast spray prior to weed emergence after planting or after harvesting plant crop (for ratoon crop). A second and third application of 2 to 4 lbs. per acre may be made as a broadcast spray over emerged cane, or by directed spray inter-row.

If weeds are emerged, add surfactant (such as Surfactant WK, Osamul 95 or Sterox SK) to the spray at the rate of 1 to 2 qts. per 100 gals. and apply as a directed spray. DO NOT SPRAY OVER TOP OF CANE.

Do not apply more than 3 treatments nor more than 10 lbs. (Puerto Rico) or 12 lbs. (Hawaii) total per acre between planting (or ratooning) and harvest. Treated areas may be planted to sugarcane or pineapple one year after last application.

Louisiana: Use on plant cane seeded on fallowed ground. Make a single application of 3 to 3¾ lbs. per acre at either of the following times: **Fall Treatment** (August through October)—Treat a 2 ft. band over the row after planting of cane, but before weeds or cane emerge. **Spring Treatment** (January through April)—if shaving and off-barring are practiced, treat a 2 ft. band over the row before weeds or cane emerge.

WHEAT, WINTER (Drill-Planted): Crop injury may result where severe winter stress, disease or insect damage follows application; winter-sensitive varieties such as McDermid and Hyslop may be less tolerant of KARMEX than winter-hardy varieties such as Gaines and Nugaines. Crop injury may also result from failure to observe the following: Do not use on sand or loamy sand soils, nor on gravelly or sandy loams low in organic matter (less than 1%) nor on thinly covered or exposed subsoil areas (clay knobs); do not treat wheat planted less than 1" deep; do not treat wheat where winter climatic conditions have caused "heaving" of plants; do not treat wheat plants are lacking in vigor due to poor emergence, insect damage, disease, high alkalinity or other causes; do not apply after wheat has reached the "boot" stage of maturity; do not use with surfactants or nitrogen solution. Do not replant treated areas to any other crop within 1 year after last treatment (except as noted), as injury to subsequent crops may result.

Idaho, Oregon and Washington—East of Cascade Range: Areas Where Average Annual Rainfall Exceeds 16 Inches: Make a single application of 1 to 1½ lbs. per acre. **Fall Treatment:** For early fall-planted wheat (seeded before September 10), apply 3 to 6 weeks after planting, but before weeds are 3" to 4" tall. Treatment after October 1 has generally given best results. Application should not be made after soil freezes in the fall. Wheat planted in late October should not be treated until the following spring. **Spring Treatment:** Apply as soon as wheat starts to grow in the spring. Treatment made prior to April 10 will usually give good results, provided weed growth is less than 4" tall. Application later than May 1 may give poor results.

Alternatively, make a single application of ½ to 1 lb. KARMEX plus ¼ lb. bromoxynil per acre as a tank mixture, either in the fall after wheat has emerged, but before soil freezes, or in the spring as soon as soil thaws; apply before weeds are 2" tall or across.

Areas Where Average Annual Rainfall is 10 to 16 Inches: After wheat is planted in the fall, make a single application of 1 to 1½ lbs. per acre, when sufficient moisture is available to germinate wheat seed. Apply before soil freezes and before weeds are 2" tall. Application later than March 1 may give poor results.

Note—If fall-planted wheat fails to grow due to winter kill or adverse conditions after fall treatment, only fields treated before November 1 may be replanted to spring wheat. Spring wheat should not be planted before April 1, and only after deep discing and plowing to a depth of 4" to 6" prior to planting. Do not retreat field with a second application during the same crop year, as injury to the crop may result.

Oregon and Washington—West of Cascade Range: Make a single application of 1½ to 2 lbs. per acre as soon as possible after planting; if wheat and weeds have emerged, apply before weeds are 3" to 4" tall. Alternatively, apply a tank mixture of KARMEX plus bromoxynil as detailed above for "East of Cascade Range".

Other Areas: Make a single application in the spring as soon as wheat (fall-planted) starts to grow, and before weeds are 2" tall. Application later than May 1 may give poor results.

Central Plains and Midwest: Use 1 to 2 lbs. per acre.

Kansas, Oklahoma and Texas: Do not use on sand or sandy loam soils. Use 1 lb. per acre on silt and silt loam soils and 1½–2 lbs. per acre on clay, clay loam, and silty clay loam soil.

Northeast: Use 1 to 1½ lbs. per acre.

FRUIT AND NUT CROPS (See Soil Limitations)

Unless otherwise directed, make a single application per year as a directed spray, avoiding contact of foliage and fruit with spray or drift. Do not graze livestock in treated orchards or groves.

APPLES: Use KARMEX alone, or apply as a tank mixture with DuPont Sinbar Herbicide.

KARMEX Alone—Use only under trees established in the orchard for at least 1 year; do not treat varieties grafted on full-dwarf root stocks. Apply 4 lbs. per acre in the spring (March through May). In the Far West, treatment may be made in winter (December through February), or apply 2 lbs. per acre as a postharvest treatment followed by 2 lbs. in the spring.

KARMEX + Sinbar—Use only under trees established in the orchard for at least 2 years. Apply either in the spring, or after harvest in the fall before weeds emerge, or during early seedling stage of weed growth.

Soil Texture	Lbs. Product Per Acre: 1 to 2% Organic Matter KARMEX + Sinbar	Lbs. Product Per Acre: More than 2% Organic Matter KARMEX + Sinbar
Sandy loam	1 + 1	1½ + 1½
Loam, silt loam, silt	1½ + 1½	2 + 2
Clay loam, clay	2 + 2	2 + 2

Where crop is grown under furrow irrigation or under raised-berm flood irrigation (trees 4" to 6" above waterline), apply only as a band treatment. Do not treat trees planted in the bottom of irrigation furrows, nor trees grown under flat flood or basin irrigation, as injury to trees may result. Where complete weed control to harvest is desired, additional weed control measures may be required during the growing season.

Georgia—Apply 2 to 3 lbs. per acre in the spring. Repeat application in the fall but do not use more than 4 lbs. per acre per year. Add Du Pont Surfactant WK at 1 pint per 25 gals. spray mixture to improve control of small, emerged weeds.

BANANAS AND PLANTAINS—New Plantings: To control annual weeds, apply 1½ to 3 lbs. per acre after planting but, before weeds emerge. Do not apply to loose soil directly over the planting material.

Established Plantings: For control of annuals and for topkill of perennials such as bermudagrass, birdseed grass and guineagrass, apply 3 to 6 lbs. per acre plus 1 pt. Surfactant WK (or suitable equivalent) per 25 gals. of spray; avoid contact of plants with spray or drift, as injury may result. When tall, dense weed growth is present, remove weed growth before application. If application is made to soil free of weeds, omit the surfactant from the spray. Repeat treatment as needed, but do not apply more often than 6-week intervals nor more than a total of 12 lbs. per acre (broadcast basis) in a 12-month period.

NOTE: Do not replant treated areas to any crop within 2 years after last application, as injury to subsequent crops may result, except that sugarcane or pineapple may be planted one year after last application.

BLUEBERRIES, CANEBERRIES AND GOOSEBERRIES: Use only in fields which have been established for at least 1 year. Do not apply to berries interplanted with fruit trees; do not apply to plants whose roots are exposed, as injury may result. Apply as a band treatment at base of canes or bushes; for spring application, apply before germination and growth of annual weeds.

Georgia Blueberries: Apply 1½ to 2 lbs. per acre in the spring and repeat treatment after harvest in the fall. Add Du Pont Surfactant WK at 1 pint per 25 gals. spray mixture to improve control of small, emerged weeds.

Indiana, Michigan and Ohio—Blueberries: Apply 2 to 4 lbs. per acre in late spring; alternatively, apply 2 lbs. per acre in the fall and repeat at the same rate in the spring. **Raspberries:** Apply 3 lbs. per acre in the spring.

Massachusetts—Blueberries: Apply 2 lbs. per acre in late spring.

New Jersey—Blueberries: For control of winter annuals, apply 2 lbs. per acre in October, November or December, or a single application of 2½ lbs. per acre in early to mid spring.

California—Raspberries, Blackberries, Boysenberries, Dewberries and Loganberries: For control of winter annuals, apply 2 lbs. per acre in October or November; repeat at same rate in late spring to control summer annuals. A single application of 3 lbs. per acre in January or February will control both winter and summer annuals in some areas, but the separate fall and spring schedule is preferred.

Western Oregon and Western Washington—Blueberries, Caneberries and Gooseberries: Use same schedule as recommended for California.

CITRUS: Use only under trees established in the grove for at least 1 year. Time the application as indicated for specific areas, except application may be made any time of the year where sprinkler or flood irrigation can be timed to activate the herbicide. Established perennial weeds require other special control procedures. Do not apply under citrus trees that have been subjected to freezing within 6 months.

Arizona (except Yuma area) and California (except Imperial and Coachella Valleys): Apply 3 to 4 lbs. per acre shortly after grove has been laid-up in final form (non-tillage program), in late fall or early winter. Alternatively, apply 2 lbs. per acre in October or November, and repeat at the same rate in March or April. Subsequent annual applications of 2 to 3 lbs. per acre will usually give adequate weed control.

Florida and Puerto Rico: Make a single application of 4 to 8 lbs. per acre, or apply 3 to 4 lbs. per acre followed by the same rate 4 to 6 months later. On bearing citrus, apply any time when seasonal rains are expected; on nonbearing trees, apply when winter banks are pulled down.

For control of guineagrass, loosestrife, maidencane, paragrass, primrose willow and seamyrtle in ditches adjacent to citrus groves, use 1 lb. per 1000 sq. ft. (40 lbs. per acre) in sufficient water (min. 4 gals. per 1000 sq. ft.) to provide thorough and uniform coverage. Apply in the spring: before weed growth starts, or after removal of vegetation. Repeat treatment on a spot basis to control hard-to-kill species, such as guineagrass. In bedded groves, do not treat water furrows between the beds, as injury to the trees may result.

Texas: Apply 2 to 4 lbs. per acre for annual weeds; use 4 to 6 lbs. per acre for control of johnsongrass seedlings. Best results accompany application in the spring; well established weeds should be eliminated by cultivation, prior to treatment.

GRAPES: Apply only to established vineyards (at least 3 years old) as a band treatment to grape rows. On soils low in clay or organic matter (1 to 2%), severe plant injury may result if heavy rainfall, or more than one inch of irrigation, occurs soon after treatment. This risk is assumed by the user.

East of the Rocky Mountains: On soils low in clay or organic matter (1 to 2%), apply 2 to 3 lbs. per acre; on soils high in clay or organic matter, apply 3 to 6 lbs. per acre. Apply in the spring, just prior to germination and growth of annual weeds.

West of the Rocky Mountains: Apply in November, December or January. For initial treatment, apply 3 to 4 lbs. per acre; subsequent annual applications of 2 lbs. per acre will usually give adequate weed control. Do not apply to vines with trunks less than 1½" in diameter as injury may result.

New York and Pennsylvania—Perennial Grasses: Use only in established vineyards (at least 4 years old) for spot control of perennial grasses such as orchardgrass, quackgrass and ryegrass. Apply in the spring as a band treatment to ridge soil (2" to 4" high) under the trellis at the rate of 8 to 12 lbs. per acre. Band width should not exceed 30". Do not apply more than once every 4 years. Use only on heavy soils such as loams, silt loams, clay loams. Do not use in areas where grape roots are shallow or exposed because of high bedrock, poor drainage, or erosion as injury to grapes may result.

MACADAMIA NUTS—Hawaii: Use only under trees established in the orchard for at least 1 year. Apply 2 to 6 lbs. per acre immediately after harvest, preferably before weeds emerge; if weeds have emerged, add 1 pt. Surfactant WK per 25 gals. of spray. Retreat as needed, but do not exceed 10 lbs. per acre per year.

OLIVES—California: Use only under trees established in the grove for at least 1 year. Apply 2 lbs. per acre after grove has been laid-up in final form, in late October or November; repeat at same rate in March or April. Remove weed growth prior to treatment.

PAPAYAS: Use only under trees established in the orchard for at least 1 year. Apply 2½ to 5 lbs. per acre, preferably before weeds emerge; if weeds have emerged, add 1 pt. Surfactant WK per 25 gals. of spray.

PEACHES: Use KARMEX alone, or apply as a tank mixture with Sinbar.

KARMEX Alone—Use only under trees established in the orchard for at least 3 years. Apply 2 to 5 lbs. per acre in the early spring, before weeds emerge or during the early seedling stage of weed growth. Do not apply within 3 months of harvest; in the Far West, do not apply within 8 months of harvest.

KARMEX + Sinbar—Use only under trees established in the orchard for at least 2 years. Apply either in the spring or after harvest in the fall before weeds emerge, or during early seedling stage of weed growth.

Soil Texture	Lbs. Product Per Acre: 1 to 2% Organic Matter KARMEX + Sinbar	Lbs. Product Per Acre: More than 2% Organic Matter KARMEX + Sinbar
Sandy loam	1 + 1	1½ + 1½
Loam, silt loam, silt	1½ + 1½	2 + 2
Clay loam, clay	2 + 2	2 + 2

Where crop is grown under furrow irrigation or under raised-berm flood irrigation (trees 4" to 6" above waterline), apply only as a band treatment. Do not treat trees planted in the bottom of irrigation furrows, nor trees grown under flat flood or basin irrigation, as injury to trees may result. Where complete weed control to harvest is desired, additional weed control measures may be required during the growing season.

Georgia—On trees established for at least 2 years, apply 2 to 3 lbs. per acre in the spring. Repeat application in the fall but do not exceed 5 lbs. per acre per year. Add Du Pont Sufactant WK at 1 pint per 25 gals. spray mixture to improve control of small, emerged weeds.

Where crop is grown under furrow irrigation or under raised-berm flood irrigation (trees 4" to 6" above waterline), apply only as a band treatment. Do not treat trees planted in the bottom of irrigation furrows, nor trees grown under flat flood or basin irrigation, as injury to trees may result. Where complete weed control to harvest is desired, additional weed control measures may be required during the growing season.

PEARS: Use only under trees established in the orchard for at least 1 year; do not treat varieties grafted on full-dwarf root stocks. Apply 4 lbs. per acre in the spring (March through May). In the Far West, treatment may be made in winter (December through February), or apply 2 lbs. per acre as a postharvest treatment followed by 2 lbs. in the spring.

PECANS: Use KARMEX alone or apply as a tank mixture with Sinbar. Make a single band or broadcast application as a directed spray, using a minimum of 30 gals. of water per acre. Apply in the spring before weeds emerge, or during the early seedling stage of growth.

	Lbs. Product Per Acre				
Soil Texture	KARMEX Alone*	OR	Tank Mixture KARMEX	+	Sinbar**
Sandy loam	2		$1\frac{1}{2}$	+	$1\frac{1}{2}$
Loam, silt loam, silt	3		$1\frac{3}{4}$	+	$1\frac{3}{4}$
Clay loam, clay	4		2	+	2

* Use only on trees established in the grove for at least 3 yrs. and on soils with at least ½% organic matter.

** Use on trees established in the grove for at least 1 yr. and on soils with at least 1% organic matter.

Note: Do not use on eroded areas where subsoil or roots are exposed, nor on trees that are diseased or lacking in vigor, or on trees planted in irrigation furrows, as injury to the trees may result.

PINEAPPLE—Hawaii and Florida: Apply 4 to 8 lbs. per acre as a broadcast spray just before or immediately after planting, but prior to weed emergence. Use 4 lbs. per acre after harvesting plant crop (for ratoon crop). For plant crop only, a second and third broadcast or interspace application may be made prior to differentiation, at the rate of 2 lbs. per acre at intervals of not less than 2 months. Additional applications to plant crop may be made as needed to interspace only, using 2 lbs. per acre. Do not apply more than 3 broadcast sprays (maximum 12 lbs. per acre) prior to differentiation, nor more than 16 lbs. total per acre per plant crop. Treated areas may be planted to pineapple or sugarcane 1 year after last application.

WALNUTS (English)—California: Use only under trees established in the orchard for at least 1 year. As an initial treatment, apply 3 to 5 lbs. per acre after the orchard has been laid-up in final form (nontillage program) in late fall or early winter; retreat annually with 2 to 3 lbs. per acre. Alternatively, apply 2 lbs. per acre in October or November and repeat at same rate in March or April.

ORNAMENTAL CROPS (See Soil Limitations)

ORNAMENTAL BULB CROPS (Bulbous Iris, Narcissus)—Western Washington: Make a single application of 4 lbs. per acre. Apply after planting but no later than 4 weeks prior to bulb emergence (usually late September or October). Do not replant treated areas to any crop within 1 year after last application, as injury to subsequent crops may result.

PLUMOSUS FERN—Florida: Hand weed and mow fern; then make a single application of 3 lbs. per acre within 3 to 5 days. Do not cultivate or disturb soil after application, as crop injury may result. Treat only established stands at least 1 year old.

TREE PLANTINGS—Colorado, Montana, Nebraska, North Dakota, South Dakota, Wyoming: Use only under established plantings (1 year or older) of American elm, caragana, cottonwood, Douglas fir, green ash, honeysuckle, Ponderosa pine, redcedar, Russian olive and Siberian elm. Use 2½ to 5 lbs. per acre; apply as a band 4 ft. wide in the tree row (2 ft. on each side of row). For example, 1 oz. KARMEX (4 level tablespoonfuls) treats 135 ft. of tree row (2 ft. on each side of row) at the rate of 5 lbs. per acre. Apply as a directed spray in early spring, before weeds emerge and before trees leaf out. Do not apply to foliage of trees, nor under trees growing in low areas, as injury to the trees may result.

NON-CROP WEED CONTROL

KARMEX is an effective herbicide for the control of many annual and perennial grasses and herbaceous weeds on noncropland areas where bare ground is desired. The degree of control and duration of effect will vary with the amount of chemical applied, soil texture, rainfall and other conditions.

KARMEX may be used as a preemergence treatment at any time of the year, except when ground is frozen, provided adequate moisture is supplied by rainfall or artificial means to activate the herbicide. Best results are obtained if application is made to the soil shortly before weed growth begins. If dense growth is present, remove tops and spray the ground.

Increased contact activity on established weeds may be obtained by the addition of DuPont Surfactant WK at the rate of 2 qts. per 100 gals. of spray mixture. Apply as a drenching spray to actively growing weeds during warm weather when daily temperature will exceed 70°F.

Except for small areas, use a fixed-boom power sprayer properly calibrated to insure a constant rate of application. Mix proper amount of KARMEX into volume of water necessary to obtain uniform coverage; if Surfactant WK is used, dilute with 10 parts of water and add as last ingredient to nearly full tank. Material must be kept in suspension at all times. Agitate by mechanical or hydraulic means in the spray tank; if bypass or return line is used, it should terminate at bottom of tank to minimize foaming. Openings in screens should be equal to or larger than 50 mesh.

General Weed Control: To control most weeds for an extended period of time on noncropland, such as utility, highway, pipeline and railroad right-of-ways, petroleum tank farms, lumberyards, storage areas, industrial plant sites and around farm buildings, apply 5 to 20 lbs. per acre. Use 20 to 60 lbs. per acre for perennial weeds; additional treatment may be required where a longer period of control is desired, or when hard-to-kill, deep-rooted perennial weeds, such as johnsongrass, are present. In low rainfall areas, KARMEX may not provide satisfactory control of deep-rooted perennial weeds.

For weed control on small areas, use one-half cupful of KARMEX per 100 sq. ft. for a dosage of approximately 50 lb. per acre.

Irrigation and Drainage Ditches: Apply 5 to 20 lbs. per acre to control most annual weeds; use 20 to 60 lbs. per acre to control both annual and perennial weeds. Apply only when water is not in the ditch. For irrigation ditches, apply during the noncrop season, and when ditch is not in use. To minimize movement of KARMEX with irrigation water (to avoid possible crop injury), it is essential that the herbicide be fixed in the soil by moisture. Apply before expected seasonal rainfall (if possible, when soil in the ditch is still moist). Following treatment, if rainfill has not totaled at least 4 inches, fill ditch with water and allow to stand for 72 hours; drain off and waste remaining water before using ditch. Do not treat any ditch into which roots of trees or other desirable plants may extend, as injury may result.

STORAGE AND DISPOSAL

STORAGE

Store product in original container only, away from other pesticides, fertilizer, food or feed.

DISPOSAL

Do not contaminate water, food or feed by storage or disposal. Wastes resulting from the use of this product may be disposed of on site, or at an approved waste disposal facility.

Completely empty bag into application equipment. Then dispose of empty bag in a sanitary landfill or by incineration, or, if allowed by State and local authorities, by burning. If burned, stay out of smoke.

SUPPLEMENTAL LABELING

KARMEX Herbicide

KARMEX
® HERBICIDE PLUS ZORIAL® COTTON HERBICIDE FOR WEED CONTROL IN COTTON

DIRECTIONS FOR USE

It is a violation of federal law to use this product in a manner inconsistent with its labeling.

DuPont KARMEX Herbicide is recommended for use as a tank mix with Zorial® 80 WP*, for preemergence control of annual weeds as shown on KARMEX® labeling, plus control of prickly sida (teaweed), anoda, spurge and goosegrass in Arkansas, Louisiana, Mississippi, the bootheel of Missouri and West Tennessee.

HOW TO USE

Apply preemergence, after planting and before cotton seedlings emerge, as follows:

Soil Texture**	lbs. 80% Per Acre*** KARMEX	+	Zorial
Sandy loam	0.5–1.0	+	0.6
Loam, silt loam, silt	0.5–1.0	+	0.9
Sandy clay loam	1.25	+	0.9
Clay loam, silty clay loam, sandy clay	1.25	+	1.25
Silty clay, clay	2.0	+	1.25

* Registered trademark of Sandoz, Inc.

** Do not use on soils with less than 1% organic matter, as crop injury may result.

*** Broadcast; for band treatment apply proportionately less.

Do not treat cotton in deep furrows, as crop injury may result; use only where cotton is planted on flat or raised seedbeds. Shallow incorporation (no deeper than 1/4") with a rotary hoe or similar equipment following planting usually improves results, especially during dry weather. A wide press wheel should be used on the planter to provide a level seedbed for subsequent early season postemergence treatments. If moisture is insufficient to activate KARMEX plus Zorial®, or if soil becomes crusted before crop emerges, a shallow rotary hoeing (no deeper than 1/4") should be made, before weeds become established.

Replanting: If initial seeding fails to produce a good stand, cotton may be replanted in soil treated preemergence with KARMEX plus Zorial. Wherever possible, avoid disturbing original bed. If necessary to rework soil before replanting, use shallow cultivation such as discing; do not relist nor move soil into the original drill area. Plant seed at least 1" deep. Do not re-treat field with a second preplant or preemergence application during the same crop year, as injury to the crop may result. The first spring following treatment, acreage may be rotated to soybeans. One year after application of the KARMEX plus Zorial, treated acreage may be rotated to peanuts.

Notes

1. Do not allow livestock to graze treated cotton.
2. Injury to cotton may occur if KARMEX plus Zorial is used in conjunction with organosphophate pesticides.
3. Cover crops planted after harvest should be plowed under and not grazed or harvested. To minimize injury to small grain cover crops, they should be broadcast planted, rather than drilled.

IMPORTANT

BEFORE USING KARMEX PLUS ZORIAL, READ AND CAREFULLY OBSERVE THE CAUTIONARY STATEMENTS AND ALL OTHER INFORMATION APPEARING ON THE PRODUCT LABELS.

This bulletin contains new or supplemental instructions for use of this product which may not appear on the package label. Follow the instructions carefully.

This labeling must be in the possession of the user at the time of pesticide application.

SPECIAL LOCAL NEED REGISTRATION [24(c)] FOR KARMEX® HERBICIDE

Crop	State	EPA SLN No.	Remarks
Apples	Georgia	GA–770013	Permits split application
Blueberries	Arkansas	AR–810037	Permits split application
	Georgia	GA–770013	Permits split application
	Maine	ME–790003	Allow Sinbar® + KARMEX® use
	New Jersey	NJ–810009	Allow Sinbar® + KARMEX® use
Cotton	Arizona	AZ–820016	Permits higher rates
Grass Seed Crops	Washington	WA–780059	Newly seeded grasses, following activated charcoal treatment.
Peaches	Georgia	GA–770013	Permits split application
Austrian Field Peas	Western Oregon	OR–770064	Permits use on solid stands
Papaya	Hawaii	HI–830007	Preplant application
Pineapple	Puerto Rico	PR–790005	Preemergence
Bermuda-grass Pastures	Arkansas	AR–770018	Established pasture and aerial
	Oklahoma	OK–830031	Permits higher rates
Sugarcane	Hawaii	HI–780017	Additional treatments, up to 20 lbs./A per crop
	Texas	TX–810036	Allow VELPAR + KARMEX use

* Copies of individual labels available upon request to nearest DuPont Agricultural Chemicals Office. For distribution and use only within the State specified.

KRENITE®

BRUSH CONTROL AGENT WATER SOLUBLE LIQUID

Contains 4 lbs. Active Per Gallon

ACTIVE INGREDIENT:
Ammonium salt of fosamine
[ethyl hydrogen (aminocarbonyl) phosphonate] 41.5%
INERT INGREDIENTS 58.5%
TOTAL 100.0%

EPA Reg. No. 352-376 U.S. Pats. 3,627,507 & 3,846,512

Keep Out of Reach of Children

PRECAUTIONARY STATEMENTS

HAZARDS TO HUMANS

CAUTION!

MAY IRRITATE EYES, NOSE, THROAT, AND SKIN.

Avoid breathing spray or mist. Avoid contact with skin, eyes, and clothing.

STATEMENT OF PRACTICAL TREATMENT

If in eyes: Flush with plenty of water.

If on skin: Wash with plenty of soap and water.

ENVIRONMENTAL HAZARDS

Do not contaminate water by cleaning of equipment or disposal of wastes. Do not apply directly to water.

IMPORTANT

Do not use on food crops. Do not allow drift or spray mist to contact desirable trees, shrubs or other plants, as injury may result. Keep from contact with fertilizers, insecticides, fungicides and seeds.

Thoroughly clean all traces of KRENITE from application equipment immediately after use. Flush tank, pump, hoses and boom with several changes of water after removing nozzle tips and screens (clean these parts separately).

GENERAL INFORMATION

Du Pont KRENITE Brush Control Agent is a water soluble liquid to be diluted with water and applied as a foliar spray for control and /or growth suppression of many woody species. A spray directed to only part of suceptible brush species will provide control of the portion sprayed, resulting in a trimming effect. It is applied to brush during the period from July to first fall coloration.

It is recommended for use on noncropland adjacent to and surrounding domestic water supply reservoirs, supply streams, lakes and ponds, as well as on drainage ditchbanks, highway right-of-ways, industrial plant sites, railroad right-of-ways, reforestation areas prior to planting, storage areas, utility and pipeline right-of-ways and other similar areas. KRENITE is nonflammable and nonvolatile.

NOTICE OF WARRANTY

Du Pont warrants that this product conforms to the chemical description on the label thereof and is reasonably fit for purposes stated on such label only when used in accordance with directions under normal use conditions. It is impossible to eliminate all risks inherently associated with use of this product. Ineffectiveness or other unintended consequences may result because of such factors as weather conditions, presence of other materials, or the manner of use or application, all of which are beyond the control of Du Pont. In no case shall Du Pont be liable for consequential, special or indirect damages resulting from the use or handling of this product. All such risks shall be assumed by the buyer. DU PONT MAKES NO WARRANTIES OF MERCHANTABILITY OR FITNESS FOR A PARTICULAR PURPOSE NOR ANY OTHER EXPRESS OR IMPLIED WARRANTY EXCEPT AS STATED ABOVE.

DIRECTIONS FOR USE

It is a violation of federal law to use this product in a manner inconsistent with its labeling.

KRENITE Brush Control Agent should be used only in accordance with recommendations on this label, or in separate published Du Pont recommendations available through local dealers.

Du Pont will not be responsible for losses or damages resulting from use of this product in any manner not specifically recommended by Du Pont. User assumes all risks associated with such nonrecommended use.

General Brush Control—Non-Cropland Areas—KRENITE at the following rates controls or suppresses undesirable brush/herbaceous plant species such as:

1½ to 6 gals. KRENITE per acre		2 to 6 gals. KRENITE per acre
Bigleaf maple*	Red Maple*	American elder
Birch	Red oak	Basswood*
Blackberry	Salmonberry	Eastern cottonwood
Black cherry*	Sassafras*	Eastern white pine
Blackgum*	Sourwood*	Field bindweed[1]
Black locust	Sumac	Multiflora rose
Bracken (fern)	Sweet gum	Slippery elm
Chinese tallow	Thimbleberry	Sycamore
Chokecherry*	Tulip tree (yellow poplar)*	Tree-of-heaven
Elm*	Vine maple	Wild grape
Hawthorn*	Virginia pine	Wild plum
Hickory*	Water oak	
Loblolly pine	White ash*	
Persimmon*	White oak	
Pin cherry	Willow*	
Quaking aspen	Winged elm*	
Red alder		

* Partial control and growth suppression.

[1] Apply KRENITE after plants begin to bloom.

Spray Preparation—Measure the proper amount of KRENITE and use agitation to thoroughly mix into the necessary volume of water. Add one of the following surfactants at the rate of 1 qt. per 100 gals. spray: Du Pont Surfactant WK, "Renex 30"[1], "Tween 20"[1], or "Triton X-100"[2]. To enhance the activity of the KRENITE treatment, the addition of ¼% (by volume) of a nonphytotoxic superior-type (60–70 second viscosity) oil as the last ingredient may be desirable in some areas (consult local sales personnel). After KRENITE has been thoroughly mixed in the spray tank, agitation of the spray solution is not required.

[1]Registered trademark of ICI Americas, Inc.

[2]Registered trademark of Rohm & Haas

Spray Application Methods—Before applying, properly calibrate spray equipment. For woody plant control, make a single foliar application of the recommended rate of KRENITE during the period from July to the first frost in the fall. For control of only a portion of a plant, as in trimming, direct the spray to thoroughly cover only the section of the plant to be controlled. Use the highest recommended rate on stands in which species partially controlled and suppressed are dominant.

For high volume ground treatment prepare a spray solution using 1½ to 3 gallons KRENITE per 100 gallons of water. Do not prepare less than a ½% solution or exceed 6 gals. KRENITE per acre. Apply a quantity of spray solution which will thoroughly and uniformly cover the target brush without causing unnecessary run-off. Spray pressure at nozzles should be 100 to 300 psi. Total rate and volume per acre will depend on plant species, height and density of growth and on type of equipment used. On tall dense brush, it is often necessary to spray from both sides to obtain adequate coverage.

For low volume ground treatment, using airblast equipment such as a "Solo backpack mist blower," prepare a spray solution using 1½ to 3 gallons KRENITE per 10 gallons of water. Apply a quantity of spray solution using equipment which will assure particle breakup and uniform coverage of the low spray volumes applied, without causing unnecessary run-off. Total rate and volume per acre will depend on plant species, height and density of growth, and on type of equipment used.

For aerial treatment, apply 1½ to 3 gallons of KRENITE per acre using 10 to 40 gallons of water to thoroughly and uniformly cover the target brush, without causing unnecessary run-off. Total rate and volume per acre will depend on plant species, height and density of growth, and on type of equipment used.

Note: Effectiveness may be reduced if, following treatment, rainfall occurs the same day. Do not cut treated brush until stems are dead, or sprouting may occur.

NOTICE TO BUYER: Purchase of this material does not confer any rights under patents of countries outside of the United States.

STORAGE AND DISPOSAL

STORAGE

Store product in original container only, away from other pesticides, fertilizer, food, or feed.

DISPOSAL

Do not contaminate water, food, or feed by storage or disposal. Wastes resulting from the use of this product may be disposed of on site or at an approved waste disposal facility.

Triple rinse (or equivalent) the container. Then offer for recycling or reconditioning, or puncture and dispose of in a sanitary landfill, or incineration, or, if allowed by state and local authorities, by burning. If burned, stay out of smoke.

SPECIAL LOCAL NEED REGISTRATION [24(c)] FOR KRENITE® BRUSH CONTROL AGENT*

Crop	State	EPA SLN No.	Remarks
Forest Land	Oregon	OR-770054	Permits lower rates for conifer release
Forest Land	Washington	WA-770045	Permits lower rates for conifer release
Land Renovation	Missouri	MO-780014	Control of mixed oak species

* Copies of individual labels available upon request to nearest Du Pont Agricultural Chemicals Office. For distribution and use only within the State specified.

KRENITE® S

BRUSH CONTROL AGENT WATER SOLUBLE LIQUID

Contains 4 Lbs. Active Per Gallon

ACTIVE INGREDIENT:	
Ammonium salt of fosamine [ethyl hydrogen (aminocarbonyl) phosphonate]	41.5%
INERT INGREDIENTS	58.5%
TOTAL	100.0%

EPA Reg. No. 352-395 U.S. Pats. 3,627,507 & 3,846,512

Keep Out of Reach of Children

PRECAUTIONARY STATEMENTS

HAZARDS TO HUMANS

WARNING!

CAUSES EYE IRRITATION.

Do not get in eyes, on skin or on clothing. Harmful if swallowed. Avoid breathing spray or mist.

STATEMENT OF PRACTICAL TREATMENT

If in eyes: flush with plenty of water for 15 minutes. Get medical attention if irritation persists.

If on skin: wash with plenty of soap and water.

ENVIRONMENTAL HAZARDS

Do not contaminate water by cleaning of equipment or disposal of wastes. Do not apply directly to water.

IMPORTANT

Do not use on food crops. Do not allow drift or spray mist to contact desirable trees, shrubs or other plants, as injury may result. Keep from contact with fertilizers, insecticides, fungicides and seeds.

Thoroughly clean all traces of KRENITE S from application equipment after use. Flush tank, pump, hoses and boom with several changes of water after removing nozzle tips and screens (clean these parts separately).

GENERAL INFORMATION

DuPont KRENITE S Brush Control Agent is a water soluble liquid to be diluted with water, and applied as a foliar spray for control and /or growth suppression of many woody species. A spray directed to only part of susceptible brush species will provide control of the portion sprayed, resulting in a trimming effect. It is applied to brush during the period from July to first fall coloration.

It is recommended for use on noncropland adjacent to and surrounding domestic water supply reservoirs, supply streams, lakes and ponds, as well as on drainage ditchbanks, highway right-of-ways, industrial plant sites, railroad right-of-ways, reforestation areas prior to planting, storage areas, utility and pipeline right-of-ways and other similar areas. KRENITE S is nonflammable and nonvolatile.

NOTICE OF WARRANTY

DuPont warrants that this product conforms to the chemical description on the label thereof and is reasonably fit for purposes stated on such label only when used in accordance with directions under normal use conditions. It is impossible to eliminate all risks inherently associated with use of this product. Ineffectiveness or other unintended consequences may result because of such factors as weather conditions, presence of other materials or the manner of use or application, all of which are beyond the control of DuPont. In no case shall DuPont be liable for consequential, special or indirect damages resulting from the use or handling of this product. All such risks shall be assumed by the Buyer. DUPONT MAKES NO WARRANTIES OF MERCHANTABILITY OR FITNESS FOR A PARTICULAR PURPOSE NOR ANY OTHER EXPRESS OR IMPLIED WARRANTY EXCEPT AS STATED ABOVE.

DIRECTIONS FOR USE

It is a violation of federal law to use this product in a manner inconsistent with its labeling.

KRENITE S Brush Control Agent should be used only in accordance with recommendations on this label, or in separate published DuPont recommendations available through local dealers.

DuPont will not be responsible for losses or damages resulting from use of this product in any manner not specifically recommended by DuPont. User assumes all risks associated with such non-recommended use.

General Brush Control—Non-Cropland Areas—KRENITE S at the following rates controls or suppresses undesirable brush/herbaceous plant species such as:

$1\frac{1}{2}$ to 6 gal. KRENITE S per acre		2 to 6 gal. KRENITE S per acre
Bigleaf maple*	Red Maple*	American elder
Birch	Red Oak	Basswood*
Blackberry	Salmonberry	Eastern cottonwood
Black cherry*	Sassafras*	Eastern white pine
Blackgum*	Sourwood*	Field bindweed[1]
Black locust	Sumac	Multiflora rose
Bracken (fern)	Sweet gum	Slippery elm
Chinese tallow	Thimbleberry	Sycamore
Chokecherry*	Tulip tree (yellow poplar)*	Tree-of-heaven
Elm*	Vine maple	Wild grape
Hawthorn*	Virginia pine	Wild plum
Hickory*	Water oak	Winged elm*
Loblolly pine	White ash*	
Persimmon*	White oak	
Pin cherry	Willow*	
Quaking aspen		
Red alder		

* Partial control and growth suppression.

[1] Apply KRENITE S after plants begin to bloom.

Spray Preparation—Measure the proper amount of KRENITE S and use agitation to thoroughly mix into the necessary volume of water. To enhance the activity of the KRENITE S treatment, the addition of $\frac{1}{4}$% (by volume) of a nonphytotoxic superior-type (60–70 second viscosity) oil as the last ingredient may be desirable in some areas (consult local sales personnel). After KRENITE S has been thoroughly mixed in the spray tank, agitation of the spray solution is not required.

Spray Application Methods—Before applying, properly calibrate spray equipment. For woody plant control, make a single foliar application of the recommended rate of KRENITE S during the period from July to the first fall coloration. For control of only a portion of a plant, as in trimming, direct the spray to thoroughly cover only the section of the plant to be controlled. Use the highest recommended rate on stands in which species partially controlled and suppressed are dominant.

Prepare a spray solution using $1\frac{1}{2}$ to 3 gallons KRENITE S per 100 gallons of water. Do not prepare less than a $1\frac{1}{2}$% solution or exceed 6 gals. KRENITE S per acre. Apply a quantity of spray solution which will thoroughly and uniformly cover the target brush without causing unnecessary run-off. Spray pressure at nozzles should be 100 to 300 psi. Total rate and volume per acre will depend on plant species, height and density of growth and on type of equipment used. On tall dense brush, it is often necessary to spray from both sides to obtain adequate coverage.

Note: Effectiveness may be reduced if, following treatment, rainfall occurs on the same day. Do not cut treated brush until stems are dead, or sprouting may occur.

NOTICE TO BUYER: Purchase of this material does not confer any rights under patents of countries outside of the United States.

STORAGE AND DISPOSAL

STORAGE

Store product in original container only, away from other pesticides, fertilizer, food or feed.

DISPOSAL

Do not contaminate water, food or feed by storage or disposal. Wastes resulting from the use of this product may be disposed of on site or at an approved waste disposal facility.

Triple rinse (or equivalent) the container. Then offer for recycling or reconditioning, or puncture and dispose of in a sanitary landfill, or incineration, or, if allowed by state or local authorities, by burning. If burned, stay out of smoke.

KROVAR® I

HERBICIDE WETTABLE POWDER

ACTIVE INGREDIENTS:	80%
Bromacil [5-bromo-3-sec-butyl-6-methyluracil] 40%	
Diuron [3-(3,4-dichlorophenyl)-1,1-dimethylurea] 40%	
INERT INGREDIENTS	20%
TOTAL	100%

EPA Reg. No. 352-352

Keep Out of Reach of Children

PRECAUTIONARY STATEMENTS

HAZARDS TO HUMANS

CAUTION!

MAY IRRITATE EYES, NOSE, THROAT, AND SKIN.

Avoid breathing dust or spray mist. Avoid contact with skin, eyes, or clothing. Wash thoroughly with soap and water after handling. Do not apply this product in such a manner as to directly or through drift expose workers or other persons. The area being treated must be vacated by unprotected persons.

STATEMENT OF PRACTICAL TREATMENT

If on skin, wash with plenty of soap and water.

If in eyes, flush with plenty of water. Get medical attention if irritation persists.

ENVIRONMENTAL HAZARDS

Do not apply directly to water. Do not contaminate water by cleaning of equipment or disposal of wastes.

IMPORTANT

Injury to or loss of desirable trees or other plants may result from failure to observe the following:

Do not apply (except as recommended for crop use), or drain or flush equipment on or near desirable trees or other plants, or on areas where their roots may extend, or in locations where the chemical may be washed or moved into contact with their roots. Do not use on lawns, walks, driveways, tennis courts or similar areas. Do not use in home fruit plantings nor in citrus orchards interplanted to other trees or desirable plants. Prevent drift of dry powder or spray to desirable plants. Do not contaminate any body of water. Keep from contact with fertilizers, insecticides, fungicides and seeds.

Thoroughly clean all traces of KROVAR I from application equipment immediately after use. Flush tank, pump, hoses, and boom with several changes of water after removing nozzle tips and screens (clean these parts separately).

GENERAL INFORMATION

DuPont KROVAR I Herbicide is a wettable powder to be mixed in water and applied as a spray for selective control of weeds in citrus and for noncrop weed control.

KROVAR I controls annual weeds such as barnyardgrass (watergrass), chickweed, crabgrass filaree, fleabane, Florida pusley, foxtail, groundsel, horseweed, johnsongrass seedlings, jungerice, lambsquarters, natalgrass (red top), nightshade (annual), pigweed, pineappleweed, puncturevine, purslane, ragweed, sandbur (sandspur), shepherdspurse, sowthistle (annual), spanishneedles, wild lettuce, and wild mustard. At higher use rates, treatment controls certain perennial weeds such as balsamapple vine seedlings, bermudagrass, heart leaf, drymary, quineagrass and milkweed (strangler) vine seedlings; and provides suppression of established balsamapple vine and milkweed (strangler) vine. Effects on perennial weeds are slow to appear, usually progressing over a period of several months.

Moisture is necessary to activate the herbicide; best results are obtained if treatment is made to moist soil, and moisture is supplied by rainfall or sprinkler irrigation within two weeks after application. The degree and duration of control will vary with the amount applied, soil texture, rainfall, and other conditions.

NOTICE OF WARRANTY

DuPont warrants that this product conforms to the chemical description on the label thereof and is reasonably fit for the purposes stated on such label only when used in accordance with the directions under normal use conditions. It is impossible to eliminate all risks inherently associated with the use of this product. Crop injury, ineffectiveness, or other unintended consequences may result because of such factors as weather conditions, presence of other materials, or the manner of use or application, all of which are beyond the control of DuPont. In no case shall DuPont be liable for consequential, special or indirect damages resulting from the use or handling of this product. All such risks shall be assumed by the buyer. DUPONT MAKES NO WARRANTIES OF MERCHANTABILITY OR FITNESS FOR A PARTICULAR PURPOSE NOR ANY OTHER EXPRESS OR IMPLIED WARRANTY EXCEPT AS STATED ABOVE.

DIRECTIONS FOR USE

It is a violation of federal law to use this product in a manner inconsistent with its labeling.

DuPont KROVAR I should be used only in accordance with recommendations on this label, or in separate published DuPont recommendations available through local dealers.

DuPont will not be responsible for losses or damages resulting from use of this product in any manner not specifically recommended by DuPont. User assumes all risk associated with such nonrecommended use.

Do not enter treated areas without protective clothing until sprays have dried. Because certain states may require more restrictive reentry intervals for various crops treated with this product, consult your State Department of Agriculture for further information.

Written or oral warnings must be given to workers who are expected to be in a treated area or in an area about to be treated with this product. When oral warnings are given, warnings shall be given in a language customarily understood by workers. Oral warnings must be given if there is reason to believe that written warnings cannot be understood by workers. Warnings must include the following information: "CAUTION. Area treated with KROVAR I on (date of application). Do not enter without appropriate protective clothing until sprays have dried. In case of contact, flush eyes with plenty of water; wash skin with plenty of soap and water. Get medical attention if irritation persists."

CITRUS

Apply any time of year provided overhead moisture (rainfall or sprinkler irrigation) is available to activate the herbicide, preferably just before or just after weeds have germinated.

Use only where trees have been established in the orchard for at least three years (except in Florida). Apply with a properly calibrated fixed-boom power sprayer as a band or broadcast treatment beneath and/or between trees. Use sufficient water (min. 40 gals. per acre) to obtain uniform coverage; avoid contact of foliage and fruit with spray or mist. Avoid overlapping, and shut off spray boom while starting, turning, slowing or stopping, or injury to trees may result. Continuous agitation in the spray tank is required to keep the material in suspension. Best results are obtained if application is made shortly before or shortly after weed growth begins; if dense growth is present, remove tops and spray ground. Do not graze cattle in treated areas.

Dosages are expressed as broadcast rates; for band treatment, use proportionately less.

ANNUAL WEEDS

California, Arizona—Best results occur when applied in late fall or early winter, but before winter annuals become well established. Application should be made after the first fall or early winter rains have settled the soil.

For the initial treatment, apply 4–5 lbs. per acre on coarse soils containing 1–2% organic matter and 5–6 lbs. per acre on fine soils, or soils with organic matter of 2½% or more and repeat annually as needed. Alternatively, apply 3–4 lbs. per acre in the fall and repeat at 2–4 lbs. per acre in the spring. Do not exceed 6 lbs. per acre per year. Use the higher range of rates where groundsel or puncturevine are known to be a problem. The higher rates will also suppress low density stands of bermudagrass and yellow nutsedge. Repeat annually.

Texas, Louisiana—Make a single application of 2 to 4 lbs. per acre on coarser soils (sands, loamy sands, sandy loams) and 4 to 6 lbs. per acre on finer soils (silt loams, clay loams, or soils with organic matter of 2½% or more); use the higher rates for maximum suppression of perennials. Alternatively, make two applications per year at rates of 2 lbs. per acre on coarser soils and 3 lbs. per acre on finer soils; make the second application when needed to maintain weed control.

Florida

Trees Established Less Than One Year—Apply 2 to 4 lbs. KROVAR I per acre as needed to maintain weed control. Do not apply at less than 60-day intervals; do not apply more than 6 lbs. per acre during any 6-month period or more than 8 lbs. per acre during the first year.

Trees Established One or Two Years—Apply 2 to 4 lbs. KROVAR I per acre. A second application may be made when needed to maintain weed control, but do not exceed 8 lbs. per acre per year.

Trees Established Three or More Years—Make one or two applications of 4 to 8 lbs. per acre. Make the second application when needed to maintain weed control. Do not apply more than 16 lbs. per acre per year.

Note: Temporary yellowing of citrus leaves may occur following treatment. Because injury to citrus trees may result, do not use on soils with less than 1% organic matter (½% in Florida), poorly drained soils, gravelly soils, nor thinly covered or exposed subsoils; do not treat trees planted in irrigation furrows; do not treat diseased trees such as those with foot rot.

Do not replant treated areas to any crop within two years after last application as injury to subsequent crops may result, except that citrus trees may be planted one year after the last application.

NON-CROP WEED CONTROL

For short-term control of annual weeds on non-cropland areas such as roadsides, storage areas, and plant sites, apply 4 to 6 lbs. per acre in 40 to 100 gals. of water. For extended control of annuals and partial control of perennials such as bermudagrass and nutsedge, apply 7 to 18 lbs. per acre. For control of hard-to-kill perennials such as bermudagrass, bouncingbet, dogbane, johnsongrass, nutsedge, and saltgrass, apply 19 to 40 lbs. per acre.

Use the higher levels of dosage rates on adsorptive soils (high in organic matter or carbon). Best results occur when application is made just before weed emergence or in early stages of weed growth.

Retreatment: Apply 4 to 6 lbs. per acre when annual weeds and grasses reappear on sites where weed growth has been controlled.

Small Areas: ¼ cupful of KROVAR I per 200 sq. ft. is approximately 15 lbs. per acre.

NOTICE TO BUYER—Purchase of this material does not confer any rights under patents of countries outside of the United States.

STORAGE AND DISPOSAL

STORAGE

Store product in original container only, away from other pesticides, fertilizer, food, or feed.

DISPOSAL

Do not contaminate water, food, or feed by storage or disposal. Wastes resulting from the use of this product may be disposed of on site or at an approved waste disposal facility.

Completely empty bag into application equipment. Then dispose of empty bag in a sanitary landfill or by incineration, or, if allowed by State and local authorities, by burning. If burned, stay out of smoke.

SUPPLEMENTAL LABELING

KROVAR® I HERBICIDE

EPA Reg. No. 352-352

KROVAR® I HERBICIDE AS A TANK MIXTURE WITH OTHER HERBICIDES FOR NON-CROP WEED CONTROL

DuPont KROVAR I Herbicide with DuPont Surfactant WK is recommended for use as a tank mixture with amitrole, amitrole T, 2, 4-D amine, dalapon, "Dowpon"* C, MSMA, paraquat, and "Phytar"** 560 (Na salt of cacodylic acid) for control of undesirable vegetation in non-crop areas such as railroad, highway, pipeline and utility right-of-ways, petroleum tank farms, lumberyards, storage areas and industrial plant sites.

*Registered trademark of Dow Chemical Company.

**Registered trademark of The Ansul Company.

DIRECTIONS

For non-cropland use, application may be made with a hand gun or boom sprayer using sufficient water (40 to 200 gals. of spray per acre) to insure uniform coverage. For smaller areas, a hand sprayer may be used. Nozzle screens should be 50 mesh or larger. Agitation may be either mechanical or hydraulic. If a bypass or return line is used, it should terminate at the bottom of the tank to minimize foaming. Do not use air agitation.

Spray Preparation—Fill spray tank half full of water. Start agitation and continue during entire mixing process. Add required amount of KROVAR I and allow to mix. If additional product is a wettable powder, add to tank and allow to mix thoroughly. If additional product is a liquid, add slowly while filling remainder of tank with water. Dilute proper amount of Surfactant WK with 10 parts of water and add to tank last and allow contents to mix thoroughly. Continuous agitation in the spray tank is required to keep the materials in suspension.

Rates for Weed Control—KROVAR I—Use KROVAR I at the following rates with Surfactant WK (1 to 2 qts. per 100 gals. spray) for non-crop weed control:

Apply 4 to 6 lbs. per acre for short-term control of annual weeds and grasses such as annual nightshade, annual sowthistle, barnyardgrass, chickweed, crabgrass, filaree, fleabane, Florida pusley, foxtail, groundsel, horseweed, johnsongrass seedlings, jungle rice, lambsquarters, natalgrass (red top), pigweed, pineappleweed, puncturevine, purslane, ragweed, sandspur, sheperdspurse, spanishneedles, wild lettuce and wild mustard.

Apply 7 to 18 lbs. per acre for extended control of annuals and partial control of perennials such as bermudagrass and nutsedge.

Apply 19 to 40 lbs. per acre for control of hard-to-kill perennials such as bermudagrass, bouncingbet, dogbane, johnsongrass, nutsedge and salt grass.

Rates for Products to be Tank Mixed with KROVAR I and Surfactant WK—Use the lower rate of KROVAR I (see note) recommended to provide control of weeds to be sprayed. Prepare a tank mixture with one of the following herbicides selected as appropriate to improve control of the weed population:

Amitrole or Amitrole T—Add 1 to 3 gals. per acre to KROVAR I to improve control of bermudagrass, bluegrass, Canada thistle, horsetail rush, nutsedge, and quackgrass.

2,4-D Amine—Add 2 to 4 lbs. active per acre to KROVAR I to improve control of blackberry, burdock, peppervine, trumpetvine, Virginia creeper and other broadleaved weeds.

Dalapon ("Dowpon" or "Dowpon" C)—Add 15 to 30 lbs. per acre to KROVAR I to improve control of bahiagrass, bermudagrass, broomsedge, bunchgrass, johnsongrass and quackgrass.

MSMA—Add 3½ to 7 lbs. active per acre to KROVAR I to aid in control of bahiagrass, broomsedge, johnsongrass, paspalum and nutsedge.

Paraquat—Add 1 to 2 qts. per acre to KROVAR I for nonselective burndown of annual weeds and top kill of perennial vines and grasses.

PHYTAR 560 (Na salt of cacodylic acid)—Add 1½ to 4 gals. per acre to KROVAR I for nonselective burndown of annual weeds and top kill of perennial vines and grasses.

MSMA + 2,4-D Amine—Add recommended rates to KROVAR I for a 3-way tank mixture to improve control of weeds listed for all three products.

Note: Use the higher rates of KROVAR I on heavier textured soils or on soils high in organic matter or carbon. Use the higher rates of other herbicides mixed with KROVAR I on heavy, dense vegetation.

IMPORTANT

BEFORE USING KROVAR I, SURFACTANT WK, AMITROLE, 2,4-D AMINE, DALAPON, MSMA, PARAQUAT OR PHYTAR 560, READ AND CAREFULLY OBSERVE THE CAUTIONARY STATEMENTS AND ALL OTHER INFORMATION APPEARING ON THE PRODUCT LABELS.

This bulletin contains new or supplemental instructions for use of these products which may not appear on package labels. Follow the instructions carefully.

SPECIAL LOCAL NEED REGISTRATION [24(c)] FOR KROVAR® I HERBICIDE

CROP STATE EPA SLN NO. REMARKS

IWC- Washington WA-820062

Use on U.S. Firebreaks Army Yakima Firing Center

KROVAR® II

HERBICIDE

WETTABLE POWDER

ACTIVE INGREDIENTS:	80%
Bromacil [5-bromo-3-sec-butyl-6-methyluracil] 53%	
Diuron [3-(3,4-dichlorophenyl)-1,1-dimethylurea] 27%	
INERT INGREDIENTS	20%
TOTAL	100%

EPA Reg. No. 352-351

Keep Out of Reach of Children

PRECAUTIONARY STATEMENTS

HAZARDS TO HUMANS

CAUTION!

MAY IRRITATE EYES, NOSE, THROAT AND SKIN.

Avoid breathing dust or spray mist. Avoid contact with skin, eyes or clothing. Wash thoroughly with soap and water after handling. Do not apply this product in such a manner as to directly or through drift expose workers or other persons. The area being treated must be vacated by unprotected persons.

STATEMENT OF PRACTICAL TREATMENT

If in eyes: Flush with plenty of water. Get medical attention if irritation persists.

If on skin: Wash with plenty of soap and water.

ENVIRONMENTAL HAZARDS

Do not apply directly to water. Do not contaminate water by cleaning of equipment or disposal of wastes.

IMPORTANT

Injury to or loss of desirable trees or other plants may result from failure to observe the following:

Do not apply (except as recommended for crop use), or drain or flush equipment on or near desirable trees or other plants, or on areas where their roots may extend, or in locations where the chemical may be washed or moved into contact with their roots. Do not use on lawns, walks, driveways, tennis courts or similar areas. Do not use in home fruit plantings nor in citrus orchards interplanted to other trees or desirable plants. Prevent drift of dry powder or spray to desirable plants. Do not contaminate any body of water. Keep from contact with fertilizers, insecticides, fungicides and seeds.

Thoroughly clean all traces of KROVAR II from application equipment immediately after use. Flush tank, pump, hoses and boom with several changes of water after removing nozzle tips and screens (clean these parts separately).

GENERAL INFORMATION

DuPont KROVAR II Herbicide is a wettable powder to be mixed in water and applied as a spray for selective control of weeds in citrus and for noncrop weed control.

KROVAR II controls certain perennial grasses and annual weeds such as barnyardgrass (watergrass), crabgrass, Florida pusley, foxtail, lambsquarters, natalgrass (red top), nightshade (annual), pigweed, purslane, ragweed, sandbur (sandspur), sowthistle (annual) and spanishneedles. Effects on perennial weeds are slow to appear, usually progressing over a period of several months. In Florida, it is also recommended for control of certain seedling vines.

Moisture is necessary to activate the herbicide; best results are obtained if treatment is made to moist soil, and moisture is supplied by rainfall or sprinkler irrigation within two weeks after application. The degree of control and duration of effect will vary with the amount applied, soil texture, rainfall and other conditions.

NOTICE OF WARRANTY

DuPont warrants that this product conforms to the chemical description on the label thereof and is reasonably fit for the purposes stated on such label only when used in accordance with the directions under normal use conditions. It is impossible to eliminate all risks inherently associated with the use of this product. Crop injury, ineffectiveness or other unintended consequences may result because of such factors as weather conditions, presence of other materials or the manner of use or application, all of which are beyond the control of DuPont. In no case shall DuPont be liable for consequential, special or indirect damages resulting from the use or handling of this product. All such risks shall be assumed by the buyer. DUPONT MAKES NO WARRANTIES OF MERCHANTABILITY OR FITNESS FOR A PARTICULAR PURPOSE NOR ANY OTHER EXPRESS OR IMPLIED WARRANTY EXCEPT AS STATED ABOVE.

DIRECTIONS FOR USE

It is a violation of federal law to use this product in a manner inconsistent with its labeling.

DuPont KROVAR II should be used only in accordance with recommendations on this label, or in separate published DuPont recommendations available through local dealers.

DuPont will not be responsible for losses or damages resulting from use of this product in any manner not specifically recommended by DuPont. User assumes all risk associated with such nonrecommended use.

Do not enter treated areas without protective clothing until sprays have dried. Because certain states may require more restrictive reentry intervals for various crops treated with this product, consult your State Department of Agriculture for further information.

Written or oral warnings must be given to workers who are expected to be in a treated area or in an area about to be treated with this product. When oral warnings are given, warnings shall be given in a language customarily understood by workers. Oral warnings must be given if there is reason to believe that written warnings cannot be understood by workers. Warnings must include the following information: "CAUTION. Area treated with KROVAR II on (date of application). Do not enter without appropriate protective clothing until sprays have dried. In case of contact, flush eyes with plenty of water; wash skin with plenty of soap and water. Get medical attention if irritation persists."

CITRUS

Use only where trees have been established in the orchard for at least four years. Apply with a properly calibrated, fixed-boom power sprayer as a band or broadcast treatment beneath and/or between trees. Use sufficient water (min. 40 gals. per acre) to obtain uniform coverage; avoid contact of foliage and fruit with spray or mist. Avoid overlapping, and shut off spray boom while starting, turning, slowing or stopping, or injury to trees may result. Continuous agitation in the spray tank is required to keep the material in suspension. Best results are obtained if application is made shortly before or shortly after weed growth begins; if dense growth is present, remove tops and spray the ground. Do not graze cattle in treated areas.

Dosages are expressed as broadcast rates; for band treatment, use proportionately less.

Do not use on citrus in Kern County, California.

Perennial Weeds: For bahiagrass, bermudagrass, nutsedge, pangolagrass, paragrass and torpedograss, make a single application per year during the period from winter to early summer at the following rates:

Soil Texture	Lbs. Per Acre
Sand, loamy sand	4 to 6
Sandy loam	5 to 7
Silt loam, clay loam	6 to 8

Alternatively, make two applications of 3 to 4 lbs. per acre per year (Florida 4 to 8 lbs.); apply in spring and summer.

Partial control of perennials usually occurs with a single treatment; repeat applications are required to control perennials. Control may be improved by cultivation prior to treatment; otherwise, avoid working the soil as long as weed control continues, or else treatment effectiveness may be reduced.

Annual Weeds: Apply 2 to 5 lbs. per acre, using the lower rates on coarse soils; apply any time of year, provided overhead moisture (rainfall or sprinkler irrigation) is available to activate the herbicide, preferably just before or just after weeds germinate.

Note: Temporary yellowing of citrus leaves may occur following treatment. Because injury to citrus trees may result: do not use on soils low in organic matter (less than 1%), poorly drained soils, gravelly soils, nor thinly covered or exposed subsoils; do not apply more than 10 lbs. per acre per year (Florida, 16 lbs.); do not treat trees planted in irrigation furrows; do not treat diseased trees, such as those with foot rot.

Do not replant treated areas to any crop within two years after last application, as injury to subsequent crops may result, except that citrus trees may be planted one year after the last application.

FLORIDA ONLY—TREES ESTABLISHED LESS THAN FOUR YEARS

Trees Established Less Than One Year—Apply 2 to 4 lbs. KROVAR II per acre as needed to maintain weed control. Do not apply at less than 60-day intervals; do not apply more than 6 lbs. per acre during any 6-month period or more than 8 lbs. per acre during the first year.

Trees Established One to Four Years—Apply 3 to 6 lbs. KROVAR II per acre; for details on timing, equipment, spray volume and other use directions, refer to product label. Use the lower rate for 1 to 2 year old trees and the higher rate for trees more than 2 years old. A second application may be made if needed, but do not apply more than 8 lbs. KROVAR II per acre per year, as injury to the trees may result.

Tank Mixtures: KROVAR II is recommended for use on citrus for weed control in tank mixtures with Paraquat. For rates of the individual products, refer to the respective product label. Do not apply tank mixture to foliage or green wood as injury may result.

Seedling Vines (balsamapple and milkweed)—Florida: Remove established vines from trees prior to treatment. Apply 6 lbs. per acre after vine growth begins in the spring (or after cultivation or chopping), but before vines reach tree limbs. For band treatment, the treated area should extend at least 1 foot beyond the dripline of the tree. Add DuPont Surfactant WK at a rate of 1 qt. per 50 gals. of spray mixture.

A repeat application may be made in late summer or early fall to control seedlings which germinate late in the season.

When used in conjunction with good mechanical cultivation practices, repeated annual applications aid in the suppression of established vines.

NONCROP WEED CONTROL

For short-term control of annual weeds on noncropland areas such as roadsides, storage areas and plant sites, apply 2 to 5 lbs. per acre in 40 to 100 gals. water. For extended control of annuals and partial control of perennials such as bermudagrass and nutsedge, apply 6 to 14 lbs. per acre. For control of hard-to-kill perennials such as bermudagrass, bouncingbet, dogbane, johnsongrass, nutsedge and saltgrass, apply 15 to 30 lbs. per acre.

Use the higher levels of dosage rates on adsorptive soils (high in organic matter or carbon). Best results occur when application is made just before weed emergence or in early stages of weed growth.

Retreatment: Apply 2 to 5 lbs. per acre when annual weeds and grasses reappear on sites where weed growth has been controlled.

Small areas: ¼ cupful of KROVAR II per 200 sq. ft. is approximately 15 lbs. per acre.

NOTICE TO BUYER: Purchase of this material does not confer any rights under patents of countries outside of the United States.

STORAGE AND DISPOSAL

STORAGE

Store product in original container only, away from other pesticides, fertilizer, food or feed.

DISPOSAL

Do not contaminate water, food or feed by storage or disposal. Wastes resulting from the use of this product may be disposed of on site or at an approved waste disposal facility. Completely empty bag into application equipment. Then dispose of empty bag in a sanitary landfill or by incineration, or, if allowed by state and local authorities, by burning. If burned, stay out of smoke.

LANNATE®

INSECTICIDE

WATER SOLUBLE POWDER

BAG CONTAINS LANNATE® IN A WATER SOLUBLE PACK—ENTIRE CONTENTS DISSOLVE IN WATER.

ACTIVE INGREDIENT:

Methomyl S-methyl-N-[(methylcarbamoyl)oxy] thioacetimidate	90%
INERT INGREDIENTS	10%
TOTAL	100%

EPA Reg. No. 352-342 U.S. Pats. 3,576,834 & 3,639,633

DANGER–POISON

Keep Out of Reach of Children

PELIGRO

PRECAUCION AL USUARIO: Si usted no lee inglés, no use este producto hasta que la etiqueta haya sido explicada ampliamente.

After opening package, immediately dump entire contents into spray tank. Do not touch inner bag; handling may cause breakage.

STATEMENT OF PRACTICAL TREATMENT

If swallowed, call a physician or Poison Control Center. Drink 1 or 2 glasses of water and induce vomiting by touching back of throat with finger. **Do not induce vomiting or give anything by mouth to an unconscious person.**

If inhaled, remove from exposure and have patient lie down and keep quiet. If patient is not breathing, start artificial respiration immediately. **Never give anything by mouth to an unconscious person.**

In case of contact, wash skin with plenty of soap and water, for eyes flush with water for 15 minutes and get medical attention; remove and wash contaminated clothing before reuse.

ATROPINE IS AN ANTIDOTE—Seek medical attention at once in all cases of suspected poisoning.

If warning symptoms appear (see **Warning Symptoms below**), get medical attention.

PRECAUTIONARY STATEMENTS

HAZARDS TO HUMANS AND DOMESTIC ANIMALS

DANGER—POISON

MAY BE FATAL IF SWALLOWED. POISONOUS IF INHALED.

Do not breathe vapors or spray mist. Do not get in eyes, on skin, or on clothing. When mixing spray, loading, applying, or otherwise handling, wear protective clothing, goggles, and a mask or respirator jointly approved by the Mine Safety and Health Administration (formerly the U.S. Bureau of Mines) and by the National Institute for Occupational Safety and Health. Wear clean clothes daily. Wash thoroughly after handling and before eating or smoking. Pilot should not assist in the mixing and loading operation.

Do not apply this product in such a manner as to directly or through drift expose workers, other persons or animals. The area being treated must be vacated by unprotected persons.

WARNING SYMPTOMS—Methomyl poisoning produces effects associated with anticholinesterase activity which may include weakness, blurred vision, headache, nausea, abdominal cramps, discomfort in the chest, constriction of pupils, sweating, slow pulse, muscle tremors.

NOTE TO PHYSICIAN

TREATMENT—Atropine sulfate should be used for treatment. Administer repeated doses, 1.2 to 2.0 mg. intravenously every 10 to 30 minutes until full atropinization is achieved. Maintain atropinization until the patient recovers. Artificial respiration or oxygen may be necessary. Allow no further exposure to any cholinesterase inhibitor until recovery is assured.

Do not use 2-PAM for exposure to LANNATE alone. However, for exposure to combinations of LANNATE and organophosphorous insecticides, 2-PAM may be used as required to supplement the atropine sulfate treatment. Do not use morphine.

ENVIRONMENTAL HAZARDS

This product is toxic to fish, birds, and other wildlife. Keep out of any body of water. Do not apply where runoff is likely to occur. Do not apply when weather conditions favor drift from areas treated. Do not contaminate water by cleaning of equipment or disposal of wastes.

This product is toxic to bees exposed to direct application. Do not apply this product while bees are actively visiting the treatment area. Time applications to coincide with periods of minimum bee activity.

PHYSICAL AND CHEMICAL HAZARDS

Keep container closed. Use with adequate ventilation.

STORAGE AND DISPOSAL

STORAGE

Handle this package carefully to prevent breakage of inner bag when stored at low temperatures. Allow to warm above 50°F for normal handling. Store product in original container only, away from other pesticides, fertilizer, food or feed. Not for use or storage in or around the home.

DISPOSAL

Do not contaminate food or feed by storage or disposal. Pesticide wastes are acutely hazardous. Improper disposal of excess pesicide, spray mixture, or rinsate is a violation of Federal Law. If these wastes cannot be disposed of by use according to label instructions contact your State Pesticide or Environmental Control Agency, or the Hazardous Waste representative at the nearest EPA regional office for guidance.

Completely empty bag into application equipment. Then dispose of empty bag in a sanitary landfill or by incineration, or, if allowed by State and local authorities, by burning. If burned, stay out of smoke.

NOTICE OF WARRANTY

Du Pont warrants that this product conforms to the chemical description on the label thereof and is reasonably fit for the purposes stated on such label only when used in accordance with the directions under normal use conditions. It is impossible to eliminate all risks inherently associated with the use of this product. Crop injury, ineffectiveness, or other unintended consequences may result because of such factors as weather conditions, presence of other materials, or the manner of use or application, all of which are beyond the control of Du Pont. In no case shall Du Pont be liable for consequential, special or indirect damages resulting from the use or handling of this product. All such risks shall be assumed by the Buyer. DUPONT MAKES NO WARRANTIES OF MERCHANTABILITY OR FITNESS FOR A PARTICULAR PURPOSE NOR ANY OTHER EXPRESS OR IMPLIED WARRANTY EXCEPT AS STATED ABOVE.

NOTICE TO BUYER

Purchase of this material does not confer any rights under patents outside of the United States.

GENERAL INFORMATION

Scouting—Monitoring insect populations to determine whether or not there is a need for application of LANNATE is recommended. More than one treatment of LANNATE may be required to control a population of pests.

Insect Predators—LANNATE at rates of 1/8 to 1/4 lb. per acre helps conserve certain beneficials, including big-eyed bugs, damsel bugs, flower bugs and spiders in cotton and soybeans. While these beneficials cannot be relied upon to control pests, they are of potential value and should be monitored along with pests in pest management programs on these crops.

Compatibility—Since formulations may be changed and new ones introduced, it is recommended that users premix a small quantity of a desired tank mix and observe for possible adverse changes (settling out, flocculation, etc.). Avoid mixtures of several materials and very concentrated spray mixtures.

Do not use LANNATE with Bordeaux mixture, "Du Ter" (triphenyltin hydroxide), lime sulfur, "Rayplex" iron nor in highly alkaline solutions. Use mildly alkaline mixtures immediately after mixing to prevent loss of insecticidal activity.

DIRECTIONS FOR USE

It is a violation of federal law to use this product in a manner inconsistent with its labeling.

Du Pont LANNATE Insecticide should be used only in accordance with recommendations on this label, or in separate published Du Pont recommendations available through local dealers.

Du Pont will not be responsible for losses or damages resulting from use of this product in any manner not specifically recommended by Du Pont. User assumes all risk associated with such nonrecommended use. LANNATE is a dry powder to be dissolved in water for application by ground or air equipment. Use only in commercial and farm plantings; do not use in home plantings.

Do not enter treated areas without appropriate protective clothing until sprays have dried. Because certain states may require more restrictive reentry intervals for various crops treated with this product, consult your State Department of Agriculture for further information.

Written or oral warnings must be given to workers who are expected to be in a treated area or in an area about to be treated with this product. When oral warnings are given, warnings shall be given in a language customarily understood by workers. Oral warnings must be given if there is reason to believe that written warnings cannot be understood by workers. Written (or oral) warnings must include the following information: "DANGER—POISON. Area treated with LANNATE on (date of application). Do not enter treated areas without appropriate protective clothing until sprays have dried. If warning symptoms appear, refer to Statement of Practical Treatment on front panel of LANNATE label and seek medical attention at once."

SPRAY PREPARATION—Fill spray tank ¼ to ½ full of water. Add LANNATE directly to spray tank. Mix thoroughly to fully dissolve the insecticide; once dissolved continued agitation is not required. Use mechanical or hydraulic means; do not use air agitation. Be sure the soluble packet is completely dissolved before adding any other ingredients.

APPLICATION—Apply at the recommended rates when insects first appear. Unless otherwise noted, continue application at 5- to 7-day intervals or as needed. Use sufficient water to obtain thorough, uniform coverage. Since LANNATE is a fast acting contact insecticide, best results follow direct spraying of the target insect. This can be accomplished with fine spray droplets at low volume. For aerial, use a minimum of 2 gals. per acre (gpa) except 10 gpa for grapes, nectarines and strawberries; 15 gpa for oranges, lemons, grapefruit, tangelos, and tangerines.

For certain crops a minimum of 1 gallon per acre may be used providing the following conditions are met.

- equipment is capable of delivering small spray droplets to insure thorough coverage,
- equipment is adjusted to distribute spray uniformly over the spray swath,
- wind conditions and other factors such as temperature humidity are such that the spray is delivered to the target area,
- local regulations do not prohibit low-volume aerial sprays,
- use rates are applied as directed on the package label or supplemental labeling for the following crops.

Alfalfa	Collards	Peppermint
Anise	Corn	Peppers
Asparagus	Cotton	Potato
Barley	Cucumber	Rye
Beans	Lettuce	Sorghum
Broccoli	Melons	Soybean
Brussels sprouts	Mint	Spinach
Cabbage	Oats	Sugar beet
Carrot	Peanuts	Summer Squash
Cauliflower	Peas (succulent)	Wheat
Celery		

Apply the low rates on small plants, small insects and light infestations of insects. Use intermediate rates on large insects and heavier infestations of insects. Use 1 to 3 applications of the highest recommended rate for controlling severe infestations. Thereafter, use the lowest rate possible to maintain control.

CROPS	INSECTS	RATES LANNATE® LBS. PER ACRE	LAST APPLICATION—DAYS TO HARVEST	LAST APPLICATION—DAYS TO LIVESTOCK GRAZING FEEDING
Alfalfa (Do not apply to dormant or semi-dormant alfalfa when min. daily temp is 50°F. or lower).	Pea Aphid, Lygus Bugs, Blotch Leafminer, Aphids, Egyptian Alfalfa Weevil Larvae	½–1		
	Fall Armyworm	½		
	Alfalfa Weevil Larvae	1		
	Loopers, Yellowstriped Armyworm, Western Yellowstriped Armyworm, Armyworm	¼–½	-	7
	Variegated Cutworm, Beet Armyworm	¼–1		
	Alfalfa Caterpillar	¼–½		
Anise (Fennel)	Cabbage Looper	1	7	.
	Beet Armyworm	½–1		
Apples **G (Do not use on Early MacIntosh & Wealthy varieties)	Apple Aphid, Rosy Apple Aphid, Tufted Apple Budmoth, Green Fruitworm, Tarnished Plant Bug	¼–½/100 gals.; up to 400 gals. per acre	8	10
	Codling Moth (10–12 day intervals)			
	Leafrollers (Fruittree, Obliquebanded, Redbanded, Variegated), Lesser Appleworm, White Apple Leafhopper, Tentiform Leafminer, Cutworm	½/100 gals.; up to 400 gals. per acre		
Asparagus	Beet Armyworm, Western Yellowstriped Armyworm, Asparagus Beetle, Spotted Asparagus Beetle, White Cutworm, Redbacked Cutworm	½–1	1	
	Variegated Cutworm	½		
Avocado	Avocado Looper, Avocado Leafroller, Omnivorous Leafroller	½–1	1	10
Barley	Armyworms, Cereal Leaf Beetle, Aphids	¼–½	7	10

CROPS	INSECTS	RATES LANNATE® LBS. PER ACRE	LAST APPLICATION—DAYS TO HARVEST	TO LIVESTOCK GRAZING FEEDING
Beans (Includes Blackeyed Peas & Cowpeas)	Leafhopper, Mexican Bean Beetle	¼–1	Succulent beans ¼ to ½ lb.—1; over ½ lb.—3; Dry beans 25	3 Vines 7 Hay
	Fall Armyworm, Variegated Cutworm	½		
	Beet Armyworm, Loopers, Corn Earworm, Saltmarsh Caterpillar, Yellowstriped Armyworm, Western Yellowstriped Armyworm, Lygus Bugs, Thrips, Aphids	½–1		
	European Corn Borer***—Initiate when moth flights first appear and continue preventive treatments at 3-4 day intervals to control eggs and larvae			
	Spotted Cucumber Beetle	¼–½		
Beets (Table)	Imported Cabbageworm, Beet Armyworm, Cabbage Looper, Diamondback Moth	¼–1	O-Roots 14-Tops	
	Cucumber Beetle, Variagated Cutworm	½		
Bermudagrass	Fall Armyworm, Armyworm	¼–1		7 Forage 3 Dehydrated Hay
Blueberries	Blueberry Leafhopper, Aphids, Tussock Moth, Weevil, Sharp Nosed Leafhopper	½	3	
	Cranberry Fruitworm, Cherry Fruitworm **G	½–1		
	Flea Beetle (larvae), Sawfly (larvae), Blueberry Leafroller	1		
	Blueberry Maggot	¼–½		
Broccoli	Loopers, Imp. Cabbageworm, Diamondback Moth	¼–1*	3	
Brussels Sprouts	Loopers, Imp. Cabbageworm, Diamondback Moth	½–1*	3	
	Variegated Cutworm	½*		
Cabbage	Loopers, Imp. Cabbageworm, Diamondback Moth	¼–1*	1	
	Fall Armyworm, Variegated Cutworm	½*		
Carrot	Beet Armyworm	⅓–1	1	
	Aster Leafhopper	¼-½–1		
	Armyworms	¼-½		
	Variegated Cutworm	¼-½		
Cauliflower	Loopers, Imp. Cabbageworm, Diamondback Moth	¼–1*	3	
	Variegated Cutworm	½*		
Celery	Beet Armyworm, Aster Leafhopper	½–1	7	
	Loopers	1		
	Variegated Cutworm	½		
	Armyworms	¼-½		
Chicory	Beet Armyworm, Variegated Cutworm, Leafhoppers	½–1	30	
Chinese Cabbage **G	Loopers, Beet Armyworm	½–1	10	
Collards (Fresh market only—do not apply when temp. is less than 50°F. Do not apply when crop is less than 10″ tall.)	Loopers, Diamondback Moth, Variegated Cutworm	½	10	
	Imported Cabbageworm, Beet Armyworm	½–1		
Corn (Sweet) **Note:** Certain hybrid varieties of sweet corn are susceptible to methomyl injury. Treat a small area to determine crop safety before full scale spraying.	Earworm—Whorl as needed	⅓–½	0 Ears	3 Forage
	Fall Armyworm, Armyworm, Earworm (Ovicide/ Larvicide) European Corn Borer—Ears 1–3 days or as needed, Corn Rootworm (adult beetles), Flea Beetles, Picnic Beetles, Aphids	¼–½		
	Variegated Cutworm, Beet Armyworm	½		
Corn (Field & Popcorn)	Earworm, Armyworm, Corn Rootworm (adult beetles), Flea Beetles, Picnic Beetles, Aphids	¼–½	0 Ears	3 Forage
	Variegated Cutworm, Beet Armyworm	½		

CROPS	INSECTS	RATES LANNATE® LBS. PER ACRE	LAST APPLICATION—DAYS TO HARVEST	LAST APPLICATION—DAYS TO LIVESTOCK GRAZING FEEDING
Cotton (Use may redden cotton)—If excessive, stop or alternate with other insecticides. U.S.	**Ovicide/Larvicide** Bollworm, Tobacco Budworm (Initiate schedule when significant numbers of eggs are present. Continue at 5- to 7-day intervals while eggs are present and larval control adequate. If significant larvae survive, use higher rates below.) Lygus Bugs (adults and numphs) start treatment on low level population for suppression.	⅛–¼ (See Predators)		
	Cotton Leafworm	¼-½		
	Aphids, Thrips	¼		
East of Rockies only	Cotton Fleahopper (as needed)	⅛-¼		
	Larvicide for worms Bollworm, Tobacco Budworm, Beet Armyworm, Cotton Leafperforator, Fall Armyworm, Lygus Bugs (adults and nymphs) use as occasional spray in regular schedule but not more often than every 10 days.	½		
	Bollworm, Tobacco Budworm, Beet Armyworm, Cotton Leafperforator, Fall Armyworm, Lygus Bugs (adults and nymphs) Up to 3 applications at 3-5 day intervals after desired boll load set on plants.	½–¾	15	Do not Graze or Feed
West of Rockies only	**Larvicide for worms** Bollworm, Fall Armyworm, Tobacco Budworm, Lygus Bugs	½–¾		
	Beet Armyworm	⅓–½		
	Cotton Leafperforator	⅓–¾		
Southwest Only	Cotton Fleahopper (as needed)	⅛-¼		
Cucumber	Loopers, Tobacco Budworm, Beet Armyworm, Yellowstriped Armyworm, Granulate Cutworm, Flea Beetles, Cucumber Beetles, Melon Aphid, Melonworm, Pickleworm	½–1	½ lb.–1 Over ½ lb.–3	
	Fall Armyworm, Variegated Cutworm	½		
Eggplant	Green Peach Aphid	¼–1	5	
	Tomato Pinworm **G, (Beet Armyworm)	½–1		
Endive, Escarole	Beet Armyworm	½-1	10	
Forest Woodlots, Plantations, and Nurseries **G	Spruce Budworm (treat when larvae are feeding on buds, re-treat after 3—4 days if needed)	¼–½		
	Forest Tent Caterpillar, Western Tent Caterpillar	½–1		
Garlic	Beet Armyworm	½*	7	
Grapefruit CA & AZ only	Thrips, Fruittree Leafroller Orange Tortrix, Western Tussock Moth			
U.S.	Beet Armyworm	½–1	1	10
FL only	Spirea Aphid			
	Whiteflies	1		
CA, AZ, & FL only	Yellow Scale, Chaff Scale, Purple Scale, Soft Brown Scale, Glover Scale	½/100 gals. up to 800 gals. per acre		
Grapes East of Rockies only	Grape Berry Moth (apply pre- & post-bloom & repeat at 10-14 day intervals or as needed), Leafhoppers **G	½–1		
U.S.	Omniverous Leafroller, Grape Leaffolder, Orange Tortrix	½-1	1 Fresh & Raisin 14 Wine Grapes	
West of Rockies only	Saltmarsh Caterpillar	½–1		
	Leafhoppers **G			
	Climbing Cutworm (Treat infestation and repeat at 7–14 day intervals as needed)	1		

CROPS	INSECTS	RATES LANNATE® LBS. PER ACRE	LAST APPLICATION—DAYS TO HARVEST	TO LIVESTOCK GRAZING FEEDING
CA only	Grape Leaf Skeletonizer **G	½–1		
Thrips	Thrips	⅛-¼		
Horseradish **G	Alphids, Thrips	½	65	
Leafy Green Vegetables: Beet (Tops), Dandelions, Kale, Mustard Greens, Parsley, Swiss Chard, Turnip Greens	Beet Armyworm, Cabbage Looper, Diamondback Moth, Imported Cabbageworm	½–1	10	
Lemon CA & AZ only	Thrips, Orange Tortrix, Western Tussock Moth	½–1	1	10
U.S.	Beet Armyworm			
FL only	Spirea Aphid			
	Whiteflies	1		
CA, AZ, & FL only	Yellow Scale, Chaff Scale, Purple Scale, Soft Brown Scale, Glover Scale	½/100 up to 800 gals. per acre		
Lentils	Western Yellowstriped Armyworm	½–1	21	
Lettuce (Head varieties in U.S. and Leaf varieties in CA and AZ)	Beet Armyworm, Loopers	¼–1	¼–½ lb.—7; over ½ lb.—10	
	Thrips, Aphids, Corn Earworm, Aster Leafhopper	½–1		
	Variegated Cutworm	½		
Melons	Loopers, Tobacco Budworm, Beet Armyworm, Yellowstriped Armyworm, Granulate Cutworm, Flea Beetles, Cucumber Beetles, Melon Aphid, Mellonworm, Pickleworm	½–1	½ lb.—1 day over ½ lb.—3 days	
	Fall Armyworm, Variegated Cutworm	½		
Mint	Variegated Cutworm, Alfalfa Looper	1	14	
	Flea Beetles	¾–1		
Nectarine CA & AZ only	Thrips (up to 3 applications)	½–1	1	10
Oats	Armyworms, Cereal Leaf Beetle, Aphids	¼–½	7	10
Onions	Beet Armyworm	½*-1	7 green and dry onions	
	Thrips, Variegated Cutworm	1		
Oranges CA & AZ only	Thrips, Western Tussock Moth, Orange Tortrix, Fruittree Leafroller, Citrus Cutworm	½–1	1	10
U.S.	Beet Armyworm			
FL only	Spirea Aphid			
	Whiteflies	1		
CA, AZ, & FL only	Yellow Scale, Chaff Scale, Purple Scale, Soft Brown Scale, Glover Scale	½/100 up to 800 gals. per acre		
Ornamentals **G (See Suppl Labeling)	Bulletin E-40477 1/81 obtainable on request	¼–½ per 100 gals.		
Peaches	Catfacing Insects (Plant Bugs and Stink Bugs)—begin at petal fall and continue in cover sprays at 7- to 10-day intervals	1–2 (or ¼–½/100 gals. up to 400 gals. per acre)	4	10
	Oriental Fruit Moth—begin at petal fall; use trapping devices and frequent field inspection to determine need for treatment. Continue treatment in cover sprays and alternate with residual-type insecticides registered for use. **G			
	Green Peach Aphid			
Peanuts	Corn Earworm***, Beet Armyworm, Potato Leafhopper—Up to 3 applications	¼–1		

CROPS	INSECTS		RATES LANNATE® LBS. PER ACRE	LAST APPLICATION—DAYS TO HARVEST	LAST APPLICATION—DAYS TO LIVESTOCK GRAZING FEEDING
	Loopers, Green Cloverworm, Velvetbean Caterpillar, Thrips, Granulate Cutworm—Up to 3 applications		½–1	21	Do not feed treated vines
	Fall Armyworm		¼–½		
Pears Northeast only	Green Fruitworm, Obliquebanded Leafroller—one application only		½/100 gals. up to 400 gals./acre	7	10
Peas (Succulent)	Armyworm, Green Cloverworm		¼–½	1	5 Forage 14 Hay
	Loopers, Pea Aphid, Beet Armyworm, Saltmarsh Caterpillar, Variegated Cutowrm		½–1		
	Alfalfa Caterpillar		¼–1		
Pecans Southeast only	Aphids		½–1	30	10
Peppers	Green Peach Aphid, Fall Armyworm, Loopers, Beet Armyworm		½	3	
	Armyworm, Variegated Cutworm		¼–½		
Pomegranates	Omniverous Leafroller		1	14	
Potato	Tuberworm, Loopers, Aphids, Beet Armyworm, Leafhoppers		½–1	6	
	Variegated Cutworm, Fall Armyworm, Flea Beetles		½		
Rye	Armyworms, Cereal Leaf Beetle, Aphids		¼–½	7	10
Sorghum	Sorghum Webworm		½	14	14
	Sorghum Midge—Apply when 50% bloom and 3–5 days later if needed		¼–½		
	Fall Armyworm (Budworm), Beet Armyworm, Corn Earworm, Armyworm				
Soybean U.S.	Green Cloverworm, Velvetbean Caterpillar, Mexican Bean Beetle	Light to moderate infestations	⅛–¼ (See Predators)	14	under ½ lb. 3 Forage 7 Hay; ½–1 lb. 10 Forage 12 Hay
		Moderate to severe infestations	¼–½		
	Corn Earworm	Moderate to severe infestations			
		Light to moderate infestations	⅛–¼ (see Predators)		
	Beet Armyworm, Salt Marsh Caterpillar, Bean Leaf Beetle, Thrips, Fall Armyworm, Silver Spotted Skipper	Light to moderate infestations	¼–⅜		
		Moderate to severe infestations	⅜–½		
AK, AL, LA, MS, TN, & TX only	Soybean Looper (worms up to ½″ long)	Use higher rate for severe infestations	½-1		
Spinach Do not apply when min. daily temp. is 32° F. or lower. Do not apply to seedlings less than 3″ diameter.	Loopers, Beet Armyworm		½-1	7	
	Fall Armyworm, Variegated Cutworm		½		
Strawberry	Beet Armyworm, Thrips, Aphids		½–1	3–Fresh 10–Processing Fruit	
	Lygus Bugs, Omniverous Tiers		1		
	Armyworms		¾–1		
Sugar Beet	Beet Webworm, Flea Beetles, Carrion Beetles, Beet Armyworm, Aphids, Western Yellowstriped Armyworm—Up to 4 applications		¼–1	7	30 Tops
	Variegated Cutworm		½		
Summer Squash	Loopers, Tobacco Budworm, Beet Armyworm, Yellowstriped Armyworm, Granulate Cutworm, Flea Beetles, Cucumber Beetles, Melon Aphid, Melonworm, Pickleworm, Fall Armyworm		½–1	½lb.—1 day, over ½lb. 3 days	

CROPS		INSECTS	RATES LANNATE® LBS. PER ACRE	LAST APPLICATION—DAYS TO HARVEST	LAST APPLICATION—DAYS TO LIVESTOCK GRAZING FEEDING
Tangelo	CA & AZ only	Thrips, Western Tussock Moth, Orange Tortrix			
	U.S.	Beet Armyworm	½–1	1	10
	FL only	Spirea Aphid			
		Whiteflies	1		
CA, AZ, & FL only		Yellow Scale, Chaff Scale, Purple Scale, Soft Brown Scale, Glover Scale	½/100 up to 800 gals. per acre		
Tangerine	CA & AZ only	Thrips, Western Tussock Moth, Orange Tortrix			
	U.S.	Beet Armyworm	½–1	1	10
	FL only	Spirea Aphid			
		Whiteflies	1		
CA, AZ, & FL only		Yellow Scale, Chaff Scale, Purple Scale, Soft Brown Scale, Glover Scale	½/100 up to 800 gals. per acre		
Tobacco (Except shade)		Flea Beetle, Hornworm	¼–½	5 Flue cured 14 Air or fire cured	
		Loopers, Aphids, Tobacco Budworm, Fall Armyworm	½		
Tomato		Tomato Fruitworm, Aphids, Hornworm, Loopers, Beet Armyworm, Southern Armyworm	½–1		
		Pinworm		1	
		Fall Armyworm, Variegated Cutworm	½		
		Armyworm	¼–½		
Turf	**G	Sod Webworm (after application, sprinkle irrigate for 15 minutes)	1–2 (⅖–⅘ oz. per 1000 sq. ft.)		Do not graze or feed
Watercress	**G	Beet Armyworm, Cabbage Looper, Diamondback Moth, Imported Cabbageworm (Use in minimum of 30 gpa of water. Water should be turned off 24 hours prior to application)	½–1	10	
Wheat		Armyworms, Cereal Leaf Beetle, Aphids	¼–½	7	10

* Add wetting agent
** All registrations for aerial and ground use unless marked G for ground use only.
*** Ovicidal and larvicidal control

NOTICE TO BUYER—Purchase of this material does not confer any rights under patents of countries outside of the United States.

SUPPLEMENTAL LABELING

LANNATE® INSECTICIDE

EPA Reg. No. 352-342

FOR LOW-VOLUME USE ON CROPS

Du Pont LANNATE Insecticide is recommended for use in low-volume aerial sprays at a minimum spray volume of 1 gallon per acre providing the following conditions are met:

- equipment is capable of delivering small spray droplets to insure thorough coverage
- equipment is adjusted to distribute spray uniformly over the spray swath
- wind conditions and other factors such as temperature and humidity are such that the spray is delivered to the target area
- local regulations do not prohibit low-volume aerial sprays
- use rates are applied as directed on the package label or supplemental labeling for the following crops:

Alfalfa	Carrot	Melons	Rye
Anise	Cauliflower	Mint	Sorghum
Asparagus	Celery	Oats	Soybean
Barley	Collards	Peanuts	Spinach
Beans	Corn	Peas (succulent)	Sugar beet
Broccoli	Cotton	Peppermint	Summer squash
Brussels sprouts	Cucumber	Peppers	Wheat
Cabbage	Lettuce	Potato	

Apply the low rates on small plants, small insects and light infestations of insects. Use intermediate rates on large insects and heavier infestations of insects. Use 1 to 3 applications of the highest recommended rate for controlling severe infestations. Thereafter, use the lowest rate possible to maintain control.

IMPORTANT

BEFORE USING LANNATE®, READ AND CAREFULLY OBSERVE THE CAUTIONARY STATEMENTS AND ALL OTHER INFORMATION APPEARING ON THE PRODUCT LABEL.

This bulletin contains new or supplemental instructions for use of this product which may not appear on package label. Follow the instructions carefully.

2/6/81

SUPPLEMENTAL LABELING

LANNATE® INSECTICIDE

EPA Reg. No. 352-342

ONIONS (GREEN & DRY)

Du Pont LANNATE Insecticide is recommended for use on onions (green & dry) for control of beet armyworm, thrips and variegated cutworm.

HOW TO USE

Apply ½ lb. LANNATE per acre in sufficient water (5 to 15 gals. per acre by air) to obtain uniform coverage. Apply when insect first appears and repeat as needed at 5–7 day intervals.

NOTE: Do not apply within 7 days of harvest.

IMPORTANT

BEFORE USING LANNATE®, READ AND CAREFULLY OBSERVE THE CAUTIONARY STATEMENTS AND ALL OTHER INFORMATION APPEARING ON THE PRODUCT LABEL.

This bulletin contains new or supplemental instructions for use of this product which may not appear on package label. Follow the instructions carefully.

4/15/83

SUPPLEMENTAL LABELING

LANNATE® INSECTICIDE

EPA Reg. No. 352-342

FOR USE ON GRAPES IN THE STATE OF CALIFORNIA

Du Pont LANNATE Insecticide is recommended for use on grapes for the control of thrips in the state of California.

HOW TO USE

Apply LANNATE at a rate of ⅛ to ¼ lbs. per acre using ground or aerial equipment. Make the first application when thrips appear in significant numbers (generally 5 to 8 thrips per leaf) but before leaf bronzing occurs. Repeat as necessary to maintain control. Do not apply within 1 day of harvest for fresh or raisin grapes. Do not apply within 14 days of harvest for wine grapes.

IMPORTANT

BEFORE USING LANNATE®, READ AND CAREFULLY OBSERVE THE CAUTIONARY STATEMENTS AND ALL OTHER INFORMATION APPEARING ON THE PRODUCT LABEL.

This bulletin contains new or supplemental instructions for use of this product which may not appear on package label. Follow the instructions carefully.

2/10/83

SUPPLEMENTAL LABELING

LANNATE® INSECTICIDE

EPA Reg. No. 352-342

FOR USE ON ENDIVE/ESCAROLE

Du Pont LANNATE Insecticide is recommended for use on endive/escarole to control beet armyworm.

HOW TO USE

Apply ½ to 1 lb. LANNATE per acre in sufficient water (5 to 15 gals per acre by air) to obtain thorough, uniform coverage. Make applications when insects first appear and repeat at 5- to 7-day intervals or as needed to maintain control. Do not apply within 10 days of harvest.

IMPORTANT

BEFORE USING LANNATE®, READ AND CAREFULLY OBSERVE THE CAUTIONARY STATEMENTS AND ALL OTHER INFORMATION APPEARING ON THE PRODUCT LABEL.

This bulletin contains new or supplemental instructions for use of this product which may not appear on package label. Follow the instructions carefully.

032384

See end of LANNATE® L listing (next) for the Summary of Current Special Local Need and Federal Tank Mix Registrations for LANNATE® and LANNATE® L Insecticide

LANNATE® L

INSECTICIDE
WATER SOLUBLE LIQUID

RESTRICTED USE PESTICIDE

For retail sale to and use only by Certified Applicators or persons under their direct supervision and only for those uses covered by the Certified Applicator's certification.

1 GALLON CONTAINS 1.8 LBS. METHOMYL

ACTIVE INGREDIENT:
Methomyl
S-methyl-N-[(methylcarbamoyl)oxy] thioacetimidate 24%
INERT INGREDIENTS 76%
TOTAL 100%

EPA Reg. No. 352-370 U.S. Pats. 3,576,834, 3,639,633 & 3,862,316

DANGER–POISON

Keep Out of Reach of Children
PELIGRO

PRECAUCION AL USUARIO: Si usted no lee ingles, no use este producto hasta que la etiqueta haya sido explicado ampliamente.

STATEMENT OF PRACTICAL TREATMENT

If swallowed, call a physician or Poison Control Center. Drink 1 or 2 glasses of water and induce vomiting by touching back of throat with finger. **Do not induce vomiting or give anything by mouth to an unconscious person.**

If inhaled, remove from exposure and have patient lie down and keep quiet. If patient is not breathing, start artificial respiration immediately. **Never give anything by mouth to an unconscious person.**

In case of contact, wash skin with plenty of soap and water; for eyes flush with water for 15 minutes and get medical attention; remove and wash contaminated clothing before reuse.

ATROPINE IS AN ANTIDOTE—Seek medical attention at once in all cases of suspected poisoning. If warning symptoms appear (see **Warning Symptoms**), get medical attention.

DO NOT SUBJECT TO TEMPERATURES BELOW 32°F

PRECAUTIONARY STATEMENTS

HAZARDS TO HUMANS AND DOMESTIC ANIMALS

DANGER—POISON

CONTAINS METHANOL.

MAY BE FATAL OR CAUSE BLINDNESS IF SWALLOWED.

POISONOUS IF INHALED. CAUSES EYE DAMAGE.

Do not breathe vapors or spray mist. Do not get in eyes, on skin or on clothing. When mixing spray, loading, applying, or otherwise handling, wear protective clothing, goggles and a mask or respirator jointly approved by the Mine Safety and Health Administration (formerly the U.S. Bureau of Mines) and by the National Institute for Occupational Safety and Health. Wear clean clothes daily. Wash thoroughly after handling and before eating and smoking. Pilot should not assist in the mixing or loading operation.

Do not apply this product in such a manner as to directly or through drift expose workers, other persons or animals. The area being treated must be vacated by unprotected persons.

WARNING SYMPTOMS—Methomyl poisoning produces effects associated with anticholinesterase activity which may include weakness, blurred vision, headache, nausea, abdominal cramps, discomfort in the chest, constriction of pupils, sweating, slow pulse, muscle tremors.

NOTE TO PHYSICIAN

TREATMENT—Atropine sulfate should be used for treatment. Administer repeated doses, 1.2 to 2.0 mg. intravenously every 10 to 30 minutes until full atropinization is achieved. Maintain atropinization until the patient recovers. Artificial respiration or oxygen may be necessary. Allow no further exposure to any cholinesterase inhibitor until recovery is assured.

Do not use 2-PAM for exposure to LANNATE L alone. However, for exposure to combinations of LANNATE L and organophosphorous insecticides, 2-PAM may be used as required to supplement the atropine sulfate treatment. [Do not use morphine.]

ENVIRONMENTAL HAZARDS

This product is toxic to fish, birds and other wildlife. Keep out of any body of water. Do not apply where runoff is likely to occur. Do not apply when weather conditions favor drift from areas treated. Do not contaminate water by cleaning of equipment or disposal of wastes.

This product is toxic to bees exposed to direct application. Do not apply this product while bees are actively visiting the treatment area. Time applications to coincide with periods of minimum bee activity.

PHYSICAL AND CHEMICAL HAZARDS

Flammable. Keep away from heat, sparks and open flame. Keep container closed. Use with adequate ventilation.

STORAGE AND DISPOSAL

STORAGE

Do not subject to temperatures below 32°F. Store product in original container only, away from other pesticides, fertilizer, food or feed. Not for use or storage in or around the home.

DISPOSAL

Do not contaminate water, food or feed by storage or disposal. Pesticide wastes are acutely hazardous. Improper disposal of excess pesticide, spray mixture or rinsate is a violation of Federal Law. If these wastes cannot be disposed of by use according to label instructions, contact your State Pesticide of Environmental Control Agency, or the Hazardous Waste representative at the nearest EPA Regional Office for guidance.

Triple rinse (or equivalent) the container. Then offer for recycling or reconditioning, or puncture and dispose of in a sanitary landfill, or, if allowed by state and local authorities, by burning. If burned, stay out of smoke.

NOTICE OF WARRANTY

DuPont warrants that this product conforms to the chemical description on the label thereof and is reasonably fit for the purposes stated on such label only when used in accordance with the directions under normal use conditions. It is impossible to eliminate all risks inherently associated with the use of this product. Crop injury, ineffectiveness or other unintended consequences may result because of such factors as weather conditions, presence of other materials or the manner of use or application, all of which are beyond the control of DuPont. In no case shall DuPont be liable for consequential, special or indirect damages resulting from the use or handling of this product. All such risks shall be assumed by the Buyer. DUPONT MAKES NO WARRANTIES OF MERCHANTABILITY OR FITNESS FOR A PARTICULAR PURPOSE NOR ANY OTHER EXPRESS OR IMPLIED WARRANTY EXCEPT AS STATED ABOVE.

GENERAL INFORMATION

Scouting—Monitoring insect populations to determine whether or not there is a need for application of LANNATE L is recommended. More than one treatment of LANNATE L may be required to control a population of pests.

Insect Predators—LANNATE L at rates of ½ to 1 pt. per acre helps conserve certain beneficials, including big-eyed bugs, damsel bugs, flower bugs and spiders in cotton and soybeans. While these beneficials cannot be relied upon to control pests, they are of potential value and should be monitored along with pests in pest management programs on these crops.

Compatibility—Since formulations may be changed, and new ones introduced, it is recommended that users premix a small quantity of a desired tank mix and observe for possible adverse changes (settling out, flocculation, etc.). Avoid mixtures of several materials and very concentrated spray mixtures.

Do not use LANNATE L with Bordeaux mixture, "Du Ter" (triphenyltin hydroxide), lime sulfur, "Rayplex" iron nor in highly alkaline solutions. Use mildly alkaline mixtures immediately after mixing to prevent loss of insecticidal activity.

DIRECTIONS FOR USE

It is a violation of federal law to use this product in a manner inconsistent with its labeling.

DuPont LANNATE L insecticide should be used only in accordance with recommendations on this label, or in separate published DuPont recommendations available through local dealers.

DuPont will not be responsible for losses or damages resulting from use of this product in any manner not specificially recommended by DuPont. User assumes all risks associated with such non-recommended use.

Do not enter treated areas without appropriate protective clothing until sprays have dried. Because certain states may require more restrictive reentry intervals for various crops treated with this product, consult your State Department of Agriculture for further information.

Written or oral warnings must be given to workers who are expected to be in a treated area or in an area about to be treated with this product. When oral warnings are given, warnings shall be given in a language customarily understood by workers. Oral warnings must be given if there is reason to believe that written warnings cannot be understood by workers. Written (or oral) warnings must include the following information: "DANGER—POISON. Area treated with LANNATE L on (date of application). Do not enter treated areas without appropriate protective clothing until sprays have dried. If warnings symptoms appear, refer to Statement of Practical Treatment on front panel of LANNATE L label and seek medical attention at once."

LANNATE L is a water soluble liquid to be diluted with water for application by ground or air equipment. Use only in commercial plantings; do not use in home plantings.

SPRAY PREPARATION—Fill spray tank ¼ to ½ full of water. Add LANNATE L directly to spray tank and mix thoroughly, using mechanical or hydraulic means; do not use air agitation.

APPLICATION—Apply at the recommended rates when insects first appear. Unless otherwise noted, continue application at 5- to 7-day intervals or as needed. Use sufficient water to obtain thorough, uniform coverage. Since LANNATE L is a fast acting contact insecticide, best results follow direct spraying of the target insect. This can be accomplished with fine spray droplets at low volume. For aerial, use a minimum of 2 gals. per acre (gpa) except: 10 gpa for grapes, nectarines and strawberries; 15 gpa for oranges and lemons, grapefruit, tangelos and tangerines.

For certain crops a minimum of 1 gallon per acre may be used providing the following conditions are met.

- equipment is capable of delivering small spray droplets to insure thorough coverage,
- equipment is adjusted to distribute spray uniformly over the spray swath,
- wind conditions and other factors such as temperature and humidity are such that the spray is delivered to the target area,
- local regulations do not prohibit low-volume aerial sprays,
- use rates are applied as directed on the package label or supplemental labeling for the following crops:

Alfalfa	Melons
Anise	Mint
Asparagus	Oats

Barley	Peanuts	Corn	Sugar beet
Beans	Peas (succulent)	Cotton	Summer Squash
Broccoli	Peppermint	Cucumber	Wheat
Brussels sprouts	Peppers	Lettuce	
Cabbage	Potato		
Carrot	Rye		
Cauliflower	Sorghum		
Celery	Soybean		
Collards	Spinach		

Apply the low rates on small plants, small insects and light infestations of insects. Use intermediate rates on large insects and heavier infestations of insects. Use 1 to 3 applications of the highest recommended rate for controlling severe infestations. Thereafter, use the lowest rate possible to maintain control.

CROPS	INSECTS	RATES: LANNATE® L PINTS PER ACRE	LAST APPLICATION—DAYS: TO HARVEST	LAST APPLICATION—DAYS: TO LIVESTOCK GRAZING/FEEDING
Alfalfa (Do not apply to dormant or semi-dormant alfalfa when min. daily temp. is 50°F. or lower.)	Pea Aphid, Lygus Bugs, Blotch Leafminer, Aphids, Egyptian Alfalfa Weevil Larvae	2–4		
	Fall Armyworm	2		
	Alfalfa Weevil Larvae	4		7
	Loopers, Yellowstriped Armyworm, Western Yellowstriped Armyworm, Armyworm	1–2		
	Variegated Cutworm, Beet Armyworm	1–4		
	Alfalfa Caterpiller	1–2		
Anise (Fennel)	Cabbage Looper	4		
	Beet Armyworm	2–4	7	
Apple (Do not use on Early MacIntosh & Wealthy varieties) **G	Apple Aphid, Rosy Apple Aphid, Tufted Apple Budmoth, Green Fruitworm, Tarnished Plant Bug	½–2/100 gals.; up to 400 gals. per acre		
	Codling Moth (10–12 day intervals)		8	10
	Leafrollers (Fruittree, Obliquebanded, Redbanded, Variegated), Lesser Appleworm, White Apple Leafhopper, Tentiform Leafminer	2/100 gals.; up to 400 gals. per acre		
	Cutworm			
Asparagus	Beet Armyworm, Western Yellowstriped Armyworm, Asparagus Beetle, Spotted Asparagus Beetle, White Cutworm, Redbacked Cutworm	2–4	1	
	Variegated Cutworm	2		
Avocado	Avocado Looper, Avocado Leafroller, Omniverous Leafroller	2–4	1	10
Barley	Armyworms, Cereal Leaf Beetle, Aphids	1–2	7	10
Beans (Includes Blackeyed Peas & Cowpeas)	Leafhopper, Mexican Bean Beetle	1–4		
	Fall Armyworm, Variegated Cutworm	2		
	Beet Armyworm, Loopers, Corn Earworm, Saltmarsh Caterpillar, Yellowstriped Armyworm, Western Yellowstriped Armyworm, Lygus Bugs, Thrips, Aphids	2–4	Succulent Beans 1–2pts—1; over 2 pts—3; Dry Beans—25	3 Vines 7 Hay
	European Corn Borrer***—Initiate when moth flights first appear and continue preventive treatments at 3–4 day intervals to control eggs and larvae			
	Spotted Cucumber Beetle	1–2		
Beets (Table)	Imported Cabbageworm, Beet Armyworm, Cabbage Looper, Diamondback Moth	1–4	0–Roots 14–Tops	
	Cucumber Beetle, Variegated Cutworm	2		
Bermudagrass	Fall Armyworm, Armyworm	1–4		7 Forage 3 Dehydrated Hay
Blueberries	Blueberry Leafhopper, Aphids, Tussock Moth, Weevil, Sharp-Nosed Leafhopper	2		
	Cranberry Fruitworm, Cherry Fruitworm **G	2–4	3	
	Flea Beetle (larvae), Sawfly (larvae), Blueberry Leafroller	4		
	Blueberry Maggot	1–2		
Broccoli	Loopers, Imp. Cabbageworm, Diamondback Moth	1–4*	3	
Brussels Sprouts	Loopers, Imp. Cabbageworm, Diamondback Moth	2–4*	3	
	Variegated Cutworm	2*		
Cabbage	Loopers, Imp Cabbageworm, Diamondback Moth	1–4*		
	Fall Armyworm, Variegated Cutworm	2*	1	
Carrot	Beet Armyworm	1⅓–4		
	Aster Leafhopper	2–4	1	

CROPS	INSECTS	RATES: LANNATE® L PINTS PER ACRE	LAST APPLICATION—DAYS: TO HARVEST	LAST APPLICATION—DAYS: TO LIVESTOCK GRAZING/ FEEDING
	Armyworms	1–2		
	Variegated Cutworm	1–2		
Cauliflower	Imp. Cabbageworm, Loopers, Diamondback Moth	1–4*		
	Variegated Cutworm	2*	3	
Celery	Beet Armyworm, Aster Leafhopper	2–4		
	Loopers	4	7	
	Variegated Cutworm	2		
	Armyworms	1–2		
Chicory	Beet Armyworm, Variegated Cutworm, Leafhoppers	2–4	80	
Chinese Cabbage **G	Loopers, Beet Armyworm	2–4	10	
Collards (Fresh market only—do not apply when temp. is less than 50°F. Do not apply when crop is less than 10″ tall)	Loopers, Diamondback Moth, Variegated Cutworm	2	10	
	Imported Cabbageworm, Beet Armyworm	2–4		
Corn (Sweet) Note: Certain hybrid varieties of sweet corn are susceptible to methomyl injury. Treat a small area to determine crop safety before full scale spraying	Earworm—Whorl as needed	1⅓–2		
	Fall Armyworm, Armyworm, Earworm (Ovicide/Larvicide), European Corn Borer—Ears 1–3 days or as needed. Corn Rootworm (adult beetles), Flea Beetles, Picnic Beetles, Aphids	1–2	0 Ears	3 Forage
	Variegated Cutworm, Beet Armyworm	2		
Corn (Field & Popcorn)	Earworm, Armyworm, Fall Armyworm, Corn Rootworm (adult beetles), Flea Beetles, Picnic Beetles, Aphids	1–2	0 Ears	3 Forage
	Variegated Cutworm, Beet Armyworm	2		
Cotton (Use may redden cotton)—If excessive, stop or alternate with other insecticides. U.S.	**Ovicide/Larvicide** Bollworm, Tobacco Budworm (Initiate schedule when significant numbers of eggs are present. Continue at 5- to 7-day intervals while eggs present and larval control adequate. If significant larvae survive, use higher rates below.) Lygus Bugs (adults and nymphs) start treatment on low level population for suppression.	½–1 (See Predators)		
	Cotton Leafworm	1–2		
	Aphids, Thrips	1		
	Cotton Fleahopper (as needed)	½–1		
East of Rockies only	**Larvicide for worms** Bollworm, Tobacco Budworm, Beet Armyworm, Cotton Leafperforator, Fall Armyworm, Lygus Bugs (adults and nymphs) use as occasional spray in regular schedule but not more often than every 10 days.	2		
	Bollworm, Tobacco Budworm, Beet Armyworm, Cotton Leafperforator, Fall Armyworm, Lygus Bugs (adult and nymphs) up to 3 applications at 3–5 day intervals after desired boll load set on plants.	2–3	15	Do not Graze or Feed
West of Rockies only	**Larvicide for worms** Bollworm, Fall Armyworm, Tobacco Budworm, Lygus Bugs	2–3		
	Beet Armyworm	1⅓–2		
	Cotton Leafperforator	1⅓–3		
Southwest only	Cotton Fleahopper (as needed)	½–2		
Cucumber	Loopers, Tobacco Budworm, Beet Armyworm, Yellow-striped Armyworm, Granulate Cutworm, Flea Beetles, Cucumber Beetles, Melon Aphid, Melonworm, Pickleworm	2–4	2 pts.—1 Over 2 pts.—3	
	Fall Armyworm, Variegated Cutworm	2		

CROPS		INSECTS	RATES: LANNATE® L PINTS PER ACRE	LAST APPLICATION—DAYS: TO HARVEST	LAST APPLICATION—DAYS: TO LIVESTOCK GRAZING/ FEEDING
Eggplant		Green Peach Aphid	1–4	5	
		Tomato Pinworm **G, Beet Armyworm	2–4		
Endive, Escarole		Beet Armyworm	2–4	10	
Forest Woodlots, Plantations, and Nurseries **G		Spruce Budworm (treat when larvae are feeding on buds, retreat after 3–4 days if needed)	1–2		
		Forest Tent Caterpillar, Western Tent Caterpillar	2–4		
Garlic		Beet Armyworm	2*	7	
Grapefruit	CA & AZ only	Thrips, Fruittree Leafroller, Orange Tortrix, Western Tossock Moth	2–4	1	10
	U.S.	Beet Armyworm			
	FL only	Spirea Aphid	4		
		Whiteflies			
	CA, AZ & Fl only	Yellow Scale, Chaff Scale, Purple Scale, Soft Brown Scale	$\frac{2}{100}$ gals. up to 800 gals per acre		
Grapes	East of Rockies only	Grape Berry Moth (apple pre- & post-bloom & repeat at 10–14 day intervals or as needed), Leafhoppers **G	2–4	1 Fresh & Raisin	
	U.S.	Omniverous Leafroller, Grape Leaffolder, Orange Tortrix	4		
	West of Rockies only	Saltmarsh Caterpillar	2–4	14 Wine Grapes	
		Leafhoppers **G			
		Climbing Cutworm (Treat infestation and repeat at 7–14 day intervals as needed)	4		
	CA only	Grape Leaf Skeletonizer **G	2–4		
		Thrips	½–1		
Horseradish **G		Aphids, Thrips	2	65	
Leafy Green Vegetables: Beet (Tops), Dandelions, Kale, Mustard Greens, Parsley, Swiss Chard, Turnip Greens		Beet Armyworm, Cabbage Looper, Diamondback Moth, Imported Cabbageworm	2–4	10	
Lemon	CA & AZ only	Thrips, Western Tussock Moth, Orange Tortrix	2–4	1	10
	U.S.	Beet Armyworm			
	FL only	Spirea Aphid	4		
		Whiteflies			
	CA, AZ, & FL only	Yellow Scale, Chaff Scale, Purple Scale, Soft Brown Scale, Glover Scale	$\frac{2}{100}$ gals. up to 800 gals. per acre		
Lentils		Western Yellowstriped Armyworm	2–4	21	
Lettuce, (head varieties in U.S. and Leaf varieties in CA and AZ)		Beet Armyworm, Loopers	1–4	1–2 pts.—7 over 2 pts.—10	
		Thrips, Aphids, Corn Earworm, Aster Leafhopper	2–4		
		Variegated Cutworm	2		
Melons		Loopers, Tobacco Budworm, Beet Armyworm, Yellow-striped Armyworm, Granulate Cutworm, Flea Bettles, Cucumber Beetles, Melon Aphid, Melonworm, Pickle-worm	2–4	2 pts.—1 day over 2 pts.—3 days	
		Fall Armyworm, Variegated Cutworm	2		
Mint		Variegated Cutworm, Alfalfa Looper	4	14	
		Flea Beetles	3–4		
Nectarine	CA & AZ only	Thrips (up to 3 applications)	2–4	1	10

CROPS	INSECTS		RATES: LANNATE® L PINTS PER ACRE	LAST APPLICATION—DAYS: TO HARVEST	LAST APPLICATION—DAYS: TO LIVESTOCK GRAZING/ FEEDING
Oats	Armyworms, Cereal Leaf Beetle, Aphids		1–2	7	10
Onions	Beet Armyworm		2–4*	7 Green and Dry Onions	
	Thrips, Variegated Cutworm		2*		
Oranges CA & AZ only	Thrips, Western Tussock Moth, Orange Tortrix, Fruit-tree Leafroller, Citrus Cutworm		2–4	1	10
U.S.	Beet Armyworm				
FL only	Spirea Aphid				
	Whiteflies		4		
CA, AZ, & FL only	Yellow Scale, Chaff Scale, Purple Scale, Soft Brown Scale, Glover Scale		²⁄₁₀₀ gals. up to 800 gals. per acre		
Ornamentals **G (See Suppl. Labeling)	Bulletin E-40477 1/81 obtainable on request		1–2 per 100 gals.		
Peaches	Catfacing Insects (Plant Bugs and Stink Bugs)—begin at petal fall and continue in cover sprays at 7- to 10-day intervals		4–8 (or 1–2/100 gals up to 400 gals. per acre	4	10
	Oriental Fruit Moth—begin at petal fall; use trapping devices and frequent field inspection to determine need for treatment. Continue treatment in cover sprays and alternate with residual-type insecticides registered for this use.**G				
	Green Peach Aphid		3–8		
Peanuts	Corn Earworm***, Beat Armyworm, Potato Leafhopper—Up to 3 applications		1–4	21	Do not feed treated vines
	Loopers, Green Cloverworm, Velvetbean Caterpillar, Thrips, Granulate Cutworm—Up to 3 applications		2–4		
	Fall Armyworm		1–2		
Pears Northeast only	Green Fruitworm, Obliquebanded Leafroller—One application only		2/100 gals. up to 400 gals./acre	7	10
Peas (Succulent)	Armyworm, Green Cloverworm		1–2	1	5 Forage 14 Hay
	Loopers, Pea Aphid, Beet Armyworm, Saltmarsh Caterpillar, Variegated Cutworm		2–4		
	Alfalfa Caterpillar		1–4		
Pecans Southeast only	Aphids		2–4	30	10
Peppers	Green Peach Aphid, Fall Armyworm, Loopers, Beet Armyworm		2	3	
	Armyworm, Variegated Cutworm		1–2		
Pomegranates	Omniverous Leafroller		4	14	
Potato	Tuberworm, Loopers, Aphids, Beet Armyworm, Leafhoppers		2–4	6	
	Variegated Cutworm, Fall Armyworm, Flea Beetles		2		
Rye	Armyworms, Cereal Leaf Beetle, Aphids		1–2	7	10
Sorghum	Sorghum Webworm		2	14	14
	Sorghum Midge—Apply when 50% bloom and 3–5 days later if needed		1–2		
	Corn earworm, Armyworm, Fall Armyworm (Budworm), Beet Armyworm				
Soybean U.S.	Green Cloverworm, Velvetbean Caterpillar Mexican Bean Beetle	Light to moderate infestations	½–1 (See Predators)	14	under 2 pts.
		Moderate to severe infestations	1–2		3 Forage 7 Hay
	Corn Earworm (worms up to ½″ long)	Moderate to severe infestations			2–4 pts.
	(worms up to ¼″ long)	Light to moderate infestations	½–1 (see Predators)		

CROPS	INSECTS		RATES: LANNATE® L PINTS PER ACRE	LAST APPLICATION—DAYS: TO HARVEST	LAST APPLICATION—DAYS: TO LIVESTOCK GRAZING/ FEEDING
	Beet Armyworm, Salt Marsh Caterpillar, Bean Leaf Beetle, Thrips, Fall Armyworm, Silver Spotted Skipper	Light to moderate infestations	1–1½	2–4	10 Forage 12 Hay
		Moderate to severe infestations	1½–2		
CA, AL, LA, MS TN, & TX only	Soybean Looper (worms up to ½″ long)		Use high rate for severe infestations		
Spinach (Do not apply when min. daily temp is 32°F or lower. Do not apply to seedlings less than 3″ diameter)	Loopers, Beet Armyworm		2–4	7	
	Fall Armyworm, Variegated Cutworm		2		
Strawberry	Beet Armyworm, Thrips, Aphids		2–4	3–Fresh 10–Processing Fruit	
	Lygus Bugs, Omniverous Leaf Tiers		4		
	Armyworms		3–4		
Sugar Beet	Beet Webworm, Flea Beetles, Carrion Beetles, Beet Armyworms, Aphids, Western Yellowstriped Armyworm—Up to 4 applications		1–4	7	30 Tops
	Variegated Cutworm		2		
Summer Squash	Loopers, Tobacco Budworm, Beet Armyworm, Yellowstriped Armyworm, Granulate Cutworm, Fleabeetles, Cucumber Beetles, Melon Aphid, Melonworm, Pickleworm, Fall Armyworm		2–4	2 pts.—1 day over 2 pts. 3 days	
Tangelo					
CA & AZ only	Thrips, Western Tussock Moth, Orange Tortrix		2–4	1	10
U.S.	Beet Armyworm				
FL only	Spirea Aphid				
	Whiteflies		4		
CA, AZ, & FL only	Yellow Scale, Chaffe Scale, Purple Scale, Soft Brown Scale, Glover Scale		2/100 gals. up to 800 gals. per acre		
Tangerine					
CA & AZ only	Thrips, Wester Tussock Moth, Orange Tortrix		2–4	1	10
U.S.	Beet Armyworm				
FL only	Spirea Aphid				
	Whiteflies		4		
CA, AZ, & FL only	Yellow Scale, Chaff Scale, Purple Scale, Soft Brown Scale, Glover Scale		2/100 gals. up to 800 gals. per acre		
Tobacco (Except shade)	Flea Beetle, Hornworm		1–2	5 Flue cured 14 Air or fire cured	
	Loopers, Aphids, Tobacco Budworm, Fall Armyworm		2		
Tomato	Tomato Fruitworm, Aphids, Hornworm, Loopers, Beet Armyworm, Southern Armyworm		2–4	1	
	Pinworm				
	Fall Armyworm, Variegated Cutworm		2		
	Armyworm		1–2		
Turf **G	Sod Webworm (after application, sprinkle irrigate for 15 minutes)		4–8 (1.5–3 fl. ozs. per 1000 sq. ft.)		Do not graze or feed
Watercress **G	Beet Armyworm, Cabbage Looper, Diamondback Moth, Imported Cabbageworm (Use in minimum of 30 gpa of water. Water should be turned off 24 hours prior to applications).		2–4	10	
Wheat	Armyworms, Cereal Leaf Beetle, Aphids		1–2	7	10

* Add wetting agent
** All registrations for aerial and ground use unless marked G for ground use only.
*** Ovicidal and larvicidal control

SUPPLEMENTAL LABELING

LANNATE® L INSECTICIDE
EPA Reg. No. 352-370

FOR LOW VOLUME USE ON CROPS

RESTRICTED USE PESTICIDE

For retail sale to and use only by Certified Applicators or persons under their direct supervision and only for those uses covered by the Certified Applicator's certification.

DuPont LANNATE® L Insecticide is recommended for use in low-volume aerial sprays at a minimum spray volume of 1 gallon per acre providing the following conditions are met:

- equipment is capable of delivering small spray droplets to insure thorough coverage
- equipment is adjusted to distribute spray uniformly over the spray swath
- wind conditions and other factors such as temperature and humidity are such that the spray is delivered to the target area
- local regulations do not prohibit low-volume aerial sprays
- use rates are applied as directed on the package label or supplemental labeling for the following crops:

Alfalfa	Cucumber
Anise	Lettuce
Asparagus	Melons
Beans	Mint
Broccoli	Peanuts
Brussels sprouts	Peas (succulent)
Cabbage	Peppermint
Carrot	Peppers
Cauliflower	Potato
Celery	Sorghum
Collards	Soybean
Corn	Spinach
Cotton	Sugar beet
	Summer squash

IMPORTANT

BEFORE USING LANNATE® L, READ AND CAREFULLY OBSERVE THE CAUTIONARY STATEMENTS AND ALL OTHER INFORMATION APPEARING ON THE PRODUCT LABEL.

This bulletin contains new or supplemental instructions for use of this product which may not appear on package label. Follow the instructions carefully.

2/6/81

SUPPLEMENTAL LABELING

LANNATE® L INSECTICIDE
EPA Reg. No. 352-370

FOR USE ON ONIONS (GREEN & DRY)

RESTRICTED USE PESTICIDE

For retail sale to and use only by Certified Applicators or persons under their direct supervision and only for those uses covered by the Certified Applicator's certification.

DuPont LANNATE L Insecticide is recommended for use on onions (green & dry) for control of beet armyworm, thrips and variegated cutworm.

HOW TO USE

Apply 2 pts. LANNATE L per acre in sufficient water (5 to 15 gals. per acre) by air to obtain uniform coverage. Apply when insect first appears and repeat as needed at 5–7 day intervals.

NOTE: Do not apply within 7 days of harvest.

IMPORTANT

BEFORE USING LANNATE L®, READ AND CAREFULLY OBSERVE THE CAUTIONARY STATEMENTS AND ALL OTHER INFORMATION APPEARING ON THE PRODUCT LABEL.

This bulletin contains new or supplemental instructions for use of this product which may not appear on package label. Follow the instructions carefully.

4/5/83

SUPPLEMENTAL LABELING

LANNATE® L INSECTICIDE
EPA Reg. No. 352-370

FOR USE ON GRAPES IN THE STATE OF CALIFORNIA

RESTRICTED USE PESTICIDE

For retail sale to and use only by Certified Applicators or persons under their direct supervision and only for those uses covered by the Certified Applicator's certification.

DuPont LANNATE L Insecticide is recommended for use on grapes for the control of thrips in the state of California.

HOW TO USE

Apply LANNATE L at the rate of ½–1 pint per acre using ground or aerial equipment. Make the first application when thrips appear in significant numbers (generally 5 to 8 thrips per leaf) but before leaf bronzing occurs. Repeat as necessary to maintain control. Do not apply within 1 day of harvest for fresh or raisin grapes. Do not apply within 14 days of harvest of wine grapes.

IMPORTANT

BEFORE USING LANNATE L®, READ AND CAREFULLY OBSERVE THE CAUTIONARY STATEMENTS AND ALL OTHER INFORMATION APPEARING ON THE PRODUCT LABEL.

This bulletin contains new or supplemental instructions for use of this product which may not appear on package label. Follow the instructions carefully.

2/10/83

SUPPLEMENTAL LABELING

LANNATE® L INSECTICIDE
EPA Reg. No. 352-370

FOR USE ON ENDIVE/ESCAROLE

RESTRICTED USE PESTICIDE

For retail sale to and use only by Certified Applicators or persons under their direct supervision and only for those uses covered by the Certified Applicator's certification.

DuPont LANNATE® L Insecticide is recommended for use on endive and escarole for the control of beet armyworm. Do not apply within 10 days of harvest.

HOW TO USE

Apply 2 to 4 pts LANNATE L per acre in sufficient water (5 to 15 gals per acre by air) to obtain thorough, uniform coverage. Make applications when insects first appear and repeat at 5–7 day intervals or as needed to maintain control.

IMPORTANT

BEFORE USING LANNATE L®, READ AND CAREFULLY OBSERVE THE CAUTIONARY STATEMENTS AND ALL OTHER INFORMATION APPEARING ON THE PRODUCT LABEL.

This bulletin contains new or supplemental instructions for use of this product which may not appear on package label. Follow the instructions carefully.

This label must be in the possession of the user at the time of pesticide application.

032384

SUMMARY OF CURRENT SPECIAL LOCAL NEED AND FEDERAL TANK MIX REGISTRATIONS FOR LANNATE® AND LANNATE® INSECTICIDE

Before using, read and carefully observe the cautionary statements and all other information appearing on product labels.

CROPS	INSECTS	RATES: LANNATE®L†† PINTS PER ACRE	RATES: LANNATE®† POUNDS PER ACRE	LAST APPLICATION—DAYS: HARVEST	LAST APPLICATION—DAYS: TO LIVESTOCK GRAZING/ FEEDING
Apple **G (Do not use on Early Macintosh & Wealthy varieties) VT, MI, PA & NY††† only	Codling Moth, Leafrollers, Tufted Apple Budmoth, Lesser Appleworm, Tentiform Leafminer, Eye Spotted Budmoth, Green Fruitworm, Leafhoppers, Plum Curculio, Tarnished Plant Bug Apple Aphid, Rosy Apple Aphid	1 + ¼ lb. Guthion® 50W per 100 gals. up to 800 gals. per acre	¼ + ¼ lb. Guthion® 50W per 100 gals. up to 800 gals. per acre	8	10
Broccoli Combination with Bacillus thuringiensis	Imported Cabbageworm Diamondback Moth	1–2 + ¼–½ # BT*	¼–½ + ¼–½ #BT*	3	
	Loopers	2 + ½ # BT*	½ + ½ # BT*		
Brussels Sprouts Combination with Bacillus thuringiensis	Imported Cabbageworm, Diamondback Moth	1–2 + ¼–½ # BT*	¼–½ + ¼–½ # BT*	3	
	Loopers	2 + ½ # BT*	½ + ½ # BT*		
Cabbage Combination with Bacillus thuringiensis	Imported Cabbageworm, Diamondback Moth	1–2 + ¼–½ # BT*	¼–½ + ¼–½ # BT*	1	
	Loopers	2 + ½ # BT*	½ + ½ # BT*		
Cauliflower Combination with Bacillus thuringiensis	Imported Cabbageworm, Loopers Diamondback Moth	1–2 + ¼–½ # BT*	¼–½ + ¼–½ # BT*	3	
	Loopers	2 + ½ # BT*	½ + ½ # BT*		
Cotton (Use may redden cotton)—If excessive, stop or alternate with other insecticides.) Combination with Bacillus thuringiensis	Bollworm*** Tobacco Budworm***	½–1 plus registered rates of BT	½–¼-plus registered rates of BT	15	Do Not Graze Or Feed
Cotton U.S. (Use may redden cotton—If excessive, stop or alternate with other insecticides) Combination with permethrin	Bollworm***, Tobacco Budworm***	½–1 + 05–2 lb. permethrin	⅛–¼ + 05–2 lb. permethrin	15	
U.S. Combination with chlordimeform	Bollworm***, Tobacco Budworm***	½ + ⅛ lb. Chlordimeform	½ + ⅛ lb. Chlordimeform	21	Do not graze or feed
TX		¼–½ + ⅛ lb. Chlordimeform	1/16–⅛ + ⅛ lb. Chlordimeform		
U.S. Combination with EPN (Do not enter field for 24 hrs. after treatment. Do not hand pick within 3 days of treatment	Bollworm***, Tobacco Budworm***, Boll Weevil	½–1 + 1–2 pts. EPN (4 lb./gal.)	⅛–¼ + 1–2 pts. EPN (4 lb./gal.)	15	
Combination with EPN + methyl parathion (Do not enter field for 48 hrs. after treatment Do not hand pick within 7 days of treatment	Bollworm***, Tobacco Budworm***, Boll Weevil	½–1 + ⅔–1⅓ pts. EPN + MP (3 lb. + 3 lb./gal.)	⅛–¼ + ⅔–1⅓ pts. EPN + MP (3 lb. + 3 lb./gal.)		
U.S. Combination with Methyl Parathion. (Do not enter field for 48 hrs. after treatment. Do not hand pick within 7 days of treatment.)	Tobacco Budworm***, Bollweevil (5–7 day intervals)	1–1¼ + 1–2 pts. Methyl Parathion (4 lb./gal.)	¼–⅓ + 1–2 pts. Methyl Parathion (4 lb./gal.)		
Mid South & Southeast only	Bollworm (5–7 day intervals)***				
TX††† only	Bollworm (5–7 day intervals)***				
Peppermint OR††† only	Flea Beetles	3–4	¾–1	14	
Radish AZ††† only	Beet Armyworm	2	½	3	

CROPS	INSECTS	RATES: LANNATE®L†† PINTS PER ACRE	RATES: LANNATE®† POUNDS PER ACRE	LAST APPLICATION—DAYS: HARVEST	LAST APPLICATION—DAYS: TO LIVESTOCK GRAZING/ FEEDING
Tomato PA†††, **G Greenhouse, NC††† & OH††† only (Wear full protective gear; apply in closed greenhouse to remain closed for 2 hrs. (or longer) after treatment, ventilate additional 1 hr. before reentry.)	White Fly	—	½–1/200 gals.	½ lb./200 gal. —7, over ½ lb. per 200–14	
LA†††	Pinworm				

* Add wetting agent
** All registrations for aerial and ground use unless marked G for ground use only
*** Ovicidal and larvicidal control

† 90% active water soluble powder or water soluble bags
†† 1.8 lb. active per gallon. Restricted use pesticide
††† FIFRA section 24(c) special local need registration

LANNATE® LV
INSECTICIDE

RESTRICTED USE PESTICIDE

For retail sale to and use only by Certified Applicators or persons under their direct supervision and only for those uses covered by the Certified Applicator's certification.

1 Gallon Contains 2.4 lb. Methomyl

ACTIVE INGREDIENT:
Methomyl
S-methyl-N-[(methylcarbamoyl)oxy] thioacetimidate 29%
INERT INGREDIENTS 71%
TOTAL 100%

EPA Reg. No. 352-384 U.S. Pats. 3,576,834 3,639,633 & 3,975,518

DANGER–POISON

Keep Out of Reach of Children
PELIGRO

PRECAUCION AL USUARIO: Si usted no lee inglés, no use este producto hasta que la etiqueta haya sido explicado ampliamente.

STATEMENT OF PRACTICAL TREATMENT

If swallowed: Call a physician or Poison Control Center. Drink 1 or 2 glasses of water and induce vomiting by touching back of throat with finger. **Do not induce vomiting or give anything by mouth to an unconscious person.**

If inhaled: Remove from exposure and have patient lie down and keep quiet. If patient is not breathing, start artificial respiration immediately. Never give anything by mouth to an unconscious person.

If in eyes: Flush eyes with plenty of water. Get medical attention.

If on skin: Wash skin with plenty of soap and water; remove and wash contaminated clothing before re-use. Get medical attention.

ATROPINE IS AN ANTIDOTE—Seek medical attention at once in all cases of suspected poisoning.

If warning symptoms appear (see **Warning Symptoms** below), get medical attention.
Do not subject to temperatures below 32°F

PRECAUTIONARY STATEMENTS

HAZARDS TO HUMANS AND DOMESTIC ANIMALS

DANGER—POISON

CONTAINS METHANOL. MAY BE FATAL OR CAUSE BLINDNESS IF SWALLOWED. POISONOUS IF INHALED. CAUSES EYE DAMAGE.

Do not breathe vapors or spray mist. Do not get in eyes, on skin, or on clothing. When mixing spray, loading, applying or otherwise handling, wear protective clothing, goggles, and a mask or respirator jointly approved by the Mine Safety and Health Administration [formerly the U.S. Bureau of Mines] and by the National Institute for Occupational Safety and Health. Wear clean clothes daily. Wash thoroughly after handling and before eating or smoking. Pilot should not assist in the mixing and loading operation. Do not apply this product in such a manner as to directly or through drift expose workers, other persons or animals. The area being treated must be vacated by unprotected persons.

WARNING SYMPTOMS—Methomyl poisoning produces effects associated with anticholinesterase activity which may include weakness, blurred vision, headache, nausea, abdominal cramps, discomfort in the chest, constriction of pupils, sweating, slow pulse, muscle tremors.

NOTE TO PHYSICIAN

TREATMENT—Altropine sulfate should be used for treatment. Administer repeated doses, 1.2 to 2.0 mg. intravenously every 10 to 30 minutes until full atropinization is achieved. Maintain atropinization until the patient recovers. Artificial respiration or oxygen may be necessary. Allow no further exposure to any cholinesterase inhibitor until recovery is assured.

Do not use 2-PAM for exposure to LANNATE LV alone. However, for exposure to combinations of LANNATE LV and organophosphorous insecticides, 2-PAM may be used as required to supplement the atropine sulfate treatment. Do not use morphine.

ENVIRONMENTAL HAZARDS

This product is toxic to fish, birds, and other wildlife. Keep out of any body of water. Do not apply where runoff is likely to occur. Do not apply when weather conditions favor drift from areas treated. Do not contaminate water by cleaning of equipment or disposal of wastes.

This product is toxic to bees exposed to direct application. Do not apply this product while bees are actively visiting the treatment area. Time applications to coincide with periods of minimum bee activity.

PHYSICAL AND CHEMICAL HAZARDS

Flammable. Keep away from heat, sparks, and open flame. Keep container closed. Use with adequate ventilation.

NOTICE TO BUYER

Purchase of this material does not confer any rights under patents outside of the United States.

STORAGE AND DISPOSAL

STORAGE

Do not subject to temperatures below 32°F. Store product in original container only, away from other pesticides, fertilizer, food, or feed. Not for use or storage in or around the home.

DISPOSAL

Do not contaminate water, food, or feed by storage or disposal. Pesticide wastes are acutely hazardous. Improper disposal of excess pesticide, spray mixture, or rinsate is a violation of Federal Law. If these wastes cannot be disposed of by use according to label instructions, contact your State Pesticide of Environmental Control Agency, or the Hazardous Waste representative at the nearest EPA Regional Office for guidance.

Triple rinse (or equivalent) the container. Then offer for recycling or reconditioning, or puncture and dispose of in a sanitary landfill, or, if allowed by state and local authorities, by burning. If burned, stay out of smoke.

NOTICE OF WARRANTY

DuPont warrants that this product conforms to the chemical description on the label thereof and is reasonably fit for the purposes stated on such label only when used in accordance with the directions under normal use conditions. It is impossible to eliminate all risks inherently associated with the use of this product. Crop injury, ineffectiveness, or other unintended consequences may result because of such factors as weather conditions, presence of other materials, or the manner of use or application, all of which are beyond the control of DuPont. In no case shall DuPont be liable for consequential, special or indirect damages resulting from the use or handling of this product. All such risks shall be assumed by the buyer. DUPONT MAKES NO WARRANTIES OF MERCHANTABILITY OR FITNESS FOR A PARTICULAR PURPOSE NOR ANY OTHER EXPRESS OR IMPLIED WARRANTY EXCEPT AS STATED ABOVE.

GENERAL INFORMATION

Scouting—Monitoring insect populations to determine whether or not there is a need for application of LANNATE LV is recommended. More than one treatment of LANNATE LV may be required to control a population of pests.

Insect Predators—LANNATE LV at rates of ½ to ¾ pt. per acre helps conserve certain beneficials, including bigeyed bugs, damsel bugs, flower bugs and spiders in cotton and soybeans. While these beneficials cannot be relied upon to control pests, they are of potential value and should be monitored along with pests in pest management programs on these crops.

Compatibility—Since formulations may be changed and new ones introduced, it is recommended that users premix a small quantity of a desired tank mix and observe for possible adverse changes (settling out, flocculation, etc.). Avoid mixtures of several materials and very concentrated spray mixtures.

Do not use LANNATE LV with Bordeaux mixture. "Du Ter" (Triphenyltin hydroxide), lime sulfur, "Rayplex" irŏn nor in highly alkaline solutions. Use mildly alkaline mixtures immediately after mixing to prevent loss of insecticidal activity.

DIRECTIONS FOR USE

It is a violation of federal law to use this product in a manner inconsistent with its labeling.

DuPont LANNATE LV Insecticide should be used only in accordance with recommendations on this label, or in separate published DuPont recommendations available through local dealers.

DuPont will not be responsible for losses or damages resulting from use of this product in any manner not specifically recommended by DuPont. User assumes all risks associated with such non-recommended use.

LANNATE LV is a water soluble liquid to be diluted with water for application by ground or air equipment. Use only in commercial plantings; do not use in home plantings.

Do not enter treated areas without appropriate protective clothing until sprays have dried. Because certain states may require more restrictive reentry intervals for various crops treated with this product, consult your State Department of Agriculture for further information.

Written or oral warnings must be given to workers who are expected to be in a treated area or in an area about to be treated with this product. When oral warnings are given, warnings shall be given in a language customarily understood by workers. Oral warnings must be given if there is reason to believe that written warnings cannot be understood by workers. Written (or oral) warnings must include the following information: "DANGER—POISON. Area treated with LANNATE LV on (date of application). Do not enter treated areas without appropriate protective clothing until sprays have dried. If warnings symptoms appear, refer to Statement of Practical Treatment amd seek medical attention at once."

Application—Apply at the recommended rates when insects first appear. Unless otherwise noted, continue application at 5–7 day intervals or as needed. Use sufficient water to obtain thorough, uniform coverage. For aerial, use a minimum of 2 gallons per acre (gpa) except: 10 gpa for grapes, nectarines, and strawberries; 15 gpa for oranges, lemons, grapefruits, tangelos and tangerines.

LANNATE LV is recommended for use as a low volume aerial spray 0.53 gpa (2L) for cotton† and soybeans† and 1 gpa for the crops listed below providing the following conditions are met:

- equipment is capable of delivering small spray droplets to insure thorough coverage
- equipment is adjusted to distribute spray uniformly over the spray swath
- wind conditions and other factors such as temperature and humidity are such that the spray is delivered to the target area
- local regulations do not prohibit low-volume aerial sprays
- use rates are applied as directed on the package label or supplemental labeling for the following crops:

Alfalfa	Collards	Peppers
Anise	Corn	Potato
Asparagus	Cotton	Rye
Barley	Cucumber	Sorghum
Beans	Lettuce	Soybean
Broccoli	Melons	Spinach
Brussels sprouts	Mint	Sugar beet
Cabbage	Oats	Summer Squash
Carrot	Peanuts	Wheat
Cauliflower	Peas (succulent)	
Celery	Peppermint	

Apply the low rates on small plants, small insects and light infestations of insects. Use intermediate rates on large insects and heavier infestations of insects. Use 1 to 3 applications of the highest recommended rate for controlling severe infestations. Thereafter, use the lowest rate possible to maintain control.

CROPS	INSECTS	RATES: LANNATE® LV PINTS PER ACRE	LAST APPLICATION—DAYS: TO HARVEST	LAST APPLICATION—DAYS: TO LIVESTOCK GRAZING/ FEEDING
Alfalfa (Do not apply to dormant or semi-dormant alfalfa when min. daily temp. is 50°F. or lower.)	Pea Aphid, Lygus Bugs, Blotch Leafminer, Aphids, Egyptian Alfalfa Weevil Larvae	1½–3		
	Fall Armyworm	1½		
	Alfalfa Weevil Larvae	3		7

CROPS	INSECTS	RATES LANNATE® LV PINTS PER ACRE	LAST APPLICATION—DAYS TO HARVEST	TO LIVESTOCK GRAZING/ FEEDING
	Loopers, Yellowstriped Armyworm, Western Yellowstriped Armyworm, Armyworm	¾–1½		
	Variegated Cutworm, Beet Armyworm	¾–3		
	Alfalfa Caterpillar	1–1½		
Anise (Fennel)	Cabbage Looper	3	7	
	Beet Armyworm	1½–3		
Apple (Do not use on Early MacIntosh & Wealthy varieties) **G	Apple Aphid	⅜–1½/100 gals.; up to 400 gals. per acre		
	Rosy Apple Aphid, Tufted Apple Budmoth, Green Fruitworm, Tarnished Plant Bug	¾–1½/100 gals.; up to 400 gals. per acre		
	Codling Moth (10-12 day intervals)		8	10
	Leafrollers (Fruittree, Obliquebanded, Redbanded, Variegated). Lesser Appleworm, White Apple Leafhoppper, Tentiform Leafminer	1½/100 gals.; up to 400 gals. per acre		
	Cutworm	1½/100 gals.		
Asparagus	Beet Armyworm, Western Yellowstriped Armyworm, Asparagus Beetle, Spotted Asparagus Beetle, White Cutworm, Redbacked Cutworm	1½–3	1	
	Variegated Cutworm	1½		
Avocado	Avocado Looper, Avocado Leafroller, Omniverous Leafroller	1½–3	1	10
Barley	Armyworms, Cereal Leaf Beetle, Aphids	¾–1½	7	10
Beans (Includes Blackeyed Peas & Cowpeas)	Leafhopper, Mexican Bean Beetle	¾–3		
	Fall Armyworm, Variegated Cutworm	1½		
	Beet Armyworm, Loopers, Corn Earworm, Saltmarsh Caterpillar, Yellowstriped Armyworm, Western Yellowstriped Armyworm, Lygus Bugs, Thrips, Aphids	1½–3	Succulent Beans ¾–1½ pts.—1 over 1½ pts.—3; Dry Beans—25	3 Vines 7 Hay
	European Corn Borer***—Initiate when moth flights first appear and continue preventive treatments at 3-4 day intervals to control eggs and larvae			
	Spotted Cucumber Beetle	¾–1½		
Beets (Table)	Imported Cabbageworm, Beet Armyworm, Cabbage Looper, Diamondback Moth	¾–3	0-Roots 14-Tops	
	Cucumber Beetle, Variegated Cutworm	1½		
Bermudagrass	Fall Armyworm, Armyworm	¾–3		7 Forage 3 Dehydrated Hay
Blueberries	Blueberry Leafhopper, Aphids, Tussock Moth, Weevil, Sharp-Nosed Leafhopper	1½		
	Cranberry Fruitworm, Cherry Fruitworm **G	1½–3	3	
	Flea Beetle (larvae), Sawfly (larvae), Blueberry Leafroller	3		
	Blueberry Maggot	¾–1½		
Broccoli	Loopers, Imp. Cabbageworm, Diamondback Moth	¾–3*	3	
Brussels Sprouts	Loopers, Imp. Cabbageworm, Diamondback Moth	1½–3*	3	
	Variegated Cutworm	1½*		
Cabbage	Loopers, Imp. Cabbageworm, Diamondback Moth	¾–3*	1	
	Fall Armyworm, Variegated Cutworm	1½*		
Carrot	Beet Armyworm	1–3		
	Aster Leafhopper	1½–3	1	
	Armyworms	¾–1½		
	Variegated Cutworm	¾–1½		
Cauliflower	Imp. Cabbageworm, Loopers, Diamondback Moth	¾–3*	3	
	Variegated Cutworm	1½*		
Celery	Beet Armyworm, Aster Leafhopper	1½–3		
	Loopers	3	7	
	Variegated Cutworm	1½		
	Armyworms	¾–1½		
Chicory	Beet Armyworm, Variegated Cutworm, Leafhoppers	1½–3	80	
Chinese Cabbage **G	Loopers, Beet Armyworm	1½–3	10	

CROPS	INSECTS	RATES: LANNATE® LV PINTS PER ACRE	LAST APPLICATION—DAYS: TO HARVEST	LAST APPLICATION—DAYS: TO LIVESTOCK GRAZING/ FEEDING
Collards (Fresh market only—do not apply when temp. is less than 50°F. Do not apply when crop is less than 10" tall)	Loopers, Diamondback Moth, Variegated Cutworm	1½	10	
	Imported Cabbageworm, Beet Armyworm	1½–3		
Corn (Sweet) Note: Certain hybrid varieties of sweet corn are susceptible to methomyl injury. Treat a small area to determine crop safety before full scale spraying	Earworm—Whorl as needed	1–1½		
	Fall Armyworm, Armyworm, Earworm (Ovicide/Larvicide), European Corn Borer—Ears 1–3 days or as needed, Corn Rootworm (adult beetles), Flea Beetles, Picnic Beetles, Aphids	¾–1½	0 Ears	3 Forage
	Variegated Cutworm, Beet Armyworm	1½		
Corn (Field & Popcorn)	Earworm, Armyworm, Fall Armyworm, Corn Rootworm (adult beetles), Flea Beetles, Picnic Beetles, Aphids	¾–1½	0 Ears	3 Forage
	Variegated Cutworm, Beet Armyworm	1½		
Cotton (Use may redden cotton)—If excessive, stop or alternate with other insecticides. U.S.	**Ovicide/Larvicide** Bollworm, Tobacco Budworm (Initiate schedule when significant number of eggs are present. Continue at 5 to 7–day intervals while eggs present and larval control adequate. If significant larvae survive, use higher rates below) Lygus Bugs (adults and nymphs) start treatment on low level population for suppresion.	2/5–3/43/4 (See Predators)		
	Cotton Leafworm	¾–1½		
	Aphids, Thrips	¾		
East of Rockies only	Cotton Fleahopper (as needed)	2/5–3/4		
	Larvicide for worms Bollworm, Tobacco Budworm, Beet Armyworm, Cotton Leafperforator, Fall Armyworm, Lygus Bugs (adults and nymphs) use as occasional spray in regular schedule but not more often than every 10 days.	1½		
	Bollworm, Tobacco Budworm, Beet Armyworm, Cotton Leafperforator, Fall Armyworm, Lygus Bugs (adults and nymphs) up to 3 applications at 3–5 day intervals after boll load set on plants.	1½–2¼	15	Do not Graze or Feed
West of Rockies only	**Larvicide for Worms** Bollworm, Fall Armyworm, Tobacco Budworm, Lygus Bugs	1½–2¼		
	Beet Armyworm	7/8–1½		
	Cotton Leafperforator (as needed)	7/8–2¼		
Southwest only	Cotton Fleahopper (as needed)	2/5–3/4		
Cucumber	Loopers, Tobacco Budworm, Beet Armyworm, Yellow-striped Armyworm, Granulate Cutworm, Flea Beetles, Cucumber Beetles, Melon Aphid, Melonworm, Pickleworm	1½–3	1½ pts.—1 Over 1½ pts.—3	
	Fall Armyworm, Variegated Cutworm	1½		
Eggplant	Green Peach Aphid	¾–3		
	Tomato Pinworm **G, Beet Armyworm	1½–3	5	
Endive, Escarole	Beet Armyworm	1½–3	10	
Forest Woodlots, Plantations, and Nurseries **G	Spruce Budworm (treat when larvae are feeding on buds, retreat after 3–4 days if needed)	¾–1½		
	Forest Tent Caterpillar, Western Tent Caterpillar	1½–3		
Garlic	Beet Armyworm	1½*	7	
Grapefruit CA & AZ only	Thrips, Fruittree Leafroller, Orange Tortrix, Western Tussock Moth			
U.S.	Beet Armyworm	1½–3	1	10
FL only	Spirea Aphid			
	Whiteflies	3		
CA, AZ & FL only	Yellow Scale, Chaff Scale, Purple Scale, Soft Brown Scale, Glover Scale	1½/100 gals. up to 800 gals./acre		
Grapes East of Rockies only	Grape Berry Moth (apply pre- & post-bloom & repeat at 10–14 day intervals or as needed), Leafhoppers **G	1½–3		

CROPS		INSECTS	RATES: LANNATE® LV PINTS PER ACRE	LAST APPLICATION—DAYS: TO HARVEST	LAST APPLICATION—DAYS: TO LIVESTOCK GRAZING/ FEEDING
	U.S.	Omniverous Leafroller, Grape Leaffolder, Orange Tortrix	1½–3	1 Fresh & Raisin	
	West of Rockies only	Saltmarsh Caterpillar	1½–3	14 Wine Grapes	
		Leafhoppers **G			
		Climbing Cutworm (Treat infestation and repeat at 7–14 day intervals as needed)	3		
	CA only	Grape Leaf Skeletonizer **G	1½–3		
		Thrips	⅜–¾		
Horseradish **G		Aphids, Thrips	1½	65	
Leafy Green Vegetables Beet (Tops), Dandelions, Kale Mustard Greens, Parsley, Swiss Chard, Turnip Greens		Beet Armyworm, Cabbage Looper, Diamondback Moth, Imported Cabbageworm	1½–3	10	
Lemon	CA & AZ only	Thrips, Western Tussock Moth, Orange Tortrix	1½–3	1	10
	U.S.	Beet Armyworm			
	FL only	Spirea Aphid			
		Whiteflies	3		
	CA, AZ & FL only	Yellow Scale, Chaff Scale, Purple Scale, Soft Brown Scale, Glover Scale	1½/100 gals. up to 800 gals./acre		
Lentils		Western Yellowstriped Armyworm	1½–3	21	
Lettuce, (head varieties in U.S. and Leaf varieties in CA and AZ)		Beet Armyworm, Loopers	¾–3	¾–1½ pts.—7 over 1½ pts.—10	
		Thrips, Aphids, Corn Earworm, Aster Leafhopper	1½–3		
		Variegated Cutworm	1½		
Melons		Loopers, Tobacco Budworm, Beet Armyworm, Yellowstriped Armyworm, Granulate Cutworm, Flea Beetles, Cucumber Beetles, Melon Aphid, Melonworm, Pickleworm	1½–3	1½ pts.—1 day; over 1½ pts.—3 days	
		Fall Armyworm, Variegated Cutworm	1½–3		
Mint		Variegated Cutworm, Alfalfa Looper	3	14	
		Flea Beetles	2¼–3		
Nectarine	CA & AZ only	Thrips (up to 3 applications)	1½–3	1	10
Oats		Armyworms, Cereal Leaf Beetle, Aphids	¾–1½	7	10
Onions		Beet Armyworm,	1½–3*	7 Green and Dry Onions	
		Thrips, Variegated Cutworm	3*		
Oranges	CA & AZ only	Thrips, Western Tussock Moth, Orange Tortrix, Fruittree Leafroller, Citrus Cutworm	1½–3	1	10
	U.S.	Beet Armyworm			
	FL only	Spirea Aphid			
		Whiteflies	3		
	CA, AZ & FL only	Yellow Scale, Chaff Scale, Purple Scale, Soft Brown Scale, Glover Scale	1½/100 gals. up to 800 gals./acre		
Peaches		Catfacing Insects (Plant Bugs and Stink Bugs)—begin at petal fall and continue in cover sprays at 7 to 10–day intervals	3–6 (or ¾–1½/100 gals. up to 400 gals. per acre)	4	10
		Oriental Fruit Moth—begin at petal fall; use trapping devices and frequent field inspection to determine need for treatment. Continue treatment in cover sprays and alternate with residual-type insecticides registered for this use. **G			
		Green Peach Aphid	2¼–6		
Peanuts		Corn Earworm***, Beet Armyworm, Potato Leafhopper—Up to 3 applications	¾–3	21	Do not feed treated vines
		Loopers, Green Cloverworm, Velvetbean Caterpillar, Thrips, Granulate Cutworm—Up to 3 applications	1½–3		
		Fall Armyworm	¾–1½		

CROPS	INSECTS		RATES LANNATE® LV PINTS PER ACRE	LAST APPLICATION—DAYS TO HARVEST	LAST APPLICATION—DAYS TO LIVESTOCK GRAZING/ FEEDING
Pears Northeast only	Green Fruitworm, Obliquebanded Leafroller—One application only		1½/100 gals. up to 400 gals./acre	7	10
Peas (Succulent)	Armyworm, Green Cloverworm		¾–1½		
	Loopers, Pea Aphid, Beet Armyworm, Saltmarsh Caterpillar, Variegated Cutworm		1½–3	1	5 Forage 14 Hay
	Alfalfa Caterpillar		¾–3		
Pecans Southeast only	Aphids		1½–3	30	10
Peppers	Green Peach Aphid, Fall Armyworm, Loopers, Beet Armyworm		1½	3	
	Armyworm, Variegated Cutworm		¾–1½		
Pomegranates	Omniverous Leafroller		3	14	
Potato	Tuperworm, Loopers, Aphids, Beet Armyworm		1½–3	6	
	Leafhoppers, Variegated Cutworm, Fall Armyworm, Flea Beetles		1½		
Rye	Armyworms, Cereal Leaf Beetle, Aphids		¾–1½	7	10
Sorghum	Sorghum Webworm		1½		
	Sorghum Midge—Apply when 50% bloom and 3–5 days later if needed.		14	14	
	Fall Armyworm (Budworm) Beet Armyworm, Corn Earworm, Armyworm		¾–1½		
†Soybeans	Green Cloverworm, Velvetbean Caterpillar, Mexican Bean Beetle	Light to moderate infestations	⅖–¾		Under 1½ pts.
		Moderate to severe infestations	¾–1½		
	Corn Earworm (worms up to ½" long)	Moderate to severe infestations			3 Forage 7 Hay
U.S.	(worms up to ¼" long)	Light to moderate infestations	⅖–¾	14	1½–3 pts.
	Beet Armyworm, Salt Marsh Caterpillar	Light to moderate infestations	¾–1		10 Forage 12 Hay
	Fall Armyworm, Bean Leaf Beetle, Thrips	Moderate to severe infestations	1–1½		
AK, AL, LA, MS, TN, TX only	Soybean Looper (worms up to ½" long)	Use higher rate for severe infestations	1½–3		
Spinach (Do not apply when min. daily temp. is 32°F. or lower. Do not apply to seedlings less than 3" diameter)	Loopers, Beet Armyworm		1½–3	7	
	Fall Armyworm, Variegated Cutworm		1½		
Strawberry	Beet Armyworm, Thrips, Aphids		1½–3	3-Fresh 10-Processing Fruit	
	Lygus Bugs, Omniverous Leaf Tiers		3		
	Armyworms		2¼–3		
Sugar Beet	Beet Webworm, Flea Beetles, Carrion Beetles, Beet Armyworm, Aphids, Western Yellowstriped Armyworm—Up to 4 applications		¾–3	7	30 Tops
	Variegated Cutworm		1½		
Summer Squash	Loopers, Tobacco Budworm, Beet Armyworm, Yellowstriped Armyworm, Granulate Cutworm, Flea Beetles, Cucumber Beetles, Melon Aphid, Melonworm, Pickleworm, Fall Armyworm		1½–3	1½ pts.—1 day over 1½ pts. 3 days	
Tangelo CA & AZ only	Thrips, Western Tussock Moth, Orange Tortrix				
U.S.	Beet Armyworm		1½–3	1	10
FL only	Spirea Aphid				
	Whiteflies		3		

CROPS	INSECTS	RATES: LANNATE® LV PINTS PER ACRE	LAST APPLICATION—DAYS: TO HARVEST	LAST APPLICATION—DAYS: TO LIVESTOCK GRAZING/ FEEDING
CA, AZ & FL only	Yellow Scale, Chaff Scale, Purple Scale, Soft Brown Scale, Glover Scale	1½/100 gals. up to 800 gals./acre		
Tangerine				
CA & AZ only	Thrips, Western Tussock Moth, Orange Tortrix			
U.S.	Beet Armyworm	1½–3	1	10
FL only	Spirea Aphid			
	Whiteflies	3		
CA, AZ & FL only	Yellow Scale, Chaff Scale, Purple Scale, Soft Brown Scale, Glover Scale	1½/100 gals. up to 800 gals./acre		
Tobacco (Except shade)	Flea Beetle, Hornworm	¾–1½	5 Flue cured 14 Air or fire cured	
	Loopers, Aphids, Tobacco Budworm, Fall Armyworm	1½		
Tomato	Tomato Fruitworm, Aphids, Hornworn, Loopers, Beet Armyworm, Southern Armyworm	1½–3		
	Pinworm		1	
	Fall Armyworm, Variegated Cutworm	1½		
	Armyworm	¾–1½		
Turf **G	Sod Webworm (after application, sprinkle irrigate for 15 minutes)	3–6 (1.1–2.3 fl. ozs. per 1000 sq. ft.)		Do not graze or feed
Watercress **G	Beet Armyworm, Cabbage Looper, Diamondback Moth, Imported Cabbageworm (Use in minimum of 30 gpa of water. Water should be turned off 24 hours prior to application.)	1½–3	10	
Wheat	Armyworms, Cereal Leaf Beetle, Aphids	¾–1½	7	10

† Not registered for aerial application in a diluted volume of less than 1 gallon per acre in California
* Add wetting agent
** All registrations for aerial and ground use unless marked G for ground use only
*** Ovicidal and larvicidal control

LEXONE® DF™

HERBICIDE DISPERSIBLE GRANULE

ACTIVE INGREDIENT:
4-Amino-6-(1,1-dimethylethyl)-3-(methylthio)-1,2,4-triazin-5(4H)-one 75%
INERT INGREDIENTS 25%
TOTAL 100%

EPA Reg. No. 352-390

Keep Out of Reach of Children

PRECAUTIONARY STATEMENTS

HAZARDS TO HUMANS

WARNING!

CAUSES EYE IRRITATION.
MAY BE HARMFUL IF INHALED OR SWALLOWED.

Avoid breathing dust or spray mist. Avoid contact with skin, eyes, and clothing. Wash thoroughly with soap and warm water after handling. Wash contaminated clothing with soap and hot water before re-use.

STATEMENT OF PRACTICAL TREATMENT

If in eyes: Flush with plenty of water. Get medical attention if irritation persists.

If on skin: Wash with plenty of soap and water. Get medical attention if irritation persists.

ENVIRONMENTAL HAZARDS

Keep out of lakes, streams, or ponds. Do not contaminate water by cleaning of equipment or disposal of wastes.

NOTICE OF WARRANTY

DuPont warrants that this product conforms to the chemical description on the label thereof and is reasonably fit for the purposes stated on such label only when used in accordance with the directions under normal use conditions. It is impossible to eliminate all risks inherently associated with the use of this product. Crop injury, ineffectiveness, or other unintended consequences may result because of such factors as weather conditions, presence of other materials, or the manner of use or application, all of which are beyond the control of DuPont. In no case shall DuPont be liable for consequential, special or indirect damages resulting from the use or handling of this product. All such risks shall be assumed by the buyer. DUPONT MAKES NO WARRANTIES OF MERCHANTABILITY OR FITNESS FOR A PARTICULAR PURPOSE NOR ANY OTHER EXPRESS OR IMPLIED WARRANTY EXCEPT AS STATED ABOVE.

STORAGE AND DISPOSAL

STORAGE

Store product in original container only, away from other pesticides, fertilizer, food, or feed.

DISPOSAL

Do not contaminate water, food, or feed by storage or disposal. Wastes resulting from the use of this product may be disposed of on site or at approved waste disposal facility.

Triple rinse (or equivalent) the container. Then offer for recycling or reconditioning, or puncture and dispose of in a sanitary landfill, or by incineration, or if allowed by state and local authorities, by burning. If burned, stay out of smoke.

GENERAL INFORMATION

DuPont LEXONE DF Herbicide is a dispersible granule to be mixed in water and applied as a spray for selective control of certain broadleaf weeds and grasses in alfalfa, asparagus, potatoes, sainfoin, soybeans, sugarcane and tomatoes. It is noncorrosive to equipment, nonflammable, and nonvolatile. For soybeans, LEXONE DF may be used in conjunction with Treflan[1], Surflan[1], Lasso[2], Roundup[2], Dual[3], Prowl[4], Paraquat[5], or Amiben[6]. Observe all cautions and limitations on labeling of all products used in mixtures.

The degree of control and duration of effect will vary with:

- Soil Organic Matter Content—High organic matter requires higher dosages than low organic matter for equivalent herbicide performance.
- Soil Texture—High clay content requires higher dosage rates than those with less clay.
- Soil pH—Activity of LEXONE DF increases as the pH increases.

Rainfall—Moisture is necessary to activate the herbicide; best results are obtained if treatment is made to moist soil and moisture is supplied by rainfall or sprinkler irrigation (1/4" to 1/2") within one week after application. If moisture is insufficient to activate the herbicide, a shallow cultivation (rotary hoe preferred) should be made after emergence of crop while weeds are small enough to be controlled by mechanical means. If heavy rains occur soon after application, injury to the crop may result.

Amount Applied—The higher recommended rate is more active than the lower rate on a given soil.

Other Conditions—Factors such as incorporation depth of LEXONE DF, competition of the crop with weeds, weed species, drought stress of weeds, high salt content of soil, crop seed quality, soil residues of herbicides previously applied, soil insecticides, and combination herbicide treatments all can affect performance.

PREEMERGENCE USE (Germinating Weeds): LEXONE DF, when used as directed, controls for an extended time weeds such as:

Broadleaves		Grasses
Beggarweed	Wild mustard	Smooth brome
Wild Buckwheat	Pepperweed	Cheatgrass
Common chickweed	Prostrate pigweed	Downy brome
Carpetweed	Redroot pigweed (careless weed)	Large crabgrass
White cockle		Smooth crabgrass
Coffeeweed	Smooth pigweed	Crowfootgrass
Hophornbeam copperleaf	Common ragweed	Giant foxtail
	Hemp sesbania (coffeebean)	Green foxtail
Flixweed		Yellow foxtail
Henbit	Shepherdspurse	Fall panicum
Jimsonweed	Sicklepod (cassia)	Signalgrass (brachiaria)
Kochia	Prickly sida (spiny sida, teaweed)	
Lambsquarters		
Venice mallow		
Meadow salsify	Pennsylvania smartweed	
Blue mustard		
Tansy mustard	Spotted spurge	
	Spurred anoda	
	Yellow rocket	
	Velvetleaf	

Partial Control	Partial Control
Cocklebur	Barnyardgrass (watergrass)
Common purslane	
Sunflower	Yellow nutsedge
Florida purslane (Florida pusley)	Wild oats
	Seedling johnsongrass

Note: Consult Treflan, Dual, Surflan, Lasso, Prowl, Paraquat, Amiben, or Roundup labels for additional weeds controlled when LEXONE DF is used in conjunction with these herbicides.

LEXONE DF applied before emergence of asparagus, potatoes, soybeans, sugarcane and weeds is an effective procedure because susceptible weeds are controlled in an early, vulnerable seedling stage before they compete with the crop. With favorable moisture conditions, LEXONE DF continues to control weeds for some time as the crop becomes better able to compete. Should weed seedlings begin to break through the preemergence treatment in significant numbers, secondary weed control procedures should be implemented; these include cultivation and postemergence herbicide application.

POSTEMERGENCE USE (Emerged Seedling or Perennial Weeds): LEXONE DF, at recommended rates, controls weeds such as Florida beggarweed, bluegrass, carpetweed, cocklebur, crabgrass (large, smooth), dayflower, fumitory, goosegrass, hemp sesbania, jimsonweed, ladysthumb, lambsquarters, Mexican weed, mustard, pigweeds (prostrate, redroot, smooth), common purslane, common ragweed, sickelpod (cassia), prickly sida (spiny sida, teaweed), Pennsylvania smartweed, toadflax, Russian thistle and velvetleaf. Treatment usually provides partial control of barnyardgrass (watergrass), curly dock, dandelion, foxtail, horsenettle, kochia, morningglory species, wild poinsettia, Florida purslane (Florida pusley), quackgrass, broadleaf signalgrass, spotted spurge and sunflower.

LEXONE DF may be sprayed over the top of certain emerged crops; they are alfalfa (established, dormant), potatoes, sainfoin (dormant), sugarcane and tomatoes (seeded and transplants) to effectively control susceptible emerged seedling weeds.

Directed postemergence applications of LEXONE DF, where spray nozzles are adjusted so that weeds are sprayed but the crop is not, may be used in emerged soybeans (southern and southeastern states) and established tomatoes to control susceptible emerged seedling weeds.

Results of postemergence treatment of emerged weeds vary with rate applied and environmental conditions; best results are obtained on succulent weeds growing under conditions of high humidity and temperatures of 70° F or higher. Addition of a surfactant such as Du Pont Surfactant WK to the spray (where recommended) increases contact effects of LEXONE DF. Application will also provide control of emerging susceptible weed seedlings for an extended period of time.

IMPORTANT

Injury to or loss of desirable trees or other plants may result from failure to observe the following:

Do not apply (except as recommended), or drain or flush equipment on or near desirable trees or other plants, or on areas where their roots may extend, or in locations where the chemical may be washed or moved into contact with their roots. Do not use on lawns, walks, driveways, tennis courts, or similar areas. Prevent drift of spray to desirable plants. Do not contaminate any body of water. Keep from contact with fertilizers, insecticides, fungicides and seeds, during storage.

Thoroughly clean all traces of LEXONE DF from application equipment immediately after use. Flush tank, pump, hoses, and boom with several changes of water after removing nozzle tips and screens (clean these parts separately).

Varietal sensitivity to LEXONE DF is known in several crops for which LEXONE DF is registered. All new crop varieties should be evaluated on a limited basis for their tolerance prior to using LEXONE DF.

DIRECTIONS FOR USE

It is a violation of federal law to use this product in a manner inconsistent with its labeling.

DuPont LEXONE DF Herbicide should be used only in accordance with recommendations on this label, or in separate published DuPont recommendations available through local dealers.

DuPont will not be responsible for losses or damages resulting from use of this product in any manner not specifically recommended by DuPont. User assumes all risks associated with such non-recommended use.

Do not use in Kern County, California.

SPRAY PREPARATION: Fill tank 1/4 full with water. Start agitation system, add LEXONE DF and continue adding water. Add separately each additional component of any tank mix while adding water. If a surfactant is used, add it last when the tank is nearly full. Continue agitation throughout. If poor mixing should occur with any component, premix the component with two parts water before adding to the spray tank; follow this procedure when mixing Treflan or Prowl/LEXONE DF plus Treflan; tank mixtures should not be stored overnight; flush sprayer with water after use and dispose of mix properly. Use spray mixture promptly.

For preemergence application to potatoes or soybeans, nonpressure nitrogen or fertilizer solution may be used in the spray mixture. Small quantities should be tested for compatibility by the following procedures before full scale mixing.

1. Put 1 pint of fertilizer solution in a quart jar.
2. Mix 2 teaspoons LEXONE DF with 2 tablespoons of water; mix thoroughly and add to fertilizer solution.
3. Close jar and shake well.
4. If other herbicides are to be used in the mixture, premix 2 teaspoons of wettable powders or 1 teaspoon of liquids with 2 tablespoons of water; add to LEXONE DF/fertilizer solution mixture.
5. Close jar and shake well.
6. Watch mixture for several seconds; check again in 30 minutes.
7. If mixture does not separate, foam, gel, or become lumpy, it may be used.
8. Mixing ability may be improved by adding compatibility agents such as Kalo Laboraties Compex or Witco Chemicals Sponto 168D. Follow directions on container.

Provided the above procedure shows the mixture to be compatible, prepare the tank mixture as follows: add the fertilizer solution to the spray tank first, with the agitator running, add the required amount of LEXONE DF and thoroughly mix. For tank mixtures with other herbicides, follow directions above.

EQUIPMENT—SPRAY VOLUMES: Apply uniformly by air (except as noted) or with a properly calibrated low pressure (20 to 40 psi) fixed-boom sprayer equipped with fan-type nozzles and screens no finer than 50 mesh. Use 10 to 40 gals. of water per acre with ground equipment (20 to 60 gals. with mixture containing Paraquat and 20 to 30 gals. with mixture containing Roundup). Use 2 to 10 gals. per acre for aerial application, except use 7 gals. per acre for sugarcane (irrigated only) in Hawaii, and 5 to 10 gals. per acre for potatoes West of Rocky Mountains and for sugarcane in Louisiana and Texas.

Continuous agitation in the spray tank is required to keep the material in suspension. Avoid overlapping, and shut off spray booms while starting, turning, slowing or stopping, or injury to the crop may result. With aerial equipment, avoid overlapping of spray swath and avoid application under conditions where excessive drift may occur.

USE RATES: All rates are expressed as broadcast rates per acre; for band treatment, use proportionately less. For example, for 14" band of 42" row, use 1/3 of broadcast rate. Where a range of dosages is given, use the lower rate on coarser soils (low in clay or organic matter) and the higher rate on finer soils (high in clay or organic matter).

REPLANTING AND CROP ROTATION RECOMMENDATIONS: Unless otherwise noted, observe the following:

Areas treated with up to 2/3 lb. LEXONE DF per acre may be replanted immediately to potatoes, soybeans and sugarcane; two months after application, asparagus, corn and forage grasses may be planted. Winter wheat may be planted 4 months after treatment in Colorado, Kansas, Montana, Nebraska, North Dakota, Oklahoma, South Dakota, Texas and Wyoming. In other areas winter wheat may be planted after 6 months.

Areas in the Midsouth and Southeast where soybeans have been treated preemergence with up to 1 lb. LEXONE DF per acre may be replanted to corn, sorghum, cotton, and rice the season following application.

Areas treated with 2/3 to 1 1/3 lbs. LEXONE DF per acre may be replanted to soybeans and sugarcane; four months after application, alfalfa, sainfoin, asparagus, corn, potatoes, tomatoes and forage grasses may be planted.

Areas treated with 1 1/3 to 2 2/3 lbs. LEXONE DF per acre may be replanted to sugarcane; two months after application, soybeans may be replanted; 10 months after application, alfalfa, sainfoin, asparagus, corn, potatoes, tomatoes and forage grasses may be planted.

Areas treated with 2 2/3 to 5 1/3 lbs. LEXONE DF per acre may be replanted to sugarcane; four months after application, soybeans may be replanted; thirteen months after application, alfalfa, sainfoin, asparagus, corn, potatoes, tomatoes and forage grasses may be planted.

Other crops: Unless otherwise directed, do not plant in treated areas during the season following treatment as crop injury may result. Cover crops may be planted anytime but stand reduction may occur; do not graze or harvest these cover crops for feed or food.

Special Conditions: For winter or drought periods, add the number of weeks of winter or drought to the above listed recropping intervals.

For soil pH 7.5 or above, add 2 months to the recropping intervals listed.

Combination Treatments: When recommended tank mix combinations are used, consult the product label for replanting and recropping instructions and observe the directions that are the most restrictive.

CROP USES

ALFALFA AND SAINFOIN (Alone or with Grasses)—Make a single application in the fall after plants become dormant or in the spring before new growth starts. Do not apply during the first growing season after seeding. Do not graze or harvest within 28 days after application. Stress which may be caused in part by low fertility, disease, insects, winter kill, overcutting, drought, or frost may increase the chance of crop injury. In Washington, Oregon, Idaho, and Utah, partial control of downy brome, flixweed, and tansy mustard will occur at the lower use rate.

In California, use only north of U.S. Interstate 80.

Pure Stands of Alfalfa or Sainfoin: Select the rate from the table below:

Soil Texture	Lbs. LEXONE DF Per Acre	
	1/2 to 2% Organic Matter	More than 2% Organic Matter
Coarse		
Loamy Sand, Sandy Loam	1/2 to 2/3	2/3 to 3/4
Medium		
Loam, Silt Loam, Silt, Sandy Clay, Sandy Clay Loam	2/3 to 3/4	3/4 to 1
Fine		
Clay, Clay Loam, Silty Clay, Silty Clay Loam	3/4 to 1	1 to 1 1/3

Do not use on sand nor on soils with less than 1/2% organic matter as crop injury may result.

Mixed Stands with Grasses: The higher range of rates for a given soil texture and organic matter classification in the table above may provide partial reduction of forage grass stands. Rates of 1 to 1 1/3 lbs. per acre may severely reduce stands of forage grasses.

ASPARAGUS—Make a single preemergence application of 1 1/3 to 2 2/3 lbs. per acre in early spring prior to spear emergence. Use the lower rate on coarse textured soils (low in clay or organic matter) and the higher rate on fine textured soils (high in clay or organic matter). If field is to be disked, apply LEXONE DF after disking, but before crop emerges.

Do not apply within 14 days of harvest; do not apply after spear emergence. Do not use on newly seeded asparagus nor on young plants during the first growing season after setting crowns.

POTATOES—Preemergence: Make a single application of 2/3 to 1 1/3 lbs. per acre after planting or after drag-off but before crop emerges. Use the lower rates on coarse textured soils and the higher rates on fine textured soils (high in clay or organic mater).

Postemergence: Do not use on early maturing smooth-skinned white or red-skinned varieties. Apply only if there have been at least 3 successive days of sunny weather prior to application. Treat before weeds are 1" tall. Make a single application of 2/3 lb. per acre. For redroot pigweed and lambsquarters only, use 1/3 to 2/3 lb. per acre.

Treatment may cause some chlorosis or minor necrosis. Do not apply within 60 days of harvest.

Application by Sprinkler Irrigation: Apply specified dosage of LEXONE DF in ¼ to ¾ inch of water per acre (¼ to ½ inch of water per acre on sandy soil) as a continuous injection in center pivot and self-propelled wheel-move systems or in the last 15-30 minutes of set-in permanent solid-set sprinkler systems.

Irrigation systems must be equipped with automatic shut-off devices which prevent back-flow to the water source. Maintain continuous agitation in injection nurse tanks during application. Treatment may be made as a single postemergence or split postemergence application. Check irrigation system to insure uniform application of water to all areas. Failure to apply LEXONE DF uniformly may result in crop injury and/or poor weed control.

Preemergence plus Postemergence: Apply as directed above. Do not use more than 1⅓ lbs. total per acre per season. Do not apply within 60 days of harvest.

Postemergence plus Postemergence—(Oregon, Washington, Idaho): Use only if LEXONE DF was not applied preemergence. Allow at least 14 days between applications. For best results, apply before weeds are 1" in height or diameter. Do not apply within 24 hours of treatment with other pesticides.

Oregon and Washington—Make 2 applications of ⅓ to ⅔ lb. per acre. On coarse (sandy) soils containing ½ to 1% organic matter do not exceed ⅓ lb. per acre per application.

Idaho—Make 2 applications of ¼ to ⅔ lb. per acre. On coarse (sandy) soils containing ½ to 1% organic matter, do not exceed ½ lb. per acre application.

CROP ROTATION RECOMMENDATIONS: In the Northwest do not plant treated areas to crops other than potatoes for one year after treatment. Cover crops planted to prevent soil erosion or for soil building purposes may be planted anytime after application of LEXONE DF but do not graze or harvest for feed or food.

See "Replanting and Crop Rotation Recommendations" for further details.

NOTE: Because potato varieties vary in their resistance to herbicides, determine tolerance to LEXONE DF prior to adoption as a field practice to prevent possible crop injury. Crop injury may occur if Norgold variety is under stress from lack of moisture, fertility, sunlight or affected by insects or disease when treated postemergence. Use of LEXONE DF on sands or loamy sands containing ½ to 1½% organic matter may result in crop injury. Do not use on any soil containing less than ½% organic matter, as crop injury may result. Preemergence use on soils containing more than 10% organic mater may result in only partial weed control. Do not apply to sweet potatoes or yams. Do not incorporate into soil.

SOYBEANS—Select one of the following herbicide treatments and apply as directed. The combination treatments control weeds that are susceptible to LEXONE DF plus those controlled by the companion products (refer to product labeling). Dosage rates listed are broadcast rates; for band applications use proportionately less.

NOTE: Soybean varieties such as Altona, Coker 102 and 156, Govan, Semmes, Tracy, Vansoy, Terra Vig 505 and 606, AP 71, and NKS 1884 are sensitive to LEXONE DF. Injury may occur if LEXONE DF is used on these varieties. Before use on any other soybean variety, tolerance to LEXCONE DF must first be determined. Injury to soybeans may occur if LEXONE DF is used on soils having a calcareous surface layer or a pH of 7.5 or higher, or if used in conjunction with soil-applied organic phosphate pesticides such as Dasanit, Disyston, Mocap, Nemacur, Thimet, Parathion, Lorsban 15G or Counter.

Note: Injury may occur if atrazine was applied on the soil the year before use of LEXONE DF. Seedling disease, cold weather, deep planting (more than 2"), excessive moisture, high soil pH (pH 7.5 or higher), high salt concentration, or drought may weaken soybean seedlings and increase possibility of crop injury.

Preplant Incorporated: Incorporate herbicide treatment into soil before planting of soybeans according to instructions on label of product used in conjunction with LEXONE DF; improper soil incorporation can result in erratic weed control or crop injury. Do not plant seed more than 2" deep. Use combination as tank mixtures in water.

LEXONE DF + Treflan—Apply within 2 weeks of planting. Treatment also controls seedling johnsongrass, goosegrass and barnyard grass.

Soil Texture	Rate Per Acre LEXONE DF (Lbs.) + Treflan (Pints)
Coarse	
Loamy sand, sandy loam	⅓ + 1
Medium	
Loam, silt loam, silt, sandy clay, sandy clay loam	½ + 1½
Fine	
Silty clay, silty clay loam, clay, clay loam	⅔ + 2

Do not use on sand. Do not use on loamy sand or sandy loam with less than 1% organic matter, nor on other soils with less than ½% organic matter as crop injury may result.

Treated vines may be grazed or fed to livestock 40 days after application.

LEXONE DF + Lasso or Lasso EC—Apply within 7 days of planting. Treatment also controls dayflower and goosegrass.

Soil Texture	Rate Per Acre: ½ to 2% Organic Matter LEXONE DF (Lbs.) + Lasso/Lasso EC (Quarts)	More than 2% Organic Matter LEXONE DF (Lbs.) + Lasso/Lasso EC (Quarts)
Coarse		
Sandy loam	⅓ + 2	½ + 2
Medium		
Loam, silt loam, silt, sandy clay, sandy clay loam	½ + 2	⅔ + 2
Fine		
Silty clay, silty clay loam, clay, clay loam	⅔ + 2	⅔ + 2½
Mississippi Delta Only—Silty clay to heavy clay	1 + 2 to 2½	1 + 2 to 2½

Do not use on soils with less than ½% organic matter, nor on sand, loamy sand or muck soils as crop injury may result. Do not apply by air.

Treated vines may be grazed or fed to livestock 40 days after application.

LEXONE DF + Dual 8E—Apply within 14 days of planting. Treatment also controls dayflower, goosegrass and barnyard grass.

Soil Texture	Rate Per Acre: ½ to 3% Organic Matter LEXONE DF (Lbs.) + Dual 8E (Pints)	More than 3% Organic Matter LEXONE DF (Lbs.) + Dual 8E (Pints)
Coarse		
Loamy sand (over 2% OM), sandy loam	⅓ + 1¼	½ + 1½
Medium		
Loam, silt loam, silt	½ + 1½	⅔ + 2
Fine		
Silty clay loam, sandy clay loam, silty clay, sandy clay, clay loam, clay	⅔ + 2	⅔ + 2 to 2½
Mississippi Delta Only—silty clay, clay	1 + 2	1 + 2 to 2½
Muck or peat soils	DO NOT USE	DO NOT USE

Do not use on any sand or loamy sand with less than 2% organic matter. Do not use on soils with less than ½% organic matter, or on alkaline soils with a pH above 7.4 as crop injury may occur. Do not use preplant incorporated on any sand, loamy sand, or sandy loam as crop injury may occur.

Treated vines may be grazed or fed to livestock 40 days after application.

LEXONE DF + Prowl—Incorporate into soil within 7 days after application and plant soybeans no later than 7 days after application.

Southern States and Eastern Coastal Plains

Soil Texture	Rate Per Acre: LEXONE DF (Lbs.) + Prowl (Pints)
Coarse	
Loamy sand, sandy loam	⅓ + 1½
Medium	
Sandy clay loam, sandy clay, loam, silt, silt loam	½ + 1½
Fine	
Clay loam, silty clay loam, clay	⅔ + 1½ to 2

Do not use on sand, muck or peat. Do not use on loamy sand or sandy loam with less than 1% organic matter, nor on other soils with less than ½% organic matter as crop injury may result.

Treated vines may be grazed or fed to livestock 40 days after application.

Northeastern and Northcentral States

Soil Texture	Rate Per Acre: ½ to 3% Organic Matter LEXONE DF (Lbs.) + Prowl (Pints)	Rate Per Acre: More than 3% Organic Matter LEXONE DF (Lbs.) + Prowl (Pints)
Coarse		
Loamy sand, sandy loam (2% or more organic matter only)	½ + 1	½ + 1½
Medium		
Sandy clay loam, sandy clay, loam, silt, silt loam	½ + 1½ to 2	½ to ⅔ + 1½ to 2
Fine		
Clay loam, silty clay loam, clay	½ to ⅔ + 1½ to 2	⅔ + 2 to 2½

Do not use on sand, muck or peat, nor on soils with less than ½% organic matter as crop injury may result.

Treated vines may be grazed or fed to livestock 40 days after application.

Preplant Non-Incorporated:

LEXONE DF + Surflan AS or Surflan 75W—Preplant Surface Application: Apply within 4 weeks of planting. Treatment also controls dayflower and goosegrass.

Soil Texture ½ to 5% Organic Matter	Rate Per Acre: LEXONE DF (Lbs.) +	Rate Per Acre: Surflan AS or Surflan 75W
Coarse		
Sandy loam (minimum 1% organic matter)	⅓ +	½ to ¾ (qt.) or ⅔ to 1 (lb.)
Medium		
Loam, silt loam, silt, sandy clay loam	½ +	¾ to 1 (qt.) or 1 to 1⅓ (lb.)
Fine		
Silty clay, silty clay loam, sandy clay, clay loam, clay	⅔ +	1 to 1¼ (qt.) or 1⅓ to 1⅔ (lb.)
Mississippi Delta Only—Silty clay to heavy clay	⅔ to 1 +	1 to 1¼ (qt.) 1⅓ to 1⅔ (lb.)

Do not use on sand or loamy sand. Do not use on sandy loam with less than 1% organic matter. Do not use on other soils with less than ½% organic matter. Do not use preemergence application on soils with more than 3% organic matter nor the preplant surface application on soils with more than 5% organic matter as poor weed control and/or crop injury may result. Do not use treated vines for feed or forage.

Preplant Incorporated (PPI) Combinations Plus Preemergence—Use combinations as tank mixes in water. Promptly incorporate PPI treatment into soil before planting of soybeans. Improper soil incorporation can result in erratic weed control or crop injury. Do not plant seed more than 2″ deep. After planting of soybeans and before emergence, apply a preemergence treatment of LEXONE DF:

LEXONE DF + Treflan (PPI) + LEXONE DF. Treatment also controls seedling johnsongrass, goosegrass and barnyardgrass.

Soil Texture	Preplant Incorporated (PPI) — LEXONE DF + Treflan, Apply within 4 weeks of planting — Rate Per Acre: LEXONE DF (Lbs.) + Treflan (Pints)	Plus Preemergence — LEXONE DF—Apply after planting but before emergence — Lbs. LEXONE DF Per Acre: Less Than 2% OM	2–4% OM	Over 4% OM
Coarse				
Loamy sand, sandy loam	⅓ + 1	⅙	⅙	⅓
Medium				
Loam, silt loam, silt, sandy clay, sandy clay loam	⅓ + 1½ or ½ + 1½	⅓ or ⅙	⅓ to ½ or ⅙ to ⅓	½ to ⅔ or ⅓ to ½
Fine				
Silty clay, silty clay loam, clay, clay loam	½ + 2 or ⅔ + 2	⅓ or ⅙	⅓ to ½ or ⅙ to ⅓	½ to ⅔ or ⅓ to ½

Higher rates in the tank mixes are recommended in low rainfall areas or where low rainfall may follow application. Use the higher rate of preemergence LEXONE DF where fields are prone to severe broadleaf weed pressure.

Do not use on sand, loamy sand, or sandy loam with less than 1% organic matter, nor on other soils with less than ½% organic matter as crop injury may result.

Treated vines may be grazed or fed to livestock 40 days after application.

LEXON DF + Lasso or Lasso EC (PPI) + LEXONE DF

Soil Texture	Preplant Incorporated (PPI) — LEXONE DF + Lasso or Lasso EC, Apply within 7 days of planting — Rate Per Acre: ½ to 2% Organic Matter LEXONE DF (Lbs.) + Lasso/Lasso EC (Qts.)	More than 2% Organic Matter LEXONE DF (Lbs.) + Lasso/Lasso EC (Qts.)	Plus Preemergence — LEXONE DF—Apply after planting but before emergence — Lbs. LEXONE DF Per Acre: Less than 2% OM	2 to 4% OM	Over 4% OM
Coarse					
Sandy loam	⅓ + 2	⅓ + 2	⅙	⅙	⅓
Medium					
Loam, silt loam, silt, sandy clay, clay loam	⅓ + 2 or ½ + 2	⅓ + 2 or ½ + 2	⅓ or ⅙	⅓ to ½ or ⅙ to ⅓	½ to ⅔ or ⅓ to ½
Fine					
Silty clay, silty clay loam, clay, clay loam	½ + 2 or ⅔ + 2	½ + 2½ or ⅔ + 2½	⅓ or ⅙	⅓ to ½ or ⅙ to ⅓	½ to ⅔ or ⅓ to ½

Higher rates in the tank mixes are recommended in low rainfall areas or where low rainfall may follow application. Use the higher rate of preemergence LEXONE DF where fields are prone to severe broadleaf weed pressure.

Do not use on soils with less than ½% organic matter, nor on sand, loamy sand or muck soils as crop injury may result. Do not apply by air.

Treated vines may be grazed or fed to livestock 40 days after application.

LEXONE DF + Dual 8E (PPI) + LEXONE DF

Preplant Incorporated (PPI)	Plus		Preemergence		
LEXONE DF + Dual 8E Apply within 2 weeks of planting			LEXONE DF—Apply after planting but before emergence		
	Rate Per Acre		Lbs. LEXONE DF Per Acre		
Soil Texture	½ to 3% Organic Matter LEXONE DF (Lbs.) + Dual 8E (Pts.)	More than 3% Organic Matter LEXONE DF (Lbs.) + Dual 8E (Pts.)	Less than 2% OM	2 to 4% OM	Over 4% OM
Coarse					
Loamy Sand (Over 2% OM), Sandy loam	⅓ + 1¼	⅓ + 1½	⅙	⅙	⅓
Medium	⅓ + 1½	⅓ + 2	⅓	⅓ to ½	½ to ⅔
Loam, silt loam, silt, sandy clay loam	or ½ + 1½	or ½ + 2	or ⅙	or ⅙ to ⅓	or ⅓ to ½
Fine					
Silty clay loam, silty clay, sandy clay, clay loam, clay	½ + 2 or ⅔ + 2	½ + 2 to 2½ or ⅔ + 2 to 2½	⅓ or ⅙	⅓ to ½ or ⅙ to ⅓	½ to ⅔ or ⅓ to ½

Higher rates in the tank mixes are recommended in low rainfall areas or where low rainfall may follow application. Use the higher rate of preemergence LEXONE DF where fields are prone to severe broadleaf weed pressure.

Do not use on any sand or loamy sand with less than 2% organic matter. Do not use on soils with less than ½% organic matter or on alkaline soils with a pH above 7.4 as crop injury may occur. Do not use preplant incorporated on any sand, loamy sand, or sandy loam as crop injury may occur.

Treated vines may be grazed or fed to livestock 40 days after application.

LEXONE DF + Prowl (PPI) + LEXONE DF

Preplant Incorporated (PPI)	Plus	Preemergence		
LEXONE DF + Prowl Incorporate into soil within 7 days after application and plant soybeans no later than 7 days after application.		LEXONE DF—Apply after planting but before emergence		
	Rate Per Acre	Lbs. LEXONE DF Per Acre		
Soil Texture	LEXONE DF (Lbs.) + Prowl (Pints)	Less Than 2% OM	2–4% OM	over 4% OM
Coarse				
Loamy sand, sandy loam	⅓ + 1½	⅙	⅙	⅓
Medium				
Loam, silt loam, silt, sandy clay, sandy clay loam	⅓ + 1½ to 2 or ½ + 1½ to 2	⅓ or ⅙	⅓ to ½ or ⅙ to ⅓	½ to ⅔ or ⅓ to ½
Fine	½ + 1½ to 2	⅓	⅓ to ½	½ to ⅔
Clay loam, silty clay loam, clay	or ⅔ + 1½ to 2	or ⅙	or ⅙ to ⅓	or ⅓ to ½

Higher rates in the tank mixes are recommended in low rainfall areas or where low rainfall may follow application. Use the higher rate of preemergence LEXONE DF where fields are prone to severe broadleaf weed pressure.

Do not use on sand, muck or peat. Do not use on loamy sand or sandy loam with less than 1% organic matter, nor on other soils with less than ½% organic matter as crop injury may result.

Treated vines may be grazed or fed to livestock 40 days after application.

Preplant Incorporation Plus Preemergence

LEXONE DF Following Dual, Treflan, or Prowl: Apply Dual, Treflan, or Prowl as a preplant incorporated treatment according to directions on product label. As a separate operation, follow with LEXONE DF using rates given under "Preemergence Application".

Preemergence Application

Apply after planting but before soybeans emerge; do not spray over top of emerged soybeans. Plant seed 1½" to 2" deep on flat or raised seedbeds only or injury to the crop may result; soil should be well prepared and as free as possible from trash and clods before planting. Use combinatons as tank mixtures in water.

Where a range of LEXONE DF dosage rates is listed for a specific soil type and organic matter, use the higher rate for difficult control problems, such as cocklebur, coffeeweed, jimsonweed, coffeebean, sicklepod, prickly sida, or velvetleaf. Use the higher rate also for minimum or no-tillage soybeans, and in the states of Georgia and South Carolina.

LEXONE DF + Lasso or Lasso EC. Treatment also controls dayflower and goosegrass.

	Rate Per Acre	
Soil Texture	½ to 2% Organic Matter LEXONE DF (Lbs.) + Lasso/Lasso EC (Quarts)	More than 2% Organic Matter LEXONE DF (Lbs.) + Lasso/Lasso EC (Quarts)
Coarse		
Sandy loam	⅙ to ⅓ + 2	½ + 2
Medium		
Loam, silt loam, silt, sandy clay, sandy clay loam	⅓ to ½ + 2	⅔ + 2
Fine		
Silty clay, silty clay loam, clay, clay loam	½ to ⅔ + 2	⅔ + 2½
Mississippi Delta Only—Silty clay to heavy clay	⅔ to 1 + 2 to 2½	1 + 2 to 2½

Do not use on soils with less than ½% organic matter, nor on sand, loamy sand or muck soils as crop injury may result. Do not apply by air.

Treated vines may be grazed or fed to livestock 40 days after application.

LEXONE DF + DUAL 8E. Treatment also controls dayflower and goosegrass.

	Rate Per Acre					
	½ to 3% Organic Matter			More than 3% Organic Matter		
Soil Texture	LEXONE DF (Lbs.)	+	DUAL 8E (Pints)	LEXONE DF (Lbs.)	+	DUAL 8E (Pints)
Coarse						
Loamy sand (over 2% OM), sandy loam	⅙ to ⅓	+	1¼	½	+	1½
Medium						
Loam, silt loam, silt	⅓ to ½	+	1½	⅔	+	2
Fine						
Silty clay loam, sandy clay loam, silty clay, sandy clay, clay loam, clay	½ to ⅔	+	2	⅔	+	2 to 2½
Mississippi Delta Only—Silty clay, clay	⅔ to 1	+	2	1	+	2 to 2½
Muck or peat soils	DO NOT USE					

Do not use on any sand or loamy sand with less than 2% organic matter. Do not use on soils with less than ½% organic matter or on alkaline soils with a pH above 7.4 as crop injury may occur. Do not use preplant incorporated on any sand, loamy sand, or sandy loam as crop injury may occur.

Treated vines may be grazed or fed to livestock 40 days after application.

LEXONE DF + Prowl

Southern States and Eastern Coastal Plains

Soil Texture	Rate Per Acre LEXONE DF (Lbs.) + Prowl (Pints)
Coarse Loamy sand, sandy loam	⅙ to ⅓ + 1½
Medium Sandy clay loam, sandy clay, loam, silt, silt loam	⅓ to ½ + 1½
Fine Clay loam, silty clay loam, clay	½ to ⅔ + 1½ to 2

Do not use on sand, muck or peat. Do not use on loamy sand or sandy loam with less than 1% organic matter, nor on other soils with less than ½% organic matter as crop injury may result.

Treated vines may be grazed or fed to livestock 40 days after application.

Northeastern and Northcentral States

Soil Texture	Rate Per Acre: ½ to 3% Organic Matter LEXONE DF (Lbs.) + Prowl (Pints)	Rate Per Acre: More than 3% Organic Matter LEXONE DF (Lbs.) + Prowl (Pints)
Coarse Loamy sand, sandy loam (2% or more organic matter only)	⅓ to ½ + 1	½ + 1½
Medium Sandy clay loam, sandy clay, loam, silt, silt loam	⅓ to ½ + 1½ to 2	½ to ⅔ + 1½ to 2
Fine Clay loam, silty clay loam, clay	½ to ⅔ + 1½ to 2	⅔ + 2 to 2½

Do not use on sand, muck or peat, nor on soils with less than ½% organic matter as crop injury may result.

Treated vines may be grazed or fed to livestock 40 days after application.

PREPLANT NON-INCORPORATED

LEXONE DF + Surflan AS or Surflan 75W—Apply within 2 days after planting. Treatment also controls dayflower and goosegrass.

Soil Texture ½ to 3% Organic Matter	Lbs. Product Per Acre: LEXONE DF (Lbs.)	+	Surflan AS or Surflan 75W
Coarse Sandy loam (minimum 1% organic matter)	⅙ to ⅓	+	½ to ¾ (qt) or ⅔ to 1 (lb)
Medium Loam, silt loam, silt, sandy clay loam	⅓ to ½	+	¾ to 1 (qt) or 1 to 1⅓ (lb)
Fine Silty clay, silty clay loam, sandy clay, clay loam, clay	½ to ⅔	+	1 to 1¼ (qt) or 1⅓ to 1⅔ (lb)
Mississippi Delta Only— Silty clay, clay	⅔ to 1	+	1 to 1¼ (qt) 1⅓ to 1⅔ (lb)

Do not use on sand or loamy sand. Do not use on sandy loam with less than 1% organic matter. Do not use on other soils with less than ½% organic matter. Do not use preemergence application on soils with more than 3% organic matter nor the preplant surface application on soils with more than 5% organic matter as poor weed control and/or crop injury may result. Do not use treated vines for feed or forage.

LEXONE DF + Amiben

Soil Texture	Rate Per Acre: ½ to 3% Organic Matters LEXONE DF (Lbs.) + Amiben (Quarts)	Rate Per Acre: More than 3% Organic Matter LEXONE DF (Lbs.) + Amiben (Quarts)
Coarse Loamy sand, sandy loam (2% or more organic matter only)	⅙ to ⅓ + 3 to 4	½ + 3 to 4
Medium Loam, silt loam, silt, sandy clay, sandy clay loam	⅓ to ½ + 3 to 4	⅔ + 4 to 5
Fine Silty clay, silty clay loam, clay, clay loam	½ to ⅔ + 4 to 5	⅔ + 4 to 5
Mississippi Delta only— silty clay to heavy clay	⅔ to 1 + 6	1 + 6

Do not use on sand or muck soils as crop injury may result.

Treated vines may be grazed or fed to livestock 40 days after application.

LEXONE DF

Soil Texture	Lbs. LEXONE DF Per Acre[3]: ½ to 2% Organic Matter	Lbs. LEXONE DF Per Acre[3]: More than 2% Organic Matter
Coarse Loamy sand, sandy loam	Do Not Use (See tank mix use directions)	½
Medium [1,2] Loam, silt loam, silt, sandy clay, sandy clay loam	⅓ to ½	⅔
Fine [1,2] Silty clay, silty clay loam, clay, clay loam	½ to ⅔	⅔
Mississippi Delta Only— Silty clay to heavy clay	⅔ to 1	1

1. **Alkaline (Calcareous) Soils:** For control of lambsquarters, redroot pigweed and wild mustard and suppression of green, yellow and giant foxtail, apply LEXONE DF at ⅓ lb. per acre on medium soils and ⅓ to ½ lb. per acre on fine textured soils containing more than ½% organic matter. Use ½ lb. per acre only where soil pH is less than 7.5 and weed pressure is heavy. The ⅓ lb. per acre rate may be used as a preemergence overlay following preplant incorporation of Treflan or Tolban.

2. **Alabama, Louisiana, Mississippi, and Tennessee:** Apply ½ to ⅔ lb. per acre on medium soils and ⅔ to ¾ lb. on fine textured soils. Use the higher range of rates for improved cocklebur control.

3. Do not use on sand nor soils with less than ½% organic matter as crop injury may result.

Treated vines may be grazed or fed to livestock 40 days after application.

DIRECTED POSTEMERGENCE TREATMENTS

Southern and Southeastern States: LEXONE DF may be applied as a directed postemergence treatment in addition to its use as a preemergence or preplant incorporated treatment. Apply when soybeans are at least 12 inches tall and when broadleaf weeds do not exceed 3 inches and grasses and common ragweed do not exceed 1 inch in height. Direct spray to cover weeds with minimal or no contact to soybean plant. Crop injury may result if spray is higher than 2 inches on soybean stem. Apply ⅓ to ⅔ lb. per acre in 25 to 40 gallons water. Use the lower rate for seedling weeds and the higher rate for larger weeds. Application of ⅓ lb. per acre controls Florida beggarweed, carpetweed, cocklebur, dayflower, Mexican weed, prostrate and smooth pigweed, redroot pigweed (carelessweed), common purslane, sicklepod (cassia), velvetleaf and crabgrass (large, smooth); ½ to ⅔ lb. per acre controls hemp sesbania, prickly sida (spiny sida, teaweed); ⅔ lb. per acre controls jimsonweed, Pennsylvania smartweed, common ragweed and gives partial control of morningglory species, horsenettle, Florida purslane (Florida pusley), spotted spurge and wild poinsettia when less than 3 inches tall and broadleaf signalgrass up to 1 inch tall. Do not use on sand, nor on soils with less than ½% organic matter as crop injury may result. Add a non-ionic surfactant such as Du Pont Surfactant WK or Ortho X-77 Spreader to the spray mixture to obtain better wetting of weed leaf surfaces. If needed, a second application may be made after 7 days.

Apply only with ground equipment designed for precise application of directed sprays such as oiling shoes or gauge wheels. Do not exceed nozzle pressure of 25 psi or use nozzle tips smaller than 8002T-Jet (or equivalent). For broadcast application, use a single flood-type spray nozzle ("K" series or equivalent) per middle, mounted on an oiling shoe or gauge wheel. For band treatment, use two nozzles per row mounted on oiling shoes or gauge wheels, one on each side of a row. Do not apply under weather conditions which favor drift.

Treated vines may be grazed or fed to livestock 40 days after application.

MINIMUM OR NO-TILLAGE SOYBEANS

LEXONE DF plus paraquat or Roundup may be used as a tank mixture for preemergence and postemergence control of many broadleaved weeds and grasses and for top kill or suppression of some perennials where soybeans will be planted directly into a preformed bed (stale seedbed), cover crop, or in previous crop residues such as corn or small grain stubble. For improved grass and volunteer small grain control, the addition of Lasso (or Lasso EC) or Surflan to the LEXONE DF plus paraquat or Roundup tank mixture is recommended. For all combinations, except Surflan, treated vines may be grazed or fed to livestock 40 days after application.

Apply with ground equipment, during or after planting but before crop emerges. Use 20 to 60 gals. of water per acre for paraquat and use the higher gallonage for dense stubble or vegetation. Use 20 gals. of water per acre for Roundup combinations.

LEXONE DF + Paraquat—Select the rate from the preemergence table for use of LEXONE DF and thoroughly mix in spray tank first; then add paraquat as directed under "Paraquat Combinations".

LEXONE DF + Lasso (or Lasso EC) + Paraquat—Select the rate from the table for use of LEXONE DF + Lasso tank mixture and thoroughly mix in spray tank first according to directions; then add paraquat as directed under "Paraquat Combinations". Do not apply by air.

LEXONE DF + Surflan + Paraquat— Select the rate from the table for use of LEXONE DF + Surflan tank mixture and thoroughly mix in spray tank first according to directions; then add paraquat as directed under "Paraquat Combinations".

Do not use treated vines for feed or forage.

Paraquat Combinations—Select the LEXONE DF or LEXONE DF combination treatment above and mix in the spray tank according to directions; then add 1 or 2 pts. paraquat (2 lb. active/gal.) per acre. Use the higher rate for weeds 4" to 6" tall. As the last ingredient, add 8 oz. Du Pont Surfactant WK or Ortho X-77 Spreader per 100 gals. of spray mixture. Maintain constant agitation.

Roundup Combinations (LEXONE DF + Lasso + Roundup): Select the rate from the table for use of LEXONE DF + Lasso tank mixture and thoroughly mix in spray tank first according to directions; then add 1 to 2 qts. Roundup per acre to the mixture and maintain constant agitation. Use the higher rates for maximum top kill of vegetation. Do not apply by air.

SUGARCANE

Hawaii—Make a single broadcast application as follows: for aerial application (irrigated cane only), use at least 7 gals. water per acre. Spot treatment may also be made at the rate of 3⅓ to 6⅔ lbs. per acre, but do not exceed 10⅔ lbs. total per acre per crop cycle. Do not apply within 17 months of harvest. Do not use treated foliage for feed or forage.

Time of Application	Lbs. LEXONE DF Per Acre	
	Irrigated	Non-Irrigated
Preemergence—apply within 2 weeks after planting —or—	5⅓ to 8	2⅔ to 5⅓
Early Postemergence—apply over cane before weeds are 3" tall —or—	5⅓ to 8	2⅔ to 5⅓
Postemergence—apply prior to "close in" time and before weeds are 3" tall	2⅔ to 5⅓	2⅔ to 5⅓

In addition to the weeds listed elsewhere on this label, treatment (non-irrigated cane) controls ageratum, spiny amaranth, wild euphoria, fireweed, Flora's paintbrush, Richardia, garden spurge, graceful spurge, tarweed, guineagrass, plushgrass, ricegrass and wiregrass (goosegrass); in irrigated cane, the foregoing weeds plus spleen amaranth, haole koa, hioloa, hila hila, rattlepod and Alexander grass.

Louisiana and Texas—Apply 2 to 4 lbs. per acre during the fall after planting but before cane emerges, or to the stubble after harvest. Make a second application in early spring prior to new cane growth. If necessary, a third application may be made in late spring at layby. Do not apply within 60 days of harvest. Do not use treated foliage for feed or forage.

In addition to the weeds listed elsewhere on this label, treatment controls curly dock, morningglory, marestail, sowthistle and winter oats.

TOMATOES

East of Rocky Mountains (Except Florida)—Apply LEXONE DF as a preplant incorporated treatment alone or in a tank mixture with Treflan EC in transplanted tomatoes, or apply as a postemergence broadcast or directed spray to established tomatoes.

Apply uniformly to seeded or transplanted tomatoes with a properly calibrated low-pressure (20 to 40 psi) fixed-boom sprayer with fan type nozzles. Use 10 to 40 gallons of water per acre with ground equipment for preplant incorporated treatment and 20 to 75 gallons for postemergence broadcast or directed spray application. Do not use air blast or other high pressure spray equipment for postemergence spray applications.

Preplant Incorporated—LEXONE DF or LEXONE DF Plus Treflan EC Tank Mix (Transplant Tomatoes): Apply ⅓ to ⅔ lb. per acre of LEXONE DF alone or as a tank mix combination with Treflan EC as a broadcast spray immediately before transplanting. Use the higher rate for heavy weed pressure or for difficult-to-control weeds. Incorporate uniformly to a depth of 2 to 4 inches. Apply Treflan EC at rates of 1 to 2 pints per acre on coarse soils; 1¼ to 2 pts. per acre on medium soils; and 1½ to 2 pts. per acre on fine soils. Consult Treflan EC label for specific use instructions and observe all cautions and limitations. When LEXONE DF is tank mixed with Treflan EC, apply immediately after mixing; do not allow tank mixture to stand for prolonged periods of time.

Postemergence Broadcast Application (Established Tomatoes): Apply ⅓ to ⅔ lb. per acre. Use the higher rate for heavy weed pressure or for difficult-to-control weeds. Treat before weeds are 1 inch tall. One or more applications may be made per crop season; allow at least 14 days between applications. Do not apply more than 1⅓ lbs. total within a 35-day period or per crop season. Do not treat seeded or transplanted tomatoes until plants have reached the 5 to 6 leaf stage or until transplants have recovered from transplant shock and new growth has started. Apply only if there have been at least 3 successive days of sunny weather prior to application, or crop injury may occur. Do not apply within 24 hours of treatment with other pesticides. Do not tank mix with other pesticides.

Postemergence Directed Application (Established Tomatoes): Apply ⅔ to 1⅓ lbs. per acre as a directed spray according to directions under "Broadcast-Postemergence Application". Do not allow spray to contact tomato foliage. This method of application is recommended for fields with heavy weed pressure or difficult-to-control weed species.

CROP ROTATION RECOMMENDATIONS: For rotation crops in fields treated with Treflan, follow instructions on product label.

See "Replanting and Crop Rotation Recommendations" for further details.

Note: Because tomato varieties vary in their resistance to LEXONE DF, determine tolerance prior to adoption as a field practice to prevent possible crop injury. Do not use hot caps on tomatoes within 7 days before or at any time after application of LEXONE DF. Crop injury or delayed maturity may result from postemergence or directed spray applications if tomatoes are growing under stress conditions caused by low fertility, disease, insects, drought, cool, wet or cloudy weather prior to application. Do not apply within 7 days of harvest.

NON-CROP USES

Apply 1⅓ to 8 lbs. LEXONE DF per acre in 20 to 60 gallons of water. For best results, apply shortly before weed growth begins.

For control of established seedling weeds add:

1 to 2 pints Paraquat (2 lbs. active per gallon) per acre in 20 to 60 gallons of water per acre using the higher spray volume for weeds 4" to 6" tall. As the last ingredient, add 8 ounces Surfactant WK or Ortho X-77 Spreader per 100 gallons of spray mixture or add 1 to 2 quarts Roundup in 20 to 30 gallons of water per acre. Use the higher rates for maximum top kill of vegetation.

[1]Registered trademarks of Elanco Products Co., Treflan contains 4 lbs. trifluralin per gal.; Surflan contains 75% oryzalin.

[2]Registered trademarks of Monsanto Co., Lasso contains 4 lbs. alachlor per gal. Roundup contains 4 lbs. glyphosate per gal.

[3]Registered trademark of Ciba-Geigy Corp., Dual 8E contains 8 lbs. metolachlor per gal.

[4]Registered trademark of American Cyanamid Co., Prowl contains 4 lbs. pendimethalin per gal.

[5]Paraquat contains 2 lbs. active per gal.; paraquat is a restricted use pesticide.

[6]Registered trademark of Amchem Products, Inc., Amiben contains 2 lbs. chloramben per gal.

SUPPLEMENTAL LABELING

LEXONE® DF™ EPA Reg. No. 352-390

Karmex® EPA Reg. No. 352-247

LEXONE ® DF™ HERBICIDE AND COMBINATIONS WITH Karmex® FOR USE ON ESTABLISHED ASPARAGUS IN THE NORTHWEST

DIRECTIONS FOR USE

It is a violation of federal law to use this product in a manner inconsistent with its labeling.

DuPont LEXONE DF™ Herbicide is recommended alone or in combination with Karmex Herbicide for use on established asparagus in the Northwest. For weeds controlled, refer to package labels for LEXONE DF™ and Karmex.

HOW TO USE

Use the following:

LEXONE DF™ Preemergence Treatment Only: Make a single preemergence application of 1⅓ or 2⅔ lbs. LEXONE DF™ per acre in early spring prior to spear emergence. Use the lower rate on coarse textured soils (low in clay or organic matter). If field is to be disked, apply LEXONE DF™ after disking but before crop emerges.

OR

LEXONE DF™ Preemergence and Post-Harvest Treatment: Make a single preemergence application of ⅔ to 1⅓ lbs. LEXONE DF™ per acre in early spring prior to spear emergence. Following harvest but before fern emerges, apply ⅔ to 2 lbs. LEXONE DF™ per acre. Do not exceed 2⅔ lbs. LEXONE DF™ per acre per growing season.

OR

LEXONE DF™ Preemergence and LEXONE DF™ Plus Karmex Post Harvest Treatment: Make a single preemergence application of ⅔ to 1⅓ lbs. LEXONE DF™ per acre in early spring prior to spear emergence. Following harvest but before fern emerges, apply a tank-mix combination of ⅔ to 1 lb. LEXONE DF™ plus 1 to 2 lbs. Karmex per acre.

Irrigated Asparagus: After the initial application of LEXONE DF™, do not exceed more than ½" of water for the first irrigation. Additional irrigations may be normal programs.

Application: Apply in 3–10 gallons per acre by air or 10–40 gallons per acre by ground equipment.

NOTE:

Do not use on newly seeded nor on young plants during the first growing season after setting crowns. Do not apply within 14 days of harvest. If field is to be disked, apply LEXONE DF™ after disking but before the crop emerges. Do not apply post-harvest application until after the last harvest of spears.

IMPORTANT

BEFORE USING LEXONE DF™ IN COMBINATIONS, READ AND CAREFULLY OBSERVE THE CAUTIONARY STATEMENTS AND ALL OTHER INFORMATION APPEARING ON THE PRODUCT LABELS.

This bulletin contains new or supplemental instructions for use of these product which may appear on these package labels. Follow the instructions carefully.

This labeling must be in the possession of the user at the time of pesticide application.

4/26/84

SUPPLEMENTAL LABELING

LEXONE® DF™ HERBICIDE

EPA Reg. No. 352-390

FOR USE ON FALLOW FOLLOWED BY SUGARCANE IN THE STATE OF LOUISIANA

DIRECTIONS FOR USE

It is a violation of federal law to use this product in a manner inconsistent with its labeling.

DuPont LEXONE DF™ Herbicide is recommended in Louisiana for use on fallow ground to be followed by sugarcane. When used as directed, LEXONE DF™ provides a short term control of seedling johnsongrass and other problem weeds as listed on the package label. LEXONE DF™ may also be used on the planted crop for weed control.

HOW TO USE

Fallow Use

Apply 1⅓ to 2 lbs. LEXONE DF™ per acre. Use the lower rate if applied less than 30 days before planting. Application should be made after the rows are built and ground should not be disturbed until planting or loss of control will result.

To aid in control of johnsongrass and other problem weeds, stubble sugarcane should be destroyed after harvest or in early spring. Land should be disked frequently in the spring to control rhizome johnsongrass. In late May to early June, rows should be built. The soil surface should be free of trash and clods. Application of LEXONE DF™ should be made at this time. Rhizome johnsongrass will not be controlled by LEXONE DF™; therefore, spot treatment with an appropriate postemergence herbicide will be necessary.

Crop Use

Apply 2 to 4 lbs. LEXONE DF™ per acre during the fall after planting but before cane emerges, or to the stubble after harvest.

Make a second application in early spring prior to new cane growth.

Note: Do not apply within 60 days of harvest. Do not use treated foliage for feed or forage.

IMPORTANT

BEFORE USING LEXONE DF™, READ AND CAREFULLY OBSERVE THE CAUTIONARY STATEMENTS AND ALL OTHER INFORMATION ON THE PRODUCT LABEL.

This bulletin contains new or supplemental directions for use of this product which may not appear on the package label. Follow the instructions carefully.

This labeling must be in the possession of the user at the time of pesticide application.

4/26/84

SUPPLEMENTAL LABELING

LEXONE® DF™ HERBICIDE

EPA Reg. No. 352-390

FOR USE ON WINTER WHEAT, BARLEY, AND FALLOW

DIRECTIONS FOR USE

It is a violation of federal law to use this product in a manner inconsistent with its labeling.

DuPont LEXONE DF™ Herbicide is recommended for weed control:

- on the winter wheat varieties TAM W101, TAM 105, Newton, Hawk, and Eagle in the states of Kansas, Oklahoma, and Texas.
- on winter wheat, winter barley, and spring barley, (except varieties Morex, Glenn, or Moravian 3) in the states of Idaho, Oregon, Utah, and Washington.
- in reduced tillage fallow where wheat will be the following crop. For use in the states of Colorado, Idaho, Kansas, Montana, Nebraska, North Dakota, Oklahoma, Oregon, South Dakota, Texas, Utah, Washington, and Wyoming.

Treatment controls or partially controls the following broadleaf weeds:

Blue mustard[1]	Filaree	Pigweed (prostrate, redroot)
Bur buttercup	Flixweed[1]	Pineappleweed
Chickweed (common mouse-ear, jagged)	Gromwell[1]	Prickly lettuce
	Henbit	Russian thistle[1]
Conical catchfly	Jacobs ladder	Shepherdspurse
Corn cockle	Jim Hill mustard	Speedwell (ivyleaf)
Dog fennel (Mayweed)	Knotweed	Tansy mustard[1]
False flax	Kochia	Tarweed (fiddleneck)
Lambsquarter	Wild buckwheat[1]	Treacle mustard
Fanweed	Miners lettuce	Wild mustard

Treatment controls or partially controls the following grasses:

Annual ryegrass (except Kansas, Oklahoma, Texas)

Downy brome[1,2]	Pacific meadow foxtail (Blackgrass)[1]
Cheatgrass[1,2]	Windgrass[1]
Green foxtail[1]	Yellow foxtail[1]

[1] Weeds and grasses partially controlled. Partial control is an observable reduction in weed competition (reduced weed population and/or vigor) as compared to an untreated area. Degree of partial control can vary with soil type, environmental factors, and size of weeds of application.

[2] Best results with late Fall treatments.

HOW TO USE ON DRYLAND WHEAT & BARLEY

Apply postemergence to the crop when wheat/barley is well tillered (at least 3 tillers) and has developed 2" secondary roots throughout the entire field. A well developed secondary root system and good crop vigor is essential to minimize injury to wheat/barley. Application must be made prior to jointing as crop injury may result.

For best weed control results:

- Apply LEXONE DF™ before broadleaf weeds emerge or shortly after emergence and before weeds are 2" tall or across.
- Use a fall treatment for winter annuals or a spring treatment for summer annual broadleaf weeds. Avoid making treatments in mid-winter in areas where severe winter weather can cause crop stress.
- Apply LEXONE DF™ before grasses emerge or shortly after emergence and before the 3 to 5 leaf stage and use the higher dosage rates listed.

Rainfall (½") within 2 to 3 weeks after application is needed to move LEXONE DF™ into the weed root zone. Lack of adequate rainfall after application will result in poor control.

USE RATES

Make a single broadcast application postemergence to the crop as follows: (Idaho, Oregon, Utah, and Washington)

Soil Texture	Pounds of LEXONE DF™ Per Acre: Less than 1% Organic Matter	1 to 2% Organic Matter	More than 25% Organic Matter
Coarse			
Loamy sand, Sandy loam	0.33 to 0.40	0.33 to 0.40	0.33 to 0.40
Medium			
Loam, silt loam, Silt, sandy clay, Sandy clay loam	0.33 to 0.40	0.40 to 0.50	0.40 to 0.60
Fine			
Silty clay, Silty clay loam, Clay, clay loam	0.33 to 0.50	0.50 to 0.67	0.50 to 0.67

Do not use on gravelly or thinly covered or exposed subsoil areas (clay knobs) as crop injury may occur.

Kansas, Oklahoma, Texas

Soil Type	Pounds of LEXONE DF™—Rate Per Acre			
	Fall Application		Spring Application	
	1 to 2% OM	More than 2% OM	1 to 2% OM	More than 2% OM
Coarse				
Loamy sand, Sandy loam	Do not Use	0.33	Do not Use	0.33 to 0.50
Medium				
Loam, silt loam, Silt, Sandy clay, Sandy clay loam	0.33 to 0.50	0.50	0.50 to 0.67	0.50 to 0.67
Fine				
Silty clay, Silty clay loam, Clay loam	0.50	0.50 to 0.67	0.50 to 0.67	0.67

Use the lower rates on coarse textured soils and the higher rates on medium or fine textured soils having more than 1% organic matter. For grass control, use the highest rate listed for your soil type.

IRRIGATED WHEAT/BARLEY

Rainfall or sprinkler irrigation is needed to activate LEXONE DF™. If adequate rainfall (½" within 2 to 3 weeks after application) does not occur, then activate via sprinkler irrigation. Do not apply more than ½" of water for the first irrigation after application. The maximum amount for each additional irrigation should not exceed 1".

SPECIFIC HERBICIDE COMBINATIONS

Winter Wheat, Spring and Winter Barley

LEXONE DF™ may be tank mixed with Igran 80W for use on winter wheat and winter barley. Do not use more than 1.0 lbs./acre Ingran 80W in tank mix. On winter wheat, spring wheat, and spring barley, LEXONE DF™ may be tank mixed with Glean, 2,4-D, Banvel, MCPA, Bronate, Buctril, Brominal Plus, or Brominal. Refer to the other product labels registered for this use for additional precautions, directions, rates, and weeds controlled.

LEXONE DF™ and Ingran 80W (Winter Wheat, Winter Barley)— Pacific NW and Utah

For added control of downy brome, apply LEXONE DF™ at 0.33 pounds per acre plus Igran 80W at 0.75 to 1.00 pound per acre. Follow LEXONE DF™ timing of application recommendation. Do not use on spring barley or irrigated wheat/barley. Refer to Igran 80W label for additional precautions and instructions.

How to Use for Reduced Tillage Fallow

Apply 0.45 to 1.0 pounds LEXONE DF™ per acre as an after harvest treatment prior to subsequent planting of winter wheat. Use the lower rate on coarse textured soils and the higher rate on medium to fine textured soil.

LEXONE DF™ can be applied in the Fall after harvest at 0.67 to 1.0 lbs./acre or in the spring at 0.45 to 0.67 lbs./acre as new weed growth begins.

If large weeds (larger than 2" tall or across) are present, or weeds other than those listed are present or anticipated, use a suitable registered companion herbicide such as Glean, Paraquat, Round-Up, or Chem-Hoe 135. Refer to labeling for these products for the additional use instructions and precautions.

When using LEXONE DF™ and Round-Up tank mix spray, volume should be 10 gpa.

Where spring application is appropriate in Colorado, Kansas, Montana, Nebraska, North Dakota, Oklahoma, South Dakota, Texas, and Wyoming improved control of downy brome, Russian thistle, and tansy mustard, tank mix 0.33 to 0.67 lbs. of LEXONE DF™ with 2 to 2.5 lbs. Igran 80 W per acre.

Do not plant spring seeded cereals following LEXONE DF™ fall fallow applications.

Do not apply LEXONE DF™ in the spring if LEXONE DF™ was used in the preceding fall or winter.

Recrop Intervals

Do not plant winter wheat within 6 months of treatment except in Colorado, Kansas, Montana, Nebraska, North Dakota, Oklahoma, South Dakota, Texas, and Wyoming where winter wheat may be planted 4 months after treatment of 0.33 to 0.67 pounds LEXONE DF™ per acre.

Rotational Crop Interval

Do not plant crops (other than wheat as listed above) within 10 months of treatment.

SPRAY PREPARATION

Mix the proper amount of LEXONE DF™ into the necessary volume of water in the spray tank with the agitator running. Other herbicides or fertilizer solutions should be added last.

Crops (as listed)

LEXONE DF™ may be mixed with liquid fertilizer. However, addition of liquid fertilizer to LEXONE DF™ alone or in a three-way tank mix with other herbicides increases the possibility of crop injury. Do not use surfactant when using LEXONE DF™ in combination with liquid fertilizer. For application of LEXONE DF™ in liquid fertilizer, where the liquid fertilizer is the carrier, slurry the LEXONE DF™ in water, then thoroughly mix the slurry into the liquid fertilizer.

Fallow: LEXONE DF™ may be used in any combination listed above.

When using tank mixes with LEXONE DF™, small quantities should be tested for compatability prior to use.

The use of mixtures with LEXONE DF™ increases the possibilities of injury or temporary yellowing to the crop, especially if 2" secondary roots are not developed throughout the field.

APPLICATION

Apply LEXONE DF™ using properly calibrated air or ground equipment. Uniform coverage is essential in order to obtain optimum weed control performance and minimize the potential for crop injury. Use a minimum spray volume of 3 gallons per acre by air or ground.

As injury to crop may result, avoid overlapping of spray swath. Shut off spray booms while starting, turning, slowing, or stopping. Continuous agitation is needed to keep LEXONE DF™ in suspension.

Note: Do not apply under conditions where excessive drift may occur.

Do not apply LEXONE DF™ alone or in combination to stressed plants.

Do not treat fields where seed had been planted less than 1" deep.

Do not treat fields that have been severely grazed.

Do not graze treated fields for 14 days following treatment.

Do not apply more than once per crop or fallow period.

Because spring barley varieties may vary in their tolerance to LEXONE DF™ limit first use to a small area to determine varietal response prior to adoption as a field practice.

Do not use on soils having a calcareous surface area or a pH of 7.5 or higher as crop injury may result.

Do not apply when soil is frozen or crop is in winter dormacy.

Do not apply when soil is saturated with moisture.

Temporary yellowing or injury of the crop may occur, especially if frost or snow occurs within a few days of application. Severe winter stress, heaving, exposed root systems, disease, or insect damage preceeding or following application may increase crop injury. Winter wheat/barley varieties differ in their degree of winter hardiness; severe injury may result from treating non-winter hardy varieties that will not withstand the combination of winter and herbicide stress. If heavy rains occur soon after application, injury to the crop may result. Extended dry weather following treatment may reduce weed control effectiveness.

IMPORTANT

BEFORE USING LEXONE DF™ OR OTHER HERBICIDES IN COMBINATION, READ AND CAREFULLY OBSERVE THE CAUTIONARY STATEMENTS AND ALL OTHER INFORMATION ON THE PRODUCT LABELS.

This bulletin contains new or supplemental directions for use of these products which may not appear on the package labels. Follow the instructions carefully.

This labeling must be in the possession of the user at the time of pesticide application.

* Registered trademark of Ciba Geigy Corporation, Igran 80W contains terbutryn.

SUPPLEMENTAL LABELING

LEXONE® DF WEED KILLER

EPA Reg. No. 352-390

FOR COMMERCIAL IMPREGNATION AND APPLICATION ON DRY BULK FERTILIZER

DIRECTIONS FOR USE

It is a violation of federal law to use this product in a manner inconsistent with its labeling.

DuPont LEXONE® DF Weed Killer may be applied with dry bulk fertilizer for preemergence weed control; dry bulk fertilizer may be impregnated or coated with LEXONE® DF application to established alfalfa and to soybeans. All recommendations, cautions and special precautions on this label must be followed along with state regulations relating to dry bulk fertilizer blending, impregnating and labeling.

HOW TO USE

Application: Uniform application of LEXONE® DF which has been impregnated on dry fertilizer is essential for satisfactory weed control. Accurate calibration of fertilizer application equipment is essential for uniform distribution to the surface. The recommended method of application is to apply ½ the recommended rate and overlap 50 percent to double apply by splitting the middles to obtain the best distribution pattern.

If fertilizer materials are excessively dusty, use diesel oil or other suitable additive to reduce dust prior to impregnation as dusty fertilizer will result in poor distribution during application. Crop injury and/or poor weed control may occur where impregnated fertilizer is not uniformly applied.

Incorporation and Combination Uses: When LEXONE® DF is to be used in combination with another herbicide, follow directions on this label for combinations, rates, crops, incorporation and special precautions.

Impregnation: To impregnate, use a system consisting of a conveyor or closed drum, which is used for dry bulk fertilizer blending. Any commonly used fertilizer can be impregnated with LEXONE® DF except ammonium nitrate or fertilizers containing ammonium nitrate, potassium nitrate or sodium nitrate. Do not use on limestone.

Apply using a minimum of 200 pounds dry bulk fertilizer per acre and up to a maximum of 450 pounds per acre. To impregnate or coat the dry bulk fertilizer with LEXONE® DF, the delivery nozzles must be directed to deliver a fine spray toward the fertilizer for thorough coverage and avoiding spray contact with mixing equipment. Uniform impregnation of LEXONE® DF to dry bulk fertilizer will vary, and if the adsorptivity is not adequate, the use of an adsorptive powder may be required to product a dry free-flowing mixture. Microcel E (Johns Manville Product Corporation) is the recommended absorbent powder. LEXONE® DF should be premixed with water to form a sprayable mixture. When another herbicide is used with LEXONE® DF, mix and impregnate immediately.

Apply immediately after impregnation unless experience has shown that impregnated fertilizer can be stored without becoming lumpy and difficult to spread.

Rates: Select the rate of LEXONE® DF to apply per acre from the appropriate section of this label and refer to the rate table to determine the amount of LEXONE® DF which is to be impregnated on a ton of dry bulk fertilizer based on the amount of fertilizer which will be distributed on one acre.

RATE CHART FOR IMPREGNATING FERTILIZER WITH LEXONE® DF

Fertilizer Rate Per Acre	LEXONE® DF Rate Per Acre			
	⅓ Pound	⅔ Pound	1 Pound	1⅓ Pounds
200 pounds	3.35 lbs/ton	6.7 lbs/ton	10 lbs/ton	13.4 lbs/ton
250 pounds	2.7 lbs/ton	5.4 lbs/ton	8 lbs/ton	10.7 lbs/ton
300 pounds	2.25 lbs/ton	4.5 lbs/ton	6.7 lbs/ton	8.9 lbs/ton
350 pounds	1.95 lbs/ton	3.9 lbs/ton	6 lbs/ton	7.7 lbs/ton
400 pounds	1.7 lbs/ton	3.4 lbs/ton	5 lbs/ton	6.7 lbs/ton
450 pounds	1.5 lbs/ton	3 lbs/ton	4.5 lbs/ton	6 lbs/ton

For rates other than those listed above, use the following formula to calculate the amount of LEXONE® DF to be impregnated per ton of dry fertilizer.

$$\text{Pounds LEXONE® DF Per Acre} \times \frac{2000}{\text{lbs. Fertilizer Per Acre}} = \text{Pounds LEXONE® DF Per ton of Fertilizer}$$

IMPORTANT

BEFORE USING LEXONE® DF, READ AND CAREFULLY OBSERVE THE CAUTIONARY STATEMENTS AND ALL OTHER INFORMATION APPEARING ON THE PRODUCT LABEL.

This bulletin contains new and supplemental instructions for use of this product which may not appear on package labels. Follow the instructions carefully.

This labeling must be in the possession of the user at the time of pesticide application.

3/22/82

ADDITIONAL EPA SUPPLEMENTAL LABELING FOR LEXONE® DF™ HERBICIDE

Crop	Title	Description	Date of Label*
Soybeans	LEXONE DF Weed Killer in combination with other Herbicides for Weed Control in Minimum or No-Tillage Soybeans	Use with other preemergence herbicides plus paraquat or Roundup for U.S.	11/9/82
Soybeans	LEXONE DF Weed Killer Plus Treflan Plus Amiben Tank Mix Preplant Incorporated for Weed Control in Soybeans	U.S.	8/12/82
Soybeans	LEXONE DF Weed Killer Plus Dual 8E Plus Amiben Tank Mix Preplant Incorporated for Weed Control in Soybeans	U.S.	8/12/82
Soybeans	LEXONE DF Weed Killer Plus Prowl Plus Amiben Tank Mix Preplant Incorporated or Preemergence for Weed Control in Soybeans in the Midwest	Midwest	9/12/82
Soybeans	LEXONE DF Weed Killer Plus Lasso Plus Amiben Tank Mix Preplant Incorporated or Preemergence for Weed Control in Soybeans	U.S.	10/3/85
Soybeans	LEXONE DF Weed Killer Tank Mixture with Lasso for Preemergence use in Soybeans in the State of Ohio	Ohio for control of black nightshade	2/22/82

* Supplement label furnished on request.

Date of
Crop Title Description Label*

SPECIAL LOCAL NEED REGISTRATIONS [24(c)] FOR LEXONE® DF™ HERBICIDE

Crop	State	EPA SLN No.	Remarks
Barley	Montana	MT-800032	Spring Barley Postemergence
Bermuda-grass	Hawaii	HI-810003	Golf Course turf
Peas and Lentils	Idaho	ID-800015	Postplant incorporated.
	Idaho	ID-810017	Postemergence weed control.
	Oregon	OR-800028	Postplant incorporated.
	Washington	WA-800012	Preemergence weed control east of Cascades.
	Washington	WA-810028	Postemergence weed control east of Cascades.
Potatoes	Idaho	ID-800055	Sprinkler injection. Change injection time.
Soybeans	Arkansas	AR-820023	Postdirect application for 8" soybeans
	Georgia	GA-810023	Preemergence weed control with center pivot irrigation

	Louisiana	LA-810032	Postdirect application. Tank mix with Butyrac or Butoxone.
Tomatoes	California	CA-800129	Fallow weed control. Fall beeded tomato field.
Wheat	Montana	MT-800012	Winter wheat, postemergence

* Copies of individual labels available upon request to nearest DuPont Agricultural Chemicals Office. For distribution and use only within the State specified.

LEXONE® 4L

HERBICIDE

ACTIVE INGREDIENT:
4-Amino-6-(1,1-dimethylethyl)-3-(methylthio)-1,2,4-triazin-5(4*H*)-one 42.1%
INERT INGREDIENTS 57.9%
TOTAL 100.0%
Contains 4 lbs. active ingredient per gallon.

EPA Reg. No. 352–382

SHAKE WELL BEFORE USING

Keep Out of Reach of Children

PRECAUTIONARY STATEMENTS

HAZARDS TO HUMANS

CAUTION!

MAY BE HARMFUL IF INHALED OR SWALLOWED.
Avoid breathing spray mist. Avoid contact with skin, eyes, and clothing. Wash thoroughly with soap and warm water after handling. Wash contaminated clothing with soap and hot water before re-use.

STATEMENT OF PRACTICAL TREATMENT

If in eyes: Flush with plenty of water. Get medical attention if irritation persists.

If on skin: Wash with plenty of soap and water. Get medical attention if irritation persists.

ENVIRONMENTAL HAZARDS
Keep out of lakes, streams, or ponds. Do not contaminate water by cleaning of equipment or disposal of wastes.

NOTICE OF WARRANTY

DuPont warrants that this product conforms to the chemical description on the label thereof and is reasonably fit for the purposes stated on such label only when used in accordance with the directions under normal use conditions. It is impossible to eliminate all risks inherently associated with the use of this product. Crop injury, ineffectiveness, or other unintended consequences may result because of such factors as weather conditions, presence of other materials, or the manner of use or application, all of which are beyond the control of DuPont. In no case shall DuPont be liable for consequential, special or indirect damages resulting from the use or handling of this product. All such risks shall be assumed by the buyer. DUPONT MAKES NO WARRANTIES OF MERCHANTABILITY OR FITNESS FOR A PARTICULAR PURPOSE NOR ANY OTHER EXPRESS OR IMPLIED WARRANTY EXCEPT AS STATED ABOVE.

STORAGE AND DISPOSAL

STORAGE
Store product in original container only, away from other pesticides, fertilizer, food, or feed.

DISPOSAL
Do not contaminate water, food, or feed by storage or disposal. Wastes resulting from the use of this product may be disposed of on site or at approved waste disposal facility.

Triple rinse (or equivalent) the container. Then offer for recycling or reconditioning, or puncture and dispose of in a sanitary landfill, or by incineration, or if allowed by state and local authorities, by burning. If burned, stay out of smoke.

GENERAL INFORMATION

DuPont LEXONE 4L Herbicide is a liquid to be mixed in water and applied as a spray for selective control of certain broadleaf weeds and grasses in alfalfa, potatoes, sainfoin, soybeans, sugarcane and tomatoes. It is noncorrosive to equipment, nonflammable, and nonvolatile. For soybeans, LEXONE 4L may be used in conjunction with Treflan[1], Surflan[1], Lasso[2], Roundup[2], Dual[3], Prowl[4], Paraquat[5], or Amiben[6]. Observe all cautions and limitations on labeling of all products used in mixtures.

The degree of control and duration of effect will vary with:

- Soil Organic Matter Content—High organic matter requires higher dosages than low organic matter for equivalent herbicide performance.
- Soil Texture—High clay content requires higher dosage rates than those with less clay.
- Soil pH—Activity of LEXONE 4L increases as the pH increases.
- Rainfall—Moisture is necessary to activate the herbicide; best results are obtained if treatment is made to moist soil and moisture is supplied by rainfall or sprinkler irrigation (¼″ to ½″) within one week after application. If moisture is insufficient to activate the herbicide, a shallow cultivation (rotary hoe preferred) should be made after emergence of crop while weeds are small enough to be controlled by mechanical means. If heavy rains occur soon after application, injury to the crop may result.

Amount Applied—The higher recommended rate is more active than the lower rate on a given soil.

Other Conditions—Factors such as incorporation depth of LEXONE 4L, competition of the crop with weeds, weed species, drought stress of weeds, high salt content of soil, crop seed quality, soil residues of herbicides previously applied, soil insecticides, and combination herbicide treatments all can affect performance.

PREEMERGENCE USE (Germinating Weeds): LEXONE 4L, when used as directed, controls for an extended time, weeds such as:

Broadleaves

Beggarweeds	Prostrate pigweed
Wild buckwheat	Reedroot pigweed (careless weed)
Common chickweed	Smooth pigweed
Carpetweed	Common ragweed
White cockle	Hemp sesbania (coffeebean)
Coffeeweed	Shepherdspurse
Hophornbeam copperleaf	Sicklepod (cassia)
Flixweed	Prickly sida (spiny sida, teaweed)
Henbit	Pennsylvania smartweed
Jimsonweed	Spotted spurge
Kochia	Spurred anoda
Lambsquarters	Yellow rocket
Venice mallow	Velvetleaf
Meadow salsify	
Blue mustard	
Tansy mustard	
Wild mustard	
Pepperweed	

Partial Control

Cocklebur	Florida purslane (Florida pusley)
Common purslane	
Sunflower	

Grasses	
Smooth brome	Giant foxtail
Cheatgrass (downy brome)	Green foxtail
	Yellow foxtail
Large crabgrass	Seedling johnsongrass
Smooth crabgrass	Fall panicum
Crowfootgrass	Signalgrass (brachiaria)
Partial Control	
Barnyardgrass (watergrass)	Wild Oats
	Seedling johnsongrass
Yellow nutsedge	

Note: Consult Treflan, Dual, Surflan, Lasso, Prowl, Paraquat, Amiben, or Roundup labels for additional weeds controlled when LEXONE 4L is used in conjunction with these herbicides.

LEXONE 4L applied before emergence of asparagus, potatoes, soybeans, sugarcane and weeds is an effective procedure because susceptible weeds are controlled in an early, vulnerable seedling stage before they compete with the crop. With favorable moisture conditions, LEXONE 4L continues to control weeds for some time as the crop becomes better able to compete. Should weed seedlings begin to break through the preemergence treatment in significant numbers, secondary weed control procedures should be implemented; these include cultivation and postemergence herbicide application.

POSTEMERGENCE USE (Emerged Seedling or Perennial Weeds): LEXONE 4L, at recommended rates, controls weeds such as Florida beggarweed, bluegrass, carpetweed, cocklebur, crabgrass (large, smooth), dayflower, fumitory, goosegrass, hemp sesbania, jimsonweed, ladysthumb, lambsquarters, Mexican weed, mustard, pigweeds (prostrate, redroot, smooth), common purslane, common ragweed, sicklepod (cassia), prickly sida (spiny sida, teaweed), Pennsylvania smartweed, toadflax, Russian thistle and velvetleaf. Treatment usually provides partial control of barnyardgrass (watergrass), curly dock, dandelion, foxtail, horsenettle, kochia, morningglory species, wild poinsettia, Florida purslane (Florida pusley), quackgrass, broadleaf signalgrass, spotted spurge and sunflower.

LEXONE 4L may be sprayed over the top of certain emerged crops; they are alfalfa (established, dormant), potatoes, sainfoin (dormant), sugarcane and tomatoes (seeded and transplants) to effectively control susceptible emerged seedling weeds.

Directed postemergence application of LEXONE 4L, where spray nozzles are adjusted so that weeds are sprayed but the crop is not, may be used in emerged soybeans (southern and southeastern states) and established tomatoes to control susceptible emerged seedling weeds.

Results of postemergence treatment of emerged weeds vary with rate applied and environmental conditions; best results are obtained on succulent weeds growing under conditions of high humidity and temperatures of 70 ° F or higher. Addition of a surfactant such as Du Pont Surfactants WK to the spray (where recommended) increases contact effects of LEXONE 4L. Application will also provide control of emerging susceptible weed seedlings for an extended period of time.

IMPORTANT—Injury to or loss of desirable trees or other plants may result from failure to observe the following:

Do not apply (except as recommended), or drain or flush equipment on or near desirable trees or other plants, or on areas where their roots may extend, or in locations where the chemical may be washed or moved into contact with their roots. Do not use on lawns, walks, driveways, tennis courts, or similar areas. Prevent drift of spray to desirable plants. Do not contaminate any body of water. Keep from contact with fertilizers, insecticides, fungicides and seeds during storage.

Thoroughly clean all traces of LEXONE 4L from application equipment immediately after use. Flush tank, pump, hoses, and boom with several changes of water after removing nozzle tips and screens (clean these parts separately).

Varietal sensitivity to LEXONE 4L is known in several crops for which LEXONE 4L is registered. All new crop varieties should be evaluated on a limited basis for their tolerance prior to using LEXONE 4L.

DIRECTIONS FOR USE

It is a violation of federal law to use this product in a manner inconsistent with its labeling.

DuPont LEXONE 4L Herbicide should be used only in accordance with recommendations on this label, or in separate published DuPont recommendations available through local dealers.

DuPont will not be responsible for losses or damages resulting from use of this product in any manner not specifically recommended by DuPont. User assumes all risk associated with such non-recommended use.

Do not use in Kern County, California.

SPRAY PREPARATION: Fill tank ¼ full with water. Start agitation system, add LEXONE 4L and continue adding water. Add separately each additional component of any tank mix while adding water. If a surfactant is used, add it last when the tank is nearly full. Continue agitation throughout. If poor mixing should occur with any component, premix the component with two parts water before adding to the spray tank: follow this procedure when mixing TREFLAN or PROWL. Use spray mixture promptly. LEXONE 4L plus TREFLAN tank mixtures should not be stored overnight; flush sprayer with water after use and dispose of mix properly.

For preemergence application to potatoes or soybeans, nonpressure nitrogen or fertilizer solution may be used in the spray mixture. Small quantities should be tested for compatibility by the following procedures before full scale mixing:

1. Put 1 pint of fertilizer solution in a quart jar.
2. Mix 2 teaspoons LEXONE 4L with 2 tablespoons of water; mix thoroughly and add to fertilizer solution.
3. Close jar and shake well.
4. If other herbicides are to be used in the mixture, premix 2 teaspoons of wettable powders or 1 teaspoon of liquids with 2 tablespoons of water; add to LEXONE 4L/fertilizer solution mixture.
5. Close jar and shake well.
6. Watch mixture for several seconds; check again in 30 minutes.
7. If mixture does not separate, foam, gel, or become lumpy, it may be used.
8. Mixing ability may be improved by adding compatibility agents such as Kalo Laboratories Compex or Witco Chemicals Sponto 168D. Follow directions on container.

Provided the above procedure shows the mixture to be compatible, prepare the tank mixture as follows: add the fertilizer solution to the spray tank first with agitator running, add the required amount of LEXONE 4L and thoroughly mix. For tank mixtures with other herbicides, follow directions above.

EQUIPMENT—SPRAY VOLUMES: Apply uniformly by air (except as noted) or with a properly calibrated low pressure (20 to 40 psi) boom sprayer equipped with fan-type nozzles and screens no finer than 50 mesh. Use 10 to 40 gals. of water per acre with ground equipment (20 to 60 gals. with mixture containing Paraquat and 20 to 30 gals. with mixture containing Roundup). Use 2 to 10 gals. per acre for aerial application, except use 7 gals. per acre for sugarcane (irrigated only) in Hawaii, and 5 to 10 gals. per acre for potatoes west of Rocky Mountains and for sugarcane in Louisiana and Texas.

Continuous agitation in the spray tank is required to keep the material in suspension. Avoid overlapping, and shut off spray booms while starting, turning, slowing or stopping, or injury to the crop may result. With aerial equipment, avoid overlapping of spray swath and avoid application under conditions where excessive drift may occur.

USE RATES: All rates are expressed as broadcast rates per acre; for band treatment, use proportionately less. For example, for 14" band on 42" row, use ⅓ of broadcast rate. Where a range of dosages is given, use the lower rate on coarser soils (low in clay or organic matter) and the higher rate on finer soils (high in clay or organic matter).

REPLANTING AND CROP ROTATION RECOMMENDATIONS: Unless otherwise noted, observe the following:

Areas treated with up to 1 pt. LEXONE 4L per acre may be replanted immediately to potatoes, soybeans and sugarcane; two months after application, asparagus, corn and forage grasses may be planted. Winter wheat may be planted 4 months after treatment in Colorado, Kansas, Montana, Nebraska, North Dakota, Oklahoma, South Dakota, Texas and Wyoming. In other areas winter wheat may be planted after 6 months.

Areas in the Midsouth and Southeast where soybeans have been treated preemergence with up to 1.5 pts. LEXONE 4L per acre may be replanted to corn, sorghum, cotton, and rice the season following application.

Areas treated with 1 to 2 pts. LEXONE 4L per acre may be replanted to soybeans and sugarcane; four months after application, alfalfa, sainfoin, asparagus, corn, potatoes, tomatoes and forage grasses may be planted.

Areas treated with 2 to 4 pts. LEXONE 4L per acre may be replanted to sugarcane; two months after application, soybeans may be replanted; 10 months after application, alfalfa, sainfoin, asparagus, corn, potatoes, tomatoes and forage grasses may be planted.

Areas treated with 4 to 8 pts. LEXONE 4L per acre may be replanted to sugarcane; four months after application, soybeans may be replanted; thirteen months after application, alfalfa, sainfoin, asparagus, corn, potatoes, tomatoes and forage grasses may be planted.

Other Crops: Unless otherwise directed, do not plant in treated areas during the season following treatment as crop injury may result. Cover crops may be planted anytime but stand reduction may occur; do not graze or harvest these cover crops for feed or food.

Special Conditions: For winter or drought periods, add the number of weeks of winter or drought to the above listed recropping intervals.

For soil pH 7.5 or above, add 2 months to the recropping intervals listed.

Combination Treatments: When recommended tank mix combinations are used, consult the product label for replanting and recropping instructions and observe the directions that are the most restrictive.

CROP USES

ALFALFA AND SAINFOIN (Alone or with Grasses)

Make a single application in the fall after plants become dormant or in the spring before new growth starts. Do not apply during the first growing season after seeding. Do not graze or harvest within 28 days after application. Stress which may be caused in part by low fertility, disease, insects, winter kill, overcutting, drought, or frost may increase the chance of crop injury. In Washington, Oregon, Idaho, and Utah, partial control of downy brome, flixweed, and tansy mustard will occur at the lower use rate.

In California, use only north of U.S. Interstate 80.

Pure Stands of Alfalfa or Sainfoin: Select the rate from the table below:

Soil Texture	Pints LEXONE 4L Per Acre: ½ to 2% Organic Matter	Pints LEXONE 4L Per Acre: More than 2% Organic Matter
Coarse: Loamy Sand, Sandy Loam	¾ to 1	1 to 1¼
Medium: Loam, Silt, Silt Loam, Sandy Clay, Sandy Clay Loam	1 to 1¼	1¼ to 1½
Fine: Clay, Clay Loam, Silty Clay, Silty Clay Loam	1¼ to 1½	1½ to 2

Do not use on sand nor on soils with less than ½% organic matter as crop injury may result.

Mixed Stands with Grasses: The higher range of rates for a given soil texture and organic matter classification in the table above may provide partial reduction of forage grass stands. Rates of 1½ to 2 pints per acre may severely reduce stands of forage grass.

POTATOES

Preemergence: Make a single application of 1 to 2 pints per acre after planting or after drag-off but before crop emerges. Use the lower rates on coarse textured soils and the higher rates on fine textured soils (high in clay or organic matter).

Postemergence: Do not use on early maturing smooth-skinned white or red-skinned varieties. Apply only if there have been at least 3 successive days of sunny weather prior to application. Treat before weeds are 1" tall. Make a single application of 1 pint per acre. For redroot pigweed and lambsquarters only, use ½ to 1 pint per acre. Treatment may cause some chlorosis or minor necrosis. Do not apply within 60 days of harvest.

Application by Sprinkler Irrigation: Apply specified dosage of LEXONE 4L in ¼ to ¾ inch of water per acre (¼ to ½ inch of water per acre on sandy soil) as a continuous injection in center pivot and self-propelled wheel-move systems or in the last 15–30 minutes of set-in permanent solid-set sprinkler systems.

Irrigation systems must be equipped with automatic shut-off devices which prevent back-flow to the water source. Maintain continuous agitation in injection nurse tanks during application. Treatment may be made as a single postemergence or split postemergence application. Check irrigation system to insure unifom application of water to all areas. Failure to apply LEXONE 4L uniformly may result in crop injury and/or poor weed control.

Preemergence plus Postemergence: Apply as directed above. Do not use more than 2 pints total per acre per season. Do not apply within 60 days of harvest.

Postemergence plus Postemergence—(Oregon, Washington, Idaho): Use only if LEXONE 4L was not applied preemergence. Allow at least 14 days between applications. For best results, apply before weeds are 1" in height or diameter. Do not apply within 24 hours of treatment with other pesticides.

Oregon and Washington—Make 2 applications of ½ to 1 pint per acre. On coarse (sandy) soils containing ½ to 1% organic matter do not exceed ½ pint per acre per application.

Idaho—Make 2 applications of ¼ to 1 pint per acre. On coarse (sandy) soils containing ½ to 1% organic matter, do not exceed ¾ pint per acre per application.

CROP ROTATION RECOMMENDATIONS: In the Northwest do not plant treated areas to crops other than potatoes for one year after treatment. Cover crops planted to prevent soil erosion or for soil building purposes may be planted anytime after application of LEXONE 4L but do not graze or harvest for feed or food.

See "Replanting and Crop Rotation Recommendations" for further details.

Note: Because potato varieties vary in their resistance to herbicides, determine tolerance to LEXONE 4L prior to adoption as a field practice to prevent possible crop injury. Crop injury may occur if Norgold variety is under stress from lack of moisture, fertility, sunlight or affected by insects or disease when treated postemergence. Use of LEXONE 4L on sands or loamy sands containing ½ to 1½% organic matter may result in crop injury. Do not use on any soil containing less than ½% organic matter, as crop injury may result. Preemergence use on soils containing more than 10% organic matter may result in only partial weed control. Postemergence use on rill irrigated potatoes may not provide adequate weed control in absence of rainfall. Do not apply to sweet potatoes or yams. Do not incorporate into soil.

SOYBEANS

Select one of the following herbicide treatments and apply as directed. The combination treatments control weeds that are susceptible to LEXONE 4L plus those controlled by the companion products (refer to product labeling). Dosage rates listed are broadcast rates; for band applications use proportionately less.

Note: Soybean varieties such as Altona, Coker 102 and 156, Govan, Semmes, Tracy, Vansoy, Terra Vig 505 and 606, AP 71, and NKS 1884 are sensitive to LEXONE 4L. Injury may occur if LEXONE 4L is used on these varieties. Before use on any other soybean variety, tolerance to LEXONE 4L must first be determined. Injury to soybeans may occur if LEXONE 4L is used on soils having a calcareous surface layer or a pH of 7.5 or higher, or if used in conjunction with soil-applied organic phosphate pesticides such as Dasanit, Disyston, Mocap, Nemacur, Thimet, Parathion, Lorsban 15G or Counter. Injury may occur if atrazine was applied on the soil the year before use of LEXONE 4L. Seedling disease, cold weather, deep planting (more than 2") excessive moisture, high soil pH (pH 7.5 or higher), high salt concentration, or drought may weaken soybean seedlings and increase possibility of crop injury.

Preplant Incorporated: Incorporate herbicide treatment into soil before planting of soybeans according to instructions on label of product used in conjunction with LEXONE 4L; improper soil incorporation can result in erratic weed control or crop injury. Do not plant seed more than 2" deep. Use combination as tank mixtures in water.

LEXONE 4L + Treflan—Apply within 2 weeks of planting. Treatment also controls seedling johnsongrass, goosegrass and barnyardgrass.

Soil Texture	Pints Per Acre LEXONE 4L + Treflan
Coarse Loamy sand, sandy loam	½ + 1
Medium Loam, silt loam, silt, sandy clay, sandy clay loam	¾ + 1½
Fine Silty clay, silty clay loam, clay, clay loam	1 + 2

Do not use on sand. Do not use on loamy sand or sandy loam with less than 1% organic matter, nor on other soils with less than ½% organic matter as crop injury may result.

Treated vines may be grazed or fed to livestock 40 days after application.

LEXONE 4L + Lasso or Lasso EC—Apply within 7 days of planting. Treatment also controls dayflower and goosegrass.

Soil Texture	Rate Per Acre: ½ to 2% Organic Matter — LEXONE 4L (Pints)	Lasso/ + Lasso EC (Quarts)	More than 2% Organic Matter — LEXONE 4L (Pints)	Lasso
Coarse Sandy loam	½	+ 2	¾	+ 2
Medium Loam, silt loam, silt, sandy clay, sandy clay loam	¾	+ 2	1	+ 2
Fine Silty clay, silty clay loam, clay, clay loam	1	+ 2	1	+ 2½
Mississippi Delta Only—Silty clay to heavy clay	1½	+ 2 to 2½	1½	+ 2 to 2½

Do not use on soils with less than ½% organic matter, nor on sand, loamy sand or muck soils as crop injury may result. Do not apply by air.

Treated vines may be grazed or fed to livestock 40 days after application.

LEXONE 4L + Dual 8E—apply within 14 days of planting. Treatment also controls dayflower, goosegrass and barnyardgrass.

Soil Texture	Pints Per Acre: ½ to 3% Organic Matter LEXONE 4L + Dual 8E	More than 3% Organic Matter LEXONE 4L + Dual 8E
Coarse Loamy Sand (Over 2% OM), Sandy Loam	½ + 1¼	¾ + 1½
Medium Loam, Silt Loam, Silt	¾ + 1½	1 + 2
Fine Silty Clay Loam, Sandy Clay Loam, Silty Clay, Sandy Clay, Clay Loam, Clay	1 + 2	1 + 2 to 2½
Mississippi Delta Only—Silty Clay, Clay	1½ + 2	1½ + 2 to 2½
Muck or Peat Soils	DO NOT USE	

Do not use on any sand or loamy sand with less than 2% organic matter. Do not use on soils with less than ½% organic matter or on alkaline soils with a pH above 7.4 as crop injury may occur. Do not use preplant incorporated on any sand, loamy sand, or sandy loam as crop injury may occur.

Treated vines may be grazed or fed to livestock 40 days after application.

LEXONE 4L + Prowl—Incorporate into soil within 7 days after application and plant soybeans no later than 7 days after application.

Southern States and Eastern Coastal Plains

Soil Texture	Pints Per Acre LEXONE 4L + Prowl
Coarse Loamy sand, sandy loam	½ + 1½
Medium Sandy clay loam, sandy clay, loam, silt, silt loam	¾ + 1½
Fine Clay loam, silty clay loam, clay	1 + 1½ to 2

Do not use on sand, muck or peat. Do not use on loamy sand or sandy loam with less than 1% organic matter, nor on other soils with less than ½% organic matter as crop injury may result.

Treated vines may be grazed or fed to livestock 40 days after application.

Northeastern and Northcentral States

Soil Texture	Pints Per Acre: ½ to 3% Organic Matter LEXONE 4L + Prowl	More than 3% Organic Matter LEXONE 4L + Prowl
Coarse Loamy sand, sandy loam (2% or more organic matter only)	¾ + 1	¾ + 1½
Medium Sandy clay loam, sandy clay, loam, silt, silt loam	¾ + 1½ to 2	¾ to 1 + 1½ to 2
Fine Clay loam, silty clay loam, clay	¾ to 1 + 1½ to 2	1 + 2 to 2½

Do not use on sand, muck or peat, nor on soils with less than ½% organic matter as crop injury may result.

Treated vines may be grazed or fed to livestock 40 days after application.

Preplant Non-Incorporated:

LEXONE 4L + Surflan AS or Surflan 75W—Preplant Surface Application: Apply within 4 weeks of planting. Treatment also controls dayflower and goosegrass.

Soil Texture ½ to 5% Organic Matter	Rate Per Acre: LEXONE 4L (Pints)	+	Rate Per Acre: Surflan AS or Surflan 75W
Coarse Sandy loam (minimum 1% organic matter)	½	+	½ to ¾ (qt) or ⅔ to 1 (lb)
Medium Loam, silt loam, silt, sandy clay loam	¾	+	¾ to 1 (qt) or 1 to 1⅓ (lb)
Fine Silty clay, silty clay loam, sandy clay, clay loam, clay	1	+	1 to 1¼ (qt) or 1⅓ to 1⅔ (lb)
Mississippi Delta Only—Silty clay to heavy clay	1 to 1¼	+	1 to 1¼ (qt) or 1⅓ to 1⅔ (lb)

Do not use on sand or loamy sand. Do not use on sandy loam with less than 1% organic matter. Do not use on other soils with less than ½% organic matter. Do not use preemergence application on soils with more than 3% organic matter nor the preplant surface application on soils with more than 5% organic matter as poor weed control and/or crop injury may result. Do not use treated vines, for feed or forage.

Preplant Incorporated (PPI) Combinations Plus Preemergence: Use combinations as tank mixes in water. Promptly incorporate PPI treatment into soil before planting of soybeans. Improper soil incorporation can result in erratic weed control or crop injury. Do not plant seed more than 2" deep. After planting of soybeans and before emergence, apply a preemergence treatment of LEXONE 4L:

LEXONE 4L + Treflan (PPI) + LEXONE 4L. Treatment also controls seedling johnsongrass, goosegrass and barnyardgrass.

	Preplant Incorporated (PPI) Plus	Preemergence		
	LEXONE 4L + Treflan Apply within 4 weeks of planting	LEXONE 4L—Apply after planting but before emergence		
	Pints per Acre	Pints LEXONE 4L Per Acre		
Soil Texture	LEXONE 4L + Treflan	Less Than 2% OM	2–4% OM	Over 4% OM
Coarse Loamy sand, sandy loam	½ + 1	¼	¼	½
Medium Loam, silt loam, silt, sandy clay, sandy clay loam	½ + 1½ or ¾ + 1½	½ or ¼	½ to ¾ or ¼ to ½	¾ to 1 or ½ to ¾
Fine Silty clay, silty clay loam, clay, clay loam	¾ + 2 or 1 + 2	½ or ¼	½ to ¾ or ¼ to ½	¾ to 1 or ½ to ¾

Higher rates in the tank mixes are recommended in low rainfall areas or where low rainfall may follow application. Use the higher rate of preemergence LEXONE 4L where fields are prone to severe broadleaf weed pressure.

Do not use on sand, loamy sand, or sandy loam with less than 1% organic matter, nor on other soils with less than ½% organic matter as crop injury may result.

Treated vines may be grazed or fed to livestock 40 days after application.

LEXONE 4L + Lasso or Lasso EC (PPI) + LEXONE 4L. Treatment also controls dayflower and goosegrass.

	Preplant Incorporated (PPI)	Plus	Preemergence		
	LEXONE 4L + Lasso or Lasso EC Apply within 7 days of planting		LEXONE 4L—Apply after planting but before emergence		
	Rate Per Acre		Pints LEXONE 4L Per Acre		
Soil Texture	½ to 2% Organic Matter LEXONE 4L (Pts.) + Lasso/Lasso EC (Qts.)	More than 2% Organic Matter LEXONE 4L (Pts.) + Lasso/Lasso EC (Qts.)	Less than 2% OM	2 to 4% OM	Over 4% OM
Coarse Sandy loam	½ + 2	½ + 2	¼	¼	½
Medium Loam, silt loam, silt, sandy clay, clay loam	½ + 2 or ¾ + 2	½ + 2 or ¾ + 2	½ or ¼	½ to ¾ or ¼ to ½	¾ to 1 or ½ to ¾
Fine Silty, clay, silty clay loam, clay, clay loam	¾ + 2 or 1 + 2	¾ + 2½ or 1 + 2½	½ or ¼	½ to ¾ or ¼ to ½	¾ to 1 or ½ to ¾

Higher rates in the tank mixes are recommended in low rainfall areas or where low rainfall may follow application. Use the higher rate of preemergence LEXONE 4L where fields are prone to severe broadleaf weed pressure.

Do not use on soils with less than ½% organic matter, nor on sand, loamy sand or muck soils as crop injury may result. Do not apply by air.

Treated vines may be grazed or fed to livestock 40 days after application.

LEXONE 4L + Dual 8E (PPI) + LEXONE 4L. Treatment also controls dayflower, goosegrass and barnyardgrass.

	Preplant Incorporated (PPI)	Plus	Preemergence		
	LEXONE 4L + Dual 8E Apply within 2 weeks of planting		LEXONE 4L—Apply after planting but before emergence		
	Pints Per Acre		Pints LEXONE 4L Per Acre		
Soil Texture	½ to 3% Organic Matter LEXONE 4L + Dual 8E	More than 3% Organic Matter LEXONE 4L + Dual 8E	Less 2% than OM	2 to 4% OM	Over 4% OM
Coarse Loamy sand (Over 2% OM), Sandy loam	½ + 1¼	½ + 1½	¼	¼	½
Medium Loam, silt loam, silt, sandy clay loam	½ + 1½ or ¾ + 1½	½ + 2 or ¾ + 2	½ or ¼	½ to ¾ or ¼ to ½	¾ to 1 or ½ to ¾
Fine Silty clay loam, silty clay, sandy clay, clay loam, clay	¾ + 2 or 1 + 2	¾ + 2 to 2½ or 1 + 2 to 2½	½ or ¼	¼ to ¾ or ¼ to ½	¾ to 1 or ½ to ¾

Higher rates in the tank mixes are recommended in low rainfall areas or where low rainfall may follow application. Use the higher rate of preemergence LEXONE 4L where fields are prone to severe broadleaf weed pressure.

Do not use on any sand or loamy sand with less than 2% organic matter. Do not use on soils with less than ½% organic matter or on alkaline soils with a pH above 7.4 as crop injury may occur. Do not use preplant incorporated on any sand, loamy sand, or sandy loam as crop injury may occur.

Treated vines may be grazed or fed to livestock 40 days after application.

LEXONE 4L + Prowl (PPI) + LEXONE 4L

Soil Texture	Preplant Incorporated (PPI) Plus: LEXONE 4L + Prowl — Incorporate into soil within 7 days after application and plant soybeans no later than 7 days after application. Pints per Acre LEXONE 4L + Prowl	Preemergence: LEXONE 4L—Apply after planting but before emergence. Pints LEXONE 4L Per Acre: Less Than 2% OM	2–4% OM	Over 4% OM
Coarse				
Loamy sand, sandy loam	½ + 1½	¼	¼	½
Medium				
Loam, silt loam, silt, sandy clay, sandy clay loam	½ + 1½ to 2 or ¾ + 1½ to 2	½ or ¼	½ to ¾ or ¼ to ½	¾ to 1 or ½ to ¾
Fine				
Silty clay, clay loam silty clay loam, clay	¾ + 1½ to 2 or 1 + 1½ to 2	½ or ¼	½ to ¾ or ¼ to ½	¾ to 1 or ½ to ¾

Higher rates in the tank mixes are recommended in low rainfall areas or where low rainfall may follow application. Use the higher rate of preemergence LEXONE 4L where fields are prone to severe broadleaf weed pressure.

Do not use on sand, muck or peat. Do not use on loamy sand or sandy loam with less than 1% organic matter, nor on other soils with less than ½% organic matter as crop injury may result.

Treated vines may be grazed or fed to livestock 40 days after application.

Preplant Incorporation Plus Preemergence:

LEXONE 4L Following Dual, Treflan, or Prowl: Apply Dual, Treflan, or Prowl as a preplant incorporated treatment according to directions on product label. As a separate operation, follow with LEXONE 4L using rates given under **"Preemergence Application."**

Preemergence Application: Apply after planting but before soybeans emerge; do not spray over top of emerged soybeans. Plant seed 1½" to 2" deep on flat or raised seedbeds only or injury to the crop may result; soil should be well prepared and as free as possible from trash and clods before planting. Use combinations as tank mixtures in water.

Where a range of LEXONE 4L dosage rates is listed for a specific soil type and organic matter, use the higher rate for difficult control problems, such as cocklebur, coffeeweed, jimsonweed, coffeebean, sicklepod, prickly sida, or velvetleaf. Use the higher rate also for minimum or no-tillage soybeans, and in the states of Georgia and South Carolina.

LEXONE 4L + Lasso or Lasso EC. Treated also controls dayflower and goosegrass.

Soil Texture	Rate Per Acre: ½ to 2% Organic Matter LEXONE 4L (Pints)	+ Lasso/Lasso EC (Quarts)	More than 2% Organic Matter LEXONE 4L (Pints)	+ Lasso/Lasso EC (Quarts)
Coarse				
Sandy loam	¼ to ½	+ 2	¾	+ 2
Medium				
Loam, silt loam, silt, sandy clay, sandy clay loam	½ to ¾	+ 2	1	+ 2
Fine				
Silty clay, silty clay loam, clay, clay loam	¾ to 1	+ 2	1	+ 2½
Mississippi Delta Only—Silty clay to heavy clay	1 to 1½	+ 2 to 2½	1½	+ 2 to 2½

Do not use on soils with less than ½% organic matter, nor on sand, loamy sand or muck soils as crop injury may result. Do not apply by air.

Treated vines may be grazed or fed to livestock 40 days after application.

LEXONE 4L + DUAL 8E. Treatment also controls dayflower and goosegrass.

Soil Texture	Rate Per Acre: ½ to 3% Organic Matter LEXONE 4L (Pints)	+ DUAL 8E (Pints)	More than 3% Organic Matter LEXONE 4L (Pints)	+ DUAL 8E (Pints)
Coarse				
Loamy sand (over 2% OM), sandy loam	¼ to ½	+ 1¼	¾	+ 1½
Medium				
Loam, silt loam, silt	½ to ¾	+ 1½	1	+ 2
Fine				
Silty clay loam, sandy clay loam, silty clay, sandy clay, clay loam, clay	¾ to 1	+ 2	1	+ 2 to 2½
Mississippi Delta Only—Silty clay, clay	1 to 1½	+ 2	1½	+ 2 to 2½
Muck or peat soils	DO NOT USE			

Do not use on any sand or loamy sand with less than 2% organic matter. Do not use on soils with less than ½% organic matter or on alkaline soils with a pH above 7.4 as crop injury may occur. Do not use preplant incorporated on any sand, loamy sand, or sandy loam as crop injury may occur.

Treated vines may be grazed or fed to livestock 40 days after application.

LEXONE 4L + Prowl

Southern States and Eastern Coastal Plains

Soil Texture	Pints Per Acre: LEXONE 4L	+	Prowl
Coarse			
Loamy sand, sandy loam	¼ to ½	+	1½
Medium			
Sandy clay loam, sandy clay, loam, silt, silt loam	½ to ¾	+	1½
Fine			
Clay loam, silty clay loam, clay	¾ to 1	+	1½ to 2

Do not use on sand, muck or peat. Do not use on loamy sand or sandy loam with less than 1% organic matter, nor on other soils with less than ½% organic matter as crop injury may result.

Treated vines may be grazed or fed to livestock 40 days after application.

Northeastern and Northcentral States

Soil Texture	Pints Per Acre: ½ to 3% Organic Matter LEXONE 4L	+ Prowl	More than 3% Organic Matter LEXONE 4L	+ Prowl
Coarse				
Loamy sand, sandy loam (2% or more organic matter only)	½ to ¾	+ 1	¾	+ 1½
Medium				
Sandy clay loam, sandy clay, loam, silt, silt loam	½ to ¾	+ 1½ to 2	¾ to 1	+ 1½ to 2

Fine Clay loam, silty clay loam, clay	3/4 to 1 + 1 1/2 to 2	1 + 2 to 2 1/2

Do not use on sand, muck or peat, nor on soils with less than 1/2% organic matter as crop injury may result.

Treated vines may be grazed or fed to livestock 40 days after application.

LEXONE 4L + Surflan AS or Surflan 75W—Apply within 2 days after planting. Treatment also controls dayflower and goosegrass.

Soil Texture 1/2 to 3% Organic Matter	Rate Per Acre		
	LEXONE 4L (Pints)	+	Surflan AS or Surflan 75W
Coarse Sandy loam (minimum 1% organic matter)	1/4 to 1/2	+	1/2 to 3/4 (qt) or 2/3 to 1 (lb)
Medium Loam, silt loam, silt, sandy clay loam	1/2 to 3/4	+	3/4 to 1 (qt) or 1 to 1 1/3 (lb.)
Fine Silty clay, silty clay loam, sandy clay, clay loam, clay	3/4 to 1	+	1 to 1 1/4 (qt) or 1 1/3 to 1 2/3 (lb)
Mississippi Delta Only—Silty clay to heavy clay	1 to 1 1/4	+	1 1/3 to 1 2/3 (lb)

Do not use on sand or loamy sand. Do not use on sandy loam with less than 1% organic matter. Do not use on other soils with less than 1/2% organic matter. Do not use preemergence application on soils with more than 3% organic matter nor the preplant surface application on soils with more than 5% organic matter as poor weed control and/or crop injury may result. Do not use treated vines for feed or forage.

LEXONE 4L + Amiben

Soil Texture	Rate Per Acre					
	1/2 to 3% Organic Matter			More than 3% Organic Matter		
	LEXONE 4L (Pints)	+	Amiben (Quarts)	LEXONE 4L (Pints)	+	Amiben (Quarts)
Coarse Loamy sand, sandy loam (2% or more organic matter only)	1/4 to 1/2	+	3 to 4	3/4	+	3 to 4
Medium Loam, silt loam, silt, sandy clay, sandy clay loam	1/2 to 3/4	+	3 to 4	1	+	4 to 5
Fine Silty clay, silty clay loam, clay, clay loam	3/4 to 1	+	4 to 5	1	+	4 to 5
Mississippi Delta only—silty clay to heavy clay	1 to 1 1/2	+	6	1 1/2	+	6

Do not use on sand or muck soils as crop injury may result.

Treated vines may be grazed or fed to livestock 40 days after application.

LEXONE 4L

Soil Texture	Pints LEXONE 4L Per Acre[3]	
	1/2 to 2% Organic Matter	More than 2% Organic Matter
Coarse Loamy sand, sandy loam	Do Not Use (See tank mix use directions on other side)	3/4
Medium[1,2] Loam, silt loam, silt, sandy clay, sandy clay loam	1/2 to 3/4	1
Fine[1,2] Silty clay, silty clay loam, clay, clay loam	3/4 to 1	1
Mississippi Delta Only—Silty clay to heavy clay	1 to 1 1/2	1 1/2

[1] Alkaline (Calcareous) Soils: For control of lambsquarters, redroot pigweed and wild mustard and suppression of green, yellow and giant foxtail, apply LEXONE 4L at 1/2 pt. per acre on medium soils and 1/2 to 3/4 pt. per acre on fine textured soils containing more than 1/2% organic matter. Use 3/4 pt. per acre only where soil pH is less than 7.5 and weed pressure is heavy. The 1/2 pt. per acre rate may be used as a preemergence overlay following preplant incorporation of Treflan or Tolban.

[2] Alabama, Louisiana, Mississippi, and Tennessee: Apply 3/4 to 1 pt. per acre on medium soils and 1 to 1 1/4 pts. on fine textured soils. Use the higher range of rates for improved cocklebur control.

[3] Do not use on sand nor on soils with less than 1/2% organic matter as crop injury may result.

Treated vines may be grazed or fed to livestock 40 days after application.

Directed Postemergence Spray Treatment (Southern and Southeastern States): LEXONE 4L may be applied as a directed postemergence treatment in addition to its use as a preemergence or preplant incorporated treatment. Apply when soybeans are at least 12 inches tall and when broadleaf weeds do not exceed 3 inches and grasses and common ragweed do not exceed 1 inch in height. Direct spray to cover weeds with minimal or no contact to soybean plant. Crop injury may result if spray is higher than 2 inches on soybean stem. Apply 1/2 to 1 pt. per acre in 25 to 40 gallons water. Use the lower rate for seedling weeds and the higher rate for larger weeds. Application of 1/2 pt. per acre controls Florida beggarweed, carpetweed, cocklebur, dayflower, Mexican weed, prostrate and smooth pigweed, redroot pigweed (carelessweed), common purslane, sicklepod (cassia), velvetleaf and crabgrass (large, smooth); 3/4 to 1 pt. per acre controls hemp sesbania, prickly sida (spiny sida, teaweed); 1 pt. per acre controls jimsonweed, Pennsylvania smartweed, common ragweed and gives partial control of morningglory species, horsenettle, Florida purslane (Florida pusley), spotted spurge and wild poinsettia when less than 3 inches tall and broadleaf signalgrass up to 1 inch tall. Do not use on sand, nor on soils with less than 1/2% organic matter as crop injury may result. Add a non-ionic surfactant such as DuPont Surfactant WK or Ortho X-77 Spreader to the spray mixture to obtain better wetting of weed leaf surfaces. If needed, a second application may be made after 7 days.

Apply only with equipment designed for precise application of directed sprays such as oiling shoes or gauge wheels. Do not exceed nozzle pressure of 25 psi or use nozzle tips smaller than 8002 T-Jet (or equivalent). For broadcast application, use a single flood-type spray nozzle ("K" series or equivalent) per middle, mounted on an oiling shoe or gauge wheel. For band treatment, use two nozzles per row mounted on oiling shoes or gauge wheels, one on each side of a row. Do not apply under weather conditions which favor drift.

Treated vines may be grazed or fed to livestock 40 days after application.

Minimum or No-Tillage Soybeans: LEXONE 4L plus paraquat or Roundup may be used as a tank mixture for preemergence and postemergence control of many broadleaved weeds and grasses and for top kill or suppression of some perennials where soybeans will be planted directly into a preformed bed (stale seedbed), cover crop, or in previous crop residues such as corn or small grain stubble. For improved grass and volunteer small grain control, the addition of Lasso (or Lasso EC) or Surflan to the LEXONE 4L plus paraquat or Roundup tank mixture is recommended. For all combinations, except Surflan, treated vines may be grazed or fed to livestock 40 days after application.

Apply with ground equipment, during or after planting but before crop emerges. Use 20 to 60 gals. of water per acre for paraquat and use the higher gallonage for dense stubble or vegetation. Use 20 to 30 gals. of water per acre for Roundup combinations.

LEXONE 4L + Paraquat—Select the rate from the preemergence table for use of LEXONE 4L and thoroughly mix in spray tank first; then add paraquat as directed under **Paraquat Combinations.**

LEXONE 4L + Lasso (or Lasso EC) + Paraquat—Select the rate from the table for use of LEXONE 4L + Lasso tank mixture and thoroughly mix in spray tank first according to directions; then add paraquat as directed under **Paraquat Combinations.** Do not apply by air.

LEXONE 4L + Surflan + Paraquat—Select the rate from the table for use of LEXONE 4L + Surflan tank mixture and thoroughly mix in spray tank first according to directions; then add paraquat as directed under **Paraquat Combinations.**

Do not use treated vines for feed or forage.

Paraquat Combinations—Select the LEXONE 4L or LEXONE 4L combination treatment above and mix in the spray tank according to directions; then add 1 or 2 pts. paraquat (2 lb. active/gal.) per acre.Use the higher rate for weeds 4" to 6" tall. As the last ingredient, add 8 oz. DuPont Surfactant WK or Ortho X-77 Spreader per 100 gals. of spray mixture. Maintain constant agitation.

Roundup Combinations (LEXONE 4L + Lasso + Roundup): Select the rate from the table for use of LEXONE 4L + Lasso tank mixture and thoroughly mix in spray tank first according to directions; then add 1 to 2 qts. Roundup per acre to the mixture and maintain constant agitation. Use the higher rates for maximum top kill of vegetation. Do not apply by air.

SUGARCANE

Hawaii—Make a single broadcast application as follows: for aerial application (irrigated cane only), use at least 7 gals. water per acre. Spot treatment may also be made at the rate of 5 to 10 pts. per acre, but do not exceed 16 pts. total per acre per crop cycle. Do not apply within 17 months of harvest. Do not use treated foliage for feed or forage.

Time of Application	Pints Per Acre	
	Irrigated	Non-Irrigated
Preemergence—apply within 2 weeks after planting	8 to 12	4 to 8
OR		
Early Postemergence—apply over cane before weeds are 3" tall	8 to 12	4 to 8
OR		
Postemergence—apply prior to "close in" time and before weeds are 3" tall	4 to 8	4 to 8

In addition to the weeds listed elsewhere on this label, treatment (non-irrigated cane) controls ageratum, spiny amaranth, wild euphoria, fireweed, Flora's paintbrush, Richardia, garden spurge, graceful spurge, tarweed, guineagrass, plushgrass, ricegrass and wiregrass (goosegrass); in irrigated cane, the foregoing weeds plus spleen amaranth, haole koa, hioloa, hila hila, rattlepod and Alexander grass.

Louisiana and Texas—Apply 3 to 6 pts. per acre during the fall after planting but before cane emerges, or to the stubble after harvest. Make a second application in early spring prior to new cane growth. If necessary, a third application may be made in late spring at layby. Do not apply within 60 days of harvest. Do not use treated foliage for feed or forage.

In addition to the weeds listed elsewhere on this label, treatment controls curly dock, morningglory, marestail, sowthistle and winter oats.

TOMATOES

East of Rocky Mountains (Except Florida)—Apply LEXONE 4L as a preplant incorporated treatment alone or in a tank mixture with Treflan EC in transplanted tomatoes, or apply as a postemergence broadcast or directed spray to established tomatoes.

Apply uniformly to seeded established or transplanted tomatoes with a properly calibrated low-pressure (20 to 40 psi) fixed-boom sprayer with fan type nozzles. Use 10 to 40 gallons of water per acre with ground equipment for preplant incorporated treatment and 20 to 75 gallons for postemergence broadcast or directed spray application. Do not use air blast or other high pressure spray equipment for postemergence spray applications.

Preplant Incorporated—LEXONE 4L or LEXONE 4L Plus Treflan EC Tank Mix (Transplant Tomatoes): Apply ½ to 1 pt. per acre of LEXONE 4L alone or as a tank mix combination with Treflan EC as a broadcast spray immediately before transplanting. Use the higher rate for heavy weed pressure or for difficult-to-control weeds. Incorporate uniformly to a depth of 2 to 4 inches. Apply Treflan EC at rates of 1 to 2 pts. per acre on coarse soils; 1¼ to 2 pts. per acre on medium soils; and 1½ to 2 pts. per acre on fine soils. Consult Treflan EC label for specific use instructions and observe all cautions and limitations. When LEXONE 4L is tank mixed with Treflan EC, apply immediately after mixing; do not allow tank mixture to stand for prolonged periods of time.

Postemergence Broadcast Application (Established Tomatoes): Apply ½ to 1 pt. per acre. Use the higher rate for heavy weed pressure or for difficult-to-control weeds. Treat before weeds are 1 inch tall. One or more applications may be made per crop season; allow at least 14 days between applications. Do not apply more than 2 pts. total within a 35-day period per crop season. Do not treat seeded or transplanted tomatoes until plants have reached the 5 to 6 leaf stage or until transplants have recovered from transplant shock and new growth has started. Apply only if there have been at least 3 successive days of sunny weather prior to application, or crop injury may occur. Do not apply within 24 hours of treatment with other pesticides. Do not tank mix with other pesticides.

Postemergence Directed Application (Established Tomatoes): Apply 1 to 2 pts. per acre as a directed spray according to directions under "Broadcast-Postemergence Application." Do not allow spray to contact tomato foliage. This method of application is recommended for fields with heavy weed pressure or difficult-to-control weed species.

CROP ROTATION RECOMMENDATIONS: For rotation crops in fields treated with Treflan, follow instructions on product label.

See "Replanting and Crop Rotation Recommendations" for further details.

Note: Because tomato varieties vary in their resistance to LEXONE 4L, determine tolerance prior to adoption as a field practice to prevent possible crop injury. Do not use hot caps on tomatoes within 7 days before or at any time after application of LEXONE 4L. Crop injury or delayed maturity may result from postemergence or directed spray applications if tomatoes are growing under stress conditions caused by low fertility, disease, insects, drought, cool, wet or cloudy weather prior to application. Do not apply within 7 days of harvest.

NON-CROP USES

Apply 2 to 12 pts. LEXONE 4L per acre in 20 to 60 gallons of water. For best results, apply shortly before weed growth begins.

For control of established seedling weeds add:

1 to 2 pints Paraquat (2 lbs. active per gallon) per acre in 20 to 60 gallons of water per acre using the higher spray volume for weeds 4" to 6" tall. As the last ingredient, add 8 ounces Surfactant WK or Ortho X-77 Spreader per 100 gallons of spray mixture or add 1 to 2 quarts Roundup in 20 to 30 gallons of water per acre. Use the higher rates for maximum top kill of vegetation.

[1] Registered trademarks of Elanco Products Co., Treflan contains 4 lbs. trifluralin per gal., Surflan contains 75% oryzalin.

[2] Registered trademarks of Monsanto Co., Lasso contains 4 lbs. alachlor per gal. Roundup contains 4 lbs glyphosate per gal.

[3] Registered trademark of Ciba-Geigy Corp., Dual 8E contains 8 lbs. metolachlor per gal.

[4] Registered trademark of American Cyanamid Co., Prowl contains 4 lbs. pendimethalin per gal.

[5] Paraquat contains 2 lbs. active per gal., paraquat is a restricted use pesticide.

[6] Registered trademark of Amchem Products, Inc., Amiben contains 2 lbs. chloramben per gal.

SUPPLEMENTAL LABELING

LEXONE® 4L HERBICIDE
EPA Reg. No. 352-382

FOR COMMERCIAL IMPREGNATION AND APPLICATION ON DRY BULK FERTILIZER

DIRECTIONS FOR USE

It is a violation of federal law to use this product in a manner inconsistent with its labeling

DuPont LEXONE 4L Herbicide may be applied with dry bulk fertilizer for preemergence weed control; dry bulk fertilizer may be impregnated or coated with LEXONE 4L for application to established alfalfa and to soybeans. All recommendations, cautions and special precautions on this label must be followed along with state regulations relating to dry bulk fertilizer blending, impregnating and labeling.

HOW TO USE

Application: Uniform application of LEXONE® 4L which has been impregnated on dry fertilizer is essential for satisfactory weed control. Accurate calibration of fertilizer application equipment is essential for uniform distribution to the soil surface. The recommended method of application is to apply $\frac{1}{2}$ the recommended rate and overlap 50 percent to double apply by splitting the middles to obtain the best distribution pattern.

If fertilizer materials are excessively dusty, use diesel oil or other suitable additive to reduce dust prior to impregnation as dusty fertilizer will result in poor distribution during application. Crop injury and/or poor weed control may occur where the impregnated fertilizer is not uniformly applied.

Incorporation and Combination Uses: When LEXONE® 4L is to be used in combination with another herbicide, follow directions on this label for combinations, rates, crops, incorporation and special precautions.

Impregnation: To impregnate, use a system consisting of a belt, conveyor or closed drum which is used for dry bulk fertilizer blending. Any commonly used fertilizer can be impregnated with LEXONE® 4L except ammonium nitrate or fertilizers containing ammonium nitrate, potassium nitrate or sodium nitrate. Do not use on limestone.

Apply using a minimum of 200 pounds dry bulk fertilizer per acre and up to a maximum of 450 pounds per acre. To impregnate or coat the dry bulk fertilizer with LEXONE® 4L, the delivery nozzles must be directed to deliver a fine spray toward the fertilizer for thorough coverage and avoiding spray contact with mixing equipment. Uniform impregnation of LEXONE® 4L to dry bulk fertilizer will vary and if the adsorptivity is not adequate, the use of an adsorptive powder may be required to produce a dry free-flowing mixture. Microcel E (JohnsManville Product Corporation) is the recommended absorbent powder. LEXONE® 4L should be premixed with water to form a sprayable mixture. When another herbicide is used with LEXONE® 4L, mix and impregnate immediately.

Apply immediately after impregnation unless experience has shown that impregnated fertilizer can be stored without becoming lumpy and difficult to spread.

Rates: Select the rate of LEXONE® 4L to apply per acre from the appropriate section of this label and refer to the rate table to determine the amount of LEXONE® 4L which is to be impregnated on a ton of dry bulk fertilizer based on the amount of fertilizer which will be distributed on one acre.

RATE CHART FOR IMPREGNATING FERTILIZER WITH LEXONE® 4L

Fertilizer Rate per Acre	LEXONE® 4L Rate Per Acre			
	$\frac{1}{2}$ Pint	1 Pint	$1\frac{1}{2}$ Pint	2 Pints
200 pounds	5 pts/ton	10 pts/ton	15 pts/ton	20 pts/ton
250 pounds	4 pts/ton	8 pts/ton	12 pts/ton	16 pts/ton
300 pounds	$3\frac{1}{3}$ pts/ton	$6\frac{2}{3}$ pts/ton	10 pts/ton	$13\frac{1}{3}$ pts/ton
350 pounds	$3\frac{5}{6}$ pts/ton	$5\frac{3}{4}$ pts/ton	$8\frac{2}{3}\frac{2}{3}$ pts/ton	$11\frac{1}{2}$ pts/ton
400 pounds	$2\frac{1}{2}$ pts/ton	5 pts/ton	$7\frac{1}{2}$ pts/ton	10 pts/ton
450 pounds	$2\frac{1}{4}$ pts/ton	$4\frac{1}{2}$ pts/ton	$6\frac{2}{3}$ pts/ton	9 pts/ton

For rates other than those listed above, use the following formula to calculate the amount of LEXONE® 4L to be impregnated per ton of dry fertilizer.

$$\text{Pounds Lexone® 4L Per Acre} \times \frac{2000}{\text{lbs. Fertilizer Per Acre}} = \text{Pounds LEXONE® 4L Per Ton of Fertilizer}$$

IMPORTANT
BEFORE USING LEXONE® 4L, READ AND CAREFULLY OBSERVE THE CAUTIONARY STATEMENTS AND ALL OTHER INFORMATION APPEARING ON THE PRODUCT LABEL.

This bulletin contains new or supplemental instructions for use of this product which may not appear on package labels. Follow the instructions carefully.

This labeling must be in the possession of the user at the time of pesticide application.

SUPPLEMENTAL LABELING

LEXONE® 4L EPA Reg. No. 352-382
KARMEX® EPA Reg. No. 352-247

LEXONE® 4L HERBICIDE AND COMBINATIONS WITH KARMEX® FOR USE ON ESTABLISHED ASPARAGUS IN THE NORTHWEST

DIRECTIONS FOR USE

It is a violation of federal law to use this product in a manner inconsistent with its labeling.

DuPont LEXONE 4L Herbicide is recommended alone or in combinations with Karmex Herbicide for use on established asparagus in the Northwest. For weeds controlled, refer to package labels for LEXONE 4L and Karmex.

HOW TO USE

Use of the following:

LEXONE 4L Preemergence Treatment Only: Make a single preemergence application of 2 to 4 pts. LEXONE 4L per acre in early spring prior to spear emergence. Use the lower rate on coarse textured soils (low in clay or organic matter). If field is to be disked, apply LEXONE 4L after disking but before crop emerges.

OR

LEXONE 4L Preemergence and Post-Harvest Treatment: Make a single preemergence application of 1 to 2 pts. LEXONE 4L per acre in early spring prior to spear emergence. Following harvest but before fern emerges, apply 1 to 3 pts. LEXONE 4L per acre. Do not exceed 4 pts. LEXONE 4L per acre per growing season.

OR

LEXONE 4L Preemergence and LEXONE 4L Plus Karmex Post Harvest Treatment: Make a single preemergence application of 1 to 2 pts. LEXONE 4L per acre in early spring to spear emergence. Following harvest but before fern emerges, apply a tank-mix combination of 1 to $1\frac{1}{2}$ pts. LEXONE 4L plus 1 to 2 lbs. Karmex per acre.

Irrigated Asparagus: After the initial application of LEXONE 4L, do not exceed more than $\frac{1}{2}''$ of water for the first irrigation. Additional irrigations may be normal programs.

Application: Apply in 3–10 gallons per acre by air, 10–40 gallons per acre by ground equipment.

Note: Do not use on newly seeded nor on young plants during the first growing season after setting crowns. Do not apply within 14 days of harvest. If field is to be disked, apply LEXONE 4L after disking but before the crop emerges. Do not apply post-harvest application until after the last harvest of spears.

IMPORTANT
BEFORE USING LEXONE 4L IN COMBINATIONS, READ AND CAREFULLY OBSERVE THE CAUTIONARY STATEMENTS AND ALL OTHER INFORMATION APPEARING ON THE PRODUCT LABELS.

This bulletin contains new or supplemental instructions for use of these products which may not appear on these package labels. Follow the instructions carefully.

This labeling must be in the possession of the user at the time of pesticide application.

SUPPLEMENTAL LABELING

LEXONE® 4L HERBICIDE
EPA Reg. No. 352-382

FOR USE ON FALLOW FOLLOWED BY SUGARCANE IN THE STATE OF LOUISIANA

DIRECTIONS FOR USE

It is a violation of federal law to use this product in a manner inconsistent with its labeling.

DuPont LEXONE 4L Herbicide is recommended in Louisiana for use on fallow ground to be followed by sugarcane. When used as directed, LEXONE 4L provides a short term control of seedling johnsongrass and other problem weeds as listed on the package label. LEXONE 4L may also be used on the planted crop for weed control.

HOW TO USE

Fallow Use

Apply 2 to 3 pts. LEXONE 4L per acre. Use the lower rate if applied less than 30 days before planting. Application should be made after the rows are built and ground should not be disturbed until planting or loss of control will result.

To aid in control of johnsongrass and other problem weeds, stubble sugarcane should be destroyed after harvest or in early spring. Land should be disked frequently in the spring to control rhizome johnsongrass. In late May to early June, rows should be built. The soil surface should be free of trash and clods. Application of LEXONE 4L should be made at this time. Rhizome johnsongrass will not be controlled by LEXONE 4L; therefore, spot treatment with an appropriate postemergence herbicide will be necessary.

Crop Use

Apply 3 to 6 lbs. LEXONE 4L per acre during the fall after planting but before cane emerges, or to the stubble after harvest. Make a second application in early spring prior to new cane growth.

Note: Do not apply within 60 days of harvest. Do not use treated foliage for feed or forage.

IMPORTANT
BEFORE USING LEXONE 4L, READ AND CAREFULLY OBSERVE THE CAUTIONARY STATEMENTS AND ALL OTHER INFORMATION ON THE PRODUCT LABEL.

This bulletin contains new or supplemental directions for use of this product which may not appear on the package label. Follow the instructions carefully.

This labeling must be in the possession of the user at the time of pesticide application.

SUPPLEMENTAL LABELING

LEXONE® 4L HERBICIDE
EPA Reg. No. 352-382

FOR USE ON WINTER WHEAT, BARLEY, AND FALLOW

DIRECTIONS FOR USE

It is a violation of federal law to use this product in a manner inconsistent with its labeling.

DuPont LEXONE 4L Herbicide is recommended for weed control:

- On the winter wheat varieties TAM W101, TAM 105, Newton, Hawk, and Eagle in the states of Kansas, Oklahoma, and Texas.
- On winter wheat, winter barley, and spring barley, (except varieties Morex, Glenn, or Moravian 3) in the states of Idaho, Oregon, Utah, and Washington.
- In reduced tillage fallow where wheat will be the following crop. For use in the states of Colorado, Idaho, Kansas, Montana, Nebraska, North Dakota, Oklahoma, Oregon, South Dakota, Texas, Utah, Washington, and Wyoming.

Treatment controls or partially controls the following broadleaf weeds:

Blue mustard[1]	Filaree	Pigweed (prostrate, redroot)
Bur buttercup	Flixweed[1]	
Chickweed (common mouse-ear, jagged)	Gromwell[1]	Pineappleweed
	Henbit	Prickly lettuce
Conical catchfly	Jacobs ladder	Russion thistle[1]
Corn cockle	Jim Hill mustard	Shepherdspurse
Dog fennel (Mayweed)	Knotweed	Speedwell (ivyleaf)
False flax	Kochia	Tansy mustard[1]
Lambsquarter	Wild buckwheat[1]	Tarweed (fiddleneck)
Fanweed	Miners lettuce	Treacle mustard
		Wild mustard

Treatment controls or partially controls the following grasses:

Annual reygrass (except Kansas, Oklahoma, Texas)

Downy brome[1, 2]	Pacific meadow foxtail (Blackgrass)[1]
Cheatgrass[1, 2]	
Green foxtail[1]	Windgrass[1]
	Yellow foxtail[1]

[1] Weeds and grasses partially controlled. Partial control is an observable reduction in weed competition (reduced weed population and/or vigor) as compared to an untreated area. Degree of partial control can vary with soil type, environmental factors, and size of weeds of application.

[2] Best results with late Fall treatments.

HOW TO USE ON DRYLAND WHEAT & BARLEY

Apply postemergence to the crop when wheat/barley is well tillered (at least 3 tillers) and has developed 2″ secondary roots throughout the entire field. A well developed secondary root system and good crop vigor is essential to minimize injury to wheat/barley. Application must be made prior to jointing as crop injury may result.

For best weed control results:

- Apply LEXONE 4L before broadleaf weeds emerge or shortly after emergence and before weeds are 2″ tall or across.
- Use a Fall treatment for winter annuals or a spring treatment for summer annual broadleaf weeds. Avoid making treatments in mid-winter in areas where severe winter weather can cause crop stress.
- Apply LEXONE 4L before grasses emerge or shortly after emergence and before the 3 to 5 leaf stage and use the higher dosage rates listed.

Rainfall (½″) within 2 to 3 weeks after application is needed to move LEXONE 4L into the weed root zone. Lack of adequate rainfall after application will result in poor control.

USE RATES

Make a single broadcast application postemergence to the crop as follows:

Idaho, Oregon, Utah and Washington.

Soil Texture	Pints of LEXONE 4L Per Acre		
	Less than 1% Organic Matter	1 to 2% Organic Matter	More than 2% Organic Matter
Coarse Loamy sand, Sandy loam	0.50 to 0.60	0.50 to 0.60	0.50 to 0.60
Medium Loam, Silt loam, Silt, Sandy clay, Sandy clay loam	0.50 to 0.60	0.60 to 0.75	0.60 to 0.90

Fine Silty clay, Silty clay loam, Clay, Clay loam	0.50 to 0.75	0.75 to 1.0	0.75 to 1.0

Do not use on gravelly or thinly covered or exposed subsoil areas (clay knobs) as crop injury may occur.

Kansas, Oklahoma, Texas

Soil Type	Pints of LEXONE 4L Per Acre			
	Fall Application		Spring Application	
	1 to 2% OM	More than 2% OM	1 to 2% OM	More than 2% OM
Coarse Loamy sand, Sandy loam	Do Not Use	0.50	Do Not Use	0.50 to 0.75
Medium Loam, Silt loam, Silt, Sandy clay, Sandy clay loam	0.50 to 0.75	0.75	0.75 to 1.0	0.75 to 1.0
Fine Silty clay, Silty clay loam, Clay loam	0.75	0.75 to 1.0	0.75 to 1.0	1.0

Do not use on gravelly or thinly covered or exposed subsoil areas (clay knobs) as crop injury may occur.

Use the lower rates on coarse textured soils and the higher rates on medium or fine textured soils having more than 1% organic matter. For grass control, use the highest rate listed for your soil type.

IRRIGATED WHEAT/BARLEY

Rainfall or sprinkler irrigation is needed to activate LEXONE 4L. If adequate rainfall (½" within 2 to 3 weeks after application) does not occur, then activate via sprinkler irrigation. Do not apply more than ½" of water for the first irrigation after application. The maximum amount for each additional irrigation should not exceed 1".

SPECIFIC HERBICIDE COMBINATIONS

Winter Wheat, Spring and Winter Barley

LEXONE 4L may be tank mixed with Igran 80W for use on winter wheat and winter barley. Do not use more than 1.0 lbs./acre Igran 80W in tank mix. On winter wheat, spring wheat, and spring barley, LEXONE 4L may be tank mixed with Glean, 2,4-D, Banvel, MCPA, Bronate, Buctril, Brominal Plus, or Brominal. Refer to the other product labels registered for this use for additional precautions, directions, rates, and weeds controlled.

LEXONE 4L and Igran 80W (Winter Wheat, Winter Barley)—Pacific NW and Utah

For added control of downy brome, apply LEXONE 4L at 0.50 pints per acre plus Igran 80W at 0.75 to 1.00 pounds per acre. Follow LEXONE 4L timing of application recommendation. Do not use on spring barley or irrigated wheat/barley. Refer to Igran 80W label for additional precautions and instructions.

How to Use for Reduced Tillage Fallow

Apply 0.67 to 1.5 pints LEXONE 4L per acre as an after harvest treatment prior to subsequent planting of winter wheat. Use the lower rate on coarse texture soils and the higher rate on medium to fine textured soil.

LEXONE 4L can be applied in the Fall after harvest at 1.0 to 1.5 pts./acre or in the spring at 0.67 to 1.0 pts./acre as new weed growth begins.

If large weeds (larger than 2" tall or across) are present, or weeds other than those listed are present or anticipated, use a suitable registered companion herbicide such as Glean, Paraquat, Roundup, or Chem-Hoe 135. Refer to labeling for these products for the additional use instructions and precautions.

When using LEXONE 4L and Roundup tank mix spray, volume should be 10 gpa.

Where spring application is appropriate in Colorado, Kansas, Montana, Nebraska, North Dakota, Oklahoma, South Dakota, Texas, and Wyoming improved control of downy brome, Russian thistle, and tansy mustard, tank mix 0.50 to 1.0 pts. of LEXONE 4L with 2 to 2.5 lbs. Igran 80 W per acre.

Do not plant spring seeded cereals following LEXONE 4L Fall fallow applications.

Do not apply LEXONE 4L in the spring if LEXONE 4L was used the preceding Fall or Winter.

Recrop Intervals

Do not plant winter wheat within 6 months of treatment except in Colorado, Kansas, Montana, Nebraska, North Dakota, Oklahoma, South Dakota, Texas, and Wyoming where winter wheat may be planted 4 months after treatment of 0.50 to 1.0 pts. LEXONE 4L per acre.

Rotational Crop Interval

Do not plant crops (other than wheat as listed above) within 10 months of treatment.

SPRAY PREPARATION

Mix the proper amount of LEXONE 4L into the necessary volume of water in the spray tank with the agitator running. Other herbicides or fertilizer solutions should be added last.

Crops (as listed)

LEXONE 4L may be mixed with liquid fertilizer. However, addition of liquid fertilizer to LEXONE 4L alone or in a three-way tank mix with other herbicides increases the possibility of crop injury. Do not use surfactant when using LEXONE 4L in combination with liquid fertilizer. For application of LEXONE 4L in liquid fertilizer, where the liquid fertilizer is the carrier, slurry the LEXONE 4L in water, then thoroughly mix the slurry into the liquid fertilizer.

Fallow: LEXONE 4L may be used in any combination listed above.

When using tank mixes with LEXONE 4L, small quantities should be tested for compatibility prior to use.

The use of mixtures with LEXONE 4L increases the possibilities of injury or temporary yellowing to the crop especially if 2" secondary roots are not developed throughout the field.

APPLICATION

Apply LEXONE 4L using properly calibrated air or ground equipment. Uniform coverage is essential in order to obtain optimum weed control performance and minimize the potential for crop injury. Use a minimum spray volume of 3 gallons per acre by air or ground.

As injury to crop may result, avoid overlapping of spray swath. Shut off spray booms while starting, turning, slowing, or stopping. Continuous agitation is needed to keep LEXONE 4L in suspension.

NOTE: Do not apply under conditions where excessive drift may occur.

Do not apply LEXONE 4L alone or in combination to stressed plants.

Do not treat fields where seed had been planted less than 1" deep.

Do not treat fields that have been severely grazed.

Do not graze treated fields for 14 days following treatment.

Do not apply more than once per crop or fallow period.

Because spring barley varieties may vary in their tolerance to LEXONE 4L limit first use to a small area to determine varietal response prior to adoption as a field practice.

Do not use on soils having a calcareous surface area or a pH of 7.5 or higher as crop injury may result.

Do not apply when soil is frozen or crop is in winter dormacy.

Do not apply when soil is saturated with moisture.

Temporary yellowing or injury of the crop may occur, especially if frost or snow occurs within a few days of application. Severe winter stress, heaving, exposed root systems, disease, or insect damage preceeding or following application may increase crop injury. Winter wheat/barley varieties differ in their degree of winter hardiness; severe injury may result from treating non-winter hardy varieties that will not withstand the combination of winter and herbicide stress. If heavy rains occur soon after application, injury to the crop may result. Extended dry weather following treatment may reduce weed control effectiveness.

IMPORTANT
BEFORE USING LEXONE 4L, IGRAN 80W*, OR OTHER HERBICIDES IN COMBINATION READ AND CAREFULLY OBSERVE THE CAUTIONARY STATEMENTS AND ALL OTHER INFORMATION ON THE PRODUCT LABELS.

This bulletin contains new or supplemental directions for use of these products which may not appear on the package labels. Follow the instructions carefully.

This labeling must be in the possession of the user at the time of pesticide application.

* Registered trademark of Ciba Geigy Corporation, "Igran" 80W contains terbutryn.

SUPPLEMENTAL LABELING

LEXONE® 4L HERBICIDE

EPA Reg. No. 352-382

ALONE; AS A TANK MIX PREPLANT INCORPORATED APPLICATION WITH TREFLAN®[1], TOLBAN®[2], PROWL®[3], OR DUAL®[2]; AS A SEQUENTIAL PREEMERGENCE APPLICATION FOLLOWING TREFLAN®, TOLBAN®, PROWL®, OR DUAL®; AS A TANK MIX PREEMERGENCE APPLICATION WITH LASSO®[4], DUAL®, PROWL®, SURFLAN®[1], OR AMIBEN®[5], FOR CONTROL OF SICKLEPOD IN SOYBEANS IN COARSE-TEXTURED, LOW ORGANIC MATTER, COASTAL PLAINS SOILS IN THE SOUTHEAST AND MIDSOUTH

DIRECTIONS FOR USE
It is a violation of state and federal law to use this product in a manner inconsistent with its labeling.

HOW TO USE
Make a single application of Lexone® 4L alone; as a tank mix preplant incorporated application with Treflan®, Tolban®, Prowl®, or Dual®; as a sequential preemergence application following Treflan®, Tolban®, Prowl®, or Dual®, preplant incorporated; or as a tank mix preemergence application with Lasso®, Dual®, Prowl®, Surflan® or Amiben® as a band or broadcast spray prior to soybean emergence.

Refer to the package label for the appropriate directions for use, recommended applications, and special precautions for use alone and in combination with other herbicides prior to use on these soils.

Apply at the following broadcast rates per acre; for band treatment, use proportionately less. For example, for 14" band on 42" rows use 1/3 of the broadcast rate.

Recommended Applications
Lexone® 4L Alone

Soil Texture	Organic Matter	Pints Lexone® 4L/Acre
Coarse		
Loamy sand, sandy loam	0.5-2.0%	1/2-3/4 pints

Tank Mixes with Other Herbicides
Coarse

Soil Texture	Organic Matter		Tank Mix Application Rates	
Loamy sand, sandy loam	0.5-2.0%			
			Lexone® 4L	1/2-3/4 pints
		plus	Treflan®	1 pint
		or	Tolban®	1 pint
		or	Lasso®	1/2-2 quarts
		or	Surflan® 75WP	2/3-1 pound
		or	Amiben®	3-4 quarts
		or	Prowl®	1 1/2 pints
		or	Dual® 8E	1 1/4 pints

Do not use on sand nor on any soils with less than 1/2% organic matter as crop injury may result.

Note: Do not use on Altona, Coker 102, Coker 156, Govan, Semmes, Tracy or Vansoy Terra Vig 505 and 606, AP 71 and NKS 1884 soybeans as these varieties are sensitive to Lexone® 4L and injury to crop may result. Injury to soybeans may occur if Lexone® 4L is used on soils having a calcareous surface layer or a pH of 7.5 of higher, or if used in conjunction with soil-applied organic phosphate pesticides such as Dasanit®, Disyston®, Mocap®, Nemacur® or Thimet®. Do not use treated vines for feed or forage.

If initial seedling fails to produce a stand, treated fields may be replanted to soybeans; do not rework soil. Do not retreat field with second application as injury to crop may result. Do not replant treated areas to any crop other than soybeans within 4 months after treatment as injury to subsequent crop may result.

IMPORTANT
BEFORE USING LEXONE® 4L READ AND CAREFULLY OBSERVE THE CAUTIONARY STATEMENTS AND ALL OTHER INFORMATION APPEARING ON THE PRODUCT LABEL.

This bulletin contains new or supplemental instructions for use of this product which may not appear on the package label. Follow the instructions carefully.

This labeling must be in the possession of the user at the time of pesticide application.

1 Registered trademark of Elanco Products Co., Treflan® contains 4 pounds trifluralin per gallon; Surflan® contains 75% oryzalin.

2 Registered trademark of Ciba-Geigy Corp., Tolban® contains 4 pounds profluralin per gallon; Dual® contains 8 pounds metolachlor per gallon.

3 Registered trademark of American Cyanamid Co., Prowl® contains 4 pounds pendimethalin per gallon.

4 Registered trademark of Monsanto Co., Lasso® contains 4 pounds alachlor per gallon.

5 Registered trademark of Amchem Products, Inc., Amiben® contains 2 pounds chloramben per gallon.

ADDITIONAL EPA SUPPLEMENTAL LABELING FOR LEXONE® 4L WEED KILLER

Crop	Title	Description	Date of Label*
Soybeans	LEXONE 4L Weed Killer in combination with other Herbicides for Weed Control in Minimum or No-Tillage Soybeans	Use with other preemergence herbicides plus paraquat or Roundup for U.S.	11/9/82
Soybeans	LEXONE 4L Weed Killer Plus Treflan Plus Amiben Tank Mix Preplant Incorporated for Weed Control in Soybeans	U.S.	8/12/82
Soybeans	LEXONE 4L Weed Killer Plus Dual 8E Plus Amiben Tank Mix Preplant Incorporated for Weed Control in Soybeans	U.S.	8/12/82
Soybeans	LEXONE 4L Weed Killer Plus Prowl Plus Amiben Tank Mix Preplant Incorporated or Preemergence for Weed Control in Soybeans in the Midwest	Midwest	9/12/82

Soybeans	LEXONE 4L Weed Killer Plus Lasso Plus Amiben Tank Mix Preplant Incorporated or Preemergence for Weed Control in Soybeans	U.S.	10/3/85
Soybeans	LEXONE 4L Weed Killer Tank Mixture with Lasso for Preemergence Use in Soybeans in the State of Ohio	Ohio for control of black nightshade	2/22/82
Tomatoes	LEXONE 4L Weed Killer for Weed Control in Established Tomatoes	Directed postemergence for U.S. (except FL)	7/21/82

* Supplemental label furnished on request.

SPECIAL LOCAL NEED REGISTRATIONS [24(c)] FOR LEXONE® 4L HERBICIDE

Crop	State	EPA SLN No.	Remarks
Alfalfa	California	CA-800132	Use only north of US Interstate 80 Dormant application Spring or fall
Barley	Idaho	ID-810031	Winter and spring barley postemergence
	Montana	MT-810001	Spring barley, postemergence
	Oregon	OR-810093	Winter and spring barley postemergence
	Utah	UT-810023	Winter and spring barley postemergence
	Washington	WA-810098	Winter and spring barley postemergence
Bermudagrass	Hawaii	HI-810004	Golf Course turf
Peas and Lentils	Idaho Oregon Washington	ID-800014 OR-80027 WA-800013	Postplant incorporated. Preemergence weed control east of Cascades.
	Idaho	ID-810016	Postemergence weed control.
	Washington	WA-810027	Postemergence weed control east of Cascades.
Potatoes	Florida Idaho N. Carolina	FL-780003 ID-780002 NC-780001	Preemergence and/or postemergence weed control.
	Idaho	ID-800054	Sprinkler irrigation injection—Change irrigation injection time.
Soybeans	Arkansas	AR-830002	Post directed application to 8″ soybeans
	N. Carolina	NC-780003	LEXONE 4L alone or with Lasso for high organic soils.
	Georgia	GA-810022	Preemergence weed control with center pivot irrigation injection.
	Louisiana	LA-810031	Post directed application. Tank mix with Butyrac 200 or Butoxone.
Tomatoes	California	CA-800128	Fallow weed control. Fall bedded tomato field.

* Copies of individual labels available upon request to nearest DuPont Agricultural Chemicals Office. For distribution and use only within the State specified.

LOROX®

HERBICIDE

ACTIVE INGREDIENT:
Linuron [3-(3,4-dichlorophenyl)-1-methoxy-1-methylurea] 50%
INERT INGREDIENTS 50%

TOTAL 100%

EPA Reg. No. 352-270

Keep Out of Reach of Children
CAUTION

PRECAUTIONARY STATEMENTS

HAZARDS TO HUMANS

HARMFUL IF SWALLOWED. MAY IRRITATE EYES, NOSE, THROAT, AND SKIN.
Avoid breathing dust or spray mist. Avoid contact with skin, eyes, and clothing.

ENVIRONMENTAL HAZARDS
Do not apply directly to water or wetlands. Do not contaminate water by cleaning of equipment or disposal of wastes.

IMPORTANT

Do not apply directly to water or wetlands. Do not contaminate water by cleaning of equipment or disposal of wastes.

IMPORTANT—When mixing/loading and spraying wear a wide-brimmed hat, mid-forearm to elbow length chemical resistant gloves, and one-piece overalls which have long sleeves and long pants. Overalls should be constructed of finely woven fabric as specified in the USDA/EPA Guide for Commercial Applicators.

Any article of clothing worn while applying product must be cleaned before reusing. Clothing which has been drenched or has otherwise absorbed concentrated pesticide must be disposed in a sanitary landfill, or incinerated, or if allowed by State and local authorities, by burning.

Instead of clothing and equipment specified above, the applicator can use an enclosed tractor cab which provides a filtered air supply (as described by Taschenberg and Bourke, 1975).

Injury to or loss of desirable trees or other plants may result from failure to observe the following:

Do not apply (except as recommended for crop use), or drain or flush equipment on or near desirable trees or other plants, or on areas where their roots may extend, or in locations where the chemical may be washed or moved into contact with their roots. Do not use on home plantings of trees, shrubs or herbaceous plants, nor on lawns, walks, driveways, tennis courts, or similar areas. Prevent drift of dry power or spray to desirable plants. Do not contaminate any body of water. Keep from contact with the fertilizers, insecticides, fungicides and seeds.

Do not apply aerially.

Thoroughly clean all traces of "Lorox" from application equipment immediately after use. Flush tank, pump, hoses, and boom with several changes of water after removing nozzle tips and screens (clean these parts separately).

NOTICE OF WARRANTY

Du Pont warrants that this product conforms to the chemical description on the label thereof and is reasonably fit for purposes stated on such label only when used in accordance with directions under normal use conditions. It is impossible to eliminate all risks inherently associated with use of this product. Crop injury, ineffectiveness or other unintended consequences may result because of such factors as weather conditions, presence of other materials, or the manner of use or application, all of which are beyond the control of Du Pont. In no case shall Du Pont be liable for consequential, special or indirect damages resulting from the use or handling of this product. All such risks shall be assumed by the Buyer. DU PONT

MAKES NO WARRANTIES OF MERCHANTABILITY OR FITNESS FOR A PARTICULAR PURPOSE NOR ANY OTHER EXPRESS OR IMPLIED WARRANTY EXCEPT AS STATED ABOVE.

STORAGE AND DISPOSAL

STORAGE

Store product in original container only, away from other pesticides, fertilizer, food, or feed.

DISPOSAL

Do not contaminate water, food, or feed by storage or disposal. Wastes resulting from the use of this product may be disposed of on site or at approved waste disposal facility.

Completely empty bag into application equipment. Then dispose of empty bag in a sanitary landfill or by incineration, or, if allowed by State and local authorities, by burning. If burned, stay out of smoke.

GENERAL INFORMATION

Du Pont LOROX Herbicide is a wettable powder to be mixed in water and applied as a spray for selective control of weeds in certain crops and for non-selective weed control on non-cropland areas. It is non-corrosive to equipment, non-flammable and non-volatile.

LOROX may be applied to soil prior to emergence of weeds to control susceptible weed seedlings for an extended period of time; the degree of control and duration of effect will vary with the amount of chemical applied, soil texture, rainfall and other conditions. Soils high in clay or organic matter require higher dosages than soil low in clay or organic matter to obtain equivalent herbicide performance. Moisture is required to activate the chemical; best results occur if rainfall (or irrigation) occurs within 2 weeks of application. In the Columbia River Basin, use LOROX only if crop is sprinkler irrigated.

LOROX may also be used to control emerged weeds. Results vary with rate applied and environmental conditions; best results are obtained on succulent weeds growing under conditions of high humidity and temperatures of 70°F or higher. Addition of a surfactant such as Du Pont Surfactant WK to the spray (where recommended) increases contact effects of LOROX.

LOROX is recommended for use in conjunction with certain other herbicides for treatment of field corn, sorghum and soybeans; observe all cautions and limitations on labeling of all products used in mixtures.

Since the effect of LOROX varies with soils, crop varieties, uniformity of application, and environmental conditions, it is suggested that growers limit their first use to small areas.

DIRECTIONS FOR USE

It is a violation of federal law to use this product in a manner inconsistent with its labeling.

LOROX Herbicide should be used only in accordance with recommendations on this label, or in separate published Du Pont recommendations available through local dealers. Do not use on any crop in Kern County, California, except carrots when a permit has been obtained from the Agricultural Commission.

Du Pont will not be responsible for losses or damages resulting from use of this product in any manner not specifically recommended by Du Pont. User assumes all risks associated with such nonrecommended use.

SELECTIVE USE IN CROPS

PREEMERGENCE USE (Germinating Weeds): LOROX, at recommended rates, controls weeds such as:

Broadleaves

Florida beggarweed	Mustard
Carpetweed	Nettleleaf goosefoot
Chickweed	Pigweed
Common dayflower	Purslane
Florida purslane (Florida pusley)	Wild radish
Galinsoga	Common ragweed
Lambsquarters	Smartweed

Grasses

Barnyardgrass (watergrass)	Foxtail (including giant)
Canarygrass	Goosegrass
Crabgrass	Fall panicum

Partial Control

Cocklebur
Annual morningglory
Prickly sida (teaweed)
Sicklepod
Velvetleaf (buttonweed)

Note: Consult Amiben[1], atrazine, Bronco[3], Dual[2] 8E, Lasso[3], Paraquat[4], propazine, Prowl[5], Ramrod[3], Roundup[3], Surflan[6] 75W, Surflan[6] AS, Tolban[2], and Treflan[6] labels for additional weeds controlled when LOROX is used in conjunction with these herbicides.

LOROX will not control established perennials such as bermudagrass, Canada thistle, field bindweed, johnsongrass and purple nutsedge. The lower dosage rates are effective on the coarser soils and the higher rates on finer soils and on the more resistant seedling weeds. Sufficient moisture (½″ to 1″ on moist soils; 1″ to 2″ on dry soils) in the form of rainfall or sprinkler irrigation is necessary after treatment to carry the chemical into the root zone of germinating weeds; best results are obtained when this occurs within two weeks after application. If heavy rainfall occurs soon after application, injury to crop may result.

A good seed bed must be prepared before application of LOROX as crop injury may result if application is made to ground which is cloddy or compacted resulting in improperly planted seed. Plant seed to depth specified. Surface of the soil should not be cultivated or disturbed after application of LOROX and before emergence of the crop as weed control may be reduced and crop injury may result. However, if moisture is insufficient to activate the herbicide, a shallow cultivation (rotary hoe preferred) should be made after emergence of row crops while weeds are small enough to be controlled by mechanical means.

POSTEMERGENCE USE (Emerged Seedling Weeds): LOROX, at recommended rates, controls weeds such as:

Broadleaves

Annual morningglory	Lambsquarters
Carpetweed	Mustard
Cocklebur	Nettleleaf goosefoot
Common dayflower	Pigweed
Common ragweed	Prickly sida (teaweed)
Dog fennel	Purslane
Fiddleneck (amsinckia)	Sesbania
Florida beggarweed	Sicklepod
Florida purslane (Florida pusley)	Smartweed
Groundsel	Velvetleaf (buttonweed)
Knawel	Wild buckwheat

Grasses

Annual ryegrass	Fall panicum
Barnyardgrass (watergrass)	Foxtail (including giant)
Broadleaf signalgrass	Goosegrass
Canarygrass	Rattail fescue
Crabgrass	Texas panicum

Note: Consult bromoxynil, Butoxone[7], Butyrac[1] 200 and Premerge 3[8] labels for additional weeds controlled when LOROX is used in conjunction with these herbicides. Best results are obtained under conditions of high humidity and temperatures over

70°F. Control of emerged weeds under drought stress is usually impractical.

EQUIPMENT—SPRAY VOLUMES AND PRESSURES: Use a tractor-mounted fixed-boom power sprayer properly calibrated to a constant speed and rate of delivery. Openings in screens should be equal to or larger than 50 mesh. Continuous agitation in the spray tank is required to keep the material in suspension. Agitate by mechanical or hydraulic means; if by-pass or return line is used, it should terminate at bottom of tank to minimize foaming. Avoid overlapping, and shut off spray booms while starting, turning, slowing or stopping, or injury to the crop may result.

For preemergence application, use 20 to 60 gals. per acre and spray pressure of 30 to 40 psi.

For postemergence application, use sufficient volume (min. 20 gals. per acre) for thorough coverage of weed foliage; use spray pressure of 20 to 25 psi to keep spray drift to a minimum.

SPRAY PREPARATION: Fill tank $\frac{1}{4}$ full with water. Start agitation system, add LOROX, and continue adding water. Add separately each additional component of any tank mix while adding water. If a surfactant is used, add it last when the tank is nearly full. Continue agitation throughout. If poor mixing should occur with any component, premix the component with two parts water before adding to the spray tank.

Fertilizer Spray Mixtures—For preemergence application, nonpressure nitrogen or fertilizer solution may be used in the spray mixture unless otherwise directed. Small quantities should be tested for compatibility by the following procedure before full scale mixing:

1. Put 1 pint of fertilizer solution in a quart jar.
2. Mix 2 teaspoons LOROX with 2 tablespoons of water; mix thoroughly and add to fertilizer solution.
3. Close jar and shake well.
4. If other herbicides are to be used in the mixture, premix 2 teaspoons of wettable powders or 1 teaspoon of liquids with 2 tablespoons of water; add to LOROX fertilizer mixture.
5. Close jar and shake well.
6. Watch mixture for several seconds; check again in 30 minutes.
7. If mixture does not separate, foam, gel, or become lumpy, it may be used.
8. Mixing ability may be improved by adding compatibility agents such as Kalo Laboratories Complex or Witco Chemicals Sponto 168D. Follow directions on container.

If the mixture is compatible, prepare spray by adding the fertilizer solution to the spray tank first; then dilute LOROX with 2 parts of water, thoroughly mix, and add slowly to spray tank with agitator running. For tank mixtures with other herbicides, follow directions above.

USE RATES: All dosages of LOROX (and tank mixtures) are expressed as broadcast rates; for band treatment, use proportionately less. For example, use $\frac{1}{3}$ of the broadcast rate when treating a 14" band where row spacing is 42". Where a range of dosages is given, use the lower rate on coarser soils (low in clay or organic matter) and the higher rate on finer soils (high in clay or organic matter); for postemergence application, use the lower rate on smaller weeds and the higher rate on larger weeds.

SOIL LIMITATIONS: Unless otherwise directed, do not use on sand, loamy sand, gravelly soils or exposed subsoils nor on soils containing less than 1% organic matter as injury to the treated or subsequent crops may result.

REPLANTING: If initial seeding fails to produce a stand, the same crop may be replanted in soil treated preemergence with LOROX (or with recommended tank mixtures). Thoroughly rework soil before replanting; do not re-treat field with a second application as injury to the crop may result. Unless otherwise directed, do not replant treated areas to any other crop within 4 months after last application as injury to subsequent crops may result. West of Rocky Mountains—carrots or celery may be planted 4 months after last application; unless otherwise directed, do not plant any other crop until 1 year after the last application. Do not rotate crops used for food or feed, which are not registered for use with LOROX, onto areas previously treated with this chemical.

ASPARAGUS—California (Includes control of dudain melon and nightshade)

Direct Seeded or Newly Planted Crowns—Do not exceed 4 lbs. total per acre per season; do not use surfactant or fertilizer solution in spray mixture.

Preemergence Application:* Make a single application of 2 to 4 lbs. per acre after planting seed $1\frac{1}{2}$" deep in coarse soil and 1" deep in fine soils. During planting operation, spray activated charcoal (Aqua Nu-Char[9] or Gro-Safe[10]) as a 1" band on soil surface directly over seed rows at the rate of 300 lbs. per acre (equivalent to 15 lbs. per acre of crop with 20" row spacing).

Postemergence Application: Make 1 or 2 applications of 1 to 2 lbs. per acre when ferns are in 6" to 18" stage and weeds are not over 4" tall. Do not reenter treated areas for 24 hours following applications unless protective clothing is worn.

Established Beds—Do not apply within 1 day of harvest; do not exceed 8 lbs. total per acre per season; do not use surfactant or fertilizer solution in spray mixture. Do not reenter treated areas for 24 hours following applications unless protective clothing is worn.

Preemergence Application:* Make a single application of 2 to 4 lbs. per acre.

Postemergence Application: Make 1 to 4 applications of 1 to 2 lbs. per acre before weeds exceed 4" in height. Apply before cutting season or immediately after cutting. Do not reenter treated areas for 24 hours following application unless protective clothing is worn.

Directed Postemergence Application (Fern Stage): Make a single application of 4 to 8 lbs. per acre as a directed spray to base of plants for control of dudain melon. Do not reenter treated areas for 24 hours following application unless protective clothing is worn.

Note: If more than 4 lbs. LOROX per acre is applied per season, do not plant any other crop until 1 year after last application.

* Preemergence weed control will be reduced in soils with high organic matter (greater than 5%) such as peat or muck.

CARROTS—Preemergence Application—Michigan and Wisconsin: Make a single application of 1 to 3 lbs. per acre after planting but before carrots emerge; plant seed at least $\frac{1}{2}$" deep. Use the lower rate on lighter soils (low in clay or organic matter) and the higher rate on heavier soils (high in clay or organic matter). Subsequent postemergence application may be made provided the total does not exceed 4 lbs. LOROX per acre per season.

Postemergence Application—U.S.: Apply $1\frac{1}{2}$ to 3 lbs. per acre as a non-directed spray after carrots are at least 3" tall. Apply before annual grasses exceed 2" in height and before broadleaf weeds exceed 6" in height. Repeat application may be made but do not exceed 4 lbs. LOROX per acre (West of Rocky Mountains, do not exceed 3 lbs. LOROX per acre). Do not exceed 40 psi spray nozzle pressure as crop injury may result.

LOROX may be applied following an application of Stoddard solvent provided treatments are at least one day apart. Stoddard solvent may be applied following an application of LOROX provided treatments are at least 2 weeks apart. Shorter time intervals between applications may result in injury to the crop. Do not apply LOROX as a tank mixture with Stoddard solvent, surfactants, nitrogen solution, other pesticides, nor when temperature exceeds 85°F as crop injury may result.

Do not reenter treated areas for 24 hours following application unless protective clothing is worn.

Note: Because carrot varieties vary in their resistance, determine tolerance to LOROX prior to adoption as a field practice to prevent possible crop injury; do not treat susceptible varieties which show an initial burning of foliage following postemergence treatment with LOROX.

CELERY—East of Rocky Mountains—Post-transplant Application: Make a single application of $1\frac{1}{2}$ to 3 lbs. per acre. Apply as a non-directed spray after celery is transplanted and established, but before celery is 8" tall. Apply before annual grasses exceed 2" in height and before broadleaf weeds exceed 6" in height. In the Northeast, use only on celery grown on muck soils.

Do not exceed 40 psi spray nozzle pressure, and do not apply when temperature exceeds 85°F nor as a tank mixture with surfactants, nitrogen solution, or other pesticides as injury to the crop may result. Do not replant to crops other than celery or carrots within 4 months after application as injury to subsequent crops may result. Do not reenter treated areas for 24 hours following application unless protective clothing is worn.

CORN (FIELD)—East of Rocky Mountains—Preemergence Application: Select one of the following herbicide treatments for application as a tank mixture. Make a single application after planting but before crop emerges. Plant seed at least 1¾" deep on flat or raised seedbeds only or injury to the crop may result. Do not spray over top of emerged corn.

LOROX + Lasso (Tank Mixture)

	Product Per Acre					
	1 to 3% Organic Matter			3 to 6% Organic Matter		
Soil Texture	LOROX (Lbs.)	+	Lasso (Qts.)	LOROX (Lbs.)	+	Lasso (Qts.)
Sandy loam	⅔ to 1¼	+	¾ to 1	1¼ to 2	+	1 to 1½
Silt loam	1 to 1⅔	+	1 to 1⅔	1⅔ to 2½	+	1½ to 2
Clay loam	1¼ to 2	+	1½ to 2	2 to 3	+	2 to 2½

Replanting: Corn or soybeans may be replanted within 4 months; after 4 months, any crop may be planted.

LOROX + Ramrod (Tank Mixture)

	Product Per Acre					
	1 to 3% Organic Matter			3 to 6% Organic Matter		
Soil Texture	LOROX (Lbs.)	+	Ramrod 65 (Lbs.)	LOROX (Lbs.)	+	Ramrod 65 (Lbs.)
Sandy loam	⅔ to 1⅓	+	1 to 2	1⅓ to 2	+	2 to 3
Silt loam	1 to 1⅔	+	1½ to 2½	1⅔ to 2½	+	2½ to 4
Clay loam	1¼ to 2	+	2 to 3	2 to 3	+	3 to 4¾

Replanting: Corn may be replanted within 4 months; after 4 months, any crop may be planted.

LOROX + atrazine (Tank Mixture)

	Product Per Acre					
	1 to 2% Organic Matter			2 to 5% Organic Matter		
Soil Texture	LOROX (Lbs.)	+	atrazine 80% (Lbs.)	LOROX (Lbs.)	+	atrazine 80% (Lbs.)
Sandy loam	⅔ to 1	+	½ to ⅔	1 to 2	+	⅔ to 1¼
Silt loam	1 to 1½	+	⅔ to 1	1½ to 2½	+	1 to 1½
Clay loam	1⅓ to 1⅔	+	¾ to 1	1⅔ to 3	+	1 to 2

Replanting: Corn may be replanted within 6 months. After 6 months, any crop may be planted except do not follow treated corn with sugar beets, tobacco, or vegetables in rotation.

CORN (FIELD AND SWEET)—Directed Postemergence Application: Make a single application as a directed spray after corn is at least 15" high (measured to the highest leaf surface on free standing plants). Do not spray over top of corn. Apply only when there is sufficient differential between height of corn and weeds so that the directed spray thoroughly covers all weed foliage without contact of upper leaves or whorl of corn by spray or drift, as such contact may cause crop injury. Early cultivation (rotary hoe or other suitable equipment) will aid in achieving proper differential between height of corn and weeds.

Use 1¼ to 3 lbs. per acre; add 1 pt. Surfactant WK for each 25 gals. spray mixture. For field corn, non-pressure nitrogen solution may be substituted for all or part of the water. Use the lower rate on lighter soils (low in clay or organic matter) and when weeds do not exceed 2" in height; use the higher rate on heavier soils (high in clay or organic matter) for weeds up to 5" in height.

Do not reenter treated areas for 24 hours following application unless protective clothing is worn.

COTTON—East of Rocky Mountains—Directed Postemergence Application: Apply as a directed spray; adjust nozzles to minimize contact of cotton leaves with spray or drift or crop injury may result. Do not spray over top of cotton.

Apply 1 to 1½ lbs. per acre when cotton is at least 8" tall and emerged weeds do not exceed 2" in height; add 1 pt. Surfactant WK for each 25 gals. spray mixture. If needed, a second application at same rate may be made one week or later after initial treatment. Alternatively, after cotton is 20" tall, make a single application of 2 to 3 lbs. per acre following last cultivation; if emerged weeds are present, add Surfactant WK as directed above. **Note:** Do not use on Pima varieties of cotton. Do not graze treated fields or feed forage from treated areas to livestock. Do not feed gin trash to livestock. Do not reenter treated areas for 24 hours following application unless protective clothing is worn.

PARSNIPS—Preemergence Application: Make a single application of 1½ to 3 lbs. per acre. Apply after planting but before crop emerges. Plant seed at least ½" deep.

POTATOES—Preemergence Application: Make a single application as a broadcast spray after planting but before crop emerges. Plant seed at least 2" deep. Do not spray over top of emerged potatoes. If beds are to be "dragged" and/or "hilled", apply after the final "dragging" or "hilling" operation. Apply before grasses are 2" tall and before broadleaf weeds are 6" tall, preferably just before or when weed seedlings emerge. If emerged weeds are present, add 1 pt. Surfactant WK for each 25 gals. spray mixture. In irrigated areas, best results are obtained when application is made to moist soil, followed within 2 weeks by 1" to 2" of sprinkler irrigation (or rainfall). On powder dry soils, irrigate prior to herbicide application and follow with sprinkler irrigation to activate the herbicide.

East of Rocky Mountains: Apply 1½ to 2½ lbs. per acre on the coarser soils (sandy loams, silt loams; 1 to 2% organic matter) and 2½ to 4 lbs. per acre on finer soils (silts, clay loams; 2 to 5% organic matter); on soils over 5% organic matter, use 4 lbs. per acre and apply to emerged weeds (before potatoes emerge).

Wisconsin—Central Sands Area: Apply 1 lb. per acre on sands and 2 lbs. per acre on loamy sands.

Northeast: For improved control of annual grasses (See Dual labeling), apply:

	Product Per Acre					
	1 to 3% Organic Matter			3 to 5% Organic Matter		
Soil Texture	LOROX (Lbs.)	+	Dual 8E (Pts.)	LOROX (Lbs.)	+	Dual 8E (Pts.)
Coarse Sandy loam	1 to 1½	+	1½	1½ to 2	+	2
Medium Silt, Loam, Silt loam	1½ to 2	+	2	2 to 2½	+	2½ to 3

SORGHUM—Preemergence Application: Select one of the following herbicide treatments for application as a tank mixture. Make a single application after planting but before crop emerges. Plant seed at least 1" deep on flat or raised seedbeds only or injury to the crop may result. Do not spray over top of emerged sorghum.

LOROX + propazine—Southwest (Tank Mixture)

	Product Per Acre					
	1 to 2% Organic Matter			2 to 4% Organic Matter		
Soil Texture	LOROX (Lbs.)	+	propazine 80% (Lbs.)	LOROX (Lbs.)	+	propazine 80% (Lbs.)
Sandy loam	⅝ to 1	+	⅜ to ⅝	1 to 1½	+	⅝ to 1
Silt loam, clay loam	1 to 1½	+	⅝ to 1	1 to 2	+	⅝ to 1¼

Replanting: Do not follow treated sorghum with any fall crop, nor with sugar beets, tobacco, vegetables or potatoes in rotation. Prior to replanting, thorough seedbed preparation including fall or spring plowing is recommended.

LOROX + Ramrod—Great Plains (Tank Mixture)

Soil Texture	Product Per Acre: 1 to 3% Organic Matter — LOROX (Lbs.) +	Ramrod 65 (Lbs.)	3 to 6% Organic Matter — LOROX (Lbs.) +	Ramrod 65 (Lbs.)
Sandy loam	⅔ to 1⅓ +	1 to 2	1⅓ to 2 +	2 to 3
Silt loam	1 to 1⅔ +	1½ to 2½	1⅔ to 2½ +	2½ to 4
Clay loam	1¼ to 2 +	2 to 3	2 to 3 +	3 to 4¾

Note: Do not graze or feed sorghum forage or silage from treated fields to dairy animals. **Replanting:** Sorghum or field corn may be replanted within 4 months; after 4 months, any crop may be planted.

Directed Postemergence Application: Make a single application of LOROX as a directed spray; add 1 pt. Surfactant WK for each 25 gals. spray mixture. If sprayer is equipped with skids, shoes or shields, apply 1 lb. per acre when sorghum is 12" tall (free standing plants) and weeds are up to 2" in height; use 1 to 2 lbs. per acre when sorghum is 15" tall and weeds are 2" to 4" in height. Apply only when there is sufficient differential between height of sorghum and weeds so that the directed spray thoroughly covers all weed foliage without contact of upper leaves or whorl of sorghum by spray or drift as such contact may cause crop injury. **Note:** Do not graze or feed plants to livestock within 3 months after postemergence application. Do not reenter treated areas for 24 hours following application unless protective clothing is worn.

SOYBEANS

Preemergence Application—Select one of the following herbicide treatments and make a single application after planting but before crop emerges. Plant seed at least 1¾" deep on flat or raised seedbeds only or injury to the crop may result. Injury to soybeans may result if application is made to fields with standing water or fields too wet to cultivate. Do not spray over top of emerged soybeans. Do not use on sand or loamy sand nor any soil containing less organic matter than listed below. Do not use on sand or loamy sand nor any soil containing less organic matter than listed below.

Preemergence Following Treflan, Prowl or Tolban Preplant—Where Treflan, Prowl or Tolban has been used as a preplant incorporated treatment (according to directions on product label), apply LOROX as a separate operation using rates recommended below for LOROX Alone. For rotation crops, follow instructions on Treflan, Prowl or Tolban labels.

LOROX Alone—If weeds have emerged, add 1 pt. Surfactant WK for each 25 gals. spray mixture.

Soil Texture	Lbs. LOROX Per Acre: ½ to 2% Organic Matter	2 to 5% Organic Matter
Sandy loam	1 to 1⅔	1⅔ to 3
Silt loam	1¼ to 2⅓	2⅓ to 4
Clay loam	1⅓ to 2⅔	2⅔ to 5

(Over 5% organic matter, use 6 lbs.)

Replanting: Soybeans or field corn may be replanted within 4 months; after 4 months, any crop may be planted.

LOROX + DuPont Lexone ® DF™ Herbicide + Lasso—For control of prickly sida, sicklepod, hemp sesbania, lambsquarters, pigweed, velvetleaf, fall panicum, foxtails, barnyard grass, broadleaf signalgrass, and goosegrass. The combination also offers partial control of cocklebur, jimsonweed and common ragweed.

Soil Texture	Product Per Acre: 1 to 3% Organic Matter — LOROX (Lbs.) +	Lexone DF (Lbs.) +	Lasso (Qts.)	3 to 6% Organic Matter — LOROX (Lbs.) +	Lexone DF (Lbs.) +	Lasso (Qts.)
Coarse Sandy loam	⅓ to ½ +	⅙ to ¼ +	¾ to 1	⅓ to ¾ +	¼ to ⅓ +	1 to 1½
		or (Lexone 4L—¼ to ⅖ pt.)			or (Lexone 4L—⅖ to ½ pt.)	
Medium Loam, Silt loam, Silt, Sandy clay, Sandy clay loam	½ to ¾ +	¼ to ⅓ +	1 to 1½	¾ to 1½ +	⅓ to ½ +	1½ to 2
		or (Lexone 4L—⅖ to ½ pt.)			or (Lexone 4L—½ to ¾ pt.)	
Fine Silty clay, Silty clay loam, Clay, Clay loam	¾ to 1½ +	⅓ to ½ +	1¼ to 2	1½ to 2 +	½ to ⅔ +	2 to 3
		or (Lexone 4L—½ to ¾ pt.)			or (Lexone 4L—¾ to 1 pt.)	

Note—Soybean varieties such as Altona, Coker 102 and 156, Govan, NKS 1884, Semmes, Tracy, Vansoy, Terra Vig 505 and 606, and AP 71 are sensitive to Lexone DF. Injury may occur if Lexone DF is used on these varieties. Before use on any other soybean variety, tolerance to Lexone DF must first be determined. Injury to soybeans may occur if Lexone DF is used on soils having calcareous surface layer or pH of 7.5 or higher, or if used in conjunction with soil applied organic phosphate pesticides such as Dasanit, Disyston, Mocap, Nemacur, Thimet, Parathion, Lorsban 15G or Counter.

Do not use treated plants for feed or forage. Injury may occur if atrazine was applied on the soil the year before. Seedling disease, cold weather, deep planting (more than 2"), excessive moisture, high soil pH (7.5 or higher), high salt concentration, or drought may weaken soybean seedlings and increase possibility of crop injury.

Cultivation after planting—Treated soil may be shallow-cultivated, rotary-hoed, or hand-hoed without reducing the weed control activity of the tank mixture. Do not cultivate deeper than the treated layer of soil since this may bring untreated soil to the surface and poor weed control may result.

Replanting: If initial seeding fails to produce a stand, treated fields may be replanted to soybeans; do not rework soil; do not retreat field with a second application as injury to the crop may result. Do not replant treated areas to any crop other than soybeans within four months after treatment as injury to subsequent crops may result.

LOROX + Lasso

Soil Texture	Product Per Acre: ½ to 3% Organic Matter — LOROX (Lbs.) +	Lasso (Qts.)	3 to 6% Organic Matter — LOROX (Lbs.) +	Lasso (Qts.)
Sandy loam	⅔ to 1¼ +	¾ to 1	1¼ to 2 +	1 to 1½
Silt loam	1 to 1⅔ +	1 to 1½	1⅔ to 2½ +	1½ to 2
Clay loam	1¼ to 2 +	1½ to 2	2 to 3 +	2 to 2½

For control of black nightshade, apply:

Soil Texture	Product Per Acre (1 to 3% Organic Matter): LOROX (Lbs.)	Lasso (Qts.)
Sandy loam	¾ to 1½	2
Silt loam	1 to 1⅔	2½ to 3
Clay loam	1¼ to 2	3

Replanting: Soybeans or field corn may be replanted within 4 months; after 4 months, any crop may be planted.

LOROX + Lexone DF + Dual 8E—For control of prickly sida, sicklepod, hemp sesbania, lambsquarters, and pigweed. The combination also offers partial control of cocklebur, jimsonweed and common ragweed.

Soil Texture	Product Per Acre: ½ to 3% Organic Matter — LOROX (Lbs.) +	Lexone DF (Lbs.) +	Dual 8E (Pts.)	3 to 6% Organic Matter — LOROX (Lbs.) +	Lexone DF (Lbs.) +	Dual 8E (Pts.)
Coarse Sandy loam	⅓ to ½	⅙ to ¼	1 to 1¼	⅓ to ¾	¼ to ⅓	1¼ to 1½
		or (Lexone 4L—¼ to ⅖ pt.)			or (Lexone 4L—⅖ to ½ pt.)	
Medium Loam, Silt loam, Silt, Sandy clay, Sandy clay loam	½ to ¾	¼ to ⅓	1¼ to 1½	¾ to ½	⅓ to ½	1½ to 2
		or (Lexone 4L—⅖ to ½ pt.)			or (Lexone 4L—½ to ¾ pt.)	

Soil Texture	Product Per Acre: ½ to 3% Organic Matter LOROX (Lbs.) +	Lexone DF (Lbs.) +	Dual 8E (Pts.)	3 to 6% Organic Matter LOROX (Lbs.) +	Lexone DF (Lbs.) +	Dual 8E (Pts.)
Fine Silty clay, Silty clay loam, Clay, Clay loam	¾ to 1½	⅓ to ½ or (Lexone 4L—½ to ¾ pt.)	1½ to 2	1½ to 2	½ to ⅔ or (Lexone 4L—¾ to 1 pt.)	2 to 2½

Note—Soybean varieties such as Altona, Coker 102 and 156, Govan, NKS 1884, Semmes, Tracy, Vansoy, Terra Vig 505 and 606, and AP 71 are sensitive to Lexone DF. Injury may occur if Lexone DF is used on these varieties. Before use on any other soybean variety, tolerance to Lexone DF must first be determined. Injury to soybeans may occur if Lexone DF is used on soils having calcareous surface layer or pH of 7.5 or higher, or if used in conjunction with soil applied organic phosphate pesticides such as Dasanit, Disyston, Mocap, Nemacur, Thimet, Parathion, Lorsban 15G or Counter.

Do not use treated plants for feed or forage. Injury may occur if atrazine was applied on the soil the year before. Seedling disease, cold weather, deep planting (more than 2″), excessive moisture, high soil pH (7.5 or higher), high salt concentration, or drought may weaken soybean seedlings and increase possibility of crop injury.

Do not use on sand, loamy sand, nor on sandy loam with less than 1% organic matter.

Cultivation after planting—Treated soil may be shallow-cultivated, rotary-hoed, or hand-hoed without reducing the weed control activity of the tank mixture. Do not cultivate deeper than the treated layer of soil since this may bring untreated soil to the surface and poor weed control may result.

Replanting: If initial seeding fails to produce a stand, treated fields may be replanted to soybeans; do not rework soil; do not retreat field with a second application as injury to the crop may result. Do not replant treated areas to any crop other than soybeans within four months after treatment as injury to subsequent crops may result.

LOROX + DUAL 8E

Soil Texture	Product Per Acre: ½ to 3% Organic Matter LOROX (Lbs.)	+ Dual 8E (Pts.)	3 to 6% Organic Matter LOROX (Lbs.)	+ Dual 8E (Pts.)
Sandy loam	⅔ to 1¼	+ 1¼	1¼ to 2	+ 1½
Silt loam	1 to 1⅔	+ 1½	1⅔ to 2½	+ 2
Clay loam	1¼ to 2	+ 2	2 to 3	+ 2 to 2½

Note: Do not graze or feed forage from treated areas to livestock.

Replanting: Soybeans or field corn may be replanted within 4 months; for rotation crops, follow instructions on Dual 8E label.

For control of black nightshade in Kentucky, Illinois, Indiana and Ohio:

Soil Texture	Product Per Acre (1 to 3% Organic Matter) LOROX (Lbs.)	Dual 8E (Pts.)
Sandy loam	¾ to 1½	1½
Silt loam	1 to 1⅔	2
Clay loam	1¼ to 2	2 to 2½

Do not use on sand or loamy sands.

Replanting: Soybeans or field corn may be replanted within four months; after four months, any crop may be planted.

LOROX + Surflan 75W or Surflan AS

Soil Texture	Product Per Acre (½ to 3% Organic Matter) LOROX (Lbs.)	+	Surflan 75W (Lbs.)	or	Surflan AS (Qts.)
Sandy loam	⅔ to 1½	+	⅔ to 1	or	½ to ¾
Silt, loam	1 to 1⅔	+	1 to 1⅓	or	¾ to 1
Clay loam	1¼ to 2	+	1⅓ to 1⅔	or	1 to 1¼

Note: Do not use treated vines for feed or forage. **Replanting:** Soybeans may be replanted within 4 months; after 4 months, any crop may be planted except do not plant root crops within 12 months.

LOROX + Prowl

Soil Texture	Product Per Acre: 1 to 3% Organic Matter LOROX (Lbs.)	+ Prowl (Pts.)	3 to 6% Organic Matter LOROX (Lbs.)	+ Prowl (Pts.)
Sandy loam	⅔ to 1¼	+ 1 to 1½	1¼ to 2	+ 1½
Silt loam	1 to 1⅔	+ 1½ to 2	1⅔ to 2½	+ 1½ to 2
Clay loam	1¼ to 2	+ 1½ to 2	2 to 3	+ 2 to 2½

Replanting: Soybeans or field corn may be replanted within 4 months; other crops may be planted the following year.

LOROX + Amiben

Soil Texture	Product Per Acre: ½ to 3% Organic Matter LOROX (Lbs.)	+ Amiben (Qts.)	3 to 6% Organic Matter LOROX (Lbs.)	+ Amiben (Qts.)
Sandy loam	⅔ to 1¼	+ 3	1¼ to 2	+ 3
Silt loam	1 to 1⅔	+ 3	1⅔ to 2½	+ 4
Clay loam	1¼ to 2	+ 4	2 to 3	+ 5

Replanting: Soybeans or field corn may be replanted within 4 months; after 4 months, any crop may be planted.

LOROX + Ramrod—Seed Crop Only

Soil Texture	Product Per Acre: 1 to 3% Organic Matter LOROX (Lbs.)	+ Ramrod 65 (Lbs.)	3 to 6% Organic Matter LOROX (Lbs.)	+ Ramrod 65 (Lbs.)
Sandy loam	⅔ to 1¼	+ 1 to 2	1¼ to 2	+ 2 to 3
Silt loam	1 to 1⅔	+ 1½ to 2½	1⅔ to 2½	+ 2½ to 4
Clay loam	1¼ to 2	+ 2 to 3	2 to 3	+ 3 to 4¾

Note: Do not graze or feed forage from treated areas to livestock; do not use seed for food, feed, or oil purposes.

Replanting: Soybeans, sorghum, or field corn may be replanted within 4 months; after 4 months, any crop may be planted.

SOYBEANS

Minimum or No-Tillage: LOROX preemergence to soybeans may be used with paraquat or Roundup in tank mix combinations for preemergence and postemergence control of many broadleaf weeds and grasses and for top kill or suppression of some perennials where soybeans will be planted directly into a preformed bed (stale seed bed), cover crop, or in previous crop residues such as corn or small grain stubble. For conventional seed bed preparation in reduced tillage systems, use rates as described for **Preemergence Application.** Addition of Lasso, Dual 8E, Surflan 75W or Surflan AS to the tank mixture improves control of grasses and volunteer small grains. Apply with ground equipment immediately before, during, or after planting but before crop emerges. Maintain constant agitation of spray mixture.

LOROX + Lasso + Paraquat or Roundup—Thoroughly mix LOROX and Lasso in spray tank first according to directions; then add paraquat or Roundup as directed under **Paraquat Combinations** or **Roundup Combinations** below.

Soil Texture	Product Per Acre: ½ to 3% Organic Matter LOROX (Lbs.)	+ Lasso (Qts.)	3 to 6% Organic Matter LOROX (Lbs.)	+ Lasso (Qts.)
Sandy loam	¾ to 1½	+ 2	1½ to 2⅖	+ 2½
Silt loam	1⅛ to 2	+ 2½	2 to 3	+ 3
Clay loam	1¼ to 2¼	+ 2½	2¼ to 3½	+ 3

For control of black nightshade, apply:

Soil Texture	Product Per Acre (1 to 3% Organic Matter) LOROX (Lbs.)	Lasso (Qts.)
Sandy loam	3/4 to 1 1/2	2
Silt loam	1 to 2	2 1/2 to 3
Clay loam	1 1/4 to 2 1/4	3

Replanting: Soybeans or field corn may be replanted within 4 months; after 4 months, any crop may be planted.

LOROX + Bronco—Do not add paraquat or Roundup to this combination.

Soil Texture[a]	Product Per Acre (1/2 to 5% Organic Matter) LOROX (Lbs.)		Bronco[b] (Lbs.)
Sandy loam	3/4 to 2	+	3 1/4 to 5
Silt loam	1 1/8 to 2 1/2	+	4 to 5
Clay loam	1 1/4 to 3	+	4 to 5

[a] Do not use on soils with less than 1/2% organic matter nor on sand, loamy sand, or muck soils as crop injury may result.

[b] Use the higher rate for dense stubble, heavy crop residue, or heavy weed population.

Replanting Soybeans: If initial seeding fails to produce a stand, treated fields may be replanted to soybeans; do not rework soil; do not retreat field with a second application as injury to the crop may result. Do not replant treated areas to any crop other than soybeans within four months after treatment as injury to subsequent crops may result.

LOROX + Dual 8E + Paraquat or Roundup—Thoroughly mix LOROX and Dual 8E in spray tank first according to directions; then add paraquat or Roundup as directed under **Paraquat Combinations** or **Roundup Combinations** below.

Soil Texture	Product Per Acre: 1/2 to 3% Organic Matter LOROX (Lbs.)	+ Dual 8E (Pts.)	3 to 6% Organic Matter LOROX (Lbs.)	+ Dual 8E (Pts.)
Sandy loam	3/4 to 1 1/2	+ 1 1/2	1 1/2 to 2 2/5	+ 1 1/2
Silt loam	1 1/8 to 2	+ 2	2 to 3	+ 2
Clay loam	1 1/4 to 2 1/4	+ 2 to 2 1/2	2 to 3 1/2	+ 2 to 2 1/2

Note: Do not graze or feed forage from treated areas to livestock.

Replanting: Soybeans or field corn may be replanted within 4 months; for rotation crops, follow instructions on Dual 8E label.

LOROX + Surflan 75W or Surflan AS + Paraquat or Roundup—Thoroughly mix LOROX and Surflan in spray tank first, according to directions, then add paraquat or Roundup as directed under **Paraquat Combinations** or **Roundup Combinations** below.

Soil Texture	Product Per Acre (1/2 to 3% Organic Matter) LOROX (Lbs.)	Surflan 75W or Surflan AS (Lbs.)		(Qts.)
Sandy loam	3/4 to 1 1/2	2/3 to 1	or	1/2 to 3/4
Silt loam	1 1/8 to 2	1 to 1 1/3	or	3/4 to1
Clay loam	1 1/4 to 2 1/4	1 1/3 to 1 2/3	or	1 to 1 1/4

Note: Do not use treated vines for feed or forage.

Replanting: Soybeans may be replanted within 4 months; after 4 months, any crop may be planted except do not plant root crops within 12 months.

LOROX + Paraquat or Roundup—Thoroughly mix LOROX and paraquat in spray tank according to directions under **Paraquat Combinations** or **Roundup Combinations** below.

Soil Texture	Lbs. LOROX Per Acre: 1/2 to 2% Organic Matter	2 to 5% Organic Matter
Sandy loam	1 to 1 2/3	1 2/3 to 3
Silt loam	1 1/4 to 2 1/3	2 1/3 to 4
Clay loam	1 1/3 to 2 2/3	2 2/3 to 5

(Over 5% organic matter, use 6 lbs.)

Replanting: Soybeans or field corn may be replanted within 4 months; after 4 months, any crop may be planted.

Paraquat Combinations—Select and tank mix in water one of the above treatments; then add 1 to 2 pts. paraquat (2 lbs. active per gal.) per acre for control of emerged weeds. Use the higher rate for weeds 4" to 6" tall. As the last ingredient, add 1/2 pint Surfactant WK or Ortho X-77 Spreader per 100 gals. of spray mixture. Maintain constant agitation. Use 20 to 60 gals. of water per acre. Use the higher gallonage for dense stubble or vegetation.

Roundup Combinations—Select and tank mix one of the above treatments. As last ingredient, add 1 1/2 quarts of Roundup per acre for control of emerged annual weeds or 2 to 4 quarts per acre for control of emerged perennial and annual weeds. Use 20 to 30 gals. of water per acre.

Bronco Combinations—Mix LOROX with water in the spray tank as directed on the package label and then add Bronco to the dilute mix. Maintain constant agitation. Use 20 to 30 gals. water per acre.

SOYBEANS

Directed Postemergence Application: Apply LOROX, alone or as a tank mixture with 2,4-DB or Premerge 3, as a directed spray to cover weed foliage with minimum contact of the soybean plant. Do not spray higher than 3" on the soybean stem or crop injury may result. Do not spray over top of soybean plants. For broadcast appplication, use a single flood-type spray nozzle ("K" series or equivalent) per middle mounted on an oiling shoe or gauge wheel. **For band treatment, use two nozzles per row mounted on oiling shoes or gauge wheels, one on each side of row. To avoid spray drift, which may cause crop injury, do not exceed nozzle pressure of 25 psi nor use nozzle tips smaller than 8002 T-Jet (or equivalent) and do not spray under windy conditions. Add 1 pint Surfactant WK for each 25 gals. spray mixture. For best results, use a preemergence treatment (such as LOROX) or cultivation to control early weed growth and to increase the differential between height of soybeans and weeds.**

Note: Do not use on soils with less than 1/2% organic matter. Do not apply more than 2 lbs. LOROX per acre per season for postemergence treatments. Do not apply within 60 days of harvest. Do not feed soybean forage or hay to livestock from fields treated postemergence. Harvested soybeans may be used for food, feed, or oil purposes. Do not reenter treated areas for 24 hours following application unless protective clothing is worn.

LOROX Alone: For soybeans at least 8" high and when weeds do not exceed 2" in height, apply 1/2 to 1 lb. per acre. Make a second application at same rate if new flush of weeds occurs.

For soybeans at least 12" high and when weeds do not exceed 4" in height, make a single application of 1 to 2 lbs. per acre. Alternatively, make a split application of 1 lb. per acre followed by a second application at the same rate after 1 week or later.

LOROX + 2,4-DB—For soybeans at least 8" high and when weeds do not exceed 4" in height, apply 1 lb. LOROX plus 1/5 lb. 2,4-DB (13 fl.oz. Butyrac 200 or 1 pint Butoxone) per acre. A second application may be made if needed, but do not make more than 2 applications per season.

LOROX + Premerge 3—For soybeans at least 8" high and when weeds do not exceed 3" in height, apply 1 lb. LOROX plus 2 qts. of Premerge 3 per acre. Add 1 pt. Surfactant WK for each 25 gals. spray mixture. A second application may be made if needed, but do not make more than 2 applications per season. Do not apply after soybeans begin to bloom.

NON-CROP WEED CONTROL

For short-term control of annual weeds on non-cropland areas such as roadsides and fence rows—apply 2 to 6 lbs. LOROX per acre in 40 to 100 gals. of water. For best results, apply shortly before weed growth begins or at early seedling stage of growth. For control of established annual weeds, add Surfactant WK at rate of 2 qts. per 100 gals. of spray mixture and apply as a thorough coverage spray during periods when daily temperatures exceed 70°F and before weed growth exceeds 8" in height. Do not reenter treated areas for 24 hours following application unless protective clothing is worn.

[1] Registered trademark of Union Carbide Agricultural Products Company, Inc.
[2] Registered trademark of Ciba-Geigy Corporation
[3] Registered trademark of Monsanto Company
[4] Paraquat contains 2 lbs. active per gal.
[5] Registered trademark of American Cyanamid Company
[6] Registered trademark of Elanco Products Company
[7] Registered trademark of Rhone-Poulenc Inc.
[8] Registered trademark of Dow Chemical Company
[9] Registered trademark of Westvaco Corporation
[10] Registered trademark of ICI United States, Inc.

SUPPLEMENTAL LABELING

LOROX HERBICIDE

EPA Registration No. 352-270

This page contains new or supplemental instructions for use of this product which may not appear on the package label. Follow instructions carefully.

LOROX

® HERBICIDE AS A TANK MIXTURE WITH DUAL 8E FOR WEED CONTROL IN POTATOES EAST OF MISSISSIPPI RIVER

DIRECTIONS FOR USE

It is a violation of federal law to use this product in a manner inconsistent with its labeling.

DuPont LOROX Herbicide plus Dual 8E is recommended as a tank mixture for application preemergence to potatoes to control annual weeds. See product labels for weeds controlled.

HOW TO USE

Make a single application as a broadcast spray after planting but before crop emerges. Plant seed at least 2″ deep. Do not spray over top of emerged potatoes. If beds are to be "dragged" and/or "hilled", apply after the final "dragging" or "hilling" operation. Apply before grasses are 2″ tall and before broadleaf weeds are 6″ tall, preferably just before or when weed seedlings emerge. If emerged weeds are present, add 1 pint Surfactant® WK for each 25 gallons of spray mixture. In irrigated areas, best results are obtained when application is made to moist soil, followed within 2 weeks by 1″ to 2″ of sprinkler irrigation (or rainfall). On powder dry soils, irrigate prior to herbicide application and follow with sprinkler irrigation to activate the herbicide.

Application

Soil Texture	Product Per Acre: 1 to 3% Organic Matter, LOROX (Lbs.) + Dual 8E (Pts.)	Product Per Acre: 3 to 5% Organic Matter, LOROX (Lbs.) + Dual 8E (Pts.)
Coarse Sandy loam	1 to 1½ + 1½	1½ to 2 + 2
Medium Silt loam, Silt loam	1½ to 2 + 2	2 to 2½ + 2½ to 3

IMPORTANT

BEFORE USING LOROX OR DUAL 8E, READ AND CAREFULLY OBSERVE THE CAUTIONARY STATEMENTS AND ALL OTHER INFORMATION APPEARING ON THE PRODUCT LABELS.

Special Local Need Registration [24(c)] for LOROX Herbicide

Crop	State	EPA SLN No.	Remarks
Asparagus	Michigan	MI-810009	General weed
	Minnesota	MN-840001	General weed control
	North Carolina	NC-810014	General weed control
	Oregon	OR-810090	General weed
	Washington	WA-790092	General weed control
Carrots	Florida	FL-780002	Preemergence— lower maximum application rate
	Ohio	OH-810027	
Hybrid Poplars	Michigan	MI-830005	General weed control

* Copies of label available upon request to nearest DuPont Agricultural Chemicals Office. For distribution and use only within the State specified.

LOROX®DF™

HERBICIDE

DISPERSIBLE GRANULE

ACTIVE INGREDIENT:
Linuron [3-(3,4-dichlorophenyl)-1-methoxy-1-methylurea] 50%
INERT INGREDIENTS 50%
TOTAL 100%

EPA Reg. No. 352-394

Keep Out of Reach of Children

PRECAUTIONARY STATEMENTS

HAZARDS TO HUMANS

CAUTION!

HARMFUL IF SWALLOWED. MAY IRRITATE EYES, NOSE, THROAT, AND SKIN.

Avoid breathing spray mist. Avoid contact with skin, eyes, and clothing.

In case of contact with eyes, immediately flush with plenty of water. Get medical attention if irritation persists.

ENVIRONMENTAL HAZARDS

Do not apply directly to water or wetlands. Do not contaminate water by cleaning of equipment or disposal of wastes.

IMPORTANT—When mixing/loading and spraying wear a wide-brimmed hat, mid-forearm to elbow length chemical resistant gloves, and one-piece overalls which have long sleeves and long pants. Overalls should be constructed of finely woven fabric as specified in the USDA/EPA Guide for Commercial Applicators.

Any article of clothing worn while applying product must be cleaned before reusing. Clothing which has been drenched or has otherwise absorbed concentrated pesticide must be disposed in a sanitary landfill, or incinerated, or if allowed by State and local authorities, by burning.

Instead of clothing and equipment specified above, the applicator can use an enclosed tractor cab which provides a filtered air supply (as described by Taschenberg and Bourke, 1975).

Injury to or loss of desirable trees or other plants may result from failure to observe the following:

Do not apply (except as recommended for crop use), or drain or flush equipment on or near desirable trees or other plants, or on areas where their roots may extend, or in locations where the chemical may be washed or moved into contact with their roots. Do not use on home plantings of trees, shrubs or herbaceous plants, nor on lawns, walks, driveways, tennis courts, or similar areas. Prevent drift of spray to desirable plants. Do not contaminate any body of water. Keep from contact with fertilizers, insecticides, fungicides and seeds.

Do not apply aerially.

Thoroughly clean all traces of LOROX from application equipment immediately after use. Flush tank, pump, hoses, and boom with several changes of water after removing nozzle tips and screens (clean these parts separately).

NOTICE OF WARRANTY

Du Pont warrants that this product conforms to the chemical description on the label thereof and is reasonably fit for purposes stated on such label only when used in accordance with directions under normal use conditions. It is impossible to eliminate all risks inherently associated with use of this product. Crop injury, ineffectiveness or other unintended consequences may result because of such factors as weather conditions, presence of other materials, or the manner of use or application, all of which are beyond the control of Du Pont. In no case shall Du Pont be liable for consequential, special or indirect damages resulting from the use or handling of this product. All such risks shall be assumed by the Buyer. DU PONT MAKES NO WARRANTIES OF MERCHANTABILITY OR FITNESS FOR A PARTICULAR PURPOSE NOR ANY OTHER EXPRESS OR IMPLIED WARRANTY EXCEPT AS STATED ABOVE.

STORAGE AND DISPOSAL

STORAGE

Store product in original container only, away from other pesticides, fertilizer, food, or feed.

DISPOSAL

Do not contaminate water, food, or feed by storage or disposal. Wastes resulting from the use of this product may be disposed of on site or at approved waste disposal facility.

Triple rinse (or equivalent) the container. Then offer for recycling or reconditioning, or puncture and dispose of in a sanitary landfill, or by incineration, or, if allowed by State and local authorities, by burning. If burned, stay out of smoke.

GENERAL INFORMATION

Du Pont LOROX DF Herbicide is a dispersible granule to be mixed in water and applied as a spray for selective control of weeds on certain crops and for non-selective weed control on non-cropland areas. It is non-corrosive to equipment, non-flammable and non-volatile.

LOROX DF may be applied to soil prior to emergence of weeds to control susceptible weed seedlings for an extended period of time; the degree of control and duration of effect will vary with the amount of chemical applied, soil texture, rainfall and other conditions. Soils high in clay or organic matter require higher dosages than soil low in clay or organic matter to obtain equivalent herbicide performance. Moisture is required to activate the chemical; best results occur if rainfall (or irrigation) occurs within 2 weeks of application. In the Columbia River Basin, use LOROX DF only if crop is sprinkler irrigated.

LOROX DF may also be used to control emerged weeds. Results vary with rate applied and environmental conditions; best results are obtained on succulent weeds growing under conditions of high humidity and temperatures of 70°F or higher. Addition of a surfactant such as Du Pont Surfactant WK to the spray (where recommended) increase contact effects of LOROX DF.

LOROX DF is recommended for use in conjunction with certain other herbicides for treatment of field corn, sorghum, and soybeans; observe all cautions and limitations on labeling of all products used in mixtures.

Since the effect of LOROX DF varies with soils, crop varieties, uniformity of application, and environmental conditions, it is suggested that growers limit their first use to small areas.

DIRECTIONS FOR USE

It is a violation of Federal law to use this product in a manner inconsistent with its labeling.

LOROX DF Herbicide should be used only in accordance with recommendations on this label, or in separate published Du Pont recommendations available through local dealers. Do not use on any crop in Kern County, California, except carrots when a permit has been obtained from the Agricultural Commission.

Du Pont will not be responsible for losses or damages resulting from use of this product in any manner not specifically recommended by Du Pont. User assumes all risks associated with such nonrecommended use.

SELECTIVE USE IN CROPS

PREEMERGENCE USE (Germinating Weeds): LOROX DF, at recommended rates, control weeds such as:

Broadleaves

Florida beggarweed	Mustard
Carpetweed	Nettleleaf goosefoot
Chickweed	Pigweed
Common dayflower	Purslane
Florida purslane (Florida pusley)	Wild radish
Galinsoga	Common ragweed
Lambsquarters	Smartweed

Grasses

Barnyardgrass (watergrass)	Foxtail (including giant)
Canarygrass	Goosegrass
Crabgrass	Fall panicum

Partial Control

Cocklebur	Sicklepod
Annual morningglory	Velvetleaf (buttonweed)
Prickly sida (teaweed)	

Note: Consult AMIBEN[1], atrazine, BRONCO[3], DUAL[2] 8E, LASSO[3], Paraquat[4], propazine, PROWL[5], RAMROD[3], ROUNDUP[3], SURFLAN[6] 75W, SURFLAN[6] AS, TOLBAN[2], and TREFLAN[6] labels for additional weeds controlled when LOROX DF is used in conjunction with these herbicides.

LOROX DF will not control established perennials such as bermudagrass, Canada thistle, field bindweed, johnsongrass and purple nutsedge. The lower dosage rates are effective on coarser soils and the higher rates on finer soils and on the more resistant seedling weeds. Sufficient moisture (½" to 1" on moist soils; 1" to 2" on dry soils) in the form of rainfall or sprinkler irrigation is necessary after treatment to carry the chemical into the root zone of germinating weeds; best results are obtained when this occurs within two weeks after application. If heavy rainfall occurs soon after application, injury to crop may result.

A good seed bed must be prepared before application of LOROX DF as crop injury may result if application is made to ground which is cloddy or compacted resulting in improperly planted seed. Plant seed to depth specified. Surface of the soil should not be cultivated or disturbed after application of LOROX DF and before emergence of the crop as weed control may be reduced and crop injury may result.

However, if moisture is insufficient to activate the herbicide, a shallow cultivation (rotary hoe preferred) should be made after emergence of row crops while weeds are small enough to be controlled by mechanical means.

POSTEMERGENCE USE (Emerged Seedling Weeds): LOROX DF, at recommended rates, controls weeds such as:

Broadleaves

Annual morningglory	Lambsquarters
Carpetweed	Mustard
Cocklebur	Nettleleaf goosefoot
Common dayflower	Pigweed
Common ragweed	Prickly sida (teaweed)
Dog fennel (amsinckia)	Purslane
Fiddleneck	Sesbania
Florida beggarweed	Sicklepod
Florida purslane (Florida pusley)	Smartweed
Groundsel	Velvetleaf (buttonweed)
Knawel	Wild buckwheat

Grasses

Annual ryegrass	Fall panicum
Barnyardgrass (watergrass)	Foxtail (including giant)
Broadleaf signalgrass	Goosegrass
Canarygrass	Rattail fescue
Crabgrass	Texas panicum

Note: Consult bromoxynil, BUTOXONE[7], BUTYRAC[1] 200, and PREMERGE 3[8] labels for additional weeds controlled when

LOROX DF is used in conjunction with these herbicides. Best results are obtained under conditions of high humidity and tempertures over 70°F. Control of emerged weeds under drought stress is usually impractical.

EQUIPMENT—SPRAY VOLUMES AND PRESSURES: Use a tractor-mounted fixed-boom power sprayer properly calibrated to a constant speed and rate of delivery. Openings in screens should be equal to or larger than 50 mesh. Continuous agitation in the spray tank is required to keep the material in suspension. Agitate by mechanical or hydraulic means; if by-pass or return line is used, it should terminate at bottom of tank to minimize foaming. Avoid overlapping, and shut off spray booms while starting, turning, slowing or stopping, or injury to the crop may result.

For preemergence application, use 20 to 60 gals. per acre and spray pressure of 30 to 40 psi.

For postemergence application, use sufficient volume (min. 20 gals. per acre) for thorough coverage of weed foliage; use spray pressure of 20 to 25 psi to keep spray drift to a minimum.

SPRAY PREPARATION: Fill tank ¼ full with water. Start agitation system, add LOROX DF, and continue adding water. Add separately each additional component of any tank mix while adding water. If a surfactant is used, add it last when the tank is nearly full. Continue agitation throughout. If poor mixing should occur with any component, premix the component with two parts water before adding to the spray tank.

Fertilizer Spray Mixtures—For preemergence application, nonpressure nitrogen or fertilizer solution may be used in the spray mixture unless otherwise directed. Small quantities should be tested for compatibility by the following procedure before full scale mixing:

1. Put 1 pint of fertilizer solution in a quart jar.
2. Mix 2 teaspoons LOROX DF with 2 tablespoons of water; mix thoroughly and add to fertilizer solution.
3. Close jar and shake well.
4. If other herbicides are to be used in the mixture, premix 2 teaspoons of wettable powders or 1 teaspoon of liquids with 2 tablespoons of water; add to LOROX DF/fertilizer mixture.
5. Close jar and shake well.
6. Watch mixture for several seconds; check again in 30 minutes.
7. If mixture does not separate, foam, gel, or become lumpy, it may be used.
8. Mixing ability may be improved by adding compatibility agents such as Kalo Laboratories COMPLEX or Witco Chemicals SPONTO 1680. Follow directions on container.

If the mixture is compatible, prepare spray by adding the fertilizer solution to the spray tank first; then dilute LOROX DF with 2 parts water, throughly mix, and add slowly to spray tank with agitator running. For tank mixtures with other herbicides, follow directions above.

USE RATES: All dosages of LOROX DF (and tank mixtures) are expressed as broadcast rates; for band treatment, use proportionately less. For example, use ⅓ of the broadcast rate when treating a 14" band where row spacing is 42". Where a range of dosages is given, use the lower rate on coarser soils (low in clay or organic matter) and the higher rate on finer soils (high in clay or organic matter); for postemergence application, use the lower rate on smaller weeds and the higher rate on larger weeds.

SOIL LIMITATIONS: Unless otherwise directed, do not use on sand, loamy sand, gravelly soils or exposed subsoils nor on soils containing less than 1% organic matter as injury to the treated or subsequent crops may result.

REPLANTING: If initial seeding fails to produce a stand, the same crop may be replanted in soil treated preemergence with LOROX DF (or with recommended tank mixtures). Thoroughly rework soil before replanting; do not re-treat field with a second application as injury to the crop may result. Unless otherwise directed, do not replant treated areas to any other crop within 4 months after last application as injury to subsequent crops may result. West of Rocky Mountains—carrots may be planted 4 months after last application; unless otherwise directed, do not plant any other crop until 1 year after the last application. Do not rotate crops used for food or feed, which are not registered for use with LOROX DF, on to areas previously treated with this chemical.

CARROTS—Preemergence Application—Michigan and Wisconsin

Make a single application of 1 to 3 lbs. per acre after planting but before carrots emerge; plant seed at least ½" deep. Use the lower rate on lighter soils (low in clay or organic matter) and the higher rate on heavier soils (high in clay or organic matter). Subsequent postemergence application may be made provided the total does not exceed 4 lbs. LOROX DF per acre per season.

Do not reenter treated areas for 24 hours following application unless protective clothing is worn.

Note: Because carrot varieties vary in their resistance, determine tolerance to LOROX DF prior to adoption as a field practice to prevent possible crop injury.

CORN (FIELD)—East of Rocky Mountains—Preemergence Application

Select one of the following herbicide treatments for application as a tank mixture. Make a single application after planting but before crop emerges. Plant seed at least 1¾" deep on flat or raised seedbeds only or injury to the crop may result. Do not spray over top of emerged corn.

LOROX DF + LASSO (Tank Mixture)

	Product Per Acre					
	1 to 3% Organic Matter			3 to 6% Organic Matter		
Soil Texture	LOROX DF (Lbs.)	+	LASSO (Quarts)	LOROX DF (Lbs.)	+	LASSO (Quarts)
Sandy loam	⅔ to 1¼	+	¾ to 1	1¼ to 2	+	1 to 1½
Silt loam	1 to 1⅔	+	1 to 1½	1⅔ to 2½	+	1½ to 2
Clay loam	1¼ to 2	+	1½ to 2	2 to 3	+	2 to 2½

REPLANTING: Corn or soybeans may be replanted within 4 months; after 4 months, any crop may be planted.

LOROX DF + RAMROD (Tank Mixture)

	Product Per Acre					
	1 to 3% Organic Matter			3 to 6% Organic Matter		
Soil Texture	LOROX DF (Lbs.)	+	RAMROD 65 (Lbs.)	LOROX DF (Lbs.)	+	RAMROD 65 (Lbs.)
Sandy loam	⅔ to 1⅓	+	1 to 2	1⅓ to 2	+	2 to 3
Silt loam	1 to 1⅔	+	1½ to 2½	1⅔ to 2½	+	2½ to 4
Clay loam	1¼ to 2	+	2 to 3	2 to 3	+	3 to 4¾

REPLANTING: Corn may be replanted within 4 months; after 4 months, any crop may be planted.

LOROX DF + atrazine (Tank Mixture)

	Product Per Acre					
	1 to 2% Organic Matter			2 to 5% Organic Matter		
Soil Texture	LOROX DF (Lbs.)	+	atrazine 80% (Lbs.)	LOROX DF (Lbs.)	+	atrazine 80% (Lbs.)
Sandy loam	⅔ to 1	+	½ to ⅔	1 to 2	+	⅔ to 1¼
Silt loam	1 to 1½	+	⅔ to 1	1½ to 2½	+	1 to 1½
Clay loam	1⅓ to 1⅔	+	¾ to 1	1⅔ to 3	+	1 to 2

REPLANTING: Corn may be replanted within 6 months. After 6 months, any crop may be planted except do not allow treated corn with sugar beets, tobacco, or vegetables in rotation.

CORN (FIELD AND SWEET)—Directed Postemergence Application:

Make a single application as a directed spray after corn is at least 15" high (measured to the highest leaf surface on free standing plants). Do not spray over top of corn. Apply only when there is sufficient differential between height of corn and weeds so that the directed spray thoroughly covers all weed foliage without contact of upper leaves or whorl of corn by spray or drift, as such contact may cause crop injury. Early cultivation (rotary hoe or other suitable equipment) will aid in achieving proper diffential between height of corn and weeds.

Use 1¼ to 3 lbs. per acre; add 1 pint Surfactant WK for each 25 gals. spray mixture. For field corn, non-pressure nitrogen solution may be substituted for all or part of the water. Use the lower rate on lighter soils (low in clay or organic matter) and when weeds do not exceed 2" in height; use the higher rate on heavier soils (high in clay or organic matter) for weeds up to 5" in height.

Do not reenter treated areas for 24 hours following application unless protective clothing is worn.

COTTON: East of Rocky Mountains—Directed Postemergence Application:
Apply as a directed spray; adjust nozzles to minimize contact to cotton leaves with spray or drift or crop injury may result. Do not spray over top of cotton.

Apply 1 to 1½ lbs. per acre when cotton is at least 8" tall and emerged weeds do not exceed 2" in height; add 1 pint Surfactant WK for each 25 gals. spray mixture. If needed, a second application at same rate may be made one week or later after inital treatment. Alternatively, after cotton is 20" tall, make a single application of 2 to 3 lbs. per acre following last cultivation; if emerged weeds are present, add Surfactant WK as directed above. Note: Do not use on Pima varieties of cotton. Do not graze treated fields or feed forage from treated areas to livetock. Do not feed gin trash to livestock. Do not reenter treated areas for 24 hours following application unless protective clothing is worn.

PARSNIPS: Preemergence Application:
Make a single application of 1½ to 3 lbs. per acre. Apply after planting but before crop emerges. Plant seed at least ½" deep.

POTATOES
Preemergence Application:
Make a single application as a broadcast spray after planting but before crop emerges. Plant seed at least 2" deep. Do not spray over top of emerged potatoes. If beds are to be "dragged" and/or "hilled", apply after the final "dragging" or "hilling" operation. Apply before grasses are 2" tall and before broadleaf weeds are 6" tall, preferably just before or when weed seedlings emerge. If emerged weeds are present, add 1 pint Surfactant WK for each 25 gals. spray mixture. In irrigated areas, best results are obtained when application is made to moist soil, followed within 2 weeks by 1" to 2" of sprinkler irrigation (or rainfall). On powder dry soils, irrigate prior to herbicide application and follow with sprinkler irrigation to activate herbicide.

East of Rocky Mountains:
Apply 1½ to 2½ lbs. per acre on the coarser soils (sandy loams, silt loams; 1 to 2% organic matter) and 2½ to 4 lbs. per acre on finer soils (silts, clay loams; 2 to 5% organic matter); on soils over 5% organic matter, use 4 lbs. per acre and apply to emerged weeds (before potatoes emerge).

Wisconsin—Central Sands Area:
Apply 1 lb. per acre on sands and 2 lbs. per acre on loamy sands.

Northeast:
For improved control of annual grasses (see "Dual" labeling), apply:

Soil Texture	Product Per Acre: 1 to 3% Organic Matter: LOROX DF (Lbs.)	+	DUAL 8E (Pints)	3 to 5% Organic Matter: LOROX DF (Lbs.)	+	DUAL 8E (Pints)
Coarse Sandy loam	1 to 1½	+	1½	1½ to 2	+	2
Medium Silt, Loam, Silt loam	1½ to 2	+	2	2 to 2½	+	2½ to 3

SORGHUM
Preemergence Application:
Select one of the following herbicide treatments for application as a tank mixture. Make a single application after planting but before crop emerges. Plant seed at least 1" deep on flat or raised seedbeds only or injury to the crop may result. Do not spray over top of emerged sorghum.

LOROX DF + propazine—Southwest (Tank Mixture)

Soil Texture	Product Per Acre: 1 to 2% Organic Matter: LOROX DF (Lbs.)	+	propazine 80% (Lbs.)	2 to 4% Organic Matter: LOROX DF (Lbs.)	+	propazine 80% (Lbs.)
Sandy loam	⅝ to 1	+	⅜ to ⅝	1 to 1½	+	⅝ to 1
Silt loam, clay loam	1 to 1½	+	⅝ to 1	1 to 2	+	⅝ to 1¼

REPLANTING: Do not follow treated sorghum with any fall crop, nor with sugar beets, tobacco, vegetables or potatoes in rotation. Prior to replanting, thorough seedbed preparation including fall or spring plowing is recommended.

LOROX DF + RAMROD—Great Plains (Tank Mixture)

Soil Texture	Product Per Acre: 1 to 3% Organic Matter: LOROX DF (Lbs.)	+	RAMROD 65 (Lbs.)	3 to 6% Organic Matter: LOROX DF (Lbs.)	+	RAMROD 65 (Lbs.)
Sandy loam	⅔ to 1⅓	+	1 to 2	1⅓ to 2	+	2 to 3
Silt loam	1 to 1⅔	+	1½ to 2½	1⅔ to 2½	+	2½ to 4
Clay loam	1¼ to 2	+	2 to 3	2 to 3	+	3 to 4¾

NOTE: Do not graze or feed sorghum forage or silage from treated fields to dairy animals. Replanting: Sorghum or field corn may be replanted within 4 months; after 4 months, any crop may be planted.

Directed Postemergence Application:
Make a single application of LOROX DF as a directed spray; add 1 pint Surfactant WK for each 25 gals. spray mixture. If sprayer is equipped with skids, shoes or shields, apply 1 lb. per acre when sorghum is 12" tall (free standing plants) and weeds are up to 2" in height; use 1 to 2 lbs. per acre when sorghum is 15" tall and weeds are 2" to 4" in height. Apply only when there is sufficient differential between height of sorghum and weeds so that the directed spray thoroughly covers all weed foliage without contact of upper leaves or whorl of sorghum by spray or drift as such contact may cause crop injury. Note: Do not graze or feed plants to livestock within 3 months after postemergence application. Do not reenter treated areas for 24 hours following application unless protective clothing is worn.

SOYBEANS
Preemergence Application
Select one of the following herbicide treatments and make a single application after planting but before crop emerges. Plant seed at least 1¾" deep on flat or raised seedbeds only or injury to the crop may result. Injury to soybeans may result if application is made to fields with standing water or fields too wet to cultivate. Do not spray over top of emerged soybeans. Do not use on sand or loamy sand nor any soil containing less organic matter than listed below.

Preemergence Following TREFLAN, PROWL or TOLBAN Preplant
Where TREFLAN, PROWL or TOLBAN has been used as a preplant incorporated treatment (according to directions on product label), apply LOROX DF as a separate operation using rates recommended below for LOROX DF Alone. For rotation crops, follow instructions on TREFLAN, PROWL, or TOLBAN labels.

LOROX DF ALONE—If weeds have emerged, add 1 pint Surfactant WK for each 25 gals. spray mixture.

Soil Texture	Lbs. LOROX DF Per Acre: ½ to 2% Organic Matter	Lbs. LOROX DF Per Acre: 2 to 5% Organic Matter
Sandy loam	1 to 1⅔	1⅔ to 3
Silt loam	1¼ to 2⅓	2⅓ to 4
Clay loam	1⅓ to 2⅔	2⅔ to 5 (Over 5% organic matter, use 6 lbs.)

REPLANTING: Soybeans or field corn may be replanted within 4 months; after 4 months; any crop may be planted.

LOROX DF + Du Pont LEXONE DF™ Herbicide + LASSO—For control of prickly sida, sicklepod, hemp sesbania, lambsquarters, pigweed, velvetleaf, fall panicum, foxtails, barnyardgrass, broadleaf signalgrass, and goosegrass. The combination also offers partial control of cocklebur, jimsonweed and common ragweed.

	Product Per Acre					
	1 to 3% Organic Matter			3 to 6% Organic Matter		
Soil Texture	LOROX DF (Lbs.) +	LEXONE DF (Lbs.) +	LASSO (Quarts)	LOROX DF (Lbs.) +	LEXONE DF (Lbs.) +	LASSO (Quarts)
Coarse Sandy loam	⅓ to ½	⅙ to ¼ or (LEXONE 4L—¼ to ⅖ pt.)	¾ to 1	⅓ to ¾	¼ to ⅓ or (LEXONE 4L—⅖ to ½ pt.)	1 to 1½
Medium Loam, Silt loam, Silt, Sandy clay, Sandy clay loam	½ to ¾	¼ to ⅓ or (LEXONE 4L—⅖ to ½ pt.)	1 to 1½	¾ to 1½	⅓ to ½ or (LEXONE 4L—½ to ¾ pt.)	1½ to 2
Fine Silty clay, Silty clay loam, Clay, Clay loam	¾ to 1½	⅓ to ½ or (LEXONE 4L—½ to ¾ pt.)	1¼ to 2	1½ to 2	½ to ⅔ or (LEXONE 4L—¾ to 1 pt.)	2 to 3

NOTE: Soybean varieties such as Altona, Coker 102 and 156, Govan, NKS 1884, Semmes, Tracy, Vansoy, Terra Vig 505 and 606, and AP 71 are sensitive to LEXONE DF. Injury may occur if LEXONE DF is used on these varieties. Before use on any other soybean variety, tolerance to LEXONE DF must first be determined. Injury to soybeans may occur if LEXONE DF is used on soils having a calcareous surface layer or pH of 7.5 or higher, or if used in conjunction with soil applied organic phosphate pesticides such as DASANIT, DISYSTON, MOCAP, NEMACUR, THIMET, PARATHION, LORSBAN 15G or Counter.

Do not use treated plants for feed or forage. Injury may occur if atrazine was applied on the soil the year before. Seedling disease, cold weather, deep planting (more than 2″), excessive moisture, high soil pH (7.5 or higher), high salt concentration, or drought may weaken soybean seedlings and increase possibility of crop injury.

Cultivation after planting: Treated soil may be shallow-cultivated, rotary-hoed, or hand-hoed without reducing the weed control activity of the tank mixture. Do not cultivate deeper than the treated layer of soil since this may bring untreated soil to the surface and poor weed control may result.

REPLANTING: If initial seeding fails to produce a stand, treated fields may be replanted to soybeans; do not rework soil; do not retreat field with a second application as injury to the crop may result. Do not replant treated areas to any crop other than soybeans within four months after treatment as injury to subsequent crops may result.

LOROX DF + LASSO

	Product Per Acre			
	½ to 3% Organic Matter		3 to 6% Organic Matter	
Soil Texture	LOROX DF (Lbs.) +	LASSO (Quarts)	LOROX DF (Lbs.) +	LASSO (Quarts)
Sandy loam	⅔ to 1¼	¾ to 1	1¼ to 2	1 to 1½
Silt loam	1 to 1⅔	1 to 1½	1⅔ to 2½	1½ to 2
Clay loam	1¼ to 2	1½ to 2	2 to 3	2 to 2½

For control of black nightshade, apply:

Soil Texture	Product Per Acre: LOROX DF (Lbs.)	Product Per Acre: LASSO (Quarts)
Sandy loam	¾ to 1½	2
Silt loam	1 to 1⅔	2½ to 3
Clay loam	1¼ to 2	3

REPLANTING: Soybeans or field corn may be replanted within 4 months; after 4 months, any crop may be planted.

LOROX DF + LEXONE DF + DUAL 8E—For control of prickly sida, sicklepod, hemp sesbania, lambsquarters, and pigweed. The combination also offers partial control of cocklebur, jimsonweed and common ragweed.

	Product Per Acre					
	½ to 3% Organic Matter			3 to 6% Organic Matter		
Soil Texture	LOROX DF (Lbs.) +	LEXONE DF (Lbs.) +	DUAL 8E (Pints)	LOROX DF (Lbs.) +	LEXONE DF (Lbs.) +	DUAL 8E (Pints)
Coarse Sandy loam	⅓ to ½	⅙ to ¼ or (LEXONE 4L—¼ to ⅖ pt.)	1 to 1¼	⅓ to ¾	¼ to ⅓ or (LEXONE 4L—⅖ to ½ pt.)	1¼ to 1½
Medium Loam, Silt loam, Silt, Sandy clay, Sandy clay loam	½ to ¾	¼ to ⅓ or (LEXONE 4L—⅖ to ½ pt.)	1¼ to 1½	¾ to ½	⅓ to ½ or (LEXONE 4L—½ to ¾ pt.)	1½ to 2
Fine Silty clay, Silty clay loam, Clay, Clay loam	¾ to 1½	⅓ to ½ or (LEXONE 4L—½ to ¾ pt.)	1½ to 2	1½ to 2	½ to ⅔ or (LEXONE 4L—¾ to 1 pt.)	2 to 2½

NOTE: Soybean varieties such as Altona, Coker 102 and 156, Govan, NKS 1884, Semmes, Tracy, Vansoy, Terra Vig 505 and 606, and AP 71 are sensitive to LEXONE DF. Injury may occur if LEXONE DF is used on these varieties. Before use on any other soybean variety, tolerance to LEXONE DF must first be determined. Injury to soybeans may occur if LEXONE DF is used on soils having a calcareous surface layer or pH of 7.5 or higher, or if used in conjunction with soil applied organic phosphate pesticides such as DASANIT, DISYSTON, MOCAP, NEMACUR, THIMET, PARATHION, LORSBAN 15G or Counter.

Do not use treated plants for feed or forage. Injury may occur if atrazine was applied on the soil the year before. Seedling disease, cold weather, deep planting (more than 2″), excessive moisture, high soil pH (7.5 or higher), high salt concentration, or drought may weaken soybean seedlings and increase possibility of crop injury.

Do not use on sand, loamy sand, nor on sandy loam with less than 1% organic matter.

Cultivation after planting: Treated soil may be shallow-cultivated, rotary-hoed, or hand-hoed without reducing the weed control activity of the tank mixture. Do not cultivate deeper than the treated layer of soil since this may bring untreated soil to the surface and poor weed control may result.

Replanting: If initial seeding fails to produce a stand, treated fields may be replanted to soybeans; do not rework soil; do not retreat field with a second application as injury to the crop may result. Do not replant treated areas to any crop other than soybeans within four months after treatment as injury to the subsequent crops may result.

LOROX DF + DUAL 8E

Soil Texture	Product Per Acre					
	½ to 3% Organic Matter			3 to 6% Organic Matter		
	LOROX DF (Lbs.)	+	DUAL 8E (Pints)	LOROX DF (Lbs.)	+	DUAL 8E (Pints)
Sandy loam	⅔ to 1¼	+	1¼	1¼ to 2	+	1½
Silt loam	1 to 1⅔	+	1½	1⅔ to 2½	+	2
Clay loam	1¼ to 2	+	2	2 to 3	+	2 to 2½

NOTE: Do not graze or feed forage from treated areas to livestock. Replanting: Soybeans or field corn may be replanted within 4 months; for rotation crops, follow instructions of DUAL 8E label.

For control of black nightshade in Kentucky, Illinois, Indiana and Ohio:

Soil Texture	Product Per Acre (1 to 3% Organic Matter)	
	LOROX DF (Lbs.)	DUAL 8E (Pints)
Sandy loam	¾ to 1½	1½
Silt loam	1 to 1⅔	2
Clay loam	1¼ to 2	2 to 2½

Do not use on sand or loamy sands.

REPLANTING: Soybeans or field corn may be replanted within four months; after four months, any crop may be planted.

LOROX DF + SURFLAN 75W or SURFLAN AS

Soil Texture	Product Per Acre (½ to 3% Organic Matter)				
	LOROX DF (Lbs.)	+	SURFLAN 75W (Lbs.)	or	SURFLAN AS (Quarts)
Sandy loam	⅔ to 1¼	+	⅔ to 1	or	½ to ¾
Silt loam	1 to 1⅔	+	1 to 1⅓	or	¾ to 1
Clay loam	1¼ to 2	+	1⅓ to 1⅔	or	1 to 1¼

NOTE: Do not use treated vines for feed or forage. Replanting: Soybeans may be replanted within 4 months; after 4 months, any crop may be planted except do not plant root crops within 12 months.

LOROX DF + PROWL

Soil Texture	Product Per Acre					
	1 to 3% Organic Matter			3 to 6% Organic Matter		
	LOROX DF (Lbs.)	+	PROWL (Pints)	LOROX DF (Lbs.)	+	PROWL (Pints)
Sandy loam	⅔ to 1¼	+	1 to 1½	1¼ to 2	+	1½
Silt loam	1 to 1⅔	+	1½ to 2	1⅔ to 2½	+	1½ to 2
Clay loam	1¼ to 2	+	1½ to 2	2 to 3	+	2 to 2½

REPLANTING: Soybeans or field corn may be replanted within 4 months; other crops may be planted the following year.

LOROX DF + AMIBEN

Soil Texture	Product Per Acre					
	½ to 3% Organic Matter			3 to 6% Organic Matter		
	LOROX DF (Lbs.)	+	AMIBEN (Quarts)	LOROX DF (Lbs.)	+	AMIBEN (Quarts)
Sandy loam	⅔ to 1¼	+	3	1¼ to 2	+	3
Silt loam	1 to 1⅔	+	3	1⅔ to 2½	+	4
Clay loam	1¼ to 2	+	4	2 to 3	+	5

REPLANTING: Soybeans or field corn may be replanted within 4 months; after 4 months, any crop may be planted.

LOROX DF + RAMROD—Seed Crop Only

Soil Texture	Product Per Acre					
	1 to 3% Organic Matter			3 to 6% Organic Matter		
	LOROX DF (Lbs.)	+	RAMROD 65 (Lbs.)	LOROX DF (Lbs.)	+	RAMROD 65 (Lbs.)
Sandy loam	⅔ to 1¼	+	1 to 2	1¼ to 2	+	2 to 3
Silt loam	1 to 1⅔	+	1½ to 2½	1⅔ to 2½	+	2½ to 4
Clay loam	1¼ to 2	+	2 to 3	2 to 3	+	3 to 4¾

NOTE: Do not graze or feed forage from treated areas to livestock; do not use seed for food, feed, or oil purposes. Replanting: Soybeans, sorghum, or field corn may be replanted within 4 months; after 4 months, any crop may be planted.

SOYBEANS

Minimum or No-Tillage

LOROX DF preemergence to soybeans may be used with paraquat, BRONCO or ROUNDUP in tank mix combinations for preemergence and postemergence control of many broadleaf weeds and grasses and for top kill or suppression of some perennials where soybeans will be planted directly into a preformed bed (stale seed bed), cover crop, or in previous crop residues such as corn or small grain stubble. For conventional seed bed preparation in reduced tillage systems, use rates as described for Preemergence Application. Addition of LASSO, DUAL 8E, SURFLAN 75W or SURFLAN AS to tank mixture improves control of grasses and volunteer small grains. Apply with ground equipment immediately before, during, or after planting but before crop emerges. Maintain constant agitation of spray mixture.

LOROX DF + LASSO + Paraquat or ROUNDUP—Thoroughly mix LOROX DF and LASSO in spray tank first according to directions; then add paraquat or ROUNDUP as directed under Paraquat Combinations or ROUNDUP Combinations below:

Soil Texture	Product Per Acre					
	½ to 3% Organic Matter			3 to 6% Organic Matter		
	LOROX DF (Lbs.)	+	LASSO (Quarts)	LOROX DF (Lbs.)	+	LASSO (Quarts)
Sandy loam	¾ to 1½	+	2	1½ to 2⅖	+	2½
Silt loam	1⅛ to 2	+	2½	2 to 3	+	3
Clay loam	1¼ to 2¼	+	2½	2¼ to 3½	+	3

For control of black nightshade, apply:

Soil Texture	Product Per Acre (1 to 3% Organic Matter)	
	LOROX DF (Lbs.)	LASSO (Quarts)
Sandy loam	¾ to 1½	2
Silt loam	1 to 2	2½ to 3
Clay loam	1¼ to 2¼	3

REPLANTING: Soybeans or field corn may be replanted within 4 months; after 4 months, any crop may be planted.

LOROX DF + BRONCO—Do not add paraquat or ROUNDUP to this combination.

Soil Texture (a)	Product Per Acre (½ to 5% Organic Matter)		
	LOROX DF (Lbs.)	+	BRONCO (b) (Lbs.)
Sandy loam	¾ to 2	+	3¼ to 5
Silt loam	1⅛ to 2½	+	4 to 5
Clay loam	1¼ to 3	+	4 to 5

(a) Do not use on soils with less than ½% organic matter nor on sand, loamy sand, or muck soils as crop injury may result.
(b) Use the higher rate for dense stubble, heavy crop residue, or heavy weed population.

REPLANTING SOYBEANS: If initial seeding fails to produce a stand, treated fields may be replanted to soybeans; do not rework soil; do not retreat field with a second application as injury to the crop may result. Do not replant treated areas to any crop other than soybeans within four months after treatment as injury to subsequent crops may result.

LOROX DF + DUAL 8E + Paraquat or ROUNDUP—Thoroughly mix LOROX DF and DUAL 8E in spray tank first according to directions; then add paraquat or ROUNDUP as directed under Paraquat Combinations or ROUNDUP Combinations below.

Soil Texture	Product Per Acre					
	$\frac{1}{2}$ to 3% Organic Matter			3 to 6% Organic Matter		
	LOROX DF (Lbs.)	+	DUAL 8E (Pints)	LOROX DF (Lbs.)	+	DUAL 8E (Pints)
Sandy loam	$\frac{3}{4}$ to $1\frac{1}{2}$	+	$1\frac{1}{2}$	$1\frac{1}{2}$ to $2\frac{2}{5}$	+	$1\frac{1}{2}$
Silt loam	$1\frac{1}{8}$ to 2	+	2	2 to 3	+	2
Clay loam	$1\frac{1}{4}$ to $2\frac{1}{4}$	+	2 to $2\frac{1}{2}$	2 to $3\frac{1}{2}$	+	2 to $2\frac{1}{2}$

NOTE: Do not graze or feed forage from treated areas to livestock. Replanting: Soybeans or field corn may be replanted within 4 months; for rotation crops, follow instructions on DUAL 8E label.

LOROX DF + SURFLAN 75W or SURFLAN AS + Paraquat or ROUNDUP—Thoroughly mix LOROX DF and SURFLAN in spray tank first, according to directions; then add paraquat or ROUNDUP as directed under Paraquat Combinations or ROUNDUP Combinations below.

Soil Texture	Product Per Acre ($\frac{1}{2}$ to 3% Organic Matter)				
	LOROX DF (Lbs.)	+	SURFLAN 75W (Lbs.)	or	SURFLAN AS (Quarts)
Sandy loam	$\frac{3}{4}$ to $1\frac{1}{2}$	+	$\frac{2}{3}$ to 1	or	$\frac{1}{2}$ to $\frac{3}{4}$
Silt loam	$1\frac{1}{8}$ to 2	+	1 to $1\frac{1}{3}$	or	$\frac{3}{4}$ to 1
Clay loam	$1\frac{1}{4}$ to $2\frac{1}{4}$	+	$1\frac{1}{3}$ to $1\frac{2}{3}$	or	1 to $1\frac{1}{4}$

NOTE: Do not use treated vines for feed or forage. Replanting: Soybeans may be replanted within 4 months; after 4 months, any crop may be planted except do not plant root crops within 12 months.

LOROX + Paraquat or ROUNDUP—Thoroughly mix LOROX DF and paraquat in spray tank according to directions under Paraquat Combinations or ROUNDUP Combinations below.

Soil Texture	Lbs. LOROX DF Per Acre	
	$\frac{1}{2}$ to 2% Organic Matter	2 to 5% Organic Matter
Sandy loam	1 to $1\frac{2}{3}$	$1\frac{2}{3}$ to 3
Silt loam	$1\frac{1}{4}$ to $2\frac{1}{3}$	$2\frac{1}{3}$ to 4
Clay loam	$1\frac{1}{3}$ to $2\frac{2}{3}$	$2\frac{2}{3}$ to 5 (Over 5% organic matter, use 6 lbs.)

REPLANTING: Soybeans or field corn may be replanted within 4 months; after 4 months, any crop may be planted.

PARAQUAT COMBINATIONS—Select and tank mix in water one of the above treatments; then add 1 to 2 pints paraquat (2 lbs. active per gal.) per acre for control of emerged weeds. Use the higher rate for weeds 4″ to 6″ tall. As the last ingredient, add $\frac{1}{2}$ pint Surfactant WK or ORTHO X-77 Spreader per 100 gals. of spray mixture. Maintain constant agitation. Use 20 to 60 gals. of water per acre. Use the higher gallonage for dense stubble or vegetation.

ROUNDUP Combinations—Select and tank mix one of the above treatments. As last ingredient, add $1\frac{1}{2}$ quarts of ROUNDUP per acre for control of emerged annual weeds or 2 to 4 quarts per acre for control of emerged perennial and annual weeds. Use 20 to 30 gals. of water per acre.

BRONCO Combinations—Mix LOROX DF with water in the spray tank as directed on the package label and then add BRONCO to the dilute mix. Maintain constant agitation. Use 20 to 30 gals. water per acre.

SOYBEANS

Directed Postemergence Application

Apply LOROX DF, alone or as a tank mixture with 2,4-DB or PREMERGE 3, as a directed spray to cover weed foliage with minimum contact of the soybean plant. Do not spray higher than 3″ on the soybean stem or crop injury may result. Do not spray over top of soybean plants. For broadcast application, use a single flood-type spray nozzle ("K" series or equivalent) per middle mounted on an oiling shoe or gauge wheel. For band treatment, use two nozzles per row mounted on an oiling shoe or gauge wheels, one on each side of row. To avoid spray drift, which may cause crop injury, do not exceed nozzle pressure of 25 psi nor use nozzle tips smaller than 8002 T-Jet (or equivalent) and do not spray under windy conditions. Add 1 pint Surfactant WK for each 25 gals. spray mixture. For best results, use a preemergence treatment (such as LOROX DF) or cultivation to control early weed growth and to increase the differential between height of soybeans and weeds. NOTE: Do not use on soils with less than $\frac{1}{2}$% organic matter. Do not apply more than 2 lbs. LOROX DF per acre per season for postemergence treatments. Do not apply within 60 days of harvest. Do not feed soybean forage or hay to livestock from fields treated postemergence. Harvested soybeans may be used for food, feed or oil purposes. Do not reenter treated areas for 24 hours following application unless protective clothing is worn.

LOROX DF Alone: For soybeans at least 8″ high and when weeds do not exceed 2″ in height, apply $\frac{1}{2}$ to 1 lb. per acre. Make a second application at same rate if new flush of weeds occurs.

For soybeans at least 12″ high and when weeds do not exceed 4″ in height, make a single application of 1 to 2 lbs. per acre. Alternatively, make a split application of 1 lb. per acre followed by a second application at the same rate after 1 week or later.

LOROX DF + 2,4-DB—For soybeans at least 8″ high and when weeds do not exceed 4″ in height, apply 1 lb. LOROX DF plus $\frac{1}{5}$ lb. 2,4-DB (13 fl. oz. BUTYRAC 200 or 1 pint BUTOXONE) per acre. A second application may be made if needed, but do not make more than 2 applications per season.

LOROX DF + PREMERGE 3—For soybeans at least 8″ high and when weeds do not exceed 3″ in height, apply 1 lb. LOROX DF plus 2 quarts of PREMERGE 3 per acre. Add 1 pint Surfactant WK for each 25 gals. spray mixture. A second application may be made if needed, but do not make more than 2 applications per season. Do not apply after soybeans begin to bloom.

NON-CROP WEED CONTROL

For short-term control of annual weeds on non-cropland areas such as roadsides and fence rows—apply 2 to 6 lbs. LOROX DF per acre in 40 to 100 gals. of water. For best results, apply shortly before weed growth begins or at early seedling stage of growth. For control of established annual weeds, add Surfactant WK at rate of 2 qts. per 100 gals. of spray mixture and apply as a thorough coverage spray during periods when daily temperatures exceed 70°F. and before weed growth exceeds 8″ in height. Do not reenter treated areas for 24 hours following application unless protective clothing is worn.

[1] Registered trademark of Union Carbide Agricultural Products Company, Inc.

[2] Registered trademark of Ciba-Geigy Corporation

[3] Registered trademark of Monsanto Company

[4] Paraquat contains 2 lbs. active per gal.

[5] Registered trademark of American Cynanamid Company

[6] Registered trademark of Elanco Products Company

[7] Registered trademark of Rhone-Poulenc Inc.

[8] Registered trademark of Dow Chemical Company

AG-475 1184 1/17/85

LOROX® L

HERBICIDE

SHAKE WELL BEFORE USING

KEEP FROM FREEZING

Contains 4 Lbs. Linuron Per Gallon

ACTIVE INGREDIENT:

Linuron [3-(3,4-dichlorophenyl)-1-methoxy-1-methylurea]	41%
INERT INGREDIENTS	59%
TOTAL	100%

EPA Reg. No. 352-391

Keep Out of Reach of Children

PRECAUTIONARY STATEMENTS

HAZARDS TO HUMANS AND DOMESTIC ANIMALS

CAUTION

HARMFUL IF SWALLOWED. MAY IRRITATE EYES, NOSE, THROAT, AND SKIN.

Avoid breathing spray mist. Avoid contact with skin, eyes, and clothing.

ENVIRONMENTAL HAZARDS

Do not apply directly to water or wetlands. Do not contaminate water by cleaning of equipment or disposal of wastes.

IMPORTANT

When mixing/loading and spraying wear a wide-brimmed hat, mid-forearm to elbow length chemical resistant gloves, and one-piece overalls which have long sleeves and long pants. Overalls should be constructed of finely woven fabric as specified in the USDA/EPA Guide for Commercial Applicators.

Any article of clothing worn while applying product must be cleaned before reusing. Clothing which has been drenched or has otherwise absorbed concentrated pesticide must be disposed in a sanitary landfill, or incinerated, or if allowed by State and local authorities, by burning.

Instead of clothing and equipment specified above, the applicator can use an enclosed tractor cab which provides a filtered air supply (as described by Taschenberg and Bourke, 1975).

Injury to or loss of desirable trees or other plants may result from failure to observe the following:

Do not apply (except as recommended for crop use), or drain or flush equipment on or near desirable trees of other plants, or on areas where their roots may extend, or in locations where the chemical may be washed or moved into contact with their roots. Do not use on home plantings of trees, shrubs or herbaceous plants, nor on lawns, walks, driveways, tennis courts, or similar areas. Prevent drift of dry powder or spray to desirable plants. Do not contaminate any body of water. Keep from contact with fertilizers, insecticides, fungicides, and seeds.

Do not apply aerially.

Thoroughly clean all traces of LOROX L from application equipment immediately after use. Flush tank, pump, hoses, and boom with several changes of water after removing nozzle tips and screens (clean these parts separately).

NOTICE OF WARRANTY

Du Pont warrants that this product conforms to the chemical description on the label thereof and is reasonably fit for purposes stated on such label only when used in accordance with directions under normal use conditions. It is impossible to eliminate all risks inherently associated with use of this product. Crop injury, ineffectiveness or other unintended consequences may result because of such factors as weather conditions, presence of other materials, or the manner of use or application, all of which are beyond the control of Du Pont. In no case shall Du Pont be liable for consequential, special or indirect damages resulting from the use or handling of this product. All such risks shall be assumed by the Buyer. DU PONT MAKES NO WARRANTIES OF MERCHANTABILITY OR FITNESS FOR A PARTICULAR PURPOSE NOR ANY OTHER EXPRESS OR IMPLIED WARRANTY EXCEPT AS STATED ABOVE.

STORAGE AND DISPOSAL

STORAGE

Keep from freezing. Store product in original container only, away from other pesticides, fertilizer, food, or feed.

DISPOSAL

Do not contaminate water, food or feed by storage or disposal. Wastes resulting from the use of this product may be disposed of on site or at approved waste disposal facility.

Triple rinse (or equivalent) the container. Then offer for recycling or reconditioning, or puncture and dispose of in a sanitary landfill, or incineration, or, if allowed by State and local authorities, by burning. If burned, stay out of smoke.

GENERAL INFORMATION

Du Pont LOROX L Herbicide is a liquid suspension to be mixed in water and applied as a spray for selective control of weeds in certain crops and for non-selective weed control on non-cropland areas. It is non-corrosive to equipment, non-flammable and non-volatile.

LOROX L may be applied to soil prior to emergence of weeds to control susceptible weed seedlings for an extended period of time; the degree of control and duration of effect will vary with the amount of chemical applied, soil texture, rainfall and other conditions. Soils high in clay or organic matter require higher dosages than soil low in clay or organic matter to obtain equivalent herbicide performance. Moisture is required to activate the chemical; best results occur if rainfall (or irrigation) occurs within 2 weeks of application. In the Columbia River Basin, use LOROX L only if crop is sprinkler irrigated.

LOROX L may also be used to control emerged weeds. Results vary with rate applied and environmental conditions; best results are obtained on succulent weeds growing under conditions of high humidity and temperatures of 70°F or higher. Addition of a surfactant such as Du Pont Surfactant WK to the spray (where recommended) increases contact effects of LOROX L.

LOROX L is recommended for use in conjunction with certain other herbicides for treatment of field corn, sorghum and soybeans; observe all cautions and limitations on labeling of all products used in mixtures.

Since the effect of LOROX L varies with soils, crop varieties, uniformity of application, and environmental conditions, it is suggested that growers limit their first use to small areas.

DIRECTIONS FOR USE

It is a violation of Federal law to use this product in a manner inconsistent with its labeling.

LOROX L Herbicide should be used only in accordance with recommendations on this label, or in separate published Du Pont recommendations available through local dealers. Do not use on any crop in Kern County, California, except carrots when a permit has been obtained from the Agricultural Commission.

Du Pont will not be responsible for losses or damages resulting from use of this product in any manner not specifically recommended by Du Pont. User assumes all risk associated with such non-recommended use.

SELECTIVE USE IN CROPS

PREEMERGENCE USE (Germinating Weeds): LOROX L, at recommended rates, controls weeds such as:

Broadleaves

Florida beggarweed	Mustard
Carpetweed	Nettleleaf goosefoot
Chickweed	Pigweed
Common dayflower	Purslane
Florida purslane (Florida pusley)	Wild radish
	Common ragweed
Galinsoga	Smartweed
Lambsquarters	

Grasses

Barnyardgrass (watergrass)	Foxtail (including giant)
Canarygrass	Goosegrass
Crabgrass	Fall panicum

Partial Control

Cocklebur	Sicklepod
Annual morningglory	Velvetleaf (buttonweed)
Prickly sida (teaweed)	

Note: Consult Amiben[1], atrazine, Bronco[3], Dual[2] 8E, Lasso[3], Paraquat[4], propazine, Prowl[5], Ramrod[3], Roundup[3], Surflan[6] 75W, Surflan[6] AS, Tolban[2], and Treflan[6] labels for additional weeds controlled when LOROX L is used in conjunction with these herbicides.

LOROX L will not control established perennials such as bermudagrass, Canada thistle, field bindweed, johnsongrass and purple nutsedge. The lower dosage rates are effective on the coarser soils and the higher rates on finer soils and on the more resistant seedling weeds. Sufficient moisture (½" to 1" on moist soils; 1" to 2" on dry soils) in the form of rainfall or sprinkler irrigation is necessary after treatment to carry the chemical into the root zone of germinating weeds; best results are obtained when this occurs within two weeks after application. If heavy rainfall occurs soon after application, injury to crop may result.

A good seed bed must be prepared before application of LOROX L as crop injury may result if application is made to ground which is cloddy or compacted resulting in improperly planted seed. Plant seed to depth specified. Surface of the soil should not be cultivated or disturbed after application of LOROX L and before emergence of the crop as weed control may be reduced and crop injury may result. However, if moisture is insufficient to activate the herbicide, a shallow cultivation (rotary hoe preferred) should be made after emergence of row crops while weeds are small enough to be controlled by mechanical means.

POSTEMERGENCE USE (Emerged Seedling Weeds): LOROX L, at recommended rates, controls weeds such as:

Broadleaves

Annual morningglory	Lambsquarters
Carpetweed	Mustard
Cocklebur	Nettleleaf goosefoot
Common dayflower	Pigweed
Common ragweed	Prickly sida (teaweed)
Dog fennel	
Fiddleneck (amsinckia)	Purslane
	Sebania
Florida beggarweed	Sicklepod
Florida purslane (Florida pusley)	Smartweed
	Velvetleaf (buttonweed)
Groundsel	
Knawel	Wild buckwheat

Grasses

Annual ryegrass	Fall panicum
Barnyardgrass (watergrass)	Foxtail (including giant)
Broadleaf signalgrass	Goosegrass
	Rattail fescue
Canarygrass	Texas panicum
Crabgrass	

Note: Consult bromoxynil, Butoxone[7], Butyrac[1] 200 and Premerge 3[8] labels for additional weeds controlled when LOROX L is used in conjunction with these herbicides.

Best results are obtained under conditions of high humidity and temperatures over 70°F. Control of emerged weeds under drought stress is usually impractical.

EQUIPMENT—SPRAY VOLUMES AND PRESSURES: Use a tractor-mounted fixed-boom sprayer properly calibrated to a constant speed and rate of delivery. Openings in screens should be equal to or larger than 50 mesh. Continuous agitation in the spray tank is required to keep the material in suspension. Agitate by mechanical or hydraulic means; if by-pass or return line is used, it should terminate at bottom of tank to minimize foaming. Avoid overlapping, and shut off spray booms while starting, turning, slowing or stopping, or injury to the crop may result.

For preemergence application, use 20 to 60 gals. per acre and spray pressure of 30 to 40 psi.

For postemergence application, use sufficient volume (min. 20 gals. per acre) for thorough coverage of weed foliage; use spray pressure of 20 to 25 psi to keep spray drift to a minimum.

Spray Preparation: Fill tank ¼ full with water. Start agitation system, add LOROX L, and continue adding water. Add separately each additional component of any tank mix while adding water. If a surfactant is used, add it last when the tank is nearly full. Continue agitation throughout. If poor mixing should occur with any component, premix the component with two parts water before adding to the spray tank.

Fertilizer Spray Mixtures—For preemergence application, nonpressure nitrogen or fertilizer solution may be used in the spray mixture unless otherwise directed. Small quantities should be tested for compatibility by the following procedure before full scale mixing:

1. Put 1 pint of fertilizer solution in a quart jar.
2. Mix 2 teaspoons LOROX L with 2 tablespoons of water; mix thoroughly and add to fertilizer solution.
3. Close jar and shake well.
4. If other herbicides are to be used in the mixture, premix 2 teaspoons of wettable powders or 1 teaspoon of liquids with 2 tablespoons of water; add to LOROX L/fertilizer mixture.
5. Close jar and shake well.
6. Watch mixture for several seconds; check again in 30 minutes.
7. If mixture does not separate, foam, gel, or become lumpy, it may be used.
8. Mixing ability may be improved by adding compatibility agents such as Kalo Laboratories Complex or Witco Chemicals Sponto 168D. Follow directions on container.

If the mixture is compatible, prepare spray by adding the fertilizer solution to the spray tank first; then dilute LOROX L with 2 parts water, thoroughly mix, and add slowly to spray tank with agitator running. For tank mixtures with other herbicides, follow directions above.

USE RATES: All dosages of LOROX L (and tank mixtures) are expressed as broadcast rates; for band treatment, use proportionately less. For example, use ⅓ of the broadcast rate when treating a 14" band where row spacing is 42". Where a range of dosages is given, use the lower rate on coarser soils (low in clay or organic matter) and the higher rate on finer soils (high in clay or organic matter); for postemergence application, use the lower rate on smaller weeds and the higher rate on larger weeds.

SOIL LIMITATIONS: Unless otherwise directed, do not use on sand, loamy sand, gravelly soils or exposed subsoils nor on soils containing less than 1% organic matter as injury to the treated or subsequent crops may result.

REPLANTING: If initial seeding fails to produce a stand, the same crop may be replanted in soil treated preemergence with LOROX L (or with recommended tank mixtures). Thoroughly rework soil before replanting; do not re-treat field with a second application as injury to the crop may result. Unless otherwise directed, do not replant treated areas to any other crop within 4 months after last application as injury to subsequent crops may result. West of Rocky Mountains—carrots or celery may be planted 4 months after last application; unless otherwise directed, do not plant any other crop until 1 year after the last application. Do not rotate crops used for food or feed, which are not registered for use with Lorox L, onto areas previously treated with this chemical.

ASPARAGUS—California (Includes control of dudain melon and nightshade):

Direct Seeded or Newly Planted Crowns—Do not exceed 4 pints total per acre per season; do not use surfactant or fertilizer solution in spray mixture.

Preemergence Application:* Make a single application of 2 to 4 pints per acre after planting seed 1½" deep in coarse soil and 1" deep in fine soils. During planting operation, spray activated charcoal (Aqua Nu-Char[9] or Gro-Safe[10]) as a 1" band on soil surface directly over seed rows at the rate of 300 lbs. per acre (equivalent to 15 lbs. per acre of crop with 20" row spacing).

Postemergence Application: Make 1 or 2 applications of 1 to 2 pints per acre when ferns are in 6" to 18" stage and weeds are not over 4" tall. Do not reenter treated areas for 24 hours following application unless protective clothing is worn.

Established Beds—Do not apply within 1 day of harvest; do not exceed 8 pints total per acre per season; do not use surfactant or fertilizer solution in spray mixture. Do not reenter treated areas for 24 hours following application unless protective clothing is worn.

Preemergence Application:* Make a single application of 2 to 4 pints per acre.

Postemergence Application: Make 1 to 4 applications of 1 to 2 pints per acre before weeds exceed 4" in height. Apply before cutting season or immediately after cutting. Do not reenter treated areas for 24 hours following application unless protective clothing is worn.

Directed Postemergence Application (Fern Stage): Make a single application of 4 to 8 pints per acre as a directed spray to base of plants for control of dudain melon. Do not reenter treated areas for 24 hours following application unless protective clothing is worn.

Note: If more than 4 pints LOROX L per acre is applied per season, do not plant any other crop until 1 year after last application.

* Preemergence weed control will be reduced in soils with high organic matter (greater than 5%) such as peat or muck.

CARROTS

Preemergence Application—Michigan and Wisconsin: Make a single application of 1 to 3 pints per acre after planting but before carrots emerge; plant seed at least ½" deep. Use the lower rate on lighter soils (low in clay or organic matter) and the higher rate on heavier soils (high in clay or organic matter). Subsequent postemergence application may be made provided the total does not exceed 4 pints LOROX L per acre per season.

Postemergence Application—U.S.: Apply 1½ to 3 pints per acre as a non-directed spray after carrots are at least 3" tall. Apply before annual grasses exceed 2" in height and before broadleaf weeds exceed 6" in height. Repeat application may be made but do not exceed 4 pints LOROX L per acre (West of Rocky Mountains, do not exceed 3 pints LOROX L per acre). Do not exceed 40 psi spray nozzle pressure as crop injury may result.

LOROX L may be applied following an application of Stoddard solvent provided treatments are at least one day apart. Stoddard solvent may be applied following an application of LOROX L provided treatments are at least 2 weeks apart. Shorter time intervals between applications may result in injury to the crop. Do not apply LOROX L as a tank mixture with Stoddard solvent, surfactants, nitrogen solution, other pesticides, nor when temperature exceeds 85°F as crop injury may result.

Do not reenter treated areas for 24 hours following application unless protective clothing is worn.

Note: Because carrot varieties vary in their resistance, determine tolerance to LOROX L prior to adoption as a field practice to prevent possible crop injury; do not treat susceptible varieties which show an initial burning of foliage following postemergence treatment with LOROX L.

CELERY

East of Rocky Mountains—Post-transplant Application: Make a single application of 1½ to 3 pints per acre. Apply as a non-directed spray after celery is transplanted and established, but before celery is 8" tall. Apply before annual grasses exceed 2" in height and before broadleaf weeds exceed 6" in height. In the Northeast, use only on celery grown on muck soils.

Do not exceed 40 psi spray nozzle pressure, and do not apply when temperature exceeds 85°F nor as a tank mixture with surfactants, nitrogen solution, or other pesticides as injury to the crop may result. Do not replant to crops other than celery or carrots within 4 months after application as injury to subsequent crops may result. Do not reenter treated areas for 24 hours following application unless protective clothing is worn.

CORN (FIELD)

East of Rocky Mountains—Preemergence Application: Select one of the following herbicide treatments for application as a tank mixture. Make a single application after planting but before crop emerges. Plant seed at least 1¾" deep on flat or raised seedbeds only or injury to the crop may result. Do not spray over top of emerged corn.

LOROX L + Lasso (Tank Mixture)

Soil Texture	Product Per Acre: 1 to 3% Organic Matter, LOROX L + (Pts.)	Lasso (Qts.)	3 to 6% Organic Matter, LOROX L + (Pts.)	Lasso (Qts.)
Sandy loam	⅔ to 1¼ +	¾ to 1	1¼ to 2 +	1 to 1½
Silt loam	1 to 1⅔ +	1 to 1½	1⅔ to 2½ +	1½ to 2
Clay loam	1¼ to 2 +	1½ to 2	2 to 3 +	2 to 2½

Replanting: Corn or soybeans may be replanted within 4 months; after 4 months, any crop may be planted.

LOROX L + Ramrod (Tank Mixture)

Soil Texture	Product Per Acre: 1 to 3% Organic Matter, LOROX L + (Pts.)	Ramrod 65 (Lbs.)	3 to 6% Organic Matter, LOROX L + (Pts.)	Ramrod 65 (Lbs.)
Sandy loam	⅔ to 1⅓ +	1 to 2	1⅓ to 2 +	2 to 3
Silt loam	1 to 1⅔ +	1½ to 2½	1⅔ to 2½ +	2½ to 4
Clay loam	1¼ to 2 +	2 to 3	2 to 3 +	3 to 4¾

Replanting: Corn may be replanted within 4 months; after 4 months, any crop may be planted.

LOROX L + atrazine (Tank Mixture)

Soil Texture	Product Per Acre: 1 to 2% Organic Matter, LOROX L + (Pts.)	atrazine 80% (Lbs.)	2 to 5% Organic Matter, LOROX L + (Pts.)	atrazine 80% (Lbs.)
Sandy loam	⅔ to 1 +	½ to ⅔	1 to 2 +	⅔ to 1¼
Silt loam	1 to 1½ +	⅔ to 1	1½ to 2½ +	1 to 1½
Clay loam	1⅓ to 1⅔ +	¾ to 1	1⅔ to 3 +	1 to 2

Replanting: Corn may be replanted within 6 months; after 6 months, any crop may be planted except do not follow treated corn with sugar beets, tobacco, or vegetables in rotation.

CORN (FIELD AND SWEET)

Directed Postemergence Application: Make a single application as a directed spray after corn is at least 15" high (measured to the highest leaf surface on free standing plants). Do not spray over top of corn. Apply only when there is sufficient differential between height of corn and weeds so that the directed spray thoroughly covers all weed foliage without contact of upper leaves or whorl of corn by spray or drift, as such contact may cause crop injury. Early cultivation (rotary hoe or other suitable equipment) will aid in achieving proper differential between height of corn and weeds.

Use $1\frac{1}{4}$ to 3 pints per acre; add 1 pint Surfactant WK for each 25 gals. spray mixture. For field corn, non-pressure nitrogen solution may be substituted for all or part of the water. Use the lower rate on lighter soils (low in clay or organic matter) and when weeds do not exceed 2″ in height; use the higher rate on heavier soils (high in clay or organic matter) for weeds up to 5″ in height.

Do not reenter treated areas for 24 hours following application unless protective clothing is worn.

COTTON

East of Rocky Mountains — Directed Postemergence Application: Apply as a directed spray; adjust nozzles to minimize contact of cotton leaves with spray or drift or crop injury may result. Do not spray over top of cotton.

Apply 1 to $1\frac{1}{2}$ pints per acre when cotton is at least 8″ tall and emerged weeds do not exceed 2″ in height; add 1 pint Surfactant WK for each 25 gals. spray mixture. If needed, a second application at same rate may be made one week or later after initial treatment. Alternatively, after cotton is 20″ tall, make a single application of 2 to 3 pints per acre following last cultivation; if emerged weeds are present, add Surfactant WK as directed above. **Note:** Do not use on Pima varieties of cotton. Do not graze treated fields or feed forage from treated areas to livestock. Do not feed gin trash to livestock. Do not reenter treated areas for 24 hours following application unless protective clothing is worn.

PARSNIPS

Preemergence Application: Make a single application of $1\frac{1}{2}$ to 3 pints per acre. Apply after planting but before crop emerges. Plant seed at least $\frac{1}{2}$″ deep.

POTATOES

Preemergence Application: Make a single application as a broadcast spray after planting but before crop emerges. Plant seed at least 2″ deep. Do not spray over top of emerged potatoes. If beds are to be "dragged" and/or "hilled", apply after the final "dragging" or "hilling" operation. Apply before grasses are 2″ tall and before broadleaf weeds are 6″ tall, preferably just before or when weed seedlings emerge. If emerged weeds are present, add 1 pint Surfactant WK for each 25 gals. spray mixture. In irrigated areas, best results are obtained when application is made to moist soil, followed within 2 weeks by 1″ to 2″ of sprinkler irrigation (or rainfall). On powder dry soils, irrigate prior to herbicide application and follow with sprinkler irrigation to activate the herbicide.

East of Rocky Mountains: Apply $1\frac{1}{2}$ to $2\frac{1}{2}$ pints per acre on the coarser soils (sandy loams, silt loams; 1 to 2% organic matter) and $2\frac{1}{2}$ to 4 pints per acre on finer soils (silts, clay loams; 2 to 5% organic matter); on soils over 5% organic matter, use 4 pints per acre and apply to emerged weeds (before potatoes emerge).

Wisconsin — Central Sands Area: Apply 1 pint per acre on sands and 2 pints per acre on loamy sands.

Northeast: For improved control of annual grasses (see Dual labeling), apply:

Soil Texture	Product Per Acre			
	1 to 3% Organic Matter		3 to 5% Organic Matter	
	LOROX L + (Pts.)	Dual 8E (Pts.)	LOROX L + (Pts.)	Dual 8E (Pts.)
Coarse Sandy loam	1 to $1\frac{1}{2}$ +	$1\frac{1}{2}$	$1\frac{1}{2}$ to 2 +	2
Medium Silt, Loam, Silt loam	$1\frac{1}{2}$ to 2 +	2	2 to $2\frac{1}{2}$ +	$2\frac{1}{2}$ to 3

SORGHUM

Preemergence Application: Select one of the following herbicide treatments for application as a tank mixture. Make a single application after planting but before crop emerges. Plant seed at least 1″ deep on flat or raised seedbeds only or injury to the crop may result. Do not spray over top of emerged sorghum.

LOROX L + propazine—Southwest (Tank Mixture)

Soil Texture	Product Per Acre			
	1 to 2% Organic Matter		2 to 4% Organic Matter	
	LOROX L + (Pts.)	propazine 80% (Lbs.)	LOROX L + (Pts.)	propazine 80% (Lbs.)
Sandy loam	$\frac{5}{8}$ to 1 +	$\frac{3}{8}$ to $\frac{5}{8}$	1 to $1\frac{1}{2}$ +	$\frac{5}{8}$ to 1
Silt loam, clay loam	1 to $1\frac{1}{2}$ +	$\frac{5}{8}$ to 1	1 to 2 +	$\frac{5}{8}$ to $1\frac{1}{4}$

Replanting: Do not follow treated sorghum with any fall crop, nor with sugar beets, tobacco, vegetables or potatoes in rotation. Prior to replanting, thorough seedbed preparation including fall or spring plowing is recommended.

LOROX L + Ramrod—Great Plains (Tank Mixture)

Soil Texture	Product Per Acre			
	1 to 3% Organic Matter		3 to 6% Organic Matter	
	LOROX L + (Pts.)	Ramrod 65 (Lbs.)	LOROX L + (Pts.)	Ramrod 65 (Lbs.)
Sandy loam	$\frac{2}{3}$ to $1\frac{1}{3}$ +	1 to 2	$1\frac{1}{3}$ to 2 +	2 to 3
Silt loam	1 to $1\frac{2}{3}$ +	$1\frac{1}{2}$ to $2\frac{1}{2}$	$1\frac{2}{3}$ to $2\frac{1}{2}$ +	$2\frac{1}{2}$ to 4
Clay loam	$1\frac{1}{4}$ to 2 +	2 to 3	2 to 3 +	3 to $4\frac{3}{4}$

Note: Do not graze or feed sorghum forage or silage from treated fields to dairy animals. **Replanting:** Sorghum or field corn may be replanted within 4 months; after 4 months, any crop may be planted.

Directed Postemergence Application: Make a single application of LOROX L as a directed spray; add 1 pint Surfactant WK for each 25 gals. spray mixture. If sprayer is equipped with skids, shoes or shields, apply 1 pint per acre when sorghum is 12″ tall (free standing plants) and weeds are up to 2″ in height; use 1 to 2 pints per acre when sorghum is 15″ tall and weeds are 2″ to 4″ in height. Apply only when there is sufficient differential between height of sorghum and weeds so that the directed spray thoroughly covers all weed foliage without contact of upper leaves or whorl of sorghum by spray or drift as such contact may cause crop injury. **Note:** Do not graze or feed plants to livestock within 3 months after postemergence application. Do not reenter treated areas for 24 hours following application unless protective clothing is worn.

SOYBEANS

Preemergence Application: Select one of the following herbicide treatments and make a single application after planting but before crop emerges. Plant seed at least $1\frac{3}{4}$″ deep on flat or raised seedbeds only or injury to the crop may result. Injury to soybeans may result if application is made to fields with standing water or fields too wet to cultivate. Do not spray over top of emerged soybeans. Do not use on sand or loamy sand nor any soil containing less organic matter than listed below:

Preemergence Following Treflan, Prowl or Tolban Preplant— Where Treflan, Prowl or Tolban has been used as a preplant incorporated treatment (according to directions on product label), apply LOROX L as a separate operation using rates recommended below for LOROX L alone. For rotation crops, follow instructions on Treflan, Prowl or Tolban labels.

LOROX L Alone—If weeds have emerged, add 1 pint Surfactant WK for each 25 gals. spray mixture.

Soil Texture	Pints LOROX L Per Acre	
	$\frac{1}{2}$ to 2% Organic Matter	2 to 5% Organic Matter
Sandy loam	1 to $1\frac{2}{3}$	$1\frac{2}{3}$ to 3
Silt loam	$1\frac{1}{4}$ to $2\frac{1}{3}$	$2\frac{1}{3}$ to 4
Clay loam	$1\frac{1}{3}$ to $2\frac{2}{3}$	$2\frac{2}{3}$ to 5 (Over 5% organic matter, use 6 pints)

Replanting: Soybeans or field corn may be replanted within 4 months; after 4 months, any crop may be planted.

LOROX L + Du Pont Lexone, DF™ Herbicide + Lasso—For control of prickly sida, sicklepod, hemp sesbania, lambsquarters, pigweed, velvetleaf, fall panicum, foxtails, barnyard grass, broadleaf signalgrass, and goosegrass. The combination also offers partial control of cocklebur, jimsonweed and common ragweed.

Soil Texture	Product Per Acre 1 to 3% Organic Matter LOROX L (Pts.) + Lexone DF (Lbs.) + Lasso (Qts.)	Product Per Acre 3 to 6% Organic Matter LOROX L (Pts.) + Lexone DF (Lbs.) + Lasso (Qts.)
Coarse Sandy loam	1/3 to 1/2 + 1/8 to 1/4 + 3/4 to 1 or (Lexone 4L—1/4 to 2/5 pt.)	1/3 to 3/4 + 1/4 to 1/3 + 1 to 1 1/2 or (Lexone 4L—2/5 to 1/2 pt.)
Medium Loam, Silt loam, Silt, Sandy clay, Sandy clay loam	1/2 to 3/4 + 1/4 to 1/3 + 1 to 1 1/2 or (Lexone 4L—2/5 to 1/2 pt.)	3/4 to 1 1/2 + 1/3 to 1/2 + 1 1/2 to 2 or (Lexone 4L—1/2 to 3/4 pt.)
Fine Silty clay, Silty clay loam, Clay, Clay loam	3/4 to 1 1/2 + 1/3 to 1/2 + 1 1/4 to 2 or (Lexone 4L—1/2 to 3/4 pt.)	1 1/2 to 2 + 1/2 to 2/3 + 2 to 3 or (Lexone 4L—3/4 to 1 pt.)

Note: Soybean varieties such as Altona, Coker 102 and 156, Govan, NKS 1884, Semmes, Tracy, Vansoy, Terra Vig 505 and 606, and AP 71 are sensitive to Lexone DF. Injury may occur if Lexone DF is used on these varieties. Before use on any other soybean variety, tolerance to Lexone DF must first be determined. Injury to soybeans may occur if Lexone DF is used on soils having a calcareous surface layer or pH of 7.5 or higher, or if used in conjunction with soil applied organic phosphate pesticides such as Dasanit, Disyton, Mocap, Nemacur, Thimet, Parathion, Lorsban 15G or Counter.

Do not use treated plants for feed or forage. Injury may occur if atrazine was applied on the soil the year before. Seedling disease, cold weather, deep planting (more than 2"), excessive moisture, high soil pH (7.5 or higher), high salt concentration, or drought may weaken soybean seedlings and increase possibility of crop injury.

Do not use on sand, loamy sand, nor on sandy loam with less than 1% organic matter.

Cultivation after planting: Treated soil may be shallow-cultivated, rotary-hoed, or hand-hoed without reducing the weed control activity of the tank mixture. Do not cultivate deeper than the treated layer of soil since this may bring untreated soil to the surface and poor weed control may result.

Replanting: If initial seeding fails to produce a stand, treated fields may be replanted to soybeans; do not rework soil; do not retreat field with a second application as injury to the crop may result. Do not replant treated areas to any crop other than soybeans within four months after treatment as injury to subsequent crops may result.

LOROX L + Lasso

Soil Texture	Product Per Acre 1/2 to 3% Organic Matter LOROX L + (Pts.)	Lasso (Qts.)	3 to 6% Organic matter LOROX L + (Pts.)	Lasso (Qts.)
Sandy loam	2/3 to 1 1/4 +	3/4 to 1	1 1/4 to 2 +	1 to 1 1/2
Silt loam	1 to 1 2/3 +	1 to 1 1/2	1 2/3 to 2 1/2 +	1 1/2 to 2
Clay loam	1 1/4 to 2 +	1 1/2 to 2	2 to 3 +	2 to 2 1/2

For control of black nightshade, apply:

Soil Texture	Product Per Acre LOROX L (Pts.)	Lasso (Qts.)
Sandy loam	3/4 to 1 1/2	2
Silt loam	1 to 1 2/3	2 1/2 to 3
Clay loam	1 1/4 to 2	3

Replanting: Soybeans or field corn may be replanted within 4 months; after 4 months, any crop may be planted.

LOROX L + Lexone DF + Dual 8E—For control of prickly sida, sicklepod, hemp sesbania, lambsquarters, and pigweed. The combination also offers partial control of cocklebur, jimsonweed and common ragweed.

Soil Texture	Product Per Acre 1/2 to 3% Organic Matter LOROX L (Pts.) +	Lexone DF (Lbs.) +	Dual 8E (Pts.)	3 to 6% Organic Matter LOROX L (Pts.) +	Lexone DF (Lbs.) +	Dual 8E (Pts.)
Coarse Sandy loam	1/3 to 1/2	1/8 to 1/4 or (Lexone 4L—1/4 to 2/5 pt.)	1 to 1 1/4	1/3 to 3/4	1/4 to 1/3 or (Lexone 4L—2/5 to 1/2 pt.)	1 1/4 to 1 1/2
Medium Loam, Silt loam, Silt Sandy clay, Sandy clay, loam	1/2 to 3/4	1/4 to 1/3 or (Lexone 4L—2/5 to 1/2 pt.)	1 1/4 to 1 1/2	3/4 to 1 1/2	1/3 to 1/2 or (Lexone 4L—1/2 to 3/4 pt.)	1 1/2 to 2
Fine Silty clay, Silty clay loam, Clay, Clay loam	3/4 to 1 1/2	1/3 to 1/2 or (Lexone 4L—1/2 to 3/4 pt.)	1 1/2 to 2	1 1/2 to 2	1/2 to 2/3 or (Lexone 4L—3/4 to 1 pt.)	2 to 2 1/2

Note: Soybean varieties such as Altona, Coker 102 and 156, Govan, NKS 1884, Semmes, Tracy, Vansoy, Terra Vig 505 and 606, and AP 71 are sensitive to Lexone DF. Injury may occur if Lexone DF is used on these varieties. Before use on any other soybean variety, tolerance to Lexone DF must first be determined. Injury to soybeans may occur if Lexone DF is used on soils having a calcareous surface layer or pH of 7.5 or higher, or if used in conjunction with soil applied organic phosphate pesticides such as Dasanit, Disyston, Mocap, Nemacur, Thimet, Parathion, Lorsban 15G or Counter.

Do not use treated plants for feed or forage. Injury may occur if atrazine was applied on the soil the year before. Seedling disease, cold weather, deep planting (more than 2"), excessive moisture, high soil pH (7.5 or higher), high salt concentration, or drought may weaken soybean seedlings and increase possibility of crop injury.

Do not use on sand, loam sand, nor on sandy loam with less than 1% organic matter.

Cultivation after planting: Treated soil may be shallow-cultivated, rotary-hoed, or hand-hoed without reducing the weed control activity of the tank mixture. Do not cultivate deeper than the treated layer of soil since this may bring untreated soil to the surface and poor weed control may result.

Replanting: If initial seeding fails to produce a stand, treated fields may be replanted to soybeans, do not rework soil, do not retreat field with a second application as injury to the crop may result. Do not replant treated areas to any crop other than soybeans within four months after treatment as injury to subsequent crops may result.

LOROX L + Dual 8E

Soil Texture	Product Per Acre 1/2 to 3% Organic Matter LOROX L + (Pts.)	Dual 8E (Pts.)	3 to 6% Organic Matter LOROX L + (Pts.)	Dual 8E (Pts.)
Sandy loam	2/3 to 1 1/4 +	1 1/4	1 1/4 to 2 +	1 1/2
Silt loam	1 to 1 2/3 +	1 1/2	1 2/3 to 2 1/2 +	2
Clay loam	1 1/4 to 2 +	2	2 to 3 +	2 to 2 1/2

Note: Do not graze or feed forage from treated areas to livestock.
Replanting: Soybeans or field corn may be replanted within 4 months; for rotation crops, follow instructions on Dual 8E label.

For control of black nightshade in Kentucky, Illinois, Indiana and Ohio:

Soil Texture	Product Per Acre (1 to 3% Organic Matter) LOROX L (Pts.)	Dual 8E (Pts.)
Sandy loam	3/4 to 1 1/2	1 1/2
Silt loam	1 to 1 2/3	2
Clay loam	1 1/4 to 2	2 to 2 1/2

Do not use on sand or loamy sands.

Replanting: Soybeans or field corn may be replanted within four months; after four months, any crop may be planted.

LOROX L + Surflan 75W or Surflan AS

Soil Texture	Product Per Acre (½ to 3% Organic Matter) LOROX L (Pts.)	+	Surflan 75W (Lbs.)	or	Surflan AS (Qts.)
Sandy loam	⅔ to 1¼	+	⅔ to 1	or	½ to ¾
Silt loam	1 to 1⅔	+	1 to 1⅓	or	¾ to 1
Clay loam	1¼ to 2	+	1⅓ to 1⅔	or	1 to 1¼

Note: Do not use treated vines for feed or forage.
Replanting: Soybeans may be replanted within 4 months; after 4 months, any crop may be planted except do not plant root crops within 12 months.

LOROX L + Prowl

Soil Texture	Product Per Acre					
	1 to 3% Organic Matter			3 to 6% Organic Matter		
	LOROX L (Pts.)	+	Prowl (Pts.)	LOROX L (Pts.)	+	Prowl (Pts.)
Sandy loam	⅔ to 1¼	+	1 to 1½	1¼ to 2	+	1½
Silt loam	1 to 1⅔	+	1½ to 2	1⅔ to 2½	+	1½ to 2
Clay loam	1¼ to 2	+	1½ to 2	2 to 3	+	2 to 2½

Replanting: Soybeans or field corn may be replanted within 4 months; other crops may be planted the following year.

LOROX L + Amiben

Soil Texture	Product Per Acre					
	½ to 3% Organic Matter			3 to 6% Organic Matter		
	LOROX L (Pts.)	+	Amiben (Qts.)	LOROX L (Pts.)	+	Amiben (Qts.)
Sandy loam	⅔ to 1¼	+	3	1¼ to 2	+	3
Silt loam	1 to 1⅔	+	3	1⅔ to 2½	+	4
Clay loam	1¼ to 2	+	4	2 to 3	+	5

Replanting: Soybeans or field corn may be replanted within 4 months; after 4 months, any crop may be planted.

LOROX L + Ramrod—Seed Crop Only

Soil Texture	Product Per Acre					
	1 to 3% Organic Matter			3 to 6% Organic Matter		
	LOROX L (Pts.)	+	Ramrod 65 (Lbs.)	LOROX L (Pts.)	+	Ramrod 65 (Lbs.)
Sandy loam	⅔ to 1¼	+	1 to 2	1¼ to 2	+	2 to 3
Silt loam	1 to 1⅔	+	1½ to 2½	1⅔ to 2½	+	2½ to 4
Clay loam	1¼ to 2	+	2 to 3	2 to 3	+	3 to 4¾

Note: Do not graze or feed forage from treated areas to livestock; do not use seed for food, feed, or oil purposes. **Replanting:** Soybeans, sorghum, or field corn may be replanted within 4 months; after 4 months, any crop may be planted.

SOYBEANS

Minimum or Non-Tillage: LOROX L preemergence to soybeans may be used with paraquat, Bronco, or Roundup in tank mix combinations for preemergence and postemergence control of many broadleaf weeds and grasses and for top kill or suppression of some perennials where soybeans will be planted directly into a preformed bed (stale seed bed), cover crop, or in previous crop residues such as corn or small grain stubble. For conventional seed bed preparation in reduced tillage systems, use rates as described for **Preemergence application.** Addition of Lasso, Dual 8E, Surflan 75W or Surflan AS, to the tank mixture improves control of grasses and volunteer small grains. Apply with ground equipment immediately before, during, or after planting but before crop emerges. Maintain constant agitation of spray mixture.

LOROX L + Lasso + Paraquat or Roundup—Thoroughly mix LOROX L and Lasso in spray tank first according to directions; then add paraquat or Roundup as directed under **Paraquat Combinations** or **Roundup Combinations** below.

Soil Texture	Product Per Acre					
	½ to 3% Organic Matter			3 to 6% Organic Matter		
	LOROX L (Pts.)	+	Lasso (Qts.)	LOROX L (Pts.)	+	Lasso (Qts.)
Sandy loam	¾ to 1½	+	2	1½ to 2⅖	+	2½
Silt loam	1⅛ to 2	+	2½	2 to 3	+	3
Clay loam	1¼ to 2¼	+	2½	2¼ to 3½	+	3

For control of black nightshade, apply:

Soil Texture	Product Per Acre (1 to 3% Organic Matter) LOROX L (Pts.)	Lasso (Qts.)
Sandy loam	¾ to 1½	2
Silt loam	1 to 2	2½ to 3
Clay loam	1¼ to 2¼	3

Replanting: Soybeans or field corn may be replanted within 4 months; after 4 months, any crop may be planted.

LOROX L + Bronco—Do not add paraquat or Roundup to this combination.

Soil Texture[a]	Product Per Acre (½ to 5% Organic Matter) LOROX L (Pts.)	+	Bronco[b] (Qts.)
Sandy loam	¾ to 2	+	3¼ to 5
Silt loam	1⅛ to 2½	+	4 to 5
Clay loam	1¼ to 3	+	4 to 5

[a] Do not use on soils with less than ½% organic matter nor on sand, loamy sand, or muck soils as crop injury may result.

[b] Use the higher rate for dense stubble, heavy crop residue, or heavy weed population.

Replanting Soybeans: If initial seeding fails to produce a stand, treated fields may be replanted to soybeans; do not rework soil; do not retreat field with a second application as injury to the crop may result. Do not replant treated areas to any crop other than soybeans within four months after treatment as injury to subsequent crops may result.

LOROX L + Dual 8E + Paraquat or Roundup—Thoroughly mix LOROX L and Dual 8E in spray tank first according to directions; then add paraquat or Roundup as directed under **Paraquat Combinations** or **Roundup Combinations** below.

Soil Texture	Product Per Acre					
	½ to 3% Organic Matter			3 to 6% Organic Matter		
	LOROX L (Pts.)	+	Dual 8E (Pts.)	LOROX L (Pts.)	+	Dual 8E (Pts.)
Sandy loam	¾ to 1½	+	1½	1½ to 2⅖	+	1½
Silt loam	1⅛ to 2	+	2	2 to 3	+	2
Clay loam	1¼ to 2¼	+	2 to 2½	2 to 3½	+	2 to 2½

Note: Do not graze or feed forage from treated areas to livestock. **Replanting:** Soybeans or field corn may be replanted within 4 months; for rotation crops, follow instructions on Dual 8E label.

LOROX L + Surflan 75W or Surflan AS + Paraquat or Roundup—Thoroughly mix Lorox L and Surflan in spray tank first according to directions; then add paraquat or Roundup as directed under **Paraquat Combinations** or **Roundup Combinations** below.

Soil Texture	Product Per Acre (½ to 3% Organic Matter) LOROX L (Pts.)	+	Surflan 75W (Lbs.)	or	Surflan AS (Qts.)
Sandy loam	¾ to 1½	+	⅔ to 1	or	½ to ¾
Silt loam	1⅛ to 2	+	1 to 1⅓	or	¾ to 1
Clay loam	1¼ to 2¼	+	1⅓ to 1⅔	or	1 to 1¼

Note: Do not use treated vines for feed or forage. **Replanting:** Soybeans may be replanted within 4 months; after 4 months, any crop may be planted except do not plant root crops within 12 months.

LOROX L + Paraquat or Roundup—Thoroughly mix LOROX L and paraquat in spray tank according to directions under **Paraquat Combinations** or **Roundup Combinations** below.

Soil Texture	Pints LOROX L Per Acre ½ to 2% Organic Matter	2 to 5% Organic Matter
Sandy loam	1 to 1⅔	1⅔ to 3
Silt loam	1¼ to 2⅓	2⅓ to 4
Clay loam	1⅓ to 2⅔	2⅔ to 5
		(Over 5% organic matter, use 6 lbs.)

Replanting: Soybeans or field corn may be replanted within 4 months; after 4 months, any crop may be planted.

Paraquat Combinations—Select and tank mix in water one of the above treatments; then add 1 to 2 pints paraquat (2 lbs. active per gal.) per acre for control of emerged weeds. Use the higher rate for weeds 4″ to 6″ tall. As the last ingredient, add ½ pint Surfactant WK or Ortho X-77 Spreader per 100 gals. of spray mixture. Maintain constant agitation. Use 20 to 60 gals. of water per acre. Use the higher gallonage for dense stubble or vegetation.

Roundup Combinations—Select and tank mix one of the above treatments. As last ingredient, add 1½ quarts of Roundup per acre for control of emerged annual weeds or 2 to 4 quarts per acre for control of emerged perennial and annual weeds. Use 20 to 30 gals. of water per acre.

Bronco Combinations—Mix LOROX L with water in the spray tank as directed on the package label and then add Bronco to the dilute mix. Maintain constant agitation. Use 20 to 30 gals. water per acre.

SOYBEANS

Directed Postemergence Application: Apply LOROX L alone or as a tank mixture with 2,4-DB or Premerge 3, as a directed spray to cover weed foliage with minimum contact of the soybean plant. Do **not spray higher than 3″ on the soybean stem or crop injury may result. Do not spray over top of soybean plants. For broadcast appplication, use a single flood-type spray nozzle ("K" series or equivalent) per middle mounted on an oiling shoe or gauge wheel. For band treatment, use two nozzles per row mounted on an oiling shoe or gauge wheels, one on each side of row. To avoid spray drift, which may cause crop injury, do not exceed nozzle pressure of 25 psi nor use nozzle tips smaller than 8002 T-Jet (or equivalent) and do not spray under windy conditions. Add 1 pint Surfactant WK for each 25 gals. spray mixture. For best results, use a preemergence treatment (such as LOROX L) or cultivation to control early weed growth and to increase the differential between height of soybeans and weeds. NOTE: Do not use on soils with less than ½% organic matter. Do not apply more than 2 pints LOROX L per acre per season for postemergence treatments. Do not apply within 60 days of harvest. Do not feed soybean forage or hay to livestock from fields treated postemergence. Harvested soybeans may be used for food, feed, or oil purposes. Do not reenter treated areas for 24 hours following application unless protective clothing is worn.**

LOROX L Alone: For soybeans at least 8″ high and when weeds do not exceed 2″ in height, apply ½ to 1 lbs. per acre. Make a second application at same rate if new flush of weeds occurs. For soybeans at least 12″ high and when weeds do not exceed 4″ in height, make a single application of 1 to 2 lbs. per acre. Alternatively, make a split application of 1 lb. per acre followed by a second application at the same rate after 1 week or later.

LOROX L + 2,4-DB: For soybeans at least 8″ high and when weeds do not exceed 4″ in height, apply 1 lb. LOROX L plus ⅕ lb. 2,4-DB (13 fl. oz. Butyrac 200 or 1 pt. Butoxone) per acre. A second application may be made if needed, but do not make more than 2 applications per season.

LOROX L + Premerge 3 (Tank Mixture): For soybeans at least 8″ high and when weeds do not exceed 3″ in height, apply 1 pint LOROX L plus 2 quarts of Premerge 3 per acre. Add 1 pint Surfactant WK for each 25 gals. spray mixture. A second application may be made if needed, but do not make more than 2 applications per season. Do not apply after soybeans begin to bloom.

NON-CROP WEED CONTROL

For short-term control of annual weeds on non-cropland areas such as roadsides and fence rows—apply 2 to 6 pints LOROX L per acre in 40 to 100 gals. of water. For best results, apply shortly before weed growth begins or at early seedling stage of growth. For control of established annual weeds, add Surfactant WK at rate of 2 qts. per 100 gals. of spray mixture and apply as a thorough coverage spray during periods when daily temperatures exceed 70°F and before weed growth exceeds 8″ in height. Do not reenter treated areas for 24 hours following application unless protective clothing is worn.

1 Registered trademark of Union Carbide Agricultural Products Company, Inc.

2 Registered trademark of Ciba-Geigy Corporation

3 Registered trademark of Monsanto Company

4 Paraquat contains 2 lbs. active per gal.

5 Registered trademark of American Cyanamid Company

6 Registered trademark of Elanco Products Company

7 Registered trademark of Rhone-Poulenc Inc.

8 Registered trademark of Dow Chemical Company

9 Registered trademark of Westvaco Corporation

10 Registered trademark of ICI United States, Inc.

AG-499 1184 1/10/85

SUPPLEMENTAL LABELING

LOROX L HERBICIDE

EPA Reg. No. 352-391

LOROX® L HERBICIDE AS A TANK MIXTURE WITH DUAL 8E FOR WEED CONTROL IN POTATOES EAST OF MISSISSIPPI RIVER

DIRECTIONS FOR USE

It is a violation of federal law to use this product in a manner inconsistent with its labeling.

DuPont Lorox® L Herbicide plus Dual 8E is recommended as a tank mixture for application preemergence to potatoes to control annual weeds. See product labels for weeds controlled.

HOW TO USE

Make a single application as a broadcast spray after planting but before crop emerges. Plant seed at least 2″ deep. Do not spray over top of emerged potatoes. If beds are to be "dragged" and/or "hilled", apply after the final "dragging" or "hilling" operation. Apply before grasses are 2″ tall and before broadleaf weeds are 6″ tall, preferably just before or when weed seedlings emerge. If emerged weeds are present, add 1 pint Surfactant® WK for each 25 gallons of spray mixture. In irrigated areas, best results are obtained when application is made to moist soil, followed within 2 weeks by 1″ to 2″ of sprinkler irrigation (or rainfall). On powder dry soils, irrigate prior to herbicide application and follow with sprinkler irrigation to activate the herbicide.

Application:

Soil Texture	1 to 3% Organic Matter LOROX L (Pts.)	1 to 3% Organic Matter Dual 8E (Pts.)	3 to 5% Organic Matter LOROX L (Pts.)	3 to 5% Organic Matter Dual 8E (Pts.)
Coarse Sandy loam	1 to 1½	1½	1½ to 2	2
Medium Silt, Loam, Silt loam	1½ to 2	2	2 to 2½	2½ to 3

Product Per Acre

IMPORTANT

BEFORE USING LOROX® L OR DUAL 8E, READ AND CAREFULLY OBSERVE THE CAUTIONARY STATEMENTS AND ALL OTHER INFORMATION APPEARING ON THE PRODUCT LABELS.

012584

This bulletin contains new or supplement instructions for use of these products which may not appear on the package labels. Follow the instructions carefully.

This labeling must be in the possession of the user at the time of pesticide application.

MANZATE® D

FUNGICIDE WITH A ZINC SALT ADDED

ACTIVE INGREDIENT:
Maneb (Manganese ethylenebisdithiocarbamate) 80%
(Manganese equivalent as metallic 16.5%)
INERT INGREDIENTS 20%

TOTAL 100%

EPA Reg. No. 352-291 U.S. Pat. 3,379,610

Keep Out of Reach of Children

PRECAUTIONARY STATEMENTS

HAZARDS TO HUMANS

CAUTION

MAY IRRITATE EYES, NOSE, THROAT, AND SKIN. MAY BE HARMFUL IF INHALED OR SWALLOWED.

Avoid breathing dust or spray mist. Avoid contact with skin, eyes, and clothing. Keep away from fire or sparks. Wear protective clothing including long pants, long sleeve shirt, gloves, hat and boots during mixing and loading operations. Do not apply this product in such a manner as to directly or through drift expose workers, other persons, or animals. The area being treated must be vacated by unprotected persons.

FIRST AID

In case of contact, flush skin or eyes with plenty of water; for eyes, get medical attention.

ENVIRONMENTAL HAZARDS

This product is toxic to fish. Do not apply directly to water. Do not apply where runoff is likely to occur. Do not apply when weather conditions favor drift from areas treated.

GENERAL INFORMATION

Du Pont MANZATE D Fungicide, a special formulation of maneb with a zinc salt added, may be used as a spray or dust for the control of many important plant diseases. MANZATE D has good adhesive properties on plants, high fungicidal efficiency and is easy to use. It is particularly effective against many of the fungus diseases of certain vegetable, fruit and ornamental crops.

MANZATE D may be used undiluted for treatment of certain crop seeds.

NOTICE OF WARRANTY

Du Pont warrants that this product conforms to the chemical description on the label thereof and is reasonably fit for purposes stated on such label only when used in accordance with the directions under normal use conditions. It is impossible to eliminate all risks inherently associated with the use of this product. Crop injury, ineffectiveness or other unintended consequences may result because of such factors as weather conditions, presence of other materials, or the manner of use or application, all of which are beyond the control of Du Pont. In no case shall Du Pont be liable for consequential, special or indirect damages resulting from the use or handling of this product. All such risks shall be assumed by the buyer. DU PONT MAKES NO WARRANTIES OF MERCHANTABILITY OR FITNESS FOR A PARTICULAR PURPOSE NOR ANY OTHER EXPRESS OR IMPLIED WARRANTY EXCEPT AS STATED ABOVE.

DIRECTIONS FOR USE

It is a violation of federal law to use this product in a manner inconsistent with its labeling.

MANZATE D should be used only in accordance with recommendations on this label, or in separate published Du Pont recommendations available through local dealers.

Du Pont will not be responsible for losses or damages resulting from use of this product in any manner not specifically recommended by Du Pont. User assumes all risks associated with such nonrecommended use.

Do not enter treated areas without protective clothing until sprays have dried or dusts have settled. Because certain states may require more restrictive reentry intervals for various crops treated with this product, consult your State Department of Agriculture for further information.

Written or oral warnings must be given to workers who are expected to be in a treated area or in an area about to be treated with this product. When oral warnings are given, warnings shall be given in a language customarily understood by workers. Oral warnings must be given if there is reason to believe that written warnings cannot be understood by workers. Warnings must include the following information: "CAUTION. Area treated with MANZATE D on (date of application). Do not enter without appropriate protective clothing until sprays have dried or dusts have settled. In case of contact, flush skin or eyes with plenty of water; for eyes, get medical attention."

SPRAYS*—Use MANZATE D at a rate of 1½ lbs. per 100 gals. of spray, unless otherwise stated. In case of severe disease outbreaks or when weather conditions favor disease, use 2 lbs. per 100 gals. of spray and shorten interval between applications. Wash required amount of product through the filler screen while tank is filling with water. Add a spreader-sticker spray adjuvant if needed.

* For small quantities, 3 to 4 level teaspoonfuls of MANZATE D per 1 gal. of water is equivalent to 1½ to 2 lbs. per 100 gals.

DUSTS—MANZATE D may be used at concentrations of 6 to 8% (4.8 to 6.4% active) unless otherwise stated, by diluting and thoroughly mixing with pyrophyllite or some other neutral diluent. Use dust mixtures as soon as possible after preparation. Commonly used insecticides may displace an equivalent amount of diluent in preparing the dust mixtures.

ALMONDS (In California): Shot-hole (fungus) and brown rot (blossom and twig blight)—Use 1½ to 2 lbs. per 100 gals. (6 to 8 lbs. per acre) at the popcorn, full-bloom and petal-fall stages (omit petal-fall spray if only brown rot is present). Add a spreader-sticker.

TOMATOES: Early blight, late blight, Septoria leaf spot, gray leaf spot (Stemphylium) and anthracnose—Apply when fruit of first clusters is well formed; repeat at 7 to 10-day intervals. (Do not use on young seedlings in greenhouses as injury may occur.) When weather conditions favor disease, more frequent applications may be necessary starting when plants are set in fields. Tomato plants treated with MANZATE D tend to develop heavy foliage and spray gallonage should be increased accordingly to assure complete coverage. In areas where anthracnose is a serious problem on canning tomatoes, apply 3 lbs. MANZATE D per acre. Do not apply within 5 days of harvest.

POTATOES: Early and late blights—Use 1 to 2 lbs. per acre. Apply as soon as plants are 2 to 6 inches high; repeat at 5 to 10-day intervals as long as blights threaten. Under severe disease conditions, use the shorter interval or the higher dosage.

APPLES: Scab, cedar apple rust, black rot (fruit and foliage), fly speck, sooty blotch, bitter rot, and white rot (Botryosphaeria)—Use 1 to 1½ lbs. per 100 gals. and apply up to 10 pounds product per acre per application (for low volume sprays, adjust accordingly). Add a spreader-sticker. Apply at 7 to 14-day intervals throughout the season. For scab, begin at green tip or delayed dormant stage; for cedar apple rust, at blossom time; for other diseases, at first cover. Do not apply within 30 days of harvest; do not graze livestock in treated areas.

CABBAGE: Alternaria leaf spot and downy mildew—Use 1½ lbs. per 100 gals. and apply up to 150 gals. per acre. Add spreader-sticker. As a dust, use up to 40 lbs. per acre. Plant beds and direct seeded fields—Apply 7 to 10 days after planting, or earlier if disease is present. Repeat at 3-day intervals until plants are set in the field or until field-seeded plants are thinned. Fields—Apply when the disease threatens; repeat weekly as needed. Shorten interval to 3 to 5 days under severe conditions. Do not apply within 7 days of harvest.

CELERY: Early blight (Cercospora) and late blight (Septoria)—Apply when plants are established in the field; repeat as frequently as required (weekly under average conditions; every 3 to 5 days under severe disease conditions). Do not apply within 14 days of harvest. Remove residues by stripping, trimming and washing.

CARROTS: Leaf spot diseases—Apply when plants are 6 weeks old (or when disease threatens); repeat every 7 to 10 days.

WATERMELONS: Downy mildew, anthracnose, Cercospora leaf spot and gummy stem blight.

CUCUMBERS: Downy mildew, anthracnose (also Pythium fruit rot in South Carolina)—Use 2 to 3 lbs. per acre. Apply as soon as plants begin to run or when disease threatens; repeat at 7 to 10-day intervals as needed. Under severe conditions, shorten intervals to 3 to 5 days. Do not apply within 5 days of harvest.

CANTALOUPES and **SQUASH** (in Texas): Downy mildew—Apply as soon as plants begin to run or when disease threatens; repeat at 7 to 10-day intervals as needed. Do not apply within 5 days of harvest.

LETTUCE and **ENDIVE:** Downy mildew—As a spray, use up to 150 gals. per acre and as a dust, up to 50 lbs. per acre. Apply when disease appears; repeat at 7 to 10-day intervals under average conditions, 3 to 5 days when disease conditions are severe. Do not apply within 10 days of harvest. Remove residues from head lettuce by stripping and trimming, and from leaf lettuce and endive by washing or other effective means.

LIMA BEANS: Downy mildew and anthracnose—use up to 200 gals. spray or 30 lbs. dust per acre. Start applications during early blooming stage, or when disease first appears; repeat at 4 to 7-day intervals as needed. Do not apply within 4 days of harvest. Do not feed treated lima bean forage or hay to livestock.

ONIONS: Downy mildew, purple blotch (blight) and blast (Botrytis leaf blight)—Apply when first spots appear; repeat weekly as needed. Add a spreader-sticker.

SPINACH: Downy mildew (blue mold) and white rust—Apply when disease first appears in area; repeat at 7 to 10-day intervals. Do not apply within 10 days of harvest. Remove residues by washing or other effective means.

PEACHES (Pacific States) and **NECTARINES** (California): Shot-hole (fungus) and peach leaf curl (California only), apply dormant and popcorn sprays. For brown rot blossom blight, apply at pink bud and 25% to 75% of full-bloom (in Pacific Northwest add a petal-fall spray). Use 1½ to 2 lbs. per 100 gals. (6 to 8 lbs. per acre). For fruit brown rot (California only), apply 2 lbs. per 100 gals. (about 10 lbs. per acre) 4 weeks before harvest, or earlier if rot is easily found on green fruit, and repeat at 7 to 14-day intervals as needed. Add a spreader-sticker. Do not apply to nectarines later than 14 days before harvest. If applied to peaches within 14 days of harvest, remove residues by brushing; do not apply later than 2 days before harvest.

APRICOTS (in California): Brown rot and shot-hole (fungus), apply at red bud, early bloom and full-bloom; for green or jacket rot, apply at full-bloom and petal-fall. Use 2 lbs. per 100 gals. (6 to 8 lbs. per acre).

GRAPES (in Great Lakes Area): Black rot—Spray just before bloom, just after bloom and 7 to 10 days later.

ROSES: Cercospora leaf spot (Texas), Rust (California), Black spot—Use 1½ lbs. spray when first leaves unfold; repeat weekly as needed. Add a spreader-sticker.

SEED TREATMENT—WHEAT, BARLEY, RYE, FLAX: Seed decay and seedling blights from many seed- and soil-borne organisms; covered smuts or bunt of wheat, barley, rye—use MANZATE D as an undiluted dust at rate of 2 oz. per bu.; for flax, use 3 to 4 oz. per bu. Apply with any suitable dust treater, or mix directly with seed in planter box by thorough stirring.

LABEL TREATED SEED—"This seed treated with maneb fungicide (MANZATE® D). Do not use for food, feed, or oil purposes."

NOTICE TO BUYER: Purchase of this material does not confer any rights under patents of countries outside of the United States.

STORAGE AND DISPOSAL

STORAGE

Important—Never allow MANZATE D to become wet during storage. This may lead to certain chemical changes which will reduce the effectiveness of MANZATE D as a fungicide and create vapors which may be flammable. Keep container closed when not in use. Store product in original container only, away from other pesticides, fertilizer, food or feed.

DISPOSAL

Do not contaminate water, food or feed by storage or disposal. Wastes resulting from the use of this product may be disposed of on site, or at an approved waste disposal facility. Completely empty bag into application equipment. Then dispose of empty bag in a sanitary landfill or by incineration, or, if allowed by state and local authorities, by burning. If burned, stay out of smoke.

SUPPLEMENTAL LABELING

MANZATE D FUNGICIDE
USDA Reg. No. 352-291

AERIAL APPLICATION OF MANZATE® D FUNGICIDE FOR DISEASE CONTROL IN CERTAIN CROPS

Du Pont MANZATE D Fungicide may be applied by aircraft at the rates shown below for control of certain crop diseases; see product labeling or listing of specific crops, details of timing, use limitations and other use instructions.

Crops	Amount Per Acre: MANZATE D (Lbs.) –in–	Amount Per Acre: Water (Gals.)
Tree Fruits and Nuts; Vineyards	5 to 7	10 to 15
Potatoes and most Vegetables; Small Fruits; Field Crops	1½ to 2	5 to 10
Tomatoes; Sweet Corn; Lettuce; Endive	1½ to 3	5 to 10

Apply with properly calibrated aerial equipment, arrange nozzles so that spray delivery is uniform over the entire spray swath. Standard whirljet nozzles may be used. Make certain spray equipment is clean (free of scale, rust, dirt, oil and pesticide deposits). MANZATE D must be kept in suspension at all times by continuous agitation.

IMPORTANT

BEFORE USING MANZATE® D, READ AND CAREFULLY OBSERVE THE CAUTIONARY STATEMENTS AND ALL

OTHER INFORMATION APPEARING ON THE PRODUCT LABEL.

This bulletin contains new or supplemental instructions for use of this product which may not appear on package label. Follow the instructions carefully.

8-6-82

SUPPLEMENTAL LABELING

MANZATE® D FUNGICIDE

FOR CONTROL OF ASPARAGUS RUST

Du Pont MANZATE D Fungicide is recommended for control of asparagus rust:

(a) In established fields after the crop has been harvested;

(b) In young fields which are not to be harvested.

HOW TO USE

Apply as a spray at the rate of 2 to 3 lbs. MANZATE D per acre, using 50 to 200 gals. of water, depending upon size of plants.

Begin application to established fields immediately after harvest and continue at 10-day intervals until disease is under control. In young fields, start when disease first appears in the area and repeat at 10-day intervals as needed.

IMPORTANT

BEFORE USING **MANZATE® D**, READ AND CAREFULLY OBSERVE THE CAUTIONARY STATEMENTS AND ALL OTHER INFORMATION APPEARING ON THE PRODUCT LABEL.

This bulletin contains new or supplemental instructions for use of this product which may not appear on package label. Follow the instructions carefully.

7/12/65

SUPPLEMENTAL LABELING

MANZATE® D FUNGICIDE

FOR CONTROL OF CROWN ROT AND SURFACE MOLDS OF BANANAS

Du Pont MANZATE D Fungicide is recommended for use on bananas for the control of crown rot and surface molds caused by organisms including: Gloeosporium musarum, Thielaviopsis paradoxa and Fusarium spp.

HOW TO USE

Apply as a spray to the crowns and fruit at the rate of 2 to 4 lbs. per 100 gallons of water. Use 5 gallons of the spray mixture to treat approximately 100 trays of fruit. Material must be kept in suspension at all times by continuous agitation. Agitate by mechanical or hydraulic means.

To prepare spray, wash required amount of MANZATE D through the filler screen while the spray tank or supply tank is being filled. Continue agitation until tank has been sprayed out.

WHEN TO SPRAY

Apply following harvest of bananas. The fruit is passed through a hydrocooler or clean water-rinse spray and the fungicide treatment should follow either of these procedures.

IMPORTANT

BEFORE USING **MANZATE® D**, READ AND CAREFULLY OBSERVE THE CAUTIONARY STATEMENTS AND ALL OTHER INFORMATION APPEARING ON THE PRODUCT LABEL.

This bulletin contains new or supplemental instructions for use of this product which may not appear on the package label. Follow the instructions carefully.

8/30/63

SUPPLEMENTAL LABELING

MANZATE® D FUNGICIDE

FOR CONTROL OF SIGATOKA DISEASE OF BANANAS

Du Pont MANZATE D Fungicide is recommended for use on bananas for the control of Sigatoka disease (Cercospora musae).

HOW TO USE

Apply as a spray at the rate of 3 to 5 lbs. MANZATE D per acre, in sufficient water (or other suitable carrier) to thoroughly cover all plant surfaces. Material must be kept in suspension at all times by continuous agitation. Agitate by mechanical or hydraulic means.

To prepare spray, wash required amount of MANZATE D through the filler screen, while spray tank or supply tank is being filled. If oil is to be used in combination with MANZATE D, add oil after MANZATE D is thoroughly mixed with water and when tank is half full. Allow mixture to become thoroughly emulsified; then add the remainder of water. Continue agitation until tank has been sprayed out.

WHEN TO APPLY

Apply at two week intervals, beginning when Sigatoka incidence is apparent.

IMPORTANT

BEFORE USING **MANZATE® D**, READ AND CAREFULLY OBSERVE THE CAUTIONARY STATEMENTS AND ALL OTHER INFORMATION APPEARING ON THE PRODUCT LABEL.

This bulletin contains new or supplemental instructions for use of this product which may not appear on the package label. Follow the instructions carefully.

8/30/63

SUPPLEMENTAL LABELING

MANZATE® D FUNGICIDE

EPA Reg. No. 352-291

FOR CONTROL OF RUST ON BEANS

Du Pont MANZATE® D Fungicide is recommended for use for control of rust on beans.

DIRECTIONS FOR USE

It is a violation of federal law to use this product in a manner inconsistent with its labeling.

HOW TO USE

Apply 1½ to 2 lbs. MANZATE® D per acre, using sufficient water to obtain thorough coverage of the plants. Begin applications when disease first appears and repeat at 7-day intervals as needed. Do not apply to string beans within 4 days of harvest.

IMPORTANT

BEFORE USING **MANZATE® D**, READ AND CAREFULLY OBSERVE THE CAUTIONARY STATEMENTS AND ALL OTHER INFORMATION APPEARING ON THE PRODUCT LABEL.

This bulletin contains new or supplemental instructions for use of this product which may not appear on package label. Follow the instructions carefully.

10/25/81

SUPPLEMENTAL LABELING

MANZATE® D FUNGICIDE

FOR CONTROL OF BEAN RUST IN FLORIDA

Du Pont MANZATE D Fungicide in combination with sulfur is recommended for use in Florida as a dust in the control of bean rust.

HOW TO APPLY

Equipment should be adjusted to give complete and uniform coverage of all plant surfaces, including undersides of leaves.

A dust containing 10% MANZATE D (8% maneb) and 90% dusting sulfur should be obtained from a commercial dust mixer. Apply at the rate of 40 lbs. of dust per acre.

WHEN TO APPLY

Begin applications when plants are small and continue at 5 to 7-day intervals, as long as disease threatens. Do not apply less than 4 days before any picking. Whenever weather conditions favor disease, shorten interval between applications. If a straight sulfur dust program has been followed at the beginning of the season, applications of MANZATE D should be started whenever rust appears in the area.

IMPORTANT

BEFORE USING MANZATE® D, READ AND CAREFULLY OBSERVE THE CAUTIONARY STATEMENTS AND ALL OTHER INFORMATION APPEARING ON THE PRODUCT LABEL.

This bulletin contains new or supplemental instructions for use of this product which may not appear on package label. Follow the instructions carefully.

SUPPLEMENTAL LABELING

MANZATE® D FUNGICIDE

FOR CONTROL OF DOWNY MILDEW OF BROCCOLI

Du Pont MANZATE D Fungicide is recommended as a protective fungicide to control downy mildew of broccoli.

HOW TO APPLY

MANZATE D should be applied at the rate of 4 lbs. per acre (3.2 lbs. active), either as a spray or in dust mixtures.

As a Spray: Mix the required amount of MANZATE D with sufficient water to obtain thorough coverage with the spray equipment used (usually 100 to 150 gals. per acre for full gallonage, hydraulic equipment; 75 gals. for airblast). Wash MANZATE D through the filler screen while the tank is filling with water. Add Du Pont Spreader-Sticker at the rate of 4 to 6 ozs. per 100 gals. of spray mixture.

As a Dust: MANZATE D may be used at a concentration of 8% (6.4% active) by diluting and thoroughly mixing with pyrophyllite or some other neutral diluent. Apply at rate of 50 lbs. per acre. Dust mixtures should be used as soon as possible after preparation.

WHEN TO APPLY

Start application when disease threatens and continue at 7 to 14-day intervals as needed. Shorten interval when disease conditions are severe. Repeat after rains. If applied less than 3 days before harvest, remove residues by trimming and washing.

IMPORTANT

BEFORE USING MANZATE® D, READ AND CAREFULLY OBSERVE THE CAUTIONARY STATEMENTS AND ALL OTHER INFORMATION APPEARING ON THE PRODUCT LABEL.

This bulletin contains new or supplemental instructions for use of this product which may not appear on package label. Follow the instructions carefully.

9/27/63

SUPPLEMENTAL LABELING

MANZATE® D FUNGICIDE

FOR CONTROL OF NEEDLE DISEASES OF CONIFERS

Du Pont MANZATE D Fungicide is recommended for the control of certain needle diseases of conifers, such as needle cast (Lophodermium pinastri) and brown spot (Scirrhia acicola).

HOW TO USE

Apply MANZATE D as a spray at the rate of 1½ lbs. per 100 gals. of water; add 4 fl. ozs. Du Pont Spreader-Sticker per 100 gals. of spray mixture.

WHEN TO APPLY

Begin applications in spring or early summer before infection occurs; repeat after heavy rains and at 2-week intervals as long as needed.

IMPORTANT

BEFORE USING MANZATE® D, READ AND CAREFULLY OBSERVE THE CAUTIONARY STATEMENTS AND ALL OTHER INFORMATION APPEARING ON THE PRODUCT LABEL.

This bulletin contains new or supplemental instructions for use of this product which may not appear on package label. Follow the instructions carefully.

5/12/69

SUPPLEMENTAL LABELING

MANZATE® D FUNGICIDE

FOR CONTROL OF CRANBERRY FRUIT ROTS IN MASSACHUSETTS

Du Pont MANZATE D Fungicide is recommended for use on cranberries in Massachusetts to control fruit rots.

HOW TO USE

For conventional spray equipment, use 3 lbs. MANZATE D per 100 gals. of water, plus 3 to 4 ozs. of Du Pont Spreader-Sticker. Apply at rate of 300 gals. of spray per acre.

For concentrate sprayers (ground equipment), use 9 lbs. MANZATE D in 20 to 35 gals. of water per acre; add 2 ozs. Du Pont Spreader-Sticker to spray mixture.

For application by aircraft, use 9 lbs. MANZATE D in 13 gals. of water per acre; add 2 ozs. Du Pont Spreader-Sticker to spray mixture.

In preparing sprays, make a thick paste with water; dilute to a smooth suspension, and pour into spray tank while adding remainder of water.

WHEN TO APPLY

Start applications as soon as the plants show 5% bloom and repeat two weeks later or at mid-bloom.

IMPORTANT

BEFORE USING MANZATE® D, READ AND CAREFULLY OBSERVE THE CAUTIONARY STATEMENTS AND ALL OTHER INFORMATION APPEARING ON THE PRODUCT LABEL.

This bulletin contains new or supplemental instructions for use of this product which may not appear on package label. Follow the instructions carefully.

9/27/63

SUPPLEMENTAL LABELING

MANZATE® D FUNGICIDE

FOR THE CONTROL OF CERTAIN CRANBERRY DISEASES IN WASHINGTON AND OREGON

Du Pont MANZATE D Fungicide is recommended for the control of cranberry twig blight and fruit rots of cranberry in Washington and Oregon.

HOW TO USE

Apply as a spray, using 2 lbs. of MANZATE D per 100 gals. of water, plus 3 to 4 ozs. of Du Pont Spreader-Sticker. Apply at the rate of 350 gals. per acre.

WHEN TO USE

First application—Immediately after blossoming.
Second application—July 10 to 20.
Third application—August 1 to 10.
Fourth application—August 20 to 25.

IMPORTANT

BEFORE USING MANZATE® D, READ AND CAREFULLY OBSERVE THE CAUTIONARY STATEMENTS AND ALL

OTHER INFORMATION APPEARING ON THE PRODUCT LABEL.

This bulletin contains new or supplemental instructions for use of this product which may not appear on package label. Follow the instructions carefully.

9/27/63

SUPPLEMENTAL LABELING

MANZATE® D FUNGICIDE

FOR THE CONTROL OF CRANBERRY FRUIT ROTS IN NEW JERSEY AND WISCONSIN

Du Pont MANZATE D Fungicide is recommended for use on cranberries in New Jersey and Wisconsin for the control of fruit rots.

HOW TO USE

Apply as a spray, using 2 lbs. of MANZATE D per 100 gals. of water, plus 3 to 4 ozs. of Du Pont Spreader-Sticker. Apply at the rate of 300 gals. per acre.

WHEN TO USE

First application—At mid-bloom.
Second application—10 to 14 days after mid-bloom.
Third application—4 weeks after mid-bloom.
Fourth application—(if conditions favor disease) 2 to 3 weeks after third application.

IMPORTANT

BEFORE USING MANZATE® D, READ AND CAREFULLY OBSERVE THE CAUTIONARY STATEMENTS AND ALL OTHER INFORMATION APPEARING ON THE PRODUCT LABEL.

This bulletin contains new or supplemental instructions for use of this product which may not appear on package label. Follow the instructions carefully.

9/27/63

SUPPLEMENTAL LABELING

MANZATE® D FUNGICIDE

USDA Reg. No. 352-291

FOR CONTROL OF CERTAIN DISEASES OF PAPAYA

Du Pont MANZATE D Fungicide is recommended for control of black spot and anthracnose of papaya.

HOW TO APPLY

Apply 2 to 3 lbs. MANZATE D per acre, using sufficient water (min. 50 gals. per acre) to provide thorough coverage; add Du Pont Spreader-Sticker at rate of 4 fl. ozs. per 100 gals. spray. Direct spray to central column of tree to thoroughly cover developing fruit.

WHEN TO APPLY

Spray every 10 to 14 days during winter and early spring months when disease incidence is high, and as often as every 7 days during periods of very wet weather.

IMPORTANT

BEFORE USING MANZATE® D, READ AND CAREFULLY OBSERVE THE CAUTIONARY STATEMENTS AND ALL OTHER INFORMATION APPEARING ON THE PRODUCT LABEL.

This bulletin contains new or supplemental instructions for use of this product which may not appear on package label. Follow the instructions carefully.

10/27/66

SUPPLEMENTAL LABELING

MANZATE® D FUNGICIDE

FOR USE ON PEPPERS

Du Pont MANZATE D Fungicide is recommended for use on peppers to control anthracnose, ripe rot and Cercospora leaf spot (frog-eye spot).

HOW TO USE

Apply as a full coverage spray, using 2 lbs. per 100 gals. of water.

WHEN TO APPLY

Cercospora leaf spot—Begin applications when disease threatens and repeat at 7 to 10-day intervals as needed.

Anthracnose and ripe rot—Begin applications when the first fruits begin to turn red. Repeat at 7 to 10-day intervals, as long as disease threatens.

IMPORTANT

BEFORE USING MANZATE® D, READ AND CAREFULLY OBSERVE THE CAUTIONARY STATEMENTS AND ALL OTHER INFORMATION APPEARING ON THE PRODUCT LABEL.

This bulletin contains new or supplemental instructions for use of this product which may not appear on package label. Follow the instructions carefully.

9/27/63 (Revised 11/23/82)

SUPPLEMENTAL LABELING

MANZATE® D FUNGICIDE

FOR CONTROL OF DOWNY MILDEW AND ANGULAR LEAF SPOT OF PUMPKINS

Du Pont MANZATE D Fungicide is recommended for use on pumpkins for the control of downy mildew and angular leaf spot.

HOW TO USE

Apply 3 lbs. MANZATE D per acre in sufficient water to wet plants thoroughly. Begin applications when disease threatens and repeat at 7 to 10-day intervals, as long as necessary.

IMPORTANT

BEFORE USING MANZATE® D, READ AND CAREFULLY OBSERVE THE CAUTIONARY STATEMENTS AND ALL OTHER INFORMATION APPEARING ON THE PRODUCT LABEL.

This bulletin contains new or supplemental instructions for use of this product which may not appear on package label. Follow the instructions carefully.

9/27/63

SUPPLEMENTAL LABELING

MANZATE® D FUNGICIDE

FOR CONTROL OF LEAF SPOT DISEASES IN NURSERY PLANTINGS OF SUMAC AND BUFFALOBERRY

Du Pont MANZATE D Fungicide is recommended for use in nursery plantings of skunkbush sumac (Rhus trilobata) and buffaloberry (Shepherdia argentea) for the control of leaf spot diseases caused by Cylindrosporium sp.

HOW TO USE

Apply MANZATE D as a spray at the rate of 1½ lbs. per 100 gals. of water. Begin applications when disease first appears, usually early June, and repeat at weekly intervals, as long as needed.

IMPORTANT

BEFORE USING MANZATE® D, READ AND CAREFULLY OBSERVE THE CAUTIONARY STATEMENTS AND ALL OTHER INFORMATION APPEARING ON THE PRODUCT LABEL.

This bulletin contains new or supplemental instructions for use of this product which may not appear on package label. Follow the instructions carefully.

5/12/69

SUPPLEMENTAL LABELING

MANZATE® D FUNGICIDE

FOR CONTROL OF HELMINTHOSPORIUM LEAF BLIGHTS OF SWEET CORN IN FLORIDA

Du Pont MANZATE D Fungicide is recommended for the control of Helminthosporium leaf blights of sweet corn in Florida, except in the Homestead area.

HOW TO APPLY

Prepare a spray mixture containing 1$^1/_2$ lbs. MANZATE D per 100 gals. of water. Apply from 100 to 200 gals. of spray per acre, depending on size of plants. Sprayers should have two nozzles over the row, in order to direct a good quantity of spray into the whorl, in addition to side nozzles.

WHEN TO APPLY

Under severe disease conditions, and when top yields of high quality sweet corn are desired, apply MANZATE D twice weekly, beginning when plants are about six inches tall and continuing as long as disease threatens. For a minimum program under mild disease conditions, spray interval may be extended to seven days.

NOTE—Do not feed treated forage to dairy animals, or to animals being finished for slaughter.

IMPORTANT

BEFORE USING MANZATE® D, READ AND CAREFULLY OBSERVE THE CAUTIONARY STATEMENTS AND ALL OTHER INFORMATION APPEARING ON THE PRODUCT LABEL.

This bulletin contains new or supplemental instructions for use of this product which may not appear on package label. Follow the instructions carefully.

9/27/63

SUPPLEMENTAL LABELING

MANZATE® D FUNGICIDE

USDA Reg. No. 352-291

FOR CONTROL OF CERTAIN DISEASES OF TURFGRASSES

Du Pont MANZATE D Fungicide is recommended for the control of rust, Rhizoctonia brown patch, and diseases incited by Helminthosporium spp. on turfgrasses.

HOW TO USE

Apply MANZATE D at the rate of 3 to 4 ozs. per 1000 sq. ft. using 3 to 5 gals. of water. Begin application when disease first appears and repeat at 7 to 10-day intervals, as long as needed. When conditions are unusually favorable for development of disease, use 6 to 8 ozs. per 1000 sq. ft. at 3 to 5-day intervals.

NOTE—Do not graze or feed clippings from treated areas to livestock.

IMPORTANT

BEFORE USING MANZATE® D, READ AND CAREFULLY OBSERVE THE CAUTIONARY STATEMENTS AND ALL OTHER INFORMATION APPEARING ON THE PRODUCT LABEL.

Special Local Need Registrations [24(c)] For MANZATE® D Fungicide*

Crop	State	EPA SLN No.	Remarks
Seed Grasses	Oregon	OR-790019	Rust Control
Tobacco	Georgia	GA-800008	Blue Mold Control
	Kentucky	KY-800018	Blue Mold Control
	Maryland	MD-800023	Blue Mold Control
	North Carolina	NC-800018	Blue Mold Control
	Pennsylvania	PA-810006	Blue Mold Control
	Puerto Rico	PR-810007	Blue Mold Control
	South Carolina	SC-800016	Blue Mold Control
	Tennessee	TN-800016	Blue Mold Control
	Virginia	VA-800033	Blue Mold Control

* Copies of label available upon request to nearest Du Pont Agricultural Chemicals Office. For distribution and use only within the State specified.

This bulletin contains new or supplemental instructions for use of this product which may not appear on package label. Follow the instructions carefully.

1/9/69

MANZATE® 200

FUNGICIDE

ACTIVE INGREDIENT:

A coordination product of zinc ion and manganese ethylenebisdithiocarbamate 80%

in which the ingredients are

Manganese ++ 16%

Zinc ++ 2%

Ethylenebisdithiocarbamate ion ($C_4H_6N_2S_4$)+ 62%

INERT INGREDIENTS 20%

TOTAL 100%

EPA Reg. No. 352-341 U.S. Pat. 3,379,610

Keep Out of Reach of Children

PRECAUTIONARY STATEMENTS

HAZARD TO HUMANS

CAUTION!

MAY IRRITATE EYES, NOSE, THROAT, AND SKIN. MAY BE HARMFUL IF INHALED OR SWALLOWED.

Avoid breathing dust or spray mist. Avoid contact with skin, eyes and clothing. Keep away from fire or sparks. Wear protective clothing including long pants, long sleeve shirt, gloves, hat and boots during mixing and loading operations.

Do not apply this product in such a manner as to directly or through drift expose workers, other persons, or animals. The area being treated must be vacated by unprotected persons.

FIRST AID:

In case of contact, flush skin or eyes with plenty of water; for eyes, get medical attention.

ENVIRONMENTAL HAZARDS

This product is toxic to fish. Do not apply directly to water. Do not apply where runoff is likely to occur. Do not apply when weather conditions favor drift from areas treated.

NOTICE OF WARRANTY

Du Pont warrants that this product conforms to the chemical description on the label thereof and is reasonably fit for purposes stated on such label only when used in accordance with the directions under normal use conditions. It is impossible to eliminate all risks inherently associated with the use of this product. Crop injury, ineffectiveness or other unintended consequences may result because of such factors as weather conditions, presence of other materials or the manner of use or application, all of which are beyond the control of Du Pont. In no case shall Du Pont be liable for consequential, special or indirect damages resulting from the use or handling of this product. All such risks shall be assumed by the Buyer. DU PONT MAKES NO WARRANTIES OF MERCHANTABILITY OR FITNESS FOR A PARTICULAR PURPOSE NOR ANY OTHER EXPRESS OR IMPLIED WARRANTY EXCEPT AS STATED ABOVE.

DIRECTIONS FOR USE

It is a violation of federal law to use this product in a manner inconsistent with its labeling.

MANZATE 200 Fungicide should be used only in accordance with recommendations on this label, or in separate published Du Pont recommendations available through local dealers.

Du Pont will not be responsible for losses or damages resulting from use of this product in any manner not specifically recommended by

Du Pont. User assumes all risks associated with such nonrecommended use.

MANZATE 200, a wettable powder containing a coordination product of zinc ion and maneb, is recommended for use as a spray or dust for the control of many important plant diseases.

Do not enter treated areas without protective clothing until sprays have dried or dusts have settled. Because certain states may require more restrictive reentry intervals for various crops treated with this product, consult your State Department of Agriculture for further information.

Written or oral warnings must be given to workers who are expected to be in a treated area or in an area about to be treated with this product. When oral warnings are given, warnings shall be given in a language customarily understood by workers. Oral warnings must be given if there is reason to believe that written warnings cannot be understood by workers. Warnings must include the following information: "CAUTION. Area treated with MANZATE 200 on (date of application). Do not enter without appropriate protective clothing until sprays have dried or dusts have settled. In case of contact, flush skin or eyes with plenty of water; for eyes, get medical attention."

As a Spray (Ground or Aerial Equipment)—Apply MANZATE 200 at the rate shown; use sufficient water to provide thorough coverage, usually 20 to 100 gals. per acre for ground equipment and 5 to 8 gals. per acre for aircraft. Add MANZATE 200 slowly to water in the spray tank with agitation, or premix thoroughly in separate holding tank for concentrate or aircraft sprayers. Continuous agitation is required to keep the product in suspension. Add a spreader-sticker spray adjuvant if needed.

As a Dust—Apply MANZATE 200 at the rate shown, using proper amount of dust containing 6% to 10% MANZATE 200 (4.8% to 8% active). Dust may be prepared by diluting and thoroughly mixing MANZATE 200 with pyrophyllite or other neutral diluent; commonly used insecticides may displace an equivalent amount of diluent. Use dust mixtures as soon as possible after preparation.

FIELD CROPS

WHEAT, BARLEY, OATS, RYE: Helminthosporium leaf blight and Septoria leaf spot—Use 2 lbs. MANZATE 200 per acre. Begin applications at tillering-to-jointing stage and repeat at 7 to 10-day intervals. Do not make more than 3 applications; do not apply within 26 days of harvest; do not graze treated area within 26 days after application.

SUGAR BEETS: Cercospora leaf spot—Use 1½ to 2 lbs. MANZATE 200 per acre. Begin applications when disease first threatens and repeat at 7 to 10-day intervals. Do not apply within 14 days of harvest. Do not feed treated sugar beet tops to livestock.

PEANUTS: Cercospora leaf spot—Use 1 to 2 lbs MANZATE 200 per acre. Begin applications when disease first appears and repeat 10 to 14-day intervals, or at 7-day intervals during humid weather. Do not feed treated peanut vine hay to livestock.

COTTON (Southwest only): Rust—Use 1¼ to 2 lbs. MANZATE 200 per acre. Begin applications when rust first appears in the area and repeat at 10 to 14-day intervals. Do not apply after bolls open; do not graze treated areas nor feed gin trash to livestock.

VEGETABLE CROPS

POTATOES: Early and late blights—Use 1 to 2 lbs. MANZATE 200 per acre. Begin applications when plants are 4 to 6 inches high; repeat 7 to 10-day intervals (more frequently under severe disease conditions) throughout the season.

POTATO TUBER TREATMENT: Fusarium seed piece decay—Dip whole or cut seed pieces in mixture of 1¼ lbs. MANZATE 200 per 50 gals. water, drain, and plant immediately; if seed pieces are to be held before planting, spread out to dry in a cool place. For dust treatment, thoroughly coat with dust containing 10% MANZATE 200 (8% active) at rate of 1 lb. per 100 lbs. of seed. Do not use treated seed pieces for food or feed purposes.

TOMATOES: Early and late blights, gray leaf spot, gray leaf mold, and anthracnose—Use 1½ to 3 lbs. MANZATE 200 per acre. Begin applications when seedlings emerge and repeat at 7-day intervals throughout the season. Do not apply with 5 days of harvest.

Bacterial leaf spot—Use 1½ lbs. MANZATE 200 combined with 4 lbs. tribasic copper sulfate, per acre. Begin applications when seedlings emerge, and repeat at 3 to 4-day intervals as needed. Do not apply within 5 days of harvest.

CUCUMBERS, MELONS, SUMMER SQUASH: Anthracnose, Alternaria and Cercospora leaf spots, downy mildew, gummy stem blight, and scab—Use 2 to 3 lbs. MANZATE 200 per acre. Begin applications at the 2-leaf stage and repeat at 5 to 7-day intervals. Do not apply within 5 days of harvest.

CORN (Sweet, Field, Popcorn): Helminthosporium leaf blight —Use 1½ lbs. MANZATE 200 per acre. Begin applications when disease first appears and repeat at 4 to 7-day intervals. Do not apply to field corn or popcorn within 40 days of harvest. Do not apply to sweet corn within 7 days of harvest; do not feed treated sweet corn forage or fodder to livestock.

ONIONS (Dry Bulb): Botrytis leaf blight, downy mildew, neck rot, and purple blotch—Use 2 to 3 lbs. MANZATE 200 per acre. Begin applications when diseases are first reported in the area and repeat at 7-day intervals throughout the season. Do not apply within 7 days of harvest; do not apply to exposed bulbs.

CELERY, FENNEL: Early and late blights—Use 2 lbs. MANZATE 200 per acre. Begin applications in plant bed as soon as plants emerge; repeat at 3 to 5-day intervals in plant bed, and at 7-day intervals after plants are set in field. Do not apply to celery within 14-days of harvest, nor to fennel within 7 days of harvest.

CARROTS: Altenaria and Cercospora leaf spots—Use 1½ to 2 lbs. MANZATE 200 per acre. Begin applications when disease first appears and repeat at 7 to 10-day intervals. Do not apply within 7 days of harvest; do not use treated tops for food or feed purposes.

ASPARAGUS: Rust—Use 2 lbs. MANZATE 200 per acre. Begin applications when rust first appears and repeat at 10-day intervals as needed. Use only on ferns after spears have been harvested.

FRUIT CROPS

APPLES, PEARS, CRABAPPLES, QUINCE: Bitter rot, black rot, brown rot, cedar apple rust, fly speck, scab, and sooty blotch — Use MANZATE 200 at 1 to 2 lbs. per 100 gals. of finished spray and apply up to 10 lbs. product per acre per application (for low volume sprays, adjust accordingly). Apply at 7 to 14-day intervals throughout the season. For scab, begin at green tip or delayed dormant stage; for cedar apple rust at blossom time; for other diseases, at first cover. Add a spreader-sticker to spray mixture for last 3 cover sprays. Do not apply within 15 days of harvest (30 days for apples); do not graze livestock in treated areas.

CRANBERRIES: Fruit Rot—Use 3 to 6 lbs. MANZATE 200 per acre. Begin applications at mid-bloom and repeat at 7 to 10-day intervals. Do not apply within 30 days of harvest.

GRAPES: Black rot, bunch rot, dead arm and downy mildew — Use 1½ to 4 lbs. MANZATE 200 per acre. Apply when shoots are ½ to 1½ inches long, when shoots are 3 to 5 inches long, and when shoots are 8 to 10 inches long. Continue at 10 to 14-day intervals but do not apply within 66 days of harvest; in California, do not apply after fruit set.

PAPAYAS: Anthrocnose and Phytophthora fruit rot—Use 2 to 2½ lbs. MANZATE 200 per acre; add a spreader-sticker to spray mixture. Begin applications at flowering time; repeat at 14-day intervals, or at 7-day intervals under severe disease conditions. Treat crown and blossom area.

SEED TREATMENT

WHEAT, BARLEY, RYE, FLAX: Seed decay and seedling blights from many seed- and soil-borne organisms; covered smuts or bunt of wheat, barley, rye—use MANZATE 200 as an undiluted dust at rate of 2 oz. per bu.; for flax, use 3 to 4 oz. per bu. Apply with any suitable dust treater, or mix directly with seed in planter box by thorough stirring.

LABEL TREATED SEED—"This seed treated with MANZATE® 200 fungicide. Do not use for food, feed or oil purposes."

NOTICE TO BUYER: Purchase of this material does not confer any rights under patents of countries outside of the United States.

STORAGE AND DISPOSAL

STORAGE
Important—Never allow MANZATE 200 to become wet during storage. This may lead to certain chemical changes which will reduce the effectiveness of MANZATE 200 as a fungicide and create vapors which may be flammable. Keep container closed when not in use. Store product in original container only, away from other pesticides, fertilizer, food or feed.

DISPOSAL
Do not contaminate water, food or feed by storage or disposal. Wastes resulting from the use of this product may be disposed of on site or at an approved waste disposal facility. Completely empty bag into application equipment. Then dispose of empty bag in a sanitary landfill or by incineration, or, if allowed by state and local authorities, by burning. If burned, stay out of smoke.

SUPPLEMENTAL LABELING

MANZATE® 200 FUNGICIDE
EPA Reg. No. 352-341

FOR TREATING WHEAT, BARLEY, RYE, OATS AND FLAX SEED

Du Pont MANZATE 200 Fungicide is recommended for use on wheat, barley, rye, oats and flax. Used according to directions MANZATE 200 will usually increase stands and yields by reducing losses from seed decay and seedling blights caused by many seed-borne and soil-borne organisms. It will usually control covered smuts or bunt of wheat, barley, rye and loose smuts of oats.

HOW TO USE
MANZATE 200 is formulated for dry or slurry treatment:

For Dry Treatment, apply at a rate of 2 ounces per bushel of wheat, barley, rye, and oats, and at a rate of 3 to 4 ounces per bushel of flax. The seed may be pretreated in a rotary drum or dust attachment for the mechanical loader or the seed may be treated in the planter box. For planter box, place ½ the seed in the planter box, but do not exceed ½ the capacity. Sprinkle ½ the total amount of MANZATE 200 over the seed surface and mix by stirring with a stick or paddle until seed is covered. Add remaining ½ of seed and fungicide and stir as outlined above.

For Slurry Treatment, MANZATE 200 should be applied by means of a slurry-type treater specifically designed and approved for this purpose. For each type of seed use the recommended slurry concentration, and the proper slurry cup and dump weight to apply the amount of slurry shown.

Prepare slurry or water suspension of MANZATE 200 by mixing the powder with water at the rates shown in the table. Weigh powder accurately and measure water carefully. Preparation of the slurry may be carried out in the slurry tank on the treater by measuring into it the desired number of gallons of water and then adding the correct amount of MANATE 200 for each gallon of water.

Crop Seeds	Treating Rate ozs./bu.	Concentration lbs./gal.	Slurry* cc/cwt. seed	Slurry cup cc	Lbs. Dump Wt.
Wheat	1.33	2.7	230*	23	10
Barley	1.33	3.6	230*	23	10
Rye	1.33	2.9	230*	23	10
Oats	1.33	4.3	288*	23	8
Flax	2.00	3.7	288*	23	8

* For Panogen and Mist-O-Matic treaters equipped to apply slurries adjust slurry cups and dump weights to apply the amount of slurry shown as cc/100 lbs. seed (cwt.). One fluid ounce is equivalent to 29.5 cc.

IMPORTANT
BEFORE USING **MANZATE 200,** READ AND CAREFULLY OBSERVE THE CAUTIONARY STATEMENTS AND ALL OTHER INFORMATION APPEARING ON THE PRODUCT LABEL.

Label Treated Seed—"This seed treated with fungicide (MANZATE 200). Do not use for food, feed or oil purposes."

This bulletin contains new or supplemental instructions for use of this product which may not appear on the package label. Follow the instructions carefully.

8/20/71

SUPPLEMENTAL LABELING

MANZATE® 200 FUNGICIDE
EPA Reg. No. 352-341

FOR CONTROL OF PEAR PSYLLA NYMPHS IN PEAR ORCHARDS IN THE PACIFIC NORTHWEST

Du Pont MANZATE 200 Fungicide is recommended for the control of pear psylla nymph in pear orchards in the Pacific Northwest.

HOW TO USE
Apply MANZATE 200 at the rate of 1.5 to 2.0 lbs. per 100 gals. for dilute spray. Apply up to 10 lbs. MANZATE 200 per acre. Use an equivalent amount per acre for concentrate sprays. Apply at any time after bloom or as a postharvest application when high populations of pear psylla nymphs are present. Repeat applications at 2 to 4-week intervals. Two to three well-timed applications per season have given the best results.

When spray oils are used in the spray program, do not apply MANZATE 200 sooner than 7 days after the last oil application as foilage injury may result.

NOTE: Do not apply within 15 days of harvest; do not graze livestock in treated areas.

IMPORTANT
BEFORE USING MANZATE 200, READ AND CAREFULLY OBSERVE THE CAUTIONARY STATEMENTS AND ALL OTHER INFORMATION APPEARING ON THE PRODUCT LABEL.

This bulletin contains new or supplemental directions for use of this product which may not appear on package label. Follow the instructions carefully.

11/24/70-A

SUPPLEMENTAL LABELING

MANZATE® 200 FUNGICIDE
EPA Reg. No. 352-341

FOR USE ON WHEAT TO CONTROL LEAF RUST

Du Pont MANZATE® 200 Fungicide is recommended for use on wheat to control leaf rust (Puccinia spp.)

HOW TO USE
Apply 2 lbs. MANZATE® 200 per acre. Use sufficient water to provide thorough coverage, usually 20 to 100 gals. per acre for ground equipment and 5 to 8 gals. per acre by air. Begin application at tillering to jointing stage and repeat at 7 to 10-day intervals.

Note: Do not make more than 3 applications. Do not apply within 26 days of harvest. Do not graze treated area within 26 days after application.

IMPORTANT
BEFORE USING MANZATE® 200, READ AND CAREFULLY OBSERVE THE CAUTIONARY STATEMENTS AND ALL OTHER INFORMATION APPEARING ON THE PRODUCT LABEL.

This bulletin contains new or supplemental instructions for use of this product which may not appear on package label. Follow the instructions carefully.

1/11/82

SUPPLEMENTAL LABELING

MANZATE® 200 FUNGICIDE
EPA Reg. No. 352-341

FOR USE ON WHEAT TO SUPPRESS GLUME BLOTCH

DuPont MANZATE 200 Fungicide is recommended for use on wheat to suppress glume blotch (Septoria spp.).

HOW TO USE

Apply 2 lbs. MANZATE 200 per acre. Use sufficient water to provide thorough coverage, usually 20 to 100 gallons per acre for ground equipment and 5 to 8 gallons per acre by air. Begin application at tillering to jointing stage and repeat at 7 to 10-day intervals.

Note: Do not make more than 3 applications. Do not apply within 26 days of harvest. Do not graze treated area within 26 days after application.

IMPORTANT

BEFORE USING MANZATE® 200, READ AND CAREFULLY OBSERVE THE CAUTIONARY STATEMENTS AND ALL OTHER INFORMATION APPEARING ON THE PRODUCT LABEL.

This bulletin contains new or supplemental instructions for use of this product which may not appear on package label. Follow the instructions carefully.

3/22/82 (Revised 5/28/82)

Special Local Need Registrations [24(c)] for Manzate 200 Fungicide*

Crop	State	EPA SLN No.	Remarks
Conifers	Oregon	OR-800081	Douglas Fir
	Washington	WA-800056	Douglas Fir
Dendrobium	Hawaii	HI-790020	Botrytis control
Seed Grasses	Oregon	OR-790020	Rust Control
Potatoes	Idaho	ID-800046	Early & late Blight Control
	Michigan	MI-810016	Early & Late Blight Control
	Oregon	OR-800092	Early Blight Control
	Washington	WA-800090	Early Blight Control
	Utah	UT-800009	Early & Late Blight Control
Tobacco	Georgia	GA-800008	Blue Mold Control
	Kentucky	KY-800010	Blue Mold Control
	Maryland	MD-800017	Blue Mold Control
	North Carolina	NC-800004	Blue Mold Control
	Pennsylvania	PA-810005	Blue Mold Control
	Puerto Rico	PR-810006	Blue Mold Control
	South Carolina	SC-800009	Blue Mold Control
	Tennessee	TN-800010	Blue Mold Control
	Virginia	VA-800017	Blue Mold Control
Tomatoes	Michigan	MI-810016	Sprinkler Irrigation

ORNAMENTAL PLANTS—DISEASES CONTROLLED	FLORIDA FL-770008	HAWAII HI-780011	ILLINOIS IL-780005	NEW JERSEY NJ-780012	NEW YORK NY-770007	TEXAS TX-780041	VIRGINIA VA-780014
Anthurium—Anthracnose spadix rot		X					
Ash, Mountain—Enthomosporium leafspot		X	X	X	X	X	X
Azalea—Cylindrocladium rot, Phytophthora, spp	X	X	X	X	X	X	X
Arborvitae—Cercospora blight	X	X	X	X	X	X	X
Carnation—Rust		X	X		X	X	X
Cedar, Red—Cercospora blight, Phomopsis blight	X	X	X	X	X	X	X
Chrysanthemum—Aschochyta blight	X	X	X	X	X	X	
—Rust		X	X		X	X	X
Cordyline—Cercospora leafspot	X	X	X	X	X	X	X
Crabapple—Sphaeropsis leafspot, Scab		X	X	X	X	X	X
Cypress, Arizona—Cercospora blight, Monochaetia canker		X	X	X	X	X	X
Dieffenbachia—Leptosphaeria brown spot	X	X	X	X	X	X	X
Dracaena—Fusarium leafspot	X	X	X	X	X	X	X
Fern—Rhizoctonia blight	X	X	X	X	X	X	X
Ficus—Cercospora leafspot	X	X	X	X	X	X	X
Firethorn—Fusicladium scab		X	X	X	X	X	X
Geranium—Puccinia rust		X	X	X	X	X	
Gladiolus—Botrytis rot, Curvularia rot	X	X	X	X	X	X	X
Hawthorn (NJ), Hawthorn, English—Fabraea leafspot		X	X	X	X	X	X
Horsechestnut—Guignardia leaf blotch		X	X	X	X	X	X
Juniper—Phomopsis blight	X	X	X	X	X	X	
Laurel, Mountain (VA), Laurel—Cercospora leafspot		X	X	X	X	X	X
Ligustrum—Cercospora leafspot	X	X	X	X	X	X	X
Oak (NJ), Oak, Red—Actinopelte leafspot, Taphrina leafspot		X	X	X	X	X	X
Peperomia—Cercospora leafspot	X	X	X	X	X	X	X
Philodendron—Daclylaria leafspot, Phytophthora leafspot	X	X	X	X	X	X	X
Pleomele—Fusarium leafspot	X	X	X	X	X	X	X
Schefflera—Alternaria blight	X	X	X	X	X	X	X
Statice—Cercospora frogeye	X	X	X	X	X	X	X
Syngonium—Cephalosporium leafspot		X	X	X	X	X	
Venus Flytrap—Anthracnose	X	X	X	X	X	X	X

* Copies of label available upon request to nearest DuPont Agricultural Chemicals Office. For distribution and use only within the State specified.

MANZATE®

FLOWABLE FUNGICIDE WITH ZINC

Contains 4 lbs. Active per Gallon

ACTIVE INGREDIENT:

Maneb (Manganese ethylenebisdithiocarbamate) (Manganese equivalent as metallic 7.6%)	37%
INERT INGREDIENTS	63%
TOTAL	100%

EPA Reg. No. 352-423 U.S. Pat. 3,379,610

Keep Out of Reach of Children

PRECAUTIONARY STATEMENTS

HAZARDS TO HUMANS AND DOMESTIC ANIMALS

CAUTION!

MAY IRRITATE EYES, NOSE, THROAT AND SKIN. MAY BE HARMFUL IF INHALED OR SWALLOWED.

Avoid breathing spray mist. Avoid contact with skin, eyes and clothing. Wear protective clothing including long pants, long sleeve shirt, gloves, hat and boots during mixing and loading operations. Do not apply this product in such a manner as to directly or through drift expose workers, other persons or animals. Area being treated must be vacated by unprotected persons.

FIRST AID

In case of contact, flush skin or eyes with plenty of water; for eyes, get medical attention.

ENVIRONMENTAL HAZARDS

This product is toxic to fish. Do not apply directly to water and wetlands. Do not apply where runoff is likely to occur. Do not apply when weather conditions favor drift from areas treated. Do not contaminate water by cleaning of equipment or disposal of wastes.

NOTICE TO BUYER

Purchase of this material does not confer any rights under patents of countries outside of the United States.

NOTICE OF WARRANTY

Du Pont warrants that this product conforms to the chemical description on the label thereof and is reasonably fit for purposes stated on such label only when used in accordance with the directions under normal use conditions. It is impossible to eliminate all risks inherently associated with the use of this product. Crop injury, ineffectiveness or other unintended consequences may result because of such factors as weather conditions, presence of other materials or the manner of use or application, all of which are beyond the control of Du Pont. In no case shall Du Pont be liable for consequential, special, or indirect damages resulting from the use or handling of this product. All such risks shall be assumed by the Buyer. DU PONT MAKES NO WARRANTIES OF MERCHANTABILITY OR FITNESS FOR A PARTICULAR PURPOSE NOR ANY OTHER EXPRESS OR IMPLIED WARRANTY EXCEPT AS STATED ABOVE.

DIRECTIONS FOR USE

It is a violation of federal law to use this product in a manner inconsistent with its labeling.

MANZATE Flowable should be used only in accordance with recommendations on this label, or in separate published Du Pont recommendations available through local dealers.

Du Pont will not be responsible for losses or damages resulting from use of this product in any manner not specifically recommended by Du Pont. User assumes all risks associated with such nonrecommended use.

Do not enter treated areas without protective clothing until sprays have dried. Because certain states may require more restrictive reentry intervals for various crops treated with this product, consult your State Department of Agriculture for further information.

Written or oral warnings must be given to workers who are expected to be in a treated area or in an area about to be treated with this product. When oral warnings are given, warnings shall be given in a language customarily understood by workers. Oral warnings must be given if there is reason to believe that written warnings cannot be understood by workers. Warnings must include the following information: "CAUTION. Area treated with MANZATE Flowable on (date of application). Do not enter without appropriate protective clothing until sprays have dried. In case of contact, flush skin or eyes with plenty of water; for eyes, get medical attention."

As a Spray: (Ground or Aerial Equipment)—Apply MANZATE Flowable at a rate of 1.2 qts. (unless otherwise stated) in sufficient water to provide thorough coverage, usually 20 to 100 gals. per acre by ground and 5 to 8 gals. per acre by air. In case of severe disease outbreaks, or when weather conditions favor disease, use 1.6 qts. and shorten the spray interval. Add MANZATE Flowable slowly to water in the spray tank with agitation, or premix thoroughly in separate holding tank for concentrate or aircraft sprayers. Continuous agitation is required to keep the product in suspension.

FRUITS AND NUTS

ALMONDS: Brown rot (blossom and twig blight). Use 1.2 to 1.6 qts. per 100 gals. (4.8 to 6.4 qts. per acre) at the popcorn, full-bloom and petal-fall stages.

Shot hole—Use the same rates, but begin applications in the dormant stage. Repeat at 7 to 10-day intervals through petal-fall.

APPLES: Scab, cedar apple rust, black rot (fruit and foliage), fly speck, sooty blotch, bitter rot, and white rot (Botryosphaeria). Use 0.8 qts. to 1.2 qts. per 100 gals. and apply up to 8 qts. product per acre per application (for low-volume sprays, adjust accordingly). Apply at 7 to 14-day intervals throughout the season. For scab, begin at green tip or delayed dormant stage; for cedar apple rust, at blossom time; for other diseases, at first cover. Do not apply within 30 days of harvest; do not graze livestock in treated areas.

APRICOTS: Brown rot (blossom, twig and fruit). Use 1.2 to 1.6 qts. per 100 gals. (4.8 to 6.4 qts. per acre) at red bud stage. Repeat at early-bloom, full-bloom, petal-fall and at 7 to 14-day intervals as necessary up to 2 weeks before harvest.

Green or jacket rot. Use 1.6 qts. per 100 gals. at full-bloom and petal-fall.

Shot-hole (fungus). Use 1.6 qts. per 100 gals. from dormant through petal-fall stages on a regular schedule.

Do not apply within 14 days of harvest.

Do not use in combination with Diazinon® on apricots.

BANANAS: Sigatoka disease (Cercospora musae). Use 2.4 to 4.0 qts. per acre in sufficient water (or other suitable carrier) to thoroughly cover all plant surfaces. Begin when disease first appears and repeat every two weeks as needed.

CRANBERRIES: Fruit rots (Massachusetts, New Jersey, Wisconsin); cranberry twig blight and fruit rot (Washington and Oregon). For conventional spray equipment, use 1.6 to 2.4 qts. per 100 gals. of water. Apply at a rate of 300 gals. of spray per acre. For concentrate sprayers (ground equipment), use 7.2 qts. in 20 to 35 gals. For aerial application, use 7.2 qts. in 13 gals. of water per acre. Begin applications at mid-bloom and repeat at 7 to 10-day intervals. Do not apply later than 4 weeks after mid-bloom.

FIGS (Kadota): Surface molds (Cladosporium) and rot (caused by Alternaria). Use 0.6 qts. per 100 gals. of water (basis high-volume spray equipment). Be sure to get uniform coverage. Make a single application 10 to 20 days before harvest (usually early September). Do not apply within 10 days of harvest.

GRAPES: Black rot (bunch rot). Use 1.2 to 3.2 qts. per acre by ground or air starting when new shoots are 8 to 10 inches long (prebloom). Repeat just after bloom and 7 to 10 days later. When fruit is present, use maximum of 1.5 qts. per acre. Do not apply within 7 days of harvest.

NECTARINES: Brown rot, shot-hole. Use 4.8 to 6.4 qts. in sufficient water per acre by ground or air. Begin at red bud and repeat in early-bloom, full-bloom, petal-fall, and at 7 to 14-day intervals. Do not apply within 14 days of harvest.

PAPAYAS: Anthracnose (Colletotricum), black spot (Cercospora). Use 1.6 to 2.4 qts. in sufficient water (min. 50 gals. per acre) to provide thorough coverage. Spray every 10 to 14 days during winter and early spring months when disease incidence is high, and as often as every 7 days during wet weather. Direct spray to center column of tree to thoroughly cover developing fruit.

PEACHES: For brown rot, scab, and shot-hole, refer to recommendations for nectarines.

For peach leaf curl, use 4.8 to 6.4 qts. in sufficient water per acre. Apply in spring dormant period.

Do not apply within 2 days of harvest. If applied within 14 days of harvest, remove residues by brushing.

VEGETABLES

ASPARAGUS: Rust. Use 1.6 to 2.4 qts. per acre and 50 to 200 gals. of water, depending upon the size of the plants. Begin application to established fields after the crop has been harvested and continue at 10-day intervals until disease is under control. In *young* fields which will be not be harvested, start when disease first appears and repeat at 10-day intervals as needed.

BEANS: Anthracnose, downy mildew, rust. Apply 1.2 to 1.6 qts. in sufficient water to obtain thorough coverage of the plants. Begin when disease first appears and repeat at 4 to 10-day intervals (7-day intervals for rust). Do not apply to succulent beans within 4 days of harvest.

Beans as a generic term include the following crops; kidney beans, lima beans, mung beans, navy beans, pinto beans, snap beans, wax beans, broad beans, fava beans, asparagus beans, blackeyed peas and cowpeas.

BROCCOLI: Alternaria leaf spot, downy mildew. Apply 1.6 to 3.2 qts. per acre in sufficient water (usually 100 to 150 gals. per acre for full gallonage hydraulic equipment, and 75 gals. for air blast) for thorough coverage. Start application when disease threatens and continue at 7 to 14-day intervals as needed. When conditions are severe, shorten interval. If applied within 3 days of harvest, remove residues by washing or trimming.

BRUSSELS SPROUTS: For Alternaria leaf spot and downy mildew, use 1.2 qts. per acre. Begin application when diseases threaten and repeat at 3-day intervals in plant beds and at 7 to 10-day intervals in the field.

CABBAGE: Alternaria leaf spot and downy mildew. Use 1.2 qts. per 100 gals. and apply up to 150 gals. per acre.

Plant beds and direct-seeded fields: Apply 7 to 10 days after planting or earlier if disease is present. Repeat at 3-day intervals until plants are set in the field or until field-seeded plants are thinned.

Fields: Apply when the disease threatens; repeat weekly as needed. Shorten intervals to 3 to 5 days under severe conditions. Do not apply within 7 days of harvest.

CANTALOUPES: Refer to recommendations for melons.

CARROTS: Alternaria and Cercospora leaf spot. Use 1.2 to 1.6 qts. per acre and start applications when plants are 6 weeks old or when disease threatens. Repeat at 7 to 10-day intervals.

CAULIFLOWER: Refer to recommendations for Brussels sprouts.

CELERY: Early blight (Cercospora) and late blight (Septoria). Use 1.2 to 1.6 qts. per acre and apply when plants are established in the field; repeat as frequently as required (weekly under average conditions or every 3 to 5 days under severe disease conditions). Do not apply within 14 days of harvest. Remove residues by stripping, trimming and washing.

COLLARDS: Alternaria leaf spot and downy mildew. Use 1.2 to 1.6 qts. per acre. Start applications when disease first threatens in the area and repeat at 7 to 10-day intervals. Do not apply within 10 days of harvest. Remove residues by washing.

CORN (sweet): Helminthosporium leaf blight. Use 1.2 qts. per acre per 100 gals. of water. Begin applications when disease first appears and repeat at 4 to 7-day intervals. Do not apply within 7 days of harvest. Do not feed treated forage to livestock.

CUCUMBERS: Alternaria (Macrosporium) leaf spot, angular leaf spot, anthracnose, downy mildew, pythium fruit rot. Use 1.6 to 2.4 qts. per acre and begin applications when disease threatens or plants begin to run. Repeat at 7 to 10-day intervals or at 3 to 5-day intervals under severe disease conditions. Do not apply within 5 days of harvest.

EGGPLANT: Anthracnose, early blight (Alternaria), Phomopsis blight, or fruit rot. Use 1.2 to 1.6 qts. per acre and begin when fruit starts to form. Repeat at 7 to 10-day intervals.

ENDIVE: Downy mildew. Use 1.2 to 1.6 qts. per acre in up to 150 gals. water. Apply when disease appears and repeat at 7 to 10-day intervals or at 3 to 5-day intervals under severe disease conditions. Do not apply within 10 days of harvest. Remove residues from head lettuce by stripping and trimming, and from leaf lettuce and endive by washing or other effective means.

HONEY BALL MELONS: Refer to recommendations for melons.

HONEY DEW MELONS: Refer to recommendations for melons.

KALE: Refer to recommendations for collards.

KOHLRABI: Refer to recommendations for Brussel sprouts.

LETTUCE: Refer to recommendations for endive.

MELONS: Anthracnose, Cercospora leaf spot, downy mildew, and gummy stem blight. Use 1.2 to 1.6 qts. per acre and start applications when runners begin to form or when disease threatens. Repeat every 7 to 10 days as needed. Under severe disease conditions, shorten spray intervals to 2 to 5 days. Do not apply within 5 days of harvest.

MUSK MELONS: Refer to recommendations for melons.

MUSTARD GREENS: Refer to recommendations for collards.

ONIONS: Blast (Botrytis leaf blight), downy mildew and purple blotch (blight). Use 1.2 to 1.6 qts. per acre, and apply when disease first threatens. Repeat every 7 days as needed.

PEPPERS: Anthracnose, Cercospora leaf spot (frogeye spot). Use 1.2 to 1.6 qts. per acre and begin applications when disease first threatens. Repeat at 7 to 10-day intervals.

PERSIAN MELONS: Refer to recommendations for melons.

POTATOES: Early and late blight. Use 0.8 to 1.6 qts. per acre. Apply as soon as plants are 2 to 6 inches high and repeat at 5 to 10-day intervals as long as blight threatens. Under severe disease conditions use the shorter interval or the higher dosage.

POTATOES (seed pieces): Fusarium seed piece decay. Use 1 qt./10 gallons of water. Dip whole or cut tubers and spread in a cool place if held before planting. This is a seed piece treatment only; do not use treated seed pieces for food or feed.

PUMPKINS: Angular leaf spot, downy mildew. Use 2.4 qts. per acre in sufficient water to wet plants thoroughly. Begin applications when disease threatens and repeat at 7 to 10-day intervals as needed.

RHUBARB: (Greenhouse, Michigan): Botrytis leaf rot. Use 1.6 qts. per 100 gals. of water and apply at the rate of 15 gals. of spray per 4,000 sq. ft. (approx. 150 gals. per acre). Make 4 applications at 7-day intervals beginning 4 weeks before harvest.

SPINACH: Downy mildew (blue mold), white rust. Use 1.2 to 1.6 qts. per acre and begin applications when disease first appears in the area. Repeat at 7 to 10-day intervals. Do not apply within 10 days of harvest. Remove residues by washing or other effective means.

SQUASH (summer and winter): Anthracnose, downy mildew. Use 1.2 to 1.6 qts. per acre and begin applications when runners form or when disease threatens. Repeat at 7 to 10-day intervals as needed.

Pythium fruit rot. Use 1.6 to 3.2 qts. per acre and begin applications when runners form or when disease threatens. Repeat at 7 to 10-day intervals as needed.

Do not apply within 5 days of harvest.

TOMATOES: Early blight, late blight, Septoria leaf spot, gray leaf spot (Stemphylium) and anthracnose. Use 1.2 to 2.4 qts. per acre for field applications, or 4.5 to 6.1 fluid oz. per 5,000 sq. ft. in greenhouses. Apply when fruit of first clusters is well formed; repeat at 7 to 10-day intervals (do not use on young seedlings in greenhouses as injury may occur). When weather conditions favor disease, more frequent applications may be necessary starting when plants are set in fields. Tomato plants treated with MANZATE Flowable tend to develop heavy foliage and spray gallonage should be increased accordingly to assure complete coverage. In areas where anthracnose is a serious problem on canning tomatoes, apply 2.4 qts. per acre. Do not apply within 5 days of harvest.

TURNIPS: Alternaria leaf spot, downy mildew. Use 1.2 to 1.6 qts. per acre and begin applications when disease threatens. Repeat at 7 to 10-day intervals.

Do not apply within 7 days of harvest. Remove residues by washing.

WATERMELONS: Refer to recommendations for melons.

FIELD CROPS

GRASSES FOR SEED: Leaf, stem, and stripe rusts. Use 1.6 to 2.4 qts. in 50 to 100 gals. of water per acre. Begin applications when first rust pustules are noted and repeat every 7 to 10 days until harvest or as long as rust threatens. Do not feed treated grass to livestock.

SUGAR BEETS: Cercospora leaf spot. Use 1.2 to 2.4 qts. per acre. Start applications when disease threatens and repeat at 7 to 10-day intervals as needed. If up to 1.6 qts. per acre are used and are not applied within 14 days of harvest, there is no restriction on feeding tops. If more than 1.6 qts. per acre are used, do not apply within 10 days of harvest and do not feed treated tops to livestock.

TOBACCO: Anthracnose. Use 1.8 to 3.2 qts. per acre and begin applications after plants are set in the field. Repeat as necessary.

Blue mold. Use 0.8 to 2.4 qts. per 100 gals. Use 2.5 to 5.0 gals. of diluted suspension per 100 square yards of plant bed. Begin applications when plants are about the size of a dime or when blue mold is first reported in the area. Repeat twice weekly, or more often if sprays are washed with rain, until plants are all in the field. Use lower rates in early applications. Increase to 2.4 qts. per 100 gals. as plants increase in size. For disease control under field conditions, use 1.6 to 2.4 qts. per acre. Repeat as necessary.

MISCELLANEOUS AGRICULTURAL USES

CONIFERS: Needle diseases such as needle cast (Lophodermium pinastri) and brown spot (Scirrhia acicola). Use 1.2 qts. per 100 gals. Begin application in spring or early summer before infection occurs. Repeat after heavy rains and at two-week intervals as long as needed.

ROSES: Cercospora leaf spot, rust, black spot. Use 1.2 qts. per acre and start applications when first leaves unfold. Repeat at 7-day intervals as needed.

STORAGE AND DISPOSAL

STORAGE

Important—Keep in a cool place but not below 32°F. Temperature extremes will affect the quality of MANZATE Flowable. Store product in original container only, away from other pesticides, fertilizer, food or feed.

DISPOSAL

Do not contaminate water, food or feed by storage or disposal. Wastes resulting from the use of this product may be disposed of on site or at approved waste disposal facility. Triple rinse (or equivalent) the container. Then offer for recycling or reconditioning, or puncture and dispose of in a sanitary landfill, or by incineration, or, if allowed by state and local authorities, by burning. If burned, stay out of smoke.

MANZATE® 200

FLOWABLE FUNGICIDE

Contains 4.0 lbs. Active per Gallon

ACTIVE INGREDIENT:
A coordination product of zinc ion and manganese ethylenebisdithiocarbamate 37.0%
in which the ingredients are
Manganese++ 7.4%
Zinc++ 0.9%
Ethylenebisdithiocarbamate ion ($C_4H_6N_2S_4$) 28.7%
INERT INGREDIENTS 63.0%
TOTAL 100.0%

EPA Reg. No. 352-398 U.S. Pat. 3,379,610

Keep Out of Reach of Children

PRECAUTIONARY STATEMENTS

HAZARDS TO HUMANS AND DOMESTIC ANIMALS

CAUTION

MAY IRRITATE EYES, NOSE, THROAT AND SKIN. MAY BE HARMFUL IF INHALED OR SWALLOWED.

Avoid contact with skin, eyes and clothing. Avoid breathing spray mist. Wear protective clothing including long pants, long sleeve shirt, gloves, hat and boots during mixing and loading operations. Do not apply in such a manner as to directly or through drift expose workers, other persons or animals. The area being treated must be vacated by unprotected persons.

FIRST AID:

In case of contact, flush skin or eyes with plenty of water; for eyes, get medical attention.

ENVIRONMENTAL HAZARDS

This product is toxic to fish. Do not apply directly to water and wetlands. Do not apply where runoff is likely to occur. Do not apply when weather conditions favor drift from areas treated. Do not contaminate water by cleaning of equipment or disposal of wastes.

NOTICE TO BUYER

Purchase of this material does not confer any rights under patents of countries outside of the United States.

NOTICE OF WARRANTY

DuPont warrants that this product conforms to the chemical description on the label thereof and is reasonably fit for purposes stated on such label only when used in accordance with the directions under normal use conditions. It is impossible to eliminate all risks inherently associated with the use of this product. Crop injury, ineffectiveness or other unintended consequences may result because of such factors as weather conditions, presence of other materials or the manner of use or application, all of which are beyond the control of DuPont. In no case shall DuPont be liable for consequential, special or indirect damages resulting from use or handling of this product. All such risks shall be assumed be the buyer. DUPONT MAKES NO WARRANTIES OF MERCHANTABILITY OR FITNESS FOR A PARTICULAR PURPOSE NOR ANY OTHER EXPRESS OR IMPLIED WARRANTY EXCEPT AS STATED ABOVE.

DIRECTIONS FOR USE

It is a violation of federal law to use this product in a manner inconsistent with its labeling.

MANZATE 200 Flowable should be used only in accordance with recommendations on this label, or in separate published DuPont recommendations available through local dealers.

DuPont will not be responsible for losses or damages resulting from use of this product in any manner not specifically recommended by DuPont. User assumes all risks associated with such nonrecommended use.

MANZATE 200 Flowable, containing a coordination product of zinc ion and maneb, is recommended for use as a spray for the control of many important plant diseases.

Do not enter treated areas without protective clothing until sprays have dried. Because certain states may require more restrictive reentry intervals for various crops treated with this product, consult your State Department of Agriculture for further information.

Written or oral warnings must be given to workers who are expected to be in a treated area or in an area about to be treated with this product. When oral warnings are given, warnings shall be given in a language customarily understood by workers. Oral warnings must be given if there is reason to believe that written warnings cannot be understood by workers. Warnings must include the following information: "CAUTION. Area treated with MANZATE 200 Flowable on (date of application). Do not enter without appropriate protective clothing until sprays have dried. In case of contact, flush skin or eyes with plenty of water; for eyes, get medical attention."

As a Spray: (Ground or Aerial Equipment)—Apply MANZATE 200 Flowable at the rate shown; use sufficient water to provide thorough coverage, usually 20 to 100 gals. per acre for ground equipment and 5 to 8 gals. per acre for aircraft. Under severe disease conditions, use the higher rate and shorter interval specified for each crop. Add MANZATE 200 Flowable slowly to water in the spray tank with agitation, or premix thoroughly in separarte holding tank for concentrate or aircraft sprayers. Continuous agitation is required to keep the product in suspension.

STORAGE AND DISPOSAL

STORAGE

Important—Keep in a cool place but not below 32°F. Temperature extremes will affect the quality of MANZATE 200 Flowable. Store product in original container only, away from other pesticides, fertilizer, food or feed.

DISPOSAL

Do not contaminate water, food or feed by storage or disposal. Wastes resulting from the use of this product may be disposed of on site or at approved waste disposal facility. Triple rinse (or equivalent) the container. Then offer for recycling or reconditioning, or puncture and dispose of in a sanitary landfill, or by incineration, or, if allowed by state and local authorities, by burning. If burned, stay out of smoke.

APPLICATION				LAST APPLICATION DAYS	
CROPS	DISEASES	QUARTS PER ACRE	TIMING/INTERVAL	TO HARVEST	TO LIVESTOCK GRAZING/FEEDING
Apples Crabapples Quince	Scab	0.8–1.7 per 100 gals. up to 8.4 qts./acre	Begin at green tip or delayed dormant/ 7–14 days		
	Cedar Apple Rust	"	Begin at blossom time/ 7–14 days	15 (30 for apples)	Do not graze or feed Note add spreader sticker during last 3 cover sprays
	Bitter Rot, Black Rot, Brown Rot, Fly Speck, Sooty Blotch	"	Begin at first cover/ 7–14 days		
Apples Tank mix with Benlate® Fungicide (Benlate DuPont registered trademark)	Scab, Powdery Mildew, Sooty Blotch, Flyspeck, Cedar Apple Rust, Quince Rust, Bitter Rot, Black Rot, Brown Rot	20 oz. plus 2 to 3 oz. Benlate† per 100 gals. May be used with or without spray oil	Apply at ½" green tip/ 7–14 days (or as needed through cover sprays)	30	Do not graze livestock in treated orchards
Asparagus	Rust	1.7	Begin when rust appears. Apply as long as rust persists— usually 4 applications/ 10 days	Apply only on ferns after spears have been harvested	
Crown (planting stock)	Crown Rot	0.8 per 100 gals.	Dip Treat for 5 min.		
Bananas	Sigatoka	1.7–3.4	Begin when leaves first appear/14–21 days		
Caprifig	Endosepsis (Fusarium Fungus), Mold	3.3 per 100 gals.	Cut figs open and dip treat for at least 15 min.		
Carrots	Alternaria And Cercospora Leaf Spot	1.3–1.7	Begin when diseases first appear/7–10 days	7	Do not use treated tops for food or feed
Cereal Grains: —Barley —Oats —Rye —Wheat	Helminthosporium Leaf Blight, Septoria Leaf Spot	1.7	Begin at tillering-to-jointing stage/7–10 days. Maximum of 3 applications	26	Do not graze livestock in treated areas within 26 days after application
Celery, Fennel	Early Blight, Late Blight	1.7	Begin at emergence/ 3–5 days in Plant bed, 7 days in field	14–Celery; 7–Fennel Remove excess residues remaining at harvest by stripping, trimming and washing	
Conifers	Lophodermium Fungi	3.3 per 100 gals.*	Nursery plantings and newly transplanted stock/21 days during sprinkler irrigation 28 days rest of year		

CROPS	DISEASES	APPLICATION QUARTS PER ACRE	APPLICATION TIMING/INTERVAL	LAST APPLICATION DAYS TO HARVEST	LAST APPLICATION DAYS TO LIVESTOCK GRAZING/FEEDING
Corn —popcorn, sweet —field, hybrid seed production	Helminthosporium Leaf Blight	1.3	Begin when disease appears/4–7 days	7 Sweet 40–field popcorn & seed prod.	Do not graze or feed sweet corn, forage, or fodder
Cotton (Southwest only)	Rust	1–1.7	Start when rust appears/ 10–14 days	Do not apply after bolls open	Do not graze or feed gin trash
Cranberries	Fruit Rot	2.5–5*	Begin at mid-bloom/ 7–10 days	30	
Cucumbers	Alternaria Leaf Spor, Anthracnose, Cercospora Leaf Spot, Downy Mildew, Gummy Blight, Scab	1.7–2.5	Begin at 2-leaf stage/ 5–7 days	5	
Grapes	Black Rot, Bunch Rot, Dead Arm, Downy Mildew	1.3–3.4	Apply when shoots are ½–1½ inches long, 3–5 inches long, 8–10 inches long. Continue at 10–14 days	66 (CA–do not apply after fruit set)	
Melons	Alternaria Leaf Spot, Anthracnose, Cercospora Leaf Spot, Downy Mildew Gummy Stem Blight, Scab	1.7–2.5	Begin at 2-leaf stage/ 5–7 days	5	
Onions (Dry Bulb)	Blast (Botrytis Leaf Blight), Downy Mildew, Purple Blotch (Blight), Neck Rot	1.7–2.5	Begin when diseases are reported in area/ 7 days	7 (Do not apply to exposed bulbs.)	
Seed— (NY only)	Onion Smut (on emerging seedlings in muck soil)	2.5	Furrow Drench at planting		
Papaya	Anthracnose, Phytophthora Fruit Rot	1.7–2.1*	Begin at flowering. Treat central column, crown, blossom area, and developing fruit of tree/ 7–14 days		
Peanuts	Cercospora Leaf Spot, Rust	0.8–1.7	Begin when disease appears/10–14 days (Normal condition) 7 (Humid conditions)	14	Do not feed treated vines to livestock
Peanuts (Tank mix with Benlate)	Cercospora Leafspot	1.3 plus 4 oz. Benlate	Begin 35–40 days after planting, or when disease first appears/10–14 days	14	Do not graze or feed treated vines, hay or hulls to livestock
	Ascochyta Web Blotch	1.3 plus 4 oz. Benlate	Begin 35–40 days after planting, or when disease first appears/ 7–14 days		
	Rust	1.3 plus 4 oz. Benlate	Begin 35–40 days after planting, or when disease first appears/7–10 days		
Pears	Scab	0.8–1.7 per 100 gals. up to 8.4 qts./acre	Begin at green tip or delayed dormant/7–14 days	15	Do not graze Note: Add spreader-sticker during last 3 cover sprays
	Cedar Apple Rust	"	Begin at blossom time/7–14 days		
	Bitter Rot, Black Rot, Brown Rot, Fly Speck, Sooty Blotch	"	Begin at first cover/7–14 days		
	Pear Psylla Nymphs (Pacific Northwest)	1.3–1.7 per 100 gals. up to 8.4 qts./acre	Begin at petal fall—thru postharvest. Use as 2–3 sprays when high populations occur/ 4–28 days	15	Do not graze
Pineapple Propagation Stock	Phytophthora Heart Rot	27 per 100 gals.	Preplant Dip		
Potatoes	Early Blight, Late Blight	0.8–1.7	Begin when plants are 4–6 inches high/ 7–10 days		
Potato seed pieces	Fusarium Seed Piece Decay, Common Scab	2.1 Per 100 gals.	Dip treat. Plant as soon as possible following treatment		Do not use treated pieces for food or feed purposes

CROPS	DISEASES	APPLICATION: QUARTS PER ACRE	APPLICATION: TIMING/INTERVAL	LAST APPLICATION DAYS: TO HARVEST	LAST APPLICATION DAYS: TO LIVESTOCK GRAZING/FEEDING
Squash (summer)	Alternaria Leaf Spot, Anthracnose, Cercospora Leaf Spot, Downy Mildew, Gummy Stem Blight, Scab	1.7–2.5	Begin when plants are in 2–leaf stage/ 5–7 days	5	
Sugar Beets	Cercospora Leaf Spot	1.3–1.7	Begin when disease first threatens/7–10 days	14	Do not feed treated tops to livestock
Tobacco					
—Plant Bed	Anthracnose, Blue Mold	0.4 per 100 gals. apply 3.5 gals./ 100 sq. yds. of plant bed	Begin when plants are about the size of a dime or when blue mold appears/3–14 days thru transplanting		
—Field	Blue Mold	0.8–1.3 per 100 gals.	Begin when blue mold is reported in the area/3–7 days		
Tomatoes	Anthracnose, Early Blight, Gray Leaf Mold, Gray Leaf Spot (Stemphyllum), Late Blight	1.3–2.5	Begin when seedlings emerge/7 days	5	
	Bacterial Spot	1.3 plus 4 lbs. tribasic copper sulphate	Begin when seedlings emerge/ 3–4 days	5	
Turf Grasses	Algae	10 fl. oz./ 1000 sq. ft.	Begin when algae begins to appear/7 days		
	Copper Spot, Fusarium Blight (F. Roseum), Red Thread, Slime Molds	7–10 fl. oz./ 1000 sq. ft.	Begin when grass greens-up in spring/7–14 days.		
		10–14 fl. oz./ 1000 sq. ft.	Use during favorable disease conditions/7 days		
	Dollar Spot (Sclerotinia)	10–14 fl. oz./ 1000 sq. ft.	Begin when grass greens-up in spring/7–14 days		
		14 fl. oz/ 1000 sq. ft.	Use during favorable disease conditions/7 days		
	Pink (Fusarium) Snow Mold	10–14 fl. oz./ 1000 sq. ft.	During winter/ 14–42 days		Do not graze treated areas or feed clippings to livestock
	Leaf Spot (Helminthosporium spp.) Rhizoctonia Brown Patch	5–7 fl. oz./ 1000 sq. ft.	Begin when disease appears/7–10 days, tees & greens—7–14 days, fairways—14–21 days, home lawns		
		10–14 fl. oz./ 1000 sq. ft.	Use during favorable disease conditions/3–5 days		
	Pythium Blight	14 fl. oz./ 1000 sq. ft.	Begin at first sign of disease/5 days or more often during favorable disease conditions		
	Leaf, Stem and Stripe Rust	5–7 fl. oz./ 1000 sq. ft.	Begin when disease appears/7–10 days		

* Add spreader sticker

† Use 3 oz. rate of "Benlate" and add spray oil for varieties more susceptible to powdery mildew, and for scab if an application is missed during an infection period in order to deactivate scab and prevent further infection

SEED TREATMENT	DISEASES	SEED TREATMENT RATE—APPLY AS A SLURRY: FLUID OZ./BU.	SEED TREATMENT RATE—APPLY AS A SLURRY: FLUID OZ./CWT.	
Barley	Seed Decay, Seedling Blights, Covered Smut, Bunt	2.2	4.5	
Corn	Damping Off, Seed Rots, Seedling Blights	2.5	4.5	
Cotton				LABEL TREATED SEED—"Do not use for food, feed or oil purposes. This seed treated with Manzate® 200 Flowable Fungicide"
Acid delinted	Damping Off, Seedling Blights	—	5.1	
Reginned		—	10.1	
Flax	Seed Decay, Seedling Blight	3.4	6.1	
Oats	Seed Decay, Seedling Blights, Loose Smut	2.2	6.7	
Potato tubers	Fusarium Seedpiece Decay, Seed Borne, Common Scab	—	25.6	
Rice	Achlya, Other Soil and Seed Borne Fungi Causing Seed Rot and Reduced Seedling Vigor	—	3.4–6.7 (of dry rice seed) (2.1–4.2 qts./ton of seed). Apply before, during, or after soaking in water	

SEED TREATMENT	DISEASES	SEED TREATMENT RATE—APPLY AS A SLURRY FLUID OZ./BU.	FLUID OZ./CWT.	
Rye	Seed Decay, Seedling Blights, Covered Smut, Bunt	2.2	4.0	Note: The incorporation of a dye into the treating slurry is recommended
Safflower	Puccinia Carthami Cda (which causes foot-and-rot disease and foliage rust disease)	—	3.4	
Sorghum	Damping Off, Seedling Blights	2.5	5.1	
Tomatoes	Damping Off, Seedling Blights	—	13.5	
Wheat	Seed Decay, Seedling Blight, Covered Smut, Bunt	2.2	3.7	

OUST®

HERBICIDE DISPERSIBLE GRANULE

ACTIVE INGREDIENT:
Sulfometuron Methyl
[Methyl 2-[[[[(4,6-dimethyl-2-pyrimidinyl) amino]-carbonyl]amino]sulfonyl]benzoate] 75%
INERT INGREDIENTS 25%
TOTAL 100%

EPA Reg. No. 352-401 EPA Est. 352-WV-1
U.S. Patent 4,394,506

Keep Out of Reach of Children

PRECAUTIONARY STATEMENTS

HAZARDS TO HUMANS

CAUTION!

MAY IRRITATE EYES, NOSE, THROAT AND SKIN.

Avoid breathing dust or spray mist. Avoid contact with skin, eyes and clothing.

FIRST AID

If in eyes, immediately flush with plenty of water and get medical attention. If on skin, immediately flush with plenty of water and get medical attention if irritation persists.

ENVIRONMENTAL HAZARDS

Do not apply directly to wetlands or any body of water.

Do not contaminate water by cleaning of equipment or disposal of wastes.

IMPORTANT

Do not use on food or feed crops. Injury to or loss of desirable trees or other plants may result from failure to observe the following: Do not apply (except as recommended), or drain or flush equipment on or near desirable trees or other plants, or on areas where their roots may extend, or in locations where the chemical may be washed or moved into contact with their roots. Do not apply where runoff water may flow onto agricultural land as injury to crops may result. In addition do not apply directly to any body of water. Do not use on lawns, walks, driveways, tennis courts or similar areas. Prevent drift of spray to desirable plants. Keep from contact with fertilizers, insecticides, fungicides and seeds.

Following an OUST application, the spray tank used should not be used for other than non-crop applications. This is extremely important as low rates of OUST can kill or severely injure most crops.

NOTICE OF WARRANTY

DuPont warrants that this product conforms to the chemical description on the label thereof and is reasonably fit for purposes stated on such label only when used in accordance with directions under normal use conditions. It is impossible to eliminate all risks inherently associated with use of this product. Ineffectiveness or other unintended consequences may result because of such factors as weather conditions, presence of other materials or the manner of use or application, all of which are beyond the control of DuPont. In no case shall DuPont be liable for consequential, special or indirect damages resulting from the use or handling of this product. All such risks shall be assumed by the Buyer. DUPONT MAKES NO WARRANTIES OF MERCHANTABILITY OR FITNESS FOR A PARTICULAR PURPOSE NOR ANY OTHER EXPRESS OR IMPLIED WARRANTY EXCEPT AS STATED ABOVE.

GENERAL INFORMATION

DuPont OUST Herbicide is a dispersible granule to be mixed in water and applied as a spray for control of many annual and perennial grasses and broadleaf weeds on non-cropland areas.

OUST may be applied preemergence or postemergence to the weeds; best results are obtained if application is made before or during early stages of weed growth. OUST may be used at any time of the year except when ground is frozen, provided adequate moisture is available for herbicide activation.

Under limited rainfall conditions, OUST may not provide satisfactory control of hard-to-kill perennials. To reduce the potential for off-target movement, do not apply OUST during periods of intense rainfall or to soils saturated with water.

OUST is absorbed by both roots and foliage of weeds, resulting in visual effects progressing from growth inhibition to reddish-purplish coloration; chlorosis, necrosis, vein discoloration and death of terminals. Initial effects are usually seen 2 or 3 weeks following application; however, the final effects are evident at about 4 to 6 weeks after application.

Since the degree of control and duration of effect will vary with the amount of chemical applied, soil texture, rainfall and other conditions, it is suggested that users limit their first use to small areas.

Notice to Buyer: Purchase of this material does not confer any rights under patents of countries outside of the United States.

DIRECTIONS FOR USE

It is a violation of federal law to use this product in a manner inconsistent with its labeling.

OUST Herbicide should be used only in accordance with recommendations on this label or in separate published DuPont recommendations available through local dealers.

DuPont will not be responsible for losses or damages resulting from the use of this product in any maner not specifically recommended by DuPont. User assumes all risks associated with such nonrecommended use.

OUST is recommended for pre- and postemergence control of many annual and perennial grasses and broadleaf weeds on noncropland areas such as airports, fence rows, roadsides and associated right-of-way, lumberyards, petroleum tank farms, pipeline and utility right-of-way, pumping installations, railroads, storage areas, plant sites and other similar areas.

SPRAY PREPARATION: Mix the proper amount of OUST in water in the spray tank with the agitator running. Continuous agitation is required to keep the product in suspension. To minimize drift, a drift control agent should be added at the recommended label rate. For nonselective postemergence applications, a nonionic surfactant such as DuPont Surfactant WK or "Ortho" X-77* may then be added at the rate of 1 quart per 100 gallons of spray to improve wetting and/or contact activity. OUST spray preparations may degrade in acid solutions if not used within 48 hours. Thoroughly reagitate before reusing.

*Registered trademark of Chevron Chemical Company.

EQUIPMENT—SPRAY VOLUME AND PRESSURE: Apply uniformly with properly calibrated ground equipment, using at least 25 gallons of spray per acre and a spray pressure of 25 to 50 psi. Extreme care must be taken to prevent drift to desirable plants or agricultural land. To minimize drift, apply at low pressures with coarse spray nozzles and booms as low as possible to ground. Keep OUST out of cropland. Do not apply by aircraft.

WEEDS CONTROLLED, RATES AND TIMING:

Early Postemergence—Selective Applications—1 Ounce Per Acre

Broadleaves		Grasses
Carolina geranium	Spotted spurge	Downy brome
Goldenrod	Sunflower	Fescue
Kochia	Sweet clover	Foxtail
Russian Thistle	Wild Carrot	

Preemergence to Early Postemergence—3 to 5 Ounces Per Acre

Broadleaves [Up to 3 to 4" in Height or Diamater]		Grasses [Up to 6 to 12" in Height]
Bouncingbet	Ox-eye daisy	Alta fescue*
Buckhorn plantain	Pigweed	Annual bluegrass
Burclover	Prickly lettuce	Annual ryegrass
Carolina geranium	Purple starthistle	Bahiagrass
Common chickweed	Ragweed	Barnyardgrass
Common dandelion	Russian thistle	Downy brome
Common speedwell	Sowthistle	Fescue
Common yarrow	St. Johnswort	Foxtail
Crimson clover	Sunflower	Indian grass
Dogfennel	Sweet clover	Italian ryegrass
Filaree	Tansy mustard	Johnsongrass (short-term control)
Hoary cress	Tansy ragwort	
Horsetail	Tumble mustard	Little barley
Kochia	Vetch	Red brome
Little mallow	Wild carrot	Red fescue
Mustard	Yellow rocket	Reed canarygrass
		Ripgut brome
		Ryegrass
		Smooth brome
		Sprangletop

Early Postemergence—6 to 12 Ounces Per Acre

Broadleaves		Grasses
Bedstraw	Jerusalem artichoke	Johnsongrass
Canada thistle	Kudzu*	(season-long control)
Curly dock	Mayweed	Yellow nutsedge
Dewberry	Musk thistle	
Fiddleneck	Poison Ivy	
Fleabane	Spanish needles	
Goldenrod	Turkey mullein	
Hemlock	Virginia pepperweed	
Honeysuckle	Wild blackberry	

*Considered hard-to-kill and use only high rate of 12 ounces per acre.

Use the higher level of the recommended dosage ranges under the following conditions: heavy weed growth; on soils containing more than 2½% organic matter; and on high soil moisture areas such as long road edges or railroad shoulders.

NONSELECTIVE WEED CONTROL

Johnsongrass Control: For short-term control (up to 3 months), apply OUST at 2 to 5 ounces per acre early postemergence to the johnsongrass. When regrowth occurs, or if further control of the actively growing johnsongrass is desired, retreat at 3 to 5 ounces per acre.

For season-long control, apply OUST at 6 to 12 ounces per acre during the stages of early postemergence to the johnsongrass through the period when the johnsongrass is actively growing.

Control of Other Weeds: Apply OUST at 3 to 5 ounces per acre for control of weeds such as Carolina geranium, foxtail, kochia, little barley, ox-eye daisy, Russian thistle, etc., during pre- to early postemergence stages of weed growth. For hard-to-control weeds such as curly dock, goldenrod, musk thistle, etc., apply OUST at 6 to 12 ounces per acre.

Tank Mixtures With Residual Herbicides (Bareground Weed Control): For improved preemergence to early postemergence control of downy brome, fescue, foxtail, johnsongrass, kochia, Russian thistle and wild blackberry add 2 to 6 ounces of OUST to the **recommended rates of one of the following DuPont products: Hyvar® X Herbicide, Karmex® Herbicide, Krovar® I Herbicide, Velpar® L Herbicide or Velpar® Herbicide.**

SELECTIVE WEED CONTROL

For use on unimproved industrial turf, on roadsides and on other noncropland areas where bermudagrass (common and coastal varieties), bahiagrass, crested wheatgrass and smooth brome are well established and are desired as ground covers.

BERMUDAGRASS RELEASE

For control of many annual and perennial broadleaf weeds and grasses use the following application timings:

Late Spring and Early Summer: Apply 1 to 2 oz. per acre for control of Carolina geranium, fescue, foxtail, goldenrod, spotted spurge, sweet clover and wild carrot after bermudagrass is well-established, usually about 30 days after initial spring flush. If regrowth of weeds occurs and additional control is desired, repeat treatment at 1 to 2 oz. during the late spring and summer. After weeds become established, best results occur when applications are made 1 to 2 weeks following mowing. Apply 2 to 4 oz. per acre for short-term control of johnsongrass after bermudagrass is well-established, usually about 30 days after initial spring flush. Retreat with 2 to 4 oz. if additional control is desired or if regrowth occurs.

Late Fall and Early Winter: Apply 1 to 4 oz. per acre for control of winter annuals such as Carolina geranium, common chickweed, fescue, little barley and wild blackberry.

BAHIAGRASS RELEASE AND SEEDHEAD SUPPRESSION

Apply 1 oz. for control of fescue, goldenrod, spotted spurge and wild carrot. Rates as low as ½ oz. per acre have reduced bahiagrass growth vigor and suppressed seedheads when applied before seedhead emergence in the spring or following mowing.

SMOOTH BROME RELEASE AND SUPPRESSION

For control of downy brome, foxtail, goldenrod, kochia, Russian thistle, sunflower and sweet clover in well established stands of smooth brome, apply 1 oz. of OUST per acre in the spring. This application may also result in smooth brome seedhead reduction, as well as growth suppression and reduced plant vigor.

CRESTED WHEATGRASS RELEASE AND SUPPRESSION

For control of downy brome, kochia, Russian thistle and sweet clover in well established stands of crested wheatgrass on noncrop sites, apply 1 oz. OUST during the rainy season or spring. This application may also result in growth suppression of the crested wheatgrass.

Note: (Selective Weed Control)

1. Temporary discoloration or topkill of bahiagrass, bermudagrass and smooth brome may result following an application of OUST. Browning and degree of stand reduction may increase in areas with warm climates.
2. Do not apply OUST to industrial turf that is under stress from drought, insects, disease, cold temperatures or late spring frost as injury may result.
3. Do not use a surfactant.
4. Follow all use precautions on label.

STORAGE AND DISPOSAL

STORAGE

Store product in original container only, away from other pesticides, fertilizer, food or feed.

DISPOSAL

Do not contaminate water, food or feed by storage or disposal. Wastes resulting from the use of this product may be disposed of on site or at approved waste disposal facility.

Triple rinse (or equivalent) the container. Then offer for recycling or reconditioning, or puncture and dispose of in a sanitary landfill, or by incineration, or, if allowed by state and local authorities, by burning. If burned, stay out of smoke.

SUPPLEMENTAL LABELING

OUST® HERBICIDE

EPA REG. NO. 352-401

FOR CONTROL OF UNDESIRABLE HERBACEOUS PLANTS IN CONIFERS

DIRECTIONS FOR USE

It is a violation of federal law to use this product in a manner inconsistent with its labeling.

Du Pont OUST® Herbicide is recommended for herbaceous weed control in commercial reforestation areas where Jack pine, loblolly pine, red pine, slash pine and Virginia pine plantings are being established. OUST is an effective herbicide providing both foliar and residual activity to control many grasses and broadleaves.

HOW TO USE

Apply 2-8 ounces of OUST per acre (2-4 ounces per acre on red pine), using appropriate ground or air (helicopter only) equipment and sufficient water, usually 25-50 gallons per acre by ground and 5-15 gallons per acre by air. Calibrate equipment to insure uniform distribution of liquid over treated area.

To reduce drift when applying by air, use precise application spray equipment (jet nozzles directed backwards [D4 or larger] or a "Microfoil"[1] boom or equivalent). When compatible with equipment, additon of a spray thickening agent, such as "Nalco-Trol"[2], is recommended to further minimize drift. Do not apply by air within 200 feet of any homestead, agricultural land or other desirable plantings.

Make application from late winter to mid-summer. Best results occur just before or after weeds have germinated. OUST may be applied preplant or postplant to young conifers.

When used as described, OUST controls many herbaceous plants such as: crabgrass, dog fennel, fescue, fireweed (willowweed), goldenrod, horseweed, Kentucky bluegrass, nutsedge (yellow), panicums (broadleaf, fall, narrow), pokeweed, ragweed, white snakeroot and yellow sweetclover.

NOTES:

1. Extreme care must be taken to prevent drift or runoff to homesteads, desirable plantings, agricultural land or any body of water. Do not apply when weather conditions favor drift from treated areas.
2. Use extreme care when applying in areas adjacent to any body of water. Keep out of lakes, streams, ponds, reservoirs or any body of water.
3. Do not apply where conifers are suffering from loss of vigor caused by insects, diseases, drought, winter damage or other stresses, as injury may result.
4. OUST is not recommended for use on poorly drained or marshy sites, but it may be used where pines have been planted in beds.
5. OUST applications may result in damage to Douglas fir, noble fir and ponderosa pine.
6. Use the lower rates on coarse textured soils (i.e. loamy sands, sandy loams) and higher rates on fine textured soils (i.e. sandy clay loams and silty clay loams). Do not use on gravelly or rocky soils, exposed subsoils or heavy clay soils (i.e. sandy clay, silty clay or clay).

IMPORTANT

BEFORE USING OUST, READ AND CAREFULLY OBSERVE THE CAUTIONARY STATEMENTS AND ALL OTHER INFORMATION APPEARING ON THE PRODUCT LABLE.

This bulletin contains new or supplemental instructions for use of this product which may not appear on the package label. Follow instructions carefully.

This labeling must be in the possession of the user at the time of pesticide application.

[1] Trademark of Union Carbide Agricultural Products Company

[2] Trademark of Nalco Chemical Company

4/23/85

SINBAR®

HERBICIDE WETTABLE POWDER

ACTIVE INGREDIENT:

Terbacil [3-tert-butyl-5-chloro-6-methyluracil]	80%
INERT INGREDIENTS	20%
TOTAL	100%

EPA Reg. No. 352-317 U.S. Pats. 3,235,357 & 3,352,862

Keep Out of Reach of Children

PRECAUTIONARY STATEMENTS

HAZARDS TO HUMANS

CAUTION!

MAY IRRITATE EYES, NOSE, THROAT AND SKIN.

Avoid breathing dust or spray mist. Avoid contact with skin, eyes, and clothing. Wash thoroughly with soap and water after handling. Do not apply this product in such a manner as to directly or through drift expose workers or other persons. The area being treated must be vacated by unprotected persons.

STATEMENT OF PRACTICAL TREATMENT

If in eyes: Flush with plenty of water; get medical attention if irritation persists.

If on skin: Wash with plenty of soap and water.

ENVIRONMENTAL HAZARDS

Keep out of lakes, streams, or ponds. Do not contaminate water by cleaning of equipment or disposal of wastes.

IMPORTANT

Injury to or loss of desirable trees or other plants may result from failure to observe the following:

Do not apply (except as recommended for crop use), or drain or flush equipment on or near desirable trees or other plants, or on areas where their roots may extend, or in locations where the chemical may be washed or moved into contact with their roots. Do not use on lawns, walks, driveways, tennis courts, or similar areas. Do not use in home planting of fruits, nuts or other crops nor in citrus, apple, peach or pecan orchards interplanted with other trees or desirable plants. Prevent drift of dry powder or spray to desirable plants. Do not contaminate any body of water. Keep from contact with fertilizers, insecticides, fungicides, and seeds. Thoroughly clean all traces of SINBAR from application equipment immediately after use. Flush tank, pump, hoses, and boom with several changes of water after removing nozzle tips and screens (clean these parts separately).

GENERAL INFORMATION

DuPont SINBAR Herbicide is a wettable powder to be mixed in water and applied as a spray for selective weed control in alfalfa, apples, peaches, blueberries, caneberries, citrus, mint, pecans, strawberries and sugarcane. It is non-volatile, non-flammable, and non-corrosive to equipment.

SINBAR controls susceptible weeds for an extended period of time; the degree of control and duration of effect will vary with the amount of chemical applied, soil texture, rainfall and other conditions. Soils high in clay or organic matter require higher dosages than soil low in clay or organic matter to obtain equivalent herbicide performance. Moisture is required to activate the chemical; best results occur if rainfall (or sprinkler irrigation) occurs within 2 weeks of application.

Observe all cautions and limitations on labeling of all products used in mixtures.

NOTICE OF WARRANTY

DuPont warrants that this product conforms to the chemical description on the label thereof and is reasonably fit for the purposes stated on such label only when used in accordance with the directions under normal use conditions. It is impossible to eliminate all risks inherently associated with the use of this product. Crop injury, ineffectiveness or other unintended consequences may result because of such factors as weather conditions, presence of other materials, or the manner of use or application, all of which are beyond the control of DuPont. In no case shall DuPont be liable for consequential, special or indirect damages resulting from the use or handling of this product. All such risks shall be assumed by the buyer. DUPONT MAKES NO WARRANTIES OF MERCHANTABILITY OR FITNESS FOR A PARTICULAR PURPOSE NOR ANY OTHER EXPRESS OR IMPLIED WARRANTY EXCEPT AS STATED ABOVE.

DIRECTIONS FOR USE

It is a violation of federal law to use this product in a manner inconsistent with its labeling.

DuPont SINBAR Herbicide should be used only in accordance with recommendations on this label, or in separate published DuPont recommendations available through local dealers.

DuPont will not be responsible for losses or damages resulting from use of this product in any manner not specifically recommended by DuPont. User assumes all risk associated with such non-recommended use.

Do not use in Kern County, California.

Do not enter treated areas without protective clothing until sprays have dried. Because certain states may require more restrictive reentry intervals for various crops treated with this product, consult your State Department of Agriculture for further information.

Written or oral warnings must be given to workers who are expected to be in a treated area or in an area about to be treated with this product. When oral warnings are given, warnings shall be given in a language customarily understood by workers. Oral warnings must be given if there is reason to believe that written warnings cannot be understood by workers. Warnings must include the following information: "CAUTION. Area treated with SINBAR on (date of application). Do not enter without appropriate protective clothing until sprays have dried. In case of contact, flush skin or eyes with plenty of water, for eyes, get medication attention."

SELECTIVE USE IN CROPS WEED CONTROL

Best results are obtained if application is made shortly before or shortly after weed growth begins; if dense growth is present, remove tops and spray the ground. Control of perennial grasses may be improved by cultivation prior to treatment; otherwise, avoid working the soils as long as weed control continues or else effectiveness of the treatment may be reduced. For alfalfa, the soil should not be disturbed. See "Crops" section for recommended use rates for specific crops.

SINBAR, at rates ½ to 1½ lbs. per acre (alfalfa), controls annual weeds including species of chickweed, henbit, lambsquarters, mustard, tansymustard, crabgrass, and foxtail, and such weeds as downy brome (cheatgrass), marestail, peppergrass, prickly lettuce, shepherdspurse, snoweed, ryegrass, wild barley, and yellow rocket. Treatment will not control established perennial bindweeds, johnsongrass and Canada thistle.

Rates of 1 to 4 lbs. per acre control johnsongrass seedlings and annual weeds including species of chickweed, clover, crabgrass, crotolaria, seedling dandelion, fiddleneck, foxtail, henbit, knotweed, lambsquarters, mustard, nightshade, panicum, plantain, pigweed, purslane, ragweed, sandbur (sandspur), signalgrass (*Brachiaria*), smartweed, and wild geranium and such weeds as American burnweed (fireweed, *Erechites hieracifolia*), barnyardgrass (watergrass), bluegrass (*Poa annua*), China lettuce, crowfootgrass, dogfennel, Flora's paintbrush, Florida pusley, guineagrass, horseweed, jimsonweed, junglerice, natalgrass, and orchardgrass. In addition, treatment usually provides partial control of species of groundsel, horsenettle, quackgrass, red sorrel, and yellow nutsedge.

Rates of 2 to 4 lbs. per acre (blueberries) control weeds including species of cinquefoil, hawkweed and such weeds as annual sedge, perennial ryegrass, quackgrass, redroot (*Lachnanthes caroliniana*), and red sorrel.

Rates of 4 to 10 lbs. per acre are usually required for control of bermudagrass and for effective control of horsenettle, quackgrass, red sorrel, and yellow nutsedge; repeated annual applications may also be necessary. Partial control of perennial weeds such as purple nutsedge, paragrass and torpedograss usually accompanies treatment at the maximum dosage rates. Effects on perennial weeds are slow to appear, usually progressing over a period of several months.

Equipment—Spray Volumes And Pressures:
Apply with a fixed-boom power sprayer properly calibrated to a constant speed and rate of delivery. Use sufficient water (minimum 20 gals. per acre) to provide thorough and uniform coverage of the ground. On sugarcane and mint, preemergence broadcast applications may be made by aircraft (5 to 10 gals. spray per acre).

Continuous agitation in the spray tank is required to keep the material in suspension. Avoid overlapping, and shut off spray booms while starting, turning, slowing or stopping, or injury to crop may result.

Spray Preparation: Mix proper amount of SINBAR into necessary volume of water.

Use Rates: All dosages of SINBAR are expressed as broadcast rates; for band treatment, use proportionately less. Where a range of dosages is given, use the lower rate on coarse textured soils (low in clay or organic matter) and the higher rate on fine textured soils (high in clay or organic matter).

Soil Limitations: Crop injury may result from failure to observe the following: Unless otherwise directed, do not use on sand, loamy sand, or gravelly soils, nor on soils low in organic matter (less than 1%).

Replanting: Unless otherwise directed, do not replant treated areas to any crop within 2 years after last application as injury to subsequent crops may result.

CROPS

ALFALFA

Treat only stands established for 1 year or more. Make a single application of ½ to 1½ lbs. per acre in the fall after plants become dormant or in the spring before new growth starts. For semi-dormant and non-dormant varieties, apply in fall or winter after last cutting or in the spring before new growth starts. Apply before or after emergence of weeds but before they are 2" tall or across.

Do not use on seedling alfalfa nor on alfalfa-grass mixtures or other mixed stands. Do not apply to established stands after new growth starts in the spring, and do not apply on snow-covered or frozen ground as injury to the crop may result. Do not use in California south of Interstate 80.

APPLES, PEACHES

Use SINBAR alone, or apply as a tank mixture with DuPont Karmex Herbicide. Make a single band or broadcast application as a directed spray; avoid contact of foliage and fruit with spray or mist. Apply either in the spring or after harvest in the fall before weeds emerge or during early seedling stage of weed growth. Where complete weed control to harvest is desired, additional weed control measures may be required during the growing season.

Where crop is grown under furrow irrigation or under raised-berm flood irrigation (trees 4" to 6" above waterline), apply only as a band treatment. Do not treat trees planted in the bottom of irrigation furrows or trees grown under flat flood or basin irrigation, and do not use on eroded areas where subsoil or tree roots are exposed, as injury to trees may result. Do not graze or feed forage from treated areas to livestock.

SINBAR Alone—Use only under trees established in the orchard for at least 3 years.

Soil Texture	Lbs. SINBAR Per Acre	
	1 to 2% Organic Matter	More than 2% Organic Matter
Sandy loam	2	3
Loam, silt loam, silt	2½	3½
Clay loam, clay	3	4

SINBAR + Karmex—Use only under trees established in the orchard for at least 2 years.

Soil Texture	Lbs. Product Per Acre	
	1 to 2% Organic Matter SINBAR + KARMEX	More than 2% Organic Matter SINBAR + KARMEX
Sandy loam	1 + 1	1½ + 1½
Loam, silt loam, silt	1½ + 1½	2 + 2
Clay loam, clay	2 + 2	2 + 2

ASPARAGUS

Direct Seeded—Plant seed 1½ inches deep in coarse soils and 1 inch deep in fine soils. During planting operation, spray activated charcoal (such as Aqua NuChar[1] or Gro-Safe[2]) as a 1 inch band on soil surface directly over rows at a rate of 300 lbs. per acre (broadcast basis equivalent to 15 lbs. per acre where row spacing is 20 inches). Follow with SINBAR as a single spray at 1 to 2 lbs. per acre. Use the lower rate on coarser soils and the higher rate on finer soils.

[1]Aqua-Nuchar—Registered trademark of Westvaco Corp.

[2]Gro-Safe—Registered trademark of ICI United States, Inc.

Established Beds—Apply 1½ to 3 lbs. SINBAR per acre prior to spear emergence. Application may be made immediately after clean cutting. Use the lower rate on coarser soils and the higher rate on finer soils. Do not apply more than 3 lbs. per acre.

Apply before weeds emerge or to small weeds (½ to 2 inches tall or across).

High organic soils absorb SINBAR so that it is substantially inactivated as a soil residual weed killer. On these soils, weed control is provided by postemergence foliar uptake only. Apply up to 3 lbs. SINBAR per acre and a maximum of 2 applications per year.

NOTE: Do not use on areas where subsoil or roots are exposed. Do not use on plants that are diseased or lacking in vigor, as injury to the crop may result. Do not harvest within 5 days after application. Treated areas may be planted to asparagus one year after application. Otherwise, do not replant to any other crop within two years after last application.

BLUEBERRIES

Treat only plantings established for 1 year or more. Make a single band or broadcast application to ground beneath bushes; avoid contact of foliage and fruit with spray or mist. Apply either in the spring or after harvest in the fall before weeds emerge or during early seedling stage of weed growth. Do not use on eroded areas where subsoil or roots are exposed, nor on plants that are diseased or lacking in vigor, as injury to the plants may result. Treated areas may be planted to blueberries one year after last application.

Soil Texture	Lbs. SINBAR in Min. 25 Gals. Water Per Acre	
	1 to 3% Organic Matter	More than 3% Organic Matter
Sand, loamy sand	Do not use	2
Sandy loam	2	3
Loam, silt loam, silt, sandy clay, sandy clay loam, silty clay, silty clay loam	3	4
clay, clay loam	4	4

CANEBERRIES

(Blackberries, Boysenberries, Dewberries, Logenberries, Raspberries, Youngberries)—Treat only plantings established for 1 year or more. Make a single band or broadcast application of 1 to 2 lbs. per acre in a minimum of 20 gals. of water to the soil beneath the canes in the fall or early spring before fruit set and before weeds emerge or during the early stage of weed growth. Do not spray foliage nor use on eroded areas where subsoil or roots are exposed, nor apply to plants that are diseased or lacking vigor as injury to the plant may result. Treated areas may be planted to mint, alfalfa, blueberries, apples, peaches or strawberries 1 year after last treatment. Do not replant to other crops within 2 years of last application. Do not apply within 70 days before harvest.

CITRUS

Apply as a band or broadcast treatment as a directed spray; avoid contact of foliage and fruit with spray or mist. Temporary yellowing of citrus leaves may occur following treatment. Treatment may be used on coarse soils but is not to be used on sands low in organic matter (less than 1%), on poorly drained soils, nor on trees planted in irrigation furrows, as injury to trees may result. Treated areas may be planted to citrus trees one year after last application. Do not graze or feed forage from treated areas to livestock.

Trees Established for One Year—Use 2 to 4 lbs. per acre for annual weeds; apply just before or just after weeds germinate.

Trees Established for Two Years or More

For annual weeds, apply 2 to 4 lbs. per acre. For perennial weeds, use as follows:

Arizona, California—For bermudagrass, make a single application of 6 to 8 lbs. per acre during the period from winter to early summer; or make a split application of 3 to 4 lbs. per acre in the fall and again in the spring.

Florida—For bermudagrass, use 5 lbs. per acre in the spring or early summer; if needed, repeat application the same season but do not apply more than 10 lbs. per acre per year as injury to trees may result.

Texas—For johnsongrass and bermudagrass, make a single application in the spring or early summer of 4 to 6 lbs. per acre on sandy loam soils and 8 to 10 lbs. per acre on clay loam soils.

MINT

(Peppermint, Spearmint)

Preemergence Application: Make a single broadcast application of 1 to 2 lbs. per acre before mint emerges.

Midwest: Apply in the spring just after the last cultivation.

Pacific Northwest:

West of Cascades Mountains—Apply in spring or fall after last cultivation.

East of Cascade Mountains—Apply any time after the first settling rain in the fall, but not when ground is frozen. If moisture is inadequate to activate the chemical, irrigation by sprinkler must follow fall or winter treatment as soon as soil thaws in the spring (or promptly after spring treatment) or poor weed control may result.

Postemergence Application (Midwest and Pacific Northwest): Apply 1 to 1½ lbs. per acre before weeds are 2" tall (or across) and grasses are 1" tall (or across). Add ½ to 1 pt. DuPont Surfactant WK (or equivalent) to each 25 gals. of spray; non-phytotoxic superior-type spray oil may be substituted at the rate of 1 gal. per acre. If preemergence treatment has not been applied, a second postemergence application may be made but do not exceed 2 lbs. total per acre per season. Do not apply within 60 days of harvest.

Preemergence + Postemergence: Apply as directed above. Do not apply more than 2 lbs. total per acre per season; do not apply within 60 days of harvest.

Note: For either newly planted roots or established mint, soil must be well prepared before preemergence application. Crop injury and/or poor weed control may result if application is made to ground which is cloddy or compacted, resulting in exposed or improperly covered roots. Do not apply to newly planted roots that are diseased or lacking in vigor, nor on thinly covered or exposed subsoil areas as injury to the crop may result. Use of insecticides in fields where SINBAR is applied may result in injury to the mint; observe use limitations on insecticide labels. Treated areas may be planted to mint one year after last application.

PECANS

Use SINBAR alone, or apply as a tank mixture with DuPont Karmex Herbicide. Use only under trees established in the grove for at least 1 year. Make a single band or broadcast application as a directed spray, using 30 gals. (min.) of water per acre. Avoid contact of foliage and nut with spray or mist. Apply in the spring before weeds emerge or during the early seedling stage of weed growth.

	Lbs. Product Per Acre	
Soil Texture	SINBAR Alone Or	Tank Mixture SINBAR + Karmex
Sandy loam	2	1½ + 1½
Loam, silt	2½	1¾ + 1¾
Clay loam, clay	3	2 + 2

Do not use on eroded areas where subsoil or roots are exposed, nor on trees that are diseased or lacking in vigor or on trees planted in irrigation furrows as injury to the trees may result. Do not graze or feed forage from treated areas to livestock.

STRAWBERRIES

(except California and Florida): Use only on plants established in the field for at least 6 months (matted row system). Make a single broadcast application of ½ to 1¼ lbs. per acre; use the lower rate on coarse-textured soils. Apply immediately after postharvest renovation (complete removal of old leaves is essential) but before new growth begins; or apply in the dormant season from fall to late winter. Treat before weeds are 2″ tall or across.

Treatment may reduce runner production or plant stand. Varieties such as Guardian, Darrow and Honeyoye may be sensitive to SINBAR; determine varietal tolerance prior to adoption as field practice to prevent possible crop injury. Do not use on soils less than 2% organic matter and do not apply over top of newly planted strawberries as crop injury may result. Alfalfa, apples, blueberries, citrus, mint, peaches, strawberries and sugar cane may be planted 1 year after last application. Do not plant any other crop within 2 years after application as injury to the crop may occur.

SUGARCANE

Because sugarcane varieties vary in their resistance to herbicides, determine tolerance to SINBAR prior to adoption as field practice to prevent possible crop injury. Do not use on varieties which are known to be susceptible to herbicide, such as 48-103 (Louisiana), 50-28 (Texas), 53-263 (Hawaii) or P.R. 1048 (Puerto Rico). Do not use where cane is grown on thinly covered sub-soils or rocky areas as injury to cane may result. Do not replant treated areas to any crop other than sugarcane or pineapple within 2 years after last application as injury to subsequent crops may result.

Louisiana—For best control of seedling johnsongrass; apply broadcast at 2 lbs. per acre in the fall after planting and before cane emerges; repeat application at same rate in the early spring before weeds emerge. Alternatively, for control of most seedling weeds in fall-planted or stubble cane not treated in the fall with "SINBAR", apply broadcast in the spring at 4 lbs. per acre. Use ⅓ of above broadcast rates when band-treating ⅓ of the area.

As a layby treatment immediately after last cultivation, apply ½ lb. per acre in a 30″ band as a directed spray to row middles; do not apply over top of cane as injury to the crop may result.

Texas—Apply broadcast at 1 to 2 lbs. per acre in the fall to stubble cane or to plant cane before emergence; repeat application at same rate in early spring. Use ⅓ of above broadcast rates when band-treating ⅓ of the area.

Hawaii—Make a single preemergence broadcast application of 1 to 2½ lbs. per acre on plant or ratoon cane before cane emerges. Longevity of control is enhanced if application is made during the relatively dry season (March through October).

Puerto Rico—Make a single preemergence broadcast application of 1 to 2½ lbs. per acre on plant cane only.

NOTICE TO BUYER

Purchase of this material does not confer any rights under patents of countries outside of the United States.

STORAGE AND DISPOSAL

STORAGE

Store product in original container only, away from other pesticides, fertilizer, food, or feed.

DISPOSAL

Do not contaminate water, food, or feed by storage or disposal. Wastes resulting from the use of this product may be disposed of on site or at an approved waste disposal facility.

Completely empty bag into application equipment. Then dispose of empty bag in a sanitary landfill, or by incineration, or, if allowed by state and local authorities, by burning. If burned, stay out of smoke.

SPECIAL LOCAL NEED REGISTRATION [24(c)] LABEL FOR SINBAR® HERBICIDE

Crop	State	EPA SLN No.	Remarks
Alfalfa	Colorado	CO-780033	Aerial application and reduced GPA
	Delaware	DE-830008	Split application
	Maryland	MD-830011	Split application
	New Jersey	NJ-820024	Allow for dormant and after cutting treatments
	New Mexico	NM-800001	Aerial application
	Oklahoma	OK-770001	Lowers spray volume to 20 GPA
	Oklahoma	OK-800003	Aerial application
	Texas	TX-800001	Aerial application
	Virginia W. Virginia	VA-830001 WV-820003	Allow for dormant and after cutting treatments
	Wyoming	WY-780004	Aerial application and reduced GPA
Blueberries	Maine Maryland New Jersey	ME-790003 MD-820018 NJ-810009	Allow SINBAR + "Karmex" use
Grass Seed Crops	Idaho Oregon Washington	ID-800009 OR-800021 WA-800010	General weed control
Strawberries	N. Carolina	NC-800021	

* Copies of individual labels available upon request to nearest DuPont Agricultural Chemicals Office. For distribution and use only within the State specified.

TELAR™

HERBICIDE
DISPERSIBLE GRANULE

ACTIVE INGREDIENT:
Chlorsulfuron
2-Chloro-N-[(4-methoxy-6-methyl-1,3,5-triazin-2-yl)aminocarbonyl]benzenesulfonamide 75%
INERT INGREDIENTS 25%

Total 100%

EPA Reg. No. 352-404 U.S. Pat. 4,127,405

Keep Out of Reach of Children

PRECAUTIONARY STATEMENTS

HAZARDS TO HUMANS

CAUTION!

MAY IRRITATE EYES, NOSE, THROAT, AND SKIN.

Avoid breathing dust or spray mist. Avoid contact with skin, eyes, and clothing. In case of contact with eyes, immediately flush with plenty of water. Get medical attention if irritation persists. Wash thoroughly after handling. Remove and wash contaminated clothing before reuse.

ENVIRONMENTAL HAZARDS

Do not apply directly to any body of water. Do not contaminate water by cleaning of equipment or disposal of wastes.

IMPORTANT

DO NOT USE ON FOOD OR FEED CROPS. Injury to or loss of desirable trees or other plants may result from failure to observe the following: Do not apply (except as recommended), or drain or flush equipment on or near desirable trees or other plants, or on areas where their roots may extend or in locations where the chemical may be washed or moved into contact with their roots. Do not apply where runoff water may flow onto agricultural land, as injury to crops may result. Do not contaminate any body of water, including irrigation water. Do not use on lawns, walks, driveways, tennis courts or similar areas. Prevent drift of spray to cropland and other desirable plants. Keep from contact with fertilizers, insecticides, fungicides and seeds. Following a TELAR application, the spray tank should not be used for other than non-crop applications. This is extremely important, as low rates of TELAR can kill or severely injure most crops.

GENERAL INFORMATION

TELAR Herbicide is a dispersible granule to be mixed in water and applied as a uniform broadcast spray for noncropland weed control. It is noncorrosive, nonflammable, nonvolatile and does not freeze.

TELAR rapidly inhibits growth of susceptible weeds. However, typical symptoms (discoloration) of dying weeds may not be noticeable for 1 to 3 weeks after application depending on growing conditions and weed susceptibility. Warm, moist conditions following treatment promote the activity of TELAR while cold, dry conditions delay the activity of TELAR. Weeds hardened off by cold weather or drought stress may not be fully controlled.

Degree of control and duration of effect depend on a) rate used, b) weed spectrum, c) weed size, d) growing conditions at and following time of treatment, e) soil pH, f) precipitation, and g) soil organic matter.

NOTICE OF WARRANTY

Du Pont warrants that this product conforms to the chemical description on the label thereof and is reasonably fit for purposes stated on such label only when used in accordance with directions under normal use conditions. It is impossible to eliminate all risks inherently associated with the use of this product. Injury to adjacent crops, ineffectiveness or other unintended consequences may result because of such factors as weather conditions, presence of other materials, or the manner of use or application, all of which are beyond the control of Du Pont. In no case shall Du Pont be liable for consequential, special or indirect damages resulting from the use or handling of this product. All such risks shall be assumed by the buyer. DU PONT MAKES NO WARRANTIES OF MERCHANTABILITY OR FITNESS FOR A PARTICULAR PURPOSE NOR ANY OTHER EXPRESS OR IMPLIED WARRANTY EXCEPT AS STATED ABOVE.

DIRECTIONS FOR USE

It is a violation of federal law to use this product in a manner inconsistent with its labeling.

TELAR Herbicide should be used only in accordance with recommendations on this label or in separate published Du Pont recommendations available through local dealers.

Du Pont will not be responsible for losses or damages resulting from the use of this product in any manner not specifically recommended by Du Pont. User assumes all risks associated with such nonrecommended use.

Du Pont TELAR Herbicide is recommended for preemergence and postemergence control of many annual and perennial broadleaf weeds. TELAR may be applied as a selective or non-selective treatment to non-cropland areas such as airports, fencerows, highways, industrial turf, lumberyards, petroleum tank farms, pipelines right-of-ways, plant sites, railroads, roadside turf, storage areas, and utility right-of-ways.

TELAR may be applied preemergence or postemergence to the weeds, but for best results, apply postemergence to young, actively growing weeds any time of the year, except when ground is frozen. Adequate moisture by rainfall is needed to activate the herbicide. Do not apply to soils saturated with moisture or during periods of intense rainfall.

SELECTIVE WEED CONTROL

Bahiagrass, bermudagrass, bluegrass, fescue and smooth brome have demonstrated good tolerance to TELAR. For selective weeding, TELAR may be applied at the rate of $\frac{1}{4}$ to 1 oz per acre per year to bahiagrass, bermudagrass and bluegrass and at the rate of $\frac{1}{4}$ to $\frac{1}{2}$ oz per acre per year to fescue and smooth brome. The higher rates and/or the addition of surfactant may result in temporary chlorosis of desirable grasses. Application of TELAR may also suppress grass growth and inhibit grass seedhead formation.

NON-SELECTIVE WEED CONTROL

Apply TELAR at the rates of $1/4$–3 oz per acre per year to control many annual and perennial broadleaf weeds in non-crop areas. The addition of a surfactant in postemergence applications enhance activity and is recommended.

Since the degree of control and duration of effect will vary with the amount of chemical applied, soil texture, soil pH, soil organic matter, weed size, rainfall and other conditions, it is suggested that users limit their first use to small areas.

WEEDS CONTROLLED AND USE RATES

$\frac{1}{4}$ oz to 1 oz per Acre

annual sowthistle**	goldenrod*
bouncingbet	groundsel
Canada thistle*	henbit**
common chickweed**	London rocket**
common lambsquarter**	sweetclover
common speedwell**	tumble mustard
common sunflower**	turkey mullein*
dandelion*	wild mustard**
fiddleneck**	wild parsnip
filaree**	

*partial control only
**controlled at $\frac{1}{4}$ to $\frac{1}{2}$ oz. per acre

1 oz to 3 oz per Acre

aster	musk thistle
bedstraw	plaintain*
black mustard	prickly lettuce
buckhorn plaintain	puncturevine*
burclover	ragweed*
Canada thistle	red clover
common cinquefoil	Russian thistle
common mallow	scouringrush
common yarrow	tansymustard
curly dock	white clover
foxtail (setaria spp)*	wild carrot
horsetail	yellow starthistle*
kochia	

* Partial control only

HOW TO USE

Spray Preparation: Fill the spray tank half full of water and with agitator running, add the proper amount of TELAR. Finish adding the required amount of water. Continuous agitation is required to keep the product in suspension. For postemergence applications, a non-ionic surfactant such as Du Pont Surfactant WK or "Ortho" X-77[1]—may be added at the rate of 1 qt. per 100 gallons of spray to improve wetting and/or contact activity. To minimize drift, a drift control agent may be added at the recommended label rate. Use TELAR spray preparation within 24 hours or product degradation may occur. Thoroughly reagitate before using.

[1]Registered trademark of Chevron Chemical Company

Application Technique: Before spraying, calibrate the equipment to determine the quantity of water necessary to uniformly cover the vegetation and soil in a measured area to be treated. Use a fixed boom, off center nozzle or boomless straight stream nozzles properly calibrated to a constant speed of travel and rate of delivery. Avoid overlapping and shut off spray booms while starting, turning, slowing or stopping when spraying industrial turf as injury to the desired species may result. Do not allow spray to drift onto adjacent crops or other valuable plants or trees, as injury may occur. Extreme care must be taken to prevent drift.

Spray Volume: Apply TELAR in ground equipment. Select a spray volume that will assure thorough coverage and uniform spray pattern. Do not use less than 10 gallons of water per acre. Spray volumes of 20 to 40 gallons per acre are preferred. Spray pressure of 25 to 50 psi is adequate.

Agitation: Use mechanical or by-pass agitation to thoroughly mix the spray solution. After initial mixing, do not use excessive agitation when using a surfactant as foaming problems may result. To avoid foaming problems, use antifoam agents.

Selection Of Use Rates: Rate selection is based on weed species, weed size and soil texture. Use the higher rates on established plants and on fine textured soils and lower rates on smaller weeds and coarse textured soils.

Note To Buyer: Purchase of this material does not confer any rights under patents of countries outside of the United States.

STORAGE AND DISPOSAL

STORAGE
Store product in original container only, away from other pesticides, fertilizer, food, or feed.

DISPOSAL
Do not contaminate water, food, or feed by storage or disposal. Wastes resulting from the use of this product may be disposed of on site or at an approved waste disposal facility. Triple rinse (or equivalent) the container. Then offer for recycling or reconditioning, or puncture and dispose of in a sanitary landfill, or incineration, or, if allowed by state and local authorities, by burning. If burned, stay out of smoke.

VELPAR®

HERBICIDE
SOLUBLE POWDER

ACTIVE INGREDIENT:
Hexazinone
3-cyclohexyl-6-(dimethylamino)-1-methyl-1,3,5-triazine-2,4 (1H,3H)-dione 90%
INERT INGREDIENTS 10%
TOTAL 100%

EPA Reg. No. 352-378 U.S. Pat. 3,902,887 & 3,983,116

Keep Out of Reach of Children

PRECAUTIONARY STATEMENTS

HAZARDS TO HUMANS
WARNING
MAY IRRITATE EYES, NOSE, THROAT AND SKIN.

Avoid breathing dust or spray mist. Avoid contact with skin, eyes and clothing.

In case of contact, flush skin or eyes with plenty of water; for eyes, get medical attention.

Do not apply this product in such a manner as to directly or through drift expose workers or other persons. The area being treated must be vacated by unprotected persons.

ENVIRONMENTAL HAZARDS
Keep out of any body of water. Do not apply where runoff is likely to occur. Do not contaminate water by cleaning of equipment or disposal of wastes.

IMPORTANT

Wear goggles when use or handling conditions favor eye exposure or if eye discomfort is encountered.

Injury to or loss of desirable trees or other plants may result from failure to observe the following: Do not apply or drain or flush equipment on or near desirable trees or other plants, or on areas where their roots may extend, or in locations where the chemical may be washed or moved into contact with their roots. Do not use on lawns, walks, driveways, tennis courts or similar areas. Prevent drift of dry powder or spray to desirable plants. Keep from contact with fertilizers, insecticides, fungicides and seeds.

Thoroughly clean all traces of VELPAR from application equipment immediately after use. Flush tank, pump, hoses and boom with several changes of water after removing nozzle tips and screens (clean these parts separately).

GENERAL INFORMATION

Du Pont VELPAR Herbicide is a water-soluble powder to be mixed in water and applied as a spray on noncropland areas. It is nonvolatile, nonflammable and noncorrosive to equipment.

VELPAR is an effective general herbicide, providing both contact and residual control of many annual and biennial weeds, woody plants and, except for johnsongrass, is effective for control of most perennial weeds. It is absorbed through the roots and foliage.

Moisture is required to activate VELPAR in the soil. Best results are obtained when weeds are young and actively growing, soil is moist at time of application and 1/4 to 1/2 inch of rainfall occurs within 2 weeks after application. Foliar application to weeds is most effective under conditions of high temperature (above 80°F), high humidity and good soil moisture. Application when vegetation is dormant or semi-dormant may not be effective.

Symptoms usually occur within 2 weeks after application under warm humid conditions, while 4 to 6 weeks may be required when weather is cool. The degree of control and duration of effect will vary with the amount of chemical applied, rainfall, temperature, weed species, soil texture and other conditions.

NOTICE OF WARRANTY

Du Pont warrants that this product conforms to the chemical description on the label thereof and is reasonably fit for purposes stated on such label only when used in accordance with the directions under normal use conditions. It is impossible to eliminate all risks inherently associated with the use of this product. Ineffectiveness or other unintended consequences may result because of such factors as weather conditions, presence of other materials or the manner of use or application, all of which are beyond the control of Du Pont. In no case shall Du Pont be liable for consequential, special or indirect damages resulting from the use or handling of this product. All such risks shall be assumed by the Buyer. DU PONT MAKES NO WARRANTIES OF MERCHANTABILITY OR FITNESS FOR A PARTICULAR PURPOSE NOR ANY OTH-

ER EXPRESS OR IMPLIED WARRANTY EXCEPT AS STATED ABOVE.

DIRECTIONS FOR USE

It is a violation of federal law to use this product in a manner inconsistent with its labeling.

Du Pont VELPAR Herbicide should be used only in accordance with recommendations on this label, or in separate published Du Pont recommendations available through local dealers.

Du Pont will not be responsible for losses or damages resulting from use of this product in any manner not specifically recommended by Du Pont. User assumes all risk associated with such nonrecommended use.

Do not enter treated areas without protective clothing until sprays have dried. Because certain states may require more restrictive reentry intervals for various crops treated with this product, consult your State Department of Agriculture for further information.

Written or oral warnings must be given to workers who are expected to be in a treated area or in an area about to be treated with this product. When oral warnings are given, warnings shall be given in a language customarily understood by workers. Oral warnings must be given if there is reason to believe that written warnings cannot be understood by workers. Warnings must include the following information: "WARNING. Area treated with VELPAR on (date of application). Do not enter without appropriate protective clothing until sprays have dried. In case of contact, flush skin or eyes with plenty of soap and water; for eyes, get medical attention."

HOW TO APPLY

Apply VELPAR as a spray just before or soon after weed emergence. Before spraying, calibrate equipment to determine the quantity of water necessary to uniformly and thoroughly cover the vegetation and soil in a measured area to be treated. Make sure the volume of water is sufficient to completely dissolve the VELPAR, using the Water Temperature/Solubility Table as a guide. Use of concentrations greater than shown in the Water Temperature/Solubility Table may result in nozzle plugging and uneven distribution.

Water Temperature/Solubility Table

Water Temperature Degrees F.	Solubility Max. Lbs. VELPAR per 100 Gals.
75	23
70	20
65	18
60	16
55	14
50	13
40	12

Add the proper amount of VELPAR into a spray tank filled with the amount of water to be used and dissolve using sufficient agitation to prevent the powder from settling. Continue agitation for approximately 10 minutes or until the powder is dissolved. Use a fixed-boom sprayer properly calibrated to a constant speed and rate of delivery. For application with a handgun sprayer, use sufficient water (50 to 200 gals. per acre) for thorough coverage. For treating small areas, a tank-type hand sprayer or sprinkling can may be used. A surfactant such as Du Pont Surfactant WK may be added at the rate of 1 quart per 100 gals. of spray to improve wetting and/or contact activity.

GENERAL WEED CONTROL—Non-Cropland Areas

Railroad, Highway, Utility and Pipeline Right-of-Ways, Petroleum Tank Farms, Storage Areas, Industrial Plant Sites, Drainage Ditchbanks and other similar areas.

Use the higher level of the recommended dosage ranges under the following conditions: on hard-to-kill species; on fine-textured soils (clay, clay loam); on soils containing more than 5% organic matter or carbon; and, on coarse-textured soils (sand, loamy sand, gravelly soils) where rainfall exceeds 40 inches per year. Use the lower levels of the dosage range if the growing season is short.

Apply 6 to 12 lbs. per acre for season-long control of many annual, biennial and perennial weeds including bahiagrass*, bermudagrass*, bluegrass, broomsedge, camphorweed, Canada thistle*, chickweed, clovers, dewberry, dogfennel, fescue*, fingergrass, foxtail, guineagrass, heath aster, honeysuckle, lantana, marestail*, natalgrass, plaintain, prickly lettuce, ragweed, smutgrass, Spanish needle, vaseygrass, wild blackberry* and wild carrot.

For short-term control (up to 3 months) of the above weeds, apply 2 to 5 lbs. per acre. Application at 2 to 5 lbs. per acre also provides short-term control of many annual, biennial and perennial weeds including barnyardgrass, bindweed*, bouncing bet*, bromegrass, buffelgrass*, burdock, cocklebur, crabgrass, crown vetch, curly dock*, dandelion*, dogbane*, fiddleneck, filaree, fleabane, goatsbeard vine, goldenrod, lespedeza, milkweed*, mustard, nutsedge*, orchardgrass*, oxalis, paragrass, pigweed, purslane, quackgrass, ryegrass, smartweed, spurge, trumpet creeper*, wild oats*, wild parsnip and wild star thistle.

* Considered hard-to-kill.

For woody plant control, apply 4 to 8 lbs. per acre as a broadcast spray for control of maple, oak, poplar, sumac, willow, hawthorn and hickory. Use at least 40 gals. of spray per acre. Apply during the period between late winter and early summer.

NOTICE TO BUYER: Purchase of this material does not confer any rights under patents of countries outside of the United States.

STORAGE AND DISPOSAL

STORAGE

Store product in original container only, away from other pesticides, fertilizer, food or feed.

DISPOSAL

Do not contaminate water, food or feed by storage or disposal. Wastes resulting from the use of this product may be disposed of on site or at an approved waste disposal facility.

Completely empty bag into application equipment. Then dispose of empty bag in a sanitary landfill or by incineration, or, if allowed by state and local authorities, by burning. If burned, stay out of smoke.

SUPPLEMENTAL LABELING

VELPAR® HERBICIDE

EPA Reg. No. 352-378

FOR SELECTIVE CONTROL OF WEEDS IN ESTABLISHED STANDS OF BERMUDAGRASS AND BAHIAGRASS IN NONCROPLAND AREAS IN SOUTHEASTERN AND SOUTHWESTERN STATES

Du Pont VELPAR Herbicide is recommended for selective weed control in established stands of bermudagrass and bahiagrass in noncropland areas in Alabama, Arkansas, Florida, Georgia, Louisiana, Mississippi, Oklahoma, North Carolina, South Carolina, Tennessee and Texas. Treatment controls weeds such as barnyardgrass, dogfennel, fescue, lespedeza, maypop passionflower, oxalis, pepperweed, pigweed, smutgrass, spiny amaranth and wild barley.

Since the effect of VELPAR on bermudagrass and bahiagrass varies with soils, varieties, uniformity of application and environmental conditions, it is suggested that initial applications be limited to small areas.

HOW TO USE

Use only on noncropland areas such as railroad, highway, utility and pipeline right-of-ways, petroleum tank farms, storage areas, industrial plant sites, drainage ditchbanks and similar areas.

Use only in stands of bermudagrass and bahiagrass established for at least one year; do not treat newly sprigged-or-sodded areas as severe injury to desirable grasses may result. Temporary injury (chlorosis and leaf burn) to bermudagrass and bahiagrass can be expected after application. Plants will recover from this injury within several weeks, if conditions are favorable for growth, while weeds will be killed.

Make a single application per year of 3/4 to 1 1/4 lbs. per acre when soil is warm and moist and weeds are actively growing. Use the lower rate on coarse-textured soils (sand to sandy loam) and the higher rate on fine-textured soils (clay loam, clay) and on soils high in organic matter (over 5%). Apply with a fixed-boom power sprayer properly calibrated to a constant speed and rate of delivery, using at least 25 gals. of spray per acre. Avoid overlapping, and shut off spray booms while starting, turning, slowing or stopping, or injury to desirable grasses may result.

Moisture is required to activate VELPAR in the soil. Best results are obtained when weeds are less than 2" in height or diameter and actively growing, soil is moist at time of application and 1/2 to 1" of rainfall occurs within 2 weeks after application. Foliar effects on weeds are most pronounced under conditions of high temperature (above 80°F) and high humidity. Symptoms usually occur within 2 weeks after application under these conditions, while 4 to 6 weeks may be required when weather is cool. If rainfall after treatment is insufficient to promote root uptake, weeds may recover from foliar effects and continue to grow. Extremely heavy rainfall after treatment may result in poor weed control and/or injury to desirable grasses, especially if application is made to dry soil.

Note: As injury to bermudagrass and bahiagrass may result, do not use on gravelly or rocky soils, thinly covered subsoils, nor on any soil with less than 1% organic matter; do not treat when desirable grasses are under drought stress or other stresses such as insect or disease injury.

As injury to desirable trees or other plants may result, do not apply or drain or flush equipment on or near desirable trees or other plants, or on areas where their roots may extend, or in locations where the chemical may be washed or moved into contact with their roots.

IMPORTANT

BEFORE USING VELPAR, READ AND CAREFULLY OBSERVE THE CAUTIONARY STATEMENTS AND ALL OTHER INFORMATION APPEARING ON THE PRODUCT LABEL.

This bulletin contains new or supplemental instructions for use of this product which may not appear on the package label. Follow the instructions carefully.

SUPPLEMENTAL LABELING

VELPAR® HERBICIDE

FOR CONTROL OF CERTAIN WEEDS IN ESTABLISHED ALFALFA

Du Pont VELPAR® Herbicide is recommended for control of certain weeds in established alfalfa.

Sufficient moisture is necessary to activate the herbicide. Best results are obtained when 1/2 to 1" of rainfall or sprinkler irrigation occurs within two weeks after application, soil is moist at time of application, and weeds have not germinated or are less than 2" tall or across.

Since the effect of VELPAR on alfalfa varies with soils, uniformity of application and environmental conditions, it is suggested that growers limit their first use to small areas.

DIRECTIONS FOR USE

Treat only stands of alfalfa established for one year or for one growing season, provided the alfalfa root system is well established and the crop is healthy, vigorous and not under stress from weather conditions, insect or disease damage. Do not use on seedling alfalfa nor on alfalfa-grass mixture or on other mixed stands.

Make a single application of VELPAR in the fall or winter after alfalfa becomes dormant or in the spring before new growth begins. **Exception:** In certain northeastern* and Midwestern states, make a single application at 1/2 to 1 1/2 lbs. per acre in the fall or winter after plants become dormant or in the spring before new growth exceeds 2" in height or to stubble after cutting following hay removal and before regrowth exceeds 2" in height.

Apply VELPAR using a fixed boom power sprayer or aerial equipment. Before spraying, calibrate equipment to determine the quantity of water necessary to uniformly and thoroughly cover the vegetation in a measured area to be treated. The sprayer must be properly calibrated to a constant speed and rate of delivery. Avoid overlapping and shut off spray booms while starting, turning, slowing or stopping to prevent crop injury.

Add the VELPAR to a water-filled tank. Agitate for approximately 10 minutes, vigorously enough to prevent settling in order to dissolve the product. Use at least 10 gallons of water for each pound of VELPAR in aerial equipment. Apply in at least 20 gallons of spray per acre in ground equipment.

USE RATES

Rate selection is to be based on soil texture and organic matter content as follows:

Soil Texture/Organic Matter Content	Pounds/Acre
Gravelly or rockly soils, exposed subsoils, hardpan, sand, poorly drained or alkali soils.	Do not use.
Sandy loam, loamy sand, loam, silt, silt loam, and sandy clay loam with less than 1% organic matter.	1/2 to 3/4
Silty clay loam, clay loam, sandy clay, silty clay and clay with 1% organic matter or higher.	3/4 to 1 1/2
Soils having more than 5% organic matter.[1]	1 to 1 1/2

[1] On soils high in organic matter, the effectiveness of VELPAR can be significantly reduced and weed control may be unsatisfactory.

WEED SPECIES CONTROLLED

VELPAR when applied preemergence or early postemergence will control the following winter annual and spring germinating weeds:

1/2 to 1 lb./acre

annual bluegrass	foxtail
blue mustard	jimsonweed
common chickweed	orchardgrass
common groundsel	purple mustard
common lambsquarter	salsify
crabgrass	shepherdspurse
fiddleneck	spurry
field pennycress	tansy mustard
filaree	yellow rocket
fleabane	

1 to 1 1/2 lbs./acre

downy brome or cheatgrass (Bromus tectorum)	annual ryegrass
ivy leaf speedwell	sweet clover
prickly lettuce	purslane speedwell

1 to 1 1/2 lbs./acre will give partial control[2] of:

common dandelion	seedling alfalfa
false dandelion	white cockle
Mexican tea	wild oats
perennial bluegrass	

[2] Partial control is an observable reduction in weed competition as compared to an untreated area. Degree of partial control can vary with soil type, environmental factors, and size of weeds at the time of application.

REPLANTING

Except in low annual rainfall areas (20" or less), corn may be planted 12 months after the last treatment provided the soil is deep disked or plowed prior to planting. Do not plant treated areas to any other crop within two years after treatment or injury to crop may result.

NOTE: Because crop injury may result, do not apply to actively growing alfalfa nor to stubble between cuttings, **except** as recommended for certain northeastern* and midwestern* states. Extremely heavy rainfall or more than 1" of irrigation soon after application may result in crop injury or nonperformance. Do not use in low desert valleys in California. Do not apply to snow covered frozen ground.

Do not graze or feed treated forage or hay to livestock within 30 days following application.

***Northeastern and midwestern states are:**

Connecticut	Missouri
Delaware	New Hampshire
Illinois	New Jersey
Indiana	New York
Iowa	Ohio
Kentucky	Pennsylvania
Main	Rhode Island
Maryland	Vermont
Massachusetts	Virginia
Michigan	West Virginia
Minnesota	Wisconsin

IMPORTANT
BEFORE USING VELPAR, READ AND CAREFULLY OBSERVE THE CAUTIONARY STATEMENTS AND ALL OTHER INFORMATION APPEARING ON THE PRODUCT LABEL.

This bulletin contains new or supplemental instructions for use of this product which may not appear on package label. Follow the instructions carefully.

SUPPLEMENTAL LABELING

VELPAR® HERBICIDE
EPA Reg. No. 352-378

FOR CONTROL OF CERTAIN WEEDS IN SUGARCANE

Du Pont VELPAR Herbicide is recommended for selective weed control in sugarcane. Since the effect of VELPAR on sugarcane varies with soils, uniformity of application and environmental conditions, it is suggested that growers limit their first use to small areas. Tolerance of new cane varieties to VELPAR should be determined prior to adoption as field practice.

Moisture is necessary to activate the herbicide. Best results are obtained when weeds are less than 2 inches in height or diameter and actively growing, soil is moist at time of application and ½ to 1 inch of rainfall occurs within 2 weeks after application. Foliar application to weeds is most effective under conditions of high temperature (above 80°F), high humidity and good soil moisture. Symptoms usually occur within 2 weeks after application under these conditions, while 4 to 6 weeks may be required when weather is cool. If rainfall after treatment is inadequate to promote root uptake, weeds may recover from foliar effects and continue to grow. Extremely heavy rainfall after application may result in poor weed control and/or crop injury, especially if the application is made to dry soil.

Note: Because injury to sugarcane may result: Do not use on cane which shows poor vigor because of insect, disease or winter injury, or shows symptoms of other stress conditions such as drought stress; do not add surfactant if applied over the top of cane; do not use on gravelly or rocky soils, thinly covered subsoils, nor on coarse-textured soils (sands to sandy loams) with less than 2% organic matter, nor on any soil with less than 1% organic matter. Temporary chlorosis of the crop may result from application over emerged cane.

Do not apply over the top of actively growing cane. Applications during active cane growth should be directed to cover the weeds and soil while minimizing crop contact. Do not use on varieties which are known to be susceptible to weed killers. Do not plant any crop other than sugarcane within 18 months of the last application of VELPAR. Do not feed sugarcane forage to livestock.

Preemergence Use (Germination Weeds): VELPAR at the recommended rates, controls alexandergrass, amaranth (slender, spiny), American burnweed (fireweed), balsam apple, barnyardgrass, Carolina geranium, castorbean, chickweed, crabgrass (hairy, large, smooth), crotalaria (showy, fuzzy), cuphea (tarweed), fingergrass (radiate, swollen), Flora's paintbrush, foxtail (bristly, yellow), goosegrass, guineagrass, henbit, jobs tears, johnsongrass (from seed), junglerice, lambsquarters, morningglory (hairy, threelobe), mustard (wild), oxalis, panicum (broadleaf, browntop, Texas, millet), paspalum (dallisgrass, ricegrass, sour, vaseygrass), pigweed (common, smooth), popolo, puslane, sandbur, sensitiveplant (Hila Hila), signalgrass (broadleaf), sowthistle, spanishneedles, sprangletop, spurge (prostrate, graceful), sunflower, waltheria (Hialoa), and provides partial (40–80%) control of ageratum, bermudagrass and itchgrass.

Post Emergence Use (Emerged Seedling Weeds): VELPAR at recommended rates controls amaranth (slender, spiny), American burnweed (fireweed), Carolina geranium, crabgrass (hairy, large, smooth), crotalaria, Flora's paintbrush, foxtail (bristly, yellow), goosegrass, guineagrass, junglerice, morningglory, mustard (wild), oxalis, panicum (blue, broadleaf, browntop, Texas), pigweed (common, smooth), popolo, purslane, sandbur, signalgrass (broadleaf), sowthistle, spurge (prostrate, graceful), sunflower, walteria (Hialoa), and provides partial (40–80%) control of ageratum, bermudagrass and itchgrass.

HOW TO USE

Apply a single treatment per year using a fixed-boom power sprayer, unless otherwise directed. Before spraying, calibrate equipment to determine the quantity of water necessary to uniformly and thoroughly cover the soil and vegetation in a measured area to be treated. The sprayer must be calibrated to a constant speed and rate of delivery. Avoid overlapping and shut off spray booms while starting, turning, slowing and stopping, or injury to the crop may result.

Apply in at least 25 gallons of spray per acre using ground equipment. Add the VELPAR powder to a water filled tank using agitation, vigorously enough to prevent settling, for approximately 10 minutes to dissolve the product. Use of concentrations greater than shown in the WATER TEMPERATURE/SOLUBILITY TABLE* may result in nozzle plugging and uneven application.

***WATER TEMPERATURE/SOLUBILITY TABLE**

Water Temperature Degrees F	Maximum Lbs. VELPAR Per 100 Gals.
75	13
70	20
65	18
60	16
55	14
50	13
40	12

Florida

Apply ½ to 1 pound VELPAR per acre on sandy soil as a pre- or directed postemergence treatment using the lower rate on soils low in organic matter and the higher rate on soils high in organic matter. On muck soils apply 1 to 2 pounds of VELPAR per acre preemergence or ½ to 1 pound per acre postemergence, using the higher rate on larger weeds. Postemergence applications should be directed to cover the weeds and soil, while minimizing crop contact.

Hawaii

Apply VELPAR pre- or postemergence at the following rates for the indicated soil texture:

Soil Texture	Pounds VELPAR Per Acre
Coarse-Textured Soils (Sandy, Loamy Sand, Sandy Loam)	½ to 1
Medium-Textured Soils (Loam, Silt Loam, Silty Clay Loam)	½ to 2
Fine-Textured Soils (Clay, Gray Hydromorphic Clay)	2 to 4

Use the higher level of the recommended dosage ranges on soils high in organic matter. A preemergence treatment may be followed by a postemergence treatment; however, do not apply more than twice the highest recommended rate for the indicated soil texture per crop cycle.

For spot treatments of emerged weeds, VELPAR may be applied with a knapsack sprayer in concentration of ½ to 2 pounds per 100 gallons of water. Apply sufficient volume to thoroughly wet weed foliage, but do not exceed 100 gallons of spray per treated acre. Use the lower concentrations on coarse-textured soils which are low in organic matter and the higher concentrations on fine-textured soils high in organic matter.

Louisiana

Apply ½ to 1 pound of VELPAR per acre broadcast in the fall before sugarcane emerges or in the spring before active cane tillering begins. Use ⅓ of the above broadcast rates when band-treating ⅓ of the area. Fall treatments of ½ to ¾ pounds per acre may be followed by a spring treatment of ½ to ¾ pounds per acre. Do not apply more than 1½ pounds per year. Use the higher levels of the recommended dosage range on fine-textured soils.

Puerto Rico

Apply 1 to 2 pounds VELPAR per acre preemergence to the cane or ½ to 1 pound per acre postemergence. Use the lower rate on coarse-textured soils and the higher rates on fine-textured soils (high in clay or organic matter).

For spot treatment of emerged weeds, VELPAR may be applied with a knapsack sprayer in concentrations of ½ to 1 pound per 100 gallons of water. Use the lower concentration on coarse-textured soils and the higher concentration on fine-textured soils. Apply a sufficient volume of spray to thoroughly wet weed foliage. Do not exceed 100 gallons of spray per treated acre.

Texas

Apply ½ to 1 pound of VELPAR per acre. On plant cane, apply before cane emerges or as a directed layby treatment. On stubble cane, apply preemergence or early postemergence (up to the 3 leaf stage) or as a directed layby treatment. A pre- or early postemergence treatment may be followed by a layby treatment, provided at least 60 days have elapsed and 3 inches of rainfall or sprinkler irrigation have occurred since the first treatment. Use the following rates for the soil texture:

Pounds VELPAR Per Acre

	Premergence		Layby
Coarse-textured soils* (Sandy Loam)	½	+	½
Medium-textured soils (Loam, Silt Loam)	¾	+	¾
Fine-textured soils (Clay Loam)	1	+	1

* With at least 2% organic matter.

SUPPLEMENTAL LABELING

VELPAR® HERBICIDE

EPA Reg. No. 352-378

FOR PRE- AND POSTEMERGENCE USE IN PINEAPPLES

DIRECTIONS FOR USE

It is a violation of federal law to use this product in a manner inconsistent with its labeling.

Du Pont VELPAR Herbicide is recommended for the control or suppression of ageratum, balsam apple, castor bean, crabgrass, crotalaria, dallisgrass, guinea grass, jungle rice, koa haole, monana loa, morning glory, oxalis, popolo, richardsonium and vaseygrass.

HOW TO USE

Apply VELPAR as directed below:

- Intercrop Period

Apply as a broadcast spray in 100 to 400 gallons water per acre at the rate of ¼ to 2 pounds per acre. For aircraft application use at least 10 gallons water per acre.

- Post mulch preplant

Apply as a broadcast spray in 100 to 400 gallons water per acre at the rate of ¼ to 2 pounds per acre.

- Post plant but before planting material starts active growth. Apply as a broadcast spray in 100 to 400 gallons water per acre at the rate of ¼ to 2 pounds per acre.

Post plant after planting material starts to grow should be made when weed growth has escaped control by other herbicide applications.

- Post plant crop harvest but prior to forcing first ratoon. Apply as a broadcast spray in 100 to 400 gallons water per acre at the rate of ¼ to 2 pounds per acre.
- Post first ratoon harvest but prior to forcing second ratoon. Apply as a broadcast spray in 100 to 400 gallons water per acre at the rate of ¼ to 2 pounds per acre.
- Directed postemergence (pineapple and weeds) interspace application.

Apply as directed spray 3 to 10 months after planting in 50 to 200 gallons water per acre (broadcast basis), at the rate of ¼ to 2 pounds per acre (broadcast basis) using a stroller boom or knapsack.

- Directed spot treatment for perennial grasses before floral induction. Spray perennial grasses postemergence to wet (50 to 200 GPA depending on size) with 1 to 2 pounds per 100 gallons water concentration as a spot treatment.

Do not exceed 6 lbs. VELPAR per acre per crop cycle.

Apply VELPAR at 2 to 4 lbs. per acre in 100 to 400 gallons of water for:

- field edges
- roadsides

Use a sprayer properly calibrated to a constant speed and rate of delivery. Mix the proper amount of VELPAR in water.

NOTE: (1) use the lower rates on coarse soils or in areas where rainfall exceeds 65 inches per year.

(2) Use the higher rates on fine soils or in areas where rain fall is less than 65 inches per year.

IMPORTANT

BEFORE USING VELPAR, READ AND CAREFULLY OBSERVE THE CAUTIONARY STATEMENTS AND ALL OTHER INFORMATION APPEARING ON THE PRODUCT LABEL.

This bulletin contains new or supplemental instructions for use of this product which may not appear on package label. Follow the instructions carefully.

This labeling must be in the possession of the user at the time of pesticide application.

SUPPLEMENTAL LABELING

VELPAR® HERBICIDE

EPA Reg. No. 352-378

FOR CONTROL OF CERTAIN WEEDS IN CHRISTMAS TREE PLANTATIONS AND REFORESTATION AREAS EAST OF THE ROCKY MOUNTAINS

Du Pont VELPAR Herbicide is recommended for control of certain weeds in Christmas tree plantations where Austrian pine, Scotch pine, Grand fir and Sitka spruce are grown. It is also recommended for use in reforestation areas where Austrian pine, Scotch pine, Loblolly pine, Ponderosa pine, Grand fir and Sitka spruce are grown.

Since the effect of VELPAR on conifers varies with soils, uniformity of application and environmental conditions, it is suggested that growers limit their first use to small areas.

Moisture is necessary to activate the herbicide. Best results are obtained when weeds are less than 2 inches in height or diameter and actively growing, soil is moist at time of application and ¼ to ½ inch of rainfall occurs within 2 weeks after application. If treatment is made during periods of active conifer growth, tender foliage may be injured.

NOTE: Because injury to conifers may result: Do not use in nurseries, seedbeds or ornamental plantings; do not use on trees which show poor vigor because of insect, disease or winter injury, or show symptoms of other stress conditions; do not add surfactant if applied over the top of conifers; do not use on poorly drained soils. In addition, crop injury may result where severe winter stress, disease or insect damage follows application, or heavy rains occur soon after application. Do not use for weed control in Christmas tree plantations in the state of Texas.

When used as directed, VELPAR controls annual bluegrass, barnyardgrass, bentgrass, bromegrass, catsear, common groundsel, common ragweed, false-dandelion, fleabane, foxtail, oxeye daisy, Pennsylvania smartweed, velvetgrass and wild carrot. Treatment provides partial control of curly dock, dandelion, willowweed (fireweed), fescue, goldenrod, horseweed, orchardgrass and ryegrass. Rates expressed are broadcast. For band application, use proportionately less; for example, use ½ of the broadcast rates when treating a 3 ft. band where row spacing is 6 ft.

HOW TO USE

Apply as a broadcast spray in the spring prior to conifer bud break. If application is made after bud break, use directional spray equipment and prevent contact of conifer foliage or injury may result.

First Year Plantations: Apply only to transplant stock that is 2 years old (2-0) or more (1 year old for loblolly pine). Allow at least two months after transplanting before applying VELPAR, then apply only if rainfall has settled the soil around the base of the transplants. Apply at the following rates for the soil texture:

Soil Texture*	Lbs. VELPAR Per Acre**
Gravelly or Rocky Soils, Exposed Subsoils, Clay Knobs, Sand, or Sandy Soils with 85% or more Sand	Do Not Use
Loamy Sand and Sandy Loam (50–85% Sand)	1
Loam, Silt Loam, Silt and Sandy Clay Loam	1 to 1¼
Silty Clay Loam, Clay Loam, Sandy Clay, Silty Clay and Clay	1¼ to 1½

Established Plantations:

Soil Texture*	Lbs. VELPAR Per Acre**
Gravelly or Rocky Soils, Exposed Subsoils, Clay Knobs, Sand, or Sandy Soils with 85% or more Sand	Do Not Use
Loamy Sand and Sandy Loam (50–85% Sand)	1 to 1¼
Loam, Silt Loam, Silt and Sandy Clay Loam	1¼ to 1¾
Silty Clay Loam, Clay Loam, Sandy Clay, Silty Clay, and Clay	1¾ to 2

* Do not use on any soil containing less than 1% organic matter; do not use on loamy sand or sandy loam with less than 2% organic matter.

** Use of rates in excess of those specified for the soil texture may result in injury to the conifers.

Before spraying, calibrate equipment to determine the quantity of water necessary to uniformly and thoroughly cover the vegetation in a measured area to be treated. Use a fixed-boom, power sprayer properly calibrated to a constant speed and rate of delivery. Avoid overlapping and shut off spray booms while starting, turning, slowing or stopping, or injury to the conifer trees may result.

Apply in at least 25 gallons of spray per acre using ground equipment. Add the VELPAR powder to a water-filled tank using agitation, vigorous enough to prevent settling, for approximately 10 minutes to dissolve the product. Use of concentrations greater than shown in the WATER/TEMPERATURE SOLUBILITY TABLE[1] may result in nozzle plugging and uneven application.

[1]WATER TEMPERATURE/SOLUBILITY TABLE

Water Temperature Degrees F	Maximum Lbs. VELPAR per 100 Gals.
75	23
70	20
65	18
60	16
55	14
50	13
40	12

IMPORTANT

BEFORE USING VELPAR, READ AND CAREFULLY OBSERVE THE CAUTIONARY STATEMENTS AND ALL OTHER INFORMATION APPEARING ON THE PRODUCT LABEL.

This bulletin contains new or supplemental instructions for use of this product which may not appear on package label. Follow the instructions carefully.

SUPPLEMENTAL LABELING

VELPAR® HERBICIDE

EPA Reg. No. 352-378B

FOR CONTROL OF CERTAIN WEEDS AND BRUSH SPECIES FOR FORESTRY SITE PREPARATION & CONIFER RELEASE WEST OF THE ROCKY MOUNTAINS

DIRECTIONS FOR USE

It is a violation of federal law to use this product in a manner inconsistent with its labeling.

Du Pont VELPAR® Herbicide is recommended for control of certain weeds and brush species in forestry site preparation and conifer release operations where Douglas Fir, Grand Fir, Noble Fir, White Fir, Ponderosa Pine, Jeffrey Pine and Sitka Spruce are grown.

Since the effect of VELPAR on conifers varies with soils, uniformity of application and environmental conditions, it is suggested that foresters limit their first use to small areas.

NOTE: Because injury to conifers may result: Do not use VELPAR in nurseries, seedbeds or ornamental plantings; do not use on trees which show poor vigor because of insect, disease or winter injury, or show symptoms of other stress conditions. Do not add surfactant if applied over the top of conifers. Do not use on poorly drained soils. Conifer injury may result where severe winter stress, disease or insect damage follows application, or if heavy rains occur soon after application.

HOW TO USE

Application Technique: Before spraying, calibrate equipment to determine the uniform quantity of water necessary to uniformly and thoroughly cover the vegetation in a measured area to be treated. Use a fixed-boom sprayer with a pressure regulator properly calibrated to a constant speed of travel and rate of delivery. Avoid overlapping and shut-off spray booms while starting, turning, slowing or stopping, as injury to conifers may result.

Spray Volume and Equipment: Apply VELPAR by air or ground equipment. Select a spray volume that will insure a thorough and uniform application, at least 8 gals. of spray per acre by air and usually 25 gals. of spray per acre by ground, but not less than 10 gals. of water for each pound of VELPAR. Use mechanical or by-pass agitation to thoroughly mix the spray solution.

Herbaceous Weed Control

When used as directed, VELPAR controls weeds such as annual bluegrass, barnyardgrass, bromegrass, false dandelion (catsear), fleabane, common groundsel, foxtail, oxeye daisy, velvet grass and wild carrot. Treatment provides partial control of dandelion, fescue, goldenrod, heath aster, orchardgrass, ryegrass, smartweed, willowweed (fireweed).

Application Rates: Apply as a preplant or postplant broadcast spray at rates of 1 to 2 lbs. per acre. Use the lower rates on coarse-textured soils and soils low in organic matter; use the higher rates on fine-textured soils and soils high in organic matter. Refer to "Use Limitations" section for specific details.

For band applications, use proportionately less; for example use ½ the broadcast rate when treating a 3 ft. band where row spacing is 6 ft.

Moisture is necessary to activate VELPAR. Best results are obtained when weeds are less than 2 inches in height or diameter, soil is moist at time of application and ½ to 1 inch of rainfall occurs within 2 weeks after application.

Timing:

West of Cascades (and other areas of high spring rainfall): For best results, apply in the spring when weeds are actively growing, but prior to bud break on conifers. If application is made after bud break, use directional spray equipment so that spray does not contact new growth, as injury may result.

East of Cascades (and other areas of low spring rainfall): For best results, apply in the fall before soil freezes, or in the spring in anticipation of rainfall. Weed control results from spring applications will be dependent on sufficient rainfall following treatment to activate VELPAR. Spring applications should be made after snow cover melts, but before bud break occurs.

Brush Control

Among the brush species controlled are: deerbrush ceonothus, squawcarpet, snow ceonothus (varnishleaf ceonothus) and greenleaf manzanita.

Application Rates: Use VELPAR for brush control in forestry site preparation and conifer release where either Ponderosa Pine or Jeffrey Pine is the primary coniferous species. Apply VELPAR at rates of 1½ to 3 lbs. per acre. Use the lower rates on coarse-textured soils and soils low in organic matter; use the higher rates on fine-textured soils and soils high in organic matter.

For best results, applications should be made to brush seedlings or sprouts when less than 18 inches in height or diameter. Applications may be made in the fall before the soil freezes or in the spring. Fall applications generally provide better results.

VELPAR effects may be slow to appear and maximum results will be seen 12–24 months following the application.

Use Limitations:

1. Do not use on gravelly or rocky soils, exposed subsoils, clay knobs, sand or sandy soils with 85% or more sand.
2. Do not use on soil containing less than 1% organic matter.
3. On soils high in organic matter (greater than 5%) effectiveness of VELPAR is reduced and results may be unsatisfactory.
4. Do not use on loamy sand or sandy loam with less than 2% organic matter (except for Ponderosa and Jeffrey Pine).
5. On first year plantations, apply VELPAR (either pre-plant or post-plant) to transplant stock that is 2 years old (2-0) or more, (1 year old (1-0) for Ponderosa Pine and Jeffrey Pine).
6. When applying VELPAR after transplanting, wait until rainfall has settled the soil around the base of the transplants before making the treatment.

IMPORTANT

BEFORE USING VELPAR, READ AND CAREFULLY OBSERVE THE CAUTIONARY STATEMENTS AND ALL OTHER INFORMATION APPEARING ON THE PRODUCT LABEL

This bulletin contains new or supplemental instructions for use of this product which may not appear on package label. Follow the instructions carefully.

SUPPLEMENTAL LABELING

VELPAR® HERBICIDE

(EPA Reg. No. 352-378)

FOR CONTROL OF CERTAIN WEEDS IN CHRISTMAS TREE PLANTATIONS WEST OF THE ROCKY MOUNTAINS

DIRECTIONS FOR USE

It is a violation of federal law to use this product in a manner inconsistent with its labeling.

Du Pont VELPAR® Herbicide is recommended for control of certain weeds in Christmas tree plantations where Douglas Fir, Grand Fir, Noble Fir, Austrian Pine, Ponderosa Pine, Scotch Pine and Sitka Spruce are grown.

Since the effect of VELPAR on conifers varies with soils, uniformity of application and environmental conditions, it is suggested that growers limit their first use to small areas.

NOTE: Because injury to conifers may result: Do not use VELPAR in nurseries, seedbeds or ornamental plantings; do not use on trees which show poor vigor because of insect, disease or winter injury, or show symptoms of other stress conditions. Do not add surfactant if applied over the top of conifers. Do not use on poorly drained soils. Conifer injury may result where severe winter stress, disease or insect damage follows application, or if heavy rains occur soon after application.

HOW TO USE

Application Technique: Before spraying, calibrate equipment to determine the quantity of water necessary to uniformly and thoroughly cover the vegetation in a measured area to be treated. Use a fixed-boom sprayer with a pressure regulator properly calibrated to a constant speed of travel and rate of delivery. Avoid overlapping and shut-off spray booms while starting, turning, slowing or stopping, as injury to conifers may result.

Spray Volume and Equipment: Apply VELPAR by air or ground equipment. Select a spray volume that will insure a thorough and uniform application, at least 8 gals. of spray per acre by air and usually 25 gals. of spray per acre by ground but not less than 10 gals. of water for every 1 lb. of VELPAR. Use mechanical or by-pass agitation to thoroughly mix the spray solution.

Herbaceous Weed Control

When used as directed, VELPAR at rates of 1 to 1½ lbs. per acre will control weeds such as annual bluegrass, barnyardgrass, bromegrass, false dandelion (catsear), fleabane, common groundsel, oxeye daisy, velvetgrass.

Treatment at 2 lbs. per acre provides partial control of crabgrass, curly dock, dandelion, fescue, goldenrod, heath aster, orchardgrass, ryegrass, smartweed, wild carrot and willowweed (fireweed).

Moisture is necessary for full activation of VELPAR. Best results are obtained when weeds are less than 2 inches in height or diameter, soil is moist at time of application and ½ to 1 inch of rainfall occurs within 2 weeks after application.

West of Cascades: For best results, apply in the spring when weeds are actively growing, but prior to bud break on conifers. If application is made after bud break, use directional spray equipment so that spray does not contact new growth, as injury may result.

East of Cascades: For best results, apply in the fall before soil freezes or in the spring in anticipation of rainfall. Weed control results from spring applications will be dependent on sufficient rainfall following treatment to activate VELPAR. Spring applications should be made after snow cover melts, but before bud break occurs.

SELECTION OF USE RATES:

Rate selection is based on soil texture. Do not use rates in excess of those listed for a particular soil texture, as injury to conifers may result. To determine soil texture and organic matter content, have soil samples classified by a reputable laboratory.

SOIL TEXTURE*	CHRISTMAS TREE PLANTATIONS 1st Year Plantings† lb. VELPAR per acre	Established Trees‡ lb. VELPAR per acre
Gravelly or rocky soils, Exposed subsoils, clay knobs, sand or sandy soils with 85% or more sand	Do Not Use	Do Not Use
Loamy sand and sandy loam (50–85% sand)**	1	1 to 1¼
Loam, silt loam, silt and sandy clay loam	1 to 1¼	1¼ to 1¾
Silty clay loam, clay loam, sandy clay, silty clay and clay, soils having more than 5% organic matter***	1¼ to 1½	1¾ to 2

Rates expressed are for broadcast application. For band application, use proportionately less; for example, use ½ the broadcast rate when treating a 3 ft. band where row spacing is 6 ft.

* Do not use on soils containing less than 1% organic matter.

** Do not use on loamy sand or sandy loam with less than 2% organic matter.

*** On soils high in organic matter, 5% or greater, the effectiveness of VELPAR can be significantly reduced and weed control may be unsatisfactory.

† Apply only to healthy transplant stock that is two years old (2–0) or more. When applying VELPAR after transplanting, wait unitl rainfall has settled the soil around the base of the transplants before making treatment.

‡ Rates recommended are for trees that have been established for 1 year or more.

IMPORTANT

BEFORE USING VELPAR, READ AND CAREFULLY OBSERVE THE CAUTIONARY STATEMENTS AND ALL OTHER INFORMATION APPEARING ON THE PRODUCT LABEL

This bulletin contains new or supplemental instructions for use of this product which may not appear on package label. Follow the instructions carefully. This labeling must be in the possession of the user at the time of pesticide application.

SPECIAL LOCAL NEED REGISTRATIONS [24(c)] FOR VELPAR® HERBICIDE

BD-9-62B O784 04/23/85

Crop	State	EPA SLN No.	Remarks
Alfalfa	California Oregon Utah Washington	CA-810083 OR-810098 UT-810026 WA-810090	Aerial application
Blueberries (lowbush)	Maine Michigan New Hampshire	ME-830002 MI-830012 NH-830007	Herbaceous and woody weeds control
Brazilian peppertree control on noncropland areas	Florida	FL-800018	Add a new local brush species as controlled
McCartney Rose control on noncrop-land areas	Texas	TX-800020	Allows higher use rates and lacing treatment
Reforestation Areas	North Carolina	NC-780015	Aerial application and blackland soils
Reforestation Areas	South Carolina	SC-790028	Aerial application
Scotch pine	Missouri	MI-770002	Broadcast
Sugarcane	Texas	TX-810036	Tank mix with "Karmex"

* Copies of individual lables available upon request to nearest Du Pont Agricultural Chemicals Office. For distribution and use only within the State specified.

VELPAR® RP
HERBICIDE

ACTIVE INGREDIENT:
Hexazione
3-cyclohexyl-6-(dimethylamino)-1-methyl-1,3,5,-triazine-2,4 (1H,3H)-dione 25%
INERT INGREDIENTS 75%
TOTAL 100%

U.S. Pat. Nos. 3,902,887 & 3,983,116

EPA Reg. No. 352-437

1 GALLON CONTAINS 2 LBS. HEXAZINONE

Keep Out of Reach of Children

PRECAUTIONARY STATEMENTS

HAZARDS TO HUMANS AND DOMESTIC ANIMALS

DANGER!

CAUSES EYE DAMAGE.

Do not get in eyes. Wear goggles or face shield and rubber gloves when handling. Avoid contact with skin and clothing. Avoid breathing spray mist. Harmful if swallowed.

In case of contact, immediately flush eyes or skin with plenty of water for at least 15 minutes. For eyes, call a physician. Get medical attention if skin irritation persists. Remove and wash contaminated clothing before re-use.

ENVIRONMENTAL HAZARDS

Do not apply directly to any body of water. Do not apply where runoff is likely to occur. Do not contaminate water by cleaning of equipment or disposal of wastes.

PHYSICAL AND CHEMICAL HAZARDS

Flammable. Keep away from heat, sparks, and open flame. Keep container closed.

IMPORTANT

Injury or loss of desirable trees or other plants may result from failure to observe the following: Do not apply except as recommended or drain or flush equipment on or near desirable trees or other plants, or on areas where their roots may extend, or in locations where the chemical may be washed or moved into contact with their roots. Do not use on lawns, walks, driveways, or tennis courts, or similar areas. Prevent drift of spray to desirable plants. Keep from contact with fertilizers, insecticides, fungicides and seeds.

Thoroughly clean all traces of VELPAR RP from application equipment immediately after use.

NOTICE OF WARRANTY

Du Pont warrants that this product conforms to the chemical description on the label thereof and is reasonably fit for the purposes stated on such label only when used in accordance with directions under normal use conditions. It is impossible to eliminate all risks inherently associated with the use of this product. Ineffectiveness or other unintended consequences may result because of such factors as weather conditions, presence of other materials or the manner of use or application, all of which are beyond the control of Du Pont. In no case shall Du Pont be liable for consequential, special or indirect damages resulting from the use or handling of this product. All such risks shall be assumed by the Buyer. DU PONT MAKES NO WARRANTIES OF MERCHANTABILITY OR FITNESS FOR A PARTICULAR PURPOSE NOR ANY OTHER EXPRESS OR IMPLIED WARRANTY EXCEPT AS STATED ABOVE.

STORAGE AND DISPOSAL

STORAGE

Do not subject to temperature below 32°F. Store product in original container only, away from other pesticides, fertilizer, food or feed.

DISPOSAL

Do not contaminate water, food, or feed by storage or disposal. Pesticide wastes are acutely hazardous. Improper disposal of excess pesticide, spray mixture, or rinsate is a violation of Federal Law. If these wastes cannot be disposed of by use according to label instructions, contact your State Pesticide or Environmental Control Agency, or the Hazardous Waste representative at the nearest EPA Regional Office for guidance.

Triple rinse (or equivalent) the container. Then offer for recycling or reconditioning, or puncture and dispose of in a sanitary landfill, or, if allowed by state and local authorities, by burning. If burned, stay out of smoke.

GENERAL INFORMATION

Du Pont VELPAR RP Herbicide is a liquid product to be applied undiluted with an exact delivery handgun applicator as an individual stem treatment for control of undesirable woody plants in rangeland. It is non-corrosive to equipment.

At recommended rates, VELPAR RP provides control of undesirable woody plants such as elm, hackberry, huisache, juniper spp. (redberry, eastern red cedar), mesquite, myrtle, oaks, osage orange (bois d' arc), tallow, whitebrush and willow. On woody plants, symptoms usually appear within 3 weeks after sufficient rainfall has carried the herbicide into the root zone where it is absorbed during periods of active plant growth. Defoliation and subsequent refoliation may occur, but susceptible plants are killed. The effect of treatment will vary with the amount applied, soil type, rainfall, timing and uniformity of application and other conditions.

DIRECTIONS FOR USE

It is a violation of federal law to use this product in a manner inconsistent with its labeling.

Du Pont will not be responsible for losses or damages resulting from use of this product in any manner not specifically recommended by Du Pont. User assumes all risk associated with such non-recommended use.

Make application just before or during period of active growth of brush species to be controlled. Apply VELPAR RP with an exact delivery handgun applicator. This equipment delivers a thin stream of predetermined volume when triggered. Direct the application to the soil within 3′ of the root collar at the rate of 2 to 4 milliliters for each 1″ of stem diameter of brush to be controlled, but not to exceed ⅓ gallon of VELPAR RP per acre per season. When treating large stems, and when more than one shot of VELPAR RP is needed per stem, make application on opposite sides of the stem. Do not treat more than 300 one-inch stem equivalents (approximately 70 brush plants) per acre per season.

NOTE: For effective brush control and prevention of damage to desirable vegetation; do not apply to brush standing in water; do not use water from treated ditches for irrigation; do not use where desirable trees or shrubs are immediately adjacent to the treated area.

Do not use on marshy or poorly drained sites nor on soils classified as clays. Do not cut hay from treated grasses. Do not use on pastures.

NOTICE TO BUYER Purchase of this material does not confer any rights under patents of countries outside of the United States.

AG-503-1 1184

VELPAR® L

HERBICIDE
WATER DISPERSIBLE LIQUID

1 Gallon contains 2 lbs. Hexazinone

ACTIVE INGREDIENT:

Hexazinone[3-cyclohexyl-6-(dimethylamino)-1-methyl-1,3,5-triazine–2,4(1H,3H)-dione]	25%
INERT INGREDIENTS	75%
TOTAL	100%

EPA Reg. No. 352-392 EPA Est. 5905-AR-1

U.S. Pats. 3,902,887 & 3,983,116

Keep Out of Reach of Children

PRECAUTIONARY STATEMENTS

HAZARDS TO HUMANS AND DOMESTIC ANIMALS

DANGER!

CAUSES EYE DAMAGE. FLAMMABLE.

Do not get in eyes. Wear goggles or face shield and rubber gloves when handling. Avoid contact with skin and clothing. Avoid breathing spray mist. Harmful if swallowed. Keep away from heat, sparks and open flame. Keep container closed.

In case of contact, immediately flush eyes or skin with plenty of water for at least 15 minutes.

For eyes, call a physician. Get medical attention if skin irritation persists. Remove and wash contaminated clothing before re-use.

ENVIRONMENTAL HAZARDS

Keep out of any body of water. Do not apply where runoff is likely to occur. Do not contaminate water by cleaning of equipment or disposal of wastes.

IMPORTANT

Injury to, or loss of, desirable trees or other plants may result from failure to observe the following: Do not apply except as recommended or drain or flush equipment, on or near desirable trees or other plants, or on areas where their roots may extend, or in locations where the chemical may be washed or moved into contact with their roots. Do not use on lawns, walks, driveways, tennis courts or similar areas. Prevent drift of spray to desirable plants. Keep from contact with fertilizers, insecticides, fungicides, and seeds.

Thoroughly clean all traces of VELPAR L from application equipment immediately after use. Flush tank, pump, hoses and boom with several changes of water after removing nozzle tips and screens (clean these parts separately).

GENERAL INFORMATION

Du Pont VELPAR L Herbicide is a water-dispersible liquid to be mixed in water and applied as a spray for selective weed control in Christmas tree plantations, for selective weed and brush control in reforestation areas, and for nonselective weed and brush control on noncropland areas. It may also be applied undiluted as a basal soil treatment for brush control on noncropland areas and in reforestation areas, or by tree injection for selective brush control. It is noncorrosive to equipment.

VELPAR L is an effective general herbicide providing both contact and residual control of many weeds. At rates recommended for noncropland areas, it provides control of many annual and biennial weeds, brush, woody vines and, except for johnsongrass, is effective for control of most perennial weeds. It is absorbed through the roots and foilage. For brush control, soil application or injection is most effective.

Moisture is required to activate VELPAR L in the soil. Best results are obtained when weeds are less than 2 inches in height or diameter, soil is moist at time of application and $^1/_4$ to $^1/_2$ inch of rainfall occurs within 2 weeks after application. Foliar application to weeds is most effective under conditions of high temperature

(80°F), high humidity and good soil moisture. Foliar application when vegetation is dormant or semi-dormant may not be effective.

On herbaceous plants, symptoms usually appear within 2 weeks after application under warm, humid conditions, while 4 to 6 weeks may be required when weather is cool. If rainfall after application is inadequate to activate VELPAR L in the soil, plants may recover from contact effects and continue to grow.

On woody plants, symptoms usually appear within 3 weeks after sufficient rainfall has carried the herbicide into the root zone, during periods of active growth. Defoliation and subsequent refoliation may occur, but susceptible plants are killed. The degree of control and duration of effect will vary with the amount of chemical applied, rainfall, temperature, weed and brush species, soil texture and other conditions.

NOTICE OF WARRANTY

Du Pont warrants that this product conforms to the chemical description on the label thereof and is reasonably fit for purposes stated on such label only when used in accordance with directions under normal use conditions. It is impossible to eliminate all risks inherently associated with the use of this product. Ineffectiveness or other unintended consequences may result because of such factors as weather conditions, presence of other materials or the manner of use or application, all of which are beyond the control of Du Pont. In no case shall Du Pont be liable for consequential, special or indirect damages resulting from the use or handling of this product. All such risks shall be assumed by the Buyer. DU PONT MAKES NO WARRANTIES OF MERCHANTABILITY OR FITNESS FOR A PARTICULAR PURPOSE NOR ANY OTHER EXPRESS OR IMPLIED WARRANTY EXCEPT AS STATED ABOVE.

DIRECTIONS FOR USE

It is a violation of federal law to use this product in a manner inconsistent with its labeling.

Du Pont VELPAR L should be used only in accordance with recommendations on this label, or in separate published Du Pont recommendations available through local dealers.

Du Pont will not be responsible for losses or damages resulting from use of this product in any manner not specifically recommended by Du Pont. User assumes all risk associated with such non-recommended use.

For broadcast treatments using ground equipment, apply VELPAR L as a spray, just before or soon after weed emergence. Do not apply to frozen or snow-covered soil. Before spraying, calibrate equipment to determine the quantity of water necessary to uniformly and thoroughly cover the vegetation and soil in an area to be treated. Use at least 5 gallons of water for each gallon of VELPAR L.

Add the proper amount of VELPAR L to a spray tank filled with the amount of water to be used and mix thoroughly to improve wetting and/or contact activity. A surfactant such as Du Pont Surfactant WK may then be added at the rate of 1 quart per 100 gals. of spray. After thoroughly mixing the spray solution, a suitable defoaming agent, such as Dow Corning "Antifoam" A or "Hodag" "Antifoam" F-1, may also be added at the rate of 1 to 5 ozs. per 1,000 gals. of spray, should foaming become a problem. Agitate thoroughly to disperse the defoaming agent.

Use a fixed boom sprayer properly calibrated to a constant speed and rate of delivery. When used in conifer plantings, avoid overlapping and shut off spray booms while starting, turning, slowing or stopping, or injury to the conifer trees may result. For application with a handgun sprayer (noncropland areas only), use sufficient water (50 to 200 gals. per acre) for thorough coverage. For treating small noncropland areas, a tank-type hand sprayer or sprinkling can may be used.

GENERAL WEED AND BRUSH CONTROL—NONCROPLAND AREAS

Railroad, Highway, Utility and Pipeline Right-of-Ways, Petroleum Tank Farms, Storage Areas, Industrial Plant Sites, Drainage Ditchbanks and other similar areas.

Use the higher levels of the recommended dosage ranges under the following conditions: on hard-to-kill species; on fine-textured soils (clay, clay loam); on soils containing more than 5% organic matter or carbon; and on coarse-textured soils (sand, loamy sand, gravelly soils) where rainfall exceeds 40 inches per year. Use the lower levels of the dosage range if the growing season is short.

Weed Control—Apply 3 to 6 gals. per acre for **season long control** of many annual, biennial and perennial weeds including bahiagrass*, bermudagrass*, bluegrass, broomsedge, camphorweed, Canada thistle*, chickweed, clovers, dewberry, dogfennel*, fescue*, fingergrass, foxtail, guineagrass, heath aster, honeysuckle, lantana, marestail*, natalgrass, plantain, prickly lettuce, ragweed, smutgrass, Spanish needle, vaseygrass, wild blackberry* and wild carrot.

For **short-term control** (up to 3 months) of the above weeds, apply 1 to 2½ gals. per acre. Application at 1 to 2½ gals. per acre also provides short-term control of many annual, biennial and perennial weeds including barnyardgrass, bindweed*, bouncing bet*, bromegrass, buffelgrass*, burdock, cocklebur, crabgrass, crown vetch, curly dock*, dandelion*, dogbane*, fiddleneck, filaree, fleabane, goatsbeard vine, goldenrod, lespedeza, milkweed*, mustard, nutsedge*, orchardgrass*, oxalis, paragrass, pigweed, purslane, quackgrass, ryegrass, smartweed, spurge, trumpet creeper*, wild oats*, wild parsnip and wild star thistle.

*Considered hard-to-kill.

For **small areas**—Three fl. ozs. VELPAR L per 250 sq. ft. is approximately 4 gals. per acre.

Brush Control—Apply to soil during the period between late winter and early summer for control of blackgum, blackcherry, elm, flowering dogweed, green ash, hawthorn, hickory, oaks (except liveoak), persimmon, sumac, sweetgum and willow. In areas where the soil normally remains frozen during the winter, and spring rains are usually inadequate for soil activation, a fall or winter treatment may be applied before the soil freezes.

Note: For effective brush control and prevention of damage to desirable vegetation: do not apply to brush standing in water; do not use water from treated ditches for irrigation; do not use where desirable trees or shrubs are immediately adjacent to the treated area.

Broadcast Treatment—Apply 2 to 4 gals. of VELPAR L per acre as a coarse spray with water. Direct the spray to the soil beneath woody plants to be controlled.

Undiluted Basal Soil Treatment—Apply VELPAR L undiluted with an exact-delivery handgun applicator. This equipment delivers a thin stream of predetermined volume when triggered. Apply at the rate of 2 to 4 milliliters for each 1" of stem diameter, at breast height. Direct the treatment to the soil within 3 ft. of the root collar of woody plants to be controlled. When treating large stems, and more than one delivery of VELPAR L is needed per stem, make applications on opposite sides of the stem.

SELECTIVE WEED CONTROL—CHRISTMAS TREE PLANTATIONS AND REFORESTATION AREAS

VELPAR L is recommended for control of certain weeds in Christmas tree plantations where Scotch Pine, Douglas Fir (west of the Rocky Mountains only), Austrian Pine, Grand Fir, Sitka Spruce and Noble Fir are grown. It is also recommended for use in reforestation areas where Scotch Pine, Douglas Fir (west of the Rocky Mountains only), Loblolly Pine, Ponderosa Pine, Austrian Pine, Grand Fir, Sitka Spruce and Noble Fir are grown.

Since the effect of VELPAR L on conifers varies with soils, uniformity of application and environmental conditions, it is suggested that growers limit their first use to small areas.

Note: Because injury to conifers may result: Do not use in nurseries, seedbeds or ornamental plantings; do not use on trees which show poor vigor because of insect, disease or winter injury, or show symptoms of other stress conditions; do not add surfactant if applied over the top of conifers; do not use on poorly drained soils. In addition, conifer injury may result where severe winter stress, disease or insect damage follows application, or heavy rains occur soon

after application. Do not use for weed control in conifers in the state of Texas.

When used as directed, VELPAR L controls annual bluegrass, barnyardgrass, bentgrass, bromegrass, catsear, common groundsel, common ragweed, false-dandelion, fleabane, foxtail, oxeye daisy, Pennsylvania smartweed, velvetgrass and wild carrot. Treatment provides partial control of curly dock, dandelion, willowweed (fireweed), fescue, goldenrod, horseweed, orchardgrass and ryegrass. Rates expressed are broadcast. For band application, use proportionately less; for example use ½ of the broadcast rates, when treating a 3 ft. band where row spacing is 6 ft.

Apply as a broadcast spray in the spring prior to conifer bud break. If application is made after bud break, use directional spray equipment and prevent contact of conifer foliage, or injury may result.

Soil Texture**	First Year Plantations† Pints VELPAR L Per Acre††	Established Plantations Pints VELPAR L Per Acre††
Gravelly or Rocky Soils, Exposed Subsoils, Clay Knobs, Sand, or Sandy Soils with 85% or more Sand	Do Not Use	Do Not Use
Loamy Sand and Sandy Loam (50 85% Sand)	4	4–5
Loam, Silt Loam, Silt and Sandy Clay Loam	4–5	5–7
Silty Clay Loam, Clay Loam, Sandy Clay, Silty Clay, and Clay	5–6	7–8

** Do not use on any soil containing less than 1% organic matter; do not use on loamy sand or sandy loam with less than 2% organic matter.

† Apply only to transplant stock that is 2 years old (2–0) or more (1 year old for loblolloy pine). Allow at least two months after transplanting before applying VELPAR L, then apply only if rainfall has settled the soil around the base of the transplants.

†† Use of rates in excess of those specified for the soil texture may result in injury to the conifers.

SELECTIVE BRUSH CONTROL—REFORESTATION AREAS

By injection—Inject 1 milliliter of undiluted VELPAR L through the bark of undesirable trees. Injections should be made at 4" intervals around the circumferences and near the ground level when using tubular injector equipment, or at waist height when using the HYPO-Hatchet® Injector or similar device. Treatments should be made in the summer.

When used as directed, VELPAR L controls black cherry, red maple, oaks (including live oak) and sweet gum.

Undiluted Basal Soil Treatment—Apply VELPAR L undiluted with an exact-delivery handgun applicator. This equipment delivers a thin stream of a predetermined volume when triggered. Apply at the rate of 2 to 4 milliliters for each 1" of stem diameter at breast height of trees to be controlled.

Direct the treatment to the soil within 3 ft. of the root collar of trees to be controlled. When treating large stems, and more than one delivery of VELPAR L is needed per stem, make applications on opposite sides of the stem.

When used as directed, VELPAR L controls aspen, elm, wild cherry, oaks (including live oak), sumac and willow.

Notice to Buyer: Purchase of this material does not confer any rights under patents of countries outside of the United States.

STORAGE AND DISPOSAL

STORAGE

Do not subject to temperatures below 32°F. Store product in original container only, away from other pesticides, fertilizer, food or feed.

DISPOSAL

Do not contaminate water, food, or feed by storage or disposal. Pesticide wastes are acutely hazardous. Improper disposal of excess pesticide, spray mixture, or rinsate is a violation of Federal Law. If these wastes cannot be disposed of by use according to label instructions, contact your State Pesticide or Environmental Control Agency, or the Hazardous Waste representative at the nearest EPA Regional Office for guidance.

Triple rinse (or equivalent) the container. Then offer for recycling or reconditioning, or puncture and dispose of in a sanitary landfill, by incineration, or, if allowed by state and local authorities, by burning. If container destroyed by burning, stay out of smoke.

SUPPLEMENTAL LABELING

VELPAR® L HERBICIDE

EPA Reg. No. 352-392

FOR CONTROL OF CERTAIN WEEDS IN ESTABLISHED ALFALFA

Du Pont VELPAR L Herbicide is recommended for control of certain weeds in established alfalfa.

Sufficient moisture is necessary to activate the herbicide. Best results are obtained when ½ to 1" of rainfall or sprinkler irrigation occurs within two weeks after application, soil is moist at time of application and weeds have not germinated or are less than 2" tall or across.

Since the effect of VELPAR L on alfalfa varies with soils, uniformity of a application and environmental conditions, it is suggested that growers limit their first use to small areas.

DIRECTIONS FOR USE

Treat only stands of alfalfa established for one year or for one growing season, provided the alfalfa root system is well established and the crop is healthy, vigorous and not under stress from weather conditions, insect or disease damage. Do not use on seedling alfalfa, nor on alfalfa-grass mixtures, or on other mixed stands.

Make a single application of VELPAR L in the fall or winter after alfalfa becomes dormant, or in the spring before new growth begins. **Exception:** In certain northeastern* and midwestern* states, make a single application of 1 to 3 qts. per acre in the fall or winter, after plants become dormant, or in the spring, before new growth exceeds 2" in height, or to stubble after cutting following hay removal and before regrowth exceeds 2" in height.

Apply VELPAR L using a fixed boom power spray or aerial equipment. Before spraying, calibrate equipment to determine the quantity of water necessary to uniformly and thoroughly cover the vegetation in an area to be treated. The sprayer must be properly calibrated to a constant speed and rate of delivery. Avoid overlapping and shut off spray booms while starting, turning, slowing or stopping to prevent crop injury.

Add the VELPAR L to a water-filled tank and mix thoroughly. Apply in at least 20 gallons of water per acre using ground equipment, and 5 to 10 gallons of water per acre by air.

USE RATES

Rate selection is to be based on soil texture and organic matter content as follows:

Soil Texture/Organic Matter	Quarts/ Acre
Gravelly or rocky soils, exposed subsoils, hardpan, sand, poorly drained or alkali soils.	Do not use
Sandy loam, loamy sand, loam, silt, silt loam, and sandy clay loam with less than 1% organic matter.	1 to 1½
Silty clay loam, clay loam, sandy clay, silty clay and clay with 1% organic matter or higher.	1½ to 3
Soils having more than 5% organic matter.[1]	2 to 3

[1] On soils high in organic matter, the effectiveness of VELPAR L can be significantly reduced, and weed control may be unsatisfactory.

WEED SPECIES CONTROLLED

VELPAR L when applied preemergence or early postemergence will control the following winter annual and spring germinating weeds:

1 to 2 qts./acre

annual bluegrass	fleabane
blue mustard	foxtail
common chickweed	jimsonweed
common groundsel	orchardgrass
common lambsquarter	purple mustard
crabgrass	salsify
fiddleneck	shepherdspurse
field pennycress	spurry
filaree	tansy mustard
	yellow rocket

2 to 3 qts./acre

downy brome or cheatgrass (Bromus tectorum)	annual ryegrass
	sweet clover
ivy leaf speedwell	purslane speedwell
prickly lettuce	

2 to 3 qts./acre will give partial control[2] of:

common dandelion	seedling alfalfa
false dandelion	white cockle
Mexican tea	wild oats
perennial bluegrass	

[2] Partial control is observable reduction in weed competition, as compared to an untreated area. Degree of partial control can vary with soil type, environmental factors and size of weeds at the time of application.

REPLANTING

Except in low annual rainfall areas (20" or less), corn may be planted 12 months after the last treatment, provided the soil is deep disked or plowed prior to planting. Do not plant treated areas to any other crop within two years after treatment, or injury to crop may result.

Note: Because crop injury may result, do not apply to actively growing alfalfa, nor to stubble between cuttings, **except** as recommended for certain northeastern* and midwestern* states. Extremely heavy rainfall, or more than 1" of irrigation soon after application, may result in crop injury or nonperformance. Do not use in low desert valleys in California. Do not apply to snow covered or frozen ground.

Do not graze or feed treated forage or hay to livestock, within 30 days following application.

*** Northeastern and midwestern states are:**

Connecticut	Missouri
Delaware	New Hampshire
Illinois	New Jersey
Indiana	New York
Iowa	Ohio
Kentucky	Pennsylvania
Maine	Rhode Island
Maryland	Vermont
Massachusetts	Virginia
Michigan	West Virginia
Minnesota	Wisconsin

IMPORTANT

BEFORE USING VELPAR® L, READ AND CAREFULLY OBSERVE THE CAUTIONARY STATEMENTS AND ALL OTHER INFORMATION APPEARING ON THE PRODUCT LABEL.

This bulletin contains new or supplemental instructions for use of this product which may not appear on package label. Follow the instructions carefully.

11/8/82

SUPPLEMENTAL LABELING

VELPAR® L HERBICIDE

EPA Reg. No. 352-392

FOR CONTROL OF UNDESIRABLE WOOD PLANTS IN NON-CROPLAND AREAS

DIRECTIONS FOR USE

It is a violation of federal law to use this product in a manner inconsistent with its labeling.

Du Pont VELPAR L Herbicide is recommended for control of undesirable woody plants, in noncropland areas, by the lacing and streaking method of application. Lacing and streaking is a method which employs a straight stream application of a concentrated solution of VELPAR L in water to the soil.

GENERAL WEED CONTROL—NONCROPLAND AREAS: Railroad, highway, utility and pipeline right-of-ways, petroleum tank farms, storage areas, industrial plant sites, drainage ditch banks and other similar areas.

Brush Control—Apply to soil during the period between late winter and early summer. Moisture is necessary to activate VELPAR L. In areas where the soil normally remains frozen during the winter, and spring rains are usually inadequate for activation, a fall or winter treatment may be applied before the soil freezes. Do not apply to frozen or snow-covered soil.

Note: For effective brush control and prevention of damage to desirable vegetation: do not apply to brush standing in water; do not use water from treated ditches for irrigation; do not use where desirable trees or shrubs are immediately adjacent to treated areas.

Brush Controlled—Alder, American elm, ash, black cherry, blackgum, Chinese elm, Chinese tallow, hackberry, hawthorne, hickory, huisache, locust, mesquite, mulberry, oak, privet, sourwood, Southern red oak, sumac, sweet bay, sweet gum, water oak, wild dogwood, wild plum, willow, winged elm and yellow poplar. Hard-to-kill species: Chinaberry, Eastern red cedar and sassafrass.

Lacing and Streaking Treatment—Apply 2 to 6 gals. VELPAR L per acre. Prepare by mixing 1 gallon of VELPAR L Herbicide with 5 or more gals. of water. Use lower rates on sands and higher rates on clays and/or hard-to-kill species. Direct the spray to the soil with a serpentine pattern using a handgun so that the swath on the soil is 6 to 12 inches wide at the base of the brush to intercept the major portion of the root system and swaths are 2 to 4 feet apart.

IMPORTANT

BEFORE USING VELPAR® L HERBICIDE, READ AND CAREFULLY OBSERVE THE CAUTIONARY STATEMENTS AND ALL OTHER INFORMATION APPEARING ON THE PRODUCT LABEL.

This bulletin contains new or supplemental instructions for use of this product which may not appear on the package label. Follow the instructions carefully.

This labeling must be in the possession of the user at the time of pesticide application.

3/9/82 (Revised 7/7/82)

SUPPLEMENTAL LABELING

VELPAR® L HERBICIDE

EPA Reg. No. 352–392

SITE PREPARATION IN REFORESTATION AREAS EAST OF THE ROCKY MOUNTAINS

GENERAL INFORMATION

VELPAR L Herbicide is recommended for use in forestry site preparation areas East of the Rocky Mountains, where Austrian Pine, Balsam fir, Black spruce, Loblolly pine, Longleaf pine, Ponderosa pine, Red pine, Red spruce, Scotch pine, Shortleaf pine, Slash pine, Spruce, Virginia pine and White spruce will be planted.

HOW TO USE

When used as directed, VELPAR L provides both contact and

residual control of many herbaceous and woody plants, including: annual bluegrass, ash, aspens (bigtooth, trembling), asters, balsam poplar, barnyardgrass, bentgrass, birch, common groundsel, common ragweed, elksedge, elm, false dandelion, fleabane, flowering dogwood*, foxtail, hawthorne, hazel, hickory, oaks, oxeye daisy, Pennsylvania smartweed, pinegrass, red maple*, sourwood*, sweet gum, velvetgrass, wild carrot, wild cherry, and willows. Treatment provides partial control of Canada thistle (suppression only), catsear, crabgrass, curly dock, dandelion, willowweed (fireweed), fescue, goldenrod, heathaster, honeysuckle, horseweed, orchardgrass and perennial grasses such as quackgrass and ryegrass.

* Considered hard-to-kill.

Apply 1 to 3 gallons of VELPAR L per acre using at least 5 gallons of water for each gallon of VELPAR L. For aerial application, apply at least 8 gallons of spray per acre, using appropriate aerial equipment to obtain thorough coverage. For ground application, use conventional spray equipment with enough water for thorough coverage, usually 25 gallons per acre over the treated area. On fine textured soils and hard-to-kill species, use the higher rates of VELPAR L, for coarser soils use a lower rate. To reduce drift, use a precise application spray boom and/or add a spray thickening agent, such as NALCO-TROL[1]. If a spray thickening agent is used in the tank mixture, follow all label directions.

[1] NALCO-TROL—Trademark of Nalco Chemical Company.

Add the VELPAR L to a water filled tank and mix thoroughly. A surfactant such as Du Pont Surfactant WK may then be added at a rate of 1 quart per 100 gallons of spray. After thorough mixing, a defoaming agent, such as Dow Corning "Antifoam A" or Hodag Antifoam F-1, may be added at the rate of 1 to 5 ounces per 1000 gallons. Agitate thoroughly to disperse the defoaming agent.

VELPAR L should be applied during the period from late spring to early summer, after the trees have reached full leaf stage or, in Maine, Michigan, Minnesota, New Hampshire, New York, Vermont and Wisconsin, after bud break. Do not burn treated sites after initial contact (browning) effect on foliage. Burn only after brush has completely defoliated at least once from sufficient root uptake of VELPAR L, usually 60 to 90 days after adequate rainfall.

IMPORTANT

BEFORE USING VELPAR® L, READ AND CAREFLLY OBSERVE THE CAUTIONARY STATEMENTS AND ALL OTHER INFORMATION APPEARING ON THE PRODUCT LABEL.

This bulletin contains new or supplemental directions for use of these products which may not appear on the package label. Follow the instructions carefully.

This labeling must be in the possession of the user at the time of pesticide application.

SUPPLEMENTAL LABELING

VELPAR® L HERBICIDE

EPA Reg. No. 352-392

FOR CONTROL OF CERTAIN WEEDS AND BRUSH SPECIES FOR FORESTRY SITE PREPARATION & CONIFER RELEASE WEST OF THE ROCKY MOUNTAINS

DIRECTIONS FOR USE

It is a violation of federal law to use this product in a manner inconsistent with its labeling.

Du Pont VELPAR® L Herbicide is recommended for control of certain weeds and brush species in forestry site preparation and conifer release operations where Douglas Fir, Grand Fir, Noble Fir, White Fir, Ponderosa Pine, Jeffrey Pine and Sitka Spruce are grown.

Since the effect of VELPAR L on conifers varies with soils, uniformity of application and environmental conditions, it is suggested that foresters limit their first use to small areas.

NOTE: Because injury to conifers may result, do not use VELPAR L in nurseries, seedbeds or ornamental plantings; do not use on trees which show poor vigor because of insect, disease or winter injury, or show symptoms of other stress conditions. Do not add surfactant if applied over the top of conifers. Do not use on poorly drained soils. Conifer injury may result where severe winter stress, disease or insect damage follows application, or if heavy rains occur soon after application.

HOW TO USE

Application Technique: Before spraying, calibrate equipment to determine the quantity of water necessary to uniformly and thoroughly cover the vegetation in a measured area to be treated. Use a fixed-boom sprayer with a pressure regulator properly calibrated to a constant speed of travel and rate of delivery. Avoid overlapping and shut-off spray booms while starting, turning, slowing or stopping, as injury to conifers may result.

Spray Volume and Equipment: Apply VELPAR L by air or ground equipment. Select a spray volume that will insure a thorough and uniform application, at least 8 gals. of spray per acre by air and usually 25 gals. of spray per acre by ground, but not less than 5 gals. of water for each gallon of VELPAR L. Use mechanical or bypass agitation to thoroughly mix the spray solution. After initial mixing, do not use excessive agitation, as foaming problems may result. To avoid foaming problems, use anti-foam agents and minimal agitation.

Herbaceous Weed Control

When used as directed, VELPAR L controls weeds such as annual bluegrass, barnyardgrass, bromegrass, false dandelion (catsear), fleabane, common groundsel, foxtail, oxeye daisy, velvet grass and wild carrot. Treatment provides partial control of dandelion, fescue, goldenrod, heath aster, orchardgrass, ryegrass, smartweed, willowweed (fireweed).

Application Rates: Apply as a pre-plant or post-plant broadcast spray at rates of 2 to 4 qts. per acre. Use the lower rates on coarse-textured soils and soils low in organic matter; use the higher rates on fine-textured soils high in organic matter. Refer to "Use Limitations" section for specific details.

For band applications, use proportionately less; for example use 1/2 the broadcast rate when treating a 3 ft band where row spacing is 6 ft.

Moisture is necessary to activate VELPAR L. Best results are obtained when weeds are less than 2 inches in height or diameter, soil is moist at time of application and 1/2 to 1 inch of rainfall occurs within 2 weeks after application.

Timing:

West of Cascades (and other areas of high spring rainfall): For best results, apply in the spring when weeds are actively growing, but prior to bud break on conifers. If application is made after bud break, use directional spray equipment so that spray does not contact new growth as injury may result.

East of Cascades (and other areas of low spring rainfall): For best results, apply in the fall before soil freezes or in the spring in anticipation of rainfall. Weed control results from spring applications will be dependent on sufficient rainfall following treatment to activate VELPAR L. Spring applications should be made after snow cover melts but before bud break occurs.

Brush Control

Among the brush species controlled are: deerbush ceonothus, squawcarpet, snowbrush ceonothus (varnishleaf ceonothus) and greenleaf manzanita.

Application Rates: Use VELPAR L for brush control in forestry site preparation and conifer release where either Ponderosa Pine or Jeffrey Pine is the primary coniferous species. Apply VELPAR L at rates of 3 to 6 qts per acre. Use the lower rates on coarse-textured soils and soils low in organic matter, use the higher rates on fine-textured soils and soils high in organic matter.

For best results, applications should be made to brush seedlings or sprouts when less than 18 inches in height or diameter. Applications may be made in the fall before the soil freezes, or in the spring. Fall applications generally provide better results.

VELPAR L effects may be slow to appear and maximum results will be seen 12–24 months following the application.

Use Limitations:

1. Do not use on gravelly or rocky soils, exposed subsoils, clay knobs, sand or sandy soils with 85% or more sand.
2. Do not use on any soil containing less than 1% organic matter.
3. On soils high in organic matter (greater than 5%) effectiveness of VELPAR L is reduced and results may be unsatisfactory.
4. Do not use on loamy sand or sandy loam with less than 2% organic matter (except for Ponderosa and Jeffrey Pine).
5. On first year plantations, apply VELPAR L (either pre-plant or post-plant) to transplant stock that is 2 years old (2-0) or more, (1 year old (1-0) for Ponderosa Pine and Jeffrey Pine).
6. When applying VELPAR L after transplanting, wait until rainfall has settled the soil around the base of the transplants before making the treatment.

IMPORTANT

BEFORE USING VELPAR® L, READ CAREFULLY. OBSERVE THE CAUTIONARY STATEMENTS AND ALL OTHER INFORMATION APPEARING ON THE PRODUCT LABEL.

This bulletin contains new or supplemental instructions for use of this product which may not appear on package label. Follow the instructions carefully.

SUPPLEMENTAL LABELING

VELPAR® L HERBICIDE

EPA Reg. No. 352-392

FOR CONTROL OF CERTAIN WEEDS IN CHRISTMAS TREE PLANTATIONS AND REFORESTATION AREAS WEST OF THE ROCKY MOUNTAINS

DIRECTION FOR USE

It is a violation of federal law to use this product in a manner inconsistent with its labeling.

Du Pont VELPAR L Herbicide is recommended for control of certain weeds in Christmas tree plantations where Douglas Fir, Grand fir, Noble Fir, Austrian Pine, Ponderosa Pine, Scotch Pine and Sitka Spruce are grown.

Since the effect of VELPAR L on conifers varies with soils, uniformity of application and environmental conditions, it is suggested that growers limit their first use to small areas.

NOTE: Because injury to conifers may result, do not use VELPAR L in nurseries, seedbeds or ornamental plantings; do not use on trees which show poor vigor because of insect, disease or winter injury, or show symptoms of other stress conditions. Do not add surfactant if applied over the top of conifers. Do not use on poorly drained soils. Conifer injury may result where severe winter stress, disease or insect damage follows application, or heavy rains occur soon after application.

HOW TO USE

Application Technique: Before spraying, calibrate equipment to determine the quantity of water necessary to uniformly and thoroughly cover the vegetation in an area to be treated. Use a fixed-boom sprayer with a pressure regulator properly calibrated to a constant speed of travel and rate of delivery. Avoid overlapping and shut-off spray booms while starting, turning, slowing or stopping, as injury to conifers may result.

Spray Volume and Equipment: Apply VELPAR L by air or ground equipment. Select a spray volume that will insure a thorough and uniform application, at least 8 gals. of spray per acre by air and usually 25 gallons of spray per acre by ground, but not less than 5 gallons of water for each gallon of VELPAR L. Use mechanical or bypass agitation to thoroughly mix the spray solution. After initial mixing, do not use excessive agitation, as foaming problems may result. To avoid foaming problems, use anti-foam agents and minimal agitation.

Herbaceous Weed Control

When used as directed, VELPAR L at rates of 2 to 3 qts per acre will control weeds such as annual bluegrass, barnyardgrass, bromegrass, false dandelion (catsear), fleabane, common groundsel, oxeye daisy, velvetgrass.

Treatment at 4 qts. per acre provides partial control of crabgrass, curly dock, dandelion, fescue, goldenrod, heath aster, orchardgrass, ryegrass, smartweed, wild carrot and willowweed (fireweed).

Moisture is necessary for full activation of VELPAR L. Best results are obtained when weeds are less than 2 inches in height or diameter, soil is moist at time of application and ½ to 1 inch of rainfall occurs within 2 weeks after application.

West of Cascades: For best results, apply in the spring when weeds are actively growing, but prior to bud break on conifers. If application is made after bud break, use directional spray equipment so that spray does not contact new growth, as injury may result.

East of Cascades: For best results, apply in the fall before soil freezes, or in the spring in anticipation of rainfall. Weed control results from spring applications will be dependent on sufficient rainfall following treatment to activate VELPAR L. Spring applications should be made after snow cover melts, but before bud break occurs.

SELECTION OF USE RATES:

Rate selection is based on soil texture. Do not use rates in excess of those listed for a particular soil texture, as injury to conifers may result. To determine soil texture and organic matter content, have soil samples classified by a reputable laboratory.

	Christmas Tree Plantations	
Soil Texture*	**1st Year Plantings† qts VELPAR L per acre**	**Established Trees†† qts VELPAR L per acre**
Gravelly or rocky soils, Exposed subsoils, clay knobs, sand or sandy soils with 85% or more sand	Do Not Use	Do Not use
Loamy sand and sandy loam (50–80% sand)**	2	2 to 2½
Loam, silt loam, silt and sandy clay loam	2 to 2½	2½ to 3½
Silty clay loam, clay loam sandy clay, silty clay and clay, soils having more than 5% organic matter***	2½ to 3	3½ to 4

Rates expressed are broadcast application. For band application, use proportionately less; for example, use ½ the broadcast rate when treating a 3 ft. band where row spacing is 6 ft.

* Do not use on soils containing less than 1% organic matter.

** Do not use on loamy sand or sandy loam with less than 2% organic matter.

*** On soils high in organic matter, 5% or greater, the effectiveness of VELPAR L can be significantly reduced, and weed control may be unsatisfactory.

† Apply only to healthy transplant stock that is two years old (2–0) or more. When applying VELPAR L after transplanting, wait until rainfall has settled the soil around the base of the transplants before making treatment.

†† Rates recommended are for trees that have been established for 1 year or more.

IMPORTANT

BEFORE USING VELPAR® L, READ AND CAREFULLY OBSERVE THE CAUTIONARY STATEMENTS AND ALL OTHER INFORMATION APPEARING ON THE PRODUCT LABEL.

This bulletin contains new or supplemental instructions for use of this product which may not appear on package label. Follow the instructions carefully. This labeling must be in the possession of the user at the time of pesticide application.

SUPPLEMENTAL LABELING

VELPAR® L HERBICIDE

EPA Reg. No. 352-392

FOR CONTROL OF UNDESIRABLE WOODY AND HERBACEOUS PLANTS FOR CONIFER RELEASE

DIRECTIONS FOR USE

It is a violation of federal law to use this product in a manner inconsistent with its labeling.

Du Pont VELPAR® L Herbicide is recommended for conifer release where balsam fir, red pine, black spruce, Norway spruce, red spruce, or white spruce are grown. VELPAR L is an effective herbicide providing both contact and residual control of many herbaceous and woody plants.

HOW TO USE

Apply ½ to 1 gallon VELPAR L per acre (1–2 gallons per acre for use in red pines) by air or by ground. For aerial application, use appropriate aerial application equipment and apply in a minimum of 8 gallons of spray per acre, but not less than 5 gallons of water for each one gallon of VELPAR L. To reduce drift, use a precise application spray boom and/or a spray thickening agent, such as NALCO-TROL[1]. Do not use a surfactant. For ground application, apply VELPAR L in enough water for thorough coverage, usually 25 gallons per acre using conventional spray equipment to insure uniform distribution of liquid over the treated area.

[1] NALCO-TROL—Trademark of Nalco Chemical Company.

Apply from late spring, at or near conifer budbreak, until early summer. For maximum contact activity on woody brush, apply after brush is in half-leaf stage. Some conifer needles may be discolored, but this usually disappears in 60–90 days. Mortality of some pine trees may occur. At least 1–2 inches of rainfall are needed for soil activation.

When used as directed, VELPAR L controls many herbaceous plants such as asters, bracken fern, goldenrod, honeysuckle, and perennial grasses, such as quackgrass. VELPAR L also suppresses the brushy growth of certain undesirable woody plants including: ash, aspen, balsam poplar, birch, blackberry, brambles, box elder, elm, hawthorns, hazel, oaks, (red, black, and white), pin cherry, red maple*, Rubus (including raspberries) and willow.

*Considered hard to kill—treatment will result in suppression only.

Notes:

1. Do not apply where conifers are suffering from loss of vigor caused by insects, diseases, winter damage or other stresses, as injury may result.
2. Application to trees less than three years old may result in injury.
3. VELPAR L is not recommended for use on poorly drained or marshy sites.
4. Severe damage may result from use on Jack pine, White pine, or Larch.
5. Use the lower rates on coarse textured soils (i.e. loamy sands, sandy loam) and higher rates on fine textured soils (i.e. sandy clay loam, silty clay loam). Do not use on gravelly or rocky soils, exposed subsoils, sand or sandy soils with more than 85% sand, or heavy clay soils (i.e. sandy clay, silty clay, or clay).
6. On tracts of land where various soil types are present and VELPAR L rate selection is difficult, conifer damage and less than expected vegetation suppression may occur.

IMPORTANT

BEFORE USING VELPAR® L, READ AND CAREFULLY OBSERVE THE CAUTIONARY STATEMENTS AND ALL OTHER INFORMATION APPEARING ON THE PRODUCT LABEL.

This bulletin contains new or supplemental instructions for use of this product which may not appear on package label. Follow the instructions carefully.

This labeling must be in the possession of the user at the time of pesticide application.

SPECIAL LOCAL NEED REGISTRATIONS [24(c)] FOR VELPAR® L HERBICIDE

Crop	State	EPA SLN No.	Remarks
Alfalfa	California Idaho Oregon Utah	CA-810084 ID-810049 OR-810099 UT-810091	Allows aerial application
Blueberries (Lowbush)	Maine Michigan N. Carolina New Hampshire	ME-830003 MI-830017 NC-380012 NH-830008	Herbaceous and woody weeds control
Rangeland	Texas New Mexico	TX-830002 NM-820023	Brush control Brush control
Reforestation Areas	Michigan Minnesota Montana	MI-820017 MN-820010 MT-820012	Site preparation Site preparation Site preparation

* Copies of individual labels available upon request to nearest Du Pont Agricultural Chemicals Office. For distribution and use only within the State specified.

VYDATE® L

INSECTICIDE/NEMATICIDE
WATER SOLUBLE LIQUID

RESTRICTED USE PESTICIDE

For retail sale to and use only by Certified Applicators or persons under their direct supervision and only for those uses covered by the Certified Applicator's certification.

1 Gallon Contains 2 lbs. Oxamyl

ACTIVE INGREDIENT:

Oxamyl

Methyl *N'N'*-dimethyl-*N*-

[(methylcarbamoyl)oxy]-1-thiooxamimidate 24%

INERT INGREDIENTS 76%

TOTAL 100%

EPA Reg. No. 352-372 U.S. Pats. 3,530,220 & 3,658,870

DANGER–POISON

Keep Out of Reach of Children

PELIGRO

PRECAUCION AL USUARIO: Si usted no lee ingles, no use este producto hasta que la etiqueta haya sido explicado ampliamente.

Do not subject to temperatures below 32°F

STATEMENT OF PRACTICAL TREATMENT

If swallowed: Call a physician or Poison Control Center. Drink 1 or 2 glasses of water and induce vomiting by touching back of throat with finger. **Do not induce vomiting or give anything by mouth to an unconscious person.**

If inhaled: Remove from exposure and have patient lie down and keep quiet. If patient is not breathing, start artificial respiration immediately. **Never give anything by mouth to an unconscious person.**

In case of contact, wash skin with plenty of soap and water; for eyes, flush with water for 15 minutes and get medical attention; remove and wash contaminated clothing before re-use.

ATROPINE IS AN ANTIDOTE—SEEK MEDICAL ATTENTION AT ONCE IN ALL CASES OF SUSPECTED POISONING.

If warning symptoms appear (see **Warning Symptoms** below), get medical attention.

PRECAUTIONARY STATEMENTS

HAZARDS TO HUMANS AND DOMESTIC ANIMALS DANGER—POISON

CONTAINS METHANOL. MAY BE FATAL OR CAUSE BLINDNESS IF SWALLOWED. MAY BE FATAL IF ABSORBED THROUGH SKIN OR INHALED.

Do not breathe vapors or spray mist. Do not get in eyes, on skin, or on clothing. When mixing spray, loading, applying or otherwise handling, wear protective clothing, goggles, and a mask or respirator jointly approved by the Mine Safety and Health Administration [formerly the U.S. Bureau of Mines] and by the National Institute for Occupational Safety and Health. Wear clean clothes daily. Wash thoroughly after handling, and before eating or smoking. Pilot should not assist in the mixing and loading operation.

Do not apply this product in such a manner as to directly, or through drift, expose workers, other persons or animals. The area being treated must be vacated by unprotected persons.

WARNING SYMPTOMS—Oxamyl poisoning produces effects associated with anticholinesterase activity which may include: weakness, blurred vision, headache, nausea, abdominal cramps, discomfort in the chest, constriction of pupils, sweating, slow pulse, muscle tremors.

NOTE TO PHYSICIAN

TREATMENT—Atropine sulfate should be used for treatment. Administer repeated doses, 1.2 to 2.0 mg. intravenously every 10 to 30 minutes until full atropinization is achieved. Maintain atropinization until the patient recovers. Artificial respiration or oxygen may be necessary. Allow no further exposure to any cholinesterase inhibitor until recovery is assured.

Do not use 2-PAM for exposure to VYDATE L alone. However, for exposure to combinations of VYDATE L and organophosphorous insecticides, 2-PAM may be used as required to supplement the atropine sulfate treatment. Do not use morphine.

ENVIRONMENTAL HAZARDS

This product is toxic to fish, birds, and other wildlife. Do not apply directly to water. Do not apply where runoff is likely to occur. Do not apply when weather conditions favor drift from areas treated. Do not contaminate water by cleaning of equipment or disposal of wastes.

This product is toxic to bees exposed to direct application. Do not apply this product while bees are actively visiting the treatment area. Time applications to coincide with periods of minimum bee activity.

PHYSICAL AND CHEMICAL HAZARDS

Flammable. Keep away from heat, sparks, and open flame. Keep container closed. Use with adequate ventilation.

NOTICE OF WARRANTY

DuPont warrants that this product conforms to the chemical description on the label thereof and is reasonably fit for the purposes stated on such label only when used in accordance with the directions under normal use conditions. It is impossible to eliminate all risks inherently associated with the use of this product. Crop injury, ineffectiveness, or other unintended consequences may result because of such factors as weather conditions, presence of other materials, or the manner of use or application, all of which are beyond the control of DuPont. In no case shall DuPont be liable for consequential, special or indirect damages resulting from the use or handling of this product. All such risks shall be assumed by the buyer. DUPONT MAKES NO WARRANTIES OF MERCHANTABILITY OR FITNESS FOR A PARTICULAR PURPOSE NOR ANY OTHER EXPRESS OR IMPLIED WARRANTY EXCEPT AS STATED ABOVE.

STORAGE AND DISPOSAL

STORAGE

Do not subject to temperatures below 32°F. Store product in original container only, away from other pesticides, fertilizer, food or feed. Not for use or storage in or around the home.

DISPOSAL

Do not contaminate water, food or feed by storage or disposal. Pesticide wastes are acutely hazardous. Improper disposal of excess pesticide, spray mixture, or rinsate is a violation of Federal Law. If these wastes cannot be disposed of by use according to label instructions, contact your State Pesticide or Environmental Control Agency, or the Hazardous Waste representative at the nearest EPA Regional Office for guidance.

Triple rinse (or equivalent) the container. Then offer for recycling or reconditioning, or puncture and dispose of in a sanitary landfill, or, if allowed by state and local authorities, by burning. If burned, stay out of smoke.

DIRECTIONS FOR USE

It is a violation of federal law to use this product in a manner inconsistent with its labeling.

Du Pont VYDATE L Insecticide/Nematicide should be used only in accordance with recommendations on the label or in separate Du Pont recommendations available through local dealers.

Du Pont will not be responsible for losses or damages resulting from use of this product in any manner not specifically recommended by Du Pont. User assumes all risks associated with such non-recommended use.

VYDATE L is a water soluble liquid to be diluted with water for application. Use only in commercial and farm plantings; do not use in home plantings. Do not rotate crops other than apples, cantaloupe, celery, citrus, cotton, cucumber, eggplant, honeydew, ornamentals, peanuts, pears, peppers, pineapple, root crop vegetables, soybeans, summer squash, tobacco, tomatoes, watermelon and winter squash, within four months after the last application.

Do not use in Suffolk and Nassau Counties, Long Island, New York.

Do not enter treated areas without appropriate protective clothing, until sprays have dried. Because certain states may require more restrictive reentry intervals for various crops treated with this product, consult your State Department of Agriculture for further information.

Written or oral warnings must be given to workers who are expected to be in a treated area or in an area about to be treated with this product. When oral warnings are given, warnings shall be given in a language customarily understood by workers. Oral warnings must be given if there is reason to believe that written warnings cannot be understood by workers. Written (or oral) warnings must include the following information: "DANGER—POISON. Area treated with VYDATE L on (date of application). Do not enter treated area without appropriate protective clothing until sprays have dried. If warning symptoms appear, refer to Statement of Practical Treatment on front panel of VYDATE L label and seek medical attention at once."

Spray Preparation—Fill spray tank ¼ to ½ full of water. Add VYDATE L directly to tank. Mix thoroughly while adding remaining water. No further agitation is necessary.

FRUITS

APPLES

Spotted Tentiform Leafminer—Apply 1 to 2 pts. VYDATE L per 100 gals. water as follows:

1st Brood Leafminer—Adults and Larvae—Make one treatment at pink (before bloom—no petals opened).

2nd Brood Leafminer—Larvae—Make an application when there are an average of two or more larvae per leaf in the sap-feeding stage. For best results, the application should be made before the larvae enter into the tissue-feeding stage.

European Red Mite and Twospotted Spider Mite—

Established Populations of Mites—Apply 1 to 2 pts. VYDATE L per 100 gals. water, as needed.

Low Populations (Preventive Control)—Apply 1/4 to 1/2 pt. VYDATE L per 100 gals. water before mites reach damaging numbers and continue every 10 to 14 days to maintain control.

White Apple Leafhopper—Apply 1 to 2 pts. VYDATE L per 100 gals. water every 10 to 14 days as needed. Begin application when significant numbers of leafhoppers are present.

Apply VYDATE L in sufficient water to obtain thorough coverage, but not in excess of 400 gals. per acre.

NOTE: Do not apply at bloom or within 30 days after bloom, or fruit thinning may occur. Do not apply to apples within 14 days of harvest. Do not graze livestock in treated orchards.

CITRUS

Citrus Rust Mite—Apply 1/4 to 1 pt. VYDATE L per 100 gals. water. Spray to run-off using up to 400 gals. of water per acre. Do not apply more than 2 qts. of product per acre.

Begin applications as soon as significant infestations are found and continue on an as-needed basis. Under light to moderate pressure, apply on 4 to 6-week intervals or as needed; whereas under moderate to heavy pressure, apply every 2 to 3 weeks, as long as the infestation continues.

DO NOT MAKE MORE THAN 6 APPLICATIONS PER YEAR

Citrus Thrips—Apply 1 to 2 qts. of VYDATE L per acre in sufficient water (100 to 500 gals. per acre) to give good coverage. Apply in the early spring before bloom when the new growth is about 3 to 4 inches long. Make additional applications as needed. Applications at petal fall may be critical to prevent fruit scarring. Applications are also recommended during mid-summer to protect new growth on young trees.

NOTE: Do not apply to citrus within 7 days of harvest. Do not graze livestock in treated orchards.

PEARS—Use not registered in California

European Red Mite, McDaniel Mite, Twospotted Spider Mite, Pear Rust Mite—Apply 3 to 4 qts. VYDATE L per acre in 100 to 600 gals. of water by ground. For best results use a dilute application.

Make application when mites first appear and repeat as needed to maintain control. Use lower rate on light infestations and higher rate on heavy infestations.

NOTE: Do not apply at bloom or within 30 days after full bloom, or fruit thinning may occur. Do not apply within 14 days of harvest. This product has been tested on Bartlett and d'Anjou varieties of pears without russeting. Its use on other varieties should be on a small scale until the possibility of russeting has been evaluated. Do not graze livestock in treated orchards.

PINEAPPLE—Use not registered in California

Reniform And Root Knot Nematodes—

Preplant or Planting Treatment—Apply 2 to 4 gals. VYDATE L per acre broadcast within 1 week prior to planting and incorporate 4 to 6 inches into the soil. Alternatively, apply 1 to 4 gals. VYDATE L per acre broadcast or apply 1/2 to 4 gals. per acre directly to the soil via drip irrigation injection, within 1 week after planting.

Subsequent Treatments—Apply 1/2 to 2 gals. VYDATE L per acre by ground equipment as a foliar spray at 2 to 4 week intervals in sufficient water to obtain thorough coverage. Begin applications when pineapple roots begin to grow following planting. Alternatively, apply 1/4 to 2 gals. VYDATE L per acre directly to the soil via drip irrigation injection at 2, 4, or 8 week intervals. For best results, subsequent applications should be made after preplant or planting application of VYDATE L or a standard fumigant. Best results occur under optimum soil moisture conditions.

NOTE: Do not apply more than 24 gals. VYDATE L per acre on the plant crop or more than 12 gals. per acre on the ratoon crop. Do not apply within 30 days of harvest.

VEGETABLES

CARROTS—Use not registered in California

Root Knot (Except Javanese), Lesion, Sting, Spiral and Stunt Nematodes—

Preplant Treatment—Apply 2 to 4 gals. of VYDATE L per acre in a minimum of 20 gals. water per acre as a broadcast treatment of the soil. Apply within one week of planting. Thoroughly incorporate to a depth of 4 to 6 inches in the soil.

In-Furrow Treatment—Apply 1 to 2 gals. of VYDATE L per acre in a minimum of 20 gals. water per acre in the seed furrow during the planting operation.

CELERY (Florida)

Serpentine Leafminers (Except Liriomyza Trifolii)—Apply 2 to 4 pts. VYDATE L per acre as a foliar spray using sufficient water (minimum 100 gals.) to obtain uniform coverage. Treat when insect first appears and repeat at 5 to 7-day intervals or as needed. Use the low rates for light infestations, the intermediate rates for heavier infestations, and the highest rates for severe infestations.

Note: Do not apply within 14 days of harvest.

CUCUMBER, CANTALOUPE, HONEYDEW MELON, WATERMELON, SQUASH

Root Knot (Except Javanese), Lesion, Sting and Stunt Nematodes—

Preplant and Planting Treatment—Apply 1 to 2 gals. VYDATE L per acre (broadcast); for band treatment, use proportionately less. Following application and before planting, incorporate VYDATE L 2 to 4 inches into the soil. Use the low rate for light infestations.

Foliar Treatment—Apply 2 to 4 pts. VYDATE L per acre in sufficient water to obtain uniform coverage of foliage. Make the first application 2 to 4 weeks after planting and repeat 2 to 3 weeks later. Use the lower rate on light infestations. The best results follow usage of VYDATE L as a soil treatment, as described above.

Leafminers (Liriomyza spp.)—Apply 2 to 4 pts. VYDATE L per acre in sufficient water to obtain uniform coverage of foliage. Make applications when insects first appear and repeat weekly or as needed. Apply the low rate for light infestations of insects and the high rate for severe infestations.

Where leafminer infestations occur annually, initiate treatment schedule 2 to 4 weeks after planting and repeat weekly or as needed.

NOTE: Do not apply within 1 day of harvest.

EGGPLANT

Aphids, Colorado Potato Beetle, Leafminers, and Mites— Apply 1 to 2 qts. VYDATE L per acre by ground equipment as a foliar spray when insects first appear. Repeat at 1 to 3 week intervals as needed.

Nematodes (except in California)—Apply 1 gal. VYDATE L per acre as a band treatment 2 to 3 weeks after transplanting and again 4 weeks later. Two to four weeks following these soil treatments, make 2 foliar applications of 2 qts. per acre by ground equipment at 1 to 2 week intervals.

NOTE: Do not apply within 1 day of harvest when using only foliar applications or within 7 days of harvest using soil with foliar applications.

PEPPER (BELL)—Use not registered in California

Root Knot (Except Javanese), Sting, Ring, Stubby Root, and Stunt Nematodes—

Transplant Water Treatment: Supplemental Foliar Treatments — Apply 2 pts. VYDATE L per acre. Add the VYDATE L to the transplant water, and use a minimum of 200 gals. per acre. Under very high nematode populations, begin with a transplant water treatment and supplement with foliar sprays. For the best results, make the first foliar application 14 days after transplanting and continue in accordance with the foliar treatment described below.

NOTE: Do not apply as a transplant water treatment during periods of slow plant growth, such as when temperatures fall below 45°F, or crop injury may result.

Green Peach Aphid, Vegetable Leafminer, and Pepper Weevil—

Foliar Treatments—Apply 2 to 4 pts. VYDATE L per acre in sufficient water (minimum 100 gals.) to obtain uniform coverage. Make applications when insects first appear and repeat at 1 to 2-week intervals or as needed. Apply low rates for light infestations of insects. Use highest recommended rates and shorter intervals for controlling severe infestations.

NOTE: Do not apply within 7 days of harvest.

POTATOES (Except in California)

Root Knot (Except Javanese), Sting, Lesion and Ring Nematodes (Except West of the Rockies); Aphids, Colorado Potato Beetle, Flea Beetles, Potato Leafhopper, and Tarnished Plant Bug—

Preplant Treatment—For control of nematodes and early season control of the above insects, apply either one of the following:

In-Furrow Treatment—Apply 1 to 2 gals. of VYDATE L per acre in a minimum of 20 gals. water per acre in the seed furrow during the planting operation.

Broadcast Treatment—Apply 2 to 4 gals. of VYDATE L per acre in a minimum of 20 gals. of water per acre as a band or broadcast treatment of the soil within one week of planting. Thoroughly incorporate to a depth of 4 to 6 inches in the soil.

Scout fields to determine when early season control diminishes and use registered foliar insecticides including VYDATE L to retain control.

Foliar Treatments—For control of the above insects, apply 2 to 4 pts. VYDATE L per acre, except in the Midwest and Northeast where 1 to 4 pts. VYDATE L per acre may be used for Colorado potato beetle. Use sufficient water (minimum 4 gals. per acre by air) to obtain uniform coverage. Make applications when insects first appear and repeat at 5 to 7-day intervals or as needed. Apply the low rates for light infestations of insects. Use the higher rates for severe infestations.

NOTE: Do not apply within 7 days of harvest.

SWEET POTATOES—Use not registered in California

Root Knot (Except Javanese) and Spiral Nematodes—

Preplant Treatment—Apply 2 to 3 gals. of VYDATE L per acre in a minimum of 20 gals. of water per acre as a broadcast treatment of the soil; for band treatment, use proportionately less. Apply within one week of planting. Thoroughly incorporate to a depth of 4 to 6 inches in the soil.

In-Furrow Treatment—As an alternate to the above, apply 1 to 2 gals. of VYDATE L per acre in a minimum of 200 gals. of water per acre in the transplant water during the planting of slips.

NOTE: Do not apply as a transplant water treatment during periods of slow plant growth, such as when temperatures fall below 45°F, or crop injury may result.

TOMATOES

Serpentine Leafminers (Except Liriomyza Trifolii)—Apply 2 to 4 pts. VYDATE L per acre in sufficient water (minimum 100 gals.) to obtain uniform coverage. Make application when insect first appears and repeat at 5 to 7-day intervals or as needed.

Apply the low rates for light infestations of insects. Use intermediate rates for heavier infestations of insects. Use the highest recommended rates for controlling severe infestation.

NOTE: Do not apply more than 6 gals. per acre per season. Do not apply within 1 day of harvest.

FIELD CROPS

COTTON

Cotton Leafperforator—Apply 1 to 4 pts. VYDATE L per acre in sufficient water to obtain thorough coverage. Make initial applications when damaging populations begin to build, and continue on a 6 to 8 day schedule for up to 4 applications.

Boll Weevil, Cotton Fleahopper and Tarnished Plant Bug—Apply ½ to 1 pt. VYDATE L per acre in sufficient water to obtain thorough coverage. Begin applications when damaging populations appear and repeat as needed to maintain control.

NOTE: Do not apply within 21 days of harvest. Do not graze or feed treated cotton to livestock.

PEANUTS—Use not registered in California

Root Knot (Except Javanese), Sting, Ring, and Lesion Nematodes, and Thrips—

Soil Treatment—Apply 2 to 6 qts. of VYDATE L in a 7" to 12" band using a minimum of 10 gals. of water per acre. Thoroughly incorporate with a rotary tiller to a depth of 3" to 5" immediately after application. Use the higher rate and wider band for severe infestations. Peanuts should be planted within 24 hours after application. Alternatively, VYDATE L may be applied in a 7" band immediately behind planting. Use 2 to 6 qts. in a minimum of 10 gals. water per acre.

Foliar Treatment—Foliar treatment is to be used only following soil fumigation, or following preplant or at planting soil application of VYDATE L or other contact nematicides. Make 2 foliar applications of VYDATE L at 2 to 4 pts. per acre in 20 to 40 gals. of water. Apply first spray 3 weeks after emergence and the second spray 6 weeks after emergence. For best results, concentrate the spray on the row using 3 cone-type nozzles positioned over and to each side of the row. Thorough spray coverage is important.

SOYBEANS—Use not registered in California

Root Knot (Except Javanese), Lesion, Ring, Stunt, Lance, and Cyst Nematodes—

Infurrow Band Treatment—Apply 2 to 4 pts. VYDATE L in 10 to 20 gals. of water per acre (based on 36" row spacing) at planting. Spray over open drill row at the juncture where the seed is covered with soil to assure mixing of VYDATE L with the soil around the seed.

Incorporated Band Treatment—Apply 2 to 4 pts. VYDATE L in 10 to 20 gals. of water per acre to a 7" to 10" band in the drill area (based on 36" row spacing) as a preplant treatment (up to 1 week before planting), or as a treatment at planting. Incorporate the VYDATE L 2" to 4" into the soil in the seed zone.

Incorporated Broadcast Treatment—Apply 1 to 2 gals. of VYDATE L in 10 to 20 gals. of water per acre as a preplant treatment (up to 1 week before planting), or as a treatment at planting. Incorporate the VYDATE L 2" to 4" into the soil in the seed zone.

NOTE: Do not cut for hay or feed treated forage to livestock. Do not apply VYDATE L to areas with severe infestations of nematodes, such as where injury to plants is manifested by severe stunting and chlorosis, as VYDATE L will not control severe infestations.

TOBACCO

Root Knot (Except Javanese) And Lesion Nematodes and Flea Beetles—Measure VYDATE L accurately as over-application may cause stunting of tobacco. VYDATE L may be used:

• As a preplant soil incorporation treatment. • Alone as a transplant water treatment for 2 to 4 weeks of nematode and flea beetle control. • As a transplant water treatment following VYDATE L or other contact nematicides, preplant fumigants or multipurpose fumigants and standard flea beetle treatments. • As a foliar treatment to extend early season nematode and flea beetle control following the use of VYDATE L in transplant water or the soil use of VYDATE L, or other contact nematicides, preplant fumigants, or multipurpose fumigants and standard flea beetle treatments.

Soil Treatment—VYDATE L may be applied to the soil as a band treatment, or it may be broadcast, disked, and bedded. For best results, the tobacco should be transplanted within 24 hours after soil treatment; however, if the transplant water treatment is used, tobacco may be transplanted within one week after the soil treatment.

Row Treatment—Apply 1 to 3 gals. of VYDATE L in an 18 to 24-inch band in a minimum of 20 gals. of water per acre of tobacco (12,000 row feet). Use the lower rates for lower infestations, the higher rates and wider band for severe infestations. Thoroughly incorporate with a rotary tiller to a depth of 4 to 6 inches.

Broadcast and Bed Treatment—Apply a broadcast spray of 1 to 3 gals. per acre in a minimum of 40 gals. of water. Use lower rates for lower infestations and the higher rates for severe infestations. Incorporate thoroughly to a depth of 4 to 6 inches, and bed the field in such a manner that only treated soil is used to form the beds.

Transplant Water Treatment—Apply VYDATE L at the following recommended rates in the transplant water depending on the number of plants per acre. Use a minimum of 100 gals. of transplant water per acre.

Plants/Acre	Rate in Pts./Acre	Rate in Fl. Oz./Acre
9,000	2	32
7,900	$1^3/_4$	28
6,800	$1^1/_2$	24
5,600	$1^1/_4$	20

For small quantities, mix 1 tablespoonful VYDATE L per 10 gals. of water, using ½ pint of solution/plant.

Foliar Treatment—Make 2 foliar applications of 4 pts. VYDATE L per acre of tobacco. Include a spreader sticker. Apply a directed spray in 20 gals. of water 2 weeks after transplanting. Apply a second directed spray in 20 to 40 gals. of water (depending on plant size) 4 to 5 weeks after transplanting.

For best results, use 3 spray nozzles positioned over and to each side of the row so that the entire plant is covered with spray.

ORNAMENTALS, NURSERY STOCK, NONBEARING FRUIT TREES

VYDATE L is recommended for use on the following ornamentals, nursery stock and nonbearing fruit trees for control of insects, mites, and nematodes:

Apple*	Dieffenbachia Picta	Pear*
Azalea	Dracaena	Peony
Bird's Nest Fern	Gardenia	Philodendron
Boxwood	Gladiolus	Rose
Cherry*	Hoya	Sansevieria
Chrysanthemum	Leatherleaf Fern	Snapdragon
Citrus*	Maranta	Strawberry
Croton	Peach*	(Nurseries)*†

* Nonbearing trees and strawberries that will not bear fruit within 12 months after application.

† NOTE: Certain varieties of strawberries are sensitive to VYDATE L, exhibiting severe phytotoxic symptoms. Known sensitive varieties that should not be treated with VYDATE L include: Bounty, Canoga (NY 1362), Cavalier, Darrow, Dunlop, Earlidawn, Earliglow, Empire, Holiday, Marlate, Midland, Midway, Mic Mac, NY 1409, Ogallalla, Redglow, Redstar, Sequoia, Stark Red, Sunrise, and Viestar; however, other varieties may also be sensitive. Therefore, VYDATE L should be used initially only on a small scale to determine crop sensitivity prior to large scale field application.

See separate Du Pont literature for additional plants.

Insects and Mites including aphids, fungus fly, fungus gnats, Japanese beetle, leafhoppers, leafminers, mealybugs, royal palm bug, scales, thrips, white fly and mites:

Foliar Treatment—Mix 2 to 4 pts. VYDATE L per 100 gals. and apply up to 200 gals. of this mixture per acre, or apply 4 to 8 pts. VYDATE L per acre in a maximum of 600 gals. water per acre. For small areas, mix 3 to 6 tablespoonsful VYDATE L per 5 to 14 gals. water and apply to 1,000 sq. ft. Apply as needed to maintain control.

Nematodes including root knot (except Javanese), sting, lesion and burrowing nematodes:

VYDATE L may be used as a pretransplant soil treatment, as a drench or as a root, corm or bulb dip. It may also be used as a foliar treatment alone, or as supplement to the above VYDATE L or other nematicide treatments. When transplanting, plants should be set into soil within 24 hours after application.

Preplant Soil Incorporation Treatment—Apply 3 to 4 gals. VYDATE L in a minimum of 20 gals. of water per acre. Thoroughly incorporate with a rotary tiller to a depth of 4 to 8 inches, immediately after application. If applied in a band, use proportionately less material.

Soil Mix Treatment—Mix $2^3/_4$ fl. ozs. of VYDATE L in 10 gals. of water. Spray $2^1/_2$ to $3^1/_3$ gals. of the dilute mix on to 1 cubic yard of soil while tumbling in a soil mixer (use 1 to $1^1/_3$ pts. for 1 bushel of soil).

Foilar Treatment—As a foliar treatment, mix 2 to 4 pts. VYDATE L per 100 gals. of water and apply as a dilute spray, not to exceed 8 pts. per acre. Apply on a 2 to 3-week schedule for 4 applications. Apply first spray at first full leaf or when plant is in active growth phase. Do not apply to plants under water stress, or to plants not actively growing. Include a spreader sticker.

Liquid Drench—Mix 1 pt. of VYDATE L in 100 gals. of water (1 ½ tablespoons/5 gals.). Apply drenches at a rate of 4 to 8 oz. per 6-inch pot or 2 to 4 oz. per 4-inch pot.

Root, Corm, or Bulb Dip—Mix 2 to 4 pts. of VYDATE L in 100 gals. of water (3 to 6 tablespoons/5 gals.). Place plant material in solution for 1 to 30 minutes. Use the higher rate and longer soaking time for more severe nematode infestation.

Note: Do not treat apple, cherry, peach or pear planting stock by root dip.

NOTE: Use only on commercial plantings; do not use on home plantings. Since varieties of ornamentals are numerous, continually changing and may respond differently to VYDATE L, test VYDATE L on a small scale before proceeding to large-scale application. Varietal response may also vary if VYDATE L is mixed with other products.

NOTICE TO BUYER

Purchase of this material does not confer any rights under patents of countries outside of the United States.

SUPPLEMENTAL LABELING

VYDATE® L INSECTICIDE/NEMATICIDE

EPA REG. NO. 352-372

FOR USE ON ORNAMENTALS

Du Pont VYDATE®L Insecticide/Nematicide is recommended for use in field or greenhouse on certain commercial and farm plantings (not home plantings) of ornamentals, nursery stock and nonbearing fruit trees for control of certain insects, mites and nematodes.

INSECTS AND MITES CONTROLLED:

Aphids	Mealybugs
Fungus Fly	Mites
Fungus Gnats	Royal Palm Bug
Japanese Beetle	Scales
Leafhoppers	Thrips
Leafminers	White Fly
Black Vine Weevil	

HOW TO USE

Large Areas—Mix 2 to 4 pts. VYDATE L per 100 gals. and apply up to 200 gals. of the mixture per acre as a foliar spray, or apply 4 to 8 pts. VYDATE L per acre in a minimum of 600 gals. water per acre.

Small Areas—Mix 3 to 6 tablespoonfuls VYDATE L per 5 to 14 gals. water and apply to 1,000 sq. ft.

Apply as needed to maintain control.

NEMATODES CONTROLLED:

Root Knot (except Javanese)	Burrowing
Sting	Stunt
Lesion	

HOW TO USE

VYDATE L may be used as a pretransplant soil treatment, as a drench or as a root, corm or bulb dip. It may also be used as a foliar treatment alone, or as a supplement to the above VYDATE L treatments or other nematicide treatments. When transplanting, plants should be set into soil within 24 hours after application.

Use the lower rates for light infestations of nematodes and the higher rates for severe infestations.

FIELD-GROWN PLANTS

Preplant Soil Incorporation Treatment—Apply 3 to 4 gals. VYDATE L in a minimum of 20 gals. of water per acre. Thoroughly incorporate with a rotary tiller to a depth of 4 to 8 inches, immediately after application. If applied in a band, use proportionately less material.

Foliar Treatment—As a foliar treatment, mix 2 to 4 pts. VYDATE L per 100 gals. of water and apply as a dilute spray, not to exceed 8 pts. per acre. Apply on a 2 to 3-week schedule for 4 applications. Apply first spray at first full leaf, or when plant is in active growth phase. Do not apply to plants under water stress, or to plants not actively growing. Include a spreader sticker.

This treatmennt works best as a supplement to a complete nematicide program following soil fumigation or use of a contact nematicide prior to transplanting.

POTTED PLANTS

Soil Mix Treatment—Mix $2^3/_4$ fluid ounces of VYDATE L in 10 gals. of water. Spray $2^1/_2$ to $3^1/_3$ gals. of the dilute mix on to 1 cubic yard of soil while tumbling in a soil mixer (use 1 to $1^1/_3$ pts. for 1 bushel of soil).

Liquid Drench—Mix 1 pt. of VYDATE L in 100 gals. of water ($1^1/_2$ tablespoons/5 gals.). Apply drenches at a rate of 4 to 8 ounces per 6-inch pot, or 2 to 4 ounces per 4-inch pot.

Root, Corm or Bulb Dip—Mix 2 to 4 pts. of VYDATE L in 100 gals. of water (3 to 6 tablespoons/5 gals.). Place plant material in solution for 1 to 30 minutes. Use the higher rate and longer soaking time for more severe nematode infestation.

NOTE: Since varieties of ornamentals are numerous, continually changing and may respond differently to VYDATE L, test VYDATE L on a small scale before proceeding to a large-scale application. Varietal response may also vary if VYDATE L is mixed with other products. Do not use in Suffolk and Nassau Counties, Long Island, New York.

IMPORTANT

BEFORE USING VYDATE® L, READ AND CAREFULLY OBSERVE THE CAUTIONARY STATEMENTS AND ALL OTHER INFORMATION APPEARING ON THE PRODUCT LABEL.

This bulletin contains new or supplemental instructions for use of this product which may not appear on package label. Follow the instructions carefully.

This label must be in the possession of the user at the time of pesticide application.

5/7/84

ORNAMENTALS FOR WHICH VYDATE® L IS RECOMMENDED

Specific Name	Common Name
Abies lasiocarpa	Fir
Acalypha wilkesiana	Copperleaf
Acer saccharum	Sugar Maple
Aesculus splendens	Buckeye
Ageratum	Flossflower, Pussy-Foot
Aglaonema commutatum	Chinese Evergreen
Aloe spp.	Medicine Plant
Antirrhinum spp.	Snapdragon
Araucaria excelsa	Norfolk Island Pine
Ardisia spp.	Ardisia
Areca sp.	Areca Palm
Asparagus sprengeri	Asparagus Fern
Asplenium nidus	Bird's Nest Fern
Aucuba spp.	Gold Dust Tree
Begonia spp.	Rieger Begonia
Berberis thunbergii	Barberry
Billbergia distachia var Straussiana	Bromeliad
Caladium spp.	Caladium
Calathea sp.	Calathea
Calendula sp.	Calendula
Callicarpa spp.	Mulberry, Beautyberry
Calycanthus floridus	Allspice
Camellia japonica	Camellia
Cattleya sp.	Orchid
Chamaecyparis spp.	Cypress (Cedar)
Chamaedorea elegans	Neanthe Bella Palm
Chrysalidocarpus lutescens	Madagascar Palm
Chrysanthemum spp.	Chrysanthemum
Citrus spp.	Citrus
Coccoloba uvifera	Sea Grape
Coleus	Flame Nettle
Cordyline terminals	Ti Plant
Cotoneaster spp.	Firethorn
Crassula spp.	Jade Plant
Cryptbergia (Billbergia X Cryptanthus)	
Cytisus purpureus	Broom plant
Dendrobium sp.	Orchid
Deutzia scabra	Mock Orange
Dianthus spp.	Carnation
Dieffenbachia amoena	Dieffenbachia
Dieffenbachia exotica	
Dieffenbachia seguine	
Dracaena spp.	Dracaena
Euonymus fortunei	Euonymus
Euonymus japonica	Japanese Euonymus
Ficus benjamina	Weeping Fig
Ficus lyrata	Fiddleleaf Fig
Ficus pumila	Creeping Fig
Ficus retusa	Indian Laurel
Fittonia verschaffeltii	Nerve Plant
Fragaria spp.	Strawberry*
Fuchsia spp.	Fuchsia, Lady's-Eardrops
Gardenia jasminoides	Gardenia
Gazania spp.	Gazania
Gernaium spp.	Geranium
Gerbera jamesonii	Transvaal Daisy
Ginkgo biloba	Ginkgo, Maidenhair Tree
Gladiolus spp.	Gladiolus
Gynura sarmentosa	Purple Passion
Gypsophila spp.	Gypsophila
Hedera helix	English Ivy
Hibiscus spp.	Hibiscus
Hoya carnosa 'Krinkle Kurl'	Hoya or Hindu Rope
Hoya carnosa 'Variegata'	Hoya
Hypericum spp.	
Hypoestes phyllostachya	Polka Dot
Hypoestes sanguinolenta	Pink Polka Dot
Ilex cornuta	Chinese Holly
Ilex crenata	Japanese Holly
Impatiens sultanii	Impatiens, Patience Plant
Iris spp.	Iris
Ixora coccinea	Red Ixora
Jasminum volubile	Wax Jasmine
Juniperus spp.	Juniper
Kalanchoe sp.	Kalanchoe
Lagerstroemia indica	Crape Myrtle
Lantana spp.	Lantana
Leucothoe fontanesiana	Doghobble, Switch Ivy
Limonium latifolium	Statice
Lobularia spp.	Lantana
Magnolia virginiana	Magnolia
Malus spp.	Apple*
Maranta kerchoviana	Prayer Plant
Mikania ternata	Plush Vine
Nephrolepis exaltata "Bostoniensis"	Boston Fern
Nerium oleander	Oleander
Osmanthus heterphyllus	Chinese Holly
Pachysandra terminalis	Japanese Pachysandra

Pachystachys spp.	
Pellionia pulchra	Satin Pellionia
Peperomia orba	Astrid Peperomia
Peperomia viridis	
Petunia spp.	Petunia
Philodendron oxycardium	Heart Leaf Philodendron
Philodendron sellium	Self-heading Philodendron
Picea spp.	Spruce
Pieris floribunda	Fetterbush
Pilea depressa	
Pilea involucrata	Frienship Plant
Pilea microphylla	Artillery Plant
Pinus spp.	Pine
Pittosporum tobira	Japanese Pittosporum
Podocarpus macrophylla	Podocarpus
Portulaca grandiflora	Rose Moss
Prunus spp.	Plum*
Pseudotsuga menziesii	Douglas Fir
Pyracantha coccinea	Pyracantha
Pyrus spp.	Pear*
Rhamnus frangla	Tallhedge
Rhododendron spp.	Azalea
Rosa multiflora 'Japonica'	Rose
Roystonea spp.	Royal Palm
Rumohra adientiformis	Leatherleaf Fern
Saintpaulia ionantha	African Violet
Salvia splendens	St. John's fire, Salvia
Sansevieria spp.	Snake Plant
Saxifraga sarmentosa	Strawberry Geranium
Schlumbergera bridgesii	Christmas Cactus
Scindapsus aureus	Devil's Ivy, Pothos
Spiraea spp.	Spiraea
Synogonium podophyllum variegatum	Nephthytis, Arrowhead Plant
Tagetes spp.	Marigold
Taxus X media 'Hicksii'	Yew
Thuja occidentalis	White Cedar
Tradescantia sp.	Wandering Jew
Tsugu canadensis	Hemlock
Ulmus parvifolia	Chinese Elm
Veitchia merrillii	Christmas Palm
Verbena X hortensis	Verbena
Viburnum suspensum	Viburnum
Viburnum japonicum	Viburnum
Viola tricolor	Pansy
Vinca rosea	Periwinkle

*Nonbearing trees and strawberries that will not bear fruit within 12 months after application.

Special Local Need Registrations [24(c)] for VYDATE® L Insecticide*

Crop	State	EPA SLN No.	Remarks
Apples	Illinois	IL-800006—	Leafhopper
	New York	NY-800005—	Aphids and leafhopper
	Vermont	VT-820003—	Aerial
	Virginia	VA-800021—	Aphids and leafhopper
	W. Virginia	WV-800001—	
Bananas/plantains	Puerto Rico	PR-820007—	Nematodes and Corm borer
Carrots	Michigan	MI-820022—	Carrot weevil
Celery	California	CA-780215—	Leafminer/aerial
	Florida	FL-760019—	Shorter PHI
	Florida	FL-770033—	Aerial
	Michigan	MI-770010—	Nematodes and carrot weevil
	Ohio	OH-780010—	Nematodes
	Pennsylvania	PA-800039—	Nematodes

Special Local Need Registrations [24(c)] for VYDATE® L Insecticide*

Crop	State	EPA SLN No.	Remarks
Citrus	Arizona	AZ-830014—	Nematodes/aerial
		AZ-830008—	Thrips/aerial
	California	CA-800063—	Thrips/aerial
	California	CA-820027—	Nematodes/irrigation
	Florida	FL-820082—	Fuller's rose beetle
	Texas	TX-800037—	Nematodes/aerial
Cotton	Texas	TX-820005—	Shorter PHI
	Texas	TX-840010	Application in oil
Cucurbits	Florida	FL-810042—	Aphids
	Indiana	IN-820004—	Ring nematodes
Ginger root	Hawaii	HI-820001—	Nematodes
Mint	Oregon	OR-840020—	Nematodes
	Washington	WA-840023—	Nematodes
Onions	Idaho	ID-830019—	Thrips
	Michigan	MI-840005—	Nematodes/thrips
	New Mexico	NM-810021—	Thrips
	Oregon	OR-830025—	Thrips
	Texas	TX-810046—	Thrips
	Washington	WA-830008—	Thrips
Onions/garlic	California	CA-820068—	Nematodes
	Oregon	OR-810033—	Nematodes
Ornamentals	California	CA-790017—	
	California	CA-790018—	Greenhouse
	Florida	FL-770043—	
	Florida	FL-770044—	Greenhouse
	Ohio	OH-780001—	Rieger begonia/nematodes
Pineapple	Hawaii	HI-810009—	Nematodes
	Puerto Rico	PR-800005—	Nematodes and mealybugs
Potatoes	New York	NY-800003—	Golden nematode
Tobacco	North Carolina	NC-810005—	Plus diazinon/fleabeetles, wireworms, and nematodes
Tomatoes	California	CA-800167—	Fleabeetles, aphids, leafminers, nematodes, drip irrigation
	Delaware	DE-810011—	Colorado potato beetle and aphids
	Florida	FL-780047—	Vegetable leafminer and aphids
	Hawaii	HI-790008—	Fleabeetles, aphids, and nematodes
	Maryland	MD-790015—	Colorado potato beetle and aphids
	New Jersey	NJ-790002—	
	Pennsylvania	PA-800024—	Colorado potato beetle and aphids
	Puerto Rico	PR-790003—	Vegetable leafminer and nematodes
	South Carolina	SC-780022—	Aerial
	Virginia	VA-800027—	Colorado potato beetle and aphids
Yams	Puerto Rico	PR-840002	Nematodes

* Copies of individual labels available upon request to nearest Du Pont Agricultural Chemicals Office. For distribution and use only within the State specified.

Elanco Products Company

A Division of Eli Lilly and Company
Lilly Corporate Center
Indianapolis, IN 46285
(317) 261-3000

BALAN® E.C.

HERBICIDE

A Selective Herbicide for the Pre-emergence Control of Annual Grasses and Broadleaf Weeds.

ACTIVE INGREDIENT:

N-butyl-N-ethyl-α,α,α-trifluoro-2,6-dinitro-p-toluidine ..	19.4%
INERT INGREDIENTS	80.6%
TOTAL	100.0%

Contains 1½ pounds active ingredient per gallon

EPA Reg. No. 1471–55

ID5023 PI 2261 AMP

BALAN—the registered trademark for Elanco Products benefin.

DANGER

Keep Out of Reach of Children

PRECAUTIONARY STATEMENTS

HAZARDS TO HUMANS AND DOMESTIC ANIMALS

Keep Out of Reach of Children.

DANGER

Corrosive. Causes eye damage. Do not get in eyes. Avoid contact with skin, eyes, or clothing. Harmful if swallowed or absorbed through the skin. Balan E.C. may cause skin sensitization reactions in certain individuals. Use eye protection or face shield and protective clothing such as coveralls, a long sleeved shirt, and impermeable gloves when handling this product.

Do not contaminate food stuffs or feeds.

First Aid: If splashed in eyes, flush thoroughly with water. In case of contact, flush with water.

ENVIRONMENTAL HAZARDS

This product is toxic to fish. Do not contaminate any body of water by direct application, cleaning of equipment or disposal of wastes.

STORAGE AND DISPOSAL

STORAGE

Avoid freezing. Store above 40°F. If frozen, poor weed control may result. Do not store near heat or open flame. Store in original container only. Tanks that have been used for phenoxy-type herbicides should not be used for Balan. In case of leak or spill, use absorbent materials to contain liquids and dispose as waste.

PESTICIDE DISPOSAL

Do not contaminate water, food, or feed by storage or disposal. Pesticide wastes are toxic. Improper disposal of excess pesticide, spray mixture, or rinsate is a violation of Federal Law. If these wastes cannot be disposed of by use according to label instructions, contact your State Pesticide or Environmental Control Agency, or the Hazardous Waste representative at the nearest EPA Regional Office for guidance.

METAL CONTAINER DISPOSAL

Triple rinse (or equivalent). Then offer for recycling, or reconditioning, or puncture and dispose of in a sanitary landfill, or by other procedures approved by state and local authorities.

PLASTIC CONTAINER DISPOSAL

Triple rinse (or equivalent). Then offer for recycling or reconditioning, or puncture and dispose of in a sanitary landfill, or, if allowed by state and local authorities, by burning. If burned, stay out of smoke.

BULK/MINI-BULK TANK CLEANING

Triple rinse (or equivalent) and wash with appropriate cleaners before reusing.

The manufacturer makes no warranties, express or implied, concerning this product or its use, which extend beyond the description on the label or in literature published by Elanco Products Company, and all statements made concerning this product apply only when used as directed. Elanco Products Company expressly disclaims all warranties, express or implied, in relation to any use of the product in combination or sequential use with any other product not specifically recommended in writing by Elanco Products Company. If used in a combination or sequential use recommended by Elanco Products Company, the liability of Elanco shall in no manner extend to any damage, loss or injury not directly caused by the inclusion of the Elanco Product in such combination or sequential use.

GENERAL USE PRECAUTIONS

Applied according to directions and under normal growing conditions, BALAN will not harm the treated crop. Over application may result in stunting, crop injury or soil residue. Uneven application or improper soil incorporation of BALAN can result in erratic weed control or crop injury. Seedling disease, cold weather, deep planting, excessive moisture, high salt concentration or drought may weaken crop seedlings and increase the possibility of damage from BALAN. Under these conditions, delayed crop development or reduced yields may result.

In the arid, irrigated areas of the Western United States—Arizona, California, Idaho, Montana, Nevada, Oregon, Utah, Washington and Wyoming—certain crops are susceptible to injury when planted in soil previously treated with BALAN. To avoid such injury, do not plant wheat, barley, rye, other grasses or onion within 10 months following a BALAN application. Milo (grain sorghum), corn, oats, spinach, red beet, sugar beet, other root crops should not be planted within 12 months following a BALAN application.

DIRECTIONS FOR USE

Read All Directions Carefully Before Applying

BALAN is a preemergence herbicide which is soil incorporated to provide season long control of many annual grasses and broadleaf weeds throughout the growing season. Incorporation of BALAN helps assure effective weed control. BALAN controls weeds as they germinate.

It is a violation of Federal Law to use this product in a manner inconsistent with its labeling.

SOIL TEXTURE GUIDE

The amount of BALAN applied will vary with the soil texture. A fine textured soil generally requires more BALAN than a coarse textured soil. BALAN is not recommended for use on muck or other soils high in organic matter content.

Refer to the following table to determine soil classification.

Soil Texture	Soil Classification
Coarse	Sand Loamy sand Sandy loam
Medium	Loam Silty clay loam* Silt loam Silt Sandy clay loam*
Fine	Clay Clay loam Silty clay loam* Silty clay Sandy clay Sandy clay loam*

* Silty clay loam and sandy clay loam soils are transitional soils and may be classified as either medium or fine textured soils. If silty clay loam or sandy clay loams are predominately sand or silt, they are usually classified as medium textured soils; if predominately clay, they are usually classified as fine textured soils.

SOIL PREPARATION

Destroy existing weeds before a BALAN application. Chop and thoroughly mix crop residues into the soil to a depth of 4 to 6 inches by deep plowing or discing before a BALAN application. Use machinery that breaks up large clods before a BALAN application.

GENERAL MIXING DIRECTIONS

Use the following instructions for mixing BALAN E.C. alone in water and tank mixed in water.

General Mixing Instructions in Water Alone:
Start with a clean sprayer. Fill the spray tank $1/3$ to $1/2$ full with clean water. Start agitation. Add the correct quantity of BALAN, continue agitation, and finish filling the tank.

General Tank Mix Instructions:
Vigorous, continuous agitation is required for all tank mixes. (Sparger pipe agitators generally provide the best agitation in spray tanks). During filling, to prevent foaming, avoid stirring or splashing air into the mixture by placing the end of the fill pipe below the surface of the water in the spray tank. Do not allow the mixture to siphon back into the water source.

Mixing order: Fill the tank $1/4$ to $1/3$ full with clean water. Start the agitation. Add dry flowables, flowables (F), liquids (L), aqueous suspensions (AS), or wettable powders (WP) to the water and agitate until the product(s) are completely dispersed in the water. Allow additional mixing and dispersion time when using dry flowable products. Continue agitation and fill tank to $3/4$ full. Add BALAN, mix thoroughly, then add any solution(S) type formulations, agitate and finish filling. Maintain agitation from filling through application. If spraying and agitation must be stopped before the tank is empty, the materials may settle to the bottom. In this case, it is important to resuspend all of the material in the bottom of the tank before continuing the spray application. Sometimes it is more difficult to resuspend settled material than it is to suspend them originally. A sparger agitator is particularly useful for this purpose. Read and carefully follow all label instructions for each material added to the tank. Premixing dry and flowable formulations with water (slurrying) and pouring the slurry through a 20 or 35 mesh wetting screen in the top of the tank will help assure good initial dispersion in the tank water. Line screens in the tank should be no finer than 50 mesh (100 mesh is finer than 50 mesh). If a buildup of material is seen on the walls of the spray tank, wash the tank with soapy water between fillings. Rinse and continue the spraying operation. Clean tank, lines, and screens thoroughly after use.

APPLICATION DIRECTIONS

Apply BALAN anytime after January 1 when the soil can be worked. Add the recommended amount of BALAN to clean water in the spray tank during the filling operation. (See General Mixing Instructions). Apply in 5 to 40 gallons of spray volume per acre (broadcast basis). Use any properly calibrated low pressure herbicide sprayer that will apply the spray uniformly. As the amount of spray volume decreases, the importance of accurate calibration and uniform application increases. Check the sprayer daily to insure proper calibration and uniform application. Apply BALAN to the soil surface and incorporate within the recommended time.

BALAN should not be applied to soils which are wet or are subject to prolonged periods of flooding as poor weed control may result.

INCORPORATION DIRECTIONS

BALAN must be incorporated one time within 8 hours after application, except in certain areas of the Western United States (see Special Note on next page). A second incorporation is required with most equipment (see under "Recommended equipment includes", for specific instructions). If BALAN is applied to a wet, warm soil surface or if the wind velocity is 10 mph or higher, variable weed control may result from delaying the first incorporation beyond 8 hours in the Eastern United States and 4 hours in the Western United States. Incorporation should place the BALAN into the top of 2 to 3 inches of the final seedbed. Place BALAN to a depth of 3 inches for more effective control of Texas panicum. Generally, incorporation equipment will mix BALAN approximately half as deep as the equipment is run. For example, a disc running 4 inches deep will incorporate BALAN into approximately the top 2 inches.

Special Note: In the Western United States, extremely high temperatures and intense sunlight may be present at the time of application. Therefore, BALAN must be incorporated into the soil within 4 hours after application to prevent loss of activity.

Incorporation Equipment
Use machinery that mixes BALAN thoroughly with the soil. Shallow incorporation with implements set to mix BALAN less than 2 inches deep in the final seedbed may result in erratic weed control. Use of incorporation equipment not listed below may result in poor or erratic weed control and/or crop injury.

Recommended equipment includes:
Disc set to cut 4 to 6 inches deep and operated in 2 different directions at 4 to 6 mph. A tandem or double disc operated one time does not provide adequate incorporation.

Power take off driven equipment (tillers, cultivators, hoes) set to incorporate BALAN in the top 2 to 3 inches of the final seedbed. P.T.O. driven equipment should be operated one time at a speed not greater than 4 mph.

Precaution: Power take off driven equipment used alone as the incorporation implement in coarse soils may cause an uneven distribution of BALAN. This occurs when the wheel track depression is below the penetrating depth of the tines on the tiller. This may result in an excessive amount of BALAN in the tractor wheel depression area and poor weed control in the diluted areas.

Where this condition exists, a tandem disc should be used on coarse soils for the incorporation of BALAN.

CULTIVATION AFTER PLANTING

Soil treated with BALAN may be shallowly cultivated or rotary hoed without reducing the weed control activity of BALAN. Do not cultivate deeper than the BALAN treated layer of soil since this may bring untreated soil to the surface and poor weed control may result.

WEEDS AND GRASSES CONTROLLED BY BALAN

BALAN will not control established weeds.

Grasses Controlled

Annual blue grass	*Poa annua*
Barnyardgrass	*Echinochloa* sp.
Watergrass	
Broadleaf signalgrass	*Brachiaria platyphylla*
(See special instructions in PEANUTS section)	
Crabgrasses	*Digitaria* spp.
Large crabgrass	
Smooth crabgrass	
Crowfootgrass	*Dactyloctenium aegyptium*
Fall panicum	*Panicum dichotomiflorum*
Spreading panicgrass	
(See special instructions in PEANUTS section)	
Foxtails	*Setaria* spp.
Bottlegrass	
Bristlegrass	
Giant foxtail	
Green foxtail	
Pigeongrass	
Robust foxtail	
Yellow foxtail	
Goosegrass	*Eleusine indica*
Silver crabgrass	
Silvergrass	
Wiregrass	
Yardgrass	
Johnsongrass	*Sorghum halepense*
Seedling only	
Junglerice	*Echinochloa colonum*

Ryegrass	*Lolium multiflorum*
Annual ryegrass	
Italian ryegrass	
Sandbur	*Cenchrus incertus*
Burgrass	
Texas panicum	*Panicum texanum*
Buffalograss	
Colorado grass	

Broadleaf Weeds Controlled

Carpetweed	*Mollugo verticillata*
Chickweed	*Stellaria media*
Florida purslane	*Richardia scabra*
Florida pusley	
Mexican clover	
Pusley	
Knotweed	*Polygonum aviculare*
Lambsquarters	*Chenopodium album*
Pigweeds	*Amaranthus* spp.
Carelessweed	
Prostrate pigweed	
Redroot	
Rough pigweed	
Spiny pigweed	
Purslane	*Portulaca oleracea*
Redmaids	*Calandrinia caulescens*

CROP RECOMMENDATIONS

These recommendations are given as the broadcast rates per acre of BALAN. For band applications, use proportionately less.

ALFALFA, BIRDSFOOT TREFOIL, CLOVER (Alsike, Ladino, Red)

Apply and incorporate BALAN before seeding at a broadcast rate per acre of 3 quarts on coarse and medium soils and 4 quarts on fine soils. Do not apply after seeding.

LETTUCE (Direct Seeded)

Apply and incorporate BALAN before seeding at a broadcast rate per acre of 3 quarts on coarse and medium soils and 4 quarts on fine soils. Do not apply after seeding.

PEANUT—BALAN ALONE:

Apply and incorporate BALAN before planting at a broadcast rate per acre of 3–4 quarts on coarse and medium soils and 4 quarts on fine soils. Use the higher rate on coarse and medium soils where the weed population may be heavy.

PEANUT—BALAN/DUAL® TANK MIX:

Dual® metolachlor, Ciba-Geigy Corporation

The BALAN/Dual tank mix controls the annual grasses and broadleaf weeds listed for BALAN alone and these additional weeds:

Foxtail millet	Southwest cupgrass
Prairie cupgrass	Witchgrass
Red rice	Yellow nutsedge

Follow recommended procedures for soil preparation, incorporation and cultivation of BALAN.

Ground application: Apply BALAN/Dual in 10 to 40 gallons of spray volume per acre.

Aerial application: Apply BALAN/Dual in a minimum of 5 gallons of spray volume per acre. BALAN/Dual may be applied and incorporated up to 14 days prior to planting.

Broadcast Rates Per Acre:

Soil Texture	BALAN	Dual 8E: Less than 3% O.M.	Dual 8E: 3% O.M. or Greater
	(quarts)	(pints)	(pints)
Coarse	3–4	1½–2	2
Medium	3–4	2–2½	2–2½
Fine	4	2–2½	2½–3

Note: Follow all cautions and precautions on the Dual label.

PEANUT—BALAN/VERNAM® TANK MIX:

Vernam®—vernolate, Stauffer Chemical Company

BALAN/Vernam controls the annual grasses and broadleaf weeds listed for BALAN alone and this additional weed:
Nutsedge (Nutgrass)

Follow recommended procedures for soil preparation and cultivation of BALAN. Apply BALAN/Vernam in 10 to 40 gallons of spray volume per acre. BALAN/Vernam can be applied up to 10 days prior to planting. Incorporate *immediately* to prevent loss of the Vernam. Follow suggested incorporation procedures for BALAN.

Broadcast Rates Per Acre:

Soil Texture	BALAN	Vernam 7E
	(quarts)	(pints)
Coarse	3–4	2⅓
Medium	3–4	2⅓
Fine	4	3

Note: Follow all cautions and precautions on the Vernam label.

PEANUT—FALL PANICUM AND BROADLEAF SIGNALGRASS CONTROL:

Fall panicum control—BALAN Alone: In the states of Alabama, Florida, Georgia, North Carolina, South Carolina and Virginia, apply BALAN at a broadcast rate of 4 quarts per acre on both coarse and medium soils.

Fall panicum control—BALAN/Vernam: For additional weed control, apply a broadcast rate of 4 quarts of BALAN and 2⅓ pints of Vernam on coarse soils and 4 quarts of BALAN and 3 pints of Vernam on medium soils. Follow incorporation directions for BALAN/Vernam.

Broadleaf signalgrass control: Apply BALAN at a broadcast rate of 4 quarts per acre on both coarse and medium soils.

TOBACCO—(TRANSPLANT BURLEY AND DARK TOBACCO)

Apply and incorporate BALAN before transplanting at a broadcast rate per acre of 3 quarts on coarse and medium soils and 4 quarts on fine soils. Set plants so that roots extend below the BALAN treated layer of soil.

BALAN is usually compatible with Ridomil®. To assure the compatibility of BALAN with Ridomil, pour the products into a small container of water in the correct proportions. After thorough mixing, let stand for five minutes. If the combination remains mixed or can be remixed readily, the mixture is compatible.

Ridomil®—metalaxyl, Ciba Geigy Corporation

Note: Read the Ridomil label carefully before using.

APPLICATION WITH DRY BULK FERTILIZER

Dry bulk fertilizer may be impregnated or coated with BALAN. Application of dry bulk fertilizer impregnated with BALAN has provided weed and grass control equal to the same rates of BALAN applied in liquid carriers.

Follow BALAN label recommendations regarding rates per acre, crops, incorporation directions, special instructions, cautions and special precautions.

Individual state regulations relating to dry bulk fertilizer blending, registration, labeling and application are the responsibility of the individual and/or company selling the fertilizer and chemical mixture.

LIMITATIONS

Apply a minimum of 200 pounds per acre of dry fertilizer impregnated with BALAN at the recommended rate. Most dry fertilizers can be used for BALAN impregnation. When coated ammonium nitrate and limestone are used alone, do not impregnate with BALAN. These materials will not absorb the herbicide. Blends containing mixtures of these materials can be impregnated.

IMPREGNATION

Use any closed drum, belt, ribbon or other commonly used dry bulk fertilizer blender. Nozzles used to spray the BALAN onto the fertilizer should be placed to provide uniform spray coverage. If less than 6 pints of BALAN are mixed per ton of fertilizer, add water to the BALAN to give a total volume of at least 6 pints per ton.

RATES

Check the crop section to determine the rate per acre of BALAN. Calculate the amount of BALAN to be impregnated on a ton of dry bulk fertilizer based on the amount of fertilizer which will be applied per acre.

$$\text{Quarts of BALAN Per Acre} \times \frac{2000}{\text{Lbs. Fertilizer}} = \text{Quarts of BALAN Per Ton of Fertilizer}$$

APPLICATION

Spread the fertilizer/chemical mixture normally with a properly calibrated applicator. Be certain the material is applied uniformly to the soil surface.

INCORPORATION

Follow recommended incorporation procedures for BALAN.

APPLICATION WITH LIQUID FERTILIZERS

BALAN may be mixed with most liquid fertilizer materials. Combination of BALAN with solutions and suspension type fertilizers has provided weed and grass control equal to the same rates of BALAN applied in water. Follow BALAN label recommendations regarding rates per acre, crops, incorporation directions, special instructions, cautions and special precautions.

Individual state regulations relating to liquid fertilizer mixing, registration, labeling and application are the responsibility of the individual and/or company selling the fertilizer and chemical mixture.

LIQUID FERTILIZER MIXING INSTRUCTIONS

BALAN E.C. in Liquid Fertilizer: Emulsifiable concentrates, such as BALAN E.C., can be mixed with liquid fertilizers. In all cases, continuous agitation is required to prevent the BALAN from rising to the surface as an oily layer. When necessary, (see Liquid Fertilizer Compatibility Test on page 17), a compatibility agent can be used to cause the BALAN to emulsify properly (i.e., have a milky appearance rather than oily layer). The use of compatibility agents is especially important when tank mixing emulsifiable concentrates (EC) with dry flowables, wettable powders (WP), flowables (F), liquids (L), aqueous suspensions (AS), or solutions (S) in liquid fertilizer. If the emulsion is not properly formed, and the E.C. rises to the surface of the fertilizer as an oil ("oils out"), the oil may combine with the wettable powder, flowable, or suspension to form oily curds (viscous phase) which is difficult to redisperse. Any one of the compatibility agents listed below is helpful in causing emulsifiable concentrates to form non-oiling mixtures with liquid fertilizers. These compatibility agents can be used at rates as low as 1½ to 2 pints per ton of liquid fertilizer and should be mixed well with the fertilizer before adding the emulsifiable concentrate. Read the label on the compatibility agent and follow the directions.

1. Sponto 168D (Witco Chemicals Co., Chicago, IL)
2. Compat (Farm Chemicals, Inc., Aberdeen, NC)
3. Unite (Hopkins Ag Chemical, Madison, WI)
4. T-Mulz 734-2 (Thompson-Hayward Chemical Co.)
5. Rigo Compatibility Agent (Rigo Company, Buckner, KY)
6. Amoco Spray Mate™ (Amoco Oil Co., Chicago, IL)
7. Kem-Link (Universal Coop, Minneapolis, MN)

Each of the above is a phosphate ester type surfactant designed to be used with liquid fertilizers. They usually do not work well as compatibility agents in tank mixtures in plain water.

TESTING FOR TANK MIX COMPATIBILITY IN LIQUID FERTILIZERS

Emulsifiable concentrates (EC) alone or in tank mixture with dry flowables, wettable powders (WP), liquids (L), flowables (F), aqueous suspensions (AS), or solutions (S) may not combine properly with some fluid fertilizer materials. Small quantities should always be tested before full-scale mixing. This will determine whether a compatibility agent is needed, and which agent does the best job. The seven agents listed above have been thoroughly tested. There are many other surfactants on the market which were not designed for use with liquid fertilizers. Use the following test to select the correct agent for your mixture.

1. Put 1 pint of the liquid fertilizer in a quart jar.
2. Add 1 to 4 teaspoons of the dry flowable, WP, L, F or AS formulation (depending on the recommended rate per acre) to the liquid fertilizer. Close jar and agitate until dispersed evenly in the fertilizer. If the materials do not disperse well, it may be necessary to slurry the chemicals in water before adding to the fertilizer.
3. After dispersing the materials (Step 2), add 3 to 4 teaspoons of the BALAN to the jar and shake well. Add solution herbicides to the mixture last and agitate. Observe the jar for about 10 minutes. If the materials rise to the surface and form a thick layer (oily curds) which will not redisperse when agitated, a compatibility agent is needed. If the mixture is easily redispersed to its original state with slight agitation, no agent is needed but good agitation must be provided in the fertilizer spray tank.
4. If the need for a compatibility agent is shown in Step 3: using a clean quart jar, start at Step 1 above, add ½ teaspoon of the compatibility agent to the liquid fertilizer, mix well, then repeat Steps 2 and 3.

An effective compatibility agent will cause the mixture to remain uniformly mixed with little or no separating or oil rising to the surface for one half hour or longer. If slight separation does occur, 2 to 3 inversions of the jar should give a uniform remix. If oily curds form which will not redisperse, more agent or another agent should be tried.

Use a clean jar for each test. The compatible mixture will have a uniform appearance and will be relatively easy to keep mixed with gentle agitation of the jar.

APPLICATION

Spread the fertilizer/chemical mixture normally with a properly calibrated applicator. Be certain the material is applied uniformly to the soil surface.

INCORPORATION

Follow recommended incorporation procedures for BALAN.

BALAN® E.C. PERFORMANCE GUARANTEE

A LIMITED WARRANTY OF BALAN WEED AND GRASS CONTROL

Elanco Products Company warrants that BALAN E.C. will control the weeds and grasses as indicated on its current label when used according to label directions and subject to the following limitations:

1. Purchaser must notify Elanco promptly if a lack of satisfactory control occurs. Such notice must be given within ninety (90) days after the first planting or application of BALAN E.C. to the treated crop, whichever occurs later.
2. An Elanco representative must be satified that Purchaser used BALAN E.C. according to the label directions and must visually observe unsatisfactory control of weeds in the field.
3. The Purchaser must provide an Elanco representative with an invoice or other satisfactory record which shows the price, quantity of BALAN E.C., and date purchased.
4. Elanco's liability will be limited to a refund of the purchase price of the BALAN E.C. applied to the acreage on which weed control was not satisfactory.
5. Void outside the U.S.A.

DISCLAIMER OF WARRANTIES

The warranty printed above is the only warranty applicable to this product. All other warranties, express or implied, including but not limited to, the implied warranties of merchantability and fitness for a particular purpose are disclaimed.

LIMITATION OF DAMAGES

Elanco's liability, whether in contract, warranty, tort, negligence, strict liability or otherwise shall not exceed the return of the amount of the purchase price of the BALAN E.C. and under no circumstances shall Elanco be liable for special, indirect or consequential damages.

INHERENT RISKS OF USE

Failure to carefully follow the directions for use of BALAN E.C. may result in unsatisfactory weed control or crop injury. Factors such as plant disease, temperature extremes, deep planting, excessive moisture, high salt concentration, drought, or immediate intense irrigation may also result in unsatisfactory weed control or crop injury.

BALAN® 2.5G

HERBICIDE

A Selective Herbicide for the Pre-emergence Control of Annual Grasses and Broadleaf Weeds in Peanuts and Tobacco

ACTIVE INGREDIENT:

N-butyl-N-ethyl-α,α,α-trifluoro-2,6-dinitro-*p*-toluidine ..	2.5%
INERT INGREDIENTS	97.5%
TOTAL	100.0%

Contains $1\frac{1}{4}$ pounds active ingredient per 50 pound bag.

EPA Reg. No. 1471-92.

ID5028

BALAN®—the registered trademark for Elanco Products benefin.

The manufacturer makes no warranties, express or implied, concerning this product or its use, which extend beyond the description of the label. All statements made concerning this product apply only when used as directed.

CAUTION
Keep Out of Reach of Children

PRECAUTIONARY STATEMENTS

HAZARDS TO HUMANS AND DOMESTIC ANIMALS

Keep out of reach of children. Harmful if swallowed or absorbed through the skin. Avoid contact with skin, eyes or clothing. In case of contact, flush with water. Do not contaminate foodstuffs or feeds.

ENVIRONMENTAL HAZARDS

This product is toxic to fish. Do not contaminate any body of water by direct application, cleaning of equipment or disposal of wastes.

STORAGE AND DISPOSAL

STORAGE

Store in original container only. In case of leak or spill, contain material and dispose as waste.

PESTICIDE DISPOSAL

Do not contaminate water, food, or feed by storage or disposal. Wastes resulting from the use of this product may be disposed of on site or at an approved waste disposal facility.

CONTAINER DISPOSAL

Completely empty bag into application equipment. Then dispose of empty bag in a sanitary landfill or by incineration, or, if allowed by State and local authorities, by burning. If burned, stay out of smoke.

SPECIAL PRECAUTIONS

Applied according to directions and under normal conditions, BALAN® will not harm the treated crop. Seedling disease, cold weather, deep planting, excessive moisture, high salt concentration or drought may weaken crop seedlings and increase the possibility of damage from BALAN. Under these conditions, delayed crop development or reduced yields may result.

DIRECTIONS FOR USE

BALAN is a pre-emergence herbicide which is mixed (incorporated) into the soil to provide long-lasting control of a wide range of annual grasses and broadleaf weeds. Incorporation of BALAN helps assure effective weed control regardless of weather conditions. BALAN controls weeds as they germinate but will not control established weeds.

It is a violation of Federal Law to use this product in a manner inconsistent with its labeling.

WEEDS AND GRASSES CONTROLLED

BALAN will not control established weeds.

Grasses Controlled

Annual bluegrass	*Poa annua*
Barnyardgrass (Watergrass)	*Echinochloa* sp.
Crabgrasses (Large crabgrass) (Smooth crabgrass)	*Digitaria* spp.
Crowfootgrass	*Dactyloctenium aegyptium*
Foxtails (Bottlegrass) (Bristlegrass) (Giant foxtail) (Green foxtail) (Pigeongrass) (Robust foxtail) (Yellow foxtail)	*Setaria* spp.
Goosegrass (Silver crabgrass) (Silvergrass) (Wiregrass) (Yardgrass)	*Eleusine indica*
Johnsongrass (Seedling only)	*Sorghum halepense*
Junglerice	*Echinochloa colonum*
Ryegrass (Annual ryegrass) (Italian ryegrass)	*Lolium multiflorum*
Sanbur (Burgrass)	*Cenchrus incertus*
Texas panicum (Buffalograss) (Colorado grass)	*Panicum texanum*

Broadleaf Weeds Controlled

Carpetweed	*Mollugo verticillata*
Chickweed	*Stellaria media*
Florida purslane (Florida pusley) (Mexican clover) (Pusley)	*Richardia scabra*
Knotweed	*Polygonum aviculare*
Lambsquarters	*Chenopodium album*
Pigweeds (Carelessweed) (Prostrate pigweed) (Redroot) (Rough pigweed) (Spiny pigweed)	*Amaranthus* spp.
Purslane	*Portulaca oleracea*
Redmaids	*Calandrinia menziesii*

SOIL TEXTURE

One key to getting good results with BALAN is to know your soil texture so that you can apply the correct rate. The amount of BALAN you apply to your soil will vary with the soil texture. A fine-textured soil requires more BALAN than a coarse-textured soil.

SOIL TEXTURE—Guide:

Refer to the following guide to determine your soil texture:

Coarse (light)* Soils	Sand Loamy sand Sandy loam
Medium Soils	Loam Silty clay loam** Silt loam Silt Sandy clay loam**
Fine (heavy)* Soils	Clay Clay loam Silty clay loam** Silty clay Sandy clay Sandy clay loam**

* Soil classification nomenclature for light and heavy-textured soils has

been changed to coarse and fine-textured soils at the direction of the Environmental Protection Agency. Throughout this label all previous references to light-textured soils have been changed to coarse-textured soils. All previous references to heavy-textured soils have been changed to fine-textured soils. These new descriptions relate more closely to the size of the soil particles in a given classification.

** Silty clay loam and sandy clay loam soils are transitional soils and may be classified as either medium or fine-textured soils. If silty clay loam or sandy clay loam soils are predominately sand or silt, they are usually classified as medium-textured soils; if predominately clay, they are usually classified as fine-textured soils.

SOIL PREPARATION

Destroy existing weeds before a BALAN application. Chop and thoroughly mix crop residues into the soil to a depth of 4 to 6 inches by deep plowing or discing before a BALAN application. Use machinery that breaks up large clods before a BALAN application.

APPLICATION DIRECTIONS

Apply BALAN with a properly calibrated herbicide or insecticide applicator that will apply the granules uniformly. Calibrate the applicator according to the equipment manufacturer's directions and check frequently during application to be certain that equipment is working properly. Avoid concentration of the material in narrow bands. Apply BALAN to the soil surface and incorporate in the same operation, if possible.

INCORPORATION DIRECTIONS

BALAN must be incorporated one time within 8 hours after application. A second incorporation is required with most equipment (see page below for specific instructions). If BALAN is applied to a wet, warm soil surface or if the wind velocity is 10 mph or higher, variable weed control may result from delaying the first incorporation beyond 8 hours.

Incorporation should place the BALAN into the top 2 to 3 inches of the final seedbed. Generally, incorporation equipment will place the chemical approximately half as deep as the equipment is run. For example, a disc running 4 inches deep will incorporate BALAN approximately 2 inches deep.

INCORPORATION EQUIPMENT

Use machinery that mixes BALAN thoroughly with the soil. Shallow incorporation with implements set to cut less than 2 inches deep may result in erratic weed control. Use of incorporation equipment not listed on the label may result in poor or erratic weed control and/or crop injury.

Recommended equipment includes:

Disc set to cut 4 to 6 inches deep and operated in 2 different directions at 4 to 6 mph. A tandem or double-disc operated one time does not provide adequate incorporation.

P.T.O.-driven equipment (tillers, cultivators, hoes) set to cut 2 to 3 inches deep with rotors spaced to provide a clean sweep of the soil and operated one time. P.T.O.-driven equipment should not be operated at a speed greater than 4 mph.

CULTIVATION AFTER PLANTING

Soil treated with BALAN may be shallow-cultivated, rotary-hoed or hand-hoed without reducing the weed control activity of BALAN. Do not cultivate deeper than the BALAN treated layer of soil since this may bring untreated soil to the surface and poor weed control may result.

CROP RECOMMENDATIONS

GENERAL

These recommendations are given as the broadcast (overall) rates of BALAN per acre. For band applications, use proportionately less. Apply BALAN anytime within 10 weeks before planting or transplanting. BALAN is not recommended for use on muck or other soils high in organic content. Do not exceed the recommended rates at any time.

PEANUTS

Apply and incorporate BALAN before planting at a broadcast rate per acre of 45 pounds on coarse and medium soils and 60 pounds on fine soils. Do not apply after planting.

TOBACCO—(Transplant burley tobacco only):

Apply and incorporate BALAN before transplanting at a broadcast rate per acre of 45 pounds on coarse and medium soils and 60 pounds on fine soils. Do not apply after transplanting. When tobacco is to be transplanted on to beds, BALAN should be applied after bedding, before transplanting and incorporated with P.T.O.-driven equipment. In all tobacco production areas, set plants in an upright position so that roots extend below the BALAN treated layer of soil.

BALAN®

DRY FLOWABLE HERBICIDE

A Selective Herbicide for the Pre-emergence Control of Annual Grasses and Broadleaf Weeds

ACTIVE INGREDIENT:	
N-butyl-*N*-ethyl-α,α,α-trifluoro-2,6 dinitro-*p* - toluidine	60%
INERT INGREDIENTS	40%
TOTAL	100%

Contains 0.6 pounds active ingredient per pound

EPA Reg. No. 1471-140

ID5529 PI 2682 AMP

* BALAN®— the registered trademark for Elanco Products benefin.

CAUTION

Keep Out of Reach of Children

PRECAUTIONARY STATEMENTS

HAZARDS TO HUMANS AND DOMESTIC ANIMALS

CAUTION

Avoid contact with skin, eyes, or clothing. Harmful if swallowed.

First Aid: In case of contact, immediately flush eyes or skin with plenty of water. If swallowed drink large quantity of water and induce vomiting by placing finger in back of throat. Never give anything by mouth to unconscious person. If accidentally swallowed, or if irritation persists, get medical attention. If inhaled, remove victim to fresh air. Give artificial respiration, and get medical attention.

ENVIRONMENTAL HAZARDS

This product is toxic to fish. Do not contaminate any body of water by direct application, cleaning of equipment or disposal of wastes.

STORAGE AND DISPOSAL

STORAGE

Store in original container only. Do not store in direct sunlight. Do not store at temperatures above 120°F. If stored above 120°F, Balan Dry Flowable could lose some of its dispersing properties, resulting in difficult application, poor weed control or crop injury. In case of leak or spill, contain material and dispose as waste.

PESTICIDE DISPOSAL

Do not contaminate water, food or feed by storage or disposal. Wastes resulting from the use of this product may be disposed of on site or at an approved waste disposal facility.

CONTAINER DISPOSAL

Triple rinse (or equivalent). Then offer for recycling or reconditioning, or puncture and dispose of in a sanitary landfill, or, if allowed by state and local authorities, by burning. If burned, stay out of smoke.

GENERAL USE PRECAUTIONS

Applied according to directions and under normal growing conditions, BALAN will not harm the treated crop. Over application may result in stunting, crop injury or soil residue. Uneven application or improper soil incorporation of BALAN can result in erratic weed control or crop injury. Seedling disease, temperature extremes, deep planting, excessive moisture, high salt concentration or drought may weaken crop seedlings and increase the possibility of damage from BALAN. Under these conditions, delayed crop development or reduced yields may result.

In the arid, irrigated areas of the Western United States—Arizona, California, Idaho, Montana, Nevada, Oregon, Utah, Washington and Wyoming—certain crops are susceptible to injury when planted in soil previously treated with BALAN. To avoid such injury, do not plant wheat, barley, rye, other grasses or onion within 10 months following a BALAN application. Milo (grain sorghum), corn, oats, spinach, red beet, sugar beet and other root crops should not be planted within 12 months following a BALAN application.

The manufacturer makes no warranties, express or implied, concerning this product or its use, which extend beyond the description on the label or in literature published by Elanco Products Company, and all statements made concerning this product apply only when used as directed. Elanco Products Company expressly disclaims all warranties, express or implied, in relation to any use of the product in combination or sequential use with any other product not specifically recommended in writing by Elanco Products Company. If used in a combination or sequential use recommended by Elanco Products Company, the liability of Elanco shall in no manner extend to any damage, loss or injury not directly caused by the inclusion of the Elanco Product in such combination or sequential use.

DIRECTIONS FOR USE

Read all direction carefully before applying.

It is a violation of Federal law to use this product in a manner inconsistent with the label.

BALAN is a preemergence herbicide which is soil incorporated to provide season long control of many annual grasses and broadleaf weeds throughout the growing season. Incorporation of BALAN helps assure effective weed control. BALAN controls weeds as they germinate.

SOIL TEXTURE GUIDE

The amount of BALAN applied will vary with the soil texture. A fine textured soil generally requires more BALAN than a coarse textured soil. BALAN is not recommended for use on muck or other soils high in organic matter content.

Refer to the following table to determine soil classification.

Soil Texture	Soil Classification
Coarse	Sand
	Loamy sand
	Sandy loam
Medium	Loam
	Silty clay loam*
	Silt loam
	Silt
	Sandy clay loam*
Fine	Clay
	Clay loam
	Silty clay loam*
	Silty clay
	Sandy clay
	Sandy clay loam*

* Silty clay loam and sandy clay loam soils are transitional soils and may be classified as either medium or fine textured soils. If silty clay loam or sandy clay loams are predominately sand or silt, they are usually classified as medium textured soils; if predominately clay, they are usually classified as fine textured soils.

SOIL PREPARATION

Destroy existing weeds before a BALAN application. Chop and thoroughly mix crop residues into the soil to a depth of four (4) to six (6) inches by deep plowing or discing before a BALAN application. Use machinery that breaks up large clods before a BALAN application.

GENERAL SPRAYER INFORMATION

Vigorous, continuous agitation is required for all dry flowables. Sparger pipe agitation generally provides the best agitation. As when using any dry flowable pesticide, the sprayer in-line strainer should be no finer than a 16 mesh (16 mesh has larger openings than 50 mesh).

1. Be sure sprayer tank is clean and not contaminated with any material, as crop injury or sprayer clogging may result.
2. Be certain that agitation system is working properly and creates a rippling or rolling action on the liquid surface.
3. Use a pump with capacity to (a) provide sufficient agitation in tank to keep mixture in suspension, and (b) to provide to the agitator bypass at all times.
4. Centrifugal pumps are preferred since they provide propeller shear action for dispersing and mixing this product.
5. Use screens to protect the pump and to prevent nozzles from clogging. Screens placed on suction side of pump should be 16 mesh or coarser. Use 50 mesh or coarser screens between the pump and boom and when required at the nozzles.
6. Empty tank as completely as possible before refilling to prevent buildup of oil or emulsifiable concentrate residue (in the case of tank mixes). If an oil or emulsifiable concentrate film starts to build up in the tank, drain and clean with strong detergent solution.
7. It is good practice to clean sprayer tank thoroughly by flushing system with water containing a detergent, then refill with clean water.

GENERAL MIXING DIRECTIONS

Use the following instructions for mixing BALAN Dry Flowable alone in water and tank mixed in water.

General Mixing Instructions in Water Alone:
Start with a clean sprayer. Fill the spray tank three-fourths ($^3/_4$) full with clean water. Start agitation and increase pump speed to operating pressure. Add the correct quantity of BALAN Dry Flowable from the original package. Do not premeasure BALAN Dry Flowable into a larger container or enlarge the container opening. Continue agitation and finish filling the tank. Maintain agitation from mixing through application. Induction systems on custom application equipment may be used to disperse BALAN Dry Flowable in water.

When the sprayer tank opening is directly above the outlet port, fill the spray tank three-fourths ($^3/_4$) full with clean water. Do not start the sprayer or agitation. Add the correct quantity of BALAN Dry Flowable from the original package and allow the BALAN Dry Flowable particles to wet as they settle to the bottom of the tank. Finish filling the tank. Start and maintain agitation until the tank is empty. If the in-line strainer becomes clogged, remove the screen and rinse in the spray tank. Replace the screen, restart the sprayer and continue agitation.

General Tank Mix Instructions:
Vigorous, continuous agitation is required for all tank mixes. Sparger pipe agitators generally provide the best agitation in spray tanks. During filling, to prevent foaming, avoid stirring or splashing air into the mixture by placing the end of the fill pipe below the surface of the water in the spray tank. Do not allow the mixture to siphon back into the water source.

Mixing order: Fill the tank three-fourths ($^3/_4$) full with clean water. Start agitation and increase pump speed to operating pressure. Add BALAN Dry Flowable, then add other dry flowables, wettable powders (WP), aqueous suspensions (AS), flowables (F), liquids (L), solutions (S), and emulsifiable or liquid concentrates (EC) in this order to the water and agitate continuously until the product(s) are completely dispersed in the water. Maintain agitation from filling through application. If spraying and agitation must be stopped before the tank is empty, the materials may settle to the bottom. In this case, it is important to suspend again all of the material in the bottom of the tank before continuing the spray application. Sometimes it is more difficult to resuspend settled material than it is to suspend it originally. A sparger agitator is particularly useful for this purpose.

Read and carefully follow all label instructions for each material added to the tank. Premixing dry and flowable formulations with water (slurrying) and pouring the slurry through a 20 or 35 mesh wetting screen in the top of the tank will help assure good initial dispersion in the tank water. In-line screens in the tank should be no finer than 16 mesh (16 mesh has bigger openings than 50 mesh).

If a buildup of material is seen on the walls of the spray tank, wash the tank with soapy water between fillings. Rinse and continue the spraying operation. Clean tank, lines, and screens thoroughly after use.

APPLICATION DIRECTIONS

Apply BALAN anytime after January 1 when the soil can be worked. Apply in at least five (5) gallons of spray volume per acre (broadcast basis). Use any properly calibrated low pressure herbicide sprayer that will apply the spray uniformly. As the amount of spray volume decreases, the importance of accurate calibration and uniform application increases. Check the sprayer daily to insure proper calibration and uniform application. Apply BALAN to the soil surface and incorporate within the recommended time.

BALAN should not be applied to soils which are wet or are subject to prolonged periods of flooding as poor weed control may result.

INCORPORATION DIRECTIONS

BALAN must be incorporated one time within eight (8) hours after application, except in certain areas of the Western United States (see Special Note below). A second incorporation is required with most equipment (refer to Incorporation Equipment below for specific instructions). If BALAN is applied to a wet, warm soil surface or if the wind velocity is 10 mph or higher, variable weed control may result from delaying the first incorporation beyond eight (8) hours in the Eastern United States and four (4) hours in the Western United States. Incorporation should place the BALAN into the top two (2) to three (3) inches of the final seedbed. Place BALAN to a depth of three (3) inches for more effective control of Texas panicum. Generaly, incorporation equipment will mix BALAN approximately half as deep as the equipment is run. For example, a disc running four (4) inches deep will incorporate BALAN into approximately the top two (2) inches.

SPECIAL NOTE: In the Western United States, extremely high temperatures and intense sunlight may be present at the time of application. Therefore, BALAN must be incorporated into the soil within four (4) hours after application to prevent loss of activity.

INCORPORATION EQUIPMENT

Use machinery that mixes BALAN thoroughly with the soil. Shallow incorporation with implements set to mix BALAN less than two (2) inches deep in the final seedbed may result in erratic weed control. Use of incorporation equipment not listed below may result in poor or erratic weed control and/or crop injury.

Recommended equipment includes:

Disc set to cut four (4) to six (6) inches deep and operated in two (2) different directions at four (4) to six (6) mph. A tandem or double disc operated one time does not provide adequate incorporation.

Power take off driven equipment (tillers, cultivators, hoes) set to incorporate BALAN in the top two (2) to three (3) inches of the final seedbed. P.T.O. driven equipment should be operated one time at a speed not greater than four (4) mph.

PRECAUTION: Power take off driven equipment used alone as the incorporation implement in coarse soils may cause an uneven distribution of BALAN. This occurs when the wheel track depression is below the penetrating depth of the tines on the tiller. This may result in an excessive amount of BALAN in the tractor wheel depression area and poor weed control in the diluted areas. Where this condition exists, a tandem disc should be used on coarse soils for the incorporation of BALAN.

CULTIVATION AFTER PLANTING

Soil treated with BALAN may be shallowly cultivated or rotary hoed without reducing the weed control activity of BALAN. Do not cultivate deeper than the BALAN treated layer of soil since this may bring untreated soil to the surface and poor weed control may result.

WEEDS AND GRASSES CONTROLLED BY BALAN

BALAN will not control established weeds.

Grasses Controlled	
Annual bluegrass	*Poa annua*
Barnyardgrass	*Echinochloa crusgalli*
Watergrass	
Broadleaf signalgrass	*Brachiaria platyphylla*
Crabgrasses	*Digitaria* spp.
Large crabgrass	
Smooth crabgrass	
Crowfootgrass	*Dactyloctenium aegyptium*
Fall panicum	*Panicum dichotomiflorum*
Spreading panicgrass	
Foxtails	*Setaria* spp.
Bottlegrass	
Bristlegrass	
Giant foxtail	
Green foxtail	
Pigeongrass	
Robust foxtail	
Yellow foxtail	
Goosegrass	*Eleusine indica*
Silver crabgrass	
Silvergrass	
Wiregrass	
Yardgrass	
Johnsongrass	*Sorghum halepense*
Seedling only	
Junglerice	*Echinochloa colonum*
Ryegrass	*Lolium multiflorum*
Annual ryegrass	
Italian ryegrass	
Sandbur	*Cenchrus incertus*
Burgrass	
Texas panicum	*Panicum texanum*
Buffalograss	
Colorado grass	
Broadleaf Weeds Controlled	
Carpetweed	*Mollugo verticillata*
Chickweed	*Stellaria media*
Florida purslane	*Richardia scabra*
Florida pusley	
Mexican clover	
Pusley	
Knotweed	*Polygonum aviculare*
Lambsquarters	*Chenopodium album*
Pigweeds	*Amaranthus* spp.
Carelessweed	
Prostrate pigweed	
Redroot	
Rough pigweed	
Spiny pigweed	
Purslane	*Portulaca oleracea*
Redmaids	*Calandrinia caulescens*

CROP RECOMMENDATIONS

These recommendations are given as the broadcast rates per acre of BALAN. For band application, use proportionately less.

ALFALFA, BIRDSFOOT TREFOIL, CLOVER (Alsike, Ladino, Red)

Apply and incorporate BALAN Dry Flowable before seeding at a broadcast rate per acre of 2 pounds on coarse and medium soils and 2½ pounds on fine soils. Do not apply after seeding.

LETTUCE (Direct Seeded)
Apply and incorporate BALAN Dry Flowable before seeding at a broadcast rate per acre of 2 pounds on coarse and medium soils and $2\frac{1}{2}$ pounds on fine soils. Do not apply after seeding.

LIQUID FERTILIZER MIXING INSTRUCTIONS

BALAN Dry Flowable in Liquid Fertilizer
BALAN Dry Flowable can be mixed with liquid fertilizer. In all cases, vigorous and continuous agitation is required to mix BALAN Dry Flowable with the liquid fertilizer. BALAN Dry Flowable may be applied alone in liquid fertilizer without a compatibility agent. The use of a compatibility agent is especially important when tank mixing BALAN Dry Flowable with emulsifiable concentrates (EC) and liquid concentrates (LC). See the liquid fertilizer compatibility test below. If the emulsion is not properly formed, the EC/LC may rise to the surface of the fertilizer as an oil ("oils out"), the oil may combine with the BALAN Dry Flowable, wettable powder, liquid, flowable or aqueous suspension to form oily curds (viscous phase) which are difficult to disperse.

Any one of the compatibility agents listed below is helpful in causing emulsifiable and liquid concentrates to form non-oiling mixtures with liquid fertilizers. These compatability agents can be used at rates as low as one and one-half ($1\frac{1}{2}$) to two (2) pints per ton of liquid fertilizer and should be mixed well with the fertilizer containing BALAN Dry Flowable before adding the emulsifiable concentrate. Read the label on the compatibility agent and follow the directions.

1. Sponto 168D (Witco Chemicals Co., Chicago, IL)
2. Compat (Farm Chemicals, Inc., Aberdeen, NC)
3. Unite (Hopkins Ag Chemical, Madison, WI)
4. T-Mulz 734-2 (Thompson-Hayward Chemical Co.)
5. Amoco Spray Mate+up,TM (Amoco Oil Co., Chicago, IL)

Each of the above is a phosphate ester type surfactant designed to be used with liquid fertilizers. They usually do not work well as compatibility agents in tank mixtures in plain water.

TESTING FOR TANK MIX COMPATIBILITY IN LIQUID FERTILIZERS:
Tank mixtures with BALAN Dry Flowable, other dry flowables, wettable powders (WP), aqueous suspensions (AS), flowables (F), liquids (L), solutions (S), or emulsifiable concentrates (EC) may not combine properly with some fluid fertilizer materials. Small quantities should always be tested before full-scale mixing. This will determine whether a compatibility agent is needed, and which agent does the best job. The five (5) agents listed on page 11 have been thoroughly tested. There are many other surfactants on the market which were not designed for use with liquid fertilizers. Use the following test to select the correct agent for your mixture.

1. Put one (1) pint of the liquid fertilizer in a quart jar.
2. Add one (1) to four (4) teaspoonful(s) of BALAN Dry Flowable with an equal amount of the WP, AS, F, L, S or EC formulation (depending on the recommended rate per acre) to the liquid fertilizer. Close jar and agitate until dispersed evenly in the fertilizer. If the materials do not disperse well, it may be necessary to slurry the chemicals in water before adding to the fertilizer.
3. Observe the jar for about 10 minutes. If the materials rise to the surface and form a thick layer (oily curds) which will not disperse when agitated, a compatibility agent is needed. If the mixture is easily dispersed to its original state with slight agitation, no agent is needed, but good agitation must be provided in the fertilizer spray tank.
4. If the need for a compatibility agent is shown in Step 3, using a clean quart jar, start at Step 1 above, add one-half ($\frac{1}{2}$) teaspoonful of the compatibility agent to the liquid fertilizer, mix well, then repeat Steps 2 and 3.

An effective compatibility agent will cause the mixture to remain uniformly mixed with little or no separating or oil rising to the surface for one half ($\frac{1}{2}$) hour or longer. If slight separation does occur, two (2) to three (3) inversions of the jar should give a uniform remix. If oily curds form which will not disperse, more agent or another agent should be tried.

Use a clean jar for each test. The compatible mixture will have a uniform appearance and will be relatively easy to keep mixed with gentle agitation of the jar.

APPLICATION

Spread the fertilizer/chemical mixture normally with a properly calibrated applicator. Be certain the material is applied uniformly to the soil surface.

INCORPORATION

Follow recommended incorporation procedures for BALAN.

BALAN DRY FLOWABLE PERFORMANCE GUARANTEE

A LIMITED WARRANTY OF BALAN WEED AND GRASS CONTROL

Elanco Products Company warrants that BALAN Dry Flowable will control the weeds and grasses as indicated on its current label when used according to label directions and subject to the following limitations:

1. Purchaser must notify Elanco promptly if a lack of satisfactory control occurs. Such notice must be given within ninety (90) days after the first planting or application of BALAN Dry Flowable to the treated crop, whichever occurs later.

2. An Elanco representative must be satisfied that Purchaser used BALAN Dry Flowable according to the label directions and must visually observe unsatisfactory control of weeds in the field.

3. The Purchaser must provide an Elanco representative with an invoice or other satisfactory record which shows the price, quantity of BALAN Dry Flowable, and date purchased.

4. Elanco's liability will be limited to a refund of the purchase price of the BALAN Dry Flowable applied to the acreage on which weed control was not satisfactory.

5. Void outside the U.S.A.

DISCLAIMER OF WARRANTIES

The warranty printed above is the only warranty applicable to this product. All other warranties, express or implied, including but not limited to the implied warranties of merchantability and fitness for a particular purpose are disclaimed.

LIMITATION OF DAMAGES

Elanco's liability, whether in contract, warranty, tort, negligence, strict liability or otherwise shall not exceed the return of the amount of the purchase price of the BALAN Dry Flowable and under no circumstances shall Elanco be liable for special indirect or consequential damages.

INHERENT RISKS OF USE

Failure to carefully follow the directions for use of BALAN Dry Flowable may result in unsatisfactory weed control or crop injury. Factors such as plant disease, temperature extremes, deep planting, excessive moisture, high salt concentration, drought, or immediate intense irrigation may also result in unsatisfactory weed control or crop injury.

PAARLAN® E.C.

HERBICIDE

A Selective Herbicide for the Pre-emergence Control of Annual Grasses and Broadleaf Weeds

ACTIVE INGREDIENT:

isopropalin* (2,6-dinitro-*N, N*-dipropylcumidine) 68.6%

INERT INGREDIENTS 31.4%

TOTAL 100.0%

Contains 6 pounds active ingredient per gallon

EPA Reg. No. 1471–79 EPA Est. No. 1471-IN-2

ID5030 PI 1618 AMP

* PAARLAN®—the registered trademark for Elanco Products isopropalin.

CAUTION

Keep Out of Reach of Children

PRECAUTIONARY STATEMENTS

HAZARDS TO HUMANS AND DOMESTIC ANIMALS

CAUTION

Harmful if swallowed or absorbed through the skin. Avoid contact with skin, eyes or clothing. In case of contact, flush with water. Do not contaminate foodstuffs or feeds.

First Aid—If swallowed: Call a physician or Poison Control Center. Drink 1 or 2 glasses of water and induce vomiting by touching back of throat with finger. Do not induce vomiting or give anything by mouth to an unconscious person. If on skin: Wash with plenty of soap and water. If in eyes: Flush with plenty of water. Get medical attention.

ENVIRONMENTAL HAZARDS

Direct contamination of any body of water with this emulsifiable concentrate may kill fish. Do not contaminate any body of water by direct application, cleaning of equipment or disposal of wastes.

STORAGE AND DISPOSAL

STORAGE

Do not store near heat or open flame. Store in original container only. In case of leak or spill, use absorbent materials to contain liquids and dispose as waste.

PESTICIDE DISPOSAL

Do not contaminate water, food or feed by storage or disposal. Pesticide wastes are toxic. Improper disposal of excess pesticide, spray mixture, or rinsate is a violation of Federal Law. If these wastes cannot be disposed of by use according to label instructions, contact your State Pesticide or Environmental Control Agency, or the Hazardous Waste representative at the nearest EPA Regional Office for guidance.

CONTAINER DISPOSAL

Triple rinse (or equivalent). Then offer for recycling or reconditioning, or puncture and dispose of in a sanitary landfill, or by other procedures approved by state and local authorities.

GENERAL USE PRECAUTIONS

Applied according to directions and under normal growing conditions, PAARLAN will not harm the treated crop. Over application, uneven application or improper soil incorporation of PAARLAN can result in erratic weed control, crop injury, delayed emergence or soil residue. Seedling disease, cold weather, deep planting, excessive moisture, high salt concentration or drought may weaken crop seedlings or transplants and increase the possibility of damage from PAARLAN. Under these conditions, delayed crop development or reduced yields may result.

The manufacturer makes no warranties, express or implied, concerning this product or its use, which extend beyond the description on the label or in literature published by Elanco Products Company, and all statements made concerning this product apply only when used as directed. Elanco Products Company expressly disclaims all warranties, express or implied, in relation to any use of the product in combination or sequential use with any other product not specifically recommended in writing by Elanco Products Company. If used in a combination or sequential use recommended by Elanco Products Company, the liability of Elanco shall in no manner extend to any damage, loss or injury not directly caused by the inclusion of the Elanco product in such combination or sequential use.

DIRECTIONS FOR USE

Read All Directions Carefully Before Applying

It is a violation of Federal law to use this product in a manner inconsistent with its labeling.

PAARLAN is a preemergence herbicide which is soil incorporated to provide season long control of many annual grasses and broadleaf weeds. Incorporation of Paarlan helps assure effective weed control. PAARLAN controls weeds as they germinate.

SOIL TEXTURE GUIDE

The amount of PAARLAN to be applied will vary with the soil texture. A fine textured soil generally requires more PAARLAN than a coarse textured soil. PAARLAN is not recommended for use on muck or other soils containing more than 10% organic matter.

Refer to the following table to determine soil texture grouping.

Soil Texture	Soil Classification
Coarse	Sand
	Loamy sand
	Sandy loam
Medium	Loam
	Silty clay loam
	Silt loam
	Silt
	Sandy clay loam*
Fine	Clay
	Clay loam
	Silty clay loam*
	Silty clay
	Sandy clay
	Sandy clay loam*

* Silty clay loam and sandy clay loam soils are transitional soils and may be classified as either medium or fine textured soils. If silty clay loam or sandy clay loams are predominately sand or silt, they are usually classified as medium textured soils; if predominately clay, they are usually classified as fine textured soils.

SOIL PREPARATION

Destroy existing weeds before a PAARLAN application. Chop and thoroughly mix crop residues into the soil to a depth of 4 to 6 inches by deep plowing or discing before a PAARLAN application. Use machinery that breaks up large clods before a PAARLAN application.

GENERAL MIXING DIRECTIONS

Use the following instructions for mixing PAARLAN E.C. alone in water and tank mixed in water.

General Mixing Instructions in Water Alone:

Start with a clean sprayer. Fill the spray tank ⅓ to ½ full with clean water. Start agitation. Add the correct quantity of PAARLAN E.C., continue agitation, and finish filling the tank.

General Tank Mix Instructions:

Vigorous, continuous agitation is required for all tank mixes. (Sparger pipe agitators generally provide the best agitation in spray tanks). During filling, to prevent foaming, avoid stirring or splashing air into the mixture by placing the end of the fill pipe below the surface of the water in the spray tank. Do not allow the mixture to siphon back into the water source.

Mixing order: Fill the tank ¼ to ⅓ full with clean water. Start the agitation. Add dry flowables, wettable powders (WP), aqueous suspensions (AS), flowables (F), and liquids (L) to the water and agitate until the product(s) are completely dispersed in the water. Allow additional mixing and dispersion time when using dry flowable products. Continue agitation and fill tank to ¾ full. Add the PAARLAN E.C., mix thoroughly, then add any solution (S) products, agitate and finish filling. Maintain agitation during filling through application. If spraying and agitation must be stopped before the tank is empty, the materials may settle to the bottom. In this case, it is important to resuspend all of the material in the bottom of the tank before continuing the spray application. Sometimes it is more difficult to resuspend settled material than it is to suspend them originally. A sparger agitator is particularly useful for this purpose.

Read and carefully follow all label instructions for each material added to the tank. Premixing dry and flowable formulations with water (slurrying) and pouring the slurry through a 20 or 35 mesh wetting screen in the top of the tank will help assure good initial dispersion in the tank water. Line screens in the tank should be no finer than 50 mesh (100 mesh is finer than 50 mesh).

If a buildup of material is seen on the walls of the spray tank, wash the tank with soapy water between fillings. Rinse and continue the spraying operation. Clean tank, lines, and screens thoroughly after use.

APPLICATION DIRECTIONS

Add the recommended amount of PAARLAN to clean water in the spray tank during the filling operation. (See General Mixing Instructions). Apply in 20 to 40 gallons of spray volume per acre (broadcast basis). Use any properly calibrated low pressure herbicide sprayer that will apply the spray uniformly. As the amount of spray volume decreases, the importance of accurate calibration and uniform application increases. Check the sprayer daily to insure proper calibration and uniform application. Apply PAARLAN to the soil surface and incorporate within the recommended time.

PAARLAN should not be applied to soils which are wet or are subject to prolonged periods of flooding as poor weed control may result.

INCORPORATION DIRECTIONS

Before planting, PAARLAN must be incorporated one time within 8 hours after application. A second incorporation is required with most equipment (see INCORPORATION EQUIPMENT, below, for specific instructions). If PAARLAN is applied to a wet, warm soil surface or if the wind velocity is 10 mph or higher, variable weed control may result from delaying the first incorporation beyond 8 hours. Incorporation should place the PAARLAN into the top 2 to 3 inches of the final seedbed. Generally, incorporation equipment will mix PAARLAN approximately half as deep as the equipment is run. For example, a disc running 4 inches deep will incorporate PAARLAN into approximately the top 2 inches.

BEDDED CULTURE

For effective weed control, PAARLAN should be incorporated into the top 2 to 3 inches of the final seedbed. Knock off beds to planting height before application of PAARLAN and incorporation of bedded ground. Avoid removal of treated soil from the seedbed before or during the planting operation.

INCORPORATION EQUIPMENT

Use machinery that mixes PAARLAN thoroughly with the soil. Shallow incorporation with implements set to mix PAARLAN less than 2 inches deep in the final seedbed may result in erratic weed control. Use of incorporation equipment not listed below may result in poor or erratic weed control and/or crop injury.

Recommended equipment includes:

Disc set to cut 4 to 6 inches deep and operated in 2 different directions at 4 to 6 mph. A tandem or double disc operated one time does not provide adequate incorporation.

Power take off driven equipment (tillers, cultivators, hoes) set to incorporate PAARLAN in the top 2 to 3 inches of the final seedbed. P.T.O. driven equipment should be operated one time at a speed not greater than 4 mph.

CULTIVATION AFTER PLANTING

Soil treated with PAARLAN may be shallowly cultivated or rotary hoed without reducing the weed control activity of PAARLAN. Do not cultivate deeper than the PAARLAN treated layer of soil. This may bring untreated soil to the surface and poor weed control may result.

In areas of bedded culture, weeds may germinate in the bottom of the furrow due to the removal of PAARLAN treated soil. These weeds should be controlled by cultivation. Shallow cultivation, rotary hoeing and hand hoeing do not reduce the weed control activity of PAARLAN.

WEEDS AND GRASSES CONTROLLED

PAARLAN will not control established weeds.

Grasses Controlled	
Barnyardgrass	*Echinochloa* sp.
(Watergrass)	
Crabgrass	*Digitaria* spp.
(Large crabgrass)	
(Small crabgrass)	
(Smooth crabgrass)	
Crowfootgrass	*Dactyloctenium aegyptium*
Foxtail	*Setaria* spp.
(Giant foxtail)	
(Green foxtail)	
(Yellow foxtail)	
(Bottlegrass)	
(Robust white)	
(Giant robust green)	
Goosegrass	*Eleusine indica*
(Silver crabgrass)	
(Silvergrass)	
(Wiregrass)	
(Yardgrass)	
Johnsongrass	*Sorghum halepense*
(seedling only)	
Ryegrass, annual	*Lolium multiflorum*
(Italian ryegrass)	
(Ryegrass)	
Field sandbur	*Cenchrus incertus*
Fall panicum	*Panicum dichotomiflorum*
Texas panicum	*Panicum texanum*
(Buffalograss)	
(Colorado grass)	
Broadleaf Weeds Controlled	
Carpetweed	*Mollugo verticillata*
Florida pusley	*Richardia scabra*
(Florida purslane)	
(Mexican clover)	
(Pusley)	
Common lambsquarters	*Chenopodium album*
Common purslane	*Portulaca oleracea*
Pigweed	*Amaranthus* spp.
Redroot pigweed	
(Carelessweed)	
(Rough pigweed)	
Prostrate pigweed	
Spiny amaranth	
Poorjoe	*Diodia teres*

These recommendations are given as broadcast rates for PAARLAN. For band application, use proportionately less. Apply PAARLAN up to 5 weeks before planting or transplanting. No waiting period is necessary. Planting or transplanting may be done the same day as a PAARLAN application.

TRANSPLANT TOBACCO

Air-cured (Burley, Maryland, Dark)—

PAARLAN Alone: Apply and incorporate PAARLAN before transplanting at a broadcast rate per acre of 2 pints on all soil textures. Do not apply PAARLAN after transplanting.

PAARLAN Tank Mix with insecticides/nematicides: PAARLAN at 2 pints per acre can be tank mixed with Dasanit®, Diazinon®, Furadan®. Mo-Cap®, and Mo-Cap Plus for effective weed, insect and/or nematode control.

Flue-cured—

PAARLAN Alone: Apply and incorporate PAARLAN before transplanting at a broadcast rate per acre of 2 pints on coarse and medium soils and 2⅔ pints on fine soils.

PAARLAN Tank Mix with insecticides/nematicides: PAARLAN at 2 pints per acre can be tank mixed with Dasanit + Di-Syston, Mo-Cap, Mo-Cap Plus, Diazinon, Di-Syston, Dasanit, and Furadan for effective weed, insect and/or nematode control.

PAARLAN is usually compatible with Ridomil®. To assure the compatibility of PAARLAN with Ridomil, pour the products into a small container of water in the correct proportion. After thorough mixing, let stand for five minutes. If the combination remains mixed or can be remixed readily, the mixture is compatible.

PRECAUTION

Read all product labels carefully before using. Note all directions, cautions, precautions, and special precautions.

NOTE: Application time before transplanting varies with different products. Read all labels carefully for this information.

APPLICATION WITH DRY FERTILIZERS IMPREGNATION/RATES

PAARLAN may be impregnated on most dry fertilizers. Individual state regulations relating to dry fertilizer mixing, registration, labeling and applications are the responsibility of the individual and/or company selling the fertilizer and PAARLAN mixture. Use any closed drum, belt, ribbon or other commonly used dry bulk fertilizer blender. The nozzle or nozzles used to spray the PAARLAN on to the fertilizer should be placed to provide uniform spray coverage. If less than 6 pints of PAARLAN are mixed per ton of fertilizer, add water to the PAARLAN to give a total volume of at least 6 pints per ton. See the following table for amounts of PAARLAN and water to be premixed based on the amount of fertilizer to be applied per acre.

PAARLAN Impregnation Volume—For PAARLAN at 2 Pints/Acre

Fertilizer Rate (Pounds/Acre)	1000	1200	1400	1600	1800	2000
PAARLAN Rate (Pints/Acre)	2	2	2	2	2	2
Pints of Water to be Premixed with PAARLAN	1	1⅔	2⅓	2⅔	3⅓	4
Total Volume PAARLAN + Water (Pints)	3	3⅔	4⅓	4⅔	5⅓	6

PAARLAN Impregnation Volume—For PAARLAN At 2⅔ Pints/Acre

Fertilizer Rate (Pounds/Acre)	1000	1200	1400	1600	1800	2000
PAARLAN Rate (Pints/Acre)	2⅔	2⅔	2⅔	2⅔	2⅔	2⅔
Pints of Water to be Premixed with PAARLAN	⅓	1	1⅔	2	2⅔	3⅓
Total Volume PAARLAN + Water (Pints)	3	3⅔	4⅓	4⅔	5⅓	6

Check the crop section to determine the rate of PAARLAN per acre. If the table above does not include the desired amount of fertilizer per acre, use the following calculation to determine the quarts of PAARLAN to be impregnated per ton of dry fertilizer.

$$\text{Pints of PAARLAN Per Acre} \times \frac{2000}{\text{Lbs. Fertilizer Per Acre}} = \text{Pints of PAARLAN Per Ton of Fertilizer}$$

APPLICATION

Spread the fertilizer/PAARLAN mixture with a properly calibrated applicator. Be certain the material is applied uniformly to the soil surface.

INCORPORATION

Follow normal PAARLAN incorporation procedures.

APPLICATION WITH LIQUID FERTILIZERS

PAARLAN may be mixed with most liquid fertilizer materials. Individual state regulations relating to liquid fertilizer mixing, registration, labeling and applications are the responsibility of the individual and/or company selling the fertilizer and PAARLAN mixture.

LIQUID FERTILIZER MIXING INSTRUCTIONS

PAARLAN E.C. in Liquid Fertilizer: Emulsifiable concentrates, such as PAARLAN E.C. can be mixed with most liquid fertilizers. In all cases, continuous agitation is required to prevent the PAARLAN from rising to the surface as an oily layer. When necessary, (see Liquid Fertilizer Compatibility Test below), a compatibility agent can be used to cause the PAARLAN E.C. to emulsify properly (i.e., have a milky appearance rather than oily layer). The use of compatibility agents is especially important when tank mixing emulsifiable concentrates (EC) with dry flowables, wettable powders (WP), aqueous suspensions (AS), flowables (F), liquids (L), aqueous suspensions (AS), or solutions (S) in liquid fertilizer.

If the emulsion is not properly formed, and the E.C. rises to the surface of the fertilizer as an oil ("oils out"), the oil may combine with the wettable powder, flowable, or suspension to form oily curds (viscous phase) which is difficult to redisperse. Any one of the compatibility agents listed below is helpful in causing emulsifiable concentrates to form non-oiling mixtures with liquid fertilizers. These compatibility agents can be used at rates as low as 1½ to 2 pints per ton of liquid fertilizer and should be mixed well with the fertilizer before adding the emulsifiable concentrate. Read the label on the compatibility agent and follow the directions.

1. Sponto 168D (Witco Chemicals Co., Chicago, IL)
2. Compat (Farm Chemicals, Inc., Aberdeen, NC)
3. Unite (Hopkins Ag Chemical, Madison, WI)
4. T-Mulz 734-2 (Thompson-Hayward Chemical Co.)
5. Rigo Compatibility Agent (Rigo Company, Buckner, KY)
6. Amoco Spray Mate (Amoco Oil Co., Chicago, IL)
7. Kem-Link (Universal Coop, Minneapolis, MN)

Each of the above is a phosphate ester type surfactant designed to be used with liquid fertilizers. They usually do not work well as compatibility agents in tank mixtures in plain water.

Testing for Tank Mix Compatibility in Liquid Fertilizers: Emulsifiable concentrates alone or in tank mixture with dry flowables, wettable powders (WP), aqueous suspensions (AS), flowables (F), liquids (L), or solutions (S) may not combine properly with some fluid fertilizer materials. Small quantities should always be tested before full-scale mixing. This will determine whether a compatibility agent is needed, and which agent does the best job. The seven agents listed above have been thoroughly tested. There are many other surfactants on the market which were not designed for use with liquid fertilizers. Use the following test to select the correct agent for your mixture.

1. Put 1 pint of the liquid fertilizer in a quart jar.
2. Add 1 to 4 teaspoonfuls of the dry flowable, WP, AS, F or L formulation (depending on the recommended rate per acre) to the liquid fertilizer. Close jar and agitate until dispersed evenly in the fertilizer. If the materials do not disperse well, it may be necessary to slurry the chemicals in water before adding to the fertilizer.
3. After dispersing the materials (Step 2), add 3 to 4 teaspoonfuls of the PAARLAN to the jar and shake well. Add solution herbicides to the mixture last and agitate. Observe the jar for about 10

minutes. If the materials rise to the surface and form a thick layer (oily curds) which will not redisperse when agitated, a compatibility agent is needed. If the mixture is easily redispersed to its original state with slight agitation, no agent is needed but good agitation must be provided in the fertilizer spray tank.

4. If the need for a compatibility agent is shown in Step 3: using a clean quart jar, start at Step 1 above, add ½ teaspoonful of the compatibility agent to the liquid fertilizer, mix well, then repeat Steps 2 and 3.

An effective compatibility agent will cause the mixture to remain uniformly mixed with little or no separating or oil rising to the surface for one half hour or longer. If slight separation does occur, 2 to 3 inversions of the jar should give a uniform remix. If oily curds form which will not redispurse, more agent or another agent should be tried. Use a clean jar for each test. The compatible mixture will have a uniform appearance and will be relatively easy to keep mixed with gentle agitation of the jar.

APPLICATION
Spread the PAARLAN/fertilizer with a properly calibrated applicator. Be certain the material is applied uniformly to the soil surface.

INCORPORATION
Follow recommended incorporation procedures for PAARLAN.

PAARLAN ® E.C. PERFORMANCE GUARANTEE

A LIMITED WARRANTY OF PAARLAN WEED AND GRASS CONTROL
Elanco Products Company warrants that PAARLAN E.C. will control the weeds and grasses as indicated on its current label when used according to label directions and subject to the following limitations:

1. Purchaser must notify Elanco promptly if a lack of satisfactory control occurs. Such notice must be given within ninety (90) days after the first planting or application of PAARLAN E.C. to the treated crop, whichever occurs later.
2. An Elanco representative must be satisfied that Purchaser used PAARLAN E.C. according to the label directions and must visually observe unsatisfactory control of weeds in the field.
3. The Purchaser must provide an Elanco representative with an invoice or other satisfactory record which shows the price, quantity of PAARLAN E.C. and date purchased.
4. Elanco's liability will be limited to a refund of the purchase price of the PAARLAN E.C. applied to the acreage on which weed control was not satisfactory.
5. Void outside the U.S.A.

DISCLAIMER OF WARRANTIES
The warranty printed above is the only warranty applicable to this product. All other warranties express or implied, including but not limited to, the implied warranties of merchantability and fitness for a particular purpose are disclaimed.

LIMITATION OF DAMAGES
Elanco's liability, whether in contract, warranty, tort, negligence, strict liability or otherwise shall not exceed the return of the amount of the purchase price of the PAARLAN E.C. and under no circumstances shall Elanco be liable for special, indirect or consequential damages.

INHERENT RISKS OF USE
Failure to carefully follow the directions for use of PAARLAN E.C. may result in unsatisfactory weed control, or crop injury factors such as plant disease, deep planting, cold weather, excessive moisture, high salt concentration, improper fertilizer placement, drought, or transplant quality may also result in unsatisfactory weed control or crop injury.

PAARLAN®— isopropalin—Elanco Products Company
Dasanit®—fensulfothion—Bayer GmbH
Diazinon®—diazinon—Ciba-Geigy
Di-Syston®—disulfoton—Bayer GmbH
Furadan®—carbofuran—F.M.C. Corporation
Mo-Cap®—ethoprop—Mobil Chemical
Ridomil®—metalaxyl—Ciba-Geigy

SONALAN® E.C.
HERBICIDE

A Selective Herbicide for Preemergence Control of Certain Annual Grasses and Broadleaf Weeds

ACTIVE INGREDIENT:
ethalfluralin* *N*-ethyl-*N*-(2-methyl-2-propenyl)-2,6-dinitro-4-(trifluoromethyl)benzenamine 35.1%
INERT INGREDIENTS 64.9%
TOTAL 100.0%
Contains 3 pounds active ingredient per gallon.

EPA Reg. No. 1471-122 EPA Est. 1471-IN-2
ID 5036 PI 2665 AMP

* SONALAN®—the registered trademark of Elanco Products ethalfluralin.

WARNING
Keep Out of Reach of Children

PRECAUTIONARY STATEMENTS

HAZARDS TO HUMANS AND DOMESTIC ANIMALS
WARNING
SONALAN E.C. formulation is irritating to the eyes and skin. Avoid breathing vapors. Avoid contact with skin, eyes, or clothing. Do not contaminate foodstuffs or feed. Harmful if swallowed or absorbed through the skin. The active ingredient, ethalfluralin, may cause skin sensitization reactions in certain individuals. Use protective clothing such as coveralls, long sleeved shirt, and impermeable gloves when handling or using this product.

First Aid: In case of contact, immediately flush eyes or skin with plenty of water. For eyes, call a physician. Remove and wash contaminated clothing before reuse.

ENVIRONMENTAL HAZARDS
This pesticide is toxic to fish. Do not apply directly to any body of water or wetlands. Do not contaminate water by cleaning of equipment or disposal of wastes. Runoff or erosion from treated areas may be hazardous to fish in neighboring areas.

STORAGE AND DISPOSAL

STORAGE
Avoid freezing. Store above 40°F (5°C). If frozen, poor weed control may result. Do not use or store near heat or open flame.

PESTICIDE DISPOSAL
Pesticide, spray mixture, or rinse water that cannot be used according to label instructions must be disposed of according to applicable federal, state or local procedures. Do not contaminate water, food or feed by storage or disposal. Waste resulting from use of this product may be disposed of on site or at an approved waste disposal facility.

Plastic Container Disposal: Triple rinse (or equivalent). Then offer for recycling or reconditioning, or puncture and dispose of in a sanitary landfill, or incineration, or, if allowed by state and local authorities, by burning. If burned, stay out of smoke.

Metal Container Disposal: Triple rinse (or equivalent). Then offer for recycling or reconditioning, or puncture and dispose of in a sanitary landfill, or by other procedures approved by state and local authorities.

The manufacturer makes no warranties, express or implied, concerning this product or its use, which extend beyond the description on the label or in literature published by Elanco Products Company, and all statements made concerning this product apply only when used as directed. Elanco Products Company expressly disclaims all warranties, express or implied, in relation to any use of the product in combination or sequential use with any other product not specifically recommended in writing by Elanco Products Company. If used in a combination or sequential use recommended by Elanco Products Company, the liability of Elanco shall in no manner extend to any damage loss or injury not directly caused by the inclusion of the Elanco product in such combination or sequential use.

GENERAL AGRICULTURAL PRECAUTION

Poor weed control and/or crop injury may result if directions are not followed. Uneven application, over application or improper incorporation can result in erratic weed control or injury to the treated or rotational crops.

Do not graze or forage crop grown in treated soil or cut for hay or silage.

Do not apply SONALAN to soils which are wet, cloddy, or subject to prolonged periods of flooding because poor weed control or crop injury may result.

Do not apply to any area not specified on the label.

SONALAN ROTATIONAL CROP RESTRICTIONS

If replanting is required, replant only crops listed on the SONALAN label. In California and Arizona, do not plant red beets, spinach or oats within eight (8) months following an application of three (3) pints per acre or more of SONALAN. In all areas do not rotate to sugarbeets the year following an application of SONALAN unless the SONALAN treated soil is moldboard plowed to a depth of at least six (6) inches prior to planting sugarbeets.

DIRECTIONS FOR USE

It is a violation of Federal Law to use this product in a manner inconsistent with its labeling.

Read all directions carefully before applying.

SONALAN is a preemergence herbicide that controls many annual grasses and broadleaf weeds. SONALAN controls weeds as they germinate. SONALAN will not control established weeds.

MIXING AND APPLICATION DIRECTIONS

SONALAN ALONE

Start with a clean spray tank. Fill sprayer one-third (1/3) to one-half (1/2) full with clean water or liquid fertilizer. Start agitation. Add correct quantity of SONALAN, continue agitation and finish filling the tank.

SONALAN TANK MIX

Vigorous, continuous agitation is required for all tank mixes. Sparger pipe agitators generally provide the best agitation in spray tanks. To prevent foaming, avoid stirring or splashing air into the mixture during filling by placing the end of the fill pipe below the surface of the water in the spray tank. Do not allow the mixture to siphon back into the water source.

Mixing order: Fill the tank one-fourth (1/4) to one-third (1/3) full with clean water or liquid fertilizer. (See page 9 for additional liquid fertilizer mixing instructions). Start the agitation. Add in the following order: dry flowables, wettable powders (WP), aqueous suspensions (AS), flowables (F), or liquids (L) to the water and agitate until the product(s) are completely dispersed. Allow additional mixing and dispersion time when using dry flowable products. Continue agitation and fill tank to three-fourths (3/4) full, add the SONALAN or other emulsifiable concentrates (EC) and mix thoroughly. Then add any solution (S) formulations, agitate and finish filling. Maintain agitation during filling and through application. If spraying and agitation must be stopped before the tank is empty, the materials may settle to the bottom. In this case, it is important to resuspend all of the material in the bottom of the tank before continuing the spray application. A sparger agitator is particularly useful for this purpose. Sometimes it is more difficult to resuspend settled material than it is to suspend it originally.

Read and carefully follow all label instructions for each material added to the tank. Premixing dry and flowable formulations with water (slurrying) and pouring the slurry through a 20 or 35 mesh wetting screen in the top of the tank will help assure good initial dispersion in the tank. Line screens in the tank should be no finer than 50 mesh (100 mesh is finer than 50 mesh).

If a buildup of material on the walls of the spray tank is observed, wash the tank with soapy water between fillings. Rinse and continue the spraying operation. Clean the tank, lines, and screens thoroughly after use. Check the sprayer daily to ensure proper calibration and uniform application. Do not apply SONALAN when the wind can cause drifting of spray particles which can result in non-uniform application.

APPLICATION

Apply SONALAN in at least five (5) gallons of water or liquid fertilizer per acre (broadcast basis), using any properly calibrated, low pressure herbicide sprayer that will apply the spray uniformly.

APPLICATION WITH LIQUID FERTILIZER

SONALAN may be mixed with most liquid fertilizer materials. Combinations of SONALAN with solutions and suspension type fertilizers have provided weed and grass control equal to the same rates of SONALAN applied in water. Follow SONALAN label recommendations regarding rates per acre, crops, incorporation directions, special instructions, cautions and special precautions.

Individual state regulations relating to liquid fertilizer mixing, registration, labeling and applications are the responsibility of the individual and/or company selling the fertilizer and chemical mixture.

LIQUID FERTILIZER MIXING INSTRUCTIONS

Emulsifiable concentrates, such as SONALAN, can be mixed with liquid fertilizers. In all cases, continuous agitation is required to prevent the SONALAN from rising to the surface as an oily layer. When necessary, (see Liquid Fertilizer Compatibility Test on page 10), a compatibility agent can be used to ensure that the SONALAN emulsifies properly (i.e., has a milky appearance rather than an oily layer). The use of compatibility agents is especially important when tank mixing emulsifiable concentrates (EC) with dry flowables, wettable powders (WP), flowables (F), liquids (L), aqueous suspensions (AS), or solutions (S) in liquid fertilizer. If the emulsion is not properly formed, and the EC rises to the surface of the fertilizer as an oil ("oils out"), the oil may combine with the wettable powder, flowable, or suspension to form oily curds (viscous phase) which are difficult to disperse.

Any one of the compatibility agents listed on page 10 is helpful in causing emulsifiable concentrates to form non-oiling mixtures with liquid fertilizers. These compatibility agents can be used at rates as low as one and one-half (1½) to two (2) pints per ton of liquid fertilizer and should be mixed well with the fertilizer before adding the emulsifiable concentrate.

Read the label on the compatibility agent and follow the directions.

1. Sponto 168D (Witco Chemicals Co., Chicago, IL)
2. Compat (Farm Chemicals, Inc., Aberdeen, NC)
3. Unite (Hopkins Ag Chemical, Madison, WI)
4. T-Mulz 734-2 (Thompson-Hayward Chemical Co.)
5. Rigo Compatibility Agent (Rigo Company, Buckner, KY)
6. Amoco Spray Mate® (Amoco Oil Co., Chicago, IL)
7. Kim-Link (Universal Coop, Minneapolis, MN)
8. Blendex (Helena Chemical Co., Cayce-West Columbia, SC)

Each of the above is a phosphate ester type surfactant designed to be used with liquid fertilizers. They usually do not work well as compatibility agents in tank mixtures in water.

Testing for Tank Mix Compatibility in Liquid Fertilizers: Emulsifiable concentrates alone or in tank mixture with dry flowables, wettable powders (WP), liquids (L), flowables (F), aqueous suspension (AS), or solutions (S), may not combine properly with some fluid fertilizer materials. Small quantities should always be tested before full-scale mixing. This will determine whether a compatibility agent is needed, and which agent does the best job. The eight (8) agents listed above have been thoroughly tested. There are many other surfactants on the market which were not designed for use with liquid fertilizers.

Use the following test to select the correct agent for your mixture:

1. Put one (1) pint of the liquid fertilizer in a quart jar.
2. Add one (1) to four (4) teaspoonful(s) of the dry flowable, WP, L, F, or AS formulations (depending on the recommended rate per acre) to the liquid fertilizer. Close jar and agitate until dispersed evenly in the fertilizer. If the materials do not disperse well, it may be necessary to slurry the chemicals in water before adding to the fertilizer.

3. After dispersing the materials (Step 2), add three (3) to four (4) teaspoonfuls of the SONALAN to the jar and shake well. Add solution herbicides to the mixture last and agitate. Observe the jar for about 10 minutes. If the materials rise to the surface and form a thick layer (oily curds) which will not disperse when agitated, a compatibility agent is needed. If the mixture is easily dispersed to its original state with slight agitation, no agent is needed, but good agitation must be provided in the fertilizer spray tank.
4. If the need for a compatibility agent is shown in Step 3, using a clean quart jar, start at Step 1 above, add one-half (½) teaspoonful of the compatibility agent to the liquid fertilizer, mix well, then repeat Steps 2 and 3.

An effective compatibility agent will cause the mixture to remain uniformly mixed with little or no separating or oil rising to the surface for one-half (½) hour or longer. If slight separation does occur, two (2) to three (3) inversions of the jar should give a uniform remix. If oily curds form which will not disperse, more agent or another agent should be tried.

Use a clean jar for each test. The compatible mixture will have a uniform appearance and will be relatively easy to keep mixed with gentle agitation of the jar.

APPLICATION WITH DRY BULK FERTILIZER

Dry bulk fertilizer may be impregnated or coated with SONALAN. Application of dry bulk fertilizer impregnated with SONALAN has provided weed and grass control equal to the same rates of SONALAN applied in liquid carriers.

Follow SONALAN label recommendations regarding rates per acre, crops, incorporation directions, special instructions, cautions and special precautions.

Individual state regulations relating to dry bulk fertilizer blending, registration, labeling and application are the responsibility of the individual and/or company selling the fertilizer and chemical mixture.

Limitations: Apply a minimum of 200 pounds per acre of dry fertilizer impregnated with SONALAN at the recommended rate. Most dry fertilizers can be used for SONALAN impregnation. When coated ammonium nitrate and/or limestone are used alone, do not impregnate with SONALAN. These materials will not absorb the herbicide. Blends containing mixtures of these materials can be impregnated.

Impregnation: Use any closed drum, belt, ribbon or other commonly used dry bulk fertilizer blender. Nozzles used to spray the SONALAN onto the fertilizer should be placed to provide uniform spray coverage. Add water to the SONALAN to give a total volume of at least six (6) pints per ton of fertilizer.

SONALAN—PREPLANT INCORPORATED APPLICATION

WEEDS CONTROLLED

SONALAN applied preplant soil incorporated will control the following weeds:

Grasses:	
Annual bluegrass	*Pao annua*
Barnyardgrass	*Echinochloa crus-galli*
Crabgrass	*Digitaria* spp.
(Large crabgrass)	
(Smooth crabgrass)	
Foxtail	*Setaria* spp.
(Giant foxtail)	
(Green foxtail)	
(Foxtail millet)	
(Pigeongrass)	
(Yellow foxtail)	
Goosegrass	*Eleusine indica*
Johnsongrass (from seed)	*Sorghum halepense*
Junglerice	*Echinochloa colonum*
Oat, wild[1]	*Avena fatua*
Panicum, fall	*Panicum dichotomiflorum*
Panicum Texas (Buffalograss) (Colorado grass)	*Panicum texanum*
Ryegrass, Italian (Annual ryegrass)	*Lolium multiflorum*
Shattercane (Wild cane)	*Sorghum bicolor*
Signalgrass, broadleaf (Brachiaria)	*Brachiaria platyphylla*
Witchgrass	*Panicum capillare*
Broadleaf Weeds:	
Buckwheat, wild	*Polygonum convolvulus*
Carpetweed	*Mollugo verticillata*
Catchfly, conical	*Silene conica*
Chickweed, common	*Stellaria media*
Fiddleneck tarweed	*Amsinckia lycopsoides*
Groundcherry[2] (annual)	
(Lanceleaf groundcherry)	*Physalis lanceifolia*
(Wright's groundcherry)	*Physalis Wrightii*
Henbit	*Lamium amplexicaule*
Kochia	*Kochia scoparia*
Lambsquarters, common	*Chenopodium album*
Nightshade[2] (annual)	*Solanum* spp.
(Black nightshade)	
(Eastern black nightshade)[3]	
(Hairy nightshade)	
Pigweed	*Amaranthus* spp.
(Prostrate pigweed)	
(Redroot pigweed)	
(Smooth pigweed)	
(Tumble pigweed)	
Purslane, common	*Portulaca oleracea*
Pusley, Florida	*Richardia scabra*
Rockpurslane, redmaids	*Caandrinia caulescens*
Thistle, Russian	*Salsola kali*

[1] (SONALAN will provide wild oat control except those germinating below the SONALAN treated zone.)
[2] (See special instructions for control in dry bean).
[3] (See special instructions for partial control in soybean).

SOIL PREPARATION—GENERAL DIRECTIONS

SONALAN should be mixed uniformly into the top two (2) to three (3) inches of soil.

Factors that influence uniform application and incorporation of SONALAN are:

Crop Residues or Existing Weeds: Ground cover, such as crop residues or existing weeds, can interfere with the incorporation of SONALAN into the soil. A manageable level of such ground cover will allow the SONALAN to be uniformly mixed into the top two (2) to three (3) inches of soil. If the level of the ground cover is such that this cannot be done, the soil must be tilled prior to the application of SONALAN.

Roughness: The soil surface should be smooth enough so that you can operate a sprayer and incorporation equipment efficiently and at speeds which ensure a uniform application and incorporation of SONALAN.

General Soil Conditions: To assure uniform incorporation of SONALAN, soil moisture conditions should be such that large clods can be broken up during the incorporation process.

SOIL MIXING DIRECTIONS

Use incorporation equipment that thoroughly and uniformly mixes SONALAN into the top two (2) to three (3) inches of the seedbed, or erratic weed control and/or crop injury may result.

Incorporation Before Planting

Apply SONALAN up to three (3) weeks prior to planting. SONALAN should be thoroughly and uniformly mixed into the top two (2) to three (3) inches of the seedbed within two (2) days after application.

Incorporation In Bedded Culture

Application prior to bedding: Apply SONALAN and mix thoroughly into the top two (2) to three (3) inches of soil. The bedding operation provides additional mixing. Do not expose untreated soil during post-bedding operations.*

Application after bedding: Knock off beds to planting height before applying SONALAN. Apply SONALAN and mix thoroughly with equipment that will conform to the bed shape. Do not leave untreated soil exposed.*

* Avoid removal of treated soil from the seedbed before or during the planting operation. This would expose untreated soil, allowing weeds to germinate in the drill row.

CULTIVATION AFTER PLANTING

Soil treated with SONALAN may be shallowly cultivated without reducing the weed control activity of SONALAN. Do not cultivate deeper than the treated soil since this may bring untreated soil to the surface, and poor weed control may result.

SOIL TEXTURE GUIDE

The amount of SONALAN applied will vary with the soil texture and organic matter. A fine textured soil will require more SONALAN per acre than a coarse soil. Choose the proper rate based on the following soil texture group and specific crop recommendations. Do not exceed recommended rates.

Soil Texture Group	Soil Classification
Coarse soils: (Light)	Sand, loamy sand, sandy loam
Medium soils:	Loam, silty clay loam*, silt loam, silt, sandy clay loam*
Fine soils: (Heavy)	Clay, clay loam, silty clay loam*, silty clay, sandy clay, sandy clay loam*

* Silty clay loam and sandy clay loam soils are transitional soils and may be classified as either medium or fine textured soils. If silty clay loam or sandy clay loam soils are predominately sand or silt, they are usually classified as medium textured soils. If they are predominately clay, they are usually classified as fine textured soils.

These recommendations are given as the broadcast rates of SONALAN per acre. Where a rate range is shown, use the lower rate for coarser textured soils or soils with lower organic matter. Use the higher rate, where a rate range is shown, on soils containing more than five percent (5%) organic matter. Where soil texture is variable within the same field, use the higher recommended SONALAN rate for the coarser textured soil. SONALAN is not recommended for use on soils containing more than 10% organic matter.

CROP RECOMMENDATIONS

SOYBEAN—SONALAN ALONE

General weed control—except groundcherry and nightshade: SONALAN should be applied as a preplant incorporated treatment. Follow recommended soil preparation, application, and incorporation procedures for SONALAN.

For broader spectrum weed control, SONALAN may be used in combination with Amiben, Basagran, Blazer, Dual, Lasso, Lexone, Lorox, Sencor and Vernam. Read the combination product label for directions, cautions and limitations prior to use.

Broadcast Rates per Acre

Soil Texture	SONALAN
	(pints)
Coarse	1½–2
Medium	2–2½
Fine	2½–3

Eastern Black Nightshade—Partial Control:
SONALAN should be applied as a preplant incorporated treatment for partial control of nightshade in medium and fine textured soils. Follow recommended soil preparation and application procedures for SONALAN.

Two incorporation passes are required for partial control of eastern black nightshade. For broader spectrum weed control, SONALAN may be used in combination with Amiben, Basagran, Blazer, Dual, Lasso, Lexone, Sencor and Vernam. Read the combination product label for directions, cautions and limitations prior to use.

Broadcast Rates per Acre

Soil Texture	SONALAN
	(pints)
Medium	3
Fine	3½

SONALAN—SOYBEAN PRECAUTION STATEMENT

Soybeans should be planted no more than two (2) inches deep after early season adverse weather conditions have passed, especially when using higher rate programs. Excessively deep planting or cool, wet weather early in the growth cycle causes additional stress to the soybean plant. These factors may result in reduced stands, delayed maturity and reduced yields.

SOYBEAN—SONALAN/AMIBEN® TANK MIX OR OVERLAY

Amiben®-chloramben, Union Carbide Agricultural Products Company, Inc.

SONALAN/Amiben should be applied as a preplant incorporated tank mix or SONALAN preplant incorporated followed by an Amiben overlay. The SONALAN/Amiben combination effectively controls the weeds listed for SONALAN alone for general weed control plus these additional weeds:
Black nightshade
Common ragweed
Pennsylvania smartweed
Velvetleaf
Wild mustard

Broadcast Rates per Acre

	SONALAN		Amiben	
Soil Texture	General Weed Control	Partial Control of Eastern Black Nightshade	Amiben 2S	Amiben DS
	(pints)	(pints)	(quarts)	(pounds)
Coarse	1¼–2	N/A	4–6**	2.4–3.6**
Medium	1¾–2½	3*	4–6	2.4–3.6
Fine	2¼–3	3½*	4–6	2.4–3.6

* Two incorporation passes are required for partial control of eastern black nightshade.
** Use the higher rate where heavy populations of smartweed, velvetleaf, ragweed, wild mustard or black nightshade are expected. Read the Amiben label for directions, cautions and limitations prior to use.

SOYBEAN—SONALAN/DUAL® TANK MIX OR OVERLAY

Dual®—melolachlor, Ciba-Geigy Corporation

SONALAN/Dual should be applied as a preplant incorporated tank mix or SONALAN preplant incorporated followed by a Dual overlay. The SONALAN/Dual combination effectively controls the weeds listed for SONALAN alone for general weed control plus these additional weeds:

Black nightshade	Southwestern cupgrass
Prairie cupgrass	Yellow nutsedge
Red rice	Galinsoga

Additional weeds partially controlled by this combination include:

Hairy nightshade	Sandbur
Volunteer sorghum	

Follow recommended soil preparation procedures for SONALAN. Apply the tank mix in at least 10 gallons of water per acre up to 14 days prior to planting. Incorporate the tank mix within two (2) days after application in the top two (2) inches of soil. When Dual is applied as an overlay to SONALAN, apply Dual after planting (behind the planter) prior to crop and weed emergence.

Broadcast Rates per Acre

Soil Texture	SONALAN General Weed Control	SONALAN Partial Control of Eastern Black Nightshade	Dual 8E Less than 3% O.M.	Dual 8E 3% O.M. or greater
	(pints)	(pints)	(pints)	(pints)
Coarse	1¼–2	NA	1½–2	2
Medium	1¾–2½	3*	2–2½	2–2½
Fine	2¼–3	3½*	2–2½	2½–3

* Two incorporation passes are required for partial control of eastern black nightshade.

Read the Dual label for directions, cautions and limitations prior to use.

SOYBEAN—SONALAN/LASSO® TANK MIX OR OVERLAY

Lasso®—alachlor, Monsanto Company

SONALAN/Lasso should be applied as a preplant incorporated tank mix or SONALAN preplant incorporated followed by a Lasso overlay. The SONALAN/Lasso combination effectively controls the weeds listed for SONALAN alone for general weed control plus these additonal weeds:

Galinsoga
Black nightshade[a]
Hairy nightshade[a]
Red rice[b]
Witchgrass
Florida pusley
Yellow nutsedge[b]

[a] or [b] Check beneath the rate tables for additional Lasso directions.

[a] Use a minimum of two and one-half (2½) quarts of Lasso to control these weeds.

[b] Use three (3) to four (4) quarts of Lasso per acre applied alone or, on medium and fine textured soils, in tank mix combination preplant incorporated only for control of these weeds.

Additional weeds partially controlled by this combination include:

Florida beggarweed
Common ragweed
Spiny sida (teaweed)
Smartweed

SONALAN/Lasso tank mix: Follow recommended soil preparation recommendations for SONALAN. Apply SONALAN/Lasso in a minimum of 15 gallons of water or liquid fertilizer per acre and incorporate thoroughly into the top two (2) inches of soil. Apply the tank mix up to seven (7) days prior to planting.

SONALAN preplant incorporated followed by Lasso overlay: Apply and incorporate SONALAN according to recommended soil preparation application and incorporation procedures.

SONALAN/LASSO TANK MIX PREPLANT INCORPORATED

Broadcast Rates per Acre

Soil Texture	SONALAN General Weed Control	SONALAN Partial Control of Eastern Black Nightshade	Lasso
	(pints)	(pints)	(quarts)
Coarse	1¼–2	N/A	2½
Medium	1¾–2½	3*	3–3½
Fine	2¼–3	3½*	3½–4

* Two incorporation passes are required for partial control of eastern black nightshade.

SONALAN PREPLANT INCORPORATED FOLLOWED BY LASSO OVERLAY

Broadcast Rates per Acre

Soil Texture	SONALAN General Weed Control	SONALAN Partial Control of Eastern Black Nightshade	Lasso Less than 3% O.M.	Lasso 3% O.M. or greater
	(pints)	(pints)	(quarts)	(quarts)
Coarse	1¼–2	N/A	2	2
Medium	1¾–2½	3*	2½	2½–3
Fine	2¼–3	3½*	3	3–4

* Two incorporation passes are required for partial control of eastern black nightshade.

Read the Lasso label for directions, cautions and limitations prior to use.

SOYBEAN—SONALAN/LEXONE® OR SONALAN/SENCOR® TANK MIX OR OVERLAY

Lexone®-metribuzin, E. I. duPont de Nemours and Company.
Sencor®-metribuzin, Bayer, GmbH

SONALAN plus Lexone or Sencor may be applied as a preplant incorporated tank mix or SONALAN preplant incorporated followed by a Sencor or Lexone overlay. Do not allow SONALAN/Sencor or SONALAN/Lexone tank mix to remain in the tank overnight. The SONALAN/Lexone or SONALAN/Sencor combination effectively controls the weeds listed for SONALAN alone for general weed control plus these additional weeds:

Jimsonweed
Venice mallow
Wild mustard
Common ragweed
Hemp sesbania
Pennsylvania smartweed
Prickly sida
Velvetleaf

Control of cocklebur, morningglory, and giant ragweed may be erratic, ranging from poor to excellent depending upon soil temperature, time of weed germination, depth of weed seed in the soil and the amount and timing of soil moisture. Control may be improved with timely cultivation.

Follow recommended soil preparation, application, and incorporation directions for SONALAN.

Broadcast Rates per Acre

Soil Texture	SONALAN General Weed Control	SONALAN Partial Control Eastern Black Nightshade	Lexone or Sencor: Lexone 4L or Sencor 4	Lexone or Sencor: Lexone Dry Flowable or Sencor Dry Flowable
	(pints)	(pints)	(pints)	(pounds)
Coarse	1¼–2	N/A	½**	⅓**
Medium	1¾–2½	3*	¾	½
Fine	2¼–3	3½*	1	⅔

* Two incorporation passes are required for partial control of eastern black nightshade.

** Do not use Sencor/Lexone on coarse soils with less than one percent (1%) organic matter.

Additional Precaution: For Sencor/Lexone tank mixes, do not plant any crop other than soybeans within four (4) months after treatment. Over application, uneven application, or improper soil incorporation may result in crop injury, soil residue, or erratic weed control. Seedling disease, cold weather, deep planting, excessive moisture, soil *p* H over seven and one-half (7.5), high salt concentration or drought may weaken crop seedlings and increase possibility of damage from the SONALAN/Sencor or SONALAN/Lexone tank mix. Under these conditions, delayed crop development or reduced yields may result. Read the Sencor and Lexone labels for directions, cautions and limitations prior to use. Do not use the foliage from soybeans treated with Sencor or Lexone for feed or forage.

SOYBEAN—SONALAN PREPLANT INCORPORATED FOLLOWED BY LOROX® OVERLAY

Lorox®-linuron, E. I. duPont de Nemours and Company

SONALAN should be applied preplant incorporated followed by a Lorox overlay. The SONALAN/Lorox combination effectively controls the weeds listed for SONALAN alone for general weed control plus these additional weeds:

Florida beggarweed
Common ragweed
Smartweed

Follow recommended soil preparaton, application, and incorporation procedures for SONALAN. When Lorox is applied as an overlay to SONALAN, apply Lorox as a preemergence application prior to crop and weed emergence. Read the Lorox label for directions, cautions and limitations prior to use.

Broadcast Rates per Acre

Soil Texture	SONALAN General Weed Control	SONALAN Partial Control of Eastern Black Nightshade	Lorox L ½ to 2% O.M.	Lorox L 2 to 5% O.M.**
	(pints)	(pints)	(pints)	(pints)
Coarse	1¼–2	N/A	1–1⅔	1⅔–3
Medium	1¾–2½	3*	1¼–2⅓	2⅓–4
Fine	2¼–3	3½*	1⅓–2⅔	2⅔–5

* Two incorporation passes are required for partial control of eastern black nightshade.
** Over 5% organic matter use 6 pints Lorox L.

SOYBEAN—SONALAN/VERNAM® TANK MIX

Vernam®-vernolate, Stauffer Chemical Company

SONALAN/Vernam effectively controls the weeds listed for SONALAN alone for general weed control plus these additional weeds:

Annual morningglory
Sicklepod
Yellow nutsedge
Purple nutsedge
Velvetleaf

Apply SONALAN/Vernam in at least 10 gallons per acre of water or liquid fertilizer. Follow recommended soil preparation procedures. Incorporate the tank mix immediately following application. Read the Vernam label for additional directions, cautions and limitations prior to use.

Broadcast Rates per Acre

Soil Texture	SONALAN General Weed Control	SONALAN Partial Control of Eastern Black Nightshade	Vernam 7E
	(pints)	(pints)	(pints)
Coarse	1¼–2	N/A	2⅓
Medium	1¾–2½	3*	3
Fine	2¼–3	3½*	3

* Two incorporation passes are required for partial control of eastern black nightshade.

SOYBEAN—SONALAN PREPLANT INCORPORATED FOLLOWED BY BASAGRAN® POSTEMERGENCE APPLICATION

Basagran®-bentazon, BASF Wyandotte Corporation

SONALAN should be applied preplant incorporated followed by a postemergence application of Basagran.

The SONALAN/Basagran combination effectively controls the weeds listed for SONALAN alone for general weed control plus these additional weeds:

Cocklebur
Jimsonweed
Prickly sida
Ragweed
Tropic croton
Velvetleaf

Consult the Basagran label for specific weed growth stage directions to achieve best results.

Broadcast Rates per Acre

Soil Texture	SONALAN General Weed Control	SONALAN Partial Control of Eastern Black Nightshade	Basagran**
	(pints)	(pints)	(pints)
Coarse	1½–2	N/A	2
Medium	2–2½	3*	2
Fine	2½–3	3½*	2

* Two incorporation passes are required for partial control of eastern black nightshade.
** Apply Basagran as follows for optimum results:

Do not treat earlier than weed stage indicated on Basagran label.
Add oil concentrate according to label use directions.
Read the Basagran label for additional weeds controlled, directions, cautions and limitations prior to use.

SOYBEAN—SONALAN PREPLANT INCORPORATED FOLLOWED BY BLAZER® POSTEMERGENCE APPLICATION (EXCEPT CALIFORNIA)

Blazer®-acifluorfen, Rohm and Haas Company

SONALAN should be applied preplant incorporated followed by a postemergence application of Blazer.

The SONALAN/Blazer combination effectively controls the weeds listed for SONALAN alone for general weed control plus these additional weeds:

Balloonvine
Tropic croton
Wooly croton
Jimsonweed
Morningglory
Black nightshade
Ragweed
Smartweed

Broadcast Rates per Acre

Soil Texture	SONALAN General Weed Control	SONALAN Partial Control of Eastern Black Nightshade	Blazer**
	(pints)	(pints)	(pints)
Coarse	1½–2	N/A	2
Medium	2–2½	3*	2
Fine	2½–3	3½*	2

* Two incorporation passes are required for partial control of eastern black nightshade.
** Apply Blazer as follows for opimium results:

1. Apply when weeds are at 2–4 leaf stage and actively growing.
2. Apply in a minimum of 20 GPA.
3. Use 40–60 PSI.
4. Do not use during periods of dry weather when crop and weeds are under stress.
5. Do not apply when temperature is below 70°F.

Read the Blazer label for additional directions, cautions and limitations prior to use.

DRY BEAN—SONALAN ALONE
(Does not include soybean)

SONALAN should be applied as a preplant incorporated treatment. Follow recommended soil preparation, application, and incorportion procedures for SONALAN.

General weed control—except groundcherry and nightshade:

Broadcast Rates per Acre

Soil Texture	SONALAN
	(pints)
Coarse	1½–2
Medium	2–2½
Fine	2½–3

For broader spectrum control, SONALAN may be used in combination with Amiben, Dual, Eptam or Lasso. Read the combination product label for directions, cautions and limitations prior to use.

Groundcherry and Nightshade control (annual)—dry bean:

Two incorporation passes are required for groundcherry and nightshade control.

Do not use these SONALAN rates for garden bean, green bean, snap bean, or string bean.

Broadcast Rates per Acre

Soil Texture	SONALAN
	(pints)
Coarse	3–3½
Medium	3½–4
Fine	4–4½

For broader spectrum control, SONALAN may be used in combination with Amiben, Dual, Eptam or Lasso. Read the combination product label for directions, cautions and limitations prior to use.

DRY BEAN—SONALAN/AMIBEN® TANK MIX OR OVERLAY

SONALAN/Amiben should be applied as a preplant incorporated tank mix or SONALAN preplant incorporated followed by an Amiben overlay.

The SONALAN/Amiben combination effectively controls the weeds listed for SONALAN alone for general weed control plus these additional weeds:

Black nightshade
Common ragweed
Pennsylvania smartweed
Velvetleaf
Wild mustard

Follow recommended soil preparation, application, and incorporation procedures for SONALAN.

Broadcast Rates per Acre

Soil Texture	SONALAN General Weed Control	SONALAN Nightshade and Groundcherry Control*	Amiben** 2S	Amiben** DS
	(pints)	(pints)	(quarts)	(pounds)
Coarse	1¼–2	3–3½	4–6	2.4–3.6
Medium	1¾–2½	3½–4	4–6	2.4–3.6
Fine	2¼–3	4–4½	4–6	2.4–3.6

* Two incorporation passes are required for groundcherry or nightshade control.
** Use the higher rate where a heavy population of smartweed, velvetleaf, ragweed, wild mustard, or black nightshade are expected. Read the Amiben label for directions, cautions and limitations prior to use.

DRY BEAN—SONALAN/DUAL® TANK MIX OR OVERLAY

SONALAN/Dual should be applied as a preplant incorporated tank mix or SONALAN preplant incorporated followed by a Dual overlay. The SONALAN/Dual combination effectively controls the weeds listed for SONALAN alone for general weed control plus these additional weeds:

Black nightshade
Prairie cupgrass
Red rice
Southwestern cupgrass
Yellow nutsedge
Galinsoga

Additional weeds partially controlled by this combination include:

Hairy nightshade
Volunteer sorghum
Sandbur

Follow recommended soil preparation procedures for SONALAN. Apply the tank mix in at least 10 gallons of water per acre up to 14 days prior to planting. Incorporate the tank mix within two (2) days after application in the top two (2) inches of soil. When Dual is applied as an overlay to SONALAN, apply Dual after planting (behind the planter) prior to crop and weed emergence.

Broadcast Rates per Acre

Soil Texture	SONALAN General Weed Control	SONALAN Nightshade and Groundcherry Control*	Dual 8E Less than 3% O.M.	Dual 8E 3% O.M. or greater
	(pints)	(pints)	(pints)	(pints)
Coarse	1¼–2	3–3½	1½–2	2
Medium	1¾–2½	3½–4	2–2½	2–2½
Fine	2¼–3	4–4½	2–2½	2½–3

* Two incorporation passes are required for groundcherry or nightshade control.

Read the Dual label for directions, cautions and limitations prior to use.

DRY BEAN—SONALAN/EPTAM®, SONALAN/EPTC TANK MIXES

Eptam®- EPTC, Stauffer Chemical Company

The SONALAN/Eptam or other SONALAN/EPTC tank mixes effectively control the weeds listed for SONALAN alone for general weed control plus these additional weeds:

Nightshade (black and hairy)
Nutsedge (purple and yellow)

Follow recommended soil preparation and application directions for SONALAN. Incorporate the tank mix immediately after application.

Broadcast Rates per Acre

Soil Texture	SONALAN General Weed Control	SONALAN Nightshade and Groundcherry Control*	Eptam 7E or other EPTC 7E formulations
	(pints)	(pints)	(pints)
Coarse	1¼–2	3–3½	3½
Medium	1¾–2½	3½–4	3½
Fine	2¼–3	4–4½	3½

* Two incorporation passes are required for groundcherry or nightshade control.

Read the Eptam or other EPTC labels for directions, cautions and limitations prior to use.

DRY BEAN—SONALAN/LASSO® TANK MIX—

(For use only west of the Mississippi River, except in Kern County, California)

SONALAN/Lasso should be applied as a preplant incorporated tank mix. The SONALAN/Lasso combination effectively controls the weeds listed for SONALAN alone for general weed control plus these additional weeds:

Galinsoga
Black nightshade[a]
Hairy nightshade[a]
Red rice[a]
Witchgrass
Florida pusley
Yellow nutsedge[a]

[a] Check beneath the rate table for additional Lasso directions.
[a] Use a minimum of two and one-half (2½) quarts of Lasso on coarse soils and three (3) quarts on medium and fine soils to control these weeds.

Additional weeds partially controlled by this combination include:

Florida beggarweed
Common ragweed
Spiny sida (teaweed)
Smartweed

SONALAN/Lasso tank mix: Follow recommended soil preparation recommendations for SONALAN. Apply SONALAN/Lasso in a minimum of 15 gallons of water per acre and incorporate thoroughly into the top two (2) inches of soil. Apply the tank mix up to seven (7) days prior to planting.

Broadcast Rates per Acre

Soil Texture	SONALAN General Weed Control	SONALAN Nightshade and Groundcherry Control*	Lasso
	(pints)	(pints)	(quarts)
Coarse	1¼–2	3–3½	2½–3
Medium	1¾–2½	3½–4	2½–3
Fine	2¼–3	4–4½	2½–3

* Two incorporation passes are required for groundcherry or nightshade control.

Read the Lasso label for directions, cautions and limitations prior to use.

DRY BEAN—SONALAN PREPLANT INCORPORATED FOLLOWED BY BASAGRAN® POSTEMERGENCE APPLICATION

SONALAN should be applied preplant incorporated followed by a postemergence application of Basagran.

The SONALAN/Basagran combination effectively controls the weeds listed for SONALAN alone for general weed control plus these additional weeds:

Cocklebur
Jimsonweed
Prickly sida
Ragweed
Tropic croton
Velvetleaf

Consult the Basagran label for specific weed growth stage directions to achieve best results.

Soil Texture	SONALAN General Weed Control	SONALAN Nightshade Groundcherry Control*	Basagran**
	(pints)	(pints)	(pints)
Coarse	1½–2	3–3½	2
Medium	2–2½	3½–4	2
Fine	2½–3	4–4½	2

* Two incorporation passes are required for groundcherry or nightshade control.

**Apply Basagran as follows for optimum results:

Do not treat earlier than weed stage indicated on Basagran label. Add oil concentrate according to label use directions.

Read the Basagran label for additional weeds controlled, directions, cautions and limitations prior to use.

SONALAN PERFORMANCE GUARANTEE

A LIMITED WARRANTY OF SONALAN WEED AND GRASS CONTROL

Elanco Products Company warrants that SONALAN will control the weeds and grasses as indicated on its current label when used according to label directions and subject to the following limitations:

1. Purchaser must notify Elanco promptly if a lack of satisfactory control occurs. Notice must be given within ninety days after the first planting or after application of SONALAN to the treated crop, whichever occurs later.
2. An Elanco representative must be satisfied that Purchaser used SONALAN according to the label directions and must visually observe unsatisfactory control of weeds in the field.
3. The Purchaser must provide an Elanco representative with an invoice or other satisfactory record which shows the price and quantity of SONALAN purchased.
4. Elanco's liability will be limited to a refund of the purchase price of the SONALAN applied to the acreage on which weed control was not satisfactory or, at the option of the Purchaser, a refund of an equivalent amount of SONALAN.
5. Void outside the U.S.A.

DISCLAIMER OF WARRANTIES

The warranty printed above is the only warranty applicable to this product. All other warranties, express or implied, including but not limited to, any implied warranties or merchantability and fitness for a particular purpose, are disclaimed.

LIMITATION OF DAMAGES

Elanco's liability, whether in contract, warranty, tort, negligence, strict liability or otherwise shall not exceed the return of the amount of the purchase price of the SONALAN and under no circumstances shall Elanco be liable for special, indirect or consequential damages.

INHERENT RISKS OF USE

Failure to carefully follow the directions for use of SONALAN, as well as other factors such as seedling disease, cold weather, deep planting, excessive moisture, high salt concentration, sodic soils or drought may result in unsatisfactory weed control or crop injury. Reduced yields may result under these conditions.

SUPPLEMENTAL LABELING

SONALAN® E.C. E.P.A. REG. NO. 1471-122

SONALAN/LASSO® AND SONALAN/LASSO/SENCOR® OR SONALAN/LASSO/LEXONE® SOIL MIXED FOR WEED CONTROL IN SOYBEANS

SONALAN may be tank mixed with Lasso or Lasso and SENCOR/ Lexone and soil mixed for weed control in soybeans.

A. SONALAN/Lasso controls those weeds listed for SONALAN alone plus these additional weeds:

Galinsoga	Nutsedge, yellow[2]
Nightshade, black[1]	Rice, red[2]
Nightshade, hairy[1]	

Additional weeds partially controlled by this combination include:

Beggarweed, Florida	Sida, spiny (teaweed)
Ragweed, common	Smartweed

[1] or [2] Check beneath rate table for additional information.

B. SONALAN/Lasso/Sencor or Lexone controls those weeds listed for SONALAN alone plus these additional weeds:

Crowfootgrass	Rice, red[2]
Galinsoga	Sesbania
Jimsonweed	(Coffeebean)
Mallow, Venice	(Coffeeweed)
Mustard, wild	Shepherdspurse
Nightshade, black[1]	Smartweed, Pennsylvania
Nightshade, hairy[1]	Spurge, spotted
Nutsedge, yellow[2]	Spurred anoda
	Velvetleaf

Additional weeds partially controlled by this combination include:

Beggarweed, Florida	Sida, spiny (teaweed)
Ragweed, common	

[1] and [2] Check beneath rate table for additional information.

Follow recommended soil preparation recommendations for SONALAN. Apply the tank mixes in a minimum of 15 gallons of water or liquid fertilizer per acre. Set equipment to mix the soil no deeper than four inches. Apply the tank mix up to seven (7) days prior to planting.

Broadcast Rates Per Acre

Use the recommended rates listed for SONALAN and Lasso for the two-way combination; add the recommended rate of Sencor or Lexone for the SONALAN/Lasso/Sencor or Lexone tank mix.

Soil Texture	SONALAN General Weed Control	SONALAN Partial Control of Eastern Black Nightshade	Lasso	Sencor[4] or Lexone 4L	Sencor Dry Flowable or Lexone Dry Flowable
	(pints)	(pints)	(quarts)	(pints)	(pounds)
Coarse	1¼–2	N/A	2½	½[4]	⅓[4]
Medium	1¾–2½	3[3]	3–3½	¾	½
Fine	2¼–3	3½[3]	3½–4	1	⅔

[1] Use a minimum of two and one-half (2½) quarts of Lasso to control these weeds.

[2] Use three (3) to four (4) quarts of Lasso per acre on medium and fine textured soils to control these weeds.

[3] Two soil mixing passes are required for partial control of eastern black nightshade.

[4] Do not use Sencor or Lexone on coarse soils with less than one (1) percent organic matter.

Read the SONALAN, Lasso, Sencor and Lexone labels for additional directions, cautions and limitations prior to use.

SONALAN®—ethalfluralin, Elanco Products Company

Lasso®—alachlor, Monsanto Company

Lexone®—metribuzin, E. I. duPont de Nemours and Company

Sencor®—metribuzin, Bayer GmbH

SUPPLEMENTAL LABELING

SONALAN® E.C.
E.P.A. REG. NO. 1471-122

SONALAN/DUAL® AND SONALAN/DUAL/SENCOR® OR SONALAN/DUAL/LEXONE® SOIL MIXED FOR WEED CONTROL IN SOYBEANS

SONALAN may be tank mixed with Dual or Dual and SENCOR/ Lexone and soil mixed for weed control in soybeans.

A. SONALAN/Dual controls those weeds listed for SONALAN alone plus these additional weeds:

Crowfootgrass	Nightshade, black
Cupgrass, prairie	Nutsedge, yellow
Cupgrass, southwestern	Rice, red
Galinsoga	

Additional weeds partially controlled by this combination include:

Nightshade, hairy	Sorghum, volunteer
Sandbur	

B. SONALAN/Dual/Sencor or Lexone controls those weeds listed for SONALAN alone plus these additional weeds:

Cupgrass, prairie
Cupgrass, southwestern
Crowfootgrass
Galinsoga
Jimsonweed
Mallow, Venice
Mustard, wild
Nightshade, black
Nutsedge, yellow
Rice, red
Sesbania
(Coffeebean)
(Coffeeweed)
Shepherdspurse
Smartweed, Pennsylvania
Spurge, spotted
Spurred anoda
Velvetleaf

Additional weeds partially controlled by this combination include:

Nightshade, hairy
Sandbur
Sorghum, volunteer

Follow recommended soil preparation recommendations for SONALAN. Apply the tank mixes in a minimum of 10 gallons of water or liquid fertilizer per acre. Mix the tank mix into the top two (2) inches of soil. Apply the tank mixes up to 14 days prior to planting.

Broadcast Rates Per Acre

Use the recommended rates listed for SONALAN and Dual for the two-way combination; add the recommended rate of Sencor or Lexone for the SONALAN/Dual/Sencor or Lexone tank mix.

	SONALAN		DUAL			
Soil Texture	General Weed Control	Partial Contrtol of Eastern Black Nightshade	Less than 3% O.M.	3% O.M. or Greater	Sencor[4] or Lexone 4L	Sencor Dry Flowable or Lexone Dry Flowable
	(pints)	(pints)	(pints)	(pints)	(pints)	(pounds)
Coarse	1¼–2	N/A	1½–2	2	½[2]	⅓[2]
Medium	1¾–2½	3[1]	2–2½	2–2½	¾	½
Fine	2¼–3	3½[3]	2-2½	2½–3	1	⅔

1 Two soil mixing passes are required for partial control of eastern black nightshade.

2 Do not use Sencor or Lexone on coarse soils with less than one (1) percent organic matter.

Read the SONALAN, Dual, Sencor and Lexone labels for additional directions, cautions and limitations prior to use.

SONALAN®—ethalfluralin, Elanco Products Company

Dual®—metolachlor, Ciba-Geigy Corporation

Lexone®—metribuzin, E. I. duPont de Nemours and Company

Sencor®—metribuzin, Bayer GmbH

SUPPLEMENTAL LABELING

SONALAN® E.C.
EPA Reg. No. 1471-122

SONALAN ALONE AND IN COMBINATION WITH AMIBEN® AND EPTAM® FOR WEED CONTROL IN SUNFLOWER

SONALAN should be applied as a preplant incorporated treatment alone or in tank mix combination with either Eptam or Amiben. SONALAN may also be applied as a preplant incorporated treatment and followed by an Amiben overlay. Check the section below for:

A. SONALAN Alone
B. SONALAN/Amiben
C. SONALAN/Eptam

A. Sunflower—SONALAN Alone

SONALAN should be applied as a preplant incorporated treatment. Follow recommended soil preparation, application and incorporation procedures for SONALAN.

For broader spectrum weed control, SONALAN may be used in combination with Amiben and Eptam. Read the combination product label for directions, cautions and limitations prior to use.

General Weed Control—Except Groundcherry and Nightshade:

Broadcast Rates Per Acre

Soil Texture	SONALAN
	(Pints)
Coarse	1½–2
Medium	2–2½
Fine	2½–3

Precaution: Do not graze or forage crop grown in treated soil or cut for hay or silage.

Read the SONALAN label for additional directions, cautions and limitations prior to use.

B. Sunflower—SONALAN/Amiben Tank Mix or Overlay

SONALAN/Amiben should be applied as a preplant incorporated tank mix, or SONALAN preplant incorporated followed by an Amiben overlay. The SONALAN/Amiben combination effectively controls the weeds listed for SONALAN alone for general weed control plus these additional weeds:

Common ragweed
Pennsylvania smartweed
Wild mustard
Black nightshade

Broadcast Rates Per Acre

Soil Texture	SONALAN +	Amiben 2S	or	Amiben DS
	(pints)	(quarts)		(pounds)
Coarse	1¼–2	4–6*		2.4–3.6*
Medium	1¾–2½	4–6		2.4–3.6
Fine	2¼–3	4–6		2.4–3.6

Precaution: Do not graze or forage crop grown in treated soil or cut for hay or silage.

*Use the higher rate where heavy populations of smartweed, velvetleaf, ragweed, wild mustard or black nightshade are expected.

Read the Amiben label for directions, cautions and limitations prior to use.

C. Sunflower—SONALAN/Eptam Tank Mix for Use Only in North Dakota, South Dakota and Minnesota

The SONALAN/Eptam tank mix effectively controls the weeds listed for SONALAN alone for general weed control plus these additional weeds:

Annual Weeds
Ryegrass (Italian)
Bermudagrass (seedling)
Rescuegrass
Volunteer grains (barley, oat, wheat)
Annual morningglory
Black nightshade
Hairy nightshade
Corn spurry
Fiddleneck
Goosefoot nettleleaf
Shepherdspurse
Tumble pigweed

Perennial Weeds
Bermudagrass
Purple nutsedge
Yellow nutsedge

Follow recommended soil preparation directions for SONALAN. Apply the tank mix in at least 10 gallons of water of liquid fertilizer per acre. Apply and immediately incorporate the tank mix prior to planting in the spring.

Broadcast Rates Per Acre

Soil Texture	SONALAN	+	Eptam 7E
	(pints)		(pints)
Coarse	1¼–2		2½–3½
Medium	1¾–2½		2½–3½
Fine	2¼–3		2½–3½

Precaution: Do not graze or forage crop grown in treated soil or cut for hay or silage.

Read the Eptam label for directions, cautions and limitations prior to use.

SONALAN®—ethalfluralin, Elanco Products Company

Amiben®—chloramben, Union Carbide Agricultural Products Company

Eptam®—EPTC, Stauffer Chemical Company

SUPPLEMENTAL LABELING

SONALAN® E.C. EPA Reg. No. 1471-122

SONALAN FOR WEED CONTROL IN DRY PEAS

SONALAN may be applied as a preplant incorporated herbicide for general weed control in dry peas. Follow recommended soil preparation, application and incorporation procedures for SONALAN.

Broadcast Rates Per Acre

Soil Texture	SONALAN (Pints)
Coarse	1½–2
Medium	2

Precaution: Do not exceed these SONALAN rates for weed control in dry peas or crop injury may occur. Do not graze or forage crop grown in treated soil or cut for hay or silage.

Read the SONALAN label for additional directions, precautions, and limitations prior to use.

SONALAN®—ethalfluralin—Elanco Products Company

SPIKE® 80W

HERBICIDE

A Preemergence and postemergence herbicide for control of brush and weeds in such areas as: Airport runways, Utility substations and rights-of-way, Tank farms, Railway roadbeds and ballast, Railroad rights-of-way, Road shoulders where no vegetation is desired, Under asphalt and concrete pavements where no future landscaping is planned, At the base of highway guardrails, signposts, and markers, At the base of transmission towers and poles, Around industrial buildings, Lumberyards, Railroad yards, Ditchbanks, Firebreaks, and Fence rows

The degree and duration of control may vary with the amount of chemical applied, soil texture, and other conditions.

ACTIVE INGREDIENT:
tebuthiuron: *N*[5-(1,1-dimethylethyl)-1,3,4-thiadiazol-2-yl]-*N,N'*-dimethylurea 80.0%
INERT INGREDIENTS: 20.0%

TOTAL 100.0%

Contains 3.2 pounds active ingredient per 4 pound bag. SPIKE®—the registered trademark for Elanco Products tebuthiuron.

EPA Reg. No. 1471-97

CAUTION

Keep Out of Reach of Children.

See additional caution statements following Individual (Spot) Treatment section.

DIRECTIONS FOR USE

Read All Directions Carefully Before Applying.

Human: Harmful if swallowed. Avoid contact with skin, eyes or clothing. Avoid inhaling dust from product. In case of contact, flush with water.

ENVIRONMENTAL

Do not contaminate any body of water, ponds or streams as death or injury may occur to vegetation irrigated by such. So not contaminate water by cleaning of equipment or disposal of wates.

STORAGE AND DISPOSAL

The herbicidal properties of SPIKE 80W require caution in handling, storage and transportation of this product. Do not contaminate food or feed by storage or disposal. Open dumping is prohibited. Do not reuse empty container. Dispose in an incinerator or land fill approved for pesticide containers or bury in a noncropland area away from desirable plants, trees and water supply.

The manufacturer makes no warranties, express or implied, concerning this product or its use which extend beyond the description on the label. All statements made concerning this product apply only when used as directed. SPIKE 80W must be applied according to Elanco's written instructions, including, but not limited to, recommended rates. Failure to do so may result in poor weed control or plant injury. Elanco expressly disclaims any warranty, either express or implied, for the use of SPIKE 80W, alone or in combination with other products, when that use is not in strict compliance with Elanco's written recommendations.

PRECAUTIONS

SPIKE 80W IS INTENDED FOR NONCROPLAND VEGETATION CONTROL. IT IS AN EXTREMELY ACTIVE HERBICIDE WHICH WILL KILL TREES, SHRUBS AND OTHER FORMS OF DESIRABLE VEGETATION HAVING ROOTS EXTENDING INTO THE TREATED AREA. FEEDER ROOTS OF MANY SPECIES OF DESIRABLE VEGETATION EXTEND MANY FEET BEYOND THE DRIP LINE OF THE BRANCHES, AND A VERY SMALL AMOUNT OF SPIKE 80W IN CONTACT WITH ONE FEEDER ROOT OF A TREE, SHRUB OR OTHER DESIRABLE VEGETATION MAY CAUSE SERIOUS INJURY OR DEATH TO THE ENTIRE PLANT.

AN ARBORICULTURIST (TREE EXPERT) SHOULD BE CONSULTED TO HELP YOU TO DETERMINE IF THE AREA OF PROPOSED APPLICATION IS FREE OF ALL ROOTS OF DESIRABLE VEGETATION. THE EFFECT OF SPIKE 80W ON DESIRABLE VEGETATION MAY BE IRREVERSIBLE AND ITS PRESENCE IN THE SOIL MAY PREVENT GROWTH OF OTHER DESIRABLE VEGETATION FOR SOME YEARS AFTER APPLICATION.

READ THE ENTIRE LABEL BEFORE USING SPIKE 80W TO DETERMINE IF THIS PRODUCT IS SUITABLE FOR THE DESIRED PURPOSE.

Do not use SPIKE 80W on areas such as walks, driveways, streets, lawns, patios, tennis courts, swimming pools, cemeteries, or other landscaped areas, or under asphalt or concrete pavement where future landscaping is planned. Do not apply on field crops. Do not apply on any area into which the roots of field crops or other desirable vegetation may extend. ROOTS OF TREES, SHRUBS, AND OTHER DESIRABLE VEGETATION MAY EXTEND FAR BEYOND THE DRIP LINE OF THE PLANT'S BRANCHES.

AVOID NONTARGET DRIFT OR PRODUCT MOVEMENT. DO NOT APPLY WHEN WINDS ARE GUSTY OR UNDER ANY OTHER CONDITIONS WHICH WILL ALLOW DRIFT OR PRODUCT MOVEMENT. DO NOT APPLY TO AREAS WHERE SOIL MOVEMENT BY WATER EROSION AND/OR NATURAL OR MECHANICAL MEANS IS LIKELY. DO NOT APPLY TO AREAS WHERE WIND IS LIKELY TO CAUSE SOIL MOVEMENT UNLESS A SOIL SEALANT IS USED. DRIFT OR ANY FORM OF PRODUCT MOVEMENT FROM TREATED AREAS MAY CAUSE DAMAGE TO ANY VEGETATION TO WHICH TREATMENT IS NOT INTENDED.

Ditchbank Usage—Do not apply SPIKE 80W to any portion of the ditchbank that will come into direct contact with water as movement of SPIKE 80W in this water to non-target plant species may result in the injury or death of those plants. Do not apply on ditches used to transport irrigation or potable water. Keep from contact with other pesticides and seeds.

Thoroughly clean all traces of SPIKE 80W from application equipment after use. DO NOT EMPTY RESIDUES CLEANED FROM APPLICATION EQUIPMENT ON AREAS WHERE THEY MAY COME IN CONTACT WITH THE ROOTS OF DESIRABLE VEGETATION OR THE WATER SOURCE FOR SUCH VEGETATION.

Woody Plant Control—Grazing is allowed in areas receiving band or individual plant treatments with 5 pounds per acre or less of SPIKE 80W. In areas receiving band or individual plant treatments with 5 pounds per acre or less of SPIKE 80W, grass may be cut for hay one year after application.

SPIKE 80W may injure or suppress certain herbaceous vegetation in the treated area. Therefore, do not apply where such injury cannot be tolerated. Do not apply broadcast applications of SPIKE 80W where forage or maintenance of a grass cover is desired. Injury to most herbaceous perennials is reduced if SPIKE 80W is applied when this vegetation is dormant.

DIRECTIONS FOR USE

Read All Directions Carefully Before Applying.

Total Vegetation Control: SPIKE 80W is a preemergence and post-emergence herbicide for total control of vegetation in such noncrop-land areas as: airport runways, utililty substations and rights-of-way, road shoulders where no vegetation is desired, under asphalt and concrete pavements where no future landscaping is planned, at the base of highway guardrails, sign posts and markers, at the base of transmission towers and poles, around industrial buildings, lum-beryards, railroad yards, ditchbanks, firebreaks, and fence rows.

For total vegetation control in areas *not treated the previous season* with SPIKE 80W or other residual herbicides, apply SPIKE 80W prior to or just after emergence of plants as follows:

At 5 pounds per acre, SPIKE 80W will control the following:

Alfalfa	Buttercup, smallflower
Aster, heath	Camphorweed
Aster, white heath	Carrot, wild
Barley, little	Catsear, spotted
Bedstraw	Cheat
Bluegrass, annual	Chickweed
Bluegrass, Kentucky	Clover, red
Bouncingbet	Cocklebur
Bromegrass, downy	Creeper, Virginia
Bromegrass, ripgut	Crowfootgrass
Bromegrass, smooth	Dock, curly
Broomsedge	Dogfennel
Buffelgrass	Fescue
Burclover	Fescue, rattail
Fiddleneck, coast	Pepperweed, Virginia
Filaree	Pigweed
Filaree, redstem	Plantain, buckhorn
Fleabane, annual	Puncturevine
Foxtail	Ragweed, giant
Gaillardia, rosering	Raspberry, red
Geranium, Carolina	Ryegrass, Italian
Goldenrod	Sedge, annual
Grape	Shepherdspurse
Gumweed	Sida, prickly
Hemlock, poison	Sowthistle, annual
Henbit	Spikeweed
Honeysuckle, Japanese	Spurge
Horseweed	Spurge, spotted
Knapweed	Starthistle, yellow
Kochia	Strawberry
Lambsquarters	Sunflower, common
Lupine	Telegraphplant
Medic, black	Timothy
Morningglory	Trumpetcreeper
Mullein, common	Velvetgrass
Nightshade, silverleaf	Vetch
Oat, wild	Witchgrass
Panicum, Texas	

At 7.5 pounds per acre, SPIKE 80W will control the following:

Awngrass, triple	Ivy, poison
Barley, foxtail	Reed, common
Bromegrass, Japanese	Sandbur, field
Canarygrass, reed	Smartweed, swamp
Carpetweed	Sowthistle, perennial
Chicory	Spurge, prostrate
Cinquefoil, common	Sumac, staghorn
	Sweetclover, white

At 10 pounds per acre, SPIKE 80W will control the following:

Barnyardgrass	Oxtongue, bristly
Crabgrass	Poorjoe
Itchgrass	Purslane, common
Johnsongrass, seedling	Ragweed, common
Lovegrass	Saltbush
Orchardgrass	Sedge, bull

At 20 pounds per acre, SPIKE 80W will control the following:

Bermudagrass	Lookingglass, Venus
Cudweed	Vaseygrass
Dallisgrass	

For the *maintenance of total vegetation control* in noncropland areas *east of the Rocky Mountains* which were treated the previous season with SPIKE 80W or other residual herbicides, apply SPIKE 80W prior to or just after emergence of plants as follows: (Some of the species listed may show erratic control depending on the time between application and weed germination.)

At 2 pounds per acre, SPIKE 80W will control the following:

Bluegrass, annual	Parsnip, wild
Bluegrass, Kentucky	Pepperweed
Carrot, wild	Pigweed
Checkweed, common	Ragweed, common
Croton	Smartweed, Pennsylvania
Fleabane, annual	Sweetclover
Horseweed	Thistle, Canada
Mullein	Woodsorrel, yellow
Panicum, fall	

At 3 pounds per acre, SPIKE 80W will control the following:

Goldenrod	Spurge

In areas of rainfall greater than 25 inches per year, the 3 pounds per acre maintenance rate should be used for all weed species listed above.

For the *maintenance of total vegetation control* in noncropland areas *west of the Rocky Mountains* which were treated the previous season with SPIKE 80W or other residual herbicides, apply SPIKE 80W prior to or just after emergence of plants as follows: (Some of the species listed may show erratic control depending on the time between application and weed germination.)

At 1.5 pounds per acre, SPIKE 80W will control the following:

Bassia, fivehook	Pigweed
Cheat	Plantain
Cudweed	Ryegrass, annual
Foxtail	Saltbush
Lettuce, prickly	Shepherdspurse
Oat, wild	Witchgrass
Oxtongue, bristly	

At 2 pounds per acre SPIKE 80W will control the following:

Buttercup	Mustard
Canarygrass, reed	Ragweed, western
Knapweed, Russian	Starthistle, yellow
Knotweed	Telegraphplant
Mallow	

At 3 pounds per acre, SPIKE 80W will control the following:

Barley	Sida, alkali
Gumweed	Smartweed, swamp
Puncturevine	

In areas of rainfall greater than 25 inches per year, the 3 pounds per acre maintenance rate should be used for all weed species listed above.

APPLICATION DIRECTIONS

Apply SPIKE 80W in 15 to 150 gallons of water per acre before or during the period of active growth of plants to be controlled. Initial control is enhanced by rainfall.

In areas of low annual rainfall (less than 15 inches per year) SPIKE 80W should be applied prior to the time of year when the predominant portion of that rainfall occurs. A minimum of 1 to $1\frac{1}{2}$ inches of rainfall is required to activate SPIKE 80W and place it in the primary weed seed germination zone.

To maximize performance under a variety of conditions, SPIKE 80W is recommended for use in tank-mix combination with any of the following products: amitrole, Atratol® 80W, Banvel® 720, Banvel® W.S., Karmex® 80W, MSMA, paraquat, Princep® 80W, Roundup®, SURFLAN® 75W, 2,4-D or Oust®. Where applications are made to existing vegetation, the contact or burning properties of paraquat or MSMA provide the rapid top kill while SPIKE 80W gives the residual long-term control desired. Applications to areas which are infested with certain difficult to control perennial weeds (such as johnsongrass, bermudagrass, quackgrass, horsetail, bindweed, dandelion or nutsedge) will benefit from tank-mixing SPIKE 80W with Roundup, amitrole, Banvel, Banvel 720, or 2,4-D. The addition of SURFLAN, Oust, Karmex, Atratol or Princep will improve SPIKE's performance on certain annual broadleaf and grass weeds such as foxtail, kochia, Russian thistle, or seedling johnsongrass.

Banvel®—dicamba, Velsicol Chemical Corporation
Karmex®—diuron, E. I. DuPont De Nemours & Company
Princep®—simazine, Ciba-Geigy
Roundup®—glyphosate, Monsanto Company
Oust®—Sulfometuron methyl, E. I. DuPont De Nemours & Company
SURFLAN®—oryzalin, Elanco Products Company
Atratol®—atrazine and prometon, Ciba-Geigy

Read the SPIKE 80W label and labels of products to be tank-mixed carefully before using. Note all warnings, cautions, precautions, and limitations of warranty on all labels.

Apply with any sprayer that will apply the spray uniformly. Check the sprayer before and during use to insure proper calibration and uniform application.

To mix, fill spray tank half full of water. Start agitation and continue during the entire mixing process. Add required amount of SPIKE 80W and allow to mix when tank mixing. If additional product is a wettable powder, add to tank and allow to mix thoroughly. If additional product is a liquid, add slowly while filling remainder of tank with water. Continuous agitation in the spray tank is required to keep the materials in suspension throughout application. Agitate by mechanical or bypass (hydraulic) means in the spray tank. If bypass or return agitation is used, it should terminate at the bottom of the tank to minimize foaming.

For treating small areas, a tank type hand sprayer or sprinkling can may be used. Before application determine the amount of water and chemical necessary to cover uniformly the area to be treated. Shake or stir frequently.

WOODY PLANT CONTROL

SPIKE 80W is an effective herbicide for the control of brush and vines. SPIKE 80W can be applied either as a broadcast spray, banded application or as an individual plant treatment depending upon the size, density and location of brush to be controlled.

SPIKE is to be applied to the soil (Not the Foliage!) where it is absorbed by the roots of plants. Effects are slow to appear and will not become apparent until sufficient moisture has carried SPIKE 80W into the root zone. The time required to achieve control is dependent on soil type, amount of rainfall and rooting depth of target species. Some species may go through several defoliations and refoliations over a period of approximately two to three years prior to dying.

SPIKE 80W can be applied anytime except when the ground is frozen or the soil is saturated with moisture. For optimum results, applications should be made just prior to the resumption of active seasonal growth in the spring and/or periods of rainfall. For applications made in the late summer or early fall in areas of average annual rainfall of greater than 25 inches, higher rates should be used and inconsistent control may result on densely infested brush areas and hard to control species.

SPIKE 80W may be used on cut brush but for optimum results time should be allowed for the brush to resprout to a height of approximately 5 feet prior to application. SPIKE requires an actively growing plant to be effective. The larger the resprouts the more SPIKE that will be taken up by the plant and the more effective and consistent the control will be.

For the control of woody plants and vines, the following rates of SPIKE 80W are recommended. These rates can vary depending upon soil type, rainfall, time of application and size/density of the woody plants.

SPIKE 80W applied at the rate of 1.25 pounds per acre will control the following species.

Haplopappus tenuisectus	Burroweed
Larrea tridentata	Creosotebush
Mimosa biuncitera	Wait-a-minute-bush

SPIKE 80W applied at the rate of 2.5 pounds per acre will control the following species.

Ailanthus altissima	Tree-of-heaven
Aloysia lycioides	Whitebrush
Artemisia tridentata	Big sagebrush
Carya glabra	Pignut hickory
Celtis occidentalis	Western hackberry
Datura discolor	Desert thornapple
Lycium berlandieri	Berlandier wolfberry
Morus rubra	Red mulberry
Pinus monticola	Western white pine
Pinus spp.	Pine
Prunus emarginata	Bitter cherry
Rhus glabra	Smooth sumac
Robinia pseudoacacia	Black locust
Rosa multiflora	Multiflora rose
Salvia leucophylla	Black sage
Symphoricarpos orbiculatus	Buckbrush
Ulmus americana	American elm
Vaccinium spp.	Blueberry
Gaylussacia spp.	Huckleberry

SPIKE 80W applied at the rate of 3.75 pounds per acre will control the following species.

Abies balsamea	Balsam fir
Acacia farnesiana	Huisache
Acer saccharum	Sugar maple
Alnus rugosa	Speckled alder
Betula populifolia	Gray birch
Carya texana	Black hickory
Celtis pallida	Granjeno
Condalia obtusifolia	Lotebush condalia
Ilex vomitoria	Yaupon
Larix laricina	Tamarack
Picea glauca	White spruce
Populus balsamifera	Balsam poplar
Populus deltoides	Eastern cottonwood
Quercus douglasii	Blue Oak
Quercus marilandica	Blackjack oak
Quercus stellata	Post oak
Salix spp.	Willow
Schaefferia cuneifolia	Desert yaupon
Spiraea tomentosa	Hardhack
Ulmus alata	Winged elm

SPIKE 80W applied at the rate of 5 pounds per acre will control the following species.

Acacia berlandieri	Guajillo
Acacia greggii	Catclaw acacia
Acacia rigidula	Blackbrush acacia
Acacia tortuosa	Twisted acacia
Acer negundo	Boxelder
Adenostoma fasciculatum	Chamise
Alnus rubra	Red alder
Campsis radicans	Trumpetcreeper
Carya ovata	Shagbark hickory
Cercocarpus betuloides	Birchleaf mountainmahogany
Colubrina texensis	Texas colubrina
Condalia obovata	Bluewood condalia
Cornus drummondii	Roughleaf dogwood

Crataegus spp.	Hawthorn
Eysenhardtia texana	Texas kidneywood
Fagus grandifolia	American beech
Jatropha dioica	Leatherstem
Leucophyllum frutenscens	Cenizo [Texas silverleaf]
Liquidambar styraciflua	Sweetgum
Parthenocissus quinquefolia	Virginia creeper
Populus grandidentata	Bigtooth aspen
Porlieria angustifolia	Guayacan
Prunus virginiana	Common chokecherry
Pseudotsuga menziesii	Douglas fir
Pueraria lobata	Kudzu
Quercus dumosa	California scrub oak
Quercus palustris	Pin oak
Quercus rubra	Red oak
Quercus virginiana	Live oak
Rhus typhina	Staghorn sumac
Rubus allegheniensis	Allegheny blackberry
Salvia ballotaeflora	Shrubby blue salvia

SPIKE 80W applied at the rate of 6.25 pounds per acre will control the following species:

Acer macrophyllum	Bigleaf maple
Acer platanoides	Norway maple
Acer saccharinum	Silver maple
Baccharis spp.	Groundsel tree
Cornus florida	Flowering dogwood
Fraxinus pennsylvanica	Green ash
Gaultheria shallon	Salal
Juniperus virginiana	Eastern redcedar
Lantana camara	Lantana
Liriodendron tulipifera	Tuliptree
Melaleuca quinquenervia	Melaleuca
Pinus banksiana	Jack pine
Pinus echinata	Shortleaf pine
Pinus resinosa	Red pine
Pinus virginiana	Virginia pine
Platanus occidentalis	American sycamore
Prunus serotina	Black Cherry
Quercus alba	White oak
Rubus laciniatus	Evergreen blackberry
Rubus occidentalis	Black raspberry
Schinus terebinthifolius	Brazilian peppertree

SPIKE 80W applied at the rate of 7.5 pounds per acre will control the following species.

Acer circinatum	Vine maple
Arctostaphylos patula	Greenleef manzanita
Ceanothus cuneatus	Wedgeleaf ceanothus
Ceanothus leucodermis	Whitethorn chaparral
Crataegus crus-galli	Cockspur hawthorn
Elaeagnus angustifolia	Russianolive
Fraxinus americana	White ash
Ligustrum spp.	Privet
Rhus laurina	Laurel sumac
Sapium sebiferum	Tallowtree
Smilax rotundifolia	Common greenbrier
Ulmus parvifolia	Chinese elm
Ulmus rubra	Slippery elm

BROADCAST APPLICATION

Apply SPIKE 80W in 15 to 150 gallons of water per acre with any properly calibrated herbicide sprayer. Check the sprayer before and during use to insure proper calibration and uniform application. Add the recommended amount of SPIKE 80W to clean water in the spray tank during the filling operation. Material must be kept in suspension at all times by constant agitation. Agitate by mechanical or bypass (hydraulic) means in the spray tank. If bypass or return agitation is used it should terminate at the bottom of the tank to minimize foaming.

For treating small areas, a tank type hand sprayer may be used. Before application detemine the amount of water and chemical necessary to cover uniformly the area to be treated. Shake or stir frequently.

Do not apply broadcast applications of SPIKE 80W where forage or maintenance of a grass cover is desired.

BANDED APPLICATION

SPIKE 80W is recommended for the control of woody plant species in noncropland areas (such as utility railroad and pipeline rights-of-way ditchbanks and fence rows) by application of a series of parallel bands to the soil surface. Individual bands should be spaced at intervals from 4 to 10 feet and at the currently labeled rate range of 2.5 to 7.5 pounds per acre depending on the woody species to be controlled. Actual herbicide bands should be kept as narrow as possible during application to achieve minimal injury or control of herbaceous vegetation. Apply SPIKE 80W to the soil surface in 5 to 75 gallons of water per acre in a series of parallel bands with spacing between bands ranging from 4 to 10 feet. In areas such as brush infested fence rows on utility rights-of-way a single band may be applied. Control is dependent upon root systems intercepting bands. Therefore larger stems should be *treated individually* when using single bands.

Band spacing should be selected based on the size of the woody plants in the area to be treated and the amount of injury or control of herbaceous vegetation that can be tolerated.

Where control of young or seedling woody plants is desired bands should be spaced closer together. This will achieve maximum exposure to their limited root systems. Where larger more mature woody species are to be controlled, bands should be spaced at the wider end of the recommended spacing range.

In addition to allowing adequate exposure of the more extensive root systems of these larger woody species for control, use of the wider spacings will further reduce injury or control of herbaceous vegetation in the area of treatment.

Within the treated band nearly all vegetation, woody and herbaceous, will be killed. Some herbaceous vegetation close to the treated band with roots extending into it may be severely injured or killed. However, since herbaceous species tend to have restricted root systems, most species outside the treated band will not be affected. Banded applications in areas of steep terrain should be applied across existing slopes in order to prevent soil erosion.

Apply with equipment designed to deliver the spray uniformly in the bands. To maintain the integrity of the individual herbicide bands, straight stream nozzles fitted with internal stabilizing vanes or their equivalent are recommended. Operating pressures should also be kept as low as will provide uniform delivery of the spray solution. Pressures in the range of 10 to 40 psi should be adequate. Pressures in excess of 40 psi will tend to cause the individual bands to break up.

When applications are made in an area where nozzles are elevated above the soil surface more than 5 feet, breakup of the individual spray streams may occur. If conditions do not permit delivery of intact spray streams to the soil surface, this method of application should not be used.

Fill the spray tank half-full of water. Start agitation and continue during entire mixing and spraying operation. Add the required amount of SPIKE 80W and allow it to mix thoroughly while completing the spray tank filling. If hand held or back pack type sprayers are used, shake vigorously after filling and periodically during application to maintain product suspension. A master shut-off switch for the entire spraying system and nozzle check valves are recommended on commercial spray equipment.

Material must be kept in suspension at all times by continuous agitation. Agitate by mechanical or bypass (hydraulic) means in the spray tank. If bypass or return agitation is used it should terminate at the bottom of the tank to minimize foaming. Check the sprayer frequently before and during use to insure proper calibration and uniform application.

INDIVIDUAL (SPOT) APPLICATION

SPIKE 80W may be applied in high or low volumes of water to selectively control individual woody plants. Recommended rates will vary depending upon site conditions with the higher rates needed for difficult to control species, large plants, heavier soils, fall applications and cut brush. Consult your local Elanco SPIKE distributor to determine the best rates for your location.

For high volume applications, mix 1 pound of SPIKE 80W in enough water to make 10 gallons of solution. Apply 10 ounces of material to the soil per every 2 to 4 inches of stem diameter.

For low volume applications, mix 1 pound of SPIKE 80W in enough water to make 1 gallon of solution. Apply 1 ounce of material to the soil per every 2 to 4 inches of stem diameter.

When treating large stems, apply the multiple treatments in even spacing around the stem.

For applying SPIKE 80W in banded or individual plant treatment, two pieces of equipment are suggested; the Solo Model 425 back pack sprayer for both banding and individual plant treatment and the Spot Gun for individual plant treatment.

The Solo sprayer is prepared for spraying by adding the pre-slurried contents of a 4 pound bag of SPIKE 80W and water to the tank. Fill to capacity with additional water and shake vigorously. Equip the Solo sprayer with a 0003-SS straight stream nozzle and the Solo pressure regulator with the green (10 psi) pressure limiting spring. To band SPIKE 80W at 5 pounds per acre, walk at 3 mph (264 feet per minute) with the Solo on continuously and space the bands 5 feet apart. Adjust the rate and walking speed according to the brush species and conditions encountered. For individual plant treatment with the Solo, apply a 1.5 second shot for every 1 to 2 inches of stem diameter at the base of unwanted woody plants.

The Spot Gun is prepared for individual plant treatment by mixing 2 pounds of SPIKE 80W in sufficient water to obtain 1 gallon of spray solution. Set the Spot Gun to deliver 8 milliliters of this solution for every 1 to 2 inches of stem diameter at the base of the unwanted woody plants. For application on steep slopes or other sensitive areas, the Spot Gun can be equipped with a soil probe to allow injection of the SPIKE 80W solution beneath the soil surface. Placement at a soil depth of 2 to 4 inches will eliminate any surface movement and reduce injury to herbaceous vegetation.

At the prescribed rates, a 4 pound bag of SPIKE 80W will treat approximately 950 stems 1 to 2 inches in diameter, because of its non-volatile nature and low potential for drift this SPIKE application technique can be used for treating unwanted woody plants growing on non-cropland areas adjacent to sensitive crops. (See label precautions.) A white spot should be visible at the base of each treated stem which should aid in inspection of the completed work.

CAUTION: DO NOT USE SPIKE 80W IN THIS MANNER IN ANY AREA WHERE DESIRABLE SPECIES ARE IN THE VICINITY OF THE PLANTS TO BE ELIMINATED. A SMALL AMOUNT OF SPIKE 80W IN CONTACT WITH THE ROOTS OF DESIRABLE TREES OR OTHER WOODY SPECIES MAY CAUSE SEVERE INJURY OR DEATH. THE ROOTS OF SUCH PLANTS MAY EXTEND FAR BEYOND THEIR DRIP LINES.

SPIKE 80W will injure or control other herbaceous vegetation in the treated area. Therefore, do not apply where such injury cannot be tolerated. See the list of herbaceous vegetation controlled by SPIKE 80W under the Total Vegetation Control section of this booklet.

SPIKE®—tebuthiuron, Elanco Products Company.

SPIKE 20P

HERBICIDE

For Control of Brush and Woody Plants in areas such as: Along highways, Ditchbanks, Fence rows, Firebreaks, Industrial sites, Pastures, Pipelines, Railraod rights-of-way, Rangeland, Utility rights-of-way and other noncropland areas for the control of woody plant species

ACTIVE INGREDIENT:
tebuthiuron*: *N*-[5-(1,1-dimethylethyl)-1,3,4-thiadiazol-2-yl]-*N,N'*-dimethylurea 20.0%
INERT INGREDIENTS: 80%
TOTAL 100.0%

Contains 5 pounds active ingredient per 25 pound bag.

* SPIKE® the registered trademark for Elanco Products tebuthiuron

SPIKE® will kill trees and shrubs. Carefully read the precautions before using.

CAUTION

Keep Out of Reach of Children

HAZARDS TO HUMANS AND DOMESTIC ANIMALS

Harmful if swallowed. Avoid contact with skin, eyes, or clothing. In case of contact, flush with water.

STATEMENTS OF PRACTICAL TREATMENT

If swallowed: Call a physician or Poison Control Center. Drink one or two glasses of water and induce vomiting by touching back of throat with finger or, if available, by administering one to two teaspoons of syrup of ipecac. Do not induce vomiting or give anything by mouth to an unconscious person.

If on skin: Wash with plenty of soap and water. Get medical attention if irritation persists.

If in eyes: Flush eyes with plenty of water. Call a physician if irritation persists.

If inhaled: Remove individual to fresh air. If breathing difficulty occurs, provide cardio-pulmonary resuscitation assistance and get medical attention.

ENVIRONMENTAL HAZARDS

Keep out lakes, ponds or streams. Do not contaminate water by cleaning of equipment or disposal of wastes.

STORAGE AND DISPOSAL

STORAGE

The herbicidal properties of SPIKE 20P require caution in handling, storage and transportation of this product. Store in original container only. In case of leak or spill, contain material and dispose as waste.

PESTICIDE DISPOSAL

Do not contaminate water, food, or feed by storage or disposal. Open dumping is prohibited. Wastes resulting from the use of this product may be disposed of on site or at an approved waste disposal facility.

CONTAINER DISPOSAL

Completely empty bag into application equipment. Do not reuse. Then dispose of empty bag in a sanitary landfill, away from desirable plants, trees and water supply, or by incineration, or, if allowed by State and local authorities, by burning. If burned, stay out of smoke.

PRECAUTIONS

SPIKE 20P IS INTENDED FOR BRUSH AND WOODY PLANT VEGETATION CONTROL. IT IS AN EXTREMELY ACTIVE HERBICIDE WHICH WILL KILL TREES, SHRUBS AND OTHER FORMS OF DESIRABLE VEGETATION HAVING ROOTS EXTENDING INTO THE TREATED AREA. FEEDER ROOTS OF MANY SPECIES OF DESIRABLE VEGETATION EXTEND MANY FEET BEYOND THE DRIP

LINE OF THE BRANCHES AND A VERY SMALL AMOUNT OF SPIKE 20P IN CONTACT WITH ONE FEEDER ROOT OF A TREE, SHRUB OR OTHER DESIRABLE VEGETATION MAY CAUSE SERIOUS INJURY OR DEATH IN THE ENTIRE PLANT. SPIKE 20P MAY SERIOUSLY INJURE DESIRABLE FORAGE LEGUMES SUCH AS LESPEDEZA OR CLOVER.

AN ARBORICULTURIST (TREE EXPERT) SHOULD BE CONSULTED TO HELP YOU TO DETERMINE IF THE AREA OF PROPOSED APPLICATION IS FREE OF ALL ROOTS OF DESIRABLE VEGETATION. THE EFFECT OF SPIKE 20P ON DESIRABLE VEGETATION MAY BE IRREVERSIBLE AND ITS PRESENCE IN THE SOIL MAY PREVENT GROWTH OF OTHER DESIRABLE VEGETATION FOR SOME YEARS AFTER APPLICATION.

READ THE ENTIRE LABEL BEFORE USING SPIKE 20P TO DETERMINE IF THIS PRODUCT IS SUITABLE FOR THE DESIRED PURPOSE.

Do not use SPIKE 20P on areas such as walks, driveways, streets, lawns, patios, tennis courts, swimming pools, cemeteries, or other landscaped areas, or under asphalt or concrete pavement where future landscaping is planned. Do not apply on field crops. Do not apply on any area into which the roots of field crops or other desirable vegetation may extend. ROOTS OF TREES, SHRUBS AND OTHER DESIRABLE VEGETATION MAY EXTEND FAR BEYOND THE DRIP LINE OF THE PLANT'S BRANCHES.

AVOID NONTARGET DRIFT OR PRODUCT MOVEMENT. DO NOT APPLY WHEN WINDS ARE GUSTY OR UNDER ANY OTHER CONDITION WHICH WILL ALLOW DRIFT OR PRODUCT MOVEMENT. DO NOT APPLY TO AREAS WHERE SOIL MOVEMENT BY WATER EROSION AND/OR NATURAL OR MECHANICAL MEANS FROM THE TARGET AREA IS LIKELY. DO NOT APPLY TO AREAS WHERE WIND IS LIKELY TO CAUSE SOIL MOVEMENT FROM THE TARGET AREA UNLESS A SOIL SEALANT IS USED. DRIFT OR ANY FORM OF PRODUCT MOVEMENT FROM TREATMENT AREAS MAY CAUSE DAMAGE TO ANY VEGETATION TO WHICH TREATMENT IS NOT INTENDED.

Ditchbank Usage—Do not apply SPIKE 20P to any portion of the ditchbank that will come into direct contact with water as movement of SPIKE 20P in this water to nontarget plant species may result in the injury or death of those plants. Do not apply on ditches used to transport irrigation or potable water. Keep from contact with other pesticides and seeds.

Thoroughly clean all traces of SPIKE 20P from application equipment after use. DO NOT EMPTY RESIDUES CLEANED FROM APPLICATION EQUIPMENT ON AREAS WHERE THEY MAY COME IN CONTACT WITH THE ROOTS OF DESIRABLE VEGETATION OR THE WATER SOURCE FOR SUCH VEGETATION.

Grazing and Haying—Grazing is only allowed in areas treated with 20 pounds per acre or less of SPIKE 20P. In areas treated with 20 pounds per acre or less of SPIKE 20P, grass may be cut for hay one year after application.

SPIKE 20P may injure or suppress certain herbaceous vegetation in the treated area. Therefore, do not apply where such injury cannot be tolerated. Injury to most herbaceous perennials is reduced if SPIKE 20P is applied when this vegetation is dormant.

DIRECTIONS FOR USE

Read All Directions Carefully Before Applying

It is a violation of Federal Law to use this product in a manner inconsistent with its labeling.

WOODY PLANT CONTROL

SPIKE 20P is an effective herbicide for the control of brush. SPIKE 20P can be applied either broadcast or as an individual plant treatment depending upon the size, density and location of brush to be controlled.

Broadcast Application: SPIKE 20P must be applied using equipment properly calibrated to provide a uniform distribution of pellets.

Use of equipment not designed to confine the spread of SPIKE 20P pellets to the target area may result in injury or death to desirable vegetation.

Spot Treatment Applications: Spot application of brush with SPIKE 20P is an excellent choice for controlling individual trees or scattered stands of brush. Apply SPIKE 20P pellets evenly on the soil over the root systems of clumps of brush at the rate of ½ ounce per 45 square feet of soil surface. For individual trees, use ½ ounce per 2 to 4 inches of stem diameter applied at the base of the tree. When treating large stems, apply the multiple treatments in even spacing around the stem.

SPIKE 20P is to be applied to the soil where it is absorbed by the roots of plants. Effects are slow to appear and will not become apparent until sufficient moisture has carried SPIKE 20P into the root zone. The time required to achieve control is dependent on soil type, amount of rainfall and rooting depth of target species. Some species may go through several defoliations and refoliations over a period of one to three years prior to dying.

SPIKE 20P can be applied anytime except when the ground is frozen or the soil is saturated with moisture. SPIKE 20P should not be used on soils containing more than five percent organic matter, more than 30% clay or on Blackland soils or less than desirable/erratic control may result. For optimum results, applications should be made just prior to the resumption of active seasonal growth in the spring and/or periods of rainfall. For applications made in the late summer or early fall in areas of average annual rainfall of greater than 25 inches, higher rates should be used and inconsistent control may result on densely infested brush areas and hard to control species.

SPIKE 20P may be used on cut brush but time should be allowed for the brush to resprout to a height of approximately 5 feet prior to application. SPIKE 20P requires an actively growing plant to be effective. The larger the resprouts the more SPIKE 20P that will be taken up by the plant and the more effective and consistent the control will be.

PASTURE/RANGELAND GENERAL USE INFORMATION

SPIKE 20P may cause temporary herbicidal symptoms to appear on perennial grasses. When more than 10 percent grass cover is present, apply SPIKE 20P when grass is dormant. Grazing should be deferred the following growing season to encourage grass response. Stands of cool season grasses may be reduced by SPIKE 20P applications. Factors contributing to the possibility of grass stand reduction include higher rates of SPIKE 20P, shallow soils and light brush density. Forage grass production usually increases as brush competition is reduced. However, increased grass production is also dependent on adequate rainfall and a sound management program. Consult the local pasture or range management specialist for details, concerning suitable overseeding species, timing, and fertilization programs.

Cool-Season Pasture Establishment: Apply SPIKE 20P at recommended rates. Do not burn treated areas for at least two growing seasons after application. Apply the recommended rate of seed and fertilizer into ash as soon as possible after the burn. Do not graze through the spring establishment period.

Warm-Season Pasture Establishment: Apply SPIKE 20P at recommended rates. Do not burn treated areas for at least two growing seasons after application. Apply seed and fertilizer into ash as soon as possible. Mix ryegrass at ⅓ the normal seeding rate along with the recommended rate of summer grasses to serve as a nurse crop. Do not graze through the establishment period.

Pasture/Rangeland Precaution: Do not cut or burn wood on treated areas for two years following a SPIKE 20P application. When more than 10 percent grass cover is present, apply SPIKE 20P when grass is dormant; and grazing should be deferred the following growing season to encourage grass response. Stands of cool season grasses such as fescue may be reduced by SPIKE 20P applications. Factors contributing to the possibility of grass and stand reduction include higher rates of SPIKE 20P, shallow soils and light brush density.

RECOMMENDATIONS FOR USE

For the control of woody plants and vines, the following rates of SPIKE 20P are recommended. These rates can very depending upon location, soil type, rainfall, time of application and size/density of the woody plants. Where a rate range is shown, the lower rate will provide for lower performance and will result in a lesser amount of grass damage; the higher rate will result in better brush control and may cause increased grass damage.

SPIKE 20P BROADCAST RATE RECOMMENDATIONS

For Rangeland Brush Control in Western United States

Apply SPIKE 20P at the rate of 3¾ to 5 pounds per acre to control the following species:

Acacia constricta	Whitethorn
Artemisia filifolia	Sand sagebrush (Partial Control)
Artemisia tridentata	Big sagebrush (Partial Control)
Cercidium spp.	Paloverde
Flourensia cernua	Tarbrush
Gutierrezia sarothrae	Perennial broomweed (Density less than 1/sq. ft.)
Larrea tridentata	Creosotebush
Leucophylium frutescens	Ceniza
Mimosa biuncifera	Catclaw mimosa

Apply SPIKE 20P at the rate of 5 to 6½ pounds per acre to control the following species:

Aloysia lycioides	Whitebrush
Arctostaphylos spp.	Manzanita
Celtis pallida	Spiny hackberry (Granjeno)
Haplopappus tenuisectus	Burroweed
Quercus turbinella	Shrub live oak
Rhus trilobata	Skunkbush sumac (Partial control)
Salvia ballotaeflora	Shrubby blue salvia

Apply SPIKE 20P at the rate of 3¾* to 7½ pounds per acre for partial control of the following species:

Quercus havardii	Sand shinnery oak (Partial control)

* NOTE: Rates for sand shinnery oak are dependent on soil conditions, rainfall and plant growth habit.

Apply SPIKE 20P at the rate of 5 to 10 pounds per acre to control the following species:

Quercus virginiana	Running live oak (Partial control)
Aloysia lycioides	Whitebrush
Quercus durandii	Bigelow oak (Partial control)
Quercus mohriana	Mohr oak (Partial control)

Apply SPIKE 20P at the rate of 10 to 15 pounds per acre to control the following species:

Acacia rigidula	Black brush
Quercus marilandica	Blackjack oak
Quercus stellata	Post oak
Ulmus alata	Winged elm

For Pasture, Rangeland, Railroad and Utiltiy Rights-of-Way, Industrial Sites, Pipelines, Fence Rows, Firebreaks, Ditchbanks and along Highways for Brush and Woody Plant Control in All Regions of United States

SPIKE 20P applied at the rate of 5 pounds per acre will control the following species:

Haplopappus tenuisectus	Burroweed
Larrea tridentata	Creosotebush
Mimosa biuncifera	Wait-a-minute-bush (Catclaw mimosa)

SPIKE 20P applied at the rate of 10 pounds per acre will control the following species:

Ailanthus altissima	Tree-of-heaven
Aloysia lycioides	Whitebrush
Artemisia tridentata	Big sagebrush
Carya glabra	Pignut hickory
Celtis occidentalis	Western hackberry
Datura discolor	Desert thornapple
Lycium berlandieri	Berlandier wolfberry
Morus rubra	Red mulberry
Pinus monticola	Western white pine
Prunus emarginata	Bitter cherry
Rhus glabra	Smooth sumac
Robinia psuedoacacia	Black locust
Rosa multiflora	Multiflora rose
Salvia leucophylla	Black sage
Symphoricarpos orbiculatus	Buckbrush
Ulmus americana	American elm
Vaccinium spp.	Blueberry
Gaylussacia spp.	Huckleberry

SPIKE 20P applied at the rate of 15 pounds per acre will control the following species:

Abies balsamea	Balsam fir
Acacia farnesiana	Huisache
Acer saccharum	Sugar maple
Alnus rugosa	Speckled alder
Betula populifolia	Gray birch
Carya texana	Black hickory
Celtis pallida	Granjeno
Condalia obtusifolia	Lotebush condalia
Ilex vomitoria	Yaupon
Larix laricina	Tamarack
Picea glauca	White spruce
Pinus spp.	Pine
Populus balsmifera	Balsam poplar
Populus deltoides	Eastern cottonwood
Quercus douglasii	Blue Oak
Quercus marilandica	Blackjack oak
Quercus stellata	Post oak
Salix spp.	Willow
Schaefferia cuneifolia	Desert yaupon
Spiraea tomentosa	Hardhack
Ulmus alata	Winged elm

SPIKE 20P applied at the rate of 20 pounds per acre will control the following species:

Acacia berlandieri	Guajillo
Acacia greggii	Catclaw acacia
Acacia rigidula	Blackbrush acacia
Acacia tortuosa	Twisted acacia
Acer negundo	Boxelder
Adenostoma fasciculatum	Chamise
Alnus rubra	Red alder
Campsis radicans	Trumpetcreeper
Carya ovata	Shagbark hickory
Cercocarpus betuloides	Birchleaf mountain-mahogany
Colubrina texensis	Texas colubrina
Condalia obovata	Bluewood condalia
Cornus drummondii	Roughleaf dogwood
Crataegus spp.	Hawthorn
Eysenhardtia texana	Texas kidneywood
Fagus grandifolia	American beech
Jatropha dioica	Leatherstem
Leucophyllum frutescens	Ceniza (Texas silverleaf)
Liquidambar styraciflua	Sweetgum
Parthenocissus quinquefolia	Virginia creeper
Populus grandidentata	Bigtooth aspen
Porlieria angustifolia	Guayacan
Prunus virginiana	Common chokecherry
Pseudotsuga menziesii	Douglas fir
Peuraria lobata	Kudzu
Quercus dumosa	California scrub oak

Quercus palustris	Pin oak
Quercus rubra	Red oak
Quercus virginiana	Live oak
Rhus typhina	Staghorn sumac
Rubus allegheniensis	Allegheny blackberry
Salvia ballotaeflora	Shrubby blue salvia

For Railroad and Utility Rights-of-Way, Industrial Sites, Pipelines, Fence Rows, Firebreaks, Ditchbanks and along Highways for Brush and Woody Plant Control in All Regions of United States

SPIKE 20P applied at the rate of 25 pounds per acre will control the following species:

Acer macrophyllum	Bigleaf maple
Acer platanoides	Norway maple
Acer saccharinum	Silver maple
Baccharis spp.	Groundsel tree
Cornus florida	Flowering dogwood
Fraxinus pennsylvanica	Green ash
Gaultheria shallon	Salal
Juniperus virginiana	Eastern redcedar
Lantana camara	Lantana
Liriodendron tulipifera	Tuliptree
Melaleuca quinquenervia	Melaleuca
Pinus banksiana	Jack pine
Pinus resinosa	Red pine
Pinus echinata	Shortleaf pine
Pinus virginiana	Virginia pine
Platanus occidentalis	American sycamore
Prunus serotina	Black cherry
Quecrus alba	White oak
Rubus laciniatus	Evergreen blackberry
Rubus occidentalis	Black raspberry
Schinus terebinthifolius	Brazilian peppertree

SPIKE 20P applied at the rate of 30 pounds per acre will control the following species:

Acer circinatum	Vine maple
Arctostaphylos patula	Greenleaf manzanita
Ceanothus cuneatus	Wedgeleaf caenothus
Ceanothus leucodermis	Whitethorn chaparral
Crataegus crus-galli	Cockspur hawthorn
Elaeagnus angustifolia	Russianolive
Fraxinus americana	White ash
Ligustrum spp.	Privet
Rhus laurina	Laurel sumac
Sapium sebiferum	Tallowtree
Smilax rotundifolia	Common greenbrier
Ulmus parvifolia	Chinese elm
Ulmus rubra	Slippery elm

PRECAUTION

SPIKE 20P WILL KILL TREES AND SHRUBS. CAREFULLY READ THE PRECAUTIONS BEFORE USING. DO NOT USE SPIKE 20P IN THIS MANNER IN ANY AREA WHERE DESIRABLE SPECIES ARE IN THE VICINITY OF THE PLANTS TO BE ELIMINATED. A SMALL AMOUNT OF SPIKE 20P IN CONTACT WITH THE ROOTS OF DESIRABLE TREES OR OTHER WOODY SPECIES MAY CAUSE SEVERE INJURY OR DEATH. THE ROOTS OF SUCH PLANTS MAY EXTEND FAR BEYOND THEIR DRIP LINES.

SPIKE 20P will injure or control other herbaceous vegetation in the treated area. Therefore, do not apply where such injury cannot be tolerated.

INHERENT RISKS OF USE

Plant injury, unsatisfactory bush or woody plant control or other unintended consequences may result from such factors as weather conditions, presence of other materials or applicaton of SPIKE 20P contrary to label instructions, all of which are beyond the control of Elanco.

DISCLAIMER OF WARRANTIES

The manufacturer makes no warranties concerning this product or its use, which extend beyond the description on the label. All other warranties, express or implied, including but not limited to, any implied warranties of merchantability and fitness for a particular purpose, are disclaimed. All statements concerning this product apply only when used as directed.

LIMITATION OF DAMAGES

Elanco's liability, whether in contract, warranty, tort, negligence, strict liability or otherwise, shall not exceed the return of the amount of the purchase price of the SPIKE 20P and under no circumstances shall Elanco be liable for special, indirect or consequential damages.

SURFLAN® A.S.
HERBICIDE

A Pre-Emergence Surface-Applied Herbicide for the Control of Most Annual Grasses and Certain Broadleaf Weeds.

ACTIVE INGREDIENT:
oryzalin* (3,5-dinitro-N^4N^4-dipropylsulfanilamide) 40.4%
INERT INGREDIENTS 59.6%
TOTAL 100.0%

Contains four pounds active ingredient per gallon.

EPA Reg. No. 1471-112 EPA Est. No. 37429-GA-01
ID 5500 PI 2652 AMP

SURFLAN®—the registered trademark for Elanco Products oryzalin.

CAUTION
Keep Out of Reach of Children

PRECAUTIONARY STATEMENTS

HAZARDS TO HUMANS AND DOMESTIC ANIMALS

Harmful if swallowed. Avoid breathing vapors. Avoid contact with skin, eyes or clothing. In case of contact, immediately flush eyes or skin with plenty of water. Get medical attention if irritation persists.

ENVIRONMENTAL HAZARDS

Keep out of lakes, ponds or streams. Do not contaminate water by cleaning of equipment or disposal of waste.

STORAGE AND DISPOSAL

PROHIBITIONS

Do not contaminate water, food or feed by storage or disposal. Open dumping is prohibited. Do not reuse empty container.

PESTICIDE DISPOSAL

Pesticide, spray mixture or rinsate that cannot be used or chemically reprocessed should be disposed of in a landfill approved for pesticides or buried in a safe place away from water supplies.

CONTAINER DISPOSAL

Shake well before using. Empty contents into spray tank. Refill container with water 1/5 to 1/4 full; shake thoroughly, pour into tank, triple rinse and drain. Dispose in an incinerator or landfill approved for pesticide containers, or bury in a safe place.

GENERAL

Consult federal, state or local disposal authorities for approved alternative procedures such as limited open burning.

The manufacturer makes no warranties, express or implied, concerning this product or its use, which extend beyond the description on the label or in literature published by Elanco Products Company, and all statements made concerning this product apply only when used as directed. Elanco Products Company expressly disclaims all warranties, express or implied, in relation to any use of the product in combination or sequential use with any other product not specifically recommended in writing by Elanco Products Company. If used in a combination or sequential use recommended by Elanco Products Company, the liability of Elanco shall in no manner extend to any damage, loss or injury not directly caused by the inclusion of the Elanco product in such combination or sequential use.

PRECAUTION

Do not feed forage from treated fields to livestock. Poor weed control or crop injury may result if directions are not followed. Over application may result in crop injury or soil residue. Do not plant any root crop for 12 months following a SURFLAN application.

WEEDS CONTROLLED IN FIELD CROPS BY SURFLAN

SURFLAN will not control established weeds.

Grasses	
Barnyardgrass (Watergrass)	*Echinochloa* sp.
Brachiaria (Signalgrass)	*Brachiaria* sp.
Browntop panicum*	*Panicum fasciculatum*
Crabgrasses (Large crabgrass) (Smooth crabgrass)	*Digitaria* spp.
Crowfootgrass	*Dactyloctenium aegyptium*
Fall panicum (Spreading panicgrass)	*Panicum dichotomiflorum*
Foxtails (Bottlegrass) (Bristlegrass) (Giant foxtail) (Green foxtail) (Pigeongrass) (Robust foxtail) (Yellow foxtail)	*Setaria* spp.
Goosegrass (Silver crabgrass) (Wiregrass) (Yardgrass)	*Eleusine indica*
Johnsongrass (Seedling only)	*Sorghum halepense*
Texas panicum* (Buffalograss) (Coloradograss)	*Panicum texanum*
Broadleaf Weeds	
Carpetweed	*Mollugo verticillata*
Florida pusley (Florida purslane) (Mexican clover) (Pusley)	*Richardia scabra*
Lambsquarters	*Chenopodium album*
Pigweeds (Carelessweed) (Prostrate pigweed) (Redroot pigweed) (Rough pigweed) (Spiny pigweed)	*Amaranthus* spp.
Purslane, common	*Portulaca oleracea*
Spurge, prostrate*	*Euphorbia supina*

* (See special instructions for Texas only)

SURFLAN also provides partial control or suppression of:

Ladysthumb	Smartweed, annual
Morningglory, annual	Spurge, spotted
Nightshade, black	Velvetleaf
Prickly sida (teaweed)	Volunteer wheat
Ragweed, common	

WEEDS CONTROLLED IN FRUIT, NUT CROPS AND VINEYARDS BY SURFLAN

SURFLAN applied in Fruit, Nut Crops and Videyards controls the following weeds:

Grasses	
Barley, little	*Hordeum pusillum*
Barnyardgrass (Watergrass)	*Echinochloa* sp.
Bluegrass, annual (Poa)	*Poa annua*
Brachiaria (Signalgrass)	*Brachiaria* sp.
Browntop panicum	*Panicum fasciculatum*
Crabgrasses (Large crabgrass) (Smooth crabgrass)	*Digitaria* spp.
Crowfootgrass	*Dactyloctenium aegyptium*
Cupgrass	*Eriochloa graclis*
Downy brome	*Bromus tectorum*
Fall panicum (Spreading panicgrass)	*Panicum dichotomiflorum*
Foxtails (Bottlegrass) (Bristlegrass) (Giant foxtail) (Green foxtail) (Pigeongrass) (Robust foxtail) (Yellow foxtail)	*Setaria* spp.
Goosegrass (Silver crabgrass) (Wiregrass) (Yardgrass)	*Eleusine indica*
Johnsongrass (Seedling only)	*Sorghum halepense*
Junglerice	*Echinochloa colonum*
Lovegrass, Mexican	*Eragrostis mexicana*
Lovegrass, orcutt	*Eragrostis orcuttiana*
Oat, wild	*Avena fatua*
Ryegrass, annual (Italia)	*Lolium multiflorum*
Texas panicum (Buffalograss) (Coloradograss)	*Panicum texanum*
Witchgrass	*Panicum capillare*
Broadleaf Weeds	
Carpetweed	*Mollugo verticillata*
Chickweed	*Stellaria media*
Cudweed	*Gnaphalium chilense*
Fiddleneck, coast	*Amsinckia intermedia*
Florida pusley (Florida purslane) (Mexican clover) (Pusley)	*Richardia scabra*
Henbit	*Lamium amplexicaule*
Knotweed, prostrate	*Polygonum aviculare*
Lambsquarters	*Chenopodium album*
Pigweeds (Carelessweed) (Prostrate pigweed) (Redroot pigweed) (Rough pigweed) (Smooth pigweed) (Spiny pigweed) (Tumble pigweed)	*Amaranthus* spp.
Puncturevine	*Tribulus terrestris*
Purslane, common	*Portulaca oleracea*
Rockpurselane, redmaids	*Calandrinia caulescens*
Sandbur, field	*Cenchrus incertus*
Shepherdspurse	*Capsella bursa-pastoris*
Sprangletop, red	*Leptochloa filiformis*

SURFLAN provides partial control or suppression of:

Filaree, red and white stem	Nightshade, black
Groundsel, common	Prickly sida (teaweed)
Ladysthumb	Ragweed, common
Lettuce, prickly	Ragweed, giant (horseweed)
Mallow, common	Rocket, London
Milkweed, climbing	Smartweed, annual
Morningglory, annual	Sowthistle, annual
Mustard, black	Spurge, prostrate
Mustard, wild	Velvetleaf

DIRECTIONS FOR USE

Read All Directions Carefully Before Applying

It is a violation of federal law to use this product in a manner inconsistent with its labeling.

SURFLAN is a preemergence, surface applied herbicide for the control of most annual grasses and certain broadleaf weeds.

General rule: The more coarse the soil or the lower the organic matter, the lower the rate.

SOIL TEXTURE GUIDE

Refer to the following guide to determine your soil texture.

Coarse (light) Soils	Sand
	Loamy sand
	Sandy loam
Medium soils	Loam
	Silt loam
	Silty clay loam*
	Silt
	Sandy clay loam*
Fine (heavy) Soils	Clay
	Clay loam
	Silty clay loam*
	Silty clay
	Sandy clay
	Sandy clay loam*

* Silty clay loam and sandy clay loam soils are transitional soils and may be classified as either medium or heavy textured soils. If silty clay loam or sandy clay loam soils are predominantly sand or silt, they are usually classified as medium textured soils. If predominantly clay, they are usually classified as fine textured soils.

MIXING DIRECTIONS

SURFLAN Alone in Water

Start with a clean spray tank. Fill sprayer one-third (1/3) to one-half (1/2) full with clean water. Start agitation. Add correct quantity of SURFLAN, continue agitation and finish filling the tank.

SURFLAN Tank Mix in Water

Vigorous, continuous agitation is required for all tank mixes. Sparger pipe agitators generally provide the best agitation in spray tanks. Avoid stirring or splashing air into the mixture during filling to prevent foaming. To do this, place the end of the fill pipe below the surface of the water in the spray tank. Do not allow the mixture to siphon back into the water source.

Mixing order: Fill the tank three-fourths (3/4) full with clean water. Start the agitation. Add dry flowables or wettable powders (WP) and agitate until the products are completely dispersed in the water. Allow additional mixing and dispersion time when using dry flowable type products. Add SURFLAN A.S., dry flowables, or other liquids and agitate until completely dispersed in the water. Next add any solution (S) products, mix thoroughly. Add emulsifiable concentrate (EC) products last, agitate and finish filling the tank. Maintain agitation during filling and through application. If spraying and agitation must be stopped before the tank is empty, the materials may settle to the bottom. In this case, it is important to again suspend all of the material in the bottom of the tank before continuing the spray application. A sparger agitator is particularly useful for this purpose. Sometimes it is more difficult to suspend settled material than it is to suspend it originally.

Read and carefully follow all label instructions for each material added to the tank. Premixing dry and flowable formulations with water (slurrying) and pouring the slurry through a 20 or 35 mesh wetting screen in the top of the tank will help assure good initial dispersion in the tank water. Line screens in the tank should be no finer than 50 mesh (100 mesh is finer than 50 mesh).

If you see a buildup of material on the walls of the spray tank, wash the tank with soapy water between fillings. Rinse and continue the spraying operation. Clean the tank, lines, and screens thoroughly after use.

SOIL PREPARATION

Destroy existing weeds before or during a SURFLAN application. The soil surface should be level enough to allow a sprayer to be operated efficiently and at speeds appropriate for uniform application.

APPLICATION

Ground Application: Apply in 20 to 40 gallons of water per acre (broadcast basis), using any properly calibrated low pressure herbicide sprayer that will apply the spray uniformly. Use herbicide tips and screens no finer than 50 mesh in nozzle and in-line strainers. As the amount of water used (spray volume) decreases, the importance of accurate calibration and uniform application increases. Check the sprayer daily to ensure proper calibration and uniform application. Agitate thoroughly before and during application. Avoid boom overlaps that will increase rates above those recommended.

Aerial Application: Use a standard aerial herbicide boom sprayer. Aerial spray equipment should be calibrated to apply the proper amount of SURFLAN A.S. alone or in tank mix combinations in two (2) to 10 gallons of spray mixture per acre. Nozzle screens and in-line strainers should be no finer than 50 mesh for tank mix combinations. SURFLAN A.S. (alone or in tank mix combinations) mixes readily with water for these concentrate aerial sprays; however, constant vigorous agitation that sweeps the contents from the bottom of the spray tank up into the main body of the liquid is required to maintain a uniform suspension until spray tank is empty. Avoid overlap of spray patterns. Do not apply when weather conditions favor drift from target area.

Chemigation: SURFLAN may be applied for weed control through center pivot* and solid set sprinkler irrigation systems. These systems should be equipped with automatic shutoff devices and valves to prevent back flow into the water source. Meter a constant flow of SURFLAN into the water during the irrigation period. Apply SURFLAN in a minimum of one inch (1") of irrigation water. Check the irrigation system before each use to ensure a uniform application.

* **Precaution:** Do not apply SURFLAN via center pivot irrigation systems to fruit and nut orchards or vineyards. In California, do not apply SURFLAN to non-dormant fruit and nut orchards or vineyards through overhead sprinkler irrigation systems.

ACTIVATION AND CULTIVATION

A single, one-half inch (1/2") rain or its equivalent in center pivot or sprinkler irrigation is necessary to activate SURFLAN when used alone or in a tank mix combination. If weeds begin to emerge, shallow cultivate one to two inches (1" to 2") to destroy existing weeds and place SURFLAN in zone of weed seed germination. If heavy rains pack the soil surface, rotary hoe to break the crust to aid crop emergence. Crops treated with SURFLAN can be shallowly cultivated without loss of activity.

FIELD CROPS

GENERAL DIRECTIONS

These recommendations are given as the broadcast (overall) rates of SURFLAN per acre. For band application, use proportionately less. SURFLAN is not recommended for use on soils containing more than five percent (5%) organic matter. Only one (1) application of SURFLAN per season should be made.

SOYBEAN—SURFLAN ALONE

Apply SURFLAN to the soil surface from four (4) weeks prior to planting up through two (2) days after planting at a broadcast rate per acre of three-fourths (3/4) quart on coarse soils; one (1) quart on medium soils and one and one-half (1 1/2) quarts on fine soils.

SOYBEAN—TANK MIX RECOMMENDATIONS

SOYBEAN—SURFLAN/SENCOR® (METRIBUZIN) OR SURFLAN/LEXONE® (METRIBUZIN) TANK MIX

Sencor®—metribuzin, Bayer GmbH
Lexone®—metribuzin, E. I. duPont de Nemours and Company

The SURFLAN/Sencor or SURFLAN/Lexone tank mix effectively controls all of the annual grasses and broadleaf weeds listed for SURFLAN plus many additional weeds including:

Browntop millet	Redweed
Hemp sesbania	Sicklepod

Jimsonweed
Prickly sida (teaweed)
Ragweed, common
Smartweed, annual
Velvetleaf

SURFLAN/Sencor and SURFLAN/Lexone also provide partial control or suppression of:

Cocklebur
Morningglory
Ragweed, giant (horseweed)

The SURFLAN/Sencor or SURFLAN/Lexone tank mix may be applied from two (2) weeks prior to planting up through two (2) days after planting, but before crop emergence with any low pressure herbicide sprayer. Do not spray over top of emerged soybeans.

Broadcast Rates Per Acre

Surface Applied Before or After Planting

Soil Texture	SURFLAN A.S.	Lexone 50 WP/4L or Sencor 50 WP/4L	Lexone Dry Flowable or Sencor Dry Flowable
	(quarts)	(pounds or pints)	(pounds)
Coarse	½–¾	½	⅓
Medium	¾–1	¾	½
Fine	1–1¼	1	⅔

Do not use the SURFLAN/Sencor or SURFLAN/Lexone tank mix on soils containing more than five percent (5%) organic matter in a preplant surface application or postplant surface application.

Special Precaution: Poor weed control and/or crop injury may result if directions are not followed. Over application may result in crop injury and/or soil residue. Do not use treated vines for feed or forage. Do not contaminate any body of water or apply to any area not specified on the label. Do not allow sprays to drift onto adjacent desirable plants. Do not plant any crop other than soybeans within four (4) months after application.

Caution: Please read the Sencor or Lexone label carefully for cautions, precautions and special precautions before applying these products.

Destroy empty containers. Do not reuse.

SOYBEAN—SURFLAN/LOROX® (LINURON) TANK MIX

Lorox®—linuron, E. I. duPont de Nemours and Company

The SURFLAN/Lorox tank mix effectively controls all of the annual grasses and broadleaf weeds listed for SURFLAN plus many additional weeds including:

Canarygrass
Chickweed, common
Cocklebur
Galinsoga
Morningglory, annual
Mustard
Nettleleaf goosefoot
Nightshade, black
Prickly sida (teaweed)
Radish, wild
Ragweed, common
Sicklepod
Smartweed, annual
Spurge, spotted
Velvetleaf

Follow SURFLAN procedures for soil preparation. The SURFLAN/Lorox tank mix may be applied within two (2) days after planting, but before crop emergence using any properly calibrated low pressure herbicide sprayer. Do not spray over top of emerged soybeans.

Broadcast Rates Per Acre

Soil Texture	SURFLAN A.S.	Lorox 50 WP/L
	(quarts)	(pounds or pints)
Coarse	½–¾	⅔–1¼
Medium	¾–1	1–1⅔
Fine	1–1¼	1¼–2

Do not use on sand or loamy sand soils.

Do not use the SURFLAN/Lorox tank mix on soils containing more than five percent (5%) organic matter.

Special Precaution: Poor weed control and/or crop injury may result if directions are not followed. Over application may result in crop injury and/or soil residue. Do not use treated vines for feed or forage. Do not contaminate any body of water or apply to any area not specified on the label. Do not allow sprays to drift onto adjacent desirable plants.

Caution: Please read the Lorox label carefully for cautions, precautions and special precautions before applying this product.

Destroy empty containers. Do not reuse.

SOYBEAN—SURFLAN/DYANAP® OR SURFLAN/KLEAN-KROP® TANK MIX

Dyanap®—naptalam + dinoseb, Uniroyal Chemical

Klean-Krop®—naptalam + dinoseb, Thompson-Hayward Company

The SURFLAN/Dyanap or SURFLAN/Klean-Krop tank mix effectively controls all of the annual grasses and broadleaf weeds listed for SURFLAN plus many additional weeds including:

Bindweed, field
Chickweed
Cocklebur
Cupgrass
Florida beggarweed
Galinsoga
Groundcherry
Morningglory
Mustard
Prickly sida (teaweed)
Ragweed, common
Sandbur
Shepherdspurse
Sprangletop
Stinkgrass
Velvetleaf
Windmillgrass

Follow SURFLAN procedures for soil preparation. SURFLAN/Dyanap or SURFLAN/Klean-Krop may be applied within two (2) days after planting but before crop emergence using any properly calibrated low pressure herbicide sprayer. Do not spray over top of emerged soybeans.

Broadcast Rates Per Acre

Soil Texture	SURFLAN A.S.	Dyanap (2 + 1)S or Klean-Krop
	(quarts)	(quarts)
Coarse	½–¾	6
Medium	¾–1	6
Fine	1–1¼	6

Do not apply SURFLAN/Dyanap or SURFLAN/Klean-Krop to any soils with over five percent (5%) organic matter. Do not use on silt loam soils of extremely fine texture.

Special Precaution: Poor weed control and/or crop injury may result if directions are not followed. Over application may result in crop injury and/or soil residue. Do not use treated vine for feed or forage.

Caution: Please read the Dyanap or Klean-Krop labels carefully for cautions, precautions and special precautions before applying these products.

Destroy empty containers. Do not reuse.

SOYBEAN—SURFLAN/AMIBEN® (CHLORAMBEN) TANK MIX

Amiben®—chloramben, Union Carbide Agricultural Products Co., Inc.

The SURFLAN/Amiben tank mix effectively controls all of the annual grasses and broadleaf weeds listed for SURFLAN plus many additional weeds including:

Chickweed
Coffeeweed
Kochia
Mustard, wild
Nightshade, black
Prickly sida (teaweed)
Ragweed, common
Smartweed, Pennsylvania
Spurge, annual
Stinkgrass
Thistle, Russian
Velvetleaf (buttonweed)

SURFLAN/Amiben may be applied within two (2) days after planting, but before crop emergence, using any properly calibrated low pressure herbicide sprayer that will apply the spray uniformly. Agitate thoroughly before and during application. Follow normal SURFLAN procedures for soil preparation. Do not spray over top of emerged soybeans.

Broadcast Rates Per Acre

Soil Texture	SURFLAN A.S.	Amiben 2S	or	Amiben DS
	(quarts)	(quarts)		(pounds)
Coarse	½–¾	4–6		2.4–3.6
Medium	¾–1	4–6		2.4–3.6
Fine	1–1¼	4–6		2.4–3.6

The higher rates of Amiben should be used on clay loams, clay or high organic (three to five percent) (3% to 5%) mineral soils. Do not use SURFLAN/Amiben on soils containing more than five percent (5%) organic matter.

Special Precaution: Poor weed control and/or crop injury may result if directions are not followed. Over application may result in crop injury and/or soil residue. Do not use treated vines for feed or forage. Do not contaminate any body of water or apply to any area not specified on the label. Do not allow sprays to drift onto adjacent desirable plants.

Caution: Please read the Amiben label carefully for cautions, precautions and special precautions before applying this product. Destroy empty containers. Do not reuse.

SURFLAN/LASSO® MICRO-TECH™, AND SURFLAN/ LASSO MICRO-TECH™/LEXONE® OR SENCOR® COMBINATIONS FOR WEED CONTROL IN SOYBEANS

SURFLAN AS may be tank-mixed with Lasso Micro-Tech™ or Lasso Micro-Tech plus Lexone or Sencor for weed control in soybeans.

SURFLAN/LASSO Micro-Tech Tank-Mix

SURFLAN/Lasso Micro-Tech controls those weeds listed for SURFLAN alone plus provides partial control or suppression of the following weeds listed on either the SURFLAN or Lasso labels.

- Ladysthumb
- Morningglory
- Nightshade, black
- Nightshade, hairy
- Prickly sida (teaweed)
- Smartweed, Pennsylvania
- Velvetleaf

The SURFLAN/Lasso Micro-Tech tank-mix may be applied to the soil surface after planting, up through two (2) days after planting and prior to crop or weed emergence. Apply the tank-mix in 20 to 40 gallons of water per acre with any low pressure herbicide sprayer. Do not spray overtop of emerged soybeans.

Broadcast Rates Per Acre

Soil Texture	SURFLAN AS (quarts)	Lasso Micro-Tech (quarts)
Coarse	1/2	1 1/2–2
Medium	1/2	2–2 1/2
Fine	1/2–3/4	2 1/2–3

Use the lower rate in the above rate ranges on soils with less than 3% organic matter and the higher rate on soils with 3% or more organic matter. Use the higher rate in the recommended range in of heavy weed infestations.

SURFLAN/LASSO MICRO-TECH/LEXONE OR SENCOR TANK-MIX

SURFLAN/Lasso Micro-Tech/Lexone or Sencor controls those weeds listed for SURFLAN alone plus these additional weeds that are listed on either the Lasso/Lexone or Sencor labels.

- Prickly sida (teaweed)
- Ragweed, common
- Smartweed, Pennsylvania
- Velvetleaf

SURFLAN/Lasso Micro-Tech/Lexone or Sencor provides partial control or suppression of the following weeds listed on either the SURFLAN, Lasso Micro-Tech, Lexone or Sencor labels:

- Cocklebur
- Jimsonweed
- Ladysthumb
- Morningglory
- Nightshade, black
- Nightshade, hairy
- Ragweed, giant
- Sicklepod

The SURFLAN/Lasso Micro-Tech/Lexone or Sencor tank-mix may be applied to the soil surface after planting, up through two (2) days after planting and prior to crop or weed emergence. Apply the tank-mix in 20 to 40 gallons of water per acre with any low pressure herbicide sprayer. Do not spray overtop of emerged soybeans.

Broadcast Rates Per Acre

Soil Texture	SURFLAN AS (Quarts)	Lasso Micro-Tech (Quarts)	Lexone 4L or Sencor 4 (Pints)	Lexone Dry Flowable or Sencor Dry Flowable (Pounds)
Coarse	1/2	1 1/2–2	1/3–1/2	1/4–1/3
Medium	1/2	2–2 1/2	1/2–3/4	1/3–1/2
Fine	1/2–3/4	2 1/2–3	3/4–1	1/2–2/3

Use the lower rate in the above rate ranges on soils with less than 3% organic matter and the high rate on soils with 3% or more organic matter. Use the higher rate in the recommended range in areas of heavy weed infestations.

Do not use Lexone or Sencor on coarse soils with less than 1% organic matter.

Special Precautions: Poor weed control and/or crop injury may result if directions are not followed. Over application may result in crop injury and/or soil residue. Do not use treated vines for feed or forage. Do not contaminate any body of water or apply to any area not specified on the label. Do not allow sprays to drift onto adjacent desirable plants. Do not plant any crop other than soybeans within four (4) months after application. Do not tank mix SURFLAN AS with Lasso EC. This combination is frequently incompatible and may cause application problems.

Caution: Read the SURFLAN, Lasso Micro-Tech, Lexone or Sencor labels carefully for directions, precautions and limitations before applying these products.

Precaution: Do not apply SURFLAN to frozen or snow covered ground.

NO-TILL SOYBEAN—SURFLAN FOR FALL OR EARLY SPRING APPLICATION

SURFLAN may be applied in the fall or early spring directly over undisturbed stubble from the previous crop. SURFLAN application can begin anytime following harvest after October 15. Spring applications should be made as early as possible in the spring prior to weed germination but at least four (4) weeks before planting. SURFLAN will not control established weeds.

The soil should not be tilled between the SURFLAN application and soybean planting.

Early applied SURFLAN effectively controls these annual grasses and broadleaf weeds:

Grasses

Barnyardgrass (Watergrass)	Foxtails (Bottlegrass) (Bristlegrass) (Giant foxtail) (Green foxtail) (Pigeongrass) (Robust foxtail) (Yellow foxtail)	Goosegrass (Silver crabgrass) (Wiregrass) (Yardgrass)
Brachiaria (Signalgrass)		Johnsongrass (Seedling only)
Browntop panicum[1]		Texas panicum[1] (Buffalograss) (Coloradograss)
Crabgrasses (Large crabgrass) (Smooth crabgrass)		
Crowfootgrass		
Fall panicum (Spreading panicgrass)		

Broadleaf Weeds

Carpetweed	Pigweeds (Carelessweed) (Prostrate pigweed) (Redroot pigweed) (Rough pigweed) (Spiny pigweed)	Purslane, common
Florida pusley (Florida purslane) (Mexican clover) (Pusley)		Ragweed, common[2]
Lambsquarters		Spurge, prostrate[1]
		Velvetleaf[2]

SURFLAN also provides partial control or suppression of:

Ladysthumb	Smartweed, annual
Morningglory, annual	Spurge, spotted
Nightshade, black	Volunteer wheat
Prickly sida (teaweed)	

SURFLAN will not control established weeds.

[1] See special instructions for Texas only.

[2] SURFLAN controls these weeds on medium and fine textured soils and provides partial control on coarse textured soils.

Broadcast Rates Per Acre

Soil Texture	SURFLAN A.S.
	(quarts)
Coarse	1
Medium	1¼–1½
Fine	1¼–1½

Use the lower rate under light crop residue conditions and under light weed pressure. Use sequential herbicide applications in the spring to control escaped broadleaved weeds and to provide control of those weeds not controlled by SURFLAN.

Precaution: Do not apply SURFLAN to frozen or snow covered ground.

NO-TILL SOYBEAN—TANK MIX RECOMMENDATIONS

NO-TILL SOYBEAN—CONTROL OF EXISTING VEGETATION WITH PARAQUAT OR ROUNDUP®

Roundup®—glyphosate, Monsanto Company

Tank mix SURFLAN with Sencor, Lexone or Lorox to give season long control of labeled annual grasses and broadleaf weeds. Add paraquat or Roundup to these tank mixes to provide postemergence control of existing vegetation.

Paraquat Combinations:

Choose SURFLAN + Sencor/Lexone or Lorox depending on your specific weed problem. Tank mix two (2) pints of paraquat 2 CL per acre to control existing annual weeds. As the last ingredient, add eight (8) ounces of Ortho X-77 Spreader (non-ionic) per 100 gallons of spray solution. Maintain constant agitation.

Caution: Check the paraquat label for additional directions, cautions and special precautions before applying this product.

Roundup Combinations:

Choose SURFLAN + Sencor/Lexone or Lorox depending on your specific weed problem. To control emerged annual weeds, tank mix one and one-half (1½) quarts of Roundup per acre. To control perennial weeds, tank mix two (2) to five (5) quarts of Roundup per acre depending on weeds to be controlled. Add Roundup as the last ingredient in the tank mix.

Caution: Check the Roundup label for additional directions, cautions, and special precautions before applying this product.

SURFLAN/2,4-D TANK MIX

SURFLAN may be tank mixed with most 2,4-D products for control of established winter weeds. When tank mixing, add SURFLAN A.S. to the tank first and agitate until completely dispersed. Maintain agitation and add the 2,4-D product.

Certain ester formulations of 2,4-D require more vigorous agitation to keep the product well dispersed when mixed with SURFLAN A.S.

To assure the compatibility of SURFLAN with 2,4-D, add the products into small containers of water in the correct proportions. After thorough mixing, let stand for five (5) minutes and check for compatibility. If the combination remains mixed or can be remixed easily, the mixture is compatible. Maintain vigorous agitation during mixing and until the spray tank is empty. Follow 2,4-D directions, precautions and limitations.

SURFLAN® A.S. (EPA Reg. No. 1471-112)

NO-TILL SOYBEAN—SURFLAN/SENCOR® OR SURFLAN/LEXONE® TANK MIX

A tank mix combination of SURFLAN/Sencor or SURFLAN/Lexone is more effective in providing preemergence weed control in non-till soybeans than any one of these herbicides used alone.

Add paraquat or Roundup as directed above to control existing vegetation if necessary.

Weeds Controlled by SURFLAN and Sencor or Lexone (See pages 5 and 13).

Apply the SURFLAN/Sencor or SURFLAN/Lexone tank mix at planting or within two (2) days after planting. Do not spray over top of emerged soybeans.

Broadcast Rates Per Acre

Soil Texture	SURFLAN A.S.	Sencor 50 WP/4L or Lexone 50 WP/4L	Lexone Dry Flowable or Sencor Dry Flowable
	(quarts)	(pounds or pints)	(pounds)
Coarse	¾	½	⅓
Medium	¾–1	¾	½
Fine	1½	1	⅔

Do not use on sandy loam or loamy sand soils with less than one percent (1%) organic matter. Not recommended on any soil containing more than five percent (5%) organic matter.

Spray Equipment: Use any properly calibrated low pressure herbicide sprayer. Agitate thoroughly before and during application. Sprayers should be accurately calibrated before applying the tank mix combination. Avoid boom overlaps that will increase rates above those recommended. Check the sprayer daily to ensure proper calibration and uniform application. As the amount of water used (spray volume) decreases, the importance of accurate calibration and uniform application increases.

Activation: A single one-half inch (½") rain or its equivalent in overhead irrigation is needed to activate the herbicide combination.

Special Precautions: Poor weed control and/or crop injury may result if directions are not followed. Over application may result in crop injury and/or soil residue. Do not use treated vines for feed or forage. Do not contaminate any body of water or apply to any area not specified on the label. Do not allow sprays to drift onto adjacent desirable plants. Do not allow SURFLAN to come into direct contact with soybean seed or crop injury may result.

Caution: Read the Sencor or Lexone labels carefully for cautions, precautions and special precautions before applying these products.

NO-TILL SOYBEAN—SURFLAN/LOROX® TANK MIX

A tank mix combination of SURFLAN/Lorox is more effective in providing preemergence weed control in no-till soybeans than either of these herbicides used alone.

Add paraquat or Roundup as directed above to control existing vegetation.

Weeds Controlled by SURFLAN and Lorox (See pages 5 and 14.)

Apply the SURFLAN/Lorox tank mix at planting or within two (2) days after planting. Do not spray over top of emerged soybeans.

Broadcast Rates Per Acre

Soil Texture	SURFLAN A.S.	Lorox 50 WP/L
	(quarts)	(pounds or pints)
Coarse	¾	⅔–1¼
Medium	¾–1	1–1⅔
Fine	1½	1¼–2

Do not use on soils containing more than five percent (5%) organic matter.

Spray Equipment: Use any properly calibrated low pressure herbicide sprayer. Sprayer should be accurately calibrated before applying the tank mix combination. Agitate thoroughly before and during application. Avoid boom overlaps that will increase rates above those recommended. Check the sprayer daily to ensure proper calibration and uniform application. As the amount of water used (spray volume) decreases, the importance of accurate calibration and uniform application increases.

Activation: A single one-half inch (½") rain or its equivalent in overhead irrigation is needed to activate the herbicide combination.

Special Precautions: Poor weed control and/or crop injury may result if directions are not followed. Over application may result in crop injury and/or soil residue. Do not use treated vines for feed or forage. Do not contaminate any body of water or apply to any area not specified on the label. Do not allow sprays to drift onto adjacent desirable plants. Do not allow SURFLAN to come into direct contact with soybean seed or crop injury may result.

SURFLAN OVER-THE-TOP SPRAY FOR WEED CONTROL IN WHEAT OR BARLEY FOLLOWED BY NO-TILL SOYBEANS

SURFLAN may be applied for weed control in wheat or barley and soybeans. Apply SURFLAN as an over-the-top spray after the small grain is fully tillered and jointing has been initiated but prior to heading. When combining wheat or barley, the cutter bar should be adjusted to leave standing stubble at a height of at least eight (8) inches. Stubble reduces wind erosion and loss of soil moisture. During combining, cut straw should be chopped and spread evenly over the field as a mulch. Straw is an important part of the stubble-crop no-till program and should not be removed from the field. Soybeans can be planted into the small grain before harvest or into the stubble immediately following harvest.

DIRECTIONS FOR USE

Follow recommended mixing and application procedures for SURFLAN.

Broadcast Rates Per Acre

Soil Texture	SURFLAN 75W	SURFLAN A.S.
	(pounds)	(quarts)
Coarse	1⅓	1
Medium	1⅓–2	1–1½
Fine	1⅔–2	1¼–1½

Note: Do not apply SURFLAN to soils containing more than five percent (5%) organic matter.

Tank Mixing Directions: SURFLAN may be tank mixed with most 2,4-D products. When tank mixing, add SURFLAN A.S. to the tank first and agitate until completely dispersed. Maintain agitation and add the 2,4-D product.

Certain ester formulations of 2,4-D require more vigorous agitation to keep the product well dispersed when mixed with SURFLAN A.S.

To assure the compatibility of SURFLAN with 2,4-D, add the products into small containers of water in the correct proportions. After thorough mixing, let stand for five (5) minutes and check for compatibility. If the combination remains mixed or can be remixed easily, the mixture is compatible. Maintain vigorous agitation during mixing and until the spray tank is empty. Follow 2,4-D application recommendations, precautions and limitations.

Precaution: Do not apply SURFLAN to winter wheat or barley prior to the fully tillered stage as crop injury may result. Do not graze or feed treated forage, hay or straw to livestock. Do not remove straw from field.

Excessive nitrogen application may result in crop lodging. Please contact your local Extension office for nitrogen recommendations.

COTTON—SURFLAN ALONE (in Texas only)

Apply SURFLAN to the soil surface at planting or within two (2) days after planting at a broadcast rate of one (1) quart on medium soils and one and one-half (1½) quarts on fine soils.

At the recommended rates for fine textured soils, SURFLAN will also control browntop panicum and prostrate spurge.

Precaution: Do not graze treated fields. If SURFLAN is not applied at or within two (2) days after planting, crop injury and poor weed control may result.

PEA (English or Green)—SURFLAN/TREFLAN TANK MIX

The SURFLAN/Treflan tank mix provides weed control and suppression of common root rot (*Aphanomyces euteiches*) in peas.

TREFLAN®-E.C., PRO-5+up,TM, M.T.F.™, trifluralin, Elanco Products Company

Mixing and Application: Apply the tank mix up to two (2) weeks prior to planting. Apply in 20 to 40 gallons of water per acre in any properly calibrated low pressure herbicide sprayer.

Broadcast Rates Per Acre

Soil Texture	SURFLAN A.S.	TREFLAN E.C. or M.T.F.™	or	TREFLAN Pro-5™
All Soil Textures	1 pint	1 pint		0.8 pint

Do not use the SURFLAN/Treflan tank mix on coarse soils with less than one and one-half percent (1½%) organic matter. Incorporate according to label directions for Treflan. See the Treflan label.

Precautions: Please read the Treflan label for additional cautions and precautions. Do not feed forage from treated fields to livestock. Poor weed control and/or crop injury may result if directions are not followed. Over application may result in crop injury and/or soil residue. Do not plant any root crop for 12 months following application.

POTATO (White)—SURFLAN ALONE

Follow SURFLAN label directions for soil preparation, activation and cultivation. SURFLAN should be applied to the soil surface at the recommended rate after planting and prior to crop emergence. Where rebedding and drag-off are standard cultural practices, applications should be made following drag-off.

Broadcast Rates Per Acre

Soil Texture	SURFLAN A.S.
	(quarts)
Coarse	¾
Medium	1

Do not use SURFLAN on soils containing more than five percent (5%) organic matter.

Precautions: Follow all directions carefully. Failure to do so may result in poor weed control. Over application may result in crop injury and/or soil residues.

POTATO (White)—SURFLAN/LEXONE® OR SURFLAN/SENCOR® TANK MIX

SURFLAN plus Sencor/Lexone effectively controls the weeds listed on the current SURFLAN and Sencor or Lexone labels.

Follow SURFLAN label directions for soil preparation, activation and cultivation. SURFLAN plus Sencor/Lexone should be applied to the soil surface at the recommended rate after planting and prior to crop emergence. Where rebedding and drag-off are standard cultural practices, applications should be made following drag-off.

Broadcast Rates Per Acre

Soil Texture	SURFLAN A.S.	Lexone 50 WP/4L or Sencor 50 WP/4L	Lexone Dry Flowable or Sencor Dry Flowable
	(quarts)	(pounds or pints)	(pounds)
Coarse	¾	½	⅓
Medium	1	⅔	½

Do not use SURFLAN/Sencor or Lexone tank mix on soils containing more than five percent (5%) organic matter.

Precautions: Follow all directions carefully. Failure to do so may result in poor weed control. Over application may result in crop injury and/or soil residues.
Read the Lexone or Sencor labels for additional warnings, cautions, precautions, and special instructions.

TOBACCO—SURFLAN ALONE

Tobacco—SURFLAN at Layby

Air-Cured (Burley, Maryland, Dark) and Flue-Cured

Soil Preparation: The last cultivation should leave the middle of the rows weed free before application. SURFLAN should be applied as a directed spray following the last normal cultivation, usually four (4) to six (6) weeks after transplanting.

Application: Applications should be made in 20 to 40 gallons of water per acre using commercially available drop nozzles that will allow treatment to be made in a 16 to 24 inch band in the middle of the row between the crop rows. SURFLAN should be applied in a manner that will not allow the spray to contact the plant. In tobacco production, in areas where spray nozzles on the ends of the spray boom pass over the same middle of the row twice, use nozzles which apply one-half (½) the normal number of gallons per acre to prevent over application. Over application may result in injury to rotational crops.

Activation and Cultivation: A one-half inch (½") rain is necessary to activate SURFLAN. If adequate rainfall is not received to activate SURFLAN, erratic weed control may occur. If weeds begin to emerge, a shallow cultivation of one to two inches (1" to 2") will destroy existing weeds and place SURFLAN in the zone of weed seed germination.

Broadcast Rates Per Acre

Soil Texture	SURFLAN A.S. (quarts)
Coarse	½–¾
Medium	¾–1
Fine	1

Precaution: Applied according to directions and under normal growing conditions, SURFLAN will not harm the treated crop. Do not spray SURFLAN in the bud of the tobacco plant as crop injury may occur. Over application may result in soil residue. Do not plant any root crop for 12 months following a SURFLAN application as crop injury may occur. Failure to follow directions may result in erratic weed control, tobacco crop injury, or soil residues. Soil residues may result in rotational crop injury.

FRUIT, NUT CROPS AND VINEYARDS (Non-bearing and Bearing)—SURFLAN ALONE

SURFLAN applied as a preemergence herbicide controls many annual grasses and broadleaf weeds in several fruit, nut crops and vineyards. It is recommended on the following crops:

Almond	Grapefruit	Peach
Apple	Guineagrass, (seedling)	Pear
Apricot		Pecan
Avocado	Kiwi	Pistachio
Cherry	Lemon	Plum
Fig	Macademia Nut	Pomegranate
Filbert	Nectarine	Prune
Grape	Olive	Walnut, English
	Orange	

Small fruits:

Blackberry	Currants	Gooseberry
Blueberry	Dewberry	Loganberry
Boysenberry	Elderberry	Raspberry

See listing of weeds controlled by SURFLAN alone.

General Directions: Existing vegetation should be destroyed either by tillage or by a tank mix with a contact or translocated herbicide. To ensure the compatibility of SURFLAN with the contact or translocated herbicide, pour the products into small containers of water in the correct proportions. After thorough mixing, let stand for five (5) minutes and check for compatibility. If the combination remains mixed or can be remixed easily, the mixture is compatible. Read the label for all directions, cautions and precautions.

Broadcast Rates Per Acre

Soil Texture		SURFLAN A.S. (quarts)
All Soil Textures	Short Term (2–4 months)	2
	Long Term (6–8 months)	4
	(8–12 months)	6

SURFLAN is not recommended for soils containing more than five percent (5%) organic matter.

Band Treatment: The amount of product needed for band treatment may be calculated by the formula:

$$\frac{\text{band width in inches}}{\text{row width}} \times \frac{\text{broadcast}}{\text{rate per acre}} = \frac{\text{amount needed per}}{\text{acre of field}}$$

Mixing and Application: Fill the spray tank three-fourths (¾) full with clean water. Add the recommended amount of SURFLAN to water. Mix thoroughly. Use 20 to 40 gallons of water per acre (broadcast basis). Apply with a properly calibrated low pressure herbicide sprayer that will uniformly apply the spray. Check the sprayer daily for proper calibration and uniform application. Provide good mixing of the spray suspension in the tank before and during application.

Chemigation: SURFLAN may be applied for weed control through center pivot* and solid set sprinkler irrigation systems. These systems should be equipped with automatic shutoff devices and valves to prevent back flow into the water source. Meter a constant flow of SURFLAN into the water during the irrigation period. Apply SURFLAN in a minimum of one inch (1") of irrigation water. Check the irrigation system before each use to ensure a uniform application.

*** Precaution:** Do not apply SURFLAN via center pivot irrigation systems to fruit and nut orchards or vineyards. In California, do not apply SURFLAN to non-dormant fruit and nut orchards or vineyards through overhead sprinkler irrigation systems.

Activation: A single one-half inch (½") to one inch (1") rainfall or sprinkler irrigation to move SURFLAN into the weed seed germination zone is required to activate the herbicide. If weeds begin to emerge, a shallow cultivation of one to two inches (1" to 2") will destroy existing weeds and place SURFLAN in the zone of weed seed germination. The higher water amount is generally required on fine textured/high organic matter soil.

Precaution: Poor weed control may result if directions are not followed. Over application may result in crop injury or excessive soil residue. Apply the spray directly to the orchard or vineyard floor.

FRUIT, NUT CROPS AND VINEYARDS—TANK MIX RECOMMENDATIONS: SURFLAN TANK MIX REFERENCE TABLE

Refer to the following table to determine which herbicides may be applied to your crop. Choose the SURFLAN combination based on your orchard or vineyard weeds to be controlled. Refer to the tank mix section for specific recommendations and precautions. Always read the label before application.* = Approved Crop.

	SURFLAN	CASORON	GOAL	HYVAR X	KARMEX	KROVAR I/II	PARAQUAT	PRINCEP	ROUNDUP	SINBAR	SOLICAM
Almond	*	*	*				*	*	*		*
Apple	*	*			*		*	*	*	*	*
Apricot	*		*				*				*
Avocado	*						*	*	*		
Cherry	*	*					*	*	*		*
Fig	*						*				
Filbert	*	*					*	*	*		*

	SURFLAN	CASORON	GOAL	HYVAR X	KARMEX	KROVAR I/II	PARAQUAT	PRINCEP	ROUNDUP	SINBAR	SOLICAM
Grape	*	*	*		*		*	*	*		
Grapefruit	*			*	*	*	*	*	*	*	*
Kiwi	*										
Lemon	*			*	*	*	*	*	*	*	*
Macadamia Nut	*				*			*			
Nectarine	*	*	*				*				*
Olive	*				*			*			
Orange	*			*	*	*	*	*	*	*	*
Peach	*	*	*		*		*	*		*	*
Pear	*	*			*		*	*	*		*
Pecan	*	*			*		*	*	*	*	
Pistachio	*								*		
Plum	*	*	*				*	*			*
Pomegranate	*										
Prune	*		*				*				*
Small Fruits											
Blackberry	*	*			*		*	*		*	
Blueberry	*	*			*		*	*		*	
Boysenberry	*				*		*	*		*	
Currants	*										
Dewberry	*				*					*	
Elderberry	*										
Gooseberry	*				*						
Loganberry	*				*			*		*	
Raspberry	*	*			*		*	*		*	
Walnut, English	*	*			*		*	*	*		*

SURFLAN/CASORON® TANK MIX

Casoron®—dichlobenil, Philips-Duphar

The SURFLAN/Casoron tank mix controls the following weeds in addition to those controlled by SURFLAN alone:

Barley, wild	Jerusalem oak	Primrose, evening
Camphorweed	Maypops	Radish, wild
Coffeeweed	Melon, citron	Ragweed
Dandelion	Milkweed vine	Rosarypea
Deadnettle, red	Miners lettuce	Smartweed
Dogfennel	Mustard, wild	Spanishneedles
Gisekia	Natalgrass	Teaweed
Goosefoot	Peppergrass	Thistle, bull
Groundsel	Pineappleweed	Thistle, Russian
Horsetail	Plantain	Woodsorrel, yellow

SURFLAN/Casoron tank mix may be used for weed control in the following crops:

Apple	Filbert	Pecan
Almond	Grape	Plum
Blackberry	Nectarine	Raspberry
Blueberry	Peach	Walnut, English
Cherry	Pear	

General Directions: Follow recommended application procedures for SURFLAN. Apply SURFLAN/Casoron in at least 20 gallons of water per acre to ensure uniform coverage. If the application is followed by one-half inch (½″) to one inch (1″) of overhead irrigation for activation, use the lower Casoron rate. SURFLAN/Casoron must be activated; check the SURFLAN label for activation directions.

Broadcast Rates Per Acre

Crop	SURFLAN A.S. (quarts)	Casoron W-50 (pounds)
Apple, Almond, Blueberry Cherry, Grape, Nectarine, Peach, Pear, Plum, Filbert, Pecan, English Walnut	2–6	8–12
Do not use on pecan seedlings within six (6) months of planting. May be used on bearing, non-bearing and nursery stock. Do not apply until four (4) weeks after transplanting. Do not make application within one (1) month of harvest of nuts or stone fruits.		
Blackberry and Raspberry	2–6	8
Use only on established plantings. Do not apply during new shoot emergence.		

Casoron Precautions: Do not use in seed beds, transplant or cutting beds or greenhouses. Do not apply until six (6) months after rooting of cuttings in the field, or after planting rooted cuttings. Do not apply on fruit stock within three (3) months prior to or following grafting or budding of root stocks or planting of new grafts. Do not apply more than the maximum recommended amount of Casoron W-50 per growing season. Do not use on "St. Lucie fine sand", "Arzell fine sand", or other light sand soil. Do not graze livestock in treated areas. Store in a tightly closed container in a dry place.

Caution: Read the SURFLAN and Casoron labels for directions, cautions and limitations prior to use.

SURFLAN/GOAL® TANK MIX

Goal®—oxyfluorfen, Rohm and Haas Company

SURFLAN/Goal provides preemergence control of the following weeds in addition to those controlled by SURFLAN alone:

Burclover	Mallow, little
Filaree, broadleaf	Sowthistle, annual
Lettuce, prickly	

The tank mix is recommended for application to dormant crops including:

Almond	Peach
Apricot	Plum
Grape	Prune
Nectarine	

General Directions: Best preemergence activity from Goal is obtained when the tank mix is applied to a relatively weed free established berm or soil surface.

Broadcast Rates Per Acre

Crop	SURFLAN A.S. (quarts)	Goal 2E (quarts)
All crops listed	2–6	4

PRECAUTIONS:

- Dormancy determines the suitability and safety of using Goal 2E on tree fruit/grapes. Should bud swell occur prior to the dates listed, Goal 2E must not be used.
- DO NOT APPLY GOAL 2E (EITHER ALONE OR IN ANY COMBINATION RECOMMENDED ON THIS LABEL) TO TREE OR VINE PLANTINGS BEFORE OCTOBER 1 OR AFTER FEBRUARY 15. DO NOT APPLY GOAL 2E (EITHER ALONE OR IN ANY COMBINATION RECOMMENDED ON THIS LABEL) TO GRAPES IN THE COACHELLA VALLEY BEFORE OCTOBER 1 OR AFTER DECEMBER 31.
- Do not apply more than four (4) quarts (two pounds active) per broadcast acre of Goal 2E in one (1) season.
- Goal 2E should not be applied to grapes established less than three (3) years.

- Do not apply Goal 2E to grapes which are not staked or trellised.
- Do not feed or graze areas treated with Goal 2E to animals.
- Do not contaminate irrigation water or water used for domestic purposes with Goal 2E.
- Goal 2E or any of the combinations recommended on this label should be applied to only healthy growing trees and vines.
- Do not apply Goal 2E or any combination containing Goal 2E, that is recommended on this label, when weather conditions favor drift.
- Apply Goal 2E only to dormant fruit or nut trees or grape vines. Do not apply when foliage, buds or fruit/nuts are present.
- Direct Goal 2E spray toward the base of tree or vines. Avoid direct plant contact.

Mixing and Application: Fill the spray tank three-fourths (3/4) full. Agitate continuously. Add the recommended amount of SURFLAN to the spray tank. Add the recommended amount of Goal. Finish filling the tank with water to proper level. Do not allow the tank mix to remain in the tank without agitation.

Activation: A single one-half inch (1/2") to one inch (1") rain or sprinkler irrigation to move SURFLAN/Goal into the weed seed germination zone is required to activate the herbicide. If weeds begin to emerge, a shallow cultivation of one to two inches (1" to 2") will destroy existing weeds and place SURFLAN in the zone of weed seed germination. The higher amount of water is generally required on fine textured/high organic matter soil.

Caution: Read the Goal label for directions, cautions and precautions before applying these products.

SURFLAN/HYVAR® X TANK MIX

Hyvar® X—bromacil, E. I. duPont de Nemours and Company

The SURFLAN/Hyvar X tank mix controls the following annual and perennial weeds in addition to those weeds controlled by SURFLAN alone:

Annual Weeds	**Perennial Weeds**
Mullein, turkey	Bermudagrass (All States)
Mustard	Johnsongrass (Texas)
Natalgrass (red top)	Nutsedge (Texas and California)
Sedge	Pangolagrass (Florida)
Sprangletop	Paragrass (Florida)
	Torpedograss (Florida)

The SURFLAN/Hyvar X tank mix is recommended on the following crops:

Citrus (Orange, Grapefuit, Lemon)

General Directions: Apply the tank mix in at least 40 gallons of water per acre. Use sufficient water to provide thorough and uniform coverage of the ground. Apply Hyvar X as a spray just before or during the period of active plant growth. If dense growth is present, results will be improved if vegetation is removed before treatment. Do not apply when ground is frozen.

Activation: Moisture is necessary to activate SURFLAN/Hyvar X tank mix; best results are obtained if moisture is supplied by rainfall or overhead sprinkler irrigation within two (2) weeks after application. Check activation instructions for SURFLAN alone.

Broadcast Rates Per Acre

Crop	SURFLAN A.S.	Hyvar X
	(quarts)	(pounds)
Citrus (Trees established for four (4) years or more) Annual Weeds	2–6	2–4

Apply any time of the year, preferably shortly before weed growth begins, when adequate moisture is available.

Broadcast Rates Per Acre

Crop	SURFLAN A.S.	Hyvar X
	(quarts)	(pounds)
Perennial Weeds	2–6	4–8

Apply before weed growth begins. If dense growth is present, remove tops and spray the ground. Note: Effects on perennial weeds from Hyvar X are slow to appear, usually progressing over a period of several months. Check the Hyvar X label for more complete directions and precautionary statements prior to use for perennial weed control.

Caution: Avoid contact of foliage and fruit with Hyvar X spray or mist. Temporary yellowing of citrus leaves may occur following Hyvar X treatment. Do not use on citrus in Kern County, California.

Because injury to citrus trees may result: do not use Hyvar X on soils low in organic matter less than one percent (1%) nor on poorly drained soils; do not apply more than eight (8) pounds per acre per year; do not treat trees planted in irrigation furrows; do not treat diseased trees such as those with foot rot. Do not use Hyvar X in citrus orchards interplanted with other trees or desirable plants, nor in home citrus plantings or in areas where roots of other valuable plants or trees may extend, as plant injury may result. Treated areas may be planted to citrus trees one (1) year after last Hyvar X application; do not replant to other crops within two (2) years after last application, as plant injury may result.

SURFLAN/KARMEX® TANK MIX

Karmex®—diuron, E. I. duPont de Nemours and Company

The SURFLAN/Karmex tank mix controls the following weeds in addition to those controlled by SURFLAN alone:

Ageratum	Kochia	Puncturevine
Buckwheat, wild	Kyllinga	Radish, wild
Buttonweed	Marigold	Ricegrass, red
Clover, Mexican	Milkweed, climbing	Ryegrass, annual
Dayflower	Morningglory, annual	Smartweed
Dogfennel	Mustard, black	Sowthistle, annual
Fescue, rattail	Mustard, wild	Spanishneedles
Gromwell	Orchardgrass	Speedwell, corn
Groundcherry, annual	Paintbrush, Flora's	Spurry, corn
Hawksbeard	Pepperweed	Tansymustard
Horseweed	Pineappleweed	Tobacco, rabbit
Knawel	Pokeweed	Velvetgrass

The SURFLAN/Karmex tank mix is recommended on the following crops:

Apple	Orange
Caneberries	Peach
Grape	Pear
Grapefruit	Pecan
Lemon	Walnut, English

Broadcast Rates Per Acre

Crop	SURFLAN A.S.	Karmex 80W
	(quarts)	(pounds)
Apple	2–6	2–4

Use Karmex only under trees established in the orchard for at least one (1) year; do not treat varieties grafted on fulldwarf root stocks.

Caneberries	2–6	2–4

(Read the Karmex label for state and crop recommendations). Use Karmex only in fields established for at least one (1) year. Do not apply Karmex to berries interplanted with fruit trees. Do not apply to plants whose roots are exposed as injury may result.

Grape

New York and Pennsylvania—Perennial Grasses:

Use Karmex only in established vineyards (at least four years old) for spot control of perennial grasses such as orchardgrass, quackgrass and ryegrass.

Apply Karmex in the spring as a band treatment to ridged soil two to four inches (2" to 4") high under the trellis at the rate of eight (8) to 12 lbs. per acre. Band width should not exceed 30 inches. Do not apply Karmex more than once every four (4) years. Use Karmex only on heavy soils such as loams, silt loams and clay loams. Do not use Karmex in areas where grape roots are shallow or exposed because of high bedrock, poor drainage, or erosion as injury to grapes may result.

U.S. (Except New York and Pennsylvania)—Perennial Grasses:

Apply Karmex only to established vineyards (at least three years old) as a band treatment to grape rows. Do not apply Karmex to vines with trunks less than one and one-half inches (1½″) in diameter as injury may result. On soils low in clay or organic matter (one to two percent) (1% to 2%), severe plant injury may result from Karmex application if heavy rainfall or more than one inch (1″) of irrigation occurs soon after treatment. This risk must be assumed by the user.

East of Rocky Mountains:		
1 to 2% organic matter	2–6	2–3
2 to 5% organic matter	2–6	3–6
West of Rocky Mountains:	2–6	2–4

(Apply Karmex in November, December, or January)

Orange, Grapefruit and Lemon

Do not apply Karmex under citrus trees that have been subjected to freezing within six (6) months. Use Karmex only under trees established in the grove for at least one (1) year.

Arizona (except Yuma area) and California (except Imperial and Coachella Valleys)	2–6	2–4
Florida (oranges and grapefruit)	2–6	4–8
Texas (oranges and grapefruit)	2–6	2–4

Peach	2–6	2–5

Use Karmex only under trees established in the orchard for at least three (3) years. Do not apply Karmex within three (3) months of harvest; in the far West, do not apply Karmex within eight (8) months of harvest.

Pear	2–6	2–4

Use Karmex only under trees established in the orchard for at least one (1) year. Do not treat varieties grafted on fulldwarf root stocks.

Pecan	2–6	2–4

Use only under trees established in the grove for at least three (3) years. Do not use on soils with less than one-half percent (½%) organic matter.

Walnut (English)		
California	2–6	2–3

Use only under trees established in the orchard for at least one (1) year.

Mixing and Application: Fill the spray tank three-fourths (¾) full with clean water and add the recommended amount of Karmex 80W. Mix thoroughly. Then add the recommended amount of SURFLAN A.S., mix thoroughly and fill the tank. Vigorous, continuous agitation is required until spray tank is empty to keep the powders in suspension. Use 20 to 40 gallons of water per acre (broadcast basis). Apply with a properly calibrated low pressure herbicide sprayer that will apply the spray uniformly. Check the sprayer daily for proper calibration and to ensure uniform application. Provide good mixing of the spray suspension in the tank before and during application.

Activation: A single one-half inch (½″) to one inch (1″) rain or sprinkler irrigation is required to activate the herbicides. If weeds begin to emerge, a shallow cultivation of one to two inches (1″ to 2″) will destroy existing weeds and place SURFLAN and Karmex in the zone of weed seed germination.

Precaution: Poor weed control may result if directions are not followed. Over application may result in crop injury or excessive soil residue. Apply the spray directly to the orchard or vineyard floor.

Caution: Read the Karmex label carefully for directions, cautions, and precautions before applying this product.

SURFLAN/KROVAR® I TANK MIX

The SURFLAN/Krovar I tank mix effectively controls the following weeds in addition to those weeds controlled by SURFLAN alone:

Filaree	Nightshade (annual)
Fleabane	Pineappleweed
Horseweed	Ragweed
Lettuce, wild	Shepherdspurse
Mustard, wild	Sowthistle (annual)
Natalgrass (red top)	Spanishneedles

The SURFLAN/Krovar I tank mix is recommended for use on the following crops:

Citrus (Orange, Lemon, Grapefruit)

General Directions: Apply any time of the year when rainfall or overhead irrigation is available to activate the tank mix, preferably just before the weeds have germinated.

Use only where trees have been established in the orchard for at least three (3) years. Use sufficient water (over 40 gallons/acre) to obtain uniform coverage.

Broadcast Rates Per Acre

Crop	SURFLAN A.S.	Krovar I
	(quarts)	(pounds)
Citrus (Trees established in the orchard for at least three (3) years)	2–6	3–6

Use for control of annual weeds in California, Arizona, Texas, Louisiana, and Florida only. Check the Krovar I label for complete directions for use and precautionary statements.

Activation and Cultivation: A single one-half inch (½″) to one inch (1″) rain or sprinkler irrigation to move the tank mix combination into the weed seed germination zone is required within two (2) weeks after application to activate the herbicides. If weeds begin to emerge, a shallow cultivation of one to two inches (1″ to 2″) will destroy existing weeds and place SURFLAN and Krovar I in the zone of weed seed germination. Citrus treated can be shallowly cultivated without loss of activity.

Caution: Avoid contact of foliage and fruit with Krovar I spray or mist. Avoid overlapping, and shut off spray boom while starting, turning, slowing or stopping, or injury to trees may result. Continuous agitation in the spray tank is required to keep the material in suspension. Best results are obtained if application is made shortly before weed growth begins; if dense growth is present, remove tops and spray ground.

Temporary yellowing of citrus leaves may occur following Krovar I treatment. Because injury to citrus trees may result: do not use Krovar I on soils with less than one percent (1%) organic matter (one and one-half percent (1½%) in Florida), poorly drained soils, gravelly soils, nor thinly covered or exposed subsoils; do not treat trees planted in irrigation furrows; do not treat diseased trees such as those with foot rot. Do not replant Krovar I treated areas to any crop within two (2) years after last application as injury to subsequent crops may result, except that citrus trees may be planted one (1) year after the last application.

SURFLAN/KROVAR® II TANK MIX

Krovar® I and II—bromacil/diuron, E. I. duPont de Nemours and Company

The SURFLAN/Krovar II tank mix effectively controls the following annual and perennial weeds in addition to those controlled by SURFLAN alone:

Annual Weeds and Vines	Perennial Weeds
Balsam apple*	Bahiagrass
Milkweed vine*	Bermudagrass
Natalgrass (red top)	Nutsedge
Nightshade (annual)	Pangolagrass
Ragweed	Paragrass
Spanishneedles	Torpedograss

* Seedling stage in Florida only.

The SURFLAN/Krovar II tank mix is recommended for weed control in the following crops:

Citrus (Orange, Lemon, Grapefuit)

General Directions: Use only where trees have been established in the orchard for at least four (4) years. Apply any time during the year when rainfall or overhead irrigation is available to activate the tank mix prior to weed germination. Moisture is necessary to activate the herbicides. Check the SURFLAN directions for activation moisture amounts. The tank mix should be activated within two (2) weeks after application. To obtain uniform coverage, use sufficient water (over 40 gallons/acre).

Broadcast Rates Per Acre

Crop	SURFLAN A.S.	Krovar II
	(quarts)	(pounds)
Citrus (Use only where trees have been established in the orchard for at least four (4) years.)	2–6	2–8

Consult the Krovar II label for complete directions for use for annual, perennial and seedling vine control, cautions and limitations prior to use.

Precautions: Avoid contact of foliage and fruit with Krovar II spray or mist. Avoid overlapping, and shut off spray boom while starting, turning, slowing or stopping, or injury to trees may result. Continuous agitation in the spray tank is required to keep the material in suspension. Best results are obtained if application is made shortly before weed growth begins; if dense growth is present, remove tops and spray the ground. Do not use Krovar II on citrus in Kern County, California.

Temporary yellowing of citrus leaves may occur following Krovar II treatment. Because injury to citrus trees may result: do not use Krovar II on soils low in organic matter (less than 1%), poorly drained soils, gravelly soils, nor thinly covered or exposed subsoils; do not apply more than 10 lbs. per acre per year (Florida, 12 lbs.); do not treat trees planted in irrigation furrows; do not treat diseased trees such as those with foot rot. Do not replant Krovar II treated areas to any crop within two (2) years after last application as injury to subsequent crops may result, except that citrus trees may be planted one (1) year after the last application.

SURFLAN/PARAQUAT TANK MIX

SURFLAN may be tank mixed with paraquat. SURFLAN provides preemergence control of certain annual grass and broadleaf weeds while paraquat provides non-selective postemergence control of annual grass and broadleaf weeds and top kill and suppression of perennial weeds.

The tank mix is recommended on the following crops:

Almond	Grape	Peach
Apple	Grapefruit	Pear
Apricot	Lemon	Pecan
Avocado	Nectarine	Plum
Cherry	Olive	Prune
Fig	Orange	Walnut, English
Filbert		

Small fruits:

Blackberry	Boysenberry
Blueberry	Raspberry

General Directions: SURFLAN/paraquat may be applied any time throughout the growing seasons. Choose an application schedule based on the specific weeds to be controlled by the SURFLAN/paraquat tank mix.

Broadcast Rates Per Acre

Crop	SURFLAN A.S.	Paraquat
	(quarts)	(quarts)
All crops listed in SURFLAN Paraquat Tank Mix section.	2–6	1–2*

Precaution: Do not allow paraquat spray to contact green stems, fruit or foliage as injury may result. Do not spray paraquat under windy conditions and use a shield for young trees or vines. Do not allow animals to graze on paraquat treated areas. Do not apply paraquat when nuts to be harvested are on the ground.

* Add a non-ionic spreader at eight (8) oz. per 100 gallons spray volume.

Mixing and Application: Add the recommended amount of SURFLAN to the spray tank according to directions for these tank mixes. Add paraquat last prior to filling the tank. Agitate continuously during filling and through application. Apply in at least 30 gallons of water per acre. Apply when the weeds and grasses are succulent and the new growth is from one inch (1″) to six inches (6″) high. Retreatment or spot treatment may be necessary for late germinating weeds and grasses.

Activation: A single one-half inch (½″) to one inch (1″) rain or sprinkler irrigation to move SURFLAN into the weed seed germination zone is required to activate the herbicide. If weeds begin to emerge, a shallow cultivation of one to two inches (1″ to 2″) will destroy existing weeds and place SURFLAN in the zone of weed seed germination.

Caution: Read the paraquat label carefully for directions, cautions and precautions before applying this product.

SURFLAN/PRINCEP® (SIMAZINE) TANK MIX

Princep®—simazine, Ciba-Geigy Corporation

The SURFLAN/Princep tank mix controls the following weeds in addition to those controlled by SURFLAN alone.

Brome, downy (cheat)	Paintbrush, Flora's
Fescue, rattail	Prickly lettuce
Fireweed	Ragweed
Hairgrass, silver	Ryegrass, annual
Knawel	Shieldscress
Milkweed, climbing	Smartweed
Morningglory, annual	Spanishneedles
Mustard, black	Tansymustard
Mustard, wild	Wiregrass
Nightshade	

The tank mix is recommended on the following crops:

Almond	Lemon
Apple	Orange
Avocado	Peach
Caneberries	Pear
Cherry	Pecan
Filbert	Plum
Grape	Walnut, English
Grapefruit	

Where a rate range of application rates for Princep is given, use the lower rate on coarser textured soil and soil lower in organic matter; use the high rate on finer textured soils and soil higher in organic matter.

Broadcast Rates Per Acre

Crop	SURFLAN A.S.	Princep 80W
	(quarts)	(pounds)
Apple, Pear and Sour Cherry	2–6	2½–5
Almond and Peach California only. Apply to two (2) to four (4) foot band on each side of free row. For peach in other states, see recommendations for peach, plum, and sweet cherry.	2–6	1¼–2½

Do not apply Princep to trees established in the grove less than three (3) years. Do not apply Princep to the Mission (Texas) variety of almonds. Do not apply Princep to almond trees propagated on plum rootstocks. Do not replant almond or peaches in soil treated with Princep for 12 months after treatment. Do not apply Princep on gravelly, sandy or loamy sand soil or on soil with less than one percent (1%) organic matter. Do not apply Princep to areas where water will accumulate.

Avocado California	2–6	2½–5

Do not apply Princep to areas where water will accumulate.

Caneberries	2–6	2½–5

Apply two and one-half (2½) to five (5) pounds Princep per acre in the spring or apply a split application of two and one-half (2½) pounds per acre in the spring plus two and one-half (2½) pounds per acre in the fall. Apply in a minimum of 40 gallons of water per acre. On plantings less than six (6) months old, use one-half (½) the above Princep rate. To control quackgrass, apply five (5) pounds Princep per acre in the fall or split the application applying two and one-half (2½) pounds per acre in the fall plus two and one-half (2½) pounds per acre in the spring, when quackgrass is actively growing. Do not apply Princep when fruit is present.

Filbert		
Washington and Oregon	2–6	2½–5

Apply Princep at two and one-half (2½) to five (5) pounds per acre in the fall or apply a split application of two and one-half (2½) pounds per acre in the spring and two and one-half (2½) pounds per acre in the fall.

Precaution: If trees are planted on a hillside, excessive soil erosion may result from the elimination of weeds. Do not apply on gravelly, sandy or loamy sand soil. Do not apply when nuts are on the ground during the harvest period.

Grape	2–6	2½–5

Apply Princep any time between harvest and early spring.

Precaution: Do not use Princep in vineyards established less than three (3) years. Do not apply Princep on gravelly, sandy or loamy sand soil.

Grapefruit, Lemon and Orange		
Arizona (lemon and orange only)	2–6	2 Spring 2 Fall
California (except Imperial, Coachella and Palo Verde Valleys)	2–6	2½–5
Florida (grapefruit and orange only)	2–6	8–12
Texas	2–6	5–6

All areas—do not use Princep in nurseries. Do not apply Princep to bedded grapefruit, lemon, or orange. To avoid possible injury, do not apply Princep to trees under stress from freeze damage for one (1) year after the freeze.

Peach Plum and Sweet Cherry	2–6	2–5

Apply Princep in late fall to early spring prior to weed emergence.

Peach: Use Princep only in Arkansas, Louisiana, Missouri, Oklahoma, Texas and the states east of the Mississippi.

Plum and Sweet Cherry: Use Princep only in Missouri and states east of the Mississippi River except Tennessee.

Pecan		
Sand and loamy sand	2–6	DO NOT USE
Sandy loam	2–6	2–5
Loam or clay soil low in organic matter	2–6	2½–3½
Clay soil high in organic matter	2–6	3½–5

Note: Do not apply Princep when nuts are on the ground. Do not allow animals to graze in Princep treated areas.

Do not use Princep in Arizona, California, New Mexico or west of the Pecos River in Texas as injury will occur. Do not make Princep applications to transplanted trees that have been established less than two (2) years in the grove as injury may occur.

Walnut, English	2–6	2½–5

Do not apply Princep to gravelly, sandy or loamy sand soils. Leveling and furrowing operations after application will lessen effectiveness of weed control. Do not apply Princep when nuts are on the ground.

Mixing and Application: Fill the spray tank three-fourths (¾) full of clean water. Follow instructions on the Princep 80W label for making a slurry of the Princep 80W in a small amount of water in a separate container, then pour the slurry into the spray tank. Mix thoroughly. Then add the recommended amount of SURFLAN A.S. Mix thoroughly and continue filling the tank. Continous, vigorous agitation is needed to keep the powders in suspension. Use 20 to 40 gallons of water per acre (broadcast basis). Apply with a properly calibrated low pressure herbicide sprayer that will apply the spray uniformly. Check the sprayer daily for proper calibration to ensure uniform application. Provide good mixing of the spray suspension in the tank before and during application.

Activation: A single one-half inch (½″) to one inch (1″) rain or sprinkler irrigation to move the tank mix combination into the weed seed germination zone is required to activate the herbicides. If weeds begin to emerge, a shallow cultivation of one to two inches (1″ to 2″) will destroy existing weeds and place SURFLAN and Princep in the zone of weed seed germination.

Precaution: Poor weed control may result if directions are not followed. Over application may result in crop injury or soil residues. Apply the spray directly to the orchard or vineyard floor.

Caution: Read the Princep label carefully for directions, cautions and precautions before applying this product.

SURFLAN/ROUNDUP® TANK MIX

SURFLAN may be tank mixed with Roundup. In this tank mix, SURFLAN provides preemergence control of certain annual grass and broadleaf weeds while the Roundup provides non-selective, postemergence control of many annual and perennial weeds. This tank mix is recommended on the following crops:

Almond	Grape	Pear
Apple	Grapefruit	Pecan
Avocado	Lemon	Pistachio
Cherry (non-bearing)	Orange	Walnut, English
Filbert		

See listing of weeds controlled by SURFLAN alone.

General Directions: SURFLAN/Roundup may be applied any time throughout the growing season. Choose an application schedule based on the specific weeds to be controlled by the SURFLAN/Roundup tank mix. Consult the Roundup label for specific weed growth stage directions to achieve best results. Karmex and Princep may be tank mixed with SURFLAN/Roundup to provide preemergence control of those weeds listed. Use Roundup with the SURFLAN/Karmex tank mix for non-selective, postemergence control of annual weeds only. Use Roundup with the SURFLAN/Princep tank mix for non-selective, postemergence control of annual and perennial weeds.

Broadcast Rates Per Acre

All Crops	SURFLAN A.S.	Roundup
	(quarts)	(quarts)
Existing annual grass and broadleaf weeds:		
Less than 6″ tall	2–6	1
Over 6″ tall	2–6	1½
Existing perennial weeds	2–6	2–5

Tree Crops	Preharvest Interval for Roundup Application
Almond	21 days
Apple	14 days
Avocado	14 days
Cherry	1 year
Filbert	21 days
Grapefruit	14 days
Lemon	14 days
Orange	14 days
Pear	14 days
Pecan	21 days
Pistachio	21 days
Walnut	21 days

Note: Repeat treatments may be necessary to control weeds regenerating from underground parts or seed. The combined total of all Roundup treatments must not exceed 10.6 quarts per acre per year. Do not feed or forage areas treated with Roundup for eight (8) weeks after application.

Precaution: EXTREME CARE MUST BE EXERCISED TO AVOID CONTACT OF ROUNDUP SPRAY, DRIFT OR MIST WITH GREEN FOLIAGE, GREEN BARK OR BARK OF

TREES ESTABLISHED LESS THAN TWO (2) YEARS, SUCKERS, OR FRUIT OF DESIRABLE TREES, CROPS, PLANTS OR OTHER DESIRABLE VEGETATION. SPRAY CONTACT WITH OTHER THAN MATURED BARK ON THE MAIN TRUNK CAN RESULT IN SERIOUS LOCALIZED OR TRANSLOCATED DAMAGE. If weeds have been mowed, allow regrowth to reach the recommended stage of growth prior to application of this product. Do not apply Roundup to weeds under drought stress.

Vines	Preharvest Interval for Roundup Application
Grape (Raisan, Table, Wine)	14 days

Note: Roundup applications should not be made when green shoots or canes or foliage are in the spray zone. (In the Northeast and Great Lakes regions, Roundup applications must be made prior to the end of bloom stage of grapes to avoid injury.)

If repeat Roundup treatments are necessary for weed control in vineyards, do not exceed a total of 10.6 quarts of Roundup per acre per year. Do not treat grapes with Roundup between 14 days before harvest to fall dormancy when no green vegetation, canes or shoots exist.

Precaution: DO NOT ALLOW ROUNDUP SPRAY, DRIFT OR MIST TO CONTACT GREEN FOLIAGE, GREEN BARK, SUCKERS OR VINES AND RENEWALS LESS THAN THREE (3) YEARS OF AGE. ROUNDUP SPRAY CONTACT OTHER THAN WITH MATURE BARK OF THE MAIN TRUNK, CAN RESULT IN SERIOUS LOCALIZED OR TRANSLOCATED DAMAGE.

Mixing and Application: Add the recommended amount of SURFLAN, or SURFLAN and other residual herbicides to the spray tank according to directions for these tank mixes. Add Roundup during the last step of the filling operation. Agitate continuously during filling and throughout application. Apply in 20 to 60 gallons of water per acre in a properly calibrated low pressure herbicide sprayer. Check the sprayer daily for proper calibration and uniform application.

Activation: A single one-half inch (½") to one inch (1") rain or sprinkler irrigation to move SURFLAN into the weed seed germination zone is required to activate the herbicide. If weeds begin to emerge, a shallow cultivation of one to two inches (1" to 2") will destroy existing weeds and place SURFLAN in the zone of weed seed germination.

Caution: Read the Roundup label carefully for directions, cautions and precautions before applying this product.

SURFLAN/SINBAR® TANK MIX

Sinbar®—terbacil, E.I. duPont de Nemours and Company

SURFLAN/Sinbar is recommended for weed control in the following crops:

Apple	Loganberry
Peach	Raspberry
Blueberry	Orange
Caneberries:	Grapefruit
Blackberry	Lemon
Boysenberry	Pecan
Dewberry	

SURFLAN/Sinbar controls weeds in addition to those controlled by SURFLAN alone. As the amount of Sinbar applied per acre increases, so does the number of additional weeds controlled. Consult the Sinbar label for application rate required for the specific additional weeds to be controlled.

General Directions: Follow recommended application procedures for SURFLAN. Apply SURFLAN/Sinbar in at least 20 gallons per acre or more to ensure a uniform application. Moisture is necessary to activate the tank mix. Consult the SURFLAN label for activation information.

Broadcast Rates Per Acre

	SURFLAN A.S. (quarts)	Sinbar 80W* (pounds)
Apple, Peach	2–6	2–4

Apply either in the spring or after harvest in the fall before weeds emerge. Where the crop is grown under furrow irrigation or under raised-berm flood irrigation (trees four to six inches (4" to 6") above water line), apply Sinbar only as a band treatment. Do not treat trees planted in the bottom of irrigation furrows or trees grown under flat flood or basin irrigation, and do not use on eroded areas where subsoil or tree roots are exposed, as injury to trees may result.

Blueberry	2–6	2–4

Apply in a minimum of 25 gallons of water per acre.

Treat only plantings established for one (1) year or more. Make a single band or broadcast application to ground beneath bushes; avoid contact of foliage and fruit with spray or mist. Apply either in the spring or after harvest in the fall before weeds emerge. Do not use Sinbar on eroded areas where subsoil or roots are exposed, nor on plants that are diseased or lacking in vigor, as injury to the plants may result. Sinbar treated areas may be planted to blueberries one (1) year after last application.

Caneberries:	2–6	1–2
Blackberry		
Boysenberry, Dewberry		
Loganberry, Raspberry		

Treat only plantings established for one (1) year or more. Make a single bank or broadcast application to the soil beneath the canes in the fall or early spring before fruit set and before weeds emerge. Do not spray foliage nor use on eroded areas where subsoil or roots are exposed, nor apply to plants that are diseased or lacking vigor as injury to the plant may result. Sinbar treated areas may be planted to caneberries one (1) year after last application. Do not apply Sinbar within 70 days before harvest.

Citrus:		
Orange, Grapefruit, Lemon Trees established for one (1) year:	2–6	2–4
Trees established for two (2) years or more:		
Arizona and California	2–6	2–8
Florida and Texas	2–6	2–10

Apply as a band or broadcast treatment as a directed spray; avoid contact of foliage and fruit with spray or mist. Temporary yellowing of citrus leaves may occur following Sinbar treatment. Sinbar may be used on coarse soils but is not to be used on sands low in organic matter (less than one percent) (1%), on poorly drained soils, or on trees planted in irrigation furrows, as injury to trees may result. Sinbar treated areas may be planted to citrus trees one (1) year after last application.

* Sinbar application rates may vary with soil texture, weeds to be controlled, length of control desired and age of crop.

Caution: Read the Sinbar label for additional directions for use, cautions and special precautions prior to use.

SURFLAN/SOLICAM® TANK MIX

Solicam®—norflurazon, Sandoz, Inc.

The SURFLAN/Solicam tank mix controls the following weeds in addition to those weeds controlled by SURFLAN alone:

Barley, wild	Filaree (red and whitestem)
Brome, downy	Fingergrass, feather
Brome, ripgut	Grama grass, sixweeks
Bursage, annual	London rocket
Cupgrass, southwestern	Pineappleweed
Dandelion, false	Thistle, Russian and sow

Weeds suppressed by this tank mix in addition to those suppressed by SURFLAN alone include:

Bermudagrass	Nutsedge, purple
Horseweed	Quackgrass
Nightshade, silverleaf	

SURFLAN/Solicam is recommended on the following crops:*

Almond	Nectarine
Apple	Peach
Apricot	Pear

Cherry	Plum
Citrus	Prune
Filbert	Walnut

* Apply Solicam on all of the above crops only in California, Oregon and Washington. In Arizona, use on citrus only. Use only on crops in the Imperial, Coachella or Palo Verde Valleys of California.

Orchard Establishment Requirements: Use SURFLAN/Solicam on crops listed above that have been established at least 18 months in the field with exception of citrus, where the trees should be established in the field for 12 months. In all cases, the soil should be settled and firm at the time of application. Do not use on nursery stock.

General Directions: Apply the tank mix to the soil surface from fall to early spring before weeds emerge, except on those crops where otherwise specified. The orchard floor must be free of weeds and must be relatively free of plant residues. Solicam does not control established weeds.

Apply as a directed spray. Avoid contact with fruit or foliage. Make only one application per year.

Activation: If no rainfall occurs within two weeks after application, the tank mix must be activated by sprinkler irrigation. Check activation instructions for SURFLAN alone.

Broadcast Rates Per Acre

Use the lower Solicam rates on coarse soils. Use the higher rates on medium to fine soils or soils having organic matter content of greater than 3%.

Crop	SURFLAN A.S. (quarts)	Solicam 80W (pounds)
Almond	2–6	2½–5
Apply from October through January. Precaution: Do not use on coarse textured soils such as sand, loamy sand, gravelly sand and fine sand.		
Apple, Apricot, Filbert, Nectarine, Peach, Pear, Plum, Prune, Walnut	2–6	2½–5
Cherry	2–6	2½–5
Precaution: Do not use on coarse textured soils such as sand, loamy sand, gravelly sand and fine sand.		
Citrus	2–6	2½–5
Apply any time during the year according to local practice. **Precaution:** Do not use where citrus is interplanted with palm trees.		

Solicam Precautions: Do not apply when nuts or fruit are on the ground at harvest. Do not graze livestock on treated areas. Do not feed or graze cover crops grown in treated orchards. Read the Solicam label for directions, cautions and limitations prior to use.

SURFLAN APPLICATION WITH LIQUID FERTILIZERS

GENERAL INSTRUCTIONS

SURFLAN may be mixed with most liquid fertilizer materials. The combination of SURFLAN with solutions and suspension type fertilizers has provided weed and grass control equal to the same rates of SURFLAN applied in water. Follow SURFLAN label recommendations regarding rates per acre, crops, application directions, special instructions, cautions and special precautions. Individual state regulations relating to liquid fertilizer mixing, registration, labeling and applications are the responsibility of the individual and/or company selling the fertilizer and chemical mixture.

Testing for Tank Mix Compatibility in Liquid Fertilizers: SURFLAN alone or in tank mixture with dry flowables, wettable powders (WP), liquids (L), flowables (F), aqueous suspensions (AS), solutions (S) or emulsifiable concentrates (EC) may not combine properly with some fluid fertilizer materials. Small quantities should always be tested before full-scale mixing. This will determine whether a compatibility agent is needed and which agent does the best job. The seven (7) agents listed on page 61 have been thoroughly tested. There are many other surfactants on the market which were not designed for use with liquid fertilizers. Use the following test to select the correct agent for your mixture.

1. Put one (1) pint of liquid fertilizer in a quart jar.
2. Add one (1) to four (4) teaspoonsful of the dry flowables, WP, L, F, or AS formulation (depending on the recommended rate per acre) to the liquid fertilizer. Close jar and agitate until the materials are dispersed evenly in the fertilizer. If the materials do not disperse well, it may be necessary to slurry chemicals in water before adding to the fertilizer.
3. After dispersing the materials (Step 2), add three (3) to four (4) teaspoonsful of any emulsifiable formulation (EC) to the jar and shake well. Add solution herbicides (S) to the mixture last and agitate. Observe the jar for about 10 minutes. If the materials rise to the surface and form a thick layer (oily curds) which will not disperse again when agitated, a compatibility agent is needed. If the mixture is easily dispersed to its original state with slight agitation, no agent is needed but good agitation must be provided in the fertilizer spray tank.
4. If the need for a compatibility agent is shown in Step 3: Using a clean quart jar, start at Step 1 above, add one-half (½) teaspoon of the compatibility agent to the liquid fertilizer, mix well, then repeat Steps 2 and 3.

An effective compatibility agent will cause the mixture to remain uniformly mixed with little or no separating or oil rising to the surface for one-half (½) hour or longer. If slight separation does occur, two (2) to three (3) inversions of the jar should give a uniform remix. If oily curds form which will not disperse, more agent or another agent should be tried.

Use a clean jar for each test. The compatibility mixture will have a uniform appearance and will be relatively easy to keep mixed with gentle agitation of the jar.

LIQUID FERTILIZER MIXING INSTRUCTIONS

SURFLAN in Liquid Fertilizer

Aqueous suspensions such as SURFLAN A.S. can be mixed with liquid fertilizers. The use of compatibility agents is especially important when tank mixing emulsifiable concentrates (EC) with dry flowables, wettable powders (WP), flowables (F), liquids (L), aqueous suspensions (AS) or solutions (S) in liquid fertilizer. These compatibility agents can be used at rates as low as one and one-half (1½) to two (2) pints per ton of liquid fertilizer and should be mixed well with the fertilizer before adding the herbicides. Read the label on the compatibility agent and follow the directions.

1. Sponto 168D (Witco Chemicals Co., Chicago, IL)
2. Compat (Farm Chemicals, Inc., Aberdeen, NC)
3. Unite (Hopkins Ag Chemical, Madison, WI)
4. T-Mulz 734-2 (Thompson-Hayward Chemical Co., Kansas City, KS)
5. Rigo Compatibility Agent (Rigo Company, Buckner, KY)
6. Amoco Spray Mate+up,TM (Amoco Oil Co., Chicago, IL)
7. Kem-Link (Universal Coop, Minneapolis, MN)

All of the above are phosphate ester-type surfactants designed to be used with liquid fertilizers. They usually do not work well as compatibility agents in tank mixtures in plain water.

APPLICATION INSTRUCTIONS

Spread the SURFLAN/fertilizer mixture normally with a properly calibrated applicator. Be certain the material is applied uniformly to the soil surface.

SURFLAN PERFORMANCE GUARANTEE (A LIMITED WARRANTY OF WEED AND GRASS CONTROL)

Elanco Products Company warrants that SURFLAN will control the weeds and grasses as indicated on its current label when used according to label directions and subject to the following limitations:

1. Purchaser must notify Elanco promptly if a lack of satisfactory control occurs. Such notice must be given within 30 days after the first planting or application of SURFLAN to the treated crop, whichever occurs later.
2. An Elanco representative must be satisfied that Purchaser used SURFLAN according to label directions and must visually observe unsatisfactory control of weeds in the field.

3. The product guarantee will extend to the following applications only:
   a) Preplant surface application (PPSA) when a light tillage is used prior to planting.
   b) Postplant surface application (SA) when a light tillage is used prior to or just following crop emergence.
   c) Any recommended application where a single one-half inch (½″) rainfall or sprinkler irrigation occurs prior to weed seedling emergence.
4. The Purchaser must provide an Elanco representative with an invoice or other satisfactory record which shows the price and quantity of SURFLAN purchased.
5. Elanco's liability will be limited to a refund of the purchase price of the SURFLAN applied to the acreage on which weed control was not satisfactory or, at the option of the Purchaser, a refund of an equivalent amount of SURFLAN.
6. Void outside the U.S.A.

DISCLAIMER OF WARRANTIES

The warranty printed above is the only warranty applicable to this product. All other warranties, express or implied, including but not limited to the implied warranties of merchantability and fitness for a particular purpose, are disclaimed.

LIMITATION OF DAMAGES

Elanco's liability, whether in contract, warranty, tort, negligence, strict liability or otherwise, shall not exceed the return of the amount of the purchase price of the SURFLAN and under no circumstances shall Elanco be liable for special, indirect or consequential damages.

INHERENT RISKS OF USE

Failure to carefully follow the directions for use of SURFLAN may result in unsatisfactory weed control or crop injury. Factors such as seedling disease, foliage disease, cold weather, deep planting, excessive moisture, high salt concentration or drought may also result in unsatisfactory weed control or crop injury.

SURFLAN®

DRY FLOWABLE HERBICIDE

A preemergence surface-applied herbicide for the control of most annual grasses and certain broadleaf weeds.

ACTIVE INGREDIENT:	
oryzalin* (3,5-Dinitro-N^4,N^4-dipropylsulfanilamide)	85.0%
INERT INGREDIENTS	15.0%
TOTAL	100.0%

Contains 0.85 pounds active ingredient per pound.

*Surflan®—the registered trademark for Elanco Products oryzalin.

EPA Reg. No. 1471-151

CAUTION
Keep Out of Reach of Children

PRECAUTIONARY STATEMENTS

HAZARDS TO HUMANS AND DOMESTIC ANIMALS

Harmful if swallowed, inhaled or absorbed through the skin. Avoid breathing dust or spray mist and contact with skin or eyes. Use eye protection and protective clothing such as coveralls, a long sleeved shirt and impermeable gloves when handling this product.

First Aid: If swallowed, call a physician or Poison Control Center. Drink one to two glasses of water and induce vomiting by touching back of throat with finger. Do not induce vomiting or give anything by mouth to an unconscious person. If on skin, wash with plenty of soap and water. Get medical attention if irritation persists. If in eyes, flush eyes with plenty of water. Get medical attention if irritation persists. If inhaled, remove individual to fresh air. If breathing difficulty occurs, get medical attention.

ENVIRONMENTAL HAZARDS

Do not apply directly to water or wetland. Do not contaminate water by cleaning of equipment or disposal of wastes. Cover or incorporate spills.

STORAGE AND DISPOSAL

STORAGE

Store in original container only. Do not store in direct sunlight. Do not store at temperatures above 120 ° F. If stored above 120 ° F, Surflan Dry Flowable could lose some of its dispersing properties, resulting in difficult application, poor weed control or crop injury. In case of leak or spill, contain material and dispose as waste.

PESTICIDE DISPOSAL

Do not contaminate water, food or feed by storage or disposal. Wastes resulting from the use of this product may be disposed of on site or at an approved waste disposal facility.

CONTAINER DISPOSAL

Triple rinse (or equivalent). Then offer for recycling or reconditioning, or puncture and dispose of in a sanitary landfill, or, if allowed by state and local authorities, by burning. If burned, stay out of smoke.

PRECAUTION

Do not feed forage from treated fields to livestock. Poor weed control or crop injury may result if directions are not followed. Over application may result in crop injury or soil residue. Do not plant any root crop for 12 months following a Surflan application.

Surflan Dry Flowable should not be impregnated on any dry bulk fertilizer. Do not impregnate nitrate containing fertilizer blends with Surflan A.S., 75W or Dry Flowable; the combination may result in explosive mixtures. Combining Surflan Dry Flowable with any dry bulk fertilizer may result in explosive mixtures.

WEEDS AND GRASSES CONTROLLED

Weeds Controlled in Field Crops by Surflan

Surflan will not control established weeds.

Grasses	
Barnyardgrass (Watergrass)	*Echinochloa sp*
Brachiaria (Signalgrass)	*Brachiaria sp.*
Browntop panicum*	*Panicum fasciculatum*
Crabgrasses (Large crabgrass) (Smooth crabgrass)	*Digitaria spp.*
Crowfootgrass	*Dactyloctenium aegyptium*
Fall panicum (Spreading panicgrass)	*Panicum dichotomiflorum*
Foxtails (Bottlegrass) (Bristlegrass) (Giant foxtail) (Green foxtail) (Pigeongrass) (Robust foxtail) (Yellow foxtail)	*Setaria spp.*
Goosegrass (Silver crabgrass) (Wiregrass) (Yardgrass)	*Eleusine indica*
Johnsongrass (Seedling only)	*Sorghum halepense*
Texas panicum (Buffalograss) (Coloradograss	*Panicum texanum*
Broadleaf Weeds	
Carpetweed	*Mollugo verticillata*
Florida pusley (Florida purslane) (Mexican clover)	*Richardia scabra*

(Pusley)	
Lambsquarters	*Chenopodium album*
Pigweeds	*Amaranthus spp.*
(Carelessweed)	
(Prostrate pigweed)	
(Redroot pigweed)	
(Rough pigweed)	
(Spiny pigweed)	
Purslane, common	*Portulaca oleracea*
Spurge, prostrate*	*Euphorbia supina*

* (See Cotton-Surflan Alone section for special instructions for Texas only)

Surflan also provides partial control or suppression of:

Ladysthumb	Smartweed, annual
Morningglory, annual	Spurge, spotted
Nightshade, black	Velvetleaf
Prickly sida (teaweed)	Volunteer wheat
Ragweed, common	

Weeds Controlled In Fruit, Nut Crops and Vineyards by Surflan

Surflan applied in Fruit, Nut Crops and Vineyards controls the following weeds:

Grasses	
Barley, little	*Hordeum pusillum*
Barnyardgrass (Watergrass)	*Echinochloa sp.*
Bluegrass, annual (Poa)	*Pos annua*
Brachiaria (Signalgrass)	*Brachiaria sp.*
Browntop panicum	*Panicum fasciculatum*
Crabgrasses	*Digitaria spp.*
(Large crabgrass)	
(Smooth crabgrass)	
Crowfootgrass	*Dactyloctenium aegyptium*
Cupgrass	*Eriochloa graclis*
Downy brome	*Bromus tectorum*
Fall panicum (Spreading panicgrass)	*Panicum dichotomiflorum*
Foxtails	*Setaria spp.*
(Bottlegrass)	
(Bristlegrass)	
(Giant foxtail)	
(Green foxtail)	
(Pigeongrass)	
(Robust foxtail)	
(Yellow foxtail)	
Goosegrass	*Eleusine indica*
(Silver crabgrass)	
(Wiregrass)	
(Yardgrass)	
Guineagrass	*Panicum maximum*
Johnsongrass (Seedling only)	*Sorghum halepense*
Junglerice	*Echinochloa colonum*
Lovegrass, Mexican	*Eragrostis mexicana*
Lovegrass, orcutt	*Eragrostis orcuttiana*
Oat, wild	*Avena fatua*
Ryegrass, annual (Italian)	*Lolium multiflorum*
Texas panicum	*Panicum texanum*
(Buffalograss)	
(Coloradograss)	
Witchgrass	*Panicum capillare*
Broadleaf Weeds	
Carpetweed	*Mollugo verticillata*
Chickweed	*Stellaria media*
Cudweed	*Gnaphalium chilense*
Fiddleneck, coast	*Amsinckia intermedia*
Florida pusley	*Richardia scabra*
(Florida purslane)	
(Mexican clover)	
(Pusley)	
Henbit	*Lamium amplexicaule*
Knotweed, prostrate	*Polygonum aviculare*
Lambsquarters	*Chenopodium album*
Pigweeds	*Amaranthus spp.*
(Carelessweed)	
(Prostrate pigweed)	
(Redroot pigweed)	
(Rough pigweed)	
(Smooth pigweed)	
(Spiny pigweed)	
(Tumble pigweed)	
Puncturevine	*Tribulus terrestris*
Purslane, common	*Portulaca oleracea*
Rockpurselane, redmaids	*Calandrinia caulescens*
Sandbur, field	*Cenchrus incertus*
Shepherdspurse	*Capsella bursa-pastoris*
Sprangletop, red	*Leptochloa filiformis*

Surflan provides partial control or suppression of:

Filaree, red and white stem	Nightshade, black
Groundsel, common	Prickly sida (teaweed)
Ladysthumb	Ragweed, common
Lettuce, prickly	Ragweed, giant (horseweed)
Mallow, common	Rocket, London
Milkweed, climbing	Smartweed, annual
Morningglory, annual	Sowthistle, annual
Mustard, black	Spurge, prostrate
Mustard, wild	Velvetleaf

DIRECTIONS FOR USE

Surflan is a preemergence, surface applied herbicide for the control of most annual grasses and certain broadleaf weeds. General rule: The more coarse the soil or the lower the organic matter, the lower the rate.

Soil Texture Guide

Refer to the following guide to determine your soil texture.

SOIL TEXTURE	SOIL CLASSIFICATION
Coarse (light) soils	Sand
	Loamy sand
	Sandy loam
	Loam
Medium soils	Silt loam
	Silty clay loam*
	Silt
	Sandy clay loam*
Fine (heavy) soils	Clay
	Clay loam
	Silty clay loam*
	Silty clay
	Sandy clay
	Sandy clay loam*

* Silty clay loam and sandy clay loam soils are transitional soils and may be classified as either medium or heavy textured soils. If silty clay loam or sandy clay loam soils are predominantly sand or silt, they are usually classified as medium textured soils. If predominantly clay, they are usually classified as fine textured soils.

MIXING DIRECTIONS

Use the following instructions for mixing Surflan Dry Flowable alone in water and tank mixed in water.

Mixing Instructions in Water Alone

1. Be sure sprayer is clean and not contaminated with any foreign material.
2. Fill tank one-fourth (¼) full with clean water and start agitation.
3. Be certain that agitation system is working properly. Vigorous, continuous agitation throughout the spray tank is required with Dry Flowable formulations. Agitation should be sufficient to create a rippling or rolling action on the liquid surface. Sparger pipe agitation generally provides the best agitation in spray tanks.
4. To prevent foaming, avoid stirring or splashing air into the

mixture by placing the end of the fill pipe below the surface of the water in the spray tank during the filling process.

5. **Pour product directly from container into tank.**
Controlled addition of the product is important to assure adequate wetting and mixing to prevent clogging of screens and outlet ports.
6. Continue filling tank with water until full.

Tank-Mix Instructions

1. Add Surflan Dry Flowable to the spray tank following the instructions outline above for Surflan alone.
2. Fill the spray tank to 90% full with water.
3. Add other formulations in this order, other dry flowables, wettable powders (WP), aqueous suspensions (A.S.), flowables (F), liquids (L), solutions (S), and emulsifiable or liquid concentrates (E.C., L.C.) in this order **ALLOWING TWO TO THREE MINUTES BETWEEN THE ADDITION OF EACH PRODUCT,** and agitate continuously until the product(s) are completely dispersed in the water. Each chemical added to a spray tank needs time to either wet or mix with water and other chemicals in the spray tank. To insure proper mixing, time and agitation is required between the addition of each of the additional products.
4. Continue filling tank with water until full.

General Instructions and Precautions

1. For either Surflan Dry Flowable alone or in tank mix combination maintain agitation from filling through application. If spraying and agitation is stopped before the tank is empty, the materials may settle to the bottom. In this case, it is important to resuspend again all the material in the bottom of the tank before continuing the spray application. Sometimes it is more difficult to resuspend settled material than it is to suspend it originally. A sparger agitator is particularly useful for this purpose.
2. In-line screens in the tank should be no finer than 16 mesh (16 mesh has bigger openings than 50 mesh).
3. If a buildup of material is seen on the walls of the spray tank, wash the tank with soapy water between fillings. Rinse and continue the spraying operation. Clean tank, lines and screens thoroughly after use.
4. Read and carefully follow all label instructions for each material added to the tank.

Read the Surflan Dry Flowable label for additional directions, precautions and limitations before use.

SOIL PREPARATION

Destroy existing weeds before or during a Surflan application. The soil surface should be level enough to allow a sprayer to be operated efficiently and at speeds appropriate for uniform application.

APPLICATION

Ground Application: Apply in 20 to 40 gallons of water per acre (broadcast basis), using any properly calibrated low pressure herbicide sprayer that will apply the spray uniformly. Use herbicide tips and screens no finer than 50 mesh in nozzle and in-line strainers. As the amount of water used (spray volume) decreases, the importance of accurate calibration and uniform application increases. Check the sprayer daily to ensure proper calibration and uniform application. Agitate thoroughly before and during application. Avoid boom overlaps that will increase rates above those recommended.

Aerial Application: Use a standard aerial herbicide boom sprayer. Aerial spray equipment should be calibrated to apply the proper amount of Surflan Dry Flowable alone or in tank mix combinations in two (2) to 10 gallons of spray mixture per acre. Nozzle screens and in-line strainers should be no finer than 50 mesh for tank mix combinations. Surflan Dry Flowable (alone or in tank mix combinations) mixes readily with water for these concentrate aerial sprays; however, constant vigorous agitation that sweeps the contents from the bottom of the spray tank up into the main body of the liquid is required to maintain a uniform suspension until spray tank is empty. Avoid overlap of spray patterns. Do not apply when weather conditions favor drift from target area.

Chemigation: Surflan Dry Flowable may be applied for weed control through center pivot and solid set sprinkler irrigation systems. Apply at rates recommended on this label. Prepare mixture with one (1) part water to one (1) part product. Injecting a larger volume of a more dilute slurry per hour will assure more accurate calibration of metering equipment. Maintain sufficient agitation to keep the herbicide in suspension. Meter slurry into irrigation water during entire period. Apply Surflan in a minimum of one inch (1") of irrigation water. The following precautions and recommendations must be followed to chemigate Surflan.

1. Apply only through properly maintained irrigation systems having continuously moving laterals and providing uniform distribution of water. Where a system does not provide uniform water distribution such as at the extreme ends of the lateral, around supports, where treated discharge water is released from water-powered motors, or where extreme changes in elevation occur, unacceptable weed control or crop injury may occur.
2. Apply the product only through irrigation systems containing anti-siphon and check valves, to prevent water source contamination and overflow and siphoning of the slurry tank, and interlocking controls between the metering device and the water pump, to insure simultaneous shut-off.
3. Inject the product with a positive displacement pump into the main line ahead of a right angle turn, to insure adequate mixing.
4. Application of more than label recommended quantities of irrigation water per acre may result in decreased product performance by removing the pesticide from the zone of effectiveness.
5. Do not apply when wind speed favors drift, when system connections or fittings leak, when nozzles do not provide uniform distribution, or when lines containing the product should be dismantled and drained. Use of low trajectory angle sprinklers or drop nozzles will decrease the likelihood of drift.
6. Greater accuracy in pesticide calibration and distribution will be achieved by injecting a larger volume of more dilute slurry per hour.
7. Constant agitation should be maintained in the slurry tank during the entire period of pesticide application.
8. Sample Calculation: One hundred thirty-five (135) acres to be covered by irrigation system.
Recommended rate of 1.2 pounds per acre times 135 acres equals 162 pounds of product.
Prepare slurry mixes using one (1) part water to one (1) part product and add to water in a 500 gallon nurse tank using the entire 162 pounds of product. Apply the entire 500 gallons to 135 acres. Irrigation system requires 20 hours to apply 0.5 inch of water to 135 acres.
Injection rate equals 500 gallons divided by 20 hours equals 25 gallons per hour or 53.33 fluid ounces to be injected per minute. Calibrate injection pump to deliver the herbicide mix into the irrigation system at the rate of 53.33 fluid ounces per minute.
9. Posting Requirements: Areas being treated with pesticides applied through irrigation systems must be posted at the usual entrances to the area, at bulletin boards where workers assemble and where the treated area lies adjacent to roads, residential areas, or any other areas such as schools, parks or other public facilities. Posted signs must be legible from 25 feet away and be printed in English and any other appropriate language such as Spanish. Signs must be posted prior to application, during application and for at least one week after application, but may remain in place indefinitely. The sign must indicate the following:

Notice: Pesticides are applied in irrigation water in this field. Do not enter this field when the irrigation system is operating. Do not drink, bathe in or play in water or in furrows, puddles, ponds, canals or ditches associated with this irrigation system.

Precaution: Do not apply Surflan via center pivot irrigation systems to fruit and nut orchards or vineyards. In California, do not apply Surflan to nondormant fruit and nut orchards or vineyards through overhead sprinkler irrigation systems.

ACTIVATION AND CULTIVATION

A single, one-half inch (½") rain or its equivalent in center pivot or sprinkler irrigation is necessary to activate Surflan when used alone or in a tank mix combination. If weeds begin to emerge, shallow cultivate one to two inches (1″ to 2″) to destroy existing weeds and place Surflan in zone of weed seed germination. If heavy rains pack the soil surface, rotary hoe to break the crust to aid crop emergence. Crops treated with Surflan can be shallowly cultivated without loss of activity.

CROP RECOMMENDATIONS

FIELD CROPS

General Directions

These recommendations are given as the broadcast (overall) rates of Surflan Dry Flowable per acre. For bank application, use proportionately less. Surflan is not recommended for use on soils containing more than five percent (5%) organic matter. Only one (1) application of Surflan per season should be made.

SOYBEAN—SURFLAN ALONE

Apply Surflan Dry Flowable to the soil surface from four (4) weeks prior to planting up through two (2) days after planting at a broadcast rate per acre of 0.9 pounds on coarse soils; 1.2 pounds on medium soils and 1.8 pounds on fine soils.

SOYBEAN—TANK MIX RECOMMENDATIONS

SOYBEAN—SURFLAN/SENCOR® (METRIBUZIN) OR SURFLAN/LEXONE® (METRIBUZIN) TANK MIX

The Surflan/Sencor or Surflan/Lexone tank mix effectively controls all of the annual grasses and broadleaf weeds listed for Surflan plus many additional weeds including:

Browntop millet	Redweed
Hemp sesbania	Sicklepod
Jimsonweed	Smartweed, annual
Prickly sida (teaweed)	Velvetleaf
Ragweed, common	

Surflan/Sencor and Surflan/Lexone also provide partial control or suppression of:

Cocklebur	Ragweed, giant (horseweed)
Morningglory	

The Surflan/Sencor or Surflan/Lexone tank mix may be applied from two (2) weeks prior to planting up through two (2) days after planting, but before crop emergence with any low pressure herbicide sprayer. Do not spray over top of emerged soybeans.

Broadcast Rates Per Acre
Surface Applied Before or After Planting

Soil Texture	Surflan Dry Flowable	Lexone 4L or Sencor 4L	Lexone Dry Flowable or Sencor Dry Flowable
	(pounds)	(pints)	(pounds)
Coarse	0.6–0.9	½	⅓
Medium	0.9–1.2	¾	½
Fine	1.2–1.5	1	⅔

Do not use the Surflan/Sencor or Surflan/Lexone tank mix on soils containing more than five percent (5%) organic matter in a preplant surface application or postplant surface application.

Special Precaution: Poor weed control and/or crop injury may result if directions are not followed. Over application may result in crop injury and/or soil residue. Do not use treated vinces for feed or forage. Do not contaminate any body of water or apply to any area not specified on the label. Do not allow sprays to drift onto adjacent desirable plants. Do not plant any crop other than soybeans within four (4) months after application.

Caution: Please read the Sencor or Lexone labels carefully for cautions, precautions and special precautions before applying these products.

Destroy empty containers. Do not reuse.

SOYBEAN—SURFLAN/LOROX® (LINURON) TANK MIX

The Surflan/Lorox tank mix effectively controls all of the annual grasses and broadleaf weeds listed for Surflan plus many additional weeds including:

Canarygrass	Nettleleaf goosefoot
Chickweed, common	Radish, wild
Galinsoga	Ragweed, common
Mustard	Smartweed, annual

Surflan/Lorox also provides partial control of:

Cocklebur	Sicklepod

Follow Surflan procedures for soil preparation. The Surflan/Lorox tank mix may be applied within two (2) days after planting, but before crop emergence using any properly calibrated low pressure herbicide sprayer. Do not spray over top of emerged soybeans.

Broadcast Rates Per Acre

Soil Texture	Surflan Dry Flowable	Lorox 50 WP/L
	(pounds)	(pounds or pints)
Coarse	0.6–0.9	⅔–1¼
Medium	0.9–1.2	1–1⅔
Fine	1.2–1.5	1¼–2

Do not use on sand or loamy sand soils.

Do not use the Surflan/Lorox tank mix on soils containing more than five percent (5%) organic matter.

Special Precaution: Poor weed control and/or crop injury may result if directions are not followed. Over application may result in crop injury and/or soil residue. Do not use treated vines for feed or forage. Do not contaminate any body of water or apply to any area not specified on the label. Do not allow sprays to drift onto adjacent desirable plants.

Caution: Please read the Lorox label carefully for cautions, precautions and special precautions before applying this product.

Destroy empty containers. Do not reuse.

SOYBEAN—SURFLAN/DYANAP® OR SURFLAN/KLEAN-KROP® TANK MIX

The Surflan/Dyanap or Surflan/Klean-Krop tank mix effectively controls all of the annual grasses and broadleaf weeds listed for Surflan plus many additional weeds including:

Bindweed, field	Groundcherry	Sandbur
Chickweed	Morningglory	Shepherdspurse
Cocklebur	Mustard	Sprangletop
Cupgrass	Prickly sida (teaweed)	Stinkgrass
Florida beggarweed	Ragweed, common	Velvetleaf
Galinsoga		Windmillgrass

Follow Surflan procedures for soil preparation. Surflan/Dyanap or Surflan/Klean-Krop may be applied within two (2) days after planting but before crop emergence using any properly calibrated low pressure herbicide sprayer. Do not spray over top of emerged soybeans.

Broadcast Rates Per Acre

Soil Texture	Surflan Dry Flowable	Dyanap (2+1)S or Klean-Krop
	(pounds)	(quarts)
Coarse	0.6–0.9	6
Medium	0.9–1.2	6
Fine	1.2–1.5	6

Do not apply Surflan/Dyanap or Surflan/Klean-Krop to any soils with over five percent (5%) organic matter. Do not use on silt loam soils of extremely fine texture.

Special Precaution: Poor weed control and/or crop injury may result if directions are not followed. Over application may result in crop injury and/or soil residue. Do not use treated vines for feed or forage.

Caution: Please read the Dyanap or Klean-Krop labels carefully for cautions, precautions and special precautions before applying these products.

Destroy empty containers. Do not reuse.

SOYBEAN—SURFLAN/AMIBEN® (CHLORAMBEN) TANK MIX

The Surflan/Amiben tank mix effectively controls all of the annual grasses and broadleaf weeds listed for Surflan plus many additional weeds including:

Chickweed	Ragweed, common
Coffeeweed	Smartweed, Pennsylvania
Kochia	Spurge, annual
Mustard, wild	Stinkgrass
Nightshade, black	Thistle, Russian
Prickly sida (teaweed)	Velvetleaf (buttonweed)

Surflan/Amiben may be applied within two (2) days after planting, but before crop emergence, using any properly calibrated low pressure herbicide sprayer that will apply the spray uniformly. Agitate thoroughly before and during application. Follow normal Surflan procedures for soil preparation. Do not spray over top of emerged soybeans.

Broadcast Rates Per Acre

Soil Texture	Surflan Dry Flowable	Amiben 2S	Amiben DS
	(pounds)	(quarts)	(pounds)
Coarse	0.6–0.9	4 to 6	2.4–3.6
Medium	0.9–1.2	4 to 6	2.4–3.6
Fine	1.2–1.5	4 to 6	2.4–3.6

The higher rates of Amiben should be used on clay loams, clay or high organic (three to five percent) (3% to 5%) mineral soils. Do not use Surflan/Amiben on soils containing more than five percent (5%) organic matter.

Special Precaution: Poor weed control and/or crop injury may result if directions are not followed. Over application may result in crop injury and/or soil residue. Do not use treated vines for feed or forage. Do not contaminate any body of water or apply to any area not specified on the label. Do not allow sprays to drfit onto adjacent desirable plants.

Caution: Please read the Amiben label carefully for cautions, precautions or special precautions before applying this product.

Destroy empty containers. Do not reuse.

SURFLAN/LASSO®, AND SURFLAN/LASSO/LEXONE® OR SENCOR® COMBINATIONS FOR WEED CONTROL IN SOYBEANS

SURFLAN Dry Flowable may be tank-mixed with Lasso or Lasso Micro-Tech™ and Lasso or Lasso Micro-Tech plus Lexone or Sencor for weed control in soybeans.

SURFLAN/LASSO TANK-MIX

SURFLAN/Lasso controls those weeds listed for SURFLAN alone plus provides partial control or suppression of the following weeds listed on either the SURFLAN or Lasso labels.

Ladysthumb
Morningglory
Nightshade, black
Nightshade, hairy
Prickly sida (teaweed)
Smartweed, Pennsylvania
Velvetleaf

The SURFLAN/Lasso tank-mix may be applied to the soil surface after planting up through two (2) days after planting and prior to crop or weed emergence. Apply the tank-mix in 20 to 40 gallons of water per acre with any low pressure herbicide sprayer. Do not spray overtop of emerged soybeans.

Broadcast Rates Per Acre

Soil Texture	SURFLAN Dry Flowable (Pounds)	Lasso or Lasso Micro-Tech (Quarts)
Coarse	0.6	1.5–2
Medium	0.6	2–2.5
Fine	0.6–0.9	2.5–3

Use the lower rate in the rate ranges on soils with less than 3% organic matter and the higher rate on soils with 3% or more organic matter. Use the higher rate in the recommended range in areas of heavy weed infestations.

SURFLAN/LASSO/LEXONE OR SENCOR TANK-MIX

SURFLAN/Lasso/Lexone or Sencor controls those weeds listed for SURFLAN alone plus these additional weeds that are listed on either the Lasso/Lexone or Sencor labels.

Prickly sida (teaweed)
Ragweed, common
Smartweed, Pennsylvania
Velvetleaf

SURFLAN/Lasso/Lexone or Sencor provides partial control or suppression of the following weeds listed on either the SURFLAN, Lasso, Lexone or Sencor labels:

Cocklebur
Jimsonweed
Ladysthumb
Morningglory
Nightshade, black
Nightshade, hairy
Ragweed, giant
Sicklepod

The SURFLAN/Lasso/Lexone or Sencor tank-mix may be applied to the soil surface after planting up through two (2) days after planting and prior to crop or weed emergence. Apply the tank-mix in 20 to 40 gallons of water per acre with any low pressure herbicide sprayer. Do not spray overtop of emerged soybeans.

Broadcast Rates Per Acre

Soil Texture	SURFLAN Dry Flowable (Pounds)	Lasso or Lasso Micro-Tech (Quarts)	Lexone 4L or Sencor 4 (Pints)	Lexone Dry Flowable or Sencor Dry Flowable (Pounds)
Coarse	0.6	1.5–2	0.33–0.5	0.25–0.33
Medium	0.6	2–2.5	0.5–0.75	0.33–0.5
Fine	0.6–0.9	2.5–3	0.75–1	0.5–0.66

Use the lower rate in the above rate ranges on soils with less than 3% organic matter and the high rate on soils with 3% or more organic matter. Use the higher rate in the recommended range in areas of heavy weed infestations.

Do not use Lexone or Sencor on coarse soils with less than 1% organic matter.

Special Precautions: Poor weed control and/or crop injury may result if directions are not followed. Over application may result in crop injury and/or soil residue. Do not use treated vines for feed or forage. Do not contaminate any body of water or apply to any area not specified on the label. Do not allow sprays to drift onto adjacent desirable plants. Do not plant any crop other than soybeans within four (4) months after application.

Caution: Read the SURFLAN, Lasso, Lexone or Sencor labels carefully for directions, precautions and limitations before applying these products.

SURFLAN®—oryzalin—Elanco Products Company
Lasso®—alachlor—Monsanto Company
Lasso® Micro-Tech™—alachlor—Monsanto Company
Lexone®—metribuzin—E.I. DuPont de Nemours and Company
Sencor®—metribuzin—Bayer GmbH

NO-TILL SOYBEAN—SURFLAN FOR FALL OR EARLY SPRING APPLICATION

Surflan may be applied in the fall or early spring directly over undisturbed stubble from the previous crop. Surflan application can begin anytime following harvest after October 15. Spring applications may be made as early as possible in the spring prior to weed germination, but at least four (4) weeks before planting. Surflan will not control established weeds.

The soil should not be tilled between the Surflan application and soybean planting.

Early applied Surflan effectively controls all of the annual grasses and broadleaf weeds listed under **Weeds & Grasses Controlled in Field Crops** by Surflan. Early applied Surflan also provides control of common ragweed and velvetleaf on medium and fine textured soils and partial control on coarse textured soils.

Surflan also provides partial control or suppression of:

Ladysthumb	Smartweed, annual
Morningglory, annual	Spurge, spotted
Nightshade, black	Volunteer wheat
Prickly sida (teaweed)	

Surflan will not control established weeds.

Broadcast Rates Per Acre

Soil Texture	Surflan Dry Flowable
	(pounds)
Coarse	1.2
Medium	1.5–1.8
Fine	1.5–1.8

Use the lower rate under light crop residue conditions and under light weed pressure. Use sequential herbicide applications in the spring to control escaped broadleaved weeds and to provide control of those weeds not controlled by Surflan.

Precaution: Do not apply Surflan to frozen or snow covered ground.

NO-TILL SOYBEAN—TANK MIX RECOMMENDATIONS

NO-TILL SOYBEAN: CONTROL OF EXISTING VEGETATION WITH PARAQUAT OR ROUNDUP®

Tank mix Surflan with Sencor, Lexone or Lorox to give season long control of labeled annual grasses and broadleaf weeds. Add paraquat or Roundup to these tank mixes to provide postemergence control of existing vegetation.

Paraquat Combinations:
Choose Surflan + Sencor/Lexone or Lorox depending on your specific weed problem. Tank mix two (2) pints of paraquat 2 CL per acre to control existing annual weeds. As the last ingredient, add eight (8) ounces of Ortho X-77 Spreader (nonionic) per 100 gallons of spray solution. Maintain constant agitation.

Caution: Check the paraquat label for additional directions, cautions and special precautions before applying this product.

Roundup Combinations:
Choose Surflan + Sencor/Lexone or Lorox depending on your specific weed problem. To control emerged annual weeds, tank mix one and one-half (1½) quarts of Roundup per acre. To control perennial weeds, tank mix two (2) to five (5) quarts of Roundup per acre depending on weeds to be controlled. Add Roundup as the last ingredient in the tank mix.

Caution: Check the Roundup label for additional directions, cautions, and special precautions before applying this product.

SURFLAN/2,4-D TANK MIX

SURFLAN may be tank mixed with most 2,4-D products for control of established winter weeds. When tank mixing, add SURFLAN A.S., 75W or Dry Flowable to the tank first and agitate until completely dispersed. Maintain agitation and add the 2,4-D product.

Certain ester formulations of 2,4-D require more vigorous agitation to keep the product well dispersed when mixed with SURFLAN A.S. or Dry Flowable. SURFLAN 75W is incompatible with some ester formulations of 2,4-D.

To assure the compatibility of SURFLAN with 2,4-D, add the products into small containers of water in the correct proportions. After thorough mixing, let stand for five (5) minutes and check for compatibility. If the combination remains mixed or can be remixed easily, the mixture is compatible. Maintain vigorous agitation during mixing and until the spray tank is empty. Follow 2,4-D directions, precautions and limitations.

Precaution: Do not apply SURFLAN to frozen or snow covered ground.

NO-TILL SOYBEAN—SURFLAN/SENCOR® OR SURFLAN/LEXONE® TANK MIX

A tank mix combination of Surflan/Sencor or Surflan/Lexone is more effective in providing preemergence weed control in no-till soybeans than any one of these herbicides used alone.

Add paraquat or Roundup as directed above to control existing vegetation if necessary.

Weeds Controlled by Surflan and Sencor or Lexone

Apply the Surflan/Sencor or Surflan/Lexone tank mix at planting or within two (2) days after planting. Do not spray over top of emerged soybeans.

Broadcast Rates Per Acre

Soil Texture	Surflan Dry Flowable	Sencor 4L or Lexone 4L	Lexone Dry Flowable or Sencor Dry Flowable
	(pounds)	(pints)	(pounds)
Coarse	0.9	½	⅓
Medium	0.9–1.2	¾	½
Fine	1.8	1	⅔

Do not use on sandy loam or loamy sand soils with less than one percent (1%) organic matter. Not recommended on any soil containing more than five percent (5%) organic matter.

Spray Equipment: Use any properly calibrated low pressure herbicide sprayer. Agitate thoroughly before and during application. Sprayers should be accurately calibrated before applying the tank mix combination. Avoid boom overlaps that will increase rates above those recommended. Check the sprayer daily to ensure proper calibration and uniform application. As the amount of water used (spray volume) decreases, the importance of accurate calibration and uniform application increases.

Activation: A single one-half inch (½″) rain or its equivalent in overhead irrigation is needed to activate the herbicide combination.

Special Precautions: Poor weed control and/or crop injury may result if directions are not followed. Over application may result in crop injury and/or soil residue. Do not use treated vines for feed or forage. Do not contaminate any body of water or apply to any area not specified on the label. Do not allow sprays to drift onto adjacent desirable plants. Do not allow Surflan to come into direct contact with soybean seed or crop injury may result.

Caution: Read the Sencor or Lexone labels carefully for cautions, precautions and special precautions before applying these products.

NO-TILL SOYBEAN—SURFLAN/LOROX® TANK MIX

A tank mix combination of Surflan/Lorox is more effective in providing preemergence weed control in no-till soybeans than either of these herbicides used alone.

Add paraquat or Roundup as directed above to control existing vegetation.

Weeds Controlled by Surflan and Lorox (See pages 6 and 21)

Apply the Surflan/Lorox tank mix at planting or within two (2) days after planting. Do not spray over top of emerged soybeans.

Broadcast Rates Per Acre

Soil Texture	Surflan Dry Flowable	Lorox 50 WP/L
	(pounds)	(pounds or pints)
Coarse	0.9	⅔–1¼
Medium	0.9–1.2	1–1⅔
Fine	1.8	1¼–2

Do not use on soils containing more than five percent (5%) organic matter.

Spray Equipment: Use any properly calibrated low pressure herbicide sprayer. Sprayer should be accurately calibrated before applying the tank mix combination. Agitate thoroughly before and during application. Avoid boom overlaps that will increase rates above those recommended. Check the sprayer daily to ensure proper calibration and uniform application. As the amount of water used (spray volume) decreases, the importance of accurate calibration and uniform application increases.

Activation: A single one-half inch (½") rain or its equivalent in overhead irrigation is needed to activate the herbicide combination.

Special Precautions: Poor weed control and/or crop injury may result if directions are not followed. Over application may result in crop injury and/or soil residue. Do not use treated vines for feed or forage. Do not contaminate any body of water or apply to any area not specified on the label. Do not allow sprays to drift onto adjacent desirable plants. Do not allow Surflan to come into direct contact with soybean seed or crop injury may result.

NO-TILL SOYBEAN—SURFLAN OVER-THE-TOP SPRAY FOR WEED CONTROL IN WHEAT OR BARLEY FOLLOWED BY NO-TILL SOYBEANS

Surflan may be applied for weed control in wheat or barley and soybeans. Apply Surflan as an over-the-top spray after the small grain is fully tillered and jointing has been initiated but prior to heading. When combining wheat or barley, the cutter bar should be adjusted to leave standing stubble at a height of at least eight (8) inches. Stubble reduces wind erosion and loss of soil moisture. During combining, cut straw should be chopped and spread evenly over the field as a mulch. Straw is an important part of the stubble-crop no-till program and should not be removed from the field. Soybeans can be planted into the small grain before harvest or into the stubble immediately following harvest.

Follow recommended mixing and application procedures for Surflan.

Broadcast Rates Per Acre

Soil texture	Surflan Dry Flowable
	(pounds)
Coarse	1.2
Medium	1.2–1.8
Fine	1.5–1.8

Precaution: Do not apply Surflan to soils containing more than five percent (5%) organic matter.

Tank Mixing Directions: Surflan may be tank mixed with most 2,4-D products. When tank mixing, add Surflan to the tank first and agitate until completely dispersed. Maintain agitation and add the 2,4-D product.

Certain ester formulations of 2,4-D require more vigorous agitation to keep the product well dispersed when mixed with Surflan Dry Flowable.

To assure the compatibility of Surflan with 2,4-D, add the products into small containers of water in the correct proportions. After thorough mixing, let stand for five (5) minutes and check for compatibility. If the combination remains mixed or can be remixed easily, the mixture is compatible. Maintain vigorous agitation during mixing and until the spray tank is empty. Follow 2,4-D application recommendations, precautions and limitations.

Precaution: Do not apply Surflan to winter wheat or barley prior to the fully tillered stage as crop injury may result. Do not graze or feed treated forage, hay or straw to livestock. Do not remove straw from field.

Caution: Read the 2,4-D labels carefully for cautions, precautions and special precautions before applying these products.

Excessive nitrogen application may result in crop lodging. Please contact your local Extension office for nitrogen recommendations.

COTTON—SURFLAN ALONE (IN TEXAS ONLY)

Apply Surflan to the soil surface at planting or within two (2) days after planting at a broadcast rate of 1.2 pounds on medium soils and 1.8 pounds on fine soils.

At the recommended rates for fine textured soils, Surflan will also control browntop panicum and prostrate spurge.

Precaution: Do not graze treated fields. If Surflan is not applied at or within two (2) days after planting, crop injury and poor weed control may result.

PEA (ENGLISH OR GREEN)—SURFLAN/TREFLAN® TANK MIX

The Surflan/Treflan tank mix provides weed control and suppression of common root rot (*Aphanomyces euteiches*) in peas.

Mixing and Application: Apply the tank mix up to two (2) weeks prior to planting. Apply in 20 to 40 gallons of water per acre in any properly calibrated low pressure herbicide sprayer.

Broadcast Rates Per Acre

Soil Texture	Surflan Dry Flowable	Treflan E.C. or M.T.F.™	Treflan 5
All soil textures	0.6 pounds	1 pint	0.8 pint

Do not use the Surflan/Treflan tank mix on coarse soils with less than one and one-half percent (1½%) organic matter. Incorporate according to label directions for Treflan. See the Treflan label.

Precautions: Please read the Treflan label for additional cautions and precautions. Do not feed forage from treated fields to livestock. Poor weed control and/or crop injury may result if directions are not followed. Over application may result in crop injury and/or soil residue. Do not plant any root crop for 12 months following application.

POTATO (WHITE)—SURFLAN ALONE

Follow Surflan label directions for soil preparation, activation and cultivation. Surflan should be applied to the soil surface at the recommended rate after planting and prior to crop emergence. Where rebedding and drag-off are standard cultural practices, applications should be made following drag-off.

Broadcast Rates Per Acre

Soil Texture	Surflan Dry Flowable
	(pounds)
Coarse	0.9
Medium	1.2

Do not use Surflan on soils containing more than five percent (5%) organic matter.

Precautions: Follow all directions carefully. Failure to do so may result in poor weed control. Over application may result in crop injury and/or soil residues.

POTATO (WHITE)—SURFLAN/LEXONE® OR SURFLAN/SENCOR® TANK MIX

Surflan plus Sencor/Lexone effectively controls the weeds listed on the current Surflan and Sencor or Lexone labels.

Follow Surflan label directions for soil preparation, activation and cultivation. Surflan plus Sencor/Lexone should be applied to the soil surface at the recommended rate after planting and prior to crop emergence. Where rebedding and drag-off are standard cultural practices, applications should be made following drag-off.

Broadcast Rates Per Acre

Soil Texture	Surflan Dry Flowable	Lexone 4L or Sencor 4L	Lexone Dry Flowable or Sencor Dry Flowable
	(pounds)	(pints)	(pounds)
Coarse	0.9	½	⅓
Medium	1.2	⅔	½

Do not use Surflan/Sencor or Lexone tank mix on soils containing more than five percent (5%) organic matter.

Precautions: Follow all directions carefully. Failure to do so may result in poor weed control. Over application may result in crop injury and/or soil residues.

Read the Lexone or Sencor labels for additional warnings, cautions, precautions, and special instructions.

TOBACCO—SURFLAN ALONE
TOBACCO—SURFLAN AT LAYBY
AIR-CURED (BURLEY, MARYLAND, DARK) AND FLUE-CURED

Soil Preparation: The last cultivation should leave the middle of the rows weed free before application. Surflan should be applied as a directed spray following the last normal cultivation, usually four (4) to six (6) weeks after transplanting.

Application: Applications should be made in 20 to 40 gallons of water per acre using commercially available drop nozzles that will allow treatment to be made in a 16 to 24 inch band in the middle of the row between the crop rows. Surflan should be applied in a manner that will not allow the spray to contact the plant. In tobacco production, in areas where spray nozzles on the ends of the spray boom pass over the same middle of the row twice, use nozzles which apply one-half (½) the normal number of gallons per acre to prevent over application. Over application may result in injury to rotational crops.

Activation and Cultivation: A one-half inch (½″) rain is necessary to activate Surflan. If adequate rainfall is not received to activate Surflan, erratic weed control may occur. If weeds begin to emerge, a shallow cultivation of one to two inches (1″ to 2″) will destroy existing weeds and place Surflan in the zone of weed seed germination.

Broadcast Rates Per Acre

Soil Texture	Surflan Dry Flowable (pounds)
Course	0.6–0.9
Medium	0.9–1.2
Fine	1.2

Precaution: Applied according to directions and under normal growing conditions, Surflan will not harm the treated crop. Do not spray Surflan in the bud of the tobacco plant as crop injury may occur. Over application may result in soil residue. Do not plant any root crop for 12 months following a Surflan application as crop injury may occur. Failure to follow directions may result in erratic weed control, tobacco crop injury, or soil residues. Soil residues may result in rotational crop injury.

FRUIT, NUT CROPS AND VINEYARDS (NON-BEARING AND BEARING—SURFLAN ALONE

Surflan applied as a preemergence herbicide controls many annual grasses and broadleaf weeds in several fruit, nut crops and vineyards. It is recommended on the following crops:

Almond	Macadamia nut	Small fruits:
Apple	Nectarine	Blackberry
Apricot	Olive	Blueberry
Avocado	Orange	Boysenberry
Cherry	Peach	Currants
Fig	Pear	Dewberry
Filbert	Pecan	Elderberry
Grape	Pistachio	Gooseberry
Grapefruit	Plum	Loganberry
Kiwi	Pomegranate	Raspberry
Lemon	Prune	Walnut, English

See listing of weeds controlled by Surflan alone.

General Directions: Existing vegetation should be destroyed either by tillage or by a tank mix with a contact or translocated herbicide. To assure the compatibility of Surflan with the contact or translocated herbicide, pour the products into small containers of water in the correct proportions. After thorough mixing, let stand for five (5) minutes and check for compatibility. If the combination remains mixed or can be remixed easily, the mixture is compatible. Read the label for all directions, cautions and precautions.

Broadcast Rates Per Acre

Soil Texture		Surflan Dry Flowable (pounds)
All Soil Textures	Short Term (2-4 months)	2.4
	Long Term (6-8 months)	4.7
	(8-12 months)	7.1

Surflan is not recommended for soils containing more than five percent (5%) organic matter.

Band Treatment: The amount of product needed for band treatment may be calculated by the formula:

$$\frac{\text{band width in inches}}{\text{row width}} \times \text{broadcast rate per acre} = \text{amount needed per acre of field}$$

Mixing and Application: Fill the spray tank three-fourths (¾) full with clean water. Add the recommended amount of Surflan to water. Mix thoroughly. Use 20 to 40 gallons of water per acre (broadcast basis). Apply with a properly calibrated low pressure herbicide sprayer that will uniformly apply the spray. Check the sprayer daily for the proper calibration and uniform application. Provide good mixing of the spray suspension in the tank before and during application.

Chemigation: Surflan may be applied for weed control through solid set sprinkler irrigation systems. Apply at rates recommended on this label. Refer to page 15 for additional chemigation directions and precautions before applying Surflan through chemigation equipment for use in fruit, nut crops and vineyards.

Precaution: Do not apply Surflan through center pivot irrigation systems to fruit, nut crops or vineyards. In California, do not apply Surflan to non-dormant fruit, nut crops or vineyards through overhead sprinkler irrigation systems.

Activation: A single one-half inch (½″) to one inch (1″) rainfall or sprinkler irrigation to move Surflan into the weed seed germination zone is required to activate the herbicide. If weeds begin to emerge, a shallow cultivation of one to two inches (1″ to 2″) will destroy existing weeds and place Surflan in the zone of weed seed germination. The higher water amount is generally required on fine textured/high organic matter soil.

Precaution: Poor weed control may result if directions are not followed. Over application may result in crop injury or excessive soil residue. Apply the spray directly to the orchard or vineyard floor.

FRUIT, NUT CROPS AND VINEYARDS—TANK MIX RECOMMENDATIONS: SURFLAN TANK MIX REFERENCE TABLE

Refer to the following table to determine which herbicides may be applied to your crop. Choose the Surflan combination based on your orchard or vineyard weeds to be controlled. Refer to the tank mix section for specific recommendations and precautions. Always read the label before application. *Approved Crop

	SURFLAN	GOAL	HYVAR X	KARMEX	KROVER I/II	PARAQUAT	PRINCAP	ROUNDUP	SINBAR	CASORON	SOLICAM
Almond	*	*				*	*	*		*	*
Apple	*			*		*	*	*	*	*	*
Apricot	*	*				*					*
Avocado	*					*	*	*			
Cherry	*					*	*	*		*	*
Fig	*					*					
Filbert	*					*	*	*		*	*
Grape	*	*		*		*	*	*		*	
Grapefruit	*		*	*	*	*	*	*	*		*
Kiwi	*										

	SURFLAN	GOAL	HYVAR X	KARMEX	KROVER I/II	PARAQUAT	PRINCAP	ROUNDUP	SINBAR	CASORON	SOLICAM
Lemon	*		*	*	*	*	*	*	*		*
Macadamia Nut	8			*			*				
Nectarine	*	*				*				*	*
Olive	*			*		*	*				
Orange	*		*	*	*	*	*	*	*		*
Peach	*	*		*		*	*			*	*
Pear	*			*		*	*	*		*	*
Pecan	*			*		*	*	*	*	*	
Pistachio	*							*			
Plum	*	*				*	*			*	*
Pomegranate	*										
Prune	*	*				*					*
Small fruits											
Blackberry	*			*		*	*		*	*	
Blueberry	*			*		*	*		*	*	
Boysenberry	*			*		*	*		*		
Currants	*										
Dewberry	*			*					*		
Elderberry	*										
Gooseberry	*			*							
Loganberry	*			*			*		*		
Raspberry	*			*		*	*		*	*	
Walnut, English	*			*		*	*	*		*	*

SURFLAN/CASARON® TANK MIX

The Surflan/Casaron tank mix controls the following weeds in addition to those controlled by Surflan alone.

Barley, wild	Jerusalem oak	Primrose, evening
Camphorweed	Maypops	Radish, wild
Coffeeweed	Melon, citron	Ragweed
Dandelion	Milkweed vine	Rosarypea
Deadnettle, red	Miners lettuce	Smartweed
Dogfennel	Mustard, wild	Spanishneedles
Gisekia	Natalgrass	Teaweed
Goosefoot	Peppergrass	Thistle, bull
Groundsel	Pineappleweed	Thistle, Russian
Horsetail	Plantain	Woodsorrel, yellow

T3 Surflan/Casaron tank mix may be used for weed control in the following crops:

Apple	Filbert	Pecan
Almond	Grape	Plum
Blackberry	Nectarine	Raspberry
Blueberry	Peach	Walnut, English
Cherry	Pear	

General Directions: Follow recommended application procedures for Surflan. Apply Surflan/Casoron in at least 20 gallons of water per acre to ensure uniform coverage. If the application is followed by one-half inch (½″) to one inch (1″) of overhead irrigation for activation, use the lower Casoron rate. Surflan/Casoron must be activated; check the Surflan label for activation directions.

Broadcast Rates Per Acre

Crop	Surflan Dry Flowable	Casoron W-50
	(pounds)	(pounds)
Apple, Almond, Blueberry, Cherry, Grape, Nectarine, Peach, Pear, Plum, Filbert, Pecan, English Walnut	2.4–7.1	8–12

Do not use on pecan seedlings within six (6) months of planting. May be used on bearing, nonbearing and nursery stock. Do not apply until four (4) weeks after transplanting. Do not make application within one (1) month of harvest of nuts or stone fruits.

Blackberry and Rasberry	2.4–7.1	8

Use only on established plantings. Do not apply during new shoot emergence.

Casoron Precautions: Do not use in seed beds, transplant or cutting beds or greenhouses. Do not apply until six (6) months after rooting of cuttings in the field, or after planting rooted cuttings. Do not apply on fruit stock within three (3) months prior to or following grafting or budding of root stocks or planting of new grafts. Do not apply more than the maximum recommended amount of Casoron W-50 per growing season. Do not use on "St. Lucie fine sand", "Arzell fine sand", or other light sand soil. Do not graze livestock in treated areas. Store in a tightly closed container in a dry place.

Caution: Read the Surflan and Casoron labels for directions, cautions and limitations prior to use.

SURFLAN/GOAL®/PARAQUAT TANK MIX

Surflan can be tank mixed with Goal 1.6E and paraquat.

Surflan/Goal provides preemergence control of the following weeds in addition to those controlled by Surflan alone:

Burclover	Lettuce, prickly
Cheeseweed (malva)	London rocket
Filaree, broadleaf	Mallow, little
Filaree, redstem	Redmaids
Groundsel, common	Sowthistle, annual

Paraquat may be tank mixed with Surflan/Goal to provide nonselective postemergence control of existing emerged annual weeds and provide top kill of existing perennial weeds.

The tank mix is recommended for application to dormant crops including:

Almond	Peach
Apricot	Plum
Grape	Prune
Nectarine	

General Directions: Best preemergence activity from Goal is obtained when the tank mix is applied to a relatively weed free established berm or soil surface.

Broadcast Rates Per Acre

Crop	Surflan Dry Flowable	Goal 1.6E	Paraquat
	(pounds)	(quarts)	(quarts)
All crops listed above	2.4–7.1	5	1

Precautions:

- Dormancy determines the suitability and safety of using Goal 1.6E on tree fruit/grapes. Should bud swell occur prior to the dates listed. Goal 1.6E must not be used.
- **Do not apply Goal 1.6E (either alone or in any combination recommended on this label) to tree or vine plantings before October 1 or after February 15. Do not apply Goal 1.6E (either alone or in any combination recommended on this label) to grapes in the Coachella Valley before October 1 or after December 31.**
- Do not apply more than five (5) quarts (two pounds active)

per broadcast acre of Goal 1.6E in one (1) season. Goal 1.6E should not be applied to grapes established less than three (3) years.

- Do not apply Goal 1.6E to grapes which are not staked or trellised.
- Do not feed or graze areas treated with Goal 1.6E to animals.
- Do not contaminate irrigation water or water used for domestic purposes with Goal 1.6E.
- Goal 1.6E or any of the combinations recommended on this label should be applied to only healthy growing trees and vines.
- Do not apply Goal 1.6E or any combination containing Goal 1.6E, that is recommended on this label, when weather conditions favor drift.
- Apply Goal 1.6E only to dormant fruit or nut trees or grape vines. Do not apply when foliage, buds or fruit/nuts are present.
- Direct Goal 1.6E spray toward the base of tree or vines. Avoid direct plant contact.

Mixing and Application: Apply Surflan/Goal in 20-40 gallons of water per acre. Fill the spray tank three-fourths (3/4) full. Agitate continuously. Add the recommended amount of Surflan to the spray tank. Add the recommended amount of Goal. Finish filling the tank with water to proper level. Do not allow the tank mix to remain in the tank without agitation.

Apply Surflan/Goal.paraquat in at least 50 gallons of water per acre. Use Ortho X-77® or Triton® AG-98 when tank mixing Surflan Dry Flowable/Goal/paraquat. Ortho X-77 and Triton AG-98 improve the compatibility of these herbicides when mixed together. Fill the spray tank one third to one half (1/3 to 1/2) full with clean water. Add either one (1) quart of X-77 or two (2) quarts of Triton AG-98 per 100 gallons of spray mixture, with the agitation running, add the recommended amount of Surflan and mix thoroughly. Continue filling the spray tank and add Goal, then paraquat. If a non-ionic spreader is being used, add it to the tank last.

Activation: A single one-half inch (1/2″) to one inch (1″) rain or sprinkler irrigation to move Surflan/Goal into the weed seed germination zone is required to activate the herbicide. If weeds begin to emerge, a shallow cultivation of one to two inches (1″ to 2″) will destroy existing weeds and place Surflan in the zone of weed seed germination. The higher amount of water is generally required on fine textured/high organic matter soil.

Caution: Read the Surflan, Goal and paraquat labels for additional limitations, directions and cautions before applying these products.

SURFLAN/HYVAR® X TANK MIX

The Surflan/Hyvar X tank mix controls the following annual and perennial weeds in addition to those weeds controlled by Surflan alone:

Annual Weeds	Perennial Weeds
Mullein, turkey	Bermudagrass (All States)
Mustard	Johnsongrass (Texas)
Natalgrass (red top)	Nutsedge (Texas and California)
Sedge	Pangolagrass (Florida)
Sprangletop	Paragrass (Florida)
	Torpedo grass (Florida)

The Surflan/Hyvar X tank mix is recommended on the following crops:

Citrus (Orange, Grapefruit, Lemon)

General Directions: Apply the tank mix in at least 40 gallons of water per acre. Use sufficient water to provide thorough and uniform coverage of the ground. Apply Hyvar X as a spray just before or during the period of active plant growth. If dense growth is present, results will be improved if vegetation is removed before treatment. Do not apply when ground is frozen.

Activation: Moisture is necessary to activate Surflan/Hyvar X tank mix; best results are obtained if moisture is supplied by rainfall or overhead sprinkler irrigation within two (2) weeks after application. Check activation instructions for Surflan alone.

Broadcast Rates Per Acre

Crop	Surflan Dry Flowable	Hyvar X
	(pounds)	(pounds)
Citrus (Trees established for four (4) years or more)		
Annual Weeds	2.4–7.1	2–4
Apply any time of the year, preferably shortly before weed growth begins, when adequate moisture is available.		
Perennial Weeds	2.4–7.1	4–8
Apply before weed growth begins. If dense growth is present, remove tops and spray the ground. Note: Effects on perennial weeds from Hyvar X are slow to appear, usually progressing over a period of several months. Check the Hyvar X label for more complete directions and precautionary statements prior to use for perennial weed control.		

Precaution: Avoid contact of foliage and fruit with Hyvar X spray or mist. Temporary yellowing of citrus leaves may occur following Hyvar X treatment. Do not use on citrus in Kern County, California.

Because injury to citrus trees may result: do not use Hyvar X on soils low in organic matter less than one percent (1%) nor on poorly drained soils; do not apply more than eight (8) pounds per acre per year; do not treat trees planted in irrigation furrows; do not treat diseased trees such as those with foot rot. Do not use Hyvar X in citrus orchards interplanted with other trees or desirable plants, nor in home citrus plantings or in areas where roots of other valuable plants or trees may extend, as plant injury may result. Treated areas may be planted to citrus trees one (1) year after last Hyvar X application; do not replant to other crops within two (2) years after last application, as plant injury may result.

SURFLAN/KARMEX® TANK MIX

The Surflan/Karmex tank mix controls the following weeds in addition to those controlled by Surflan alone:

Ageratum	Kochia	Puncturevine
Buckwheat, wild	Kyllinga	Radish, wild
Buttonweed	Marigold	Ricegrass, red
Clover, Mexican	Milkweed, climbing	Ryegrass, annual
Dayflower	Morningglory, annual	Smartweed
Dogfennel	Mustard, black	Sowthistle, annual
Fescue, rattail	Mustard, wild	Spanishneedles
Gromwell	Orchardgrass	Speedwell, corn
Groundcherry, annual	Paintbrush, Flora's	Spurry, corn
Hawksbeard	Pepperweed	Tansymustard
Horseweed	Pineappleweed	Tobacco, rabbit
Knawel	Pokeweed	Velvetgrass

The Surflan/Karmex tank mix is recommended on the following crops:

Apple	Orange
Caneberries	Peach
Grape	Pear
Grapefruit	Pecan
Lemon	Walnut, English

Broadcast Rates Per Acre

Crop	Surflan Dry Flowable	+	Karmex 80W
	(pounds)		(pounds)
Apple	2.4–7.1		2–4
Use Karmex only under trees established in the orchard for at least one (1) year; do not treat varieties grafted on full-dwarf root stocks.			
Caneberries	2.4–7.1		2–4

(Read the Karmex label for state and crop recommendations). Use Karmex only in fields established for at least one (1) year. Do not apply Karmex to berries interplanted with fruit trees. Do not apply to plants whose roots are exposed as injury may result.

Grape

New York and Pennsylvania—Perennial Grasses:

Use Karmex only in established vineyards (at least four years old) for spot control of perennial grasses such as orchardgrass, quackgrass and ryegrass. Apply Karmex in the spring as a band treatment to ridged soil two to four inches (2" to 4") high under the trellis at the rate of eight (8) to 12 lbs. per acre. Band width should not exceed 30 inches. Do not apply Karmex more than once every four (4) years. Use Karmex only on heavy soils such as loams, silt loams and clay loams. Do not use Karmex in areas where grape roots are shallow or exposed because of high bedrock, poor drainage, or erosion as injury to grapes may result.

U.S. (Except New York and Pennsylvania)—Perennial Grasses:

Apply Karmex only to established vineyards (at least three years old) as a band treatment to grape rows. Do not apply Karmex to vines with trunks less than one and one-half inches (1½") in diameter as injury may result. On soils low in clay or organic matter (one to two percent) (1% to 2%), severe plant injury may result from Karmex application if heavy rainfall or more than one inch (1") of irrigation occurs soon after treatment. This risk must be assumed by the user.

Broadcast Rates Per Acre

Crop	Surflan Dry Flowable	Karmex 80W
East of Rocky Mountains:	(pounds)	(pounds)
1 to 2% organic matter	2.4–7.1	2–3
2 to 5% organic matter	2.4–7.1	3–6
West of Rocky Mountains:	2.4–7.1	2–4
(Apply Karmex in November, December, or January)		
Orange, Grapefruit and Lemon		
Do not apply Karmex under citrus trees that have been subjected to freezing within six (6) months.		
Use Karmex only under trees established in the grove for at least one (1) year:		
Arizona (except Yuma area) and California (except Imperial and Coachella Valleys)	2.4–7.1	2–4
Florida (oranges and grapefruit)	2.4–7.1	4–8
Texas (oranges and grapefruit)	2.4–7.1	2–4
Peach	2.4–7.1	2–5
Use Karmex only under trees established in the orchard for at least three (3) years. Do not apply Karmex within three (3) months of harvest; in the far West, do not apply Karmex within eight (8) months of harvest.		
Pear	2.4–7.1	2–4
Use Karmex only under trees established in the orchard for at least one (1) year. Do not treat varieties grafted on full-dwarf root stocks.		
Pecan	2.4–7.1	2–4
Use only under trees established in the grove for at least three (3) years. Do not use on soils with less than one-half percent (½%) organic matter.		
Walnut (English) California	2.4–7.1	2–3
Use only under trees established in the orchard for at least one (1) year.		

Mixing and Application: Fill the spray tank three-fourths (¾) full with clean water and add the recommended amount of Karmex 80W. Mix thoroughly. Then add the recommended amount of Surflan, mix thoroughly and fill the tank. Vigorous, continuous agitation is required until spray tank is empty to keep the powders in suspension. Use 20 to 40 gallons of water per acre (broadcast basis). Apply with a properly calibrated low pressure herbicide sprayer that will apply the spray uniformly. Check the sprayer daily for proper calibration and to ensure uniform application. Provide good mixing of the spray suspension in the tank before and during application.

Activation: A single one-half inch (½") to one inch (1") rain or sprinkler irrigation is required to activate the herbicides. If weeds begin to emerge, a shallow cultivation of one to two inches (1" to 2") will destroy existing weeds and place Surflan and Karmex in the zone of weed seed germination.

Precaution: Poor weed control may result if directions are not followed. Over application may result in crop injury or excessive soil residue. Apply the spray directly to the orchard or vineyard floor.

Caution: Read the Karmex label carefully for directions, cautions, and precautions before applying this product.

SURFLAN/KROVAR® I TANK MIX

The Surflan/Krovar I tank mix effectively controls the following weeds in addition to those weeds controlled by Surflan alone:

Filaree	Nightshade (annual)
Fleabane	Pineappleweed
Horseweed	Ragweed
Lettuce, wild	Shepherdspurse
Mustard, wild	Sowthistle (annual)
Natalgrass (red top)	Spanishneedles

The Surflan/Krovar I tank mix is recommended for use on the following crops:

Citrus (Orange, Lemon, Grapefruit)

General Directions: Apply any time of the year when rainfall or overhead irrigation is available to activate the tank mix, preferably just before the weeds have germinated.

Use only where trees have been established in the orchard for at least three (3) years. Use sufficient water (over 40 gallons/acre) to obtain uniform coverage.

Broadcast Rates Per Acre

Crop	Surflan Dry Flowable	Krovar I
	(pounds)	(pounds)
Citrus	2.4–7.1	3–6
(Trees established in the orchard for at least three (3) years)		
Use for control of annual weeds in California, Arizona, Texas, Louisiana, and Florida only. Check the Krovar I label for complete directions for use and precautionary statements.		

Activation and Cultivation: A single one-half inch (½") to one inch (1") rain or sprinkler irrigation to move the tank mix combination into the weed seed germination zone is required within two (2) weeks after application to activate the herbicides. If weeds begin to emerge, a shallow cultivation of one to two inches (1" to 2") will destroy existing weeds and place Surflan and Krovar I in the zone of weed seed germination. Citrus treated can be shallowly cultivated without loss of activity.

Precaution: Avoid contact of foliage and fruit with Krovar I spray or mist. Avoid overlapping, and shut off spray boom while starting, turning, slowing or stopping, or injury to trees may result. Continuous agitation in the spray tank is required to keep the material in suspension. Best results are obtained if application is made shortly before weed growth begins; if dense growth is present, remove tops and spray ground.

Temporary yellowing of citrus leaves may occur following Krovar I treatment. Because injury to citrus trees may result, do not use Krovar I on soils with less than one percent (1%) organic matter (one and one-half percent (1½%) in Florida), poorly drained soils, gravelly soils, nor thinly covered or exposed subsoils; do not treat trees planted in irrigation furrows; do not treat diseased trees such as those with foot rot.

Do not replant Krovar I treated areas to any crop within two (2) years after last application as injury to subsequent crops may result, except that citrus trees may be planted one (1) year after the last application.

SURFLAN/KROVAR® II TANK MIX

The Surflan/Krovar II tank mix effectively controls the following annual and perennial weeds in addition to those controlled by Surflan alone:

Annual Weeds and Vines	Perennial Weeds
Balsam apple*	Bahiagrass
Milkweed vine*	Bermudagrass
Natalgrass (red top)	Nutsedge
Nightshade (annual)	Pangolagrass
Ragweed	Paragrass
Spanishneedles	Torpedograss

* Seedling stage in Florida only.

The Surflan/Krovar II tank mix is recommended for weed control in the following crops:

Citrus (Orange, Lemon, Grapefruit)

General Directions: Use only where trees have been established in the orchard for at least four (4) years. Apply any time during the year when rainfall or overhead irrigation is available to activate the tank mix prior to weed germination. Moisture is necessary to activate the herbicides. Check the Surflan directions for activation moisture amounts. The tank mix should be activated within two (2) weeks after application. To obtain uniform coverage, use sufficient water (over 40 gallons/acre).

Broadcast Rates Per Acre

Crop	Surflan Dry Flowable	Krovar II
	(pounds)	(pounds)
Citrus (Use only where trees have been established in the orchard for at least four (4) years.)	2.4–7.1	2–8

Consult the Krovar II label for complete directions for use for annual, perennial and seedling vine control, cautions and limitations prior to use.

Precautions: Avoid contact of foliage and fruit with Krovar II spray or mist. Avoid overlapping, and shut off spray boom while starting, turning, slowing or stopping, or injury to trees may result. Continuous agitation in the spray tank is required to keep the material in suspension. Best results are obtained if application is made shortly before weed growth begins; if dense growth is present, remove tops and spray the ground. Do not use Krovar II on citrus in Kern County, California.

Temporary yellowing of citrus leaves may occur following Krovar II treatment. Because injury to citrus trees may result, do not use Krovar II on soils low in organic matter (less than 1%), poorly drained soils, gravelly soils, nor thinly covered or exposed subsoils; do not apply more than 10 lbs. per acre per year (Florida, 12 lbs.); do not treat trees planted in irrigation furrows; do not treat diseased trees such as those with foot rot.

Do not replant Krovar II treated areas to any crop within two (2) years after last application as injury to subsequent crops may result, except that citrus trees may be planted one (1) year after the last application.

SURFLAN/PARAQUAT TANK MIX

Surflan may be tank mixed with paraquat. Surflan provides preemergence control of certain annual grass and broadleaf weeds while paraquat provides nonselective postemergence control of annual grass and broadleaf weeds and top kill and suppression of perennial weeds.

The tank mix is recommended on the following crops:

Almond	Grape	Peach
Apple	Grapefruit	Pear
Apricot	Lemon	Pecan
Avocado	Nectarine	Plum
Cherry	Olive	Prune
Fig	Orange	Walnut, English

Filbert
Small fruits:

Blackberry	Boysenberry
Blueberry	Raspberry

General Directions: Surflan/paraquat may be applied any time throughout the growing season. Choose an application schedule based on the specific weeds to be controlled by the Surflan/paraquat tank mix.

Broadcast Rates Per Acre

Crop	Surflan Dry Flowable	Paraquat
	(pounds)	(quarts)
All crops listed above	2.4–7.1	1–2*

Precaution: Do not allow paraquat spray to contact green stems, fruit or foliage as injury may result. Do not spray paraquat under windy conditions and use a shield for young trees or vines. Do not allow animals to graze on paraquat treated areas. Do not apply paraquat when nuts to be harvested are on the ground

* Add a non-ionic spreader at eight (8) oz. per 100 gallons spray volume

Mixing and Application: Add the recommended amount of Surflan to the spray tank according to directions for these tank mixes. Add paraquat last prior to filling the tank. Agitate continuously during filling and through application. Apply in at least 30 gallons of water per acre. Apply when the weeds and grasses are succulent and the new growth is from one inch (1″) to six inches (6″) high. Retreatment or spot treatment may be necessary for late germinating weeds and grasses.

Activation: A single one-half inch (½″) to one inch (1″) rain or sprinkler irrigation to move Surflan into the weed seed germination zone is required to activate the herbicide. If weeds begin to emerge, a shallow cultivation of one to two inches (1″ to 2″) will destroy existing weeds and place Surflan in the zone of weed seed germination.

Caution: Read the paraquat label carefully for directions, cautions and precautions before applying this product.

SURFLAN/PRINCEP® (SIMAZINE) TANK MIX

The Surflan/Princep tank mix controls the following weeds in addition to those controlled by Surflan alone.

Brome, downy (cheat)	Paintbrush, Flora's
Fescue, rattail	Prickly lettuce
Fireweed	Ragweed
Hairgrass, silver	Ryegrass, annual
Knawel	Shieldscress
Milkweed, climbing	Smartweed
Morningglory, annual	Spanishneedles
Mustard, black	Tansymustard
Mustard, wild	Wiregrass
Nightshade	

The tank mix is recommended on the following crops:

Almond	Lemon
Apple	Orange
Avocado	Peach
Caneberries	Pear
Cherry	Pecan
Filbert	Plum
Grape	Walnut, English
Grapefruit	

Where a rate range of application rates for Princep is given, use the lower rate on coarser textured soil and soil lower in organic matter; use the high rate on finer textured soils and soil higher in organic matter.

Broadcast Rates per Acre

Crop	Surflan Dry Flowable	Princep 80W
	(pounds)	(pounds)
Apple, Pear and Sour Cherry	2.4–7.1	2½–5

Crop	Surflan Dry Flowable	Princep 80W
	(pounds)	(pounds)
Almond and Peach		
California only		
Apply two (2) to four (4) foot band on each side of tree row. For peach in other states, see recommendations for peach, plum, and sweet cherry.	2.4–7.1	1¼–2½

Do not apply Princep to trees established in the grove less than three (3) years. Do not apply Princep to the Mission (Texas) variety of almonds. Do not apply Princep to almond trees propagated on plum rootstocks. Do not replant almond or peaches in soil treated with Princep for 12 months after treatment. Do not apply Princep on gravelly, sandy or loamy sand soil or on soil with less than one percent (1%) organic matter. Do not apply Princep to areas where water will accumulate.

Avocado		
California	2.4–7.1	2½–5

Do not apply Princep to areas where water will accumulate.

Caneberries	2.4–7.1	2½–5

Apply two and one-half (2½) to five (5) pounds of Princep per acre in the spring or apply a split application of two and one-half (2½) pounds per acre in the spring plus two and one-half (2½) pounds per acre in the fall. Apply in a minimum of 40 gallons of water per acre. On plantings less than six (6) months old, use one-half (½) the above Princep rate. To control quackgrass, apply five (5) pounds Princep per acre in the fall or split the application applying two and one-half (2½) pounds per acre in the fall plus two and one-half (2½) pounds per acre in the spring, when quackgrass is actively growing. Do not apply Princep when fruit is present.

Filbert		
Washington and Oregon	2.4–7.1	2½–5

Apply Princep at two and one-half (2½) to five (5) pounds per acre in the fall or apply a split application of two and one-half (2½) pounds per acre in the spring and two and one-half (2½) pounds per acre in the fall.

Precaution: If trees are planted on a hillside, excessive soil erosion may result from the elimination of weeds. Do not apply on gravelly, sandy or loamy sand soil. Do not apply when nuts are on the ground during the harvest period.

Grape	2.4–7.1	2½–5

Apply Princep any time between harvest and early spring.

Precaution: Do not use Princep in vineyard established less than three (3) years. Do not apply Princep on gravelly, sandy or loamy sand soil.

Grapefruit, Lemon and Orange		
Arizona (lemon and orange only)	2.4–7.1	2 Spring 2 Fall
California (except Imperial, Coachella and Palo Verde Valleys)	2.4–7.1	2½–5
Florida (grapefruit and orange only)	2.4–7.1	8–12
Texas	2.4–7.1	5-6

All areas: do not use Princep in nurseries. Do not apply Princep to bedded grapefruit, lemon, or orange. To avoid possible injury, do not apply Princep to trees under stress from freeze damage for one (1) year after the freeze.

Peach Plum and Sweet Cherry	2.4–7.1	2–5

Apply Princep in late fall to early spring prior to weed emergence.

Peach: Use Princep only in Arkansas, Louisiana, Missouri, Oklahoma, Texas and the states east of the Mississippi.

Plum and Sweet Cherry: Use Princep only in Missouri and states east of the Mississippi River except Tennessee.

Pecan		
Sand and loamy sand	2.4–7.1	Do Not Use
Sandy loam	2.4–7.1	2–5
Loam or clay soil low in organic matter	2.4–7.1	2½–3½
Clay soil high in organic matter	2.4–7.1	3½–5

Note: Do not apply Princep when nuts are on the ground. Do not allow animals to graze in Princep treated areas.

Do not use Princep in Arizona, California, New Mexico or west of the Pecos River in Texas as injury will occur. Do not make Princep applications to transplanted trees that have been established less than two (2) years in the grove as injury may occur.

Walnut, English	2.4–7.1	2½–5

Do not apply Princep to gravelly, sandy or loamy sand soils. Leveling and furrowing operations after application will lessen effectiveness of weed control. Do not apply Princep when nuts are on the ground.

Mixing and Application: Fill the spray tank three-fourths (¾) full of clean water. Follow instructions on the Princep 80W label for making a slurry of the Princep 80W in a small amount of water in a separate container, then pour the slurry into the spray tank. Mix thoroughly. Then add the recommended amount of Surflan. Mix thoroughly and continue filling the tank. Continuous, vigorous agitation is needed to keep the powders in suspension. Use 20 to 40 gallons of water per acre (broadcast basis). Apply with a properly calibrated low pressure herbicide sprayer that will apply the spray uniformly. Check the sprayer daily for proper calibration to ensure uniform application. Provide good mixing of the spray suspension in the tank before and during application.

Activation: A single one-half inch (1/2″) to one inch (1″) rain or sprinkler irrigation to move the tank mix combination into the weed seed germination zone is required to activate the herbicides. If weeds begin to emerge, a shallow cultivation of one to two inches (1″ to 2″) will destroy existing weeds and place Surflan and Princep in the zone of weed seed germination.

Precaution: Poor weed control may result if directions are not followed. Over application may result in crop injury or soil residues. Apply the spray directly to the orchard or vineyard floor.

Caution: Read the Princep label carefully for directions, cautions and precautions before applying this product.

SURFLAN/ROUNDUP® TANK MIX

Surflan may be tank mixed with Roundup. In this tank mix, Surflan provides preemergence control of certain annual grass and broadleaf weeds while the Roundup provides non-selective, postemergence control of many annual and perennial weeds. This tank mix is recommended on the following crops:

Almond	Grape	Pear
Apple	Grapefruit	Pecan
Avocado	Lemon	Pistachio
Cherry (non-bearing)	Orange	Walnut, English
Filbert		

See listing of weeds controlled by Surflan alone.

General Directions: Surflan/Roundup may be applied any time thoughout the growing season. Choose an application schedule based on the specific weeds to be controlled by the Surflan/Roundup tank mix. Consult the Roundup label for specific weed growth stage directions to achieve best results. Karmex and Princep may be tank mixed with Surflan/Roundup to provide preemergence control of those weeds listed. Use Roundup with the Surflan/Karmex tank mix for non-selective, postemergence control of annual weeds only. Use Roundup with the Surflan/Princep tank mix for non-selective, postemergence control of annual and perennial weeds.

Broadcast Rates Per Acre

All Crops	Surflan Dry Flowable	Roundup
	(pounds)	(quarts)
Existing annual grass and broadleaf weeds:		
Less than 6″ tall	2.4–7.1	1
Over 6″ tall	2.4–7.1	1½
Existing perennial weeds	2.4–7.1	2–5

Tree Crops	Preharvest Interval for Roundup Application
Almond	21 days
Apple	14 days
Avocado	14 days
Cherry	1 year
Filbert	21 days
Grapefruit	14 days
Lemon	14 days
Orange	14 days
Pear	14 days
Pecan	21 days
Pistachio	21 days
Walnut	21 days

Note: Repeat treatments may be necessary to control weeds regenerating from underground parts or seed. The combined total of all Roundup treatments must not exceed 10.6 quarts per acre per year. Do not feed or forage areas treated with Roundup for eight (8) weeks after application.

PRECAUTION: *Extreme care must be exercised to avoid contact of Roundup spray, drift or mist with green foliage, green bark or bark of trees established less than two (2) years, suckers, or fruit of desirable trees, crops, plants or other desirable vegetation. Spray contact with other than matured bark on the main trunk can result in serious localized or translocated damage.* If weeds have been mowed, allow regrowth to reach the recommended stage of growth prior to application of this product. Do not apply Roundup to weeds under drought stress.

Vines	Preharvest Interval for Roundup Application
Grape (Raisin, Table, Wine)	14 days

Note: Roundup applications should not be made when green shoots or canes or foliage are in the spray zone. (In the Northeast and Great Lakes regions, Roundup applications must be made prior to the end of bloom stage of grapes to avoid injury.)

If repeat Roundup treatments are necessary for weed control in vineyards, do not exceed a total of 10.6 quarts of Roundup per acre per year. Do not treat grapes with Roundup between 14 days before harvest to fall dormancy when no green vegetation, canes or shoots exist.

PRECAUTION: *Do not allow Roundup spray, drift or mist to contact green foliage, green bark, suckers or vines and renewals less than three (3) years of age. Roundup spray contact other than with mature bark of the main trunk, can result in serious localized or translocated damage.*

Mixing and Application: Add the recommended amount of Surflan, or Surflan and other residual herbicides to the spray tank according to directions for these tank mixes. Add Roundup during the last step of the filling operation. Agitate continuously during filling and throughout application. Apply in 20 to 60 gallons of water per acre in a properly calibrated low pressure herbicide sprayer. Check the sprayer daily for proper calibration and uniform application.

Activation: A single one-half inch (½″) to one inch (1″) rain or sprinkler irrigation to move Surflan into the weed seed germination zone is required to activate the herbicide. If weeds begin to emerge, a shallow cultivation of one to two inches (1″ to 2″) will destroy existing weeds and place Surflan in the zone of weed seed germination.

Caution: Read the Roundup label carefully for directions, cautions and precautions before applying this product.

SURFLAN/SINBAR® TANK MIX

Surflan/Sinbar is recommended for weed control in the following crops:

Apple
Peach
Blueberry
Orange
Grapefruit
Lemon
Pecan

Caneberries:
Blackberry
Boysenberry
Dewberry
Loganberry
Raspberry

Surflan/Sinbar controls weeds in addition to those controlled by Surflan alone. As the amount of Sinbar applied per acre increases, so does the number of additional weeds controlled. Consult the Sinbar label for application rate required for the specific additional weeds to be controlled.

General Directions: Follow recommended application procedures for Surflan. Apply Surflan/Sinbar in at least 20 gallons per acre or more to ensure a uniform application. Moisture is necessary to activate the tank mix. Consult the Surflan label for activation information.

Broadcast Rates Per Acre

	Surflan Dry Flowable (pounds)	Sinbar 80W* (pounds)
Apple, Peach	2.4–7.1	2–4

Apply either in the spring or after harvest in the fall before weeds emerge. Where the crop is grown under furrow irrigation or under raised-berm flood irrigation (trees four to six inches (4″ to 6″) above water line), apply Sinbar only as a band treatment. Do not treat trees planted in the bottom of irrigation furrows or trees grown under flat flood or basin irrigation, and do not use on eroded areas where subsoil or tree roots are exposed, as injury to trees may result.

	Surflan Dry Flowable (pounds)	Sinbar 80W* (pounds)
Blueberry	2.4–7.1	2–4

Apply in a minimum of 25 gallons of water per acre.

Treat only plantings established for one (1) year or more. Make a single band or broadcast application to ground beneath bushes: avoid contact of foliage and fruit with spray or mist. Apply either in the spring or after harvest in the fall before weeds emerge. Do not use Sinbar on eroded areas where subsoil or roots are exposed, nor on plants that are diseased or lacking in vigor, as injury to the plants may result. Sinbar treated areas may be planted to blueberries one (1) year after last application.

	Surflan Dry Flowable (pounds)	Sinbar 80W* (pounds)
Caneberries: Blackberry, Boysenberry, Dewberry, Loganberry, Raspberry	2.4–7.1	1–2

Treat only plantings established for one (1) year or more. Make a single band or broadcast application to the soil beneath the canes in the fall or early spring before fruit set and before weeds emerge. Do not spray foliage nor use on eroded areas where subsoil or roots are exposed, nor apply to plants that are diseased or lacking vigor as injury to the plant may result. Sinbar treated areas may be planted to caneberries one (1) year after last application. Do not apply Sinbar within 70 days before harvest.

	Surflan Dry Flowable (pounds)	Sinbar 80W* (pounds)
Citrus: Orange, Grapefruit, Lemon		
Trees established for one (1) year:	2.4–7.1	2–4
Trees established for two (2) years or more:		
Arizona and California	2.4–7.1	2–8
Florida and Texas	2.4–7.1	2–10

Apply as a band or broadcast treatment as a directed spray: avoid contact of foliage and fruit with spray or mist. Temporary yellowing of citrus leaves may occur following Sinbar treatment. Sinbar may be used on coarse soils but is not to be used on sands low in organic matter (less than one percent) (1%), on poorly drained soils, or on trees planted in irrigation furrows, as injury to trees may result. Sinbar treated areas may be planted to citrus trees one (1) year after last application.

* Sinbar application rates may vary with soil texture, weeds to be controlled, length of control desired and age of crop.

Caution: Read the Sinbar label for additional directions for use, cautions and special precautions prior to use.

SURFLAN/SOLICAM® TANK MIX

The Surflan/Solicam tank mix controls the following weeds in addition to those weeds controlled by Surflan alone:

Barley, wild	Filaree (red and whitestem)
Brome, downy	Fingergrass, feather
Brome, ripgut	Grama grass, sixweeks
Bursage, annual	London rocket
Cupgrass, southwestern	Pineappleweed
Dandelion, false	Thistle, Russian and sow

Weeds suppressed by this tank mix in addition to those suppressed by Surflan alone include:

Bermudagrass	Nutsedge, purple
Horseweed	Quackgrass
Nightshade, silverleaf	

Surflan/Solicam is recommended on the following crops:*

Almond	Nectarine
Apple	Peach
Apricot	Pear
Cherry	Plum
Citrus	Prune
Filbert	Walnut

* Apply Solicam on all of the above crops only in California, Oregon and Washington. In Arizona, use on citrus only. Use only on crops in the Imperial, Coachella or Palo Verde Valleys of California.

Orchard Establishment Requirements: Use Surflan/Solicam on crops listed above that have been established at least 18 months in the field with exception of citrus, where the trees should be established in the field for 12 months. In all cases, the soil should be settled and firm at the time of application. Do not use on nursery stock.

General Directions: Apply the tank mix to the soil surface from fall to early spring before weeds emerge, except on those crops where otherwise specified. The orchard floor must be free of weeds and must be relatively free of plant residues. Solicam does not control established weeds.

Apply as a directed spray. Avoid contact with fruit or foliage. Make only one application per year.

Activation: If no rainfall occurs within two weeks after application, the tank mix must be activated by sprinkler irrigation. Check activation instructions for Surflan alone.

Broadcast Rates Per Acre

Use the lower Solicam rates on coarse soils. Use the higher rates on medium to fine soils or soils having organic matter content of greater than 3%.

Crop	Surflan Dry Flowable	Solicam 80W
	(pounds)	(pounds)
Almond	2.4–7.1	2½–5
Apply from October through January		
Precaution: Do not use on coarse textured soils such as sand, loamy sand, gravelly sand and fine sand.		
Apple, Apricot, Filbert, Nectarine, Peach, Pear, Plum, Prune, Walnut	2.4–7.1	2½–5
Cherry	2.4–7.1	2½–5
Precaution: Do not use on coarse textured soils such as sand, loamy sand, gravelly sand and fine sand.		
Citrus	2.4–7.1	2½–5
Apply any time during the year according to local practice.		
Precaution: Do not use where citrus is interplanted with palm trees.		

Solicam Precautions: Do not apply when nuts or fruit are on the ground at harvest. Do not graze livestock on treated areas. Do not feed or graze cover crops grown in treated orchards. Read the Solicam label for directions, cautions and limitations prior to use.

SURFLAN APPLICATION WITH LIQUID FERTILIZERS

GENERAL INSTRUCTIONS

Surflan Dry Flowable may be mixed with most liquid fertilizer materials. The combination of Surflan with solutions and suspension-type fertilizers has provided weed and grass control equal to the same rates of Surflan applied in water. Follow Surflan label recommendations regarding rates per acre, crops, application directions, special instructions, cautions and special precautions. Individual state regulations relating to liquid fertilizer mixing, registration, labeling and applications are the responsibility of the individual and/or company selling the fertilizer and chemical mixture. In all cases, vigorous and continuous agitation is required to mix Surflan Dry Flowable with the liquid fertilizer. Surflan Dry Flowable may be applied alone in liquid fertilizer without a compatibility agent. The use of a compatibility agent is especially important when tank mixing Surflan Dry Flowable with emulsifiable concentrates (EC) and liquid concentrates (LC). See the liquid fertilizer compatibility test below. If the emulsion is not properly formed, the EC/LC may rise to the surface of the fertilizer as an oil ("oils out"), the oil may combine with the Surflan Dry Flowable, wettable powder, liquid, flowable or aqueous suspension to form oily curds (viscous phase) which are difficult to disperse.

Any one of the compatibility agents listed below is helpful in causing emulsifiable and liquid concentrates to form non-oiling mixtures with liquid fertilizers. These compatibility agents can be used at rates as low as one and one-half (1½) to two (2) pints per ton of liquid fertilizer and should be mixed well with the fertilizer containing Surflan Dry Flowable before adding the emulsifiable concentrate. Read the label on the compatibility agent and follow the directions.

1. Sponto 168D (Witco Chemicals Co., Chicago, IL)
2. Compat (Farm Chemicals, Inc., Aberdeen, NC)
3. Unite (Hopkins Ag Chemical, Madison, WI)
4. T-Mulz 734-2 (Thompson-Hayward Chemical Co., Kansas City, KS)
5. Rigo Compatibility Agent (Rigo Co., Buckner, KY)
6. Amoco Spray Mate™ (Amoco Oil Co., Chicago, IL)
7. Kem-Link (Universal Coop, Minneapolis, MN)
8. Blendex™ (Helena Chemical Co., Cayce-West, Columbia, SC)
9. Spray-Aide™ (Miller Chemical and Fertilizer Corp., Hanover, PA)

Each of the above is a phosphate ester type surfactant designed to be used with liquid fertilizers. They usually do not work well as compatibility agents in tank mixtures in plain water.

TESTING FOR TANK MIX COMPATIBILITY IN LIQUID FERTILIZERS:

Tank mixtures with Surflan Dry Flowable, other dry flowables, wettable powders (WP), aqueous suspensions (AS), flowables (F), liquids (L), solutions (S), or emulsifiable concentrates (EC) may not combine properly with some fluid fertilizer materials. Small quantities should always be tested before full-scale mixing. This will determine whether a compatibility agent is needed, and which agent does the best job. The nine (9) agents listed above have been thoroughly tested. There are many other surfactants on the market which were not designed for use with liquid fertilizers. Use the following test to select the correct agent for your mixture:

1. Put one (1) pint of liquid fertilizer in a quart jar.
2. Add one (1) to four (4) teaspoonful(s) of Surflan Dry Flowable with an equal amount of the WP, AS, F, L, S, or EC formulation (depending on the recommended rate per acre) to the liquid fertilizer. Close jar and agitate until dispersed evenly in the fertilizer. If the materials do not disperse well, it may be necessary to slurry the chemicals in water before adding to the fertilizer.
3. Observe the jar for about 10 minutes. If the materials rise to the surface and form a thick layer (oil curds), which will not disperse when agitated, a compatibility agent is needed. If the mixture is easily dispersed to its original state with slight agitation, no agent is needed, but good agitation must be provided in the fertilizer spray tank.
4. If the need for a compatibility agent is shown in Step 3, using a clean quart jar, start at Step 1 above, add one-half (1/2) teaspoonful of the compatibility agent to the liquid fertilizer, mix well, then repeat Steps 2 and 3.

An effective compatibility agent will cause the mixture to remain uniformly mixed with little or no separating or oil rising to the surface for one-half (1/2) hour or longer. If slight separation does occur, two (2) to three (3) inversions of the jar should give a uniform remix. If oily curds form which will not disperse, more agent or another agent should be tried.

Use a clean jar for each test. The compatible mixture will have a uniform appearance and will be relatively easy to keep mixed with gentle agitation of the jar.

SURFLAN PERFORMANCE GUARANTEE

(A LIMITED WARRANTY OF WEED AND GRASS CONTROL)

Elanco Products Company **warrants** that Surflan will control the weeds and grasses as indicated on its current label when used according to label directions and subject to the following limitations:

1. Purchaser must notify Elanco promptly if a lack of satisfactory control occurs. Such notice must be given within 30 days after the first planting or application of Surflan to the treated crop, whichever occurs later.
2. An Elanco representative must be satisfied that Purchaser used Surflan according to label directions and must visually observe unsatisfactory control of weeds in the field.
3. The product guarantee will extend to the following applications only:
   a) Preplant surface application (PPSA) when a light tillage is used prior to planting.
   b) Postplant surface application (SA) when a light tillage is used prior to or just following crop emergence.
   c) Any recommended application where a single one-half inch (1/2") rainfall or sprinkler irrigation occurs prior to weed seedling emergence.
4. The Purchaser must provide an Elanco representative with an invoice or other satisfactory record which shows the price and quantity of Surflan purchased.
5. Elanco's liability will be limited to a refund of the purchase price of the Surflan applied to the acreage on which weed control was not satisfactory or, at the option of the Purchaser, a refund of an equivalent amount of Surflan.
6. Void outside the U.S.A.

DISCLAIMER OF WARRANTIES

The warranty printed above is the only warranty applicable to this product. All other warranties, express or implied, including but not limited to the implied warranties of merchantability and fitness for a particular purpose, are disclaimed.

LIMITATION OF DAMAGES

Elanco's liability, whether in contract, warranty, tort, negligence, strict liability or otherwise, shall not exceed the return of the amount of the purchase price of the Surflan and under no circumstances shall Elanco be liable for special, indirect or consequential damages.

INHERENT RISKS OF USE

Failure to carefully follow the directions for use of Surflan may result in unsatisfactory weed control or crop injury. Factors such as seedling disease, foliage disease, cold weather, deep planting, excessive moisture, high salt concentration or drought may also result in unsatisfactory weed control or crop injury.

Amiben®-chloramben, Union Carbide Agricultural Products Co., Inc.
Casoron®-dichlobenil, Philips-Duphar
Dyanap®-naptalam + dinoseb, Uniroyal Chemical
Goal®-oxyfluorfen, Rohm and Haas Company
Hyvar® X-bromacil, E. I. duPont de Nemours and Company
Karmex®-diuron, E. I. duPont de Nemours and Company
Klean-Krop®-naptalam + dinoseb, Thompson-Hayward Company
Krovar® I and II-bromacil/diuron, E. I. duPont de Nemours and Company
Lasso®-alachlor, Monsanto Company
Lasso®-Micro-Tech™-alachlor, Monsanto Company
Lexone®-metribuzin, E. I. duPont de Nemours and Company
Lorox®-linuron, E. I. duPont de Nemours and Company
Princep®-simazine, Ciba-Geigy Corporation
Roundup®-glyphosate, Monsanto Company
Sencor®-metribuzin, Bayer GmbH
Sinbar®-terbacil, E. I. duPont de Nemours and Company
Solicam®-norflurazon, Sandoz, Inc.
Treflan®-E.C., 5, M.T.F.™, trifluralin, Elanco Products Company

TREFLAN® E.C.

HERBICIDE

A Selective Herbicide for the Pre-Emergence Control of Annual Grasses and Broadleaf Weeds

ACTIVE INGREDIENT:

trifluralin* (*a,a,a*–trifluoro–2,6–dinitro–*N*, *N*–dipropyl–*p*–toluidine)	44.5%
INERT INGREDIENTS	55.5%
TOTAL	100.0%

Contains 4 pounds active ingredient per gallon.
TREFLAN®—the registered trademark for Elanco Products trifluralin.
U.S. Patent No. Re. 31227
EPA Reg. No. 1471-35

WARNING
Keep Out of Reach of Children

PRECAUTIONARY STATEMENTS

HAZARDS TO HUMANS AND DOMESTIC ANIMALS

WARNING

Do not get in eyes. Avoid contact with skin and clothing. Harmful if swallowed or absorbed through the skin. Treflan E.C. may cause skin sensitization reactions in certain individuals. Use eye protection and protective clothing such as coveralls, a long sleeved shirt, and impermeable gloves when handling this product

First Aid: In case of contact, immediately flush eyes or skin with plenty of water. For eyes, call a physician. Remove and wash contaminated clothing before reuse.

ENVIRONMENTAL HAZARDS

Direct contamination of any body of water with this emulsifiable concentrate may kill fish and other aquatic organisms. Do not contaminate any body of water by direct application, cleaning of equipment or disposal of wastes.

STORAGE AND DISPOSAL

STORAGE

Store in original container only. Avoid freezing. Store above 40°F. If frozen, poor weed control may result. Do not store near heat or flame. In case of leak or spill, use absorbent materials to contain liquids and dispose as waste.

DISPOSAL

Do not contaminate water, food or feed by storage or disposal. Wastes resulting from the use of this product may be disposed of on site or at an approved waste disposal facility.

Plastic Container Disposal: Triple rinse (or equivalent). Then offer for recycling or reconditioning, or puncture and dispose of in a sanitary landfill, or incineration, or, if allowed by state and local authorities, by burning. If burned, stay out of smoke.

Metal Container Disposal: Triple rinse (or equivalent). Then offer for recycling, or reconditioning, or puncture and dispose of in a sanitary landfill, or by other procedures approved by state and local authorities.

Special Precautions

Applied according to directions and under normal growing conditions, Treflan will not harm the treated crop. Overapplication may result in crop injury or a soil residue. Uneven application or improper soil incorporation of Treflan can result in erratic weed control or crop injury. Seedling disease, cold weather, deep planting, excessive moisture, high salt concentration or drought may weaken crop seedlings and increase the possibility of damage from Treflan. Under these conditions, delayed crop development or reduced yields may result.

In Arizona, Colorado, California, Idaho, Montana, Nevada, New Mexico, Oregon, Utah, Washington and Wyoming, sugarbeets, red beets or spinach should not be planted for 12 months after a spring application or for 14 months after a fall application of Treflan. Plow the land to a depth of 12 inches prior to planting sugar beets to prevent the possibility of crop injury. Sorghum (milo), proso millet, corn or oats should not be planted for 14 months after a spring application or for 16 months after a fall application of Treflan to avoid crop injury. If land has not been irrigated, do not plant any of these crops for 18 months after a spring application or 20 months after a fall application of Treflan.

In those portions of Kansas, Nebraska, North Dakota, Oklahoma, South Dakota and Texas where at least 20 inches of irrigation and/or rainfall (total) was used to produce the crop, sorghum or oats should not be planted for 12 months after an application of Treflan.

If less than 20 inches of total water was used to produce the crop, do not plant sorghum, proso millet, or oats for 18 months after an application of Teflan. Cool, wet weather conditions during the early stage of growth may increase the possibility of injury to sorghum.

In all other areas receiving greater than 20″ rainfall/year: Moldboard plow before planting sugar beets where a spring application of Teflan was made the previous season. Also note planting restrictions listed in the section on control of rhizome johnsongrass and other higher rate programs.

Vegetable Growing Areas: Vegetable crops other than those listed on this label should not be planted within 5 months following the application of Treflan.

DIRECTIONS FOR USE

TREFLAN is a preemergence herbicide which is incorporated into the soil to provide long-lasting control of many annual grasses and broadleaf weeds. TREFLAN controls weeds as they germinate. TREFLAN will not control established weeds.

WEEDS AND GRASSES CONTROLLED BY TREFLAN

Grasses

Annual bluegrass	*(poa annua)*
Barnyardgrass (Watergrass)	*(Echinochloa sp.)*
Brachiaria (Signalgrass)	*(Brachiaria sp.)*
Bromegrass	*(Bromus tectorum)*
(Cheatgrass)	
(Downy brome)	
Cheat (Chess)	*(Bromus secalinus)*
Crabgrass	*(Digitaria spp.)*
(Large crabgrass)	
(Smooth crabgrass)	
Foxtail	*(Setaria spp.)*
(Bottlegrass)	
(Bristlegrass)	
(Giant foxtail)	
(Green foxtail)	
(Foxtail millet)	
(Pigeongrass)	
(Robust foxtail)	
(Yellow foxtail)	
Goosegrass	*(Eleusine indica)*
(Silver crabgrass)	
(Silvergrass)	
(Wiregrass)	
(Yardgrass)	
Johnsongrass (from seed)	*(Sorghum halepense)*
(Rhizome—see special instructions for control in cotton and special instructions for control in soybeans)	
Junglerice	*(Echinochloa colonum)*
Panicum	
Fall panicum	*(Panicum dichotomiflorum)*
(Spreading panicgrass—see special instructions in cotton and see special instructions in soybeans)	
Guineagrass	*(Panicum maximum)*
(See special instructions in sugarcane)	
Texas panicum	*(Panicum texanum)*
(Buffalograss)	
(Coloradograss)	
Itchgrass (Raoulgrass)	*(Rottboellia exaltata)*
(See special instructions in sugarcane)	
Red rice	*(Oryza sativa)*
(See special instructions in soybean)	
Sandbur (Burgrass)	*(Cenchrus incertus)*
Sprangletop	*(Leptochloa filiformis)*
Stinkgrass (Lovegrass)	*(Eragrostis cilianensis)*
Wild cane (Shattercane)	*(Sorghum bicolor)*
(See special instructions in soybean)	
Woolly cupgrass	*(Eriochloa villosa)*

Broadleaf Weeds

Carpetweed	*(Mollugo verticillata)*
Chickweed	*(Stellaria media)*
Field bindweed	*(Convolvulus arvensis)*
(See special instructions in Fruit and Nut Crops and Vineyards)	
Florida pusley	*(Richardia scabra)*
(Florida purslane)	
Mexican clover)	
(Pusley)	
Goosefoot	*(Chenopodium hybridum)*
Henbit (fall application only)	*(Lamium amplexicaule)*
Knotweed	*(Polygonum aviculare)*
Kochia	*(Kochia scoparia)*
(Fireweed)	
(Mexican fireweed)	
Lambsquarters	*(Chenopodium album)*
Pigweed	*(Amaranthus spp.)*
(Carelessweed)	
(Prostrate pigweed)	
(Redroot)	
(Rough pigweed)	
(Spiny pigweed)	
Puncturevine (Western U.S. only)	*(Tribulus terrestris)*
(Caltrop)	
(Goathead)	
Purslane	*(Portulaca oleracea)*
Russian thistle (Tumbleweed)	*(Salsola kali)*
Stinging nettle (Nettle)	*(Urtica dioica)*

SOIL PREPARATION

Crop Residues or Existing Weeds: Ground cover, such as crop residues or existing weeds, can interfere with the incorporation of TREFLAN into the soil. A manageable level of such ground cover will allow the TREFLAN to be uniformly incorporated into the top 2 to 3 inches of soil. If the level of the ground cover is such that this cannot be done, you must till the soil prior to the application of TREFLAN.

Roughness: The soil surface should be smooth enough so that you can operate a sprayer and incorporation equipment efficiently and at speeds which insure a uniform application and incorporation of TREFLAN.

General Soil Conditions: To assure uniform incorporation of TREFLAN, soil moisture conditions should be such that large clods can be broken up during the incorporation process.

SOIL TEXTURE GUIDE

The amount of TREFLAN you apply will vary with the soil texture and organic matter. A fine textured soil will require more TREFLAN per acre than a coarse soil. Choose the proper rate for each application based on the following soil texture group and specific crop recommendations. Do not exceed recommended rates.

Soil Texture	Soil Classification
Coarse Soils: (Light)	Sand, loamy sand, sandy loam
Medium Soils:	Loam, silty clay loam*, silt loam, silt, sandy clay loam*
Fine Soils:	Clay, clay loam, silty clay loam*, silty clay, sandy clay, sandy clay loam*

* Silty clay loam and sandy clay loam soils are transitional soils and may be classified as either medium or fine textured soils. If silty clay loam or sandy clay loam soils are predominately sand or silt, they are usually classified as medium textured soils. If they are predominately clay, they are usually classified as fine textured soils.

MIXING AND APPLICATION DIRECTIONS

TREFLAN E.C. Alone in Water

Start with a clean spray tank. Fill sprayer ⅓ to ½ full with clean water. Start agitation. Add correct quantity of TREFLAN, continue agitation and finish filling the tank.

TREFLAN E.C. Tank Mix in Water

Vigorous, continuous agitation is required for all tank mixes. (Sparger pipe agitators generally provide the best agitation in spray tanks). Avoid stirring or splashing air into the mixture during filling to prevent foaming. To do this, place the end of the fill pipe below the surface of the water in the spray tank. Do not allow the mixture to siphon back into the water source.

Mixing order: Fill the tank ¼ to ⅓ full with clean water. Start the agitation. Add dry flowables, wettable powders (WP), aqueous suspensions (AS), flowables (F), and liquids (L) to the water and agitate until the product(s) are completely dispersed in the water. Allow additional mixing and dispersion time when using dry flowable products. Continue agitation and fill tank to ¾ full, add the TREFLAN E.C., mix thoroughly. Then add any solution (S) formulations, agitate and finish filling. Maintain agitation during filling and through application. If spraying and agitation must be stopped before the tank is empty, the materials may settle to the bottom. In this case, it is important to resuspend all of the material in the bottom of the tank before continuing the spray application. A sparger agitator is particularly useful for this purpose. Sometimes it is more difficult to resuspend settled material than it is to suspend it originally.

Read and carefully follow all label instrucions for each material added to the tank. Premixing dry and flowable formulations with water (slurrying) and pouring the slurry through a 20 or 35 mesh wetting screen in the top of the tank will help assure good initial dispersion in the tank water. Line screens in the tank should be no finer than 50 mesh (100 mesh is finer than 50 mesh).

If you see a buildup of material on the walls of the spray tank, wash the tank with soapy water between fillings. Rinse and continue the spraying operation. Clean the tank, lines, and screens thoroughly after use.

As the spray volume decreases, the importance of accurate calibration and uniform application increases. Check the sprayer daily to insure proper calibration and uniform application. Do not apply TREFLAN when the wind can cause drifting of spray particles which can result in non-uniform application. TREFLAN should not be applied to soils which are wet or are subject to prolonged periods of flooding as poor weed control may result.

Ground application: Apply TREFLAN in 5 to 40 gallons of water or liquid fertilizer per acre (broadcast basis), using any properly calibrated, low pressure herbicide sprayer that will apply the spray uniformly.

Aerial application: Apply TREFLAN in 5 to 10 gallons of water or liquid fertilizer per acre. Adjust pump pressure, nozzle arrangements, speed and height to provide a uniform application to the soil surface. Use swath markers or flagmen to assure proper application spray widths.

INCORPORATION DIRECTIONS

Incorporation Equipment—General Directions

Use incorporation equipment that mixes TREFLAN into the top 2 to 3 inches of the final seedbed, or erratic weed control and/or crop injury may result. Incorporation equipment such as a disc will mix TREFLAN approximately half as deep as the equipment is set to operate. For example, a disc set to cut 4 inches deep will incorporate most of the TREFLAN within the top 2 inches of soil.

Incorporation Before Planting

TREFLAN must be incorporated one time within 24 hours after application. Then any time prior to planting, a second incorporation is necessary, this time running the equipment in a different direction from the first. You should incorporate the TREFLAN uniformly into the top 2 to 3 inches of the final seedbed.

Incorporation After Planting

Check specific crop for incorporation directions after planting.

Incorporation in Bedded Culture

For effective weed control, TREFLAN needs to be incorporated into the top 2 to 3 inches of the final seedbed.

Application prior to bedding: Apply TREFLAN and incorporate it one time with recommended equipment. The bedding operation serves as the second incorporation. Do not expose untreated soil during post-bedding operations.*

Application after bedding: Knock off beds to planting height before applying TREFLAN. Apply TREFLAN and incorporate it with recommended equipment that will conform to the bed shape. Do not leave untreated soil exposed.*

* Avoid removal of untreated soil from the seedbed before or during the planting operation. This would expose untreated soil, allowing weeds to germinate in the drill row.

Recommended Equipment

Any recommended incorporation tool may be used alone or in combination with any other recommended tool. Two incorporation passes are required unless specifically stated. The second incorporation should not be deeper than the first.

Disc: set to cut 4 to 6 inches deep and operate at 4 to 6 mph.

Field Cultivator: set to cut 3 to 4 inches deep and operate at 5 mph or more. A field cultivator is defined as an implement with 3 to 4 rows of sweeps, spaced at intervals of 7 inches or less and staggered so that no soil is left unturned. Chisel points should not be used.

Combination Seedbed Conditioners: set to cut 3 to 4 inches deep and operate at a speed of at least 5 mph. These implements are defined as three or more tillage devices combined and used as a single tool. For example, 2 to 3 rows of field cultivator C- or S-shaped shanks with an effective sweep spacing of 6 to 9 inches (staggered so that no soil is left unturned), followed by a spike-tooth or flextine harrow, followed by a ground-driven reel or basket.

Rolling Cultivator: set to cut 2 to 4 inches deep and operate at 6 to 8 mph. Rolling cultivators are adequate for use on coarse and medium textured soils only, except when used in sugarcane where the rolling cultivator may be used on fine textured soils.

Bed Conditioner (Do-All): set to cut 2 to 4 inches deep and operate at 4 to 6 mph. The do-all is adequate for use on coarse and medium textured soils only. When using the do-all in bedded culture, only one incorporation pass is required. However, two passes with a do-all are required in flat planted culture.

Mulch Treader (other similar disc-type implements): set to cut 3 to 4 inches deep and operate at 5 to 8 mph.

P.T.O. Driven Equipment (tillers, cultivators, hoes): adjust to incorporate TREFLAN into the top 2 to 3 inches of the seedbed with rotors spaced to provide a clean sweep of the soil. Only one incorporation is necessary. P.T.O. driven equipment should not be operated at a speed greater than 4 mph.

Other equipment, including the flexible tine-tooth harrow (Flex-tine, Melroe), is also recommended but only for the special programs for which it is specified in this label.

CULTIVATION AFTER PLANTING

Soil treated with TREFLAN may be shallow cultivated without reducing the weed control activity of TREFLAN. Do not cultivate deeper than the treated soil since this may bring untreated soil to the surface, and poor weed control may result.

CROP RECOMMENDATIONS

These recommendations are given as the broadcast rates of TREFLAN per acre. For band applications, decrease the amount of TREFLAN in proportion to the amount of surface treated per acre. Apply any time after January 1 when the soil can be worked and is suitable for good incorporation. TREFLAN can be applied in the fall—see specific crop for recommendations. For general fall application directions where specific recommendations are not given, see page 58. Where a rate range is shown, use the lower rate for coarser soils or soils with lower organic matter. TREFLAN should not be used on soils containing more than 10% organic matter.

COTTON—TREFLAN ALONE

TREFLAN can be applied and incorporated before or at planting, immediately after planting, or at layby.

COTTON—Preemergence:

Broadcast Rates Per Acre

	TREFLAN E.C.	
Soil Texture	**Areas receiving less than 20" average annual rainfall***	**Areas receiving greater than 20" average annual rainfall***
	(pints)	(pints)
Coarse	1	1
Medium	1¼–1½	1½
Fine	1½	2

* Use 1½ pints per acre on coarse and medium textured soils and 2 pints on fine soils with 2–5% organic matter; use 2 to 2½ pints on all soils with 5–10% organic matter.

COTTON—Postplant:
When incorporating TREFLAN after planting (postplant), be careful not to disturb the seed.

COTTON—Layby:
Apply and incorporate TREFLAN any time up to layby, but not less than 90 days before harvest. Direct the layby applications onto the soil between the rows and beneath emerged cotton plants. Use the same rates as for a premergence application.

COTTON—Fall Application:
Apply and incorporate TREFLAN any time between October 15 and December 31. The ground may be left flat or bedded-up over winter. On bedded ground, knock beds down to desired height before planting, moving some treated soil from beds into furrows. Where soil is left flat over winter, be careful not to turn up untreated soil during spring bedding operations. Destroy established weeds during seedbed preparation. If weeds become established in furrows due to uncovering of untreated soil during bedding, destroy these weeds before planting. In the fall, do not apply TREFLAN to soils which are wet or subject to prolonged periods of flooding.

Broadcast Rates per Acre—Fall application only:
In Alabama, Arkansas, northern Florida, Georgia, Louisiana, Mississippi, southeastern Missouri bootheel, North Carolina, New Mexico, Oklahoma, South Carolina, Tennessee and Texas, apply and incorporate TREFLAN at a broadcast rate of 2 pints per acre on coarse and medium soils and 2½ pints on fine soils.

In Arizona, California and Nevada, apply and incorporate TREFLAN at a broadcast rate of 1½ pints per acre on course soils; 2 pints on medium soils; and 2½ pints on fine soils.

For cotton grown in other states, apply and incoporate TREFLAN at a broadcast rate of 1 pint per acre on coarse soils; 1½ pints on medium soils; 2 pints on fine soils, 1½ pints on coarse soils with 2–5% organic matter; and 2 to 2½ pints on soils with 5–10% organic matter.

COTTON—SPECIAL USE DIRECTIONS

COTTON—Fall panicum:
Apply and incorporate TREFLAN at the broadcast rate of 2 pints per acre on both coarse and medium soils.

COTTON—Rhizome johnsongrass:
In all cotton-producing states except Arizona and California, you can obtain commercially acceptable control of rhizome johnsongrass with a double rate program which you apply for 2 consecutive years in accordance with the following directions.

Soil Preparation—Proper preparation of the soil before application is very important for satisfactory results. Use a chisel plow or similar implement to bring rhizomes to the top of the soil. Then follow with a disc two times before application to cut the rhizomes into small (2 to 3 inch) pieces. This should also destroy any emerged johnsongrass.

Application—Choose the one application program that best fits your cultural practices:

Spring Application—Apply TREFLAN any time before planting in the spring for 2 years in a row. Use a broadcast rate of 2 pints per acre on coarse soils; 3 pints on medium soils; and 4 pints on fine soils,

OR

Fall Application—Apply TREFLAN between October 15 and December 31 for 2 years in a row at the same rates as a spring application for the control of rhizome johnsongrass.

Incorporation—Deep incorporation is essential for good rhizome johnsongrass control. Incorporate TREFLAN thoroughly with a disc set to cut 4 to 6 inches deep and operate at 4 to 6 mph. Two passes are necessary, with the second pass in a different direction from the first.

Cultivation—Some johnsongrass plants will escape. Timely cultivations during the crop season to remove escaped plants are necessary to obtain commercially acceptable control. You cannot obtain commercially acceptable control with only 1 year of double rate TREFLAN use.

Crop Rotation—In the season following a double rate treatment, plant only rice and those crops for which TREFLAN can be applied as a preplant treatment or injury may result.

COTTON—Pigweed and seedling johnsongrass control: In Alabama, Arkansas, Florida, Georgia, Louisiana, Mississippi, southeastern Missouri, North Carolina, South Carolina, Tennessee and southern Virginia, TREFLAN may be applied preplant at a broadcast rate of 1 to 1½ pints per acre on coarse soils; 1½ to 2 pints on medium soils; and 2 pints on fine soils. Exception: in State of Louisiana, 3 pints per acre are recommended on fine soils.

COTTON—Additional weed and grass control (Texas Gulf Coast):
In the Texas Gulf Coast counties of Brazoria, Calhoun, Chambers, Fort Bend, Galveston, Harris, Jackson, Jefferson, Liberty, Matagorda, Orange, Victoria, Waller and Wharton, TREFLAN may be applied up to 2 weeks before planting at a broadcast rate of 1½ pints

per acre on coarse soils; 2 pints on medium soils; and 3 pints on fine soils.

COTTON PRECAUTIONS: Cotton should be planted after early season adverse weather conditions have passed, especially when using higher rate programs. Cool, wet weather early in the growth cycle causes additional stress to the cotton plant. This may result in reduced stands, delayed maturity and reduced yields.

COTTON—TANK MIXES AND OVERLAYS

Follow recommended soil preparation and incorporation procedures for TREFLAN.

COTTON—TREFLAN/Caparol tank mix for cotton grown in California, Arizona, New Mexico and west Texas:

The TREFLAN/Caparol combination will control certain grasses and broadleaf weeds listed for TREFLAN alone plus the following weeds:

Annual morningglory	Prickly sida (Teaweed)
Groundcherry (Annual)	Ragweed
Malva	Smartweed
Mustard	Wild oat

The tank mix also controls shallow germinating seedlings of cocklebur and coffeeweed.

Broadcast Rates Per Acre

Soil Texture	TREFLAN E.C.	Caparol 80W
	(pints)	(pounds)
Coarse	1	2 *
Medium	1¼–1½	2½
Fine	2	2½

* Do not use on sands and loamy sands. For band applications use proportionately less.

Mixing Directions: Carefully follow the procedures on the Caparol 80W label for making a slurry and adding it to a partially filled tank of water. After the Caparol is thoroughly mixed with the partially filled tank of water, add TREFLAN E.C. and continue filling. Agitate continuously during the filling and spraying operation.

Avoid leaving the spray mixture in the tank without constant agitation. If bypass agitation is used, the bypass line should stop at the bottom of the tank to minimize foaming.

Additional Precautions: The combination of TREFLAN/Caparol should not be used in the cut areas of newly leveled fields, in areas of excess salt, or where flooding over the beds is likely to happen. Do not plant cotton in tractor wheel depressions. These conditions may cause crop injury. On mulch-planted cotton, water back only after cotton seedlings are well-established.

Crop Rotations: Cabbage, okra, onions and peas may be planted in the fall after a spring application of TREFLAN plus Caparol. Winter barley, winter rye and winter wheat can be planted in the fall also, if they are plowed down and not used for food or feed. Refer to the Caporal label for directions, cautions and precautions.

COTTON—TREFLAN/Cotoran tank mix (except in Arizona and California):

Follow recommended spoil preparation and incorporation procedures for TREFLAN.

The TREFLAN/Cotoran tank mix effectively controls all the annual grasses and broadleaf weeds listed for TREFLAN alone plus these additional weeds:

Buttonweed	Ragweed
Cocklebur	Ryegrass
Groundcherry, Wright	Sesbania
Jimsonweed	Sicklepod
Morningglory	Smartweed
Prickly sida (Teaweed)	Tumbleweed

Broadcast Rates Per Acre

Soil Texture	TREFLAN E.C.	Cotoran 80W
	(pints)	(pounds)
Coarse	1	1¼
Medium	1½	2
Fine	2	2½

Mixing Directions: Carefully follow the procedures on the Cotoran label for making a Cotoran slurry and adding it to a partially filled tank of water. After the Cotoran is thoroughly mixed with the partially filled tank of water, add the TREFLAN and continue filling. Agitate continuously throughout the filling and application operations. Do not leave spray mixture in tank without constant agitation. If bypass agitation is used, the bypass line should stop at the bottom of the tank to minimize foaming. Apply in 15–40 gallons of water per acre.

Additional Precautions: Do not plant crops other than cotton on the treated land within 6 months after the application of TREFLAN plus Cotoran, or crop injury may result. Do not feed foliage from treated cotton plants or gin trash to livestock. Do not mix TREFLAN plus Cotoran with liquid fertilizer.

West Texas Only: Do not use the tank mix of TREFLAN plus Cotoran on sandy, loamy sand or fine sandy loam soils. Do not use on cotton planted in furrows.

Arkansas, Louisiana and Mississippi Only: Use 1 pound Cotoran 80W in tank mix with TREFLAN on sandy loam soils low in organic matter.

New Mexico: Do not plant treated land with crops other than cotton until 1 year after the last application. Do not use on sandy or coarse textured soils of less than 1% organic matter.

COTTON—TREFLAN preplant followed by Cotoran overlay:

Apply and incorporate TREFLAN as recommended for the specific soil texture. Then apply Cotoran as a preemergence surface treatment at 1¼ to 2½ pounds per acre. On light silt and sandy soils low in organic matter, use the lower rate.

Additional Precaution: Refer to the Cotoran label for all cautions and precautions when using this preemergence surface treatment.

COTTON—TREFLAN preplant followed by Karmex overlay:

For cotton grown east of the Mississippi River plus Arkansas, southeastern Missouri, Louisiana and Eastern Texas, incorporate TREFLAN before planting. Then follow with a preemergence application of Karmex. This will effectively control all the weeds controlled by TREFLAN alone plus these additional weeds:

Annual morningglory	Shepherdspurse
Annual groundcherry	Velvetgrass
Dogfennel	Wild Lettuce
Pennycress	Wild Mustard
Ragweed	

Broadcast Rates Per Acre

Soil Texture	TREFLAN E.C.	Karmex 80W
	(pints)	(pounds)
Coarse	1	⅓
Medium	1½	⅔
Fine	2	1

Additional Precautions: Do not use Karmex on soils with less than 1% organic matter as crop injury may result. Do not allow livestock to graze on cotton treated with Karmex. Consult the Karmex label for additional instructions, cautions and precautions.

SOYBEAN—TREFLAN ALONE

SOYBEAN—Preemergence:

Follow recommended soil preparation, application, and incorporation procedures for TREFLAN.

Broadcast Rates Per Acre

Soil Texture	TREFLAN E.C.*
	(pints)
Coarse	1
Medium	1½
Fine	2

* Use 1½ pints per acre on coarse and medium textured soils and 2 pints on fine soils with 2–5% organic matter, use 2–2½ pints on all soils with 5–10% organic matter.

SOYBEAN—Fall Application:

Apply and incorporate TREFLAN anytime between October 15 and December 31. Ground may be left flat or bedded-up over winter.

On bedded ground, knock beds down to desired height before planting, moving some treated soil from tops into furrows. Where soil is left flat over winter, take care during spring bedding operations to prevent turning up untreated soil. Destroy established weeds during seedbed preparation. If weeds become established in furrows due to uncovering of untreated soil during listing, destroy these weeds before planting. Do not fall apply TREFLAN to soils which are wet, are subject to prolonged periods of flooding, or where rice was grown the previous year.

For soybeans grown in Alabama, Arkansas, northern Florida, Georgia, Louisiana, Mississippi, southeastern Missouri Bootheel, North Carolina, Oklahoma, South Carolina, Tennessee and Texas, apply and incorporate TREFLAN at a broadcast rate of 2 pints per acre on coarse and medium soils and 2½ pints on fine soils.

For soybeans grown in states other than those listed above, apply and incorporate TREFLAN at a broadcast rate of 1 pint per acre on coarse soils; 1½ pints on medium soils; 2 pints on fine soils; 1½ pints on coarse soils with 2–5% organic matter; and 2 to 2½ pints on soils with 5–10% organic matter.

SOYBEAN—SPECIAL USE DIRECTIONS

SOYBEAN—Fall panicum:

Apply TREFLAN at the broadcast rate of 2 pints per acre on both coarse and medium soils.

SOYBEAN—Pigweed and seedling johnsongrass:

In Alabama, Arkansas, Florida, Georgia, Kansas, Louisiana, Mississippi, southeastern Missouri, North Carolina, Oklahoma, South Carolina, Tennessee and southern Virginia, TREFLAN may be applied at a broadcast rate of 1 to 1½ pints per acre on coarse soils; 1½ to 2 pints on medium soils; and 2 pints on fine soils. Exception: in state of Louisiana, 3 pints per acre are recommended on fine soils.

SOYBEAN—Additional weed and grass control (Texas Gulf Coast):

In the Texas Gulf Coast counties of Brazoria, Calhoun, Chambers, Fort Bend, Galveston, Harris, Jackson, Jefferson, Liberty, Matagorda, Orange, Victoria, Waller and Wharton, TREFLAN may be applied up to 2 weeks before planting at a broadcast rate of 1½ pints per acre on coarse soils; 2 pints on medium soils; and 3 pints on fine soils.

SOYBEAN—Charcoal soils in Arkansas, Lousiana and Mississippi:

Newly cleared land often contains high organic matter (5–10%) and charcoal from burning debris. This charcoal and/or organic matter tends to bind TREFLAN and reduce its weed control activity. Under these conditions, higher rates of TREFLAN are necessary for weed control. Increased rates, however, can cause crop injury if charcoal or organic matter is not present to bind some of the TREFLAN. In the burn row a high level of charcoal is present; consequently, poor weed control may result even with an increased rate of TREFLAN.

Apply TREFLAN at the broadcast rate of 1½ to 2½ pints per acre on coarse soils; 2½ pints on medium soils; and 3 pints on fine textured soils. Follow recommended soil preparation, application and incorporation procedures for TREFLAN.

SOYBEAN—Red rice in Arkansas, Louisiana, Mississippi and Texas only:

You can obtain suppression or partial control of red rice when you apply TREFLAN at the following recommended rates. Follow recommended soil preparation and incorporation procedures for TREFLAN.

Apply and incorporate TREFLAN in the spring before planting.

Broadcast Rates Per Acre

Soil Texture	TREFLAN E.C. Application Year 1	TREFLAN E.C. Application Year 2
	(pints)	(pints)
Coarse	2	1
Medium	3	1½
Fine	4	2
Coarse soils with 2 to 5% organic matter	3	1½
Soils with 5 to 10% organic matter	4	2–2½

If a combination of high organic matter (5–10%) and charcoal are present in the soil, apply TREFLAN the second year at the following rates for charcoal soils in Arkansas, Louisiana and Mississippi:

Soil Texture	TREFLAN E.C.
	(pints)
Coarse	1½–2½
Medium	2½
Fine	3

For more information on charcoal soils see page 33. Charcoal soils heading.

Crop Rotation: The program for red rice control in soybeans is a 2-year program. Use the rates listed for first year application and plant soybeans. The second year use the normal TREFLAN rates listed for your soil type and charcoal level and plant only those crops for which TREFLAN has been registered as a preplant treatment, or crop injury may result. Do not plant rice the second year. Rice may be planted the third year.

SOYBEAN—Rhizome johnsongrass (Eastern United States and the state of Texas):

You can obtain commercially acceptable control of rhizome johnsongrass with a double rate program applied for 2 consecutive years in accordance with the following directions.

Soil Preparation—Proper preparation of the soil before application is very important for satisfactory results. Use a chisel plow or similar implement to bring rhizomes to the top of the soil. Then follow with a disc two times before application to cut the rhizomes into small (2 to 3 inch) pieces and to destroy any emerged johnsongrass.

Application—choose the one application program that best fits your cultural practices:

Spring Application—Apply TREFLAN anytime in the spring before planting for 2 years in a row. Use a broadcast rate of 2 pints per acre on coarse soils; 3 pints on medium soils; 4 pints on fine soils; 3 pints on coarse soils with 2–5% organic matter; and 4 pints on soils with 5–10% organic matter.

OR

Fall Application—Apply TREFLAN between October 15 and December 31 for 2 years in a row at the same rates as a spring application for the control of rhizome johnsongrass.

OR

Split Application—Apply TREFLAN at the same rate in both the spring and fall for 2 years in a row using the rates in the following table.

Broadcast Rates Per Acre

Soil Texture	TREFLAN E.C. Spring and Fall
	(pints)
Coarse	1
Medium	1½
Fine	2
Coarse soils with 2–5% organic matter	1½
Soils with 5–10% organic matter	2

Incorporation—Deep incorporation is essential for good rhizome johnsongrass control. Incorporate TREFLAN thoroughly with a disc set to cut 4 to 6 inches deep and operate at 4 to 6 mph. Two passes are necessary, with the second pass in a different direction from the first.

Cultivation—Some johnsongrass plants will escape. Timely cultivations during the crop season to remove escaped plants are necessary to obtain commercially acceptable control.

Crop Rotation: In the season following a double rate treatment, plant only rice and those crops for which TREFLAN can be applied as a preplant treatment or injury may result.

SOYBEAN—Rhizome johnsongrass—TREFLAN/Sencor or TREFLAN/ Lexone tank mix:

TREFLAN/Sencor or TREFLAN/Lexone may be used for rhizome johnsongrass control and for the control of those weeds listed for TREFLAN alone. For the additional weeds controlled by Sencor or Lexone in tank mix, see below. Follow procedures for soil preparation, incorporation, and cultivation recommended in the Soybean—Rhizome johnsongrass section.

Apply TREFLAN/Sencor or Lexone up to two weeks before planting for two consecutive years at the following broadcast rates per acre:

Soil Texture	TREFLAN E.C.	+	Lexone 50WP/4L or Sencor 50WP/4	or	Lexone (dry flowable) or Sencor (dry flowable)
	(pints)		(pounds/pints)		(pounds)
Coarse*	2		1/2		1/3
Medium	3		3/4		1/2
Fine	4		1		2/3

*Do not use on coarse soils with less than 1% organic matter.

Read and follow all additional precautions listed for the TREFLAN/Sencor or TREFLAN/Lexone tank mix which follow at the end of the Sencor and Lexone section.

SOYBEAN—Wild cane (shattercane):

Follow recommended soil preparation and application procedures for TREFLAN.

Wild cane (shattercane) can germinate throughout the growing season and from greater soil depth than most other weed seeds. Commercially acceptable control of wild cane can be obtained with the following increased rates of TREFLAN.

Apply TREFLAN at the broadcast rate of 1 pint per acre on a coarse soil; 2 pints on a medium soil; and 2½ pints on a fine textured soil.

Incorporation—Deep incorporation is essential for good wild cane control. Incorporate TREFLAN thoroughly with a disc set to cut 4 to 6 inches deep and operate at 4 to 6 mph. Two passes are necessary, with the second pass in a different direction from the first.

Cultivation—Cultivations during the crop season will also contribute to control.

SOYBEAN PRECAUTIONS: Soybeans should be planted after early season adverse weather conditions have passed, especially when using higher rate programs. Cool, wet weather early in the growth cycle causes additional stress to the soybean plant. This may result in reduced stands, delayed maturity and reduced yields.

SOYBEAN—TANK MIXES AND OVERLAYS

SOYBEAN—TREFLAN/Sencor or TREFLAN/Lexone tank mix:

The TREFLAN/Sencor or TREFLAN/Lexone tank mix effectively controls the annual grasses and broadleaf weeds controlled by TREFLAN alone plus these additional weeds:

Jimsonweed	Ragweed, common
Mallow, Venice (Flower-of-an-hour)	Sesbania, hemp
Mustard, wild	Smartweed, Pennsylvania
Prickly sida	Velvetleaf

Control of cockelbur, morningglory and giant ragweed (horseweed) may be erratic. Control may be improved with timely cultivation. Where cocklebur is a serious problem, an overlay of Sencor or Lexone may be preferred to the TREFLAN/Sencor or TREFLAN/Lexone tank mix.

Follow recommended procedures for soil preparation, incorporation, and cultivation of TREFLAN. Mix according to instructions in Mixing and Application Directions. The TREFLAN/Sencor or TREFLAN/Lexone tank mix can be applied from 2 weeks before planting up to planting.

Broadcast Rates Per Acre

Soil Texture	TREFLAN E.C.	+	Lexone 50WP/4L or Sencor 50WP/4	or	Lexone (dry flowable) or Sencor (dry flowable)
	(pints)		(pounds/pints)		(pounds)
Coarse*	1		1/2		1/3
Medium	1½		3/4		1/2
Fine	2		1		2/3

*Do not use Sencor/Lexone on coarse soils with less than 1% organic matter.

Additional Precautions: Do not plant any crop other than soybeans within 4 months after treatment. Overapplication, uneven application, or improper soil incorporation may result in crop injury, herbicide residue, or erratic weed control. Additional stress factors are seedling disease, cold weather, deep planting, excessive moisture, soil *p* H over 7.5, high salt concentration, or drought. Any of these may weaken crop seedlings and increase possibility of damage from the tank mix. These additional factors may also delay crop development or reduce yields when Sencor or Lexone is applied. Observe all cautions and limitations on the Sencor and Lexone labels. Do not use the foliage from soybeans treated with the TREFLAN/Sencor or TREFLAN/Lexone tank mix for feed or forage.

SOYBEAN—TREFLAN preplant followed by Sencor or Lexone as an overlay:

Apply TREFLAN as a preplant incorporated herbicide. As a separate operation, make a single application of Sencor or Lexone as either a band or broadcast spray during planting or as a separate operation after planting, but before the soybeans emerge. Do not spray Sencor or Lexone over the top of emerged soybeans, or crop injury may result.

Use Directions—Follow directions on the Sencor or Lexone labels for specific instructions regarding each herbicide.

Broadcast Rates Per Acre

Soil Texture	TREFLAN E.C.	Postplant/Preemergence: Sencor 50WP/4 or [Lexone 50WP/4L]*	or	Sencor (dry flowable) or [Lexone (dry flowable)]*
	(pints)	(pounds or pints)		(pounds)
Coarse**	1	3/4–1 [3/4]		1/2–2/3 [1/2]
Medium	1½	3/4–1½ [3/4–1]		1/2–1 [1/2–2/3]
Fine	2	1–1¾ [1]		2/3–1⅙ [2/3]

* Lexone rates for each soil texture are enclosed in brackets "[ ]".
** Do not apply Lexone to sand or soils with less than ½% organic matter. Do not apply Sencor to coarse soils (sandy loam and loamy sand) containing less than 2% organic matter.

Additional Sencor and Lexone precautions: Do not use Lexone or Sencor on Tracy, Semmes, Altona, Vansoy or Coker102 soybeans. These varieties are sensitive to Lexone or Sencor, and crop injury may result. Seed must be planted at least 1½ inches but not more than 2 inches below the soil surface before a Sencor or Lexone application. Do not apply Sencor or Lexone at these rates more than once per season. Do not replant areas treated with Sencor or Lexone with any crop other than soybeans within 4 months after treatment. Injury to soybeans may occur if you use Lexone or Sencor on soils having a calcarous surface or *p* H of 7.5 or higher, or if you use them in conjunction with soil-applied organic phosphate pesticides. Do not use the foliage from treated soybeans for feed or forage.

SOYBEAN—TREFLAN/Amiben tank mix or overlay:

Tank mix—Amiben may be applied several days prior to planting as a broadcast tank mix with TREFLAN. Weeds controlled by this tank mix, in addition to those controlled by TREFLAN alone, are smartweed, velvetleaf and ragweed. The tank mixture should be used as a spring preplant incorporated treatment.

Overlay—Amiben may be applied broadcast or in a band over the soybean row at planting time in fields where TREFLAN has been preplant incorporated. Weeds controlled by Amiben when surface applied, in addition to those controlled by TREFLAN alone, are:

Coffeeweed (Sesbania)	Spurge, annual
Mustard, wild	Smartweed, Pennsylvania
Nightshade, black	Stinkgrass
Prickly sida (Teaweed)	Velvetleaf
Ragweed, common	

Apply TREFLAN as a tank mix with Amiben, or apply and incorporate TREFLAN alone followed by an overlay application of Amiben at these rates:

Broadcast Rates Per Acre

Soil Texture	TREFLAN E.C.	Amiben 2S
	(pints)	(quarts)
Coarse	1	4–6*
Medium	1½	4–6
Fine	2	4–6

* Use the higher rate where you expect heavy populations of smartweed, velvetleaf, ragweed, wild mustard or black nightshade. Do not use on muck or charcoal soils. Read and observe directions and cautions on the Amiben label.

SOYBEAN—TREFLAN/Amiben/Sencor or Lexone tank mix:
The TREFLAN/Amiben/Sencor or Lexone tank mix effectively controls all weeds listed for TREFLAN/Amiben and TREFLAN/Sencor or Lexone tank mixes.

Follow recommended soil preparation, application, and incorporation procedures for TREFLAN. The TREFLAN/Amiben/Sencor or Lexone tank mix may be applied from several days prior to planting in 10 to 40 gallons of water per acre. Use screens no finer than 50 mesh.

Apply the TREFLAN/Amiben/Sencor or Lexone tank mix at the following broadcast rate per acre:

Soil Texture	TREFLAN E.C. +	Amiben 2S +	Lexone 50WP/4L or Sencor 50WP/4 or	Lexone (dry flowable) or Sencor (dry flowable)
	(pints)	(quarts)	(pounds/pints)	(pounds)
Coarse[a]	1	3–4[b]	½	⅓
Medium	1½	3–4[b]	½–¾[c]	⅓–½[c]
Fine	2	4–5	¾[c]	½[c]

[a] Do not use Sencor or Lexone on coarse soils with less than 1% organic matter.
[b] Use the higher rate of Amiben when velvetleaf or black nightshade is a problem.
[c] On Clarion/Webster soils in Minnesota and Iowa or in similar alkaline (calcareous) soils with a *p*H of 7.5 or above, apply Sencor or Lexone at the rates listed below:

Soil Texture	Lexone 50WP/4L or Sencor 50WP/4 or	Lexone (dry flowable) or Sencor (dry flowable)
	(pounds/pints)	(pounds)
Medium	½	⅓
Fine	½–¾[d]	⅓–½[d]

[d] Use the higher rate only where soil *p*H is less than 7.5 and where weed pressure is heavy.

Additional precautions: The TREFLAN/Amiben/Sencor or Lexone tank mix will not harm the treated crop when you apply it according to directions and under normal growing conditions. However, overapplication, uneven application or improper soil incorporation of the tank mix can result in erratic weed control or crop injury. Additional stress factors are seedling disease, cold weather, deep planting, excessive moisture, soil *p* H over 7.5, high salt concentration, or drought. These additional factors may weaken crop seedlings and increase the possibility of damage from the tank mix. These additional factors may also delay crop development or reduce yields. Observe all cautions and limitations of all products used in mixtures. Do not use the foliage from soybeans treated with the TREFLAN/Amiben/Sencor or Lexone tank mix for feed or forage.

SOYBEAN—TREFLAN/Vernam tank mix:
TREFLAN/Vernam tank mix effectively controls those weeds listed for TREFLAN alone plus these additional weeds:

Purple nutsedge (nutgrass)	Coffeeweed
Yellow nutsedge (nutgrass)	Velvetleaf
Annual morningglory	

Broadcast Rates Per Acre

Soil Texture	TREFLAN E. C.	Vernam 7E
	(pints)	(pints)
Coarse	1	1¾–2⅓
Medium	1½	2⅓–3*
Fine	2	3–3½

* For nutsedge, wild cane and velvetleaf control, use the higher rate of 3 pints per acre on medium textured soils.

ALFALFA—ESTABLISHED

In areas receiving less than 20″ average annual rainfall per year, apply TREFLAN to established alfalfa stands at a broadcast rate of 1½ pints per acre on coarse soils and 2 pints on medium and fine soils. Use incorporation equipment that will insure thorough soil mixing with minimum damage to the established alfalfa.

ASPARAGUS—ESTABLISHED

Follow recommended soil preparation, application and incorporation procedures for TREFLAN.

TREFLAN can be applied to established asparagus as a single or a split application. In the winter or early spring, apply TREFLAN to asparagus after ferns are removed but before spear emergence. Or, apply after harvest in the late spring or early summer before ferning begins. TREFLAN will suppress volunteer seedling asparagus and field bindweed if you use the following recommended rates and application schedules.

Follow recommended soil preparation, application and incorporation procedures for TREFLAN.

Broadcast Rate Per Acre

Soil Texture	TREFLAN* Split Application: Before Harvest +	After Harvest	TREFLAN* Single Application: Before Harvest or	After Harvest
	(pints)		(pints)	
Coarse	1 +	1	2 or	2
Medium	1½ +	1½	3 or	3
Fine	2 +	2	4 or	4

* In any single calendar year, the maximum TREFLAN to apply is 2 pints per acre on coarse soils; 3 pints on medium soils; and 4 pints on fine soils.

FOR THE FOLLOWING CROP GROUPING, USE THE RATE TABLE BELOW.

CARROT

CASTOR BEAN

CELERY—(Direct seeded and transplant in areas receiving less than 20″ average annual rainfall.)

COLE CROPS—TRANSPLANT
Apply and incorporate prior to transplanting only. (Broccoli, Brussels Sprout, Cabbage and Cauliflower) See next section for direct seeded.

OKRA

PEPPER—TRANSPLANT
Apply an incorporate prior to transplanting only.

SOUTHERN PEA—(Before planting only.)

Apply and incorporate TREFLAN before planting, at planting or immediately after planting, unless otherwise indicated.

Broadcast Rates Per Acre

Soil Texture	TREFLAN E.C. Areas receiving less than 20″ average annual rainfall*	TREFLAN E.C. Areas receiving greater than 20″ average annual rainfall*
	(pints)	(pints)
Coarse	1	1
Medium	$1\frac{1}{4}$–$1\frac{1}{2}$	$1\frac{1}{2}$
Fine	$1\frac{1}{2}$	2

* Use $1\frac{1}{2}$ pints per acre on coarse and medium textured soils and 2 pints on fine soils with 2–5% organic matter; use 2 pints on all soils with 5–10% organic matter.

COLE CROP—DIRECT SEEDED (Broccoli, Brussels Sprout, Cabbage and Cauliflower) See above section for transplant.

For direct-seeded cole crops, apply and incorporate TREFLAN before planting at a broadcast rate of 1 pint per acre on coarse and medium soils and $1\frac{1}{2}$ pints on fine soils and soils with 2–5% organic matter. Direct-seeded cole crops have exhibited marginal tolerance to recommended rates of TREFLAN. Stunting or reduced stands may occur.

CUCURBITS—POSTPLANT EMERGED (Cantaloupe, Cucumber and Watermelon)

Western United States including Texas:
Apply TREFLAN as a directed spray to the soil between the rows and beneath plants which are in the 3 to 4 true leaf stage.

Broadcast Rates Per Acre

Soil Texture	TREFLAN E.C. Areas receiving less than 20″ average annual rainfall*	TREFLAN E.C. Areas receiving greater than 20″ average annual rainfall*
	(pints)	(pints)
Coarse	1	1
Medium	$1\frac{1}{4}$–$1\frac{1}{2}$	$1\frac{1}{2}$
Fine	$1\frac{1}{2}$	2

* Use $1\frac{1}{2}$ pints per acre on coarse and medium textured soils and 2 pints on fine soils with 2–5% organic matter; use 2 pints on all soils with 5–10% organic matter.

Set incorporation equipment to throw treated soil around the plants during incorporation.

DRY BEAN—TREFLAN ALONE

Apply and incorporate TREFLAN before planting using the following rates:

Broadcast Rates Per Acre

Soil Texture	TREFLAN E.C. Areas receiving less than 20″ average annual rainfall*	TREFLAN E.C. Areas receiving greater than 20″ average annual rainfall*
	(pints)	(pints)
Coarse	1	1
Medium	$1\frac{1}{4}$–$1\frac{1}{2}$	$1\frac{1}{2}$
Fine	$1\frac{1}{2}$	2

* Use $1\frac{1}{2}$ pints per acre on coarse and medium textured soils and 2 pints on fine soils with 2–5% organic matter; use 2 pints on all soils with 5–10% organic matter.

DRY BEAN—TREFLAN/Eptam tank mix:
The TREFLAN/Eptam tank mix effectively controls all the following weeds in addition to those weeds listed for TREFLAN:

Henbit (spring applications)	Oat, wild
Nightshade, black	Ragweed, common
Nightshade, hairy	Smartweed, Pennsylvania
Nutsedge	Velvetleaf (Buttonweed)

Follow recommended soil preparation and incorporation procedures for TREFLAN. The TREFLAN/Eptam tank mix should be applied from 2 days before planting up to planting. Incorporate immediately after application.

Broadcast Rates Per Acre

Soil Texture	TREFLAN E.C. Areas receiving less than 20″ average annual rainfall*	TREFLAN E.C. Areas receiving greater than 20″ average annual rainfall*	Eptam 7E
	(pints)	(pints)	(pints)
Coarse	1	1	$2\frac{1}{2}$–$3\frac{1}{2}$**
Medium	$1\frac{1}{4}$–$1\frac{1}{2}$	$1\frac{1}{2}$	$2\frac{1}{2}$–$3\frac{1}{2}$
Fine	$1\frac{1}{2}$	2	$2\frac{1}{2}$–$3\frac{1}{2}$

* Use $1\frac{1}{2}$ pints per acre on coarse and medium textured soils and 2 pints on fine soils with 2–5% organic matter; use 2 pints on all soils with 5–10% organic matter.

** Use Eptam 7E at a rate of $2\frac{1}{2}$ pints per acre to control annual grasses; $3\frac{1}{2}$ pints to control nutsedge and additional broadleaf weeds.

Precaution: Read the Eptam label before using. Observe all cautions and limitations of all products used in mixtures. The combination of TREFLAN and Eptam should not be used on soybeans, black-eyed peas (beans), lima beans and other flatpodded beans, except Romano. Do not use the foliage from a crop treated with the TREFLAN/Eptam tank mix for feed or for grazing.

DRY BEAN—Fall application for dry bean grown in Idaho, Oregon, and Washington:
Apply and incorporate TREFLAN any time between October 15 and December 31 at a broadcast rate of 1 pint per acre on coarse soils; $1\frac{1}{4}$ to $1\frac{1}{2}$ pints on medium soils; and $1\frac{1}{2}$ pints on fine soils. Destroy established weeds during seedbed preparation.

FOR THE FOLLOWING CROP GROUPING, USE THE RATE LISTED BELOW

BEANS—(Guar and Mungbean)

GREENS—Turnip greens grown for processing, Collard, Kale and Mustard greens.

MUSTARD—Grown for seed or processing for food in Minnesota, Montana and North Dakota.

Apply and incorporate TREFLAN before planting at 1 pint per acre on coarse soils and $1\frac{1}{2}$ pints on medium and fine soils.

BEANS—(Lima Bean and Snap Bean):
Apply and incorporate TREFLAN before planting at a broadcast rate of 1 pint per acre on coarse and medium soils and $1\frac{1}{2}$ pints on fine soils.

CORN (Field Corn) and GRAIN SORGHUM (Milo)

Apply TREFLAN to field corn or grain sorghum (8 inches or taller) as an over-the-top or directed spray to effectively control weeds listed for TREFLAN.

Soil Preparation—Cultivate before a TREFLAN application to insure loose, friable soil, to remove established weeds, and to cover the base of plants with soil.

Application Directions—TREFLAN should be applied and incorporated at the recommended rates for the soil texture when the crop is well established (8 inches or taller). TREFLAN may be applied either as an over-the-top spray or as a directed spray. Drop nozzles should be used if foliage prevents uniform coverage of soil surface. Soil incorporation may be accomplished with only one pass of a sweep-type cultivator or a properly adjusted rolling cultivator. The sweep-type cultivator should have 3 to 5 sweeps per row middle and be operated at 6 to 8 mph. Set the middle sweeps so as to avoid exposing untreated soil. Adjust the incorporation tools to prevent crop injury.

Broadcast Rates Per Acre

Soil Texture	TREFLAN E.C.
	(pints)
Coarse	$\frac{3}{4}$–1*
Medium	1–$1\frac{1}{2}$
Fine	$1\frac{1}{2}$–2

Use the lower rates when you anticipate light weed pressure and the higher rates when you anticipate heavy weed pressure.

* Corn only: Apply 1 to 1½ pints per acre in Alabama, Florida, Georgia, North Carolina, South Carolina and Virginia to control fall panicum and Texas panicum.

Precaution: Do not apply TREFLAN to corn grown for seed. Do not apply TREFLAN to corn or sorghum as a preplant or preemergence treatment, or crop injury may occur.

HOPS

Apply and incorporate TREFLAN while the crop is dormant. Use a broadcast rate of 1 pint per acre on coarse soils; 1¼ to 1½ pints on medium soils; and 1½ pints on fine soils and soils with 2–10% organic matter.

MINT—(Established Peppermint and Spearmint)

Apply TREFLAN at a rate of 1 pint per acre on coarse soils; 1¼ pints on medium soils; and 1½ pints on fine soils. Use in incorporation equipment that will insure thorough soil mixing with minimum damage to the crop.

PEA—(Dry and English)—TREFLAN ALONE

Apply and incorporate TREFLAN before planting at a rate of 1 pint per acre on coarse and medium soils and 1½ pints on fine soils.

PEA—TREFLAN/Far-Go tank mix for pea in Idaho, Oregon and Washington.
The tank mix combination of TREFLAN plus Far-Go will provide control of wild oat in addition to other annual grasses and broadleaf weeds controlled by TREFLAN.

Application Rates: Broadcast ¾ pint of TREFLAN per acre on coarse and medium soils; 1 pint of TREFLAN on fine soils. Use 1¼ quarts of Far-Go per acre for all soil textures.

Incorporation Directions: Apply and incorporate up to 3 weeks before planting. Follow recommended incorporation procedures for TREFLAN.

Precaution: Do not apply to lentils. Leaf crinkling and delayed maturity of peas may occur, particularly on clay points in the northwest; but this is usually more than offset by a reduction of wild oat. Do not use foliage from treated peas for feed or forage. Refer to the cautions, precautions and directions on the Far-Go label.

PEA—Fall application to dry pea and English pea in Idaho, Oregon and Washington:
Apply and incorporate TREFLAN any time between October 15 and December 31 at a broadcast rate of 1 pint per acre on coarse soils; 1¼ to 1½ pints on medium soils; and 1½ pints on fine soils. Destroy established weeds during seedbed preparation. Do not apply TREFLAN in the fall to soils which are wet or are subject to prolonged periods of flooding.

PEANUT—(Spanish Peanut in Texas and Oklahoma)

Apply and incorporate TREFLAN before planting, at planting or immediately after planting at a broadcast rate of 1 pint per acre on coarse soils. When incorporating after planting, take care not to disturb the seed.

PEANUT—TREFLAN/Vernam tank mix (Spanish Peanut in Texas and Oklahoma):
TREFLAN/Vernam tank mix effectively controls those weeds listed for TREFLAN alone plus the additional weeds:

Purple nutsedge (nutgrass)	Coffeeweed
Yellow nutsedge (nutgrass)	Velvetleaf
Annual morningglory	

Follow recommended soil preparation procedures for TREFLAN. You may apply the TREFLAN/Vernam tank mix up to 10 days prior to planting. Incorporate the tank mix immediately after application. Apply TREFLAN/Vernam at these rates:

Broadcast Rates Per Acre

Soil Texture	TREFLAN E.C.	Vernam 7E
	(pints)	(pints)
Coarse	1	2⅓

POTATO—(All states except Maine)

Apply and incorporate TREFLAN after planting, before emergence, or immediately following dragoff or after the potato plants have fully emerged.

Broadcast Rates Per Acre

Soil Texture	TREFLAN E.C. Areas receiving less than 20″ average annual rainfall*	TREFLAN E.C. Areas receiving greater than 20″ average annual rainfall*
	(pints)	(pints)
Coarse	1	1
Medium	1¼–1½	1½
Fine	1½	2

* Use 1½ pints per acre on coarse and medium soils with 2–5% organic matter; use 2 pints on all soils with 5–10% organic matter.

Set incorporation equipment so that the bed and furrow will be uniformly covered with a layer of treated soil. If the layer of treated soils is not uniform and the herbicide is concentrated over the bed, potato emergence may be retarded, and stem brittleness can occur. When applying and incorporating TREFLAN after potato plants have fully emerged, do not completely cover the foliage with treated soil. Likewise, do not completely cover foliage at subsequent cultivations. Be careful that incorporation machinery does not damage potato seed pieces or elongating sprouts.

POTATO—Split application in Idaho, Oregon and Washington:
On all soils, apply and incorporate ¾ pint of TREFLAN per acre before planting and ¾ pint after planting when potato plants have fully emerged. Do not apply to soils containing 2% or more organic matter. Follow incorporation directions listed above for application to potato after planting.

POTATO—TREFLAN/Eptam tank mix for potatoes grown in Kansas, Minnesota, Nebraska, North Dakota, Oklahoma, South Dakota and Texas:
The TREFLAN/Eptam tank mix effectively controls the following weeds in addition the those weeds controlled by TREFLAN.

Henbit (Spring applications)	Oat, wild
Nightshade, black	Ragweed, common
Nightshade, hairy	Smartweed, Pennsylvania
Nutsedge	Velvetleaf (Buttonweed)

Follow recommended soil preparation and application procedures for TREFLAN. The TREFLAN/Eptam tank mix may be applied after planting, but prior to crop emergence. In areas where potatoes are normally dragged off, the TREFLAN/Eptam tank mix should be applied and incorporated up to or immediately following drag off.

Broadcast Rates Per Acre

Soil Texture	TREFLAN E.C. Areas receiving less than 20″ average annual rainfall*	TREFLAN E.C. Areas receiving greater than 20″ average annual rainfall	Eptam 7E
	(pints)	(pints)	(pints)
Coarse	1	1	1¾–7**
Medium	1–1½	1–1½	1¾–7
Fine	1–1½	1–2	1¾–7

* Use 1½ pints per acre on coarse and medium soils with 2–5% organic matter; use 2 pints on all soils with 5–10% organic matter.
** Use the higher rate of Eptam 7E for nutsedge control.

Precaution: Read the Eptam label before using. Observe cautions and limitations of products used in mixtures. Do not graze or feed forage to livestock from fields treated with TREFLAN/Eptam tank mix.

POTATO—TREFLAN/Eptam application before planting in Washington, Idaho and Oregon:
TREFLAN/Eptam may be applied before planting at a broadcast rate of ¾ pint of TREFLAN per acre and 3½ pints of Eptam 7E per acre on all soil textures. Incorporate immediately.

Precaution: Do not use this tank mix both before and after planting in the same season. Read the Eptam label before using. Observe all cautions and limitations on labeling of all products used in mixtures. Do not use foliage from treated crops for feed or forage.

SAFFLOWER

Follow recommended soil preparation, application and incorporation procedures for TREFLAN.

Apply and incorporate TREFLAN in the spring before planting or in the fall between October15 and December 31.

Broadcast Rates Per Acre

Soil Texture	TREFLAN E.C. Areas receiving less than 20" average annual rainfall* (pints)	TREFLAN E.C. Areas receiving greater than 20" average annual rainfall* (pints)
Coarse	1	1
Medium	1¼–1½	1½
Fine	1½	2

* Use 1½ pints per acre on coarse and medium textured soils and 2 pints on fine soils with 2–5% organic matter; use 2 to 2½ pints on all soils with 5–10% organic matter.

SAFFLOWER—Fall application in Arizona, California, Idaho, Montana, Nevada, Oregon, Utah, Washington and Wyoming:
Apply and incorporate TREFLAN any time between October 15 and December 31. Ground may be left flat or bedded-up over winter. On bedded ground, knock beds down to desired height before planting, moving some treated soil from tops into furrows. Where soil is left flat over winter, take care during spring bedding operations to prevent turning up untreated soil. Destroy established weeds during seedbed preparation. If weeds become established in furrows due to uncovering of untreated soil during listing, destroy these weeds before planting. Apply and incorporate TREFLAN at a broadcast rate of 1½ pints per acre on coarse soils; 2 pints on medium soils; and 2½ pints on fine soils.

Do not apply TREFLAN in the fall to soils which are wet or are subject to prolonged periods of flooding.

SUGAR BEET

Apply TREFLAN as a broadcast, overtop spray when plants are between 2 and 6 inches tall at a rate of 1 pint per acre on coarse soils and 1¼ to 1½ pints on medium and fine soils. Use the higher rate for medium and fine soils in areas receiving more than 20" average annual rainfall. Set incorporation machinery to throw treated soil toward the plants in the row. Be careful that incorporation machinery does not damage the sugar beet taproot.

Precaution: Exposed beet roots should be covered with soil before a TREFLAN application to reduce the possibility of girdling.

SUGAR BEET—Incorporation with a tine-tooth harrow in California, Colorado, Idaho, Montana, Nebraska, Oregon, Texas, Utah, Washington and Wyoming:
A properly operated tine-tooth harrow (Flextine or Melroe) can incorporate TREFLAN for effective weed control in sugar beet. Operate the tine-tooth harrow 2 times over the field in opposing directions at a speed of 3 to 6 mph. Set the harrow to cut 1 or 2 inches deep. Be careful that the tine-tooth harrow does not damage the sugar beet taproot. Follow recommended application procedures and broadcast rates per acre for sugar beet (see preceding paragraph).

SUGARCANE—(Plant Cane)

Apply and incorporate TREFLAN twice a year at a broadcast rate of 2 to 4 pints per acre for all soil textures. Make the first TREFLAN application in the fall on firmly packed beds immediately after the seed pieces are planted. Make the second TREFLAN application in the spring before or shortly after the cane emerges. Loosen rain-packed beds 2 to 3 inches deep before the spring application. Take care that incorporation machinery does not damage the seed pieces or emerging shoots.

SUGARCANE—Postplant in Hawaii for control of most annual grasses, including guineagrass:
Surface apply TREFLAN after planting (for plant cane) or after harvesting (for ratoon cane), before weeds and cane emerge. Use a broadcast rate of 6 to 8 pints per acre for all soil textures. In plant cane, the beds should be formed or rolled before application. In ratoon cane, the crop residue should be removed before application. If large amounts of crop residues are present, TREFLAN will not be effective. Apply TREFLAN just before anticipated rainfall or sprinkle irrigate immediately after application.

SUGARCANE—Applications up to layby for plant cane or ratoon cane grown in Lousiana or Texas:
Apply and incorporate TREFLAN at a broadcast rate of 2 to 4 pints per acre for all soil textrues. Do this in the spring from before or shortly after the cane emerges up to layby. Apply the TREFLAN after the beds have been shaved or false shaved. Loosen rain-packed beds 2 to 3 inches deep before application. Be careful that incorporation machinery does not damage seed pieces or emerging shoots. You may use a rolling cultivator or bed chopper to incorporate TREFLAN layby applications in sugarcane on all soil textures. Follow normal incorporation directions for the rolling cultivator. Set bed chopper to cut 3 to 4 inches deep and operate at 4 to 6 mph. Two incorporation passes are necessary.

SUGARCANE—Itchgrass (Raoulgrass) control in Louisiana:
Apply and incorporate TREFLAN on either plant or ratoon cane at a broadcast rate of 4 pints per acre for all soil textures. Follow the directions above for sugarcane layby application in Louisana and Texas.

SUNFLOWER—TREFLAN ALONE

Apply and incorporate TREFLAN in the spring or in the fall between October 15 and December 31. Follow recommended soil preparation, application and incorporation procedures for TREFLAN.

Broadcast Rates Per Acre

Soil Texture	TREFLAN E.C. Areas receiving less than 20" average annual rainfall* (pints)	TREFLAN E.C. Areas receiving greater than 20" average annual rainfall* (pints)
Coarse	1	1
Medium	1¼–1½	1½
Fine	1½	2

* Use 1½ to 2 pints per acre on coarse and medium soils with 2–5% organic matter and 2 pints on all soils with 5–10% organic matter.

SUNFLOWER—TREFLAN/Amiben tank mix or overlay:
Follow recommended soil preparation, application and incorporation procedures for TREFLAN.

Amiben may be applied in a band or broadcast over sunflowers at planting in fields where TREFLAN has been incorporated prior to planting. Or, the TREFLAN/Amiben tank mix may be incorporated prior to planting. TREFLAN/Amiben tank mix improves mustard, smartweed, velvetleaf and ragweed control in addition to those weeds controlled by TREFLAN alone.

Apply TREFLAN/Amiben or TREFLAN with an Amiben overlay at the following broadcast rates per acre:

Soil Texture	TREFLAN E.C. (pints)	Amiben 2S (quarts)
Coarse	1	4
Medium	1½	4–6*
Fine	2	4–6*

* For best control of mustard, common ragweed or black nightshade, use the 6 quart rate.

In coarse textured soils, heavy rains on the incorporated Amiben may move it below the weed seed germination zone, and erratic weed control may result. If sufficient rain does not fall within 7 days after a preemergence application of Amiben, but there is enough soil moisture to germinate weeds and grasses, a light cultivation with a rotary hoe or similar tool will uproot these small broadleaf weeds and grasses. The shallow mixing of Amiben in the surface soil will not interfere with the action of Amiben when rains come.

TOMATO

For direct-seeded tomato, apply TREFLAN at blocking or thinning as a directed spray to the soil between rows and beneath the plants, and incorporate. For transplant tomato, apply and incorporate TREFLAN before transplanting. Do not apply TREFLAN after transplanting.

Broadcast Rates Per Acre

Soil Texture	TREFLAN E.C. Areas receiving less than 20″ average annual rainfall*	TREFLAN E.C. Areas receiving greater than 20″ average annual rainfall*
	(pints)	(pints)
Coarse	1	1
Medium	1¼–1½	1½
Fine	1½	2

* Use 1½ pints per acre on coarse and medium textured soils and 2 pints on fine soils with 2–5% organic matter; use 2 pints on all soils with 5–10% organic matter.

FRUIT AND NUT CROPS AND VINEYARDS

For areas receiving more than 20″ average annual rainfall.
For new plantings of vineyards, citrus and pecan trees, apply and incorporate TREFLAN before planting at a broadcast rate of 1 pint per acre on coarse soils; 1½ pints on medium soils; 2 pints on fine soils; 1½ pints on fine soils with 2–5% organic matter; and 2 pints on soils with 5–10% organic matter.

For non-bearing, established plantings of citrus and pecan trees and bearing plantings of grapefruit, lemon, orange, pecan, tangelo and tangerine trees, apply TREFLAN at a broadcast rate of 2 to 4 pints per acre for all soil textures.

For areas receiving less than 20″ average annual rainfall.
For new plantings of almond, apricot, citrus, nectarine, peach, pecan and walnut trees, apply and incorporate TREFLAN before planting at a broadcast rate of 1 pint per acre on coarse soils; 1¼ to 1½ pints on medium soils; 1½ pints on fine soils; 1½ to 2 pints on soils with 2–5% organic matter; and 2 pints on soils with 5–10% organic matter.

For new plantings of vineyards, apply and incorporate TREFLAN before planting at a broadcast rate of 1 to 1½ pints per acre on coarse soils; 1½ to 3 pints on medium soils; and 3 to 4 pints on fine soils or soils with 2–10% organic matter. Do not use more than 2 pints per acre on heat-treated grape rootings.

For postplant applications on bearing or non-bearing, established plantings of vineyards and almond, apricot, grapefruit, lemon, nectarine, orange, peach, plum, prune, tangelo, tangerine and walnut trees, apply TREFLAN at a broadcast rate of 2 to 4 pints per acre for all soil textures. Do not apply to vineyards within 60 days of harvest.

In established plantings, apply TREFLAN as a directed spray to the soil and use incorporation methods not injurious to the trees or vines.

FRUIT AND NUT CROPS AND VINEYARDS—Rhizome johnsongrass control:

For areas receiving less than 20″ average annual rainfall.
You can obtain commercially acceptable control of rhizome johnsongrass with postplant applications in bearing and non-bearing, established plantings of vineyards and almond, apricot, grapefruit, lemon, nectarine, orange, peach, pecan, tangelo, tangerine and walnut trees with a TREFLAN program when applied for 2 years in a row.

Soil Preparation—Work the soil thoroughly to bring the rhizomes nearer the surface.

Application—Apply TREFLAN at a broadcast rate of 2 quarts per acre on all soil textures each year for 2 years in a row. Do not apply to vineyards within 60 days of harvest.

Incorporation—Incorporate TREFLAN thoroughly with a disc set to cut 4 to 6 inches deep and operate at 4 to 6 mph. Two incorporation passes are necessary, with the second pass in a different direction from the first.

Cultivation—Some johnsongrass plants will escape. Timely cultivations are necessary to obtain commercially acceptable control. You cannot obtain commercially acceptable control with only 1 year of TREFLAN use.

Precautions: Do not use the 2 quart rate on new plantings, or crop injury may result. Do not interplant orchards or vineyards with other crops. If the treated vineyards and orchards are diverted to other crop uses, plant only those crops for which TREFLAN has been registered as a preplant treatment for the next cropping season.

FRUIT AND NUT CROPS AND VINEYARDS—Bindweed control in California:
TREFLAN can be used for the control of field bindweed in vineyards and for almond, apricot, grapefruit, lemon, nectarine, orange, peach, pecan, tangelo, tangerine and walnut trees.

Apply TREFLAN at a broadcast rate of 4 pints per acre on all soil textures. TREFLAN must be applied in the spring with a specially designed spray blade which applies a thin, concentrated layer at a soil depth of 4 to 6 inches. This layer of TREFLAN prevents bindweed shoots from emerging.

Land Preparation—Destroy all weeds and grasses with soil tillage before applying TREFLAN. This tillage is necessary to prevent trash from interfering with the operation of the spray blade.

Equipment—This operation requires a spray blade capable of running 4 to 6 inches below the surface of the soil. The spray blade should be equipped with nozzles located under the blade and directed so that the TREFLAN spray will be trapped under the soil which is flowing over the blade as it is pulled through the soil. Use a sufficient number of nozzles with spacing that will uniformly apply the TREFLAN underground in a thin, horizontal layer.

Application—Apply TREFLAN in 40 to 80 gallons of water per acre. Operate the spray blade at a depth of 4 to 6 inches.

Precaution: Some soils develop cracks as they dry after rainfall or irrigation. Field bindweed may emerge if the cracks extend through the TREFLAN layer. Prevent or eliminate cracks by shallow discing or other tillage. Avoid deep tillage which disturbs the subsurface layer. Cultivation or tillage also aids the control of germinating seeds.

WHEAT (WINTER)—(Idaho, Montana, Oregon and Washington)

TREFLAN may be applied for preplant preemergence control of cheatgrass and other annual grasses and broadleaf weeds controlled by TREFLAN. The growth, development and yield of winter wheat will not be adversely affected, provided the seed is placed below the zone of soil treated with TREFLAN.

Apply TREFLAN any time during a period from 3 weeks up to immediately prior to planting. Broadcast TREFLAN at a rate of 1½ pints per acre on coarse and medium soils and 2 pints on fine soils.

Incorporation Directions—Incorporate TREFLAN into the soil with a flexible tine-tooth harrow (Flextine, Melroe) set to cut 1 to 2 inches deep and operate at 3 to 6 mph. Incorporate one time within 24 hours after application, followed by a second incorporation in a different direction from the first prior to planting. Do not till the soil with a disc after the TREFLAN has been incorporated with a flexible tine harrow.

Seeding Directions—Use only a deep furrow or semi-deep furrow drill that will place the seed below the zone of soil into which TREFLAN has been incorporated.

Precaution: Wheat planted in direct contact with treated soil may suffer crop injury in the form of delayed emergence and development.

WHEAT (WINTER)—Fallow soil application in Washington and Oregon:

TREFLAN applied and shallowly incorporated into fallow soil up to four months ahead of planting will control cheatgrass and certain annual grasses and broadleaf weeds. The growth, development or yield will not be adversely affected as long as the seed is placed below the zone of soil treated with TREFLAN. Use deep or semi-deep furrow drills.

Broadcast rates are $1\frac{1}{2}$ pints per acre on coarse and medium soils and 2 pints on the fine soils. Apply TREFLAN any time from May to September prior to the fall planting of winter wheat.

Incorporation—Incorporate TREFLAN with a flexible tine-tooth harrow (Flextine or Melroe) set to cut to 1 to 2 inches deep and operated at 3 to 6 mph. For thorough incorporation, two passes of the equipment in different directions over the field are necessary. Incorporate one time within 24 hours after application followed by a second incorporation prior to seeding. Do not till the soil with a disc after TREFLAN has been applied with a flexible tine harrow.

Precaution: Use only deep furrow or semi-deep furrow drills. Place seed below the zone of soil into which TREFLAN has been incorporated. Do not plant wheat directly into the zone of soil treated with TREFLAN; injury to the crop or a delay in its emergence and development may occur.

WHEAT (SPRING), DURUM AND BARLEY—TREFLAN ALONE

TREFLAN is recommended as a postplant incorporated treatment to control foxtail (pigeongrass).

Apply TREFLAN at a broadcast rate of 1 pint per acre on coarse and medium soils and $1\frac{1}{2}$ pints on fine soils.

Plant 2 to 3 inches deep in a well-tilled seedbed. Apply TREFLAN after seeding but before the crop emerges. To incorporate, use flex-tine or diamond harrows operated two times in different directions, at speeds of at least 5 mph. Incorporate by operating equipment 1 to $1\frac{1}{2}$ inches deep. Application and the first incorporation should be done in the same operation if possible. Both incorporations must be done within 24 hours.

WHEAT (SPRING), DURUM AND BARLEY—TREFLAN/ Far-Go tank mix:

TREFLAN/Far-Go applied as a postplant incorporated treatment will control foxtail (pigeongrass) and wild oat.

Plant 2 to 3 inches deep in a well-tilled seedbed. Apply TREFLAN/ Far-Go after seeding but before crop emerges. To incorporate, use flex-tine or diamond harrows two times, operated in different directions, at speeds of at least 5 mph. Incorporate by operating equipment 1 to $1\frac{1}{2}$ inches deep. Application and the first incorporation should be done in the same operation if possible. If not, incorporate immediately after application.

Broadcast Rates Per Acre

	TREFLAN E.C.	Far-Go	
Soil Texture	Barley Durum Spring Wheat	Durum Spring Wheat	Barley
	(pints)	(pints)	(pints)
Coarse	1	2	$2\frac{1}{2}$
Medium	1	2	$2\frac{1}{2}$
Fine	$1\frac{1}{2}$	2	$2\frac{1}{2}$

Precaution: Overapplication may result in crop injury. Read the Far-Go label carefully before using.

FALL APPLICATION

General: (Areas receiving more than 20" average annual rainfall)

See specific crop for recommendations. For all crops for which there are no specific fall application instructions and for which TREFLAN is recommended as a preemergence application, use the rates listed for spring applications. Do not apply TREFLAN in the fall for sugarbeets, potatoes and direct-seeded tomatoes.

Apply and incorporate TREFLAN any time between October 15 and December 31. Ground may be left flat or bedded-up over winter. On bedded ground, knock beds down to desired height before planting, moving some treated soil from beds into furrows. Where soil is left over winter, be careful not to turn up untreated soil during spring bedding operations. Destroy established weeds during seedbed preparation. If weeds become established in furrows due to uncovering of untreated soil during bedding, destroy these weeds before planting. Do not apply TREFLAN in the fall to soils which are wet, are subject to prolonged periods of flooding, or where rice was grown the previous year.

FERTILIZER USE DIRECTIONS APPLICATION WITH LIQUID FERTILIZERS

TREFLAN may be mixed with most liquid fertilizer materials. The combination of TREFLAN with solutions and suspension-type fertilizers has provided weed and grass control equal to the same rates of TREFLAN applied in water. Follow TREFLAN label recommendations regarding rates per acre, crops, incorporation directions, special instructions, cautions and special precautions. Individual state regulations relating to liquid fertilizer mixing, registration, labeling and applications are the responsibility of the individual and/or company selling the fertilizer and chemical mixture.

Testing for Tank Mix Compatibility in Liquid Fertilizers: TREFLAN alone or in tank mixture with dry flowables, wettable powders (WP), aqueous suspensions (AS), flowables, (F), liquids (L), or solutions (S) may not combine properly with some fluid fertilizer materials. Small quantities should always be tested before full-scale mixing. This will determine whether a compatibility agent is needed, and which agent does the best job. The seven agents listed below have been thoroughly tested. There are many other surfactants on the market which were not designed for use with liquid fertilizers. Use the following test to select the correct agent for your mixture.

1. Put 1 pint of the liquid fertilizer in a quart jar.
2. Add 1 to 4 teaspoonful(s) of the dry flowable, WP, AS, F or L formulation (depending on the recommended rate per acre) to the liquid fertilizer. Close jar and agitate until the materials are dispersed evenly in the fertilizer. If the materials do not disperse well, it may be necessary to slurry the chemicals in water before adding to the fertilizer.
3. After dispersing the materials (Step 2), add 3 to 4 teaspoonsful of TREFLAN to the jar and shake well. Add solution herbicides to the mixture last and agitate. Observe the jar for about 10 minutes. If the materials rise to the surface and form a thick layer (oily curds) which will not redisperse when agitated, a compatibility agent is needed. If the mixture is easily redispersed to its original state with slight agitation, no agent is needed but good agitation must be provided in the fertilizer spray tank.
4. If the need for a compatibility agent is shown in Step 3: Using a clean quart jar, start at Step 1 above, add $\frac{1}{2}$ teaspoon of the compatibility agent to the liquid fertilizer, mix well, then repeat Steps 2 and 3.

An effective compatibility agent will cause the mixture to remain uniformly mixed with little or no separating or oil rising to the surface for one half hour or longer. If slight separation does occur, 2 or 3 inversions of the jar should give a uniform remix. If oily curds form which will not redisperse, more agent or another agent should be tried.

Use a clean jar for each test. The compatible mixture will have a uniform appearance and will be relatively easy to keep mixed with gentle agitation of the jar.

LIQUID FERTILIZER MIXING INSTRUCTIONS TREFLAN E.C. in Liquid Fertilizer

General—Emulsifiable concentrates, such as TREFLAN E.C., can be mixed with liquid fertilizers. In all cases, continuous agitation is required to prevent the TREFLAN from rising to the surface as on oily layer. When necessary, (see Testing for Tank Mix Compatibility in Liquid Fertilizers, above) a compatibility agent can be used to cause the TREFLAN to emulsify properly (i.e., have a milky appearance rather than oily layer). The use of compatibility agents is especially important when tank mixing emulsifiable concentrates (E.C.) with dry flowables, wettable powders (WP), aqueous suspensions (AS), flowables (F), liquids (L), or solutions (S) in liquid fertilizer. If the emulsion is not properly formed and the TREFLAN rises to the surface of the fertilizer as an oil ("oils out"), the oil may combine with the wettable powder, flowable, or suspension to form oily curds (viscous phase) which is difficult to redisperse. Any one of the compatibility agents listed below is helpful in causing liquid concentrates to form non-oiling mixtures with liquid fertilizers. These compatibility agents can be used at rates as low as 1½ to 2 pints per ton of liquid fertilizer and should be mixed well with the fertilizer before adding the liquid concentrate. Read the label on the compatibility agent and follow the directions.

1. Sponto 168D (Witco Chemicals Co., Chicago, IL)
2. Compat (Farm Chemicals, Inc., Aberdeen, NC)
3. Unite (Hopkins Ag Chemical, Madison, WI)
4. T-Mulz 734-2 (Thompson-Hayward Chemical Co.)
5. Rigo Compatibility Agent (Rigo Company, Buckner, KY)
6. Amoco Spray Mate+up,TM (Amoco Oil Co., Chicago, IL)
7. Kem-Link (Universal Coop, Minneapolis, MN)

All of the above are phosphate, ester-type surfactants designed to be used with liquid fertilizers. They usually do not work as compatibility agents in tank mixtures in plain water.

APPLICATION
Spread the fertilizer/pesticide mixture with a properly calibrated applicator. Be certain the material is applied uniformly to the soil surface.

INCORPORATION
Follow normal TREFLAN incorporation procedures.

TREFLAN APPLICATION WITH DRY BULK FERTILIZERS

General—Dry bulk fertilizers may be impregnated or coated with TREFLAN. Application of dry bulk fertilizers impregnated with TREFLAN has provided weed and grass control equal to the same rates of TREFLAN applied in water.

All TREFLAN label recommendations regarding rates per acre, approved crops, incorporation, special instructions, cautions and special precautions must be followed.

Apply aminimum of 200 pounds per acre of dry fertilizer impregnated with TREFLAN at the recommended rates. Any commonly used dry fertilizers can be used for TREFLAN impregnation except coated ammonium nitrate and straight limestone. These materials will not absorb the herbicide. Blends containing mixtures of these materials can be impregnated.

Impregnation—Use any closed drum, belt, ribbon or other commonly used dry bulk fertilizer blender. Apply TREFLAN uniformly to the fertilizer.

Rates—Check specific crop recommendations for the rate of TREFLAN per acre. See the rate table which follows to determine the amount of TREFLAN to be impregnated into a ton of dry bulk fertilizer based on the amount of fertilizer which will be applied per acre. (See rate chart).

Application—Spread the fertilizer/chemical mixture with a properly calibrated applicator. Be certain the material is applied uniformly to the soil surface.

Incorporation—Follow TREFLAN incorporation procedures.

RATE CHART FOR IMPREGNATING FERTILIZER WITH TREFLAN E.C. (TREFLAN Added to a TON of Fertilizer)

Fertilizer Rate Per Acre	TREFLAN Rate Per Acre				
	1 pint	1½ pints	2 pints	3 pints	4 pints
200 pounds	5 qts. per ton	7½ qts. per ton	10 qts. per ton	15 qts. per ton	20 qts. per ton
250 pounds	4 qts. per ton	6 qts. per ton	8 qts. per ton	12 qts. per ton	16 qts. per ton
300 pounds	3⅓ qts. per ton	5 qts. per ton	6⅔ qts. per ton	10 qts. per ton	13⅓ qts. per ton
350 pounds	2¾ qts. per ton	4¼ qts. per ton	5¾ qts. per ton	8½ qts. per ton	11½ qts. per ton
400 pounds	2½ qts. per ton	3¾ qts. per ton	5 qts. per ton	7½ qts. per ton	10 qts. per ton
450 pounds	2¼ qts. per ton	3⅓ qts. per ton	4½ qts. per ton	6⅔ qts. per ton	9 qts. per ton

For rates other than those listed above, use the following formula to calculate the amount of TREFLAN to be impregnated on a ton of dry bulk fertilizer:

$$\text{Pints TREFLAN Per Acre} \times \frac{1000}{\text{Lbs. Fertilizer Per Acre}} = \text{Quarts TREFLAN Per Ton of Fertilizer}$$

All individual state regulations relating to dry bulk fertilizer blending, registration, labeling and application are the responsibility of the individual and/or company selling the fertilizer and chemical mixture.

TREFLAN PERFORMANCE GUARANTEE—A LIMITED WARRANTY OF TREFLAN WEED AND GRASS CONTROL

Elanco Products Company warrants that TREFLAN will control the weeds and grasses as indicated on its current label when used according to label directions and subject to the following limitations:

1. Purchaser must notify Elanco promptly if a lack of satisfactory control occurs. Notice must be given within ninety days after the first planting or after application of TREFLAN to the treated crop, whichever occurs later.
2. An Elanco representative must be satisfied that Purchaser used TREFLAN according to the label directions and must visually observe unsatisfactory control of weeds in the field.
3. The Purchaser must provide Elanco representative with an invoice or other satisfactory record which shows the price and quantity of TREFLAN purchased.
4. Elanco's liability will be limited to a refund of the purchase price of the TREFLAN applied to the acreage on which weed control was not satisfactory or, at the option of the Purchaser, a refund of an equivalent amount of TREFLAN.
5. Void outside the USA.

DISCLAIMER OF WARRANTIES
The warranty printed above is the only warranty applicable to this product. All other warranties, expressed or implied, including, but not limited to, any implied warranties of merchantability and fitness for a particular purpose, are disclaimed.

LIMITATION OF DAMAGES
Elanco's liability, whether in contract, warranty, tort, negligence, strict liability or otherwise, shall not exceed the return of the amount of the purchase price of the TREFLAN and under no circumstances shall Elanco be liable for special, indirect or consequential damages.

INHERENT RISKS OF USE
Failure to carefully follow the directions for use of TREFLAN, as well as other factors such as seedling disease, cold weather, deep planting, excessive moisture, high salt concentration or drought, may result in unsatisfactory weed control or crop injury. Reduced yields may result under these conditions.

AMIBEN®—chloramben, Union Carbide Agricultrual Products Co., Inc.
CAPAROL®—prometryn, Ciba-Geigy Corporation.
COTORAN®—fluometuron, Ciba-Geigy Corporation.
EPTAM®—EPTC, Stauffer Chemical Company.
FAR-GO®—triallate, Monsanto Agricultural Products Company.
KARMEX®—diuron, E. I. duPont de Nemours and Company.
LEXONE®—metribuzin, E. I. duPont de Nemours and Company.
SENCOR®—metribuzin, Bayer, GmbH.
VERNAM®—vernolate, Stauffer Chemical Company.

SUPPLEMENTAL LABELING

TREFLAN® E.C. (EPA Reg. No. 1471-35)
TREFLAN® M.T.F.™ (EPA Reg. No. 1471-116)
TREFLAN® 5 (EPA Reg. No. 1471-120)

TREFLAN FALL APPLICATION FOR WEED CONTROL IN FLAX

TREFLAN may be applied and incorporated in the fall for weed control in spring seeded flax. For TREFLAN E.C., M.T.F., and 5, ground cover from existing weeds or previous crop should be at a manageable level so that there is no interference with incorporation.

Broadcast Rates Per Acre

Soil Texture	TREFLAN E.C./M.T.F.	TREFLAN 5
	(pints)	(pints)
Coarse	1	0.8
Medium	1½	1.2
Fine	2	1.6

APPLICATION DIRECTIONS
Follow recommended application directions for TREFLAN.

INCORPORATION DIRECTIONS
Incorporate TREFLAN one time within 24 hours after application. The second incorporation should be performed in the spring prior to seeding flax. The incorporation operations should result in a thorough mixing of TREFLAN with the soil. Otherwise, erratic weed control may result.

INCORPORATION EQUIPMENT
Use machinery that mixes TREFLAN thoroughly with the soil. Recommended equipment includes:

Disc—set to cut 3-4 inches deep and operated in 2 different directions at 4 to 6 mph. A tandem or double-disc operated one time does not provide adequate incorporation.

Field cultivator—set to cut 3-4 inches deep and operated at 5 mph or more. The field cultivator used alone or in combination with the double-disc will provide effective incorporation providing the following instructions are used:

1. Two passes over the field with a field cultivator with second pass running at an angle to the first. Do not set cultivator to cut deeper than 4 inches. Be sure a depth of not greater than 4 inches is maintained on the second pass since untreated soil may be turned up.
2. Field cultivator used for the first pass and the double-disc used for the second pass.
3. Double-disc used for the first pass and the field cultivator used for the second pass.

NOTE: A field cultivator is defined as an implement with 3 to 4 rows of sweeps spaced at intervals of 7 inches or less and staggered so that no soil is left unturned. Chisel points should not be used.

Mulch treader and other similar disc-type implements—set to cut 3-4 inches deep and operated at 5 to 8 mph in two different directions.

SPECIAL INSTRUCTIONS FOR FLAX

1. Incorporation operations or any other tillages performed in the spring prior to seeding should be relatively shallow so as to maintain a firm seedbed, and the seedbed should be packed just prior to seeding.
2. Seeding should be done with a press drill or hoe drill. Seed into a moist seedbed no more than 1.5 inches deep.
3. Flax should not be seeded until the seedbed has warmed up.
4. Refer to the special precautions statement on the TREFLAN label regarding stresses that can lead to crop injury or yield reduction.

NOTE: Do not delay the first incorporation more than 24 hours after application.

CAUTION
Refer to the TREFLAN label for additional warnings, cautions, precautions, and special precautions before use.

TREFLAN® E.C., M.T.F.™, 5™—trifluralin, Elanco Products Company

SUPPLEMENTAL LABELING

TREFLAN® M.T.F.™ (EPA Reg. No. 1471-116)
TREFLAN® 5 (EPA Reg. No. 1471-120)

TREFLAN PREPLANT INCORPORATED FOR FOXTAIL (PIGEONGRASS) CONTROL IN SPRING WHEAT, DURUM, AND BARLEY

TREFLAN may be fall applied for foxtail/pigeongrass (Setaria spp.) control in spring wheat, durum, and barley planted the following spring. TREFLAN M.T.F. and 5 may be applied to ground that has a manageable trash level, been fallow, or pre-tilled. The first incorporation is required within 24 hours after application. A second incorporation is required prior to planting to destroy emerged weeds and insure an even distribution of TREFLAN treated soil.

Broadcast Rates Per Acre

Soil Texture	TREFLAN M.T.F.	TREFLAN 5
	(pints/A)	(pints/A)
Coarse/Medium	1	0.8
Fine	1.5	1.2

APPLICATION DIRECTIONS
Mix and/or apply TREFLAN according to labelled directions.

INCORPORATION DIRECTIONS
Any of the following tools are recommended for fall incorporation. The disc or field cultivator may be used for the spring incorporation pass. Care should be taken to operate the tool at a more shallow depth than the fall incorporation.

1. Chisel plow: *May be used for first pass only.* Operate at 4-5 inches deep at 4-6 mph. A chisel plow is defined as having 3 rows of up to 18-inch sweeps on no greater than 12-inch centers. Stagger sweeps so that no soil is left unturned.
2. Tandem disc: Operate at 3-4 inches deep at 4-6 mph.
3. Field cultivator: Operate at 3-4 inches deep at 5 mph or more. A field cultivator is defined as having 3 to 4 rows of sweeps with "c" or "s" shaped shanks, spaced 7 inches or less and staggered so that no soil is left unturned.

PLANTING DIRECTIONS
Wheat, durum, or barley should be seeded approximately 2 inches deep.

PRECAUTIONS
While use of this practice may result in a stand reduction, slight stand reductions do not normally affect yield.
Check the TREFLAN label for additional directions, cautions, and precautions prior to use.

TREFLAN® M.T.F.™ and 5™—trifluralin, Elanco Products Company

SUPPLEMENTAL LABELING

TREFLAN® E.C. (EPA Reg. No. 1471-35)
TREFLAN® M.T.F.™ (EPA Reg. No. 1471-116)
TREFLAN® 5 (EPA Reg. No. 1471-120)

TREFLAN FOR WEED CONTROL IN RAPESEED

TREFLAN as a broadcast application will control certain annual grasses and broadleaf weeds in rapeseed.

DIRECTIONS FOR USE
For use in all states except Alaska: Follow recommended procedures for soil preparation and application according to the TREFLAN label. TREFLAN may be applied in the fall or early spring prior to seeding. Set incorporation equipment to incorporate TREFLAN to a depth of 3 to 4 inches with equipment specified in the TREFLAN label.

Broadcast Rates Per Acre

Soil Texture	TREFLAN E.C. or M.T.F.	TREFLAN 5
	(pints)	(pints)
Coarse	1	0.8
Medium	1½	1.2
Fine	2	1.6

Read and follow all directions for use, cautions and precautions on the TREFLAN label.

TREFLAN®, M.T.F.™, 5™ —trifluralin, Elanco Products Company

SUPPLEMENTAL LABELING

TREFLAN® E.C. (EPA Reg. No. 1471-35)
TREFLAN® M.T.F,™ (EPA Reg. No. 1471-116)
TREFLAN® 5 (EPA Reg. No. 1471-120)

TREFLAN/DUAL® TANK MIX PREPLANT INCORPORATED FOR WEED CONTROL IN SOYBEANS AND TREFLAN PREPLANT INCORPORATED WITH DUAL SURFACE APPLIED

TREFLAN/Dual effectively controls those weeds listed on the label for TREFLAN plus these additional weeds:

Black nightshade
Prairie cupgrass
Red rice
Southwestern cupgrass
Yellow nutsedge

Additional weeds partially controlled by this combination include:
Hairy nightshade
Volunteer sorghum

Follow recommended soil preparation and incorporation procedures for TREFLAN (see the TREFLAN label).

Apply TREFLAN/Dual in 10-40 gallons of water per acre with conventional low pressure sprayers and in a minimum of 5 gallons of water per acre by air. TREFLAN/Dual tank mix can be applied up to 14 days prior to planting. TREFLAN can also be applied preplant incorporated and followed by Dual at planting or after planting as a banded or broadcast application.

Broadcast Rates Per Acre

Soil Texture	TREFLAN E.C./M.T.F.™	TREFLAN 5	Dual 8E Less than 3% O.M.	Dual 8E 3% O.M. or Greater
	(pints)	(pints)	(pints)	(pints)
Coarse	1	0.8	1½–2	2
Medium	1½	1.2	2–2½	2–2½
Fine	2	1.6	2–2½	2½–3

Check the TREFLAN and Dual labels for additional directions for use, cautions, and precautions prior to application.

TREFLAN® E.C., M.T.F.™, 5™—trifluralin, Elanco Products Company
Dual®—metolachlor, Ciba-Geigy

SUPPLEMENTAL LABELING

TREFLAN® E.C. (EPA Reg. No. 1471–35)
TREFLAN® M.T.F.™ (EPA Reg. No. 1471–116)
TREFLAN® 5 (EPA Reg. No. 1471-120)

TREFLAN FOR SHATTERCANE CONTROL IN FIELD CORN/GRAIN SORGHUM AND TREFLAN/ATRAZINE TANK MIX FOR ADDITIONAL WEED CONTROL IN FIELD CORN/GRAIN SORGHUM

TREFLAN applied as an over-the-top spray or as a directed spray in field corn/grain sorghum is now labeled for shattercane (*Sorghum bicolor*) control in addition to those other weeds listed on the label for TREFLAN.

TREFLAN can be tank mixed with atrazine for additional weed control in field corn/grain sorghum.

Broadcast Rates Per Acre

Soil Texture	TREFLAN E.C./M.T.F.	TREFLAN 5 +	Atrazine 4L*
	(pints)	(pints)	(pints)
Coarse (Sandy loam only)	¾–1	0.6–0.8	2.4
Medium	1–1½	0.8–1.2	4¾
Fine	1½–2	1.2–1.6	6

* When using AAtrex® 4L, use the rates listed above. For other atrazine formulations, use equivalent rates. When using AAtrex NineO®, 1 pint 4L = 0.55 pound of Nine-O. One pint of 4L equals 0.62 pound of atrazine 80W.

Apply and incorporate the TREFLAN/atrazine tank mix as directed on the TREFLAN label for field corn and grain sorghum.

Check the TREFLAN and atrazine labels for additional directions, cautions, and precautions prior to use.

TREFLAN® E.C., M.T.F.™, 5™—trifluralin, Elanco Products Company
AAtrex®—atrazine, Ciba-Geigy
Nine-O®—Ciba-Geigy

SUPPLEMENTAL LABELING

TREFLAN® E.C. (EPA Reg. No. 1471–35)
TREFLAN® 5 (EPA Reg. No. 1471–120)

TREFLAN/ZORIAL® COMBINATION FOR WEED CONTROL IN COTTON

For distribution and use only within those states where Zorial is registered.

DIRECTIONS FOR USE

It is a violation of Federal law to use this product in a manner inconsistent with the label.

A tank mix combination of TREFLAN and Zorial is recommended as a preplant soil incorporated treatment for broad spectrum control of grass and broadleaf weeds in cotton. TREFLAN/Zorial may also be tank mixed and incorporated using one-half the recommended rate of Zorial with the remaining one-half rate of Zorial surface applied after planting. TREFLAN/Zorial will effectively control or suppress the following weeds in addition to those controlled by TREFLAN alone:

Control
Cocklebur*
Morningglory*
Pennsylvania smartweed
Prickly sida
Sickle pod
Spotted spurge
Spurred anoda
Velvetleaf

Suppression
Bermudagrass
Johnsongrass (rhizome)
Cocklebur
Hemp sesbania
Morningglory
Nutsedge
Ragweed
Other smartweeds

* Controlled by Zorial split application only.

APPLICATION

Follow soil preparation procedures normally used with TREFLAN. The tank mix of TREFLAN/Zorial may be applied up to 30 days before planting using any properly calibrated, low pressure herbicide sprayer that will apply the spray uniformly.

Fill the spray tank three-fourths (¾) full with clean water and start agitation. Premix the required amount of Zorial with a small amount of water and add to the spray tank. Continue agitation and add the required amount of TREFLAN to the spray tank. Agitate continuously during filling and throughout the spraying operation. Agitation, such as a sparger system, should be used to sweep bottom contents of the tank up into the main body of the spray mixture for thorough mixing and uniform suspension. If a bypass line is used, discharge at the bottom of the tank to minimize foaming. Apply in from 10–20 gallons of water per acre with a sprayer equipped with herbicide tips and screens sized 50 mesh or larger. Do not allow the spray mixture to remain in the tank overnight as settling could occur and resuspension would be difficult.

Apply TREFLAN/Zorial at the following broadcast rates per acre:

Soil Texture	TREFLAN E.C. Pints per acre	TREFLAN 5 Pints per acre	Zorial 80WP** lbs. per acre
Coarse	1	0.8	1.25
Medium	1.5	1.2	1.9
Fine	2	1.6	2.5

** Split application—Apply and incorporate the full rate of TREFLAN with half the recommended rate of Zorial. Surface apply the remaining half-rate of Zorial preemergence.

INCORPORATION DIRECTIONS
Incorporate TREFLAN plus Zorial one time within 24 hours after application. A second incorporation is required with most equipment. Consult the complete label for TREFLAN for recommendations regarding specific incorporation tools.

PRECAUTIONS
Follow all directions carefully. Failure to do so may result in poor weed conrol or crop injury. Overapplication may result in crop injury. TREFLAN M.T.F.™ is not compatible with Zorial 80W. Do not tank mix TREFLAN M.T.F. with Zorial 80W. Read the labels for TREFLAN and Zorial for additional warnings, cautions, precautions, and special precautions.

TREFLAN® E.C., M.T.F.™, 5™—trifluralin, Elanco Products Company
Zorial®—norflurazon, Sandoz

SUPPLEMENTAL LABELING
TREFLAN® E.C. (EPA Reg. No. 1471–35)
TREFLAN® M.T.F.™ (EPA Reg. No. 1471–116)
TREFLAN® 5 (EPA Reg. No. 1471–120)

TREFLAN POST EMERGENCE SOIL INCORPORATED FOR WEED CONTROL IN COTTON

DIRECTIONS FOR USE
TREFLAN is recommended as a post emergence soil incorporated application in cotton. TREFLAN may now be applied after final bed preparation any time from the four true leaf stage up to layby, but not less than 90 days before harvest.

APPLICATION DIRECTIONS
TREFLAN may be applied to cotton from the four true leaf stage as either a broadcast application or as a post directed spray. Drop nozzles should be used if cotton foliage prevents uniform coverage of the soil surface. Refer to the TREFLAN label for ground and aerial aplication directions.

APPLICATION RATES
Apply TREFLAN to cotton at the following broadcast rates per acre:

Soil Texture	TREFLAN E.C./M.T.F.™	TREFLAN 5
	(pints)	(pints)
Coarse	¾–1	0.6–0.8
Medium	1–1½	0.8–1.2
Fine	1½–2	1.2–1.6

INCORPORATION DIRECTIONS
Final beds should be established before the TREFLAN application is made. The soil srface should be free of existing weeds and excessive trash or clods. TREFLAN should be incorporated within 24 hours following application and may be accomplished in one pass with a sweep type or a rolling cultivator. The sweep type cultivator should have 3 to 5 sweeps per row middle and be operated at 6 to 8 mph. Set sweeps to provide maximum soil mixture and to insure movement of treated soil into the crop row. One pass with the sweep cultivator properly set will provide adequate soil incorporation. The rolling cultivator should be set to provide maximum soil mixing and to insure movement of treated soil into the crop row. The rolling cultivator should be operated at 6 to 8 mph. Care should be taken to insure that middle sweeps do not expose untreated soil. One pass with the rolling cultivator properly set will provide adequate incorporation.

WEEDS CONTROLLED
For weeds controlled, refer to the TREFLAN label.

ROTATIONAL CROPS
Wheat or barley can be planted in the fall following this application at recommended rates.

SPECIAL PRECAUTIONS
Do not apply to cotton before the four true leaf stage or crop injury may result.

Poor weed control may result if directions are not followed.

Soil treated with TREFLAN may be shallow cultivated, rotary hoed, or hand hoed without reducing the weed control activity of TREFLAN. Do not cultivate deeper than the TREFLAN treated layer of soil (1 to 2 inches) since this may bring untreated soil to the surface and poor weed control may result.

Refer to the TREFLAN label for additional information, cautions, precautions, and special precautions before use.

WARRANTY
Elanco Products Company guarantees that TREFLAN will control the weeds and grasses as indicated on its current label when used according to label directions. Refer to the TREFLAN label for specific information on the TREFLAN Performance Guarantee.

TREFLAN® E.C., M.T.F.™, 5™—trifluralin, Elanco Products Company.

SUPPLEMENTAL LABELING
TREFLAN® E.C. (EPA Reg. No. 1471–35)
TREFLAN® M.T.F.™ (EPA Reg. No. 1471–116)
TREFLAN® 5 (EPA Reg. No. 1471–120)

TREFLAN/EPTAM® TANK MIX APPLIED AS AN OVERTOP SPRAY AND INCORPORATED FOR WEED CONTROL IN SUGAR BEETS

TREFLAN may be tank mixed with Eptam and used as an overtop spray and incorporated for weed control in sugar beets. Apply the tank mix anytime after the first true leaves have formed until the beets are six (6) inches tall.

Broadcast Rates Per Acre

Soil Texture	TREFLAN E.C./M.T.F.™	TREFLAN 5	Eptam 7E**
	(pints)	(pints)	(pints)
Coarse	1	0.8	2¼–3½
Medium/Fine	1¼–1½*	1–1.2*	2¼–3½

* Use the higher rate of TREFLAN for medium and fine soils in areas receiving more than 20" average annual rainfall.
** Check the Eptam label for proper rate for your area.

Read the TREFLAN and Eptam labels for additional weeds controlled, incorporation directions, additional directions, cautions and limitations prior to use.

TREFLAN® E.C., M.T.F.™, 5™—trifluralin, Elanco Products Company
Eptam®—E.P.T.C.—Stauffer Chemical Company

SUPPLEMENTAL LABELING
TREFLAN® E.C. (EPA Reg. No. 1471–35)
TREFLAN® M.T.F.™ (EPA Reg. No. 1471–116)
TREFLAN® 5 (EPA Reg. No. 1471–120)

TREFLAN FOR WEED CONTROL IN ONIONS GROWN FOR DRY BULBS ONLY

TREFLAN effectively controls certain annual grasses and broadleaf weeds in onions grown for dry bulbs only.

APPLICATION DIRECTIONS
Apply TREFLAN as a directed spray to the soil between the onion rows. Avoid spraying the onion tops or exposed bulbs.

PREHARVEST INTERVAL
Do not apply TREFLAN within 60 days of harvest.

Broadcast Rates Per Acre

Soil Texture	TREFLAN E.C. or M.T.F.	TREFLAN 5
	(pints)	(pints)
Coarse	¾–1	0.6–0.8
Medium	1–1¼	0.8–1

For band applications, use proportionately less TREFLAN. Use the lower rates where light weed pressure is anticipated.

INCORPORATION DIRECTIONS

Soil incorporation may be accomplished by operating sweep-type or rolling cultivators 2 to 4 inches deep at 6 to 8 mph. Incorporation equipment must mix TREFLAN uniformly in the soil. Incorporate with two passes. The first pass must be within 24 hours of application or erratic weed control may result. Avoid covering exposed onion bulbs with treated soil during incorporation as injury to the crop may occur. Care should be taken to avoid injury to the roots during incorporation.

PRECAUTIONS

Applied according to these directions and under normal growing conditions, TREFLAN will not adversely affect onions. Diseases, improper incorporation depth, excessive moisture, high salt concentration or drought may weaken the crop and increase the possiblity of damage from TREFLAN. Under these conditions, delayed crop development or reduced yields may result. Refer to the TREFLAN label for additional warnings and cautions before use.

TREFLAN® E.C., M.T.F.™, PRO-5™—trifluralin, Elanco Products Company

SUPPLEMENTAL LABELING

TREFLAN® E.C. (EPA Reg. No. 1471–35)
TREFLAN® M.T.F.™ (EPA Reg. No. 1471–116)
TREFLAN® 5 (EPA Reg. No. 1471–120)

WILD OAT CLAIM ADDED TO TREFLAN LABEL

TREFLAN applied as a preplant incorporated (PPI) treatment will control those wild oats (*Avena fatua*) that germinate in the treated zone. This claim is for all PPI uses except fall applications for spring seeded cereals.

TREFLAN® E.C., M.T.F.™, 5™—trifluralin, Elanco Products Company

SUPPLEMENTAL LABELING

TREFLAN® E.C. (EPA Reg. No. 1471-35)
TREFLAN® M.T.F.™ (EPA Reg. No. 1471-116)
TREFLAN® 5 (EPA Reg. No. 1471-120)

TREFLAN/EPTAM® TANK MIX FOR WEED CONTROL IN SUNFLOWER IN THE STATES OF MINNESOTA, NORTH DAKOTA AND SOUTH DAKOTA

TREFLAN/Eptam (EPTC) may be tank mixed and preplant incorporated for weed control in sunflower.

The TREFLAN/Eptam tank mix controls the following annual and perennial weeds in addition to those controlled by TREFLAN alone:

Annual Weeds

Ryegrass (Italian)	Nightshade, black and hairy
Bermudagrass (seedling)	Spurry, corn
Rescuegrass	Fiddleneck
Volunteer grains (barley, oats, wheat)	Nettleleaf, goosefoot
Wild Oats	Shepherdspurse
Morningglory, annual	Pigweed, tumble

Perennial Weeds

Bermudagrass
Nutsedge, purple and yellow

DIRECTIONS FOR USE

Follow recommended soil preparation directions for TREFLAN. Apply the tank mix in at least ten (10) gallons of water or liquid fertilizer per acre. Apply and immediately incorporate the tank mix prior to planting in the spring. TREFLAN/Eptam may also be applied in the late fall before the ground freezes. Follow recommended incorporation directions for TREFLAN.

Broadcast Rates Per Acre

	TREFLAN				EPTAM 7E	
	Areas Receiving less than 20" average annual rainfall		Areas Receiving greater than 20" average annual rainfall			
	E.C./ M.T.F.	PRO-5	E.C./ M.T.F.	TREFLAN 5	Spring	Fall
	(pints)	(pints)	(pints)	(pints)	(pints)	(pints)
Coarse	1	0.8	1	0.8	3½	4½
Medium	1¼–1½	1–1.2	1½	1.2	3½	5¼
Fine	1½	1.2	2	1.6	3½	5¼

Check the TREFLAN and Eptam labels for additional directions for use, cautions, and precautions prior to use.

TREFLAN® E.C., M.T.F.™, 5™—trifluralin, Elanco Products Company
Eptam®—EPTC, Stauffer Chemical Company

SUPPLEMENTAL LABELING

TREFLAN® E.C. (EPA Reg. No. 1471-35)
TREFLAN® M.T.F.™ (EPA Reg. No. 1471-116)
TREFLAN® 5 (EPA Reg. No. 1471-120)

TREFLAN INCORPORATION DIRECTIONS UPDATE

TREFLAN can be incorporated with one pass when using combination seedbed conditioners. These tools should be set to cut three (3) to four (4) inches deep and operate at a minimum speed of six (6) mph. Combination seedbed conditioners are defined as three (3) or more tillage devices combined into a single tool. Only one incorporation is necessary when soil conditions allow for optimum soil mixing. Soil conditions such as excessive trash, roughness, cloddiness, moisture extremes, or high clay content which prevent optimum soil mixing action will require two (2) incorporations.

Refer to the TREFLAN label for other recommended incorporation tools and directions for use.

TREFLAN® E.C., M.T.F.™, 5™—trifluralin, Elanco Products Company

SUPPLEMENTAL LABELING

TREFLAN® E.C. (EPA Reg. No. 1471-35)
TREFLAN® M.T.F.™ (EPA Reg. No. 1471-116)
TREFLAN® 5 (EPA Reg. No. 1471-120)

TREFLAN/LASSO® TANK MIX PREPLANT INCORPORATED OR TREFLAN PREPLANT INCORPORATED WITH LASSO SURFACE APPLIED FOR WEED CONTROL IN SOYBEANS

TREFLAN/Lasso® tank mixed and TREFLAN preplant incorporated followed by Lasso surface applied control those weeds listed on the TREFLAN label plus these additional weeds:

Galinsoga	Witchgrass
Nightshade, black*	Pusley, Florida
Nightshade, hairy*	Nutsedge, yellow**
Rice, red**	

Additional weeds partially controlled by this combination include:

Beggarweed, Florida	Sida, spiny (teaweed)
Ragweed, common	Smartweed

* or ** Check rate table for additional information.

For TREFLAN/Lasso tank mix: Follow recommended soil preparation recommendations for TREFLAN. Apply TREFLAN/Lasso in a minimum of 15 gallons of water or liquid fertilizer per acre. Set incorporation equipment to work the soil no deeper than four (4) inches. Apply the tank mix up to seven (7) days prior to planting.

For TREFLAN preplant incorporated followed by Lasso surface applied: Apply and incorporate TREFLAN according to recommended soil preparation application and incorporation procedures for TREFLAN. Refer to the Lasso label for additional application directions, cautions and precautions prior to use.

Soil Texture	TREFLAN E.C./M.T.F.	or	TREFLAN 5	+	Lasso
	(pints)				(quarts)
Coarse	1		0.8		2½
Medium	1½		1.2		3–3½
Fine	2		1.6		3½–4

TREFLAN preplant incorporated followed by Lasso surface applied:

Soil Texture	TREFLAN E.C./M.T.F.	TREFLAN 5	Lasso Less than 3% Organic Matter	Lasso 3% or More Organic Matter
	(pints)	(pints)	(quarts)	(quarts)
Coarse	1	0.8	2	2
Medium	1½	1.2	2½	2½–3
Fine	2	1.6	3	3–4

* Use a minimum of two and one-half (2½) quarts of Lasso to control these weeds.

** Use three (3) to four (4) quarts of Lasso per acre applied alone or, on medium and fine textured soils, in tank mix combination preplant incorporated only for control of these weeds.

Check the TREFLAN and Lasso labels for additional directions for use, precautions and limitations prior to use.

TREFLAN® E.C., M.T.F.™, 5™—trifluralin, Elanco Products Company
Lasso®—alachlor, Monsanto Company

SUPPLEMENTAL LABELING

TREFLAN® E.C. (EPA Reg. No. 1471-35)
TREFLAN® M.T.F.™ (EPA Reg. No. 1741-116)
TREFLAN® 5 (EPA Reg. No. 1741-120)

INCORPORATION OF TREFLAN® WITH BEDDING IMPLEMENTS FOR WEED CONTROL IN COTTON

Bedding implements (listers and hippers) may be used to incorporate TREFLAN into the soil for weed control in cotton. Since bedding equipment does not provide thorough incorporation under all conditions, it is important to closely follow the directions below to optimize weed control performance. Weed control resulting from single pass incorporation with bedding equipment will be reduced compared to conventional double pass incorporation. Apply the amount of TREFLAN recommended in the label.

SOIL PREPARATION

Crop Residues or Existing Weeds: Ground cover, such as crop residues or existing weeds, can interfere with the incorporation of TREFLAN into the soil. A manageable level of such ground cover will allow the TREFLAN to be uniformly incorporated into the top 2 to 3 inches of soil. If the level of the ground cover is such that this cannot be done, you must till the soil prior to the application of TREFLAN.

Roughness: The soil surface should be smooth enough so that you can operate a sprayer and incorporation equipment efficiently and at speeds which ensure a uniform application and incorporation of TREFLAN.

General Soil Conditions: To assure uniform incorporation of TREFLAN, soil moisture conditions should be such that large clods can be broken up during the incorporation process.

BEDDER USE DIRECTIONS

Either a lister or disk bedder may be used. Operate the bedder in the manner recommended by the equipment manufacturer in order to bed to the desired height. A ripper shank, sweep or chisel shank should be mounted on the bedder behind the spray nozzles and ahead of the bedder tool to help distribute TREFLAN in the center of the bed. The use of bed tillage equipment such as rolling cultivators, P.T.O. driven rod weeders or bed conditioners after the bedding operation will provide additional mixing of the soil.

Weather conditions, cultural practices, bed tillage and planting procedures will determine location of the TREFLAN in the soil. Weed control obtained will be dependent upon location of the TREFLAN at the time of planting.

If TREFLAN is moved during bed tillage or planting, a band application of TREFLAN at planting or a postemergence application of TREFLAN may be required to ensure good weed control.

PRECAUTIONS

Do not incorporate with the bedder if the soil is too wet for good mixing.

Read the TREFLAN label for additional directions, precautions and limitations prior to use. Do not exceed recommended rates.

TREFLAN® E.C., M.T.F.™, 5™—trifluralin, Elanco Products Company

SUPPLEMENTAL LABELING

TREFLAN® E.C. (EPA Reg. No. 1471-35)
TREFLAN® M.T.F.™ (EPA Reg. No. 1471-116)
TREFLAN® 5 (EPA Reg. No. 1471-120)

TREFLAN FOR WEED CONTROL IN FORAGE LEGUMES USED AS COVER CROPS IN THE ACREAGE CONSERVATION RESERVE PROGRAM

TREFLAN can be used as a preplant incorporated broadcast applied treatment for preemergence control of many annual grasses and broadleaf weeds in direct seeded forage legumes.

Apply and incorporate TREFLAN following recommended procedures in the label. Apply TREFLAN at the following broadcast rates per acre:

Soil Texture	TREFLAN E.C./M.T.F.™	TREFLAN 5
	(pints)	(pints)
Coarse	1	0.8
Medium	1–1½	0.8–1.2
Fine	1½	1.2

Follow the most severe grazing restrictions imposed either by the pesticide label or by the USDA Conservation Use Program, whichever is longest. Consult the local ASC committee or other State Agency to determine the period of the USDA grazing restriction.

Precaution: Some stand reductions may occur with this use; however, excellent weed control will allow time for establishment of a quality stand.

TREFLAN® E.C., M.T.F.™, 5™—trifluralin, Elanco Products Company

SUPPLEMENTAL LABELING

TREFLAN® E.C. (EPA Reg. No. 1471-35)
TREFLAN® M.T.F.™ (EPA Reg. No. 1471-116)
TREFLAN® 5 (EPA Reg. No. 1471-120)

SEPTEMBER 1 FALL APPLICATION FOR TREFLAN IN CALIFORNIA, NORTH DAKOTA, SOUTH DAKOTA, MINNESOTA AND MONTANA

TREFLAN can now be applied beginning September 1 for all recommended fall applications.

Follow soil preparation, application, broadcast rate per acre, and incorporation procedures listed on the TREFLAN label for fall application.

Check the TREFLAN label for additional directions for use, cautions, and precautions prior to use.

TREFLAN® E.C., M.T.F.™, 5—trifluralin, Elanco Products Company

TREFLAN®
M.T.F.™
HERBICIDE

A Selective Herbicide for the Pre-emergence Control of Annual Grasses and Broadleaf Weeds

ACTIVE INGREDIENT:

trifluralin* (*α,α,α*-trifluro-2,6-dinitro-*N,N*-dipropyl-*p*-toluidine)	41.2%
INERT INGREDIENTS	58.8%
TOTAL	100.0%

Contains 4 pounds active ingredient per gallon

EPA Reg. No. 1471-116
U.S. Patent No. Re. 31227
ID 5560 PI 2693 AMP

* TREFLAN®, M.T.F.™ —Trademarks for Elanco Products multiple temperature formulation of trifluralin

CAUTION
Keep Out of Reach of Children

PRECAUTIONARY STATEMENTS

HAZARDS TO HUMANS AND DOMESTIC ANIMALS

CAUTION

Avoid contact with skin, eyes and clothing. Avoid breathing vapors. Harmful if swallowed, inhaled, or absorbed through the skin. Treflan M.T.F. may cause skin sensitization reactions in certain individuals. Use eye protection and protective clothing such as coveralls, a long sleeved shirt, and impermeable gloves when handling this product. Do not contaminate foodstuffs or feeds.

First Aid: In case of contact, immediately flush eyes or skin with plenty of water. For eyes, call a physician. Remove and wash contaminated clothing before reuse.

ENVIRONMENTAL HAZARDS

Direct contamination of any body of water with this emulsifiable concentrate may kill fish and other aquatic organisms. Do not contaminate any body of water by direct application, cleaning of equipment or disposal of wastes.

STORAGE AND DISPOSAL

STORAGE

Store in original container only. May be stored in unheated facilities. Do not store near heat or open flame. In case of leak or spill, use absorbent materials to contain liquids and dispose as waste.

DISPOSAL

Do not contaminate water, food or feed by storage or disposal. Wastes resulting from the use of this product may be disposed of on site or at an approved waste disposal facility.

PLASTIC CONTAINER DISPOSAL

Triple rinse (or equivalent). Then offer for recycling or reconditioning, or puncture and dispose of in a sanitary landfill, or incineration, or, if allowed by state and local authorities, by burning. If burned, stay out of smoke.

METAL CONTAINER DISPOSAL

Triple rinse (or equivalent). Then offer for recycling, or reconditioning, or puncture and dispose of in a sanitary landfill, or by other procedures approved by state and local authorities.

SPECIAL PRECAUTIONS

Applied according to directions and under normal growing conditions, TREFLAN will not harm the treated crop. Overapplication may result in crop injury or a soil residue. Uneven application or improper soil incorporation of TREFLAN can result in erratic weed control or crop injury. Seedling disease, cold weather, deep planting, excessive moisture, high salt concentration or drought may weaken crop seedlings and increase the possibility of damage from TREFLAN. Under these conditions, delayed crop development or reduced yields may result.

In Arizona, Colorado, California, Idaho, Montana, Nevada, New Mexico, Oregon, Utah, Washington and Wyoming, sugarbeets, red beets or spinach should not be planted for 12 months after a spring application or for 14 months after a fall application of TREFLAN. Plow the land to a depth of 12 inches prior to planting sugar beets to prevent the possibility of crop injury. Sorghum (milo), proso millet, corn or oats should not be planted for 14 months after a spring application or for 16 months after a fall application of TREFLAN to avoid crop injury. If land has not been irrigated, do not plant any of these crops for 18 months after a spring application or 20 months after a fall application of TREFLAN.

In those portions of Kansas, Nebraska, North Dakota, Oklahoma, South Dakota and Texas where at least 20 inches of irrigation and/or rainfall (total) was used to produce the crop, sorghum or oats should not be planted for 12 months after an application of TREFLAN.

If less than 20 inches of total water was used to produce the crop, do not plant sorghum, proso millet, or oats for 18 months after an application of TREFLAN. Cool, wet weather conditions during the early stage of growth may increase the possibility of injury to sorghum.

In all other areas receiving greater than 20″ rainfall/year: Moldboard plow before planting sugar beets where a spring application of TREFLAN was made the previous season. Also note planting restrictions listed in the section on control of rhizome johnsongrass and other higher rate programs.

Vegetable Growing Areas: Vegetable crops other than those listed on this label should not be planted within 5 months following the application of TREFLAN.

DIRECTIONS FOR USE

TREFLAN is a preemergence herbicide which is incorporated into the soil to provide long-lasting control of many annual grasses and broadleaf weeds. TREFLAN controls weeds as they germinate. TREFLAN will not control established weeds.

WEEDS AND GRASSES CONTROLLED BY TREFLAN

Grasses

Annual bluegrass	*Poa annua*
Barnyardgrass (Watergrass)	*Echinochloa* sp.
Brachiaria (Signalgrass)	*Brachiaria* sp.
Bromegrass (Cheatgrass) (Downy brome)	*Bromus tectorum*
Cheat (Chess)	*Bromus secalinus*
Crabgrass (Large crabgrass) (Smooth crabgrass)	*Digitaria* spp.
Foxtail (Bottlegrass) (Bristlegrass) (Giant foxtail) (Green foxtail) (Foxtail millet) (Pigeongrass) (Robust foxtail) (Yellow foxtail)	*Setaria* spp.
Goosegrass (Silver crabgrass) (Silvergrass) (Wiregrass) (Yardgrass)	*Eleusine indica*
Johnsongrass (from Seed)	*Sorghum halepense*

(Rhizome—see special instructions for control in cotton and see special instructions for control in soybeans)

Junglerice — *Echinochloa colonum*
Panicum
Fall panicum — *Panicum dichotomiflorum*
(Spreading panicgrass—see special instructions in cotton and see special instructions in soybeans)
Guineagrass — *Panicum maximum*
(See special instructions in sugarcane)
Texas panicum — *Panicum texanum*
(Buffalograss)
(Coloradograss)
Itchgrass — *Rottboellia exaltata*
(Raoulgrass)
(See special instructions in sugarcane)
Red rice — *Oryza sativa*
(See special instructions in soybean)
Sandbur — *Cenchrus incertus*
(Burgrass)
Sprangletop — *Leptochloa filiformis*
Stinkgrass — *Eragrostis cilianensis*
(Lovegrass)
Wild cane — *Sorghum bicolor*
(Shattercane)
(See special instructions in soybean)
Woolly cupgrass — *Eriochloa villosa*

Broadleaf Weeds

Carpetweed — *Mollugo verticillata*
Chickweed — *Stellaria media*
Field bindweed — *Convolvulus arvensis*
(See special instructions in Fruit and Nut Crops and Vineyards)
Florida pusley — *Richardia scabra*
(Florida purslane)
(Mexican clover)
(Pusley)
Goosefoot — *Chenopodium hybridum*
Henbit (fall application only) — *Lamium amplexicaule*
Knotweed — *Polygonum aviculare*
Kochia — *Kochia scoparia*
(Fireweed)
(Mexican fireweed)
Lambsquarters — *Chenopodium album*
Pigweed — *Amaranthus* spp.
(Carelessweed)
(Prostrate pigweed)
(Redroot)
(Rough pigweed)
(Spiny pigweed)
Puncturevine (Western U.S. only) — *Tribulus terrestris*
(Caltrop)
(Goathead)
Purslane — *Portulaca oleracea*
Russian thistle — *Salsola kali*
(Tumbleweed)
Stinging nettle — *Urtica dioica*
(Nettle)

SOIL PREPARATION

Crop Residues or Existing Weeds: Ground cover, such as crop residues or existing weeds, can interfere with the incorporation of TREFLAN into the soil. A manageable level of such ground cover will allow the TREFLAN to be uniformly incorporated into the top 2 to 3 inches of soil. If the level of the ground cover is such that this cannot be done, you must till the soil prior to the application of TREFLAN.

Roughness: The soil surface should be smooth enough so that you can operate a sprayer and incorporation equipment efficiently and at speeds which insure a uniform application and incorporation of TREFLAN.

General Soil Conditions: To assure uniform incorporation of TREFLAN, soil moisture conditions should be such that large clods can be broken up during the incorporation process.

SOIL TEXTURE GUIDE

The amount of TREFLAN you apply will vary with the soil texture and organic matter. A fine textured soil will require more TREFLAN per acre than a coarse soil. Choose the proper rate for each application based on the following soil texture group and specific crop recommendations. Do not exceed recommended rates.

Soil Texture	Soil Classification
Coarse Soils: (Light)	Sand, loamy sand, sandy loam
Medium Soils:	Loam, silty clay loam*, silt loam, silt, sandy clay loam*
Fine Soils:	Clay, clay loam, silty clay loam*, silty clay, sandy clay, sandy clay loam*

* Silty clay loam and sandy clay loam soils are transitional soils and may be classified as either medium or fine textured soils. If silty clay loam or sandy clay loam soils are predominately sand or silt, they are usually classified as medium textured soils. If they are predominately clay, they are usually classified as fine textured soils.

MIXING AND APPLICATION DIRECTIONS

TREFLAN M.T.F. Alone in Water

Start with a clean spray tank. Fill sprayer ⅓ to ½ full with clean water. Start agitation. Add correct quantity of TREFLAN, continue agitation and finish filling the tank.

TREFLAN M.T.F. Tank Mix in Water

Vigorous, continuous agitation is required for all tank mixes. (Sparger pipe agitators generally provide the best agitation in spray tanks). Avoid stirring or splashing air into the mixture during filling to prevent foaming. To do this, place the end of the fill pipe below the surface of the water in the spray tank. Do not allow the mixture to siphon back into the water source.

Mixing order: Fill the tank ¼ to ⅓ full with clean water. Start the agitation. Add dry flowables, wettable powders (WP), aqueous suspensions (AS), flowables (F), and liquids (L) to the water and agitate until the product(s) are completely dispersed in the water. Allow additional mixing and dispersion time when using dry flowable products. Continue agitation and fill tank to ¾ full, add the TREFLAN M.T.F., mix thoroughly. Then add any solution (S) formulations, agitate and finish filling. Maintain agitation during filling and through application. If spraying and agitation must be stopped before the tank is empty, the materials may settle to the bottom. In this case, it is important to resuspend all of the material in the bottom of the tank before continuing the spray application. A sparger agitator is particularly useful for this purpose. Sometimes it is more difficult to resuspend settled material than it is to suspend it originally.

Read and carefully follow all label instructions for each material added to the tank. Premixing dry and flowable formulations with water (slurrying) and pouring the slurry through a 20 or 35 mesh wetting screen in the top of the tank will help assure good initial dispersion in the tank water. Line screens in the tank should be no finer than 50 mesh (100 mesh is finer than 50 mesh).

If you see a buildup of material on the walls of the spray tank, wash the tank with soapy water between fillings. Rinse and continue the spraying operation. Clean the tank, lines, and screens thoroughly after use.

As the spray volume decreases, the importance of accurate calibration and uniform application increases. Check the sprayer daily to insure proper calibration and uniform application. Do not apply TREFLAN when the wind can cause drifting of spray particles which can result in non-uniform application. TREFLAN should not be applied to soils which are wet or are subject to prolonged periods of flooding as poor weed control may result.

Ground application: Apply TREFLAN in 5 to 40 gallons of water or liquid fertilizer per acre (broadcast basis), using any properly calibrated, low pressure herbicide sprayer that will apply the spray uniformly.

Aerial application: Apply TREFLAN in 5 to 10 gallons of water or liquid fertilizer per acre. Adjust pump pressure, nozzle arrangements, speed and height to provide a uniform application to the soil surface. Use swath markers or flagmen to assure proper application spray widths.

INCORPORATION DIRECTIONS

Incorporation Equipment—General Directions

Use incorporation equipment that mixes TREFLAN into the top 2 to 3 inches of the final seedbed, or erratic weed control and/or crop injury may result. Incorporation equipment such as a disc will mix TREFLAN approximately half as deep as the equipment is set to operate. For example, a disc set to cut 4 inches deep will incorporate most of the TREFLAN within the top 2 inches of soil.

Incorporation Before Planting

TREFLAN must be incorporated one time within 24 hours after application. Then any time prior to planting, a second incorporation is necessary, this time running the equipment in a different direction from the first. You should incorporate the TREFLAN uniformly into the top 2 to 3 inches of the final seedbed.

Incorporation After Planting

Check specific crop for incorporation directions after planting.

Incorporation in Bedded Culture

For effective weed control, TREFLAN needs to be incorporated into the top 2 to 3 inches of the final seedbed.

Application prior to bedding: Apply TREFLAN and incorporate it one time with recommended equipment. The bedding operation serves as the second incorporation. Do not expose untreated soil during post-bedding operations.*

Application after bedding: Knock off beds to planting height before applying TREFLAN. Apply TREFLAN and incorporate it with recommended equipment that will conform to the bed shape. Do not leave untreated soil exposed.*

* Avoid removal of treated soil from the seedbed before or during the planting operation. This would expose untreated soil, allowing weeds to germinate in the drill row.

Recommended Equipment

Any recommended incorporation tool may be used alone or in combination with any other recommended tool. Two incorporation passes are required unless specifically stated. The second incorporation should not be deeper than the first.

Disc: set to cut 4 to 6 inches deep and operate at 4 to 6 mph.

Field Cultivator: set to cut 3 to 4 inches deep and operate at 5 mph or more. A field cultivator is defined as an implement with 3 to 4 rows of sweeps, spaced at intervals of 7 inches or less and staggered so that no soil is left unturned. Chisel points should not be used.

Combination Seedbed Conditioners: set to cut 3 to 4 inches deep and operate at a speed of at least 5 mph. These implements are defined as three or more tillage devices combined and used as a single tool. For example, 2 to 3 rows of field cultivator C- or S-shaped shanks with an effective sweep spacing of 6 to 9 inches (staggered so that no soil is left unturned), followed by a spike-tooth or flex-tine harrow, followed by a ground-driven reel or basket.

Rolling Cultivator: set to cut 2 to 4 inches deep and operate at 6 to 8 mph. Rolling cultivators are adequate for use on coarse and medium textured soils only, except when used in sugarcane where the rolling cultivator may be used on fine textured soils.

Bed Conditioner (Do-All): set to cut 2 to 4 inches deep and operate at 4 to 6 mph. The Do-All is adequate for use on coarse and medium textured soils only. When using the Do-All in bedded culture, only one incorporation pass is required. However, two passes with a Do-All are required in flat planted culture.

Mulch Treader (other similar disc-type implements): set to cut 3 to 4 inches deep and operate at 5 to 8 mph.

P.T.O. Driven Equipment (tillers, cultivators, hoes): adjust to incorporate TREFLAN into the top 2 to 3 inches of the seedbed with rotors spaced to provide a clean sweep of the soil. Only one incorporation is necessary. P.T.O. driven equipment should not be operated at a speed greater than 4 mph.

Other equipment, including the flexible tine-tooth harrow (Flex-tine, Melroe), is also recommended but only for the special programs for which it is specified in this label.

CULTIVATION AFTER PLANTING

Soil treated with TREFLAN may be shallow cultivated without reducing the weed control activity of TREFLAN. Do not cultivate deeper than the treated soil since this may bring untreated soil to the surface, and poor weed control may result.

CROP RECOMMENDATIONS

These recommendations are given as the broadcast rates of TREFLAN per acre. For band applications, decrease the amount of TREFLAN in proportion to the amount of surface treated per acre. Apply any time after January 1 when the soil can be worked and is suitable for good incorporation. TREFLAN can be applied in the fall—see specific crop for recommendations. For general fall application directions where specific recommendations are not given, see page 74. Where a rate range is shown, use the lower rate for coarser soils or soils with lower organic matter. TREFLAN should not be used on soils containing more than 10% organic matter.

COTTON—TREFLAN ALONE

TREFLAN can be applied and incorporated before or at planting, immediately after planting, or at layby.

COTTON—Preemergence:

Broadcast Rates Per Acre

Soil Texture	TREFLAN M.T.F. Areas receiving less than 20" average annual rainfall*	TREFLAN M.T.F. Areas receiving greater than 20" average annual rainfall*
	(pints)	(pints)
Coarse	1	1
Medium	1¼–1½	1½
Fine	1½	2

* Use 1½ pints per acre on coarse and medium textured soils and 2 pints on fine soils with 2–5% organic matter; use 2 to 2½ pints on all soils with 5–10% organic matter.

COTTON—Postplant:

When incorporating TREFLAN after planting (postplant), be careful not to disturb the seed.

COTTON—Layby:

Apply and incorporate TREFLAN any time up to layby, but not less than 90 days before harvest. Direct the layby applications onto the soil between the rows and beneath emerged cotton plants. Use the same rates as for a preemergence application.

COTTON—Fall Application:

Apply and incorporate TREFLAN any time between October 15 and December 31. The ground may be left flat or bedded-up over winter. On bedded ground, knock beds down to desired height before planting, moving some treated soil from beds into furrows. Where soil is left flat over winter, be careful not to turn up untreated soil during spring bedding operations. Destroy established weeds during seedbed preparation. If weeds become established in furrows due to uncovering of untreated soil during bedding, destroy these weeds before planting. In the fall, do not apply TREFLAN to soils which are wet or subject to prolonged periods of flooding.

Broadcast Rates per Acre—Fall application only:

In Alabama, Arkansas, northern Florida, Georgia, Louisiana, Mississippi, southeastern Missouri bootheel, North Carolina, New Mexico, Oklahoma, South Carolina, Tennessee and Texas, apply and incorporate TREFLAN at a broadcast rate of 2 pints per acre on coarse and medium soils and 2½ pints on fine soils.

In Arizona, California and Nevada, apply and incorporate TREFLAN at a broadcast rate of 1½ pints per acre on coarse soils; 2 pints on medium soils; and 2½ pints on fine soils.

For cotton grown in other states, apply and incorporate TREFLAN at a broadcast rate of 1 pint per acre on coarse soils; 1½ pints on medium soils; 2 pints on fine soils, 1½ pints on coarse soils with 2–5% organic matter; and 2 to 2½ pints on soils with 5–10% organic matter.

COTTON—SPECIAL USE DIRECTIONS

COTTON—Fall panicum:

Apply and incorporate TREFLAN at the broadcast rate of 2 pints per acre on both coarse and medium soils.

COTTON—Rhizome johnsongrass:

In all cotton-producing states except Arizona and California, you can obtain commercially acceptable control of rhizome johnsongrass with a double rate program which you apply for 2 consecutive years in accordance with the following directions.

Soil Preparation—Proper preparation of the soil before application is very important for satisfactory results. Use a chisel plow or similar implement to bring rhizomes to the top of the soil. Then follow with a disc two times before application to cut the rhizomes into small (2 to 3 inch) pieces. This should also destroy any emerged johnsongrass.

Application—Choose the one application program that best fits your cultural practices:

Spring Application—Apply TREFLAN any time before planting in the spring for 2 years in a row. Use a broadcast rate of 2 pints per acre on coarse soils; 3 pints on medium soils; and 4 pints on fine soils.

OR

Fall Application—Apply TREFLAN between October 15 and December 31 for 2 years in a row at the same rates as a spring application for the control of rhizome johnsongrass.

Incorporation—Deep incorporation is essential for good rhizome johnsongrass control. Incorporate TREFLAN thoroughly with a disc set to cut 4 to 6 inches deep and operate at 4 to 6 mph. Two passes are necessary, with the second pass in a different direction from the first.

Cultivation—Some johnsongrass plants will escape. Timely cultivations during the crop season to remove escaped plants are necessary to obtain commercially acceptable control. You cannot obtain commercially acceptable control with only 1 year of double rate TREFLAN use.

Crop Rotation—In the season following a double rate treatment, plant only rice and those crops for which TREFLAN can be applied as a preplant treatment or injury may result.

COTTON—Pigweed and seedling johnsongrass control: In Alabama, Arkansas, Florida, Georgia, Louisiana, Mississippi, southeastern Missouri, North Carolina, South Carolina, Tennessee and southern Virginia, TREFLAN may be applied preplant at a broadcast rate of 1 to 1½ pints per acre on coarse soils; 1½ to 2 pints on medium soils; and 2 pints on fine soils. Exception: in the State of Lousiana, 3 pints per acre are recommended on fine soils.

COTTON—Additional weed and grass control (Texas Gulf Coast):

In the Texas Gulf Coast counties of Brazoria, Calhoun, Chambers, Fort Bend, Galveston, Harris, Jackson, Jefferson, Liberty, Matagorda, Orange, Victoria, Waller and Wharton, TREFLAN may be applied up to 2 weeks before planting at a broadcast rate of 1½ pints per acre on coarse soils; 2 pints on medium soils; and 3 pints on fine soils.

COTTON PRECAUTIONS: Cotton should be planted after early season adverse weather conditions have passed, especially when using higher rate programs. Cool, wet weather early in the growth cycle causes additional stress to the cotton plant. This may result in reduced stands, delayed maturity and reduced yields.

COTTON—TANK MIXES AND OVERLAYS

Follow recommended soil preparation and incorporation procedures for TREFLAN.

COTTON—TREFLAN/Caparol tank mix for cotton grown in California, Arizona, New Mexico and west Texas:

The TREFLAN/Caparol combination will control certain grasses and broadleaf weeds listed for TREFLAN alone plus the following weeds:

Annual morningglory	Prickly sida (Teaweed)
Groundcherry (Annual)	Ragweed
Malva	Smartweed
Mustard	Wild oat

The tank mix also controls shallow germinating seedlings of cocklebur and coffeeweed.

Broadcast Rates Per Acre

Soil Texture	TREFLAN M.T.F.	Caparol 80W
	(pints)	(pounds)
Coarse	1	2*
Medium	1¼ - 1½	2½
Fine	2	2½

* Do not use on sands and loamy sands. For band applications use proportionately less.

Mixing Directions: Carefully follow the procedures on the Caparol 80W label for making a slurry and adding it to a partially filled tank of water. After the Caparol is thoroughly mixed with the partially filled tank of water, add the TREFLAN M.T.F. and continue filling. Agitate continuously during the filling and spraying operation.

Avoid leaving the spray mixture in the tank without constant agitation. If bypass agitation is used, the bypass line should stop at the bottom of the tank to minimize foaming.

Additional Precautions: The combination of TREFLAN/Caparol should not be used in the cut areas of newly leveled fields, in areas of excess salt, or where flooding over the beds is likely to happen. Do not plant cotton in tractor wheel depressions. These conditions may cause crop injury. On mulch-planted cotton, water back only after cotton seedlings are well established.

Crop Rotations: Cabbage, okra, onions and peas may be planted in the fall after a spring application of TREFLAN plus Caparol. Winter barley, winter rye and winter wheat can be planted in the fall also, if they are plowed down and not used for food or feed. Refer to the Caparol label for directions, cautions and precautions.

COTTON—TREFLAN/Cotoran tank mix (except in Arizona and California):

Follow recommended soil preparation and incorporation procedures for TREFLAN.

The TREFLAN/Cotoran tank mix effectively controls all the annual grasses and broadleaf weeds listed for TREFLAN alone plus these additional weeds:

Buttonweed	Ragweed
Cocklebur	Ryegrass
Groundcherry, Wright	Sesbania
Jimsonweed	Sicklepod
Morningglory	Smartweed
Prickly sida (Teaweed)	Tumbleweed

Broadcast Rates Per Acre

Soil Texture	TREFLAN M.T.F.	Cotoran 80W
	(pints)	(pounds)
Coarse	1	1¼
Medium	1½	2
Fine	2	2½

Mixing Directions: Carefully follow the procedures on the Cotoran label for making a Cotoran slurry and adding it to a partially filled tank of water. After the Cotoran is thoroughly mixed with the partially filled tank of water, add the TREFLAN and continue filling. Agitate continuously throughout the filling and application operations. Do not leave spray mixture in tank without constant agitation. If bypass agitation is used, the bypass line should stop at the bottom of the tank to minimize foaming. Apply in 15–40 gallons of water per acre.

Additional Precautions: Do not plant crops other than cotton on the treated land within 6 months after the application of TREFLAN plus Cotoran, or crop injury may result. Do not feed foliage from treated cotton plants or gin trash to livestock. Do not mix TREFLAN plus Cotoran with liquid fertilizer.

West Texas Only: Do not use the tank mix of TREFLAN plus Cotoran on sandy, loamy sand or fine sandy loam soils. Do not use on cotton planted in furrows.

Arkansas, Louisiana, and Mississippi Only: Use 1 pound Cotoran 80W in tank mix with TREFLAN on sandy loam soils low in organic matter.

New Mexico: Do not plant treated land with crops other than cotton until 1 year after the last application. Do not use on sandy or coarse textured soils of less than 1% organic matter.

COTTON—TREFLAN preplant followed by Cotoran overlay:
Apply and incorporate TREFLAN as recommended for the specific soil texture. Then apply Cotoran as a preemergence surface treatment at 1¼ to 2½ pounds per acre. On light silt and sandy soils low in organic matter, use the lower rate.

Additional Precaution: Refer to the Cotoran label for all cautions and precautions when using this preemergence surface treatment.

COTTON—TREFLAN preplant followed by Karmex overlay:
For cotton grown east of the Mississippi River plus Arkansas, southeastern Missouri, Louisiana and Eastern Texas, incorporate TREFLAN before planting. Then follow with a preemergence application of Karmex. This will effectively control all the weeds controlled by TREFLAN alone plus these additional weeds:

Annual groundcherry	Shepherdspurse
Annual morningglory	Velvetgrass
Dogfennel	Wild lettuce
Pennycress	Wild mustard
Ragweed	

Broadcast Rates Per Acre

Soil Texture	TREFLAN M.T.F.	Karmex 80W
	(pints)	(pounds)
Coarse	1	⅓
Medium	1½	⅔
Fine	2	1

Additional Precautions: Do not use Karmex on soils with less than 1% organic matter as crop injury may result. Do not allow livestock to graze on cotton treated with Karmex. Consult the Karmex label for additional instructions, cautions and precautions.

SOYBEAN—TREFLAN ALONE

SOYBEAN—Preemergence:
Follow recommended soil preparation, application, and incorporation procedures for TREFLAN.

Broadcast Rates Per Acre

Soil Texture	Treflan M.T.F.*
	(pints)
Coarse	1
Medium	1½
Fine	2

* Use 1½ pints per acre on coarse and medium textured soils and 2 pints on fine soils with 2–5% organic matter; use 2–2½ pints on all soils with 5–10% organic matter.

SOYBEAN—Fall Application:
Apply and incorporate TREFLAN anytime between October 15 and December 31. Ground may be left flat or bedded-up over winter. On bedded ground, knock beds down to desired height before planting, moving some treated soil from tops into furrows. Where soil is left flat over winter, take care during spring bedding operations to prevent turning up untreated soil. Destroy established weeds during seedbed preparation. If weeds become established in furrows due to uncovering of untreated soil during listing, destroy these weeds before planting. Do not fall apply TREFLAN to soils which are wet, are subject to prolonged periods of flooding, or where rice was grown the previous year.

For soybeans grown in Alabama, Arkansas, northern Florida, Georgia, Louisiana, Mississippi, southeastern Missouri bootheel, North Carolina, Oklahoma, South Carolina, Tennessee and Texas, apply and incorporate TREFLAN at a broadcast rate of 2 pints per acre on coarse and medium soils and 2½ pints on fine soils.

For soybeans grown in states other than those listed above, apply and incorporate TREFLAN at a broadcast rate of 1 pint per acre on coarse soils; 1½ pints on medium soils; 2 pints on fine soils; 1½ pints on coarse soils with 2–5% organic matter; and 2 to 2½ pints on soils with 5–10% organic matter.

SOYBEAN—SPECIAL USE DIRECTIONS

SOYBEAN—Fall panicum:
Apply TREFLAN at the broadcast rate of 2 pints per acre on both coarse and medium soils.

SOYBEAN—Pigweed and seedling johnsongrass:
In Alabama, Arkansas, Florida, Georgia, Kansas, Louisiana, Mississippi, southeastern Missouri, North Carolina, Oklahoma, South Carolina, Tennessee and southern Virginia, TREFLAN may be applied at a broadcast rate of 1 to 1½ pints per acre on coarse soils; 1½ to 2 pints on medium soils; and 2–2½ pints on fine soils. Exception: in the State of Louisiana, 3 pints per acre are recommended on fine soils.

SOYBEAN—Additional weed and grass control (Texas Gulf Coast):
In the Texas Gulf Coast counties of Brazoria, Calhoun, Chambers, Fort Bend, Galveston, Harris, Jackson, Jefferson, Liberty, Matagorda, Orange, Victoria, Waller and Wharton, TREFLAN may be applied up to 2 weeks before planting at a broadcast rate of 1½ pints per acre on coarse soils; 2 pints on medium soils; and 3 pints on fine soils.

SOYBEAN—Charcoal soils in Arkansas, Louisiana and Mississippi:
Newly cleared land often contains high organic matter (5–10%) and charcoal from burning debris. This charcoal and/or organic matter tends to bind TREFLAN and reduce its weed control activity. Under these conditions, higher rates of TREFLAN are necessary for weed control. Increased rates, however, can cause crop injury if charcoal or organic matter is not present to bind some of the TREFLAN. In the burn row a high level of charcoal is present; consequently, poor weed control may result even with an increased rate of TREFLAN.

Apply TREFLAN at the broadcast rate of 1½ to 2½ pints per acre on coarse soils; 2½ pints on medium soils; and 3 pints on fine textured soils. Follow recommended soil preparation, application and incorporation procedures for TREFLAN.

SOYBEAN—Red rice in Arkansas, Louisiana, Mississippi and Texas only:
You can obtain suppression or partial control of red rice when you apply TREFLAN at the following recommended rates. Follow recommended soil preparation and incorporation procedures for TREFLAN.

Apply and incorporate TREFLAN in the spring before planting.

Broadcast Rates Per Acre

Soil Texture	TREFLAN M.T.F. Application Year 1	TREFLAN M.T.F. Application Year 2
	(pints)	(pints)
Coarse	2	1
Medium	3	1½
Fine	4	
Coarse soils with 2 to 5% organic matter	3	2 1½
Soils with 5 to 10% organic matter	4	2–2½

If a combination of high organic matter (5–10%) and charcoal are present in the soil, apply TREFLAN the second year at the following rates for charcoal soils in Arkansas, Louisiana and Mississippi:

Soil Texture	TREFLAN M.T.F.
	(pints)
Coarse	1½–2½
Medium	2½
Fine	3

For more information on charcoal soils see under SOYBEAN—charcoal soils heading.

Crop Rotation: The program for red rice control in soybeans is a 2-year program. Use the rates listed for first year application and plant soybeans. The second year use the normal TREFLAN rates listed for your soil type and charcoal level and plant only those crops for which TREFLAN has been registered as a preplant treatment, or crop injury may result. Do not plant rice the second year. Rice may be planted the third year.

SOYBEAN—Rhizome johnsongrass (Eastern United States and the State of Texas):

You can obtain commercially acceptable control of rhizome johnsongrass with a double rate program applied for 2 consecutive years in accordance with the following directions.

Soil Preparation—Proper preparation of the soil before application is very important for satisfactory results. Use a chisel plow or similar implement to bring rhizomes to the top of the soil. Then follow with a disc two times before application to cut the rhizomes into small (2 to 3 inch) pieces and to destroy any emerged johnsongrass.

Application—Choose the one application program that best fits your cultural practices:

Spring Application—Apply TREFLAN anytime in the spring before planting for 2 years in a row. Use a broadcast rate of 2 pints per acre on coarse soils; 3 pints on medium soils; 4 pints on fine soils; 3 pints on coarse soils with 2–5% organic matter; and 4 pints on soils with 5–10% organic matter.

OR

Fall Application—Apply TREFLAN between October 15 and December 31 for 2 years in a row at the same rates as a spring application for the control of rhizome johnsongrass.

OR

Split Application—Apply TREFLAN at the same rate in both the spring and fall for 2 years in a row using the rates in the following table.

Broadcast Rates Per Acre

Soil Texture	TREFLAN M.T.F. Spring and Fall
	(pints)
Coarse	1
Medium	1½
Fine	2
Coarse soils with 2–5% organic matter	1½
Soils with 5–10% organic matter	2

Incorporation—Deep incorporation is essential for good rhizome johnsongrass control. Incorporate TREFLAN thoroughly with a disc set to cut 4 to 6 inches deep and operate at 4 to 6 mph. Two passes are necessary, with the second pass in a different direction from the first.

Cultivation—Some johnsongrass plants will escape. Timely cultivations during the crop season to remove escaped plants are necessary to obtain commercially acceptable control.

Crop Rotation: In the season following a double rate treatment, plant only rice and those crops for which TREFLAN can be applied as a preplant treatment or injury may result.

SOYBEAN—Rhizome johnsongrass—TREFLAN/Sencor or TREFLAN/Lexone tank mix:

Lexone®—metribuzin, E. I. duPont de Nemours and Company.
Sencor®—metribuzin, Bayer, GmbH.

TREFLAN/Sencor or TREFLAN/Lexone may be used for rhizome johnsongrass control and for the control of those weeds listed for TREFLAN alone. For the additional weeds controlled by Sencor or Lexone in tank mix, see SOYBEAN—Tank Mixes and Overlays section. Follow procedures for soil preparation, incorporation, and cultivation recommended in the Soybean—Rhizome johnsongrass section.

Apply TREFLAN/Sencor or Lexone up to two weeks before planting for two consecutive years at the following broadcast rates per acre:

Soil Texture	TREFLAN M.T.F.	+	Lexone 50WP/4L or Sencor 50WP/4	or	Lexone (dry flowable) or Sencor (dry flowable)
	(pints)		(pounds/pints)		(pounds)
Coarse*	2		½		⅓
Medium	3		¾		½
Fine	4		1		⅔

* Do not use on coarse soils with less than 1% organic matter.

Read and follow all additional precautions listed for the TREFLAN/Sencor or TREFLAN/Lexone tank mix (See page 39).

SOYBEAN—Wild cane (shattercane):

Follow recommended soil preparation and application procedures for TREFLAN.

Wild cane (shattercane) can germinate throughout the growing season and from greater soil depth than most other weed seeds. Commercially acceptable control of wild cane can be obtained with the following increased rates of TREFLAN.

Apply TREFLAN at the broadcast rate of 1 pint per acre on a coarse soil; 2 pints on a medium soil; and 2½ pints on a fine textured soil.

Incorporation—Deep incorporation is essential for good wild cane control. Incorporate TREFLAN thoroughly with a disc set to cut 4 to 6 inches deep and operate at 4 to 6 mph. Two passes are necessary, with the second pass in a different direction from the first.

Cultivation—Cultivations during the crop season will also contribute to control.

SOYBEAN PRECAUTIONS: Soybeans should be planted after early season adverse weather conditions have passed, especially when using higher rate programs. Cool, wet weather early in the growth cycle causes additional stress to the soybean plant. This may result in reduced stands, delayed maturity and reduced yields.

SOYBEAN—TANK MIXES AND OVERLAYS

SOYBEAN—TREFLAN/Sencor or TREFLAN/Lexone tank mix:

The TREFLAN/Sencor or TREFLAN/Lexone tank mix effectively controls the annual grasses and broadleaf weeds controlled by TREFLAN alone plus these additional weeds:

Jimsonweed	Ragweed, common
Mallow, Venice (Flower-of-an-hour)	Sesbania, hemp
	Smartweed, Pennsylvania
Mustard, wild	Velvetleaf
Prickly sida	

Control of cocklebur, morningglory and giant ragweed (horseweed) may be erratic. Control may be improved with timely cultivation. Where cocklebur is a serious problem, an overlay of Sencor or Lexone may be preferred to the TREFLAN/Sencor or TREFLAN/Lexone tank mix.

Follow recommended procedures for soil preparation, incorporation, and cultivation of TREFLAN. Mix according to instructions under Mixing and Application Directions. The TREFLAN/Sencor or TREFLAN/Lexone tank mix can be be applied from 2 weeks before planting up to planting.

Broadcast Rates Per Acre

Soil Texture	TREFLAN M.T.F.	+	Lexone 50WP/4L or Sencor 50WP/4	or	Lexone (dry flowable) or Sencor (dry flowable)
	(pints)		(pounds/pints)		(pounds)
Coarse*	2		½		⅓
Medium	1½		¾		½
Fine	2		1		⅔

* Do not use Sencor/Lexone on coarse soils with less than 1% organic matter.

Additional Precautions: Do not plant any crop other than soybeans within 4 months after treatment. Overapplication, uneven application, or improper soil incorporation may result in crop injury, herbicide residue, or erratic weed control. Additional stress factors are seedling disease, cold weather, deep planting, excessive moisture, soil pH over 7.5, high salt concentration, or drought. Any of these may weaken crop seedlings and increase possibility of damage from the tank mix. These additional factors may also delay crop development or reduce yields when Sencor or Lexone is applied. Observe all cautions and limitations on the Sencor and Lexone labels. Do not use the foliage from soybeans treated with the TREFLAN/Sencor or TREFLAN/Lexone tank mix for feed or forage.

SOYBEAN—TREFLAN preplant followed by Sencor or Lexone as an overlay:

Apply TREFLAN as a preplant incorporated herbicide. As a separate operation, make a single application of Sencor or Lexone as either a band or broadcast spray during planting or as a separate operation after planting, but before the soybeans emerge. Do not spray Sencor or Lexone over the top of emerged soybeans, or crop injury may result.

Use Directions—Follow directions on the Sencor or Lexone labels for specific instructions regarding each herbicide.

Broadcast Rates Per Acre

Soil Texture	TREFLAN M.T.F.	Postplant/Preemergence: Sencor 50WP/4 or [Lexone 50WP/4L]*	or Sencor (dry flowable) or [Lexone (dry flowable)]*
	(pints)	(pounds or pints)	(pounds)
Coarse**	1	¾–1 [¾]	½–⅔ [½]
Medium	1½	¾–1½ [¾–1]	½–1 [½–⅔]
Fine	2	1–1¾ [1]	⅔–1⅙ [⅔]

* Lexone rates for each soil texture are enclosed in brackets "[ ]".

** Do not apply Lexone to sand or soils with less than ½% organic matter. Do not apply Sencor to coarse soils (sandy loam and loamy sand) containing less than 2% organic matter.

Additional Sencor and Lexone precautions: Do not use Lexone or Sencor on Tracy, Semmes, Altona, Vansoy or Coker 102 soybeans. These varieties are sensitive to Lexone or Sencor, and crop injury may result. Seed must be planted at least 1½ inches but not more than 2 inches below the soil surface before a Sencor or Lexone application. Do not apply Sencor or Lexone at these rates more than once per season. Do not replant areas treated with Sencor or Lexone with any crop other than soybeans within 4 months after treatment. Injury to soybeans may occur if you use Lexone or Sencor on soils having a calcarous surface or pH of 7.5 or higher, or if you use them in conjunction with soil-applied organic phosphate pesticides. Do not use the foliage from treated soybeans for feed or forage.

SOYBEAN—TREFLAN/Amiben tank mix or overlay:

Tank mix—Amiben may be applied several days prior to planting as a broadcast tank mix with TREFLAN. Weeds controlled by this tank mix, in addition to those controlled by TREFLAN alone, are smartweed, velvetleaf and ragweed. The tank mixture should be used as a spring preplant incorporated treatment.

Overlay—Amiben may be applied broadcast or in a band over the soybean row at planting time in fields where TREFLAN has been preplant incorporated. Weeds controlled by Amiben when surface applied, in addition to those controlled by TREFLAN alone, are:

Coffeeweed (Sesbania)	Spurge, annual
Mustard, wild	Smartweed, Pennsylvania
Nightshade, black	Stinkgrass
Prickly sida (Teaweed)	Velvetleaf
Ragweed, common	

Apply TREFLAN as a tank mix with Amiben, or apply and incorporate TREFLAN alone followed by an overlay application of Amiben at these rates:

Broadcast Rates Per Acre

Soil Texture	TREFLAN M.T.F.	Amiben 2S
	(pints)	(quarts)
Coarse	1	4–6*
Medium	1½	4–6
Fine	2	4–6

* Use the higher rate where you expect heavy populations of smartweed, velvetleaf, ragweed, wild mustard or black nightshade. Do not use on muck or charcoal soils. Read and observe all directions and cautions on the Amiben label.

SOYBEAN—TREFLAN/Amiben/Sencor or Lexone tank mix: The TREFLAN/Amiben/Sencor or Lexone tank mix effectively controls all weeds listed for TREFLAN/Amiben and TREFLAN/Sencor or Lexone tank mixes.

Follow recommended soil preparation, application, and incorporation procedures for TREFLAN. The TREFLAN/Amiben/Sencor or Lexone tank mix may be applied from several days prior to planting up to planting in 10 to 40 gallons of water per acre. Use screens no finer than 50 mesh.

Apply the TREFLAN/Amiben/Sencor or Lexone tank mix at the following broadcast rates per acre:

Soil Texture	TREFLAN M.T.F.	+ Amiben 2S +	Lexone 50WP/4L or Sencor 50WP/4	or Lexone (dry flowable) or Sencor (dry flowable)
	(pints)	(quarts)	(pounds/pints)	(pounds)
Coarse[a]	1	3–4[b]	½	⅓
Medium	1½	3–4[b]	½–¾[c]	⅓–½[c]
Fine	2	4–5	¾[c]	½[c]

[a] Do not use Sencor or Lexone on coarse soils with less than 1% organic matter.

[b] Use the higher rate of Amiben when velvetleaf or black nightshade is a problem.

[c] On Clarion/Webster soils in Minnesota and Iowa or on similar alkaline (calcareous) soils with a pH of 7.5 or above, apply Sencor or Lexone at the rates listed below:

Soil Texture	Lexone 50WP/4L or Sencor 50WP/4	or Lexone (dry flowable) or Sencor (dry flowable)
	(pounds/pints)	(pounds)
Medium	½	⅓
Fine	½–¾[d]	⅓–½[d]

d Use the higher rate only where soil pH is less than 7.5 and where weed pressure is heavy.

Additional precautions: The TREFLAN/Amiben/Sencor or Lexone tank mix will not harm the treated crop when you apply it according to directions and under normal growing conditions. However, overapplication, uneven application or improper soil incorporation of the tank mix can result in erratic weed control or crop injury. Additional stress factors are seedling disease, cold weather, deep planting, excessive moisture, soil pH over 7.5, high salt concentration, or drought. These additional factors may weaken crop seedlings and increase the possibility of damage from the tank mix. These additional factors may also delay crop development or reduce yields. Observe all cautions and limitations of all products used in mixtures. Do not use the foliage from soybeans treated with the TREFLAN/ Amiben/Sencor or Lexone tank mix for feed or forage.

SOYBEAN—TREFLAN/Vernam tank mix:

Vernam®—vernolate, Stauffer Chemical Company

TREFLAN/Vernam tank mix effectively controls those weeds listed for TREFLAN alone plus these additional weeds:

Annual morningglory	Velvetleaf
Coffeeweed	Yellow nutsedge (nutgrass)
Purple nutsedge (nutgrass)	

Follow recommended soil preparation procedures for TREFLAN. You may apply the TREFLAN/Vernam tank mix up to 10 days prior to planting. Incorporate the tank mix immediately after application. Apply TREFLAN/Vernam at these rates:

Broadcast Rates Per Acre

Soil Texture	TREFLAN M.T.F.	Vernam 7E
	(pints)	(pints)
Coarse	1	1¾–2⅓
Medium	1½	2⅓–3*
Fine	2	3–3½

* For nutsedge, wild cane and velvetleaf control, use the higher rate of 3 pints per acre on medium textured soils.

ALFALFA—ESTABLISHED

In areas receiving less than 20" average annual rainfall per year, apply TREFLAN to established alfalfa stands at a broadcast rate of 1½ pints per acre on coarse soils and 2 pints on medium and fine soils. Use incorporation equipment that will insure thorough soil mixing with minimum damage to the established alfalfa.

ASPARAGUS—ESTABLISHED

Follow recommended soil preparation, application and incorporation procedures for TREFLAN.

TREFLAN can be applied to established asparagus as a single or as a split application. In the winter or early spring, apply TREFLAN to asparagus after ferns are removed but before spear emergence. Or, apply after harvest in the late spring or early summer before ferning begins. TREFLAN will suppress volunteer seedling asparagus and field bindweed if you use the following recommended rates and application schedules. Follow recommended soil preparation, application and incorporation procedures for TREFLAN.

Broadcast Rates Per Acre

Soil Texture	TREFLAN*					
	Split Application			Single Application		
	Before Harvest	+	After Harvest	Before Harvest	or	After Harvest
	(pints)			(pints)		
Coarse	1	+	1	2	or	2
Medium	1½	+	1½	3	or	3
Fine	2	+	2	4	or	4

In any single calendar year, the maximum TREFLAN to apply is 2 pints per acre on coarse soils; 3 pints on medium soils; and 4 pints on fine soils.

FOR THE FOLLOWING CROP GROUPING, USE THE RATE TABLE BELOW.

CARROT

CASTOR BEAN

CELERY—(Direct seeded and transplant in areas receiving less than 20" average annual rainfall.)

COLE CROPS—TRANSPLANT

Apply and incorporate prior to transplanting only. (Broccoli, Brussels Sprout, Cabbage and Cauliflower) See next section for direct seeded.

OKRA

PEPPER—TRANSPLANT

Apply and incorporate prior to transplanting only.

SOUTHERN PEA—(Before planting only.)

Apply and incorporate TREFLAN before planting, at planting or immediately after planting, unless otherwise indicated.

Broadcast Rates Per Acre

Soil Texture	TREFLAN M.T.F.	
	Areas receiving less than 20" average annual rainfall*	Areas receiving greater than 20" average annual rainfall*
	(pints)	(pints)
Coarse	1	1
Medium	1¼–1½	1½
Fine	1½	2

* Use 1½ pints per acre on coarse and medium textured soils and 2 pints on fine soils with 2–5% organic matter; use 2 pints on all soils with 5–10% organic matter.

COLE CROPS—DIRECT SEEDED (Broccoli, Brussels Sprout, Cabbage and Cauliflower) See above section for transplant.

For direct-seeded cole crops, apply and incorporate TREFLAN before planting at a broadcast rate of 1 pint per acre on coarse and medium soils and 1½ pints on fine soils and soils with 2–5% organic matter. Direct-seeded cole crops have exhibited marginal tolerance to recommended rates of TREFLAN. Stunting or reduced stands may occur.

CUCURBITS—POSTPLANT EMERGED (Cantaloupe, Cucumber and Watermelon)

Western United States including Texas:
Apply TREFLAN as a directed spray to the soil between the rows and beneath plants which are in the 3 to 4 true leaf stage.

Broadcast Rates Per Acre

Soil Texture	TREFLAN M.T.F.	
	Areas receiving less than 20" average annual rainfall*	Areas receiving greater than 20" average annual rainfall*
	(pints)	(pints)
Coarse	1	1
Medium	1¼–1½	1½
Fine	1½	2

* Use 1½ pints on coarse and medium textured soils and 2 pints on fine soils with 2–5% organic matter; use 2 pints on all soils with 5–10% organic matter.

Set incorporation equipment to throw treated soil around the plants during incorporation.

DRY BEAN—TREFLAN ALONE

Apply and incorporate TREFLAN before planting using the following rates:

Broadcast Rates Per Acre

Soil Texture	TREFLAN M.T.F.	
	Areas receiving less than 20" average annual rainfall*	Areas receiving greater than 20" average annual rainfall*
	(pints)	(pints)
Coarse	1	1
Medium	1¼–1½	1½
Fine	1½	2

* Use 1½ pints per acre on coarse and medium textured soils and 2 pints on fine soils with 2–5% organic matter; use 2 pints on all soils with 5–10% organic matter.

DRY BEAN—TREFLAN/Eptam tank mix:

The TREFLAN/Eptam tank mix effectively controls all the following weeds in addition to those weeds listed for TREFLAN:

Henbit (Spring applications)	Oat, wild
Nightshade, black	Ragweed, common
Nightshade, hairy	Smartweed, Pennsylvania
Nutsedge	Velvetleaf (Buttonweed)

Follow recommended soil preparation and incorporation procedures for TREFLAN. The TREFLAN/Eptam tank mix should be applied from 2 days before planting up to planting. Incorporate immediately after application.

Broadcast Rates Per Acre

Soil Texture	TREFLAN M.T.F. Areas receiving less than 20" average annual rainfall*	TREFLAN M.T.F. Areas receiving greater than 20" average annual rainfall*	Eptam 7E
	(pints)	(pints)	(pints)
Coarse	1	1	2½–3½**
Medium	1¼–1½	1½	2½–3½
Fine	1½	2	2½–3½

* Use 1½ pints per acre on coarse and medium textured soils and 2 pints on fine soils with 2–5% organic matter; use 2 pints on all soils with 5–10% organic matter.

** Use Eptam 7E at a rate of 2½ pints per acre to control annual grasses; 3½ pints to control nutsedge and additional broadleaf weeds.

Precaution: Read the Eptam label before using. Observe all cautions and limitations of all products used in mixtures. The combination of TREFLAN and Eptam should not be used on soybeans, black-eyed peas (beans), lima beans and other flatpodded beans, except Romano. Do not use the foliage from a crop treated with the TREFLAN/Eptam tank mix for feed or for grazing.

DRY BEAN—Fall application for dry bean grown in Idaho, Oregon and Washington:

Apply and incorporate TREFLAN any time between October 15 and December 31 at a broadcast rate of 1 pint per acre on coarse soils; 1¼ to 1½ pints on medium soils; and 1½ pints on fine soils. Destroy established weeds during seedbed preparation.

FOR THE FOLLOWING CROP GROUPING, USE THE RATE LISTED BELOW.

BEANS—(Guar and Mungbean)

GREENS—Turnip greens grown for processing, Collard, Kale and Mustard greens.

MUSTARD—Grown for seed or processing for food in Minnesota, Montana and North Dakota:

Apply and incorporate TREFLAN before planting at 1 pint per acre on coarse soils and 1½ pints on medium and fine soils.

BEANS—(Lima Bean and Snap Bean):

Apply and incorporate TREFLAN before planting at a broadcast rate of 1 pint per acre on coarse and medium soils and 1½ pints on fine soils.

CORN (Field Corn) and GRAIN SORGHUM (Milo)

Apply TREFLAN to field corn or grain sorghum (8 inches or taller) as an over-the-top or directed spray to effectively control weeds listed for TREFLAN.

Soil Preparation—Cultivate before a TREFLAN application to insure loose, friable soil, to remove established weeds, and to cover the base of plants with soil.

Application Directions—TREFLAN should be applied and incorporated at the recommended rates for the soil texture when the crop is well established (8 inches or taller). TREFLAN may be applied either as an over-the-top spray or as a directed spray. Drop nozzles should be used if foliage prevents uniform coverage of soil surface. Soil incorporation may be accomplished with only one pass of a sweep-type cultivator or a properly adjusted rolling cultivator.

The sweep-type cultivator should have 3 to 5 sweeps per row middle and be operated at 6 to 8 mph. Set the middle sweeps so as to avoid exposing untreated soil. Adjust the incorporation tools to prevent crop injury.

Broadcast Rates Per Acre

Soil Texture	TREFLAN M.T.F.
	(pints)
Coarse	¾–1*
Medium	1–1½
Fine	1½–2

Use the lower rates when you anticipate light weed pressure and the higher rates when you anticipate heavy weed pressure.

* Corn only: Apply 1 to 1½ pints per acre in Alabama, Florida, Georgia, North Carolina, South Carolina and Virginia to control fall panicum and Texas panicum.

Precaution: Do not apply TREFLAN to corn grown for seed. Do not apply TREFLAN to corn or sorghum as a preplant or preemergence treatment, or crop injury may occur.

HOPS

Apply and incorporate TREFLAN while the crop is dormant. Use a broadcast rate of 1 pint per acre on coarse soils; 1¼ to 1½ pints on medium soils; and 1½ pints on fine soils and soils with 2–10% organic matter.

MINT—(Established Peppermint and Spearmint)

Apply TREFLAN at a rate of 1 pint per acre on coarse soils; 1¼ pints on medium soils; and 1½ pints on fine soils. Use incorporation equipment that will insure thorough soil mixing with minimum damage to the crop.

PEA—(Dry and English)—TREFLAN ALONE

Apply and incorporate TREFLAN before planting at a rate of 1 pint per acre on coarse and medium soils and 1½ pints on fine soils.

PEA—TREFLAN/Far-Go tank mix for pea in Idaho, Oregon and Washington:

The tank mix combination of TREFLAN plus Far-Go will provide control of wild oat in addition to other annual grasses and broadleaf weeds controlled by TREFLAN.

Application Rates: Broadcast ¾ pint of TREFLAN per acre on coarse and medium soils; 1 pint of TREFLAN on fine soils. Use 1¼ quarts of Far-Go per acre for all soil textures.

Incorporation Directions: Apply and incorporate up to 3 weeks before planting. Follow recommended incorporation procedures for TREFLAN.

Precaution: Do not apply to lentils. Leaf crinkling and delayed maturity of peas may occur, particularly on clay points in the northwest; but this is usually more than offset by a reduction of wild oat. Do not use foliage from treated peas for feed or forage. Refer to the cautions, precautions and directions on the Far-Go label.

PEA—Fall application to dry pea and English pea in Idaho, Oregon and Washington:

Apply and incorporate TREFLAN any time between October 15 and December 31 at a broadcast rate of 1 pint per acre on coarse soils; 1¼ to 1½ pints on medium soils; and 1½ pints on fine soils. Destroy established weeds during seedbed preparation. Do not apply TREFLAN in the fall to soils which are wet or are subject to prolonged periods of flooding.

PEANUT—(Spanish Peanut in Texas and Oklahoma)

Apply and incorporate TREFLAN before planting, at planting or immediately after planting at a broadcast rate of 1 pint per acre on coarse soils. When incorporating after planting, take care not to disturb the seed.

PEANUT—TREFLAN/Vernam tank mix (Spanish Peanut in Texas and Oklahoma):

TREFLAN/Vernam tank mix effectively controls those weeds listed for TREFLAN alone plus these additional weeds:

Annual morningglory	Velvetleaf
Coffeeweed	Yellow nutsedge (nutgrass)
Purple nutsedge (nutgrass)	

Follow recommended soil preparation procedures for TREFLAN. You may apply the TREFLAN/Vernam tank mix up to 10 days prior to planting. Incorporate the tank mix immediately after application. Apply TREFLAN/Vernam at these rates:

Broadcast Rates Per Acre

Soil Texture	TREFLAN M.T.F.	Vernam 7E
	(pints)	(pints)
Coarse	1	2⅔

POTATO—(All states except Maine)

Apply and incorporate TREFLAN after planting, before emergence, or immediately following dragoff or after the potato plants have fully emerged.

Broadcast Rates Per Acre

Soil Texture	TREFLAN M.T.F.	
	Areas receiving less than 20″ average annual rainfall*	Areas receiving greater than 20″ average annual rainfall*
	(pints)	(pints)
Coarse	1	1
Medium	1¼–1½	1½
Fine	1½	2

* Use 1½ pints per acre on coarse and medium soils with 2–5% organic matter; use 2 pints on all soils with 5–10% organic matter.

Set incorporation equipment so that the bed and furrow will be uniformly covered with a layer of treated soil. If the layer of treated soils is not uniform and the herbicide is concentrated over the bed, potato emergence may be retarded, and stem brittleness can occur. When applying and incorporating TREFLAN after potato plants have fully emerged, do not completely cover the foliage with treated soil. Likewise, do not completely cover foliage at subsequent cultivations. Be careful that incorporation machinery does not damage potato seed pieces or elongating sprouts.

POTATO—Split application in Idaho, Oregon and Washington: On all soils, apply and incorporate ¾ pint of TREFLAN per acre before planting and ¾ pint after planting when potato plants have fully emerged. Do not apply to soils containing 2% or more organic matter. Follow incorporation directions listed above for application to potato after planting.

POTATO—TREFLAN/Eptam tank mix for potatoes grown in Kansas, Minnesota, Nebraska, North Dakota, Oklahoma, South Dakota and Texas:

The TREFLAN/Eptam tank mix effectively controls the following weeds in addition to those weeds controlled by TREFLAN.

Henbit (Spring applications)	Oat, wild
Nightshade, black	Ragweed, common
Nightshade, hairy	Smartweed, Pennsylvania
Nutsedge	Velvetleaf (Buttonweed)

Follow recommended soil preparation and application procedures for TREFLAN. The TREFLAN/Eptam tank mix may be applied after planting, but prior to crop emergence. In areas where potatoes are normally dragged off, the TREFLAN/Eptam tank mix should be applied and incorporated up to or immediately following drag off.

Broadcast Rates Per Acre

Soil Texture	TREFLAN M.T.F.		
	Areas receiving less than 20″ average annual rainfall*	Areas receiving greater than 20″ average annual rainfall*	Eptam 7E
	(pints)	(pints)	(pints)
Coarse	1	1	1¾–7**
Medium	1–1½	1–1½	1¾–7
Fine	1–1½	1–2	1¼–7

* Use 1½ pints per acre on coarse and medium soils with 2–5% organic matter; use 2 pints on all soils wih 5–10% organic matter.

** Use the higher rate of Eptam 7E for nutsedge control.

Precaution: Read the Eptam label before using. Observe cautions and limitations of products used in mixtures. Do not graze or feed forage to livestock from fields treated with TREFLAN/Eptam tank mix.

POTATO—TREFLAN/Eptam application before planting in Washington, Idaho and Oregon:

TREFLAN/EPTAM may be applied before planting at a broadcast rate of ¾ pint of TREFLAN per acre and 3½ pints of Eptam 7E per acre on all soil textures. Incorporate immediately.

Precaution: Do not use this tank mix both before and after planting in the same season. Read the Eptam label before using. Observe all cautions and limitations on labeling of all products used in mixtures. Do not use foliage from treated crops for feed or forage.

SAFFLOWER

Follow recommended soil preparation, application and incorporation procedures for TREFLAN.

Apply and incorporate TREFLAN in the spring before planting or in the fall between October 15 and December 31.

Broadcast Rates Per Acre

Soil Texture	TREFLAN M.T.F.	
	Areas receiving less than 20″ average annual rainfall*	Areas receiving greater than 20″ average annual rainfall*
	(pints)	(pints)
Coarse	1	1
Medium	1¼–1½	1½
Fine	1½	2

* Use 1½ pints per acre on coarse and medium textured soils and 2 pints on fine soils with 2–5% organic matter; use 2 to 2½ pints on all soils with 5–10% organic matter.

SAFFLOWER—Fall application in Arizona, California, Idaho, Montana, Nevada, Oregon, Utah, Washington and Wyoming:

Apply and incorporate TREFLAN any time between October 15 and December 31. Ground may be left flat or bedded-up over winter. On bedded ground, knock beds down to desired height before planting, moving some treated soil from tops into furrows. Where soil is left flat over winter, take care during spring bedding operations to prevent turning up untreated soil. Destroy established weeds during seedbed preparation. If weeds become established in furrows due to uncovering of untreated soil during listing, destroy these weeds before planting. Apply and incorporate TREFLAN at a broadcast rate of 1½ pints per acre on coarse soils; 2 pints on medium soils; and 2½ pints on fine soils.

Do not apply TREFLAN in the fall to soils which are wet or are subject to prolonged periods of flooding.

SUGAR BEET

Apply TREFLAN as a broadcast, overtop spray when plants are between 2 and 6 inches tall at a rate of 1 pint per acre on coarse soils and 1¼ to 1½ pints on medium and fine soils. Use the higher rate for medium and fine soils in areas receiving more than 20″ average annual rainfall. Set incorporation machinery to throw treated soil toward the plants in the row. Be careful that incorporation machinery does not damage the sugar beet taproot.

Precaution: Exposed beet roots should be covered with soil before a TREFLAN application to reduce the possibility of girdling.

SUGAR BEET—Incorporation with a tine-tooth harrow in California, Colorado, Idaho, Montana, Nebraska, Oregon, Texas, Utah, Washington and Wyoming:

A properly operated tine-tooth harrow (Flextine or Melroe) can incorporate TREFLAN for effective weed control in sugar beet. Operate the tine-tooth harrow 2 times over the field in opposing directions at a speed of 3 to 6 mph. Set the harrow to cut 1 or 2 inches deep. Be careful that the tine-tooth harrow does not damage the sugar beet taproot. Follow recommended application procedures and broadcast rates per acre for sugar beet (see preceding paragraph).

SUGARCANE—(Plant Cane)

Apply and incorporate TREFLAN twice a year at a broadcast rate of 2 to 4 pints per acre for all soil textures. Make the first TREFLAN application in the fall on firmly packed beds immediately after the seed pieces are planted. Make the second TREFLAN application in the spring before or shortly after the cane emerges. Loosen rain-packed beds 2 to 3 inches deep before the spring application. Take care that incorporation machinery does not damage the seed pieces or emerging shoots.

SUGARCANE—Postplant in Hawaii for control of most annual grasses, including guineagrass:

Surface apply TREFLAN after planting (for plant cane) or after harvesting (for ratoon cane), before weeds and cane emerge. Use a broadcast rate of 6 to 8 pints per acre for all soil textures. In plant cane, the beds should be formed or rolled before application. In ratoon cane, the crop residue should be removed before application. If large amounts of crop residues are present, TREFLAN will not be effective. Apply TREFLAN just before anticipated rainfall or sprinkle irrigate immediately after application.

SUGARCANE—Applications up to layby for plant cane or ratoon cane grown in Louisiana or Texas:

Apply and incorporate TREFLAN at a broadcast rate of 2 to 4 pints per acre for all soil textures. Do this in the spring from before or shortly after the cane emerges up to layby. Apply the TREFLAN after the beds have been shaved or false shaved. Loosen rain-packed beds 2 to 3 inches deep before application. Be careful that incorporation machinery does not damage seed pieces or emerging shoots. You may use a rolling cultivator or bed chopper to incorporate TREFLAN layby applications in sugarcane on all soil textures. Follow normal incorporation directions for the rolling cultivator. Set bed chopper to cut 3 to 4 inches deep and operate at 4 to 6 mph. Two incorporation passes are necessary.

SUGARCANE—Itchgrass (Raoulgrass) control in Louisiana:

Apply and incorporate TREFLAN on either plant or ratoon cane at a broadcast rate of 4 pints per acre for all soil textures. Follow the directions above for sugarcane layby application in Louisiana and Texas.

SUNFLOWER—TREFLAN ALONE

Apply and incorporate TREFLAN in the spring or in the fall between October 15 and December 31. Follow recommended soil preparation, application and incorporation procedures for TREFLAN.

Broadcast Rates Per Acre

Soil Texture	TREFLAN M.T.F. Areas receiving less than 20″ average annual rainfall*	TREFLAN M.T.F. Areas receiving greater than 20″ average annual rainfall*
	(pints)	(pints)
Coarse	1	1
Medium	1¼–1½	1½
Fine	1½	2

* Use 1½ to 2 pints per acre on coarse and medium soils with 2–5% organic matter and 2 pints on all soils with 5–10% organic matter.

SUNFLOWER—TREFLAN/Amiben tank mix or overlay:

Follow recommended soil preparation, application and incorporation procedures for TREFLAN.

Amiben may be applied in a band or broadcast over sunflowers at planting in fields where TREFLAN has been incorporated prior to planting. Or, the TREFLAN/Amiben tank mix may be incorporated prior to planting. TREFLAN/Amiben tank mix improves mustard, smartweed, velvetleaf and ragweed control in addition to those weeds controlled by TREFLAN alone.

Apply TREFLAN/Amiben or TREFLAN with an Amiben overlay at the following broadcast rates per acre:

Soil Texture	TREFLAN M.T.F.	Amiben 2S
	(pints)	(quarts)
Coarse	1	4
Medium	1½	4–6*
Fine	2	4–6*

*For best control of mustard, common ragweed or black nightshade, use the 6 quart rate.

In coarse textured soils, heavy rains on the incorporated Amiben may move it below the weed seed germination zone, and erratic weed control may result. If sufficient rain does not fall within 7 days after a preemergence application of Amiben, but there is enough soil moisture to germinate weeds and grasses, a light cultivation with a rotary hoe or similar tool will uproot these small broadleaf weeds and grasses. The shallow mixing of Amiben in the surface soil will not interfere with the action of Amiben when rains come.

TOMATO

For direct-seeded tomato, apply TREFLAN at blocking or thinning as a directed spray to the soil between rows and beneath the plants, and incorporate. For transplant tomato, apply and incorporate TREFLAN before transplanting. Do not apply TREFLAN after transplanting.

Broadcast Rates Per Acre

Soil Texture	TREFLAN M.T.F. Areas receiving less than 20″ average annual rainfall*	TREFLAN M.T.F. Areas receiving greater than 20″ average annual rainfall*
	(pints)	(pints)
Coarse	1	1
Medium	1¼–1½	1½
Fine	1½	2

* Use 1½ pints per acre on coarse and medium textured soils and 2 pints on fine soils with 2–5% organic matter; use 2 pints on all soils with 5–10% organic matter.

FRUIT AND NUT CROPS AND VINEYARDS

For areas receiving more than 20″ average annual rainfall.

For new plantings of vineyards, citrus and pecan trees, apply and incorporate TREFLAN before planting at a broadcast rate of 1 pint per acre on coarse soils; 1½ pints on medium soils; 2 pints on fine soils; 1½ pints on fine soils with 2–5% organic matter; and 2 pints on soils with 5–10% organic matter.

For non-bearing, established plantings of citrus and pecan trees and bearing plantings of grapefruit, lemon, orange, pecan, tangelo and tangerine trees, apply TREFLAN at a broadcast rate of 2 to 4 pints per acre for all soil textures.

For areas receiving less than 20″ average annual rainfall.

For new plantings of almond, apricot, citrus, nectarine, peach, pecan and walnut trees, apply and incorporate TREFLAN before planting at a broadcast rate of 1 pint per acre on coarse soils; 1¼ to 1½ pints on medium soils; 1½ pints on fine soils; 1½ to 2 pints on soils with 2–5% organic matter; and 2 pints on soils with 5–10% organic matter.

For new plantings of vineyards, apply and incorporate TREFLAN before planting at a broadcast rate of 1 to 1½ pints per acre on coarse soils; 1½ to 3 pints on medium soils; and 3 to 4 pints on fine soils or soils with 2–10% organic matter. Do not use more than 2 pints per acre on heat-treated grape rootings.

For postplant applications on bearing or non-bearing, established plantings of vineyards and almond, apricot, grapefruit, lemon, nectarine, orange, peach, plum, prune, tangelo, tangerine and walnut trees, apply TREFLAN at a broadcast rate of 2 to 4 pints per acre for all soil textures. Do not apply to vineyards within 60 days of harvest.

In established plantings, apply TREFLAN as a directed spray to the soil and use incorporation methods not injurious to the trees or vines.

FRUIT AND NUT CROPS AND VINEYARDS—Rhizome johnsongrass control:

For areas receiving less than 20" average annual rainfall.

You can obtain commercially acceptable control of rhizome johnsongrass with postplant applications in bearing and non-bearing, established plantings of vineyards and almond, apricot, grapefruit, lemon, nectarine, orange, peach, pecan, tangelo, tangerine and walnut trees with a TREFLAN program when applied for 2 years in a row.

Soil Preparation—Work the soil thoroughly to bring the rhizomes nearer the surface.

Application—Apply TREFLAN at a broadcast rate of 2 quarts per acre on all soil textures each year for 2 years in a row. Do not apply to vineyards within 60 days of harvest.

Incorporation—Incorporate TREFLAN thoroughly with a disc set to cut 4 to 6 inches deep and operate at 4 to 6 mph. Two incorporation passes are necessary, with the second pass in a different direction from the first.

Cultivation—Some johnsongrass plants will escape. Timely cultivations are necessary to obtain commercially acceptable control. You cannot obtain commercially acceptable control with only 1 year of TREFLAN use.

Precautions: Do not use the 2 quart rate on new plantings, or crop injury may result. Do not interplant orchards or vineyards with other crops. If the treated vineyards and orchards are diverted to other crop uses, plant only those crops for which TREFLAN has been registered as a preplant treatment for the next cropping season.

FRUIT AND NUT CROPS AND VINEYARDS—Bindweed control in California:

TREFLAN can be used for the control of field bindweed in vineyards and for almond, apricot, grapefruit, lemon, nectarine, orange, peach, pecan, tangelo, tangerine and walnut trees.

Apply TREFLAN at a broadcast rate of 4 pints per acre on all soil textures. TREFLAN must be applied in the spring with a specially designed spray blade which applies a thin, concentrated layer at a soil depth of 4 to 6 inches. This layer of TREFLAN prevents bindweed shoots from emerging.

Land Preparation—Destroy all weeds and grasses with soil tillage before applying TREFLAN. This tillage is necessary to prevent trash from interfering with the operation of the spray blade.

Equipment—This operation requires a spray blade capable of running 4 to 6 inches below the surface of the soil. The spray blade should be equipped with nozzles located under the blade and directed so that the TREFLAN spray will be trapped under the soil which is flowing over the blade as it is pulled through the soil. Use a sufficient number of nozzles with spacing that will uniformly apply the TREFLAN underground in a thin, horizontal layer.

Application—Apply TREFLAN in 40 to 80 gallons of water per acre. Operate the spray blade at a depth of 4 to 6 inches.

Precaution: Some soils develop cracks as they dry after rainfall or irrigation. Field bindweed may emerge if the cracks extend through the TREFLAN layer. Prevent or eliminate cracks by shallow discing or other tillage. Avoid deep tillage which disturbs the subsurface layer. Cultivation or tillage also aids the control of germinating seeds.

WHEAT (WINTER)—(Idaho, Montana, Oregon and Washington)

TREFLAN may be applied for preplant preemergence control of cheatgrass and other annual grasses and broadleaf weeds controlled by TREFLAN. The growth, development and yield of winter wheat will not be adversely affected, provided the seed is placed below the zone of soil treated with TREFLAN.

Apply TREFLAN any time during a period from 3 weeks up to immediately prior to planting. Broadcast TREFLAN at a rate of 1½ pints per acre on coarse and medium soils and 2 pints on fine soils.

Incorporation Directions—Incorporate TREFLAN into the soil with a flexible tine-tooth harrow (Flextine, Melroe) set to cut 1 to 2 inches deep and operate at 3 to 6 mph. Incorporate one time within 24 hours after application, followed by a second incorporation in a different direction from the first prior to planting. Do not till the soil with a disc after the TREFLAN has been incorporated with a flexible tine harrow.

Seeding Directions—Use only a deep furrow or semi-deep furrow drill that will place the seed below the zone of soil into which TREFLAN has been incorporated.

Precaution: Wheat planted in direct contact with treated soil may suffer crop injury in the form of delayed emergence and development.

WHEAT (WINTER)—Fallow soil application in Washington and Oregon:

TREFLAN applied and shallowly incorporated into fallow soil up to four months ahead of planting will control cheatgrass and certain annual grasses and broadleaf weeds. The growth, development, or yield will not be adversely affected as long as the seed is placed below the zone of soil treated with TREFLAN. Use deep or semi-deep furrow drills.

Broadcast rates are 1½ pints per acre on coarse and medium soils and 2 pints on fine soils. Apply TREFLAN any time from May to September prior to the fall planting of winter wheat.

Incorporation—Incorporate TREFLAN with a flexible tine-tooth harrow (Flextine or Melroe) set to cut 1 to 2 inches deep and operated at 3 to 6 mph. For thorough incorporation, two passes of the equipment in different directions over the field are necessary. Incorporate one time within 24 hours after application followed by a second incorporation prior to seeding. Do not till the soil with a disc after TREFLAN has been applied with a flexible tine harrow.

Precaution: Use only deep furrow or semi-deep furrow drills. Place seed below the zone of soil into which TREFLAN has been incorporated. Do not plant wheat directly into the zone of soil treated with TREFLAN; injury to the crop or a delay in its emergence and development may occur.

WHEAT (SPRING), DURUM AND BARLEY—TREFLAN ALONE

TREFLAN is recommended as a postplant incorporated treatment to control foxtail (pigeongrass).

Apply TREFLAN at a broadcast rate of 1 pint per acre on coarse and medium soils and 1½ pints on fine soils.

Plant 2 to 3 inches deep in a well-tilled seedbed. Apply TREFLAN after seeding but before the crop emerges. To incorporate, use flextine or diamond harrows operated two times in different directions, at speeds of at least 5 mph. Incorporate by operating equipment 1 to 1½ inches deep. Application and the first incorporation should be done in the same operation if possible. Both incorporations must be done within 24 hours.

WHEAT (SPRING), DURUM, BARLEY (FALL APPLICATION)—Foxtail/Pigeongrass control:

TREFLAN may be fall applied for foxtail/pigeongrass control in spring wheat, durum and barley planted the following spring. TREFLAN may be applied to ground that has a manageable trash level, has been fallowed or pre-tilled. The first incorporation is required within 24 hours after application. A second incorporation is required prior to planting to destroy emerged weeds and to ensure an even distribution of TREFLAN treated soil.

Broadcast Rates Per Acre

Soil Texture	TREFLAN M.T.F.
	(pints)
Coarse	1
Medium	1
Fine	1½

Incorporation Directions—Any of the following tools are recommended for fall incorporation. The disc or field cultivator may be used for the spring incorporation pass. Care should be taken to operate the tool at a more shallow depth than the fall incorporation.
1. Chisel plow: *May be used for the first pass only.* Operate at 4–5

inches deep at 4–6 mph. A chisel plow is defined as having 3 rows of up to 18-inch sweeps on no greater than 12-inch centers. Stagger sweeps so that no soil is left unturned.

2. Tandem disc: Operate at 3–4 inches deep at 4–6 mph.
3. Field cultivator: Operate at 3–4 inches deep at 5 mph or more. A field cultivator is defined as having 3 to 4 rows of sweeps with "c" or "s" shaped shanks, spaced 7 inches or less and staggered so that no soil is left unturned.

Planting Directions—Wheat, durum, or barley should be approximately 2 inches deep.

Precaution—While use of this practice may result in a stand reduction, slight stand reductions do not normally affect yield.

WHEAT (SPRING), DURUM AND BARLEY—TREFLAN/Far-Go tank mix:

TREFLAN/Far-Go applied as a postplant incorporated treatment will control foxtail (pigeongrass) and wild oat.

Plant 2 to 3 inches deep in a well-tilled seedbed. Apply TREFLAN/Far-Go after seeding but before crop emerges. To incorporate, use flex-tine or diamond harrows two times, operated in different directions, at speeds of at least 5 mph. Incorporate by operating equipment 1 to 1½ inches deep. Application and the first incorporation should be done in the same operation if possible. If not, incorporate immediately after application.

Broadcast Rates Per Acre

Soil Texture	TREFLAN M.T.F. Barley Durum Spring Wheat	Far-Go Durum Spring Wheat	Far-Go Barley
	(pints)	(pints)	(pints)
Coarse	1	2	2½
Medium	1	2	2½
Fine	1½	2	2½

Precaution: Overapplication may result in crop injury. Read the Far-Go label carefully before using.

FALL APPLICATION

General: (Areas receiving more than 20″ average annual rainfall)

See specific crop for recommendations. For all crops for which there are no specific fall application instructions and for which TREFLAN is recommended as a preemergence application, use the rates listed for spring applications. Do not apply TREFLAN in the fall for sugarbeets, potatoes and direct-seeded tomatoes.

In most states apply and incorporate TREFLAN any time between October 15 and December 31. In Minnesota, Montana, North and South Dakota, apply and incorporate TREFLAN anytime between September 1 and December 31. Ground may be left flat or bedded-up over winter. On bedded ground, knock beds down to desired height before planting, moving some treated soil from beds into furrows. Where soil is left over winter, be careful not to turn up untreated soil during spring bedding operations. Destroy established weeds during seedbed preparation. If weeds become established in furrows due to uncovering of untreated soil during bedding, destroy these weeds before planting. Do not apply TREFLAN in the fall to soils which are wet, are subject to prolonged periods of flooding, or where rice was grown the previous year.

FERTILIZER USE DIRECTIONS—APPLICATION WITH LIQUID FERTILIZERS

TREFLAN may be mixed with most liquid fertilizer materials. The combination of TREFLAN with solutions and suspension-type fertilizers has provided weed and grass contol equal to the same rates of TREFLAN applied in water. Follow TREFLAN label recommendations regarding rates per acre, crops, incorporation directions, special instructions, cautions and special precautions.

Individual state regulations relating to liquid fertilizer mixing, registration, labeling and applications are the responsibility of the individual and/or company selling the fertilizer and chemical mixture.

Testing for Tank Mix Compatibility in Liquid Fertilizers: TREFLAN alone or in tank mixture with dry flowables, wettable powders (WP), aqueous suspensions (AS), flowables (F), liquids (L), or solutions (S) may not combine properly with some fluid fertilizer materials. Small quantities should always be tested before full-scale mixing. This will determine whether a compatibility agent is needed, and which agent does the best job. The seven agents listed below have been thoroughly tested. There are many other surfactants on the market which were not designed for use with liquid fertilizers. Use the following test to select the correct agent for your mixture.

1. Put 1 pint of the liquid fertilizer in a quart jar.
2. Add 1 to 4 teaspoonful(s) of the dry flowable, WP, AS, F or L formulation (depending on the recommended rate per acre) to the liquid fertilizer. Close jar and agitate until the materials are dispersed evenly in the fertilizer. If the materials do not disperse well, it may be necessary to slurry the chemicals in water before adding to the fertilizer.
3. After dispersing the materials (Step 2), add 3 to 4 teaspoonsful of TREFLAN to the jar and shake well. Add solution herbicides to the mixture last and agitate. Observe the jar for about 10 minutes. If the materials rise to the surface and form a thick layer (oily curds) which will not redisperse when agitated, a compatibility agent is needed. If the mixture is easily redispersed to its original state with slight agitation, no agent is needed but good agitation must be provided in the fertilizer spray tank.
4. If the need for a compatibility agent is shown in Step 3: Using a clean quart jar, start at Step 1 above, add ½ teaspoon of the compatibility agent to the liquid fertilizer, mix well, then repeat Steps 2 and 3.

An effective compatibility agent will cause the mixture to remain uniformly mixed with little or no separating or oil rising to the surface for one half hour or longer. If slight separation does occur, 2 to 3 inversions of the jar should give a uniform remix. If oily curds form which will not redisperse, more agent or another agent should be tried.

Use a clean jar for each test. The compatible mixture will have a uniform appearance and will be relatively easy to keep mixed with gentle agitation of the jar.

LIQUID FERTILIZER MIXING INSTRUCTIONS—TREFLAN M.T.F. in Liquid Fertilizer

General—Emulsifiable concentrates, such as TREFLAN M.T.F., can be mixed with liquid fertilizers. In all cases, continuous agitation is required to prevent the TREFLAN from rising to the surface as an oily layer. When necessary, (see Testing for Tank Mix Compatibility in Liquid Fertilizers, above) a compatibility agent can be used to cause the TREFLAN to emulsify properly (i.e., have a milky appearance rather than oily layer). The use of compatibility agents is especially important when tank mixing emulsifiable concentrates (E.C.) with dry flowables, wettable powders (WP), aqueous suspensions (AS), flowables (F), liquids (L), or solutions (S) in liquid fertilizer. If the emulsion is not properly formed and the TREFLAN rises to the surface of the fertilizer as an oil ("oils out"), the oil may combine with the wettable powder, flowable, or suspension to form oily curds (viscous phase) which is difficult to redisperse. Any one of the compatibility agents listed below is helpful in causing liquid concentrates to form non-oiling mixtures with liquid fertilizers. These compatibility agents can be used at rates as low as 1½ to 2 pints per ton of liquid fertilizer and should be mixed well with the fertilizer before adding the liquid concentrate. Read the label on the compatibility agent and follow the directions.

1. Sponto 168D (Witco Chemicals Co., Chicago, IL)
2. Compat (Farm Chemicals, Inc., Aberdeen, NC)
3. Unite (Hopkins Ag Chemical, Madison, WI)
4. T-Mulz 734-2 (Thompson-Hayward Chemical Co.)
5. Rigo Compatibility Agent (Rigo Company, Buckner, KY)
6. Amoco Spray Mate™ (Amoco Oil Co., Chicago, IL)
7. Kem-Link (Universal Coop, Minneapolis, MN)

All of the above are phosphate ester-type surfactants designed to be used with liquid fertilizers. They usually do not work as compatibility agents in tank mixtures in plain water.

APPLICATION

Spread the fertilizer/pesticide mixture with a properly calibrated applicator. Be certain the material is applied uniformly to the soil surface.

INCORPORATION

Follow normal TREFLAN incorporation procedures.

TREFLAN APPLICATION WITH DRY BULK FERTILIZERS

GENERAL

Dry bulk fertilizers may be impregnated or coated with TREFLAN. Application of dry bulk fertilizers impregnated with TREFLAN has provided weed and grass control equal to the same rates of TREFLAN applied in water.

All TREFLAN label recommendations regarding rates per acre, approved crops, incorporation, special instructions, cautions and special precautions must be followed.

Apply a minimum of 200 pounds per acre of dry fertilizer impregnated with TREFLAN at the recommended rates. Any commonly used dry fertilizers can be used for TREFLAN impregnation except coated ammonium nitrate and straight limestone. These materials will not absorb the herbicide. Blends containing mixtures of these materials can be impregnated.

Impregnation—Use any closed drum, belt, ribbon or other commonly used dry bulk fertilizer blender. Apply TREFLAN uniformly to the fertilizer.

Rates—Check specific crop recommendations for the rate of TREFLAN per acre. See the rate table which follows to determine the amount of TREFLAN to be impregnated into a ton of dry bulk fertilizer based on the amount of fertilizer which will be applied per acre. (See rate chart below).

Application—Spread the fertilizer/chemical mixture with a properly calibrated applicator. Be certain the material is applied uniformly to the soil surface.

Incorporation—Follow TREFLAN incorporation procedures.

RATE CHART FOR IMPREGNATING FERTILIZER WITH TREFLAN M.T.F. (TREFLAN Added to a TON of Fertilizer)

Fertilizer Rate Per Acre	TREFLAN Rate Per Acre				
	1 pint	1½ pints	2 pints	3 pints	4 pints
200 pounds	5 qts. per ton	7½ qts. per ton	10 qts. per ton	15 qts. per ton	20 qts. per ton
250 pounds	4 qts. per ton	6 qts. per ton	8 qts. per ton	12 qts. per ton	16 qts. per ton
300 pounds	3⅓ qts. per ton	5 qts. per ton	6⅔ qts. per ton	10 qts. per ton	13⅓ qts. per ton
350 pounds	2¾ qts. per ton	4¼ qts. per ton	5¾ qts. per ton	8½ qts. per ton	11½ qts. per ton
400 pounds	2½ qts. per ton	3¾ qts. per ton	5 qts. per ton	7½ qts. per ton	10 qts. per ton
450 pounds	2¼ qts. per ton	3⅓ qts. per ton	4½ qts. per ton	6⅔ qts. per ton	9 qts. per ton

For rates other than those listed above, use the following formula to calculate the amount of TREFLAN to be impregnated on a ton of dry bulk fertilizer:

$$\text{Pints TREFLAN Per Acre} \times \frac{1000}{\text{Lbs. Fertilizer Per Acre}} = \text{Quarts TREFLAN Per Ton of Fertilizer}$$

All individual state regulations relating to dry bulk fertilizer blending, registration, labeling and application are the responsibility of the individual and/or company selling the fertilizer and chemical mixture.

TREFLAN PERFORMANCE GUARANTEE—A LIMITED WARRANTY OF TREFLAN WEED AND GRASS CONTROL

Elanco Products Company warrants that TREFLAN will control the weeds and grasses as indicated on its current label when used according to label directions and subject to the following limitations:

1. Purchaser must notify Elanco promptly if a lack of satisfactory control occurs. Notice must be given within ninety days after the first planting or after application of TREFLAN to the treated crop, whichever occurs later.
2. An Elanco representative must be satisfied that Purchaser used TREFLAN according to the label directions and must visually observe unsatisfactory control of weeds in the field.
3. The Purchaser must provide Elanco representative with an invoice or other satisfactory record which shows the price and quantity of TREFLAN purchased.
4. Elanco's liability will be limited to a refund of the purchase price of the TREFLAN applied to the acreage on which weed control was not satisfactory or, at the option of the Purchaser, a refund of an equivalent amount of TREFLAN.
5. Void outside the USA.

DISCLAIMER OF WARRANTIES

The warranty printed above is the only warranty applicable to this product. All other warranties, express or implied, including, but not limited to, any implied warranties of merchantability and fitness for a particular purpose, are disclaimed.

LIMITATION OF DAMAGES

Elanco's liability, whether in contract, warranty, tort, negligence, strict liability or otherwise, shall not exceed the return of the amount of the purchase price of the TREFLAN and under no circumstances shall Elanco be liable for special, indirect or consequential damages.

INHERENT RISKS OF USE

Failure to carefully follow the directions for use of TREFLAN, as well as other factors such as seedling disease, cold weather, deep planting, excessive moisture, high salt concentration or drought, may result in unsatisfactory weed control or crop injury. Reduced yields may result under these conditions.

SUPPLEMENTAL LABELING

See Treflan E. C. for Supplemental Labeling pertaining to this product.

TREFLAN® 5

CONCENTRATED FORMULA HERBICIDE

A Selective Herbicide for the Preemergence Control of Annual Grasses and Broadleaf Weeds

ACTIVE INGREDIENT:
trifluralin* (*α,α,α* -trifluoro-2,6-dinitro-*N*, *N* -dipropyl-*p* -toluidine) 50.8%
INERT INGREDIENTS 49.2%
TOTAL 100.0%

Contains 5 pounds active ingredient per gallon.

EPA Reg. No. 1471-120 EPA Est. No. 1471-IN-2
U.S. Patent No. Re. 31227
ID 5555 PI 2700 AMP

* Treflan® 5—the registered trademark for Elanco Products trifluralin.

CAUTION
Keep Out of Reach of Children

PRECAUTIONARY STATEMENTS

HAZARDS TO HUMANS AND DOMESTIC ANIMALS

CAUTION

Causes eye irritation. Do not get in eyes. Avoid contact with skin and clothing. Harmful if swallowed or absorbed through the skin. Treflan 5 may cause skin sensitization reactions in certain individuals. Use eye protection and protective clothing such as coveralls, a long sleeved shirt, and impermeable gloves when handling this product.

Do not contaminate foodstuffs or feeds.

First Aid: In case of contact, immediately flush eyes or skin with plenty of water. For eyes, call a physician. Remove and wash contaminated clothing before reuse.

ENVIRONMENTAL HAZARDS

Direct contamination of any body of water with this emulsifiable concentrate may kill fish and other aquatic organisms.

Do not contaminate any body of water by direct application, cleaning of equipment or disposal of wastes.

PHYSICAL OR CHEMICAL HAZARDS

Do not use or store near heat or open flame.

STORAGE AND DISPOSAL

STORAGE

Avoid freezing. Store above 40°F. If frozen, poor weed control may result. Do not store near heat or open flame. Store in original container only. In case of leak or spill, use absorbent materials to contain liquids and dispose as waste.

PESTICIDE DISPOSAL

Do not contaminate water, food or feed by storage or disposal. Wastes resulting from the use of this product may be disposed of on site or at an approved waste disposal facility.

PLASTIC CONTAINER DISPOSAL

Triple rinse (or equivalent). Then offer for recycling or reconditioning, or puncture and dispose of in a sanitary landfill, or incineration, or, if allowed by state and local authorities, by burning. If burned, stay out of smoke.

METAL CONTAINER DISPOSAL

Triple rinse (or equivalent). Then offer for recycling, or reconditioning, or puncture and dispose of in a sanitary landfill, or by other procedures approved by state and local authorities.

BULK/MINI-BULK TANK HANDLING

Tanks that have been used for phenoxy-type herbicides should not be used for Treflan 5. To clean, triple rinse (or equivalent) and wash with appropriate cleaners before reusing.

The manufacturer makes no warranties, express or implied, concerning this product or its use, which extend beyond the description on the label or in literature published by Elanco Products Company, and all statements made concerning this product apply only when used as directed. Elanco Products Company expressly disclaims all warranties, express or implied, in relation to any use of the product in combination or sequential use with any other product not specifically recommended in writing by Elanco Products Company. If used in a combination or sequential use recommended by Elanco Products Company, the liability shall in no manner extend to any damage, loss or injury not directly caused by the inclusion of the Elanco product in such combination or sequential use.

SPECIAL PRECAUTIONS

Applied according to directions and under normal growing conditions, TREFLAN will not harm the treated crop. Overapplication may result in crop injury or a soil residue. Uneven application or improper soil incorporation of TREFLAN can result in erratic weed control or crop injury. Seedling disease, cold weather, deep planting, excessive moisture, high salt concentration or drought may weaken crop seedlings and increase the possibility of damage from TREFLAN. Under these conditions, delayed crop development or reduced yields may result.

In Arizona, Colorado, California, Idaho, Montana, Nevada, New Mexico, Oregon, Utah, Washington and Wyoming, sugarbeets, red beets or spinach should not be planted for 12 months after a spring application or for 14 months after a fall application of TREFLAN. Plow the land to a depth of 12 inches prior to planting sugar beets to prevent the possibility of crop injury. Sorghum (milo), proso millet, corn or oats should not be planted for 14 months after a spring application or for 16 months after a fall application of TREFLAN to avoid crop injury. If land has not been irrigated, do not plant any of these crops for 18 months after a spring application or 20 months after a fall application of TREFLAN.

In those portions of Kansas, Nebraska, North Dakota, Oklahoma, South Dakota and Texas where at least 20 inches of irrigation and/or rainfall (total) was used to produce the crop, sorghum or oats should not be planted for 12 months after an application of TREFLAN.

If less than 20 inches of total water was used to produce the crop, do not plant sorghum, proso millet, or oats for 18 months after an application of TREFLAN. Cool, wet weather conditions during the early stage of growth may increase the possibility of injury to sorghum.

In all other areas receiving greater than 20″ rainfall/year: Moldboard plow before planting sugar beets where a spring application of TREFLAN was made the previous season. Also note planting restrictions listed in the section on control of rhizome johnsongrass and other higher rate programs.

Vegetable Growing Areas:

Vegetable crops other than those listed on this label should not be planted within 5 months following the application of TREFLAN.

DIRECTIONS FOR USE

TREFLAN 5 is a preemergence herbicide which is incorporated into the soil to provide long-lasting control of many annual grasses and broadleaf weeds. TREFLAN controls weeds as they germinate. TREFLAN will not control established weeds.

WEEDS AND GRASSES CONTROLLED BY TREFLAN 5

Grasses

Annual bluegrass	*Poa annua*
Barnyardgrass (Watergrass)	*Echinochloa* sp.
Brachiaria (Signalgrass)	*Brachiaria* sp.
Bromegrass (Cheatgrass) (Downy brome)	*Bromus tectorum*
Cheat (Chess)	*Bromus secalinus*
Crabgrass (Large crabgrass) (Smooth crabgrass)	*Digitaria* spp.
Foxtail: Bottlegrass (Bristlegrass) (Giant foxtail) (Green foxtail) (Foxtail millet) (Pigeongrass) (Robust foxtail) (Yellow foxtail)	*Setaria* spp.
Goosegrass (Silver crabgrass) (Silvergrass) (Wiregrass) (Yardgrass)	*Eleusine indica*
Johnsongrass (from seed) (Rhizome—See special instructions for control in cotton and see special instructions for control in soybeans)	*Sorghum halepense*
Junglerice	*Echinochloa colonum*
Panicum	
Fall panicum	*Panicum dichotomiflorum*

(Spreading panicgrass—see special instructions in cotton and see special instructions in soybeans)

Guineagrass	*Panicum maximum*
(See Sugarcane for special instructions)	
Texas panicum	*Panicum texanum*
(Buffalograss)	
(Coloradograss)	
Itchgrass	*Rottboellia exaltata*
(Raoulgrass)	
(See special instructions in Sugarcane Itchgrass section)	
Red rice	*Oryza sativa*
(See Soybean—Special Use Directions for special instructions)	
Sandbur	*Cenchrus incertus*
(Burgrass)	
Sprangletop	*Leptochloa filiformis*
Stinkgrass	*Eragrostis cilianesis*
(Lovegrass)	
Wild cane	*Sorghum bicolor*
(Shattercane)	
(See special instructions in Soybean—Wild Cane section)	
Woolly cupgrass	*Eriochloa villosa*
Broadleaf Weeds	
Carpetweed	*Mollugo verticillata*
Chickweed	*Stellaria media*
Field bindweed	*Convolvulus arvensis*
(See special instructions in Fruit and Nut Crops and Vineyards)	
Florida pusley	*Richardia scabra*
(Florida purslane)	
(Mexican clover)	
(Pusley)	
Goosefoot	*Chenopodium hybridum*
Henbit (fall application only)	*Lamium amplexicaule*
Knotweed	*Polygonum aviculare*
Kochia	*Kochia scoparia*
(Fireweed)	
(Mexican fireweed)	
Lambsquarters	*Chenopodium album*
Pigweed	*Amaranthus* spp.
(Carelessweed)	
(Prostrate pigweed)	
(Redroot)	
(Rough pigweed)	
(Spiny pigweed)	
Puncturevine (Western U.S. only)	*Tribulus terrestris*
(Caltrop)	
(Goathead)	
Purslane	*Portulaca oleracea*
Russian thistle	*Salsola kali*
(Tumbleweed)	
Stinging nettle	*Urtica dioica*
(Nettle)	

SOIL PREPARATION

Crop Residues or Existing Weeds: Ground cover, such as crop residues or existing weeds, can interfere with the incorporation of TREFLAN into the soil. A manageable level of such ground cover will allow the TREFLAN to be uniformly incorporated into the top 2 to 3 inches of soil. If the level of the ground cover is such that this cannot be done, you must till the soil prior to the application of TREFLAN.

Roughness: The soil surface should be smooth enough so that you can operate a sprayer and incorporation equipment efficiently and at speeds which insure a uniform application and incorporation of TREFLAN.

General Soil Conditions: To assure uniform incorporation of TREFLAN, soil moisture conditions should be such that large clods can be broken up during the incorporation process.

SOIL TEXTURE GUIDE

The amount of TREFLAN you apply will vary with the soil texture and organic matter. A fine textured soil will require more TREFLAN per acre than a coarse soil. Choose the proper rate for each application based on the following soil texture group and specific crop recommendations. Do not exceed recommended rates.

Soil Texture	Soil Classification
Coarse Soils: (Light)	Sand, loamy sand, sandy loam
Medium Soils:	Loam, silty clay loam*, silt loam, silt, sandy clay loam*
Fine Soils:	Clay, clay loam, silty clay loam*, silty clay, sandy clay, sandy clay loam*

* Silty clay loam and sandy clay loam soils are transitional soils and may be classified as either medium or fine textured soils. If silty clay loam or sandy clay loam soils are predominately sand or silt, they are usually classified as medium textured soils. If they are predominately clay, they are usually classified as fine textured soils.

MIXING AND APPLICATION DIRECTIONS

TREFLAN 5 Alone in Water

Start with a clean spray tank. Fill sprayer ⅓ to ½ full with clean water. Start agitation. Add correct quantity of TREFLAN, continue agitation and finish filling the tank.

TREFLAN 5 Tank Mix in Water

Vigorous, continuous agitation is required for all tank mixes. (Sparger pipe agitators generally provide the best agitation in spray tanks). Avoid stirring or splashing air into the mixture during filling to prevent foaming. To do this, place the end of the fill pipe below the surface of the water in the spray tank. Do not allow the mixture to siphon back into the water source.

Mixing order: Fill the tank ¼ to ⅓ full with clean water. Start the agitation. Add dry flowables, wettable powders (WP), aqueous suspensions (AS), flowables (F), and liquids (L) to the water and agitate until the product(s) are completely dispersed in the water. Allow additional mixing and dispersion time when using dry flowable products. Continue agitation and fill tank to ¾ full, add the TREFLAN 5, mix thoroughly. Then add any solution (S) formulations, agitate and finish filling. Maintain agitation during filling and through application. If spraying and agitation must be stopped before the tank is empty, the materials may settle to the bottom. In this case, it is important to resuspend all of the material in the bottom of the tank before continuing the spray application. A sparger agitator is particularly useful for this purpose. Sometimes it is more difficult to resuspend settled material than it is to suspend it originally.

Read and carefully follow all label instructions for each material added to the tank. Premixing dry and flowable formulations with water (slurrying) and pouring the slurry through a 20 or 35 mesh wetting screen in the top of the tank will help assure good initial dispersion in the tank water. Line screens in the tank should be no finer than 50 mesh (100 mesh is finer than 50 mesh).

If you see a buildup of material on the walls of the spray tank, wash the tank with soapy water between fillings. Rinse and continue the spraying operation. Clean the tank, lines, and screens thoroughly after use.

As the spray volume decreases, the importance of accurate calibration and uniform application increases. Check the sprayer daily to insure proper calibration and uniform application. Do not apply TREFLAN when the wind can cause drifting of spray particles which can result in non-uniform application. TREFLAN should not be applied to soils which are wet or are subject to prolonged periods of flooding as poor weed control may result.

Ground application: Apply TREFLAN in 5 to 40 gallons of water or liquid fertilizer per acre (broadcast basis), using any properly calibrated, low pressure herbicide sprayer that will apply the spray uniformly.

Aerial application: Apply TREFLAN in 5 to 10 gallons of water or liquid fertilizer per acre. Adjust pump pressure, nozzle arrangements, speed and height to provide a uniform application to the soil surface. Use swath markers or flagmen to assure proper application spray widths.

INCORPORATION DIRECTIONS

Incorporation Equipment—General Directions
Use incorporation equipment that mixes TREFLAN into the top 2 to 3 inches of the final seedbed, or erratic weed control and/or crop injury may result. Incorporation equipment such as a disc will mix TREFLAN approximately half as deep as the equipment is set to operate. For example, a disc set to cut 4 inches deep will incorporate most of the TREFLAN within the top 2 inches of soil.

Incorporation Before Planting
TREFLAN must be incorporated one time within 24 hours after application. Then any time prior to planting, a second incorporation is necessary, this time running the equipment in a different direction from the first. You should incorporate the TREFLAN uniformly into the top 2 to 3 inches of the final seedbed.

Incorporation After Planting
Check specific crop for incorporation directions after planting.

Incorporation In Bedded Culture
For effective weed control, TREFLAN needs to be incorporated into the top 2 to 3 inches of the final seedbed.

Application prior to bedding: Apply TREFLAN and incorporate it one time with recommended equipment. The bedding operation serves as the second incorporation. Do not expose untreated soil during post-bedding operations.*

Application after bedding: Knock off beds to planting height before applying TREFLAN. Apply TREFLAN and incorporate it with recommended equipment that will conform to the bed shape. Do not leave untreated soil exposed.*

* Avoid removal of untreated soil from the seedbed before or during the planting operation. This would expose untreated soil, allowing weeds to germinate in the drill row.

RECOMMENDED EQUIPMENT

Any recommended incorporation tool may be used alone or in combination with any other recommended tool. Two incorporation passes are required unless specifically stated. The second incorporation should not be deeper than the first.

Disc: set to cut 4 to 6 inches deep and operate at 4 to 6 mph.

Field Cultivator: set to cut 3 to 4 inches deep and operate at 5 mph or more. A field cultivator is defined as an implement with 3 to 4 rows of sweeps, spaced at intervals of 7 inches or less and staggered so that no soil is left unturned. Chisel points should not be used.

Combination Seedbed Conditioners: set to cut 3 to 4 inches deep and operate at a speed of at least 5 mph. These implements are defined as three or more tillage devices combined and used as a single tool. For example, 2 to 3 rows of field cultivator C- or S-shaped shanks with an effective sweep spacing of 6 to 9 inches (staggered so that no soil is left unturned), followed by a spike-tooth or flextine harrow, followed by a ground-driven reel or basket.

Rolling Cultivator: set to cut 2 to 4 inches deep and operate at 6 to 8 mph. Rolling cultivators are adequate for use on coarse and medium textured soils only, except when used in sugarcane where the rolling cultivator may be used on fine textured soils.

Bed Conditioner (Do-All): set to cut 2 to 4 inches deep and operate at 4 to 6 mph. The do-all is adequate for use on coarse and medium textured soils only. When using the do-all in bedded culture, only one incorporation pass is required. However, two passes with a do-all are required in flat planted culture.

Mulch Treader (other similar disc-type implements): set to cut 3 to 4 inches deep and operate at 5 to 8 mph.

P.T.O. Driven Equipment (tillers, cultivators, hoes): adjust to incorporate TREFLAN into the top 2 to 3 inches of the seedbed with rotors spaced to provide a clean sweep of the soil. Only one incorporation is necessary. P.T.O. driven equipment should not be operated at a speed greater than 4 mph.

Other equipment, including the flexible tine-tooth harrow (Flex-tine, Melroe), is also recommended but only for the special programs for which it is specified in this label.

CULTIVATION AFTER PLANTING

Soil treated with TREFLAN may be shallow cultivated without reducing the weed control activity of TREFLAN. Do not cultivate deeper than the treated soil since this may bring untreated soil to the surface, and poor weed control may result.

CROP RECOMMENDATIONS

These recommendations are given as the broadcast rates of TREFLAN per acre. For band applications, decrease the amount of TREFLAN in proportion to the amount of surface treated per acre. Apply any time after January 1 when the soil can be worked and is suitable for good incorporation. TREFLAN can be applied in the fall—see specific crop for recommendations. For general fall application directions where specific recommendations are not given, see page 74. Where a rate range is shown, use the lower rate for coarser soils or soils with lower organic matter. TREFLAN should not be used on soils containing more than 10% organic matter.

COTTON—TREFLAN ALONE

TREFLAN can be applied and incorporated before or at planting, immediately after planting, or at layby.

COTTON—Preemergence:

Broadcast Rates per Acre

Soil Texture	TREFLAN 5: Areas receiving less than 20" average annual rainfall*	TREFLAN 5: Areas receiving greater than 20" average annual rainfall*
	(pints)	(pints)
Coarse	0.8	0.8
Medium	1.0–1.2	1.2
Fine	1.2	1.6

* Use 1.2 pints per acre on coarse and medium textured soils and 1.6 pints on fine soils with 2–5% organic matter; use 1.6 to 2 pints on all soils with 5–10% organic matter.

COTTON—Postplant:
When incorporating TREFLAN after planting (postplant), be careful not to disturb the seed.

COTTON—Layby:
Apply and incorporate TREFLAN any time up to layby, but not less than 90 days before harvest. Direct the layby applications onto the soil between the rows and beneath emerged cotton plants. Use the same rates as for a preemergence application.

COTTON—Fall Application:
Apply and incorporate TREFLAN any time between October 15 and December 31. The ground may be left flat or bedded-up over winter. On bedded ground, knock beds down to desired height before planting, moving some treated soil from beds into furrows. Where soil is left flat over winter, be careful not to turn up untreated soil during spring bedding operations. Destroy established weeds during seedbed preparation. If weeds become established in furrows due to uncovering of untreated soil during bedding, destroy these weeds before planting. In the fall, do not apply TREFLAN to soils which are wet or subject to prolonged periods of flooding.

Broadcast Rates per Acre—Fall application only:
In Alabama, Arkansas, northern Florida, Georgia, Louisiana, Mississippi, southeastern Missouri bootheel, North Carolina, New Mexico, Oklahoma, South Carolina, Tennessee and Texas, apply and incorporate TREFLAN at a broadcast rate of 1.6 pints per acre on coarse and medium soils and 2 pints on fine soils.

In Arizona, California and Nevada, apply and incorporate TREFLAN at a broadcast rate of 1.2 pints per acre on coarse soils; 1.6 pints on medium soils; and 2 pints on fine soils.

For cotton grown in other states, apply and incorporate TREFLAN at a broadcast rate of 0.8 pint per acre on coarse soils; 1.2 pints on medium soils; 1.6 pints on fine soils, 1.2 pints on coarse soils with

2–5% organic matter; and 1.6 to 2 pints on soils with 5–10% organic matter.

COTTON—SPECIAL USE DIRECTIONS

COTTON—Fall panicum:

Apply and incorporate TREFLAN at the broadcast rate of 1.6 pints per acre on both coarse and medium soils.

COTTON—Rhizome johnsongrass:

In all cotton-producing states except Arizona and California, you can obtain commercially acceptable control of rhizome johnsongrass with a double rate program which you apply for 2 consecutive years in accordance with the following directions.

Soil Preparation—Proper preparation of the soil before application is very important for satisfactory results. Use a chisel plow or similar implement to bring rhizomes to the top of the soil. Then follow with a disc two times before application to cut the rhizomes into small (2 to 3 inch) pieces. This should also destroy any emerged johnsongrass.

Application—Choose the one application program that best fits your cultural practices:

Spring Application—Apply TREFLAN any time before planting in the spring for 2 years in a row. Use a broadcast rate of 1.6 pints per acre on coarse soils; 2.4 pints on medium soils; and 3.2 pints on fine soils.

OR

Fall Application—Apply TREFLAN between October 15 and December 31 for 2 years in a row at the same rates as a spring application for the control of rhizome johnsongrass.

Incorporation—Deep incorporation is essential for good rhizome johnsongrass control. Incorporate TREFLAN thoroughly with a disc set to cut 4 to 6 inches deep and operate at 4 to 6 mph. Two passes are necessary, with the second pass in a different direction from the first.

Cultivation—Some johnsongrass plants will escape. Timely cultivations during the crop season to remove escaped plants are necessary to obtain commercially acceptable control. You cannot obtain commercially acceptable control with only 1 year of double rate TREFLAN use.

Crop Rotation—In the season following a double rate treatment, plant only rice and those crops for which TREFLAN can be applied as a preplant treatment or injury may result.

COTTON—Pigweed and seedling johnsongrass control:

In Alabama, Arkansas, Florida, Georgia, Louisiana, Mississippi, southeastern Missouri, Nebraska, North Carolina, South Carolina, Tennessee and southern Virginia, TREFLAN may be applied preplant at a broadcast rate of 0.8 to 1.2 pints per acre on coarse soils; 1.2 to 1.6 pints on medium soils; and 1.6-2 pints on fine soils. Exception: in the State of Louisiana, 2.4 pints per acre are recommended on fine soils.

COTTON—Additional weed and grass control (Texas Gulf Coast):

In the Texas Gulf Coast counties of Brazoria, Calhoun, Chambers, Fort Bend, Galveston, Harris, Jackson, Jefferson, Liberty, Matagorda, Orange, Victoria, Waller and Wharton, TREFLAN may be applied up to 2 weeks before planting at a broadcast rate of 1.2 pints per acre on coarse soils; 1.6 pints on medium soils; and 2.4 pints on fine soils.

COTTON PRECAUTIONS: Cotton should be planted after early season adverse weather conditions have passed, especially when using higher rate programs. Cool, wet weather early in the growth cycle causes additional stress to the cotton plant. This may result in reduced stands, delayed maturity and reduced yields.

COTTON—TANK MIXES AND OVERLAYS

Follow recommended soil preparation and incorporation procedures for TREFLAN.

COTTON—TREFLAN/Caparol tank mix for cotton grown in California, Arizona, New Mexico and west Texas:

The TREFLAN/Caparol combination will control certain grasses and broadleaf weeds listed for TREFLAN alone plus the following weeds:

Annual morningglory	Prickly sida (Teaweed)
Groundcherry (Annual)	Ragweed
Malva	Smartweed
Mustard	Wild oat

The tank mix also controls shallow germinating seedlings of cocklebur and coffeeweed.

Broadcast Rates Per Acre

Soil Texture	TREFLAN 5	Caparol 80W
	(pints)	(pounds)
Coarse	0.8	2*
Medium	1.0 to 1.2	2½
Fine	1.6	2½

* Do not use on sands and loamy sands. For band applications use proportionately less.

Mixing Directions: Carefully follow the procedures on the Caparol 80W label for making a slurry and adding it to a partially filled tank of water. After the Caparol is thoroughly mixed with the partially filled tank of water, add the TREFLAN 5 and continue filling. Agitate continuously during the filling and spraying operation.

Avoid leaving the spray mixture in the tank without constant agitation. If bypass agitation is used, the bypass line should stop at the bottom of the tank to minimize foaming.

Additional Precautions: The combination of TREFLAN/Caparol should not be used in the cut areas of newly leveled fields, in areas of excess salt, or where flooding over the beds is likely to happen. Do not plant cotton in tractor wheel depressions. These conditions may cause crop injury. On mulch-planted cotton, water back only after cotton seedlings are well established.

Crop Rotations: Cabbage, okra, onions and peas may be planted in the fall after a spring application of TREFLAN plus Caparol. Winter barley, winter rye and winter wheat can be planted in the fall also if they are plowed down and not used for food or feed. Refer to the Caparol label for directions, cautions and precautions.

COTTON—TREFLAN/Cotoran tank mix (except in Arizona and California):

Follow recommended soil preparation and incorporation procedures for TREFLAN.

The TREFLAN/Cotoran tank mix effectively controls all the annual grasses and broadleaf weeds listed for TREFLAN alone plus these additional weeds:

Buttonweed	Ragweed
Cocklebur	Ryegrass
Groundcherry, Wright	Sesbania
Jimsonweed	Sicklepod
Morningglory	Smartweed
Prickly sida (Teaweed)	Tumbleweed

Broadcast Rates Per Acre

Soil Texture	TREFLAN 5	Cotoran 80W
	(pints)	(pounds)
Coarse	0.8	1¼
Medium	1.2	2
Fine	1.6	2½

Mixing Directions: Carefully follow the procedures on the Cotoran label for making a Cotoran slurry and adding it to a partially filled tank of water. After the Cotoran is thoroughly mixed with the partially filled tank of water, add the TREFLAN and continue filling. Agitate continuously throughout the filling and application operations. Do not leave spray mixture in tank without constant agitation. If bypass agitation is used, the bypass line should stop at the bottom of the tank to minimize foaming. Apply in 15–40 gallons of water per acre.

Additional Precautions: Do not plant crops other than cotton on the treated land within 6 months after the application of TREFLAN plus Cotoran, or crop injury may result. Do not feed foliage from treated cotton plants or gin trash to livestock. Do not mix TREFLAN plus Cotoran with liquid fertilizer.

West Texas Only: Do not use the tank mix of TREFLAN plus Cotoran on sandy, loamy sand or fine sandy loam soils. Do not use on cotton planted in furrows.

Arkansas, Louisiana, and Mississippi Only: Use 1 pound Cotoran 80W in tank mix with TREFLAN on sandy loam soils low in organic matter.

New Mexico: Do not plant treated land with crops other than cotton until 1 year after the last application. Do not use on sandy or coarse textured soils of less than 1% organic matter.

COTTON—TREFLAN preplant followed by Cotoran overlay:
Apply and incorporate TREFLAN as recommended for the specific soil texture. Then apply Cotoran as a preemergence surface treatment at 1¼ to 2½ pounds per acre. On light silt and sandy soils low in organic matter, use the lower rate.

Additional Precaution: Refer to the Cotoran label for all cautions and precautions when using this preemergence surface treatment.

COTTON—TREFLAN preplant followed by Karmex overlay:

For cotton grown east of the Mississippi River plus Arkansas, southeastern Missouri, Louisiana and Eastern Texas, incorporate TREFLAN before planting. Then follow with a preemergence application of Karmex. This will effectively control all the weeds controlled by TREFLAN alone plus these additional weeds:

Annual groundcherry	Shepherdspurse
Annual morningglory	Velvetgrass
Dogfennel	Wild lettuce
Pennycress	Wild mustard
Ragweed	

Broadcast Rates Per Acre

Soil Texture	TREFLAN 5	Karmex 80W
	(pints)	(pounds)
Coarse	0.8	⅓
Medium	1.2	⅔
Fine	1.6	1

Additional Precautions: Do not use Karmex on soils with less than 1% organic matter as crop injury may result. Do not allow livestock to graze on cotton treated with Karmex. Consult the Karmex label for additional instructions, cautions and precautions.

SOYBEAN—TREFLAN ALONE

SOYBEAN—Preemergence:
Follow recommended soil preparation, application, and incorporation procedures for TREFLAN.

Broadcast Rates Per Acre

Soil Texture	TREFLAN 5*
	(pints)
Coarse	0.8
Medium	1.2
Fine	1.6

* Use 1.2 pints per acre on coarse and medium textured soils and 1.6 pints on fine soils with 2–5% organic matter; use 1.6-2 pints on all soils with 5–10% organic matter.

SOYBEAN—Fall Application:
Apply and incorporate TREFLAN anytime between October 15 and December 31. Ground may be left flat or bedded-up over winter. On bedded ground, knock beds down to desired height before planting, moving some treated soil from tops into furrows. Where soil is left flat over winter, take care during spring bedding operations to prevent turning up untreated soil. Destroy established weeds during seedbed preparation. If weeds become established in furrows due to uncovering of untreated soil during listing, destroy these weeds before planting. Do not fall apply TREFLAN to soils which are wet, are subject to prolonged periods of flooding or where rice was grown the previous year.

For soybeans grown in Alabama, Arkansas, northern Florida, Georgia, Louisiana, Mississippi, southeastern Missouri bootheel, North Carolina, Oklahoma, South Carolina, Tennessee and Texas, apply and incorporate TREFLAN at a broadcast rate of 1.6 pints per acre on coarse and medium soils and 2 pints on fine soils.

For soybeans grown in states other than those listed above, apply and incorporate TREFLAN at a broadcast rate of 0.8 pint per acre on coarse soils; 1.2 pints on medium soils; 1.6 pints on fine soils; 1.2 pints on coarse soils with 2–5% organic matter; and 1.6 to 2 pints on soils with 5–10% organic matter.

SOYBEAN—SPECIAL USE DIRECTIONS

SOYBEAN—Fall panicum:
Apply TREFLAN at the broadcast rate of 1.6 pints per acre on both coarse and medium soils.

SOYBEAN—Pigweed and seedling johnsongrass:
In Alabama, Arkansas, Florida, Georgia, Kansas, Louisiana, Mississippi, southeastern Missouri, North Carolina, Oklahoma, South Carolina, Tennessee and southern Virginia, TREFLAN may be applied at a broadcast rate of 0.8 to 1.2 pints per acre on coarse soils; 1.2 to 1.6 pints on medium soils; and 1.6 pints on fine soils. Exception: in the State of Louisiana, 2.4 pints per acre are recommended on fine soils.

SOYBEAN—Additional weed and grass control (Texas Gulf Coast):
In the Texas Gulf Coast counties of Brazoria, Calhoun, Chambers, Fort Bend, Galveston, Harris, Jackson, Jefferson, Liberty, Matagorda, Orange, Victoria, Waller and Wharton, TREFLAN may be applied up to 2 weeks before planting at a broadcast rate of 1.2 pints per acre on coarse soils; 1.6 pints on medium soils; and 2.4 pints on fine soils.

SOYBEAN—Charcoal soils in Arkansas, Louisiana and Mississippi:
Newly cleared land often contains high organic matter (5–10%) and charcoal from burning debris. This charcoal and/or organic matter tends to bind TREFLAN and reduce its weed control activity. Under these conditions, higher rates of TREFLAN are necessary for weed control. Increased rates, however, can cause crop injury if charcoal or organic matter is not present to bind some of the TREFLAN. In the burn row a high level of charcoal is present; consequently, poor weed control may result even with an increased rate of TREFLAN.
Apply TREFLAN at the broadcast rate of 1.6 to 2 pints per acre on coarse soils; 2 pints on medium soils; and 2.4 pints on fine textured soils. Follow recommended soil preparation, application and incorporation procedures for TREFLAN.

SOYBEAN—Red rice in Arkansas, Louisiana, Mississippi and Texas only:
You can obtain suppression or partial control of red rice when you apply TREFLAN at the following recommended rates. Follow recommended soil preparation and incorporation procedures for TREFLAN.

Apply and incorporate TREFLAN in the spring before planting.

Broadcast Rates Per Acre

Soil Texture	TREFLAN 5	
	Application Year 1	Application Year 2
	(pints)	(pints)
Coarse	1.6	0.8
Medium	2.4	1.2
Fine	3.2	1.6
Coarse soils with 2 to 5% organic matter	2.4	1.2
Soils with 5 to 10% organic matter	3.2	1.6–2

If a combination of high organic matter (5–10%) and charcoal are present in the soil, apply TREFLAN the second year at the following rates for charcoal soils in Arkansas, Louisiana and Mississippi:

Soil Texture	TREFLAN 5*
	(pints)
Coarse	1.2–2
Medium	2
Fine	2.4

For more information on charcoal soils see SOYBEAN—charcoal soils heading.

Crop Rotation: The program for red rice control in soybeans is a 2-year program. Use the rates listed for first year application and plant soybeans. The second year use the normal TREFLAN rates listed for your soil type and charcoal level and plant only those crops for which TREFLAN has been registered as a preplant treatment, or crop injury may result. Do not plant rice the second year. Rice may be planted the third year.

SOYBEAN—Rhizome johnsongrass (Eastern United States and the State of Texas):

You can obtain commercially acceptable control of rhizome johnsongrass with a double rate program applied for 2 consecutive years in accordance with the following directions.

Soil Preparation—Proper preparation of the soil before application is very important for satisfactory results. Use a chisel plow or similar implement to bring rhizomes to the top of the soil. Then follow with a disc two times before application to cut the rhizomes into small (2 to 3 inch) pieces and to destroy any emerged johnsongrass.

Application—Choose the one application program that best fits your cultural practices.

Spring Application—Apply TREFLAN anytime in the spring before planting for 2 years in a row. Use a broadcast rate of 1.6 pints per acre on coarse soils; 2.4 pints on medium soils; 3.2 pints on fine soils; 2.4 pints on coarse soils with 2–5% organic matter; and 3.2 pints on soils with 5–10% organic matter.

OR

Fall Application—Apply TREFLAN between October 15 and December 31 for 2 years in a row at the same rates as a spring application for the control of rhizome johnsongrass.

OR

Split Application—Apply TREFLAN at the same rate in both the spring and fall for 2 years in a row using the rates in the following table.

Broadcast Rates Per Acre

Soil Texture	TREFLAN 5 Spring and Fall
	(pints)
Coarse	0.8
Medium	1.2
Fine	1.6
Coarse soils with 2–5% organic matter	1.2
Soils with 5–10% organic matter	1.6

Incorporation—Deep incorporation is essential for good rhizome johnsongrass control. Incorporate TREFLAN thoroughly with a disc set to cut 4 to 6 inches deep and operate at 4 to 6 mph. Two passes are necessary, with the second pass in a different direction from the first.

Cultivation—Some johnsongrass plants will escape. Timely cultivations during the crop season to remove escaped plants are necessary to obtain commercially acceptable control.

Crop Rotation: In the season following a double rate treatment, plant only rice and those crops for which TREFLAN can be applied as a preplant treatment or injury may result.

SOYBEAN — Rhizome johnsongrass — TREFLAN/Sencor or TREFLAN/Lexone tank mix:

TREFLAN/Sencor or TREFLAN/Lexone may be used for rhizome johnsongrass control and for the control of those weeds listed for TREFLAN alone. For the additional weeds controlled by Sencor or Lexone in tank mix, see Soybean—Tank Mixes and Overlays section. Follow procedures for soil preparation, incorporation, and cultivation recommended in the Soybean—Rhizome johnsongrass section.

Apply TREFLAN/Sencor or Lexone up to two weeks before planting for two consecutive years at the following broadcast rates per acre.

Soil Texture	TREFLAN 5	+ Lexone 50WP/4L or Sencor 50WP/4	or Lexone (dry flowable) or Sencor (dry flowable)
	(pints)	(pounds/pints)	(pounds)
Coarse*	1.6	1/2	1/3
Medium	2.4	3/4	1/2
Fine	3.2	1	2/3

* Do not use on coarse soils with less than 1% organic matter.

Read and follow all additional precautions listed for the TREFLAN/ Sencor or TREFLAN/Lexone tank mix.

SOYBEAN—Wild cane (shattercane):

Follow recommended soil preparation and application procedures for TREFLAN.

Wild cane (shattercane) can germinate throughout the growing season and from greater soil depth than most other weed seeds. Commercially acceptable control of wild cane can be obtained with the following increased rates of TREFLAN.

Apply TREFLAN at the broadcast rate of 0.8 pint per acre on a coarse soil; 1.6 pints on a medium soil; and 2 pints on a fine textured soil.

Incorporation—Deep incorporation is essential for good wild cane control. Incorporate TREFLAN thoroughly with a disc set to cut 4 to 6 inches deep and operate at 4 to 6 mph. Two passes are necessary, with the second pass in a different direction from the first.

Cultivation—Cultivation during the crop season will also contribute to control.

SOYBEAN PRECAUTIONS: Soybeans should be planted after early season adverse weather conditions have passed, especially when using higher rate programs. Cool, wet weather early in the growth cycle causes additional stress to the soybean plant. This may result in reduced stands, delayed maturity and reduced yields.

SOYBEAN—TANK MIXES AND OVERLAYS

SOYBEAN—TREFLAN/Sencor or TREFLAN/Lexone tank mix:

The TREFLAN/Sencor or TREFLAN/Lexone tank mix effectively controls the annual grasses and broadleaf weeds controlled by TREFLAN alone plus these additional weeds:

Jimsonweed	Ragweed, common
Mallow, Venice (Flower-of-an-hour)	Sesbania, hemp
	Smartweed, Pennsylvania
Mustard, wild	Velvetleaf
Prickly sida	

Control of cocklebur, morningglory and giant ragweed (horseweed) may be erratic. Control may be improved with timely cultivation. Where cocklebur is a serious problem, an overlay of Sencor or Lexone may be preferred to the TREFLAN/Sencor or TREFLAN/Lexone tank mix.

Follow recommended procedures for soil preparation, incorporation, and cultivation of TREFLAN. Mix according to instructions in Mixing and Application Directions section. The TREFLAN/Sencor or TREFLAN/Lexone tank mix can be be applied from 2 weeks before planting up to planting.

Broadcast Rates Per Acre

Soil Texture	TREFLAN 5	Lexone 50WP/4L or Sencor 50WP/4	or	Lexone (dry flowable) or Sencor (dry flowable)
	(pints)	(pounds/pints)		(pounds)
Coarse*	0.8	1/2		1/3
Medium	1.2	3/4		1/2
Fine	1.6	1		2/3

* Do not use Sencor/Lexone on coarse soils with less than 1% organic matter.

Additional Precautions: Do not plant any crop other than soybeans within 4 months after treatment. Overapplication, uneven application, or improper soil incorporation may result in crop injury, herbicide residue, or erratic weed control. Additional stress factors are seedling disease, cold weather, deep planting, excessive moisture, soil *p*H over 7.5, high salt concentration, or drought. Any of these may weaken crop seedlings and increase possibility of damage from the tank mix. These additional factors may also delay crop development or reduce yields when Sencor or Lexone is applied. Observe all cautions and limitations on the Sencor and Lexone labels. Do not use the foliage from soybeans treated with the TREFLAN/Sencor or TREFLAN/Lexone tank mix for feed or forage.

SOYBEAN—TREFLAN preplant followed by Sencor or Lexone as an overlay:
Apply TREFLAN as a preplant incorporated herbicide. As a separate operation, make a single application of Sencor or Lexone as either a band or broadcast spray during planting or as a separate operation after planting, but before the soybeans emerge. Do not spray Sencor or Lexone over the top of emerged soybeans, or crop injury may result.

Use Directions—Follow directions on the Sencor or Lexone labels for specific instructions regarding each herbicide.

Broadcast Rates Per Acre

Soil Texture	TREFLAN 5	Postplant/Preemergence: Sencor 50WP/4 or [Lexone 50WP/4L]*	or	Postplant/Preemergence: Sencor (dry flowable) or [Lexone (dry flowable)]*
	(pints)	(pounds or pints)		(pounds)
Coarse**	0.8	3/4–1 [3/4]		1/2–2/3 [1/2]
Medium	1.2	3/4–1 1/2 [3/4–1]		1/2–1 [1/2–2/3]
Fine	1.6	1–1 3/4 [1]		2/3–1 1/6 [2/3]

* Lexone rates for each soil texture are enclosed in brackets "[ ]".
** Do not apply Lexone to sand or soils with less than 1/2% organic matter. Do not apply Sencor to coarse soils (sandy loam or loamy sand) containing less than 2% organic matter.

Additional Sencor and Lexone precautions: Do not use Lexone or Sencor on Tracy, Semmes, Altona, Vansoy or Coker 102 soybeans. These varieties are sensitive to Lexone or Sencor, and crop injury may result. Seed must be planted at least 1 1/2 inches but not more than 2 inches below the soil surface before a Sencor or Lexone application. Do not apply Sencor or Lexone at these rates more than once per season. Do not replant areas treated with Sencor or Lexone with any crop other than soybeans within 4 months after treatment. Injury to soybeans may occur if you use Lexone or Sencor on soils having a calcarous surface or *p*H of 7.5 or higher, or if you use them in conjunction with soil-applied organic phosphate pesticides. Do not use the foliage from treated soybeans for feed or forage.

SOYBEAN—TREFLAN/Amiben tank mix or overlay:

Tank mix—Amiben may be applied several days prior to planting as a broadcast tank mix with TREFLAN. Weeds controlled by this tank mix, in addition to those controlled by TREFLAN alone, are smartweed, velvetleaf and ragweed. The tank mixture should be used as a spring preplant incorporated treatment.

Overlay—Amiben may be applied broadcast or in a band over the soybean row at planting time in fields where TREFLAN has been preplant incorporated. Weeds controlled by Amiben when surface applied, in addition to those controlled by TREFLAN alone, are:

Coffeeweed (Sesbania)	Spurge, annual
Mustard, wild	Smartweed, Pennsylvania
Nightshade, black	Stinkgrass
Prickly sida (Teaweed)	Velvetleaf
Ragweed, common	

Apply TREFLAN as a tank mix with Amiben, or apply and incorporate TREFLAN alone followed by an overlay application of Amiben at these rates:

Broadcast Rates Per Acre

Soil Texture	TREFLAN 5	Amiben 2S
	(pints)	(quarts)
Coarse	0.8	4–6*
Medium	1.2	4–6
Fine	1.6	4–6

* Use the higher rate where you expect heavy populations of smartweed, velvetleaf, ragweed, wild mustard or black nightshade. Do not use on muck or charcoal soils. Read and observe all directions and cautions on the Amiben label.

SOYBEAN—TREFLAN/Amiben/Sencor or Lexone tank mix:
The TREFLAN/Amiben/Sencor or Lexone tank mix effectively controls all weeds listed for TREFLAN/Amiben and TREFLAN/Sencor or Lexone tank mixes.

Follow recommended soil preparation, application, and incorporation procedures for TREFLAN. The TREFLAN/Amiben/Sencor or Lexone tank mix may be applied from several days prior to planting up to planting in 10 to 40 gallons of water per acre. Use screens no finer than 50 mesh.

Apply the TREFLAN/Amiben/Sencor or Lexone tank mix at the following broadcast rates per acre:

Soil Texture	TREFLAN 5	Amiben 2S +	Lexone 50WP/4L or Sencor 50WP/4	or	Lexone (dry flowable) or Sencor (dry flowable)
	(pints)	(quarts)	(pounds/pints)		(pounds)
Coarse[a]	0.8	3–4[b]	1/2		1/3
Medium	1.2	3–4[b]	1/2–3/4[c]		1/3–1/2[c]
Fine	1.6	4–5	3/4[c]		1/2[c]

[a] Do not use Sencor or Lexone on coarse soils with less than 1% organic matter.
[b] Use the higher rate of Amiben when velvetleaf or black nightshade is a problem.
[c] On Clarion/Webster soils in Minnesota and Iowa or on similar alkaline (calcareous) soils with a *p*H of 7.5 or above, apply Sencor or Lexone at the rates listed below:

Soil Texture	Lexone 50WP/4L or Sencor 50WP/4	or	Lexone (dry flowable) or Sencor (dry flowable)
	(pounds/pints)		(pounds)
Medium	1/2		1/3
Fine	1/2–3/4[d]		1/3–1/2[d]

[d] Use the higher rate only where soil pH is less than 7.5 and where weed pressure is heavy.

Additional precautions: The TREFLAN/Amiben/Sencor or Lexone tank mix will not harm the treated crop when you apply it according to directions and under normal growing conditions. However, overapplication, uneven application or improper soil incorporation of the tank mix can result in erratic weed control or crop injury. Additional stress factors are seedling disease, cold weather, deep planting, excessive moisture, soil *p*H over 7.5, high salt concentration, or drought. These additional factors may weaken crop seedlings and increase the possibility of damage from the tank mix.

These additional factors may also delay crop development or reduce yields. Observe all cautions and limitations of all products used in mixtures. Do not use the foliage from soybeans treated with the TREFLAN/Amiben/Sencor or Lexone tank mix for feed or forage.

SOYBEAN—TREFLAN/Vernam tank mix:

TREFLAN/Vernam tank mix effectively controls those weeds listed for TREFLAN alone plus these additional weeds:

Annual morningglory	Velvetleaf
Coffeeweed	Yellow nutsedge (nutgrass)
Purple nutsedge (nutgrass)	

Follow recommended soil preparation procedures for TREFLAN. You may apply the TREFLAN/Vernam tank mix up to 10 days prior to planting. Incorporate the tank mix immediately after application. Apply TREFLAN/Vernam at these rates:

Broadcast Rates Per Acre

Soil Texture	TREFLAN 5	Vernam 7E
	(pints)	(pints)
Coarse	0.8	1¾–2⅓
Medium	1.2	2⅓–3*
Fine	1.6	3–3½

* For nutsedge, wild cane and velvetleaf control, use the higher rate of 3 pints per acre on medium textured soils.

ALFALFA—ESTABLISHED

In areas receiving less than 20″ average annual rainfall per year, apply TREFLAN to established alfalfa stands at a broadcast rate of 1.2 pints per acre on coarse soils and 1.6 pints on medium and fine soils. Use incorporation equipment that will insure thorough soil mixing with minimum damage to the established alfalfa.

ASPARAGUS—ESTABLISHED

Follow recommended soil preparation, application and incorporation procedures for TREFLAN.

TREFLAN can be applied to established asparagus as a single or as a split application. In the winter or early spring, apply TREFLAN to asparagus after ferns are removed but before spear emergence. Or, apply after harvest in the late spring or early summer before ferning begins. TREFLAN will suppress volunteer seedling asparagus and field bindweed if you use the following recommended rates and application schedules.

Follow recommended soil preparation, application and incorporation procedures for TREFLAN.

Broadcast Rates Per Acre

	TREFLAN 5*					
	Split Application			Single Application		
Soil Texture	Before Harvest	+	After Harvest	Before Harvest	or	After Harvest
	(pints)			(pints)		
Coarse	0.8	+	0.8	1.6	or	1.6
Medium	1.2	+	1.2	2.4	or	2.4
Fine	1.6	+	1.6	3.2	or	3.2

*In any single calendar year, the maximum TREFLAN 5 to apply is 1.6 pints per acre on coarse soils; 2.4 pints on medium soils; and 3.2 pints on fine soils.

FOR THE FOLLOWING CROP GROUPING, USE THE RATE TABLE BELOW.

CARROT

CASTOR BEAN

CELERY—(Direct seeded and transplant in areas receiving less than 20″ average annual rainfall.)

COLE CROPS—TRANSPLANT

Apply and incorporate prior to transplanting only. (Broccoli, Brussels Sprout, Cabbage and Cauliflower) See next section for direct seeded.

OKRA

PEPPER—TRANSPLANT

Apply and incorporate prior to transplanting only.

SOUTHERN PEA—(Before planting only.)

Apply and incorporate TREFLAN before planting, at planting or immediately after planting, unless otherwise indicated.

Broadcast Rates Per Acre

	TREFLAN 5	
Soil Texture	Areas receiving less than 20″ average annual rainfall*	Areas receiving greater than 20″ average annual rainfall*
	(pints)	(pints)
Coarse	0.8	0.8
Medium	1.0–1.2	1.2
Fine	1.2	1.6

* Use 1.2 pints per acre on coarse and medium textured soils and 1.6 pints on fine soils with 2–5% organic matter; use 1.6 pints on all soils with 5–10% organic matter.

COLE CROPS—DIRECT SEEDED (Broccoli, Brussels Sprout, Cabbage and Cauliflower) See above section for transplant.

For direct-seeded cole crops, apply and incorporate TREFLAN before planting at a broadcast rate of 0.8 pint per acre on coarse and medium soils and 1.2 pints on fine soils and soils with 2–5% organic matter. Direct-seeded cole crops have exhibited marginal tolerance to recommended rates of TREFLAN. Stunting or reduced stands may occur.

CUCURBITS—POSTPLANT EMERGED (Cantaloupe, Cucumber and Watermelon)

Western United States including Texas:

Apply TREFLAN as a directed spray to the soil between the rows and beneath plants which are in the 3 to 4 true leaf stage.

Broadcast Rates Per Acre

	TREFLAN 5	
Soil Texture	Areas receiving less than 20″ average annual rainfall*	Areas receiving greater than 20″ average annual rainfall*
	(pints)	(pints)
Coarse	0.8	0.8
Medium	1.0–1.2	1.2
Fine	1.2	1.6

* Use 1.2 pints on coarse and medium textured soils and 1.6 pints on fine soils with 2–5% organic matter; use 1.6 pints on all soils with 5–10% organic matter.

Set incorporation equipment to throw treated soil around the plants during incorporation.

DRY BEAN—TREFLAN ALONE

Apply and incorporate TREFLAN before planting using the following rates:

Broadcast Rates Per Acre

	TREFLAN 5	
Soil Texture	Areas receiving less than 20″ average annual rainfall*	Areas receiving greater than 20″ average annual rainfall*
	(pints)	(pints)
Coarse	0.8	0.8
Medium	1.0–1.2	1.2
Fine	1.2	1.6

* Use 1.2 pints per acre on coarse and medium textured soils and 1.6 pints on fine soils with 2–5% organic matter; use 1.6 pints on all soils with 5–10% organic matter.

DRY BEAN—TREFLAN/Eptam tank mix:

The TREFLAN/Eptam tank mix effectively controls all the following weeds in addition to those weeds listed for TREFLAN:

Henbit (Spring applications)	Oat, wild
Nightshade, black	Ragweed, common
Nightshade, hairy	Smartweed, Pennsylvania
Nutsedge	Velvetleaf (Buttonweed)

Follow recommended soil preparation and incorporation procedures for TREFLAN. The TREFLAN/Eptam tank mix should be applied from 2 days before planting up to planting. Incorporate immediately after application.

Broadcast Rates Per Acre

Soil Texture	TREFLAN 5: Areas receiving less than 20" average annual rainfall*	TREFLAN 5: Areas receiving greater than 20" average annual rainfall*	Eptam 7E
	(pints)	(pints)	(pints)
Coarse	0.8	0.8	2½-3½**
Medium	1.0-1.2	1.2	2½-3½
Fine	1.2	1.6	2½-3½

* Use 1.2 pints per acre on coarse and medium textured soils and 1.6 pints on fine soils with 2–5% organic matter; use 1.6 pints on all soils with 5–10% organic matter.

** Use Eptam 7E at a rate of 2½ pints per acre to control annual grasses; 3½ pints in control nutsedge and additional broadleaf weeds.

Precaution: Read the Eptam label before using. Observe all cautions and limitations of all products used in mixtures. The combination of TREFLAN and Eptam should not be used on soybeans, black-eyed peas (beans), lima beans and other flatpodded beans, except Romano. Do not use the foliage from a crop treated with the TREFLAN/Eptam tank mix for feed or for grazing.

DRY BEAN—Fall application for dry bean grown in Idaho, Oregon, and Washington:

Apply and incorporate TREFLAN any time between October 15 and December 31 at a broadcast rate of 0.8 pint per acre on coarse soils; 1 to 1.2 pints on medium soils; and 1.2 pints on fine soils. Destroy established weeds during seedbed preparation.

FOR THE FOLLOWING CROP GROUPING, USE THE RATE LISTED BELOW.

BEANS—(Guar and Mungbean)

GREENS—Turnip greens grown for processing, Collard, Kale and Mustard greens.

MUSTARD—Grown for seed or processing for food in Minnesota, Montana and North Dakota:

Apply and incorporate TREFLAN before planting at 0.8 pint per acre on coarse soils and 1.2 pints on medium and fine soils.

BEANS—(Lima Bean and Snap Bean):

Apply and incorporate TREFLAN before planting at a broadcast rate of 0.8 pint per acre on coarse and medium soils and 1.2 pints on fine soils.

CORN (Field Corn) and GRAIN SORGHUM (Milo)

Apply TREFLAN to field corn or grain sorghum (8 inches or taller) as an over-the-top or directed spray to effectively control weeds listed for TREFLAN.

Soil Preparation—Cultivate before a TREFLAN application to insure loose, friable soil, to remove established weeds, and to cover the base of plants with soil.

Application Directions—TREFLAN should be applied and incorporated at the recommended rates for the soil texture when the crop is well established (8 inches or taller). TREFLAN may be applied either as an over-the-top spray or as a directed spray. Drop nozzles should be used if foliage prevents uniform coverage of soil surface. Soil incorporation may be accomplished with only one pass of a sweep-type cultivator or a properly adjusted rolling cultivator.

The sweep-type cultivator should have 3 to 5 sweeps per row middle and be operated at 6 to 8 mph. Set the middle sweeps so as to avoid exposing untreated soil. Adjust the incorporation tools to prevent crop injury.

Broadcast Rates Per Acre

Soil Texture	TREFLAN 5
	(pints)
Coarse	0.6-0.8*
Medium	0.8-1.2
Fine	1.2-1.6

Use the lower rates when you anticipate light weed pressure and the higher rates when you anticipate heavy weed pressure.

* Corn only: Apply 0.8 to 1.2 pints per acre in Alabama, Florida, Georgia, North Carolina, South Carolina and Virginia to control fall panicum and Texas panicum.

Precaution: Do not apply TREFLAN to corn grown for seed. Do not apply TREFLAN to corn or sorghum as a preplant or preemergence treatment, or crop injury may occur.

HOPS

Apply and incorporate TREFLAN while the crop is dormant. Use a broadcast rate of 0.8 pint per acre on coarse soils; 1 to 1.2 pints on medium soils; and 1.2 pints on fine soils and soils with 2–10% organic matter.

MINT—(Established Peppermint and Spearmint)

Apply TREFLAN at a rate of 0.8 pint per acre on coarse soils; 1.0 pint on medium soils; and 1.2 pints on fine soils. Use incorporation equipment that will insure thorough soil mixing with minimum damage to the crop.

PEA—(Dry and English)—TREFLAN ALONE

Apply and incorporate TREFLAN before planting at a rate of 0.8 pint per acre on coarse and medium soils and 1.2 pints on fine soils.

PEA—TREFLAN/Far-Go tank mix for pea in Idaho, Oregon and Washington:

The tank mix combination of TREFLAN plus Far-Go will provide control of wild oat in addition to other annual grasses and broadleaf weeds controlled by TREFLAN.

Application Rates: Broadcast 0.6 pint of TREFLAN per acre on coarse and medium soils; 0.8 pint of TREFLAN on fine soils. Use 1¼ quarts of Far-Go per acre for all soil textures.

Incorporation Directions: Apply and incorporate up to 3 weeks before planting. Follow recommended incorporation procedures for TREFLAN.

Precaution: Do not apply to lentils. Leaf crinkling and delayed maturity of peas may occur, particularly on clay points in the northwest; but this is usually more than offset by a reduction of wild oat. Do not use foliage from treated peas for feed or forage. Refer to the cautions, precautions and directions on the Far-Go label.

PEA—Fall application to dry pea and English pea in Idaho, Oregon and Washington:

Apply and incorporate TREFLAN any time between October 15 and December 31 at a broadcast rate of 0.8 pint per acre on coarse soils; 1 to 1.2 pints on medium soils; and 1.2 pints on fine soils. Destroy established weeds during seedbed preparation. Do not apply TREFLAN in the fall to soils which are wet or are subject to prolonged periods of flooding.

PEANUT—(Spanish Peanut in Texas and Oklahoma)

Apply and incorporate TREFLAN before planting, at planting or immediately after planting at a broadcast rate of 0.8 pint per acre on coarse soils. When incorporated after planting, take care not to disturb the seed.

PEANUT—TREFLAN/Vernam tank mix (Spanish Peanut in Texas and Oklahoma):

TREFLAN/Vernam tank mix effectively controls those weeds listed for TREFLAN alone plus these additional weeds:

Annual morningglory	Velvetleaf
Coffeeweed	Yellow nutsedge (nutgrass)
Purple nutsedge (nutgrass)	

Follow recommended soil preparation procedures for TREFLAN. You may apply the TREFLAN/Vernam tank mix up to 10 days prior to planting. Incorporate the tank mix immediately after application. Apply TREFLAN/Vernam at these rates:

Broadcast Rates Per Acre

Soil Texture	TREFLAN 5	Vernam 7E
	(pints)	(pints)
Coarse	0.8	2⅓

POTATO—(All states except Maine)

Apply and incorporate TREFLAN after planting before emergence, or immediately following dragoff or after the potato plants have fully emerged.

Broadcast Rates Per Acre

Soil Texture	TREFLAN 5: Areas receiving less than 20″ average annual rainfall*	TREFLAN 5: Areas receiving greater than 20″ average annual rainfall
	(pints)	(pints)
Coarse	0.8	0.8
Medium	1.0–1.2	1.2
Fine	1.2	1.6

* Use 1.2 pints per acre on coarse and medium soils with 2–5% organic matter; use 1.6 pints on all soils with 5–10% organic matter.

Set incorporation equipment so that the bed and furrow will be uniformly covered with a layer of treated soil. If the layer of treated soil is not uniform and the herbicide is concentrated over the bed, potato emergence may be retarded, and stem brittleness can occur. When applying and incorporating TREFLAN after potato plants have fully emerged, do not completely cover the foliage with treated soil. Likewise, do not completely cover foliage at subsequent cultivations. Be careful that incorporation machinery does not damage potato seed pieces or elongating sprouts.

POTATO—Split application in Idaho, Oregon and Washington:
On all soils, apply and incorporate 0.6 pint of TREFLAN per acre before planting and 0.6 pint after planting when potato plants have fully emerged. Do not apply to soils containing 2% or more organic matter. Follow incorporation directions listed above for application to potato after planting.

POTATO—TREFLAN/Eptam tank mix for potatoes grown in Kansas, Minnesota, Nebraska, North Dakota, Oklahoma, South Dakota and Texas:
The TREFLAN/Eptam tank mix effectively controls the following weeds in addition to those weeds controlled by TREFLAN.

Henbit (Spring applications)	Oat, wild
	Ragweed, common
Nightshade, black	Smartweed, Pennsylvania
Nightshade, hairy	Velvetleaf (Buttonweed)
Nutsedge	

Follow recommended soil preparation and application procedures for TREFLAN. The TREFLAN/Eptam tank mix may be applied after planting, but prior to crop emergence. In areas where potatoes are normally dragged off, the TREFLAN/Eptam tank mix should be applied and incorporated up to or immediately following drag off.

Broadcast Rates Per Acre

Soil Texture	TREFLAN 5: Areas receiving less than 20″ average annual rainfall*	TREFLAN 5: Areas receiving greater than 20″ average annual rainfall*	Eptam 7E
	(pints)	(pints)	(pints)
Coarse	0.8	0.8	1¾–7**
Medium	0.8–1.2	0.8–1.2	1¾–7
Fine	0.8–1.2	0.8–1.6	1¾–7

* Use 1.2 pints per acre on coarse and medium soils with 2–5% organic matter; use 1.6 pints on all soils with 5–10% organic matter.
** Use the higher rate of Eptam 7E for nutsedge control.

Precaution: Read the Eptam label before using. Observe cautions and limitations of products used in mixtures. Do not graze or feed forage to livestock from fields treated with TREFLAN/Eptam tank mix.

POTATO—TREFLAN/Eptam application before planting in Washington, Idaho and Oregon:
TREFLAN/Eptam may be applied before planting at a broadcast rate of 0.6 pint of TREFLAN per acre and 3½ pints of Eptam 7E per acre on all soil textures. Incorporate immediately.

Precaution: Do not use this tank mix both before and after planting in the same season. Read the Eptam label before using. Observe all cautions and limitations on labeling of all products used in mixtures. Do not use foliage from treated crops for feed or forage.

SAFFLOWER

Follow recommended soil preparation, application and incorporation procedures for TREFLAN.

Apply and incorporate TREFLAN in the spring before planting or in the fall between October 15 and December 31.

Broadcast Rates Per Acre

Soil Texture	TREFLAN 5: Areas receiving less than 20″ average annual rainfall*	TREFLAN 5: Areas receiving greater than 20″ average annual rainfall
	(pints)	(pints)
Coarse	0.8	0.8
Medium	1.0–1.2	1.0–1.2
Fine	1.2	1.6

* Use 1.2 pints per acre on coarse and medium textured soils and 1.6 pints on fine soils with 2–5% organic matter; use 1.6 to 2 pints on all soils with 5–10% organic matter.

SAFFLOWER—Fall application in Arizona, California, Idaho, Montana, Nevada, Oregon, Utah, Washington and Wyoming:
Apply and incorporate TREFLAN any time between October 15 and December 31. Ground may be left flat or bedded-up over winter. On bedded ground, knock beds down to desired height before planting, moving some treated soil from tops into furrows. Where soil is left flat over winter, take care during spring bedding operations to prevent turning up untreated soil. Destroy established weeds during seedbed preparation. If weeds become established in furrows due to uncovering of untreated soil during listing, destroy these weeds before planting. Apply and incorporate TREFLAN at a broadcast rate of 1.2 pints per acre on coarse soils; 1.6 pints on medium soils; and 2 pints on fine soils.

Do not apply TREFLAN in the fall to soils which are wet or are subject to prolonged periods of flooding.

SUGAR BEET

Apply TREFLAN as a broadcast, overtop spray when plants are between 2 and 6 inches tall at a rate of 0.8 pint per acre on coarse soils and 1 to 1.2 pints on medium and fine soils. Use the higher rate for medium and fine soils in areas receiving more than 20″ average annual rainfall. Set incorporation machinery to throw treated soil toward the plants in the row. Be careful that incorporation machinery does not damage the sugar beet taproot.

Precaution: Exposed beet roots should be covered with soil before a TREFLAN application to reduce the possibility of girdling.

SUGAR BEET—Incorporation with a tine-tooth harrow in California, Colorado, Idaho, Montana, Nebraska, Oregon, Texas, Utah, Washington and Wyoming:
A properly operated tine-tooth harrow (Flextine or Melroe) can incorporate TREFLAN for effective weed control in sugar beet. Operate the tine-tooth harrow 2 times over the field in opposing directions at a speed of 3 to 6 mph. Set the harrow to cut 1 or 2 inches deep. Be careful that the tine-tooth harrow does not damage the sugar beet taproot. Follow recommended application procedures and broadcast rates per acre for sugar beet (see preceding paragraph).

SUGARCANE—(Plant Cane)

Apply and incorporate TREFLAN twice a year at a broadcast rate of 1.6 to 3.2 pints per acre for all soil textures. Make the first TREFLAN application in the fall on firmly packed beds immediately after the seed pieces are planted. Make the second TREFLAN application in the spring before or shortly after the cane emerges. Loosen rain-packed beds 2 to 3 inches deep before the spring application. Take care that incorporation machinery does not damage the seed pieces or emerging shoots.

SUGARCANE—Postplant in Hawaii for control of most annual grasses, including guineagrass:

Surface apply TREFLAN after planting (for plant cane) or after harvesting (for ratoon cane), before weeds and cane emerge. Use a broadcast rate of 4.8 to 6.4 pints per acre for all soil textures. In plant cane, the beds should be formed or rolled before application. In ratoon cane, the crop residue should be removed before application. If large amounts of crop residues are present, TREFLAN will not be effective. Apply TREFLAN just before anticipated rainfall or sprinkle irrigate immediately after application.

SUGARCANE—Applications up to layby for plant cane or ratoon cane grown in Louisiana or Texas:

Apply and incorporate TREFLAN at a broadcast rate of 1.6 to 3.2 pints per acre for all soil textures. Do this in the spring from before or shortly after the cane emerges up to layby. Apply the TREFLAN after the beds have been shaved or false shaved. Loosen rain-packed beds 2 to 3 inches deep before application. Be careful that incorporation machinery does not damage seed pieces or emerging shoots. You may use a rolling cultivator or bed chopper to incorporate TREFLAN layby applications in sugarcane on all soil textures. Follow normal incorporation directions for the rolling cultivator. Set bed chopper to cut 3 to 4 inches deep and operate at 4 to 6 mph. Two incorporation passes are necessary.

SUGARCANE—Itchgrass (Raoulgrass) control in Louisiana:

Apply and incorporate TREFLAN on either plant or ratoon cane at a broadcast rate of 3.2 pints per acre for all soil textures. Follow the directions above for sugarcane layby application in Louisiana and Texas.

SUNFLOWER—TREFLAN ALONE

Apply and incorporate TREFLAN in the spring or in the fall between October 15 and December 31. Follow recommended soil preparation application and incorporation procedures for TREFLAN.

Broadcast Rates Per Acre

Soil Texture	TREFLAN 5: Areas receiving less than 20" average annual rainfall*	TREFLAN 5: Areas receiving greater than 20" average annual rainfall*
	(pints)	(pints)
Coarse	0.8	0.8
Medium	1.0–1.2	1.2
Fine	1.2	1.6

* Use 1.2 to 1.6 pints per acre on coarse and medium soils with 2–5% organic matter and 1.6 pints on all soils with 5–10% organic matter.

SUNFLOWER—TREFLAN/Amiben tank mix or overlay:

Follow recommended soil preparation, application and incorporation procedures for TREFLAN.

Amiben may be applied in a band or broadcast over sunflowers at planting in fields where TREFLAN has been incorporated prior to planting. Or, the TREFLAN/Amiben tank mix may be incorporated prior to planting. TREFLAN/Amiben tank mix improves mustard, smartweed, velvetleaf and ragweed control in addition to those weeds controlled by TREFLAN alone.

Apply TREFLAN/Amiben or TREFLAN with an Amiben overlay at the following broadcast rates per acre:

Soil Texture	TREFLAN 5	Amiben 2S
	(pints)	(quarts)
Coarse	0.8	4
Medium	1.2	4–6*
Fine	1.6	4–6*

* For best control of mustard, common ragweed or black nightshade, use the 6 quart rate.

In coarse textured soils, heavy rains on the incorporated Amiben may move it below the weed seed germination zone, and erratic weed control may result. If sufficient rain does not fall within 7 days after a preemergence application of Amiben, but there is enough soil moisture to germinate weeds and grasses, a light cultivation with a rotary hoe or similar tool will uproot these small broadleaf weeds and grasses. The shallow mixing of Amiben in the surface soil will not interfere with the action of Amiben when rains come.

TOMATO

For direct-seeded tomato, apply TREFLAN at blocking or thinning as a directed spray to the soil between rows and beneath the plants, and incorporate. For transplant tomato, apply and incorporate TREFLAN before transplanting. Do not apply TREFLAN after transplanting.

Broadcast Rates Per Acre

Soil Texture	TREFLAN 5: Areas receiving less than 20" average annual rainfall*	TREFLAN 5: Areas receiving greater than 20" average annual rainfall*
	(pints)	(pints)
Coarse	0.8	0.8
Medium	1.0–1.2	1.2
Fine	1.2	1.6

* Use 1.2 pints per acre on coarse and medium textured soils and 1.6 pints on fine soils with 2–5% organic matter; 1.6 pints on all soils with 5–10% organic matter.

FRUIT AND NUT CROPS AND VINEYARDS

For areas receiving more than 20" average annual rainfall.

For new plantings of vineyards, citrus and pecan trees, apply and incorporate TREFLAN before planting at a broadcast rate of 0.8 pint per acre on coarse soils; 1.2 pints on medium soils; 1.6 pints on fine soils; 1.2 pints on fine soils with 2–5% organic matter; and 1.6 pints on soils with 5–10% organic matter.

For non-bearing, established plantings of citrus and pecan trees and bearing plantings of grapefruit, lemon, orange, pecan, tangelo and tangerine trees, apply TREFLAN at a broadcast rate of 1.6 to 3.2 pints per acre for all soil textures.

For areas receiving less than 20" average annual rainfall.

For new plantings of almond, apricot, citrus, nectarine, peach, pecan and walnut trees, apply and incorporate TREFLAN before planting at a broadcast rate of 0.8 pint per acre on coarse soils; 1 to 1.2 pints on medium soils; 1.2 pints on fine soils; 1.2 to 1.6 pints on soils with 2–5% organic matter; and 1.6 pints on soils with 5–10% organic matter.

For new plantings of vineyards, apply and incorporate TREFLAN before planting at a broadcast rate of 0.8 to 1.2 pints per acre on coarse soils; 1.2 to 2.4 pints on medium soils; and 2.4 to 3.2 pints on fine soils or soils with 2–10% organic matter. Do not use more than 1.6 pints per acre on heat-treated grape rootings.

For postplant applications on bearing or non-bearing, established plantings of vineyards and almond, apricot, grapefruit, lemon, nectarine, orange, peach, plum, prune, tangelo, tangerine and walnut trees, apply TREFLAN at a broadcast rate of 1.6 to 3.2 pints per acre for all soil textures. Do not apply to vineyards within 60 days of harvest.

In established plantings, apply TREFLAN as a directed spray to the soil and use incorporation methods not injurious to the trees or vines.

FRUIT AND NUT CROPS AND VINEYARDS—Rhizome johnsongrass control:

For areas receiving less than 20" average annual rainfall.
You can obtain commercially acceptable control of rhizome johnsongrass with postplant applications in bearing and non-bearing, established plantings of vineyards and almond, apricot, grapefruit, lemon, nectarine, orange, peach, pecan, tangelo, tangerine and walnut trees with a TREFLAN program when applied for 2 years in a row.

Soil Preparation—Work the soil thoroughly to bring the rhizomes nearer the surface.

Application—Apply TREFLAN at a broadcast rate of 1.6 quarts per acre on all soil textures each year for 2 years in a row. Do not apply to vineyards within 60 days of harvest.

Incorporation—Incorporate TREFLAN thoroughly with a disc set to cut 4 to 6 inches deep and operate at 4 to 6 mph. Two incorporation passes are necessary, with the second pass in a different direction from the first.

Cultivation—Some johnsongrass plants will escape. Timely cultivations are necessary to obtain commercially acceptable control. You cannot obtain commercially acceptable control with only 1 year of TREFLAN use.

Precautions: Do not use the 1.6 quart rate on new plantings, or crop injury may result. Do not interplant orchards or vineyards with other crops. If the treated vineyards and orchards are diverted to other crop uses, plant only those crops for which TREFLAN has been registered as a preplant treatment for the next cropping season.

FRUIT AND NUT CROPS AND VINEYARDS—Bindweed control in California:
TREFLAN can be used for the control of field bindweed in vineyards and for almond, apricot, grapefruit, lemon, nectarine, orange, peach, pecan, tangelo, tangerine and walnut trees.
Apply TREFLAN at a broadcast rate of 3.2 pints per acre on all soil textures. TREFLAN must be applied in the spring with a specially designed spray blade which applies a thin, concentrated layer at a soil depth of 4 to 6 inches. This layer of TREFLAN prevents bindweed shoots from emerging.

Land Preparation—Destroy all weeds and grasses with soil tillage before applying TREFLAN. This tillage is necessary to prevent trash from interfering with the operation of the spray blade.

Equipment—This operation requires a spray blade capable of running 4 to 6 inches below the surface of the soil. The spray blade should be equipped with nozzles located under the blade and directed so that the TREFLAN spray will be trapped under the soil which is flowing over the blade as it is pulled through the soil. Use a sufficient number of nozzles with spacing that will uniformly apply the TREFLAN underground in a thin, horizontal layer.

Application—Apply TREFLAN in 40 to 80 gallons of water per acre. Operate the spray blade at a depth of 4 to 6 inches.

Precaution: Some soils develop cracks as they dry after rainfall or irrigation. Field bindweed may emerge if the cracks extend through the TREFLAN layer. Prevent or eliminate cracks by shallow discing or other tillage. Avoid deep tillage which disturbs the subsurface layer. Cultivation or tillage also aids the control of germinating seeds.

WHEAT (WINTER)—(Idaho, Montana, Oregon and Washington)

TREFLAN may be applied for preplant preemergence control of cheatgrass and other annual grasses and broadleaf weeds controlled by TREFLAN. The growth, development and yield of winter wheat will not be adversely affected provided the seed is placed below the zone of soil treated with TREFLAN.
Apply TREFLAN any time during a period from 3 weeks up to immediately prior to planting. Broadcast TREFLAN at a rate of 1.2 pints per acre on coarse and medium soils and 1.6 pints on fine soils.

Incorporation Directions—Incorporate TREFLAN into the soil with a flexible tine-tooth harrow (Flextine, Melroe) set to cut 1 to 2 inches deep and operate at 3 to 6 mph. Incorporate one time within 24 hours after application, followed by a second incorporation in a different direction from the first prior to planting. Do not till the soil with a disc after the TREFLAN has been incorporated with a flexible tine harrow.

Seeding Directions—Use only a deep furrow or semi-deep furrow drill that will place the seed below the zone of soil into which TREFLAN has been incorporated.

Precaution: Wheat planted in direct contact with treated soil may suffer crop injury in the form of delayed emergence and development.

WHEAT (WINTER)—Fallow soil application in Washington and Oregon:
TREFLAN applied and shallowly incorporated into fallow soil up to four months ahead of planting will control cheatgrass and certain annual grasses and broadleaf weeds. The growth, development, or yield will not be adversely affected as long as the seed is placed below the zone of soil treated with TREFLAN. Use deep or semi-deep furrow drills.

Broadcast rates are 1.2 pints per acre on coarse and medium soils and 1.6 pints on fine soils. Apply TREFLAN any time from May to September prior to the fall planting of winter wheat.

Incorporation—Incorporate TREFLAN with a flexible tine-tooth harrow (Flextine or Melroe) set to cut 1 to 2 inches deep and operated at 3 to 6 mph. For thorough incorporation, two passes of the equipment in different directions over the field are necessary. Incorporate one time within 24 hours after application followed by a second incorporation prior to seeding. Do not till the soil with a disc after TREFLAN has been applied with a flexible tine harrow.

Precaution: Use only deep furrow or semi-deep furrow drills. Place seed below the zone of soil into which TREFLAN has been incorporated. Do not plant wheat directly into the zone of soil treated with TREFLAN. Injury to the crop or a delay in its emergence and development may occur.

WHEAT (SPRING), DURUM AND BARLEY—TREFLAN ALONE

TREFLAN is recommended as a postplant incorporated treatment to control foxtail (pigeongrass).

Apply TREFLAN at a broadcast rate of 0.8 pint per acre on coarse and medium soils and 1.2 pints on fine soils.

Plant 2 to 3 inches deep in a well-tilled seedbed. Apply TREFLAN after seeding but before the crop emerges. To incorporate, use flextine or diamond harrows operated two times in different directions, at speeds of at least 5 mph. Incorporate by operating equipment 1 to $1\frac{1}{2}$ inches deep. Application and the first incorporation should be done in the same operation if possible. Both incorporations must be done within 24 hours.

WHEAT (SPRING), DURUM, BARLEY (FALL APPLICATION)—Foxtail/Pigeongrass control:
TREFLAN may be fall applied for foxtail/pigeongrass control in spring wheat, durum and barley planted the following spring. TREFLAN may be applied to ground that has a manageable trash level, has been fallowed or pre-tilled. The first incorporation is required within 24 hours after application. A second incorporation is required prior to planting to destroy emerged weeds and to ensure an even distribution of TREFLAN treated soil.

Broadcast Rates Per Acre

Soil Texture	TREFLAN 5
	(pints)
Coarse	0.8
Medium	0.8
Fine	1.2

Incorporation Directions—Any of the following tools are recommened for fall incorporation. The disc or field cultivator may be used for the spring incorporation pass. Care should be taken to operate the tool at a more shallow depth than the fall incorporation.

1. Chisel plow: *May be used for the first pass only.* Operate at 4–5 inches deep at 4–6 mph. A chisel plow is defined as having 3 rows of up to 18-inch sweeps on no greater than 12-inch centers. Stagger sweeps so that no soil is left unturned.
2. Tandem disc: Operate at 3–4 inches deep at 4–6 mph.
3. Field cultivator: Operate at 3–4 inches deep at 5 mph or more. A field cultivator is defined as having 3 to 4 rows of sweeps with "c" or "s" shaped shanks, spaced 7 inches or less and staggered so that no soil is left unturned.

Planting Directions—Wheat, durum, or barley should be approximately 2 inches deep.

Precaution—While use of this practice may result in a stand reduction, slight stand reductions do not normally affect yield.

WHEAT (SPRING), DURUM AND BARLEY—TREFLAN/ Far-Go tank mix:

TREFLAN/Far-Go applied as a postplant incorporated treatment will control foxtail (pigeongrass) and wild oat.

Plant 2 to 3 inches deep in a well-tilled seedbed. Apply TREFLAN/ Far-Go after seeding but before crop emerges. To incorporate, use flex-tine or diamond harrows two times, operated in different directions, at speeds of at least 5 mph. Incorporate by operating equipment 1 to 1½ inches deep. Application and the first incorporation should be done in the same operation if possible. If not, incorporate immediately after application.

Broadcast Rates Per Acre

	TREFLAN 5	Far-Go	
Soil Texture	**Barley Durum Spring Wheat**	**Durum Spring Wheat**	**Barley**
	(pints)	(pints)	(pints)
Coarse	0.8	2	2½
Medium	0.8	2	2½
Fine	1.2	2	2½

Precaution: Overapplication may result in crop injury. Read the Far-Go label carefully before using.

FALL APPLICATION

General: (Areas receiving more than 20" average annual rainfall) See specific crop for recommendations. For all crops for which there are no specific fall application instructions and for which TREFLAN is recommended as a preemergence application, use the rates listed for spring applications. Do not apply TREFLAN in the fall for sugarbeets, potatoes and direct-seeded tomatoes.

In most states apply and incorporate TREFLAN any time between October 15 and December 31. In Minnesota, Montana, North and South Dakota, apply and incorporate TREFLAN anytime between September 1 and December 31. Ground may be left flat or bedded-up over winter. On bedded ground, knock beds down to desired height before planting, moving some treated soil from beds into furrows. Where soil is left over winter, be careful not to turn up untreated soil during spring bedding operations. Destroy established weeds during seedbed preparation. If weeds become established in furrows due to uncovering of untreated soil during bedding, destroy these weeds before planting. Do not apply TREFLAN in the fall to soils which are wet, are subject to prolonged periods of flooding, or where rice was grown the previous year.

FERTILIZER USE DIRECTIONS
APPLICATION WITH LIQUID FERTILIZERS

TREFLAN 5 may be mixed with most liquid fertilizer materials. The combination of TREFLAN with solutions and suspension-type fertilizers has provided weed and grass control equal to the same rates of TREFLAN applied in water. Follow TREFLAN label recommendations regarding rates per acre, crops, incorporation directions, special instructions, cautions and special precautions. Individual state regulations relating to liquid fertilizer mixing, registration, labeling and applications are the responsibility of the individual and/or company selling the fertilizer and chemical mixture.

Testing for Tank Mix Compatibility in Liquid Fertilizers: TREFLAN alone or in tank mixture with dry flowables, wettable powders (WP), aqueous suspensions (AS), flowables, (F), liquids (L), or solutions (S) may not combine properly with some fluid fertilizer materials. Small quantities should always be tested before full-scale mixing. This will determine whether a compatibility agent is needed, and which agent does the best job. The seven agents listed below have been thoroughly tested. There are many other surfactants on the market which were not designed for use with liquid fertilizers. Use the following test to select the correct agent for your mixture.

1. Put 1 pint of the liquid fertilizer in a quart jar.
2. Add 1 to 4 teaspoonful(s) of the dry flowable, WP, AS, F or L formulation (depending on the recommended rate per acre) to the liquid fertilizer. Close jar and agitate until the materials are dispersed evenly in the fertilizer. If the materials do not disperse well, it may be necessary to slurry the chemicals in water before adding to the fertilizer.
3. After dispersing the materials (Step 2), add 3 to 4 teaspoonsful of TREFLAN to the jar and shake well. Add solution herbicides to the mixture last and agitate. Observe the jar for about 10 minutes. If the materials rise to the surface and form a thick layer (oily curds) which will not redisperse when agitated, a compatibility agent is needed. If the mixture is easily redispersed to its original state with slight agitation, no agent is needed but good agitation must be provided in the fertilizer spray tank.
4. If the need for a compatibility agent is shown in Step 3: Using a clean quart jar, start at Step 1 above, add ½ teaspoon of the compatibility agent to the liquid fertilizer, mix well, then repeat Steps 2 and 3.

An effective compatibility agent will cause the mixture to remain uniformly mixed with little or no separating or oil rising to the surface for one half hour or longer. If slight separation does occur, 2 to 3 inversions of the jar should give a uniform remix. If oily curds form which will not redisperse, more agent or another agent should be tried.

Use a clean jar for each test. The compatible mixture will have a uniform appearance and will be relatively easy to keep mixed with gentle agitation of the jar.

LIQUID FERTILIZER MIXING INSTRUCTIONS
TREFLAN 5 in Liquid Fertilizer

General—Emulsifiable concentrates, such as TREFLAN 5, can be mixed with liquid fertilizers. In all cases, continuous agitation is required to prevent the TREFLAN from rising to the surface as an oily layer. When necessary, (see Testing for Tank Mix Compatibility in Liquid Fertilizers) a compatibility agent can be used to cause the TREFLAN to emulsify properly (i.e., have a milky appearance rather than oily layer). The use of compatibility agents is especially important when tank mixing emulsifiable concentrates (E.C.) with dry flowables, wettable powders (WP), aqueous suspensions (AS), flowables (F), liquids (L), or solutions (S) in liquid fertilizer. If the emulsion is not properly formed and the TREFLAN rises to the surface of the fertilizer as an oil ("oils out"), the oil may combine with the wettable powder, flowable, or suspension to form oily curds (viscous phase) which is difficult to redisperse. Any one of the compatibility agents listed below is helpful in causing liquid concentrates to form non-oiling mixtures with liquid fertilizers. These compatibility agents can be used at rates as low as 1½ to 2 pints per ton of liquid fertilizer and should be mixed well with the fertilizer before adding the liquid concentrate. Read the label on the compatibility agent and follow the directions.

1. Sponto 168D (Witco Chemicals Co., Chicago, IL)
2. Compat (Farm Chemicals, Inc., Aberdeen, NC)
3. Unite (Hopkins Ag Chemical, Madison, WI)
4. T-Mulz 734-2 (Thompson-Hayward Chemical Co.)
5. Rigo Compatibility Agent (Rigo Company, Buckner, KY)

6. Amoco Spray Mate™ (Amoco Oil Co., Chicago, IL)
7. Kem-Link Universal Coop., Minneapolis, MN)

All of the above are phosphate, ester-type surfactants designed to be used with liquid fertilizers. They usually do not work as compatibility agents in tank mixtures in plain water.

APPLICATION

Spread the fertilizer/pesticide mixture with a properly calibrated applicator. Be certain the material is applied uniformly to the soil surface.

INCORPORATION

Follow normal TREFLAN incorporation procedures.

TREFLAN 5 APPLICATION WITH DRY BULK FERTILIZERS

GENERAL

Dry bulk fertilizers may be impregnated or coated with TREFLAN. Application of dry bulk fertilizers impregnated with TREFLAN has provided weed and grass control equal to the same rates of TREFLAN applied in water.

All TREFLAN label recommendations regarding rates per acre, approved crops, incorporation, special instructions, cautions and special precautions must be followed.

Apply a minimum of 200 pounds per acre of dry fertilizer impregnated with TREFLAN at the recommended rates. Any commonly used dry fertilizers can be used for TREFLAN impregnation except coated ammonium nitrate and straight limestone. These materials will not absorb the herbicide. Blends containing mixtures of these materials can be impregnated.

Impregnation—Use any closed drum, belt, ribbon or other commonly used dry bulk fertilizer blender. Apply TREFLAN uniformly to the fertilizer.

Rates—Check specific crop recommendations for the rate of TREFLAN per acre. See the rate table which follows to determine the amount of TREFLAN to be impregnated into a ton of dry bulk fertilizer based on the amount of fertilizer which will be applied per acre. (See rate chart below).

Application—Spread the fertilizer/chemical mixture with a properly calibrated applicator. Be certain the material is applied uniformly to the soil surface.

Incorporation—Follow TREFLAN incorporation procedures.

Rate Chart for Impregnating Fertilizer With TREFLAN 5 (TREFLAN Added to a Ton of Fertilizer)

Fertilizer Rate Per Acre	TREFLAN Rate Per Acre				
	0.8 pint	1.2 pints	1.6 pints	2.4 pints	3.2 pints
200 pounds	4 qts. per ton	6 qts. per ton	8 qts. per ton	12 qts. per ton	16 qts. per ton
250 pounds	3.2 qts. per ton	4.8 qts. per ton	6.4 qts. per ton	9.6 qts. per ton	12.8 qts. per ton
300 pounds	2.7 qts. per ton	4 qts. per ton	5.6 qts. per ton	8 qts. per ton	10.8 qts. per ton
350 pounds	2.3 qts. per ton	3.6 qts. per ton	4.8 qts. per ton	6.8 qts. per ton	9.2 qts. per ton
400 pounds	2 qts. per ton	3 qts. per ton	4 qts. per ton	6 qts. per ton	8 qts. per ton
450 pounds	1.8 qts. per ton	2.7 qts. per ton	3.6 qts. per ton	5.4 qts. per ton	7.2 qts. per ton

For rates other than those listed above, use the following formula to calculate the amount of TREFLAN to be impregnated on a ton of dry bulk fertilizer:

$$\text{Pints TREFLAN Per Acre} \times \frac{1000}{\text{Lbs. Fertilizer Per Acre}} = \text{Quarts TREFLAN Per Ton of Fertilizer}$$

All individual state regulations relating to dry bulk fertilizer blending, registration, labeling and application are the responsibility of the individual and/or company selling the fertilizer and chemical mixture.

TREFLAN 5 PERFORMANCE GUARANTEE— A LIMITED WARRANTY OF TREFLAN WEED AND GRASS CONTROL

Elanco Products Company warrants that TREFLAN will control the weeds and grasses as indicated on its current label when used according to label directions and subject to the following limitations:

1. Purchaser must notify Elanco promptly if a lack of satisfactory control occurs. Notice must be given within ninety days after the first planting or after application of TREFLAN to the treated crop, whichever occurs later.
2. An Elanco representative must be satisfied that Purchaser used TREFLAN according to the label directions and must visually observe unsatisfactory control of weeds in the field.
3. The Purchaser must provide Elanco representative with an invoice or other satisfactory record which shows the price and quantity of TREFLAN purchased.
4. Elanco's liability will be limited to a refund of the purchase price of the TREFLAN applied to the acreage on which weed control was not satisfactory or, at the option of the Purchaser, a refund of an equivalent amount of TREFLAN.
5. Void outside the USA.

DISCLAIMER OF WARRANTIES

The warranty printed above is the only warranty applicable to this product. All other warranties, expressed or implied, including, but not limited to, any implied warranties of merchantability and fitness for a particular purpose, are disclaimed.

LIMITATION OF DAMAGES

Elanco's liability, whether in contract, warranty, tort, negligence, strict liability or otherwise, shall not exceed the return of the amount of the purchase price of the TREFLAN and under no circumstances shall Elanco be liable for special, indirect or consequential damages.

INHERENT RISKS OF USE

Failure to carefully follow the directions for use of TREFLAN, as well as other factors such as seedling disease, cold weather, deep planting, excessive moisture, high salt concentration or drought, may result in unsatisfactory weed control or crop injury. Reduced yields may result under these conditions.

SUPPLEMENTAL LABELING

See TREFLAN E.C. for Supplemental Labeling pertaining to this product.

TREFLAN® TR-10

GRANULES HERBICIDE

A Selective Herbicide for the Preemergence Control of Annual Grasses and Broadleaf Weeds

ACTIVE INGREDIENT:

trifluralin* (α,α,α-trifluoro-2,6-dinitro-*N*,*N*-dipropyl-*p*-toluidine) 10.0%

INERT INGREDIENTS 90.0%

TOTAL 100.0%

Contains 5 pounds active ingredient per 50 pound bag.

*TREFLAN®, TR-10™ —Trademarks for Elanco Products granular formulation of trifluralin containing 10 percent active ingredient.

EPA Reg. No. 1471-143 U.S. Patent. No. Re. 31227
ID 5535 PI 2883 AMP

CAUTION

Keep Out of Reach of Children

PRECAUTIONARY STATEMENTS

HAZARDS TO HUMANS AND DOMESTIC ANIMALS

CAUTION

Harmful if swallowed or absorbed through the skin. Avoid breathing dust. Avoid contact with skin, eyes or clothing. In case of skin or eye contact, flush with plenty of water. If irritation persists, get medical attention.

ENVIRONMENTAL HAZARD

This product is toxic to fish. Use with care when applying in areas adjacent to any body of water. Do not apply directly to water. Do not contaminate water by cleaning of equipment or disposal of wastes.

STORAGE AND DISPOSAL

STORAGE

Store in original container only. In case of leak or spill, contain material and dispose as waste.

PESTICIDE DISPOSAL

Do not contaminate water, food or feed by storage or disposal. Waste resulting from the use of this product may be disposed of on site at an approved waste disposal facility.

CONTAINER DISPOSAL

Completely empty bag by shaking or tapping sides and bottom to loosen clinging particles. Empty residue into application equipment. Then dispose of bags in a sanitary landfill or by incineration if allowed by State and local authorities. If burned, stay out of smoke.

SPECIAL PRECAUTIONS

Applied according to directions and under normal growing conditions, TREFLAN will not harm the treated crop. Overapplication may result in crop injury or a soil residue. Uneven application or improper soil incorporation of TREFLAN can result in erratic weed control or crop injury. Seedling disease, cold weather, deep planting, excessive moisture, high salt concentration or drought may weaken crop seedlings and increase the possibility of damage from TREFLAN. Under these conditions, delayed crop development or reduced yields may result.

In Arizona, Colorado, California, Idaho, Montana, Nevada, New Mexico, Oregon, Utah, Washington and Wyoming, sugar beets, red beets or spinach should not be planted for 12 months after a spring application or for 14 months after a fall application of TREFLAN. Plow the land to a depth of 12 inches prior to planting sugar beets to prevent the possibility of crop injury. Sorghum (milo), proso millet, corn or oats should not be planted for 14 months after a spring application or for 16 months after a fall application of TREFLAN to avoid crop injury. If land has not been irrigated, do not plant any of these crops for 18 months after a spring application or 20 months after a fall application of TREFLAN.

In those portions of Kansas, Nebraska, North Dakota, Oklahoma, South Dakota and Texas where at least 20 inches of irrigation and/or rainfall (total) was used to produce the crop, sorghum or oats should not be planted for 12 months after an application of TREFLAN.

If less than 20 inches of total water was used to produce the crop, do not plant sorghum, proso millet, or oats for 18 months after an application of TREFLAN. Cool, wet weather conditions during the early stage of growth may increase the possibility of injury to sorghum.

In all other areas receiving greater than 20″ rainfall per year: Moldboard plow before planting sugar beets where a spring application of TREFLAN was made the previous season.

Vegetable Growing Areas: Vegetable crops other than those listed on this label should not be planted within 5 months folowing the application of TREFLAN.

DIRECTIONS FOR USE

TREFLAN is a preemergence herbicide which is incorporated into the soil to provide long-lasting control of many annual grasses and broadleaf weeds. TREFLAN controls weeds as they germinate. TREFLAN will not control established weeds.

It is a violation of Federal law to use this product in a manner inconsistent with its labeling.

WEEDS AND GRASSES CONTROLLED BY TREFLAN

Grasses	
Annual bluegrass	*Poa annua*
Barnyardgrass (Watergrass)	*Echinochloa* sp.
Brachiaria (Signalgrass)	*Brachiaria* sp.
Bromegrass (Cheatgrass) (Downy brome)	*Bromus tectorum*
Cheat (Chess)	*Bromus secalinus*
Crabgrass (Large crabgrass) (Smooth crabgrass)	*Digitaria* spp.
Foxtail (Bottlegrass) (Bristlegrass) (Giant foxtail) (Green foxtail) (Foxtail millet) (Pigeongrass) (Robust foxtail) (Yellow foxtail)	*Setaria* spp.
Goosegrass (Silver crabgrass) (Silvergrass) (Wiregrass) (Yardgrass)	*Eleusine indica*
Johnsongrass (from seed)	*Sorghum halepense*
Junglerice	*Echinochloa colonum*
Sandbur (Burgrass)	*Cenchrus incertus*
Sprangletop	*Leptochloa filiformis*
Stinkgrass (Lovegrass)	*Eragrostis cilianensis*
Wild cane (Shattercane)	*Sorghum bicolor*
Wild oat[1]	*Avena fatua*
Woolly cupgrass	*Eriochloa villosa*
Broadleaf weeds	
Carpetweed	*Mollugo verticillata*
Chickweeed	*Stellaria media*
Florida pusley (Florida purslane) (Mexican clover) (Pusley)	*Richardia scabra*
Goosefoot	*Chenopodium hybridum*
Henbit (fall application only)	*Lamium amplexicaule*
Knotweed	*Polygonum aviculare*
Lambsquarters	*Chenopodium album*
Pigweed (Carelessweed) (Prostrate pigweed) (Redroot pigweed) (Rough pigweed) (Spiny pigweed)	*Amaranthus* spp.
Puncturevine (Western U.S. only) (Caltrop) (Goathead)	*Tribulus terrestris*

Purslane	*Portulaca oleracea*
Stinging nettle (Nettle)	*Urtica dioica*

[1] TREFLAN applied as a preplant incorporated (PPI) treatment will provide partial control of wild oats. This claim is for all PPI uses except fall applications for spring seeded cereals

SOIL PREPARATION

Crop Residues or Existing Weeds: Apply TREFLAN to soil that has a manageable trash level, been fallowed, pre-tilled or to soil that has standing stubble. A manageable level of such ground cover would allow the TREFLAN to be uniformly incorporated into the top 2 to 3 inches of the final seedbed.

Roughness: The soil surface should be smooth enough so that application and incorporation equipment can be operated efficiently and at speeds which ensure a uniform application and incorporation of TREFLAN.

General Soil Conditions: To assure uniform incorporation of TREFLAN, soil moisture conditions should be such that large clods can be broken up during the incorporation process.

SOIL TEXTURE GUIDE

The amount of TREFLAN applied will vary with the soil texture and organic matter. A fine textured soil will require more TREFLAN per acre than a coarse soil. Choose the proper rate for each application based on the following soil texture group and specific crop recommendations. Do not exceed recommended rates.

Soil Texture	Soil Classification
Coarse soils: (Light)	Sand, loamy sand, sandy loam
Medium Soils:	Loam, silty clay loam*, silt loam, silt, sandy clay loam*
Fine Soils (Heavy)	Clay, clay loam, silty clay loam*, silty clay, sandy clay, sandy clay loam*

* Silty clay loam and sandy clay loam soils are transitional soils and may be classified as either medium or fine textured soils. If silty clay loam or sandy clay loam soils are predominately sand or silt, they are usually classified as medium textured soils. If they are predominately clay, they are usually classified as fine textured soils.

APPLICATION DIRECTIONS

Apply TREFLAN with a properly calibrated herbicide applicator that will apply the granules uniformly. Calibrate the applicator according to the manufacturer's directions and check frequently during application to be certain that equipment is working properly. Avoid concentration of the material in narrow bands.

Freezing will not adversely affect this product. If product is frozen at time of application, agitate or thaw to produce a free-flowing granule.

INCORPORATION DIRECTIONS

Incorporation Equipment—General Directions

Use incorporation equipment that mixes TREFLAN into the top 2 to 3 inches of the final seedbed, or erratic weed control and/or crop injury may result. Incorporation equipment such as a disc will mix TREFLAN approximately half as deep as the equipment is set to operate. For example, a disc set to cut 4 inches deep will incorporate most of the TREFLAN within the top 2 inches of the final seedbed.

Incorporation Before Planting

TREFLAN must be incorporated one time within 24 hours after application. **The second incorporation must be delayed at least five (5) days after the first,** completed prior to planting, and should be run in a different direction from the first. Incorporate the TREFLAN uniformly into the top 2 to 3 inches of the final seedbed.

Incorporation After Planting

Check specific crop for incorporation directions after planting.

Incorporation in Bedded Culture

For effective weed control, TREFLAN needs to be incorporated into the top 2 to 3 inches of the final seedbed.

Application prior to bedding: Apply TREFLAN and incorporate it one time with recommended equipment. The bedding operation serves as the second incorporation. Do not expose untreated soil during post-bedding operations.*

Application after bedding: Knock off beds to planting height before applying TREFLAN. Apply TREFLAN and incorporate it with recommended equipment that will conform to the bed shape. Do not leave untreated soil exposed.*

* Avoid removal of treated soil from the seedbed before or during the planting operation. This would expose untreated soil, allowing weeds to germinate in the drill row.

Recommended Equipment

Any recommended incorporation tool may be used alone or in combination with any other recommended tool. Two incorporation passes are required unless specifically stated. The second incorporation should not be deeper than the first.

Disc: Set to cut 4 to 6 inches deep and operate at 4 to 6 mph.

Field Culivator: Set to cut 3 to 4 inches deep and operate at 5 mph or more. A field cultivator is defined as an implement with 3 to 4 rows of sweeps, spaced at intervals of 7 inches or less and staggered so that no soil is left unturned. Chisel points should not be used.

Chisel Plow: (For use in northern Great Plains). The chisel plow may be used for the first incorporation pass only. Any other recommended tool may be used for the second pass for row crops. The chisel plow may be used for any tillage or incorporation pass in the Summer Fallow program. Operate at 4–5 inches deep at 4–6 mph. A chisel plow is defined as having 3 rows of 14 to 18 inch sweeps on no greater than 12 inch center. Stagger sweeps so that no soil is left unturned. Chisel points should not be used.

Combination Seedbed Conditioners: Set to cut 3 to 4 inches deep and operate at a speed of at least 5 mph. These implements are defined as three or more tillage devices combined and used as a single tool. For example, 2 to 3 rows of field cultivator "c" or "s" shaped shanks with an effective sweep spacing of 6 to 9 inches (staggered so that no soil is left unturned), followed by a spike-tooth or flextine harrow, followed by a ground-driven reel or basket.

Rolling Cultivator: Set to cut 2 to 4 inches deep and operate at 6 to 8 mph. Rolling cultivators are adequate for use on coarse and medium textured soils only.

Mulch treader (other similar disc-type implements): Set to cut 3 to 4 inches deep and operate at 5 to 8 mph.

P.T.O. Driven Equipment (tillers, cultivators, hoes): Adjust to incorporate TREFLAN into the top 2 to 3 inches of the seedbed with rotors spaced to provide a clean sweep of the soil. Only one incorporation is necessary. P.T.O. driven equipment should not be operated at a speed greater than 4 mph.

Other equipment, including the flexible tine-tooth harrow, (Flextine, Melroe), is also recommended but only for the special programs for which it is specified in this label.

CULTIVATION AFTER PLANTING

Soil treated with TREFLAN may be shallow cultivated without reducing the weed control activity of TREFLAN. Do not cultivate deeper than the treated soil since this may bring untreated soil to the surface, and poor weed control may result.

CROP RECOMMENDATIONS

These recommendations are given as the broadcast rates of TREFLAN per acre. Apply any time after January 1 when the soil can be worked and is suitable for good incorporation—see specific crop for recommendations. Where a rate range is shown, use the lower rate for coarser soils or soils with lower organic matter. TREFLAN should not be used on soils containing more than 10% organic matter or on muck soils.

ALFALFA—ESTABLISHED
(For Barnyardgrass, Crabgrass and Foxtail Control)

Treflan TR-10 can be activated for barnyardgrass, crabgrass and foxtail control in established alfalfa either by rainfall/overhead sprinkler irrigation or by mechanical incorporation.

Apply Treflan TR-10 to established alfalfa stands prior to weed emergence at a broadcast rate of 20 pounds per acre for all soil textures. Treflan TR-10 may be applied using properly calibrated ground or aerial application equipment. A single rainfall or overhead sprinkler irrigation of 0.5 inches or more after application is required to activate Treflan. If this does not occur within three (3) days after application, Treflan should be activated using incorporation equipment that will ensure thorough soil mixing with minimum damage to the established alfalfa.

Precaution: In the season following a 20 pound per acre Treflan TR-10 treatment, where established alfalfa is to be rotated to another crop, plant only those crops for which Treflan can be applied as a preplant treatment or injury may result.

FOR THE FOLLOWING CROP GROUPING, USE THE RATE TABLE BELOW.

CASTOR BEAN

COLE CROPS—TRANSPLANT

(Broccoli, Brussels Sprout, Cabbage and Cauliflower). See Direct Seeded below.

Apply and incorporate prior to transplanting only.

OKRA

PEPPER—TRANSPLANT

Apply and incorporate prior to transplanting only.

SOUTHERN PEA—(Before planting only.)

Apply and incorporate TREFLAN before planting, at planting or immediately after planting, unless otherwise indicated.

Broadcast Rates Per Acre

Soil Texture	TREFLAN TR-10: Areas receiving less than 20″ average annual rainfall*	TREFLAN TR-10: Areas receiving greater than 20″ average annual rainfall*
	(pounds)	(pounds)
Coarse	5	5
Medium	6¼–7½	7½
Fine	7½	10

* Use 7½ pounds per acre on coarse and medium textured soils and 10 pounds on fine soils with 2–5% organic matter; use 10 pounds on all soils with 5–10% organic matter.

COLE CROPS—DIRECT SEEDED

(Broccoli, Brussels Sprout, Cabbage and Cauliflower). See above for transplant.

For direct-seeded cole crops, apply and incorporate TREFLAN before planting at a broadcast rate of 5 pounds per acre on coarse and medium soils and 7½ pounds on fine soils and soils with 2–5% organic matter. Direct-seeded cole crops have exhibited marginal tolerance to recommended rates of TREFLAN. Stunting or reduced stands may occur.

CUCURBITS—POSTPLANT EMERGED

(Cantaloupe, Cucumber and Watermelon)

Western United States including Texas:

Apply TREFLAN as a broadcast granule to the soil between the rows and beneath plants which are in the 3 to 4 true leaf stage.

Broadcast Rates Per Acre

Soil Texture	TREFLAN TR-10: Areas receiving less than 20″ average annual rainfall*	TREFLAN TR-10: Areas receiving greater than 20″ average annual rainfall*
	(pounds)	(pounds)
Coarse	5	5
Medium	6¼–7½	7½
Fine	7½	10

* Use 7½ pounds per acre on coarse and medium textured soils and 10 pounds on fine soils with 2–5% organic matter; use 10 pounds on all soils with 5–10% organic matter.

Set incorporation equipment to throw treated soil around the plants during incorporation.

FRUIT AND NUT CROPS AND VINEYARDS

For areas receiving less than 20″ average annual rainfall.

For new plantings of almond, apricot, citrus, nectarine, peach, pecan and walnut trees, apply and incorporate TREFLAN before planting at a broadcast rate of 5 pounds per acre on coarse soils; 6¼ to 7½ pounds on medium soils; 7½ pounds on fine soils; 7½ to 10 pounds on soils with 2–5% organic matter; and 10 pounds on soils with 5–10% organic matter.

For new plantings of vineyards, apply and incorporate TREFLAN before planting at a broadcast rate of 5 to 7½ pounds per acre on coarse soils; 7½ to 15 pounds on medium soils; and 15 to 20 pounds on fine soils or soils with 2 to 10% organic matter. Do not use more than 10 pounds per acre on heat-treated grape rootings.

For postplant applications on bearing or non-bearing, established plantings of vineyards and almond, apricot, grapefruit, lemon, nectarine, orange, peach, pecan, plum, prune, tangelo, tangerine and walnut trees, apply TREFLAN at a broadcast rate of 10 to 20 pounds per acre for all soil textures. Do not apply to vineyards within 60 days of harvest.

In established plantings, apply TREFLAN to the soil and use incorporation methods not injurious to the trees or vines.

COTTON—TREFLAN ALONE

TREFLAN can be applied and incorporated before or at planting, immediately after planting, or at layby.

COTTON—Preemergence:

Broadcast Rates Per Acre

Soil Texture	TREFLAN TR-10: Areas receiving less than 20″ average annual rainfall*	TREFLAN TR-10: Areas receiving greater than 20″ average annual rainfall*
	(pounds)	(pounds)
Coarse	5	5
Medium	6¼–7½	7½
Fine	7½	10

* Use 7½ pounds per acre on coarse and medium textured soils and 10 pounds on fine soils with 2–5% organic matter; use 10 pounds on all soils with 5–10% organic matter.

COTTON—Postplant:

When incorporating TREFLAN after planting (postplant), be careful not to disturb the seed.

COTTON—Layby:

Apply and incorporate TREFLAN any time up to layby, but not less than 90 days before harvest. Direct the layby applications onto the soil between the rows and beneath emerged cotton plants. Use the same rates as for a preemergence application.

COTTON—Fall Application:

Apply and incorporate TREFLAN any time between October 15 and December 31. The ground may be left flat or bedded-up over winter. On bedded ground, knock beds down to desired height before planting, moving some treated soil from beds into furrows. Where soil is left flat over winter, be careful not to turn up untreated soil during spring bedding operations. Destroy established weeds during seedbed preparation. If weeds become established in furrows due to uncovering of untreated soil during bedding, destroy these weeds before planting. In the fall, do not apply TREFLAN to soils which are wet or subject to prolonged periods of flooding.

Broadcast Rates per Acre—Fall Application Only:
In Alabama, Arkansas, northern Florida, Georgia, Louisiana, Mississippi, southeastern Missouri bootheel, North Carolina, New Mexico, Oklahoma, South Carolina, Tennessee and Texas, apply and incorporate TREFLAN at a broadcast rate of 10 pounds per acre on coarse and medium soils and 12½ pounds on fine soils.

In Arizona, California and Nevada, apply and incorporate TREFLAN at a broadcast rate of 7½ pounds per acre on coarse soils; 10 pounds on medium soils; and 12½ pounds on fine soils.

For cotton grown in other states, apply and incorporate TREFLAN at a broadcast rate of 5 pounds per acre on coarse soils; 7½ pounds on medium soils; 10 pounds on fine soils; 7½ pounds on coarse soils with 2–5% organic matter; and 10 to 12½ pounds on soils with 5–10% organic matter.

DRY BEAN—TREFLAN ALONE

Apply and incorporate TREFLAN before planting using the following rates:

Broadcast Rates Per Acre

Soil Texture	TREFLAN TR-10 Areas receiving less than 20" average annual rainfall*	TREFLAN TR-10 Areas receiving greater than 20" average annual rainfall*
	(pounds)	(pounds)
Coarse	5	5
Medium	6¼–7½	7½
Fine	7½	10

* Use 7½ pounds per acre on coarse and medium textured soils and 10 pounds on fine soils with 2–5% organic matter; use 10 pounds on all soils with 5–10% organic matter.

DRY BEAN—Fall Application
Apply and incorporate TREFLAN any time between October 15 and December 31 at a broadcast rate of 5 pounds per acre on coarse soils; 6¼ to 7½ pounds on medium soils; and 7½ pounds on fine soils. Destroy established weeds during seedbed preparation.

FLAX—FALL APPLICATION

Apply and incorporate TREFLAN in the fall from September 1 to December 31.

Broadcast Rates Per Acre

Soil Texture	TREFLAN TR-10
	(pounds)
Coarse	5
Medium	7½
Fine	10

Special Instructions for Flax
1. Incorporation operations or any other tillages performed in the spring prior to seeding should be relatively shallow so as to maintain a firm seedbed, and the seedbed should be packed just prior to seeding.
2. Seeding should be done with a press drill or hoe drill. Seed into a moist seedbed no more than 1½ inches deep.
3. Flax should not be seeded until the seedbed has warmed up.

BEANS—(Lima Bean and Snap Bean)

Apply and incorporate TREFLAN before planting at a broadcast rate of 5 pounds per acre on coarse and medium soils and 7½ pounds on fine soils.

MUSTARD—Grown for seed or processing for food in Minnesota, Montana and North Dakota

Apply and incorporate TREFLAN before planting at a broadcast rate of 5 pounds per acre on coarse soils and 7½ pounds on medium and fine soils.

PEANUT—(Spanish Peanut in Texas and Oklahoma)

Apply and incorporate TREFLAN before planting, at planting or immediately after planting at a broadcast rate of 5 pounds per acre on coarse soils. When incorporating after planting, take care not to disturb the seed.

RAPE

Apply and incorporate TREFLAN in the spring before planting or in the fall after September 1.

Broadcast Rates Per Acre

Soil Texture	TREFLAN TR-10
	(pounds)
Coarse	5
Medium	7½
Fine	10

SAFFLOWER

Follow recommended soil preparation, application and incorporation procedures for TREFLAN.
Apply and incorporate TREFLAN in the spring before planting or in the fall between October 15 and December 31.

Broadcast Rates Per Acre

Soil Texture	TREFLAN TR-10 Areas receiving less than 20" average annual rainfall*	TREFLAN TR-10 Areas receiving greater than 20" average annual rainfall*
	(pounds)	(pounds)
Coarse	5	5
Medium	6¼–7½	7½
Fine	7½	10

* Use 7½ pounds per acre on coarse and medium textured soils and 10 pounds on fine soils with 2–5% organic matter; use 10 to 12½ pounds on all soils with 5–10% organic matter.

SAFFLOWER—Fall Application—(Arizona, California, Idaho, Montana, Nevada, Oregon, Utah, Washington and Wyoming):

Apply and incorporate TREFLAN any time between October 15 and December 31. Ground may be left flat or bedded-up over winter. On bedded ground, knock beds down to desired height before planting, moving some treated soil from tops into furrows. Where soil is left flat over winter, take care during spring bedding operations to prevent turning up untreated soil. Destroy established weeds during seedbed preparation. If weeds become established in furrows due to uncovering of untreated soil during listing, destroy these weeds before planting.

Apply and incorporate TREFLAN at a broadcast rate of 7½ pounds per acre on coarse soils; 10 pounds on medium soils; and 12½ pounds on fine soils.

Do not apply TREFLAN in the fall to soils which are wet or are subject to prolonged periods of flooding.

SOYBEAN—TREFLAN ALONE

SOYBEAN—Preemergence:
Follow recommended soil preparation, application, and incorporation procedures for TREFLAN.

Broadcast Rates Per Acre

Soil Texture	TREFLAN TR-10*
	(pounds)
Coarse	5
Medium	7½
Fine	10

* Use 7½ pounds per acre on coarse and medium textured soils and 10 pounds on fine soils with 2–5% organic matter; use 10–12½ pounds on all soils with 5–10% organic matter.

SOYBEAN—Fall Application:
Apply and incorporate TREFLAN any time between October 15 and December 31. Ground may be left flat or bedded-up over winter. On bedded ground, knock beds down to desired height before planting, moving some treated soil from tops into furrows. Where soil is left flat over winter, take care during spring bedding operations to prevent turning up untreated soil. Destroy established

weeds during seedbed preparation. If weeds become established in furrows due to uncovering of untreated soil during listing, destroy these weeds before planting. Do not apply TREFLAN in the fall to soils which are wet, are subject to prolonged periods of flooding, or where rice was grown the previous year.

For soybeans grown in Alabama, Arkansas, northern Florida, Georgia, Louisiana, Mississippi, southeastern Missouri Bootheel, North Carolina, Oklahoma, South Carolina, Tennessee and Texas, apply and incorporate TREFLAN at a broadcast rate of 10 pounds per acre on coarse and medium soils and 12½ pounds on fine soils.

For soybeans grown in states other than those listed above, apply and incorporate TREFLAN at a broadcast rate of 5 pounds per acre on coarse soils; 7½ pounds on medium soils; 10 pounds on fine soils; 7½ pounds on coarse soils with 2–5% organic matter; and 10–12½ pounds on soils with 5–10% organic matter.

SUGAR BEET

Broadcast TREFLAN overtop when plants are between 2 and 6 inches tall at a rate of 5 pounds per acre on coarse soils and 6¼ to 7½ pounds on medium and fine soils. Use the higher rate for medium and fine soils in areas receiving more than 20" average annual rainfall. Set incorporation machinery to throw treated soil toward the plants in the row. Be careful that incorporation machinery does not damage the sugar beet taproot.

Precaution: Exposed beet roots should be covered with soil before TREFLAN application to reduce the possibility of girdling.

SUNFLOWER—TREFLAN ALONE

Apply and incorporate TREFLAN in the spring or in the fall between October 15 and December 31. Follow recommended soil preparation, application and incorporation procedures for TREFLAN.

Broadcast Rates Per Acre

	TREFLAN TR-10	
Soil Texture	Areas receiving less than 20" average annual rainfall*	Areas receiving greater than 20" average annual rainfall*
	(pounds)	(pounds)
Coarse	5	5
Medium	6¼–7½	7½
Fine	7½	10

* Use 7½ to 10 pounds per acre on coarse and medium textured soils with 2–5% organic matter and 10 pounds on all soils with 5–10% organic matter.

TOMATO

For direct seeded tomato, apply TREFLAN at blocking or thinning to the soil between rows and beneath the plants, and incorporate. For transplant tomato, apply and incorporate TREFLAN before transplanting. Do not apply TREFLAN after transplanting.

Broadcast Rates Per Acre

	TREFLAN TR-10	
Soil Texture	Areas receiving less than 20" average annual rainfall*	Areas receiving greater than 20" average annual rainfall*
	(pounds)	(pounds)
Coarse	5	5
Medium	6¼–7½	7½
Fine	7½	10

* Use 7½ pounds per acre on coarse and medium textured soils and 10 pounds on fine soils with 2–5% organic matter; use 10 pounds on all soils with 5–10% organic matter.

WHEAT (SPRING), DURUM AND BARLEY—SUMMER FALLOW

TREFLAN may be applied to coarse, medium and fine textured soils for control of labeled weeds in the summer fallow period and for pigeongrass (foxtail) control in wheat, durum, and barley seeded the following spring.

TREFLAN TR-10 may be applied to ground that has a manageable trash level, been fallow, pre-tilled, or to soil left with a stubble cover. The first incorporation is required within 24 hours after application. The second incorporation and subsequent tillage may be done whenever necessary to destroy resistant weed growth during the remainder of the fallow year. During the fallow year, susceptible weeds may not be controlled until after the second incorporation.

Broadcast Rates Per Acre (All Soil Textures)

	TREFLAN TR-10	
Application Date	Area Subject to Less Than 10 Inches Annual Rainfall	All Other Areas
April 15–April 30	8¾	10
May 1–May 31	8¾–7½	10–8¾
June 1–June 30	7½–6¼	8¾–7½
July 1–July 31	6¼–5	7½–6¼
August 1–August 31	5	6¼–5

Where a rate range is shown, use the higher rate per acre during the early part of an application period and the lower rate per acre during the latter part of an application period.

Incorporation Directions: Incorporate TREFLAN one (1) time within twenty-four (24) hours after application. Care should be taken to ensure that the second incorporation and subsequent tillages are more shallow than the first. Incorporate and till the TREFLAN treated ground with the following equipment:

1. **Chisel plow:**
   Operate at 4–5 inches deep at 4–6 mph. A chisel plow is defined as having 3 rows of 14 to 18 inch sweeps on no greater than 12 inch center. Stagger sweeps so that no soil is left unturned. Chisel points should not be used.
2. **Tandem disc:**
   Operate at 3–4 inches deep at 4–6 mph.
3. **Field cultivator:**
   Operate at 3–4 inches deep at 5 mph or more. A field cultivator is defined as having 3 to 4 rows of sweeps with "c" or "s" shaped shanks, spaced 7 inches or less and staggered so that no soil is left unturned.

Seeding Directions: Wheat, durum, or barley should be seeded approximately two (2) inches deep.

Precautions: While use of this practice may result in a stand reduction, slight stand reductions do not normally affect yield.

Read the TREFLAN label for additional directions, precautions and limitations prior to use.

WHEAT (SPRING), DURUM AND BARLEY—FALL APPLICATION

For the control of foxtail (pigeongrass), apply TREFLAN at a broadcast rate of 5 pounds per acre on coarse and medium soils and 7½ pounds on fine soils.

Incorporation Directions: Any of the following tools are recommended for fall incorporation. The disc or field cultivator may be used for the spring incorporation pass. Care should be taken to operate the tool at a more shallow depth than the fall incorporation.

1. **Chisel plow:**
   May be used for first pass only. Operate at 4–5 inches deep at 4–6 mph. A chisel plow is defined as having 3 rows of 14 to 18 inch sweeps on no greater than 12 inch center. Stagger sweeps so that no soil is left unturned. Chisel points should not be used.
2. **Tandem disc:**
   Operate at 3–4 inches deep at 4–6 mph.
3. **Field cultivator:**
   Operate at 3–4 inches deep at 5 mph or more. A field cultivator is defined as having 3 to 4 rows of sweeps with "c" or "s" shaped shanks, spaced 7 inches or less and staggered so that no soil is left unturned.

Seeding Directions: Wheat, durum, or barley should be seeded approximately two (2) inches deep.

Precautions: While use of this practice may result in a stand reduction, slight stand reductions do not normally affect yield.

WHEAT (WINTER)—(Idaho, Montana, Oregon and Washington):

TREFLAN may be applied for preplant preemergence control of cheatgrass and other annual grasses and broadleaf weeds controlled by TREFLAN.

Apply TREFLAN any time up to three (3) weeks prior to planting. Broadcast TREFLAN at a rate of 7½ pounds per acre on coarse and medium soils and 10 pounds on fine soils.

Incorporation Directions: Incorporate TREFLAN into the soil with a flexible tine-tooth harrow, (Flextine or Melroe), set to cut one (1) to two (2) inches deep and operate at 3 to 6 mph. Incorporate one (1) time within 24 hours after application, followed by a second incorporation at least five (5) days after the first pass in a different direction prior to planting. Do not till the soil with a disc after the TREFLAN has been incorporated with a flexible tine harrow.

Seeding Directions: Use only a deep furrow or semi-deep furrow drill that will place the seed below the zone of soil into which TREFLAN has been incorporated.

Precaution: Do not plant wheat directly into the zone of soil treated with TREFLAN as injury to the crop (delay in emergence or stand reduction) may occur. Delayed emergence or slight stand reductions do not normally affect yield.

WHEAT (WINTER)—Fallow Soil Application (Idaho, Montana, Oregon and Washington)

TREFLAN applied and shallowly incorporated into fallow soil up to four (4) months ahead of planting will control cheatgrass and certain annual grasses and broadleaf weeds.

Broadcast rates are 7½ pounds per acre on coarse and medium soils and 10 pounds on fine soils. Apply TREFLAN any time from May to September prior to the fall planting of winter wheat.

Incorporation Directions: Incorporate TREFLAN with a flexible tine-tooth harrow, (Flextine or Melroe), set to cut one (1) to two (2) inches deep and operated at three (3) to six (6) mph. For thorough incorporation, two (2) passes of the equipment in different directions over the field are necessary. Incorporate one (1) time within 24 hours after application followed by a second incorporation at least five (5) days after the first prior to seeding. Do not till the soil with a disc after TREFLAN has been incorporated with a flexible tine harrow.

Seeding Directions: Use only a deep furrow or semi-deep furrow drill that will place the seed below the zone of soil into which TREFLAN has been incorporated.

Precaution: Do not plant wheat directly into the zone of soil treated with TREFLAN as injury to the crop (delay in emergence or stand reduction) may occur. Delayed emergence or slight stand reductions do not normally affect yield.

FALL APPLICATION

General: (Areas receiving more than 20" average annual rainfall). See specific crop for recommendations. For all crops for which there are no specific fall application instructions and for which TREFLAN is recommended as a preemergence application, use the rates listed for spring applications. Do not apply TREFLAN in the fall for sugar beets, potatoes and direct-seeded tomatoes.

In most states: Apply and incorporate TREFLAN any time between October 15 and December 31.

In California, Minnesota, Montana, North and South Dakota: Apply and incorporate TREFLAN any time between September 1 and December 31. Ground may be left flat or bedded-up over winter. On bedded ground, knock beds down to desired height before planting, moving some treated soil from beds into furrows. Where soil is left over winter, be careful not to turn up untreated soil during spring bedding operations. Destroy established weeds during seedbed preparation.

If weeds become established in furrows due to uncovering of untreated soil during bedding, destroy these weeds before planting. Do not apply TREFLAN in the fall to soils which are wet, are subject to prolonged periods of flooding, or where rice was grown the previous year.

TREFLAN PERFORMANCE GUARANTEE
A LIMITED WARRANTY OF TREFLAN WEED AND GRASS CONTROL

Elanco Products Company warrants that TREFLAN will control the weeds and grasses as indicated on its current label when used according to label directions and subject to the following limitations:

1. Purchaser must notify Elanco promptly if a lack of satisfactory control occurs. Notice must be given within ninety days after the first planting or after application of TREFLAN to the treated crop, whichever occurs later.
2. An Elanco representative must be satisfied that Purchaser used TREFLAN according to the label directions and must visually observe unsatisfactory control of weeds in the field.
3. The Purchaser must provide an Elanco representative with an invoice or other satisfactory record which shows the price and quantity of TREFLAN purchased.
4. Elanco's liability will be limited to a refund of the purchase price of the TREFLAN applied to the acreage on which weed control was not satisfactory or, at the option of the Purchaser, a refund of an equivalent amount of TREFLAN.
5. Void outside the USA.

DISCLAIMER OF WARRANTIES

The warranty printed above is the only warranty applicable to this product. All other warranties, expressed or implied, including, but not limited to, any implied warranties of merchantability and fitness for a particular purpose, are disclaimed.

LIMITATION OF DAMAGES

Elanco's liability, whether in contract, warranty, tort, negligence, strict liability or otherwise, shall not exceed the return of the amount of the purchase price of the TREFLAN and under no circumstances shall Elanco be liable for special, indirect or consequential damages.

INHERENT RISKS OF USE

Failure to carefully follow the directions for use of TREFLAN, as well as other factors such as seedling disease, cold weather, deep planting, excessive moisture, high salt concentration or drought, may result in unsatisfactory weed control or crop injury. Reduced yields may result under these conditions.

FMC Corporation

Agricultural Chemical Group
2000 Market Street
Philadelphia, PA 19103
(215) 299-6000

AMMO® 2.5 EC

INSECTICIDE

RESTRICTED USE PESTICIDE

For retail sale to, and use only by, Certified Applicators or persons under their direct supervision, and only for those uses covered by the Certified Applicator's certification.

ACTIVE INGREDIENT:	
*Cypermethrin**	30.6%
INERT INGREDIENTS:	69.4%
TOTAL	100.0%

Contains xylene range aromatic solvents.

* (±) Cyano (3-phenoxyphenyl) methyl (±) *cis/trans* 3-(2,2-dichloroethenyl)-2,2 dimethylcyclopropane carboxylate

** *Cis/trans* ratio: Max 55% (±) cis and min. 45% (±) trans

Contains 2.5 pounds cypermethrin per gallon.

EPA Reg.No. 279-3027 EPA Est. No. 279-
U.S. Patent No. 4,024,163
Ammo—FMC Trademark

CAUTION
Keep Out of Reach of Children

For Emergency Assistance Call 716–735-3765
See other panels for additional precautionary information.

FIRST AID:

If swallowed: Do not cause vomiting. Call a physician.

If on skin: Wash skin thoroughly with soap and water.

If in eyes: Flush thoroughly with water. If signs of irritation persist, contact physician.

If inhaled: Exposure to vapor or dust may cause tingling or numbness of exposed areas, especially the face, but this is temporary and requires no treatment. Lung irritation from breathing vapors is possible, and medical attention should be obtained if it is severe.

NOTE TO PHYSICIAN:
Like the natural pyrethrins, the synthetic derivative is expected to have relatively minor toxicity in humans; in fact, any significant acute toxic effects are more likely from a carrier by hydrocarbon solvent. Consequently, induction of vomiting may increase the likelihood of the most important toxic potential, chemical pneumonia, and so should either be avoided or done only under medical supervision. Ingestion of a large amount calls for gastric lavage, with care (Trendelenburg position, suction available, cuffed endotracheal tube if patient is unconscious) to avoid intrapulmonary aspiration. A saline cathartic (sodium or magnesium sulfate), 15–30 gm. dissolved in water should be given, as should 15–20 gm. activated charcoal as a slurry in water. Digestible fats, oils or alcohol may increase absorption and so should be avoided. Skin contact (vapor or powder) may be followed by transient tingling or numbness, usually of the face, but this subsides without treatment.

PRECAUTIONARY STATEMENTS

HAZARDS TO HUMANS (AND DOMESTIC ANIMALS)

CAUTION
Harmful if absorbed through skin, inhaled or swallowed. Causes eye irritation. Avoid breathing vapor or spray mist. Avoid contact with skin, may cause sensitization reaction in some individuals. Avoid contact with eyes or clothing. Wash thoroughly with soap and water after handling. Remove contaminated clothing and wash before reuse.

Applicators must wear protective clothing. Mixers and loaders must wear a full face shield, impermeable gloves, a rubber apron, boots and protective clothing.

ENVIRONMENTAL HAZARDS
This product is highly toxic to bees exposed to direct treatment or blooming crops or weeds. Do not apply this product or allow it to drift to blooming crops or weeds while bees are actively visiting the treatment area.

This product is extremely toxic to fish. Use with care when applying in areas adjacent to any body of water. Do not apply directly to water. Do not apply when weather conditions favor drift from treated areas. Do not contaminate water by cleaning of equipment, or disposal of wastes.

PHYSICAL OR CHEMICAL HAZARDS
Do not use or store near heat or open flame.

DIRECTIONS FOR USE

It is a violation of Federal law to use this product in a manner inconsistent with its labeling.

RE-ENTRY STATEMENTS
Do not treat areas while unprotected humans or domestic animals are present in the treatment area.

Do not allow entry into treated areas without protective clothing until sprays have dried.

State re-entry regulations will apply if they are more restrictive than those stated on this label. Consult with State regulatory officials.

ROTATIONAL CROPS
Rotational crops may be planted within 30 days after last application.

STORAGE AND DISPOSAL

PESTICIDE STORAGE
Do not store below 10°F, (−12°C).

Keep out of reach of children and animals. Store in original containers only. Store in a cool, dry place and avoid excessive heat. Carefully open containers. After partial use, replace lids and close tightly. Do not put concentrate or dilute material into food or drink containers. Do not contaminate other pesticides, fertilizers, water, food, or feed by storage or disposal.

In case of spill, avoid contact, isolate area and keep out animals and unprotected persons. Confine spills. Call FMC collect: (716) 735-3765.

To confine spill: If liquid, dike surrounding area or absorb with sand, cat litter or commercial clay. If dry material, cover to prevent dispersal. Place damaged package in a holding container. Identify contents.

PESTICIDE DISPOSAL
Pesticide wastes are toxic. Improper disposal of excess pesticide, spray mixture, or rinsate is a violation of Federal Law. If these wastes cannot be disposed of by use according to label instructions, contact your State Pesticide or Environmental Control Agency, or the Hazardous Waste representative at the nearest EPA Regional Office for guidance.

CONTAINER DISPOSAL
Glass Containers: Triple rinse (or equivalent). Then dispose of in a sanitary landfill or by other approved State and local procedures.
Metal Containers: Triple rinse (or equivalent). Then offer for recycling or reconditioning, or puncture and dispose of in a sanitary landfill, or by other procedures approved by state and local authorities. Do not cut or weld metal containers.

Use low rate under light to moderate infestation. Higher rates should be used under heavy insect pressure. The rate of application is variable according to insect pressure, timing or spray and field scouting.

Crop	Insect	Dosage lb ai/A	Dosage fl oz/A	Remarks
Cotton	Cutworms Tobacco Thrips Soybean (banded) Thrips	0.025–0.1	1.3–5	Use low rate under light to moderate infestation. Higher rates should be used under heavy insect pressure. The rate of application is variable according to insect pressure, timing of spray and field scouting.
	Beet Armyworm Fall Armyworm Yellow Striped Armyworm Cotton Bollworm Pink Bollworm Tobacco Budworm Tarnished Plant Bug Lygus Bugs Cotton Fleahopper Cotton leafperforator Cabbage Looper Boll Weevil other plant bugs White Flies* Saltmarsh Caterpillar	0.04–0.1	2–5	AMMO 2.5EC may be applied in water or refined vegetable oil. If water is used, apply a minimum of one gallon of finished spray per acre by aircraft or five gallons of finished spray per acre with ground equipment. When applying in water by aircraft, one quart of emulsified oil may be substituted for one quart of water in the finished spray. When using oil, use a minimum of one quart per acre in the finished spray. Ammo 2.5 EC may be injected into overhead sprinkler irrigation water provided 1) an anti-backflow check valve is present between the injection port and the water source, 2) a check valve is present in the line to prevent irrigation water from entering the chemical supply tank and 3) the irrigation injection system has interlocking on-off switches. For boll weevil control, apply AMMO 2.5EC at a 3-4 day interval until pest numbers are reduced to acceptable levels. A maximum of 1.0 lb. may be applied per acre per season. Ammo 2.5 EC may be applied within 14 days of harvest. Do not graze or feed cotton for forage.

*Aids in control

TANK-MIXTURE

Ammo 2.5 EC may be applied in tank mixtures with other products approved for use on cotton. Observe all restrictions and precautions which appear on the labels of these products.

For improved control of aphids and mites, Ammo 2.5 EC may be tank-mixed with Curacron® 6E. The tank-mixture is most effective when pest populations are first observed. Subsequent applications may be needed to maintain control.

DEALERS SHOULD SELL IN ORIGINAL PACKAGES ONLY.

TERMS OF SALE OR USE: On purchase of this product buyer and user agree to the following conditions:

WARRANTY: FMC makes no warranty, expressed or implied, concerning the use of this product other than indicated on the label. Except as so warranted, the product is sold as is. Buyer and user assume all risk of use and/or handling and/or storage of this material when such use and/or handling and/or storage is contrary to label instructions.

DIRECTIONS AND RECOMMENDATIONS: Follow directions carefully. Timing and method of application, weather and crop conditions, mixture with other chemicals not specifically recommended and other influencing factors in the use of this product are beyond the control of the seller and are assumed by the buyer at his own risk.

USE OF PRODUCT: FMC's recommendations for the use of this product are based upon tests believed to be reliable. The use of this product being beyond the control of the manufacturer, no guarantee, expressed or implied, is made as to the effect of such or the results to be obtained if not used in accordance with directions or established safe practice.

DAMAGES: Buyer's or user's exclusive remedy for damages for breach of warranty or negligence shall be limited to direct damages not exceeding the purchase price paid and shall not include incidental or consequential damages.

AMMO® 2.5 OIL

INSECTICIDE

RESTRICTED USE PESTICIDE

For retail sale to, and use only by, Certified Applicators or persons under their direct supervision, and only for those uses covered by the Certified Applicator's certification.

ACTIVE INGREDIENT:
*Cypermethrin** 30.6%
INERT INGREDIENTS 69.4%
TOTAL 100.0%

Contains xylene range aromatic solvents.

* (±) Cyano (3-phenoxyphenyl) methyl (±) *cis/trans* 3-(2,2-dichloroethenyl)-2,2 dimethylcyclopropane carboxylate

** *Cis/trans* ratio: Max. 55% (±) cis and min. 45% (±) trans

Contains 2.5 pounds cypermethrin per gallon.

EPA Reg. No. 279-3044 EPA Est. No. 279-
U.S. Patent No. 4,024,163 Ammo-FMC Trademark

CAUTION

Keep Out of Reach of Children

For Emergency Assistance Call 716-735-3765.

See other panels for additional precautionary information.

FIRST AID:

If swallowed: Do not cause vomiting. Call a physician.

If on skin: Wash thoroughly with soap and water.

If in eyes: Flush thoroughly with water. If signs of irritation persist, contact physician.

If inhaled: Exposure to vapor or dust may cause tingling or numbness of exposed areas, especially the face, but this is temporary and requires no treatment. Lung irritation from breathing vapors is possible, and medical attention should be obtained if it is severe.

Note to Physician: Like the natural pyrethrins, the synthetic derivative is expected to have relatively minor toxicity in humans; in fact, any significant acute toxic effects are more likely from a carrier hydrocarbon solvent. Consequently, induction of vomiting may increase the likelihood of the most important toxic potential, chemical pneumonia, and so should either be avoided or done only under medical supervision. Ingestion of a large amount calls for gastric lavage, with care (Trendelenburg position, suction available, cuffed endotracheal tube if patient is unconscious) to avoid intrapulmonary aspiration. A saline cathartic (sodium or magnesium sulfate), 15–30 gm. dissolved in water should be given, as should 15–20 gm. activated charcoal as a slurry in water. Digestible fats, oils or alcohol may increase absorption and so should be avoided. Skin contact (vapor or powder) may be followed by transient tingling or numbness, usually of the face, but this subsides without treatment.

PRECAUTIONARY STATEMENTS

HAZARDS TO HUMANS (AND DOMESTIC ANIMALS)
CAUTION

Harmful if absorbed through skin, inhaled or swallowed. Causes eye irritation. Avoid breathing vapor or spray mist. Avoid contact with skin, may cause sensitization reaction in some individuals. Avoid contact with eyes or clothing. Wash thoroughly with soap and water after handling. Remove contaminated clothing and wash before reuse.

Applicators must wear protective clothing. Mixers and loaders must wear full face shield, impermeable gloves, a rubber apron, boots and protective clothing.

ENVIRONMENTAL HAZARDS

This product is highly toxic to bees exposed to direct treatment or blooming crops or weeds. Do not apply this product or allow it to drift to blooming crops or weeds while bees are actively visiting the treatment area.

This product is extremely toxic to fish. Use with care when applying in areas adjacent to any body of water. Do not apply directly to water. Do not apply when weather conditions favor drift from treated areas. Do not contaminate water by cleaning of equipment, or disposal of wastes.

PHYSICAL OR CHEMICAL HAZARDS

Do not use or store near heat or open flame.

DIRECTIONS FOR USE

It is a violation of Federal law to use this product in a manner inconsistent with its labeling.

RE-ENTRY STATEMENTS

Do not treat areas while unprotected humans or domestic animals are present in the treatment area.

Do not allow entry into treated areas without protective clothing until sprays have dried.

State re-entry regulations will apply if they are more restrictive than those stated on this label. Consult with State regulatory officials.

STORAGE AND DISPOSAL

PESTICIDE STORAGE

Do not store below 10°F, (−12°C).

Keep out of reach of children and animals. Store in original containers only. Store in a cool, dry place and avoid excess heat. Carefully open containers. After partial use, replace lids and close tightly. Do not put concentrate or dilute material into food or drink containers. Do not contaminate other pesticides, fertilizers, water, food, or feed by storage or disposal.

In case of spill, avoid contact, isolate area and keep out animals and unprotected persons. Confine spills. Call FMC collect: (716) 735-3765.

To confine spill: If liquid, dike surrounding area or absorb with sand, cat litter or commercial clay. If dry material, cover to prevent dispersal. Place damaged package in a holding container. Identify contents.

PESTICIDE DISPOSAL

Pesticide wastes are toxic. Improper disposal of excess pesticide, spray mixture, or rinse is a violation of Federal law. If these wastes cannot be disposed of by use according to label instructions, contact your State Pesticide or Environmental Control Agency, or the Hazardous Waste representative at the nearest EPA Regional Office for guidance.

CONTAINER DISPOSAL

Glass Containers: Triple rinse (or equivalent). Then dispose of in a sanitary landfill or by other approved State and local procedures. **Metal Containers:** Triple rinse (or equivalent). Then offer for recycling or reconditioning, or puncture and dispose of in a sanitary landfill, or by other procedures approved by State and local authorities. Do not cut or weld metal containers.

Use low rate under light to moderate infestation. Higher rates should be used under heavy insect pressure. The rate of application is variable according to insect pressure, timing of spray and field scouting.

Crop	Insect	Dosage Lb AI/A	Fl Oz/A	Remarks
Cotton	Cutworms Tobacco Thrips Soybean (banded) Thrips	0.025 to 0.1	1.3 to 5	Ammo 2.5 Oil may be applied in water or refined vegetable oil. When water is used, apply a minimum of one gallon of finished spray per acre by air or five gallons of finished spray with ground equipment. When applying in water by air, one quart of emulsified oil may be substituted for one quart of water in the finished spray. When using oil, use a minimum of one quart per acre in the finished spray.
	Beet Armyworm Fall Armyworm Yellow Striped Armyworm Cotton Bollworm Pink Bollworm Tobacco Budworm Tarnished Plant Bug Lygus Bugs Saltmarsh Caterpillar Cotton Fleahopper Cotton Leaf-perforator Cabbage Looper Boll Weevil Other Plant Bugs White Flies*	0.04 to 0.1	2 to 5	Ammo 2.5 Oil may be injected into overhead sprinkler irrigation water provided 1) an anti-backflow check valve is present between the injection port and the water source, 2) a check valve is present in the line to prevent irrigation water from entering the chemical supply tank and 3) the irrigation injection system has interlocking on-off switches. For boll weevil control, apply Ammo 2.5 Oil at a 3 to 4 day interval until pest numbers are reduced to acceptable levels. A maximum of 1.0 lb. may be applied per acre per season. Ammo 2.5 Oil may be applied within 14 days of harvest. Do not graze or feed cotton for forage.

* Aids in control.

Tank-Mixtures

Ammo 2.5 Oil may be applied in tank mixtures with other products approved for use on cotton. Observe all restrictions and precautions which appear on the labels of these products.

For improved control of aphids and mites, Ammo 2.5 Oil may be tank-mixed with Curacron® 6E. The tank-mixture is most effective when pest populations are first observed. Subsequent applications may be needed to maintain control.

DEALERS SHOULD SELL IN ORIGINAL PACKAGES ONLY.

TERMS OF SALE OR USE: On purchase of this product buyer and user agree to the following conditions:

WARRANTY: FMC makes no warranty, expressed or implied, concerning the use of this product other than indicated on the label. Except as so warranted, the product is sold as is. Buyer and user assume all risk of use and/or handling and/or storage of this material when such use and/or handling and/or storage is contrary to label instructions.

DIRECTIONS AND RECOMMENDATIONS: Follow directions carefully. Timing and method of application, weather and crop conditions, mixture with other chemicals not specifically recommended and other influencing factors in the use of this product are beyond the control of the seller and are assumed by buyer at his own risk.

USE OF PRODUCT: FMC's recommendations for the use of this product are based upon tests believed to be reliable. The use of this product being beyond the control of the manufacturer, no guarantee, expressed or implied, is made as to the effects of such or the results to be obtained if not used in accordance with directions or established safe practice.

DAMAGES: Buyer's or user's exclusive remedy for damages for breach of warranty or negligence shall be limited to direct damages not exceeding the purchase price paid and shall not include incidental or consequential damages.

AQUA ETHION 8

INSECTICIDE

For Agricultural or Commercial Use Only.

ACTIVE INGREDIENTS:	BY WT.
Ethion (0, 0, 0′, 0′-Tetraethyl S, S′-methylene bisphosphorodithioate)	81.0%
INERT INGREDIENTS:	19.0%
TOTAL	100.0%

Contains 8 lbs. Ethion per gallon

EPA Reg. No. 279-2102 EPA Est.,

DANGER–POISON

Keep Out of Reach of Children

PELIGRO

PRECAUCION AL USUARIO: Si usted no lee ingles, no use este producto hasta que la etiqueta le haya sido explicada ampliamente.

FIRST AID TREATMENT

If swallowed, drink one or two glasses of water and induce vomiting by touching back of throat with finger or blunt object. Do not induce vomiting or give anything by mouth to an unconscious person. Get medical attention.

If on skin or clothing, immediately flush skin or eyes with water for at least 15 minutes while removing contaminated clothing and shoes.

If in eye, flush eye for 15 minutes with plenty of water. Call a physician immediately.

If inhaled, remove victim to fresh air. If not breathing, give artificial respiration, preferably mouth-to-mouth. Get medical attention.

Note to Physician: Active ingredient is a cholinesterase inhibitor. Patient should be kept under observation for 48 hours. Atropine is antidotal. Give 4 mg. atropine sulfate intramuscularly or intravenously. Repeat every 15 minutes until patient is fully atropinized. Do not give morphine.

For Emergency Assistance Call 716–735-3765.

PRECAUTIONARY STATEMENTS

HAZARDS TO HUMANS (& DOMESTIC ANIMALS)

DANGER

Poisonous if swallowed, inhaled, or absorbed through skin! Do not breathe vapor or spray mist. Do not get in eyes, on skin or on clothing. In case of contact wash immediately with water; for eyes flush with plenty of water and get medical attention. Wear a mask or respirator of a type approved by the Mining Enforcement and Safety Administration (formerly U.S. Bureau of Mines) and by the National Institute for Occupational Safety and Health under the provisions of 30 CFR. Part II for Ethion protection. Wash hands, arms and face thoroughly with soap and water before eating or smoking. Wash all contaminated clothing with soap and hot water before reuse.

ENVIRONMENTAL HAZARDS

This product is toxic to fish, birds and other wildlife. Birds feeding on treated areas may be killed. Keep out of lakes, ponds and streams. Do not apply where runoff is likely to occur. Do not apply when weather conditions favor drift from areas treated. Do not contaminate water by cleaning of equipment, or disposal of wastes. Apply this product only as specified on this label.

This product is toxic to bees and should not be applied when bees are actively visiting the area.

PHYSICAL OR CHEMICAL HAZARDS

Do not use, pour, spill or store near heat or open flame.

DIRECTIONS FOR USE

It is a violation of Federal Law to use this product in a manner inconsistent with its labeling.

Do not apply this product in such manner as to directly or through drift expose workers or other persons. The area being treated must be vacated by unprotected persons.

Do not enter treated areas for 24 hours unless appropriate protective clothing is worn.

Protective clothing means, at least, a hat or other suitable head covering, a long sleeved shirt and long legged trousers or a coverall type garment (all of closely woven fabric covering the body, including the arms and legs), shoes and socks.

Because certain states may require more restrictive reentry intervals for various crops treated with this product, consult your State Department of Agriculture for further information.

Written or oral warnings must be given to workers who are expected to be in a treated area or in an area about to be treated with this product. Inform workers of areas or fields that may not be entered without specific protective clothing, period of time field must be vacated and appropriate actions to take in case of accidental exposure. An example of such information is given under written warnings. When oral warnings are given, warnings shall be given in a language customarily understood by workers. Oral warnings must be given if there is reason to believe that written warnings cannot be understood by workers. Written warnings must include the following information:

DANGER

Area treated with AQUA ETHION 8 on (insert date). Do not enter without appropriate protective clothing for 24 hours (insert State Department of Agriculture's re-entry interval, if more restrictive.)

In case of accidental exposure to pesticide spray or dust, wash the skin thoroughly with soap and water. Remove contaminated clothing and wash before reuse. If in eyes, flush with plenty of water. If inhaled, go to an area where the pesticide has not been applied. Get medical attention if needed.

STORAGE AND DISPOSAL

PESTICIDE STORAGE

Do not store below 20°F, (−7°C).

Do not use or store near heat, open flame or hot surfaces.

Keep out of reach of children and animals. Store in original containers only. Store in a cool, dry place and avoid excess heat. Carefully open containers. After partial use, replace lids and close tightly. Do not put concentrate or dilute material into food or drink containers. Do not contaminate other pesticides, fertilizers, water, food, or feed by storage or disposal.

In case of spill, avoid contact, isolate area and keep out animals and unprotected persons. Confine spills. Call FMC collect: (716) 735-3765.

To confine spill: If liquid, dike surrounding area and absorb with sand, cat litter or commercial clay. If dry material, cover to prevent dispersal. Place damaged package in a holding container. Identify contents.

PESTICIDE DISPOSAL

Pesticide wastes are acutely hazardous. Improper disposal of excess pesticide, spray mixture, or rinsate is a violation of Federal Law. If these wastes cannot be disposed of by use according to label instructions, contact your State Pesticide or Environmental Control Agency, or the Hazardous Waste representative at the nearest EPA Regional Office for guidance.

CONTAINER DISPOSAL

Triple rinse (or equivalent). Then offer for recycling or reconditioning, or puncture and dispose of in a sanitary landfill, or by other procedures approved by state and local authorities. Do not cut or weld metal containers.

Apply when insects first appear and repeat as required to maintain effective control. Coverage of both upper and lower leaf surfaces is essential for good control. Where amount of water is not specified, use sufficient water for thorough coverage. For aircraft application to field crops and vegetables, mix recommended amounts with sufficient water to provide a MINIMUM of 1 gallon of finished spray per acre. Observe use limitations. If insect control is required beyond the following use patterns, supplement the control program with other suitable pesticides. Observe days interval between last application and harvest indicated by number in parentheses ( ) following crop. Do not use in undiluted form.

APPLE (60): Codling Moth—½ pint per 100 gals. Do not exceed 4 applications during fruiting period.

APPLE (40): Codling Moth—½ pint per 100 gals. Do not exceed 2 applications during fruiting period. Clover Mite, European Red Mite, Twospotted Spider Mite—¼ pint per 100 gals. Do not exceed 4 applications during fruiting period.

APPLE (20): Clover Mite, European Red Mite, Twospotted Spider Mite—¼ pint per 100 gals. Do not exceed 2 applications during fruiting period.

Do not use on Astrachen, Davey, Duchess, Margaret Pratt, Melba, Wealthy, Williams Red (Red Williams), Yellow Transparent or on Crimson and N.J. #3 (Brite Mac), or any varieties maturing with or before Early McIntosh, as injury is likely to occur.

BEANS (Succulent and Dry) (2): Bean Leaf Beetle, Mexican Bean Beetle, Twospotted Spider Mite—¼ to ½ pt. per acre. Potato Leafhopper—½ pt. per acre. Do not feed bean foliage to livestock.

BEANS (Succulent and Dry) (4): Lygus Bugs—1 pt. per acre. Do not feed treated bean foliage to livestock.

BERMUDA GRASS (Lawns and Seed Crop): Eriophyid Mites—½ tablespoon per gal. of water to cover 200 sq. ft. of lawn or use 1½ fl. ozs. per 6 gals. of water per 1200 sq. ft. Apply as needed to maintain effective control. Keep children and animals off treated areas until the spray has completely dried. Do not apply to Bermuda grass to be used for grazing or forage to be fed to livestock.

CHESTNUT, FILBERT, PECAN (0): Aphids, Mites—2½ pints per acre. Do not apply more than twice during the fruiting period. On chestnuts and pecans, do not apply after husk split. On filberts, do not apply when nut is visible. Do not graze livestock in treated groves.

CORN, FIELD (50): Pacific Mite, Twospotted Spider Mite—1 pint per acre. Do not make more than 1 application after ear formation. Do not feed treated fodder to livestock.

COTTON: Aphids—½ to 1 pt. per acre. Spider Mites—¼ to ½ pt. per acre. Cotton Leafworm—½ pint per acre. Lygus Bugs, Spider Mites (WEST COAST AREA ONLY)—1 to 1½ pints per acre. Treat when an infestation reaches the point where control is required and repeat as necessary. Do not apply after the bolls open. Do not graze dairy or meat animals in treated fields.

CUCUMBER (0), SUMMER SQUASH (0): Serpentine Leafminer, Twospotted Spider Mite—½ pt. per acre.

EGGPLANT (0): Twospotted Spider Mite—¼ to ½ pt. per acre.

MELON (0): Melon Leafminer, Serpentine Leafminer, Twospotted Spider Mite—½ pt. per acre. Melon Leafhopper—¾ to 1 pint per acre. Petrobia Mite (California only)—½ to 1 pint per acre.

NECTARINE (30): Clover Mite, European Red Mite, Twospotted Spider Mite—¼ pint per 100 gals., or 1¼ to 1¾ pints per acre. Do not exceed 2 applications during fruiting period.

ORNAMENTALS, GREENHOUSE (Commercial Use): Millipedes—SOIL TREATMENT: Use 5 fl. ozs. AQUA ETHION 8 in 75 gals. of water and apply to 2,500 sq. ft. of bench space. Wear mask or respirator passed for Ethion protection by Mining Enforcement and Safety Administration (formerly the U.S. Bureau of Mines) and by the National Institute for Occupational Safety and Health under the provisions of 30 CFR Part II when applying in greenhouse.

ORNAMENTALS, OUT-OF-DOORS (Commercial Use): Twospotted Spider Mite—¼ pt. per 100 gals. Slight needle drop has occurred on some varieties of Taxus following application of this material.

PEAR (30): Clover Mite, European Red Mite, Twospotted Spider Mite—¼ pint to ⅓ pint per 100 gals. Pear Psylla—⅓ to ½ pint per 100 gals. Do not exceed 3 applications after first cover spray.

PEPPER (0), PIMENTO (0): Serpentine Leafminer—½ pt. per acre. Twospotted Spider Mite—¼ to ½ pt. per acre.

STRAWBERRY (2): Strawberry Leafroller, Twospotted Spider Mite—½ pt. per 100 gals., ½ to 1 pt. per acre. Spray plants thoroughly. Continue application after harvest as required.

SORGHUM (Grain & Forage): Sorghum Midge, Sorghum Mite—¼ to ½ pint per acre. Do not exceed 3 applications per season. Do not harvest or feed forage within 30 days after last application. For midge, apply at first bloom and repeat 4 to 10 days later. Some leaf burning may occur on phosphate-sensitive varieties.

TOMATO (2): Tropical Spider Mite, Twospotted Spider Mite, Leafminer—½ pt. in sufficient water to treat 1 acre.

TURF (Commercial Use): Chinch Bugs, Sod Webworm—Use ⅙ to ¼ pt. per 1000 sq. ft. of lawn applied in 20 gals. of water or ¾ to 1¼ pts. in 100 gals. of water to treat 5000 sq. ft. To treat one acre, use 3¾ to 5 qts. applied in 800 gals. of water. It is important that the spray penetrates through the mat of turf to reach the insects. It helps to have the lawn well irrigated prior to treatment.

CAUTION: Keep children and animals off treated areas until this material has been thoroughly watered into soil and the grass has completely dried.

DEALERS SHOULD SELL IN ORIGINAL PACKAGES ONLY.

TERMS OF SALE OR USE: On purchase of this product buyer and user agree to the following conditions:

WARRANTY: FMC makes no warranty, expressed or implied, concerning the use of this product other than indicated on the label. Except as so warranted, the product is sold as is. Buyer and user assume all risk of use and/or handling and/or storage of this material when such use and/or handling and/or storage is contrary to label instructions.

DIRECTIONS AND RECOMMENDATIONS: Follow directions carefully. Timing and method of application, weather and crop conditions, mixture with other chemicals not specifically recommended and other influencing factors in the use of this product are beyond the control of the seller and are assumed by the buyer at his own risk.

USE OF PRODUCT: FMC's recommendations for the use of this product are based upon tests believed to be reliable. The use of this product being beyond the control of the manufacturer, no guarantee, expressed or implied, is made as to the effects of such or the results to be obtained if not used in accordance with directions or established safe practice.

DAMAGES: Buyer's or user's exclusive remedy for damages for breach of warranty or negligence shall be limited to direct damages not exceeding the purchase price paid and shall not include incidental or consequential damages.

AQUA 8 PARATHION

INSECTICIDE

RESTRICTED USE PESTICIDE

For retail sale to and use only by certified applicators or persons under their direct supervision, and only for those uses covered by the certified applicator's certification.

ACTIVE INGREDIENTS:	BY WT.
Parathion	76.35%
Related Compounds	3.65%
INERT INGREDIENTS:	20.00%
TOTAL	100.00%

Contains 8 pounds of Parathion per gallon

EPA Reg. No. 279-1611 EPA Est., 279-

STOP! READ THE LABEL

CAN KILL YOU

DANGER–POISON

Keep Out of Reach of Children

See side panels for antidote and precautions

PELIGRO

PRECAUCION AL USUARIO: Si usted no lee ingles, no use este producto hasta que la etiqueta le haya sido explicada ampliamente.

DANGER-POISON-PRECAUTIONS

POISONOUS IF SWALLOWED

This product can kill you if swallowed even in small amounts; spray mist or dust may be fatal if swallowed.

POISONOUS BY SKIN CONTACT

Poisonous if touched by hands or spilled or splashed on skin, in eyes or on clothing (liquid goes through clothes).

POISONOUS IF BREATHED

Breathing vapors, spray mist or dust may be fatal.

ENVIRONMENTAL HAZARDS

This product is toxic to fish, shrimp, crab, birds and other wildlife. Birds and other wildlife in treated areas may be killed. Keep out of lakes, streams, ponds, tidal marshes and estuaries. Shrimp and crab may be killed at application rates recommended on this label. Do not apply where these are important resources. Do not apply where run-off is likely to occur. Do not apply when weather conditions favor drift from the areas treated. Do not contaminate water by cleaning of equipment or disposal of wastes. Apply this product only as specified on this label.

This product is highly toxic to bees exposed to direct treatment or residues on crops. Protective information may be obtained from your Cooperative Agricultural Extension Service.

POISONOUS TO FISH AND WILDLIFE

POISON SIGNS (Symptoms)

PARATHION is a very dangerous poison. It rapidly enters the body on contact with all skin surfaces and eyes. Clothing wet with this material must be removed immediately. Exposed persons must receive prompt medical treatment or they may die.

Some of the signs and symptoms of poisoning are: Headache, nausea, vomiting, cramps, weakness, blurred vision, pin-point pupils, tightness in chest, labored breathing, nervousness, sweating, watering of eyes, drooling or frothing of mouth and nose, muscle spasms and coma.

USE ONLY WHEN WEARING THE FOLLOWING PROTECTIVE EQUIPMENT AND CLOTHING

(1) Wear water-proof pants, coat, hat, rubber boots or rubber overshoes. (2) Wear safety goggles. (3) Wear a mask or a pesticide respirator jointly approved by the Mining Enforcement and Safety Administration (formerly the U.S. Bureau of Mines) and by the National Institute for Occupational Safety and Health under the provisions of 30 CFR Part II. (4) Wear heavy duty, natural rubber gloves.

WORK SAFETY RULES

Keep all unprotected persons and children away from treated area or where there is danger of drift.

Do not rub eyes or mouth with hands. If you feel sick in any way, STOP work and get help right away. Call a doctor (physician), clinic or hospital—immediately. Explain that the victim has been exposed to parathion and describe his condition. After first aid is given (see First Aid Treatment Section) and if a doctor cannot come, take victim to clinic or hospital.

IMPORTANT! Before removing gloves, wash them with soap and water. Always wash hands, face and arms with soap and water before smoking, eating or drinking.

AFTER WORK, take off all work clothes and shoes. Shower, using soap and water. Wear only clean clothes when leaving job. Do not wear contaminated clothing. Wash protective clothing and protective equipment with soap and water after each use. Respirator should be cleaned and filter replaced according to instructions included with respirator.

FIRST AID TREATMENT

Call a doctor (physician), clinic or hospital immediately. Explain that the victim has been exposed to parathion and describe his condition.

If breathing has stopped, start artificial respiration immediately and maintain until doctor sees victim.

If swallowed, call a physician or Poison Control Center. Drink 1 or 2 glasses of water and induce vomiting by touching back of throat with finger, or, if available by administering syrup of ipecac. Do not induce vomiting or give anything by mouth to an unconscious person.

In case of contact, immediately flush eyes or skin with plenty of water for at least 15 minutes while removing contaminated clothing and shoes. See doctor immediately.

NOTE TO PHYSICIAN

Antidote—administer atropine sulfate in large doses, TWO to FOUR mg. intravenously or intramuscularly as soon as cyanosis is overcome. Repeat at 5 to 10 minute intervals until signs of atropinization appear. 2-PAM chloride is also antidotal and may be administered in conjunction with atropine. DO NOT GIVE MORPHINE

OR TRANQUILIZERS. Parathion is a strong cholinesterase inhibitor affecting the central and peripheral nervous systems and producing cardiac and respiratory depression. At first sign of pulmonary edema, the patient should be given supplemental oxygen and treated symptomatically. Continued absorption of the poison may occur and fatal relapses have been reported after initial improvement; VERY CLOSE SUPERVISION OF THE PATIENT IS INDICATED FOR AT LEAST 48 HOURS.

FOR EMERGENCY ASSISTANCE CALL 716–735-3765.

POST TREATED AREA

Consult your State Agricultural Extension Service or Experiment Station regarding posting treated areas.

PHYSICAL OR CHEMICAL HAZARDS

FLAMMABLE. Keep away from heat and open flame.

DIRECTIONS FOR USE

It is a violation of Federal Law to use this product in a manner inconsistent with its labeling.

Do not apply this product in such a manner as to directly or through drift expose workers or other persons. The area being treated must be vacated by unprotected persons.

Do not enter treated areas for 48 hours unless appropriate protective clothing is worn.

Protective clothing means, at least, a hat or other suitable head covering, a long sleeved shirt and long legged trousers or a coverall type garment (all of closely woven fabric covering the body, including the arms and legs), shoes and socks.

Because certain states may require more restrictive reentry intervals for various crops treated with this product, consult your State Department of Agriculture for further information.

Written or oral warnings must be given to workers who are expected to be in a treated area or in an area about to be treated with this product. Inform workers of areas or fields that may not be entered without specific protective clothing, period of time field must be vacated and appropriate actions to take in case of accidental exposure. An example of such information is given under written warnings. When oral warnings are given, warnings shall be given in a language customarily understood by workers. Oral warnings must be given if there is reason to belief that written warnings cannot be understood by workers. Written warnings must include the following information:

DANGER

Area treated with AQUA 8 PARATHION on (insert date). Do not enter without appropriate protective clothing for 48 hours (insert State Department of Agriculture's re-entry interval, if more restrictive).

In case of accidental exposure to pesticide spray or dust, wash the skin thoroughly with soap and water. Remove contaminated clothing and wash before reuse. If in eyes, flush with plenty of water. If inhaled, go to an area where the pesticide has not been applied. Get medical attention if needed.

STORAGE AND DISPOSAL

PESTICIDE STORAGE

Do not store in or around the home.

Do not store below 20°F, (−7°C).

Do not use or store near heat, open flame or hot surfaces.

Keep out of reach of children and animals. Store in original containers only. Store in a cool, dry place and avoid excess heat. Carefully open containers. After partial use, replace lids and close tightly. Do not put concentrate or dilute material into food or drink containers. Do not contaminate other pesticides, fertilizers, water, food, or feed by storage or disposal.

In case of spill, avoid contact, isolate area and keep out animals and unprotected persons. Confine spills. Call FMC collect: (716) 735-3765.

To confine spill: If liquid, dike surrounding area or absorb with sand, cat litter or commercial clay. If dry material, cover to prevent dispersal. Place damaged package in a holding container. Identify contents.

PESTICIDE DISPOSAL

Pesticide wastes are acutely hazardous. Improper disposal of excess pesticide, spray mixture, or rinsate is a violation of Federal Law. If these wastes cannot be disposed of by use according to label instructions, contact your State Pesticide or Environmental Control Agency, or the Hazardous Waste representative at the nearest EPA Regional Office for guidance.

CONTAINER DISPOSAL

Triple rinse (or equivalent). Then offer for recycling or reconditioning, or puncture and dispose of in a sanitary landfill, or incineration, or, if allowed by state and local authorities, by burning. If burned, stay out of smoke.

Not for use or storage in or around the house. Do not use in undiluted form.

Do not allow this material to drift onto neighboring crops or non-crop areas or use in a manner or at a time other than in accordance with label directions because animal, plant or crop injury, illegal residues or other undesirable results may occur.

Do not use in undiluted form. Apply as a full coverage (dilute) spray when insects first appear and repeat as required to maintain control. May be applied by ground or airplane application equipment. When applying by air, mix the recommended amount with sufficient water to provide a minimum of 1 gallon finished spray per acre. For ground application, mix recommended amounts with a minimum of 10 gallons of water to cover one acre. Observe days interval between last application and harvest indicated by number in [ ] following crop.

FIELD CROPS

ALFALFA [15]: Aphids, Armyworms, Alfalfa Caterpillar, Grasshoppers, Webworms, Alfalfa Weevil—Use ½ pint. **California and Nevada:** Use no more than 6 ounces.

BARLEY (see Clover).

CLOVER [15], SMALL GRAINS (Barley, Oats, Wheat) [15], VETCH [15]: Aphids, Armyworms, Grasshoppers—Use ½ pint. Stink Bug, Petrobia Mite, Black Grass Bug—Use ½ to ¾ pint.

HOPS [15]: Aphids—Use ½ to ¾ pint.

OATS (see Clover).

SUGAR BEETS [15]: Alfalfa Looper, Aphids, Armyworms, Leafhoppers, Blister Beetle, Flea Beetles, Leafminer, Lygus Bugs, Stink Bug, Webworm—Use ½ to ¾ pint.

TOBACCO (Field): Aphids, Suckfly—Use ½ pint. Do not apply within 5 days of priming or 15 days of cutting. Avoid getting wet with plant juices when cutting **(Seed Bed):** Aphids—Use ⅙ ounce in 2 gallons of water per 100 square yards of seed.

VETCH (see Clover).

WHEAT (see Clover).

FRUITS

For fruit trees mix the recommended amount with 100 gallons of water full coverage (dilute) spray.

APPLES [14]: Codling Moth; Plum Curculio, Orange Tortrix; San Jose, Forbes, Olive Parlatoria and Scurfy Scales; Redbanded Leafroller (second and third broods); Woolly Apple Aphid; Mealybug —Use ⅓ pint (2⅔ pints per acre). Eyespotted Bud Moth, Apple Red Bug (Chigger Mites), Fruitree Leafroller, Leafhoppers, Rosy and Green Apple Aphids, Redbanded Leafroller (first brood), Grasshoppers, European Red, Twospotted, Pacific, Schoene, Willamette and Clover Mites—Use ¼ pint (2 pints per acre). European Sawfly—Use ⅜ pint (3 pints per acre) at petal fall. Apply at petal fall and every 10 to 14 days thereafter until control is achieved. Parathion may injure the fruit and foliage of Golden Delicious, Jonathan and McIntosh apples as well as related varieties such as Cortland, Kendall, Macoun and Melba. For greater safety, the use of as low dosages as possible is recommended on apples.

CHERRY [14]: Aphids, Mites, Shothole Borer (following petal fall and in late summer), Red Spider Mites, Leafminer, Plum Curculio, Oriental Fruit Moth, San Jose Scale Crawlers (apply after emergence)—Use $\frac{1}{4}$ pint ($1\frac{1}{2}$ pints per acre). Fruitfly—Use $\frac{1}{4}$ to $\frac{1}{3}$ pint ($1\frac{1}{2}$ to 2 pints per acre). Apply within 5 days after first fly emerges and repeat every 10 days. Repeat immediately after every rain. For additional protection use the same dosage in a post-harvest spray. Cherry Fruitworm, Eyespotted Bud Moth, Cankerworm — Use $\frac{1}{3}$ pint (2 pints per acre).

CITRUS (Grapefruit, Lemons, Oranges) [14]: Florida— Purple, Florida Red, Cottony-cushion and Snow Scales, and Mealybug —Use $\frac{1}{3}$ pint between June and September or use 3 ozs. in two sprays, the first in the spring and a second between June and September. Do not apply more than 2 quarts of this material per acre.

Fig [30]: Use $\frac{1}{4}$ pint (2 pints per acre).

GRAPEFRUIT (see Citrus).

LEMONS (see Citrus).

NECTARINES (see Peaches).

ORANGES (see Citrus).

PEACHES [21], NECTARINES [21], In California: Fruittree Leafroller, Catfacing Insects—Use $\frac{1}{4}$ pint (minimum of 1 quart per acre). Aphids, Oriental Fruit Moth, Thrips—Use $\frac{1}{4}$ to $\frac{1}{2}$ pint (minimum of 1 pint per acre). Peach Twig Borer—Use $\frac{1}{2}$ pint per 100 gallons of water (4 pints per acre). For Peach Tree Borer, make applications during appearance of the moths in early to late summer. Apply 2 or 3 sprays to trunk from ground to scaffold limbs timed with moth flight. Do not apply more than once after bloom. Do not apply more than 5 pounds actual parathion per acre between January 1 and harvest.

PEACHES [14], NECTARINES [14]: Areas Other Than California: Fruittree Leafroller, Catfacing Insects, Cottony Peach Scale, Green Peach Aphid, Spider Mites, Shothole Borer (apply during peak of adult beetle activity), Redbanded Leafroller—Use $\frac{1}{4}$ pint. Lecanium Scale, (apply after all eggs have hatched), San Jose Scale—Use $\frac{3}{8}$ pint. Plum Curculio, Oriental Fruit Moth, Peachtree Borer—Use $\frac{1}{3}$ pint. Apply at petal fall and every 10 to 14 days thereafter until control is achieved. For Peach Tree Borer, make applications during apperance of the moths in early to late summer. Apply 2 or 3 sprays to trunk from ground to scaffold limbs timed with moth flight. Do not apply more than 5 pounds actual parathion per acre per year.

PEARS [14]: Pearleaf Blister Mite—Use $\frac{1}{3}$ pint ($2\frac{2}{3}$ pints per acre) in preblossom spray. Pear Psylla—Use $\frac{1}{4}$ pint (2 pints per acre) in pre-blossom or post-blossom sprays. Codling Moth, Mealybug, Woolly Aphid and Spider Mites—Use $\frac{1}{4}$ pint (2 pints per acre). Injury may occur on Bosc variety.

PLUMS [14], PRUNES [14]: Plum Curculio—Use $\frac{3}{8}$ pint (2 pints per acre). Eyespotted Bud Moth, Leafroller, Leafhoppers, Aphids, Spider Mites, Orange Tortrix, Codling Moth, Thrips—Use $\frac{1}{4}$ to $\frac{1}{3}$ pint ($1\frac{1}{4}$ to $1\frac{2}{3}$ pints per acre). San Jose Scale Crawlers, Tussock Moth—Use $\frac{1}{3}$ pint ($1\frac{2}{3}$ pints per acre) when crawlers emerge. Peach Twig Borer—Use $\frac{1}{2}$ pint (4 pints per acre).

PRUNES (see Plums).

SMALL FRUITS

Use amounts listed in sufficient water, 10 gallons minimum by ground or 1 gallon minimum by air, to cover 1 acre.

BLACKBERRIES [15], BOYSENBERRIES [15], DEWBERRIES [15], LOGANBERRIES [15], RASPBERRIES [15]: Aphids, Twospotted Spider Mite, Red Spider Mite—Use $\frac{1}{4}$ pint.

BLUEBERRIES [14]: Maggots, Thrips—Use $\frac{1}{2}$ pint.

BOYSENBERRIES (see Blackberries).

CRANBERRY [15]: Sparganothis Fruitworm, Girdler Tipworms, Cranberry Fruitworm, Fireworm, Leafhopper—Use $\frac{3}{4}$ pint. When applied by aircraft use $\frac{3}{4}$ in a minimum of 5 gallons of water per acre.

CURRANT [30]: Aphids—Use $\frac{1}{8}$ to $\frac{1}{4}$ pint as leaf buds open in spring. Red Spider Mites—Use $\frac{1}{4}$ to $\frac{1}{2}$ pint before bloom.

DEWBERRIES (see Blackberries).

GRAPES [14]: Hoplia Beetle, Mealybug, Thrips, Mites, Grape Berry Moth—Use $\frac{1}{2}$ pint per acre. Leaffolder—Use $\frac{1}{4}$ pint per acre. For Grape Berry Moth, apply just before bloom, at petal fall, 7 to 14 days after petal fall, 35 to 45 days after bloom and 14 days later. Apply 200 gallons of water per acre with ground equipment only. **Dormant Application:** Use $1\frac{1}{4}$ pints plus 2 gallons Dormant Oil Emulsion per 100 gallons of water. Apply 200 gallons per acre as a dormant spray.

LOGANBERRIES (see Blackberries).

RASPBERRIES (see Blackberries).

STRAWBERRIES [14]: Aphids, Lygus Bugs, Leafhoppers, Leafroller, Red Spider Mites, Root Weevil, Whitefly—Use $\frac{1}{4}$ to $\frac{1}{2}$ pint. Repeat at 7 to 10 day intervals.

VEGETABLES

Use amounts listed in sufficient water, 10 gallons minimum by ground or 1 gallon minimum by air, to cover 1 acre.

ARTICHOKES [7]: Artichoke Plume Moth—Use $\frac{1}{2}$ to 1 pint. Aphids—Use $\frac{1}{2}$ pint.

BEANS [7]*: Aphids, Armyworms, Leafhoppers, Leafminer, Leafroller, Mexican Bean Beetle, Red Spider Mite—Use $\frac{1}{3}$ to $\frac{1}{2}$ pint. Corn Earworm, Diabrotica Beetles, Cabbage Looper—Use $\frac{1}{4}$ to $\frac{3}{4}$ pint. Whitefly—Use $\frac{1}{2}$ pint. Repeat applications at 7 to 10 day intervals as required. *At rates exceeding $\frac{1}{2}$ pint, do not apply within 15 days of harvest.

BLACKEYED PEAS [15]: Aphids, Leafminer—Use $\frac{1}{2}$ pint.

BEETS [15]: Aphids, Webworms, Blister Beetles, Leafhoppers —Use $\frac{1}{2}$ pint. Do not apply within 21 days of harvest if tops are to be used for feed or food.

BROCCOLI [7], BRUSSELS SPROUTS [7], CABBAGE [10], CAULIFLOWER [7], COLLARDS [10], KALE [10], KOHLRABI [7], MUSTARD GREENS [10], TURNIPS [10]: Aphids, Armyworms, Cabbage Looper, Climbing Cutworms, Diamondback Moth, Imported Cabbageworm, Thrips, Vegetable Weevil—Use $\frac{1}{2}$ pint.

BRUSSELS SPROUTS (see Broccoli).

CABBAGE (see Broccoli).

CARROTS [15]: Aphids, Petrobia Mite, Vegetable Weevil—Use $\frac{1}{2}$ pint. Do not use treated carrot tops for food or feed.

CAULIFLOWER (see Broccoli).

CELERY [21]*: Aphids, Climbing Cutworms, Cabbage Loopers, Leafhoppers—Use $\frac{1}{2}$ pint. Leafminer, Whitefly—Use $\frac{3}{4}$ pint. *At rates exceeding $\frac{1}{2}$ pint do not apply within 30 days of harvest.

COLLARDS (see Broccoli).

CORN: Armyworms, Budworm, European Corn Borer, Mites—Use $\frac{1}{2}$ to 1 pint. Do not apply within 12 days of harvest or cutting forage.

CORN [12]: Armyworm, Aphids—Use $\frac{1}{4}$ pint. European Corn Borer, Spider Mite, Adult Corn Rootworm—Use $\frac{1}{2}$ pint. Corn Earworm, Chinch Bugs—Use $\frac{3}{4}$ pint. For Corn Earworm, apply to silks as soon as insects appear. Make 3 to 4 applications at 2 to 3 day intervals as infestation warrants. Reapply immediately after rain. Climbing Cutworms, Grasshoppers, Japanese Beetles—Use 6.08 ounces per acre.

COTTON [7]: Cotton Leafworms, Spider Mites, Fleahoppers, Thrips —Use $\frac{1}{4}$ to $\frac{1}{2}$ pint per acre. Cotton Aphids, Cabbage Loopers —Use $\frac{1}{2}$ pint per acre. Cotton Leafperforators, Saltmarsh Caterpillars, Stink Bugs, False Chinch Bugs, Lygus Bugs—Use $\frac{1}{2}$ to 1 pint per acre. Workers entering fields within 24 hours after application should wear protective clothing. Do not hand pick within 7 days of application.

CUCUMBERS [15], MELONS [7], PUMPKINS [10], SQUASH [15]: Cucumber Beetle, Aphids, Stink Bugs, Squash Vine Borer —Use $\frac{1}{2}$ pint. Do not apply before plants start to vine and unless plants are dry.

EGGPLANTS [15]: Aphids, Whitefly, Leafminer—Use ½ pint.

ENDIVE (Escarole) [21]: Aphid, Leafhopper—Use ½ pint.

ESCAROLE (see Endive).

GARLIC [15]: Leafminer, Thrips—Use ½ pint.

KALE (see Broccoli).

KOHLRABI (see Broccoli).

LETTUCE [Leaf-21, Head-7]: Aphids, Leafhoppers, Cabbage Loopers, Armyworms—Use ½ pint.

MELONS (see Cucumbers).

MUSTARD GREENS (see Broccoli).

OKRA [21]: Aphids, Leafminers (including Serpentine) Use ½ pint.

ONIONS [15]: Leafminer, Petrobia Mite, Thrips—Use ½ pint. Repeat weekly as required.

PEAS [10]: Aphids, Pea Weevil, Thrips, Climbing Cutworms, Armyworms, Leafminer—Use ½ pint.

PEPPERS [15]: Aphids, Serpentine Leafminer—Use ½ pint.

POTATOES [5], TOMATOES [10]: Colorado Potato Beetle, Flea Beetle, Leafhoppers, Leafminer, Grasshoppers, Whitefly, Aphids, Thrips, Potato Psyllid (Potato only), Blister Beetle, Vegetable Weevil, Stink Bug, Armyworms, Tomato Russet Mite, Plant Bugs* Use ½ pint. Tomato Pinworm*—Use ¾ to 1 pint. *Tomato only.

PUMPKINS (see Cucumbers).

RICE, California Only [1]: Tadpole Shrimp, Rice Leafminer — Use 1/10 pint per acre. Apply when eggs and larvae of Rice Leafminer are abundant on seedling rice. Restrict spill from rice fields for 3 days following application. Do not spray over canals or laterals. Do not apply within 14 days before or after application of the herbicide propanil (Stam F-34 or Rogue) as injury may occur. Shrimp, crabs and crayfish may be killed. De not apply where these are important resources.

RUTABAGAS [7]: Aphids, Armyworms, Diamondback Moth, Imported Cabbageworm—Use ½ pint. Cabbage Looper, Climbing Cutworm, Vegetable Weevil—Use ½ to ¾ pint.

SPINACH [14], SWISS CHARD [21]: Aphids, Leafminers, Crown Mite, Vegetable Weevil, Seed Corn Maggot in crowns—Use ½ pint.

SQUASH (see Cucumbers).

SWISS CHARD (see Spinach).

TOMATO (see Potato).

TURNIP (see Broccoli).

MOSQUITOES; IRRIGATED PASTURES, RICE [1], ALFALFA FIELDS [15]: Apply 1.6 fluid ounces per acre in sufficient water to cover (minimum of 1 gallon of finished spray per acre). Application should only be made under supervision of Mosquito Abatement Districts and other official agencies. Do not graze livestock on irrigated pastures within 7 days of application. Do not apply to water drainage areas where run-off drainage will contaminate lakes, ponds or streams.

ORNAMENTALS (Outdoor Only): Aphids, Whitefly, Spider Mites, Leafhoppers, Thrips, Lacebug, Bagworm, Cottonycushion Scale—Use ½ pint. For Spider Mites and Leafhoppers repeat at 7 to 10 days. Juniper Webworm, Lecanium Scale—Use ⅔ pint. Mealybugs, Scales, Leaftiers—Use ¾ pint against young or crawler stages. Repeat in 3 weeks if necessary.

ALMONDS: Fruittree Leafroller, Tent Caterpillar, Peach Twig Borer—Use ½ pint. Do not exceed 1½ quarts per acre. Do not apply after hulls open. Do not feed treated hulls to livestock.

WALNUTS: Walnut Aphids, Walnut Husk Flies—Use ¼ pint (minimum ½ pint per acre. Do not apply after husks are open. Apply in 200 to 400 gallons of finished spray per acre).

DEALERS SHOULD SELL IN ORIGINAL PACKAGES ONLY.

TERMS OF SALE OR USE: On purchase of this product buyer and user agree to the following conditions:

WARRANTY: FMC makes no warranty, expressed or implied, concerning the use of this product other than indicated on the label. Except as so warranted, the product is sold as is. Buyer and user assume all risk of use and/or handling and/or storage of this material when such use and/ or handling and/or storage is contrary to label instructions.

DIRECTIONS AND RECOMMENDATIONS: Follow directions carefully. Timing and method of application, weather and crop conditions, mixture with other chemicals not specifically recommended, and other influencing factors in the use of this product are beyond the control of the seller and are assumed by buyer at his own risk.

USE OF PRODUCT: FMC's recommendations for the use of this product are based upon tests believed to be reliable. The use of this product being beyond the control of the manufacturer, no guarantee, expressed or implied, is made as to the effects of such or the results to be obtained if not used in accordance with directions or established safe practice.

DAMAGES: Buyer's or user's exclusive remedy for damages for breach of warranty or negligence shall be limited to direct damages not exceeding the purchase price paid and shall not include incidental or consequential damages.

CARBAMATE

FUNGICIDE

ACTIVE INGREDIENT:	BY WT.
Ferbam (Ferric dimethyl dithiocarbamate)	76.0%
INERT INGREDIENTS:	24.0%
TOTAL	100.0%

EPA Reg. No. 279-388AA EPA Est., 279-

CAUTION

Keep Out of Reach of Children

For Emergency Assistance Call 716–735-3765.

See other panels for additional precautionary information.

PRECAUTIONARY STATEMENTS

HAZARDS TO HUMANS (& DOMESTIC ANIMALS)

CAUTION

May cause irritation of eyes, nose, throat and skin! Avoid breathing dust or spray mist. Avoid contact with eyes, skin or clothing. In case of contact, flush with plenty of water; for eyes, get medical attention.

ENVIRONMENTAL HAZARDS

This product is toxic to fish. Keep out of lakes, streams, ponds, tidal marshes and estuaries. Do not apply where runoff is likely to occur. Do not apply when weather conditions favor drift from areas treated. Do not contaminate water by cleaning of equipment, or disposal of wastes. Apply this product only as specified on this label.

Do not allow this material to drift onto neighboring crops or noncrop areas or use in a manner or at a time other than in accordance with label directions because animal, plant or crop injury, illegal residues or other undesirable results may occur.

PHYSICAL OR CHEMICAL HAZARDS

Keep packages or dust away from sparks, flames or hot surfaces.

DIRECTIONS FOR USE

It is a violation of Federal Law to use this product in a manner inconsistent with its labeling.

Do not apply this product in such a manner as to directly or through drift expose workers or other persons. The area being treated must be vacated by unprotected persons.

Do not enter treated areas without protective clothing until sprays have dried.

Protective clothing means, at least, a hat or other suitable head covering, a long sleeved shirt and long legged trousers or a coverall type garment (all of closely woven fabric covering the body, including the arms and legs), shoes and socks.

Because certain states may require more restrictive reentry intervals for various crops treated with this product, consult your State Department of Agriculure for further information.

Written or oral warnings must be given to workers who are expected to be in a treated area or in an area about to be treated with this product. Inform workers of areas or fields that may not be entered without specific protective clothing, period of time field must be vacated and appropriate actions to take in case of accidental exposure. An example of such information is given under written warnings. When oral warnings are given, warnings shall be given in a language customarily understood by workers. Oral warnings must be given if there is reason to believe that written warnings cannot be understood by workers. Written warnings must include the following information:

CAUTION

Area treated with CARBAMATE on (insert date). Do not enter without appropriate protective clothing until sprays have dried (insert State Department of Agriculture's re-entry interval, if more restrictive).

In case of accidental exposure to pesticide spray or dust, wash the skin thoroughly with soap and water. Remove contaminated clothing and wash before reuse. If in eyes, flush with plenty of water. If inhaled, go to an area where the pesticide has not been applied. Get medical attention if needed.

STORAGE AND DISPOSAL

PESTICIDE STORAGE

Keep away from fire and sparks.

Keep out of reach of children and animals. Store in original containers only. Store in a cool, dry place and avoid excess heat. Carefully open containers. After partial use, fold and roll back bags, clamp and close tightly. Do not put concentrate or dilute material into food or drink containers. Do not contaminate other pesticides, fertilizers, water, food or feed by storage or disposal.

In case of spill, avoid contact, isolate area and keep out animals and unprotected persons. Confine spills. Call FMC collect: (716) 735-3765.

To confine spill: If liquid, dike surrounding area or absorb with sand, cat litter or commercial clay. If dry material, cover to prevent dispersal. Place damaged package in a holding container. Identify contents.

PESTICIDE DISPOSAL

Pesticide wastes are toxic. Improper disposal of excess pesticide, spray mixture, or rinsate is a violation of Federal Law. If these wastes cannot be disposed of by use according to label instructions, contact your State Pesticide or Environmental Control Agency, or the Hazardous Waste representative at the nearest EPA Regional Office for guidance.

CONTAINER DISPOSAL

Completely empty bag into application equipment. Then dispose of empty bag in a sanitary landfill or by incineration, or, if allowed by State and local authorities, by burning. If burned, stay out of smoke.

To mix this material, slowly sift the desired amount into the partially filled tank of water with the agitator running and when thoroughly mixed apply at once with continuous agitation. Where desired this material may first be made to a thick paste by adding a little water. This should then be diluted and added to the tank while it is filling.

Where this material is used with Lead Arsenate, no Lime need be used as a safener, since this material is itself a good safener for Lead Arsenate. Observe all cautions and limitations on labeling of all products used in mixtures.

Observe days interval between last application and harvest indicated by number in ( ) following crop.

APPLES (7): For early season control of Scab, Cedar Rust, Blotch —Use 1 to 1½ pounds of this material, or ½ pound plus 3 pounds Kolospray or Kolofog, per 100 gallons of water in prebloom, calyx and early cover sprays. For Rust control, make an additional bloom application using ½ pound per 100 gallons of water. For late season control of Scab, Black Rot, Bitter Rot, Sooty Blotch, Fly Speck, and Brooks' Spot—Use ¾ pound per 100 gallons of water. Where Bitter Rot and Brooks' Spot are severe problems increase dosage to 1½ to 2 pounds per 100 gallons of water. On Golden Delicious, Jonathan, Baldwin and other varieties of Apples which are especially apt to Russet, some injury may result from the use of this material if used in the early season or bloom applications. The best time for use is in the cover sprays.

When used in combination with sulfur, do not apply within 10 days to 2 weeks of an oil spray nor just prior to or during periods of excessively high temperatures.

CABBAGE (Plant Beds Only): Downy Mildew—Use 1½ pounds per 100 gallons of water, making 2 or 3 applications each week.

CHERRIES (0): Brown Rot, Leaf Spot—Use 1½ pounds per 100 gallons of water in petal fall and cover sprays.

CITRUS (0) (Grapefruit, Lemons, Limes, Oranges, Tangerines, Tangelos): Scab—Use 1½ pounds per 100 gallons of water. Do not apply more than 1000 gallons of spray per acre. Apply during prebloom period and when ⅔ of petals have fallen. May be applied in late summer or early fall if a heavy flush of growth appears.

CRANBERRIES: Fruit Rots—Use 2 pounds per 100 gallons of water. Begin applications early in blossoming period and repeat at two week intervals for a total of 5 applications. Do not apply within 28 days after mid-bloom. Fairy Ring—Use 9 pounds per 100 gallons of water and apply 1 gallon of this mixture to a one foot square area. Treat an area 3 feet beyond the advancing line of dead vines and 2 feet within this line. Apply in the fall immediately after harvest.

GRAPES (7): Black Rot—Use 2 pounds per 100 gallons of water, applying just before bloom, just after bloom and two weeks later.

ORNAMENTALS: Black Spot and Rust on Roses; Rust on Snapdragons and Carnations; Leaf Spot on Chrysanthemums; Botrytis Blight on Peonies and Tulips—Use 1 to 1½ pounds per 100 gallons of water, applying at 7 to 10 day intervals during growing season.

PEACHES (21): Leaf Curl—Use 1½ pounds per 100 gallons of water. Apply in the fall after leaves drop or in the spring before buds begin to swell.

PEARS (7): Scab, Sooty Blotch—Use 1½ pounds per 100 gallons of water in pink, calyx, first and second cover sprays, and 1 pound in summer sprays.

PINES: Fusiform Rust of Seedlings—Use 1 to 2 pounds per 100 gallons of water. Apply at 3 to 7 day intervals as needed, beginning when plants emerge. Southern Cone Rust (Cronortium Strobilinum) in Slash and Long Leaf Pines—Use 2 pounds per 100 gallons of water plus 4 to 8 ounces of suitable spreader-sticker. Begin at time of flowering when strobili are emerging from bud scales. Repeat at 5 day intervals until conelet stage. Direct spray to flower-bearing portions of tree crowns.

TOBACCO (Plant Beds): Blue Mold—Use 2 to 3 pounds per 100 gallons of water. Apply at the rate of 3 gallons per 100 square yards when plants are small, increasing to 6 gallons when plants are ready for transplanting. Begin applications when plants are the size of a dime, or when Blue Mold is reported in the area, and repeat twice weekly until plants are transplanted.

TOMATOES: Gray Mold (Botrytis)—Use 3 pounds per 100 gallons of water. Spray for complete coverage. Make seed bed application just before seedlings break the soil surface and repeat at 7 day intervals. Make field application when disease threatens and repeat at 7 to 10 day intervals as necessary to maintain control.

DEALERS SHOULD SELL IN ORIGINAL PACKAGES ONLY.

TERMS OF SALE OR USE: On purchase of this product buyer and user agree to the following conditions:

WARRANTY: FMC makes no warranty, expressed or implied, concerning the use of this product other than indicated on the label. Except as so warranted, the product is sold as is. Buyer and user assume all risk of use and/or handling and/or storage of this material when such use and/or handling and/or storage is contrary to label instructions.

DIRECTIONS AND RECOMMENDATIONS: Follow directions carefully. Timing and method of application, weather and crop conditions, mixture with other chemicals not specifically recommended and other influencing factors in the use of this product are beyond the control of the seller and are assumed by buyer at his own risk.

USE OF PRODUCT: FMC's recommendations for the use of this product are based upon tests believed to be reliable. The use of this product being beyond the control of the manufacturer, no guarantee, expressed or implied, is made as to the effects of such or the results to be obtained if not used in accordance with directions or established safe practice.

DAMAGES: Buyer's or user's exclusive remedy for damages for breach of warranty or negligence shall be limited to direct damages not exceeding the purchase price paid and shall not include incidental or consequential damages.

Kolofog, Kolospray,—FMC Trademarks

DICHLONE 50 WP

FUNGICIDE

ACTIVE INGREDIENT:

*Dichlone	50.0%
INERT INGREDIENTS:	50.0%
TOTAL	100.0%

*2,3-Dichloro-1,4-naphthoquinone

EPA Reg. No. 279-491 EPA Est., 279-

CAUTION

Keep Out of Reach of Children

May cause skin irritation. Avoid contact with skin, eyes, or clothing. Avoid inhalation of dust or spray mist. Wear protective clothing. Wash with soap and water after using. Avoid storage near feed and food products.

For Emergency Assistance Call 716–735-3765

See other panels for additional precautionary information.

ENVIRONMENTAL CAUTIONS

This product is toxic to fish. Avoid use in shallow farm ponds which contain desirable fish. Do not use in potable water. Do not apply where runoff is likely to occur. Do not apply when weather conditions favor drift from areas treated. Do not contaminate water by cleaning of equipment or disposal of wastes. Apply this product only as specified on this label.

DIRECTIONS FOR USE

It is a violation of Federal Law to use this product in a manner inconsistent with its labeling.

Do not apply this product in such a manner as to directly or through drift expose workers or other persons. The area being treated must be vacated by unprotected persons.

Do not enter treated areas without protective clothing until sprays have dried.

Protective clothing means, at least, a hat or other suitable head covering, a long sleeved shirt and long legged trousers or a coverall type garment (all of closely woven fabric covering the body, including the arms and legs), shoes and socks.

Because certain states may require more restrictive reentry intervals for various crops treated with this product, consult your State Department of Agriculture for further information.

Written or oral warnings must be given to workers who are expected to be in a treated area or in an area about to be treated with this product. Inform workers of areas or fields that may not be entered without specific protective clothing, period of time field must be vacated and appropriate actions to take in case of accidental exposure. An example of such information is given under written warnings. When oral warnings are given, warnings shall be given in a language customarily understood by workers. Oral warnings must be given if there is reason to believe that written warnings cannot be understood by workers. Written warnings must include the following information:

CAUTION

Area treated with DICHLONE 50 WP on (insert date). Do not enter without appropriate protective clothing until sprays have dried (insert State Department of Agriculture's re-entry interval, if more restrictive.)

In case of accidental exposure to pesticide spray or dust, wash the skin thoroughly with soap and water. Remove contaminated clothing and wash before reuse. If in eyes, flush with plenty of water. If inhaled, go to an area where the pesticide has not been applied. Get medical attention if needed.

STORAGE AND DISPOSAL

PESTICIDE STORAGE

Keep out of reach of children and animals. Store in original containers only. Store in a cool, dry place and avoid excess heat. Carefully open containers. After partial use, fold and roll back bags, clamp and close tightly. Do not put concentrate or dilute material into food or drink containers. Do not contaminate other pesticides, fertilizers, water, food or feed by storage or disposal.

In case of spill, avoid contact, isolate area and keep out animals and unprotected persons. Confine spills. Call FMC collect: (716) 735-3765.

To confine spill: If liquid, dike surrounding area and absorb with sand, cat litter or commercial clay. If dry material, cover to prevent dispersal. Place damaged package in a holding container. Identify contents.

PESTICIDE DISPOSAL

Wastes resulting from the use of this product may be disposed of on site or at an approved waste disposal facility.

CONTAINER DISPOSAL

Completely empty bag into application equipment. Then dispose of empty bag in a sanitary landfill or by incineration, or, if allowed by State and local authorities, by burning. If burned, stay out of smoke.

Do not use DICHLONE with mixtures containing oils, Bordeaux mixture, lime sulfur or organic mercuries.

DO NOT USE IN UNDILUTED FORM

APPLE: Apple Scab—Use 1/4 pound per 100 gallons of water on a protective schedule at 5 to 7 day intervals from delayed dormant through the early cover sprays. Use 1/2 pound per 100 gallons of water, in applications up to 50 hours after the start of a rain, for after rain infection period. Use a full coverage spray but do not exceed 8 pounds per acre. For late cover applications, use a suitable summer fungicide. DICHLONE may cause injury to apples during periods of high temperatures.

CHERRY: Brown Rot Blossom Blight—Use 1/2 pound per 100 gallons of water at popcorn, full bloom and petal fall stages of growth. When cool, wet weather occurs, resulting in a prolonged bloom period, apply every 4 days. Brown Rot of Fruit—Use 1/2 pound per 100 gallons of water at 7 to 10 day intervals beginning at petal fall. Cherry Leaf Spot—Use 1/2 to 3/4 pound per 100 gallons of water beginning immediately after petal fall. Make two additional treatments at 10 day intervals. Use a full coverage spray but do not exceed 10 pounds per acre. Do not apply within 3 days of harvest.

PEACH: Brown Rot—For blossoms and twig brown rot blossom blight, use 1/2 to 3/4 pound per 100 gallons of water at pink bud stage, full bloom and petal fall. When cool, wet weather occurs, resulting in a prolonged bloom period, apply every 4 days. Do not apply after petal fall. Peach California Blight (Coryneum Blight)—Use 3/4 pound per 100 gallons of water as a post-harvest spray between September 15th and September 30th in Oregon and Washington. Peach Leaf Curl—Use 1 pound per 100 gallons of water as a dormant spray just before the buds begin to swell.

Use a full coverage spray but do not exceed 5 pounds per acre.

PLUM, PRUNE: Brown Rot Blossom Blight—Use ½ to ¾ pound per 100 gallons of water at green bud, popcorn, full bloom and pre-shuck fall stages of growth. When cool, wet weather occurs, resulting in a prolonged bloom period, apply every 4 days. Do not apply after the petal fall period. Prune Russet Scab (Lacy Scab) (California Only)—Use ½ to ¾ pound per 100 gallons of water. Apply in popcorn to full bloom stage of growth. Do not apply after the petal fall period.

Use a full coverage spray but do not exceed 2.5 pounds per acre.

STRAWBERRY: Fruit Rot (Botrytis)—Use ¾ pound in 200 gallons of water per acre. Make the first application as new growth starts and repeat applications at full bloom, calyx, and every 10 to 14 days until 3 days before first picking.

BEANS: Anthracnose—Use 1 to 1½ pounds per 100 gallons of water at weekly intervals in a series of 4 to 5 applications, starting just before bloom. Use 100 to 150 gallons of spray per acre. Do not apply within 7 days of harvest.

CELERY: Early and Late Blight—Use 1 pound per acre in 200 gallons of water. Begin applications 7 to 10 days after plants are set and repeat every 5 to 7 days during infection period.

TOMATOES: Early and Late Blight, Gray Mold—Use 2 to 2½ pounds per acre in sufficient water for thorough coverage. Apply at 4 to 7 day intervals during infection periods. For control of gray mold, late blight, phoma rot and leaf spot in plant beds: Use 1 pound per 100 gallons of water (3 level teaspoons per gallon) per acre. Begin when plants emerge. Repeat at 5 to 7 day intervals until transplanting time. For control of soil borne diseases at transplanting: Mix 1 pound per 100 gallons. Use as a dip after pulling and before planting.

Note: DICHLONE may injure tomato plants if applied when temperature reaches 85°F.

RECREATIONAL LAKES, FARM PONDS: For control of Certain Bloom-Producing Blue-Green Algae—Use 0.05 ppm or 2.2 ounces of DICHLONE 50 WP per acre-foot of water (325,960) gallons). Application in excess of 0.07 ppm will kill fish. If extremely heavy growths of algae are treated, apply to about one third of the pond at a time to allow the fish to move to an area more favorable for oxygen supply. Retreat the segment if necessary in 2 to 3 days. The interval between treatment of segments of the pond or lake should be sufficient to allow for decomposition of the algae before adjacent areas are treated or retreated. Once under control, retreat only when it is apparent that re-growth of algae is occurring rapidly. Use one of the following methods of treatment:

Consult your State Fish and Game Agency before applying this product to public waters.

Spray Method: Mix required amount of DICHLONE 50 WP in 5 or more gallons of water and spray as uniformly as possible over the water surface. Any sprayer equipped with agitation may be used.

For isolated groups of these algae, thoroughly spray the surface of floating mats with ½ to 1 ounce of DICHLONE 50 WP in 3 gallons of water. Apply at the rate of 1 oz. of DICHLONE 50 WP per 20,000 sq. ft. where a 1 foot depth is intended to be treated. A knapsack sprayer may be used for spot treatment.

Drag Method: Place the required amount of DICHLONE 50 WP in a finely woven cloth bag—such as a flour bag—and drag it through the area to be treated. Attach the bag to the end of a boat or tie it on a long pole. Make applications as uniformly as possible.

Swimming Pools: For control of Certain Bloom-Producing Blue-Green Algae—Use 1 ounce of DICHLONE 50 WP per 50,000 gallons of water (0.15 ppm of DICHLONE 50 WP). Begin applications when algae first appear at the bottom of the water area and repeat when necessary at 1 to 2 week intervals. Use one of the methods of treatment listed above for recreational lakes and farm ponds. Three (3) level teaspoonfuls per gallon equal 1 lb. per 100 gallons.

DEALERS SHOULD SELL IN ORIGINAL PACKAGES ONLY.

TERMS OF SALE OR USE: On purchase of this product buyer and user agree to the following conditions:

WARRANTY: FMC makes no warranty, expressed or implied, concerning the use of this product other than indicated on the label. Except as so warranted, the product is sold as is. Buyer and user assume all risk of use and/or handling and/or storage of this material when such use and/or handling and/or storage is contrary to label instructions.

DIRECTIONS AND RECOMMENDATIONS: Follow directions carefully. Timing and method of application, weather and crop conditions, mixture with other chemicals not specifically recommended, and other influencing factors in the use of this product are beyond the control of the seller and are assumed by the buyer at his own risk.

USE OF PRODUCT: FMC's recommendations for use of this product are based upon tests believed to be reliable. The use of this product being beyond the control of the manufacturer, no guarantee, expressed or implied, is made as to the effects of such or the results to be obtained if not used in accordance with directions or established safe practice.

DAMAGES: Buyer's or user's exclusive remedy for damages for breach of warranty or negligence shall be limited to direct damages not exceeding the purchase price paid and shall not include incidental or consequential damages.

ETHION 4 MISCIBLE

MITICIDE–INSECTICIDE

For Agricultural or Commercial Use Only

ACTIVE INGREDIENTS:	BY WT.
Ethion (0,0,0′, 0′-tetraethyl S,S′-methylene bisphosphorodithiote)	46.5%
Petroleum Distillate	32.0%
INERT INGREDIENTS:	21.5%
TOTAL	100.0%

Contains 4 pounds Ethion per gallon.

EPA Reg. No. 279-1254 EPA Est., 279-

WARNING

Keep Out of Reach of Children

AVISO

PRECAUCION AL USUARIO: Si usted no lee ingles, no use este producto hasta que la etiqueta le haya sido explicada ampliamente.

For Emergency Assistance Call 716–735-3765.

See other panels for additional precautionary information.

STATEMENT OF PRACTICAL TREATMENT

If swallowed—Call a physician or poison control center. Give one or two glasses of water and induce vomiting by touching back of throat with finger. Do not induce vomiting or give anything by mouth to an unconscious or convulsing person.

If inhaled—If not breathing, give artificial respiration. Get medical attention.

If in eyes—Flush with plenty of water. Get medical attention if irritation persists.

If on skin—Wash off with soap and water and remove contaminated clothing and shoes.

Antidote

NOTE TO PHYSICIAN:

Active ingredient is a cholinesterase inhibitor. Patient should be kept under observation for 48 hours. Atropine is antidotal. Give 4 mg. Atropine sulfate intramuscularly or intravenously as soon as cyanosis is overcome. Repeat every 15 minutes until patient is fully atropanized. Pralidoxime (2-PAM) is also antidotal and should be administered in conjunction with Atropine. Never give morphine.

PRECAUTIONARY STATEMENTS

HAZARDS TO HUMANS (& DOMESTIC ANIMALS)

May be fatal if swallowed or absorbed through skin. Do not breathe vapor or spray mist. Avoid contact with skin, eyes and clothing. Wash thoroughly after handling.

ENVIRONMENTAL HAZARDS

This product is toxic to fish and wildlife. Birds feeding on treated areas may be killed. Do not contaminate water by cleaning of equipment or disposal of wastes and containers. Apply this product only as specified on this label.

This product is toxic to bees and should not be applied when bees are actively visiting the area.

Do not allow this material to drift onto neighboring crops or non-crop areas or use in a manner or at a time other than in accordance with label directions because animal, plant or crop injury, illegal residues or other undesirable results may occur.

Do not apply directly to water.

PHYSICAL OR CHEMICAL HAZARDS

Do not use or store near heat or open flame.

DIRECTIONS FOR USE

It is a violation of Federal Law to use this product in a manner inconsistent with its labeling.

Do not apply this product in such a manner as to directly or through drift expose workers or other persons. The area being treated must be vacated by unprotected persons.

Do not enter treated areas for 24 hours unless appropriate protective clothing is worn.

Protective clothing means, at least, a hat or other suitable head covering, a long sleeved shirt and long legged trousers or a coverall type garment (all of closely woven fabric covering the body, including the arms and legs), shoes and socks.

Because certain states may require more restrictive reentry intervals for various crops treated with this product, consult your State Department of Agriculture for further information.

Written or oral warnings must be given to workers who are expected to be in a treated area or in an area about to be treated with this product. Inform workers of areas or fields that may not be entered without specific protective clothing, period of time field must be vacated and appropriate actions to take in case of accidental exposure. An example of such information is given under written warnings. When oral warnings are given, warnings shall be given in a language customarily understood by workers. Oral warnings must be given if there is reason to believe that written warnings cannot be understood by workers. Written warnings must include the following information:

WARNING

Area treated with ETHION 4 MISCIBLE on (insert date). Do not enter without appropriate protective clothing for 24 hours (insert State Department of Agriculture's re-entry interval, if more restrictive).

In case of accidental exposure to pesticide spray or dust, wash the skin thoroughly with soap and water. Remove contaminated clothing and wash before reuse. If in eyes, flush with plenty of water. If inhaled, go to an area where the pesticide has not been applied. Get medical attention if needed.

STORAGE AND DISPOSAL

PESTICIDE STORAGE

Do not use or store near heat, open flame or hot surfaces.

Keep out of reach of children and animals. Store in original containers only. Store in a cool, dry place and avoid excess heat. Carefully open containers. After partial use, replace lids and close tightly. Do not put concentrate or dilute material into food or drink containers. Do not contaminate other pesticides, fertilizers, water, food, or feed by storage or disposal.

In case of spill, avoid contact, isolate area and keep out animals and unprotected persons. Confine spills. Call FMC collect: (716) 735-3765.

To confine spill: If liquid, dike surrounding area or absorb with sand, cat litter or commercial clay. If dry material, cover to prevent dispersal. Place damaged package in a holding container. Identify contents.

PESTICIDE DISPOSAL

Pesticide wastes are toxic. Improper disposal of excess pesticide, spray mixture, or rinsate is a violation of Federal Law. If these wastes cannot be disposed of by use according to label instructions, contact your State Pesticide or Environmental Control Agency, or the Hazardous Waste representative at the nearest EPA Regional Office for guidance.

CONTAINER DISPOSAL

Triple rinse (or equivalent). Then offer for recycling or reconditioning, or puncture and dispose of in a sanitary landfill, or by other procedures approved by state and local authorities. Do not cut or weld metal containers.

Do not use in undiluted form.

Do not use if material does not emulsify.

Apply the listed amount per acre when insects first appear and repeat as required to maintain effective control. Use in sufficient water for thorough coverage, unless otherwise noted for listed crops. Coverage of both upper and lower leaf surfaces is essential for good control. Observe use limitations. If insect control is required beyond the following use patterns, supplement the control program with other suitable pesticides. When applying this material by aircraft, mix the recommended amount with sufficient water to provide a minimum of 1 gallon finished spray per acre. Where more than 2 quarts are recommended, mix with sufficient water to provide a minimum of finished spray equal to twice the amount of ETHION MISCIBLE used. Observe days interval between last application and harvest indicated by number in ( ) following the crop.

ALMONDS: Clover (Brown Almond) Mite, European Red Mite—Use 0.75 pt. plus 1 to 2 gals. dormant oil per 100 gals. of water. Olive (Parlatoria) Scale—Use 1 pt. plus 3 gals. dormant oil per 100 gals. of water. San Jose Scale—Use 0.5 to 0.75 pt. plus 2 gals. dormant oil per 100 gals. of water. Apply during dormant or delayed dormant period.

APPLES: Apple Aphid, European Red Mite, Rosy Apple Aphid, San Jose Scale—Use 0.6 pt. plus 2 gals. dormant oil per 100 gals. water. Apply during dormant or delayed dormant period ("Superior" type oil of 60 to 70 second viscosity preferred for delayed dormant use.)

BEANS (Succulent and Dry) (2): Bean Leaf Beetle, Mexican Bean Beetle, Twospotted Spider Mite—0.5 to 1 pt. per acre. Potato Leafhopper—1 pt. per acre. Do not feed treated foliage to livestock.

BEANS (Succulent and Dry) (4): Lygus Bugs—1 pt. per acre. Do not feed treated foliage to livestock.

BERMUDA GRASS (Lawns and Seed Crop): Eriophyid Mites—1 tablespoon per gal. of water to cover 200 sq. ft. of lawn or use 3 fluid ozs. per 6 gals. of water per 1200 sq. ft. Apply as needed to maintain effective control. Keep children and animals off treated areas until the spray has completely dried. Do not apply to Bermuda grass to be used for grazing or forage to be fed to livestock.

CHERRIES: Black Cherry Aphid, San Jose Scale—0.5 to .75 pt. plus 2 gals. dormant oil per 100 gals. water. Apply during dormant or delayed dormant period ("Superior" type oil of 60 to 70 second viscosity preferred for delayed dormant use).

CHESTNUTS, FILBERTS, PECANS: Aphids, Mites—2.5 qts. per acre. Do not apply more than twice during fruiting period. Do not apply after husk split on Chestnuts and Pecans, nor when nut is visible on Filberts. Do not graze livestock on treated groves.

CORN, FIELD (50): Pacific Spider Mite, Twospotted Spider Mite—1 qt. per acre. Do not make more than 1 application after ear formation. Do not feed treated forage to livestock.

COTTON: Aphids—1 to 2 pts. per acre. Cotton Leafworm—1 pt. per acre. Spider Mites—0.5 to 1 pt. per acre. Spider Mites, Lygus Bugs and other Mirids **(West Coast Only)**—2 to 3 pts. per acre. Do not apply after bolls open. Do not graze dairy or meat animals in treated fields.

CUCUMBERS (0), SUMMER SQUASH (0): Serpentine Leafminer, Twospotted Spider Mite—1 pt. per acre.

EGGPLANT (0): Twospotted Spider Mite—0.5 to 1 pt. per acre.

GRAPEFRUIT (0), LEMONS (21), LIMES (21), ORANGES (0), TANGELOS (0), TANGERINES (0): To control Citrus Red Mite, Citrus Rust Mite, Sixspotted Mite, Texas Citrus Mite, Black Scale, Brown Soft Scale, California Red Scale, Chaff Scale, Florida Red Scale, Glover Scale, Purple Scale, Snow Scale, Yellow Scale, Citrus Whitefly, Mites — Use 0.5 to 0.75 pt. in 100 gals. water. Mites, Scales, and Citrus Whitefly—Use 0.75 pt. per 100 gals. water combined with 0.5 to 0.7% summer oil **(Florida and Texas areas)** or 0.5 to 0.75 pt. per 100 gals. water combined with 1 to 2 gals. summer oil **(West Coast)**. Where Snow Scale is a problem, spray trunk, limbs and twigs thoroughly. For Whitefly, apply shortly after Whiteflies have laid their eggs. Do not apply more than 2000 gals. spray per acre. Do not apply more than once on Lemons and Limes; on other citrus crops listed, do not apply within 90 days of previous application. On Tangerines, do not apply more than twice per crop. Do not apply oil within 3 weeks of a sulfur application. Do not use oil on Naval Oranges.

GRAPES (California only) (30): To control Grape Leafhopper, Twospotted Spider Mite, Williamette Mite—Use 1.5–2.5 pints per acre. Do not apply more than twice during fruiting period.

HOME GROUNDS (For Use By Trained Pest Control Operators) For outside use only: Clover (Brown Almond) Mite infesting houses—1 pt. per 2500 square feet area. Mix well with sufficient water to cover 2500 sq. ft. Spray areas of mite infestation from the bottom of first floor windows down to the ground and 15 to 20 ft. away from building. Keep children and animals off treated areas until spray has completely dried.

MELONS (0): Melon Leafminer, Serpentine Leafminer, Twospotted Spider Mite—1 pt. per acre. Melon Leafhopper—1.5 to 2 pts. per acre. Petrobia Mite **(California only)**, Cucumber Beetle—1 to 2 pts. in a minimum of 10 gals. water per acre.

NECTARINES, PEACHES, PLUMS, PRUNES: Clover (Brown Almond) Mite, European Red Mite and including Leafcurl Plum Aphid and Mealy Plum Aphid on Plum and Prune—.75 pt. plus 1 to 2 gals. dormant oil per 100 gals. water. Olive (Parlatoria) Scale—1 pt. plus 3 gals. dormant oil per 100 gals. water. San Jose Scale—0.5 to .75 pt. plus 2 gals. dormant oil per 100 gals. water. Apply during dormant or delayed dormant period. ("Superior" type oil of 60 or 70 second viscosity preferred for delayed dormant use).

NECTARINES (30), PEACHES (30): Clover (Brown Almond) Mite, European Red Mite, Twospotted Spider Mite—0.5 pt. per 100 gals. water or 2.5 to 3.5 pts. per acre. Do not apply more than twice during fruiting period.

ORNAMENTALS (Greenhouse) SOIL TREATMENT: Millipedes—Use 10 fl. ozs. in 75 gals. of water and apply to 2,500 sq. ft. of bench space. Wear mask or respirator approved by Mining Enforcement and Safety Administration (formerly the U.S. Bureau of Mines) and by the National Institute for Occupational Safety and Health under the provisions of 30 CFR Part II for Ethion protection.

ORNAMENTALS (Out of Doors): Twospotted Spider Mite—Use 0.5 pint per 100 gals. May cause some needle drop on Anderson Yew.

PEARS: Clover (Brown Almond) Mite, European Red Mite, San Jose Scale—.75 pt. plus 1 to 2 gals. dormant oil per 100 gals. water. Apply during dormant or delayed dormant period. ("Superior" type oil of 60 or 70 second viscosity preferred for delayed dormant use).

PEPPERS (0), PIMENTOS (0): Serpentine Leafminer—1 pt. per acre, Twospotted Spider Mite—0.5 to 1 pt. per acre.

SORGHUM (30): Sorghum Midge, Sorghum Mite—0.5 to 1.0 pt. per acre. For Midge, apply at first bloom and repeat 4 to 10 days later. Do not harvest or feed forage within 30 days after last application. Do not apply more than 3 times per season. Some leaf burning may occur on phosphate-sensitive varieties.

STRAWBERRIES (2): Strawberry Leafroller, Twospotted Spider Mite—1 pt. per 100 gals.; 1 to 2 pts. per acre. Spray plants thoroughly. Continue application after harvest as required.

TOMATOES (2): Leafminer, Tropical Spider Mite, Twospotted Spider Mite—1 pt. per acre.

TURF: (Commercial Use)—Chinch Bugs, Sod Webworm—Use 0.33 to 0.5 pt. per 1000 sq. ft. of turf applied in 20 gals. of water or 1.67 to 2.5 pts. in 100 gals. of water to treat 5000 sq. ft. To treat one acre, use 1.75 to 2.5 gals applied in 800 gals. of water. It is important that the spray penetrate through the mat of turf to reach the insects. It helps to have the turf well irrigated prior to treatment. **Caution:** Keep children and animals off treated areas until this material has been thoroughly watered into soil and the grass has completely dried.

DEALERS SHOULD SELL IN ORIGINAL PACKAGES ONLY.

TERMS OF SALE OR USE: On purchase of this product buyer and user agree to the following conditions:

WARRANTY: FMC makes no warranty, expressed or implied, concerning the use of this product other than indicated on the label. Except as so warranted, the product is sold as is. Buyer and user assume all risk of use and/or handling and/or storage of this material when such use and/or handling and/or storage is contrary to label instructions.

DIRECTIONS AND RECOMMENDATIONS: Follow directions carefully. Timing and method of application, weather and crop conditions, mixture with other chemicals not specifically recommended and other influencing factors in the use of this product are beyond the control of the seller and are assumed by buyer at his own risk.

USE OF PRODUCT: FMC's recommendations for the use of this product are based upon tests believed to be reliable. The use of this product being beyond the control of the manufacturer, no guarantee, expressed or implied, is made as to the effects of such or the results to be obtained if not used in accordance with directions or established safe practice.

DAMAGES: Buyer's or user's exclusive remedy for damages for breach of warranty or negligence shall be limited to direct damages not exceeding the purchase price paid and shall not include incidental or consequential damages.

ETHION 8 EC

INSECTICIDE–MITICIDE

For Agricultural or Commercial Use Only

ACTIVE INGREDIENTS:	
Ethion; O,O,O′,O′-tetraethyl S,S′-methylene bisphosphorodithioate	81.90%
INERT INGREDIENTS	18.10%
TOTAL	100.00%

Contains 8 pounds Ethion per gallon.

EPA Reg. No. 279-2820 EPA Est.

DANGER–POISON

Keep Out of Reach of Children

PELIGRO

PRECAUCION AL USUARIO: Si usted no lee ingles, no use este producto hasta que la etiqueta le haya sido explicada ampliamente.

FIRST AID

If swallowed, drink 1 or 2 glasses of water and induce vomiting by touching back of throat with finger. Never give anything by mouth to an unconscious person. Get medical attention. If breathing has stopped, start artificial respiration immediately and maintain until a doctor sees victim.

If in eye, flush with water for 15 minutes and get immediate medical attention.

If on skin or clothing, immediately flush skin or eyes with water for 15 minutes while removing contaminated clothing and shoes.

Antidote

Note To Physician: Active ingredient is a cholinesterase inhibitor. Patient should be kept under observation for 48 hours. Atropine is antidotal. Give 4 mg. atropine sulfate intramuscularly or intravenously as soon as cyanosis is overcome. Repeat every 15 minutes until patient is fully atropinized. 2-PAM is also antidotal and may be administered in conjunction with atropine. Never give morphine.

For Emergency Assistance Call 716–735-3765.

See Side Panel For Additional Precautionary Information.

PRECAUTIONARY STATEMENTS

HAZARDS TO HUMANS (AND DOMESTIC ANIMALS)

Danger

Poisonous if swallowed, inhaled or absorbed through skin. Do not breathe vapor or spray mist. Do not get in eyes, on skin or on clothing. Wear a mask or pesticide respirator jointly approved by the Mining Enforcement and Safety Administration (formerly the U.S. Bureau of Mines) and the National Institute for Occupational Safety and Health. Wash hands, face and arms thoroughly with soap and water after handling and before eating or smoking. Remove contaminated clothing and wash with soap and hot water before reuse.

ENVIRONMENTAL HAZARDS

This product is toxic to fish and wildlife. Keep out of lakes, ponds and streams. Do not contaminate water by cleaning of equipment or disposal of wastes and containers. Shrimp and crab may be killed at application rates recommended on this label. Do not apply where these are important resources. Do not apply when weather conditions favor drift from the areas treated. Do not apply where run-off is likely to occur. Apply this product only as specified on this label.

This product is toxic to bees and should not be applied when bees are actively visiting the area.

PHYSICAL OR CHEMICAL HAZARDS

Do not use or store near heat or open flame.

DIRECTIONS FOR USE

Not for use or storage in or around the home.

It is a violation of Federal law to use this product in a manner inconsistent with its labeling.

Do not apply this product in such a manner as to directly or through drift expose workers or other persons. The area being treated must be vacated by unprotected persons.

Do not enter treated areas for 24 hours unless appropriate protective clothing is worn.

Protective clothing means, at least, a hat or other suitable head covering, a long-sleeved shirt and long-legged trousers or a coverall type garment (all of closely woven fabric covering the body, including the arms and legs), shoes and socks.

Because certain states may require more restrictive reentry intervals for various crops treated with this product, consult your State Department of Agriculture for further information.

Written or oral warnings must be given to workers who are expected to be in a treated area or in an area about to be treated with this product. Inform workers of areas or fields that may not be entered without specific protective clothing, period of time field must be vacated and appropriate actions to take in case of accidental exposure. An example of such information is given under written warnings. When oral warnings are given, warnings shall be given in a language customarily understood by workers. Oral warnings must be given if there is reason to believe that written warnings cannot be understood by workers. Written warnings must include the following information:

DANGER

Area treated with ETHION 8 EC on (insert date). Do not enter without appropriate protective clothing for 24 hours (insert State Department of Agriculture's re-entry interval, if more restrictive).

In case of accidental exposure to pesticide spray or dust, wash the skin thoroughly with soap and water. Remove contaminated clothing and wash before reuse. If in eyes, flush with plenty of water. If inhaled, go to an area where the pesticide has not been applied. Get medical attention if needed.

STORAGE AND DISPOSAL

PESTICIDE STORAGE

Do not store below 0°F, (—18°C).

Do not use or store near heat, open flame or hot surfaces.

Keep out of reach of children and animals. Store in original containers only. Store in a cool, dry place and avoid excess heat. Carefully open containers. After partial use, replace lids and close tightly. Do not put concentrate or dilute material into food or drink containers. Do not contaminate other pesticides, fertilizers, water, food, or feed by storage or disposal.

In case of spill, avoid contact, isolate area and keep out animals and unprotected persons. Confine spills. Call FMC collect: (716) 735-3765.

To confine spill: If liquid, dike surrounding area or absorb with sand, cat litter or commercial clay. If dry material, cover to prevent dispersal. Place damaged package in a holding container. Identify contents.

PESTICIDE DISPOSAL

Pesticide wastes are acutely hazardous. Improper disposal of excess pesticide, spray mixture, or rinsate is a violation of Federal law. If these wastes cannot be disposed of by use according to label instructions, contact your State Pesticide or Environmental Control Agency, or the Hazardous Waste representative at the nearest EPA Regional Office for guidance.

CONTAINER DISPOSAL

Triple rinse (or equivalent). Then offer for recycling or reconditioning, or puncture and dispose of in a sanitary landfill, or by other procedures approved by State and local authorities. Do not cut or weld metal containers.

Not for use or storage in or around the home.

Do not use if material does not emulsify.

Apply the listed amount per acre when insects first appear and repeat as required to maintain effective control. Use in sufficient water for thorough coverage with dilute or concentrate sprayers, unless otherwise noted for listed crops. Coverage of upper and lower leaf surfaces is essential for good control. Observe use limitations. If insect control is required beyond the following use patterns, supplement the control program with other suitable pesticides. When applying this material by aircraft, mix the recommended amount with sufficient water to provide a minimum of 1 gallon finished spray per acre. Where more than 2 quarts are recommended mix with sufficient water to provide a minimum of finished spray equal to twice the amount of Ethion 8 EC used. Do not use in undiluted form. Observe days interval between last application and harvest indicated by the number in ( ) following the crop.

ALMONDS (Dormant or Delayed Dormant): Brown Mite, European Red Mite—6 fluid ounces plus 1 to 2 gallons dormant oil per 100 gallons water. Olive Scale—8 fluid ounces plus 3 gallons dormant oil per 100 gallons water. San Jose Scale—4 to 6 fluid ounces plus 2 gallons dormant oil per 100 gallons water. Apply during dormant or delayed dormant period. **Foliar:** Brown Mite, European Red Mite and Twospotted Spider Mite—Use 1½ to 2½ pints per acre. By ground, apply a minimum of 200 gallons of water per acre; by air, a minimum of 15 gallons of water per acre. At 1½ to 2 pints per acre, do not apply more than once during the fruiting period nor within 30 days of harvest. When 2 to 2½ pints per acre are used, do not apply after husks split. Do not graze livestock in treated groves.

APPLES (Dormant or Delayed Dormant): Apple Aphid, European Red Mite, Rosy Apple Aphid, San Jose Scale—4.8 fluid ounces plus 2 gallons dormant oil per 100 gallons water. (Superior type oil of 60 to 70 second viscosity preferred for delayed dormant use.) **Summer:** Brown Mite, European Red Mite, Twospotted Spider Mite—Use ¼ pint per 100 gallons of water (maximum of 2½ pints per acre). Apply during the fruiting period. If making no more than 4 applications, wait 40 days before harvest; if making no more than 3 applications, wait 30 days before harvest; if making no more than 2 applications, wait 20 days before harvest. Codling Moth—Use ½ pint per 100 gallons of water (maximum of 5 pints per acre). If making no more than 4 applications, wait 60 days before harvest; if making no more than 2 applications, wait 40 days before harvest. Apply at petal fall and every 10 to 14 days thereafter until control is achieved. Do not use ethion on Astrachen, Pratt, Melba, N.J. No. 3 (Brite Mac), Wealthy, Williams Red (Red Williams), Yellow Transparent or any variety maturing with or before Early McIntosh as injury is likely to occur.

BEANS (Succulent and Dry) (2): Bean Leaf Beetle, Mexican Bean Beetle, Twospotted Spider Mites—4 to 8 fluid ounces per acre. Potato Leafhopper—8 fluid ounces per acre. **Beans (4):** Lygus Bugs—1 pint per acre. Do not feed treated foliage to livestock.

CHERRIES: Black Cherry Aphid, San Jose Scale—4 to 6 fluid ounces plus 2 gallons dormant oil per 100 gallons water. Apply during dormant or delayed dormant period.

CHESTNUTS: Aphids, Mites—Use 2½ pints per acre. Do not apply more than twice during the fruiting period. Do not apply after husks split. Do not graze livestock in treated groves.

CORN, FIELD (50): Pacific Spider Mite, Twospotted Spider Mite—1 pint per acre. Do not make more than 1 application after ear formation. Do not feed treated foliage to livestock.

COTTON: Aphids—Use ½ to 1 pint per acre. Cotton Leafworm—Use ½ pint per acre. Spider Mites—Use ¼ to ½ pint per acre. Spider Mites, Lygus Bugs and other Mirids on **West Coast only**—Use 1 to 1½ pints per acre. Do not apply after bolls open. Do not graze dairy or meat animals in treated fields. Workers entering treated fields within 24 hours following application should wear protective clothing.

CUCUMBERS (0), SUMMER SQUASH (0): Serpentine Leafminer, Twospotted Spider Mite—8 fluid ounces per acre.

EGGPLANT (0): Two Spotted Spider Mite—Use 4 to 8 fluid ounces per acre.

FILBERTS: Aphids, Mites—1¼ quarts per acre. Do not apply more than twice during fruiting period. Do not apply when nut is visible. Do not graze livestock in treated groves.

GRAPEFRUIT (0), LEMONS (21), LIMES (21), ORANGES (0), TANGELOS (0), TANGERINES (0): Citrus Red Mite, Citrus Rust Mite, Sixspotted Mite, Scales (Black, Brown Soft, California Red, Purple, Yellow Chaff, Glover), and Citrus Whitefly. For Mites use 4 to 6 pints per acre. For Mites, Scales and Citrus Whitefly use 4 to 6 fluid ounces per 100 gallons of water combined with 1 to 2 gallons summer oil (West Coast). For Whitefly, apply shortly after eggs have been laid. Do not apply more than 2000 gallons spray per acre. With concentrate sprayers apply 50 to 100 gallons of spray per acre. Citrus Thrips **California and Arizona only**—Use 2½ to 3 pints per acre. Apply when Thrips first appear and before they exceed 2 to 3 Thrips per leaf. Do not apply more than once per season on lemons and limes. On grapefruit, oranges, tangelos and tangerines do not apply within 90 days of previous application. On tangerines do not apply more than twice per season. Do not apply oil within 3 weeks of a sulphur application.

GRAPES (30): Grape Leafhopper, Twospotted Spider Mite, Willamette Mite (West Coast insects)—Use 1 to 2¼ pints per acre. When used at rates of 1¼ pints per acre or less, do not apply within 30 days of harvest or more than twice during the fruiting period; or do not apply within 60 days of harvest with no limit on the number of applications per season. At rates above 1¼ pints per acre, make only 1 application per season. Do not apply to Salvador, Muscat or Fehr-Szago varieties of grapes.

MELONS (0): Melon Leafminer, Serpentine Leafminer, Twospotted Spider Mite—Use ½ pint per acre. Melon Leafhopper—Use ¾ to 1 pint per acre in a minimum of 5 gallons of water per acre by ground. Petrobia Mite (California only)—Use ½ to 1 pint in a minimum of 10 gallons of water per acre.

NECTARINES, PEACHES (Dormant or Delayed Dormant): Brown Mite, European Red Mite—6 fluid ounces plus 1 to 2 gallons dormant oil per 100 gallons water. Olive Scale—8 fluid ounces plus 3 gallons dormant oil per 100 gallons water. San Jose Scale—4 to 6 fluid ounces plus 2 gallons dormant oil per 100 gallons water. Do not apply on Nectarines after foliage or edible parts start to form.

NECTARINES (30), PEACHES (30): Brown Mite, European Red Mite, Twospotted Spider Mites—4 fluid ounces per 100 gallons water or 1¼ to 1¾ pints per acre. Do not apply more than twice during fruiting period.

ORNAMENTALS (Greenhouse): Millipedes—Soil Treatment: Use 5 fluid ounces in 75 gallons of water and apply to 2500 square feet of bench space. Wear mask or respirator approved by the Mining Enforcement and Safety Administration and the National Institute for Occupational Safety and Health for ethion protection. **(Out-of-doors):** Twospotted Spider Mite—Use 4 fluid ounces per 100 gallons of water. May cause some needle drop on Anderson Yew.

PEARS: (Dormant or Delayed Dormant): Brown Mite, European Red Mite, San Jose Scale—6 fluid ounces plus 1 to 2 gallons dormant oil per 100 gallons water. **Summer:** Brown Mite, European Red Mite, Twospotted Spider Mites—Use ¼ to ⅓ pint per 100 gallons of water (maximum of 2½ pints per acre). Do not make more than 3 applications after the first cover spray nor apply within 30 days of harvest; or do not make more than 2 applications after the first cover spray nor apply within 20 days of harvest. Pear Psylla—Use ⅓ to ½ pint per 100 gallons of water (maximum of 5 pints per acre). Apply just before bloom if insects are present. Repeat 30 days later if necessary. Do not make more than 3 applications after the first cover spray nor apply within 60 days of harvest.

PECANS: Aphids and Mites—Use 2½ pints per acre. Do not apply more than twice during the fruiting period. Do not apply after husks split. Do not graze livestock in treated groves.

PEPPER (0), PIMENTO (0): Serpentine Leafminer—Use ½ pint per acre. Twospotted Spider Mite—Use ¼ to ½ pint per acre.

PLUMS, PRUNES (Dormant or Delayed Dormant): Brown Mite, European Red Mite, Leaf Curl Plum Aphids, Mealy Plum Aphid—6 fluid ounces plus 1 to 2 gallons dormant oil per 100 gallons water. Olive Scale—8 fluid ounces plus 3 gallons dormant oil per 100 gallons water. San Jose Scale—4 to 6 fluid ounces plus 2 gallons dormant oil per 100 gallons water. **Summer:** Brown Mite, European Red Mite, Twospotted Spider Mite—Use ¼ pint per 100 gallons of water (maximum of 1¾ pints per acre). Do not make more than 3 applications during the fruiting period or apply within 21 days of harvest. European Fruit Lecanium—Use ⅓ pint per 100 gallons of water (maximum of 2½ pints per acre). Do not make more than 2 applications during the fruiting period or apply within 45 days of harvest.

SORGHUM (30): Sorghum Midge, Sorghum Mite—Use 4 to 8 fluid ounces per acre. For Midge, apply at first bloom and repeat 4 to 10 days later. Do not utilize for forage within 30 days of treatment. Do not apply more than 3 times per season. Some leaf burning may occur on phosphate-sensitive varieties.

STRAWBERRIES (2): Strawberry Leafroller and Spider Mites—Use ½ to 1 pint per acre in sufficient water to cover upper and lower leaf surfaces. Spray plants thoroughly. Continue applications after harvest as required.

TOMATOES (2): Leafminer, Tropical Spider Mite, Twospotted Spider Mite—Use ½ pint per acre.

TURF (Commercial Use): Chinch Bugs, Sod Webworm—Use ¼ pint per 1000 square feet of turf applied in 20 gallons of water or 1¼ pints in 100 gallons of water to treat 5000 square feet. To treat one acre, use 7 to 9 pints applied in 800 gallons of water. It is important that the spray penetrates through the mat of turf to reach the insects. It helps to have the turf well–irrigated prior to treatment. CAUTION: Keep children and animals off treated areas until this material has been thoroughly watered into the soil and the turf has completely dried.

WALNUTS (Dormant or Delayed Dormant): Overwintering eggs of European Red Mite, Brown Mite, San Jose Scale and Soft Scale (West Coast only)—Use 2 pints per acre with 2 gallons of a dormant oil in a minimum of 100 gallons of water. This treatment should only be used on trees that have not suffered from lack of moisture at anytime during the growing season. Oil should not be applied to orchards when soil moisture is low. Do not graze livestock in treated groves. **Summer:** European Red Mite, Twospotted Spider Mite, Walnut Aphid, and Walnut Husk Fly (California only)—Use 2½ pints per acre. Do not make more than 2 applications during the fruiting period. Do not apply after husks split. Do not graze livestock in treated groves.

DEALERS SHOULD SELL IN ORIGINAL PACKAGES ONLY.

TERMS OF SALE OR USE: On purchase of this product buyer and user agree to the following conditions:

WARRANTY: FMC makes no warranty, expressed or implied, concerning the use of this product other than indicated on the label. Except as so warranted, the product is sold as is. Buyer and user assume all risk of use and/or handling and/or storage of this material when such use and/or handling and/or storage is contrary to label instructions.

DIRECTIONS AND RECOMMENDATIONS: Follow directions carefully. Timing and method of application, weather and crop conditions, mixture with other chemicals not specifically recommended, and other influencing factors in the use of this product are beyond the control of the seller and are assumed by buyer at his own risk.

USE OF PRODUCT: FMC's recommendations for the use of this product are based upon tests believed to be reliable. The use of this product being beyond the control of the manufacturer, no guarantee, expressed or implied, is made as to the effects of such or the results to be obtained if not used in accordance with directions or established safe practice.

DAMAGES: Buyer's or user's exclusive remedy for damages for breach of warranty or negligence shall be limited to direct damages not exceeding the purchase price paid and shall not include incidental or consequential damages.

ETHION SUPERIOR 70 OIL

INSECTICIDE

ACTIVE INGREDIENTS:	BY WT.
*Ethion	2.0%
Petroleum Oil	94.0%
INERT INGREDIENTS:	4.0%
TOTAL	100.0%

*000′,0′-Tetraethyl S,S′-methylene bisphosphorodithioate

Contains 0.15 lbs. Ethion per gallon.

Petroleum Oil Specifications:

Unsulfonated Residue (Min.)	92%
Viscosity	70 Seconds

EPA Reg. No. 279-1731 EPA Est., 279-

CAUTION

Keep Out of Reach of Children

For Emergency Assistance Call: 716–735–3765

See other panels for additional precautionary information.

PRECAUTIONARY STATEMENTS

HAZARDS TO HUMANS (& DOMESTIC ANIMALS)

CAUTION

Harmful if swallowed. May be absorbed through skin. Avoid breathing vapor or spray mist. Avoid contact with skin, eyes and clothing. Wash thoroughly after handling. Avoid storage near feed or food products.

ENVIRONMENTAL HAZARDS

This product is toxic to fish, birds and other wildlife. Birds feeding on treated areas may be killed. Keep out of lakes, ponds and streams. Do not apply where runoff is likely to occur. Do not apply when weather conditions favor drift from areas treated. Do not contaminate water by cleaning of equipment, or disposal of wastes. Apply this product only as specified on this label.

DIRECTIONS FOR USE

It is a violation of Federal law to use this product in a manner inconsistent with its labeling.

STORAGE AND DISPOSAL

PESTICIDE STORAGE

Do not store below 32°F, (0°C).

Do not use or store near heat, open flame or hot surfaces.

Keep out of reach of children and animals. Store in original containers only. Store in a cool, dry place and avoid excess heat. Carefully open containers. After partial use, replace lids and close tightly. Do not put concentrate or dilute material into food or drink containers. Do not contaminate other pesticides, fertilizers, water, food, or feed by storage or disposal.

In case of spill, avoid contact, isolate area and keep out animals and unprotected persons. Confine spills. Call FMC collect: (716) 735-3765.

To confine spill: If liquid, dike surrounding area or absorb with sand, cat litter or commercial clay. If dry material, cover to prevent dispersal. Place damaged package in a holding container. Identify contents.

PESTICIDE DISPOSAL

Pesticide wastes are toxic. Improper disposal of excess pesticide, spray mixture, or rinsate is a violation of Federal Law. If these wastes cannot be disposed of by use according to label instructions, contact your State Pesticide or Environmental Control Agency, or the Hazardous Waste representative at the nearest EPA Regional Office for guidance.

CONTAINER DISPOSAL

Triple rinse (or equivalent). Then offer for recycling or reconditioning, or puncture and dispose of in a sanitary landfill, or by other procedures approved by state and local authorities. Do not cut or weld metal containers.

Spray on warm days in late Winter or early Spring, and before the buds separate into the clusters. Do not spray in freezing weather or apply strong dosages after the buds have burst to any extent. Do not apply when buds have separated in the clusters.

Do not apply within 2 weeks of sulfur spray or within 7 days of Captan.

Do not use if this spray oil does not emulsify. Correct emulsification is indicated if, when this product is added to the water in the spray tank, a uniform milk-white spray mixture results provided the spray mixture is being agitated. If the spray mixture, after adding the oil to the spray tank with the agitation running, is not milk-white, do not use the resulting spray or the remainder of this product.

Fill the spray tank about one-third full of water, start the agitator and pour in the required amounts of ETHION SUPERIOR 70 OIL. If other spray materials are to be used with the oil, mix the other materials thoroughly with the water in the spray tank first and then add the ETHION SUPERIOR 70 OIL as the filling of the tank is being completed. Do not use in undiluted form.

When a gun has been shut off or the outfit stopped for any length of time, spray on the ground for a few moments until the spray appears creamy white. This is desirable also where there are dead ends in the pipe lines of stationary outfits. All the material in stationary outfits should be sprayed out each day.

ALMONDS, PEACHES: Clover (Brown Almond) Mite, European Red Mite—To control overwintering Mite eggs use 2 gallons per 100 gallons of water. Oystershell Scale, San Jose Scale—Use 2 gallons per 100 gallons of water. Parlatoria (Olive) Scale—Use 3 gallons per 100 gallons of water. Apply during dormant or delayed dormant period. Do not graze livestock in treated groves.

APPLES: Apple Aphid, European Red Mite, Oystershell Scale, Rosy Apple Aphid, San Jose Scale—Use 2 gallons per 100 gallons of water. Make a single application during the delayed dormant period. Do not apply later than the delayed dormant period.

APRICOTS: Oystershell Scale—Use 2 gallons per 100 gallons of water. Apply during dormant or delayed dormant period.

CHERRIES (Sweet & Tart): Black Cherry Aphid, European Red Mite, Forbes Scale, Putnam Scale, San Jose Scale—Use 2 gallons per 100 gallons of water. Apply during delayed dormant period.

NECTARINES: Clover (Brown Almond) Mite, European Red Mite—To control overwintering Mite eggs use 2 gallons per 100 gallons of water. San Jose Scale—Use 2 gallons per 100 gallons of water. Parlatoria (Olive) Scale—Use 3 gallons per 100 gallons of water. Apply during dormant or delayed dormant period.

PEARS: Clover (Brown Almond) Mite, European Red Mite—To control overwintering Mite eggs use 2 gallons per 100 gallons of water. Oystershell Scale, San Jose Scale, Pear Psylla—Use 2 gallons per 100 gallons of water. Apply during dormant or delayed dormant period.

PLUMS, PRUNES: Clover (Brown Almond) Mite, European Red Mite, Leaf Curl Plum Aphid, Mealy Plum Aphid—To control overwintering Mite and Aphid eggs use 2 gallons per 100 gallons of water. Oystershell Scale, San Jose Scale—Use 2 gallons per 100 gallons of water. Parlatoria (Olive) Scale—Use 3 gallons per 100 gallons of water. Apply during dormant or delayed dormant period.

ORNAMENTALS: (Arborvitae, Ash, Birch, Boxwood, Catalpa, Dogwood, Elm, Euonymus, Juniper, Lilac, Magnolia, Maple, Mock Orange, Oak, Pine, Poplar, Privet, Taxus, Tulip Tree, Willow): Soft Scales including Cottony Maple, European Elm, European Fruit Lecanium, Magnolia and Tulip Tree Scale. Hard or Armored Scales including Euonymus, Juniper, Oystershell, Pine Needle, Scurfy and San Jose Scale—For deciduous ornamentals, use 2 to 3 gallons per 100 gallons of water. For evergreen ornamentals, use 2 gallons per 100 gallons of water. Apply as dormant spray after most danger of freezing nights has passed but before buds have begun to open. Spray on a clear mild day and do not spray in late afternoon. Be cautious with oil sprays. Do not use on easily injured plants such as Beech, Butternut, Hickory, Maple (Sugar & Japanese), or Walnut. Oil will injure Douglas Fir flower buds and remove blue color from Colorado Blue Spruce.

DEALERS SHOULD SELL IN ORIGINAL PACKAGES ONLY.

TERMS OF SALE OR USE: On purchase of this product buyer and user agree to the following conditions:

WARRANTY: FMC makes no warranty, expressed or implied, concerning the use of this product other than indicated on the label. Except as so warranted, the product is sold as is. Buyer and user assume all risk of use and/or handling and/or storage of this material when such use and/or handling and/or storage is contrary to label instructions.

DIRECTIONS AND RECOMMENDATIONS: Follow directions carefully. Timing and method of application, weather and crop conditions, mixture with other chemicals not specifically recommended and other influencing factors in the use of this product are beyond the control of the seller and are assumed by buyer at his own risk.

USE OF PRODUCT: FMC's recommendations for the use of this product are based upon tests believed to be reliable. The use of this product being beyond the control of the manufacturer, no guarantee, expressed or implied, is made as to the effects of such or the results to be obtained if not used in accordance with directions or established safe practice.

DAMAGES: Buyer's or user's exclusive remedy for damages for breach of warranty or negligence shall be limited to direct damages not exceeding the purchase price paid and shall not include incidental or consequential damages.

FUNGINEX® 1.6 EC

FUNGICIDE

ACTIVE INGREDIENT:

Triforine (N, N'-[1,4-piperazinediyl-bis-(2,2,2-trichloroethylidene)]-bis-[formamide])	18.2%
INERT INGREDIENTS:	81.8%
TOTAL	100.0%

This product contains 1.6 lb. of Triforine per gallon.

EPA Reg. No. 21137-4-279 EPA Est. 279-FL-1

Funginex® is a registered trademark of Celamerck GMbh & Company KG, Ingelheim/Germany.

DANGER

Keep Out of Reach of Children

STATEMENT OF PRACTICAL TREATMENT

If in eyes, flush with plenty of water for 15 minutes and get medical attention.

If swallowed, drink 1 or 2 glasses of water and induce vomiting by touching back of throat with finger or blunt object. Do not induce vomiting or give anything by mouth to an unconscious person. Get medical attention.

If inhaled, remove patient from contaminated area and get medical attention.

If on skin, remove contaminated clothing and wash skin with soap and water.

For Emergency Assistance Call 716–735-3765.

PRECAUTIONARY STATEMENTS

HAZARDS TO HUMANS AND DOMESTIC ANIMALS

DANGER

Corrosive, causes eye damage. Do not get in eyes. Wear goggles or face shield. Harmful if swallowed. Do not inhale spray mist. Avoid contact with skin. After handling, wash thoroughly with soap and water.

ENVIRONMENTAL HAZARDS

Do not apply directly to water. Do not contaminate water by cleaning of equipment or disposal of wastes. Apply this product only as specified on this label.

USE PRECAUTIONS

Do not graze animals in treated orchards.

As timing of fungicide applications for disease control varies due to climatic and other conditions, consult Agricultural Experiment Station or State Extension Service specialist.

Do not use this material if it cannot be applied according to the use pattern on this label.

Do not mix FUNGINEX with wetting agents, spreader-slickers or other adjuvants.

Do not let spray mixture stand in tank overnight.

DIRECTIONS FOR USE

It is a violation of Federal law to use this product in a manner inconsistent with its labeling. For use as a fungicide for the control of certain important plant diseases.

STORAGE AND DISPOSAL

Store in a cool, dry place. Do not store below 32°F.
Do not contaminate water, food or feed by storage or disposal. Open dumping is prohibited. Do not re-use empty container.

PESTICIDE DISPOSAL
Pesticide, spray mixture, or rinse water that cannot be used according to label instructions, must be disposed of according to applicable Federal, State or local procedures.

CONTAINER DISPOSAL
Dispose of in an incinerator or landfill approved for pesticide containers or bury in a safe place. Consult Federal, State or local disposal authorities for approved alternative procedures such as limited open burning.

APPLES: Scab (Venturia Inaequalis), Powdery Mildew (Podosphaera leucotricha), and Rust (Gymnosporangium spp.) For full coverage spray only, mix 10 fl. oz. of FUNGINEX per 100 gallons and apply to run-off. For low volume sprayers, apply 36 to 40 fl. oz. of undiluted FUNGINEX per acre per application in sufficient water (50-200 gallons of water per acre). For aerial application, apply 36 to 40 fl. oz. of FUNGINEX in a minimum of 20 gallons of water per acre. Complete coverage is essential to insure adequate control. Make first application at ½ inch green tip and repeat every 7 days for a preventive control program. Do not apply after petal fall. Do not exceed a total of 5 applications. Consult Agricultural Experiment Station or State Extension Service specialist for use of FUNGINEX in apple scab monitoring control program.

APRICOTS, CHERRIES, NECTARINES, PEACHES, PLUMS and PRUNES: Brown Rot Blossom Blight (Monilinia spp.) For full coverage spray only, mix 12 to 16 fl. oz. of FUNGINEX per 100 gallons and apply to run-off. For low volume sprayers apply 36 to 48 fl. oz. of undiluted FUNGINEX per acre per application in sufficient water (50 to 200 gallons of water per acre). For aerial application, apply 36 to 48 fl. oz. of FUNGINEX in a minimum of 20 gallons of water per acre. Complete coverage is essential to insure adequate control. Make first application at early bloom (peaches, nectarines: pink bud; apricots: red bud; cherries, plums, prunes: white bud or popcorn). Repeat after 50% bloom. If necessary, depending upon the length of the bloom period and conditions favoring brown rot blossom blight development, make a third application at early petal fall. Alternately, if warm, wet weather prevails, apply the two or three applications at 2 to 4 day intervals beginning at early bloom, since blossom period will be shortened. Do not exceed three sprays of FUNGINEX during the blossom period. The higher rate of FUNGINEX is only necessary under conditions of severe disease pressure:

California Only: For full coverage spray only, mix 12 fl. oz. of FUNGINEX per 100 gallons and apply to run-off. For low volume sprayers, apply 36 to 48 fl. oz. of undiluted FUNGINEX per acre per application in sufficient water (50 to 200 gallons of water per acre). For aerial application, apply 36 to 48 fl. oz. of FUNGINEX in a minimum of 20 gallons of water per acre. Complete coverage is essential to insure adequate control. Make first application on peaches and nectarines at pink bud to 5% bloom, on apricots at red bud, on cherries, plums and prunes at popcorn or white bud, followed by a second application at 50-100% bloom. Do not exceed two sprays of FUNGINEX during the blossom period.

APRICOTS, NECTARINES, PEACHES: Brown Rot, Fruit Rot (Monilinia spp.) For full coverage spray only mix 12 to 16 fl. oz. of FUNGINEX per 100 gallons and apply to run-off. For low volume sprayers, apply 36 to 48 fl. oz. of undiluted FUNGINEX per acre per application in sufficient water (50 to 200 gallons of water per acre). For aerial application, apply 36 to 48 fl. oz. of FUNGINEX in a minimum of 20 gallons of water per acre. Complete coverage is essential to insure adequate control. Make the first application 2 or 3 weeks before harvest and repeat in 5 to 10 days. Make a third application just prior to harvest. Do not exceed three sprays of FUNGINEX during the pre-harvest period. The higher rate of FUNGINEX is only necessary under conditions of severe disease pressure.

California Only: For full coverage spray only, mix 12 fl. oz. of FUNGINEX per 100 gallons and apply to run-off. For low volume sprayers, apply 36 to 48 fl. oz. of undiluted FUNGINEX per acre per application in sufficient water (50 to 200 gallons of water per acre). For aerial application, apply 36 to 48 fl. oz. of FUNGINEX in a minimum of 20 gallons of water per acre. Complete coverage is essential to insure adequate control. Make first application 2 or 3 weeks before harvest and repeat in 5 to 10 days. Do not exceed two sprays of FUNGINEX during the pre-harvest period.

HIGHBUSH BLUEBERRIES: Mummyberry Disease (Monilinia vaccinii-corymbosi), Pacific and Mid-Western States Apply 24 fl. oz. of FUNGINEX per acre in 20 to 50 gallons of water for ground application or in 5 gallons of water for aerial application. Make the first application at leaf bud break and repeat in 7 to 10 days. Make the third application at pink bud stage and repeat in 7 to 10 days at early bloom. For the last application, apply 16 fl. oz. of FUNGINEX per acre in 20 to 50 gallons of water for ground application or in 5 gallons of water for aerial application. Make the last application between full bloom and early petal fall. Do not make more than five applications from leaf bud break to early petal fall.

Eastern Seaboard States (For primary Infection only) Apply 24 fl. oz. of FUNGINEX per acre in 20 to 50 gallons of water for ground application or in 5 gallons of water for aerial application. Make the first application at leaf bud break and repeat in 7 to 10 days. Make the last application at pink bud stage. Do not make more than three applications from leaf bud break to pink bud stage. Application of FUNGINEX during or beyond early bloom may result in fruit russetting.

DEALERS SHOULD SELL IN ORIGINAL PACKAGES ONLY.

TERMS OF SALE OR USE On purchase of this product buyer and user agree to the following conditions:

WARRANTY FMC makes no warranty, expressed or implied, concerning the use of this product other than indicated on the label. Except as so warranted, the product is sold as is. Buyer and user assume all risk of use and/or handling and/or storage of this material when such use and/or handling and/or storage is contrary to label instructions.

DIRECTIONS AND RECOMMENDATIONS Follow directions carefully. Timing and method of application, weather and crop conditions, mixture with other chemicals not specifically recommended and other influencing factors in the use of this product are beyond the control of the seller and are assumed by buyer at his own risk.

USE OF PRODUCT FMC's recommendations for the use of this product are based upon tests believed to be reliable. The use of this product being beyond the control of the manufacture, no guarantee, expressed or implied, is made as to the effects of such or the results to be obtained if not used in accordance with directions or established safe practice.

DAMAGES Buyer's or user's exclusive remedy or damages for breach of warranty or negligence shall be limited to direct damages not exceeding the purchase price paid and shall not include incidental or consequential damages.

FURADAN® 3G

INSECTICIDE

RESTRICTED USE PESTICIDE

For retail sale to and application only by Certified Applicators or personnel under their direct supervision, and only for those uses covered by the Certified Applicator's certification.

For Agricultural or Commercial Use Only

ACTIVE INGREDIENT:

* Carbofuran	3.00%
INERT INGREDIENTS	97.00%
TOTAL	100.00%

*2,3-Dihydro-2,2-dimethyl-7-benzofuranyl methylcarbamate

EPA Reg. No. 279-2792 EPA Est., No. 279
U.S. Patent No. 3,474,171 Furadan—FMC Trademark

WARNING

Keep Out of Reach of Children

STATEMENT OF PRACTICAL TREATMENT

If swallowed, drink 1 to 2 glasses of water and induce vomiting by touching back of throat with finger or blunt object. Do not induce vomiting or give anything by mouth to an unconscious person. Get medical attention.

If in eyes, flush with plenty of water for at least 15 minutes. Get medical attention.

If on skin: wash immediately with plenty of soap and water. Remove contaminated clothing and wash before reuse.

Antidote

Note to Physician: FURADAN is a reversible cholinesterase inhibitor. Do not use oximes such as 2-PAM. Start by giving 2 mg. atropine intramuscularly. According to clinical response, continue until signs of atropinization occur (dry mouth or dilated pupils). If in eye, instill one drop of homatropine.

For Emergency Assistance Call 716-735-3765.

PRECAUTIONARY STATEMENTS

HAZARDS TO HUMANS (& DOMESTIC ANIMALS)

WARNING

Poisonous if swallowed. May be fatal or harmful as a result of skin or eye contact or by breathing dust. Causes cholinesterase inhibition. Warning symptoms of poisoning include weakness, headache, sweating, nausea, vomiting, diarrhea, tightness in chest, blurred vision, pinpoint eye pupils, abnormal flow of saliva, abdominal cramps, and unconsciousness. Atropine sulfate is antidotal.

Wear long-sleeved clothing and protective gloves when handling. Wash hands and face before eating or smoking. Bathe at the end of the work day. Change clothing daily and wash before reuse.

ENVIRONMENTAL HAZARDS

This pesticide is toxic to fish, birds and other wildlife. Birds feeding on treated area may be killed. Birds killed by carbofuran pose a hazard to hawks and other birds-of-prey; bury or otherwise dispose of dead birds to prevent poisoning of other wildlife. Cover or incorporate granules in spill areas. Runoff from treated areas may be hazardous to fish in neighboring areas. Fish may be killed at recommended rates. Do not contaminate wells, wetlands or any body of water by cleaning of equipment or disposal of waste.

Do not allow this material to drift onto neighboring crops or non-crop areas or use in a manner or at a time other than in accordance with label directions because animal, plant or crop injury, illegal residues or other undesirable results may occur.

Carbofuran is a chemical which can travel (seep or leach) through soil and can contaminate ground water which may be used as drinking water. Carbofuran has been found in ground water as a result of agricultural use. Users are advised not to apply carbofuran where the water table (ground water) is close to the surface and where the soils are very permeable, i.e., well-drained soils such as loamy sands. Your local agricultural agencies can provide further information on the type of soil in your area and the location of ground water.

DIRECTIONS FOR USE

It is a violation of Federal law to use this product in a manner inconsistent with its labeling.

Do not re-enter treated fields less than 24 hours following application.

Do not use this product on Long Island, N.Y.

STORAGE AND DISPOSAL

PESTICIDE STORAGE

Do not store in or around the home.

Keep out of reach of children and animals. Store in original containers only. Store in a cool, dry place and avoid excess heat. Carefully open containers. After partial use, fold and roll back bags, clamp and close tightly. Do not put concentrate or dilute material into food or drink containers. Do not contaminate other pesticides, fertilizers, water, food or feed by storage or disposal.

In case of spill, avoid contact, isolate area and keep out animals and unprotected persons. Confine spills. Call FMC collect: (716) 735-3765.

To confine spill: If liquid, dike surrounding area or absorb with sand, cat litter or commercial clay. If dry material, cover to prevent dispersal. Place damaged package in a holding container. Identify contents.

PESTICIDE DISPOSAL

Pesticide wastes are acutely hazardous. Improper disposal of excess pesticide, spray mixture, or rinsate is a violation of Federal Law. If these wastes cannot be disposed of by use according to label instructions, contact your State Pesticide or Environmental Control Agency, or the Hazardous Waste representative at the nearest EPA Regional Office for guidance.

CONTAINER DISPOSAL

Completely empty bag into application equipment. Then dispose of empty bag in a sanitary landfill or by incineration, or, if allowed by State and local authorities, by burning. If burned, stay out of smoke.

RICE: (Arkansas, Louisiana, Mississippi, Texas only) Rice Water Weevil and Mosquitoes—Apply 15 to 20 pounds of FURADAN 3G per acre one day before or within 21 days after permanent flooding. Do not make more than one application per season. Apply by air or ground equipment. The subsequent use of propanil herbicide may result in crop injury. For effective larval control of the dark rice field mosquito, application must be made within 2 to 4 days after the field has been flooded.

DEALERS SHOULD SELL IN ORIGINAL PACKAGES ONLY.

TERMS OF SALE OR USE: On purchase of the product buyer and user agree to the following conditions:

WARRANTY: FMC makes no warranty, expressed or implied, concerning the use of this product other than indicated on the label. Except as so warranted, the product is sold as is. Buyer and user assume all risk of use and/or handling and/or storage of this material when such use and/or handling and/or storage is contrary to label instructions.

DIRECTIONS AND RECOMMENDATIONS: Follow directions carefully. Timing and method of application, weather and crop conditions, mixture with other chemicals not specifically recommended, and other influencing factors in the use of this product are beyond the control of the seller and are assumed by buyer at his own risk.

USE OF PRODUCT: FMC's recommendations for the use of this product are based upon tests believed to reliable. The use of this product being beyond the control of the manufacturer, no guarantee, expressed or implied, is made as to the effects of such or the results to be obtained if not used in accordance with directions or established safe practice.

DAMAGES: Buyer's or user's exclusive remedy for damages for breach of warranty or negligence shall be limited to direct damages not exceeding the purchase price paid and shall not include incidental or consequential damages.

FURADAN® 4 F

INSECTICIDE–NEMATICIDE

RESTRICTED USE PESTICIDE

For retail sale to and use only by Certified Applicators or persons under their direct supervision, and only for those uses covered by the Certified Applicator's certification.

ACTIVE INGREDIENT:	
*Carbofuran	44.90%
INERT INGREDIENTS:	55.10%
TOTAL	100.00%

* 2,3-Dihydro-2,2-dimethyl-7-benzofuranyl methylcarbamate.
This product contains 4 lbs. of carbofuran per gallon.

EPA Reg. No. 279-2876 ZA　　EPA Est., 279

U.S. Patent No. 3,474,171　　Furadan—FMC Trademark

DANGER–POISON

Keep Out of Reach of Children

For Emergency Assistance Call 716–735-3765.

STATEMENT OF PRACTICAL TREATMENT

If swallowed: drink 1 or 2 glasses of water and induce vomiting by touching back of throat with finger. Do not induce vomiting or give anything by mouth to an unconscious person. Get medical attention.

If inhaled: remove to fresh air. Call a physician immediately.

If in eyes: flush with plenty of water for at least 15 minutes. Get medical attention.

If on skin: wash skin immediately with soap and water. Remove contaminated clothing and wash before reuse.

NOTE TO PHYSICIAN: FURADAN (carbofuran) is a reversible cholinesterase inhibitor. Do not use oximes such as 2-PAM. Start by giving 2 mg. atropine intramuscularly. According to clinical response, continue until signs of atropinization occur (dry mouth or dilated pupils). If in eye, instill one drop of homatropine.

PRECAUTIONARY STATEMENTS

HAZARDS TO HUMANS (& DOMESTIC ANIMALS)

DANGER

Poisonous if swallowed or inhaled. May be fatal or harmful as a result of skin or eye contact or by breathing spray mist. Causes cholinesterase inhibition. Warning symptoms of poisoning include weakness, headache, sweating, nausea, vomiting, diarrhea, tightness in chest, blurred vision, pinpoint eye pupils, abnormal flow of saliva, abdominal cramps, and unconsciousness. Atropine sulfate is antidotal.

Wear a pesticide respirator jointly approved by the Mining Enforcement and Safety Administration (formerly the U.S. Bureau of Mines) and by the National Institute for Occupational Safety and Health under the provisions of 30 CFR Part II. Wear goggles or face shield for eye protection when measuring or mixing. Wear long sleeved clothing and protective gloves when handling. Wash hands and face before eating or smoking. Bathe at the end of the work day. Change clothing daily and wash before reuse.

Keep all unprotected persons out of the operating area or vicinity where there may be danger of drift.

ENVIRONMENTAL HAZARDS

This product is toxic to fish, birds and other wildlife. Birds feeding on treated areas may be killed. For waterfowl protection, do not apply immediately before or during irrigation, or on fields in proximity of waterfowl nesting areas, or on fields where waterfowl are known to repeatedly feed. Drift and runoff from treated areas may be hazardous to fish in neighboring areas. Do not apply directly to water.

Notice: It is a Federal offense to use any pesticide in a manner that results in the death of a member of an endangered species.

The use of FURADAN 4 F may pose a hazard to the following Federally designated endangered/threatened species known to be found in certain areas within the named locations:

Attwater's Greater Prairie Chicken—Texas counties including: Aransas, Austin, Brazoria, Colorado, Galveston, Goliad, Harris, Refugio and Victoria

Aleutian Canada Goose—California counties including: Colusa, Merced, Stanislaus and Sutter

Kern Primrose Sphinx Moth—Walker Basin of Kern County, California

This product may not be used in areas where adverse impact on the Federally designated endangered/threatened species noted above is likely. Prior to making applications, the user of this product must determine that no such species are located in or immediately adjacent to the area to be treated. If the user is in doubt whether or not the above named endangered species may be affected, he should contact either the regional U.S. Fish and Wildlife Service office (Endangered Species Specialist) or personnel of the State Fish and Game office.

This product is highly toxic to bees exposed to direct treatment or residues on crops. Do not apply this product or allow it to drift to blooming crops or weeds if bees are visiting the treatment area. Protective information may be obtained from your Cooperative Agricultural Extension Service.

Carbofuran is a chemical which can travel (seep or leach) through soil and can contaminate ground water which may be used as drinking water. Carbofuran has been found in ground water as a result of agricultural use. Users are advised not to apply carbofuran where the water table (ground water) is close to the surface and where the soils are very permeable, i.e., well-drained soils such as loamy sands. Your local agricultural agencies can provide further information on the type of soil in your area and the location of ground water.

DIRECTIONS FOR USE

It is a violation of Federal law to use this product in a manner inconsistent with its labeling.

Do not use this product on Long Island, N.Y.

Do not apply this product in such a manner as to directly or through drift expose workers or other persons. The area being treated must be vacated by unprotected persons.

If prolonged intimate contact with corn and or sorghum foliage will result, do not re-enter treated fields within 14 days of application without wearing proper protective clothing. For all other situations, do not re-enter fields less than 24 hours following application, unless appropriate protective clothing is worn.

Protective clothing means, at least, a hat or other suitable head covering, a long sleeved shirt and long legged trousers or a coverall type garment (all of closely woven fabric covering the body, including the arms and legs), shoes and socks.

Because certain states may require more restrictive re-entry intervals for various crops treated with this product, consult your State Department of Agriculture for further information.

Written or oral warnings must be given to workers who are expected to be in treated area or in an area about to be treated with this product. Inform workers of areas or fields that may not be entered without specific protective clothing, period of time field must be vacated and appropriate actions to take in case of accidental exposure. An example of such information is given under written warnings. When oral warnings are given, warnings shall be given in a language customarily understood by workers. Oral warnings must be given if there is reason to believe that written warnings cannot be understood by workers. Written warnings must include the following information:

DANGER
Area treated with Furadan ® 4 F insecticide-nematicide on (insert date). Do not enter without appropriate protective clothing for 24 hours (insert State Department of Agriculture's reentry interval, if more restrictive).

In case of accidental exposure to pesticide spray, wash the skin thoroughly with the soap and water. Remove contaminated clothing and wash before reuse. If in eyes, flush with plenty of water. If inhaled, go to an area where the pesticide has not been applied. Get medical attention if needed.

SHAKE WELL BEFORE USING

STORAGE AND DISPOSAL

PESTICIDE STORAGE

Not for use or storage in or around the house.

Do not store below 35°F, (2°C).

Keep out of reach of children and animals. Store in original containers only. Store in cool, dry place and avoid excess heat. Carefully open containers. After partial use, replace lids and close tightly. Do not put concentrate or dilute material into food or drink containers. Do not contaminate other pesticides, fertilizers, water, food or feed by storage or disposal.

In case of spill, avoid contact, isolate area and keep out animals and unprotected persons. Confine spills. Call FMC collect: (716) 735-3765.

To confine spill: If liquid, dike surrounding area or absorb with sand, cat litter or commercial clay. If dry material, cover to prevent dispersal. Place damaged package in a holding container. Identify contents.

PESTICIDE DISPOSAL

Pesticide wastes are acutely hazardous. Improper disposal of excess pesticide, spray mixture, or rinsate is a violation of Federal Law. If these wastes cannot be disposed of by use according to label instructions, contact your State Pesticide or Environmental Control Agency, or Hazardous Waste representative at the nearest EPA Regional Office for guidance.

CONTAINER DISPOSAL

Metal Containers: Triple rinse (or equivalent). Then offer for recycling or reconditioning, or puncture and dispose of in a sanitary landfill, or by other procedures approved by state and local authorities. Do not cut or weld metal containers.

Plastic Containers: Triple rinse (or equivalent). Then offer for recycling or reconditioning, or puncture and dispose of in a sanitary landfill, or incineration, or, if allowed by state and local authorities, by burning. If burned, stay out of smoke.

Do not apply FURADAN 4 F through overhead irrigation systems. Do not rotate to a succeeding crop other than alfalfa, corn (field, pop or sweet), cotton, cucurbits (cucumbers, melons, pumpkins or squash), grapes, peanuts, peppers, potatoes, rice, small grains (barley, oats, wheat), sorghum, soybeans, strawberries, sugar beets, sugarcane and tobacco. Tomatoes, cabbage, peas, succulent beans and dry beans may be planted the following season provided the prior season's application did not exceed 8.7 pounds/acre of FURADAN 15 G, 13 pounds/acre of FURADAN 10 G or 2 pints 10 fluid ounces of FURADAN 4 F. Any other crop may be planted if it is not harvested or grazed.

ALFALFA: Alfalfa Weevil Larvae, Egyptian Alfalfa Weevil, Pea Aphids, and in N.Y. State for Snout Beetle Control—Apply the amount of FURADAN 4 F, indicated in the chart, when feeding is noticed or when insects appear. Alfalfa Weevil Adult—Apply 1 to 2 pints per acre when insects appear. Alfalfa Blotch Leafminer (Northeastern States only) and Potato Leafhopper. Apply 1 to 2 pints per acre when insects appear. Lygus Bugs—Apply 2 pints per acre prior to bloom. Grasshoppers—Apply ¼ to ½ pint per acre when grasshopper feeding is noticed. For control of Blue Alfalfa Aphid (nymphs and wingless adults)—Apply FURADAN 4 F at ½ to 1 pint per acre when insect feeding is noticed or when insects appear. Do not apply more than twice per season. Do not apply more than once per cutting. Do not use more than 1 pint per acre in the second application. Apply only to fields planted to pure stands of alfalfa. Do not move bees to alfalfa fields within 7 days of application. Observe the indicated number of days after application before cutting or grazing.

Pints of FURADAN 4 F Per Acre	Do not cut or graze within
½	7 days
1	14 days
2	28 days

Minimum gallonage requirements. Ten gallons of finished spray per acre with ground equipment, two gallons per acre with aircraft, except Blue Alfalfa Aphid use 30 gallons of finished spray per acre with ground equipment, 5 gallons per acre with aircraft.

For waterfowl protection, do not apply on fields in proximity of waterfowl nesting areas and/or fields where waterfowl are known to repeatedly feed.

COTTON: Thrips—Use Furadan 4 F at 2.5 fluid ounces per 1,000 linear feet of row (1 quart per acre with 40 inch row spacing). Apply in the seed furrow at planting. Furadan 4 F may be mixed with water or liquid fertilizer. When Furadan 4 F is used with liquid fertilizers, premix one part of Furadan 4F with two parts of water. Check physical compatibility before mixing large quantities. Add this premix to the tank of fertilizer along with the rinsings from the premixing container. Maintain agitation in the tank after mixing and during application. Do not mix until ready to use. Do not feed cotton forage.

FIELD CORN, POPCORN, SWEET CORN—AT PLANTING: Corn Rootworms, Flea Beetles, and to aid in the control of first generation European Corn Borer—Use 2.5 fluid ounces of FURADAN 4 F per 1,000 linear feet of row (1 quart per acre with 40 inch row spacing). Apply at planting, as a 7 inch band over the row or inject on each side of the row by mixing with water or liquid fertilizers.

Southern Corn Rootworm, Flea Beetles, Seed Corn Maggot, Wireworms, and to aid in the control of first generation European Corn Borer and Armyworm for approximately 4 to 6 weeks after planting—Use 2.5 fluid ounces of Furadan 4 F per 1,000 linear feet of row (1 quart per acre with 40 inch row spacing). Apply at planting directly into the seed furrow. If FURADAN 4 F is used with liquid fertilizers, premix 1 part of FURADAN 4 F with two parts of water, making certain that the mixture is physically compatible. (Do not prepare the premix until immediately before use.) Add the premix and rinsate from the "premix" container to the tank of fertilizer. Maintain agitation in the tank after mixing and during application. Do not mix until ready to use. Do not feed forage within 30 days of last application.

FIELD CORN, POPCORN—POST PLANT: Corn rootworms (northern, southern and western)—use 2.5 fluid ounces of FURADAN 4 F per 1,000 linear feet of row (1 quart per acre with 40 inch row spacing). Apply as a post emergent spray by banding over the row, or by side dressing or basal spraying both sides of the row after corn emerges. Control will generally be improved if the treatment is cultivated into the soil. Do not feed forage within 30 days of last application.

FIELD CORN—FOLIAR APPLICATION
European Corn Borer—Use 1½–2 pints of FURADAN 4 F per acre as a foliar spray when corn borer eggs begin to hatch. Use the higher rate for heavier pest infestations. For treatment with aerial equipment, apply as a broadcast spray using a minimum of 1 gallon of

finished spray per acre. For treatment with ground equipment, direct the spray into the corn whorl for first brood and into the ear zone for second brood, using a minimum of 10 gallons of finished spray per acre. Repeat if necessary. Observe all precautions listed below.

Southwestern Corn Borer—Use FURADAN 4 F at 1–2 pints per acre as a broadcast foliar spray when eggs begin to hatch. Rate used will depend on desired residual activity. If infestation continues, retreat in 7 days after a 1 pint application and within 14 days after a 2 pint application. Apply in sufficient water for thorough coverage using a minimum of 1 gallon of finished spray per acre with air equipment or 10 gallons of finished spray per acre with ground equipment. Observe all precautions listed below.

Banks Grass Mites (suppression)—FURADAN 4 F when applied at 2 pints per acre for the control of European or Southwestern Corn Borers will also suppress Banks Grass Mites.

Grasshoppers—Use FURADAN 4 F as a foliar spray at ¼–½ pint per acre when insects appear or feeding is noticed.

—Use the ¼ pint rate under light to moderate population levels (0–14 grasshoppers/sq. yd.).

—Use the ½ pint rate under more severe population levels (15 or more grasshoppers/sq. yd.).

Apply in sufficient water for thorough coverage using a minimum of 2 gallons of finished spray per acre with air equipment or 10 gallons of finished spray per acre with ground equipment. Observe all precautions listed below.

Do not make more than two applications per season at the 1½–2 pint use rate. Do not make more than four applications per season at the 1 pint use rate. Do not forage, cut or harvest within 30 days of last application. Do not apply on seed corn less than 14 days prior to detasseling or rogueing. If prolonged intimate contact will result, do not reenter treated field within 14 days of application without wearing proper protective clothing.

SWEET CORN—FOLIAR APPLICATION (Machine Harvested Only): European Corn Borers—For control of second generation borers apply 1 pint FURADAN 4 F per acre. Make first application just prior to first silking and repeat at weekly intervals. Do not make more than four (4) applications per season. Do not apply within seven (7) days of harvest. If prolonged, intimate contact will result, do not re-enter treated field within 14 days of application without wearing proper protective clothing. Do not graze or harvest stalks within 21 days of last application.

Do not make a foliar application if more than 12 ozs. of FURADAN 10 G per 1,000 linear feet of row (10 lbs. per acre with 40″ row spacing) or 8 ozs. of FURADAN 15 G per 1,000 linear feet of row (6.7 lbs. per acre with 40″ row spacing) or 2.5 fluid ounces of FURADAN 4 F per 1,000 linear feet of row (1 quart per acre with 40 inch row spacing) were used in an at-planting application.

Minimum gallonage requirements: 10 gallons of finished spray per acre with ground equipment, 2 gallons per acre with aircraft.

GRAPES (For use in California): Nematodes (Root-Knot and Dagger) and Grape Phylloxera—Apply FURADAN 4 F at 2½ gallons per acre as a broadcast treatment only, to the soil surface between the vine rows and immediately incorporate by mechanical means. Do not apply within 200 days of harvest. Remove dense weed growth prior to treatment. Do not use on soils of pH 8.0 or greater. Do not make more than one application per crop year.

Wear protective clothing including rubber gloves and impervious foot covering at time of application.

ORNAMENTALS—CONTAINER GROWN

Root Weevil Larvae—Prepare a stock solution by thoroughly mixing 1 to 2 fluid ounces of Furadan 4F per 100 gallons of water. Apply as a soil drench in sufficient volume to saturate the entire soil profile within each container. The following guidelines give approximate amounts to use on various sized containers. Make a single application when larvae are present (usually from July to mid-October in outdoor growing areas). Later application may be less effective due to lower temperatures and/or the presence of more mature larvae.

Container Size	Amount of Stock Solution Per Container
6 inches diameter	1 pint
8 inches diameter	1 to 2 pints
10 inches diameter	2 to 4 pints
12 inches diameter	3 to 6 pints
16 inches diameter	4 to 8 pints

Early marginal necrosis or leaf drop may occur on Hydrangea or Birch. Not all species or varieties of ornamentals have been tested. Before treating large numbers of plants of a particular variety, treat a few plants and observe prior to full scale application.

Do not handle treated plants, soil or containers that are wet with spray.

Application of Furadan 4 F through overhead sprinkler equipment is prohibited.

PINE SEEDLINGS: Pales Weevil and Pitch-Eating Weevil in pine plantations—Apply a 1% (W/W) active FURADAN clay slurry (see following for preparation) to the roots of pine seedlings prior to transplanting. Treat seedlings by dipping roots or use any other suitable means which allows thorough coating. Keep roots moist until transplanted. Prepare the slurry as follows: Add 1.6 ounces (2½ tablespoons) of FURADAN 4 F to ½ gallon of water. Mix thoroughly. Add 2 pounds of pulverized kaolin clay (pH 4.5) to this suspension. Mix thoroughly. This is sufficient to treat the roots of 150 to 200 seedlings. Adequate ventilation is required for indoor treatment. Wear protective unlined rubber gloves at time of application. Users should avoid contamination of the inside of the gloves. If contamination of the inside of the gloves is suspected, the gloves should be immediately discarded.

POTATOES: Potato Tuberworm (Virginia only); Colorado Potato Beetle, European Corn Borer, Potato Flea Beetle, Potato Leafhopper—Use 1 to 2 pints of FURADAN 4 F in sufficient water to treat one acre. Apply when insects first appear and repeat as necessary to maintain control. Do not make more than 8 foliar applications per season. Do not apply more than 6 pints to the foliage if either FURADAN 10 G or FURADAN 15 G was used at planting. Do not apply more than 1 quart per application. Do not apply within 14 days of harvest. Minimum gallonage requirements: 10 gallons of finished spray per acre with ground equipment and 3 gallons per acre with aircraft.

Do not use this product on Long Island, New York.

For waterfowl protection do not apply:

—immediately before or during furrow irrigation

—on fields in proximity of waterfowl nesting areas

—on fields where waterfowl are known to repeatedly feed

SMALL GRAINS (Wheat, Oats, Barley): Grasshoppers—Use FURADAN 4 F at ¼ to ½ pint per acre when insects appear. Cereal Leaf Beetle—Use ½ pint per acre when insects appear. Apply before heads emerge from boot. Do not make more than two applications per season. Do not feed treated forage to livestock.

Minimum gallonage requirements: Ten gallons of finished spray per acre with ground equipment, two gallons per acre with aircraft.

For waterfowl protection do not apply on fields in proximity of waterfowl nesting areas and/or on fields where waterfowl are known to repeatedly feed.

SOYBEANS—AT PLANTING APPLICATION: Nematodes (Root-Knot, Cyst, Stunt, Ring, Sting, Spiral, Lesion, Lance, Dagger, Stubby-Root)—Use 3.75–5 fluid ounces per 1,000 linear feet of row (3–4 pints per acre with 40-inch row spacing). Apply at planting as a 12-inch surface band. Soybeans may be grazed or cut for forage 30 days or later following an at planting application.

SOYBEANS—FOLIAR APPLICATION: Mexican Bean Beetle—Use FURADAN 4 F at ½–1 pint per acre when insects appear. Grasshoppers—Use FURADAN 4 F at ¼–½ pint per acre when insects appear. Do not use FURADAN 4 F as a foliar application if FURADAN 10 G, FURADAN 15G, FURADAN 4 F was applied to soybeans at planting time. Do not make more than 2 foliar applications per season. Do not apply within 21 days of harvest. Do not graze or feed foliar-treated forage to livestock or cut for silage or hay. Minimum gallonage requirements: 20 gallons of finished spray per acre with ground equipment, 1½ gallons per acre with aircraft.

STRAWBERRY: (For use in Washington & Oregon) Root Weevils —Use FURADAN 4 F at 2 to 4 pints per acre (2.6 to 5.1 fluid ounces per 1000 linear feet of row for 42 inch row spacing). Apply as a 10 to 12 inch band over the row after last harvest but before October 1st. Do not make more than one application per season. Do not apply if berries are present.

SUGARCANE: Sugarcane Borer—Apply 1 to 1½ pints FURADAN 4 F per acre using ground or aerial equipment. Check sugarcane fields weekly, beginning in early June and continuing through August. Make first application only after visible joints form and 5% or more of the plants are infested with young larvae feeding in or under the leaf sheath and which have not bored into the stalks. Repeat whenever field checks indicate the infestation exceeds 5 percent. Do not apply within 17 days of harvest. Do not use in Hawaii.

SUNFLOWERS (Confectionary and Oil)—At Planting
Sunflower Stem Weevil, Sunflower Beetle, Grasshoppers—Use 2.5 to 5.0 fluid ounces of Furadan 4 F per 1000 linear feet of row (1.4 to 2.8 quarts per acre with 30 inch row spacing). Apply directly into the seed furrow—OR—Apply in a 7 inch band. Furadan 4 F may be mixed with water or liquid fertilizer. When Furadan 4 F is used with liquid fertilizers, premix one part of Furadan 4 F with two parts of water. Check physical compatibility before mixing large quantities. Add this premix to the tank of fertilizer along with the rinsings from the premixing container. Maintain agitation in the tank after mixing and during application. Do not mix until ready to use.

Use the higher rate when heavier insect infestations are anticipated.

SUNFLOWERS (Confectionary and Oil)—Foliar Application
Sunflower Moth, Banded Sunflower Moth, Stem Weevils, Seed Weevils—Use Furadan 4 F at 1 pint per acre. Grasshoppers—Use Furadan 4 F at ¼ to 1 pint per acre. Sunflower Beetle—Use Furadan 4 F at ¼ to ½ pint per acre. Apply as a foliar spray when insects appear. When a rate range is indicated, use higher rate for heavier insect infestations. Repeat applications as necessary but not more than four times per season. Apply a minimum of two gallons of finished spray per acre with aircraft and 10 gallons by ground equipment. If prolonged, intimate contact will result, do not re-enter treated field within 14 days of application without wearing proper protective clothing. Do not harvest crop within 28 days of last application.

TOBACCO—BURLEY: Flea Beetles—Use 1 gallon of FURADAN 4 F per acre. Apply as a broadcast spray over the soil surface prior to transplanting and incorporate into the top 3 inches of soil with a suitable device.

TOBACCO—FLUE-CURED: Flea Beetles, Wireworms, and to aid in the control of Budworms and Root-Knot Nematodes—Use 1½ gallons of FURADAN 4 F per acre—OR—Use 1 gallon of FURADAN 4 F per acre for control of Flea Beetles only. Apply as a broadcast spray over the soil surface prior to forming beds. Incorporate into the top 3 inches of soil. Form beds and plant as usual. This product may induce flecking of the bottom or lower leaves.

DEALERS SHOULD SELL IN ORIGINAL PACKAGES ONLY.

TERMS OF SALE OR USE: On purchase of this product buyer and user agree to the following conditions:

WARRANTY: FMC Makes no warranty, expressed or implied, concerning the use of this product other than indicated on the label. Except as so warranted, this product is sold as is. Buyer and user assume all risk of use and/or handling and/or storage of this material when such use and/or handling and/or storage is contrary to label instructions.

DIRECTIONS AND RECOMMENDATIONS: Follow directions carefully. Timing and method of application, weather and crop conditions, mixture with other chemicals not specifically recommended, and other influencing factors in the use of this product are beyond the control of the seller and are assumed by buyer at his own risk.

USE OF PRODUCT: FMC's recommendations for the use of this product are based upon tests believed to be reliable. The use of this product being beyond the control of the manufacturer, no guarantee, expressed or implied, is made as to the effects of such or the results to be obtained if not used in accordance with directions or established safe practice.

DAMAGES: Buyer's or user's exclusive remedy for damages for breach of warranty or negligence shall be limited to direct damages not exceeding the purchase price paid and shall not include incidental or consequential damages.

FURADAN® 5G

INSECTICIDE

RESTRICTED USE PESTICIDE

For retail sale to and application only by Certified Applicators or personnel under their direct supervision, and only for those uses covered by the Certified Applicator's certification.

For Agricultural or Commercial Use Only

ACTIVE INGREDIENTS:	
*Carbofuran	5.0%
INERT INGREDIENTS:	95.0%
TOTAL	100.0%

* 2,3-Dihydro-2, 2-dimethyl-7-benzofuranyl methylcarbamate

EPA Reg. No. 279-2874 AA EPA Est., 279-
U.S. Patent No. 3,474,171 Furadan—FMC Trademark

WARNING

Keep Out of Reach of Children

STATEMENT OF PRACTICAL PURPOSE

If swallowed, drink 1 or 2 glasses of water and induce vomiting by touching back of throat with finger. Do not induce vomiting or give anything by mouth to an unconscious person. Get medical attention.

If in eyes, flush with plenty of water for at least 15 minutes. Get medical attention.

NOTE TO PHYSICIAN:
FURADAN is a reversible cholinesterase inhibitor. Do not use oximes such as 2-PAM.
Start by giving 2 mg. atropine intramuscularly. According to clinical response, continue until signs of atropinization occur (dry mouth or dilated pupils).
If in eye, instill one drop of homatropine.

For Emergency Assistance Call 716–735-3765.

See side panel for additional precautionary information.

PRECAUTIONARY STATEMENTS

HAZARDS TO HUMANS (AND DOMESTIC ANIMALS)

WARNING

May be fatal if swallowed or absorbed through eyes. Do not get in eyes or on clothing. Avoid contact with skin. Wash thoroughly with soap and water after handling and before eating or smoking. Remove contaminated clothing and wash before reuse.

ENVIRONMENTAL HAZARDS

This pesticide is toxic to fish, birds and other wildlife. Birds feeding on treated areas may be killed. Birds killed by carbofuran pose a hazard to hawks and other birds-of-prey; bury or otherwise dispose of dead birds to prevent poisoning of other wildlife. Cover or incorporate granules in spill areas. Runoff from treated areas may be hazardous to fish in neighboring areas. Fish may be killed at recommended rates. Do not contaminate wells, wetlands or any body of water by cleaning of equipment or disposal of waste.

Carbofuran is a chemical which can travel (seep or leach) through soil and can contaminate ground water which may be used as drinking water. Carbofuran has been found in ground water as a result of agricultural use. Users are advised not to apply carbofuran where the water table (ground water) is close to the surface and where the soils are very permeable, i.e., well-drained soils such as loamy sands. Your local agricultural agencies can provide further information on the type of soil in your area and the location of ground water.

DIRECTIONS FOR USE

It is a violation of Federal law to use this product in a manner inconsistent with its labeling.

Do not re-enter treated fields less than 24 hours following application.

Do not use this product on Long Island, N.Y.

STORAGE AND DISPOSAL

PESTICIDE STORAGE
Do not store in or around the home.

Keep out of reach of children and animals. Store in original containers only. Store in a cool, dry place and avoid excess heat. Carefully open containers. After partial use, fold and roll back bags, clamp and close tightly. Do not put concentrate or dilute material into food or drink containers. Do not contaminate other pesticides, fertilizers, water, food or feed by storage or disposal.

In case of spill, avoid contact, isolate area and keep out animals and unprotected persons. Confine spills. Call FMC collect: (716) 735-3765.

To confine spill: If liquid, dike surrounding area or absorb with sand, cat litter or commercial clay. If dry material, cover to prevent dispersal. Place damaged package in a holding container. Identify contents.

PESTICIDE DISPOSAL
Pesticide wastes are acutely hazardous. Improper disposal of excess pesticide, spray mixture, or rinsate is a violation of Federal Law. If these wastes cannot be disposed of by use according to label instructions, contact your State Pesticide or Environmental Control Agency, or the Hazardous Waste representative at the nearest EPA Regional Office for guidance.

CONTAINER DISPOSAL
Completely empty bag into application equipment. Then dispose of empty bag in a sanitary landfill or by incineration, or, if allowed by State and local authorities, by burning. If burned, stay out of smoke.

RICE: To control Rice Water Weevil and Dark Rice Field Mosquito in Arkansas, Louisiana, Mississippi and Texas apply 10 pounds of FURADAN 5 G per acre one day before or within 21 days after flooding. Do not make more than one application per season. The subsequent use of propanil herbicide may result in crop injury.

For effective control of the dark rice field mosquito, application must be made within 2 to 4 days after flooding.

To control Rice Water Weevil in California apply 10 pounds of FURADAN 5 G per acre prior to flooding and planting. Do not make more than one application per season. The subsequent use of propanil herbicide may result in crop injury.

SUGARCANE: Sugarcane borer—Apply 10 to 15 pounds FURADAN 5 G per acre using ground or aerial equipment. Check sugarcane fields weekly, beginning in early June and continuing through August. Make first application only after visible joints form and 5% or more of the plants are infested with young larvae feeding in or under the leaf sheath and which have not bored into the stalks. Repeat whenever field checks indicate the infestation exceeds 5 per cent. Do not apply within 17 days of harvest. Do not use in Hawaii.

DEALERS SHOULD SELL IN ORIGINAL PACKAGES ONLY.

TERMS OF SALES OR USE On purchase of this product buyer and user agree to the following conditions:

WARRANTY: FMC makes no warranty, expressed or implied, concerning the use of this product other than indicated on the label. Except as so warranted, the product is sold as is. Buyer and user assume all risk of use and/or handling and/or storage of this material when such use and/or handling and/or storage is contrary to label instructions.

DIRECTIONS AND RECOMMENDATIONS: Follow directions carefully. Timing and method of application, weather and crop conditions, mixture with other chemicals not specifically recommended, and other influencing factors in the use of this product are beyond the control of the seller and are assumed by buyer at his own risk.

USE OF PRODUCT: FMC's recommendations for the use of this product are based upon tests believed to be reliable. The use of this product being beyond the control of the manufacturer, no guarantee, expressed or implied, is made as to the effects of such or the results to be obtained if not used in accordance with directions or established safe practice.

DAMAGES: Buyer's or user's exclusive remedy for damages for breach of warranty or negligence shall be limited to direct damages not exceeding the purchase price paid and shall not include incidental or consequential damages.

FURADAN® 10G

INSECTICIDE–NEMATICIDE

For Agricultural or Commercial Use Only

ACTIVE INGREDIENT:	BY WT.
*Carbofuran	10.0%
INERT INGREDIENTS:	90.0%
TOTAL	100.0%

* 2,3-Dihydro-2,2-dimethyl-7-benzofuranyl methylcarbamate

EPA Reg. No. 279-2712 AA EPA Est., 279-
U.S. Patent No. 3,474,171 Furadan—FMC Trademark

WARNING

Keep Out of Reach of Children

STATEMENT OF PRACTICAL TREATMENT

If swallowed, drink 1 or 2 glasses of water and induce vomiting by touching back of throat with finger. Do not induce vomiting or give anything by mouth to an unconscious person. Get medical attention.

If in eyes, flush with plenty of water for at least 15 minutes. Get medical attention.

NOTE TO PHYSICIAN: CARBOFURAN is a reversible cholinesterase inhibitor. Do not use oximes such as 2-PAM. Start by giving 2 mg. atropine intramuscularly. According to clinical response, continue until signs of atropinization occur (dry mouth or dilated pupils). If in eye, instill one drop of homatropine.

For Emergency Assistance Call 716–735-3765.

See other panels for additional precautionary information.

PRECAUTIONARY STATEMENTS

HAZARDS TO HUMANS (& DOMESTIC ANIMALS)

WARNING
May be fatal if swallowed. Do not get in eyes or on clothing. Avoid contact with skin. Wash thoroughly with soap and water after handling and before eating or smoking. Remove contaminated clothing and wash before reuse.

ENVIRONMENTAL HAZARDS
This pesticide is toxic to fish and wildlife. Use with care when applying in areas frequented by wildlife or adjacent to any body of water. Birds feeding on treated areas may be killed. Keep out of lakes, ponds or streams. Do not contaminate water by cleaning of equipment or disposal of wastes. Cover or incorporate granules

which are spilled during loading and which are visible on soil surfaces in turn areas.

Notice: It is a Federal offense to use any pesticide in a manner that results in the death of a member of an endangered species.

The use of FURADAN 10 G may pose a hazard to the following Federally designated endangered/threatened species known to be found in certain areas within the named locations:

Attwater's Greater Prairie Chicken—Texas counties including: Aransas, Austin, Brazoria, Colorado, Galveston, Goliad, Harris, Refugio and Victoria

Aleutian Canada Goose—California counties including: Colusa, Merced, Stanislaus and Sutter

Kern Primrose Sphinx Moth—Walker Basin of Kern County, California

This product may not be used in areas where adverse impact on the Federally designated endangered/threatened species, noted above, is likely. Prior to making applications, the user of this product must determine that no such species are located in or immediately adjacent to the area to be treated. If the user is in doubt whether or not the above named endangered species may be affected, he should contact either the regional U.S. Fish and Wildlife Service office (Endangered Species Specialist) or personnel of the State Fish and Game office.

Carbofuran is a chemical which can travel (seep or leach) through soil and can contaminate ground water which may be used as drinking water. Carbofuran has been found in ground water as a result of agricultural use. Users are advised not to apply carbofuran where the water table (ground water) is close to the surface and where the soils are very permeable, i.e., well-drained soils such as loamy sands. Your local agricultural agencies can provide further information on the type of soil in your area and the location of ground water.

DIRECTIONS FOR USE

It is a violation of Federal Law to use this product in a manner inconsistent with its labeling.

Do not use this product on Long Island, N.Y.

If prolonged intimate contact with corn and/or sorghum foliage will result, do not re-enter treated fields within 14 days of application without wearing proper protective clothing. For all other situations, do not re-enter fields less than 24 hours following application.

Cover or incorporate granules in spill areas.

STORAGE AND DISPOSAL

PESTICIDE STORAGE

Do not store in or around the home.

Keep out of reach of children and animals. Store in original containers only. Store in a cool, dry place and avoid excess heat. Carefully open containers. After partial use, fold and roll back bags, clamp and close tightly. Do not put concentrate or dilute material into food or drink containers. Do not contaminate other pesticides, fertilizers, water, food or feed by storage or disposal.

In case of spill, avoid contact, isolate area and keep out animals and unprotected persons. Confine spills. Call FMC collect: (716) 735-3765.

To confine spill: If liquid, dike surrounding area or absorb with sand, cat litter or commercial clay. If dry material, cover to prevent dispersal. Place damaged package in a holding container. Identify contents.

PESTICIDE DISPOSAL

Pesticide wastes are acutely hazardous. Improper disposal of excess pesticide, spray mixture, or rinsate is a violation of Federal Law. If these wastes cannot be disposed of by use according to label instructions, contact your State Pesticide or Environmental Control Agency, or the Hazardous Waste representative at the nearest EPA Regional Office for guidance.

CONTAINER DISPOSAL

Completely empty bag into application equipment. Then dispose of empty bag in a sanitary landfill or by incineration, or, if allowed by State and local authorities, by burning. If burned, stay out of smoke.

USE CAUTION

Do not rotate to a succeeding crop other than alfalfa, corn (field, pop or sweet), cotton, cucurbits (cucumbers, melons, pumpkins or squash), grapes, peanuts, peppers, potatoes, rice, small grains (barley, oats, wheat), sorghum, soybeans, strawberries, sugar beets, sugarcane and tobacco. Tomatoes, cabbage, peas, succulent beans and dry beans may be planted the following season provided the prior season's application did not exceed 8.7 lbs. per acre of Furadan® 15 G insecticide-nematicide, 13 lbs. per acre of Furadan® 10 G insecticide-nematicide or 2 pt. 10 fl. oz. per acre, of Furadan® 4 F insecticide-nematicide. Any other crop may be planted if it is not harvested or grazed.

FIELD CORN, POPCORN, SWEET CORN—CONVENTIONAL, MINIMUM & NO TILLAGE

Corn Rootworms (Northern, Southern & Western), Flea Beetles, Wireworms, aid in the control of White Grubs and first generation European Corn Borer, reduce losses due to Stalk Rot by reducing the incidence of insect wounds which permit entry by the Stalk Rot Fungi; Armyworm control for approximately 4 to 6 weeks after planting and suppression of cutworms—Use 12 ounces of FURADAN 10 G per 1000 linear feet of row (10 lbs. per acre with 40 inch row spacing). Early season suppression of Common Stalk Borer — Use 24–36 ounces per 1000 linear feet of row (20 to 30 lbs. per acre with 40 inch row spacing). Apply in a 7 inch band ahead of the planter press wheel and incorporate into the top 1 inch of soil by use of special covering devices or by dragging a short length of chain. —OR—Direct the granules into the planter shoe with the seed. —OR—Place the applicator tube directly behind the planter shoe so that the granules drop into the furrow and mix with the covering soil.

Lesion Nematodes (Northeast and Midwest Corn States including Colorado and Wyoming)—Use 12–24 ounces per 1000 linear feet of row (10–20 lbs. per acre with 40 inch row spacing). Apply "at planting" as above (in a covered band or in the seed furrow).

European Corn Borer (first generation)—Use 24–36 ounces per 1000 linear feet of row (20–30 lbs. per acre with 40 inch row spacing) "at planting" as above (in a covered band or in the seed furrow).

Seed Corn Maggot—Use 12 ounces per 1000 linear feet of row (10 lbs. per acre with 40 inch row spacing). Southwestern Corn Borer (second to third generation)—Use 18–36 ounces per 1000 linear feet of row (15 to 30 lbs. per acre with 40 inch row spacing). Southern Corn Billbug (Southeastern States), and to aid in the control of Leafhoppers and thereby reduce losses due to Maize Chlorotic Dwarf Virus and Corn Stunt diseases—Use 24–36 ounces per 1000 linear feet of row (20 to 30 lbs. per acre with 40 inch row spacing). Apply into the seed furrow as noted above.

Nematodes (Sting, Stunt, Stubby Root, Root Knot, Dagger, Lesion, Lance, and Spiral)—Use 18–24 ounces per 1000 linear feet of row (15–20 lbs. per acre with 40 inch row spacing). Apply at planting time in a 7 to 15 inch band and incorporate into the top 3 inches of soil.

FIELD CORN, POPCORN—POST PLANT

Northern & Western Corn Rootworm—Use 12 ounces of FURADAN 10 G per 1000 linear feet of row (10 lbs. per acre with 40 inch row spacing) per acre by banding over the row or by side dressing both sides of the row and cultivating into the soil.

FIELD CORN—FOLIAR APPLICATION

Southwestern Corn Borers and European Corn Borers—Use 10 lbs. per acre. Apply with aircraft by broadcasting over the corn plants or with ground equipment by directing the granules into the corn whorls. Apply when eggs begin to hatch. Do not make a foliar application if FURADAN 10 G was applied at more than 12 ounces per 1000 linear feet of row (10 lbs. per acre with 40 inch row spacing) at planting. Do not make more than two foliar applications per season.

CORN INSECTICIDE APPLICATOR SETTINGS

FURADAN 10 G is a dense, free-flowing insecticide requiring accurate equipment calibration to apply the recommended amount per acre. Always calibrate your planter at the speed which you plan to travel during the planting operation. The following are approximate settings for guidance in calibration:

Applicator Gauge Setting

Application Rate	10 lbs./A			15 lbs./A			20 lbs./A		
Planter Speed (MPH)	4	6	8	4	6	8	4	6	8
Equipment									
Allis Chalmers	5	5	5	8	8	8	10	10	10
Gandy 901 Series									
Snap on	18	21	24	21	26	29	24	29	34
Bolt on	17	21	24	21	26	30	24	30	34
IHC Cyclo Planter									
Gate	1	1	2	1	2	3	2	3	4
Dial	6	9	4	9	6	3	4	3	2
John Deere									
Max Emerge (even numbered notches —0-10-20-30-40)	11	14	17	14	19	22	17	22	26
Max Emerge (odd numbered notches —5-15-25-35-45)	13	19	24	19	26	31	24	31	35
John Deere									
Range	1	1	2	2	2	2	2	2	3
Notch	26	33	10	1	12	24	8	24	6
Noble									
Round Opening	8	11	13	11	15	18	13	18	23
Rectangular Opening	11	15	20	15	22	27	20	27	33

CORN APPLICATION RATES

When the granular application equipment is set to deliver FURADAN 10 G at a rate of 12 ounces per 1000 linear feet, the total amount applied per acre will vary depending on row spacing. Use the table below to determine your FURADAN needs for each acre.

Row Spacing (Inches)	40″	38″	36″	30″	28″
FURADAN 10 G (lbs./acre)	10	10.5	11	13	14

COTTON

Thrips—Use FURADAN 10 G at 10 pounds per acre (40 inch row spacing); 12 ounces per 1000 linear feet. Apply in the seed furrow at planting. Do not feed cotton forage.

CRANBERRIES

For control of Black Vine Weevils—Use FURADAN 10 G at 20 pounds per acre. For control of adult weevils and larvae, spread granules uniformly over soil surface using a rotary spreader. Make one application about June 15–25 and repeat one month later. Incorporate granules by following each application with sprinkler irrigation. For use only on cranberry bogs that are sprinkler irrigated and are not flooded for purposes of irrigation or harvest. Do not apply within 60 days of harvest. For use only in the State of Washington.

CUCURBITS (Cucumber, Melons, Squash, Pumpkins)

Nematodes (Root-Knot, Sting, Lance)—For use in Southeastern States—Use FURADAN 10 G at the rate of 2¼ pounds per 1000 linear feet of row (20 pounds per acre based on 60 inch row spacing). Apply in a 12 to 15 inch band and incorporate into the top 3 inches of soil.

GRAPES (For use in California)

For control of Nematodes (Root-Knot, Dagger and Lesion) and Grape Phylloxera—Broadcast FURADAN 10 G at 100 pounds per acre over the soil surface between the vine rows and immediately incorporate. Do not apply within 200 days of harvest. Remove dense weed growth prior to treatment. Do not use on soils of pH 8.0 or greater. Retreat as necessary but do not make more than one application per crop year.

PEANUTS

Do not feed treated peanut forage to dairy animals or animals being finished for slaughter.

In-furrow at planting application method:

Thrips, Potato Leafhopper and as an aid in the control of Southern Corn Rootworm in Southeastern States—Use in-furrow at planting, 5 to 10 lbs. per acre (36 inch row spacing); ⅓ to ⅔ lb. per 1000 linear feet of row.

Row banding application method:

Nematodes (Root Knot, Sting, Stunt, Lesion, Ring) in Southeastern States and in the Southwestern States of Texas and Oklahoma; and in the Southeastern States to aid in the control of Potato Leafhopper and Southern Corn Rootworm and for control of Thrips—Apply FURADAN 10 G in a band over the row prior to planting and incorporate 3 inches into the soil. Use the rate indicated below for various band widths. Additional control measures may be needed late in the season for Potato Leafhopper.

Band Width	Rate/Acre (36 inch rows)	Rate/1000 linear feet
12 inches	20 to 40 lbs.	1⅓ to 2¾ lbs.
18 inches	30 to 50 lbs.	2 to 3½ lbs.

Banded Pegging—Time Application Method

For control of Southern Corn Rootworm in the Southeastern States —Use FURADAN 10 G at 11 to 22 ounces per 1000 linear feet (10–20 pounds per acre with 36 inch row spacing). Apply on a 14 to 18 inch band over the row when pegs begin to form or up to 30 days after pegging begins (approximately 60 to 90 days after planting). Incorporate granules immediately if vine growth permits. Do not apply FURADAN 10 G at pegging *if FURADAN 10 G was used in a band at planting.*

PEPPERS

Green Peach Aphid—(Delmarva Peninsula and South Jersey only). European Corn Borer—Make two applications of FURADAN 10 G by sidedressing on either one or both sides of the row. Make a first application two to four weeks after transplanting using 20 lbs. per acre (equivalent to ¾ lb. per 500 linear feet with 38 inch rows). Make the second application four to six weeks later using 30 lbs. per acre (1 lb. per 500 linear feet). Do not make the second application within 21 days of harvest.

POTATOES

Colorado Potato Beetle, European Corn Borer, Potato Flea Beetle, Potato Leafhopper, Tuber Flea Beetle; Green Peach Aphid (Western and Northeastern States including the Delmarva Peninsula); Potato Tuberworm (Virginia only); and to aid in the control of Wireworm —Use 2¼ lbs. of FURADAN 10 G per 1000 linear feet of row (30 lbs. per acre 38-inch row spacing). Apply directly into the bottom of the furrow during the planting operation.

SORGHUM

Chinch Bug, Corn Leaf Aphid, Flea Beetle, Greenbug, Lesion Nematode, Seedcorn Maggot, Southern Corn Rootworm, Wireworm, Yellow Sugarcane Aphid and aids in the control of White Grubs and Cutworms—Use only on grain sorghum or sorghum grown for forage. Do not graze or harvest sorghum grown for forage within 75 days of application.

For Corn Leaf Aphid, Greenbug, Yellow Sugarcane Aphid—Use 9 to 12 ounces of Furadan® 10 G insecticide-nematicide per 1000 linear feet of row (7.5 to 10 lbs. per acre with 40 inch row spacing) applied directly into the seed furrow—OR—Use 12 ounces of Furadan 10G per 1000 linear feet of row (10 lbs. per acre with 40 inch row spacing) applied in a 7 inch band over the row and incorporate the granules into the top 1 inch of soil by the use of a press wheel, dragging a short length of chain or suitable device.

For Chinch bug, Flea Beetle, Southern Corn Rootworm, Lesion Nematode, Wireworm, Seedcorn Maggot (in furrow application only), and to aid in the control of White Grubs and Cutworms—Use 12 ounces of Furadan 10 G per 1000 linear feet of row (10 lbs. per acre with 40 inch row spacing). Apply at planting as noted above.

Equivalent pounds of FURADAN 10 G per acre for various row spacings are as follows:

	Amount of FURADAN 10 G per 1000 Linear Feet of Row	Pounds of FURADAN 10 G per Acre for Various Row Spacings				
		40"	36"	34"	30"	20"
In Furrow	9 Ounces	7.5	8.25	8.75	10	15
	12 Ounces	10	11	11.50	13	20
7" Band	12 Ounces	10	11	11.50	13	20

When a rate range is indicated, use the higher rate when heavier pest infestations are anticipated.

SOYBEANS

Nematodes (Rootknot, Cyst, Stunt, Ring, Sting, Spiral, Lesion, Lance, Dagger, Stubby-Root)—Use FURADAN 10 G at 18–24 ounces/1000 linear feet (equivalent to an application of 15–20 lbs./acre on a 40-inch row spacing) at planting in a 7-inch band and mix with the covering soil.

—OR—

Apply FURADAN 10 G at 18 ounces/1000 linear feet (equivalent to an application of 15 lbs./acre on a 40-inch row spacing). Apply at planting by directing the granules into the planter shoe with the seed or by placing the applicator tube directly behind the planter shoe so that the granules drop into the furrow and mix with the covering soil.

Mexican Bean Beetle (larvae)—Use FURADAN 10 G at 18 ounces/1000 linear feet (equivalent to 15 lbs./acre on a 40-inch row spacing) at planting in a 7-inch band and mix with the covering soil.

—OR—

Apply FURADAN 10 G at 12–18 ounces/1000 linear feet (equivalent to an application of 10–15 lbs./acre on a 40-inch row spacing). Apply at planting by directing the granules into the planter shoe with the seed or by placing the applicator tube directly behind the planter shoe so that the granules drop into the furrow and mix with the covering soil.

SUGAR BEETS

Sugar Beet Root Maggot—Use 13 ounces of Furadan 10 G per 1,000 linear feet of row (20 lbs. per acre using 22 inch row spacing). Apply granules in a 6 to 7 inch band and incorporate into the top 1 inch of soil using a suitable device or use 6.5 ounces of Furadan 10 G per 1,000 linear feet of row (10 lbs. per acre using 22 inch row spacing) by applying granules in a "modified in-furrow" application. (Drop granules from a tube in a narrow band of 1 to 2 inches in front of the planter disks and allow some granules to fall into seed furrow with seed.)

SUGAR BEET APPLICATION RATES

When the granular application equipment is set to deliver Furadan 10 G at rates of 6.5, or 13 ounces per 1,000 linear feet, the total amount applied per acre will vary depending on row spacing. Use the table below to determine your Furadan needs for each acre.

Ounces of Furadan 10G per 1,000 Linear Feet of Row	Pounds of Furadan 10G per Acre for Various Row Spacings 20"	22"	24"	30"
6.5	10.5	9.5	9	7
13.0	21	19	17.5	14

SUGARCANE

Apply FURADAN 10 G at the rate of $2\frac{1}{2}$ to 5 lbs. per 1000 linear feet of row (equivalent to 20 to 40 lbs. per acre based on 60 inch row spacings). Do not use in Hawaii.

Plant cane

Nematodes (Root Knot and Stunt), Wireworm—Apply the above rate as a 15 inch band directly over the planted cane and cover with soil.

Stubblecane

Wireworm (Florida only), Nematodes (Root Knot & Stunt) — Apply the above rate as a 15 inch band over the stubble row. Use any time after harvest until regrowth reaches 18 inches. Cover with a 1 to 2 inch layer of soil. Covering with more than 2 inches of soil may reduce stand.

SUNFLOWERS (Confectionary and Oil)

Sunflower Stem Weevils, Grasshoppers—Use 12–24 ounces of FURADAN 10 G per 1,000 linear feet of row (13 to 26 lbs. per acre with 30 inch row spacing). Apply at planting by directing the granules into the planter shoe with the seed or by placing the applicator tube directly behind the planter shoe so that the granules drop into the furrow and mix with the covering soil—OR—apply at planting in a 7 inch band and mix with the covering soil. Sunflower Beetle—Use 12 to 24 ounces of Furadan 10 G per 1,000 linear feet of row (13 to 26 lbs. per acre with 30 inch row spacing) and apply in furrow as noted above—OR—use 24 ounces of Furadan 10 G per 1,000 linear feet of row (26 lbs. per acre with 30 inch row spacing) and apply at planting in a 7 inch band as noted above. When a rate range is indicated, use the higher rate when heavier insect infestation is anticipated.

TOBACCO—ALL TYPES EXCEPT FLUE-CURED

For control of Flea Beetles, Hornworms and to aid in the control of Budworms—Use FURADAN 10 G at 60 pounds per acre. Broadcast granules over the soil surface prior to transplanting and incorporate into the top 3 inches with a suitable device.

For full season control of Flea Beetles only—Use FURADAN 10 G at 30–40 pounds per acre—OR—for full season control of Flea Beetles and approximately 60 days control of Hornworms—Use FURADAN 10 G at 40 pounds per acre. Apply broadcast as noted above.

TOBACCO—FLUE-CURED

For control of Stunt Nematodes, Flea Beetles, Hornworms, Wireworms and to aid in the control of Root Knot Nematodes and Budworms—Use FURADAN 10 G at 60 pounds per acre. Broadcast granules over the soil surface prior to transplanting and incorporate into the top 3 inches with a suitable device. Form beds and plant as usual. This product may induce flecking of the bottom or lower leaves.

For full season control of Flea Beetles and approximately 60 days control of Hornworms—Use FURADAN 10 G at 40 pounds per acre. Apply broadcast as noted above. Note leaf flecking caution listed above.

COTTONWOOD

Cottonwood Twig Borer, Cottonwood Leaf Beetle, Clearwing Borer—**In Commercial Plantations** use 10 to 20 lbs. per acre (12 ft. row spacing) or $2\frac{3}{4}$ to $5\frac{1}{2}$ lbs. per 1000 linear feet of row. Apply during June or July to the root zone of the cutting by the use of a granular subsoil applicator as a sidedress 10 to 12 inches from the trees in a continuous band on both sides of the trees. **In Nursery Plantings** use 40 lbs. per acre (40 inch row spacing) or 3 lbs. per 1000 linear feet of row. Apply during May or June to the root zone of the cutting, sidedressing 10 to 12 inches from the trees in a continuous band on both sides of the trees.

PINE SEEDLINGS

Pales Weevil and Pitch-Eating Weevil control in pine plantations — Use 10 grams (1 teaspoon) per seedling at transplanting. Distribute the granules in a 6 inch radius on the soil around each seedling. Cover granules with soil. Do not use at planting if seedlings were treated with FURADAN 4 F slurry root dip.

SOUTHERN PINE SEED ORCHARDS

Seedbugs, Coneworms, Coneborers—Use 4 to 8 ounces per inch of tree diameter. Cone Beetles—Use 8 ounces per inch of tree diameter. Broadcast within the drip area of the tree and incorporate with a suitable device. Make one (1) application in winter or early spring. Do not graze or feed orchard plants or plant parts.

SIBERIAN ELM

Elm Leaf Beetle control by commercial applicators only and in the States of Colorado, Nebraska, Kansas, New Mexico, Arizona, Utah and Wyoming only—Apply FURADAN 10 G, before insects appear in the spring, at the rate of $\frac{1}{2}$ ounce per inch of trunk circumference. Measure the circumference (girth) of the tree 3 to 4 feet above the ground. Dig one hole for each inch of girth, 3 to 9 inches deep under the drip line of the tree. Thoroughly soak area under the tree. Place granules required in each hole. Replace soil plug immediately and compress soil. Keep treated areas soaked with water for 14 days.

If Girth of Tree is	Locate holes Away from Trunk	Amount of FURADAN 10 G
1 to 10 inches	1 ft.	0.5 to 5 ounces
14 to 20 inches	3 ft.	7 to 10 ounces
24 to 80 inches	6 ft. & 12 ft. (alternating)	12 to 20 ounces
44 to 80 inches	10 ft. & 12 ft. (alternating)	22 to 40 ounces
80 inches and larger	12 ft. & 20 ft. (alternating)	40 Ounces

DEALERS SHOULD SELL IN ORIGINAL PACKAGES ONLY.

TERMS OF SALE OR USE: On purchase of this product buyer and user agree to the following conditions:

WARRANTY: FMC makes no warranty, expressed or implied, concerning the use of this product other than indicated on the label. Except as so warranted, the product is sold as is. Buyer and user assume all risk of use and/or handling and/or storage of this material when such use and/or handling and/or storage is contrary to label instructions.

DIRECTIONS AND RECOMMENDATIONS: Follow directions carefully. Timing and method of application, weather and crop conditions, mixture with other chemicals not specifically recommended, and other influencing factors in the use of this product are beyond the control of the seller and are assumed by buyer at his own risk.

USE OF PRODUCT: FMC's recommendations for the use of this product are based upon tests believed to be reliable. The use of this product being beyond the control of the manufacturer, no guarantee, expressed or implied, is made as to the effects of such or the results to be obtained if not used in accordance with directions or established safe practice.

DAMAGES: Buyer's or user's exclusive remedy for damages for breach of warranty or negligence shall be limited to direct damages not exceeding the purchase price paid and shall not include incidental or consequential damages.

FURADAN® 15G

INSECTICIDE–NEMATICIDE

RESTRICTED USE PESTICIDE

For retail sale to and application only by Certified Applicators or personnel under their direct supervision, and only for those uses covered by the Certified Applicator's certification.

For Agricultural or Commercial Use Only

ACTIVE INGREDIENT

*Carbofuran	15.0%
INERT INGREDIENTS	85.0%
TOTAL	100.0%

*2,3-Dihydro-2,2-dimethyl-7-benzofuranyl methylcarbamate

EPA Reg. No. 279-3023 EPA Est.,279-

Furadan—FMC Trademark U.S. Patent No. 3,474,171

WARNING
Keep Out of Reach of Children

PRECAUTIONARY STATEMENTS

HAZARDS TO HUMANS (& DOMESTIC ANIMALS)

WARNING

Poisonous if swallowed. May be fatal or harmful as a result of skin or eye contact or by breathing dust. Causes cholinesterase inhibition. Warning symptoms of poisoning include weakness, headache, sweating, nausea, vomiting, diarrhea, tightness in chest, blurred vision, pinpoint eye pupils, abnormal flow of saliva, abdominal cramps, and unconsciousness. Atropine sulfate is antidotal.

Wear long-sleeved clothing and protective gloves when handling. Wash hands and face before eating or smoking. Bathe at the end of the work day. Change clothing daily and wash before reuse.

ENVIRONMENTAL HAZARDS

This pesticide is toxic to fish, birds and other wildlife. Birds feeding on treated areas may be killed. Birds killed by carbofuran pose a hazard to hawks and other birds-of-prey; bury or otherwise dispose of dead birds to prevent poisoning of other wildlife. Cover or incorporate granules in spill areas. Runoff from treated areas may be hazardous to fish in neighboring areas. Do not apply directly to water. Do not contaminate wells, wetlands or any body of water by cleaning of equipment or disposal of waste.

Notice: It is a Federal offense to use any pesticide in a manner that results in the death of a member of an endangered species.

The use of FURADAN 15G may pose a hazard to the following Federally designated endangered/threatened species known to be found in certain areas within the named locations.

Attwater's Greater Prairie Chicken—Texas counties including: Aransas, Austin, Brazoria, Colorado, Galveston, Goliad, Harris, Refugio and Victoria

Aleutian Canada Goose—California counties including Colusa, Merced, Stanislaus and Sutter

Kern Primrose Sphinx Moth—Walker Basin of Kern County, California

This product may not be used in areas where adverse impact on the Federally designated endangered/threatened species, noted above, is likely. Prior to making applications, the user of this product must determine that no such species are located in or immediately adjacent to the area to be treated. If the user is in doubt whether or not the above named endangered species may be affected, he should contact either the regional U.S. Fish and Wildlife Service office (Endangered Species Specialist) or personnel of the State Fish and Game office.

Carbofuran is a chemical which can travel (seep or leach) through soil and can contaminate ground water which may be used as drinking water. Carbofuran has been found in ground water as a result of agricultural use. Users are advised not to apply carbofuran where the water table (ground water) is close to the surface and where the soils are very permeable, i.e., well-drained soils such as loamy sands. Your local agricultural agencies can provide further information on the type of soil in your area and the location of ground water.

STATEMENT OF PRACTICAL TREATMENT

If swallowed, drink 1 to 2 glasses of water and induce vomiting by touching back of throat with finger. Do not induce vomiting or give anything by mouth to an unconscious person. Get medical attention.

If in eyes, flush with plenty of water for at least 15 minutes. Get medical attention.

If on skin, wash with soap and water. Remove contaminated clothing and wash before reuse.

Note to Physician: FURADAN (Carbofuran) is a reversible cholinesterase inhibitor. Do not use oximes such as 2-PAM. Start by giving 2 mg. atropine intramuscularly. According to clinical response, continue until signs of atropinization occur (dry mouth or dilated pupils). If in eye, instill one drop of homatropine.

For Emergency Assistance Call 716–735-3765.

DIRECTIONS FOR USE

It is a violation of Federal law to use this product in a manner inconsistent with its labeling.

Do not use this product on Long Island, N.Y.

If prolonged intimate contact with corn and/or sorghum foliage will result, do not re-enter treated fields within 14 days of application without wearing proper protective clothing. For all other situations, do not re-enter fields less than 24 hours following application.

Cover or incorporate granules in spill ares.

STORAGE AND DISPOSAL

PESTICIDE STORAGE

Do not store in or around the home.

Keep out of reach of children and animals. Store in original containers only. Store in a cool, dry place and avoid excess heat. Carefully open containers. After partial use, fold and roll back bags, clamp and close tightly. Do not put concentrate or dilute material into food or drink containers. Do not contaminate other pesticides, fertilizers, water, food or feed by storage or disposal.

In case of spill, avoid contact, isolate area and keep out animals and unprotected persons. Confine spills. Call FMC collect: (716) 735-3765.

To confine spill: If liquid, dike surrounding area or absorb with sand, cat litter or commercial clay. If dry material, cover to prevent dispersal. Place damaged package in a holding container. Identify contents.

PESTICIDE DISPOSAL

Pesticide wastes are acutely hazardous. Improper disposal of excess pesticide, spray mixture, or rinsate is a violation of Federal Law. If these wastes cannot be disposed of by use according to label instructions, contact your State Pesticide or Environmental Control Agency, or the Hazardous Waste representative at the nearest EPA Regional Office for guidance.

CONTAINER DISPOSAL

Completely empty bag into application equipment. Then dispose of empty bag in a sanitary landfill or by incineration, or, if allowed by State and local authorities, by burning. If burned, stay out of smoke.

USE CAUTION

Do not rotate to a succeeding crop other than alfalfa, corn (field, pop or sweet), cotton, cucurbits (cucumbers, melons, pumpkins or squash), grapes, peanuts, peppers, potatoes, rice, small grains (barley, oats, wheat), sorghum, soybeans, strawberries, sugar beets, sugarcane and tobacco. Tomatoes, cabbage, peas, succulent beans and dry beans may be planted the following season provided the prior season's application did not exceed 8.7 lbs/A of FURADAN 15G, 13 lbs/A of FURADAN 10G or 2 pt. 10 fl. oz./A of FURADAN 4F. Any other crop may be planted if it is not harvested or grazed.

FIELD CORN, POPCORN, SWEET CORN—Conventional, Minimum & No Tillage

Corn Rootworms (Northern, Southern & Western), Flea Beetles, Wireworms, aid in the control of White Grubs and first generation European Corn Borer, reduce losses due to Stalk Rot by reducing the incidence of insect wounds which permit entry by the Stalk Rot Fungi; Armyworm control for approximately 4 to 6 weeks after planting and suppression of Cutworms—Use 8 ounces of FURADAN 15G per 1,000 linear feet of row (6.7 lbs. per acre with 40 inch row spacing). Early season suppression of Common Stalk Borer — Use 16–24 ounces per 1,000 linear feet of row (13.3–20 lbs. per acre with 40 inch row spacing). Apply in a 7 inch band ahead of the planter press wheel and incorporate into the top 1 inch of soil by use of special covering devices or by dragging a short length of chain. —OR—Direct the granules into the planter shoe with the seed. —OR—Place the applicator tube directly behind the planter shoe so that the granules drop into the furrow and mix with the covering soil.

Lesion Nematodes (Northeast and Midwest Corn States including Colorado and Wyoming—Use 8–16 ounces per 1,000 linear feet of row (6.7–13.3 lbs. per acre with 40 inch row spacing). Apply "at planting," as above, (in a covered band or in the seed furrow).

European Corn Borer (first generation)—Use 16–24 ounces per 1,000 linear feet of row (13.3–20 lbs. per acre with 40 inch row spacing) "at planting" as above (in a covered band or in the seed furrow).

Seed Corn Maggot—Use 8 ounces per 1,000 linear feet of row (6.7 lbs. per acre with 40 inch row spacing). Southwestern Corn Borer (second to third generation)—Use 12–24 ounces per 1,000 linear feet of row (10–20 lbs. per acre with 40 inch row spacing). Southern Corn Billbug (Southeastern states), and to aid in the control of Leafhoppers and thereby reduce losses due to Maize Chlorotic Dwarf Virus and Corn Stunt disease—Use 16–24 ounces per 1,000 linear feet of row (13.3–20 lbs. per acre with 40 inch row spacing). Apply into the seed furrow as noted above.

Nematodes (Sting, Stunt, Stubby Root, Root Knot, Dagger, Lesion, Lance and Spiral)—Use 12–16 ounces per 1,000 linear feet of row (10–13.3 lbs. per acre with 40 inch row spacing). Apply at planting time a 7 to 15 inch band and incorporate into the top 3 inches soil.

FIELD CORN, POPCORN—Post Plant

Northern and Western Corn Rootworm—Use 8 ounces of FURADAN 15G per 1,000 linear feet of row (6.7 lbs. per acre with 40 inch row spacing) by banding over the row or by side dressing both sides of the row and cultivating into the soil.

FIELD CORN—Foliar Application

Southwestern Corn Borers and European Corn Borers—Use 6.7 lbs. per acre. Apply with aircraft by broadcasting over the corn plants or with ground equipment by directing the granules into the corn whorls. Apply when eggs begin to hatch. Do not make a foliar application if FURADAN 15G was applied at more than 8 ounces per 1,000 linear feet of row (6.7 lbs. per acre with 40 inch row spacing) at planting. Do not make more than two foliar applications per season.

Corn Insecticide Application Setting

FURADAN 15G is a dense, free flowing insecticide requiring accurate equipment calibration to apply the recommended amount per acre. Always calibrate your planter at the speed which you plan to travel during the planting operation. The following are approximate settings for guidance in calibration:

Applicator Gauge Setting

Application Rate	6.7 lbs./A			10 lbs./A			13.3 lbs./A		
Planter Speed (MPH)	4	6	8	4	6	8	4	6	8
Equipment									
Allis Chalmers	3	3	3	5	5	5	7	7	7
Gandy	14	18	21	18	22	25	22	25	29
IHC Cyclo Planter									
Gate	1	1	2	1	2	2	2	2	3
Dial	7	9	2	9	3	8	2	8	5
John Deere									
Range	1	2	2	1	2	2	2	2	2
Notch	18	1	2	20	2	4	1	4	8
John Deere									
Max Emerge (even number notches —0–10–20–30–40)	10	14	19	14	21	25	19	25	30
Max Emerge (odd number notches —5–15–25–35)	9	12	14	12	15	18	14	18	21
Noble									
Round Opening	8	12	16	12	17	22	16	22	28
Rectangular Opening	5	7	9	7	10	13	9	13	17

Corn Application Rates

When the granular application equipment is set to deliver FURADAN 15G at a rate of 8 ounces per 1,000 linear feet, the total amount applied per acre will vary depending on row spacing. Use the table below to determine your FURADAN needs for each acre.

Row Spacing (Inches)	40″	38″	36″	30″	28″
FURADAN 15G (lbs./acre)	6.7	7.0	7.3	8.7	9.3

COTTON

Thrips—Use FURADAN 15G at 8 ounces per 1,000 linear feet of row (6.7 lbs. per acre with 40 inch row spacing). Apply in the seed furrow at planting. Do not feed cotton forage.

CRANBERRIES

For control of Black Vine Weevils—Use FURADAN 15G at 13.3 lbs. per acre. For control of adult weevils and larvae, spread granules uniformly over soil surface using a rotary spreader. Make one application about June 15–25 and repeat one month later. Incorporate granules by following each application with sprinkler irrigation. For use only on cranberry bogs that are sprinkler irrigated and are not flooded for purposes of irrigation or harvest. Do not apply within 60 days of harvest. For use only in the State of Washington.

CUCURBITS (Cucumber, Melons, Squash, Pumpkins)
Nematodes (Root Knot, Sting, Lance)—For use in Southeastern States—Use FURADAN 15G at the rate of 1½ lbs. per 1,000 linear feet of row (13.3 lbs. per acre based on 60 inch row spacing). Apply in a 12 to 15 inch band and incorporate into the top 3 inches of soil.

GRAPES (For Use in California)
For control of Nematodes (Root-Knot, Dagger and Lesion) and Grape Phylloxera—Broadcast FURADAN 15G at 66.7 lbs. per acre over the soil surface between the vine rows and immediately incorporate. Do not apply within 200 days of harvest. Remove dense weed growth prior to treatment. Do not use on soils of pH 8.0 or greater. Do not make more than one application per crop year.

PEANUTS
Do not feed treated peanut forage to dairy animals or animals being finished for slaughter.

In-furrow at planting application method:
Thrips, Potato Leafhopper and as an aid in the control of Southern Corn Rootworm in Southeastern States—Use in–furrow at planting, 3.3 to 6.7 lbs. per acre (36 inch row spacing); 3.5 to 6.7 ounces per 1,000 linear feet of row.

Row banding application method:
Nematodes (Root Knot, Sting, Stunt, Lesion, Ring) in Southeastern States and in the Southwestern States of Texas and Oklahoma; and in the Southeastern States to aid in the control of Potato Leafhopper and Southern Corn Rootworm and for control of Thrips—Apply FURADAN 15G in a band over the row prior to planting and incorporate 3 inches into the soil. Use the rate indicated below for various band widths. Additional control measures may be needed late in the season for Potato Leafhopper.

Band Width	Rate/Acre (36 inch rows)	Rate/1,000 linear feet
12 inches	13.3 to 26.7 lbs.	0.9 to 1.8 lbs.
18 inches	20 to 33.3 lbs.	1.33 to 2.3 lbs.

Banded Pegging—Time Application Method
For control of Southern Corn Rootworm in the Southeastern States —Use FURADAN 15G at 7.3 to 14.7 ounces per 1,000 linear feet (6.7–13.3 lbs. per acre with 36 inch row spacing). Apply on a 14 to 18 inch band over the row when pegs begin to form or up to 30 days after pegging begins (approximately 60 to 90 days after planting). Incorporate granules immediately if vine growth permits. Do not apply FURADAN 15G at pegging *if FURADAN 15G was used in a band at planting.*

PEPPERS
Green Peach Aphid—(Delmarva Peninsula and South Jersey only). European Corn Borer—Make two applications of FURADAN 15G by sidedressing on either one or both sides of the row. Make a first application two to four weeks after transplanting using 13.3 lbs. per acre (equivalent to ½ lb. per 500 linear feet with 38 inch rows). Make the second application four to six weeks later using 20 lbs. per acre (0.7 lb. per 500 linear feet). Do not make the second application within 21 days of harvest.

POTATOES
Colorado Potato Beetle, European Corn Borer, Potato Flea Beetle, Potato Leafhopper, Tuber Flea Beetle; Green Peach Aphid (Western and Northeastern states including the Delmarva Peninsula); Potato Tuberworm (Virginia only); and to aid in the control of Wireworm —Use 1½ lbs. of FURADAN 15G per 1,000 linear feet of row (20 lbs. per acre 38-inch row spacing). Apply directly into the bottom of the furrow during the planting operation.

SORGHUM
Chinch Bug, Corn Leaf Aphid, Flea Beetle, Greenbug, Lesion Nematode, Seedcorn Maggot, Southern Corn Rootworm, Wireworm, Yellow Sugarcane Aphid and aids in the control of White Grubs and Cutworms—Use only on grain sorghum or sorghum grown for forage. Do not graze or harvest sorghum grown for forage within 75 days of application.

For Corn Leaf Aphid, Greenbug, Yellow Sugarcane Aphid—Use 6 to 8 ounces of Furadan 15 G per 1000 linear feet of row (5 to 6.7 lbs. per acre with 40 inch row spacing) applied directly into the seed furrow—OR—Use 8 ounces of Furadan 15 G per 1000 linear feet of row (6.7 lbs. per acre with 40 inch row spacing) applied in a 7 inch band over the row and incorporate the granules into the top 1 inch of soil by the use of a press wheel, dragging a short length of chain or suitable device.

For Chinch Bug, Flea Beetle, Southern Corn Rootworm, Lesion Nematode, Wireworm, Seedcorn Maggot (in furrow application only), and to aid in the control of White Grubs and Cutworms—Use 8 ounces of Furadan 15 G per 1000 linear feet of row (6.7 lbs. per acre with 40 inch row spacing). Apply at planting as noted above.

Equivalent pounds of FURADAN 15G per acre for various row spacings are as follows:

	Amount of FURADAN 15G per 1,000 Linear Feet of Row	Pounds of FURADAN 15G per Acre for Various Row Spacings				
		40″	36″	34″	30″	20″
In Furrow	6 Ounces	5	5.5	5.8	6.7	10
	8 Ounces	6.7	7.3	7.7	8.7	13.3
7″ Band	8 Ounces	6.7	7.3	7.7	8.7	13.3

When a rate range is indicated, use the higher rate when heavier pest infestations are anticipated.

SOYBEANS
Nematodes (Root Knot, Cyst, Stunt, Ring, Sting, Spiral Lesion, Lance, Dagger, Stubby-Root)—Use FURADAN 15G at 12–16 ounces/1,000 linear feet (equivalent to an application of 10–13.3 lbs./A on a 40-inch row spacing) at planting in a 7-inch band and mix with the covering soil.

—OR—

Apply FURADAN 15G at 12 ounces/1,000 liner feet (equivalent to an application of 10 lbs./A on a 40-inch row spacing). Apply at planting by directing the granules into the planter shoe with the seed or by placing the applicator tube directly behind the planter shoe so that the granules drop into the furrow and mix with the covering soil.

Mexican Bean Beetle (larvae)—Use FURADAN 15G at 12 ounces/1,000 linear feet (equivalent to 10 lbs./A on a 40-inch row spacing) at planting in a 7-inch band and mix with the covering soil.

—OR—

Apply FURADAN 15G at 8–12 ounces/1,000 linear feet (equivalent to an application of 6.7–10 lbs./A on a 40-inch row spacing). Apply at planting by directing the granules into the planter shoe with the seed or by placing the applicator tube into the furrow and mix with the covering soil.

SUGAR BEETS
Sugar Beet Root Maggot—Use 9 ounces of Furadan 15G per 1,000 linear feet of row (13.3 lbs. per acre using 22 inch row spacing). Apply granules in a 6 to 7 inch band and incorporate into the top 1 inch of soil using a suitable device or use 4.5 ounces of Furadan 15G per 1,000 linear feet of row (6.7 lbs. per acre using 22 inch row spacing) by applying granules in a "modified in-furrow" application. (Drop granules from a tube in a narrow band of 1 to 2 inches in front of the planter disks and allow some granules to fall into seed furrow with seed.)

Sugar Beet Application Rates
When the granular application equipment is set to deliver Furadan 15G at rates of 4.5, or 9 ounces per 1,000 linear feet, the total amount applied per acre will vary depending on row spacing. Use the table below to determine your Furadan needs for each acre.

Ounces of Furadan 15G per 1,000 Linear Feet of Row	Pounds of Furadan 15G per Acre for Various Row Spacings			
	20″	22″	24″	30″
4.5	7.5	6.5	6	5
9.0	15	13.5	12	10

SUGARCANE

Apply FURADAN 15G at the rate of 1.7 lbs. to 3.3 lbs. per 1,000 linear feet of row (equivalent to 13.3 to 26.7 lbs. per acre based on 60 inch row spacings). Do not use in Hawaii.

Plant Cane:

Nematodes (Root Knot and Stunt), Wireworm—Apply the above rate as a 15 inch band directly over the planted cane and cover with soil.

Stubblecane:

Wireworm (Florida only), Nematodes (Root Knot and Stunt) —Apply the above rate as a 15 inch band over the stubble row. Use any time after harvest until regrowth reaches 18 inches. Cover with a 1 to 2 inch layer of soil. Covering with more than 2 inches of soil may reduce stand.

SUNFLOWERS (Confectionary and Oil)

Sunflower Stem Weevils, Grasshoppers—Use 8–16 ounces of FURADAN 15G per 1,000 linear feet of row (8.7 to 17.4 lbs. per acre with 30 inch row spacing). Sunflower beetle—Use 8 to 16 ounces of FURADAN 15G per 1,000 linear feet of row (8.7 to 17.4 lbs. per acre with 30 inch row spacing) and apply in furrow as noted above—OR—use 16 ounces of Furadan 15 G per 1,000 linear feet of row (17.4 lbs. per acre with 30 inch row spacing) and apply at planting in a 7 inch band as noted above. When a rate is indicated, use the higher rate when heavier insect infestation is anticipated. Apply at plant by directing the granules into the planter shoe with the seed or by placing the applicator tube directly behind the planter shoe so that the granules drop into the furrow and mix with the covering soil.—OR—Apply at planting in a 7 inch band and mix with the covering soil. When a rate range is indicated, use higher rate when heavier insect infestation is anticipated.

TOBACCO—ALL TYPES EXCEPT FLUE-CURED

For control of Flea Beetles, Hornworms and to aid in the control of Budworms—Use FURADAN 15G at 40 pounds per acre. Broadcast granules over the soil surface prior to transplanting and incorporate into the top 3 inches with a suitable device.

For full season control of Flea Beetles only—Use FURADAN 15G at 20–26.7 lbs. per acre—OR—for full season control of Flea Beetles and approximately 60 days control of Hornworms—Use FURADAN 15G at 26.7 lbs. per acre. Apply broadcast as noted above.

TOBACCO—FLUE CURED

For control of Stunt Nematodes, Flea Beetles, Hornworms, Wireworms and to aid in the control of Root Knot Nematodes and Budworms—Use FURADAN 15G at 40 lbs. per acre. Broadcast granules over the soil surface prior to transplanting and incorporate into the top 3 inches with a suitable device. Form beds and plant as usual. This product may induce flecking of the bottom or lower leaves.

For full season control of Flea Beetles and approximately 60 days control of Hornworms—Use FURADAN 15G at 26.7 lbs. per acre. Apply broadcast as noted above. Note leaf flecking caution listed above.

COTTONWOOD

Cottonwood Twig Borer, Cottonwood Leaf Beetle, Clearwing Borer—**In Commercial Plantations**—use 6.7–13.3 lbs. per acre (12 foot row spacing) or 1.8 to 3.5 lbs. per 1,000 linear feet of row. Apply during June or July to the root zone of the cutting by the use of a granular subsoil applicator as a sidedress 10 to 12 inches from the trees in a continuous band on both sides of the trees. **In Nursery plantings**—use 26.7 lbs. per acre (40 inch row spacing) or 2 lbs. per 1,000 linear feet of row. Apply during May or June to the root zone of the cutting, side-dressing 10 to 12 inches from the trees in a continuous band on both sides of the trees.

PINE SEEDLINGS IN PINE PLANTATIONS

Pales Weevil and Pitch-Eating Weevil, Nantucket Pine Tip Moth and Pitch Pine Tip Moth—Use 7 grams (3/4 teaspoon) per seedling at transplanting. Distribute the granules in a 6 inch radius on the soil around each seedling. Cover granules with soil. Do not use at planting if seedlings were treated with FURADAN 4F slurry root dip.

PINE SEED ORCHARDS

Seedbugs, Coneworms, Coneborers—Use 2.7 to 5.3 oz. per inch of tree diameter. Cone beetles—Use 5.3 ounces per inch of tree diameter. Broadcast within the drip area of the tree and incorporate with a suitable device. Make one (1) application in winter or early spring. Do not graze or feed orchard plants or plant parts.

SIBERIAN ELM

Elm Leaf Beetle control by commercial applicators only and in the States of Colorado, Nebraska, Kansas, New Mexico, Arizona, Utah and Wyoming only—Apply FURADAN 15G before insects appear in the spring, at the rate of 1/3 ounce per inch of trunk circumference. Measure the circumference (girth) of the tree 3 to 4 feet above the ground. Dig one hole for each inch of girth, 3 to 9 inches deep under the drip line of the tree. Thoroughly soak areas under the tree. Place granules required in each hole. Replace soil plug immediately and compress soil. Keep treated areas soaked with water for 14 days.

If Girth of tree is	Locate Holes away from trunk	Amount of FURADAN 15G
1 to 10 inches	1 ft.	1/3–3 1/3 oz.
14 to 20 inches	3 ft.	4.7–6.7 oz.
24 to 80 inches	6 ft. and 12 ft. (alternating)	8–13.3 oz.
44 to 80 inches	10 ft. and 12 ft. (alternating)	14.7–26.7 oz.
80 inches and larger	12 ft. and 20 ft. (alternating)	26.7 oz.

DEALERS SHOULD SELL IN ORIGINAL PACKAGES ONLY.

TERMS OF SALE OR USE: On purchase of this product buyer and user agree to the following conditions:

WARRANTY: FMC makes no warranty, expressed or implied, concerning the use of this product other than indicated on the label. Except as so warranted, this product is sold as is. Buyer and user assume all risk of use and/or handling and/or storage of this material when such use and/or handling and/or storage is contrary to label instructions.

DIRECTIONS AND RECOMMENDATIONS: Follow directions carefully. Timing and method of application, weather and crop conditions, mixture with other chemicals not specifically recommended, and other influencing factors in the use of this product are beyond the control of the seller and are assumed by buyer at his own risk.

USE OF PRODUCT: FMC's recommendations for the use of this product are based on tests believed to be reliable. The use of this product being beyond the control of the manufacturer, no guarantee, expressed or implied, is made as to the effects of such or the results to be obtained if not used in accordance with directions or established safe practice.

DAMAGES: Buyer's or user's exclusive remedy for damages for breach of warranty or negligence shall be limited to direct damages not exceeding the purchase price paid and shall not include incidental or consequential damages.

KOLO®-100

FUNGICIDE

ACTIVE INGREDIENTS:	BY WT.
Sulfur	75.4%
Dichlone (2,3-Dichloro-1,4-naphthoquinone)	3.5%
INERT INGREDIENTS:	21.1%
TOTAL	100.0%

This product contains Bentonite Sulfur.
Sulfur Fineness 99.7% thru 325 mesh.

EPA Reg. No. 279-2925 EPA Est., 279-FL-1
Kolo—FMC Trademark

CAUTION

Keep Out of Reach of Children

STATEMENT OF PRACTICAL TREATMENT

If swallowed: Drink one or two glasses of water and induce vomiting by touching back of throat with finger. Do not induce vomiting or give anything by mouth to an unconscious person.

If in eyes: Flush with plenty of water. Get medical attention if irritation persists.

If on skin: Wash with plenty of soap and water.

For Emergency Assistance Call 716-735-3765
See other panels for additional precautionary information.

PRECAUTIONARY STATEMENTS

HAZARDS TO HUMAN (& DOMESTIC ANIMALS)

CAUTION

Harmful if swallowed. Avoid breathing dust or spray mist. May cause irritation of nose, throat or skin. Avoid contact with eyes, skin and clothing.

ENVIRONMENTAL HAZARDS

This product is toxic to fish. Keep out of any lakes, ponds, or streams. Do not apply where runoff is likely to occur. Do not apply when weather conditions favor drift from areas treated. Do not contaminate water by cleaning of equipment or disposal of wastes. Apply this product only as specified on this label.

DIRECTIONS FOR USE

It is a violation of Federal Law to use this product in a manner inconsistent with its labeling.

Do not apply this product in such a manner as to directly or through drift expose workers or other persons. The area being treated must be vacated by unprotected persons.

Do not enter treated areas without protective clothing until sprays have dried.

Protective clothing means, at least, a hat or other suitable head covering, a long sleeved shirt and long legged trousers or a coverall type garment (all of closely woven fabric covering the body, including the arms and legs), shoes and socks.

Because certain states may require more restrictive reentry intervals for various crops treated with this product, consult your State Department of Agriculture for further information.

Written or oral warnings must be given to workers who are expected to be in a treated area or in an area about to be treated with this product. Inform workers of areas or fields that may not be entered without specific protective clothing, period of time field must be vacated and appropriate actions to take in case of accidental exposure. An example of such information is given under written warnings. When oral warnings are given, warnings shall be given in a language customarily understood by workers. Oral warnings must be given if there is reason to believe that written warnings cannot be understood by workers. Written warnings must include the following information:

CAUTION

Area treated with KOLO®–100 on (insert date). Do not enter without appropriate protective clothing until sprays have dried (insert State Department of Agriculture's re-entry interval, if more restrictive).

In case of accidental exposure to pesticide spray or dust, wash the skin thoroughly with soap and water. Remove contaminated clothing and wash before reuse. If in eyes, flush with plenty of water. If inhaled, go to an area where the pesticide has not been applied. Get medical attention if needed.

STORAGE AND DISPOSAL

PESTICIDE STORAGE

Keep away from fire and sparks.

Keep out of reach of children and animals. Store in original containers only. Store in a cool, dry place and avoid excess heat. Carefully open containers. After partial use, fold and roll back bags, clamp and close tightly. Do not put concentrate or dilute material into food or drink containers. Do not contaminate other pesticides, fertilizers, water, food or feed by storage or disposal.

In case of spill, avoid contact, isolate area and keep out animals and unprotected persons. Confine spills. Call FMC collect: (716) 735-3765.

To confine spill: If liquid, dike surrounding area or absorb with sand, cat litter or commercial clay. If dry material, cover to prevent dispersal. Place damaged package in a holding container. Identify contents.

PESTICIDE DISPOSAL

Wastes resulting from the use of this product may be disposed of on site or at an approved waste disposal facility.

CONTAINER DISPOSAL

Completely empty bag into application equipment. Then dispose of empty bag in a sanitary landfill or by incineration, or, if allowed by State and local authorities, by burning. If burned, stay out of smoke.

DIRECTIONS FOR MIXING: Fill the spray tank with water until the agitator blades are covered. Start the engine, and with the agitator in motion sift in slowly the required amount of material, holding the bag well into the tank opening to avoid drifting of dust into the face of the operator. Finish filling the tank and apply immediately. Keep the agitator running at all times.

Apply the following recommended rates in sufficient water to thoroughly cover one acre. (With ground application equipment use a minimum of 300 gallons of diluted spray per acre on pome fruits and a minimum of 200 gallons of diluted spray per acre on stone fruits.)

APPLE: Scab, Powdery Mildew—Use 3½ pounds per 100 gallons of water. Apply on a protective schedule at 5 to 7 day intervals from delayed dormant through the 4th cover spray. May be applied on apples up to 1 day of harvest; however, for late cover applications another suitable fungicide should be used.

PEACH: Brown Rot—For blossom and twig brown rot blossom blight use 3½ pounds per 100 gallons of water at pink bud stage, full bloom and just before shuck fall. When cool, wet weather occurs, resulting in a prolonged bloom period, apply every 4 days. As a preharvest spray use 3½ pounds per 100 gallons of water at 14 and 7 days before harvest. Do not apply within 7 days of harvest.

CHERRY: Brown Rot—Use 3½ pounds per 100 gallons of water. For Twig Brown Rot begin at pink bud. Repeat at full bloom and before shuck shed or every 4 days in cool, wet weather. For Blossom Blight, apply at popcorn, full bloom and petal fall. For Fruit Rot apply at 7 to 10 day intervals from petal fall to 3 days from harvest. Do not apply within 3 days of harvest.

PLUM, PRUNE: Brown Rot, Blossom Blight, Coryneum Blight—Use 3½ pounds per 100 gallons of water at pink, full bloom and pre-shuck fall or at 4 day intervals during cool, wet weather. Brown Rot Fruit Rot—Apply in cover sprays as needed. Do not apply within 3 days of harvest.

NOTE: Do not apply within 10 to 14 days of an oil spray. Do not apply prior to or during periods of excessively high temperatures.

DEALERS SHOULD SELL IN ORIGINAL PACKAGES ONLY.

TERMS OF SALE OR USE: On purchase of this product buyer and user agree to the following conditions:

WARRANTY: FMC makes no warranty, expressed or implied, concerning the use of this product other than indicated on the label. Except as so warranted, the product is sold as is. Buyer and user assume all risk of use and/or handling and/or storage of this material when such use and/or handling and/or storage is contrary to label instructions.

DIRECTIONS AND RECOMMENDATIONS: Follow directions carefully. Timing and method of application, weather and crop conditions, mixture with other chemicals not specifically recommended, and other influencing factors in the use of this product are beyond the control of the seller and are assumed by buyer at his own risk.

USE OF PRODUCT: FMC's recommendations for the use of this product are based upon tests believed to be reliable. The use of this product being beyond the control of the manufacturer, no guarantee, expressed or implied, is made as to the effects of such or the results to be obtained if not used in accordance with directions or established safe practice.

DAMAGES: Buyer's or user's exclusive remedy for damages for breach of warranty or negligence shall be limited to direct damages not exceeding the purchase price paid and shall not include incidental or consequential damages.

KOLODUST® XTRA DUST OR SPRAY

FUNGICIDE

ACTIVE INGREDIENT:	BY WT.
* Sulfur	53.0%
INERT INGREDIENTS:	47.0%
TOTAL	100.0%

*Sulfur fineness 99.7% through 325 mesh.

This product contains bentonite sulfur.

EPA Reg. No. 279-391 EPA Est., 279-

Kolodust—FMC Trademark

CAUTION

Keep Out of Reach of Children

For Emergency Assistance Call 716–735-3765.

See other panels for additional precautionary information.

ENVIRONMENTAL HAZARDS

Keep out of lakes, ponds and streams. Do not contaminate water by cleaning of equipment or disposal of wastes.

Do not allow this material to drift onto neighboring crops or non-crop areas or use in a manner or at a time other than in accordance with label directions because animal, plant or crop injury, illegal residues or other undesirable results may occur.

DIRECTIONS FOR USE

It is a violation of Federal Law to use this product in a manner inconsistent with its labeling.

Do not apply this product in such a manner as to directly or through drift expose workers or other persons. The area being treated must be vacated by unprotected persons.

Do not enter treated areas without protective clothing until sprays have dried or dusts have settled.

Protective clothing means, at least, a hat or other suitable head covering, a long sleeved shirt and long legged trousers or a coverall type garment (all of closely woven fabric covering the body, including the arms and legs), shoes and socks.

Because certain states may require more restrictive reentry intervals for various crops treated with this product, consult your State Department of Agriculture for further information.

Written or oral warnings must be given to workers who are expected to be in a treated area or in an area about to be treated with this product. Inform workers of areas or fields that may not be entered without specific protective clothing, period of time field must be vacated and appropriate actions to take in case of accidental exposure. An example of such information is given under written warnings. When oral warnings are given, warnings shall be given in a language customarily understood by workers. Oral warnings must be given if there is reason to believe that written warnings cannot be understood by workers. Written warnings must include the following information:

CAUTION

Area treated with KOLODUST® XTRA Dust or Spray on (insert date). Do not enter without appropriate protective clothing until sprays have dried (insert State Department of Agriculture's re-entry interval, if more restrictive).

In case of accidental exposure to pesticide spray or dust, wash the skin thoroughly with soap and water. Remove contaminated clothing and wash before reuse. If in eyes, flush with plenty of water. If inhaled, go to an area where the pesticide has not been applied. Get medical attention if needed.

STORAGE AND DISPOSAL

PESTICIDE STORAGE

Keep out of reach of children and animals. Store in original containers only. Store in a cool, dry place and avoid excess heat. Carefully open containers. After partial use, fold and roll back bags, clamp and close tightly. Do not put concentrate or dilute material into food or drink containers. Do not contaminate other pesticides, fertilizers, water, food or feed by storage or disposal.

In case of spill, avoid contact, isolate area and keep out animals and unprotected persons. Confine spills. Call FMC collect: (716) 735-3765.

To confine spill: If liquid, dike surrounding area or absorb with sand, cat litter or commercial clay. If dry material, cover to prevent dispersal. Place damaged package in a holding container. Identify contents.

PESTICIDE DISPOSAL

Wastes resulting from the use of this product may be disposed of on site or at an approved waste disposal facility.

CONTAINER DISPOSAL

Completely empty bag into application equipment. Then dispose of empty bag in a sanitary landfill or by incineration, or, if allowed by State and local authorities, by burning. If burned, stay out of smoke.

Do not use within two weeks before or after the application of an oil spray. Applications of spraying and dusting materials containing sulfur should not be made prior to or during periods of excessively high summer temperatures.

FOR USE AS A DUST in the control of Scab on **APPLES, PEARS,** and **PEACHES;** Brown Rot and Leaf Spot on **CHERRIES;** Brown Rot on **PEACHES, PLUMS,** and **PRUNES;** and Leaf Spot on **STRAWBERRIES.** This material may be applied to dry foliage ahead of infection. It may also be applied in the rain when combating Scab on Apples during prolonged rain periods. When using this material, dust in such a manner that all parts of the tree or plant will be covered with a thin film of dust. Cover the top and upper part of each tree as thoroughly as you do the lower half. Make repeated applications frequently enough to keep new growth protected.

Where desired this material may be used as a spray. General recommendations for its use in this way are given below.

DIRECTIONS FOR MIXING: Fill the spray tank with water until the agitator blades are covered. Start the engine, and with the agitator in motion sift in slowly the required amount of this material. If desired, water may be added while the material is sifted in. Finish filling the tank with the agitator running and apply immediately.

FOR USE AS A SPRAY ON APPLE AND PEAR TREES for Scab control, the usual practice is to use 5 to 8 pounds of this material per 100 gallons of water. Apply in the delayed dormant (silver tip) and early prepink (green tip) stages, during the bloom period and in the calyx and early cover applications.

Under severe Scab conditions, up to 8 pounds per 100 gallons of water may be used up to and including the Calyx application. After the Calyx application if Scab is under control, and ahead of and during periods of high temperature, the dosage may be reduced to 2 to 4 pounds per 100 gallons of water depending on the temperature and severity of the disease to be controlled. Sulfur may injure certain apple varieties. Check plant tolerance prior to use.

FOR USE AS A SPRAY ON APPLE TREES for Powdery Mildew control—Use 2 to 4 pounds per 100 gallons of water. Apply 3 to 4 pounds during the delayed dormant (silver tip) and early prepink (green tip) stages of plant growth. Reduce rates to 2 to 3 pounds from petal fall through to harvest. Sulfur may injure certain apple varieties. Check plant tolerance prior to use.

In the later cover applications (beginning about three weeks after the Calyx) many Experiment Stations recommend the use of one pound of Hydrated Spray Lime to each pound of Lead Arsenate used in the spray mixture. This practice is recommended when using this material and Lead Arsenate in the later cover sprays.

When using Nicotine Sulfate as a contact spray in combination with this materal, always use 5 pounds of Hydrated Spray Lime per 100 gallons of water. When used in combination with Lead Arsenate or Zinc Sulfate also add 6 to 8 pounds of Hydrated Spray Lime per each 100 gallons of water.

When used in combination with Zinc Sulfate and Lead Arsenate, mix with water in the spray tank in the following order: this material, Zinc Sulfate, Hydrated Spray Lime and Lead Arsenate.

FOR USE AS A SPRAY ON PEACH TREES for Brown Rot and Scab control —Use 5 to 8 pounds of this material per 100 gallons of water. Apply in the pink, bloom and petal fall stages and at 10 to 14 day intervals as long as necessary.

DEALERS SHOULD SELL IN ORIGINAL PACKAGES ONLY.

TERMS OF SALE OR USE: On purchase of this product buyer and user agree to the following conditions:

WARRANTY: FMC makes no warranty, expressed or implied, concerning the use of this product other than indicated on the label. Except as so warranted, the product is sold as is. Buyer and user assume all risk of use and/or handling and/or storage of this material when such use and/or handling and/or storage is contrary to label instructions.

DIRECTIONS AND RECOMMENDATIONS: Follow directions carefully. Timing and method of application, weather and crop conditions, mixture with other chemicals not specifically recommended and other influencing factors in the use of this product are beyond the control of the seller and are assumed by buyer at his own risk.

USE OF PRODUCT: FMC's recommendations for the use of this product are based upon tests believed to be reliable. The use of this product being beyond the control of the manufacturer, no guarantee, expressed or implied, is made as to the effects of such or the results to be obtained if not used in accordance with directions or established safe practice.

DAMAGES: Buyer's or user's exclusive remedy for damages for breach of warranty or negligence shall be limited to direct damages not exceeding the purchase price paid and shall not include incidental or consequential damages.

KOLOSPRAY®

FUNGICIDE–INSECTICIDE

ACTIVE INGREDIENT:	BY WT.
Sulfur	81.25%
INERT INGREDIENTS:	18.75%
TOTAL	100.00%

This product contains bentonite sulfur. Sulfur fineness 99.7% thru 325 mesh.

EPA Reg. No. 279–47 EPA Est., 279-

Kolospray—FMC Trademark

CAUTION

Keep Out of Reach of Children

For Emergency Assistance Call 716–735-3765.

See other panels for additional precautionary information.

ENVIRONMENTAL HAZARDS

Keep out of lakes, ponds and streams. Do not contaminate water by cleaning of equipment, or disposal of wastes. Do not apply where runoff is likely to occur. Do not apply when weather conditions favor drift from areas treated. Apply this product only as specified on this label.

DIRECTIONS FOR USE

It is a violation of Federal Law to use this product in a manner inconsistent with its labeling.

Do not apply this product in such a manner as to directly or through drift expose workers or other persons. The area being treated must be vacated by unprotected persons.

Do not enter treated areas without protective clothing until sprays have dried.

Protective clothing means, at least, a hat or other suitable head covering, a long sleeved shirt and long legged trousers or a coverall type garment (all closely woven fabric covering the body, including the arms and legs), shoes and socks.

Because certain states may require more restrictive reentry intervals for various crops treated with this product, consult your State Department of Agriculture for further information.

Written or oral warnings must be given to workers who are expected to be in a treated area or in an area about to be treated with this product. Inform workers of areas or fields that may not be entered without specific protective clothing, period of time field must be vacated and appropriate actions to take in case of accidental exposure. An example of such information is given under written warnings. When oral warnings are given, warnings shall be given in a language customarily understood by workers. Oral warnings must be given if there is reason to believe that written warnings cannot be understood by workers. Written warnings must include the following information:

CAUTION

Area treated with KOLOSPRAY® on (insert date). Do not enter without appropriate protective clothing until sprays have dried (insert State Department of Agriculture's re-entry interval, if more restrictive).

In case of accidental exposure to pesticide spray or dust, wash the skin thoroughly with soap and water. Remove contaminated clothing and wash before reuse. If in eyes, flush with plenty of water. If inhaled, go to an area where the pesticide has not been applied. Get medical atention if needed.

STORAGE AND DISPOSAL

PESTICIDE STORAGE

Keep away from fire and sparks.

Keep out of reach of children and animals. Store in original containers only. Store in a cool, dry place and avoid excess heat. Carefully open containers. After partial use, fold and roll back bags, clamp and close tightly. Do not put concentrate or dilute material into food or drink containers. Do not contaminate other pesticides, fertilizers, water, food or feed by storage or disposal.

In case of spill, avoid contact, isolate area and keep out animals and unprotected persons. Confine spills. Call FMC collect: (716) 735-3765.

To confine spill: If liquid, dike surrounding area or absorb with sand, cat litter or commercial clay. If dry material, cover to prevent dispersal. Place damaged package in a holding container. Identify contents.

PESTICIDE DISPOSAL

Wastes resulting from the use of this product may be disposed of on site or at an approved waste disposal facility.

CONTAINER DISPOSAL

Completely empty bag into application equipment. Then dispose of empty bag in a sanitary landfill or by incineration, or, if allowed by State and local authorities, by burning. If burned, stay out of smoke.

Application of dusting and spraying materials containing sulfur should not be made just prior to or during periods of excessively high temperatures. Do not make applications within two weeks of an oil spray.

Do not use in undiluted form.

Fill the spray tank about two-thirds full of water. Start the engine so that the agitator is in motion and sift the KOLOSPRAY into the tank while the tank is being filled. Add other spray materials that are required after the KOLOSPRAY has been dispersed in the spray tank. Observe all cautions and limitations on labeling of all products used in mixtures.

ALMONDS: Brown Rot, Scab—Use 6 pounds per 100 gallons of water in pre-bloom, bloom and post-bloom periods. Apply when disease threatens during rain periods and repeat at 7 to 10 day intervals.

APPLES, PEARS: Scab and Powdery Mildew—Spray in the delayed dormant (silver tip), early pre-pink (green tip) and early cover stages of growth. Use 6 to 8 pounds of KOLOSPRAY or use 4 to 6 pounds of KOLOSPRAY plus 2 to 3 pounds of Polysulfide Compound per 100 gallons of water. Do not apply on pears after the fruit begins to turn down. Sulfur may injure certain apple varieties. Check plant tolerance prior to use.

APPLES ONLY: Powdery Mildew—Use 1 to 3 pounds per 100 gallons of dilute spray. Apply 3 pounds per 100 gallons at the delayed dormant (silver tip) through the early pre-pink (green tip) stages and 2 to 3 pounds per 100 gallons during full bloom, and 1 to 3 pounds per 100 gallons from petal fall through to harvest. Sulfur may injure certain apple varieties. Check plant tolerance prior to use.

PEARS ONLY: As a fall clean-up spray for control of Blister Mite, European Red Mite and Twospotted Spider Mite—Use 6 pounds of KOLOSPRAY plus 6 pounds of Polysulfide Compound or 5 gallons of Liquid Lime Sulfur Solution per 100 gallons of dilute spray.

BEANS: Powdery Mildew, Leaf Spot, Rust—Use 4 to 7 pounds per 100 gallons of water. Begin applications when first true leaves form or at first sign of disease. Repeat at 7 to 10 day intervals. Sulfur may injure certain varieties of beans.

BLACKBERRIES, CURRANTS, DEWBERRIES, GOOSEBERRIES, HUCKLEBERRIES, RASPBERRIES, STRAWBERRIES: Powdery Mildew—Use 2 pounds per 100 gallons of water. Apply when disease first appears and repeat as necessary.

CITRUS: Citrus Flat Mite, Yuma Spider Mite, Sixspotted Mite, and Citrus Rust Mite—Use 4 to 5 pounds per 100 gallons of water (a minimum of 60 pounds per acre). Apply when mites appear and repeat as necessary. Do not apply sulfur within 60 days of an oil application. Do not apply sulfur just prior to or during periods of excessively hot weather.

GRAPES: Powdery Mildew—Use 2 to 5 pounds per 100 gallons. Begin applications when shoots are 6 to 8 inches long and repeat at 7 to 10 day intervals. Do not apply to Concord or other sulfur sensitive varieties.

PEACHES, PLUMS, PRUNES, NECTARINES, CHERRIES: Powdery Mildew, Cherry Leaf Spot, Brown Rot, Prune Rust—Use 6 to 10 pounds per 100 gallons of water. Peach Silver Mite—Use 6 pounds per 100 gallons of water. (By ground, apply a minimum of 200 gallons of water per acre.) Apply in the pre-bloom, bloom and post-bloom periods. Apply when mites appear and when diseases threaten during rain periods and repeat at 7 to 10 day intervals.

PEAS: Powdery Mildew, Rust—Use 4 to 7 pounds per 100 gallons of water. Begin early in season before disease appears. Repeat at 7 to 10 day intervals.

POTATOES: Powdery Mildew—Use 3 to 4 pounds of KOLOSPRAY in sufficient water to cover one acre. Applying by aircraft, use a minimum of 8 gallons of spray per acre. With ground equipment, use a minimum of 10 gallons of spray per acre. Apply at the first sign of infection and repeat at three week intervals.

TOMATOES: Russet Mite—Use 7 pounds per acre in a minimum of 25 gallons of water by ground applicator or in a minimum of 5 gallons of water by aircraft. Apply when mites first appear and repeat as necessary.

DEALERS SHOULD SELL IN ORIGINAL PACKAGES ONLY.

TERMS OF SALE OR USE: On purchase of this product buyer and user agree to the following conditions:

WARRANTY: FMC makes no warranty, expressed or implied, concerning the use of this product other than indicated on the label. Except as so warranted, the product is sold as is. Buyer and user assume all risk of use and/or handling and/or storage of this material when such use and/or handling and/or storage is contrary to label instructions.

DIRECTIONS AND RECOMMENDATIONS: Follow directions carefully. Timing and method of application, weather and crop conditions, mixture with other chemicals not specifically recommended and other influencing factors in the use of this product are beyond the control of the seller and are assumed by buyer at his own risk.

USE OF PRODUCT: FMC's recommendations for the use of this product are based upon tests believed to be reliable. The use of this product being beyond the control of the manufacturer, no guarantee, expressed or implied, is made as to the effects of such or the results to be obtained if not used in accordance with directions or established safe practice.

DAMAGES: Buyer's or user's exclusive remedy for damages for breach of warranty or negligence shall be limited to direct damages not exceeding the purchase price paid and shall not include incidental or consequential damages.

PHOS KIL® SPRAY

INSECTICIDE
Contains 15% Parathion

RESTRICTED USE PESTICIDE

For retail sale to and use only by certified applicators or persons under their direct supervision, and only for those uses covered by the certified applicator's certification.

ACTIVE INGREDIENT:	
*Parathion	15.0%
INERT INGREDIENTS:	85.0%
TOTAL	100.0%

*O, O-Diethyl O-p-nitrophenyl thiophosphate

EPA Reg. No. 279-336 EPA Est., 279-

Phos kil Spray—FMC Trademark

STOP! READ THE LABEL

CAN KILL YOU

DANGER–POISON

Keep Out of Reach of Children
PELIGRO

PRECAUCION AL USUARIO: Si usted no lee ingles, no use este producto hasta que la etiqueta le haya sido explicada ampliamente.

USE ONLY WHEN WEARING THE FOLLOWING PROTECTIVE EQUIPMENT AND CLOTHING

(1) Wear water-proof pants, coat, hat, rubber boots or rubber overshoes. (2) Wear safety goggles. (3) Wear a mask or a pesticide respirator jointly approved by the Mining Enforcement and Safety Administration (formerly the U.S. Bureau of Mines) and by the

National Institute for Occupational Safety and Health under the provisions of 30 CFR Part II. (4) Wear heavy duty, natural rubber gloves.

DANGER—POISON—PRECAUTIONS

POISONOUS IF SWALLOWED

This product can kill you if swallowed even in small amounts; spray mist or dust may be fatal if swallowed.

POISONOUS BY SKIN CONTACT

Poisonous if touched by hands or spilled or splashed on skin, in eyes or on clothing (liquid goes through clothes).

POISONOUS IF BREATHED

Breathing vapors, spray mist or dust may be fatal.

POISONOUS TO FISH AND WILDLIFE

POISON SIGNS (Symptoms)

PARATHION is a very dangerous poison. It rapidly enters the body on contact with all skin surfaces and eyes. Clothing wet with this material must be removed immediately. Exposed persons must receive prompt medical treatment or they may die.

Some of the signs and symptoms of poisoning are: Headache, nausea, vomiting, cramps, weakness, blurred vision, pin-point pupils, tightness in chest, labored breathing, nervousness, sweating, watering of eyes, drooling or frothing of mouth and nose, muscle spasms and coma.

WORK SAFETY RULES

Keep all unprotected persons and children away from treated area or where there is danger of drift.

Do not rub eyes or mouth with hands. If you feel sick in any way, STOP work and get help right away. Call a doctor (physician), clinic or hospital—immediately. Explain that the victim has been exposed to parathion and describe his condition. After first aid is given (see First Aid Treatment Section) and if a doctor cannot come, take victim to clinic or hospital.

IMPORTANT! Before removing gloves, wash them with soap and water. Always wash hands, face and arms with soap and water before smoking, eating or drinking.

AFTER WORK, take off all work clothes and shoes. Shower, using soap and water. Wear only clean clothes when leaving job. Do not wear contaminated clothing. Wash protective clothing and protective equipment with soap and water after each use. Respirator should be cleaned and filter replaced according to instructions included with respirator.

FIRST AID TREATMENT

Call a doctor (physician), clinic or hospital immediately. Explain that the victim has been exposed to parathion and describe his condition.

If breathing has stopped, start artificial respiration immediately and maintain until doctor sees victim.

If swallowed: Call a physician or Poison Control Center. Drink 1 or 2 glasses of water and induce vomiting by touching back of throat with finger, or, if available, by administering syrup of ipecac. Do not induce vomiting or give anything by mouth to an unconscious person.

In case of contact, immediately flush eyes or skin with plenty of water for at least 15 minutes while removing contaminated clothing and shoes. See doctor immediately.

NOTE TO PHYSICIAN

Antidote—administer atropine sulfate in large doses, TWO to FOUR mg. intravenously or intramuscularly as soon as cyanosis is overcome. Repeat at 5 to 10 minute intervals until signs of atropinization appear. 2-PAM chloride is also antidotal and may be administered in conjunction with atropine. DO NOT GIVE MORPHINE OR TRANQUILIZERS. Parathion is a strong cholinesterase inhibitor affecting the central and peripheral nervous systems and producing cardiac and respiratory depression. At first sign of pulmonary edema, the patient should be given supplemental oxygen and treated symptomatically. Continued absorption of the poison may occur and fatal relapses have been reported after initial improvement; VERY CLOSE SUPERVISION OF THE PATIENT IS INDICATED FOR AT LEAST 48 HOURS.

FOR EMERGENCY ASSISTANCE CALL 716–735-3765.

POST TREATED AREA

Consult your State Agricultural Extension Service or Experiment Station regarding posting treated areas.

ENVIRONMENTAL HAZARDS

This product is toxic to fish and wildlife. Birds and other wildlife in treated areas may be killed. Keep out of any body of water. Do not apply when weather conditions favor drift from treated areas. Do not apply where runoff is likely to occur.

Do not contaminate water by cleaning of equipment or disposal of wastes.

This product is highly toxic to bees exposed to direct treatment or residues on crops. Protective information may be obtained from your Cooperative Agricultural Extension Service.

DIRECTIONS FOR USE

It is a violation of Federal Law to use this product in a manner inconsistent with its labeling.

Do not apply this product is such a manner as to directly or through drift expose workers or other persons. The area being treated must be vacated by unprotected persons.

Do not enter treated areas for 48 hours unless appropriate protective clothing is worn.

Protective clothing means, at least, a hat or other suitable head covering, a long sleeved shirt and long legged trousers or a coverall type garment (all of closely woven fabric covering the body, including the arms and legs), shoes and socks.

Because certain states may require more restrictive reentry intervals for various crops treated with this product, consult your State Department of Agriculture for further information.

Written or oral warnings must be given to workers who are expected to be in a treated area or in an area about to be treated with this product. Inform workers of areas or fields that may not be entered without specific protective clothing, period of time field must be vacated and appropriate actions to take in case of accidental exposure. An example of such information is given under written warnings. When oral warnings are given, warnings shall be given in a language customarily understood by workers. Oral warnings must be given if there is reason to believe that written warnings cannot be understood by workers. Written warnings must include the following information:

DANGER

Area treated with Phos Kil® Spray (insert date). Do not enter without appropriate protective clothing for 48 hours (insert State Department of Agriculture's re-entry interval, if more restrictive).

In case of accidental exposure to pesticide spray or dust, wash the skin thoroughly with soap and water. Remove contaminated clothing and wash before reuse. If in eyes, flush with plenty of water. If inhaled, go to an area where the pesticide has not been applied. Get medical attention if needed.

STORAGE AND DISPOSAL

PESTICIDE STORAGE

Do not store in or around the home.

Keep out of reach of children and animals. Store in original containers only. Store in a cool, dry place and avoid excess heat. Carefully open containers. After partial use, fold and roll back bags, clamp and close tightly. Do not put concentrate or dilute material into food or drink containers. Do not contaminate other pesticides, fertilizers, water, food or feed by storage or disposal.

In case of spill, avoid contact, isolate area and keep out animals and unprotected persons. Confine spills. Call FMC collect: (716) 735-3765.

To confine spill: If liquid, dike surrounding area or absorb with sand, cat litter or commercial clay. If dry material, cover to prevent dispersal. Place damaged package in a holding container. Identify contents.

PESTICIDE DISPOSAL

Pesticide wastes are acutely hazardous. Improper disposal of excess pesticide, spray mixture, or rinsate is a violation of Federal Law. If these wastes cannot be disposed of by use according to label instructions, contact your State Pesticide or Environmental Control Agency, or the Hazardous Waste representative at the nearest EPA Regional Office for guidance.

CONTAINER DISPOSAL

Completely empty bag into application equipment. Then dispose of empty bag in a sanitary landfill or by incineration, or, if allowed by State and local authorities by burning. If burned, stay out of smoke.

DO NOT USE IN UNDILUTED FORM.

Use a knife to slit open the package containing this material. Hold the cut end of the bag closed while introducing the bag into the spray tank opening or slurry bucket to minimize the amount of powder floating in the air. Do not shake bag until it is deep within the tank opening. It is recommended that the concentrated wettable powder be emptied into water in a slurry bucket and then emptied into the tank. This procedure is preferred to dumping the dry material into the tank or onto a screen and then washing it into the tank.

Avoid use of this material in combination with strongly alkaline materials such as Bordeaux, Lime Sulfur or Hydrated Lime. OBSERVE DAYS INTERVAL BETWEEN LAST APPLICATION AND HARVEST INDICATED BY NUMBER IN [ ] FOLLOWING CROP.

APPLES [14]: Mites—Use 1 pound per 100 gallons of water. Apply in 2 or 3 applications at 7 to 10 day intervals. Make first application when mites first become troublesome. Aphids—Use 1 pound per 100 gallons of water when aphids first appear, and make a second application if necesssary. Mealybugs—Use 1 to 1½ pounds per 100 gallons of water in 1 or 2 applications when necessary. Plum Curculio—Use 1 to 1½ pounds per 100 gallons of water in the calyx and first cover spray for the control of early season injury. Where Curculio feeding later in the season causes fruit damage, one or more additional applications may be desirable starting when the first damage appears. Bud Moth—Use 1 pound per 100 gallons of water in an application ahead of bloom where bud moth is a serious problem and satisfactory control is not resulting from the use of a regular program. Redbanded Leafroller—Use 2 lbs. per 100 gallons applied in the calyx and 10 days later for control of first brood larvae. For later broods, use 2 pounds per 100 gallons in 2 or 3 cover sprays when the larvae are present.

CAUTION: For greater safety the use of as low dosages as possible is advised. Slow drying weather increases the danger of injury. Russeting has been reported on Jonathans. Applications of this product may cause injury to fruit and foliage of McIntosh and related varieties.

PEARS [14]: Mites—Use 1 pound per 100 gallons of water. Apply in 2 or 3 applications at 7 to 10 day intervals. Make first application when mites first become troublesome. Mealybugs—Use 1 to 1½ pounds per 100 gallons of water in 1 or 2 applications when necessary. Pear Psylla—Use 1 pound per 100 gallons of water early in the season and later if needed when Psylla becomes troublesome. More than one application may be needed.

PEACHES (Except California) [14], PLUMS [14], PRUNES [14]: Plum Curculio—Use 1½ to 2 pounds per 100 gallons of water in 2 or more applications at 8 to 10 day intervals. Begin when adults are present and continue throughout the period when overwintering Curculio are active. Additional applications will be necessary for later broods. Do not apply more than 5 pounds actual Parathion per acre per year on peaches.

PEACHES (Except California) [14]: Oriental Fruit Moth—Use 1½ to 2 pounds per 100 gallons of water. Time the applications with egg hatch of the broods. Mites—Use 1 pound per 100 gallons of water in 2 or 3 applications at 7 to 10 day intervals. Make applications when mites become troublesome. Cottony-cushion Scale—Use 2 pounds per 100 gallons of water in two applications 10 days apart beginning when the crawlers first appear. Do not apply more than 5 pounds actual Parathion per acre per year on peaches.

CITRUS [14]: Including Grapefruit, Lemons, Limes, Oranges, Tangelos, Tangerines: Florida Red and Purple Scale, Mealybugs, Aphids, Cottony-cushion Scale—Use 2 pounds per 100 gallons of water. Begin when insects first appear and repeat as needed. Do not make more than 3 applications per season. Do not apply within 30 days of harvest if exceed 1300 gallons of water per acre at listed rate.

TOBACCO: Aphids—Use 1 pound per 100 gallons of water as needed. Make no applications within 5 days of priming or 15 days of cutting. Avoid getting wet with plant juices when cutting. Gloves should be worn while transplanting plants or harvesting leaves if an application has been made within 30 days.

TOMATOES [10]: Aphids—Use 2 pounds per acre in sufficient water to cover. Begin when aphids first appear and repeat as needed.

CABBAGE [10], BROCCOLI [7], BRUSSELS SPROUTS [7], CAULIFLOWER [7]: Aphids, Diamond-back Moth Larvae, Imported Cabbageworm, Cabbage Loopers, Armyworms—Use 2 to 3 pounds per acre in sufficient water to cover. Begin when insects first appear, and repeat as needed.

CARROTS [15]: Leafminer, Leafhoppers, Vegetable Weevils—Use 2 to 3 pounds per acre in sufficient water to cover. Aphids, Stink Bugs —Use 3⅓ pounds per acre. Do not use treated carrot tops for food or feed.

TABLE BEETS [15] [21 days if tops are to be used for food or feed]: Flea Beetle, Leaf Miners, Aphids, and Webworms—Use 3 pounds per acre in sufficient water to cover. Blister Beetle—Use 3⅓ pounds per acre.

TURNIPS [10]: Aphids, Cabbage Webworms, Cutworms, Vegetable Weevils, False Chinch Bugs and Harlequin Bugs—Use 3 pounds per acre in sufficient water to cover. Cabbage Loopers—Use 3⅓ pounds per acre. If greens are used for food, do not apply within 21 days of harvest.

POTATOES [5]: Aphids—Use 2 pounds per acre in sufficient water to cover. Begin when aphids first appear and repeat as needed.

DEALERS SHOULD SELL IN ORIGINAL PACKAGES ONLY.

TERMS OF SALE OR USE:

On purchase of this product buyer and user agree to the following conditions:

WARRANTY: FMC makes no warranty, expressed or implied, concerning the use of this product other than indicated on the label. Except as so warranted, the product is sold as is. Buyer and user assume all risk of use and/or handling and/or storage of this material when such use and /or handling and/or storage is contrary to label instructions.

DIRECTIONS AND RECOMMENDATIONS: Follow directions carefully. Timing and method of application, weather and crop conditions, mixture with other chemicals not specifically recommended, and other influencing factors in the use of this product are beyond the control of the seller and are assumed by buyer at his own risk.

USE OF PRODUCT: FMC's recommendations for the use of this product are based upon tests believed to be reliable. The use of this product being beyond the control of the manufacturer, no guarantee, expressed or implied, is made as to the effects of such or the results to be obtained if not used in accordance with directions or established safe practice.

DAMAGES: Buyer's or user's exclusive remedy for damages for breach of warranty or negligence shall be limited to direct damages not exceeding the purchase price paid and shall not include incidental or consequential damages.

POLYRAM® 80 WP

FUNGICIDE

ACTIVE INGREDIENTS:
A mixture of 5.2 parts by weight (83.9%) of ammoniates of [ethylenebis(dithiocarbamato)]-zinc with 1 part by weight (16.1%) ethylene-bis-[dithiocarbamic acid] bimolecular and trimolecular cyclic anhydrosulfides and disulfides 80.0%
INERT INGREDIENTS: 20.0%
TOTAL 100.0%
Zinc expressed as metallic—14.0%

EPA Reg. No. 279-2032AA EPA Est., 279-
Polyram is a Registered Trademark of BASF Aktiengesellschaft

CAUTION
Keep Out of Reach of Children

PRECAUTIONARY STATEMENTS

HAZARDS TO HUMANS (& DOMESTIC ANIMALS)

CAUTION
May be Harmful if Swallowed or Inhaled. May Cause Irritation of Eyes, Nose, Throat and Skin! In case of contact, flush with plenty of water; for eyes, get medical attention. During the mixing and loading process, wear protective clothing (long pants, long sleeve shirt, impermeable gloves, hat and boots).

ENVIRONMENTAL HAZARDS
The product is toxic to fish. Do not apply directly to water. Do not contaminate water by cleaning of equipment or disposal of wastes. Apply this product only as specified on the label.

Do not allow this material to drift onto neighboring crops or non-crop areas or use in a manner or at a time other than in accordance with label directions because animal, plant or crop injury, illegal residues or other undesirable results may occur.

PHYSICAL OR CHEMICAL HAZARDS
Protect from excessive heat.

For Emergency Assistance Call 716–735-3765.
See other panels for precautionary information.

DIRECTIONS FOR USE

It is a violation of Federal Law to use this product in a manner inconsistent with its labeling.

Do not apply this product in such a manner as to directly or through drift expose workers or other persons. The area being treated must be vacated by unprotected persons.

Do not enter treated areas without protective clothing until sprays have dried.

Protective clothing means, at least, a hat or other suitable head covering, a long sleeved shirt and long legged trousers or a coverall type garment (all of closely woven fabric covering the body, including the arms and legs), shoes and socks.

Because certain states may require more restrictive reentry intervals for various crops treated with this product, consult your State Department of Agriculture for further information.

Written or oral warnings must be given to workers who are expected to be in a treated area or in an area about to be treated with this product. Inform workers of areas or fields that may not be entered without specific protective clothing, period of time field must be vacated and appropriate actions to take in case of accidental exposure. An example of such information is given under written warnings. When oral warnings are given, warnings shall be given in a language customarily understood by workers. Oral warnings must be given if there is reason to believe that written warnings cannot be understood by workers. Written warnings must include the following information:

CAUTION
Area treated with POLYRAM® 80 WP on (insert date). Do not enter without appropriate protective clothing until sprays have dried (insert State Department of Agriculture's re-entry interval, if more restrictive).

In case of accidental exposure to pesticide spray or dust, wash the skin thoroughly with soap and water. Remove contaminated clothing and wash before reuse. If in eyes, flush with plenty of water. If inhaled, go to an area where the pesticide has not been applied. Get medical attention if needed.

STORAGE AND DISPOSAL

PESTICIDE STORAGE
Keep out of reach of children and animals. Store in original containers only. Store in a cool, dry place and avoid excess heat. Carefully open containers. After partial use, fold and roll back bags, clamp and close tightly. Do not put concentrate or dilute material into food or drink containers. Do not contaminate other pesticides, fertilizers, water, food or feed by storage or disposal.

In case of spill, avoid contact, isolate area and keep out animals and unprotected persons. Confine spills. Call FMC collect: (716) 735-3765.

To confine spill: If liquid, dike surrounding area or absorb with sand, cat litter or commercial clay. If dry material, cover to prevent dispersal. Place damaged package in a holding container. Identify contents.

PESTICIDE DISPOSAL
Pesticide wastes are toxic. Improper disposal of excess pesticide, spray mixture, or rinsate is a violation of Federal Law. If these wastes cannot be disposed of by use according to label instructions, contact your State Pesticide or Environmental Control Agency, or the Hazardous Waste representative at the nearest EPA Regional Office for guidance.

CONTAINER DISPOSAL
Completely empty bag into application equipment. Then dispose of empty bag in a sanitary landfill or by incineration, or, if allowed by State and local authorities by burning. If burned, stay out of smoke.

Observe days interval between last application and harvest indicated by number in parentheses following crop.

TANK MIXTURES: POLYRAM is compatible with most of the pesticides with which it might normally be used, including spray oils. POLYRAM is compatible with parathion, but should not be used in combination with parathion-oil. When used with oil, Karathane, or Diazinon, POLYRAM should be thoroughly mixed with water in the spray tank and the tank filled nearly full of water before the other material is added. Do not combine POLYRAM with fixed coppers or highly alkaline materials such as Bordeaux mixture or lime sulfur. Observe all cautions and limitations on labeling of all products used in mixtures.

Karathane is a registered Trademark of Rohm and Haas Company

Benlate is a registered Trademark of E. I. Du Pont De Nemours & Co. (Inc.)

APPLES (30*): Cedar Apple Rust, Scab, Sooty Blotch, Fly Speck, Botryosphaeria Rot, Bitter Rot, Black Rot, Brown Rot, and as an aid in reduction of European Red Mite—Use only on a protective schedule. POLYRAM 80 WP must be applied before infection occurs. Do not graze livestock on treated areas. For ground application: Use 2 pounds per 100 gallons of water starting when green tissue is first exposed and continue applications as needed through petal fall. In the cover sprays use 1½ pounds per 100 gallons of water. Use a maximum of 12 to 16 pounds in sufficient water to treat one acre. For aerial application: Use 8 pounds in a minimum of 6 gallons of water per acre starting when green tissue is first exposed and continue applications as needed through first cover. For second and later cover sprays use 3 to 6 pounds in a minimum of 6 gallons of water per acre depending upon weather conditions and severity of listed diseases in the area. Apple Scab, Powdery Mildew— POLYRAM 80 WP can be tank mixed with Benlate 50 WP. Use 1 lb. of POLYRAM 80 WP and 2 oz. of Benlate 50 WP per 100 gallons of water. Apply when green tissue is first exposed and continue as needed through cover sprays.

* Observe a 15 day interval between the last application and harvest in the following states: Arkansas, Delaware, Illinois, Indiana, Kansas, Kentucky, Maryland, Missouri, New Jersey, North Carolina, Ohio, Pennsylvania, South Carolina, Tennessee, Virginia, and West Virginia.

Observe a 30 day interval between the last application and harvest in all other states.

ASPARAGUS (Ferns): Asparagus Rust—Use 2 to 3 pounds per 50 gallons of water per acre by ground application or 3 pounds per acre in 5 to 10 gallons of water by aircraft application. Apply in early August when first signs of disease appear and repeat at 7 to 10 day intervals as required to maintain control.

CELERY (14): Early Blight, Late Blight—Apply 2 pounds in 100 to 150 gallons of water per acre by ground or in 5 to 10 gallons of water per acre by air. Apply at weekly intervals after plants are set in the field. Use the 150 gallons of water per acre rate as plants increase in size or when threat of disease becomes serious. If necessary reduce spray interval to 3 to 5 days when conditions are especially favorable for disease development. Remove excess residues by stripping, trimming and washing.

CUCUMBERS (5), CANTALOUPE (5): Alternaria Leaf Blight, Downy Mildew, Gummy Stem Blight—Use 2 pounds per acre in 100 gallons of water by ground, or in 5 to 10 gallons of water by air. Apply when disease is first observed and repeat at 7 to 10 day intervals as necessary to maintain control.

PEANUTS (0): Cercospora Leaf Spot—Apply 1 to 2 pounds in 25 to 30 gallons of water per acre by ground applicators or 1½ to 2 lbs. in 4 gallons of water per acre by aircraft application. Apply when spots first appear on old leaves at the base of the plants. If a heavy rain occurs within 24 hours after the first application, repeat in 3 to 7 days; otherwise repeat at 10 to 14 day intervals. Do not feed treated forage to dairy animals or animals being finished for slaughter.

PECANS: Pecan Scab—Ground Applications: Use 2 pounds per 100 gallons of water. Airplane or Mist Blower Application: Use 8 pounds per acre in a minimum of 20 gallons of water. Begin applications when buds are bursting and first leaves are showing. Repeat in 10 to 14 days or when first leaves are half grown. Apply a third application when the tips of the small nuts have turned brown. Repeat as required to maintain control. Do not apply after shuck-split. Do not graze livestock on treated areas.

POTATOES (0): Early Blight, Late Blight—Use 1 to 2 pounds in 100 gallons of water per acre by ground application or 2 pounds per acre in 3 to 6 gallons of water per acre by aircraft application. Start applications when plants are small (2 to 6 inches). Apply at 5 to 10 day intervals or more often if weather conditions are conducive for rapid blight development. **Sprinkler Application, Western States only (except California)**—Early Blight Control—Apply two pounds POLYRAM 80 WP per acre using solid set, wheel move, or center pivot spinkler irrigation equipment. Begin application when plants are small (2 to 4 inches). Apply at 5 to 10 day intervals or more often if weather conditions are conducive to rapid blight development.

Do not run irrigation system without check valves or other devices to prevent backup of POLYRAM 80 WP into water source. Irrigation water treated with POLYRAM 80 WP should be maintained on the treated area until the water is absorbed by the soil. Tank containing POLYRAM 80 WP should be connected to the suction side of the irrigation pump or other pressurized equipment attached to the irrigation line.

ROSES (0): Blackspot—Use 1½ pounds per 100 gallons of water (6 teaspoonfuls per 1 gallon). Apply when new growth first appears and repeat at weekly intervals. During periods of continuously rainy weather, shorten application intervals to twice a week.

SUGAR BEETS (0): Cercospora Leaf Spot—Use 1½ to 3 pounds per acre in 100 to 150 gallons of water by ground, or in 5 to 10 gallons of water by air. Apply when disease first appears in area (usually late July or early August). Repeat at 7 to 10 day intervals as required to maintain control. Do not feed treated tops to meat or dairy animals.

SWEET CORN (for fresh vegetables use) (1): For Florida Use Only. Not to be used on sweet corn for processing. Helminthosporium Leaf Spot—Use 1½ to 2 pounds per 100 gallons of water or 3 to 4 pounds in sufficient water to treat one acre. Apply when first signs of disease appear and repeat as required. Do not feed forage to livestock.

TOBACCO (Plant Beds): Anthracnose, Blue Mold—Use 1½ to 2 pounds per 100 gallons of water. Apply 3 to 6 gallons of spray per 100 square yards of plant bed when plants reach size of a dime. Repeat 2 to 3 times per week depending upon the weather and disease severity.

TOMATOES (5): Anthracnose, Early Blight, Gray Leaf Spot, Late Blight—Use 1½ to 3 pounds in 100 to 150 gallons of water per acre by ground, or in 5 to 10 gallons of water by air. Apply as a full coverage dilute spray. Start applications 10 to 14 days after field seeded tomatoes emerge or soon after transplanting. Repeat at 7 to 10 day intervals. In areas where growth is rapid and conditions are conducive for disease development, it may be necessary to shorten the treatment intervals.

DEALERS SHOULD SELL IN ORIGINAL PACKAGES ONLY.

TERMS OF SALE OR USE: On purchase of this product buyer and user agree to the following conditions:

WARRANTY: FMC makes no warranty, expressed or implied, concerning the use of this product other than indicated on the label. Except as so warranted, the product is sold as is. Buyer and user assume all risk of use and/or handling and/or storage of this material when such use and/or handling and/or storage is contrary to label instructions.

DIRECTIONS AND RECOMMENDATIONS: Follow directions carefully. Timing and method of application, weather and crop conditions, mixture with other chemicals not specifically recommended and other influencing factors in the use of this product are beyond the control of the seller and are assumed by buyer at his own risk.

USE OF PRODUCT: FMC's recommendations for the use of this product are based upon tests believed to be reliable. The use of this product being beyond the control of the manufacturer, no guarantee, expressed or implied, is made as to the effects of such or the results to be obtained if not used in accordance with directions or established safe practice.

DAMAGES: Buyer's or user's exclusive remedy for damages for breach of warranty or negligence shall be limited to direct damages not exceeding the purchase price paid and shall not include incidental or consequential damages.

POUNCE ® 1.5 G
INSECTICIDE

RESTRICTED USE PESTICIDE

For retail sale to and use only by certified applicators or persons under their direct supervision, and only for those uses covered by the certified applicator's certification.

ACTIVE INGREDIENT:	BY WT.
*Permethrin	1.50%
INERT INGREDIENTS:	98.50%
TOTAL	100.00%

*(3-Phenoxyphenyl) methyl (±) cis-trans 3-(2,2-dichloroethenyl)-2,2-dimethylcyclopropanecarboxylate cis/trans ratio: Max. 55% (±) cis and min. 45% (±) trans
U.S. Patent No. 4,024,163

EPA Reg. No. 279-3059 EPA Est. 279-
Pounce—FMC trademark

CAUTION
Keep Out of Reach of Children

PRECAUTIONARY STATEMENTS

HAZARDS TO HUMANS (AND DOMESTIC ANIMALS)
CAUTION

Harmful if swallowed or absorbed through skin. Causes moderate eye irritation. Avoid contact with eyes, skin or clothing. Wash thoroughly after handling. Remove and wash clothing before reuse.

STATEMENT OF PRACTICAL TREATMENT

If swallowed, drink 1 or 2 glasses of water and induce vomiting by touching back of throat with finger. Do not induce vomiting or give anything by mouth to an unconscious person. Get medical attention.

If in eyes, flush with plenty of water for at least 15 minutes. Get medical attention.

If on skin, wash with soap and water.

ENVIRONMENTAL HAZARDS

This product is extremely toxic to fish. Use with care when applying in areas adjacent to any body of water. Do not apply directly to water. Keep out of lakes, ponds or streams. Do not contaminate water by cleaning of equipment, or disposal of wastes.

PHYSICAL OR CHEMICAL HAZARDS

Do not use or store near heat or open flame.

For Emergency Assistance Call (716)735-3765

DIRECTIONS FOR USE

It is a violation of Federal Law to use this product in a manner inconsistent with its labeling.

STORAGE AND DISPOSAL

PESTICIDE STORAGE

Do not store in or around the home.

Keep out of reach of children and animals. Store in original containers only. Store in a cool, dry place and avoid excess heat. Carefully open containers. After partial use, fold and roll back bags, clamp and close tightly. Do not put concentrate or dilute material into food or drink containers. Do not contaminate other pesticides, fertilizers, water, food or feed by storage or disposal.

In case of spill, avoid contact, isolate area and keep out animals and unprotected persons. Confine spills. Call FMC collect: (716)735-3765.

To confine spill: If liquid, dike surrounding area or absorb with sand, cat litter or commercial clay. If dry material cover to prevent dispersal. Place damaged package in a holding container. Identify contents.

PESTICIDE DISPOSAL

Wastes resulting from the use of this product may be disposed of on site or at an approved waste disposal facility.

CONTAINER DISPOSAL

Completely empty bag into application equipment. Then dispose of empty bag in a sanitary landfill or by incineration, or, if allowed by State and local authorities, by burning. If burned, stay out of smoke.

FIELD CORN, POPCORN

Preemergent Use: To control Armyworm and Cutworms—Use Pounce 1.5G at a rate of 6.7 to 13.3 pounds (0.1 to 0.2 pounds active) per acre in the time period from five days prior to planting up to emergence of the crop. Apply as a broadcast granule using ground equipment or by aircraft.

Foliar Use: For control of Armyworm, Cutworms, European Corn Borer and Stalk Borer, apply Pounce 1.5G at a rate of 6.7 to 13.3 pounds (0.1 to 0.2 pounds active) per acre prior to ear formation. When treating for armyworms and European corn borer, apply when eggs begin to hatch. For treatment of stalk borer, Pounce 1.5G must be applied when or shortly before the stalk borer larvae are moving into the corn from the surrounding weeds and grasses. Mowing or burn-down herbicide is suggested to initiate movement of stalk borer larvae. Apply by aircraft by broadcasting over corn plants or with ground equipment by directing the granule into the whorl.

DEALERS SHOULD SELL IN ORIGINAL PACKAGES ONLY.

TERMS OF SALE OR USE: On purchase of this product buyer and user agree to the following conditions:

WARRANTY: FMC makes no warranty, expressed or implied, concerning the use of this product other than indicated on the label. Except as so warranted, the product is sold as is. Buyer and user assume all risk of use and/or handling and/or storage of this material when such use and/or handling and or storage is contrary to label instructions.

DIRECTIONS AND RECOMMENDATIONS: Follow directions carefully. Timing and method of application, weather and crop conditions, mixture with other chemicals not specifically recommended and other influencing factors in the use of this product are beyond the control of the seller and are assumed by buyer at his own risk.

USE OF PRODUCT: FMC's recommendations for the use of this product are based upon tests believed to be reliable. The use of this product being beyond the control of the manufacturer, no guarantee, expressed or implied, is made as to the effects of such or the results to be obtained if not used in accordance with directions or established safe practice.

DAMAGES: Buyer's or user's exclusive remedy for damages for breach of warranty or negligence shall be limited to direct damages not exceeding the purchase price paid and shall not include incidental or consequential damages.

POUNCE® 3.2 EC
INSECTICIDE

RESTRICTED USE PESTICIDE

For retail sale to and use only by certified applicators or persons under their direct supervision, and only for those uses covered by the certified applicator's certification.

ACTIVE INGREDIENT:	
*Permethrin**	38.4%
INERT INGREDIENTS:***	61.6%
TOTAL	100.0%

* (3-Phenoxyphenyl) methyl (±) *cis-trans* 3-(2,2-dichloroethenyl)-2,2-dimethylcyclopropanecarboxylate

** *cis/trans* ratio: Max. 55% (±) *cis* and min. 45% (±) *trans*

*** Contains xylene range aromatic solvents.

Contains 3.2 pounds permethrin per gallon.

EPA Reg. No. 279-3014 EPA Est., 279-FL-1
U.S. Patent No. 4,024,163 Pounce—FMC Trademark

CAUTION
Keep Out of Reach of Children

For Emergency Assistance Call 716–735-3765.
See other panels for additional precautionary information.

FIRST AID:

If in eye: flush eyes with water for 15 minutes. Call a physician.

If swallowed: do not induce vomiting. Call a physician.

If on skin: remove contaminated clothing. Wash with plenty of soap and water. Get medical attention if irritation persists.

Note to Physician: Vomiting should be supervised by a physician or the professional staff because of the possible pulmonary damages by aspiration of the solvent.

PRECAUTIONARY STATEMENTS

HAZARDS TO HUMANS (& DOMESTIC ANIMALS)

CAUTION
Harmful if swallowed or absorbed through skin. Causes moderate eye irritation. Avoid contact with eyes, skin or clothing. Wash thoroughly after handling. Remove and wash clothing before reuse.

ENVIRONMENTAL HAZARDS
This product is highly toxic to bees exposed to direct treatment or residues on crops or weeds. Do not apply this product or allow it to drift to crops or weeds on which bees are actively foraging. Additional information may be obtained from your Cooperative Extension Service.

This product is extremely toxic to fish. Use with care when applying in areas adjacent to any body of water. Do not apply directly to water. Do not apply when weather conditions favor drift from treated areas. Do not contaminate water by cleaning of equipment, or disposal of wastes.

PHYSICAL OR CHEMICAL HAZARDS
Do not use or store near heat or open flame.

DIRECTIONS FOR USE

It is a violation of Federal Law to use this product in a manner inconsistent with its labeling.

Do not apply this product in such a manner as to directly or through drift expose workers or other persons. The area being treated must be vacated by unprotected persons.

Do not enter treated areas without protective clothing until sprays have dried.

Protective clothing means, at least, a hat or other suitable head covering, a long sleeved shirt and long legged trousers or a coverall type garment (all of closely woven fabric covering the body, including the arms and legs), shoes and socks.

Because certain states may require more restrictive reentry intervals for various crops treated with this product, consult your State Department of Agriculture for further information.

Written or oral warnings must be given to workers who are expected to be in a treated area or in an area about to be treated with this product. Inform workers of areas or fields that may not be entered without specific protective clothing, period of time field must be vacated and appropriate actions to take in case of accidental exposure. An example of such information is given under written warnings. When oral warnings are given, warnings shall be given in a language customarily understood by workers. Oral warnings must be given if there is reason to believe that written warnings cannot be understood by workers. Written warnings must include the following information:

CAUTION
Area treated with POUNCE 3.2 EC on (insert date). Do not enter without appropriate protective clothing until sprays have dried (insert State Department of Agriculture's re-entry interval, if more restrictive).
In case of accidental exposure to pesticide spray or dust, wash the skin thoroughly with soap and water. Remove contaminated clothing and wash before reuse. If in eyes, flush with plenty of water. If inhaled, go to an area where the pesticide has not been applied. Get medical attention if needed.

STORAGE AND DISPOSAL

PESTICIDE STORAGE
Do not store below 10°F, (−12°C).

Do not use or store near heat, open flame or hot surfaces.

Keep out of reach of children and animals. Store in original containers only. Store in a cool, dry place and avoid excess heat. Carefully open containers. After partial use, replace lids and close tightly. Do not put concentrate or dilute material into food or drink containers. Do not contaminate other pesticides, fertilizers, water, food, or feed by storage or disposal.

In case of spill, avoid contact, isolate area and keep out animals and unprotected persons. Confine spills. Call FMC collect: (716) 735-3765.

To confine spill: If liquid, dike surrounding area or absorb with sand, cat litter or commercial clay. If dry material, cover to prevent dispersal. Place damaged package in a holding container. Identify contents.

PESTICIDE DISPOSAL
Pesticide wastes are toxic. Improper disposal of excess pesticide, spray mixture, or rinsate is a violation of Federal Law. If these wastes cannot be disposed of by use according to label instructions, contact your State Pesticide or Environmental Control Agency, or the Hazardous Waste representative at the nearest EPA Regional Office for guidance.

CONTAINER DISPOSAL
Metal Containers: Triple rinse (or equivalent). Then offer for recycling or reconditioning, or puncture and dispose of in a sanitary landfill, or by other procedures approved by state and local authorities. Do not cut or weld metal containers.
Glass Containers: Triple rinse (or equivalent). Then dispose of in a sanitary landfill or by other approved State and local procedures.

This label must be in the possession of the user at the time of application.

GENERAL INSTRUCTIONS

POUNCE 3.2 EC is a 3.2 pounds per gallon formulation of the insecticide permethrin. Apply POUNCE when insects appear or feeding is noticed. The higher rate should be used as pest populations increase. Repeat the application as necessary to maintain control. POUNCE may be applied by both ground and aerial equipment. Use sufficient water to obtain full coverage. With the exception of crops listed below, rotational crops should not be planted within 60 days of last application.

ALMONDS: To control Navel Orangeworm, Peach Twig Borer—Use POUNCE 3.2 EC at a rate of 8–16 ounces (0.2 to 0.4 pounds active) per acre. Apply when insects appear. Do not make more than two applications during hull split. Do not make more than 5 applications per season. Apply in a minimum of 15 gallons of finished spray per acre by aircraft or 20 gallons of finished spray per acre with ground equipment. Do not apply within seven days of harvest. Do not graze livestock in treated orchards.

APPLES: To control Plum Curculio, Redbanded Leafroller, Rosy Apple Aphid, Spotted Tentiform Leafminer, Tarnished Plant Bug, White Apple Leafhopper, Oblique Banded Leafroller and Green Fruitworm—Use POUNCE 3.2 EC at a rate of 4–8 ounces (0.1 to 0.2 pounds active) per acre with ground equipment only. Apply in 25 to 400 gallons of finished spray per acre when insects appear and repeat as required to maintain control. Do not make more than three applications per season. Do not apply after petal fall.

ASPARAGUS: To control Cutworms and Asparagus Beetle—Use Pounce 3.2 EC at the rate of 2 to 4 ounces (0.05 to 0.1 lbs. active) per acre. Apply by ground equipment using sufficient water to obtain uniform coverage. Apply as needed. Do not exceed 4 applications. Do not apply within 3 days of harvest.

For post harvest application, use Pounce 3.2 EC at a rate of 4 ounces (0.1 lbs. active) per acre. Apply to the fern stage of the asparagus plant after spear harvest when larval and adult stage of the Asparagus Beetle, Tarnished Plant Bug and Lygus Bugs and the adult stage of the Japanese beetle are present.

BROCCOLI, BRUSSELS SPROUTS, CAULIFLOWER: To control Beet Armyworm, Cabbage Looper, Imported Cabbageworm and Diamondback Moth—Use POUNCE 3.2 EC at a rate of 2 to 4 ounces (0.05 to 0.1 pound active) per acre. Apply in a minimum of 3 gallons of finished spray per acre by aircraft and 20 gallons per acre by ground equipment. Do not make more than 8 applications per season. Do not apply within one day of harvest.

CABBAGE: To control Beet Armyworm, Cabbage Looper, Diamondback Moth, and Imported Cabbageworm—Use POUNCE 3.2 EC at a rate of 2 to 4 ounces (0.05 to 0.1 pound active) per acre. Apply in a minimum of 1 gallon of finished spray per acre by aircraft and 20 gallons per acre with ground equipment. Do not make more than 10 applications per season. Do not apply within one day of harvest.

CELERY: To control Leafminer, Cutworms and Southern Armyworm—Use POUNCE 3.2 EC at a rate of 4 to 8 ounces (0.1–0.2 pound active) per acre on a 3 to 5 day schedule, or as needed to maintain control. Apply in a minimum of 5 gallons of finished spray per acre by aircraft and 10 gallons per acre by ground equipment. Do not make more than 10 applications per season in California. Do not apply within 3 days of harvest. Do not graze celery forage nor feed trimmings from treated fields to livestock.

CHRYSANTHEMUMS (Greenhouse): For control of Liromyza Leafminer flies, use POUNCE 3.2 EC at a rate of 20 fluid ounces (0.5 pound active) per 100 gallons (1 teaspoon per gallon). Avoid spraying the blooms. POUNCE may be applied on a weekly schedule.

Caution: Pounce has demonstrated excellent plant safety; however, not all cultivars have been tested. Before treating large numbers of plants of a particular cultivar, treat a few plants and observe prior to full scale applications.

CONIFERS (Container and Field Grown): For control of Nantucket Pine Tip Moth use POUNCE 3.2 EC at a rate of 4 to 8 fluid ounces (0.1 to 0.2 pound active) per acre. POUNCE may be diluted in a non-volatile vegetable oil or water in a minimum of 1 gallon of finished spray per acre using equipment calibrated to give adequate coverage. Begin application when the adults appear and repeat at 5–7 day intervals or as needed throughout the season.

COTTON: To control Beet Armyworm (California and Arizona only), Bollworm, Cabbage Looper, Cotton Fleahopper, Cotton Leaf Perforator, Cutworm, Lygus Bugs, Pink Bollworm, Tarnished Plant Bug and Tobacco Budworm—Use POUNCE 3.2 EC at a rate of 4 to 8 ounces (0.1 to 0.2 pound active) per acre. To control Boll Weevil, alternate POUNCE 3.2 EC at 4 ounces (0.1 pound active) per acre with POUNCE 3.2 EC at 2 ounces (0.05 pound active) per acre every 3–4 days until control is achieved.

Apply the above rates in a minimum of one gallon of water per acre by aircraft and 5 gallons of water per acre with ground equipment. POUNCE 3.2 EC may also be applied using refined non-volatile vegetable oil for control of the above named insects. POUNCE 3.2 EC should be diluted with oil and applied in a minimum of one quart total volume/acre using equipment calibrated to give adequate coverage. To control Bollworm and Tobacco Budworm use 2 to 8 ounces (0.05 to 0.2 pounds active) per acre. The 2 ounce rate per acre should be used only when insect infestation levels are light and the cotton is closely scouted before and after application.

To control cutworms—Use POUNCE 3.2 EC at a rate of 4 to 8 ounces (0.1 to 0.2 pounds active) per acre. Apply prior to planting or pre-emergent to crop. Use a minimum of 1 gallon of water per acre by aircraft and 5 gallons of water per acre by ground equipment. POUNCE 3.2 EC has been found to be effective and compatible in tank mixtures with herbicides and liquid fertilizers.

Do not make more than 15 applications per season. Do not apply within 14 days of harvest. Do not graze or feed cotton forage.

Cotton Tank-Mix: POUNCE® 3.2 EC Plus Tank Mixtures (all cotton growing states). Mix POUNCE 3.2 EC at 4.0 to 8.0 ounces (0.1 to 0.2 pound active) per acre with other insecticides recommended on this label for control of cotton insect pests. POUNCE 3.2 EC has been found to be effective and compatible in tank mixtures with the following products:

- Methyl Parathion
- EPN plus Methyl Parathion
- Methomyl
- Guthion
- Azinphos-Methyl
- Malathion
- Chlordimeform

When using POUNCE 3.2 EC in tank mixtures, observe all restrictions and precautions which appear on the labels of these products. Apply tank mixtures immediately after mixing.

EGGPLANTS: To control Colorado Potato Beetle—Use POUNCE 3.2 EC at the rate of 8 ounces (0.2 lbs. active) per acre. To control Vegetable Leafminer—Use POUNCE 3.2 EC at the rate of 4 to 8 ounces (0.1 to 0.2 lbs. active) per acre. Apply using sufficient water to obtain uniform coverage. Apply as needed. Do not exceed 16 applications. Do not apply within 3 days of harvest.

FIELD CORN, POPCORN

Preemergent Use: To control Armyworm and Curworms—Use POUNCE 3.2 EC at a rate of 4 to 8 ounces (0.1 to 0.2 pounds active) in the time period from five days prior to planting up to emergence of the crop. Apply as a broadcast spray in a minimum of 20 gallons of finished spray per acre with ground equipment. Pounce has been found compatible and may be tank mixed with the following herbicides:

AAtrex®	Lasso®	Prowl®
Banvel®	Lorox®	Ramrod®
Bladex®	Paraquat CL®	Roundup®
Dual®	Princep®	

When using Pounce 3.2 EC in tank mixtures, observe all restrictions and precautions which appear on the labels of these products. Provide constant agitation during mixing and application to keep the mixture in suspension. Label dosage rates should not be exceeded.

Foliar Use: For control of Cutworms, European Corn Borer, Armyworm and Stalk Borer—Apply POUNCE 3.2 EC at a rate of 4 to 8 ounces (0.1 to 0.2 pounds active) prior to brownsilk stage. When treating for stalk borer, Pounce 3.2 EC must be applied when or shortly before the stalk borer larvae are moving into the corn from surrounding weeds and grasses. Mowing or burndown herbicide are suggested to initiate movement. Apply a minimum of 1 gallon of finished spray per acre by air and 10 gallons per acre with ground equipment. POUNCE may also be injected into overhead sprinkler irrigation water provided (1) an anti-backflow check valve is present between the injection point and the water source, (2) the injection system has a check valve in the line to prevent irrigation water from entering the chemical supply tank and (3) the irrigation injection system must have interlocking on-off switches.

HEAD LETTUCE: To control Alfalfa Looper and Cabbage Looper —Use POUNCE 3.2 EC at a rate of 2 to 8 ounces (0.05 to 0.2 pound active) per acre. To control Beet Armyworm, Corn Earworm and Tobacco Budworm—Use POUNCE 3.2 EC at a rate of 4 to 8 ounces (0.1 to 0.2 pounds active) per acre. Apply a minimum of 5 gallons of finished spray per acre by aircraft and 15 gallons per acre with ground equipment. Do not make more than 15 applications per season.

HORSERADISH: To control Imported Crucifer Weevil (*Baris lepidii*).

As a spring preplant dip—Use a 0.1% active solution (2 pints, 1 ounce of POUNCE 3.2 EC per 100 gallons). Soak sets for 30 minutes and air-dry before planting.

For foliar application, use POUNCE 3.2 EC at a rate of 1 pint (0.2 pound active) per acre. Apply by ground equipment in a minimum spray volume of 20 gallons per acre. Make up to 3 foliar applications as needed to control weevil adults during oviposition.

Do not apply within 22 days of harvest.

Leafy Vegetables (except Brassica): Amaranth; arrugula; celtuce; chervil; corn salad; chrysanthemum (edible-leaved and garland); cress, garden; cress, upland; dandelion; dock; endive; fennel; orach; parsley; purslane, garden and winter; rhubarb; Swiss chard: To control Vegetable Leafminer, Cabbage Looper, Cutworms, Fall Armyworm, Beet Armyworm, Corn Earworm, Green Cloverworm, Leafhoppers and European Corn Borer—Use Pounce 3.2 EC at a rate of 4 to 8 ounces (0.1 to 0.2 pounds active) per acre. Apply every 3 to 5 days or as needed. Use sufficient water to obtain full coverage of foilage. Do not apply within one day of harvest. Do not graze treated areas or field crop refuse to livestock.

ORNAMENTAL NURSERY STOCK (Field Grown): To control Beet Armyworm, Cabbage Looper, Citrus Thrips, *Heliothis spp.*, Leafhoppers, Leafminers and Whiteflies—Use 4 to 8 ounces per 100 gallons of water. POUNCE 3.2 EC may be used to control specified pests on non-edible ornamentals and non-bearing plants of fruiting species.

Caution: Marginal leaf burn may occur on Salvia, Dieffenbachia and Pteris Fern. Application to blooming plants may cause browning of petals. POUNCE has demonstrated excellent plant safety; however, not all species and varieties have been tested. Before treating large numbers of plants of a particular variety, treat a few plants and observe prior to full scale application.

PEACHES: To control Lesser Peach Tree Borer, Peach Twig Borer, Rose Chafer, Plum Curculio, Tarnished Plant Bug, Green Fruitworm, Oriental Fruit Moth—Use POUNCE 3.2 EC at a rate of 2–4 ounces (0.05–0.1 pound active) per 100 gallons of water. Do not exceed 400 gallons per acre. Apply when insects appear. Make applications with ground equipment, using sufficient water to obtain uniform coverage. Do not apply within 7 days of harvest. Do not make more than 8 applications per season. Do not graze livestock in treated orchards.

PEARS (Dormant through Delayed Dormant): To control Pear Psylla—Apply POUNCE 3.2 EC at a rate of 8 to 16 ounces (0.2 to 0.4 pound active) per acre. POUNCE 3.2 EC may be combined with 2 to 8 gallons of spray oil per acre. Apply during the dormant through delayed dormant growth periods only. Apply in 3 to 20 gallons of finished spray per acre by aircraft and 25 to 400 gallons per acre by ground equipment. Do not make more than two applications per season.

PEARS (Summer): To control Pear Psylla, Green Fruitworm, Codling Moths—Use POUNCE 3.2 EC at a rate of 8–16 ounces (0.2–0.4 pound active) per acre. Use in sufficient water to achieve runoff and apply by ground equipment. Do not make more than 3 applications per season. Do not apply within 14 days of harvest.

PEPPERS (Bell): To control Vegetable Leafminer, Cabbage Looper, Flea Beetle, Corn Earworm, Pepper Weevil—Use POUNCE 3.2 EC at a rate of 4 to 8 ounces (0.1 to 0.2 lbs. active) per acre. To control European Corn Borer—Use POUNCE 3.2 EC at the rate of 8 ounces (0.2 lbs. active) per acre. Apply using sufficient water to obtain uniform coverage. Apply as needed. Do not exceed 8 sprays per season. Do not apply within 3 days of harvest.

PINE SEED ORCHARDS: To control Coneworms, Seed Bugs—Use POUNCE 3.2 EC at the following rates:

13 ounces in 100 gallons of water (0.025% dilution by weight) for high volume sprayers;

68 ounces in 100 gallons of water (0.125% dilution by weight) for low volume sprayers;

48 ounces in not less than 10 gallons of water per acre for aerial application.

Apply first application within 1 week of female flower closure or peak pollen flight for webbing coneworm control. For other coneworms and seed bugs, apply first application within 30 days following female flower closure. Repeat applications at 4 week intervals, but do not apply more than 6 applications. Apply approximately 5–10 gallons of the 0.025% dilution per tree with high volume sprayers. With low volume sprayers, apply 100 gallons of the 0.125% dilution per acre. Do not graze or harvest cover crop. Avoid contact with open water.

PISTACHIOS: To control Peach Twig Borer, Navel Orangeworm and Chinch Bug—Use Pounce 3.2 EC at a rate of 8 to 16 ounces (0.2 to 0.4 lbs. active) per acre. Use sufficient water to obtain full coverage of foliage. Do not make more than two applications per season. Do not apply after 10% hull split. Nuts may be harvested on the day of application. Do not graze livestock in treated orchards.

POTATOES: To control Beet Armyworm, Cabbage Looper, Colorado Potato Beetle, Cutworms, European Corn Borer, Potato Flea Beetle, Potato Leafhopper, Potato Psyllid, Potato Tuberworm, Aster Leafhopper and Tarnished Plant Bug. Use POUNCE 3.2 EC at a rate of 4 to 8 ounces (0.1 to 0.2 pound active) per acre. Apply in a minimum of 3 gallons of finished spray per acre by air or 20 gallons of finished spray per acre using ground equipment. Pounce 3.2 EC may also be applied using refined non-volatile vegetable oil for control of the above-named insects. Pounce 3.2 EC should be diluted with oil and applied in a minimum of one quart total volume per acre using equipment calibrated to give adequate coverage. Do not make more than 12 applications per season. Do not apply within 7 days of harvest. Do not graze or feed potato forage.

RANGE GRASS: To control Range Caterpillar—Use POUNCE 3.2 EC at a rate of 0.4 ounces (0.01 lb. active) per acre. Apply using sufficient water to obtain uniform coverage. Do not apply more than once per year. Cattle may be present during application. Do not harvest or feed hay to livestock.

ROSES (Field Grown): For control of *Heliothis sp.*, use 4 to 8 fluid ounces (0.1 to 0.2 pound active) per acre. POUNCE may be applied in 5 gallons of finished spray per acre with ground equipment and 1 gallon per acre by air.

ROSES (Greenhouse): For control of Beet Armyworm, Cabbage Looper and Omniverous Leafroller, use POUNCE at a rate of 8 fluid ounces (0.2 pound active) per 100 gallons of water.

Caution: POUNCE is not phytotoxic to the following varieties of greenhouse roses: Ballena, Bettina, Cara Mia, Coquette, Excitement, Forever Yours, G. Wave, Jack Frost, Jr. Bridesmaid, Matador, Paul's Pink, Samantha, Seventeen, Sonia, Town Crier, Tropicana and Visa. Other varieties may vary in their sensitivity to POUNCE, and a small number of plants should be treated under local conditions to determine plant safety prior to commercial use.

SOYBEANS: To control Beet Armyworm, Cabbage Looper, Corn Earworm, Green Cloverworm, Mexican Bean Beetle, Soybean Looper, and Velvetbean Caterpillar, Bean Leaf Beetle and Potato Leafhopper—Use POUNCE 3.2 EC at a rate of 2–4 ounces (0.05 to 0.1 pound active) per acre. To control Beet Armyworm—Use

POUNCE 3.2 EC at a rate of 4–8 oz. (0.1 to 0.2 lb. active) per acre. Use higher dosage rate as population pressure increases. Apply a minimum of one gallon of finished spray per acre by aircraft and 5 gallons with ground equipment.

POUNCE 3.2 EC may also be applied using refined non-volatile vegetable oil for control of the above named insects. POUNCE 3.2 EC should be diluted with oil and applied in a minimum of one quart total volume/acre using equipment calibrated to give adequate coverage.

Do not make more than two applications per season. Do not apply within 60 days of harvest. Do not graze or feed soybean forage. Do not plant rotational crops within 60 days of last application.

SPINACH: To control Beet Armyworm, Corn Earworm, Cabbage Looper, Cutworms, European Corn Borer, Green Cloverworm, Leafhoppers—Use POUNCE 3.2 EC at a rate of 4 ounces (0.1 pound active) per acre. Apply when insects appear. Apply in sufficient water to obtain uniform coverage. Do not make more than 7 applications per season. Do not apply within 7 days of harvest.

SWEET CORN: To control Corn Earworm, European Corn Borer, Fall Armyworm—Use POUNCE 3.2 EC at a rate of 4–8 ounces (0.1–0.2 pound active) per acre. Apply every 5 days or as needed. Do not apply more than 6 applications. Do not apply within 1 day of harvest of ears or forage or livestock grazing.

SWEET CORN (Florida only) fresh market use only: To control Corn Earworm, European Corn Borer, Fall Armyworm, Southern Armyworm—Use 4–10 ounces (0.1–0.25 pound active) per acre. Apply every 5 days or as needed.

Do not apply more than 3.25 pounds active ingredient per acre per crop season. Do not graze livestock in treated areas or harvest for forage.

TOMATOES (ONLY FOR APPLICATION TO TOMATOES GROWN IN FLORIDA FOR FINAL MARKETING AS FRESH TOMATOES)

To control Beet Armyworm, Cabbage Looper, Colorado Potato Beetle, Granulate Cutworm, Hornworms, Vegetable Leafminers, Southern Armyworm, Tomato Fruitworm and Tomato Pinworm—Use Pounce 3.2 EC at a rate of 2 to 8 ounces (0.05 to 0.2 pound active) per acre. Apply by aerial or ground equipment in sufficient water to obtain uniform coverage. Do not apply more than 1.2 pound (active) per acre per season. Can be applied up to the day of harvest. Do not apply to cherry tomatoes or other varieties which produce fruit less than one inch in diameter.

Treatment of Preconstruction Lumber and Logs

General Information

Pounce® 3.2 EC insecticide should be diluted with water. To prepare the spray, dilute Pounce 3.2 EC as shown in the following spray dilution chart:

Spray Dilution Chart

Gallons of Spray Mixture Desired	Gallons of Pounce 3.2 EC To Use		
	0.5% Solution	0.75% Solution	1.0% Solution
40	1/2	3/4	1
80	1	1 1/2	2
200	2 1/2	3 3/4	5
400	5	7 1/2	10
800	10	15	20

Directions for Application

To protect unseasoned lumber and logs from wood destroying insects, such as Termites, Carpenter Ants and Beetles (Ambrosia, Powder-post, Old house borers and others), totally treat wood with a 0.5% to 1.0% solution of Pounce 3.2 EC. This solution can be applied by various methods including spraying, brushing, dipping, and pressure treatment. Frequent monitoring of dip and pressure systems are necessary to insure that the desired level of Pounce 3.2 EC is maintained. Wood can be handled after treatment when dry.

1. For dip treatments, the wood should be totally submersed in the solution until thoroughly wet and then allowed to dry in a suitable location. Dipping solutions to which Pounce 3.2 EC has been added should be agitated before use if left unused for long periods of time. Sediment, debris and other deposits should be periodically cleaned from the tank.

2. For pressure treatments, the wood should be placed in the treatment chamber, the Pounce 3.2 EC solution added and the system pressurized up to 250 psi for up to one hour depending on the density and type of wood treated. After the pressure is released and the system drained, the wood should be placed in a suitable location for drying.

3. For spray treatments, the wood should be sprayed thoroughly including back and ends.

4. For brush treatments, all parts of wood surfaces should be thoroughly treated.

DEALERS SHOULD SELL IN ORIGINAL PACKAGES ONLY.

TERMS OF SALE OR USE: On purchase of this product buyer and user agree to the following conditions:

WARRANTY: FMC makes no warranty, expressed or implied, concerning the use of this product other than indicated on the label. Except as so warranted, the product is sold as is. Buyer and user assume all risk of use and/or handling and/or storage of this material when such use and/or handling and/or storage is contrary to label instructions.

DIRECTIONS AND RECOMMENDATIONS: Follow directions carefully. Timing and method of application, weather and crop conditions, mixture with other chemicals not specifically recommended, and other influencing factors in the use of this product are beyond the control of the seller and are assumed by buyer at his own risk.

USE OF PRODUCT: FMC's recommendations for the use of this product are based upon tests believed to be reliable. The use of this product being beyond the control of the manufacturer, no guarantee, expressed or implied, is made as to the effects of such or the results to be obtained if not used in accordance with directions or established safe practice.

DAMAGES: Buyer's or user's exclusive remedy for damages for breach of warranty or negligence shall be limited to direct damages not exceeding the purchase price paid and shall not include incidental or consequential damages.

POUNCE® MUSHROOM DUST

INSECTICIDE

For Agricultural or Commercial Use Only

ACTIVE INGREDIENT:
*Permethrin** 2%
INERT INGREDIENTS: 98%
TOTAL 100%

* (3-Phenoxyphenyl) methyl (±) *cis-trans* 3-(2,2-dichloroethenyl)-2,2-dimethycyclopropanecarboxylate

** *cis/trans* ratio: Max. 55% (±) *cis* and min. 45% (±) *trans*

EPA Reg. No. 279-3040 EPA Est., 279-FL-1
U.S. Patent No. 4,024,163 Pounce—FMC Trademark

CAUTION

Keep Out of Reach of Children

FIRST AID

If swallowed: Call a physician or Poison Control Center. Drink 1 or 2 glasses of water and induce vomiting by touching back of throat with finger, or, if available, by administering syrup of ipecac. Do not induce vomiting or give anything by mouth to an unconscious person.

If on skin: Wash with plenty of soap and water. Get medical attention.

If in eyes: Flush with plenty of water. Get medical attention if irritation persists.

For Emergency Assistance Call 716–735-3765

See other panels for additional precautionary information.

PRECAUTIONARY STATEMENTS

HAZARDS TO HUMANS (& DOMESTIC ANIMALS)

CAUTION

Harmful if swallowed or absorbed through skin. Causes moderate eye irritation. Avoid contact with eyes, skin or clothing. Wash thoroughly after handling. Remove and wash clothing before reuse.

ENVIRONMENTAL HAZARDS

Do not contaminate water by cleaning of equipment, or disposal of wastes.

DIRECTIONS FOR USE

It is a violation of Federal law to use this product in a manner inconsistent with its labeling.

STORAGE AND DISPOSAL

PESTICIDE STORAGE

Keep out of reach of children and animals. Store in original containers only. Store in a cool, dry place and avoid excess heat. Carefully open containers. After partial use, fold and roll back bags, clamp and close tightly. Do not put concentrate or dilute material into food or drink containers. Do not contaminate other pesticides, fertilizers, water, food or feed by storage or disposal.

In case of spill, avoid contact, isolate area and keep out animals and unprotected persons. Confine spills. Call FMC collect: (716) 735-3765.

To confine spill: If liquid, dike surrounding area or absorb with sand, cat litter or commercial clay. If dry material, cover to prevent dispersal. Place damaged package in a holding container. Identify contents.

PESTICIDE DISPOSAL

Wastes resulting from the use of this product may be disposed of on site or at an approved waste disposal facility.

CONTAINER DISPOSAL

Completely empty bag into application equipment. Then dispose of empty bag in a sanitary landfill or by incineration, or, if allowed by State and local authorities, by burning. If burned, stay out of smoke.

This label must be in the possession of the user at the time of application.

Mushroom Houses and Adjacent Premise Areas

For control of mushroom flies (Sciarid and Phorid adults):

Preparation of the Building Prior to Fogging:

(1) Close all doors, windows and ventilators
(2) Lock or barricade all entrances, turn off pilot lights, post warning signs, and take precautions to prevent persons and animals from entering the area.

APPLICATION:

Apply POUNCE MUSHROOM DUST at the rate of 3 pounds (0.06 lb. a.i.) of POUNCE MUSHROOM DUST per standard double (35,000 cu. ft.: 8000 sq. ft.).

Use POUNCE MUSHROOM DUST prior to filling house, during cool-down, during spawning, up to pinning and between breaks. Do not use when mushrooms are present. Treat once daily or as needed when flies appear. Do not make more than 20 applications prior to pinning of first break. Apply no more than two applications between each break. Do not apply more than 30 applications total per crop of 5 breaks.

Do not apply within 3 days of harvest.

Length of exposure time should be limited to 1 hour, then ventilate the house. Use fans to ventilate in houses that do not have forced air circulation. Wear full-faced gas mask with canister type recommended for general insecticide protection.

DEALERS SHOULD SELL IN ORIGINAL PACKAGES ONLY.

TERMS OF SALE OR USE: On purchase of this product buyer and user agree to the following conditions:

WARRANTY: FMC makes no warranty, expressed or implied, concerning the use of this product other than indicated on the label. Except as so warranted, the product is sold as is. Buyer and user assume all risk of use and/or handling and/or storage of this material when such use and/or handling and/or storage is contrary to label instructions.

DIRECTIONS AND RECOMMENDATIONS: Follow directions carefully. Timing and method of application, weather and crop conditions, mixture with other chemicals not specifically recommended, and other influencing factors in the use of this product are beyond the control of the seller and are assumed by buyer at his own risk.

USE OF PRODUCT: FMC's recommendations for the use of this product are based upon tests believed to be reliable. The use of this product being beyond the control of the manufacturer, no guarantee, expressed or implied, is made as to the effects of such or the results to be obtained if not used in accordance with directions or established safe practice.

DAMAGES: Buyer's or user's exclusive remedy for damages for breach of warranty or negligence shall be limited to direct damages not exceeding the purchase price paid and shall not include incidental or consequential damages.

POUNCE® MUSHROOM SPRAYMIST

INSECTICIDE

For Agricultural or Commercial Use Only

ACTIVE INGREDIENT:

*Permethrin**	38.4%
Xylene range aromatic solvent	50.2%
INERT INGREDIENTS***	11.4%
TOTAL	100.0%

* (3-Phenoxyphenyl) methyl (±) *cis-trans 3* -(2,2-dichloroethenyl)-2,2-dimethylcyclopropanecarboxylate

** *cis/trans* ratio: Max. 55% (±) *cis* and min. 45% (±) *trans*

Contains 3.2 lbs. permethrin per gallon.

EPA Reg. No. 279-3039 EPA Est., 279-FL-1
U.S. Patent No. 4,024,163
Pounce—FMC Trademark

CAUTION

Keep Out of Reach of Children

FIRST AID:

If in eye, flush eyes with plenty of water. Get medical attention if irritation persists.

If swallowed, do not induce vomiting. Call a physician.

If on skin, remove contaminated clothing. Wash with plenty of soap and water. Get medical attention if irritation persists.

Note to Physician: Vomiting should be supervised by a physician or the professional staff because of the possible pulmonary damages by aspiration of the solvent.

For Emergency Assistance Call 716–735-3765

See other panels for additional precautionary information.

PRECAUTIONARY STATEMENTS

HAZARDS TO HUMANS (& DOMESTIC ANIMALS)

CAUTION

Harmful if swallowed or absorbed through skin. Causes moderate eye irritation. Avoid contact with eyes, skin or clothing. Wash thoroughly after handling. Remove and wash clothing before reuse.

ENVIRONMENTAL HAZARDS

Do not contaminate water by cleaning of equipment, or disposal of wastes.

PHYSICAL OR CHEMICAL HAZARDS

Do not use or store near heat or open flame.

DIRECTIONS FOR USE

It is a violation of Federal law to use this product in a manner inconsistent with its labeling.

STORAGE AND DISPOSAL

PESTICIDE STORAGE

Do not store below 10°F, (−12°C).

Do not use or store near heat, open flame or hot surfaces.

Keep out of reach of children and animals. Store in original containers only. Store in a cool, dry place and avoid excess heat. Carefully open containers. After partial use, replace lids and close tightly. Do not put concentrate or dilute material into food or drink containers. Do not contaminate other pesticides, fertilizers, water, food, or feed by storage or disposal.

In case of spill, avoid contact, isolate area and keep out animals and unprotected persons. Confine spills. Call FMC collect: (716) 735-3765.

To confine spill, if liquid, dike surrounding area or absorb with sand, cat litter or commercial clay. If dry material, cover to prevent dispersal. Place damaged package in a holding container. Identify contents.

PESTICIDE DISPOSAL

Pesticide wastes are toxic. Improper disposal of excess pesticide, spray mixture, or rinsate is a violation of Federal Law. If these wastes cannot be disposed of by use according to label instructions, contact your State Pesticide or Environmental Control Agency, or the Hazardous Waste representative at the nearest EPA Regional Office for guidance.

CONTAINER DISPOSAL

Glass Container Triple rinse (or equivalent). Then dispose of in a sanitary landfill or by other approved State and local procedures.

Metal Container Triple rinse (or equivalent). Then offer for recycling or reconditioning, or puncture and dispose of in a sanitary landfill, or by other procedures approved by state and local authorities. Do not cut or weld metal containers.

This label must be in the possession of the user at the time of application.

Mushroom Houses and Adjacent Premise Areas

For control of mushroom flies (Sciarid and Phorid adults):

Preparation of the Building Prior to Fogging:

(1) Close all doors, windows and ventilators
(2) Lock or barricade all entrances, turn off pilot lights, post warning signs, and take precautions to prevent persons and animals from entering the area.

APPLICATION:

Apply POUNCE MUSHROOM SPRAYMIST as a fogging or aerosol treatment at the rate of 2 to 2.5 ounces (0.05–0.0625 lb. a.i.) POUNCE MUSHROOM SPRAYMIST per 30 ounces of water or suitable diluent. Use 1 quart of solution per standard double (35,000 cu. ft.: 8,000 sq. ft.).

Use POUNCE MUSHROOM SPRAYMIST prior to filling house, during cooldown, during spawning up to pinning and between breaks. Do not use when mushrooms are present. Treat once daily or as needed when flies appear. Do not make more than 20 applications prior to pinning of first break. Apply no more than two applications between each break. Do not apply more than 30 applications total per crop of 5 breaks.

Do not apply within 3 days of harvest.

Length of exposure time should be limited to 1 hour, then ventilate the house. Use fans to ventilate in houses that do not have forced air circulation. Wear full-faced gas mask with canister type recommended for general insecticide protection.

DEALERS SHOULD SELL IN ORIGINAL PACKAGES ONLY.

TERMS OF SALE OR USE: On purchase of this product buyer and user agree to the following conditions:

WARRANTY: FMC makes no warranty, expressed or implied, concerning the use of this product other than indicated on the label. Except as so warranted, the product is sold as is. Buyer and user assume all risk of use and/or handling and/or storage of this material when such use and/or handling and/or storage is contrary to label instructions.

DIRECTIONS AND RECOMMENDATIONS: Follow directions carefully. Timing and method of application, weather and crop conditions, mixture with other chemicals not specifically recommended, and other influencing factors in the use of this product are beyond the control of the seller and are assumed by buyer at his own risk.

USE OF PRODUCT: FMC's recommendations for the use of this product are based upon tests believed to be reliable. The use of this product being beyond the control of the manufacturer, no guarantee, expressed or implied, is made as to the effects of such or the results to be obtained if not used in accordance with directions or established safe practice.

DAMAGES: Buyer's or user's exclusive remedy for damages for breach of warranty or negligence shall be limited to direct damages not exceeding the purchase price paid and shall not include incidental or consequential damages.

STIK®

PLANT GROWTH REGULATOR

ACTIVE INGREDIENT:

1-Naphthaleneacetic Acid, Sodium Salt	7.1%
INERT INGREDIENTS:	92.9%
TOTAL	100.0%

1-Naphthaleneacetic Acid equivalent 6.34%.
Contains 4 grams 1-Naphthaleneacetic Acid Sodium Salt per 2 oz.

EPA Reg. No. 279–1735 EPA Est.,-
Stik—FMC Trademark

CAUTION

Keep Out of Reach of Children

STATEMENT OF PRACTICAL TREATMENT

If in eyes: flush with water for at least 15 minutes.

If on skin: wash with soap and plenty of water. Remove and wash contaminated clothing before reuse.

PRECAUTIONARY STATEMENTS

HAZARDS TO HUMANS AND DOMESTIC ANIMALS

CAUTION

Harmful if swallowed. Avoid contact with skin, eyes and clothing. Avoid breathing spray mist.

For Emergency Assistance Call 716–735–3765.

ENVIRONMENTAL HAZARDS

Keep out of lakes, streams and ponds. Do not contaminate water by cleaning of equipment or disposal of wastes.

See other panels for additional precautionary information.

DIRECTIONS FOR USE

It is a violation of Federal Law to use this product in a manner inconsistent with its labeling.

Do not apply this product in such a manner as to directly or through drift expose workers or other persons. The area being treated must be vacated by unprotected persons.

Do not enter treated areas without protective clothing until sprays have dried.

Protective clothing means, at least, a hat or other suitable head covering, a long sleeved shirt and long legged trousers or a coverall type garment (all of closely woven fabric covering the body, including the arms and legs), shoes and socks.

Because certain states may require more restrictive reentry intervals for various crops treated with this product, consult your State Department of Agriculture for further information.

Written or oral warnings must be given to workers who are expected to be in a treated area or in an area about to be treated with this product. Inform workers of areas or fields that may not be entered without specific protective clothing, period of time field must be vacated and appropriate actions to take in case of accidental exposure. An example of such information is given under written warnings. When oral warnings are given, warnings shall be given in a language customarily understood by workers. Oral warnings must be given if there is reason to believe that written warnings cannot be understood by workers. Written warnings must include the following information:

CAUTION

Area treated with STIK® on (insert date). Do not enter without appropriate protective clothing until sprays have dried (insert State Department of Agriculture's re-entry interval, if more restrictive).

In case of accidental exposure to pesticide spray or dust, wash the skin thoroughly with soap and water. Remove contaminated clothing and wash before reuse. If in eyes, flush with plenty of water. If inhaled, go to an area where the pesticide has not been applied. Get medical attention if needed.

STORAGE AND DISPOSAL

PESTICIDE STORAGE

Keep out of reach of children and animals. Store in original containers only. Store in a cool, dry place and avoid excess heat. Carefully open containers. After partial use, replace lids and close tightly. Do not put concentrate or dilute material into food or drink containers. Do not contaminate other pesticides, fertilizers, water, food, or feed by storage or disposal.

In case of spill, avoid contact, isolate area and keep out animals and unprotected persons. Confine spills. Call FMC collect: (716) 735-3765.

To confine spill: If liquid, dike surrounding area or absorb with sand, cat litter or commercial clay. If dry material, cover to prevent dispersal. Place damaged package in a holding container. Identify contents.

PESTICIDE DISPOSAL

Wastes resulting from the use of this product may be disposed of on site or at an approved waste disposal facility.

CONTAINER DISPOSAL

Triple rinse (or equivalent). Then offer for recycling or reconditioning, or puncture and dispose of in a sanitary landfill, or incineration, or, if allowed by state and local authorities, by burning. If burned, stay out of smoke.

This material, when used as directed, will aid in preventing seed pod and nut formation on certain varieties of shade and ornamental trees.

It is important to observe usage cautions as noted in "Caution" statement below, as injury may result from improper use.

CAUTION: This formulation and its application on shade trees is intended for use by professional arborists only. Maximum dosages should not exceed 40 ppm or 8 oz. by weight of STIK per 100 gals. of dilute spray. Avoid application or drift onto flowering plants and shrubs other than those intended for treatment as injury may result. Avoid spraying or permitting drift onto food crops. Some wilting of leaves may follow use of this material—this will vary with plant vigor, the less vigorous trees being affected the greatest. Avoid use on weak trees or under slow drying conditions.

Do not allow this material to drift onto neighboring crops or non-crop areas or use in a manner or at a time other than in accordance with label directions because animal, plant or crop injury, illegal residues or other undesirable results may occur.

When the foliage has been seriously injured by insects, diseases, mineral deficiencies, frost or extreme drought conditions exist, little or no control can be expected.

SHADE TREES: (Catalpa, Horsechestnut, Elm, and Maple): To prevent seed pod and nut formation—Use 4 to 8 oz. (20–40 ppm) per 100 gals. of spray. Apply sufficient spray to thoroughly wet tree. Make application at the full-bloom stage of tree development.

DIRECTIONS FOR THINNING APPLES

Apply as a post bloom spray at the rate of 1 to 4 oz. per 100 gals. of water (5 to 20 ppm) depending upon the apple variety. Do not use on summer or early fall varieties later than 1 week following full bloom. Other varieties may be treated 10 to 21 days after full bloom.

DIRECTIONS FOR PREVENTING PREMATURE DROP

APPLES: Use 2 to 4 oz. per 100 gals. of water. Do not apply more than 425 gals. of spray per acre. Apply spray 7 to 14 days before harvest. Do not make more than 2 applications. Do not apply within 2 days of harvest.

PEARS: Use 2 oz. per 100 gals. of water. Do not apply more than 425 gals. of spray per acre. Apply to fruit 5 to 7 days before harvest. Do not make more than 2 applications. Do not apply within 2 days of harvest.

DEALERS SHOULD SELL IN ORIGINAL PACKAGES ONLY.

TERMS OF SALE OR USE: On purchase of this product buyer and user agree to the following conditions:

WARRANTY: FMC makes no warranty, expressed or implied, concerning the use of this product other than indicated on the label. Except as so warranted, the product is sold as is. Buyer and user assume all risk of use and/or handling and/or storage of this material when such use and/or handling and/or storage is contrary to label instructions.

DIRECTIONS AND RECOMMENDATIONS: Follow directions carefully. Timing and method of application, weather and crop conditions, mixture with other chemicals not specifically recommended and other influencing factors in the use of this product are beyond the control of the seller and are assumed by buyer at his own risk.

USE OF PRODUCT: FMC's recommendations for the use of this product are based upon tests believed to be reliable. The use of this product being beyond the control of the manufacturer, no guarantee, expressed or implied, is made as to the effects of such or the results to be obtained if not used in accordance with directions or established safe practice.

DAMAGES: Buyer's or user's exclusive remedy for damages for breach of warranty or negligence shall be limited to direct damages not exceeding the purchase price paid and shall not include incidental or consequential damages.

THIODAN® 3 E.C.

INSECTICIDE

For Agricultural or Commercial Use Only.

ACTIVE INGREDIENTS:	BY WT.
*Endosulfan (Hexachlorohexahydromethano-2,4,3-benzodioxathiepin oxide)	33.7%
Xylene Range Aromatic Solvent	57.7%
INERT INGREDIENTS:	8.6%
TOTAL	100.0%

* Thiodan

Thiodan 3 E.C. contains 3 pounds of Endosulfan per gallon.

EPA Reg. No. 279-2924 EPA Est., 279-

Thiodan is a registered trademark of Hoechst Ag.

Keep Out of Reach of Children

DANGER–POISON

PELIGRO

PRECAUCION AL USUARIO: Si usted no lee ingles, no use este producto hasta que la etiqueta le haya sido explicada ampliamente.

STATEMENT OF PRACTICAL TREATMENT

If swallowed. Call a physician or Poison Control Center immediately. If possible, vomiting should be induced under medical supervision. Drink one or two glasses of water and induce vomiting by giving one (1) ounce of syrup of ipecac, if available, or by touching the back of the throat with a finger. Do not induce vomiting or give anything by mouth to a person who is unconscious or convulsing.

If inhaled: Remove victim to fresh air. Apply artificial respiration if indicated.

If on skin: Remove contaminated clothing and wash skin with soap and water.

If in eyes: Flush eyes with plenty of water. Call a physician immediately.

Note to Physician: Endosulfan is a central nervous system stimulant absorbable by mouth, inhalation or through contact with skin. There is no antidote, but, excitant effects can usually be controlled by barbiturates or diazepam. There also is hazard of chemical pneumonia from aspiration, because of the petroleum hydrocarbon content.

See other panels for additional precautionary information.

PRECAUTIONARY STATEMENTS

HAZARDS TO HUMANS (& DOMESTIC ANIMALS)

DANGER

Fatal if Swallowed, Hazardous if Inhaled or Absorbed Through Skin! Do not breathe spray mist. Do not get in eyes, on skin or on clothing. Wash thoroughly with soap and water after handling and before eating or smoking; wear clean clothing. During commercial exposure in spray-mixing and loading operations, wear clean synthetic rubber gloves and a pesticide respirator jointly approved by the MSHA (Mines Safety and Health Administration) and by the National Institute for Occupational Safety and Health under the provisions of 30 CFR Part II.

ENVIRONMENTAL HAZARDS

This product is toxic to fish, birds, and other wildlife. Birds feeding on treated areas may be killed. Keep out of lakes, streams, ponds, tidal marshes, and estuaries. Do not apply to water that will be used for recreational purposes and human and livestock consumption. Shrimp and crab may be killed at application rates recommended on this label. Do not apply where fish, shrimp, crab and other aquatic life are important resources. This product must not be used in areas where impact on threatened endangered species is likely. Contact your State Fish and Game Agency before applying this product. Apply this product only as specified on this label.

This pesticide is toxic to bees exposed to direct application. Applications should be timed to coincide with periods of minimum bee activity, usually between late evening and early morning.

PHYSICAL OR CHEMICAL HAZARDS

Do not use, pour, spill or store near heat or open flame.

DIRECTIONS FOR USE

It is a violation of Federal law to use this product in a manner inconsistent with its labeling.

Do not apply this product in such a manner as to directly or through drift expose workers or other persons. The area being treated must be vacated by unprotected persons.

Do not enter treated areas without protective clothing for 24 hours.

Protective clothing means, at least, a hat or other suitable head covering, a long sleeved shirt and long legged trousers or a coverall type garment (all of closely woven fabric covering the body, including the arms and legs), shoes and socks.

Because certain states may require more restrictive reentry intervals for various crops treated with this product, such as the 48 hour re-entry interval established by California, consult your State Department of Agriculture for further information.

Written or oral warnings must be given to workers who are expected to be in a treated area or in an area about to be treated with this product. Inform workers of areas or fields that may not be entered without specific protective clothing, period of time field must be vacated and appropriate actions to take in case of accidental exposure. An example of such information is given under written warnings. When oral warnings are given, warnings shall be given in a language customarily understood by workers. Oral warnings must be given if there is reason to believe that written warnings cannot be understood by workers. Written warnings must include the following information:

DANGER

Area treated with THIODAN® 3 EC. on (insert date). Do not enter without appropriate protective clothing for 24 hours (insert State Department of Agriculture's re-entry interval, if more restrictive).

In case of accidental exposure to pesticide spray or dust, wash the skin thoroughly with soap and water. Remove contaminated clothing and wash before reuse. If in eyes, flush with plenty of water. If inhaled, go to an area where the pesticide has not been applied. Get medical attention if needed.

STORAGE AND DISPOSAL

PESTICIDE STORAGE

Do not store in or around the home.

Do not store below 20°F, (−7°C).

Do not use or store near heat, open flame or hot surfaces.

Keep out of reach of children and animals. Store in original containers only. Store in a cool, dry place and avoid excess heat. Carefully open containers. After partial use, replace lids and close tightly. Do not put concentrate or dilute material into food or drink containers. Do not contaminate other pesticides, fertilizers, water, food, or feed by storage or disposal.

In case of spill, avoid contact, isolate area and keep out animals and unprotected persons. Confine spills. Call FMC collect: (716) 735-3765.

To confine spill: If liquid, dike surrounding area or absorb with sand, cat litter or commercial clay. If dry material, cover to prevent dispersal. Place damaged package in a holding container. Identify contents.

PESTICIDE DISPOSAL

Pesticide wastes are acutely hazardous. Improper disposal of excess pesticide, spray mixture, or rinsate is a violation of Federal Law. If these wastes cannot be disposed of by use according to label instructions, contact your State Pesticide or Environmental Control Agency, or the Hazardous Waste representative at the nearest EPA Regional Office for guidance.

CONTAINER DISPOSAL
Triple rinse (or equivalent). Then offer for recycling or reconditioning, or puncture and dispose of in a sanitary landfill, or by other procedures approved by state and local authorities. Do not cut or weld metal containers.

Not for use or storage in or around the home.
Do not use in undiluted form.

Apply the listed amount per acre when insects first appear and repeat as required to maintain effective control. Use in sufficient water for thorough coverage of listed crops, unless otherwise noted. Coverage of upper and lower leaf surfaces is essential for good control. Observe use limitations. If insect control is required beyond the following use patterns, supplement the control program with other suitable pesticides. When applying this material by aircraft, mix the recommended amount with sufficient water to provide a minimum of 1 gallon finished spray per acre on vegetable and field crops and a minimum of 20 gallons of finished spray per acre on deciduous trees and vines. Where more than 2 quarts of this material are recommended, mix with sufficient water to provide a minimum of finished spray equal to twice the amount of THIODAN 3 E.C. used. Do not plant root crops other than carrots, potatoes, sweet potatoes, and sugar beets as follow-up crops. Observe days interval between last application and harvest indicated by number in ( ) following the crop.

ALFALFA, GROWN FOR FORAGE (21): Meadow Spittlebug—2/3 pt. per acre. Apply 1 week after eggs begin to hatch or when first small masses of froth are produced in the crowns of the plant which is usually about the time the alfalfa is 4 to 6 inches tall. Apply in a minimum of 10 gallons of water per acre by ground or 2 gallons of water per acre by air. Do not apply more than once per cutting. Do not graze or cut forage within 21 days of treatment. **Seed Crop Only:** Aphids—1 1/3 pints. Lygus Bugs—2 to 2 2/3 quarts. Meadow Spittlebug—2/3 pints.

Do not feed treated forage to dairy or meat animals. Do not use on forage to be sold commercially or shipped interstate.

ALMONDS: Peach Twig Borer—2 1/2 to 3 1/3 qts. per acre for control of early season infestation. Apply during popcorn, pink or petal fall. Use in a minimum of 200 gallons of water per acre dilute or in 40 gallons per acre concentrate. For areas less than one acre, use 1 qt. per 100 gallons of water. Do not exceed one application per season. Treated hulls may be fed to livestock and dairy animals.

APPLES (30): Apple Aphid, Rosy Apple Aphid, Woolly Aphid—2/3 qt. per 100 gallons or 2 2/3 to 3 1/3 qts. per acre. Apple Rust Mite—1/3 to 2/3 qts. per 100 gallons or 4 to 5 1/3 qts. per acre. Do not feed pomace from treated apples to livestock. Do not exceed 2 applications during the fruiting period at rate above 3 1/3 quarts per acre.

APPLES (21): Apple Aphid, Rosy Apple Aphid—2/3 qt. per 100 gallons or 2 2/3 to 3 1/3 qts. per acre. Do not exceed 3 applications during fruiting period. Do not feed pomace to livestock.

APPLES (30) Western Use Only: Woolly Apple Aphid—2/3 qt. per 100 gallons or 5 1/3 qts. per acre. Do not exceed 2 applications during fruiting period. Do not feed pomace from treated apples to livestock.

APRICOTS (21), NECTARINES (21), PEACHES (21): Peach Tree Borer, Lesser Peach Tree Borer—Use 1 qt. per 100 gallons. West Coast—2/3 qt. per 100 gallons. Southeastern States—Use 2 to 3 1/3 qts. per 100 gallons. Best control is obtained with a single application during the first week of September. Spray all bark areas from ground level to scaffold limbs.

—OR—

APRICOTS (30), NECTARINES (30), PEACHES (30): Aphids (Black Cherry, Black Peach, Green Peach, Rusty Plum), Catfacing Insects, Peach Twig Borer—2/3 qt. per 100 gallons or 2 2/3 to 3 1/3 qts. per acre. Peach Silver Mite—1/3 qt. per 100 gallons or 2 2/3 to 3 1/3 qts. per acre. Do not exceed 2 applications during the fruiting period.

ARTICHOKES (7): Artichoke Aphid, Green Peach Aphid—1 to 1 1/3 qts. per acre. Artichoke Plume Moth—1 1/3 qts. per acre.

BARLEY, OATS, RYE, WHEAT: Army Cutworm—1 1/3 pt. per acre. For aerial application, apply in 2 gallons diesel fuel oil per acre. Cereal Leaf Beetle (Illinois, Indiana, Michigan, and Ohio only)—Use 2/3 to 1 1/3 pints per acre. Apply when small larvae are readily found in the field. For aerial application use a minimum of 1 to 2 gallons of water per acre. Do not apply after heads begin to form. Do not feed treated forage to livestock.

BEANS, SUCCULENT AND DRY (Except Lima Beans) (3): Black Bean Aphid, Bean Leafskeletonizer, Cucumber Beetles, Green Stink Bug, Mexican Bean Beetle—Use 1 1/3 to 2 2/3 pts. Western Bean Cutworm —Use 1 1/3 qts. Do not exceed 3 applications per season. Do not feed threshings or allow livestock to graze in treated fields.

BLUEBERRIES: Blueberry Bud Mite—2 qts. per acre in 300 gallons of water. Apply immediately after harvest and repeat 6 to 8 weeks later. Do not apply after buds are well formed.

BROCCOLI (7), BRUSSELS SPROUTS (14), CABBAGE (7), CAULIFLOWER (14): Cabbage Aphid, Cabbage Looper, Cross-Striped Cabbageworm, Diamondback Moth Larvae, Flea Beetle, Harlequin Bug, Imported Cabbageworm, Stink Bug—1 to 1 1/3 qts. Armyworms, Cutworms—1 1/3 qts. per acre.

CARROTS (7): Green Peach Aphid—1 1/3 to 2 2/3 pts. Do not exceed 2 applications. Do not use tops for food or feed.

CELERY (7): Green Peach Aphid—1 1/3 pts.—or—**Celery (4):** Green Peach Aphid, Cabbage Looper—2 2/3 pts. Do not exceed 3 applications per season.

CHERRIES (21): Black Cherry Aphid, Plum Rust (Nursery) Mite—2/3 qt. per 100 gallons or 2 2/3 to 3 1/3 qts. per acre. Do not exceed 2 applications after shuck split. **(Michigan only)**—Mineola Moth—1 1/3 qts. per 100 gallons applied in delayed dormant period.

CHERRIES (21)—Pacific Northwest Only: Eyespotted Bud Moth, Fruittree Leafroller—2/3 qt. per 100 gallons or 2 2/3 to 3 1/3 qts. per acre. For Eyespotted Bud Moth, apply during popcorn stage of growth. For Fruittree Leafroller, apply during pre-pink stage of growth. Do not exceed 2 applications after shuck split.

CHERRY, PEACH, PLUM NURSERY STOCK DIP: Peachtree Borer— Thoroughly mix 2/3 gallon per 40 gallons of water. Immerse the bundles of trees so that the roots and crowns are covered well above the grafting bud scar. Plant immediately, or dry before returning to storage. Wear rubber gloves when dipping.

CITRUS (Nonbearing Trees and Nursery Stock): Citrus Aphid—2/3 pt. per 100 gallons or 3 1/3 qts. per acre. Do not apply to bearing trees or trees that will bear fruit within 12 months. Shrimp and crab may be killed at application rates recommended. Do not apply where these are important resources.

COLLARDS (21): Cabbage Looper, Diamondback Moth Larvae, Harlequin Bug, Imported Cabbageworm—1 qt. per acre. Apply only once per season.

COTTON: Thrips—2 qts. per acre. Do not apply after bolls open. Do not graze dairy or meat animals in treated fields.

COTTON—Eastern Use Only: Boll Weevil—2/3 to 1 2/3 qts. per acre. Bollworm. Cabbage Looper, Cotton Leafperforator, Lygus Bugs, Stink Bugs—1 1/3 qts. per acre. Do not apply after bolls open. Do not graze dairy or meat animals in treated fields.

COTTON—Western Use Only: Boll Weevil—2/3 to 2 qts. per acre. Bollworm, Cabbage Looper, Cotton Leafperforator, Lygus Bugs, Stink Bugs—1 1/3 to 2 qts. per acre. Do not apply after bolls open. Do not graze dairy or meat animals in treated fields.

CUCUMBERS (0), MELONS (0), SUMMER & WINTER SQUASH (0): Aphids, Cucumber Beetles, Striped Flea Beetle, Melonworm, Pickleworm, Rindworm (on Watermelons), Squash Beetle, Squash Bug, Squash Vine Borer—1 1/3 to 2 2/3 pts. For Vine Borer control, apply weekly to flowerbuds, stems and vines beginning when moths first appear. For Cabbage Looper and Omnivorous Leafroller on Melons use 1 1/3 qts.

EGGPLANTS (1): Colorado Potato Beetle, Flea Beetle, Green Peach Aphid, Whitefly, Blister Beetle, Green Stink Bug—$1\frac{1}{3}$ pts. per acre.

GRAPES (7): Grape Leafhopper—$\frac{2}{3}$ quart per 100 gallons or $1\frac{1}{3}$ to 2 qts. per acre. Rose Chafer—$\frac{2}{3}$ qt. per 100 gallons. Do not apply to Concord variety as severe injury is likely to occur. Ground application preferred.

KALE (21): Cabbage Flea Beetle, Harlequin Bug, Imported Cabbageworm—1 qt. per acre. Apply only once per season.

LETTUCE (14): Cabbage Looper, Green Peach Aphid, Diamondback Moth Larvae, Imported Cabbageworm—1 to $1\frac{1}{3}$ qts. Do not exceed 3 applications after thinning head lettuce and remove outside leaves at harvest. Do not exceed 2 applications per season on leaf lettuce. Do not feed crop refuse to livestock.

MACADAMIA NUTS (0): Southern Green Stink Bug—$1\frac{1}{3}$ qts. per 100 gallons of water (maximum of 5 gallons per acre). Do not graze livestock on orchard crops or grasses in treated areas.

MUSTARD GREENS (21): Aphids, Cabbage Looper, Imported Cabbageworm—1 qt. per acre. Do not exceed 1 application per season.

PEARS (7)*: Consperse Stink Bug—**Ground Treatment,** use $\frac{2}{3}$ qt. per 100 gallons, 200 to 400 gallons spray per acre applied to orchard floor and ground around trees prior to bloom. Do not graze livestock in treated orchards. **Foliar Treatment,** use $\frac{2}{3}$ qt. per 100 gallons or $2\frac{2}{3}$ to $3\frac{1}{3}$ qts. per acre. Stink Bugs must be wet to obtain control. Pear Rust Mite—$\frac{2}{3}$ qt. per 100 gallons or $2\frac{2}{3}$ to $3\frac{1}{3}$ qts. per acre. Pearleaf Blister Mite—$\frac{1}{3}$ to $\frac{2}{3}$ qt. per 100 gallons applied as a post harvest or dormant treatment. Pear Psylla—$3\frac{1}{3}$ qts. in sufficient water to treat 1 acre. (Minimums of 10 gallons per acre for aerial application, 40 gallons per acre for semi-concentrate, and 300 gallons per acre for dilute). Apply when adults are first observed or nymphs are small and repeat to maintain control. Make no more than 2 applications during the fruiting period.

*(21): Make no more than 5 applications during the fruiting period.

NOTE: Aerial application does not result in satisfactory control and should only be employed if impossible to apply by ground. Do not feed cull fruits to animals nor allow livestock to graze in treated fields.

PEAS, SEED CROP ONLY (0): Pea Aphid, Pea Weevil—$\frac{2}{3}$ to 1 qt. per acre. Do not exceed 2 applications during the fruiting period. Use only on peas to be harvested by combine. Do not feed treated vines to livestock or allow livestock to graze in treated fields.

PEAS, SUCCULENT (0): Pea Aphid, Pea Weevil—$1\frac{1}{3}$ to $2\frac{2}{3}$ pts. per acre. Do not exceed 2 applications during the fruiting period. Use only on peas to be harvested by combine. Do not feed treated vines to livestock or allow to graze in treated fields.

PEAS, SOUTHERN (Black-eyed Peas, Crowder Peas, Southern Peas) (3): Cowpea Curculio—$\frac{2}{3}$ to $1\frac{1}{3}$ qt. per acre. Do not feed treated threshings to livestock or allow to graze in treated fields. Make 3 applications at 5 day intervals starting when pods are $\frac{1}{2}$ inch long.

PECANS: Black Pecan Aphid—1 qt. per 100 gallons. Pecan Leaf Phylloxera—$\frac{2}{3}$ to 1 qt. per 100 gallons when nymphs appear and before they are enclosed in plant tissue. For high populations, use the higher rate. Do not apply after shuck split. Do not graze livestock in treated groves. Do not graze livestock on orchard crops or grasses in treated areas.

PEPPERS (4)*: Green Peach Aphid, Flea Beetle, Hornworms, Pepper Maggot—$1\frac{1}{3}$ to $2\frac{2}{3}$ pts. *1 day if maximum of $1\frac{1}{3}$ pts. per acre is applied.

PINEAPPLES (for Fresh Market only) (7): Pineapple Fruit Mite—$\frac{2}{3}$ gallon per acre. Apply at intervals of 7 to 10 days if necessary, particularly during the 40 days period of blooming. Do not exceed 4 applications per season. Do not feed treated forage or pineapple by-products to livestock.

PLUMS (7), Prunes (7): Hop Aphid, Leafcurl Plum Aphid, Thistle Aphid—$\frac{2}{3}$ qt. per 100 gallons or $2\frac{2}{3}$ to $3\frac{1}{3}$ qts. per acre. Apply when eggs hatch during pre-bloom or petal fall. Summer applications should be made before leaves curl. Plum Rust (Plum Nursery) Mite—$\frac{2}{3}$ qt. per 100 gallons or $2\frac{2}{3}$ to $3\frac{1}{3}$ qts. per acre. Lesser Peachtree Borer—1 qt. per 100 gallons or $2\frac{2}{3}$ to $3\frac{1}{3}$ qts. per acre. Apply spray to thoroughly wet trunk and main branches. Peachtree Borer—1 qt. per 100 gallons or $2\frac{2}{3}$ to $3\frac{1}{3}$ qts. per acre. Spray thoroughly to cover at bark areas from ground to scaffold limbs. Do not graze livestock on orchard crops or grasses in treated areas.

PLUMS (7), Prunes (7)—Eastern Use Only: Peachtree Borer—1 qt. per 100 gallons. Spray thoroughly to cover all bark areas from ground to scaffold limbs. Do not graze livestock on orchard crops or grasses in treated areas.

PLUMS (7), Prunes (7)—Pacific Northwest Only: Fruittree Leafroller—$\frac{2}{3}$ qt. per 100 gallons or $2\frac{2}{3}$ to $3\frac{1}{3}$ qts. per acre. Apply during pre-pink stage of growth. Do not graze livestock on orchard crops or grasses in treated areas.

PLUMS (7), Prunes (7)—Western Use Only: Peachtree Borer—$\frac{2}{3}$ to 1 qt. per 100 gallons. Spray thoroughly to cover all bark areas from ground to scaffold limbs. Peach Twig Borer—$\frac{2}{3}$ qt. per 100 gallons or $2\frac{2}{3}$ to $3\frac{1}{3}$ qts. per acre. Do not graze livestock on orchard crops or grasses in treated areas.

POTATOES (0): Aphids, Armyworms, Colorado Potato Beetle, Green Stink Bug, Leaffooted Bug, Potator Flea Beetles, Potato Leafhopper, Plant Bugs, Potato Tuberworm, Whitefly—$1\frac{1}{3}$ to $2\frac{2}{3}$ pts. per acre. European Corn Borer, Potato Psyllid—1 to $1\frac{1}{3}$ qts. per acre. False Chinch Bug—$1\frac{1}{3}$ qts. per acre. Threelined Potato Beetle—$1\frac{1}{3}$ pts. per acre.

POTATOES (0)—Eastern Use Only: Threelined Potato Beetle—$\frac{2}{3}$ qt. per acre.

PUMPKINS (1): Aphids, Cucumber Beetles, Squash Vine Borer—$1\frac{1}{3}$ to $2\frac{2}{3}$ pts. per acre. For Vine Borer control, apply weekly to flower buds, stems and vines beginning when moths first appear.

SAFFLOWER: Green Peach Aphid—$1\frac{1}{3}$ qts. per acre. Do not apply after flower heads open.

SPINACH (21): Crown Mite, Green Peach Aphid—1 qt. Do not exceed one application per season.

STRAWBERRIES (4): Cyclamen Mite—$2\frac{2}{3}$ qts. per acre in 400 gallons of water. Do not reapply within 35 days when fruit is present. Meadow Spittlebug, Strawberry Aphid, Tarnished Plant Bug—$1\frac{1}{3}$ qts. per acre in 200 gallons of water. Do not reapply within 15 days or more than twice within a 35 day period when fruit is present.

STRAWBERRIES (4)—Eastern Use Only: Cyclamen Mite—$2\frac{2}{3}$ qts. per acre in 400 gallons of water. Do not apply at intervals of less than 35 days when fruit is present.

STRAWBERRIES—Northwest Use Only: Garden Symphylan (aids in reducing damage)—$1\frac{1}{3}$ qts. per 100 gallons (plant dip). Mix thoroughly. Dip entire plant. When immersing bundles of plants, make certain any trapped air is forced out to assure thorough wetting of entire plant. Drain and allow plants to dry before setting them out in the field. Wear rubber gloves during dipping operation.

STRAWBERRIES (4)—Western Use Only: Cyclamen Mite—8 qts. per acre in 800 gallons of water. Do not apply at intervals of less than 35 days when fruit is present.

SUGAR BEETS (30): Green Peach Aphid—$\frac{2}{3}$ to $1\frac{1}{3}$ quarts. Meadow Spittlebug—$\frac{2}{3}$ qt. Sugar Beet Webworm—1 to $1\frac{1}{3}$ qts. Do not feed treated tops to livestock.

SUGAR BEETS (30)—Eastern Use Only: Meadow Spittlebug—$\frac{2}{3}$ qt. per acre. Do not feed treated tops to livestock.

SUGAR BEETS (30)—Western Use Only: Meadow Spittlebug—$\frac{2}{3}$ to $1\frac{1}{3}$ qts. per acre. Do not feed treated tops to livestock.

SUNFLOWER (0): Sunflower Moth—$1\frac{1}{3}$ qts. per acre. Do not exceed 3 applications. Make first application at onset of bloom and make two more applications at 4 to 7 day intervals. Do not feed treated forage to livestock.

SWEET CORN, FRESH VEGETABLE USE ONLY (0): Corn Leaf Aphid—$1\frac{1}{3}$ qts. Corn Earworm—2 qts. Apply at 2 to 3 day intervals beginning when silks first appear and continuing until they begin to dry. Do not exceed 5 applications. Do not apply to sweet corn to be processed. Do not feed treated forage or ensilage to livestock or allow to graze in treated fields.

SWEET POTATOES (0): Sweet Potato Flea Beetle—2/3 qts. per acre. Begin applications shortly after transplanting or as soon as Flea Beetles appear. Banded Cucumber Beetle Larvae (aids in control of, and for use in **South Central States only**)—1 1/3 to 2 2/3 qts. per acre broadcast. Work into the soil to a depth of approximately 3 inches making a single treatment just prior to plant set. Do not feed treated potatoes to livestock.

TOBACCO (5): Budworm, Cabbage Looper, Flea Beetle, Green Peach Aphid, Hornworm—**Field:** 2/3 to 1 1/3 qts. **Seed Bed:** Mix 2/3 qt. per 100 gals. of water and apply 6 gals. spray per 100 sq. yds. Green June Bug Larvae—1/3 qt. per 100 gals. of water. Drench plant bed with 1 gal. spray per sq. yd. Stink Bug—Use field application of 2/3 qt per 100 gals.

TOMATOES, FIELD AND GREENHOUSE (1): Blister Beetle, Colorado Potato Beetle, Flea Beetle, Green Peach Aphid—2/3 qt. Hornworms—2/3 to 1 1/3 qts. Tomato Fruitworm, Tomato Russet Mite, Yellowstriped Armyworm—1 1/3 qts. Stink Bugs—1 to 1 1/3 qts. Cabbage Looper—1 qt. Whitefly—2/3 qt. per 100 gals. in 100 to 200 gals. water. In greenhouse applications, wear approved mask or respirator approved by MSA and OSHA for protection against endosulfan.

TOMATOES (Field & Greenhouse) (1)—California Only: Webworm—2/3 to 1 1/3 qts. per acre. In greenhouse applications, wear approved mask or respirator approved by MSA and OSHA for protection against endosulfan.

TOMATOES (Field & Greenhouse) (1)—Eastern Use Only: Blister Beetles, Colorado Potato Beetle, Flea Beetles, Green Peach Aphid—2/3 qt. per acre. Cabbage Looper—1 qt. per acre. In greenhouse applications, wear approved mask or respirator approved by MSA and OSHA for protection against endosulfan.

TOMATOES (Field & Greenhouse) (1)—Western Use Only: Blister Beetles, Colorado Potato Beetle, Flea Beetles, Green Peach Aphid—2/3 to 1 1/3 qts. per acre. Cabbage Looper—1 to 1 1/3 qts. per acre. In greenhouse applications, wear approved mask or respirator approved by MSA and OSHA for protection against endosulfan.

WALNUTS: Walnut Aphid—2 to 3 1/3 qts. per acre. Do not apply after husk split. Do not graze livestock on orchard and crops or grasses in treated areas.

ORNAMENTALS

ORNAMENTAL PLANTS

LEATHERLEAF FERN: Leatherleaf Fern Borer—2/3 qt. per 100 gallons. Begin treatment when first larval feeding is observed in the mid-vein area at the base of the leaflets. Repeat at intervals of 2 to 3 weeks as necessary.

ORNAMENTALS (Greenhouse, and Out-of-Doors): Aphids, Cyclamen Mite, Whitefly, Rose Chafer—2/3 qt. per 100 gallons. On chrysanthemums, best results will be obtained if applied before plants flower. Do not apply to "Bonnafon Deluxe," "Fred Shoesmith," and "White Knight" chrysanthemums as injury may result. Do not use on birch trees. In greenhouse applications, wear a mask or respirator approved by MSA and OSHA for protection against endosulfan.

ORNAMENTAL TREES AND SHRUBS

BOXELDER: Boxelder Bug—2 qts. per 100 gallons (1 1/3 tablespoons per gallon). Spray foliage and trunks when insects first appear. Repeat as required. In late summer or fall spray tree trunks, outside walls, fences and other places where bugs congregate.

DOGWOOD: Dogwood Borer. **Lilac:** Lilac Borer—2/3 to 1 1/3 qts. per 100 gallons. Apply in early June and repeat in 10 to 14 days. Drench all bark areas down to the ground level.

DOUGLAS FIR (Grown for Ornamentals)—Pacific Northwest Only: Cooley Spruce Gall Aphid, Douglas Fir Needle Midge—2/3 qt. per 100 gallons or 1 1/3 tsps. per gallon. To control Gall Aphid, apply when white, cottony tufts appear. To control Needle Midge, apply in late April or early May just before buds open.

PINES (Austrian, Jack, Red, Scotch, White): Zimmerman Pine Moth—1 qt. per 100 gallons. Apply in mid-April and again if necessary in late fall. Thoroughly wet bark and main stem, especially where branches join main stem.

SHADE TREES (except Birch), SHRUBS: Aphids—2/3 qt. per 100 gallons.

SPRUCE: Spruce Gall Aphid (**New England area only**)—2/3 qt. per 100 gallons. Apply in late April or early May when Aphids are present but before galls are formed.

TAXUS: Taxus Bud Mite—2/3 quart per 100 gallons. Make 3 to 5 applications beginning in mid-May. Thorough treatment of foliage, twigs, and bark is essential. Black Vine Weevil—1 1/3 qt. per 100 gallons. Spray thoroughly and drench soil under plants. Apply when beetles first appear. Repeat in 14 days.

DEALERS SHOULD SELL IN ORIGINAL PACKAGES ONLY.

TERMS OF SALE OR USE: On purchase of this product buyer and user agree to the following conditions:

WARRANTY: FMC makes no warranty, expressed or implied, concerning the use of this product other than indicated on the label. Except as so warranted, the product is sold as is. Buyer and user assume all risk of use and/or handling and/or storage of this material when such use and/or handling and/or storage is contrary to label instructions.

DIRECTIONS AND RECOMMENDATIONS: Follow directions carefully. Timing and method of application, weather and crop conditions, mixture with other chemicals not specifically recommended and other influencing factors in the use of this product are beyond the control of the seller and are assumed by buyer at his own risk.

USE OF PRODUCT: FMC's recommendations for the use of this product are based upon tests believed to be reliable. The use of this product being beyond the control of the manufacturer, no guarantee, expressed or implied, is made as to the effects of such or the results to be obtained if not used in accordance with directions or established safe practice.

DAMAGES: Buyer's or user's exclusive remedy for damages for breach of warranty or negligence shall be limited to direct damages not exceeding the purchase price paid and shall not include incidental or consequential damages.

THIODAN® 50 WP

INSECTICIDE

For Agricultural or Commercial Use Only

ACTIVE INGREDIENT:	BY WT.
*Endosulfan: (Hexachlorohexahydromethano-2,4,3-benzodioxathiepin oxide)	50.0%
INERT INGREDIENTS:	50.0%
TOTAL	100.0%

* Thiodan

EPA Reg. No. 279-1380AA EPA Est., 279-

Thiodan is a registered trademark of Hoechst Ag.

WARNING

Keep Out of Reach of Children

AVISO

PRECAUCION AL USUARIO: Si usted no lee ingles, no use este producto hasta que la etiqueta le haya sido explicada ampliamente.

STATEMENT OF PRACTICAL TREATMENT

If swallowed: Call a physician or Poison Control Center immediately. If possible, vomiting should be induced under medical supervision. Drink one or two glasses of water and induce vomiting by giving one (1) ounce of syrup of ipecac, if available, or by touching the back of the throat with a finger. Do not induce vomiting or give anything by mouth to a person who is unconscious or convulsing.

If inhaled: Remove victim to fresh air. Apply artificial respiration if indicated.

If on skin: Remove contaminated clothing and wash skin with soap and water.

If in eyes: Flush eyes with plenty of water. Call a physician immediately.

Note to Physicians: Endosulfan is a central nervous system stimulant absorbable by mouth, inhalation or through contact with skin. There is no antidote, but excitant effects can usually be controlled by barbituates or diazepam.

For Emergency Assistance Call 716–735-3765.

See other panels for additional precautionary information.

PRECAUTIONARY STATEMENTS

HAZARDS TO HUMANS (& DOMESTIC ANIMALS)

WARNING

Hazardous if swallowed, inhaled, or absorbed through skin. Do not breathe dust or spray mist. Do not get in eyes, on skin, or on clothing. Wash thoroughly with soap and water after handling and before eating or smoking; wear clean clothing. During commercial or prolonged exposure in spray-mixing and loading operations, wear clean synthetic rubber gloves and a pesticide respirator jointly approved by MSHA (Mine, Safety and Health Administration) and by the National Institute for Occupational Safety and Health under the provisions of 30 CFR Part II. Do not apply or allow to drift to areas occupied by unprotected humans or beneficial animals.

ENVIRONMENTAL HAZARDS

This product is toxic to fish, birds, and other wildlife. Birds feeding on treated areas may be killed. Keep out of lakes, streams, ponds, tidal marshes, and estuaries. Do not apply to water that will be used for recreational purposes and human and livestock consumption. Shrimp and crab may be killed at application rates recommended on this label. Do not apply where fish, shrimp, crab, and other aquatic life are important resouces. This product must not be used in areas where impact on threatened endangered species is likely. Contact your State Fish and Game Agency before applying this product. Apply this product only as specified on this label.

This pesticide is toxic to bees exposed to direct application. Applications should be timed to coincide with periods of minimum bee activity, usually between late evening and early morning.

DIRECTIONS FOR USE

It is a violation of Federal Law to use this product in a manner inconsistent with its labeling.

Do not apply this product in such a manner as to directly or through drift expose workers or other persons. The area being treated must be vacated by unprotected persons.

Do not enter treated areas without appropriate protective clothing for 24 hours.

Protective clothing means, at least, a hat or other suitable head covering, a long sleeved shirt and long legged trousers or a coverall type garment (all of closely woven fabric covering the body, including the arms and legs), shoes and socks.

Because certain states may require more restictive re-entry intervals for various crops treated with this product, consult your State Department of Agriculture for further information.

Written or oral warnings must be given to workers who are expected to be in a treated area or in an area about to be treated with this product. Inform workers of areas or fields that may not be entered without specific protective clothing, period of time field must be vacated and appropriate actions to take in case of accidental exposure. An example of such information is given under written warnings. When oral warnings are given, warnings shall be given in a language cutomarily understood by workers. Oral warnings must be given if there is reason to believe that written warnings cannot be understood by workers. Written warnings must include the following information:

WARNING

Area treated with THIODAN® 50 WP on (insert date). Do not enter without appropriate protective clothing for 24 hours (insert State Department of Agriculture's re-entry interval, if more restrictive).

In case of accidental exposure to pesticide spray or dust, wash the skin thoroughly with soap and water. Remove contaminated clothing and wash before reuse. If in eyes, flush with plenty of water. If inhaled, go to an area where the pesticide has not been applied. Get medical attention if needed.

STORAGE AND DISPOSAL

PESTICIDE STORAGE

Do not store in or around the home.

Keep out of reach of children and animals. Store in original containers only. Store in a cool, dry place and avoid excess heat. Carefully open containers. After partial use, fold and roll back bags, clamp and close tightly. Do not put concentrate or dilute material into food or drink containers. Do not contaminate other pesticides, fertilizers, water, food or feed by storage or disposal.

In case of spill, avoid contact, isolate area and keep out animals and unprotected persons. Confine spills. Call FMC collect: (716) 735-3765.

To confine spill: If liquid, dike surrounding area or absorb with sand, cat litter or commercial clay. If dry material, cover to prevent dispersal. Place damaged package in a holding container. Identify contents.

PESTICIDE DISPOSAL

Pesticide wastes are acutely hazardous. Improper disposal of excess pesticide, spray mixture, or rinsate is a violation of Federal Law. If these wastes cannot be disposed of by use according to label instructions, contact your State Pesticide or Environmental Control Agency, or the Hazardous Waste representative at the nearest EPA Regional Office for guidance.

CONTAINER DISPOSAL

Completely empty bag into application equipment. Then dispose of empty bag in a sanitary landfill or by incineration, or, if allowed by State and local authorities, by burning. If burned, stay out of smoke.

Not for use or storage in or around the home.

Apply the listed amount per acre when insects first appear and repeat as required to maintain effective control. Depending upon the type of equipment used, apply in sufficient water for thorough coverage, unless otherwise noted, for listed crops. Coverage of upper and lower leaf surfaces is essential for good control. Observe use limitations. If insect control is required beyond the following use patterns, supplement the control program with other suitable pesticides. Observe days interval between last application and harvest indicated by number in [ ] following the crop.

ALFALFA (Seed Crop Only): Aphids, Meadow Spittlebug—Use 1 to 4 lbs. per acre. Lygus Bug, Stink Bug—Use 3 to 4 lbs. per acre. Apply in sufficient water for thorough coverage. By ground, apply a minimum of 20 gals.; by air, apply a minimum of 7½ gals. of spray per acre. Do not feed treated forage to meat or dairy animals. Do not use on forage sold commercially or shipped interstate. Crop to be used for seed production only.

ALFALFA (Grown for Forage) [21]: Meadow Spittlebug—½ lb. per acre. Apply 1 week after eggs begin to hatch or when first small masses of froth are produced in the crowns of the plant which is usually about the time the Alfalfa is 4 to 6 inches tall. Apply in a minimum of 10 gallons of water per acre by ground or 2 gallons of water per acre by air. Do not apply more than once per cutting. Do not graze or cut forage within 21 days of treatment.

ALMONDS: Peach Twig Borer—Use 4 to 5 lbs. per acre for control of early season infestation. Apply during popcorn, pink or petal fall in a minimum of 200 gals. dilute or 40 gals. per acre concentrate. (For areas less than one acre use 1 lb. per 100 gals.) Do not exceed one application per season. Treated hulls may be fed to livestock and dairy animals.

APPLES [21]: Apple Aphid, Rosy Apple Aphid, Green Fruitworm, Tarnished Plant Bug—Use 1 lb. per 100 gals. of water or a maximum of 4 to 5 lbs. per acre. Applications made at pink and/or petal fall provide best control of Tarnished Plant Bug and Green Fruitworms. If necessary, use 1½ lbs. per 100 gals. prior to petal fall to control Apple Aphids and Rosy Apple Aphids. Do not exceed three applications during the fruiting period. White Apple Leafhopper—For control of first generation use 1 lb per 100 gals. of water; 4 to 5 lbs. per acre. Apply when nymphs first appear on the leaves.—**OR—APPLES [30]** Apple Rust Mite, Woolly Apple Aphid—Use 1 lb. per 100 gals. or a maximum of 8 lbs. per acre during the fruiting period. **For State of Washington** dosage may be reduced to ¾ lb. per 100 gals. of water or a maximum of 6 lbs. per acre. Do not exceed two applications during the fruiting period. Do not feed pomace from treated apples to livestock.

APRICOTS [21], NECTARINES [21], PEACHES [21]: Peach Tree Borer, Lesser Peach Tree Borer—Use 1½ lbs. per 100 gals. **West Coast** 1 lb. per 100 gals. **Southeastern States,** apply 3 to 5 lbs. per 100 gals. post-harvest. In the Southeastern States, best control is obtained with a single application during the first week of September. Spray all bark areas from ground level to lower scaffold limbs.**—OR— APRICOTS [30], NECTARINES [30], PEACHES [30]:** Catfacing insects (Stinkbug type), Green Peach Aphid, Peach Twig Borer, Green Fruitworm—Use 1 lb. per 100 gals.; 4 to 5 lbs. per acre. **West Coast** 8 lbs. per acre. Black Cherry Aphid, Black Peach Aphid, Rusty Plum Aphid—Use 1 lb. per 100 gals.; 4 to 5 lbs. per acre. Peach Silver Mite—Use ½ lb. per 100 gals.; 4 to 5 lbs. per acre. Do not exceed two applications during the fruiting season.

ARTICHOKES [7]: Artichoke Aphid, Green Peach Aphid—Use 1½–2 lbs. per acre. Artichoke Plume Moth—Use 2 lbs. per acre. By ground, apply a minimum of 20 gals.; by air apply a minimum of 10 gals. of spray per acre.

BARLEY, OATS, RYE, WHEAT: Army Cutworm—1 lb. per acre. For aerial applications, apply in 2 gallons diesel fuel oil per acre. Cereal Leaf Beetle (Illinois, Indiana, Michigan, and Ohio only)—Use ½–1 lb. per acre. Apply when small larvae are readily found in the field. For aerial application use a minimum of 1 to 2 gallons water per acre. Do not apply after heads begin to form. Do not feed treated forage to livestock.

BEANS, [Succulent and Dry], SOUTHERN FIELD PEAS [3]: Black Bean Aphid, Bean Leafskeletonizer, Cucumber Beetle, Green Stink Bug, Mexican Bean Beetle, Cowpea Curculio—Use 1 to 2 lbs. per acre. Do not exceed 3 applications per season. Do not feed treated threshings to livestock or allow to graze in treated fields. Do not use on Lima Beans that are to be removed from the field for processing. For control of Cowpea Curculio make 3 applications at 5-day intervals, starting when the pods are ½ inch long.

BLUEBERRIES: Blueberry Bud Mite—3 lbs. per acre in 300 gallons of water. Apply immediately after harvest and repeat 6 to 8 weeks later. Do not apply after buds are well formed.

BOXELDER: Boxelder Bug—Use 3 lbs. per 100 gals. (3 tablespoons per gal.) Spray foliage and trunks beginning in May when insects first appear and continue through June. In late summer or fall, spray tree trunks, outside walls, fences and other places where bugs are congregating.

BROCCOLI [7], BRUSSELS SPROUTS [14], CABBAGE [7], CAULIFLOWER [14]: Cabbage Aphid, Cabbage Looper, Imported Cabbage Worm, Diamondback Moth Larvae, Cross-striped Cabbage Worm, Flea Beetles, Harlequin Bugs, Stink Bugs—Use 1½ to 2 lbs. per acre. Armyworm, Cutworm—Use 2 lbs. per acre.

CARROTS [7]: Green Peach Aphid—Use 1 to 2 lbs. per acre. Do not use tops for food or feed. Do not exceed 2 applications per season. Do not plant root crops other than carrots, potatoes, sugar beets, and sweet potatoes as follow-up crops.

CELERY [4]: Green Peach Aphid, Cabbage Looper—Use 1 to 2 lbs. per acre. Do not exceed 3 applications.**—OR—CELERY [7];** Green Peach Aphid—Use 1 lb. per acre. Unlimited number of applications.

CHERRIES [21]: Peach Tree Borer, Lesser Peach Tree Borer—Use 1½ lbs. per 100 gals. **West Coast** 1 to 1½ lbs. per 100 gals. Thoroughly wet all bark from ground to scaffold limbs. Black Cherry Aphid, Plum Rust Nursery Mite, Green Fruitworm—Use 1 lb. per 100 gals.; 4 to 5 lbs. per acre. **Pacific Northwest Only**—Eyespotted Bud Moth, Fruit Tree Leafroller—Use 1 lb. per 100 gals., at "popcorn" stage for Bud Moth control and during pre-pink stage of growth for leafroller control. **Michigan Only**—For control of Mineola Moth—Use 2 lbs. per 100 gals. in the delayed dormant period. Do not exceed two applications after shuck split.

CHERRY, PEACH, PLUM NURSERY STOCK: Peachtree Borer—Use 4 lbs. per 40 gals. water. Mix thoroughly. Wear rubber gloves during dipping operation. Immerse trees so that roots and crowns are covered well above grafting bud scar. Plant immediately or dry before returning stock to storage.

CITRUS, NURSERY AND NON-BEARING STOCK: Citrus Aphid—Use ½ lb. per 100 gals.; 5 lbs. per acre. Do not apply to bearing trees or to trees that will bear fruit within twelve months.

COLLARDS [21]: Cabbage Looper, Diamondback Moth Larvae, Harlequin Bug, Imported Cabbageworm—Use 1½ lbs. per acre. Do not exceed one application.

COTTON: Cotton Leafperforator—Use 2 lbs. per acre. Thrips—3 lbs. per acre. Do not apply after bolls open. Do not graze meat or dairy animals in treated fields.

COTTON—Eastern Use Only: Boll Weevil—1 lb. to 2½ lbs. per acre. Bollworm, Cabbage Looper, Cotton Leafperforator, Lygus Bugs, Stink Bugs—2 lbs. per acre. Do not apply after bolls open. Do not graze dairy or meat animals in treated fields.

COTTON—Western Use Only: Boll Weevil—1 lb. to 3 lbs. per acre. Bollworm, Cabbage Looper, Cotton Leafperforator, Lygus Bugs, Stink Bugs—2 lbs. to 3 lbs. per acre. Do not apply after bolls open. Do not graze dairy or meat animals in treated fields.

CUCUMBERS [0], MELONS [0], SUMMER AND WINTER SQUASH [0]: Aphids, Cucumber Beetles, Striped Flea Beetle, Melonworm, Pickleworm, Rindworm (on Watermelons), Squash Beetle, Squash Bug, Squash Vine Borer—Use 1 to 2 lbs. per acre. For Vine Borer control, apply weekly to flower buds, stems, and vines beginning when moths first appear. For Cabbage Looper and Omnivorous Leafroller on Melons use 2 lbs. per acre.

DOGWOOD, LILACS: Dogwood Borer, Lilac Borer—Use 1 to 2 lbs. per 100 gals. Apply in early June and repeat in 10 to 14 days. Drench all bark areas down to the ground level.

DOUGLAS FIR (Grown for Ornamentals)—Pacific Northwest only: Cooly Spruce Gall Aphid, Douglas Fir Needle Midge—1 lb. per 100 gallons or ⅙ oz. per gallon. To control Gall Aphid, apply when white cottony tufts appear. To control Needle Midge, apply in late April or early May just before buds open.

EGGPLANT [1]: Blister Beetle, Colorado Potato Beetle, Flea Beetle, Green Peach Aphid, Green Stink Bug, Whitefly—Use 1 lb. per acre.

FILBERTS [0]: Filbert Aphid—Use 1 lb. per 100 gals. or 3 to 4 lbs. per acre. Filbert Leafroller—Use 3 to 4 lbs. per acre. Do not graze livestock in treated groves.

GRAPES [7]: Grape Leafhopper, Grape Phylloxera (leaf form), Rose Chafer—Use 1 lb. per 100 gals., 2 to 3 lbs. per acre. Do not use on Baco No. 1, Baco Noir, Concord, Seibel 7053 (Chancellor), Seibel 8357 (Colobel), Seibel 13053 (Cascade) varieties of grapes as severe injury is likely to occur.

KALE [21]: Cabbage Flea Beetle, Imported Cabbageworm, Harlequin Bug—Use 1½ lbs. per acre. Do not exceed one application per season.

LEATHERLEAF FERN: Leatherleaf Fern Borer—Use 1 lb. per 100 gals. Begin treatment when first larvae feeding is observed in the mid-vein area at the base of the leaflets. Repeat at intervals of 2 to 3 weeks as necessary.

LETTUCE [14]: Cabbage Looper, Diamondback Moth Larvae, Green Peach Aphid, Imported Cabbageworm—Use 1½ to 2 lbs. per acre. On Head Lettuce, do not exceed 3 applications after thinning. Remove wrapper leaves at harvest. Do not feed crop refuse to livestock. On Leaf Lettuce, do not exceed 2 applications per season.

MACADAMIA NUTS [0]: Southern Green Stink Bug—Use 2 lbs. per 100 gals. of water (maximum of 30 lbs. per acre). Do not graze livestock in treated groves.

MUSTARD GREENS [21]: Aphids, Cabbage Looper, Imported Cabbageworm—Use 1½ lbs. per acre. Do not exceed one application per season.

ORNAMENTALS [Greenhouse & Out of Doors]: Aphids, Cyclamen Mite, Rose Chafer, Whitefly—Use 1 lb. per 100 gals. On Chrysanthemums, best results will be obtained if applied before plants flower. Do not use on Birch Trees. In greenhouse wear mask or respirator. Do not apply to "Bonnafon Deluxe," "Fred Shoesmith," and "White Knight" chrysanthemums as injury may result.

PEARS [7]: Green Fruitworm, Tarnished Plant Bug—Use 1 lb. per 100 gals.; 4 to 5 lbs. per acre. Apply at white bud or petal fall.

Consperse Stink Bug (Ground Treatment)—Use 1 lb. per 100 gals.; 200 to 400 gals. per acre applied to orchard floor around trees prior to bloom. Do not graze livestock in treated orchards.

Consperse Stink Bug (Foliar Treatment), Pear Rust Mite—Use 1 lb. per 100 gals.; 4 to 5 lbs. per acre. Do not exceed 2 applications during the fruiting period. Stink Bugs must be wet by spray to obtain control. Pear Leaf Blister Mite—Apply ½ to 1 lb. per 100 gals. as a post harvest or dormant treatment.

Pear Psylla—Use 5 lbs per acre in a minimum of 10 gals. of water by aerial application or in 40 gals. of water for semi-concentrate or in 300 gals. of water for dilute application. Apply when adults are first observed or nymphs are small and repeat to maintain control. Do not exceed 5 applications during the fruiting period. Do not apply within 21 days of harvest except if only 2 applications are made during the fruiting period, may be applied up to 7 days prior to harvest.

PEAS [0], Seed: Pea Aphid, Pea Weevil—Use 1 to 1½ lbs. per acre. Do not feed treated vines or threshings to dairy or meat animals. **Succulent:** Pea Aphid, Pea Weevil—Use 1 to 2 lbs. per acre. Do not exceed 2 applications during the fruiting period. Use only on peas to be harvested by combine. Do not feed treated vines to livestock or allow to graze in treated fields.

PECANS: Black Pecan Aphid, Pecan Nut Casebearer, Spittlebug—Use 1½ lbs. per 100 gals. Do not apply after shuck split. Do not graze livestock in treated groves. For Casebearer, apply when eggs of first generation appear on tips of the young nuts. Another application may be required after the second generation of eggs is deposited. For Spittlebug, apply when first leaves are half-grown and repeat as required. Pecan Leaf Phylloxera—1 to 1½ lbs. per 100 gallons when nymphs appear and before they are enclosed in plant tissue. For high populations, use the higher rate. Do not apply after Shuck Split. Do not graze livestock in treated groves.

PEPPERS [4]*: Green Peach Aphid, Flea Beetles, Hornworms, Pepper Maggot—Use 1 to 2 lbs. per acre. *1 day if no more than 1 lb. per acre is used.

PINEAPPLES [for fresh market only] [7]: Pineapple Fruit Mite—4 lb. per acre. Apply at intervals of 7 to 10 days if necessary, particularly during the 40 day period of blooming. Do not exceed 4 applications per season. Do not feed treated forage or pineapple byproducts to livestock.

PINES [Austrian, Jack, Red, Scotch, White]: Zimmerman Pine Moth—Use 1½ lbs. per 100 gals. Apply in mid-April and again, if necessary, in late fall. Thoroughly wet bark and main stem especially where branches join main stem.

PLUMS [7], PRUNES [7]: Fruittree Leafroller (**Pacific Northwest Only**)—Use 1 lb. per 100 gals.; 4 to 5 lbs. per acre. Apply during pre-pink stage of growth. Hop Aphid, Leafcurl Plum Aphid, Plum Rust (Plum Nursery) Mite, Thistle Aphid—Use 1 lb. per 100 gals.; 4 to 5 lbs. per acre. For Aphids, apply when eggs hatch during pre-bloom or petal fall. Summer applications should be made before leaves curl. Lesser Peachtree Borer—Use 1½ lbs. per 100 gals.; 4 to 5 lbs. per acre. Apply spray to thoroughly wet trunk and main branches. Peachtree Borer—Use 1½ lbs. per 100 gals.; 4 to 5 lbs. per acre except **West Coast** use 1 to 1½ lbs. per 100 gals. Spray thoroughly to cover all bark areas from ground to scaffold limbs. Peach Twig Borer—Use 4 to 5 lbs. per acre.

POTATOES [0]: Aphids, Colorado Potato Beetle, Green Stink Bug, Leaffooted Bug, Potato Flea Beetle, Potato Leafhopper, Potato Tuberworm, Southern Armyworm, Whitefly—Use 1 to 2 lbs. per acre. European Corn Borer, Potato Psyllid—Use 1½ to 2 lbs. per acre. False Chinch Bug—Use 2 lbs. per acre. Threelined Potato Beetle—Use 1 lb. per acre. Do not plant root crops other than carrots, potatoes, sugar beets, and sweet potatoes as follow-up crops.

PUMPKINS [0]: Aphids, Cucumber Beetle, Squash Vine Borer—Use 1 to 2 lbs. per acre. For Vine Borer, apply weekly to flower buds, stems and vines, beginning when moths first appear.

SAFFLOWER: Green Peach Aphid—Use 2 lbs. per acre. Do not apply after flower heads open.

SHADE TREES [except Birch], SHRUBS: Aphids—1 lb. per 100 gallons. Do not exceed one application per season.

SPINACH [21]: Green Peach Aphid, Crown Mite—Use 1½ lbs. per acre. Do not exceed one application per season.

SPRUCE: Spruce Gall Aphid (New England area only) 1 lb. per 100 gallons. Apply in late April or early May when Aphids are present but before galls are formed.

STRAWBERRIES [4]: Tarnished Plant Bug, Meadow Spittlebug, Strawberry Aphid—Use 2 lbs. per acre. Do not re-apply within 15 days or more than twice during any 35 day period when fruit is present. Cyclamen Mite—Thoroughly wet the foliage, stem, and crowns of the plant. For multiple applications, do not apply at intervals less than 35 days during the fruiting period. **West Coast**—Use 8 lbs. in 800 gals. water per acre. Other areas—Use 1 lb. per 100 gals.; 4 lbs. per acre in 400 gals. of water.

STRAWBERRIES—Northwest Use Only: Garden Symphylan (aids in reducing damage)—2 lbs. per 100 gallons (plant dip). Mix thoroughly. Dip entire plant. When immersing bundles of plants, make certain any trapped air is forced out to assure thorough wetting of entire plant. Drain and allow plants to dry before setting them out in the field. Wear rubber gloves during dipping operation.

SUGAR BEETS [30]: Green Peach Aphid, Meadow Spittlebug—1 to 2 lbs. per acre. Sugar Beet Webworm—1½ to 2 lbs. per acre. Make 1 or 2 applications as needed for control. Do not feed treated tops to livestock. **Eastern Use Only:** Meadow Spittlebug—1 lb. per acre. Do not feed treated tops to livestock. **Western Use Only:** Meadow Spittlebug—1 to 2 lbs. per acre. Do not feed treated tops to livestock. Do not plant root crops other than carrots, potatoes, sugar beets and sweet potatoes as follow-up crops.

SUNFLOWERS [0]: Sunflower Moth—Use 2 lbs. per acre. Do not exceed 3 applications. Make the first application at the onset of bloom and make two more applications at 4 to 7 day intervals. Do not feed treated forage to livestock.

SWEET CORN (Fresh Vegetable Use Only) [0]: Corn Leaf Aphid—2 lbs. Corn Earworm—3 lbs. Apply at 2 to 3 day intervals beginning when silks first appear and continuing until they begin to dry. Do not exceed 5 applications. Do not apply to sweet corn to be processed. Do not feed treated forage or ensilage to livestock or allow to graze in treated fields.

SWEET POTATOES [0]: Sweet Potatoe Flea Beetle—1 lb. per acre. Begin applications shortly after transplanting or as soon as flea beetles appear. **South Central States Only:** Aids in control of Banded Cucumber Beetle Larvae—Apply 2 to 4 lbs. per acre broadcast, or ⅔ to 1⅓ lbs. per acre in a 16 inch band over the row (48 inch spacing). Work into the soil to a depth of approximately 3 inches. A single treatment should be made just prior to plant set. Do not feed treated potatoes to livestock. Do not plant root crops other than carrots, potatoes, sugar beets, and sweet potatoes as follow-up crops.

TAXUS: Taxus Bud Mite—Use 1 lb. per 100 gals. Apply 3 to 5 applications beginning in mid-May. Spray foliage, twigs and bark, Black Vine Weevil—Use 2 lbs. per 100 gals. Spray thoroughly and drench soil under plants. Apply when weevils first appear. Repeat in 14 days.

TOBACCO [5]: Budworm, Cabbage Looper, Flea Beetle, Green Peach Aphid, Hornworm—SEED BED: Use 1 lb. per 100 gals. Apply about 6 gallons of finished spray per 100 square yards. FIELD: Use 1 to 2 lbs. per acre to control above insets. For Stink Bugs—Use 2 to 3 lbs. per 100 gals. PLANT BED: Green June Bug Larvae—Use ½ lb. per 100 gals. drenching at rate of 1 gal. per sq. yd.

TOMATOES [Field and Greenhouse] [1]: Blister Beetle, Colorado Potato Beetle, Flea Beetle, Green Peach Aphid—Use 1 lb. per acre. Hornworm—Use 1 to 2 lbs. per acre. Tomato Fruitworm, Yellowstriped Armyworm—Use 2 lbs. per acre. Cabbage Looper, Stink Bugs—Use 1½ to 2 lbs. per acre. Whitefly—Use 1 lb. per 100 gals.; 100 to 200 gals. per acre. In greenhouse applications, wear mask or respirator.

WALNUT: Walnut Aphid—Use 3 to 4 lbs. per acre. Do not apply after husk split. Do not graze livestock in treated groves.

DEALERS SHOULD SELL IN ORIGINAL PACKAGES ONLY.

TERMS OF SALE OR USE: On purchase of this product buyer and user agree to the following conditions:

WARRANTY: FMC makes no warranty, expressed or implied, concerning the use of this product other than indicated on the label. Except as so warranted, the product is sold as is. Buyer and user assume all risk of use and/or handling and/or storage of this material when such use and/or handling and/or storage is contrary to label instructions.

DIRECTIONS AND RECOMMENDATIONS: Follow directions carefully. Timing and method of application, weather and crop conditions, mixture with other chemicals not specifically recommended and other influencing factors in the use of this product are beyond the control of the seller and are assumed by buyer at his own risk.

USE OF PRODUCT: FMC's recommendations for the use of this product are based upon tests believed to be reliable. The use of this product being beyond the control of the manufacturer, no guarantee, expressed or implied, is made as to the effects of such or the results to be obtained if not used in accordance with directions or established safe practice.

DAMAGES: Buyer's or user's exclusive remedy for damages for breach of warranty or negligence shall be limited to direct damages not exceeding the purchase price paid and shall not include incidental or consequential damages.

WETTABLE SULPHUR

FUNGICIDE-INSECTICIDE

ACTIVE INGREDIENT:	BY WT.
Sulphur	92.0%
INERT INGREDIENTS:	8.0%
TOTAL	100.0%

The material in this bag is guaranteed to pass 95% through a 325 mesh screen.

EPA Reg. No. 279-387AA EPA Est., 279-

CAUTION

Keep Out of Reach of Children

For Emergency Assistance Call 716–735-3765.

See other panels for additional information.

DIRECTIONS FOR USE

It is a violation of Federal Law to use this product in a manner inconsistent with its labeling.

Do not apply this product in such a manner as to directly or through drift expose workers or other persons. The area being treated must be vacated by unprotected persons.

Do not enter treated areas without protective clothing until sprays have dried.

Protective clothing means, at least, a hat or other suitable head covering, a long sleeved shirt and long legged trousers or a coverall type garment (all of closely woven fabric covering the body, including the arms and legs), shoes and socks.

Because certain states may require more restrictive reentry intervals for various crops treated with this product, consult your State Department of Agriculture for further information.

Written or oral warnings must be given to workers who are expected to be in a treated area or in an area about to be treated with this product. Inform workers of areas or fields that may not be entered without specific protective clothing, period of time field must be vacated and appropriate actions to take in case of accidental exposure. An example of such information is given under written warnings. When oral warnings are given, warnings shall be given in a language customarily understood by workers. Oral warnings must be given if there is reason to believe that written warnings cannot be understood by workers. Written warnings must include the following information:

CAUTION

Area treated with WETTABLE SULPHUR on (insert date). Do not enter treated areas for 24 hours unless appropriate protective clothing until sprays have dried (insert State Department of Agriculture's re-entry interval, if more restrictive).

In case of accidental exposure to pesticide spray or dust, wash the skin thoroughly with soap and water. Remove contaminated clothing and wash before reuse. If in eyes, flush with plenty of water. If inhaled, go to an area where the pesticide has not been applied. Get medical attention if needed.

STORAGE AND DISPOSAL

PESTICIDE STORAGE

Keep away from fire and sparks.

Keep out of reach of children and animals. Store in original containers only. Store in a cool, dry place and avoid excess heat. Carefully open containers. After partial use, fold and roll back bags, clamp and close tightly. Do not put concentrate or dilute material into food or drink containers. Do not contaminate other pesticides, fertilizers, water, food or feed by storage or disposal.

In case of spill, avoid contact, isolate area and keep out animals and unprotected persons. Confine spills. Call FMC collect: (716) 735-3765.

To confine spill: If liquid, dike surrounding area or absorb with sand, cat litter or commercial clay. If dry material, cover to prevent dispersal. Place damaged package in a holding container. Identify contents.

PESTICIDE DISPOSAL

Wastes resulting from the use of this product may be disposed of on site or at an approved waste disposal facility.

CONTAINER DISPOSAL

Completely empty bag into application equipment. Then dispose of empty bag in a sanitary landfill or by incineration, or, if allowed by State and local authorities, by burning. If burned, stay out of smoke.

ENVIRONMENTAL HAZARDS

Keep out of lakes, ponds and streams. Do not contaminate water by cleaning of equipment or disposal of wastes. Do not apply when weather conditions favor drift from the areas treated. Do not apply where runoff is likely to occur. Do not use in a manner or at a time other than in accordance with label directions because animal, plant or crop injury, or other undesirable results may occur.

Applications of materials containing sulphur should not be made prior to or during periods of excessively high summer temperatures or within 2 weeks of an oil spray.

Fill the spray tank with water until the agitator blades are covered. With the agitator running, sift in the required amount of WETTABLE SULPHUR. Continue filling and keep agitator running.

ALMONDS: Brown Rot, Scab—Use 6 pounds per 100 gallons of water in pre-bloom, bloom and post-bloom periods. Apply when disease threatens during rain periods and repeat at 7 to 10 day intervals.

APPLES AND PEARS: For Scab or Powdery Mildew control, use as a spray in pre-blossom, Calyx and Early cover sprays. Use 6 to 8 pounds of WETTABLE SULPHUR or 4 to 6 pounds of WETTABLE SULPHUR plus 2 to 3 pounds Polysulphide Compound or 2 gallons of Liquid Lime Sulphur Solution per 100 gallons of dilute spray. Sulphur may injure certain varieties of Apples. Check specific varietal tolerance to sulphur prior to using. **Pears only:** Do not apply after fruit begins to turn down. **Fall Clean-up Spray:** Blister Mites, European Red Mites and Twospotted Spider Mites—Use 6 pounds of WETTABLE SULPHUR plus 6 pounds of Polysulphide Compound or 5 gallons of Liquid Lime Sulphur Solution to each 100 gallons of dilute spray.

BLACKBERRIES, BOYSENBERRIES, GOOSEBERRIES, LOGANBERRIES, RASPBERRIES, STRAWBERRIES: Powdery Mildew—Use 2 pounds per 100 gallons of water. Apply when disease first appears and repeat as necessary.

CITRUS: Citrus Flat Mite, Yuma Spider Mite, Sixspotted Mite and Citrus Rust Mite—Use 5 to 12 pounds per 100 gallons of water (75 to 125 pounds per acre). Apply when mites appear and repeat as necessary. Do not make an oil application within 60 days prior to or after a sulphur application.

GRAPES: Powdery Mildew—Use 2 to 5 pounds per 100 gallons of dilute spray. Use as a spray beginning when shoots are 6 to 8 inches long and repeat at 7 to 10 day intervals. Ground application preferred. Concord and other Labrusa type grapes may be injured by sulphur.

PLUMS, PRUNES, PEACHES, NECTARINES, CHERRIES: Powdery Mildew, Silver Mite, Cherry Leaf Spot, Prune Rust and Brown Rot—Use 5 to 10 pounds per 100 gallons of water. (By ground apply a minimum of 200 gallons of water per acre.) Apply in the pre-bloom, bloom and post-bloom periods. Apply when disease threatens during rain periods and repeat at 7 to 10 day intervals.

TOMATOES: Tomato Russet Mite—Use 12 to 25 pounds per acre. Apply with ground equipment at the rate of 60 to 250 gallons of water per acre. Use only application equipment which has paddle type agitation. Continuous and complete agitation is necessary.

DEALERS SHOULD SELL IN ORIGINAL PACKAGES ONLY.

TERMS OF SALE OR USE: On purchase of this product buyer and user agree to the following conditions:

WARRANTY: FMC makes no warranty, expressed or implied, concerning the use of this product other than indicated on the label. Except as so warranted, the product is sold as is. Buyer and user assume all risk of use and/or handling and/or storage of this material when such use and/or handling and/or storage is contrary to label instructions.

DIRECTIONS AND RECOMMENDATIONS: Follow directions carefully. Timing and method of application, weather and crop conditions, mixture with other chemicals not specifically recommended and other influencing factors in the use of this product are beyond the control of the seller and are assumed by buyer at his own risk.

USE OF PRODUCT: FMC's recommendations for the use of this product are based upon tests believed to be reliable. The use of this product being beyond the control of the manufacturer, no guarantee, expressed or implied, is made as to the effects of such or the results to be obtained if not used in accordance with directions or established safe practice.

DAMAGES: Buyer's or user's exclusive remedy for damages for breach of warranty or negligence shall be limited to direct damages not exceeding the purchase price paid and shall not include incidental or consequential damages.

ZINEB 75 WETTABLE POWDER

FUNGICIDE

ACTIVE INGREDIENT:	BY WT.
*Zineb (zinc ethylene bisdithiocarbomate)	75.0%
INERT INGREDIENTS:	25.0%
TOTAL	100.0%
* Total Zinc as Metallic	17.7%

EPA Reg. No. 279-1387 EPA Est.,

CAUTION

Keep Out of Reach of Children

PRECAUTIONARY STATEMENTS

HAZARDS TO HUMANS (& DOMESTIC ANIMALS)

CAUTION

May Cause Irritation of Nose and Throat! Do not breathe dust or spray mist. During the mixing and loading process, wear protective clothing (long pants, long sleeved shirt, impermeable gloves, hat and boots).

ENVIRONMENTAL HAZARDS

Keep out of any body of water. Do not contaminate water by cleaning of equipment or disposal of wastes. Do not allow this material to drift onto neighboring crops or non-crop areas or use in a manner or at a time other than in accordance with label directions because animal, plant or crop injury, illegal residues or other undesirable results may occur.

PHYSICAL OR CHEMICAL HAZARDS

Keep away from fire and sparks.

Never allow this product to become wet during storage. This may lead to certain chemical changes which will reduce its effectiveness as a fungicide and create vapors which may be inflammable. Keep container closed when not in use.

For Emergency Assistance Call 716–735-3765.

See other panels for additional information.

DIRECTIONS FOR USE

It is a violation of Federal Law to use this produce in a manner inconsistent with its labeling.

Do not apply this product in such a manner as to directly or through drift expose workers or other persons. The area being treated must be vacated by unprotected persons.

Do not enter treated areas without protective clothing until sprays have dried.

Protective clothing means, at least, a hat or other suitable head covering, a long sleeved shirt and long legged trousers or a coverall type garment (all of closely woven fabric covering the body, including the arms and legs), shoes and socks.

Because certain states may require more restrictive reentry intervals for various crops treated with this product, consult your State Department of Agriculture for further information.

Written or oral warnings must be given to workers who are expected to be in a treated area or in an area about to be treated with this product. Inform workers of areas or fields that may not be entered without specific protective clothing, period of time field must be vacated and appropriate actions to take in case of accidental exposure. An example of such information is given under written warnings. When oral warnings are given, warnings shall be given in a language customarily understood by workers. Oral warnings must be given if there is reason to believe that written warnings cannot be understood by workers. Written warnings must include the following information:

CAUTION

Area treated with ZINEB 75 Wettable Powder on (insert date). Do not enter without appropriate protective clothing until sprays have dried (insert State Department of Agriculture's re-entry interval, if more restrictive).

In case of accidental exposure to pesticide spray or dust, wash the skin thoroughly with soap and water. Remove contaminated clothing and wash before reuse. If in eyes, flush with plenty of water. If inhaled, go to an area where the pesticide has not been applied. Get medical attention if needed.

STORAGE AND DISPOSAL

PESTICIDE STORAGE

Keep away from fire and sparks.

Keep out of reach of children and animals. Store in original containers only. Store in a cool, dry place and avoid excess heat. Carefully open containers. After partial use, fold and roll back bags, clamp and close tightly. Do not put concentrate or dilute material into food or drink containers. Do not contaminate other pesticides, fertilizers, water, food or feed by storage or disposal.

In case of spill, avoid contact, isolate area and keep out animals and unprotected persons. Confine spills. Call FMC collect: (716) 735-3765.

To confine spill: If liquid, dike surrounding area or absorb with sand, cat litter or commercial clay. If dry material, cover to prevent dispersal. Place damaged package in a holding container. Identify contents.

PESTICIDE DISPOSAL

Pesticide wastes are toxic. Improper disposal of excess pesticide, spray mixture, or rinsate is a violation of Federal Law. If these wastes cannot be disposed of by use according to label instructions, contact your State Pesticide or Environmental Control Agency, or the Hazardous Waste representative at the nearest EPA Regional Office for guidance.

CONTAINER DISPOSAL

Completely empty bag into application equipment. Then dispose of empty bag in a sanitary landfill or by incineration, or, if allowed by State and local authorities, by burning. If burned, stay out of smoke.

Unless otherwise indicated use 1½ to 2 pounds per 100 gallons of water. Start filling the tank with water. Sift in the required amount of product when the agitator paddles are covered or the jet or recirculating pipe from pump is covered. Fill the tank when the product is thoroughly mixed. In regular spray equipment use 100 to 125 gallons of spray per acre for low growing crops. Use 150 to 200 gallons per acre for heavy tomato vines.

Dilution Directions for Small Quantities: Eight teaspoonfuls of this product per 1 gallon of water is equivalent to 2 pounds per 100 gals.

APPLES: Summer control of Scab, Sooty Blotch, Fly Speck, Brook's Spot, Black Rot—Use 1 to 1½ lbs. per 100 gals water in cover sprays. Apply full coverage spray. Do not apply within 15 days of harvest in the following states: Arkansas, Delaware, Illinois, Indiana, Kansas, Kentucky, Maryland, Missouri, New Jersey, North Carolina, Ohio, Pennsylvania, South Carolina, Tennessee, Virginia, and West Virginia. In all other states, do not apply within 30 days of harvest.

BEANS (Field and Snap): Rust, Anthracnose, **LIMA BEANS:** Anthracnose, Downy Mildew—Start applications when disease first appears and repeat at 4- to 7-day intervals as long as disease threatens. Do not apply to Snap, Lima and String Beans later than 7 days before picking.

BEETS, CARROTS: Cercospora Leaf Spot, Alternaria Leaf Spot—Apply when disease threatens and continue at 7- to 10-day intervals as needed. Do not apply within 7 days of harvest if tops are used for food or feed.

CABBAGE, CAULIFLOWER, BROCCOLI, BRUSSELS SPROUTS: Downy Mildew—Apply when disease threatens and continue at 7- to 10-day intervals as needed. Do not apply within 7 days of harvest.

CELERY: Early blight (Cercospora) and late blight (Septoria)—Start applications in the plant bed and continue in the field as frequently as required (weekly under average conditions; every 3 to 5 days under severe conditions). Remove residues by stripping, trimming and washing. Do not apply within 14 days of harvest.

CORN: Helminthosporium Blight—Apply when disease threatens and continue at 7- to 10-day intervals as needed. Do not feed forage to dairy animals or animals being finished for slaughter. Do not apply within 40 days of harvest.

CRANBERRIES: Fruit Rots—at 5% bloom stage, apply 2½ pounds in 100 gallons of water using 300 gallons of spray per acre. Apply a second application 2 weeks later. For aircraft application, apply 8 pounds in 13 or more gallons of water per acre. Do not apply after mid-bloom.

CUCUMBERS, MELONS, PUMPKINS, SQUASH: Downy Mildew, Anthracnose—Apply as soon as plants appear above ground or when disease threatens and continue at 7- to 10-day intervals as long as necessary to maintain control. Under severe conditions, shorten interval to 3 to 5 days. Do not apply within 5 days of harvest.

EGGPLANT: Alternaria, Anthracnose, Phomopsis—Apply when disease threatens and continue at 7- to 10-day intervals as needed.

GRAPES: Downy Mildew, Black Rot—Use 1½ lbs. per 100 gallons of water. Apply just before bloom, just after bloom, 10 days after bloom and 2 to 3 weeks later. Cover thoroughly. Do not apply within 7 days of harvest.

GRAPEFRUIT, LEMONS, LIMES, ORANGES, TANGELOS, TANGERINES: Fruit Russet—Use ½ pound per 100 gallons of water. May be applied when 10% of the leaves are infected with Rust Mites; however, the best time to apply this material is in the post-bloom period and in the summer application (June or July). Greasy Spot—Use 1 lb. per 100 gallons of water. Make two spray applications at 4 to 8 week intervals during July and August. If only one application can be made, use 2 lbs. and apply in July. Best results are expected with the split schedule.

HOPS: Downy Mildew—Apply soon after training vines and repeat at 7- to 10-day intervals as long as needed. Do not apply within 14 days of harvest.

LETTUCE: Downy Mildew—in plant bed use 2 lbs. as drenching spray starting when seed is planted and continue at 7- to 10-day intervals until transplanting. In field, use 2 lbs. as a spray; begin applications when disease appears and continue at 7- to 10-day intervals as long as needed (under severe conditions, shorten interval to 3 to 5 days) but do not apply later than 7 days before harvest on leaf lettuce (5 days on head lettuce).

MUSHROOMS: Verticillium (Brown Spot, Dry Bubble), Dactyllum (Mildew, Soft Decay, Cobweb), Mycogone (Bubbles) and Trichoderma (Green Mold)—Use 10 gallons of spray (1 lb. per 100 gals.) per 4000 sq. ft. of bed. Begin applications soon after casing soil is applied. Apply at 3- to 7-day intervals until buttons begin to form, also between "breaks" as required.

ONIONS (dry and green): Downy Mildew, Purple Blotch—Use 2 to 3 lbs. per 100 gallons of water per acre. Apply when disease threatens and continue at 7- to 10-day intervals as needed. Do not apply within 7 days of harvest on green onions. May be applied to day of harvest on dry onions.

ORNAMENTALS: Septoria Leaf Spot of Chrysanthemums; Rust of Snapdragons; Petal Blight of Azaleas and Camelias—Apply at 3- to 5-day intervals as needed.

PEPPERS: Cercospora Leaf Spot, Anthracnose—Apply when disease threatens and continue at 7-to 10-day intervals as needed.

SPINACH: Downy Mildew (Blue Mold), White Rust—Start applications when diseases first appear in area and continue at 7- to 10-day intervals. Do not apply within 7 days of harvest.

STRAWBERRIES: To control Leaf Spot. Apply when new growth starts in the spring and repeat as required until fruit begins to form. Do not apply within 7 days of harvest.

TOBACCO: Blue Mold—Apply spray directly to the young plants. For each 100 sq. yds. of plant bed, apply 3 to 6 gals. of spray depending upon size of plants. Start applications when leaves are about the size of a dime, or when Blue Mold first appears in the vicinity. Apply at 3- to 4-day intervals until transplanting time.

TOMATOES: Late and early blights, Septoria Leaf Spot, Gray Leaf Spot (Stemphylium), Anthracnose—Start applications when fruit of first-clusters are well formed; repeat at 7- to 10-day intervals. When weather conditions favor late blight, more frequent applications may be necessary starting when plants are set in the fields.

WHEAT: Rust—Apply $1\frac{1}{2}$ lbs. per acre in 25 to 40 gals. of water. Make first application when the plants are in the boot or early heading stage or when Rust is first reported in the area. Repeat at weekly or 10-day intervals until the milk stage.

DEALERS SHOULD SELL IN ORIGINAL PACKAGES ONLY.

TERMS OF SALE OR USE: On purchase of this product buyer and user agree to the following conditions:

WARRANTY: FMC makes no warranty, expressed or implied, concerning the use of this product other than indicated on the label. Except as so warranted, the product is sold as is. Buyer and user assume all risk of use and/or handling and/or storage of this material when such use and/or handling and/or storage is contrary to label instructions.

DIRECTIONS AND RECOMMENDATIONS: Follow directions carefully. Timing and method of application, weather and crop conditions, mixture with other chemicals not specifically recommended and other influencing factors in the use of this product are beyond the control of the seller and are assumed by buyer at his own risk.

USE OF PRODUCT: FMC's recommendations for the use of this product are based upon tests believed to be reliable. The use of this product being beyond the control of the manufacturer, no guarantee, expressed or implied, is made as to the effects of such or the results to be obtained if not used in accordance with directions or established safe practice.

DAMAGES: Buyer's or user's exclusive remedy for damages for breach of warranty or negligence shall be limited to direct damages not exceeding the purchase price paid and shall not include incidental or consequential damages.

ZIRAM SPRAY

FUNGICIDE

ACTIVE INGREDIENT:	
Ziram (zinc dimethyl dithiocarbamate)	76.0%
INERT INGREDIENTS:	24.0%
TOTAL	100.0%

Zinc expressed as metallic 16.25%

EPA Reg. No. 279–268 EPA Est. 279-NY-1

CAUTION

Keep Out of Reach of Children

May be harmful if swallowed or inhaled. May cause irritation of eyes, nose, throat and skin. Avoid contact with eyes, skin or clothing. Avoid breathing dust or spray mist. In case of contact, flush with plenty of water, for eyes, get medical attention.

For Emergency Assistance Call 716–735-3765.

See side panel for additional precautionary statements.

PRECAUTIONARY STATEMENTS

ENVIRONMENTAL HAZARDS

This product is toxic to fish and wildlife. Do not apply directly to water. Do not apply where runoff is likely to occur. Do not apply when weather conditions favor drift from areas treated. Do not contaminate water by cleaning of equipment or disposal of wastes. Apply this product only as specified on this label.

PHYSICAL OR CHEMICAL HAZARDS

Keep away from fire and sparks.

DIRECTIONS FOR USE

It is a violation of Federal Law to use this product in a manner inconsistent with its labeling.

Do not apply this product in such a manner as to directly or through drift expose workers or other persons. The area being treated must be vacated by unprotected persons.

Do not enter treated areas without protective clothing until sprays have dried.

Protective clothing means, at least, a hat or other suitable head covering, a long sleeved shirt and long legged trousers or a coverall type garment (all of closely woven fabric covering the body, including the arms and legs), shoes and socks.

Because certain states may require more restrictive reentry intervals for various crops treated with this product, consult your State Department of Agriculture for further information.

Written or oral warnings must be given to workers who are expected to be in a treated area or in an area about to be treated with this product. Inform workers of areas or fields that may not be entered without specific protective clothing, period of time field must be vacated and appropriate actions to take in case of accidental exposure. An example of such information is given under written warnings. When oral warnings are given, warnings shall be given in a language customarily understood by workers. Oral warnings must be given if there is reason to believe that written warnings cannot be understood by workers. Written warnings must include the following information:

CAUTION

Area treated with ZIRAM SPRAY on (insert date). Do not enter without appropriate protective clothing until sprays have dried (insert State Department of Agriculture's reentry interval, if more restrictive).

In case of accidental exposure to pesticide spray or dust, wash the skin thoroughly with soap and water. Remove contaminated clothing and wash before reuse. If in eyes, flush with plenty of water. If inhaled, go to an area where the pesticide has not been applied. Get medical attention if needed.

STORAGE AND DISPOSAL

PESTICIDE STORAGE

Keep away from fire and sparks.

Keep out of reach of children and animals. Store in original containers only. Store in a cool, dry place and avoid excess heat. Carefully open containers. After partial use, fold and roll back bags, clamp and close tightly. Do not put concentrate or dilute material into food or drink containers. Do not contaminate other pesticides, fertilizers, water, food or feed by storage or disposal.

In case of spill, avoid contact, isolate area and keep out animals and unprotected persons. Confine spills. Call FMC collect: (716) 735-3765.

To confine spill: If liquid, dike surrounding area or absorb with sand, cat litter or commercial clay. If dry material, cover to prevent dispersal. Place damaged package in a holding container. Identify contents.

PESTICIDE DISPOSAL

Wastes resulting from the use of this product may be disposed of on site or at an approved waste disposal facility.

CONTAINER DISPOSAL

Completely empty bag into application equipment. Then dispose of empty bag in a sanitary landfill or by incineration, or, if allowed by State and local authorities, by burning. If burned, stay out of smoke.

Use the recommended rates for concentrate, dilute and aerial applications. Ground application is preferred, especially when full canopy of foliage is developed.

For deciduous fruits and almonds, apply the recommended rate in a minimum of 50 gallons per acre by ground. By air, use 20 gallons per acre, or sufficient water to provide complete coverage.

For cucumbers, melons, beans and tomatoes, apply in a minimum of 10 gallons per acre by ground and 3 gallons per acre by air.

To mix, fill the spray tank about ⅔ full of water, and with agitator running, slowly sift in the required amount of ZIRAM SPRAY. Finish filling the tank with water and let mixing continue in the tank until thoroughly mixed before applying. Continue agitation and apply immediately.

ALMONDS: Brown Rot, Shot-Hole (Coryneum Blight), Leaf Blight —Use 8 pounds per acre in pre-bloom, blossom and petal fall periods. Scab—Use 8 pounds per acre at petal fall and 5 weeks after petal fall. Do not use later than 5 weeks after petal fall.

APPLES: Scab—Use 6 to 12 pounds per acre in pre-blossom, blossom and early cover stages. Bull's Eye Rot (Oregon and Washington)—Use 6 to 12 pounds per acre beginning 12 to 14 days after calyx stage and repeat as necessary in summer sprays. Perennial Canker (Oregon)—Use 6 to 12 pounds per acre in pre-blossom and calyx sprays.

APRICOTS: Brown Rot, Shot-Hole (Coryneum Blight)—Use 8 to 10 pounds per acre in pre-bloom, blossom and petal fall through early cover sprays.

CHERRIES: Brown Rot, Blossom Blight—Use 4 to 5 pounds per acre in popcorn, full blossom, petal fall and shuck stages and approximately 2 weeks after shuck fall. Do not apply within 7 days of harvest.

NECTARINES, PEACHES: Peach Blight (Coryneum) and Leaf Curl —Use 8 to 12 pounds per acre in the fall before heavy rains. For Leaf Curl only, use 8 to 12 pounds per acre as a dormant application prior to bud swell. Brown Rot and Peach Blight (Coryneum)—Use 8 to 10 pounds per acre in pre-bloom, blossom and petal fall sprays. May be used in early cover and before harvest. Remove excess residue by washing or wiping.

PEARS: Scab—Use 6 to 12 pounds per acre in pre-blossom, blossom and early cover sprays. Storage Rots (Oregon)—Use 6 to 12 pounds per acre in pre-harvest spray.

CUCUMBERS, MELONS: Scab, Anthracnose—Use 2 pounds per acre at 7 to 10 day intervals when plants are small. Increase rate to 3 pounds per acre after plants vine.

BEANS (Snap, Lima): Anthracnose—Use 2 pounds per acre in sufficient water for thorough coverage when plants are small. Increase to three pounds per acre on large plants. Make applications at 7 day intervals starting when true leaves appear. Do not apply to snap beans later than 4 days before picking.

TOMATOES: Anthracnose, Early Blight—Use 2 to 3 pounds per acre at 7 to 10 day intervals starting 2 to 4 weeks after plants are set. Under severe conditions, apply 3 pounds per acre at 7 to 10 day intervals.

DEALERS SHOULD SELL IN ORIGINAL PACKAGES ONLY.

TERMS OF SALE OR USE: On purchase of this product buyer and user agree to the following conditions:

WARRANTY: FMC makes no warranty, expressed or implied, concerning the use of this product other than indicated on the label. Except as so warranted, the product is sold as is. Buyer and user assume all risk of use and/or handling and/or storage of this material when such use and/or handling and/or storage is contrary to label instructions.

DIRECTIONS AND RECOMMENDATIONS: Follow directions carefully. Timing and method of application, weather and crop conditions, mixture with other chemicals not specifically recommended, and other influencing factors in the use of this product are beyond the control of the seller and are assumed by buyer at his own risk.

USE OF PRODUCT: FMC's recommendations for the use of this product are based upon tests believed to be reliable. The use of this product being beyond the control of the manufacturer, no guarantee, expressed or implied, is made as to the effects of such or the results to be obtained if not used in accordance with directions or established safe practice.

DAMAGES: Buyer's or user's exclusive remedy for damages for breach of warranty or negligence shall be limited to direct damages not exceeding the purchase price paid and shall not include incidental or consequential damages.

Hoechst-Roussel Agri-Vet Company

Agricultural Chemicals
Route 202–206 North
Somerville, NJ 08876
(201) 231-2000

HOELON® 3EC

HERBICIDE

RESTRICTED USE PESTICIDE

For retail sale to and use only by certified applicators or persons under their direct supervision and only for those uses covered by the certified applicator's certification.

For Control of Wild Oats and Other Annual Grasses in Wheat (Including Durum), Barley, Lentils, Soybeans and Acreage Conservation Reserve (Set Aside).

ACTIVE INGREDIENT*
Diclofop-methyl: methyl 2-
[4-(2,4-dichlorophenoxy) phenoxy] propanoate 35.49%
INERT INGREDIENTS 64.51%
TOTAL 100.0%

* Equivalent to 3.0 pounds of active ingredient per gallon

EPA Reg. No. 8340-20-54382

The name and logo HOECHST are registered trademarks of Hoechst AG
The name and logo ROUSSEL are registered trademarks of Roussel Uclaf S.A.

Keep Out of Reach of Children

DANGER

Directions for Use:
This product is registered for use in the following states:

Wheat, Barley, Soybeans and Acreage Conservation Reserve	Wheat and Barley Only		Lentils
Delaware	Arizona	Oklahoma	Idaho
Kansas	California	Oregon	Washington
Maryland	Colorado	South Carolina	
Minnesota	Idaho	Texas	
Nebraska	Montana	Utah	
North Dakota	Nevada	Washington	
South Dakota	New Mexico	West Virginia	
Virginia	North Carolina	Wyoming	

Wheat Only	Soybeans & Acreage Conservation Reserve		Wheat, Soybeans & Acreage Conservation Reserve
Alabama	Illinois	Missouri	Louisiana
Arkansas	Indiana	New Jersey	
Florida	Iowa	Ohio	
Georgia	Kentucky	Pennsylvania	
Mississippi	Michigan	Wisconsin	
Tennessee			

GENERAL INFORMATION

HOELON® 3EC Herbicide is a highly selective herbicide for the control of a broad spectrum of annual grassy weeds in wheat (including durum), barley, lentils, soybeans and acreage conservation reserve. HOELON® 3EC Herbicide is a three pound active ingredient per gallon emulsifiable concentrate. For best results applications should be made when the majority of annual grassy weeds are in the one-to-three leaf stage of growth. Larger weeds, three-to-four leaves, will require a higher dosage rate. The stage of weed growth is as important as the number of weeds. In high populations, thorough uniform spray coverage of weeds is critical and use of flat fan nozzles is recommended for optimum coverage with ground equipment.

Special Notes:
1. HOELON® 3EC Herbicide does NOT control broadleaf weeds or perennial grassy weeds.
2. HOELON® 3EC Herbicide is a slow acting product; therefore, visual results will not be evident immediately following application.
3. DO NOT exceed 3⅓ pints per acre of HOELON® 3EC Herbicide in a growing season. Do not apply beyond the specified crop growth stages.
4. DO NOT allow livestock to graze treated fields. DO NOT harvest forage, hay or straw from treated fields prior to grain harvest.

Equipment Notes
Thorough uniform spray coverage of weeds is essential. Ground applications should be in at least 10 gallons of carrier per acre using a minimum of 40 psi and use of flat fan or hollow cone nozzles is recommended. Aerial applications must be applied in at least 5 gallons of carrier per acre. Use spray nozzle tips and spraying pressures which provide a small uniform droplet size (200-350 micron range).

Do not make aerial applications when wind is above 5 mph or within 100 feet of a lake, stream, drainage basin, tidal marsh, or estuary.
Only clean water, free of suspended matter or other contaminants, should be used when applying HOELON® 3EC Herbicide.

DIRECTIONS FOR THE USE OF HOELON® 3EC HERBICIDE ON WHEAT AND BARLEY

(The general information section must also be read for complete use instructions.)
When using HOELON® 3EC Herbicide to control grassy weeds in the North Central, Southwest and Western U.S. and the grassy weed(s) do not include annual ryegrass, apply the 2 pint rate when the annual grassy weeds are in the 1-3 leaf stage of growth and growing conditions are optimal. Higher rates should be used (see chart) when the majority of the grassy weeds is not clearly in the specified stage of growth or if growth is retarded due to lack of moisture. Thorough spray coverage is essential.
Cold (lower than 40°F) and/or prolonged wet weather conditions increase the sensitivity of barley to HOELON® 3EC Herbicide, and *should not* be used under these adverse growing conditions.
DO NOT exceed 2⅔ pints per acre of HOELON® 3EC Herbicide on barley.

APPLY HOELON® 3EC HERBICIDE AS A BROADCAST POSTEMERGENCE SPRAY AT THE RATES LISTED BELOW

SUSCEPTIBLE ANNUAL GRASSY WEEDS	AMOUNT OF HOELON® 3EC HERBICIDE PER ACRE (PINTS)* RELATIVE TO GROWTH STAGE OF ANNUAL GRASSY WEEDS			
	1-3 LEAF	3-4 LEAF	4-5 LEAF	5 LEAF—2 TILLERS
Annual Ryegrass (Italian) *(Lolium multiflorum)*[1]	1⅓ pints	1⅓ pints	2 pints	2⅔ pints
Broadleaf Signalgrass (Suppression) *Brachiaria platylphylla)*	2⅔-3⅓ pints	Do Not Apply	Do Not Apply	
Crabgrass *(Digitaria sanguinalis)*	2⅔-3⅓ pints	Do Not Apply		
Fall Panicum *(Panicum dichotomiflorum)* Barnyardgrass (watergrass) *(Echinochloa crusgalli)* Giant Foxtail *(Setaria faberii)* Green Foxtail (pigeongrass) *(Setaria viridis)* Persian Darnel *(Lolium persicum)* Wild Oats *(Avena fatua)* Yellow Foxtail (pigeongrass) *(Setaria lutescens)* Witchgrass (Suppression) *(Panicum capillare)*	2-2⅔ pints	2⅔-3⅓ pints		
Itchgrass (Raoulgrass) *(Rottboellia exaltata)* Volunteer Corn *(Zea mays)*	See specific control recommendations for Itchgrass and Volunteer Corn in this labeling			

[1] Annual ryegrass can also be controlled preemergence in wheat. See appropriate recommendation section in this labeling.

* When controlling mixed populations of grassy weeds, always use the rate that will control the least susceptible species.

Special Notes:

For best results, application should be made when the majority of annual grassy weeds is in the one-to-four leaf stage of growth. *For best control of crabgrass and yellow foxtail, application should be made before the second leaf fully emerges.* The stage of weed growth is more important than the number of weeds. **Thorough uniform spray coverage of weeds is essential.**

RATE RANGE FOR EACH CROP/REGION

REGISTERED CROPS	RATE RANGE FOR EACH CROP/REGION	CROP GROWTH STAGE	
		Minimum	Maximum
1. Wheat, including durum (All Regions)	1⅓-3⅓ pints	None	Before one node (jointing)
2. Barley (North Central, Southwest and Western Regions of U.S.)	1⅓-2⅔ pints	After emergence	Before one node (jointing)
3. Barley (Southeast and Midatlantic Regions of U.S.)*	1⅓-2 pints	After Tiller (Stooling) Initiation	Before one node (jointing)
4. Soybeans, All registered states	2-3⅓ pints	None	Before formation of sixth trifoliate leaf
5. Lentils	2-2⅔ pints	After emergence	Before bloom

* HOELON 3EC Herbicide can only be used on the following varieties of barley in the Southeast and Midatlantic Regions: Boone, Milton, Henry and Sussex. Applications of this product to other varieties may result in crop damage.

Barley must be developed to tiller initiation in the Southeast and Midatlantic Regions prior to application or increased crop response may be noted.

TANK MIXES

Broadleaf Herbicides

All broadleaf herbicides should be used in accordance with the label limitations and precautions for each product. No label dosage rates should be exceeded. For best results, tank mixes should be used only when growing conditions (air temperature and soil moisture) are optimum, and grasses are in the 2 to 3 leaf stage of growth with light to moderate infestations.

ME 4 Brominal® or Buctril® (2EC)

HOELON® 3EC Herbicide can be tank mixed with ME 4 Brominal or Buctril (2EC) herbicide for broadleaf weed control in wheat and barley. HOELON® 3EC Herbicide should be used at the 2 to 3⅓ pints per acre rate as specified on the label. If ME 4 Brominal is used, it should be mixed at a rate of ¾ to 1 pint per acre according to the directions specified on the label. If Buctril (2EC) is used, it should be mixed at a rate of 1½ to 2 pints per acre as specified on the label.

ME 4 Brominal or Buctril (2EC) Plus MCPA (ester) (4EC)

Low rates of MCPA (ester) (4EC) can be added to a tank mix of HOELON® 3EC Herbicide plus ME 4 Brominal or Buctril (2EC). HOELON® 3EC Herbicide should be used at a rate of 2⅔ to 3⅓ pints per acre. If Buctril (2EC) is used, a rate of 1 pint per acre should be tank mixed with MCPA (ester) (4EC) at 1½ ounces per acre. If ME 4 Brominal is used, a rate of ½ pint per acre should be tank mixed with MCPA (ester) (4EC) at 1½ ounces per acre. MCPA rates in excess of 1½ ounces per acre may reduce the grassy weed control from HOELON® 3EC Herbicide. No other formulations of MCPA should be substituted for the 4.0 pound active ingredient ester formation.

Glean® Herbicide

HOELON® 3EC Herbicide can be tank mixed with Glean for broadleaf weed control in winter wheat only. HOELON® 3EC Herbicide should be used at a rate of 2⅔ to 3⅓ pints per acre. The rate of Glean should not exceed ¼ ounce per acre rate. Refer to the Glean label for crop rotation restrictions and weed species controlled at the ¼ ounce per acre rate. Prior to the use of Glean, determine the soil pH by laboratory analysis using a 1:1 soil:water suspension. Do not use Glean on soils with a pH greater than 7.5. In all cases, ground applications should be made with a minimum of 10 gallons of water per acre and air applications with a minimum of 5 gallons of water per acre. If other broadleaf herbicides are applied within 5 days, reduced grassy weed control may occur.

Fertilizers

HOELON® 3EC Herbicide can be tank mixed with liquid nitrogen fertilizers (e.g., 28-32%) if the timing for use is compatible. The concentration of liquid nitrogen fertilizer to water should not exceed a 1:1 ratio of total carrier in the spray tank. When mixing with liquid nitrogen fertilizers, as a partial carrier, crop burn may occur. This condition is intensified under higher temperatures (greater than 75° F daytime) and/or low soil moisture. This mixture should not be used when these environmental conditions are present. Thorough spray coverage is essential when using HOELON® 3EC Herbicide and fan-type nozzles are recommended for ground applications. Do not use less than 2 pints per acre of HOELON® 3EC Herbicide when tank mixing with liquid nitrogen fertilizers.

Crop Oil Concentrates

In spring wheat only, at the lower labeled usage rates, it may be helpful to use 1 pint to 1 quart of crop oil concentrate, approved for use on growing crops. Use a crop oil concentrate containing a blend of 80% (minimum) petroleum or vegetable base oil and the remaining composed of a tolerance exempt surfactant. Some slight wheat yellowing may be noted when crop oil concentrates are added to HOELON® 3EC Herbicide. DO NOT apply crop oil concentrate or other additives with HOELON® 3EC Herbicide for use on barley.

For more detailed information on determining the low use rates in combination with crop oil concentrates, refer to the Hoechst-Roussel Agri-Vet Company Technical Bulletin No. 110, "Low Rate Technology". Contact your local dealer or Hoechst-Roussel Agri-Vet Company representative for this Technical Bulletin.

Preemergence Use to Control Annual Ryegrass in Fall Planted Wheat (Winter Wheat)

Apply HOELON® 3EC Herbicide preemergence (at planting) to wheat to control annual ryegrass. HOELON® 3EC Herbicide should be applied at a rate of 2 to 2⅔ pints depending upon soil type and annual ryegrass density. In coarse textured soils with less than 2% organic matter, use the 2 pint per acre rate. In fine textured soils with greater than 2% organic matter, use the 2⅔ pint per acre rate. In fields that have a history of heavy annual ryegrass pressures, always use the 2⅔ pint rate. If rainfall does not occur within seven (7) days after application, reduced control may occur.

DO NOT apply HOELON® 3EC Herbicide preemergence to barley as barley damage will occur. HOELON® 3EC Herbicide applied preemergence will not control wild oats.

DIRECTIONS FOR THE USE OF HOELON® 3EC HERBICIDE ON LENTILS

(The general information section must also be read for complete use instructions.)

HOELON® 3EC Herbicide should be applied at a rate of 2 to 2⅔ pints per acre. Apply the 2 pint per acre rate only when the annual grassy weeds are in the 1 to 2 leaf stage of growth and growing conditions are normal. The higher rate should be used when the development of the annual grassy weeds are not clearly in the specified stages of growth or if the growth is retarded due to adverse growing conditions. Thorough uniform spray coverage of weeds is essential. Ground applications should be made in at least 10 gallons of water per acre using a minimum of 40 psi and the use of fan type or hollow cone nozzles is recommended.

Special Notes for Lentils:

1. DO NOT apply more than one application per growing season.
2. DO NOT tank mix with any broadleaf herbicides.
3. Only ground applications should be made on lentils.
4. Apply prior to the bloom stage of the lentils.

DIRECTIONS FOR THE USE OF HOELON® 3EC HERBICIDE ON SOYBEANS

(The general information section must also be read for complete use instructions.)

Apply HOELON® 3EC Herbicide at the higher labeled rates when the development of the majority of grassy weeds is not clearly in the specified stages of growth or if growth is retarded due to adverse growing conditions.

Under these conditions, the addition of 1 pint to 1 quart of crop oil concentrate may result in improved weed control. Use a crop oil concentrate containing a blend of 80% (minimum) petroleum or vegetable base oil with the remaining composed of a tolerance exempt surfactant.

Banded applications may be made using the following formula to calculate applications:

$$\frac{\text{Band width in inches}}{\text{Row width in inches}} \times \text{Rate/acre broadcast} = \text{Amount needed for treatment}$$

Volunteer Corn and Itchgrass

Heavy infestations of volunteer corn (especially when uneven emergence is occurring from ears buried at varying soil depths) should be treated as a special weed problem. **Apply after ALL of the volunteer corn has emerged but before the first emerged corn grows to a height where thorough coverage (including the whorl) is not possible.** Spray boom height and pressures should be adjusted to obtain total coverage of the entire volunteer corn plant. Thorough coverage of the volunteer corn foliage and penetration of clumps is essential. Use 2 to 3⅓ pints of HOELON® 3EC Herbicide in 10 to 40 gallons of water per acre at not less than 40 psi.

At the lower labeled use rate, use 1 pint to 1 quart of crop oil concentrate approved for use on growing crops. Use an oil concentrate containing a blend of 80% (minimum) petroleum or vegetable base oil with the remaining composed of a tolerance exempt surfactant. Spot treatments of clumps may be necessary.

For treatment of individual clumps of volunteer corn use 1⅓ fluid ounces per one gallon of spray/acre (3⅓ pints per 40 gallons). **In mixed annual grassy weed/volunteer corn infestations, the delayed broadcast application of HOELON® 3EC Herbicide to control heavy infestations of volunteer corn could result in reduced control of any annual grassy weeds that have grown beyond the recommended leaf stages. Do not apply more than one application in a growing season.**

For postemergence control of itchgrass, apply to emerged itchgrass before it has grown beyond 10-12 inches in height. Use 2⅔ pints of HOELON® 3EC Herbicide in at least 10 gallons of water per acre. Flat fan nozzles and a pressure of at least 40 pounds per square inch have proven most effective. HOELON® 3EC Herbicide may be used on a broadcast basis sprayed directly over the top of the soybeans. Complete coverage of itchgrass is essential. Delayed application to control heavy infestations of itchgrass may give reduced control of other susceptible annual grasses. HOELON® 3EC Herbicide is most effective when the grasses are actively growing. It is less effective when used under drought stress.

Special Notes:

1. DO NOT tank mix HOELON® 3EC Herbicide with any other herbicide or apply any other herbicide within one week of a HOELON® 3EC Herbicide application.
2. Banvel®, 2,4-D and other phenoxy herbicide injury sometimes occurs in soybean fields treated with HOELON® 3EC Herbicide even though these herbicides were not recently used in the sprayer. This situation is not caused by HOELON® 3EC Herbicide and is often caused by sprayer contamination. If these products have been used in your sprayer, refer to Hoechst-Roussel Agri-Vet Company Technical Bulletin No. 100, "How to Avoid Sprayer Contamination," available through your HOELON® 3EC Herbicide supplier.

DIRECTIONS FOR THE USE OF HOELON® 3EC HERBICIDE ON ACREAGE CONSERVATION RESERVE (SET ASIDE)

(The general information section must also be read for complete use instructions.)

HOELON® 3EC Herbicide may be used to control the annual grassy weeds in acreage conservation reserve or set-aside programs. This acreage is often seeded to wheat, clover or alfalfa as a cover crop. These cover crops have excellent tolerance to HOELON® 3EC Herbicide. Use 2 pints to 3⅓ pints per acre when the annual grassy weeds are in the one to four leaf stage as noted in the use directions section of this labeling. In most situations, 2 pints of HOELON® 3EC Herbicide plus 2 pints of crop oil concentrate will offer commercially acceptable control of annual grasses in acreage conservation reserve programs. Apply early to young actively growing grassy weeds. DO NOT harvest or graze cover crops treated with HOELON® 3EC Herbicide. DO NOT apply to cover crops such as oats, sorghum, and sudan grass as injury may occur.

Keep Out of Reach of Children

DANGER

PRECAUTIONARY STATEMENTS

HAZARDS TO HUMANS AND DOMESTIC ANIMALS

Causes eye and skin damage. Do not get in eyes or on skin or clothing. Wear goggles and respirator when handling. Avoid contact or inhalation of spray mist. Wash thoroughly after using and change clothing. Do not take internally.

The following protective apparel must be worn during all loading and mixing operations and when applied by ground equipment: gloves, waterproof boots, impermeable pants and shirts, goggles and a cartridge type respirator. Aerial pilots are not to be involved in the mixing and loading unless the aforementioned equipment is worn. Pilots must wear a cartridge type respirator during application unless the aircraft is equipped with an air filter system.

The Environmental Protection Agency has concluded that diclofop-methyl, the active ingredient in this product, produces tumors in laboratory mice. This effect was not observed in other species tested. The user must read and follow all precautionary statements on this label. Give particular attention to protective clothing requirements.

ENVIRONMENTAL HAZARDS

This pesticide is toxic to fish. Use with care when applying in areas adjacent to any body of water. Keep out of lakes, streams, ponds, drainage basins, tidal marshes and estuaries. Do not apply when weather conditions favor runoff or drift. Avoid direct application or drift of spray material to water surfaces. Do not apply within 100 feet of aquatic habitat. Do not contaminate arable land and/or water by cleaning of equipment and/or disposal of waste.

FIRST AID

In case of skin or eye contact, immediately flush with plenty of water for at least 15 minutes.

If swallowed, DO NOT INDUCE VOMITING but seek medical attention immediately.

Note to Physician: Aspiration hazard may contraindicate the use of gastric lavage.

For emergency assistance call (201) 231-2244.

STORAGE AND DISPOSAL

Do not contaminate water, food, or feed by storage or disposal.

Pesticide washes are toxic. Improper disposal of excess pesticide, spray mixture, or rinsate is a violation of Federal Law. If these wastes cannot be disposed of by use according to label instructions, contact your State Pesticide or Environmental Control Agency, or the Hazardous Waste representative at the nearest EPA Regional Office for guidance.

CONTAINER DISPOSAL

Triple rinse (or equivalent). Then offer for recycling or reconditioning, or puncture and dispose of in a sanitary landfill, or by other procedures approved by state and local authorities.

Do not store below 20°F.

IMPORTANT NOTICE: DISCLAIMER

Read "IMPORTANT NOTICE: DISCLAIMER" before buying or using. If terms are not acceptable, return at once unopened. HOECHST-ROUSSEL AGRI-VET COMPANY warrants only that the product conforms to the chemical description on the label and is reasonably fit for the purpose stated on the label when used in accordance with the directions under normal conditions of use. This warranty does not extend to the use of this product contrary to label instructions or under abnormal conditions, or under conditions not reasonably foreseeable to HOECHST-ROUSSEL AGRI-VET COMPANY and user assumes the risk of any such use. HOECHST-ROUSSEL AGRI-VET COMPANY MAKES NO OTHER WARRANTY, EXPRESS OR IMPLIED, INCLUDING ANY IMPLIED WARRANTY OF FITNESS FOR A PARTICULAR PURPOSE OR OF MERCHANTABILITY. In no case shall HOECHST-ROUSSEL AGRI-VET COMPANY be liable for consequential, special, indirect or incidental damages resulting from the use or handling of this product. The foregoing conditions of sale and warranty can be varied only by an agreement in writing signed by a duly authorized representative of HOECHST-ROUSSEL AGRI-VET COMPANY.

HOELON® 3EC Herbicide is a registered trademark of Hoechst AG.

ME 4 Brominal® Bromoxynil Broadleaf Herbicide is a registered trademark of Union Carbide Agricultural Products Company, Inc.

Buctril® Herbicide is a registered trademark of Rhône-Poulenc Inc., Agrochemical Division.

Glean® Herbicide is a registered trademark of E.I. Dupont deNemours and Company, Inc.

SCOUT®

INSECTICIDE

RESTRICTED USE PESTICIDE

For Retail Sale to and Application Only by Certified Applicators or Persons Under their Direct Supervision

ACTIVE INGREDIENTS:

[1R[1(S*)3(RS*)]]-2,2-dimethyl-3-(1,2,2,2-tetrabromoethyl) cyclopropanecarboxylic acid α-cyano-(3-phenoxyphenyl) methyl ester[1]	3.75%
Xylene Range Aromatic Solvent	12.48%
INERT INGREDIENTS:	83.77%
TOTAL	100.00%

[1]Equivalent to 0.3 lbs. active ingredient per gallon

EPA Reg. No. 34147-2-54382 EPA Est. No. 42545—MO-1

SCOUT® is a registered trademark of Roussel Uclaf

Keep Out of Reach of Children

DANGER

PRECAUTIONARY STATEMENTS

HAZARDS TO HUMANS AND DOMESTIC ANIMALS

Corrosive, causes irreversible eye damage. Wear goggles while handling. Can cause irritation to skin such as burning and tingling. Do not get in eyes, or on skin or on clothing. Wear protective gloves while handling. Harmful if swallowed, inhaled or absorbed through the skin. If material gets on skin, the affected area should be thoroughly washed with plenty of soap and water. Remove and wash clothing before reuse.

ENVIRONMENTAL HAZARDS

This pesticide is extremely toxic to fish and other aquatic organisms, including crab and shrimp. Use with care when applying in areas adjacent to any body of water. Do not apply directly to water or wetlands. Do not apply when weather conditions favor drift from treated area. Drift or runoff from treated fields may kill fish and other aquatic organisms. Do not contaminate water by cleaning of equipment or disposal of wastes.

This pesticide is highly toxic to bees exposed to direct treatment on blooming crops or weeds. Do not apply this product or allow it to drift to blooming crops or weeds while bees are actively visiting the treatment area.

Notify State and/or Federal authorities and HOECHST-ROUSSEL AGRI-VET COMPANY immediately if you observe any adverse environmental effects due to the use of SCOUT Insecticide.

NOTICE: It is a Federal offense to use any pesticide in a manner that results in the death of a member of an endangered species.

The use of SCOUT Insecticide may pose a hazard to the following Federally designated endangered/threatened species known to be found in certain areas within the named locations:

SLACKWATER DARTER—Tennessee counties including Lawrence and Wayne. Alabama counties including Limestone, Lauderdale and Madison. FRESHWATER MUSSELS—Headwaters on small rivers in the middle and upper drainage of the Cumberland and Tennessee Rivers in the Tennessee counties of Claiborne, Hanc, Hawking and Sullivan and the Virginia counties of Smyth, Washington, Scott, Lee and Grayson. Any tributary of the White or St. Francis rivers in the Arkansas counties of Independence, Jackson, Woodruff, Monroe, Arkansas, Lee, St. Francis, Cross, Poinsett, Greene and Craighead. PECOS GAMBUSIA—Texas counties including Reeves, Jefferson Davis and Pecos. New Mexico counties including Chaves and Eddy. COMANCHE SPRINGS PUPFISH—Texas counties including Reeves and Jefferson Davis. DEXTER NATIONAL FISH HATCHERY—Chaves county, New Mexico.

This product may not be used in these areas where adverse impact on the Federally designated endangered/threatened species is likely.

FIRST AID

In case of eye or skin contact, immediately flush with plenty of water. SCOUT Insecticide contains petroleum distillates. If swallowed, do not induce vomiting. Call a physician. Vomiting should be supervised by a physician because of the possible pulmonary damage via aspiration of the solvent. If possible, bring the product container and labeling to the attending physician.

For emergency assistance, call (201)231-2244.

Physical Hazards

Do not use or store near heat or open flame.

STORAGE AND DISPOSAL

Do not contaminate water, food or feed by storage or disposal. Do not store over 100°F or below 10°F.

Pesticide wastes are toxic. Improper disposal of excess pesticide, spray mixture, or rinstate is a violation of Federal Law. If these wastes cannot be disposed of by use according to label instructions, contact your State Pesticide or Environmental Control Agency, or the hazardous waste representative at the nearest EPA Regional Office for guidance.

Container Disposal: Empty containers should be triple rinsed into the spray tank during the spray operation. Then offer for recycling or reconditioning, or puncture and dispose of in a sanitary landfill, or by other procedures approved by state and local authorities.

DIRECTIONS FOR USE

It is a violation of Federal Law to use this product in a manner inconsistent with this label.

SCOUT INSECTICIDE IS RESTRICTED FOR USE IN THE FOLLOWING STATES: Alabama, Arkansas, Arizona, Georgia, Louisiana, Mississippi, Missouri, New Mexico, North Carolina, Oklahoma, South Carolina, Tennessee and Texas.

Dilute SCOUT Insecticide in water or at least once refined crop oil according to the directions and use as a foliar spray on cotton at the rates listed. Uniform coverage is essential. Begin applications when economic infestations occur based on local information and repeat as needed.

Do not enter treated field before spray is dried.

Crop	Insect	Fluid* Ounces Per Acre	Dilution Requirements
Cotton	Cotton Leaf Perforator Cotton Bollworm Pink Bollworm Tobacco Budworm	6.4 to 8	Minimum gallonage requirements: Four gallons of finished spray per acre with ground equipment, or no less than one gallon per acre by air (minimum of three gallons per acre in Arizona) or one quart of at least once refined crop oil per acre.
	Bollweevil**	8	

* Apply as required by scouting reports, usually at intervals of five to seven days. Under normal conditions, the rate of 6.4 fluid ounces will give adequate control.

** Use the 8 fluid ounces per acre rate for optimal suppression on a five-day interval schedule.

Conversion Table For SCOUT Insecticide

Fluid Ounces Per Acre	Pints Per Acre	Acres Treated Per Gallon	Quarts Per 5 Acres
6.4	0.4000	20.0	1.00
7.0	0.4375	18.3	1.10
8.0	0.5000	16.0	1.25

Note:

1. DO NOT graze or feed cotton foliage. 2. DO NOT apply more than 81 fluid ounces per acre (0.19 pounds active ingredient per acre) in one growing season. 3. DO NOT apply within 28 days of harvest. 4. DO NOT plant rotational crops other than cotton within four months of last application.

IMPORTANT NOTICE: DISCLAIMER

Read IMPORTANT NOTICE: DISCLAIMER before buying or using. If terms are not acceptable, return at once unopened. HOECHST-ROUSSEL AGRI-VET COMPANY warrants only that the product conforms to the chemical description on the label and is reasonably fit for the purpose stated on the label when used in accordance with the directions under normal conditions of use. This warranty does not extend to the use of this product contrary to label instructions or under abnormal conditions, or under conditions not reasonably foreseeable to HOECHST-ROUSSEL AGRI-VET COMPANY, and user assumes the risk of any such use. HOECHST-ROUSSEL AGRI-VET COMPANY MAKES NO OTHER WARRANTY, EXPRESSED OR IMPLIED, INCLUDING ANY IMPLIED WARRANTY OF FITNESS FOR A PARTICULAR PURPOSE OR OF MERCHANTABILITY. In no case shall HOECHST-ROUSSEL AGRI-VET COMPANY be liable for consequential, special, indirect or incidental damages resulting from the use or handling of this product. The foregoing conditions of sale and warranty can be varied only by an agreement in writing signed by a duly authorized representative of HOECHST-ROUSSEL AGRI-VET COMPANY.

ICI Americas Inc.

Agricultural Chemicals Division
Wilmington, DE 19897
(302) 575-3000

ACTELLIC® 7E

INSECTICIDE

For Control of Stored Grain Insects

ACTELLIC May Only be Used to Treat Corn, Rice, Wheat, and Grain Sorghum Intended for Export Only.

ACTIVE INGREDIENT:
O-[2-(diethylamino)-6-methyl-4-pyrimidinyl]
O,O-dimethyl phosphorothioate* 74%
INERT INGREDIENTS 26%
TOTAL 100%

* Pirimiphos-methyl

ACTELLIC® Insecticide contains 7 pounds active ingredient per gallon.

EPA Reg. No. 10182-87

ACTELLIC® is a registered trademark of Imperial Chemical Industries PLC

Keep Out of Reach of Children
WARNING

See Side Panel For Additional Precautionary Statements and Statement of Practical Treatment.

PRECAUTIONARY STATEMENTS

HAZARDS TO HUMANS AND DOMESTIC ANIMALS
KEEP OUT OF REACH OF CHILDREN
WARNING

CAUSES EYE IRRITATION: Do not get in eyes. Wear goggles or face shield when handling.

HARMFUL IF SWALLOWED, INHALED OR ABSORBED THROUGH SKIN: Avoid breathing vapors or spray mist. Keep container closed. Avoid contact with skin or clothing. Wear rubber gloves when handling concentrate. Wash thoroughly after handling.

FIRST AID: In case of contact with eyes, immediately flush with plenty of water for at least 15 minutes. Call a physician if irritation persists.

IF SWALLOWED: Drink 1 or 2 glasses of water and induce vomiting by touching back of throat with finger or blunt object. Do not induce vomiting or give anything by mouth to an unconscious person. Call a physician.

IN CASE OF CONTACT WITH SKIN: Flush with plenty of water. Wash clothing before reuse.

IF INHALED: Remove to fresh air. If not breathing, give artificial respiration, preferably mouth-to-mouth. If breathing is difficult, give oxygen. Call a physician.

NOTE TO PHYSICIAN: Atropine is antidotal. Give 2 mg. of atropine intramuscularly or subcutaneously and repeat if symptoms of poisoning reappear. Pralidoxime (2 PAM) may be given if available. Give 1 gram intramuscularly or intravenously and repeat as necessary.

For emergency assistance, call ICI Americas Inc., (302) 575-3000.

ENVIRONMENTAL HAZARDS

Keep out of lakes, streams and ponds. Do not contaminate water by cleaning of equipment or disposal of waste.

In case of a significant spill, call CHEMTREC, (800) 424-9300.

PHYSICAL AND CHEMICAL HAZARDS

Do not use or store near heat or open flame.
(Combustible Liquid—Class III A: DOL Reg. 29 CFR 1910.106)

STORAGE AND DISPOSAL

PROHIBITIONS

Do not contaminate water, food or feed by storage or disposal. Open dumping is prohibited. Do not reuse empty container.

STORAGE

Store in a cool place and protect from freezing. If accidently frozen, reconstitute by gently warming to room temperature and thoroughly mixing contents. Keep container closed when not in use.

PESTICIDE DISPOSAL

Pesticide wastes are toxic. Improper disposal of excess pesticide, spray mixture, or rinsate is a violation of federal law. If these wastes cannot be disposed of by use according to label instructions, contact your State Pesticide or Environmental Control Agency, or the Hazardous Waste representative at the nearest EPA Regional Office for guidance.

CONTAINER DISPOSAL

Triple rinse (or equivalent). Then offer for recycling or reconditioning, or puncture and dispose of in a sanitary landfill, or by other procedures approved by state and local authorities.

DIRECTIONS FOR USE

It is a violation of federal law to use this product in a manner inconsistent with its labeling.

ACTELLIC 7E must be applied in accordance with the written requirements and specifications of the foreign purchaser (see Remarks and Precautions).

The following countries have approved ACTELLIC for use on stored grains:

Belgium	Peru
Brazil*	Philippines
Dominican Republic	Portugal
Indonesia	Spain
Italy	United Kingdom
Mexico	U.S.S.R.
Morocco	West Germany

* wheat only

Apply the specified amount of ACTELLIC in 5 gallons of water to each 60,000 pounds of grain.

Apply spray solution uniformly with suitable applicating equipment as grain enters storage or export container. Select nozzle size and operating pressure to adjust the application rate with the flow of grain. Apply the spray solution as a coarse, uniform, low pressure spray in order to minimize drift. Shield nozzle if necessary.

INSECTS CONTROLLED WITH ACTELLIC

Coleoptera

Cigarette beetle	Hairy fungus bettle
Confused flour beetle	Red flour beetle
Corn sap beetle	Sawtoothed grain beetle
Flat grain beetle	Granary weevil
Maize weevil	Merchant grain beetle
	Rice weevil

Lepidoptera

Angoumois grain moth
Indian meal moth
Almond moth

REMARKS AND PRECAUTIONS

ICI research has concluded that application rates of 0.4 to 0.55 pints of ACTELLIC 7E in 5 gallons of water per 30 tons (60,000 pounds) of wheat, corn, or grain sorghum (6–8 ppm) or 0.65 to 1.0 pints of ACTELLIC in 5 gallons of water per 30 tons of rice (9–15 ppm) are necessary for consistent and lasting control of insects. Application of rates less than the aforementioned may affect efficacy and may result in reduced length of control. Use at less than these rates shall be at the sole risk of the buyer and user.

Shipping documents accompanying ACTELLIC treated commodities must state that the treated commodity is intended for export only and is not to be sold for use in the U.S.A.

NOTICE TO BUYER AND USER

Seller warrants that this product conforms to the chemical description on the label and is reasonably fit for the purposes stated on the label when used in accordance with directions under normal conditions of use. This warranty does not extend to the use of this product contrary to label instructions, or under abnormal use conditions, or under conditions not reasonably foreseeable to Seller and Buyer and User assumes the risk of any such use. SELLER DISCLAIMS ALL OTHER WARRANTIES, EXPRESSED OR IMPLIED, INCLUDING ANY WARRANTY OF FITNESS OR MERCHANTABILITY. SELLER SHALL NOT BE LIABLE FOR CONSEQUENTIAL, SPECIAL OR INDIRECT DAMAGES RESULTING FROM THE USE OR HANDLING OF THIS PRODUCT AND SELLER'S SOLE LIABILITY AND BUYER'S AND USER'S EXCLUSIVE REMEDY SHALL BE LIMITED TO THE REFUND OF THE PURCHASE PRICE.

AMBUSH®

INSECTICIDE

RESTRICTED USE PESTICIDE

For retail sale to and use only by Certified Applicators or persons under their direct supervision and only for those uses covered by the Certified Applicator's certification.

ACTIVE INGREDIENT:
Permethrin—
(3–Phenoxyphenyl)methyl (±)–cis, trans–3–(2,2–dichloroethenyl)–2,2–dimethyl cyclopropanecarboxylate* 25.6%
INERT INGREDIENTS 74.4%
TOTAL 100.0%

* Cis/trans ratio: Min. 35% (±) cis and max. 65% (±)trans.

AMBUSH contains 2 pounds active ingredient per gallon

EPA Reg. No. 10182-18

U.S. Patent No. 4,024,163

AMBUSH® Insecticide is a registered trademark of ICI Americas Inc.

Keep Out of Reach of Children

WARNING

AVISO

PRECAUCION AL USUARIO: Si usted no lee ingles, no use este producto hasta que la etiqueta le haya sido explicado ampliamente.

PRECAUTIONARY STATEMENTS

HAZARDS TO HUMANS AND DOMESTIC ANIMALS

WARNING

HARMFUL IF SWALLOWED OR ABSORBED THROUGH SKIN. MAY CAUSE EYE IRRITATION.

Avoid contact with eyes, skin or clothing. Avoid breathing vapor or spray mist. Wash thoroughly after handling.

STATEMENT OF PRACTICAL TREATMENT

IF SWALLOWED: Do not induce vomiting. Call a physician. Vomiting should by supervised by a physician or the professional staff because of possible pulmonary damages via aspiration of the solvent.

IF IN EYES: Flush with plenty of water. Get medical attention if irritation persists.

IF ON SKIN: Wash with plenty of soap and water.

For 24-hour emergency assistance, call ICI AMERICAS INC., 302/575-3000.

ENVIRONMENTAL HAZARDS

This pesticide is extremely toxic to fish. Use with care when applying in areas adjacent to any body of water. Do not apply directly to water. Do not apply when weather conditions favor drift from treated areas. Do not contaminate water by cleaning of equipment or disposal of wastes. This product is highly toxic to bees exposed to direct treatment or residues on crops or weeds. Do not apply AMBUSH, or allow it to drift to crops or weeds on which bees are actively foraging. Additional information may be obtained from your Cooperative Extension Service. Apply this product only as specified on this label.

In case of a significant spill, call CHEMTREC 800/424-9300.

PHYSICAL OR CHEMICAL HAZARDS

Do not use or store near heat or open flame.

STORAGE AND DISPOSAL

PROHIBITIONS

Do not contaminate water, food or feed by storage or disposal. Open dumping is prohibited. Do not reuse empty container.

STORAGE

Keep container closed when not in use. Do not store near food or feed. Protect from freezing. In case of spill or leak on floor or paved surfaces, soak up with sand, earth or synthetic absorbent. Remove to chemical waste area.

PESTICIDE DISPOSAL

Pesticide wastes are toxic. Improper disposal of excess pesticide, spray mixture or rinsate is a violation of federal law. If these wastes cannot be disposed of by use according to label instructions, contact your State Pesticide or Environmental Control Agency, or the Hazardous Waste representative at the nearest EPA Regional Office for guidance.

CONTAINER DISPOSAL

Metal Containers: Triple rinse (or equivalent). Then offer for recycling or reconditioning, or puncture and dispose of in a sanitary landfill, or by other procedures approved by state and local authorities.

Plastic Containers: Triple rinse (or equivalent). Then offer for recycling or reconditioning, or puncture and dispose of in a sanitary landfill, or incineration, or, if allowed by state and local authorities, by burning. If burned, stay out of smoke.

Glass Containers: Triple rinse (or equivalent). Then dispose of in a sanitary landfill or by other approved state and local procedures.

DIRECTIONS FOR USE

It is a violation of federal law to use this product in a manner inconsistent with its labeling.

GENERAL INSTRUCTIONS: Apply AMBUSH as shown in "Directions for Application". Use sufficient water to obtain full coverage of foliage. Timing and frequency of applications should be based upon insect populations reaching economic thresholds. The higher rates may be required under conditions of heavy infestation. Do not apply more than 12 pints per acre per season (3 lbs ai/acre). Crops in this label may be planted immediately after last treatment. Do not plant other crops within 60 days after last application. Any other crop may be planted if it is not harvested or grazed.

REENTRY STATEMENTS

Do not apply this product in such a manner as to directly or through drift expose workers or other persons. The area being treated must be vacated by unprotected persons.

Do not enter treated areas without protective clothing until sprays have dried.

Protective clothing means, at least, a hat or other suitable head covering, a long sleeved shirt and long-legged trousers or a coverall type garment (all of closely woven fabric covering the body, including the arms and legs), shoes and socks.

Because certain states may require more restrictive reentry intervals for various crops treated with this product, consult your State Department of Agriculture for further information.

Written or oral warnings regarding use of protective clothing and accidental exposure must be given to workers who are expected to be in treated areas or in areas about to be treated. Inform workers of areas or fields that may not be entered without specific protective clothing, period of time field must be vacated and appropriate actions to take in case of accidental exposure. An example of such information is given under written warnings. When oral warnings are given, warnings shall be given in a language customarily understood by workers. Oral warnings must be given if there is reason to believe that written warnings cannot be understood by workers. Written warnings must include the following information:

WARNING

Area treated with AMBUSH 2E on (insert date). Do not enter without appropriate protective clothing until sprays have dried (if more restrictive, insert State Department of Agriculture reentry interval).

In case of accidental exposure to pesticide spray, wash the skin thoroughly with soap and water. Remove contaminated clothing and wash before reuse. If in eyes, flush with plenty of water. Get medical attention if needed.

Rate Conversion Chart

Pounds Active/ Acre	Fluid Ounces/ Acre	Pints/ Acre	Acres Treated/ Gallon
0.05	3.2	0.2	40.0
0.10	6.4	0.4	20.0
0.125	8.0	0.5	16.0
0.15	9.6	0.6	13.3
0.175	11.2	0.7	11.4
0.20	12.8	0.8	10.0

NOTICE TO BUYER AND USER

Seller warrants that this product conforms to the chemical description on the label and is reasonably fit for the purposes stated on the label when used in accordance with directions under normal conditions of use. This warranty does not extend to the use of this product contrary to label instructions, or under abnormal use conditions, or under conditions not reasonably foreseeable to Seller and Buyer and User assumes the risks of any such use. SELLER DISCLAIMS ALL OTHER WARRANTIES, EXPRESSED OR IMPLIED, INCLUDING ANY WARRANTY OF FITNESS OR MERCHANTABILITY. SELLER SHALL NOT BE LIABLE FOR CONSEQUENTIAL, SPECIAL OR INDIRECT DAMAGES RESULTING FROM THE USE OR HANDLING OF THIS PRODUCT AND SELLER'S SOLE LIABILITY AND BUYER'S AND USER'S EXCLUSIVE REMEDY SHALL BE LIMITED TO THE REFUND OF THE PURCHASE PRICE.

DIRECTIONS FOR APPLICATION

CROP	TARGET PESTS	RATE (LBS AI/A)	REMARKS
Almonds	Navel Orangeworm, Peach Twig Borer	0.2–0.4	Apply by ground equipment using sufficient water to obtain uniform coverage. Can be applied up to 7 days before harvest. Do not apply more than 2.0 lbs. a.i. per acre per season. Do not graze livestock in treated orchards.
Apples	Apple Aphid, Redbanded Leafroller, Obliquebanded Leafroller, Plum Curculio, White Apple Leafhopper, Spotted Tentiform Leafminer, Tarnished Plant Bug	0.1–0.4	Apply by ground equipment only, using 25–400 gallons of spray per acre. Spray to wet all foliage. Do not apply after petal fall.
Asparagus	Cutworm complex Asparagus beetle	0.05–0.1	Apply by ground equipment using sufficient water to obtain uniform coverage. Apply as needed. Do not apply more than 0.4 lbs. a.i. per acre per season. Do not apply within 3 days of harvest.
Asparagus ferns	Japanese Beetle Asparagus Beetle Lygus Bugs	0.1	Apply to the fern stage of the asparagus plant after spear harvest when larval and adult stage of the asparagus beetle and lygus bugs and the adult stage of the Japanese beetle are present.
Celery	(Refer to Leafy Vegetables for application directions)		
Chrysanthemums (Florida only)	Vegetable Leafminer	0.1–0.2	Apply by ground.
COLE CROPS Broccoli	Cabbage Looper, Imported Cabbageworm, Diamondback Moth, Cabbage Aphid*	0.05–0.1	Apply every 5 days or as needed by air or ground. Use sufficient water to obtain full coverage of foliage. Do not apply more than 0.8 lbs. a.i. per acre per season. Do not apply within 1 day of harvest.
Brussels Sprouts	Cabbage Looper, Imported Cabbageworm, Diamondback Moth, Cabbage Aphid*	0.05–0.1	Apply as needed by air or ground. Use sufficient water to obtain full coverage of foliage. Do not apply more than 0.8 lbs. a.i. per acre per season. Do not apply within 1 day of harvest.
Cabbage	Cabbage Looper, Imported Cabbageworm, Diamondback Moth, Cabbage Aphid*	0.05–0.1	Apply every 5–10 days or as needed by air or ground. Use sufficient water to obtain full coverage of foliage. Do not apply more than 1.0 lb. a.i. per acre per season. Do not apply within 1 day of harvest.
Cauliflower	Cabbage Looper, Imported Cabbageworm, Diamondback Moth, Cabbage Aphid*	0.05–0.1	Apply every 5 days or as needed by air or ground. Use sufficient water to obtain full coverage of foliage. Do not apply more than 0.8 lbs. a.i. per acre per season. Do not apply within 1 day of harvest.
Conifers (container and field grown)	Nantucket Pine Tip Moth	0.1–0.2	Apply first application when the adults appear and repeat at 5–7 day intervals or as needed throughout the season. AMBUSH may be diluted in a nonvolatile vegetable oil or in water in a minimum of one gallon of finished spray per acre using equipment calibrated to give adequate coverage.

CROP	TARGET PESTS	RATE (LBS AI/A)	REMARKS
Cotton	Bollworm, Tobacco Budworm, Pink Bollworm, Lygus Bugs, Cabbage Looper, Cotton Leafperforator, Boll Weevil, Cotton Fleahopper, Saltmarsh Caterpillar, Whitefly**, Cotton Aphid**, Cutworms	0.1–0.2	Apply as required by scouting reports, usually at intervals of 5 to 7 days. Apply by air or ground. Use sufficient water to obtain full coverage of foliage. AMBUSH may also be applied by air diluted with a vegetable oil such as cottonseed oil or soybean oil. Agricultural mineral oils (crop oils) are not recommended. Apply using a minimum of 1 quart of finished spray per acre. For control of boll weevils, spray on a 3 to 5 day schedule or as needed. Do not apply within 14 days of harvest. The higher rates of AMBUSH (0.15 to 0.2 lbs. a.i. per acre) may be required under conditions of heavy lepidopterous worm infestation. AMBUSH aids in the reduction of lepidopterous egg hatch by contact action. Do not graze livestock in treated areas. Do not apply more than 3.0 lbs. a.i. per acre per season.
Eggplant	Colorado Potato Beetle Vegetable Leafminer, Cabbage Looper, Flea Beetle	0.2 0.1–0.2	Apply by ground or air equipment using sufficient water to obtain uniform coverage. Apply as needed. Do not apply more than 3.0 lbs. a.i. per acre per season. Do not apply within 3 days of harvest.
Field Corn/ Popcorn/ Field Corn Grown for Seed (Preemergent Use)	Armyworm, Cutworms	0.1–0.2	Apply in the time period from 5 days prior to planting up to emergence of the crop. Apply as a broadcast spray in a minimum of 20 gallons of finished spray/acre with ground equipment.***
Field Corn/ Popcorn/Field Corn Grown for Seed (Foliar Use)	Cutworms, European Corn Borer, Armyworm, Southwestern Corn Borer, Flea Beetle Corn Earworm,† Corn Rootworm Beetle,† Tobacco Budworm,† Stalk Borer	0.1–0.2	Apply *prior to the brown silk stage.* Apply a minimum of 1 gallon of finished spray/acre by air and 10 gallons/acre with ground equipment.*** When treating for stalk borer, AMBUSH must be applied when or shortly before the stalk borer larvae are moving into the corn from surrounding weeds and grasses. Mowing or burndown herbicide are suggested to initiate movement.***
Horseradish	Imported Crucifer Weevil (*Baris lepidii*)	0.1% ai solution‡ in a spring preplant dip and 0.2 lb ai/A as a foliar spray.	Soak sets for 30 min. and air-dry before planting; make up to 3 foliar applications as needed to control weevil adults during oviposition. Apply by ground equipment in a minimum spray volume of 20 gallons per acre. Do not apply within 22 days of harvest.
Leafy Vegetables¶ (except *Brassica*)	Vegetable Leafminer, Cabbage Looper, Cutworm Complex, Fall Armyworm, Beet Armyworm, Corn Earworm, Green Cloverworm, Leafhoppers, European Corn Borer	0.1–0.2	Apply every 3–5 days or as needed by air or ground. Use sufficient water to obtain full coverage of foliage. Do not apply more than 2.0 lbs. a.i. per acre per season. Can be applied up to 1 day of harvest. Do not graze treated areas or feed crop refuse to livestock.
Lettuce	(Refer to Leafy Vegetables for application directions)		
Mushrooms (Mushroom houses & adjacent premise areas)	Mushroom Flies (Sciarid and Phorid Adults)	Apply as a fogging or aerosol treatment at the rate of 3.2 to 4.0 ounces (0.05–0.0625 lb. ai) per 30 ounces of water or suitable diluent. Use 1 quart of solution per standard double house (35,000 cu. ft.: 8,000 sq. ft.)	Preparation of the building prior to fogging: (1) Close all doors, windows and ventilators. (2) Lock or barricade all entrances, turn off pilot lights, post warning signs, and take precautions to prevent persons and animals from entering the area. Use prior to filling house, during cool-down, during spawning, up to pinning and between breaks. Do not use when mushrooms are present. Treat once daily or as needed when flies appear. Do not make more than 20 applications prior to pinning of first break; apply no more than two applications between each break. Do not apply more than 30 applications total per crop of 5 breaks. Do not apply within 3 days of harvest. Length of exposure time should be limited to 1 hour; then ventilate the house. Use fans to ventilate in houses that do not have forced air circulation. Wear full-faced gas mask with canister type recommended for general insecticide protection.
Peaches	Peach Twig Borer, Oriental Fruit Moth, Green Fruitworm, Lesser Peachtree Borer, Plum Curculio, Tarnished Plantbug, Rose Chafer	0.1–0.4	Apply by ground equipment using 25–400 gallons of spray per acre. Spray to wet all foliage. Can be applied up to 7 days before harvest. Do not apply more than 3.0 lbs. a.i. per acre per season. Do not graze livestock in treated orchards.
Pears (Dormant)	Pear Psylla	0.2–0.4	Apply at a rate of 0.8 to 1.6 pints (0.2 to 0.4 lbs. active) per acre. May be combined with 2 to 8 gallons of spray oil per acre. Apply during the dormant through delayed dormant growth periods only. Apply in 3 to 20 gallons of finished spray per acre by aircraft and 25 to 400 gallons per acre by ground equipment. Do not apply more than 0.8 lbs. a.i. per acre per season. Do not graze orchard floor.

CROP	TARGET PESTS	RATE (LBS AI/A)	REMARKS
Pears (Pre-Bloom/ Seasonal)	Pear Psylla, Codling Moth, Green Fruitworm	0.2–0.4	Apply two pre-bloom sprays followed by three seasonal sprays. Pre-bloom sprays can be applied from dormant through bud burst stages. Seasonal sprays can be applied from petal fall up to 14 days of harvest. Do not apply more than 2.0 lbs. a.i. per acre per season.
Peppers (Bell)	Vegetable Leafminer, Cabbage Looper, Flea Beetle, Pepper Weevil	0.1–0.2	Apply by ground or air equipment using sufficient water to obtain uniform coverage. Apply as needed. Do not apply more than 1.6 lbs. a.i. per acre per season. Do not apply within 3 days of harvest.
	European Corn Borer	0.2	
Pine Seed Orchard	Coneworms, Seed Bugs	13 fl oz/100 gals water (0.025% dilution by wt) for high volume sprayers; 68 fl oz/100 gals water (0.125% dilution by wt) for low volume sprayers; 48 fl oz/A/ application in not less than 10 gals of water for aerial applications	Apply first application within 1 week of female flower closure or peak pollen flight for webbing coneworm control. For other coneworms and seed bugs, apply first application within 30 days following female flower closure. Repeat applications at intervals of 4 weeks, but do not apply more than 6 applications. Apply approximately 5–10 gals of the 0.025% dilution per tree with high volume sprayers. With low volume sprayers, apply 100 gallons of the 0.125% dilution per acre. Do not graze or harvest cover crop. Avoid contact with open water.
Pistachios	Peach Twig Borer, Navel Orangeworm, Chinch Bug	0.2–0.4	Apply at a rate of 0.8 to 1.6 pints (0.2–0.4 lb active) per acre. Use sufficient water to obtain full coverage of foliage. Do not apply more than 0.8 lbs. a.i. per acre per season. Do not apply after 10 percent hull split. Nuts may be harvested on the day of application. Do not graze livestock in treated orchards.
Potatoes	Colorado Potato Beetle, Potato Leafhopper, Aster Leafhopper, Potato Flea Beetle, Potato Aphid, Potato Tuberworm, Potato Psyllid, Cabbage Looper	0.05–0.2	Apply as needed by air or ground. Use sufficient water to obtain full coverage of foliage. Do not apply more than 2.4 lbs a.i. per acre per season. Can be applied up to 7 days prior to harvest.
Range Grass	Range Caterpillar	0.01	Apply by ground or air using sufficient volume to obtain uniform coverage. Do not apply more than once per year. Cattle may be present during application. Do not harvest or feed hay to livestock.
Soybeans	Green Cloverworm, Mexican Bean Beetle, Bean Leaf Beetle, Soybean Looper, Cabbage Looper, Potato Leafhopper, Velvetbean Caterpillar, Saltmarsh Caterpillar	0.05–0.1	Apply by air or ground. Do not apply more than 0.4 lbs. a.i. per acre per season. Can be applied up to 60 days before harvest. Do not graze treated areas or harvest for forage or hay.
	Corn Earworm, Beet Armyworm	0.1–0.2	
Spinach	(Refer to Leafy Vegetables for application directions)		
Sweet Corn	Corn Earworm, European Corn Borer, Fall Armyworm, Corn Rootworm Adults, Western Bean Cutworm, Aster Leafhopper	0.1–0.2	Apply every 5 days or as needed by air or ground. Do not apply more than 1.2 lbs. a.i. per acre per season. Do not apply within 1 day of harvest of ears or forage or livestock grazing. AMBUSH may be mixed with LANNATE® (methomyl) in accordance with label limitations and precautions. Label dosage rates should not be exceeded
Sweet Corn (Florida only—corn grown for the fresh market)	Corn Earworm, European Corn Borer, Fall Armyworm, Southern Armyworm, Corn Rootworm Adults, Western Bean Cutworm, Aster Leafhopper	0.1–0.25	Apply every 5 days or as needed by air or ground. Do not apply more than 3.25 lbs. a.i. per acre per crop season. Do not graze livestock in treated areas or harvest for forage. AMBUSH may be mixed with LANNATE® (methomyl) in accordance with label limitations and precautions. Label dosage rates should not be exceeded.

* Aids in suppression.

** Only suppression can be expected when utilized on a 5–7 day schedule.

*** AMBUSH may be applied in tank-mixtures with the following herbicides in accordance with label limitations and precautions. Label dosage rates should not be exceeded. PARAQUAT + PLUS™; GRAMOXONE®Paraquat; BICEP®; LASSO®/Atrazine Pre-mix; LASSO® 4E & ME; DUAL® 8E; Atrazine 4L & 80W; BLADEX® 80W & 4L; PRINCEP®; PRINCEP® CALIBER 90®; 2.4-D; ROUNDUP®, BRONCO®, BANVEL®; EXTRAZINE®.

† Not for control of this insect in California.

‡ 0.1% ai solution = $3\frac{1}{3}$ pt formulation per 100 gallons of water.

¶ These are the crops of the leafy vegetables crop group that AMBUSH can be used on: Amaranth (leafy amaranth, Chinese spinach, tampala); arrugula (Roquette); celery; celtuce; chervil; corn salad; chrysanthemum, edible-leaved; chrysanthemum, garland; cress, garden; cress, upland (yellow rocket, winter cress); dandelion; dock (sorrel); endive (escarole); fennel, Florence; lettuce (head and leaf); orach; parsley; purslane, garden; purslane, winter; rhubarb; spinach, fine; spinach, New Zealand; Swiss chard.

AMBUSH® 25W

INSECTICIDE
WETTABLE POWDER

RESTRICTED USE PESTICIDE

For retail sale to and use only by Certified Applicators, or persons under their direct supervision, and only for those uses covered by the Certified Applicator's certification.

ACTIVE INGREDIENT:
Permethrin—
(3-Phenoxyphenyl)methyl (±) *cis, trans*-3-(2,2-dichloroethenyl)-2,2-dimethylcyclopropanecarboxylate* 25%
INERT INGREDIENTS 75%
TOTAL 100%

* Cis/trans ratio: Min. 35% (±) cis and max. 65% (±) trans

EPA Reg. No. 10182-35
US PATENT No 4,024,163

Keep Out of Reach of Children
WARNING
AVISO

PRECAUCION AL USUARIO: Si usted no lee ingles, no use este producto hasta que la etiqueta le haya sido explicado ampliamente.

PRECAUTIONARY STATEMENTS

HAZARDS TO HUMANS AND DOMESTIC ANIMALS
WARNING
MAY CAUSE EYE IRRITATION
HARMFUL IF SWALLOWED
Do not get in eyes, on skin, or on clothing. Wash thoroughly after handling.

STATEMENT OF PRACTICAL TREATMENT
IF IN EYES: Flush with water for 15 minutes. Get medical attention if irritation persists.

IF SWALLOWED: Drink 1 or 2 glasses of water and induce vomiting by touching back of throat with finger. Do not induce vomiting or give anything by mouth to an unconscious person.

For 24-hour emergency assistance call ICI AMERICAS INC., 302/575-3000.

ENVIRONMENTAL HAZARDS
This pesticide is extremely toxic to fish. Use with care when applying in areas adjacent to any body of water. Do not apply directly to water. Do not apply when weather conditions favor drift from treated areas. Do not contaminate water by cleaning of equipment or disposal of wastes. This product is highly toxic to bees exposed to direct treatment or residues on crops or weeds. Do not apply AMBUSH 25W or allow it to drift to crops or weeds on which bees are actively foraging. Additional information may be obtained from your local Cooperative Extension Service. Apply this product only as specified on this label.

In case of a significant spill, call CHEMTREC, 800-424-9300.

STORAGE AND DISPOSAL

PROHIBITIONS
Do not contaminate water, food or feed by storage or disposal. Do not reuse empty container.

STORAGE
Mix as needed. Do not store diluted material. Keep container closed when not in use. Do not store near food or feed. In case of spill on floor or paved surfaces, sweep up and remove to chemical waste area.

PESTICIDE DISPOSAL
Wastes resulting from the use of this product may be disposed of on site or at an approved waste disposal facility.

CONTAINER DISPOSAL
Completely empty bag into application equipment. Then dispose of empty bag in a sanitary landfill or by incineration, or, if allowed by State and local authorities, by burning. If burned, stay out of smoke.

NOTICE TO BUYER AND USER

Seller warrants that this product conforms to the chemical description on the label and is reasonably fit for the purposes stated on the label when used in accordance with directions under normal conditions of use. This warranty does not extend to the use of this product contrary to label instructions, or under abnormal use conditions, or under conditions not reasonably foreseeable to Seller and Buyer and User assumes the risk of any such use. SELLER DISCLAIMS ALL OTHER WARRANTIES, EXPRESSED OR IMPLIED, INCLUDING ANY WARRANTY OF FITNESS OR MERCHANTABILITY. SELLER SHALL NOT BE LIABLE FOR CONSEQUENTIAL, SPECIAL OR INDIRECT DAMAGES RESULTING FROM THE USE OR HANDLING OF THIS PRODUCT AND SELLER'S SOLE LIABILITY AND BUYER'S AND USER'S EXCLUSIVE REMEDY SHALL BE LIMITED TO THE REFUND OF THE PURCHASE PRICE.

DIRECTIONS FOR USE

It is a violation of federal law to use this product in a manner inconsistent with its labeling.

GENERAL INSTRUCTIONS
Apply AMBUSH 25W as shown in "Directions for Application". Use sufficient water to obtain full coverage of foliage. Timing and frequency of applications should be based upon insect populations reaching economic thresholds. The higher rates may be required under conditions of heavy infestation. Do not apply more than 12 lbs. per acre per season (3 lbs ai/acre). Crops in this label may be planted immediately after last treatment. Do not plant other crops within 60 days after last application. Any other crop may be planted if it is not harvested or grazed.

REENTRY STATEMENTS
Do not apply this product in such a manner as to directly or through drift expose workers or other persons. The area being treated must be vacated by unprotected persons.

Do not enter treated areas without protective clothing until sprays have dried.

Protective clothing means, at least, a hat or other suitable head covering, a long sleeved shirt and long-legged trousers or a coverall type garment (all of closely woven fabric covering the body, including the arms and legs), shoes and socks.

Because certain states may require more restrictive reentry intervals for various crops treated with this product, consult your State Department of Agriculture for further information.

Written or oral warnings regarding use of protective clothing and accidental exposure must be given to workers who are expected to be in treated areas or in areas about to be treated. Inform workers of areas or fields that may not be entered without specific protective clothing, period of time field must be vacated and appropriate actions to take in case of accidental exposure. An example of such information is given under written warnings. When oral warnings are given, warnings shall be given in a language customarily understood by workers. Oral warnings must be given if there is reason to believe that written warnings cannot be understood by workers. Written warnings must include the following information.

WARNING
Area treated with AMBUSH 25W on (insert date). Do not enter without appropriate protective clothing until sprays have dried (if more restrictive, insert State Department of Agriculture reentry interval).

In case of accidental exposure to pesticide spray, wash the skin thoroughly with soap and water. Remove contaminated clothing and wash before reuse. If in eyes, flush with plenty of water. Get medical attention if needed.

DIRECTIONS FOR APPLICATION

CROP	TARGET PESTS	RATE (lbs ai/A)	REMARKS
Almonds	Navel Orangeworm, Peach Twig Borer	0.2–0.4	Apply by ground equipment using sufficient water to obtain uniform coverage. Can be applied up to 7 days before harvest. Do not apply more than 2.0 lbs. a.i. per acre per season. Do not graze livestock in treated orchards.
Apples	Apple Aphid, Redbanded Leafroller, Obliquebanded Leafroller, Plum Curculio, White Apple Leafhopper, Spotted Tentiform Leafminer, Tarnished Plant Bug	0.1–0.4	Apply by ground equipment only, using 25–400 gallons of spray per acre. Spray to wet all foliage. Do not apply after petal fall.
Asparagus	Cutworm Complex, Asparagus Beetle	0.05–0.1	Apply by ground equipment using sufficient water to obtain uniform coverage. Apply as needed. Do not apply more than 0.4 lbs. a.i. per acre per season. Do not apply within 3 days of harvest.
Asparagus Ferns	Japanese Beetle, Asparagus Beetle, Lygus Bugs	0.1	Apply to the fern stage of the asparagus plant after spear harvest when larval and adult stage of the asparagus beetle, lygus bugs and the adult stage of the Japanese beetle are present.
Celery (Refer to Leafy Vegetables for application directions)			
Chrysanthemums (Florida only)	Vegetable Leafminer	0.1–0.2	Apply by ground.
COLE CROPS			
Broccoli	Cabbage Looper, Imported Cabbageworm, Diamondback Moth, Cabbage Aphid*	0.05–0.1	Apply every 5 days or as needed by air or ground. Use sufficient water to obtain full coverage of foliage. Do not apply more than 0.8 lbs. a.i. per acre per season. Do not apply within 1 day of harvest.
Brussels Sprouts	Cabbage Looper, Imported Cabbageworm, Diamondback Moth, Cabbage Aphid*	0.05–0.1	Apply as needed by air or ground. Use sufficient water to obtain full coverage of foliage. Do not apply more than 0.8 lbs. a.i. per acre per season. Do not apply within 1 day of harvest.
Cabbage	Cabbage Looper, Imported Cabbageworm, Diamondback Moth, Cabbage Aphid*	0.05–0.1	Apply every 5-10 days or as needed by air or ground. Use sufficient water to obtain full coverage of foliage. Do not apply more than 1.0 lb. a.i. per acre per season. Do not apply within 1 day of harvest.
Cauliflower	Cabbage Looper, Imported Cabbageworm, Diamondback Moth, Cabbage Aphid*	0.05–0.1	Apply every 5 days or as needed by air or ground. Use sufficient water to obtain full coverage of foliage. Do not apply more than 0.8 lbs. a.i. per acre per season. Do not apply within 1 day of harvest.
Cotton	Bollworm, Tobacco Budworm, Pink Bollworm, Lygus Bugs, Cabbage Looper, Cotton Leafperforator, Boll Weevil, Cotton Fleahopper, Saltmarsh Caterpillar, Whitefly,** Cotton Aphid,** Cutworms	0.1–0.2	Apply as required by scouting reports, usually at intervals of 5 to 7 days. Apply by air or ground. Use sufficient water to obtain full coverage of foliage. For control of boll weevils spray on a 3 to 5 day schedule or as needed. Do not apply within 14 days of harvest. The higher rates of AMBUSH 25W (0.15 to 0.2 lbs. a.i. per acre) may be required under conditions of heavy lepidopterous worm infestation. AMBUSH 25W aids in the reduction of lepidopterous egg hatch by contact action. Do not graze livestock in treated areas. Do not apply more than 3.0 lbs. a.i. per acre per season.
Eggplant	Colorado Potato Beetle	0.2	Apply by ground or air equipment using sufficient water to obtain uniform coverage. Apply as needed. Do not apply more than 3.0 lbs. a.i. per acre per season. Do not apply within 3 days of harvest.
	Vegetable Leafminer, Cabbage Looper, Flea Beetle	0.1–0.2	
Field Corn/ Popcorn/Field Corn Grown for Seed (Preemergent Use)	Armyworm, Cutworms	0.1–0.2	Apply in the time period from 5 days prior to planting up to emergence of the crop. Apply as a broadcast spray in a minimum of 20 gallons of finished spray/acre with ground equipment.***
Field Corn/ Popcorn/Field Corn Grown for Seed (Foliar Use)	Cutworms, European Corn Borer, Armyworm, Southwestern Corn Borer, Flea Beetle, Corn Earworm,‡ Corn Rootworm Beetle,‡ Tobacco Budworm‡ Stalk Borer	0.1–0.2	Apply *prior to the brown silk stage.* Apply a minimum of 1 gallon of finished spray/acre by air and 10 gallons/acre with ground equipment.*** When treating for stalk borer, AMBUSH 25W must be applied when or shortly before the stalk borer larvae are moving into the corn from surrounding weeds and grasses. Mowing or burndown herbicide are suggested to initiate movement.***

CROP	TARGET PESTS	RATE (lbs ai/A)	REMARKS
Leafy Vegetables¶ (except *Brassica*)	Vegetable Leafminer, Cabbage Looper, Cutworm Complex, Fall Armyworm, Beet Armyworm, Corn Earworm, Green Cloverworm, Leafhoppers, European Corn Borer	0.1–0.2	Apply every 3-5 days or as needed by air or ground. Use sufficient water to obtain full coverage of foliage. Do not apply more than 2.0 lbs. a.i. per acre per season. Can be applied up to 1 day of harvest. Do not graze treated areas or feed crop refuse to livestock.
Lettuce (Refer to Leafy Vegetables for application directions)			
Mushrooms§ (Mushroom houses & adjacent premise areas)	Mushroom Flies (Sciarid and Phorid Adults)	Apply 6 oz.[1] to 11 gal. water or 8 level tablespoons to 3 gal. water. As a guide, use 1 gal. spray per 750 sq. ft. (400-1, 400 sq. ft.)	Spray directly to walls and ceilings as residual surface treatment only. Spray to point of runoff. Use AMBUSH 25W prior to filling house, during cool-down, during spawning, up to pinning and between breaks. Do not use when mushrooms are present. Treat as needed when flies appear. Do not make more than 20 applications prior to pinning of first break; apply no more than two applications between each break. Do not apply more than 30 applications total per crop of 5 breaks. Do not apply within 3 days of harvest.
Peaches	Peach Twig Borer, Oriental Fruit Moth, Green Fruitworm, Lesser Peachtree Borer, Plum Curculio, Tarnished Plantbug, Rose Chater	0.1–0.4	Apply by ground equipment using 25–400 gallons of spray per acre. Spray to wet all foliage. Can be applied up to 7 days before harvest. Do not apply more than 3.0 lbs. a.i. per acre per season. Do not graze livestock in treated orchards.
Pears (Dormant)	Pear Psylla	0.2–0.4	Apply at a rate of 0.8 to 1.6 lbs. (0.2 to 0.4 lbs active) per acre. May be combined with 2 to 8 gallons of spray oil per acre. Apply during the dormant through delayed dormant growth periods only. Apply in 3 to 20 gallons of finished spray per acre by aircraft and 25 to 400 gallons per acre by ground equipment. Do not apply more than 0.8 lbs. a.i. per acre per season. Do not graze orchard floor.
Pears (Pre-Bloom/ Seasonal)	Pear Psylla, Codling Moth, Green Fruitworm	0.2–0.4	Apply two pre-bloom sprays followed by three seasonal sprays. Pre-bloom sprays can be applied from dormant through bud burst stages. Seasonal sprays can be applied from petal fall up to 14 days of harvest. Do not apply more than 2.0 lbs. a.i. per acre per season.
Peppers (Bell)	Vegetable Leafminer, Cabbage Looper, Flea Beetle, Pepper Weevil	0.1–0.2	Apply by ground or air equipment using sufficient water to obtain uniform coverage. Apply as needed. Do not apply more than 1.6 lbs. a.i. per acre per season. Do not apply within 3 days of harvest.
	European Corn Borer	0.2	
Pistachios	Peach Twig Borer, Navel Orangeworm, Chinch Bug	0.2–0.4	Apply at a rate of 0.8 to 1.6 lbs. (0.2–0.4 lb active) per acre. Use sufficient water to obtain full coverage of foliage. Do not apply more than 0.8 lbs. a.i. per acre per season. Do not apply after 10 percent hull split. Nuts may be harvested on the day of application. Do not graze livestock in treated orchards.
Potatoes	Colorado Potato Beetle, Potato Leafhopper, Aster Leafhopper, Potato Flea Beetle, Potato Aphid, Potato Tuberworm, Potato Psyllid, Cabbage Looper	0.05–0.2	Apply as needed by air or ground. Use sufficient water to obtain full coverage of foliage. Do not apply more than 2.4 lbs. a.i. per acre per season. Can be applied up to 7 days prior to harvest.
Soybeans	Green Cloverworm, Mexican Bean Beetle, Bean Leaf Beetle, Soybean Looper, Cabbage Looper, Potato Leafhopper, Velvetbean Caterpillar, Saltmarsh Caterpillar	0.05–0.1	Apply by air or ground. Do not apply more than 0.4 lbs. a.i. per acre per season. Can be applied up to 60 days before harvest. Do not graze treated areas or harvest for forage or hay.
	Corn Earworm, Beet Armyworm	0.1–0.2	
Spinach (Refer to Leafy Vegetables for application directions)			
Sweet Corn	Corn Earworm, European Corn Borer, Fall Armyworm, Corn Rootworm Adults, Western Bean Cutworm, Aster Leafhopper	0.1–0.2	Apply every 5 days or as needed by air or ground. Do not apply more than 1.2 lbs. a.i. per acre per season. Do not apply within 1 day of harvest of ears or forage or livestock grazing. AMBUSH 25W may be mixed with LANNATE® (methomyl) in accordance with label limitations and precautions. Label dosage rates should not be exceeded.
Sweet Corn (Florida only—corn grown for the fresh market)	Corn Earworm, European Corn Borer, Fall Armyworm, Southern Armyworm, Corn Rootworm Adults, Western Bean Cutworm, Aster Leafhopper	0.1–0.25	Apply every 5 days or as needed by air or ground. Do not apply more than 3.25 lbs. a.i. per acre per crop season. Do not graze livestock in treated areas or harvest for forage. AMBUSH 25W may be mixed with LANNATE® (methomyl) in accordance with label limitations and precautions. Label dosage rates should not be exceeded.

* Aids in suppression.

** Only suppression can be expected when utilized on a 5-7 day schedule.

*** AMBUSH 25W may be applied in tank mixtures with the following herbicides in accordance with label limitations and precautions. Label dosage rates should not be exceeded LASSO®/Atrazine Pre-mix; LASSO® 4E & ME; DUAL® 8E; Atrazine 4L & 80W; BLADEX® 80W & 4L; PRINCEP®; PRINCEP® CALIBER® 90, 2,4-D; ROUNDUP®; BRONCO®; BANVEL®; EXTRAZINE®.
NOTE: Do not tank-mix AMBUSH® 25W Insecticide with BICEP® GRAMOXONE® Paraquat or PARAQUAT + PLUS™

‡ Not for control of this insect in California.

¶ These are the crops of the leafy vegetables crop group that AMBUSH 25W can be used on: Amaranth (leafy amaranth, Chinese spinach, tampala); arrugula (Roquette); celery; celtuce; chervil; corn salad; chrysanthemum, edible-leaved; chrysanthemum, garland; cress, garden; cress, upland (yellow rocket, winter cress); dandelion; dock (sorrel); endive (escarole); fennel, Florence; lettuce (head and leaf); orach; parsley; purslane, garden; purslane, winter; rhubarb; spinach; spinach, fine; spinach, New Zealand; Swiss chard.

§ Not registered for use in the state of California.

[1] 1 oz. of this powder equals 5 level teaspoons.

AMBUSH® Insecticide is a registered trademark of ICI Americas Inc.
LANNATE® is a registered trademark of E.I. du Pont de Nemours & Co.
PARAQUAT + PLUS™ is a registered trademark of Chevron Chemical Company.
LASSO; ROUNDUP; and BRONCO are registered trademarks of Monsanto Company.
BICEP; DUAL; PRINCEP; and CALIBER are registered trademarks of Ciba-Geigy Corp.
BLADEX; ATRAZINE; and EXTRAZINE are registered trademarks of Shell Chemical Company.
BANVEL is a registered trademark of Velsicol Chemical Company

RATE CONVERSION CHART

Pounds Active/ Acre	Formulation Ounces/ Acre	Formulation Lbs/ Acre
0.05	3.2	0.2
0.10	6.4	0.4
0.125	8.0	0.5
0.15	9.6	0.6
0.175	11.2	0.7
0.20	12.8	0.8

CYMBUSH® 3E

INSECTICIDE

RESTRICTED USE PESTICIDE

For retail sale to and use only by Certified Applicators, or persons under their direct supervision, and only for those uses covered by the Certified Applicator's certification.

ACTIVE INGREDIENT:
Cypermethrin
(±)α-cyano-(3-phenoxyphenyl) methyl (±)-*cis, trans*-3-(2,2-dichloroethenyl)-2, 2-dimethylcyclopropanecarboxylate* 35.6%
INERT INGREDIENTS 64.4%
TOTAL 100.0%

* *Cis/trans* ratio: 45/55 ± 10

CYMBUSH 3E contains 3 pounds active ingredient per gallon.

EPA Reg. No. 10182-65
U.S. Patent No. 4,024,163

Keep Out of Reach of Children
DANGER

See Side Panel For Additional Precautionary Statements and Statement of Practical Treatment.

PRECAUTIONARY STATEMENTS

HAZARDS TO HUMANS AND DOMESTIC ANIMALS

DANGER

Corrosive. Causes eye burns and skin irritation. May be fatal if swallowed. Harmful if inhaled or absorbed through skin. Do not get in eyes, on skin or on clothing. Wear protective clothing when spraying. Wear protective clothing, impermeable gloves, and full face shield when handling or mixing concentrate. Avoid breathing vapors or spray mist. Wash thoroughly after handling. May cause allergic skin reactions.

STATEMENT OF PRACTICAL TREATMENT:

IF SWALLOWED: Do not induce vomiting. Call a physician immediately.

IF IN EYES: Immediately flush eyes with plenty of water for at least 15 minutes. Get medical attention.

IF ON SKIN: Immediately remove contaminated clothing and wash with plenty of soap and water. Get medical attention if irritation persists.

NOTE TO PHYSICIAN

Inducing vomiting as first aid for this substance may result in increased risk of chemical pneumonia or pulmonary edema caused by aspiration of the hydrocarbon solvent. Vomiting should be induced only under professional supervision.

For 24-hour emergency assistance, call ICI AMERICAS INC. 302/575-3000.

ENVIRONMENTAL HAZARDS

This pesticide is extremely toxic to fish. Use with care when applying in areas adjacent to any body of water. Do not apply directly to water. Do not apply when weather conditions favor drift from treated areas. Do not contaminate water by cleaning of equipment or disposal of wastes. This product is highly toxic to bees exposed to direct treatment on crops or weeds. Do not apply CYMBUSH or allow it to drift to crops or weeds on which bees are actively foraging. Additional information may be obtained from your Cooperative Extention Service. Apply this product only as specified on this label. In case of a significant spill, call CHEMTREC, 800/424-9300.

PHYSICAL OR CHEMICAL HAZARDS

Do not use or store near heat or open flame.

STORAGE AND DISPOSAL

PROHIBITIONS

Do not contaminate water, food or feed by storage or disposal. Open dumping is prohibited. Do not reuse empty container.

STORAGE

Keep container closed when not in use. Do not store near food or feed. Protect from freezing. In case of spill or leak on floor or paved surfaces, soak up with sand, earth or synthetic absorbent. Remove to chemical waste area.

PESTICIDE DISPOSAL

Pesticide wastes are acutely hazardous. Improper disposal of excess pesticide, spray mixture or rinsate is a violation of federal law. If these wastes cannot be disposed of by use according to label instructions, contact your State Pesticide or Environmental Control Agency, or the hazardous waste representative of the nearest EPA Regional Office for guidance.

CONTAINER DISPOSAL

Metal Containers: Triple rinse (or equivalent). Then offer for recycling or reconditioning, or puncture and dispose of in a sanitary landfill, or by other procedures approved by state and local authorities.

Plastic Containers: Triple rinse (or equivalent). Then offer for recycling or reconditioning, or puncture and dispose of in a sanitary landfill, or incineration, or, if allowed by state and local authorities, by burning. If burned, stay out of smoke.

Glass Containers: Triple rinse (or equivalent). Then dispose of in a sanitary landfill or by other approved state and local procedures.

DIRECTIONS FOR USE

It is a violation of federal law to use this product in a manner inconsistent with its labeling. Apply CYMBUSH 3E as shown in the following chart, using sufficient spray to obtain full coverage of foliage. Apply every 5 to 7 days or as needed. Timing and frequency of applications should be based upon insect populations reaching locally determined economic thresholds. The use of less than recommended rates should only occur under conditions of light to moderate insect infestations.

Cotton Spray Recommendations

Target Pests	Rate (lbs ai/A)	Remarks
Cotton Bollworm Tobacco Budworm Boll Weevil Pink Bollworm Lygus Bug Cabbage Looper Cotton Leafperforator Cutworm spp. Thrips Beet Armyworm Flea Beetle Cotton Fleahopper Saltmarsh Caterpillar Whitefly* Cotton Aphid*	0.06–0.10	Apply as required by scouting reports, usually at intervals of 5 to 7 days. For boll weevil control, spray on a 3 to 4 day schedule. Apply by air or ground. Use sufficient water to obtain full coverage of foliage. Do not apply more than 1.0 lb a.i. per acre per season. In areas of the country where insects have a history of being highly susceptible to insecticides, available data indicate rates as low as 0.04 lb a.i./acre will control light infestations of bollworm, budworm, pink bollworm, cabbage looper, and lygus. NOTE: 1) The use of rates lower than 0.06 lb a.i./A should be used only in conjunction with intense field monitoring. 2) Do not use less than 0.06 lb a.i./A on initial (first) application because of the variation in size and age of insects present. CYMBUSH 3E may also be applied by air diluted with at least a once refined vegetable oil such as cottonseed oil or soybean oil. Agricultural mineral oils are not recommended. Apply using 1–2 quarts of finished spray per acre. CYMBUSH may be applied in tank mixture with the following insecticides in accordance with label limitations and precautions. Label dosage rate should not be exceeded. Do not tank mix in low volume dilutions of vegetable oil. Chlordimeform — Malathion Methyl Parathion — LANNATE® GUTHION® — LORSBAN™ Do not apply within 14 days of harvest. Do not graze livestock in treated areas.

* Aids in Suppression

CYMBUSH® is a registered trademark of ICI Americas Inc.
GUTHION® is a registered trademark of the Parent Company of Farbenfabriken Bayer GmbH. Leverkusen.
LANNATE® is a registered trademark of the E.I. du Pont de Nemours & Co.
LORSBAN™ is a trademark of the Dow Chemical Company.

Rate Conversion Chart

Pounds Active/ Acre	Fluid Ounces/ Acre	Pints/ Acre	Acres Treated/ Gallon
0.06	2.56	0.16	50.0
0.08	3.41	0.21	37.5
0.10	4.27	0.27	30.0

NOTICE TO BUYER AND USER

Seller warrants that this product conforms to the chemical description on the label and is reasonably fit for the purposes stated on the label when used in accordance with directions under normal conditions of use. This warranty does not extend to the use of this product contrary to label instructions, or under abnormal use conditions, or under conditions not reasonably foreseeable to Seller and Buyer and User assumes the risk of any such use. SELLER DISCLAIMS ALL OTHER WARRANTIES, EXPRESSED OR IMPLIED, INCLUDING ANY WARRANTY OF FITNESS OR MERCHANTABILITY. SELLER SHALL NOT BE LIABLE FOR CONSEQUENTIAL, SPECIAL OR INDIRECT DAMAGES RESULTING FROM THE USE OR HANDLING OF THIS PRODUCT AND SELLER'S SOLE LIABILITY AND BUYER'S AND USER'S EXCLUSIVE REMEDY SHALL BE LIMITED TO THE REFUND OF THE PURCHASE PRICE.

CYMBUSH® OL

INSECTICIDE

RESTRICTED USE PESTICIDE

For retail sale to and use only by Certified Applicators, or persons under their direct supervision, and only for those uses covered by the Certified Applicator's certification.

FOR USE IN OIL ONLY

DO NOT MIX WITH WATER

ACTIVE INGREDIENT:
Cypermethrin
(±)α-cyano-(3-phenoxyphenyl) methyl (±)-*cis, trans*-3-(2,2-dichloroethenyl)-2, 2-dimethylcyclopropanecarboxylate* 36.6%
INERT INGREDIENTS 63.4%
TOTAL 100.0%

* *Cis/trans* ratio: 45/55 ± 10

CYMBUSH OL contains 3 pounds active ingredient per gallon.

EPA Reg. No. 10182-80

U.S. Patent No. 4,024,163

DANGER

Keep Out of Reach of Children

See Side Panel For Additional Precautionary Statements and Statement of Practical Treatment.

PRECAUTIONARY STATEMENTS

HAZARDS TO HUMANS AND DOMESTIC ANIMALS

DANGER

Corrosive. Causes eye burns and skin irritation. May be fatal if swallowed. Harmful if inhaled or absorbed through skin. Do not get in eyes, on skin or on clothing. Wear protective clothing when spraying. Wear protective clothing, impermeable gloves, full face shield, rubber apron and boots when handling or mixing concentrate. Avoid breathing vapors or spray mist. Wash thoroughly after handling. May cause allergic skin reactions.

STATEMENT OF PRACTICAL TREATMENT

IF SWALLOWED: Do not induce vomiting. Call a physician immediately.

IF IN EYES: Immediately flush eyes with plenty of water for at least 15 minutes. Get medical attention.

IF ON SKIN: Immediately remove contaminated clothing and wash with plenty of soap and water. Get medical attention if irritation persists.

NOTE TO PHYSICIAN

Inducing vomiting as first aid for this substance may result in increased risk of chemical pneumonia or pulmonary edema caused by aspiration of the hydrocarbon solvent. Vomiting should be induced only under professional supervision.

For 24-hour emergency assistance, call ICI AMERICAS INC. 302/575-3000.

ENVIRONMENTAL HAZARDS

This pesticide is extremely toxic to fish. Use with care when applying in areas adjacent to any body of water. Do not apply directly to water. Do not apply when weather conditions favor drift from treated areas. Do not contaminate water by cleaning of equipment or disposal of wastes. This product is highly toxic to bees exposed to direct treatment or residues on crops or weeds. Do not apply CYMBUSH or allow it to drift to crops or weeds on which bees are actively foraging. Additional information may be obtained from your Cooperative Extension Service. Apply this product only as specified on this label.

In case of a significant spill, call CHEMTREC, 800/424-9300.

PHYSICAL OR CHEMICAL HAZARDS
Do not use or store near heat or open flame.

STORAGE AND DISPOSAL

PROHIBITIONS
Do not contaminate water, food or feed by storage or disposal. Open dumping is prohibited. Do not reuse empty container.

STORAGE
Keep container closed when not in use. Do not store near food or feed. Protect from freezing. In case of spill or leak on floor or paved surfaces, soak up with sand, earth or synthetic absorbent. Remove to chemical waste area.

PESTICIDE DISPOSAL
Pesticide wastes are acutely hazardous. Improper disposal of excess pesticide, spray mixture or rinsate is a violation of federal law. If these wastes cannot be disposed of by use according to label instructions, contact your State Pesticide or Environmental Control Agency, or the hazardous waste representative of the nearest EPA Regional Office for guidance.

CONTAINER DISPOSAL
Metal Containers: Triple rinse (or equivalent). Then offer for recycling or reconditioning, or puncture and dispose of in a sanitary landfill, or by other procedures approved by state and local authorities.
Plastic Containers: Triple rinse (or equivalent). Then offer for recycling or reconditioning, or puncture and dispose of in a sanitary landfill, or incineration, or, if allowed by state and local authorities, by burning. If burned, stay out of smoke.
Glass Containers: Triple rinse (or equivalent). Then dispose of in a sanitary landfill or by other approved state and local procedures.

DIRECTIONS FOR USE

It is a violation of federal law to use this product in a manner inconsistent with its labeling. Apply CYMBUSH OL as shown in the following chart, using sufficient spray to obtain full coverage of foliage. Apply every 5 to 7 days or as needed. Timing and frequency of applications should be based upon insect populations reaching locally determined economic thresholds. The use of less than recommended rates should only occur under conditions of light to moderate insect infestations.

Cotton Spray Recommendations

Target Pests	Rate (lbs ai/A)	Remarks
Cotton Bollworm Tobacco Budworm Boll Weevil Pink Bollworm Lygus Bug Cabbage Looper Cotton Leafperforator Cutworm spp. Thrips Beet Armyworm Flea Beetle Cotton Fleahopper Saltmarsh Caterpillar Whitefly* Cotton Aphid*	0.06-0.10	Dilute CYMBUSH OL with at least a once refined non-phytotoxic crop oil such as cottonseed or soybean oil. Apply using a finished spray volume of 1–2 quarts per acre. Agricultural mineral oils are not recommended. Do not apply more than 1.0 lb. a.i. per acre per season. Apply as required by scouting reports, usually at intervals of 5 to 7 days. For boll weevil control, spray on a 3 to 4 day schedule. In areas of the country where insects have a history of being highly susceptible to insecticides, available data indicate rates as low as 0.04 lb a.i./acre will control light infestation of bollworm, budworm, pink bollworm, cabbage looper, and lygus. NOTE: 1). The use of rates lower than 0.06 lb a.i./A should be used only in conjunction with intense field monitoring. 2). Do not use less than 0.06 lb a.i./A on initial (first) application because of the variation in size and age of insects present. Do not apply within 14 days of harvest. Do not graze livestock in treated areas.

* Aids in Suppression.

Rate Conversion Chart

Pounds Active/ Acre	Fluid Ounces/ Acre	Pints/ Acre	Acres Treated/ Gallon
0.06	2.56	0.16	50.0
0.08	3.41	0.21	37.5
0.10	4.27	0.27	30.0

NOTICE TO BUYER AND USER

Seller warrants that this product conforms to the chemical description on the label and is reasonably fit for the purposes stated on the label when used in accordance with directions under normal conditions of use. This warranty does not extend to the use of this product contrary to label instructions, or under abnormal use conditions, or under conditions not reasonably foreseeable to Seller and Buyer and User assumes the risk of any such use. SELLER DISCLAIMS ALL OTHER WARRANTIES, EXPRESSED OR IMPLIED, INCLUDING ANY WARRANTY OF FITNESS OR MERCHANTABILITY. SELLER SHALL NOT BE LIABLE FOR CONSEQUENTIAL, SPECIAL OR INDIRECT DAMAGES RESULTING FROM THE USE OR HANDLING OF THIS PRODUCT AND SELLER'S SOLE LIABILITY AND BUYER'S AND USER'S EXCLUSIVE REMEDY SHALL BE LIMITED TO THE REFUND OF THE PURCHASE PRICE.

FUSILADE® 4E
HERBICIDE

Postemergence Herbicide for Control of Perennial and Annual Grass Weeds for Cotton, Soybeans and Nonfood Uses

ACTIVE INGREDIENT:
Fluazifop-butyl-
butyl (*R S*)-2-[4-[[5-trifluoromethyl)-2-pyridinyl]oxy]phenoxy]propanoate 46.5%
INERT INGREDIENTS 53.5%
TOTAL 100.0%

FUSILADE 4E contains 4 pounds active ingredient per gallon.

EPA Reg. No. 10182-67

Keep Out of Reach of Children
WARNING

PRECAUTIONARY STATEMENTS

HAZARDS TO HUMANS AND DOMESTIC ANIMALS
WARNING
CAUSES EYE IRRITATION. HARMFUL IF SWALLOWED, INHALED OR ABSORBED THROUGH SKIN. Do not get in eyes, on skin or on clothing. Wear protective clothing and PVC gloves when spraying. Wear protective clothing, PVC gloves and apron and full face shield when handling or mixing concentrate. Avoid breathing vapors or spray mist. Wash thoroughly with soap and water after handling and before eating or smoking. Remove contaminated clothing and wash before reuse.

STATEMENT OF PRACTICAL TREATMENT
IF SWALLOWED: Do not induce vomiting. Call a physician immediately.

IF IN EYES: Immediately flush eyes with plenty of water for at least 15 minutes. Get medical attention if irritation persists.

IF ON SKIN: Immediately remove contaminated clothing and wash thoroughly with plenty of soap and water. Get medical attention if irritation persists.

NOTE TO PHYSICIAN: Inducing vomiting as first aid for this substance may result in increased risk of chemical pneumonia or pulmonary edema caused by aspiration of the hydrocarbon solvent. Vomiting should be induced only under professional supervision. For 24-hour emergency assistance, call ICI Americas Inc. (302) 575–3000.

ENVIRONMENTAL HAZARDS

This product is toxic to fish. Do not apply directly to water. Do not contaminate water by cleaning of equipment or disposal of waste. Do not apply when weather conditions favor drift from target area.

In case of a significant spill, call CHEMTREC (800) 424–9300.

PHYSICAL OR CHEMICAL HAZARDS

Do not use or store near heat or open flame.

STORAGE AND DISPOSAL

PROHIBITIONS

Do not contaminate water, food or feed by storage or disposal. Open dumping is prohibited. Do not reuse empty container.

STORAGE

Store in original container only. Keep container closed when not in use. Do not store near food or feed. In case of spill or leak on floor or paved surfaces, soak up with sand, earth or synthetic absorbent. Remove to chemical waste area.

PESTICIDE DISPOSAL

Pesticide wastes are toxic. Improper disposal of excess pesticide, spray mixture or rinsate is a violation of federal law. If these wastes cannot be disposed of by use according to label instructions, contact your State Pesticide or Environmental Control Agency, or the Hazardous Waste Representative at the nearest EPA Regional Office for guidance.

CONTAINER DISPOSAL

Metal Containers: Triple rinse (or equivalent). Then offer for recycling by or reconditioning, or puncture and dispose of in a sanitary landfill, or by other procedures approved by state and local authorities.

Plastic Containers: Triple rinse (or equivalent). Then offer for recycling or reconditioning, or puncture and dispose of in a sanitary landfill, by incineration, or, if allowed by state and local authorities, by burning. If burned, stay out of smoke.

Glass Containers: Triple rinse (or equivalent). Then dispose of in a sanitary landfill or by other approved state and local procedures.

DIRECTIONS FOR USE

It is a violation of federal law to use this product in a manner inconsistent with its labeling. This labeling must be in the possession of the user at the time of application.

GENERAL INFORMATION

FUSILADE 4E is a selective postemergence herbicide for control of annual and perennial grass weeds. FUSILADE 4E will provide effective control of grass weeds in conventional tillage, minimum tillage, and no-till plantings. FUSILADE 4E does not control broadleaf weeds or sedges (nutgrass).

CONTROL SYMPTOMS

FUSILADE 4E is a systemic herbicide which moves from the treated foliage into the shoots, roots, rhizomes, stolons, and growing points (meristematic regions) of treated grass weeds. Excellent control of a wide range of annual and perennial grass weeds will be obtained with FUSILADE 4E when applied as recommended on this label.

Growth of treated grass weeds stops soon after application. Symptoms include loss of vigor, yellowing and/or reddening, and eventual death to the treated grass plant. Symptoms are generally observed within one to three weeks, depending on grass weed species and environmental conditions.

APPLICATION DIRECTIONS

Timing

Apply to actively growing grass weeds before they exceed the recommended growth stages shown on this label.

Do not apply FUSILADE 4E if rainfall is expected within 1 hour.

SPRAY ADDITIVES (CLEARED FOR USE ON GROWING CROPS)

ALWAYS ADD ONE OF THE FOLLOWING:

- Crop Oil Concentrate

Add a nonphytotoxic crop oil or a once refined vegetable oil concentrate containing 15–20% approved emulsifier; at 1% (1 quart per 25 gallons) of the finished spray volume.

- Nonionic Surfactant

Add nonionic surfactant containing at least 75% surface active agent, at 0.25% (1/2 pint per 25 gallons) of the finished spray volume.

Ground Application—Use 5–40 gallons of spray volume per acre and 40–60 psi pressure. A sufficient spray volume and pressure should be used to provide complete coverage of the treated grass weed species.

Use higher volumes and increase the pressure to 60 psi if the grass foliage is dense to ensure complete coverage.

DO NOT USE FLOOD TYPE OR OTHER SPRAY NOZZLE TIPS WHICH DELIVER COARSE SPRAYS.

DO NOT APPLY FUSILADE 4E WITH RECIRCULATING SPRAYERS, ROPEWICK APPLICATORS, CONTROLLED DROPLET APPLICATORS (CDA) OR ANY SIMILAR DEVICES.

Band Applications

Band application equipment should be adjusted to provide maximum coverage of grasses in the planted row. Thorough coverage can be obtained with a nozzle directed to each side of the row toward the grass weeds in the planted row. Application with a single nozzle over the top of the row is not recommended.

The use of band applications on perennial grasses may result in reinfestation of the treated band from the untreated middle and is not recommended.

Aerial Application—Use a minimum of 5 gallons per acre. If the grass foliage is dense, increase the spray volume to 10 gallons per acre to improve spray coverage.

Soil Residual Activity

Under conditions where rainfall is adequate following FUSILADE 4E application, soil residual activity may be observed. This may result in control or suppression of new flushes of annual grasses. The amount of residual control is dependent on the FUSILADE 4E rate used, grass density, and the soil type, with more activity being seen on coarse-textured soils low in organic matter.

SPOT TREATMENTS

Mix FUSILADE 4E and crop oil concentrate or nonionic surfactant with water according to the amounts shown below. Spray to obtain thorough coverage, but do not spray to runoff. Retreat if necessary. **Use only nonionic surfactant on ornamentals.**

In cotton, applications must be made prior to boll set. Do not apply a total of more than 1 quart of FUSILADE 4E per acre per season to cotton.

In soybeans, applications must be made prior to bloom. Do not apply a total of more than 1 pint of FUSILADE 4E per acre per season to soybeans.

Spot Spray Recommendations

To Make This Spray Volume	Add These Amounts: FUSILADE 4E	Crop/Vegetable Oil Concentrate*	or	Nonionic Surfactant
1 gallon	½ fl. oz. (1 tbsp.)	1½ fl. oz.	or	½ fl. oz.
25 gallons	¾ pint	1 quart		½ pint
50 gallons	1½ pints	2 quarts		1 pint
100 gallons	1½ quarts	1 gallon		1 quart

* Do not use a crop oil concentrate on ornamentals. Use only nonionic surfactant on ornamentals.

Cultivation—Cultivation no sooner than 7 days after application will often assist in grass weed control. Cultivation of treated grasses 7 days prior to or within 7 days after application of FUSILADE 4E may reduce weed control.

DIRECTIONS FOR USES

ANNUAL AND PERENNIAL GRASS WEED CONTROL FOR COTTON AND SOYBEANS

REGION A

- FOR ALL STATES EXCEPT THE AREAS OF ARIZONA, CALIFORNIA, NEW MEXICO, WEST OKLAHOMA, WEST TEXAS, UTAH, AND COLORADO AS OUTLINED ON THE MAP
- APPLY FUSILADE 4E TO ACTIVELY GROWING GRASSES
- APPLY FUSILADE 4E AT THE RECOMMENDED RATE TO GRASSES AT THE RECOMMENDED GROWTH STAGES AS OUTLINED
- DO NOT APPLY FUSILADE 4E TO GRASSES WHICH ARE STRESSED DUE TO MOISTURE, TEMPERATURE, HERBICIDE OR MECHANICAL INJURY
- DO NOT APPLY FUSILADE 4E TO GRASSES WHICH HAVE TILLERED, FORMED SEED HEADS, OR EXCEED RECOMMENDED GROWTH STAGES
- APPLY FUSILADE 4E WHEN THE FIRST GRASS WEED SPECIES IN A MIXED GRASS WEED POPULATION REACHES THE RECOMMENDED GROWTH STAGE FOR TREATMENT

ANNUAL GRASS CONTROL RECOMMENDATIONS FOR REGION A

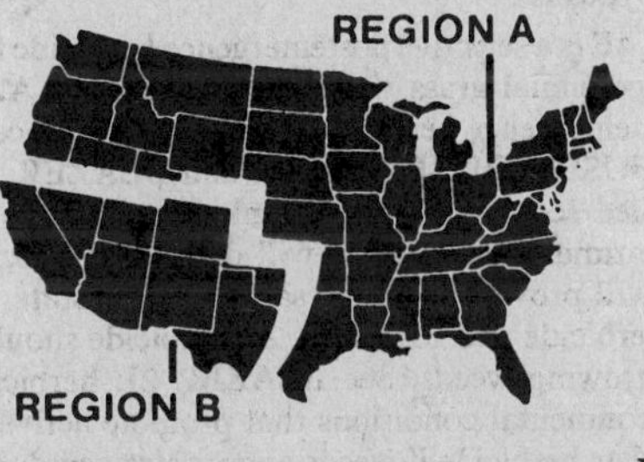

ANNUALS — **REGION A**

Grass Species	Height (inches)	Rate Pints/A
Group I[1,2]		
Shattercane/Wildcane/Volunteer milo (*Sorghum Bicolor*)	6–12	¼
Volunteer corn (*Zea mays*)	12–18	¼
Wild Proso Millet (*Panicum miliaceum*)	4–8	¼
Group II[1,2]		
Seedling johnsongrass (*Sorghum halepense*)	2–4	⅜
Goosegrass (*Eleusine indica*)	2–4	⅜
Group III[1,2]		
Barnyardgrass (*Echinochloa crusgalli*)	2–4	½
Broadleaf signalgrass (*Brachiaria platyphylla*)	2–4	½
Fall panicum (*Panicum dichotomiflorum*)	2–4	½
Foxtail, giant (*Setaria faberi*)	4–6	½
Foxtail, green (*Setaria virdis*)	2–4	½
Foxtail, yellow (*Setaria lutescens*)	2–4	½
Itchgrass (*Rottboellia exaltata*)	2–4	½
Junglerice (*Echinochloa colonum*)	2–4	½
Large crabgrass (*Digitaria sanguinalis*)	2–4	½
Smooth crabgrass (*Digitaria ischaemum*)	2–4[3]	½
Southern sandbur (*Cenchrus echinatus*)	2–4	½
Southwestern cupgrass (*Eriochloa gracilis*)	2–4	½
Texas panicum (*Panicum texanum*)	2–4	½
Wild oats (*Avena fatua*)		
Witchgrass (*Panicum capillare*)	2–4	½
Red rice (*Oryza sativa*)	1–2[4]	½
Volunteer Cereals		
Barley (*Hordeum vulgare*)	2–4	½
Oats (*Avena sativa*)	2–4	½
Rye (*Secale cereale*)	2–4	½
Wheat (*Triticum aestivum*)	2–4	½

[1] When grass populations include grasses from all groups above, use the ½ pt/A rate of FUSILADE 4E for broad-spectrum control.
[2] Retreat at the recommended rate if needed to control late-germinating grasses.
[3] Length of runners.
[4] 2–3 leaves.

PERENNIAL GRASS CONTROL RECOMMENDATIONS FOR REGION A

PERENNIALS

Grass Species[5,6]	Height (inches)	Rate Pints/A
Rhizome johnsongrass[7] (*Sorghum halepense*)	12–18	½
Bermudagrass[8] (*Cynodon dactylon*)	4–8 (runner length)	½
Quackgrass[9] (*Agropyron repens*)	less than 10 (minimum of 4 leaves)	½
Wirestem muhly[10] (*Muhlenbergia frondosa*)	4–12 (3–10 leaves)	½

[5] Best results can be obtained if rhizomes or stolons are cut up by preplan tillage practices (discing, plowing, etc.).
[6] Two applications of the recommended rate of FUSILADE 4E are generally more effective for full season control than a single application of the same total amount.
[7] Apply before the boot stage of growth. If new shoots emerge or regrowth occurs, make a second application of ⅜ pt/A when the johnsongrass is 6–12 inches tall.
[8] If regrowth occurs, retreat at recommended height and rate. Bermudagrass is a low-growing perennial grass, and the crop canopy may shield it from herbicide sprays. Improve herbicide spray coverage by directing the spray under the crop canopy.
[9] If regrowth occurs, retreat at the recommended rate when 3–5 quackgrass leaves are present.
[10] If regrowth occurs, retreat at the recommended rate two to six weeks after the first application.

DIRECTIONS FOR USE

ANNUAL AND PERENNIAL GRASS WEED CONTROL FOR COTTON AND SOYBEANS REGION B

- FOR THE AREAS OF ARIZONA, CALIFORNIA, NEW MEXICO, WEST OKLAHOMA, WEST TEXAS, UTAH, AND COLORADO AS OUTLINED ON THE MAP
- APPLY FUSILADE 4E TO ACTIVELY GROWING GRASSES
- FOR BEST RESULTS APPLY FUSILADE 4E WITHIN 7 DAYS AFTER IRRIGATION
- APPLY FUSILADE 4E AT THE RECOMMENDED RATE TO GRASSES AT THE RECOMMENDED GROWTH STAGES AS OUTLINED IN THE TABLE
- DO NOT APPLY FUSILADE 4E TO GRASSES WHICH ARE STRESSED DUE TO MOISTURE, TEMPERATURE, HERBICIDE OR MECHANICAL INJURY
- DO NOT APPLY FUSILADE 4E TO GRASSES WHICH HAVE TILLERED, FORMED SEED HEADS, OR EXCEED RECOMMENDED GROWTH STAGES
- APPLY FUSILADE 4E WHEN THE FIRST GRASS WEED SPECIES IN A MIXED GRASS WEED POPULATION REACHES THE RECOMMENDED GROWTH STAGE FOR TREATMENT

ANNUAL GRASS CONTROL RECOMMENDATIONS FOR REGION B

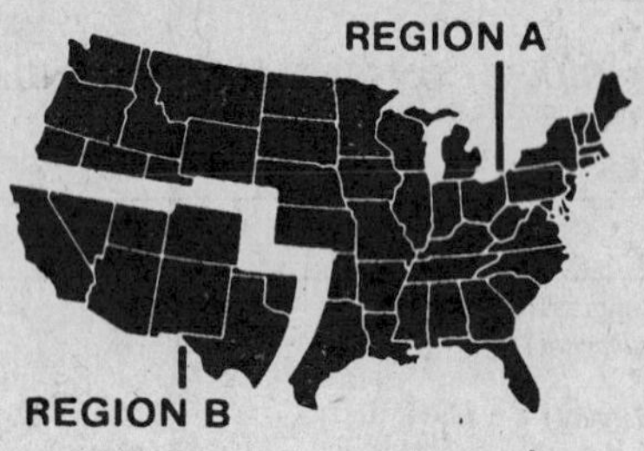

ANNUALS — **REGION B**

Grass Species	Height (inches)	Rate Pints/A
Group I[11,12]		
Seedling johnsongrass (*Sorghum halepense*)	2–4	½
Volunteer corn (*Zea mays*)*	12–24	½
Group II[11,12]		
Texas panicum (*Panicum texanum*)*	2–4	¾
Volunteer cereals	2–4	¾
Barley (*Hordeum vulgare*)		
Milo (*Sorghum bicolor*)		
Oats (*Avena sativa*)		
Wheat (*Triticum aestivum*)		
Group III[11,12]		
Barnyardgrass (*Echinochloa crusgalli*)	2–4	1
Large crabgrass (*Digitaria sanguinalis*)*	2–4	1
Yellow foxtail (*Setaria lutescens*)*	2–4	1
Field sandbur (*Cenchrus pauciflorus*)*	2–4	1
Junglerice (*Echinochloa colonum*)*	2–4	1

[11] In cotton, retreat at the recommended rate if needed to control late germinating grasses or regrowth.

[12] In soybeans, do not apply more than one pint per acre per season.

*Not recommended for control of this grass in Arizona and California.

PERENNIALS

Grass Species[13]	Height (inches)	Rate Pints/A
Rhizome johnsongrass[14] (*Sorghum halepense*)	12–18	3/4 to 1
Bermudagrass[15] (*Cynodon dactylon*)	3 or 4–8 runner (stolon) length	3/4 to 1

[13] Best results can be obtained if rhizomes and stolons are cut up by preplant tillage practices (discing, plowing, etc.)

[14] Apply before the boot stage. In cotton, a second application is necessary for season long control. Retreat at 12–18 inches.

[15] For Arizona and California, 1 pint per acre is recommended. In cotton, retreat at recommended height and rate for season long control.

LABELED CROPS—Remarks and Precautions

COTTON—Do not apply a total of more than 1 quart of FUSILADE 4E per acre per season to cotton.

Do not harvest cotton within 100 days of application.

Do not apply after boll set.

Do not graze or harvest for forage or hay.

Do not plant rotational grass crops such as corn, sorghum and cereals within 60 days after last application.

SOYBEANS—Do not apply a total of more than 1 pint of FUSILADE 4E per acre per season to soybeans.

Make last application before bloom.

Do not graze or harvest for forage or hay.

Do not plant rotational grass crops such as corn, sorghum and cereals within 60 days after last application.

TANK-MIX–SOYBEANS FUSILADE 4E HERBICIDE/BLAZER 2L HERBICIDE

POSTEMERGENCE CONTROL OF GRASS AND BROADLEAF WEEDS

FUSILADE 4E herbicide and BLAZER 2L herbicide may be used together in a postemergence program for broad-spectrum weed control in soybeans.

FUSILADE 4E is a selective postemergence herbicide for control of annual and perennial grass weeds in soybeans. BLAZER 2L is a selective postemergence herbicide for the control of certain broadleaf weeds. FUSILADE 4E herbicide and BLAZER 2L herbicide may be applied sequentially or in tank-mix. The growth stage of weeds at the time of application will determine which method of application will provide the most satisfactory results. Both FUSILADE 4E herbicide and BLAZER 2L herbicide should be applied to actively growing weeds. See BLAZER 2L herbicide label for defined environmental conditions that promote active growth. Do not apply either herbicide if weeds appear stressed due to unfavorable temperatures, drought and/or low soil fertility. Do not apply BLAZER 2L herbicide or FUSILADE 4E/BLAZER 2L herbicide tank-mix if soybeans show injury from prior herbicide applications. Read and observe all applicable label directions and limitations for both FUSILADE 4E herbicide and BLAZER 2L herbicide before using.

METHOD 1: Sequential Application—FUSILADE 4E Herbicide Followed by BLAZER Herbicide

(Annual and/or perennial grass weeds at proper growth stage for treatment, prior to broadleaf weed treatment)

Apply FUSILADE 4E herbicide and an approved adjuvant at the recommended rate and growth stage for the grass weeds being treated, as per the FUSILADE 4E herbicide label directions. FUSILADE 4E is a systemic herbicide, therefore when treating annual grasses allow at least 3 days, and when treating perennial grasses allow at least 5 days to elapse prior to a BLAZER 2L herbicide application. This will enable FUSILADE 4E herbicide to be adequately translocated to the meristematic regions of the plant. After the appropriate time interval has elapsed, apply BLAZER 2L herbicide with Triton AG-98 or an approved non-ionic surfactant to susceptible broadleaf weeds as outlined on the BLAZER 2L herbicide label.

METHOD 2: Sequential Application—BLAZER 2L Herbicide Followed By FUSILADE 4E Herbicide

(Broadleaf weeds at proper growth stage for treatment, prior to annual and/or perennial grass weed treatment)

Apply BLAZER 2L herbicide with Triton AG-98 or an approved non-ionic surfactant at the recommended rate and growth stage to susceptible broadleaf weeds, as per the BLAZER 2L herbicide label directions. Broadleaf herbicides can stress and affect the growth of grass weed species. FUSILADE 4E herbicide must be applied to actively growing grasses for best results. A sequential application of FUSILADE 4E herbicide may be applied following a BLAZER 2L herbicide application when grasses resume active growth with the development of a new leaf. Follow the recommended rates and growth stage for the grass weeds being treated, as per the FUSILADE 4E herbicide label.

METHOD 3: Tank-Mix Applications of FUSILADE 4E Herbicide And BLAZER 2L Herbicide

(Annual grasses only and broadleaf weeds are at the proper growth stage for treatment, as per the respective labels)

A tank-mix of FUSILADE 4E herbicide and BLAZER 2L herbicide may be applied at the recommended rates and growth stages to susceptible annual grass and broadleaf weed species in a manner as described on the respective labels. Use only an approved non-ionic surfactant, as per the FUSILADE 4E herbicide label recommendations. Follow water volume and spray pressure recommendations, as per the BLAZER 2L herbicide label. Load the spray tank with half the amount of required water and add the recommended amounts of FUSILADE 4E herbicide, BLAZER 2L herbicide, and non-ionic surfactant while the agitator is running, and then add the remaining quantity of water. Do not make more than one application of this tank-mix per season.

Note: Tank-mix applications sometimes have resulted in reduced grass weed control and possible increases in crop injury as compared to either product used alone. If grass regrowth occurs following an application of the tank-mix or an additional flush of grasses emerge, make a second application of FUSILADE 4E herbicide to actively growing grass weeds, as per the label recommendations. A tank-mix application is not recommended if perennial grass weeds are the predominant grasses to be controlled.

Postemergence Use Restrictions (Partial List)

- Always read and follow the restrictions and limitations for all products whether used alone or in a tank-mix. The most restrictive labeling of any product used applies in tank mixtures.
- Do not apply a total of more than 1 pint of FUSILADE 4E herbicide per acre per season to soybeans.
- Do not apply more than 4 pints (1.0 lb active) per acre of BLAZER 2L herbicide per growing season for soybeans.
- Make the last FUSILADE 4E herbicide application before soybean bloom. Do not apply within 100 days of harvest for soybeans.
- Do not apply BLAZER 2L herbicide within 50 days of harvest for soybeans.
- Do not use treated plants for feed or forage.
- Do not plant rotational crops other than cotton and soybeans within 60 days after the last FUSILADE 4E herbicide application.
- Crop Rotation Restriction: Root crops (such as carrots, turnips, sweet potatoes, etc.) must not be planted in fields that have been treated with BLAZER 2L herbicide for a period of 18 months following the last treatment.
- In the case of crop failure, only soybeans may be immediately replanted.
- Avoid drift to all other crops and non-target areas.
- Thoroughly flush spray tanks with water before and after each use.
- BLAZER 2L herbicide requires a six hour rain-free period for best results. Do not apply BLAZER 2L herbicide or FUSILADE 4E/ BLAZER 2L herbicide tank-mix if rain is threatening.
- Do not make more than one application of the FUSILADE 4E/ BLAZER 2L herbicide tank-mix in a single season.
- Do not use during dry weather conditions when crop and weeds are under stress or wilt and not actively growing, as weed control may be reduced.
- Do not apply BLAZER 2L herbicide or FUSILADE 4E/ BLAZER 2L herbicide tank-mix when crop and weeds are under stress from cold temperature, when maximum day temperature is below 70°F, or when soil temperature is below 60°F, as weeds will not be actively growing and control may be reduced.
- Do not apply BLAZER 2L herbicide or FUSILADE 4E/ BLAZER 2L herbicide tank-mix when crops are exhibiting injury from previously applied pesticides or are exhibiting stress symptoms from disease, nematodes or insects, as increased crop response may result.
- Do not apply BLAZER 2L herbicide or FUSILADE 4E/ BLAZER 2L herbicide tank-mix when crop and weeds are under stress from excessive fertilizer or soil salts, wind injury, frost damage, or high temperature stress or wilt, as increased crop response may result.
- Do not mix FUSILADE 4E herbicide, or FUSILADE 4E/ BLAZER 2L herbicide tank-mix with other pesticides or liquid fertilizers except as specified on this label or other approved ICI Americas supplemental labeling.

COTTON TANK-MIX WITH AMBUSH INSECTICIDE OR WITH CYMBUSH 3E INSECTICIDE

Apply FUSILADE 4E, AMBUSH or CYMBUSH 3E in the tank mix when required according to the weed and insect problems present. Follow rate recommendations indicated in the tables below.

Region A[1]

Application Rate Table

FUSILADE 4E + AMBUSH

Product	Rate[2,3]
FUSILADE 4E + AMBUSH	¼ to ½ pint (0.125-0.25 lb ai)/A 0.4-0.8 pints (0.1-0.2 lb ai)/A

FUSILADE 4E + CYMBUSH 3E

Product	Rate[2,3]
FUSILADE 4E + CYMBUSH 3E	¼ to ½ pint (0.125-0.25 lb ai)/A 0.16-0.27 pints (0.06-0.10 lb ai)/A

[1] Region A—All states except the areas of Arizona, California, New Mexico, West Oklahoma, West Texas, Utah and Colorado as outlined in the FUSILADE 4E label.

[2] Follow the recommended rate and growth stage for the grass weeds being treated, as per the FUSILADE 4E label.

[3] Follow the recommended rate for the target insect pest, as per the AMBUSH or CYMBUSH 3E label.

Region B[4]

Application Rate Table

FUSILADE 4E + AMBUSH

Product	Rate[5,6]
FUSILADE 4E + AMBUSH	½ to 1 pint (0.25-0.5 lb ai)/A 0.4-0.8 pints (0.1-0.2 lb ai)/A

FUSILADE 4E + CYMBUSH 3E

Product	Rate[5,6]
FUSILADE 4E + CYMBUSH 3E	½ to 1 pint (0.25-0.5 lb ai)/A 0.16-0.27 pints (0.06-0.10 lb ai)/A

[4] Region B—For the areas of Arizona, California, New Mexico, West Okla-

homa, West Texas, Utah and Colorado as outlined in the FUSILADE 4E label.

[5] Follow the recommended rate and growth stage for the grass weeds being treated, as per the FUSILADE 4E label.

[6] Follow the recommended rate for the target insect pest, as per the AMBUSH or CYMBUSH 3E label.

Timing—Apply to actively growing grasses before they exceed the recommended growth stages shown on the FUSILADE 4E label. Apply as required for insect control based on scouting reports. Subsequent applications of AMBUSH or CYMBUSH 3E may be required at intervals of 5 to 7 days. New growth or regrowth of grasses may be treated when subsequent growth reaches recommended growth stages shown on label.

Do not apply if rainfall is expected within 1 hour.

Spray Additives (Cleared for use on growing crops)
Always add one of the following:

- Crop oil concentrate—Add a nonphytotoxic crop oil or a once refined vegetable oil concentrate containing 15-20% approved emulsifier; at 1% (1 qt. per 25 gallons) of the finished spray volume.
- Nonionic surfactant—Add nonionic surfactant containing at least 75% surface active agent, at 0.25% (½ pint per 25 gallons) of the finished spray volume.

Ground Application—Use 5-40 gallons of total spray mixture per acre and 40-60 psi pressure. Sufficient spray volume and pressure should be provided to provide complete coverage of the foliage.

Use higher volumes and increase the pressure to 60 psi if foliage is dense to ensure complete coverage.

Band Applications—Band application equipment should be adjusted to provide maximum coverage of foliage in the planted row. Thorough coverage can be obtained with a nozzle directed to each side of the row towards the foliage in the planted row. Applications with a single nozzle over the top of the row are not recommended.

Do not use band applications when the grass is a perennial. Reinfestation of the treated band can result from the untreated middle.

Aerial Application—Use a minimum of 5 gallons of finished spray volume per acre. If the foliage is dense, increase the spray volume to 10 gallons per acre to improve coverage.

Mixing—Fill the spray tank half-full with water and add the recommended amounts of FUSILADE 4E + AMBUSH or CYMBUSH 3E while the agitator is running. Then add crop oil concentrate or nonionic surfactant and the remaining quantity of water and mix thoroughly.

USE RESTRICTIONS

- AMBUSH AND CYMBUSH ARE RESTRICTED USE PESTICIDES FOR RETAIL SALE TO AND USE ONLY BY CERTIFIED APPLICATORS, OR PERSONS UNDER THEIR DIRECT SUPERVISION, AND ONLY FOR THOSE USES COVERED BY THE CERTIFIED APPLICATOR'S CERTIFICATION.
- Always read and follow the restrictions and limitations for all products whether used alone or in tank mix. The most restrictive labeling of any product used applies in tank mixtures.
- Do not apply a total of more than 1 quart of FUSILADE 4E herbicide per acre per season to cotton (1 lb ai/acre).
- Do not apply a total of more than 12 pints of AMBUSH per acre per season (3 lbs ai/acre).
- Do not apply a total of more than 1.0 lb ai of CYMBUSH 3E per acre per season.
- Do not harvest cotton within 100 days of FUSILADE 4E application.
- Do not apply FUSILADE 4E after boll set.
- Do not graze or harvest for forage or hay.
- Do not plant rotational crops such as corn, sorghum and cereals within 60 days after the last tank mix application containing FUSILADE 4E.
- Avoid drift to all other crops and nontarget areas.
- Thoroughly flush spray tanks with water before and after each use.
- Apply to foliage when grasses are actively growing. When weeds are under stress use AMBUSH and CYMBUSH alone to control insects.

DIRECTIONS FOR USE
NONFOOD USES

Not for use or storage in or around the home.

FALLOW LAND

FUSILADE 4E herbicide can be used to control grass weeds during fallow periods.

Do not plant rotational crops other than cotton and soybeans within 60 days after last application.

NONBEARING FOOD CROPS

FUSILADE 4E herbicide can be used to control grass weeds in nonbearing groves, orchards or vineyards or for site preparation prior to transplanting.

APPLY AS A DIRECTED SPRAY TO PREVENT CONTACT OF SPRAY WITH FOLIAGE OF TREES OR VINES.

DO NOT APPLY TO TREES, VINES OR OTHER LISTED CROPS THAT WILL BE HARVESTED WITHIN ONE YEAR AFTER APPLICATION.

DO NOT APPLY TO ASPARAGUS FROM WHICH SPEARS WILL BE HARVESTED WITHIN ONE YEAR OF TREATMENT.

FUSILADE 4E is recommended on the following nonbearing crops:

Almonds	Macadamia
Apples	Mango*
Apricots	Nectarines
Asparagus	Olives
Avocados	Oranges
Berries	Papaya*
Cherries	Peaches
Dates	Pears
Figs	Pecans
Filberts	Pistachios
Ginseng*	Plums
Grapes	Pomegranates
Grapefruit	Prunes
Guava*	Tangelos
Jojoba	Tangerines
Lemons	Walnuts
Limes	West Indian Cherry*

* Not applicable in California.

ORNAMENTALS

FUSILADE 4E herbicide can be used to control grass weeds in many newly transplanted and established nongrassy ornamentals, trees, shrubs and ground covers. See the following Tables I through IV for specific recommended uses.

Application of FUSILADE 4E may be made over-the-top to ornamentals listed in Table I. Plants listed in Tables II through IV should have a directed spray applied to prevent contact of spray with foliage. If the growth habit of the plant or other factors do not allow a directed spray, an over-the-top application can be made to plants in Tables II-IV. However, there is the possibility of damage in the range listed for each table when applied at label rates.

DO NOT APPLY TO ORNAMENTALS THAT MAY BE HARVESTED FOR FOOD WITHIN ONE YEAR AFTER APPLICATION.

NOTICE TO BUYER AND USER: It is impossible to test every species and variety or cultivar of ornamental or nursery plants under all conditions. Plant tolerance of pesticides vary as conditions vary. Plant tolerance of FUSILADE 4E at label rates has been found to be acceptable within the ranges specified for the indicated genera and species. Neither the manufacturer nor the seller has determined whether or not FUSILADE 4E can safely be used on plants not specified on this label. The professional user should determine if FUSILADE 4E can be used safely prior to use.

APPLICATION DIRECTIONS

Timing—Apply to actively growing grass weeds before they exceed the recommended growth stages shown on this label.

Do not apply FUSILADE 4E if rainfall is expected within 1 hour.

Spray Additives (Cleared for use on growing crops)

ALWAYS ADD ONE OF THE FOLLOWING:

- **Crop Oil Concentrate**—Add a nonphytotoxic crop oil or a once refined vegetable oil concentrate containing 15-20% emulsifier; at 1% (1 quart per 25 gallons) of the finished spray volume.
- **Nonionic Surfactant**—Add nonionic surfactant containing at least 75% surface active agent, at 0.25% (½ pint per 25 gallons) of finished spray volume. **Use only nonionic surfactant on ornamentals.**

DIRECTIONS FOR USE

ANNUAL AND PERENNIAL GRASS CONTROL RECOMMENDATIONS FOR NONFOOD USES

REGIONS A & B

- Use **1 pint** of FUSILADE 4E per acre and 1% crop oil concentrate (1 quart per 25 gallons) or 0.25% (½ pint per 25 gallons) of a nonionic surfactant. **Use only nonionic surfactant on ornamentals.**
- Spray to obtain thorough coverage of grass foliage, but not to runoff. Repeat treatments may be necessary to provide full-season control.
- Use standard pesticide hollow cone or flat fan nozzles.
- Apply FUSILADE 4E to actively growing grasses before they exceed the recommended growth stages shown below.
- Do not apply FUSILADE 4E to grasses which are stressed due to moisture, temperature, herbicide or mechanical injury.
- Do not apply FUSILADE 4E to grasses which have tillered, formed seed heads, or exceeded recommended growth stages for treatment.
- **Do not use a crop oil concentrate with FUSILADE 4E on ornamentals.**

ANNUAL GRASS CONTROL RECOMMENDATIONS FOR REGIONS A & B

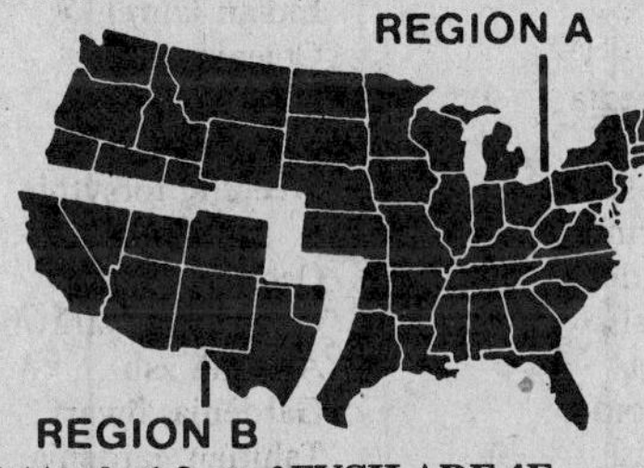

ANNUALS (Apply 1.0 pt of FUSILADE 4E per acre.)

Grass Species

Growth Stage: 2-8" tall, before tillering and/or heading

Barnyardgrass (*Echinochloa crusgalli*)
Broadleaf signalgrass (*Brachiaria platyphylla*)
Fall panicum (*Panicum dichotomiflorum*)
Field sandbur (*Cenchrus pauciflorus*)
Foxtail, green (*Setaria viridis*)
Foxtail, giant (*Setaria faberi*)
Foxtail, yellow (*Setaria lutescens*)
Goosegrass (*Eleusine indica*)
Guineagrass, seedling (*Panicum maximum*)
Itchgrass (*Rottboellia exaltata*)
Johnsongrass, seedling (*Sorghum halepense*)
Junglerice (*Echinochloa colonum*)
Large crabgrass (*Digitaria sanguinalis*)
Prairie cupgrass (*Eriochloa contracta*)
Red rice (*Oryza sativa*)
Shattercane/Wildcane/Volunteer milo (*Sorghum bicolor*)
Smooth crabgrass (*Digitaria ischaemum*)
Southern sandbur (*Cenchrus echinatus*)
Southwestern cupgrass (*Eriochloa gracilis*)
Texas panicum (*Panicum texanum*)
Volunteer barley (*Hordeum vulgare*)
Volunteer oats (*Avena fatua*)
Volunteer rye (*Secale cereale*)
Volunteer corn (*Zea mays*)
Volunteer wheat (*Triticum aestivum*)
Wild oats (*Avena fatua*)
Wild proso millet (*Panicum miliaceum*)
Witchgrass (*Panicum capillare*)

PERENNIAL GRASS CONTROL RECOMMENDATIONS FOR REGIONS A & B

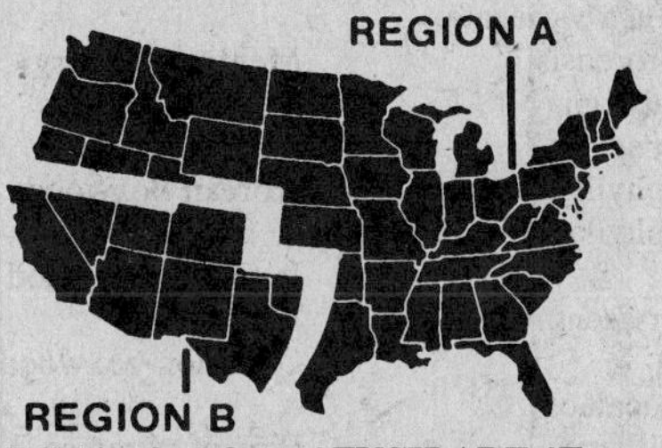

PERENNIALS (Apply 1.0 pt of FUSILADE 4E per acre.)

Grass Species

Bermudagrass (*Cynodon dactylon*)[16]
Guineagrass (*Panicum maximum*)[17]
Johnsongrass (*Sorghum halepense*)[18]
Quackgrass (*Agropyron repens*)[19]
Wirestem muhly (*Muhlenbergia frondosa*)[20]

[16] 3" tall or with 4-8" runners
[17] 6-12" tall and before seedhead initiation
[18] 12-18" tall and before boot stage
[19] 3-5 leaves, not more than 10" tall
[20] 4-12" tall and before seedhead initiation

TABLE I

OVER-THE-TOP APPLICATIONS
may be applied to the following ornamentals:
Use Only Nonionic Surfactant on Ornamentals

Scientific Name	Common Name/ Variety
Abelia grandiflora	Glossy abelia
Abies concolor	Concolor fir
Abies procera	Noble fir
Acacia redolens	Acacia ongerops
Acacia salacina	Willow-leafed acacia
Acacia saligna	Willow acacia
Acacia schafnerii	Jim wheat acacia
Acacia stenophylla	Shoe-string acacia
Acer palmatum	Maple/japanese
Acer platanoides	Norway maple
Acer saccharinum	Silver maple
Achillea filipendulina	Coronation gold, fernleaf yarrow
Achillea millefolium	Common yarrow
Acrecastrum romanzoffianum	Queen palm
Agapanthus africanus	Peter pan lily of the Nile
Agave americana	Century plant
Ageratum sp.	
Aglanoema commutatum	Silver Queen evergreen, Fransher evergreen, Treubii Ribbon evergreen
Aglanoema crispum	Painted evergreen
Alcea rosea	Hollyhock

Scientific Name	Common Name/ Variety
Allium schoenoprasum	Chives
Aloe barbadensis	Barbados aloe
Aloe vera	
Aloe zanzibarica	
Alyssum sp.	
Araucaria heterophylla	Norfolk Island pine
Arctostaphylos uva-ursi	red bearberry
Arctotheca calendula	Cape weed
Arenaria verna	sandwort
Asparagus densiflorus	Sprengeri, Myres asparagus, Sprenger asparagus
Astilbe x arendsii	False spirea
Atriplex sp.	Saltbrush
Aucuba japonica	Aucuba
Aucuba japonica variegata	
Berberis mentorensis	Mentor barberry
Berberis thunbergii	Crimson pygmy
Bougainvillea sp.	Raspberry ice
Buxus microphylla koreana	Korean boxwood
Buxus microphylla var. japonica	Japanese boxwood
Buxus sempervirens	Welleri, Common boxwood
Caesalpinia cacalaco	
Caesalpinia gilliesii	Bird of paradise
Calendula sp.	Marigold
Callistemon lanceolatus	Bottle-brush
Calluna vulgaris	Scotch heather
Camellia japonica	Camellia
Camellia sasanqua	Sasanqua camellia
Campanula carpatica	bellflower
Carissa grandiflora	Natal plum
Carissa tuttlei	
Carnegiea gijantea	Saguaro cactus
Cassia artemisioides	Feathery cassia
Cassia didymobotrya	African cassia
Cassia sturdii	
Centaurea cineraria	Dusty miller
Cerastium tomentosum	Snow in summer
Ceratonia siliqua	Carob tree
Cercis canadensis	Red bud
Chamaecyparis obtusa	Cripps hinoki falsecypress
Chamaecyparis lawsoniana	Allum lawson cypress
Chamaecyparis pisifera	Plumosa
Chamaerops excelsa	Windmill palm
Chamaerops humilis	Mediterranean fan palm
Chrysalidocarpus lutescens (small)	Golden fruited palm
Chrysanthemum x superbum	Shasta daisy
Cissus rhombifloia	Ellen Danica, grape ivy
Cistus hybridus	Rock rose
Citrus aurantium	Sour orange
Citrus paradisi	Grapefruit
Clerodendrum speciosum	Pagoda flower
Cleyera spp.	Cleyera
Clivia miniata	Kaffir lily
Cocos plumosa	Queen palm
Codiaeum variegatum	Croton
Coffea arabica	Coffee
Coleus x hybridus	Coleus
Convolvulus oneorum	Bush morning glory
Coprosma baueri	Mirror plant
Coprosma repens	Varigated mirror plant
Cordyline terminalis	Ti
Coreopsis verticillata	Threadleaf coreopsis
Cornus florida	Flowering dogwood
Cornus mas	Cornelian cherry dogwood
Cornus sericea	Red twig dogwood, Flaviramea
Cotoneaster apiculata	
Cotoneaster dammeri	Coral beauty, Royal beauty
Cotoneaster divaricatus	Spreading cotoneaster
Cotoneaster microphyllus	
Cotoneaster repens	Cotoneaster
Cotoneaster salicifolius franch	Willowleaf cotoneaster
Crossandra nilotica	Crossandra
Cupressus sempervirens	Italian cypress
Cycus revoluta	Sago palm
Deutzia gracilis	slender deutzia
Dianthus barbatus	Sweet William
Dieffenbachia amoena	Giant dumbcane
Diervilla lonicera	Bush honeysuckle
Dodonea viscosa purplurea	Purple hopseed bush
Echinocatus sp.	Hedgehog cactus
Elaeagnus angustrifolia	Russian olive
Encelia farinosa	Bittle brush
Erythrina fusca	Fastadiata, Swamp immortella
Escallonia fradessii	
Escallonia rubra	
Eucalyptus microtheca	Gum-barked coolibah
Eucalyptus polyanthemos	Red box gum
Eucalyptus rostrata	Red gum
Eucalyptus spathulata	Narrow-leaf gimlet
Eucalyptus rudis	Desert gum
Eugenia myrtifolia	Brush cherry
Euonymus alata	Siebold
Euonymus fortunei	
Euonymus japonica	Silver King
Euonymus kiautschovicus	Spreading Euonymus
Euphorbia lactea	Candelabra plant
Euryops pectinatus	
Fatshedera lizei	Fatshedera
Ferocactus sp.	Barrel cactus
Ficus benjamina	Exotica weeping fig
Ficus elastica decora	Rubber tree
Ficus microcarpa nitida	Indian laurel
Ficus nitida	Indian laurel
Ficus repens	Creeping fig
Forsythia intermedia	
Forsythia sp.	
Forsythia suspensa	Weeping forsythia
Forsythia x intermedia	
Fouquieria splendens	Ocotillo
Fragaria chiloensis	Ornamental strawberry
Fraxinus velutina	Arizona ash
Gardenia jasminoides	Gardenia, dwarf
Gardenia taitensis	Tahitian gardenia
Gazania splendens	Gazania gold rush
Gazania uniflora leucoleana	
Geijera parviflora	Australia willow
Gelsemium sempervirens	Carolina jessamine
Geranium pratense	Cranesbill
Gladiolus x hortulanus	Debbie, Jennie, Mahogany, Stargazer
Gleditsia triancanthos var. inermis	Honey locust/shade master
Graptophyllum pictum	Caricature plant
Grevillea robusta	Silk oak
Grevillea rosmarinifolia	Spider flower
Hedera canariensis	Algerian ivy
Hedera helix	English ivy
Hedera helix hahnii	Hahn's ivy
Hemerocallis hybrids	Daylily
Hemigraphis sp.	Waffle plant/Metallic plant
Hesperaloe parviflora	

Scientific Name	Common Name/ Variety
Hibiscus rosa-sinensis	Chinese hibiscus
Hibiscus syriacus	Althea
Hosta lancifolia	Variegated hosta
Hydrangea panicultata	Panicle hydrangea
Hydrangea quercifolia	Oakleaf hydrangea
Iberis sempervirens	Candytuff
Ilex cornuta	Dwarf buford holly
Ilex crenata	Japanese holly
Ilex glabra	Compact inkberry
Ilex opaca	American holly
Ilex vomitoria	Dwarf yaupon/Tall yaupon
Ilex x Meserveae	Meserve holly
Ilex x attenuata	Fosteri holly
Iris sp.	
Isotoma sp.	Blue star creeper
Ixora coccinea	Ixora
Jacaranda acutifolia	Jacaranda
Jacobinia ghiesbreghtiana	
Juniperius horizontalis	Admiral
Juniperius scopulorum	Cologreen
Juniperus chinensis	Hetzi, San Jose, Torulosa, Gold tip, Nicks compact juniper, Keteleeri, Hollywood, Blue Point, Blue vase juniper, Parsoni, Gold coast juniper, Sargent juniper
Juniperus conferta	Shore juniper
Juniperus horizontalis	Plumosa
Juniperus procumbens	
Juniperus sabina	Pepin, Skandia, Buffalo
Juniperus scopulorum	Witchita blue
Juniperus virginiana	Red ceder
Justicia spicigera	Firecracker, Mexican indigo
Kochia scoparia f. trychophylla	Compact burningbush
Lagerstroemia indica	Crepe myrtle
Lantana camera	Bush lantana
Lantana sellowiana	Purple lantana (trailing)
Leptospermum laevigatum	
Leucophyllum frutescens	Texas sage
Liatris spicata	Gay feather
Ligustrum amurense	Amur River
Ligustrum lucidum	Wax ligustrum
Ligustrum ovalifolium	Privet/california
Ligustrum texanum	Texas privet
Ligustrum x Vicari	Vicari
Liquidambar styraciflua	American sweetgum
Liriope spicata	Liriope
Liriope muscari	Green liriope/Variegated liriope
Livistona chinenis	Chinese fan palm
Lonicera x morrowii	Morrow honeysuckle
Lonicera xylosteum	Emerald mound
Lysiloma thornberii	Desert fern tree
Lysimachia nummularia	Moneywort/Wandering Sally
Magnolia sp.	Oregon Grapeholly
Magnolia stellata	Star magnolia
Mahonia aquifolium	Mahonia
Malus floribunda	Showy crabapple
Mesembryanthemum drosanhemum hispidum	Rose ice plant
Mesembryanthemum lampranthus prductus	Purple trailing ice plant
Mesembryanthemum lampranthus spectabilis	Red spike ice plant
Metrosideros excelsus	New Zealand christmas tree
Musa maurelli	Ethiopia banana
Myoponum parvifolium	Prostrate myoponum
Myrica cerifera	Wax myrtle
Nephrolepsis exaltata	Sword fern
Nerium oleander	Pink, variegated, petite oleander
Odocanthus sp.	
Olea europaea	Olive tree
Olneya tesota	Ironwood
Opuntia Cholla	Cholla cactus
Osmanthus fragrans	Osmanthus, tea olive
Osteospermum fruticosum alba	White african daisy
Pachysandra terminalis	Japanese pachysandra
Parkensonia aculeata	Green palo verde
Passiflora pfordtiii	Passion vine
Pelargonium domesticum	Geranium
Pelargonium peltatum	Ivy geranium
Pennisetum setaceum	Red fountain grass
Peperomia scandens	Radiator plant
Petunia sp.	
Philadelphus coronarius	Aurea
Philodendron oxycardium	"Micans" velvetleaf philodendron
Philodendron selloum	
Phoenix canariensis	Canary Island date palm
Phoenix roebelenii	Pygmy date palm
Photinia x fraseri	Photinia
Phyllostachys aurea	Golden bamboo
Physocarpus fruticosa	Gold drop, Jackmanni, Abbotswood
Physocarpus opulifolius	Dwarf Ninebark, Nanus
Picea abies	Norway spruce
Picea glauca	Dwarf Alberta, Black Hills, Densata
Picea omarika	Serbian spruce
Picea pungens	Blue Spruce
Pieris japonica	Lily-of-the-valley bush
Pilea nummulariifolia	Creeping Charlie
Pinus canariensis	Canary Island pine
Pinus mugo	Dwarf Swiss mountain pine
Pinus nigra	Austrian pine/Black pine
Pinus ponderosa	Western/Ponderosa pine
Pinus resinosa	Red pine
Pinus strobiformus	Mexican border pine
Pinus strobus	Eastern white pine
Pinus sylvestris	Scotch pine
Pittosporum Tobira	Wheelers dwarf, variegated
Pittosporum phillyraeoides	Desert willow
Plumbago capensis	Cane plumbago
Podocarpus gracilior	African fern pine
Podocarpus macrophylla	Yew pine
Polygonum capitatum	Pink clover
Polyscias fruticosa	Parsley panax
Polystichum capense	
Potentilla fructosa	Gold drop, Primrose beauty
Prosopis chilensis	Chilean mesquite
Prumus trialoba	Flowering almond
Prunus caroliniana ompacta	Carolina cherry
Pseuderanthemum atropurpureum	Purple false eranthemum
Pseudotsuga mensiessi	Douglas fir
Pyracantha koidzumii	Lodense pyracantha
Pyracantha angustifolia	Variegated firethorn
Pyracantha coccinea	Scarlet firethorn, Lalandei
Pyracantha graberi	

Scientific Name	Common Name/ Variety
Pyracantha koidzumii x coccinea	Mojave firethorn
Pyrus calleryana	Bradford pear
Quercus virginiana	Oak, live
Raphiolepis indica	Pink lady
Raphiolepis umbellata	Yedda/Indian hawthorn
Rhamnus frangula	Tallhedge buckthorn
Rhododendron "Gable Hybrid"	Elizabeth Gable, Purple splendor, Rose greely
Rhododendron catawbiense	Lees dark purple, Roseum elegans, Chionoides rhododendron, White catawba, Roseum elegans, Purple elegans, Roseum superbum, English roseum
Rhododendron formosa	
Rhododendron obtusum	Amoenum, Coral bells
Rhododendron sp.	Blaauw's pink, Rutherfordiana Constance, White cascase, Prize, Kluis sensation, Sweetheart, Kaempo, Royalty, Hahm red, Red wing, Gumpo white, Salmon spray, Road runner, Tradition, Gloria, Gumpo pink, Greeting, Pink pearl, Pericat, Tabor, Mother's Day, Snow, Masasoit, Red ruffle, Delaware Valley white, Purple Gem, H.H. Hume, Stewartstonian, Hino red, Rosebud, Fashion, Herbert, Gibralter, Girard's rose, Boule de neige, President lincoln
Rhododendron yedoense	Korean azalea/ Poukhanense
Rhuellia californica	
Rhus lancea	African sumar standard
Rosa Banksiae	Banksia
Rosa hybrida	Hybrid tea rose
Rosa sp.	Rose
Rosmarinus officinalis prostratus	Rosemary dwarf
Salix alba	Weeping willow
Salix matsudana	Tortuosa corkscrew willow
Salvia greggii	
Sansevieria sp.	Moon Glow
Sansevieria trifasciata	Hahaii/Mother-in-law's tongue
Santolina chanaecy parissus	Lavender cotton
Schefflera arboricola	Manila Ripple
Schinus molle	California pepper
Schinus terebinthifolius	Brazilean pepper
Sedum brevifolium	Green stone crop
Sedum quatemalense	Brown beam
Sedum spectabile	Sedum
Sedum x rubrotinctum	
Sempervivum tectorum	Hen and chickens
Simmiondsia chinensii	Jojoba
Spiraea nipponica	Snowmound spirea
Spiraea x billiardii	Billiard spirea
Spiraea x bumalda	Froebelii, Crispa spirea, Gold flame, Anthony Waterer
Statice sinuata	Annual statice
Stenocarpus sinuatus	Firewheel tree
Strelitzia nicolai	Giant bird of paradise
Strelitzia reginae	Bird of paradise
Syringa meyeri	Palibin
Syringa patula	Korean lilac
Syringa villosa	James McFarlane lilac
Syzgium paniculatum	Australian bush cherry
Tagetes sp.	Marigold
Taxus cuspidata	Japanese yew
Taxus x media	Thayeri yew, Hicks yew, Dense yew
Tecoma stans angustate	Yellow bells
Tecomaria capensis	Cape honeysuckle
Ternstroemia gymnanthera	Cleyera
Thevetia peruvianaa	Yellow oleander tree
Thuga occidentalis	White cedar, Emerald green, Woodwardii, Techny, Pyramidalis, Globosa, Techny american arborvitae
Thuja orientalis	Berkman's
Trachelospermum asiaticum	
Trachelospermum asiaticum	Asiatic jasmine
Trachelospermum jasminoides	Star jasmine
Trifolium fragiferum	O'Conners legume
Tsuga canadensis	Eastern hemlock
Viburnum dentatum	Arrowwood viburnum
Viburnum japonicum	Viburnum/japanese snowball
Viburnum odoratissium	Viburnum
Viburnum plicatum	Doublefile viburnum/ tomentosum
Viburnum suspensum	Sandankwa viburnum
Viburnum trilobum	Compact cranberrybush viburnum
Viburnum x juddi	Judd
Viburnum x rhytidophylloides	Viburnum/willowwood
Vicia sp.	Crown Vetch
Vinca major	Periwinkle
Vinca minor	Myrtle, Dwarf periwinkle
Washingtonia robusta	Mexican fan palm
Weigela florida	Newport red, Pink weigelia
Xylosma senticosa	
Yucca filamentosa	Yucca

Scientific Name	Common Name/ Variety
Yucca gloriosa	Spanish dagger
Yucca pendula	Weeping dagger
Zinnia sp.	

TABLE II

DIRECTED APPLICATIONS

to avoid ornamental foliage are recommended for the list below:

- When possible and plant growth habit allows, applications should be made as a directed spray to the plants listed below.
- Limited testing of the same plants has shown phytotoxicity of up to 20% when FUSILADE 4E is applied over-the-top at label rates.

Scientific Name	Common Name/ Variety
Abies fraseri	Fraser Fir
Acer rubrum	Maple/red
Ajuga variegata	Bugle weed
Ajuga reptans	Varigated ajuga
Aquilegia hybrida	Columbine
Arbutus unedo	Strawberry tree
Bigonia tweediana	Cats claw, Yellow trumpet
Callistemom viminalis	Weeping bottle-brush
Ceonothus griseus	
Cupressocyparis leylandi	Leyland cypress
Dracaena fragans	Massangeana, dracaena
Dracaena marginata	Tricolor dracaena
Eucalyptus citriodora	Lemon-scented gum
Feijoa sellowiana	Pineapple guava
Gamolepsis chrysanthemoides	
Gazania ringens	
Gladiolus x hortulanus	Gallery
Grewia caffra	Lavender star plant
Herniaria glabra	Green carpet
Juniperus chinensis	Pfitzeriana, Nana, Spearmint, Old Gold, Prostrata, Robusta, Maney
Juniperus conferta	Blue Pacific
Juniperus horizontalis	Wiltonii, Hughes, Webberi, Variagata, Blue Rug, Youngtown Compacta
Juniperus japonica	San Jose
Juniperus sabina	Tamariseifolia, Arcadia, Scandia, Broadmoor
Juniperus scopulorum	Welchii, Pathfinder
Juniperus virginiana	Sky Rocket, Grey owl
Lantana monteuidensis	White lantana
Leucothoe axillaris	Fetterbush
Ligustrum japonicum	Japanese privet
Lonicera japonica	Japanese honeysuckle
Nandina domestica	Heavenly bamboo
Nerium oleander standard	Oleander
Ophiopogon japonicum	Mondo grass
Opuntia sp.	Prickly pear cactus
Pennisetum sectaceum	Green fountain grass
Philodendron sp.	
Plumeria acuminata	Temple tree
Podocarpus macrophyllus	Maki
Potentilla verna	Spring cinquefoil
Rhododendron indicum	Indica
Rhododendron obtusum	Kurume Eureka
Rhododendron sp.	P.J.M.
Rhoeo spathacea	Oyster plant
Ruellia ciliosa	Ruellia
Salix caroliniana	Willow
Spirea x vanhouteii	Vanhoutte spirea
Syringa chinensis	Lilac
Viburnum macrocephalum	Chinese snowball

TABLE III

DIRECTED APPLICATIONS

to avoid ornamental foliage are recommended for the list below:

- When possible and plant growth habit allows, applications should be made as a directed spray to the plants listed below.
- Limited testing of the same plants has shown phytotoxicity of up to 50% when FUSILADE 4E is applied over-the-top at label rates.

Scientific Name	Common Name/ Variety
Acacia farnesiana	Sweet acacia
Acacia latifolia	
Aptenia cordifolia	Hearts and flowers
Asystasia gangetica	Coromandel
Baccharis sarothroides	Desert broom
Brachychiton populneum	Bottle tree
Cassia condolioma	
Cercidium praecox	Sonoran palo verde
Citharexylum spinosum	Fiddlewood
Cordyline stricta	Cordyline
Cupaniopsis anacardioides	Carrot wood
Crassula argentea	Jade plant
Cyperus alternifolius	Umbrella plant
Dicentra spectabilis	Bleeding heart
Dracaena deremensis	Janet Craig/Warneckii
Epipremnum aureum	Marble Queen pothos
Eucalyptus nicholii	
Exococaria cochichinensis	Chinese crenate croton
Hibiseus lepenk	
Juniperus chinensis	Armstrongii
Juniperus scopulorum	Spiny Greek, Excelsa Strieta
Juniperus virginiana	Burkii
Justicia betonia	White shrimp plant
Justicia brandegeana	Shrimp plant
Justicia carnea	Kings crown
Magnolia gradiflora	Southern magnolia
Mesembryanthemum delosperma alba	White ice plant (trailing)
Odontonema strictum	Red justicia
Oenothera berlandier	Mexican evening primrose
Pachystachys lutea	Yellow shrimp plant
Pedilanthus tithymaloides	Slipper flower
Peperomia obtusifolia	Baby rubber plant
Plectranthus australis	Swedish ivy
Polygonum capitatum	Pinkhead knotweed
Rhododendron indicum	Formosa
Rhododendron obtusum	Hersey red, Hino pink, Hinodegeri
Rhododendron poukhanensis	Karen
Rhus aromatica	Fragrant sumac
Thunbergia laurifloia	Laurel-leaved thunbergia
Vaccinum ashei	Tifblue Blueberry

TABLE IV
DIRECTED APPLICATIONS

to avoid ornamental foliage are recommended for the list below:

- When possible and plant growth habit allows, applications should be made as a directed spray to the plants listed below.
- Limited testing of the same plants has shown phytotoxicity greater than 50% when FUSILADE 4E is applied over-the-top at label rates.

Scientific Name	Common Name/ Variety
Acalypha godsefeiana hertophylla	
Alsophia australis	Australia tree fern
Betula nigra	Birch/River
Catharanthus roseus	Madagascar periwinkle
Chamaecyparis pisifera	Boulevard falsecypress
Chlorophytum comosum	Spider plant
Cortederia selloana	Pampas grass
Euonymus alata	Compacta
Juniperus chinensis	Sea Green
Juniperus horizontalis	Blue Chip, Prince of Wales, Bar Harbor
Juniperus scopulorum	Blue Haven
Kalanchoe fedtschenkoi	Lavender-scallops
Kalanchoe tubiflora	Chandelier plant
Physocarpus fruticosa	Katherine Dykes
Rhododendron obtusum	Hinocrimson
Setcreasea purpurea	Purple heart
Zebrina pendula	Wandering jew

GENERAL REMARKS AND PRECAUTIONS

Clean sprayer thoroughly before and after the application of FUSILADE 4E. Failure to clean sprayer thoroughly after application may result in injury to any grass crop subsequently sprayed, including corn, sorghum, small grains, sugarcane, rice and turfgrass.

Clean sprayer with water and a commercial sprayer cleaner or surfactant/adjuvant as per the recommended use directions of the respective products. Circulate the cleaning solution through the entire sprayer system. Spray approximately half of the tank solution through the hoses and nozzles to clean these parts. Properly drain the tank, as per procedures approved by state and local authorities, and rinse the total system several times with clean water.

Avoid drift to all other crops and nontarget areas. Grass crops are highly susceptible to FUSILADE 4E.

Do not plant rotational grass crops such as corn, sorghum and cereals within 60 days after last application.

Do not tank mix FUSILADE 4E with herbicides or other pesticides, with liquid fertilizers or with any other additives except as specified on this label or other supplemental labeling. Sequential applications of other herbicides except as specified on supplemental labeling within five days before or after FUSILADE 4E application has sometimes resulted in crop injury and reduced grass control.

Conversion Table

FUSILADE 4E Rate to be Applied

lb ai/A	pints/A	fl oz/A	A/gal
1/8	1/4	4	32
3/16	3/8	6	21.3
1/4	1/2	8	16
3/8	3/4	12	10.7
1/2	1	16	8

FUSILADE 4E contains 4 pounds of active ingredient per gallon.

Always add crop oil concentrate or nonionic surfactant (cleared for use on growing crops).

BLAZER 2L is a registered trademark of Rohm and Haas Company.

NOTICE TO BUYER AND USER

Seller warrants that this product conforms to the chemical description on the label and is reasonably fit for the purposes stated on the label when used in accordance with directions under normal conditions of use. This warranty does not extend to the use of this product contrary to label instructions, or under abnormal use conditions, or under conditions not reasonably foreseeable to Seller and Buyer and User assumes the risk of any such use. SELLER DISCLAIMS ALL OTHER WARRANTIES EXPRESSED OR IMPLIED INCLUDING ANY WARRANTY OF FITNESS OR MERCHANTABILITY. SELLER SHALL NOT BE LIABLE FOR CONSEQUENTIAL, SPECIAL OR INDIRECT DAMAGES RESULTING FROM THE USE OR HANDLING OF THIS PRODUCT AND SELLER'S SOLE LIABILITY AND BUYER'S AND USER'S EXCLUSIVE REMEDY SHALL BE LIMITED TO THE REFUND OF THE PURCHASE PRICE.

FUSILADE® 2000

1E HERBICIDE

For use in Cotton, Soybeans, Nonbearing Crops, Ornamentals and Fallow Land

ACTIVE INGREDIENT:
fluazifop-P-butyl—
butyl(**R**)-2-[4-[[5-(trifluoromethyl)-2-pyridinyl]
oxy]phenoxy]propanoate* 13.0%
INERT INGREDIENTS 87.0%
TOTAL 100.0%

*FUSILADE 2000 1E contains 1 pound (+) isomer (fluazifop-P-butyl) per gallon

EPA Reg. No. 10182-88

Keep Out of Reach of Children
DANGER

PRECAUTIONARY STATEMENTS

HAZARDS TO HUMANS AND DOMESTIC ANIMALS
CORROSIVE, CAUSES EYE DAMAGE AND SKIN BURNS. HARMFUL IF SWALLOWED, INHALED OR ABSORBED THROUGH SKIN. MAY CAUSE ALLERGIC SKIN REACTIONS. Do not get in eyes, on skin or on clothing. Wear protective clothing and PVC gloves when spraying. Wear protective clothing, PVC gloves and apron and full face shield when handling or mixing concentrate. Avoid breathing vapors or spray mist. Wash thoroughly with soap and water after handling and before eating or smoking. Remove contaminated clothing and wash before reuse.

STATEMENT OF PRACTICAL TREATMENT

IF SWALLOWED: Do not induce vomiting. Call a physician immediately.

IF IN EYES: Immediately flush eyes with plenty of water for at least 15 minutes. Get medical attention.

IF ON SKIN: Immediately remove contaminated clothing and wash thoroughly with plenty of soap and water. Get medical attention.

NOTE TO PHYSICIAN: Inducing vomiting as first aid for this substance may result in increased risk of chemical pneumonia or pulmonary edema caused by aspiration of the hydrocarbon solvent. Vomiting should be induced only under professional supervision.

For 24-hour emergency assistance, call ICI Americas Inc., (302) 575-3000.

ENVIRONMENTAL HAZARDS
This product is toxic to fish. Do not apply directly to water. Do not contaminate water by cleaning of equipment or disposal of waste. Do not apply when weather conditions favor drift from target area.

In case of a significant spill, call CHEMTREC (800) 424-9300.

PHYSICAL AND CHEMICAL HAZARDS

Do not use or store near heat or open flame.

STORAGE AND DISPOSAL

PROHIBITIONS

Do not contaminate water, food or feed by storage or disposal. Open dumping is prohibited. Do not reuse empty container.

STORAGE

Store in original container only. Keep container closed when not in use. Do not store near food or feed. In case of spill or leak on floor or paved surfaces, soak up with sand, earth or synthetic absorbent. Remove to chemical waste area.

PESTICIDE DISPOSAL

Pesticide wastes are acutely hazardous. Improper disposal of excess pesticide, spray mixture or rinsate is a violation of federal law. If these wastes cannot be disposed of by use according to label instructions, contact your State Pesticide or Environmental Control Agency or the Hazardous Waste Representative at the nearest EPA Regional Office for guidance.

CONTAINER DISPOSAL:

Metal Containers:

Triple rinse (or equivalent). Then offer for recycling or reconditioning, or puncture and dispose of in a sanitary landfill, or by other procedures approved by state and local authorities.

Plastic Containers:

Triple rinse (or equivalent). Then offer for recycling or reconditioning, or puncture and dispose of in a sanitary landfill, or incineration, or, if allowed by state and local authorities, by burning. If burned, stay out of smoke.

Glass Containers:

Triple rinse (or equivalent). Then dispose of in a sanitary landfill or by other approved state and local procedures.

DIRECTIONS FOR USE

It is a violation of federal law to use this product in a manner inconsistent with its labeling.

This labeling must be in the possession of the user at the time of application.

GENERAL INFORMATION

FUSILADE 2000 1E is a selective post-emergence herbicide for control of annual and perennial grass weeds. FUSILADE 2000 1E will provide effective control of grass weeds in conventional tillage, minimum tillage, and no-till plantings. FUSILADE 2000 1E does not control broadleaf weeds or sedges (nutgrass).

Control Symptoms—FUSILADE 2000 1E is a systemic herbicide which moves from the treated foliage into the shoots, roots, rhizomes, stolons, and growing points (meristematic regions) of treated grass weeds. Excellent control of a wide range of annual and perennial grass weeds will be obtained with FUSILADE 2000 1E when applied as recommended on this label.

Growth of treated grass weeds stops soon after application. Symptoms include loss of vigor, yellowing and/or reddening, and eventual death to the treated grass plant. Symptoms are generally observed within one to three weeks, depending on grass weed species and environmental conditions.

APPLICATION DIRECTIONS

Timing—Apply to actively growing grass weeds before they exceed the recommended growth stages shown on this label.

Do not apply FUSILADE 2000 1E if rainfall is expected within 1 hour.

Spray Additives (Cleared for use on growing crops)

ALWAYS ADD ONE OF THE FOLLOWING:

- **Crop Oil Concentrate**—Add a non-phytotoxic crop oil concentrate or a once refined vegetable oil concentrate containing 15-20% approved emulsifier, at 1% (1 quart per 25 gallons) of the finished spray volume. For aerial applications, add one pint of crop oil concentrate per acre.
- **Nonionic Surfactant**—Add nonionic surfactant containing at least 75% surface active agent, at 0.25% (1/2 pint per 25 gallons) of the finished spray volume for ground sprays. For aerial application add 1 pint of surfactant per acre.

Ground Application—Use sufficient spray volume and pressure to ensure complete coverage. Use 5-40 gallons of spray volume per acre and 40-60 psi pressure. A sufficient spray volume and pressure should be used to provide complete coverage of the treated grass weed species. Use higher volumes and pressure if the grass foliage is dense to ensure complete coverage.

DO NOT USE FLOOD TYPE OR OTHER SPRAY NOZZLE TIPS WHICH DELIVER COARSE, LARGE DROPLET SPRAYS.

DO NOT APPLY FUSILADE 2000 1E WITH RECIRCULATING SPRAYERS, ROPEWICK APPLICATORS, CONTROLLED DROPLET APPLICATORS (CDA) OR ANY SIMILAR DEVICES.

Band Applications—Band application equipment should be adjusted to provide maximum coverage of grasses in the planted row. Thorough coverage can be obtained with a nozzle directed to each side of the row toward the grass weeds in the row. Application with a single nozzle directly over the row is not recommended.

The use of band applications on perennial grasses may result in reinfestation of the treated band from the untreated middle and is not recommended.

Aerial Application—Use a minimum of 5 gallons per acre. If the grass foliage is dense, increase the spray volume to 10 gallons per acre to improve spray coverage.

Cultivation—Cultivation no sooner than 7 days after application will often assist in grass weed control. Cultivation of treated grasses 7 days prior to or within 7 days after application of FUSILADE 2000 1E may reduce weed control.

Spot Treatments

Mix FUSILADE 2000 1E and crop oil concentrate or nonionic surfactant with water according to the amounts shown below. Spray to obtain thorough coverage, but do not spray to runoff. Re-treat if necessary. **Use only nonionic surfactant on ornamentals.**

In cotton, applications must be made prior to boll set. Do not apply a total of more than 96 oz. of FUSILADE 2000 1E per acre per season to cotton.

In soybeans, applications must be made prior to bloom. Do not apply a total of more than 64 oz. of FUSILADE 2000 1E per acre per season to soybeans.

Spot Spray Recommendations

To Make This Spray Volume	Add These Amounts: FUSILADE 2000 1E	Crop/Vegetable Oil Concentrate*	or	Nonionic Surfactant
1 gallon	1 1/2 fl. oz. (3 tbsp)	1 1/2 fl. oz.	or	1/2 fl. oz.
10 gallons	13 fl. oz.	13 fl. oz.		3 fl. oz.
25 gallons	32 fl. oz.	1 quart		1/2 pint
50 gallons	64 fl. oz.	2 quarts		1 pint

* Do not use a crop oil concentrate on ornamentals. Use only nonionic surfactant on ornamentals.

DIRECTIONS FOR USE

ANNUAL AND PERENNIAL GRASS WEED CONTROL FOR COTTON AND SOYBEANS

[1] Alabama, Arkansas, California (Northern; see regional map), Connecticut, Delaware, Florida, Georgia, Idaho, Indiana, Iowa, Kentucky, Louisiana, Maine, Maryland, Massachusetts, Michigan, Minnesota, Missouri, Montana, Nebraska, Nevada (Northern; see regional map), New Hampshire, New Jersey, New York, North Carolina, North Dakota, Ohio, Oklahoma (Eastern; broadly defined as east of Interstate 35), Oregon, Pennsylvania, Rhode Island, South Carolina, South Dakota, Tennessee, Texas (Eastern; broadly defined as east of Interstate 35), Utah (Northern; see regional map), Vermont, Washington, West Virginia, Wisconsin, and Wyoming.

[2] Arizona, California (Southern; see regional map), Colorado, Oklahoma (Western; broadly defined as west of Interstate 35), Nevada (Southern; see regional map), New Mexico, Utah (Southern; see regional map), and Texas (Western; broadly defined as west of Interstate 35).

REGIONS A & B

- Apply to actively growing grasses.
- Apply at the recommended rate to grasses at the recommended growth stages as outlined.
- Do not apply to grasses which are stressed due to moisture, temperature, low soil fertility, mechanical or chemical injury.
- Do not apply to grasses which have formed seed heads, or exceeded recommended growth stages.
- Apply when the first grass weed species in a mixed grass weed population reaches the recommended growth stage for treatment.
- For mixed grass populations, use the highest recommended rate for grasses in that group.
- Where irrigation is used as part of normal cropping practice best results are usually obtained when FUSILADE 2000 1E is applied within 7 days after irrigation.

REGION A—Annual Grass Control

- Rates for annual grass control in Region A are 12-32 oz. of FUSILADE 2000 1E per acre.
- Re-treat at the recommended rate if needed to control late-germinating grasses.

REGION B—Annual Grass Control

- Rates for annual grass control in Region B are 16-48 oz. of FUSILADE 2000 1E per acre.
- In cotton, re-treat at the recommended rate if needed to control late germinating grasses or re-growth.
- In soybeans, do not apply more than 64 oz. of FUSILADE 2000 1E per acre per season.

GRASS WEED CONTROL

ANNUALS **REGION A**

Annual Grass Species	Height (inches)	No. of Leaves Not to Exceed	Rate(s) fl. oz./A	Special[1] Rate fl. oz./A
Barnyardgrass (*Echinochloa crusgalli*)	2–3	3	24	20
Broadleaf signalgrass (*Brachiaria platyphylla*)	2–4	5	24	20
Goosegrass (*Eleusine indica*)	2–4	6	16	
Crabgrass				
Large crabgrass (*Digitaria sanguinalis*)	1–2	4	24	20
Smooth crabgrass (*Digitaria ischaemum*)	1–2	4	24	20
Southern Crabgrass (*Digitaria ciliaris*)	1–2	4	24	20
Tropical Crabgrass (*Digitaria bicornis*)	1–2	4	24	20
Fall Panicum (*Panicum dichotomiflorum*)	2–6	6	24	20
Field Sandbur (*Cenchrus incertus*)	2–4	4	24	20
Foxtails				
Giant foxtail (*Setaria faberi*)	2–6	4	24	20
Green foxtail (*Setaria viridis*)	2–4	4	24	20
Yellow foxtail (*Setaria lutescens*)	2–4	4	24	20
Italian Ryegrass (*Lolium multiflorum*)	2–4	4	24	20
Itchgrass (*Rottboellia exaltata*)	4–24	6	16	
Johnsongrass, seedling (*Sorghum halepense*)	2–8	4	12	
Junglerice (*Echinochloa colonum*)	2–3	3	24	20
Red rice (*Oryza sativa*)	½–1	2	32	20[2]
Shattercane (*Sorghum bicolor*)	6–12	8	12	
Sorghum almum (*Sorghum almum*)	6–12	8	16	
Southern sandbur (*Cenchrus echinatus*)	2–6	6	20	
Texas panicum (*Panicum texanum*)	2–4	6	24	20
Volunteer Cereals				
V. Barley (*Hordeum vulgare*)	2–6	6	16	
V. Milo (*Sorghum bicolor*)	6–12	4	12	
V. Oats (*Avena sativa*)	2–6	6	16	
V. Rye (*Secale cereale*)	2–6	6	16	
V. Wheat (*Triticum aestivum*)	2–6	6	16	
V. Corn (*Zea mays*)	12–24	10	12	
Wild Proso Millet (*Panicum miliaceum*)	4–8	6	12	
Witchgrass (*Panicum capillare*)	2–4	6	24	20
Wild Oats (*Avena fatua*)	2–6	6	16	
Woolly cupgrass (*Eriochloa villosa*)	2–4	6	24	20

[1] SPECIAL USE DIRECTIONS
Apply 20 oz. of FUSILADE 2000 1E herbicide/acre under these additional conditions
- Apply under favorable soil moisture and humidity which normally exists within a few days after rainfall or irrigation. Avoid extreme air temperatures.
- Apply at earliest growth stages as indicated on rate tables.
- Apply in highly competitive crop stands, e.g. narrow row or drilled soybeans, or where cultivation is planned.
- Apply under conditions where light to moderate weed densities exist.
- Always use 1% V/V crop oil concentrate.
- Do not tank mix with other pesticides.
- Spray coverage and equipment calibration are critical.
- Scout on a field-to-field, day-to-day basis to insure that the previously described conditions are met.

[2] Most infestations require two applications at 20 oz. of FUSILADE 2000 1E herbicide/acre to achieve satisfactory control.

ANNUALS			REGION B
Grass Species	Height (inches)	No. of Leaves Not to Exceed	Rate fl. oz./A
Barnyardgrass (*Echinochloa crusgalli*)	1–2	3	24*
Crabgrass, Large (*Digitaria sanquinalis*)	1–2	3	24**
Johnsongrass, seedling (*Sorghum halepense*)	2–4	3	16
Junglerice (*Echinochloa colonum*)	2–3	3	24**
Volunteer Cereals			
V. Barley (*Hordeum vulgare*)	2–4	3	24
V. Milo (*Sorghum bicolor*)	2–4	4	24
V. Oats (*Avena sativa*)	2–4	3	24
V. Wheat (*Triticum aestivum*)	2–4	3	24
Volunteer Corn (*Zea mays*)	12–18	6	24**

* Use 48 oz. in California and Arizona

** Not labeled for control of this grass in Arizona and California

PERENNIAL GRASS CONTROL FOR REGIONS A & B

- Best results can be obtained if rhizomes or stolons are cut up by preplant tillage practices (discing, plowing, etc.) to stimulate maximum emergence of grass shoots.
- A second application may be needed for season-long control, particularly in no-till and reduced tillage plantings.

PERENNIALS			REGION A
Grass Species	Appli.	Height (inches)	Rate fl. oz./A
Bermudagrass (*Cynodon dactylon*)	1st	4–8 (runner length)	24 oz.

Remarks
If re-growth occurs, re-treat at recommended height and rate. Bermudagrass is a low-growing perennial, and the crop canopy may shield it from herbicide sprays. Improve herbicide spray coverage by directing the spray under the crop canopy. A minimum spray volume of 15 gallons/acre is suggested to improve coverage.

	2nd	4–8	16 oz.
Quackgrass (*Agropyron repens*)	1st	6–10	24 oz.

Remarks
For best results, make second application 2-3 weeks after the first but before quackgrass exceeds 10 inches. Always use 1% v/v crop oil concentrate. Spot treatment is not recommended. In no-till soybeans only, where rhizomes cannot be cut up by tillage, a preplant application of GRAMOXONE® paraquat is recommended. Follow all directions, restrictions and precautions on the GRAMOXONE paraquat labeling.

	2nd	up to 10 in.	16 oz.
Rhizome johnsongrass (*Sorghum halepense*)	1st	8–18*	24 oz.

Remarks
Apply before the boot stage of growth. If new shoots emerge or regrowth occurs, make a second application of 16 oz./A when the johnsongrass is 6-12 inches tall.

	2nd	6–12*	16 oz.

* Apply at 8-12 inches in Eastern Oklahoma, the Brazos bottoms, the Blacklands, Coastal Bend and Rio Grande areas of Eastern Texas. Apply second application at 4-6 inches.

Grass Species	Appli.	Height (inches)	Rate fl. oz./A
Wirestem muhly (*Muhlenbergia frondosa*)	1st	4–12	24 oz.

Remarks
If re-growth occurs, re-treat at the recommended rate and height.

PERENNIALS — REGION B

For best results, apply FUSILADE 2000 1E 3 days before to 7 days after irrigation.

Grass Species	Appli.	Height (inches)	Rate fl. oz./A
Bermudagrass (*Cynodon dactylon*)	1st	4–8 (runner length)	32–48 oz.

Remarks
In cotton re-treat at recommended rate and runner length (usually occurs about 4 weeks after first application). In soybeans, do not apply more than 64 oz. in one season. Bermudagrass is a low-growing perennial, and the crop canopy may shield it from herbicide sprays. Improvement in coverage may be achieved by directing the spray under the crop canopy. A minimum spray volume of 20 gallons/acre is suggested to improve coverage.

	2nd	4–8	32 to 48 oz.
Rhizome johnsongrass (*Sorghum halepense*)	1st	12–18	32 oz.

Remarks
Apply before the boot stage of growth. The second application should be made at the recommended rate to 12-18 inch tall johnsongrass (usually occurs about 4 weeks after first application).

	2nd	12–18	24 to 32 oz.

LABELED CROPS—Remarks and Precautions

COTTON—Do not apply a total of more than 96 oz. of FUSILADE 2000 1E herbicide per acre per season to cotton.

Do not harvest cotton within 90 days of application.

Do not apply after boll set.

Do not graze or harvest for forage or hay.

Do not plant rotational grass crops such as corn, sorghum and cereals within 60 days after last application.

SOYBEANS—Do not apply a total of more than 64 oz. of FUSILADE 2000 1E herbicide per acre per season to soybeans.

Make last application before bloom.

Do not graze or harvest for forage or hay.

Do not plant rotational grass crops such as corn, sorghum and cereals within 60 days after last application.

SOYBEAN TANK-MIX
FUSILADE 2000 1E HERBICIDE/BLAZER 2L HERBICIDE

Postemergence Control of Grass and Broadleaf Weeds
FUSILADE 2000 1E herbicide and BLAZER 2L herbicide may be used together in a postemergence program for broad spectrum weed control in soybeans.

FUSILADE 2000 1E herbicide is a selective postemergence herbicide for control of annual and perennial grass weeds in soybeans. BLAZER 2L herbicide is a selective postemergence herbicide for the control of certain broadleaf weeds.

FUSILADE 2000 1E herbicide and BLAZER 2L herbicide may be applied sequentially or in a tank-mix. The growth stage of weeds at the time of application will determine which application will provide the most satisfactory results.

Both FUSILADE 2000 1E herbicide and BLAZER 2L herbicide should be applied to actively growing weeds. See BLAZER 2L herbicide label for defined environmental conditions that promote active growth. Do not apply either herbicide if weeds appear stressed due to unfavorable temperatures, drought and/or low soil fertility. Do not apply BLAZER 2L herbicide or FUSILADE 2000 1E herbicide tank-mix if soybeans show injury from prior herbicide applications.

Read and observe all applicable label directions and limitations for both FUSILADE 2000 1E herbicide and BLAZER 2L herbicide before using.

METHOD 1: Sequential Application

FUSILADE 2000 1E Herbicide Followed by

BLAZER 2L Herbicide

(Annual and/or perennial grass weeds at proper growth stage for treatment, prior to broadleaf weed treatment)

Apply FUSILADE 2000 1E herbicide and an approved adjuvant at the recommended rate and growth stage for the grass weeds being treated, as per the FUSILADE 2000 1E herbicide label directions.

FUSILADE 2000 1E herbicide is a systemic herbicide, therefore when treating annual grasses allow at least 3 days, and when treating perennial grasses allow at least 5 days to elapse prior to a BLAZER 2L herbicide application. This will enable FUSILADE 2000 1E herbicide to be adequately translocated to the meristematic regions of the plant.

After the appropriate time interval has elapsed, apply BLAZER 2L herbicide with Triton® AG-98 or an approved non-ionic surfactant to susceptible broadleaf weeds as outlined on the BLAZER 2L herbicide label.

METHOD 2: Sequential Application

BLAZER 2L Herbicide Followed by

FUSILADE 2000 1E Herbicide

(Broadleaf weeds at proper growth stage for treatment, prior to annual and/or perennial grass weed treatment)

Apply BLAZER 2L herbicide with Triton AG-98 or an approved non-ionic surfactant at the recommended rate and growth stage to susceptible broadleaf weeds, as per the BLAZER 2L herbicide label directions. Broadleaf herbicides can stress and affect the growth of grass weed species.

FUSILADE 2000 1E herbicide must be applied to actively growing grasses for best results. A sequential application of FUSILADE 2000 1E herbicide may be applied following a BLAZER 2L herbicide application when grasses resume active growth with the development of a new leaf. Follow the recommended rates and growth stage for the grass weeds being treated, as per the FUSILADE 2000 1E herbicide label.

METHOD 3: Tank-Mix Applications

FUSILADE 2000 1E Herbicide and BLAZER 2L Herbicide

(Annual grasses only and broadleaf weeds are at the proper stage of growth for treatment, as per the respective labels)

A tank-mix of FUSILADE 2000 1E herbicide and BLAZER 2L herbicide may be applied at the recommended rates and growth stages to susceptible annual grass and broadleaf weed species in a manner as described on the respective labels.

Use only an approved non-ionic surfactant, as per the FUSILADE 2000 1E herbicide label recommendations. Follow water volume and spray pressure recommendations, as per the BLAZER 2L herbicide label.

Load the spray tank with half the amount of required water and add the recommended amounts of FUSILADE 2000 1E herbicide, BLAZER 2L herbicide, and non-ionic surfactant while the agitator is running, and then add the remaining quantity of water. Do not make more than one application of this tank-mix per season.

Note: Tank-mix applications sometimes have resulted in reduced grass weed control and possible increases in crop injury as compared to either product used alone. If grass regrowth occurs following an application of the tank-mix or an additional flush of grasses emerge, make a second application of FUSILADE 2000 1E herbicide to actively growing grass weeds, as per the label recommendations. A tank-mix application is not recommended if perennial grass weeds are the predominant grasses to be controlled.

Postemergence Use Restrictions

- Always read and follow the restrictions and limitations for all products whether used alone or in a tank mix. The most restrictive labeling of any product used applies in tank mixtures.
- Do not apply a total of more than 64 oz. of FUSILADE 2000 1E per acre per season to soybeans.
- Do not apply more than 4 pints (1.0 lb active) per acre of BLAZER 2L herbicide per growing season for soybeans.
- Make the last FUSILADE 2000 1E herbicide application before soybean bloom.
- Do not apply BLAZER 2L herbicide within 50 days of harvest for soybeans.
- Do not use treated plants for feed or forage.
- Do not plant rotational grass crops such as corn, sorghum and cereals within 60 days after the last FUSILADE 2000 1E herbicide application.
- Crop Rotation Restriction: Root crops (such as carrots, turnips, sweet potatoes, etc.) must not be planted in fields that have been treated with BLAZER 2L herbicide for a period of 18 months following the last treatment.
- In the case of crop failure, only soybeans may be immediately replanted.
- Avoid drift to all other crops and non-target areas.
- Thoroughly flush spray tanks with water before and after each use.
- BLAZER 2L herbicide requires a six hour rain-free period for best results. Do not apply BLAZER 2L herbicide or FUSILADE 2000 1E/BLAZER 2L herbicide tank mix if rain is threatening.
- Do not make more than one application of the FUSILADE 2000 1E/BLAZER 2L herbicide tank mix in a single season.
- Do not use during dry weather conditions when crop and weeds are under stress or wilt and not actively growing, as weed control may be reduced.
- Do not apply BLAZER 2L herbicide or FUSILADE 2000 1E/BLAZER 2L herbicide tank mix when crop and weeds are under stress from cold temperature, when maximum day temperature is below 70°F, or when soil temperature is below 60°F, as weeds will not be actively growing and control may be reduced.
- Do not apply BLAZER 2L herbicide or FUSILADE 2000 1E/BLAZER 2L herbicide tank mix when crops are exhibiting injury from previously applied pesticides or are exhibiting stress symptoms from disease, nematodes or insects, as increased crop response may result.
- Do not apply BLAZER herbicide or FUSILADE 2000 1E/BLAZER 2L herbicide tank mix when crop and weeds are under stress from excessive fertilizer or soil salts, wind injury, frost damage, or high temperature stress or wilt, as increased crop response may result.
- Do not mix FUSILADE 2000 1E herbicide, or FUSILADE 2000 1E/BLAZER 2L herbicide tank mix with other pesticides or liquid fertilizers except as specified on this label or other approved ICI Americas supplemental labeling.

COTTON TANK-MIX WITH AMBUSH INSECTICIDE OR WITH CYMBUSH 3E INSECTICIDE

Apply FUSILADE 2000 1E, AMBUSH or CYMBUSH 3E in the tank mix when required according to the weed and insect problems present. Follow rate recommendations indicated in the tables below.

REGION A[1]

APPLICATION RATE TABLE

FUSILADE 2000 1E + AMBUSH

Product	Rate[2,3]
FUSILADE 2000 1E	12–24 fl. oz. (.094–0.188 lb ai)/A
+	
AMBUSH	0.4–0.8 pints (0.1–0.2 lb ai)/A

FUSILADE 2000 1E + CYMBUSH 3E

Product	Rate[2,3]
FUSILADE 2000 1E +	12–24 fl. oz. (.094–0.188 lb ai)/A
CYMBUSH 3E	0.16–0.27 pints (0.06–0.10 lb ai)/A

[1] Region A—All states except the areas of Arizona, California, New Mexico, West Oklahoma, West Texas, Utah and Colorado as outlined in the FUSILADE 2000 1E label.

[2] Follow the recommended rate and growth stage for the grass weeds being treated, as per the FUSILADE 2000 1E label.

[3] Follow the recommended rate for the target insect pest, as per the AMBUSH or CYMBUSH 3E label.

REGION B[4]

APPLICATION RATE TABLE

FUSILADE 2000 1E + AMBUSH

Product	Rate[5,6]
FUSILADE 2000 1E +	16–48 fl. oz. (0.125–0.375 lb ai)/A
AMBUSH	0.4–0.8 pints (0.1–0.2 lb ai)/A

FUSILADE 2000 1E + CYMBUSH 3E

Product	Rate[5,6]
FUSILADE 2000 1E +	16–48 fl. oz. (0.125–0.375 lb ai)/A
CYMBUSH 3E	0.16–0.27 pints (0.06–0.10 lb ai)/A

[4] Region B—For the areas of Arizona, California, New Mexico, West Oklahoma, West Texas, Utah and Colorado as outlined in the FUSILADE 2000 1E label.

[5] Follow the recommended rate and growth stage for the grass weeds being treated, as per the FUSILADE 2000 1E label.

[6] Follow the recommended rate for the target insect pest, as per the AMBUSH or CYMBUSH 3E label.

Timing—Apply to actively growing grasses before they exceed the recommended growth stages shown on the FUSILADE 2000 1E label. Apply as required for insect control based on scouting reports. Subsequent applications of AMBUSH or CYMBUSH 3E may be required at intervals of 5 to 7 days. New growth or regrowth of grasses may be treated when subsequent growth reaches recommended growth stages shown on label.

Do not apply if rainfall is expected within 1 hour.

Spray Additives (Cleared for use on growing crops)
Always add one of the following:

- Crop oil concentrate—Add a nonphytotoxic crop oil or a once refined vegetable oil concentrate containing 15-20% approved emulsifier; at 1% (1 qt. per 25 gallons) of the finished spray volume. For aerial applications, add one pint of crop oil concentrate per acre.
- Nonionic surfactant—Add nonionic surfactant containing at least 75% surface active agent, at 0.25% (½ pint per 25 gallons) of the finished spray volume. For aerial applications, add one pint of surfactant per acre.

Ground Application—Use 5-40 gallons of total spray mixture per acre and 40-60 psi pressure. Sufficient spray volume and pressure should be provided to provide complete coverage of the foliage.

Use higher volumes and increase the pressure to 60 psi if foliage is dense to ensure complete coverage.

Band Applications—Band application equipment should be adjusted to provide maximum coverage of foliage in the planted row. Thorough coverage can be obtained with a nozzle directed to each side of the row towards the foliage in the planted row. Applications with a single nozzle over the top of the row are not recommended.

Do not use band applications when the grass is a perennial. Reinfestation of the treated band can result from the untreated middle.

Aerial Application—Use a minimum of 5 gallons of finished spray volume per acre. If the foliage is dense, increase the spray volume to 10 gallons per acre to improve coverage.

Mixing—Fill the spray tank half-full with water and add the recommended amounts of FUSILADE 2000 1E + AMBUSH or CYMBUSH 3E while the agitator is running. Then add crop oil concentrate or nonionic surfactant and the remaining quantity of water and mix thoroughly.

USE RESTRICTIONS

- AMBUSH AND CYMBUSH ARE RESTRICTED USE PESTICIDES FOR RETAIL SALE TO AND USE ONLY BY CERTIFIED APPLICATORS, OR PERSONS UNDER THEIR DIRECT SUPERVISION, AND ONLY FOR THOSE USES COVERED BY THE CERTIFIED APPLICATOR'S CERTIFICATION.
- Always read and follow the restrictions and limitations for all products whether used alone or in tank mix. The most restrictive labeling of any product used applies in tank mixtures.
- Do not apply a total of more than 96 fl. oz. of FUSILADE 2000 1E herbicide per acre per season to cotton (1 lb ai/acre).
- Do not apply a total of more than 12 pints of AMBUSH per acre per season (3 lbs ai/acre).
- Do not apply a total of more than 1.0 lb ai of CYMBUSH 3E per acre per season.
- Do not harvest cotton within 90 days of FUSILADE 2000 1E application.
- Do not apply FUSILADE 2000 1E after boll set.
- Do not graze or harvest for forage or hay.
- Do not plant rotational crops such as corn, sorghum and cereals within 60 days after the last tank mix application containing FUSILADE 2000 1E.
- Avoid drift to all other crops and nontarget areas.
- Thoroughly flush spray tanks with water before and after each use.
- Apply to foliage when grasses are actively growing. When weeds are under stress use AMBUSH and CYMBUSH alone to control insects.

DIRECTIONS FOR USE

NONFOOD USES

Not for use or storage in or around the home.

FALLOW LAND

FUSILADE 2000 1E herbicide can be used to control grass weeds during fallow period.

Do not plant rotational crops other than cotton and soybeans within 60 days after last application.

NONBEARING CROPS

FUSILADE 2000 1E herbicide can be used to control grass weeds in nonbearing groves, orchards or vineyards or for site preparation prior to transplanting.

APPLY AS A DIRECTED SPRAY TO PREVENT CONTACT OF SPRAY WITH FOLIAGE OF TREES OR VINES.

DO NOT APPLY TO TREES, VINES OR OTHER LISTED CROPS THAT WILL BE HARVESTED WITHIN ONE YEAR AFTER APPLICATION.

DO NOT APPLY TO ASPARAGUS FROM WHICH SPEARS WILL BE HARVESTED WITHIN ONE YEAR OF TREATMENT.

FUSILADE 2000 1E herbicide is recommended on the following nonbearing crops:

Acerola (West Indian Cherry)	Lemons
Almonds	Limes
Apples	Macadamia
Apricots	Mango*
Asparagus	Nectarines
Avocados	Olives
Berries	Oranges
Cherries	Papaya*
Coffee*	Peaches
Dates	Pears
Figs	Pecans
Filberts	Pistachios

Ginseng*	Plums
Grapes	Pomegranates
Grapefruit	Prunes
Guava*	Tangelos
Jojoba	Tangerines
Kiwi*	Walnuts

* Not applicable in California.

ORNAMENTALS

FUSILADE 2000 1E herbicide can be used to control grass weeds in many newly transplanted and established non-grassy ornamentals, trees, shrubs and ground covers. See the following Tables I through IV for specific recommended uses.

Application of FUSILADE 2000 1E may be made over-the-top to ornamentals listed in Table I. Plants listed in Tables II through IV should have a directed spray applied to prevent contact of spray with foliage. If the growth habit of the plant or other factors do not allow a directed spray, an over-the-top application can be made to plants in Table II-IV. However, there is the possibility of damage in the range listed for each table when applied at label rates.

DO NOT APPLY TO ORNAMENTALS THAT MAY BE HARVESTED FOR FOOD WITHIN ONE YEAR AFTER APPLICATION.

NOTICE TO BUYER AND USER: It is impossible to test every species and variety or cultivar of ornamental or nursery plants under all conditions. Plant tolerance of pesticides vary as conditions vary. Plant tolerance of FUSILADE 2000 1E at label rates has been found to be acceptable within the ranges specified for the indicated genera and species. Neither the manufacturer nor the seller has determined whether or not FUSILADE 2000 1E can safely be used on plants not specified on this label. The professional user should determine if FUSILADE 2000 1E can be used safely prior to use.

APPLICATION DIRECTIONS—NON-FOOD USES

Timing—For non-food use apply to actively growing grass weeds before they exceed the recommended growth stages shown on this page.

Do not apply FUSILADE 2000 1E herbicide if rainfall is expected within 1 hour.

Spray Additives (Cleared for use on growing crops) ALWAYS ADD ONE OF THE FOLLOWING:

- **Crop Oil Concentrate**—Add a nonphytotoxic crop oil or a once refined vegetable oil concentrate containing 15-20% emulsifier, at 1% (1 quart per 25 gallons) of the finished spray volume.
- **Nonionic Surfactant**—Add nonionic surfactant containing at least 75% surface wetting agent, at 0.25% (½ pint per 25 gallons) of finished spray volume. **Use only nonionic surfactant on ornamentals.**

DIRECTIONS FOR USE

ANNUAL AND PERENNIAL GRASS CONTROL RECOMMENDATIONS FOR NONFOOD USES

- Use **32-48 oz.** of FUSILADE 2000 1E herbicide per acre and 1% crop oil concentrate (1 quart per 25 gallons) or 0.25% (½ pint per 25 gallons) of a nonionic surfactant. **Use only nonionic surfactant on ornamentals.**
- Spray to obtain thorough coverage of grass foliage, but not to runoff. Repeat treatments may be necessary to provide full season control.
- Use standard pesticide hollow cone or flat fan nozzles.
- Apply FUSILADE 2000 1E herbicide to actively growing grasses before they exceed the recommended growth stages shown below.
- Do not apply FUSILADE 2000 1E herbicide to grasses which are stressed due to moisture, temperature, low soil fertility, mechanical or chemical injury.
- Do not apply to grasses which have tillered, formed seed heads, or exceeded recommended growth stages.
- Apply FUSILADE 2000 1E herbicide when the first grass species in a mixed grass weed population reaches the recommended growth stage for treatment.
- **Do not use a crop oil concentrate with FUSILADE 2000 1E on ornamentals.**

ANNUAL & PERENNIAL GRASS CONTROL FOR NONFOOD USES

ANNUALS

Grass Species Growth Stage: 2-8″ tall, before tillering and/or heading

Barnyardgrass (*Echinochloa crusgalli*)
Broadleaf signalgrass (*Brachiaria platyphylla*)
Crabgrass, Large (*Digitaria sanquinalis*)
Crabgrass, smooth (*Digitaria ischaemum*)
Fall Panicum (*Panicum dichotomiflorum*)
Field sandbur (*Cenchrus pauciflorus*)
Foxtail, giant (*Setaria faberi*)
Foxtail, green (*Setaria viridis*)
Foxtail, yellow (*Setaria lutescens*)
Goosegrass (*Eleusine indica*)
Guineagrass, seedling (*Panicum maximum*)
Itchgrass (*Rottboellia exaltata*)
Johnsongrass, seedling (*Sorghum halepense*)
Junglerice (*Echinochloa colonum*)
Prairie cupgrass (*Eriochloa contracta*)
Red rice (*Oryza sativa*)
Shattercane/Wildcane/Volunteer milo (*Sorghum bicolor*)
Southern sandbur (*Cenchrus echinatus*)
Southwestern cupgrass (*Eriochloa gracilis*)
Texas panicum (*Panicum texanum*)
Volunteer barley (*Hordeum vulgare*)
Volunteer oats (*Avena fatua*)
Volunteer rye (*Secale cereale*)
Volunteer corn (*Zea mays*)
Volunteer wheat (*Triticum aestivum*)
Wild oats (*Avena fatua*)
Wild proso millet (*Panicum miliaceum*)
Witchgrass (*Panicum capillare*)
Woolly cupgrass (*Eriochloa villosa*)

PERENNIALS

Grass Species	Growth Stage:
Bermudagrass (*Cynodon dactylon*)	4-8″ runners
Guineagrass (*Panicum maximum*)	6-12″ tall before seedhead initiation
Quackgrass (*Agropyron repens*)	6-10″
Rhizome johnsongrass (*Sorghum halepense*)	8-18″ tall and before boot stage
Torpedograss* (*Panicum repens*)	3-6″ tall (1st application) 2-3″ tall (subsequent application)
Wirestem muhly (*Muhlenbergia frondosa*)	4-12″ tall and before seedhead initiation

* Use 48 fl. oz. per acre per application. Three applications will be needed.

TABLE 1

OVER-THE-TOP APPLICATIONS

May be applied to the following ornamentals:

Use Only Nonionic Surfactant on Ornamentals

Scientific Name	Common Name/ Variety
Abelia grandiflora	Glossy abelia
Abies concolor	Concolor fir
Abies procera	Noble fir
Acacia redolens	Acacia ongerops

Scientific Name	Common Name/ Variety
Acacia salacina	Willow-leafed acacia
Acacia saligna	Willow acacia
Acacia schafnerii	Jim wheat acacia
Acacia stenophylla	Shoe-string acacia
Acer palmatum	Maple/japanese
Acer platanoides	Norway maple
Acer saccharinum	Silver maple
Achillea filipendulina	Coronation gold, fernleaf yarrow
Achillea millefolium	Common yarrow
Acrecastrum romanzoffianum	Queen palm
Agapanthus africanus	Peter pan lily of the Nile
Agave americana	Century plant
Ageratum sp.	
Aglanoema commutatum	Silver Queen evergreen, Fransher evergreen, Treubii Ribbon evergreen
Aglanoema crispum	Painted evergreen
Alcea rosea	Hollyhock
Allium schoenoprasum	Chives
Aloe barbadensis	Barbados aloe
Aloe vera	
Aloe zanzibarica	
Alyssum sp.	
Araucaria heterophylla	Norfolk Island pine
Arctostaphylos uva-ursi	red bearberry
Arctotheca calendula	Cape weed
Arenaria verna	sandwort
Asparagus densiflorus	Sprengeri, Myres asparagus, Sprenger asparagus
Astilbe x arendsii	False spirea
Atriplex sp.	Saltbrush
Aucuba japonica	Aucuba
Aucuba japonica variegata	
Berberis mentorensis	Mentor barberry
Berberis thunbergii	Crimson pygmy
Bougainvillea sp.	Raspberry ice
Buxus microphylla koreana	Korean boxwood
Buxus microphylla var. japonica	Japanese boxwood
Buxus sempervirens	Welleri, Common boxwood
Caesalpinia cacalaco	
Caesalpinia gilliesii	Bird of paradise
Calendula sp.	Marigold
Callistemon lanceolatus	Bottle-brush
Calluna vulgaris	Scotch heather
Camellia japonica	Camellia
Camellia sasanqua	Sasanqua camellia
Campanula carpatica	bellflower
Carissa grandiflora	Natal plum
Carissa tuttlei	
Carnegiea gijantea	Saguaro cactus
Cassia artemisioides	Feathery cassia
Cassia didymobotrya	African cassia
Cassia sturdii	
Centaurea cineraria	Dusty miller
Cerastium tomentosum	Snow in summer
Ceratonia siliqua	Carob tree
Cercis canadensis	Red bud
Chamaecyparis obtusa	Cripps hinoki falsecypress
Chamaecyparis lawsoniana	Allum lawson cypress
Chamaecyparis pisifera	Plumosa
Chamaerops excelsa	Windmill palm
Chamaerops humilis	Mediterranean fan palm
Chrysalidocarpus lutescens (small)	Golden fruited palm
Chrysanthemum x superbum	Shasta daisy
Cissus rhombifloia	Ellen Danica, grape ivy
Cistus hybridus	Rock rose
Citrus aurantium	Sour orange
Citrus paradisi	Grapefruit
Clerodendrum speciosum	Pagoda flower
Cleyera spp.	Cleyera
Clivia miniata	Kaffir lily
Cocos plumosa	Queen palm
Codiaeum variegatum	Croton
Coffea arabica	Coffee
Coleus x hybridus	Coleus
Convolvulus oneorum	Bush morning glory
Coprosma baueri	Mirror plant
Coprosma repens	Varigated mirror plant
Cordyline terminalis	Ti
Coreopsis verticillata	Threadleaf coreopsis
Cornus florida	Flowering dogwood
Cornus mas	Cornelian cherry dogwood
Cornus sericea	Red twig dogwood, Flaviramea
Cotoneaster apiculata	
Cotoneaster dammeri	Coral beauty, Royal beauty
Cotoneaster divaricatus	Spreading cotoneaster
Cotoneaster microphyllus	
Cotoneaster repens	Cotoneaster
Cotoneaster salicifolius franch	Willowleaf cotoneaster
Crossandra nilotica	Crossandra
Cupressus sempervirens	Italian cypress
Cycus revoluta	Sago palm
Deutzia gracilis	slender deutzia
Dianthus barbatus	Sweet William
Dieffenbachia amoena	Giant dumbcane
Diervilla lonicera	Bush honeysuckle
Dodonea viscosa purplurea	Purple hopseed bush
Echinocatus sp.	Hedgehog cactus
Elaeagnus angustrifolia	Russian olive
Encelia farinosa	Bittle brush
Erythrina fusca	Fastadiata, Swamp immortella
Escallonia fradessii	
Escallonia rubra	
Eucalyptus microtheca	Gum-barked coolibah
Eucalyptus polyanthemos	Red box gum
Eucalyptus rostrata	Red gum
Eucalyptus spathulata	Narrow-leaf gimlet
Eucalytus rudis	Desert gum
Eugenia myrtifolia	Brush cherry
Euonymus alata	Siebold
Euonymus fortunei	
Euonymus japonica	Silver King
Euonymus kiautschovicus	Spreading Euonymus
Euphorbia lactea	Candelabra plant
Euryops pectinatus	
Fatshedera lizei	Fatshedera
Ferocactus sp.	Barrel cactus
Ficus benjamina	Exotica weeping fig
Ficus elastica decora	Rubber tree
Ficus microcarpa nitida	Indian laurel
Ficus nitida	Indian laurel
Ficus repens	Creeping fig
Forsythia intermedia	
Forsythia sp.	
Forsythia suspensa	Weeping forsythia
Forsythia x intermedia	
Fouquieria splendens	Ocotillo
Fragaria chiloensis	Ornamental strawberry
Fraxinus velutina	Arizona ash

Scientific Name	Common Name/ Variety
Gardenia jasminoides	Gardenia, dwarf
Gardenia taitensis	Tahitian gardenia
Gazania splendens	Gazania gold rush
Gazania uniflora leucoleana	
Geijera parviflora	Australia willow
Gelsemium sempervirens	Carolina jessamine
Geranium pratense	Cranesbill
Gladiolus x hortulanus	Debbie, Jennie, Mahogany, Stargazer
Gleditsia triancanthos var. inermis	Honey locust/shade master
Graptophyllum pictum	Caricature plant
Grevillea robusta	Silk oak
Grevillea rosmarinifolia	Spider flower
Hedera canariensis	Algerian ivy
Hedera helix	English ivy
Hedera helix hahnii	Hahn's ivy
Hemerocallis hybrids	Daylily
Hemigraphis sp.	Waffle plant/Metallic plant
Hesperaloe parviflora	
Hibiscus rosa-sinensis	Chinese hibiscus
Hibiscus syriacus	Althea
Hosta lancifolia	Variegated hosta
Hydrangea panicultata	Panicle hydrangea
Hydrangea quercifolia	Oakleaf hydrangea
Iberis sempervirens	Candytuff
Ilex cornuta	Dwarf buford holly
Ilex crenata	Japanese holly
Ilex glabra	Compact inkberry
Ilex opaca	American holly
Ilex vomitoria	Dwarf yaupon/Tall yaupon
Ilex x Meserveae	Meserve holly
Ilex x attenuata	Fosteri holly
Iris sp.	
Isotoma sp.	Blue star creeper
Ixora coccinea	Ixora
Jacaranda acutifolia	Jacaranda
Jacobinia ghiesbreghtiana	
Juniperius horizontalis	Admiral
Juniperius scopulorum	Cologreen
Juniperus chinensis	Hetzi, San Jose, Torulosa, Gold tip, Nicks compact juniper, Keteleeri, Hollywood, Blue Point, Blue vase juniper, Parsoni, Gold coast juniper, Sargent juniper
Juniperus conferta	Shore juniper
Juniperus horizontalis	Plumosa
Juniperus procumbens	
Juniperus sabina	Pepin, Skandia, Buffalo
Juniperus scopulorum	Witchita blue
Juniperus virginiana	Red ceder
Justicia spicigera	Firecracker, Mexican indigo
Kochia scoparia f. trychophylla	Compact burningbush
Lagerstroemia indica	Crepe myrtle
Lantana camera	Bush lantana
Lantana sellowiana	Purple lantana (trailing)
Leptospermum laevigatum	
Leucophyllum frutescens	Texas sage
Liatris spicata	Gay feather
Ligustrum amurense	Amur River
Ligustrum lucidum	Wax ligustrum
Ligustrum ovalifolium	Privet/california
Ligustrum texanum	Texas privet
Ligustrum x Vicari	Vicari
Liquidambar styraciflua	American sweetgum
Leriope spicata	Liriope
Liriope muscari	Green liriope/Variegated liriope
Livistona chinenis	Chinese fan palm
Lonicera x morrowii	Morrow honeysuckle
Lonicera xylosteum	Emerald mound
Lysiloma thornberii	Desert fern tree
Lysimachia nummularia	Moneywort/Wandering Sally
Magnolia sp.	Oregon Grapeholly
Magnolia stellata	Star magnolia
Mahonia aquifolium	Mahonia
Malus floribunda	Showy crabapple
Mesembryanthemum drosanhemum hispidum	Rose ice plant
Mesembryanthemum lampranthus prductus	Purple trailing ice plant
Mesembryanthemum lampranthus spectabilis	Red spike ice plant
Metrosideros excelsus	New Zealand christmas tree
Musa maurelli	Ethiopia banana
Myoponum parvifolium	Prostrate myoponum
Myrica cerifera	Wax myrtle
Nephrolepsis exaltata	Sword fern
Nerium oleander	Pink, variegated, petite oleander
Odocanthus sp.	
Olea europaea	Olive tree
Olneya tesota	Ironwood
Opuntia Cholla	Cholla cactus
Osmanthus fragrans	Osmanthus, tea olive
Osteospermum fruticosum alba	White african daisy
Pachysandra terminalis	Japanese pachysandra
Parkensonia aculeata	Green palo verde
Passiflora pfordtiii	Passion vine
Pelargonium domesticum	Geranium
Pelargonium peltatum	Ivy geranium
Pennisetum setaceum	Red fountain grass
Peperomia scandens	Radiator plant
Petunia sp.	
Philadelphus coronarius	Aurea
Philodendron oxycardium	"Micans" velvetleaf philodendron
Philodendron selloum	
Phoenix canariensis	Canary Island date palm
Phoenix roebelenii	Pygmy date palm
Photinia x fraseri	Photinia
Phyllostachys aurea	Golden bamboo
Physocarpus fruticosa	Gold drop, Jackmanni, Abbotswood
Physocarpus opulifolius	Dwarf Ninebark, Nanus
Picea abies	Norway spruce
Picea glauca	Dwarf Alberta, Black Hills, Densata
Picea omarika	Serbian spruce
Picea pungens	Blue Spruce
Pieris japonica	Lily-of-the-valley bush
Pilea nummulariifolia	Creeping Charlie
Pinus canariensis	Canary Island pine
Pinus mugo	Dwarf Swiss mountain pine
Pinus nigra	Austrian pine/Black pine
Pinus ponderosa	Western/Ponderosa pine
Pinus resinosa	Red pine
Pinus strobiformus	Mexican border pine

Scientific Name	Common Name/ Variety
Pinus strobus	Eastern white pine
Pinus sylvestris	Scotch pine
Pittosporum Tobira	Wheelers dwarf, variegated
Pittosporum phillyraeoides	Desert willow
Plumbago capensis	Cane plumbago
Podocarpus gracilior	African fern pine
Podocarpus macrophylla	Yew pine
Polygonum capitatum	Pink clover
Polyscias fruticosa	Parsley panax
Polystichum capense	
Potentilla fructosa	Gold drop, Primrose beauty
Prosopis chilensis	Chilean mesquite
Prumus trialoba	Flowering almond
Prunus caroliniana ompacta	Carolina cherry
Pseuderanthemum atropurpureum	Purple false eranthemum
Pseudotsuga mensiessi	Douglas fir
Pyracantha koidzumii	Lodense pyracantha
Pyracantha angustifolia	Variegated firethorn
Pyracantha coccinea	Scarlet firethorn, Lalandei
Pyracantha graberi	
Pyracantha koidzumii x coccinea	Mojave firethorn
Pyrus calleryana	Bradford pear
Quercus virginiana	Oak, live
Raphiolepis indica	Pink lady
Raphiolepis umbellata	Yedda/Indian hawthorn
Rhamnus frangula	Tallhedge buckthorn
Rhododendron "Gable Hybrid"	Elizabeth Gable, Purple splendor, Rose greely
Rhododendron catawbiense	Lees dark purple, Roseum elegans, Chionoides rhododendron, White catawba, Roseum elegans, Purple elegans, Roseum superbum, English roseum
Rhododendron formosa	
Rhododendron obtusum	Amoenum, Coral bells
Rhododendron sp.	Blaauw's pink, Rutherfordiana Constance, White cascase, Prize, Kluis sensation, Sweetheart, Kaempo, Royalty, Hahm red, Red wing, Gumpo white, Salmon spray, Road runner, Tradition, Gloria, Gumpo pink, Greeting, Pink pearl, Pericat, Tabor, Mother's Day, Snow, Masasoit, Red ruffle, Delaware Valley white, Purple Gem,
	H. H. Hume, Stewartstonian, Hino red, Rosebud, Fashion, Herbert, Gibralter, Girard's rose, Boule de neige, President Lincoln
Rhododendron vedoense	Korean azalea/ Poukhanense
Rhuellia californica	
Rhus lancea	African sumar standard
Rosa Banksiae	Banksia
Rosa hybrida	Hybrid tea rose
Rosa sp.	Rose
Rosmarinus officinalis prostratus	Rosemary dwarf
Salix alba	Weeping willow
Salix matsudana	Tortuosa corkscrew willow
Salvia greggii	
Sansevieria sp.	Moon Glow
Sansevieria trifasciata	Hahaii/Mother-in-law's tongue
Santolina chanaecy parissus	Lavender cotton
Schefflera arboricola	Manila Ripple
Schinus molle	California pepper
Schinus terebinthifolius	Brazilean pepper
Sedum brevifolium	Green stone crop
Sedum quatemalense	Brown beam
Sedum spectabile	Sedum
Sedum x rubrotinctum	
Sempervivum tectorum	Hen and chickens
Simmiondsia chinensii	Jojoba
Spiraea nipponica	Snowmound spirea
Spiraea x billiardii	Billiard spirea
Spiraea x bumalda	Froebelii, Crispa spirea, Gold flame, Anthony Waterer
Statice sinuata	Annual statice
Stenocarpus sinuatus	Firewheel tree
Strelitzia nicolai	Giant bird of paradise
Strelitzia reginae	Bird of paradise
Syringa meyeri	Palibin
Syringa patula	Korean lilac
Syringa villosa	James McFarlane lilac
Syzgium paniculatum	Australian bush cherry
Tagetes sp.	Marigold
Taxus cuspidata	Japanese yew
Taxus x media	Thayeri yew, Hicks yew, Dense yew
Tecoma stans angustate	Yellow bells
Tecomaria capensis	Cape honeysuckle
Ternstroemia gymnanthera	Cleyera
Thevetia peruvianaa	Yellow oleander tree
Thuga occidentalis	White cedar, Emerald green, Woodwardii, Techny, Pyramidalis, Globosa, Techny american arborvitae
Thuja orientalis	Berkman's
Trachelospermum asiaticum	
Trachelospermum asiaticum	Asiatic jasmine
Trachelospermum jasminoides	Star jasmine
Trifolium fragiferum	O'Conners legume

Scientific Name	Common Name/ Variety
Tsuga canadensis	Eastern hemlock
Viburnum dentatum	Arrowwood viburnum
Viburnum japonicum	Viburnum/japanese snowball
Viburnum opulus	Nanum
Viburnum odoratissium	Viburnum
Viburnum plicatum	Doublefile viburnum/ tomentosum
Viburnum suspensum	Sandankwa viburnum
Viburnum trilobum	Compact cranberrybush viburnum
Viburnum x juddi	Judd
Viburnum x rhytidophylloides	Viburnum/willowwood
Vicia sp.	Crown Vetch
Vinca major	Periwinkle
Vinca minor	Myrtle, Dwarf periwinkle
Washingtonia robusta	Mexican fan palm
Weigela florida	Newport red, Pink weigelia
Xylosma senticosa	
Yucca filamentosa	Yucca
Yucca gloriosa	Spanish dagger
Yucca pendula	Weeping dagger
Zinnia sp.	

TABLE II

DIRECTED APPLICATIONS

to avoid ornamental foliage are recommended for the list below:

- When possible and plant growth habit allows, applications **should be made as a directed spray to the plants** listed below.
- Limited testing of the same plants has shown phytotoxicity of up to 20% when FUSILADE 2000 1E is applied over-the-top at label rates.

Scientific Name	Common Name/ Variety
Abies fraseri	Fraser Fir
Acer rubrum	Maple/red
Ajuga variegata	Bugle weed
Ajuga reptans	Varigated ajuga
Aquilegia hybrida	Columbine
Arbutus unedo	Strawberry tree
Bigonia tweediana	Cats claw, Yellow trumpet
Callistemom viminalis	Weeping bottle-brush
Ceonothus griseus	
Cupressocyparis leylandi	Leyland cypress
Dracaena fragans	Massangeana, dracaena
Dracaena marginata	Tricolor dracaena
Eucalyptus citriodora	Lemon-scented gum
Feijoa sellowiana	Pineapple guava
Gamolepsis chrysanthemoides	
Gazania ringens	
Gladiolus x hortulanus	Gallery
Grewia caffra	Lavender star plant
Herniaria glabra	Green carpet
Juniperus chinensis	Pfitzeriana Nana, Spearmint, Old Gold, Prostrata, Robusta, Maney
Juniperus conferta	Blue Pacific
Juniperus horizontalis	Wiltonii, Hughes, Webberi, Variagata, Blue Rug, Youngtown Compacta
Juniperus japonica	San Jose
Juniperus sabina	Tamariseifolia, Arcadia, Scandia, Broadmoor
Juniperus scopulorum	Welchii, Pathfinder
Juniperus virginiana	Sky Rocket, Grey owl
Lantana monteuidensis	White lantana
Leucothoe axillaris	Fetterbush
Ligustrum japonicum	Japanese privet
Lonicera japonica	Japanese honeysuckle
Nandina domestica	Heavenly bamboo
Nerium oleander standard	Oleander
Ophiopogon japonicum	Mondo grass
Opuntia sp.	Prickly pear cactus
Pennisetum sectaceum	Green fountain grass
Philodendron sp.	
Plumeria acuminata	Temple tree
Podocarpus macrophyllus	Maki
Potentilla verna	Spring cinquefoil
Rhododendron indicum	Indica
Rhododendron obtusum	Kurume Eureka
Rhododendron sp.	P.J.M.
Rhoeo spathacea	Oyster plant
Ruellia ciliosa	Ruellia
Salix caroliniana	Willow
Spirea x vanhouteii	Vanhoutte spirea
Syringa chinensis	Lilac
Viburnum macrocephalum	Chinese snowball

TABLE III

DIRECTED APPLICATIONS

to avoid ornamental foliage are recommended for the list below:

- When possible and plant growth habit allows, applications should be made as a directed spray to the plants listed below.
- Limited testing of the same plants has shown phytotoxicity of up to 50% when FUSILADE 2000 1E is applied over-the-top at label rates.

Scientific Name	Common Name/ Variety
Acacia farnesiana	Sweet acacia
Acacia latifolia	
Aptenia cordifolia	Hearts and flowers
Asystasia gangetica	Coromandel
Baccharis sarothroides	Desert broom
Brachychiton populneum	Bottle tree
Cassia condolioma	
Cercidium praecox	Sonoran palo verde
Citharexylum spinosum	Fiddlewood
Cordyline stricta	Cordyline
Cupaniopsis anacardioides	Carrot wood
Crassula argentea	Jade plant
Cyperus alternifolius	Umbrella plant
Dicentra spectabilis	Bleeding heart
Dracaena deremensis	Janet Craig/Warneckii
Epipremnum aureum	Marble Queen pothos
Eucalyptus nicholii	
Exococaria cochichinensis	Chinese crenate croton
Hibiseus lepenk	
Juniperus chinensis	Armstrongii
Juniperus scopulorum	Spiny Greek, Excelsa Strieta
Juniperus virginiana	Burkii
Justicia betonia	White shrimp plant
Justicia brandegeana	Shrimp plant
Justicia carnea	Kings crown
Magnolia gradiflora	Southern magnolia

Scientific Name	Common Name/ Variety
Mesembryanthemum delosperma alba	White ice plant (trailing)
Odontonema strictum	Red justicia
Oenothera berlandier	Mexican evening primrose
Pachystachys lutea	Yellow shrimp plant
Pedilanthus tithymaloides	Slipper flower
Peperomia obtusifolia	Baby rubber plant
Plectranthus australis	Swedish ivy
Polygonum capitatum	Pinkhead knotweed
Rhododendron indicum	Formosa
Rhododendron obtusum	Hersey red, Hino pink, Hinodegeri
Rhododendron poukhanensis	Karen
Rhus aromatica	Fragrant sumac
Thunbergia laurifloia	Laurel-leaved thunbergia
Vaccinum ashei	Tifblue Blueberry

TABLE IV

DIRECTED APPLICATIONS

to avoid ornamental foliage are recommended for the list below:

- When possible and plant growth habit allows, applications should be made as a directed spray to the plants listed below.
- Limited testing of the same plants has shown phytotoxicity greater than 50% when FUSILADE 2000 1E is applied over-the-top at label rates.

Scientific Name	Common Name Variety
Acalypha godsefeiana hertophylla	
Alsophia australis	Australia tree fern
Betula nigra	Birch/River
Catharanthus roseus	Madagascar periwinkle
Chamaecyparis pisifera	Boulevard falsecypress
Chlorophytum comosum	Spider plant
Cortederia selloana	Pampas grass
Euonymus alata	Compacta
Juniperus chinensis	Sea Green
Juniperus horizontalis	Blue Chip, Prince of Wales, Bar Harbor
Juniperus scopulorum	Blue Haven
Kalanchoe fedtschenkoi	Lavender-scallops
Kalanchoe tubiflora	Chandelier plant
Physocarpus fruticosa	Katherine Dykes
Rhododendron obtusum	Hinocrimson
Setcreasea purpurea	Purple heart
Zebrina pendula	Wandering jew

GENERAL REMARKS AND PRECAUTIONS

Do not apply to ornamentals that may be harvested for food within one year after application.

Clean sprayer thoroughly before and after the application of FUSILADE 2000 1E herbicide. Failure to clean sprayer thoroughly after application may result in injury to any grass crop subsequently sprayed, including corn, sorghum, small grains, sugarcane, rice and turfgrass.

Clean sprayer with water and a commercial sprayer cleaner or surfactant/adjuvant as per the recommended use directions of the respective products. Circulate the cleaning solution through the entire sprayer system. Spray approximately half of the tank solution through the hoses and nozzles to clean these parts. Properly drain the tank, as per procedures approved by state and local authorities, and rinse the total system several times with clean water.

Avoid drift to all other crops and nontarget areas. Grass crops are highly susceptible to FUSILADE 2000 1E herbicide.

Do not plant rotational grass crops such as corn, sorghum and cereals within 60 days after last application.

Do not tank-mix FUSILADE 2000 1E herbicide with herbicides or other pesticides, with liquid fertilizers or with any other additives except as specified on this label or other supplemental labeling.

Sequential applications of other herbicides except as specified on supplemental labeling within five days before or after FUSILADE 2000 1E herbicide application has sometimes resulted in crop injury and reduced grass control.

CONVERSION TABLE
FUSILADE 2000 1E Rate to be Applied

lb ai/A	pints/A	fl oz/A	A/gal
0.094	3/4	12	10.7
0.125	1	16	8
0.156	1 1/4	20	6.4
0.188	1 1/2	24	5.3
0.250	2	32	4
0.375	3	48	2.7

FUSILADE 2000 1E contains 1 pound of active ingredient per gallon.

Always add crop oil concentrate or nonionic surfactant (cleared for use on growing crops).

BLAZER 2L is a registered trademark of Rohm and Haas Company.

NOTICE TO BUYER AND USER

Seller warrants that this product conforms to the chemical description on the label and is reasonably fit for the purposes stated on the label when used in accordance with directions under normal conditions of use. This warranty does not extend to the use of this product contrary to label instructions, or under abnormal use conditions, or under conditions not reasonably foreseeable to Seller and Buyer and User assumes the risk of any such use. SELLER DISCLAIMS ALL OTHER WARRANTIES EXPRESSED OR IMPLIED INCLUDING ANY WARRANTY OF FITNESS OR MERCHANTABILITY. SELLER SHALL NOT BE LIABLE FOR CONSEQUENTIAL, SPECIAL OR INDIRECT DAMAGES RESULTING FROM THE USE OR HANDLING OF THIS PRODUCT AND SELLER'S SOLE LIABILITY AND BUYER'S AND USER'S EXCLUSIVE REMEDY SHALL BE LIMITED TO THE REFUND OF THE PURCHASE PRICE.

Registration pending in the state of California.

GRAMOXONE®
PARAQUAT

RESTRICTED USE PESTICIDE

For retail sale to and use only by Certified Applicators or persons under their direct supervision and only for those uses covered by the Certified Applicator's certification.

A Weed and Grass Killer—Harvest Aid Chemical

ACTIVE INGREDIENT:	BY WT.
Paraquat dichloride (1,1′-dimethyl-4,4′-bipyridinium dichloride)	29.1%
INERT INGREDIENTS	70.9%
TOTAL	100.0%

Contains 2 lbs. Paraquat cation per gallon.

EPA Reg. No. 239-2186-10182

Form 10418-D

Keep Out of Reach of Children

DO NOT USE OR STORE IN OR AROUND THE HOME

DANGER–POISON

DANGER:

ONE SWALLOW CAN KILL.

SYMPTOMS ARE PROLONGED AND PAINFUL.

ONSET OF SYMPTOMS MAY BE DELAYED FOR UP TO 3 DAYS AFTER SWALLOWING.

PELIGRO

PRECAUCION AL USUARIO: Si usted no lee ingles, no use este producto hasta que la etiqueta le haya sido explicado ampliamente.

POISON

DO NOT REMOVE CONTENTS EXCEPT FOR IMMEDIATE USE.

TO PREVENT ACCIDENTS, NEVER PUT INTO FOOD, DRINK, OR OTHER CONTAINERS.

READ DANGER STATEMENTS AND WORKER SAFETY RULES ON SIDE PANEL

DANGER

Can Kill If Swallowed. May Be Harmful or Fatal If Absorbed Through Skin or Inhaled. Harmful to Skin and Eyes.

STATEMENT OF PRACTICAL TREATMENT

IF SWALLOWED

IMMEDIATELY induce vomiting by inserting finger in throat. Administer fluids and induce further vomiting. Get to a hospital or physician fast. PROMPT TREATMENT IS ESSENTIAL and must be initiated before signs and symptoms of injury appear. **Note to Physicians:** Call the Chevron Poison Information Center (415) 233-3737 at any hour to obtain toxicology and medical management consultation, a supply of bentonite clay, and paraquat analysis. Symptoms following ingestion may be delayed up to 3 days. To be effective, treatment for paraquat poisoning must begin **IMMEDIATELY.** Treatment consists of binding paraquat in the gut with suspensions of bentonite clay or charcoal and removal of paraquat from the blood by prolonged charcoal hemoperfusion or continuous hemodialysis.

SKIN CONTACT

IMMEDIATELY wash with soap and water. Remove and wash all contaminated clothing with soap and hot water before reuse. Prolonged contact will cause severe irritation. Contact with irritated skin, or a cut, or repeated contact with intact skin may result in poisoning.

EYE CONTACT

IMMEDIATELY wash eyes with water for at least 15 minutes and get medical attention. Contact may cause eye injury.

INHALATION

IMMEDIATELY get away from spray mist. Exposure may cause irritation, nose bleeds and may lead to poisoning. Stop and check spray procedure. Follow Worker Safety Rules on side panel.

DIRECTIONS FOR USE

It is a violation of Federal law and may endanger the applicator or third parties to use this product without complying with all label directions and warnings.

READ ENTIRE LABEL: USE STRICTLY IN ACCORDANCE WITH DANGER STATEMENTS AND DIRECTIONS, AND WITH APPLICABLE STATE AND FEDERAL REGULATIONS.

READ DANGER AND FIRST AID TREATMENT ON INSIDE FRONT COVER AND WORKER SAFETY RULES.

NOTE: Do Not Use Around Home Gardens, Schools, Recreational Parks, or Playgrounds.

Do not combine GRAMOXONE paraquat with liquid fertilizer solutions except as specified below.

DO NOT combine with suspension type liquid fertilizers as the activity of paraquat will be reduced.

Use ORTHO X-77®* Spreader (non-ionic). ORTHO X-77 Spreader is recommended, but a similar and suitable non-ionic surfactant may be substituted.

Do not use dirty or muddy water for GRAMOXONE paraquat dilution as GRAMOXONE paraquat will be inactivated.

DIRECTIONS

PREPLANT OR PREEMERGENCE USES

PREPLANT OR PREEMERGENCE USE—For Kill of Emerged Annual Broadleaf Weeds and Grasses and for Top Kill Suppression of Perennials—Apply when the weeds and grasses are succulent and growth is from 1″ to 6″ high (larger plants are less affected by this treatment). Note: To prevent injury to germinating crop seedlings, do not apply preplant or preemergence to soils lacking clay minerals, i.e., peat, muck, pure sand, artificial planting media.

ALFALFA, CORN, LETTUCE, MELONS, PEPPERS, SORGHUM, SUGAR BEETS, TOMATOES, SAFFLOWER AND SUNFLOWER: Preplant or Preemergence Treatment—Apply 1 to 2 qts. GRAMOXONE paraquat per sprayed acre as a band treatment over the crop row, or as a broadcast treatment prior to, during, or after planting—but before emergence of the crop. **Aerial Application**—5 to 10 gals. water per acres. **Ground Application**—Use 20 to 60 gals. diluted spray. Add ORTHO X-77 Spreader (non-ionic) at 8 to 32 fl. oz. per 100 gals. of diluted spray. Seedbeds should be formed as far ahead of planting and treatment as possible to permit maximum weed and grass emergence. Seeding should be done with a minimum amount of soil disturbance. Weeds and grasses emerging after application will not be controlled. Crop plants emerged at time of application will be killed.

ASPARAGUS: Preplant or Preemergence—Apply 1 to 2 qts. GRAMOXONE paraquat per sprayed acre as a band treatment over the row or as a broadcast treatment prior to, during, or after planting, but before emergence of the crop. Add ORTHO X-77 Spreader (non-ionic) at 8 to 32 fl. oz. per 100 gals. spray. Apply 50 to 100 gals. spray per acre. Allow maximum weeds and grass emergence prior to treatment. Seeding or transplanting should be done with a minimum of soil disturbance. Weeds and grasses emerging after treatment will not be controlled.

ASPARAGUS (Plantings Established 2 Years or More): Preemergence—Apply 1 to 2 qts. GRAMOXONE paraquat per sprayed acre in 25 to 100 gals. of spray mix. Apply as a band treatment over the row or as a broadcast application prior to emergence of the crop and/or after the last harvest. Add ORTHO X-77 Spreader (non-ionic) at 8 to 32 fl. oz. per 100 gals. of spray mix. For residual control add $2\frac{1}{2}$ to 5 lbs. PRINCEP®* 80 W as a tank mix. Refer to PRINCEP 80 W label for specific recommendations and cautions. Crop plants emerged at time of application will be severely injured. Do not apply within 6 days of harvest.

BARLEY, WHEAT: Preplant or Preemergence Treatment—Apply 1 to 4 pts. GRAMOXONE paraquat per sprayed acre as a band treatment over the crop row, or as a broadcast treatment prior to, during, or after planting—but before emergence of the crop. **Aerial Application**—5 to 10 gals. water per acre. **Ground Application**—Use 20 to 60 gals. diluted spray. Add ORTHO X-77 Spreader (non-ionic) at 8 to 32 fl. oz. per 100 gals. of diluted spray. Seedbeds should be formed as far ahead of planting and treatment as possible to permit maximum weed and grass emergence. Seeding should be done with a minimum amount of soil disturbance. Weeds and grasses emerging after application will not be controlled. Crop plants emerged at time of application may be killed. Do not graze livestock in treated area.

BEANS (LIMA, SNAP), BROCCOLI, CABBAGE, CARROTS, CAULIFLOWER, CHINESE CABBAGE, COLLARDS, CUCURBITS (WATERMELON, CANTALOUPE/MUSKMELON, PUMPKIN, SQUASH, CUCUMBER), EGGPLANT, PEAS, TURNIP: Preplant and Preemergence Treatment—Apply 1 to 2 qts. GRAMOXONE paraquat per sprayed acre as a band treatment over the crop row, or as a broadcast treatment before, during or after planting—but before emergence of the crop. Use sufficient water to give thorough coverage. Add ORTHO X-77 Spreader (non-ionic) at 8 to 32 fl. oz. per 100 gals. diluted spray. Use the higher rate with heavier weed infestations. Seedbeds should be formed as far ahead of planting and treatment as possible to permit maximum weed and grass emergence. Seeding should be done with a minimum amount of soil disturbance. Weeds and grasses emerging after application will not be controlled. Crop plants emerged at time of application will be killed.

COTTON: Preplant Treatment—Ground Application—Apply 1 to 2 qts. GRAMOXONE paraquat per acre as a single broadcast application in 20 to 60 gals. water. Beds should be preformed to permit maximum weed and grass emergence prior to treatment. Seeding should be done with a minimum amount of soil disturbance. Weeds and grasses emerging after application will not be controlled. Add ORTHO X-77 Spreader (non-ionic) at 8 to 32 fl. oz. per 100 gals. spray.

ONIONS (Seeded) For Green or Dry Bulb Harvest: Preplant and Preemergence—Apply 1 to 2 qts. GRAMOXONE paraquat per sprayed acre as a broadcast treatment prior to, during or after planting, but before emergence of the crop. Use the high rate for heavier weed infestations and for control of Wild Oats. Add ORTHO X-77 Spreader (non-ionic) at 8 to 32 fl. oz. per 100 gals. spray. Apply in 20 to 100 gals. water per acre. Allow maximum weed and grass emergence prior to treatment. Seeding should be done with a minimum of soil disturbance. Weeds and grasses emerging after treatment will not be controlled. Crop plants emerged at time of application will be injured. Maximum of one application at the 2 qts. per acre dosage. Do not apply within 60 days of harvest. **In California**—Do not apply within 200 days of harvest.

POTATOES (Preplant or Preemergence Treatment)—Apply 1 to 2 pts. GRAMOXONE paraquat per acre as a broadcast application in 30 to 100 gals. water by ground application, or in 5 to 10 gals. by air. Add ORTHO X-77 Spreader (non-ionic) at 8 to 32 fl. oz. per 100 gals. of diluted spray. The application should be delayed to provide maximum weed and grass emergence but should be applied not later than ground cracking, before potatoes have emerged. Application made after crop emergence has reduced yields of certain sensitive varieties. Weeds and grasses emerging after application will not be controlled. Postemergence management should be performed as required to control subsequent weed and grass emergence.

FOR KILL OF EMERGED VOLUNTEER BARLEY (California Only)—Preplant in Cotton, Potato, Lettuce, Melon, Safflower, Sugar Beet, and Tomato Fields—Apply $\frac{1}{2}$ to 1 pt. GRAMOXONE paraquat per acre broadcast as a preplant treatment to preformed seedbeds. **Aerial Application**—Use 5 to 10 gals. water per acre. Add ORTHO X-77 Spreader (non-ionic) at 8 to 32 fl. oz. per 100 gals. diluted spray.

SEEDBED PREPARATION FOR ESTABLISHING GRASSES FOR SEED PRODUCTION: To Control Grasses and Broadleaf Weeds—Prepare the seedbed and allow weed seeds to germinate. When broadleaf weeds and annual grasses are at the 3 to 5 leaf stage apply 1 qt. GRAMOXONE paraquat in 20 to 40 gals. water per acre (larger weeds will be less affected by this treatment). Add ORTHO X-77 Spreader (non-ionic) at 8 to 32 fl. oz. per 100 gals. diluted spray. This treatment can be repeated prior to or on day of seeding. Do not allow animals to graze on treated areas. Do not use seed from treated areas for animal feed. Do not use straw from treated areas for animal bedding or feed.

CONSERVATION TILLAGE USES

GRAMOXONE paraquat and GRAMOXONE paraquat tank mixes may be applied in liquid nitrogen and/or complete liquid fertilizer solutions (follow label recommendations for crop involved) when 20 to 60 gals. of dilute solution are applied per acre in combination with 1 to 2 pts. of ORTHO X-77 Spreader (non-ionic) per 100 gals. Equipment must be adjusted and calibrated to obtain thorough coverage of undesirable vegetation. Inadequate coverage and/or the absence of ORTHO X-77 Spreader (non-ionic) will result in a severe reduction of GRAMOXONE paraquat activity. Phosphate containing liquid fertilizer solutions will reduce the activity of GRAMOXONE paraquat. When liquid fertilizer solutions containing phosphorus are used with GRAMOXONE paraquat, and a more rapid kill is desired, use the high GRAMOXONE paraquat label rate.

CORN ONLY: For Kill of Existing Vegetation and Residual Weed Control Where Corn Will Be Planted Directly Into a Cover Crop, Established Sod, or in Previous Crop Residues. Tank-Mix with Atrazine:

Apply 1 to 2 pts. GRAMOXONE paraquat plus $2\frac{1}{2}$ to $3\frac{3}{4}$ lbs. Atrazine 80 Wettable per acre, as a broadcast spray. Refer to Atrazine 80 Wettable label for directions, limitations and cautions. Apply 20 to 60 gals. diluted spray per acre. Add ORTHO X-77 Spreader (non-ionic) at 8 fl. oz. per 100 gals. diluted spray. Add Atrazine to spray tank first, mixing thoroughly in water. Then add GRAMOXONE paraquat and surfactant. Constant agitation recommended at all times.

Tank Mix with AATREX®* 80 W and PRINCEP 80 W:

Apply 1 to 2 pts. GRAMOXONE paraquat plus $1\frac{1}{4}$ to $2\frac{1}{2}$ lbs. PRINCEP 80 W plus $1\frac{1}{4}$ to $2\frac{1}{2}$ lbs. AATREX 80 W per sprayed acre prior to, during, or after planting—but before crop emergence. Use the low rates of PRINCEP 80 W and AATREX 80 W on light soils, and the higher rates on heavy clay soils. Use 20 to 60 gals. of diluted spray per sprayed acre. Add ORTHO X-77 Spreader (non-ionic) at 8 fl. oz. per 100 gals. diluted spray. Add the AATREX 80 W to the spray tank while agitating, then add the PRINCEP 80 W, GRAMOXONE paraquat, and ORTHO X-77 Spreader (non-ionic). Continue agitation during application. Refer to the AATREX 80 W and PRINCEP 80 W labels for specific precautionary statements.

Tank Mix with BLADEX®* 80 W:

Apply 1 to 2 pts. GRAMOXONE paraquat plus $1\frac{1}{2}$ to 5 lbs. BLADEX 80 W, depending on soil type, as a broadcast spray. Apply 20 to 60 gals. diluted spray per acre. Add ORTHO X-77 Spreader (non-ionic) at 1 qt. per 100 gals. diluted spray. Add BLADEX 80 W to spray tank first, mixing thoroughly in water. Then add GRAMOXONE paraquat and surfactant. Constant agitation recommended at all times. Do not mix with liquid fertilizer. Refer to BLADEX 80 W label for directions, limitations and cautions.

Tank Mix with LASSO®* and Atrazine:
Apply 1 to 2 pts. GRAMOXONE paraquat (0.25 to 0.5 lb. active) plus 4 to 6 pts. LASSO (2.0 to 3.0 lbs. active), plus 1.25 to 2.5 lbs. Atrazine 80 Wettable (1.0 to 2.0 lbs. active) per treated acre prior to, during, or after planting, but before crop emergence. Use 20 to 60 gals. of diluted spray per treated acre. Add ORTHO X-77 Spreader (non-ionic) at 8 oz. per 100 gals. of diluted spray. Add the Atrazine to the spray tank while agitating, then add the LASSO, GRAMOXONE paraquat and ORTHO X-77 Spreader (non-ionic). Continue agitation during application. Refer to Atrazine and LASSO labels for cautions and for specific rates for a given soil type.

Tank Mix with DUAL®* and AATREX: Apply 1 to 2 pts. GRAMOXONE paraquat per acre plus DUAL plus AATREX as a broadcast ground spray before, during spray before, during, or after planting but before crop emergence. Apply in 20 to 60 gals. of diluted spray per acre. Add ORTHO X-77 Spreader (non-ionic) at 8 fl. oz. per 100 gals. of diluted spray. Refer to DUAL and AATREX labels for rates of application, directions, limitations, cautions and for a list of weeds controlled by preemergence activity of DUAL and AATREX. Add AATREX to spray tank first, mixing thoroughly in carrier. Then add DUAL, GRAMOXONE paraquat and ORTHO X-77 Spreader (non-ionic). Constant agitation recommended at all times.

Tank Mix with DUAL and BLADEX: Apply 2 pts. GRAMOXONE paraquat per acre plus DUAL plus BLADEX as a broadcast ground spray before, during or after planting but before crop emergence. Apply in 20 to 60 gals. of diluted spray per acre. Add ORTHO X-77 Spreader (non-ionic) at 1 to 2 qts. per 100 gals. of diluted spray. Refer to DUAL and BLADEX labels for rates of application, directions, limitations, cautions and for a list of weeds controlled by preemergence activity of DUAL and BLADEX. Add BLADEX to spray tank first, mixing thoroughly in carrier. Then add DUAL, GRAMOXONE paraquat and ORTHO X-77 Spreader (non-ionic). Constant agitation recommended at all times.

GRAIN SORGHUM: For Desiccation and Residual Control of Annual Broadleaf Weeds and Grasses Where Sorghum Will Be Planted Directly Into Previous Crop Residues—Apply 1 to 2 pts. GRAMOXONE paraquat plus 2 to 3 lbs. of Atrazine 80 Wettable per acre as a broadcast ground spray before, during, or after planting but before crop emergence. Refer to Atrazine 80 Wettable label for directions, limitations, cautions and for a list of weeds controlled by preemergence activity of Atrazine. Apply 20 to 60 gals. diluted spray per acre. Add ORTHO X-77 Spreader (non-ionic) at 16 to 32 fl. oz. per 100 gals. diluted spray. Water or nitrogen solutions may be used as a carrier. Add Atrazine to spray tank first, mixing thoroughly in carrier. Then add GRAMOXONE paraquat and ORTHO X-77 Spreader (non-ionic). Constant agitation recommended at all times. This herbicide tank mixture will not control volunteer sorghum or shattercane which emerges after application.

Tank Mix with Atrazine and IGRAN®* 80 W: This tank mix should be used instead of GRAMOXONE paraquat + Atrazine when better preemergence control of barnyardgrass, crabgrass, and fall panicum is desired). Apply 1 to 2 pts. GRAMOXONE paraquat plus 1 to 2 lbs. Atrazine 80 W or 1.6 to 3.2 pts. of Atrazine 4L plus 2 to 2½ lbs. IGRAN 80 W per acre as a broadcast ground spray before, during, or after planting but before crop emergence. Weed control is often superior from application made approximately one week prior to planting. Use low rates of Atrazine and IGRAN 80 W on medium soils, and the higher rates on heavy soils. Refer to Atrazine and IGRAN 80 W labels for directions, limitations, cautions and for a list of weeds controlled by preemergence activity of Atrazine and IGRAN 80 W. The addition of 2,4-D (ester) at 0.5 lb. active per sprayed acre may aid in the suppression of perennial broadleaf weeds and annual broadleaf weeds that have made excessive growth before application. Refer to the 2,4-D label for directions, limitations, and cautions. Apply 20 to 40 gals. diluted spray per acre. Add ORTHO X-77 Spreader (non-ionic) at 16 to 32 fl. oz. per 100 gals. diluted spray. Use only water or nitrogen solutions as a carrier. Add IGRAN 80 W to spray tank first, mixing thoroughly in carrier. Then add Atrazine, 2,4-D (if used), GRAMOXONE paraquat, and ORTHO X-77 Spreader (non-ionic). Constant agitation recommended at all times.

SOYBEANS: Preplant or Preemergence Treatment—Apply 1 to 4 pts. GRAMOXONE paraquat per sprayed acre as a band treatment over the crop row, or as a broadcast treatment, prior to, during, or after planting—but before emergence of the crop. Use the 1.0 pt. rate when spring annual broadleaf weeds and grasses are less then 4 inches high or when a directed spray or cultivation will be used within 3 weeks after planting. **Aerial Application**—5 to 10 gals. spray mix per acre. **Ground Application**—20 to 60 gals. spray mix per acre. Add ORTHO X-77 Spreader (non-ionic) at 8 fl. oz. per 100 gals. of diluted spray. Seeding should be done with a minimum amount of soil disturbance. Weeds and grasses emerging after application will not be controlled. Crop plants emerged at time of application may be killed and yields reduced.

SOYBEANS: Minimum Tillage System. Tank Mix with LASSO and LOROX®*: Use this tank mixture for control of most emerged annual weeds, suppression of many emerged perennial weeds and preemergence control of many annual weeds when soybeans will be planted directly into a cover crop, stale seed bed, or in previous crop residues such as wheat stubble. This tank mixture will not control regrowth from perennial weeds. Do not apply by air. For weeds controlled preemergence, see the LASSO and LOROX labels. Refer to the LASSO and LOROX labels for precautionary statements, crop rotation restrictions, rates which are dependent upon soil textures, directions, limitations, and cautions.

Broadcast Treatment—Apply 1 to 2 pts. GRAMOXONE paraquat in this tank mixture in 20 to 60 gals. water per acre immediately before, during or after planting but before crop emergence. Use the 2 pt. rate if any weeds are 4 to 6 inches tall. This mixture may not control weeds taller than 6 inches. As density of stubble, crop residue or weeds increases, spray gallonage should be increased within the recommended range for complete coverage. Add ORTHO X-77 Spreader (non-ionic) at 8 fl. oz. per 100 gals. of diluted spray. First, add the LOROX to the spray tank while agitating, then add the LASSO, GRAMOXONE paraquat and ORTHO X-77 Spreader (non-ionic). Continue agitation during application.

SOYBEANS (Preemergence Treatment)—For Kill of Emerged Annual Broadleaf Weeds and Grasses and for Top Kill and Suppression of Emerged Perennials Where Soybeans Will be Planted Directly into a Preformed Bed, Cover Crop, or in Previous Crop Residues. Tank Mix with LOROX: Apply 1 pt. GRAMOXONE paraquat in this tank mixture as a band treatment over the crop row or as a broadcast treatment after planting—but before emergence of the crop. Refer to the LOROX labels for precautionary statements, rates which are dependent upon soil textures, directions, limitations and cautions. **Aerial Application**—Apply 5 to 10 gals. spray mix per acre. **Ground Application**—Use 20 to 60 gals. diluted spray. Add ORTHO X-77 Spreader (non-ionic) at 8 fl. oz. per 100 gals. of diluted spray. Seeding should be done with a minimum amount of soil disturbance. Crop plants emerged at time of application will be killed. For band applications, compute amount of Paraquat and LOROX needed per acre by dividing the band width by the inches between crop rows and then multiplying the answer by the recommended broadcast rates. Continue agitation during application. This herbicide tank mixture WILL NOT CONTROL Rhizome Johnsongrass.

Tank Mix with LASSO and SENCOR®*: For Contact Action on Emerged Weeds and Preemerged Control of Smartweed, Pigweed and Cocklebur, Foxtail and Crabgrass. (This tank mix should be used instead of Paraquat + LOROX or Paraquat + SENCOR when better preemergence control of crabgrass and foxtail is desired). Apply 1 to 2 pts. GRAMOXONE paraquat in this tank mixture per acre as a preemergence broadcast ground spray. Refer to LASSO and SENCOR labels for rates which are dependent upon soil textures, directions, limitations and cautions. Apply in 20 to 60 gals. spray mix per acre. Use the 2 pt. rate of Paraquat if weeds are 4 to 6 inches tall. This mixture will not control weeds taller than six inches. Add ORTHO X-77 Spreader (non-ionic) at 8 fl. oz. per 100 gals. of diluted spray. Add SENCOR to the spray tank first and mix thoroughly. Then add LASSO followed by GRAMOXONE paraquat and ORTHO X-77 Spreader (non-ionic). Constant agitation is recommended at all times. Do not use treated vines for feed or forage.

Tank Mix with Prowl®* and SENCOR or LEXONE®* and/or LOROX (Except CA): Use this tank mixture for control of most emerged annual weeds, suppression of many emerged perennial weeds and preemergence control of many annual weeds when soybeans will be planted directly into a cover crop, stale seed bed, or in previous crop residues such as wheat stubble. This tank mixture will not control regrowth from perennial weeds. Apply 1 to 4 pts. GRAMOXONE paraquat per acre plus PROWL plus SENCOR or LEXONE and/or LOROX as a broadcast ground spray before, during, or no more than 2 days after planting (before crop emergence). The higher Paraquat rates will be required under dry conditions or when weeds are 3-6 inches tall. Apply in 20 to 60 gals. of diluted spray per acre. Add ORTHO X-77 Spreader (non-ionic) at 2 to 4 pts. per 100 gals. of diluted spray. Refer to PROWL, SENCOR, LEXONE and LOROX labels for rates of application, directions, limitations, cautions and for a list of weeds controlled preemergence by each herbicide. Fill spray tank one-half to three-quarters full with water. Add SENCOR or LEXONE or LOROX to spray tank first, mixing thoroughly in water. Refer to respective labels for mixing directions. Then add PROWL, GRAMOXONE paraquat and ORTHO X-77 Spreader (non-ionic). Constant agitation recommended at all times.

Tank Mix with SENCOR: For Contact Action on Emerged Weeds and Preemergence Control of Smartweed, Pigweed, and Cocklebur. (This tank mix should be used instead of Paraquat + LOROX when infestations of emerged smartweed, pigweed and cocklebur are present at application and when better preemergence control of these species is desired).

Apply 1 to 2 pts. GRAMOXONE paraquat in this tank mixture per acre as a broadcast ground spray before, during or after planting but before emergence of the crop. Use the 2 pt. rate of GRAMOXONE paraquat if weeds are 4 to 6 inches tall. This mixture will not control weeds taller than six inches. Apply in 20 to 60 gals. spray mix per acre. Add ORTHO X-77 Spreader (non-ionic) at 1 qt. per 100 gals. of diluted spray. Do not use treated vines for feed or forage. Refer to the SENCOR label(s) for cautions, limitations and recommended rates for specific soil textures. In using minimum tillage equipment, it is important to observe the proper planting depth caution stated on the SENCOR labels so application is not made directly onto the soybean seed.

Tank Mix with SURFLAN®* and SENCOR or with SURFLAN and LEXONE: For Contact Action on Emerged Weeds and Preemergence Control of Smartweed, Pigweed, Lambsquarters, Velvetleaf, Common Ragweed, Foxtail and Fall Panicum. (This tank mix should be used instead of Paraquat + LASSO + SENCOR, Paraquat + SENCOR or Paraquat + LOROX when longer residual control of foxtail and fall panicum is desired).

Apply 1 qt. GRAMOXONE paraquat in this tank mixture per acre at planting or within 2 days after planting. Refer to SURFLAN and SENCOR or LEXONE labels for rates which are dependent upon soil textures, directions, limitations and cautions. Apply in 20 to 40 gals. spray mix per acre. Add ORTHO X-77 Spreader (non-ionic) at 8 fl. oz. per 100 gals. of diluted spray. Add SENCOR or LEXONE to the spray tank first and mix thoroughly. Then add SURFLAN followed by GRAMOXONE paraquat and ORTHO X-77 Spreader (non-ionic). Constant agitation is recommended at all times. Do not use treated vines for feed or forage.

Tank Mix with DUAL and SENCOR or LEXONE or LOROX: Apply 1 to 2 pts. GRAMOXONE paraquat per acre plus DUAL plus SENCOR or LEXONE or LOROX as a broadcast ground spray before, during, or after planting but before crop emergence. Apply in 20 to 60 gals. of diluted spray per acre. Add ORTHO X-77 Spreader (non-ionic) at 8 fl. oz. per 100 gals. of diluted spray. Refer to DUAL, SENCOR, LEXONE and LOROX labels for rates of application, directions, limitations, cautions, and for a list of weeds controlled by preemergence activity of DUAL, SENCOR, LEXONE and LOROX. Add SENCOR or LEXONE or LOROX to spray tank first, mixing thoroughly in carrier. Then add DUAL, GRAMOXONE paraquat and ORTHO X-77 Spreader (non-ionic). Constant agitation recommended at all times.

WHEAT: Chemical Fallow

For Desiccation of Most Emerged Annual Broadleaf Weeds and Grasses and Suppression of Many Emerged Perennial Weeds During the Fallow Period(s) of the Wheat-Fallow, Wheat-Annual Crop-Fallow, Wheat-Annual Crop and Continuous Wheat Cropping Systems—Apply 1 to 2 pts. GRAMOXONE paraquat per acre as a broadcast spray. **Aerial Application**—Apply 5 to 10 gals. spray mix per acre. **Ground Application**—Use 20 to 60 gals. spray mix per acre. As density of stubble, crop residue or weeds increases, spray gallonage should be increased within the recommended range for complete coverage. Add ORTHO X-77 Spreader (non-ionic) at 1 to 4 pts. per 100 gals. of diluted spray. Use only clean water or nitrogen solutions as a carrier. Weeds and grasses emerging after application and regrowth from perennial weeds will not be controlled. Weeds taller than 6 inches at application may not be controlled. The addition of 2,4-D or BANVEL®* may aid in the suppression of perennial broadleaf weeds and annual broadleaf weeds that have made excessive growth before application. Refer to 2,4-D and BANVEL labels for directions, limitations and cautions. For extended weed control during the fallow period, tank mix with a registered residual herbicide or herbicide combination. Refer to residual herbicide label(s) for directions, limitations, cautions and for a list of weeds controlled by preemergence activity of the residual herbicide(s). Add wettable powder herbicide(s) to spray tank first, mixing thoroughly in carrier. Then add liquid herbicide(s), GRAMOXONE paraquat and ORTHO X-77 Spreader (non-ionic). Constant agitation recommended at all times.

PASTURE AND RANGE RESEEDING

PASTURE AND RANGE RESEEDING: Suppression of Existing Sod and Undesirable Emerged Broadleaf Weeds and Grasses to Permit Pasture and Range Reseeding—Apply recommended rates in water to make 18 to 75 gals. diluted mix per sprayed acre. Add 8 to 32 fl. oz. ORTHO X-77 Spreader (non-ionic) per 100 gals. spray. Apply broadcast or in bands over the drill row prior to, or at time of seeding desired grasses and/or forage legumes. Apply as a band treatment when recovery of the sprayed sod is desired, such as in the case with desirable fescue grass. Apply as a broadcast treatment when existing grass species is undesirable and should be totally suppressed, such as in the case with run-down bluegrass sod.

West of Cascade and Sierra Nevada Mountains—1 to 2 pts. GRAMOXONE paraquat per sprayed acre prior to, or at time of seeding grasses such as Harddinggrass and Palestine Orchardgrass. Apply in October through December after first fall rains and after weeds have emerged and sod has started new growth. For best seeding results, apply on moderate to heavily grazed areas. Do not use in areas with heavy sod and weed growth. Do not pasture treated areas until growth of newly planted seedlings is 3 to 6 inches high.

East of Rocky Mountains—1 to 2 pts. GRAMOXONE paraquat per sprayed acre during the growing season. Use the 2 pts. rate to suppress vigorous and coarse sod species such as Bromegrass. See specific directions for use in the South and Southeast for suppression of Bermudagrass or Bahiagrass sods. Apply prior to, or at time of seeding grasses or forage legumes such as alfalfa, clover and birdsfoot trefoil. Apply only to grazed or mowed pastures which are not more than 2 to 3 inches in height at time of treatment. Do not graze treated areas until the newly planted seedlings have reached their recommended grazing height. Under good seedling emergence and growing conditions, allow a minimum of one month between treatment and first grazing. Seedling grasses, and forage legumes should be at least 6″ tall; Sudangrass must be 18″ tall and Sorghum-sudan 24″ tall at first grazing. Under late fall or winter conditions, it may take 3 to 5 months for seeded grasses and/or legumes to reach recommended grazing height.

South and Southeast—Bermudagrass or Bahiagrass Sods—Late Summer or Early Fall Application for Sod Suppression when Seeding Winter Annuals: Apply 1 pt. GRAMOXONE paraquat per acre in 20 to 30 gals. water in late summer or early fall to sod that does not exceed 3 inches in height. Apply prior to, or at time of seeding winter annuals. Add 1 qt. ORTHO X-77 Spreader (non-ionic) per 100 gals. spray. Do not pasture in treated areas until 60 days after treatment or until winter annual seedlings are 9 inches tall.

BERMUDAGRASS PASTURES: For the Suppression of Emerged Annual Broadleaf and Grassy Weeds Including Little Barley, Bromes, Buttercup, and Carolina Geranium in Dormant Bermudagrass Pastures—Apply 1 pt. per acre GRAMOXONE paraquat in 20 to 30 gals. water with ground equipment or 3 to 5 gals. water per acre with fixed wing aircraft during March. For control of little barley, apply before the mid boot stage. Add 1 qt. of ORTHO X-77 Spreader (non-ionic) per 100 gals. spray mix. Do not pasture or mow for hay until 40 days after treatment.

ALFALFA, CLOVER

ALFALFA, CLOVER—Washington, Oregon, Colorado, Idaho, Montana, Utah, Nevada, Wyoming, and California (counties of Del Norte, Siskiyou, Modoc, Shasta, Lassen, Plumas, Sierra and Nevada): Apply for Desiccation of Ryegrass, Bluegrass, Cheatgrass, Dogfennel, Chickweed, and Tansymustard in Dormant Alfalfa and Clover, Except Annual Clover—Apply 2 to 3 pts. GRAMOXONE paraquat per acre (use high rate for cheatgrass) in 20 to 100 gals. spray mix by ground as a broadcast applicaton. Apply by air in 5 to 10 gals. spray mix per acre. Apply to established stands after the last fall cutting when the crop is dormant but before spring growth starts. Do not apply if regrowth following the last fall cutting is more than 2" tall. Alfalfa and clover foliage present at the time of application will be burned. Apply when the broadleaf weeds and grasses are succulent and growth is from 1" to 6" tall. Add ORTHO X-77 Spreader (non-ionic) at 8 to 32 fl. oz./100 gals. spray mix. Do not pasture animals in treated fields before first cutting. Do not graze, cut or harvest within 60 days of application for clover or 42 days of applicaton for alfalfa. Do not apply more than once per season.

ALFALFA: For Use in States East of the Rockies and North of the Southern Borders of Colorado, Kansas, Missouri, Kentucky, and Virginia—For control of Bluegrass, Chickweed, Henbit, and Downy Brome and suppression of perennial grasses including Orchard Grass, Timothy and Smooth Brome. Apply 2 to 3 pts. GRAMOXONE paraquat per acre in 20 to 60 gals. water by ground equipment as a broadcast application. Apply by air in 5 to 10 gals. spray mix per acre. Use 3 pt. rate for kill or suppression of harder to control weeds and grasses, such as the perennial species. Do not apply if fall regrowth following the last fall cutting is more than 6" tall. Apply to well-established stands (at least one year old) after the last fall cutting when the crop is dormant or before spring growth reaches one inch. Alfalfa foliage present at the time of application will be burned which may reduce the yield of the first cutting. Weeds and grasses should be succulent and growing at the time of application. Weeds that germinate after application will not be controlled. Add ORTHO X-77 Spreader (non-ionic) at 8 to 32 fl. oz./100 gals. spray mix. Do not graze, cut or harvest within 60 days of application. Do not apply more than once per season.

Tank Mix with PRINCEP 80 W Simazine Weed Killer:

ALFALFA only—For Control of Filaree, Jim Hill Mustard, and Shepherdspurse, and for improved Control of Bluegrass, Cheatgrass, Chickweed, and Tansymustard. Apply 1 to 3 pts. GRAMOXONE paraquat plus 1 to 2 lbs. of PRINCEP 80 W in 20 to 100 gals. spray mix per acre with ground equipment or 5 to 10 gals. spray mix per acre by air. Add ORTHO X-77 Spreader (non-ionic) at 8 to 32 fl. oz. per 100 gals. spray mix. Use the 1 pt. rate of GRAMOXONE paraquat when weeds and grasses are less than 4" tall. Refer to PRINCEP 80 W label for cautions, limitations, and recommended rates for specific soil types. Do not pasture animals in treated fields before first cutting. Do not graze, cut or harvest within 90 days of application. Do not apply more than once per season.

ALFALFA only—Suppression and Control of Broadleaf Weeds and Grasses in New Alfalfa Seedlings Grown for Hay (California Only).

For Control of:	Use Rate Directions—Pint/Acre For Supression	For Control
Spikeweed (4 in. tall or less)	1/2 pt.	1 to 2 pts.
Volunteer Small Grain (8 in. tall or less)	1/2 to 1 pt.	2 pts.
Fiddleneck (6 in. tall or less)	1/2 to 1 pt.	2 pts.
Shepherdspurse	1 to 2 pts.	—
Annual Bluegrass	—	1 to 2 pts.
Chickweed	—	1 to 2 pts.
Red Maids (6 in. tall or less)	—	1 to 2 pts.

Apply in 25 to 50 gals. water per acre by ground or in 5 to 10 gals. by air as a broadcast application during late winter or early spring. Add ORTHO X-77 Spreader (non-ionic) at 8 to 32 fl. oz. per 100 gals. spray mix. Do not use the 1/2 pt. rate unless the alfalfa has at least 3 trifoliate leaves; the 1 pt. rate unless the alfalfa has 6 trifoliate leaves; or rates over 1 pt. unless there are 9 trifoliate leaves. Alfalfa foliage present at time of application will be burned. Do not pasture animals in treated fields before first cutting. Do not graze, cut or harvest within 70 days of application. Do not apply more than once during the first growing season.

CAUTION: SEEDLING ALFALFA STANDS WILL BE REDUCED AND REPLANTING MAY BE NECESSARY. NOT RECOMMENDED IN SEEDLING ALFALFA GROWN FOR SEED.

HARVEST—AID USES

COTTON HARVEST AID

TO AID IN OPENING OF MATURE BOLLS AND TO DESICCATE GREEN LEAVES: 1/2 pt. of GRAMOXONE paraquat in combination with 1 pt. of phosphate or 1 gal. of chlorate defoliant per acre. Apply when 80% or more of the bolls are open and the remaining bolls to be harvested are mature. Development of immature bolls will be inhibited.

TO AID IN DEFOLIATION AND OPENING OF MATURE BOLLS OF WESTERN COTTON (ARIZONA, CALIFORNIA): First Application—1/3 to 1/2 pt. GRAMOXONE paraquat plus 1 to 2 pts. phosphate defoliant or 1 to 2 gals. chlorate defoliant per acre. Use higher dosages on rank cotton. (Caution—Do not use more than 1/2 pt. GRAMOXONE in initial application, as excessive desiccation may occur.) Apply when 60% or more of the bolls are open and remaining bolls to be harvested are mature. Development of immature bolls will be inhibited. Do not use more than 4.0 lbs. actual sodium chlorate per acre or more than 1.5 lbs. actual phosphate type defoliants per acre.

Second Application—To Aid in Opening of Mature Bolls at Second or Later Harvest Aid Treatment and to Desiccate Green Leaves—1/2 to 1 pt. GRAMOXONE paraquat per acre (in desert cotton areas or on rank vigorous cotton, use the 1 pt. rate only) or 1/2 to 1 pt. GRAMOXONE paraquat plus 1 1/2 to 2 pts. phosphate defoliant or 1 1/2 to 2 gals. chlorate defoliant. Apply when 75% or more of the bolls are open and the remaining bolls to be harvested are mature. Development of immature bolls will be inhibited.

TO DESSICATE STRIPPER COTTON: 1 to 2 pts. GRAMOXONE paraquat per acre. When foliage is dense, use 2 applications of 1 pt. if necessary. Apply when 85% or more of the bolls are open and the remaining bolls to be harvested are mature. Development of immature bolls will be inhibited.

AERIAL APPLICATION—Apply 3 to 10 gals. spray mix per acre. Use ORTHO X-77 Spreader (non-ionic) at 8 to 32 fl. oz. per 100 gals. spray mix. Do not apply during periods of thermal inversion to avoid drift.

GROUND APPLICATION—Apply 10 to 30 gals. spray mix per acre. Arrange nozzles to provide thorough coverage of the foliage. Use ORTHO X-77 Spreader (non-ionic) at 8 to 32 fl. oz. per 100 gals. spray mix.

May apply as split application. Do not pasture lactating dairy animals. Do not pasture livestock in treated fields within 15 days after treatment. Remove livestock from treated area 30 days before slaughter. Do not feed gin trash to livestock. Do not apply cotton within 3 days before harvest. Repeat application if necessary. Do not make more than 2 applications or exceed a total of 2 pts. per acre. When combined with phosphate defoliants, observe livestock cautions listed on the phosphate label. When combined with chlorate defoliants, do not pasture livestock on treated areas or feed treated foliage or gin trash. Do not apply within 7 days when used with phosphate or chlorate defoliant.

GUAR—For Use as a Preharvest Crop Desiccant—Apply 1 qt. GRAMOXONE paraquat in 20 to 30 gals. of water per acre. Add ORTHO X-77 Spreader (non-ionic) at 1 to 2 qts. per 100 gals. water. Apply after pods are fully matured. Do not harvest until at least 4 days after application. Do not graze treated areas or feed treated forage to livestock.

POTATO VINES: PREHARVEST VINE KILLING—1 to 2 pts. GRAMOXONE paraquate per acre. Use the 2 pt. rate where quick vine kill is desired or 2 applications of 1 pt. when vine growth is dense. Apply in 50 to 100 gals. water per acre with thorough coverage. Add 8 to 32 fl. oz. of ORTHO X-77 Spreader (non-ionic) per 100 gals. water. Begin application when leaves are beginning to turn yellow. Immature potato foliage and drought stressed potato foliage are tolerant to GRAMOXONE paraquat and desiccation will not be complete. Do not make more than 2 applications with a minimum of 5 days between applications. Do not apply GRAMOXONE paraquat to potatoes within 3 days before harvest. Do not pasture livestock in treated potato fields. To avoid injury to subsequent crops, do not use on muck or peat soils. Do not use GRAMOXONE paraquat for desiccating potato vines when the potatoes are to be stored or used for seed as this use may result in the tuber decomposition and failure of seed pieces to germinate and grow normally.

SOYBEAN HARVEST AID—Apply ½ to 1 pt. GRAMOXONE paraquat per acre on broadleaf weeds and grasses. Use high rate on cocklebur. Add 1 qt. of ORTHO X-77 Spreader (non-ionic) per 100 gals. spray. **Aerial Application**—2 to 5 gals. spray per acre. **Ground Application**—20 to 40 gals. spray per acre. **Indeterminate Varieties:** Apply when at least 65% of the seed pods have reached a mature brown color or when seed moisture is 30% or less. **Determinate Varieties:** Apply when soybean plants are mature, i.e., beans are fully developed, at least ½ of leaves have dropped, and leaves left on plants are turning yellow. Immature soybeans will be injured. Mature cockleburs, especially drought-stressed plants, are tolerant to GRAMOXONE paraquat and desiccation will not be complete. Do not pasture livestock within 15 days of treatment. Remove livestock from treated fields at least 30 days before slaughter.

SUNFLOWER (OILSEED VARIETIES ONLY)—For Desiccation of Sunflower Plants and Broadleaf Weeds and Grasses to Facilitate Harvesting: Apply 1 to 2 pts. GRAMOXONE paraquat per acre in 5 gals. spray mix by air or in 20 to 40 gals. by ground application equipment. Use the higher rate when crop stands or weed infestations are heavy. Apply when the sunflower seeds reach physiological maturity (when seed moisture is 35 percent or lower). For many varieties, this corresponds to the time when the back of the heads are yellow and the bracts are turning brown. Harvest 7 to 21 days after application. Add ORTHO X-77 Spreader (non-ionic) at 1 qt. per 100 gals. water. Do not graze treated areas or feed treated forage to livestock.

SUGARCANE—Preharvest Crop Desiccation (Florida and Texas Only)—Apply ½ to 1 pt. GRAMOXONE paraquat per acre. Apply to mature plant or stubble cane using 4 to 10 gals. water per acre. Use higher rate under cool, cloudy weather conditions. ORTHO X-77 Spreader (non-ionic) should be added to all sprays at the rate of 1 pt. per 100 gals. water. Apply with aircraft sprayer 3 to 14 days before burning and harvest. For rapid results, apply when weather is clear and warm. Apply as a coarse spray at low pressure when air is calm to prevent spray drift.

DIRECTED SPRAYS

ORCHARDS, VINEYARDS, WINDBREAK, SHADE AND ORNAMENTAL TREES: For Kill of Annual Broadleaf Weeds and Grasses and For Top Kill and Suppression of Perennials and Green Suckers under Filbert Trees, in the Interspaces and Around the Base of Trees and Vines—Acerola, Almonds, Apples, Apricots, Avocados, Bananas, Cherries, Citrus, Coffee, Figs, Filberts, Grapes, Kiwifruit, Macadamia Nuts, Nectarines, Olives, Papayas, Peaches, Pears, Pecans, Pistachios, Plums, Prunes, Walnuts and other Trees such as Arborvitae, Ash, Elm, Fir, Oak and Pine—Apply as a directed spray at 1 to 2 qts. GRAMOXONE paraquat per sprayed acre in 30 to 150 gals. water (use 30 to 50 gals. water for pecans) to thoroughly cover the weeds and grasses. Add ORTHO X-77 Spreader (non-ionic) at 8 to 32 fl. oz. per 100 gals. Apply when the weeds and grasses are succulent and the new growth is from 1 to 6 inches high. For mature woody weeds, green suckers, late germinating weeds and grasses and for perennials—retreatment or spot treatment may be necessary. **CAUTION:** Do not allow spray to contact green stems, fruit or foliage as injury may result. Do not spray under windy conditions and use a shield for young trees or vines. Do not allow animals to graze on treated areas. Do not apply when nuts to be harvested are on the ground. For Kiwifruit, do not treat more than 3 times per year and do not apply within 14 days of harvest. For Pistachios, do not harvest within 7 days of treatment and do not exceed two applications after shells split.

ALMONDS, GRAPES, AND WALNUTS (California Only); ORANGES (California and Florida Only); SOUR CHERRIES (Michigan Only); APPLES, PEACHES, PEARS: Tank Mix with PRINCEP 80 W Simazine Weed Killer—For Top Kill and Residual Control of Annual Broadleaf Weeds and Grasses and for Suppression of Perennial Weeds—For broadcast application apply 1 to 2 qts. GRAMOXONE paraquat per acre and PRINCEP 80 W at the following rates per acre:

Crop	Rate PRINCEP 80 W Per Sprayed Acre	Precautions
Almonds and Peaches (California Only)	1¼-2½ lbs. in 2-4 ft. band on each side of tree row. 1¼ lbs. provides residual control of burclover, common chickweed, wild mustard and sheperdspurse.	Do not treat trees established less than 3 years. Do not treat Mission variety of almonds. Do not apply to almond trees propagated on plum root stalks. Do not replant almonds or peaches in treated soil. Do not apply on gravelly sand or loamy sand soil or on soil with less than 1% organic matter. Do not treat areas where water will accumulate.
Apples Pears Sour Cherries (Michigan Only) Walnuts (California Only)	2½-5 lbs.	Do not apply to sandy soil. Leveling and furrowing operations after application will lessen effectiveness of weed control.
Peaches (Arkansas, Louisiana, Missouri, Oklahoma, Texas and states East of the Mississippi River)	2-5 lbs.	Do not apply to sandy or gravelly soil.

Grapes (California Only)	$2\frac{1}{2}$-6 lbs.	Do not use in vineyards established less than 3 years.
Oranges (California Only) Except Imperial, Coachella and Palo Verde Valleys (Florida Only)—Bedded	$2\frac{1}{2}$-5 lbs. 4 lbs.	Do not use in nurseries. To avoid possible injury, do not apply to trees under stress from freeze damage for one year following the freeze.

Use the low rate of PRINCEP 80 W on coarse textured soils and low organic matter soils.

Use the high rate on fine textured soils and high organic matter soils. Apply as a tank mix in 50 to 200 gals. of water per acre to the orchard floor avoiding contact with fruit, foliage or stems. Add ORTHO X-77 Spreader (non-ionic) at 8 to 32 fl. oz. per 100 gals. diluted spray. Apply when weeds and grasses are succulent and the new growth is from 1 to 6 inches tall. For band applications or spot applications around trees, reduce the broadcast rate and the amount of water in proportion to the area actually sprayed.

Make only one application per year. Apply in orchards where trees have been established one year or more. Do not spray under windy conditions and use a shield for young trees. Do not apply to almonds and walnuts when nuts are on the ground during the harvest period. Do not allow animals to graze treated areas.

Refer to the PRINCEP 80 W Simazine label for general information, caution and warning statements.

HOPS (Washington, Oregon, and Idaho Only): For Kill of Annual Bluegrass, Ryegrass, Barnyardgrass, Pigweed and Chickweed— Apply as a directed spray in the interspaces and around the base of hop plantings using 1 qt. GRAMOXONE paraquat in 20 to 100 gals. water per acre to thoroughly cover weeds and grasses. Add ORTHO X-77 Spreader (non-ionic) at 8 to 32 fl. oz. per 100 gals. spray mix. Do not apply around hop vines less than 10 ft. tall. Apply when weeds and grasses are succulent and growth is from 1 to 6 inches tall. Retreatment or spot treatment may be necessary. **For Suckering and Stripping (Cascade, Yakima Cluster and Bullion Varieties Only):** Spray only basal 2 ft. of vines using 1 qt. in 20 to 100 gals. water per acre. Add ORTHO X-77 Spreader (non-ionic) at 8 to 32 fl. oz. per 100 gals. spray mix. Do not apply to hop vines less than 10 ft. tall. Repeat as necessary. **CAUTION:** Do not apply more than 3 times per season. Do not apply within 14 days of harvest. Do not allow spray to contact green stems, foliage, flowers, or cones, as injury may result. Do not allow animals to graze in treated hopyards. Hop vine refuse and silage may be fed to livestock.

GUAVA—Apply as directed spray at 2 qts. GRAMOXONE paraquat per sprayed acre in 50 to 200 gals. to thoroughly cover the weeds and grasses. Add ORTHO X-77 Spreader (non-ionic) at 8 to 32 fl. oz. per 100 gals. Apply when the weeds and grasses are succulent and the new growth is from 1 to 6 inches high. For mature woody weeds, late germinating weeds and grasses and for perennials—retreatment or spot treatment may be necessary.

CAUTION: Do not allow spray to contact green stems, fruit or foliage as injury may result. Do not spray under windy conditions and use a shield for young trees. Do not allow animals to graze on treated areas.

PASSION FRUIT—Apply as directed spray at 2 qts. GRAMOXONE paraquat per sprayed acre in 50 to 200 gals. to thoroughly cover the weeds and grasses. Add ORTHO X-77 Spreader (non-ionic) at 8 to 32 fl. oz. per 100 gals. Apply when weeds and grasses are succulent and the new growth is from 1 to 6 inches high. For mature woody weeds, late germinating weeds and grasses and for perennials—retreatment or spot treatment may be necessary.

CAUTION: Do not allow spray to contact green stems, fruit or foliage as injury may result. Do not spray under windy conditions and use a shield for young vines. If bark is still green, wrap vine prior to application to prevent injury. If application is to be made during harvesting season, apply GRAMOXONE paraquat only after picking fruits off the ground. Do not allow animals to graze on treated areas.

PINEAPPLES: For Kill of Annual Broadleaf Weeds and Grasses and Top Kill and Suppression of Perennials—Apply as a directed spray using 1 to 2 qts. GRAMOXONE paraquat in 100 to 200 gals. water per sprayed acre to thoroughly cover the weeds and grasses. Add ORTHO X-77 Spreader (non-ionic) at 8 to 32 fl. oz. per 100 gals. Apply when weeds and grasses are succulent and new growth is from 1 to 6 inches high. Retreatment may be necessary on more mature weeds. Do not apply within 20 days of harvest.

SMALL FRUITS (Blackberries, Boysenberries, Raspberries, Blueberries)—For Kill of Annual Broadleaf Weeds and Grasses and for Top Kill and Suppression of Perennial Weeds in the Interspaces and Around the Base of Bushes—Apply as a coarse directed spray at 1 to 2 qts. GRAMOXONE paraquat per sprayed acre in 50 to 200 gals. water to thoroughly wet the weeds and grasses. Add ORTHO X-77 Spreader (non-ionic) at 8 to 32 fl. oz. per 100 gals. water. Apply before emergence of new canes or shoots as injury to those canes or shoots can result. Apply as a coarse spray to avoid drift injury from fine spray mist.

STRAWBERRIES: Postemergence Directed Spray—For Control of Emerged Annual Broadleaf Weeds and Grasses and for Top Kill and Suppression of Emerged Perennial Weeds Between the Rows After Crop Plant Emergence or Establishment Apply 1 qt. GRAMOXONE paraquat per sprayed acre in 20 to 60 gals. spray mix. Apply with conventional ground equipment directing the spray between the rows and using shields to prevent spray contact with crop plants. Add ORTHO X-77 Spreader (non-ionic) at 8 fl. oz. per 100 gals. spray mix. For best results, apply when weeds and grasses are succulent and weed growth is 1 to 6 inches high. Weeds and grasses emerging after application will not be controlled. Do not allow spray to contact strawberry plants as injury or excessive residues may result. Do not apply more than 3 times per crop season. Do not apply within 21 days of harvest. Do not graze livestock in treated areas.

GRAIN SORGHUM: Postemergence Directed Spray—For Kill of Emerged Crabgrass and Pigweed 3 Inches or Less in Height 1 to 2 pts. GRAMOXONE paraquat in 20 to 40 gals. water per sprayed acre. Use higher gallonage of water for larger weeds. For band applications, 1 pt. GRAMOXONE paraquat will cover 2 land or crop acres when a 20 inch band is treated in 40 inch rows. Treat when grain sorghum is naturally standing 12 inches tall or more. Add ORTHO X-77 Spreader (non-ionic) at 1 qt. per 100 gals. of spray. Do not exceed 30 psi; high pressure contributes to spray drift. For best results, apply GRAMOXONE paraquat at 1 pt. per sprayed acre to grass and weeds less than 2 inches tall. Grass and weeds 2 to 3 inches tall should be sprayed with 2 pts. of GRAMOXONE paraquat per sprayed acre. Use precision application equipment with spray nozzles mounted on skid shoes, oiling shoes, fenders, or cultivators with gauge wheels and adjusted so that only the lower 3 inches or less of the sorghum stalk is sprayed. For best coverage on single rows, use 4 flat fan nozzles per row, 2 nozzles on each side of the row. The 2 forward nozzles should point forward and downward, and rear nozzles should point to the rear and downward. For double row sorghum use one skid between the two rows with one nozzle spraying backward and downward. Some sheath and lower leaf spotting and desiccation will occur. Basal spotting does not penetrate beneath outer sheath. Degree of injury is related to precision of application and wind velocity. Do not spray under windy conditions.

FIELD CORN: Postemergence Directed Spray Using GRAMOXONE paraquat Alone or GRAMOXONE paraquat + Atrazine Tank Mix—For Control of Emerged Annual Broadleaf Weeds and Annual Grasses Less than 4 Inches Tall—Apply 1 pt. GRAMOXONE paraquat per acre in 20 to 40 gals. of spray mix. Apply when corn plants are at least 10 inches tall. Corn plants shorter than 10 inches may be injured and not recover. Corn height measured from soil surface to top of whorl. Apply with directed spray application equipment. Arrange nozzles to spray no higher than the lower 3 inches of corn stalks to provide for maximum contact of weeds, with minimum contact of corn plants to prevent excessive sheath and lower leaf spotting and desiccation. Add 1 qt. of ORTHOX-77 Spreader (non-ionic) per 100 gals. of spray. Do not mix with liquid fertilizer. **For Improved Control of Weeds, Especially Broadleaves, Tank Mix GRAMOXONE paraquat and AATREX:** Add 5/8 to 1 1/4 lbs. of AATREX 80 W or 1 to 2 pts. of AATREX 4L in combination with 1 pt. GRAMOXONE paraquat per acre. Add AATREX 80 W or AATREX 4L to spray tank first, mixing thoroughly in water. Then add GRAMOXONE paraquat and surfactant. Constant agitation recommended at all times. Refer to AATREX labels for limitation and cautions.

Do not graze treated areas or feed treated forage to livestock.

CORN: For use in the USDA Witchweed eradication program to kill grassy weeds. Apply 1 qt. GRAMOXONE paraquat in 10 to 20 gals. spray per acre. Add ORTHO X-77 Spreader (non-ionic) at 8 fl. oz. per 100 gals. spray mix. Apply as a spray directed onto the grassy weeds at the base of the corn stalks in the hills or row. Initiate sprays in early July and repeat in early August if regrowth of grassy weeds occurs. Do not graze livestock in treated area.

SOYBEANS: Postemergence Directed Spray—For Control of Seedling Johnsongrass, Crabgrass, Goosegrass, Brachiaria, Echinochloa spp. (Texas Millet), and Pigweed less than 2 inches tall apply GRAMOXONE paraquat at 4 fl. oz. (.063 lb. active) per acre. Use proportionately less for band treatment . . . for example on 40 inch rows and 16 inch band apply 1 gal. GRAMOXONE paraquat to 80 acres. For 2 to 4 inch grass and 2 to 3 inch pigweed apply GRAMOXONE paraquat at 8 fl. oz. (.125 lb. active) per acre . . . for example on 40 inch rows and 16 inch band apply 1 gal. GRAMOXONE paraquat to 40 acres. Soybeans should be 8 inches tall at first application. If needed, make second application 7 to 14 days later. Do not treat more than twice.

Apply 20 gals. spray mix per treated acre. Use proportionately less for band treatments . . . for example 8 gals. spray mix per 16 inch band on 40 inch rows.

For best coverage, use 4 flat fan nozzles per row, 2 nozzles on each side of row. The two forward nozzles should point forward and downward, and rear nozzles should point to the rear and downward. Do not exceed 30 psi; high pressure contributes to drift. Adjust nozzles to spray no higher than the lower three inches of the soybean plant. Do not treat if soybeans are below 8 inches tall; soybeans treated topically will be killed. Add ORTHO X-77 Spreader (non-ionic) at 1 to 2 qts. per 100 gals. of spray. Do not graze treated areas or feed treated forage to livestock.

SUGARCANE—Directed Spray: For Kill of Emerged Annual Broadleaf Weeds and Grasses and for Top Kill of Perennials—Apply as a shielded or directed spray when weeds are 2 to 6 inches high. Avoid contact with cane foliage to prevent serious leaf burn and reduced yields. Make a second and final application, if necessary, when new weed growth is 2 to 6 inches high. Add ORTHO X-77 Spreader (non-ionic) at 8 fl. oz. per 100 gals. spray mix.

Florida Only—Apply 1 qt. GRAMOXONE paraquat in 50 to 100 gals. of water per acre. For optimum results, apply early in the season (March-April when weeds are small. Do not apply after June 1st as cane growth may be stunted and yields reduced.

Hawaii Only—Apply 1 qt. GRAMOXONE paraquat in 20 to 200 gals. of water per acre. Do not apply after cane rows have closed in.

Louisiana Only—Apply 1/2 to 1 qt. GRAMOXONE paraquat in 20 to 200 gals. of water per acre. For tiller control, apply when tillers are less than 18" high. Use the higher rate of GRAMOXONE paraquat with heavier weed infestations or tiller growth. Do not apply within 30 days of harvest.

TOMATOES AND PEPPERS (Middle Atlantic, Southeast, South Central Regions, and Southwest to the Western Boundary of Texas): Postemergence Directed Spray—For Kill of Emerged Annual Broadleaf Weeds and Grasses and for Top Kill and Suppression of Emerged Perennial Weeds between Plastic Mulch Covered Rows after Crop Plant Emergence or Establishment—Apply as directed spray using 1 qt. GRAMOXONE paraquat per sprayed acre in 20 to 100 gals. spray mix. Apply with conventional ground equipment directing spray between the rows and using shields to prevent spray contact with crop plants. Add ORTHO X-77 Spreader (non-ionic) at 8 fl. oz. per 100 gals. spray mix. Apply when weeds and grasses are succulent and weed growth is 1 to 6 inches high. Weeds and grasses emerging after application will not be controlled. Do not allow spray to contact tomato or pepper plants as injury or excessive residues may result. Do not apply more than 3 applications per crop season. Do not allow animals to graze on treated areas.

DORMANT USE

PEPPERMINT, SPEARMINT: For Suppression of Weeds Such as Italian Ryegrass, Prickly Lettuce, Groundsel, Chickweed, Downy Brome and Bluegrasses During the Dormant Season—Apply 1 1/2 to 3 pts. per acre as a broadcast spray in 25 to 65 gals. spray mix. Use the higher rate with heavier weed infestations. Add ORTHO X-77 Spreader (non-ionic) at 8 fl. oz. per 100 gals. spray mix. Apply when peppermint or spearmint is dormant and before spring growth begins. Do not apply to weeds which are greater than 6" tall. Do not apply more than 3 pts. per acre per dormant season as injury or excessive residues may result. **For Improved Contact Activity and Residual Control of Italian Ryegrass, Prickly Lettuce and Groundsel**—Add 1 to 2 lbs. of SINBAR®* (Terbacil) Weed Killer as a tank mix. Use the higher tank mix rate with heavier weed infestations. Apply tank mix treatment no more than once per season. Use a minimum of 25 gals. spray mix per acre. Refer to the SINBAR label for specific recommendations and cautions.

RHUBARB: For Kill of Emerged Annual Broadleaved Weeds and Grasses and for Top Kill and Suppression of Emerged Perennial Weeds in Dormant Rhubarb—Apply 1 to 2 qts GRAMOXONE paraquat per sprayed acre by ground application equipment. Use 50 to 150 gals. spray mix per acre. Add ORTHO X-77 Spreader (non-ionic) at 8 to 32 fl. oz. per 100 gals. spray mix. Use the higher rate with heavier weed infestations. Apply during the dormant season before buds in crown begin to grow. Do not make more than two applications per season.

NONCROP USES

NONCROP AREAS (Public Airports, Electric Transformer Stations and Substations, Pipeline Pumping Stations, Around Commercial Buildings, Storage Yards and Other Installations, Fence Lines, and Similar Noncrop Areas)—For the Control of Annual Broadleaved Weeds and Grasses (such as Burclover, Chickweed, Filaree, Groundsel, Nettle, Pigweed, Plantain, Puncturevine, Purslane, Red Clover, Shepherdspurse, Thistle, Wild Mustard, Wild Radish, Wild Oats, Bluegrass, Cheatgrass, Crabgrass) and Top Kill and Suppression of Perennial Weeds (such as Bermudagrass, Johnsongrass, and Morningglory—1 qt. GRAMOXONE paraquat per acre. Apply for full coverage and thorough weed contact (50 to 100 gals. diluted spray per acre).

Best results are obtained when applications are made to young, succulent weeds and grasses. Mature, woody weeds are less susceptible. Repeat as needed. Add ORTHO X-77 Spreader (non-ionic) at 8 oz. per 100 gals. GRAMOXONE paraquat is compatible with many residual herbicides. Avoid spray contact with foliage or fruit of food crops and ornamentals.

RESIN SOAKING

RESIN SOAKING IN PINES (Loblolly, Shortleaf, Longleaf, Slash, Virginia, Pond, Pitch, and Spruce Pines):

Tree Selection—Select trees to be treated from stands on sites not subject to periods of extreme drought stress as the desiccating effect of GRAMOXONE paraquat to pines is accentuated during such periods, causing a reduction in the amount of oleoresin deposited in the xylem. Select trees to be treated from vigorous, non-stagnated stands, either natural or planted. In stagnated stands or commercial timber stands, plan treating with GRAMOXONE paraquat not sooner than three years after a commercial thinning.

Application Directions: Apply GRAMOXONE paraquat diluted in water (0.2% to 5% cation, wt./wt. basis, depending on the type of treatment method) to a suitable wound in the tree trunk to bring the treatment into contact with the xylem (sapwood). One of the following methods may be used:

Bark Streaks or Cuts: This type of wound is made using a standard or rotary bark hack or a chainsaw chipping tool employed in naval stores work to remove a single one-inch wide streak of bark about 1 to 2 feet from ground level. The total length should not exceed 1/3 of the tree circumference. Multiple streaks or cuts can result in serious girdling of the trunk and premature death of the tree. A coarse spray (about 2 to 6 ml.) of GRAMOXONE paraquat solution (1% to 5% cation, wt./wt. basis) should be applied to runoff to the exposed xylem, using a plastic acid bottle or a low-pressure sprayer. Wear rubber gloves when using the acid bottle. The amount of spray required per cut depends on tree circumference and the length of cut or streak (1/3 of circumference). For a 9-inch diameter tree, 3 ml. of spray will cover the 1 inch wide streak. Using 3 ml. of a 2% or 4% GRAMOXONE paraquat solution will result in application of 60 or 120 mg. GRAMOXONE paraquat per streak.

Tree Injection: Pressure injections of 1% to 4% GRAMOXONE paraquat solution (cation, wt./wt. basis) should be applied beneath the bark and just beneath the cambium layer. Injections should be made 1 to 2 inches apart around 1/3 of the tree circumference or no closer than 3 inches apart all around the tree. From 0.2 ml. to 0.4 ml. of GRAMOXONE paraquat solution is applied per injection. Using 2% to 4% GRAMOXONE paraquat solution, this results in 36 to 144 mg. GRAMOXONE paraquat applied per 9-inch diameter tree, for holes spaced 1 inch apart around 1/3 of the tree or 3 inches apart all around the tree.

Bore Holes: 3/8 to 5/8 inch diameter holes 2 to 6 inches deep, depending on tree diameter, are bored sloping slightly downward and tangentially inward rather than directly towards the center of the tree. Solutions of 0.3% to 5% GRAMOXONE paraquat can be used. Lower concentrations (0.3%) are usually applied at higher volumes of 15 to 35 ml. per tree, resulting in application of 45 to approximately 100 mg. GRAMOXONE paraquat per tree. Using 2% to 4% solutions and applying 5 ml. per single hole per tree results in application of 100 or 200 mg. GRAMOXONE paraquat per tree. Bore hole applications in the range of 45 to 100 mg. GRAMOXONE paraquat per tree have resulted in effective oleoresin induction.

Time of Treatment: Resin soaking can occur from treatments made any time of the year; however, cool season treatments under non-drouth conditions usually result in less severe pine beetle infestations and longer tree life.

Interval Between Treatment and Tree Harvest: The interval between application of GRAMOXONE paraquat and tree harvest should be a minimum of 6 months and preferably from 12 to 24 months. Intervals of over 6 months may not be possible under conditions of drouth or serious pine beetle attacks, which may make early harvest necessary. The GRAMOXONE paraquat treatment may encourage beetle attack, or may cause premature death of the tree. Desiccation of the xylem tissue, rather than the desired resin soaking, may occur, and is more likely at higher dosage rates.

Effect on Stem Growth:

GRAMOXONE paraquat treatment can result in reduced stem growth during the interval between treatment and tree harvest.

Dilution Table for GRAMOXONE paraquat (2lbs. cation per gallon)

Concentration of Cation Desires (Wt./Wt. Basis)	To 1 Gal. of GRAMOXONE paraquat, add the following No. Gals. of Water:
0.2%	118.8
0.5%	46.8
1.0%	22.9
2.0%	10.9
3.0%	6.9
4.0%	4.9
5.0%	3.7

WORKER SAFETY RULES/REENTRY STATEMENTS

USE STRICTLY IN ACCORDANCE WITH DANGER STATEMENTS AND DIRECTIONS, AND WITH APPLICABLE STATE AND FEDERAL REGULATIONS.

DO NOT get on skin, eyes or clothing.
DO NOT inhale spray mist.
WASH SPLASHES from skin and eyes immediately.
REMOVE and wash contaminated clothing.
WASH before eating, smoking and drinking.
WEAR full face shield, rubber gloves and apron and waterproof footwear when handling or mixing concentrate.
WEAR waterproof footwear and clothing when spraying or when contacting vegetation wet with spray.
DO NOT enter treated areas without protective clothing until sprays have dried.
AVOID working in spray mist. If there is risk of exposure wear goggles and approved face mask capable of filtering spray droplets.

KEEP all unprotected persons out of operating areas or vicinity where there may be danger of drift.

Certain states may require more restrictive reentry intervals; consult your State Department of Agriculture for further information.

Written or oral warnings regarding use of protective clothing and accidental exposure must be given to workers who are expected to be in treated areas or in areas about to be treated.

IMPORTANT: The effect of swallowing Paraquat is far more severe than that which usually results from skin contact or inhalation of spray mist. No opportunity for mist or product to cause serious injury or death should arise when used in strict compliance with these rules but the opportunity may arise in case of gross violations so strictly follow all these rules **as if your life depends on it.**

ENVIRONMENTAL SAFETY CAUTIONS

WILDLIFE

This product is toxic to wildlife. Birds and other wildlife in treated areas may be killed. Do not apply directly to water.

DRIFT

GRAMOXONE paraquat is a contact herbicide that kills all green plant tissue. Do not apply under conditions involving possible drift to food, forage, or other plantings that might be damaged or the crops thereof rendered unfit for sale, use, or consumption. Do not apply when weather conditions favor drift from areas treated. For aerial applicators use, do not apply during periods of thermal inversion to avoid drift.

CHEMICAL HAZARD

This product reacts with aluminum to produce hydrogen gas which may form a highly combustible gas mixture. Such gas could flash or explode, causing serious personal injury, if ignited by open flame, spark, welder's torch, lighted cigarette, or other ignition source. Do not mix or store in containers, spray tanks, nurse tanks, or such systems made of aluminum or having aluminum fittings. After dilution with water to spray concentrations, spray systems made of, or containing aluminum parts, may be used but should not be used for extended (overnight) storage.

EQUIPMENT/CONTAINERS

Flush all spray equipment with water after use each day. GRAMOXONE paraquat is corrosive to aluminum. Aluminum spray equipment and aluminum aircraft structures that are exposed to spray solution or spray drift should be flushed thoroughly with water immediately after use.

STORAGE AND DISPOSAL

PROHIBITIONS

Do not contaminate water, food or feed by storage, disposal or cleaning of equipment. Open dumping is prohibited.

STORAGE

Store at temperatures above 32° F. Do not contaminate feed, foodstuffs or drinking water. Do not store next to feed or food, or transport in or on vehicles containing foodstuffs or feeds. For help with any spill, leak, fire or exposure involving this material, call day or night (415) 233-3737.

PESTICIDE DISPOSAL

This product is acutely hazardous. Improper disposal of excess pesticide, spray mixture, or rinsate is a violation of Federal Law. If these wastes cannot be disposed of by use according to label instructions, contact your State Pesticide or Environmental Control Agency, or the Hazardous Waste representative at the nearest EPA Regional Office for guidance.

CONTAINER DISPOSAL

See Container Label For Container Disposal Directions.

CONDITIONS OF SALE

1. ICI Americas Inc. (ICIA) warrants that this material conforms to the chemical description on the label and is reasonably fit for use as directed hereon. ICIA neither makes, nor authorizes any agent or representative to make, any other warranty of FITNESS or MERCHANTABILITY, guarantee or representation, express or implied, concerning this material.

2. Critical and unforeseeable factors beyond ICIA's control prevent it from eliminating all risks in connection with the use of chemicals. Such risks include, but are not limited to, damage to plants and crops to which the material is applied, lack of complete control, and damage caused by drift to other plants or crops. Such risks occur even though the product is reasonably fit for the uses stated hereon and even though label directions are followed. Buyer and user acknowledge and assume all risks and liability (except those assumed by ICIA under 1 above) resulting from handling, storage and use of this material.

Throughout this supplemental label, a number of chemicals manufactured by companies other than ICI Americas Inc. are mentioned, they are indicated by an asterisk and include:

AATREX®—Reg. TM of CIBA-GEIGY Corp.;

BANVEL®—Reg. TM of Velsicol Chemical Corp.;

BLADEX®—Reg. TM of Shell Chemical Company;

DUAL®—Reg. TM of CIBA-GEIGY Corp.;

IGRAN®—Reg. TM of CIBA-GEIGY Corp.;

LASSO®—Reg. TM of Monsanto Company;

LEXONE®—Reg. TM of E. I. du Pont de Nemours & Co., Inc.;

LOROX®—Reg. TM of E. I. du Pont de Nemours & Co., Inc.;

PRINCEP®—Reg. TM of CIBA-GEIGY Corp.;

PROWL®—Reg. TM of American Cyanamid Company;

SENCOR®—Reg. TM of the Parent Company of Farbenfabriken Bayer GmbH, Leverkusen;

SINBAR®—Reg. TM of E. I. du Pont de Nemours & Co., Inc.;

SURFLAN®—Reg. TM of Elanco Products Company;

X-77®—Reg. TM of Chevron Chemical Company.

Please consult each appropriate label for specific use recommendations.

TALON®-G

RODENTICIDE PELLETS

For Effective Control of Commensal Mice and Rats

Kills warfarin-resistant Norway Rats and house mice. Rodents may consume a lethal dose in one feeding with first dead rodents appearing four or five days after treatment begins.

ACTIVE INGREDIENT:	
Brodifacoum-3-[3-(4′-bromo-[1,1′-biphenyl]-4-yl)-1,2,3,4-tetrahydro-1-naphthalenyl]-4-hydroxy-2H-1-benzopyran-2-one	0.005%
INERT INGREDIENTS	99.995%
TOTAL	100.000%

EPA Reg. No. 10182-38

Keep Out of Reach of Children

CAUTION

May Be Harmful or Fatal if Swallowed. Read Additional Precautionary Statements on Side Panel.

PRECAUTIONARY STATEMENTS

HAZARDS TO HUMANS AND DOMESTIC ANIMALS

CAUTION

MAY BE HARMFUL OR FATAL IF SWALLOWED. KEEP AWAY FROM HUMANS, DOMESTIC ANIMALS, AND PETS. WASH HANDS AFTER HANDLING BAIT.

IF BAIT IS EATEN BY HUMANS, CALL A PHYSICIAN AT ONCE. For 24-hour emergency assistance, call ICI Americas Inc. 302/575-3000.

IF BAIT IS EATEN BY ANIMALS OR PETS, call the National Animal Poison Control Center 217/333-3611 or ICI Americas Inc. 302/575-3000 for 24-hour emergency assistance.

In case of significant spill, call CHEMTREC 800/424-9300.

NOTE TO PHYSICIAN OR VETERINARIAN

This product may reduce the clotting ability of the blood and cause hemorrhaging. If poisoning occurs, intramuscular and oral administration of vitamin K_1 are indicated, as in poisoning from overdose of bishydroxy coumarin. For human cases, vitamin K_1 is antidotal at doses of 10–20 mg (not mg/kg). For animal cases, vitamin K_1 is antidotal at 2–5 mg/kg. Repeated doses may need to be given up to two weeks (based on monitoring of prothrombin times). In severe cases, blood transfusions may be necessary.

ENVIRONMENTAL HAZARDS

This product is toxic to fish, birds and wildlife. This product can pose a secondary hazard to birds of prey and mammals. Keep out of any body of water.

DIRECTIONS FOR USE

GENERAL CLASSIFICATION

It is a violation of federal law to use this product in a manner inconsistent with its labeling.

USE RESTRICTIONS

For control of Norway Rats, Roof rats and House mice in and around homes, industrial, commercial, agricultural and public buildings. TALON may also be used in and around transport vehicles (ships, trains, aircraft) and related port or terminal buildings. Do not use in sewers. This product must be placed in tamper-proof bait boxes or in locations not accessible to children, pets, domestic animals or wildlife. Do not place bait in areas where there is a possibility of contaminating food or surfaces that come in direct contact with food.

Authorized by the USDA for use in federally inspected meat, poultry and egg processing plants.

SELECTION OF TREATMENT AREAS

Determine areas where rats and/or mice will most likely find and consume the bait. Generally, these areas are along walls, by gnawed openings, in or beside burrows, in corners and concealed places, between floors and walls, or in locations where rodents or their signs have been observed. Remove as much food as possible.

APPLICATION DIRECTIONS

Norway and Roof Rats:

Apply 4 to 16 ounces of bait (usually at intervals of 15 to 30 feet) per placement. Maintain an uninterrupted supply of fresh bait for 10 days or until signs of rat activity cease.

House Mice:

Apply $\frac{1}{4}$ to $\frac{1}{2}$ ounce of bait at intervals of 8 to 12 feet per placement. Larger placements (up to 2 ounces) may be needed at points of very high mouse activity. Maintain an uninterrupted supply of fresh bait for 15 days or until signs of mouse activity cease.

Rats & Mice:

Replace contaminated or spoiled bait immediately. Collect and dispose of all dead animals and unconsumed bait according to "Disposal" paragraph. To prevent reinfestation, eliminate food, water, and harborage as much as possible. If reinfestation does occur, repeat treatment. Where a continuous source of infestation is present, establish permanent bait stations and replenish bait as needed.

STORAGE AND DISPOSAL

Do not contaminate water, food or feed by storage or disposal.

STORAGE

Store in original container only in a dry place inaccessible to children and pets.

PESTICIDE DISPOSAL

Wastes resulting from the use of this product may be disposed of on site or at an approved waste disposal facility.

CONTAINER DISPOSAL

Dispose of bait container in a sanitary landfill or by incineration if allowed by state and local authorities.

NOTICE TO BUYER AND USER

Seller warrants that this product conforms to the chemical description on the label and is reasonably fit for the purposes stated on the label when used in accordance with directions under normal conditions of use. This warranty does not extend to the use of this product contrary to label instructions, or under abnormal use conditions, or under conditions not reasonably foreseeable to Seller and Buyer and User assumes the risk of any such use. SELLER DISCLAIMS ALL OTHER WARRANTIES, EXPRESSED OR IMPLIED, INCLUDING ANY WARRANTY OF FITNESS OF MERCHANTABILITY. SELLER SHALL NOT BE LIABLE FOR CONSEQUENTIAL, SPECIAL OR INDIRECT DAMAGES RESULTING FROM THE USE OR HANDLING OF THIS PRODUCT AND SELLER'S SOLE LIABILITY AND BUYER'S AND USER'S EXCLUSIVE REMEDY SHALL BE LIMITED TO THE REFUND OF THE PURCHASE PRICE.

Mobay Corporation

Agricultural Chemicals Division
Box 4913
Kansas City, MO 64120
(816) 242-2000

®BAYLETON

50% WETTABLE POWDER
FUNGICIDE

For Control of Specified Diseases on Certain Field, Fruit, and Vegetable Crops.

Only for agricultural use, or for sale to, use, and storage by commercial applicators, and other servicepersons.

ACTIVE INGREDIENT:

1-(4-Chlorophenoxy)-3,3-dimethyl-1-(1*H*-1,2,4-triazol-1-yl)-2-butanone	50%
INERT INGREDIENTS	50%
TOTAL	100%

EPA Reg. No. 3125-320 U.S. Patent No. 3,912,752

BAYLETON is a Reg. TM of the Parent Company of Farbenfabriken Bayer GmbH, Leverkusen.

CAUTION
Keep Out of Reach of Children
AVISO

PRECAUCION AL USARIO: Si usted no puede leer o entender inglés, no use este producto hasta que la etiqueta le haya sido explicada ampliamente.

(TO THE USER: If you cannot read or understand English, do not use this product until the label has been fully explained to you.)

RECOMMENDED USES

Almonds	Wheat
Apples	Barley
Cucurbits	Pine Seedlings
Grapes	Sugar Beets
Pears	and Grasses grown for seed
Pineapple	

RESTRICTIONS ON WAREHOUSE STORAGE

Protect from rain and moisture. Store in a cool, dry place. Do not store next to herbicides. Do not store near feed or food products.

STATEMENTS OF PRACTICAL TREATMENT

If swallowed, vomiting should be induced. Administer water freely and induce vomiting by giving one dose (1/2 oz. or 15 ml.) of syrup of ipecac. If vomiting does not occur within 10 to 20 minutes, administer second dose. If syrup of ipecac is not available, induce vomiting by sticking finger down throat. Repeat until vomit fluid is clear. Never give anything by mouth to an unconscious person. Professional medical assistance should be secured immediately.

If on skin, wash skin immediately with soap and water.

If eyes are contaminated, wash with plenty of water for at least 15 minutes. Get medical attention.

If inhaled, remove to fresh air. If not breathing, give artificial respiration, preferably mouth to mouth. Get medical attention.

SYMPTOMS OF POISONING: The compound does not cause any definite symptoms that would be diagnostic. Poisoning is accompanied by hyperactivity followed by sedation.

NOTE TO PHYSICIAN: No specific antidote. Treat symptomatically.

PRECAUTIONARY STATEMENTS

HAZARDS TO HUMANS AND DOMESTIC ANIMALS

WARNING

Harmful if swallowed, inhaled, or absorbed through the skin. Causes eye irritation. Do not breathe dust or allow to get in eyes, on skin, or on clothing.

Wash hands, arms and face thoroughly with soap and warm water after use and before eating or smoking. Wash all contaminated clothing with soap and hot water before reuse. Do not contaminate feed or food items. Keep out of reach of children.

ENVIRONMENTAL HAZARDS

Do not use on other crops grown for food or forage. Keep out of lakes, streams, and ponds. Do not contaminate water by cleaning of equipment or disposal of wastes. Apply this product only as specified on this label.

Do not make applications when weather conditions favor drift from target area.

ROTATIONAL CROPS

Small grains, corn, sorghum, soybeans, beans, peas and cucurbits may be planted 35 days after the last application of BAYLETON, however forage or vines from these crops may not be used for food or feed. Root crops may be planted 120 days after the last application but tops must not be used for food or feed. All crops may be planted 12 months or later after the last application of BAYLETON without any restrictions.

RE-ENTRY STATEMENTS

Do not apply this product in such a manner as to directly or through drift expose workers or other persons. The area being treated must be vacated by unprotected persons. Do not enter treated areas without protective clothing until sprays have dried. Because certain states may require more restrictive re-entry intervals for various crops treated with this product, consult your State Department of Agriculture for further information.

Written or oral warnings must be given to workers who are expected to be in a treated area or in an area about to be treated with this product. Oral warnings should include those statements found under the label heading "Precautionary Statements" and "Re-entry Statements" included on this label. When oral warnings are given, warnings shall be given in a language customarily understood by workers. Oral warnings must be given if there is reason to believe that written warnings cannot be understood by workers. Written warnings must include the following information: WARNING, area treated with BAYLETON 50% Wettable Powder on (date of application). Do not enter without appropriate protective clothing until sprays have dried. **If on skin,** remove contaminated clothing and wash skin immediately with soap and warm water. **If eyes are contaminated,** wash with flowing water for at least 15 minutes. **If swallowed,** vomiting should be induced. (See Precautionary Statements on label).

STORAGE AND DISPOSAL

PESTICIDE DISPOSAL

Do not contaminate water, food or feed by storage or disposal.

Wastes resulting from the use of this product may be disposed of on site or at an approved waste disposal facility.

CONTAINER DISPOSAL

Completely empty bag into application equipment. Then dispose of empty bag in a sanitary landfill or by incineration, or, if allowed by State and local authorities, by burning. If burned, stay out of smoke.

Store in a cool, dry place and in such a manner as to prevent cross contamination with other pesticides, fertilizers, food, and feed. Store in original container and out of the reach of children, preferably in a locked storage area.

Handle and open container in a manner as to prevent spillage. If the container is leaking or material spilled for any reason or cause, carefully sweep material into a pile. Refer to Precautionary Statements on label for hazards associated with the handling of this material. Do not walk through spilled material. Dispose of pesticide as directed above. In spill or leak incidents, keep unauthorized people away. You may contact the Mobay Emergency Response Team for decontamination procedures or any other assistance that may be necessary. The Mobay Kansas City Emergency Response Telephone No. is 816-242-2582, or contact Chemtrec at 800-424-9300.

IMPORTANT

This bulletin is not intended to provide adequate information for application. Before using any product, read and carefully observe directions, cautionary statements and other information appearing on the product label. This product is sold subject to the Conditions of Sale set forth on the container label.

READ THE LABEL BEFORE USE

DIRECTIONS FOR USE

It is a violation of Federal Law to use this product in a manner inconsistent with its labeling.

MIXING

Add the recommended amount of BAYLETON 50% Wettable Powder into the spray tank while filling with water to the desired level. Operate the agitator while mixing. If other materials are added to the spray tank, the BAYLETON 50% Wettable Powder should be thoroughly dispersed prior to the addition of other materials.

COMPATIBILITY

BAYLETON is physically compatible with many registered insecticides, miticides, and fungicides. To determine the compatibility of BAYLETON with other products, the following procedure should be followed: Pour the recommended proportions of the products into a suitable container of water, mix thoroughly and allow to stand at least five (5) minutes. If the combination remains mixed or can be re-mixed readlily, the mixture is considered physically compatible. For futher information contact your local Mobay representative.

NOTE TO USER

For worker protection during mixing, loading, and during application, wear a hat, long sleeve shirt, and long legged trousers or overalls. In addition, during mixing and loading, wear rubber or neoprene gloves and a dust mask. Protective clothing should be laundered separately following application.

RECOMMENDED APPLICATIONS

BAYLETON 50% Wettable Powder may be applied in a minimum of 20 gallons of spray solution per acre by ground or in a minimum of 5 gallons of spray solution per acre by aircraft spray equipment except as noted under "Remarks" for each crop. Check equipment calibration frequently. Complete coverage and uniform application are essential for the most effective results, especially when lower spray volumes are applied. If necessary, increase the spray volume per acre for complete crop coverage.

Consult your State Agricultural Experiment Station or Extension Service Specialist or Mobay representative for guidance on the selection of specific dosage rates and timing of applications appropriate for your specific crop and disease. State regulations may contain additional restrictions or requirements. Check the Remarks and Restrictions sections of this label for additional directions.

CROP	DISEASE		RATE OF BAYLETON 50% WETTABLE POWDER		REMARKS (ALSO REFER TO DIRECTIONS FOR USE)
FRUITS			oz./A	oz./100 gal[1]	
Apples	Powdery mildew		2 to 8	½ to 2	Make the first application at the green-tip stage and continue applications at 7 to 14 day intervals, or as needed until terminal growth ceases. Use the higher label rates under heavy disease pressure, with extended spray intervals, and/or on highly disease susceptible varieties.
	Cedar apple rust	East of the Rocky Mts.	2 to 8	½ to 2	A maximum of 24 ounces of BAYLETON 50% Wettable Powder may be applied per acre per season. BAYLETON 50% Wettable Powder may be applied up to the day of harvest. Do not graze livestock in treated orchards.
Pears	Powdery mildew		2 to 8	½ to 2	Make the first application at the budburst stage and continue applications at 7 to 14 day intervals, or as needed. A maximum of 24 ounces of BAYLETON 50% Wettable Powder may be applied per acre per season. BAYLETON 50% Wettable Powder may be applied up to the day of harvest. Do not graze livestock in treated orchards.
Grapes					**GENERAL COMMENTS** Use the higher label rates under heavy disease pressure, with extended spray intervals, and/or on highly disease susceptible varieties. Fields should be observed closely for early disease symptoms when susceptible varieties are planted and/or under prolonged conditions favorable for disease development. A maximum of 18 ounces of BAYLETON 50% Wettable Powder may be applied per acre per crop season. BAYLETON 50% Wettable Powder may be applied up to 14 days before harvest.
	Powdery mildew		2 to 6		**POWDERY MILDEW** Make the first application pre-bloom and continue applications at 14 to 21 day intervals, or as needed.
	Black rot	Protective Schedule	2 to 3		**BLACK ROT** **Protective Schedule** Make the first application at 10-inch green shoot and continue applications at 7 to 14 day intervals through 5° Brix stage or until veraison (berry coloring) is complete. Under severe disease conditions and/or on highly susceptible varieties, make applications at 7 day intervals.
		Post-infection	3 to 4		**Post-Infection Schedule** During bloom apply within 72 hours after the beginning of an infection period. Do not make spray applications any closer than 7 day intervals. After bloom, continue applications at rates of 2 to 3 ounces of BAYLETON 50% W.P. at 7 to 14 day intervals, or as needed to maintain control.

[1] Rates of BAYLETON 50% Wettable Powder are based on a standard of 400 gallons of dilute spray per acre, or the equivalent amount of product per acre in a concentrate spray.

CROP	DISEASE	RATE OF BAYLETON 50% WP	REMARKS (ALSO REFER TO DIRECTIONS FOR USE)
Pineapples (Fresh market only)	Fresh fruit rot *(Ceratocystis paradoxa)*	13⅓ oz.	**Post-Harvest Treatment:** Apply specified dosage in 100 gallons of water to fruit after harvest. Fruit should be dipped or sprayed for thorough coverage and allowed to drain. Limit dipping time to not more than 3 minutes.
NUTS Almonds	Blossom blight	oz./A oz./100 gal[1] 24 to 32 6 to 8	Make applications at pink bud and full bloom in aerial or ground equipment. Concentrate sprays may be applied provided the amount of BAYLETON 50% WP applied per acre is the same as that which would be applied per acre in a full coverage spray. Do not apply more than 64 ozs. (4 lbs.) BAYLETON 50% WP per acre per crop season. The last application can be made up to full bloom.
SMALL GRAINS	Powdery mildew	2 to 6 oz/A	Fields should be observed closely for early disease symptoms when susceptible varieties are planted and/or under prolonged conditions favorable for disease development.
Wheat Barley	Rusts (stripe, leaf, and stem) Leaf blotch *(Septoria tritici)* Barley scald (Except California)	4 to 8 oz/A	Apply specified dosage per acre in a minimum of 10 gallons of water per acre by ground equipment or 5 gallons of water per acre by air when disease symptoms first appear on leaves or stems. Additional applications should be made if new disease symptoms appear. The total amount of BAYLETON 50% W.P. should not exceed 16 oz. per acre per crop season and the last application should not be made within 21 days of harvest. **Powdery mildew:** In areas having severe powdery mildew infection apply 4 to 6 oz. of BAYLETON 50% W.P. per acre. **Rusts:** In areas having severe rust infection apply 6 to 8 oz. of BAYLETON 50% W.P. per acre. **Leaf blotch:** BAYLETON 50% W.P., applied at first sign of infection, will suppress light to moderate leaf blotch infections. For control of severe leaf blotch infections it is recommended that a registered fungicide with good activity against this disease be tank mixed with BAYLETON. **Barley scald:** Apply 4 to 8 ozs. of BAYLETON 50% W.P. at the first sign of scald symptoms.
GRASSES GROWN FOR SEED			Apply specified dosage in 20 gallons of water or more per acre with ground sprayers, or in 7 or more gallons of water per acre with aircraft. Thorough coverage is important for optimum disease control. Disease control may be enhanced by the addition of a wetting agent to spray mixtures of BAYLETON 50% W.P. A maximum of 2 lbs. of BAYLETON 50% W.P. may be applied per acre per year. The last application may be made up to 5 days before harvest. Chaff and straw from treated area may be used for feed purposes, however, do not forage or cut green crop or use seed for feed purposes. Regrowth may be grazed 11 weeks after last application.
Bermudagrass Perennial ryegrass Bluegrass	Rusts (Puccinia spp.)	**PREVENTATIVE RATE** 4 to 8 ozs. per acre.	**RUSTS** **Preventative Schedule** For optimum disease control, apply specified dosage of BAYLETON 50% W.P. as soon as weather conditions are favorable for rust development. Apply before rust pustules are present, and continue applications at 14 to 21 day intervals depending on disease pressure.
		CURATIVE RATE 8 to 16 ozs. per acre.	**RUSTS** **Curative Schedule** Apply the specified dosage if rust pustules are already present. A second application 14 to 21 days later may be necessary. On highly susceptible grass varieties and/or under prolonged conditions favorable for disease development use rates of 12 to 16 ozs. per acre and a shorter interval between applications. Rust control is especially important during the period from early head emergence through last flowering. Disease control may not be adequate and yield loss may occur if rust infection is allowed to reach 5 percent or greater before treating. BAYLETON 50% W.P. will prevent young, developing pustules from producing spores. However, pustules and other symptoms caused by the rust fungus prior to the application of BAYLETON 50% W.P. will remain visible.
	Powdery mildew	2 to 4 ozs. per acre.	**POWDERY MILDEW:** One or two applications at the 2 to 4 oz. rate applied at the first signs of disease and when new powdery mildew symptoms reappear are generally sufficient for disease control. In cases where severe mildew infections are present, use the higher rate.

[1] Rates of BAYLETON 50% Wettable Powder are based on a standard of 400 gallons of dilute spray per acre, or the equivalent amount of product per acre in a concentrate spray.

CROP	DISEASE	RATE OF BAYLETON 50% WP	REMARKS (ALSO REFER TO DIRECTIONS FOR USE)
Sugar beets	Powdery mildew	**BROADCAST** 8 to 16 ozs./A. **BAND** 16 ozs./A.	**GENERAL COMMENTS** Apply the specified dosage per acre in a minimum of 10 gallons of water by ground and 5 gallons of water by air. Under severe disease conditions, use of the higher recommended rate may be necessary. Do not apply more than 16 ozs. BAYLETON 50% WP per acre per season. **BROADCAST APPLICATIONS:** Apply specified dosage per acre as a broadcast foliar spray when disease symptoms first appear. To maintain control or if reinfection occurs, additional applications may be made at 10 to 14 day intervals. Broadcast applications may be made up to 15 days before harvest. **BAND APPLICATIONS:** Apply specified dosage per acre as a spray directed into the leaf whorl in a 2 inch band by turning a flat fan spray nozzle sideways to achieve a narrow band. Band application may be made up to 30 days before harvest.
Cucurbits [Balsam pear (bitter melon), Chinese waxgourd (Chinese preserving melon), citron melon, cucumber, gherkin, gourds, edible melons including hybrids (including cantaloupe, casaba, crenshaw, honeydew melons, honey balls mango melon, muskmelon, Persian melon), pumpkin, squash, summer squash, winter watermelon, including hybrids]	Powdery mildew (Erysiphe cichoracearum)*	2 to 4 ozs. per acre.	Apply specified dosage per acre in a minimum of 10 gallons of water by ground and 5 gallons of water by air as a foliar spray when disease symptoms first appear. Repeat at 14 day intervals to maintain control if new disease symptoms appear. Under severe disease conditions use the 4 oz. rate. Applications may be made up to day of harvest. A maximum of 16 ozs. of BAYLETON 50% WP may be applied per acre per season. BAYLETON 50% WP will not give commercially acceptable control of Sphaerotheca fuliginea.

TANK MIXES: The following tank mix recommendations are made for the improved control of specific diseases. Before mixing BAYLETON 50% Wettable Powder with any other fungicide, read and carefully observe the precautionary statements, directions for use, and other information appearing on the EPA registered product labels.

CROP	CO-APPLIED FUNGICIDES AND RATE	FOR IMPROVED CONTROL OF:		BAYLETON 50% WP RATE
Apples	Dithane® M-45 @ 1 to 2 lbs./100 gals.	Bitter rot Black rot Brown rot	Fly speck Scab Sooty blotch	½ to 2 ozs./100 gals.
Barley	Refer to wheat			
Cantaloupe Cucumbers	Refer to melons Refer to melons			
Grapes	Dithane M-45 @ 1½ to 4 lbs./A	Bunch rot	Deadarm Downy mildew	2 to 6 ozs./A
Melons	Dithane M-45 @ 2 to 3 lbs./A	Downy mildew Alternaria leaf spot Anthracnose Cercospora leaf spot	Gummy stem blight Scab	2 to 4 ozs./A
Pears (Except California)	Refer to apples			
Pumpkins	Refer to melons			
Squash, summer	Refer to melons			
Sugar beets	Dithane M-45 @ 1½ to 2 lbs./A	Cercospora leaf spot		4 to 16 ozs./A
Watermelons	Refer to melons			
Wheat	Dithane M-45 @ 2 lbs./A	Helminthosporium leaf spots Septoria leaf spots (leaf blotch)	Septoria glume blotch Tan spot	2 to 8 ozs./A

SPECIAL LOCAL NEED REGISTRATIONS [SECTION 24(c)] FOR BAYLETON 50% WETTABLE POWDER*

CROP	DISEASE	STATE EPA SLN NO.	DOSAGE	REMARKS
Non-bearing fruit trees and ornamentals (orchards and nurseries)	Powdery mildew	Idaho ID 80-0031 Oregon OR 80-0030 Utah UT 80-0001 Washington WA 80-0020	2 ounces per 100 gallons of water.	Apply as a uniform dilute spray to the point of drip.
Non-bearing apple trees	Powdery mildew	New Mexico NM 82-0011	2 to 8 ounces per acre or ½ to 2 ounces per 100 gallons.	First application at green-tip and continue at 7- to 14-day intervals.
Pine seed (Nurseries)	Fusiform rust *(Cronartium quercuum)*	Alabama AL 83-0009 North Carolina NC 83-0015 Oklahoma OK 83-0012 Tennessee TN 83-0001 Texas TX 83-0040	20 ounces.	SEED TREATMENT Mix specified dosage in 100 gals. of water. Cover seeds with the mixture and soak them, at room temperature (75°F), for 24 hours, stirring occasionally.

* For distribution and use only within the state specified. These Special Local Need registrations approved as of date of this publication are, however, subject to cancellation by the state. Contact your nearest Mobay Agricultural Chemicals Division Office for the current status and a copy of the complete individual supplemental Special Local Need labeling. This labeling must be in the possession of the user at time of pesticide application.

®BAYLETON

50% WETTABLE POWDER FUNGICIDE IN WATER SOLUBLE PACKETS

For Control of Specified Diseases on Certain Field, Fruit and Vegetable Crops.

Only for agricultural use, or for sale to, use, and storage by commercial applicators, and other servicepersons.

ACTIVE INGREDIENT:
1-(4-Chlorophenoxy)-3,3-dimethyl-1-(1*H*-1,2,4-triazol-1-yl)-2-butanone 50%
INERT INGREDIENTS 50%
TOTAL 100%

EPA Reg. No. 3125-340 U.S. Patent No. 3,912,752

BAYLETON is a Reg. TM of the Parent Company of Farbenfabriken Bayer GmbH, Leverkusen.

CAUTION

Keep Out of Reach of Children

AVISO

PRECAUCION AL USUARIO: Si usted no puede leer o entender inglés, no use este producto hasta que la etiqueta le haya sido explicada ampliamente.

(TO THE USER: If you cannot read or understand English, do not use this product until the label has been fully explained to you.)

RECOMMENDED USES

Almonds	Wheat
Apples	Barley
Cucurbits	Pine Seedlings
Grapes	Sugar Beets
Pears	and Grasses grown for seed

RESTRICTIONS ON WAREHOUSE STORAGE

Protect from rain and moisture. Store in a cool, dry place above 32°F (freezing). Do not store next to herbicides. Do not store near feed or food products.

STATEMENTS OF PRACTICAL TREATMENT

If swallowed, vomiting should be induced. Administer water freely and induce vomiting by giving one dose (½ oz. or 15 ml.) of syrup of ipecac. If vomiting does not occur within 10 to 20 minutes, administer second dose. If syrup of ipecac is not available, induce vomiting by sticking finger down throat. Repeat until vomit fluid is clear. Never give anything by mouth to an unconscious person. Professional medical assistance should be secured immediately.

If on skin, wash skin immediately with soap and water.

If eyes are contaminated, wash with plenty of water for at least 15 minutes. Get medical attention.

If inhaled, remove to fresh air. If not breathing, give artificial respiration, preferably mouth to mouth. Get medical attention.

SYMPTOMS OF POISONING: The compound does not cause any definite symptoms that would be diagnostic. Poisoning is accompanied by hyperactivity followed by sedation.

NOTE TO PHYSICIAN: No specific antidote. Treat symptomatically.

PRECAUTIONARY STATEMENTS

HAZARDS TO HUMANS AND DOMESTIC ANIMALS

WARNING

Harmful if swallowed, inhaled, or absorbed through the skin. Causes eye irritation. Do not allow to get in eyes, on skin, or on clothing.

Wash hands, arms and face thoroughly with soap and warm water after use and before eating or smoking. Wash all contaminated clothing with soap and hot water before reuse. Do not contaminate feed or food items. Keep out of reach of children.

ENVIRONMENTAL HAZARDS

Do not use on other crops grown for food or forage. Do not apply directly to water. Do not contaminate water by cleaning of equipment or disposal of wastes. Apply this product only as specified on this label.

Do not make applications when weather conditions favor drift from target area.

ROTATIONAL CROPS

Small grains, corn, sorghum, soybeans, beans, peas, and cucurbits may be planted 35 days after the last application of BAYLETON, however forage or vines from these crops may not be used for food or feed. Root crops may be planted 120 days after the last application but tops must not be used for food or feed. All crops may be planted 12 months or later after the last application of BAYLETON without any restrictions.

RE-ENTRY STATEMENTS

Do not apply this product in such a manner as to directly or through drift expose workers or other persons. The area being treated must be vacated by unprotected persons. Do not enter treated areas without protective clothing until sprays have dried. Because certain states may require more restrictive re-entry intervals for various crops treated with this product, consult your State Department of Agriculture for further information.

Written or oral warnings must be given to workers who are expected to be in a treated area or in an area about to be treated with this product. Oral warnings should include those statements found under the label heading "Precautionary Statements" and "Re-entry Statements" included on this label. When oral warnings are given, warnings shall be given in a language customarily understood by workers. Written warnings must include the following information: WARNING, area treated with BAYLETON 50% Wettable Powder on (date of application). Do not enter without appropriate protective clothing until sprays have dried. **If on skin,** remove contaminated clothing and wash skin immediately with soap and warm water. **If eyes are contaminated,** wash with flowing water for at least 15 minutes. **If swallowed,** vomiting should be induced. (See Precautionary Statements on label).

STORAGE AND DISPOSAL

PESTICIDE DISPOSAL

Do not contaminate water, food or feed by storage or disposal.

Wastes resulting from the use of this product may be disposed of on site or at an approved waste disposal facility.

CONTAINER DISPOSAL

Completely empty bag into application equipment. Then dispose of empty bag in a sanitary landfill or by incineration, or, if allowed by State and local authorities, by burning. If burned, stay out of smoke.

Store in a cool, dry place and in such a manner as to prevent cross contamination with other pesticides, fertilizers, food and feed. Do not store below freezing (32°F). Exposure to moisture or excessive handling of water soluble packets may cause breakage. Store packets in original container and out of the reach of children, preferably in a locked storage area.

Handle and open container in a manner as to prevent spillage. If the container is leaking or material spilled for any reason or cause, carefully sweep material into a pile. Refer to Precautionary Statements on label for hazards associated with the handling of this material. Do not walk through spilled material. Dispose of pesticide as directed above. In spill or leak incidents, keep unauthorized people away. You may contact the Mobay Emergency Response Team for decontamination procedures or any other assistance that may be necessary. The Mobay Kansas City Emergency Response Telephone No. is 816-242-2582, or contact Chemtrec at 800-424-9300.

IMPORTANT: This bulletin is not intended to provide adequate information for application. Before using any product, read and carefully observe directions, cautionary statements and other information appearing on the product label. This product is sold subject to the Conditions of Sale set forth on the container label.

READ THE LABEL BEFORE USE

DIRECTIONS FOR USE

It is a violation of Federal Law to use this product in a manner inconsistent with its labeling.

MIXING: The enclosed packets are **water soluble.** Do not allow packets to become wet prior to mixing spray solution. Do not handle with wet hands. Reseal outer bag to protect remaining packets.

Ground Application Equipment: The order of mixing is important and should be followed.

1. The spray tank should be filled $1/4$ to $1/3$ full of water.
2. While recirculating and with the agitator running, add the required number of PVA packets, as determined under "Recommended Applications," into the spray tank as you continue filling the tank to the desired level of water. One packet contains 8 ounces of wettable powder. If it is determined, under "Recommended Applications," that 2 pounds of BAYLETON 50% Wettable Powder should be added to the spray tank, add four 8-ounce packets. The PVA packets should be completely dissolved in 5 to 10 minutes. If an adjuvant is needed, add prior to addition of PVA packets to the spray tank.
3. The addition of other pesticides should be after the PVA packets are completely dissolved, but prior to completion of filling the tank.
4. As with any pesticide, maintain adequate agitation prior to and during spraying.

Aerial Application Equipment: It is important to pre-mix the PVA packets and water before loading into the aerial spray tank. The mixing process should be made according to the same directions for ground application equipment.

CAUTION: Do not use PVA packets in a tank-mix with products that contain Boron or release free chlorine. The resultant reaction of PVA and Boron or free chlorine is a plastic which is not soluble in water or solvents such as diesel oils, kerosene, gasoline or alcohol.

Do not attempt to use the PVA packets directly in diesel oils or summer spray type oils as in ULV or LV uses. PVA packets are water soluble not oil soluble.

NOTICE TO USER: For worker protection during application, wear a hat, long sleeve shirt, and long legged trousers or overalls. Protective clothing should be laundered separately following application.

RECOMMENDED APPLICATIONS

BAYLETON 50% Wettable Powder may be applied in a minimum of 20 gallons of spray solution per acre by ground or in a minimum of 5 gallons of spray solution per acre by aircraft spray equipment except as noted under "Remarks" for each crop. Check equipment calibration frequently. Complete coverage and uniform application are essential for the most effective results, especially when lower spray volumes are applied. If necessary, increase the spray volume per acre for complete crop coverage.

Consult your State Agricultural Experiment Station or Extension Service Specialist or Mobay representative for guidance on the selection of specific dosage rates and timing of applications appropriate for your specific crop and disease. State regulations may contain additional restrictions or requirements. Check the Remarks and Restrictions sections of this label for additional directions.

CROP	DISEASE		RATE OF BAYLETON 50% WETTABLE POWDER OZ./A	RATE OF BAYLETON 50% WETTABLE POWDER OZ./100 GAL[1]	REMARKS (ALSO REFER TO DIRECTIONS FOR USE)
FRUITS					
Apples	Cedar apple rust Powdery mildew		2 to 8	$1/2$ to 2	Make the first application at the green-tip stage and continue applications at 7 to 14 day intervals, or as needed until terminal growth ceases. Use the higher label rate under heavy disease pressure, with extended spray intervals and/or on highly disease susceptible varieties. A maximum of 24 ounces of BAYLETON 50% Wettable Power may be applied per acre per season. BAYLETON 50% Wettable Powder may be applied up to the day of harvest. Do not graze livestock in treated orchards.

CROP	DISEASE	RATE OF BAYLETON 50% WETTABLE POWDER	OZ./A	OZ./100 GAL[1]	REMARKS (ALSO REFER TO DIRECTIONS FOR USE)
Pears	Powdery mildew		2 to 8	½ to 2	Make the first application at the budburst stage and continue applications at 7 to 14 day intervals, or as needed. Use the higher label rate under heavy disease pressure, with extended spray intervals and/or on highly disease susceptible varieties. A maximum of 24 ounces of BAYLETON 50% Wettable Power may be applied per acre per season. BAYLETON 50% Wettable Powder may be applied up to the day of harvest. Do not graze livestock in treated orchards.
Grapes	Powdery mildew		2 to 6	—	**GENERAL COMMENTS** Use the higher label rates under heavy disease pressure, with extended spray intervals, and/or on highly disease susceptible varieties. Fields should be observed closely for early disease symptoms when susceptible varieties are planted and/or under prolonged conditions favorable for disease development. A maximum of 18 ounces of BAYLETON 50% Wettable Powder may be applied per acre per crop season. BAYLETON 50% Wettable Powder may be applied up to 14 days before harvest. **POWDERY MILDEW** Make the first application pre-bloom and continue applications at 14 to 21 day intervals, or as needed. **BLACK ROT** **Protective Schedule** Make the first application at 10-inch green shoot and continue applications at 7 to 14 day intervals through 5° Brix stage or until veraison (berry coloring) is complete. **Post-infection Schedule** During bloom apply within 48 to 72 hours after the beginning of an infection period. Do not make spray applications any closer than 7 day intervals. After bloom, continue applications at rates of 2 to 3 ounces of BAYLETON 50% W.P. at 7 to 14 day intervals, or as needed to maintain control.
	Black rot	Protective Schedule	2 to 3	—	
		Post-infection	3 to 4		
NUTS Almonds	Blossom blight		24 to 32	6 to 8	Make applications at pink bud and full bloom in aerial or ground equipment. Concentrate sprays may be applied provided the amount of BAYLETON 50% WP applied per acre is the same as that which would be applied per acre in a full coverage spray. Do not apply more than 64 ozs. (4 lbs) BAYLETON 50% WP per acre per crop season. The last application can be made up to full bloom.
SMALL GRAINS	Powdery mildew		2 to 6 oz./A		Fields should be observed closely for early disease symptoms when susceptible varieties are planted and/or under prolonged conditions favorable for disease development. Apply specified dosage per acre when disease symptoms first appear on leaves or stems. Additional applications should be made if new disease symptoms appear. The total amount of BAYLETON 50% W.P. should not exceed 16 oz. per acre per crop season and the last application should not be made within 21 days of harvest. **Powdery mildew:** In areas having severe powdery mildew infection apply 4 to 6 oz. of BAYLETON 50% W.P. per acre. **Rusts:** In areas having severe rust infection apply 6 to 8 oz. of BAYLETON 50% W.P. per acre. **Leaf blotch:** BAYLETON 50% W.P., applied at first sign of infection, will suppress light to moderate leaf blotch infections. For control of severe leaf blotch infections it is recommended that a registered fungicide with good activity against this disease be tank mixed with BAYLETON. **Barley scald:** Apply 4 to 8 ozs. of BAYLETON 50% W.P. at the first sign of scald symptoms.
Wheat Barley	Rusts (stripe, leaf, Leaf blotch and stem) *(Septoria tritici)* (Except California) Barley scald (Except California)		4 to 8 oz./A		

CROP	DISEASE	RATE OF BAYLETON 50% WETTABLE POWDER	REMARKS (ALSO REFER TO DIRECTIONS FOR USE)
GRASSES GROWN FOR SEED Bermudagrass Perennial ryegrass Bluegrass	Rusts *(puccinia spp.)*	**PREVENTATIVE RATE** 4 to 8 ozs. per acre. (Except California) 4 to 8 ozs. per acre. **CURATIVE RATE** 8 to 16 ozs. per acre.	Apply the specified dosage in 20 gallons of water or more per acre with ground sprayers, or in 7 or more gallons of water per acre with aircraft. Thorough coverage is important for optimum disease control. Disease control may be enhanced by the addition of a wetting agent to spray mixtures of BAYLETON 50% W.P. A maximum of 2 lbs. of BAYLETON 50% W.P. may be applied per acre per year. The last application may be made up to 5 days of harvest. Chaff and straw from treated area may be used for feed purposes, however, do not forage or cut green crop or use seed for feed purposes. Regrowth may be grazed 11 weeks after last application. **RUSTS** Preventative Schedule For optimum disease control, apply specified dosage of BAYLETON 50% W.P. as soon as weather conditions are favorable for rust development. Apply before rust pustules are present, and continue applicatons at 14 to 21 day intervals depending on disease pressure. RUSTS **Curative Schedule** Apply the specified dosage if rust pustules are already present. A second application 14 to 21 days later may be necessary. On highly susceptible grass varieties and/or under prolonged conditions favorable for disease development use rates of 12 to 16 oz. per acre and a shorter interval between applications. Rust control is especially important during the period from early head emergence through leaf flowering. Disease control may not be adequate and yield loss may occur if rust infection is allowed to reach 5 percent or greater before treating. BAYLETON 50% W.P. will prevent young, developing pustules from producing spores. However, pustules and other symptoms caused by the rust fungus prior to the application of BAYLETON 50% W.P. will remain visible.
	Powdery mildew	2 to 4 ozs. per acre.	**POWDERY MILDEW:** One or two applications at the 2 to 4 oz. rate applied at the first signs of disease and when new powdery mildew symptoms reappear are generally sufficient for disease control. In cases where severe mildew infections are present, use the higher rate.
Pine (Seedlings) (Except California)	Pine rust (Fusiform rust)	4 to 16 oz.	Begin application prior to infection period, repeat as necessary at 2 to 3 week intervals depending upon disease pressure. Use lower rates in areas of low disease incidence and higher rates in areas of severe disease incidence. A maximum of 64 ozs. of BAYLETON 50% W.P. may be applied per acre per season. A spreader-sticker is needed to help adhere spray solution to the pine trees.
Sugar beets	Powdery mildew	**BROADCAST** 4 to 16 ozs./A **BAND** 8 to 16 ozs./A	**GENERAL COMMENTS** Apply the specified dosage per acre in a minimum of 10 gallons of water by ground and 5 gallons of water by air. Under severe disease conditions, use of the higher recommended rate may be necessary. Do no apply more than 16 ozs. of BAYLETON 50% WP per acre per season. **BROADCAST APPLICATIONS:** Apply specified dosage per acre as a broadcast foliar spray when disease symptoms first appear. To maintain control or if reinfection occurs, additional applications may be made in 10 to 14 day intervals. **BAND APPLICATIONS:** Apply specified dosage per acre as a spray directed into the leaf whorl in a 2-inch band by turning a flat fan spray nozzle sideways to achieve a narrow band. Band applications may be made up to 30 days before harvest.

CROP	DISEASE	RATE OF BAYLETON 50% WETTABLE POWDER	REMARKS (ALSO REFER TO DIRECTIONS FOR USE)
Cucurbitis [Balsam pear (bitter melon), Chinese waxgourd (Chinese preserving melon), citron melon, cucumber, gherkin gourds, edible melons including hybrids (including cantaloupe casaba, crenshaw, honeydew melons, honey balls mango melon, muskmelon, Persian melon, pumpkin, squash, summer squash, winter watermelon, including hybrids]	Powdery mildew	2 to 4 ozs. per acre	Apply specified dosage per acre in a minimum of 10 gallons of water by ground and 5 gallons of water by air as a foliar spray when disease symptoms first appear. Repeat at 14 day intervals to maintain control if new disease symptoms appear. Under severe disease conditions use the 4 oz. rate. Applications may be made up to day of harvest. A maximum of 16 oz. per acre of BAYLETON 50% W.P. may be applied per acre per season. Bayleton 50% WP will not give commercially acceptable control of Sphaerotheca fulisinea.

[1] Rates of BAYLETON 50% Wettable Powder are based on a standard of 400 gallons of dilute spray per acre, or the equivalent amount of product per acre in a concentrate spray.

TANK MIXES

The following tank mix recommendations are made for the improved control of specific diseases. Before mixing BAYLETON 50% Wettable Powder with any other fungicide, read and carefully observe the precautionary statements, directions for use, and other information appearing on the EPA registered product labels.

CROP	CO-APPLIED FUNGICIDE AND RATE	FOR IMPROVED CONTROL OF:		BAYLETON 50% WETTABLE POWDER RATE
Apples	Dithane® M-45 @ 1 to 2 lbs/100 gals.	Bitter rot Black rot Brown rot	Fly speck Scab Sooty blotch	½ to 2 ozs./100 gals.
Barley	Refer to wheat			
Cantaloupe Cucumbers	Refer to melons Refer to melons			
Grapes	Dithane M-45 @ 1½ to 4 lbs./A	Bunch rot Deadarm Downy mildew		2 to 6 ozs./A
Melons	Dithane M-45 @ 2 to 3 lbs./A	Alternaria leaf spot Anthracnose Cercospora leaf spot Downy mildew Gummy stem blight Scab		2 to 4 ozs./A
Pears (except California)	Refer to apples			
Pumpkins	Refer to melons			
Squash, summer	Refer to melons			
Sugar beets	Dithane M-45 @ 1½ to 2 lbs./A	Cercospora leaf spot		4 to 16 ozs./A
Watermelons	Refer to melons			
Wheat	Dithane M-45 @ 2 lbs./A	Helminthosporium leaf spots Septoria leaf spots (leaf blotch) Septoria glume blotch Tan spot		2 to 8 ozs./A

®BOLSTAR 6

EMULSIFIABLE INSECTICIDE

RESTRICTED USE PESTICIDE

For retail sale to and use only by Certified Applicators or persons under their direct supervision, and only for those uses covered by the Certified Applicator's certification.

For the Control of Certain Insects on Cotton and Soybeans.

ACTIVE INGREDIENTS:

O-Ethyl *O*-[4-(methylthio)phenyl] *S*-propyl phosphorodithioate	64%
Aromatic Petroleum Distillate	15%
INERT INGREDIENTS	21%
TOTAL	100%

Contains 6 lbs. *O*-Ethyl *O*-[4-(methylthio)phenyl]S-propyl phosphorodithioate per gallon.

EPA Reg. No. 3125-321 U.S. Patent No. 3,947,529

BOLSTAR and GUTHION are Reg. TMs of the Parent Company of Farbenfabriken Bayer GmbH, Leverkusen.

CAUTION

Keep Out of Reach of Children

AVISO

PRECAUCION AL USUARIO: Si usted no puede leer o entender inglés, no use este producto hasta que la etiqueta le haya sido explicada ampliamente.

(TO THE USER: If you cannot read or understand English, do not use this product until the label has been fully explained to you.)

RECOMMENDED USES

COTTON AND SOYBEANS

RESTRICTIONS ON WAREHOUSE STORAGE

Do not use, pour, spill or store near heat or open flame.

STATEMENTS OF PRACTICAL TREATMENT

If in eyes, immediately flush eyes with plenty of water for at least 15 minutes. For further treatment, get medical attention.

If swallowed, vomiting should be induced. Administer water freely, and induce vomiting by giving one dose (1/2 ounce or 15 ml.) of syrup of ipecac. If vomiting does not occur within 10 to 20 minutes, administer second dose. If syrup of ipecac is not available, induce vomiting by sticking finger down throat. Repeat until vomit fluid is clear. Never give anything by mouth to an unconscious person. Professional medical assistance should be secured immediately.

If on skin, wash thoroughly with soap and warm water.

ENVIRONMENTAL HAZARDS

This pesticide is toxic to fish and wildlife. Use with care when applying in areas frequented by wildlife or adjacent to any body of water. When BOLSTAR 6 is applied by air at rates of 1 1/2 pints or greater per acre, applications should be made no closer than 100 feet of shorelines of fish bearing waters. Keep out of lakes, streams and ponds. Do not apply when weather conditions favor runoff or drift from target areas. Do not contaminate water by cleaning of equipment or disposal of wastes.

This product is toxic to bees exposed to direct treatment or residues on crops or weeds. Do not apply this product, or allow it to drift to crops or weeds on which bees are foraging. Additional information may be obtained from your Cooperative Extension Service.

Do not use on other crops used for food or forage. Use only according to label directions. Application at rates above those shown may result in illegal crop residues. Do not treat food crops grown in the greenhouse.

Crops to be grown in rotation with cotton should not be planted within 60 days after last application of BOLSTAR 6.

PHYSICAL OR CHEMICAL HAZARDS

Do not use or store near heat or open flame.

RE-ENTRY STATEMENTS

Do not apply this product in such a manner as to directly or through drift expose workers or other persons. The area being treated must be vacated by unprotected persons. Do not enter treated areas without protective clothing until sprays have dried. Because certain states may require more restrictive re-entry intervals for various crops treated with this product, consult your State Department of Agriculture for further information.

STORAGE AND DISPOSAL

PESTICIDE DISPOSAL

Do not contaminate water, food or feed by storage or disposal.

Pesticide wastes are toxic. Improper disposal of excess pesticide, spray mixture, or rinsate is a violation of Federal Law. If these wastes cannot be disposed of by use according to label instructions, contact your State Pesticide or Environmental Control Agency, or the Hazardous Waste representative at the nearest EPA Regional Office for guidance.

CONTAINER DISPOSAL

Triple rinse (or equivalent). Then offer for recycling or reconditioning, or puncture and dispose of in a sanitary landfill, or by other procedures approved by State and local authorities.

Store in a cool, dry place, away from open flame and extreme heat, and in such a manner as to prevent cross contamination with other pesticides, fertilizers, food, and feed. Store in original container and out of the reach of children, preferably in a locked storage area.

Handle and open container in a manner as to prevent spillage. If container is leaking invert to prevent leakage. If the container is leaking or material spilled for any reason or cause, carefully dam up spilled material to prevent runoff. Refer to Precautionary Statements on label for hazards associated with the handling of this material. Do not walk through spilled material. Absorb spilled material with absorbing type compounds and dispose of as directed for pesticides above. In spill or leak incidents, keep unauthorized people away. You may contact the Mobay Emergency Response Team for decontamination procedures or any other assistance that may be necessary. The Mobay Kansas City Emergency Response Telephone No. is 816-242-2582, or contact Chemtrec at 800-424-9300.

IMPORTANT

This bulletin is not intended to provide adequate information for application. Before using the product, read and carefully observe directions, cautionary statements and other information appearing on the product label. This product is sold subject to the Conditions of Sale set forth on the container label.

READ THE LABEL BEFORE USE

DIRECTIONS FOR USE

It is a violation of Federal Law to use this product in a manner inconsistent with its labeling.

BOLSTAR 6 emulsifiable insecticide may be used for the control of susceptible as well as resistant *Heliothis virescens* (tobacco budworm) and *Heliothis zea* (cotton bollworm) on cotton. Control or suppression of additional insects as noted under "RECOMMENDED APPLICATIONS" may also be obtained. Bolstar 6 emulsifiable insecticide may be used for the control of certain insects infesting soybeans.

RECOMMENDED APPLICATIONS			
CROP	INSECT	PINTS BOLSTAR 6	REMARKS
FIELD CROPS Cotton	Tobacco budworm Cotton bollworm Lygus (nymphs)	$\frac{2}{3}$ to 2	Apply specified dosage per acre by air or ground equipment alone or in sufficient water or refined vegetable oil for complete coverage. For best results during mid-to-late season, or under hot, dry conditions, use a higher gallonage per acre for better coverage. For example, by air use up to 5 gallons of water per acre. When using oil, one (1) quart of total volume is recommended. Pre-mix material. Oil mixture should be sprayed within 12 hours to avoid compatibility problems. Always run a small jar compatibility test before large scale mixing as each batch of vegetable oil tends to differ greatly in composition. For **light infestations**, apply $\frac{2}{3}$ to 1 pint ($\frac{1}{2}$ to $\frac{3}{4}$ lb. active) per acre, or for **moderate infestations** apply 1 to $1\frac{1}{3}$ pints ($\frac{3}{4}$ to 1 lb. active) per acre. Apply as needed usually at 3 to 7 day intervals. **Heavy infestations, or infestations of large larvae:** Apply 1 or 2 applications of the 2-pint ($1\frac{1}{2}$ lb. active) rate per acre as needed, or until infestations are brought under control, then reduce application rates to $\frac{2}{3}$ to $1\frac{1}{3}$ pints per acre and apply as needed usually at 3 to 7 day intervals between applications. In California and Arizona the minimum use rate is $1\frac{1}{3}$ pints (1 lb. active) per acre, regardless of degree of infestation of tobacco budworm or cotton bollworm. **NOTE:** In areas of frequent rainfall, the higher rates and closer spray intervals may be needed for adequate control, however, in **any case** do not apply more than a total of 20 pints per acre with the last application no closer to harvest than 21 days in California and Arizona, or 14 days in all other states. Do not graze treated fields.
	Fall armyworm Beet armyworm Pink bollworm	1 to 2	
	Fleahoppers	$\frac{2}{3}$	
	Suppression of: Lygus (adults) Whiteflies Spider mites (Carmine and two-spotted)	$\frac{2}{3}$ to 2	
	Thrips Fleahoppers Plant bugs (For early to mid-season use in Southern cotton growing areas)	$\frac{1}{3}$	Apply specified dosage per acre by air or ground equipment alone, in water, or in refined vegetable oil as stated above. Apply as needed during early to mid-season when thrips, plant bug or fleahoppers reach economic threshold. **In California**, do not make ULV applications of BOLSTAR 6 alone or in vegetable oil.
Soybeans (Except California)	Bean thrips Potato leafhopper Velvet bean caterpillar	$\frac{1}{3}$ to 1	Apply specified dosage per acre in sufficient diluent for complete coverage on a 10 to 14 day interval. Apply high rates for heaviest infestations. **NOTE:** In areas of frequent rainfall the higher rates and closer spray intervals may be needed for adequate control. Do not apply more than a total of **4** pints per acre within one season. The last application should be no closer to harvest than **60** days. Grazing or ensilage should be no earlier than 35 days from the last application.
	Plant bugs *Lygus* spp. Green cloverworm *Heliothis* spp. (podworm, bollworm, corn earworm)	$\frac{2}{3}$ to $1\frac{1}{3}$	
	Beet armyworm Mexican bean beetle	1 to $1\frac{1}{3}$	Apply specified dosage per acre in sufficient diluent for complete coverage on a 10 to 14 day interval. Apply high rates for heaviest infestations. **NOTE:** In areas of frequent rainfall the higher rates and closer spray intervals may be needed for adequate control. Do not apply more than a total of **4** pints per acre within one season. The last application should be no closer to harvest than **60** days. Grazing or ensilage should be no earlier than 35 days from the last application.
	Three-cornered alfalfa hopper (suppression) Mexican bean beetle adults (suppression)		

NOTE: BOLSTAR 6 when used on cotton may be tank-mixed with ®GUTHION 2L or 2S when GUTHION is used for control of insects at recommended rates of GUTHION.
Do not apply GUTHION alone within 14 days of harvest of cotton after applying tank mixes of BOLSTAR plus GUTHION. In California and Arizona, GUTHION alone should not be applied within 21 days of harvest of cotton after applying this tank mix.

SPECIAL LOCAL NEED REGISTRATIONS [SECTION 24(c)] FOR BOLSTAR 6*

FIELD CROPS	INSECT	STATE EPA SLN NO.	DOSAGE	REMARKS
Chrysanthemums (Grown for stock plants)	Beet armyworms	Florida FL 82-0046	2 pints per acre	Apply in 100 gallons of water as a dilute full coverage spray.
Cotton	Tobacco budworm Cotton bollworm Lygus (nymphs)	Arkansas AR 82-0027 Georgia GA 82-0017 Louisiana LA 81-0042 Mississippi MS 82-0034 South Carolina SC 82-0022 Texas TX 82-0044	$\frac{2}{3}$ to 2 pints per acre	ULV application in vegetable oil.
	Fall armyworm Beet armyworm Pink bollworm		1 to 2 pints per acre	
	Fleahoppers		$\frac{2}{3}$ pint per acre	
	Suppression of: Lygus (adults) Whiteflies Spider mites (Carmine and two-spotted)		$\frac{2}{3}$ to 2 pints per acre	
	Thrips Fleahoppers Plant bugs (For early to mid-season use)		$\frac{1}{3}$ pint per acre	

* For distribution and use only within the state specified. These Special Local Need registrations approved as of date of this publication are, however, subject to cancellation by the state. Contact your nearest Mobay Agricultural Chemicals Division Office for the current status and a copy of the complete individual supplemental Special Local Need labeling. This labeling must be in the possession of the user at time of pesticide application.

®DASANIT

15% GRANULAR INSECTICIDE—NEMATICIDE

RESTRICTED USE PESTICIDE

Acute Oral Toxicity.

For retail sale to, and use only by Certified Applicators or persons under the direct supervision of a Certified Applicator, and only for those uses covered by the Certified Applicator's certification.

For Effective Control of Soil Insects and Suppression or Control of Parasitic Nematodes on Certain Crops, and Commercial Turf Grasses and Ornamentals.

ACTIVE INGREDIENT:
O, O-Diethyl *O*-[4-(methylsulfinyl)phenyl] phosphorothioate 15%
INERT INGREDIENTS 85%
TOTAL 100%

EPA Reg. No. 3125-213-AA

DASANIT is a Reg. TM of the Parent Company of Farbenfabriken Bayer GmbH, Leverkusen.

DANGER–POISON

Keep Out of Reach of Children

PELIGRO

PRECAUCION AL USUARIO: Si usted no puede leer o entender inglés, no use este producto hasta que la etiqueta le haya sido explicada ampliamente.

TO THE USER: If you cannot read or understand English, do not use this product until the label has been fully explained to you.)

RECOMMENDED USES

Corn (field, sweet, popcorn), Cotton (SE, Gulf states only), Onions (dry bulb), Ornamentals (shrubs, flowers, foliage plants), Peanuts, Potatoes, Rutabagas, Soybeans, Sugar beets, Sugarcane (Florida and Texas only), Sweet potatoes, Tobacco, Tomatoes, Bananas and Plantain (Puerto Rico only), Turf, Grain Sorghum (Texas and Oklahoma only).

RESTRICTIONS ON WAREHOUSE STORAGE

Do not store adjacent to or near proximity of foodstuffs, drugs, feeds or any other material intended for consumption by humans or animals. Do not store adjacent to toys, cosmetics, dishes, furniture or clothing. Do not store next to herbicides. Protect from rain and moisture. In case of puncture, use overslip bag and securely close. Not for use or storage in or around the home.

STATEMENTS OF PRACTICAL TREATMENT

In case of poisoning call a physician immediately. Have patient lie down and keep quiet.

If swallowed, vomiting should be induced. Administer water freely and induce vomiting by giving one dose ($\frac{1}{2}$ oz. or 15 ml.) of syrup of ipecac. If vomiting does not occur within 10 to 20 minutes, administer second dose. If syrup of ipecac is not available, induce vomiting by sticking finger down throat. Repeat until vomit fluid is clear. Never give anything by mouth to an unconscious person. Professional medical assistance should be secured immediately.

If on skin, remove contaminated clothing and wash skin immediately with soap and warm water.

If eyes are contaminated, wash with flowing water for at least 15 minutes.

TO PHYSICIAN: Antidote—Administer atropine sulfate in large therapeutic doses. Repeat as necessary to the point of tolerance. 2-PAM is also antidotal and may be administered in conjunction with atropine.

Compound inhibits cholinesterase resulting in stimulation of the central nervous system, the parasympathetic nervous system, and the somatic motor nerves. Do not give morphine. Watch for pulmonary edema, which may develop in serious cases of poisoning even after 12 hours. At first sign of pulmonary edema, the patient should be placed in an oxygen tent and treated symptomatically.

PRECAUTIONARY STATEMENTS

HAZARDS TO HUMANS AND DOMESTIC ANIMALS

Poisonous if swallowed, inhaled, or absorbed through the skin. Rapidly absorbed through the skin. Do not get in eyes, on skin or on clothing. Do not breathe dust. Wear protective clothing and cotton or natural rubber gloves. If cotton gloves are used, they must be laundered or discarded after each day's use. Do not wear the same gloves for other work. Rubber gloves should be washed with soap and water after each use.

When using DASANIT 15% Granular in enclosed areas, wear a mask or respirator jointly approved by the Mining Enforcement and Safety Administration (formerly the U.S. Bureau of Mines) and by the National Institute for Occupational Safety and Health under the provisions of 30 CFR Part 11.

Do not contaminate feed or food. Keep out of reach of children. Keep all unprotected persons out of the operating area or vicinity. Wash hands, arms, and face thoroughly with soap and warm water before eating or smoking. Wash all contaminated clothing with soap and hot water before reuse.

SYMPTOMS OF POISONING: A sense of "tightness" in the chest. Sweating. Contracted pupils. Stomach pains. Vomiting and diarrhea.

ENVIRONMENTAL HAZARDS

This product is toxic to fish and extremely toxic to wildlife. Use with care when applying in areas frequented by wildlife. Birds feeding on treated areas may be killed. Cover, disc, or incorporate spill areas. Drift and runoff from treated areas may be hazardous to aquatic organisms in neighboring areas. Do not apply directly to water or wetlands. Do not contaminate water by cleaning of equipment or disposal of wastes. This product is highly toxic to bees exposed to direct treatment on blooming crops or weeds. Do not apply this product or allow it to drift to blooming crops or weeds while bees are actively visiting the treatment area.

RESTRICTIONS

Do not use in or near human dwellings. Do not use on plants other than those recommended. Do not apply to plant foliage other than commercial turf. Application of rates above those shown may result in illegal crop residues.

Do not treat crops grown in the greenhouse. Do not cut sod or handle treated sod for 30 days after treatment.

Do not apply this product with a knapsack or similar equipment that is placed on the user's body.

RE-ENTRY STATEMENTS

Do not apply this product in such a manner as to directly or through drift expose workers or other persons. The area being treated must be vacated by unprotected persons. Unprotected workers should not re-enter treated fields until 24 hours after application. Unprotected workers should not re-enter fields where the soil is wet until 7 days after soil application. Because certain states may require more restrictive re-entry intervals for various crops treated with this product, consult your State Department of Agriculture for further information.

Written or oral warnings must be given to workers who are expected to be in a treated area or in an area about to be treated with this product. Oral warnings should include those statements found under the label heading "Precautionary Statements" and "Re-entry Statements" included on this label. When oral warnings are given, warnings shall be given in a language customarily understood by workers. Oral warnings must be given if there is reason to believe that written warnings cannot be understood by workers. Written warnings must include the following information: DANGER, area treated with DASANIT 15% Granular on (date of application). Unprotected workers should not re-enter treated fields until 24 hours after application. Unprotected workers should not re-enter fields where the soil is wet until 7 days after soil application.

If on skin, remove contaminated clothing and wash skin immediately with soap and warm water.

If eyes are contaminated, wash with flowing water for at least 15 minutes.

If swallowed, vomiting should be induced (See Precautionary Statements on label).

STORAGE AND DISPOSAL

Do not contaminate water, food, or feed by storage or disposal. Pesticide wastes are acutely hazardous. Improper disposal of excess pesticide or rinsate is a violation of Federal Law. If these wastes cannot be disposed of by use according to label instructions, contact your State Pesticide or Environmental Control Agency, or the Hazardous Waste representative at the nearest EPA Regional Office for guidance.

Do not use container in connection with food, feed, or drinking water. Completely empty bag into application equipment. Then dispose of empty bag in a sanitary landfill or by incineration, or, if allowed by State and local authorities, by burning. If burned, stay out of smoke.

Store in a cool, dry place and in such a manner as to prevent cross contamination with other pesticides, fertilizers, food, and feed. Store in original container and out of reach of children, preferably in a locked storage area.

Handle and open container in a manner as to prevent spillage. If the container is leaking or material spilled for any reason or cause, carefully sweep material into a pile. Refer to Precautionary Statements on label for hazards associated with the handling of this material. Do not walk through spilled material. Dispose of pesticide as directed above. In spill or leak incidents, keep unauthorized people away. You may contact the Mobay Emergency Response Team for decontamination procedures or any other assistance that may be necessary. The Mobay Kansas City Emergency Response Telephone No. is 816-242-2582, or contact Chemtrec at 800-424-9300.

IMPORTANT

This bulletin is not intended to provide adequate information for application. Before using the product, read and carefully observe directions, cautionary statements and other information appearing on the label. This product is sold subject to the Conditions of Sale set forth on the container label.

READ THE LABEL BEFORE USE

DIRECTIONS FOR USE

It is a violation of Federal law to use this product in a manner inconsistent with its labeling.

AGRICULTURAL CROPS: DASANIT 15% Granular is highly effective for control of those insects noted under "RECOMMENDED APPLICATIONS," including those resistant to some chlorinated hydrocarbon insecticides. Applications should be made with granular equipment capable of distributing the required amount of granules evenly as prescribed in the recommended application for each specific crop.

DASANIT 15% Granular is effective for early-season suppression of nematodes attacking corn, cotton, tomatoes, peanuts, soybeans and sweet potatoes. For best results, it is essential that the granules be thoroughly mixed into the top 2 to 6 inches of soil by disking or using other suitable tilling equipment.

ORNAMENTALS: DASANIT 15% Granular kills plant parasitic nematodes by contact action. When applied to soil, it will not harm plants if it is thoroughly worked into the soil before transplanting. Do not direct-seed into treated areas until safety to germinating seeds has been established. One application will protect plants from nematode injury for 4 to 6 months.

COMMERCIAL TURF GRASSES: DASANIT has proven to be very effective for the control of nematodes in established commercial turf, such as sod farms, golf courses and cemeteries. When applied as recommended, one application will control nematodes for periods up to 9 months, without injury to the grass. Thorough watering of treated areas will leach the material into the root zone, resulting in maximum kill. DASANIT 15% Granular is also recommended for control of white grubs in commercial turf grasses.

Consult your local Agricultural Experiment Station or Extension authorities for information regarding proper timing of application to turf.

RECOMMENDED APPLICATIONS

CROP	PEST	DOSAGE DASANIT 15% GRAN.	REMARKS
FIELD CROPS	Seed corn beetle Seed corn maggot (Midwest states only) Corn rootworms Wireworms	4 to 8 ounces per 1000 feet of row (for any row spacing*), or 3.3 to 6.7 lbs. per acre (40 inch row spacing)	**FOR APPLICATION AT PLANTING:** Place the granules in a 4 to 6 inch band directly behind the planter shoe in front of the press wheel. Apply the high rate only for control of the seed corn beetle, seed corn maggot and wire worms. In fields with a history of moderate to severe rootworm infestations, apply 6 to 8 ounces per 1000 feet of row (for any row spacing) or 5 to 6.7 pounds per acre (40 inch row spacing). Light to moderate infestations of rootworms can be controlled with the lower rates of 4 oz. per 1000 feet of row, or 3.3 lbs. per acre for 40 inch row spacing.
Corn Field Sweet Popcorn	Corn rootworms	6 to 8 ounces per 1000 feet of row (for any row spacing*) or 5 to 6.7 pounds per acre (40 inch row spacing)	**FOR APPLICATION AT CULTIVATION (BASAL):** Distribute the granules in a uniform 4 to 6 inch band along the row around the base of the stalks. For maximum incorporation into the soil, drop the granules directly in front of the cultivating equipment. Consult your local Agricultural Extension Service for the proper timing of cultivation applications. **NOTE:** Cultivation application may be made in addition to an at-planting application, but not within 40 days of harvest for food, feed or forage.
	European corn borers (second brood) (Midwest states only)	5 to 6.7 pounds per acre	**POST-PLANT TOPICAL BROADCAST APPLICATION:** Apply specified dosage per acre into whorls about 10 days following maximum moth activity when egg masses average 1 or more per plant. A second application may be necessary in 7 to 10 days. Do not apply more than 2 topicals per season nor within 30 days of harvest for food, feed or forage. **NOTE:** Topical applications may be made in addition to soil applications described above or on the label for DASANIT Spray Concentrate.
Corn Field Sweet Popcorn (Southeastern) states only	Stubby root and sting nematodes (Early-season suppression)	14.5 oz. per 1000 feet of row (for any row spacing*), or 13½ pounds per acre (36 inch row spacing)	**FOR APPLICATION AT PLANTING:** Place the granules in a 12 inch band over the row in front of the planter shoe. Incorporate thoroughly to a depth of 3 inches to insure adequate mixing of granules in the soil.
Sugar beets	Sugar beet root maggot	4½ to 9 ounces per 1000 feet of row (for any row spacing*), or 6⅔ to 13⅓ lbs. per acre (22 inch row spacing)	Apply specified dosage in a 4 to 6 inch band in one of the following ways: (1) In front of the planter. Incorporate the granules lightly into the soil with shallow tilling equipment. Plant seed into the center of the treated band of soil. (2) Directly behind the planter shoe in front of the press wheel. Place the granules 1 inch or more above seed. Do not allow the granules to come into direct contact with the seed. NOTE: For light sandy soils, use only the lowest dosage (4½ oz. per 1000 feet of row, or 6⅔ lbs. per acre). Use higher dosages (up to the maximum of 9 oz. per 1000 feet of row, or 13⅓ lbs. per acre) on heavier soils, or soils that are high in organic matter.
Sugarcane	Nematodes (Florida only)	1.5 to 3.8 lbs. per 1000 feet of row regardless of row spacing*, or 13.3 to 33.3 lbs. per acre (5 foot row spacing)	Apply specified dosage in a 12 to 18 inch band in open furrow over the seed pieces at planting.
	White grubs (*Phyllophaga crinita*) (Texas only)	13.3 to 26.6 lbs. per acre	Apply specified dosage, using ground or air equipment, as a broadcast treatment at layby. Incorporate mechanically or with irrigation water. Do not harvest within 150 days of treatment. Do not use treated foliage as animal feed.
Tobacco	Wireworms	20 ounces per 1000 feet of row regardless of row spacing*, or 13.3 lbs. per acre (broadcast)	Apply as a broadcast application or in a 10 to 12 inch band over the row before planting. After the granules are uniformly applied, disk or till thoroughly to a depth of 2 to 3 inches to insure adequate mixing in the soil. Plant crop in the usual manner.
FRUITS Bananas Plantain (Puerto Rico only)	Nematodes (burrowing, spiral, reniform)	1 to 1½ oz.	Distribute the specified dosage uniformly around the base of each plant in a 2 to 2½ foot band, using suitable granular applicator equipment such as a shaker can. Protective clothing and cotton or natural rubber gloves as described in the precautionary statements on this label should be worn while making applications. Apply at planting, and at 6 month intervals. Do not apply within 60 days of harvest. The granules should be mixed with the covering soil at planting. During dry or wet periods, at planting and post planting applications should be incorporated into the soil to a depth of approximately 2 inches.
VEGETABLES Dry bulb onions	Onion maggot	3.7 oz. per 1000 feet of row (for any row spacing*), or 6.7 lbs. per acre (18 inch row spacing)	Distribute the granules evenly in the seed furrow at planting time. Do not apply to green-bunch onions. Under certain conditions, application may result in reduced yields.
Cotton (Southeastern and Gulf states only)	Nematodes (Early-season suppression)	26⅔ to 53⅓ oz./ 1000 feet of row (for any row spacing*), or 21⅓ to 43⅓ lbs. per acre (40 inch row spacing)	Apply uniformly in a 12 inch band at planting. Thoroughly mix the granules into the soil to a depth of 4 to 6 inches. Plant seed down the center of the treated band. Do not use foliage for feed or forage.
Grain Sorghum (Texas and Oklahoma only)	White grubs (*Phyllophaga crinita*) (Suppression of feeding damage)	8 oz. per 1000 feet of row (for any row spacing*), or 6.5 pounds per acre (40 inch row spacing)	Apply the granules in the seed furrow at planting as a 2 inch band, approximately ½ inch above the seed. Do not apply directly to the seed.

CROP	PEST	DOSAGE DASANIT 15% GRAN.	REMARKS
Peanuts	Thrips (Early-season reduction) (Southeastern states only) Nematodes (Early season suppression)	$14\frac{2}{3}$ to $29\frac{1}{3}$ oz. per 1000 feet of row (for any row spacing*), or $13\frac{1}{3}$ to $26\frac{2}{3}$ lbs. per acre (36 inch row spacing)	**BAND APPLICATION:** Apply uniformly in a 12 to 18 inch band. Thoroughly mix the granules into the soil to a depth of 4 to 6 inches. Plant seed down the center of the treated band. Do not feed vines or hay. Do not exceed the stated broadcast rate per acre, no matter how the pesticide is applied.
		20 to 40 pounds per acre	**BROADCAST APPLICATION:** Apply specified dosage uniformly per acre. Disk or till the soil thoroughly to a depth of 4 to 6 inches to insure adequate mixing in the soil. Plant crop in the usual manner. Do not feed vines or hay.
	Lesser cornstalk borer (Alabama and Georgia, and on irrigated peanuts in Texas and Oklahoma.) Southern corn rootworm	$14\frac{2}{3}$ to $29\frac{1}{3}$ oz. per 1000 feet of row (for any row spacing*), or $13\frac{1}{3}$ to $26\frac{2}{3}$ lbs. per acre (36 inch row spacing)	Apply uniformly in a 12 to 18 inch band over the row at pegging. On irrigated peanuts, follow application with irrigation. For maximum control of corn rootworm, mix the granules into the soil to a depth of 1 to 2 inches on each side of the row. For moderate to heavy infestations of lesser cornstalk borer, a second pegging-time application should be made 18 to 21 days later, but not within 65 days of harvest. Applications at pegging should not exceed a total of 40 pounds per acre. Do not exceed the stated broadcast rate per acre, no matter how the pesticide is applied. **NOTE:** Do not apply a total of more than $46\frac{2}{3}$ pounds of DASANIT 15% Granular in any single crop year. Do not feed vines or hay.
	Lesion nematode (Early-season suppression)	16.5 to 22 oz./ 1000 feet of row (for any row spacing*), 15 to 20 lbs. per acre (36 inch row spacing)	Apply in a 12 to 14 inch band over the row at pegging. Mix granules into the soil to a depth of 1 to 2 inches on each side of the row if plant size permits. For maximum performance, irrigation or rainfall should follow soon after application. Application may be made up to 65 days of harvest. **NOTE:** Where at planting or other applications of DASANIT 15% Granular are made, do not apply a total of more than $46\frac{2}{3}$ pounds of product in any crop year. Do not feed vines or hay.
Soybeans	Nematodes (Except cyst nematode) (Early-season suppression)	$6\frac{2}{3}$ to $13\frac{1}{3}$ oz. per 1000 feet of row (for any row spacing*), or 6 to 12 lbs. per acre (36 inch row spacing)	Apply uniformly in a 12 inch band at planting. Thoroughly mix the granules into the soil to a depth of 4 to 6 inches. Plant seed down the center of the treated band.
Potatoes	Tobacco wireworms Southern potato wireworms	13.3 pounds per acre	Apply specified dosage as a preplant broadcast application. Mix the insecticide into the top 3 to 4 inches of soil by disking or with other suitable tilling equipment. Plant crop in the usual manner.
	Tuber flea beetle larvae Wireworms (*Ctenicera* spp.)	33.3 pounds per acre	
Rutabagas	Cabbage maggot	6.7 to 13.3 ounces per 1000 feet of row (for any row spacing*), but not more than 16.7 lbs. total per acre	Apply specified dosage in a 4 to 6 inch band prior to seeding. Cultivate granules into top 1 inch of soil, then seed into the center of treated band. Make a second application approximately 4 weeks after planting, using DASANIT Spray Concentrate in water applied as a drench to wet the crown of the plant and 3 inches of soil on both sides of the row. One or two additional drench-type sprays may be applied as required, but not closer than 40 days from harvest. Do not apply DASANIT more than 4 times per season.
Sweet potatoes	Nematodes (Earlyseason suppression)	20 to 46.7 lbs. per acre	**BROADCAST APPLICATION:** Apply specified dosage as a preplant broadcast application. Incorporate into the soil 4 to 6 inches deep. Plant crop in the usual manner. Use the higher dosage for fields with a high population of nematodes, or fields with a history of serious nematode damage.
	Southern potato wireworms (aids in the reduction of feeding damage) (Southeastern states only)	46.7 lbs. per acre	
Tomatoes	Nematodes (Early-season suppression)	66.7 to 134 pounds per acre	**BROADCAST APPLICATION:** Apply specified dosage as a uniform broadcast application, and incorporate into the soil 4 to 6 inches deep by disking or with other suitable tilling equipment. Plant crop in usual manner.
		1.5 to 3 pounds per 1000 feet of row (for any row spacing*), or 21.1 to 42.2 pounds per acre (38 inch row spacing)	**BAND APPLICATION:** Apply specified dosage uniformly in a 12 inch band over the row before planting. Disk or till thoroughly to a depth of 4 to 6 inches to insure adequate mixing of chemical in the soil. Plant crop in usual manner.

CROP	PEST	DOSAGE DASANIT 15% GRAN.	REMARKS
Commercial turf grasses	Nematodes	$1\frac{1}{2}$ to 3 lbs. per 1000 sq. ft. (approximately 68 to 134 lbs. per acre)	Distribute specified dosage evenly over turf area to be treated. Immediately following treatment, apply $\frac{1}{4}$ to $\frac{1}{2}$ inch of water to drench the insecticide into the soil. Do not treat newly-seeded areas. Do not use on grasses other than those specified. Treated turf should not be cut for sod, or sod handled for 30 days after treatment.
	White grubs (including Japanese beetle, *Ataenius spretulus*, Chafers, *Phyllophaga* spp. Asiatic garden beetle larvae) (Except California)	6.25 oz./1000 sq. ft. approximately 17 lbs. per acre)	
ORNAMENTALS Flowering Shrubs: Azaela Blue mist Camellia Chinese hibiscus Gardenia Honeysuckle Jasmine Jungleflame Metrosideros Pyracantha Surinam cherry Roses Viburnum Perennial Flowers: Aloe Butterfly iris Century plant Chrysanthemum Cyclamen Easter lily Gladiolus Gloxinia Vines: Bougainvillea Other Shrubs:	Nematodes	6.7 pounds per 1000 sq. ft.	**Preplant Treatment of Beds and Benches (Outdoor only):** Apply specified dosage evenly over bed or bench area. Work thoroughly into soil to a depth of 4 to 6 inches. Do not apply to newly-seeded areas.
Ardisia Boxwood Croton Dwarf yaupon Dwarf Japanese holly Euonymus Eurya Gallberry Holly Holly leaf osmanthus Japanese yew Juniper Privet Natal plum Mahonia (California barberry) Red leaf barberry Yew Ornamental Foliage Plants	Nematodes	(See Remarks)	**Potting Soil (Outdoor only):** Treat potting soil by thoroughly mixing $\frac{1}{3}$ ounce of DASANIT 15% Granular with 1 cubic foot of soil ($7\frac{1}{2}$ one-gallon cans of soil = 1 cubic foot). 5-28-85
	IMPORTANT: Do not treat peony, begonia, hydrangea or *Aucuba* spp. because of possible plant injury.		

* WHERE ROW SPACING IS EXTREMELY NARROW, DO NOT PLACE TREATED ZONES (BAND OR FURROW) CLOSER TOGETHER THAN 6 INCHES.

SPECIAL LOCAL NEED REGISTRATIONS [SECTION 24(c)] FOR DASANIT 15% GRANULAR*

CROP	INSECT	STATE EPA SLN NO.	DOSAGE	REMARKS
Commercial seed grass	Billbugs	Oregon OR 76-0001	26.7 pounds per acre.	Broadcast application.
Potatoes	Southern potato wireworm Tobacco wireworms	Idaho ID 78-0012 Oregon OR 81-0084 Washington WA 79-0049	13.3 pounds per acre.	Apply as a preplant broadcast application.
	Tuber flea beetle larvae Wireworms (*Ctenicera* spp.) (*Limonius* spp.)		33.3 pounds per acre.	
	Nematodes (suppression of low to moderate populations.)	Idaho ID 78-0012		
	Colorado potato beetle Wireworms (*Ctenicera* spp.) (*Limonius* spp.)	Idaho ID 78-0012 Oregon OR 81-0084 Washington WA 79-0049	20 pounds based on 36 inch rows or 23 ounces per 1000 foot of row for any row spacing.	Apply at-planting in a 4 to 6 inch band.

* For distribution and use only within the state specified. These Special Local Need registrations approved as of date of this publication are, however, subject to cancellation by the state. Contact your nearest Mobay Agricultural Chemicals Division Office for the current status and a copy of the complete individual supplemental Special Local Need labeling. This labeling must be in the possession of the user at time of pesticide application.

®DASANIT

SPRAY CONCENTRATE INSECTICIDE—NEMATICIDE

RESTRICTED USE PESTICIDE

For retail sale to and use only by Certified Applicators or persons under the direct supervision of a Certified Applicator, and only for those uses covered by the Certified Applicator's certification.

For Effective Control of Soil Insects and Suppression of Parasitic Nematodes.

For Use Only by Commercial Growers and Applicators

ACTIVE INGREDIENT:

O, O-Diethyl *O*-[4-(methylsulfinyl)phenyl] phosphorothioate 63%

INERT INGREDIENTS 37%

TOTAL 100%

Contains 6 lbs. *O, O*-Diethyl *O*-[4-(methylsulfinyl)phenyl] phosphorothioate per gallon.

EPA Reg. No. 3125-163-AA

DASANIT is a Reg. TM of the Parent Company of Farbenfabriken Bayer GmbH, Leverkusen.

Tillam is a Reg. TM of Stauffer Chemical Company.

DANGER–POISON

Keep Out of Reach of Children

PELIGRO

PRECAUCION AL USUARIO: Si usted no puede leer o entender inglés, no use este producto hasta que la etiqueta le haya sido explicada ampliamente.

(TO THE USER: If you cannot read or understand English, do not use this product until the label has been fully explained to you.)

RECOMMENDED USES

Tobacco	Tomatoes
Citrus seedlings	Grain sorghum (Texas and Oklahoma only)
Corn	
Dry bulb onions	Sugarcane (Texas only)
Peanuts	
Rutabagas	Sweet potatoes

RESTRICTIONS ON WAREHOUSE STORAGE

Do not store adjacent to or near proximity of foodstuffs, drugs, feeds or any other material intended for consumption by humans or animals. Do not store adjacent to toys, cosmetics, dishes, furniture or clothing. Do not store next to herbicides. Store below 140°F. Do not store in front of space heater.

STATEMENTS OF PRACTICAL TREATMENT

In case of poisoning, call a physician immediately. Have patient lie down and keep quiet.

If swallowed, vomiting should be induced. Administer water freely, and induce vomiting by giving one dose (½ oz. or 15 ml.) of syrup of ipecac. If vomiting does not occur within 10 to 20 minutes, administer second dose. If syrup of ipecac is not available, induce vomiting by sticking finger down throat. Repeat until vomit fluid is clear. Never give anything by mouth to an unconscious person. Professional medical assistance should be secured immediately.

If on skin, remove contaminated clothing and wash skin immediately with soap and warm water.

If eyes are contaminated, wash with flowing water for at least 15 minutes.

TO PHYSICIAN: Antidote—Administer atropine sulfate in large therapeutic doses. Repeat as necessary, to the point of tolerance. 2-PAM is also antidotal and may be administered in conjunction with atropine.

Compound inhibits cholinesterase resulting in stimulation of the central nervous system, the parasympathetic nervous system, and the somatic motor nerves. Do not give morphine. Watch for pulmonary edema, which may develop in serious cases of poisoning even after 12 hours. At first sign of pulmonary edema, the patient should be placed in an oxygen tent and treated symptomatically.

PRECAUTIONARY STATEMENTS

HAZARDS TO HUMANS AND DOMESTIC ANIMALS

Poisonous if swallowed, inhaled or absorbed through the skin. Rapidly absorbed through the skin. Do not get in eyes or on skin. Wear protective clothing, natural rubber gloves and goggles.

Do not breathe fumes or spray mist. For protection against DASANIT insecticide-nematicide, wear a mask or respirator jointly approved by the Mining Enforcement and Safety Administration (formerly the U.S. Bureau of Mines) and by the National Institute for Occupational Safety and Health under the provisions of 30 CFR Part 11.

Do not contaminate feed or food. Keep out of reach of children. Keep all unprotected persons out of the operating area until DASANIT Spray Concentrate has been thoroughly drenched into or covered by soil, and all mist has dissipated. Wash hands, arms and face thoroughly with soap and warm water before eating or smoking.

HANDLING OF CONTAMINATED ARTICLES: Clothing and rags receiving undiluted DASANIT Spray Concentrate as a result of leaks, spills or mishaps should be removed immediately and not reused. These contaminated articles should be buried in the same manner as the empty container; see container disposal section for directions. (Contaminated articles which should be discarded include: hats, gloves, aprons, coats, boots, etc. not made of rubber or coated with rubber or other similar materials.) Clothing receiving spray mist or droplets from the mixed, diluted DASANIT Spray Concentrate should be removed and decontaminated before reuse. This clothing can be decontaminated by machine washing separately from other items with soap or detergent and bleach in hot water twice before reuse. Protective clothing and equipment should be washed down with detergent or soap and bleach in water. Wash-water from the cleaning of protective clothing or equipment should not be allowed to run off or otherwise enter water supplies.

SYMPTOMS OF POISONING: A sense of "tightness" in the chest. Contracted pupils. Stomach pains. Vomiting and diarrhea.

ENVIRONMENTAL HAZARDS

This product is toxic to fish and extremely toxic to wildlife. Use with care when applying in areas frequented by wildlife. Birds feeding on treated areas may be killed. Cover, disc, or incorporate spill areas. Drift and runoff from treated areas may be hazardous to aquatic organisms in neighboring areas. Do not apply directly to water or wetlands. Do not contaminate water by cleaning of equipment or disposal of wastes. This product is highly toxic to bees exposed to direct treatment on blooming crops or weeds. Do not apply this product or allow it to drift to blooming crops or weeds while bees are actively visiting the treatment area.

READ THE LABEL BEFORE USE

RESTRICTIONS

Do not use in or near human dwellings. Do not use on plants other than those recommended. Use only according to label directions. Applications at rates above those shown may result in illegal crop residues. Do not apply to plant foliage, or treat food crops grown in the greenhouse.

RE-ENTRY STATEMENTS

Do not apply this product in such a manner as to directly or through drift expose workers or other persons. The area being treated must be vacated by unprotected persons. Unprotected workers should not re-enter treated fields until 24 hours after application. Unprotected workers should not re-enter fields where the soil is wet until 7 days after soil application. Because certain states may require more restrictive re-entry intervals for various crops treated with this product, consult your State Department of Agriculture for further information.

Written or oral warnings must be given to workers who are expected to be in a treated area or in an area about to be treated with this product. Oral warnings should include those statements found under the label heading "Precautionary Statements" and "Re-entry Statements" included on this label. When oral warnings are given, warnings shall be given in a language customarily understood by workers. Oral warnings must be given if there is reason to believe that written warnings cannot be understood by workers. Written warnings must include the following information: DANGER, area treated with DASANIT Spray Concentrate on (date of application). Unprotected workers should not re-enter treated fields until 24 hours after application. Unprotected workers should not re-enter fields where the soil is wet until 7 days after soil application. **If on skin,** remove contaminated clothing and wash skin immediately with soap and warm water. **If eyes are contaminated,** wash with flowing water for at least 15 minutes. **If swallowed,** vomiting should be induced. (See Precautionary Statements on label).

STORAGE AND DISPOSAL

PESTICIDE DISPOSAL

Do not contaminate water, food or feed by storage or disposal.

Pesticide wastes are acutely hazardous. Improper disposal of excess pesticide, spray mixture, or rinsate is a violation of Federal Law. If these wastes cannot be disposed of by use according to label instructions, contact your State Pesticide or Environmental Control Agency for guidance.

CONTAINER DISPOSAL

Triple rinse (or equivalent). Then offer for recycling or reconditioning, or puncture and dispose of in a sanitary landfill, or by other procedures approved by State and local authorities.

Store in a cool, dry place and in such a manner as to prevent cross contamination with other pesticides, fertilizers, food, and feed. Store away from open flame and extreme heat. Store in original container and out of the reach of children, preferably in a locked storage area.

Handle and open container in a manner as to prevent spillage. If container is leaking invert to prevent leakage. If the container is leaking or material spilled for any reason or cause, carefully dam up spilled material to prevent runoff. Refer to Precautionary Statements on label for hazards associated with the handling of this material. Do not walk through spilled material. Absorb spilled material with absorbing type compounds and dispose of as directed for pesticides above. In spill or leak incidents, keep unauthorized people away. You may contact the Mobay Emergency Response Team for decontamination procedures or any other assistance that may be necessary. The Mobay Kansas City Response Telephone No. is 816-242-2582, or contact Chemtrec at 800-424-9300.

IMPORTANT: This bulletin is not intended to provide adequate information for application. Before using the product, read and carefully observe directions, cautionary statements and other information appearing on the label. This product is sold subject to the Conditions of Sale set forth on the container label.

DIRECTIONS FOR USE

It is a violation of Federal Law to use this product in a manner inconsistent with its labeling.

DASANIT Spray Concentrate is highly effective for control of those insects noted under "Recommended Applications", including those resistant to some chlorinated hydrocarbon insecticides. Applications should be made with spray equipment capable of distributing the required amount of product evenly as described in the recommended application for each crop.

DASANIT Spray Concentrate is effective for early season suppression of nematodes attacking peanuts, tomatoes and sweet potatoes. For best results it is essential that DASANIT Spray Concentrate be thoroughly mixed into the top 2 to 6 inches of soil by disking or using other suitable tilling equipment.

RECOMMENDED APPLICATIONS

CROP	PEST	DOSAGE DASANIT S.C.	REMARKS
FIELD CROPS		0.8 to 1.6 fl. ounces per 1000 feet of row (for any row spacing*), or ⅔ to 1⅓ pints per acre (40 inch row spacing)	**FOR APPLICATION AT PLANTING:** Apply as a water-emulsion in a 4 to 6 inch band directly behind the planter shoe in front of the press wheel. **FOR SPLIT-BOOT APPLICATION AT PLANTING WITH LIQUID FERTILIZER:** Apply in combination with liquid fertilizer† as a split-boot application (bands 2 inches away from the seed on both sides of the planted row). † Where sufficient agitation is not available, certain liquid high phosphorus fertilizers may require the use of Compex* compatibility agent at the rate of 1 quart per 100 gals. of mixture to assure proper mixing.
Corn Field Sweet Popcorn	Corn rootworm	1.2 to 1.6 fl. ounces per 1000 feet of row (for any row spacing*), or 1 to 1⅓ pints per acre (40 inch row spacing)	**FOR APPLICATION AT CULTIVATION:** Apply as a water-emulsion evenly in a 4 to 6 inch band along the row around the base of the stalks. For maximum incorporation into the soil, apply directly in front of the cultivating equipment. Consult your local Agricultural Extension Service for proper timing of cultivation applications. **NOTE:** Cultivation application may be made in addition to an at-planting application, but not within 40 days of harvest for food, feed or forage.
Peanuts	Thrips (Early-season reduction)	3 to 6 fl. oz. per 1000 ft. of row (regardless of row spacing*), or 2.7 to 5.4 pints per acre (36 inch row spacing)	**BAND APPLICATION:** Apply uniformly as a water-emulsion spray in a 12 to 18 inch band. Disk or till the soil thoroughly to a depth of 4 to 6 inches to insure adequate mixing in the soil. Plant seed into the center of the treated band of soil. Do not feed vines or hay. Do not exceed the stated broadcast rate per acre, no matter how the pesticide is applied.
	Nematodes (Early-season suppresion)	4 to 8 pints (½ to 1 gallon) per acre	**BROADCAST APPLICATION:** Apply specified dosage uniformly as a water-emulsion spray. Disk or till the soil thoroughly to a depth of 4 to 6 inches to insure adequate mixing in the soil. Plant crop in the usual manner. Do not feed vines or hay.
	Lesser cornstalk borer (Texas and Oklahoma only)	1½ fl. oz. per 1000 ft. of row (regardless of row spacing*), or 1⅓ pints per acre (36 inch row spacing)	**POST-PLANTING BAND APPLICATIONS:** Apply specified dosage as a basal-directed spray, using 20 to 30 gallons of water per acre. Apply as needed, up to a maximum of 4 times per crop. Do not apply within 30 days before harvest. Do not feed vines or hay. Do not exceed the stated broadcast rate per acre, no matter how the pesticide is applied. **NOTE:** If soil applications of DASANIT 15% Granular or DASANIT Spray Concentrate have been made at planting or at pegging, do not apply these basal-directed sprays.
Sugarcane (Texas only)	White grubs *(Phyllophaga crinita)*	⅓ to ⅔ gallons per acre	Apply specified dosage, as a layby treatment. Apply in 5 to 15 gallons total spray volume per acre as a broadcast interline spray and incorporate mechanically or with irrigation water. Do not spray sugarcane foliage. Do not harvest within 150 days of treatment. Do not use treated foliage as animal feed.
Tobacco	Wireworms	4 fl. ounces per 1000 feet of row (regardless of row spacing*), or ⅓ gallon per acre (broadcast)	Apply in a water-emulsion as a broadcast spray, or in a 10 to 12 inch band over the row before planting. After the material is uniformly applied, disk or till thoroughly to a depth of 2 to 3 inches to insure adequate mixing in the soil. Plant crop in the usual manner. Do not exceed the stated broadcast rate per acre, no matter how the pesticide is applied.
Grain sorghum (Texas and Oklahoma only)	White grubs *(Phyllophaga crinita)* (Suppression of feeding damage)	1.7 fl. oz. per 1000 feet of row (for any row spacing*), or 22 fl. oz. per acre (40 inch row spacing)	Apply specified dosage in 20 gallons of water as a 3 to 4 inch band in the seed furrow at planting. Apply approximately ½ inch above the seed. Do not apply directly to the seed.
FRUIT		⅓ to ½ pint	**Bare Root Dip:** Mix specified dosage per 50 gallons of water. This is equivalent to approximately 575 to 860 ppm active ingredient. Remove soil from roots of plants to be treated, and immerse only the root systems into the solution. Allow plants to soak for 30 minutes, then remove for planting or shipment. Treated plants to be moved into commercial trade within one week after dipping must bear warning tags containing the same information as in the warning section on this label. These tags are available from your dealer. Treated plants to be held in storage for more than one week do not require tags.
Citrus seedlings	Nematodes	(See Remarks)	**Potted Plants:** Add 3 to 6 fluid ounces of DASANIT Spray Concentrate to 25 gallons of water. Apply ⅓ pint of this solution per 8-inch pot. For best results apply pot drenches when the soil is damp. Do not apply to dry pots, or to plants in a wilted condition.
VEGETABLES Onions (dry bulb)	Onion maggot	0.7 fl. oz. per 1000 ft. of row (regardless of row spacing*), or 1.3 pints per acre (18 inch row spacing)	Mix the required amount of DASANIT Spray Concentrate in sufficient water (approximately 100 gallons of water per acre) to distribute the chemical evenly in the seed furrow at planting time. Do not apply to green-bunch onions.
Rutabagas	Cabbage maggot	1⅓ to 2⅔ fl. ounces in not less than 3½ gallons of water per 1000 feet of row (for any row spacing), but not more than 3 pints total per acre	Apply specified dosage in a 6 inch band sprayed over the row immediately after seeding. Make a second application at the same rate in a 6 inch band sprayed over the row approximately 4 weeks after planting. One or two additional sprays may be applied as required, but not closer than 40 days from harvest. Apply as drenches to wet the crown of the plant and 3 inches of soil on both sides of the row. Do not apply DASANIT more than 4 times per season.
		1⅔ to 3⅓ gallons per acre	**Broadcast Application:** Apply specified dosage uniformly as a preplant broadcast spray. Mix insecticide into the top 4 to 6 inches of soil by disking or with other suitable tilling equipment. Plant crop in the usual manner.

CROP	PEST	DOSAGE DASANIT S.C.	REMARKS
Tomatoes	Nematodes (Early-season suppression)	5 to $9\frac{1}{2}$ fluid ounces per 1000 feet of row (any row spacing*), or $4\frac{1}{3}$ to $8\frac{1}{3}$ pints per acre (38 inch row spacing)	**Band Application:** Apply in a water-emulsion in a uniform 12 inch band over the row before planting. Disk or till thoroughly to a depth of 4 to 6 inches to insure adequate mixing in the soil. Plant crop in the usual manner. Do not exceed the stated broadcast rate per acre, no matter how the pesticide is applied.
Sweet potatoes	Nematodes: *Meloidogyne-* (Root knot) *Rotylenchu-lus* Reniform (Except California) *Helicotylenchus* Spiral (Early-season suppression) (Except California)	4 to 9.3 pints per acre (.5 to $1\frac{1}{6}$ gal)	**Broadcast Application:** Apply specified dosage as a preplant broadcast application. Incorporate into the soil 4 to 6 inches deep. Plant crop in the usual manner. Use the higher dosage for fields with a high population of neomatodes or fields with a history of serious nematode damage. 5-28-85
	Southern potato wireworms (aids in the reduction of feeding damage. Southeastern states only.)	9.3 pints per acre ($1\frac{1}{6}$ gal)	

* WHERE ROW SPACING IS EXTREMELY NARROW, DO NOT PLACE TREATED ZONES (BAND OR FURROW) CLOSER TOGETHER THAN 6 INCHES.

DASANIT Spray Concentrate in Combination with Tillam® 6-E for Use on Tobacco:

For control of wireworms, and certain weeds in tobacco (flue-cured and light air-cured), DASANIT Spray Concentrate may be tank-mixed with Tillam 6-E. For complete list of weeds controlled refer to the label for Tillam 6-E.

Soil Preparation: The soil should be well-worked and dry enough to permit good soil incorporation (mixing into the soil).

Mixing: Add to the spray tank $2\frac{2}{3}$ qts. Tillam 6-E and 4 qts. DASANIT Spray Concentrate in 20 to 50 gal. of water per acre. Use a properly calibrated low-pressure boom-type sprayer having good agitation.

Soil Incorporation: The tank mix must be incorporated (mixed) into the soil immediately after application to prevent loss of the herbicide by evaporation. Whenever possible, application and incorporation should be done in the same operation.

Use power-driven cultivation equipment set to cut to a depth of 3″ in light soil, and 3 to 4″ in heavy soil, or a tandem disc set to cut to a depth of 4 to 6″.

Cross-disc at right angles for thorough mixing. The bedding operation should follow the incorporation as soon as practical. Planting may be done immediately after bedding or at a later date; however, to insure long-term weed control, the crop should be planted as soon as possible after application and bedding.

NOTE: Refer to the label for Tillam 6-E for pertinent directions and precautions for use, and restrictions.

SPECIAL LOCAL NEED REGISTRATIONS [SECTION 24(c)] FOR DASANIT SPRAY CONCENTRATE*

CROP	PEST	STATE EPA SLN NO.	DOSAGE	REMARKS
Potatoes	Wireworms (*Ctenicera spp.*) (*Limonius spp.*) Nematodes (suppression of low to moderate populations.)	Idaho ID 79-0016	$6\frac{2}{3}$ pints per acre.	Apply specified dosage in a water-emulsion or with liquid fertilizer as a preplant broadcast application.
Tobacco (Flue-cured)	Root-knot nematodes Wireworms	Georgia GA 78-0012 North Carolina NC 78-0005 South Carolina SC 79-0006 Virginia VA 78-0012	Tank-mix $2\frac{2}{3}$ quarts DASANIT Spray Concentrate and $2\frac{2}{3}$ quarts NEMACUR 3 in 10 to 50 gallons of water per acre	Uniformly broadcast and incorporate thoroughly followed by bedding.
Tomatoes	Nematodes	California CA 78-0201	5 to $9\frac{1}{2}$ fluid ounces per 1000 feet of row (any row spacing**) or $4\frac{1}{3}$ to $8\frac{1}{3}$ pints per acre (38 inch row spacing).	Soil injection in a water-emulsion or with liquid fertilizer.

* For distribution and use only within the state specified. These Special Local Need registrations approved as of date of this publication are, however, subject to cancellation by the state. Contact your nearest Mobay Agricultural Chemicals Division Office for the current status and a copy of the complete individual supplemental Special Local Need labeling. This labeling must be in the possession of the user at time of pesticide application.

** Where row spacing is extremely narrow, do not place treated zones (band or furrow) closer together than 6 inches.

®DASANIT
®DI-SYSTON

4-2 SPRAY CONCENTRATE INSECTICIDE—NEMATICIDE

RESTRICTED USE PESTICIDE

Acute human oral, dermal, and inhalation toxicity.

For retail sale to and use only by Certified Applicators or persons under their direct supervision, and only for those uses covered by the Certified Applicator's certification.

Three-Way Protection For Tobacco Transplants.

For Use Only by Commercial Growers and Applicators.

ACTIVE INGREDIENTS:

O, O-Diethyl *O*-[4-(methylsulfinyl) phenyl] phosphorothioate	43%
O, O-Diethyl *S*-[2-(ethylthio)ethyl] phosphorodithioate	21%
INERT INGREDIENTS	36%
TOTAL	100%

Contains 4 lbs. *O, O*-Diethyl *O*-[4-(methylsulfinyl) pheynl] phosphorothioate and 2 lbs. *O, O*-Diethyl *S*-[2-(ethylthio) ethyl] phosphorodithioate per gallon.

EPA Reg. No. 3125-299

DASANIT and DI-SYSTON are Reg. TMs of the Parent Company of Farbenfabriken Bayer GmbH, Leverkusen.

DANGER–POISON

Keep Out of Reach of Children
PELIGRO

PRECAUCION AL USUARIO: Si usted no puede leer o entender inglés, no use este producto hasta que la etiqueta le haya sido explicada ampliamente.

(TO THE USER: If you cannot read or understand English, do not use this product until the label has been fully explained to you.)

RECOMMENDED USES

Tobacco transplants—Registered for use only in Delaware, Maryland, Virginia, West Virginia, North Carolina, South Carolina, and Georgia.

RESTRICTIONS ON WAREHOUSE STORAGE

Do not store adjacent to or near proximity of foodstuffs, drugs, feeds or any other material intended for consumption by humans or animals. Do not store adjacent to toys, cosmetics, dishes, furniture or clothing. Do not store next to herbicides. Store below 140°F. Do not store in front of space heater.

STATEMENTS OF PRACTICAL TREATMENT

In case of poisoning, call a physician immediately. Have patient lie down and keep quiet.

If swallowed, vomiting should be induced. Administer water freely, and induce vomiting by giving one dose (½ oz. or 15 ml.) of syrup of ipecac. If vomiting does not occur within 10 to 20 minutes, administer second dose. If syrup of ipecac is not available, induce vomiting by sticking finger down throat. Repeat until vomit fluid is clear. Never give anything by mouth to an unconscious person. Professional medical assistance should be secured immediately.

If on skin, remove contaminated clothing and wash skin immediately with soap and warm water.

If eyes are contaminated, wash with flowing water for at least 15 minutes.

TO PHYSICIAN: Antidote—Administer atropine sulfate in large therapeutic doses. Repeat as necessary, to the point of tolerance. 2-PAM is also antidotal, and may be administered in conjunction with atropine.

Compound inhibits cholinesterase, resulting in stimulation of the central nervous system, the parasympathetic nervous system and the somatic motor nerves. Do not give morphine. Watch for pulmonary edema, which may develop in serious cases of poisoning, even after 12 hours. At first sign of pulmonary edema, the patient should be placed in an oxygen tent and treated symptomatically.

PRECAUTIONARY STATEMENTS

HAZARDS TO HUMANS AND DOMESTIC ANIMALS

Poisonous if swallowed, inhaled or absorbed through the skin. Rapidly absorbed through the skin. Do not get in eyes, on skin, or on clothing. Wear waterproof protective clothing, natural rubber gloves and goggles.

Do not breathe fumes or spray mist. For protection against DASANIT and DI-SYSTON insecticides wear a mask or respirator jointly approved by the Mining Enforcement and Safety Administration (formerly the U.S. Bureau of Mines) and by the National Institute for Occupational Safety and Health under the provisions of 30 CFR Part 11.

Do not contaminate feed or food. Keep out of reach of children. Keep all unprotected persons out of the operating area or vicinity. Wash hands, arms and face thoroughly with soap and warm water before eating or smoking.

HANDLING OF CONTAMINATED ARTICLES: Clothing and rags receiving undiluted DASANIT + DI-SYSTON 4-2 Spray Concentrate as a result of leaks, spills or mishaps should be removed immediately and not reused. These contaminated articles should be buried in the same manner as the empty container (see container disposal section for directions). Contaminated articles which should be discarded include: hats, gloves, aprons, coats, boots, etc. not made of rubber or other similar materials. Clothing receiving spray mist or droplets from the mixed, diluted DASANIT + DI-SYSTON 4-2 Spray Concentrate should be removed and decontaminated before reuse. This clothing can be decontaminated by machine washing separately from other items with soap or detergent and bleach in hot water twice before reuse. Protective clothing and equipment should be washed down with detergent or soap and bleach in water. Wash-water from the cleaning of protective clothing or equipment should not be allowed to run off or otherwise enter water supplies.

SYMPTOMS OF POISONING: A sense of "tightness" in the chest. Sweating. Contracted pupils. Stomach pains. Vomiting and diarrhea.

ENVIRONMENTAL HAZARDS

This product is toxic to fish and extremely toxic to wildlife. Use with care when applying in areas frequented by wildlife. Birds feeding on treated areas may be killed. Cover, disc, or incorporate spill areas. Drift and runoff from treated areas may be hazardous to aquatic organisms in neighboring areas. Do not apply directly to water or wetlands. Do not contaminate water by cleaning of equipment or disposal of wastes. This product is highly toxic to bees exposed to direct treatment on blooming crops or weeds. Do not apply this product or allow it to drift to blooming crops or weeds while bees are actively visiting the treatment area.

RESTRICTIONS

Not for home garden use. Do not use on other crops used for food or forage. Use only according to label directions. Application at rates above those shown may result in illegal crop residues. Do not treat food crops. Do not apply in any manner not specified on the label.

CROP ROTATION STATEMENT

Do not plant any food or feed crop in rotation after a field treatment with disulfoton unless it is a registered use for disulfoton.

RE-ENTRY STATEMENTS

Do not apply this product in such a manner as to directly or through drift expose workers or other persons. The area being treated must be vacated by unprotected persons. Do not allow re-entry into treated fields within 24 hours of application unless appropriate protective clothing is worn. Unprotected workers should not re-enter fields where the soil is wet until 7 days after soil application. Because certain states may require more restrictive re-entry intervals for various crops treated with this product, consult your State Department of Agriculture for further information.

Written or oral warnings must be given to workers who are expected to be in a treated area or in an area about to be treated with this product. Oral warnings should include those statements found under the label heading "Precautionary Statements" and "Re-entry Statements" included on this label. When oral warning are given, warnings shall be given in a language customarily understood by workers. Oral warnings must be given if there is reason to believe that written warnings cannot be understood by workers. Written warnings must include the following information: DANGER, area treated with DASANIT + DI-SYSTON 4-2 Spray Concentrate on (date of application). Do not allow re-entry into treated fields within 24 hours of application unless appropriate protective clothing is worn. Unprotected workers should not re-enter fields where soil is wet until 7 days after soil application. **If on skin,** remove contaminated clothing and wash skin immediately with soap and warm water. **If eyes are contaminated,** wash with flowing water for at least 15 minutes. **If swallowed,** vomiting should be induced. (See Precautionary Statements on label).

STORAGE AND DISPOSAL

PESTICIDE DISPOSAL

Do not contaminate water, food or feed by storage or disposal.

Pesticide wastes are acutely hazardous. Improper disposal of excess pesticide, spray mixture, or rinsate is a violation of Federal Law. If these wastes cannot be disposed of by use according to label instructions, contact your State Pesticide or Environmental Control Agency, or the Hazardous Waste representative at the nearest EPA Regional Office for guidance.

CONTAINER DISPOSAL

Triple rinse (or equivalent). Then offer for recycling or reconditioning, or puncture and dispose of in a sanitary landfill, or by other procedures approved by State and local authorities.

Store in a cool, dry place and in such a manner as to prevent cross contamination with other pesticides, fertilizers, food, and feed. Store in original container and out of the reach of children, preferably in a locked storage area.

Handle and open container in a manner as to prevent spillage. If container is leaking invert to prevent leakage. If the container is leaking or material spilled for any reason or cause, carefully dam up spilled material to prevent runoff. Refer to Precautionary Statements on label for hazards associated with the handling of this material. Do not walk through spilled material. Absorb spilled material with absorbing type compounds and dispose of as directed for pesticides above. In spill or leak incidents, keep unauthorized people away. You may contact the Mobay Emergency Response Team for decontamination procedures or any other assistance that may be necessary. The Mobay Kansas City Emergency Response Telephone No. is 816-242-2582, or contact Chemtrec at 800-424-9300.

IMPORTANT

This bulletin is not intended to provide adequate information for application. Before using the product, read and carefully observe directions, cautionary statements and other information appearing on the label. This product is sold subject to the Conditions of Sale set forth on the container label.

READ THE LABEL BEFORE USE

DIRECTIONS FOR USE

It is a violation of Federal law to use this product in a manner inconsistent with its labeling.

A single pre-plant application of DASANIT + DI-SYSTON 4-2 Spray Concentrate insecticide-nematicide provides excellent **systemic control of foliar insects, control of soil insects, and suppression of parasitic nematodes.** This **triple-action** crop protection is obtained with one application made within two weeks prior to transplanting.

DASANIT + DI-SYSTON 4-2 Spray Concentrate is recommended for early-season suppression of low nematode populations. When nematode populations (particularly of root-knot nematodes) are moderate to high, use of more efficacious products such as ®NEMACUR 3 or ®NEMACUR - ®DASANIT 2 + 4 Spray Concentrate are recommended as per Directions for Use found on Special Local Need [Section 24(c)] labeling for these products issued by certain states.

RECOMMENDED APPLICATIONS

CROP	PEST	DOSAGE DASANIT + DI-SYSTON 4-2 S.C.	REMARKS
Tobacco (South-eastern states only*)	Aphids Flea beetles Wireworms	18 fl. oz./1000 ft. of row (for 48 inch row spacing), or 1½ gallons per acre (broadcast)	Apply in a water emulsion, or in liquid fertilizer, in a 10 to 12 inch band over the row, or as a broadcast spray before planting. Disk or till thoroughly to a depth of 4 to 6 inches to insure adequate mixing in the soil. Plant crop in the usual manner. Apply only once per season.
	Nematodes (Early-season suppression)		
			7-16-85

* Delaware, Maryland, Virginia, West Virginia, North Carolina, South Carolina and Georgia.

SPECIAL LOCAL NEED REGISTRATIONS [SECTION 24(C)] FOR DASANIT + DI-SYSTON 4-2 SPRAY CONCENTRATE*

CROP	PEST	STATE EPA SLN NO.	DOSAGE	REMARKS
Tobacco	Aphids Flea beetles Wireworms	Georgia GA 81–0005 South Carolina SC 81–0015	18 fluid ounces per 1,000 feet of row (for 48 inch row spacing), or 1½ gallons per acre (broadcast).	May be tank-mixed with herbicides, and with Ridomil® to control black shank and blue mold.
	Nematodes (Early-season suppression)	North Carolina NC 81–0035 Virginia VA 82–0014		May be tank-mixed with Ridomil to control black shank and blue mold.

* For distribution and use only within the state specified. These Special Local Need registrations approved as of date of this publication are, however, subject to cancellation by the state. Contact your nearest Mobay Agricultural Chemicals Division Office for the current status and a copy of the complete individual supplemental Special Local Need labeling. This labeling must be in the possession of the user at time of pesticide application.

®DEF 6

EMULSIFIABLE DEFOLIANT

For Cotton

Fast clean leaf drop for better yields of clean, high-quality lint.

ACTIVE INGREDIENT:

S,S,S-Tributyl phosphorotrithioate	70.5%
INERT INGREDIENTS	29.5%
TOTAL	100.0%

Contains 6 lbs. *S,S,S*-Tributyl phosphorotrithioate per gallon.

EPA Reg. No. 3125-282-AA

DEF is a Reg. TM of the Parent Company of Farbenfabriken Bayer GmbH, Leverkusen.

CAUTION
Keep Out of Reach of Children
AVISO

PRECAUCION AL USUARIO: Si usted no puede leer o entender inglés, no use este producto hasta que la etiqueta le haya sido explicada ampliamente.

(TO THE USER: If you cannot read or understand English, do not use this product until the label has been fully explained to you.)

RECOMMENDED USES

For bottom and complete cotton defoliation. May be used in all power-operated ground and aircraft equipment.

RESTRICTIONS ON WAREHOUSE STORAGE

Do not store near fertilizers, seed, insecticides and fungicides. Do not store near heat or open flame. Store in a cool well-ventilated place.

DIRECTIONS FOR USE

It is a violation of Federal law to use this product in a manner inconsistent with its labeling.

Under favorable conditions, DEF 6 gives effective defoliation of cotton within four to seven days after application. DEF 6 drops the leaves in a green condition, and does not appear effective until defoliation actually begins. When continued low temperatures prevail at night (below 60°F.), complete defoliation may require nine to fourteen days. DEF 6 may be used whether or not moisture is present on cotton, and the presence or absence of dew or rain, during or following application, does not adversely affect the performance. DEF 6 does not suppress second growth, especially where rainfall follows defoliation. Schedule all applications in accordance with local recommendations. Consult your State Agricultural Experiment Station or Extension Service for specific use information.

COMPLETE DEFOLIATION

PROPER TIMING: DEF 6 should be applied to mature cotton plants when 50% or more of the bolls are open, and 7 to 10 days prior to anticipated picking. Plants are considered suitably mature if the youngest bolls (1) cannot be dented by pressure between thumb and forefinger or (2) cannot easily be cut through with a sharp knife.

THOROUGH COVERAGE: All leaves must be treated to accomplish complete defoliation.

RAINFALL: DEF 6 will provide efficient complete defoliation when properly applied, whether rain or dew is present during or after application, but does not suppress second growth, especially in areas where rainfall follows treatment sufficient to delay picking.

BOTTOM DEFOLIATION

Much seed cotton is lost due to rot and exposure of the bottom bolls which are the first to mature. An application of DEF 6, using shielded drop nozzles (four 8001 fans or two $\frac{1}{8}$ K 1 or 1.5 flooding) to the lower leaves where all the bolls are mature, will cause the treated leaves to drop, exposing the mature bolls to air movement and sunlight. By removing the top 8 to 12 rows of spindles from a picker, the exposed bolls may be harvested. DEF 6 leaves the plant green and flexible so that no damage is done by the picker to the untreated top leaves and immature bolls. Complete defoliation and picking can follow, usually in 3 to 4 weeks. Other additional benefits gained by bottom defoliation, even without picking the exposed bolls, are:

Reduction of boll rot (up to $\frac{1}{4}$ to $\frac{1}{2}$ bale per acre).

Faster rate of opening of lower mature bolls.

Reduced lodging.

Higher grade of cotton.

More cotton from fewer pickings.

MIXING: DEF 6 forms an emulsion when diluted with water, and is suitable for use in all power-operated ground and aircraft sprayers. It is non-corrosive and non-clogging to spray equipment, and does not constitute a fire hazard. DEF 6 may be added to diesel oil, in accordance with local recommendations, to increase effectiveness, particularly where night temperatures drop to below 60°F., where plants have had insufficient moisture, and on storm-proof varieties of cotton. The use of 3 to 5 gallons of oil per acre is generally required for aircraft spraying, while 5 to 10 gallons is recommended for ground applications. Where oils are used, particular care should be exercised to avoid spray drift to adjacent crops.

DOSAGE: Use specified dosage of DEF 6 per acre in the amount of water or diesel oil necessary to give complete coverage. The quantity of water or oil to be used per acre is dependent somewhat upon the equipment used. In all cases, however, sufficient spray should be applied to treat all leaves. Larger droplets more effectively reach the lower leaves. The gallons of water ordinarily required for adequate coverage are recommended in the table below for both air and ground spray applications. Make sure the spray pattern is such that a uniform wetting of leaf surfaces results.

PRECAUTIONS

If chlorate type chemicals have been used in the same spray equipment, thoroughly wash spray tank, and particularly dried deposits on aircraft fabric.

RESTRICTIONS

Do not use on other crops used for food or forage.

Avoid spray drift to susceptible plants other than cotton, as this product may injure or defoliate other crops. (Coarse sprays are less likely to drift.)

Thoroughly clean the spray equipment with a chemical cleaner before using for other products.

Do not store near fertilizers, seeds, insecticides or fungicides.

STATEMENT OF PRACTICAL TREATMENT

If swallowed, vomiting should be induced. Administer water freely and induce vomiting by giving one dose ($\frac{1}{2}$ oz. or 15 ml.) of syrup of ipecac. If vomiting does not occur within 10 to 20 minutes, administer second dose. If syrup of ipecac is not available, induce vomiting by sticking finger down throat. Repeat until vomit fluid is clear. Never give anything by mouth to an unconscious person. Have patient lie down and keep quiet. Professional medical assistance should be secured immediately.

In case of contact, remove contaminated clothing and flush skin and eyes with plenty of water. Wash contaminated clothing with soap and hot water before reuse. If poisoning occurs, obtain prompt medical aid.

TO PHYSICIAN: This product inhibits cholinesterase resulting in stimulation of the central nervous system, the parasympathetic nervous system and the somatic motor nerves. Poisoning with this product also results in cardiovascular and respiratory symptoms which must be treated as separate pathological entities apart from the cholinergic effects.

Use atropine sulfate to reverse cholinergic symptoms; and maintain a systematic, symptomatic treatment of the cardiovascular and respiratory effects, even though the cholinergic symptoms have ceased. Do not give central nervous system depressants. Watch for pulmonary edema, which may develop in serious cases of poisoning even after 12 hours. The use of oxygen is recommended in case of respiratory distress.

PRECAUTIONARY STATEMENTS

HAZARDS TO HUMANS AND DOMESTIC ANIMALS

WARNING

May be fatal if swallowed, inhaled, or absorbed through the skin. Do not breathe spray mist. Do not get in eyes, on skin, or on clothing. Wash with soap and warm water after handling and before eating or smoking.

For protection during application wear a hat, long sleeved shirt and long legged trousers. In addition, mixer-loaders must wear rubber or neoprene gloves.

Do not contaminate feed or food. Keep out of reach of children.

ENVIRONMENTAL HAZARDS

This product is toxic to fish. Keep out of lakes, streams, or ponds. Do not apply where runoff is likely to occur. Do not apply when weather conditions favor drift from areas treated. Do not contaminate water by cleaning of equipment, or disposal of wastes. Apply this product only as specified on the label.

IMPORTANT

This bulletin is not intended to provide adequate information for application. Before using the product, read and carefully observe directions, cautionary statements and other information appearing on the product label. This product is sold subject to the Conditions of Sale set forth on the container label.

READ THE LABEL BEFORE USE

STORAGE AND DISPOSAL

PESTICIDE DISPOSAL

Do not contaminate water, food or feed by storage or disposal. Pesticide wastes are toxic. Improper disposal of excess pesticide, spray mixture, or rinsate is a violation of Federal Law. If these wastes cannot be disposed of by use according to label instructions, contact your State Pesticide or Environmental Control Agency, or the Hazardous Waste representative at the nearest EPA Regional Office for guidance.

CONTAINER DISPOSAL

Triple rinse (or equivalent). Then offer for recycling or reconditioning, or puncture and dispose of in a sanitary landfill, or by other procedures approved by State and local authorities.

Store in a cool, dry place, away from open flame and extreme heat, and in such a manner as to prevent cross contamination with other pesticides, fertilizers, food, and feed. Store in original container and out of the reach of children, preferably in a locked storage area.

Handle and open container in a manner as to prevent spillage. If container is leaking invert to prevent leakage. If the container is leaking or material spilled for any reason or cause, carefully dam up spilled material to prevent runoff. Refer to Precautionary Statements on label for hazards associated with the handling of this material. Do not walk through spilled material. Absorb spilled material with absorbing type compounds and dispose of as directed for pesticides above. In spill or leak incidents, keep unauthorized people away. You may contact the Mobay Emergency Response Team for decontamination procedures or any other assistance that may be necessary. The Mobay Kansas City Emergency Response Telephone No. is 816-242-2582, or contact Chemtrec at 800-424-9300.

RECOMMENDED APPLICATIONS

CROP	PINTS DEF 6	GALLONS OF WATER AIR	GALLONS OF WATER GROUND	DIRECTIONS
Cotton	1⅓ to 2	5 to 10	15 to 25	**COMPLETE DEFOLIATION** Apply specified dosage per acre in recommended gallons of water to give thorough coverage to mature cotton when 50% or more of the bolls are open. Applications can be made until first frost, but should not be used within 7 days of picking. For rank cotton, see recommendation below.
	1 to 1½	—	10 to 15	**BOTTOM DEFOLIATION** Apply specified dosage* per acre in recommended gallons of water with spray directed only on the lower part of the plant in the area where only mature bolls are found. Do not use within 7 days of picking. * The bottom defoliation rate should be proportional to the fraction of the plant to be defoliated.
Rank cotton	3	5 to 10	15 to 25	**COMPLETE DEFOLIATION OF RANK COTTON** Apply specified dosage per acre in the recommended amount of water to give thorough coverage of mature cotton when 50% or more of the bolls are open. Schedule 2 applications of the specified dosage two to six weeks apart for complete defoliation of rank cotton. Applications can be made until first frost, but not within 7 days of picking.

1-10-85

SPECIAL LOCAL NEED REGISTRATION [SECTION 24(c)] FOR DEF 6*

CROP	USE	STATE EPA SLN NO.	DOSAGE	REMARKS
Cotton & Rank cotton	Defoliation	Texas TX 81-0045	1⅓ to 3 pints per acre.	Apply in not less than 1½ pints of once refined vegetable oil.

* For distribution and use only within the state specified. This Special Local Need registration approved as of date of this publication is, however, subject to cancellation by the state. Contact your nearest Mobay Agricultural Chemicals Division Office for the current status and a copy of the complete individual supplemental Special Local Need labeling. This labeling must be in the possession of the user at time of pesticide application.

®DI-SYSTON

15% GRANULAR SYSTEMIC INSECTICIDE

Better crop protection for higher crop yields.

RESTRICTED USE PESTICIDE

Acute human, oral, dermal, and inhalation toxicity.

For retail sale to and use only by Certified Applicators or persons under their direct supervision and only for those uses covered by the Certified Applicator's certification.

For Use by Commercial Growers or Professional Applicators.

Not for Home Garden Use.

ACTIVE INGREDIENT:

O,O-Diethyl *S*-[2-(ethylthio) ethyl] phosphorodithioate .. 15%

INERT INGREDIENTS 85%

TOTAL 100%

EPA Reg. No. 3125-172-AA

DI-SYSTON is a Reg. TM of the Parent Company of Farbenfabriken Bayer GmbH, Leverkusen.

DANGER–POISON

Keep Out of Reach of Children

PELIGRO

PRECAUCION AL USUARIO: Si usted no puede leer o entender inglés, no use este producto hasta que la etiqueta le haya sido explicada ampliamente.

(TO THE USER: If you cannot read or understand English, do not use this product until the label has been fully explained to you.)

Recommended Uses

Alfalfa	Peanuts
Barley	Peas
Beans	Pecans**
Broccoli	Potatoes
Brussels sprouts	Raspberries***
Cabbage	Sorghum
Cauliflower	Soybeans
Coffee*	Strawberries (propagating plants only)
Cotton	Tobacco
Fruit trees (non-bearing)	Wheat (Fall)
Hops	
Lettuce	
Ornamental flowers	
Ornamental trees and shrubs	

* Puerto Rico only.
** South Central and Southwestern states.
*** Nursery stock in Northeast states only.

RESTRICTIONS ON WAREHOUSE STORAGE

Do not store adjacent to or near proximity of foodstuffs, drugs, feeds or any other material intended for consumption by humans or animals. Do not store adjacent to toys, cosmetics, dishes, furniture or clothing. Do not store next to herbicides. Protect from rain and moisture. Store in a cool, well-ventilated place. In case of puncture, use overslip bag and securely close.

STATEMENTS OF PRACTICAL TREATMENT

TREATMENT: **In case of poisoning,** call a physician immediately. Have patient lie down and keep quiet.

If swallowed, vomiting should be induced. Administer water freely and induce vomiting by giving one dose ($\frac{1}{2}$ oz. or 15 ml.) of syrup of ipecac. If vomiting does not occur within 10 to 20 minutes, administer second dose. If syrup of ipecac is not available, induce vomiting by sticking finger down throat. Repeat until vomit fluid is clear. Never give anything by mouth to an unconscious person. Professional medical assistance should be secured immediately. **If on skin,** remove contaminated clothing, and wash skin immediately with soap and warm water.

If eyes are contaminated, wash with flowing water for at least 15 minutes.

TO PHYSICIAN: Antidote—Administer atropine sulfate in large therapeutic doses. Repeat as necessary to the point of tolerance. 2-PAM is also antidotal, and may be administered in conjunction with atropine.

Compound inhibits cholinesterase, resulting in stimulation of the central nervous system, the parasympathetic nervous system and the somatic motor nerves. Do not give morphine. Watch for pulmonary edema, which may develop in serious cases of poisoning even after 12 hours. At first sign of pulmonary edema, the patient should be placed in an oxygen tent, and treated symptomatically.

PRECAUTIONARY STATEMENTS

HAZARDS TO HUMANS AND DOMESTIC ANIMALS

DANGER–POISON

Poisonous if swallowed, inhaled, or absorbed through the skin. Rapidly absorbed through the skin. Do not get in eyes or on skin. Do not breathe dust. Wear protective clothing and cotton or natural rubber gloves. If cotton gloves are used, they must be laundered or discarded after each day's use. Do not wear the same gloves for other work. Rubber gloves should be washed with soap and water after each use.

When using DI-SYSTON 15% Granular insecticide in enclosed areas, wear a mask or respirator jointly approved by the Mining Enforcement and Safety Administration (formerly the U.S. Bureau of Mines) and by the National Institute for Occupational Safety and Health under the provisions of 30 CFR Part 11.

Do not contaminate feed or food. Keep out of reach of children. Keep all unprotected persons out of the operating area or vicinity. Wash hands, arms and face thoroughly with soap and warm water before eating or smoking. Wash all contaminated clothing with soap and hot water before reuse.

SYMPTOMS OF POISONING: A sense of "tightness" in the chest. Sweating. Contracted pupils. Stomach pains. Vomiting and diarrhea.

ENVIRONMENTAL HAZARDS

This pesticide is toxic to fish and wildlife. Do not apply directly to water or wetlands. Drift and runoff from treated areas may be hazardous to fish in neighboring areas. Do not contaminate water by cleaning of equipment or disposal of wastes. Cover or incorporate spills.

This pesticide is toxic to bees exposed to direct application. Applications should be timed to coincide periods of minimum bee activity, usually between late evening and early morning.

CROP ROTATION STATEMENT

Do not plant any food or feed crop in rotation after a field treatment with disulfoton unless it is a registered use for disulfoton.

RESTRICTIONS

Not for home garden use. In commercial topping of flowers by hand, clean gloves should be worn. Do not use on other crops used for food or forage. Use only according to label directions. Application at rates above those shown may result in illegal crop residues. Do not treat crops grown in the greenhouse.

Do not apply this product with a knapsack or other equipment that is placed on the user's body.

RE-ENTRY STATEMENT

Do not apply this product in such a manner as to directly or through drift expose workers or other persons. The area being treated must be vacated by unprotected persons. Do not allow re-entry into treated fields within 24 hours of application unless appropriate protective clothing is worn. Because certain states may require more restrictive re-entry intervals for various crops treated with this product, consult your State Department of Agriculture for further information.

Written or oral warnings must be given to workers who are expected to be in a treated area or in an area about to be treated with this product. Oral warnings should include those statements found under the label heading "Precautionary Statements" and "Re-entry Statements" included on this label. When oral warnings are given, warnings shall be given in a language customarily understood by workers. Oral warnings must be given if there is reason to believe that written warnings cannot be understood by workers. Written warnings must include the following information: DANGER, area treated with DI-SYSTON 15% Granular on (date of application). Do not allow re-entry into treated fields within 24 hours of application unless appropriate protective clothing is worn. **If on skin,** remove contaminated clothing and wash skin immediately with soap and warm water. **If eyes are contaminated,** wash with flowing water for at least 15 minutes. **If swallowed,** vomiting should be induced. (See Precautionary Statements on label).

STORAGE AND DISPOSAL

PESTICIDE DISPOSAL

Do not contaminate water, food or feed by storage or disposal.

Pesticide wastes are acutely hazardous. Improper disposal of excess pesticide is a violation of Federal Law. If these wastes cannot be disposed of by use according to label instructions, contact your State Pesticide or Environmental Control Agency, or the Hazardous Waste representative at the nearest EPA Regional Office for guidance.

CONTAINER DISPOSAL

Completely empty bag into application equipment. Then dispose of empty bag in a sanitary landfill or by incineration, or, if allowed by State and local authorities, by burning. If burned, stay out of smoke.

Store in a cool, dry place and in such a manner as to prevent cross contamination with other pesticides, fertilizers, food, and feed. Store in original container and out of the reach of children, preferably in a locked storage area.

Handle and open container in a manner as to prevent spillage. If the container is leaking or material spilled for any reason or cause, carefully sweep material into a pile. Refer to Precautionary Statements on label for hazards associated with the handling of this material. Do not walk through spilled material. Dispose of pesticide as directed above. In spill or leak incidents, keep unauthorized people away. You may contact the Mobay Emergency Response Team for decontamination procedures or any other assistance that may be necessary. The Mobay Kansas City Emergency Response Telephone No. is 816-242-2582, or contact Chemtrec at 800-424-9300.

IMPORTANT: This bulletin is not intended to provide adequate information for application. Before using the product, read and carefully observe directions, cautionary statements and other information appearing on the product label. This product is sold subject to the Conditions of Sale set forth on the container label.

READ THE LABEL BEFORE USE

DIRECTIONS FOR USE

It is a violation of Federal Law to use this product in a manner inconsistent with its labeling.

DI-SYSTON 15% Granular systemic insecticide enters the plant through the root system. The insecticide, when placed near the roots, is picked up and moves in the sap stream to all parts of the plant, where it kills insects as they feed on the plant.

This formulation of DI-SYSTON 15% Granular is highly versatile, and may be applied by various methods. It can be drilled in the furrow with seed, broadcast or banded on each side of the seed furrow. Foliage broadcast application may be applied by ground or air equipment. Time and method of application and limits for use are discussed in detail for each specific crop under "Remarks" below.

As a preplant treatment for ornamental flowers, shrubs and trees, apply granules evenly to seed bed and mix into soil; seeds or transplants are then placed in the normal manner. Established plants may be treated by distributing granules uniformly under foliage canopy. Mixing the granules into top soil and watering will accelerate entry into the plant, as the chemical will leach into the root zone and be taken up by the root system.

RECOMMENDED APPLICATIONS

IMPORTANT: When used on cotton, beans or lettuce at the maximum recommended rates under adverse conditions, such as extremely cool or wet or extremely dry weather, DI-SYSTON systemic insecticide may cause some delay in emergence, stunting of seedlings or reduction of stand. Such damage may be more pronounced in light, sandy soils. Plant injury may also occur, or may be enhanced when certain preemergence herbicides are used in conjunction with DI-SYSTON.

CROP	INSECT	DOSAGE DI-SYSTON 15% GRAN.	REMARKS
FIELD CROPS Alfalfa Clover	Aphids Mites	6.7 pounds per acre	Apply only one broadcast treatment but not within 28 days of grazing or cutting for forage. Make only one application per crop season.
Barley	Aphids Mites	6.7 pounds per acre	Apply by drilling or broadcast at planting time, or as a broadcast after emergence. Do not apply more than twice per season, nor within 60 days of harvest for grain. Allow a minimum of 21 days between applications. Do not graze or cut for forage within 30 days of treatment.
Cotton	Aphids Mites Thrips	5 to 8 oz. per 1000 feet of row (for any row spacing*), or 4 to 6.7 pounds per acre (40 inch row spacing)	**At Planting:** Place the granules in the seed furrow, or in a band on each side of the furrow. **RATE CHART FOR PLANTING TIME APPLICATION** **Heavy Soils** — 8 oz. 15% Gran./1000 ft. of row; **6.7 lbs. 15% Gran./acre** **Medium Soils** — 6.6 oz. 15% Gran./1000 ft. of row; **5.3 lbs. 15% Gran./acre** **Light Soils** — 5 oz. 15% Gran./1000 ft. of row; **4 lbs. 15% Gran./acre** **Post-Plant:** (Irrigated Cotton Only) If an additional treatment is necessary, make one side-dress application, or a single broadcast application over the foliage. Rates up to 13.3 pounds per acre can be applied to first-squaring cotton (90 days before harvest), and rates up to 6.7 pounds per acre can be applied to within 28 days of harvest. Do not apply more than twice per season. Allow a minimum of 21 days

CROP	INSECT	DOSAGE DI-SYSTON 15% GRAN.	REMARKS
			between applications. Read "IMPORTANT" statement under "RECOMMENDED APPLICATIONS" for additional information before use on this crop.
		2.5 to 3.3 ounces per 1000 feet of row (for any row spacing*), or 2 to 2.7 pounds per acre (40 inch row spacing)	**Hill-Dropped Cotton:** Apply specified dosage with the seed at planting. Use the lower rate where hill spacing is greater than 14 inches apart, or where applications are made to light, sandy soils. Read "IMPORTANT" statement under "RECOMMENDED APPLICATIONS" for additional information before use on this crop.
Hops	Aphids Mites	13.3 to 26.7 pounds per acre (See Remarks)	Uniformly spread the granules in a 6 inch square area at the base of each plant. For standard 7 ft. by 7 ft. hopyard spacing, the recommended dosage is equal to 0.24 oz. (2 level teaspoonfuls) to 0.48 oz. (4 level teaspoonfuls) per plant. Apply in mid to late June, just prior to hilling. Hilling operation should cover granules with soil. Irrigate as normal crop culture requires. Do not graze or feed crop refuse.
Peanuts	Aphids Leafhoppers Mites Thrips	7.4 to 14.7 oz. per 1000 feet of row (for any row spacing*), or 6.7 to 13.3 pounds per acre (36 inch row spacing)	Place the granules in a band on each side of the seed furrow at planting, or as a side-dressing after plants are established. Do not apply directly on the seed. Where necessary, granules may be applied in a band over the row, or as a side-dressing at pegging time. Do not feed treated vines.
Sorghum	Aphids (Greenbugs) (See "NOTE" below)	6 to 8 oz. per 1000 ft. of row (for any row spacing*), or 5 to 6.7 pounds per acre (40 inch row spacing)	**Application at Planting: In-Furrow**—Place the granules in the seed furrow approximately ½ inch above the seed at planting. Do not apply directly on the seed. See "NOTE" below.
		8 oz. per 1000 feet of row (for any row spacing*), or 6.7 pounds per acre (40 inch row spacing)	**Application at Planting: Band**—Place the granules in a 4 to 6 inch band directly behind the planter shoe, in front of the press wheel. Do not apply directly on the seed. See "NOTE" below. **Post-planting Application: Whorl**—Apply specified dosage by broadcasting into the whorl. Do not apply more than 3 times per season, nor within 30 days of harvest of grain, or 45 days of harvest for forage or fodder. See "NOTE" below.
Soybeans	Thrips (Except flower thrips) Mexican bean beetle	6.7 pounds per acre (36 inch row spacing)	Place the granules in a 4 inch band over the seed furrow behind the planter shoe and in front of the press wheel, or in a band on each side of the seed furrow at planting time. Do not apply directly on the seed.
	NOTE: (1) Post-planting broadcast applications into the whorl may be made following an at-planting application. (2) Due to greenbug resistance in certain areas, DI-SYSTON used alone may not provide satisfactory control. Consult your local agricultural advisor or Extension Service for recommendations.		
Tobacco	Aphids Flea beetles	**Transplant Beds:** 1 oz. per 100 square feet of bed	Apply evenly over bed area. Work the granules into the top 2 to 3 inches of soil and seed beds in normal manner. An alternative treatment may be made by broadcasting the granules evenly over the plants after emergence, and watering thoroughly. Do not make more than one application to transplant beds.
		Field Application: 20 to 40 oz. per 1000 ft. of row (for any row spacing*), or 13.3 to 26.7 lbs. per acre (broadcast, or with 48 inch row spacing)	Apply in a 6 to 12 inch band over the row, or as a broadcast treatment immediately before transplanting. Incorporate the granules by working them into the top 2 to 3 inches of soil. Plant crop in the usual way. Do not make more than 1 field application per crop season, regardless of DI- SYSTON formulation used.
	Southern potato wireworm	13.3 to 26.7 lbs. per acre	When applied as described for "Field Application" above, DI-SYSTON 15% Granular will aid in control of this wireworm.
Wheat (Fall)	Hessian fly Aphids (Greenbugs) (Oat bird-cherry) Grasshoppers	6.7 pounds per acre	Apply by drilling or broadcast at planting time in the fall. Make only one application, and do not graze or cut for forage within 75 days of treatment. **NOTE:** Due to greenbug resistance in certain areas, DI-SYSTON used alone may not provide satisfactory control. Consult your local agricultural advisor or Extension Service for recommendations.
SEED FIELD CROPS Soybeans	Aphids Leafhoppers Mites Thrips	8 ounces per 1000 feet of row (for any row spacing*), or 6.7 pounds per acre (40 inch row spacing)	Place the granules in a band on each side of the seed furrow at planting or as a side-dressing after plants are established. Do not pasture or use treated crop for feed, food or forage. Do not apply directly to seed.
FRUIT Coffee (Puerto Rico only)	Leafminer	2 to 4 grams per foot of tree height	Distribute granules uniformly under the tree canopy. Do not apply more than twice per year. Make no more than one pre-harvest and one post-harvest application. Do not make the pre-harvest application within 90 days of harvest.

CROP	INSECT	DOSAGE DI-SYSTON 15% GRAN.	REMARKS
NON-BEARING FRUIT Fruit trees: Apples Apricots Cherries Crab apples Peaches Pears Plums Prunes	Aphids Lace bug Leafhoppers Mites Thrips Whiteflies	2.5 to 27 oz. per tree, or 2.5 ounces per inch of trunk diameter	Spread granules uniformly from trunk to drip line on all sides. Work into soil, and water thoroughly. The higher dosage is recommended for heavy organic or muck soils. For transplanting, equivalent quantities should be applied into planting hole. Do not apply to trees that will bear fruit during current crop year.
Strawberries (Propagating plants only)	Aphids Mites (Except cyclamen mite)	17 to 34.4 oz. per 1000 ft. of row (for any row spacing*), or 13.3 to 26.7 lbs. per acre (42 inch row spacing)	Apply by side-dressing on each side of the row, or by broadcasting over the row as needed. Do not use fruit from treated plants for food purposes.
Raspberries (Nursery stock in Northeast states only)	Aphids	53.3 lbs. per acre	**Band Application:** Apply specified dosage in a 24 inch band over the row to new or established plantings. Work granules 2 to 3 inches into the soil, using a cultivator or other suitable equipment. A second application of granules may be applied to the soil later in the season. Consult your State Agricultural Experiment Station or Extension Service for proper timing of applications.
NUTS Pecans (South Central and Southwestern states) (see **Note 1**)	Aphids	10 to 20 pounds per acre	Apply by treating 6 foot bands of soil on both sides of the trees. The bands should be located under the main drip area of the trees. Work the granules into the upper 2 to 3 inches of soil. On irrigated pecans, apply water as soon as possible after treatment. Make only 1 application per season. Early spring treatment is recommended. Do not harvest nuts within 80 days of treatment. Do not graze grass under treated trees. Do not mix this product with Phosalone.
VEGETABLES Beans	Aphids Leafhoppers Mexican bean beetle (larvae) Mites Thrips	6 to 12 oz. per 1000 ft. of row (for any row spacing*), or 6.7 to 13.3 lbs. per acre (30 inch row spacing)	For snap or green Lima beans, place the granules in a band on each side of the seed furrow at planting time only. For dry beans, apply as above at planting time or as a side-dressing on each side of the row after plants become established, but not within 60 days of harvest. Do not apply directly on the seed. Do not apply more than once per season. Read "IMPORTANT" statement under "RECOMMENDED APPLICATIONS" for additional information before use on this crop. Do not feed vines or hay.
Brussels sprouts	Aphids Flea beetles Leafhoppers Mites Thrips	7.4 ounces per 1000 ft. of row (for any row spacing*),	**Transplant Seed Beds:** Apply granules evenly over bed area at the rate of 6.7 pounds per acre. Work into the top 2 to 3 inches of soil and seed beds in the normal manner. An alternative seed bed treatment may be made by broadcasting the granules evenly over the plants after emergence, and watering thoroughly. Do not apply more than one seed bed treatment. **Field Applications:** Place the granules in a band on each side of the seed furrow or transplanted row at planting, and where necessary, as a side-dressing after plants become established. Do not apply directly on the seed. Do not apply more than two field applications per season, nor within 30 days of harvest for Brussels sprouts, or 40 days of harvest for cauliflower. Allow a minimum of 21 days between applications.
Cauliflower	Aphids Flea beetles Leafhoppers	or 6.7 pounds per acre (36 inch row spacing)	
Broccoli	Aphids Flea beetles Leafhoppers	7.4 ounces per 1000 ft. of row (for any row spacing*), or 6.7 pounds per acre	**Transplant Seed Beds:** Apply granules evenly over bed area at the rate of 6.7 pounds per acre. Work into the top 2 to 3 inches of soil, and seed beds in normal manner. An alternative seed bed treatment may be made by broadcasting the granules evenly over the plants after emergence, and watering thoroughly. Do not apply more than one seed bed treatment.
Cabbage	Aphids Flea beetles	(36 inch row spacing)	
	Root aphid	11.3 oz. per 1000 feet of row (for any row spacing*)	**Field Applications:** Place the granules as a band on each side of the seed furrow or transplanted row at planting, or as a side-dressing after plants become established. Do not apply directly on the seed. Do not apply more than one field application per season. Do not use within 14 days of harvest for broccoli, or 42 days of harvest for cabbage.
Lettuce	Aphids Leafhoppers Mites Thrips	4 to 8 oz. per 1000 ft. of row (for any row spacing*), or 6.7 to 13.3 lbs. per acre (20 inch row spacing)	**Spring- or Summer-Seeded Lettuce:** Use the higher rate on heavy organic soils. Place the granules in a band on each side of the row at seeding time only. Do not apply directly on the seed. Do not apply to transplanted lettuce. **Fall- or Winter-Seeded Lettuce:** Apply as a side-dressing at thinning time. Do not apply within 60 days of harvest. Read "IMPORTANT" statement under "RECOMMENDED APPLICATIONS" for additional information before use on this crop.
Peas	Aphids Leafhoppers	6.7 to 16.7 lbs. per acre	Apply by drilling or broadcasting at planting time, or as a side-dress on each side of the row after energence. Do not apply more than once per season, nor within 50 days of harvest. Do not feed vines or hay.

CROP	INSECT	DOSAGE DI-SYSTON 15% GRAN.	REMARKS
Potatoes	Aphids Flea beetles (Except tuber flea beetles) Leafhoppers Potato psyllid Southern potato wireworm Colorado potato beetle (Early-season reduction)	15 to 23 oz. per 1000 ft. of row (for any row spacing*), or 13.5 to 20.5 lbs. per acre (38 inch row spacing)	**Row Treatments:** Place the granules in the seed furrow or in a band on each side of the seed furrow at planting. An additional application may also be made as a side-dressing after plants become established. Do not make more than two applications per season, nor apply within 75 days of harvest. Higher rate to be used in organic soils. See NOTE below for "foliar applications." Do not mix this product with Phosalone.
		20 to 26.7 lbs. per acre	**Broadcast Application:** Apply specified dosage per acre as a preplant broadcast application. Work granules into soil to a depth of 2 to 3 inches, using a disc or other suitable equipment. Plant crop in usual manner. A postemergence side-dress may also be made at rates of 15 to 23 ounces per 1000 feet of row (for any row spacing*), after the plants become established. Do not make more than two applications per season, nor apply within 75 days of harvest. See NOTE below for "foliar applications." Do not mix this product with Phosalone.
	NOTE: Foliar applications of DI-SYSTON 8 may be made in addition to an **at-planting** soil treatment. Refer to the label for DI-SYSTON 8.		
ORNAMENTALS			
Flowers (See **Note 1**)	Aphids Lace bug Leafhoppers Mites Thrips Whiteflies	10.7 oz. per 150 to 300 feet of row	Apply in a band on each side of the seed furrow, or as a band over the row. Work into soil, and water thoroughly. For treatment of Easter lilies and gladiolus, apply the specified dosage in the furrow over the bulbs after they are set, but before covering with soil. Do not make in-furrow applications to garden lily bulbs. The higher dosage is recommended for heavy organic or muck soils.
Shrubs Trees Forest plantings (See **Note 1**)	Aphids Birch leafminer Camellia scale Elm leaf beetle European elm scale Holly leafminer Lace bug Leafhoppers Mimosa webworm Mites Pine tip moth Soft scale *(Coccus hesperidium)* Tea scale Thrips Whiteflies	2.5 to 27 oz. per tree, or 2.5 oz. per inch of trunk diameter	Spread granules uniformly from trunk to drip line on all sides. Work into soil and water thoroughly. The higher dosage is recommended for heavy organic or muck soils. For transplanting, equivalent quantities should be applied into planting hole or by dipping seedlings in a puddling clay slurry made up of 1 part by weight of DI-SYSTON 15% Granular in 6 to 14 parts by weight of clay. 4–26–85

*** ON ROW CROPS WHERE ROW SPACING IS EXTREMELY NARROW, DO NOT PLACE TREATED ZONES (BAND OR FURROW) CLOSER TOGETHER THAN 6 INCHES. Note 1:** For use by commercial growers or professional applicators only.

SPECIAL LOCAL NEED REGISTRATIONS [SECTION 24(c)] FOR DI-SYSTON 15% GRANULAR*

CROP	INSECT	STATE EPA SLN NO.	DOSAGE	REMARKS
Alfalfa	Aphids Mites	Arizona AZ 77–0018 California CA 77–0083	6.7 pounds per acre	Apply once per cutting as a broadcast application.
Barley	Hessian fly Aphids **(Greenbugs) (Oat birdcherry) Grasshoppers	Oregon OR 80-0034 Washington WA 81-0087	6.7 pounds per acre	Apply by drilling or broadcast at planting time.
	Aphids **(Greenbugs)			Broadcast postemergence by air or ground. Irrigated only.
Potatoes (sprinkler irrigated only)	Green peach aphids	Oregon OR 79–0042 Washington WA 79-0067	20 pounds per acre	Apply as a postemergence aerial broadcast application.
Potatoes (sprinkler irrigated only)	Aphids	Idaho ID 83–0035 Oregon OR 83–0057 Washington WA 84–0012	20 lbs. per acre	Apply suggested rate per acre as a soil application and/or a foliar broadcast treatment.
Wheat	Aphids	California CA 76–0019	6.7 pounds per acre	Apply postemergence as an aerial or ground broadcast application.

CROP	INSECT	STATE EPA SLN NO.	DOSAGE	REMARKS
Wheat (Fall) (Spring)	Hessian fly Aphids **(Greenbugs) (Oat birdcherry) Grasshoppers	Oregon OR 80–0034 Washington WA 81–0087	6.7 pounds per acre	Apply by drilling or broadcast at planting time.
	Aphids **(Greenbugs)			Broadcast postemergence by air or ground. Irrigated only.
Wheat (Fall) (Spring) Barley	Hessian fly Aphids **(Greenbugs (Oat birdcherry) Grasshoppers	Montana MT 80–0004	6.7 pounds per acre	Apply by drilling or broadcast at planting time. Make only one application and do not graze or harvest for grain or forage within 60 days of treatment.

* For distribution and use only within the state specified. These Special Local Need registrations approved as of date of this publication are, however, subject to cancellation by the state. Contact your nearest Mobay Agricultural Chemicals Division Office for the current status and a copy of the complete individual supplemental Special Local Need labeling. This labeling must be in the possession of the user at time of pesticide application.

** Due to greenbug resistance in certain areas, DI-SYSTON used alone may not provide satisfactory control. Consult your local agricultural advisor or Extension Service for recommendations.

®DI-SYSTON 8

EMULSIFIABLE SYSTEMIC INSECTICIDE

RESTRICTED USE PESTICIDE

Acute human oral, dermal, and inhalation toxicity.

For retail sale to and use only by Certified Applicators or persons under their direct supervision, and only for those uses covered by the Certified Applicator's certification.

For Effective Crop Protection.

8 Pound Per Gallon Formulation.

ACTIVE INGREDIENTS:

O,O-Diethyl *S*-[2-(ethylthio)ethyl] phosphorodithioate ..	85%
Aromatic Petroleum Distillate	3%
INERT INGREDIENTS	12%
TOTAL	100%

Contains 8 lbs. *O,O*-Diethyl *S*-[2-(ethylthio)ethyl] phosphorodithioate per gallon.

EPA Reg. No. 3125-307

DI-SYSTON is a Reg. TM of the Parent Company of Farbenfabriken Bayer GmbH, Leverkusen.

DANGER–POISON

Keep Out of Reach of Children

PELIGRO

PRECAUCION AL USUARIO: Si usted no puede leer o entender inglés, no use este producto hasta que la etiqueta le haya sido explicada ampliamente.

(**TO THE USER:** If you cannot read or understand English, do not use this product until the label has been fully explained to you.)

RECOMMENDED USES

Barley	Peas
Beans	Pecans
Broccoli	Potatoes
Brussels sprouts	Sorghum
Cabbage	Strawberries (propagating plants only)
Cauliflower	Sugarcane†
Corn	Tobacco
Cotton	Wheat
Lettuce	

† Puerto Rico only

RESTRICTIONS ON WAREHOUSE STORAGE

Do not store adjacent to or near proximity of foodstuffs, drugs, feeds or any other material intended for consumption by humans or animals. Do not store adjacent to toys, cosmetics, dishes, furniture or clothing. Do not store next to herbicides.

STATEMENTS OF PRACTICAL TREATMENT

In case of poisoning, call a physician immediately. Have patient lie down and keep quiet.

If swallowed, vomiting should be induced. Administer water freely and induce vomiting by giving one dose (½ oz. or 15 ml.) of syrup of ipecac. If vomiting does not occur within 10 to 20 minutes, administer second dose. If syrup of ipecac is not available, induce vomiting by sticking finger down throat. Repeat until vomit fluid is clear. Never give anything by mouth to an unconscious person. Professional medical assistance should be secured immediately.

If on skin, remove contaminated clothing and wash skin immediately with soap and warm water.

If eyes are contaminated, wash with flowing water for at least 15 minutes.

TO PHYSICIAN: Antidote—Administer atropine sulfate in large therapeutic doses. Repeat as necessary to the point of tolerance. 2-PAM is also antidotal, and may be administered in conjunction with atropine.

Compound inhibits cholinesterase, resulting in stimulation of the central nervous system, the parasympathetic nervous system and the somatic motor nerves. Do not give morphine. Watch for pulmonary edema, which may develop in serious cases of poisoning, even after 12 hours. At first sign of pulmonary edema, the patient should be placed in an oxygen tent and treated symptomatically.

PRECAUTIONARY STATEMENTS

HAZARDS TO HUMANS AND DOMESTIC ANIMALS

DANGER–POISON

Poisonous if swallowed, inhaled or absorbed through the skin. Rapidly absorbed through the skin. Do not get in eyes, on skin or on clothing. Wear protective clothing, natural rubber gloves and goggles.

Do not breathe fumes or spray mist. Wear a pesticide respirator jointly approved by the Mining Enforcement and Safety Administration (formerly the U.S. Bureau of Mines) and by the National Institute for Occupational Safety and Health under the provisions of 30 CFR Part 11.

Do not contaminate feed or food. Keep out of reach of children. Keep all unprotected persons out of the operating area or vicinity. Wash hands, arms and face thoroughly with soap and warm water before eating or smoking.

HANDLING OF CONTAMINATED ARTICLES: Clothing and rags receiving undiluted DI-SYSTON 8 as a result of leaks, spills or mishaps should be removed immediately and not reused. These contaminated articles should be buried in the same manner as the empty container; see container disposal section for directions. (Contaminated articles which should be discarded would include: hats, gloves, aprons, coats, boots, etc. not made of rubber, or coated with rubber or other similar materials). Clothing receiving spray mist or droplets from the mixed, diluted DI-SYSTON 8 should be removed and decontaminated before reuse. This clothing can be decontaminated by machine washing separately from other items with soap or detergent and bleach in hot water twice before reuse. Protective clothing and equipment should be washed down with detergent or soap and bleach in water. Wash water from the cleaning of protective clothing or equipment should not be allowed to run off or otherwise enter water supplies.

SYMPTOMS OF POISONING: A sense of "tightness" in the chest. Sweating. Contracted pupils. Stomach pains. Vomiting and diarrhea.

ENVIRONMENTAL HAZARDS

This product is toxic to fish, and wildlife. Do not apply directly to water or wetlands. Drift and runoff from treated areas may be hazardous to fish in neighboring areas. Do not contaminate water by cleaning of equipment or disposal of wastes. Cover or incorporate spills.

This pesticide is toxic to bees exposed to direct application. Applications should be timed to coincide with periods of minimum bee activity, usually between late evening and early morning.

CROP ROTATION STATEMENT

Do not plant any food or feed crop in rotation after a field treatment with disulfoton unless it is a registered use for disulfoton.

RESTRICTIONS

Not for home garden use. Do not use on other crops used for food or forage. Use only according to label directions. Application at rates above those shown may result in illegal crop residues. Do not treat food crops grown in the greenhouse. Do not apply in any manner not specified on the label.

RE-ENTRY STATEMENTS

Do not apply this product in such a manner as to directly or through drift expose workers or other persons. The area being treated must be vacated by unprotected persons. Do not allow re-entry into treated fields within 24 hours of application unless appropriate protective clothing is worn. Because certain states may require more restrictive re-entry intervals for various crops treated with this product, consult your State Department of Agriculture for further information.

Written or oral warnings must be given to workers who are expected to be in a treated area or in an area about to be treated with this product. Oral warnings should include those statements found under the label heading "Precautionary Statements" and "Re-entry Statements" included on this label. When oral warnings are given, warnings shall be given in a language customarily understood by workers. Oral warnings must be given if there is reason to believe that written warnings cannot be understood by workers. Written warnings must include the following information: DANGER, area treated with DI-SYSTON 8 on (date of application). Do not allow re-entry into treated fields within 24 hours of application unless appropriate protective clothing is worn. **If on skin,** remove contaminated clothing and wash skin immediately with soap and warm water. **If eyes are contaminated,** wash with flowing water for at least 15 minutes. **If swallowed,** vomiting should be induced. (See Precautionary Statements on label).

STORAGE AND DISPOSAL

PESTICIDE DISPOSAL

Do not contaminate water, food or feed by storage or disposal.

Pesticide wastes are acutely hazardous. Improper disposal of excess pesticide, spray mixture, or rinsate is a violation of Federal Law. If these wastes cannot be disposed of by use according to label instructions, contact your State Pesticide or Environmental Control Agency, or the Hazardous Waste representative at the nearest EPA Regional Office for guidance.

CONTAINER DISPOSAL

Triple rinse (or equivalent). Then offer for recycling or reconditioning, or puncture and dispose of in a sanitary landfill, or by other procedures approved by State and local authorities.

Store in a cool, dry place and in such a manner as to prevent cross contamination with other pesticides, fertilizers, food, and feed. Store in original container and out of the reach of children, preferably in a locked storage area.

Handle and open container in a manner as to prevent spillage. If container is leaking invert to prevent leakage. If the container is leaking or material spilled for any reason or cause, carefully dam up spilled material to prevent runoff. Refer to Precautionary Statements on label for hazards associated with the handling of this material. Do not walk through spilled material. Absorb spilled material with absorbing type compounds and dispose of as directed for pesticides above. In spill or leak incidents, keep unauthorized people away. You may contact the Mobay Emergency Response Team for decontamination procedures or any other assistance that may be necessary. The Mobay Kansas City Emergency Response Telephone No. is 816-242-2582, or contact Chemtrec at 800-424-9300.

DIRECTIONS FOR USE

It is a violation of Federal Law to use this product in a manner inconsistent with its labeling.

DI-SYSTON systemic insecticide enters the plant by absorption through the roots. The insecticide placed near the roots is translocated to all parts of the plant, and kills sucking insects feeding on the plant.

The formulation of DI-SYSTON 8 is highly versatile, and offers many forms of application. It can be sprayed or injected into the furrow at planting time or soil-injected as a side-dress treatment at planting time, or after plants become established. Applications may be made in water emulsion, with liquid fertilizer, or undiluted in specially designed equipment for low volume soil injection, but never on crops for which less than 1 quart liquid concentrate per acre is the recommended dosage. When mixing with certain liquid high-phosphorous fertilizers, the use of Compex® compatibility agent, or other equivalent agents at the rate of 1 quart per 100 gals. of mixture may be needed to assure proper mixing. Time and method of application and limits for use are discussed in detail for each specific crop under "Remarks" below.

RECOMMENDED APPLICATIONS

IMPORTANT: When used on cotton, beans or lettuce at the maximum recommended rates, under adverse conditions, such as extremely cool or wet or extremely dry weather, DI-SYSTON may cause some delay in emergence, stunting of seedlings or reduction of stand. Such damage may be more pronounced in light sandy soils. Plant injury may also occur or may be enhanced when certain preemergence herbicides are used in conjunction with DI-SYSTON.

CROP	INSECT	DOSAGE—DI-SYSTON 8	REMARKS
FIELD CROPS Barley	Aphids Mites	1 pint per acre	Apply as a soil injection in a water emulsion, or with liquid fertilizer at planting time. Do not apply within 60 days of harvest. Do not graze or cut for forage within 30 days of treatment.

CROP	INSECT	DOSAGE—DI-SYSTON 8	REMARKS
FIELD CROPS (Cont'd) Barley (Foliar Application Spring or Fall)	Aphids (Including greenbugs) (See **NOTE** in Remarks)	½ to 1 pint per acre	For application by aircraft or ground equipment: Apply specified dosage per acre in sufficient water to obtain thorough coverage but not less than 1 gallon total volume per acre. Use the lower rates on plants up to tillering (stooling). Higher rates should be used after plants begin to tiller. For application in liquid fertilizer: Mix ½ to 1 pint of DI-SYSTON 8 in the amount of liquid fertilizer to be applied per acre. Apply this mixture as a top dressing with suitable ground equipment. Foliar applications may be made following an at-planting application, however, regardless of method of application do not apply more than a total of 2 pints per acre of DI-SYSTON 8 per crop season. Do not make any applications within 30 days of grain harvest. **NOTE:** (1) Do not graze treated fields after **any** application of DI-SYSTON 8. (2) Due to greenbug resistance in certain areas, DI-SYSTON used alone may not provide satisfactory control. Consult your local agricultural advisor or Extension Service for recommendations.
Corn Field, Popcorn	Aphids Leafhoppers Mites Thrips	1.2 fl. ounces per 1000 feet of row for any row spacing* (for 40 inch row spacing use 1 pint per acre.)	**Treatment at Planting Time:** As a water emulsion, spray in a 4 to 6 inch band directly behind the planter shoe in front of the press wheel, **or** apply with **liquid fertilizer** in split bands on each side of the corn row. Do not apply directly to the seed. Do not forage or cut treated corn for food, feed or ensilage within 28 days of treatment. **Post-plant Side-dress Treatment:** An alternative treatment for early-season control of aphids, leafhoppers, mites and thrips may be made in a water emulsion, or with liquid fertilizer as a soil-injection on each side of the furrow after emergence. Do not forage or cut treated corn for food, feed or ensilage within 28 days of treatment.
	Aphids Mites	8 to 16 fl. ounces **NOTE:** Apply only one soil treatment and one foliar treatment per crop season.	**Foliar Treatment:** Apply specified dosage per acre as a foliar spray in sufficient water for thorough coverage, but not less than 1 gallon total volume per acre. Do not forage or cut treated corn for food, feed or ensilage within 28 days of treatment.
Cotton	Aphids Mites Thrips	.75 to 1.2 fluid ounces per 100 feet of row (for any row spacing*), or 9 to 16 fluid ounces per acre (40 inch row spacing)	**Preplant:** As a preplant soil incorporated treatment, DI-SYSTON 8 may be applied in approximately a 16 inch band as a water emulsion either alone or with a preemergence herbicide. Request Bulletin No. DS 6720 for specific recommendations on a herbicide to be used with DI-SYSTON. **Planting:** Apply as a water emulsion, or with liquid fertilizer. Spray in the furrow on either side of the seed, or soil-inject on each side of the furrow. Do not apply directly on the seed. **RATE CHART FOR PREPLANT OR PLANTING-TIME APPLICATION** **Heavy Soil**: 1.2 fl. oz./1000 ft. of row, or 1 pt./acre **Medium Soil**: 1 fl. oz./1000 ft. of row, or 12 fl. oz./acre **Light Soil**: .75 fl. oz./1000 ft. of row, or 9 fl. oz./acre **Post-plant:** If an additional treatment is needed, apply in a water emulsion, or with liquid fertilizer as a soil-injection on each side of the row. A single post-plant application at rates up to 2 pints/acre may be applied up to first-squaring (90 days before harvest), or a single post-plant application at rates up to 1 pint/acre may be applied up to 28 days of harvest. Do not apply more than twice per season. Allow a minimum of 21 days between applications. Read "IMPORTANT" statement under "RECOMMENDED APPLICATIONS" for additional information before use on this crop.
	Aphids Thrips	3 to 9 fluid ounces **NOTE:** If foliar applications are made, do not apply soil treatments within the same crop year.	**FOLIAR—Prior to Bloom:** Apply specified dosage per acre as a foliar spray, using aircraft or ground equipment, in sufficient water for thorough coverage, but not less than 1 gallon total volume per acre. Repeat as necessary, but do not apply more than 3 times per season, with the last application no later than bloom.
Sorghum	Aphids (Greenbugs) (See **NOTE** below)	.9 to 1.2 fl. oz. per 1000 feet of row (for any row spacing*), (for 40 inch row spacing use 12 to 16 fl. ozs. per acre)	**AT-PLANTING APPLICATION: Furrow**—Apply as a water emulsion, or in liquid fertilizer in the seed furrow approximately ½ inch above the seed. Do not apply directly on the seed.
		1.2 fl. oz. per 1000 feet of row (for any row spacing*) (for 40 inch row spacing use 1 pint per acre)	**AT-PLANTING APPLICATION: Band**—Apply as a water emulsion, or in liquid fertilizer in a 4 to 6 inch band directly behind the planter shoe, in front of the press wheel. Do not apply directly on the seed.
		NOTICE: The following side-dress and/or foliar applications may be made in addition to granular or liquid at-planting applications. **A total of 5 applications** may be made to a single sorghum crop (2 soil applications, granular or liquid, plus 3 foliar applications of rates 8 fluid ounces or less/acre).	

CROP	INSECT	DOSAGE—DI-SYSTON 8	REMARKS
FIELD CROPS (Cont'd) Sorghum (Cont'd)	Aphids (Greenbugs) (See **NOTE** below)	1.2 fl. oz. per 1000 feet of row (for any row spacing*) (for 40 inch row spacing use 1 pint per acre)	**POST-PLANTING APPLICATION: Side-dress**—Apply as a water emulsion, or with liquid fertilizer injected into the row on both sides of the furrow at any time up to boot stage, but not within 45 days of use for forage or fodder.
	Aphids (Greenbugs) (see **NOTE** below) Sorghum midge (Arkansas, Louisiana, Oklahoma, Texas)	4 to 8 fluid ounces per acre	**POST-PLANTING APPLICATION: Foliar Spray—A maximum of 3 foliar applications may be made of rates 8 fluid ounces or less/acre.** Apply specified dosage per acre in sufficient water to obtain thorough coverage. For aerial application, use not less than 1 gallon of water per acre. Use not less than 5 gallons of water per acre with ground equipment. **For Midge:** Make one application when 50% of the seed heads have emerged from the boot. Make a second application 3 to 5 days later.
	Banks grass mites (in Texas, excluding Trans-Pecos area, and in Oklahoma)	8 to 16 fluid ounces per acre	See table below. **For Banks Grass Mites:** The 8 fluid ounce rate may be applied as directed above for aphids. Rates above 8 fluid ounces per acre may not be applied if **any** soil applications have been made. If rates above 8 fluid ounces per acre are used, do not apply more than twice per crop season, nor within 7 days of harvest of grain, or 45 days of use as forage or fodder.

DAYS FROM LAST APPLICATION TO HARVEST FOR FOLIAR RATES OF 4 to 8 FLUID OUNCES/ACRE

APPLICATION SCHEDULE	FOR GRAIN	FOR FORAGE OR FODDER
Any soil **plus** any foliar	34	45
1 or 2 foliars	7	45
3 foliars	34	60

NOTE: Due to greenbug resistance in certain areas, DI-SYSTON used alone may not provide satisfactory control. Consult your local agricultural advisor or Extension Service for recommendations.

CROP	INSECT	DOSAGE—DI-SYSTON 8	REMARKS
Sugarcane (Puerto Rico only)	Aphids	6 to 9 fluid ounces per acre	Apply specified dosage per acre as a foliar spray, in sufficient water for thorough coverage, utilizing either an aircraft or ground sprayer. Approximately 4 gallons of water per acre should be used when making aircraft applications. Begin applications when aphids first appear. Repeat as necessary, but not more than 4 times per season, nor within 28 days of harvest.
Tobacco	Aphids Flea beetles	4 pints (1/2 gallon) per acre	Apply as a water emulsion broadcast spray in sufficient water for thorough coverage. Disk or till the soil to a depth of 2 to 3 inches before setting plants. Do not make more than one field application per crop season, regardless of the formulation of DI-SYSTON used.
Wheat (Foliar application Spring or Fall)	Aphids (Greenbugs) Mites (See **NOTE** in Remarks)	4 to 12 fluid ounces per acre	**For application by aircraft or ground equipment:** Apply specified dosage per acre in sufficient water to obtain thorough coverage, but not less than 1 gallon total volume per acre. Use the lower rates on plants up to tillering (stooling). Higher rates should be used after plants begin to tiller. **For application in liquid fertilizer:** Mix 6 to 12 fluid ounces of DI-SYSTON 8 in the amount of liquid fertilizer to be applied per acre. Apply this mixture as a top-dressing, with suitable ground equipment. **Fall Applications:** Two applications may be made in the Fall. Allow 30 days between applications. **Spring Applications:** Application at green-up may be made following Fall applications. A second application may be made 30 days following the first Spring application. Do not make **any** applications within 30 days of grain harvest. **NOTE** (1) Do **not** graze treated fields after **any** application of DI-SYSTON 8. (2) Due to greenbug resistance in certain areas, DI-SYSTON used alone may not provide satisfactory control. Consult your local agricultural advisor or Extension Service for recommendations.
Wheat (Fall)	Hessian fly Aphids (Greenbugs) and oat bird cherry—Except California)	0.25 fluid ounces per 1000 feet of row (see chart below) or up to 1 pint per acre	Apply as a soil-injection in water emulsion or with liquid fertilizer at planting time in the Fall. **NOTE:** Due to greenbug resistance in certain areas, DI-SYSTON used alone may not provide satisfactory control. Consult your local agricultural advisor or Extension Service for recommendations.

Row Spacing	Ounces/Acre	Ounces DI-SYSTON 8 per 1000 feet of row based on 3 inch band
7 or 8 inch rows	15.5	.25
10 inch rows	13.25	.25
12 inch rows	10.75	.25
14 inch rows	9.5	.25
16 inch rows	8.5	.25
18 inch rows	7.0	.25

CROP	INSECT	DOSAGE—DI-SYSTON 8	REMARKS
NON-BEARING FRUIT Strawberries (Propagating plants only)	Aphids Mites (Except cyclamen mite)	2.8 to 5 fluid ounces per 1000 feet of row (for any row spacing*), or 2 to 4 pints per acre (42 inch row spacing)	Apply in a water emulsion, or with liquid fertilizer as a soil injection on each side of the row as needed. Do not use fruit from treated plants for food purposes.
NUTS Pecans (South Central and Southwestern states only)	Aphids	1.5 to 3 pints per acre	Apply by spraying 6 foot bands of soil on both sides of the trees. The bands should be located under the main drip area of the trees. Work the treated soil to a depth of 2 to 3 inches. On irrigated pecans, apply water as soon as possible after treatment. Make only 1 application per season. Early Spring treatment is recommended. Do not harvest nuts within 80 days of treatment. Do not graze grass under treated trees. Do not tank mix this product with Phosalone.
Pecans	Aphids	4 to 6 fluid ounces	**Foliar Applications:** Apply specified dosage in 100 gallons of water as a full-coverage spray, using not more than 400 gallons of finished spray per acre. Repeat as necessary, up to 3 times per season. Do not apply within 30 days of harvest. Do not graze grass under treated trees. Do not tank mix this product with Phosalone.
		¾ to 1 pint	**Application by Aircraft:** Apply specified dosage per acre in sufficient water to obtain thorough coverage, but not less than 5 gallons total volume per acre. Do not apply more than 3 times per season, nor within 30 days of harvest. Do not graze grass under treated trees. Do not tank mix this product with Phosalone.
VEGETABLES Beans	Aphids Leafhoppers Mexican bean beetle (larvae) Mites Thrips	.9 to 1.9 fluid ounces per 1000 feet of row (for any row spacing*), or 1 to 2 pints per acre (30 inch row spacing) (For extremely narrow row spacing, do not exceed the 2 pints per acre rate)	For snap or green Lima beans, apply in a water emulsion, or with liquid fertilizer as a soil injection on each side of the seed furrow at planting time only. For dry beans, apply as above at planting time, or inject as a side-dressing on each side of the row after plants become established, but not within 60 days of harvest. Do not apply directly on the seed. Do not apply more than once per season. Read "IMPORTANT" statement under "RECOMMENDED APPLICATIONS" for additional information before use on this crop. Do not feed vines or hay.
Brussels sprouts	Aphids Flea beetles Leafhoppers Mites Thrips	1.1 fluid ounces per 1000 feet of row (for any row spacing*), or 1 pint per acre (36 inch row spacing)	**Transplant Seed Beds:** Apply evenly over bed area in a water emulsion as a broadcast spray prior to seeding, at the rate of 1 pint per acre. Work into the top 2 to 3 inches of soil, and plant seed in normal manner. **Field Applications:** Apply specified dosage in a water emulsion, or with liquid fertilizer as an injection on each side of the seed furrow or transplanted row at planting, and where necessary, injected as a side-dressing after plants are established. Do not apply directly on the seed. Do not apply more than two field applications per season, nor within 30 days of harvest for Brussels sprouts, or 40 days of harvest for cauliflower. Allow a minimum of 21 days between applications.
Cauliflower	Aphids Flea beetles Leafhoppers		
Broccoli	Aphids Flea beetles Leafhoppers	1.1 fluid ounces per 1000 feet of row (for any row spacing*), or 1 pint per acre (36 inch row spacing*)	**Transplant Seed Beds:** Apply evenly over bed area in a water emulsion as a broadcast spray prior to seeding, at the rate of 1 pint per acre. Work into the top 2 to 3 inches of soil, and plant seed in normal manner. **Field Applications:** Apply specified dosage in a water emulsion, or with liquid fertilizer as an injection on each side of the seed furrow or transplanted row at planting, or injected as a side-dressing after plants are established. Do not apply directly on the seed. Do not apply more than one field application per season. Do not use within 14 days of harvest for broccoli, or 42 days of harvest for cabbage.
Cabbage	Aphids Flea beetles		
	Root aphids	1.7 fluid ounces per 1000 feet of row (for any row spacing*)	
Lettuce	Aphids Leafhoppers Mites Thrips	.6 to 1.2 fluid ounces per 1000 feet of row (for any row spacing*), or 1 to 2 pints per acre (20 inch row spacing)	**Spring or Summer Seeded Lettuce:** Apply in a water emulsion, or with liquid fertilizer as an injection on each side of the seed furrow at planting time only. Use the higher rate on heavy organic soils. Do not apply directly on the seed. Do not apply to transplanted lettuce. **Fall or Winter Seeded Lettuce:** Apply specified dosage in a water emulsion, or as a soil-injected side-dressing with liquid fertilizer at thinning time. Do not apply within 60 days of harvest. Read "IMPORTANT" statement under "RECOMMENDED APPLICATIONS" for additional information before use on this crop.
Peas	Aphids Leafhoppers	1 to 2.5 pints per acre	Apply in a water emulsion, or with liquid fertilizer as an in-furrow spray or as an injection on each side of the seed furrow at planting, or injected as a side-dressing after plants are established. Do not apply more than once per season, nor within 50 days of harvest. Do not feed vines or hay.
Potatoes	Aphids	3 to 4 pints per acre	**Soil Application—Preplant Broadcast:** Apply specified dosage per acre as a preplant broadcast application. Incorporate to a depth of 2 to 3 inches. Plant crop in usual manner. Do not make more than two **soil** applications of DI-SYSTON per crop season, regardless of method of application or formulation used. Do not tank mix this product with Phosalone.
	Aphids Flea beetles (Except tuber flea beetles) Leafhoppers Potato psyllid Southern potato wireworm	2.25 to 3.5 fl. oz. per 1000 feet of row (for any row spacing*), or 2 to 3 pints per acre (38 inch row spacing)	**Soil Application—Furrow:** Apply undiluted in specifically-designed equipment for low volume soil injection, or apply in a water emulsion, or with liquid fertilizer as an in-furrow spray, or inject on each side of the seed furrow at planting. An **additional** side-dress application may be made as a water emulsion injection after plants become established. Do not make more than two **soil** applications of DI-SYSTON per crop season, regardless of method of application or formulation used. Do not apply second soil application within 75 days of harvest. Higher rate to be used in organic soils. Do not tank mix this product with Phosalone.

CROP	INSECT	DOSAGE—DI-SYSTON 8	REMARKS
	Colorado potato beetle (Early-season reduction)		
	Green peach aphids	6 to 16 fluid ounces per acre	**Foliar Applications:** Apply specified dosage per acre in sufficient water to obtain thorough coverage, but not less than 1 gallon total volume per acre. Begin applications when pest first appears. Use the higher rates for heavy infestations. Repeat as necessary, but not more than 3 times per season, nor within 30 days of harvest. In Arizona and California, do not make foliar applications of DI-SYSTON 8 to the potato variety, Centennial. Do not tank mix this product with Phosalone.
	NOTE: Foliar applications may be made following **at-planting** soil treatment as recommended above, or on the label for DI-SYSTON 15% Granular. 3–28–85		

* ON ROW CROPS WHERE ROW SPACING IS EXTREMELY NARROW, DO NOT PLACE TREATED ZONES (BAND OR FURROW) CLOSER TOGETHER THAN 6 INCHES.

SPRINKLER IRRIGATION APPLICATION TO POTATOES

For control of green peach aphid, and Colorado potato beetle on potatoes in the states of Oregon, Washington, Idaho, and Utah, apply DI-SYSTON 8 at the rate of 1 to 3 pints per acre through sprinkler irrigation systems. Use only in solid set or center pivot systems equipped with automatic shutoff devices and valves to prevent backflow into the water source. Do not apply when wind velocities exceed 10 mph.

Up to 3 applications of the 1 pint per acre rate may be made but not within 30 days of harvest, or, only 1 application of rates above 1 pint per acre may be made but not within 60 days of harvest. Applications may be made following **at-planting** soil treatments as recommended on this label or on the label for DI-SYSTON 15% Granular. Do not tank mix this product with Phosalone.

Solid Set Systems: Apply specified dosage for the entire length of the irrigation period **or** for a 30 to 60 minute period at the end of a regular irrigation set **or** as a 30 to 60 minute injection as a separate application not associated with a regular irrigation. Allow time for all lines to flush the pesticide through all nozzles before turning off irrigation water. To ensure the lines are flushed and free of remaining pesticide, a dye indicator may be injected into the lines to mark the end of the application period. See **NOTE.**

Circular Systems: Inject the specified dosage per acre continuously for one complete revolution of the system. See **NOTE** below.

SAMPLE CALCULATIONS:

Center Pivot Calibrations:

(1) To apply DI-SYSTON 8 at the rate of 1 pint per acre to 130 acres you will need 130 × 1 pint or 130 pints (16¼ gallons) of product.

(2) If your injection pump delivers 15 gallons per hour and your system takes 20 hours for 1 revolution, you will need to mix enough water with the DI-SYSTON 8 to obtain 300 gallons of emulsion.

(3) Mix the required amounts of DI-SYSTON 8 and water in a nurse tank equipped with mechanical agitation, start the irrigation system and begin the injection.

(4) Check frequently to ensure all systems are functioning properly.

Solid Set of Calibration:

(1) Determine the area to be treated per irrigation set. If one set irrigates four ¼ mile long lines on a 50′ spacing, 4 × 1,320′ × 50 feet = 264,000 square feet or approximately 6 acres.

(2) If 1 pint of DI-SYSTON 8 is to be applied per acre, 6 × 1 or 6 pints of DI-SYSTON 8 is needed.

(3) Dilute with sufficient water so injection time will require approximately 30 minutes or more per set. Under no circumstances should injection be less than 10 minutes.

(4) Repeat applications until the entire field is treated.

NOTE: Apply the product only through systems containing anti-siphon and check valves which will prevent water source contamination and overflow of the slurry tank, and containing interlocking controls between the metering device and the water pump to ensure simultaneous shutoff. Constant agitation must be maintained in the chemical supply tank during the entire period of insecticide application. Inject the product with a positive displacement pump into the main line ahead of a right angle turn to ensure adequate mixing.

Application of more than label recommended quantities of irrigation water per acre may result in decreased product performance by removing the chemical from the zone of effectiveness. Do not apply when wind speed favors drift, when systems connections or fittings leak, when nozzles do not provide uniform distribution or when lines containing the product must be dismantled or drained.

Greater accuracy in calibration and distribution will be achieved by injecting a larger volume of a more dilute slurry per hour. Where sprinkler distribution patterns do not overlap sufficiently, unacceptable insect control may result. Where sprinkler distribution patterns overlap excessively, crop injury may result.

Allow foliage to dry thoroughly before entering field. If there is a need to enter fields earlier wear protective clothing.

SPECIAL LOCAL NEED REGISTRATIONS [SECTION 24(c)] FOR DI-SYSTON 8*

CROP	INSECT	STATE EPA SLN NO.	DOSAGE	REMARKS
Asparagus	Asparagus aphids Thrips	California CA 84-0192	1 pint per acre.	Apply specified dosage to fern stage of asparagus when aphid or thrip infestation appears. A maximum of 3 applications may be made per year. Do not apply within 4 months of harvest.
		Oregon OR 84-0032 Washington WA 84-0036		Apply specified dosage to fern stage of asparagus when aphid or thrip infestation appears. A maximum of 3 applications may be made per year. Do not apply within 6 months of harvest.
Barley	Aphids	California CA 79-0109	4 to 12 fluid ounces per acre*	Application by ground or air.
Bermuda grass (Seed crop only)	Thrips Whiteflies Mealybugs	Arizona AZ 79-0044	1 pint per acre.	Application by ground or air.

CROP	INSECT	STATE EPA SLN NO.	DOSAGE	REMARKS
Lettuce	Lettuce root aphids	California CA 81-0044	1.2 fluid ounce per 1,000 ft. of row (for any row spacing) or 2 pints per acre (20 inch row spacing)*	Incorporate by sprinkler injection. Do not use within 60 days of harvest.
Potatoes (Sprinkler Irrigated Only)	Aphids	Idaho ID 83-0034 Oregon OR 83-0058 Washington WA 84-0013	3 pints	Apply suggested rate per acre as a soil application and/or a foliar treatment as a sprinkler injection.
Sugarcane	Yellow sugarcane aphids	Florida Fl 80-0003	6 to 9 fluid ounces per acre*	Application by ground or air.
Tobacco	Weed, insect and nematode control	Virginia VA 79-0001	Refer to Special Local Need Label*	Paarlan EC, DI-SYSTON 8, DASANIT Spray Concentrate†.

* For distribution and use only within the state specified. These Special Local Need registrations approved as of date of this publication are, however, subject to cancellation by the state. Contact your nearest Mobay Agricultural Chemicals Division Office for the current status and a copy of the complete individual supplemental Special Local Need labeling. This labeling must be in the possession of the user at the time of pesticide application.
† Contact Elanco Products Company for individual labels.

IMPORTANT:
This bulletin is not intended to provide adequate information for application. Before using the product, read and carefully observe directions, cautionary statements and other information appearing on the product label. This product is sold subject to the Conditions of Sale set forth on the container label. READ THE LABEL BEFORE USE

®DYLOX

80% SOLUBLE POWDER INSECTICIDE

A Selective Insecticide For Effective Control Of Insect Pests On A Variety Of Field And Vegetable Crops.

Only for agricultural use, or for sale to, use, and storage by commercial applicators, and other servicepersons.

Not for Sale or Use in the State of Maine.

ACTIVE INGREDIENT:
Dimethyl (2,2,2-trichloro-1-hydroxyethyl)phosphonate ... 80%
INERT INGREDIENTS ... 20%
TOTAL ... 100%

EPA Reg. No. 3125-184-AA
DYLOX is a Reg. TM of the Parent Company of Farbenfabriken Bayer GmbH, Leverkusen.

CAUTION
Keep Out of Reach of Children
AVISO

PRECAUCION AL USUARIO: Si usted no puede leer o entender inglés, no use este producto hasta que la etiqueta le haya sido explicada ampliamente.
(TO THE USER: If you cannot read or understand English, do not use this product until the label has been fully explained to you.)

RECOMMENDED USES

Alfalfa	Flowers, Shrubs and Trees
Barley	Grasses (including rangeland and pastures)
Beans (dry and succulent)	Lawns and turf
Blueberries	Lettuce
Brussels sprouts	Lima beans
Cabbage	Narcissus
Carrots	Oats
Cauliflower	Peppers
Citrus (California only)	Pumpkin
Clover	Safflower
Collards	Soybeans (seed)
Corn	Sugar beets
Cotton	Table beets
Cowpeas (Southern peas, Black-eyed peas, Crowder peas)	Tobacco
Flax	Tomatoes
	Wheat

Keep away from space heaters, or walls that sweat. Store below 120°F. Protect from rain and moisture. Store in a cool, dry place. Do not store next to herbicides.

STATEMENTS OF PRACTICAL TREATMENT

If swallowed, vomiting should be induced. Administer water freely and induce vomiting by giving one dose (1/2 oz. or 15 ml.) of syrup of ipecac. If vomiting does not occur within 10 to 20 minutes, administer second dose. If syrup of ipecac is not available, induce vomiting by sticking finger down throat. Repeat until vomit fluid is clear. Never give anything by mouth to an unconscious person. Have patient lie down and keep quiet. Professional medical assistance should be secured immediately.
Do not contaminate feed or food. Keep out of reach of children. Should poisoning occur, obtain prompt medical aid. Prolonged exposure will result in cholinesterase depression.
TO PHYSICIAN: Atropine sulfate is antidotal. 2-PAM is also antidotal and may be administered in conjunction with atropine.

PRECAUTIONARY STATEMENTS

HAZARDS TO HUMANS AND DOMESTIC ANIMALS
WARNING
May be fatal if swallowed. Harmful if inhaled, or absorbed through the skin. Do not get in eyes or on skin. Do not breathe dust or spray mist. Always wash thoroughly with soap and warm water after handling. Wash contaminated clothing with soap and hot water before reuse.

ENVIRONMENTAL HAZARDS
Protect Wildlife!!
This pesticide is toxic to fish and wildlife and is extremely toxic to aquatic invertebrates. Apply the product in strict accordance with label directions. Do not apply when weather conditions favor drift from areas treated. Do not apply to water or wetlands. Do not apply where runoff is likely to occur. Runoff from treated areas may be hazardous to aquatic organisms in neighboring areas. In cleaning of equipment or disposal of wastes, do not contaminate water.

RE-ENTRY STATEMENTS
Do not apply this product in such a manner as to directly or through drift expose workers or other persons. The area being treated must be vacated by unprotected persons. Do not reenter treated fields within 24 hours unless protective clothing is worn. Because certain states may require more restrictive re-entry intervals for various crops treated with this product, consult your State Department of Agriculture for further information.

Written or oral warnings must be given to workers who are expected to be in a treated area or in an area about to be treated with this product. Oral warnings should include those statements found under the label heading "Precautionary Statements" and "Re-entry Statements" included on this label. When oral warnings are given, warnings shall be given in a language customarily understood by workers. Oral warnings must be given if there is reason to believe that written warnings cannot be understood by workers. Written warnings must include the following information: WARNING, area treated with DYLOX 80% Soluble Powder on (date of application). Do not reenter treated fields within 24 hours unless protective clothing is worn.

If on skin, remove contaminated clothing and wash skin immediately with soap and warm water.

If eyes are contaminated, wash with flowing water for at least 15 minutes.

If swallowed, vomiting should be induced. (See Precautionary Statements on label).

STORAGE AND DISPOSAL

PESTICIDE DISPOSAL

Do not contaminate water, food or feed by storage or disposal. Wastes resulting from the use of this product may be disposed of on site or at an approved waste disposal facility.

CONTAINER DISPOSAL

Do not use container in connection with food, feed, or drinking water. Completely empty bag into application equipment. Then dispose of empty bag in a sanitary landfill or by incineration, or, if allowed by State and local authorities, by burning. If burned, stay out of smoke.

Store in a cool, dry place and in such a manner as to prevent cross contamination with other pesticides, fertilizers, food, and feed. Store in original container and out of the reach of children, preferably in a locked storage area.

Handle and open container in a manner as to prevent spillage. If the container is leaking or material spilled for any reason or cause, carefully sweep material into a pile. Refer to Precautionary Statements on label for hazards associated with the handling of this material. Do not walk through spilled material. Dispose of pesticide as directed above. In spill or leak incidents, keep unauthorized people away. You may contact the Mobay Emergency Response Team for decontamination procedures or any other assistance that may be necessary. The Mobay Kansas City Emergency Response Telephone No. is 816-242-2582, or contact Chemtrec at 800-424-9300.

IMPORTANT

This bulletin is not intended to provide adequate information for application. Before using the product, read and carefully observe directions, cautionary statements and other information appearing on the product label. This product is sold subject to the Conditions of Sale set forth on the container label.

READ THE LABEL BEFORE USE

DIRECTIONS FOR USE

It is a violation of Federal Law to use this product in a manner inconsistent with its labeling.

DYLOX insecticide offers the unique advantage of not significantly affecting beneficial insects (parasites, predators and pollinators), especially when applied at the minimum recommended rates per acre. Thus, DYLOX supplements the natural control factors, instead of destroying them. This advantage is lost when DYLOX is used in conjunction with parathion or other non-selective insecticides. Accordingly, for the most effective and economical control, DYLOX should not be used in conjunction with, or alternated in, a spray schedule with pesticides destroying parasites, predators or other beneficial insects. This applies to alfalfa and cotton, in particular. While DYLOX is not recommended for the control of aphids, mites and cabbage looper, some suppression of these pests may result from recommended dosages.

MIXING: DYLOX 80% Soluble Powder dissolves readily in water, and is suitable for use in all power-operated ground and aircraft sprayers. To dissolve in water, pour the required amount of DYLOX 80% Soluble Powder into full amount of water, and then agitate. Use immediately after mixing. It is recommended that DYLOX 80% Soluble Powder be used in sprayers equipped with 50-mesh screens. If 100-mesh screens are used, some screen clogging may occur.

DOSAGE: Use specified dosage of DYLOX 80% Soluble Powder in the amount of water necessary to give complete coverage of foliage. The type of equipment used will determine the concentration required.

SPRAYING: Complete coverage and thorough application are essential for most effective results. Schedule applications in accordance with local conditions. Consult your State Agricultural Extension Service or Experiment Station for specific timing of applications in your area.

NOTE: Spray droplets from applications of DYLOX 80% Soluble Powder may cause spotting of paint finishes on aluminum and steel siding, automobiles, trucks and tractors. Do not allow spray droplets to contact metal painted surfaces. If accidental exposure does occur, the painted surface should be thoroughly washed immediately after exposure.

RECOMMENDED APPLICATIONS

CROP	INSECT	OUNCES DYLOX 80% S.P.	REMARKS
FIELD CROPS	Alfalfa caterpillar	7½ to 10	Apply specified dosage per acre, using sufficient water for complete coverage, but not less than 1 gallon per acre. Three applications may be made per cutting, with the last application up to the day of cutting.
	Alfalfa webworm	5 to 20	
Alfalfa	Western yellow-striped armyworm	10	
Clover (including mixed stands with grasses)	Beet armyworm Leafhoppers Variegated cutworm	10 to 20	
	Armyworm Alfalfa plant bug Lygus bugs Stink bugs Tarnished plant bug	20	
Barley Flax Oats Wheat	Armyworms Beet webworm Variegated cutworm	10 to 20	Apply specified dosage per acre, using sufficient water for complete coverage, but not less than 1 gallon per acre. Three applications may be made per season, but not within 21 days of harvest. **NOTE:** Applications may be made without removal of grazing livestock.
	Bertha armyworm Diamondback moth	20	

CROP	INSECT	OUNCES DYLOX 80% S.P.	REMARKS
Corn Field, Sweet Popcorn	Armyworms Cutworms	10 to 20	Apply specified dosage per acre, using sufficient water for complete coverage, but not less than 1 gallon per acre. Three applications may be made per season. For early application, when plants are 3 to 12 inches high, direct the spray to the lower portions of plants and to the soil around the base of the plants. Later applications may be made as full-coverage spray up to day of harvest.
Cotton	Cotton fleahopper	5 to 20	Apply specified dosage per acre, using sufficient water for complete coverage, but not less than 1 gallon per acre. Do not pasture treated fields within 14 days of application. Do not apply within 7 days of picking. DYLOX application may occasionally cause marginal leaf burn. **NOTE:** For control of light to moderate infestation of cotton fleahoppers, use 5 ounces per acre. Higher rates may be required if heavy infestations or migrations occur.
	Cotton leafworm Darkling ground beetle Western yellow-striped armyworm	10 to 20	
	Beet armyworm Southern garden leafhopper	20	
	Black fleahopper complex Cotton leaf perforator Leaf roller Lygus bugs Stink bugs	20 to 30	
	Salt-marsh caterpillar	30	
Grasses (including rangeland and pastures)	Leafhoppers	10 to 20	Apply specified dosage per acre, using sufficient water for complete coverage, but not less than 1 gallon per acre. Three applications may be made per cutting, with the last application up to the day of cutting. Do not apply more than 3 times per season if crop is not cut for hay. **NOTE:** Applications may be made without removal of grazing livestock.
	Armyworm Fall armyworm Plant bugs	20	
Safflower	Armyworms Lygus bugs Thrips Variegated cutworm	20 to 30	Apply specified dosage per acre in each of two applications, using sufficient water for complete coverage, but not less than 1 gallon per acre. Do not apply more than twice per year, or **later than bloom stage.** Make first application approximately 2 weeks prior to bloom—second application at onset of bloom.
Sugar beets	Beet webworm Variegated cutworm	10 to 20	Apply specified dosage per acre, using sufficient water for complete coverage, but not less than 1 gallon per acre. Repeat as necessary. Do not apply within 14 days of harvest.
	Dipterous leafminers	20	
	Alfalfa webworm Beet armyworm	20 to 30	
	Salt-marsh caterpillar	30	
Tobacco	Budworm Hornworm	20	Apply specified dosage per acre, using sufficient water for complete coverage, but not less than 1 gallon per acre. For budworm control, make application directly into bud. Repeat as necessary. Do not treat within 3 days of harvest.
	Green June beetle larvae	10	Apply specified dosage in 100 gallons of water per 100 square yards to uprooted areas of plant beds.
SEED FIELD CROPS			
Alfalfa Clover	Armyworms Lygus bugs Stink bugs Variegated cutworm	20 to 30	Apply specified dosage per acre, using sufficient water for complete coverage, but not less than 1 gallon per acre. Repeat as necessary, up to 7 days before harvest. Chaff from alfalfa and clover seed crop may be used for feed or forage; however, **do not** cut green crop for these purposes. Do not pasture or use treated soybeans for feed, food, forgae or oil purposes. **NOTE:** Due to lygus bug resistance in certain areas of alfalfa and clover DYLOX used alone may not provide satisfactory control. Consult your local agricultural advisor or extension service for recommendation.
Soybeans	Armyworms Dipterous leafminers Lygus bugs Stink bugs Variegated cutworm		
FRUIT			
Blueberries	Spanworm	16	Apply specified dosage per acre using sufficient water for complete coverage, but not less than 1 gallon per acre. Make a prebloom application as soon as insects appear. A second application may be made during bloom but not less than 7 days after first application. Do not apply after bloom stage.
Citrus (California only)	Fruit tree leaf roller Omnivorous leaf roller Black swallowtail	40	Apply specified dosage per acre, in sufficient water for complete coverage, but not less than 1 gallon per acre. One application may be made by aircraft up to 7 days before harvest. Two applications with ground equipment may be made in addition to one aerial application. However, if more than one application is made, do not apply last application within 21 days of harvest.
	Avocado leaf roller Orange tortrix Citrus cutworm Citrus looper	80	

CROP	INSECT	OUNCES DYLOX 80% S.P.	REMARKS
VEGETABLES	Western bean cutworm	10 to 20	Apply specified dosage per acre, using sufficient water for complete coverage, but not less than 1 gallon per acre. Repeat as necessary. Do not apply within 14 days of harvest
Beans (dry and succulent)*	Armyworms Dipterous leafminers Lygus bugs Mexican bean beetle Stink bugs Variegated cutworm	20 to 30	*Identified as *Phaseolus* sp. (including kidney beans, mung beans, navy beans, pinto beans, snap beans, and wax beans); *Vicia* sp. (including broad beans, fava beans); *Vigna* sp. (including asparagus beans).
Lima beans	Western bean cutworm	10 to 20	Apply specified dosage per acre, using sufficient water for complete coverage, but not less than 1 gallon per acre. Up to 5 applications may be made per season. Do not apply within 3 days of harvest.
	Armyworms Dipterous leafminers Lygus bugs Mexican bean beetle Stink bugs Variegated cutworm	20 to 30	
Cowpeas (Southern peas, Black-eyed peas, Crowder peas)	Armyworms Dipterous leafminers Imported cabbageworm Lygus bugs Mexican bean beetle Stink bugs Variegated cutworm	20 to 30	Apply specified dosage per acre, using sufficient water for complete coverage, but not less than 1 gallon per acre. Do not apply more than twice per season, nor within 14 days of harvest.
Brussels sprouts Cabbage Cauliflower	Western yellow-striped armyworm	10	Apply specified dosage per acre, using sufficient water for complete coverage, but not less than 1 gallon per acre. Repeat as necessary. Do not apply within 21 days of harvest.
	Imported cabbageworm Variegated cutworm	10 to 20	
	Diamonback moth	20	
Carrots	Western yellow-striped armyworm	10	Apply specified dosage per acre, using sufficient water for complete coverage, but not less than 1 gallon per acre. Repeat as necessary. Do not apply within 28 days of harvest. Do not use tops for food or feed.
	Imported cabbageworm Variegated cutworm	10 to 20	
	Diamondback moth Dipterous leafminers	20	
	Beet armyworm Lygus bugs	20 to 30	
	Salt-marsh caterpillar	30	
Collards Lettuce	Thrips Variegated cutworm	10 to 20	Apply specified dosage per acre, using sufficient water for complete coverage, but not less than 1 gallon per acre. Repeat as necessary, until heads begin to form. Do not apply within 28 days of harvest.
	Armyworms Beet webworm Diamondback moth Dipterous leafminers Salf-marsh caterpiller	20	
Peppers	Dipterous leafminers Pepper maggot Serpentine leafminers	20	Apply specified dosage per acre, using sufficient water for complete coverage, but not less than 1 gallon per acre. Repeat as necessary. Do not apply within 21 days of harvest.
Pumpkin	Variegated cutworm	10 to 20	Apply specified dosage per acre, in sufficient water for complete coverage, but not less than 1 gallon per acre. Do not apply more than 3 times per season, nor within 3 days of harvest.
	Squash bug	20	
Table beets	Variegated cutworm	10 to 20	Apply specified dosage per acre, using sufficient water for complete coverage, but not less than 1 gallon per acre. Repeat as necessary. Do not apply within 28 days of harvest. Do not eat tops.
	Alfalfa webworm Beet webworm Dipterous leafminers Salt-marsh caterpillar	20	
	Beet armyworm Lygus bugs	20 to 30	
Tomatoes	Serpentine leafminers Tomato hornworms Dipterous leafminers	20	Apply specified dosage per acre, using sufficient water for complete coverage, but not less than 1 gallon per acre. Repeat as necessary. Do not apply within 21 days of harvest.
ORNAMENTALS	Nantucket pine tip moth Zimmerman pine moth	20	Mix specified dosage in 100 gallons of water and spray plants until they are thoroughly wet. Repeat as necessary. **CAUTION:** Phytotoxicity has occurred on certain varieties of carnations and zinnias. When large plantings of these flowers are to be sprayed for the first time, treat only a few plants and observe for 4 to 5 days before spraying the entire planting.
Flowers, Shrubs, & Trees	Armyworms Bagworms Climbing cutworms Dipterous leafminers Lygus bugs Stink bugs	20 to 30	

CROP	INSECT	OUNCES DYLOX 80% S.P.	REMARKS
	Tarnished plant bug Tobacco budworm Webworms		Injury from spray droplets may occur on leaves of certain trees.
Narcissus	Narcissus bulb fly	20	Apply specified dosage in 100 gallons of water as a drench per 1,000 feet of row. Direct stream to base of plants at beginning of adult fly activity (early May to June). Repeat treatments annually.
Lawns and Turf	Cutworms Sod webworms (Lawn moths)	2½ to 3¾	Mix specified dosage (approximately ⅓ to ½ cup) of DYLOX 80% Soluble Powder in 15 to 30 gallons of water and apply uniformly to 1,000 sq. ft. of lawn or turf area by means of a watering can, compressed air sprayer, or any commercial power sprayer. For best results, mow the lawn and rake dead grass from damaged spots. Water the lawn and allow grass to dry before spraying. For sod webworm and cutworm control do not irrigate following application. Three applications at monthly intervals may be required for maximum control. In most areas, the first of these applications should be made in May or June.
	White grubs	3¾	For white grub control thoroughly irrigate following application. Consult your State Agricultural Experiment Station or State Extension Service Specialists, for more specific information regarding timing of applications. Do not use DYLOX 80% Soluble Powder in hose proportioner sprayers. **NOTE:** Do not use treated areas or clippings from treated areas for feed or forage. 7–29–85

PREMISE SPRAY FOR FLY CONTROL: For Fly Control in and around Farm Buildings (Except Dairy Barns and Poultry Houses): Mix 5 pounds of DYLOX 80% Soluble Powder in 40 gallons of water (equivalent to 5 level tablespoonfuls per gallon). Thoroughly wet all areas to be treated by spraying 1 gallon per 500 square feet of surface area, or to the point of runoff. Apply to floors, walls, ceiling, around windows, doors and other areas where flies congregate. Also spray stock pens (after removal of animals), garbage areas and other locations where refuse has collected. Repeat applications every 1 to 2 weeks, or as fly populations increase. For longer residual control, add 1 pound of sugar to each gallon of spray solution.

RESTRICTIONS (Premise Spray): Remove animals before spraying barn. Do not treat portions of buildings where poultry feed. Direct application to walls, floors or other surfaces previously treated with lime, whitewash or other alkaline materials may be ineffective. Do not contaminate feed, water or milk-handling equipment.

Rate/Acre in Ounces of DYLOX 80% S.P.	Pounds of DYLOX 80% S.P. to Treat: 5 Acres	10 Acres	15 Acres
5	1½	3⅛	4⅔
7½	2⅓	4⅔	7
8	2½	5	7½
10	3⅛	6¼	9⅜
16	5	10	15
20	6¼	12½	18¾
30	9⅜	18¾	28⅛

NOTE: When using a more convenient whole number or fraction of pounds of formulated material, do not exceed maximum registered rate.

SPECIAL LOCAL NEED REGISTRATIONS [SECTION 24(c)] FOR DYLOX 80% SOLUBLE POWDER*

CROP	INSECT	STATE EPA SLN NO.	DOSAGE	REMARKS
Alfalfa	Army cutworm	Kansas KS 76-0002	20 ounces per acre	Three applications may be made per cutting with last application up to day of cutting.
Birdsfoot trefoil (seed crop only)	Armyworms Lygus bugs Stink bugs Variegated cutworm	New York NY 81-0020	20 to 30 ounces per acre	Repeat as necessary up to 7 days before harvest.
Blueberry	Gypsy moth	New Jersey NJ 84-0006	1.5 pounds	Apply specified dosage with aerial spray equipment or with ground spray equipment as a broadcast spray just before or during bloom stage when gypsy moth populations reach damaging levels. Do not apply more than one time per season nor within 45 days of harvest.
Cotton	Pink bollworm (adults)	California CA 82-0094	1.25–1.75 pounds per acre.	Apply specified dosage per acre, using sufficient water for complete coverage. Do not apply within 7 days of picking.

* For distribution and use only within the state specified. These Special Local Need registrations approved as of date of this publication are, however, subject to cancellation by the state. Contact your nearest Mobay Agricultural Chemicals Division Office for the current status and a copy of the complete individual supplemental Special Local Need labeling. This labeling must be in the possession of the user at time of pesticide application.

RESTRICTIONS

Do not use on other crops used for food or forage. Use only according to label directions. Application at rates above those shown may result in illegal crop residues. Do not treat food crops grown in the greenhouse.

DO NOT ALLOW DYLOX SPRAY TO DRIFT ONTO SORGHUM. TO DO SO MAY CAUSE BURNING OF THE SORGHUM OR MILO. Keep children and pets off treated areas until spray deposit has dried.

®DYRENE

50% WETTABLE POWDER
FOLIAGE FUNGICIDE

For Use on a Wide Variety of Vegetables and Other Crops.

ACTIVE INGREDIENT:	
4,6-Dichloro-N-(2-chlorophenyl)-1, 3, 5-triazin-2-amine ...	50%
INERT INGREDIENTS	50%
TOTAL	100%

EPA Reg. No. 3125-50-AA

DYRENE is a Reg. TM of Mobay Chemical Corporation.

CAUTION
Keep Out of Reach of Children
AVISO

PRECAUCION AL USUARIO: Si usted no puede leer o entender inglés, no use este producto hasta que la etiqueta le haya sido explicada ampliamente.

(TO THE USER: If you cannot read or understand English, do not use this product until the label has been fully explained to you.)

RECOMMENDED USES

Celery	Potatoes
Garlic	Strawberries
Gladioli	Tomatoes
Onions	

RESTRICTIONS ON WAREHOUSE STORAGE

Protect from rain and moisture. Keep away from space heaters, and walls that sweat. Do not store next to herbicides. Store in a cool, dry place.

STATEMENTS OF PRACTICAL TREATMENT

If on skin, wash immediately with soap and warm water. If irritation occurs, get medical attention.

PRECAUTIONARY STATEMENTS

HAZARDS TO HUMANS AND DOMESTIC ANIMALS

CAUTION

Harmful if swallowed or inhaled. Avoid contact with eyes, skin and clothing. Avoid breathing of dust or spray mist. Wash thoroughly with soap and warm water after handling. Wash contaminated clothing with soap and hot water before reuse.

NOTE: Protective clothing should be worn during periods of exposure, such as during application, or when contacting treated foliage. DYRENE may cause an allergic skin reaction in some susceptible persons. This condition should be treated as an allergic dermatitis.

Do not contaminate feed or food. Keep out of reach of children.

ENVIRONMENTAL HAZARDS

This product is toxic to fish. Keep out of lakes, streams or ponds. Do not apply when weather conditions favor drift from areas treated. Apply this product only as specified on the label. Do not contaminate water by cleaning of equipment or disposal of wastes.

RESTRICTIONS

Do not exceed the maximum dosage. Do not use on other crops used for food or feed. Use only according to label directions. Do not treat food crops grown in the greenhouse. Do not graze treated areas.

RE-ENTRY STATEMENTS

Do not apply this product in such a manner as to directly or through drift expose workers or other persons. The area being treated must be vacated by unprotected persons. Do not enter treated areas without protective clothing until sprays have dried. Because certain states may require more restrictive re-entry intervals for various crops treated with this product, consult your State Department of Agriculture for further information.

Written or oral warnings must be given to workers who are expected to be in a treated area or in an area about to be treated with this product. Oral warnings should include those statements found under the label heading "Precautionary Statements" and "Re-entry Statements" included on this label. When oral warnings are given, warnings shall be given in a language customarily understood by workers. Oral warnings must be given if there is reason to believe that written warnings cannot be understood by workers. Written warnings must include the following information: WARNING, area treated with DYRENE 50% Wettable Powder on (date of application). Do not enter without appropriate protective clothing until sprays have dried. **If on skin,** remove contaminated clothing and wash skin immediately with soap and warm water.

STORAGE AND DISPOSAL

PESTICIDE DISPOSAL

Do not contaminate water, food or feed by storage or disposal.

Wastes resulting from the use of this product may be disposed of on site or at an approved waste disposal facility.

CONTAINER DISPOSAL

Do not use container in connection with food, feed, or drinking water. Completely empty bag into application equipment. Then dispose of empty bag in a sanitary landfill or by incineration, or, if allowed by State and local authorities, by burning. If burned, stay out of smoke.

Store in a cool, dry place and in such a manner as to prevent cross contamination with other pesticides, fertilizers, food, and feed. Store in original container and out of the reach of children, preferably in a locked storage area.

Handle and open container in a manner as to prevent spillage. If the container is leaking or material spilled for any reason or cause, carefully sweep material into a pile. Refer to Precautionary Statements on label for hazards associated with the handling of this material. Do not walk through spilled material. Dispose of pesticide as directed above. In spill or leak incidents, keep unauthorized people away. You may contact the Mobay Emergency Response Team for decontamination procedures or any other assistance that may be necessary. The Mobay Kansas City Emergency Response Telephone No. is 816-242-2582, or contact Chemtrec at 800-424-9300.

IMPORTANT

This bulletin is not intended to provide adequate information for application. Before using the product, read and carefully observe directions, cautionary statements and other information appearing on the product label. This product is sold subject to the Conditions of Sale set forth on the container label.

READ THE LABEL BEFORE USE

DIRECTIONS FOR USE

It is a violation of Federal Law to use this product in a manner inconsistent with its labeling.

DYRENE 50% Wettable Powder effectively protects crops from various destructive plant diseases. Complete and thorough coverage of foliage is necessary for satisfactory control.

CROP APPLICATIONS: Begin applications when disease appears, and repeat if needed at intervals of 7 to 10 days, and during extended periods of wet weather, every 4 or 5 days. Observe time limitations for each use for interval between last application and harvest. Schedule applications in accordance with local conditions. Consult your State Agricultural Experiment Station or Extension Service for specific use information, and for state regulations which may contain additional restrictions or requirements.

RECOMMENDED APPLICATIONS

CROP	DISEASE	POUNDS DYRENE 50% W.P.	REMARKS
FRUIT Strawberries	Gray mold *(Botrytis)* Leaf spot Leaf scorch	2 to 6	Apply specified dosage, using 100 to 200 gallons of water per acre for good coverage*. Do not apply within 5 days of harvest for strawberries.
VEGETABLES Celery	Early blight Late blight *Rhizoctonia*	2 to 6	Apply specified dosage, using 75 to 200 gallons of water per acre for good coverage*. Treated celery should be washed and trimmed.
Onions (bulb or dry) Garlic	*Botrytis* Purple blotch *(Alternaria)*	2 to 6	Apply specified dosage, using 75 to 200 gallons of water per acre for good coverage*. Not recommended for use in Florida.
Onions (green)	*Botrytis* Purple blotch *(Alternaria)*	2 to 3	Apply specified dosage, using 75 to 100 gallons of water per acre for good coverage*. Not recommended for use in Florida.
Potatoes	Early blight Late blight	2 to 5	Apply specified dosage, using 75 to 175 gallons of water per acre for good coverage*.
Tomatoes	Anthracnose *Botrytis* Early blight Gray leaf spot Late blight *Septoria* leaf spot	2 to 5	Apply specified dosage, using 75 to 175 gallons of water per acre for good coverage*.
ORNAMENTALS Gladioli	Leaf spot *(Stemphylium)*	2	Apply specified dosage in 100 gallons of water as a full-coverage spray*. Repeat twice weekly during periods of infection. Recommended for use in Florida only. 7–29–85

* This dosage is calculated for conventional hydraulic-type sprayers. When lower volumes of spray are applied per acre, increase the concentration of DYRENE in the spray mixture in order to apply amount of DYRENE per acre equivalent to a full-coverage spray.

SPECIAL LOCAL NEED REGISTRATIONS [SECTION 24(c)] FOR DYRENE 50% WETTABLE POWDER*

CROP	DISEASE	STATE EPA SLN NO.	DOSAGE	REMARKS
Fruits Strawberries Vegetables Celery Garlic Onions (bulb or dry) Onions (green) Potatoes Tomatoes	Same as Federal label. Black mold added for tomatoes.	California CA 77–0066	Same as Federal label.	This Special Local Need registration is for a product packaged in water soluble packets. (PVA) Aerial applications permitted in not less than 10 gallons of water.
Tomatoes	Anthracnose *Botrytis* Early blight Gray leaf spot Late blight *Septoria* leaf spot Black mold *(Alternaria tenuis)*	California CA 76–0103	2 to 5 pounds per acre.	Apply specified dosage using 75 to 175 gallons of water per acre for ground application or apply by air in 5 to 15 gallons of water.

* For distribution and use only within the state specified. These Special Local Need registrations approved as of date of this publication are, however, subject to cancellation by the state. Contact your nearest Mobay Agricultural Chemicals Division Office for the current status and a copy of the complete individual supplemental Special Local Need labeling. This labeling must be in the possession of the user at time of pesticide application.

FURADAN® 3
GRANULES INSECTICIDE

RESTRICTED USE PESTICIDE

Due to acute inhalation toxicity

For retail sale to and application only by Certified Applicators or personnel under their direct Supervision.

For Control of Mosquitoes and Rice Water Weevil on Rice.

For Agricultural or Commercial Use Only.

ACTIVE INGREDIENT:

carbofuran (2,3-dihydro-2,2-dimethyl-7-benzofuranyl methylcarbamate)	3%
INERT INGREDIENTS	97%
TOTAL	100%

EPA Reg. No. 279-2792-3125 U.S. Patent No. 3,474,171

FURADAN—FMC trademark

CAUTION
Keep Out of Reach of Children
AVISO

PRECAUCION AL USUARIO: Si usted no puede leer o entender inglés, no use este producto hasta que la etiqueta le haya sido explicada ampliamente.

(**TO THE USER:** If you cannot read or understand English, do not use this product until the label has been fully explained to you.)

RECOMMENDED USES

Rice water weevil and Mosquitoes in Arkansas, Louisiana, Mississippi, Texas.

RESTRICTIONS ON WAREHOUSE STORAGE

The package will not withstand stacking two pallets high. Do not store next to herbicides. In case of puncture, use overslip bag and securely close.

STATEMENTS OF PRACTICAL TREATMENT

If swallowed, vomiting should be induced. Administer water freely, and induce vomiting by giving one dose (½ oz. or 15 ml.) of syrup of ipecac. If vomiting does not occur within 10 to 20 minutes, administer second dose. If syrup of ipecac is not available, induce vomiting by sticking finger down throat. Repeat until vomit fluid is clear. Professional medical assistance should be secured immediately. Never give anything by mouth to an unconscious person. Get medical attention.

If in eyes, immediately flush eyes with plenty of water for at least 15 minutes.

If on skin, in case of contact, remove contaminated clothing and immediately wash skin with soap and water.

NOTE TO PHYSICIAN: Antidote—FURADAN is a reversible cholinesterase inhibitor. Do not use oximes, such as 2-PAM. Start by giving 2 mg. of atropine intramuscularly. According to clinical response, continue until signs of atropinization occur (dry mouth or dilated pupils). If in eye, instill one drop of homatropine.

PRECAUTIONARY STATEMENTS

HAZARDS TO HUMANS (AND DOMESTIC ANIMALS)

WARNING

Poisonous if swallowed. May be fatal or harmful as a result of skin or eye contact or by breathing dust. Causes cholinesterase inhibition. Warning symptoms of poisoning include weakness, headache, sweating, nausea, vomiting, diarrhea, tightness in chest, blurred vision, pinpoint eye pupils, abnormal flow of saliva, abdominal cramps, and unconsciousness.

Atropine sulfate is antidotal.

Wear long-sleeved clothing and protective gloves when handling. Wash hands and face before eating or smoking. Bathe at the end of the work day. Change clothing daily and wash before reuse.

ENVIRONMENTAL HAZARDS

This pesticide is toxic to fish, birds and other wildlife. Birds feeding on treated areas may be killed. Birds killed by carbofuran pose a hazard to hawks and other birds of prey; bury or otherwise dispose of dead birds to prevent poisoning of other wildlife. Cover or incorporate granules in spill areas. Runoff from treated areas may be hazardous to fish in neighboring areas. Fish may be killed at recommended rates. Do not contaminate wells, wetlands or any body of water by cleaning of equipment or disposal of waste.

Carbofuran is a chemical which can travel (seep or leach) through soil and can contaminate ground water which may be used as drinking water. Carbofuran has been found in ground water as a result of agricultural use. Users are advised not to apply carbofuran where the water table (ground water) is close to the surface and where the soils are very permeable, i.e., well-drained soils as loamy sands. Your local agricultural agencies can provide further information on the type of soil in your area and the location of ground water.

Do not allow this material to drift onto neighboring crops or non-crop areas, or use in a manner or at a time other than in accordance with label directions because animal, plant or crop injury, illegal residues or other undesirable results may occur.

STORAGE AND DISPOSAL

PESTICIDE STORAGE

Do not store in or around the home.

Keep out of reach of children and animals. Store in original containers only. Store in a cool, dry place and avoid excess heat. Carefully open containers. After partial use, fold and roll back bags, clamp and close tightly. Do not put concentrate or dilute material into food or drink containers. Do not contaminate other pesticides, fertilizers, water, food or feed by storage or disposal.

In case of spill, avoid contact, isolate area and keep out animals and unprotected persons. Confine spills. Call Mobay Kansas City Emergency Response Telephone No. 816-242-2582, or contact Chemtrec at 800-424-9300.

To confine spill: If liquid, dike surrounding area or absorb with sand, cat litter or commercial clay. If dry material, cover to prevent dispersal. Place damaged package in a holding container. Identify contents.

PESTICIDE DISPOSAL

Pesticide wastes are acutely hazardous. Improper disposal of excess pesticide, spray mixture, or rinsate is a violation of Federal Law. If these wastes cannot be disposed of by use according to label instructions, contact your State Pesticide or Environmental Control Agency, or the Hazardous Waste representative at the nearest EPA Regional Office for guidance.

CONTAINER DISPOSAL

Completely empty bag into application equipment. Then dispose of empty bag in a sanitary landfill or by incineration, or, if allowed by State and local authorities, by burning. If burned, stay out of smoke.

IMPORTANT

This bulletin is not intended to provide adequate information for application. Before using the product, read and carefully observe directions, cautionary statements and other information appearing on the product label. This product is sold subject to the Conditions of Sale set forth on the container label.

READ THE LABEL BEFORE USE

DIRECTIONS FOR USE

It is a violation of Federal law to use this product in a manner inconsistent with its labeling.

Do not re-enter treated fields less than 24 hours following application. Do not use this product on Long Island, N.Y.

RICE: Arkansas, Louisiana, Mississippi, and Texas—For control of **rice water weevil and mosquitoes,** apply 15 to 20 pounds of FURADAN 3 Granules per acre 1 day before, or within 21 days after permanent flooding. Do not make more than one application per season. Apply by air or ground equipment. The subsequent use of propanil herbicide may result in crop injury.

For effective larval control of dark rice field mosquito, application must be made within 2 to 4 days after the field has been flooded.

8-29-85

SPECIAL LOCAL NEED REGISTRATIONS [SECTION 24(c)] FOR FURADAN 3 GRANULES*

Crop	Pest	State EPA SLN No.	Dosage	Remarks
Rice	Rice water weevil Mosquitoes	Arkansas AR 81-0045 Louisiana LA 81-0020 Mississippi MS 81-0051	15 to 20 pounds per acre.	One application as needed but not within 60 days of harvest.

* For distribution and use only within the state specified. These Special Local Need registrations approved as of date of this publication are, however, subject to cancellation by the state. Contact your nearest Mobay Agricultural Chemicals Division Office for the current status and a copy of the complete individual supplemental Special Local Need labeling. This labeling must be in the possession of the user at time of pesticide application.

FURADAN® 4F

INSECTICIDE

RESTRICTED USE PESTICIDE

Due to acute oral, dermal, and inhalation toxicity.

For retail sale to and application only by Certified Applicators or personnel under their direct supervision, and only for those uses covered by the Certified Applicator's certification.

For Contact and Systemic Control of Pests on Certain Field and Vegetable Crops.

ACTIVE INGREDIENT:
carbofuran (2,3-Dihydro-2,2-dimethyl-7-benzofuranyl methylcarbamate) 44.90%
INERT INGREDIENTS 55.10%
TOTAL 100.00%

EPA Reg. No. 279-2876-ZC-3125

U.S. Patent No. 3,474,171

This product contains 4 lbs. carbofuran per gallon.

FURADAN—FMC trademark.

DANGER–POISON

Keep Out of Reach of Children
PELIGRO

PRECAUCION AL USUARIO: Si usted no puede leer o entender inglés, no use este producto hasta que la etiqueta le haya sido explicada ampliamente.

(TO THE USER: If you cannot read or understand English, do not use this product until the label has been fully explained to you.)

RECOMMENDED USES

Alfalfa	Small grains (wheat, oats, barley)
Cotton	Soybeans
Field Corn	Strawberries (Washington and Oregon)
Grapes	Sugarcane
Sweet Corn	Sunflowers*
Ornamentals	Tobacco
Pine seedlings	
Potatoes	

* See label restrictions.

STATEMENTS OF PRACTICAL TREATMENT

If swallowed, vomiting should be induced. Administer water freely, and induce vomiting by giving one dose (½ oz. or 15 ml.) of syrup of ipecac. If vomiting does not occur within 10 to 20 minutes, administer second dose. If syrup of ipecac is not available, induce vomiting by sticking finger down throat. Repeat until vomit fluid is clear. Do not induce vomiting or give anything by mouth to an unconscious person. Professional medical assistance should be secured immediately.

If inhaled, remove to fresh air. Call a physician immediately.

If on skin: In case of contact, remove contaminated clothing and immediately wash skin with soap and water.

If in eyes, get medical attention; in the meantime, flush eyes with plenty of water for at least 15 minutes.

ANTIDOTE

Note to Physician: Carbofuran is a reversible cholinesterase inhibitor. Do not use oximes such as 2-PAM. Start by giving 2 mg. atropine intramuscularly. According to clinical response, continue until signs of atropinization occur (dry mouth or dilated pupils). If in eye, instill one drop of homatropine.

PRECAUTIONARY STATEMENTS

HAZARDS TO HUMANS AND DOMESTIC ANIMALS

DANGER

Poisonous if swallowed or inhaled. May be fatal or harmful as a result of skin or eye contact or by breathing spray mist. Causes cholinesterase inhibition. Warning symptoms of poisoning include weakness, headache, sweating, nausea, vomiting, diarrhea, tightness in chest, blurred vision, pinpoint eye pupils, abnormal flow of saliva, abdominal cramps, and unconsciousness. Atropine Sulfate is antidotal.

Wear a pesticide respirator jointly approved by the Mining Enforcement and Safety Administration (formerly the U.S. Bureau of Mines) and by the National Institute for Occupational Safety and Health under the provisions of 30 CFR Part 11. Wear goggles or face shield for eye protection when measuring or mixing. Wear long sleeved clothing and protective gloves when handling. Wash hands and face before eating or smoking. Bathe at the end of the work day. Change clothing daily and wash before reuse.

Keep all unprotected persons out of the operating area or vicinity where there may be danger of drift.

ENVIRONMENTAL HAZARDS

This product is toxic to fish, birds and other wildlife. Birds feeding on treated areas may be killed. For waterfowl protection do not apply immediately before or during irrigation or on fields in proximity of waterfowl nesting areas, or on fields where waterfowl are known to repeatedly feed. Drift & runoff from treated areas may be hazardous to fish in neighboring areas. Do not apply directly to water.

Notice: It is a Federal offense to use any pesticide in a manner that results in the death of a member of an endangered species.

The use of FURADAN 4F may pose a hazard to the following Federally designated endangered/threatened species known to be found in certain areas within the named locations:

Attwater's Greater Prairie Chicken—Texas counties including: Aransas, Austin, Brazoria, Colorado, Galveston, Goliad, Harris, Refugio and Victoria

Aleutian Canada Goose—California counties including: Colusa, Merced, Stanislaus and Sutter

Kern Primrose Sphinx Moth—Walker Basin of Kern County, California

This product may not be used in areas where adverse impact on the Federally designated endangered/threatened species noted above is likely. Prior to making applications, the user of this product must determine that no such species are located in or immediately adjacent to the area to be treated. If the user is in doubt whether or not the above named endangered species may be affected, he should contact either the regional U.S. Fish and Wildlife Service office

(Endangered Species Specialist) or personnel of the State Fish and Game office.

This product is highly toxic to bees exposed to direct treatment or residues on crops. Do not allow it to drift to blooming crops or weeds if bees are visiting the treatment area. Protective information may be obtained from your Cooperative Agricultural Extension Service.

Carbofuran is a chemical which can travel (seep or leach) through soil and can contaminate ground water which may be used as drinking water. Carbofuran has been found in ground water as a result of agricultural use. Users are advised not to apply carbofuran where the water table (ground water) is close to the surface and where the soils are very permeable, i.e., well-drained soils such as loamy sands. Your local agricultural agencies can provide further information on the type of soil in your area and the location of ground water.

STORAGE AND DISPOSAL

PESTICIDE STORAGE

Not for use or storage in or around the house.

Do not store below 35°F, (2°C).

Keep out of reach of children and animals. Store in original containers only. Store in a cool, dry place and avoid excess heat. Carefully open containers. After partial use, replace lids and close tightly. Do not put concentrate or dilute material into food or drink containers. Do not contaminate other pesticides, fertilizers, water, food, or feed by storage or disposal.

In case of spill, avoid contact, isolate area and keep out animals and unprotected persons. Confine spills. Call Mobay Kansas City Emergency Response Telephone No. (816) 242-2582 or contact Chemtrec at (800) 424-9300.

To confine spill: If liquid, dike surrounding area or absorb with sand, cat litter or commercial clay. If dry material, cover to prevent dispersal. Place damaged package in a holding container. Identify contents.

PESTICIDE DISPOSAL

Pesticide wastes are acutely hazardous. Improper disposal of excess pesticide, spray mixture, or rinsate is a violation of Federal Law. If these wastes cannot be disposed of by use according to label instructions, contact your State Pesticide or Environmental Control Agency, or the Hazardous Waste representative at the nearest EPA Regional Office for guidance.

CONTAINER DISPOSAL

Metal Containers: Triple rinse (or equivalent). Then offer for recycling or reconditioning, or puncture and dispose of in a sanitary landfill, or by other procedures approved by State and local authorities. Do not cut or weld metal containers.

Plastic Containers: Triple rinse (or equivalent). Then offer for recycling or reconditioning, or puncture and dispose of in a sanitary landfill, or incineration, or, if allowed by State and local authorities, by burning. If burned, stay out of smoke.

IMPORTANT: This bulletin is not intended to provide adequate information for application. Before using the product, read and carefully observe directions, cautionary statements and other information appearing on the product label. This product is sold subject to the Conditions of Sale set forth on the container label.

READ THE LABEL BEFORE USE

DIRECTIONS FOR USE

It is a violation of Federal Law to use this product in a manner inconsistent with its labeling.

FURADAN 4F is effective for the control of corn rootworms and sugarcane borer. FURADAN has a long residual life against alfalfa weevil and Egyptian alfalfa weevil and can be applied earlier than other insecticides for the best control of this pest.

IMPORTANT: Read these entire Directions and Conditions of Sale before using FURADAN 4F.

CONDITIONS OF SALE

THE DIRECTIONS ON THIS LABEL WERE DETERMINED THROUGH RESEARCH TO BE THE DIRECTIONS FOR CORRECT USE OF THIS PRODUCT. THIS PRODUCT HAS BEEN TESTED FOR A RANGE OF WEATHER CONDITIONS SIMILAR TO THOSE WEATHER CONDITIONS THAT ARE ORDINARY AND CUSTOMARY IN THE GEOGRAPHIC AREA WHERE THE PRODUCT IS USED. INSUFFICIENT CONTROL OF PESTS AND/OR INJURY TO THE CROP TO WHICH THE PRODUCT IS APPLIED MAY RESULT FROM THE OCCURRENCE OF EXTRAORDINARY OR UNUSUAL WEATHER, OR FROM FAILURE TO FOLLOW LABEL DIRECTIONS. IN ADDITION, FAILURE TO FOLLOW LABEL DIRECTIONS MAY CAUSE INJURY TO OTHER CROPS, ANIMALS, MAN, OR THE ENVIRONMENT. MOBAY OFFERS, AND THE BUYER ACCEPTS AND USES, THIS PRODUCT SUBJECT TO THE CONDITIONS THAT EXTRAORDINARY OR UNUSUAL WEATHER, OR FAILURE TO FOLLOW LABEL DIRECTIONS ARE BEYOND THE CONTROL OF MOBAY AND ARE, THEREFORE, THE RESPONSIBILITY OF THE BUYER.

RECOMMENDED APPLICATIONS

Do not apply FURADAN 4F through overhead irrigation systems.

Do not rotate to a succeeding crop other than alfalfa, corn (field, popcorn or sweet), cotton, cucurbits (cucumbers, melons, pumpkins or squash), grapes, peanuts, peppers, potatoes, rice, small grains (barley, oats, wheat), sorghum, soybeans, strawberries, sugar beets, sugarcane and tobacco. Tomatoes, cabbage, peas, succulent beans and dry beans may be planted the following season provided the prior season's application did not exceed 8.7 lb/A of FURADAN 15G, 13 lb/A of FURADAN 10 Granules or 2 pt. 10 fl. oz. of FURADAN 4F. Any other crop may be planted if it is not harvested or grazed.

ALFALFA: Alfalfa Weevil Larvae, Egyptian Alfalfa Weevil, Pea Aphid, and in New York State for Snout Beetle control—Apply the amount of FURADAN 4F indicated in the chart when feeding is noticed or when insects appear. Alfalfa Weevil Adult—Apply 1 to 2 pints per acre when insects appear. Alfalfa Blotch Leafminer (Northeastern states only) and Potato Leafhopper—Apply 1 to 2 pints per acre when insects appear. Lygus Bugs—Apply 2 pints per acre prior to bloom. Grasshoppers—Apply 1/4 to 1/2 pint per acre when grasshopper feeding is noticed. For control of Blue Alfalfa Aphid (nymphs and wingless adults)—apply FURADAN 4F at 1/2 to 1 pint per acre when insect feeding is noticed or when insects appear.

Observe the indicated number of days after application before cutting or grazing. Do not apply more than twice per season. Do not apply more than once per cutting. Do not use more than 1 pint per acre in the second application. Apply only to fields planted to pure stands of alfalfa. Do not move bees into alfalfa fields within 7 days of application.

Pints of FURADAN 4F per acre	Do not cut or graze within:
1/2	7 days
1	14 days
2	28 days

Minimum gallonage requirements: Ten gallons of finished spray per acre with ground equipment, two gallons per acre with aircraft, except Blue Alfalfa Aphid use 30 gallons of finished spray per acre with ground equipment, 5 gallons per acre with aircraft.

For waterfowl protection, do not apply on fields in proximity of waterfowl nesting areas, and/or on fields where waterfowl are known to repeatedly feed.

COTTON Thrips—Use FURADAN 4F at 2.5 fluid ounces per 1,000 linear feet of row (1 quart per acre with 40 inch row spacing). Apply in the seed furrow at planting. FURADAN 4F may be mixed with water or liquid fertilizer. When FURADAN 4F is used with liquid fertilizers, premix one part of FURADAN 4F with two parts of water. Check physical compatibility before mixing large quantities. Add this premix to the tank of fertilizer along with the rinsings from the premixing container. Maintain agitation in the tank after mixing and during application. Do not mix until ready to use. Do not feed cotton forage.

FIELD CORN, POPCORN, SWEET CORN—At Planting: Corn Rootworms, Flea Beetles, and to aid in the control of first generation European Corn Borer—Use 2.5 fluid ounces of FURADAN 4F per 1,000 linear feet of row (1 quart per acre with 40 inch row spacing). Apply at planting, as a 7 inch band over the row or inject on each side of the row by mixing with water or liquid fertilizers.

Southern Corn Rootworm, Flea Beetles, Seed Corn Maggot, Wireworms, and to aid in the control of first generation European Corn Borer and Armyworm for approximately 4 to 6 weeks after planting—Use 2.5 fluid ounces of FURADAN 4F per 1,000 linear feet of row (1 quart per acre with 40 inch row spacing). Apply at planting directly into the seed furrow.

FURADAN 4F may be mixed with water or liquid fertilizers. If FURADAN 4F is used with liquid fertilizers, premix one part of FURADAN 4F with two parts of water. Check physical compatibility before mixing large quantities. Add this premix to the tank of fertilizer along with the rinsings from the premixing container. Maintain agitation in the tank after mixing and during application. Do not mix until ready to use. Do not feed forage within 30 days of last application.

FIELD CORN, POPCORN—Post Plant: Corn Rootworms (northern, southern and western)—use 2.5 fluid ounces of FURADAN 4F per 1,000 linear feet of row (1 quart per acre with 40 inch row spacing). Apply as a post emergent spray by banding over the row, or by side dressing or basal spraying both sides of the row after corn emerges. Control will generally be improved if the treatment is cultivated into the soil. Do not feed forage within 30 days of last application.

FIELD CORN—Foliar Application: European Corn Borer—Use 1½ to 2 pints of FURADAN 4F per acre as a foliar spray when corn borer eggs begin to hatch. Use the higher rate for heavier pest infestations. For treatment with aerial equipment, apply as broadcast spray using a minimum of 1 gallon of finished spray per acre. For treatment with ground equipment, direct the spray into the corn whorl for first brood and into the ear zone for second brood, using a minimum of 10 gallons of finished spray per acre. Repeat if necessary. Observe all precautions listed below.

Southwestern Corn Borer—Use FURADAN 4F at 1 to 2 pints per acre as a broadcast foliar spray when eggs begin to hatch. Rate used will depend on desired residual activity. If infestation continues, retreat in 7 days after a 1 pint application and within 14 days after a 2 pint application. Apply in sufficient water for thorough coverage using a minimum of 1 gallon of finished spray per acre with air equipment or 10 gallons of finished spray per acre with ground equipment. Observe all precautions listed below.

Banks Grass Mites (suppression)—FURADAN 4F when applied at 2 pints per acre for the control of European or Southwestern Corn Borers will also suppress Banks Grass Mites.

Grasshoppers—Use FURADAN 4F as a foliar spray at ¼ to ½ pint per acre when insects appear or feeding is noticed.

— Use the ¼ pint rate under light to moderate population levels (0-14 grasshoppers/sq. yd.).
— Use the ½ pint rate under more severe population levels (15 or more grasshoppers/sq. yd.).

Apply in sufficient water for thorough coverage using a minimum of 2 gallons of finished spray per acre with air equipment or 10 gallons of finished spray per acre with ground equipment. Observe all precautions listed below.

Do not make more than two applications per season at the 1½ to 2 pint use rate. Do not make more than four applications per season at the 1 pint use rate. Do not forage, cut or harvest within 30 days of last application. Do not apply on seed corn less than 14 days prior to detasseling or rogueing. If prolonged intimate contact will result, do not re-enter treated field within 14 days of application without wearing proper protective clothing.

SWEET CORN FOLIAR APPLICATION (MACHINE HARVESTED ONLY): European Corn Borer—For control of second generation borers apply 1 pint of FURADAN 4F per acre. Make first application just prior to first silking and repeat at weekly intervals. Do not make more than four (4) applications per season. Do not apply within seven (7) days of harvest. If prolonged intimate contact will result, do not reenter treated field within 14 days of application without wearing proper protective clothing. Do not graze or harvest stalks within 21 days of last application.

Minimum gallonage requirements: 10 gallons of finished spray per acre with ground equipment, 2 gallons per acre with aircraft.

Do not make a foliar application if more than 12 ozs. of FURADAN 10G per 1,000 linear feet of row (10 lbs. per acre with 40" row spacing) or 8 ozs. of FURADAN 15G per 1,000 linear feet of row (6.7 lbs. per acre with 40 inch row spacing) or 2.5 fluid ounces of FURADAN 4F per 1,000 linear feet of row (1 quart per acre with 40 inch row spacing) were used in an at-planting application. At-planting application not permitted in California.

GRAPES (For use in California): For control of Nematodes (root-knot and dagger) and Grape Phylloxera. Apply FURADAN 4F insecticide-nematicide at 2½ gallons per acre as a broadcast treatment only, to the soil surface between the vine rows and immediately incorporate by mechanical means. Do not apply within 200 days of harvest. Remove dense weed growth prior to treatment. Do not use on soils of pH 8.0 or greater. Do not make more than one application per crop year. Wear protective clothing including rubber gloves and impervious foot covering at time of application.

PINE SEEDLINGS: Pales Weevil and Pitch-eating Weevil in pine plantations—Apply a 1% (W/W) active FURADAN clay slurry (see following for preparation) to the roots of pine seedlings prior to transplanting. Treat seedlings by dipping roots, or use any other suitable means which allows thorough coating. Keep roots moist until transplanted.

Prepare the slurry as follows: Add 1.6 ounces (2½ tablespoons) of FURADAN 4F to ½ gallon of water. Mix thoroughly. Add 2 pounds of pulverized kaolin clay (pH 4.5) to this suspension. Mix thoroughly. This is sufficient to treat the roots of 150 to 200 seedlings.

Adequate ventilation is required for indoor treatment. Wear protective unlined rubber gloves at time of application. Users should avoid contamination of the inside of the gloves. If contamination of the inside of the gloves is suspected, the gloves should be immediately discarded.

POTATOES: Potato Tuberworm (Virginia only); Colorado Potato Beetle, European Corn Borer, Potato Flea Beetle, Potato Leafhopper—Use 1 to 2 pints of FURADAN 4F in sufficient water to treat one acre. Apply when insects first appear and repeat as necessary to maintain control. Do not make more than 8 foliar applications per season. Do not apply more than 6 pints to the foliage if FURADAN 10 Granules or Furadan 15G was used at planting. Do not apply more than 2 pints per application. Do not apply within 14 days of harvest. Minimum gallonage requirements: 10 gallons of finished spray per acre with ground equipment and 3 gallons per acre with aircraft.

Do not use this product in Long Island, New York.

For waterfowl protection do not apply:

— immediately before or during furrow irrigation
— on fields in proximity of waterfowl nesting areas
— on fields where waterfowl are known to repeatedly feed

SMALL GRAINS (WHEAT, OATS, BARLEY): Grasshoppers—Use FURADAN 4F at ¼ to ½ pint per acre when insects appear. Cereal Leaf Beetle—Use ½ pint per acre when insects appear. Apply before heads emerge from boot. Do not make more than two applications per season. Do not feed treated forage to livestock.

Minimum gallonage requirements: Ten gallons of finished spray per acre with ground equipment, two gallons per acre with aircraft.

For waterfowl protection do not apply on fields in proximity of waterfowl nesting areas and/or on fields where waterfowl are known to repeatedly feed.

SOYBEANS—At planting Application: Nematodes (Root-knot, Cyst, Stunt, Ring, Sting, Spiral, Lesion, Lance, Dagger, Stubby-root)—use 3.75 to 5 fluid ounces per 1,000 linear feet of row (3 to 4 pints per acre with 40-inch row spacing). Apply at planting as a 12-inch surface band. Soybeans may be grazed or cut for forage 30 days or later following an at planting application.

SOYBEANS—Foliar Application: Mexican Bean Beetle—Use FURADAN 4F at ½ to 1 pint per acre when insects appear. Grasshoppers—Use FURADAN 4F at ¼ to ½ pint per acre when insects appear. Do not use FURADAN 4F as a foliar application if FURADAN 10G, 15G or 4F was applied to soybeans at planting time. Do not make more than 2 foliar applications per season. Do not apply within 21 days of harvest. Do not graze or feed foliar-treated forage to livestock or cut for silage or hay. Minimum gallonage requirements: 20 gallons of finished spray per acre with ground equipment, 1½ gallons per acre with aircraft.

STRAWBERRY (For use in Washington and Oregon): Root Weevils—Use FURADAN 4F at 2 to 4 pints per acre (2.6 to 5.1 fluid ounces per 1,000 linear feet of row for 42 inch row spacing). Apply as a 10- to 12-inch band over the row after last harvest but before October 1. Do not make more than one application per season. Do not apply if berries are present.

SUNFLOWERS (CONFECTIONARY AND OIL)—At Planting: Sunflower Stem Weevil, Sunflower Beetle, Grasshoppers—Use 2.5 to 5.0 fluid ounces of FURADAN 4F per 1000 linear feet of row (1.4 to 2.8 quarts per acre with 30 inch row spacing). Apply directly into the seed furrow-OR-Apply in a 7 inch band. FURADAN 4F may be mixed with water or liquid fertilizer. When FURADAN 4F is used with liquid fertilizers, premix one part of FURADAN 4F with two parts of water. Check physical compatibility before mixing large quantities. Add this premix to the tank of fertilizer along with the rinsings from the premixing container. Maintain agitation in the tank after mixing and during application. Do not mix until ready to use.

Use the higher rate when heavier insect infestations are anticipated.

SUNFLOWERS (CONFECTIONARY AND OIL)—Foliar Application: Sunflower Moth, Banded Sunflower Moth, Stem Weevils, Seed Weevils—Use FURADAN 4F at 1 pint per acre. Grasshoppers—Use FURADAN 4F at ¼ to 1 pint per acre. Sunflower Beetle—Use FURADAN 4F at ¼ to ½ pint per acre. Apply as a foliar spray when insects appear. When a rate range is indicated, use higher rate for heavier insect infestations. Repeat applications as necessary but not more than four times per season. Apply a minimum of two gallons of finished spray per acre with aircraft and 10 gallons by ground equipment. If prolonged, intimate contact will result, do not re-enter treated field within 14 days of application without wearing proper protective clothing. Do not harvest crop within 28 days of last application.

SUGARCANE: Sugarcane Borer—Apply 1 to 1½ pints FURADAN 4F per acre using ground or aerial equipment. Check sugarcane fields weekly, beginning in early June and continuing through August. Make first application only after visible joints form and 5% or more of the plants are infested with young larvae feeding in or under the leaf sheath, and which have not bored into the stalks. Repeat whenever field checks indicate the infestation exceeds 5%. Do not apply within 17 days of harvest. Do not use in Hawaii.

TOBACCO—BURLEY: Flea Beetles—Use 1 gallon of FURADAN 4F per acre. Apply as a broadcast spray over the soil surface prior to transplanting and incorporate into the top 3 inches of soil with a suitable device.

TOBACCO—FLUE-CURED: Flea Beetles, Wireworms, and to aid in the control of Budworms and Root-knot Nematodes—Use 1½ gallons of FURADAN 4F per acre—or—use 1 gallon of FURADAN 4F per acre for control of Flea Beetles only. Apply as a broadcast spray over the soil surface prior to forming beds. Incorporate into the top 3 inches of soil. Form beds and plant as usual. This product may induce flecking of the bottom or lower leaves.

ORNAMENTALS—Container Grown (Except California): Root Weevil Larvae—Prepare a stock solution by thoroughly mixing 1 to 2 fluid ounces of FURADAN 4F per 100 gallons of water. Apply as a soil drench in sufficient volume to saturate the entire soil profile within each container. The following guidelines give approximate amounts to use on various sized containers. Make a single application when larvae are present (usually from July to mid-October in outdoor growing areas). Later application may be less effective due to lower temperatures and/or the presence of more mature larvae.

Container Size	Amount of Stock Solution Per Container
6 inches diameter	1 pint
8 inches diameter	1 to 2 pints
10 inches diameter	2 to 4 pints
12 inches diameter	3 to 6 pints
16 inches diameter	4 to 8 pints

Early marginal necrosis or leaf drop may occur on Hydrangea or Birch. Not all species or varieties of ornamentals have been tested. Before treating large numbers of plants of a particular variety, treat a few plants and observe prior to full scale application.

Do not handle treated plants, soil or containers that are wet with spray.

Application of FURADAN 4F through overhead sprinkler equipment is prohibited.

SPECIAL LOCAL NEED REGISTRATIONS [SECTION 24(c)] FOR FURADAN 4F*

CROP	INSECT	STATE EPA SLN NO.	DOSAGE	REMARKS
Alfalfa	Blue alfalfa aphid	Nevada NV 82-0007	1 to 2 pints per acre.	Application by ground or air.
		Oklahoma OK 81-0008	1 to 2 pints per acre.	
	Spotted alfalfa aphid	Oklahoma OK 82-0003		
Alfalfa (Pure seeded)	Alfalfa blotch leafminer Potato leafhopper Clover root curculio	Pennsylvania PA 78-0012	1 to 2 quarts per acre.	At-planting application.
	Alfalfa blotch leafminer Potato leafhopper Clover root curculio Nematodes	New Jersey NJ 81-0014 New York NY 81-0017		
	Alfalfa blotch leafminer Potato leafhopper	Maryland MD 80-0025 Virginia VA 80-0037		

CROP	INSECT	STATE EPA SLN NO.	DOSAGE	REMARKS
	Potato leafhopper Clover root curculio	Michigan MI 81-0011		
	Alfalfa blotch leafminer (suppression) Clover root curculio Potato leafhopper (suppression) Nematodes	Connecticut CT 81-0007	1 to 2 quarts per acre.	At-planting application.
	Clover root curculio Potato leafhopper Crickets Grasshoppers	North Carolina NC 82-0031		
Alfalfa (Pure seeded) (no-till seedings)	Potato leafhopper	Illinois IL 84-0003		
Bermuda grass (grown for seed)	Thrips Fulgorids (leafhoppers)	Arizona AZ 77-0017 California CA 80-0168	1 quart per acre in 5 gallons of water by air, or 20 gallons of water by ground application.	Repeat treatment as necessary to maintain control.
Corn (field)	Corn rootworm larvae	Indiana IN 81-0023 Michigan MI 82-0006	2 pints per 13,000 feet of row.	Apply evenly in a 2- to 4-inch band on both sides of row after crop emergence.
	Grasshopper	Colorado CO 79-0011 Missouri MO 80-0004 Nebraska NE 78-0022 South Dakota SD 79-0016	¼ to ½ pint per acre.	Application by ground or air.
	Grasshopper nymphs	Kansas KS 78-0015		
	European corn borer	Delaware DE 78-00020	1 to 2 pints per acre.	Foliar application.
		Alabama AL 81-0030 Kansas KS 79-0008 Nebraska NE 80-0027 Tennessee TN 81-0010	1 quart per acre.	Application by ground or air.
		Maryland MD 79-0011 Texas TX 81-0016 Virginia VA 78-0022	1 to 2 pints per acre.	
	Southwestern corn borer	Alabama AL 81-0030	1 to 2 pints per acre.	Maximum 2 applications.
		Arizona AZ 81-0024 Colorado CO 79-0012	2 pints per acre.	
		Kansas KS 79-0008	1 quart per acre.	Maximum 2 applications.
		New Mexico NM 78-0017 Tennessee TN 81-0010 Texas TX 81-0016	1 to 2 pints per acre.	Maximum 4 applications at 1 pint rate and 2 applications at 2 pint rate. Do not apply to seed corn prior to rogueing or detasseling. Workers should not enter treated field within 14 days of application unless protective equipment is used.
	Southwestern corn borer Mites	New Mexico NM 79-0025	1 to 2 pints of FURADAN 4F tank-mixed with 1½ to 2 pints ®METASYSTOX-R Spray Concentrate.	Apply tank-mix when both pests are present at the same time.
	Southwestern corn borer Sugarcane borer	Louisiana LA 79-0031	1½ to 2 pints per acre.	Broadcast foliar spray.

CROP	INSECT	STATE EPA SLN NO.	DOSAGE	REMARKS
	Southern corn billbug	North Carolina NC 80-0019 South Carolina SC 80-0026	1 to 2 quarts per acre.	Postemergent spray.
Corn (sweet)	European corn borer	Connecticut CT 82-0007 Delaware DE 82-0009 New Jersey NJ 82-0010 Pennsylvania PA 82-0018 Virginia VA 82-0028	1 pint per acre.	Make first application prior to first silking and repeat at weekly intervals. Maximum four applications per season. Application by air. (MACHINE HARVESTED ONLY)
	European corn borer Corn earworm	Michigan MI 82-1019 Ohio OH 82-0015		Application by air or ground. (MACHINE HARVESTED ONLY)
Grain sorghum	Greenbug Corn leaf aphid	New Mexico NM 78-0011 Oklahoma OK 81-0012 Texas TX 78-0027	0.5 to 1 pint per acre.	Application by ground or air.
	Chinch bug	Mississippi MS 82-0032 Missouri MO 80-0011 Nebraska NE 80-0002		Ground application only.
		Kansas KS 78-0014	1 pint per acre.	
		Kansas KS 81-0027	2 pints per 13,000 feet of row.	In furrow application.
	Chinch bug (for early plant protection and stand establishment) Greenbugs	Missouri MO 81-0030	2 pints per 13,000 linear feet of row (40″ row spacing)	Apply directly into seed furrow.
	Greenbug	Colorado CO 79-0013 Kansas KS 78-0014 Missouri MO 80-0011	1 pint per acre.	Application by ground or air.
		Kansas KS 81-0027	2 pints per 13,000 feet of row.	In furrow application.
Grapes	Black vine weevil	Washington WA 82-0075	1 quart per acre.	Ground application after adult weevils have emerged in Spring.
Non-bearing peach and apple trees	Nematodes	West Virginia WV 78-0008	2.5 gallons per acre (10.0 pounds active).	Apply in the Spring when tree growth begins.
Non-bearing peach, nectarine and apple trees		Delaware DE 81-0008 Maryland MD 80-0019 Virginia VA 80-0039	1.5 gallons per acre. (6.0 pounds active).	Preplant or post-plant applications.
Non-bearing peach, nectarine, apple, apricot, cherry, and plum trees		Ohio OH 81-0033 Michigan MI 82-0002	1.5 to 2.5 gallons per acre (6.0 to 10.0 pounds per (acre).	
Potatoes	Colorado potato beetle Potato leafhopper Potato flea beetle	Delaware DE 76-0001 Delaware DE 78-0019 North Carolina NC 80-0027 Ohio OH 81-0037 Virginia VA 77-0014	1 to 2 pints per acre.	Ground application. Aerial application in 2 gallons finished spray per acre.

CROP	INSECT	STATE EPA SLN NO.	DOSAGE	REMARKS
		Oregon OR 80-0088 Utah UT 80-0011		Application by ground or air. For aerial use 3 to 5 gallons finished spray per acre.
	Colorado potato beetle European corn borer	Idaho ID 81-0042 Washington WA 81-0021 Ohio OH 81-0037		Application by ground or air.
	Nematodes and soil pests	Nevada NV 82-0013	3 quarts per acre.	By air. Immediately sprinkler irrigate after application.
Rotation crops (wheat or barley)		Oregon OR 81-0047		May be planted as rotation crops 4 months following the last application of FURADAN 4F or FURADAN 10 Granules provided straw is not harvested or grazed.
Sorghum	Grasshopper	Nebraska NE 78-0022	1/4 to 1/2 pint per acre.	Application by ground or air.
	Greenbug	Nebraska NE 78-0021	1 pint (0.5 pounds active) per acre.	
Sorghum (grain)	Greenbug Corn leaf aphid	Oklahoma OK 81-0012	1 pint per acre.	Application by ground or air.
	Chinch bug		1/2 to 1 pint per acre.	Ground application only.
Sorghum (grown for grain or forage)	Greenbug Corn leaf aphid Southern corn rootworm Yellow Sugarcane aphid	Texas TX 81-0016	1 1/2 to 2 pints per 13,000 linear fee or row.	Apply in-furrow at-planting.
	Chinch bug		1/2 to 1 pint in 20 to 30 gallons of water.	Ground application only.
Sorghum (forage and silage)	Chinch bug (for early plant protection and stand establishment) Greenbugs	Missouri MO 81-0030	2 pints per 13,000 linear feet of row (40″ row spacing).	Apply directly into seed furrow.
Soybeans	Grasshopper	Nebraska NE 78-0023	1/4 to 1/2 pint per acre.	Application by ground or air.
Soybeans	Nematodes	Arkansas AR 81-0016 Louisiana LA 82-0020 Mississippi MS 82-0017 Missouri MO 82-0017 Oklahoma OK 82-0020 Tennessee TN 82-0013	3 to 4 pints per 13,000 linear feet of row (one acre in 40 inch row spacing)	At planting.
		Ohio OH 82-0019	4 pints per 13,000 linear feet of row.	
Soybeans	Mexican bean beetle	Arkansas AR 82-0016 Mississippi MS 82-0017 Missouri MO 82-0017 Tennessee TN 82-0013	2 to 4 pints per 13,000 linear feet of row (one acre with 40 inch row spacing)	At planting.
Strawberries	Root weevils	California CA 82-0099 New Jersey NJ 81-0013 New York NY 81-0009 Vermont VT 81-0001	4 pints per acre.	Apply after harvest.
Sugarcane	Diaprepes abbreviatuas	Puerto Rico PR 80-0002	1 pint per acre.	Application by ground or air.

CROP	INSECT	STATE EPA SLN NO.	DOSAGE	REMARKS
Sunflowers (grown for seed)	Stem weevil	North Dakota ND 80-0014	1 pint per acre.	Application by ground or air.
Tobacco (burley)	Flea beetles	Indiana IN 81-0003 Kentucky KY 78-0019 Ohio OH 80-0017	1 gallon per acre in 20 to 40 gallons of water.	Broadcast spray over soil surface.
Tobacco (dark air-cured, dark fire-cured)		Kentucky KY 79-0012		
Tobacco (dark-fired)	Flea beetles	Tennessee TN 81-0026	¾ to 1 gallon per acre.	Broadcast spray over soil surface.
Tobacco (flue-cured)	Budworms Hornworms Flea beetles Nematodes (including cyst nematode) Wireworms	Virginia VA 79-0021	1 to 1½ gallons per acre (4 to 6 pounds active)	
		Virginia VA 78-0021		Tank mix with Ridomil 2E for black shank and blue mold.
	Budworms Flea Beetles Nematodes Wireworms	North Carolina NC 77-0021 South Carolina SC 81-0012		Broadcast over soil surface prior to forming beds.
	Green peach aphid (aids in control)	South Carolina SC 81-0012		
Tobacco (fire-cured)*	Hornworms Flea beetles Blue mold	Virginia VA 81-0021	1 gallon per acre.	Tank mix with Ridomil 2E.
Tobacco (seedleaf)	Flea beetles Hornworms Wireworms Nematodes Aphids (early-season)	Pennsylvania PA 80-0042	1 to 1½ gallons per acre.	Broadcast spray over soil surface.
Tobacco (Maryland type)		Maryland MD 81-0020		
Winter wheat	Grasshopper	Nebraska NE 78-0023	¼ to ½ pint per acre.	Application by ground or air.

* For distribution and use only within the state specified. These Special Local Need registrations approved as of date of this publication are, however, subject to cancellation by the state. Contact your nearest Mobay Agricultural Chemicals Division Office for the current status and a copy of the complete individual supplemental Special Local Need labeling. This labeling must be in the possession of the user at time of pesticide application.

FURADAN® 15G

INSECTICIDE–NEMATICIDE

RESTRICTED USE PESTICIDE

For retail sale to and application only by Certified Applicators or personnel under their direct supervision, and only for those uses covered by the Certified Applicator's certification.

For Agricultural or Commercial Use Only.

ACTIVE INGREDIENT:
Carbofuran (2,3-Dihydro-2,2-dimethyl-7-benzofuranyl methylcarbamate) 15%
INERT INGREDIENTS 85%
TOTAL 100%

EPA Registration No. 279-3023-3125 U.S. Patent No. 3,474,171

FURADAN—FMC trademark

Keep Out of Reach of Children

CAUTION—AVISO

PRECAUCION AL USUARIO: Si usted no puede leer o entender inglés, no use este producto hasta que la etiqueta le haya sido explicada ampliamente.
(TO THE USER: If you cannot read or understand English, do not use this product until the label has been fully explained to you.)

RECOMMENDED USES

Field corn	Grapes*	Sugar beets
Cotton	Peanuts	Sunflowers*
Cranberries	Peppers*	Cottonwood
Cucurbits (cucumber, melons, squash pumpkins)	Potatoes*	Pine seedling
	Tobacco	Siberian elm*
	Sugarcane	Southern pine seed orchards
	Sorghum	
	Soybeans	

* See label for restrictions

RESTRICTIONS ON WAREHOUSE STORAGE

Do not store next to herbicides. In case of puncture, use overslip bag and securely close.

STATEMENTS OF PRACTICAL TREATMENT

If swallowed, vomiting should be induced. Administer water freely and induce vomiting by giving one dose (½ oz. or 15 ml.) of syrup of ipecac. If vomiting does not occur within 10 to 20 minutes, administer second dose. If syrup of ipecac is not available, induce vomiting by sticking finger down throat. Repeat until vomit fluid is clear. Professional medical assistance should be secured immediately. Never give anything by mouth to an unconscious person. Get medical attention.

If in eyes, immediately flush eyes with plenty of water for at least 15 minutes. Get medical attention.

If on skin, wash with soap and water.

ANTIDOTE: Note to physician: Carbofuran is a reversible cholinesterase inhibitor. Do not use oximes such as 2-PAM. Start by giving 2 mg. atropine intramuscularly. According to clinical response, continue until signs of atropinization occur (dry mouth or dilated pupils). If in eye, instill one drop of homatropine.

PRECAUTIONARY STATEMENTS

HAZARD TO HUMANS (AND DOMESTIC ANIMALS)

WARNING

Poisonous if swallowed. May be fatal or harmful as a result of skin or eye contact or by breathing dust. Causes cholinesterase inhibition. Warning symptoms of poisoning include weakness, headache, sweating, nausea, vomiting, diarrhea, tightness in chest, blurred vision, pinpoint eye pupils, abnormal flow of saliva, abdominal cramps, and unconsciousness. Atropine sulfate is antidotal.

Wear long sleeved clothing and protective gloves when handling. Wash hands and face before eating or smoking. Bathe at the end of the work day. Change clothing daily and wash before reuse.

ENVIRONMENTAL HAZARDS

This pesticide is toxic to fish, birds and other wildlife. Birds feeding on treated areas may be killed. Birds killed by carbofuran pose a hazard to hawks and other birds-of-prey; bury or otherwise dispose of dead birds to prevent poisoning of other wildlife. Cover or incorporate granules in spill areas. Runoff from treated area may be hazardous to fish in neighboring areas. Do not apply directly to water. Do not contaminate wells, wetlands or any body of water by cleaning of equipment or disposal of wastes.

NOTICE: It is a Federal offense to use any pesticide in a manner that results in the death of a member of an endangered species.

The use of FURADAN 15G may pose a hazard to the following Federally designated endangered/threatened species known to be found in certain areas within the named locations.

Attwater's Greater Prairie Chicken—Texas counties including: Aransas, Austin, Brazoria, Colorado, Galveston, Goliad, Harris, Refugio and Victoria.

Aleutian Canada Goose—California counties including Colusa, Merced, Stanislaus and Sutter.

Kern Primrose Sphinx Moth—Walker Basin of Kern County, California.

This product may not be used in areas where adverse impact on the Federally designated endangered/threatened species, noted above, is likely. Prior to making applications, the user of this product must determine that no such species are located in or immediately adjacent to the area to be treated. If the user is in doubt whether or not the above named endangered species may be affected, he should contact either the regional U.S. Fish and Wildlife Service office (Endangered Species Specialist) or personnel of the State Fish and Game office.

Use Caution

Do not rotate to a succeeding crop other than alfalfa, corn (field, pop or sweet), cotton, cucurbits (cucumbers, melons, pumpkins or squash), grapes, peanuts, peppers, potatoes, rice, small grains (barley, oats, wheat), sorghum, soybeans, strawberries, sugarbeets, sugarcane and tobacco. Tomatoes, cabbage, peas, succulent beans and dry beans may be planted the following season provided the prior season's application did not exceed 8.7 lbs./A of FURADAN 15G, 13 lbs./A of FURADAN 10G or 2 pts. 10 fl. ozs., of FURADAN 4F. Any other crop may be planted if it is not harvested or grazed.

STORAGE AND DISPOSAL

PESTICIDE STORAGE

Do not store in or around the home.

Keep out of reach of children and animals. Store in original containers only. Store in a cool, dry place and avoid excess heat. Carefully open containers. After partial use, fold and roll back bags, clamp and close tightly. Do not put concentrate or dilute material into food or drink containers. Do not contaminate other pesticides, fertilizers, water, food or feed by storage or disposal.

In case of spill, avoid contact, isolate area and keep out animals and unprotected persons. Confine spills. Call Mobay Kansas City Emergency Response Telephone No. (816) 242-2582 or contact Chemtrec at (800) 424-9300.

To confine spill: If liquid, dike surrounding area or absorb with sand, cat litter or commercial clay. If dry material, cover to prevent dispersal. Place damaged package in a holding container. Identify contents.

PESTICIDE DISPOSAL

Pesticide wastes are acutely hazardous. Improper disposal of excess pesticide, spray mixture, or rinsate is a violation of Federal Law. If these wastes cannot be disposed of by use according to label instructions, contact your State Pesticide or Environmental Control Agency, or the Hazardous Waste representative at the nearest EPA Regional Office for guidance.

CONTAINER DISPOSAL

Completely empty bag into application equipment. Then dispose of empty bag in a sanitary landfill or by incineration, or, if allowed by State and local authorities, by burning. If burned, stay out of smoke.

IMPORTANT

This bulletin is not intended to provide adequate information for application. Before using the product, read and carefully obseve directions, cautionary statements and other information appearing on the product label. This product is sold subject to the Conditions of Sale set forth on the container label.

READ THE LABEL BEFORE USE

DIRECTIONS FOR USE

It is a violation of Federal law to use this product in a manner inconsistent with its labeling.

SWEET CORN, FIELD CORN, POPCORN—Conventional, Minimum and No-Tillage

CORN ROOTWORMS (Northern, Southern & Western), FLEA BEETLES, WIREWORMS, aids in the control of WHITE GRUBS and first generation EUROPEAN CORN BORER, reduce losses due to STALK ROT by reducing the incidence of insect wounds which permit entry by the Stalk Rot Fungi; ARMYWORM control for approximately 4 to 6 weeks after planting, and suppression of CUTWORMS—Use 8 ounces of FURADAN 15G per 1,000 linear feet of row (6.7 lbs. per acre with 40 inch row spacing). Early season suppression of COMMON STALK BORER—Use 16 to 24 ounces per 1,000 linear feet of row (13.3 to 20 lbs. per acre with 40 inch row spacing). Apply in a 7 inch band ahead of the planter press wheel and incorporate into the top 1 inch of soil by use of special covering devices or by dragging a short length of chain.—OR—Direct the granules into the planter shoe with the seed.—OR—Place the applicator tube directly behind the planter shoe so that the granules drop into the furrow and mix with the covering soil.

LESION NEMATODES (Northeast and Midwest Corn States including Colorado and Wyoming)—Use 8 to 16 ounces per 1,000 linear feet of row (6.7 to 13.3 lbs. per acre with 40 inch row spacing). Apply "at planting" as above (in a covered band or in the seed furrow).

SEED CORN MAGGOT—Use 8 ounces per 1,000 linear feet of row (6.7 lbs. per acre with 40 inch row spacing). SOUTHWESTERN CORN BORER (second to third generation)—Use 12 to 24 ounces per 1,000 linear feet of row (10 to 20 lbs. per acre with 40 inch row spacing). SOUTHERN CORN BILLBUG (Southeastern states), and to aid in the control of LEAFHOPPERS and thereby reduce losses due to Maize Chlorotic Dwarf Virus and Corn Stunt diseases—Use 16 to 24 ounces per 1,000 linear feet of row (13.3 to 20 lbs. per acre with 40 inch row spacing). Apply into the seed furrow as noted above.

EUROPEAN CORN BORER (first generation)—Use 16 to 24 ounces per 1,000 linear feet of row (13.3 to 20 lbs. per acre with 40 inch row spacing) and apply "at planting" as above (in a covered band or in the seed furrow).

NEMATODES (sting, stunt, stubby root, root knot, dagger, lesion, lance, and spiral)—Use 12 to 16 ounces per 1,000 linear feet of row (10 to 13.3 lbs. per acre with 40 inch row spacing). Apply at planting time in a 7 to 15 inch band and incorporate into the top 3 inches of soil.

FIELD CORN, POPCORN—Post Plant

NORTHERN & WESTERN CORN ROOTWORM—Use 8 ounces of FURADAN 15G per 1,000 linear feet of row (6.7 lbs. per acre with 40 inch row spacing) by banding over the row or by side dressing both sides of the row and cultivating into the soil.

FIELD CORN—Foliar Application

SOUTHWESTERN CORN BORERS (second to third generation), AND EUROPEAN CORN BORERS—Use 6.7 lbs. per acre. Apply with aircraft by broadcasting over the corn plants or with ground equipment by directing the granules into the corn whorls. Apply when eggs begin to hatch. Do not make a foliar application if FURADAN 15G were applied at more than 8 ounces per 1000 linear feet of row (6.7 lbs. per acre with 40 inch row spacing) at planting. Do not make more than two foliar applications per season.

CORN APPLICATION RATES

When the granular application equipment is set to deliver FURADAN 15G at a rate of 8 ounces per 1,000 linear feet, the total amount applied per acre will vary depending on row spacing. Use the table below to determine your FURADAN needs for each acre.

Row Spacing (Inches)	40″	38″	36″	30″	28″
FURADAN 15G (lbs./acre)	6.7	7.0	7.3	8.7	9.3

CORN INSECTICIDE APPLICATOR SETTINGS

FURADAN 15G is a dense, free flowing insecticide requiring accurate equipment calibration to apply the recommended amount per acre. Always calibrate your planter at the speed which you plan to travel during the planting operation. The following are approximate settings for guidance in calibration.

	Applicator Gauge Setting								
Application Rate	**6.7 lbs./A**			**10 lbs./A**			**13.3 lbs./A**		
Planter Speed (MPH)	**4**	**6**	**8**	**4**	**6**	**8**	**4**	**6**	**8**
Equipment									
Allis Chalmers	3	3	3	5	5	5	7	7	7
Gandy	14	18	21	18	22	25	22	25	29
IHC Cyclo Planter									
Gate	1	1	2	1	2	2	2	2	3
Dial	7	9	2	9	3	8	2	8	5
John Deere									
Range	1	2	2	1	2	2	2	2	2
Notch	18	1	2	20	2	4	1	4	8
John Deere									
Max. Emerge (even number notches-0-10-20-30-40)	10	14	19	14	21	25	19	25	30
Max. Emerge (odd number notches-5-15-25-35)	9	12	14	12	15	18	14	18	21
Noble									
Round opening	8	12	16	12	17	22	16	22	28
Rectangular Opening	5	7	9	7	10	13	9	13	17

SORGHUM

Chinch Bug, Corn Leaf Aphid, Flea Beetle, Greenbug, Lesion Nematode, Seedcorn Maggot, Southern Corn Rootworm, Wireworm, Yellow Sugarcane Aphids and aids in control of White Grubs and Cutworms—Use only on grain sorghum or sorghum grown for forage. Do not graze or harvest sorghum grown for forage within 75 days of application.

For Corn Leaf Aphid, Greenbug, Yellow Sugarcane Aphid—Use 6 to 8 ounces of FURADAN 15G per 1000 linear feet of row (5 to 6.7 lbs. per acre with 40 inch row spacing) applied directly into the seed furrow—OR—use 8 ounces of FURADAN 15G per 1000 linear feet of row (6.7 lbs per acre with 40 inch row spacing) applied in a 7 inch band over the row and incorporate the granules into the top 1 inch of soil by use of a press wheel, dragging a short length of chain or other suitable device.

For Chinch Bug, Flea Beetle, Southern Corn Rootworm, Lesion Nematode, Wireworm, Seedcorn Maggot (in furrow application only), and to aid in the control of White Grubs and Cutworms—Use 8 ounces of FURADAN 15G per 1000 linear feet of row (6.7 lbs per acre with 40 inch row spacing). Apply at planting as noted above.

Equivalent pounds of FURADAN 15G per acre for various row spacings are as follows:

	Amount of FURADAN 15G per 1000 Linear Feet of Row	Pounds of FURADAN 15G per Acre for Various Row Spacings				
		40″	36″	34″	30″	20″
In Furrow	6 Ounces	5	5.5	5.83	6.7	10
	8 Ounces	6.7	7.3	7.7	8.7	13.3
7″ Band	8 Ounces	6.7	7.3	7.7	8.7	13.3

When a rate range is indicated, use a higher rate when heavier pest infestations are anticipated.

COTTON

Thrips—Use FURADAN 15G at 8 ounces per 1,000 linear feet of row (6.7 lbs. per acre with 40 inch row spacing). Apply in the seed furrow at planting. Do not feed cotton forage.

CUCURBITS (CUCUMBER, MELONS, SQUASH, PUMPKINS)

Nematodes (Root Knot, Sting, Lance)—For use in Southeastern States—Use FURADAN 15G at the rate of 1½ lbs. per 1,000 linear feet of row (13.3 lbs. per acre based on 60 inch row spacing). Apply in a 12 to 15 inch band and incorporate into the top 3 inches of soil.

GRAPES (FOR USE IN CALIFORNIA)

For control of Nematodes (Root Knot, Dagger and Lesion) and Grape Phylloxera—Broadcast FURADAN 15G at 66.7 lbs. per acre over the soil surface between the vine rows and immediately incorporate. Do not apply within 200 days of harvest. Remove dense weed growth prior to treatment. Do not use on soils of pH 8.0 or greater. Retreat as necessary but do not make more than one application per crop year.

SOYBEANS

Nematodes (Root Knot, Cyst, Stunt, Ring, Sting, Spiral, Lesion, Lance, Dagger, Stubby-Root)—Use FURADAN 15G at 12 to 16 ounces per 1,000 linear feet (equivalent to an application of 10 to 13.3 lbs. per acre on a 40 inch row spacing) at planting time in a 7 inch band and mix with the covering soil.

—OR—

Apply FURADAN 15G at 12 ounces/1000 linear feet (equivalent to an application of 10 lbs./Acre on a 40 inch row spacing). Apply at planting by directing the granules into the planter shoe with the seed or by placing the applicator tube directly behind the planter shoe so that the granules drop into the furrow and mix with the covering soil.

Mexican Bean Beetle (larvae)—Use FURADAN 15G at 12 ounces per 1,000 linear feet (equivalent to 10 lbs. per acre on a 40 inch row spacing) at planting in a 7 inch band and mix with the covering soil.

—OR—

Apply FURADAN 15G at 8 to 12 ounces/1000 linear feet (equivalent to an application of 6.7 to 10 lbs./Acre on a 40 inch row spacing). Apply at planting by directing the granules into the planter shoe with the seed or by placing the applicator tube directly behind the planter shoe so that the granules drop into the furrow and mix with the covering soil.

SUGAR BEETS

Sugar Beet Root Maggot—Use 9 ounces of FURADAN 15G per 1,000 linear feet of row (13.3 lbs. per acre using 22 inch row spacing). Apply granules in a 6 to 7 inch band and incorporate into the top 1 inch of soil using a suitable device or use 4.5 ounces of FURADAN 15G per 1,000 linear feet of row (6.7 lbs. per acre using

22 inch row spacing) by applying granules in a "modified in-furrow" application. (Drop granules from a tube in a narrow band of 1 to 2 inches in front of the planter disks and allow some granules to fall into seed furrow with seed.)

Sugar Beet Application Rates

When the granular application equipment is set to deliver FURADAN 15G at rates of 4.5, or 9 ounces per 1,000 linear feet, the total amount applied per acre will vary depending on row spacing. Use the table below to determine your FURADAN needs for each acre.

Ounces of FURADAN 15G per 1,000 Linear Feet of Row	Pounds of FURADAN 15G per Acre for Various Row Spacings			
	20″	22″	24″	30″
4.5	7.5	6.5	6	5
9.0	15	13.5	12	10

PEANUTS

Do not feed treated peanut forage to dairy animals or animals being finished for slaughter.

In-furrow at-planting application method:

Thrips, Potato Leafhopper and as an aid in the control of Southern Corn Rootworm in Southeastern States—Use in-furrow at planting, 3.3 to 6.7 lbs. per acre (36 inch row spacing); 3.5 to 6.7 ounces per 1,000 linear feet of row.

Row banding application method:

Nematodes (Root Knot, Sting, Stunt, Lesion, Ring) in Southeastern States and in the Southwestern States of Texas and Oklahoma; and in the Southeastern States to aid in the control of Potato Leafhopper and Southern Corn Rootworm and for the control of Thrips—apply FURADAN 15G in a band over the row prior to planting and incorporate 3 inches into the soil. Use the rate indicated below for various band widths. Additional control measures may be needed late in the season for Potato Leafhopper.

Band Width	Rate/Acre (36 inch rows)	Rate/1,000 linear feet
12 inches	13.3 to 26.7 lbs.	0.9 to 1.8 lbs.
18 inches	20 to 33.3 lbs.	1.33 to 2.3 lbs.

Banded pegging—time application method:

For control of Southern Corn Rootworm in the Southeastern States—Use FURADAN 15G at 7.3 to 14.7 ounces per 1,000 linear feet (6.7-13.3 lbs. per acre with 36 inch row spacing). Apply on a 14 to 18 inch band over the row when pegs begin to form or up to 30 days after pegging begins (approximately 60 to 90 days after planting). Incorporate granules immediately if vine growth permits. Do not apply FURADAN 15G at pegging **if FURADAN 15G was used in a band at planting.**

SUGARCANE

Apply FURADAN 15G at the rate of 1.7 to 3.3 lbs. per 1,000 linear feet of row (equivalent to 13.3 to 26.7 lbs. per acre based on 60 inch row spacing). Do not use in Hawaii.

Plant cane. Nematodes (Root Knot and Stunt), Wireworm—Apply the above rate as a 15 inch band directly over the planted cane and cover with soil.

Stubblecane: Wireworm (Florida only), Nematodes (Root Knot and Stunt)—Apply the above rate as a 15 inch band over the stubble row. Use anytime after harvest until regrowth reaches 18 inches. Cover with a 1 to 2 inch layer of soil. Covering with more than 2 inches of soil may reduce stand.

SUNFLOWERS (CONFECTIONARY AND OIL) (Except California)

Sunflower Stem Weevils, Grasshoppers—Use 8 to 16 ounces of FURADAN 15G per 1,000 linear feet of row (8.7 to 17.4 lbs. per acre with 30 inch row spacing). Apply at planting by directing the granules into the planter shoe with the seed or by placing the applicator tube directly behind the planter shoe so that the granules drop into the furrow and mix with the covering soil—OR—apply at planting in a 7 inch band and mix with the covering soil.

Sunflower Beetle—Use 8 to 16 ounces of FURADAN 15G per 1,000 linear feet of row (8.7 to 17.4 lbs. per acre with 30 inch row spacing) and apply in furrow as noted above—OR—use 16 ounces of FURADAN 15G per 1,000 linear feet of row (17.4 lbs. per acre with 30 inch row spacing) and apply at planting in a 7 inch band as noted above. When a rate range is indicated, use the higher rate when heavier insect infestation is anticipated.

PEPPERS

Green Peach Aphids (Delmarva Peninsula and South Jersey only); European Corn Borer: Make two applications of FURADAN 15G by side dressing on either one or both sides of the row. Make a first application two to four weeks after transplanting using 13.3 lbs. per acre (equivalent to ½ lb. per 500 linear feet with 38 inch rows). Make the second application four to six weeks later using 20 lbs. per acre (0.7 lbs. per 500 linear feet). Do not make the second application within 21 days of harvest.

TOBACCO (Flue-cured)

For control of Stunt Nematodes, Wireworms, Hornworms, Flea Beetles and to aid in the control of Budworms and Root Knot Nematodes—Use FURADAN 15G at 40 lbs. per acre. Broadcast granules over the soil surface and incorporate into top 3 inches with a suitable device. Form beds and plant as usual. This product may induce flecking of the bottom or lower leaves.

For full season control of Flea Beetles and approximately 60 days control of Hornworms use FURADAN 15G at 26.7 lbs. per acre. Apply broadcast as noted above. Note leaf flecking caution listed above.

TOBACCO (All Types Except Flue-Cured)

For control of Flea Beetles, Hornworms, and to aid in the control of Budworms, use FURADAN 15G at 40 lbs. per acre. Broadcast granules over the soil surface prior to transplanting and incorporate into the top 3 inches with a suitable device.

For full season control of Flea Beetles only, use 20 to 26.7 lbs. of FURADAN 15G per acre,—OR—for full season control of Flea Beetles and approximately 60 days control of Hornworms use FURADAN 15G at 26.7 lbs. per acre. Apply broadcast as noted above.

POTATOES

Colorado Potato Beetle, European Corn Borer, Potato Flea Beetle, Potato Leafhopper, Tuber Flea Beetle, Green Peach Aphid (Western and Northeastern states including Delmarva Peninsula), Potato Tuberworm (Virginia only); and to aid in the control of Wireworm—Use 1½ lbs. of FURADAN 15G per 1,000 linear feet of row (20 lbs. per acre on 38″ row spacing). Apply directly into the bottom of the furrow during the planting operation.

CRANBERRIES

For control of Black Vine Weevils—Use FURADAN 15G at 13.3 lbs. per acre. For control of adult weevils and larvae, spread granules uniformly over soil surface using a rotary spreader. Make one application about June 15 to 25 and repeat one month later. Incorporate granules by following each application with sprinkler irrigation. For use only on cranberry bogs that are sprinkler irrigated and are not flooded for purposes of irrigation or harvest. Do not apply within 60 days of harvest. **For use only in the State of Washington.**

COTTONWOOD

Cottonwood Twig Borer, Cottonwood Leaf Beetle, Clearwing Borer—**In commercial plantations:** Use 6.7 to 13.3 pounds per acre (12 ft. row spacing) or 1.8 to 3.5 pounds per 1,000 linear feet of row. Apply during June or July to the root zone of the cutting by the use of a granular subsoil applicator as a sidedress 10 to 12 inches from the trees in a continuous band on both sides of the trees. **In Nursery plantings:** Use 26.7 pounds per acre (40 inch row spacing) or 2 pounds per 1,000 linear feet of row. Apply during May or June to the root zone of the cutting, sidedressing 10 to 12 inches from the trees in a continuous band on both sides of the trees.

PINE SEEDLINGS

Pales Weevil and Pitch-Eating Weevil control in pine plantations—Use 7 grams (¾teaspoonful) of FURADAN 15G per seedling at transplanting. Distribute the granules in a 6-inch radius on the soil around each seedling. Cover granules with soil. **NOTE:** Do not use at planting if seedlings were treated with FURADAN 4F slurry root dip.

SOUTHERN PINE SEED ORCHARDS
Seedbugs, Coneworms, Coneborers—Use 2.7 to 5.3 ounces per inch of tree diameter. Cone Beetles—Use 5.3 ounces per inch of tree diameter. Broadcast within the drip area of the tree and incorporate with a suitable device. Make one (1) application in winter or early spring. Do not graze or feed orchard plants or plant parts.

SIBERIAN ELM
Elm Leaf Beetle control by commercial applicators only and in the States of Colorado, Nebraska, Kansas, New Mexico, Arizona, Utah and Wyoming only—Apply FURADAN 15G before insects appear in the spring, at the rate of 1/3 ounce per inch of trunk circumference. Measure the circumference (girth) of the tree 3 to 4 feet above the ground. Dig one hole for each inch of girth, 3 to 9 inches deep under the drip line of the tree. Thoroughly soak area under the tree. Place granules required in each hole. Replace soil plug immediately and compress soil. Keep treated areas soaked with water for 14 days.

If Girth of tree is:	Locate Holes away from trunk	Amount of FURADAN 15G
1 to 10 inches	1 ft.	1/3 to 3 1/3 ounces
14 to 20 inches	3 ft.	4.7 to 6.7 ounces
24 to 80 inches	6 ft. & 12 ft. (alternating)	8 to 13.3 ounces
44 to 80 inches	10 ft. & 12 ft. (alternating)	14.7 to 26.7 ounces
80 inches and larger	12 ft. & 20 ft. (alternating)	26.7 ounces

®GUTHION 2S
EMULSIFIABLE INSECTICIDE

RESTRICTED USE PESTICIDE

For retail sale to and use only by Certified Applicators or persons under their direct supervision, and only for those uses covered by the Certified Applicator's certification.

For Control of a Wide Variety of Agricultural Pests.

ACTIVE INGREDIENTS:

0.0-Dimethyl *S*-[(4-oxo-1,2,3-benzotriazin-3 (4*H*)-yl) methyl] phosphorodithioate	22%
Aromatic Petroleum Distillate	52%
INERT INGREDIENTS	26%
TOTAL	100%

Contains 2 lbs. 0,0-Dimethyl-*S*-[(4-oxo-1,2,3-benzotriazin-3 (4*H*)-yl)methyl] phosphorodithioate per gallon.

EPA Reg. No. 3125-123-ZC

GUTHION is a Reg. TM of the Parent Company of Farbenfabriken Bayer GmbH, Leverkusen.

DANGER–POISON

Keep Out of Reach of Children
PELIGRO

PRECAUCION AL USUARIO: Si usted no puede leer o entender inglés, no use este producto hasta que la etiqueta le haya sido explicada ampliamente.

(TO THE USER: If you cannot read or understand English, do not use this product until the label has been fully explained to you.)

RECOMMENDED USES

Almonds
Apples‡
Apricots
Artichokes
Barley
Beans (dry and snap)
Blackberries
Blackeyed peas
Blueberries**
Boysenberries
Broccoli
Brussels sprouts
Cabbage
Cauliflower
Celery
Citrus fruits
Cotton
Crab apples‡
Cranberries
Cucumbers
Eggplant
Filberts*
Forest and shade trees
Grapes
Loganberries
Melons
Nectarines
Nursery plants
Oats
Onions (green and dry)
Ornamentals
Pasture grass†
Peaches
Pecans
Peppers
Plums
Potatoes
Prunes
Quinces‡
Raspberries
Rye
Southern pine seed orchards
Soybeans
Spinach
Strawberries
Tobacco
Tomatoes
Wheat
Walnuts

* Pacific Northwest only
** Eastern and North Central states only
‡ Washington, Oregon, Idaho and Utah
† States east of the Mississippi River

RESTRICTIONS ON WAREHOUSE STORAGE
Do not store in front of space heaters. Store below 120°F. Do not store adjacent to or near proximity of foodstuffs, drugs, feed or any other material intended for consumption by humans or animals. Do not store adjacent to toys, cosmetics, dishes, furniture or clothing. Do not store next to herbicides. If product freezes-out, it may be reconstituted by warming and agitating. Do not store below 25°F. Protect from heat. Keep away from open flame. Do not heat.

STATEMENTS OF PRACTICAL TREATMENT
In case of poisoning, call a physician immediately. Have patient lie down and keep quiet. **If swallowed,** vomiting should be induced. Administer water freely, and induce vomiting by giving one dose (1/2 oz. or 15 ml.) of syrup of ipecac. If vomiting does not occur within 10 to 20 minutes administer second dose. If syrup of ipecac is not available, induce vomiting by sticking finger down throat. Repeat until vomit fluid is clear. Never give anything by mouth to an unconscious person. Professional medical assistance should be secured immediately. **If on skin,** remove contaminated clothing, and wash skin immediately with soap and warm water. **If eyes are contaiminated,** wash with flowing water for at least 15 minutes.

TO PHYSICIAN: Antidote—Administer atropine sulfate in large therapeutic doses. Repeat as necessary, to the point of tolerance. 2-PAM is also antidotal, and may be administered in conjunction with atropine.

Compound inhibits cholinesterase, resulting in stimulation of the central nervous system, the parasympathetic nervous system and the somatic motor nerves. Do not give morphine. Watch for pulmonary edema, which may develop in serious cases of poisoning, even after 12 hours. At first sign of pulmonary edema, the patient should be placed in an oxygen tent, and treated symptomatically.

PRECAUTIONARY STATEMENTS

HAZARDS TO HUMANS AND DOMESTIC ANIMALS
Poisonous if swallowed, inhaled or absorbed through the skin. Do not get in eyes or on skin. Do not breathe fumes or spray mist. Spray operator should work to windward to stay out of drift or mist.

When loading spray tank, or handling the concentrate at any other time, wear protective clothing, natural rubber gloves and goggles. Wear a pesticide respirator jointly approved by the Mining Enforcement and Safety Administration (formerly the U.S. Bureau of Mines) and by the National Institute for Occupational Safety and Health under the provisions of 30 CFR Part 11.

Do not contaminate feed or foodstuffs. Keep out of reach of children and domestic animals. Keep all unprotected persons out of the operating area or vicinity where there may be danger of drift. Wash hands, arms and face thoroughly with soap and warm water before eating or smoking.

HANDLING OF CONTAMINATED ARTICLES: Clothing and rags receiving undiluted GUTHION 2S as a result of leaks, spills or mishaps should be removed immediately, and not reused. These contaminated articles should be buried in the same manner as the empty container; see container disposal section for directions. (Contaminated articles which should be discarded include: hats, gloves, aprons, coats, boots, etc. not made of rubber, or coated with rubber or other similar materials.) Clothing receiving spray mist or droplets from the mixed, diluted GUTHION 2S should be removed and decontaminated before reuse. This clothing can be decontaminated by machine-washing, separately from other items, with soap or detergent and bleach in hot water twice before reuse. Protective clothing and equipment should be washed down with detergent or soap and bleach in water. Washwater from the cleaning of protective clothing or equipment should not be allowed to runoff or otherwise enter water supplies.

SYMPTOMS OF POISONING: A sense of "tightness" in the chest. Sweating. Contracted pupils. Stomach pains. Vomiting and diarrhea.

ENVIRONMENTAL HAZARDS

This product is toxic to fish, birds and other wildlife. Keep out of any body of water. Do not apply where runoff is likely to occur. Do not apply when weather conditions favor drift from areas treated. Do not contaminate water by cleaning of equipment, or disposal of wastes.

Apply this product only as specified on this label.

PHYSICAL AND CHEMICAL HAZARDS

Keep away from open flame. Do not heat.

ROTATIONAL CROPS

Treated areas may be replanted with any crop specified on this label as soon as practical after last application. Root crops may be planted 180 days after last application. All other crops must not be planted within 30 days of last application.

RESTRICTIONS

Do not use on other crops used for food or forage. Use only according to label directions. Application at rates above those shown may result in illegal crop residues. Do not graze livestock in treated orchards or groves for 21 days after treatment. Do not treat food crops grown in the greenhouse. Do not apply when weather conditions favor drift from areas treated.

RE-ENTRY STATEMENTS

Do not apply this product in such a manner as to directly or through drift expose workers or other persons. The area being treated must be vacated by unprotected persons. Do not enter treated areas without protective clothing until 24 hours after application. Because certain states may require more restrictive re-entry intervals for various crops treated with this product, consult your State Department of Agriculture for further information.

Written or oral warnings must be given to workers who are expected to be in a treated area or in an area about to be treated with this product. Oral warnings should include those statements found under the label heading "Precautionary Statements" and "Re-entry Statements" included on this label. When oral warnings are given, warnings shall be given in a language customarily understood by workers. Oral warnings must be given if there is reason to believe that written warnings cannot be understood by workers. Written warnings must include the following information: DANGER, area treated with GUTHION 2S on (date of application). Do not enter without appropriate protective clothing until 24 hours after application. **If on skin,** remove contaminated clothing and wash skin immediately with soap and warm water. **If eyes are contaminated,** wash with flowing water for at least 15 minutes. **If swallowed,** vomiting should be induced. (See Precautionary Statements on label).

STORAGE AND DISPOSAL

PESTICIDE DISPOSAL

Do not contaminate water, food or feed by storage or disposal.

Pesticide wastes are acutely hazardous. Improper disposal of excess pesticide, spray mixture, or rinsate is a violation of Federal Law. If these wastes cannot be disposed of by use according to label instructions, contact your State Pesticide or Environmental Control Agency, or the Hazardous Waste representative at the nearest EPA Regional Office for guidance.

CONTAINER DISPOSAL

Triple rinse (or equivalent). Then offer for recycling or reconditioning, or puncture and dispose of in a sanitary landfll, or by other procedures approved by State and local authorities.

Store in a cool dry place and in such a manner as to prevent cross contamination with other pesticides, fertilizers, food, and feed. Do not store below 25°F. Keep way from open flame and extreme heat. Store in original container and out of the reach of children, preferably in a locked storage area.

Handle and open container in a manner as to prevent spillage. If container is leaking invert to prevent leakage. If the container is leaking or material spilled for any reason or cause, carefully dam up spilled material to prevent runoff. Refer to Precautionary Statements on label for hazards associated with the handling of this material. Do not walk through spilled material. Absorb spilled material with absorbing type compounds and dispose of as directed for pesticides above. In spill or leak incidents, keep unauthorized people away. You may contact the Mobay Emergency Response Team for decontamination procedures or any other assistance that may be necessary. The Mobay Kansas City Emergency Response Telephone No. is 816-242-2582, or contact Chemtrec at 800-424-9300.

READ THE LABEL BEFORE USE

DIRECTIONS FOR USE

It is a violation of Federal law to use this product in a manner inconsistent with its labeling.

MIXING: GUTHION 2S emulsifiable insecticide forms an emulsion when diluted with water, and is suitable for use in all power-operated ground and aircraft sprayers. To mix with water, pour the required amount of GUTHION 2S into full amount of water, and then agitate. If cold water is to be used, and nozzle or screen plugging is experienced, 16 mesh slotted screens (from Spraying Systems Co.) are recommended.

DOSAGE: Use specified dosage of GUTHION 2S in the amount of water necessary to give complete coverage of foliage. The type of equipment used will determine the concentration required.

SPRAYING: Work to windward. Protect sprayer operators from drift or mist. When low volumes of spray are applied, complete coverage and thorough application are essential for most effective results. Schedule applications in accordance with local conditions. Consult your State Agricultural Experiment Station or Extension Service for specific use information in your area.

RECOMMENDED APPLICATIONS

CROP	INSECT		PINTS GUTHION 2S	REMARKS
FIELD CROPS	Boll weevil		½ to 1	Apply specified dosage per acre, by air or ground equipment, in sufficient water for complete coverage, but not less than 1 gallon per acre. For early-season control of bollworm and pink bollworm, in states west of the Rocky Mts. use 2 to 3 pints and for mid-to-late season applications, use 3 to 4 pints. Repeat as necessary. Do not apply within 1 day of picking at rates up to 2 pints per acre or within 17 days of picking at rates above 2 pints per acre. Cotton receiving late-season applications should not be pastured.
Cotton	Brown cotton leafworm Cotton fleahopper	Cotton leafworm Lygus bugs Thrips	1	
	Rapid plant bug	Tarnished plant bug	1 to 2	
	Stink bug		2	

CROP	INSECT		PINTS GUTHION 2S	REMARKS
	Bollworm	Pink bollworm	East of Rocky Mts. 2 West of Rocky Mts. 2 to 4	
Cotton[1] (Ultra-low-volume spray)	Boll weevil		½ to 1	GUTHION 2S may be used undiluted in any ground or aerial spray equipment that has been adapted and calibrated for ultra-low-volume spraying. Spray machines must be equipped with accepted low-volume devices that will produce droplets within the range of 30 to 100 microns in size. ULV aerial applications should be made at altitudes of 10 to 20 feet. Repeat applications as necessary, but not within 2 days of hand-picking. Cotton may be machine-harvested any time after application. Do not graze livestock in treated areas. **Early- and Mid-Season Control:** Apply specified dosage per acre in accordance with local recommendations.
Cotton[1] (Ultra low-volume spray)	Boll weevil		1	**Diapause Weevil Control:** The one-pint-per-acre rate only is recommended for control of diapausing boll weevils. Schedule applications in accordance with local recommendations.
Barley Oats Rye Wheat	Cereal leaf beetle		1½ to 2	Apply specified dosage per acre, by air or ground equipment, in sufficient water for complete coverage, but not less than 1 gallon per acre. Do not apply more than once per season. Do not harvest for food, feed, forage or graze within 30 days of treatment.
Pasture grasses (States east of Mississippi River)	Grasshoppers Meadow spittlebugs (adults)		2 to 3	Apply specified dosage per acre by air or ground equipment in sufficient water for complete coverage but not less than 1 gallon per acre. Do not apply more than once per cutting. Do not graze or harvest within 16 days of application at the 2 pint rate or 21 days at rates above 2 pints.
Soybeans	Aphids Bean leaf beetle Green cloverworm Leafhoppers Leafminers	Leaf rollers Stink bugs Velvet bean caterpillar	1½ to 2	Apply specified dosage per acre, by air or ground equipment, in sufficient water to give complete coverage, but not less than 1 gallon per acre. Repeat as necessary. Do not apply within 45 days of harvest. Do not graze or feed treated vines to livestock.
	Mexican bean beetle		2 to 3	
Tobacco	Aphids Grasshoppers Tobacco hornworm Tobacco flea beetle		2 to 3	Apply specified dosage per acre, by air or ground equipment, in sufficient water for complete coverage, but not less than 1 gallon per acre. Treat hornworms as soon as egg masses or worms are first found. Repeat as necessary, but not within 6 days of harvest. Prime before treating.
	Tobacco budworm		3	
FRUIT Apples Crab apples Quinces (Washington) Oregon, Idaho, and Utah)	Apple maggot Codling moth European apple sawfly Eye-spotted bud moth Fruit tree leaf roller Green fruitworm Orange tortrix Plum curculio Red-banded leaf roller		2 to 4	Apply specified dosage per acre undiluted by helicopter or fixed wing aircraft, using equipment capable of applying ultra-low-volume sprays.[1]. Use Beecomist nozzles with 20-micron sleeves. Do not apply more than 5 times per season, nor within 7 days of harvest.
Apricots Nectarines Peaches	Aphids Cottony peach scale European fruit lecanium scale Forbes scale Lesser peach tree borer Mites Peach tree borer	*Platynota flavedana* leaf roller Plum curculio Red-banded leaf roller San Jose scale** Stink bug Tarnished plant bug Terrapin scale Walnut scale White peach scale	1 to 1¼	Apply specified dosage in 100 gallons of water as a full-coverage spray,* using not more than 800 gallons of finished spray per acre. Repeat as necessary. Do not apply more than 8 times per season, within 21 days of harvest for apricots, nectarines and peaches, nor within 15 days of harvest for plums and prunes. Apply the 1½-pint rate for Oriental fruit moth or peach twig borer on apricots, nectarines and peaches only before mid-season. For control of peach tree borer, apply 2 or 3 sprays to trunk from ground to scaffold limbs, timed with moth flight. For scale control, apply when crawlers are present. **NOTE:** It is suggested that when treating nectarines during bloom period, beekeepers should be warned well in advance to remove hives a safe distance from orchards to be treated. ****GUTHION** used alone may not proivde satisfactory control. Consult your local agricultural advisor or Extension Service for recommendations.
	Oriental fruit moth	Peach twig borer	1 to 1½	
Nectarines	Thrips		1 to 1¼	
Plums Prunes	Aphids Eye-spotted bud moth Forbes scale Fruit tree leaf roller	Peach tree borer Peach twig borer Plum curculio Red-banded leaf roller San Jose scale**	1 to 2	

CROP	INSECT	PINTS GUTHION 2S	REMARKS
	Lesser peach tree borer Mites Orange tortrix Stink bug Tarnished plant bug Tussock moth		
	American plum borer	2	
Blackberries Boysenberries Loganberries Raspberries	Leafhoppers Leaf rollers	1	Apply specified dosage per acre to foliage, using approximately 200 gallons of water for good coverage. Where ground conditions dictate an air application, use specified rate in minimum of 1 gallon of water per acre. Repeat as necessary, but not within 14 days of harvest.
	Leafminers	$1\frac{1}{4}$	
	Aphids	$1\frac{1}{4}$ to 2	
	Obscure root weevil	2	
	Raspberry crown (root) borer	4 to 8	For control of root weevils and borers prior to harvest, apply specified dosage per acre to lower portion of canes, and to the soil beneath the plants, using approximately 200 gallons of water. Do not apply more than twice per season. Do not make applications within 3 days of harvest at rates up to 4 pints per acre. Rates above 4 pints per acre should be applied only before fruit-set, or after crop is harvested.
	Obscure root weevil	2	
Blueberries (Eastern and North Central states only)	Blueberry maggot Fruitworms Lecanium scale Plum curculio	2 to 3	Apply specified dosage per acre, using approximately 200 gallons of water for good coverage. Where ground conditions dictate an air application, use specified rate in a minimum of 1 gallon of water per acre. Repeat as necessary, but not more than 4 times per season, nor within 3 days of harvest. If multiple late-season applications of dust formulations containing GUTHION are to be made for blueberry maggot control, do not make more than two applications of GUTHION 2S per season, with the last spray no later than June 15.
Cranberries	Cranberry fruitworm *Sparganothis sulfureana* Tipworm	2 to 4	Apply specified dosage per acre, using approximately 200 gallons of water for good coverage. Where ground conditions dictate an air application, use specified rate in a minimum of 1 gallon of water per acre. Repeat as necessary. Do not apply within 21 days of harvest.
	Fireworms	4	
Citrus fruits	Aphids Black scale Brown soft scale Chaff scale Citricola scale Citrus mealybug Citrus rust mite Citrus thrips Cottony-cushion scale European brown snail Florida red scale Fruit tree leaf rollers Fuller rose beetle Glover scale Orange tortrix Purple scale Snow scale Western tussock moth Whiteflies	1 to $1\frac{1}{2}$	Apply specified dosage in 100 gallons of water as a full-coverage spray,* using not more than 2000 gallons of finished spray per acre. A single application per year may be applied up to within 7 days of picking. Where 2 applications are required, the second spray should not be applied within 28 days of harvest. Do not apply more than twice per fruit year. Do not pick fruit, or do other work involving contact with the tree, such as pruning, within 7 days of treatment.
	California red scale Texas citrus mite Yellow scale	$1\frac{1}{2}$	
Grapes	Grape berry moth Grape cane girdlers Grape mealybug Leafhoppers Mites Red-banded leaf rollers Thrips	1 to 2	Apply specified dosage in 100 gallons of water as a full-coverage spray* using not more than 300 gallons of finished spray per acre. Repeat as necessary, but not more than 3 times per season. Minimum dosage specified may be applied up to harvest. Higher rates up to $1\frac{1}{2}$ pints require a 10-day interval to harvest, and above $1\frac{1}{2}$ pints a 28-day interval to harvest.
Strawberries	Aphids Meadow spittlebug Oblique-banded leaf roller Obscure root weevil Omnivorous leaf tier Pea leaf weevil Small black (grass) weevil Strawberry leaf rollers Whitefly	2	Apply specified dosage in 200 gallons of water per acre as a full-coverage spray.* Where ground conditions dictate an air application, use specified rate in a minimum of 1 gallon of water per acre. Repeat as necessary. Do not apply within 5 days of harvest.
NUTS Almonds	Peach twig borer	$1\frac{1}{2}$ to 2	Apply specified dosage in 100 gallons of water as a full-coverage spray,* using not more than 500 gallons of finished spray per acre. Do not apply more than twice per season, nor within 60 days of harvest. Allow 30 days between applications.
Filberts (Pacific Northwest only)	Apple mealybug Filbert aphid Filbert leaf roller Filbertworm	1 to 3	Apply specified dosage in 100 gallons of water as a full-coverage spray,* using not more than 1000 gallons of finished spray per acre. Repeat as necessary. Do not apply within 30 days of harvest. Do not graze livestock in treated groves for 21 days after treatment.
Pecans	Aphids Fall webworm Hickory shuckworm Leafminers May beetles Mites Pecan casebearer Southern green stink bug Spittlebug Twig girdlers Walnut caterpillar	$1\frac{1}{2}$ to $2\frac{1}{4}$	Apply specified dosage in 100 gallons of water as a full-coverage spray,* using not more than 1800 gallons of finished spray per acre. Repeat as necessary. Do not apply after husks split. Where more than $22\frac{1}{2}$ pints of GUTHION 2S are applied per acre in a single application, do not graze livestock in treated groves. Where $22\frac{1}{2}$ pints or less of GUTHION 2S are applied per acre, livestock may be grazed in treated groves after a 21-day post- treatment interval.

CROP	INSECT	PINTS GUTHION 2S	REMARKS
Walnuts	Codling moth; Filbertworm	1½ to 3¼	Apply specified dosage in 100 gallons of water as a full-coverage spray,* using not more than 1100 gallons of finished spray per acre. Repeat as necessary. Do not apply after husks split. Do not graze livestock in treated groves for 21 days after treatment.
VEGETABLES Artichokes	Plum moth	6	Apply specified dosage per acre, by air or ground equipment, in sufficient water for complete coverage, but not less than 1 gallon per acre. Repeat as necessary. Do not apply within 30 days of harvest. Do not feed or ensile treated forage.
Beans (Snap and dry)	Leaf rollers; Mites	1 to 2	Apply specified dosage per acre, by air or ground equipment, in sufficient water for complete coverage, but not less than 1 gallon per acre. Repeat as necessary. Do not apply to snap beans within 7 days of harvest, nor to dry beans within 30 days of harvest. Do not exceed 4 applications on dry beans. Do not feed or ensile treated forage.
	Aphids; Bean leaf beetle; Green cloverworm; Leafhoppers; Leafminers; Stink bugs; Velvet bean caterpillar	1½ to 2	
	Mexican bean beetle; Spotted cucumber beetle; Striped cucumber beetle; Tarnished plant bug; Western-striped cucumber beetle	2	
Blackeyed peas (Southern peas, Crowder peas)	Corn earworm; Cowpea curculio	3 to 4	Apply specified dosage per acre, by air or ground equipment, in sufficient water for complete coverage, but not less than 1 gallon per acre. Do not apply more than 4 times per season, not within 7 days of harvest. Do not use vines for feed or forage, nor pasture treated areas.
	Leafminers; Stink bugs	1½ to 2	
Broccoli, Brussels sprouts, Cabbage, Cauliflower	Aphids; Cabbage looper; Diamondback moth; Imported cabbageworm	2 to 3	Apply specified dosage per acre, by air or ground equipment, in sufficient water for complete coverage, but not less than 1 gallon per acre. Repeat as necessary. Do not apply within 7 days of harvest for Brussels sprouts, 15 days of harvest for broccoli and cauliflower, nor within 21 days of harvest for cabbage.
	Cabbage maggot	½	Mix specified dosage in 50 gallons of water. Apply 4 to 6 ounces of this emulsion per plant immediately after transplanting.
	Cabbage maggot (Transplant fields in California only)	3	Apply specified dosage in 300 to 400 gallons of water per acre as a soil drench in the rows when damage first appears. Additional applications may be necessary. Do not apply within 21 days of harvest.
	Cabbage maggot (Direct seeded fields in California only)	3	Apply specified dosage per acre, in sufficient water for uniform distribution. Mix in the upper 2 inches of soil prior to seeding, or spray in the seed row at planting time. Usually, 2 to 3 additional sprays are necessary during the growing season, depending upon time of year and maggot population. Do not apply within 21 days of harvest.
Celery	Aphids; Leafminers; Leafhoppers; Spittlebugs; Tarnished plant bug	2	Apply specified dosage in 100 gallons of water as a full-coverage spray,* using not more than 200 gallons of finished spray per acre. Repeat as necessary, but not within 14 days of harvest.
Cucumbers	Spotted cucumber beetle; Striped cucumber beetle; Western-striped cucumber beetle	2	Apply specified dosage per acre, by air or ground equipment, in sufficient water for complete coverage, but not less than 1 gallon per acre. Do not apply more than 3 times per season, nor within 1 day of harvest.
Eggplant	Leafminers	1½ to 2	Apply specified dosage per acre, by air or ground equipment, in sufficient water to give complete coverage, but not less than 1 gallon per acre. Repeat as necessary. Do not apply after fruit set.
	European corn borer; Flea beetles	2	
Onions (green and dry)	Thrips	2 to 3	Apply specified dosage per acre, by air or ground equipment, in sufficient water to give complete coverage, but not less than 1 gallon per acre. Do not apply more than 3 times per season, nor within 28 days of harvest of dry onions, or 7 days of harvest of green onions.
Peppers	Leafminers	1½ to 2	Apply specified dosage per acre, by air or ground equipment in sufficient water to give complete coverage, but not less than 1 gallon per acre. A maximum of 4 treatments may be made up to within 3 days of harvest. Where more than 4 applications are required, do not apply last spray within 14 days of harvest.
	European corn borer; Flea beetles	2	
Potatoes	Colorado potato beetle	1½	Apply specified dosage per acre, by air or ground equipment, in sufficient water to give complete coverage, but not less than 1 gallon per acre. Repeat as necessary. Do not apply within 7 days of harvest. NOTE: Resistance to Colorado potato beetle has occurred in some areas of the Northeast. Consult your local Extension Service or Mobay sales representative for details.
	Banded cucumber beetle; Leafminers	1½ to 2	
	European corn borer; Flea beetle; Leafhoppers; Spittlebugs; Tarnished plant bug	2 to 3	
	Tuberworm	2¼ to 3	

CROP	INSECT	PINTS GUTHION 2S	REMARKS
Spinach	Aphids Leafminers Mites	1½ to 2	Apply specified dosage per acre, by air or ground equipment, in sufficient water to give complete coverage, but not less than 1 gallon per acre. Repeat as necessary. Do not apply within 14 days of harvest.
Tomatoes	Colorado potato beetle	1½	Apply specified dosage per acre, by air or ground equipment, in sufficient water for complete coverage, but not less than 1 gallon per acre. Repeat as necessary. The high rates should be used where heavy infestations of late-instar lepidopterous larvae (large worms) and pinworms are present. Rates of pints or less per acre may be applied up to day of harvest. Rates above 3 pints per acre require an interval of 14 days between application and harvest.
	Banded cucumber beetle *Drosophila* Green stink bug Leafminers Whitefly	1½ to 2	
	Aphids European corn borer Flea beetles Grasshoppers Leafhoppers Thrips	2 to 3	Apply specified dosage per acre, by air or ground equipment, in sufficient water for complete coverage, but not less than 1 gallon per acre. Repeat as necessary. The high rates should be used where heavy infestations of late-instar lepidopterous larvae (large worms) and pinworms are present. Rates of 3 pints or less per acre may be applied up to day of harvest. Rates above 3 pints per acre require an interval of 14 days between application and harvest. NOTE: Resistance to Colorado potato beetle has occurred in some areas of the Northeast. Consult your local Extension Service or Mobay sales representative for details.
	Tuberworm	2¼ to 3	
	Corn earworm Fruitworm Hornworm Pinworm Yellow-striped armyworm	3 to 6	
MELONS Honeydew melons Muskmelons (cantaloupe) Watermelons Other melons	Leafhoppers Leafminers	1½ to 2	Apply specified dosage per acre, by air or ground equipment, in sufficient water to give complete coverage, but not less than 1 gallon per acre. Repeat as necessary, up to the day of harvest, but not more than 4 times per season.
	melons Rindworms Spotted cucumber beetle Striped cucumber beetle Western-striped cucumber beetle	2	
ORNAMENTALS Ornamentals, nursery plants, forest and shade trees	Aphids Black vine weevil *Cerococcus* scale Euonymus scale Juniper scale Lace bugs Leafhoppers Mites Olive scale Oystershell scale *Pulvinaria* scale Thrips	1½ to 2	Apply specified dosage per 100 gallons of water (2 teaspoonfuls per gallon). Spray all foliage surfaces, including the underside of leaves, for complete coverage. For control of black pine leaf, brown soft, European elm and Putnam scales, use 1 tablespoonful per gallon. Repeat as necessary.
	Brown soft scale Putnam scale	4	
	Black pine leaf scale European elm scale	3 to 4	
	Cone midge Cone moth	8 to 16	Apply specified dosage per 100 gallons of water. Time applications to coincide with moth flight, when cones are open for pollination. Thorough coverage of cones is necessary for maximum control. Repeat as necessary.
	European pine shoot moth Nantucket pine tip moth	1½ to 3	Apply specified dosage per acre, in sufficient water for good coverage. Time applications to coincide with moth flight. For application to individual trees, use 1 tablespoonful of GUTHION 2S per gallon of water.
	Injury to hawthorn or American linden may occur under some conditions. Do not allow children or pets on treated area until material has been washed into soil, and treated area is dry.		
Southern pine seed orchards	Coneworm Seedworm	(See Remarks)	Use a maximum of 6 pints per 100 gallons of water (0.2% dilution) as a high volume spray.* Use a maximum of 3 pints per 10 gallons of water (1% dilution) for low volume applications. Apply 5 to 10 gallons of the 0.2% dilution or 1 to 2 gallons of the 1% dilution per tree. Make the first application within 30 days of conelet closure. Apply either dilution as needed with no more than 6 applications per season. Applications may be made more frequently at lower concentrations but do not exceed the quantity allowed at the highest dilution. Thorough coverage is necessary for maximum control.

3-5-85

[1] **NOTE:** This formulation, when used undiluted, may cause spotting of automobile finishes if prolonged exposure is permitted. Do not spray directly over automobiles. If accidental exposure does occur, automobiles should be washed immediately.

* This concentration is calculated for conventional hydraulic-type sprayers. When lower volumes of spray are applied per acre with concentrate sprayers, increase the concentration of GUTHION 2S in the spray mixture in order to apply amount of GUTHION 2S per acre equivalent to a full coverage spray. Where conditions dictate an air application, apply dosage per acre equivalent to a full coverage ground spray, in not less than one gallon of water per acre.

SPECIAL LOCAL NEED REGISTRATIONS [SECTION 24(c)] FOR GUTHION 2S*

CROP	INSECT	STATE EPA SLN NO.	DOSAGE	REMARKS
Blueberries	Blueberry maggot	Maine ME 79-0006	1 pint per acre.	Aerial application. Follow state recommendations for timing of sprays.
Celery	Carrot weevil	Ohio OH81-0018	2 pints per acre.	Do not apply within 14 days of harvest.
Cranberries	Cranberry weevil Cranberry fruitworm *Sparganothis sulfureana* Tipworm	Massachusetts MA 78-0002	2 to 4 pints per acre.	Application by ground or air.
	Fireworms		4 pints per acre.	
Eggplant	Leafminers	New Jersey NJ 78-0008	$1\frac{1}{2}$ to 2 pints per acre.	Application by ground or air.
	European corn borer Flea beetles		2 pints per acre.	
Parsley (root and moss-curled)	Carrot weevil	Ohio OH 81-0018	2 pints per acre.	Apply up to 21 days of harvest.
		Maryland MD 81-0005 New Jersey NJ 81-0003	1 quart per acre.	
Walnuts	Aphids Codling moth European red mite Filbertworm Walnut huskfly	California CA 75-0001	$1\frac{1}{2}$ to $3\frac{3}{4}$ pints per acre.	Application by ground or air.

* For distribution and use only within the state specified. These Special Local Need registrations approved as of date of this publication are, however, subject to cancellation by the state. Contact your nearest Mobay Agricultural Chemicals Division Office for the current status and a copy of the complete individual supplemental Special Local Need labeling. This labeling must be in the possession of the user at time of pesticide application.

®GUTHION 2L

EMULSIFIABLE INSECTICIDE

RESTRICTED USE PESTICIDE

For retail sale to and use only by Certified Applicators or persons under their direct supervision, and only for those uses covered by the Certified Applicator's certification.

For control of a wide variety of agricultural pests.

ACTIVE INGREDIENTS:

O,O-Dimethyl *S*-[(4-oxo-1,2,3-benzotriazin-3(4*H*)-yl) methyl]phosphorodithioate	22.2%
Aromatic Petroleum Distillate	59.8%
INERT INGREDIENTS	18.0%
TOTAL	100.0%

Contains 2 lbs. *O,O*-Dimethyl *S*-[(4-oxo-1,2,3-benzotriazin-3(4*H*)-yl) methyl] phosphorodithioate per gal.

EPA Reg. No. 3125-102

BOLSTAR, GUTHION and DEF are Reg. TM's of the Parent Company of Farbenfabriken Bayer GmbH, Leverkusen.

DANGER–POISON

Keep Out of Reach of Children

PELIGRO

PRECAUCION AL USUARIO: Si usted no puede leer o entender inglés, no use este producto hasta que la etiqueta le haya sido explicada ampliamente.

(**TO THE USER:** If you cannot read or understand English, do not use this product until the label has been fully explained to you.)

RECOMMENDED USES

Almonds
Apricots
Artichokes
Nectarines
Nursery plants
Oats
Barley
Beans (dry and snap)
Blackberries
Blackeyed peas
Blueberries**
Boysenberries
Broccoli
Brussels sprouts
Cabbage
Cauliflower
Celery
Citrus fruits
Cotton
Cranberries
Cucumbers
Eggplant
Filberts*
Forest and shade trees
Grapes
Loganberries
Melons
Onions (green and dry)
Ornamentals
Pasture grass‡
Peaches
Pecans
Peppers
Plums
Potatoes
Prunes
Raspberries
Rye
Southern pine seed orchards
Soybeans
Spinach
Strawberries
Sugarcane†
Tobacco
Tomatoes
Wheat
Walnuts

† Florida, Louisiana and Texas only
* Pacific Northwest only
** Eastern and North Central states only
‡ States East of the Mississippi River

RESTRICTIONS ON WAREHOUSE STORAGE

Do not store in front of space heater. Do not store in draft from an open door. Do not store below 45°F. Keep away from open flame. Do not heat. Store below 120°F. Difficult to reconstitute formulation after exposure to cold. Do not store adjacent to, or near proximity of foodstuffs, drugs, feeds or any other material intended for consumption by humans or animals. Do not store adjacent to toys, cosmetics, dishes, furniture or clothing. Do not store next to herbicides. In case of freeze-out, do not sell until reconstituted.

STATEMENTS OF PRACTICAL TREATMENT

In case of poisoning, call a physician immediately. Have patient lie down and keep quiet.

If swallowed, vomiting should be induced. Administer water freely, and induce vomiting by giving one dose ($\frac{1}{2}$ oz. or 15 ml.) of syrup of ipecac. If vomiting does not occur within 10 to 20 minutes, administer second dose. If syrup of ipecac is not available, induce vomiting by sticking finger down throat. Repeat until vomit fluid is clear. Never give anything by mouth to an unconscious person. Professional medical assistance should be secured immediately.

If on skin, remove contaminated clothing, and wash skin immediately with soap and warm water.

If eyes are contaminated, wash with flowing water for at least 15 minutes.

TO PHYSICIAN: Antidote—Administer atropine sulfate in large therapeutic doses. Repeat as necessary, to the point of tolerance. 2-PAM is also antidotal, and may be administered in conjunction with atropine. Compound inhibits cholinesterase, resulting in stimulation of the central nervous system, the parasympathetic nervous system and the somatic motor nerves. Do not give morphine. Watch for pulmonary edema, which may develop in serious cases of poisoning, even after 12 hours. At first sign of pulmonary edema, the patient should be placed in an oxygen tent, and treated symptomatically.

PRECAUTIONARY STATEMENTS

HAZARDS TO HUMANS AND DOMESTIC ANIMALS

Poisonous if swallowed, inhaled or absorbed through the skin. Do not get in eyes or on skin. Do not breathe fumes or spray mist. Spray operator should work to windward to stay out of drift or mist.

When loading spray tank, or handling the concentrate at any other time, wear protective clothing, natural rubber gloves and goggles. When handling the concentrate, wear a pesticide respirator jointly approved by the Mining Enforcement and Safety Administration (formerly the U.S. Bureau of Mines) and by the National Institute for Occupational Safety and Health under the provisions of 30 CFR Part 11.

Do not contaminate feed or foodstuffs. Keep out of reach of children and domestic animals. Keep all unprotected persons out of the operating area or vicinity where there may be danger of drift. Wash hands, arms and face thoroughly with soap and warm water before eating or smoking.

HANDLING OF CONTAMINATED ARTICLES: Clothing and rags receiving undiluted GUTHION 2L as a result of leaks, spills or mishaps should be removed immediately, and not reused. These contaminated articles should be buried in the same manner as the empty container (see container disposal section for directions). Contaminated articles which should be discarded include: hats, gloves, aprons, coats, boots, etc. not made of rubber, or coated with rubber or other similar materials. Clothing receiving spray mist or droplets from the mixed, diluted GUTHION 2L should be removed, and decontaminated before reuse. This clothing can be decontaminated by machine-washing, separately from other items, with soap or detergent and bleach in hot water twice before reuse. Protective clothing and equipment should be washed-down with detergent or soap and bleach in water. Wash-water from the cleaning of protective clothing or equipment should not be allowed to run-off, or otherwise enter water supplies.

SYMPTOMS OF POISONING: A sense of "tightness" in the chest. Sweating. Contracted pupils. Stomach pains. Vomiting and diarrhea.

ENVIRONMENTAL HAZARDS

Highly toxic to fish and wildlife. Do not contaminate any body of water nor apply to any area not specified on the label.

PHYSICAL AND CHEMICAL HAZARDS

Keep away from open flame. Do not heat.

ROTATIONAL CROPS

Treated areas may be replanted with any crop specified on this label as soon as practical after last application. Root crops may be planted 180 days after last application. All other crops must not be planted within 30 days of last application.

RESTRICTIONS

Do not use on other crops used for food or forage. Use only according to label directions. Application at rates above those shown may result in illegal crop residues. Do not graze livestock in treated orchards or groves for 21 days after treatment. Do not treat food crops grown in the greenhouse.

RE-ENTRY STATEMENTS

Do not apply this product in such a manner as to directly or through drift expose workers or other persons. The area being treated must be vacated by unprotected persons. Do not enter treated areas without protective clothing until 24 hours after application. Because certain states may require more restrictive re-entry intervals for various crops treated with this product, consult your State Department of Agriculture for further information.

Written or oral warnings must be given to workers who are expected to be in a treated area or in an area about to be treated with this product. Oral warnings should include those statements found under the label heading "Precautionary Statements" and "Re-entry Statements" included on this label. When oral warnings are given, warnings shall be given in a language customarily understood by workers. Oral warnings must be given if there is reason to believe that written warnings cannot be understood by workers. Written warnings must include the following information: DANGER, area treated with GUTHION 2L on (date of application). Do not enter without appropriate protective clothing until 24 hours after application.

If on skin, remove contaminated clothing and wash skin immediately with soap and warm water.

If eyes are contaminated, wash with flowing water for at least 15 minutes.

If swallowed, vomiting should be induced. (See Precautionary Statements on label).

STORAGE AND DISPOSAL

PESTICIDE DISPOSAL

Do not contaminate water, food or feed by storage or disposal.

Pesticide wastes are acutely hazardous. Improper disposal of excess pesticide, spray mixture, or rinsate is a violation of Federal Law. If these wastes cannot be disposed of by use according to label instructions, contact your State Pesticide or Environmental Control Agency, or the Hazardous Waste representative at the nearest EPA Regional Office for guidance.

CONTAINER DISPOSAL

Triple rinse (or equivalent). Then offer for recycling or reconditioning, or puncture and dispose of in a sanitary landfill, or by other procedures approved by State and local authorities.

Store in a cool, dry place and in such a manner as to prevent cross contamination with other pesticides, fertilizers, food, and feed. Keep away from open flame and extreme heat. Do not store below 45°F. Store in original container and out of the reach of children, preferably in a locked storage area.

Handle and open container in a manner as to prevent spillage. If container is leaking invert to prevent leakage. If the container is leaking or material spilled for any reason or cause, carefully dam up spilled material to prevent runoff. Refer to Precautionary Statements on label for hazards associated with the handling of this material. Do not walk through spilled material. Absorb spilled material with absorbing type compounds and dispose of as directed for pesticides above. In spill or leak incidents, keep unauthorized people away. You may contact the Mobay Emergency Response Team for decontamination procedures or any other assistance that may be necessary. The Mobay Kansas City Emergency Response Telephone No. is 816-242-2582, or contact Chemtrec at 800-424-9300.

Galecron is a trademark of CIBA-GEIGY for chlordimeform.

Fundal is a trademark of Schering AG Berlin/Bergamen.

IMPORTANT

This bulletin is not intended to provide adequate information for application. Before using the product, read and carefully observe directions, cautionary statements and other information appearing on the product label. This product is sold subject to the Conditions of Sale set forth on the container label.

DIRECTIONS FOR USE

It is a violation of Federal Law to use this product in a manner inconsistent with its labeling.

MIXING: GUTHION 2L emulsifiable insecticide forms an emulsion when diluted with water, and is suitable for use in all power-operated ground and aircraft sprayers. To mix with water, pour the required amount of GUTHION 2L into full amount of water, and then agitate. GUTHION 2L may also be applied undiluted as an ultra-low-volume spray with either ground or aircraft equipment that has been adapted and calibrated for ultra-low-volume spraying as described below under "Recommended Applications" for those crops specified.

DOSAGE: Use specified dosage of GUTHION 2L in the amount of water necessary to give complete coverage of foliage. The type of equipment used will determine the concentration required.

SPRAYING: Work to windward. Protect sprayer operators from drift or mist. When low volumes of spray are applied, complete coverage and thorough application are essential for most effective results. Schedule applications in accordance with local conditions. Consult your State Agricultural Extension Service or Experiment Station for specific use information in your area.

RECOMMENDED APPLICATIONS

CROP	INSECT	PINTS GUTHION 2L	REMARKS
FIELD CROPS	Boll Weevil	½ to 1	Apply specified dosage per acre by air or ground equipment in sufficient water for complete coverage but not less than 1 gallon per acre. For early-season control of bollworm and pink bollworm in states west of the Rocky Mts. use 2 to 3 pints and for mid-to-late season applications, use 3 to 4 pints. Repeat as necessary. Do not apply within 1 day of picking at rates up to 2 pints per acre or within 17 days of picking at rates above 2 pints per acre. Cotton receiving late-season applications should not be pastured.
	Brown cotton leafworm, Cotton leafhopper, Cotton leafworm, Lygus bugs, Thrips	1	
Cotton (Conventional or low-volume spray)	Rapid plant bug, Tarnished plant bug	1 to 2	
	Stink bug	2	
	Bollworm, Pink bollworm	East of Rocky Mts. 2; West of Rocky Mts. 2 to 4	GUTHION 2L at rates up to 1 pint per acre may be tank mixed with Galecron 4E or Fundal 4EC for ovicidal control of bollworms and tobacco budworms. For control of the larval stages of resistant and non-resistant bollworm and/or tobacco budworm, BOLSTAR® 6 may be included in GUTHION 2L tank mixes. When using GUTHION 2L in a tank mix with Galecron 4E or Fundal 4EC, observe all limitations, precautions and rate recommendations which appear on the labels for these products. **See NOTE below.**
Cotton (Ultra-low-volume spray)	Boll weevil	½ to 1	GUTHION 2L may be tank mixed in vegetable oil in any ground or aerial spray equipment that has been adapted and calibrated for ultra-low-volume spraying. Spray machines must be equipped with accepted low-volume devices that will produce droplets within the range of 30 to 100 microns in size. ULV aerial applications should be made at altitudes of 10 to 20 feet. Repeat applications as necessary, but not within 2 days of hand-picking. Cotton may be machine-harvested any time after application. Do not graze livestock in treated areas. **See NOTE below.** **Early-and Mid-season Control:** Apply specified dosage per acre in accordance with local recommendations.
		1	**Diapause Weevil Control:** The one-pint-per-acre rate only is recommended for control of diapausing boll weevils. Schedule applications in accordance with local recommendations. **See NOTE below.**
Barley, Oats, Rye, Wheat	Cereal leaf beetle	1½ to 2	Apply specified dosage per acre, by air or ground equipment, in sufficient water for complete coverage, but not less than 1 gallon per acre. Do not apply more than once per season. Do not harvest for food, feed, forage or graze within 30 days of treatment.
Pasture Grasses (States east of Mississippi River)	Grasshoppers, Meadow spittlebugs (adults)	2 to 3	Apply specified dosage per acre by air or ground equipment in sufficient water for complete coverage but not less than 1 gallon per acre. Do not apply more than once per cutting. Do not graze or harvest within 16 days of application at the 2 pint rate or 21 days at rates above 2 pints.
Soybeans	Aphids, Bean leaf beetle, Green cloverworm, Leafhoppers, Leafminers, Leaf rollers, Stink bugs, Velvet bean caterpillar	1½ to 2	Apply specified dosage per acre, by air or ground equipment in sufficient water to give complete coverage, but not less than 1 gallon per acre. Repeat as necessary. One or 2 applications of rates up to 2 pints per acre may be made up to 14 days of harvest. For additional applications or for use of rates above 2 pints per acre, do not apply within 45 days of harvest.
	Mexican bean beetle	2 to 3	Do not graze or feed treated vines to livestock.
Sugarcane (Conventional spray) (Florida, Louisiana and Texas only)	Sugarcane borer	3	Apply specified dosage per acre by air application, using a minimum of 2 gallons of water per acre. For best results, applications should be made at approximately cane-top level, and not more than 10 feet above cane-top level. Do not apply more than 5 times per season, nor within 30 days of harvest. Bagasse from sugarcane treated with GUTHION may be used for feed of cattle, goats and sheep. Consult your State Agricultural Extension Service or Experiment Station for specific use information.

CROP	INSECT	PINTS GUTHION 2L	REMARKS
Sugarcane[1] (Ultra-low-volume spray) (Florida, Louisiana and Texas only)	Sugarcane borer	3	Apply specified dosage per acre undiluted. GUTHION 2L may be used undiluted in any aerial spray equipment that has been adapted and calibrated for ultra-low-volume spraying. Planes must be equipped with accepted low-volume devices that will produce droplets within the range of 30 to 100 microns in size. For best results, applications should be made at approximately cane-top level, and not more than 10 feet above cane-top level. Do not apply more than 5 times per season, nor within 30 days of harvest. Bagasse from sugarcane treated with GUTHION may be used for feed of cattle, sheep and goats. Consult your State Agricultural Extension Service or Experiment Station for specific use information.
Tobacco	Aphids Grasshoppers Tobacco hornworm Tobacco flea beetle	2 to 3	Apply specified dosage per acre, by air or ground equipment, in sufficient water for complete coverage, but not less than 1 gallon per acre. Treat for hornworms as soon as egg masses or worms are first found. Repeat as necessary, but not within 6 days of harvest. Prime before treating.
	Tobacco budworm	3	
FRUIT			
Apricots Nectarines Peaches	Aphids Cottony peach scale European fruit lecanium scale Forbes scale Lesser peach tree borer Mites Peach tree borer *Platynota flavedana* leaf roller Plum curculio Red-banded leaf roller San Jose scale** Stink bug Tarnished plant bug Terrapin scale Walnut scale White peach scale	1 to 1¼	Apply specified dosage in 100 gallons of water as a full-coverage spray*, using not more than 800 gallons of finished spray per acre. Repeat as necessary. Do not apply more than 8 times per season, within 21 days of harvest for apricots, nectarines and peaches, nor within 15 days of harvest for plums and prunes. Apply the 1½ pint rate for oriental fruit moth and peach twig borer on apricots, nectarines and peaches only before mid-season. For control of peach tree borer, apply 2 or 3 sprays to trunk from ground to scaffold limbs, timed with moth flight. For scale control, apply when crawlers are present.
	Oriental fruit moth Peach twig borer	1 to 1½	
Nectarines	Thrips	1 to 1¼	
Plums Prunes	Aphids Eye-spotted bud moth Forbes scale Fruit tree leaf roller Lesser peach tree borer Mites Orange tortrix Peach tree borer Peach twig borer Plum cucrculio Red-banded leaf roller San Jose scale** Stink bug Tarnished plant bug Tussock moth	1 to 2	NOTE: It is suggested that when treating nectarines during bloom period, beekeepers should be warned well in advance to remove hives a safe distance from orchards to be treated. **GUTHION used alone may not provide satisfactory control. Consult your local agricultural adviser or Extension Service for recommendations.
	American plum borer		
Blackberries Boysenberries Loganberries Raspberries	Leafhoppers Leaf rollers	1	Apply specified dosage per acre to foliage, using approximately 200 gallons of water for good coverage. Where ground conditions dictate an air application of GUTHION 2L, use specified rate in a minimum of 1 gallon of water per acre. Repeat as necessary, but not within 14 days of harvest.
	Leafminers	1¼	
	Aphids	1¼ to 2	
	Obscure root weevil	2	
	Raspberry crown (root) borer	4 to 8	For control of root weevils and borers prior to harvest, apply specified dosage per acre to lower portion of canes, and to the soil beneath the plants, using approximately 200 gallons of water. Do not apply more than twice per season. Do not make applications within 3 days of harvest at rates up to 4 pints per acre. Rates above 4 pints per acre should be applied only before fruit-set, or after crop is harvested.
	Obscure root weevil	2	
Blueberries (Eastern and North Central states only)	Blueberry maggot Fruitworms Lecanium scale Plum curculio	2 to 3	Apply specified dosage per acre, using approximately 200 gallons of water for good coverage. Where ground conditions dictate an air application of GUTHION 2L, use specified rate in a minimum of 1 gallon of water per acre. Repeat as necessary, but not more than 4 times per season, nor within 3 days of harvest. If multiple late-season applications of dust formulations containing GUTHION are to be made for blueberry maggot control, do not make more than two spray applications of GUTHION 2L per season, with the last spray no later than June 15.

CROP	INSECT	PINTS GUTHION 2L	REMARKS
Cranberries	Cranberry fruitworm *Sparganothis sulfureana* Tipworm	2 to 4	Apply specified dosage per acre, using approximately 200 gallons of water for good coverage. Where ground conditions dictate an air application of GUTHION 2L, use specified rate in a minimum of 1 gallon of water per acre. Repeat as necessary. Do not apply within 21 days of harvest.
	Fireworms	4	
Citrus fruits	Aphids Black scale Brown soft scale Chaff scale Citricola scale Citrus mealybug Citrus rust mite Citrus thrips Cottony-cushion scale European brown snail Florida red scale Fruit tree leaf rollers Fuller rose beetle Glover scale Orange tortrix Purple scale Snow scale Western tussock moth Whiteflies	1 to 1½	Apply specified dosage in 100 gallons of water as a full-coverage spray*, using not more than 2000 gallons of finished spray per acre. A single application per year may be applied up to within 7 days of picking. Where 2 applications are required, the second spray should not be applied within 28 days of harvest. Do not apply more than twice per fruit year. Do not pick fruit, or do other work involving contact with the tree, such as pruning, within 7 days of treatment.
	California red scale Texas citrus mite Yellow scale	1½	
Grapes	Grape berry moth Grape bud beetle Grape cane girdlers Grape mealybug Leafhoppers Mites Red-banded leaf roller Thrips	1 to 2	Apply specified dosage in 100 gallons of water as a full-coverage spray*, using not more than 300 gallons of finished spray per acre. Repeat as necessary, but not more than 3 times per season. Minimum dosage specified may be applied up to harvest. Higher rates up to 1½ pints require a 10-day interval to harvest, and above 1½ pints a 28-day interval to harvest.
Strawberries	Aphids Meadow spittlebug Oblique-banded leaf roller Obscure root weevil Omnivorous leaf tier Pea leaf weevil Small black (grass) weevil Strawberry leaf rollers Whitefly	2	Apply specified dosage in 200 gallons of water per acre as a full-coverage spray*. Where ground conditions dictate an air application, use specified rate in a minimum of 1 gallon of water per acre. Repeat as necessary. Do not apply within 5 days of harvest.
NUTS Almonds	Peach twig borer	1½ to 2	Apply specified dosage in 100 gallons of water as a full-coverage spray*, using not more than 500 gallons of finished spray per acre. Do not apply more than twice per season, nor within 60 days of harvest. Allow 30 days between applications.
Filberts (Pacific Northwest only)	Apple mealybug Filbert aphid Filbert leaf roller Filbertworm	1 to 3	Apply specified dosage in 100 gallons of water as a full-coverage spray*, using not more than 1000 gallons of finished spray per acre. Repeat as necessary. Do not apply within 30 days of harvest. Do not graze livestock in treated groves for 21 days after treatment.
Pecans	Aphids Fall webworm Hickory shuckworm Leafminers May beetles Mites Pecan casebearer Southern green stink bug Spittlebug Twig girdlers Walnut caterpillar	1½ to 2¼	Apply specified dosage in 100 gallons of water as a full-coverage spray*, using not more than 1800 gallons of finished spray per acre. Repeat as necessary. Do not apply after husks split. Where more than 22½ pints of GUTHION 2L are applied per acre in a single application, do not graze livestock in treated groves. Where 22½ pints or less of GUTHION 2L are applied per acre, livestock may be grazed in treated groves after a 21-day post-treatment interval.
Walnuts	Codling moth Filbertworm	1½ to 3¾	Apply specified dosage in 100 gallons of water as a full-coverage spray*, using not more than 1100 gallons of finished spray per acre. Repeat as necessary. Do not apply after husks split. Do not graze livestock in treated groves for 21 days after treatment.
VEGETABLES Artichokes	Plum moth	6	Apply specified dosage per acre, by air or ground equipment, in sufficient water for complete coverage, but not less than 1 gallon per acre. Repeat as necessary. Do not apply within 30 days of harvest. Do not feed or ensile treated forage.
Beans (snap and dry)	Leaf rollers Mites	1 to 2	Apply specified dosage per acre, by air or ground equipment, in sufficient water to give complete coverage, but not less than 1 gallon per acre. Repeat as necessary. Do not apply to snap beans within 7 days of harvest, nor to dry beans within 30 days of harvest. Do not exceed 4 applications on dry beans. Do not feed or ensile treated forage.
	Aphids Bean leaf beetle Green cloverworm Leafhoppers Leafminers Stink bugs Velvet bean caterpillar	1½ to 2	
	Mexican bean beetle Spotted cucumber beetle Striped cucumber beetle Tarnished plant bug Western-striped cucumber beetle	2	

CROP	INSECT	PINTS GUTHION 2L	REMARKS
Broccoli Brussels sprouts Cabbage Cauliflower	Aphids, Cabbage looper, Diamondback moth, Imported cabbageworm	2 to 3	Apply specified dosage per acre, in sufficient water for complete coverage, but not less than 1 gallon per acre. Repeat as necessary. Do not apply within 7 days of harvest for Brussels sprouts, 15 days of harvest for broccoli and cauliflower, nor within 21 days of harvest for cabbage.
	Cabbage maggot	½	Mix specified dosage in 50 gallons of water. Apply 4 to 6 ounces of this emulsion per plant immediately after transplanting.
	Cabbage maggot (Transplant fields in California only)	3	Apply specified dosage in 300 to 400 gallons of water per acre as a soil drench in the rows when damage first appears. Additional applications may be necessary. Do not apply within 21 days of harvest.
	Cabbage maggot (Direct-seeded fields in California only)	3	Apply specified dosage per acre, in sufficient water for uniform distribution. Mix in the upper 2 inches of soil prior to seeding, or spray in the seed row at planting time. Usually 2 to 3 additional sprays are necessary during the growing season depending upon time of year and maggot population. Do not apply within 21 days of harvest.
Celery	Aphids, Leafminers, Leafhoppers, Spittlebugs, Tarnished plant bug	2	Apply specified dosage in 100 gallons of water as a full-coverage spray*, using not more than 200 gallons of finished spray per acre. Repeat as necessary, but not within 14 days of harvest.
Blackeyed peas (Southern peas, Crowder peas)	Corn earworm, Cowpea curculio	3 to 4	Apply specified dosage per acre, by air or ground equipment, in sufficient water for complete coverage, but not less than 1 gallon per acre. Do not apply more than 4 times per season, nor within 7 days of harvest. Do not use vines for feed or forage, nor pasture treated areas.
	Leafminers, Stink bugs	1½ to 2	Apply specified dosage per acre, by air or ground equipment, in sufficient water for complete coverage, but not less than 1 gallon per acre. Do not apply more than 4 times per season, nor within 7 days of harvest. Do not use vines for feed or forage, nor pasture treated areas.
Cucumbers	Spotted cucumber beetle, Striped cucumber beetle, Western-striped cucumber beetle	2	Apply specified dosage per acre, by air or ground equipment, in sufficient water for complete coverage, but not less than 1 gallon per acre. Do not apply more than 3 times per season, nor within 1 day of harvest.
Eggplant	Leafminers	1½ to 2	Apply specified dosage per acre, by air or ground equipment, in sufficient water to give complete coverage, but not less than 1 gallon per acre. Repeat as necessary. Do not apply after fruit set.
	European corn borer, Flea beetles	2	Apply specified dosage per acre, by air or ground equipment, in sufficient water to give complete coverage, but not less than 1 gallon per acre. Repeat as necessary. Do not apply after fruit set.
Onions (green and dry)	Thrips	2 to 3	Apply specified dosage per acre, by air or ground equipment, in sufficient water to give complete coverage, but not less than 1 gallon per acre. Do not apply more than 3 times per season, nor within 28 days of harvest of dry onions, or 7 days of harvest of green onions.
Peppers	Leafminers	1½ to 2	Apply specified dosage per acre, by air or ground equipment, in sufficient water to give complete coverage, but not less than 1 gallon per acre. A maximum of 4 treatments may be made up to within 3 days of harvest. Where more than 4 applications are required, do not apply last spray within 14 days of harvest.
	European corn borer, Flea beetles	2	Apply specified dosage per acre, by air or ground equipment, in sufficient water to give complete coverage, but not less than 1 gallon per acre. A maximum of 4 treatments may be made up to within 3 days of harvest. Where more than 4 applications are required, do not apply last spray within 14 days of harvest.
Potatoes	Colorado potato beetle	1½	Apply specified dosage per acre, by air or ground equipment, in sufficient water for complete coverage, but not less than 1 gallon per acre. Repeat as necessary. Do not apply within 7 days of harvest.
	Banded cucumber beetle, Leafminers	1½ to 2	Apply specified dosage per acre, by air or ground equipment, in sufficient water for complete coverage, but not less than 1 gallon per acre. Repeat as necessary. Do not apply within 7 days of harvest.
	European corn borer, Flea beetle, Leafhoppers, Spittlebugs, Tarnished plant bug	2 to 3	Apply specified dosage per acre, by air or ground equipment, in sufficient water for complete coverage, but not less than 1 gallon per acre. Repeat as necessary. Do not apply within 7 days of harvest.
	Tuberworm	2¼ to 3	Apply specified dosage per acre, by air or ground equipment, in sufficient water for complete coverage, but not less than 1 gallon per acre. Repeat as necessary. Do not apply within 7 days of harvest.
Spinach	Aphids, Leafminers, Mites	1½ to 2	Apply specified dosage per acre, by air or ground equipment, in sufficient water for complete coverage, but not less than 1 gallon per acre. Repeat as necessary. Do not apply within 14 days of harvest.
Tomatoes	Colorado potato beetle	1½	Apply specified dosage per acre, by air or ground equipment, in sufficient water for complete coverage, but not less than 1 gallon per acre. Repeat as necessary. The high rates should be used where heavy infestations of late-instar lepidopterous larvae (large worms) and pinworms are present. Rates of 3 pints per acre or less may be applied up to day of harvest. Rates above 3 pints per acre require an interval of 14 days between application and harvest.
	Banded cucumber beetle, *Drosophila*, Green stink bug, Leafminers, Whitefly	1½ to 2	Apply specified dosage per acre, by air or ground equipment, in sufficient water for complete coverage, but not less than 1 gallon per acre. Repeat as necessary. The high rates should be used where heavy infestations of late-instar lepidopterous larvae (large worms) and pinworms are present. Rates of 3 pints per acre or less may be applied up to day of harvest. Rates above 3 pints per acre require an interval of 14 days between application and harvest.
	Aphids, European corn borer, Flea beetles, Grasshoppers, Leafhoppers, Thrips	2 to 3	Apply specified dosage per acre, by air or ground equipment, in sufficient water for complete coverage, but not less than 1 gallon per acre. Repeat as necessary. The high rates should be used where heavy infestations of late-instar lepidopterous larvae (large worms) and pinworms are present. Rates of 3 pints per acre or less may be applied up to day of harvest. Rates above 3 pints per acre require an interval of 14 days between application and harvest.
	Tuberworm	2¼ to 3	Apply specified dosage per acre, by air or ground equipment, in sufficient water for complete coverage, but not less than 1 gallon per acre. Repeat as necessary. The high rates should be used where heavy infestations of late-instar lepidopterous larvae (large worms) and pinworms are present. Rates of 3 pints per acre or less may be applied up to day of harvest. Rates above 3 pints per acre require an interval of 14 days between application and harvest.
	Corn earworm, Fruitworm, Hornworm, Pinworm, Yellow-striped armyworm	3 to 6	Apply specified dosage per acre, by air or ground equipment, in sufficient water for complete coverage, but not less than 1 gallon per acre. Repeat as necessary. The high rates should be used where heavy infestations of late-instar lepidopterous larvae (large worms) and pinworms are present. Rates of 3 pints per acre or less may be applied up to day of harvest. Rates above 3 pints per acre require an interval of 14 days between application and harvest.

CROP	INSECT		PINTS GUTHION 2L	REMARKS
MELONS Honeydew melons Muskmelons (Cantaloupe) Watermelons Other melons	Leafhoppers	Leafminers	1½ to 2	Apply specified dosage per acre, by air or ground equipment, in sufficient water to give complete coverage, but not less than 1 gallon per acre. Repeat as necessary, up to the day of harvest, but not more than 4 times per season.
	Ringworms Spotted cucumber beetle	Striped cucumber beetle Western-striped cucumber beetle	2	
ORNAMENTALS	Aphids *Cerococcus* scale Euonymus scale Juniper scale Lace bugs Leafhoppers	Mites Olive scale Oystershell scale *Pulvinaria* scale Thrips	1½ to 2	Apply specified dosage per 100 gallons of water (2 teaspoonfuls per gallon). Spray all foliage surfaces, including the underside of leaves, for complete coverage. For control of black pine leaf, brown soft, European elm and Putnam scales, use 1 tablespoonful per gallon. Repeat as necessary.
	Brown soft scale		4	
	Black pine leaf scale	European elm scale Putnam scale	3 to 4	
Ornamentals nursery plants, forest and shade trees	Cone midge Cone moth		8 to 16	Apply specified dosage per 100 gallons of water. Time applications to coincide with moth flight, when cones are open for pollination. Thorough coverage of cones is necessary for maximum control. Repeat as necessary.
	European pine shoot moth Nantucket pine tip moth		1½ to 3	Apply specified dosage per acre in sufficient water for good coverage. Time applications to coincide with moth flights. For application to individual trees, use 1 tablespoonful of GUTHION 2L per gallon of water.
	Injury to hawthorn or American linden may occur under some conditions. Do not allow children or pets on treated area until material has been washed into soil, and treated area is dry.			
Southern pine seed orchards	Coneworm Seedworm		(See Remarks)	Use a maximum of 6 pints per 100 gallons of water (0.2% dilution) as a high volume spray*. Use a maximum of 3 pints per 10 gallons of water (1% dilution) for low volume applications. Apply 5 to 10 gallons of the 0.2% dilution or 1 to 2 gallons of the 1% dilution per tree. Make the first application within 30 days of conelet closure. Apply either dilution as needed with no more than 6 applications per season. Applications may be made more frequently at lower concentrations but do not exceed the quantity allowed at the highest dilution. Thorough coverage is necessary for maximum control. 3–5–85

NOTE: GUTHION 2L may be tank mixed in one application with ®DEF 6 defoliant. Refer to label for DEF 6 for recommended rates, precautions, and restrictions. Preceding the use of this tank mix, GUTHION 2L may be applied in a maximum of 9 times at 1 pint per acre. The maximum rate of DEF 6 in a tank mix combination is 2 pints per acre.

[1] **NOTE:** This formulation may cause spotting of automobile finishes if prolonged exposure is permitted. Do not spray directly over automobiles. If accidental exposure does occur, automobiles should be washed immediately.

* This concentration is calculated for conventional hydraulic-type sprayers. When lower volumes of spray are applied per acre with concentrate sprayers, increase the concentration of GUTHION 2L in the spray mixture in order to apply amount of GUTHION 2L per acre equivalent to a full-coverage spray. Where conditions dictate an air application, apply dosage per acre equivalent to a full-coverage ground spray, in not less than one gallon of water per acre.

SPECIAL LOCAL NEED REGISTRATIONS [SECTION 24(c)] FOR GUTHION2L*

CROP	INSECT	STATE EPA SLN NO.	DOSAGE	REMARKS
Cotton	Boll weevil	Louisiana LA 82–0028 South Carolina SC 82–0021 Texas TX 84–0005	½ to 2 pints per acre	Ultra Low Volume application in vegetable oil.

* For distribution and use only within the state specified. These Special Local Need registrations approved as of date of this publication are, however, subject to cancellation by the state. Contact your nearest Mobay Agricultural Chemicals Division Office for the current status and a copy of the complete individual supplemental Special Local Need labeling. This labeling must be in the possession of the user at time of pesticide application.

®GUTHION

50% WETTABLE POWDER INSECTICIDE IN WATER SOLUBLE PACKETS

For Control Of A Wide Variety Of Agricultural Insect Pests.

Only for agricultural use, or for sale to, use, and storage by commercial applicators, and other servicepersons.

ACTIVE INGREDIENT:	
O,O-Dimethyl *S*-[(4-oxo-1,2,3-benzotriazin-3(4*H*)-yl)methyl]phosphorodithioate	50%
INERT INGREDIENTS	50%
TOTAL	100%

EPA Reg. No. 3125–301–AA

GUTHION is a Reg. TM of the Parent Company of Farbenfabriken Bayer GmbH, Leverkusen.

DANGER–POISON

Keep Out of Reach of Children

PELIGRO

PRECAUCION AL USUARIO: Si usted no puede leer o entender ingles, no use este productor hasta que la etiqueta le haya sido explicada ampliamente.

(TO THE USER: If you cannot read or understand English, do not use this product until the label has been fully explained to you.)

RECOMMENDED USES

Alfalfa†
Almonds
Apples
Apricots
Artichokes
Barley
Beans (snap and dry)
Birdsfoot trefoil**
Blackberries
Blackeyed peas
Boysenberries
Broccoli
Brussels sprouts
Cabbage
Cantaloupe
Cauliflower
Celery
Cherries***
Citrus fruits
Clover
Crab apples
Cranberries
Cucumbers
Eggplant
Filberts*
Forest and shade trees
Grapes
Honeydew melons
Loganberries
Nectarines
Nursery plants
Oats
Onions (green and dry)
Other melons
Ornamentals
Peaches
Pears
Pecans
Peppers
Plums
Potatoes
Prunes
Quinces
Raspberries
Rye
Slash pine
Soybeans
Spinach
Strawberries
Tobacco
Tomatoes
Walnuts
Watermelons
Wheat

* Pacific Northwest only
** East of the Mississippi River
*** Except California
† GUTHION is currently not registered by the state of California for use on alfalfa.

RESTRICTIONS ON WAREHOUSE STORAGE

Do not store in front of space heater. Protect from heat. Keep away from open flame. Do not heat. Store in a cool dry place above 32°F. (freezing) and below 120°F. Do not store adjacent to or near proximity of foodstuffs, drugs, feeds or any other material intended for consumption by humans or animals. Do not store adjacent to toys, cosmetics, dishes, furniture or clothing. Do not store next to herbicides.

STATEMENTS OF PRACTICAL TREATMENT

In case of poisoning, call a physician immediately. Have patient lie down and keep quiet.

If swallowed, vomiting should be induced. Administer water freely and induce vomiting by giving one dose (½ oz. or 15 ml.) of syrup of ipecac. If vomiting does not occur within 10 to 20 minutes, administer second dose. If syrup of ipecac is not available, induce vomiting by sticking finger down throat. Repeat until vomit fluid is clear. Never give anything by mouth to an unconscious person. Professional medical assistance should be secured immediately.

If on skin, remove contaminated clothing and wash skin immediately with soap and warm water.

If eyes are contaminated, wash with flowing water for at least 15 minutes.

TO PHYSICIAN: ANTIDOTE—Administer atropine sulfate in large therapeutic doses. Repeat as necessary to the point of tolerance. 2-PAM is also antidotal and may be administered in conjunction with atropine.

Compound inhibits cholinesterase resulting in stimulation of the central nervous system, the parasympathetic nervous system, and the somatic motor nerves. Do not give morphine. Watch for pulmonary edema, which may develop in serious cases of poisoning even after 12 hours. At first sign of pulmonary edema, the patient should be placed in an oxygen tent and treated symptomatically.

PRECAUTIONARY STATEMENTS

HAZARDS TO HUMANS AND DOMESTIC ANIMALS

Poisonous if swallowed, inhaled, or absorbed through the skin. Do not get in eyes or on skin. Do not breathe dust or spray mist. Spray operator should work to windward to stay out of drift or mist.

When loading spray tank wear protective clothing, natural rubber gloves and goggles.

Do not contaminate feed or foodstuffs. Keep out of reach of children and domestic animals. Keep all unprotected persons out of the operating area or vicinity where there may be danger of drift. Wash hands, arms, and face thoroughly with soap and warm water before eating or smoking. Wash all contaminated clothing with soap and hot water before reuse.

SYMPTOMS OF POISONING: A sense of "tightness" in the chest. Sweating. Contracted pupils. Vomiting and diarrhea.

ENVIRONMENTAL HAZARDS

This product is toxic to fish, birds, and other wildlife. Keep out of lakes, streams, or ponds, tidal marshes and estuaries. Do not apply where runoff is likely to occur. Do not apply when weather conditions favor drift from areas treated. Do not contaminate water by cleaning of equipment, or disposal of wastes. Apply this product only as specified on the label.

This product is highly toxic to bees exposed to direct treatment or residues on crops. Protective information may be obtained from your Cooperative Agricultural Extension Service.

RESTRICTIONS

Do not use on other crops used for food and forage. Use only according to label directions. Application at rates above those shown may result in illegal crop residues. Do not graze livestock in treated orchards or groves for 21 days after treatment. Do not treat food crops grown in the greenhouse.

RE-ENTRY STATEMENTS

Do not apply this product in such a manner as to directly or through drift expose workers or other persons. The area being treated must be vacated by unprotected persons. Do not enter treated areas without protective clothing until 24 hours after treatment. Because certain states may require more restrictive re-entry intervals for various crops treated with this product, consult your State Department of Agriculture for futher information.

Written or oral warnings must be given to workers who are expected to be in a treated area or in an area about to be treated with this product. Oral warnings should include those statements found under the label heading "Precautionary Statements" and "Re-entry Statements" included on this label. When oral warnings are given, warnings shall be given in a language customarily understood by workers. Oral warnings must be given if there is reason to believe that written warnings cannot be understood by workers. Written warnings must include the following information: DANGER, area treated with GUTHION 50% Wettable Powder on (date of application). Do not enter without appropriate protective clothing until 24 hours after treatment. **If on skin,** remove contaminated clothing and wash skin immediately with soap and warm water. **If eyes are contaminated,** wash with flowing water for at least 15 minutes. **if swallowed,** vomiting should be induced (see Precautionary Statements on label).

STORAGE AND DISPOSAL

PESTICIDE DISPOSAL

Do not contaminate water, food or feed by storage disposal.

Pesticide wastes are acutely hazardous. Improper disposal of excess pesticide, spray mixture, or rinsate is a violation of Federal Law. If these wastes cannot be disposed of by use according to label instructions, contact your State Pesticide or Environmental Control Agency, or the Hazardous Waste representative at the nearest EPA Regional Office for guidance.

CONTAINER DISPOSAL

Do not use container in connection with food, feed, or drinking water. Completely empty bag into application equipment. Then dispose of empty bag in a sanitary landfill or by incineration, or, if allowed by State and local authorities, by burning. If burned, stay out of smoke.

Store in a cool, dry place and in such a manner as to prevent cross contamination with other pesticides, fertilizers, food, and feed. Do not store below freezing (32°F). Exposure to moisture or excessive handling of water soluble packets may cause breakage. Store packets in original container and out of the reach of children, preferably in a locked storage area.

Handle and open container in a manner as to prevent spillage. If the container is leaking or material spilled for any reason or cause, carefully sweep material into a pile. Refer to Precautionary Statements on label for hazards associated with the handling of this material. Do not walk through spilled material. Dispose of pesticide as directed above. In spill or leak incidents, keep unauthorized people away. You may contact the Mobay Emergency Response Team for decontamination procedures or any other assistance that may be necessary. The Mobay Kansas City Emergency Response Telephone No. is 816-242-2582, or contact Chemtrec at 800-424-9300.

READ THE LABEL BEFORE USE

DIRECTIONS FOR USE

It is a violation of Federal law to use this product in a manner inconsistent with its labeling.

MIXING: Packets containing GUTHION 50% Wettable Powder insecticide are **water soluble.** Do not allow packet to become wet prior to adding to the spray tank. Do not handle with wet hands. Reseal outer bag to protect remaining packets.

To prepare the spray mixture, drop the required number of unopened packets, as determined under "Recommended Applications," into the spray tank while filling with water to the desired level. Operate the agitator while mixing. Depending on the water temperature and the degree of agitation, the packets should be completely dissolved with approximately 5 minutes from the time they were added to the water.

DOSAGE: Use specified dosage of GUTHION 50% Wettable Powder in the amount of water necessary to give complete coverage of foliage. Determine the **total amount** of wettable powder to be added to the spray tank based on the rates under "Recommended Applications." For each **one pound** of wettable powder to be added to the spray tank use one 1 pound packet. For example, if it is determined that 2 pounds of GUTHION 50% Wettable Powder should be added to the spray tank, add two 1 pound packets. The type of equipment used will determine the concentration required; however, use of these packets is not recommended for making highly concentrated mixtures such as used in aircraft spraying.

SPRAYING: Work to windward. Protect sprayer operators from drift or mist. When low volumes of spray are applied complete coverage and thorough application are essential for most effective results. Schedule applications in accordance with local conditions. Consult your State Agricultural Experiment Station or Extension Service for specific use information in your area.

RECOMMENDED APPLICATIONS

CROP	PEST	POUNDS GUTHION 50% W.P.	REMARKS
FIELD CROPS Alfalfa Birdsfoot trefoil (East of the Mississippi River only) Clover **(Not for use in California)**	**Leafhoppers** Mites	½ to 1	Apply specified dosage per acre, using a minimum of 10 gallons of water per acre. It may be necessary to use 20 to 25 gallons of water per acre on heavy growth for control of alfalfa weevil or Egyptian alfalfa weevil with ground equipment. Do not apply within 14 days of harvest at the rates of ½ to ¾ pounds per acre, or within 16 days at the rate of 1 pound per acre. Do not apply rates above 1 pound per acre within 21 days of harvest. Do not apply more than once per cutting at any rate.
	Alfalfa weevil Egyptian alfalfa weevil	¾ to 1½	
	Alfalfa plant bug Fleahopper Grasshoppers Lygus bugs Spittlebug	1 to 1½	

CROP	PEST	POUNDS GUTHION 50% W.P.	REMARKS
Barley Oats Rye Wheat	Cereal leaf beetle	$\frac{3}{4}$ to 1	Apply specified dosage per acre, in sufficient water for complete coverage. Do not apply more than once per season. Do not harvest for food, feed, forage or graze within 30 days of treatment.
Soybeans	Aphids Bean leaf beetle Green cloverworm Leafhoppers Leafminers Leaf rollers Stink bugs Velvet bean caterpillar	$\frac{3}{4}$ to 1	Apply specified dosage per acre, in sufficient water to give complete coverage. Repeat as necessary. Do not apply within 45 days of harvest. Do not graze or feed treated vines to livestock.
	Mexican bean beetle	1 to $1\frac{1}{2}$	
Tobacco	Aphids Grasshoppers Tobacco hornworm Tobacco flea beetle	1 to $1\frac{1}{2}$	Apply specified dosage per acre, in sufficient water for complete coverage. Treat for hornworms as soon as egg masses or worms are first found. Repeat as necessary, but not within 6 days of harvest. Prime before treating.
	Tobacco budworm	$1\frac{1}{2}$	
FRUIT Apples Crab apples Pears Quinces	Aphids Apple maggot Codling moth European apple sawfly Eye-spotted bud moth Forbes scale Fruit tree leaf roller Green fruitworm Leafhoppers Mealybug Orange tortrix Plum curculio Putnam scale Red-banded leaf roller San Jose scale† Stink bug Tarnished plant bug	$\frac{1}{2}$ to $\frac{5}{8}$	Apply specified dosage in 100 gallons of water as a full-coverage spray,* using not more than 1000 gallons of finished spray per acre. Repeat as necessary. Do not apply more than 8 times per season, nor within 15 days of harvest. An alternate spray schedule may be used by making 6 applications at rates of $\frac{1}{2}$ to $\frac{5}{8}$ pound, followed by 2 applications at the $\frac{1}{2}$ pound rate, but not within 7 days of harvest. GUTHION is compatible with dormant and summer oils, which may be added to apple and pear sprays in accordance with local recommendations. Do not use oils within 30 days of sulfur or lime-sulfur treatments. †GUTHION used alone may not provide satisfactory control. Consult your local agricultural advisor or extension service for recommendations.
Cherries (Except California)	Eye-spotted bud moth Forbes scale Fruit flies Fruit tree leaf roller Lesser peach tree borer Mites Plum curculio San Jose scale†	$\frac{1}{2}$ to 1	Apply specified dosage in 100 gallons of water as a full-coverage spray,* using not more than 800 gallons of finished spray per acre. Repeat as necessary. Do not apply more than 8 times per season, nor within 15 days of harvest. †GUTHION used alone may not provide satisfactory control. Consult your local agricultural advisor or extension service for recommendations.
	Cherry leafminer Mineola moth	$\frac{1}{2}$	
Apricots Nectarines Peaches	Aphids Cottony peach scale European fruit lecanium scale Forbes scale Lesser peach tree borer Mites Peach tree borer *Platynota flavedena* leaf roller Plum curculio Red-banded leaf roller San Jose scale† Stink bug Tarnished plant bug Terrapin scale Walnut scale White peach scale	$\frac{1}{2}$ to $\frac{5}{8}$	Apply specified dosage in 100 gallons of water as a full-coverage spray,* using not more than 800 gallons of finished spray per acre. Repeat as necessary. Do not apply more than 8 times per season, within 21 days of harvest for apricots, nectarines and peaches, nor within 15 days of harvest for plums and prunes. Apply the $\frac{3}{4}$-pound rate for oriental fruit moth and peach twig borer on apricots, nectarines and peaches only before mid-season. For control of peach tree borer, apply 2 or 3 sprays to trunk from ground to scaffold limbs, timed with moth flight.
	Oriental fruit moth Peach twig borer	$\frac{1}{2}$ to $\frac{3}{4}$	
Nectarines	Thrips	$\frac{1}{2}$ to $\frac{5}{8}$	For scale control, apply when crawlers are present. GUTHION is compatible with dormant and summer oils, which may be added to peach and prune sprays in accordance with local recommendations. Do not use oils within 30 days of sulfur or lime-sulfur treatments. **NOTE:** It is suggested that when treating nectarines during bloom period, beekeepers should be warned well in advance to remove hives a safe distance from orchards to be treated. †GUTHION used alone may not provide satisfactory control. Consult your local agricultural advisor or extension service for recommendations.
Plums Prunes	Aphids Codling moth Eye-spotted bud moth Forbes scale Fruit tree leaf roller Lesser peach tree borer Mites Orange tortrix Peach tree borer Peach twig borer Plum curculio Red-banded leaf roller San Jose scale† Stink bug Tarnished plant bug Tussock moth	$\frac{1}{2}$ to 1	
	American plum borer	1	

CROP	PEST	POUNDS GUTHION 50% W.P.	REMARKS
Blackberries Boysenberries Loganberries Raspberries	Leafhoppers Leaf rollers	$\frac{1}{2}$	Apply specified dosage per acre to foliage, using approximately 200 gallons of water for good coverage. Repeat as necessary, but not within 14 days of harvest.
	Leafminers	$\frac{5}{8}$	
	Aphids	$\frac{5}{8}$ to 1	
	Obscure root weevil	1	
	Raspberry crown (root) borer	2 to 4	For control of root weevils and borers prior to harvest, apply specified dosage per acre to lower portion of canes, and to the soil beneath the plants, using approximately 200 gallons of water. Do not apply more than twice per season. Do not make applications within 3 days of harvest at rates up to 2 pounds per acre. Rates above 2 pounds per acre should be applied only before fruit set, or after crop is harvested.
	Obscure root weevil	1	
Blueberries (Eastern and North Central states only)	Blueberry maggot Fruitworms Lecanium scale Plum curculio	1 to $1\frac{1}{2}$	Apply specified dosage per acre, using approximately 200 gallons of water for good coverage. Repeat as necessary, but not more than 4 times per season, nor within 3 days of harvest. If multiple late season applications of dust formulations of GUTHION are to be made for blueberry maggot control, do not make more than two spray applications of GUTHION per season, with the last spray no later than June 15.
Cranberries	Cranberry fruitworm *Sparganothis sulfureana* Tipworm	1 to 2	Apply specified dosage per acre, using approximately 200 gallons of water for good coverage. Repeat as necessary. Do not apply within 21 days of harvest.
	Fireworms	2	
Citrus fruits	Aphids Black scale Brown soft scale Chaff scale Citricola scale Citrus mealybug Citrus rust mite Citrus thrips Cottony-cushion scale European brown small Florida red scale Fruit tree leaf rollers Fuller rose beetle Glover scale Orange tortrix Purple scale Snow scale Western tussock moth White flies	$\frac{1}{2}$ to $\frac{3}{4}$	Apply specified dosage in 100 gallons of water as a full-coverage spray,* using not more than 2000 gallons of finished spray per acre. A single application per year may be applied up to within 7 days of picking. Where 2 applications are required, the second spray should not be applied within 28 days of harvest. Do not apply more than twice per fruit year. Do not pick fruit or do other work involving contact with the tree, such as pruning, within 7 days of treatment.
	California red scale Texas citrus mite Yellow scale	$\frac{3}{4}$	
Grapes	Grape berry moth Grape bud beetle (Except California) Grape cane girdlers Grape mealybug Leafhoppers Mites Red-banded leaf rollers Thrips	$\frac{1}{2}$ to 1	Apply specified dosage in 100 gallons of water as a full-coverage spray,* using not more than 300 gallons of finished spray per acre. Repeat as necessary, but not more than 3 times per season. Minimum dosage specified may be applied up to harvest. Higher rates up to $\frac{3}{4}$ pound require a 10-day interval to harvest, and above $\frac{3}{4}$ pound a 28-day interval to harvest.
Strawberries	Aphids Meadow spittlebug Oblique-banded leaf roller Obscure root weevil Omnivorous leaf tier Pea leaf weevil Small black (grass) weevil Strawberry leaf rollers Whitefly	1	Apply specified dosage in 200 gallons of water per acre as a full-coverage spray.* Repeat as necessary. Do not apply within 5 days of harvest.
NUTS Almonds	Peach twig borer	$\frac{3}{4}$ to 1	Apply specified dosage in 100 gallons of water as a full-coverage spray,* using not more than 500 gallons of finished spray per acre. Do not apply more than twice per season, nor within 60 days of harvest. Allow 30 days between applications.
Filberts (Pacific Northwest only)	Apple mealybug Filbert aphid Filbert leaf roller Filbertworm	$\frac{1}{2}$ to $1\frac{1}{2}$	Apply specified dosage in 100 gallons of water as a full-coverage spray,* using not more than 1000 gallons of finished spray per acre. Repeat as necessary. Do not apply within 30 days of harvest. Do not graze livestock in treated groves for 21 days after treatment.

CROP	PEST	POUNDS GUTHION 50% W.P.	REMARKS
Pecans	Aphids Fall webworm Hickory shuckworm Leafminers May beetles Mites Pecan casebearer Southern green stink bug Spittlebugs Twig girdlers Walnut caterpillars	$\frac{3}{4}$ to $1\frac{1}{8}$	Apply specified dosage in 100 gallons of water as a full-coverage spray,* using not more than 1800 gallons of finished spray per acre. Repeat as necessary. Do not apply after husks split. Where more than $11\frac{1}{4}$ pounds of GUTHION 50% Wettable Powder are applied per acre in single application, do not graze livestock in treated groves. Where $11\frac{1}{4}$ pounds or less of GUTHION 50% Wettable Powder are applied per acre, livestock may be grazed in treated groves after a 21-day post-treatment interval.
Walnuts	Codling moth Filbertworm Navel orangeworm	$\frac{3}{4}$ to $1\frac{7}{8}$	Apply specified dosage in 100 gallons of water as a full-coverage spray,* using not more than 1100 gallons of finished spray per acre. Repeat as necessary. Do not apply after husks split. Do not graze livestock in treated groves for 21 days after treatment.
VEGETABLES Artichokes	Plum moth	3	Apply specified dosage per acre in sufficient water for complete coverage. Repeat as necessary. Do not apply within 30 days of harvest. Do not feed or ensile treated forage.
Beans (dry and snap)	Leaf rollers Mites	$\frac{1}{2}$ to 1	Apply specified dosage per acre, in sufficient water for complete coverage. Repeat as necessary. Do not apply to snap beans within 7 days of harvest, nor to dry beans within 30 days of harvest. Do not exceed 4 applications on dry beans. Do not feed or ensile treated forage.
	Aphids Bean leaf beetle Green cloverworm Leafhoppers Leafminers Stink bugs Velvet bean caterpillar	$\frac{3}{4}$ to 1	
	Mexican beans beetle Spotted cucumber beetle Striped cucumber beetle Tarnished plant bug Western-striped cucumber beetle	1	
Broccoli Brussels sprouts Cabbage Cauliflower	Aphids Cabbage looper Diamondback moth Imported cabbageworm	1 to $1\frac{1}{2}$	Apply specified dosage per acre, in sufficient water for complete coverage. Repeat as necessary. Do not apply within 7 days of harvest for Brussels sprouts, 15 days of harvest for broccoli and cauliflower, nor within 21 days of harvest for cabbage.
	Cabbage maggot	$\frac{1}{4}$ to $\frac{3}{8}$	Mix specified dosage in 50 gallons of water. Apply 4 to 6 ounces of this emulsion per plant at or immediately after transplanting.
	Cabbage maggot (Direct-seeded fields in California only)	$1\frac{1}{2}$	Apply specified dosage per acre, in sufficient water for uniform distribution. Mix in upper 2 inches of soil prior to seeding, or spray in the seed row at planting time. Usually, 2 to 3 additional sprays are necessary during the growing season, depending on time of year and maggot population. Do not apply within 21 days of harvest.
	Cabbage maggot (Transplant fields in California only)	$1\frac{1}{2}$	Apply specified dosage in 300 to 400 gallons of water per acre as a soil drench in rows when damage first appears. Additional applications may be necessary. Do not apply within 21 days of harvest.
Celery	Aphids Leafminers Leafhoppers Spittlebug Tarnished plant bug	1	Apply specified dosage in 100 gallons of water as a full-coverage spray,* using not more than 200 gallons of finished spray per acre. Repeat as necessary, but not within 14 days of harvest.
Blackeyed peas (Southern peas, Crowder peas)	Corn earworm Cowpea curculio	$1\frac{1}{2}$ to 2	Apply specified dosage per acre, in sufficient water for complete coverage. Do not apply more than 4 times per season nor within 7 days of harvest. Do not use vines for feed or forge nor pasture treated areas.
	Leafminers Stink bugs	$\frac{3}{4}$ to 1	
Cucumbers	Spotted cucumber beetle Striped cucumber beetle Western-striped cucumber beetle	1	Apply specified dosage per acre, in sufficient water for complete coverage. Do not apply more than 3 times per season, nor within 1 day of harvest.
Eggplant	Leafminers	$\frac{3}{4}$ to 1	Apply specified dosage per acre, in sufficient water to give complete coverage. Repeat as necessary. Do not apply after fruit set.
	European corn borer Flea beetles	1	

CROP	PEST		POUNDS GUTHION 50% W.P.	REMARKS
Onions (green and dry)	Thrips		1 to 1½	Apply specified dosage per acre, in sufficient water to give complete coverage. Do not apply more than 3 times per season, nor within 28 days of harvest of dry onions, or 7 days of harvest of green onions.
Peppers	Leafminers		¾ to 1	Apply specified dosage per acre, in sufficient water to give complete coverage. A maximum of 4 treatments may be made up to within 3 days of harvest. Where more than 4 applications are required, do not apply last spray within 14 days of harvest.
	European corn borer Flea beetles		1	
Potatoes	Colorado potato beetle		¾	Applied specified dosage per acre, in sufficient water for complete coverage. Repeat as necessary. Do not apply within 7 days of harvest. **NOTE:** Resistance to Colorado potato beetle has occurred in some areas of the Northeast. Consult your local Extension Service or Mobay sales representative for details.
	Banded cucumber beetle Leafminers		¾ to 1	
	European corn borer Flea beetle	Leafhoppers Spittlebugs Tarnished plant bug	1 to 1½	
	Tuberworm		1⅛ to 1½	
Spinach	Aphids Leafminers Mites		¾ to 1	Apply specified dosage per acre, in sufficient water for complete coverage. Repeat as necessary. Do not apply within 14 days of harvest.
Tomatoes	Colorado potato beetle		¾	Apply specified dosage per acre, in sufficient water for complete coverage. Repeat as necessary. The high rates should be used where heavy infestations of late-instar lepidopterous larvae (large worms) and pinworms are present. Rates of 1½ pounds per acre or less may be applied up to day of harvest. Rates above 1½ pounds per acre require an interval of 14 days between application and harvest. **NOTE:** Resistance to Colorado potato beetle has occurred in some areas of the Northeast. Consult your local Extension Service or Mobay sales representative for details.
	Banded cucumber beetle *Drosophila*	Green stink bug Leafminers Whitefly	¾ to 1	
	Aphids European corn borer Flea beetles Grasshoppers Leafhoppers Thrips		1 to 1½	
	Tuberworm		1⅛ to 1½	
	Corn earworm Fruitworm Pinworm	Yellow-striped armyworm Hornworm	1½ to 3	
MELONS Honeydew melons Muskmelon (cantaloupe) Watermelons Other melons	Leafhoppers Leafminers		¾ to 1	Apply specified dosage per acre, in sufficient water to give complete coverage. Repeat as necessary, up to the day of harvest, but not more than 4 times per season.
	Rindworms Spotted cucumber beetle	Striped cucumber beetle Western-striped cucumber beetle	1	
ORNAMENTALS				
Ornamentals, nursery plants, forest and shade trees	Aphids *Cerococcus* scale Euonymus scale Juniper scale Lace bugs Leafhoppers	Mites Olive scale Oystershell scale *Pulvinaria* scale Thrips	¾ to 1	Apply specified dosage per 100 gallons of water (2 level teaspoonfuls per gallon). Spray all foliage surfaces, including the underside of leaves for complete coverage. For control of black pine leaf, brown soft, European elm and Putnam scales, use 4 level teaspoonfuls per gallon. Repeat as necessary.
	Brown soft scale Putnam scale		2	
	Black pine leaf scale European elm scale		1½ to 2	
	Cone midge Cone moth		4 to 8	Apply specified dosage per 100 gallons of water. Time applications to coincide with moth flight, when cones are open for pollination. Thorough coverage of cones is necessary for maximum control. Repeat as necessary.
	European pine shoot moth Nantucket pine tip moth		¾ to 1½	Apply specified dosage per acre, in sufficient water for good coverage. Time applications to coincide with moth flights. For application to individual trees use 1 level tablespoonful of GUTHION 50% Wettable Powder per gallon of water.
	Do not allow children or pets on treated area until material has been washed into soil, and treated area is dry.			
Slash pine	Coneworm Seedworm		3	Apply specified dosage per 100 gallons of water. Thorough coverage of cones is necessary for maximum control. Repeat as necessary. 1-15-85

* This concentration is calculated for conventional hydraulic-type sprayers. When lower volumes of spray are applied per acre with concentrate sprayers, increase the concentration of GUTHION 50% W.P. in the spray mixture in order to apply amount of GUTHION 50% W.P. per acre equivalent to a full-coverage spray.

SPECIAL LOCAL NEED REGISTRATIONS [SECTION 24(c)] FOR GUTHION 50% WETTABLE POWDER (PVA)*

CROP	PEST	STATE EPA SLN NO.	DOSAGE	REMARKS
Almonds	Navel orangeworm Peach twig borer	California CA 82-0045	3 to 4 pounds per acre.	Ground or aerial application. 28 day pre-harvest interval.
Cotton	Boll weevil	Arizona AZ 81-0021	1/4 to 1/2 pound per acre.	Application by ground or air.
	Pink bollworm	Arizona AZ 81-0021 California CA 81-0074	1 to 2 pounds per acre.	
Grapes	Omnivorous leaf roller Grape leaf roller	California CA 80-0146	1 to 2 pounds per acre	Up to 3 applications per season
Pistachios	Navel orangeworm	California CA 79-0149	5 pounds per acre.	Apply at onset of hull splitting but prior to 10% hullsplit.

* For distribution and use only within the state specified. These Special Local Need registrations approved as of date of this publication are, however, subject to cancellation by the state. Contact your nearest Mobay Agricultural Chemicals Division Office for the current status and a copy of the complete individual supplemental Special Local Need labeling. This labeling must be in the possession of the user at time of pesticide application.

®GUTHION

50% WETTABLE POWDER INSECTICIDE

For Control of a Wide Variety of Agricultural Insect Pests.

Only for agricultural use, or for sale to, use, and storage by commercial applicators, and other servicepersons.

ACTIVE INGREDIENT:
O,O-Dimethyl *S*-[(4-oxo-1,2,3-benzotriazin-3(4*H*)-yl)methyl] phosphorodithioate 50%
INERT INGREDIENTS 50%
TOTAL 100%

EPA Reg. No. 3125-193-AA

GUTHION is a Reg. TM of the Parent Company of Farbenfabriken Bayer GmbH, Leverkusen.

DANGER–POISON

Keep Out of Reach of Children
PELIGRO

PRECAUCION AL USUARIO: Si usted no puede leer o entender inglés, no use este producto hasta que la etiqueta le haya sido explicada ampliamente.
(TO THE USER: If you cannot read or understand English, do not use this product until the label has been fully explained to you.)

RECOMMENDED USES

Alfalfa†
Almonds
Apples
Apricots
Artichokes
Barley
Beans (snap and dry)
Birdsfoot trefoil***
Blackberries
Blackeyed peas
Blueberries**
Boysenberries
Broccoli
Brussels sprouts
Cabbage
Cantaloupe
Cauliflower
Celery
Honeydew melons
Loganberries
Nectarines
Nursery plants
Oats
Onions (green and dry)
Other melons
Ornamentals
Pasture grasses***
Peaches
Pears
Pecans
Peppers
Plums
Potatoes
Prunes
Quinces
Cherries****
Citrus fruits
Clover
Crab apples
Cranberries
Cucumbers
Eggplant
Filberts*
Forest and shade trees
Grapes
Raspberries
Rye
Slash pine
Soybeans
Spinach
Strawberries
Tobacco
Tomatoes
Walnuts
Watermelons
Wheat

* Pacific Northwest only
** Eastern and North Central states only
*** East of the Mississippi River
**** Except California
† GUTHION is currently not registered by the state of California for use on alfalfa.

RESTRICTIONS ON WAREHOUSE STORAGE

Do not store in front of space heater. Protect from heat. Keep away from open flame. Do not heat. Store below 120°F. Do not store adjacent to or near proximity of foodstuffs, drugs, feeds or any other material intended for consumption by humans or animals. Do not store adjacent to toys, cosmetics, dishes, furniture or clothing. Do not store next to herbicides. Store in a cool dry place.

STATEMENTS OF PRACTICAL TREATMENT

In case of poisoning, call a physician immediately. Have patient lie down and keep quiet.

If swallowed, vomiting should be induced. Administer water freely, and induce vomiting by giving one dose (1/2 oz. or 15 ml.) of syrup of ipecac. If vomiting does not occur within 10 to 20 minutes, administer second dose. If syrup of ipecac is not available, induce vomiting by sticking finger down throat. Repeat until vomit fluid is clear. Never give anything by mouth to an unconscious person. Professional medical assistance should be secured immediately. **If on skin,** remove contaminated clothing and wash skin immediately with soap and warm water. **If eyes are contaminated,** wash with flowing water for at least 15 minutes.

TO PHYSICIAN: Antidote—Administer atropine sulfate in large therapeutic doses. Repeat as necessary to the point of tolerance. 2-PAM is also antidotal and may be administered in conjunction with atropine.

Compound inhibits cholinesterase, resulting in stimulation of the central nervous system, the parasympathetic nervous system, and the somatic motor nerves. Do not give morphine. Watch for pulmonary edema, which may develop in serious cases of poisoning even after 12 hours. At first sign of pulmonary edema, the patient should be placed in an oxygen tent and treated symptomatically.

PRECAUTIONARY STATEMENTS

HAZARDS TO HUMANS AND DOMESTIC ANIMALS

DANGER–POISON

Poisonous if swallowed, inhaled, or absorbed through the skin. Do not get in eyes or on skin. Do not breathe dust or spray mist. Spray operator should work to windward to stay out of drift or mist.

When loading spray tank, or handling the powder at any other time, wear protective clothing, natural rubber gloves and goggles. For protection against GUTHION, when handling the powder, wear a pesticide mask or respirator jointly approved by the Mining Enforcement and Safety Administration (formerly the U.S. Bureau of Mines) and by the National Institute for Occupational Safety and Health under the provisions of 30 CFR Part 11.

Do not contaminate feed or foodstuffs. Keep out of reach of children and domestic animals. Keep all unprotected persons out of the operating area or vicinity where there may be danger of drift. Wash hands, arms, and face thoroughly with soap and warm water before eating or smoking. Wash all contaminated clothing with soap and hot water before reuse.

SYMPTOMS OF POISONING: A sense of "tightness" in the chest. Sweating. Contracted pupils. Stomach pains. Vomiting and diarrhea.

ENVIRONMENTAL HAZARDS

This product is toxic to fish, birds, and other wildlife. Keep out of lakes, streams or ponds. Do not apply where runoff is likely to occur. Do not apply when weather conditions favor drift from areas treated. Do not contaminate water by cleaning of equipment, or disposal of wastes. Apply this product only as specified on the label.

This product is highly toxic to bees exposed to direct treatment or residues on crops. Protective information may be obtained from your Cooperative Agricultural Extension Service.

ROTATIONAL CROPS

Treated areas may be replanted with any crop specified on this label as soon as practical after last application. Root crops may be planted 180 days after last application. All other crops must not be planted within 30 days of last application.

RE-ENTRY STATEMENTS

Do not apply this product in such a manner as to directly or through drift expose workers or other persons. The area being treated must be vacated by unprotected persons. Do not enter treated areas without protective clothing until 24 hours after treatment. Because certain states may require more restrictive re-entry intervals for various crops treated with this product, consult your State Department of Agriculture for further information.

Written or oral warnings must be given to workers who are expected to be in a treated area or in an area about to be treated with this product. Oral warnings should include those statements found under the label heading "Precautionary Statements" and "Re-entry Statements" included on this label. When oral warnings are given, warnings shall be given in a language customarily understood by workers. Oral warnings must be given if there is reason to believe that written warnings cannot be understood by workers. Written warnings must include the following information: DANGER, area treated with GUTHION 50% Wettable Powder on (date of application). Do not enter without appropriate protective clothing until 24 hours after treatment. **If on skin,** remove contaminated clothing and wash skin immediately with soap and warm water. **If eyes are contaminated,** wash with flowing water for at least 15 minutes. **If swallowed,** vomiting should be induced. (See Precautionary Statements on label).

STORAGE AND DISPOSAL

PESTICIDE DISPOSAL

Do not contaminate water, food or feed by storage or disposal.

Pesticide wastè are acutely hazardous. Improper disposal of excess pesticide, spray mixture, or rinsate is a violation of Federal Law. If these wastes cannot be disposed of by use according to label instructions, contact your State Pesticide or Environmental Control Agency, or the Hazardous Waste representative at the nearest EPA Regional Office for guidance.

CONTAINER DISPOSAL

Do not use container in connection with food, feed, or drinking water. Completely empty bag into application equipment. Then dispose of empty bag in a sanitary landfill or by incineration, or, if allowed by State and local authorities, by burning. If burned, stay out of smoke.

Store in a cool, dry place and in such a manner as to prevent cross contamination with other pesticides, fertilizers, food, and feed. Store in original container and out of the reach of children, preferably in a locked storage area.

Handle and open container in a manner as to prevent spillage. If the container is leaking or material spilled for any reason or cause, carefully sweep material into a pile. Refer to Precautionary Statements on label for hazards associated with the handling of this material. Do not walk through spilled material. Dispose of pesticide as directed above. In spill or leak incidents, keep unauthorized people away. You may contact the Mobay Emergency Response Team for decontamination procedures or any other assistance that may be necessary. The Mobay Kansas City Emergency Response Telephone No. is 816-242-2582, or contact Chemtrec at 800-424-9300.

READ THE LABEL BEFORE USE

RESTRICTIONS

Do not use on other crops used for food or forage. Use only according to label directions. Application a rates above those shown may result in illegal crop residues. Do not graze livestock in treated orchards or groves for 21 days after treatment. Do not treat food crops grown in the greenhouse.

DIRECTIONS FOR USE

It is a violation of Federal Law to use this product in a manner inconsistent with its labeling.

MIXING: GUTHION 50% Wettable Powder insecticide mixes easily with water to form a suspension which is readily usable in all conventional ground spray equipment. Mix the required amount of GUTHION 50% Wettable Powder with a small quantity of water. Add this premix through the screen while filling the spray tank, or fill the tank to the required level and add the premix. Operate the agitator while mixing.

DOSAGE: Use specified dosage of GUTHION 50% Wettable Powder in the amount of water necessary to give complete coverage of foliage. The type of equipment used will determine the concentration required.

SPRAYING: Work to windward. Protect sprayer operators from drift or mist. When low volumes of spray are applied, complete coverage and thorough application are essential for most effective results. Schedule applications in accordance with local conditions. Consult your State Agricultural Experiment Station or Extension Service for specific use information in your area.

RECOMMENDED APPLICATIONS

CROP	PEST	POUNDS GUTHION 50% W.P.	REMARKS
FIELD CROPS Alfalfa Clover (Including mixed stands with grasses, States east of the Mississippi River only) Birdsfoot trefoil (East of the Mississippi River only)	Leafhoppers Mites	½ to 1	Apply specified dosage per acre, using a minimum of 10 gallons of water per acre. It may be necessary to use 20 to 25 gallons of water per acre on heavy growth for control of alfalfa weevil or Egyptian alfalfa weevil with ground equipment. Do not apply within 14 days of harvest or grazing at the rates of ½ to ¾ pounds per acre, or within 16 days at the rate of 1 pound per acre. Do not apply rates above 1 pound per acre within 21 days of harvest or grazing. Do not apply more than once per cutting at any rate.
	Alfalfa weevil Egyptian alfalfa weevil	¾ to 1½	
	Alfalfa plant bug Fleahopper Grasshoppers Lygus bugs Spittlebug	1 to 1½	
Barley Oats Rye Wheat	Cereal leaf beetle	¾ to 1	Apply specified dosage per acre, in sufficient water for complete coverage. Do not apply more than once per season. Do not harvest for food, feed, forage or graze within 30 days of treatment.
Pasture grasses (States east of the Mississippi River)	Grasshoppers Meadow spittlebugs (adult)	1 to 1½	Apply specified dosage per acre using a minimum of 10 gallons of water per acre. Do not apply more than once per cutting. Do not graze or harvest within 16 days of application at the 1 pound rate or 21 days at rates above 1 pound.
Soybeans	Aphids Bean leaf beetle Green cloverworm Leafhoppers Leafminers Leaf rollers Stink bugs Velvet bean caterpillar	¾ to 1	Apply specified dosage per acre, in sufficient water to give complete coverage. Repeat as necessary. Do not apply within 45 days of harvest. Do not graze or feed treated vines to livestock.
	Mexican bean beetle	1 to 1½	
Tobacco	Aphids Grasshoppers Tobacco hornworm Tobacco flea beetle	1 to 1½	Apply specified dosage per acre, in sufficient water for complete coverage. Treat for hornworms as soon as egg masses or worms are first found. Repeat as necessary, but not within 6 days of harvest. Prime before treating.
	Tobacco budworm	1½	
FRUIT Apples Crab apples Pears Quinces	Aphids Apple maggot Codling moth European apple sawfly Eye-spotted bud moth Forbes scale Fruit tree leaf roller Green fruit-worm Leafhoppers Mealybug Orange tortrix Plum curculio Putnam scale Red-banded leaf roller San Jose scale† Stink bug Tarnished plant bug	½ to ⅝	Apply specified dosage in 100 gallons of water as a full-coverage spray,* using not more than 1000 gallons of finished spray per acre. Repeat as necessary. Do not apply more than 8 times per season, nor within 15 days of harvest. An alternate spray schedule may be used by making 6 applications at rates of ½ to ⅝ pound, followed by 2 applications at the ½ pound rate, but not within 7 days of harvest. GUTHION 50% Wettable Powder is compatible with dormant and summer oils, which may be added to apple and pear sprays in accordance with local recommendations. Do not use oils within 30 days of sulfur or lime-sulfur treatments. † GUTHION used alone may not provide satisfactory control. Consult your local agricultural advisor or extension service for recommendations.
		2 to 3	**Aerial Applications:** Apply specified dosage in a minimum of 3 gallons of water per acre. Apply by fixed-wing or helicopter aircraft. Where aphids are a problem, aerial applications may not provide satisfactory control. Do not apply more than 5 times at the 2 pound rate, nor 3 times at rates above 2 pounds. Do not apply within 7 days of harvest.
Cherries (Except California)	Eye-spotted bud moth Forbes scale Fruit flies Fruit tree leaf roller Lesser peach tree borer Mites Plum curculio San Jose scale†	½ to 1	Apply specified dosage in 100 gallons of water as a full-coverage spray,* using not more than 800 gallons of finished spray per acre. Repeat as necessary. Do not apply more than 8 times per season, nor within 15 days of harvest.
	Cherry leafminer Mineola moth	½	† GUTHION used alone may not provide satisfactory control. Consult your local agricultural advisor or extension service for recommendations.
Apricots Nectarines Peaches	Aphids Cottony peach scale European fruit lecanium scale Forbes scale Lesser peach tree borer Mites Plum curculio Red-banded leaf roller San Jose scale† Stink bug Tarnished plant bug Terrapin scale	½ to ⅝	Apply specified dosage in 100 gallons of water as a full-coverage spray,* using not more than 800 gallons of finished spray per acre. Repeat as necessary. Do not apply more than 8 times per season nor within 21 days of harvest for apricots, nectarines and peaches, nor within 15 days of harvest for plums and prunes.

CROP	PEST	POUNDS GUTHION 50% W.P.	REMARKS
FRUIT (Cont'd) Apricots Nectarines Peaches	Peach tree borer *Platynota flavedana* leaf roller Walnut scale White peach scale	½ to ⅝	Apply the ¾-pound rate for oriental fruit moth and peach twig borer on apricots, nectarines and peaches only before mid-season. For control of peach tree borer, apply 2 or 3 sprays to trunk from ground to scaffold limbs, timed with moth flight.
	Oriental fruit moth Peach twig borer	½ to ¾	For scale control, apply when crawlers are present.
Nectarines	Thrips	½ to ⅝	
Plums Prunes	Aphids Codling moth Eye-spotted bud moth Forbes scale Fruit tree leaf roller Lesser peach tree borer Mites Orange tortrix Peach tree borer Peach twig borer Plum curculio Red-banded leaf roller San Jose scale† Stink bug Tarnished plant bug Tussock moth	½ to 1	GUTHION 50% Wettable Powder is compatible with dormant and summer oils, which may be added to peach and prune sprays in accordance with local recommendations. Do not use oils within 30 days of sulfur or lime-sulfur treatments. **NOTE:** It is suggested that when treating nectarines during bloom period, beekeepers should be warned well in advance to remove hives a safe distance from orchards to be treated. †GUTHION used alone may not provide satisfactory control. Consult your local agricultural advisor or extension service for recommendations.
	American plum borer	1	
Blackberries Boysenberries Loganberries Raspberries	Leafhoppers Leaf rollers	½	Apply specified dosage per acre to foliage, using approximately 200 gallons of water for good coverage. Repeat as necessary, but not within 14 days of harvest.
	Leafminers	⅝	
	Aphids	⅝ to 1	
	Obscure root weevil	1	
	Raspberry crown (root) borer	2 to 4	For control of root weevils and borers prior to harvest, apply specified dosage per acre to lower portion of canes, and to the soil beneath the plants, using approximately 200 gallons of water. Do not apply more than twice per season. Do not make applications within 3 days of harvest at rates up to 2 pounds per acre. Rates above 2 pounds per acre should be applied only before fruit set, or after crop is harvested.
	Obscure root weevil	1	
Blueberries (Eastern and North Central states only)	Blueberry maggot Fruitworms Lecanium scale Plum curculio	1 to 1½	Apply specified dosage per acre, using approximately 200 gallons of water for good coverage. Repeat as necessary, but not more than 4 times per season, nor within 3 days of harvest. If multiple late season applications of dust formulations of GUTHION are to be made for blueberry maggot control, do not make more than two spray applications of GUTHION per season, with the last spray no later than June 15.
Cranberries	Cranberry fruitworm *Sparganothis sulfureana* Tipworm	1 to 2	Apply specified dosage per acre, using approximately 200 gallons of water for good coverage. Repeat as necessary. Do not apply within 21 days of harvest.
	Fireworms	2	
Citrus fruits	Aphids Black scale Brown soft scale Chaff scale Citricola scale Citrus mealybug Citrus rust mite Citrus thrips Cottony-cushion scale European brown snail Florida red scale Fruit tree leaf rollers Fuller rose beetle Glover scale Orange tortrix Purple scale Snow scale Western tussock moth White flies	½ to ¾ ½ to ¾	Apply specified dosage in 100 gallons of water as a full-coverage spray,* using not more than 2000 gallons of finished spray per acre. A single application per year may be applied up to within 7 days of picking. Where 2 applications are required, the second spray should not be applied within 28 days of harvest. Do not apply more than twice per fruit year. Do not pick fruit or do other work involving contact with the tree, such as pruning, within 7 days of treatment.
	California red scale Texas citrus mite Yellow scale	¾	
Grapes	Grape berry moth Grape bud beetle Grape cane girdlers Grape mealybug Leafhoppers Mites Red-banded leaf rollers Thrips	½ to 1	Apply specified dosage in 100 gallons of water as a full-coverage spray,* using not more than 300 gallons of finished spray per acre. Repeat as necessary, but not more than 3 times per season. Minimum dosage specified may be applied up to harvest. Higher rates up to ¾ pound require a 10-day interval to harvest, and above ¾ pound, a 28-day interval to harvest.

CROP	PEST	POUNDS GUTHION 50% W.P.	REMARKS
Strawberries	Aphids Meadow spittle-bug Oblique-banded leaf roller Obscure root weevil Omnivorous leaf tier Pea leaf weevil Small black (grass) weevil Strawberry leaf rollers Whitefly	1	Apply specified dosage in 200 gallons of water per acre as a full-coverage spray.* Repeat as necessary. Do not apply within 5 days of harvest.
NUTS Almonds	Peach twig borer	$\frac{3}{4}$ to 1	Apply specified dosage in 100 gallons of water as a full-coverage spray,* using not more than 500 gallons of finished spray per acre. Do not apply more than twice per season, nor within 60 days of harvest. Allow 30 days between applications.
Filberts (Pacific Northwest only)	Apple mealybug Filbert aphid Filbert leaf roller Filbertworm	$\frac{1}{2}$ to $1\frac{1}{2}$	Apply specified dosage in 100 gallons of water as a full-coverage spray,* using not more than 1000 gallons of finished spray per acre. Repeat as necessary. Do not apply within 30 days of harvest. Do not graze livestock in treated groves for 21 days after treatment.
Pecans	Aphids Fall webworm Hickory shuckworm Leafminers May beetles Mites Pecan casebearer Southern green stink bug Spittlebugs Twig girdlers Walnut caterpillars	$\frac{3}{4}$ to $1\frac{1}{8}$	Apply specified dosage in 100 gallons of water as a full-coverage spray,* using not more than 1800 gallons of finished spray per acre. Repeat as necessary. Do not apply after husks split. Where more than $11\frac{1}{4}$ pounds of GUTHION 50% Wettable Powder are applied per acre in a single application, do not graze livestock in treated groves. Where $11\frac{1}{4}$ pounds or less of GUTHION 50% Wettable Powder are applied per acre, livestock may be grazed in treated groves after a 21-day post-treatment interval.
Walnuts	Codling moth Filbertworm Navel orangeworm	$\frac{3}{4}$ to $1\frac{7}{8}$	Apply specified dosage in 100 gallons of water as a full-coverage spray,* using not more than 1100 gallons of finished spray per acre. Repeat as necessary. Do not apply after husks split. Do not graze livestock in treated groves for 21 days after treatment.
VEGETABLES Artichokes	Plum moth	3	Apply specified dosage per acre in sufficient water for complete coverage. Repeat as necessary. Do not apply within 30 days of harvest. Do not feed or ensile treated forage.
Beans (Dry and snap)	Leaf rollers Mites	$\frac{1}{2}$ to 1	Apply specified dosage per acre, in sufficient water for complete coverage. Repeat as necessary. Do not apply to snap beans within 7 days of harvest, nor to dry beans within 30 days of harvest. Do not exceed 4 applications on dry beans. Do not feed or ensile treated forage.
	Aphids Bean leaf beetle Green cloverworm Leafhoppers Leafminers Stink bugs Velvet bean caterpillar	$\frac{3}{4}$ to 1	
	Mexican bean beetle Spotted cucumber beetle Striped cucumber beetle Tarnished plant bug Western-striped cucumber beetle	1	
	Aphids Cabbage looper Diamondback moth Imported cabbageworm	1 to $1\frac{1}{2}$	Apply specified dosage per acre, in sufficient water for complete coverage. Repeat as necessary. Do not apply within 7 days of harvest for Brussels sprouts, 15 days of harvest for broccoli and cauliflower, nor within 21 days of harvest for cabbage.
Broccoli Brussels sprouts Cabbage Cauliflower	Aphids Cabbage looper Diamondback moth Imported cabbageworm	1 to $1\frac{1}{2}$	Apply specified dosage per acre, in sufficient water for complete coverage. Repeat as necessary. Do not apply within 7 days of harvest for Brussels sprouts, 15 days of harvest for broccoli and cauliflower, nor within 21 days of harvest for cabbage.
	Cabbage maggot	$\frac{1}{4}$ to $\frac{3}{8}$	Mix specified dosage in 50 gallons of water. Apply 4 to 6 ounces of this emulsion per plant at or immediately after transplanting.
	Cabbage maggot (Direct-seeded fields in California only)	$1\frac{1}{2}$	Apply specified dosage per acre, in sufficient water for uniform distribution. Mix in upper 2 inches of soil prior to seeding, or spray in the seed row at planting time. Usually, 2 to 3 additional sprays are necessary during the growing season, depending on time of year and maggot population. Do not apply within 21 days of harvest.
	Cabbage maggot (Transplant fields in California only)	$1\frac{1}{2}$	Apply specified dosage in 300 to 400 gallons of water per acre as a soil drench in rows when damage first appears. Additional applications may be necessary. Do not apply within 21 days of harvest.

CROP	PEST	POUNDS GUTHION 50% W.P.	REMARKS
Celery	Aphids Leafminers Leafhoppers Spittlebug Tarnished plant bug	1	Apply specified dosage in 100 gallons of water as a full-coverage spray,* using not more than 200 gallons of finished spray per acre. Repeat as necessary, but not within 14 days of harvest.
Blackeyed peas (Southern peas, Crowder peas)	Corn earworm Cowpea curculio	1½ to 2	Apply specified dosage per acre, in sufficient water for complete coverage. Do not apply more than 4 times per season nor within 7 days of harvest. Do not use vines for feed or forage nor pasture treated areas.
	Leafminers Stink bugs	¾ to 1	
Cucumbers	Spotted cucumber beetle Striped cucumber beetle Western-striped cucumber beetle	1	Apply specified dosage per acre, in sufficient water for complete coverage. Do not apply more than 3 times per season, nor within 1 day of harvest.
Eggplant	Leafminers	¾ to 1	Apply specified dosage per acre, in sufficient water to give complete coverage. Repeat as necessary. Do not apply after fruit set.
	European corn borer Flea beetles	1	
Onions (green and dry)	Thrips	1 to 1½	Apply specified dosage per acre, in sufficient water to give complete coverage. Do not apply more than 3 times per season, nor within 28 days of harvest of dry onions, or 7 days of harvest of green onions.
Peppers	Leafminers	¾ to 1	Apply specified dosage per acre, in sufficient water to give complete coverage. A maximum of 4 treatments may be made up to within 3 days of harvest. Where more than 4 applications are required, do not apply last spray within 14 days of harvest.
	European corn borer Flea beetles	1	
Potatoes	Colorado potato beetle	¾	Apply specified dosage per acre, in sufficient water for complete coverage. Repeat as necessary. Do not apply within 7 days of harvest.
	Banded cucumber beetle Leafminers	¾ to 1	
	European corn borer Flea beetle Leafhoppers Spittlebugs Tarnished plant bug	1 to 1½	
	Tuberworm	1⅛ to 1½	
Spinach	Aphids Leafminers Mites	¾ to 1	Apply specified dosage per acre, in sufficient water for complete coverage. Repeat as necessary. Do not apply within 14 days of harvest.
Tomatoes	Colorado potato beetle	¾	Apply specified dosage per acre, in sufficient water for complete coverage. Repeat as necessary. The high rates should be used where heavy infestations of late-instar lepidopterous larvae (large worms) and pinworms are present. Rates of 1½ pounds per acre or less may be applied up to day of harvest. Rates above 1½ pounds per acre require an interval of 14 days between application and harvest.
	Banded cucumber beetle *Drosophila* Green stink bug Leafminers Whitefly	¾ to 1	
	Aphids European corn borer Flea beetles Grasshoppers Leafhoppers Thrips	1 to 1½	
	Tuberworm	1⅛ to 1½	
	Corn earworm Fruitworm Pinworm Yellow-striped armyworm Hornworm	1½ to 3	
MELONS Honeydew melons Muskmelon (cantaloupe) Watermelons Other melons	Leafhoppers Leafminers	¾ to 1	Apply specified dosage per acre, in sufficient water to give complete coverage. Repeat as necessary, up to the day of harvest, but not more than 4 times per season.
	Rindworms Spotted cucumber beetle Striped cucumber beetle Western-striped cucumber beetle	1	
ORNAMENTALS Ornamentals, nursery plants, forest and shade trees	Aphids Black vine weevil *Cerococcus* scale Euonymus scale Leafhoppers Mites Olive scale Oystershell scale	¾ to 1	Apply specified dosage per 100 gallons of water (2 level

CROP	PEST	POUNDS GUTHION 50% W.P.	REMARKS
Ornamentals nursery plants, forest and shade trees	Juniper scale Lace bugs *Pulvinaria* scale Thrips	¾ to 1	teaspoonfuls per gallon). Spray all foliage surfaces, including the underside of leaves for complete coverage. For control of black pine leaf, brown soft, European elm and Putnam scales, use 4 level teaspoonfuls per gallon. Repeat as necessary.
	Brown soft scale Putnam scale	2	
	Black pine leaf scale European elm scale	1½ to 2	
	Cone midge Cone moth	4 to 8	Apply specified dosage per 100 gallons of water. Time applications to coincide with moth flight, when cones are open for pollination. Thorough coverage of cones is necessary for maximum control. Repeat as necessary.
	European pine shoot moth Nantucket pine tip moth	¾ to 1½	Apply specified dosage per acre, in sufficient water for good coverage. Time applications to coincide with moth flights. For application to individual trees use 1 level tablespoonful of GUTHION 50% Wettable Powder per gallon of water.
	Do not allow children or pets on treated area until material has been washed into soil, and treated area is dry.		
Slash pine	Coneworm Seedworm	3	Apply specified dosage per 100 gallons of water. Thorough coverage of cones is necessary for maximum control. Repeat as necessary. 1-15-85

* This concentration is calculated for conventional hydraulic-type sprayers. When lower volumes of spray are applied per acre with concentrate sprayers, increase the concentration of GUTHION 50% W.P. in the spray mixture in order to apply amount of GUTHION 50% Wettable Powder per acre equivalent to a full-coverage spray.

SPECIAL LOCAL NEED REGISTRATIONS [SECTION 24(c)] FOR GUTHION 50% WETTABLE POWDER*

CROP	PEST	STATE EPA SLN NO.	DOSAGE	REMARKS
Apples	Various insects	New York NY 81-0004	1½ to 2 pounds per acre	Aerial application in at least 3 gallons of water.
		Vermont VT 78-0004	1½ pounds per acre.	Low volume aerial application.
Border treatment around alkali bee beds	Black blister beetle (Melone niger) Blister beetle (Zonitis atripennis flavida)	Idaho ID 80-0021 Oregon OR 80-0036 Washington WA 80-0018	¼ pound in 2 gallons of water per 1000 feet of ditch 4 to 6 inches wide and 6 to 10 inches deep.	Apply prior to emergence of overwintering beetles.
Celery	Carrot weevil	Ohio OH 81-0017	1 pound per acre.	Do not apply within 14 days of harvest.
Cotton	Insects	Alabama Al 81-0011 Arkansas AR 80-0020 Louisiana LA 80-0029 New Mexico NM 80-0018 Tennessee TN 80-0023 Texas TX 80-0036	¼ to 2 pounds per acre.	Application by ground or air.
	Boll weevil	South Carolina SC 81-0023	¼ to ½ pound per acre.	
Parsley	Carrot weevil	Maryland MD 81-0006 New Jersey NJ 81-0004 Ohio OH 81-0017	1 pound per acre.	Up to 3 applications per season.

* For distribution and use only within the state specified. These Special Local Need registrations approved as of date of this publication are, however, subject to cancellation by the state. Contact your nearest Mobay Agricultural Chemicals Division Office for the current status and a copy of the complete individual supplemental Special Local Need labeling. This labeling must be in the possession of the user at time of pesticide application.

®MESUROL

75% WETTABLE POWDER INSECTICIDE—MOLLUSCICIDE

FOR CONTROL OF CERTAIN INSECTS AND MOLLUSKS.

Only for agricultural use, or for sale to, use, and storage by commercial applicators, and other servicepersons.

ACTIVE INGREDIENT:	
3,5-Dimethyl-4-(methylthio)phenol methylcarbamate	75%
INERT INGREDIENTS	25%
TOTAL	100%

U.S. Patent No. 3,313,684

EPA Reg. No. 3125-288-AA

MESUROL is a Reg. TM of the Parent Company of Farbenfabriken Bayer GmbH, Leverkusen.

CAUTION
Keep Out of Reach of Children
AVISO

PRECAUCION AL USUARIO: Si usted no puede leer o entender inglés, no use este producto hasta que la etiqueta le haya sido explicada ampliamente.

(**TO THE USER:** If you cannot read or understand English, do not use this product until the label has been fully explained to you.)

RESTRICTIONS ON WAREHOUSE STORAGE

Do not store in front of space heater. Store below 120°F. Do not store adjacent to, or near foodstuffs, drugs, feeds or any other material intended for consumption by humans or animals. Do not store adjacent to toys, cosmetics, dishes, furniture or clothing. Do not store next to herbicides. Store in a cool dry place.

RECOMMENDED USES

Non-bearing fruit trees, ornamentals

STATEMENTS OF PRACTICAL TREATMENT

If swallowed, vomiting should be induced. Administer water freely, and induce vomiting by giving one dose (1/2 oz. or 15 ml.) of syrup of ipecac. If vomiting does not occur within 10 to 20 minutes, administer second dose. If syrup of ipecac is not available, induce vomiting by sticking finger down throat. Repeat until vomit fluid is clear. Never give anything by mouth to an unconscious person. Professional medical assistance should be secured immediately. In case of poisoning, call a physician immediately. Have patient lie down and keep quiet.

If on skin, remove contaminated clothing and wash skin immediately with soap and warm water.

If eyes are contaminated, wash with flowing water for at least 15 minutes.

PRECAUTIONARY STATEMENTS

HAZARDS TO HUMANS AND DOMESTIC ANIMALS

WARNING

Poisonous if swallowed, inhaled, or absorbed through the skin. Rapidly absorbed through the skin. Do not get in eyes or on skin. Wear protective clothing, natural rubber gloves, and goggles.

Do not breathe dust or spray mist. Wear a mask or respirator similar to those used for protection against organocarbamate insecticides jointly approved by the Mining Enforcement and Safety Administration (formerly the U.S. Bureau of Mines) and by the National Institute for Occupational Safety and Health under the provisions of 30 CFR Part 11.

Do not contaminate feed or food. Keep out of reach of children. Keep all unprotected persons out of the operating area, or vicinity where there may be danger of drift. Wash hands, arms and face thoroughly with soap and warm water before eating or smoking. Wash all contaminated clothing with soap and hot water before reuse.

SYMPTOMS OF POISONING: A sense of "tightness" in the chest. Sweating. Contracted pupils. Stomach pains. Vomiting and diarrhea.

TO PHYSICIAN: Antidote—Administer atropine sulfate in large therapeutic doses. Repeat as necessary to the point of tolerance.

Compound inhibits cholinesterase resulting in stimulation of the central nervous system, the parasympathetic nervous system, and the somatic motor nerves. Do not give morphine. Watch for pulmonary edema, which may develop in serious cases of poisoning even after 12 hours. At first sign of pulmonary edema, the patient should be placed in an oxyen tent and treated symptomatically.

ENVIRONMENTAL HAZARDS

Do not use on other crops used for food or forage. Use only according to label directions. Applications at rates above those shown may result in illegal crop residues.

This product is toxic to fish and wildlife. Use with care when applying in areas adjacent to any body of water. Keep out of any body of water. Do not apply where runoff is likely to occur. Do not apply when weather conditions favor drift from areas treated. Do not contaminate water by cleaning of equipment, or disposal of wastes.

This product is highly toxic to bees exposed to direct treatment or residues on crops. Protective information may be obtained from your Cooperative Agricultural Extension Service.

RE-ENTRY STATEMENTS

Do not apply this product in such a manner as to directly or through drift expose workers or other persons. The area being treated must be vacated by unprotected persons. Do not enter treated areas without protective clothing until sprays have dried. Because certain states may require more restrictive re-entry intervals for various crops treated with this product, consult your State Department of Agriculture for further information.

Written or oral warnings must be given to workers who are expected to be in a treated area or in an area about to be treated with this product. Oral warnings should include those statements found under the label heading "Precautionary Statements" and "Re-entry Statements" included on this label. When oral warnings are given, warnings shall be given in a language customarily understood by workers. Oral warnings must be given if there is reason to believe that written warnings cannot be understood by workers. Written warnings must include the following information: WARNING, area treated with MESUROL 75% Wettable Powder on (date of application). Do not enter without appropriate protective clothing until sprays have dried. **If on skin,** remove contaminated clothing and wash skin immediately with soap and warm water. **If eyes are contaminated,** wash with flowing water for at least 15 minutes. **If swallowed,** vomiting should be induced. (See Precautionary Statements on label).

STORAGE AND DISPOSAL

PESTICIDE DISPOSAL

Do not contaminate water, food or feed by storage or disposal.

Wastes resulting from the use of this product may be disposed of on site or at an approved waste disposal facility.

CONTAINER DISPOSAL

Do not use container in connection with food, feed, or drinking water. Completely empty bag into application equipment. Then dispose of empty bag in a sanitary landfill or by incineration, or, if allowed by State and local authorities, by burning. If burned, stay out of smoke.

Store in a cool, dry place and in such a manner as to prevent cross contamination with other pesticides, fertilizers, food, and feed. Store in original container and out of the reach of children, preferably in a locked storage area.

Handle and open container in a manner as to prevent spillage. If the container is leaking or material spilled for any reason or cause, carefully sweep material into a pile. Refer to Precautionary Statements on label for hazards associated with the handling of this material. Do not walk through spilled material. Dispose of pesticide as directed above. In spill or leak incidents, keep unauthorized people away. You may contact the Mobay Emergency Response Team for decontamination procedures or any other assistance that may be necessary. The Mobay Kansas City Emergency Response Telephone No. is 816-242-2582, or contact Chemtrec at 800-424-9300.

IMPORTANT: This bulletin is not intended to provide adequate information for application. Before using the product, read and carefully observe directions, cautionary statements and other information appearing on the label. This product is sold subject to the Conditions of Sale set forth on the container label.

DIRECTIONS FOR USE

It is a violation of Federal Law to use this product in a manner inconsistent with its labeling.

FOR INSECT CONTROL: MESUROL 75% Wettable Powder effectively controls certain specified destructive pests on non-bearing fruit trees, and ornamentals. Use specific quantities of MESUROL 75% Wettable Powder in sufficient water to give complete coverage of foliage. The type of equipment used will determine the concentration required. Thorough coverage is necessary for maximum insect control.

The product mixes readily with water to form a suspension in all conventional ground spray equipment. Maintain adequate agitation in the spray tank to keep the wettable powder in suspension. Do not apply with oil.

Schedule applications in accordance with local conditions.

RECOMMENDED APPLICATIONS

CROP	PEST	POUNDS MESUROL 75% WP	REMARKS
Blueberries (Except California and Massachusetts)	Blueberry maggot (control of adult flies)	1 to 2	Apply specified dosage per acre by air or ground equipment in a minimum of 5 gallons of water per acre. Repeat as necessary with a minimum of 7 days between applications. A maximum of 8 pounds of product may be applied per season for maggot control and/or bird repellency. Applications may be made up to the day of harvest. **NOTE:** Since timing of maggot control sprays is very important, contact your State Agricultural Experiment Station or Extension Service for accurate timing. For maximum repellency of birds, apply as berries first begin to ripen and repeat as necessary to maintain repellency with at least 7 days between applications.
	Repellency of: Robins, Starlings, Grackles, Redwings, Baltimore orioles, Bluejays, Finches and other pest birds	2 to 2⅔	
Cherries	Cherry fruit flies	1 to 1⅓	Apply specified dosage in 100 gallons of water as a full-coverage spray, using not more than 400 gallons of finished spray per acre. Repeat as necessary. For cherries, do not apply more than 3 times per season, nor within 7 days of harvest; for peaches do not apply more than 4 times per season, nor with 21 days of harvest. Use the high rate when an eradicative treatment is made to control heavy insect populations. The low rate may be used in seasonal spray programs. Do not graze or feed cover crops grown in treated areas. **NOTE:** Spray timing is very important to control plum curculio and cherry fruit flies. Consult you local state extension representative for number and timing of sprays in seasonal spray programs.
Cherries Peaches	European red mite McDaniel mite		
	Plum curculio	1⅓	
Cherries	Robins, Starlings, Finches, Grackles, Sparrows, Bluejays, Cedar waxwings	2⅔ to 5⅓	**FOR BIRD REPELLENCY:** Apply specified dosage per acre with conventional sprayer as a full-coverage spray as cherries begin to ripen. The rate used is dependent on the size and density of the trees. Do not apply more than a total of 3 times per season, whether applied for insect control, or bird repellency, or a combination of both. Do not apply within 7 days of harvest. Do not graze grass under treated trees.
NON-BEARING FRUITS Deciduous Fruit Trees Citrus	Aphids Mites	½ to 1	Apply specified dosage per acre in 50 gallons of water. Spray plants and foliage until thoroughly wet. Repeat as necessary applying up to 4 applications per season (8 pounds of product per acre). Do not apply to trees, other than cherries and peaches, which will bear fruit within 1 year of application.
	Snails Slugs	2	
ORNAMENTALS (Field and Greenhouse) Flowers Shrubs Trees	Aphids Mites	½ to 1	Apply specified dosage per acre in 50 gallons of water. Spray plants and foliage until thoroughly wet. Repeat as necessary applying up to 4 applications per season (8 pounds of product per acre.) **NOTE:** Do not apply in conjunction with foliar fertilizer applications. 3-25-85
	Snails Slugs	2	

SPECIAL LOCAL NEED REGISTRATIONS [SECTION 24(c)] FOR MESUROL 75% WETTABLE POWDER*

CROP	PEST	STATE EPA SLN NO.	DOSAGE	REMARKS
Corn	Snails and slugs	Delaware DE 84-0004 Indiana IN 84-0007 Maryland MD 84-0002 Michigan MI 84-0010 New York NY 84-0002 Ohio OH 84-0002 Pennsylvania PA 84-0007 Virginia VA 84-0003 West Virginia WV 84-0002	20 pounds of prepared bait per acre.	Prepare bait by applying ⅔ pound of MESUROL 75% Wettable Powder to 100 pounds of cracked corn. Broadcast to the soil surface when damage becomes apparent.
Tree seeds (non-edible) for Nursery Plantings	Cowbirds Crows Grackles Mourning doves Sparrows	Maine ME 84-0002 New Hampshire NH 84-0003	10⅔ ounces per 100 pounds of seed	Use equipment capable of uniformly mixing MESUROL 75% WP and seed. The addition of a sticking agent (maximum 1 pint/1000 pounds of seed) such as milk, cola or vegetable oil will enhance adherence to the seed. Do not use treated seed for food or feed.

* For distribution and use only within the state specified. These Special Local Need registrations approved as of date of this publication are, however, subject to cancellation by the state. Contact your nearest Mobay Agricultural Chemicals Division Office for the current status and a copy of the complete individual supplemental Special Local Need labeling. This labeling must be in the possession of the user at time of pesticide application.

®METASYSTOX-R

SPRAY CONCENTRATE
SYSTEMIC INSECTICIDE

For Control of Aphids, Mites and Other Pests on a Variety of Food Crops and Field Crops.

Only for agricultural use, or for sale to, use and storage by commercial applicators, and other servicepersons.

ACTIVE INGREDIENTS:

S-[2-(Ethylsulfinyl)ethyl] *O,O* -dimethyl phosphorothioate	25%
Aromatic Petroleum Distillate	14%
INERT INGREDIENTS	61%
TOTAL	100%

Contains 2 lbs. *S* -[2-(Ethylsulfinyl)ethyl] *O,O* -dimethyl phosphorothioate per gallon.

EPA Reg. No. 3125-111-AA

METASYSTOX is a Reg. TM of the Parent Company of Farbenfabriken Bayer GmbH, Leverkusen.

CAUTION
Keep Out of Reach of Children
AVISO

PRECAUCION AL USUARIO: Si usted no puede leer o entender inglés, no use este producto hasta que la etiqueta le haya sido explicada ampliamente.

(TO THE USER: If you cannot read or understand English, do not use this product until the label has been fully explained to you.)

RECOMMENDED USES

Alfalfa (seed)
Apples*
Apricots*
Blackberries
Beans (snap and lima)
Broccoli
Brussels sprouts
Cabbage
Cauliflower
Cherries*
Oranges
Other melons
Peaches*
Pears
Peas†
Peppermint
Peppers
Plums**
Potatoes
Prunes**
Citrus*
Clover (seed)
Cotton
Corn
Crab apples*
Cucumbers
Eggplant
Filberts‡
Grapefruit
Head lettuce
Lemons
Muskmelon (cantaloupe)
Nectarines*
Pumpkin
Quinces*
Raspberries
Safflower
Spearmint
Sorghum
Spanish onions†‡
Squash
Sugar beets
Turnips
Walnuts
Watermelon

* Non-bearing only
** Non-bearing and bearing
† North Central states
‡ Oregon and Washington only
†‡ West of the Mississippi River only

RESTRICTIONS ON WAREHOUSE STORAGE

Flammable. Keep away from heat and open flame.

STATEMENTS OF PRACTICAL TREATMENT

If splashed in eyes, immediately flush eyes with plenty of water for at least 15 minutes. For further treatment, call a physician.

If swallowed, vomiting should be induced. Administer water freely, and induce vomiting by giving one dose (½ oz. or 15 ml.) of syrup of ipecac. If vomiting does not occur within 10 to 20 minutes, administer second dose. If syrup of ipecac is not available, induce vomiting by sticking finger down throat. Repeat until vomit fluid is clear. Never give anything by mouth to an unconscious person. Professional medical assistance should be secured immediately.

If poisoning occurs, obtain prompt medical aid.

NOTE TO PHYSICIAN: May cause cholinesterase depression. Atropine sulfate is antidotal. 2-PAM is also antidotal and may be administered in conjunction with atropine.

PRECAUTIONARY STATEMENTS

HAZARDS TO HUMANS AND DOMESTIC ANIMALS

WARNING

Causes eye irritation. May be fatal if swallowed, inhaled or absorbed through the skin. Do not get in eyes, on skin, or on clothing. Do not breathe spray mist. Wash thoroughly with soap and warm water after handling. Wash clothing with soap and hot water before reuse.

Avoid contamination of feed or food. Keep out of reach of children.

ENVIRONMENTAL HAZARDS

This product is toxic to fish and wildlife. Birds and other wildlife in treated areas may be killed. This product is toxic to bees and should not be applied when bees are actively visiting the area. Keep out of lakes, streams, ponds, tidal marshes and estuaries. Do not apply where runoff is likely to occur. Do not apply when weather conditions favor drift from areas treated. Do not contaminate water by cleaning of equipment, or disposal of wastes. Apply this product only as specified on the label.

PHYSICAL OR CHEMICAL HAZARDS

Flammable. Keep away from heat and open flame.

RESTRICTIONS

Do not use on other crops used for food or forage. Use only according to label directions. Application at rates above those shown may result in illegal crop residues. Do not treat food crops grown in the greenhouse.

RE-ENTRY STATEMENTS

Do not apply this product in such a manner as to directly or through drift expose workers or other persons. The area being treated must be vacated by unprotected persons. Do not enter treated areas without protective clothing until 48 hours after application. Because certain states may require more restrictive re-entry intervals for various crops treated with this product, consult your State Department of Agriculture for further information.

Written or oral warnings must be given to workers who are expected to be in a treated area or in an area about to be treated with this product. Oral warnings should include those statements found under the label heading "Precautionary Statements" and "Re-entry Statements" included on this label. When oral warnings are given, warnings shall be given in a language customarily understood by workers. Oral warnings must be given if there is reason to believe that written warnings cannot be understood by workers. Written warnings must include the following information: WARNING, area treated with METASYSTOX-R Spray Concentrate on (date of application). Do not enter without appropriate protective clothing until 48 hours after application.

If on skin, remove contaminated clothing and wash skin immediately with soap and water.

If eyes are contaminated, wash with flowing water for at least 15 minutes.

If swallowed, vomiting should be induced. (See Precautionary Statements on label).

STORAGE AND DISPOSAL

PESTICIDE DISPOSAL

Do not contaminate water, food or feed by storage or disposal.

Pesticide wastes are toxic. Improper disposal of excess pesticide, spray mixture, or rinsate is a violation of Federal Law. If these wastes cannot be disposed of by use according to label instructions, contact your State Pesticide or Environmental Control Agency, or the Hazardous Waste representative at the nearest EPA Regional Office for guidance.

CONTAINER DISPOSAL

Triple rinse (or equivalent). Then offer for recycling or reconditioning, or puncture and dispose of in a sanitary landfill, or by other procedures approved by state and local authorities.

Store in a cool, dry place and in such a manner as to prevent cross contamination with other pesticides, fertilizers, food, and feed. Store away from heat and open flame. Store in original container and out of the reach of children, preferably in a locked storage area.

Handle and open container in a manner as to prevent spillage. If container is leaking invert to prevent leakage. If the container is leaking or material spilled for any reason or cause, carefully dam up spilled material to prevent runoff. Refer to Precautionary Statements on label for hazards associated with the handling of this material. Do not walk through spilled material. Absorb spilled material with absorbing type compounds and dispose of as directed for pesticides above. In spill or leak incidents, keep unauthorized people away. You may contact the Mobay Emergency Response Team for decontamination procedures or any other assistance that may be necessary. The Mobay Kansas City Emergency Response Telephone No. is 816-242-2582, or contact Chemtrec at 800-424-9300.

IMPORTANT

This bulletin is not intended to provide adequate information for application. Before using the product read and carefully observe directions, cautionary statements and other information appearing on the product label. This product is sold subject to the Conditions of Sale set forth on the container label.

READ THE LABEL BEFORE USE

DIRECTIONS FOR USE

It is a violation of Federal law to use this product in a manner inconsistent with its labeling.

METASYSTOX-R systemic insecticide penetrates by absorption, and is translocated in the plant. The best time to apply METASYSTOX-R is early morning or evening, especially when crops such as alfalfa and cotton are in bloom. This practice will minimize the possibility of beneficial pollinating insects, including honeybees, from coming into contact with the spray.

MIXING: METASYSTOX-R Spray Concentrate forms an emulsion when diluted with water, and is suitable for use in all power-operated ground and aircraft sprayers. To mix with water, pour the required amount of METASYSTOX-R Spray Concentrate into full amount of water, and then agitate.

DOSAGE: Use specified dosage of METASYSTOX-R Spray Concentrate in the amount of water necessary to give complete coverage of foliage, but in no case use less than one gallon of water per acre. The type of equipment used will determine the concentration required.

SPRAYING: Work to windward. Protect sprayer operators from drift or mist. When low volumes of spray are applied, complete coverage and thorough application are essential for most effective results. Schedule applications in accordance with local conditions. Consult your State Agricultural Experiment Station or Extension Service for specific use information and for state regulations, which may contain additional restrictions or requirements.

RECOMMENDED APPLICATIONS

CROP	PEST	PINTS MSR S.C.	REMARKS
FIELD CROPS			
Cotton (Except California)	Aphids Thrips Fleahoppers	½ to 1	Apply specified dosage per acre in sufficient water for complete coverage but not less than one gallon per acre. Do no apply more than 3 times per season nor within 14 days of picking. Do not graze or feed gin trash to dairy or meat animals.
	Mites	1½ to 2	
Cotton (California only)	Aphids Mites Lygus bugs	1½ to 2	

CROP	PEST	PINTS MSR S.C.	REMARKS
Corn Field, Sweet, Popcorn	Aphids Corn rootworm beetles Leafhoppers Mites Thrips	1½ to 2	Apply specified dosage per acre in sufficient water for complete coverage but not less than one gallon per acre. For control of corn rootworm beetles apply as needed in accordance with established economic thresholds in your locality. When using one application per season, do not apply within 7 days of harvest of corn, or within 7 days of foraging or harvest of corn fodder. When using two or three applications per season, do not make the last application within 21 days of harvest of corn, or within 21 days of foraging or harvest of corn fodder. Do not apply more than three times per season.
Peppermint Spearmint	Aphids Mites	3	Apply specified dosage, in at least 20 gallons of water per acre, with ground equipment. FOR established infestations, make 2 applications 10 to 14 days apart. Do not apply within 14 days of harvest.
Safflower	Aphids	2	Apply specified dosage per acre, in sufficient water for complete coverage, but not less than 10 gallons per acre. Do not apply more than 3 times per season, within 7 days of harvest.
	Lygus bugs	4	
Sorghum	Greenbugs Corn leaf aphids Yellow sugarcane aphid	1 to 2	Apply specified dosage per acre, by air or ground equipment, in sufficient water for complete coverage, but not less than one gallon per acre. Use the higher dosage for control of mites. Do not apply more than 3 times per season, nor within 45 days of harvest of grain, or 21 days of grazing or cutting for forage. Do not use on sweet sorghum. **CAUTION:** Since certain sorghum varieties may be sensitive to organophosphate insecticides, METASYSTOX-R Spray Concentrate should be applied to a small area of the field, and observed for a few days, to determine if any crop injury will occur. **NOTE:** Due to greenbug resistance in certain areas METASYSTOX-R used alone may not provide satisfactory control. Consult your local agricultural advisor or Extension Service for recommendations.
	Banks grass mites (Southwest only, including Kansas, Oklahoma, Texas, Colorado, New Mexico and Arizona)		
Sugar beets	Aphids Leafhoppers Mites	1½ to 3	Apply specified dosage per acre, in sufficient water for complete coverage, but not less than one gallon per acre. Do not apply more than 6 times per season. Do not harvest beets or use beet tops for feed or forage within 30 days of last treatment.
SEED FIELD CROPS Alfalfa Clover	Aphids Leafhoppers Mites Thrips	1½ to 2	Apply specified dosage per acre, in sufficient water for complete coverage, but not less than one gallon per acre. Repeat as necessary, up to 21 days before seed harvest. Chaff from seed crop may be used for feed or forage; but do not cut green crop for these purposes.
	Lygus bugs (California only)		
FRUITS Blackberries Raspberries	Aphids Mites	1 to 1½	Apply specified dosage, in 100 gallons of water, as full-coverage spray,* applying not more than 600 gallons dilute spray per acre. Do not apply more than once per season, nor within 7 days of harvest.
Grapefruit Lemons Oranges	Aphids Texas citrus mite	1 to 1½	Apply specified dosage, in 100 gallons of water, as a full-coverage spray.* Do not apply more than twice per season, nor within 7 days of harvest.
Pears	Aphids Mites	1	Apply specified dosage, in 100 gallons of water, as a full-coverage spray.* Do not apply more than once per season, nor within 30 days of harvest.
Plums Prunes	Aphids Mites	1	Apply specified dosage, in 100 gallons of water, as a full-coverage spray.* Do not apply more than twice per season, nor within 35 days of harvest.
NON-BEARING FRUITS Apples Apricots Cherries Crab Apples Grapes Nectarines Peaches Plums Prunes Quinces	Aphids Mites	1 to 1½	Apply specified dosage, in 100 gallons of water, as a full-coverage spray * on non-bearing trees. Repeat as necessary. Do not apply to trees that will bear fruit within the next 12 months.
Citrus	Aphids Texas citrus mite		
VEGETABLES Beans (Snap and lima)	Leafhoppers Mites	2	Apply specified dosage, in at least 4 gallons of water per acre. Three applications may be made per picking season, with the last application no later than 21 days before harvest. Do not graze or cut treated vines for feed or forage within 21 days of application.
	Lygus bugs (California only)		

CROP	PEST	PINTS MSR S.C.	REMARKS
Broccoli Brussels sprouts Cauliflower	Aphids	1½ to 2	Apply specified dosage per acre, in sufficient water for complete coverage, but not less than one gallon per acre. Do not apply more than 3 times per season, nor within 10 days of harvest for brussels sprouts, or 7 days of harvest for broccoli and cauliflower.
Cabbage	Aphids	1½ to 3	Apply specified dosage per acre, in sufficient water for complete coverage, but not less than one gallon per acre. Do not apply more than 3 times per season, nor within 7 days of harvest.
Cucumbers	Aphids Mites	1½ to 2	Apply specified dosage per acre, in sufficient water for complete coverage, but not less than one gallon per acre. Do not apply more than twice per season.
Eggplant	Aphids Mites	2	Apply specified dosage per acre, in sufficient water for complete coverage, but not less than one gallon per acre. Do not apply more than 3 times per season, nor within 7 days of harvest.
Head lettuce	Aphids Mites	1½ to 2	Apply specified dosage per acre, in sufficient water for complete coverage, with a minimum of one gallon per acre. A maximum of 3 applications may be made per season. The following preharvest intervals based on number of applications applies only to California's spring and summer crops: 1 application .... 14 days 2 applications .... 21 days 3 applications .... 28 days All crops in Arizona, and California's fall and winter crops require a 28 day PHI (regardless of the number of applications.) All other states require a 21 day PHI.
Spanish onions (bulb) (West of the Mississippi River only)	Thrips (Population suppression only)	3	Apply specified dosage per acre, in sufficient water for complete coverage, but not less than ten gallons per acre. For best results, make 3 applications at 10 to 14 day intervals, with the last application no later than 30 days before harvest.
Peas (North Central)	Aphids	1½	Apply specified dosage in at least 4 gallons of water per acre. Up to 3 applications may be made per season, with the last application no later than 21 days before harvest. Do not graze or cut treated vines for feed or forage within 21 days of application.
Peppers	Aphids	2	Apply specified dosage per acre, in sufficient water for complete coverage, but not less than one gallon per acre. Do not apply more than twice per season.
Potatoes	Aphids Flea beetles Leafhoppers	1½ to 2	Apply specified dosage per acre, in sufficient water for complete coverage, but not less than one gallon per acre. Repeat as necessary. Do not apply within 7 days of harvest.
Pumpkin Squash (Winter and Summer)	Aphids	1½ to 2	Apply specified dosage per acre, in sufficient water for complete coverage, but not less than one gallon per acre. Do not apply more than once per season. Do not apply within 14 days of harvest of pumpkins or Winter squash, nor within 1 day of harvest of Summer squash.
Turnips	Aphids Mites	1½ to 2	Apply specified dosage per acre, in sufficient water for complete coverage, but not less than one gallon per acre. Do not apply more than twice per season. Do not apply within 21 days of harvest if tops are to be used for food, nor within 7 days of harvest if tops are not to be utilized.
MELONS Muskmelon (Cantaloupe) Other melons	Aphids Mites	1½ to 2	Apply specified dosage per acre, in sufficient water for complete coverage, but not less than one gallon per acre. Do not apply more than 3 times per season nor within 14 days of harvest.
Watermelons	Aphids Mites	1½ to 2	Apply specified dosage per acre, in sufficient water for complete coverage, but not less than one gallon per acre. Do not apply more than twice per season, nor within 7 days of harvest.
NUTS Filberts (Oregon and Washington only)	Aphids	1 pint per 10 trees (average 4½ in. diameter)	Apply either undiluted or diluted with an equal volume of water. For undiluted application, apply ⅓ fluid ounce (2 teaspoonfuls) of product per inch of trunk diameter. For diluted application apply ⅔ fluid ounce of mixture per inch of trunk diameter. One pint of product is sufficient to treat 10 trees averaging 4½ inches in diameter, when applied diluted or undiluted. Apply specified dosage with a paint brush or low pressure sprayer in a band completely around the trunk. On trees with extremely short trunks, it may be necessary to treat bases of scaffold limbs in order to apply the correct dosage. Width of the treated band will vary with trunk diameter. Do not apply more than once per season, nor within 105 days of harvest. Do not use on heavily stressed trees, or on young trees with trunk diameters less than 2 inches.
Walnuts	Aphids Mites	1½	Apply specified dosage per 100 gallons of water as a full-coverage spray.* Do not apply more than once per season, nor within 30 days of harvest.

* This METASYSTOX-R dosage is calculated for conventional sprayers. When lower volumes of spray are applied per acre, as with low pressure, low volume, airplane or mist-type equipment, increase the concentration of METASYSTOX-R in the spray mixture in order to apply amount of METASYSTOX-R per acre equivalent to a full-coverage spray, but in no case use less than one gallon of water per acre.

SPECIAL LOCAL NEED REGISTRATIONS [SECTION 24(c)] FOR METASYSTOX-R SPRAY CONCENTRATE*

CROP	PEST	STATE EPA SLN NO.	DOSAGE	REMARKS
Alfalfa (seed crop)	Aphids Leafhoppers Mites Thrips Lygus bugs	Idaho ID 78-0001 Oregon OR 77-0067 Nevada NV 77-0014 Utah UT 77-0012 Washington WA 77-0058	$1\frac{1}{2}$ to 2 pints per acre.	Application by ground or air. Lygus bugs is the basis for the SLN registration
Bluegrass (Commercial seed crop)	Thrips	Idaho ID 77-0017 Oregon OR 80-0070 Washington WA 77-0018	4 pints per acre.	Repeat as necessary.
Corn (Field)	Mites	Kansas KS 80-0010 Texas TX 80-0024	$1\frac{1}{2}$ to 2 pints per acre.	Maximum 3 applications, with a 14 day pre-harvest interval.
	Southwestern corn borer Mites	New Mexico NM 79-0025	$1\frac{1}{2}$ to 2 pints MSR Spray Concentrate tank-mixed with 1 to 2 pints of FURADAN® 4F per acre.	Apply by air in a minimum of 3 gallons total volume when both pests are present at the same time.
Douglas fir seed & breeding orchards	Cone and seed insects	Oregon OR 80-0071 Washington WA 80-0069	$3\frac{1}{8}$ to $4\frac{1}{8}$ gallons per acre	Application by ground.
Filberts	Filbert aphid	Oregon OR 80-0079 Washington WA 80-0077	2 pints per acre.	Foliar spray.
Melons	Leafhopper	California CA 82-0104	2 pints per acre.	Application by air at vining stage.
Non-bearing peaches and apricots	Green peach aphid	Idaho ID 77-0014 Oregon OR 79-0014 Washington WA 79-0013	$\frac{1}{2}$ to 1 fluid ounce per inch of trunk diameter.	Soil drench application.
Peppermint	Aphids Mites	Idaho ID 77-0006 Oregon OR 77-0019 Washington WA 77-0014	3 pints per acre.	Application by ground or air.
Peppers	Leafhopper	California CA 82-0103	2 pints per acre.	Application by air after crop is established. Do not apply more than twice per season.
Spearmint	Aphids Mites	Idaho ID 77-0006 Oregon OR 77-0019 Washington WA 77-0014	3 pints per acre.	Application by ground or air.
Strawberries (Only pre-bloom & post-final harvest)	Strawberry aphid	Oregon OR 79-0067 Washington WA-85-0003	2 pints	Apply by air or ground. Do not apply to fruit. Three applications per year, one application pre-bloom and two post-final harvest.

* For distribution and use only within the state specified. These Special Local Need registrations approved as of date of this publication are, however, subject to cancellation by the state. Contact your nearest Mobay Agricultural Chemicals Division Office for the current status and a copy of the complete individual supplemental Special Local Need labeling. This labeling must be in the possession of the user at time of pesticide application.

MONITOR® 4

LIQUID INSECTICIDE

RESTRICTED USE PESTICIDE

For retail sale to and use only by Certified Applicators or persons under their direct supervision, and only for those uses covered by the Certified Applicator's certification.

For Effective Insect Control on Cole Crops, Cotton, Potatoes, and Sugar Beets.

ACTIVE INGREDIENT:

O,S-Dimethylphosphoramidothioate*	40%
INERT INGREDIENTS	60%
TOTAL	100%

Contains 4 lbs. O,S-Dimethylphosphoramidothioate per gallon.

*U.S. Patent Nos. 3,309,266; 3,689,604; 6,676,555

EPA Reg. No. 3125-280-AA

MONITOR is a Reg. TM of Chevron Chemical Company.

DANGER–POISON

Keep Out of Reach of Children

PELIGRO

PRECAUCION AL USUARIO: Si usted no puede leer o entender inglés, no use este producto hasta que la etiqueta le haya sido explicada ampliamente.

(TO THE USER: If you cannot read or understand English, do not use this product until the label has been fully explained to you.)

RECOMMENDED USES

Cotton	Cauliflower
Broccoli	Celery (Florida only)
Brussels sprouts	Potatoes
Cabbage	Sugar beets (Arizona and California only)

RESTRICTIONS ON WAREHOUSE STORAGE

Do not store in front of space heaters. Store below 120°F. Do not store adjacent to or near foodstuffs, drugs, feed or any other material intended for consumption by humans or animals. Do not store adjacent to toys, cosmetics, dishes, furniture or clothing. Do not store next to herbicides.

STATEMENTS OF PRACTICAL TREATMENT

If swallowed, vomiting should be induced. Administer water freely and induce vomiting by giving one dose (½ oz. or 15 ml.) of syrup of ipecac. If vomiting does not occur within 10 to 20 minutes, administer second dose. If syrup of ipecac is not available, induce vomiting by sticking finger down throat. Repeat until vomit fluid is clear. Never give anything by mouth to an unconscious person. Have patient lie down and keep quiet. Professional medical assistance should be secured immediately.

If inhaled, remove victim to fresh air. If not breathing give artifical respiration, preferably mouth to mouth. Get medical assistance immediately.

If on skin, In case of contact, remove contaminated clothing and immediately wash with soap and water.

If in eyes, wash with plenty of water for at least 15 minutes. Get medical attention.

PRECAUTIONARY STATEMENTS

HAZARDS TO HUMANS AND DOMESTIC ANIMALS

Keep out of reach of children. Do not inhale. Do not get on skin. Do not take internally.

Poisonous if swallowed, inhaled or absorbed through skin. Rapidly absorbed through skin. Do not get in eyes, on skin or on clothing. Wear a mask or respirator jointly approved by the Mining Enforcement and Safety Administration (formerly the U.S. Bureau of Mines) and by the National Institute for Occupational Safety and Health under the provisions of 30 CFR Part 11. Wear natural rubber gloves, protective clothing and goggles. In case of contact, wash immediately with soap and water. Wash hands, arms and face thoroughly with soap and water before eating or smoking. Keep all unprotected persons out of the operating areas or vicinity where there may be danger of drift. Do not store or transport with feed or food.

ANTIDOTE: Atropine is antidotal. 2-PAM is also antidotal, and may be administered in conjunction with atropine.

Handling of Contaminated Articles: Clothing and rags receiving undiluted MONITOR 4 as a result of leaks, spills or mishaps should be removed immediately and not reused. These contaminated articles should be disposed of in the same manner as the empty container; see container disposal section for directions. (Contaminated articles which should be discarded include: hats, gloves, aprons, coats, boots, etc. not made of rubber, or coated with rubber or other similar materials). Clothing receiving spray mist or droplets from the mixed, diluted MONITOR 4 should be removed and decontaminated before reuse. This clothing can be decontaminated by machine washing, separately from other items, with soap or detergent and bleach in hot water twice before reuse. Protective clothing and equipment should be washed down with detergent or soap and bleach water. Wash water from the cleaning of protective clothing or equipment should not be allowed to run off or otherwise enter water supplies.

ENVIRONMENTAL HAZARDS

This product is extremely toxic to birds and other wildlife. Birds and other wildlife in treated areas may be killed. Do not apply directly to water bodies or wetlands (eg., lakes, streams, ponds, canals). Do not contaminate water by cleaning of equipment or disposal of wastes. This product is highly toxic to bees exposed to direct treatment or residues on blooming crops or weeds. Do not apply this product or allow it to drift to blooming crops or weeds if bees are visiting the treatment area.

RESTRICTIONS

Do not apply under conditions involving possible drift to food, forage or other planting that might be damaged or the crops thereof rendered unfit for sale, use or consumption. Not for use or storage in or around home.

RE-ENTRY STATEMENTS

Do not apply this product in such a manner as to directly or through drift expose workers or other persons. The area being treated must be vacated by unprotected persons. Do not enter treated areas without protective clothing until sprays have dried. Because certain states may require more restrictive re-entry intervals for various crops treated with this product, consult your State Department of Agriculture for further information.

Written or oral warnings must be given to workers who are expected to be in a treated area or in an area about to be treated with this product. Oral warnings should include those statements found under the label heading "Precautionary Statements" and "Re-entry Statements" included on this label. When oral warnings are given, warnings shall be given in a language customarily understood by workers. Oral warnings must be given if there is reason to believe that written warnings cannot be understood by workers. Written warnings must include the following information: DANGER, area treated with MONITOR 4 on (date of application). Do not enter without appropriate protective clothing until sprays have dried. **If on skin,** remove contaminated clothing and wash skin immediately with soap and warm water. **If eyes are contaminated,** wash with flowing water for at least 15 minutes. **If swallowed,** vomiting should be induced. (See Precautionary Statements on label).

STORAGE AND DISPOSAL

PESTICIDE DISPOSAL

Do not contaminate water, food or feed by storage or disposal.

Pesticide wastes are acutely hazardous. Improper disposal of excess pesticide, spray mixture, or rinsate is a violation of Federal Law. If these wastes cannot be disposed of by use according to label instructions, contact your State Pesticide or Environmental Control Agency, or the Hazardous Waste representative at the nearest EPA Regional Office for guidance.

CONTAINER DISPOSAL

Triple rinse (or equivalent). Then offer for recycling or reconditioning, or puncture and dispose of in a sanitary landfill, or by other procedures approved by state and local authorities.

Store in a cool, dry place and in such a manner as to prevent cross contamination with other pesticides, fertilizers, food, and feed. Store in original container and out of the reach of children, preferably in a locked storage area.

Handle and open container in a manner as to prevent spillage. If container is leaking, invert to prevent leakage. If the container is leaking or material spilled for any reason or cause, carefully dam up spilled material to prevent runoff. Refer to Precautionary Statements on label for hazards associated with the handling of this material. Do not walk through spilled material. Absorb spilled material with absorbing type compounds and dispose of as directed for pesticides above. In spill or leak incidents, keep unauthorized people away. You may contact the Mobay Emergency Response Team for decontamination procedures or any other assistance that may be necessary. The Mobay Kansas City Emergency Response Telephone No. is 816-242-2582, or contact Chemtrec at 800-424-9300.

READ THE LABEL BEFORE USE

DIRECTIONS FOR USE

It is a violation of Federal law to use this product in a manner inconsistent with its labeling.

Use the designated amount of product in the following gallons of water per acre, unless otherwise noted:

Ground Application: 25 to 125 gallons of water
Air Application: 3 to 10 gallons of water

RECOMMENDED APPLICATIONS

CROP	INSECT	PINTS MONITOR 4	REMARKS
FIELD			
Cotton	Aphids Thrips (Early-season)	$\frac{1}{5}$ to $\frac{2}{5}$ (0.1 to 0.2 lbs. A1/A)	Apply specified dosage per acre as needed up to 50 days before harvest and before bolls open. For whiteflies, fleahoppers and low rates for aphids and thrips, a second application may be needed for heavy infestations. Do not feed gin trash to livestock or allow animals to graze on treated areas.
	Fleahoppers	$\frac{2}{5}$ (0.2 lb. A1/A)	
	Whiteflies	$\frac{1}{2}$ to 1	
	Beet armyworms Cabbage looper Lygus Mites	1 to 2	
	Aphids Thrips (Mid-to late-season)		
	Stink bugs (Arizona and California only)		
VEGETABLES Broccoli Brussels sprouts Cabbage Cauliflower	Cabbage looper Imported cabbageworm Diamondback moth larvae Aphids	1 to 2	**In a Preventive Program:** Beginning when the plants are small, use the 1 pint rate (0.5 pound active) per acre, and apply at 7-day intervals. **For Clean-up of Existing Populations:** Use $1\frac{1}{2}$ to 2 pints (0.75 to 1.0 pound active) per acre. Do not apply to Brussels sprouts later than 14 days before the first harvest, or to cauliflower later than 28 days before the first harvest, or to cabbage later than 35 days before first harvest. For broccoli, the dosage rate of 1 pint per acre may be made up to 14 days before first harvest, but do not apply rates above 1 pint per acre within 21 days of first harvest.
Celery (Florida only)	Dipterous leafminers	1 to 2 (0.5 to 1.0 lb. active)	Apply specified dosage per acre with ground equipment only. Up to 10 applications may be made per season at 7 to 10 day intervals. Do not apply within 21 days of harvest. All celery must be trimmed in field and no trimming may be fed to livestock. Treated celery must not be dried for celery flakes.
Potatoes	Aphids Potato tubeworm Cabbage looper Colorado potato beetle (Except Long Island, New York) Flea beetles Potato leafhopper Cutworms (Foliage feeding) (Except California) Armyworms (Except California) European corn borer (Except California) Lygus bug (Except California)	$1\frac{1}{2}$ to 2	Apply specified dosage (0.75 to 1.0 pound active) per acre. Apply in a 7- to 10-day preventive program, or as necessary. Do not apply later than 14 days before harvest. Resistance to Colorado potato beetle has occurred in other areas of New York. Consult your local extension service for details.
Sugar beets (Arizona and California only)	Green peach aphid	1 to 2	Apply specified dosage per acre as a full coverage spray as needed, allowing 14 days between applications. Do not make more than 6 applications per crop season and not within 30 days of harvest. Sprays may be directed to obtain full foliar coverage.
	Beet armyworm	2	

SPRINKLER IRRIGATION APPLICATION TO POTATOES: For control of green peach aphid on potatoes in the states of Oregon, Washington, Idaho, and Utah apply MONITOR 4 at the rate of 1½ to 2 pints per acre through sprinkler irrigation systems.

Solid Set Systems: Apply specified dosage for the entire length of the irrigation period **or** for a 30 to 60 minute period at the end of a regular irrigation set **or** as a 30 to 60 minute injection as a separate application not associated with a regular irrigation. Allow time for all lines to flush the pesticide through all nozzles before turning off irrigation water. To ensure the lines are flushed and free of remaining pesticide, a dye indicator may be injected into the lines to mark the end of the application period. See **NOTE.**

Solid Set Calibration Sample Calculations:

(1) Determine the area to be treated per irrigation set. If one set irrigates four ¼ mile long lines on a 50′ spacing, 4 × 1,320′ × 50 feet = 264,000 square feet or approximately 6 acres.
(2) If 1½ pints of MONITOR 4 are to be applied per acre, 6 × 1½ or 9 pints of MONITOR 4 are needed.
(3) Dilute with sufficient water so injection time will require approximately 30 minutes or more per set. Under no circumstances should injection be less than 10 minutes.
(4) Repeat applications until the entire field is treated.

Center Pivot Systems: Inject the specified dosage per acre continuously for one complete revolution of the system. See **NOTE.**

Center Pivot Calibrations Sample Calculations:

(1) To apply MONITOR 4 at the rate of 1.5 pints per acre to 130 acres you still need 130 × 1.5 pints or 195 pints or (24⅜ gallons) of product.
(2) If your injection pump delivers 15 gallons per hour and your system takes 20 hours for 1 revolution, you will need to mix enough water with the MONITOR 4 to obtain 300 gallons of emulsion.
(3) Mix the required amounts of MONITOR 4 and water in a nurse tank equipped with mechanical agitation, start the irrigation system and begin the injection.
(4) Check frequently to insure all systems are functioning properly.

NOTE: Apply the product only through systems containing anti-siphon and check valve which will prevent water source contamination and overflow of the slurry tank, and containing interlocking controls between the metering device and the water pump to insure simultaneous shut-off. Constant agitation must be maintained in the chemical supply tank during the entire period of insecticide application. Inject the product with a positive displacement pump into the main line ahead of a right angle turn to insure adequate mixing.

Application of more than label recommended quantities of irrigation water per acre may result in decreased product performance by removing the chemical from the zone of effectiveness. Do not apply when wind speed favors drift, when system connections or fittings leak, when nozzles do not provide uniform distribution or when lines containing the product must be dismantled or drained.

Greater accuracy in calibration and distribution will be achieved by injecting a larger volume of a more dilute slurry per hour. Where sprinkler distribution patterns do not overlap sufficiently, unacceptable insect control may result. Where sprinkler distribution patterns overlap excessively, crop injury may result. 1-9-85

SPECIAL LOCAL NEED REGISTRATIONS [SECTION 24(c)] FOR MONITOR 4*

CROP	INSECT	STATE EPA SLN NO.	DOSAGE	REMARKS
Alfalfa (Seed crop)	Lygus Stinkbug	Arizona AZ 82-0013	1½ to 2 pints per acre.	Application by air.
Beets (Table—grown for seed use only)	Green peach aphid	Washington WA 79-0057	2 pints per acre.	Application during the early bloom to maturation period.
Bermudagrass (Seed crop)	Fulgorids Banks grass mite	Arizona AZ 82-0014	1 quart per acre.	Application by air or ground.
Cotton	Aphids Beet armyworm Cabbage looper Mites Thrips Lygus	California CA 79-0188	1 to 2 pints per acre.	Do not apply after 65% of the bolls are open.
	Cabbage Looper	Arizona AZ 99-0037	1 to 2 pints per acre.	
	Whiteflies	Arizona AZ 79-0037 Arkansas AR 81-0044 Mississippi MS 81-0055 Tennessee TN 81-0028	½ to 1 pint per acre.	
	Thrips (early-season)	Arkansas AR 81-0044	⅕ to ⅖ pint per acre.	
	Fleahoppers		⅖ per acre.	
	Aphids (early-season) Thrips (early-season)	Mississippi MS 81-0055	⅕ to ⅖ pint per acre.	
	Fleahoppers	Tennessee TN 81-0028	⅖ pint per acre.	
	Beet armyworms Cabbage looper Mites Aphids Thrips (mid-to late-season)	Arkansas AR 81-0044 Mississippi MS 81-0055 Tennessee TN 81-0028	1 to 2 pints per acre.	
Cucumbers	Aphids Pickleworms Melonworms Leafminer	Florida FL 81-0009	1 to 2 pints per acre.	Maximum 12 pints per acre per crop season. Apply up to 7 days of harvest.

CROP	INSECT	STATE EPA SLN NO.	DOSAGE	REMARKS
Eggplant	Spidermites Aphids Leafminers Lepidopterous larvae	Florida FL 81-0033	1 to 2 pints per acre.	Maximum 14 pints per acre per crop season. Apply up to 7 days of harvest.
Head lettuce	Aphids Cabbage looper Vegetable leafminer	Florida FL 81-0012	1 to 2 pints per acre.	Apply when insects first appear.
Head lettuce (crisphead only)	Tobacco budworm (suppression)	Arizona AZ 81-0005	1 to 2 pints per acre.	Application by ground or air.
	Aphids Armyworms Cabbage looper Leafminers Thrips	California CA 80-0186		Ground application.
Melons	Leafminer Pickleworm Melonworm Rindworm Complex (cabbage looper, tobacco budworm, granulate cutworm)	Florida FL 81-0034	1 to 2 pints per acre.	Apply at 7 day intervals. Do not apply more than 12 pints per acre per crop season.
	Melon aphid Leaf miner Pickleworm Melonworm Rindworm Complex (cabbage looper, corn earworm, tobacco budworm, granulate cutworm) Green stink bug Potato leafhopper Spotted cucumber beetle Two-spotted mite	Texas TX 84-0020	1 to 2 pints per acre.	Apply specified dosage (0, .5 to 1.0 pound active) in 20-100 gallons of water per acre at 7 day intervals. Do not apply more than 5 times per season. Applications may be made up to 14 days of harvest for watermelon, cantaloupe and honeydew melons.
Peppers	Green peach aphids Leafminers Thrips Flea beetles	New Mexico NM 82-0008 Texas TX 82-0019	1 pint per acre.	Application by air or ground up to 7 days before harvest.
Seed alfalfa Seed clover	Aphids Lygus Thrips Armyworm Spider mite suppression	California CA 83-0064	2 pints per acre.	Apply as needed by ground or air.
Seed carrots	Lygus	Idaho ID 76-0005 Oregon OR 80-0068 Washington WA 76-0026	2 pints per acre.	Apply prior to introduction of pollinating insects into the seed field.
Sugar beets & table beets (grown for seed only)	Green peach aphid	Oregon OR 81-0069	2 pints per acre.	Ground or air application.
Tomatoes	Aphids Beet armyworm[1] Black cutworm[1] Cabbage looper Colorado potato beetle Fruitworm Hornworm Leafminers Variegated cutworm [1]Not for Indiana SLN	California CA 78-0163 Indiana IN 79-0001 Michigan MI 78-0016 Ohio OH 79-0008	$1\frac{1}{2}$ to 2 pints per acre.	Applications may be made up to 7 days of harvest.

CROP	INSECT	STATE EPA SLN NO.	DOSAGE	REMARKS
Tomatoes (fresh fruit only)	Leafminers (adults)	South Carolina SC 78-0007	$1\frac{1}{2}$ to 2 pints per acre.	Ground application up to 14 days of harvest.
	Aphids Fruitworms Leafminers Tomato pinworm	Florida FL 80-0046		Application by ground or air, up to 7 days of harvest.

* For distribution and use only within the state specified. These Special Local Need registrations approved as of date of this publication are, however, subject to cancellation by the state. Contact your nearest Mobay Agricultural Chemicals Division Office for the current status and a copy of the complete individual supplemental Special Local Need labeling. This labeling must be in the possession of the user at time of pesticide application.

®MORESTAN

25% WETTABLE POWDER MITICIDE—FUNGICIDE—INSECTICIDE

For Control of Mites, Powdery Mildew, Pear Psylla and Aphids.

ACTIVE INGREDIENT:

6-Methyl-1,3-dithiolo [4,5-*b*] quinoxalin-2-one	25%
INERT INGREDIENTS	75%
TOTAL	100%

EPA Reg. No. 3125-117-AA

MORESTAN is a Reg. TM of the Parent Company of Farbenfabriken Bayer GmbH, Leverkusen.

CAUTION

Keep Out of Reach of Children

AVISO

PRECAUCION AL USUARIO: Si usted no puede leer o entender inglés, no use este producto hasta que la etiqueta le haya sido explicada ampliamente.

(TO THE USER: If you cannot read or understand English, do not use this product until the label has been fully explained to you.)

RECOMMENDED USES

Bearing Fruit

Apples	Pears
Apricots	Citrus (CA, AZ, TX, FL only)

Non-Bearing Fruit

Apples	Peaches
Apricots	Pears
Cherries	Plums
Citrus	Prunes
Nectarines	Strawberries

Other Crops

Macadamia nuts	Flowers
Walnuts	Shrubs

RESTRICTIONS ON WAREHOUSE STORAGE

Protect from rain and moisture. Do not store next to herbicides.

STATEMENTS OF PRACTICAL TREATMENT

If swallowed, administer water freely and induce vomiting by giving one dose ($\frac{1}{2}$ oz. or 15 ml.) of syrup of ipecac. If vomiting does not occur within 10 to 20 minutes, administer second dose. If syrup of ipecac is not available, induce vomiting by sticking finger down throat. Repeat until vomit fluid is clear. Never give anything by mouth to an unconscious person. Get medical attention immediately.

If on skin, wash skin immediately with soap and warm water. If irritation occurs, get medical attention.

If eyes are contaminated, wash with plenty of water for at least 15 minutes. Get medical attention.

If inhaled, remove to fresh air. If not breathing, give artificial respiration, preferably mouth to mouth. Get medical attention.

TO PHYSICIAN: No specific antidote is available. Treat symptomatically.

PRECAUTIONARY STATEMENTS

HAZARDS TO HUMANS AND DOMESTIC ANIMALS

CAUTION

Harmful if swallowed, inhaled, or absorbed through the skin. Avoid breathing of spray mist. Wash thoroughly with soap and warm water after handling.

Under conditions of bright sunlight, particularly if wind is present, some persons spraying MORESTAN experience a skin reaction on exposed parts of the body similar to sunburn. When applying sprays containing more than $\frac{1}{2}$ lb. MORESTAN 25% Wettable Powder per 100 gal. of water it is recommended that sufficient clothing be worn to cover most body areas and that a suitable sun-screen lotion such as Uval (Dome Laboratories) or Solbar (Person and Covey), be applied to body areas necessarily left uncovered such as the neck and face.

Do not contaminate feed or food. Keep out of reach of children.

ENVIRONMENTAL HAZARDS

MORESTAN is toxic to fish. Do not apply where runoff is likely to occur. In cleaning of equipment or disposal of wastes, do not contaminate streams, lakes, or ponds with this material.

RESTRICTIONS

Under certain adverse weather conditions, particularly those that create slow drying of spray deposits, injury may occur to foliage of strawberries and deciduous fruit trees.

Do not use summer oils if MORESTAN is used in the cover spray program.

Do not use spreader-sticker materials with MORESTAN.

Do not allow sprays to drift onto other crops, including pastures. Do not graze treated areas.

Do not use on other crops used for food and forage. Use only according to label directions. Applications at rates above those shown may result in illegal crop residues. Do not treat food crops grown in the greenhouse.

Phytotoxicity has occurred on the foliage and flowers of several varieties of roses, especially during periods of high temperatures. Treat roses only after cutting, and before new growth appears. Before treating large plantings for the first time, treat a few plants and observe for 4 to 5 days. Under certain conditions, some injury to other ornamentals may occur.

RE-ENTRY STATEMENTS

Do not apply this product in such a manner as to directly or through drift expose workers or other persons. The area being treated must be vacated by unprotected persons. Do not enter treated areas without protective clothing until sprays have dried. Because certain states may require more restrictive re-entry intervals for various crops treated with this product, consult your State Department of Agriculture for further information.

Written or oral warnings must be given to workers who are expected to be in a treated area or in an area about to be treated with this product. Oral warnings should include those statements found under the label heading "Precautionary Statements" and "Re-entry Statements" included on this label. When oral warnings are given, warnings shall be given in a language customarily understood by workers. Oral warnings must be given if there is reason to believe that written warnings cannot be understood by workers. Written warnings must include the following information: WARNING, area treated with MORESTAN 25% Wettable Powder on (date of application). Do not enter without appropriate protective clothing until sprays have dried. **If on skin,** remove contaminated clothing and wash skin immediately with soap and warm water. **If eyes are contaminated,** wash with flowing water for at least 15 minutes. **If swallowed.** vomiting should be induced. (See Precautionary Statements on label).

STORAGE AND DISPOSAL

PESTICIDE DISPOSAL

Do not contaminate water, food or feed by storage or disposal.

Wastes resulting from the use of this product may be disposed of on site or at an approved waste disposal facility.

CONTAINER DISPOSAL

Do not use container in connection with food, feed, or drinking water. Completely empty bag into application equipment. Then dispose of empty bag in a sanitary landfill or by incineration, or, if allowed by State and local authorities, by burning. If burned, stay out of smoke.

Store in a cool, dry place and in such a manner as to prevent cross contamination with other pesticides, fertilizers, food, and feed. Store in original container and out of the reach of children, preferably in a locked storage area.

Handle and open container in a manner as to prevent spillage. If the container is leaking or material spilled for any reason or cause, carefully sweep material into a pile. Refer to Precautionary Statements on label for hazards associated with the handling of this material. Do not walk through spilled material. Dispose of pesticide as directed above. In spill or leak incidents, keep unauthorized people away. You may contact the Mobay Emergency Response Team for decontamination procedures or any other assistance that may be necessary. The Mobay Kansas City Emergency Response Telephone No. is 816-242-2582, or contact Chemtrec at 800-424-9300.

IMPORTANT: This bulletin is not intended to provide adequate information for application. Before using the product, read and carefully observe directions, cautionary statements and other information appearing on the product label. This product is sold subject to the Conditions of Sale set forth on the container label.

READ THE LABEL BEFORE USE

DIRECTIONS FOR USE

It is a violation of Federal Law to use this product in a manner inconsistent with its labeling.

MORESTAN 25% Wettable Powder is recommended for residual control of aphids, mites, mite eggs and powdery mildew. The formulation mixes readily with water to form a suspension, which is usable in all conventional hand- or machine-operated sprayers. For best results, apply a thorough coverage before mites or mite eggs are present in large numbers, and before mites cause bronzing of the foliage. Apply during good drying conditions (see "RESTRICTIONS"). Consult your State Agricultural Extension Service or Experiment Station for detailed information on timing of applications for most effective control.

RECOMMENDED APPLICATIONS

CROP	PEST	Pounds MORESTAN 25% W.P.	REMARKS
BEARING FRUIT Pre-Bloom and Post-Harvest	Powdery mildew (see NOTE in Remarks	½ to 1	Apply specified dosage per 100 gallons of water in pre-bloom sprays (delayed-dormant, pre-pink or pink). Do not apply after first bloom. Additional applications may be made to tree foliage after harvest.
Apples Pears	Mites Mite eggs		**NOTE:** Pre-bloom sprays with MORESTAN 25% W.P. will provide effective control of early-season powdery mildew infections. No additional fungicides are needed for powdery mildew control during the pre-bloom period. Use other recommended fungicides in post-bloom spray programs for seasonal control.
	Pear psylla	1½	
Apricots (Post-harvest only)	Mites (Except cyclamen mite) Mite eggs	4	Apply specified dosage per acre in a minimum of 10 gallons of water per acre by air or 40 gallons of water per acre by ground. For best control thorough coverage of all surfaces is necessary. Make only one application between harvest and up to the time trees break dormancy. Do not graze cover crops.
BEARING CITRUS TREES Citrus (California, Arizona, Texas and Florida only)	Citrus red mite Citrus rust mite Texas citrus mite Mite eggs	5	Apply specified dosage per acre by concentrate or dilute ground spray using 20 to 500 gallons of water per acre. Schedule applications no closer than 90 days apart in Texas and Florida, or 180 days apart in California or Arizona. Allow at least 60 days between last application and harvest.
NON-BEARING FRUITS Apples	Powdery mildew Mites Mite eggs	½ to 1	Apply specified dosage per 100 gallons of water.* Spray all foliage surfaces, including the underside of leaves, for complete coverage. Repeat as necessary. For control of powdery mildew, begin preventive applications before disease appears, and apply at 7- to 14-day intervals. During heavy infection periods, make more frequent applications. Do not apply to fruit trees that will bear fruit within one year of applicaton. Apply to strawberries, as required, in propagating beds, and in the field during the year of setting new plants, but not after first bloom appears in the following year.
Apricots, Pears, Cherries, Plums, Nectarines, Prunes, Peaches, Strawberries	Mites (Except cyclamen mite) Mite eggs	½ to 1	
Citrus	Citrus red mite Citrus rust mite Texas citrus mite Mite eggs	1 to 1½	Apply specified dosage per 100 gallons of water. Spray all surfaces, including the underside of leaves, for complete coverage. Repeat as necessary. Do not apply to trees that will bear fruit within one year of application.

CROP	PEST	Pounds MORESTAN 25% W.P.	REMARKS
NUTS Macadamia nuts	Mites	1 to 2	Apply specified dosage per 100 gallons of water as a full-coverage spray*, using up to 400 gallons of finished spray per acre, depending upon the number and size of trees. Spray all surfaces, including the underside of leaves, for complete coverage. Three applications per season may be made, up to the day of harvest
Walnuts	Aphids Mites Mite eggs	1 to 1½	Apply specified dosage per 100 gallons of water as full-coverage spray*, using up to 250 gallons of finished spray per acre, depending upon the number and size of trees. Spray all foliage surfaces, including the underside of leaves, for complete coverage. Do not apply more than twice per season, or within 30 days of harvest.
ORNAMENTALS Flowers Shrubs	Mites Mite eggs	½ to 1	Apply specified dosage per 100 gallons of water (½ pounds per 100 gallons = 1 level teaspoonful per 1 gallon). Spray all foliage surfaces, including the underside of leaves, for complete coverage. Repeat as necessary. 3–12–84

* This MORESTAN dosage is calculated for conventional sprayers. When lower volumes of spray are applied per acre, as with low pressure, low volume, airplane or mist-type equipment, increase the concentration of MORESTAN in the spray mixture in order to apply amount of MORESTAN per acre equivalent to a full-coverage spray.

SPECIAL LOCAL NEED REGISTRATIONS [Section 24(c)] FOR MORESTAN 25% WETTABLE POWDER*

CROP	INSECT	STATE EPA SLN NO.	DOSAGE	REMARKS
Bearing apples and pears (Pre-bloom and Post-harvest)	Mites Mite eggs Powdery mildew*	Oregon OR 76–0017 Washington WA 85-0009	½ to 1 pound per 100 gallons of water	Permits low volume full-coverage spray. *Washington only.
	Pear psylla		1½ pounds per 100 gallons of water.	

* For distribution and use only within the state specified. These Special Local Need registrations approved as of date of this publication are, however, subject to cancellation by the state. Contact your nearest Mobay Agricultural Chemicals Division Office for the current status and a copy of the complete individual supplemental Special Local Need labeling. This labeling must be in the possession of the user at time of pesticide application.

®MORESTAN

25% WETTABLE POWDER
MITICIDE–FUNGICIDE
INSECTICIDE
IN WATER SOLUBLE PACKETS

For Control of Mites, Powdery Mildew, Pear Psylla and Aphids.

ACTIVE INGREDIENT:	
6-Methyl-1,3-dithiolo [4,5-*b*] quinoxalin-2-one	25%
INERT INGREDIENTS	75%
TOTAL	100%

EPA Reg. No. 3125-302-AA

MORESTAN is a Reg. TM of the Parent Company of Farbenfabriken Bayer GmbH, Leverkusen.

CAUTION
Keep Out of Reach of Children
AVISO

PRECAUCION AL USUARIO: Si usted no puede leer o entender inglés, no use este producto hasta que la etiqueta le haya sido explicada ampliamente.

(TO THE USER: If you cannot read or understand English, do not use this product until the label has been fully explained to you.)

RECOMMENDED USES

Bearing Fruit

Apples	Pears
Apricots (Post-harvest only)	Citrus (CA, AZ, TX, FL only)

Non-Bearing Fruit

Apples	Peaches
Apricots	Pears
Cherries	Plums
Citrus	Prunes
Nectarines	Strawberries

Other Crops

Macadamia nuts	Flowers
Walnuts	Shrubs

RESTRICTIONS ON WAREHOUSE STORAGE

Protect from rain and moisture. Store in a cool dry place but not below 32°F. Do not store next to herbicides.

STATEMENTS OF PRACTICAL TREATMENT

If swallowed, administer water freely and induce vomiting by giving one dose (½ oz. or 15 ml.) of syrup of ipecac. If vomiting does not occur within 10 to 20 minutes, administer second dose. If syrup of ipecac is not available, induce vomiting by sticking finger down throat. Repeat until vomit fluid is clear. Never give anything by mouth to an unconscious person. Get medical attention immediately.

If on skin, wash skin immediately with soap and warm water. If irritation occurs, get medical attention.

If eyes are contamined, wash with plenty of water for at least 15 minutes. Get medical attention.

If inhaled, remove to fresh air. If not breathing, give artificial respiration, preferably mouth to mouth. Get medical attention.

TO PHYSICIAN: No specific antidote is available. Treat symptomatically.

PRECAUTIONARY STATEMENTS

HAZARDS TO HUMANS AND DOMESTIC ANIMALS

CAUTION

Harmful if swallowed, inhaled or absorbed through the skin. Avoid breathing of spray mist. Wash thoroughly with soap and warm water after handling.

Under conditions of bright sunlight, particularly if wind is present, some persons spraying MORESTAN experience a skin reaction on exposed parts of the body similar to sunburn. When applying sprays containing more than ½ lb. MORESTAN 25% Wettable Powder per 100 gal. of water, it is recommended that sufficient clothing be worn to cover most body areas, and that a suitable sun-screen lotion, such as Uval (Dome Laboratories) or Solbar (Person and Covey), be applied to body areas necessarily left uncovered such as the neck and face.

Do not contaminate feed and food. Keep out of reach of children.

ENVIRONMENTAL HAZARDS

MORESTAN is toxic to fish. Do not apply where runoff is likely to occur. In cleaning of equipment or disposal of wastes, do not contaminate streams, lakes, or ponds with this material.

RE-ENTRY STATEMENTS

Do not apply this product in such a manner as to directly or through drift expose workers or other persons. The area being treated must be vacated by unprotected persons. Do not enter treated areas without protective clothing until sprays have dried. Because certain states may require more restrictive re-entry intervals for various crops treated with this product, consult your State Department of Agriculture for further information.

Written or oral warnings must be given to workers who are expected to be in a treated area or in an area about to be treated with this product. Oral warnings should include those statements found under the label heading "Precautionary Statements" and "Re-entry Statements" included on this label. When oral warnings are given, warnings shall be given in a language customarily understood by workers. Oral warnings must be given if there is reason to believe that written warnings cannot be understood by workers. Written warnings must include the following information: WARNING, area treated with MORESTAN 25% Wettable Powder on (date of application). Do not enter without appropriate protective clothing until sprays have dried.

If on skin, remove contaminated clothing and wash skin immediately with soap and warm water.

If eyes are contaminated, wash with flowing water for at least 15 minutes.

If swallowed, vomiting should be induced. (See Precautionary Statements on label).

STORAGE AND DISPOSAL

PESTICIDE DISPOSAL

Do not contaminate water, food or feed by storage or disposal.

Wastes resulting from the use of this product may be disposed of on site or at an approved waste disposal facility.

CONTAINER DISPOSAL

Do not use container in connection with food, feed, or drinking water. Completely empty bag into application equipment. Then dispose of empty bag in a sanitary landfill or by incineration, or if allowed by State and local authorities, by burning.

If burned, stay out of smoke.

Store in a cool, dry place and in such a manner as to prevent cross contamination with other pesticides, fertilizers, food, and feed. Do not store below freezing (32°F). Exposure to moisture or excessive handling of water soluble packets may cause breakage. Store packets in original container and out of the reach of children, preferably in a locked storage area.

Handle and open container in a manner as to prevent spillage. If the container is leaking or material spilled for any reason or cause, carefully sweep material into a pile. Refer to Precautionary Statements on label for hazards associated with the handling of this material. Do not walk through spilled material. Dispose of pesticide as directed above. In spill or leak incidents, keep unauthorized people away. You may contact the Mobay Emergency Response Team for decontamination procedures or any other assistance that may be necessary. The Mobay Kansas City Emergency Response Telephone No. is 816-242-2582, or contact Chemtrec at 800-424-9300.

IMPORTANT

This bulletin is not intended to provide adequate information for application. Before using the product, read and carefully observe directions, cautionary statements and other information appearing on the product label. This product is sold subject to the Conditions of Sale set forth on the container label.

READ THE LABEL BEFORE USE

DIRECTIONS FOR USE

It is a violation of Federal Law to use this product in a manner inconsistent with its labeling.

MORESTAN 25% Wettable Powder miticide-fungicide-insecticide is recommended for residual control of aphids, mites, mite eggs, and powdery mildew. The formulation mixes readily with water to form a suspension which is usable in all conventional hand or machine-operated sprayers. For best results, apply a thorough coverage before mites or mite eggs are present in large numbers and before mites cause bronzing of the foliage. Apply during good drying conditions. (See Restrictons). Consult your State Agricultural Extension Service or Experiment Station for detailed information on timing of applications for most effective control.

MIXING: The enclosed packets containing MORESTAN 25% Wettable Powder are **water soluble.** Do not allow packets to become wet prior to adding to the spray tank. Do not handle with wet hands. Reseal outer bag to protect remaining packets.

To prepare the spray mixture, drop the required number of unopened packets, as determined under "Recommended Applications," into the spray tank while filling with water to the desired level. Operate the agitator while mixing. Depending on the water temperature and the degree of agitation, the packets should be completely dissolved within approximately 5 minutes from the time they were added to the water.

DOSAGE: Use specified dosage of MORESTAN 25% Wettable Powder in the designated amount of water necessary to give complete coverage of foliage. Determine the **total amount** of wettable powder to be added to the spray tank based on the rates under "Recommended Applications." **For each 1 pound (16 ounces)** of wettable powder to be added to the spray tank use one 1 pound packet. For example, if it is determined that 2 pounds (32 ounces) of MORESTAN 25% Wettable Powder should be added to the spray tank, add two 1 pound packets. The type of equipment used will determine the concentration required; however, use of these packets is not recommended for making highly concentrated mixtures such as used in aircraft spraying.

RECOMMENDED APPLICATIONS

CROP	PEST	Pounds MORESTAN 25% W.P.	REMARKS
BEARING FRUIT Pre-Bloom and Post-Harvest Apples Pears	Powdery mildew (see NOTE in Remarks) Mites Mite eggs	½ to 1	Apply specified dosage per 100 gallons of water in pre-bloom sprays (delayed-dormant, pre-pink or pink). Do not apply after first bloom. Additional applications may be made to tree foliage after harvest. NOTE: Pre-Bloom sprays with MORESTAN 25% W.P. will provide effective control of early-season powdery mildew infections. No additional fungicides are needed for powdery mildew control during the pre-bloom period. Use other recommended fungicides in post-bloom spray programs for seasonal control.
	Pear psylla	1½	
Apricots (Post-harvest only)	Mites (Except cyclamen mite) Mite eggs	4	Apply specified dosage per acre in a minimum of 10 gallons of water per acre by air or 40 gallons of water per acre by ground. For best control through coverage of all surfaces is necessary. Make only one application between harvest and up to the time trees break dormancy. Do not graze cover crops.

CROP	PEST	Pounds MORESTAN 25% W.P.	REMARKS
BEARING CITRUS TREES Citrus (California only)	Citrus red mite Citrus rust mite Texas citrus mite Mite eggs	1 to 1½	Apply specified dosage per 100 gallons of water as a full-coverage spray* using 200 to 1000 gallons of finished spray per acre, depending upon the number and size of trees. Spray all surfaces, including the underside of leaves for complete coverage. Schedule applications no closer than 180 days apart. Do not harvest within 60 days of last application.
Citrus (Arizona Texas, and Florida only)	Citrus red mite Citrus rust mite Texas citrus mite Mite eggs	5	Apply specified dosage per acre by concentrate or dilute ground spray using 20 to 500 gallons of water per acre. Schedule applications no closer than 90 days apart in Texas and Florida, or 180 days apart in Arizona. Allow at least 60 days between last application and harvest.
NON-BEARING FRUITS Apples	Powdery mildew Mites Mite eggs	½ to 1	Apply specified dosage per 100 gallons of water.* Spray all foliage surfaces, including the underside of leaves, for complete coverage. Repeat as necessary. For control of powdery mildew, begin preventive applications before disease appears, and apply at 7 to 14-day intervals. During heavy infection periods make more frequent applications. Do not apply to fruit trees that will bear fruit within one year of application. Apply to strawberries, as required, in propagating beds, and in the field during the year of setting new plants, but not after first bloom appears in the following year.
Apricots Pears Cherries Plums Nectarines Prunes Peaches Strawberries	Mites (Except cyclamen mite) Mite eggs	½ to 1	
Citrus	Citrus red mite Citrus rust mite Texas citrus mite Mite eggs	1 to 1½	Apply specified dosage per 100 gallons of waste. Spray all surfaces, including the underside of leaves, for complete coverage. Repeat as necessary. Do not apply to trees that will bear fruit within one year of application.
NUTS Macadamia nuts	Mites	1 to 2	Apply specified dosage per 100 gallons of water as a full-coverage spray*, using up to 400 gallons of finished spray per acre, depending upon the number and size of trees. Spray all surfaces, including the underside of leaves, for complete coverage. Three applications per season may be made, up to the day of harvest.
Walnuts	Aphids Mites Mite eggs	1 to 1½	Apply specified dosage per 100 gallons of water as a full-coverage spray*, using up to 250 gallons of finished spray per acre, depending upon the number and size of trees. Spray all foliage surfaces, including the underside of leaves, for complete coverage. Do not apply more than twice per season, or within 30 days of harvest.
ORNAMENTALS Flowers Shrubs	Mites Mite eggs	½ to 1	Apply specified dosage per 100 gallons of water (½ pound of 25% Wettable Powder per 100 gallons = 1 level teaspoonful per 1 gallon). Spray all foliage surfaces, including the underside of leaves, for complete coverage. Repeat as necessary. 10-5-83

* This concentration is calculated for conventional sprayers. When lower volumes of spray are applied per acre, as with low pressure, low volume, airplane or mist-type equipment, increase the concentration of MORESTAN in the spray mixture in order to apply amount of MORESTAN per acre equivalent to a full-coverage spray. Refer also to "DOSAGE" paragraph.

SPECIAL LOCAL NEED REGISTRATIONS [SECTION 24(c)] FOR MORESTAN 25% WETTABLE POWDER (PVA)*

CROP	PEST	STATE EPA SLN NO.	DOSAGE	REMARKS
Bearing citrus trees	Mites Mite eggs	California CA 79-0059	5 pounds per acre.	By concentrate or dilute ground spray.

* For distribution and use only within the state specified. These Special Local Need registrations approved as of date of this publication are, however, subject to cancellation by the state. Contact your nearest Mobay Agricultural Chemicals Divisions Office for the current status and a copy of the complete individual supplement Special Local Need labeling. This labeling must be in the possession of the user at time of pesticide application.

RESTRICTIONS

Under certain adverse weather conditions, particularly those that create slow drying of spray deposits, injury may occur to foliage of strawberries and deciduous fruit trees.

Do not use summer oils if MORESTAN is used in the cover spray program.

Do not use spreader-sticker materials with MORESTAN.

Do not allow sprays to drift onto other crops, including pastures. Do not graze treated areas.

Do not use on other crops used for food and forage. Use only according to label directions. Applications at rates above those shown may result in illegal crop residues. Do not treat food crops grown in the greenhouse.

Phytotoxicity has occured on the foliage and flowers of several varities of roses, especially during periods of high temperatures. Treat roses only after cutting and before new growth appears. Before treating large plantings for the first time, treat a few plants and observe for 4 to 5 days. Under certain conditions, some injury to other ornamentals may occur.

®NEMACUR

15% GRANULAR SYSTEMIC INSECTICIDE–NEMATICIDE

For Effective Control of Nematodes in Certain Field, Fruit, and Vegetable Crops.

Only for agricultural use, or for sale to, use, and storage by commercial applicators, and other servicepersons.

ACTIVE INGREDIENT:	
Ethyl 3-methyl-4-(methylthio)phenyl (1-methylethyl)phosphoramidate	15%
INERT INGREDIENTS	85%
TOTAL	100%

EPA Reg. No. 3125-236-AA

NEMACUR is a Reg. TM of the Parent Company of Farbenfabriken Bayer GmbH, Leverkusen.

DANGER–POISON

Keep Out of Reach of Children

PELIGRO

PRECAUCION AL USUARIO: Si usted no puede leer o entender inglés, no use este producto hasta que la etiqueta le haya sido explicada ampliamente.

(**TO THE USER:** If you cannot read or understand English, do not use this product until the label has been fully explained to you.)

RECOMMENDED USES

Apples, Brussels sprouts, Cabbage, Cherries, Citrus, Cotton, Grapes, Okra, Peaches, Peanuts, Pineapple, Raspberry, Soybeans, Tobacco

RESTRICTIONS ON WAREHOUSE STORAGE

Do not store in front of space heater. Store below 120°F. Do not store adjacent to or near foodstuffs, drugs, feed or any other material intended for consumption by humans or animals. Do not store adjacent to toys, cosmetics, dishes, furniture, or clothing. Protect from rain and moisture. Do not store next to herbicides. In case of puncture, use overslip bag and securely close.

STATEMENTS OF PRACTICAL TREATMENT

In case of poisoning, call a physician immediately. Have patient lie down and keep quiet. **If swallowed, vomiting should be induced.** Administer water freely and induce vomiting by giving one dose (½ oz. or 15 ml.) of syrup of ipecac. If vomiting does not occur within 10 to 20 minutes, administer second dose. If syrup of ipecac is not available, induce vomiting by sticking finger down throat. Repeat until vomit fluid is clear. Never give anything by mouth to an unconscious person. Professional medical assistance should be secured immediately. **If on skin,** remove contaminated clothing and wash skin immediately with soap and warm water. **If eyes are contaminated,** wash with flowing water for at least 15 minutes. **If inhaled,** remove victim to fresh air. If not breathing, give artificial respiration, preferably mouth-to-mouth.

TO PHYSICIAN: Antidote—Administer atropine sulfate in large therapeutic doses. Repeat as necessary to the point of tolerance. 2-PAM is also antidotal and may be administered in conjunction with atropine.

Compound inhibits cholinesterase resulting in stimulation of the central nervous system, the parasympathetic nervous system and the somatic motor nerves. Do not give morphine. Watch for pulmonary edema, which may develop in serious cases of poisoning even after 12 hours. At first sign of pulmonary edema, the patient should be placed in an oxygen tent and treated symptomatically.

PRECAUTIONARY STATEMENTS

HAZARDS TO HUMANS AND DOMESTIC ANIMALS

DANGER–POISON

Fatal if swallowed or absorbed through the skin. Causes eye irritation. May be fatal if inhaled. Do not get in eyes, on skin, or on clothing. Do not breathe dust. Wear protective clothing, rubber gloves, and goggles or face shield when handling. Wash thoroughly with soap and water after handling and before eating or smoking. Remove and wash contaminated clothing before reuse. Fatal if swallowed or absorbed through the skin. Causes eye irritation. May be fatal if inhaled. Do not get in eyes, on skin, or on clothing. Do not breathhould be secured immediately. **If on skin,** remove contaminated clothing and wash skin immediately with soap and warm water. **If eyes are contaminated,** wash wit Fatal if swallowed or absorbed through the skh with flowing water for at least 15 minutes. **If inhaled,** remove victim to fresh air. If not breathing, give artificial respiration, preferably mouth-to-mouth.

Do not contaminate feed or food. Keep all unprotected persons out of the operating area or vicinity. Wash hands, arms, and face thoroughly with soap and warm water before eating or smoking. Wash all contaminated clothing with soap and hot water before reuse.

Keep out of the reach of children.

Not for use or storage in or around the home.

SYMPTOMS OF POISONING: A sense of "tightness" in the chest. Sweating. Contracted pupils. Stomach pains. Vomiting and diarrhea.

ENVIRONMENTAL HAZARDS

This product is toxic to fish and wildlife. Keep out of lakes, streams, or ponds. Birds feeding on treated areas may be killed.

IMPORTANT: This bulletin is not intended to provide adequate information for application. Before using the product, read and carefully observe directions, cautionary statements and other information appearing on the label. This product is sold subject to the Conditions of Sale set forth on the container label.

READ THE LABEL BEFORE USE

RESTRICTIONS

Any food crop not specified on this label may be planted into treated areas 120 days after the last application. Any cover crops that are planted during the 120 day period must be plowed under and not grazed.

Do not use on other crops. Use only according to label directions. Do not allow livestock or poultry to feed on grass or grass clippings. Do not apply this product with a knapsack or similar equipment that is placed on the user's body.

RE-ENTRY STATEMENTS

Do not apply this product in such a manner as to directly or through drift expose workers or other persons. The area being treated must be vacated by unprotected persons. Mixer-loaders must wear protective coveralls, rubber or neoprene boots, rubber or neoprene gloves, and goggles or face shields when handling this product. Any person coming in direct contact with treated soil prior to incorporation or irrigation, or with treated wet soil after the initial irrigation, must wear protective coveralls, rubber or neoprene boots, and rubber or neoprene gloves. Wash thoroughly with soap and water after handling and before eating or smoking.

Because certain states may require more restrictive re-entry intervals for various crops treated with this product, consult your State Department of Agriculture for further information.

Written or oral warnings must be given to workers who are expected to be in a treated area or in an area about to be treated with this product. Oral warnings should include those statements found under the label heading "Precautionary Statements" and "Re-entry Statements" included on this label. When oral warnings are given, warnings shall be given in a language customarily understood by workers. Oral warnings must be given if there is reason to believe that written warnings cannot be understood by workers. Written warnings must include the following information: DANGER, area treated with NEMACUR 15% Granular on (date of application). See "Re-entry Statements" for use of proper protective clothing. **If on skin,** remove contaminated clothing and wash skin immediately with soap and warm water. **If eyes are contaminated,** wash with flowing water for at least 15 minutes. **If swallowed,** vomiting should be induced. (See Precautionary Statements on label).

STORAGE AND DISPOSAL

PESTICIDE DISPOSAL

Do not contaminate water, food or feed by storage or disposal.

Pesticide wastes are acutely hazardous. Improper disposal of excess pesticide, spray mixture, or rinsate is a violation of Federal Law. If these wastes cannot be disposed of by use according to label instructions, contact your State Pesticide or Environmental Control Agency, or the Hazardous Waste representative at the nearest EPA Regional Office for guidance.

CONTAINER DISPOSAL

Do not use container in connection with food, feed, or drinking water. Completely empty bag into application equipment. Then dispose of empty bag in a sanitary landfill or by incineration, or, if allowed by State and local authorities, by burning. If burned, stay out of smoke.

Store in a cool, dry place and in such a manner as to prevent cross contamination with other pesticides, fertilizers, food, and feed. Store in original container and out of the reach of children, preferably in a locked storage area.

Handle and open container in a manner as to prevent spillage. If the container is leaking or material spilled for any reason or cause, carefully sweep material into a pile. Refer to Precautionary Statements on label for hazards associated with the handling of this material. Do not walk through spilled material. Dispose of pesticide as directed above. In spill or leak incidents, keep unauthorized people away. You may contact the Mobay Emergency Response Team for decontamination procedures or any other assistance that may be necessary. The Mobay Kansas City Emergency Response Telephone No. is 816-242-2582, or contact Chemtrec at 800-424-9300.

DIRECTIONS FOR USE

It is a violation of Federal Law to use this product in a manner inconsistent with its labeling.

AGRICULTURAL CROPS

NEMACUR 15% Granular has shown excellent activity against the major genera of nematodes attacking the following crops. NEMACUR kills nematodes by contact action.

COTTON AND PEANUTS: When applied according to directions on this label NEMACUR 15% Granular will provide early season reduction of thrips infesting peanuts and control of thrips infesting cotton.

RECOMMENDED APPLICATIONS

CROP	PEST	DOSAGE NEMACUR 15% GRANULAR IN FURROW OR BAND OZS./1000 FT. OF ROW		REMARKS
FIELD CROPS Cotton	Thrips (early-season reduction)	6 to 8 oz. per 1000 ft. of row for any row spacing. (5 to 6.5 lbs. per acre on 40 inch row).		**In furrow/covering soil:** Apply specified dosage per 1000 ft. of row in furrow or band in covering soil behind the seed drop and in front of covering devices.
	Nematodes Thrips	8 to 18 oz. per 1000 ft. of row for any row spacing. (6.5 to 14.7 lbs. per acre on 40 inch row spacing).		**In furrow/covering soil:** Apply specified dosage per 1000 ft. of row in furrow or band in covering soil behind the seed drop and in front of covering devices. **Band:** Apply specified dosage per 1000 ft. of row in a 6 to 12 inch band over the row. Incorporate thoroughly to insure uniform distribution.[1]
				Suggested use rates of NEMACUR 15% G in oz./1000 feet of row based on nematode infestation levels:* (see below)

Band width	Low	Moderate	High*
In furrow	8	10	12
In covering soil	8	10	12
6 inch band	8	10	12
8 inch band	10	12	14
10 inch band	11	13	16
12 inch band	12	14	18

* Use the highest recommended rates in fields with a history of nematode problems.

CROP	PEST	DOSAGE (ozs./1000 ft. of row)	DOSAGE (pounds)	REMARKS
Peanuts	Nematodes Thrips (early-season reduction)	11 to 18.7 ozs. for any row spacing (or 10 to 17 pounds per acre on 36 inch rows)	20 to 33.5 pounds	**FOR BAND OR BROADCAST APPLICATION** **BAND:** Apply specified dosage in ozs. per 1000 ft. of row in front of the planter shoe as a 12 to 18 inch band. Incorporate the granules into the soil to a depth of 2 to 6 inches. Use the high rate in fields with high populations of nematodes or in fields having a history of serious nematode damage. WHEN BAND APPLICATIONS ARE USED ON NARROW ROW CROPS, DO NOT USE BAND WIDTHS THAT WILL ALLOW TREATED AREAS TO OVERLAP
Soybeans (California only)	Nematodes	6.7 to 26.7 ozs. for any row spacing (or 6 to 24 lbs. per acre on 36 inch rows)	26.7 to 40 pounds	BROADCAST: Distribute the granules uniformly over the entire area to be treated and immediately incorporate to a depth of 2 to 6 inches by disking or tilling. Use the high rate in fields with high populations of nematodes or in fields having a history of serious nematode damage. **NOTE:** When NEMACUR is applied to soybeans, crop injury from Propanil drift may be intensified. For both band and broadcast applications do not feed or graze soybean vines, green peanut vines or peanut vine hay. Do not hog down treated peanut fields.

CROP	PEST	DOSAGE NEMACUR 15% GRANULAR IN FURROW OR BAND OZS./1000 FT. OF ROW		REMARKS
Soybeans (Except California)	Nematodes	8 to 22 ozs. per 1000 ft. of row for any row spacing. (7.3 to 20 lbs. per acre on 36 inch row spacing)		**In furrow/covering soil:** Apply specified dosage in ozs. per 1000 ft. of row in furrow or band in covering soil behind the seed drop in front of the covering devices. **Band:** Apply specified dosage per 1000 ft. of row in a 6 to 12 inch band over the row. Incorporate thoroughly to a depth of 2 to 3 inches to insure uniform distribution.[1] **Suggested use rate of NEMACUR 15% G in oz./1000 ft. of row based on nematode infestation levels:** **Nematode Infestation Levels** **Application — Low — Moderate — High*** In furrow — 8 — 11 — 14 In covering soil — 8 — 11 — 14 6 inch band — 8 — 12 — 15 8 inch band — 10 — 12 — 16 10 inch band — 12 — 14 — 18 12 inch band — 14 — 16 — 22 * Includes fields with a history of nematode problems. **NOTE:** Do not overlap bands on narrow row crops. Do not feed or graze soybean vines. When NEMACUR is applied to soybeans, crop injury from Propanil drift may be intensified. Use NEMACUR only as a band application when metribuzin is used for control of weeds.
Tobacco (not for use on shade-grown tobacco)	Nematodes		26.7 to 40 pounds	Distribute the granules uniformly over the entire area to be treated and immediately incorporate to a depth of 2 to 4 inches by disking or tilling to insure uniform distribution. Where a range in rates is recommended use the high rate in fields with high populations of nematodes or in fields having a history of serious nematode damage.
FRUIT CROPS Apple Cherry Peach (Except California)	Nematodes	66.7 to 133.3 pounds		**BROADCAST APPLICATION:** Apply specified dosage per acre with equipment properly calibrated to insure uniform distribution. Immediately incorporate granules into the top 2 to 4 inches of soil following application.
		66.7 to 133.3 pounds		**BAND APPLICATION:** Apply specified dosage per **treated** acre with equipment properly calibrated to apply the product uniformly in a band covering the feeder root system of the tree. Immediately incorporate granules into the top 2 to 4 inches of soil following application.
	Incorporation of this product may be accomplished by mechanical incorporation 2 to 4 inches deep, or by sprinkler irrigation applied immediately following application (1–2 inches of water). Control of nematodes is best obtained when there is adequate rainfall or irrigation after application to move the product into the root zone. Do not apply within 72 days of harvest of apples, or within 45 days of harvest of peaches or within 45 days of harvest of cherries. Do not apply more than 133.3 pounds/acre per year per planting site. Do not graze livestock in treated orchards. Do not feed cover crops grown in treated orchards to livestock.			
Citrus (Arizona and Texas only)	Nematodes	67 to 133 pounds		**BROADCAST:** Apply specified dosage per acre as a broadcast application. Immediately after application, incorporate the granules into the soil by light raking or tilling. **BAND:** Apply proportionally less product per acre based on band width. (Example: to treat trees planted on 20 foot centers with a 10 foot band centered on the row, apply one-half of the broadcast rate). Center the treated band on the tree row using a minimum band width of "to the drip line." Incorporate immediately after application. Under dry conditions, follow with ½ to 1 inch of overhead or flood irrigation. Do not apply more than once per season nor within 180 days of harvest. Do not graze livestock on treated areas.
Grapes (West of the Rocky Mountains-Except California)	Nematodes	120 pounds		**Broadcast:** With ground injection equipment, apply specified dosage per acre 2 or more inches below soil surface.*
		60 pounds (120 pounds per **treated** acre)		**Band:** With ground injection equipment, apply specified dosage per acre 2 or more inches below the soil surface.* Center the treated band on the vine row using a minimum band width equal to 50% of the row spacing.

CROP	PEST	DOSAGE NEMACUR 15% GRANULAR	REMARKS
Grapes (East of the Rocky Mountains)	Nematodes	120 pounds	**Broadcast:** Apply specified dosage broadcast to the soil* with suitable ground equipment and incorporate 2 to 4 inches deep.*
		60 pounds (120 pounds per **treated** acre)	**Band:** Use proportionally less NEMACUR 15% Granular per acre. Treated band should center on the vine row with a minimum width of 50% of row spacing. Incorporate immediately.* Example: To treat a 5-foot band on rows 10 feet apart (50% of the row spacing), use one-half the broadcast rate.
	*NOTE: Control of nematodes is best obtained when there is adequate rainfall or irrigation after application to move the product into the root zone. **RESTRICTIONS:** Up to 3 applications to grapes can be made per season, but do not apply more than 120 pounds of NEMACUR 15% Granular (18 pounds active ingredient) per acre in one season. Do not graze or feed treated crop to livestock.		
Pineapple* (Puerto Rico)	Nematodes	67 to 133	**BROADCAST APPLICATION:** Apply specified dosage (10 to 20 lbs. active ingredient) per acre as a preplant broadcast treatment. Thoroughly incorporate the granules into the soil using suitable tillage equipment.
		(See Remarks)	**BAND APPLICATION:** Use proportionally less NEMACUR 15% Granular per acre. Apply in a band 2 feet wide before planting. Thoroughly incorporate the granules into the soil using suitable tillage equipment.
		NOTE: Postplanting applications of NEMACUR 3 can be made in addition to a preplant band or broadcast application of NEMACUR 15% Granular. Refer to the labeling for NEMACUR 3 for directions and dosage rates. Do not apply more than a total of 20 lbs. active ingredient per acre per crop season regardless of the formulation or method of application. Do not use forage or fodder for animal feed.	
		*NEMACUR 15% Granular nematicide has shown excellent activity against the major genera of nematodes infesting pineapple plants. When applied as a preplant soil treatment, NEMACUR protects young developing root systems.	

CROP	PEST	DOSAGE NEMACUR 15% GRANULAR BAND: OZS./ 1000 FT. OF ROW	BROADCAST: LBS./ACRE	REMARKS
VEGETABLES Cabbage (Direct seeded and transplanted)	Nematodes	11 to 22 ozs. for any row spacing (or 10 to 20 lbs. per acre on 36 inch rows)	26.7 to 40 pounds	**FOR BAND OR BROADCAST APPLICATION** **BAND:** Apply specified dosage in ozs. per 1000 ft. of row in front of the planter shoe as a 12 to 15 inch band. Incorporate the granules into the soil to a depth of 2 to 6 inches. Use the high rate in fields with high populations of nematodes or in fields having a history of serious nematode damage. WHEN BAND APPLICATIONS ARE USED ON NARROW ROW CROPS, DO NOT USE BAND WIDTHS THAT WILL ALLOW TREATED AREAS TO OVERLAP. **BROADCAST:** Distribute the granules uniformly over the entire area to be treated and immediately incorporate to a depth of 2 to 6 inches by disking or tilling. Use the high rate in fields with high populations of nematodes or in fields having a history of serious nematode damage.
Brussels sprouts (Transplanted)	Nematodes	11 to 22 ozs. for any row spacing (or 10 to 20 lbs. per acre on 36 inch rows)	26.7 to 40 pounds	
Okra (Except California)	Nematodes	14.7 to 18.4 ozs. (or 13.4 to 16.7 pounds per acre on 36 inch rows)	40 pounds	**FOR BAND OR BROADCAST APPLICATION** **BAND:** Apply specified dosage in ounces per 1000 ft. of row in front of the planter shoe as a 12 to 15 inch band. Incorporate the granules into the soil to a depth of 2 to 6 inches. Use low rate in 12 inch band. Use high rate in 15 inch band. **BAND WIDTH (36 inch row spacing)** **12 inch**: 14.7 oz./1000 ft. of row or 13.4 lb./acre **15 inch**: 18.4 oz./1000 ft. of row or 16.7 lb./acre WHEN BAND APPLICATIONS ARE USED ON NARROW ROW CROPS, DO NOT USE BAND WIDTHS THAT WILL ALLOW TREATED AREAS TO OVERLAP. **BROADCAST:** Distribute the granules uniformly over the entire area to be treated and immediately incorporate to a depth of 2 to 6 inches by disking or tilling. 8-1-85

[1] The action of some planters during planting operation may provide adequate incorporation. When using planters which do not disturb enough soil for incorporation or under dry conditions, additional incorporation can be achieved with scratchers, paddlewheels, pickwheels, ground driven rototillers or similar equipment behind the planter.

SPECIAL LOCAL NEED REGISTRATIONS [SECTION 24(c)] FOR NEMACUR 15% GRANULAR*

CROP	PEST	STATE EPA SLN NO.	DOSAGE	REMARKS
Grapes	Nematodes (Dagger and root-knot)	California CA 83–0062	120 pounds	**Broadcast:** With ground injection equipment, apply specified dosage per acre 2 or more inches below soil surface.
			60 pounds (120 pounds per **treated** acre)	**Band:** With ground injection equipment, apply specified dosage per acre 2 or more inches below the soil surface.

CROP	PEST	STATE EPA SLN NO.	DOSAGE	REMARKS
Non-bearing citrus, deciduous fruit and nut trees, vineyards and commercial tree and vine nurseries	Nematodes	California CA 82–0012	120 pounds per acre.	**Broadcast** application with ground injection equipment.
			60 pounds per acre.	**Band** application with ground injection equipment.
Raspberries	Nematodes	Oregon OR 84-0034 Washington WA 84-0055	40 to 80 pounds	**Broadcast Application:** Apply specified dosage per acre with equipment properly calibrated to insure uniform distribution. Immediately incorporate granules into the top 2 to 4 inches of soil following application. Use the high rate for heavy nematode infested areas. **Band Application:** Apply specified dosage per **treated** acre with equipment properly calibrated to apply the product uniformly in a band covering the feeder root system of the plant. Immediately incorporate granules into the top 2 to 4 inches of soil following application. Use the high rate for heavy nematode infested areas.

Incorporation of this product may be accomplished by mechanical incorporation 2 to 4 inches deep, by sprinkler irrigation or rainfall applied immediately following application (1-2 inches of water).
Control of nematodes is best obtained when there is adequate rainfall or irrigation after application to move the product into the root zone (October through December 31).
Do not apply more than 80 pounds of NEMACUR 15% Granular (12 pounds active ingredient) per acre per year per planting site.
Do not apply within 6 months of harvest.

* For distribution and use only within the state specified. These Special Local Need registrations approved as of date of this publication are, however, subject to cancellation by the state. Contact your nearest Mobay Agricultural Chemicals Division Office for the current status and a copy of the complete individual supplemental Special Local Need labeling. This labeling must be in the possession of the user at time of pesticide application.

®NEMACUR

15% GRANULAR
SYSTEMIC NEMATICIDE

For Effective Control of Nematodes in Citrus Crops (Except Arizona, California and Texas).

Only for agricultural use, or for sale to, use, and storage by commercial applicators, and other servicepersons.

ACTIVE INGREDIENT:
Ethyl 3-methyl-4-(methylthio)phenyl
(1-methylethyl)phosphoramidate 15%
INERT INGREDIENTS 85%
TOTAL 100%

EPA Reg. No. 3125-236-AA

NEMACUR is a Reg. TM of the Parent Company of Farbenfabriken Bayer GmbH, Leverkusen.

DANGER–POISON

Keep Out of Reach of Children
PELIGRO

PRECAUCION AL USUARIO: Si usted no puede leer o entender inglés, no use este producto hasta que la etiqueta le haya sido explicada ampliamente.

(TO THE USER: If you cannot read or understand English, do not use this product until the label has been fully explained to you.)

RESTRICTIONS ON WAREHOUSE STORAGE

Do not store in front of space heater. Store below 120°F. Do not store adjacent to or near foodstuffs, drugs, feed or any other material intended for consumption by humans or animals. Do not store adjacent to toys, cosmetics, dishes, furniture, or clothing. Protect from rain and moisture. Do not store next to herbicides. In case of puncture, use overslip bag and securely close.

RECOMMENDED USES

Citrus (Except Arizona, California and Texas).

STATEMENTS OF PRACTICAL TREATMENT

In case of poisoning, call a physician immediately. Have patient lie down and keep quiet. **If swallowed, vomiting should be induced.** Administer water freely, and induce vomiting by given one dose (½ oz. or 15 ml.) of syrup of ipecac. If vomiting does not occur within 10 to 20 minutes, administer second dose. If syrup of ipecac is not available, induce vomiting by sticking finger down throat. Repeat until vomit fluid is clear. Never give anything by mouth to an unconscious person. Professional medical assistance should be secured immediately. **If on skin,** remove contaminated clothing and wash skin immediately with soap and warm water. **If eyes are contaminated,** wash with flowing water for at least 15 minutes. **If inhaled,** remove victim to fresh air. If not breathing give artificial respiration, preferably mouth to mouth.

TO PHYSICIAN: Antidote—Administer atropine sulfate in large therapeutic doses. Repeat as necessary to the point of tolerance. 2-PAM is also antidotal and may be administered in conjunction with atropine.

Compound inhibits cholinesterase, resulting in stimulation of the central nervous system, the parasympathetic nervous system and the somatic motor nerves. Do not give morphine. Watch for pulmonary edema, which may develop in serious cases of poisoning even after 12 hours. At first sign of pulmonary edema, the patient should be placed in an oxygen tent and treated symptomatically.

PRECAUTIONARY STATEMENTS

HAZARDS TO HUMANS AND DOMESTIC ANIMALS

Fatal if swallowed or absorbed through the skin. Causes eye irritation. May be fatal if inhaled. Do not get in eyes, on skin, or on clothing. Do not breathe dust. Wear protective clothing, rubber gloves, and goggles or face shield when handling. Wash thoroughly with soap and water after handling and before eating or smoking. Remove and wash contaminated clothing before reuse.

Do not contaminate feed or food. Keep all unprotected persons out of the operating area or vicinity. Wash hands, arms, and face thoroughly with soap and warm water before eating or smoking. Wash all contaminated clothing with soap and hot water before reuse. Keep out of the reach of children.

Not for use or storage in or around the home.

SYMPTOMS OF POISONING: A sense of "tightness" in the chest. Sweating. Contracted pupils. Stomach pains. Vomiting and diarrhea.

ENVIRONMENTAL HAZARDS

This product is toxic to fish and wildlife. Keep out of lakes, streams or ponds. Birds feeding on treated areas may be killed.

RE-ENTRY STATEMENTS

Do not apply this product in such a manner as to directly or through drift expose workers or other persons. The area being treated must be vacated by unprotected persons. Mixer-loaders must wear protective coveralls, rubber or neoprene boots, rubber or neoprene gloves, and goggles or face shields when handling this product. Any person coming in direct contact with treated soil prior to incorporation or irrigation, or with treated wet soil after the initial irrigation, must wear protective coveralls, rubber or neoprene boots, and rubber or neoprene gloves. Wash thoroughly with soap and water after handling and before eating or smoking.

Because certain states may require more restrictive re-entry intervals for various crops treated with this product, consult your State Department of Agriculture for further information.

Written or oral warnings must be given to workers who are expected to be in a treated area or in an area about to be treated with this product. Oral warnings should include those statements found under the label heading "Precautionary Statements" and "Re-entry Statements" included on this label. When oral warnings are given, warnings shall be given in a language customarily understood by workers. Oral warnings must be given if there is reason to believe that written warnings cannot be understood by workers. Written warnings must include the following information: DANGER, area treated with NEMACUR 15% Granular on (date of application). See "Re-entry Statements" for use of proper protective clothing. **If on skin,** remove contaminated clothing and wash skin immediately with soap and warm water. **If eyes are contaminated,** wash with flowing water for at least 15 minutes. **If swallowed,** vomiting should be induced. (See Precautionary Statements on label).

STORAGE AND DISPOSAL

PESTICIDE DISPOSAL

Do not contaminate water, food or feed by storage or disposal.

Pesticide wastes are acutely hazardous. Improper disposal of excess pesticide, spray mixture, or rinstate is a violation of Federal Law. If these wastes cannot be disposed of by use according to label instructions, contact your State Pesticide or Environment Control Agency, or the Hazardous Waste representative at the nearest EPA Regional Office for guidance.

CONTAINER DISPOSAL

Do not use container in connection with food, feed, or drinking water. Completely empty bag into application equipment. Then dispose of empty bag in a sanitary landfill or by incineration, or, if allowed by State and local authorities, by burning. If burned, stay out of smoke.

Store in a cool, dry place and in such a manner as to prevent cross contamination with other pesticides, fertilizers, food, and feed. Store in original container and out of the reach of children, preferably in a locked storage area.

Handle and open container in a manner as to prevent spillage. If the container is leaking or material spilled for any reason or cause, carefully sweep material into a pile. Refer to Precautionary Statements on label for hazards associated with the handling of this material. Do not walk through spilled material. Dispose of pesticide as directed above. In spill or leak incidents, keep unauthorized people away. You may contact the Mobay Emergency Response Team for decontamination procedures and any other assistance that may be necessary.

The Mobay Kansas City Emergency Response Telephone No. is 816-242-2582, or contact Chemtrec at 800-424-9300.

IMPORTANT

This bulletin is not intended to provide adequate information for application. Before using the product, read and carefully observe directions, cautionary statements and other information appearing on the label. This product is sold subject to the Conditions of Sale set forth on the container label.

READ THE LABEL BEFORE USE

DIRECTIONS FOR USE

It is a violation of Federal law to use this product in a manner inconsistent with its labeling.

NEMACUR 15% Granular has shown excellent activity against the major genera of nematodes attacking citrus. NEMACUR kills nematodes by contact action.

Recommended Applications

Crop	Pest	Dosage NEMACUR 15% G	Remarks
Citrus (Except Arizona, California and Texas).	Nematodes	67 to 133 pounds	**BROADCAST:** Apply specified dosage per acre as a broadcast application. Immediately after application, incorporate the granules into the soil by light raking or tilling. **BAND:** Apply proportionally less product per acre based on band width. (Example: to treat trees planted on 20 foot centers with a 10 foot band centered on the row, apply one-half of the broadcast rate). Center the treated band on the tree row using a minimum band width of "to the drip line." Incorporate immediately after application. Under dry conditions, follow with ½ to 1 inch of overhead or flood irrigation. Do not apply more than once per season nor within 180 days of harvest. Do not graze livestock in treated areas. 7–29–85

RESTRICTIONS

Any food crop not specified on the label may be planted into treated areas 120 days after the last application. Any cover crops that are planted during the 120 day period must be plowed under and not grazed.

Do not use on other crops. Use only according to label directions.

Do not apply this product with a knapsack or similar equipment that is placed on the user's body.

®NEMACUR 3

EMULSIFIABLE SYSTEMIC INSECTICIDE—NEMATICIDE

RESTRICTED USE PESTICIDE

For retail sale to and use only by Certified Applicators or persons under their direct supervision, and only for those uses covered by the Certified Applicator's certification.

For Effective Control of Nematodes in Certain Field and Fruit Crops.

ACTIVE INGREDIENT:

Ethyl 3-methyl-4-(methylthio)phenyl (1-methylethyl) phosphoramidate	35%
INERT INGREDIENTS	65%
TOTAL	100%

Contains 3 pounds Ethyl-3-methyl-4-(methylthio)phenyl (1-methylethyl) phosphoramidate per gallon.

EPA Reg. No. 3125-283

NEMACUR is a Reg TM of the Parent Company of Farbenfabriken Bayer GmbH, Leverkusen.

DANGER–POISON

Keep Out of Reach of Children

PELIGRO

PRECAUCION AL USUARIO: Si usted no puede leer o entender inglés, no use este producto hasta que la etiqueta le haya sido explicada ampliamente.

(**TO THE USER:** If you cannot read or understand English, do not use this product until the label has been fully explained to you.)

RECOMMENDED USES

Apples	Peanuts
Cherries	Pineapples
Citrus	Raspberries
Cotton	Soybeans
Grapes	Tobacco
Peaches	

RESTRICTIONS ON WAREHOUSE STORAGE

Do not store in front of space heater. Store below 120°F, but not below 32°F. Do not store adjacent to, or near foodstuffs, drugs, feed or any other material intended for consumption by humans or animals. Do not store adjacent to toys, cosmetics, dishes furniture, or clothing. Protect from rain and moisture. Do not store next to herbicides.

STATEMENTS OF PRACTICAL TREATMENT

In case of poisoning, call a physician immediately. Have patient lie down and keep quiet.

If swallowed, vomiting should be induced. Administer water freely and induce vomiting by giving one dose (½ oz. or 15 ml.) of syrup of ipecac. If vomiting does not occur within 10 to 20 minutes, administer second dose. If syrup of ipecac is not available, induce vomiting by sticking finger down throat. Repeat until vomit fluid is clear. Never give anything by mouth to an unconscious person. Professional medical assistance should be secured immediately.

If on skin, remove contaminated clothing and wash skin immediately with soap and warm water.

If eyes are contaminated, wash with flowing water for at least 15 minutes.

If inhaled, remove victim to fresh air. If not breathing, give artificial respiration, preferably mouth to mouth.

TO PHYSICIAN: ANTIDOTE Administer atropine sulfate in large therapeutic doses. Repeat as necessary to the point of tolerance. 2-PAM is also antidotal and may be administered in conjunction with atropine.

Compound inhibits cholinesterase, resulting in stimulation of the central nervous system, the parasympathetic nervous system, and the somatic motor nerves. Do not give morphine. Watch for pulmonary edema, which may develop in serious cases of poisoning even after 12 hours. At first sign of pulmonary edema, the patient should be placed in an oxygen tent and treated symptomatically.

PRECAUTIONARY STATEMENTS

HAZARDS TO HUMANS AND DOMESTIC ANIMALS

Store in a cool, dry place and in such a manner as to prevent cross contamination with other pesticides, fertilizers, food and feed. Do not store below 32°F. Do not store near heat or open flame. Store in original container and out of the reach of children, preferably in a locked storage area.

Handle and open container in a manner as to prevent spillage. If container is leaking invert to prevent leakage. If the container is leaking or material spilled for any reason or cause, carefully dam up spilled material to prevent runoff. Refer to Precautionary Statements on label for hazards associated with the handling of this material. Do not walk through spilled material. Absorb spilled material with absorbing type compounds and dispose of as directed for pesticides above. In spill or leak incidents, keep unauthorized people away. You may contact the Mobay Emergency Response Team for decontamination procedures or any other assistance that may be necessary. The Mobay Kansas City Emergency Response Telephone No. is 816-242-2582, or contact Chemtrec at 800-424-9300.

Fatal if swallowed or absorbed through the skin. Causes eye damage. May be fatal if inhaled. Do not get in eyes, on skin or on clothing. Wear protective clothing, rubber gloves and goggles or face shield when handling. Wash thoroughly with soap and water after handling and before eating or smoking.

Do not breathe fumes or spray mist. For protection against organophosphorous insecticides, wear a mask or respirator jointly approved by the Mining Enforcement and Safety Administration (formerly the U.S. Bureau of Mines) and by the National Institute for Occupational Safety and Health under the provisions of 30 CFR Part 11.

Do not contaminate feed or food. Keep out of reach of children. Keep all unprotected persons out of the operating area or vicinity. Wash hands, arms, and face thoroughly with soap and warm water before eating or smoking.

HANDLING OF CONTAMINATED ARTICLES: Clothing and rags receiving undiluted NEMACUR 3 as a result of leaks, spills, or mishaps should be removed immediately and not reused. These contaminated articles should be buried in the same manner as the empty container: see container disposal section for directions. (Contaminated articles which should be discarded would include: hats, gloves, aprons, coats, boots, etc. not made of rubber or coated with rubber or other similar materials). Clothing receiving spray mist or droplets from the mixed, diluted NEMACUR 3 should be removed and decontaminated before reuse. The clothing can be decontaminated by machine washing separately from other items with soap or detergent and bleach in hot water twice before reuse. Protective clothing and equipment should be washed down with detergent or soap and bleach in water. Wash water from the cleaning of protective clothing or equipment should not be allowed to run off or otherwise enter water supplies.

SYMPTOMS OF POISONING: A sense of "tightness" in the chest. Sweating. Contracted pupils. Stomach pains. Vomiting and diarrhea.

ENVIRONMENTAL HAZARDS

This product is extremely toxic to fish, birds and other wildlife. Keep out of lakes, ponds or streams. Do not apply when weather conditions favor drift from target area. Do not contaminate water by cleaning of equipment or disposal of wastes.

RESTRICTIONS
Do not use on other crops. Use only according to label directions. Any food crop not specified on this label may be planted into treated areas 120 days after the last application. Any cover crops that are planted during the 120 day period must be plowed under and not grazed.

RE-ENTRY STATEMENTS
Do not apply this product in such a manner as to directly or through drift expose workers or other persons. The area being treated must be vacated by unprotected persons.

Mixer-loaders must wear protective coveralls, rubber or neoprene boots, rubber or neoprene gloves, goggles or face shields when handling this product. Any person coming in direct contact with treated water, and/or treated soil prior to incorporation or irrigation, or with treated wet soil after the initial irrigation, must wear protective coveralls, rubber or neoprene boots, and rubber or neoprene gloves. Wash thoroughly with soap and water after handling and before eating or smoking.

Because certain states may require more restrictive re-entry intervals for various crops treated with this product, consult your State Department of Agriculture for further information.

Written or oral warnings must be given to workers who are expected to be in a treated area or in an area about to be treated with this product. Oral warnings should include those statements found under the label heading "Precautionary Statements" and "Re-Entry Statements" included on this label. When oral warnings are given, warnings shall be given in a language customarily understood by workers. Oral warnings must be given if there is reason to believe that written warnings cannot be understood by workers. Written warnings must include the following information: DANGER, area treated with NEMACUR 3 on (date of application). See "Re-entry Statements" for use of proper protective clothing. **If on skin,** remove contaminated clothing and wash skin immediately with soap and warm water. **If eyes are contaminated,** wash with flowing water for at least 15 minutes. **If swallowed,** vomiting should be induced. (See Precautionary Statements on label).

STORAGE AND DISPOSAL

PESTICIDE DISPOSAL
Do not contaminate water, food or feed by storage or disposal.

Pesticide wastes are acutely hazardous. Improper disposal of excess pesticide, spray mixture, or rinsate is a violation of Federal Law. If these wastes cannot be disposed of by use according to label instructions, contact your State Pesticide or Environmental Control Agency, or the Hazardous Waste representative at the nearest EPA Regional Office for guidance.

CONTAINER DISPOSAL
Triple rinse (or equivalent). Then offer for recycling or reconditioning, or puncture and dispose of in a sanitary landfill, or by other procedures approved by State and local authorities.

IMPORTANT

This bulletin is not intended to provide adequate information for application. Before using the product, read and carefully observe directions, cautionary statements and other information appearing on the label. This product is sold subject to the Conditions of Sale set forth on the container label.

READ THE LABEL BEFORE USE

DIRECTIONS FOR USE

It is a violation of Federal law to use this product in a manner inconsistent with its labeling.

NEMACUR 3 nematicide has shown excellent activity against the major genera of nematodes attacking the following crops. NEMACUR kills nematodes by contact action.

COTTON AND PEANUTS: When applied according to directions on the label, NEMACUR 3 will provide early season reduction of thrips infesting peanuts and control of thrips infesting cotton.

RECOMMENDED APPLICATIONS

CROP	PEST	DOSAGE NEMACUR 3 — IN FURROW OR BAND FL. OZS./1000 FT. OF ROW	REMARKS
FIELD CROPS Cotton	Thrips (early-season reduction)	2.4 to 3.3 fl. ozs. per 1,000 ft. of row for any row spacing. (1 to 1⅓ quarts per acre, based on 40 inch rows).	**In furrowing/covering soil:** Apply specified dosage per 1,000 ft. of row as a spray in furrow or band in covering soil behind the seed drop, and in front of covering devices.
	Nematodes Thrips	3.3 to 7.1 fl. ozs. per 1,000 ft. of row for any spacing. (1⅓ to 2.9 quarts per acre, based on 40 inch rows).	**In furrow/covering soil:** Apply specified dosage per 1,000 ft. of row as a spray in furrow or band in covering soil behind the seed drop, and in front of covering devices. **Band:** Apply specified dosage per 1,000 ft. of row as a spray in a 6 to 12 inch band over the row. Incorporate thoroughly to insure uniform distribution.[1] Suggested use rates of NEMACUR 3 in fluid oz. 1,000 feet of row based on nematode infestation levels:* (see below) * Use the highest recommended rates in fields with a history of nematode problems.

Band width	Low	Moderate	High*
In furrow	3.3	4.0	4.8
In covering soil	3.3	4.0	4.8
6 inch band	3.3	4.0	4.8
8 inch band	4.0	4.8	5.7
10 inch band	4.4	5.2	6.4
12 inch band	4.8	5.7	7.1

CROP	PEST	DOSAGE NEMACUR 3 — BAND: FL. OZS./ 1,000 FT. OF ROW	DOSAGE NEMACUR 3 — BROADCAST: GALLONS/ACRE	REMARKS
Peanuts	Nematodes Thrips (early season reduction)	4.5 to 7.3 fl. ozs. for any row spacing (or 2 to 3.3 quarts per acre on 36 inch rows).	1 to 1.7 gallons.	**FOR BAND OR BROADCAST APPLICATION-BAND:** Apply specified dosage in fl. oz. per 1,000 ft. of row as a water emulsion spray in a 12 to 18 inch band over the row in front of the planter shoe. Use sufficient water and incorporate thoroughly to insure uniform distribution. Use the high rate in fields with high populations of nematodes or in fields having a history of serious nematode damage.

<table>
<tr><th rowspan="2">CROP</th><th rowspan="2">PEST</th><th colspan="2">DOSAGE NEMACUR 3</th><th rowspan="2">REMARKS</th></tr>
<tr><th>BAND: FL. OZS./ 1,000 FT. OF ROW</th><th>BROADCAST: GALLONS/ACRE</th></tr>
<tr><td>Soybeans (California only)</td><td>Nematodes</td><td>3 to 11 fl. ozs. for any row spacing (or 1.3 to 5 quarts per acre on 36 inch rows).</td><td>1.3 to 2 gallons.</td><td>WHEN BAND APPLICATIONS ARE USED ON NARROW ROW CROPS, DO NOT USE BAND WIDTHS THAT WILL ALLOW TREATED AREAS TO OVERLAP.
BROADCAST: Apply as a water emulsion spray over the entire area to be treated, using sufficient water and thorough incorporation to insure uniform distribution. Use the high rate in fields with high populations of nematodes or in fields having a history of serious nematode damage. Plant crop in the usual manner.
NOTE: When NEMACUR is applied to soybeans, crop injury from Propanil drift may be intensified. For both band and broadcast applications do not feed or graze soybean vines, green peanut vines or peanut vine hay. Do not hog down treated peanut fields.</td></tr>
<tr><td>Soybeans (Except California)</td><td>Nematodes</td><td>3.2 to 8.7 fl. ozs. per 1,000 ft. of row for any row spacing. ($1\frac{1}{2}$ to 4 quarts of NEMACUR 3 per acre based on 36 inch rows).</td><td></td><td>In furrow/covering soil: Spray specified dosage in fluid ounces per 1,000 ft. of row in furrow or band in covering soil behind the seed drop in front of the covering devices. Band: Spray specified dosage per 1,000 ft. of row in a 6 to 12 inch band over the row. Incorporate thoroughly to insure uniform distribution.[1]
Suggested use rate of NEMACUR 3 in fluid ounces/1,000 ft. row based on nematode infestation levels:
<table>
<tr><th></th><th colspan="3">Nematode Infestation Levels</th></tr>
<tr><th>Application</th><th>Low</th><th>Moderate</th><th>High*</th></tr>
<tr><td>In furrow</td><td>3.2</td><td>4.4</td><td>5.6</td></tr>
<tr><td>In covering soil</td><td>3.2</td><td>4.4</td><td>5.6</td></tr>
<tr><td>6 inch band</td><td>3.2</td><td>4.8</td><td>5.9</td></tr>
<tr><td>8 inch band</td><td>4.0</td><td>4.8</td><td>6.4</td></tr>
<tr><td>10 inch band</td><td>4.8</td><td>5.6</td><td>7.1</td></tr>
<tr><td>12 inch band</td><td>5.6</td><td>6.4</td><td>8.7</td></tr>
</table>
* Includes fields with a history of nematode problems.
NOTE: Do not overlap bands on narrow row crops. Do not feed or graze soybean vines. When NEMACUR is applied to soybeans, crop injury from Propanil drift may be intensified. Use NEMACUR only as a band application when metribuzin is used for control of weeds.</td></tr>
</table>

<table>
<tr><th>CROP</th><th>PEST</th><th>DOSAGE NEMACUR 3</th><th>REMARKS</th></tr>
<tr><td>Tobacco (Not for use on shade-grown tobacco)</td><td>Nematodes</td><td>$1\frac{1}{3}$ to 2 gallons</td><td>Apply as a water emulsion spray over the entire area to be treated using a minimum of 20 gallons of water per acre to insure uniform distribution. Incorporate to a depth of 2 to 4 inches by disking or tilling. Where a range in rates is recommended use the high rate in fields with high populations of nematodes or in fields having a history of serious nematode damage. Plant crop in the usual manner.</td></tr>
<tr><td>FRUIT
Apple
Cherry
Peach
(Except California)</td><td>Nematodes</td><td>$3\frac{1}{3}$ to $6\frac{2}{3}$ gallons.</td><td>BROADCAST APPLICATION: Apply specified dosage in 20 to 40 gallons of water per acre as a spray to the soil surface. Incorporate immediately 2 to 4 inches deep.
BAND APPLICATION: Apply specified dosage in 20 to 40 gallons of water per treated acre as a spray to the soil surface with equipment properly calibrated to apply the product in a band covering the feeder root system of the tree. Incorporate immediately 2 to 4 inches deep. Incorporation of this product may be accomplished by mechanical incorporation 2 to 4 inches deep, or by sprinkler irrigation applied immediately following application (1-2 inches of water).
Control of nematodes is best obtained when there is adequate rainfall or irrigation after application to move the product into the root zone.
NEMACUR 3 may be tank mixed with herbicides registered and recommended for orchard weed control. Observe all cautions and limitations on labeling of all products used in mixtures.
Do not apply within 72 days of harvest of apples, or within 45 days of harvest of peaches or within 45 days of harvest of cherries. Do not apply more than $6\frac{2}{3}$ gallons/acre per year per planting site. Do not graze livestock in treated orchards. Do not feed cover crops grown in treated orchards to livestock.
Special Precautions: Avoid contacting tree foliage with the spray mixture. Do not add oils or surfactants to the spray mixture. Do not allow spray or drift to contact operator.</td></tr>
<tr><td>Citrus (Arizona and Texas only)</td><td>Nematodes</td><td>$3\frac{1}{3}$ to $6\frac{2}{3}$ gallons</td><td>BROADCAST: Apply specified dosage in 20 to 40 gallons of water per acre as a water emulsion broadcast spray to the soil surface.
Band: Apply proportionally less product per acre based on width. (Example: to treat trees planted on 20 foot centers with a 10 foot band centered on the row, apply one-half of the broadcast rate). Center the treated band on the tree row using a minimum band width of "to the drip line".
Immediately after application, incorporate the product into the soil to a depth of 2 to 3 inches or apply sufficient irrigation for incorporation. Under dry conditions, follow with $\frac{1}{2}$ to 1 inch of irrigation.
Do not apply more than once per season nor within 180 days of harvest. Do not graze livestock in treated areas.</td></tr>
</table>

CROP	PEST	DOSAGE NEMACUR 3	REMARKS
Grapes (West of the Rocky Mountains —Except California)	Nematodes	6 gallons.	**Broadcast:** With ground injection equipment, apply specified dosage per acre in 20 to 40 gallons of solution 2 or more inches below soil surface.*
		3 gallons (6 gallons per **treated** acre).	**Band:** With ground injection equipment, apply specified dosage per acre in 20 to 40 gallons of solution 2 or more inches below the soil surface.* Center the treated band on the vine row using a minimum band width equal to 50% of the row spacing.
Grapes (East of the Rocky Mountains)	Nematodes	6 gallons.	**Broadcast:** Apply specified dosage per acre in 20 to 40 gallons of spray solution per acre to the soil * with suitable ground equipment and incorporate 2 to 4 inches deep.
		3 gallons (6 gallons per **treated** acre).	**Band.** Use proportionally less NEMACUR 3 per acre. Treated band should center on the vine row with a minimum width of 50% of the row spacing. Incorporate immediately. Example: To treat a 5-foot band on rows 10 feet apart (50% of the row spacing), use one-half the broadcast rate.
	* NOTE: Control of nematodes is best obtained when there is adequate rainfall or irrigation after application to move the product into the root zone. **RESTRICTIONS:** Up to 3 applications can be made per season, but do not apply more than 6 gallons of NEMACUR 3(18 pounds active ingredient) per acre in one season. Do not graze or feed treated crop to livestock.		
Pineapple (Hawaii only)	Nematodes (*Rotylenchulus* and *Meloidogyne* species)	$1\frac{1}{3}$ to 8 pints.	**POSTPLANT APPLICATION:** **PLANT CROP:** Apply specified dosage in 50 to 250 gallons of water per acre as a broadcast spray or through drip irrigation. Applications may begin immediately after planting. Make additional applications at intervals of 1 to 3 months as needed. Do not apply within 30 days before harvest. Do not use green forage or green fodder (cannery waste, such as cull fruit, fruit skin or shells, crowns, cores and basal ends of fruit may be fed) for animal feed. Do not apply more than a total of 13.3 gallons of NEMACUR 3 per acre per plant crop. **RATOON CROP.** Apply specified dosage in 50 to 250 gallons of water per acre as a broadcast spray or through drip irrigation. The first application may be made immediately following crop harvest. Make additional applications at intervals of 1 to 3 months as needed. Do not apply within 30 days before harvest. Do not use green forage or green fodder (cannery waste, such as cull fruit, fruit skin or shells, crowns, cores and basal ends of fruit may be fed) for animal feed. Do not apply more than a total of 6.7 gallons of NEMACUR 3 to each ratoon crop.
Pineapple* (Puerto Rico)	Nematodes	$13\frac{1}{3}$ to $26\frac{2}{3}$ pints	**POSTPLANT BROADCAST APPLICATION: PLANT CROP—** Apply specified dosage (5 to 10 lbs. active ingredient) in 50 to 250 gallons of water per acre as a broadcast spray. Begin applications 1 to 3 months after planting. Make additional applications at intervals of 3 to 6 months as needed. Do not apply within $7\frac{1}{2}$ months before harvest. NOTE: A preplant soil application using NEMACUR 15% Granular or a soil fumigant according to label directions should be made in addition to post-plant broadcast applications of NEMACUR 3. **FIRST RATOON CROP.** Apply specified dosage in 50 to 250 gallons of water per acre as a broadcast spray. Make the first application immediately following crop harvest. Make additional applications at intervals of 3 to 6 months as needed. Do not apply within $7\frac{1}{2}$ months before harvest.
		Do not apply more than a total of 20 lbs. active ingredient per acre per crop season regardless of the formulation or method of application. Do not use for forage or fodder for animal feed. *NEMACUR 3 emulsifiable nematicide has shown excellent activity against the major genera of nematodes infesting pineapple plants. When applied as a broadcast spray after planting, NEMACUR protects young developing root systems. For best results, applications should be made when plants are actively growing.	
Raspberry (Except California)	Nematodes	2 to 4 gallons	Apply as a broadcast spray to established plantings during the period (October 1 to December 31) when adequate rainfall can be expected. Apply in sufficient water to insure uniform distribution (15 to 100 gal. of spray mixture per acre). Direct spray to the soil surface and the base of the plant. Do not apply within six months of harvest. Do not apply more than once a year. Use the higher rate in fields with high populations of nematodes or in fields having a history of serious nematode damage.

[1] The action of some planters during planting operation may provide adequate incorporation. When using planters which do not disturb enough soil for incorporation or under dry conditions, additional incorporation can be achieved with scratchers, paddlewheels, ground driven rototillers or similar equipment behind the planter.

SPECIAL LOCAL NEED REGISTRATIONS [SECTION 24(c)] FOR NEMACUR 3

CROP	PEST	STATE EPA SLN NO.	DOSAGE	REMARKS
Bulbous iris	Nematodes	Oregon OR 80-0063	2 to 4 gallons per acre.	Apply in-furrow at time of planting.
Citrus: Grapefruit Lemons Limes Oranges Tangerines	Nematodes	California CA 85-0007	6 gallons	**Broadcast:** With ground injection equipment, apply specified dosage per acre in 20 to 40 gallons of solution 2 or more inches below the soil surface**.
			3 gallons (6 gallons per **treated** acre)	**Band:** With ground injection equipment, apply specified dosage per acre in 20 to 40 gallons of solution 2 or more inches below the soil surface**. Center the treated band in the tree row using a minimum band width equal to 50% of the row spacing. **LOW PRESSURE IRRIGATION** (including drip, mini-sprinklers, and misters): Apply specified dosage per acre in 1 to 6 applications per year using specified irrigation equipment. When this product is being applied through the specified irrigation systems, at the entry point to the field, or if there are no obvious entry points, at every 600 feet, there shall be signs posted that are readable at 25 feet that state (in English and Spanish): "DANGER! Pesticides are being applied in the water through the irrigation system. Do not drink water or walk on wet soil." Instead of posting treated groves, the operator of the property may have a person present during the application period to keep bystanders, etc., from entering the application site.
			NOTE: Control of nematodes is best obtained when there is adequate rainfall or irrigation after application to move the product into the root zone.	
Cotton	Nematodes Thrips (early-season reduction)	New Mexico NM 79-0021 Texas TX 79-0035	2.6 to 5.3 fluid ounces per 1000 feet of row for any row spacing. (1.1 to 2.2 quarts per acre, based on 40-inch rows.)	In a water emulsion or as a tank mix with Treflan EC and/or liquid fertilizer.
	Nematodes Thrips	California CA 82-0101	9.8 fluid ounces per 1,000 ft. of row or 4.0 quarts per acre based on 40 inch rows.	Application by soil injection in water emulsion or liquid fertilizer.
Grapes	Nematodes (Dagger and Root-knot)	California CA 83-0061	6 gallons	**Broadcast:** With ground injection equipment, apply specified dosage per acre in 20 to 40 gallons of solution 2 or more inches below soil surface.
			3 gallons (6 gallons per **treated** acre)	**Band:** With ground injection equipment, apply specified dosage per acre in 20 to 40 gallons of solution 2 or more inches below the soil surface. **Drip Irrigation.** Apply specified dosage using drip irrigation equipment.
Iris and narcissus bulbs	Nematodes	Washington WA 76-0034	2 to 4 gallons per acre.	Apply in-furrow at time of planting.
Non-bearing citrus, deciduous fruit and nut trees, vineyards, and commercial tree and vine nurseries	Nematodes	California CA 82-0011	6 gallons per acre.	Broadcast application with ground injection equipment.
			3 gallons per acre. (6 gallons per **treated** acre)	Band application with ground injection equipment.
Pineapple (Hawaii only)	Nematodes	Hawaii HI 85-31	1⅓ to 8 pints.	Post-plant application. Ratoon crop application.
Tobacco (not for use on shade-grown tobacco)	Nematodes	Georgia GA 78-0011 South Carolina SC 79-0004	2 gallons per acre.	Broadcast application. May be tank mixed with Ridomil and tobacco herbicides.
		Georgia GA 78-0012 Virginia VA 78-0012	2⅔ quarts per acre.	Tank mix with DASANIT S.C.
	Root-knot and tobacco cyst nematodes	Virginia VA 78-0010	2 gallons per acre.	Broadcast application. May be tank mixed with Ridomil.

CROP	PEST	STATE EPA SLN NO.	DOSAGE	REMARKS
	Nematodes	Maryland MD 82-0005		Broadcast application.
	Root knot nematodes Black shank Blue mold	North Carolina NC 78-0006	2 gallons per acre.	Broadcast application. May be tank mixed with Prowl, Paarlan, Ridomil or Tillam plus Devrinol.
Transplant Cabbage	Nematodes	Florida FL 84-0019	5 fl. oz. per 1,000 ft. of row spacing or $2\frac{1}{4}$ quarts per acre on 36-inch row spacing.	**Post-plant Drench Application**** Apply specified dosage in a 10-inch band directly over the row within 24 hours of transplanting using 62 gallons of water per 1,000 ft. of row, or 800 to 1,000 gallons of water per acre depending on width of row spacing. **Restriction: Application must be separate from hand transplanting.

* For distribution and use only within the state specified. These Special Local Need registrations approved as of date of this publication are, however, subject to cancellation by the state. Contact your nearest Mobay Agricultural Chemicals Division Office for the current status and a copy of the complete individual supplemental Special Local Need labeling. This labeling must be in the possession of the user at time of pesticide application.

®NEMACUR 3

EMULSIFIABLE SYSTEMIC NEMATICIDE

RESTRICTED USE PESTICIDE

For retail sale to and use only by Certified Applicators or persons under their direct supervision, and only for those uses covered by the Certified Applicator's certification.

For Effective Control of Nematodes in Citrus (except Arizona, California and Texas).

ACTIVE INGREDIENT:
Ethyl 3-methyl-4-(methylthio)phenyl (1-methylethyl) phosphoramidate 35%
INERT INGREDIENTS 65%
TOTAL 100%

EPA Reg. No. 3125-283

NEMACUR is a Reg. TM of the Parent Company of Farbenfabriken Bayer GmbH, Leverkusen.

DANGER–POISON

Keep Out of Reach of Children
PELIGRO

PRECAUCION AL USUARIO: Si usted no puede leer o entender inglés, no use este producto hasta que la etiqueta le haya sido explicada ampliamente.

(TO THE USER): If you cannot read or understand English, do not use this product until the label has been fully explained to you.)

RESTRICTIONS ON WAREHOUSE STORAGE

Do not store in front of space heater. Store below 120°F but not below 32°F. Do not store adjacent to, or near foodstuffs, drugs, feed or any other material intended for consumption by humans or animals. Do not store adjacent to toys, cosmetics, dishes, furniture, or clothing. Protect from rain and moisture. Do not store next to herbicides. In case of puncture, use overslip bag and securely close.

RECOMMENDED USES

Citrus (except Arizona, California and Texas)

PRECAUTIONARY STATEMENTS

HAZARDS TO HUMANS AND DOMESTIC ANIMALS

Fatal if swallowed or absorbed through the skin. Causes eye damage. May be fatal if inhaled. Do not get in eyes, on skin, or on clothing. Wear protective clothing, rubber gloves and goggles or face shield when handling. Wash thoroughly with soap and water after handling and before eating or smoking.

Do not breathe fumes or spray mist. For protection against organophosphorous insecticides, wear a mask or respirator jointly approved by the Mining Enforcement and Safety Administration (formerly the U.S. Bureau of Mines) and by the National Institute for Occupational Safety and Health under the provisions of 30 CFR Part 11.

Do not contaminate feed or food. Keep out of reach of children. Keep all unprotected persons out of the operating area or vicinity. Wash hands, arms, and face thoroughly with soap and warm water before eating or smoking.

HANDLING OF CONTAMINATED ARTICLES: Clothing and rags receiving undiluted NEMACUR 3 as a result of leaks, spills, or mishaps should be removed immediately and not reused. These contaminated articles should be buried in the same manner as the empty container: see container disposal section for directions. (Contaminated articles which should be discarded would include: hats, gloves, aprons, coats, boots, etc. not made of rubber, or coated with rubber or other similar materials.) Clothing receiving spray mist or droplets from the mixed, diluted NEMACUR 3 should be removed and decontaminated before reuse. This clothing can be decontaminated by machine washing separately from other items with soap or detergent and bleach in hot water twice before reuse. Protective clothing and equipment should be washed down with detergent or soap and bleach in water. Wash water from the cleaning of protective clothing or equipment should not be allowed to run off or otherwise enter water supplies.

SYMPTOMS OF POISONING: A sense of "tightness" in the chest. Sweating. Contracted pupils. Stomach pains. Vomiting and diarrhea.

STATEMENTS OF PRACTICAL TREATMENT

In case of poisoning, call a physician immediately. Have patient lie down and keep quiet.

If swallowed, vomiting should be induced. Administer water freely, and induce vomiting by giving one dose ($\frac{1}{2}$ oz. or 15 ml.) of syrup of ipecac. If vomiting does not occur within 10 to 20 minutes, administer second dose. If syrup of ipecac is not available, induce vomiting by sticking finger down throat. Repeat until vomit fluid is clear. Never give anything by mouth to an unconscious person. Professional medical assistance should be secured immediately.

If on skin, remove contaminated clothing and wash skin immediately with soap and warm water.

If eyes are contaminated, wash with flowing water for at least 15 minutes.

If inhaled, remove victim to fresh air. If not breathing, give artificial respiration, preferably mouth to mouth.

TO PHYSICIAN: Antidote—Administer atropine sulfate in large therapeutic doses. Repeat as necessary to the point of tolerance. 2-PAM is also antidotal and may be administered in conjunction with atropine.

Compound inhibits cholinesterase, resulting in stimulation of the central nervous system, the parasympathetic nervous system and the somatic motor nerves. Do not give morphine. Watch for pulmonary edema, which may develop in serious cases of poisoning even after 12 hours. At first sign of pulmonary edema, the patient should be placed in an oxygen tent and treated symptomatically.

ENVIRONMENTAL HAZARDS

This product is extremely toxic to fish, birds and other wildlife. Keep out of lakes, ponds or streams. Do not apply when weather conditions favor drift from target area. Do not contaminate water by cleaning of equipment or disposal of wastes.

RESTRICTIONS

Do not use on other crops. Use only according to label directions.

Any food crop not specified on the label may be planted into treated areas 120 days after the last application. Any cover crops that are planted during the 120 day period must be plowed under and not grazed.

RE-ENTRY STATEMENTS

Do not apply this product in such a manner as to directly or through drift expose workers or other persons. The area being treated must be vacated by unprotected persons.

Mixer-loaders must wear protective coveralls, rubber or neoprene boots, rubber or neoprene gloves, and goggles or face shields when handling this product. Any person coming in direct contact with treated water, and/or treated soil prior to incorporation or irrigation, or with treated wet soil after the initial irrigation, must wear protective coveralls, rubber or neoprene boots, and rubber or neoprene gloves. Wash thoroughly with soap and water after handling and before eating or smoking.

Because certain states may require more restrictive re-entry intervals for various crops treated with this product, consult your State Department of Agriculture for further information.

Written or oral warnings must be given to workers who are expected to be in a treated area or in an area about to be treated with this product. Oral warnings should include those statements found under the label heading "Precautionary Statements" and "Worker Protection Statements" included on this label. When oral warnings are given, warnings shall be given in a language customarily understood by workers. Oral warnings must be given if there is reason to believe that written warnings cannot be understood by workers. Written warnings must include the following information: DANGER, area treated with NEMACUR 3 on (date of application). See "Re-entry Statements" for use of proper protective clothing.

If on skin, remove contaminated clothing and wash skin immediately with soap and warm water.

If eyes are contaminated, wash with flowing water for at least 15 minutes.

If swallowed, vomiting should be induced. (See Precautionary Statements on label).

STORAGE AND DISPOSAL

PESTICIDE DISPOSAL

Do not contaminate water, food or feed by storage or disposal.

Pesticide wastes are acutely hazardous. Improper disposal of excess pesticide, spray mixture, or rinsate is a violation of Federal Law. If these wastes cannot be disposed of by use according to label instructions, contact your State Pesticide or Environmental Control Agency, or the Hazardous Waste representative at the nearest EPA Regional Office for guidance.

CONTAINER DISPOSAL

Triple rinse (or equivalent). Then offer for recycling or reconditioning, or puncture and dispose of in a sanitary landfill, or by other procedures approved by State and local authorities.

Store in a cool, dry place and in such a manner as to prevent cross contamination with other pesticides, fertilizers, food, and feed. Do not store below 32°F. Do not store near heat or open flame. Store in original container and out of the reach of children, preferably in a locked storage area.

Handle and open container in a manner as to prevent spillage. If container is leaking invert to prevent leakage. If the container is leaking or material spilled for any reason or cause, carefully dam up spilled material to prevent runoff. Refer to Precautionary Statements on label for hazards associated with the handling of this material. Do not walk through spilled material. Absorb spilled material with absorbing type compounds and dispose of as directed for pesticides above. In spill or leak incidents, keep unauthorized people away. You may contact the Mobay Emergency Response Team for decontamination procedures or any other assistance that may be necessary. The Mobay Kansas City Emergency Response Telephone No. is 816-242-2582, or contact Chemtrec at 800-424-9300.

IMPORTANT

This bulletin is not intended to provide adequate information for application. Before using the product, read and carefully observe directions, cautionary statements and other information appearing on the label. This product is sold subject to the Conditions of Sale set forth on the container label.

READ THE LABEL BEFORE USE

DIRECTIONS FOR USE

It is a violation of Federal law to use this product in a manner inconsistent with its labeling.

NEMACUR 3 nematicide has shown excellent activity against the major genera of nematodes attacking citrus. NEMACUR kills nematodes by contact action.

Recommended Applications

Crop	Pest	Dosage of NEMACUR 3	Remarks
Citrus (Except California)	Nematodes	3⅓ to 6⅔ gallons	**BROADCAST:** Apply specified dosage in 20 to 40 gallons of water per acre as a water emulsion broadcast spray to the soil surface. **BAND:** Apply proportionally less product per acre based on band width. (Example: To treat trees planted on 20 foot centers with a 10 foot band centered on the row, apply one-half of the broadcast rate). Center the treated band on the tree row using a minimum band width of "to the drip line." Immediately after application, incorporate the product into the soil to a depth of 2 to 3 inches or apply sufficient irrigation for incorporation. Under dry conditions, follow with ½ to 1 inch or irrigation. Do not apply more than once per season nor within 180 days of harvest. Do not graze livestock in treated areas.

7-29-85

®SENCOR DF

75% DRY FLOWABLE HERBICIDE

For Control of Certain Grasses and Broadleaf Weeds

ACTIVE INGREDIENT:	
4–Amino–6–(1,1–dimethylethyl)–3–(methylthio)–1,2,4–triazin–5(4*H*)–one†	75%
INERT INGREDIENTS:	25%
TOTAL	100%

† U.S. Patent No. 3,671,523

EPA Reg. No. 3125-325 EPA Est. 3125-MO-1

SENCOR is a Reg. TM and SPLIT-SHOT and EXTENDED SPLIT-SHOT are TMs of the Parent Company of Farbenfabriken Bayer GmbH, Leverkusen.

Keep Out of Reach of Children
CAUTION
AVISO

PRECAUCION AL USUARIO: Si usted no puede leer o entender inglés, no use este producto hasta que la etiqueta le haya sido explicada ampliamente.

(**TO THE USER:** If you cannot read or understand English, do not use this product until the label has been fully explained to you.)

RECOMMENDED USES

Soybeans, Potatoes, Dormant Established Alfalfa and Sainfoin, Established Asparagus, Sugarcane, Tomatoes, Wheat/Fallow/Wheat Rotation, Winter Wheat, Barley, Field Corn, Barley and Wheat/Fallow/Barley and Wheat Rotation.

RESTRICTIONS ON WAREHOUSE STORAGE

Protect from rain and moisture. Store in a cool, dry place. Do not store in front of a space heater.

DIRECTIONS FOR USE

It is a violation of Federal law to use this product in a manner inconsistent with its labeling.

IMPORTANT: Read these entire Directions and Conditions of Sale before using SENCOR DF 75% Dry Flowable herbicide.

GENERAL INFORMATION

MIXING

When using SENCOR, make sure the sprayer is completely clean, free of rust or corrosion which occurs from Winter storage. Examine strainers and screens to be sure the sprayer is clean from previously used pesticides.

Any tank mix containing SENCOR should be kept agitated and sprayed out immediately. Do not allow tank mixes to stand for prolonged periods of time.

The proper mixing procedure for SENCOR alone or in tank mix combinations with other herbicides is:

1. Fill the spray tank ¼ to ⅓ full with clean water.
2. Add recommended rate of SENCOR while recirculating and with agitator running.
3. Follow the triple rinse procedure described under "STORAGE AND DISPOSAL" to insure that all product is removed from the container.
4. Mix thoroughly and add clean water to fill spray tank to desired level.
5. Add the other herbicide to tank last and agitate thoroughly.
6. Continue agitation during application and until sprayer tank is empty.

Application

Use a standard low pressure (20 to 40 psi.) herbicide boom sprayer equipped with suitable nozzles and screens no finer than 50-mesh in-nozzle and in-line strainers. Agitate thoroughly before and during application with bypass agitation.

GROUND APPLICATION: Apply the proper rate of SENCOR in a minimum 10 to 40 gallons of spray mixture per acre broadcast.

Banded Application:

Use proportionally less SENCOR per acre in a band versus a broadcast application. For band application use ¼ to 1 gallon of spray mix per inch of band width regardless of row spacing.

EXAMPLES: To treat a 15-inch band on rows 30 inches apart, use one-half of the broadcast rate of SENCOR.

To treat a 14-inch band on rows 42 inches apart, use one-third of the broadcast rate of SENCOR.

AERIAL APPLICATION: Where permitted, apply specified rate in a minimum of 2 to 10 gallons of spray mixture per acre. Do not apply aerially when wind speed is greater than 10 MPH. Do not apply aerially when SENCOR is tank mixed with Lasso.

For All Applications of SENCOR: Sprayer must be accurately calibrated before applying SENCOR. Check sprayer during application to be sure it is working properly and delivering a uniform spray pattern. As the volume of spray mixture decreases per acre, the importance of accurate calibration and uniform application increases. Avoid over application, misapplication, and boom and spray swath overlapping that will increase spray dosage. (Crop injury may occur as a result). Avoid spray skips and gaps which allow weeds to grow in untreated soil. Do not apply when weather conditions favor spray drift and/or when sensitive or cool season crops, such as cole crops, onions, peas, or strawberries are present in adjacent fields or in areas where wheat is growing in coarse textured soils.

SPRAYER CLEANUP

Spray equipment must be thoroughly cleaned to remove remaining traces of herbicide that might injure other crops to be sprayed. Drain any remaining spray solution of SENCOR from the spray tank and discard in non-crop areas away from water supplies. Rinse the spray tank and refill with water, adding a heavy-duty detergent at the rate of one cup per 20 gallons of water. Recycle this mixture through the equipment for 5 minutes and spray out. Repeat this procedure twice. Fill the spray tank with clean water, recycle for 5 minutes, and spray out. Clean pump and nozzle screens thoroughly. Wash away any spray mixture from the outside of spray tank, nozzles or spray rig. All rinse water must be disposed of in compliance with local, State and Federal guidelines.

APPLICATION OF SENCOR IN FLUID FERTILIZERS

SENCOR may be applied in fluid fertilizer solutions to alfalfa and soybeans by following the appropriate mixing procedures and compatibility check. When using tank mix combinations, be sure all components are compatible.

Compatibility checks of SENCOR and tank mix combinations which include SENCOR should be made for each batch of fluid fertilizer because of the variability of these fertilizers.

Compatibility Check:

1. Pre-mix 2 teaspoonfuls of SENCOR with 8 teaspoonfuls of water (1:4 ratio) in a quart jar by adding the water first and follow with SENCOR. Mix thoroughly. If a second herbicide is to be used, double the amount of water (1:8 ratio) and add the second herbicide after mixing SENCOR first.

2. Then pour 1 pint of fluid fertilizer into the quart jar and shake well.

3. Allow to stand for 5 minutes.

THIS COMPATIBILITY CHECK SHOULD ONLY BE USED WHEN MIXING WITH FLUID FERTILIZERS.

Interpretation of Results:
If the solution in the jar appears to be uniform, without signs of agglomeration, or without a separation of an oily film on top of the fertilizer, the mixture may be used. If not, repeat the compatibility check using twice the amount of water or add a compatibility agent to the water. If separation occurs, but the mixture can be resuspended by shaking, then application is possible with good agitation in the spray tank.

TANK MIXING GUIDELINES

1. Add the required amount of water and compatibility agent (if required) to the tank. Start agitation system while adding SENCOR and follow by adding the fluid fertilizer and agitate.
2. If a second herbicide is to be used, follow as above in 1, but use twice the amount of water. Start agitation and add SENCOR and follow by adding the second herbicide, and then continue filling the tank with fluid fertilizer.
3. Maintain continuous agitation to assure uniform spray mixture until the tank is emptied.

COMMERCIAL IMPREGNATION AND APPLICATION OF SENCOR ON DRY BULK FERTILIZER

Dry bulk fertilizer may be impregnated or coated with SENCOR for application to established alfalfa and to soybeans. All recommendations, cautions, and special precautions on this label must be followed along with State regulations relating to dry bulk fertilizer blending, impregnating and labeling.

Impregnation: To impregnate, use a system consisting of a belt, conveyor, or closed drum which is used for dry bulk fertilizer blending. Any commonly used fertilizer can be impregnated with SENCOR except ammonium nitrate, or fertilizers containing ammonium nitrate, potassium nitrate, or sodium nitrate. Do not use on powder limestone.

Apply using a minimum of 200 pounds dry bulk fertilizer per acre and up to a maximum of 450 pounds per acre. To impregnate or coat the dry bulk fertilizer with SENCOR, the delivery nozzles must be directed to deliver a fine spray toward the fertilizer for thorough coverage while avoiding spray contact with mixing equipment. Uniform impregnation of SENCOR to dry bulk fertilizer will vary and if the absorptivity is not adequate, an absorptive powder may be added to produce a dry, free-flowing mixture. Microcel E (Johns-Manville Product Corporation) is the recommended absorbent powder. SENCOR should be pre-mixed with water to form a sprayable mixture. When another herbicide is used with SENCOR, mix and impregnate immediately.

Apply immediately after impregnation unless experience has shown that impregnated fertilizer can be stored without becoming lumpy and difficult to spread.

Rates: Select the recommended rate of SENCOR per acre from the appropriate section of this label and refer to the formula below to determine the amount of SENCOR which is to be impregnated on a ton of dry bulk fertilizer based on the amount of fertilizer which will be distributed on one acre.

$$\text{Pounds SENCOR DF Per Acre} \times \frac{2000}{\text{lbs. Fertilizer Per Acre}} = \text{Pounds SENCOR DF Per Ton of Fertilizer}$$

APPLICATION: Uniform application is essential for satisfactory weed control. Accurate calibration of fertilizer application equipment is essential for uniform distribution to the soil surface. The recommended method of application is to apply ½ the recommended rate and overlap 50 percent or to double apply by splitting the middles to obtain the best distribution pattern.

If fertilizer materials are excessively dusty, use diesel oil or other suitable additive to reduce dust prior to impregnation as dusty fertilizer will result in poor distribution during application. Crop injury and/or poor weed control may occur where the impregnated fertilizer is not uniformly applied.

INCORPORATION AND COMBINATION USES: When SENCOR is to be used in combination with another herbicide, follow directions on this label for combinations, rates, crops, incorporation, and special precautions.

SOYBEANS (EXCEPT CALIFORNIA)

SENCOR herbicide tank mix combinations are recommended for preplant incorporated applications, preemergence surface applications, and SPLIT-SHOT™, and EXTENDED SPLIT-SHOT™ application. SENCOR is also recommended as an overlay application following a preplant incorporated application of a recommended grass herbicide and alone as a preemergence surface application. All these applications can be applied with ground equipment, and some can be applied with aerial spray equipment. In addition, SENCOR can be applied as a postemergence directed spray to soybeans in certain states.

SPECIAL PRECAUTIONS (SOYBEANS)

Injury to soybeans may occur when SENCOR is used under the following conditions:

1. When soils have a calcareous surface area or a pH of 7.5 or higher.
2. Due to the sensitivity of certain soybean varieties, SENCOR is not recommended for use on sensitive varieties, such as, Tracy, Semmes, Altona, Vansoy, Coker 102, Coker 156, Portage, Maple Amber, Govan, NKS-1884, Terra-Vig 505, Terra-Vig 606, Vinton 81, Asgrow 6520, Agri Pro 55 and Agri Pro 71. Consult your Mobay Representative or your seed supplier for information on the tolerance to SENCOR of newly released soybean varieties, prior to use of SENCOR.
3. When applied in conjunction with soil-applied organic phosphate pesticides.
4. Over application or boom overlapping may result in stand loss and soil residues.
5. Uneven application or improper incorporation can decrease the level of weed control and/or increase the level of injury.
6. When applied to any soil with less than ½% organic matter.
7. Soil incorporation deeper than recommended.
8. When sprayers are not calibrated accurately.
9. When heavy rains occur soon after application, especially in poorly drained areas where water may stand for several days.
10. When soybeans are planted less than 1½" deep, particularly in preemergence application.

Activation

A minimum amount of soil moisture is required to activate SENCOR. In areas of low rainfall, preemergence applications to dry soil should be followed with light irrigation of ¼ acre inch of water. Do not apply heavy irrigation immediately after application. As with many surface-applied herbicides, weed control and crop tolerance may vary with rainfall and/or soil texture.

Grazing and Feeding Treated Vines

Treated vines may be grazed or fed to livestock 40 days after application when SENCOR is applied alone or with: Treflan®, Dual®, Prowl®, Lasso®, Amiben®, Paraquat, Lasso plus Paraquat.

Do not use treated vines for feed or forage when SENCOR is applied with: Surflan®, Dyanap®, Dyanap plus Lasso, Surflan plus Paraquat, Sonalan®, linuron plus Lasso, linuron plus Dual.

Rate Ranges

Where a rate range is shown, use a lower rate on soils that are coarse-textured or low in organic matter. Use a higher rate on soils that are relatively fine-textured or high in organic matter.

REPLANTING

If replanting is necessary in fields treated with SENCOR as directed on the label, the field may be replanted to soybeans. When replanting, a minimum of tillage is recommended. Do not apply a second treatment as injury to soybeans may occur.

WEEDS CONTROLLED BY SENCOR AND SENCOR TANK-MIX COMBINATIONS

C = Control S = Suppression or Erratic Control P = Poor or No Control
• = No information (Control may range from poor to excellent)

1 = SENCOR Alone
2 = SENCOR SPLIT-SHOT
3 = SENCOR plus Treflan
4 = SENCOR plus Dual
5 = SENCOR plus Prowl
6 = SENCOR plus Lasso
7 = SENCOR plus Surflan
8 = SENCOR plus Amiben
9 = SENCOR plus Dyanap
10 = SENCOR plus Lasso and Dyanap
11 = SENCOR plus Paraquat
12 = SENCOR plus Lasso and Paraquat
13 = SENCOR plus Surflan and Paraquat
14 = EXTENDED SPLIT-SHOT
15 = SENCOR plus Sonalan
16 = SENCOR plus linuron plus (Lasso or Dual)

ANNUAL BROADLEAF WEEDS	1	2	3	4	5	6	7	8	9	10	11	12	13	14	15	16
Black Nightshade *(Solanum nigrum)*	P	P	P	C	P	C	S	S	P	P	P	C	P	C	P	S
Bristly Starbur *(Acanthospermum hispidum)*	C	C	C	C	C	C	C	C	C	C	C	C	C	C	C	C
Buffalobur *(Solanum rostratum)*	C	C	P	P	P	P	P	P	C	P	P	P	P	C	P	•
Carpetweed *(Mollugo verticillata)*	C	C	C	C	C	C	C	C	C	C	C	C	C	C	C	C
Cocklebur *(Xanthium pensylvanicum)*	C	C	S	S	S	S	S	S	C	S	C	C	S	C	S	S
Cooperleaf, Hophornbeam *(Acalypha ostryaefolia)*	C	C	C	C	C	C	C	C	C	C	C	C	C	C	C	C
Florida Beggarweed *(Desmodium tortuosum)*	C	C	C	C	C	C	C	C	C	C	C	C	C	C	C	C
Florida Pusley *(Richardia scabra)*	C	C	C	C	C	C	C	C	C	C	C	C	C	C	C	C
Galinsoga *(Galinsoga spp.)*	C	C	C	C	C	C	C	C	C	C	C	C	C	C	C	C
Horseweed Marestail *(Conyza canadensis)*	•	•	•	•	•	•	•	•	•	•	C	C	C	C	•	•
Jimsonweed *(Datura stramonium)*	C	C	C	C	C	C	C	S	C	C	C	C	C	C	C	S
Knotweed *(Polygonum spp.)*	C	C	C	C	C	C	C	C	C	C	C	C	C	C	C	C
Kochia *(Kochia scoparia)*	C	C	C	C	C	C	C	C	C	C	C	C	C	C	C	C
Lambsquarters *(Chenopodium spp.)*	C	C	C	C	C	C	C	C	C	C	C	C	C	C	C	C
Morningglory, Ivyleaf *(Ipomoea hederacea)*	P	P	S	P	P	P	P	P	S	S	P	P	P	P	P	P
Morningglory, Pitted *(Ipomoea lacunosa)*	P	P	S	P	P	P	P	P	P	P	P	P	P	P	P	P
Morningglory, Smallflower *(Jacquemontia tamnifolia)*	P	P	C	P	P	P	P	P	P	P	P	P	P	P	P	P
Morningglory, Tall *(Ipomoea purpurea)*	P	P	S	P	P	P	S	P	S	S	P	P	P	P	P	P
Pigweeds *(Amaranthus spp.)*	C	C	C	C	C	C	C	C	C	C	C	C	C	C	C	C
Prickly Sida/Teaweed *(Sida spinosa)*	C	C	C	C	C	C	C	C	C	C	C	C	C	C	C	C
Purslane *(Portulaca oleracea)*	C	C	C	C	C	C	C	C	C	C	C	C	C	C	C	C
Ragweed, Common *(Ambrosia artemisiifolia)*	C	C	C	C	C	C	C	C	C	C	C	C	C	C	C	C
Redweed *(Melochia corchorifolia)*	C	C	C	C	C	C	C	C	C	C	C	C	C	C	C	C
Russian Thistle *(Salsola kali)*	C	C	C	C	C	C	C	C	C	C	C	C	C	C	C	C
Sesbania *(Sesbania spp.)*	C	C	C	C	C	C	C	C	C	C	C	C	C	C	C	C

WEEDS CONTROLLED BY SENCOR AND SENCOR TANK-MIX COMBINATIONS

C = Control
• = No information
S = Suppression or Erratic Control (Control may range from poor to excellent)
P = Poor or No Control

1 = SENCOR Alone
2 = SENCOR SPLIT-SHOT
3 = SENCOR plus Treflan
4 = SENCOR plus Dual
5 = SENCOR plus Prowl
6 = SENCOR plus Lasso
7 = SENCOR plus Surflan
8 = SENCOR plus Amiben
9 = SENCOR plus Dyanap
10 = SENCOR plus Lasso and Dyanap
11 = SENCOR plus Paraquat
12 = SENCOR plus Lasso and Paraquat
13 = SENCOR plus Surflan and Paraquat
14 = EXTENDED SPLIT-SHOT
15 = SENCOR plus Sonalan
16 = SENCOR plus linuron plus (Lasso or Dual)

ANNUAL BROADLEAF WEEDS	1	2	3	4	5	6	7	8	9	10	11	12	13	14	15	16
Shepherdspurse (*Capella bursa-pastoris*)	C	C	C	C	C	C	C	C	C	C	C	C	C	C	C	C
Sicklepod (*Cassia obtusifolia*)	C	C	C	C	S	C	S	C	C	C	C	C	C	C	S	S
Smartweeds (*Polygonum* spp.)	C	C	C	C	C	C	C	C	C	C	C	C	C	C	C	S
Spotted Spurge (*Euphorbia maculata*)	C	C	P	C	P	C	S	P	C	C	C	C	C	C	P	•
Spurred Anoda (*Anoda cristata*)	C	C	C	C	C	C	C	C	C	C	C	C	C	C	C	•
Sunflower (*Helianthus* spp.)	C	C	S	S	S	S	S	S	C	S	S	S	S	C	S	P
Velvetleaf (*Abutilon theophrasti*)	C	C	C	C	C	C	C	C	C	C	C	C	C	C	C	C
Venice Mallow (*Hibiscus trionum*)	C	C	C	C	C	C	C	C	C	C	C	C	C	C	C	C
Wild Mustards (*Brassica* spp.)	C	C	C	C	C	C	C	C	C	C	C	C	C	C	C	C
ANNUAL GRASSES	**1**	**2**	**3**	**4**	**5**	**6**	**7**	**8**	**9**	**10**	**11**	**12**	**13**	**14**	**15**	**16**
Barnyardgrass (*Echinochloa crus-galli*)	S	C	C	C	C	C	C	C	C	C	C	C	C	C	C	C
Bluegrass (*Poa annua*)	C	C	C	C	C	C	•	C	C	C	C	C	C	C	C	C
Broadleaf Signalgrass (*Brachiaria platyphylla*)	C	C	C	C	C	C	C	C	C	C	C	C	C	C	C	•
Browntop Millet (*Panicum ramosum*)	C	C	C	P	C	S	C	P	C	P	C	P	C	C	•	•
Crabgrass (*Digitaria* spp.)	C	C	C	C	C	C	C	C	C	C	C	C	C	C	C	C
Crowfootgrass (*Dactyloctenium aegyptium*)	C	C	C	C	C	C	C	C	C	C	C	C	C	•	•	
Cupgrass (*Eriochloa gracilis*)	P	C	P	P	P	P	P	P	P	P	P	P	P	C	•	•
Foxtails (*Setaria* spp.)	S	C	C	C	C	C	C	C	C	C	C	C	C	C	C	C
Goosegrass (*Eleusine indica*)	C	C	C	C	C	C	C	C	C	C	C	C	C	C	C	C
Johnsongrass, Seedling (*Sorghum halepense*)	C	C	C	C	C	C	C	P	C	C	C	C	C	C	C	•
Junglerice (*Echinochloa colonum*)	C	C	C	C	C	C	•	C	C	C	C	C	C	C	C	•
Nutsedge, Yellow (*Cyperus esculentus*)	P	P	P	C	P	C	P	P	P	P	P	C	P	C	P	•
Panicum, Fall (*Panicum dichotomiflorum*)	P	C	C	C	C	C	C	P	P	P	P	C	P	C	C	C
Panicum, Texas (*Panicum texanum*)	P	C	C	P	C	S	C	P	P	P	P	S	P	S	C	•
Red Rice (*Oryza sativa*)	P	C	C	C	P	C	•	P	P	P	P	C	P	C	•	•

WEEDS CONTROLLED BY SENCOR AND SENCOR TANK-MIX COMBINATIONS

C = Control
S = Suppression or Erratic Control (Control may range from poor to excellent)
P = Poor or No Control
• = No information
1 = SENCOR Alone
2 = SENCOR SPLIT-SHOT
3 = SENCOR plus Treflan
4 = SENCOR plus Dual
5 = SENCOR plus Prowl
6 = SENCOR plus Lasso
7 = SENCOR plus Surflan
8 = SENCOR plus Amiben
9 = SENCOR plus Dyanap
10 = SENCOR plus Lasso and Dyanap
11 = SENCOR plus Paraquat
12 = SENCOR plus Lasso and Paraquat
13 = SENCOR plus Surflan and Paraquat
14 = EXTENDED SPLIT-SHOT
15 = SENCOR plus Sonalan
16 = SENCOR plus linuron plus (Lasso or Dual)

ANNUAL GRASSES	1	2	3	4	5	6	7	8	9	10	11	12	13	14	15	16
Sandbur *(Cenchrus* spp.*)*	P	C	C	P	C	S	•	P	P	C	P	C	P	S	•	•
Shattercane *(Sorghum bicolor)*	P	C	C	P	P	P	P	P	P	P	P	P	P	P	C	•
Sorghum, Volunteer *(Sorghum* spp.*)*	P	C	C	P	P	P	P	P	P	P	P	P	P	P	•	P
Sprangletop *(Leptochloa* spp.*)*	P	C	C	P	P	P	•	P	P	P	P	P	P	P	•	P
Stinkgrass *(Eragrostis* spp.*)*	P	C	C	P	P	P	•	P	P	P	P	P	P	P	•	P
Wheat, Volunteer *(Triticum* spp.*)*	P	P	P	P	P	P	S	P	P	P	P	C	C	P	•	P
Witchgrass *(Panicum capillare)*	P	C	C	C	C	C	•	P	P	C	P	C	C	C	C	•

SENCOR ALONE

SENCOR (Alone) Preemergence Application

SENCOR can be applied broadcast or banded. This application may be made during planting or as a separate operation after planting but before crop emergence.

See the "APPLICATION" section under "GENERAL INFORMATION" in the front of this label.

Do not apply to sand soils; or to sandy loam or loamy sand soils containing less than 2% organic matter. Do not incorporate into soil or apply more than once per season.

Pounds of SENCOR DF Per Acre

SOIL TEXTURE	ORGANIC MATTER		
	Less than 2%	2 to 4%	Over 4%
Coarse Soils (Sandy loam, loamy sand)	DO NOT USE[3]	½	⅔
Medium Soils[1] (Loam, silt loam, silt, sandy clay, sandy clay loam)	½ to ⅔	⅔ to ⅚	⅚ to 1
Fine Soils[1] (Silty clay, silty clay loam[2], clay, clay loam)	⅔ to ⅚	⅚ to 1	1 to 1⅙
Mississippi Delta Only	1	1⅙	1⅓

[1] For control of lambsquarters, redroot pigweed and wild mustard, and for suppression of green, yellow and giant foxtails on alkaline (calcareous) soils in Nebraska, Minnesota, South Dakota and North Dakota only, apply SENCOR at rates of ⅓ pound/acre on medium soils and ⅓ to ½ pound/acre on fine soils regardless of soil organic matter percentage (use ½ pound only where soil pH is less than 7.5 and weed pressure is heavy). The ⅓ pound/acre rate of SENCOR alone can be applied regardless of soil pH. For control of other weeds listed on the label use SENCOR at full rates recommended in the table above, **but note that crop injury may occur on soils having a calcareous surface area or a pH of 7.5 or higher.**

[2] Silty clay loam soils are transitional soils and may be classified as medium textured soils in some regions of the U.S.

[3] Refer to the appropriate section of the label for use of SENCOR on soybeans in coarse soils with 0.5 to 2.0% organic matter in certain states.

USES OF SENCOR IN COMBINATIONS WITH OTHER HERBICIDES
SPLIT-SHOT™ APPLICATION

A preplant incorporated application of SENCOR tank mixed with either Treflan, Lasso, Dual, Prowl or Sonalan and followed by a preemergence surface application of SENCOR alone, after planting but prior to soybean emergence, will control more broadleaf and grass weeds in soybeans than when either herbicide is used alone.

Refer to the Treflan, Lasso, Dual, Prowl or Sonalan labels, and to appropriate sections of the SENCOR label, for directions on soil preparation, herbicide application, incorporation techniques, herbicide rates, weed species controlled, and restrictions for using tank mix combinations of SENCOR. Carefully observe the "Special Precautions" section concerning the use of SENCOR in tank mix combinations on soybeans.

When a SPLIT-SHOT application of SENCOR with Prowl, Treflan or Sonalan is used, the preplant incorporated tank mix may be applied up to 21 days prior to planting soybeans; with Dual or Lasso, the preplant incorporated tank mix may be applied up to 14 days prior to planting.

On medium and fine textured soils with greater than 2% organic matter, a rate range is recommended for the SENCOR preemergence overlay application. The higher rate should be used (a) in fields with a history of severe broadleaf weed pressure, (b) when the time between preplant incorporated tank mix and preemergence overlay applications approaches the maximum stated above, and/or (c) when the organic matter content of the soil is at the upper end of the indicated range.

NOTE: Tank mix combinations which include Lasso cannot be applied with aerial spray equipment.

For black nightshade control, refer to the appropriate section of the Lasso, Dual, or Sonalan labels.

SPLIT-SHOT APPLICATION

	PREPLANT INCORPORATED TANK MIX APPLICATION			—FOLLOWED BY—	PREEMERGENCE OVERLAY APPLICATION		
					RATE OF SENCOR DF POUND/ACRE ORGANIC MATTER		
SOIL TEXTURE(1)	RATE OF COMBINATION PRODUCT/ACRE	PLUS	RATE OF SENCOR DF POUND/ACRE		LESS THAN 2%	2 TO 4%	OVER 4%
Coarse (Light) (Sand, loamy sand, sandy loam)	Treflan 1 pt or Lasso 2–2½ qt or Dual 1¼–1½ pt or Prowl 1½ pt or Sonalan 1¼–2 pt	plus	⅓	—FOLLOWED BY—	⅙	⅙	⅙–⅓
Medium (Loam, silt loam, sandy clay loam, silt, sandy clay)	Treflan 1½ pt or Lasso 2½—3 qt or Dual 1½ pt or Prowl 1½ pt or Sonalan 1¾–2½ pt	plus	½	—FOLLOWED BY—	⅙	⅙–⅓	⅓–½
			or				
			⅓[2]	—FOLLOWED BY—	⅓	⅓–½	(½–⅔)[3]
Fine (Heavy) (Silty clay loam*, clay loam, silty clay, clay)	Treflan 2 pt or Lasso 2½–3 qt or Dual 2–2½ pt or Prowl 1½–2 pt or Sonalan 2¼–3 pt	plus	⅔	—FOLLOWED BY—	⅙	⅙–⅓	⅓–½
			or				
			½[2]	—FOLLOWED BY—	⅓	⅓–½	(½–⅔)[3]

* Silty clay loam soils are transitional soils and may be classified as medium textured soils in some regions of the U.S.

1 On coarse textured soils, do not use on sand soils with less than 1% organic matter, or on loamy sand or sandy loam soils with less than 0.5% organic matter. However, on coarse textured soils **with a calcareous surface area or a pH of 7.5 or higher,** do not use on sand soils with less than 2% organic matter, or on loamy sand or sandy loam soils with less than 1% organic matter.

2 Use this lower rate of SENCOR in the preplant incorporated tank mix **on soils having a calcareous surface area or a pH of 7.5 or higher,** and in those situations where soils within a field vary extremely in texture or organic matter content.

3 Reduce this preemergence overlay rate of SENCOR by ⅙ pound/acre when using SPLIT-SHOT application **on soils with over 4% organic matter which have a calcareous surface area or a pH of 7.5 or higher.**

EXTENDED SPLIT-SHOT APPLICATION

(Includes No-Till, Reduced Till, Ridge Till, Strip Till, Mulch Till)

An early preplant (surface-applied or shallow incorporated) application of SENCOR tank mixed with either Dual or Lasso, followed by a preemergence surface application of SENCOR tank mixed with Dual or Lasso after planting but prior to soybean emergence, will control more broadleaf and grass weeds in soybeans than either herbicide used alone.

An EXTENDED SPLIT-SHOT application will decrease the need for tillage and/or contact herbicides for the control of existing vegetation prior to planting, while providing residual control of weeds after planting.

When an EXTENDED SPLIT-SHOT application of SENCOR with Dual or Lasso is used, the preplant tank mix combination may be applied 15 to 30 days prior to planting soybeans. Follow directions on the label accompanying the product for SPLIT-SHOT applications from 0 to 14 days before planting.

Where a rate range is recommended, the higher rates should be used (a) in fields with a history of severe weed pressure, (b) when the time between early preplant tank mix and preemergence overlay applications approaches the maximum 30 days, (c) when the organic matter content of the soil is at the upper end of the indicated range, (d) when heavy crop residues are present on the soil surface, and/or (e) when the early preplant tank mix application is shallow incorporated (e.g. use 2 to 2½ qts Lasso in the early preplant tank mix when surface applied and use 2½ to 3 qts Lasso when the tank mix is to be lightly incorporated).

When weeds exceed 1 to 1½ inches in height or diameter at application, use a contact herbicide, such as Roundup or Paraquat.

Refer to the Dual or Lasso label, and to appropriate sections of the SENCOR label for additional information on soil preparation, herbicide application, weeds controlled, precautions, restrictions, limitations and sprayer clean-up.

EXTENDED SPLIT-SHOT APPLICATION

SOIL TEXTURE[1]	Early Preplant Tank Mix Application (Surface-Applied or Shallow Incorporated): Rate of Combination Product/acre	Plus	Rate of SENCOR DF pound/acre	followed by	Preemergence Overlay Application: Rate of Combination Product/acre	Plus	Rate of SENCOR DF pound/acre ORGANIC MATTER ½ to 2%	2 to 4%	Over 4%
COARSE (Light) Sand, loamy sand, sandy loam	Dual 1⅓ pt or Lasso 1½-2 qt	plus	⅓-½	followed by	Dual ⅔ pt or Lasso 1½ qt	plus	⅙	⅙-⅓	⅓
MEDIUM Loam, silt loam, sandy clay loam, silt, sandy clay	Dual 1¾ pt or Lasso 2-3 qt	plus	[2]½-⅔	followed by	Dual ¾ pt or Lasso 1-2 qt	plus	⅓	⅓-½	½-⅔
FINE (Heavy) Silty clay loam*, clay loam, silty clay, clay	Dual 2 pt or Lasso 2-3 qt	plus	[2]⅔-⅚	followed by	Dual 1 pt or Lasso 1-2 qt	plus	⅓	⅓-½	½-⅔

* Silty clay loam soils are transitional soils and may be classified as medium textured soils in some regions of the U.S.

[1] On **coarse textured** soils, do not use on sand soil with less than 1% organic matter. However, on coarse textured soils **with a calcareous surface area or a pH of 7.5 or higher,** do not use on sand soils with less than 2% organic matter, or on loamy sand or sandy loam soils with less than 1% organic matter.

[2] Use the lower rate of SENCOR in the early preplant tank mix **on soils having a calcareous surface area or a pH of 7.5 or higher,** and in those rare situations where soils within a field vary extremely in texture or organic matter content.

SENCOR PLUS SONALAN

SENCOR PLUS SONALAN OVERLAY APPLICATION

SENCOR may be applied as a preemergence overlay application following a preplant incorporated application of Sonalan 3 EC. Consult the Sonalan label for specific directions on use, recommendations, restrictions and any additional weeds not specified on the SENCOR label.

SENCOR PLUS SONALAN TANK MIX APPLICATION

Incorporate the tank mixture into the top 1 to 2 inches of soil within 21 days before planting according to label directions for Sonalan.

Apply SENCOR plus Sonalan preplant incorporated if furrow irrigation is used or when a period of dry weather after application is expected. If soybeans are planted on beds, apply and incorporate the tank mixture after bed formation.

Mixing: Refer to "Mixing" under the "GENERAL INFORMATION" section in the front of the SENCOR label.

Application: Sonalan should be uniformly applied and thoroughly mixed into the soil within 2 days after application. For specific application information, refer to "Application" under the "GENERAL INFORMATION" section in the front of the SENCOR label.

BROADCAST RATES

SOIL TEXTURE	SENCOR DF lbs/acre	Sonalan 3 EC pts/acre
COARSE[1] (Sandy loam, loamy sand)	⅓	1¼-2
MEDIUM[2] (Loam, silt loam, silt, sandy clay, sandy clay loam)	½	1¾-2½
FINE[3] (Silty clay, silty clay loam[2], clay, clay loam)	⅔	2¼-3

[1] Do not use on coarse soils with less than 1% organic matter.

[2] Silty clay loam soils are transitional soils and may be classified as medium textured soils in some regions of the U.S.

[3] For control of lambsquarters, redroot pigweed, wild mustard, and green and yellow foxtails on alkaline (calcareous) soils in Minnesota, Nebraska, South Dakota, and North Dakota only, apply SENCOR at rates of ⅓ pound/acre on medium soils and ⅓ to ½ pound/acre on fine soils regardless of soil organic matter percentage (use ½ pound only where soil pH is less than 7.5 and weed pressure is heavy). The ⅓ pound rate of SENCOR in tank mix combination with Sonalan can be applied regardless of soil pH. For control of other weeds not listed on the label, use SENCOR at full rates recommended in the table above, **but note that crop injury may occur on soils having a calcareous surface area or a pH of 7.5 or higher.**

SPECIAL PRECAUTIONS (SENCOR PLUS SONALAN)

For additional precautions, restrictions, limitations, incorporation, and sprayer cleanup information, refer to the appropriate sections of the SENCOR label and the Sonalan label.

For black nightshade control, refer to the Sonalan label for specific rates and application instructions.

SENCOR PLUS TREFLAN

SENCOR PLUS TREFLAN OVERLAY APPLICATION

SENCOR may be applied as a preemergence broadcast or band overlay application following a preplant incorporated treatment of Treflan. Consult the Treflan label for specific directions for use, recommendations, restrictions and any additional weeds not specified on the SENCOR label.

SENCOR PLUS TREFLAN TANK MIX APPLICATION

A single application of a tank mix combination of SENCOR and Treflan EC will control more broadleaf and grass weeds in soybeans than when either herbicide is used alone.

Prepare the soil surface by deep plowing, offset disking or tandem disking prior to the application of the herbicide combination. The soil surface should be well prepared and free of clods and trash.

This SENCOR plus Treflan tank mix combination may be applied and incorporated into the soil up to 10 days before planting.

Mixing

Refer to "Mixing" under the "General Information" section in the front of the SENCOR label.

Application:

For specific application information refer to "Application" under the "General Information" section in the front of the SENCOR label.

Apply SENCOR plus Treflan to the soil surface and incorporate in the same operation, if possible. Variable weed control may result from delayed incorporation if SENCOR plus Treflan are applied to a wet, warm soil surface or if the wind velocity is 10 miles per hour or higher. Use machinery that mixes SENCOR plus Treflan thoroughly with the soil. Incorporation may be delayed up to 24 hours after application. Shallow incorporation with implements set to cut less than 2 inches deep may result in erratic weed control. Do not use spike or spring-tooth harrows alone for incorporation.

Broadcast Rates

Soil Texture	SENCOR DF Pounds per Acre	Treflan EC Pints per Acre
Coarse[1] (Sandy loam, loamy sand)	⅓	1
Medium (Loam, silt loam, silt, sandy clay, sandy clay loam)[3]	½	1½
Fine (Silty clay, silty clay loam,[2] clay, clay loam)[3]	⅔	2

[1] Do not use on coarse soils with less than 1% organic matter.
[2] Silty clay loam soils are transitional soils and may be classified as medium textured soils in some regions of the U.S.
[3] For control of lambsquarters, redroot pigweed, wild mustard, and green and yellow foxtails on alkaline (calcareous) soils in Minnesota, Nebraska, South Dakota, and North Dakota only, apply SENCOR at rates of ⅓ pound/acre on medium soils and ⅓ to ½ pound/acre on fine soils regardless of soil organic matter percentage (use ½ pound only where soil pH is less than 7.5 and weed pressure is heavy). The ⅓ pound rate of SENCOR in tank mix combination with Treflan can be applied regardless of soil pH. For control of other weeds listed on the label use SENCOR at full rates recommended in the table above, **but note that crop injury may occur on soils having a calcareous surface area or a pH of 7.5 or higher.**

Incorporation Equipment:

1. Set PTO-driven equipment (tillers, cultivators, hoes) to cut 2 to 3 inches deep and space rotors to provide a clean sweep of the soil. PTO equipment should not be operated at a speed greater than 4 miles per hour.
2. Set disk to cut 4 to 6 inches deep and operate twice in different directions at 4 to 6 miles per hour.
3. Set mulch treader and other similar disk-type implements to cut 3 to 4 inches deep and operate twice in different directions at 5 to 8 miles per hour.

For Coarse and Medium Textured Soils Only

4. Set rolling cultivator to cut 2 to 4 inches deep and operate twice at 6 to 8 miles per hour. Set bed conditioner (Do-All) to cut 2 to 4 inches deep and operate at 4 to 6 miles per hour.

SPECIAL PRECAUTIONS (SENCOR PLUS TREFLAN)

Seedling disease, cold weather, excessive moisture, high salt concentration or drought may weaken soybean seedlings and increase the possibility of damage from the tank mix. Do not plant soybeans deeper than 2 inches.

In the Central United States, do not plant sorghum or oats for 12 months where the tank mix has been applied unless 20 inches or more of irrigation and/or rainfall (total) was used to produce the crop. If less than 20 inches total water was used to produce the crop during the year, do not plant either crop for 18 months after the tank mix application. Cool wet weather conditions during the early stage of growth may increase the possibility of injury to sorghum.

For additional precautions, restrictions, limitations and sprayer clean-up information refer to the appropriate section of the SENCOR label. Do not use this tank mix combination on soils containing charcoal in Arkansas, Louisiana and Mississippi.

SENCOR PLUS DUAL

SENCOR PLUS DUAL OVERLAY APPLICATION

Apply a preplant incorporated treatment of Dual 8E as directed on that product label for use on soybeans. Follow with a preemergence treatment of SENCOR as directed on this label for use on soybeans.

SENCOR PLUS DUAL TANK MIX APPLICATION

Preplant Incorporated Application: Incorporate the tank mixture into the top 2 inches of soil within 14 days before planting using a disk, harrow, rolling cultivator, or similar implement.

Apply SENCOR plus Dual preplant incorporated if furrow irrigation is used or when a period of dry weather after application is expected. If soybeans are planted on beds, apply and incorporate the tank mixture after bed formation.

Preemergence Application: Dry weather following preemergence application of SENCOR plus Dual tank mixture may reduce effectiveness. If weeds develop, cultivate uniformly with shallow tilling equipment such as a rotary hoe that will not damage soybeans.

Mixing Instructions: Refer to "Mixing" under the "General Information" section in the front of the SENCOR label.

For information on applying SENCOR in fluid or dry fertilizer refer to the "Application Of SENCOR In Fluid Fertilizers" or "Commercial Impregnation And Application Of SENCOR On Dry Bulk Fertilizer" under the "General Information" section in the front of the SENCOR label.

Broadcast Rates
SENCOR plus Dual Tank Mix Preemergence Applications

Soil Texture	Organic Matter			
	0.5% to 3%		Over 3%	
	SENCOR DF Pounds/Acre	Dual 8EC Pints/Acre	SENCOR DF Pounds/Acre	Dual 8EC Pints/Acre
Coarse[1] (Loamy sand, Sandy loam)	⅓	1¼	½	1½
Medium (Loam, silt loam, silt)	½	1½	⅔	2
Fine (Silty clay loam[2], sandy clay loam, silty clay, sandy clay, clay loam, clay)	⅔	2	⅔ to ⅚	2 to 2½
Mississippi Delta Only (Silty clay, clay)	1	2	1	2 to 2½

[1] Do not use on sand soils. Do not apply SENCOR and Dual overlay or tank mix preemergence on loamy sand with less than 2% organic matter.
[2] Silty clay loam soils are transitional soils and may be classified as medium textured soils in some regions of the U.S.

Broadcast Rates
SENCOR plus Dual Tank Mix Preplant Incorporated Applications

Soil Texture	Organic Matter			
	0.5% to less than 3%		3% or greater	
	SENCOR DF Pounds/Acre	Dual 8EC Pints/Acre	SENCOR DF Pounds/Acre	Dual 8EC Pints/Acre
Coarse[1] (Loamy sand, sandy loam)	⅓	1¼	⅓	1½
Medium (Loam, silt loam, silt)	½	1½	½	2
Fine (Silty clay loam[2], sandy clay loam, silty clay, sandy clay, clay loam, clay)	⅔	2	⅔	2 to 2½
Mississippi Delta Only: (Silty clay, clay)	⅔ to ⅚	2	⅔ to ⅚	2 to 2½

[1] Do not use on sand soils. Do not apply SENCOR plus Dual tank mix preplant incorporated on sand or loamy sand with less 2% organic matter or crop injury may occur.
[2] Silty clay loam soils are transitional soils and may be classified as medium textured soils in some regions of the U.S.

SPECIAL PRECAUTIONS (SENCOR PLUS DUAL)

For additional precautions, restrictions, limitations, and sprayer clean-up information refer to the appropriate sections of the SENCOR and the Dual label.

SENCOR PLUS PROWL

SENCOR PLUS PROWL OVERLAY APPLICATION

Apply a preplant incorporated treatment of Prowl as directed on that product label for use on soybeans. Follow with a preemergence treatment of SENCOR as directed on the SENCOR label for use on soybeans.

SENCOR PLUS PROWL TANK MIX APPLICATION

Preplant Incorporated Application: Prepare the soil by plowing or disking to mix previous crop residues into the soil to a depth of 4 to 6 inches.

For specific application information refer to "Application" under the "General Information" section in the front of the SENCOR label.

Incorporate the tank mixture into the top 1 or 2 inches of soil within 7 days after application according to label directions for Prowl. Mechanical incorporation is not required if a rain of one-quarter inch or more occurs within 7 days after application. Soybeans must be planted no later than 7 days after application of the tank mixture.

Preemergence Application: Except for minimum and no-tillage systems, the seed bed should be firm and free of trash and clods.

For specific application information refer to "Application" under the "General Information" section in the front of the SENCOR label.

This application must be made after planting and before crop emergence. Do not incorporate.

If cultivation is necessary because of soil crusting, soil compaction or weed germination before rain or irrigation, use shallow tilling equipment such as a rotary hoe that does not damage soybeans.

Mixing Instructions: Refer to "Mixing" under the "General Information" section in the front of the SENCOR label.

For information on applying SENCOR in fluid or dry fertilizer refer to the "Application Of SENCOR In Fluid Fertilizers" or "Commercial Impregnation And Application Of SENCOR On Dry Bulk Fertilizer" under the "General Information" section in the front of the SENCOR label.

SOUTHERN STATES AND EASTERN COASTAL PLAINS

For use only in Alabama, Arkansas, Florida, Georgia, Louisiana, Mississippi, North Carolina, Oklahoma, Southeastern Missouri "Bootheel" Region and Coastal Plains of Delaware*, Maryland*, New Jersey* and Virginia*.

* SENCOR plus Prowl is not recommended for use on soils with less than 2% organic matter in the coastal plain of New Jersey or the Delmarva Peninsula.

BROADCAST RATES
SENCOR plus Prowl Tank Mix Applications

Soil Texture	SENCOR DF Pounds/Acre	Prowl Pints/Acre
Coarse[1] (Sandy loam, loamy sand)	⅓	1½
Medium (Loam, silt loam, silt, sandy clay, sandy clay loam)	½	1½
Fine (Silty clay, silty clay loam[2], clay, clay loam)	⅔	1½ to 2

[1] Do not use on sand soils. Do not use on loamy sand or sandy loam containing less than 1% organic matter. Where a range of rates is shown for medium and fine soils, use the higher rate if heavy weed infestations are anticipated.

[2] Silty clay loam soils are transitional soils and may be classified as medium textured soils in certain regions of the U.S.

Do not use on muck or peat soils.

NORTHEASTERN AND NORTH CENTRAL STATES

For use only in Illinois, Indiana, Iowa, Kansas, Kentucky, Michigan, Minnesota, Nebraska, New York, North Dakota, Ohio, Pennsylvania, South Dakota, Wisconsin and Missouri (except the "Bootheel" Region).

Broadcast Rates
SENCOR plus Prowl Tank Mix Applications

Soil Texture	Organic Matter ½ to 3% SENCOR DF Pounds/Acre	½ to 3% Prowl Pints/Acre	Over 3% SENCOR DF Pounds/Acre	Over 3% Prowl Pints/Acre
Coarse (Sandy loam, loamy sand)	⅓	1	½	1½
Medium (Loam, silt loam, sandy clay, sandy clay loam)	½	1½ to 2	½ to ⅔	1½ to 2
Fine (Silty clay, silty clay loam[2], clay, clay loam)	½ to ⅔	1½ to 2	⅔ to ⅚	2 to 2½

[1] Do not use on sand soils. Do not use on loamy sand or sandy loam containing less than 1% organic matter. Where a range of rates is shown for medium and fine soils, use the higher rate if heavy weed infestations are anticipated.

[2] Silty clay loam soils are transitional soils and may be classified as medium textured soils in certain areas of the U.S.

Do not use on muck or peat soils.

SPECIAL PRECAUTIONS (SENCOR PLUS PROWL)

Soil incorporation deeper than recommended will reduce weed control and can result in crop injury.

For additional precautions, restrictions, limitations, and sprayer clean-up information, refer to the appropriate sections of the SENCOR and the Prowl label.

SENCOR PLUS LASSO

SENCOR PLUS LASSO TANK MIX APPLICATION

PREEMERGENCE: SENCOR may be used in a tank mix combination with Lasso as a preemergence band or broadcast application to soybeans in accordance with the specified soil types and dosages recommended.

For specific information regarding spray equipment, dilution rates, mixing, directions for use, methods of application, limitations and restrictions refer to the appropriate section of the SENCOR label.

Refer to the Lasso label for pertinent recommendations, directions for use, restrictions and any additional weeds not specified on the SENCOR label.

Recommended Applications
SENCOR plus Lasso Tank Mix Preemergence Application
(Broadcast rates)

Soil Texture	½ to 3% Organic Matter SENCOR DF Lbs./Acre	+	½ to 3% Organic Matter Lasso Qts./Acre	Greater than 3% Organic Matter SENCOR DF Lbs./Acre	+	Greater than 3% Organic Matter Lasso Qts./Acre
Coarse[1] (Sandy loam)	⅓	+	1½ to 2	½	+	1½ to 2
Medium[2] (Loam, silt loam, silt, sandy clay, sandy clay loam)	½	+	1½ to 2	⅔	+	1½ to 2
Fine[2] (Silty clay, silty clay loam,[3] clay, clay loam)	⅔	+	2	⅔ to ⅚	+	2 to 2½
Mississippi Delta Only (Silty clay to heavy clay)	SENCOR DF Lbs./Acre 1⅓	+	Lasso Qts./Acre 2 to 2½			

[1] Do not use SENCOR plus Lasso on sand or loamy sand soils with less than 2% organic matter.

[2] For control of lambsquarters, redroot pigweed, wild mustard, green and yellow foxtails on alkaline (calcareous) soils in Minnesota, Nebraska, South Dakota, and North Dakota only, apply SENCOR DF at rates of ⅓ pound/acre on medium soils and ⅓ to ½ pound/acre on fine soils regardless of soil organic matter percentage (use ½ pound only where soil pH is less than 7.5 and weed pressure is heavy). The ⅓ pound/acre rate of SENCOR in tank mix combination with Lasso can be applied regardless of soil pH. For control of other weeds use SENCOR at full rates recommended in the table above **but note that crop injury may occur on soils having a calcareous surface area or a pH of 7.5 or higher.**

[3] Silty clay loam soils are transitional soils and may be classified as medium textured soils in some regions of the U.S.

DO NOT USE ON MUCK SOILS.

PREPLANT INCORPORATED: For specific application information refer to "Application" under the "General Information" section in the front of the SENCOR label.

Apply SENCOR plus Lasso preplant incorporated if furrow irrigation is used or when a period of dry weather after application is expected. If soybeans are planted on beds, apply and incorporate the tank mixture after bed formation. Apply within 7 days prior to planting and shallowly incorporate into the upper 1 to 2 inches of soil.

RECOMMENDED APPLICATIONS
SENCOR plus Lasso Tank Mix Preplant Incorporated Applications (Broadcast Rates)

Soil Texture	SENCOR DF Pounds/Acre	Lasso Quarts/Acre
Coarse[1] (Loamy sand [over 2% organic matter], sandy loam)	1/3	2 to 2 1/2
Medium (Loam, silt loam, silt)	1/2	2 1/2 to 3
Fine (Silty clay loam[2], sandy clay loam, silty clay, sandy clay, clay loam, clay)	2/3	2 1/2 to 3
Mississippi Delta Only: (Silty clay, clay)	2/3 to 5/6	2 1/2 to 3

[1] Do not use SENCOR plus Lasso on sand or loamy sand soils with less than 2% organic matter.
[2] Silty clay loam soils are transitional soils and may be classified as medium textured soils in some regions of the U.S.

DO NOT USE ON MUCK SOILS

SPECIAL PRECAUTIONS (SENCOR PLUS LASSO)

For additional precautions, restrictions, limitations and sprayer clean-up information refer to the appropriate sections of the SENCOR and the Lasso label.

SENCOR PLUS SURFLAN

SENCOR PLUS SURFLAN TANK MIX APPLICATION

SENCOR may be used in a tank mix combination with Surflan A.S. as a preemergence application to control weeds in soybeans in accordance with the specified soil types and dosages recommended. Consult the Surflan label for specific directions for use, recommendations, restrictions and any additional weeds not specified on the SENCOR label.

Mixing Instructions: Refer to "Mixing" under the "GENERAL INFORMATION" section in the front of the SENCOR label.

Application: Apply within 2 days after planting but before crop emergence.

A one-half inch rain or more or its equivalent in overhead irrigation is needed to activate the herbicide combination. If a one-half inch rain does not occur within 7 days after application, incorporate the herbicide combination into the top one inch of the soil with a ground-driven rotary hoe, rolling cultivator, or similar implement that will not damage soybean plants. Cultivation will control emerging annual weeds and grasses and improve herbicidal effectiveness. If heavy rains pack the soil surface, rotary hoe to break the crust and to aid soybean emergence.

For specific information on application, refer to "Application" under the "GENERAL INFORMATION" section in the front of the SENCOR label.

BROADCAST RATES

Soil Texture	SENCOR DF Pounds Per Acre	Surflan A.S. Quarts Per Acre
Coarse[1] (Sand, sandy loam, loamy sand)	1/3	1/2 to 3/4
Medium (Loam, silt loam, silt, sandy clay, sandy clay loam)	1/2	3/4 to 1
Fine (Silty clay, silty clay loam[2], clay, clay loam)	2/3	1 to 1 1/4
Mississippi Delta Only	2/3 to 5/6	1 to 1 1/4

[1] Not recommended on sand with less than 1% organic matter. Do not use on soils containing more than 5% organic matter.
[2] Silty clay loam soils are transitional soils and may be classified as medium textured soils in some regions of the U.S.

SPECIAL PRECAUTIONS (SENCOR PLUS SURFLAN)

Do not plant root crops for 12 months following application.

Do not contaminate any body of water nor apply to any area not specified on the label.

For additional precautions, restrictions, limitations and sprayer clean-up information, refer to the appropriate section of this label and the Surflan A.S. labels.

SENCOR PLUS AMIBEN

SENCOR PLUS AMIBEN PREEMERGENCE TANK MIX APPLICATION

SENCOR may be applied in a tank mix combination with Amiben 2 as a band or broadcast application to soybeans.

SENCOR plus Amiben Preemergence Tank Mix Application Broadcast Rates

Soil Texture	1/2 to 3% Organic Matter: SENCOR DF Lbs./Acre	+	Amiben 2 Gal./Acre	Greater than 3% Organic Matter: SENCOR DF Lbs./Acre	+	Amiben 2 Gal./Acre
Coarse[1] (Sandy loam, loamy sand)	1/3	+	3/4 to 1	1/2	+	3/4 to 1
Medium (Loam, silt loam, silt, sandy clay, sandy clay loam)	1/2	+	3/4 to 1	2/3	+	1 to 1 1/4
Fine (Silty clay, silty clay loam[2], clay, clay loam)	2/3	+	1 to 1 1/4	2/3 to 5/6	+	1 to 1 1/4
Mississippi Delta Only (Silty clay to heavy clay)	SENCOR DF Lbs./Acre: 1 1/3	+	Amiben Gals./Acre: 1 1/2			

[1] Do not use on sandy loam or loamy sand soil with less than 2% organic matter. Do not use on sand soils.
[2] Silty clay loam soils are transitional soils and may be classified as medium textured soils in some regions of the U.S.

Do not use on muck soils.

Where a rate range is specified, use the higher rate in fields with a history of intense weed pressure. See "Rate Ranges" in the front of the "Soybeans" section on SENCOR label.

For additional precautions and information regarding spray equipment, mixing, application, limitations and restrictions refer to the appropriate section of SENCOR label and the Amiben label.

SENCOR PLUS DYANAP

SENCOR PLUS DYANAP TANK MIX APPLICATION

A tank mix combination of SENCOR and Dyanap preemergence, band or broadcast, may be applied up to cracking stage of soybeans. Apply SENCOR at the rates recommended on the label for use alone on soybeans. Refer to the Dyanap label for correct rates of that product and for additional information on weed control, mixing, application, restrictions and precautionary statements.

SENCOR PLUS LASSO AND DYANAP

SENCOR PLUS LASSO AND DYANAP TANK MIX APPLICATION

SENCOR may be applied in a tank mix combination with Lasso and Dyanap as a preemergence band or broadcast application up to the cracking stage of soybeans. For additional information concerning weeds controlled, mixing, application, limitations or restrictions, refer to the Lasso and Dyanap labels under Lasso plus SENCOR and Dyanap plus Lasso tank mixes and appropriate sections of SENCOR label.

SENCOR DF (Pounds) plus Lasso 4EC (Quarts) and Dyanap (Quarts)

Soil Texture	Broadcast Rates Per Acre		
	SENCOR (Lbs.)	Lasso (Qts.)	Dyanap (Qts.)
Coarse[1] (Sandy loam, loamy sand)	⅓	1½	2 to 4
Medium (Loam, silt loam, silt, sandy clay, sandy clay loam)	½ to ⅔	1½ to 2	3 to 4
Fine (Silty clay, silty clay loam[2], clay, clay loam)	½ to ⅔	2	4

[1] Not recommended on sand soil with less than 1% organic matter.
[2] Silty clay loam soils are transitional soils and may be classified as medium textured soils in some regions of the U.S.

SENCOR PLUS LINURON PLUS (LASSO OR DUAL)

SENCOR PLUS LINURON PLUS (LASSO OR DUAL) TANK MIX APPLICATION

SENCOR may be applied in combination with linuron and Lasso 4 or Dual 8EC as a preemergence application for the control of certain weeds in soybeans. Consult the linuron, Lasso, or Dual labels for specific directions for use, recommendations, restrictions and any additional weeds not specified on the SENCOR label.

Mixing: Refer to "Mixing" under the "GENERAL INFORMATION" section in the front of the SENCOR label.

Application: Applications can be made only with ground spray equipment in accordance with specific soil types and dosage ranges. For specific application information, refer to "Application" under the "GENERAL INFORMATION" section in the front of the SENCOR label.

BROADCAST RATES
(0.5 to 3% Organic Matter Only

Soil Texture	SENCOR DF Pounds/Acre	linuron 50 WP or 4 L (Pounds/Acre) or (Pints/Acre)	Lasso 4 Quarts/Acre	or	Dual 8EC Pints/Acre
Coarse[1] (Sand, loamy sand, sandy loam)	⅙ to ¼	⅓ to ½	¾ to 1	or	1 to 1¼
Medium (Loam, silt loam, silt, sandy clay, sandy clay loam)	¼ to ⅓	½ to ¾	1 to 1½	or	1¼ to 1½
Fine (Silty clay, silty clay loam[2], clay, clay loam)	⅓ to ½	¾ to 1½	1¼ to 2	or	1½ to 2

[1] DO NOT USE SENCOR plus linuron plus (Lasso or Dual) on sand soils with less than 1% organic matter.
[2] Silty clay loam soils are transitional soils and may be classified as medium textured soils in some regions of the U.S.

SPECIAL PRECAUTIONS SENCOR PLUS LINURON PLUS (LASSO OR DUAL)

For additional precautions, restrictions, limitations, and sprayer cleanup information, refer to the appropriate sections of the SENCOR, linuron, Lasso, or Dual labels.

FOR USE IN COARSE (LIGHT) SOILS IN Alabama, Arkansas, Florida, Georgia, Louisiana, Mississippi, Missouri, North Carolina, Oklahoma, South Carolina, Tennessee, Texas and Virginia.

SENCOR herbicide is recommended alone or in combination with Treflan, Lasso, Dual, Surflan or Amiben for use in coarse-textured, low organic matter soils in the states listed above for the control of certain weeds in soybeans. Refer to the appropriate sections of the SENCOR, Treflan, Lasso, Dual, Surflan or Amiben labels for specific directions for use, recommendations, restrictions and any additional weeds not specified on the SENCOR label.

Mixing: Refer to "Mixing" under the "GENERAL INFORMATION" section in the front of the SENCOR label.

Application: For specific application information, refer to "Application" under the "GENERAL INFORMATION" section in the front of the SENCOR label.

SENCOR (ALONE) PREEMERGENCE APPLICATION

BROADCAST RATES

Soil Texture	Organic Matter	SENCOR DF Pounds/Acre
Coarse (Light) Soils (Sand[1], loamy sand, sandy loam)	0.5 to 2.0%	⅓ to ½[2]

[1] Not recommended for use on sand with less than 1% organic matter.
[2] Use the higher rate under heavy weed pressures and/or on soils higher in organic matter.

SENCOR IN COMBINATION WITH OTHER HERBICIDES

SENCOR is recommended in a tank mix combination with Treflan as a preplant incorporated application or as a preemergence overlay application following a preplant incorporated application of Treflan. SENCOR is also recommended for use as a preemergence application in combination with Lasso, Dual, Surflan or Amiben.

BROADCAST RATES

Soil Texture	Organic Matter	Combination Product/Acre	Plus SENCOR DF Pounds/Acre
Coarse (Light) Soils (Sand[1], loamy sand sandy loam)	0.5 to 2.0%	Preplant Incorporated Treflan 4EC 1 pt	plus ⅓ to ½[2]
		-or- Preemergence Lasso 4EC 1½ to 2 qt -or- Dual 8EC 1¼ to 1½ pt -or- Surflan A.S. ½ to ¾ qt -or- Amiben 2 ¾ to 1 gal	plus ⅓ to ½[2]

[1] Not recommended for use on sand with less than 1% organic matter.
[2] Use the higher rate under heavy weed pressures and/or on soils higher in organic matter.

SPECIAL PRECAUTIONS

For additional precautions, restrictions, limitations and sprayer cleanup information, refer to the appropriate sections of the SENCOR, Treflan, Lasso, Dual, Surflan or Amiben labels.

Do not use on sand soils with less than 1% organic matter, or on sandy loam or loamy sand soils with less than 0.5% organic matter.

FOR CONTACT ACTION AND RESIDUAL CONTROL

For Control of Emerged Annual Broadleaf Weeds and Grasses and For Suppression of Emerged Perennials Where Soybeans Will Be Planted Directly Into a Pre-formed Bed, Cover Crop or in Previous Crop Residues as in "Minimum Till" or "No Till" Crop Production.

SENCOR PLUS PARAQUAT

SENCOR PLUS PARAQUAT TANK MIX APPLICATION

Apply 1 to 2 pints Paraquat + Plus plus 1/2 to 1⅓ pounds SENCOR DF according to the recommendations for SENCOR on the label as a broadcast ground spray before, during or after planting but before emergence of the crop. Soybeans exposed at time of application will be killed. Use the higher rate of Paraquat + Plus for heavier infestations or for taller weeds (4 to 6 inches tall). Apply in 20 to 60 gallons spray mix per acre. Use the higher volume in fields with heavy weed infestations or with large amounts of previous crop residues. Add Ortho X-77® Spreader or similar non-ionic spreader at 1 quart per 100 gallons of diluted spray. This mixture will not control weeds taller than 6 inches.

SENCOR PLUS LASSO AND PARAQUAT

SENCOR PLUS LASSO AND PARAQUAT TANK MIX APPLICATION

Apply 1 to 2 pints Paraquat + Plus plus 2 to 2½ quarts Lasso 4 E.C. plus ⅓ to 1⅓ pounds SENCOR DF as a preemergence broadcast ground spray. Use the higher rate of Paraquat + Plus for weeds 4 to 6 inches tall. Soybeans exposed at time of application will be killed. Refer to the appropriate sections of this label and the Paraquat and Lasso labels for rates which are dependent upon soil textures, directions, limitations and cautions.

Apply in 20 to 60 gallons spray mix per acre. Use the higher volume in fields with heavy weed infestations or with large amounts of previous crop residues. Add Ortho X-77 Spreader or similar non-ionic spreader at 8 fl. ounces per 100 gallons of diluted spray. Add SENCOR to the spray tank first and mix thoroughly. Then add Lasso followed by Paraquat + Plus and Ortho X-77 Spreader. Constant agitation is recommended at all times. This tank mix will not control shattercane which emerges after treatment or other weeds taller than 6 inches.

SENCOR PLUS SURFLAN AND PARAQUAT

SENCOR PLUS SURFLAN AND PARAQUAT TANK MIX APPLICATION

The combination provides control of weeds and grasses which are emerged at the time of application plus many weeds and grasses which may germinate after planting. Soybeans exposed at time of application will be killed.

Weeds Controlled by Paraquat: Paraquat will kill emerged annual broadleaf weeds and grasses and provide top kill of perennials.

When to Apply: Apply the tank mix at planting or within 2 days after planting. Do not spray over top of emerged soybeans.

SENCOR PLUS SURFLAN AND PARAQUAT PREEMERGENCE TANK MIX APPLICATION

	Broadcast Rates Per Acre		
Soil Texture:	SENCOR DF	Surflan A.S.	Paraquat + Plus
Coarse[1]	⅓ pound	¾ quart	1 quart
Medium	½ pound	1 quart	1 quart
Fine	⅔ pound	1½ quart	1 quart

[1] Not recommended on sand soils with less than 1% organic matter. Do not use on soils containing more than 5% organic matter.

Mixing: Refer to "Mixing" under the "General Information" section in the front of the SENCOR label.

Add Ortho X-77 or similar non-ionic spreader at 8 fl. ozs. per 100 gallons of diluted spray mixture.

APPLICATION: For specific application information refer to "Application" under the "General Information" section in the front of the SENCOR label.

A one-half inch rain or more or its equivalent in overhead irrigation is needed to activate the herbicide combination.

SPECIAL PRECAUTIONS (SENCOR PLUS SURFLAN AND PARAQUAT): Poor weed control and/or crop injury may result if directions are not followed.

Do not use a rib-type press wheel on your no-till planter or crop injury may result.

For additional precautions, special precautions, restrictions, limitations and sprayer cleanup information, refer to the appropriate section of the SENCOR, Surflan and Paraquat labels.

SOUTHERN AND SOUTHEASTERN STATES ONLY

POSTEMERGENCE DIRECTED SPRAY APPLICATIONS

SENCOR can be applied in postemergence directed sprays to soybeans for control of certain weeds which escape preplant or preemergence herbicide applications and for control of additional flushes of weeds that may occur after soybeans have emerged. Postemergence directed sprays of SENCOR can be applied to soybeans in addition to a preemergence or preplant application of SENCOR according to label directions.

WEEDS CONTROLLED

SENCOR, applied postemergence to soybeans as a directed spray according to directions on this label, will control the following at rates shown (broadcast basis) when grasses and common ragweed are less than 1 inch tall and other broadleaves are less than 3 inches tall:

⅓ pound/acre	
Florida beggarweed	*Desmodium tortuosum*
Pigweeds	*Amaranthus* spp.
Carpetweed	*Mollugo verticillata*
Cocklebur	*Xanthium pensylvanicum*
Dayflower	*Commelina* spp.
Mexicanweed	*Caperonia castaniifolia*
Purslane	*Portulaca oleracea*
Sicklepod	*Cassia obtusifolia*
Velvetleaf	*Abutilon theophrasti*
Crabgrass	*Digitari* spp.
⅓ to ⅔ pound/acre	
Sesbania	*Sesbania* spp.
Prickly sida/Teaweed	*Sida spinosa*
⅔ pound/acre	
Ragweed, common	*Ambrosia artemisiifolia*

At the rate of ⅔ pound/acre morningglory species, (*Ipomoea* spp.) horsenettle, (*Solanum* spp.) Florida pusley, (*Richardia scabra*) spotted spurge (*Euphorbia maculata*) and wild poinsettia (*Euphorbia heterophylla*) are suppressed when SENCOR is applied before these weeds are 3 inches tall. The ⅔ pound/acre rate will suppress broadleaf signalgrass (*Brachiaria platyphylla*) up to 1 inch tall.

SENCOR POSTEMERGENCE DIRECTED SPRAY APPLICATIONS
Recommended Applications

Crop	SENCOR DF Pounds Per Acre	Remarks
Soybeans (Alabama, Arkansas, Florida, Georgia, Kentucky, Louisiana, Mississippi, Missouri, North Carolina, Oklahoma, South Carolina, Tennessee and Texas)	⅓ to ⅔ (broadcast basis)	Apply proper dosage using 10 to 40 gallons of water per acre as a directed spray in a 6 to 8 inch band on each side of the row after soybeans are 8 inches tall and before broadleaf weeds are 3 inches tall and before grasses and common ragweed are 1 inch tall. For best results the spray must cover weed foliage with minimum or no contact with soybean foliage. Add a non-ionic surfactant such as Ortho X-77® to the spray mixture to obtain better wetting of weed leaf surfaces. To determine the correct dosage of SENCOR DF

Crop	SENCOR DF Pounds Per Acre	Remarks
		for a band application see "Banded Application" under "Application" in the "General Information" section in the front of the label. If necessary, a second postemergence directed spray application can be made after 7 days. Do not feed or graze green soybean vines. Do not harvest soybeans or use dry soybean vines for feed or forage within 70 days of last application.

SPECIAL PRECAUTIONS (DIRECTED POSTEMERGENCE)

Do not apply directly to soybeans or serious crop injury will occur. Do not allow spray to contact more than the lower ¼ to ⅓ of soybean plants. Soybean leaves contacted by the spray will be killed.

Do not apply SENCOR postemergence to the sensitive soybean varieties. See "Soybean Precautions" on the front of the SENCOR label.

To avoid injury to other crops or desirable plants from spray drift, sprayer pressure must not exceed 30 psi and the sprayer must be fitted with nozzles no smaller than 8002 T-Jet (or equivalent). Do not apply under weather conditions which favor drift.

POTATOES

SENCOR herbicide is recommended for use in ground or aircraft spray equipment as a preemergence and/or postemergence application to potatoes. Postemergence applications may be made only on white-skinned potato varieties that are not early maturing. For optimum control of weeds in potatoes apply SENCOR DF before weeds are 1-inch tall. Refer to special precautions for potatoes for additional information.

Ground Application: SENCOR is recommended for use with ground spray equipment, except air blast sprayers, as a preemergence and/or postemergence application for control of certain grass and broadleaf weeds in potatoes. Apply as a broadcast spray after planting and before crop emergence or after drag-off if this operation is part of the normal cultural practice. Postemergence applications should be applied after crop emergence.

Aerial Application: SENCOR is recommended for use in aerial spray equipment as a preemergence and/or postemergence application for control of certain grass and broadleaf weeds in potatoes. Apply as a broadcast spray after planting and before crop emergence or after drag-off if this operation is part of the normal cultural practice. Postemergence applications should be applied after crop emergence. Do not apply aerially when SENCOR is tank mixed with Lasso.

PREEMERGENCE

WEEDS CONTROLLED

SENCOR, applied preemergence to potatoes according to directions, will effectively control:

Broadleaves

Carpetweed	*Mollugo verticillata*
Cocklebur	*Xanthium pensylvanicum*
Jimsonweed	*Datura stramonium*
Lambsquarters	*Chenopodium album*
Pigweeds	*Amaranthus* spp.
Ragweed, Common	*Ambrosia artemisiifolia*
Shepherdspurse	*Capella bursa-pastoris*
Sicklepod	*Cassia obtusifolia*
Smartweed, Pennsylvania	*Polygonum pensylvanicum*
Wild Mustards	*Brassica* spp.

Grasses

Broadleaf Signalgrass	*Brachiaria platyphylla*
Crabgrass	*Digitaria* spp.
Foxtails	*Setaria* spp.
Johnsongrass, Seedling	*Sorghum halepense*
Panicum, Fall	*Panicum dichotomiflorum*

For Control of Wild Mustard (Brassica spp.) in Potatoes (Except California)—Apply at the rate of ⅓ to ⅔ pound per acre as a preemergence broadcast spray using aerial or ground spray equipment in accordance with directions on this label. Application should be made at or near ground crack and/or before the mustard is one inch tall. For control of mustard larger than 2 inches in diameter or other weeds listed on the label, a minimum of ⅔ pound per acre must be used. When less than ⅔ pound per acre is used, a postemergence application in accordance with directions on the SENCOR label may be required for control of other weeds or later germinating weed species.

Hard To Control Weeds: Although SENCOR applied preemergence may not provide commercially acceptable control in every instance, it will suppress growth of the following weeds and reduce their competition with potato plants.

Broadleaf Weeds

Common Purslane	*Portulaca oleracea*

Grasses

Barnyardgrass	*Echinochloa crus-galli*
Nutsedge	*Cyperus* spp.

POSTEMERGENCE

WEEDS CONTROLLED

SENCOR, applied early postemergence to potatoes according to directions will effectively control:

Broadleaves

Cocklebur	*Xanthium pensylvanicum*
Lambsquarters	*Chenopodium album*
Pigweed, Redroot	*Amaranthus retroflexus*
Ragweed, Common	*Ambrosia artemisiifolia*
Russian thistle	*Salsola kali*
Smartweed, Pennsylvania	*Polygonum pensylvanicum*

Grasses

Foxtails	*Setaria* spp.

On medium and fine soils in Idaho, Oregon and Washington, two postemergence applications, each at the rate of ⅔ pound per acre, will control barnyardgrass (*Echinochloa crus-galli*), kochia (*Kochia scoparia*), sunflower (*Helianthus* spp.), and Jim Hill mustard (*Sisymbrium altissimum*) in addition to weeds listed above.

Hard To Control Weeds: Although SENCOR applied early postemergence may not provide commercially acceptable control in every instance, it will suppress growth of the following weeds and reduce their competition with potato plants.

Broadleaves:

Kochia	*Kochia scoparia*
Sunflower	*Helianthus spp.*

Grasses:

Barnyardgrass	*Echinochloa crus-galli*

Recommended Broadcast Applications

Crop	SENCOR DF Pounds Per Acre	Remarks
Potatoes	⅔ to 1⅓	**PREEMERGENCE:** Apply specified dosage per acre as a broadcast spray to the soil surface after planting and before crop emergence or after drag-off if this operation is part of normal cultural practice.* Distribute the chemical evenly over the area to be treated. Do not incorporate into soil. Use the lower rate on sandy soils.
Potatoes (white-skinned varieties that are not early maturing)	⅓ to ⅔	**POSTEMERGENCE:** Apply specified dosage per acre as a broadcast spray over the tops of potato plants.* Use rates of ⅓ to ⅔ lb. for control of redroot pigweed and lambsquarters only. Apply the ⅔ lb. rate for control of other weeds listed on this label. Do not apply within 60 days of harvest.

		SPLIT APPLICATIONS: This product may be applied once preemergence and once post-emergence as directed above.* Do not apply postemergence within 60 days of harvest.
		IDAHO, OREGON AND WASHINGTON ONLY: Two postemergence applications can be made as broadcast sprays over the tops of potato plants if SENCOR is not applied preemergence. Use $\frac{1}{3}$ to $\frac{2}{3}$ pound/acre for control of redroot pigweed and lambsquarters only. On coarse (sandy) soils with low organic matter do not exceed $\frac{1}{3}$ pound/acre per application in Oregon and Washington or $\frac{1}{2}$ pound/acre per application in Idaho. On medium and heavy soils only use $\frac{2}{3}$ pound/acre for control of other weeds listed on the label. Make the first application early in the season while weeds are still small. Allow at least 14 days before the second application. For optimum control apply before weeds are 1 inch tall. Do not apply after June 30 if treated land is to be planted to crops other than potatoes. Do not apply within 60 days of harvest.

* Refer to Special Precautions—Potatoes below.

SPRINKLER IRRIGATION APPLICATIONS TO POTATOES

SENCOR may be applied early postemergence to potatoes through center pivot, permanent solid set or self-propelled wheel move irrigation systems equipped with automatic shutoff devices and valves to prevent backflow into the water source. Interlocking controls should be installed between the water pump and pesticide pump so that both pumps stop if either one should fail. Maintain continuous agitation in injection nurse tanks during application. Inject the product with a positive displacement pump in to the main line ahead of a right angle turn to insure adequate mixing. Greater accuracy in calibration and distribution will be achieved by injecting a larger volume of more dilute mixture of product and water per hour. Application of more than the quantity of irrigation water recommended on this label may result in decreased product performance by removing the chemical from the zone of effectiveness. Where sprinkler distribution patterns do not overlap sufficiently unacceptable weed control may result. Where sprinkler distribution patterns overlap excessively crop injury may result. Allow sufficient time for pesticide to be flushed through all lines and all nozzles before turning off irrigation water. To ensure that lines are flushed and free of remaining pesticide, an indicator dye may be injected into the lines to mark the end of the application period.

Calibration: (Center Pivot and Self-Propelled Wheel Move Systems)

1. Determine number of minutes required to make one complete revolution while applying $\frac{1}{4}$ to $\frac{3}{4}$ inch of water per acre.
2. With the system at operating pressure determine the exact number of minutes required to inject one gallon of water.
3. Divide the time required for one revolution (step 1) by the time required to inject one gallon (step 2). This gives total gallons of product-water mixture to be added to nurse tank.
4. Add required amount of water to nurse tank and start the agitation system. Then add sufficient SENCOR herbicide at the recommended rate (see RECOMMENDED BROADCAST APPLICATIONS) to the nurse tank.

EXAMPLE

If 20 hours (1200 minutes) were required for one revolution and if 2 minutes were required to inject one gallon then a total of 600 gallons of product-water mixture are required ($1200 \div 2 = 600$); to treat 135 acres at $\frac{2}{3}$ lb/acre, 90.5 pounds of SENCOR DF are required.

Recommended Broadcast Applications

Crop	Pounds/Acre SENCOR DF	Remarks
Potatoes (white-skinned varieties that are not early maturing)	$\frac{1}{3}$ to $\frac{2}{3}$	**POSTEMERGENCE IN SPRINKLER IRRIGATION:** Apply specified dosage in $\frac{1}{4}$ to $\frac{3}{4}$ inch of water per acre as a continuous injection in center pivot and self-propelled wheel move systems or in the last 15 to 30 minutes of set in permanent solid set sprinkler systems. On sandy soil apply in $\frac{1}{4}$ to $\frac{1}{2}$ inch of water. Use rates of $\frac{1}{3}$ to $\frac{2}{3}$ pounds for control of redroot pigweed and lambsquarters only. Apply the $\frac{2}{3}$ pound rate for control of other weeds listed on the label. Do not apply within 60 days of harvest.

Do not apply when wind speed favors drift, when system connections leak, when nozzles do not provide uniform distribution or when lines containing the product must be dismantled or drained.

SENCOR PLUS DUAL TANK MIX APPLICATION (Except California)

SENCOR may be applied in a tank mix combination with Dual 8 EC as a preemergence broadcast application to potatoes. Apply SENCOR at $\frac{1}{2}$ to $1\frac{1}{3}$ pounds and Dual 8 EC at 1.5 to 3 pints per acre according to the respective labels for use of each product alone on potatoes. Refer to appropriate sections of the SENCOR and Dual labels for precautionary statements, restrictions, application information and weeds controlled.

SENCOR PLUS PROWL TANK MIX APPLICATION (Except California)

SENCOR may be applied in a tank mix combination with Prowl as a preemergence broadcast application to potatoes. Apply SENCOR DF at $\frac{2}{3}$ to $1\frac{1}{3}$ pounds and Prowl at 1 to 3 pints per acre according to the respective labels for use of each product alone on potatoes. Refer to the appropriate sections of the SENCOR and the Prowl labels for precautionary statements, restrictions, application information and weeds controlled.

SPECIAL PRECAUTIONS (Potatoes)

Do Not Use SENCOR DF on Potatoes in Kern County, California

Do not apply more than $1\frac{1}{3}$ pounds total SENCOR DF per acre in a single crop season regardless of the method of application. **Do not apply within 24 hours of application of other pesticides.** Do not apply postemergence within 3 days after periods of cool, wet or cloudy weather or crop injury may occur. Postemergence applications may cause some chlorosis or minor necrosis.

In addition to early maturing smooth-skinned white, or red-skinned varieties of potatoes, postemergence applications of SENCOR are not recommended on Atlantic, Shepody, Chip Bell, Bellchip, and Centennial varieties. Preemergence applications to these varieties may cause crop injury, especially under adverse weather conditions and when higher SENCOR rates are used.

Do not apply to sweet potatoes or yams.

Do not plant sensitive crops such as onions, lettuce, cole crops and cucurbits during the next growing season following applications of SENCOR or injury may occur.

ALFALFA AND SAINFOIN

SENCOR herbicide is labelled for use in alfalfa and sainfoin in the following areas:

1. Alfalfa and sainfoin (Including mixed stands with grasses) (all areas except California)
2. Alfalfa and sainfoin (Including mixed stands with grasses) (California only)
3. Alfalfa—Post-dormant application of SENCOR impregnated on dry fertilizer only
4. Alfalfa—Non-dormant, non-winter hardy varieties (Arizona Only)

SENCOR is recommended for use in aerial or ground spray equipment as a broadcast surface application to dormant established crops of alfalfa and sainfoin for the control of certain grass and broadleaf weeds.

APPLICATION: Refer to the specific "Application" section under "General Information" in the front of the label for detailed information on the application of SENCOR. For information on applying SENCOR in fluid or on dry fertilizer refer to the "Application Of SENCOR In Fluid Fertilizers" or "Commercial Impregnation And Application Of SENCOR On Dry Bulk Fertilizer" under the "General Information" section in the front of the SENCOR label.

SPECIAL PRECAUTIONS:

—Do not apply SENCOR after growth begins in the spring or before growth ceases in the fall.

—Treat only dormant established alfalfa.

—Do not graze or harvest within 28 days after application.

—For best weed control, apply SENCOR when weeds are less than 2 inches tall or before weeds foliage is 2 inches in diameter.

—Reduced weed control may occur when extended dry conditions follow application of SENCOR.

—Crop injury may occur if alfalfa or sainfoin are:

1. Under stress conditions such as diseases, insect infestations, poorly drained soils, drought or winter injury when SENCOR is applied, or
2. When SENCOR is applied during the first growing season after seeding or if applied earlier than 12 months after seeding.

ALFALFA AND SAINFOIN (All Areas except California)

Recommended Broadcast Applications

Crop	Pounds/Acre SENCOR DF	Remarks
Alfalfa and Sainfoin	½ to 1⅓	**ALL AREAS:** (Including Idaho, Oregon, and Washington) Select the proper dosage according to weeds known to be present in field to be treated. Use the lower rate of SENCOR on sandy loam or loamy sand soils in areas other than Idaho, Oregon, and Washington.
	⅓ to ½	**IDAHO, OREGON AND WASHINGTON ONLY:** On sandy loam or loamy sand soils apply SENCOR DF at rates of ⅓ to ½ pound per acre for control of common chickweed only.

FOR USE ON MIXED STANDS OF ALFALFA AND GRASSES:

Rates of ⅔ to 1 pound of SENCOR DF per acre will provide partial reduction of forage grass stands. These rates may be used to reduce forage grass stands to prevent crowding out of alfalfa. Higher rates will severely reduce forage grass stands.

SENCOR should not be used on sand soils. In areas West of the Rocky Mountains, avoid using SENCOR on soils with calcareous surface area, high levels of lime or sodium, or a pH greater than 7.5.

ALFALFA AND SAINFOIN

WEEDS CONTROLLED

⅓ to ½ pound SENCOR DF/acre (Idaho, Oregon, and Washington Only)

Chickweed, Common — *Cerastium vulgatum*

½ to ⅔ pound SENCOR DF/acre

Cheat — *Bromus secalinus*
Downy brome — *Bromus tectorum*
Japanese brome — *Bromus japonicus*
Rescuegrass — *Bromus catharticus*

⅔ to 1⅓ pounds SENCOR DF/acre

Broadleaves

Fleabane, Rough — *Erigeron strigosus*
Flixweed — *Descurainia sophia*
Henbit — *Lamium amplexicaule*
Kochia — *Kochia scoparia*
Lambsquarters, Common — *Chenopodium album*
Marestail (Horseweed) — *Hippuris vulgaris*
Meadow Salsify — *Tragopogon pratensis*
Mustard, Blue — *Chorispora tenella*
Mustard, Jim Hill (tumble) — *Sisymbrium altissimum*
Mustard, Tansy — *Descurainia pinnata*
Pepperweed, Virginia — *Lepidium virginicum*
Pennycress, Field — *Thlaspi arvense*
Pigweed, Redroot — *Amaranthus retroflexus*
Prickly Lettuce — *Lactuca serriola*
Russian Knapweed — *Centaurea repens*
Shepherdspurse — *Capsella bursa-pastoris*
White Cockle — *Melandrium album*
Wild Buckwheat — *Polygonum convolvulus*
Yellow Rocket — *Barbarea vulgaris*

Grasses

Foxtail, Green — *Setaria viridis*
Little Barley — *Hordeum pusillum*
Smooth Brome — *Bromus inermis*
Wild Oats — *Avena fatua*

1⅓ pounds SENCOR DF/acre

Broadleaves

Chickweed, Mousear — *Cerastium vulgatum*
Dandelion — *Taraxacum officinale*
Ragweed, Common — *Ambrosia artemisiifolia*

Grasses

Barnyardgrass — *Echinochloa crus-galli*
Bluegrass — *Poa annua*
Foxtail Barley — *Hordeum jubatum*

Weeds Partially Controlled: At the rate of 1⅓ pound/acre SENCOR DF may be used to reduce the competition from curly dock *(Rumex crispus)*.

ALFALFA AND SAINFOIN (California Only)

(Including Mixed Stands With Grasses)

SENCOR DF is recommended for use in aerial or ground spray equipment as a broadcast surface application to dormant established crops of alfalfa and sainfoin.

APPLICATION: SENCOR DF is recommended for use in aerial or ground spray equipment as a broadcast surface application to dormant established crops of alfalfa and sainfoin for control of certain grass and broadleaf weeds. Do not apply SENCOR DF after growth begins in the spring or before growth ceases in the fall. Do not apply to either alfalfa or sainfoin during the first growing season after seeding.

For information on applying SENCOR in fluid fertilizer solutions to alfalfa, refer to the appropriate section of the SENCOR label.

For information on Commercial impregnation and application of SENCOR on dry bulk fertilizer, refer to the appropriate section of the SENCOR label.

WEEDS CONTROLLED

½ to ⅔ pounds SENCOR DF/acre

Cheatgrass (downy brome)

⅔ to 1⅓ pounds SENCOR DF/acre

Broadleaves

Chickweed, Common — *Cerastium vulgatum*
Flixweed — *Descurainia sophia*
Henbit — *Lamium amplexicaule*
Kochia — *Kochia scoparia*
Meadow Salisfy — *Tragopogon pratensis*
Mustard, Blue — *Chlorispora tenella*
Mustard, Tansy — *Descurainia pinnata*
Pepperweed, Virginia — *Lepidium virginicum*
Shepherdspurse — *Capsella bursa-pastoris*
White Cockle — *Melandrium album*
Wild Buckwheat — *Polygonum convolvulus*
Yellow Rocket — *Barbarea vulgaris*

Grasses

Smooth Brome — *Bromus inermis*
Wild Oats — *Avena fatua*

1⅓ pounds SENCOR DF/Acre

Broadleaves	
Dandelion	*Taraxacum officinale*
Grasses	
Barnyardgrass	*Echinochloa crus-galli*
Bluegrass	*Poa annua*
Foxtail Barley	*Hordeum jubatum*

Recommended Broadcast Application

Crop	SENCOR DF Pounds/Acre	Remarks
Alfalfa and Sainfoin	½ to 1⅓	Select the proper dosage according to weeds known to be present in the field to be treated. Apply specified dosage in 20 to 40 gallons of water per acre with ground spray equipment or 3 to 10 gallons of water per acre with aerial spray equipment fitted with nozzles suitable for broadcast applications of herbicides. Treat only dormant established crops of alfalfa and sainfoin. Injury may occur to alfalfa if SENCOR is applied earlier than 12 months after seeding. Do not apply after Spring growth begins or before growth ceases in the Fall. Do not graze or harvest within 28 days after application. At the 1⅓ pound/acre rate, SENCOR DF may be used for suppression of curly dock.

FOR USE ON MIXED STANDS OF ALFALFA AND GRASSES: Rates of ⅔ to 1⅓ pounds of SENCOR DF per acre will provide partial reduction of forage grass stands. These rates may be used to reduce forage grass stands to prevent crowding out of alfalfa. Higher rates will severely reduce forage grass stands.

Do not apply with aerial spray equipment when wind speed is greater than 10 mph. Do not apply when weather conditions favor spray drift and/or when sensitive cool season crops, such as cole crops, onions, peas, or strawberries, are present in adjacent fields. Applications should not be made when weather conditions favor spray drift, especially in areas where wheat is growing on coarse textured soils in adjacent fields, or injury may occur.

ALFALFA

POST DORMANT APPLICATION OF SENCOR IMPREGNATED ON DRY FERTILIZER ONLY

SENCOR DF may be applied after dormancy has broken, but prior to three inches of new alfalfa shoot growth, only when impregnated on dry fertilizer in Connecticut, Illinois, Indiana, Iowa, Kansas, Kentucky, Michigan, Minnesota, Nebraska, New York, Ohio, Pennsylvania, South Dakota and Wisconsin. Apply at rates of 1 to 1⅓ pounds per acre as directed on the label for application during dormancy. Apply only when alfalfa foliage is dry or crop injury may occur. when using this application method, do not harvest or graze treated alfalfa for 60 days after application.

ALFALFA

NON-DORMANT, NON-WINTER HARDY VARIETIES

(Arizona only)

SENCOR is recommended as a broadcast surface application to established crops of non-dormant alfalfa varieties for preemergence and postemergence control of certain winter annual weeds following either a Fall or Winter sheep grazing/green-chop harvest.

Weeds Controlled:	London rocket (mustard)
Mouse barley	Shepheardspurse
Nettleleaf goosefoot	Littleseed canarygrass
Prickly lettuce	Lambsquarters
Field pepperweed	Little mallow (cheeseweed) (Malva)
Silversheath knotweed	Spiny sowthistle

Recommended Broadcast Applications

Crop	SENCOR DF Pounds/Acre	Remarks
Alfalfa Non-Dormant Non-Winter hardy varieties	½ to ⅔	Apply specified dosage by aerial or ground spray equipment in 7 to 40 gallons of water per acre. Treat established alfalfa stubble after Fall or Winter sheep grazing or green-chop harvest and prior to the time regrowth is 2″ tall. Alfalfa foliage present at time of application can exhibit yellowing. Injury may occur to alfalfa in areas of high salt concentration where the crop is stunted and/or has a poorly developed root system, or if alfalfa is under stressed growing conditions such as diseases, insect infestations, or drought. For most effective postemergence weed control, treatment should be made before weeds are 2″ tall or before leaf rosettes are 2″ wide. For maximum control, rainfall (¼″ or more) or irrigation is necessary within 30 days of treatment, however, do not flood irrigate within 2 days after treatment. Use ½ pound SENCOR on sand soil when only mustard, goosefoot, lambsquarters, or canarygrass are the weeds to be controlled. Do not apply earlier than 6 months after seeding. Do not graze or harvest within 28 days after application.

SPECIAL PRECAUTIONS

Maintain continuous mechanical agitation in the spray tank to insure a uniform spray mixture.

Do not apply with aerial spray equipment when wind speed is greater than 10 mph. Do not apply when weather conditions favor spray drift and/or when sensitive cool season crops, such as cole crops, onions, peas or strawberries are present in adjacent fields. Applications should not be made when weather conditions favor spray drift especially in areas where wheat is growing on coarse textured soils in adjacent fields, or injury may occur.

ASPARAGUS (Established)

SENCOR is recommended for use in aerial or ground spray equipment as a single preemergence broadcast application or as a split application consisting of a preemergence broadcast application followed by a post harvest broadcast application.
For specific application information see under the "General Information" section "Application" of SENCOR in the front of the SENCOR label.

Weeds Controlled
SENCOR DF, applied to established asparagus according to directions, will effectively control:

Broadleaves	
Chickweed, Common	*Stellaria media*
Jimsonweed	*Datura stramonium*
Lambsquarters	*Chenopodium album*
Pigweed, Redroot	*Amaranthus retroflexus*
Ragweed, Common	*Ambrosia artemisiifolia*
Smartweed, Pennsylvania	*Polygonum pensylvanicum*
Sorrel, Red	*Rumex acetosella*
Velvetleaf	*Abutilon theophrasti*
Grasses	
Crabgrass	*Digitaria* spp.
Foxtails	*Setaria* spp.
Sandbur, Field	*Cenchrus pauciflorus*

Recommended Broadcast Applications

Crop	SENCOR DF Dosage/Acre	Remarks
Asparagus	1⅓ to 2⅔	**PREEMERGENCE APPLICATION ONLY:** Make a single surface application in early Spring before asparagus spears or ferns emerge. If the field is to be disked, apply SENCOR DF after disking but before the crop emerges. Use the lower rate for control of the broadleaf weeds listed above. Use the higher rate in fields with a history of severe infestations of grasses and for maximum residual control. Do not apply within 14 days of harvest.
	⅔ to 1⅓ preemergence plus 1⅓ to 2 post harvest	**SPLIT APPLICATION PREEMERGENCE AND POST HARVEST** **Preemergence Application:** Apply before asparagus spears or ferns emerge. If the field is to be disked, apply after disking but prior to crop emergence. Do not apply within 14 days of harvest. **Post Harvest Application:** Apply after last harvest of the season but prior to emergence. The lower combination rates may be used for control of common ragweed, lambsquarters, redroot pigweed, and red sorrel. Use the higher combination rates for other weeds listed or in fields with severe grass infestations or for maximum post harvest control of emerged weeds.

IMPORTANT: The total amount of SENCOR DF applied in one crop season may not exceed 2⅔ pounds per acre.

SPECIAL PRECAUTIONS (Asparagus)
Do not use on newly seeded asparagus nor on young plants during the first growing season after setting crowns.

DO NOT APPLY POST HARVEST APPLICATIONS UNTIL AFTER THE LAST HARVEST OF SPEARS.

FIELD CORN

(IOWA, KANSAS, MISSOURI, NEBRASKA, AND SOUTH DAKOTA ONLY)

SENCOR herbicide is recommended for the added effect on the control of certain broadleaf weeds in corn when applied preemergence in a tank mix combination with labeled rates of Lasso plus atrazine, Lasso plus Bladex®, Dual plus atrazine, or Dual plus Bladex. Refer to the appropriate product labels for weeds controlled by these combinations.

SENCOR in each of the tank mix combinations will control the following weeds:

Lambsquarters	*Chenopodium album*
Pigweed, Redroot (carelessweed)	*Amaranthus retroflexus*
Smartweed, Pennsylvania	*Polygonum pensylvanicum*
Velvetleaf	*Abutilon theophrasti*

MIXING: Select proper dosages based on soil type as shown below. Fill the spray tank ¼ to ⅓ full with clean water and add SENCOR DF to the spray tank first. Mix thoroughly and then add atrazine or Bladex to the spray tank. Add water to fill spray tank to desired level. Mix thoroughly and then add Lasso or Dual to the spray tank. Maintain sufficient agitation during mixing and spraying operations to insure a uniform spray mixture.

APPLICATION: For specific application information see under the "General Information" section "Application" of SENCOR in the front of this label.

Recommended Broadcast Applications

Rates Per Acre:

Soil Texture	SENCOR DF	Lasso 4 E.C.	or Dual 8 E.C.	Atrazine 4L or 80 WP	or Bladex 4L or 80 WP
Medium (Loam, silt loam, silt, sandy clay, sandy clay loam)	⅓ lb.	1.5 to 2.0 qts.	1.5 to 2.0 pts.	0.75 to 1.5 qts. of 4L or 0.95 to 1.9 lbs. of 80 WP	1.25 to 2.5 qts. of 4L or 1.5 to 3.1 lbs. of 80 WP
Fine (Silty clay, silty clay loam[1], clay, clay loam)	⅓ lb.	2.0 to 2.5 qts.	2.0 to 2.5 pts.	0.75 to 1.5 qts. of 4L or 0.95 to 1.9 lbs. of 80 WP	1.5 to 3.0 qts. of 4L or 1.75 to 3.75 lbs. of 80 WP

[1] Silty clay loam soils are transitional soils and may be classified as medium textured soils in some regions of the U.S.

Follow the Lasso, Dual, Bladex, or atrazine labels for use rates on specific soil textures and organic matter content. Not recommended for use on muck soils as reduced weed control may result.

SPECIAL PRECAUTIONS: (FIELD CORN)
Not for use on coarse textured soils or soils containing less than 2% organic matter content.

Do not use on soils having a calcareous surface or high in lime or sodium and/or a pH of 7.0 or higher.

Corn seed should be planted 1½ inches deep.

Use in hybrid seed corn production fields only if both parents are known to be tolerant to SENCOR and other herbicides used.

Observe all cautions and limitations on labeling of all products used in mixtures.

SUGARCANE (Hawaii Only)

SENCOR, a selective herbicide, is effective as a preemergence and an early postemergence broadcast application for control of certain grass and broadleaf weeds. When applied as a spot treatment, it also provides excellent control of perennial grasses and broadleaves.

Ground Application: SENCOR should be mixed by filling the spray tank half full of clean water. Then add the recommended amount of SENCOR DF to suit the total tank capacity and the rate of application per acre (preferably 25 to 35 gallons per acre). Complete filling the tank and maintain sufficient agitation during mixing and spraying to insure a uniform spray mixture.

Aerial Application: SENCOR is recommended for use in aerial spray equipment as a preemergence or postemergence application to irrigated sugarcane. Aerial spray equipment should be calibrated to apply the proper amount of SENCOR DF in 7 to 10 gallons of spray mixture per acre.

SENCOR applied preemergence or postemergence to the sugarcane as a broadcast spray or spot treatment will effectively control the following when weeds are less than 3 inches in height.

WEEDS CONTROLLED IN IRRIGATED AND NON-IRRIGATED SUGARCANE

Broadleaves	
Amaranth, Spiny	*Amaranthus spinosus*
Euphorbia, Wild	*Euphorbia* sp.
Fireweed	*Erechtites hieracifolius*
Floras paintbrush	*Emilia sonochifolia*
Spurge, Garden	*Euphorbia hirta*
Spurge, Graceful	*Euphorbia glomerifera*
Grasses	
Crabgrass	*Digitaria* spp.
Guineagrass	*Panicum maximum*
Plushgrass	*Chloris radiata*
Ricegrass	*Oryzopsis hymenoides*
Wiregrass	*Eleusine indica*

WEEDS CONTROLLED IN IRRIGATED SUGARCANE ONLY

Broadleaves

Amaranth, Spleen	*Amaranthus dubius*
Haole Koa	*leucaena leucocephala*
Hialoa	*Waltheria americana*
Hilahila	*Mimosa pudica*
Purslane, Common	*Portulaca oleracea*
Rattlepod	*Crotalaria spectabilis*

Grasses

Alexandergrass	*Brachiaria plantaginea*
Bristly foxtail	*Setaria verticillata*

WEEDS CONTROLLED IN NON-IRRIGATED SUGARCANE ONLY

Broadleaves

Ageratum	*Ageratum conyzoides*
Richardia	*Richardia brasiliensis*
Tarweed	*Cuphea carthagenesis*

Recommended Broadcast Applications

Crop	SENCOR DF Pounds/Acre	Remarks
Sugarcane (Hawaii only)	2⅔ to 5⅓ (non-irrigated) 5⅓ to 8 (irrigated)	**PREEMERGENCE:** (Irrigated and non-irrigated sugarcane) Apply specified dosage per acre as a broadcast spray to the soil surface. Applications should be made within two weeks after planting prior to cane emergence or shortly after emergence (spike stage). OR **EARLY POSTEMERGENCE:** (Irrigated and non-irrigated sugarcane) Apply specified dosage per acre as a broadcast spray over the cane. Application may be delayed as long as 4 to 6 weeks after planting provided weeds are less than 3 inches in height.
	2⅔ to 5⅓	**POSTERMERGENCE:** Apply specified dosage per acre as a broadcast spray to control weeds prior to "close in" time when cane shades out the weed growth.
	3⅓ to 6⅔	**SPOT TREATMENT:** Apply specified dosage in 30 to 50 gallons of finishing spray per acre. Spot treatments may be used to control weeds in missed areas, corners of fields, or areas of hard to control weeds.

NOTE: Do not apply more than 10⅔ pounds of SENCOR DF (8 lb. active ingredient)/acre per crop cycle regardless of the method of application. The last application may be made up to 17 months of harvest.

SUGARCANE (Louisiana and Texas Only)

Preemergence and postemergence applications of SENCOR DF with aerial or ground spray equipment are recommended for control of the following weeds in sugarcane in Louisiana and Texas:

Broadleaves

Amaranth, Spiny	*Amaranthus spinosus*
Bindweed, Field	*Convolvulus arvensis*
Chickweed	*Cerastium vulgatum*
Henbit	*Lamium amplexicaule*
Lambsquarters	*Chenopodium album*
London rocket	*Sisymbrium irio*
Marestail	*Conyza canadensis*
Mustard, Wild	*Brassica kaber*
Pigweeds	*Amaranthus* spp.
Purslane	*Portulaca oleracea*
Sowthistle	*Sonchus* spp.

Grasses

Broadleaf Signalgrass	*Brachiaria platyphylla*
Crabgrass	*Digitaria* spp.
Foxtails	*Setaria* spp.
Johnsongrass, Seedling	*Sorghum halepense*
Oats, Winter	*(Avena spp.)*

Recommended Applications

Crop	SENCOR DF Pounds/Acre	Remarks
Sugarcane (Louisiana and Texas only)	2 to 4	**BROADCAST:** Apply specified dosage per acre using 20 to 30 gallons of water with ground equipment or 5 gallons of water with aircraft spray equipment. Apply as a broadcast spray during the Fall after planting or to the stubble after harvest. Make a second application early in the Spring.
	1 to 2	**BAND:** Apply specified dosage in 10 to 20 gallons of water per acre in a 30 to 36 inch band over the row during the Fall after planting or to the stubble after harvest. Make a second application early in the Spring.

SPECIAL PRECAUTIONS: (Louisiana and Texas only)

Do not use treated foliage for feed or forage.

Use the higher rate on heavy clay soil and soil with a high percentage of organic matter. If necessary, a third application may be made in late Spring at layby. Do not apply within 60 days of harvest.

SUGARCANE (Florida Only)

Postemergence directed spray applications of SENCOR are recommended for control of the following weeds in sugarcane in Florida.

Broadleaves

Amaranth, Spiny (seedling)	*Amaranthus spinosus*
Butterweed Cressleaf groundsel	*Senecio glabellus*

Grasses

Broadleaf panicum	*Brachiaria adspersa*
Goosegrass	*Eleusine indica*

Recommended Broadcast Applications

Crop	SENCOR DF Pounds/Acre	Remarks
Sugarcane	1⅓ to 2⅔	Apply specified dosage in 10 to 40 gallons of water per acre as a postemergence directed spray at the base of the plant after sugarcane is 12 inches tall but before row closing. Avoid spray contact with sugarcane foliage. For best weed control apply before weeds are 6 inches tall. Do not apply more than 2⅔ pounds per acre in one season.

SPECIAL PRECAUTIONS (FLORIDA SUGARCANE):

Do not use on sugarcane grown on sand.

Spray contact with sugarcane foliage may result in minor leaf margin chlorosis and/or necrosis.

Do not apply within 60 days of harvest

Do not use treated crop for feed or forage.

Care should be taken to insure that spray adequately covers the foliage of weeds.

Avoid spray overlaps or variations in application speed that may result in insufficient or excessive rates of application.

TOMATOES

Apply SENCOR herbicide with ground equipment to seeded and transplanted tomatoes as specified below under "Recommended Applications" on the label.

For effective control of grasses and broadleaf weeds with postemergence applications, apply SENCOR before weeds are 1-inch tall. Thorough spray coverage on weed foliage is essential for adequate control with postemergence applications.

Do not use air blast or other high pressure spray equipment to make postemergence applications of SENCOR. Refer to the appropriate section of this label for additional information regarding spray equipment, dilution rates, mixing, sprayer cleanup, restrictions, container disposal and cautions.

Refer to "Mixing" under the "General Information" section in the front of the label.

For specific application information see under the "General Information" section "Application" of SENCOR in the front of the label.

Weeds Controlled:

PREPLANT INCORPORATED APPLICATIONS—TRANSPLANT TOMATOES ONLY

Broadleaves	
Galinsoga	*Galinsoga* spp.
Lambsquarters	*Chenopodium album*
*Pigweed, Redroot	*Amaranthus retroflexus*
*Purslane, Common	*Portulaca oleracea*
Grasses	
*Goosegrass	*Eleusine indica*

Preplant incorporated applications applied as directed will suppress foxtails, panicums and barnyardgrass.

SENCOR/TREFLAN TANK MIX

This tank mix combination applied preplant incorporated as directed on this label will control the weeds listed above plus those weeds listed on the Treflan label.

POSTEMERGENCE APPLICATIONS—ESTABLISHED TOMATOES

For effective control of weeds with postemergence applications, apply SENCOR before weeds are 1-inch tall.

Broadcast Sprays—⅓ to ½ lb. of SENCOR DF/acre

Directed Sprays ⅔ to 1⅓ lbs. of SENCOR DF/acre

Broadleaves	
Carpetweed	*Mollugo verticillata*
Fumitory	*Fumaria officinalis*
Galinsoga	*Galinsoga* spp.
*Jimsonweed	*Datura stramonium*
*Ladysthumb	*Polygonum persicaria*
Lambsquarters	*Chenopodium album*
Mustard, Wild	*Brassica kaber*
Pigweeds	*Amaranthus* spp.
Purslane	*Portulaca oleracea*
*Ragweed, Common	*Ambrosia artemisiifolia*
*Smartweed, Pennsylvania	*Polygonum pensylvanicum*
Toadflax	*Linaria* spp.
*Velvetleaf	*Abutilon theophrasti*
Grasses	
*Foxtail, Yellow	*Setaria glauca*
Goosegrass	*Eleusine indica*
(Plus Weeds Listed Under "Broadcast Sprays")	

* For optimum control of these weeds, use the highest rate recommended on the label for the type of application to be made. Repeat postemergence applications may be needed for best control.

Postemergence applications as directed on this label will suppress barnyardgrass and crabgrass when these weeds are less than 1-inch tall.

Recommended Broadcast Applications

Crop	SENCOR DF *Pounds/Acre	Remarks
Tomatoes	⅓ to ⅔	**PREPLANT INCORPORATED—TRANSPLANT TOMATOES ONLY (EXCEPT CALIFORNIA):** Apply specified dosage in 10 to 40 gallons of water per acre as a broadcast spray to the soil surface immediately before transplanting. Incorporate to a depth of 2 to 4 inches with equipment capable of uniformly mixing the chemical into the soil. This application may be made alone or in a tank mix combination with Treflan E.C. Refer to the Treflan label for specific rate of application and for additional precautions and restrictions for tomatoes.
	⅓ to ½	**POSTEMERGENCE BROADCAST SPRAY—ESTABLISHED TOMATOES:** Apply specified dosage in 20 to 75 gallons of water per acre as a broadcast spray over the plants in single or multiple applications with a minimum of 14 days between treatments. For transplanted tomatoes do not apply within 14 days of transplanting. Transplants **must** have recovered from transplant shock and new growth evident. Do not apply to established tomatoes within 24 hours of application of other pesticides. Do not tank mix with other pesticides. (See "Special Precautions" below.)
	⅔ to 1⅓	**POSTEMERGENCE DIRECTED SPRAY—ESTABLISHED TOMATOES:** Apply specified dosage in 20 to 75 gallons of water per acre as a directed spray in single or multiple applications with a minimum of 14 days between treatments. Avoid contacting tomato foliage with spray. This method of treatment is recommended for use in fields with a history of severe weed pressure or in fields infested with hard to control weeds. For transplanted tomatoes, do not apply within 14 days of transplanting. Transplants **must** have recovered from transplant shock and new growth evident. Do not apply to established tomatoes within 24 hours of application of other pesticides. (See "Special Precautions" below.) When banding see the appropriate section in the front of the label.

* Use the higher rate in fields with a history of severe weed pressure and for maximum residual weed control.

SPECIAL PRECAUTIONS (TOMATOES)

Do not apply more than a total of 1⅓ lbs. SENCOR DF per crop season.

Do not apply the total amount of 1⅓ lbs. SENCOR DF within a time span of less than 35 days, except in the case of directed sprays.

Allow at least 14 days between applications, regardless of dosage or method of application.

Do not apply within 7 days of harvest.

Do not apply within 3 days after periods of cool, wet or cloudy weather, or crop injury will occur.

Do not use hot caps on tomatoes within 7 days before or at any time after application of SENCOR.

Do not treat seeded tomatoes until plants have reached the 5 to 6 leaf stage.

Crop injury or delayed maturity may result from broadcast or directed spray applications if tomatoes are growing under stress conditions such as periods of drought or cool, wet and cloudy weather preceding application.

For newly introduced tomato varieties with unknown tolerance to SENCOR, treat only a small area to determine if SENCOR can be used without injury to the crop.

DO NOT USE SENCOR ON TOMATOES IN KERN COUNTY, CALIFORNIA.

CEREALS (BARLEY AND WHEAT)

For the control of certain grass and broadleaf weeds, SENCOR herbicide is recommended as a single post emergence application in barley (winter and spring) and winter wheat (see the appropriate section of this label for geographic limitations).

SENCOR may be used in cereals (barley and wheat) as follows:

1. Barley (spring and winter)—Only East of the Cascade Mountains in Idaho, Oregon, Montana, and Washington.

2. Winter wheat in Idaho, Oregon, Utah, & Washington.

3. Dryland (non-irrigated) Winter wheat in Kansas, Oklahoma, & Texas.

4. Dryland (non-irrigated) Winter wheat only (Arkansas, Louisiana, Mississippi, & Tennessee)

Mixing

See the "Mixing" section under "General Information" in the front of the label for specific mixing procedures. Do not tank mix SENCOR with pesticides other than those specified below. When tank mixing, carefully follow the instructions on this label. Refer to the other product label registered for the specific use intended for directions, rates, and weeds controlled.

Application

Refer to the specific "Application" section under "General Information" in the front of the label for detailed information on the application of SENCOR.

Apply SENCOR when wheat stand is uniform in height. For best results apply SENCOR when weeds are small (less than 1 to 1½ inches in diameter or height). **In the Fall** apply after wheat has a minimum of 3 tillers and 4 secondary roots at least 2 inches in length, but before winter dormancy and wheat jointing occurs. **In the Spring** apply after wheat has recovered from winter dormancy and after wheat has fully tillered and has developed secondary roots at least 2 inches in length, but before jointing.

Grazing Treated Wheat

Wheat fields treated with SENCOR may be grazed 14 days following application.

Apply SENCOR only once per crop season for wheat or barley.

SPECIAL PRECAUTIONS (CEREALS)

Crop injury may occur on barley or wheat when SENCOR is used under the following conditions:

1. When applied while the crop is under stress such as winter kill, frost damage, disease, drought or excessive moisture, or when these conditions follow the application.

2. When applied in combination with fluid fertilizer.

3. When applications are made prior to the growth stage specified on this label.

4. When soils are high in lime or sodium, have a pH greater than 7.7, or are calcareous soils.

5. When proper application instructions are not followed.

Weed control may not be observed for 2 to 4 weeks after application; under very dry conditions, weed control may not be apparent for 4 to 6 weeks. Moisture (at least ½") is needed within 2 to 3 weeks after application to move SENCOR into the weed root zone. Lack of adequate moisture after application may result in poor control.

Do not exceed the rates specified on this label.

BARLEY (Spring and Winter)

FOR USE ONLY EAST OF THE CASCADE MOUNTAINS IN IDAHO, OREGON, UTAH, MONTANA AND WASHINGTON ON THE VARIETIES COMPANA, HANNCHEN, HECTOR, HUDSON, LUTHER, KAMIAK, KLAGES, OLYMPIC, PIROLINE AND STEPTOE.

A single postemergence application of SENCOR is recommended for control of certain grass and broadleaf weeds in Spring and Winter barley.

Recommended Broadcast Applications

Soil Texture	SENCOR DF Pounds Per Acre — % Organic Matter[1]: 1 to 2	SENCOR DF Pounds Per Acre — % Organic Matter[1]: Over 2	Remarks
Coarse (Sandy loam, loamy sand)	⅓	⅓ to ½	Barley should be seeded to obtain a uniform crop stand at application time. Apply after barley has fully tillered and developed secondary roots but before jointing and before weeds are 2 inches tall or have 2 inch diameter rosettes. Do not graze or harvest treated barley for feed before crop maturity.
Medium (Loam, silt loam, silt, sandy clay, sandy clay loam)	⅓ to ½	½	
Fine (Silty clay, silty clay loam,[2] clay, clay loam)	½	½ to ⅔	

[1] Do not use on any soils containing less than 1% organic matter.
[2] Silty clay loam soils are transitional soils and may be classified as medium textured soils in some regions of the U.S.

WEEDS CONTROLLED IN BARLEY

Broadleaves	
Chickweed, Common	*Stellaria media*
Fiddleneck, Tarweed	*Amsinckia lycopsoides*
Henbit	*Lamium amplexicaule*
*Kochia	*Kochia scoparia*
Lambsquarters	*Chenopodium album*
*Mustard, Blue or Purple	*Chorispora tenella*
*Mustard, Jim Hill	*Sisymbrium altissimum*
*Mustard, Wild	*Brassica kaber*
Pigweeds	*Amaranthus* spp.
Grasses	
*Cheatgrass	*Bromus secalinus*
*Downy Brome	*Bromus tectorum*
*Foxtails	*Setaria* spp.
*Ryegrass, Annual	*Lolium multiflorum*

* Use the highest rate recommended for soil texture and percent organic matter for the control of these weeds.

Hard to Control Weed: Wild oats (*Avena fatua*)

NOTE: Control of wild oats may be variable, ranging from poor to excellent depending on timing of application, stage of growth of wild oats at time of application, temperatures and soil moisture conditions following treatment.

SENCOR IN COMBINATION WITH OTHER HERBICIDES

SENCOR may be applied as directed above or may be tank mixed with 2,4-D, Banvel®, Bronate®, Brominal®, Brominal Plus, or Buctril®. Refer to the other product label registered for this use for additional directions, rates, and weed species controlled.

WINTER WHEAT

FOR USE ON WINTER WHEAT IN IDAHO, OREGON, UTAH AND WASHINGTON ON THE VARIETIES DAWS, GAINES, HATTON, HILL 81, HYSLOP, LUKE, McDERMID, MORO, NuGAINES, PAHA, PECK, SPRAGUE, STEPHENS, WANSER, FARO, TYEE, AND WESTON AND IN MONTANA ON THE VARIETIES CENTURK, CHEYENNE AND WINALTA.

A single post-emergence application of SENCOR is recommended for control of certain grass and broadleaf weeds in Winter wheat.

Recommended Broadcast Applications[1]

Soil Texture	SENCOR DF Pounds/Acre: 0.75% to 1% Organic Matter[2]	SENCOR DF Pounds/Acre: 1% to 2% Organic Matter	SENCOR DF Pounds/Acre: More than 2% Organic Matter
Coarse loamy sand, sandy loam	⅓ to ⅖	⅓ to ⅖	⅓ to ⅖

Soil Texture	SENCOR DF Pounds/Acre 0.75% to 1% Organic Matter[2]	1% to 2% Organic Matter	More than 2% Organic Matter
Medium loam, silt loam, silt, sandy clay, sandy clay loam	⅓ to ⅖	⅖ to ½	⅖ to ⅗
Fine silty clay, silty clay loam[3], clay, clay loam	⅓ to ½	½ to ⅔	½ to ⅔

[1] Do not use on gravelly or thinly covered or exposed subsoil areas (clay knobs) as crop injury may occur.
[2] Do not use on any soils containing less than 0.75% organic matter.
[3] Silty clay loam soils are transitional soils and may be classified as medium textured soils in some regions of the U.S.

WEEDS CONTROLLED IN WHEAT

Broadleaves

Annual Polemonium (Jacob's Ladder)	*Polemonium micranthum*
Chickweed, Common	*Stellaria media*
Conical Catchfly	*Silene conica*
Corncockle	*Agrostemma githago*
Fiddleneck, Tarweed	*Amsinckia lycopsoides*
Gromwell	*Lithospermum officinale*
Henbit	*Lamium amplexicaule*
Ivyleaf Speedwell	*Veronica hederifolia*
Knotweed, Prostrate	*Polygonum aviculare*
*Kochia	*Kochia scoparia*
Lambsquarters	*Chenopodium album*
Madwort (catchweed)	*Asperugo procumbens*
Mayweed (dog fennel)	*Eupatorium capillifoium*
Miners Lettuce	*Claytonia perfoliata*
*Mustard, Blue or Purple	*Chorispora tenella*
*Mustard, Jim Hill (tumble)	*Sisymbrium altissimum*
*Mustard, Wild	*Brassica kaber*
Pennycress, Field	*Thlaspi arvense*
Pineappleweed	*matricaria matricarioides*
Prickly Lettuce	*Lactuca serriola*
Pigweeds	*Amaranthus* spp.
Redstem Filaree	*Erodium cicutarium*
*Russian Thistle	*Salsola kali*
Shepherdspurse	*Capsella bursa-pastoris*
Grasses	
*Blackgrass (Black twitchgrass)	*Alopecurus myosuroides*
*Bluegrass, bulbous	*Poa bulbosa*
*Cheatgrass	*Bromus secalinus*
*Downy Brome	*Bromus tectorum*
*Foxtails	*Setaria* spp.
*Rescuegrass	*Bromus catharticus*
*Ryegrass, Italian	*Lolium multiflorum*
*Wild Barley	*Hordeum leporinum*
*Windgrass	*Aspera spica-venti*

*Suppression only may occur

For suppression of these weeds, use the highest rate recommended for soil texture and percent organic matter. Suppression is a reduction in weed size and growth compared to a non-treated area in the same field.

HARD-TO-CONTROL WEEDS

Broadleaves

Cowcockle	*Vaccaria pyamidata*
Mustard, Tansy	*Descurainia pinnata*
Wild Buckwheat	*Polygonum convolvulus*
Grasses	
Ripgut Brome	*Bromus diandrus*
Wild Oats	*Avena fatua*

NOTE: Control of these weeds may be variable, ranging from poor to excellent depending on timing of application, stage of growth at the time of application, temperatures, and soil moisture conditions following treatment.

SENCOR IN COMBINATION WITH OTHER HERBICIDES

SENCOR may be applied as a tank-mix with Glean®, 2,4-D, MCPA, Igran®, Banvel®, Bronate®, Brominal®, ME4 Brominal, 3+3 Brominal, Brominal Plus or Buctril®. Refer to the other product labels registered for this use for additional directions, rates and weed species controlled. Observe all cautions and limitations on labeling of all products used in mixtures.

IRRIGATED WHEAT/BARLEY

Follow timing directions for dryland wheat/barley found at the beginning of this cereals section. After application of SENCOR, do not apply more than ½″ of water for the first irrigation; the maximum amount for each additional irrigation should not exceed 1″.

SENCOR PLUS IGRAN TANK-MIX APPLICATION

Apply SENCOR DF at ⅓ to ½ pound per acre and Igran 80WP at ½ to ¾ pound per acre according to label directions. The rate of each product depends on soil type. Do not use Igran on sand or loamy sand soils. Use lower rate of Igran on sandy loam soils.

BROADLEAF WEED CONTROL ONLY

Where only broadleaf control is needed, SENCOR applied at the rates listed below in a tankmix combination with Brominal, Brominal Plus, ME4 Brominal, 3+3 Brominal, Bronate or Buctril can be used to control wild buckwheat, in addition to all the weeds listed on these labels.

Recommended Broadcast Applications

SENCOR DF Pounds/Acre	Remarks
¼ to ⅓	Apply SENCOR in a tank mix combination with Brominal, Brominal Plus, ME4 Brominal, 3+3 Brominal, Buctril or Bronate. Use higher rate of SENCOR for control of larger broadleaf weeds that are 2 inches in diameter or height. Apply before weeds are greater than 3 inches in diameter or height. Not for use on sand soils.

SPECIAL PRECAUTIONS: Refer to the "General Mixing and Application", and "Precautions" in the front of the cereal section.

Temporary chlorosis or injury may occur after application of SENCOR to wheat, especially if frost or snow occurs within a few days after application. Heavy rainfall (1″ or more) within 4 days following application may result in crop injury.

DRYLAND (NON-IRRIGATED) WINTER WHEAT

FOR USE ONLY ON DRYLAND WINTER WHEAT IN KANSAS, OKLAHOMA AND TEXAS ON THE VARIETIES TAM W101, TAM 105, AND NEWTON AND IN OKLAHOMA AND TEXAS ON HAWK VARIETY.

WEEDS CONTROLLED IN WHEAT

Broadleaf Weeds

Annual Polemonium (Jacob's Ladder)	*Polemonium micranthum*
Chickweed, Common	*Stellaria media*
Fiddleneck, Tarweed	*Amsinckia lycopsoides*
Henbit	*Lamium amplexicaule*
*Kochia	*Kochia scoparia*
Lambsquarters	*Chenopodium album*
*Mustard, Blue	*Chorispora tenella*
*Mustard, Jim Hill	*Sisymbrium altissimum*
*Mustard, Wild	*Brassica kaber*
Pigweeds	*Amaranthus* spp.
Grasses	
*Cheatgrass	*Bromus secalinus*
*Downy Brome	*Bromus tectorum*
*Foxtails	*Setaria* spp.
*Rescuegrass	*Bromus catharticus*

* For control of these weeds, use the highest rate recommended for soil texture and percent organic matter.

Hard to Control Weeds: *Italian Ryegrass *(Lolium multiflorum)*, *Wild Oats *(Avena fatua)*, and *Wild Buckwheat *(Polygonum convolvulus)*.

NOTE: Control of these weeds may be variable, ranging from poor to excellent depending on timing of application, stage of growth at time of application, temperatures and soil moisture conditions following treatment.

Recommended Broadcast Applications

Soil Texture	SENCOR DF Pounds Per Acre — % Organic Matter[1] — Fall — 1 to 2	Over 2	Spring — 1 to 2	Over 2	Remarks
Coarse (Sandy loam, loamy sand)	⅓	⅓	⅓	⅓ to ½	Apply specified dosage per acre with aerial or ground spray equipment as a broadcast postemergence spray. For Fall and Spring application refer to "Application" in the beginning of this Cerals section for information on timing of application and growth restrictions.
Medium (Loam, silt loam, silt, sandy clay, sandy clay loam)	⅓ to ½	½	½ to ⅔	½ to ⅚	
Fine (Silty clay, silty clay loam[2], clay, clay loam)	½	½ to ⅔	½ to ⅚	⅔ to 1	

[1] Do not use on any soils containing less than 1% organic matter.
[2] Silty clay loam soils are transitional soils and may be classified as medium textured soils in some regions of the U.S.

SENCOR IN COMBINATION WITH OTHER HERBICIDES

SENCOR may be applied as directed above or may be tank mixed with 2,4-D, Banvel, Bronate, Brominal, or Buctril. Refer to the other product label registered for additional directions, rates, and weed species controlled.

DRYLAND (NON-IRRIGATED) WINTER WHEAT ONLY (Arkansas, Louisiana, Mississippi and Tennessee)

WEEDS CONTROLLED

Broadleaves

Bittercress	*Cardamine* spp.
*Buttercup	*Ranunculus* spp.
Carolina Geranium	*Geranium carolinianum*
Chickweed, Common	*Stellaria media*
Chickweed, Mouseear	*Cerastium vulgatum*
Cutleaf Evening Primrose	*Oenothera laciniata*
Henbit	*Lamium amplexicaule*
*Kochia	*Kochia scorparia*
Lambsquarters, Common	*Chenopodium album*
Mayweed (dog fennel)	*Eupatorium capillifolium*
Mustard, Wild	*Brassica kaber*
Pepperweed, Virginia	*Lepidium virginicum*
Pigweeds	*Amaranthus* spp.
Shepherdspurse	*Capsella bursa-pastoris*
Wild Turnip	*Brassica rapa*
Grasses	
Bluegrass, Annual	*Poa annua*
*Cheatgrass	*Bromus secalinus*
*Downy Brome	*Bromus tectorum*
Little Barley	*Hordeum pusillum*
*Rescuegrass	*Bromus catharticus*
*Ryegrass, Italian	*Lolium multiflorum*
Vernal Withlograss	*Drababerna* spp.

* For conrol of these weeds on fine soils, use SENCOR DF at ⅔ pound/acre.

Note: Control of Wild Buckwheat (*Polygonum convolvulus*) and Winter vetch (*vicia spp.*) may be variable with SENCOR, ranging from poor to excellent depending on growth at time of application, temperatures and moisture conditions following application.

Recommended for use only on the following wheat varieties:
Abe, Arthur 71, Caldwell, Coker 747, Coker 762, Coker 916, Coker 68-15, Doublecrop, Hunter, McNair 1003, McNair 1813, Nelson, Pike, Southern Belle and Tyler.

Recommended Broadcast Applications

Soil Texture	SENCOR DF[1] Pounds/A	Remarks
Coarse (Sandy loam, loamy sand)	⅓	Apply specified dosage per acre as a broadcast postemergence spray. For wheat plants that are coming out of Winter dormancy, allow at least two weeks for plants to recover before application. If wheat has been grazed during the Winter, allow at least two weeks for recovery before application. Refer to the beginning of the Cereals section for information on timing and growth stages for application.
Medium (Loam, silt loam, silt, sandy clay, sandy clay loam)	½	
Fine (Silty clay, silty clay loam[2], clay, clay loam)	½–⅔[3]	

[1] Do not use on Coarse soils containing less than 1% organic matter or on Medium or Fine textured soils containing less than 0.5% organic matter.
[2] Silty clay loam soils are transitional soils and may be classified as Medium textured soils in some regions of the U.S.
[3] Use ⅔ pound/acre for control of weed species designated with an asterisk under "Weeds Controlled".

SENCOR IN COMBINATION WITH OTHER HERBICIDES

SENCOR may be applied as directed above and may be tank mixed with 2,4-D, Banvel, Bronate, Brominal, or Buctril. Refer to the other registered product label for additional directions, rates, and weed species controlled.

Special Precautions

Temporary chlorosis may occur after application of SENCOR to wheat, especially if frost occurs.

FOR WEED CONTROL IN A WHEAT/FALLOW/WHEAT ROTATION (Idaho, Oregon, Utah and Washington Only)

SENCOR may be applied to provide weed control during the fallow period after wheat harvest or in the Spring before winter wheat is planted. Winter wheat can be seeded 4 months (120 days) after Spring application. Mechanical tillage or the application of a contact herbicide may be required to control weeds germinating prior to seeding of winter wheat. Best results will be obtained where straw and chaff are evenly distributed across the field.

Refer to "Mixing" on the "General Information" section in the front of the label.

For specific application information see under the "General Information" section "Application" of SENCOR in the front of the label.

Where weed growth is present at application time, SENCOR should be applied with Ortho Paraquat or other contact herbicide. Refer to the other product label registered for additional directions, rates, and weed species controlled.

WEEDS CONTROLLED

Broadleaves

Chickweed, Common	*Stellaria media*
Henbit	*Lamium amplexicaule*
*Kochia	*Kochia scoparia*
Lambsquarters	*Chenopodium album*
Mustard, Blue or Purple	*Chorispora tenella*
Mustard, Jim Hill	*Sisymbrium altissimum*
Mustard, Tansy	*Descuainia pinnata*
Mustard, Treacle	*Eyrsimum repandum*
Mustard, Wild	*Brassica kaber*
Pennycress, Field (Fanweed)	*Thlaspi arvense*
Pigweeds	*Amaranthus* spp.
*Russian thistle	*Salsola kali*
*Wild Sunflower	*Helianthus* spp.
Grasses	
Cheatgrass	*Bromus secalinus*

Downy Brome	*Bromus tectorum*
*Wheat, Volunteer	*Triticum* spp.

* Note: Since control of these weeds may be variable depending on moisture following application, the higher labelled rate is recommended.

After Harvest Application (Fall Fallow): SENCOR may be applied to wheat stubble after harvest in the Fall. Apply ⅔ to ⅚ pound per acre broadcast before weeds emerge. Use higher rate for longer weed control or for weeds designated as requiring the higher rate for control. Rainfall (½ inch or more) is necessary for herbicide activation.

Do not plant crops in treated areas for at least 10 months following fall applications.

SENCOR may be applied at ⅔ to ⅚ pound per acre as directed above for a Fall application and tank mixed with 2⅔ to 5⅓ quarts Chem-Hoe® 135 FL3 per acre. Apply 4 to 5⅓ quarts Chem-Hoe 135 per acre if volunteer grain or grass weeds have 4 or more leaves. If other vegetation is present at the time of application use a contact herbicide. Refer to the Chem-Hoe 135 FL3 label for additional directions, rates, weed species controlled and precautions.

Spring Application (Summer Fallow): SENCOR may be applied to wheat stubble in the Spring. Apply ½ to ⅔ pound per acre broadcast before weeds emerge in the Spring. Use higher rate for longer weed control or weeds designated as requiring higher rate for control. Rainfall (½ inch or more) is necessary for herbicide activation.

SENCOR may be applied at ½ to ⅔ pound after January 1 and before March 1 as a tank mix with 2⅔ to 5⅓ quarts Chem-Hoe 135 FL3 per acre. Apply 4 to 5⅓ quarts Chem-Hoe 135 per acre if volunteer grain or grass weeds have 4 or more leaves. A contact herbicide is recommended if other vegetation is present at the time of application. Refer to the Chem-Hoe 135 FL3 label for additional directions, rates, weed species controlled and precautions.

PRECAUTIONS AND RESTRICTIONS

Do not graze treated fields.

Do not plant Spring seeded cereals following Fall fallow applications of SENCOR.

Where SENCOR was applied in the Fall, do not apply SENCOR in the Spring.

FOR WEED CONTROL IN A FALLOW ROTATION WITH BARLEY AND WHEAT
(Colorado, Kansas, Montana, Nebraska, and Wyoming only)

SENCOR may be applied to provide weed control during the fallow period after wheat or barley harvest or in the Spring before planting of Winter wheat or barley. Mechanical tillage or the application of a contact herbicide may be required to control weeds germinating prior to seeding of Winter wheat or barley.

Refer to "Mixing" on the "General Information" section in the front of the label.

For specific application information see under the "General Information" section "Application" of SENCOR in the front of this label.

Where weed growth is present at application time, SENCOR should be applied with Paraquat + Plus, Roundup, or other contact herbicide. Refer to the other product label registered for additional directions, rates, and weed species controlled. Do not plant crops in treated areas earlier than 10 months following Fall applications.

WEEDS CONTROLLED

Broadleaves	
Chickweed, Common	*Stellaria media*
Cowcockle	*Vaccaria pyramidata*
Henbit	*Lamium amplexicaule*
*Kochia	*Kochia scoparia*
Lambsquarters	*Chenopodium album*
Mustard, Blue or Purple	*Chorispora tenella*
Mustard, Jim Hill	*Sisymbrium altissimum*
Mustard, Tansy	*Descurainia pinnata*
Mustard, Treacle	*Eyrsimum repandum*
Mustard, Wild	*Brassica kaber*
Pennycress, Field (fanweed)	*Thlaspi arvense*
Pigweeds	*Amaranthus* spp.
Russian thistle	*Salsola kali*
Wild Sunflower	*Helianthus* spp.
Grasses	
Cheatgrass	*Bromus secalinus*
Downy Brome	*Bromus tectorum*
*Foxtail, Green	*Setaria viridis*
*Wheat, Volunteer	*Triticum* spp.
*Wild Oats	*Avena fatua*

* Note: Since control of these weeds may vary depending on moisture following application, use the higher rate recommended below.

AFTER HARVEST APPLICATION (Fall Fallow)—SENCOR may be applied to the stubble after harvest in the Fall. Apply ⅚ to 1 pound per acre broadcast before weeds emerge. Use the higher rate for longer weed control or for weeds designated as requiring the higher rate for control. Rainfall (½ inch or more) is necessary for herbicide activation.

SPRING APPLICATION (Summer Fallow)—SENCOR may be applied to the stubble in the Spring. Apply ½ to ⅔ pound per acre broadcast before weeds emerge in the Spring. Use the higher rate for longer weed control or weeds designated as requiring the higher rate for control. Rainfall (½ inch or more) is necessary for herbicide activation. Wheat or barley can be seeded 120 days after Spring application.

PRECAUTIONS AND RESTRICTIONS

Do not graze treated fields.

Do not plant Spring seeded cereals following Fall applications for fallow.

Where SENCOR was applied in the Fall, do not apply SENCOR in the Spring.

CROP ROTATION RECOMMENDATIONS

Waiting Period After SENCOR Herbicide Application[1]

4 Months	8 Months	12 Months	18 Months
Alfalfa	Barley	Other Crops not listed (Except root crops)	Sugar Beets
Barley[2]	Cotton		Onions (And other root crops)
Asparagus	Lentils		
Corn	Peas		
Forage Grasses	Rice		
Potatoes	Wheat		
Sainfoin			
Soybeans			
Sugarcane			
Tomatoes			
Wheat[2]			

[1] Cover crops for soil building or erosion control may be planted any time, but do not graze or harvest for food or feed. Stand reductions may occur in some areas.

[2] Following peas, lentils or soybeans.

SPECIAL LOCAL NEED REGISTRATIONS [SECTION 24(c)] FOR SENCOR DF HERBICIDE*

CROP	PEST	STATE EPA SLN No.	DOSAGE	REMARKS
Alfalfa	Various broadleaf weeds and grasses	Washington WA 85-0005	⅔ to 1⅓ pounds	Application by ground or air. Do not graze or harvest within 28 days of application.
Alfalfa (Established) (Including mixed stands with grasses)	Various weeds	Utah UT 81-0018 Washington WA 81-0081 Oregon OR 81-0077	⅓ to 1 pound per acre.	Tank mix with Paraquat + Plus.
Alfalfa (Established) Non-dormant (non-Winter hardy varieties)	Various weeds	Arizona AZ 81-0027	½ to ⅔ pound per acre.	Following either a Fall or Winter sheep grazing/green-chop harvest.
Asparagus (Established)	Various weeds	Oregon OR 81-0108 Washington WA 81-0105	⅔ to 2⅔ pounds per acre.	Application through sprinkler irrigation systems.
Barley (Spring)	Various weeds	Alaska AK 83-0003	⅙ to ⅓ pounds per acre.	Postemergence ground application.
Bermudagrass (grown for seed)	Various weeds	Arizona AZ 83-0007	⅔ pound per acre.	Aerial or ground application.
Peas (Dry field)	Broadleaf weeds	Idaho ID 82-0010 Oregon OR 81-0014 Washington WA 80-0095	⅓ to ½ pound per acre.	Preemergence application.
Dry field peas	Chickweed Dog Fennel Henbit Lambsquarters Wild Mustard Field Pennycress (Fanweed) Shepherdspurse	Washington WA 85-0002 Idaho ID 84-0015	⅕ to ⅓ pound per acre.	One postemergence application may be made per season. Do not apply within 75 days of harvest. Do not apply over very moist soils or wet crop foliage.
Lentils	Broadleaf weeds	Idaho ID 82-0010 Oregon OR 81-0014 Washington WA 80-0095	⅓ to ½ pound per acre.	Preemergence application.
		Idaho ID 82-0011 Oregon OR 82-0016 Washington WA 82-0017	⅙ to ¼ pound per acre.	Early post application.
Potatoes	Weeds	Idaho ID 81-0046 Oregon OR 81-0040 Washington WA 81-0042	See label rates.	Tank mix with Eptam 7-E.
Soybeans	Weeds	Michigan MI 81-0012	⅓ to ½ pounds per acre.	For use on alkaline (calcareous) soils.
	Red rice	Louisiana LA 81-0007	See label rates.	Overlay application following Treflan PPI or tank mixed with Lasso.
	Sickelpod or coffeeweed (Cassia obtusifolia)	South Carolina SC 81-0017	⅓ to ½ pound per acre.	Low organic sandy soils.

CROP	PEST	STATE EPA SLN No.	DOSAGE	REMARKS
	Sicklepod	Arkansas AR 81-0053 Georgia GA 84-0006 Louisiana LA 81-0047	$\frac{1}{3}$ to $\frac{1}{2}$ pound per acre.	Low organic sandy soils.
		Mississippi MS 81-0057 Missouri MO 81-0029 Florida FL 84-0022 Oklahoma OK 84-0003	$\frac{1}{3}$ to $\frac{1}{2}$ pound per acre.	Low organic sandy soils, alone and in combination with other herbicides.
		North Carolina NC 84-0005	$\frac{1}{3}$ lb. per acre.	SENCOR Overlay application following Treflan applied preplant incorporated.
Sugarcane	Various grass and broadleaf weeds	Florida FL 83-0011	$1\frac{1}{3}$ to $2\frac{2}{3}$ pounds per acre.	Aerial or postemergence ground application.
Sugarcane (Fallow)	Various grass and broadleaf weeds	Louisiana LA 84-0018	$1\frac{1}{3}$ to 2 pounds per acre.	Broadcast (Summer fallow). Application by ground or air in late May to early June.
Tomatoes (chemical fallow)	Common chickweed London rocket Shepherdspurse Wild mustard Lambsquarters Redroot pigweed Purslane	California CA 79-0234	$\frac{2}{3}$ to 1 pound per acre.	Broadcast by air or ground at planting. Not for use in Kern County, California.
Wheat (Winter)	Ivyleaf speedwell	Oregon OR 81-0013	$\frac{1}{3}$ pound per acre.	Postemergence. Application by ground or air.
Wheat (Dryland Winter)	Broadleaf weeds and grasses	Kansas KS 82-0022 Oklahoma OK 80-0017	$\frac{1}{3}$ to 1 pound per acre.	Use on Eagle, TAM-101, TAM-105, and Newton varieties. On variety TAM W-101, TAM-105, Newton, and Hawk.

* For distribution and use only within the state specified. These Special Local Need registrations approved as of date of this publication are, however, subject to cancellation by the state. Contact your nearest Mobay Agricultural Chemicals Division Office for the current status and a copy of the complete individual supplemental Special Local Need labeling. This labeling must be in the possession of the user at time of pesticide application.

PRECAUTIONARY STATEMENTS

HAZARDS TO HUMANS AND DOMESTIC ANIMALS

CAUTION

May be harmful if swallowed or inhaled. Avoid contact with eyes, skin or clothing. Avoid breathing spray mist. Wash thoroughly with soap and warm water after handling. Wash contaminated clothing with soap and hot water before reuse.

Do not contaminate feed or food. Keep out of reach of children.

Obtain prompt medical aid if poisoning should occur.

To Physician: Treat the patient symptomatically.

Symptoms of Poisoning: The compound does not cause any definite symptoms that would be diagnostic. Poisoning is accompanied by breathing difficulties and sedation.

STATEMENTS OF PRACTICAL TREATMENT

If in eyes, immediately flush eyes with plenty of water for at least 15 minutes. For further treatment get medical attention.

If swallowed, vomiting should be induced. Administer water freely and induce vomiting by giving one dose ($\frac{1}{2}$ oz. or 15 ml.) of syrup of ipecac. If vomiting does not occur within 10 to 20 minutes, administer second dose. If syrup of ipecac is not available, induce vomiting by sticking finger down throat. Repeat until vomit fluid is clear. Never give anything by mouth to an unconscious person. Professional medical assistance should be secured immediately.

If on skin, wash thoroughly with soap and warm water.

ENVIRONMENTAL HAZARDS

Do not use on other crops grown for food or forage. Keep out of lakes, streams, ponds, tidal marshes and estuaries. Do not apply when weather conditions favor drift from areas treated. Do not contaminate water by cleaning of equipment or disposal of wastes. Apply this product only as specified on the label.

Do not allow sprays to drift on to adjacent desirable plants.

Observe all cautions and limitations on labeling of all products used in mixtures.

RE-ENTRY STATEMENTS

Do not apply this product in such a manner as to directly or through drift expose workers or other persons. The area being treated must be vacated by unprotected persons. Do not enter treated areas without protective clothing until sprays have dried. Because certain states may require more restrictive reentry intervals for various crops treated with this product, consult your State Department of Agriculture for further information.

Written or oral warnings must be given to workers who are expected to be in a treated area or in an area about to be treated with this product. Oral warnings should include those statements found under the label heading "Precautionary Statements" and "Re-entry

recommended grass herbicide and alone as a preemergence surface application. All these applications can be applied with ground equipment, and some can be applied with aerial spray equipment. In addition, SENCOR can be applied as a postemergence directed spray to soybeans in certain states.

SPECIAL PRECAUTIONS (SOYBEANS)

Injury to soybeans may occur when SENCOR is used under the following conditions:

1. When soils that have a calcareous surface area or a pH of 7.5 or higher.
2. Due to the sensitivity of certain soybean varieties, SENCOR is not recommended for use on sensitive varieties, such as, Tracy, Semmes, Altona, Vansoy, Coker 102, Coker 156, Portage, Maple Amber, Govan, NKS-1884, Terra-Vig 505, Terra-Vig 606, Vinton 81, Asgrow 6520, Agri Pro 55 and Agri Pro 71. Consult your Mobay Representative or your seed supplier for information on the tolerance to SENCOR of newly released soybean varieties, prior to use of SENCOR.
3. When applied in conjunction with soil-applied organic phosphate pesticides.
4. Over application or boom overlapping may result in stand loss and soil residues.
5. Uneven application or improper incorporation can decrease the level of weed control and/or increase the level of injury.
6. When applied to any soil with less than ½% organic matter.
7. Soil incorporation deeper than recommended.
8. When sprayers are not calibrated accurately.
9. When heavy rains occur soon after application, especially in poorly drained areas where water may stand for several days.
10. When soybeans are planted less than 1½" deep, particularly in preemergence application.

Activation

A minimum amount of soil moisture is required to activate SENCOR. In areas of low rainfall, preemergence applications to dry soil should be followed with light irrigation of ¼ acre inch of water. Do not apply heavy irrigation immediately after application. As with many surface-applied herbicides, weed control and crop tolerance may vary with rainfall and/or soil texture.

Grazing and Feeding Treated Vines

Treated vines may be grazed or fed to livestock 40 days after application when SENCOR is applied alone or with: Treflan®, Dual®, Prowl®, Lasso®, Amiben®, Paraquat, Lasso plus Paraquat.

Do not use treated vines for feed or forage when SENCOR is applied with: Surflan®, Dyanap®, Dyanap plus Lasso, Surflan plus Paraquat, Sonalan, linuron plus Lasso, linuron plus Dual.

Rate Ranges

Where a rate range is shown, use a lower rate on soils that are coarse-textured or low in organic matter. Use a higher rate on soils that are relatively fine-textured or high in organic matter.

REPLANTING

If replanting is necessary in fields treated with SENCOR as directed on this label, the field may be replanted to soybeans. When replanting a minimum of tillage is recommended. Do not apply a second treatment as injury to soybeans may occur.

WEEDS CONTROLLED BY SENCOR AND SENCOR TANK MIX COMBINATIONS

C = Control S = Suppression or Erratic Control (Control may range from poor to excellent) P = Poor or No Control • = No Information

1 = SENCOR Alone
2 = SENCOR SPLIT-SHOT
3 = SENCOR plus Treflan
4 = SENCOR plus Dual
5 = SENCOR plus Prowl
6 = SENCOR plus Lasso
7 = SENCOR plus Surflan
8 = SENCOR plus Amiben
9 = SENCOR plus Dyanap
10 = SENCOR plus Lasso and Dyanap
11 = SENCOR plus Paraquat
12 = SENCOR plus Lasso and Paraquat
13 = SENCOR plus Surflan and Paraqut
14 = Extended SPLIT-SHOT
15 = SENCOR plus Sonalan
16 = SENCOR plus linuron plus (Lasso or Dual)

ANNUAL BROADLEAF WEEDS	1	2	3	4	5	6	7	8	9	10	11	12	13	14	15	16
Black Nightshade *Solanum nigrum*	P	P	P	C	P	C	S	S	P	P	P	C	P	C	P	S
Bristly Starbur *Acanthospermum hispidum*	C	C	C	C	C	C	C	C	C	C	C	C	C	C	C	C
Buffalobur *Solanum rostratum*	C	C	P	P	P	P	P	P	C	P	P	P	P	C	P	•
Carpetweed *Mollugo verticillata*	C	C	C	C	C	C	C	C	C	C	C	C	C	C	C	C
Cocklebur *Xanthium pensylvanicum*	C	C	S	S	S	S	S	S	C	S	C	C	S	C	S	S
Copperleaf, Hophornbeam *Acalypha ostryaefolia*	C	C	C	C	C	C	C	C	C	C	C	C	C	C	C	C
Florida Beggarweed *Desmodium tortuosum*	C	C	C	C	C	C	C	C	C	C	C	C	C	C	C	C
Florida Pusley *Richardia scabra*	C	C	C	C	C	C	C	C	C	C	C	C	C	C	C	C
Galinsoga *Galinsoga* spp.	C	C	C	C	C	C	C	C	C	C	C	C	C	C	C	C
Horseweed Marestail *Conyza canadensis*	•	•	•	•	•	•	•	•	•	•	C	C	C	C	•	•
Jimsonweed *Datura stramonium*	C	C	C	C	C	C	C	S	C	C	C	C	C	C	C	S
Knotweed *Polygonum* spp.	C	C	C	C	C	C	C	C	C	C	C	C	C	C	C	C
Kochia *Kochia scoparia*	C	C	C	C	C	C	C	C	C	C	C	C	C	C	C	C
Lambsquarters *Chenopodium* spp.	C	C	C	C	C	C	C	C	C	C	C	C	C	C	C	C
Morningglory, Ivyleaf *Ipomoea hederacea*	P	P	S	P	P	P	P	P	S	S	P	P	P	P	P	P

WEEDS CONTROLLED BY SENCOR AND SENCOR TANK MIX COMBINATIONS

C = Control S = Suppression or Erratic Control (Control may range from poor to excellent) P = Poor or No Control • = No Information

1 = SENCOR Alone
2 = SENCOR SPLIT-SHOT
3 = SENCOR plus Treflan
4 = SENCOR plus Dual
5 = SENCOR plus Prowl
6 = SENCOR plus Lasso
7 = SENCOR plus Surflan
8 = SENCOR plus Amiben
9 = SENCOR plus Dyanap
10 = SENCOR plus Lasso and Dyanap
11 = SENCOR plus Paraquat
12 = SENCOR plus Lasso and Paraquat
13 = SENCOR plus Surflan and Paraqut
14 = Extended SPLIT-SHOT
15 = SENCOR plus Sonalan
16 = SENCOR plus linuron plus (Lasso or Dual)

ANNUAL BROADLEAF WEEDS	1	2	3	4	5	6	7	8	9	10	11	12	13	14	15	16
Morningglory, Pitted *Ipomoea lacunosa*	P	P	S	P	P	P	P	P	P	P	P	P	P	P	P	P
Morningglory, Smallflower *Jacquemontia tamnifolia*	P	P	C	P	P	P	P	P	P	P	P	P	P	P	P	P
Morningglory, Tall *Ipomoea purpurea*	P	P	S	P	P	P	S	P	S	S	P	P	P	P	P	P
Pigweeds *Amaranthus* spp.	C	C	C	C	C	C	C	C	C	C	C	C	C	C	C	C
Prickly Sida/Teaweed *Sida spinosa*	C	C	C	C	C	C	C	C	C	C	C	C	C	C	C	C
Purslane *Portulaca oleracea*	C	C	C	C	C	C	C	C	C	C	C	C	C	C	C	C
Ragweed, Common *Ambrosia artemisiifolia*	C	C	C	C	C	C	C	C	C	C	C	C	C	C	C	C
Redweed *Melochia corchorifolia*	C	C	C	C	C	C	C	C	C	C	C	C	C	C	C	C
Russian Thistle *Salsola kali*	C	C	C	C	C	C	C	C	C	C	C	C	C	C	C	C
Sesbania *Sesbania* spp.	C	C	C	C	C	C	C	C	C	C	C	C	C	C	C	C
Shepherdspurse *Capsella bursa-pastoris*	C	C	C	C	C	C	C	C	C	C	C	C	C	C	C	C
Sicklepod *Cassia obtusifolia*	C	C	C	C	S	C	S	C	C	C	C	C	C	C	S	S
Smartweeds *Polygonum* spp.	C	C	C	C	C	C	C	C	C	C	C	C	C	C	C	S
Spotted Spurge *Euphorbia maculata*	C	C	P	C	P	C	S	P	C	C	C	C	C	C	P	•
Spurred Anoda *Anoda cristata*	C	C	C	C	C	C	C	C	C	C	C	C	C	C	C	•
Sunflower *Helianthus* spp.	C	C	S	S	S	S	S	S	C	S	S	S	S	C	S	P
Velvetleaf *Abutilon theophrasti*	C	C	C	C	C	C	C	C	C	C	C	C	C	C	C	C
Venice Mallow *Hibiscus trionum*	C	C	C	C	C	C	C	C	C	C	C	C	C	C	C	C
Wild Mustards *Brassica* spp.	C	C	C	C	C	C	C	C	C	C	C	C	C	C	C	C
ANNUAL GRASSES	**1**	**2**	**3**	**4**	**5**	**6**	**7**	**8**	**9**	**10**	**11**	**12**	**13**	**14**	**15**	**16**
Barnyardgrass *Echinochloa crus-galli*	S	C	C	C	C	C	C	C	C	C	C	C	C	C	C	C
Bluegrass *Poa annua*	C	C	C	C	C	C	•	C	C	C	C	C	C	C	C	C
Broadleaf Signalgrass *Brachiaria platyphylla*	C	C	C	C	C	C	C	C	C	C	C	C	C	C	C	•
Browntop Millet *Panicum ramosum*	C	C	C	P	C	S	C	P	C	P	C	P	C	C	•	•
Crabgrass *Digitaria* spp.	C	C	C	C	C	C	C	C	C	C	C	C	C	C	C	C
Crowfootgrass *Dactyloctenium aegyptium*	C	C	C	C	C	C	C	C	C	C	C	C	C	C	•	•
Cupgrass *Eriochloa gracilis*	P	C	P	P	P	P	P	P	P	P	P	P	P	C	•	•
Foxtails *Setaria* spp.	S	C	C	C	C	C	C	C	C	C	C	C	C	C	C	C
Goosegrass *Eleusine indica*	C	C	C	C	C	C	C	C	C	C	C	C	C	C	C	C

WEEDS CONTROLLED BY SENCOR AND SENCOR TANK MIX COMBINATIONS

C = Control
S = Suppression or Erratic Control (Control may range from poor to excellent)
P = Poor or No Control
• = No Information

1 = SENCOR Alone
2 = SENCOR SPLIT-SHOT
3 = SENCOR plus Treflan
4 = SENCOR plus Dual
5 = SENCOR plus Prowl
6 = SENCOR plus Lasso
7 = SENCOR plus Surflan
8 = SENCOR plus Amiben
9 = SENCOR plus Dyanap
10 = SENCOR plus Lasso and Dyanap
11 = SENCOR plus Paraquat
12 = SENCOR plus Lasso and Paraquat
13 = SENCOR plus Surflan and Paraquat
14 = Extended SPLIT-SHOT
15 = SENCOR plus Sonalan
16 = SENCOR plus linuron plus (Lasso or Dual)

ANNUAL GRASSES	1	2	3	4	5	6	7	8	9	10	11	12	13	14	15	16
Johnsongrass, Seeding *Sorghum halepense*	C	C	C	C	C	C	C	P	C	C	C	C	C	C	C	•
Junglerice *Echinochloa colonum*	C	C	C	C	C	C	•	C	C	C	C	C	C	C	C	•
Nutsedge, Yellow *Cyperus esculentus*	P	P	P	C	P	C	P	P	P	P	P	C	P	C	P	•
Panicum, Fall *Panicum dichotomiflorum*	P	C	C	C	C	C	C	P	P	P	P	C	P	C	C	C
Panicum, Texas *Panicum texanum*	P	C	C	P	C	S	C	P	P	P	P	S	P	S	C	•
Red Rice *Oryza sativa*	P	C	C	C	P	C	•	P	P	P	P	C	P	C	•	•
Sandbur *Cenchrus* spp.	P	C	C	P	C	S	•	P	P	C	P	C	P	S	•	•
Shattercane *Sorghum bicolor*	P	C	C	P	P	P	P	P	P	P	P	P	P	P	C	•
Sorghum, Volunteer *Sorghum* spp.	P	C	C	P	P	P	P	P	P	P	P	P	P	P	•	P
Sprangletop *Leptochloa* spp.	P	C	C	P	P	P	•	P	P	P	P	P	P	P	•	P
Stinkgrass *Eragrostis* spp.	P	C	C	P	P	P	•	P	P	P	P	P	P	P	•	P
Wheat, Volunteer *Triticum* spp.	P	P	P	P	P	P	S	P	P	P	P	C	C	P	•	P
Witchgrass *Panicum capillare*	P	C	C	C	C	C	•	P	P	C	P	C	C	C	C	•

SENCOR ALONE

SENCOR (Alone) Preemergence Application

SENCOR can be applied broadcast or banded. This application may be made during planting or as a separate operation after planting but before crop emergence.

See the "APPLICATION" section under "GENERAL INFORMATION" in the front of this label.

Do not apply to sand soils; or to sandy loam or loamy sand soils containing less than 2% organic matter. Do not incorporate into soil or apply more than once per season.

PINTS OF SENCOR 4 PER ACRE

SOIL TEXTURE	Organic Matter: Less than 2%	2 to 4%	Over 4%
COARSE SOILS (Sandy loam, loamy sand)	DO NOT USE[3]	3/4	1
MEDIUM SOILS[1] (Loam, silt loam, silt sandy clay, sandy clay loam)	3/4 to 1	1 to 1 1/4	1 1/4 to 1 1/2
FINE SOILS[1] (Silty clay, silty clay loam[2], clay, clay loam)	1 to 1 1/4	1 1/4 to 1 1/2	1 1/2 to 1 3/4
Mississippi Delta Only	1 1/2	1 3/4	2

[1] For control of lambsquarters, redroot pigweed and wild mustard, and for suppression of green, yellow and giant foxtails on alkaline (calcareous) soils in Nebraska, Minnesota, South Dakota and North Dakota only, apply SENCOR at rates of 1/2 pint/acre on medium soils and 1/2 to 3/4 pint/acre on fine soils regardless of soil organic matter percentage (use 3/4 pint only where soil pH is less than 7.5 and weed pressure is heavy). The 1/2 pint/acre rate of SENCOR alone can be applied regardless of soil pH. For control of other weeds listed on the label use SENCOR at full rates recommended in the table above, **but note that crop injury may occur on soils having a calcareous surface area or a pH of 7.5 or higher.**

[2] Silty clay loam soils are transitional soils and may be classified as medium textured soils in some regions of the U.S.

[3] Refer to the appropriate section of the label for use of SENCOR on soybeans in coarse soils with 0.5 to 2.0% organic matter in certain states.

USES OF SENCOR IN COMBINATIONS WITH OTHER HERBICIDES

SPLIT-SHOT,™ APPLICATION

A preplant incorporated application of SENCOR tank mixed with either Treflan, Lasso, Dual, Prowl, or Sonalan and followed by a preemergence surface application of SENCOR alone, after planting but prior to soybean emergence, will control more broadleaf and grass weeds in soybeans than when either herbicide is used alone.

Refer to the Treflan, Lasso, Dual, Prowl, or Sonalan labels, and to appropriate sections of the SENCOR label, for directions on soil preparation, herbicide application, incorporation techniques, herbicide rates, weed species controlled, and restrictions for using tank mix combinations of SENCOR. Carefully observe the "Special Precautions" section concerning the use of SENCOR in tank mix combinations on soybeans.

When a SPLIT-SHOT application of SENCOR with Prowl, Treflan, or Sonalan is used, the preplant incorporated tank mix be applied up to 21 days prior to planting soybeans; with Dual or Lasso, the preplant incorporated tank mix may be applied up to 14 days prior to planting.

On medium and fine textured soils with greater than 2% organic matter, a rate range is recommended for the SENCOR preemergence overlay application. The higher rate should be used (a) in fields with a history of severe broadleaf weed pressure, (b) when the time between preplant incorporated tank mix and preemergence overlay applications approaches the maximum stated above, and/or (c) when the organic matter content of the soil is at the upper end of the indicated range.

NOTE: Tank mix combinations which include Lasso cannot be applied with aerial spray equipment.

For black nightshade control, refer to the appropriate section of the Lasso, Dual or Sonalan labels.

SPLIT-SHOT APPLICATION

	PREPLANT INCORPORATED TANK MIX APPLICATION			—FOLLOWED BY—	PREEMERGENCE OVERLAY APPLICATION		
SOIL TEXTURE[1]	RATE OF COMBINATION PRODUCT/ACRE	PLUS	RATE OF SENCOR 4 PINT/ACRE		RATE OF SENCOR 4 PINT/ACRE ORGANIC MATTER LESS THAN 2%	2 TO 4%	OVER 4%
COARSE (Light) (Sand, loamy sand, sandy loam)	Treflan 1 pt or Lasso 2–2½ qt or Dual 1¼–1½ pt or Prowl 1½ pt or Sonalan 1¼–2 pt	plus—	½	—FOLLOWED BY—	¼	¼	¼–½
MEDIUM (Loam, silt loam, sandy clay loam, silt, sandy clay)	Treflan 1½ pt or Lasso 2½–3 qt or Dual 1½ pt or Prowl 1½ pt or Sonalan 1¾–2½ pt	plus—	¾	—FOLLOWED BY—	¼	¼–½	½–¾
				or			
			½[2]	—FOLLOWED BY—	½	½–¾	(¾–1)[3]
FINE (Heavy) (Silty clay loam*, clay loam, silty clay, clay)	Treflan 2 pt or Lasso 2½–3 qt or Dual 2–2½ pt or Prowl 1½–2 pt or Sonalan 2¼–3 pt	plus—	1.0	—FOLLOWED BY—	¼	¼–½	½–¾
				or			
			¾[2]	—FOLLOWED BY—	½	½–¾	(¾–1)[3]

* Silty clay loam soils are transitional soils and may be classified as medium textured soils in some regions of the U.S.

[1] On **coarse textured** soils, do not use on sand soils with less than 1% organic matter, or on loamy sand or sandy loam soils with less than 0.5% organic matter. However, on coarse textured soils **with a calcareous surface area or a pH of 7.5 or higher**, do not use on sand soils with less than 2% organic matter, or on loamy sand or sandy loam soils with less than 1% organic matter.

[2] Use this lower rate of SENCOR in the preplant incorporated tank mix **on soils having a calcareous surface area or a pH of 7.5 or higher**, and in those situations where soils within a field vary extremely in texture or organic matter content.

[3] Reduce this preemergence overlay rate of SENCOR by ¼ pint/acre when using SPLIT-SHOT application **on soils with over 4% organic matter and which have a calcareous surface area or a pH of 7.5 or higher**.

EXTENDED SPLIT-SHOT™ APPLICATION

(Includes No-Till, Reduced Till, Ridge Till, Strip Till, Mulch Till)

An early preplant (surface-applied or shallow incorporated) application of SENCOR tank mixed with either Dual or Lasso, followed by a preemergence surface application of SENCOR tank mixed with Dual or Lasso after planting but prior to soybean emergence, will control more broadleaf and grass weeds in soybeans than either herbicide used alone.

An EXTENDED SPLIT-SHOT application will decrease the need for tillage and/or contact herbicides for the control of existing vegetation prior to planting, while providing residual control of weeds after planting.

When an EXTENDED SPLIT-SHOT application of SENCOR with Dual or Lasso is used, the preplant tank mix combination may be applied 15 to 30 days prior to planting soybeans. Follow directions on the label accompanying the product for SPLIT-SHOT applications from 0-14 days before planting.

Where a rate range is recommended, the higher rates should be used (a) in fields with a history of severe weed pressure, (b) when the time between early preplant tank mix and preemergence overlay applications approaches the maximum 30 days, (c) when the organic matter content of the soil is at the upper end of the indicated range, (d) when heavy crop residues are present on the soil surface, and/or (e) when the early preplant tank mix application is shallow incorporated (e.g. use 2 to 2½ qts Lasso in the early preplant tank mix when surface applied and use 2½ to 3 qts Lasso when the tank mix is to be lightly incorporated).

When weeds exceed 1 to 1½ inches in height or diameter at application, use a contact herbicide, such as Round-Up or Paraquat.

Refer to the Dual or Lasso label, and to appropriate sections of the SENCOR label for additional information on soil preparation, herbicide application, weeds controlled, precautions, restrictions, limitations, and sprayer clean-up.

EXTENDED SPLIT-SHOT APPLICATION

EARLY PREPLANT TANK MIX APPLICATION (SURFACE-APPLIED OR SHALLOW INCORPORATED)				FOLLOWED BY	PREEMERGENCE OVERLAY APPLICATION				
SOIL TEXTURE(1)	RATE OF COMBINATION PRODUCT/ACRE	PLUS	RATE OF SENCOR 4 PINT/ ACRE		RATE OF COMBINATION PRODUCT/ ACRE	PLUS	RATE OF SENCOR 4 PINT/ ACRE ORGANIC MATTER ½ TO 2%	2 TO 4%	OVER 4%
COARSE (Light) Sand, loamy sand, sandy loam	Dual 1⅓ pt or Lasso 1½-2 qt	plus	½-¾	followed by	Dual ⅔ pt or Lasso 1½ qt	plus	¼	¼-½	½
MEDIUM Loam, silt loam, sandy clay loam, silt, sandy clay	Dual 1¾ pt or Lasso 2-3 qt	plus	(2)¾-1	followed by	Dual ¾ pt or Lasso 1-2 qt	plus	½	½-¾	*¾-1
FINE (Heavy) Silty clay loam*, clay loam, silty clay, clay	Dual 2 pt or Lasso 2-3 qt	plus	(2)1-1¼	followed by	Dual 1 pt or Lasso 1-2 qt	plus	½	½-¾	¾-1

* Silty clay loam soils are transitional soils and may be classified as medium textured soils in some regions of the U.S.

(1) On coarse textured soils, do not use on sand soil with less than 1% organic matter. However, on coarse textured soils with a calcareous surface area or a pH of 7.5 or higher, do not use on sand soils with less than 2% organic matter, or on loamy sand or sandy loam soils with less than 1% organic matter.

(2) Use the lower rate of SENCOR in the early preplant tank mix on soils having calcareous surface area or a pH of 7.5 or higher, and in those rare situations where soils within a field vary extremely in texture or organic matter content.

SENCOR PLUS SONALAN

SENCOR PLUS SONALAN OVERLAY APPLICATION

SENCOR may be applied as a preemergence overlay application following a preplant incorporated application of Sonalan 3 EC. Consult the Sonalan label for specific directions on use, recommendations, restrictions and any additional weeds not specified on the SENCOR label.

SENCOR PLUS SONALAN TANK MIX APPLICATION

Incorporate the tank mixture into the top 1 to 2 inches of soil within 21 days before planting according to label directions for Sonalan.

Apply SENCOR plus Sonalan preplant incorporated if furrow irrigation is used or when a period of dry weather after application is expected. If soybeans are planted on beds, apply and incorporate the tank mixture after bed formation.

Mixing: Refer to "Mixing" under the "General Information" section in the front of the SENCOR label.

Application: Sonalan should be uniformly applied and thoroughly mixed into the soil within 2 days after application. For specific application information, refer to "Application" under the "General Information" section in the front of the SENCOR label.

Broadcast Rates

Soil Texture	SENCOR 4 pts/acre	Sonalan 3 EC pts/acre
COARSE[1] (Sandy loam, loamy sand)	½	1¼-2
MEDIUM[3] (Loam, silt loam, silt, sandy clay, sandy clay loam)	¾	1¾-2½
FINE[3] (Silty clay, silty clay loam[2], clay, clay loam)	1	2¼-3

[1] Do not use on coarse soils with less than 1% organic matter.

[2] Silty clay loam soils are transitional soils and may be classified as medium textured soils in some regions of the U.S.

[3] For control of lambsquarters, redroot pigweed, wild mustard, and green and yellow foxtails on alkaline (calcareous) soils in Minnesota, Nebraska, South Dakota, and North Dakota only, apply SENCOR at rates of ½ pint/acre on medium soils and ½ to ¾ pint/acre on fine soils regardless of soil organic matter percentage (use ¾ pint only where soil pH is less than 7.5 and weed pressure is heavy). The ½ pint rate of SENCOR in tank mix combination with Sonalan can be applied regardless of soil pH. For control of other weeds not listed on the label, use SENCOR at full rates recommended in the table above, **but note that crop injury may occur on soils having a calcareous surface area or a pH of 7.5 or higher.**

SPECIAL PRECAUTIONS (SENCOR PLUS SONALAN)

For additional precautions, restrictions, limitations, incorporation, and sprayer clean-up information, refer to the appropriate sections of the SENCOR label and the Sonalan label.

For black nightshade control, refer to the Sonalan label for specific rates and application instructions.

SENCOR PLUS TREFLAN

SENCOR PLUS TREFLAN OVERLAY APPLICATION

SENCOR may be applied as a preemergence broadcast or band overlay application following a preplant incorporated treatment of Treflan. Consult the Treflan label for specific directions for use, recommendations, restrictions and any additional weeds not specified on the SENCOR label.

SENCOR PLUS TREFLAN TANK MIX APPLICATION

A single application of a tank mix combination of SENCOR and Treflan EC will control more broadleaf and grass weeds in soybeans than when either herbicide is used alone.

Prepare the soil surface by deep plowing, offset disking or tandem disking prior to the application of the herbicide combination. The soil surface should be well prepared and free of clods and trash.

This SENCOR plus Treflan tank mix combination may be applied and incorporated into the soil up to 10 days before planting.

Mixing Refer to "Mixing" under the "General Information" section in the front of the SENCOR label.

Application: For specific application information refer to "Application" under the "General Information" section in the front of the SENCOR label.

Apply SENCOR plus Treflan to the soil surface and incorporate in the same operation, if possible. Variable weed control may result from delayed incorporation if SENCOR plus Treflan are applied to a wet, warm soil surface or if the wind velocity is 10 miles per hour or higher. Use machinery that mixes SENCOR plus Treflan thoroughly with the soil. Incorporation may be delayed up to 24 hours after application. Shallow incorporation with implements set to cut less than 2 inches deep may result in erratic weed control. Do not use spike or spring-tooth harrows alone for incorporation.

Broadcast Rates

Soil Texture	SENCOR 4 Pints per Acre	Treflan EC Pints per Acre
Coarse[1] (Sandy loam, loamy sand)	½	1
Medium (Loam, silt loam, silt, sandy clay, sandy clay loam)[3]	¾	1½
Fine (Silty clay, silty clay loam,[2] clay, clay loam)[3]	1	2

[1] Do not use on coarse soils with less than 1% organic matter.

[2] Silty clay loam soils are transitional soils and may be classified as medium textured soils in some regions of the U.S.

[3] For control of lambsquarters, redroot pigweed, wild mustard, and green and yellow foxtails on alkaline (calcareous) soils in Minnesota, Nebraska, South Dakota, and North Dakota only, apply SENCOR at rates of ½ pint/acre on medium soils and ½ to ¾ pint/acre on fine soils regardless of soil organic matter percentage (use ¾ pint only where soil pH is less than 7.5 and weed pressure is heavy). The ½ pint rate of SENCOR in tank mix combination with Treflan can be applied regardless of soil pH. For control of other weeds listed on the label use SENCOR at full rates recommended in the table above, **but note that crop injury may occur on soils having a calcareous surface area or a pH of 7.5 or higher.**

Incorporation Equipment:

1. Set PTO-driven equipment (tillers, cultivators, hoes) to cut 2 to 3 inches deep and space rotors to provide a clean sweep of the soil. PTO equipment should not be operated at a speed greater than 4 miles per hour.

2. Set disk to cut 4 to 6 inches deep and operate twice in different directions at 4 to 6 miles per hour.

3. Set mulch treader and other similar disk-type implements to cut 3 to 4 inches deep and operate twice in different directions at 5 to 8 miles per hour.

For Coarse and Medium Textured Soils Only

4. Set rolling cultivator to cut 2 to 4 inches deep and operate twice at 6 to 8 miles per hour. Set bed conditioner (Do-All) to cut 2 to 4 inches deep and operate at 4 to 6 miles per hour.

SPECIAL PRECAUTIONS (SENCOR PLUS TREFLAN)

Seedling disease, cold weather, excessive moisture, high salt concentration or drought may weaken soybean seedlings and increase possibility of damage from the tank mix. Do not plant soybeans deeper than 2 inches.

In the Central United States, do not plant sorghum or oats for 12 months where the tank mix has been applied unless 20 inches or more of irrigation and/or rainfall (total) was used to produce the crop. If less than 20 inches total water was used to produce the crop during the year, do not plant either crop for 18 months after the tank mix application. Cool wet weather conditions during the early stage of growth may increase the possibility of injury to sorghum.

For additional precautions, restrictions, limitations and sprayer clean-up information refer to the appropriate section of this label. Do not use this tank mix combination on soils containing charcoal in Arkansas, Louisiana and Mississippi.

SENCOR PLUS DUAL

SENCOR PLUS DUAL OVERLAY APPLICATION

Apply a preplant incorporated treatment of Dual 8E as directed on that product label for use on soybeans. Follow with a preemergence treatment of SENCOR as directed on this label for use on soybeans.

SENCOR PLUS DUAL TANK MIX APPLICATIONS

Preplant Incorporated Application: Incorporate the tank mixture into the top 2 inches of soil within 14 days before planting using a disk, harrow, rolling cultivator, or similar implement.

Apply SENCOR plus Dual preplant incorporated if furrow irrigation is used or when a period of dry weather after application is expected. If soybeans are planted on beds, apply and incorporate the tank mixture after bed formation.

Preemergence Application: Dry weather following preemergence application of SENCOR plus Dual tank mixture may reduce effectiveness. If weeds develop, cultivate uniformly with shallow tilling equipment such as a rotary hoe that will not damage soybeans.

Mixing Instructions: Refer to "Mixing" under the "General Information" section in the front of the SENCOR label.

For information on applying SENCOR in fluid or dry fertilizer refer to the "Application of SENCOR In Fluid Fertilizers" or "Commercial Impregnation And Application Of SENCOR On Dry Bulk Fertilizer" under the "General Information" section in the front of the SENCOR label.

Broadcast Rates:
SENCOR plus Dual Tank Mix Preemergence Applications

Soil Texture	Organic Matter 0.5% to 3% SENCOR 4 Pints/Acre	0.5% to 3% Dual 8E Pints/Acre	Over 3% SENCOR 4 Pints/Acre	Over 3% Dual 8E Pints/Acre
COARSE[1] (Loamy sand, sandy loam)	½	1¼	¾	1½
MEDIUM (Loam silt loam, silt)	¾	1½	1	2
FINE (Silty clay loam[2], sandy clay loam, silty clay, sandy clay, clay loam, clay)	1	2	1 to 1¼	2 to 2½
MISSISSIPPI DELTA ONLY (Silty clay, clay)	1½	2	1½	2 to 2½

[1] Do not use on sand soils. Do not apply SENCOR and Dual overlay or tank mix preemergence on loamy sand with less than 2% organic matter.
[2] Silty clay loam soils are transitional soils and may be classified as medium textured soils in some regions of the U.S.

Broadcast Rates
SENCOR plus Dual Tank Mix Preplant Incorporated Applications

Soil Texture	Organic Matter 0.5% to less than 3% SENCOR 4 Pints/Acre	0.5% to less than 3% Dual 8E Pints/Acre	3% or greater SENCOR 4 Pints/Acre	3% or greater Dual 8E Pints/Acre
COARSE[1] (Loamy sand, sandy loam)	½	1¼	½	1½
MEDIUM (Loam silt loam, silt)	¾	1½	¾	2
FINE (Silty clay loam[2], sandy clay loam, silty clay, sandy clay, clay loam, clay)	1	2	1	2 to 2½
MISSISSIPPI DELTA ONLY (Silty clay, clay)	1 to 1¼	2	1 to 1¼	2 to 2½

[1] Do not use on sand soils. Do not apply SENCOR plus Dual tank mix preplant incorporated on sand or loamy sand with less than 2% organic matter or crop injury may occur.
[2] Silty clay loam soils are transitional soils and may be classified as medium textured soils in some regions of the U.S.

SPECIAL PRECAUTIONS (SENCOR AND DUAL)

For additional precautions, restrictions, limitations, and sprayer clean-up information refer to the appropriate sections of the SENCOR label and the Dual label.

SENCOR PLUS PROWL

SENCOR PLUS PROWL OVERLAY APPLICATION

Apply a preplant incorporated treatment of Prowl as directed on that product label for use on soybeans. Follow with a preemergence treatment of SENCOR as directed on the SENCOR label for use on soybeans.

SENCOR PLUS PROWL TANK MIX APPLICATION

Preplant Incorporated Application: Prepare the soil by plowing or disking to mix previous crop residues into the soil to a depth of 4 to 6 inches.

For specific application information refer to "Application" under the "General Information" section in the front of the SENCOR label.

Incorporate the tank mixture into the top 1 or 2 inches of soil within 7 days after application according to label directions for Prowl. Mechanical incorporation is not required if a rain of one-quarter inch or more occurs within 7 days after application. Soybeans must be planted no later than 7 days after application of the tank mixture.

Preemergence Application: Except for minimum and no-tillage systems, the seed bed should be firm and free of trash and clods.

For specific application information refer to "Application" under the "General Information" section in the front of the SENCOR label.

This application must be made after planting and before crop emergence. Do not incorporate.

If cultivation is necessary because of soil crusting, soil compaction or weed germination before rain or irrigation, use shallow tilling equipment such as a rotary hoe that does not damage soybeans.

Mixing Instructions: Refer to "Mixing" under the "General Information" section in the front of the SENCOR label.

For information on applying SENCOR in fluid or dry fertilizer refer to the "Application Of SENCOR In Fluid Fertilizers" or "Commercial Impregnation and Application Of SENCOR On Dry Bulk Fertilizer" under the "General Information" section in the front of the SENCOR label.

SOUTHERN STATES AND EASTERN COASTAL PLAINS

For use only in Alabama, Arkansas, Florida, Georgia, Louisiana, Mississippi, North Carolina, Oklahoma, Southeastern Missouri "Bootheel" Region and Coastal Plains of Delaware*, Maryland*, New Jersey* and Virginia*.

* SENCOR plus Prowl is not recommended for use on soils with less than 2% organic matter in the coastal plain of New Jersey or the Delmarva Peninsula.

Broadcast Rates
SENCOR plus Prowl Tank Mix Applications

Soil Texture	SENCOR 4 Pints/Acre	Prowl Pints/Acre
COARSE[1] (Sandy loam, loamy sand)	½	1½
MEDIUM (Loam, silt loam, silt, sandy clay, sandy clay loam)	¾	1½
FINE (Silty clay, silty clay loam[2], clay, clay loam)	1	1½ to 2

[1] Do not use on sand soils. Do not use on loamy sand or sandy loam containing less than 1% organic matter.

[2] Silty clay loam soils are transitional soils and may be classified as medium textured soils in certain regions of the U.S.

Do not use on muck or peat soils.

NORTHEASTERN AND NORTH CENTRAL STATES

For use only in Illinois, Indiana, Iowa, Kansas, Kentucky, Michigan, Minnesota, Nebraska, New York, North Dakota, Ohio, Pennsylvania, South Dakota, Wisconsin and Missouri (except the "Bootheel" Region).

Broadcast Rates
SENCOR plus Prowl Tank Mix Applications

Soil Texture	Organic Matter ½ to 3% SENCOR 4 Pints/Acre	½ to 3% Prowl Pints/Acre	Over 3% SENCOR 4 Pints/Acre	Over 3% Prowl Pints/Acre
COARSE[1] (Sandy loam, loamy sand)	½	1	¾	1½
MEDIUM (Loam, silt loam sandy clay, sandy clay loam)	¾	1½ to 2	¾ to 1	1½ to 2
FINE (Silty clay, salty clay loam,[2] clay, clay loam)	¾ to 1	1½ to 2	1 to 1¼	2 to 2½

[1] Do not use on sand soils. Do not use on loamy sand or sandy loam containing less than 1% organic matter. Where a range of rates is shown for medium and fine soils, use the higher rate if heavy weed infestations are anticipated.

[2] Silty clay loam soils are transitional soils and may be classified as medium textured soils in certain areas of the U.S.

Do not use on muck or peat soils.

SPECIAL PRECAUTIONS (SENCOR AND PROWL)

Soil incorporation deeper than recommended will reduce weed control and can result in crop injury.

For additional precautions, restrictions, limitations, and sprayer clean-up information, refer to the appropriate sections of the SENCOR label and the Prowl label.

SENCOR PLUS LASSO

SENCOR PLUS LASSO TANK MIX APPLICATION

PREEMERGENCE: SENCOR may be used in a tank mix combination with Lasso as a preemergence band or broadcast application to soybeans in accordance with the specified soil types and dosages recommended.

For specific information regarding spray equipment, dilution rates, mixing, directions for use, methods of application, limitations and restrictions refer to the appropriate section of the SENCOR label.

Refer to the Lasso label for pertinent recommendations, directions for use, restrictions and any additional weeds not specified on the SENCOR label.

RECOMMENDED APPLICATIONS
SENCOR plus Lasso Tank Mix Preemergence Application (Broadcast rates)

Soil Texture	½ to 3% Organic Matter SENCOR 4 Pts./Acre	+	½ to 3% Organic Matter Lasso Qts./Acre	Greater than 3% Organic Matter SENCOR 4 Pts./Acre	+	Greater than 3% Organic Matter Lasso Qts./Acre
COARSE[1] (Sandy loam)	½	+	1½ to 2	¾	+	1½ to 2
MEDIUM[2] (Loam, silt loam, silt, sandy clay, sandy clay loam)	¾	+	1½ to 2	1	+	1½ to 2
FINE[2] (Silty clay, silty clay loam,[3] clay, clay loam)	1	+	2	1 to 1¼	+	2 to 2½
Mississippi Delta Only (Silty clay to heavy clay)	SENCOR 4 Pts./Acre: 2	+	Lasso Qts./Acre: 2 to 2½			

[1] Do not use SENCOR plus Lasso on sand or loamy sand soils with less than 2% organic matter.

[2] For control of lambsquarters, redroot pigweed, wild mustard, green and yellow foxtails on alkaline (calcareous) soils in Minnesota, Nebraska, South Dakota, and North Dakota only, apply SENCOR 4 at rates of ½ pint/acre on medium soils and ½ to ¾ pint/acre on fine soils regardless of

soil organic matter percentage (use 3/4 pint only where soil pH is less than 7.5 and weed pressure is heavy). The 1/2 pint/acre rate of SENCOR in tank mix combination with Lasso can be applied regardless of soil pH. For control of other weeds use SENCOR at full rates recommended in the table above, **but note that crop injury may occur on soils having a calcareous surface area or a pH of 7.5 or higher.**

3 Silty clay loam soils are transitional soils and may be classified as medium textured soils in some regions of the U.S

DO NOT USE ON MUCK SOILS.

PREPLANT·INCORPORATED: For specific application information refer to "Application" under the "General Information" section in the front of the SENCOR label.

Apply SENCOR plus Lasso preplant incorporated if furrow irrigation is used or when a period of dry weather after application is expected. If soybeans are planted on beds, apply and incorporate the tank mixture after bed formation. Apply within 7 days prior to planting and shallowly incorporate into the upper 1 to 2 inches of soil.

Recommended Applications
SENCOR plus Lasso Tank Mix Preplant Incorporated Applications (Broadcast Rates)

Soil Texture	SENCOR 4 Pints/Acre	Lasso Quarts/Acre
COARSE[1] (Loamy sand [over 2% organic matter], sandy loam)	1/2	2 to 2 1/2
MEDIUM (Loam, silty loam, silt)	3/4	2 1/2 to 3
FINE (Silty clay loam[2], sandy clay loam, silty clay, sandy clay, clay loam, clay)	1	2 1/2 to 3
Mississippi Delta Only (Silty clay, clay)	1 to 1 1/4	2 1/2 to 3

1 Do not use SENCOR plus Lasso on sand or loamy sand soils with less than 2% organic matter.

2 Silty clay loam soils are transitional soils and may be classified as medium textured soils in some regions of the U.S.

DO NOT USE ON MUCK SOILS.

SPECIAL PRECAUTIONS (SENCOR PLUS LASSO)

For additional precautions, restrictions, limitations and sprayer clean-up information refer to the appropriate sections of the SENCOR label and the Lasso label.

SENCOR PLUS SURFLAN®

SENCOR PLUS SURFLAN TANK MIX APPLICATION

SENCOR may be used in a tank mix combination with Surflan A.S. as a preemergence application to control weeds in soybeans in accordance with the specified soil types and dosages recommended. Consult the Surflan label for specific directions for use, recommendations, restrictions and any additional weeds not specified on the SENCOR label.

Mixing Instructions: Refer to "Mixing" under the "GENERAL INFORMATION" section in the front of the SENCOR label.

Application: Apply within 2 days after planting but before crop emergence.

A one-half inch rain or more or its equivalent in overhead irrigation is needed to activate the herbicide combination. If a one-half inch rain does not occur within 7 days after application, incorporate the herbicide combination into the top one inch of the soil with a ground-driven rotary hoe, rolling cultivator, or similar implement that will not damage soybean plants. Cultivation will control emerging annual weeds and grasses and improve herbicidal effectiveness. If heavy rains pack the soil surface, rotary hoe to break the crust and to aid soybean emergence.

For specific information on application, refer to "Application" under the "GENERAL INFORMATION" section in the front of the SENCOR label.

Broadcast Rates

Soil Texture	SENCOR 4 Pints Per Acre	Surflan A.S. Quarts Per Acre
COARSE[1] (Sand, sandy loam, loamy sand)	1/2	1/2 to 3/4
MEDIUM (Loam, silt loam, silt, sandy clay, sandy clay loam)	3/4	3/4 to 1
FINE (Silty clay, silty clay loam,[2] clay, clay loam)	1	1 to 1 1/4
Mississippi Delta Only	1 to 1 1/4	1 to 1 1/4

1 Not recommended on sand with less than 1% organic matter. Do not use on soils containing more than 5% organic matter.

2 Silty clay loam soils are transitional soils and may be classified as medium textured soils in some regions of the U.S.

SPECIAL PRECAUTIONS (SENCOR PLUS SURFLAN)

For additional precautions, restrictions, limitations and sprayer cleanup information, refer to the appropriate section of the SENCOR and the Surflan A.S. labels.

SENCOR PLUS AMIBEN®

SENCOR PLUS AMIBEN PREEMERGENCE TANK MIX APPLICATION

SENCOR may be applied in a tank mix combination with Amiben 2 as a band or broadcast application to soybeans.

SENCOR plus AMIBEN Preemergence Tank Mix Application
Broadcast Rates

Soil Texture	1/2 to 3% Organic Matter: SENCOR 4 Pts./Acre	+	Amiben 2 Gal./Acre	Greater than 3% Organic Matter: SENCOR 4 Pts./Acre	+	Amiben 2 Gal./Acre
COARSE[1] (Sandy loam, loamy sand)	1/2	+	3/4 to 1	3/4	+	3/4 to 1
MEDIUM (Loam, silt loam, silt, sandy clay, sandy clay loam)	3/4	+	3/4 to 1	1	+	1 to 1 1/4
FINE (Silty clay, silty clay loam[2], clay, clay loam)	1	+	1 to 1 1/4	1 to 1 1/4	+	1 to 1 1/4
Mississippi Delta Only	SENCOR 4 Pts./Acre	+	Amiben Gal./Acre			
(Silty clay to heavy clay)	2	+	1 1/2			

1 Do not use on sandy loam or loamy sand soil with less than 2% organic matter. Do not use on sand soils.

2 Silty clay loam soils are transitional soils and may be classified as medium textured soils in some regions of the U.S.

Do not use on muck soils.

Where a rate range is specified, use the higher rate in fields with a history of intense weed pressure. See "Rate Ranges" in the front of the "Soybeans" section on the SENCOR label.

For additional precautions and information regarding spray equipment, mixing, application, limitations and restrictions refer to the appropriate section of the SENCOR label and the Amiben label.

SENCOR PLUS DYANAP®

SENCOR PLUS DYANAP TANK MIX APPLICATION

A tank mix combination of SENCOR and Dyanap preemergence, band or broadcast, may be applied up to cracking stage of soybeans. Apply SENCOR at the rates recommended on the label for use alone on soybeans. Refer to the Dyanap label for correct rates of that product and for additional information on weed control, mixing, application, restrictions and precautionary statements.

SENCOR PLUS LASSO AND DYANAP

SENCOR PLUS LASSO AND DYANAP TANK MIX APPLICATION

SENCOR may be applied in a tank mix combination with Lasso and Dyanap as a preemergence band or broadcast application up to the cracking stage of soybeans. For additional information concerning weeds controlled, mixing, application, limitations or restrictions, refer to the Lasso and Dyanap labels under Lasso plus SENCOR and Dyanap plus Lasso tank mixes and appropriate sections of the SENCOR label.

SENCOR 4 (Pints) plus Lasso 4EC (Quarts) and Dyanap (Quarts)

	Broadcast Rates Per Acre		
Soil Texture	SENCOR (Pts.)	Lasso (Qts.)	Dyanap (Qts.)
COARSE[1] (Sandy loam, loamy sand)	½	1½	2 to 4
MEDIUM (Loam, silt loam, silt, sandy clay, sandy clay loam)	¾ to 1	1½ to 2	3 to 4
FINE (Silty clay, silty clay loam[2], clay, clay loam)	¾ to 1	2	4

[1] Not recommended on sand soil with less than 1% organic matter.
[2] Silty clay loam soils are transitional soils and may be classified as medium textured soils in some regions of the U.S.

SENCOR PLUS LINURON PLUS (LASSO OR DUAL)

SENCOR PLUS LINURON PLUS (LASSO OR DUAL) TANK MIX APPLICATION

SENCOR may be applied in combination with linuron and Lasso 4 or Dual 8 EC as a preemergence application for the control of certain weeds in soybeans. Consult the linuron, Lasso, or Dual label for specific directions on use, recommendations, restrictions and any additional weeds not specified on the SENCOR label.

Mixing: Refer to "Mixing" under the "GENERAL INFORMATION" section in the front of the SENCOR label.

Application: Applications can be made only with ground spray equipment in accordance with specified soil types and dosage rates. For specific application information, refer to "Application" under the "GENERAL INFORMATION" section in the front of the label.

Broadcast Rates
(0.5 to 3% Organic Matter Only)

Soil Texture	SENCOR 4 Pts/Acre	linuron 50 WP or 4 L (Lbs/Acre) or (Pts/Acre)	Lasso 4 Qts/Acre	or	Dual 8 EC Pts/Acre
Coarse[1] (Sand, loamy sand, sandy loam)	¼ to ⅜	⅓ to ½	¾ to 1	or	1 to 1¼
Medium Loam, silt loam, silt, sandy clay, sandy clay loam)	⅜ to ½	½ to ¾	1 to 1½	or	1¼ to 1½
Fine (Silty clay, silty clay loam[2], clay, clay loam)	½ to ¾	¾ to 1½	1¼ to 2	or	1½ to 2

[1] DO NOT USE SENCOR plus linuron plus (Lasso or Dual) on sand soils with less than 1% organic matter.
[2] Silty clay loam soils are transitional soils and may be classified as medium textured soils in some regions of the U.S.

SPECIAL PRECAUTIONS SENCOR PLUS LINURON PLUS (LASSO OR DUAL)

For additional precautions, restrictions, limitations, and sprayer cleanup information, refer to the appropriate sections of the SENCOR label and the linuron label and the Lasso or Dual labels.

SOYBEANS

FOR USE IN COARSE (LIGHT) SOILS in Alabama, Arkansas, Florida, Georgia, Louisiana, Mississippi, Missouri, North Carolina, Oklahoma, South Carolina, Tennessee, Texas, Virginia.

SENCOR herbicide is recommended alone or in combination with Treflan, Lasso, Dual, Surflan or Amiben for use in coarse-textured, low organic matter soils in the states listed above for the control of certain weeds in soybeans. Refer to the appropriate sections of the SENCOR, Treflan, Lasso, Dual, Surflan or Amiben label for specific directions on use, recommendations, restrictions and any additional weeds not specified on the SENCOR label.

Mixing: Refer to "Mixing" under the "GENERAL INFORMATION" section in the front of the SENCOR label.

Application: For specific application infornation, refer to "Application" under the "GENERAL INFORMATION" section in the front of the SENCOR label.

SENCOR (ALONE) PREEMERGENCE APPLICATION

Broadcast Rates

Soil Texture	Organic Matter	SENCOR 4 Pts/Acre
Coarse (Light) Soils Sand[1], loamy sand, sandy loam	0.5 to 2.0%	½ to ¾[2]

[1] Not recommended for use on sand with less than 1% organic matter.
[2] Use the higher rate under heavy weed pressures and/or on soils higher in organic matter.

SENCOR IN COMBINATION WITH OTHER HERBICIDES

SENCOR is recommended in a tank mix combination with Treflan as a preplant incorporated application or as a preemergence overlay application following a preplant incorporated application of Treflan. SENCOR is also recommended for use as a preemergence application in combination with Lasso, Dual, Surflan or Amiben.

Broadcast Rates

Soil Texture	Organic Matter	Combination Product/Acre		Plus SENCOR 4 Pts/Acre
Coarse (Light) Soils sand[1], loamy sand, sandy loam	0.5 to 2.0%	Preplant Incorporated Treflan 4EC	1 pt	plus ½ to ¾[2]
		–or–		
		Preemergence Lasso 4 EC	1½ to 2 qts	
		–or–		
		Dual 8EC	1¼ to 1½ pts	
		–or–		plus ½ to ¾[2]
		Surflan A.S.	½ to ¾ qt	
		–or–		
		Amiben 2	¾ to 1 gal	

[1] Not recommended for use on sand with less than 1% organic matter.
[2] Use the higher rate under heavy weed pressures and/or on soils higher in organic matter.

SPECIAL PRECAUTIONS

For additional precautions, restrictions, limitations and sprayer cleanup information, refer to the appropriate sections of the SENCOR, Treflan, Lasso, Dual, Surflan or Amiben labels.

Do not use on sand soils with less than 1% organic matter, or on sandy loam or loamy sand soils with less than 0.5% organic matter.

FOR CONTACT ACTION AND RESIDUAL CONTROL

For Control of Emerged Annual Broadleaf Weeds and Grasses and For Suppression of Emerged Perennials Where Soybeans Will Be Planted Directly Into a Pre-formed Bed, Cover Crop or in Previous Crop Residues as in "Minimum Till" or "No Till" Crop Production.

SENCOR PLUS PARAQUAT

SENCOR PLUS PARAQUAT TANK MIX APPLICATION

Apply 1 to 2 pints Ortho Paraquat + Plus plus 3/4 to 2 pints SENCOR 4 according to the recommendations for SENCOR on this label as a broadcast ground spray before, during or after planting but before emergence of the crop. Soybeans exposed at time of application will be killed. Use the higher rate of Paraquat + Plus for heavier infestations or for taller weeds (4 to 6 inches tall). Apply in 20 to 60 gallons spray mix per acre. Use the higher volume in fields with heavy weed infestations or with large amounts of previous crop residues. Add Ortho X-77 Spreader or similar non-ionic spreader at 1 quart per 100 gallons of diluted spray. This mixture will not control weeds taller than 6 inches.

SENCOR PLUS LASSO AND PARAQUAT

SENCOR PLUS LASSO AND PARAQUAT TANK MIX APPLICATION

Apply 1 to 2 pints Ortho Paraquat + Plus plus 2 to 2½ quarts Lasso 4 E.C. plus ½ to 2 pints SENCOR 4 as a preemergence broadcast ground spray. Use the higher rate of Paraquat + Plus for weeds 4 to 6 inches tall. Soybeans exposed at time of application will be killed. Refer to the appropriate sections of this label and the Paraquat and Lasso labels for rates which are dependent upon soil textures, directions, limitations and cautions.

Apply in 20 to 60 gallons spray mix per acre. Use the higher volume in fields with heavy weed infestations or with large amounts of previous crop residues. Add Ortho X-77 Spreader or similar non-ionic spreader at 8 fl. ounces per 100 gallons of diluted spray. Add SENCOR to the spray tank first and mix thoroughly. Then add Lasso followed by Paraquat + Plus and Ortho X-77 Spreader. Constant agitation is recommended at all times. This tank mix will not control shattercane which emerges after treatment or other weeds taller than 6 inches.

SENCOR PLUS SURFLAN AND PARAQUAT

SENCOR PLUS SURFLAN AND PARAQUAT TANK MIX APPLICATION

The combination provides control of weeds and grasses which are emerged at the time of application plus many weeds and grasses which may germinate after planting. Soybeans exposed at time of application will be killed.

Weeds Controlled by Paraquat: Paraquat will kill emerged annual broadleaf weeds and grasses and provide top kill of perennials.

When to Apply: Apply the tank mix at planting or within 2 days after planting. Do not spray over top of emerged soybeans.

SENCOR PLUS SURFLAN AND PARAQUAT PREEMERGENCE TANK MIX APPLICATION

	Broadcast Rates Per Acre		
Soil Texture	**SENCOR 4**	**Surflan A.S.**	**Paraquat + Plus**
Coarse[1]	½ pint	¾ quart	1 quart
Medium	¾ pint	1 quart	1 quart
Fine	1 pint	1½ quarts	1 quart

[1] Not recommended on sand soils with less than 1% organic matter. Do not use on soils containing more than 5% organic matter.

Do not use on soils containing more than 5% organic matter.

Mixing: Refer to "Mixing" under the "General Information" section in the front of the SENCOR label.

Add Ortho X-77 Spreader or similar non-ionic spreader at 8 fl. ozs. per 100 gallons of diluted spray mixture.

APPLICATION: For specific application information refer to "Application" under the "General Information" section in the front of the SENCOR label.

A one-half inch rain or more or its equivalent in overhead irrigation is needed to activate the herbicide combination.

Special Precautions (SENCOR plus Surflan and Paraquat): Poor weed control and/or crop injury may result if directions are not followed.

Do not use a rib-type press wheel on your no-till planter or crop injury may result.

For additional precautions, special precautions, restrictions, limitations and sprayer cleanup information, refer to the approriate section of the SENCOR, Surflan and Paraquat labels.

SOUTHERN AND SOUTHEASTERN STATES ONLY

POSTEMERGENCE DIRECTED SPRAY APPLICATIONS

SENCOR can be applied in postemergence directed sprays to soybeans for control of certain weeds which escape preplant or preemergence herbicide applications and for control of additional flushes of weeds that may occur after soybeans have emerged. Postemergence directed sprays of SENCOR can be applied to soybeans in addition to a preemergence or preplant application of SENCOR according to label directions.

WEEDS CONTROLLED

SENCOR, applied postemergence to soybeans as a directed spray according to directions on this label, will control the following at rates shown (broadcast basis) when grasses and common ragweed are less than 1 inch tall and other broadleaves are less than 3 inches tall:

½ pint/acre	
Florida beggarweed	*Desmodium tortuosum*
Pigweeds	*Amaranthus* spp.
Carpetweed	*Mollugo verticillata*
Cocklebur	*Xanthium pensylvanicum*
Dayflower	*Commelina* spp.
Mexicanweed	*Caperonia castaniifolia*
Purslane	*Portulaca oleracea*
Sicklepod	*Cassia obtusifolia*
Velvetleaf	*Abutilon theophrasti*
Crabgrass	*Digitaria* spp.
½ to 1 pint/acre	
Sesbania	*Sesbania* spp.
Prickly sida/Teaweed	*Sida spinosa*
1 pint/acre	
Ragweed, common	*Ambrosia artemisiifolia*

At the rate of 1 pint/acre morningglory species, (*Impomoea* spp.) horsenettle, (*Solanum* spp.) Florida pusley, (*Richardia scabra*) spotted spurge (*Euphorbia maculata*) and wild poinsettia (*Euphorbia heterophylla*) are suppressed when SENCOR is applied before these weeds are 3 inches tall. The 1 pint/acre rate will suppress broadleaf signalgrass (*Brachiaria platyphylla*) up to 1 inch tall.

SENCOR Postemergence Directed Spray Applications Recommended Applications

Crop	SENCOR 4 Pints Per Acre	Remarks
Soybeans (Alabama, Arkansas, Florida Georgia, Kentucky, Louisiana, Mississippi, Missouri, North Carolina, Oklahoma, South Carolina, Tennessee and Texas)	½ to 1 (broadcast basis)	Apply proper dosage using 10 to 40 gallons of water per acre as a directed spray in a 6 to 8 inch band on each side of the row after soybeans are 8 inches tall and before broadleaf weeds are 3 inches tall and before grasses and common ragweed are 1 inch tall. For best results the spray must cover weed foliage with minimum or no contact with soybean foliage. Add a non-ionic surfactant such as Ortho X-77[R] to the spray mixture to obtain better wetting of weed leaf surfaces. To determine the correct dosage of SENCOR 4 for a Band application see "Banded Application" under "Application" in the "General Information" section in front of the label. If necessary, a second postemergence directed spray application can be made after 7 days. Do not feed or graze green soybean vines. Do not harvest soybeans or use dry soybean vines for feed or forage within 70 days of last application.

SPECIAL PRECAUTIONS (DIRECTED POSTEMERGENCE)

Do not apply directly to soybeans or serious crop injury will occur. Do not allow spray to contact more than the lower ¼ to ⅓ of soybean plants. Soybean leaves contacted by the spray will be killed.

Do not apply SENCOR postemergence to the sensitive soybean varieties. See "Soybean Precautions" on the front of the SENCOR label.

To avoid injury to other crops or desirable plants from spray drift, sprayer pressure must not exceed 30 psi and the sprayer must be fitted with nozzles no smaller than 8002 T-Jet (or equivalent). Do not apply under weather conditions which favor drift.

POTATOES

SENCOR herbicide is recommended for use in ground or aircraft spray equipment as a preemergence and/or postemergence application to potatoes. Postemergence applications may be made only on white-skinned potato varieties that are not early maturing. For optimum control of weeds in potatoes apply SENCOR 4 before weeds are 1-inch tall. Refer to special precautions for potatoes for additional information.

Ground Application: SENCOR is recommended for use with ground spray equipment, except air blast sprayers, as a preemergence and/or postemergence application for control of certain grass and broadleaf weeds in potatoes. Apply as a broadcast spray after planting and before crop emergence or after drag-off if this operation is part of the normal cultural practice. Postemergence applications should be applied after crop emergence.

Aerial Application: SENCOR is recommended for use in aerial spray equipment as a preemergence and/or postemergence application for control of certain grass and broadleaf weeds in potatoes. Apply as a broadcast spray after planting and before crop emergence or after drag-off if this operation is part of the normal cultural practice. Postemergence applications should be applied after crop emergence. Do not apply aerially when SENCOR is tank mixed with Lasso.

PREEMERGENCE WEEDS CONTROLLED

SENCOR, applied preemergence to potatoes according to directions, will effectively control:

Broadleaves

Carpetweed	*Mollugo verticillata*
Cocklebur	*Xanthium pensylvanicum*
Jimsonweed	*Datura stramonium*
Lambsquarters	*Chenopodium album*
Pigweeds	*Amaranthus* spp.
Ragweed, Common	*Ambrosia artemisiifolia*
Shepherdspurse	*Capsella bursa-pastoris*
Sicklepod	*Cassia obtusifolia*
Smartweed, Pennsylvania	*Polygonum pensylvanicum*
Wild Mustards	*Brassica* spp.

Grasses

Broadleaf Signalgrass	*Brachiaria platyphylla*
Crabgrass	*Digitaria* spp.
Foxtails	*Setaria* spp.
Johnsongrass, Seedling	*Sorghum halepense*
Panicum, Fall	*Pancium dichotomiflorum*

For Control of Wild Mustard (Brassica spp.) in Potatoes (Except California) - Apply at the rate of 1/2 to 1 pint per acre as a preemergence broadcast spray using aerial or ground spray equipment in accordance with directions on the label. Application should be made at or near ground crack and/or before the mustard is one inch tall. For control of mustard larger than 2 inches in diameter or other weeds listed on the label, a minimum of 1 pint per acre must be used. When less than 1 pint per acre is used a postemergence application in accordance with directions on the SENCOR label may be required for control of other weeds or later germinating weed species.

Hard To Control Weeds: Although SENCOR applied preemergence may not provide commercially acceptable control in every instance, it will suppress growth of the following weeds and reduce their competition with potato plants.

Broadleaf Weeds

Common purslane	*Portulaca oleracea*

Grasses

Barnyardgrass	*Echinochloa crus-galli*
Nutsedge	*Cyperus* spp.

POSTEMERGENCE WEEDS CONTROLLED

SENCOR, applied early postemergence to potatoes according to directions, will effectively control:

Broadleaves

Cocklebur	*Xanthium pensylvanicum*
Lambsquarters	*Chenopodium album*
Pigweed, Redroot	*Amaranthus retroflexus*
Ragweed, Common	*Ambrosia artemisiifolia*
Russian thistle	*Salsola kali*
Smartweed, Pennsylvania	*Polygonum pensylvanicum*

Grasses

Foxtails	*Setaria* spp.

On medium and fine soils in Idaho, Oregon and Washington, two postemergence applications, each at the rate of 1 pint per acre, will control barnyardgrass (*Echinochloa crus-galli*), kochia (*Kochia scoparia*), sunflower (*Helianthus* spp.), and Jim Hill mustard (*Sisymbrium altissimum*) in addition to weeds listed above.

Hard To Control Weeds: Although SENCOR applied early postemergence may not provide commercially acceptable control in every instance, it will suppress growth of the following weeds and reduce their competition with potato plants.

Broadleaves

Kochia	*Kochia scoparia*
Sunflower	*Helianthus* spp.

Grasses

Barnyardgrass	*Echinochloa crus-galli*

Recommended Broadcast Applications

Crop	SENCOR 4 Pints Per Acre	Remarks
Potatoes	1 to 2	**PREEMERGENCE:** Apply specified dosage per acre as a broadcast spray to the soil surface after planting and before crop emergence or after drag-off if this operation is part of normal cultural practice.* Distribute the chemical evenly over the area to be treated. Do not incorporate into soil. Use the lower rate on sandy soils.
		POSTEMERGENCE: Apply specified dosage per acre as a broadcast spray over the tops of potato plants.* Use rate of 1/2 to 1 pt. for control of redroot pigweed and lambsquarters only. Apply the 1 pt. rate for control of other weeds listed on this label. Do not apply within 60 days of harvest.
Potatoes (white-skinned varieties that are not early maturing)	1/2 to 1	**SPLIT APPLICATIONS:** This product may be applied once preemergence and once postemergence as directed above.* **Do not apply postemergence within 60 days of harvest. IDAHO, OREGON AND WASHINGTON ONLY:** Two postemergence applications can be made as broadcast sprays over the tops of potato plants if SENCOR is not applied preemergence. Use 1/2 to 1 pint/acre for control of redroot pigweed and lambsquarters only. On coarse (sandy) soils with low organic matter do not exceed 1/2 pint/acre per application in Oregon and Washington or 3/4 pint/acre per application in Idaho. On medium and heavy soils only use 1 pint/acre for control of other weeds listed on the label. Make the first application early in the season while weeds are still small. Allow at least 14 days before the second application. For optimum control apply before weeds are 1 inch tall. Do not apply after June 30 if treated land is to be planted to crops other than potatoes. Do not apply within 60 days of harvest.

* Refer to Special Precautions—Potatoes below.

SPRINKLER IRRIGATION APPLICATION TO POTATOES

SENCOR may be applied early postemergence to potatoes through center pivot, permanent solid set or self-propelled wheel move irrigation systems equipped with automatic shutoff devices and valves to prevent backflow into the water source. Interlocking controls should be installed between the water pump and pesticide pump so that both pumps stop if either one should fail. Maintain continuous agitation in injection nurse tanks during application. Inject the product with a positive displacement pump in to the main line ahead of a right angle turn to insure adequate mixing. Greater accuracy in calibration and distribution will be achieved by injecting a larger volume of more dilute mixture of product and water per hour. Application of more than the quantity of irrigation water recommended on this label may result in decreased product performance by removing the chemical from the zone of effectiveness. Where sprinkler distribution patterns do not overlap sufficiently unacceptable weed control may result. Where sprinkler distribution patterns overlap excessively crop injury may result. Allow sufficient time for pesticide to be flushed through all lines and all nozzles before turning off irrigation water. To ensure that lines are flushed and free of remaining pesticide, an indicator dye may be injected into the lines to mark the end of application period.

Calibration: (Center Pivot and Self-Propelled Wheel Move Systems)

1. Determine number of minutes required to make one complete revolution while applying $\frac{1}{4}$ to $\frac{3}{4}$ inch of water per acre.
2. With the system at operating pressure determine the exact number of minutes required to inject one gallon of water.
3. Divide the time required for one revolution (step 1) by the time required to inject one gallon (step 2). This gives total gallons of product-water mixture to be added to nurse tank.
4. Add required amount of water to nurse tank and start the agitation system. Then add sufficient SENCOR herbicide at the recommended rate (see RECOMMENDED BROADCAST APPLICATIONS) to the nurse tank.

EXAMPLE:

If 20 hours (1200 minutes) were required for one revolution and if 2 minutes were required to inject one gallon then a total of 600 gallons of product-water mixture are required ($1200 \div 2 = 600$); to treat 135 acres at 1 pt/acre, 135 pints (16 gallons and 7 pints) of SENCOR 4 are required.

Recommended Broadcast Applications

Crop	Pints/Acre SENCOR 4	Remarks
Potatoes (white skinned varieties that are not early maturing)	$\frac{1}{2}$ to 1	**POSTEMERGENCE IN SPRINKLER IRRIGATION:** Apply specified dosage in $\frac{1}{4}$ to $\frac{3}{4}$ inch of water per acre as a continuous injection in center pivot and self-propelled wheel move systems or in the last 15 to 30 minutes of set in permanent solid set sprinkler systems. On sandy soil apply in $\frac{1}{4}$ to $\frac{1}{2}$ inch of water. Use rates of $\frac{1}{2}$ to 1 pint for control of redroot pigweed and lambsquarters only. Apply the 1 pint rate for control of other weeds listed on the label. Do not apply within 60 days of harvest.

Do not apply when wind speed favors drift, when system connections leak, when nozzles do not provide uniform distribution or when lines containing the product must be dismantled or drained.

SENCOR PLUS DUAL TANK MIX APPLICATION (Except California)

SENCOR may be applied in a tank mix combination with Dual 8 EC as a preemergence broadcast application to potatoes. Apply SENCOR at 0.75 to 2 pints and Dual 8 EC at 1.5 to 3 pints per acre according to the respective labels for use of each product alone on potatoes. Refer to appropriate sections of the SENCOR and Dual labels for precautionary statements, restrictions, application information and weeds controlled.

SENCOR PLUS PROWL TANK MIX APPLICATION (Except California)

SENCOR may be applied in a tank mix combination with Prowl as a preemergence braodcast application to potatoes. Apply SENCOR at 1 to 2 pints and Prowl at 1 to 3 pints per acre according to the respective labels for use of each product alone on potatoes. Refer to the appropriate sections of the SENCOR and Prowl labels for precautionary statements, restrictions, applicable information and weeds controlled.

SPECIAL PRECAUTIONS (Potatoes)

Do Not Use SENCOR 4 on Potatoes in Kern County, California

Do not apply more than 2 pints total SENCOR 4 per acre in a single crop season regardless of the method of application. **Do not apply within 24 hours of application of other pesticides.** Do not apply postemergence within 3 days after periods of cool, wet or cloudy weather or crop injury may occur. Postemergence applications may cause some chlorosis or minor necrosis.

In addition to early maturing smooth-skinned white, or red-skinned varieties of potatoes, postemergence applications of SENCOR are not recommended on Atlantic, Shepody, Chip Bell, Bellchip, and Centennial varieties. Preemergence applications to these varieties may cause crop injury, especially under adverse weather conditions and when higher SENCOR rates are used.

Do not apply to sweet potatoes or yams.

Do not plant sensitive crops such as onions, lettuce, cole crops and cucurbits during the next growing season following applications of SENCOR or injury may occur.

ALFALFA AND SAINFOIN (Except California)

SENCOR herbicide is labelled for use in alfalfa and sainfoin in the following areas:

1. Alfalfa and sainfoin (Including mixed stands with grasses) (all areas except California)
2. Alfalfa and sainfoin (Including mixed stands with grasses) (California only)
3. Alfalfa—Post-dormant application of SENCOR impregnated on dry fertilizer only.
4. Alfalfa—Non-dormant, non-winter hardy varieties. (Arizona only)

SENCOR is recommended for use in aerial or ground spray equipment as a broadcast surface application to dormant established crops of alfalfa and sainfoin for the control of certain grass and broadleaf weeds.

APPLICATION: Refer to the specific "Application" section under "General Information" in the front of this label for detailed information on the application of SENCOR. For information on applying SENCOR in fluid or on dry fertilizer refer to the "Application Of SENCOR In Fluid Fertilizers" or "Commercial Impregnation And Application Of SENCOR On Dry Bulk Fertilizer" under the "General Information" section in the front of the SENCOR label.

SPECIAL PRECAUTIONS:

—Do not apply SENCOR after growth begins in the spring or before growth ceases in the fall.

—Treat only dormant established alfalfa.

—Do not graze or harvest within 28 days after application.

—For best weed control, apply SENCOR when weeds are less than 2 inches tall or before weeds foilage is 2 inches in diameter.

—Reduced weed control may occur when extended dry conditions follow application of SENCOR.

—Crop injury may occur if alfalfa or sainfoin are:

1. Under stress conditions such as diseases, insect infestations, poorly drained soils, drought or winter injury when SENCOR is applied, or
2. When SENCOR is applied during the first growing season after seeding or if applied earlier than 12 months after seeding.

ALFALFA and SAINFOIN (All Areas except California)

Recommended Broadcast Applications

Crop	Pints/Acre SENCOR 4	Remarks
Alfalfa and Sainfoin	¾ to 2	ALL AREAS (Including Idaho, Oregon, and Washington) Select the proper dosage according to weeds known to be present in field to be treated. Use the lower rate of SENCOR on sandy loam or loamy sand soils in areas other than Idaho, Oregon, and Washington.
	½ to ¾	IDAHO, OREGON AND WASHINGTON ONLY: On sandy loam or loamy sand soils apply SENCOR 4 at rates of ½ to ¾ pint per acre for control of common chickweed only.

FOR USE ON MIXED STANDS OF ALFALFA AND GRASSES:

Rates of 1 to 1½ pints of SENCOR 4 per acre will provide partial reduction of forage grass stands. These rates may be used to reduce forage grass stands to prevent crowding out of alfalfa. Higher rates will severely reduce forage grass stands.

SENCOR should not be used on sand soils. In areas West of the Rocky Mountains, avoid using SENCOR on soils with calcareous surface area, high levels of lime or sodium, or a pH greater than 7.5.

ALFALFA AND SAINFOIN

WEEDS CONTROLLED

½ to ¾ pint SENCOR 4/acre
(Idaho, Oregon, and Washington Only)

Chickweed, Common	*Cerastium vulgatum*

¾ to 1 pint SENCOR 4/acre

Cheat	*Bromus secalinus*
Downy brome	*Bromus tectorum*
Japanese brome	*Bromus japonicus*
Rescuegrass	*Bromus catharticus*

1 to 2 pints SENCOR 4/acre

Broadleaves

Fleabane, Rough	*Erigeron strigosus*
Flixweed	*Descurainia sophia*
Henbit	*Lamium amplexicaule*
Kochia	*Kochia scoparia*
Lambsquarters, Common	*Chenopodium album*
Marestail (Horseweed)	*Hippuris vulgaris*
Meadow Salsify	*Tragopogon pratensis*
Mustard, Blue	*Chorispora tenella*
Mustard, Jim Hill (tumble)	*Sisymbrium altissimum*
Mustard, Tansy	*Descurainia pinnata*
Pepperweed, Virginia	*Lepidium virginicum*
Pennycress, Field	*Thlaspi arvense*
Pigweed, Redroot	*Amaranthus retroflexus*
Prickly Lettuce	*Lactuca serriola*
*Russian Knapweed	*Centaurea repens*
Shepherdspurse	*Capsella bursa-pastoris*
White Cockle	*Melandrium album*
Wild Buckwheat	*Polygonum convolvulus*
Yellow Rocket	*Barbarea vulgaris*

Grasses:

Foxtail, Green	*Setaria viridis*
Little Barley	*Hordeum pusillum*
Smooth Brome	*Bromus inermis*
Wild Oats	*Avena fatua*

2 pints SENCOR 4/Acre

Broadleaves

Chickweed, Mousear	*Cerastium vulgatum*
Dandelion	*Taraxacum officinale*
Ragweed, Common	*Ambrosia artemisiifolia*

Grasses

Barnyardgrass	*Echinochloa crus-galli*
Bluegrass	*Poa annua*
Foxtail Barley	*Hordeum jubatum*

Weeds Partially Controlled: At the rate of 2 pints/acre SENCOR 4 may be used to reduce the competition from curly dock (*Rumex crispus*).

ALFALFA AND SAINFOIN (California Only)

(Including Mixed Stands With Grasses)

SENCOR 4 is recommended for use in aerial or ground spray equipment as a broadcast surface application to dormant established crops of alfalfa and sainfoin.

APPLICATION: SENCOR 4 is recommended for use in aerial or ground spray equipment as a broadcast surface application to dormant established crops of alfalfa and sainfoin for control of certain grass and broadleaf weeds. Do not apply SENCOR 4 after growth begins in the spring or before growth ceases in the fall. Do not apply to either alfalfa or sainfoin during the first growing season after seeding.

For information on applying SENCOR in fluid fertilizer solutions to alfalfa, refer to the appropriate section of the SENCOR label.

For information on Commercial impregnation and application of SENCOR on dry bulk fertilizer, refer to the appropriate section of the SENCOR label.

WEEDS CONTROLLED

¾ to 1 pint SENCOR 4/acre
Cheatgrass (downy brome)

1 to 2 pints SENCOR 4/acre

Broadleaves

Chickweed, Common	*Cerastium vulgatum*
Flixweed	*Descurainia sophia*
Henbit	*Lamium amplexicaule*
Kochia	*Kochia scoparia*
Meadow Salsify	*Tragopogon pratensis*
Mustard, Blue	*Chlorispora tenella*
Mustard, Tansy	*Descurainia pinnata*
Pepperweed, Virginia	*Lepidium virginicum*
Shepherdspurse	*Capsella bursa-pastoris*
White Cockle	*Melandrium album*
Wild Buckwheat	*Polygonum convolvulus*
Yellow Rocket	*Barbarea vulgaris*

Grasses

Smooth Brome	*Bromus inermis*
Wild Oats	*Avena fatua*

2 pints SENCOR 4/Acre

Broadleaves

Dandelion	*Taraxacum officinale*

Grasses

Barnyardgrass	*Echinochloa crus-galli*
Bluegrass	*Poa annua*
Foxtail Barley	*Hordeum jubatum*

Recommended Broadcast Application

Crop	SENCOR 4 Pints/Acre	Remarks
Alfalfa and Sainfoin	¾ to 2	Select the proper dosage according to weeds known to be present in the field to be treated. Apply specified dosage in 20 to 40 gallons of water per acre with ground spray equipment or 3 to 10 gallons of water per acre with aerial spray equipment fitted with nozzles suitable for broadcast applications of herbicides. Treat only dormant established crops of alfalfa and sainfoin. Injury may occur to alfalfa if SENCOR is applied earlier than 12 months after seeding. Do not apply after Spring growth begins or before growth ceases in the Fall. Do not graze or harvest within 28 days after application. At the 2 pint/acre rate, SENCOR 4 may be used for suppression of curly dock.

FOR USE ON MIXED STANDS OF ALFALFA AND GRASSES: Rates of 1 to 1½ pints of SENCOR 4 per acre will provide partial reduction of forage grass stands. These rates may be used to reduce forage grass stands to prevent crowding out of alfalfa. Higher rates will severely reduce forage grass stands.

Do not apply with aerial spray equipment when wind speed is greater than 10 mph. Do not apply when weather conditions favor spray drift and/or when sensitive cool season crops, such as cole crops, onions, peas, or strawberries, are present in adjacent fields. Applications should not be made when weather conditions favor spray drift, especially in areas where wheat is growing on coarse textured soils in adjacent fields, or injury may occur.

ALFALFA

Post Dormant Application of SENCOR Impregnated on Dry Fertilizer Only—SENCOR 4 may be applied after dormancy has broken, but prior to three inches of new alfalfa shoot growth, only when impregnated on dry fertilizer in Connecticut, Illinois, Indiana, Iowa, Kansas, Kentucky, Michigan, Minnesota, Nebraska, New York, Ohio, Pennsylvania, South Dakota and Wisconsin. Apply at rates of 1½ to 2 pints per acre as directed on the label for application during dormancy. Apply only when alfalfa foliage is dry or crop injury may occur. When using this application method, do not harvest or graze treated alfalfa for 60 days after application.

ALFALFA

Non-Dormant, Non-Winter Hardy Varieties

(Arizona Only)

SENCOR is recommended as a broadcast surface application to established crops of non-dormant alfalfa varieties for preemergence and postemergence control of certain winter annual weeds following either a Fall or Winter sheep grazing/green-chop harvest.

Weeds Controlled:

Mouse barley	Silversheaf knotweed	Lambsquarters
Nettleleaf goosefoot	London rocket (mustard)	Little mallow (cheeseweed) (Malva)
Prickly lettuce	Shepherdspurse	Spiny sowthistle
Field pepperweed	Littleseed canarygrass	

Crop	SENCOR 4 Pints/Acre	Remarks
Alfalfa Non-Dormant, Non-Winter hardy varieties	¾ to 1	Apply specified dosage by aerial or ground spray equipment in 7 to 40 gallons of water per acre. Treat established alfalfa stubble after Fall or Winter sheep grazing or green-chop harvest and prior to the time regrowth is 2″ tall. Alfalfa foliage present at time of application can exhibit yellowing. Injury may occur to alfalfa in areas of high salt concentration where the crop is stunted and/or has a poorly developed root system, or if alfalfa is under stressed growing conditions such as diseases, insect infestations, or drought. For most effective postemergence weed control, treatment should be made before weeds are 2″ tall or before leaf rosettes are 2″ wide. For maximum control, rainfall (¼″ or more) or irrigation is necessary within 30 days of treatment, however, do not flood irrigate within 2 days after treatment. Use ¾ pint SENCOR on sand soil when only mustard, goosefoot, lambsquarters, or canarygrass are the weeds to be controlled. Do not apply earlier than 6 months after seeding. Do not graze or harvest within 28 days after application.

SPECIAL PRECAUTIONS: Maintain continuous mechanical agitation in the spray tank to insure a uniform spray mixture. Do not apply with aerial spray equipment when wind speed is greater than 10 mph. Do not apply when weather conditions favor spray drift and/or when sensitive cool season crops, such as cole crops, onions, peas or strawberries, are present in adjacent fields. Applications should not be made when weather conditions favor spray drift especially in areas where wheat is growing on coarse textured soils in adjacent fields, or injury may occur.

ASPARAGUS (Established)

SENCOR is recommended for use in aerial or ground spray equipment as a single preemergence broadcast application or as a split application consisting of a preemergence broadcast application followed by a post harvest broadcast application.

For specific application information see under the "General Information" section "Application" of SENCOR in the front of the SENCOR label.

WEEDS CONTROLLED

SENCOR 4, applied to established asparagus according to directions, will effectively control:

Broadleaves

Chickweed, Common	*Stellaria media*
Jimsonweed	*Datura stramonium*
Lambsquarters	*Chenopodium album*
Pigweed, Redroot	*Amaranthus retroflexus*
Ragweed, Common	*Ambrosia artemisiifolia*
Smartweed, Pennsylvania	*Polygonum pensylvanicum*
Sorrel, Red	*Rumex acetosella*
Velvetleaf	*Abutilon theopharasti*

Grasses

Crabgrass	*Digitaria* spp.
Foxtails	*Setaria* spp.
Sandbur, Field	*Cenchrus pauciflorus*

Recommended Broadcast Applications

Crop	SENCOR 4 Dosage/Acre	Remarks
	2 to 4	**PREEMERGENCE APPLICATION ONLY:** Make a single surface application in early Spring before asparagus spears or ferns emerge. If the field is to be disked, apply SENCOR 4 after disking but before the crop emerges. Use the lower rate for control of the broadleaf weeds listed above. Use the higher rate in fields with a history of severe infestations of grasses and for maximum residual control. Do not apply within 14 days of harvest.
Asparagus	1 to 2 pre-emergence plus 2 to 3 post harvest	SPLIT APPLICATION PREEMERGENCE AND POST HARVEST **Preemergence Application:** Apply before asparagus spears or ferns emerge. If the field is to be disked, apply after disking but prior to crop emergence. Do not apply within 14 days of harvest. **Post Harvest Application:** Apply after last harvest of the season but prior to emergence. The lower combination rates may be used for control of common ragweed, lambsquarters, redroot pigweed, and red sorrel. Use the higher combination rates for other weeds listed or in fields with severe grass infestations or for maximum post harvest control of emerged weeds.

IMPORTANT: The total amount of SENCOR 4 applied in one crop season may not exceed 4 pints per acre.

SPECIAL PRECAUTIONS (ASPARAGUS)

Do not use on newly seeded asparagus nor on young plants during the first growing season after setting crowns.

DO NOT APPLY POST HARVEST APPLICATIONS UNTIL AFTER THE LAST HARVEST OF SPEARS

FIELD CORN

(IOWA, KANSAS, MISSOURI, NEBRASKA, AND SOUTH DAKOTA ONLY)

SENCOR herbicide is recommended for the added effect on the control of certain broadleaf weeds in corn when applied preemergence in a tank mix combination with labeled rates of Lasso plus atrazine, Lasso plus Bladex®, Dual plus atrazine, or Dual plus Bladex. Refer to the appropriate product labels for weeds controlled by these combinations.

SENCOR in each of the tank mix combinations will control the following weeds:

Lambsquarters	*Chenopodium album*
Pigweed, Redroot (carelessweed)	*Amaranthus retroflexus*
Smartweed, Pennsylvania	*Polygonum pensylvanicum*
Velvetleaf	*Abutilon theophrasti*

MIXING: Select proper dosages based on soil type as shown below. Fill the spray tank ¼ to ⅓ full with clean water and add SENCOR 4 to the spray tank first. Mix thoroughly and then add atrazine or Bladex to the spray tank. Add water to fill spray tank to desired level. Mix thoroughly and then add Lasso or Dual to the spray tank. Maintain sufficient agitation during mixing and spraying operations to insure a uniform spray mixture.

APPLICATION: For specific application information see under the "General Information" section "Application" of SENCOR in the front of the label.

Recommended Broadcast Applications

Rates Per Acre

Soil Texture	SENCOR 4	Lasso 4 E.C. or	Dual 8 E.C.	Atrazine 4L or 80 WP or	Bladex 4L or 80 WP
MEDIUM (Loam, silt loam, silt, sandy clay, sandy clay loam)	½ pt.	1.5 to 2.0 qts.	1.5 to 2.0 pts.	0.75 to 1.5 qts. of 4L or 0.95 to 1.9 lbs. of 80 WP	1.25 to 2.5 qts. of 4L or 1.5 to 3.1 lbs. of 80 WP
FINE (Silty clay, silty clay loam[1], clay, clay loam)	½ pt.	2.0 to 2.5 qts.	2.0 to 2.5 pts.	0.75 to 1.5 qts. of 4L or 0.95 to 1.9 lbs. of 80 WP	1.5 to 3.0 qts. of 4L or 1.75 to 3.75 lbs. of 80 WP

[1] Silty clay loam soils are transitional soils and may be classified as medium textured soils in some regions of the U.S.

Follow the Lasso, Dual, Bladex, or atrazine labels for use rates on specific soil textures and organic matter content. Not recommended for use on muck soils as reduced weed control may result.

SPECIAL PRECAUTIONS: (FIELD CORN)

Not for use on coarse textured soils or soils containing less than 2% organic matter content.

Do not use on soils having a calcareous surface or high in lime or sodium and/or a pH of 7.0 or higher.

Corn seed should be planted 1½ inches deep.

Use in hybrid seed corn production fields only if both parents are known to be tolerant to SENCOR and other herbicides used.

Observe all cautions and limitations on labeling of all products used in mixtures.

SUGARCANE (Hawaii Only)

SENCOR, a selective herbicide, is effective as a preemergence and an early postemergence broadcast application for control of certain grass and broadleaf weeds. When applied as a spot treatment, it also provides excellent control of perennial grasses and broadleaves.

Ground Application: SENCOR should be mixed by filling the spray tank half full of clean water. Then add the recommended amount of SENCOR 4 to suit the total tank capacity and the rate of application per acre (preferably 25 to 35 gallons per acre). Complete filling the tank and maintain sufficient agitation during mixing and spraying to insure a uniform spray mixture.

Aerial Application: SENCOR is recommended for use in aerial spray equipment as a a preemergence or postemergence application to irrigated sugarcane. Aerial spray equipment should be calibrated to apply the proper amount of SENCOR 4 in 7 to 10 gallons of spray mixture per acre.

SENCOR applied preemergence or postemergence to the sugarcane as a broadcast spray or spot treatment will effectively control the following when weeds are less than 3 inches in height.

WEEDS CONTROLLED IN IRRIGATED AND NON-IRRIGATED SUGARCANE

Broadleaves	
Amaranth, Spiny	*Amaranthus spinosus*
Euphorbia, Wild	*Euphorbia* spp.
Fireweed	*Erechtites hieracifolius*
Floras paintbrush	*Emilia sonochifolia*
Spurge, Garden	*Euphorbia hirta*
Spurge, Graceful	*Euphorbia glomerifera*
Grasses	
Crabgrass	*Digitaria* spp.
Guineagrass	*Panicum maximum*
Plushgrass	*Chloris radiata*
Ricegrass	*Oryzopsis hymenoides*
Wiregrass	*Eleusine indica*

Weeds Controlled In Irrigated Sugarcane Only:

Broadleaves	
Amaranth, Spleen	*Amaranthus dubius*
Haole Koa	*Leucaena leucocephala*
Hialoa	*Waltheria americana*
Hilahila	*Mimosa pudica*
Purslane, Common	*Portulaca oleracea*
Rattlepod	*Crotalaria spectabilis*
Grasses	
Alexandergrass	*Brachiaria plantaginea*
Bristly foxtail	*Setaria verticillata*

Weeds Controlled In Non-Irrigated Sugarcane Only:

Broadleaves	
Ageratum	*Ageratum conyzoides*
Richardia	*Richardia brasiliensis*
Tarweed	*Cuphea carthagenesis*

Recommended Broadcast Applications

Crop	SENCOR 4 Pints/Acre	Remarks
Sugarcane (Hawaii only)	4 to 8 (½ to 1 gal) (non-irrigated) 8 to 12 (1 to 1½ gal) (irrigated)	**PREEMERGENCE:** (Irrigated and non-irrigated sugarcane) Apply specified dosage per acre as a broadcast spray to the soil surface. Applications should be made within two weeks after planting prior to cane emergence or shortly after emergence (spike stage). OR **EARLY POSTEMERGENCE:** (Irrigated and non-irrigated sugarcane) Apply specified dosage per acre as a broadcast spray over the cane. Application may be delayed as long as 4 to 6 weeks after planting provided weeds are less than 3 inches in height.
	4 to 8 (½ to 1 gal)	OR **POSTEMERGENCE:** Apply specified dosage per acre as broadcast spray to control weeds prior to "close in" time when cane shades out the weed growth.
	5 to 10 (1⅛ to 1¼ gal)	**SPOT TREATMENT:** Apply specified dosage in 30 to 50 gallons of finished spray per acre. Spot treatments may be used to control weeds in missed areas, corners of fields, or areas of hard to control weeds.

NOTE: Do not apply more than 16 pints (2 gal) of SENCOR 4 (8 lb.

active ingredient)/acre per crop cycle regardless of the method of application. The last application may be made up to 17 months of harvest.

SUGARCANE (Louisiana and Texas Only)

Preemergence and postemergence applications of SENCOR 4 with aerial or ground spray equipment are recommended for control of the following weeds in sugarcane in Louisiana and Texas:

Broadleaves	
Amaranth, Spiny	*Amaranthus spinosus*
Bindweed, Field	*Convolvulus arvensis*
Chickweed	*Cerastium vulgatum*
Henbit	*Lamium amplexicaule*
Lambsquarters	*Chenopodium album*
London rocket	*Sisymbrium irio*
Marestail	*Conyza canadensis*
Mustard, Wild	*Brassica kaber*
Oats, Winter	*Avena* spp.
Pigweeds	*Amaranthus* spp.
Purslane	*Portulaca oleracea*
Sowthistle	*Sonchus* spp.
Grasses	
Broadleaf Signalgrass	*Brachiaria platyphylla*
Crabgrass	*Digitaria* spp.
Foxtails	*Setaria* spp.
Johnsongrass, Seedling	*Sorghum halepense*

Recommended Broadcast Applications

Crop	SENCOR 4 Pints/Acre	Remarks
Sugarcane (Louisiana and Texas only)	3 to 6	**BROADCAST:** Apply specified dosage per acre using 20 to 30 gallons of water with ground equipment or 5 gallons of water with aircraft spray equipment. Apply as a broadcast spray during the Fall after planting or to the stubble after harvest. Make a second application early in the Spring.
	1½ to 3	**BAND:** Apply specified dosage in 10 to 20 gallons of water per acre in a 30 to 36 inch band over the row during the Fall after planting or to the stubble after harvest. Make a second application early in the Spring.

SPECIAL PRECAUTIONS: (Louisiana and Texas only)

Do not use treated foliage for feed or forage.

Use the higher rate on heavy clay soil and soil with a high percentage of organic matter. If necessary, a third application may be made in late Spring at layby. Do not apply within 60 days of harvest.

SUGARCANE (Florida Only)

Postemergence directed spray applications of SENCOR are recommended for control of the following weeds in sugarcane in Florida.

Broadleaves	
Amaranth, Spiny (seedling)	*Amaranthus spinosus*
Butterweed (Cressleaf groundsel)	*Senecio glabellus*
Grasses	
Broadleaf panicum	*Brachiaria adspersa*
Goosegrass	*Eleusine indica*

Recommended Broadcast Applications

Crop	SENCOR 4 Pints/Acre	Remarks
Sugarcane	2 to 4	Apply specified dosage in 10 to 40 gallons of water per acre as a postemergence directed spray at the base of the plant after sugarcane is 12 inches tall but before row closing. Avoid spray contact with sugarcane foliage. For best weed control apply before weeds are 6 inches tall. Do not apply more than 4 pints per acre in one season.

SPECIAL PRECAUTIONS (FLORIDA SUGARCANE):

Do not use on sugarcane grown on sand.

Spray contact with sugarcane foliage may result in minor leaf margin chlorosis and/or necrosis.

Do not apply within 60 days of harvest

Do not use treated crop for feed or forage.

Care should be taken to insure that spray adequately covers the foliage of weeds.

Avoid spray overlaps or variations in application speed that may result in insufficient or excessive rates of application.

TOMATOES

Apply SENCOR herbicide with ground equipment to seeded and transplanted tomatoes as specified below under "Recommended Applications" on the label.

For effective control of grasses and broadleaf weeds with postemergence applications, apply SENCOR before weeds are 1-inch tall. Thorough spray coverage on weed foliage is essential for adequate control with postemergence applications.

Do not use air blast or other high pressure spray equipment to make postemergence applications of SENCOR. Refer to the appropriate section of this label for additional information regarding spray equipment, dilution rates, mixing, sprayer cleanup, restrictions, container disposal and cautions.

Refer to "Mixing" under the "General Information" section in the front of the label.

For specific application information see under the "General Information" section "Application" of SENCOR in the front of the label.

WEEDS CONTROLLED

PREPLANT INCORPORATED APPLICATIONS—TRANSPLANT TOMATOES ONLY

Broadleaves	
Galinsoga	*Galinsoga* spp.
Lambsquarters	*Chenopodium album*
*Pigweed, Redroot	*Amaranthus retroflexus*
*Purslane, Common	*Portulaca oleracea*
Grasses	
*Goosegrass	*Eleusine indica*

Preplant incorporated applications applied as directed will suppress foxtails, panicums and barnyardgrass.

SENCOR/TREFLAN TANK MIX

This tank mix combination applied preplant incorporated as directed on this label will control the weeds listed above plus those weeds listed on the Treflan label.

POSTEMERGENCE APPLICATIONS—ESTABLISHED TOMATOES

For effective control of weeds with postemergence applications, apply SENCOR before weeds are 1-inch tall.

Broadcast Sprays—½ to ¾ pt. of SENCOR 4/acre

Directed Sprays 1 to 2 pts. of SENCOR 4/acre

Broadleaves	
Carpetweed	*Mollugo verticillata*
Fumitory	*Fumaria officinalis*
Galinsoga	*Galinsoga* spp.
*Jimsonweed	*Datura stramonium*
*Ladysthumb	*Polygonum persicaria*
Lambsquarters	*Chenopodium album*
Mustard, Wild	*Brassica kaber*
Pigweeds	*Amaranthus* spp.
Purslane	*Portulaca oleracea*
*Ragweed, Common	*Ambrosia artemisiifolia*
*Smartweed, Pennsylvania	*Polygonum pensylvanicum*
Toadflax	*Linaria* spp.
*Velvetleaf	*Abutilon theophrasti*
Grasses	
*Foxtail, Yellow	*Setaria glauca*
Goosegrass	*Eleusine indica*

(Plus Weeds Listed Under "Broadcast Sprays")

* For optimum control of these weeds, use the highest rate recommended on the label for the type of application to be made. Repeat postemergence applications may be needed for best control.

Postemergence applications as directed on this label will suppress barnyardgrass and crabgrass when these weeds are less than 1-inch tall.

Recommended Broadcast Applications

Crop	SENCOR 4 *Pints/Acre	Remarks
	½ to 1	**PREPLANT INCORPORATED—TRANSPLANT TOMATOES ONLY (EXCEPT CALIFORNIA):** Apply specified dosage in 10 to 40 gallons of water per acre as a broadcast spray to the soil surface immediately before transplanting. Incorporate to a depth of 2 to 4 inches with equipment capable of uniformly mixing the chemical into the soil. This application may be made alone or in a tank mix combination with Treflan E.C. Refer to the Treflan label for specific rate of application and for additional precautions and restrictions for tomatoes.
Tomatoes	½ to ¾	**POSTEMERGENCE BROADCAST SPRAY—ESTABLISHED TOMATOES:** Apply specified dosage in 20 to 75 gallons of water per acre as a broadcast spray over the plants in single or multiple applications with a minimum of 14 days between treatments. For transplanted tomatoes do not apply within 14 days of transplanting. Transplants **must** have recovered from transplant shock and new growth evident. Do not apply to established tomatoes within 24 hours of application of other pesticides. Do not tank mix with other pesticides. (See "Special Precautions" below.)
	1 to 2	**POSTEMERGENCE DIRECTED SPRAY—ESTABLISHED TOMATOES:** Apply specified dosage in 20 to 75 gallons of water per acre as a directed spray in single or multiple applications with a minimum of 14 days between treatments. Avoid contacting tomato foliage with spray. This method of treatment is recommended for use in fields with a history of severe weed pressure or in fields infested with hard to control weeds. For transplanted tomatoes, do not apply within 14 days of transplanting. Transplants **must** have recovered from transplant shock and new growth evident. Do not apply to established tomatoes within 24 hours of application of other pesticides. (See "Special Precautions" below.) When banding see the appropriate section in the front of the label.

* Use the higher rate in fields with a history of severe weed pressure and for maximum residual weed control.

SPECIAL PRECAUTIONS (TOMATOES)

Do not apply more than a total of 2 pts. SENCOR 4 per crop season.

Do not apply the total amount of 2 pts. SENCOR 4 within a time span of less than 35 days, except in the case of directed sprays.

Allow at least 14 days between applications, regardless of dosage or method of application.

Do not apply within 7 days of harvest.

Do not apply within 3 days after periods of cool, wet or cloudy weather, or crop injury will occur.

Do not use hot caps on tomatoes within 7 days before or at any time after application of SENCOR.

Do not treat seeded tomatoes until plants have reached the 5 to 6 leaf stage.

Crop injury or delayed maturity may result from broadcast or directed spray applications if tomatoes are growing under stress conditions such as periods of drought or cool, wet and cloudy weather preceding application.

For newly introduced tomato varieties with unknown tolerance to SENCOR, treat only a small area to determine if SENCOR can be used without injury to the corp.

DO NOT USE SENCOR ON TOMATOES IN KERN COUNTY, CALIFORNIA.

CEREALS (BARLEY AND WHEAT)

For the control of certain grass and broadleaf weeds, SENCOR herbicide is recommended as a single post emergence application in barley (winter and spring) and winter wheat (see the appropriate section of this label for geographic limitations).

SENCOR may be used in cereals (barley and wheat) as follows:

1. Barley (spring and winter)—Only East of the Cascade Mountains in Idaho, Oregon, Montana, and Washington.
2. Winter Wheat in Idaho, Oregon, Utah, & Washington.
3. Dryland (Non-irrigated) Winter wheat in Kansas, Oklahoma, & Texas.
4. Dryland (Non-irrigated) winter wheat only (Arkansas, Louisiana, Mississippi, & Tennessee)

Mixing

See the "Mixing" section under "General Information" in the front of the label for specific mixing procedures. Do not tank mix SENCOR with pesticides other than those specified below. When tank-mixing, carefully follow the instructions on this label. Refer to the other product label registered for the specific use intended for directions, rates, and weeds controlled.

Application

Refer to the specific "Application" section under "General Information" in the front of the label for detailed information on the application of SENCOR.

Apply SENCOR when wheat stand is uniform in height. For best results apply SENCOR when weeds are small (less than 1 to 1½ inches in diameter or height). **In the Fall** apply after wheat has a minimum of 3 tillers and 4 secondary roots at least 2 inches in length, but before winter dormancy and wheat jointing occurs. **In the Spring** apply after wheat has recovered from winter dormancy and after wheat has fully tillered and has developed secondary roots at least 2 inches in length, but before jointing.

Grazing Treated Wheat

Wheat fields with SENCOR may be grazed 14 days following application.

Apply SENCOR only once per crop season for wheat or barley.

SPECIAL PRECAUTIONS (CEREALS)

Crop injury may occur on barley or wheat when SENCOR is used under the following conditions:

1. When applied while the crop is under stress such as winter kill, frost damage, disease, drought or excessive moisture, or when these conditions follow the application.
2. When applied in combination with fluid fertilizer.
3. When applications are made prior to the growth stage specified on this label.
4. When soils are high in lime or sodium, have a pH greater than 7.7, or are calcareous soils.
5. When proper application instructions are not followed.

Weed control may not be observed for 2 to 4 weeks after application; under very dry conditions, weed control may not be apparent for 4 to 6 weeks. Moisture (at least ½") is needed within 2 to 3 weeks after application to move SENCOR into the weed root zone. Lack of adequate moisture after application may result in poor control.

Do not exceed the rates specified on this label.

BARLEY (Spring and Winter)

FOR USE ONLY EAST OF THE CASCADE MOUNTAINS IN IDAHO, OREGON, UTAH, MONTANA AND WASHINGTON ON THE VARIETIES COMPANA, HANNCHEN, HECTOR, HUDSON, LUTHER, KAMIAK, KLAGES, OLYMPIC, PIROLINE AND STEPTOE.

A single postemergence application of SENCOR is recommended for control of certain grass and broadleaf weeds in Spring and Winter barley.

Recommended Broadcast Applications

Soil Texture	SENCOR 4 Pints Per Acre % Organic Matter[1] 1 to 2	Over 2	Remarks
COARSE (Sandy loam, loamy sand)	½	½ to ¾	Barley should be seeded to obtain a uniform crop stand at application time. Apply after barley has fully tillered and developed secondary roots but before jointing and before weeds are 2 inches tall or have 2 inch diameter rosettes. Do not graze or harvest treated barley for feed before crop maturity.
MEDIUM (Loam, silt loam, silt, sandy clay, sandy clay loam)	½ to ¾	¾	
FINE (Silty clay, silty clay loam,[2] clay, clay loam	¾	¾ to 1	

[1] Do not use on any soils containing less than 1% organic matter.
[2] Silty clay loam soils are transitional soils and may be classified as medium textured soils in some regions of the U.S.

Weeds Controlled In Barley

Broadleaves

Chickweed, Common	*Stellaria media*
Fiddleneck, Tarweed	*Amsinckia lycopsoides*
Henbit	*Lamium amplexicaule*
* Kochia	*Kochia scoparia*
Lambsquarters	*Chenopodium album*
* Mustard, Blue or Purple	*Chorispora tenella*
* Mustard, Jim Hill	*Sisymbrium altissimum*
* Mustard, Wild	*Brassica kaber*
Pigweeds	*Amaranthus* spp.
Grasses	
* Cheatgrass	*Bromus secalinus*
* Downey Brome	*Bromus tectorum*
* Foxtails	*Setaria* spp.
* Ryegrass, Annual	*Lolium multiflorum*

* Use the highest rate recommended for soil texture and percent organic matter for the control of these weeds.

Hard to Control Weed

Wild oats — *Avena fatua*

NOTE: Control of wild oats may be variable, ranging from poor to excellent depending on timing of application, stage of growth of wild oats at time of application, temperatures and soil moisture conditions following treatment.

SENCOR in Combination with Other Herbicides

SENCOR may be applied as directed above or may be tank mixed with 2, 4–D, Banvel®, Bronate®, Brominal®, Brominal Plus, or Buctril®. Refer to the other product label registered for this use for additional directions, rates, and weed species controlled.

WINTER WHEAT

FOR USE ON WINTER WHEAT IN IDAHO, OREGON, UTAH AND WASHINGTON ON THE VARIETIES DAWS, GAINES, HATTON, HILL 81, HYSLOP, LUKE, McDERMID, MORO, NuGAINES, PAHA, PECK, SPRAGUE, STEPHENS, WANSER, FARO, TYEE, AND WESTON AND IN MONTANA ON THE VARIETIES CENTURK, CHEYENNE AND WINALTA.

A single post-emergence application of SENCOR is recommended for control of certain grass and broadleaf weeds in Winter wheat.

Recommended Broadcast Applications[1]

Soil Texture	SENCOR[4] Pints/Acre 0.75% to 1% Organic Matter[2]	1% to 2% Organic Matter	More than 2% Organic Matter
COARSE Loamy sand, sandy loam	0.5 to 0.6	0.5 to 0.6	0.5 to 0.6
MEDIUM Loam, silt loam, silt, sandy clay, sandy clay loam	0.5 to 0.6	0.6 to 0.75	0.6 to 0.9
FINE Silty clay, silty clay loam[3] clay, clay loam	0.5 to 0.75	0.75 to 1.0	0.75 to 1.0

[1] Do not use on gravelly or thinly covered or exposed subsoil areas (clay knobs) as crop injury may occur.
[2] Do not use on any soils containing less than 0.75% organic matter.
[3] Silty clay loam soils are transitional soils and may be classified as medium textured soils in some regions of the U.S.

WEEDS CONTROLLED IN WHEAT

Broadleaves

Annual Polemonium (Jacob's Ladder)	*Polemonium micranthum*
Chickweed, Common	*Stellaria media*
Conical Catchfly	*Silene conica*
Corncockle	*Agrostemma githago*
Fiddleneck, Tarweed	*Amsinckia lycopsoides*
Gromwell	*Lithospermum officinale*
Henbit	*Lamium amplexicaule*
Ivyleaf Speedwell	*Veronica hederifolia*
Knotweed, Prostrate	*Polygonum aviculare*
* Kochia	*Kochia scoparia*
Lambsquarters	*Chenopodium album*
Madwort (catchweed)	*Asperugo procumbens*
Mayweed (dog fennel)	*Eupatorium capillifoium*
Miners Lettuce	*Claytonia perfoliata*
* Mustard, blue or purple	*Chorispora tenella*
* Mustard, Jim Hill (tumble)	*Sisymbrium altissimum*
* Mustard, Wild	*Brassica kaber*
Pennycress, Field	*Thlaspiarvense*
Pineappleweed	*Matricaria matricarioides*
Prickly Lettuce	*Lactuca serriola*
Pigweeds	*Amaranthus* spp.
Redstem filaree	*Erodium cicutarium*
* Russian Thistle	*Salsola kali*
Shepherdspurse	*Capsella bursa-pastoris*
Grasses	
* Blackgrass (Black twitch-grass)	*Alopecurus myosuroides*
* Bluegrass, bulbous	*Poa bulbosa*
* Cheatgrass	*Bromus secalinus*
* Downy Brome	*Bromus tectorum*
* Foxtails	*Setaria* spp.
* Rescuegrass	*Bromus catharticus*
* Ryegrass, Italian	*Lolium multiflorum*
* Wild Barley	*Hordeum leporinum*
* Windgrass	*Apera spica-venti*

* Suppression only may occur.

For suppression of these weeds, use the highest rate recommended for soil texture and percent organic matter. Suppression is a reduction in weed size and growth compared to a non-treated area in the same field.

HARD-TO-CONTROL WEEDS

Broadleaves

Cowcockle	*Vaccaria pyamidata*
Mustard, Tansy	*Descurainia pinnata*
Wild Buckwheat	*Polygonum convolvulus*

Grasses

Ripgut Brome	*Bromus diandrus*
Wild Oats	*Avena fatua*

NOTE: Control of these weeds may be variable, ranging from poor to excellent depending on timing of application, stage of growth at the time of application, temperatures, and soil moisture conditions following treatment.

SENCOR IN COMBINATION WITH OTHER HERBICIDES

SENCOR may be applied as a tank-mix with Glean®, 2,4–D, MCPA, Igran®, Banvel, Bronate, Brominal, ME4 Brominal, 3+3 Brominal, Brominal Plus or Buctril. Refer to the other product labels registered for this use for additional directions, rates and weed species controlled. Observe all cautions and limitations on labeling of all products used in mixtures.

IRRIGATED WHEAT/BARLEY

Follow timing directions for dryland wheat/barley found at the beginning of this cereals section. After application of SENCOR, do not apply more than ½ inch of water for the first irrigation; the maximum amount for each additional irrigation should not exceed 1 inch.

SENCOR PLUS IGRAN TANK-MIX APPLICATION

Apply SENCOR 4 at ½ to ¾ pint per acre and Igran 80WP at ½ to ¾ pound per acre according to label directions. The rate of each product depends on soil type. Do not use Igran on sand or loamy sand soils. Use lower rate of Igran on sandy loam soils.

BROADLEAF WEED CONTROL ONLY:

Where only broadleaf control is needed, SENCOR applied at the rates listed below in a tank-mix combination with Brominal, Brominal Plus, ME4 Brominal, 3+3 Brominal, Bronate or Buctril can be used to control wild buckwheat, in addition to all the weeds listed on these labels.

Recommended Broadcast Application

SENCOR 4 Pints/Acre	Remarks
⅜ to ½	Apply SENCOR in a tank mix combination with Brominal, Brominal Plus, ME4 Brominal, 3+3 Brominal, Buctril or Bronate. Use higher rate of SENCOR for control of larger broadleaf weeds that are 2 inches in diameter or height. Apply before weeds are greater than 3 inches in diameter or height. Not for use on sand soils.

SPECIAL PRECAUTIONS: Refer to the "General Mixing and Application", and "Precautions" in the front of the cereal section.

Temporary chlorosis or injury may occur after application of SENCOR to wheat, especially if frost or snow occurs within a few days after application. Heavy rainfall (1 inch or more) within 4 days following application may result in crop injury.

DRYLAND (NON-IRRIGATED) WINTER WHEAT

FOR USE ONLY ON DRYLAND WINTER WHEAT IN KANSAS, OKLAHOMA AND TEXAS ON THE VARIETIES TAM W101, TAM 105, AND NEWTON AND IN OKLAHOMA AND TEXAS ON HAWK VARIETY.

WEEDS CONTROLLED IN WHEAT

Broadleaf Weeds

Annual Polemonium (Jacob's Ladder)	*Polemonium micranthum*
Chickweed, Common	*Stellaria media*
Fiddleneck, Tarweed	*Amsinckia lycopsoides*
Henbit	*Lamium amplexicaule*
*Kochia	*Kochia scoparia*
Lambsquarters	*Chenopodium album*
*Mustard, Blue	*Chorispora tenella*
*Mustard, Jim Hill	*Sisymbrium altissimum*
*Mustard, Wild	*Brassica kaber*
Pigweeds	*Amaranthus* spp.

Grasses

*Cheatgrass	*Bromus secalinus*
*Downy Brome	*Bromus tectorum*
*Foxtails	*Setaria* spp.
*Rescuegrass	*Bromus catharticus*

* For control of these weeds, use the highest rate recommended for soil texture and percent organic matter.

Hard to Control Weeds: *Italian ryegrass (*Lolium multiflorum*), *Wild Oats (*Avena fatua*), and *Wild Buckwheat (*Polygonum convolvulus*).

NOTE: Control of these weeds may be variable, ranging from poor to excellent depending on timing of application, stage of growth at time of application, temperatures and soil moisture conditions following treatment.

Recommended Broadcast Applications

	SENCOR 4 Pints Per Acre				
	% Organic Matter[1]				
	Fall		Spring		
Soil Texture	1 to 2	Over 2	1 to 2	Over 2	Remarks
COARSE (Sandy loam, loamy sand)	½	½	½	½ to ¾	Apply specified dosage per acre with aerial or ground spray equipment as a broadcast post-emergence spray. For Fall and Spring application, refer to "Application" in the beginning of this Cereals section for information on timing of application and growth restrictions.
MEDIUM (Loam, silt loam, silt, sandy clay, sandy clay loam)	½ to ¾	¾	¾ to 1	¾ to 1¼	
FINE (Silty clay, silty clay loam[2], clay, clay loam)	¾	¾ to 1	¾ to 1¼	1 to 1½	

[1] Do not use on any soils containing less than 1% organic matter.
[2] Silty clay loam soils are transitional soils and may be classified as medium textured soils in some regions of the U.S.

SENCOR IN COMBINATION WITH OTHER HERBICIDES

SENCOR may be applied as directed above or may be tank mixed with 2,4-D, Banvel, Bronate, Brominal, or Buctril. Refer to the other product labels registered for additional directions, rates, and weed species controlled.

DRYLAND (NON-IRRIGATED) WINTER WHEAT ONLY (Arkansas, Louisiana, Mississippi and Tennessee)

WEEDS CONTROLLED

Broadleaves

Bittercress	*Cardamine* spp.
*Buttercup	*Ranunculus* spp.
Carolina Geranium	*Geranium carolinianum*
Chickweed, Common	*Stellaria media*
Chickweed, Mousear	*Cerastium vulgatum*
Cutleaf Evening Primrose	*Oenothera laciniata*
Henbit	*Lamium amplexicaule*
*Kochia	*Kochia scorparia*
Lambsquarters, Common	*Chenopodium album*
Mayweed (dog fennel)	*Eupatorium capillifolium*
Mustard, Wild	*Brassica kaber*
Pepperweed, Virginia	*Lepidium virginicum*
Pigweeds	*Amaranthus* spp.
Shepherdspurse	*Capsella bursa-pastoris*
Wild Turnip	*Brassica rapa*

Grasses

Bluesgrass, Annual	*Poa annua*
*Cheatgrass	*Bromus secalinus*
*Downy Brome	*Bromus tectorum*
Little Barley	*Hordeum pusillum*
*Rescuegrass	*Bromus catharticus*
*Ryegrass, Italian	*Lolium multiflorum*
Vernal Withlograss	*Drababerna* spp.

* For control of these weeds on fine soils, use SENCOR 4 at 1 pint/acre.

Note: Control of wild buckwheat (Polygonum convolvulus) and Winter Vetch (Vicia spp.) may be variable with SENCOR, ranging from poor to excellent depending on growth at time of application, temperatures and moisture conditions following application.

Recommended for use only on the following varieties:

Abe, Arthur 71, Caldwell, Coker 747, Coker 762, Coker 916, Coker 68-15, Doublecrop, Hunter, McNair 1003, McNair 1813, Nelson, Pike, Southern Belle, & Tyler.

Recommended Broadcast Application

Soil Texture	SENCOR 4[1] Pints/A	Remarks
Coarse (Sandy loam, loamy sand)	½	Apply specified dosage per acre as a broadcast postemergence spray. For wheat plants that are coming out of Winter Dormancy, allow at least two weeks for plants to recover before application. If wheat has been grazed during the Winter, allow at least two weeks for recovery before application. Refer to the beginning of this Cereals section for information on timing and growth stages for application.
Medium (Loam, silt loam, silt, sandy clay, sandy clay loam)	¾	
Fine (Silty clay, silty clay loam[2], clay, clay loam)	¾–1.0[3]	

[1] Do not use on Coarse soils containing less than 1% organic matter or on Medium or Fine textured soils containing less than 0.5% organic matter.
[2] Silty clay loam soils are transitional soils and may be classified as Medium textured soils in some regions of the U.S.
[3] Use 1 pint/acre for control of weed species designated with an asterick under "Weeds Controlled".

SENCOR IN COMBINATION WITH OTHER HERBICIDES

SENCOR may be applied as directed above and may be tank mixed with 2,4-D, Banvel, Bronate, Brominal, or Buctril. Refer to the other registered product labels for additional directions, rates, and weed species controlled.

Special Precautions

Temporary chlorosis may occur after application of SENCOR to wheat, especially if frost occurs.

FOR WEED CONTROL IN A WHEAT/FALLOW/WHEAT ROTATION

(Idaho, Oregon, Utah and Washington Only)

SENCOR may be applied to provide weed control during the fallow period after wheat harvest or in the Spring before winter wheat is planted. Winter wheat can be seeded 4 months (120 days) after Spring application. Mechanical tillage or the application of a contact herbicide may be required to control weeds germinating prior to seeding of winter wheat. Best results will be obtained where straw and chaff are evenly distributed across the field.

Refer to "Mixing" on the "General Information" section in the front of the label.

For specific application information see under the "General Information" section "Application" of SENCOR in the front of the label.

Where weed growth is present at application time, SENCOR should be applied with Ortho Paraquat or other contact herbicide. Refer to the other product label registered for additional directions, rates, and weed species controlled.

WEEDS CONTROLLED

Broadleaves

Chickweed, Common	*Stellaria media*
Henbit	*Lamium amplexicaule*
* Kochia	*Kochia scoparia*
Lambsquarters	*Chenopodium album*
Mustard, Blue or Purple	*Chorispora tenella*
Mustard, Jim Hill	*Sisymbrium altissium*
Mustard, Tansy	*Descuainia pinnata*
Mustard, Treacle	*Eyrsimum repandum*
Mustard, Wild	*Brassica kaber*
Pennycress, Field (Fanweed)	*Thlaspi arvense*
Pigweeds	*Amaranthus* spp.
* Russian thistle	*Salsola kali*
Wild Sunflower	*Helianthus* spp.

Grasses

Cheatgrass	*Bromus secalinus*
Downy Brome	*Bromus tectorum*
* Wheat, Volunteer	*Triticum* spp.

* Note: Since control of these weeds may be variable depending on moisture following application, the higher labelled rate is recommended.

After Harvest Application (Fall Fallow): SENCOR may be applied to wheat stubble after harvest in the Fall. Apply 1 to 1-¼ pints per acre broadcast before weeds emerge. Use higher rate for longer weed control or for weeds designated as requiring the higher rate for control. Rainfall (½ inch or more) is necessary for herbicide activation.

Do not plant crops in treated areas for at least 10 months following fall applications.

SENCOR may be applied at 1 to 1¼ pints per acre as directed above for a Fall application and tank mixed with 2⅔ to 5⅓ quarts Chem-Hoe® 135 FL3 per acre. Apply 4 to 5⅓ quarts Chem-Hoe 135 per acre if volunteer grain or grass weeds have 4 or more leaves. If other vegetation is present at the time of application use a contact herbicide. Refer to the Chem-Hoe 135 FL3 label for additional directions, rates, weed species controlled and precautions.

Spring Application (Summer Fallow): SENCOR may be applied to wheat stubble in the Spring. Apply ¾ to 1 pint per acre broadcast before weeds emerge in the Spring. Use higher rate for longer weed control or weeds designated as requiring higher rate for control. Rainfall (½ inch or more) is necessary for herbicide activation.

SENCOR may be applied at ¾ to 1 pint after January 1 and before March 1 as a tank mix with 2⅔ to 5⅓ quarts Chem-Hoe 135 FL3 per acre. Apply 4 to 5⅓ quarts Chem-Hoe 135 per acre if volunteer grain or grass weeds have 4 or more leaves. A contact herbicide is recommended if other vegetation is present at the time of application. Refer to the Chem-Hoe 135 FL3 label for additional directions, rates, weed species controlled and precautions.

PRECAUTIONS AND RESTRICTIONS

Do not graze treated fields.

Do not plant Spring seeded cereals following Fall fallow applications of SENCOR.

Where SENCOR was applied in the Fall, do not apply SENCOR in the Spring.

FOR WEED CONTROL IN A FALLOW ROTATION WITH BARLEY AND WHEAT (Colorado, Kansas, Montana Nebraska, and Wyoming only)

SENCOR may be applied to provide weed control during the fallow period after wheat or barley harvest or in the Spring before planting of Winter wheat or barley. Mechanical tillage or the application of a contact herbicide may be required to control weeds germinating prior to seeding of Winter wheat or barley.

Refer to "Mixing" on the "General Information" section in the front of this label.

For specific application information see under the "General Information" section "Application" of SENCOR in the front of the label.

Where weed growth is present at application time, SENCOR should be applied with Paraquat + Plus, Roundup, or other contact herbicide. Refer to the other product labels registered for additional directions, rates, and weed species controlled. Do not plant crops in treated areas earlier than 10 months following Fall applications.

WEEDS CONTROLLED

Broadleaves

Chickweed, Common	*Stellaria media*
Cowcockle	*Vaccaria pyramidata*
Henbit	*Lamium amplexicaule*
* Kochia	*Kochia scoparia*
Lambsquarters	*Chenopodium album*
Mustard, Blue or Purple	*Chorispora tenella*
Mustard, Jim Hill	*Sisymbrium altissimum*
Mustard, Tansy	*Descurainia pinnata*
Mustard, Treacle	*Eyrsimum repandum*
Mustard, Wild	*Brassica kaber*
Pennycress, Field (fanweed)	*Thlaspi arvense*
Pigweeds	*Amaranthus* spp.

Russian thistle	*Salsola kali*
Wild Sunflower	*Helianthus* spp.
Grasses	
Cheatgrass	*Bromus secalinus*
Downy Brome	*Bromus tectorum*
* Foxtail, Green	*Setaria viridis*
* Wheat, Volunteer	*Triticum* spp.
* Wild Oats	*Avena fatua*

* Note: Since control of these weeds may vary depending on moisture following application, use the higher rate recommended below.

AFTER HARVEST APPLICATION (Fall Fallow)—SENCOR may be applied to the stubble after harvest in the Fall. Apply 1¼ to 1½ pints per acre broadcast before weeds emerge. Use the higher rate for longer weed control or for weeds designated as requiring the higher rate for control. Rainfall (½ inch or more) is necessary for herbicide activation.

SPRING APPLICATION (Summer Fallow)—SENCOR may be applied to the stubble in the Spring. Apply ¾ to 1 pint per acre broadcast before weeds emerge in the Spring. Use the higher rate for longer weed control or weeds designated as requiring the higher rate for control. Rainfall (½ inch or more) is necessary for herbicide activation. Wheat or barley can be seeded 120 days after Spring application.

PRECAUTIONS AND RESTRICTIONS

Do not graze treated fields.

Do not plant Spring seeded cereals following Fall applications for fallow.

Where SENCOR was applied in the Fall, do not apply SENCOR in the Spring.

SPECIAL LOCAL NEED REGISTRATIONS [SECTION 24(c)] FOR SENCOR 4*

CROP	PEST	STATE EPA SLN NO.	DOSAGE	REMARKS
Alfalfa	Various broadleaf weeds and grasses	Washington WA 85-0006	¾ to 2 pts.	Application by ground or air. Do not graze or harvest within 28 days of application.
Alfalfa Established (Including mixed stands with grasses)	Various weeds	Oregon OR 81-0075 Utah UT 81-0017 Washington WA 81-0083	SENCOR 4: ½ to 1½ pints per acre Paraquat + Plus: 1½ to 2 pints per acre.	Tank mix with Paraquat + Plus.
Asparagus (Established)	Weeds	Oregon OR 81-0109 Washington WA 81-0107	1 to 2 pints preemergence in combination with 2 to 3 pints postharvest.	Sprinkler irrigation.
Barley (Spring)	Various weeds	Alaska AK 83-0002	¼ to ½ pint/acre.	Postemergence ground application.
Corn (Chemical fallow)	Various weeds	California CA 84-0007	1 pint per acre.	Apply by ground or air in the Fall.
Lentils	Suppression of certain broadleaf weeds	Idaho ID 82-0012 Montana MT 82-0005 Oregon OR 82-0017 Washington WA 82-0015	¼ to ⅜ pints per acre.	Application by ground early postemergence.
Lentils and dry field peas	Suppression of certain broadleaf weeds	Idaho ID 80-0020	½ to ¾ pint per acre.	Apply before or after planting but before crop emergence.
		Montana MT 80-0007		Northern Idaho only.
		Oregon OR 80-0038		In Oregon, east of Cascade Mountains only.
		Washington WA 80-0025		Eastern Washington only.
Dry field peas	Chickweed Dog Fennel Henbit Lambsquarters Wild Mustard Field Pennycress (Fanweed) Shepherdspurse	Idaho ID 84-0014 Washington WA 85-0001	⅓ to ½ pints per acre.	One postemergence application may be made per season. Do not apply within 75 days of harvest. Do not apply over very moist soils or wet crop foliage.
Potatoes	Weeds	Idaho ID 81-0095 Oregon OR 81-0039 Washington WA 81-0041	Label rates.	Tank mix with Eptam 7-E.

CROP	PEST	STATE EPA SLN NO.	DOSAGE	REMARKS
	Various broadleaf and grass weeds including quackgrass	Maine ME 79-0009	1 to 2 pints per acre.	May be tank mixed with 7 to 10 pounds DOWPON M as a preemergence broadcast spray.
Soybeans	Red rice	Louisiana LA 79-0009	See SENCOR, Lasso and Treflan labels for specific dosage.	Apply as a preemergence broadcast or band overlay.
	Sicklepod	Arkansas AR 81-0054 Florida FL 78-0058 Georgia GA 78-0034 Louisiana LA 81-0045 Mississippi MS 81-0059 Missouri MO 81-0027	½ to ¾ pints per acre.	Use on low organic coarse soils.
		Oklahoma OK 84-0004		Low organic sandy soils, alone and in combination with other herbicides.
		North Carolina NC79-0009	See SENCOR and Treflan or Tolban labels for specific dosage.	Use on low organic soils.
	Sicklepod Coffeeweed *(Cassia obtusifolia L.)*	South Carolina SC 78-0031	½ to ¾ pints per acre.	Use on low organic coarse soils.
	Various weeds	Michigan MI 79-0021		Use on alkaline soils.
Sugarcane	Various grass and broadleaf weeds	Florida FL 83-0010	2 to 4 pints per acre ground. 2 to 3 pints per acre aerial.	Postemergence application by ground or air
Sugarcane	Various grass and broad-leaf weeds	Louisiana LA 84-0017	2 to 3 pints	BROADCAST (Summer Fallow): Apply specified dosage per acre using 20 to 30 gallons of water with ground equipment or 5 gallons of water with aircraft spray equipment. Apply in late May to early June after spring tillage and rows have been rebuilt. Use the lower rate if the application is made less than 5 weeks prior to planting. Apply to a soil surface which is free of weeds, trash and clods.
Tomatoes (Chemical fallow)	Various weeds	California CA 79-0235	1 to 1.5 pints per acre.	Application by ground or air. Not for use in Kern County.
Wheat (Dryland Winter only)	Ivyleaf speedwell	Oregon OR 80-0017	½ pint per acre.	Single postemergence application by ground or air
	Various grass and broad-leaf weeds	Kansas KS 82-0021	½ to 1½ pints per acre.	Use on Eagle, Tam W-101, Tam-105 and Newton variety wheat.
		Oklahoma OK 79-0017		On variety TAM W-101, TAM 105, Newton, and Hawk.
		Missouri MO 84-0003	½ to 1½ pints per acre.	On Caldwell, Centurk 78, Coker 747, Hart, Newton, Pike, Pioneer 876, Pioneer 578, TAM-105, and TAM W-101 wheat variety.

* For distribution and use only within the state specified. These Special Local Need registrations approved as of date of this publication are, however, subject to cancellation by the state. Contact your nearest Mobay Agricultural Chemicals Division Office for the current status and a copy of the complete individual supplemental Special Local Need labeling. This labeling must be in the possession of the user at time of pesticide application.

Crop Rotation Recommendations

Waiting Period After SENCOR herbicide application[1]			
4 Months	**8 Months**	**12 Months**	**18 Months**
Alfalfa	Barley	Other Crops not listed (Except root crops)	Sugar Beets
Barley[2]	Cotton		Onions (And other root crops)
Asparagus	Lentils		
Corn	Peas		
Forage Grasses	Rice		
Potatoes	Wheat		
Sainfoin			
Soybeans			
Sugarcane			
Tomatoes			
Wheat[2]			

[1] Cover crops for soil building or erosion control may be planted any time, but do not graze or harvest for food or feed. Stand reductions may occur in some areas.
[2] Following peas, lentils or soybeans.

PRECAUTIONARY STATEMENTS

HAZARDS TO HUMANS AND DOMESTIC ANIMALS

CAUTION

May be harmful if swallowed or inhaled. Avoid contact with eyes, skin or clothing. Avoid breathing spray mist. Wash thoroughly with soap and warm water after handling. Wash contaminated clothing with soap and hot water before reuse.

Do not contaminate feed or food. Keep out of reach of children.

Obtain prompt medical aid if poisoning should occur.

To Physician: Treat the patient symptomatically.

Symptoms of Poisoning: The compound does not cause any definite symptoms that would be diagnostic. Poisoning is accompanied by breathing difficulties and sedation.

STATEMENTS OF PRACTICAL TREATMENT

If in eyes, immediately flush eyes with plenty of water for at least 15 minutes. For further treatment get medical attention.

If swallowed, vomiting should be induced. Administer water freely and induce vomiting by giving one dose (1/2 oz. or 15 ml.) of syrup of ipecac. If vomiting does not occur within 10 to 20 minutes, administer second dose. If syrup of ipecac is not available, induce vomiting by sticking finger down throat. Repeat until vomit fluid is clear. Never give anything by mouth to an unconscious person. Professional medical assistance should be secured immediately.

If on skin, wash thoroughly with soap and warm water.

ENVIRONMENTAL HAZARDS

Do not use on other crops grown for food or forage. Keep out of lakes, streams, ponds, tidal marshes and estuaries. Do not apply when weather conditions favor drift from areas treated. Do not contaminate water by cleaning of equipment or disposal of wastes. Apply this product only as specified on the label.

Do not allow sprays to drift on to adjacent desirable plants.

Observe all cautions and limitations on labeling of all products used in mixtures.

RE-ENTRY STATEMENTS

Do not apply this product in such a manner as to directly or through drift expose workers or other persons. The area being treated must be vacated by unprotected persons. Do not enter treated areas without protective clothing until sprays have dried. Because certain states may require more restrictive reentry intervals for various crops treated with this product, consult your State Department of Agriculture for further information.

Written or oral warnings must be given to workers who are expected to be in a treated area or in an area about to be treated with this product. Oral warnings should include those statements found under the label heading "Precautionary Statements" and "Re-entry Statements" included on this label. When oral warnings are given, warnings shall be given in a language customarily understood by workers. Oral warnings must be given if there is reason to believe that written warnings cannot be understood by workers. Written warnings must include the following information: WARNING, area treated with SENCOR 4 flowable herbicide on (date of application). Do not enter without appropriate protective clothing until sprays have dried. **If on skin,** remove contaminated clothing and wash skin immediately with soap and warm water. **If eyes are contaminated,** wash with flowing water for at least 15 minutes. **If swallowed,** vomiting should be induced. (See Precautionary Statements on label).

STORAGE AND DISPOSAL

PESTICIDE DISPOSAL

Do not contaminate water, food or feed by storage or disposal.

Wastes resulting from the use of this product may be disposed of on site or at an approved waste disposal facility.

CONTAINER DISPOSAL

Triple rinse (or equivalent). Then offer for recycling or reconditioning, or puncture and dispose of in a sanitary landfill, or incineration, or, if allowed by state and local authorities, by burning. If burned, stay out of smoke.

Store in a cool, dry place and in such a manner as to prevent cross contamination with other pesticides, fertilizers, food, and feed. Store in original container and out of the reach of children, preferably in a locked storage area.

Handle and open container in a manner as to prevent spillage. If container is leaking invert to prevent leakage. If the container is leaking or material spilled for any reason or cause, carefully dam up spilled material to prevent runoff. Refer to Precautionary Statements on label for hazards associated with the handling of this material. Do not walk through spilled material. Absorb spilled material with absorbing type compounds and dispose of as directed for pesticides above. In spill or leak incidents, keep unauthorized people away. You may contact the Mobay Emergency Response Team for decontamination procedures or any other assistance that may be necessary. The Mobay Kansas City Emergency Response Telephone No. is 816-242-2582, or contact Chemtrec at 800-424-9300.

SENCOR is a Reg. TM, SPLIT-SHOT and EXTENDED SPLIT-SHOT are TMs of the Parent Company of Farbenfabriken Bayer GmbH, Leverkusen.

Amiben and Brominal are Reg. TM's of Union Carbide Agricultural Products Co., Inc.

Banvel is a Reg. TM of Velsicol Chemical Corp.

Bladex is a Reg. TM of Shell Chemical Co.

Bronate and Buctril are Reg. TM's of Rhone-Poulenc Chemical Co.

Chem-Hoe is a Reg. TM of PPG Industries, Inc.

Dual and Igran are Reg. TM's of Ciba-Geigy.

Dyanap is a Reg. TM of Uniroyal Chemical Co.

Glean is a Reg. TM of E. I. duPont de Nemours and Co.

Lasso and Roundup are Reg. TM's of Monsanto Co.

Paraquat + Plus is a TM of Chevron Chemical Co.

Prowl is a Reg. TM of American Cyanamid Co.

Surflan, Sonalan and Treflan are Reg. TM's of Elanco Products Co.

Ortho X-77 is a Reg. TM of Kalo Laboratories, Inc.

IMPORTANT

This bulletin is not intended to provide adequate information for application. Before using any product, read and carefully observe directions, cautionary statements and other information appearing on the product label. This product is sold subject to the Conditions of Sale set forth on the container label.

®SYSTOX 2

EMULSIFIABLE SYSTEMIC INSECTICIDE

RESTRICTED USE PESTICIDE

For retail sale to and use only by Certified Applicators or persons under their direct supervision, and only for those uses covered by the Certified Applicator's certification.

For Control of Aphids, Mites and Leafhoppers on a Wide Variety of Crops.

ACTIVE INGREDIENTS:	
Demeton	
O,O-Diethyl *O*-[2-(ethylthio) ethyl] phosphorothioate mixture with *O,O*-Diethyl *S*-[2-(ethylthio) ethyl] phosphorothioate	25%
Aromatic Petroleum Distillate	68%
INERT INGREDIENTS	7%
TOTAL	100%

Contains 2 lbs. Demeton per gallon

EPA Reg. No. 3125-4-ZA

SYSTOX is a Reg. TM of the Parent Company of Farbenfabriken Bayer GmbH, Leverkusen.

DANGER–POISON

Keep Out of Reach of Children

PELIGRO

PRECAUCION AL USUARIO: Si usted no puede leer o entender inglés, no use este producto hasta que la etiqueta le haya sido explicada ampliamente.

(TO THE USER: If you cannot read or understand English, do not use this product until the label has been fully explained to you.)

RECOMMENDED USES

Almonds	Nectarines
Apples	Oats
Apricots	Oranges
Beans	Ornamentals
Barley	Peas
Blackberries	Peaches
Broccoli	Pears
Brussels sprouts	Pecans
Cabbage	Peppers
Cauliflower	Pineapples*
Celery	Plums
Cotton	Potatoes
Dewberries	Prunes (fresh)
Eggplant	Raspberries
Filberts	Seed alfalfa
Gooseberries	Seed clover
Grapefruit	Sorghum
Grapes	Stone fruits*
Hops	Strawberries
Lemons	Sugar beets
Lettuce	Tomatoes
Loganberries	Walnuts
Muskmelons	Wheat

* Non-bearing

RESTRICTIONS ON WAREHOUSE STORAGE

Do not store adjacent to or near foodstuffs, drugs, feeds or any other material intended for consumption by humans or animals. Do not store adjacent to toys, cosmetics, dishes, furniture or clothing. Do not store next to herbicides. Do not use, pour, spill or store near heat or open flame. Do not store in or around the home.

STATEMENTS OF PRACTICAL TREATMENT

In case of poisoning call a physician immediately. Have patient lie down and keep quiet.

If swallowed, vomiting should be induced. Administer water freely and induce vomiting by giving one dose (1/2 oz. or 15 ml.) of syrup of ipecac. If vomiting does not occur within 10 to 20 minutes, administer second dose. If syrup of ipecac is not available, induce vomiting by sticking finger down throat. Repeat until vomit fluid is clear. Never give anything by mouth to an unconscious person. Professional medical assistance should be secured immediately.

If on skin, remove contaminated clothing and wash skin immediately with soap and warm water.

If eyes are contaminated, wash with flowing water for at least 15 minutes.

TO PHYSICIAN: Antidote—Administer atropine sulfate in large therapeutic doses. Repeat as necessary to the point of tolerance. 2-PAM is also antidotal and may be administered in conjunction with atropine.

Compound inhibits cholinesterase resulting in stimulation of the central nervous system, the parasympathetic nervous system, and the somatic motor nerves. Do not give morphine. Watch for pulmonary edema, which may develop in serious cases of poisoning even after 12 hours. At first sign of pulmonary edema, the patient should be placed in an oxygen tent and treated symptomatically.

PRECAUTIONARY STATEMENTS

HAZARDS TO HUMANS AND DOMESTIC ANIMALS

Poisonous if swallowed, inhaled or absorbed through the skin. Rapidly absorbed through the skin. Do not get in eyes or on skin. Wear protective clothing, natural rubber gloves, and goggles.

Do not breathe fumes or spray mist. For protection against demeton insecticide wear a mask or respirator jointly approved by the Mining Enforcement and Safety Administration (formerly the U.S. Bureau of Mines) and by the National Institute for Occupational Safety and Health under the provisions of 30 CFR Part 11.

Do not contaminate or store near feed or foodstuffs. Keep out of reach of children. Keep all unprotected persons out of the operating area or vicinity where there may be danger of drift. Wash hands, arms, and face thoroughly with soap and warm water before eating or smoking.

HANDLING OF CONTAMINATED ARTICLES: Clothing and rags receiving undiluted SYSTOX 2 as a result of leaks, spills, or mishaps should be removed immediately and not reused. These contaminated articles should be buried in the same manner as the empty container; see container disposal section for directions. (Contaminated articles which should be discarded would include: hats, gloves, aprons, coats, boots, etc. not made of rubber or coated with rubber or other similar materials). Clothing receiving spray mist or droplets from the mixed, diluted SYSTOX 2 should be removed and decontaminated before reuse. This clothing can be decontaminated by machine washing separately from other items with soap or detergent and bleach in hot water twice before reuse. Protective clothing and equipment should be washed down with detergent or soap and bleach in water. Wash water from the cleaning of protective clothing or equipment should not be allowed to run off or otherwise enter water supplies.

SYMPTOMS OF POISONING: A sense of "tightness" in the chest. Contracted pupils. Stomach pains. Vomiting and diarrhea.

ENVIRONMENTAL HAZARDS

This product is toxic to fish, shrimp, birds, and other wildlife. Keep out of lakes, streams, or ponds, tidal marshes and estuaries. Do not apply where runoff is likely to occur. Do not apply when weather conditions favor drift from areas treated. Do not contaminate water by cleaning of equipment, or disposal of wastes. Apply this product only as specified on this label.

This product is toxic to bees and should not be applied when bees are actively visiting the area.

PHYSICAL AND CHEMICAL HAZARDS

Do not use, pour, spill or store near heat or open flame.

RESTRICTIONS

Do not use on other crops used for food or forage. Use only according to label directions. Application at rates above those shown may result in illegal crop residues. Do not graze livestock in treated orchards or groves. Do not treat food crops grown in the greenhouse.

RE-ENTRY STATEMENTS

Do not apply this product in such a manner as to directly or through drift expose workers or other persons. The area being treated must be vacated by unprotected persons. Do not enter treated areas without protective clothing until 48 hours after application. Because certain states may require more restrictive re-entry intervals for various crops treated with this product, consult your State Department of Agriculture for further information.

Written or oral warnings must be given to workers who are expected to be in a treated area or in an area about to be treated with this product. Oral warnings should include those statements found under the label heading "Precautionary Statements" and "Re-entry Statements" included on this label. When oral warnings are given, warnings shall be given in a language customarily understood by workers. Oral warnings must be given if there is reason to believe that written warnings cannot be understood by workers. Written warnings must include the following information: DANGER, area treated with SYSTOX 2 on (date of application). Do not enter without appropriate protective clothing until 48 hours after application. **If on skin,** remove contaminated clothing and wash skin immediately with soap and warm water. **If eyes are contaminated,** wash with flowing water for at least 15 minutes. **If swallowed,** vomiting should be induced. (See Precautionary Statements on label).

STORAGE AND DISPOSAL

PESTICIDE DISPOSAL

Do not contaminate water, food or feed by storage or disposal.

Pesticide wastes are acutely hazardous. Improper disposal of excess pesticide, spray mixture, or rinsate is a violation of Federal Law. If these wastes cannot be disposed of by use according to label instructions, contact your State Pesticide or Environmental Control Agency, or the Hazardous Waste representative at the nearest EPA Regional Office for guidance.

CONTAINER DISPOSAL

Triple rinse (or equivalent). Then offer for recycling or reconditioning, or puncture and dispose of in a sanitary landfill, or by other procedures approved by State and local authorities.

Store in a cool, dry place, away from open flame and extreme heat, and in such a manner as to prevent cross contamination with other pesticides, fertilizers, food, and feed. Store in original container and out of the reach of children, preferably in a locked storage area.

Handle and open container in a manner as to prevent spillage. If container is leaking invert to prevent leakage. If the container is leaking or material spilled for any reason or cause, carefully dam up spilled material to prevent runoff. Refer to Precautionary Statements on label for hazards associated with the handling of this material. Do not walk through spilled material. Absorb spilled material with absorbing type compounds and dispose of as directed for pesticides above. In spill or leak incidents, keep unauthorized people away. You may contact the Mobay Emergency Response Team for decontamination procedures or any other assistance that may be necessary. The Mobay Kansas City Emergency Response Telephone No. is 816-242-2582, or contact Chemtrec 800-424-9300.

IMPORTANT: This bulletin is not intended to provide adequate information for application. Before using the product, read and carefully observe directions, cautionary statements and other information appearing on the product label. This product is sold subject to the Conditions of Sale set forth on the container label.

READ THE LABEL BEFORE USE

DIRECTIONS FOR USE

It is a violation of Federal Law to use this product in a manner inconsistent with its labeling.

SYSTOX 2 emulsifiable systemic insecticide penetrates by absorption and is translocated in the plant. The best time to apply SYSTOX 2 is early morning or evening, especially when crops such as alfalfa and cotton are in bloom. By observing this practice, beneficial pollinating insects, including honey bees, will not be brought into contact with spray, and so not adversely affected.

MIXING: SYSTOX 2 forms an emulsion when diluted with water and is suitable for use in all power-operated ground sprayers and aircraft sprayers. To mix with water, pour the required amount of SYSTOX 2 into full amount of water and then agitate. For ground application, apply in sufficient water for good coverage. For airplane application, use the recommended per acre rate in at least 1 gallon of water.

DOSAGE: Use specified dosage of SYSTOX 2 in the amount of water necessary to give complete coverage of foliage. The type of equipment used will determine the concentration required.

SPRAYING: Work to windward. Protect sprayer operators from drift or mist. When low volumes of spray are applied, complete coverage and thorough application are essential for most effective results. Schedule applications in accordance with local conditions. Consult your State Agricultural Experiment Station or Extension Service for specific use information consistent with label directions and cautions.

RECOMMENDED APPLICATIONS

CROP	PEST	PINTS SYSTOX 2	REMARKS
FIELD CROPS Barley Oats Wheat	Aphids	½ to 1	Apply specified dosage per acre by air or ground equipment in sufficient water for complete coverage, but not less than 1 gallon per acre. Repeat as necessary, but do not apply more than twice per season, nor within 14 days of previous application. Do not apply within 45 days of harvest for grain. Do not graze treated fields.
	Mites	1	
Cotton	Thrips	1	Apply specified dosage per acre by air or ground equipment in sufficient water for complete coverage, but not less than 1 gallon per acre. Repeat as necessary. Do not apply within 21 days of picking. Do not graze dairy or meat animals on treated fields. Do not feed gin waste to livestock.
	Aphids Mites	1½	
Sorghum	Aphids Mites	1	Apply specified dosage per acre by air or ground equipment in sufficient water for complete coverage, but not less than 1 gallon per acre. Do not apply more than once per season nor within 35 days of harvest of grain or of grazing or cutting for forage.
Sugar beets	Aphids Mites	2	Apply specified dosage per acre by air or ground equipment in sufficient water for complete coverage, but not less than 1 gallon per acre. Repeat as necessary, but not more than 3 times per season nor within 30 days of harvest.
Hops	Aphids	1	Apply specified dosage in 100 gallons of water as a full coverage spray* using not more than 200 gallons of dilute spray per acre. Do not apply within 21 days of harvest.

CROP	PEST	PINTS SYSTOX 2	REMARKS
SEED FIELD CROPS Alfalfa Clover	Aphids Mites	1 to 2	Apply specified dosage per acre, using a minimum of 10 gallons of water per acre when applying with ground equipment, or a minimum of 2 gallons per acre for aerial applications. Repeat as necessary up to 15 days before harvest. Chaff from seed crop may be used for feed or forage, however, **do not** cut green crop for these purposes. **NOTE:** Due to spotted alfalfa aphid resistance to organophosphate products in certain areas, SYSTOX used alone may not provide satisfactory control. Consult your local agricultural advisor or Extension Service for recommendations.
FRUITS Apples	Aphids	½ to 2	Apply specified dosage in 100 gallons of water as a full-coverage spray using not more than 1200 gallons of finished spray per acre.* Do not apply the 2 pint rate more than once nor the lower rates more than 3 times per season. Do not apply within 21 days of harvest. Lower rates are effective when used in regular preventive spray programs. To control established mite or aphid populations, use the high rate. †NOTE: SYSTOX used alone may not provide satisfactory control of mites. Consult your local Agricultural Advisor, Extension Service or Mobay representative for recommendation.
	Mites†	¾ to 2	
	Unspotted tentiform leafminer	1	
	White apple leafhopper	½ to 1	
Apricots Nectarines Peaches	Aphids Mites	1	Apply specified dosage in 100 gallons of water as a full-coverage spray* using not more than 800 gallons of finished spray per acre. Do not apply more than 3 times per season nor within 30 days of harvest.
Pears	Aphids Mites	¾	Apply specified dosage in 100 gallons of water as a full-coverage spray* using not more than 900 gallons of finished spray per acre. Do not apply more than 3 times per season nor within 21 days of harvest.
	Unspotted tentiform leafminer	1	
Blackberries Dewberries Gooseberries Loganberries Raspberries	Aphids Leafhoppers Mites	1 to 2	Apply as a post harvest treatment, in the late Summer or Fall months only, for reduction of insect infestation and to permit further plant growth. Use specified dosage in 100 gallons of water as a full coverage spray,* applying not more than 600 gallons dilute spray per acre at the 1-pint rate nor more than 300 gallons per acre at the 2-pint rate.
Grapefruit Lemons Oranges	Aphids Citrus red mite	1	Apply specified dosage in 100 gallons of water as a full-coverage spray* using not more than 1000 gallons of finished spray per acre. Do not apply within 21 days of harvest.
Grapes	Leafhoppers Mites	1½	Apply specified dosage per acre in sufficient water for complete coverage. Where ground conditions dictate an air application, use specified rate in a minimum of 1 gallon of water per acre. Do not apply within 21 days of harvest.
Strawberries	Aphids Mites (Except cyclamen mite)	1½	
Plums Prunes (fresh)	Aphids Mites	1 to 1½	Apply specified dosage in 100 gallons of water as a full-coverage spray* using not more than 800 gallons of finished spray per acre. Do not apply more than three times per season or within 30 days of harvest at the low rate nor more than twice per season or within 60 days of harvest at the high rate.
NON-BEARING FRUITS Stone fruits	Aphids Mites	1 to 2	Apply specified dosage in 100 gallons of water as a full-coverage spray* on non-bearing stone fruits. Repeat as necessary.
Pineapple	Mites	(See Remarks)	For temporary protection of pineapple planting stock, dip stock in emulsion containing 2 pints of SYSTOX 2 per 100 gallons of water. Do not handle treated stock with bare hands.
NUTS Almonds Walnuts	Aphids Mites	3	Apply specified dosage per acre, once a season in sufficient water for complete coverage. When ground conditions indicate an air application, use specified rate in a minimum of 1 gallon of water per acre. May be combined with codling moth spray. Do not apply within 21 days of harvest.
Filberts	Aphids	1	Apply specified dosage in 100 gallons of water as a full-coverage spray,* using not more than 1000 gallons dilute spray per acre on filberts or 400 gallons per acre on pecans. Repeat as necessary. Do not apply within 40 days of harvest for filberts nor 21 days of harvest for pecans.
Pecans	Aphids Mites	1	
VEGETABLES Beans	Aphids Mites	1½ to 2	Apply specified dosage per acre by air or ground equipment in sufficient water for complete coverage, but not less than 1 gallon per acre. Repeat as necessary. Do not apply within 21 days of harvest.
Broccoli Brussels sprouts Cabbage Cauliflower	Aphids	2	Apply specified dosage per acre by air or ground equipment in sufficient water for complete coverage, but not less than 1 gallon per acre. Repeat as necessary. Do not apply within 21 days of harvest. First application may be made at specified dosage per 100 gallons of transplant water.
Celery	Aphids Mites	1	Apply specified dosage in 100 gallons of water as a full-coverage spray* using not more than 200 gallons of dilute spray per acre at the 1-pint rate, nor more than 100 gallons per acre at the 2-pint rate. Do not apply within 21 days of harvest for lettuce and peas nor within 28 days of harvest for celery.
Lettuce	Aphids	1 to 2	
Peas	Aphids	1	

CROP	PEST	PINTS SYSTOX 2	REMARKS
Eggplant	Aphids	1 to $1\frac{1}{2}$	Apply specified dosage per acre by air or ground equipment in sufficient water for complete coverage, but not less than 1 gallon per acre. Do not apply more than twice per season nor within 7 days of harvest.
Peppers	Aphids	$1\frac{1}{2}$	Apply specified dosage per acre by air or ground equipment in sufficient water for complete coverage, but not less than 1 gallon per acre. Repeat as necessary. Do not apply within 3 days of harvest.
Potatoes	Aphids Leafhoppers	$\frac{3}{4}$ to 2	Apply specified dosage per acre by air or ground equipment in sufficient water for complete coverage, but not less than 1 gallon per acre. Repeat as necessary. Do not apply with 21 days of harvest.
Tomatoes	Aphids Mites	1 to $1\frac{1}{2}$	Apply specified dosage in 100 gallons of water as a full-coverage spray,* using not more than 150 gallons of dilute spray per acre at the 1-pint rate nor more than 100 gallons per acre at the $1\frac{1}{2}$-pint rate. Do not apply within 3 days of harvest.
MELONS Muskmelons	Aphids Beet leafhopper Mites	$1\frac{1}{2}$	Apply specified dosage per acre in sufficient water for complete coverage. If band spray applications are used, reduce the specified dosage in proportion to the amount of acreage actually sprayed. Repeat as necessary. Do not apply within 21 days of harvest.
ORNAMENTALS Flowers Shrubs and trees	Aphids Mealybugs Mites Whiteflies	$\frac{1}{2}$ to $\frac{3}{4}$	**FOLIAR SPRAY:** Mix specified dosage in 50 gallons of water and spray plants until they are thoroughly wet. **SOIL DRENCH:** Mix specified dosage in 50 gallons of water to form an emulsion (1 pint SYSTOX 2 per 50 gallons is equivalent to 2 teaspoonsfuls per gallon of water). Apply 1 pint of this emulsion per each square foot of bench or bed area to be treated. For potted plants apply $\frac{1}{4}$ pint of emulsion per 6-inch pot. For treatment of shrubs and trees up to 6 feet in height, apply 1 to 2 gallons of the water emulsion around the base of each plant. Use proportionately more for taller plants. Repeat applications as required. Do not apply within 5 days of hand-picking, pruning, transplanting, or other handling. Caution: Phytotoxicity has occurred on African violets, cibotium ferns, and Croft lilies. Caution should be exercised when treating these plants.
	Birch leafminer Boxwood leafminer Holly leafminer Lace bugs Camellia scale Rhododendron scale Soft brown scale Tea scale	$\frac{3}{4}$ to 1	
Chrysanthemums Daffodils Lilies Primroses	Foliar nematodes *(Ditylenchus spp.)* *(Aphelenchoides spp.)*	1 to 2	Mix specified dosage in 50 gallons of water and spray plants until they are thoroughly wet (1 pint SYSTOX 2 per 50 gallons is equivalent to 2 teaspoonfuls per gallon of water). For use as a soil drench, prepare an emulsion as above and apply 1 pint of the emulsion to each square foot of bench or bed area to be treated. For potted plants, apply $\frac{1}{4}$ pint of the water emulsion per 6-inch pot. Apply foliar sprays or soil drenches at 1 to 2 week intervals. Three applications may be required for maximum control of nematodes. The intervals between applications may be extended to 2 to 3 weeks at the 2 pint rate. Do not apply within 5 days of hand-picking, pruning, transplanting, or other handling. Caution: Phytotoxicity has occurred on Croft lilies. Caution should be exercised when treating these plants. 3-15-85

* This concentration is calculated for conventional hydraulic-type sprayers. When lower volumes of spray are applied per acre with concentrate sprayers, increase the concentration of SYSTOX 2 in the spray mixture in order to apply amount of active ingredient per acre equivalent to a full coverage spray. Where conditions dictate an air application, apply dosage per acre equivalent to a full-coverage ground spray in not less than one gallon of water per acre.

SPECIAL LOCAL NEED REGISTRATIONS [SECTION 24(c)] FOR SYSTOX 2*

CROP	INSECT	STATE EPA SLN NO.	DOSAGE	REMARKS
Alfalfa (Forage crop)	Aphids	Arizona AZ 78-0029 California CA 77-0093 Colorado CO 81-0001	$\frac{1}{2}$ pint per acre.	Application by ground or air.
		Idaho ID 79-0014		Summer crop only. Application by ground or air.
		Montana MT 77-0004 Nevada NV 77-0008 Oregon OR 77-0051		Application by ground or air.
		Utah UT 79-0003		Summer crop only. Application by ground or air.
		Washington WA 77-0022		Application by ground or air.

* For distribution and use only within the states specified. These Special Local Need registrations approved as of date of this publication are, however, subject to cancellation by the state. Contact your nearest Mobay Agricultural Chemicals Division Office for the current status and a copy of the complete individual supplemental Special Local Need labeling. This labeling must be in the possession of the user at time of pesticide application.

®SYSTOX 6

EMULSIFIABLE SYSTEMIC INSECTICIDE

RESTRICTED USE PESTICIDE

For retail sale to and use only by Certified Applicators or persons under their direct supervision, and only for those uses covered by the Certified Applicator's certification.

For Control of Aphids, Mites and Leafhoppers on a Wide Variety of Crops.

ACTIVE INGREDIENTS:

Demeton	
O,O-Diethyl *O*-[2-(ethylthio) ethyl] phosphorothioate mixture with *O,O*-Diethyl *S*-[2-(ethylthio) ethyl] phosphorothioate	66%
Aromatic Petroleum Distillate	26%
INERT INGREDIENTS	8%
TOTAL	100%

Contains 6 lbs. Demeton per gallon.

EPA Reg. No. 3125-221-ZA

SYSTOX is a Reg. TM of the Parent Company of Farbenfabriken Bayer GmbH, Leverkusen.

DANGER–POISON

Keep Out of Reach of Children

PELIGRO

PRECAUCION AL USUARIO: Si usted no puede leer o entender inglés, no use este producto hasta que la etiqueta le haya sido explicada ampliamente.

(TO THE USER: If you cannot read or understand English, do not use this product until the label has been fully explained to you.)

RECOMMENDED USES

Almonds	Nectarines
Apples	Oats
Apricots	Oranges
Beans	Ornamentals
Barley	Peas
Blackberries	Peaches
Broccoli	Pears
Brussels sprouts	Pecans
Cabbage	Peppers
Cauliflower	Pineapples*
Celery	Plums
Cotton	Potatoes
Dewberries	Prunes (fresh)
Eggplant	Raspberries
Filberts	Seed alfalfa
Gooseberries	Seed clover
Grapefruit	Sorghum
Grapes	Stone fruits*
Hops	Strawberries
Lemons	Sugar beets
Lettuce	Tomatoes
Loganberries	Walnuts
Muskmelons	Wheat

* Non-bearing

RESTRICTIONS ON WAREHOUSE STORAGE

Do not store adjacent to or near foodstuffs, drugs, feeds or any other material intended for consumption by humans or animals. Do not store adjacent to toys, cosmetics, dishes, furniture or clothing. Do not store next to herbicides. Do not use, pour, spill or store near heat or open flame. Do not store in or around the home.

STATEMENTS OF PRACTICAL TREATMENT

In case of poisoning call a physician immediately. Have patient lie down and keep quiet.

If swallowed, vomiting should be induced. Administer water freely, and induce vomiting by giving one dose (½ oz. or 15 ml.) of syrup of ipecac. If vomiting does not occur within 10 to 20 minutes, administer second dose. If syrup of ipecac is not available, induce vomiting by sticking finger down throat. Repeat until vomit fluid is clear. Never give anything by mouth to an unconscious person. Professional medical assistance should be secured immediately.

If on skin, remove contaminated clothing and wash skin immediately with soap and warm water.

If eyes are contaminated, wash with flowing water for at least 15 minutes.

TO PHYSICIAN: ANTIDOTE—Administer atropine sulfate in large therapeutic doses. Repeat as necessary to the point of tolerance. 2-PAM is also antidotal and may be administered in conjunction with atropine.

Compound inhibits cholinesterase resulting in stimulation of the central nervous system, the parasympathetic nervous system, and the somatic motor nerves. Do not give morphine. Watch for pulmonary edema, which may develop in serious cases of poisoning even after 12 hours. At first sign of pulmonary edema, the patient should be placed in an oxygen tent and treated symptomatically.

PRECAUTIONARY STATEMENTS

HAZARDS TO HUMANS AND DOMESTIC ANIMALS

Poisonous if swallowed, inhaled, or absorbed through the skin. Rapidly absorbed through the skin. Do not get in eyes or on skin. Wear protective clothing, natural rubber gloves and goggles.

Do not breathe fumes or spray mist. For protection against demeton insecticide, wear a mask or respirator jointly approved by the Mining Enforcement and Safety Administration (formerly the U.S. Bureau of Mines) and by the National Institute for Occupational Safety and Health under the provisions of 30 CFR Part 11.

Do not contaminate or store near feed or foodstuffs. Keep out of reach of children. Keep all unprotected persons out of the operating area or vicinity where there may be danger of drift. Wash hands, arms and face thoroughly with soap and warm water before eating or smoking.

HANDLING OF CONTAMINATED ARTICLES: Clothing and rags receiving undiluted SYSTOX 6 as a result of leaks, spills or mishaps should be removed immediately and not reused. These contaminated articles should be buried in the same manner as the empty container (see container disposal section for directions). Contaminated articles which should be discarded include: hats, gloves, aprons, coats, boots, etc. not made of rubber or coated with rubber or other similar materials. Clothing receiving spray mist or droplets from the mixed, diluted SYSTOX 6 should be removed and decontaminated before reuse. This clothing can be decontaminated by machine washing separately from other items with soap or detergent and bleach in hot water twice before reuse. Protective clothing and equipment should be washed down with detergent and soap and bleach in water. Wash water from the cleaning of protec-

CROP	PEST	PINTS SYSTOX 2	REMARKS
Eggplant	Aphids	1 to $1\frac{1}{2}$	Apply specified dosage per acre by air or ground equipment in sufficient water for complete coverage, but not less than 1 gallon per acre. Do not apply more than twice per season nor within 7 days of harvest.
Peppers	Aphids	$1\frac{1}{2}$	Apply specified dosage per acre by air or ground equipment in sufficient water for complete coverage, but not less than 1 gallon per acre. Repeat as necessary. Do not apply within 3 days of harvest.
Potatoes	Aphids Leafhoppers	$\frac{3}{4}$ to 2	Apply specified dosage per acre by air or ground equipment in sufficient water for complete coverage, but not less than 1 gallon per acre. Repeat as necessary. Do not apply with 21 days of harvest.
Tomatoes	Aphids Mites	1 to $1\frac{1}{2}$	Apply specified dosage in 100 gallons of water as a full-coverage spray,* using not more than 150 gallons of dilute spray per acre at the 1-pint rate nor more than 100 gallons per acre at the $1\frac{1}{2}$-pint rate. Do not apply within 3 days of harvest.
MELONS Muskmelons	Aphids Beet leafhopper Mites	$1\frac{1}{2}$	Apply specified dosage per acre in sufficient water for complete coverage. If band spray applications are used, reduce the specified dosage in proportion to the amount of acreage actually sprayed. Repeat as necessary. Do not apply within 21 days of harvest.
ORNAMENTALS Flowers Shrubs and trees	Aphids Mealybugs Mites Whiteflies	$\frac{1}{2}$ to $\frac{3}{4}$	**FOLIAR SPRAY:** Mix specified dosage in 50 gallons of water and spray plants until they are thoroughly wet. **SOIL DRENCH:** Mix specified dosage in 50 gallons of water to form an emulsion (1 pint SYSTOX 2 per 50 gallons is equivalent to 2 teaspoonsfuls per gallon of water). Apply 1 pint of this emulsion per each square foot of bench or bed area to be treated. For potted plants apply $\frac{1}{4}$ pint of emulsion per 6-inch pot. For treatment of shrubs and trees up to 6 feet in height, apply 1 to 2 gallons of the water emulsion around the base of each plant. Use proportionately more for taller plants. Repeat applications as required. Do not apply within 5 days of hand-picking, pruning, transplanting, or other handling. Caution: Phytotoxicity has occurred on African violets, cibotium ferns, and Croft lilies. Caution should be exercised when treating these plants.
	Birch leafminer Boxwood leafminer Holly leafminer Lace bugs Camellia scale Rhododendron scale Soft brown scale Tea scale	$\frac{3}{4}$ to 1	
Chrysanthemums Daffodils Lilies Primroses	Foliar nematodes *(Ditylenchus spp.)* *(Aphelenchoides spp.)*	1 to 2	Mix specified dosage in 50 gallons of water and spray plants until they are thoroughly wet (1 pint SYSTOX 2 per 50 gallons is equivalent to 2 teaspoonfuls per gallon of water). For use as a soil drench, prepare an emulsion as above and apply 1 pint of the emulsion to each square foot of bench or bed area to be treated. For potted plants, apply $\frac{1}{4}$ pint of the water emulsion per 6-inch pot. Apply foliar sprays or soil drenches at 1 to 2 week intervals. Three applications may be required for maximum control of nematodes. The intervals between applications may be extended to 2 to 3 weeks at the 2 pint rate. Do not apply within 5 days of hand-picking, pruning, transplanting, or other handling. Caution: Phytotoxicity has occurred on Croft lilies. Caution should be exercised when treating these plants. 3-15-05

* This concentration is calculated for conventional hydraulic-type sprayers. When lower volumes of spray are applied per acre with concentrate sprayers, increase the concentration of SYSTOX 2 in the spray mixture in order to apply amount of active ingredient per acre equivalent to a full coverage spray. Where conditions dictate an air application, apply dosage per acre equivalent to a full-coverage ground spray in not less than one gallon of water per acre.

SPECIAL LOCAL NEED REGISTRATIONS [SECTION 24(c)] FOR SYSTOX 2*

CROP	INSECT	STATE EPA SLN NO.	DOSAGE	REMARKS
Alfalfa (Forage crop)	Aphids	Arizona AZ 78-0029 California CA 77-0093 Colorado CO 81-0001	$\frac{1}{2}$ pint per acre.	Application by ground or air.
		Idaho ID 79-0014		Summer crop only. Application by ground or air.
		Montana MT 77-0004 Nevada NV 77-0008 Oregon OR 77-0051		Application by ground or air.
		Utah UT 79-0003		Summer crop only. Application by ground or air.
		Washington WA 77-0022		Application by ground or air.

* For distribution and use only within the states specified. These Special Local Need registrations approved as of date of this publication are, however, subject to cancellation by the state. Contact your nearest Mobay Agricultural Chemicals Division Office for the current status and a copy of the complete individual supplemental Special Local Need labeling. This labeling must be in the possession of the user at time of pesticide application.

®SYSTOX 6

EMULSIFIABLE SYSTEMIC INSECTICIDE

RESTRICTED USE PESTICIDE

For retail sale to and use only by Certified Applicators or persons under their direct supervision, and only for those uses covered by the Certified Applicator's certification.

For Control of Aphids, Mites and Leafhoppers on a Wide Variety of Crops.

ACTIVE INGREDIENTS:
Demeton
O,O-Diethyl *O*-[2-(ethylthio) ethyl] phosphorothioate mixture with
O,O-Diethyl *S*-[2-(ethylthio) ethyl] phosphorothioate 66%
Aromatic Petroleum Distillate 26%
INERT INGREDIENTS 8%

TOTAL 100%

Contains 6 lbs. Demeton per gallon.

EPA Reg. No. 3125-221-ZA

SYSTOX is a Reg. TM of the Parent Company of Farbenfabriken Bayer GmbH, Leverkusen.

DANGER–POISON

Keep Out of Reach of Children

PELIGRO

PRECAUCION AL USUARIO: Si usted no puede leer o entender inglés, no use este producto hasta que la etiqueta le haya sido explicada ampliamente.

(TO THE USER: If you cannot read or understand English, do not use this product until the label has been fully explained to you.)

RECOMMENDED USES

Almonds
Apples
Apricots
Beans
Barley
Blackberries
Broccoli
Brussels sprouts
Cabbage
Cauliflower
Celery
Cotton
Dewberries
Eggplant
Filberts
Gooseberries
Grapefruit
Grapes
Hops
Lemons
Lettuce
Loganberries
Muskmelons
Nectarines
Oats
Oranges
Ornamentals
Peas
Peaches
Pears
Pecans
Peppers
Pineapples*
Plums
Potatoes
Prunes (fresh)
Raspberries
Seed alfalfa
Seed clover
Sorghum
Stone fruits*
Strawberries
Sugar beets
Tomatoes
Walnuts
Wheat

* Non-bearing

RESTRICTIONS ON WAREHOUSE STORAGE

Do not store adjacent to or near foodstuffs, drugs, feeds or any other material intended for consumption by humans or animals. Do not store adjacent to toys, cosmetics, dishes, furniture or clothing. Do not store next to herbicides. Do not use, pour, spill or store near heat or open flame. Do not store in or around the home.

STATEMENTS OF PRACTICAL TREATMENT

In case of poisoning call a physician immediately. Have patient lie down and keep quiet.

If swallowed, vomiting should be induced. Administer water freely, and induce vomiting by giving one dose (½ oz. or 15 ml.) of syrup of ipecac. If vomiting does not occur within 10 to 20 minutes, administer second dose. If syrup of ipecac is not available, induce vomiting by sticking finger down throat. Repeat until vomit fluid is clear. Never give anything by mouth to an unconscious person. Professional medical assistance should be secured immediately.

If on skin, remove contaminated clothing and wash skin immediately with soap and warm water.

If eyes are contaminated, wash with flowing water for at least 15 minutes.

TO PHYSICIAN: ANTIDOTE—Administer atropine sulfate in large therapeutic doses. Repeat as necessary to the point of tolerance. 2-PAM is also antidotal and may be administered in conjunction with atropine.

Compound inhibits cholinesterase resulting in stimulation of the central nervous system, the parasympathetic nervous system, and the somatic motor nerves. Do not give morphine. Watch for pulmonary edema, which may develop in serious cases of poisoning even after 12 hours. At first sign of pulmonary edema, the patient should be placed in an oxygen tent and treated symptomatically.

PRECAUTIONARY STATEMENTS

HAZARDS TO HUMANS AND DOMESTIC ANIMALS

Poisonous if swallowed, inhaled, or absorbed through the skin. Rapidly absorbed through the skin. Do not get in eyes or on skin. Wear protective clothing, natural rubber gloves and goggles.

Do not breathe fumes or spray mist. For protection against demeton insecticide, wear a mask or respirator jointly approved by the Mining Enforcement and Safety Administration (formerly the U.S. Bureau of Mines) and by the National Institute for Occupational Safety and Health under the provisions of 30 CFR Part 11.

Do not contaminate or store near feed or foodstuffs. Keep out of reach of children. Keep all unprotected persons out of the operating area or vicinity where there may be danger of drift. Wash hands, arms and face thoroughly with soap and warm water before eating or smoking.

HANDLING OF CONTAMINATED ARTICLES: Clothing and rags receiving undiluted SYSTOX 6 as a result of leaks, spills or mishaps should be removed immediately and not reused. These contaminated articles should be buried in the same manner as the empty container (see container disposal section for directions). Contaminated articles which should be discarded include: hats, gloves, aprons, coats, boots, etc. not made of rubber or coated with rubber or other similar materials. Clothing receiving spray mist or droplets from the mixed, diluted SYSTOX 6 should be removed and decontaminated before reuse. This clothing can be decontaminated by machine washing separately from other items with soap or detergent and bleach in hot water twice before reuse. Protective clothing and equipment should be washed down with detergent and soap and bleach in water. Wash water from the cleaning of protec-

CROP	INSECT	PINTS SYSTOX 6	REMARKS
VEGE-TABLES Beans	Aphids Mites	½ to ⅔	Apply specified dosage per acre by air or ground equipment in sufficient water for complete coverage, but not less than 1 gallon per acre. Repeat as necessary. Do not apply within 21 days of harvest.
Broccoli Brussels sprouts Cabbage Cauliflower	Aphids	⅔	Apply specified dosage per acre by air or ground equipment in sufficient water for complete coverage, but not less than 1 gallon per acre. Repeat as necessary. Do not apply within 21 days of harvest. First application may be made at specified dosage per 100 gallons of transplant water.
Celery	Aphids Mites	⅓	Apply specified dosage in 100 gallons of water as a full-coverage spray* using not more than 200 gallons of dilute spray per acre at the ⅓ pint rate, nor more than 100 gallons per acre at the ⅔ pint rate. Do not apply within 21 days of harvest for lettuce and peas nor within 28 days of harvest for celery.
Lettuce	Aphids	⅓ to ⅔	
Peas	Aphids	⅓	
Eggplant	Aphids	⅓ to ½	Apply specified dosage per acre by air or ground equipment in sufficient water for complete coverage, but not less than 1 gallon per acre. Do not apply more than twice per season nor within 7 days of harvest.
Peppers	Aphids	½	Apply specified dosage per acre by air or ground equipment in sufficient water for complete coverage, but not less than 1 gallon per acre. Repeat as necessary. Do not apply within 3 days of harvest.
Potatoes	Aphids Leafhoppers	¼ to ⅔	Apply specified dosage per acre by air or ground equipment in sufficient water for complete coverage, but not less than 1 gallon per acre. Repeat as necessary. Do not apply within 21 days of harvest.
Tomatoes	Aphids Mites	⅓ to ½	Apply specified dosage in 100 gallons of water as a full-coverage spray,* using not more than 150 gallons of dilute spray per acre at the ⅓ pint rate nor more than 100 gallons per acre at the ½ pint rate. Do not apply within 3 days of harvest.
MELONS Muskmelons	Aphids Beet leafhoppers Mites	½	Apply specified dosage per acre in sufficient water for complete coverage, but not less than 1 gallon per acre. If band spray applications are used, reduce the specified dosage in proportion to the amount of acreage actually sprayed. Repeat as necessary. Do not apply within 21 days of harvest.

* This concentration is calculated for conventional hydraulic-type sprayers. When lower volumes of spray are applied per acre with concentrate sprayers, increase the concentration SYSTOX 6 in the spray mixture in order to apply amount of active ingredient per acre equivalent to a full-coverage spray. Where conditions dictate an air application, apply dosage per acre equivalent to a full-coverage ground spray in not less than one gallon of water per acre.

ORNAMENTALS: FOR USE BY COMMERCIAL APPLICATORS AND COMMERCIAL GROWERS ONLY

CROP	INSECT	PINTS SYSTOX 6	REMARKS
Flowers Shrubs and trees	Aphids Mealybugs Mites Whiteflies	⅙ to ¼	**FOLIAR SPRAY:** Mix specified dosage in 50 gallons of water and spray plants until they are thoroughly wet. **SOIL DRENCH:** Mix specified dosage in 50 gallons of water to form an emulsion (⅓ pint SYSTOX 6 per 50 gallons is equivalent to ⅔ teaspoonful per gallon of water). Apply 1 pint of this emulsion per each square foot of bench or bed area to be treated. For potted plants apply ¼ pint of emulsion per 6 inch pot. For treatment of shrubs and trees up to 6 feet in height, apply 1 to 2 gallons of the water emulsion around the base of each plant. Use proportionately more for taller plants. Repeat applications as required. Do not apply within 5 days of hand-picking, pruning, transplanting, or other handling. Caution: Phytotoxicity has occurred on African violets, cibotium ferns and Croft lilies. Caution should be exercised when treating these plants.
	Birch leafminer Boxwood leafminer Holly leafminer Lace bugs	¼ to ⅓	
	Camellia scale Rhododendron scale Soft brown scale Tea scale	⅓	
Chrysanthemums Daffodils Lilies Primroses	Foliar nematodes (*Ditylenchus* sp.) (*Aphelenchoides* sp.)	⅓ to ⅔	Mix specified dosage in 50 gallons of water and spray plants until they are thoroughly wet (⅓ pint SYSTOX 6 per 50 gallons is equivalent to ⅔ teaspoonful per gallon of water). For use as a soil drench, prepare an emulsion as above and apply 1 pint of the emulsion to each square foot of bench or bed area to be treated. For potted plants, apply ¼ pint of the water emulsion per 6 inch pot. Apply foliar sprays or soil drenches at 1 to 2 week intervals. Three applications may be required for maximum control of nematodes. The intervals between applications may be extended to 2 to 3 weeks at the ⅔ pint rate. Do not apply within 5 days of hand-picking, pruning, transplanting, or other handling. Caution: Phytotoxicity has occurred on Croft lilies. Caution should be exercised when treating these plants. 3–15–85

SPECIAL LOCAL NEED REGISTRATIONS [SECTION 24(c)] FOR SYSTOX 6*

CROP	INSECT	STATE EPA SLN NO.	DOSAGE	REMARKS
Alfalfa (Forage crop)	Aphids	Arizona AZ 78-0030	$2\frac{2}{3}$ ounces per acre.	Application by ground or air.
		California CA 77-0094		
		Colorado CO 81-0002		
		Idaho ID 79-0013		Summer crop only. Application by ground or air.
		Montana MT 77-0005		Application by ground or air.
		Nevada NV 77-0007		
		Oregon OR 77-0050		
		Utah UT 79-0002		Summer crop only. Application by ground or air.
		Washington WA 77-0021		Application by ground or air.

* For distribution and use only within the state specified. These Special Local Need registrations approved as of date of this publication are, however, subject to cancellation by the state. Contact your nearest Mobay Agricultural Chemicals Division Office for the current status and a copy of the complete individual supplemental Special Local Need labeling. This labeling must be in the possession of the user at time of pesticide application.

Monsanto Company

Agricultural Products
800 North Lindbergh Boulevard
St. Louis, MO 63167
(314) 694-1000

AVADEX®
HERBICIDE

RESTRICTED USE PESTICIDE

For retail sale to and use only by Certified Applicators or persons under their direct supervision and only for those uses covered by the Certified Applicator's certification.

Emulsifiable Herbicide for Fall or Spring Soil Treatment to Control Wild Oats in Certain Spring Planted Crops.

ACTIVE INGREDIENT:	
*S-(2,3-dichloroallyl) diisopropylthiocarbamate	45.7%
INERT INGREDIENTS:	54.3%
TOTAL	100.0%

*Contains 4 pounds S-(2,3-dichloroallyl) diisopropylthiocarbamate per gallon.

EPA Reg. No. 524-119-AA EPA Est. 37507-MT-1
84351.00024T

In case of an emergency involving this product, Call Collect, day or night, (314) 694–4000.

WARNING
Keep Out of Reach of Children
Read the entire label before using this product.

Use only according to label instructions.

Read "LIMIT OF WARRANTY AND LIABILITY" before buying or using. If terms are not acceptable, return at once unopened.

CAUSES EYE AND SKIN IRRITATION.

HARMFUL IF SWALLOWED.

Do not get in eyes, on skin or clothing.

When handling or spraying wear rubber gloves, goggles, long trousers and long sleeved shirt or jacket of closely knit material along with leather or rubber boots high enough to cover the ankle.

Do not use or store near heat or flame.

FIRST AID:

IF IN EYES, immediately flush with plenty of water for at least 15 minutes. Call a physician.

IF ON SKIN, wash with plenty of soap and water. Remove contaminated clothing. Wash clothing before reuse.

LIMIT OF WARRANTY AND LIABILITY

This company warrants that this product conforms to the chemical description on the label and is reasonably fit for the purposes set forth in the Complete Directions for Use ("Directions"), when used in accordance with those Directions under the conditions described therein. NO OTHER EXPRESS OR IMPLIED WARRANTY OF FITNESS FOR PARTICULAR PURPOSE OR MERCHANTABILITY IS MADE. This warranty is also subject to the conditions and limitations stated herein.

Buyer and all users shall promptly notify this company of any claims whether based in contract, negligence, strict liability, other tort or otherwise.

Buyer and all users are responsible for all loss or damage from use or handling which results from conditions beyond the control of this company, including, but not limited to, incompatibility with products other than those set forth in the Directions, unusual weather (i. weather conditions which are outside the range considered normal at the application site and for the time period when the product is applied with the normal range being determined on the basis of the average range for the prior 40 years computed from the best available information, and ii. weather perils, including but not limited to hurricanes, tornadoes and floods) as well as weather considerations set forth in the Directions, application in any manner not explicitly set forth in the Directions, moisture conditions outside the moisture range specified in the Directions, or the presence of products other than those set forth in the Directions in or on the soil or crop.

THE EXCLUSIVE REMEDY OF THE USER OR BUYER, AND THE LIMIT OF THE LIABILITY OF THIS COMPANY OR ANY OTHER SELLER FOR ANY AND ALL LOSSES, INJURIES OR DAMAGES RESULTING FROM THE USE OR HANDLING OF THIS PRODUCT (INCLUDING CLAIMS BASED IN CONTRACT, NEGLIGENCE, STRICT LIABILITY, OTHER TORT OR OTHERWISE) SHALL BE THE PURCHASE PRICE PAID BY THE USER OR BUYER FOR THE QUANTITY OF THIS PRODUCT INVOLVED, OR, AT THE ELECTION OF THIS COMPANY OR ANY OTHER SELLER, THE REPLACEMENT OF SUCH QUANTITY OR, IF NOT ACQUIRED BY PURCHASE, REPLACEMENT OF SUCH QUANTITY. IN NO EVENT SHALL THIS COMPANY OR ANY OTHER SELLER BE LIABLE FOR ANY INCIDENTAL OR CONSEQUENTIAL DAMAGES.

The buyer and all users are deemed to have accepted the terms of this LIMIT OF WARRANTY AND LIABILITY which may not be varied by any verbal or written agreement.

NOT FOR REFORMULATION OR REPACKAGING.

STORAGE AND DISPOSAL

Do not contaminate water, food or feed by storage and disposal.

STORAGE

STORE ABOVE 32° F TO KEEP FROM FREEZING.

Freezing will result in crystals which settle to the bottom of the can. If allowed to freeze, place in a warm room and roll and/or shake the container frequently for several days to redissolve before using.

DISPOSAL

Wastes of this pesticide may cause eye and skin irritation and may be dangerous. Improper disposal of excess pesticide, spray mixtures, or rinsate is a violation of Federal Law. If these wastes cannot be disposed of according to label use instructions, contact your State Pesticide or Environmental Control Agency, or the Hazardous Waste representative at the nearest EPA Regional Office for guidance.

Empty container retains vapor and product residue. Observe all labeled safeguards until container is cleaned, reconditioned or destroyed. DO NOT CUT OR WELD ON OR NEAR THIS CONTAINER.

Triple rinse container. Then offer for recycling or reconditioning, or puncture and dispose of in a sanitary landfill, or by other procedures approved by state and local authorities.

ENVIRONMENTAL HAZARDS

This product is toxic to fish and wildlife. Keep out of lakes, ponds and streams. Do not contaminate water by cleaning of equipment or disposal of wastes.

PHYSICAL OR CHEMICAL HAZARD

COMBUSTIBLE

Do not use or store near heat or flame.

Use only with adequate ventilation.

In case of:
FIRE, use water spray, foam, dry chemical or CO_2.
SPILL or LEAK, flush area with water spray.

HOW TO MIX AND APPLY AVADEX

Dilute AVADEX with 5 or more gallons of water per acre for an overall treatment. Higher volumes approaching 10 gallons per acre will usually result in more uniform spraying, less loss of AVADEX by evaporation, and slightly better weed control.

Spraying during winds above 10 miles per hour will reduce the amount of AVADEX below the amount required for good weed control.

For best mixing of AVADEX with water, proceed as follows:

1. Fill the tank approximately three-quarters full with water.
2. Carefully mix while adding the proper amount of AVADEX.
3. Fill the remaining portion of the tank with water.
4. Mix thoroughly again if the solution has been left in the tank for any length of time.
5. The sprayer by-pass valve should be in operation prior to mixing and through time of application.

Promptly change clothing which becomes wet by accidental spillage of AVADEX or drift from spray. Wash clothing before reuse.

Broadcast Application—Before spraying, check all hose connections to see that they are properly secured. Adjust the height of the spray boom in the field to ensure an even distribution of the chemical on the soil. Avoid overlapping of the spray. Pressure should not be above 30 p.s.i. and just high enough to provide a uniform spray pattern. Boomless sprayers should not be used because distribution of AVADEX will be too uneven.

A field marker or chain should be attached to both ends of the sprayer boom to serve as a marker, thus avoiding overlapping or untreated strips. Overlapping will double the recommended rate and crop injury may result.

Aerial applications should not be made on hilly fields as this results in uneven distribution which can cause crop injury or poor weed control. Care should be taken to use recommended volumes of water and to avoid delays in incorporation of AVADEX into the soil.

Band Application—The amount needed for various band width and row spacing can be calculated as follows:

$$\frac{\text{Band width}}{\text{Inches between crop rows}} \times \text{Recommended overall rate in quarts} = \text{Quarts of AVADEX needed per acre for band treatment}$$

For example, if 2 quarts per acre are recommended for overall treatment and you wish to treat a band 8 inches wide in crop rows spaced 24 inches apart:

$\frac{8}{24} \times 2 = \frac{2}{3}$ quarts of AVADEX needed per acre of crop for the band treatment.

For best results, mix the calculated rate of AVADEX in from 5 to 10 gallons of water per acre for band applications.

GENERAL AVADEX INFORMATION

Keep AVADEX container closed to prevent spills, causing contamination of other materials and to prevent evaporation which will increase the strength of AVADEX.

AVADEX is recommended for use in the crops barley, corn (grain and silage), flax, forage legumes, lentils, peas, potatoes, soybeans and sugar beets only. Other crops should not be treated with AVADEX because injury may occur. Spraying equipment must be properly calibrated—application of too much AVADEX may injure the crop—application of too little AVADEX may result in poor weed control.

AVADEX should be sprayed on the soil before weeds germinate. If weeds have emerged, tillage prior to or following spraying must be relied upon for their control.

Before applying AVADEX, be sure the soil is in good working condition. This applies particularly to clay soils.

AVOID APPLICATION TO A FIELD LEFT IN A RIDGED CONDITION. Application to a field which is wet, lumpy, rough or ridged will result in reduced weed control and increased crop thinning.

The incorporation of the chemical into the soil must be completed as soon as possible on the day of spraying. The incorporation should not be deeper than 2 inches because deeper incorporation tends to dilute the chemical, decreasing weed control and increasing the risk of crop thinning.

When summerfallow or plowed ground is being treated, only a single disking (incorporation) is required. If stubble ground with a heavy trash cover is being treated, or if the soil is wet or lumpy (making proper mixing difficult), a second tillage operation either before or after spraying is recommended.

Recommended rates should be followed exactly in order to: 1. Avoid crop injury. 2. Control wild oats.

Wild oat control may be evaluated by removing the surface inch or two of soil at the time of germination to inspect the number of wild oats that were killed before emergence. Wild oats are usually killed before emergence, but occasionally and particularly under dry conditions, plants may emerge before they die.

NOTE

Domestic oats should not be seeded where AVADEX was used at 1½ quarts per acre or more the previous year. There is a possibility that AVADEX may carry over sufficiently to injure this crop.

Under conditions of prolonged high temperature at the time of weed germination, or drought in spring, AVADEX may not maintain the usual high standard of control.

Seeding should be done either before or after spraying and incorporation, depending upon the crop that is to be sown. If seeding is delayed, shallow reworking of the treated area prior to seeding will not destroy the effects of the chemical.

Rates of application given on this label are general. Consult the Agricultural Experiment Station or Extension Service Weed Specialists for specific recommendations regarding dosage and method of application. Use only in the manner recommended on the label to avoid possible injury to persons, animals or crops, or other unintended consequences.

DIRECTIONS FOR USE

It is a violation of Federal Law to use this product in any manner inconsistent with its labeling.

Before using, be sure to read and understand the "Warning", "General AVADEX Information" and "How to Mix and Apply AVADEX" sections of this label.

FALL TREATMENT

For fall use on spring planted sugar beets, flax and barley only.

When used in the fall, AVADEX should be applied within 2 to 3 weeks of soil freeze-up. Applications made more than 3 weeks before freeze-up may result in reduced effectiveness of the chemical.

1. Apply the recommended rate of AVADEX shown under "Specific Crop Use" section. (See "How to Mix and Apply AVADEX" section.)
2. On summerfallow ground, which is loose and free of lumps and trash, incorporate AVADEX to a depth of 2 inches or less with a disk plus harrow or 2 harrowings at right angles as soon as possible after spraying. On stubble or lumpy ground, disk before AVADEX application and double disk at right angles to a depth of 2 inches or less as soon as possible after spraying. (See "General AVADEX Information" and "Specific Crop Use" sections.)
3. If soil must be ridged after incorporating AVADEX to prevent soil erosion by high winds, the depth of ridging should be kept to a minimum. Deep ridging may decrease the effectiveness of the chemical or result in crop injury in the spring.
4. In preparing the AVADEX-treated area for seeding in the spring, care must be taken to avoid working the soil any deeper than 2 inches. Deep tillage, greater than 2 inches in depth, tends to dilute the chemical, decreasing the wild oat control and increasing the risk of crop injury.

SPRING TREATMENT

For use on barley, corn, flax, forage legumes, lentils, peas, potatoes, soybeans and sugar beets only.

AVADEX can be applied and incorporated either before seeding or after seeding on certain crops in the spring. (See "Specific Crop Use" section.) When AVADEX is applied before seeding in the spring, do not exceed 3 weeks between the time of application and crop planting. When application is made after planting of the crop, apply and incorporate AVADEX before crop seeds begin to sprout, generally within 7 days after planting.

Before Seeding Incorporation—Incorporate with a disk-type implement followed by a harrow. Two diskings at right angles may be required under heavy or trashy soil conditions. Under light and easily worked soil conditions, 2 harrowings at right angles may be sufficient.

After Seeding Incorporation—Drill crop seed to a depth of 2 to 3 inches when possible. Apply the recommended rate of AVADEX and incorporate with a harrow only. Two harrowings at right angles will be necessary. This method of incorporation is not effective if compact soil prevents penetration of harrow teeth, or trash accumulates in the harrow sections.

Band Application Incorporation—Use hooded power-driven rotary tillers or ground-driven tillers. Consult a local agricultural authority for specific recommendations on suitable band application and incorporation equipment for use under specific soil and weather conditions.

SPECIFIC CROP USE

BARLEY—Application of AVADEX at 1¼ quarts per acre is recommended after seeding of barley for the control of wild oats. Seed must be at a depth of at least 2 inches to provide a safety zone of untreated soil above the seed. Application and incorporation must be done before kernels sprout to prevent crop injury. Incorporate with a double harrowing at right angles immediately after application.

CORN (North Dakota, Minnesota and Montana only)—Apply and incorporate AVADEX at the rate of 1½ quarts per acre before or after planting for the control of wild oats. For band application, reduce the amount used per acre in proportion to the area treated. Calculate this amount according to the formula given on this label.

FLAX—Apply and incorporate Avadex at the rate of 1½ to 2 quarts per acre before or after seeding for the control of wild oats. Under favorable spring moisture conditions, moderate control of green foxtail and wild millet will be obtained with an application of 2 quarts per acre of AVADEX.

FORAGE LEGUMES (Alfalfa, sweet, red and alsike clovers) — Apply and incorporate AVADEX at the rate of 1½ to 2 quarts per acre before seeding for the control of wild oats, and for the control of henbit in alfalfa in Oklahoma only.

Where forage legumes are underseeded with flax, use 1½ to 2 quarts per acre before seeding for the control of wild oats. Where legumes are underseeded with barley, DO NOT EXCEED an application of 1¼ quarts per acre after planting of the crop.

PEAS AND LENTILS—The application and incorporation of AVADEX at 1¼ quarts per acre for peas and 1½ quarts per acre for lentils can be made up to 3 weeks before planting or immediately after planting into a well prepared seed bed for the control of wild oats. Leaf crinkling and delayed maturity of peas may occur on clay points in the Northwest. This is usually more than offset by increased yield resulting from the control of wild oats.

POTATOES (North Dakota, Minnesota and Montana only)—Apply and incorporate AVADEX at the rate of 1½ to 2 quarts per acre before planting for the control of wild oats.

SOYBEANS (North Dakota and Minnesota only)—Apply and incorporate AVADEX at the rate of 1½ to 2 quarts per acre before or after planting for the control of wild oats. For band application, reduce the amount used per acre in proportion of the area treated. Calculate this amount according to the formula given on this label.

SUGAR BEETS—Apply and incorporate AVADEX before seeding. For wild oat control, 1½ to 2 quarts per acre overall is recommended for both light and heavy clay loam soils. For band application, reduce the amount per acre in proportion to the area treated. Calculate this amount according to the formula given on this label.

NOTE—When AVADEX is used in combination with other registered sugar beet herbicides to control weeds other than wild oats, DO NOT EXCEED a broadcast application rate of 2 quarts of AVADEX per acre on heavy soils or 1½ quarts of AVADEX per acre on light sandy soils.

AVADEX CROP USE SUMMARY

Use this table in combination with detailed label instructions shown for specific crops.

Crop	Recommended Rate (quarts per acre)		When to Use
Barley	1¼	Fall— Spring—	Before Seeding After Seeding
Corn	1½	Spring—	Before or After Seeding
Flax	1½—2	Fall— Spring—	Before Seeding Before or After Seeding
Forage Legumes (alfalfa; sweet, red, alsike clover)	1½—2	Spring—	Before Seeding
Lentils	½	Spring—	Before or After Seeding
Peas	1¼	Spring—	Before or After Seeding
Potatoes	1½—2	Spring—	Before Seeding
Soybeans	1½—2	Spring—	Before or After Seeding
Sugar Beets	1½—2	Fall— Spring—	Before Seeding Before Seeding

AVADEX® GRANULAR
HERBICIDE

RESTRICTED USE PESTICIDE

For retail sale to and use only by Certified Applicators or persons under their direct supervision and only for those uses covered by the Certified Applicator's certification.

For Control of Wild Oats in Sugar Beets in North Dakota and Minnesota Only.

Do Not Use on Other Crops.

ACTIVE INGREDIENT:
S-(2,3 Dichloroallyl)-
diisopropylthiocarbamate 10%
INERT INGREDIENTS 90%

TOTAL 100%

EPA Reg. No. 524-151-AA EPA Est. 524-1A-1
843.56-000.07M

In case of emergency involving this product, Call Collect, day or night, (314) 694-4000.

WARNING
Keep Out of Reach of Children

Read the entire label.
Use only according to label instructions.
Read "LIMIT OF WARRANTY AND LIABILITY" before buying or using. If terms are not acceptable, return at once unopened.
CAUSES EYE IRRITATION
MAY CAUSE SKIN IRRITATION
Do not get in eyes, on skin or on clothing.

When mixing, loading or applying, wear rubber gloves, goggles, long trousers and long sleeved shirt or jacket of closely knit material along with leather or rubber boots high enough to cover the ankle.

Wash thoroutly after handling.

FIRST AID: IF IN EYES, immediatley flush with plenty of water for at least 15 minutes. Call a physician.

IF ON SKIN: wash with plenty of soap and water. Remove contaminated clothing. Wash clothing before reuse.

LIMIT OF WARRANTY AND LIABILITY

This company warrants that this product conforms to the chemical description on the label and is reasonably fit for the purposes set forth in the Complete Directions for Use ("Directions"), when used in accordance with those Directions under the conditions described therein. NO OTHER EXPRESS OR IMPLIED WARRANTY OF FITNESS FOR PARTICULAR PURPOSE OR MECHANTABILITY IS MADE. This warranty is also subject to the conditions and limitations stated herein.

Buyer and all users shall promptly notify this company of any claims whether based in contract, negligence, strict liability, other tort or otherwise.

Buyer and all users are responsible for all loss or damage from use or handling which results from conditions beyond the control of this company, including, but not limited to, incompatibility with products other than those set forth in the Directions, unusual weather (i. weather conditions which are outside the range considered normal at the application site and for the time period when the product is applied with the normal range being determined on the basis of the average range for the prior 40 years computed from the best available information, and ii. weather perils, including but not limited to hurricanes, tornadoes and floods) as well as weather considerations set forth in the Directions, application in any manner not explicity set forth in the Directions, moisture conditions outside the moisture range specified in the Directions, or the presence of products other than those set forth in the Directions in or on the soil or crop.

THE EXCLUSIVE REMEDY OF THE USER OR BUYER, AND THE LIMIT OF THE LIABILITY OF THIS COMPANY OR ANY OTHER SELLER FOR ANY AND ALL LOSSES, INJURIES OR DAMAGES RESULTING FROM THE USE OR HANDLING OF THIS PRODUCT (INCLUDING CLAIMS BASED IN CONTRACT, NEGLIGENCE, STRICT LIABILITY, OTHER TORT OR OTHERWISE) SHALL BE THE PURCHASE PRICE PAID BY THE USER OR BUYER FOR THE QUANTITY OF THIS PRODUCT INVOLVED, OR, AT THE ELECTION OF THIS COMPANY OR ANY OTHER SELLER, THE REPLACEMENT OF SUCH QUANTITY OR, IF NOT ACQUIRED BY PURCHASE, REPLACEMENT OF SUCH QUANTITY. IN NO EVENT SHALL THIS COMPANY OR ANY OTHER SELLER BE LIABLE FOR ANY INCIDENTAL OR CONSEQUENTIAL DAMAGES.

The buyer and all users are deemed to have accepted the terms of this LIMIT OF WARRANTY AND LIABILITY which may not be varied by any verbal or written agreement.

NOT FOR REFORMULATION OR REPACKAGING.

APPLICATION EQUIPMENT AND CALIBRATION

This granular herbicide must be applied through equipment capable of applying small quantities of granules evenly. It is important that the equipment be calibrated properly to deliver the desired amount of product. For even distribution, a scattering plate must be attached under each delivery tube or outlet in such a manner to give overall coverage. To calibrate, attach a bag over each delivery tube or a catch pan under the scattering plate. Operate on normal terrain at a constant speed. Four to 5 mph is suggested. Collect the granules from all outlets while traveling the desired distance. Check to see that each outlet dispenses the same amount of granules. Combine all samples and weigh.

For example: If a 20 foot applicator is being used, stake off a distance of 436 feet or $\frac{1}{5}$ of an acre in the field to be treated. For other applicator widths, distances are:

Applicator Width in feet	Distance for $\frac{1}{5}$ of an acre
12	726
14	622
16	545
20	436
28	311
36	242

The quantities of granules collected for $\frac{1}{5}$ acre of broadcast treatment are:

	Recommended rate per acre		
	20 lb	25 lb	30 lb
Amount for $\frac{1}{5}$ acre	4 lb	5 lb	6 lb

Continue equipment adjustment and calibration until the proper amount is collected.

For band application, the amount of granules needed for band width and row spacing can be calculated as follows:

$$\frac{\text{Band Width in inches}}{\text{Inches Between Crop Rows}} \times \text{Recommended Broadcast Rate} = \text{Pounds of Granules Needed Per Acre For Band Treatment}$$

For example: If 30 pounds per acre are recommended for overall treatment for a band 7 inches wide in crop rows spaced 20 inches apart:

$$\frac{7}{20} \times 30 = 10.5 \text{ pounds of granules needed per acre for band treatment}$$

GENERAL INFORMATION

This granular herbicide should be used only for the control of wild oats in sugar beets and only at recommended rates.

Incorporation of this product should be done with 48 hours after application.

If soil moisture is low before planting time, a post-planting irrigation will aid herbicidal action.

Use of this product in any manner not consistent with this label may result in injury to persons, animals or crops, or other unintended consequences.

DIRECTIONS FOR USE

It is a violation of federal law to use this product in any manner inconsistent with its labeling.

SUGAR BEETS

FALL TREATMENT

In North Dakota And Minnesota Only.

When used in the fall, this herbicide should be applied within 3 weeks before soil freeze-up. It is suggested that the soil temperature at 2 inches be 40°F or less. Do not apply on snow cover. Applications not made in accordance with the above may reduce wild oat control.

Broadcast Treatment—Apply 25 pounds per acre on fields known to have a light to medium wild oat infestation, and 30 pounds per acre on fields with heavy infestation.

Granules may be applied to soil that has an established cover crop (weeds, fall germinated wild oats, barley, etc.) to prevent wind erosion. Lightly incorporate the chemical within 48 hours after application either with a harrow or rotary hoe. Incorporation with this equipment will not disturb the established cover crop.

In preparing the treated area for seeding in the spring, care must be taken to avoid working the soil any deeper than 2 inches. Tillage deeper than 2 inches will dilute the chemical and reduce wild oat control.

SPRING TREATMENT

The seed bed should be free of clods. Any existing weed growth should be thoroughly worked into the soil before treatment. Improper incorporation or dry conditions may reduce wild oat control.

Thorough mixing of granules in the top 2 inches of soil is required for best wild oat control.

Broadcast Treatment—Use equipment that can apply small amounts of granules evenly. Applications may be made anytime within 7 days before seeding Incorporation can be accomplished by cross discing or harrowing at least twice; or with hooded power driven rotary tillers. Use the following recommended rates for soil textures:

SOIL TEXTURE	AMOUNT PER ACRE Broadcast Treatment
LIGHT	
Sandy Loams	20 lb
HEAVY	
Clay Loams and Clays	30 lb

Band Treatment—Apply over the row. Satisfactory results have been obtained with equipment such as the rotary hoe, Eversman tillers-incorporators, by-hoes, or similar mechanical equipment. Application can be made at planting time by mounting a granular applicator on such equipment as the "Tiller-incorporator."

STORAGE AND DISPOSAL

Do not contaminate water, seed, feed and foodstuffs by storage or disposal.

Keep bag dry. Product that cannot be used according to label instructions, due to spillage or otherwise, must be disposed of according to Federal, State or local procedures.

Completely empty bag into application equipment. Then dispose of bags in a sanitary landfill or by incineration if allowed by State and local authorities.

BRONCO

HERBICIDE

For Weed Control in Corn, Soybeans and Grain Sorghum (Milo)

Avoid Contact With Desirable Plants And Trees Since Severe Injury Or Destruction May Result.

ACTIVE INGREDIENTS:	
alachlor	27.6%
Isopropylamine salt of glyphosate	14.8%
INERT INGREDIENTS	57.6%
TOTAL	100.0%

Contains 2.6 pounds of 2-chloro-2′-6′diethyl-N-(methoxymethyl) acetanilide and 1.4 pounds* of the isopropylamine salt of N-(phosphonomethyl) glycine per U.S. gallon. *Equivalent to 1.04 pounds per U.S. gallon of the acid, glyphosate.

EPA Reg. No. 524-341-AA

U.S. Patent No. 3,547,620 U.S. Pat. No. 3,799,758 covers use.

Other patents pending.

®BRONCO is a trademark of Monsanto Company.

1984-4 817.37-000.12/53

DANGER

Keep Out of Reach of Children

Complete Directions for Use.

SHAKE WELL BEFORE USING

The use of this product may be hazardous to your health. This product contains alachlor, which has been determined to cause tumors in laboratory animals.

Users are urged to obtain information on careful handling and use of pesticides and to participate in Monsanto Company's education program on potential hazards and safer use of this product. For additional information, contact Monsanto Agricultural Products Co. at 1-800-453-4000. [NOTE: This number is for user safety information only. *NOT* for emergencies. For emergencies, call collect day or night (314)694-4000.]

Read the entire label before using this product.

Use only according to label instructions.

NOT FOR REFORMULATION OR REPACKAGING.

Read "LIMIT OF WARRANTY AND LIABILITY" before buying or using. If terms are not acceptable, return at once unopened.

LIMIT OF WARRANTY AND LIABILITY

This company warrants that this product conforms to the chemical description on the label and is reasonably fit for the purposes set forth in the Complete Directions for Use label booklet ("Directions") when used in accordance with those Directions under the conditions described therein. NO OTHER EXPRESS OR IMPLIED WARRANTY OF FITNESS FOR PARTICULAR PURPOSE OR MERCHANTABILITY IS MADE. This warranty is also subject to the conditions and limitations stated herein.

Buyer and all users shall promptly notify this company of any claims whether based in contract, negligence, strict liability, other tort or otherwise. Buyer and all users are responsible for all loss or damage from use or handling which results from conditions beyond the control of this company, including but not limited to incompatibility with products other than those set forth in the Directions, application to or contact with desirable vegetation, unusual weather (i. weather conditions which are outside the range considered normal at the application site and for the time period when the product is applied with the normal range being determined on the basis of the average range for the prior 40 years computed from the best available information, and ii. weather perils, including but not limited to hurricanes, tornadoes and floods) as well as weather considerations set forth in the Directions, application in any manner not explicitly set forth in the Directions, moisture conditions outside the moisture range specified in the Directions, or the presence of products other than those set forth in the Directions in or on the soil, crop or treated vegetation.

THE EXCLUSIVE REMEDY OF THE USER OR BUYER, AND THE LIMIT OF THE LIABILITY OF THIS COMPANY OR ANY OTHER SELLER FOR ANY AND ALL LOSSES, INJURIES OR DAMAGES RESULTING FROM THE USE OR HANDLING OF THIS PRODUCT (INCLUDING CLAIMS BASED IN CONTRACT, NEGLIGENCE, STRICT LIABILITY, OTHER TORT OR OTHERWISE) SHALL BE THE PURCHASE PRICE PAID BY THE USER OR BUYER FOR THE QUANTITY OF THIS PRODUCT INVOLVED, OR, AT THE ELECTION OF THIS COMPANY OR ANY OTHER SELLER, THE REPLACEMENT OF SUCH QUANTITY OR, IF NOT ACQUIRED BY PURCHASE, REPLACEMENT OF SUCH QUANTITY. IN NO EVENT SHALL THIS COMPANY OR ANY OTHER SELLER BE LIABLE FOR ANY INCIDENTAL OR CONSEQUENTIAL DAMAGES.

The buyer and all users are deemed to have accepted the terms of this LIMIT OF WARRANTY AND LIABILITY which may not be varied by any verbal or written agreement.

In case of emergency involving this product, call collect, day or night, (314) 694-4000.

PRECAUTIONARY STATEMENTS

HAZARDS TO HUMANS AND DOMESTIC ANIMALS

Keep out of reach of children

DANGER!

CAUSES EYE BURNS AND SKIN IRRITATION.

HARMFUL IF SWALLOWED.

MAY CAUSE ALLERGIC SKIN REACTION.

Do not get in eyes, on skin, or on clothing.

Wear goggles or face shield, rubber gloves, long trousers, long sleeved shirt or jacket of tightly woven material along with boots high enough to cover ankles when transferring and mixing, and when adjusting, repairing or cleaning equipment. Wear rubber boots when pouring from open containers. Clothing which comes in contact with Bronco must be washed before reuse. Clothing or other materials which become drenched with the concentrated pesticide must be disposed of in a sanitary landfill, by incineration or, if allowed by state and local authorities, by burning. If burned, stay out of smoke.

Wash thoroughly with soap and water after handling.

FIRST AID: IF IN EYES, immediately flush with plenty of water for at least 15 minutes. Call a physician.

IF ON SKIN, immediately flush with plenty of water while removing contaminated clothing. Call a physician. Wash clothing before reuse. Sensitized persons should avoid further contact and reuse of clothing.

IF SWALLOWED, induce vomiting immediately by giving two glasses of water and sticking finger down throat. Call a physician. NEVER GIVE ANYTHING BY MOUTH TO AN UNCONSCIOUS PERSON.

PHYSICAL OR CHEMICAL HAZARD

COMBUSTIBLE.

Do not use or store near heat or flame.

Use only with adequate ventilation.

In case of:

FIRE, use water spray, foam, dry chemical or CO_2.

SPILL OR LEAK, flush area with water spray.

Spray solutions of this product should be mixed, stored and applied only in stainless steel, aluminum, fiberglass, plastic and plastic-lined steel containers.

DO NOT MIX, STORE OR APPLY THIS PRODUCT OR SPRAY SOLUTIONS OF THIS PRODUCT IN GALVANIZED STEEL OR UNLINED STEEL (EXCEPT STAINLESS STEEL) CONTAINERS OR SPRAY TANKS. This product or spray solutions of this product react with such containers and tanks to produce hydrogen gas which may form a highly combustible gas mixture. This gas mixture could flash or explode, causing serious personal injury, if ignited by open flame, spark, welder's torch, lighted cigarette or other ignition source.

ENVIRONMENTAL HAZARD

This chemical has been identified in limited ground water sampling and there is the possibility that it can leach through the soil to ground water, especially where soils are coarse and ground water is near the surface. Following application, and during rainfall events that cause runoff, this chemical may reach surface water bodies including streams, rivers and reservoirs.

Care must be taken when using this product to prevent back siphoning into wells, spills or improper disposal of excess pesticide, spray mixtures or rinsates.

Check valves or antisiphoning devices must be used on all mixing and/or irrigation equipment. Disposal of excess pesticide, spray mixtures, or rinsate should be according to label use instructions or according to the state pesticide or environmental control agency or the hazardous waste representative at the nearest EPA regional office.

Do not apply directly to water or wetlands. Do not contaminate water by cleaning of equipment or disposal of wastes.

STORAGE AND DISPOSAL

Do not contaminate water, foodstuffs, seed or feed by storage and disposal.

STORAGE

Store above 40°F. to keep product in solution.

Below 40°F. crystals may form and settle to the bottom. If crystals form, place in a warm room and shake the container frequently until crystals dissolve. For bulk containers, see the container label for alternate storage information.

DISPOSAL

Wastes of this pesticide may cause eye burns and skin irritation and may be dangerous. Improper disposal of excess pesticide, spray mixtures, or rinsate is a violation of Federal law. If these wastes cannot be disposed of according to label use instructions, contact your State Pesticide or Environmental Control Agency, or the Hazardous Waste representative at the nearest EPA Regional Office for guidance.

Empty container retains vapor and product residue. Observe all labeled safeguards until container is cleaned, reconditioned or destroyed.

(See the individual container label for disposal information.)

GENERAL INFORMATION

BRONCO, applied alone or in tank mixtures, is recommended for the preemergence and postemergence control of many annual grasses and broadleaf weeds in soybeans and corn grown under NO-TILL or other minimum tillage practices. This product will also provide reduced competition from certain emerged perennial weeds.

MOISTURE CONDITIONS

Rainfall or sprinkler irrigation within 6 hours after application may reduce control of emerged weeds.

Moisture in the form of rainfall or irrigation is required after application to move this product into the weed germination zone for preemergence weed control. The amount of moisture required depends on the soil moisture content at the time of application, soil type, and plant residues. Performance is improved when ⅓ to ¾ inch of moisture is received within one week after application and prior to weed emergence. High intensity or excessive rainfall or irrigation after application may reduce preemergence weed control on coarse soils.

READ AND CAREFULLY OBSERVE PRECAUTIONARY STATEMENTS AND ALL OTHER INFORMATION APPEARING ON THE LABELING OF ALL PRODUCTS USED IN MIXTURES.

ATTENTION

AVOID DRIFT. EXTREME CARE MUST BE USED WHEN APPLYING THIS PRODUCT TO PREVENT INJURY TO DESIRABLE PLANTS, TREES AND CROPS.

Do not allow the herbicide solution to mist, drip, drift, or splash onto desirable vegetation since minute quantities of this herbicide can cause severe damage or destruction to crops, plants, or other areas on which treatment was not intended. The likelihood of injury occurring from the use of this product is greatest when winds are gusty or in excess of 5 miles per hour or when other conditions, including lesser wind velocities, will allow spray drift to occur. When spraying avoid combinations of pressure and nozzle type that will result in splatter or fine particles (mist) which are likely to drift. AVOID APPLYING AT EXCESSIVE SPEED OR PRESSURE.

DO NOT MIX THIS PRODUCT WITH LIQUID FERTILIZERS, HERBICIDE OILS OR ANY MATERIALS OTHER THAN WATER, EXCEPT THOSE LISTED ON THIS LABEL.

NOTE: Use of this product in any manner not consistent with this label may result in injury to persons, animals or crops, or other unintended consequences. Keep container closed to prevent spills and contamination.

DO NOT APPLY THIS PRODUCT TO EMERGED CORN OR SOYBEANS.

DO NOT FEED FORAGE FROM OR GRAZE TREATED AREAS WITHIN 8 WEEKS AFTER APPLICATION.

Flush sprayer with clean water after use.

SOIL TEXTURE

Applicators should evaluate soil and weed conditions carefully to assure that they choose the lowest effective label rate.

The recommended use rates of this product and the other herbicides labeled for use in tank mixtures with this product generally vary with soil texture. Unless soil texture is specifically named, rate tables throughout this label refer to only three soil texture groups: COARSE, MEDIUM and FINE. The following is a complete listing of soil textures included in each of these three soil texture groups:

Soil Texture Group	Soil Texture
COARSE:	sand, loamy sand, sandy loam
MEDIUM:	loam, silt loam, silt, sandy clay loam
FINE:	silty clay loam, clay loam, sandy clay, silty clay, clay

Refer to the above table to determine the corresponding soil texture group for the soil to be treated.

WEEDS CONTROLLED

When applied as directed under conditions described, this product alone or in tank mixtures with other products listed will control or reduce competition from the weeds indicated. Refer to the "DIRECTIONS FOR USE" section to determine rates for each crop.

Reduced control of emerged weeds may occur when weeds are under poor growing conditions such as from drought stress, when application follows recent mowing or when excessive amounts of dust cover the plants.

This product applied alone or in tank mixtures with other products listed will CONTROL the following EMERGED WEEDS:

Annual Grasses

Bluegrass (annual)	*Poa annua*
Brome (downy)	*Bromus tectorum*
Corn (volunteer)	*Zea mays*
Crabgrass	*Digitaria ischaemum* *Digitaria sanguinalis*
Foxtail (giant)	*Setaria faberi*
Foxtail (green)	*Setaria viridis*
Foxtail (yellow)	*Setaria lutescens*
*Panicum (fall)	*Panicum dichotomiflorum*
Sandbur (field)	*Cenchrus* spp.
Shattercane	*Sorghum bicolor*
Wheat (volunteer)	*Triticum aestivum*

Annual Broadleaves

Fleabane	*Erigeron* spp.
Kochia	*Kochia scoparia*
Lambsquarters	*Chenopodium album*
Lettuce (prickly)	*Lactuca serriola*
Marestail	*Eragrostis canadensis*
Pigweed, Carelessweed (redroot)	*Amaranthus retroflexus*
Pigweed (smooth)	*Amaranthus hybridus*
Ragweed (common)	*Ambrosia artemisiifolia*
Ragweed (giant)	*Ambrosia trifida*
Smartweed	*Polygonum pensylvanicum*
Sunflowers	*Helianthus annuus*
Thistle (Russian)	*Salsola kali*
Velvetleaf, Buttonweed	*Abutilon theophrasti*

* 5 quarts of this product are required for satisfactory postemergence control of fall panicum.

This product applied alone or in tank mixtures with other products listed will REDUCE COMPETITION from the following EMERGED WEEDS:

Perennial Grasses

Bluegrass (Kentucky)	*Poa* spp.
Bromegrass (smooth)	*Bromus inermis*
Fescues	*Festuca* spp.
Muhly (wirestem)	*Muhlenbergia frondosa*
Johnsongrass	*Sorghum halepense*
Orchardgrass	*Dactylis glomerata*
Quackgrass	*Agropyron repens*
Ryegrass (perennial)	*Lolium perenne*
Timothy	*Phleum pratense*

Perennial Broadleaves

Alfalfa	*Medicago sativa*
Clover (red)	*Trifolium pratense*
Clover (white)	*Trifolium repens*
Dock (curly)	*Rumex crispus*
Milkweed	*Asclepias* spp.
Mullein (common)	*Verbascum thapsus*
Smartweed (swamp)	*Polygonum coccineum*
Thistle (Canada)	*Cirsium arvense*

This product applied alone or in tank mixuture with other products will provide PREEMERGENCE CONTROL of the following weeds:

Annual Grasses

Barnyardgrass	*Echinochloa crus-galli*
Crabgrass	*Digitaria ischaemum* *Digitaria sanguinalis*
Foxtail (giant)	*Setaria faberi*
Foxtail (green)	*Setaria viridis*
Foxtail (yellow)	*Setaria lutescens*
Goosegrass	*Elusine indica*
Panicum (fall)	*Panicum dichotomiflorum*
Signalgrass (broadleaf)	*Brachiaria* spp.
Witchgrass	*Panicum capillare*

Annual Broadleaves

When applied as directed under conditions described, BRONCO® alone or in tank mixtures with products listed in the following table will provide preemergence control or reduced competition from the additional grass and broadleaf weeds as indicated. Refer to the crop on which treatment is intended in the "DIRECTIONS FOR USE" section of this label to determine the rate of BRONCO alone and of tank mixtures recommended for use in each crop.

NOTE: C=Control R=Reduced Competition

		Tank Mixtures				
	BRONCO®	Atrazine™	Bladex™	Lorox™	Princep™	Sencor™ Lexone™
Carpetweed Mollugo verticillata	C	C	C	C	C	C
Cocklebur Xanthium pensylvanicum	•	R	R	R	R	R
Jimsonweed Datura stramonium	•	C	R	R	R	R
Kochia Kochia scoparia	•	C	•	•	•	•
Lambsquarters Chenopodium album	R	C	C	C	C	C
Morningglory (annual) Ipomeoea purpurea	•	R	R	R	R	R
Mustard Brassica spp.	•	C	C	C	C	C
Nightshade (black) Solanum nigrum	C*	C	C	C	C	C*
Nightshade (hairy) Solanum sarachoides	C*	C	C	C	C	C*
Pigweed, Carelessweed (redroot) Amaranthus retroflexus	C	C	C	C	C	C
Pigweed (smooth) Amaranthus hybridus	C	C	C	C	C	C
Purslane Portulaca oleracea	C	C	C	C	C	C
Pusley (Florida) Richardia scabra	C	C	C	C	C	C

	Tank Mixtures					
	BRONCO®	Atrazine™	Bladex™	Lorox™	Princep™	Sencor™ Lexone™
Ragweed (common)						
Ambrosia artemisiifolia	R	C	C	C	C	C
Ragweed (giant)						
Ambrosia trifida	•	C	R	R	R	R
Sandbur (field)						
Cenchrus spp.	•	C	•	•	•	•
Sida (spiny), teawood						
Sida spinosa	R	C	C	C	C	C
Smartweed						
Polygonum pensylvanicum						
Polygonum persicaria	R	C	C	C	C	C
Velvetleaf, Buttonweed						
Abutilon theophrasti	•	R	R	R	R	R

* Use the maximum labeled rates to achieve control.

™Bladex is a registered trademark of the Shell Chemical Company.

™Lorox is a trademark of E. I. duPont de Nemours and Company.

™Princep is a registered trademark of Ciba-Geigy Corporation.

™Lexone is a trademark of E. I. duPont de Nemours and Company.

™Sencor is a trademark of the parent company of Farbenfabriken Bayer GmbH, Leverkusen.

REFER TO SPECIFIC PRODUCT LABELS FOR ADDITIONAL WEEDS CONTROLLED, CROP ROTATION RESTRICTIONS AND CAUTIONARY STATEMENTS.

APPLICATION INFORMATION

SHAKE CONTAINER VIGOROUSLY PRIOR TO MIXING AND USE.

Broadcast Treatment—Apply this product and labeled tank mixture products in 10 to 30 gallons of water, or 10 to 50 gallons of liquid nitrogen (28% or 32% solutions only), per acre during or after planting, but BEFORE CROP EMERGENCE. Application of this product in fluid fertilizer should only be used for treatment of annual weeds less than 6 inches tall. Spray coverage should be uniform and complete under conditions of dense stubble or plant residue or very high weed populations, higher rates within the recommended ranges may be needed for best results.

DO NOT APPLY BY AIR.

MIX ONLY WITH CLEAN WATER.

MIXING, SPRAYING AND HANDLING INSTRUCTIONS

NOTE

Direct contact or exposure to Bronco or spray mixtures of Bronco should be minimized. The following instructions for transfer, mixing, cleaning or repairing equipment should be followed in order to minimize this exposure. Review the protective clothing requirements as listed in the Precautionary Statements section of this label (front panel) and do not use this product until you have the necessary protective clothing.

2.5 Gallon Containers

Open pouring from these containers can result in exposure from splashing or spilling. Special care in lifting and pouring are strongly recommended.

Bulk Containers

Open pouring from these containers can result in exposure from splashing or spilling and is not recommended. This product should be transferred from these containers to the mix or spray tank using pumps or transfer probes. The probe or pump should not be removed from the container or disconnected until the container is emptied and rinsed. Use the pump or probe system to rinse the empty container and transfer the rinsate directly to the mix or spray tank.

Equipment Cleaning & Repair

Cleaning and repair of transfer systems and application equipment is a source of exposure to this product. Care should be taken to minimize exposure during cleaning and repair of transfer systems and application equipment. Whenever possible, these systems or equipment should be rinsed before being cleaned or repaired. When repairs must be made during transfer or application, the equipment should be shutdown, and special care taken to avoid contact with the pesticide.

Always predetermine the compatibility of BRONCO with other labeled herbicides by mixing small proportional quantities in advance. NOTE: Reduced results may occur if muddy water is used, such as water from ponds or ditches.

Mix this product or labeled tank mixtures with clean water as follows:

1. Place a 20- to 35-mesh screen or wetting basket over filling port.
2. Through the screen, fill the sprayer tank one-half full with water.
3. Add tank mix components. If a wettable powder is used, make a slurry with water, and add it SLOWLY through the screen into the tank. Continue agitation.

If flowable herbicides are used, pre-mix one part flowable with one part water. Add diluted mixture SLOWLY through the screen into the tank.

4. Pre-mix one part of BRONCO with one part water. Add diluted mixture SLOWLY through the screen into the tank.
5. Complete filling the sprayer tank with water. Remove hose from tank immediately to avoid siphoning back into the water source. Maintain good agitation at all times until the contents of the tank are sprayed.

NOTE: If spray mixture is allowed to settle at any time, thorough agitation is required to resuspend the mixture before spraying is resumed. Keep bypass line on or near bottom of tank to minimize foaming. Screen size in nozzle or line strainers should be no finer than 50 mesh. Carefully select proper nozzle to avoid spraying a fine mist. Check for even distribution of spray droplets. For best results with ground application equipment, use flat fan or whirl-chamber nozzle. To reduce loss of chemical due to drift of a fine mist, apply at pressures less than 40 psi.

DIRECTIONS FOR USE

It is a violation of Federal Law to use this product in any manner inconsistent with this labeling.

SHAKE WELL BEFORE USING

No-Till and Minimum Tillage Systems
CORN, SOYBEANS, AND GRAIN SORGHUM (MILO)

When applied as directed under the conditions described, this product or labeled tank mixtures will control many emerged annual weeds, suppress many emerged perennial weeds and give preemergence control of many annual weeds when corn, soybeans, or grain sorghum will be planted directly into a cover crop, or in previous crop residues. This product or labeled tank mixtures will not control regrowth from perennial weeds or unemerged perennial weeds.

For use on grain sorghum, apply only to grain sorghum seed that has been properly treated with a protectant containing the active ingredient, flurazole.

For mixing instructions, see the "Mixing and Spraying Instructions" section of this label.

The higher recommended rates listed for each soil texture grouping should be used when:

- PREVIOUS CROP RESIDUES ARE EXTREMELY DENSE
- EMERGED WEEDS ARE OVER 6 INCHES TALL
- WHEN HEAVY INFESTATIONS OF ANNUAL GRASSES OR BROADLEAVES ARE EXPECTED
- WHEN SOILS CONTAIN MORE THAN 3% ORGANIC MATTER

Soil Texture Group	Broadcast Rate Of BRONCO® (Qts./A.)
COARSE	3 to 5
MEDIUM	4 to 5
FINE	4 to 5

Double Crop Systems
SOYBEANS AND GRAIN SORGHUM (Milo)

Bronco or Bronco plus a labeled tank mixture will control the weed species listed below in medium and fine texture soils at the 3 to 5 quarts per acre rate. For additional information for weeds controlled on coarse textured soils consult the "WEEDS CONTROLLED" section of this label.

Annual Grasses	Annual Broadleaves*
Bluegrass (annual)	Lambsquarters
Brome (downy)	Marestail**
Corn (volunteer)	Pigweed (redroot)
Crabgrass	Pigweed (smooth)
Foxtail (giant)	Ragweed (common)
Foxtail (green)	
Foxtail (yellow)	
Wheat (volunteer)	

* A tank mixture of Bronco plus Lorox or Lexone or Sencor is recommended to improve control of emerged labeled weeds in double crop soybeans. Tank mixtures of Bronco plus atrazine are recommended to improve weed control in double crop grain sorghum. Consult the soybean and grain sorghum tank mixture sections of this label for recommended rates.

** A bronco rate of 3 quarts per acre plus tank mixture with Lorox or Lexone or Sencor will control light infestations of this broadleaf weed. When medium to heavy infestations exist, use the 4 to 5 quart per acre rate of Bronco alone or in tank mixture with Lorox or Lexone or Sencor. Consult the tank mixtures section of this label for recommended use rates.

When using the 3 quart per acre rate of Bronco or Bronco tank mixtures, flat fan nozzles or low volume flood jet nozzles and application volumes of 20 gallons per acre or less are recommended for optimum performance. Refer to the "GENERAL INFORMATION" section of this label for additional information.

The higher Bronco rates (4 to 5 quarts per acre) should be used in double crop systems when:

- Weeds not listed under double crop systems (Note: A 5 quart per acre Bronco rate must be used to control emerged fall panicum).
- Weeds are drought stressed.
- Heavy infestations of annual grasses or broadleaves are present.

TANK MIXTURES

BRONCO® PLUS Roundup® AND/OR Lasso®

In cases of heavy weed infestation or hard-to-control weed species found on this label. Roundup OR Lasso herbicides OR both can be added to the recommended treatment rates to obtain the following improved types of control:

Roundup	Postemergence
Lasso	Preemergence
Roundup and Lasso	Postemergence and Preemergence

Refer to the respective labels for Roundup and Lasso for specific recommendations regarding the timing of application.

The following table shows the maximum amounts of Roundup OR Lasso herbicides, OR both that can be added to specific treatment rates of this product.

BRONCO®	Recommended Rates (Qts./A.) Roundup®	And/Or	Lasso®
3.0	2.0		2.2
3.5	1.8		1.7
4.0	1.6		1.4
4.5	1.4		1.1
5.0	1.25		0.75

TANK MIXTURES—CORN

BRONCO® plus Atrazine

OR

BRONCO plus Bladex

OR

BRONCO plus Princep

Apply BRONCO® alone or in tank mixture with atrazine, OR Bladex OR Princep herbicides for control of grasses and broadleaf weeds listed in the "Weeds Controlled" section of this label when CORN is planted directly into a cover crop or in previous crop residues.

Refer to specific product labels for crop rotation restrictions and cautionary statements of all products used in these tank mixtures.

See the following table for recommended rates of atrazine, OR Bladex OR Princep in this tank mixture applied on various soil types. To determine the recommended rates for BRONCO refer to the rate table for this product.

Soil Texture Group*	Broadcast Rates (Qts./A.)** Atrazine 4L	Bladex 4L	Princep 4L
Coarse-Sandy Loam Only	1 to 1.2	1 to 1.6	1 to 1.2
Medium	1.2 to 1.6	1.2 to 1.6	1.2 to 1.6
Fine	1.6 to 2	1.6 to 2.2	1.6 to 2

* Refer to the "Soil Texture" section of this label to determine the corresponding soil texture group for the soil to be treated.

** When using 4L formulations use the rates listed.

- Use equivalent rates when using 80W or 90% dry flowable formulations. One quart of the 4L formulation of these products equals 1.25 pounds of the 80W formulation. One quart of atrazine 4L or Princep 4L equals 1.1 pounds of the 90% dry flowable formulations of these products.
- Use the higher rates when recommended for each product listed when soils contain more than 3% organic matter.

Tank Mixtures—CORN

BRONCO® plus Atrazine plus Princep

OR

BRONCO® plus Atrazine plus Bladex

For additional preemergence control of heavy weed infestation and/or hard to control weeds, a three-way tank mixture of BRONCO® herbicide plus atrazine plus Princep OR BRONCO plus atrazine plus Bladex may be applied. Refer to the following rate tables and the rate table for BRONCO to determine the recommended rates for these tank mixtures on various soil types.

Soil Texture Group*	Broadcast Rates (QTS./A)** Atrazine 4L	+	Princep 4L
COARSE	0.75 to 1		0.75 to 1
MEDIUM	1 to 1.25		1 to 1.5
FINE	1 to 1.25		1 to 1.5
	Atrazine 4L	**+**	**Bladex 4L**
COARSE	0.75 to 1		1 to 1.25
MEDIUM	1 to 1.25		1.25 to 2
FINE	1 to 1.25		1.25 to 2

* Refer to the "Soil Texture" section of this label to determine the corresponding soil texture group for the soil to be treated.

** When using 4L formulations use the rates listed.

- Use equivalent rates when using 80W or 90% dry flowable formulations. One quart of the 4L formulation of these products equals 1.25 pounds of the 80W formulation. One quart of atrazine 4L or Princep 4L equals 1.1 pounds of the 90% dry flowable formulations of these products.

- USE THE LOWER RATES RECOMMENDED FOR EACH PRODUCT LISTED WHEN SOILS CONTAIN LESS THAN 3% ORGANIC MATTER.

NOTE: PLANT ONLY CORN, PEANUTS, SORGHUM OR SOYBEANS* the year following use of this mixture.

* There is a possibility of crop injury due to carryover of atrazine if soybeans are planted the following year. DO NOT plant soybeans the year following use of these tank mixtures on furrow irrigated corn. DO NOT graze treated area or feed treated forage to livestock for 21 days following application of these tank mixtures.

Tank Mixtures—SOYBEANS

BRONCO® plus Lorox

OR

BRONCO® plus LEXONE or SENCOR

Apply BRONCO herbicide alone or in tank mixtures with Lorox, OR Lexone, OR Sencor herbicides for control of grasses and broadleaf weeds listed in the "Weeds Controlled" section of this label when SOYBEANS are planted directly into a cover crop, stale seedbed, or in previous residues such as wheat stubble.

Refer to specific product labels for crop rotation restrictions and cautionary statements of all products used in these tank mixtures.

See the following table for recommended rates of Lorox, OR Lexone OR Sencor in tank mixtures applied on various soil types. To determine rates for BRONCO refer to the rate table for this product.

	Broadcast Rates (Qts./A)	
Soil Texture Group*	**Lorox L**†**	**LEXONE 4L OR SENCOR 4F†**
COARSE—Sandy Loam Only	0.5 to 0.75	0.25 to 0.375
MEDIUM	0.75 to 1	0.375 to 0.5
FINE	1 to 1.5	0.5 to 0.75

* Refer to the "Soil Texture" section of this label to determine the corresponding soil texture group for the soil to be treated.

** Use the higher rate of Lorox L in the recommended ranges on soils containing more than 3% organic matter.

† Use rates listed in this label when using Lorox L, Lexone 4L or Sencor 4F. Use equivalent rates when using the 50 WP or DF formulations of these products. One pint of Lorox L, Lexone 4L or Sencor 4F equals 1 pound of the corresponding 50 WP formulation or ⅔ pound of the corresponding DF formulation.

THE FOLLOWING INFORMATION APPLIES ONLY TO USE OF LEXONE OR SENCOR.

On the silty clay or heavy clay soils of the Mississippi Delta, use 0.75 to 1 quart of Lexone or Sencor per acre.

Use the higher rate of Lexone or Sencor in the recommended ranges on soils with greater than 2% organic matter.

Do not use on muck soils.

Do not apply on alkaline soils with a pH higher than 7.4.

Crop injury may occur if any atrazine was applied on the soil the year before use of this Lexone or Sencor tank mixture.

DO NOT REPLANT CROPS OTHER THAN SOYBEANS FOR 120 DAYS AFTER APPLICATION.

Tank Mixtures
GRAIN SORGHUM (Milo)

BRONCO® PLUS ATRAZINE

Apply BRONCO alone or in tank mixture with atrazine herbicide for control of grasses and broadleaf weeds listed in the "WEEDS CONTROLLED" section of this label when grain sorghum is planted directly into a cover crop or in previous crop residues.

Refer to specific product labels for crop rotation restrictions and cautionary statements of all products used in these tank mixtures.

See the following table for recommended rates of atrazine in this tank mixture applied on various soil types. To determine the recommended rates for BRONCO refer to the rate table for this product.

	Broadcast Rates (Qts./A)**
Soil Texture Group*	**ATRAZINE 4L**
COARSE (Except Sand)	0.75 to 1
MEDIUM	1 to 1.25
FINE	1.2 to 1.5

* Refer to the "SOIL TEXTURE" section of this label to determine the corresponding soil texture group for the soil to be treated.

** When using 4L formulations use the rates listed.

- Use equivalent rates when using 80W or 90% dry flowable formulations. One quart of the 4L formulation of this product equals 1.25 pounds of the 80W formulation. One quart of atrazine 4L equals 1.1 pound of the 90% dry flowable formulation of this product.
- Use the higher rates when recommended for each product listed when soils contain more than 3% organic matter.

®ROUNDUP and ®LASSO are registered trademarks of Monsanto Company.

FAR-GO®
HERBICIDE

Emulsifiable Herbicide for Control of Wild Oats in Winter Wheat, Spring and Durum Wheat, Barley, Green Peas, Field Dried Peas and Lentils.

ACTIVE INGREDIENT:
* S-(2,3,3-Trichloroallyl)-diisopropylthiocarbamate 46.3%
INERT INGREDIENTS: 53.7%
TOTAL 100.0%

* Contains 4 pounds S-(2,3,3-Trichloroallyl)-diisopropylthiocarbamate per gallon.

EPA Reg. No. 524–145–AA 815.15-000.48/53

® FAR-GO is a registered trademark of Monsanto Company.

WARNING
Keep Out of Reach of Children

Read the entire label before using this product.
Use only according to label instructions.
NOT FOR REFORMULATION OR REPACKAGING
Read "LIMIT OF WARRANTY AND LIABILITY" before buying or using. If terms are not acceptable, return at once unopened.

LIMIT OF WARRANTY AND LIABILITY

This company warrants that this product conforms to the chemical description on the label and is reasonably fit for the purposes set forth in the complete Directions for Use label booklet ("Directions") when used in accordance with those Directions under the conditions described therein. NO OTHER EXPRESS OR IMPLIED WARRANTY OF FITNESS FOR PARTICULAR PURPOSE OR MERCHANTABILITY IS MADE. This warranty is also subject to the conditions and limitations stated herein.

Buyer and all users shall promptly notify this company of any claims whether based in contract, negligence, strict liability, other tort or otherwise.

Buyer and all users are responsible for all loss or damage from use or handling which results from conditions beyond the control of this company, including, but not limited to, incompatibility with products other than those set forth in the Directions, unusual weather, weather conditions which are outside the range considered normal at the application site and for the time period when the product is applied, as well as weather considerations set forth in the Directions, application in any manner not explicity set forth in the Directions, moisture conditions outside the moisture range specified in the Directions, or the presence of products other than those set forth in the Directions in or on the soil or crop.

THE EXCLUSIVE REMEDY OF THE USER OR BUYER, AND THE LIMIT OF THE LIABILITY OF THIS COMPANY

OR ANY OTHER SELLER FOR ANY AND ALL LOSSES, INJURIES OR DAMAGES RESULTING FROM THE USE OR HANDLING OF THIS PRODUCT (INCLUDING CLAIMS BASED IN CONTRACT, NEGLIGENCE, STRICT LIABILITY, OTHER TORT OR OTHERWISE) SHALL BE THE PURCHASE PRICE PAID BY THE USER OR BUYER FOR THE QUANTITY OF THIS PRODUCT INVOLVED, OR, AT THE ELECTION OF THIS COMPANY OR ANY OTHER SELLER, THE REPLACEMENT OF SUCH QUANTITY, OR, IF NOT ACQUIRED BY PURCHASE, REPLACEMENT OF SUCH QUANTITY. IN NO EVENT SHALL THIS COMPANY OR ANY OTHER SELLER BE LIABLE FOR ANY INCIDENTAL OR CONSEQUENTAL DAMAGES.

The buyer and all users are deemed to have accepted the terms of this LIMIT OF WARRANTY AND LIABILITY which may not be varied by any verbal or written agreement.

PRECAUTIONARY STATEMENTS

HAZARD TO HUMANS AND DOMESTIC ANIMALS

Keep Out of Reach of Children

WARNING!

CAUSES SKIN AND EYE IRRITATION

HARMFUL IF SWALLOWED

Do not get in eyes, on skin or on clothing.

Avoid breathing vapors or spray mist.

Wash thoroughly after handling.

To avoid contact with skin or eyes, wear rubber gloves and goggles when handling or spraying.

FIRST AID:

IF IN EYES, immediately flush with plenty of water for at least 15 minutes. Call a physician.

IF ON SKIN, wash with plenty of soap and water. Remove contaminated clothing. Wash clothing before reuse. If irritation persists, get medical attention.

IF SWALLOWED, do NOT induce vomiting. Give large quantities of water. Call a physician immediately. NEVER GIVE ANYTHING BY MOUTH TO AN UNCONSCIOUS PERSON.

In case of emergency involving this product, Call Collect, day or night. (314) 694-4000.

ENVIRONMENTAL HAZARDS

Avoid direct application to any body of water. Do not contaminate water by disposal of waste or cleaning of equipment.

PHYSICAL OR CHEMICAL HAZARDS

COMBUSTIBLE.

Do not use, pour, spill or store near heat or flame.

Use only with adequate ventilation.

In case of:

FIRE, use water spray, foam, dry chemical or CO_2.

SPILL or LEAK, flush area with water spray.

STORAGE AND DISPOSAL

Do not contaminate water, foodstuffs, seed or feed by storage and disposal.

STORAGE

STORE ABOVE 32° F TO KEEP FROM FREEZING

Freezing will result in crystals which settle to the bottom of the can. If allowed to freeze, place in a warm room (72°F.) and roll and shake the container frequently for several days to redissolve before using.

Keep this container closed to prevent spills, evaporation and contamination.

DISPOSAL

Wastes of this pesticide may cause skin and eye irritation and may be dangerous. Improper disposal of excess pesticide, spray mixtures, or rinsate is a violation of Federal law. If these wastes cannot be disposed of by use according to label instructions, contact your State Pesticide or Environmental Control Agency, or the Hazardous Waste representative at the nearest EPA Regional Office for guidance.

Emptied container retains vapor and product residue. Observe all labeled safeguards until container is cleaned, reconditioned or destroyed.

(See individual container label for disposal information.)

GENERAL INFORMATION

This herbicide or any labeled tank mixture of this herbicide should be sprayed on the soil and incorporated before weeds germinate. If weeds, including wild oats, have emerged, prior to treating and/or planting they must be controlled.

Before applying this herbicide or any labeled tank mixture of this herbicide, the soil should be in good working condition. This applies particularly to clay soils. Avoid application to a field left in a ridged condition. Application to a field which is wet, lumpy, rough or ridged, will result in reduced wild oat control and increased crop thinning.

This product or any labeled tank mixture of this herbicide must be incorporated into the soil immediately after spraying.

Seeding may be done either before or after spraying and incorporation, depending upon the crop that is to be sown. See "Directions for Use" section of this label. If seeding is delayed, shallow reworking of the treated area before seeding will not destroy the effects of the chemical.

Applied according to label directions and under normal growing conditions, Far-Go herbicide plus Treflan will not adversely affect spring wheat and barley. Seedling disease, cold weather, improper seeding or incorporation depth, excessive moisture, high salt concentration or drought may weaken crop seedlings and increase the possibility of crop damage from Far-Go herbicide plus Treflan. Under these conditions, delayed crop development or reduced yields may result.

Weed control may be evaluated by removing a surface inch of soil at the time of germination to inspect the number of wild oats that were killed before emergence.

This herbicide is primarily absorbed by wild oat shoots from the treated layer of the soil. Wild oats are usually controlled before they emerge through the soil. Occasionally, and particularly under dry conditions and cold soil temperatures, some wild oats may emerge and reach the 3 to 4-leaf stage before dying. Under conditions of prolonged high temperature at the time of wild oat germination or drought in the spring, this product may not provide the usual high standard of wild oat control.

ATTENTION

DELAYED EMERGENCE, STAND REDUCTION, STUNTING AND YIELD LOSS MAY RESULT DUE TO COLD OR WET CONDITIONS, IRRIGATION DURING GERMINATION OR EMERGENCE, DEEP PLANTING, SOIL CRUSTING, DISEASE, INSECTS, OR SELECTION OF A VARIETY GENETICALLY SUSCEPTIBLE TO STRESS. THESE ARE CONDITIONS BEYOND THE CONTROL OF MONSANTO, AND THE GROWER SHOULD CONSULT THE STATE UNIVERSITY EXTENSION SPECIALIST OR SEED PRODUCER OR SUPPLIER FOR LOCAL VARIETAL INFORMATION AND RECOMMENDED TILLAGE AND PLANTING PRACTICES AND DATES.

Domestic oats should not be seeded where Far-Go was used the previous year.

Use of this product in any manner not consistent with this label may result in injury to persons, animals or crops, or other unintended consequences.

STANDARD SPRAYABLE FLUID FERTILIZER COMPATIBILITY TEST

Herbicides may not always mix evenly throughout a sprayable fluid fertilizer or the compounds may separate too quickly to make their combined use of practical value. This may be due to certain characteristics of the different fluid fertilizers. A simple test using small quantities of the components is suggested to provide compatibility potential.

The test follows:

A. Materials Required For A Compatibility Test

1. Two one-quart jars with lid or stopper (marked "with" and "without").
2. Teaspoons (for a more exacting test, a five to ten milliliter (ml.) pipette or graduated cylinder is desirable).
3. A minimum of one quart of the sprayable fluid fertilizer to be tested.
4. The herbicide to be mixed.
5. A compatibility agent (The purpose of the adjuvant is to help keep the fertilizer and crop protection chemical in suspension, if this assistance is needed).

B. Procedure

1. Add one pint of the sprayable fluid fertilizer that will be used or other herbicide carrier to each jar marked "with" and "without"

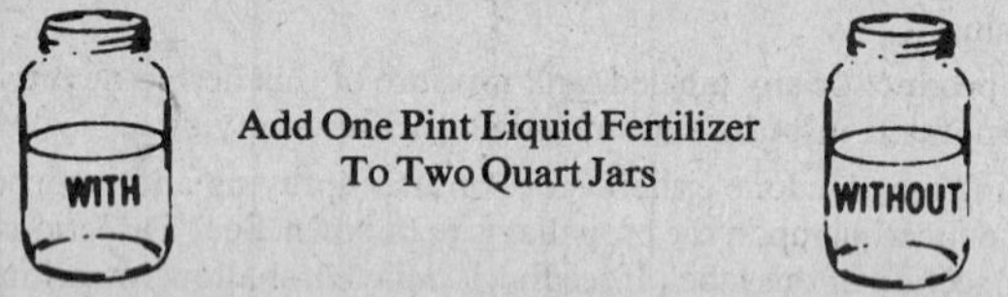

2. To the jar marked "with", add 1/4 TEAspoon or 1.2 milliliters of a suitable compatibility agent; shake gently for five to ten seconds to mix. (1/4 TEAspoon is the equivalent of two pints per 100 gallons of liquid fertilizer.)

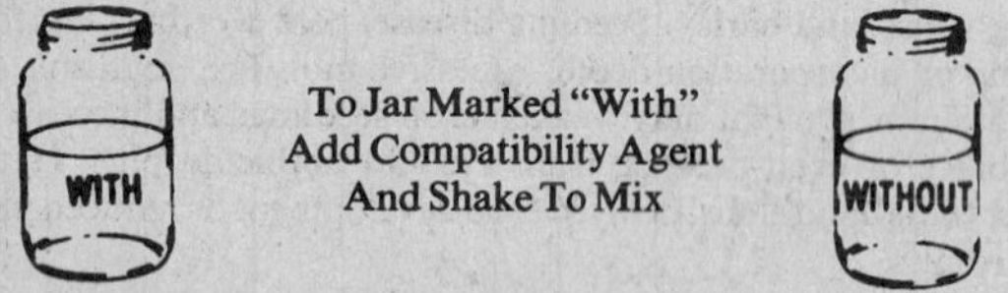

3. To each jar add the appropriate amount of herbicide(s). Shake gently for five to ten seconds after each addition.

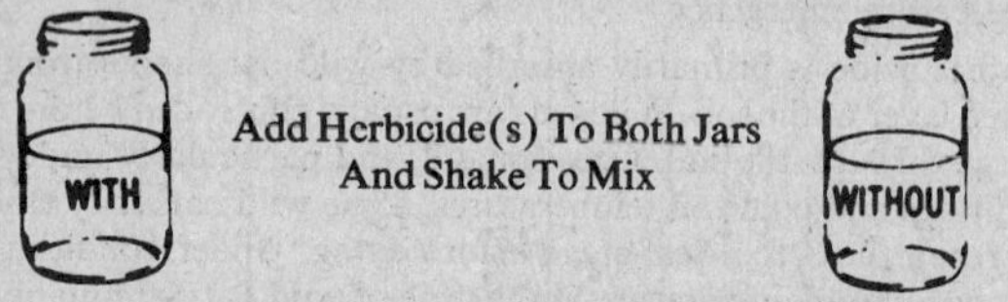

Herbicide	Rate/A		Amount To Be Added Per Pint Of Sprayable Fluid Fertilizer (Assuming Volume is 25 gal/A) Level Teaspoons		Milliliters
FAR-GO®	1 qt.	=	0.9	or	4.7
	1.25 qt.	=	1.2	or	5.9

This compatibility test is designed for 25 gallons of spray per acre with the maximum labeled rate of herbicide. For changes in spray volume, make appropriate changes in the ingredients of the test. Regardless of spray volume, the amount of compatibility agent should be equal to two or three pints (two pints = 1/4 TEAspoon or 1.2. milliliters, three pints = 3/8 TEAspoon or 1.8 milliliters per pint of sprayable fluid fertilizer) per 100 gallons of liquid fertilizer.

C. Observations And Decisions

1. If the herbicide(s) and the sprayable fluid fertilizer are compatible.
2. If a compatibility agent is necessary.

Five minutes after the final addition and mixing, observe both jars for the formation of large flakes, sludge, gels or other precipitates. Observe if the herbicide(s) cannot be physically mixed with the liquid fertilizer but remains as small oily particles in the solution.

If incompatibility in any form described above occurs in the jar "with" the compatibility agent added, the liquid fertilizer and the herbicides(s) should not be used together in the same spray tank.

If incompatibility as described above occurs in the jar "without" the adjuvant but not in the jar "with" adjuvant, the use of a compatibility adjuvant is recommended.

Both jars should be allowed to stand and to be observed periodically for one-half hour. If the separate layers of liquid fertilizer and additives can be resuspended by shaking, commercial application is possible. This product normally will go to the top after standing.

If the herbicide(s) is compatible with fluid fertilizer in the foregoing test without having to use a compatibility agent, fluid fertilizer may be used for the premixing. If it is not compatible, the herbicide(s) should be premixed with water before adding to the spray tank.

MIXING INSTRUCTIONS

Always predetermine the compatibility of this herbicide or labeled tank mix of this herbicide with water carrier or sprayable fluid fertilizer carrier by mixing small proportional quantities in advance. See the compatibility test section of the label for instructions as to how to determine the compatibility of this herbicide with sprayable fluid fertilizer.

NOTE: This herbicide can be applied in sprayable fluid fertilizer in the spring to barley, spring and Durum wheat only.

Mix FAR-GO or labeled tank mix of FAR-GO with the appropriate carrier as follows:

1. Fill the tank approximately three-quarters full with water. Check for adequate agitation.
2. Carefully add the proper amount of herbicide(s) into the tank while mixing.
3. Fill the remaining portion of the tank with water.
4. Maintain good agitation at all times until the contents of the tank are sprayed.
5. Mix thoroughly again if the solution has been left in the tank for any length of time.
6. The sprayer by-pass valve should be in operation before and during application.

APPLICATION INFORMATION

Before spraying, check all hose connections to see that they are properly secured. Spraying equipment must be properly calibrated. Adjust the height of the spray boom in the field to ensure an even distribution of the chemical on the soil. Pressure should not be above 30 p.s.i. and just high enough to provide a uniform spray pattern. Boomless sprayers should not be used because the distribution of this product will be too uneven.

A field marker or chain should be attached to both ends of the sprayer boom to serve as a marker, thus avoiding overlapping or untreated strips. Overlapping will double the recommended rate and crop injury may result.

Apply in 10 or more gallons of water or sprayable fluid fertilizer per acre. Higher volumes may result in more uniform spraying and less evaporation.

Spraying during winds above 10 miles per hour may result in a loss of herbicide below the amount required for good wild oat control.

WEEDS CONTROLLED WITH FAR-GO®

When applied as directed under conditions described on this label, FAR-GO CONTROLS the following weeds:

Oats (wild)	Ryegrass (annual)
Avena fatua	Lolium multiflorum (Oregon only)

DIRECTIONS FOR USE

It is a violation of Federal Law to use this product in any manner inconsistent with its labeling.

FIELD PREPARATION

Before applying this herbicide or any labeled tank mixture, the soil should be in good working condition. All deep tillage must be completed prior to application of this herbicide. If plowed ground is being treated, a single cultivation is recommended prior to spraying. If stubble ground with a heavy trash cover is being treated, or if the soil is wet or lumpy (making proper mixing difficult), two cultivations before spraying are recommended.

HOW TO INCORPORATE

For proper incorporation of Far-Go, set the incorporation implement to work the soil no deeper than 3 to 4 inches. Incorporation should be performed immediately after herbicide application. Do not use disk implements for incorporation.

CROPPING SYSTEMS

FAR-GO®

FALL TREATMENT: Winter Wheat

Crop	FAR-GO Rate (Qts/A)
Winter Wheat* Apply just before seeding to soon after seeding before wild oats germinate.	1.25

* DO NOT use on clay knobs.

For applications to Winter Wheat after seeding, apply and shallowly incorporate with a spike tooth, or spring tooth harrow followed with a second incorporation at right angles. When incorporating after seeding Winter Wheat, adjust cultivation depth so as not to disturb wheat seed.

FALL TREATMENT: Barley, Durum Wheat, and Spring Wheat

Crop	FAR-GO Rate (Qts/A)
Spring or Durum Wheat	1
Barley—Apply within 3 weeks of soil freezeup.	1.25

Breakdown of this product is minimal at 40 F. and below, therefore very limited herbicide will be lost after soil temperatures drop below 40°F.

For applications to provide wild oat control in spring planted Spring Wheat, Durum Wheat and Barley, apply on summer fallow ground which is free of lumps and trash, and single pass incorporate with equipment similar to a culti-harrow or a duckfoot cultivator multi-weeder, or spring tooth harrow.

For application to stubble ground with heavy trash cover, fields should be worked twice with a field cultivator or chisel plow to remove as much surface trash as possible. When a chisel plow is used it should be followed by a field cultivator before application. Apply this herbicide and single pass incorporate with a field cultivator or culti-harrow.

When incorporating with a single pass in the fall, a second incorporation must be performed in the spring during seedbed preparation. If no spring work is anticipated, then both incorporations should be done in the fall.

If soil must be ridged after incorporating to prevent soil erosion by high winds, the depth of ridging should be kept to a minimum. Deep ridging may decrease the effectiveness of the herbicide or result in crop injury in the spring.

In preparing the herbicide-treated area for seeding in the spring, care must be taken to avoid working the soil any deeper than 3 to 4 inches.

SPRING TREATMENT: Spring Wheat, Barley, Green Peas, Field Dried Peas and Lentils

BEFORE SEEDING APPLICATION

Crop	FAR-GO® Rate (Qts/A)
Spring and Durum Wheat	1
Barley—Apply just before seeding.	1.25
Green Peas*	1.25
Field Dried Peas*	1.25
Lentils—Apply up to 3 weeks before seeding.	1.25

* Leaf crinkling and delayed maturity of peas may occur, particularly on clay points in the Northwest, but this is usually more than offset by a reduction of wild oat competition and increased yields.

BEFORE SEEDING INCORPORATION: For applications to stubble ground, work fields twice with a disk, field cultivator or chisel plow to provide soil in a good working condition and to remove surface trash. Apply and incorporate with equipment such as a culti-harrow, springtooth cultivator or duckfoot cultivator. A second incorporation at right angles should provide best results.

SPRING TREATMENT: Barley, Green Peas, Field Dried Peas, Lentils, Durum Wheat and Spring Wheat

AFTER SEEDING INCORPORATION

Crop	FAR-GO® Rate (Qts/A)
Spring and Durum Wheat	1
Barley	1.25
Green Peas*	1.25
Field Dried Peas*	1.25
Lentils	1.25

* Leaf crinkling and delayed maturity of peas may occur, particularly on clay points in the Northwest, but this is usually more than offset by a reduction of wild oat competition and increased yields

Apply after seeding before crop emergence and before wheat and barley kernel sprouts exceed 1/2 inch in length and before green peas, field dried peas, and lentils shoot sprouts exceed 1/4 inch in length.

AFTER SEEDING INCORPORATION: Apply immediately after seeding and shallowly incorporate with equipment such as a flex multiweeder or harrow, and follow with a second incorporation at a right angle. Adjust incorporation equipment to a depth so as not to disturb the seed.

FAR-GO® PLUS TREFLAN™ (TANK MIXTURE)

Barley and Wheat

When applied as directed under conditions described on this label, this tank mixture CONTROLS these additional weeds:

Foxtail (green)	Foxtail (yellow)
Setaria viridis	Setaria lutescens

Green and yellow foxtail are commonly referred to as pigeon grass.

™ Treflan is a trademark of Elanco Products Co.

SPRING TREATMENT: Barley, Spring Wheat, and Durum Wheat

AFTER SEEDING INCORPORATION

Crop	FAR-GO® Qts/A +	TREFLAN Qts/A
Spring Wheat, Durum Wheat and Barley	1	0.5

AFTER SEEDING INCORPORATION: Apply immediately after seeding and shallowly incorporate with equipment such as a flex multiweeder or harrow, and follow with a second incorporation at a right angle. Adjust incorporation equipment to a depth so as not to disturb the seed.

NOTE: Do not apply the FAR-GO herbicide plus Treflan tank mix to soils which are subject to prolonged periods of flooding or excessive irrigation.
If less than 25 inches of total water is used to produce the crop, do not plant sorghum or oats for 18 months after application.

Refer to Treflan label for crop rotation restrictions and cautionary statements.

ANNUAL CANARYGRASS SEED PRODUCTION

FALL TREATMENT: Apply the herbicide at a rate of 1 quart per acre within 3 weeks of soil freezeup. The seeding depth of the crop should be 1 to 2 inches.

SPRING TREATMENT: Incorporate this product according to the BEFORE SEEDING INCORPORATION instructions in the

SPRING TREATMENT section of this label. Apply the herbicide at a rate of 1 quart per acre just before seeding. The seeding depth of the crop should be 1 to 2 inches.

FERTILIZER IMPREGNATION WITH FAR-GO®

FOR USE IN ALL STATES EXCEPT CALIFORNIA

The application of FAR-GO impregnated on fertilizer may be made as follows:

Crops: Winter Wheat, Spring Wheat and Barley

Time of Application and Rate

- Fall Applied and Fall Incorporated

—Spring wheat and barley
1.25 lbs. ai/A of FAR-GO (The 1.5 lb. rate may be used only on fine textured soil.)

—Winter wheat
1.25 lbs. ai/A of FAR-GO
Preplant incorporated

- Spring Applied and Preplant Incorporated

—Spring barley only
1.25 lbs. ai/A of FAR-GO

APPLICATION WITH DRY BULK FERTILIZER

FOR USE IN ALL STATES EXCEPT CALIFORNIA

The herbicide/fertilizer impregnation process must be done at commercial fertilizer or chemical dealerships that are properly equipped for this procedure.

Dry bulk fertilizer may be impregnated with FAR-GO herbicide. FAR-GO must be impregnated on a minimum of 150 pounds per acre of dry fertilizer. Applications of dry bulk fertilizer impregnated with FAR-GO herbicide may be made in the fall or spring, but only preplant incorporated treatments to barley may be made in the spring. Spring planted wheat or barley may be seeded into fall FAR-GO herbicide applications in accordance with label directions. Applications should be made as soon as possible after blending.

The following table provides a reference to determine the amount of liquid herbicide to be mixed per ton of dry bulk fertilizer.

Dry Fertilizer (Lbs/Acre)	Acres Covered (Per Ton)	FAR-GO® Herbicide Rate (Quarts Per Acre) 1.25	1.50
		(Quarts Per Ton)	
150	13.3	16.5	20.0
160	12.5	15.8	18.8
170	11.7	14.8	17.5
180	11.1	14.0	16.8
190	10.5	13.0	15.8
200	10.0	12.5	15.0
210	9.5	12.0	14.3
220	9.1	11.5	13.5
230	8.7	11.0	13.0
240	8.3	10.5	12.5
250	8.0	10.0	12.0
260	7.7	9.6	11.6
270	7.4	9.3	11.1

Mix and blend the dry fertilizer and herbicide in a closed rotary drum-type mixer allowing sufficient time to ensure uniform coverage. The nozzle used to spray the herbicide treatment must be placed inside the mixer to provide uniform spray coverage of the tumbling fertilizer.

Spread the fertilizer/herbicide mixture with a properly calibrated applicator. Be certain the material is applied uniformly to the soil surface. When using fan spreaders, a 100% overlap is recommended. Fan spreaders should be calibrated to put out one-half the desired rate per acre and application overlapped by doubling back to cover one-half of the previous swath. Dribble spreaders must be calibrated with the fertilizer/herbicide mixture, since the flow pattern will generally be slower than straight dry fertilizer. Non-uniform spreading of the fertilizer/herbicide mixture may result in unsatisfactory weed control or crop injury.

Follow normal FAR-GO herbicide incorporation procedures to incorporate the fertilizer/herbicide mixture.

A partial list of fertilizers which may be impregnated with FAR-GO are:

Diammonium phosphate	18–46–0
Treble super-phosphate	0–46–0
Urea	45–0–0
Ammonium nitrate	33–0–0
Ammonium sulfate	21–0–0
Potassium chloride	0–0–60
Potassium sulfate	0–0–52

NOTE: FAR-GO CAN BE IMPREGNATED ON ANY COMMONLY USED DRY FERTILIZER EXCEPT THOSE CONTAINING POTASSIUM NITRATE OR SODIUM NITRATE. WHEN MIXING FAR-GO WITH AMMONIUM NITRATE, IT IS IMPERATIVE THAT FAR-GO BE EVENLY MIXED, THAT CONCENTRATIONS ABOVE 5 PERCENT BY WEIGHT OF FAR-GO BE AVOIDED, AND THAT THE MIXTURE BE UTILIZED WITHIN 24 HOURS OF MIXING WITH IMMEDIATE APPLICATION RECOMMENDED. CONCENTRATIONS IN EXCESS OF 5 PERCENT BY WEIGHT OF FAR-GO WITH AMMONIUM NITRATE OR LONG-TERM STORAGE OF FAR-GO AND AMMONIUM NITRATE COULD RESULT IN EXCESSIVE CONCENTRATIONS OF FAR-GO IN THE MIXTURE DUE TO DRAINING OF FAR-GO. THESE EXCESSIVE CONCENTRATIONS OF FAR-GO MAY CAUSE PERSONAL INJURY AND/OR PROPERTY DAMAGE FROM FIRE AND/OR EXPLOSION IF IGNITED OR DETONATED. EXCESSIVE CONCENTRATIONS OF FAR-GO IN THE MIXTURE MAY ALSO CAUSE CROP INJURY. AMMONIUM NITRATE IS AN OXIDIZING AGENT AND ALL REGULATIONS RELATING TO ITS STORAGE, BLENDING, LABELING, OR TRANSPORTATION ARE THE RESPONSIBILITY OF THE ENTITY MIXING AND/OR SELLING THE FERTILIZER/FAR-GO HERBICIDE MIXTURE.

If you have any questions concerning this Label Pamphlet or the proper application or use of Far-Go herbicide, contact your local Monsanto Sales Representative.

In case of an emergency involving this product, Call Collect, day or night, (314) 694-4000.

FAR-GO® GRANULAR

HERBICIDE

Granular Selective Herbicide for Fall or Spring Application to Control Wild Oats in Certain Fall or Spring Planted Crops.

ACTIVE INGREDIENT:
S-(2,3,3′-Trichloroallyl)diisopropyl-thiocarbamate 10%

INERT INGREDIENTS 90%

TOTAL 100%

EPA Reg. No. 524-292-AA EPA Est. 524-1A-1

843.53-000.32M

CAUTION

Keep Out of Reach of Children

In case of an emergency involving this product, Call Collect, day or night, (314) 694-4000.

Read the entire label before using this product.
Use only according to label instructions.

NOT FOR REFORMULATION OR REPACKAGING.

Read "LIMIT OF WARRANTY AND LIABILITY" before buying or using. If terms are not acceptable, return at once unopened.

PRECAUTIONARY STATEMENTS

HAZARD TO HUMANS AND DOMESTIC ANIMALS

CAUTION!

Keep out of reach of children.

MAY CAUSE SKIN AND EYE IRRITATION.
HARMFUL IF SWALLOWED.
Avoid contact with skin, eyes, or clothing.
Wash thoroughly with soap and water after handling.

FIRST AID: IF IN EYES: immediately flush with plenty of water for at least 15 minutes. Call a physician.

IF ON SKIN: immediately flush with plenty of water while removing contaminated clothing. Wash clothing before reuse. If irritation persists, get medical attention.

IF SWALLOWED: do NOT induce vomiting. Give large quantities of water. Call a physician immediately. NEVER GIVE ANYTHING BY MOUTH TO AN UNCONSCIOUS PERSON.

ENVIRONMENTAL HAZARDS

Avoid direct application to any body of water. Do not contaminate water by disposal of waste or cleaning of equipment.

STORAGE AND DISPOSAL

Do not contaminate water, foodstuffs, seed or feed by storage and disposal.

Keep bag closed to prevent spills and contamination.

DISPOSAL

Wastes resulting from the use of this product that cannot be used or chemically reprocessed should be disposed of in a landfill approved for pesticide disposal or buried on site in a safe place away from water supplies. All disposal should be in accordance with applicable Federal, state or local procedures.

Completely empty bag into application equipment. Then dispose of empty bag in a sanitary landfill or by incineration, or, if allowed by State and local authorities, by burning. If burned, stay out of smoke.

If you have any questions concerning this Label or the proper application or use of Far-Go herbicide, contact your local Monsanto Sales Representative.

In case of emergency involving this product, Call collect, day or night, (314) 694-4000.

DIRECTIONS FOR USE

It is a violation of Federal Law to use this product in any manner inconsistent with its labeling.

This herbicide is recommended for wild oat control in barley, lentils, durum, spring and winter wheat only.

Other crops should not be treated with this product because injury may occur. For barley, durum and spring wheat, this product may be applied on the soil either in the fall or in the spring before wild oats germinate. For lentils, this product may only be applied in the spring before wild oats germinate. For winter wheat, this product may only be applied in the fall before wild oats germinate.

Application equipment must be properly calibrated—application of too much herbicide may injure the crop—application of too little may result in poor wild oat control. Recommended rates should be followed in order to:

1. Avoid crop injury
2. Avoid crop residue at harvest
3. Control wild oats

Use of this product not consistent with this label may result in injury to persons, animals or crops, or other unintended consequences. Keep container closed to prevent spills and contamination.

Application to a field which is wet, lumpy, rough or ridged will result in reduced wild oat control and promote crop thinning. Incorporation must be completed within 48 hours after application and before germination of the wild oats. If weeds, including wild oats, have emerged prior to treating and/or planting, they must be controlled. All deep tillage by cultivators or double disc implements must be completed prior to application. Do not plow following application of this herbicide.

Seeding may be done either before or after application, depending upon the crop that is to be sown. If seeding is delayed, shallow reworking of the treated area before seeding will not destroy the effects of the chemical.

When using this herbicide, a strip should be left untreated for proof of results. Weed control may be evaluated also by removing a surface inch or two of the soil at the time of germination to inspect the number of wild oats that were killed before emergence.

Wild oats are usually killed before emergence, but occasionally and particularly under dry conditions, plant may reach the 3 to 4 leaf stage before they die.

Under conditions of prolonged high temperature at the time of germination or extreme drought in the spring, this product may not maintain the usual high standard of wild oat control.

ATTENTION

Delayed emergence, stand reduction, stunting and yield loss may result due to cold or wet conditions, irrigation during germination or emergence, deep planting soil crusting, disease, insects or selection of a variety genetically susceptible to stress. These are conditions beyond the control of Monsanto, and the grower should consult the state university extension specialist or seed producer or supplier for local varietal information and recommended tillage and planting practices and dates.

Domestic oats should not be seeded where Far-Go was used the previous year.

Do not graze livestock on treated crops.

Use in any manner not consistent with label directions can result in serious injury to the crop.

HOW TO APPLY AND CALIBRATE EQUIPMENT

This granular herbicide must be applied through a specially designed ground applicator or airplane capable of applying small quantities of granules evenly.

GROUND EQUIPMENT: It is important that the applicator be calibrated properly to deliver the desired amount of this product to avoid applying too little or too much material. To give even distribution, scatter plates (similar to those used for applying granules in a band) must be attached to each delivery tube or outlet in such a manner to give overall coverage. To calibrate, attach a collector pan, or strong plastic bag, over each spreader plate or delivery tube. Operate over normal terrain to be treated at 4 to 5 miles per hour. Collect the granules from all outputs after covering the desired distance.

For example: If a 14-foot applicator is being used, stake off a distance of 324 feet in the field to be treated (distance equal to $\frac{1}{10}$ acre treated). After attaching a bag, or collecting pan, to each outlet, collect the granules while driving the desired distance. Check to see that each outlet disperses the same amount of granules. Combine all samples and weigh. For the above set of conditions, the quantity of granules that should be collected for the following recommended rates for broadcast treatment are:

	RECOMMENDED RATE PER ACRE		
	10 lbs.	**12.5 lbs.**	**15 lbs.**
Proper amount to be collected	1.0 lbs.	1.25 lbs.	1.5 lbs.

If more or less than the desired quantity is collected, adjust opening accordingly and again collect the granules while driving the staked-off distance. Continue this procedure until the proper amount is delivered.

AIRPLANE: For aerial applications, attachments designed for applying low volumes of granules must be used. In order to ensure uniform aerial application, it is recommended that the field distribution pattern is checked, and any necessary gate and vent modifications are made to ensure an even pattern distribution. In order to ensure uniform application and to avoid overlapping and possible crop injury, it is recommended that two flagmen, one at each end of the field, be used.

FIELD PREPARATION

Before applying this product, be sure the soil is in good working condition. All deep tillage by cultivation, or double disc implements, must be completed prior to application of this herbicide. If stubble ground is being treated, one or two passes with a field cultivator may be required before application and incorporation.

INCORPORATED APPLICATIONS

When summer fallow, or plowed ground is being treated, a field cultivator will provide adequate incorporation. For proper incorporation of Far-Go, set incorporation implement to work the soil no deeper than 3 to 4 inches. Do not use disk implements for incorporation.

FALL—On summer fallow ground which is loose and free of lumps and trash, incorporate this product using equipment similar to a culti-harrow or a duckfoot with rod weeder attachment.

For applications to fields or standing stubble, fields may be worked once or twice with a field cultivator or chisel plow. Apply granules and incorporate with a field cultivator or culti-harrow.

If soil must be ridged after incorporating to prevent soil erosion by high winds, the depth of ridging should be kept to a minimum.

In preparing the herbicide treated area for seeding in the spring, care must be taken to avoid working the soil any deeper than fall tillage.

When incorporating with a single pass in the fall, a second incorporation must be performed in the spring during seed bed preparation. If no spring work is anticipated, then both incorporations should be done in the fall.

For applications to winter wheat after seeding, apply and shallow incorporate with a spike tooth, or spring tooth harrow set shallow as not to disturb wheat seed.

SPRING—Before Seeding Incorporation: For applications to fields of standing stubble, work fields once or twice with a disc, field cultivator or chisel plow, to provide soil in a good working condition. Apply granules and incorporate with equipment such as a culti-harrow or duckfoot cultivator. A second incorporation at right angeles should provide best results.

FOR SPRING WHEAT AND BARLEY IN MONTANA ONLY. For suppression of Persian Darnel (Lolium temulentum) and Downy Brome (Bromus tectorum), apply this product in the spring before seeding and shallowly incorporate.

After Seeding Incorporation: Apply granules immediately after seeding and shallow incorporate at right angles with equipment such as a flex multiweeder or harrow. Adjust incorporation equipment to a depth so as not to disturb the seed.

WHEN TO USE

FALL TREATMENT INCORPORATED

Crop	Rate (Lbs./A)	When to Apply
Spring and Durum Wheat and Barley	12.5 to 15	*Within 3 weeks of normal freezeup or until snow cover occurs.
Winter Wheat	12.5 to 15	Before or after seeding only.

Apply lower rates on light soils and apply higher rates on heavy soils.

Incorporate within 48 hours. For methods, see "FIELD PREPARATION" and "HOW TO INCORPORATE" sections.

* Average soil temperature at the 2-inch depth must be 40°F. or less.

SPRING TREATMENT INCORPORATED

Crop	Rate (Lbs./A)	When to Apply
Spring and Durum Wheat	10.0 to 12.5	After seeding before wild oats germinate.
Spring and Durum Wheat*	12.5	Just before seeding with HOEDRILLS, before wild oats germinate.
Barley	12.5 to 15.0	Just before or after seeding before wild oats germinate.
Lentils	12.5 to 15.0	Just before or immediately after seeding before wild oats germinate.
Peas	12.5 to 15.0	Just before or immediately after seeding before wild oats germinate.

Apply lower rates on light soils and apply higher rates on heavy soils.

Incorporate within 48 hours. For methods, see "FIELD PREPARATION" and "HOW TO INCORPORATE" sections.

* For press drills use the 10.0 lb. rate. For HOEDRILLS use the 12.5 lb. rate.

Best spring and Durum wheat results will be obtained with at least a 3-day delay between herbicide incorporation and seeding.

FALL SURFACE APPLICATION

USE ONLY IN IDAHO, MINNESOTA, NORTH DAKOTA, MONTANA AND UTAH

The fall surface application rate for spring planting of spring wheat, durum wheat and barley is 15 pounds per acre for all soil types.

Fall surface application must be incorporated in the spring: two passes are recommended. If surface applications are made because soil conditions don't permit incorporation and these conditions change making a fall incorporation possible, incorporate even if several weeks after application.

Fall application of this product may be made in standard stubble or to fields with surface residue beginning 3 weeks prior to normal soil freeze-up. (Average solid temperature at the 2-inch depth must be 40°F. or less.) Surface applications should not be made to fields covered with snow or with excessive crop residue, which will not allow granule contact with the soil.

Fall surface applications may not provide the same level of wild oat control as fall incorporated surface applications. Wild oat control with fall surface applications may be reduced if drought or abnormally warm temperatures occur between application and crop emergence in the spring. To achieve the most reliable wild oat control from fall application, this product should be incorporated.

LIMIT OF WARRANTY

This company warrants that this product conforms to the chemical description on the label and is reasonably fit for the purposes set forth in the Complete Directions for Use ("Directions"), when used in accordance with those Directions under the conditions described therein. NO OTHER EXPRESS OR IMPLIED WARRANTY OF FITNESS FOR PARTICULAR PURPOSE OR MERCHANTABILITY IS MADE. This warranty is also subject to the conditions and limitations stated herein.

Buyer and all users shall promptly notify this company of any claims whether based in contract, negligence, strict liability, other tort or otherwise.

Buyer and all users are responsible for all loss or damage from use or handling which results from conditions beyond the control of this company, including but not limited to depth of incorporation and seeding except as set forth in the Directions, incompatibility with products other than those set forth in the Directions, unusual weather, weather conditions which are outside the range considered normal at the application site and for the time period when the product is applied, as well as weather considerations, application in any manner not explicitly set forth in the Directions, moisture conditions outside the moisture range specified in the Directions, or the presence of products other than those set forth in the Directions in or on the soil or crop.

THE EXCLUSIVE REMEDY OF THE USER OR BUYER AND THE LIMIT OF THE LIABILITY OF THIS COMPANY OR ANY OTHER SELLER FOR ANY AND ALL LOSSES, INJURIES OR DAMAGES RESULTING FROM THE USE OR HANDLING OF THIS PRODUCT (INCLUDING CLAIMS BASED IN CONTRACT, NEGLIGENCE, STRICT LIABILITY, OTHER TORT OR OTHERWISE) SHALL BE THE PURCHASE PRICE PAID BY THE USER OR BUYER FOR THE QUANTITY OF THIS PRODUCT INVOLVED, OR, AT THE ELECTION OF THIS COMPANY OR ANY OTHER SELLER, THE REPLACEMENT OF SUCH QUANTITY OR, IF NOT ACQUIRED BY PURCHASE, REPLACEMENT OF SUCH QUANTITY. IN NO EVENT SHALL THIS COMPANY OR ANY OTHER SELLER BE LIABLE FOR ANY INCIDENTAL OR CONSEQUENTIAL DAMAGES.

The buyer and all users are deemed to have accepted the terms of this LIMIT OF WARRANTY AND LIABILITY which may not be varied by any verbal or written agreement.

LANDMASTER®

REDUCED TILLAGE HERBICIDE

For control or suppression of emerged weeds in fallow and reduced tillage small grain systems in the states of Colorado, Idaho, Kansas, Montana, Nebraska, New Mexico, North Dakota, Oklahoma, Oregon, South Dakota, Texas, Utah, Washington and Wyoming.

AVOID CONTACT WITH DESIRABLE VEGETATION SINCE SEVERE INJURY OR DESTRUCTION MAY RESULT.

ACTIVE INGREDIENTS*:	
Glyphosate [N-phosphonomethyl glycine] as the isopropylamine salt	12.9%
2,4-D (2.4-dichlorophenoxyacetic acid) as the isopropylamine salt	21.5%
INERT INGREDIENTS	65.6%
TOTAL	100.0%

* Contains 0.9 pound/gallon of glyphosate and 1.6 pounds/gallon of 2,4-D.

The glyphosate component of Landmaster herbicide is protected by U.S. Patent No. 3,799,758, and U.S. Patent No. 4,405,531. Other patents are pending.

EPA Reg. No. 524-351

®Landmaster is a trademark of Monsanto Company.

1985-1 897.14-000.01/53

DANGER

Keep Out of Reach of Children

Complete Directions for Use

Read the entire label before using product.

Use only according to label instructions.

NOT FOR REFORMULATION OR REPACKAGING.

Read "LIMIT OF WARRANTY AND LIABILITY" before buying or using. If terms are not acceptable, return at once unopened.

LIMIT OF WARRANTY AND LIABILITY

This company warrants that this product conforms to the chemical description on the label and is reasonably fit for the purposes set forth in the complete Directions for Use label booklet ("Directions") when used in accordance with those Directions under the conditions described therein. NO OTHER EXPRESS OR IMPLIED WARRANTY OF FITNESS FOR PARTICULAR PURPOSE OR MERCHANTABILITY IS MADE. This warranty is also subject to the conditions and limitations stated therein.

Buyer and all users shall promptly notify this company of any claims whether based in contract, negligence, strict liability, other tort or otherwise.

Buyer and all users are responsible for all loss or damage from use or handling which results from conditions beyond the control of this company, including but not limited to incompatibility with products other than those set forth in the Directions, application to or contact with desirable vegetation, unusual weather (i. weather conditions which are outside the range considered normal at the application site and for the time period when the product is applied with the normal range being determined on the basis of the average range for the prior 40 years computed from the best available information, and ii. weather perils, including but not limited to hurricanes, tornadoes and floods) as well as weather considerations set forth in the Directions, application in any manner not explicity set forth in the Directions, moisture conditions outside the moisture range specified in the Directions, or the presence of products other than those set forth in the Directions in or on the soil, crop or treated vegetation.

THE EXCLUSIVE REMEDY OF THE USER OR BUYER, AND THE LIMIT OF THE LIABILITY OF THIS COMPANY OR ANY OTHER SELLER FOR ANY AND ALL LOSSES, INJURIES OR DAMAGES RESULTING FROM THE USE OR HANDLING OF THIS PRODUCT (INCLUDING CLAIMS BASED IN CONTRACT, NEGLIGENCE, STRICT LIABILITY, OTHER TORT OR OTHERWISE) SHALL BE THE PURCHASE PRICE PAID BY THE USER OR BUYER FOR THE QUANTITY OF THIS PRODUCT INVOLVED, OR, AT THE ELECTION OF THIS COMPANY OR ANY OTHER SELLER, THE REPLACEMENT OF SUCH QUANTITY OR, IF NOT ACQUIRED BY PURCHASE, REPLACEMENT OF SUCH QUANTITY. IN NO EVENT SHALL THIS COMPANY OR ANY OTHER SELLER BE LIABLE FOR ANY INCIDENTAL OR CONSEQUENTIAL DAMAGES.

The buyer and all users are deemed to have accepted the terms of this LIMIT OF WARRANTY AND LIABILITY which may not be varied by any verbal or written agreement.

PRECAUTIONARY STATEMENTS

HAZARDS TO HUMANS AND DOMESTIC ANIMALS

Keep out of reach of children.

DANGER!

CAUSES EYE BURNS. HARMFUL IF SWALLOWED. MAY CAUSE ALLERGIC SKIN REACTION.

Do not get in eyes, on skin or on clothing.

Wear goggles or face shield and rubber gloves in transferring and mixing and when adjusting, repairing or cleaning equipment.

Wash thoroughly with soap and water after handling.

FIRST AID: IF IN EYES, immediately flush with plenty of water for at least 15 minutes. Call a physician.

IF ON SKIN, immediately flush with plenty of water for at least 15 minutes while removing contaminated clothing. Wash clothing before reuse. Sensitized persons should avoid further contact and reuse of clothing.

IF SWALLOWED, induce vomiting immediately by either giving two glasses of water and sticking finger down throat or, when available, administering syrup of ipecac as described on its container. Call a physician. NEVER GIVE ANYTHING BY MOUTH TO AN UNCONSCIOUS PERSON.

In case of an emergency involving this product, Call Collect, day or night, (314) 694-4000.

ENVIRONMENTAL HAZARDS

Avoid direct application to any body of water. Do not contaminate water by cleaning of equipment or disposal of wastes.

PHYSICAL OR CHEMICAL HAZARDS

Spray solutions of this spray should be mixed, stored and applied only in stainless steel aluminum, fiberglass, plastic and plastic-lined steel containers.

DO NOT MIX, STORE OR APPLY THIS PRODUCT OR SPRAY SOLUTIONS OF THIS PRODUCT IN GALVANIZED STEEL OR UNLINED STEEL (EXCEPT STAINLESS STEEL) CONTAINERS OR SPRAY TANKS. This product or spray solutions of this product react with such containers and tanks to produce hydrogen gas which may form a highly combustible gas mixture. This gas mixture could flash or explode, causing serious personal injury, if ignited by open flame, spark, welder's torch, lighted cigarette or other ignition source.

STORAGE AND DISPOSAL

Do not contaminate water, food, or feed by storage or disposal. Do not store near fertilizers, seeds, insecticides or fungicides.

STORAGE

Store above 40°F. to keep product in solution. If crystals form, place in a warm room (72°F.), allow the product to reach room temperature and roll or shake periodically until crystals have redissolved. Keep container closed to prevent spills and contamination.

DISPOSAL

Wastes of this pesticide may cause eye burns and may be dangerous. Improper disposal of excess pesticide, spray mixtures, or rinsate is a violation of Federal law. If these waste cannot be disposed of by use according to label instructions, contact your State Pesticide or Environmental Control Agency, or the Hazardous Waste representative at the nearest EPA Regional Office for guidance.

Empty container retains vapor and product residue. Observe all labeled safeguards until container is destroyed.

Triple rinse container. Then puncture and dispose of in a sanitary landfill, or by incineration, or, if allowed by state and local authorities, by burning. If burned, stay out of smoke.

GENERAL INFORMATION

Landmaster is a postemergence herbicide for control or suppression of emerged weeds in fallow and reduced tillage systems, prior to planting or emergence of wheat, barley, rye and oats.

This product enters the plant through the foliage and moves throughout the plant. Visual effects of control are a gradual wilting or yellowing of the plant, which advances to complete browning of above-ground growth and deterioration of affected underground plant parts. Visible symptoms will usually develop on labeled weeds within 2 to 4 days after application, but may not occur for 7 or more days. Extremely cool or cloudy weather following treatment may slow activity of this product and delay the visual effects of control.

APPLICATION PRECAUTIONS AND WARNINGS

- Do not plant any crop other than wheat, barley, oats or rye for 3 months after treatment or until this product has disappeared from the soil.
- Do not feed or forage treated vegetation from treated areas within 8 weeks after application.
- Do not apply in the vicinity of sensitive crops such as cotton, grapes, tomatoes and other desirable vegetation.
- The likelihood of injury occurring to adjacent crops from the use of this product is greatest when winds are gusty or in excess of 5 miles per hour or when other conditions including lesser wind velocities will favor spray drift.
- Movement of this product on soil particles during violent windstorms may cause damage to susceptible plants that are contacted. This hazard is reduced if rainfall occurs shortly after application.
- Buyer and all users are responsible for all loss or damage in connection with the use or handling of mixtures of this herbicide or other materials that are not expressly recommended in this labeling. Mixing this product with herbicides or other materials not recommended on this label may result in reduced performance. Use of this product in any manner not consistent with this label may result in injury to persons, animals or crops, or other unintended consequences.

Landmaster is subject to all state and county regulations for 2,4-D amine.

DIRECTIONS FOR USE

TIMING OF APPLICATION

This product should be applied postemergence to vigorously growing weeds when they have reached the recommended size given in the "RECOMMENDED RATES AND WEEDS CONTROLLED" section of this label. Application should be delayed until maximum emergence of the target weeds, but before weeds exceed the maximum size recommended. For annual weeds, allow at least 3 days after treatment before tillage. For field bindweed, allow at least 7 days after treatment before tillage.

Reduced control may result if treatments are made during poor growing conditions such as drought stress, disease or insect damage or if weeds have been mowed, grazed or cut. Heavy dust on foliage or an overstory canopy covering targeted weeds may also reduce control.

Rainfall or irrigation occurring within 6 hours after application may reduce effectiveness. Heavy rainfall or irrigation within 2 hours after application will wash this product off the foliage and a repeat treatment will be required.

RECOMMENDED RATES AND WEEDS CONTROLLED

When applied as directed, Landmaster herbicide will provide control or suppression of the grass and broadleaf weed species listed below:

NOTE: C = Control, S = Suppression

		Landmaster Rate Per Acre	
Weed Species	**Weed Height**	**40 fluid oz.**	**54 fluid oz.**
ANNUAL WEEDS			
Barley Hordeum vulgare		C	C
Downy brome* Bromus tectorum		C	C
Foxtail Setaria sp.		C	C
Kochia Kochia scoparia		C	C
Lambsquarters Chenopodium album		C	C
Prickly lettuce Lactuca serriola	2 to 6 inches	C	C
Redroot pigweed Amaranthus retroflexus		C	C
Russian thistle Salsola kali		C	C
Rye Secale cereale		C	C
Tansy mustard Descurainia pinnata		C	C
Tumble mustard Sisymbrium altissimum		C	C
Wheat Triticum aestivum		C	C
Wild oats Avena fatua		—	C
PERENNIAL WEEDS			
Field bindweed Convolvulus arvensis	6 to 18" (Length of vines)	—	S

* For control in no-till systems, use 54 fluid ounces per acre.

The addition of 2% ammonium sulfate by weight or 17 pounds per 100 gallons of water may increase the performance of Landmaster. The improvement in performance may be apparent where environmental stress is a concern. Dissolve dry ammonium sulfate in water and filter to remove undissolved material prior to addition to the spray tank. Lower quality ammonium sulfate may contain foreign material which could result in nozzle tip plugging. Thoroughly rinse the spray system with clean water after use to reduce corrosion.

APPLICATION EQUIPMENT
This product may be applied using either ground or aerial spray equipment. Use extreme care to avoid misting or drifting of herbicide solution onto foliage, green stems or fruit of desirable crops, trees, or plants during both growing and dormant periods since even very small quantities of spray can cause severe plant injury.

GROUND APPLICATION: Apply recommended rates of this product in 3 to 10 gallons of water per acre as a broadcast spray. For optimum spray distribution and coverage, use flat fan or low volume food nozzles. When using flood nozzles, space them no more than 40 inches apart and ensure double overlap of spray pattern. Refer to the manufacturers recommendations for correct pressure and nozzle height above the target canopy. Avoid pressure and nozzles which produce fine droplets or mist.

Use appropriate marking devices to ensure uniform spray coverage and best results from Landmaster herbicide.

AERIAL APPLICATION: Apply the recommended rates of this product in 3 to 5 gallons of water per acre as a broadcast spray. DO NOT APPLY DURING INVERSION CONDITIONS, WHEN WINDS ARE GUSTY, OR WHEN OTHER CONDITIONS WILL FAVOR DRIFT. DRIFT MAY CAUSE DAMAGE TO ANY VEGETATION CONTACTED TO WHICH TREATMENT WAS NOT INTENDED. TO PREVENT INJURY TO ADJACENT DESIRABLE VEGETATION, APPROPRIATE BUFFER ZONES MUST BE MAINTAINED.

Coarse sprays are less likely to drift, therefore, do not use nozzles or nozzle configurations which dispense spray as fine spray droplets. Do not angle nozzles forward into the airstream and do not increase spray volume by increasing nozzle pressure.

MIXING INSTRUCTIONS
Fill the spray tank to about 3/4 of the desired volume with clean water. Add the recommended amount of Landmaster herbicide, then complete the filling process while maintaining agitation. Remove the hose from the mix tank immediately after filling to avoid siphoning back into the carrier source. During mixing and application, foaming of the spray solution may occur. To prevent or minimize foam, terminate by-pass and return lines at the tank bottom and/or use an agriculturally approved anti-foam or defoaming agent.

NOTE: Reduced results may occur if water containing soil is used, such as water from ponds and unlined ditches.

SPRAYER CLEANUP
CLEAN THE ENTIRE SPRAYER AFTER APPLICATION OF THIS PRODUCT. Failure to clean the sprayer thoroughly may result in injury to desirable crops which are subsequently sprayed. First, add clean water to the tank and thoroughly rinse the entire sprayer system. Secondly, fill the tank with water and ammonia. Add 1 quart of household ammonia per 25 gallons of water. Pump enough solution through the hoses, boom and nozzles to fill these parts completely. Then fill the tank, close and leave for 24 hours before draining and rinsing thoroughly with water.

Application or use of other agricultural chemicals with the equipment used for this product may result in injury to desirable vegetation.

Thoroughly wash aircraft, especially landing gear, after each day of spraying to remove residues of this product accumulated during spraying or from spills. PROLONGED EXPOSURE OF THIS PRODUCT TO UNCOATED STEEL SURFACES MAY RESULT IN CORROSION AND POSSIBLE FAILURE OF THE PART. LANDING GEAR ARE MOST SUSCEPTIBLE. The maintenance of an organic coating (Paint) which meets aerospace specification MIL-C-38413 may prevent corrosion.

In case of an emergency involving this product, Call Collect, day or night, (314) 694-4000.

LASSO®
HERBICIDE

Emulsifiable herbicide for weed control in corn, soybeans, peanuts, grain sorghum (milo), and crops listed.

ACTIVE INGREDIENT:	
*Alachlor	45.1%
INERT INGREDIENTS	54.9%
TOTAL	100.0%

* Contains 480 grams/litre or 4 pounds/gallon of 2-chloro-2'-6'-diethyl-N-(methoxymethyl) acetanilide.

EPA Reg. No. 524-314-AA U.S. Pat. No. 3,547,620

For use of this product in California see page 103.

843.41-001.37/53 1984-2

DANGER
Keep Out of Reach of Children

The use of this product may be hazardous to your health. This product contains alachlor, which has been determined to cause tumors in laboratory animals.

Users are urged to obtain information on careful handling and use of pesticides and to participate in Monsanto Company's education program on potential hazards and safer use of this product. For additional information, contact Monsanto Agricultural Products Co. at 1-800-453-4000. [NOTE: This number is for user safety information only. *NOT* for emergencies. For emergencies, call collect day or night (314) 694-4000.]

Complete Directions for Use

Read the entire label before using this product.

Use only according to label instructions.

Read "LIMIT OF WARRANTY AND LIABILITY" before buying or using. If terms are not acceptable, return at once unopened.

LIMIT OF WARRANTY AND LIABILITY

This company warrants that this product conforms to the chemical description on the label and is reasonably fit for the purposes set forth in the Complete Directions for Use label booklet ("Directions") when used in accordance with those Directions under the conditions described therein. NO OTHER EXPRESS OR IMPLIED WARRANTY OF FITNESS FOR PARTICULAR PURPOSE OR MERCHANTABILITY IS MADE. This warranty is also subject to the conditions and limitations stated herein.

Buyer and all users shall promptly notify this company of any claims whether based in contract, negligence, strict liability, other tort or otherwise.

Buyer and all users are responsible for all loss or damage from use or handling which results from conditions beyond the control of this company, including but not limited to incompatibility with products other than those set forth in the Directions, unusual weather (i. weather conditions which are outside the range considered normal at the application site and for the time period when the product is applied with the normal range being determined on the basis of the average range for the prior 40 years computed from the best available information, and ii. weather perils, including but not limited to hurricanes, tornadoes and floods) as well as weather considerations set forth in the Directions, application in any manner not explicitly set forth in the Directions, moisture conditions outside the moisture range specified in the Directions, or the presence of products other than those set forth in the Directions in or on the soil or crop.

THE EXCLUSIVE REMEDY OF THE USER OR BUYER, AND THE LIMIT OF THE LIABILITY OF THIS COMPANY OR ANY OTHER SELLER FOR ANY AND ALL LOSSES, INJURIES OR DAMAGES RESULTING FROM THE USE OR HANDLING OF THIS PRODUCT (INCLUDING CLAIMS BASED IN CONTRACT, NEGLIGENCE, STRICT LIABILITY, OTHER TORT OR OTHERWISE) SHALL BE THE PURCHASE PRICE PAID BY THE USER OR BUYER FOR THE

QUANTITY OF THIS PRODUCT INVOLVED, OR, AT THE ELECTION OF THIS COMPANY OR ANY OTHER SELLER, THE REPLACEMENT OF SUCH QUANTITY OR, IF NOT ACQUIRED BY PURCHASE, REPLACEMENT OF SUCH QUANTITY. IN NO EVENT SHALL THIS COMPANY OR ANY OTHER SELLER BE LIABLE FOR ANY INCIDENTAL OR CONSEQUENTIAL DAMAGES.

The buyer and all users are deemed to have accepted the terms of this LIMIT OF WARRANTY AND LIABILITY which may not be varied by any verbal or written agreement.

PRECAUTIONARY STATEMENTS

HAZARD TO HUMANS AND DOMESTIC ANIMALS

Keep Out of Reach of Children

DANGER!

CAUSES EYE BURNS AND SKIN IRRITATION.

HARMFUL IF SWALLOWED.

MAY CAUSE ALLERGIC SKIN REACTION.

Do not get in eyes, on skin or on clothing.

Wear goggles or face shield, rubber gloves, long trousers, long sleeved shirt or jacket of tightly woven material along with boots high enough to cover the ankles when transferring and mixing, and when adjusting, repairing or cleaning equipment. Wear rubber boots when pouring from open containers or when reentry is made into fields following application of this product through center pivot irrigation systems and the field is still wet. Clothing which comes in contact with Lasso must be washed before reuse. Clothing or other materials which become drenched with the concentrated pesticide must be disposed of in a sanitary landfill, by incineration or, if allowed by state and local authorities, by burning. If burned, stay out of smoke.

Wash thoroughly with soap and water after handling.

FIRST AID: IF IN EYES, immediately flush with plenty of water for at least 15 minutes. Call a physician.

IF ON SKIN, immediately flush with plenty of water while removing contaminated clothing. Call a physician. Wash clothing before reuse. Sensitized persons should avoid further contact and reuse of clothing.

IF SWALLOWED, induce vomiting immediately by giving two glasses of water and sticking finger down throat. Call a physician. NEVER GIVE ANYTHING BY MOUTH TO AN UNCONSCIOUS PERSON.

In case of an emergency involving this product, Call Collect, day or night, (314) 694-4000.

ENVIRONMENTAL HAZARDS

This chemical has been identified in limited ground water sampling and there is the possibility that it can leach through the soil to ground water, especially where soils are coarse and ground water is near the surface. Following application, and during rainfall events that cause runoff, this chemical may reach surface water bodies including streams, rivers and reservoirs.

Care must be taken when using this product to prevent back siphoning into wells, spills or improper disposal of excess pesticide, spray mixtures or rinsates.

Check valves or antisiphoning devices must be used on all mixing and/or irrigation equipment. Disposal of excess pesticide, spray mixtures, or rinsate should be according to label use instructions or according to the state pesticide or environmental control agency or the hazardous waste representative at the nearest EPA regional office.

Do not apply directly to water or wetlands. Do not contaminate water by cleaning of equipment or disposal of wastes.

PHYSICAL OR CHEMICAL HAZARD

COMBUSTIBLE.

Do not use or store near heat or flame.

Use only with adequate ventilation.

In case of:

FIRE, use water spray, foam, dry chemical or CO_2.

SPILL or LEAK, flush area with water spray.

STORAGE AND DISPOSAL

Do not contaminate water, foodstuffs, seed or feed by storage and disposal.

STORE ABOVE 32°F. TO KEEP PRODUCT IN SOLUTION.

STORAGE

Below 32°F. crystals may form and settle to the bottom. If crystals form, place in a warm room and roll and/or shake the container frequently to redissolve before using. For bulk containers, see the container label for alternate storage information.

DISPOSAL

Wastes of this pesticide may cause eye burns and skin irritation and may be dangerous. Improper disposal of excess pesticide, spray mixtures, or rinsate is a violation of Federal law. If these wastes cannot be disposed of according to label use instructions, contact your State Pesticide or Environmental Control Agency, or the Hazardous Waste representative at the nearest EPA Regional Office for guidance.

Empty container retains vapor and product residue. Observe all labeled safeguards until container is cleaned, reconditioned or destroyed.

See the individual container label for disposal information.

GENERAL INFORMATION

LASSO® herbicide is recommended for control of yellow nutsedge and the annual grasses and broadleaf weeds listed in the "WEEDS CONTROLLED" section of this label. This product may be applied either as a surface application after planting or shallowly incorporated prior to planting to blend the herbicide treatment into the upper 1 to 2 inches of soil. Except for minimum or conservation tillage systems, the seedbed should be fine, firm and free of clods and trash.

®LASSO is a registered trademark of Monsanto Company.

Read and carefully observe precautionary statements and all other information appearing on the labeling of all products used in mixtures and sequential treatments.

NOTE THAT DYANAP™ PARAQUAT AND PREMERGE™ (dinoseb) HERBICIDES ARE POISONS WHICH MAY BE FATAL OR HARMFUL TO HUMANS AND DOMESTIC ANIMALS.

Do not flood irrigate following application of this product or tank mixtures. This product will not control emerged seedlings.

Do not apply when conditions favor drift.

DO NOT APPLY WITH AERIAL EQUIPMENT.

Use of this product not consistent with this label may result in injury to persons, animals or crops, or other unintended consequences. Keep container closed to prevent spills and contamination.

Flush sprayer with clean water after use.

™Dyanap is a trademark of Uniroyal, Inc.

™ Premerge is a trademark of the Dow Chemical Company.

SOIL TEXTURE

Applicators should evaluate soil and weed conditions carefully to assure that they choose the lowest effective label rate.

The recommended use rates of this product and the other herbicides labeled for use in tank mixtures with this product generally vary with soil texture. Unless soil texture is specifically named, rate tables throughout this label refer to only three soil texture groups: coarse, medium and fine. The following is a complete listing of soil textures included in each of these three soil texture groups:

Soil Texture Group	Soil Texture
Coarse:	sand, lomy sand, sandy loam
Medium:	loam, silt loam, silt, sandy clay loam

Fine:	silty clay loam, clay loam, sandy clay, silty clay, clay

Refer to the above table to determine the corresponding soil texture group for the soil to be treated.

MIXING, SPRAYING AND HANDLING INSTRUCTIONS

NOTE

Direct contact or exposure to Lasso or spray mixtures of Lasso should be minimized. The following instructions for transfer, mixing, cleaning or repairing equipment should be followed in order to minimize this exposure. Review the protective clothing requirements as listed in the Precautionary Statements section of this label (front panel) and do not use this product until you have the necessary protective clothing.

2.5 Gallon Containers

Open pouring from these containers can result in exposure from splashing or spilling. Special care in lifting and pouring are strongly recommended.

55 Gallon and Bulk Containers

Open pouring from these containers can result in exposure from splashing or spilling and is not recommended. This product should be transferred from these containers to the mix or spray tank using pumps or transfer probes. The probe or pump should not be removed from the container or disconnected until the container is emptied and rinsed. Use the pump or probe system to rinse the empty container and transfer the rinsate directly to the mix or spray tank.

Equipment Cleaning & Repair

Cleaning and repair of transfer systems and application equipment is a source of exposure to this product. Care should be taken to minimize exposure during cleaning and repair of transfer systems and application equipment. Whenever possible, these systems or equipment should be rinsed before being cleaned or repaired. When repairs must be made during transfer or application, the equipment should be shutdown, and special care taken to avoid contact with the pesticide.

Always predetermine the compatibility of LASSO or labeled tank mixtures of LASSO with water carrier or sprayable fluid fertilizer carrier by mixing small proportional quantities in advance. See the "Standard Sprayable Fluid Fertilizer Compatibility Test" section in this label to determine the compatibility of this product and the labeled tank mixtures recommended for use with sprayable fluid fertilizer carrier. NOTE: LASSO plus Banvel™ herbicide may be applied in sprayable fluid fertilizers only on field corn and silage corn. LASSO and the tank mixtures of LASSO plus atrazine and LASSO plus Bladex™ herbicide may be applied in sprayable fluid fertilizers on all corn. For soybeans, only LASSO and the tank mixtures of LASSO plus Lorox™ herbicide and LASSO plus Sencor™ or Lexone™ herbicides may be applied in sprayable fluid fertilizers. DO NOT apply LASSO plus Dyanap or LASSO plus Premerge herbicide in sprayable fluid fertilizers. DO NOT apply LASSO or labeled tank mixtures of LASSO in sprayable fluid fertilizer on any other crops.

Mix LASSO or labeled tank mixture of LASSO with the appropriate carrier as follows:

1. Place a 20 to 35 mesh screen or wetting basket over filling port.
2. Through the screen, fill the sprayer tank one-half full with the appropriate carrier.
3. If a compatibility agent is necessary to improve mixing or to prevent the formation of undesirable and unsprayable gels or precipitates, while agitating add it to the carrier already in the tank. Use only compatibility agents, cleared by FDA for this use. Read and follow all directions for use, cautionary statements and all other information appearing on the selected compatibility agent label. Check for adequate agitation.
4. If a wettable powder or dry flowable formulation is used, make a slurry with water and add it slowly through the screen into the tank. Continue agitation.
5. If a flowable formulation is used, add slowly through screen into the tank. Mixing and compatibility may be improved when flowable is pre-mixed one part flowable with one part water and added to the tank in diluted form.
6. Add LASSO slowly through the screen into the tank. Mixing and compatibility may be improved when this product is pre-diluted with two parts of water and added to the tank in diluted form.
7. Complete filling the sprayer tank with carrier. If Roundup or paraquat is used, add the required amount near the end of the filling process. Remove hose from tank immediately after filling to avoid siphoning back into the carrier source.

Maintain good agitation at all times until the contents of the tank are sprayed. NOTE: If spray mixture is allowed to settle at any time thorough agitation is required to resuspend the mixture before spraying is resumed.

Keep by-pass line on or near bottom of tank to minimize foaming. Screen size in nozzle or line strainers should be no finer than 50 mesh. Carefully select proper nozzle to avoid spraying a fine mist. Check for even distribution of spray droplets. To reduce loss of the chemical due to drift of a fine mist, apply at nozzle pressures below 40 psi.

™Banvel is a trademark of Velsicol Chemical Corp.

™ Bladex is a trademark of the Shell Chemical Company.

™ Lorox is a trademark of E. I. duPont de Nemours and Company.

™ Lexone is a trademark of E. I. duPont de Nemours and Company.

™ Sencor is a trademark of the parent company of Farbenfabriken Bayer GmbH, Leverkusen.

STANDARD SPRAYABLE FLUID FERTILIZER COMPATIBILITY TEST

Herbicides may not always mix evenly throughout a sprayable fluid fertilizer or the components may separate too quickly to make their combined use of practical value. This may be due to certain characteristics of the different fluid fertilizers. A simple test using small quantities of the components is suggested to provide compatibility potential.

The test follows:

A. Materials Required For A Compatibility Test

1. Two one-quart jars with lid or stopper (marked "with" and "without").
2. Teaspoons (for a more exacting test, a five to ten milliliter (ml.) pipette or graduated cylinder is desirable).
3. Sprayable fluid fertilizer to be tested.
4. The herbicide chemicals to be mixed.
5. A compatibility agent (the purpose of the adjuvant is to help keep the fertilizer and crop protection chemical in suspension, if this assistance is needed).

B. Procedure

1. Add one pint of the sprayable fluid fertilizer that will be used or other herbicide carrier to each jar marked "with" and "without".

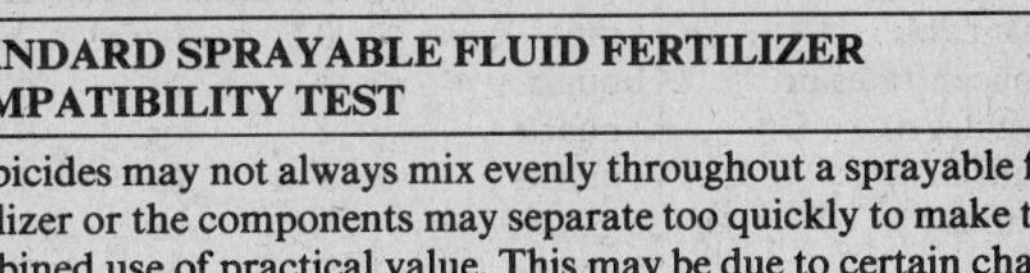

2. To the jar marked "with", add 1/4 teaspoon or 1.2 milliliters of a suitable compatibility agent; shake gently for five to ten seconds to mix. (1/4 teaspoon in one pint is the equivalent of two pints per 100 gallons of liquid fertilizer.)

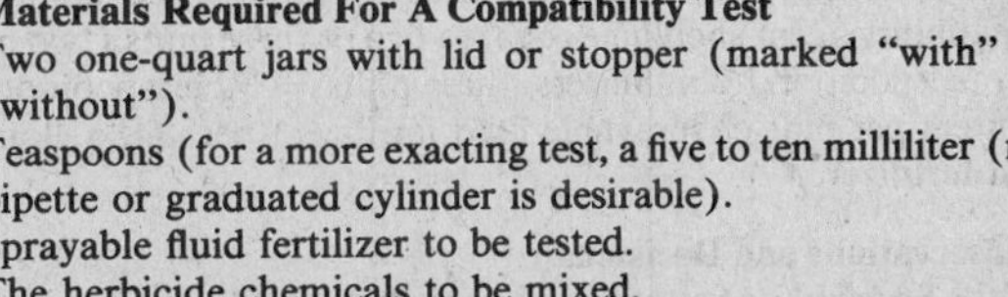

3. To each jar add the appropriate amount of herbicide(s). If more than one is used, add them separately with the wettable powders or dry flowables added first, flowables second and liquid last. Shake gently five to ten seconds after each addition.

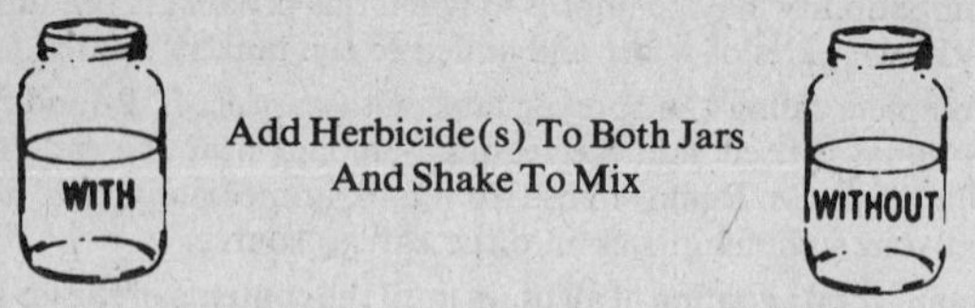

Herbicide	Rate/A		Amount To Be Added Per Pint Of Sprayable Fluid Fertilizer (Assuming Volume Is 25 Gal/A) Level Teaspoons
Wettable	1 lb.	=	1.5
Powders	2 lb.	=	3
or	3 lb.	=	4.5
Dry Flowables	4 lb.	=	6
	5 lb.	=	7.5

Herbicide	Rate/A		Level Teaspoons		Milliliters
Emulsifiable	1 pint	=	0.5	or	2.4
Concentrates or	1 quart	=	1	or	4.7
Flowables or	2 quarts	=	2	or	9.5
Liquids or	3 quarts	=	3	or	14.2
Solutions	1 gallon	=	4	or	19.0
	5 quarts	=	5	or	23.8

This compatibility test is designed for 25 gallons of spray per acre with the maximum labeled rate of herbicide. For changes in spray volume or herbicide concentration, make appropriate changes in the ingredients of the test. Regardless of spray volume, the amount of compatibility agent should be equal to two or three pints (two pints = $\frac{1}{4}$ teaspoon or 1.2 milliliters, three pints = $\frac{3}{8}$ teaspoon or 1.8 milliliters per pint of sprayable fluid fertilizer) per 100 gallons of liquid fertilizer.

C. Observations and Decisions

1. If the herbicide(s) and the sprayable fluid fertilizer are compatible.
2. If a compatibility agent is necessary.

Five minutes after the final addition and mixing, observe both jars for the formation of large flakes, sludge, gels or other precipitates. Observe if the herbicide(s) cannot be physically mixed with the liquid fertilizer but remains as small oily particles in the solution.

If incompatibility in any form described above occurs in the jar "with" the compatibility agent added, the liquid fertilizer and the herbicide(s) should not be used together in the same spray tank.

If incompatibility as described above occurs in the jar "without" the adjuvant but not in the jar "with" adjuvant, the use of a compatibility adjuvant is recommended.

Both jars should be allowed to stand and be observed periodically for one-half hour. If the separate layers of liquid fertilizer and additives can be resuspended by shaking, commercial application is possible. An emulsifiable concentrate normally will go to the top after standing; wettable powders will either settle to the bottom of the tank or jar, or float to the top, depending upon the density of the fertilizers.

If the herbicide(s) is compatible with fluid fertilizer in the foregoing test without having to use a compatibility agent, fluid fertilizer may be used for the premixing. If it is not compatible without the compatibility agent, the herbicide(s) should be premixed with water before adding to the spray tank.

APPLICATION INFORMATION

APPLICATION SYSTEMS

Ground Broadcast Treatment—Apply this product and the labeled tank mixtures in 10 or more gallons of solution per acre using broadcast boom equipment. The carrier may be either water or sprayable fluid fertilizer as specified for the crop to be treated in the "DIRECTIONS FOR USE" section of this label. Do not apply during periods of gusty winds, when winds are in excess of 15 mph or when other conditions favoring drift exist.

Band Treatment—Apply a broadcast equivalent rate and volume per acre. To determine these:

$$\frac{\text{Band width in inches}}{\text{Row width in inches}} \times \text{Broadcast RATE per acre} = \text{Band RATE per acre}$$

$$\frac{\text{Band width in inches}}{\text{Row width in inches}} \times \text{Broadcast VOLUME per acre} = \text{Band VOLUME per acre}$$

Center Pivot Irrigation Treatment—Center pivot is the only irrigation equipment through which this product may be used. Refer to the Monsanto booklet "How to Apply Lasso Herbicide with Center Pivot Irrigation" before use.

Use only in systems that apply uniformly. Do not apply when wind speed favors drift, when system connections or fittings leak, when nozzles do not provide uniform distribution or when the line containing the product must be dismantled and drained. Meter diluted LASSO herbicide or a labeled tank mixture of LASSO into the center pivot irrigation system after planting and before crop or weed emergence. Herbicide applications should be made in $\frac{1}{3}$ to $\frac{3}{4}$ inch of water per acre. Do not apply in more than $\frac{3}{4}$ inch water per acre under any conditions or reduced performance may occur. On very sandy soils (more than 60% sand and less than 1% organic matter), apply these herbicides in maximum of $\frac{1}{2}$ inch water per acre. Good agitation must be maintained during the entire application period. Have appropriate check valves or other suitable devices in the system to insure that the herbicide solution cannot siphon back into the water supply. Flush the system with water when application is complete. If spills of the herbicide concentrate occur, immediately wash off the system.

NOTE: Applications can be made by injection through center pivot irrigation equipment only when specified for the crop to be treated in the "DIRECTIONS FOR USE" section of this label. Do not apply this product or any labeled tank mixture through center pivot irrigation unless the treatment is specifically recommended in this label for the crop to be treated.

Pumps, injection equipment, agitation equipment, hoses and connections between nurse tank and the point of injection must be constructed of materials which are highly resistant to LASSO. Calibration and resistant materials information is available in the booklet "How to Apply Lasso Herbicide with Center Pivot Irrigation" at your local farm chemical dealer or by writing Monsanto Company.

Application with Dry Bulk Fertilizer—The herbicide/fertilizer impregnation process must be done only at commercial fertilizer or chemical dealerships properly equipped for this procedure.

Dry bulk fertilizer may be impregnated with LASSO or the tank mixtures of LASSO plus atrazine or LASSO plus Bladex on corn; and LASSO or the tank mixtures of LASSO plus Lexone or Sencor on soybeans. LASSO and these tank mixtures must be applied with 200 to 450 pounds of dry bulk fertilizer per acre and shallowly incorporated within 7 days prior to planting. The proper amount of herbicide must be applied per acre in the blend as recommended in this label for the crop, weed and soil type treated. Refer to the table for broadcast rate per acre in the appropriate crop section of this label to determine the recommended rate per acre for the herbicide treatment to be applied.

The following table provides a reference to determine the amount of LIQUID herbicide to be mixed per ton of dry bulk fertilizer for a range of herbicide recommendations for fertilizer rates per acre:

Fertilizer Rate (Lb./Acre)	Acres Covered (Per Ton)	Recommended Quarts Liquid Herbicide/Acre 1.5	2	2.5	3
		(Quarts of Herbicide/Ton Dry Bulk Fertilizer)			
200	10	15	20	25	30
250	8	12	16	20	24
300	6.7	10	13.3	16.7	20
350	5.7	8.6	11.4	14.3	17.1
400	5	7.5	10	12	15
450	4.5	6.7	8.9	11.1	13.3

To determine the amount of herbicide needed for rates not included in the preceding table, use the following formula:

$$\frac{\text{Recommended Herbicide Rate (lb. or qt.)/Acre} \times 2000}{\text{Pounds Fertilizer/Acre}} = \text{Pounds or Quarts of herbicide per ton of dry bulk fertilizer}$$

Mix and blend the dry fertilizer and herbicide mixture in a closed rotary-drum type mixer allowing sufficient time to ensure uniform coverage. Use at least one ton of dry fertilizer per mixing operation. Inject the herbicide into the drum over a minimum of a 2-minute period and allow at least 2 additional minutes mixing time to ensure uniformity. The nozzle used to spray the herbicide treatment must be placed inside the mixer to provide uniform spray coverage of the tumbling fertilizer.

If the dry fertilizer used has inadequate adsorptive capacity, use a highly adsorptive powder such as Microcel E to provide a free-flowing mixture. The adsorptive powder must be added separately and uniformly to the prepared herbicide-fertilizer mixture. Generally, the weight of Microcel E is no more than 2% of the total weight of the herbicide. The following table provides a partial list of dry fertilizers which may be impregnated with LASSO or tank mixtures of LASSO with other herbicides:

TM Microcel is a trademark of Manville Products Corporation.

Fertilizer	LASSO®	LASSO + atrazine	LASSO + Bladex or	LASSO + Lexone Sencor
Ammonium sulfate (21-0-0)	Yes	Yes	Yes	Yes
Ammonium phosphate-sulfate (16-20-0)	Yes	Yes	Yes	Yes
Diammonium phosphate (18-46-0)	Yes	Yes	Yes	Yes
Potassium chloride (0-0-60)	Yes	Yes	Yes	Yes
Potassium sulfate (0-0-52)	Yes	Yes	Yes	Yes
Single super-phosphate (0-20-0)	Yes	No	No	No
Treble super-phosphate (9-46-0)	Yes	No	No	No
*Urea (45-0-0)	Yes	Yes	Yes	Yes

* Some ureas may be phytotoxic when applied on corn. Use only ureas known to be safe to corn. Blends suitable for use in soybeans, such as 0-26-26 are also approved for use with LASSO and the tank mixtures listed.

NOTE: DO NOT impregnate LASSO or tank mixtures of LASSO with other herbicides on fertilizers containing ammonium nitrate, potassium nitrate or sodium nitrate.

Spread the herbicide-dry fertilizer mixture uniformly with a properly calibrated applicator. Fertilizers impregnated with LASSO or tank mixtures of LASSO with other herbicides must be spread at half-rate and overlapped 100 percent to obtain full rate and uniform distribution. Non-uniform spreading of the fertilizer-herbicide mixture may result in unsatisfactory weed control or crop injury.

APPLICATION TIMING AND METHODS

Preemergence Surface Application—This product and all labeled tank mixtures except LASSO plus Treflan™ may be applied to the soil surface after planting and prior to either crop or weed emergence. Apply within 5 days of last preplant tillage. If weeds emerge after treatment, or if treatment is applied more than 5 days after last preplant tillage, rotary hoe or shallowly cultivate immediately to improve performance.

TM Treflan is a trademark of Elanco Products Company.

Precipitation or overhead sprinkler irrigation is required after application to move the herbicide treatment into the weed germination zone. The amount of precipitation or overhead sprinkler irrigation required depends on existing soil moisture, soil type and precent organic matter content, but ⅓ to ¾ inches is normally adequate. Performance is improved when moisture is received within 7 days after application and prior to weed emergence. High intensity or excessive rainfall or excessive irrigation after application may reduce control.

Postemergence Surface Application—This product and certain tank mixtures may be applied to specific crops at the ground cracking stage of the crop or after crop emergence. Refer to the specific treatment and crop intended in the "DIRECTIONS FOR USE" section of the label to determine if applications at cracking or postemergence to the crop are recommended and to determine the proper weed and crop growth stage limitations. Precipitation or overhead sprinkler irrigation is required after application to move the herbicide treatment into the weed germination zone to control unemerged weeds. If weeds emerge after treatment, rotary hoe or shallowly cultivate immediately to improve performance.

Preplant Incorporation—This product and many of the labeled tank mixtures may be mixed into the soil using shallow incorporation equipment any time within 7 days prior to planting. Refer to the crop to be treated in the "DIRECTIONS FOR USE" section of this label to determine if preplant incorporation is recommended. Apply the recommended treatment rate to the soil surface as a broadcast application. Either existing soil moisture or subsequent precipitation or irrigation is required to bring incorporated herbicide treatments into contact with germinating weed seedlings. If weeds emerge after treatment, rotary hoe or shallowly cultivate immediately to improve performance.

NOTE: Do not incorporate unless specifically recommended in this label for the crop and herbicide treatment selected. Incorporation of this product and its labeled tank mixtures on coarse soil types may reduce length of weed control due to leaching with rainfall or irrigation. Incorporation is not recommended on coarse soils in the Southeastern States.

1-Pass Incorporation (Surface Blend™): Shallowly incorporate (Surface Blend) the treatment into the upper 1 to 2 inches of the soil. Equipment should be operated at manufacturer's designed speed for incorporation to ensure adequate mixing and distribution of the herbicide treatment in the soil. Equipment design including any drag attachments must be adequate to avoid soil ridging which may result in streaked or reduced weed control. Soil conditions, including moisture content and crop residue levels, must be suitable to allow thorough and uniform mixing with the equipment used for 1-pass incorporation.

Contact Monsanto Company for additional information regarding recommended equipment and methods for Surface Blend application.

TM Surface Blend is a Trademark of Monsanto Company.

2-Pass Incorporation: When 2-pass incorporation is used, shallowly incorporate the herbicide treatment into the upper 1 to 2 inches of the soil with equipment set to work the soil NO DEEPER THAN 4 INCHES. The second pass must be made at an angle to and no deeper than the first pass to ensure proper distribution of the herbicide treatment in the soil.

Cultivation Information—Delay cultivation after application for as long as possible unless weeds or grasses emerge. Shallowly cultivate or rotary hoe immediately if weeds or grasses emerge. If cultivation is necessary because of soil crusting or compaction, set equipment shallow and minimize lateral soil movement to avoid dilution or displacement of the herbicide treatment. Only after weeds emerge in the treated band, set cultivator to throw soil into the row covering the band. On peanuts, do not throw soil into the row.

WEEDS CONTROLLED

Applicators should evaluate soil and weed conditions carefully to assure that they choose the lowest effective label rate.

When applied as directed under conditions described, yellow nutsedge and the annual grasses and broadleaf weeds listed are CONTROLLED by Lasso:

LASSO®

Annual Grasses	
Barnyardgrass	*Echinochloa crus-galli*
Crabgrass	*Digitaria ischaemum*
	Digitaria sanguinalis
Foxtail (giant)	*Setaria faberi*
Foxtail (green, robust purple, robust white)	*Setaria viridis*
Foxtail (yellow)	*Setaria lutescens*
Goosegrass	*Eleusine indica*
Panicum (fall)	*Panicum dichotomiflorum*
Rice (red)**	*Oryza sativa*
Signalgrass (broadleaf)	*Brachiaria platyphylla*
Witchgrass	*Panicum capillare*
Annual Broadleaves	
Carpetweed	*Mollugo verticillata*
Galinsoga	*Galinsoga* spp.
Groundcherry (cutleaf)*	*Physalis angulata*
Nightshade (black)*	*Solanum nigrum*
Nightshade (hairy)*	*Solanum sarachoides*
Pigweed, Carelessweed	*Amaranthus* spp.
Purslane	*Pourtulaca oleracea*
Pusley (Florida)	*Richadia scabra*
Sedge	
Nutsedge (yellow)**	*Cyperus esculentus*

* Use a minimum of 2.5 quarts of this product per acre on coarse soils, and 3 to 4 quarts alone or in tank mix combinations on medium and fine textured soils for control of these weeds.

** Use 3 to 4 quarts of this product per acre applied alone or in tank mix combinations listed on this label. Preplant incorporate only for control of these weeds.

When applied preplant incorporated, or preemergence immediately after planting and within 5 days of last tillage, LASSO at a rate of 2.5 to 4 quarts per acre on a broadcast basis applied alone or in a labeled tank mixture REDUCES COMPETITION from the following hard to control annual grasses and broadleaf weeds.

Annual Grasses	
Cupgrass (woolly)*****	*Eriochloa villosa*
Johnsongrass (seedling)	*Sorghum halepense*
Millet (proso)****	*Panicum miliaceum*
Panicum (Texas)*	*Panicum texanum*
Sandbur, Grassbur	*Cenchrus pauciflorus*
Shattercane (wild cane)**	*Sorghum bicolor*
Annual Broadleaves	
Beggarweed (Florida)	*Desmodium tortuosum*
Lambsquarters	*Chenopodium album*
Ragweed (common)	*Ambrosia artemisiifolia*
Sicklepod******	*Cassia obtusifolia*
Sida (prickly), Teaweed	*Sida spinosa*
Smartweed	*Polygonum pensylvanicum*
	Polygonum persicaria
Sedge	
Nutsedge (yellow)***	*Cyperus esculentus*

* For reduced competition from this weed, use a minimum of 3 quarts per acre and apply through center pivot irrigation or under sprinkler irrigation systems.

** Use 4 quarts per acre of this product applied alone or in tank mix combinations listed in this label. Preplant incorporate only to reduce competition from this weed. In areas of heavy weed infestations, apply an additional 2 quarts per acre of LASSO any time from planting until corn in 5 inches tall or any time before soybeans exceed unifoliate stage (first two true leaves). This sequential treatment must be made prior to emergence of shattercane.

*** Use 3 to 4 quarts per acre as a preemergence surface application on coarse textured soils with less than 3% organic matter to reduce competition from this weed.

**** For reduced competition from this weed, use 4 quarts per acre of this product applied preplant incorporated. Improved control may be obtained by applying subsequent labeled treatments of registered postemergence herbicides.

***** Use 4 quarts of LASSO per acre plus 2 quarts of Bladex to reduce competition from this weed.

****** For reduced competition, use LASSO in a tank mix with Sencor, Lexone, or Dyanap.

When applied as directed under conditions described, tank mixtures of LASSO with products listed in the following table will control or reduce competition from the additional broadleaf weeds as indicated. Refer to the crop on which treatment is intended in the "DIRECTIONS FOR USE" section of this label to determine the tank mixtures and rates recommended for use in each crop.

NOTE: C = Control R = Reduced Competition

	Amiben™	atrazine	Banvel™	Bladex™	Dyanap™	linuron	Premerge™	propazine	Sencor™/Lexone™	Treflan™
Cocklebur Xanthium pensylvanicum	•	R	R	R	R	•	R	R	R	•
Coffeeweed (Hemp sesbania) Sesbania exaltata	•	•	•	•	R	•	•	•	•	•
Copperleaf (hophornbeam) Acalypha spp.	•	•	•	•	•	•	C	•	•	•
Cupgrass (wooly)* Eriochloa villosa	•	•	•	R	•	•	•	•	•	•
Groundcherry (annual) Physalis spp.	•	C	•	•	•	C	•	C	C	•
Jimsonweed Datura Stramonium	•	C	•	•	•	•	R	C	R	•
Kochia Kochia scoparia	•	C	•	•	•	•	•	C	•	C
Lambsquarters Chenopodium album	C	C	C	C	•	C	C	C	C	•
Morningglory (tall) Ipomea purpurea	•	R	R	R	R	•	R	R	•	•
Mustard Brassica spp.	R	C	C	C	•	C	C	C	•	•
Nightshade (black) Solanum nigrum	•	C	•	C	•	•	•	C	•	•
Nightshade (hairy) Solanum sarachoides	•	C	•	C	•	•	•	C	•	•
Ragweed (common) Ambrosia artemisiifolia	C	C	C	C	•	C	C	C	C	•
Sicklepod Cassia obtusifolia	•	C	•	•	R	•	R	C	R	•
Sida (prickly), Teaweed Sida spinosa	•	•	•	•	C	C	C	•	•	•

Smartweed										
Polygonum pensylvanicum										
Polygonum persicaria	C	C	C	C	•	C	•	C	C	•
Starbur (bristly)										
Acanthospermum hispidum	•	•	•	•	C	•	•	•	•	•
Velvetleaf, Buttonweed										
Abutilon theophrasti	R	R	R	R	•	R	R	R	R	•
Waterhemp										
Amaranthus tuberculatos	•	•	C	•	•	•	•	•	•	•

* Use 4 quarts of LASSO per acre plus 2 quarts of Bladex to reduce competition from this weed.

NOTE: The 3-way tank mixture of LASSO plus atrazine plus Bladex controls or reduces competition as specified for the respective weeds listed under the LASSO plus atrazine and LASSO plus Bladex recommendations.

The 3-way tank mixture of LASSO plus Amiben plus Lexone or Sencor controls or reduces competition as specified for the respective weeds listed under the LASSO plus Amiben and LASSO plus Lexone or Sencor recommendations.

Low Volatile Ester and Amine formulations of 2,4-D (2,4-dichlorophenoxy) acetic acid only, may be applied sequentially to LASSO in field and silage corn only for control of additional emerged broadleaf weeds listed on the 2,4-D label.

Basagran™ may be applied sequentially to LASSO in soybeans for control of additional emerged weeds listed on the Basagran label such as cocklebur and velvetleaf.

Certain tank mixtures of LASSO and broadleaf herbicides may be combined with applications of Roundup® herbicide or paraquat for control of many emerged weeds prior to corn or soybean emergence. Refer to the Roundup or paraquat label for a list of emerged weeds controlled by tank mixtures with these products.

™ Basagran is a trademark of BASF Aktiengesellschaft Ludwigshafen am Rhein.

®Roundup is a registered trademark of Monsanto Company.

CROPS
DIRECTIONS FOR USE

It is a violation of federal law to use this product in any manner inconsistent with its labeling.

CORN
(ALL TYPES)

LASSO®

Apply this product in water or sprayable fluid fertilizer solution for control of yellow nutsedge and the annual grasses and broadleaf weeds listed in the "WEEDS CONTROLLED" section of this label.

APPROVED APPLICATION SYSTEMS

Ground—Broadcast Boom; banded
Center Pivot Irrigation
Dry Bulk Fertilizer Impregnation

Reference: The "APPLICATION SYSTEMS" section of this label provides detailed information on the proper application of this product using the system selected.

APPROVED APPLICATION METHODS

Preplant Incorporated—Apply this product within 7 days prior to planting and shallowly incorporate or Surface Blend into the upper 1 to 2 inches of soil. This product must be incorporated when treating furrow irrigated corn. Performance may be increased by packing the soil following incorporation if furrow irrigation is to be used. Irrigation within 10 days following planting may improve weed control.

Preemergence Surface—Apply this product after planting, before crop and weed emergence and within 5 days after last preplant tillage operation.

Postemergence Surface—This product may be applied after crop emergence to reduce competition from weeds listed as controlled on this label. Apply before corn exceeds 5 inches in height.

Minimize herbicide contact with the corn foliage since leaf burn may occur. Subsequent growth or yield should not be affected.

Weeds which are emerged at time of application are not controlled by this product. If weeds are emerged at application, shallowly cultivate or rotary hoe immediately to improve performance. DO NOT apply in sprayable fluid fertilizer after crop emergence.

NOTE: The highest levels of weed control will be obtained when applications are made prior to the emergence of the crop and weeds.

Reference: The "APPLICATION TIMING AND METHODS" section of this label provides detailed information and procedures for the application timing and method selected.

Sequential Treatment—This product may be applied as two sequential treatments for improved control of hard-to-control weeds. The combined total application rate for the two treatments must not exceed 8 quarts per acre per year. The second treatment must be applied before the corn exceeds 5 inches in height.

RECOMMENDED RATES: Refer to the following tables for the recommended broadcast treatment rates for this product. Applications which are not consistent with recommendations in this label may result in unsatisfactory weed control or crop injury.

Surface Application Rates

Soil Texture Group*	Broadcast Rate Per Acre (Quarts)** — Less than 3% organic matter	Broadcast Rate Per Acre (Quarts)** — 3% or more organic matter
Coarse***	2 to 3	2 to 3
Medium	2.5 to 3	2.5 to 3.5
Fine	3 to 3.5	3 to 4

* Refer to the "SOIL TEXTURE" section of this label to determine the corresponding texture group for the soil to be treated

** Use the higher rate in the recommended range in areas of heavy weed infestations. Use a minimum of 2.5 quarts of this product per acre on coarse soils and 3 to 4 quarts on medium or fine textured soils to control black or hairy nightshade or to reduce competition from hard-to-control weeds listed in this label.

*** When applied through CENTER PIVOT IRRIGATION or under sprinkler irrigation systems on coarse soils, use a minimum of 3 quarts per acre of this product.

NOTE: For postemergence surface treatment—When applied following LASSO or tank mixtures of LASSO, the total amount of LASSO applied alone or in combination must not exceed 8 quarts per acre per year. Do not apply in sprayable fluid fertilizer.

On coarse soils with 10% or more organic matter use 4 to 6 quarts of LASSO herbicide per acre. On muck or peat soils use 6 to 8 quarts per acre. Use the higher rate of LASSO herbicide in the recommended ranges in areas of heavy weed infestations.

Preplant Incorporated Application Rates

Soil Texture Group*	Broadcast Rate Per Acre (Quarts)** — Less than 3% organic matter	Broadcast Rate Per Acre (Quarts)** — 3% or more organic matter
Coarse	2.5 to 3	2.5 to 3
Medium	3 to 3.5	3 to 4
Fine	3 to 4	3.5 to 4

* Refer to the "SOIL TEXTURE" section of this label to determine the corresponding texture group for the soil to be treated.

** Use the higher rate in the recommended range in areas of heavy weed infestations. Use 2.5 to 3.5 quarts per acre when furrow irrigation is used. Use a minimum of 3 quarts on coarse textured soils for control of yellow nutsedge or red rice. Rates may be increased to a maximum of 4 quarts on any soil type when heavy infestations of yellow nutsedge or red rice are present, or to reduce competition from the hard-to-control weeds listed in this label.

On coarse soils with 10% or more organic matter use 4 to 6 quarts of LASSO herbicide per acre. On muck or peat soils use 6 to 8 quarts per acre. Use the higher rate of LASSO herbicide in the recommended ranges in areas of heavy weed infestations.

LASSO® PLUS ATRAZINE (TANK MIXTURE)

Apply this tank mixture in water or sprayable fluid fertilizer solutions for control of yellow nutsedge and the annual grasses and broadleaf weeds listed in the "WEEDS CONTROLLED" section of this label.

APPROVED APPLICATION SYSTEMS
Ground—Broadcast Boom; banded
Center Pivot Irrigation
Dry Bulk Fertilizer Impregnation

Reference: The "APPLICATION SYSTEMS" section of this label provides detailed information on the proper application of this tank mixture using the system selected.

APPROVED APPLICATION METHODS

Preplant Incorporated—Apply this tank mixture within 7 days prior to planting and shallowly incorporate or Surface Blend into the upper 1 to 2 inches of soil.

Preemergence Surface—Apply this tank mixture after planting, before crop and weed emergence and within 5 days after last preplant tillage operation.

Postemergence Surface—Apply this tank mixture after crop emergence until weeds reach the 2-leaf stage and the crop is no more than 5 inches tall. Applications made beyond the 2-leaf stage will not provide satisfactory weed control. Some leaf burn may occasionally occur when applied early postemergence to corn, but subsequent growth or yield should not be affected.

DO NOT apply to sweet corn after crop emergence.

DO NOT apply in sprayable fluid fertilizer after crop emergence since crop injury may result.

Reference: The "APPLICATION TIMING AND METHODS" section of this label provides detailed information and procedures for the application timing and method selected.

NOTE: PLANT ONLY CORN, PEANUTS, SORGHUM OR SOYBEANS* the year following use of this mixture.

* There is a possibility of crop injury due to carryover of atrazine if soybeans are planted the following year. DO NOT plant soybeans the year following use of this tank mixture on furrow irrigated corn. DO NOT graze treated area or feed treated forage to livestock for 21 days following application of this tank mixture.

RECOMMENDED RATES: Refer to the following tables for the recommended broadcast treatment rates for this tank mixture. Applications which are not consistent with recommendations in this label may result in unsatisfactory weed control or crop injury.

Surface Application Rates

Soil Texture Group*	Broadcast Rate Per Acre (Quarts)**			
	Less than 3% organic matter		3% or more organic matter	
	LASSO® +	atrazine 4L***	LASSO +	atrazine 4L***
Coarse	1.5 to 2.5	1	2 to 3	1
Medium	1.75 to 2.5	1 to 1.2	2 to 3	1.2 to 1.4
Fine	2.25 to 3	1.2 to 1.6	3 to 4	1.2 to 1.6

* Refer to the "SOIL TEXTURE" section of this label to determine the corresponding texture group for the soil to be treated.
** Use the higher rates in the recommended ranges in areas of heavy weed infestations. Use a minimum of 2.5 quarts of LASSO per acre in this tank mixture on coarse soils and 3 to 4 quarts on medium or fine textured soils to reduce competition from hard-to-control weeds listed in this label.
*** Use rates listed in this label when using atrazine 4L. Use equivalent rates when using atrazine 80W or 90% dry flowable formulations. One quart of atrazine 4L equals 1.25 pounds of atrazine 80W or 1.1 pounds of atrazine 90% dry flowable.

NOTE: When applied through CENTER PIVOT IRRIGATION or under sprinkler irrigation systems on coarse soils, use 1 quart per acre of atrazine 4L and a minimum of 2 quarts per acre of LASSO in this tank mixture.

When applied following LASSO or tank mixtures of LASSO, the total amount of LASSO applied alone or in combination must not exceed 8 quarts per acre per year. The total amount of atrazine 4L must not exceed 3.2 quarts per acre per year. Do not apply in sprayable fluid fertilizer. Minimize herbicide contact with the corn foliage since leaf burn may occur. Subsequent growth or yield should not be affected. Do not graze or feed treated forage to live stock for 21 days following application.

Preplant Incorporated Application Rates

Soil Texture Group*	Broadcast Rate Per Acre (Quarts)**			
	Less than 3% organic matter		3% or more organic matter	
	LASSO* +	atrazine 4L***	LASSO +	atrazine 4L***
Coarse	2 to 2.5	1.25	2.5 to 3	1.25
Medium	2.25 to 3	1.25 to 1.45	2.5 to 3.5	1.45 to 1.65
Fine	2.5 to 3.5	1.45 to 1.85	3 to 4	1.45 to 1.85

* Refer to the "SOIL TEXTURE" section of this label to determine the corresponding texture group for the soil to be treated.
** Use the higher rates in the recommended ranges in areas of heavy weed infestations. Use 3 to 4 quarts of LASSO per acre for control of yellow nutsedge or to reduce competition from the hard-to-control weeds listed in this label.
*** Use rates listed in this label when using atrazine 4L. Use equivalent rates when using atrazine 80W or 90% dry flowable formulations. One quart of atrazine 4L equals 1.25 pounds of atrazine 80W or 1.1 pounds of atrazine 90% dry flowable.

LASSO® plus BANVEL (TANK MIXTURE)

For use in ILLINOIS, IOWA, MINNESOTA and WISCONSIN on level or flat-planted field corn on fine textured (silty clay loam, clay loam, sandy clay, silty clay or clay) soils with more than 2.5% organic matter.

Apply in water or sprayable fluid fertilizer solutions for control of the annual grasses and broadleaf weeds listed in the "WEEDS CONTROLLED" section of this label.

NOTE: Use on course or medium textured soils or on fine textured soils with 2.5% or less organic matter may result in crop injury and/or destruction.

APPROVED APPLICATION SYSTEMS
Ground—Broadcast Boom; banded

NOTE: DO NOT apply by injection through center pivot systems.

Reference: The "APPLICATION SYSTEMS" section of this label provides detailed information on the proper application of this tank mixture using ground spray equipment.

APPROVED APPLICATION METHODS

Preemergence Surface—Apply this tank mixture after planting, before crop and weeds emerge and within 5 days of last preplant tillage operation. Corn seeds must be planted 1½ inches or deeper beneath the soil surface. Direct chemical contact with corn seed must be avoided since crop injury may result. Apply far enough behind planter equipment to avoid any incorporation by the planter wheel or other covering device. If corn seeds are planted less than 1½ inches beneath the soil surface, delay application until corn has spiked.

Postemergence Surface—This tank mixture may be applied after crop emergence until weeds reach the 2-leaf stage and the crop is no more than 3 inches tall. Applications made beyond the 2-leaf stage will not provide satisfactory grass control. If rain does not occur within 3 days after application, rotary hoe immediately to control emerged weeds.

Some leaf burn may occasionally occur when applied early postemergence to corn, but this should not affect subsequent growth or yield. DO NOT apply in sprayable fluid fertilizer after crop emergence.

Reference: The "APPLICATION TIMING AND METHODS" section of this label provides detailed information and procedures for the application timing and method selected.

NOTE: PREVENT DRIFT TO SOYBEANS OR OTHER DESIRABLE PLANTS. Do not use on furrow irrigated corn, or when corn is planted at the bottom of a furrow, utilizing lister, till or other similar planting methods. DO NOT incorporate prior to planting or corn emergence. If it is necessary to drag for leveling or rotary hoe to break soil crust, DO NOT disturb the soil more than 1/2 inch deep.

RECOMMENDED RATES:
Refer to the following table for the recommended broadcast treatment rates for this tank mixture. Applications which are not consistent with recommendations in this label may result in unsatisfactory weed control or crop injury and/or destruction.

Soil Texture Group	Broadcast Rate Per Acre (Quarts) More than 2.5% organic matter LASSO®	Banvel
Fine—Silty Clay Loam through Clay	2.5	0.5

LASSO® plus BLADEX (TANK MIXTURE)
Apply this tank mixture in water or sprayable fluid fertilizer solutions for control of yellow nutsedge, and the annual grasses and broadleaf weeds listed in the "WEEDS CONTROLLED" section of this label.

APPROVED APPLICATION SYSTEMS
Ground—Broadcast Boom; banded
Dry Bulk Fertilizer Impregnation

Reference: The "APPLICATION SYSTEMS" section of this label provides information on the proper application of this tank mixture using the system selected.

APPROVED APPLICATION METHODS

Preplant Incorporated—Apply this tank mixture within 7 days prior to planting and shallowly incorporate or Surface Blend into the upper 1 to 2 inches of soil.

Preemergence Surface—Apply this tank mixture after planting, before crop and weed emergence and within 5 days after last preplant tillage operation.

Reference: The "APPLICATION TIMING AND METHODS" section of this label provides detailed information and procedures for the application timing and method selected.

RECOMMENDED RATES: Refer to the following tables for the recommended broadcast treatment rates for this tank mixture. Applications which are not consistent with recommendations in this label may result in unsatisfactory weed control or crop injury.

Surface Application Rates

Soil Texture Group*	Broadcast Rate Per Acre (Quarts)**** 1% to 3% organic matter LASSO® +	Bladex 4L***	3% or more organic matter LASSO® +	Bladex 4L***
Coarse**	1.5 to 2.5	1 to 1.2	2 to 3	1.2 to 1.6
Medium	1.75 to 2.5	1.2 to 1.4	2 to 3	1.6 to 2
Fine	2.25 to 3	1.4 to 1.8	3 to 4	1.8 to 2.2

* Refer to the "SOIL TEXTURE" section of this label to determine the corresponding texture group for the soil to be treated.
** On coarse textured soils containing less than 2% organic matter, use LASSO plus Bladex only on sandy loam. Do not use on sand and loamy sand with less than 2% organic matter.
*** Use rates listed in this label using Bladex 4L. Use equivalent rates when using Bladex 80W. One quart of Bladex 4L equals 1.25 lbs. of Bladex 80W.
**** Use a minimum of 2.5 quarts of LASSO per acre on coarse soils and 3 to 4 quarts on medium and fine textured soils to reduce competition from hard-to-control weeds listed in this label.

Preplant Incorporated Application Rates

Soil Texture Group*	Broadcast Rate Per Acre (Quarts)**** 1% to 3% organic matter LASSO® +	Bladex 4L***	3% or more organic matter LASSO® +	Bladex 4L***
Coarse**	2 to 2.5	1.2 to 1.4	2.5 to 3	1.4 to 1.8
Medium	2.25 to 3	1.4 to 1.6	2.5 to 3.5	1.8 to 2.2
Fine	2.5 to 3.5	1.6 to 2	3 to 4	2 to 2.4

NOTE: See "Incorporation" instructions.

* Refer to the "SOIL TEXTURE" section of this label to determine the corresponding texture group for the soil to be treated.
** On coarse textured soils containing less than 2% organic matter, use LASSO plus Bladex only on sandy loam. Do not use on sand and loamy sand with less than 2% organic matter.
*** Use rates listed in this label using Bladex 4L. Use equivalent rates when using Bladex 80W. One quart of Bladex 4L equals 1.25 lbs. of Bladex 80W.
**** Use 3 to 4 quarts of LASSO per acre to control yellow nutsedge or to reduce competition from hard-to-control weeds listed in this label.

LASSO® PLUS ATRAZINE PLUS BLADEX (TANK MIXTURE)
Apply this tank mixture in water or sprayable fluid fertilizer solutions for control of yellow nutsedge and the annual grasses and broadleaf weeds listed in the "WEEDS CONTROLLED" section of this label.

APPROVED APPLICATION SYSTEMS
Ground—Broadcast Boom; banded

Reference: The "APPLICATION SYSTEMS" section of this label provides detailed information on the proper application of this tank mixture using the system selected.

APPROVED APPLICATION METHODS

Preemergence Surface—Apply this tank mixture after planting, before crop and weed emergence and within 5 days after last preplant tillage operation.

Preplant Incorporated—Apply this tank mixture within 7 days prior to planting and shallowly incorporate or Surface Blend into the upper 1 to 2 inches of soil.

Reference: The "APPLICATION TIMING AND METHODS" section of this label provides detailed information and procedures for the application timing and method selected.

NOTE: PLANT ONLY CORN, PEANUTS, SORGHUM OR SOYBEANS* the year following use of this mixture.

* There is a possibility of crop injury due to carryover of atrazine if soybeans are planted the following year. DO NOT plant soybeans the year following use of this tank mixture on furrow irrigated corn. DO NOT graze treated area or feed treated forage to livestock for 21 days following application of this tank mixture.

RECOMMENDED RATES: Refer to the following tables for the recommended broadcast treatment rates for this tank mixture. Applications which are not consistent with recommendations in this label may result in unsatisfactory weed control or crop injury.

Surface Application Rates

Soil Texture Group*	Broadcast Rate Per Acre (Quarts) Less than 3% organic matter LASSO +	atrazine 4L** +	Bladex 4L***
Coarse****	1.5 to 2	0.75	1
Medium	1.75 to 2	0.75 to 1	1 to 1.25
Fine	2.25 to 2.5	1 to 1.25	1.25 to 1.5

Soil Texture Group*	Broadcast Rate Per Acre (Quarts) 3% or more organic matter LASSO +	atrazine 4L** +	Bladex 4L***
Coarse****	1.5 to 2	0.75 to 1	1 to 1.25
Medium	2 to 2.5	1 to 1.25	1.5 to 2
Fine	2.5 to 3	1 to 1.25	1.5 to 2

* Refer to the "SOIL TEXTURE" section of this label to determine the corresponding texture group for the soil to be treated.
** Use rates listed in this label when using atrazine 4L. Use equivalent rates when using atrazine 80W or 90% dry flowable formulations. One quart of atrazine 4L equals 1.25 pounds of atrazine 80W or 1.1 pounds of atrazine 90% dry flowable.
*** Use rates listed in this label using Bladex 4L. Use equivalent rates when using Bladex 80W. One quart of Bladex 4L equals 1.25 pounds of Bladex 80W.
**** Do not use this mixture on sand or loamy sand soils with less than 2% organic matter.

NOTE: Use the higher rates in the recommended ranges in areas of heavy weed infestations and for preplant incorporation. Use 3 to 4 quarts of LASSO per acre preplant incorporated for control of yellow nutsedge.

Use a minimum of 2.5 quarts of LASSO on coarse soils and 3 to 4 quarts in this tank mixture on medium or fine textured soils when hard-to-control weeds are present.

2,4-D SEQUENTIAL TREATMENT

Field And Silage Corn Only

Only Low Volatile Ester and Amine formulations of 2,4-D (2,4-dichlorophenoxyacetic acid) may be applied for control of emerged broadleaf weeds following an application of LASSO. Apply 2,4-D as a post directed spray only. Read and follow the precautionary statements, directions for use and all other information appearing on the 2,4-D label.

RECOMMENDED RATES: Low Volatile Ester formulations—0.25 pounds acid equivalent per acre.

Amine formulations—0.50 pounds acid equivalent per acre.

Conservation or Minimum Tillage Systems
TANK MIXTURES

FIELD CORN OR SILAGE CORN ONLY

When applied as directed under the conditions described, these tank mixtures control many emerged annual weeds, suppress many emerged perennial weeds and give preemergence control of many annual grasses and weeds when corn will be planted directly into a cover crop, established sod or in previous crop residues. These tank mixtures will not control regrowth from perennial weeds.

For mixing instructions, see the "MIXING AND SPRAYING INSTRUCTIONS" section of this label.

LASSO® plus ROUNDUP®

or

LASSO plus ATRAZINE plus ROUNDUP

or

LASSO plus ATRAZINE plus PARAQUAT

or

LASSO plus BLADEX plus ROUNDUP

or

LASSO plus PRINCEP™ plus ROUNDUP

Apply these tank mixtures with Roundup in 10 to 40 gallons of water, or the tank mixtures with paraquat in 15 to 40 gallons of water spray solution per acre immediately before, during or after planting, but BEFORE CROP EMERGENCE. As density of stubble, crop residue or weeds increases, spray gallonage and rate should be increased within the recommended ranges to insure complete coverage. In the absence of emerged vegetation, delete the Roundup or paraquat portion of these tank mixtures.

™ Princep is a trademark of Ciba-Geigy Corporation.

CONTROL OR SUPPRESSION OF EMERGED WEEDS

ROUNDUP®

Annual Weeds—Apply 1 quart of Roundup herbicide per acre in these tank mixtures if weeds are less than 6 inches tall, and use 1.5 quarts of Roundup if weeds are more than 6 inches tall for control of many emerged annual grasses and broadleaf weeds. For emerged annual weeds controlled, see the "WEEDS CONTROLLED" section of the label for Roundup.

Perennial Weeds—At normal application dates in minimum tillage systems, perennial weeds may not be at the proper stage of growth for control. See the "GENERAL INFORMATION" section of the label for Roundup for the proper stage of growth for perennial weeds. Use of 2 to 4 quarts of Roundup per acre in the above mixtures under these conditions provides top kill and reduces competition from many emerged perennial grasses and broadleaf weeds.

For emerged perennial weeds controlled see the "WEEDS CONTROLLED" section of the label for Roundup. To obtain control of perennial weeds, follow recommendations on the label for Roundup for stage of growth and rate of application.

USE OF THIS MIXTURE FOR BERMUDAGRASS OR JOHNSONGRASS CONTROL IS NOT RECOMMENDED.

NOTE: When using these tank mixtures, do not exceed 4 quarts of Roundup herbicide per acre.

PARAQUAT

When used as directed, paraquat in tank mixture with LASSO and atrazine controls many emerged annual weeds and suppresses many emerged perennial weeds.

Broadcast Treatment—Apply 1 to 2 pints of paraquat (0.25 to 0.5 lb. active) in this tank mixture immediately before, during or after planting but BEFORE CROP EMERGENCE. Use the 2 pint rate if any weeds are 4 to 6 inches tall. This mixture may not control weeds taller than 6 inches. As density of stubble, crop residue or weeds increases, spray gallonage should be increased within the recommended range for complete coverage. Add Ortho X-77™ Spreader at 8 oz. per 100 gallons of diluted spray. Refer to the paraquat label for precautionary statements.

™ Ortho X-77 is a trademark of Chevron Chemical Company.

PREEMERGENCE WEED CONTROL

LASSO®

For weeds controlled preemergence, see the sections of this label on LASSO. See the following table for recommended rates of LASSO in this tank mixture on various soil types:

Soil Texture Group*	Broadcast Rate Per Acre (Quarts) LASSO®
Coarse	2.5 to 3
Medium	3
Fine	3 to 4

* Refer to the "SOIL TEXTURE" section of the label to determine the corresponding texture group for the soil to be treated.

Use the higher rate of LASSO in the recommended ranges in areas of heavy grass infestation, or when organic matter content is 3 percent or more. The rate of LASSO may be increased to a maximum of 4 quarts per acre on any soil type to reduce competition from the hard-to-control weeds listed in this label.

LASSO® plus ATRAZINE

For weeds controlled preemergence, see the sections of this label for LASSO and LASSO plus atrazine.

See the following table for recommended rates of LASSO plus atrazine in this tank mixture on various soil types.

	Broadcast Rate Per Acre (Quarts)		
Soil Texture Group*	**LASSO®**	+	**atrazine 4L****
Coarse—Sandy loam only	2 to 2.5		1 to 1.2
Medium	2.5 to 3		1.2 to 1.6
Fine	2.5 to 3		1.6 to 2

* Refer to the "SOIL TEXTURE" section of the label to determine the corresponding texture group for the soil to be treated.

** Use rates listed in this label when using atrazine 4L. Use equivalent rates when using atrazine 80W or 90% dry flowable formulations. One quart of atrazine 4L equals 1.25 pounds of atrazine 80W or 1.1 pounds of atrazine 90% dry flowable.

Use the higher rate of LASSO in the recommended ranges in areas of heavy grass infestation or when fall panicum or crabgrass will be present. The rate of LASSO may be increased to 4 quarts per acre on any soil type to reduce competition from the hard-to-control weeds listed in this label.

Use the higher rate of atrazine in the recommended ranges on soils with greater than 3% organic matter.

NOTE: PLANT ONLY CORN, PEANUTS, SORGHUM, OR SOYBEANS† the year following use of this mixture.

† There is a possibility of crop injury due to carryover of atrazine if soybeans are planted the following year. DO NOT plant soybeans the year following use of this tank mixture on furrow irrigated corn.

DO NOT graze treated area or feed treated forage to livestock for 21 days following application of this tank mixture.

LASSO® plus BLADEX

For weeds controlled preemergence, see the "WEEDS CONTROLLED" section of the label for LASSO and this tank mixture.

See the following table for recommended rates of LASSO plus Bladex in this tank mixture with Roundup on various soil types.

Soil Texture Group*	Broadcast Rate Per Acre (Quarts) LASSO®	+	Bladex 4L**
Coarse—Sandy loam only	2 to 2.5		1 to 1.6
Medium	2.5 to 3		1.2 to 1.6
Fine	2.5 to 3		1.6 to 2.2

* Refer to the "SOIL TEXTURE" section of the label to determine the corresponding texture group for the soil to be treated.

** When using Bladex 80W use equivalent rates. One quart of Bladex 4L equals 1.25 pounds of Bladex 80W.

Use the higher rate of LASSO in the recommended ranges in areas of heavy grass infestation or when fall panicum or crabgrass will be present. The rate of LASSO may be increased to a maximum of 4 quarts per acre in this tank mixture on any soil type to reduce competition from the hard-to-control weeds listed in this label.

Use the higher rate of Bladex in the recommended ranges on soils with greater than 3% organic matter.

NOTE: Do not use this mixture on sand or loamy sand soils with less than 2% organic matter.

LASSO® plus PRINCEP

For weeds controlled preemergence see the "WEEDS CONTROLLED" section of the labels for LASSO and Princep herbicides.

See the following table for recommended rates of LASSO plus Princep in this tank mixture on various soil types.

Soil Texture Group*	Broadcast Rate Per Acre (Quarts) LASSO®	+	Princep 4L**
Coarse—Sandy loam only	2 to 2.5		1 to 1.2
Medium	2.5 to 3		1.2 to 1.6
Fine	2.5 to 3		1.6 to 2.2

* Refer to the "SOIL TEXTURE" section of the label to determine the corresponding texture group for the soil to be treated.

** When using Princep 80W use equivalent rates. One quart of Princep 4L equals 1.25 pounds of Princep 80W.

Use the higher rate of LASSO in the recommended ranges in areas of heavy grass infestation or when fall panicum or crabgrass will be present.

Use the higher rate of Princep in the recommended ranges on soils with greater than 3% organic matter.

NOTE: LAND TREATED WITH PRINCEP SHOULD NOT BE PLANTED TO ANY CROP EXCEPT CORN FOR ONE YEAR FOLLOWING TREATMENT AS CROP INJURY MAY OCCUR. AFTER HARVEST OF TREATED CROP, PLOW AND THOROUGHLY TILL THE SOIL IN THE FALL OR SPRING TO MINIMIZE POSSIBLE INJURY TO SPRING SEEDED ROTATIONAL CROPS.

LASSO® plus ATRAZINE plus PRINCEP

For additional preemergence control of heavy weed infestation and/or hard-to-control weeds, a three-way tank mixture of LASSO herbicide plus atrazine plus Princep may be applied. Refer to the following rate table to determine the recommended rates for these tank mixtures on various soil types.

Soil Texture Group*	Broadcast Rates (Qts/A)** LASSO®	+	atrazine 4L	+	Princep 4L
Coarse	2 to 3		0.75 to 1		0.75 to 1
Medium	2.5 to 3.5		1 to 1.25		1 to 1.5
Fine	2.5 to 4		1 to 1.25		1 to 1.5

* Refer to the "SOIL TEXTURE" section of this label to determine the corresponding texture group for the soil to be treated.

** When using 4L formulations use the rates listed. Use equivalent rates when using 80W or 90% dry flowable formulations. One quart of the 4L formulation of these products equals 1.25 pounds of the 80W formulation. One quart of atrazine 4L or Princep 4L equals 1.1 pound of the 90% dry flowable formulations of these products.

Use the higher rates of atrazine and Princep in the recommended ranges on soils with greater than 3% organic matter.

In the areas of heavy grass infestation, use the higher rate of LASSO in the recommended ranges when fall panicum, crabgrass, or soils with greater than 3% organic matter will be present. The rate of LASSO may be increased to 4 quarts per acre on any soil type to reduce competition from the hard-to-control weeds listed in this label.

NOTE: LAND TREATED WITH PRINCEP SHOULD NOT BE PLANTED TO ANY CROP EXCEPT CORN FOR ONE YEAR FOLLOWING TREATMENT AS CROP INJURY MAY OCCUR. AFTER HARVEST OF TREATED CROP, PLOW AND THOROUGHLY TILL THE SOIL IN THE FALL OR SPRING TO MINIMIZE POSSIBLE INJURY TO SPRING SEEDED ROTATIONAL CROPS. PLANT ONLY CORN, PEANUTS, SORGHUM OR SOYBEANS† THE YEAR FOLLOWING USE OF THIS MIXTURE.

† There is a possibility of crop injury due to carryover of atrazine if soybeans are planted the following year.

DO NOT plant soybeans the year following use of this tank mixture on furrowed irrigated corn.

DO NOT graze treated area or feed treated forage to livestock for 21 days following application of this tank mixture.

GRAIN SORGHUM (MILO)

LASSO®

Apply this product in water or sprayable fluid fertilizer solution for control of yellow nutsedge and the annual grasses and broadleaf weeds listed in the "WEEDS CONTROLLED" section of this label. Apply only to grain sorghum planted with seed that has been properly treated with a protectant containing the active ingredient, flurazole.

APPROVED APPLICATION SYSTEMS

Center Pivot Irrigation

Ground—Broadcast Boom; banded

Reference: The "APPLICATION SYSTEMS" section of this label provides detailed information on the proper application of this product using the system selected.

APPROVED APPLICATION METHODS

Preplant Incorporated—Apply this product within 7 days prior to planting and shallowly incorporate or Surface Blend into the upper 1 to 2 inches of soil. This product must be incorporated when treating furrow irrigated milo. Performance may be increased by packing the soil following incorporation if furrow irrigation is to be used. Irrigation within 10 days following planting may improve weed control.

Preemergence Surface—Apply this product after planting, before crop and weed emergence, and within 5 days after last preplant tillage operation.

Reference: The "APPLICATION TIMING AND METHODS" sections of this label provides detailed information and procedures for the application timing and method selected.

RECOMMENDED RATES: Refer to the following table for the recommended broadcast treatment rates for this product. Applications which are not consistent with recommendations in this label may result in unsatisfactory weed control or crop injury.

Surface Application Rates

Soil Texture Group*	Broadcast Rate Per Acre (Quarts)**	
	Less than 3% Organic Matter	3% or more Organic Matter
Coarse and Medium	2.5 to 3	2.5 to 3
Fine	3.5 to 4	3.5 to 4

* Refer to the "SOIL TEXTURE" section of this label to determine the corresponding texture group for the soil to be treated.

** Use the higher rates in the recommended ranges in areas of heavy weed infestation. Use a minimum of 3 quarts on coarse and medium soils, and 4 quarts on fine textured soils to reduce competition from hard-to-control weeds listed in this label, or for fields under irrigation.

NOTE: When applied through CENTER PIVOT IRRIGATION or under sprinkler irrigation systems, use a minimum of 3 quarts per acre of this product on coarse soils.

Preplant Incorporated Application Rates—Use 3 quarts on medium* soils, and 3.5 to 4 quarts on fine** soils. Incorporation is not recommended on coarse soils.

* Refer to the "SOIL TEXTURE" section of this label to determine the corresponding texture group for the soil to be treated.

** Rates may be increased to a maximum of 4 quarts on any soil type when heavy infestation of yellow nutsedge is present, or to reduce competition from the hard-to-control weeds listed in this label.

LASSO® PLUS ATRAZINE (TANK MIXTURE)

Apply this tank mixture in water or sprayable fluid fertilizer solutions for control of yellow nutsedge and the annual grasses and broadleaf weeds listed in the "WEEDS CONTROLLED" section of this label. Apply only to grain sorghum seed that has been properly treated with a protectant containing the active ingredient, flurazole.

APPROVED APPLICATION SYSTEMS

Center Pivot Irrigation

Ground—Broadcast Boom; banded

Reference: The "APPLICATION SYSTEMS" section of this label provides detailed information on the proper application of this tank mixture using the system selected.

APPROVED APPLICATION METHODS

Preplant Incorporated—Apply this tank mixture within 7 days prior to planting and shallowly incorporate or Surface Blend into the upper 1 to 2 inches of soil.

Preemergence Surface—Apply this tank mixture after planting, before crop and weed emergence and within 5 days after last preplant tillage operation.

Reference: The "APPLICATION TIMING AND METHODS" section of this label provides detailed information and procedures for the application timing and method selected.

NOTE: PLANT ONLY CORN, PEANUTS, SORGHUM OR SOYBEANS* the year following use of this mixture.

* There is a possibility of crop injury due to carry-over of atrazine if soybeans are planted the following year. DO NOT plant soybeans the year following use of this tank mixture on furrow irrigated milo. On the Gulf Coast of Texas, use for grain sorghum on fine textured soils only. DO NOT graze treated area or feed treated forage to livestock for 21 days following application of this tank mixture.

RECOMMENDED RATES: Refer to the following tables for the recommended broadcast treatment rates for this tank mixture. Applications which are not consistent with recommendations in this label may result in unsatisfactory weed control or crop injury.

Surface Application Rates

Soil Texture Group*	Broadcast Rate Per Acre (Quarts)**			
	Less than 3% organic matter		3% or more organic matter	
	LASSO® +	atrazine 4L***	LASSO +	atrazine 4L***
Coarse (Except Sand)	2	0.75 to 1	2	1 to 1.5
Medium	2	1 to 1.25	2	1.25 to 1.5
Fine	2.25	1.2 to 1.5	2.5	1.25 to 1.75

* Refer to the "SOIL TEXTURE" section of this label to determine the corresponding texture group for the soil to be treated.

** Use the higher rates in the recommended ranges in areas of heavy weed infestations, or for fields under irrigation. Use a minimum of 2.5 quarts of LASSO per acre in this tank mixture on coarse soils and 3 to 4 quarts on medium or fine textured soils to reduce competition from hard-to-control weeds listed in this label.

*** Use rates listed in this label when using atrazine 4L. Use equivalent rates when using atrazine 80W or 90% dry flowable formulations. One quart of atrazine 4L equals 1.25 pounds of atrazine 80W or 1.1 pounds of atrazine 90% dry flowable.

NOTE: When applied through CENTER PIVOT IRRIGATION or under sprinkler irrigation systems use a minimum of 2 quarts per acre of LASSO and 1 quart per acre of atrazine 4L in this tank mixture on coarse soils.

PREPLANT INCORPORATED APPLICATION RATES

Apply 2.5 quarts of LASSO per acre on medium* soils and 2.75 quarts on fine* soils, and use the higher rates of atrazine in the recommended ranges for surface application rates. Incorporation is not recommended on coarse soils.

* Refer to the "SOIL TEXTURE" section of this label to determine the corresponding texture group for the soil to be treated.

NOTE: Use 3 to 4 quarts of LASSO per acre for control of yellow nutsedge or to reduce competition from the hard-to-control weeds listed on this label.

LASSO® PLUS MODOWN™ 4F FLOWABLE (TANK MIXTURE)

Apply this tank mixture in water or sprayable fluid fertilizer solution for control of the annual grasses and broadleaf weeds listed in the "WEEDS CONTROLLED" section of this label. Apply only to grain sorghum seed that has been properly treated with a protectant containing the active ingredient, flurazole.

APPROVED APPLICATION SYSTEMS

Ground—Broadcast Boom; banded

Reference: The "APPLICATION SYSTEMS" section of this label provides detailed information on the proper application of this tank mixture using the equipment selected.

NOTE: Do not apply this tank mixture after milo has begun to crack the soil surface just prior to emergence. Rainfall or irrigation which splashes treated soil onto the crop at time of emergence may result in crop injury from Modown. Do not rotate to crops other than corn, soybeans, grain sorghum, small grains or rice for one year following treatment with this tank mixture.

APPROVED APPLICATION METHOD

Preemergence Surface—Apply this tank mixture after planting, before crop and weed emergence and within 5 days after last tillage operation.

Reference: The "APPLICATION TIMING AND METHODS" section of this label provides detailed information and procedures for the application method and timing selected.

RECOMMENDED RATES: Refer to the following tables for the recommended broadcast treatment rates for this tank mixture. Applications which are not consistent with recommendations in this label may result in unsatisfactory weed control or crop injury.

Surface Application Rates

Soil Texture Group*	Broadcast Rate Per Acre (Quarts)**		
	LASSO®	+	Modown 4F
Coarse	2		1.25
Medium	2 to 2.5		1.25 to 1.6
Fine	2 to 3		1.25 to 1.6

* Refer to the "SOIL TEXTURE" section of this label to determine the corresponding texture group for the soil to be treated.

** Use the higher rates in the recommended ranges in areas of heavy weed infestations, for fields under irrigation, or when soil organic matter is greater than 3 percent. Use a minimum of 2.5 quarts of LASSO per acre in this tank mixture on coarse soils and 3 to 4 quarts on medium or fine textures soils to reduce competition from hard-to-control weeds listed in this label.

™ Modown is a trademark of Mobil Chemical Company

LASSO® PLUS PROPAZINE (TANK MIXTURE)

Apply this tank mixture in water or sprayable fluid fertilizer solution for control of yellow nutsedge and the annual grasses and broadleaf weeds listed in the "WEEDS CONTROLLED" section of this label. Apply only to grain sorghum seed that has been properly treated with a protectant containing the active ingredient, flurazole.

APPROVED APPLICATION SYSTEMS

Ground—Broadcast Boom; banded

Reference: The "APPLICATION SYSTEMS" section of this label provides detailed information on the proper application of this tank mixture using the equipment selected.

APPROVED APPLICATION METHOD

Preemergence Surface—Apply this tank mixture after planting, before crop and weed emergence and within 5 days after last tillage operation.

Reference: The "APPLICATION TIMING AND METHODS" section of this label provides detailed information and procedures for the application method and timing selected.

NOTE: Applications of this tank mixture to alkali soils where cuts, fills, or erosions have exposed calcareous or alkali subsoils may result in crop injury from propazine. Do not plant crops other than corn or soybeans for 18 months following application of this tank mixture due to possibility of crop injury from propazine carryover.

RECOMMENDED RATES: Refer to the following tables for the recommended broadcast treatment rates for this tank mixture. Applications which are not consistent with recommendations in this label may result in unsatisfactory weed control or crop injury.

Surface Application Rates

Soil Texture Group*	Broadcast Rate Per Acre (Quarts)**	
	LASSO®	Propazine 4L***
Coarse	2	0.75 to 1****
Medium	2 to 2.5	1
Fine	2 to 3	1

* Refer to the "SOIL TEXTURE" section of this label to determine the corresponding texture group for the soil to be treated.

** Use the higher rates in the recommended ranges in areas of heavy weed infestation, for fields under irrigation, or where soil organic matter is greater than 3 percent. Use 3 to 4 quarts of LASSO per acre for control of yellow nutsedge or to reduce competition from the hard-to-control weeds listed in this label.

*** Use rates listed in this label when using propazine 4L. Use equivalent rates when using propazine 80W or 90% dry flowable formulations. One quart of propazine 4L equals 1.25 pounds of propazine 80W or 1.1 pounds of propazine 90% dry flowable.

**** In coarse soils with 1% or less organic matter, use the following reduced rates of propazine. When using reduced rates, duration of weed control may be lessened or weed control may be reduced if either an irrigation or a rainfall in excess of 1 inch occurs after application.

Percent Organic Matter	Propazine 4L* (Quarts)
Less than 0.75%	0.4
0.75 to 1%	0.6

* See reference *** just preceding this table.

SOYBEANS

LASSO®

Apply this product in water or sprayable fluid fertilizer solution for control of yellow nutsedge, and the annual grasses and broadleaf weeds listed in the "WEEDS CONTROLLED" section of this label.

APPROVED APPLICATION SYSTEMS

Ground—Broadcast Boom; banded
Center Pivot Irrigation
Dry Bulk Fertilizer Impregnation

Reference: The "APPLICATION SYSTEMS" section of this label provides detailed information on the proper application of this product using the system selected.

APPROVED APPLICATION METHODS

Preplant Incorporated—Apply this product within 7 days prior to planting and shallowly incorporate or Surface Blend into the upper 1 to 2 inches of soil.

Preemergence Surface—Apply this product after planting, before crop and weed emergence and within 5 days after last preplant tillage operation.

Postemergence Surface—This product may be applied after crop emergence to reduce competition from weeds listed as controlled on this label. Apply before soybeans exceed the unifoliate stage (first two true leaves).

Weeds which are emerged at time of application are not controlled by this product. If weeds are emerged at application, shallowly cultivate or rotary hoe immediately to improve performance. DO NOT apply in sprayable fluid fertilizer after crop emergence.

NOTE: The highest levels of weed control will be obtained when applications are made prior to the emergence of the crop and weeds.

DO NOT graze treated area or feed treated forage to livestock following postemergence application of this product.

Reference: The "APPLICATION TIMING AND METHODS" section of this label provides detailed information and procedures for the application timing and method selected.

Sequential Treatment—This product may be applied as two sequential treatments for improved control of hard-to-control weeds. The combined total application rate for the two treatments must not exceed 8 quarts per acre per year. The second treatment must be applied before soybeans exceed unifoliate stage (first two true leaves).

RECOMMENDED RATES: Refer to the following tables for the recommended broadcast treatment rates for this product. Applications which are not consistent with recommendations in this label may result in unsatisfactory weed control or crop injury.

Surface Application Rates

Soil Texture Group*	Broadcast Rate Per Acre (Quarts)**	
	Less than 3% organic matter	3% or more organic matter
Coarse***	2 to 3	2 to 3
Medium	2.5 to 3	2.5 to 3.5
Fine	3 to 3.5	3 to 4

* Refer to the "SOIL TEXTURE" section of this label to determine the corresponding texture group for the soil to be treated.

** Use the higher rate in the recommended range in areas of heavy weed infestations. Use a minimum of 2.5 quarts of this product per acre on coarse soils and 3 to 4 quarts on medium or fine textured soils to control black or hairy nightshade or to reduce competition from hard-to-control weeds listed in this label.

*** When applied through CENTER PIVOT IRRIGATION or under sprinkler irrigation systems on coarse soils, use a minimum of 3 quarts per acre of this product.

On coarse soils with 10% or more organic matter use 4 quarts of LASSO herbicide per acre.

Preplant Incorporated Application Rates

Soil Texture Group*	Broadcast Rate Per Acre (Quarts)**	
	Less than 3% organic matter	3% or more organic matter
Coarse	2.5 to 3	2.5 to 3
Medium	3 to 3.5	3 to 4
Fine	3 to 4	3.5 to 4

* Refer to the "SOIL TEXTURE" section of this label to determine the corresponding texture group for the soil to be treated.

** Use the higher rate in the recommended range in areas of heavy weed infestations. Use a minimum of 3 quarts per acre on coarse textured soils for control of yellow nutsedge or red rice. Rates may be increased to a maximum of 4 quarts on any soil type when heavy infestations of yellow nutsedge or red rice are present or to reduce competition from the hard-to-control weeds listed in this label.

On coarse soils with 10% or more organic matter use 4 quarts of LASSO herbicide per acre.

LASSO® plus AMIBEN (TANK MIXTURE)

Apply this tank mixture in water solution for control of the annual grasses and broadleaf weeds listed in the "WEEDS CONTROLLED" section of this label.

DO NOT use this tank mixture on sand or loamy sand soils as crop injury may result.

APPROVED APPLICATION SYSTEMS

Ground—Broadcast Broom; banded

Reference: The "APPLICATION SYSTEMS" section of this label provides detailed information on the proper application of this tank mixture using the system selected.

APPROVED APPLICATION METHODS

Preplant Incorporated— Apply this tank mixture within 7 days prior to planting and shallowly incorporate or Surface Blend into the upper 1 to 2 inches of soil.

Preemergence Surface—Apply after planting, before crop and weed emergence and within 5 days after last preplant tillage operation.

Reference: The "APPLICATION TIMING AND METHODS" section of this label provides detailed information and procedures for the application timing and method selected.

RECOMMENDED RATES:

Preplant Incorporated—Apply 2.5 to 3 quarts of LASSO plus 4 quarts of Amiben per acre.

NOTE: Use the 3 quart per acre rate of LASSO in areas of heavy weed infestation. Use 3 to 4 quarts of LASSO when black or hairy nightshade, yellow nutsedge or red rice is present or to reduce competition from the hard-to-control weeds listed in this label.

Preemergence Surface—2 quarts LASSO plus 4 quarts Amiben per acre.

NOTE: Use a minimum of 2.5 quarts of LASSO on coarse soils and 3 to 4 quarts in this tank mixture on medium or fine textured soils to control black or hairy nightshade or to reduce competition from hard-to-control weeds listed in this label.

LASSO®plus DYANAP (TANK MIXTURE)

Apply this tank mixture in water solution for control of the annual grasses and broadleaf weeds listed in the "WEEDS CONTROLLED" section of this label.

APPROVED APPLICATION SYSTEMS

Ground—Broadcast Boom; banded

Reference: The "APPLICATION SYSTEMS" section of this label provides detailed information on the proper application of this tank mixture using the system selected.

APPROVED APPLICATION METHODS

Preemergence Surface—Apply this tank mixture after planting, before crop and weed emergence and within 5 days after last preplant tillage operation.

Reference: The "APPLICATION TIMING AND METHODS" section of this label provides detailed information and procedures for the application timing and method selected.

NOTE: DO NOT apply this tank mixture in sprayable fluid fertilizer.

RECOMMENDED RATES: Refer to the following table for the recommended broadcast treatment rates for this tank mixture. Applications which are not consistent with recommendations in this label may result in unsatisfactory weed control or crop injury.

Surface Application Rates

Soil Texture Groups*	Broadcast Rate Per Acre (Quarts) LASSO	+	Dyanap
Coarse	2		4.5
Medium and Fine	2		6

* Refer to the "SOIL TEXTURE" section of this label to determine the corresponding texture group for the soil to be treated.

LASSO® plus LEXONE or LASSO plus SENCOR (TANK MIXTURE)

Apply this tank mixture in water or sprayable fluid fertilizer solution for control of yellow nutsedge and the annual grasses and broadleaf weeds listed in the "WEEDS CONTROLLED" section of this label.

NOTE: Crop injury may result if atrazine was applied the year preceding use of this tank mixture.

DO NOT REPLANT CROPS OTHER THAN SOYBEANS FOR 120 DAYS AFTER APPLICATION OF THESE TANK MIXTURES.

APPROVED APPLICATION SYSTEMS

Ground—Broadcast Boom; banded

Dry Bulk Fertilizer Impregnation

Reference: The "APPLICATION SYSTEMS" section of this label provides detailed information on the proper application of this tank mixture using the system selected.

APPROVED APPLICATION METHODS

Preplant Incorporated—Apply this tank mixture within 7 days prior to planting and shallowly incorporate or Surface Blend into the upper 1 to 2 inches of soil. Preplant incorporation may cause slight soybean injury early, but soybeans should recover with no subsequent effect on growth or yield.

Preemergence Surface—Apply this tank mixture after planting, before crop and weed emergence and within 5 days after last preplant tillage operation.

Reference: The "APPLICATION TIMING AND METHODS" section of this label provides detailed information and procedures for application timing and method selected.

RECOMMENDED RATES: Refer to the following tables for the recommended broadcast treatment rates for this tank mixture. Applications which are not consistent with recommendations in this label may result in unsatisfactory weed control or crop injury.

Surface Application Rates

Soil Texture Group*	Broadcast Rates Per Acre: Less than 3% organic matter LASSO® (quarts)	+	Lexone 4L** or Sencor 4F** (pint)	3% or more organic matter LASSO® (quarts)	+	Lexone 4L** or Sencor 4F** (pints)
Coarse—Sandy Loam only	1.5 to 2.5		0.5	2 to 3		0.75
Medium	1.75 to 2.5		0.75	2 to 3		1
Fine	2.25 to 3		1	3 to 4		1

* Refer to the "SOIL TEXTURE" section of this label to determine the corresponding texture group for the soil to be treated.

** Use the rates listed in this label when using Lexone 4L or Sencor 4F. Use the equivalent rates when using the 50WP or DF formulations of these products. One pint of Lexone 4L or Sencor 4F equals one pound of the corresponding 50 WP formulation or ⅔ pound of the corresponding DF formulation.

NOTE: DO NOT use on sand or loamy sand soil. Use 1.5 to 2 pints of Lexone 4L or Sencor 4F per acre in this tank mixture when applying on silty clay and heavy clay soils of the Mississippi Delta. Use a minimum of 2.5 quarts of LASSO on coarse soils and 3 to 4 quarts in this tank mixture on medium or fine textured soils to control black or hairy nightshade or to reduce competition from heavy infestations of the hard-to-control weeds listed in this label.

Preplant Incorporated Application Rates

Surface Application Rates

Soil Texture Group*	Broadcast Rates Per Acre (Quarts)***			
	Less than 3% organic matter		3% or more organic matter	
	LASSO® + (quarts)	Lexone 4L** or Sencor 4F** (pints)	LASSO® + (quarts)	Lexone 4L* or Sencor 4F** (pints)
Medium	2.25 to 3	0.75	2.5 to 3.5	0.75
Fine	2.5 to 3.5	1	3 to 4	1

* Refer to the "SOIL TEXTURE" section of this label to determine the corresponding texture group for the soil to be treated.

** Use the rates listed in this label when using Lexone 4L or Sencor 4F. Use the equivalent rates when using the 50WP or DF formulations of these products. One pint of Lexone 4L or Sencor 4F equals one pound of the corresponding 50WP formulation or ⅔ pound of the corresponding DF formulation.

*** Use the 3 quart per acre rate of LASSO in areas of heavy weed infestation. Use 3 to 4 quarts of LASSO when black or hairy nightshade, yellow nutsedge or red rice is present or to reduce competition from the hard-to-control weeds listed in this label.

NOTE: Do not use on coarse textured soils. Do not use on medium or fine textured soils with less than 1% organic matter. These mixtures, preplant incorporated, may cause slight soybean injury early, but the soybeans should recover and subsequent growth or yield should not be affected.

LASSO® PLUS AMIBEN PLUS LEXONE OR SENCOR (TANK MIXTURES)

Apply this tank mixture in water solution for control of the annual grasses and broadleaf weeds listed in the "WEEDS CONTROLLED" section of this label.

NOTE: DO NOT use on sand, loamy sand, or on soils with less than 1% organic matter. DO NOT use the foliage from soybeans treated with LASSO plus Amiben plus Lexone or Sencor for feed or forage.

APPROVED APPLICATION SYSTEMS

Ground— Broadcast Boom

Reference: The "APPLICATION SYSTEMS" section of this label provides detailed information on the proper application of this tank mixture using the system selected.

APPROVED APPLICATIONS METHODS

Preplant Incorporation—Apply this tank mixture within 7 days prior to planting and shallowly incorporate or Surface Blend into the upper 1 to 2 inches of soil. Preplant incorporation may cause slight soybean injury early but soybeans should recover with no subsequent effect on growth or yield.

Preemergence Surface— Apply this tank mixture after planting, before crop and weed emergence and within 5 days after last preplant tillage operation.

Reference: The "APPLICATION TIMING AND METHODS" section of this label provides detailed information and procedures for application timing methods selected.

RECOMMENDED RATES: Refer to the following tables for recommended broadcast treatment rates. Applications which are not consistent with recommendations in this label may result in unsatisfactory weed control or crop injury.

Soil Texture Group*	Broadcast Rate Per Acre		
	Less than 3% organic matter		
	LASSO® (Quarts)	Amiben (Quarts)	Lexone 4L** Sencor 4F (Pints)
Coarse Sandy Loam Only	2	3	0.5
Medium	2 to 2.5	3 to 4	0.5
Fine	2.5	4 to 5	0.75

Soil Texture Group*	Broadcast Rate Per Acre		
	More than 3% organic matter		
	LASSO® (Quarts)	Amiben (Quarts)	Lexone 4L** Sencor 4F (Pints)
Coarse Sandy Loam Only	2	3 to 4	0.75
Medium	2.5 to 3	4 to 5	0.5 to 0.75
Fine	2.5 to 3	4 to 6	0.75 to 1

* Refer to the "SOIL TEXTURE" section of this label to determine the corresponding texture group for the soil to be treated. On alkaline (calcareous) soils with a pH of 7.4 or greater, apply Lexone 4L or Sencor 4F at .5 pint/A on medium soils and .5 to .75 pint or corresponding equivalent of WP or DF formulations on fine soils. Use the higher rates of LASSO in the recommended ranges in areas of heavy weed infestations and for preplant incorporation applications. Use a minimum of 2.5 quarts of LASSO on coarse soils and 3 to 4 quarts in this tank mixture on medium or fine soils, when hard-to-control weeds or black or hairy nightshade are present. Use 3 to 4 quarts of LASSO per acre preplant incorporated for control of yellow nutsedge.

** Use rates listed in this label when using Lexone 4L or Sencor 4F. Use equivalent rates when using the 50 WP or DF formulations of these products. One pint of Lexone 4L or Sencor 4F equals 1 pound of the corresponding 50 WP formulation or ⅔ pound of the corresponding DF formulation.

LASSO® plus LINURON (TANK MIXTURE)

Apply this tank mixture in water or sprayable fluid fertilizer solution for control of the annual grasses and broadleaf weeds listed in the "WEEDS CONTROLLED" section of this label.

NOTE: Plant seed at least 1¾ inches deep and DO NOT use on sand or loamy sand or on soils with less than 1% organic matter as crop injury may occur. DO NOT spray over the top of emerged soybeans.

APPROVED APPLICATION SYSTEMS

Ground—Broadcast Boom; banded

Reference: The "APPLICATION SYSTEMS" section of this label provides detailed information on the proper application of this tank mixture using the system selected.

APPROVED APPLICATION METHODS

Preemergence Surface—Apply after planting, before crop and weed emergence and within 5 days after last preplant tillage operation.

Reference: The "APPLICATION TIMING AND METHODS" section of this label provides detailed information and procedures for application timing and method selected.

RECOMMENDED RATES: Refer to the following table for the recommended broadcast treatment rates for this tank mixture. Applications which are not consistent with recommendations in this label may result in unsatisfactory weed control or crop injury.

Surface Application Rates

Soil Texture Group*	Broadcast Rates Per Acre (Quarts)			
	1 to 3% organic matter		3% or more organic matter	
	LASSO® +	linuron** 4L	LASSO® +	linuron** 4L
Coarse—Sandy Loam only	1.5 to 2.5	0.5	2 to 3	0.5 to 0.75
Medium	1.75 to 2.5	0.5 to 0.75	2 to 3	0.75 to 1
Fine	2.25 to 3	1	3 to 4	1.25 to 1.5

* Refer to the "SOIL TEXTURE" section of this label to determine the corresponding texture group for the soil to be treated.

** Use the rates listed on this label when using linuron 4L. Use equivalent rates when using linuron 50WP. One quart of linuron 4L equals 2 pounds of linuron 50WP.

NOTE: Use a minimum of 2.5 quarts per acre of LASSO on coarse textured soils, and use 3 to 4 quarts on medium and fine textured soils to control black or hairy nightshade, or to reduce competition from the hard-to-control weeds listed in this label.

LASSO® plus PREMERGE (TANK MIXTURE)

Soybeans

Apply this tank mixture in water solution for contol of the annual grasses and broadleaf weeds listed in the "WEEDS CONTROLLED" section of this label.

NOTE: DO NOT apply in sprayable fluid fertilizer or on sand or loamy sand soils as crop injury may occur.

APPROVED APPLICATIONS SYSTEMS

Ground—Broadcast Boom; banded

Reference: The "APPLICATION SYSTEMS" section of this label provides detailed information on the proper application of this tank mixture using the system selected.

APPROVED APPLICATION METHODS

Preemergence Surface—Apply this tank mixture anytime after planting and before crop emergence.

Early Postemergence—This tank mixture may be applied after crop and weeds emerge, but before first soybean leaves expand to expose buds. Some soybean injury may occur but effects are usually temporary and do not result in yield reduction.

NOTE: For reduced competition from cocklebur, jimsonweed or morningglory, better results can be obtained by early postemergence treatment.

Reference: The "APPLICATION TIMING AND METHODS" section of this label provides detailed information and procedures for application timing and method selected.

NOTE: DO NOT apply early postemergence if soil surface is very wet, if first soybean cotyledon leaves have expanded to expose the bud, or if air temperature is expected to exceed 90°F. within 24 hours after application.

RECOMMENDED RATES: Refer to the following tables for the recommended broadcast treatment rates for this tank mixture. Applications which are not consistent with recommendations in this label may result in unsatisfactory weed control or crop injury.

Surface Application Rates (Preemergence)

Soil Texture Group*	Broadcast Rate Per Acre (Quarts)		
	LASSO®	+	Premerge
Coarse—Sandy Loam only	2		4
Medium	2 to 2.5		6
Fine	2.5 to 3		6

* Refer to the "SOIL TEXTURE" section of this label to determine the corresponding texture group for the soil to be treated.

Surface Applications Rates (Early Postemergence)

Soil Texture Group*	Broadcast Rate Per Acre (Quarts)		
	LASSO®	+ Premerge**	
		Up to 75°F.	75°F. to 90°F
Coarse—Sandy Loam Only	1.5	3	2
Medium	2	3	2
Fine	2.5 to 3	3	2

* Refer to the "SOIL TEXTURE" section of this label to determine the corresponding texture group for the soil to be treated.

** Select the rate of Premerge (dinoseb) corresponding to the maximum air temperature expected at spraying and within the next 24 hours after spraying as indicated in this rate table.

BASAGRAN SEQUENTIAL TREATMENT

Soybeans Only

Basagran may be applied following preemergence applications of LASSO or tank mixtures of LASSO plus Lorox, LASSO plus Dyanap, or LASSO plus metribuzin (Sencor or Lexone) for control of emerged weeds such as cocklebur, velvetleaf and other weed species listed on the label for Basagran. Read and follow the cautionary statements, directions for use and all other information appearing on the label for Basagran.

RECOMMENDED RATES: Apply 0.75 to 1.5 quarts of Basagran as directed on the label for Basagran.

Conservation or Minimum Tillage Systems
TANK MIXTURES

SOYBEANS ONLY

When applied as directed under the conditions described, these tank mixtures control many emerged annual weeds, suppress many emerged perennial weeds and give preemergence control of many annual grasses and weeds when soybeans will be planted directly into a cover crop, stale seedbed, or in previous crop residues such as wheat stubble.

These tank mixtures will not control regrowth from perennial weeds.

For mixing instuctions, see the "MIXING AND SPRAYING INSTRUCTIONS" section of this label.

LASSO® plus ROUNDUP®

or

LASSO plus LINURON plus ROUNDUP

or

LASSO plus LINURON plus PARAQUAT

or

LASSO plus LEXONE plus ROUNDUP

or

LASSO plus SENCOR plus ROUNDUP

Apply these tank mixtures in 15 to 40 gallons of water per acre immediately before, during or after planting, but BEFORE CROP EMERGENCE. As density of stubble, crop residue or weeds increases, spray gallonage and rate should be increased within the recommended ranges to insure complete coverage. In the absence of emerged vegetation delete the Roundup or paraquat portion of these mixes.

CONTROL OR SUPPRESSION OF EMERGED WEEDS

ROUNDUP®

Annual Weeds—Apply 1 quart of Roundup per acre in these tank mixtures if weeds are less than 6 inches tall, and use 1.5 quarts of Roundup if weeds are more than 6 inches tall for control of many emerged annual grasses broadleaf weeds. For emerged annual weeds controlled, see the "WEEDS CONTROLLED" section of the label for Roundup.

Perennial Weeds—At normal application dates in minimum tillage systems, perennial weeds may not be at the proper stage of growth for control. See the "GENERAL INFORMATION" section of the label for Roundup for the proper stage of growth for perennial weeds. Use of 2 to 4 quarts of Roundup per acre in this tank mixture under these conditions provides top kill and reduces competition from many emerged perennial grass and broadleaf weeds. For perennial weeds controlled, see the "WEEDS CONTROLLED" section of the label for Roundup.

To obtain control of perennial weeds, follow recommendations on the label for Roundup for stage of growth and rate of application.

USE OF THIS TANK MIXTURE FOR BERMUDAGRASS OR JOHNSONGRASS CONTROL IS NOT RECOMMENDED.

NOTE: When using these tank mixtures, do not exceed 4 quarts of Roundup herbicide per acre.

PARAQUAT

When used as directed, paraquat in tank mixture with LASSO plus Lorox controls many emerged annual weeds and suppresses many emerged perennial weeds.

Broadcast Treatment—Apply 1 to 2 pints of paraquat (0.25 to 0.5 lb. active) in this tank mixture immediately before, during or after planting but BEFORE CROP EMERGENCE. Use the 2 pint rate if any weeds are 4 to 6 inches tall. This mixture may not control weeds taller than 6 inches. As density of stubble, crop residue or weeds increases, spray gallonage should be increased within the recommended range for complete coverage. Add Ortho X-77 Spreader at 8 oz. per 100 gallons of diluted spray. Refer to the paraquat label for precautionary statements.

PREEMERGENCE WEED CONTROL

LASSO®

For weeds controlled preemergence, see the "WEEDS CONTROLLED" section of this label.

See the following table for recommended rates of LASSO on various soil types.

Soil Texture Group*	Broadcast Rate Per Acre (Quarts) LASSO®
Coarse	2.5 to 3
Medium	3 to 3.5
Fine	3 to 4

* Refer to the "SOIL TEXTURE" section of the label to determine the corresponding texture group for the soil to be treated.

Use the higher rate in the recommended ranges in areas of heavy grass infestation, or when organic matter content is 3 percent or more. The rate of LASSO may be increased to a maximum of 4 quarts per acre on any soil type to reduce competition from the hard-to-control weeds listed in this label.

LASSO® plus LINURON

For weeds controlled preemergence, see the "WEEDS CONTROLLED" section of this label for LASSO and LASSO plus linuron.

See the following table for recommended rates of LASSO and linuron in this tank mixture.

Soil Texture Group*	Broadcast Rate Per Acre (Quarts) LASSO®	+	linuron 4L**
Coarse—Sandy Loam only	2 to 2.5		0.5 to 0.75
Medium	2.5 to 3		0.75 to 1
Fine	2.5 to 3		1 to 1.5

* Refer to the "SOIL TEXTURE" section of this label to determine the corresponding texture group for the soil to be treated.

** Use the rates listed on this label when using linuron 4L. Use equivalent rates when using linuron 50WP. One quart of linuron 4L equals 2 pounds of linuron 50WP.

Use the higher rate of LASSO in the recommended ranges in areas of heavy grass infestation or when fall panicum or crabgrass will be present. The rate of LASSO may be increased to a maximum of 4 quarts per acre on any soil type to reduce competition from the hard-to-control weeds listed in this label.

Use the higher rate of linuron in the recommended ranges on soils with greater than 3% organic matter.

DO NOT use this tank mixture on sand or loamy sand or on soil with less than 1% organic matter as crop injury from linuron may occur.

LASSO® plus LEXONE
or
LASSO plus SENCOR

For weeds controlled preemergence, see the "WEEDS CONTROLLED" section of this label for LASSO and LASSO plus Lexone 4L or Sencor 4F.

See the following table for recommended rates of LASSO plus Lexone 4L or LASSO plus Sencor 4F in this tank mixture on various soil types.

Soil Texture Group*	Broadcast Rate Per Acre LASSO® (quarts)	+	Lexone 4L** or Sencor 4F** (pints)
Coarse—Sandy Loam only	2 to 2.5		0.5 to 0.75
Medium	2.5 to 3		0.75 to 1
Fine	2.5 to 3		1 to 1.5***

* Refer to the "SOIL TEXTURE" section of this label to determine the corresponding texture group for the soil to be treated.

** Use the rates listed in this label when using Lexone 4L or Sencor 4F. Use the equivalent rates when using the 50WP or DF formulations of these products. One pint of Lexone 4L or Sencor 4F equals one pound of the corresponding 50WP formulation or 2/3 pound of the corresponding DF formulation.

*** On the silty clay or heavy clay soils of the Mississippi Delta, use 1.5 to 2 pints of Lexone 4L or Sencor 4F per acre.

Use the higher rate of LASSO in the recommended ranges in areas of heavy grass infestations or when fall panicum or crabgrass will be present. The rate of LASSO may be increased to a maximum of 4 quarts per acre on any soil type to reduce competition from the hard-to-control weeds listed in this label.

Use the higher rate of Lexone or Sencor in the recommended ranges on soils with greater than 3% organic matter.

DO NOT use this mixture on sand or loamy sand soils as crop injury from Lexone or Sencor may occur.

DO NOT use on muck soils.

DO NOT apply on alkaline soils with a pH greater than 7.4.

Crop injury may occur if atrazine was applied the year before use of this tank mixture.

DO NOT REPLANT CROPS OTHER THAN SOYBEANS FOR 120 DAYS AFTER APPLICATION OF THESE TANK MIXTURES.

PEANUTS

LASSO®

Apply this product in water solution for control of yellow nutsedge and the annual grasses and broadleaf weeds listed in the "WEEDS CONTROLLED" section of this label.

APPROVED APPLICATION SYSTEMS

Ground—Broadcast Boom; banded
Center Pivot Irrigation

Reference: The "APPLICATION SYSTEMS" section of this label provides detailed information on the proper application of this product using the system selected.

APPROVED APPLICATION METHODS

Preplant Incorporated*—Apply within 7 days prior to planting and shallowly incorporate or Surface Blend into the upper 1 to 2 inches of soil.

Postplant Incorporated*—Apply after planting and shallowly incorporate or Surface Blend into the upper 1 to 2 inches of soil prior to peanut germination. Incorporation depth after planting must be kept above the seed to avoid mechanical injury.

Preemergence Surface—Apply after planting, before crop and weed emergence and within 5 days after last preplant tillage operation.

Reference: The "APPLICATION TIMING AND METHODS" section of this label provides detailed information and procedures for application timing and method selected.

* Incorporation on coarse soils may result in reduced length of weed control due to leaching with rainfall or irrigation. Incorporation is not recommended on coarse soils in the Southeastern States.

RECOMMENDED RATES: Refer to the following table for recommended broadcast treatment rates for this product. Applications which are not consistent with recommendations in this label may result in unsatisfactory weed control or crop injury.

Application Method	Lasso®* (Quarts)
Preplant or Postplant Incorporated	3.5 to 4
Preemergence Surface	3 to 4

* Use the higher rate in the recommended range if hard-to-control weeds or heavy weed infestations are present.

LASSO® plus DYANAP (TANK MIXTURE)

Peanuts

Apply this tank mixture in water solution for control of the annual grasses and broadleaf weeds listed in the "WEEDS CONTROLLED" section of this label.

NOTE: DO NOT cultivate before applying this treatment. DO NOT graze or feed forage from treated peanuts to livestock prior to crop maturity.

APPROVED APPLICATION SYSTEMS

Ground—Broadcast Boom; banded

Reference: The "APPLICATION SYSTEMS" section of this label provides detailed information on the proper application of this tank mixture using the system selected.

APPROVED APPLICATION METHODS

Preemergence Surface*—Apply this tank mixture after planting, before crop and weed emergence and within 5 days after last preplant tillage operation.

Ground-crack Surface*—Apply this tank mixture at the groundcrack stage of peanuts.

Reference: The "APPLICATION TIMING AND METHODS" section of this label provides detailed information and procedures for application timing and method selected.

* Weed control performance is frequently improved when application is delayed until the peanut ground-cracking stage.

NOTE: This tank mixture may be applied as a sequential treatment after application of any labeled preplant incorporated herbicide when applied as directed.

RECOMMENDED RATES: Refer to the following table for recommended broadcast treatment rates for this tank mixture. Applications which are not consistent with recommendations in this label may result in unsatisfactory weed control or crop injury.

Surface Application Rates

	Broadcast Rate Per Acre (Quarts)		
	LASSO®	+ Dyanap*	
Soil Texture		**up to 90°F.**	**above 90°F.**
Sand and Loamy Sand	2 to 4	4	4
Sand Loam and Loam	2 to 4	6	4

* Use the amount of Dyanap indicated for maximum air temperature at application time or expected within 24 hours after application.

NOTE: When using 4 quarts per acre of LASSO, DO NOT exceed 4 quarts per acre of Dyanap. Use the higher recommended rate of LASSO in areas where heavy weed infestations or hard-to-control weed species exist.

LASSO® plus PREMERGE (TANK MIXTURE)

Peanuts

Apply this tank mixture in water solution for control of the annual grasses and broadleaf weeds listed in the "WEEDS CONTROLLED" section of this label.

NOTE: DO NOT add wetting agents or other adjuvants to this tank mixture.

APPROVED APPLICATION SYSTEMS

Ground—Broadcast Boom; banded

Reference: The "APPLICATION SYSTEMS" section of this label provides detailed information on the proper application of this tank mixture using the system selected.

APPROVED APPLICATION METHODS

Ground-crack Surface—Apply this tank mixture at the groundcrack stage of peanuts.

Reference: The "APPLICATION TIMING AND METHODS" section of this label provides detailed information and procedures for application timing and method selected.

RECOMMENDED RATES: Refer to the following table for recommended broadcast treatment rates for this tank mixture. Applications which are not consistent with recommendations in this label may result in unsatisfactory weed control or crop injury.

Surface Application Rates

Broadcast Rate Per Acre (Quarts)		
LASSO®*	+ Premerge**	
	up to 90°F.	**above 90°F.**
2 to 3	2	1.5

* Use the higher rate of LASSO in the recommended range in areas of heavy weed infestation or to reduce competition from the hard-to-control weeds listed in this label.

** Use the amount of Premerge indicated for maximum air temperature at time of spraying or expected within 24 hours after application.

DRY BEANS

Dry Beans (All Types)

FOR USE ONLY WEST OF THE MISSISSIPPI RIVER, EXCEPT IN KERN COUNTY, CALIFORNIA

LASSO®

Apply this product in water solution for control of yellow nutsedge and the annual grasses and broadleaf weeds listed in the "WEEDS CONTROLLED" section of this label.

NOTE: DO NOT APPLY ON DRY BEANS AFTER PLANTING AS CROP INJURY MAY OCCUR. This product may delay crop maturity and/or reduce yield if cold, wet soil conditions occur after planting.

APPROVED APPLICATION SYSTEMS

Ground—Broadcast Boom

Reference: The "APPLICATION SYSTEMS" section of this label provides detailed information on the proper application of this product using the system selected.

APPROVED APPLICATION METHODS

Preplant Incorporated—Apply this product within 7 days prior to planting and shallowly incorporate or Surface Blend into the upper 1 to 2 inches of soil.

Reference: The "APPLICATION TIMING AND METHODS" section of this label provides detailed information and procedures for application timing and method selected.

RECOMMENDED RATES: Apply 2.5 to 3 quarts of this product per acre and shallowly incorporate prior to planting. Use the higher rate in areas of heavy weed infestations or when hard-to-control weeds, black or hairy nightshade or yellow nutsedge are present.

Applications which are not consistent with recommendations in this label may result in unsatisfactory weed control or crop injury.

LASSO®

Lima Beans (Green)

FOR USE IN ALL STATES, EXCEPT IN KERN COUNTY, CALIFORNIA

Apply this product in water solution for control of yellow nutsedge and the annual grasses and broadleaf weeds listed in the "WEEDS CONTROLLED" section of this label.

NOTE: This product may delay crop maturity and/or reduce yield if cold, wet soil conditions occur after planting.

APPROVED APPLICATION SYSTEMS

Ground—Broadcast Boom

Reference: The "APPLICATION SYSTEMS" section of this label provides detailed information on the proper application of this product using the system selected.

APPROVED APPLICATION METHODS

Preplant incorporated—Apply this product within 7 days prior to planting and shallowly incorporate or Surface Blend into the upper 1 to 2 inches of soil.

Preemergence Surface—Apply this product after planting and before crop and weed emergence and within 5 days after last preplant tillage operation.

Reference: The "APPLICATION TIMING AND METHODS" section of this label provides detailed information and procedures for application timing and method selected.

RECOMMENDED RATES: Refer to the following table for recommended broadcast treatment rates.

Preplant Incorporated Application Rates

Soil Texture Group*	Broadcast Rates per Acre (Quarts) LASSO®
Coarse	2
Medium and Fine**	2.5 to 3

* Refer to the "SOIL TEXTURE" section of this label to determine the corresponding texture group for the soil to be treated.

** Use the higher rate in the recommended range in areas of heavy weed infestation or when hard-to-control weeds, black or hairy nightshade or yellow nutsedge are present.

RECOMMENDED RATES: Refer to the following table for recommended broadcast treatment rates.

Surface Application Rates

Soil Texture Group*	Broadcast Rates Per Acre (Quarts) LASSO®
Coarse	1.5 to 2
Medium and Fine**	2.5 to 3

* Refer to the "SOIL TEXTURE" section of this label to determine the corresponding texture group for the soil to be treated.

** Use the higher rate in the recommended ranges in areas of heavy weed infestation or when hard-to-control weeds, black or hairy nightshade are present.

NOTE: Preemergence applications are not recommended for use in Illinois or Wisconsin as crop injury may occur.

LASSO® FOLLOWED BY AMIBEN or PREMERGE

Dry Beans (All Types)

FOR USE IN MICHIGAN ONLY

Apply Amiben or Premerge as a sequential preemergence application to LASSO preplant incorporated.

APPROVED APPLICATION SYSTEMS
Ground—Broadcast Boom

Reference: The "APPLICATION SYSTEMS" section of this label provides detailed information on the proper application of these products using the system selected.

APPROVED APPLICATION METHODS

LASSO®

Preplant Incorporated—Apply this product within 7 days prior to planting and shallowly incorporate or Surface Blend into the upper 1 to 2 inches of soil.

Reference: The "APPLICATION TIMING AND METHODS" section of this label provides detailed information and procedures for application timing and method selected.

AMIBEN or PREMERGE

Preemergence Surface—Apply Amiben or Premerge after planting and before crop and weed emergence. Read and carefully follow all use instructions and precautionary statements listed on the respective labels for these products.

Reference: The "APPLICATION, TIMING AND METHODS" section of this label provides detailed information and procedures for application timing and method selected.

RECOMMENDED RATES:
LASSO—Apply 2 quarts of this product per acre and shallowly incorporate or Surface Blend prior to planting.

NOTE: DO NOT USE ON SAND OR LOAMY SANDS. DO NOT APPLY ON DRY BEANS AFTER PLANTING AS CROP INJURY MAY OCCUR. This product may delay crop maturity and/or reduce yield if cold, wet soil conditions occur after planting.

AMIBEN—Apply 4 quarts of this product per acre.

PREMERGE—Apply 5 to 6 quarts of this product per acre.

NOTE: When LASSO followed by Amiben or Premerge are used on soils with 6% or more organic matter, weed control may be reduced.

LASSO® plus EPTAM™ (TANK MIXTURE)

Dry Beans (All Types)

FOR USE ONLY IN MICHIGAN AND WEST OF THE MISSISSIPPI RIVER, EXCEPT IN CALIFORNIA

Apply this tank mixture in water solution for control of yellow nutsedge and the annual grasses and broadleaf weeds listed in the "WEEDS CONTROLLED" section of this label.

NOTE: Read the label for Eptam thoroughly for rotational crop restrictions. This tank mixture may delay crop maturity and/or reduce yield if cold, wet soil conditions occur after planting.

APPROVED APPLICATION SYSTEMS
Ground—Broadcast Boom

Reference: The "APPLICATION SYSTEMS" section of this label provides detailed information on the proper application of this tank mixture using the system selected.

APPROVED APPLICATION METHODS

Preplant Incorporated—Apply this tank mixture within 7 days prior to planting and shallowly incorporate into the upper 2 inches of soil immediately.

Reference: The "APPLICATION TIMING AND METHODS" section of this label provides detailed information and procedures for application timing and method selected.

RECOMMENDED RATES: Apply 2 to 3 quarts of LASSO plus 2 to 3 pints of Eptam 7-E per acre. Use the higher recommended rate of LASSO for this tank mixture when heavy weed infestations, hard-to-control weeds, black or hairy nightshade or yellow nutsedge are present.

USE ONLY THE 2 QUART RATE OF LASSO IN MICHIGAN. Applications which are not consistent with recommendations in this label may result in unsatisfactory weed control or crop injury.

™ Eptam is a trademark of the Stauffer Chemical Company.

LASSO® plus TREFLAN (TANK MIXTURE)

Dry Beans (All Types)

FOR USE ONLY WEST OF THE MISSISSIPPI RIVER, EXCEPT IN KERN COUNTY, CALIFORNIA

Apply this tank mixture in water solution for control of yellow nutsedge and the annual grasses and broadleaf weeds listed in the "WEEDS CONTROLLED"section of this label.

NOTE: Read the label for Treflan thoroughly for rotational crop restrictions. This tank mixture may delay crop maturity and/or reduce yield if cold, wet soil conditions occur after planting.

APPROVED APPLICATION SYSTEMS
Ground—Broadcast Boom

Reference: The "APPLICATION SYSTEMS" section of this label provides detailed information on the proper application of this tank mixture using the system selected.

APPROVED APPLICATION METHODS

Preplant Incorporated—Apply this tank mixture within 7 days prior to planting and shallowly incorporate into the upper 2 inches of soil within 8 hours after application.

Reference: The "APPLICATION TIMING AND METHODS" section of this label provides detailed information and procedures for application timing and method selected.

RECOMMENDED RATES: Apply 2.5 to 3 quarts of this product plus ½ quart of Treflan per acre. Use the higher rate of this product for this tank mixture in areas of heavy weed infestation or when hard-to-control weeds, black or hairy nightshade or yellow nutsedge are present. Applications which are not consistent with recommendations in this label may result in unsatisfactory weed control or crop injury.

LASSO®

Red Kidney Bean Types Only

FOR USE IN ILLINOIS, INDIANA AND WISCONSIN ONLY

Apply this product in water solution for control of yellow nutsedge and the annual grasses and broadleaf weeds listed in the "WEEDS CONTROLLED" section of this label.

NOTE: This product may delay crop maturity and/or reduce yield if cold, wet soil conditions occur after planting.

APPROVED APPLICATION SYSTEMS

Ground—Broadcast Boom; banded

Reference: The "APPLICATION SYSTEMS" section of this label provides detailed information on the proper application of this tank mixture using the system selected.

APPROVED APPLICATION METHODS

Preplant Incorporated—Apply this product within 7 days prior to planting and shallowly incorporate or Surface Blend into the upper 1 to 2 inches of soil.

Preemergence Surface—Apply this product after planting and before crop and weed emergence and within 5 days after last preplant tillage operation.

Reference: The "APPLICATION TIMING AND METHODS" section of this label provides detailed information and procedures for application timing and method selected.

RECOMMENDED RATES: Apply 2.5 to 3 quarts of this product per acre. Use the higher recommended rate when heavy weed infestations, hard-to-control weeds, black or hairy nightshade or yellow nutsedge are present. Preplant incorporate only for control of yellow nutsedge.

LASSO®

Mung Beans

FOR USE IN OKLAHOMA ONLY

Apply this product in water solution for control of the annual grasses and broadleaf weeds listed in the "WEEDS CONTROLLED" section of this label.

APPROVED APPLICATION SYSTEMS

Ground—Broadcast Boom; banded

Reference: The "APPLICATION SYSTEMS" section of this label provides detailed information on the proper application of product using the system selected.

APPROVED APPLICATION METHODS

Preemergence Surface—Apply this product after planting and before crop and weed emergence and within 5 days after last preplant tillage operation.

Reference: The "APPLICATION TIMING AND METHODS" section of this label provides detailed information and procedures for application timing and method selected.

RECOMMENDED RATES: Refer to the following table for recommended broadcast treatment rates.

Surface Application Rates

	Broadcast Rate Per Acre (Quarts)
Soil Texture	**LASSO®**
Coarse*	1.5
Medium and Fine	2.0

*DO NOT USE ON SAND AS CROP INJURY MAY OCCUR.

PEAS

LASSO®

Peas, Green (for processing)

FOR USE IN MINNESOTA ONLY

Apply this product in water solution for control of the annual grasses and broadleaf weeds listed in the "WEEDS CONTROLLED" section of this label.

NOTE: This product may delay crop maturity and/or reduce yield if cold, wet soil conditions occur after application.

APPROVED APPLICATION SYSTEMS

Ground—Broadcast Boom

Reference: The "APPLICATION SYSTEMS" section of this label provides detailed information on the proper application of this tank mixture using the system selected.

Preemergence Surface—Apply this product after planting before crop and weed emergence and within 5 days after last preplant tillage operation.

Reference: The "APPLICATION TIMING AND METHODS" section of this label provides detailed information and procedures for application timing and method selected.

RECOMMENDED RATES: Refer to the following table for the recommended broadcast treatment rate for this product. Applications which are not consistent with recommendations in this label may result in unsatisfactory weed control or crop injury.

Surface Application Rates

	Broadcast Rate Per Acre (Quarts)
Soil Texture Group	**LASSO®**
Medium	2
Fine	2.5

* Refer to the "SOIL TEXTURE" section of this label to determine the corresponding texture group for the soil to be treated.

SUNFLOWERS

LASSO®

Apply this product in water or sprayable fluid fertilizer solution for control of yellow nutsedge, and the annual grasses and broadleaf weeds listed in the "WEEDS CONTROLLED" section of this label.

APPROVED APPLICATION SYSTEMS

Ground—Broadcast Boom; banded

Reference: The "APPLICATION SYSTEMS" section of this label provides detailed information on the proper application of this product using the system selected.

APPROVED APPLICATION METHODS

Preplant Incorporated—Apply this product within 7 days prior to planting and shallowly incorporate or Surface Blend into the upper 1 to 2 inches of soil.

Preemergence Surface—Apply this product after planting before crop and weed emergence and within 5 days after last preplant tillage operation.

Reference: The "APPLICATION TIMING AND METHODS" section of this label provides detailed information and procedures for the application timing and method selected.

NOTE: DO NOT graze treated area or feed treated forage to livestock following application of this product.

RECOMMENDED RATES: Refer to the following tables for the recommended broadcast treatment rates for this product. Applications which are not consistent with recommendations in this label may result in unsatisfactory weed control or crop injury.

Surface Application Rates

	Broadcast Rate Per Acre (Quarts)
Soil Texture Group*	**LASSO®**
Coarse	2
Medium	2.5 to 3
Fine	3 to 4

* Refer to the "SOIL TEXTURE" section of this label to determine the corresponding texture group for the soil to be treated.

Preplant Incorporated Application Rates

	Broadcast Rate Per Acre (Quarts)
Soil Texture Group*	**LASSO®**
Coarse	2.5
Medium	3 to 3.5
Fine	3.5 to 4

* Refer to the "SOIL TEXTURE" section of this label to determine the

corresponding texture group for the soil to be treated.

LASSO PLUS AMIBEN (TANK MIXTURE)

Apply this tank mixture in water solution for control of the annual grasses and broadleaf weeds listed in the "WEEDS CONTROLLED" section of this label.

APPROVED APPLICATION SYSTEMS

Ground—Broadcast Boom; banded

Reference: The "APPLICATION SYSTEMS" section of this label provides detailed information on the proper application of this tank mixture using the system selected.

APPROVED APPLICATION METHODS

Preplant Incorporated—Apply this tank mixture within 7 days prior to planting and shallowly incorporate or Surface Blend into the upper 1 to 2 inches of soil.

Preemergence Surface—Apply after planting, before crop and weed emergence and within 5 days after last preplant tillage operation.

Reference: The "APPLICATION TIMING AND METHODS" section of this label provides detailed information and procedures for the application timing and method selected.

NOTE: DO NOT graze treated area or feed treated forage to livestock following application of this tank mixture.

RECOMMENDED RATES:

Preplant Incorporated—Apply to 2.5 to 3 quarts of LASSO plus 4 quarts of Amiben per acre.

Preemergence Surface—Refer to the following table for the recommended broadcast treatment rates for LASSO plus Amiben. Applications which are not consistent with recommendations in this label may result in unsatisfactory weed control or crop injury.

Surface Application Rates

Soil Texture Group*	Broadcast Rate Per Acre (Quarts) LASSO®	Amiben
Coarse	2	4
Medium	2 to 2.5	4
Fine	2.5 to 3	4

* Refer to the "SOIL TEXTURE" section of this label to determine the corresponding texture group for the soil to be treated.

COTTON

LASSO®

FOR USE ONLY IN TEXAS, OKLAHOMA, LOUISIANA, NEW MEXICO, ARIZONA AND CALIFORNIA

Apply this product in water solution for control of the annual grasses and broadleaf weeds listed in the "WEEDS CONTROLLED" section of this label.

NOTE: DO NOT use this product on sand or loamy sand soils. This treatment may cause stunting and/or reduce yield if seedling disease, cold weather, excessive moisture or drought stress occurs after application. If the crop must be replanted into treated soil the seedbed should be tilled lightly prior to replanting. DO NOT plant in furrows (use only in flat planted and cotton planted on top of beds).

APPROVED APPLICATION SYSTEMS

Ground—Broadcast Boom; banded

Reference: The "APPLICATION SYSTEMS" section of this label provides detailed information on the proper application of this product using the system selected.

NOTE: If a band treatment is used then the spray nozzles must be located behind the drags or the covering mechanism of the planter to avoid the hazard of cotton injury and reduced yield.

APPROVED APPLICATION METHODS

Preplant Incorporated (Except Louisiana)—Apply this product within 7 days prior to planting and shallowly incorporate into the upper 1 to 2 inches of soil.

Preemergence Surface—Apply this product after planting, before crop or weed emergence and within 5 days after last preplant tillage operation.

Reference: The "APPLICATION TIMING AND METHODS" section of this label provides detailed information and procedures for application timing and method selected.

RECOMMENDED RATES: Refer to the following tables for the recommended broadcast treatment rates for this product.

Surface Application Rates*

Soil Texture Group**	Broadcast Rate Per Acre (Quarts) LASSO®
Coarse (sandy loam only)	2
Medium	2.5
Fine	3

* FOR LOUISIANA ONLY, apply 1 quart of this product per acre on sandy loam, 1.5 quarts per acre on medium soils, and 2 quarts per acre on fine soils.
NOTE: FOR PLANTING DATES AND AGRONOMIC PRACTICES, REFER TO THE LOCAL USE GUIDE FOR LOUISIANA.

** Refer to the "SOIL TEXTURE" section of this label to determine the corresponding texture group for the soil to be treated.

Preplant Incorporated Application Rates*

Soil Texture Group**	Broadcast Rate Per Acre (Quarts) LASSO®
Coarse (sandy loam only)	2.5 to 3
Medium	3 to 3.5
Fine	3.5 to 4

* DO NOT USE PREPLANT INCORPORATION IN LOUISIANA.

** Refer to the "SOIL TEXTURE" section of this label to determine the corresponding texture group for the soil to be treated.

WOODY ORNAMENTALS

LASSO®

FOR USE ON JUNIPER (JUNIPERUS SPP.) AND YEW (TAXUS SPP.)

Apply this product in water solution for control of the annual grasses and broadleaf weeds listed in the "WEEDS CONTROLLED" section of this label.

NOTE: Topical application or contact of this product with green foliage of the woody ornamental may cause injury and is not recommended. In case green foliage is accidentally sprayed, overhead irrigation within 1 hour after application will minimize injury symptoms (i.e. leaf tip burn, discoloration, distortion, etc.). This product is recommended for this use for application by certified, professional applicators only.

DO NOT USE THIS PRODUCT ON SEEDBEDS, UNROOTED CUTTINGS OR PRIOR TO TRANSPLANTING. DO NOT USE THIS PRODUCT IN GREENHOUSES OR OTHER ENCLOSED STRUCTURES, APPLIED THROUGH IRRIGATION SYSTEMS OR MIXED WITH SPRAYABLE FLUID FERTILIZERS. APPLICATIONS DURING PERIODS WHEN TEMPERATURE EXCEEDS 90°F MAY CAUSE INJURY TO ORNAMENTALS.

APPROVED APPLICATION SYSTEMS

Ground—Broadcast Boom; banded

Reference: The "APPLICATION SYSTEMS" section of this label provides detailed information on the proper application of this tank mixture using the system selected.

APPROVED APPLICATION METHODS

Post Transplant Directed—Apply this product as a directed spray after transplanting or to established plantings prior to weed emergence.

Reference: The "APPLICATION TIMING AND METHODS" section of this label provides detailed information and procedures for application timing and method selected.

RECOMMENDED RATES: Apply 4 quarts of this product per acre as a broadcast treatment. Applications which are not consistent with the recommendations in this label may result in unsatisfactory weed control or injury to ornamental plantings.

NOTE: Repeat applications of this product may be required for sustained weed control. DO NOT exceed 3 applications per year or retreat within 21 days.

CALIFORNIA

LASSO® herbicide has been approved by the U.S. Environmental Protection Agency for the uses, crops and sites listed in this label. Approval of the items listed below is pending under the State of California registration requirements.

These use conditions, crops and sites may not be treated with this product in California until approval is received:

- LASSO, and LASSO plus Amiben on Sunflowers
- LASSO plus atrazine plus Bladex on Corn
- LASSO plus atrazine plus Princep on field corn or silage corn in conservation or minimum tillage systems
- LASSO plus Amiben plus Lexone or Sencor on Soybeans.
- Center Pivot Irrigation Application on Soybeans and Peanuts
- LASSO plus Amiben™ PPI on Soybeans
- LASSO and LASSO tank mixes on Grain Sorghum (Milo)
- Sequential treatment on Corn and Soybeans
- Tank mixes with atrazine 90% DF formulations except AAtrex™ NINE-O™
- Tank mixes with linuron 4L
- Tank mixes with Sencor+up,TM DF
- Control of **Acanthospermum hispidum**
- Control of **Physalis** spp.
- Reduced competition from **Cassia obtusifolia**
- Reduced competition from **Cyperus esculentus**
- Reduced competition from **Eriochloa villosa**
- Reduced competition of **Panicum miliaceum**
- Reduced competition from **Sorghum bicolor** (shattercane)

™AAtrex NINE-O is a trademark of the Ciba-Geigy Corporation.

In case of an emergency involving this product, Call Collect, day or night, (314) 694-4000.

LASSO ATRAZINE

FLOWABLE PRE-MIX HERBICIDE

Selective Herbicide for Preemergence Weed Control in Corn (All Types), and Grain Sorghum (milo).

SHAKE WELL BEFORE USING

ACTIVE INGREDIENTS:*	
Alachlor, [2-chloro-2'6'-diethyl-N-(Methoxymethyl) acetanilide]	27.2%
Atrazine, [2-chloro-4-(ethylamino)-6-(isopropylamino)-s-triazine]	15.5%
Related atrazine compounds	0.8%
INERT INGREDIENTS:	56.5%
TOTAL	100.0%

* Contains 300 grams/litre or 2.5 pounds/gallon of alachlor and 180 grams/litre or 1.5 pounds/gallon of atrazine and related compounds.

EPA Reg. No. 524-329-AA 1694.24/53

WARNING

Keep Out of Reach of Children

The use of this product may be hazardous to your health. This product contains alachlor, which has been determined to cause tumors in laboratory animals.

Users are urged to obtain information on careful handling and use of pesticides and to participate in Monsanto Company's education program on potential hazards and safer use of this product. For additional information, contact Monsanto Agricultural Products Co. at 1-800-453-4000. [NOTE: This number is for user safety information only. *NOT* for emergencies. For emergencies, call collect day or night (314)694-4000.]

Complete Directions for Use.

Read the entire label before using this product.

Use only according to label instructions.

Read "LIMIT OF WARRANTY AND LIABILITY" before buying or using. If terms are not acceptable, return at once unopened.

LIMIT OF WARRANTY AND LIABILITY

This company warrants that this product conforms to the chemical description on the label and is reasonably fit for the purposes set forth in the complete Directions for Use label booklet ("Directions") when used in accordance with those Directions under the conditions described therein. NO OTHER EXPRESS OR IMPLIED WARRANTY OF FITNESS FOR PARTICULAR PURPOSE OR MERCHANTABILITY IS MADE. This warranty is also subject to the conditions and limitations stated herein.

Buyer and all users shall promptly notify this company of any claims whether based in contract, negligence, strict liability, other tort or otherwise.

Buyer and all users are responsible for all loss or damage from use or handling which results from conditions beyond the control of this company, including, but not limited to, incompatibility with products other than those set forth in the Directions, unusual weather (i, weather conditions which are outside the range considered normal at the application site and for the time period when the product is applied with the normal range being determined on the basis of the average range for the prior 40 years computed from the best available information, and ii, weather perils, including but not limited to hurricanes, tornadoes and floods) as well as weather considerations set forth in the Directions, application in any manner not explicitly set forth in the Directions, moisture conditions outside the moisture range specified in the Directions, or the presence of products other than those set forth in the Directions in or on the soil or crop.

THE EXCLUSIVE REMEDY OF THE USER OR BUYER AND THE LIMIT OF THE LIABILITY OF THIS COMPANY OR ANY OTHER SELLER FOR ANY AND ALL LOSSES, INJURIES OR DAMAGES RESULTING FROM THE USE OR HANDLING OF THIS PRODUCT (INCLUDING CLAIMS BASED IN CONTRACT, NEGLIGENCE, STRICT LIABILITY, OTHER TORT OR OTHERWISE) SHALL BE THE PURCHASE PRICE PAID BY THE USER OR BUYER FOR THE QUANTITY OF THIS PRODUCT INVOLVED, OR, AT THE ELECTION OF THIS COMPANY OR ANY OTHER SELLER, THE REPLACEMENT OF SUCH QUANTITY, OR, IF NOT ACQUIRED BY PURCHASE, REPLACEMENT OF SUCH QUANTITY. IN NO EVENT SHALL THIS COMPANY OR ANY OTHER SELLER BE LIABLE FOR ANY INCIDENTAL OR CONSEQUENTIAL DAMAGES.

The buyer and all users are deemed to have accepted the terms of this LIMIT OF WARRANTY AND LIABILITY which may not be varied by any verbal or written agreement.

In case of an emergency involving this product, Call Collect, day or night, (314) 694-4000

PRECAUTIONARY STATEMENTS

HAZARDS TO HUMANS AND DOMESTIC ANIMALS

Keep out of reach of children.

WARNING!

HARMFUL IF SWALLOWED OR ABSORBED THROUGH SKIN.

CAUSES EYE AND SKIN IRRITATION.

MAY CAUSE ALLERGIC SKIN REACTION.

Do not get in eyes, on skin, or on clothing.

Wear goggles or face shield and rubber gloves, long trousers, long sleeved shirt or jacket of tightly woven material along with boots high enough to cover ankles when transferring and mixing, and when adjusting, repairing or cleaning equipment. Wear rubber boots when pouring from open containers or when reentry is made into fields following application of this product through center pivot irrigation systems and the field is still wet. Clothing which comes in contact with this product must be washed before reuse. Clothing or other materials which become drenched with the concentrated pesticide must be disposed of in a sanitary landfill, by incineration or, if allowed by state and local authorities, by burning. If burned, stay out of smoke.

Wash thoroughly with soap and water after handling.

FIRST AID: IF IN EYES, immediately flush with plenty of water for at least 15 minutes. Call a physician.

IF ON SKIN, immediately flush with plenty of water while removing contaminated clothing. Call a physician. Wash clothing before reuse. Sensitized persons should avoid further contact and reuse of clothing.

IF SWALLOWED, induce vomiting immediately by giving two glasses of water and sticking finger down throat. Call a physician. NEVER GIVE ANYTHING BY MOUTH TO AN UNCONSCIOUS PERSON.

ENVIRONMENTAL HAZARD

This chemical has been identified in limited ground water sampling and there is the possibility that it can leach through the soil to ground water, especially where soils are coarse and ground water is near the surface. Following application, and during rainfall events that cause runoff, this chemical may reach surface water bodies including streams, rivers and reservoirs.

Care must be taken when using this product to prevent back siphoning into wells, spills or improper disposal of excess pesticide, spray mixtures or rinsates.

Check valves or antisiphoning devices must be used on all mixing and/or irrigation equipment. Disposal of excess pesticide, spray mixtures, or rinsate should be according to label use instructions or according to the state pesticide or environmental control agency or the hazardous waste representative at the nearest EPA regional office.

PHYSICAL OR CHEMICAL HAZARD

COMBUSTIBLE.

Do not use or store near heat or flame.

Use only with adequate ventilation.

In case of:

FIRE, use water spray, foam, dry chemical or CO_2.

SPILL or LEAK, soak up, remove and keep away from humans and domestic animals.

STORAGE AND DISPOSAL

Do not contaminate water, foodstuffs, seed or feed by storage and disposal.

STORAGE

STORE ABOVE 40°F.

Below 32°F. this product may freeze. If frozen, place in a warm room and roll and/or shake the container frequently to redissolve before using. For bulk containers, see the container label for alternate storage information.

DISPOSAL

Wastes of this pesticide may cause eye and skin irritation and may be dangerous. Improper disposal of excess pesticide, spray mixtures, or rinsate is a violation of Federal law. If these wastes cannot be disposed of according to label use instructions, contact your State Pesticide or Environmental Control Agency, or the Hazardous Waste representative at the nearest EPA Regional Office for guidance.

Empty container retains vapor and product residue. Observe all labeled safeguards until container is cleaned, reconditioned or destroyed.

(See the individual container label for disposal information.)

GENERAL INFORMATION

This product is recommended for the control of many annual grasses and broadleaf weeds. Except for minimum and conservation tillage systems, the seed bed should be fine, firm and free of clods and trash.

Performance is improved when moisture occurs within one week after application. The amount required depends primarily upon the amount of soil moisture in the weed germination zone at the time of application, soil type and organic matter content, but $1/3$ to $3/4$ inch is generally sufficient for performance of this product. High intensity or excessive rainfall, or excessive irrigation, after application may reduce control.

Read and carefully observe the cautionary statements and all other information appearing on this label and on the labels of paraquat and Roundup® herbicides before using in combination with this product.

Use of this product in any manner not consistent with this label may result in injury to persons, animals or crops, or other unintended consequences. Keep container closed to prevent spills and contamination.

Do not apply when weather conditions favor drift.

DO NOT APPLY WITH AERIAL EQUIPMENT.

Flush sprayer with clean water after use.

Do not graze treated area or feed treated forage to livestock for 21 days following application.

NOTE: Use this product for weed control in corn and grain sorghum (milo) only. PLANT ONLY CORN, PEANUTS, SORGHUM OR SOYBEANS THE YEAR FOLLOWING THE USE OF THIS MIXTURE. IF SOYBEANS ARE TO BE PLANTED, THERE IS A POSSIBILITY OF CROP INJURY DUE TO CARRYOVER OF ATRAZINE. Do not plant soybeans in areas where furrow irrigation is practiced. In Texas, on grain sorghum, use only on the fine textured soils of the Gulf Coast.

®Trademark of Monsanto Company

MIXING, SPRAYING AND HANDLING INSTRUCTIONS

NOTE

Direct contact or exposure to this product or spray mixtures of this product should be minimized. The following instructions for transfer, mixing, cleaning or repairing equipment should be followed in order to minimize this exposure. Review the protective clothing requirements as listed in the Precautionary Statements section of this label (front panel) and do not use this product until you have the necessary protective clothing.

2.5 Gallon Containers

Open pouring from these containers can result in exposure from splashing or spilling. Special care in lifting and pouring are strongly recommended.

Bulk Containers

Open pouring from these containers can result in exposure from splashing or spilling and is not recommended. This product should be transferred from these containers to the mix or spray tank using pumps or transfer probes. The probe or pump should not be removed from the container or disconnected until the container is emptied and rinsed. Use the pump or probe system to rinse the empty container and transfer the rinsate directly to the mix or spray tank.

Equipment Cleaning & Repair

Cleaning and repair of transfer systems and application equipment is a source of exposure to this product. Care should be taken to minimize exposure during cleaning and repair of transfer systems and application equipment. Whenever possible, these systems or equipment should be rinsed before being cleaned or repaired. When repairs must be made during transfer or application, the equipment should be shutdown, and special care taken to avoid contact with the pesticide.

Always predetermine the compatibility of this product with water carrier or sprayable fluid fertilizer or other labeled herbicide by mixing small proportional quantities in advance.

Mix this product with the appropriate carrier as follows:

1. Place a 20 to 35 mesh screen or wetting basket over filling port.
2. Through the screen, fill the sprayer tank one-half full with the carrier.
3. If a compatibility agent is necessary to improve mixing or to prevent the formation of undesirable and unsprayable gels or precipitates, while agitating add it to the carrier already in the tank.

   Use only compatibility agents, cleared by the Federal Food and Drug Adminsitration for this use. Follow the directions for use of the compatibility agent label. Check for adequate agitation.
4. Pre-mix one part of this product with one part water. Add diluted mixture SLOWLY through the screen into the tank. When necessary, this product may be diluted with a fertilizer carrier instead of water, after compatibility has been confirmed.
5. With minimum tillage systems, add the required amounts of Roundup or paraquat near the end of the filing process.
6. Complete filling the sprayer tank with carrier. Remove hose from tank immediately to avoid siphoning back into the carrier source. Maintain good agitation at all times until the contents of the tank are sprayed. NOTE: If spray mixture is allowed to settle at any time, thorough agitation is required to resuspend the mixture before spraying is resumed.

Keep by-pass line on or near the bottom of tank to minimize foaming. Screen size in nozzle or line strainers should be no finer than 50 mesh. Carefully select proper nozzle to avoid spraying a fine mist. For best results, use flat fan or flood jet nozzles. Check for even distribution of spray droplets. To reduce loss of the chemical due to drift of a fine mist, apply at pressures from 20 to 40 psi.

STANDARD SPRAYABLE FLUID FERTILIZER COMPATIBLITY TEST

Herbicides may not always mix evenly throughout a sprayable fluid fertilizer or the components may separate too quickly to make their combined use of practical value. This may be due to certain characteristics of the different fluid fertilizers. A simple test using small quantities of the components is suggested to provide compatibility potential.

The test follows:

A. Materials Required For A Compatibility Test

1. Two one-quart jars with lid or stopper (marked "with" and "without").
2. Teaspoons (for a more exacting test, a five to ten milliliter (ml.) pipette or graduated cylinder is desirable).
3. A minimum of one quart of the sprayable fluid fertilizer to be tested.
4. The herbicide(s) to be mixed.
5. A compability agent (the purpose of the adjuvant is to help keep the fertilizer and herbicide(s) in suspension, if this assistance is needed).

B. Procedure

1. Add one pint of the sprayable fluid fertilizer that will be used or other herbicide carrier to each jar marked "with" and "without".

Add One Pint Liquid Fertilizer
To Two Quart Jars

2. To the jar marked "with", add 1/4 TEAspoon or 1.2 milliliters of a suitable compatibility agent; shake gently for five to ten seconds to mix. (1/4 TEAspoon is the equivalent of two pints per 100 gallons of liquid fertilizer.)

To Jar Marked "With"
Add Compatibility Agent
And Shake To Mix

3. To each jar add the appropriate amount of herbicide(s). If more than one is used, add them separately with the wettable powders added first, flowables second and liquid last. Shake gently five to ten seconds after each addition.

Add Herbicide(s) To Both Jars
And Shake To Mix

Amount To Be Added Per Pint Of Sprayable Fluid Fertilizer (Assuming Volume is 25 gal/A)

Herbicide	**Rate/A**		**Level TEAspoons**		**Milliliters**
This Product	3.0 qts	=	2.9	or	14.2
	3.5 qts.	=	3.4	or	16.6
	4.0 qts.	=	3.8	or	19.0
	4.5 qts.	=	4.3	or	21.4
	5.0 qts.	=	4.8	or	23.8

This compatibility test is designed for 25 gallons of spray per acre with the maximum labeled rate of herbicide. For changes in spray volume or herbicide concentration, make appropriate changes in the ingredients of the test. Regardless of spray volume, the amount of compatibility agent should be equal to two or three pints (two pints = 1/4 TEAspoon or 1.2 milliliters, three pints = 3/8 TEAspoon or 1.8 milliliters per pint of sprayable fluid fertilizer) per 100 gallons of liquid fertilizer.

C. Observations And Decisions

1. If the herbicide(s) and the sprayable fluid fertilizer are compatible.
2. If a compatibility agent is necessary.

Five minutes after the final addition and mixing, observe both jars for the formation of large flakes, sludge, gels or other precipitates. Observe if the herbicide(s) cannot be physically mixed with the liquid fertilizer but remains as small oily particles in the solution.

If incompatibility in any form described above occur in the jar "with" the compatibility agent added, the liquid fertilizer and the herbicide(s) should not be used together in the same spray tank.

If incompatibility as described above occurs in the jar "without" the adjuvant but not in the jar "with" adjuvant, the use of a compatibility adjuvant is recommended.

Both jars should be allowed to stand and be observed periodically for one-half hour. If the separate layers of liquid fertilizer and additives can be resuspended by shaking, commercial application is possible. Wettable powders will either settle to the bottom of the tank or jar, or float to the top, depending upon the density of the fertilizers.

If the herbicide(s) is compatible with fluid fertilizer in the foregoing test without having to use a compatibility agent, fluid fertilizer may be used for the premixing. If it is not compatible, the herbicide(s) should be premixed with water before adding to the spray tank.

WEEDS CONTROLLED

Applicators should evaluate soil and weed conditions carefully to assure that they choose the lowest effective label rate.

When applied as directed this product will CONTROL the following weeds:

Annual Grasses	
Barnyardgrass	*Echinochloa crus-galli*
Crabgrass	*Digitaria ischaemum*
	Digitaria sanguinalis
Foxtail (giant)	*Setaria faberi*
Foxtail (green)	*Setaria viridis*
Foxtail (yellow)	*Setaria lutescens*
Goosegrass	*Eleusine indica*
Panicum (fall)	*Panicum dichotomiflorum*
Signalgrass (broadleaf)	*Brachiaria* spp.
Witchgrass	*Panicum capillare*
Annual Broadleaves	
Carpetweed	*Mollugo verticillata*
Jimsonweed	*Datura stramonium*
Kochia	*Kochia scoparia*
Lambsquarters	*Chenopodium album*
Mustard	*Brassica* spp.
Nighshade (black)	*Solanum nigrum*
Nightshade (hairy)	*Solanum sarachoides*
Pigweed, Carelessweed	*Amaranthus* spp.
Purslane	*Portulaca oleracea*
Pusley (Florida)	*Richardia scabra*
Ragweed (common)	*Ambrosia artemisiifolia*
Smartweed	*Polygonum Pensylvanicum*
	Polygonum persicaria
Sedges	
Nutsedge (yellow)*	*Cyperus esculentus*

* Use 5 quarts per acre of this product on medium and fine textured soils and preplant incorporate only for control of this weed.

When applied preplant incorporated or immediately after planting and within 5 days of last tillage, this product at rate of 4 to 5 quarts per acre on a broadcast basis will reduce competition from the following HARD TO CONTROL weeds:

Annual Grasses	
Johnsongrass (seedling)	*Sorghum halepense*
Rice (red)	*Oryza sativa*
Sandbur, Grassbur	*Cenchrus pauciflorus*
Annual Broadleaves	
Beggarweed (Florida)	*Desmodium tortuosum*
Cocklebur	*Xanthium pensylvanicum*
Morningglory (annual)	*Ipomoea purpurea*
Sida (spiny), Teaweed	*Sida spinosa*
Velvetleaf, Buttonweed	Abutilon theophrasti
Sedges	
Nutsedge (yellow)	*Cyperus esculentus*

DIRECTIONS FOR USE

It is a violation of Federal Law to use this product in any manner inconsistent with its labeling.

GROUND APPLICATION

Apply as a broadcast treatment in 10 or more gallons of water or sprayable fluid fertilizer per acre. Do not apply during periods of gusty winds or winds in excess of 15 mph.

Band Treatment—Apply a broadcast equivalent rate and volume per acre to determine these:

$$\frac{\text{Band width in inches}}{\text{Row width in inches}} \times \text{Broadcast RATE per acre} = \text{Band RATE per acre}$$

$$\frac{\text{Band width in inches}}{\text{Row width in inches}} \times \text{Broadcast VOLUME per acre} = \text{Band VOLUME per acre}$$

Mount spray equipment on the planter with nozzles behind press wheel and apply this product at planting time.

CENTER PIVOT IRRIGATION APPLICATION

Center pivot is the only irrigation equipment through which this product may be used. Refer to the Monsanto booklet "How to Apply Lasso Herbicide with Center Pivot Irrigation" before use. Apply the recommended ground application rates with center pivot irrigation systems. Use only in systems that apply uniformly. Do not apply when wind speed favors drift, when system connections or fittings leak, when nozzles do not provide uniform distribution or when the line containing the product must be dismantled and drained. Meter this product diluted into the center pivot irrigation system after seeding and before crop or weed emergence. Herbicide applications should be made in $\frac{1}{3}$ to $\frac{3}{4}$ inch of water per acre. Do not apply in more than $\frac{3}{4}$ inch water per acre under any conditions or reduced performance may occur. On very sandy soils (more than 60% sand and less than 1% organic matter), apply this product in a maximum of $\frac{1}{2}$ inch water per acre. Good agitation must be maintained during the entire application period. Have appropriate check valves in the system to insure that this product cannot siphon back into the water supply. Flush the system with water when application is complete. If spills of the herbicide concentrate occur, immediately wash off the system.

CULTIVATION INFORMATION

If annual grass or broadleaf weed seedlings emerge, rotary hoe immediately to improve weed control.

If cultivation is necessary because of soil crusting or compaction, tillage should be shallow to minimize dilution of this herbicide.

Except for the above, as long as the grasses and weeds are controlled, delay cultivation as long as possible.

If band application is used, avoid throwing soil during cultivation on to the band before weeds emerge in the band.

CORN

Surface Application—Apply this product in 10 or more gallons of water or sprayable fluid fertilizer anytime from immediately after planting until weeds or grasses reach the 2-leaf stage and the crop is no more than 5 inches tall. Performance may be improved when application is made prior to weed emergence. Applications made beyond the 2-leaf stage of weed growth will not provide satisfactory weed control. Occasionally, some leaf burn may occur, but this should not affect subsequent crop growth or yield. If applied in sprayable fluid fertilizer, application MUST be made prior to crop emergence. On sweet corn use this product before crop or weeds emerge. See the following table for recommended broadcast rates per acre for surface applications:

	Broadcast Rate Per Acre	
Soil Texture	Less than 3% Organic Matter Quarts	3% or More Organic Matter Quarts
COARSE Sand thru sandy loam	(See Information Below)	
MEDIUM Loam thru clay loam	3 to 3.5	3.5 to 4
FINE Silty clay loam thru clay	3.5 to 4	4 to 4.5

Use the higher rate of this product in the recommended ranges in areas where heavy grass infestations will be present.

Use 4 to 5 quarts of this product per acre to reduce competition from the hard-to-control weeds listed on this label.

COARSE SOILS (NON-IRRIGATED): Apply 2.5 to 3.5 quarts of this product per acre.

COARSE SOILS (SPRINKLER IRRIGATED INCLUDING CENTER PIVOT INJECTION): Apply 3 to 4 quarts of this product per acre. Do not exceed 1/3 to 3/4 inches of water per acre with the first irrigation as reduced effectiveness may result.

Preplant Incorporation—Broadcast spray this product in 10 or more gallons of water or sprayable fluid fertilizer and incorporate thoroughly into the top 2 inches of soil within 7 days prior to planting. When SURFACE BLENDING using one-pass incorporation, the equipment used must provide uniform and shallow blending to avoid streaking the treatment. To insure shallow incorporation of 2 inches or less, pull-type tillage equipment should be set to work the soil NO DEEPER THAN 4 INCHES.

This product is diluted when incorporated into the soil. Therefore, SHALLOW AND UNIFORM incorporation of the herbicide treatment is essential.

See the following table for recommended broadcast rates per acre for preplant incorporated applications:

Soil Texture	Broadcast Rate Per Acre (Quarts) Less than 3% Organic Matter	3% or more Organic Matter
COARSE*		
Sand thru sandy loam	3.5	3.5
MEDIUM		
Loam thru clay loam	3.75	4.0
FINE		
Silty clay loam thru clay	4.25	4.5

NOTE: TO CONTROL yellow nutsedge apply 5 quarts of this product per acre on medium and fine textured soils and preplant incorporate only.

* Performance may be reduced when this product is incorporated on coarse soils with less then 3% organic matter content.

Use 4 to 5 quarts of this product per acre to reduce competition from the hard-to-control weeds listed on this label.

Equivalent quarts of LASSO and quarts of atrazine 4L in the recommended rates of this product are as follows:

This Product	=	LASSO	+	Atrazine
2.5 qt.		1.6 qt.		0.9 qt.
3.0 qt.		1.9 qt.		1.1 qt.
3.5 qt.		2.2 qt.		1.3 qt.
4.0 qt.		2.5 qt.		1.5 qt.
4.5 qt.		2.8 qt.		1.7 qt.
5.0 qt.		3.1 qt.		1.9 qt.

BLADEX™ PLUS THIS PRODUCT (TANK MIXTURE)

Apply this tank mixture in water or sprayable fluid fertilizer solutions for control of annual grasses and broadleaf weeds listed in the "WEEDS CONTROLLED" section of this label. Refer to the rate table for this product and to the Bladex rate table below for either surface application or preplant incorporation.

Surface or Preplant Incorporated Application of Bladex

Soil Texture Group	Broadcast Rate Per Acre (Quarts) Less than 3% Organic Matter	3% or More Organic Matter
COARSE* Sand thru sandy loam	1	1 to 1.25
MEDIUM Loam thru clay loam	1 to 1.25	1.5 to 2
FINE Silty clay loam thru clay	1.25 to 1.5	1.5 to 2

* Do not use this mixture on sand or loamy sand soils with less than 2% organic matter.

™ Bladex is a trademark of the Shell Chemical Company.

MINIMUM TILLAGE SYSTEMS

TANK MIXTURES OF THIS PRODUCT PLUS ROUNDUP® OR PARAQUAT

When applied as directed under the conditions described, these tank mixtures control many emerged annual weeds, suppress many emerged perennial weeds and give preemergence control of many annual weeds when corn will be planted directly into a cover crop, established sod or in previous crop residues. These tank mixtures will not control regrowth from perennial weeds. Do not apply by air.

Refer to specific product labels for crop rotation restrictions and cautionary statements of all products used in the tank mixture.

For mixing instructions, see the "Mixing and Application Instructions" section of this label.

Apply these tank mixtures in 10 to 30 gallons of water per acre immediately before, during or after planting but BEFORE CROP EMERGENCE. As previous crop residue or vegetation increases, spray gallonage should be increased within the recommended range to insure complete coverage.

ATTENTION

AVOID DRIFT—EXTREME CARE MUST BE USED WHEN APPLYING THESE TANK MIXTURES TO PREVENT INJURY TO DESIRABLE PLANTS AND CROPS. Do not allow spray mist to drift since even minute quantities of spray can cause severe damage or destruction to nearby crops, plants or other areas on which treatment is not intended. Do not apply when winds are gusty or in excess of 5 miles per hour or when other conditions, including lesser wind velocities, will allow drift to occur. When spraying avoid combinations of pressure and nozzle type that will result in fine particles (mist) which are more likely to drift.

For weeds controlled preemergence see the "Weeds Controlled" section of this label. For control of emerged weeds, see the "Weeds Controlled" sections of the Roundup or paraquat labels. PLEASE NOTE THAT PARAQUAT IS A HIGHLY TOXIC SUBSTANCE WHICH CAN BE FATAL IF MISUSED. READ THE ENTIRE PARAQUAT LABEL AND USE ONLY AS DIRECTED.

ROUNDUP® PLUS THIS PRODUCT (TANK MIXTURE)

Annual Weeds—Use 1.5 quarts of Roundup herbicide per acre with the amount of this product recommended in the Surface Application section of this label.

Perennial Weeds—At normal application dates in minimum tillage systems, perennial weeds may not be at the proper stage of growth for control. See the "General Information" section of the label for Roundup for the proper stage of growth for perennial weeds. Use of 2 to 4 quarts of Roundup herbicide in tank mixture with the amount of this product recommended in the rate table of this label under these conditions provides top kill and reduces competition from many emerged perennial grass and broadleaf weeds. To obtain control, follow label recommendations for stage of growth and rate of application for specific perennial weeds. To obtain the desired stage of growth, it may be necessary to apply Roundup herbicide alone in the late summer or fall and then follow with this product at planting.

NOTE: When using this tank mixture, do not exceed 4 quarts of Roundup herbicide per acre.

USE OF THIS TANK MIXTURE FOR BERMUDAGRASS OR JOHNSONGRASS CONTROL IS NOT RECOMMENDED.

®Trademark of Monsanto Company

PARAQUAT PLUS THIS PRODUCT (TANK MIXTURE)

When used as directed, paraquat in tank mixture with this product controls many emerged annual weeds and suppresses many emerged perennial weeds. Apply 1 to 2 pints of paraquat (0.25 to 0.5 lb active) with the amount of this product recommended in the Surface Application section of this label per acre immediately before, during or after planting but BEFORE CROP EMERGENCE. Use the 2 pint paraquat rate if weeds are 4 to 6 inches tall. This mixture may not control weeds taller than 6 inches. Add Ortho™ X-77 Spreader at 8 oz. per 100 gallons of diluted spray.

™Ortho is a registered trademark of Chevron Chemical Company.

APPLICATION WITH DRY BULK FERTILIZER

The herbicide/fertilizer impregnation process must be done only at commercial fertilizer or chemical dealerships properly equipped for this procedure.

Dry bulk fertilizer may be impregnated with this product for use on corn. These tank mixtures must be applied with 200 to 450 pounds of dry bulk fertilizer per acre and shallowly incorporated within 7 days prior to planting. The proper amount of herbicide must be applied per acre in the blend as recommended in this label for the crop, weed and soil type treated. Refer to the table for broadcast rate per acre in the Preplant Incorporated section of this label to determine the recommended broadcast rate per acre for the herbicide treatment to be applied.

The following table provides a reference to determine the amount of LIQUID herbicide to be mixed per ton of dry bulk fertilizer for a range of herbicide recommendations for fertilizer rates per acre:

Fertilizer Rate (Lb./Acre)	Acres Covered (Per Ton)	Recommended Quarts Liquid Herbicide/Acre 3	3.5	4.0	4.5
		(Quarts of Herbicide/Ton Dry Bulk Fertilizer)			
200	10	30	35	40	45
250	8	24	28	32	36
300	6.7	20	23.3	26.6	30
350	5.7	17.1	20	22.8	25.7
400	5	15	17.5	20	22.5
450	4.5	13.3	15.6	17.8	20

To determine the amount of herbicide needed for rates not included in the preceding table, use the following formula:

$$\frac{\text{Recommended Herbicide Rate (lb. or qt.)/Acre} \times 2000}{\text{Pounds Fertilizer/Acre}} = \text{Pounds or Quarts of herbicide per ton of dry bulk fertilizer}$$

Mix and blend the dry fertilizer and herbicide mixture in a closed rotary-drum type mixer allowing sufficient time to ensure uniform coverage. Use at least one ton of dry fertilizer per mixing operation. Inject the herbicide into the drum over a minimum of a 2-minute period and allow at least 2 additional minutes mixing time to ensure uniformity. The nozzle used to spray the herbicide treatment must be placed inside the mixer to provide uniform spray coverage of the tumbling fertilizer.

If the dry fertilizer used has inadequate adsorptive capacity, use a highly adsorptive powder such as Microcel E to provide a free-flowing mixture. The adsorptive powder must be added separately and uniformly to the prepared herbicide-fertilizer mixture. Generally 2 percent by weight of Microcel E is required. The following is a partial list of dry fertilizers which may be impregnated with this product.

Ammonium sulfate	21–0–0
Ammonium phosphate-sulfate	16–20–0
Diammonium phosphate	18–46–0
Potassium chloride	0–0–60
Potassium sulfate	0–0–52
*Urea	45–0–0

* Some ureas may be phytotoxic when applied on corn. Use only ureas known to be safe to corn.

NOTE: DO NOT impregnate tank mixtures of this product with other herbicides on fertilizers containing ammonium nitrate, potassium nitrate or sodium nitrate.

APPLICATION

Spread the fertilizer/chemical mixture with a properly calibrated applicator. Be certain the material is applied uniformly to the soil surface. When using spin spreaders, a 100% overlap is recommended to ensure even distribution. Spin applicators should be calibrated to put one-half the desired rate per acre and application overlapped by doubling back to cover one-half of the previous swath. Dribble spreaders must be calibrated with the fertilizer/chemical mixture since the flow pattern for the LASSO herbicide-coated fertilizer will generally be slower than straight dry fertilizer.

GRAIN SORGHUM (MILO)

Surface Application—Apply this product in 10 or more gallons of water or sprayable fluid fertilizer any time from immediately after planting until weeds reach the 2-leaf stage. Applications with water may be made until the crop reaches 5 inches in height. Occasionally, some leaf burn may occur but this should not affect subsequent crop growth or yield. DO NOT APPLY IN SPRAYABLE FLUID FERTILIZER AFTER CROP EMERGENCE SINCE CROP INJURY MAY RESULT.

See the following table for recommended broadcast rates per acre for surface application. Apply only to grain sorghum seed that has been properly treated with a protectant containing the active ingredient flurazole.

NOTE: Applications made in milo growing on alkali soils or where cuts, fills or erosion have exposed calcareous or alkali subsoils may result in crop injury.

Soil Texture	Broadcast Rate Per Acre (Quarts) Less than 3% organic matter	3% or more organic matter
COARSE—Sand thru sandy loam	Do not use	3
MEDIUM—Loam thru clay loam	3 to 3.5	3.5 to 4
FINE—Silty clay loam thru clay	3.5 to 4	4 to 4.5

Use the higher rate of this product in the recommended ranges in areas where heavy weed infestations are present.

Use 4 to 5 quarts of this product per acre on medium or fine textured soils only to reduce competition from the hard-to-control weeds listed on this label.

NOTE: When applied through center pivot irrigation or under sprinkler irrigation systems on medium or fine textured soils only, use 4 to 5 quarts of this product. So not exceed ¾ inches of water per acre with the first irrigation as reduced effectiveness may result.

Preplant Incorporation—Broadcast spray this product in 10 or more gallons of water or sprayable fluid fertilizer and incorporate thoroughly into the top 1 to 2 inches of soil within 7 days prior to planting. When SURFACE BLENDING, using one-pass incorporation, the equipment used must provide uniform and shallow blending to avoid streaking. To insure shallow incorporation of 2 inches or less, pull-type tillage equipment should be set to work the soil NO DEEPER THAN 4 INCHES.

This product is diluted when incorporated into the soil. Therefore, SHALLOW AND UNIFORM incorporation of the herbicide treatment is essential.

See the following table for recommended broadcast rates per acre for preplant incorporation applications:

Soil Texture	Broadcast Rate Per Acre (Quarts) Less than 3% organic matter	3% or more organic matter
COARSE—Sand thru sandy loam	Do not use	Do not use
MEDIUM—Loam thru clay loam	3.5 to 3.75	4
FINE—Silty clay loam thru clay	4 to 4.25	4.5

On medium soil, use 4 to 5 quarts of this product per acre to reduce competition from the hard-to-control weeds listed on this label. On fine soils, use 5 quarts of this product per acre.

LASSO® and/or ATRAZINE (TANK MIXTURE)

For hard-to-control weeds, additional amounts of LASSO and/or atrazine may be added to the recommended treatment rates for this product. The following table shows the amounts of LASSO and/or atrazine that can be added to specific treatment rates of this product.

Recommended Rate (Quarts)	Additional Product (Quarts)	
LASSO® & atrazine flowable	LASSO®	Atrazine (maximum)
2.5	1.25 to 2.25	1.1
3	1 to 2	0.9
3.5	.75 to 1.75	0.7
4	.5 to 1.5	0.5

In case of an emergency involving this product, Call Collect, day or night (314) 694-4000

LASSO II

HERBICIDE

POWER GRANULAR

Selective Herbicide for Preemergence Weed Control in Corn, Soybeans, Peanuts, Sunflowers and Woody Ornamentals.

ACTIVE INGREDIENT:
Alachlor:2-chloro-2′, 6′-diethyl-N-(methoxymethyl) acetanilide 15%
INERT INGREDIENTS 85%
TOTAL 100%

EPA Reg. No. 524-296-AA EPA Est. No. 524-IA-1
U.S. Patent 3,547,620
®Lasso is a registered trademark of Monsanto Company.
©MONSANTO COMPANY

WARNING

Keep Out of Reach of Children

The use of this product may be hazardous to your health. This product contains alachlor, which has been determined to cause tumors in laboratory animals.

Users are urged to obtain information on careful handling and use of pesticides and to participate in Monsanto Company's education program on potential hazards and safer use of this product. For additional information, contact Monsanto Agricultural Products Co. at 1-800-453-4000. [NOTE: This number is for user safety information only. *NOT* for emergencies. For emergencies, call collect day or night (314) 694-4000.]

Completely empty bag into application equipment. Then dispose of empty bag in a sanitary landfill or by incineration, or, if allowed by state and local authorities, by burning. If burned, stay out of smoke. Read the entire label before using this product.

Use only according to label instructions.

Read "LIMIT OF WARRANTY AND LIABILITY" before buying or using. If terms are not acceptable, return at once unopened.

LIMIT OF WARRANTY AND LIABILITY

This company warrants that this product conforms to the chemical description on the label and is reasonably fit for the purposes set forth in the Complete Directions for Use, when used in accordance with those directions under the conditions described therein. NO OTHER EXPRESS OR IMPLIED WARRANTY OF FITNESS FOR PARTICULAR PURPOSE OR MERCHANTABILITY IS MADE. This warranty is also subject to the conditions and limitations stated herein.

Buyer and all users shall promptly notify this company of any claims whether based on contract, negligence, strict liability, other tort or otherwise.

Buyer and all users are responsible for all loss or damage from use or handling which results from conditions beyond the control of this company, including but not limited to incompatibility with products other that those set forth in the Directions, unusual weather (i. weather conditions which are outside the range considered normal at the application site and for the time period when the product is applied with the normal range being determined on the basis of the average range for the prior 40 years computed from the best available information, and ii. weather perils, including but not limited to hurricanes, tornadoes and floods) as well as weather considerations set forth in the Directions, application in any manner not explicitly set forth in the Directions, moisture conditions outside the moisture range specified in the Directions, or the presence of products other than those set forth in the Directions in or on the soil or crop.

THE EXCLUSIVE REMEDY OF THE USER OR BUYER, AND THE LIMIT OF THE LIABILITY OF THIS COMPANY OR ANY OTHER SELLER FOR ANY AND ALL LOSSES, INJURIES OR DAMAGES RESULTING FROM THE USE OR HANDLING OF THIS PRODUCT (INCLUDING CLAIMS BASED IN CONTRACT, NEGLIGENCE, STRICT LIABILITY, OTHER TORT OR OTHERWISE) SHALL BE THE PURCHASE PRICE PAID BY THE USER OR BUYER FOR THE QUANTITY OF THIS PRODUCT INVOLVED, OR, AT THE ELECTION OF THIS COMPANY OR ANY OTHER SELLER, THE REPLACEMENT OF SUCH QUANTITY OR, IF NOT ACQUIRED BY PURCHASE, REPLACEMENT OF SUCH QUANTITY. IN NO EVENT SHALL THIS COMPANY OR ANY OTHER SELLER BE LIABLE FOR ANY INCIDENTAL OR CONSEQUENTIAL DAMAGES.

The buyer and all users are deemed to have accepted the terms of this LIMIT OF WARRANTY AND LIABILITY which may not be varied by any verbal or written agreement.

PRECAUTIONARY STATEMENTS

HAZARDS TO HUMANS AND DOMESTIC ANIMALS

Keep out of reach of children.

WARNING!

IRRITATING TO EYES.
MAY CAUSE ALLERGIC SKIN REACTION.

Do not get in eyes, on skin or on clothing.

Avoid breathing dust.

Do not take internally.

Wash thoroughly with soap and water after handling.

Wear rubber gloves, long trousers, long sleeved shirt or jacket of tightly woven material along with boots high enough to cover ankles when loading and when adjusting, repairing or cleaning equipment. Clothing which comes in contact with Lasso must be washed before reuse.

FIRST AID:

IF IN EYES, immediately flush with plenty of water for at least 15 minutes. Call a physician.

IF ON SKIN, immediately flush with plenty of water. Remove contaminated clothing. If irritation persists, get medical attention.

ENVIRONMENTAL HAZARDS

This chemical has been identified in limited ground water sampling and there is the possibility that it can leach through the soil to ground water, especially where soils are coarse and ground water is near the surface. Following application, and during rainfall events that cause runoff, this chemical may reach surface water bodies including streams, rivers and reservoirs.

Do not apply directly to water or wetlands. Do not contaminate water by cleaning of equipment or disposal of wastes.

STORAGE AND DISPOSAL

Do not contaminate water, foodstuffs, seed or feed by storage and disposal.

Wastes resulting from the use of this product that cannot be used or chemically reprocessed should be disposed of in a landfill approved for pesticide disposal or buried on site in a safe place away from water supplies. All disposal should be in accordance with applicable Federal, state or local procedures.

Completely empty bag into application equipment. Then dispose of empty bag in a sanitary landfill or by incineration, or, if allowed by state and local authorities, by burning. If burned, stay out of smoke.

In case of emergency involving this product,
Call Collect, day or night (314) 694-4000

APPLICATION EQUIPMENT, CALIBRATION AND HANDLING INSTRUCTIONS

NOTE

Direct contact or exposure to Lasso should be minimized. The following instructions for loading and repairing equipment should be followed in order to minimize this exposure. Review the protective clothing requirements as listed in the Precautionary Statements section of this label and do not use this product until you have the necessary protective clothing.

Equipment Cleaning and Repair

Cleaning and repair of application equipment is a source of exposure to this product. Care should be taken to minimize exposure during cleaning and repair of application equipment. Whenever possible, this equipment should be rinsed before being cleaned or repaired. When repairs must be made during application, special care should be taken to avoid contact with the pesticide.

Broadcast Treatment—Use broadcast applicators that can apply these granules evenly.

Band Treatment (Behind Press Wheel only)—Use applicators designed for this purpose. Calculate the amount of granules per acre needed for band treatment as follows:

$$\frac{\text{Band width in inches}}{\text{Row width in inches}} \times \text{Broadcast Rate per acre} = \text{Band Rate per acre}$$

For example: If 26 pounds per acre are recommended for broadcast treatment and you wish to treat a band of 14 inches wide in crop rows spaced 38 inches apart:

$$\frac{14}{38} \times 26 = 9.6 \text{ pounds of granules needed for the band treatment}$$

NOTE: For application of this product do not rely on settings used for other granular herbicides.

Calibration—The following settings are only guidelines for calibration; they do not allow for equipment variation or wear. Check settings after initial calibration so that the proper rate is applied.

	Granular Applicator Setting			
	Broadcast Rate Per Acre			
	16 Pounds		20 Pounds	
Applicator	4 mph	5 mph	4 mph	5 mph
Allis Chalmers	2	2	3	3
Gandy (bolt-on)	12	13	13	15
(snap-on)	12	13	14	15
I.H. before 1981	1/9	2/2	2/2	2/6
1981—	1/9	2/1	2/1	2/4
John Deere—				
metal hopper	1/27	1/31	1/31	2/1
plastic hopper.				
—10, 20 etc. gauge	13	15	15	19
—5, 15 etc. gauge	11	13	13	15
Noble-hopper bottom hole ●	10	13	13	15
Noble and White–bottom hole ■	7	8	9	10
Buffalo				
Fluted-row feed	2.0	2.0	2.5	2.5
Plateless planter	8	10	10	13

	Granular Applicator Setting			
	Broadcast Rate Per Acre			
	23 Pounds		26 Pounds	
Applicator	4 mph	5 mph	4 mph	5 mph
Allis Chalmers	4	4	4	4
Gandy (bolt-on)	14	17	16	18
(snap on)	15	16	16	17
I.H. before 1981	2/4	2/9	2/6	3/5
1981—	2/3	2/6	2/5	2/8
John Deere				
metal hopper	1/33	2/4	2/2	2/6
plastic hopper.				
—10, 20 etc. gauge	17	21	19	24
—5, 15 etc. gauge	14	16	15	17
Noble-hopper bottom hole ●	14	17	16	19
Noble and White bottom hole ■	10	12	11	13
Buffalo				
Fluted-row feed	3.0	3.0	3.5	3.5
Plateless planter	12	15	13	18

It is important that the applicator be calibrated properly to deliver the desired amount of this product. To calibrate, set the applicator and attach a bag over the spreader plate or remove the delivery tube and attach a plastic nursing bottle over the spout. Operate at a constant speed on terrain to be planted. Four to 5 mph is suggested. For a 14-inch band, collect the granules for 500 linear feet (583 sq. ft. area) and measure the VOLUME or WEIGHT collected. Tap plastic nursing bottle or other measuring container gently to settle the granules when taking a reading.

Refer to the following table to determine the broadcast rate per acre for the amount collected. If necessary adjust the application equipment so that each spout dispenses the proper amount.

Broadcast Rate Per Acre	Volume Collected in 500 linear feet for a 14 inch band	Weight Collected in 500 linear feet for a 14 inch band
16 lb.	4.4 fl. oz.	3.4 oz.
20 lb.	5.6 fl. oz.	4.3 oz.
23 lb.	6.4 fl. oz.	4.9 oz.
26 lb.	7.2 fl. oz.	5.6 oz.

For other widths calculate the calibration distance as follows:

$$\frac{14 \text{ inches}}{\text{Band width in inches}} \times 500 = \text{Calibration distance in linear feet.}$$

CULTIVATION INFORMATION

If grass or weed seedlings emerge before it rains rotary hoeing will improve performance.

If cultivation is necessary because of soil crusting or compaction, tillage should be shallow to minimize dilution of this herbicide.

Except for the above, as long as the grasses and weeds are controlled, delay cultivation as long as possible.

Only after weeds emerge in the treated band, set cultivator to throw soil into the row covering the band. On peanuts, do not throw soil into the row.

WEED CONTROL

When applied as directed under the conditions described, this product CONTROLS the following weeds:

Annual Grasses

Barnyardgrass	*Echinochloa crus-galli*
Crabgrass	*Digitaria sanguinalis*
Foxtail (giant)	*Setaria faberi*
Foxtail (green)	*Setaria viridis*
Foxtail (yellow)	*Setaria lutescens*
Goosegrass	*Pleusine indica*
Panicum (fall)	*Panicum dichtomiflorum*

Signalgrass (broadleaf)	*Brachiaria* spp.
Witchgrass	*Panicum capillare*

Annual Broadleaves

Carpetweed	*Mollugo verticillata*
Pigweed, Carelessweed	*Amaranthus* spp.
Purslane	*Portulaca oleracea*
Pusley (Florida)	*Richardia scabria*

When applied immediately after planting, under the conditions described, this product at a rate of 16 to 26 pounds per acre on a broadcast basis REDUCES COMPETITION from the following hard to control weeds:

Sedges

Nutsedge (yellow)	*Cyperus esculentus*

Annual Broadleaves

Beggarweed (Florida)	*Desmodium tortuosum*
Lambsquarters	*Chenopodium album*
Nightshade (black)	*Solanum nigrum*
Ragweed (common)	*Ambrosia artemisiifolia*
Smartweed	*Polygonum pensylvancium*
	Polygonum persicaria
Sida (spiny), Teaweed	*Sida spinosa*

This product WILL NOT CONTROL Rhizome Bermudagrass, Field Bindweed, Bull Nettle, Canada Thistle, Rhizome Johnsongrass, Morningglory and Quackgrass.

DIRECTIONS FOR USE

This product is recommended for preemergence control of most annual grasses and certain broadleaf weeds. It should be applied to the soil surface prior to crop or weed emergence. The soil surface should be freshly worked and free of weed growth. The seed bed should be fine, firm and free of clods and trash.

Better results are obtained when moisture occurs within one week after application. High intensity or excessive rainfall, or excessive irrigation, after application may reduce control.

Use of this product not consistent with this label may result in injury to persons, animals or crops, or other unintended consequences. Keep bag closed to prevent spills and contamination.

FIELD CORN, HYBRID SEED CORN, SILAGE CORN, SWEET CORN AND SOYBEANS

Applicators should evaluate soil and weed conditions carefully to assure that they choose the lowest effective label rate.

For control of listed annual grasses and broadleaf weeds apply after planting and before crop or weeds emerge. For band treatment apply granules behind the press wheel. See table below for recommended rates on various soil textures.

Soil Texture	Broadcast Rate Per Acre	
	Less than 3% organic matter	More than 3% organic matter
MEDIUM		
Loams	16 lb	20 lb
HEAVY		
Silty Clay Loam thru Clay	20 lb	26 lb

FOR LISTER OR DEEP FURROW PLANTED CORN

Broadcast Treatment—Apply at the rate of 16 pounds per acre.

Band Treatment (Behind Press Wheel)—For a 10 inch band treatment over the row (40 inch spacing) apply 4 pounds per acre. For a 7 inch band treatment over the row (40 inch spacing), apply 2.8 pounds per acre.

If heavy rains follow application, weed control on the top and sides of the furrow may not be as good as that obtained in the bottom of the furrow.

Where deep lister planting is practiced, only partial weed control can be expected on steep furrow sides. It is suggested that a 7 inch band be applied at the bottom of the furrow. Better results can be expected where the granules lie on an even flat surface which is not subject to shifting by wind or rain.

FOR FURROW IRRIGATED CORN AND SOYBEANS

This product must be incorporated into the soil by harrow, finger weeder or rotary hoe. Shallow incorporation of at least ½ inch but not more than 2 inches is most effective for weed control. Incorporation deeper than 2 inches may result in unsatisfactory weed control. Incorporation depth must be kept above the seed. Soil moisture at application must be sufficient for crop and weed emergence. Irrigation within 10 days following planting may improve weed control. See the table above for recommended rates.

PEANUTS

Broadcast—Apply 20 pounds of this product per acre from immediately after planting and until either crop or weeds emerge. NOTE: Hard-to-control weeds require 23 to 26 pounds per acre.

SUNFLOWERS

For control of listed annual grasses and broadleaf weeds apply after planting and before crops or weeds emerge. For band treatment apply granules behind the press wheel.

Broadcast Treatment—Apply at the rate of 20 to 26 pounds per acre.

Band Treatment (Behind Press Wheel)—For a 10 inch band treatment over the row (40 inch spacing) apply 5 to 6.5 pounds per acre. For a 7 inch band treatment over the row (40 inch spacing), apply 3.5 to 4.6 pounds per acre. Use the higher rate in areas of heavy weed infestations, to reduce competition from hard-to-control weeds listed in this label, or on soil with 3 percent or more organic matter.

WOODY ORNAMENTALS

Field and Liner Grown Ornamentals

Use in **Cotoneaster** (*Cotoneaster* spp.), **Crabapple** (*Malus* spp.), **Dogwood** (*Cornus* spp.), **Euonymus** (*Euonymus* spp. *shrub* spp. *only*), **Holly** (*Ilex* spp.), **Juniper** (*Juniper* spp.), and **Yew** (*Taxus* spp.).

CONTAINER GROWN ORNAMENTALS

Use in Holly (*Ilex* spp.) and Juniper (*Juniperus* spp.).

DO NOT USE IN GREENHOUSES OR OTHER ENCLOSED STRUCTURES.

NOT RECOMMENDED FOR DOMESTIC APPLICATION, EXCEPT BY CERTIFIED APPLICATORS.

When applied as directed under the conditions described, this product controls the weeds listed on this label in newly planted or established field and container grown woody ornamentals.

DO NOT USE THIS PRODUCT ON SEEDBEDS, UNROOTED CUTTINGS OR PRIOR TO TRANSPLANTING. DO NOT APPLY THIS PRODUCT TO PLANTS WHOSE FOLIAGE IS MOIST OR WET AS GRANULES MAY STICK TO THIS FOLIAGE AND CROP INJURY MAY RESULT. OVERHEAD IRRIGATION APPLIED WITHIN ONE HOUR AFTER APPLICATION WILL WASH GRANULES OFF THE FOLIAGE AND INJURY WILL BE MINIMIZED.

Broadcast Treatment—Apply this product at a broadcast rate of 27 pounds per acre anytime after transplanting and before weeds emerge or after existing weeds have been removed. See the "Weed Control" and "Cultivation Information" sections of this label for further information. Repeat applications of this product may be required for sustained long-term weed control. Do not exceed three (3) applications of this product per year.

NOTE: Plant injury symptoms (leaf tip burn, discoloration, distortion, etc.) may occur with applications of this product to actively growing woody ornamentals. Do not retreat within twenty-one (21) days.

In case of an emergency involving this product, Call Collect, day or night, (314) 694-4000.

MICRO-TECH™ LASSO®

HERBICIDE

Herbicide for weed control in Soybeans and crops listed.

ACTIVE INGREDIENTS*

Alachlor, [2-chloro-2′ 6′-diethyl-N-(methoxymethyl) acetanilide]	41.5%
INERT INGREDIENTS:	58.5%
TOTAL	100.0%

*Contains 480 grams/litre or 4 pounds/gallon.

EPA Reg. No. 524-344 U.S. Pat. No. 3,547,620

1984-2 843.11-000.23/53

CAUTION

Keep Out of Reach of Children

The use of this product may be hazardous to your health. This product contains alachlor, which has been determined to cause tumors in laboratory animals.

Users are urged to obtain information on careful handling and use of pesticides and to participate in Monsanto Company's education program on potential hazards and safer use of this product. For additional information, contact Monsanto Agricultural Products Co. at 1-800-453-4000. [Note: This number is for user safety information only. *NOT* for emergencies. For emergencies, call collect day or night (314)694-4000.]

Complete Directions for Use

*LASSO® MICRO-TECH™ is referred to throughout these instructions as "LASSO".

Read the entire label before using this product.

Use only according to label instructions.

Read "LIMIT OF WARRANTY AND LIABILITY" before buying or using. If terms are not acceptable, return at once unopened.

LIMIT OF WARRANTY AND LIABILITY

This company warrants that this product conforms to the chemical description on the label and is reasonably fit for the purposes set forth in the Complete Directions for Use label booklet ("Directions") when used in accordance with those Directions under the conditions described therein. NO OTHER EXPRESS OR IMPLIED WARRANTY OF FITNESS FOR PARTICULAR PURPOSE OR MERCHANTABILITY IS MADE. This warranty is also subject to the conditions and limitations stated herein.

Buyer and all users shall promptly notify this company of any claims whether based in contract, negligence, strict liability, other tort or otherwise.

Buyer and all users are responsible for all loss or damage from use or handling which results from conditions beyond the control of this company, including but not limited to incompatibility with products other than those set forth in the Directions, unusual weather (i. weather conditions which are outside the range considered normal at the application site and for the time period when the product is applied with the normal range being determined on the basis of the average range for the prior 40 years computed from the best available information, and ii. weather perils, including but not limited to hurricanes, tornadoes and floods) as well as weather considerations set forth in the Directions, application in any manner not explicitly set forth in the Directions, moisture conditions outside the moisture range specified in the Directions, or the presence of products other than those set forth in the Directions in or on the soil or crop.

THE EXCLUSIVE REMEDY OF THE USER OR BUYER, AND THE LIMIT OF THE LIABILITY OF THIS COMPANY OR ANY OTHER SELLER FOR ANY AND ALL LOSSES, INJURIES OR DAMAGES RESULTING FROM THE USE OR HANDLING OF THIS PRODUCT (INCLUDING CLAIMS BASED IN CONTRACT, NEGLIGENCE, STRICT LIABILITY, OTHER TORT OR OTHERWISE) SHALL BE THE PURCHASE PRICE PAID BY THE USER OR BUYER FOR THE QUANTITY OF THIS PRODUCT INVOLVED, OR, AT THE ELECTION OF THIS COMPANY OR ANY OTHER SELLER. THE REPLACEMENT OF SUCH QUANTITY OR, IF NOT ACQUIRED BY PURCHASE, REPLACEMENT OF SUCH QUANTITY. IN NO EVENT SHALL THIS COMPANY OR ANY OTHER SELLER BE LIABLE FOR ANY INCIDENTAL OR CONSEQUENTIAL DAMAGES.

The buyer and all users are deemed to have accepted the terms of this LIMIT OF WARRANTY AND LIABILITY which may not be varied by any verbal or written agreement.

PRECAUTIONARY STATEMENTS

HAZARD TO HUMANS AND DOMESTIC ANIMALS

Keep Out of Reach of Children.

CAUTION!

MAY CAUSE EYE AND SKIN IRRITATION.
HARMFUL IF ABSORBED THROUGH SKIN.
MAY CAUSE ALLERGIC SKIN REACTION.

Avoid contact with eyes, skin and clothing.

Wear goggles or face shield, rubber gloves, long trousers, long sleeved shirt or jacket of tightly woven material along with boots high enough to cover ankles when transferring and mixing, and when adjusting, repairing or cleaning equipment. Wear rubber boots when pouring from open containers or when reentry is made into fields following application of this product through center pivot irrigation systems and the field is still wet. Clothing which comes in contact with Lasso must be washed before reuse. Clothing or other materials which become drenched with the concentrated pesticide must be disposed of in a sanitary landfill, by incineration or, if allowed by state and local authorities, by burning. If burned, stay out of smoke.

Wash thoroughly with soap and water after handling.

FIRST AID: IF IN EYES, immediately flush with plenty of water for at least 15 minutes. Call a physician.

IF ON SKIN, immediately flush with plenty of water while removing contaminated clothing. Call a physician. Wash clothing before reuse. Sensitized persons should avoid further contact and reuse of clothing.

ENVIRONMENTAL HAZARDS

This chemical has been identified in limited ground water sampling and there is the possibility that it can leach through the soil to ground water, especially where soils are coarse and ground water is near the surface. Following application, and during rainfall events that cause runoff, this chemical may reach surface water bodies including streams, rivers and reservoirs.

Care must be taken when using this product to prevent back siphoning into wells, spills or improper disposal of excess pesticide, spray mixtures or rinsates.

Check valves or antisiphoning devices must be used on all mixing and/or irrigation equipment. Disposal of excess pesticide, spray mixtures, or rinsate should be according to label use instructions or according to the state pesticide or environmental control agency or the hazardous waste representative at the nearest EPA regional office.

Do not apply directly to water or wetlands. Do not contaminate water by cleaning of equipment or disposal of wastes.

PHYSICAL OR CHEMICAL HAZARD

In case of:

SPILL or LEAK, soak up, remove and keep away from humans and domestic animals.

STORAGE AND DISPOSAL

Do not contaminate water, foodstuffs, seed or feed by storage and disposal.

STORAGE

STORE ABOVE −4°F. TO KEEP PRODUCT FROM FREEZING. If frozen, thaw before use.

DISPOSAL

Wastes resulting from the use of this product that cannot be used or chemically reprocessed should be disposed of in a landfill approved for pesticide disposal or buried on site in a safe place away from water supplies. All disposal should be in accordance with applicable Federal, state or local procedures.

Emptied container retains product residue. Observe all labeled safeguards until container is cleaned, reconditioned or destroyed.

(See individual container label for disposal information.)

GENERAL INFORMATION

LASSO® herbicide is recommended for control of yellow nutsedge and the annual grasses and broadleaf weeds listed in the "WEEDS CONTROLLED" section of this label. This product may be applied either as a surface application after planting or shallowly incorporated prior to planting to blend the herbicide treatment into the upper 1 to 2 inches of soil. Except for minimum or conservation tillage systems, the seedbed should be fine, firm and free of clods and trash.

®LASSO is a registered trademark of Monsanto Company.

Read and carefully observe precautionary statements and all other information appearing on the labeling of all products used in mixtures and sequential treatments.

NOTE THAT PARAQUAT HERBICIDE IS A POISON WHICH MAY BE FATAL OR HARMFUL TO HUMANS AND DOMESTIC ANIMALS.

Do not flood irrigate following application of this product or tank mixtures. This product will not control emerged seedlings.

Do not apply when conditions favor drift.

DO NOT APPLY WITH AERIAL EQUIPMENT.

Use of this product not consistent with this label may result in injury to persons, animals or crops, or other unintended consequences. Keep containers closed to prevent spills and contamination.

Flush sprayer with clean water after use.

SOIL TEXTURE

Applicators should evaluate soil and weed conditions carefully to assure that they choose the lowest effective label rate.

The recommended use rates of this product and the other herbicides labeled for use in tank mixtures with this product generally vary with soil texture. Unless soil texture is specifically named, rate tables throughout this label refer to only three soil texture groups: coarse, medium and fine. The following is a complete listing of soil textures included in each of these three soil texture groups:

SOIL TEXTURE GROUP	SOIL TEXTURE
COARSE:	sand, loamy sand, sandy loam
MEDIUM:	loam, silt loam, silt, sandy clay loam
FINE:	silty clay loam, clay loam, sandy clay, silty clay, clay

Refer to the above table to determine the corresponding soil texture group for the soil to be treated.

MIXING, SPRAYING AND HANDLING INSTRUCTIONS

NOTE

Direct contact or exposure to Lasso or spray mixtures of Lasso should be minimized. The following instructions for transfer, mixing, cleaning or repairing equipment should be followed in order to minimize this exposure. Review the protective clothing requirements as listed in the Precautionary Statements section of this label (front panel) and do not use this product until you have the necessary protective clothing.

2.5 Gallon Containers

Open pouring from these containers can result in exposure from splashing or spilling. Special care in lifting and pouring are strongly recommended.

55 Gallon and Bulk Containers

Open pouring from these containers can result in exposure from splashing or spilling and is not recommended. This product should be transferred from these containers to the mix or spray tank using pumps or transfer probes. The probe or pump should not be removed from the container or disconnected until the container is emptied and rinsed. Use the pump or probe system to rinse the empty container and transfer the rinsate directly to the mix or spray tank.

Equipment Cleaning & Repair

Cleaning and repair of transfer systems and application equipment is a source of exposure to this product. Care should be taken to minimize exposure during cleaning and repair of transfer systems and application equipment. Whenever possible, these systems or equipment should be rinsed before being cleaned or repaired. When repairs must be made during transfer or application, the equipment should be shutdown, and special care taken to avoid contact with the pesticide.

Always predetermine the compatibility of LASSO or labeled tank mixtures of LASSO with water carrier or sprayable fluid fertilizer carrier by mixing small proportional quantities in advance. See the "Standard Sprayable Fluid Fertilizer Compatibility Test" section in this label to determine the compatibility of this product and the labeled tank mixtures recommended for use with sprayable fluid fertilizer carrier. For soybeans, only LASSO and the tank mixtures of LASSO plus Lorox™ herbicide and LASSO plus Sencor™ or Lexone™ herbicides may be applied in sprayable fluid fertilizers. DO NOT apply LASSO or labeled tank mixtures of LASSO in sprayable fluid fertilizer on any other crops.

Mix LASSO or labeled tank mixture of LASSO with the appropriate carrier as follows:

1. Place a 20 to 35 mesh screen or wetting basket over filling port.
2. Through the screen, fill the sprayer tank one-half full with the appropriate carrier.
3. If a compatibility agent is necessary to improve mixing or to prevent the formation of undesirable and unsprayable gels or precipitates, while agitating add it to the carrier already in the tank. Use only compatibility agents, cleared by FDA for this use. Read and follow all directions for use, cautionary statements and all other information appearing on the selected compatibility agent label. Check for adequate agitation.
4. If a wettable powder or dry flowable formulation is used, make a slurry with water and add it slowly through the screen into the tank. Continue agitation.
5. If a flowable formulation is used, add slowly through screen into the tank. Mixing and compatibility may be improved when flowable is pre-mixed one part flowable with one part water and added to the tank in diluted form.
6. Add LASSO slowly through the screen into the tank. Mixing and compatibility may be improved when this product is pre-diluted with two parts of water and added to the tank in diluted form.
7. Complete filling the sprayer tank with carrier. If Roundup or paraquat is used, add the required amount near the end of the filling process. Remove hose from tank immediately after filling to avoid siphoning back into the carrier source.

Maintain good agitation at all times until the contents of the tank are sprayed. NOTE: If spray mixture is allowed to settle at any time thorough agitation is required to resuspend the mixture before spraying is resumed.

Keep by-pass line on or near bottom of tank to minimize foaming. Screen size in nozzle or line strainers should be no finer than 50 mesh. Carefully select proper nozzle to avoid spraying a fine mist. Check for even distribution of spray droplets. To reduce loss of the chemical due to drift of a fine mist, apply at nozzle pressures below 40 psi.

™Banvel is a trademark of Velsicol Chemical Corp.

™Bladex is a trademark of the Shell Chemical Company.

™Lorox is a trademark of E.I. duPont de Nemours and Company.

TMLexone is a trademark of E.I. duPont de Nemours and Company.

TMSencor is a trademark of the parent company of Farbenfabriken Bayer GmbH, Leverkusen.

STANDARD SPRAYABLE FLUID FERTILIZER COMPATIBILITY TEST

Herbicides may not always mix evenly throughout a sprayable fluid fertilizer or the components may separate too quickly to make their combined use of practical value. This may be due to certain characteristics of the different fluid fertilizers. A simple test using small quantities of the components is suggested to provide compatibility potential.

The test follows:

A. Materials Required For A Compatibility Test

1. Two one-quart jars with lid or stopper (marked "with" and "without").
2. Teaspoons (for a more exacting test, a five to ten milliliter (ml.) pipette or graduated cylinder is desirable).
3. Sprayable fluid fertilizer to be tested.
4. The herbicide chemicals to be mixed.
5. A compatibility agent (the purpose of the adjuvant is to help keep the fertilizer and crop protection chemical in suspension, if this assistance is needed).

B. Procedure

1. Add one pint of the sprayable fluid fertilizer that will be used or other herbicide carrier to each jar marked "with" and "without".

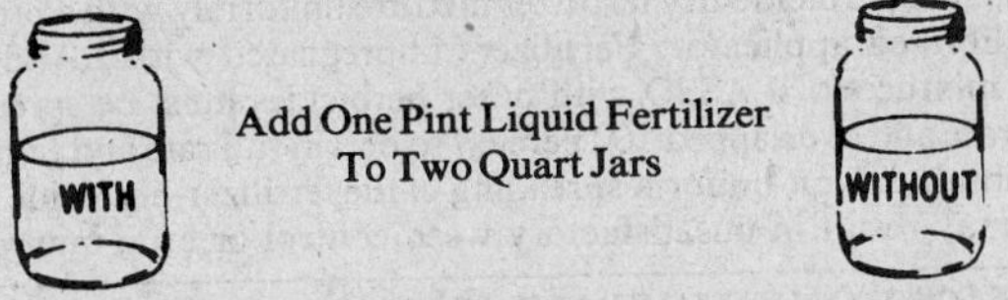

2. To the jar marked "with", add ¼ teaspoon or 1.2 milliliters of a suitable compatibility agent; shake gently for five to ten seconds to mix. (¼ teaspoon in one pint is the equivalent of two pints per 100 gallons of liquid fertilizer.)

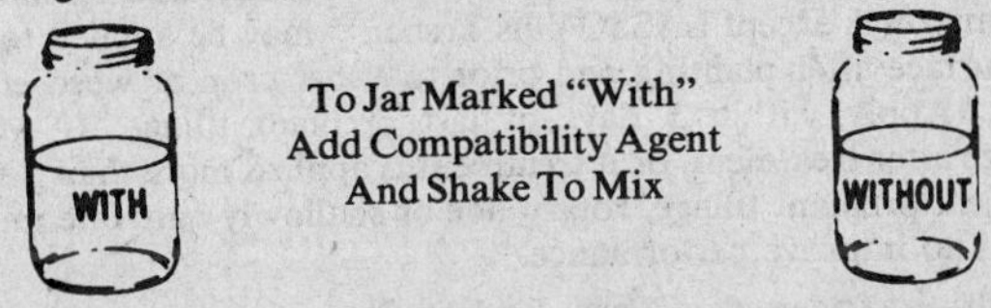

3. To each jar add the appropriate amount of herbicide(s). If more than one is used, add them separately with the wettable powders or dry flowables added first, flowables second and liquid last. Shake gently five to ten seconds after each addition.

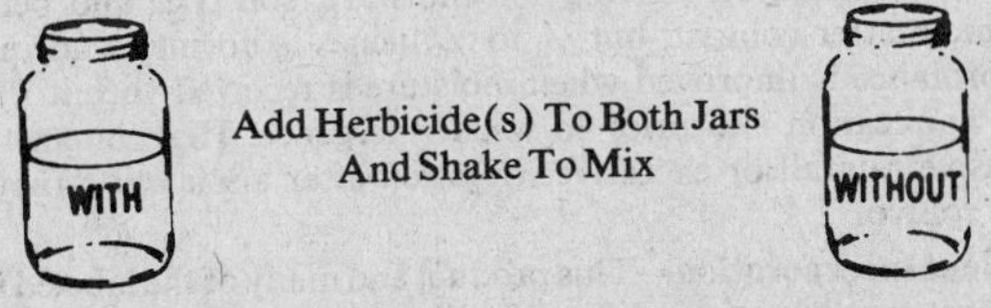

Amount to Be Added Per Pint Of Sprayable Fluid Fertilizer (Assuming Volume is 25 gal/A)

Herbicide	Rate/A		Level Teaspoons
Wettable Powders or Dry Flowables	1 lb.	=	1.5
	2 lb.	=	3
	3 lb.	=	4.5
	4 lb.	=	6
	5 lb.	=	7.5

Herbicide	Rate/A		Level Teaspoons		Milliliters
Emulsifiable	1 pint	=	0.5	or	2.4
Concentrates or	1 quart	=	1	or	4.7
Flowables or	2 quarts	=	2	or	9.5
Liquids or	3 quarts	=	3	or	14.2
Solutions	1 gallon	=	4	or	19.0
	5 quarts	=	5	or	23.8

This compatibility test is designed for 25 gallons of spray per acre with the maximum labeled rate of herbicide. For changes in spray volume or herbicide concentration, make appropriate changes in the ingredients of the test. Regardless of spray volume, the amount of compatibility agent should be equal to two or three pints (two pints = ¼ teaspoon or 1.2 milliliters, three pints = ⅜ teaspoon or 1.8 milliliters per pint of sprayable fluid fertilizer) per 100 gallons of liquid fertilizer.

C. Observations and Decisions

1. If the herbicide(s) and the sprayable fluid fertilizer are compatible.
2. If a compatibility agent is necessary.

Five minutes after the final addition and mixing, observe both jars for the formation of large flakes, sludge, gels or other precipitates. Observe if the herbicide(s) cannot be physically mixed with the liquid fertilizer but remains as small oily particles in the solution.

If incompatibility in any form described above occurs in the jar "with" the compatibility agent added, the liquid fertilizer and the herbicide(s) should not be used together in the same spray tank.

If incompatibility as described above occurs in the jar "without" the adjuvant but not in the jar "with" adjuvant, the use of a compatibility adjuvant is recommended.

Both jars should be allowed to stand and be observed periodically for one-half hour. If the separate layers of liquid fertilizer and additives can be resuspended by shaking, commercial application is possible. An emulsifiable concentrate normally will go to the top after standing; wettable powders will either settle to the bottom of the tank or jar, or float to the top, depending upon the density of the fertilizers.

If the herbicide(s) is compatible with fluid fertilizer in the foregoing test without having to use a compatibility agent, fluid fertilizer may be used for the premixing. If it is not compatible without the compatibility agent, the herbicide(s) should be premixed with water before adding to the spray tank.

APPLICATION INFORMATION

APPLICATION SYSTEMS

Ground Broadcast Treatment—Apply this product and the labeled tank mixtures in 15 or more gallons of solution per acre using broadcast boom equipment. The carrier may be either water or sprayable fluid fertilizer as specified for the crop to be treated in the "DIRECTIONS FOR USE" section of this label. Do not apply during periods of gusty winds, when winds are in excess of 15 mph or when other conditions favoring drift exist.

Band Treatment—Apply a broadcast equivalent rate and volume per acre. To determine these:

$$\frac{\text{Band width in inches}}{\text{Row width in inches}} \times \text{Broadcast RATE per acre} = \text{Band RATE per acre}$$

$$\frac{\text{Band width in inches}}{\text{Row width in inches}} \times \text{Broadcast VOLUME per acre} = \text{Band VOLUME per acre}$$

Center Pivot Irrigation Treatment—Center pivot is the only irrigation equipment through which this product may be used. Refer to the Monsanto booklet "How to Apply Lasso Herbicide with Center Pivot Irrigation" before use.

Use only in systems that apply uniformly. Do not apply when wind speed favors drift, when system connections or fittings leak, when nozzles do not provide uniform distribution or when the line containing the product must be dismantled and drained. Meter diluted LASSO herbicide or a labeled tank mixture of LASSO into the center pivot irrigation system after planting and before crop or weed emergence. Herbicide applications should be made in 1/3 to 3/4 inch of water per acre. Do not apply in more than 3/4 inch water per acre under any conditions or reduced performance may occur. On very sandy soils (more than 60% sand and less than 1% organic matter), apply these herbicides in a maximum of 1/2 inch water per acre. Good agitation must be maintained during the entire application period. Have appropriate check valves or other suitable devices in the system to insure that the herbicide solution cannot siphon back into the water supply. Flush the system with water when application is complete. If spills of the herbicide concentrate occur, immediately wash off the system.

NOTE: Applications can be made by injection through center pivot irrigation equipment only when specified for the crop to be treated in the "DIRECTIONS FOR USE" section of this label. Do not apply this product or any labeled tank mixture through center pivot irrigation unless the treatment is specifically recommended in this label for the crop to be treated.

Equipment must be constructed of materials which are resistant to LASSO. Calibration information is available in the booklet "How to Apply LASSO Herbicide with Center Pivot Irrigation" at your local farm chemical dealer or by writing Monsanto Company.

Application with Dry Bulk Fertilizer—The herbicide/fertilizer impregnation process must be done only at commercial fertilizer or chemical dealerships properly equipped for this procedure.

Dry bulk fertilizer may be impregnated with LASSO or the tank mixtures of LASSO plus Lexone or Sencor on soybeans. LASSO and these tank mixtures must be applied with 200 to 450 pounds of dry bulk fertilizer per acre and shallowly incorporated within 7 days prior to planting. The proper amount of herbicide must be applied per acre in the blend as recommended in this label for the crop, weed and soil type treated. Refer to the table for broadcast rate per acre in the appropriate crop section of this label to determine the recommended rate per acre for the herbicide treatment to be applied.

The following table provides a reference to determine the amount of LIQUID herbicide to be mixed per ton of dry bulk fertilizer for a range of herbicide recommendations for fertilizer rates per acre:

Fertilizer Rate (Lb./Acre)	Acres Covered (Per Ton)	Recommended Quarts Liquid Herbicide/Acre 1.5	2	2.5	3
		(Quarts of Herbicide/Ton Dry Bulk Fertilizer)			
200	10	15	20	25	30
250	8	12	16	20	24
300	6.7	10	13.3	16.7	20
350	5.7	8.6	11.4	14.3	17.1
400	5	7.5	10	12	15
450	4.5	6.7	8.9	11.1	13.3

To determine the amount of herbicide needed for rates not included in the preceding table, use the following formula:

$$\frac{\text{Recommended Herbicide Rate (lb. or qt.)/Acre} \times 2000}{\text{Pounds Fertilizer/Acre}} = \text{Pounds or Quarts of herbicide per ton of dry bulk fertilizer}$$

Mix and blend the dry fertilizer and herbicide mixture in a closed rotary-drum type mixer allowing sufficient time to ensure uniform coverage. Use at least one ton of dry fertilizer per mixing operation. Inject the herbicide into the drum over a minimum of a 2-minute period and allow at least 2 additional minutes mixing time to ensure uniformity. The nozzle used to spray the herbicide treatment must be placed inside the mixer to provide uniform spray coverage of the tumbling fertilizer.

If the dry fertilizer used has inadequate adsorptive capacity, use a highly adsorptive powder such as Microcel E to provide a free-flowing mixture. The adsorptive powder must be added separately and uniformly to the prepared herbicide-fertilizer mixture. Generally, less than 2 percent by weight of Microcel E is required. The following table provides a partial list of dry fertilizers which may be impregnated with LASSO or tank mixtures of LASSO with other herbicides:

Fertilizer	LASSO®	LASSO + Lexone Sencor
Ammonium sulfate (21–0–0)	Yes	Yes
Ammonium phosphate-sulfate (16–20–0)	Yes	Yes
Diammonium phosphate (18–46–0)	Yes	Yes
Potassium chloride (0–0–60)	Yes	Yes
Potassium sulfate (0–0–52)	Yes	Yes
Single super-phosphate (0–20–0)	Yes	No
Treble super-phosphate (0–46–0)	Yes	No
*Urea (45–0–0)	Yes	Yes

* Blends suitable for use in soybeans, such as 0–26–26 are also approved for use with LASSO and the tank mixtures listed.

NOTE: DO NOT impregnate LASSO or tank mixtures of LASSO with other herbicides on fertilizers containing ammonium nitrate, potassium nitrate or sodium nitrate.

Spread the herbicide-dry fertilizer mixture uniformly with a properly calibrated applicator. Fertilizers impregnated with LASSO or tank mixtures of LASSO with other herbicides must be spread at half-rate and overlapped 100 percent to obtain full rate and uniform distribution. Non-uniform spreading of the fertilizer-herbicide mixture may result in unsatisfactory weed control or crop injury.

APPLICATION TIMING AND METHODS

NOTE: Unless specifically recommended, only one treatment of LASSO may be applied to a labeled crop.

Preemergence Surface Application—This product and all labeled tank mixtures except LASSO plus Treflan™ may be applied to the soil surface after planting and prior to either crop or weed emergence. Apply within 5 days of last preplant tillage. If weeds emerge after treatment, or if treatment is applied more than 5 days after last preplant tillage, rotary hoe or shallowly cultivate immediately to improve performance.

™ Treflan is a trademark of Elanco Products Company.

Precipitation or overhead sprinkler irrigation is required after application to move the herbicide treatment into the weed germination zone. The amount of precipitation or overhead sprinkler irrigation required depends on existing soil moisture, soil type and percent organic matter content, but 1/3 to 3/4 inches is normally adequate. Performance is improved when moisture is received within 7 days after application and prior to weed emergence. High intensity or excessive rainfall or excessive irrigation after application may reduce control.

Preplant Incorporation—This product and many of the labeled tank mixtures may be mixed into the soil using shallow incorporation equipment any time within 7 days prior to planting. Refer to the crop to be treated in the "DIRECTIONS FOR USE" section of this label to determine if preplant incorporation is recommended. Apply the recommended treatment rate to the soil surface as a broadcast application. Either existing soil moisture or subsequent precipitation or irrigation is required to bring incorporated herbicide treatments into contact with germinating weed seedlings. If weeds emerge after treatment, rotary hoe or shallowly cultivate immediately to improve performance.

NOTE: Do not incorporate unless specifically recommended in this label for the crop and herbicide treatment selected. Incorporation of this product and its labeled tank mixtures on coarse soil types may reduce length of weed control due to leaching with rainfall or irrigation. Incorporation is not recommended on coarse soils in the Southeastern States.

1-Pass Incorporation (Surface Blend™): Shallowly incorporate (Surface Blend) the treatment into the upper 1 to 2 inches of the soil. Equipment should be operated at manufacturer's designed speed for incorporation to ensure adequate mixing and distribution of the herbicide treatment in the soil. Equipment design including any drag attachments must be adequate to avoid soil ridging which may result in streaked or reduced weed control. Soil conditions, including moisture content and crop residue levels, must be suitable to allow thorough and uniform mixing with the equipment used for 1-pass incorporation.

Contact Monsanto Company for additional information regarding recommended equipment and methods for Surface Blending.

2-Pass Incorporation: When 2-pass incorporation is used, shallowly incorporate the herbicide treatment into the upper 1 to 2 inches of the soil with equipment set to work the soil NO DEEPER THAN 4 INCHES. The second pass must be made at an angle to and no deeper than the first pass to ensure proper distribution of the herbicide treatment in the soil.

Cultivation Information—Delay cultivation after application for as long as possible unless weeds or grasses emerge. Shallowly cultivate or rotary hoe immediately if weeds or grasses emerge. If cultivation is necessary because of soil crusting or compaction, set equipment shallow and minimize lateral soil movement to avoid dilution or displacement of the herbicide treatment. If a band application is used and weeds have emerged in the treated band, set cultivator to throw soil into the row, covering the band.

™Surface Blend is a Trademark of Monsanto Company

WEEDS CONTROLLED

Applicators should evaluate soil and weed conditions carefully to assure that they choose the lowest effective label rate.

When applied as directed under conditions described, yellow nutsedge and the annual grasses and broadleaf weeds listed are CONTROLLED by LASSO:

LASSO®

Annual Grasses	
Barnyardgrass	*Echinochloa crus-galli*
Crabgrass	*Digitaria ischaemum*
	Digitaria sanguinalis
Foxtail (giant)	*Setaria faberi*
Foxtail (green, robust purple, robust white)	*Setaria virdis*
Foxtail (yellow)	*Setaria lutescens*
Goosegrass	*Eleusine indica*
Panicum (fall)	*Panicum dichotomiflorum*
Rice (red)**	*Oryza sativa*
Signalgrass (broadleaf)	*Brachiaria platyphylla*
Witchgrass	*Panicum capillare*
Annual Broadleaves	
Carpetweed	*Mollugo verticillata*
Galinsoga	*Galinsoga* spp.
Nightshade (black)*	*Solanum nigrum*
Nightshade (hairy)*	*Solanum sarachoides*
Pigweed, Carelessweed	*Amaranthus* spp.
Purslane	*Portulaca oleracea*
Pusley (Florida)	*Richardia scabra*
Sedge	
Nutsedge (yellow)**	*Cyperus esculentus*

* Use a minimum of 2.5 quarts of this product per acre on coarse soils, and 3 to 4 quarts alone or in tank mix combinations on medium and fine textured soils for control of these weeds.

** Use 3 to 4 quarts of this product per acre applied alone or in tank mix combinations listed on this label. Preplant incorporate only for control of these weeds.

When applied preplant incorporated, or preemergence immediately after planting and within 5 days of last tilage, LASSO at a rate of 2.5 to 4 quarts per acre on a broadcast basis applied alone or in a labeled tank mixture REDUCES COMPETITION from the following hard to control annual grasses and broadleaf weeds.

Annual Grasses	
Johnsongrass (seedling)	*Sorghum halepense*
Panicum (Texas)*	*Panicum texanum*
Sandbur, Grassbur	*Cenchrus pauciflorus*
Shattercane (wild cane)**	*Sorghum bicolor*
Annual Broadleaves	
Beggarweed (Florida)	*Desmodium tortuosum*
Lambsquarters	*Chenopodium album*
Ragweed (common)	*Ambrosia artemisiifolia*
Sida (prickly), Teaweed	*Sida spinosa*
Smartweed	*Polygonum pensylvanicum*
	Polygonum persicaria
Sedge	
Nutsedge (yellow)***	*Cyperus esculentus*

* For reduced competition from this weed use a minimum of 3 quarts per acre and apply through center pivot irrigation or under sprinkler irrigation systems.

** Use 4 quarts per acre of this product applied alone or in tank mix combinations listed in this label. Preplant incorporate only to reduce competition from this weed.

*** Use 3 to 4 quarts per acre as a preemergence surface application on coarse textured soils with less than 3% organic matter to reduce competition from this weed.

When applied as directed under conditions described, tank mixtures of LASSO with products listed in the following table will control or reduce competition from the additional broadleaf weeds as indicated. Refer to the crop on which treatment is intended in the "DIRECTIONS FOR USE" section of this label to determine the tank mixtures and rates recommended for use in each crop.

NOTE: C = Control R = Reduced Competition

	Amiben™	Lorox™	Sencor™/Lexone™	Treflan™
Cocklebur Xanthium pensylvanicum	•	•	R	•
Jimsonweed Datura stramonium	•	•	R	•
Kochia Kochia scoparia	•	•	•	C
Lambsquarters Cheopodium album	C	C	C	•
Mustard Brassica spp	R	C	•	•
Ragweed (common) Ambrosia artemisiifolia	C	C	C	•
Sicklepod Cassia obstusifolia	•	•	R	•
Sida (prickly), Teaweed Sida spinosa	•	C	•	•
Smartweed Polygonum pensylvanicum Polygonum persicaria	C	C	C	•
Velvetleaf, Buttonweed Abutilon theophrasti	R	R	R	•

Basagran™ may be applied sequentially to LASSO in soybeans for control of additional emerged weeds listed on the Basagran label such as cocklebur and velvetleaf.

Certain tank mixtures of LASSO and broadleaf herbicides may be combined with applications of paraquat for control of many emerged weeds prior to soybean emergence. Refer to the paraquat label for a list of emerged weeds controlled by tank mixtures with these products.

™Basagran is a trademark of BASF Aktiengesellschaft Ludwigshafen am Rhein.

CROPS
DIRECTIONS FOR USE

It is a violation of federal law to use this product in any manner inconsistent with its labeling.

SOYBEANS

LASSO®

Apply one treatment of this product in water or sprayable fluid fertilizer solution for control of yellow nutsedge, and the annual grasses and braodleaf weeds listed in the "WEEDS CONTROLLED" section of this label.

APPROVED APPLICATION SYSTEMS

Ground—Broadcast Boom; banded

Center Pivot Irrigation

Dry Bulk Fertilizer Impregnation

Reference: The "APPLICATION SYSTEMS" section of this label provides detailed information on the proper application of this product using the system selected.

APPROVED APPLICATION METHODS

Preplant Incorporated—Apply this product within 7 days prior to planting and shallowly incorporate or Surface Blend into the upper 1 to 2 inches of soil.

Preemergence Surface—Apply this product after planting, before crop and weed emergence and within 5 days after last preplant tillage operation.

Reference: The "APPLICATION TIMING AND METHODS" section of this label provides detailed information and procedures for the application timing and method selected.

RECOMMENDED RATES: Refer to the following tables for the recommended broadcast treatment rates for this product. Applications which are not consistent with recommendations in this label may result in unsatisfactory weed control or crop injury.

Surface Application Rates

Soil Texture Group*	Broadcast Rate per Acre (Quarts)** Less than 3% organic matter	3% or more organic matter
Coarse***	2	2
Medium	2.5	2.5 to 3
Fine	3	3 to 4

* Refer to the "SOIL TEXTURE" section of this label to determine the corresponding texture group for the soil to be treated.

** Use the higher rate in the recommended range in areas of heavy weed infestations. Use a minimum of 2.5 quarts of this product per acre on coarse soils and 3 to 4 quarts on medium or fine textured soils to control black or hairy nightshade or to reduce competition from hard-to-control weeds listed in this label.

*** When applied through CENTER PIVOT IRRIGATION or under sprinkler irrigation systems on coarse soils, use a minimum of 3 quarts per acre of this product.

On coarse soils with 10% or more organic matter use 4 quarts of LASSO herbicide per acre.

Preplant Incorporated Application Rates

Soil Texture Group*	Broadcast Rate per Acre (Quarts)**
Coarse***	2.5
Medium	3 to 3.5
Fine	3.5 to 4

* Refer to the "SOIL TEXTURE" section of this label to determine the corresponding texture group for the soil to be treated.

** Use the higher rate in the recommended range in areas of heavy weed infestations. Use a minimum of 3 quarts per acre on coarse textured soils for control of yellow nutsedge on red rice. Rates may be increased to a maximum of 4 quarts on any soil type when heavy infestations of yellow nutsedge or red rice are present or to reduce competition from the hard-to-control weeds listed in this label.

*** On coarse soils with 10% or more organic matter use 4 quarts of LASSO herbicide per acre.

NOTE: Do not feed soybean hay from treated crop.

LASSO® plus AMIBEN (TANK MIXTURE)

Apply one treatment of this tank mixture in water solution for control of the annual grasses and broadleaf weeds listed in the "WEEDS CONTROLLED" section of this label.

DO NOT use this tank mixture on sand or loamy sand soils as crop injury may result.

APPROVED APPLICATION SYSTEMS

Ground—Broadcast Boom; banded

Reference: The "APPLICATION SYSTEMS" section of this label provides detailed information on the proper application of this tank mixture using the system selected.

APPROVED APPLICATION METHODS

Preplant Incorporated—Apply this tank mixture within 7 days prior to planting and shallowly incorporate or Surface Blend into the upper 1 to 2 inches of soil.

Preemergence Surface—Apply after planting, before crop and weed emergence and within 5 days after last preplant tilage operation.

Reference: The "APPLICATION TIMING AND METHODS" section of this label provides detailed information and procedures for the application timing and method selected.

RECOMMENDED RATES:

Preplant Incorporated—Apply 2.5 to 3 quarts of LASSO plus 4 quarts of Amiben per acre.

NOTE: Use the 3 quart per acre rate of LASSO in areas of heavy weed infestation. Use 3 to 4 quarts of LASSO when black or hairy nightshade, yellow nutsedge or red rice is present or to reduce competition from the hard-to-control weeds listed in this label.

Preemergence Surface—2 quarts LASSO plus 4 quarts Amiben per acre.

NOTE: Use a minimum of 2.5 quarts of LASSO on coarse soils and 3 to 4 quarts in this tank mixture on medium or fine textured soils to control black or hairy nightshade or to reduce competition from hard-to-control weeds listed in this label.

NOTE: Do not feed soybean hay from treated crop.

LASSO® plus LEXONE
or
LASSO plus SENCOR
(TANK MIXTURE)

Apply one treatment of this tank mixture in water or sprayable fluid fertilizer solution for control of yellow nutsedge and the annual grasses and broadleaf weeds listed in the "WEEDS CONTROLLED" section of this label.

NOTE: DO NOT use on muck soils or on alkaline soils with greater than 7.4 pH. Crop injury may result if atrazine was applied the year preceding use of this tank mixture.

DO NOT REPLANT CROPS OTHER THAN SOYBEANS FOR 120 DAYS AFTER APPLICATION OF THESE TANK MIXTURES.

APPROVED APPLICATION SYSTEMS

Ground—Broadcast Boom; banded

Dry Bulk Fertilizer Impregnation

Reference: The "APPLICATION SYSTEMS" section of this label provides detailed information on the proper application of this tank mixture using the system selected.

APPROVED APPLICATION METHODS

Preplant Incorporated—Apply this tank mixture within 7 days prior to planting and shallowly incorporate or Surface Blend into the upper 1 to 2 inches of soil. Preplant incorporation may cause slight soybean injury early, but soybeans should recover with no subsequent effect on growth or yield.

Preemergence Surface—Apply this tank mixture after planting, before crop and weed emergence and within 5 days after last preplant tillage operation.

Reference: The "APPLICATION TIMING AND METHODS" section of this label provides detailed information and procedures for application timing and method selected.

RECOMMENDED RATES: Refer to the following tables for the recommended broadcast treatment rates for this tank mixture. Applications which are not consistent with recommendations in this label may result in unsatisfactory weed control or crop injury.

Surface Application Rates

Soil Texture Group*	Broadcast Rates per Acre** Less than 3% organic matter: LASSO® (quarts) +	Lexone 4L** or Sencor 4F** (pints)	3% or more organic matter: LASSO (quarts) +	Lexone 4L** or Sencor 4F** (pints)
Coarse—Sandy Loam only	2	0.5	2	0.75
Medium	2	0.75	2	1
Fine	2	1	2.5	1

* Refer to the "SOIL TEXTURE" section of this label to determine the corresponding texture group for the soil to be treated.

** Use the rates listed in this label when using Lexone 4L or Sencor 4F. Use the equivalent rates when using the 50 WP or DF formulations of these products. One pint of Lexone 4L or Sencor 4F equals one pound of the corresponding 50 WP formulation or ⅔ pound of the corresponding DF formulation.

NOTE: DO NOT use on sand or loamy sand soil. Use 1.5 to 2 pints of Lexone 4L or Sencor 4F per acre in this tank mixture when applying on silty clay and heavy clay soils of the Mississippi Delta. Use a minimum of 2.5 quarts of LASSO on coarse soils and 3 to 4 quarts in this tank mixture on medium or fine textured soils to control black or hairy nightshade or to reduce competition from heavy infestations of the hard-to-control weeds listed in this label.

Preplant Incorporated Application Rates

Soil Texture Group*	Broadcast Rate per Acre*** LASSO® (quarts) +	Lexone 4L** or Sencor 4F** (pints)
Medium	2.5 to 3	0.75
Fine	2.5 to 3	1

* Refer to the "SOIL TEXTURE" section of this label to determine the corresponding texture group for the soil to be treated.

** Use the rates listed in this label when using Lexone 4L or Sencor 4F. Use the equivalent rates when using the 50 WP or DF formulations of these products. One pint of Lexone 4L or Sencor 4F equals one pound of the corresponding 50 WP formulation or ⅔ pound of the corresponding DF formulation.

*** Use the 3 quart per acre rate of LASSO in areas of heavy weed infestation. Use 3 to 4 quarts of LASSO when black or hairy nightshade, yellow nutsedge or red rice is present or to reduce competition from the hard-to-control weeds listed in this label.

NOTE: Do not use on coarse textured soils. Do not use on medium or fine textured soils with less than 1% organic matter. These mixtures, preplant incorporated, may cause slight soybean injury early, but the soybeans should recover and subsequent growth or yield should not be affected.

NOTE: Do not feed soybean hay from treated crop.

LASSO® plus LOROX (TANK MIXTURE)

Apply one treatment of this tank mixture in water or sprayable fluid fertilizer solution for control of the annual grasses and broadleaf weeds listed in the "WEEDS CONTROLLED" section of this label.

NOTE: Plant seed at least 1¾ inches deep and DO NOT use on sand or loamy sand or on soils with less than 1% organic matter as crop injury may occur. DO NOT spray over the top of emerged soybeans.

APPROVED APPLICATION SYSTEMS

Ground—Broadcast Boom; banded

Reference: The "APPLICATION SYSTEMS" section of this label provides detailed information on the proper application of this tank mixture using the system selected.

APPROVED APPLICATION METHODS

Preemergence Surface— Apply after planting, before crop and weed emergence and within 5 days after last preplant tillage operation.

Reference: The "APPLICATION TIMING AND METHODS" section of this label provides detailed information and procedures for application timing and method selected.

RECOMMENDED RATES: Refer to the following table for the recommended broadcast treatment rates for this tank mixture. Applications which are not consistent with recommendations in this label may result in unsatisfactory weed control or crop injury.

Surface Application Rates

Soil Texture Group*	Broadcast Rates per Acre 1 to 3% organic matter: LASSO® (quarts) +	Lorox 50WP (pounds)	More than 3% organic matter: LASSO (quarts) +	Lorox 50WP (pounds)
Coarse—Sandy Loam only	1.5	1	1.5	1 to 1.5
Medium	2	1 to 1.5	2.25	1.5 to 2
Fine	2.25	2	2.5 to 3	2.5 to 3

* Refer to the "SOIL TEXTURE" section of this label to determine the corresponding texture group for the soil to be treated.

NOTE: Use a minimum of 2.5 quarts per acre of LASSO on coarse textured soils, and use 3 to 4 quarts on medium and fine textured soils to control black or hairy nightshade, or to reduce competition from the hard-to-control weeds listed in this label.

NOTE: Do not feed soybean hay from treated crop.

BASAGRAN SEQUENTIAL TREATMENT

Soybeans Only

Basagran may be applied following preemegence applications of LASSO or tank mixtures of LASSO plus Lorox, or LASSO plus Sencor or Lexone for control of emerged weeds such as cocklebur, velvetleaf and other weed species listed on the label for Basagran.

RECOMMENDED RATES: Apply 0.75 to 1.5 quarts of Basagran as directed on the label for Basagran.

Conservation or Minimum Tillage Systems TANK MIXTURES

SOYBEANS ONLY

When applied as directed under the conditions described, these tank mixtures control many emerged annual weeds, suppress many emerged perennial weeds and give preemergence control of many annual grasses and weeds when soybeans will be planted directly into a cover crop, stale seedbed, or in previous crop residues such as wheat stubble.

These tank mixtures will not control regrowth from perennial weeds.

For mixing instructions, see the "MIXING AND SPRAYING INSTRUCTIONS" section of this label.

LASSO plus LOROX plus Paraquat

Apply one treatment of these tank mixtures in 15 to 40 gallons of water per acre immediately before, during or after planting, but BEFORE CROP EMERGENCE. As density of stubble, crop residue or weeds increases, spray gallonage and rate should be increased within the recommended ranges to insure complete coverage. In the absence of emerged vegetation delete the paraquat portion of these mixes.

CONTROL OR SUPPRESSION OF EMERGED WEEDS

PARAQUAT

When used as directed, paraquat in tank mixture with LASSO plus Lorox controls many emerged annual weeds and suppresses many emerged perennial weeds.

Broadcast Treatment—Apply 1 to 2 pints of paraquat (0.25 to 0.5 lb. active) in this tank mixture immediately before, during or after planting but BEFORE CROP EMERGENCE. Use the 2 pint rate if any weeds are 4 to 6 inches tall. This mixture may not control weeds taller than 6 inches. As density of stubble, crop residue or weeds increases, spray gallonage should be increased within the recommended range for complete coverage. Add Ortho X-77 Spreader at 8 oz. per 100 gallons of diluted spray. Refer to the Paraquat label for precautionary statements.

PREEMERGENCE WEED CONTROL

LASSO® plus LOROX

For weeds controlled preemergence, see the "WEEDS CONTROLLED" section of this label for LASSO and LASSO plus Lorox.

See the following table for recommended rates of LASSO and Lorox 50WP in this tank mixture.

Soil Texture Group*	Broadcast Rate per Acre: LASSO® (quarts)	+	Lorox 50WP (pounds)
Coarse—Sandy Loam Only	2 to 2.5		1 to 1.5
Medium	2.5 to 3		1.5 to 2
Fine	2.5 to 3		2 to 3

* Refer to the "SOIL TEXTURE" section of this label to determine the corresponding texture group for the soil to be treated.

Use the higher rate of LASSO in the recommended ranges in areas of heavy grass infestation or when fall panicum or crabgrass will be present. The rate of LASSO may be increased to a maximum of 4 quarts per acre on any soil type to reduce competition from the hard-to-control weeds listed in this label.

Use the higher rate of Lorox 50WP in the recommended ranges on soils with greater than 3% organic matter.

DO NOT use this tank mixture on sand or loamy sand or on soil with less than 1% organic matter as crop injury from Lorox may occur.

NOTE: Do not feed soybean hay from treated crop.

DRY BEANS

DRY BEANS (ALL TYPES)

FOR USE ONLY WEST OF THE MISSISSIPPI RIVER, EXCEPT IN CALIFORNIA

LASSO®

Apply one treatment of this product in water solution for control of yellow nutsedge and the annual grasses and broadleaf weeds listed in the "WEEDS CONTROLLED" section of this label.

NOTE: DO NOT APPLY ON DRY BEANS AFTER PLANTING AS CROP INJURY MAY OCCUR. This product may delay crop maturity and/or reduce yield if cold, wet soil conditions occur after planting.

APPROVED APPLICATION SYSTEMS

Ground—Broacast Boom

Reference: The "APPLICATION SYSTEMS" section of this label provides detailed information on the proper application of this product using the system selected.

APPROVED APPLICATION METHODS

Preplant Incorporated—Apply this product within 7 days prior to planting and shallowly incorporate or Surface Blend into the upper 1 to 2 inches of soil.

Reference: The "APPLICATION TIMING AND METHODS" section of this label provides detailed information and procedures for application timing and method selected.

RECOMMENDED RATES: Apply 2.5 to 3 quarts of this product per acre and shallowly incorporate prior to planting. Use the higher rate in areas of heavy weed infestations or when hard-to-control weeds, black or hairy nightshade or yellow nutsedge are present. Applications which are not consistent with recommendations in this label may result in unsatisfactory weed control or crop injury.

LASSO®

Lima Beans (Green)

FOR USE IN ALL STATES, EXCEPT IN CALIFORNIA

Apply one treatment of this product in water solution for control of yellow nutsedge and the annual grasses and broadleaf weeds listed in the "WEEDS CONTROLLED" section of this label.

NOTE: This product may delay crop maturity and/or reduce yield if cold, wet soil conditions occur after planting.

APPROVED APPLICATION SYSTEMS

Ground—Broadcast Boom

Reference: The "APPLICATION SYSTEMS" section of this label provides detailed information on the proper application of this product using broadcast boom equipment.

APPROVED APPLICATION METHODS

Preplant Incorporated—Apply this product within 7 days prior to planting and shallowly incorporate or Surface Blend into the upper 1 to 2 inches of soil.

Preemergence Surface—Apply this product after planting and before crop and weed emergence and within 5 days after last preplant tillage operation.

Reference: The "APPLICATION TIMING AND METHODS" section of this label provides detailed information and procedures for the application timing and method selected.

RECOMMENDED RATES: Refer to the following table for the recommended broadcast treatment rates.

Surface Application Rates

Soil Texture Group*	Broadcast Rate per Acre (Quarts) LASSO®
Coarse	1.5 to 2
Medium and Fine**	2.5 to 3

* Refer to the "SOIL TEXTURE" section of this label to determine the corresponding texture group for the soil to be treated.

** Use the higher rates in the recommended ranges in areas of heavy weed infestations or when hard-to-control weeds or black or hairy nightshade are present.

NOTE: Preemergence applications are not recommended for use in Illinois or Wisconsin as crop injury may occur.

Preplant Application Rates

Soil Texture*	Broadcast Rate per Acre (Quarts) LASSO®
Coarse	2
Medium and Fine**	2.5 to 3

* Refer to the "SOIL TEXTURE" section of this label to determine the corresponding texture group for the soil to be treated.

** Use the higher rates in the recommended ranges in areas of heavy weed infestations, or when hard-to-control weeds, black or hairy nightshade or yellow nutsedge are present.

LASSO® FOLLOWED BY AMIBEN or PREMERGE

Dry Beans (All Types)

FOR USE IN MICHIGAN ONLY

Apply Amiben or Premerge as a sequential preemergence application to LASSO preplant incorporated.

APPROVED APPLICATION SYSTEMS

Ground—Broadcast Boom

Reference: The "APPLICATION SYSTEMS" section of this label provides detailed information on the proper application of these products using the system selected.

APPROVED APPLICATION METHODS

LASSO®

Preplant Incorporated—Apply this product within 7 days prior to planting and shallowly incorporate or Surface Blend into the upper 1 to 2 inches of soil.

Reference: The "APPLICATION, TIMING AND METHODS" section of this label provides detailed information and procedures for application timing and method selected.

AMIBEN or PREMERGE

Preemergence Surface—Apply Amiben or Premerge after planting and before crop and weed emergence. Read and carefully follow all use instructions and precautionary statements listed on the respective labels for these products.

Reference: The "APPLICATION, TIMING AND METHODS" section of this label provides detailed information and procedures for application timing and method selected.

RECOMMENDED RATES:

LASSO—Apply 2 quarts of this product per acre and shallowly incorporate or Surface Blend prior to planting.

NOTE: DO NOT USE ON SAND OR LOAMY SANDS. DO NOT APPLY ON DRY BEANS AFTER PLANTING AS CROP INJURY MAY OCCUR. This product may delay crop maturity and/or reduce yield if cold, wet soil conditions occur after planting.

AMIBEN—Apply 4 quarts of this product per acre.

PREMERGE—Apply 5 to 6 quarts of this product per acre.

NOTE: When LASSO followed by Amiben or Premerge are used on soils with 6% or more organic matter, weed control may be reduced.

LASSO® plus TREFLAN (TANK MIXTURE)

Dry Beans (All Types) and Lima Beans

FOR USE ONLY WEST OF THE MISSISSIPPI RIVER, EXCEPT IN CALIFORNIA

Apply one treatment of this tank mixture in water solution for control of yellow nutsedge and the annual grasses and broadleaf weeds listed in the "WEEDS CONTROLLED" section of this label.

NOTE: Read the label for Treflan thoroughly for rotational crop restrictions. This tank mixture may delay crop maturity and/or reduce yield if cold, wet soil conditions occur after planting.

APPROVED APPLICATION SYSTEMS

Ground—Broadcast Boom

Reference: The "APPLICATION SYSTEMS" section of this label provides detailed information on the proper application of this tank mixture using the system selected.

APPROVED APPLICATION METHODS

Preplant Incorporated—Apply this tank mixture within 7 days prior to planting and shallowly incorporate into the upper 2 inches of soil within 8 hours after application.

Reference: The "APPLICATION TIMING AND METHODS" section of this label provides detailed information and procedures for application timing and method selected.

RECOMMENDED RATES: Apply 2.5 to 3 quarts of this product plus ½ quart of Treflan per acre. Use the higher rate of this product for this tank mixture in areas of heavy weed infestation or where hard-to-control weeds, black or hairy nightshade or yellow nutsedge are present. Applications which are not consistent with recommendations in this label may result in unsatisfactory weed control or crop injury.

LASSO®

Red Kidney Bean Types Only

FOR USE IN ILLINOIS AND WISCONSIN ONLY

Apply one treatment of this product in water solution for control of yellow nutsedge and the annual grasses and broadleaf weeds listed in the "WEEDS CONTROLLED" section of this label.

NOTE: This product may delay crop maturity and/or reduce yield if cold wet soil conditions occur after planting.

APPROVED APPLICATION SYSTEMS

Ground—Broadcast Boom; banded

Reference: The "APPLICATION SYSTEMS" section of this label provides detailed information on the proper application of this tank mixture using the system selected.

APPROVED APPLICATION METHODS

Preplant Incorporated—Apply this product within 7 days prior to planting and shallowly incorporate or Surface Blend into the upper 1 to 2 inches of soil.

Preemergence Surface—Apply this product after planting and before crop and weed emergence and within 5 days after last preplant tillage operation.

Reference: The "APPLICATION TIMING AND METHODS" section of this label provides detailed information and procedures for application timing and method selected.

RECOMMENDED RATES: Apply 2.5 to 3 quarts of this product per acre. Use the higher recommended rate when heavy weed infestations, hard-to-control weeds, black or hairy nightshade or yellow nutsedge are present. Preplant incorporate only for control of yellow nutsedge.

LASSO®

Mung Beans

FOR USE IN OKLAHOMA ONLY

Apply one treatment of this product in water solution for control of the annual grasses and broadleaf weeds listed in the "WEEDS CONTROLLED" section of this label.

APPROVED APPLICATION SYSTEMS

Ground—Broadcast Boom; banded

Reference: The "APPLICATION SYSTEMS" section of this label provides detailed information on the proper application of product using the system selected.

APPROVED APPLICATION METHODS

Preemergence Surface—Apply this product after planting and before crop and weed emergence and within 5 days after last preplant tillage operation.

Reference: The "APPLICATION TIMING AND METHODS" section of this label provides detailed information and procedures for application timing and method selected.

RECOMMENDED RATES: Refer to the following table for recommended broadcast treatment rates.

Surface Application Rates

Soil Texture	Broadcast Rate per Acre (Quarts) LASSO®
Coarse*	1.5
Medium and Fine	2.0

* DO NOT USE ON SAND AS CROP INJURY MAY OCCUR.

PEAS

LASSO®

Peas, Green (for processing)

FOR USE IN MINNESOTA ONLY

Apply one treatment of this product in water solution for control of the annual grasses and broadleaf weeds listed in the "WEEDS CONTROLLED" section of this label.

NOTE: This product may delay crop maturity and/or reduce yield if cold, wet soil conditions occur after application.

APPROVED APPLICATION SYSTEMS

Ground—Broadcast Boom

Reference: The "APPLICATION SYSTEMS" section of this label provides detailed information on the proper application of this tank mixture using the system selected.

Preemergence Surface—Apply this product after planting before crop and weed emergence and within 5 days after last preplant tillage operation.

Reference: The "APPLICATION TIMING AND METHODS" section of this label provides detailed information and procedures for application timing and method selected.

RECOMMENDED RATES: Refer to the following table for the recommended broadcast treatment rate for this product. Applications which are not consistent with recommendations in this label may result in unsatisfactory weed control or crop injury.

Surface Application Rates

Soil Texture Group*	Broadcast Rate per Acre (Quarts) LASSO®
Medium	2
Fine	2.5

* Refer to the "SOIL TEXTURE" section of this label to determine the corresponding texture group for the soil to be treated.

WOODY ORNAMENTALS

LASSO®

FOR USE ON JUNIPER (JUNIPERUS SPP.) AND YEW (TAXUS SPP.)

Apply one treatment of this product in water solution for control of the annual grasses and broadleaf weeds listed in the "WEEDS CONTROLLED" section of this label.

NOTE: Topical application or contact of this product with green foliage of the woody ornamental may cause injury and is not recommended. In case green foliage is accidentally sprayed, overhead irrigation within 1 hour after application will minimize injury symptoms (i.e. leaf tip burn, discoloration, distortion, etc.). This product is recommended for this use for application by certified, professional applicators only.

DO NOT USE THIS PRODUCT ON SEEDBEDS, UNROOTED CUTTINGS OR PRIOR TO TRANSPLANTING. DO NOT USE THIS PRODUCT IN GREENHOUSES OR OTHER ENCLOSED STRUCTURES, APPLIED THROUGH IRRIGATION SYSTEMS OR MIXED WITH SPRAYABLE FLUID FERTILIZERS. APPLICATIONS DURING PERIODS WHEN TEMPERATURE EXCEEDS 90°F MAY CAUSE INJURY TO ORNAMENTALS.

APPROVED APPLICATION SYSTEMS

Ground—Broadcast Boom; banded

Reference: The "APPLICATION SYSTEMS" section of this label provides detailed information on the proper application of this tank mixture using the system selected.

APPROVED APPLICATION METHODS

Post Transplant Directed—Apply this product as a directed spray after transplanting or to established plantings prior to weed emergence.

Reference: The "APPLICATION TIMING AND METHODS" section of this label provides detailed information and procedures for application timing and method selected.

RECOMMENDED RATES: Apply 4 quarts of this product per acre as a broadcast treatment. Applications which are not consistent with the recommendations in this label may result in unsatisfactory weed control or injury to ornamental plantings.

NOTE: Repeat applications of this product may be required for sustained weed control. DO NOT exceed 3 applications per year or retreat within 21 days.

In case of an amergency involving this product, Call Collect, day or night, (314) 694-4000

NIRAN® E-4

PARATHION INSECTICIDE

RESTRICTED USE PESTICIDE

For retail sale to and use only by Certified Applicators or persons under their direct supervision and only for those uses covered by the Certified Applicator's certification.

Emulsifiable Insecticide for Controlling Certain Insects on the Listed Field, Forage, Fruit and Vegetable Crops.

ACTIVE INGREDIENTS:

*Parathion: O,O-diethyl O-p-nitrophenyl phosphorothioate	44.3%
*Related compounds	2.3%
Xylene range aromatic solvents	47.9%
INERT INGREDIENTS:	5.5%
TOTAL	100.0%

*Parathion and related compounds equivalent to 4.0 lbs. per gallon.

EPA Reg. No. 524-132 896.09-000.27

DANGER–POISON

Keep Out of Reach of Children

NOT FOR HOME USE

Complete Directions for Use

Read the entire label before using this product.

Use only according to these label instructions. Read "LIMIT OF WARRANTY AND LIABILITY" before buying or using. If terms are not acceptable return at once unopened.

LIMIT OF WARRANTY AND LIABILITY

This company warrants that this product conforms to the chemical description on the label and is reasonably fit for the purposes set forth in the Complete Directions for Use label booklet ("Directions") when used in accordance with those Directions under the conditions described therein. NO OTHER EXPRESS OR IMPLIED WARRANTY IS MADE AND THE IMPLIED WARRANTY OF MERCHANTABILITY AND FITNESS FOR PARTICULAR PURPOSE IS EXCLUDED. This warranty is also subject to the conditions and limitations stated herein.

Buyer and all users shall promptly notify this company of any claims whether based in contact, negligence, strict liability, other tort or otherwise.

Buyer and all users are responsible for all loss or damage from use or handling which results from conditions beyond the control of this company, including, but not limited to, incompatibility with products other than those set forth in the Directions, and application in any manner not explicitly set forth in the Directions.

THE EXCLUSIVE REMEDY OF THE USER OR BUYER, AND THE LIMIT OF THE LIABILITY OF THIS COMPANY OR ANY OTHER SELLER FOR ANY AND ALL LOSSES, INJURIES OR DAMAGES RESULTING FROM THE USE OR HANDLING OF THIS PRODUCT (INCLUDING CLAIMS BASED IN CONTRACT, NEGLIGENCE, STRICT LIABILITY, OTHER TORT OR OTHERWISE) SHALL BE THE PURCHASE PRICE PAID BY THE USER OR BUYER FOR THE QUANTITY OF THIS PRODUCT INVOLVED, OR, AT THE ELECTION OF THIS COMPANY OR ANY OTHER SELLER, THE REPLACEMENT OF SUCH QUANTITY OR, IF NOT ACQUIRED BY PURCHASE, REPLACEMENT OF SUCH QUANTITY. IN NO EVENT SHALL THIS COMPANY OR ANY OTHER SELLER BE LIABLE FOR ANY INCIDENTAL OR CONSEQUENTIAL DAMAGES.

The buyer and all users are deemed to have accepted the terms of this LIMIT OF WARRANTY AND LIABILITY which may not be varied by any verbal or written agreement.

PRECAUTIONARY STATEMENTS

HAZARDS TO HUMANS AND DOMESTIC ANIMALS

Keep Out of Reach of Children

DANGER! PELIGRO!

PRECAUCION AL USUARIO: Si usted no lee ingles, no use este producto hasta que la etiqueta haya sido explicada ampliamente.

POISONOUS IF SWALLOWED, INHALED OR ABSORBED THROUGH SKIN.

RAPIDLY ABSORBED THROUGH SKIN.

REPEATED EXPOSURE MAY, WITHOUT SYMPTOMS, BE INCREASINGLY HAZARDOUS.

Do not breathe vapors or spray mist.

Do not get in eyes, on skin or on clothing.

POISON SIGNS (Symptoms)

Parathion is a very dangerous poison. It rapidly enters the body on contact with all skin surfaces and eyes. Clothing wet with this material must be removed immediately. Exposed persons must receive prompt medical treatment or they may die.

Some of the signs and symptoms of poisoning are: Headache, nausea, vomiting, cramps, weakness, blurred vision, pin-point pupils, tightness in chest, labored breathing, nervousness, sweating, watering of eyes, drooling or frothing of mouth and nose, muscle spasms and coma.

POISON

ATROPINE IS AN ANTIDOTE. CONSULT PHYSICIAN FOR EMERGENCY SUPPLY.

CALL A PHYSICIAN AT ONCE IN ALL CASES OF SUSPECTED PARATHION POISONING.

If symptoms or signs of poisoning include blurred vision, abdominal cramps, and tightness in the chest, do not wait for a doctor but give two atropine tablets (each 1/100 grain or 0.65 milligrams) at once. (One tablet to children under five years of age.)

FIRST AID TREATMENT

STATEMENT OF PRACTICAL TREATMENT

Call a doctor (physician), clinic or hospital immediately. Explain that the victim has been exposed to Parathion and describe his condition. If doctor cannot come, take victim to clinic or hospital.

Move patient immediately from the area where Parathion is present.

IF INHALED, remove victim to fresh air. If not breathing, immediately give artificial respiration, preferably mouth-to-mouth, and maintain until doctor sees victim. If breathing is difficult, give oxygen.

IF IN EYES OR ON SKIN, immediately flush with plenty of water for at least 15 minutes while removing contaminated personal clothing and shoes to avoid continued possible exposure to victim or persons in attendance.

IF SWALLOWED, induce vomiting immediately by giving two glasses of water and by touching back of throat with finger. DO NOT INDUCE VOMITING OR GIVE ANYTHING BY MOUTH TO AN UNCONSCIOUS PERSON. Have victim lie down and keep quiet.

NOTE TO PHYSICIAN

Antidote—administer atropine sulfate in large doses, TWO to FOUR mg. intravenously or intramuscularly as soon as cyanosis is overcome. Repeat at 5 to 10 minute intervals until signs of atropinization appear. 2-PAM chloride is a pharmacologic antidote and may be administered as an adjunct to, but not a substitute for, atropine which is a symptomatic and often life-saving antidote. DO NOT GIVE MORPHINE OR TRANQUILIZERS BECAUSE THESE ANTIDOTES MAY PRODUCE ADVERSE EFFECTS ACCENTUATING THE PHARMACOLOGIC EFFECT OF THIS PRODUCT. This product is a strong cholinesterase inhibitor affecting the central and peripheral nervous systems and producing cardiac and respiratory depression. At first sign of pulmonary edema, the patient should be given supplemental oxygen and treated symptomatically. Continued absorption of the poison may occur and fatal relapses have been reported after initial improvements; VERY CLOSE SUPERVISION OF THE PATIENT IS INDICATED FOR AT LEAST 48 HOURS.

WORK SAFETY RULES

REPEATED EXPOSURES TO CHOLINESTERASE INHIBITORS SUCH AS ARE CONTAINED IN THIS PRODUCT MAY, WITHOUT WARNING, CAUSE PROLONGED SUSCEPTIBILITY TO VERY SMALL DOSES OF ANY CHOLINESTERASE INHIBITOR.

Persons working with this product should have frequent blood tests of their cholinesterase levels. If the cholinesterase level falls below a critical point, no further exposure should be allowed until it has been determined by means of blood tests that the cholinesterase level has returned to normal. Before using this product, consult the State Agricultural Extension Service or Experimental Station for specific recommendations regarding such blood tests.

If handled indoors, provide mechanical exhaust ventilation.

Keep all unprotected persons and animals away from treated area or where there is danger of drift.

Do not rub eyes or mouth with hands. If you feel sick in any way, STOP work and get help right away. See First Aid (Practical Treatment) section.

USE ONLY WHEN WEARING THE FOLLOWING PROTECTIVE EQUIPMENT AND CLOTHING.

(1) Wear water-proof pants, coat, hat, rubber boots or rubber overshoes. (2) Wear safety goggles. (3) Wear a pesticide respirator jointly approved by the Mining Enforcement and Safety Administration (formerly the U.S. Bureau of Mines) and the National Institute for Occupational Safety and Health under the provisions of 30 CFR part II. (4) Wear heavy duty, natural rubber gloves.

IMPORTANT! Before removing gloves, wash them with soap and water. Always wash hands, face and arms with soap and water before smoking, eating or drinking.

AFTER WORK, take off all work clothes, and shoes. Shower, using soap and water. Wear only clean clothes when leaving job. Do not wear contaminated clothing. Wash protective clothing and protective equipment with soap and water after each use. Respirator should be cleaned and filter replaced according to instructions included with respirator.

REENTRY STATEMENT

Do not apply this product in such a manner as to directly or through drift expose workers or other persons. The area being treated must be vacated by unprotected persons.

Do not enter treated areas for at least 48 hours unless appropriate protective clothing is worn.

Because certain states may require more restrictive reentry intervals for various crops treated with this product, consult your State Department of Agriculture for further information.

Written or oral warnings must be given to workers who are expected to be in a treated area or in an area about to be treated with this product.

ORAL WARNINGS shall inform workers of areas or fields that may not be entered, without protective equipment and clothing specified on this label, during treatment or for 48 hours following treatment. Follow First Aid Treatment instructions shown on this label in case of accidental exposure.

When oral warnings are given, warnings shall be given in a language customarily understood by workers. Oral warnings must be given if there is reason to believe that written warnings cannot be understood by workers.

WRITTEN WARNINGS must include the following information.

DANGER—Area treated with Parathion on (date). Do not enter without appropriate protective clothing for 48 hours. Follow First Aid Treatment instructions shown on this product label in case of accidental exposure.

ATTENTION

To avoid excessive residues of parathion on food or forage crops always observe the statements found under "Directions for Use," limiting the time before harvest when parathion may be applied.

Because this material is poisonous by skin contact, inhalation or swallowing it should not be used in such a manner or under weather conditions as will permit drift of the spray onto areas not intended to be sprayed.

Do not apply or allow drift to areas occupied by unprotected humans or beneficial animals.

ENVIRONMENTAL HAZARD

This product is toxic to fish, shrimp, crab, birds, and other wildlife. Birds and other wildlife in treated areas may be killed. Keep out of lakes, streams, ponds, tidal marshes and estuaries. Shrimp and crab may be killed at application rates recommended on this label. Do not apply where these are important resources. Do not apply where runoff is likely to occur. Do not apply when weather conditions favor drift from areas treated. Do not contaminate water by cleaning of equipment, or disposal of wastes. Apply this product only as specified on this label.

This product is highly toxic to bees exposed to direct treatment or residues on crops. Protective information may be obtained from your Cooperative Agricultural Extension Service.

PHYSICAL OR CHEMICAL HAZARD

Combustible: Do not use, pour, spill or store near heat or flame.

In case of:
FIRE, use water spray, foam, dry chemical or CO_2.
SPILL or LEAK keep all unprotected persons away.
Cover with absorbent such as soda ash, lime, clay or saw dust. Sweep up and bury. Wash area thoroughly with strong lye solution.

STORAGE AND DISPOSAL

Do not contaminate water, food, or feed by storage or disposal.

STORAGE AND HANDLING

1. Handle container carefully so as to avoid damage and prevent leakage.
2. Do not use or store in or around the home.
3. Store container in a well ventilated place.
4. Empty container retains vapor and product residue. Observe all labeled safeguards until container is cleaned, reconditioned or destroyed.
5. Keep out of reach of children and domestic animals.

DISPOSAL

Wastes of this pesticide are acutely hazardous. Improper disposal of excess pesticide, spray mixtures, or rinsate is a violation of federal law. If these wastes cannot be disposed of according to label use instructions, contact your state pesticide or Environmental Control Agency, or the Hazardous Waste Representative at the nearest EPA Regional Office for guidance.

Triple rinse empty container with 5% caustic soda solution. Flush with water. Then offer for recycling or reconditioning, or puncture to prevent other use and dispose of in an approved landfill, or by other procedures approved by state and local authorities.

DIRECTIONS FOR USE

It is a violation of Federal Law to use this product in any manner inconsistent with its labeling.

Be sure to read the precautionary statements before using!

This product is designed for application after dilution with water and for use by trained operators using airplane or power ground equipment. The hazards and precautions for handling the product in this container are equally applicable to it after dilution with water for spray application. Add the concentrate to the spray tank while filling with water, and mix thoroughly either by means of a tank agitator or pump by-pass. For best results, thoroughly cover all surfaces to be treated with spray. Rates of application given below should not be exceeded. Never apply later than indicated to assure residue levels at harvest are below tolerances established by the Environmental Protection Agency.

Consult the State Agricultural Extension Service or Experiment Station for specific recommendations regarding application, dosage and timing of sprays. For application by ground equipment, add the desired amount of concentrate to sufficient water to apply at least 3 gallons of water per acre. For application by aircraft, add the amount of concentrate desired per acre to ½ to 3 gallons of water consistent with crop growth and good coverage. Greater quantities of water may be required to give sufficient coverage of orchard trees.

CEREAL

DO NOT APPLY WITHIN 15 DAYS OF HARVEST, CUTTING OR FORAGE USE ON THE FOLLOWING CROPS.

RICE (California)—To control rice leaf miners and tadpole shrimp, use 3/16 pint per acre. Shrimp, crabs and crayfish may be killed. Do not apply where these are important resources.

DO NOT APPLY WITHIN 12 DAYS OF HARVEST, CUTTING OR FORAGE USE ON THE FOLLOWING CROPS.

CORN—For control of European corn borers, use 1 pint per acre. Apply the first spray when 75% of the corn plants show "shot-hole" feeding and follow with two additional sprays at 5 to 7 day intervals. Use sufficient water per acre to provide complete coverage and be certain whorls of plant are well treated. For corn leaf aphids and grasshoppers, use ½ to 1 pint per acre. For fall armyworms, corn earworms, corn rootworm adults, armyworms, climbing cutworms, grasshoppers and Japanese beetles, use ¾ pint per acre. To control stink bugs and spider mites, use 1 pint per acre. To control chinch bugs, use 1½ pints per acre.

SORGHUM—To control sorghum midge, apply at rate of 1 pint to 1 quart per acre, 2 applications 3 to 5 days apart when approximately 90% of the heads have completely emerged from the boot or not later than start of blooming. For corn leaf aphids, grasshoppers, greenbugs and mites, use ½ pint per acre. For sorghum webworms, fall armyworms, armyworms, and corn earworms, use ¾ to 1 pint per acre. To control chinch bugs, use 1½ pints per acre. Leaf injury may occur on some hybrid varieties of sorghum. Spray a few rows a week or so before booting to test effects on plants.

DO NOT APPLY WITHIN 15 DAYS OF HARVEST, CUTTING OR FORAGE USE ON THE FOLLOWING CROPS.

SMALL GRAINS (Wheat, Oats, Barley)—To control armyworms, aphids (greenbugs) and winter grain mites, use ½ pint per acre. For thrips, use ½ to ¾ pint per acre. For Say's plant bugs, use ¾ pint per acre. For black grass bugs, stink bugs, white spider mites, leafhoppers, climbing cutworms, grasshoppers and brown wheat mites, use 1 pint per acre. For chinch bugs, false chinch bugs and banks grass mites, use 1½ pints per acre.

FIELD AND FORAGE CROPS

DO NOT APPLY WITHIN 20 DAYS OF HARVEST, CUTTING OR FORAGE USE ON THE FOLLOWING CROPS.

SOYBEANS—To control webworms, use ½ pint per acre. To control velvet bean caterpillars, green cloverworms, two-spotted mites and stink bugs, use 1 pint per acre. To control corn earworms and fall armyworms, use 1 to 1⅗ pints per acre. To control white grubs and wireworms, broadcast 1 gallon per acre just prior to planting and thoroughly incorporate into upper 4 to 6 inches of soil.

DO NOT APPLY WITHIN 15 DAYS OF HARVEST, CUTTING OR FORAGE USE ON THE FOLLOWING CROPS.

ALFALFA, CLOVER, VETCH AND GRASS—For sweet clover aphids, three-cornered alfalfa hoppers, alfalfa caterpillars, and spittlebugs, use ½ pint per acre. For aphids, alfalfa weevil larvae, and adult weevils, armyworms, clover leaf weevils, climbing cutworms, webworms, grasshoppers, crickets, spotted alfalfa aphids, leafhoppers, Lygus bugs, thrips, and tortricid moths, use ½ to 1 pint per acre.

For control of range caterpillar, use 1 pint per acre. For alfalfa seed chalcids control on alfalfa grown for seed, use ½ to 1 pint per acre. California and Nevada regulations limit the use of this material to not more than $^3/_4$ pint per acre. For clover head weevils, spider mites, blister beetles, Asiatic garden beetles, sweet clover weevils, and green June beetles, use 1 pint per acre. For beet armyworms and corn earworms use 1 to $1^1/_2$ pints per acre. Do not spray legumes during bloom period to avoid injury to honey bees.

PEANUTS—To control fall armyworms, climbing cutworms, corn earworm, grasshoppers, leafhoppers and red-necked peanutworms, saltmarsh caterpillar, three-cornered alfalfa hopper, and webworm, use 1 pint per acre. To control lesser cornstalk borers, use 1 pint to 1 quart per acre, direct spray to soil surface and base of plants.

SUGAR BEETS—For alfalfa loopers, aphids, armyworms, leafhoppers, blister beetles, flea beetles, leaf miners, Lygus bugs, stink bugs, webworms, climbing cutworms and grasshoppers, use 1 pint per acre. For false celery leaf tiers, use $1^1/_2$ pints per acre. For beet crown borers, use $1^1/_2$ pints per acre, ground application over the row during seedling stage. To control white grubs and wireworms, broadcast 1 gallon per acre just prior to planting and thoroughly incorporate into upper 4 to 6 inches of soil.

SUGARCANE—To control wireworms, use 2 quarts in 10 to 12 inch band in the open furrow at time of planting.

DO NOT APPLY WITHIN 7 DAYS OF HARVEST ON THE FOLLOWING CROPS

COTTON—To control aphids, mites, cotton leafworms, cotton fleahoppers, garden webworms and thrips, use $^1/_2$ pint per acre. For some spider mites, use $^1/_4$ to $^1/_2$ pint per acre. For cabbage loopers, use $^3/_4$ to 1 pint per acre. For boll weevils and stink bugs, use 1 to $1^1/_2$ pints per acre. For salt-marsh caterpillars, use 1 to 2 pints per acre. For bollworms, cotton leaf perforators, Lygus bugs, false chinch bugs, serpentine leaf miners and southern garden leafhoppers, use 1 quart per acre. Use enough water for complete coverage. Make first application when insects appear and repeat at 7 day intervals if required. If desired, this formulation may be combined with other insecticides in a complete cotton spray program.

DO NOT APPLY WITHIN 5 DAYS OF HARVEST ON THE FOLLOWING CROPS.

TOBACCO—For control of aphids, stink bugs and tobacco suckflies, use $^1/_2$ pint per acre. Do not apply within 5 days of priming or 15 days of cutting. Avoid plant juices coming in contact with the skin or other parts of the body of those who are engaged in cutting the crop.

FRUIT

DO NOT USE TREATED CITRUS PEEL FOR FOOD PURPOSES.

CITRUS (California)—Grapefruits, Kumquats, Lemons, Limes, Oranges, Tangeloes and Tangerines.

Scale—For purple, black, brown soft, California red, citricola, cottony-cushion and yellow scales, use $^1/_3$ to $^2/_3$ pint in 100 gallons of water applied at petal fall to prevent fruit scattering.

Other insects—Use 1 to 2 quarts in 100 gallons of water for control of the following additional insects infesting citrus: climbing cutworms, fruit tree leaf rollers, katydids, omnivorous leaf rollers, Fuller rose beetles, pink scavenger caterpillars, orange tortrix, or angeworms and Western tussock moths. Do not use more than $2^1/_2$ gallons of this product per acre up to 30 days of harvest. Do not use more than 1 gallon of this product per acre from 30 days up to 15 days of harvest. Consult agricultural experimental authorities for specific recommendations in your area.

CITRUS (Areas other than California)—Grapefruits, Kumquats, Lemons, Limes, Oranges, Tangeloes and Tangerines.

Treat for mealybugs; chaff, cottony-cushion, Glover's, purple, Florida red, yellow, snow scales; aphids; orange dog and plant bugs, using $^1/_2$ to $^5/_8$ pint in 100 gallons of water. For control of mites and whiteflies, use $^1/_2$ to $^3/_4$ pint with 1 gallon of emulsive oil concentrate in 100 gallons of water. For controlling grasshoppers, use 1 pint per acre. Thorough coverage is essential for best results. Do not use more than $2^1/_2$ gallons of this product per acre up to 30 days of harvest. Do not use more than 1 gallon of this product per acre from 30 days up to 15 days of harvest.

DO NOT APPLY TO CURRANTS AND FIGS WITHIN 30 DAYS OF HARVEST.

CURRANTS—For control of currant aphids, use $^1/_2$ to $1^3/_5$ pints per acre. For control of two-spotted spider mites, use 1 to $1^3/_5$ pints per acre. For currant borers, use $1^3/_5$ pints per acre.

FIGS—For two-spotted and Pacific mites, use $^1/_2$ to 1 pint per 100 gallons of water. For fig scales, use 1 pint per 100 gallons of water. Do not use more than $2^1/_2$ quarts of this product per acre.

DO NOT APPLY TO AVOCADO WITHIN 21 DAYS OF HARVEST.

AVOCADO—To control banded cucumber beetles, grasshoppers, citrus root weevils, red-banded thrips, avocado lace bugs, pyriform scales, webbing worms, blossom anomala, little fire ants, greenhouse thrips and tortricids, use $^3/_8$ pint in 100 gallons of water. To control whiteflies, use $^3/_8$ pint with 1 gallon of emulsive oil concentrate in 100 gallons of water. To control Florida red scales, Florida wax scales, dictyospermum scales and avocado leafhoppers, use $^1/_2$ pint in 100 gallons of water. To control pumpkin bugs and mealybugs, use $^1/_2$ to $^3/_4$ pint in 100 gallons of water. To control latania scales, use $^3/_4$ pint in 100 gallons of water. Do not use more than $2^1/_2$ quarts of this product per acre.

DO NOT APPLY TO THE FRUITS LISTED BELOW WITHIN 15 DAYS OF HARVEST.

CANEBERRIES (Raspberries, Loganberries, Boysenberries and Blackberries)—For control of two-spotted spider mites, use $^3/_4$ pint per acre. For control of obsure and woods weevils, use at 1 quart per acre as a post harvest application to the soil or ground cover over roots of plants. For crown borers, use at 1 quart per acre but apply to crown area and lower canes.

CRANBERRIES—For control of fireworms, fruitworms, tipworms and lecanium scales, use $1^1/_2$ pints per acre.

GOOSEBERRIES—For control of currant aphids, use $^1/_2$ to $1^3/_5$ pints per acre. For control of two-spotted spider mites, use 1 to $1^3/_5$ pints per acre. For currant borers, use $1^3/_5$ pints per acre.

DO NOT APPLY TO THE FRUITS LISTED BELOW WITHIN 14 DAYS OF HARVEST.

APPLES—For control of European sawflies, San Jose, Forbes or scurfy scales, mealybugs, European red and two-spotted mites, bagworms, Japanese beetles, shot-hole borers, orange tortrix and apple lace bugs, dilute $^1/_2$ pint in 100 gallons of water and spray to cover foliage thoroughly. For codling moths, use $^1/_2$ pint in 100 gallons of water, 3 to 4 applications, 10 to 14 days apart, starting 10 to 14 days after petal fall; for second and third broods, spray 1 to 3 times at 10 to 14 day intervals. For fruit tree leaf rollers, use $^1/_2$ pint per 100 gallons of water at petal fall and for red-banded leaf rollers, apply $^1/_2$ pint per 100 gallons of water at petal fall and at first, fifth and sixth cover spray. For plum curculio, apply at $^1/_2$ pint per 100 gallons of water at petal fall and 1 or 2 additional times each 7 to 10 days apart. For grasshoppers, use $^3/_4$ pint in 100 gallons. For the following insects, $^3/_8$ pint per 100 gallons of water is adequate: bud moths; clover, Pacific, Willamette or Schoenii mites; flea weevils; rosy, wooly and green apple aphids; leafhoppers; leaf miners; and red bugs. Certain insects, such as two-spotted and Willamette mites, may require repeat treatments at 7 to 10 day intervals during the summer months. Parathion sprays may injure the foliage and fruit of McIntosh apples and related varieties, such as Cortland, Kendall, Macoun, Melba, etc., and Golden Delicious or Jonathan. Consult the State Agricultural Extension Service or Experiment Station for advice on possibility of injury and safening the spray by using activated carbon. Do not use more than $1^1/_2$ gallons of this product per acre.

APRICOTS—To control aphids, mites, bud moths, peach tree borers, Japanese beetles and leaf rollers, use $^3/_4$ pint per 100 gallons of water. Control of codling moths, lesser peach tree borers, grasshoppers, and tortrix requires $^1/_2$ to $^7/_8$ pint per 100 gallons. To control Oriental fruit moths, use $^1/_2$ to $^7/_8$ pint per 100 gallons of water at shuck split, 10 to 12 days later and if needed 6 and 3 weeks before harvest. For peach tree borers and lesser peach tree borers, apply 2 or 3 sprays to trunk from ground to scaffold limbs timed with moth emergence. Use $^3/_8$ pint in 100 gallons of water for control of Pandemis moths. Avoid injury to bees by delaying spray till after full bloom. Do not use more than $3^1/_2$ quarts of this product per acre.

BLUEBERRIES—For thrips, maggots, curculio and tip borers, use $^1/_2$ pint in 100 gallons of water. For lecanium scales, use 1 pint per 100 gallons of water. Use before fruit sets or after harvest. Use from 100 to 300 gallons of diluted spray per acre, but do not apply more than $1^1/_5$ pints of this product to one acre of blueberries at any application.

CHERRIES—For aphids and mites, mix $^3/_8$ pint in 100 gallons of water. For sawflies, use $^3/_8$ to $^1/_2$ pint in 100 gallons of water. Use $^1/_2$ pint per 100 gallons for thrips, cherry fruitworms, pear slugs, Pandemis moths, bud moths, cankerworms, rose chafers, San Jose scale crawlers, fruit flies and tortrix. For fruit tree leaf rollers, use $^1/_2$ pint per 100 gallons of water at petal fall or shuck split; for plum curculio, use $^1/_2$ pint per 100 gallons of water, 2 or 3 applications, 8 to 10 days apart, beginning at petal fall or shuck split; for Oriental fruit moths, use $^1/_2$ pint in 100 gallons of water at shuck split and 10 to 12 days later. For Japanese beetles, use $^3/_4$ to 1 pint per 100 gallons. Do not use more than $2^1/_{10}$ quarts of this product per acre.

GRAPES—For mites, aphids, mealybugs and berry moths, use $^3/_8$ pint per 100 gallons of water. For leaf rollers, Japanese beetles and leaf folders, use $^1/_2$ pint per 100 gallons of water. For false chinch bugs, use 1 pint in 100 gallons of water per acre by ground equipment or in 10 gallons of water by aircraft. For consperse stink bugs, use $1^1/_2$ quarts per acre. For grape leafhoppers, use $1^1/_2$ to $2^1/_2$ quarts per acre. For black vine weevils, use $2^1/_2$ quarts per acre. Do not use more than $1^1/_2$ quarts of this product per acre after the fruit is the size of buckshot. Use 300 to 500 gallons of water per acre depending on age of vineyard and stage of plant growth.

PEACHES AND NECTARINES (Areas other than California)—For control of green peach aphids, use $^3/_8$ pint in 100 gallons of water. For peach tree borers, leaf rollers, mites, catfacing insects, tarnished plant bugs, shot-hole borers, peach bark beetles, scales and bud moths, mix $^1/_2$ pint per 100 gallons of water, and repeat if re-infestation occurs. For Oriental fruit moths, see under apricots. For plum curculio, use $^1/_2$ pint per 100 gallons of water. In the South, treat at petal fall, 10 days later and repeat at 7 to 10 day intervals up to 3 weeks before harvest. In the North, treat 3 to 4 times, 7 to 10 days apart, beginning at shuck-off. For lesser peach tree and American plum borers and grasshoppers, use $^3/_4$ to 1 pint per 100 gallons. For peach tree borers and lesser peach tree borers, apply 2 or 3 sprays to trunk from ground to scaffold limbs timed with moth emergence. Do not apply more than 4 quarts of this material per acre at any application, and do not use more than 5 quarts per acre per year.

PEACHES AND NECTARINES (California)—Use as shown for other areas except do not apply within 21 days of harvest. Do not apply more than once after bloom. Do not apply more than $2^1/_2$ quarts of this product per acre at any application, and do not use more than 5 quarts per acre between January 1 and harvest.

PEARS—For control of leaf miners, aphids, leaf rollers, grasshoppers, scales, mealybugs and certain mites, use the dosage described for those insects on apples. For pear psylla, use $^3/_8$ pint per 100 gallons of water. For pear blister mites, pear slugs, green fruitworms and plant bugs, use $^1/_2$ pint per 100 gallons of water. For codling moths, use $^1/_2$ pint in 100 gallons of water in 2 to 4 cover sprays, beginning with the first cover. For plum curculio, apply $^1/_2$ pint in 100 gallons of water at petal fall and 10 days later. Some injury may occur on Bosc pears, under some conditions. Do not use more than $3^1/_2$ quarts of this product per acre.

PLUMS AND PRUNES—Apply $^1/_2$ to $^5/_8$ pint per 100 gallons of water for control of these insects: pear thrips, flower thrips, mites, aphids, leafhoppers, leaf rollers, peach tree borers, shot-hole borers, bud moths, tortrix, mealy plum lice and scales. Apply scale treatment when crawlers emerge. For plum curculio make 3 to 4 applications, beginning at petal fall, at rate of $^1/_2$ pint in 100 gallons of water. For codling moths, use $^1/_2$ to 1 pint per 100 gallons of water at petal fall and a summer application timed with moth emergence. For peach twig borers, use 1 pint per 100 gallons of water. Do not use more than 4 quarts of this product per acre.

STRAWBERRIES—To control flower thrips, crickets, flea beetles, red spider mites, aphids, Lygus bugs, leafhoppers, whiteflies and leaf rollers, use $^1/_2$ to 1 pint in 100 to 150 gallons of water per acre. To control weevils, crown borers and strawberry leaf beetles, use $^3/_4$ pint in 100 to 150 gallons of water per acre. As a foliar treatment, do not use more than $1^3/_5$ pints of this product per acre. To control garden symphylans, use 5 quarts in 40 gallons of water per acre as a preplant soil treatment.

DO NOT APPLY TO THE FRUITS LISTED BELOW WITHIN 7 DAYS OF HARVEST.

PINEAPPLES—For control of crickets and mealybugs, use $^1/_2$ pint per 100 gallons of water, and apply 300 gallons of spray per acre.

DO NOT USE PARATHION ON OLIVES AFTER AUGUST 1.

OLIVES—For black, oleander and parlatoria scales, use 1 pint in $1^1/_2$ gallons light-medium grade summer oil emulsion, or 1 gallon light-medium grade summer emulsive oil in 100 gallons of water, post-bloom.

NUTS

DO NOT APPLY AFTER HULLS OR HUSKS BEGIN TO OPEN. DO NOT FEED TREATED HULLS OR HUSKS TO LIVESTOCK.

ALMONDS—To control fruit tree leaf rollers, tent caterpillars and peach twig borers, use 1 pint per 100 gallons of water. As a dormant spray for parlatoria and San Jose scales, use 1 pint with 3 gallons dormant oil emulsion or $2^1/_2$ gallons dormant emulsive oil in 100 gallons of water. Do not use more than 3 quarts of this product per acre.

FILBERTS—For apple mealybugs, filbert aphids, bud moths and spider mites, use $^1/_2$ pint per 100 gallons of water. Do not use more than 3 quarts of this product per acre.

PECANS—For control of aphids, use $^1/_2$ to $^3/_4$ pint in 100 gallons of water. To control mites, pecan nut casebearers and pecan leaf casebearers, use $^3/_4$ pint in 100 gallons of water. To control black and yellow pecan aphids, fall webworms and twig girdlers, use 1 quart per 100 gallons of water. Do not use more than $5^3/_8$ quarts of this product per acre.

WALNUTS—To control aphids, lecanium scales and walnut husk flies, use $^1/_2$ pint in 100 gallons of water. Do not use more than $2^1/_2$ gallons of this product per acre.

VEGETABLES

DO NOT APPLY WITHIN 21 DAYS OF HARVEST ON THE FOLLOWING CROPS UNLESS OTHERWISE SPECIFIED.

CELERY—To control aphids, mites, celery worms and tarnished plant bugs, use 1 pint per acre. To control leaf miners, whiteflies and leafhoppers, use $1^1/_2$ pints per acre, but do not use within 30 days of harvest.

ENDIVE—To control green peach aphids and alfalfa loopers, use 1 pint per acre.

LETTUCE (Leaf and Bibb)—For aphids, armyworms, cabbage loopers, imported cabbageworms, banded cucumber beetles and Lygus bugs, use $^1/_2$ to 1 pint per acre. To control six-spotted leafhoppers, use $^3/_4$ pint per acre. For harlequin bugs and vegetable weevils, use 1 pint per acre. At the $^1/_2$ pint rate, harvest can be made within 14 days of application.

OKRA—To control leaf miners and spider mites, use $^1/_2$ to 1 pint per acre. For aphids and stink bugs, use 1 pint per acre.

SWISS CHARD—To control aphids and serpentine leaf miners, use 1 pint per acre.

DO NOT APPLY WITHIN 15 DAYS OF HARVEST ON THE FOLLOWING CROPS UNLESS OTHERWISE SPECIFIED.

BEANS—For control of bean leaf beetles and two-spotted mites, use $^1/_2$ pint per acre. Use $^1/_2$ to $^3/_4$ pint per acre to control thrips and lima pod borers. To control stink bugs, plant bugs, Mexican bean beetles, leaf rollers, leaf miners, leafhoppers, aphids, red spider mites and armyworms use 1 pint per acre.

BEETS—To control flea beetles and leaf miners, use $^3/_5$ pint per acre. For aphids, blister beetles and webworms, use 1 pint per acre. If greens are used for food, do not use within 21 days of harvest.

BLACKEYED PEAS—To control aphids, leaf miners, bean leaf rollers and stink bugs, use 1 pint per acre.

CARROTS—To control leaf miners, use $^1/_2$ to $^3/_4$ pint per acre. To control leafhoppers, use $^3/_4$ pint per acre. Use 1 pint per acre to control aphids, vegetable weevils, stink bugs and petrobia mites. To control rust fly maggots (first brood), mix 1 pint with 100 gallons per acre and dribble into furrow at planting time. To control rust fly maggots (second brood), use 1 pint per acre as a foliage spray. Do not feed tops.

CUCUMBERS—For squash vine borers, aphids, cucumber beetles, leaf miners, pickleworms, mites and thrips, use $^1/_2$ to 1 pint per acre. For squash bugs, stink bugs, flea beetles and leafhoppers, use 1 pint per acre. Do not apply prior to vining.

EGGPLANT—To control thrips, leaf miners, blister beetles, and flea beetles, use $^1/_2$ to $^3/_4$ pint per acre. To control Colorado potato beetles, use $^3/_4$ pint per acre. To control spider mites and lace bugs, use $^3/_4$ to 1 pint per acre. To control aphids, whiteflies and stink bugs, use 1 pint per acre.

GARLIC—To control onion thrips, use $^1/_2$ pint per acre. To control leaf miners and petrobia mites, use 1 pint per acre.

ONIONS—To control onion thrips, use $^1/_2$ pint per acre. To control onion maggot flies, use $^3/_4$ pint per acre. To control aphids, stink bugs, leaf miners, and petrobia mites, use 1 pint per acre. To control brown wheat mites, use $1^1/_4$ pints per acre.

PEPPERS—To control thrips, use $^7/_{16}$ to $^1/_2$ pint per acre. To control aphids, leaf miners and western potato flea beetles, use 1 pint per acre.

RADISHES—To control aphids, false chinch bugs and harlequin bugs, use $^1/_2$ to 1 pint per acre. To control cabbage loopers and diamondback moths, use 1 pint per acre.

SPINACH—To control aphids, leaf miners, armyworms, cabbage loopers, vegetable weevils, harlequin bugs, seed corn maggots, crown mites and leafhoppers, use 1 pint per acre.

SQUASH—To control cucumber beetles, aphids, stink bugs, melonworms, pickleworms, climbing cutworms, serpentine leaf miners and squash vine borers, use $^1/_2$ to 1 pint per acre. To control squash bugs, flea beetles and leafhoppers, use 1 pint per acre.

SWEET POTATOES—To control aphids, spider mites, leafhoppers and stink bugs, use 1 pint per acre. To control serpentine leaf miners and morningglory leaf miners, use 1 to $1^1/_2$ pints per acre.

DO NOT APPLY WITHIN 12 DAYS OF HARVEST ON THE FOLLOWING CROPS.

SWEET CORN—To control corn earworms feeding in the bud, fall armyworms, aphids and silkflies, use $^1/_2$ pint per acre. To control sap beetles and spider mites, use 1 pint per acre. To control chinch bugs, use $1^1/_2$ pints per acre.

DO NOT APPLY WITHIN 10 DAYS OF HARVEST ON THE FOLLOWING CROPS UNLESS OTHERWISE SPECIFIED.

PEAS—To control aphids, pea weevils, spider mites, stink bugs, thrips, armyworms, climbing cutworms, leaf miners, alfalfa loopers and celery loopers, use 1 pint per acre. If vines are to be used for forage, do not harvest for 15 days after treatment.

PUMPKINS—To control aphids, cucumber beetles, climbing cutworms, squash bugs and squash vine borers, use $^1/_2$ to 1 pint per acre.

COLLARDS, KALE AND MUSTARD GREENS—To control aphids, leaf miners, armyworms, cabbage loopers, vegetable weevils, harlequin bugs, seed corn maggots, crown mites and leafhoppers, use 1 pint per acre. On fall and winter crops, do not use within 15 days of harvest.

TOMATOES—To control blister beetles, use $^1/_2$ pint per acre. To control hornworms, leafhoppers and psyllids, use $^3/_4$ pint per acre. For aphids, leaf miners, whiteflies, armyworms, grasshoppers, mites, tomato russet mites, leaf-footed bugs, stink bugs, loopers and plant bugs, use 1 pint per acre.

TURNIPS—To control aphids, cabbage webworms, climbing cutworms, vegetable weevils, false chinch bugs and harlequin bugs, use $^1/_2$ to 1 pint per acre. To control cabbage loopers, use 1 pint per acre. If greens are used for food, do not apply within 21 days of harvest.

DO NOT APPLY WITHIN 7 DAYS OF HARVEST ON THE FOLLOWING CROPS UNLESS OTHERWISE SPECIFIED.

ARTICHOKES—To control artichoke plume moths, use 1 quart per acre.

CABBAGE AND COLE CROPS (Broccoli, Brussels Sprouts, Cauliflower)—To control aphids, thrips, diamondback moth larvae, imported cabbageworms, cabbage loopers and armyworms, use $^1/_2$ to 1 pint per acre. To control harlequin bugs, vegetable weevils, climbing cutworms and flea beetles, use 1 pint per acre. Rates above $^1/_2$ pint should not be applied to cabbage closer than 10 days until harvest.

KOHLRABI—To control aphids, use 1 pint per acre.

LETTUCE (Head)—To control aphids, cabbage loopers, imported cabbageworms, banded cucumber beetles, Lygus bugs, webworms and armyworms up to third instar, use $^1/_2$ to 1 pint per acre. To control six-spotted leafhoppers, use $^3/_4$ pint per acre. For harlequin bugs, vegetable weevils and leaf miners, use 1 pint per acre. To control garden symphylans, broadcast $1^1/_4$ gallons per acre just prior to planting and thoroughly incorporate into upper 6 to 9 inches of soil.

MELONS—For squash vine borers, leaf miners and false chinch bugs, use $^1/_2$ pint per acre. For aphids, leafhoppers, cucumber beetles, pickleworms and mites, use $^1/_2$ to 1 pint per acre. To control thrips, squash bugs and stink bugs, use 1 pint per acre.

RUTABAGAS—To control aphids, cabbage loopers and climbing cutworms, use 1 pint per acre.

DO NOT APPLY WITHIN 5 DAYS OF HARVEST ON THE FOLLOWING CROP.

POTATOES—To control aphids, blister beetles, Colorado potato beetles, leaf miners, mites, plant bugs, potato psyllid, thrips, vegetable weevils and grasshoppers, use $^1/_2$ to 1 pint per acre. For armyworms, cabbage loopers, and climbing cutworms, use $^3/_4$ pint per acre. For leafhoppers, stink bugs and flea beetles, use 1 pint per acre.

MISCELLANEOUS

CABBAGE—For application to cabbage grown for seed only to control cabbage seed pod weevils, use 1 quart per acre.

CHRISTMAS TREES—To control aphids and mites, use $^1/_2$ pint per 100 gallons of water.

HOPS—For control of hop aphids, use 1 to $1^3/_5$ pints per acre. For spider mites, use $1^3/_5$ pints per acre. Do not apply within 15 days of harvest.

SAFFLOWER—To control aphids, Lygus bugs and grasshoppers, use 1 pint per acre. Do not use parathion after flowering.

SOIL INSECTS

Keep all persons and animals out of treated areas for 48 hours.

WIREWORMS—To control on:

Beans	Corn	Rutabagas
Beets	Endive	Soybeans
Broccoli	Egg Plant	Sugar Beets
Brussels Sprouts	Escarole	Sugarcane
Cabbage	Kale	Sweet Corn
Cantaloupe	Lettuce	Sweet Potatoes
Carrots	Onions	Tomatoes
Cauliflower	Peas	Turnips
Celery	Peppers	Watermelon

Broadcast 3 qt. to 1 gal. per acre on soil before planting and thoroughly work into upper 4 to 9 inches.

WIREWORMS—To control on: Tobacco

Broadcast 2 qt. per acre on soil at least 3 weeks before planting and work into top 6 to 9 inches.

WIREWORMS—To control on: Potatoes

Broadcast 1 to 1½ gal. per acre on soil before planting and thoroughly work into upper 4 to 9 inches.

CORN ROOTWORMS—To control on: Peanuts

Apply 2 to 2½ qt. per acre as a row soil treatment at planting or pegging time, work lightly into soil.

GARDEN SYMPHYLAN—To control on:
Beans Lettuce Sugar Beets
Corn Potatoes Tomatoes

Apply 5 qt. per acre to soil surface before planting time and thoroughly work into upper 6 to 9 inches.

CUTWORMS—To control on: Corn, Cucumbers

Broadcast 3 qt. to 1 gal. per acre before planting and thoroughly work into upper 1 to 3 inches.

WHITE GRUBS—To control on: Corn, Soybeans, Sugar Beets

Broadcast 3 qt. to 1 gal. per acre before planting and thoroughly work into upper 4 to 6 inches.

NOTE: Consult the State Agricultural Extension Service or Experiment Station concerning specific usage, dosages and methods of application.

NIRAN® M-4

INSECTICIDE

RESTRICTED USE PESTICIDE

For retail sale to and use only by Certified Applicators or persons under their direct supervision and only for those uses covered by the Certified Applicator's certification.

Emulsifiable Insecticide for Controlling Certain Insects on the Listed Field, Forage, Fruit and Vegetable Crops.

ACTIVE INGREDIENTS:	
* O,O-dimethyl O-p-nitrophenyl phosphorothioate	43.3%
* Related compounds	2.3%
Aromatic petroleum derivative solvent	47.9%
INERT INGREDIENTS	6.5%
TOTAL	100.0%

* Methyl parathion and related compounds equivalent to 4.0 lbs. per gallon.

EPA Est. 524.AL-1 EPA Reg. No. 524-128

869.11-000.34/53 AN

DANGER–POISON

Keep Out of Reach of Children

NOT FOR HOME USE

COMPLETE DIRECTIONS FOR USE

Read the entire label before using this product.

Use only according to label instructions. Read "LIMIT OF WARRANTY AND LIABILITY" before buying or using. If terms are not acceptable return at once unopened.

LIMIT OF WARRANTY AND LIABILITY

This company warrants that this product conforms to the chemical description on the label and is reasonably fit for the purposes set forth in the Complete Directions for Use label booklet ("Directions") when used in accordance with those Directions under the conditions described herein. NO OTHER EXPRESS OR IMPLIED WARRANTY IS MADE AND THE IMPLIED WARRANTY OF MERCHANTABILITY AND FITNESS FOR PARTICULAR PURPOSE IS EXCLUDED. This warranty is also subject to the conditions and limitations stated herein.

Buyer and all users shall promptly notify this company of any claims whether based in contract, negligence, strict liability, other tort or otherwise.

Buyer and all users are responsible for all loss or damage from use or handling which results from conditions beyond the control of this company, including, but not limited to, incompatibility with products other than those set forth in the Directions and application in any manner not explicitly set forth in the Directions.

THE EXCLUSIVE REMEDY OF THE USER OR BUYER, AND THE LIMIT OF THE LIABILITY OF THIS COMPANY OR ANY OTHER SELLER FOR ANY AND ALL LOSSES, INJURIES OR DAMAGES RESULTING FROM THE USE OR HANDLING OF THIS PRODUCT (INCLUDING CLAIMS BASED IN CONTRACT, NEGLIGENCE, STRICT LIABILITY, OTHER TORT OR OTHERWISE) SHALL BE THE PURCHASE PRICE PAID BY THE USER OR BUYER FOR THE QUANTITY OF THIS PRODUCT INVOLVED, OR, AT THE ELECTION OF THIS COMPANY OR ANY OTHER SELLER, THE REPLACEMENT OF SUCH QUANTITY OR, IF NOT ACQUIRED BY PURCHASE, REPLACEMENT OF SUCH QUANTITY. IN NO EVENT SHALL THIS COMPANY OR ANY OTHER SELLER BE LIABLE FOR ANY INCIDENTAL OR CONSEQUENTIAL DAMAGES.

The buyer and all users are deemed to have accepted the terms of this LIMIT OF WARRANTY AND LIABILITY which may not be varied by any verbal or written agreement.

PRECAUTIONARY STATEMENTS

HAZARDS TO HUMANS AND DOMESTIC ANIMALS

Keep Out of Reach of Children.

DANGER! PELIGRO!

PRECAUCION AL USUARIO: Si usted no lee inglés, no use este producto hasta que la etiqueta haya sido explicada ampliamente.

POISONOUS IF SWALLOWED, INHALED OR ABSORBED THROUGH SKIN. RAPIDLY ABSORBED THROUGH SKIN.

REPEATED EXPOSURE MAY, WITHOUT SYMPTOMS, BE INCREASINGLY HAZARDOUS.

Do not breathe vapors or spray mist.

Do not get in eyes, on skin, or on clothing.

POISON SIGNS (Symptoms)

Methyl parathion is a very dangerous poison. It rapidly enters the body on contact with all skin surfaces and eyes. Clothing wet with this material must be removed immediately. Exposed persons must receive prompt medical treatment or they may die.

Some of the signs and symptoms of poisoning are: Headaches, nausea, vomiting, cramps, weakness, blurred vision, pin-point pupils, tightness in chest, labored breathing, nervousness, sweating, watering of eyes, drooling or frothing of mouth and nose, muscle spasms and coma.

POISON

ATROPINE IS AN ANTIDOTE. CONSULT PHYSICIAN FOR EMERGENCY SUPPLY.

CALL A PHYSICIAN AT ONCE IN ALL CASES OF SUSPECTED METHYL PARATHION POISONING.

If symptoms or signs of poisoning include blurred vision, abdominal cramps, and tightness in the chest, do not wait for a doctor but give two atropine tablets (each 1/100 grain or 0.65 miligrams) at once. (One tablet to children under five years of age.)

FIRST AID TREATMENT

STATEMENT OF PRACTICAL TREATMENT

Call a doctor (physician), clinic or hospital immediately. Explain that the victim has been exposed to Methyl parathion and describe his condition. If doctor cannot come, take victim to clinic or hospital.

Move victim immediately from the area where Methyl parathion is present.

IF INHALED, remove victim to fresh air. If not breathing, immediately give artificial respiration, preferably mouth-to-mouth, and maintain until doctor sees victim. If breathing is difficult, give oxygen.

IF IN EYES OR ON SKIN, immediately flush with plenty of water for at least 15 minutes while removing contaminated personal clothing and shoes to avoid continued possible exposure to victim or persons in attendence.

IF SWALLOWED, induce vomiting immediately by giving two glasses of water and by touching back of throat with finger. DO NOT INDUCE VOMITING OR GIVE ANYTHING BY MOUTH TO AN UNCONSCIOUS PERSON. Have victim lie down and keep quiet.

NOTE TO PHYSICIAN

Antidote—administer atropine sulfate in large doses, TWO to FOUR mg. intravenously or intramuscularly as soon as cyanosis is overcome. Repeat at 5 to 10 minute intervals until signs of atropinization appear. 2-PAM chloride is a pharmacologic antidote and may be administered as an adjunct to, but not a substitute for, atropine which is a symptomatic and often life-saving antidote. DO NOT GIVE MORPHINE OR TRANQUILIZERS BECAUSE THESE ANTIDOTES MAY PRODUCE ADVERSE EFFECTS ACCENTUATING THE PHARMACOLOGIC EFFECT OF THIS PRODUCT. This product is a strong cholinesterase inhibitor affecting the central and peripheral nervous systems and producing cardiac and respiratory depression. At first sign of pulmonary edema, the patient should be given supplemental oxygen and treated symptomatically. Continued absorption of the poison may occur and fatal relapses have been reported after initial improvements; VERY CLOSE SUPERVISION OF THE PATIENT IS INDICATED FOR AT LEAST 48 HOURS.

WORK SAFETY RULES

REPEATED EXPOSURES TO CHOLINESTERASE INHIBITORS SUCH AS ARE CONTAINED IN THIS PRODUCT MAY, WITHOUT WARNING, CAUSE PROLONGED SUSCEPTIBILITY TO VERY SMALL DOSES OF ANY CHOLINESTERASE INHIBITOR.

Persons working with this product should have frequent blood tests of their cholinesterase levels. If the cholinesterase level falls below a critical point, no further exposure should be allowed until it has been determined by means of blood tests that the cholinesterase level has returned to normal. Before using this product, consult the State Agriculture Extension Service or Experimental Station for specific recommendations regarding such blood tests.

If handled indoors, provide mechanical exhaust ventilation.

Keep all unprotected persons and children away from treated area or where there is danger of drift.

Do not rub eyes or mouth with hands. If you feel sick in any way, STOP work and get help right away. See First Aid (Practical treatment) section.

USE ONLY WHEN WEARING THE FOLLOWING PROTECTIVE EQUIPMENT AND CLOTHING

(1) Wear water-proof pants, coat, hat, rubber boots or rubber overshoes. (2) Wear safety goggles. (3) Wear a pesticide respirator jointly approved by the Mining Enforcement and Safety Administration (formerly the U.S. Bureau of Mines) and by the National Institute for Occupational Safety and Health under the provisions of 30 CFR part II. (4) Wear heavy duty, natural rubber gloves.

IMPORTANT! Before removing gloves, wash them with soap and water. Always wash hands, face and arms with soap and water before smoking, eating or drinking.

AFTER WORK, take off all work clothes, and shoes. Shower, using soap and water. Wear only clean clothes when leaving job. Do not wear contaminated clothing. Wash protective clothing and protective equipment with soap and water after each use. Respirator should be cleaned and filler replaced according to instructions included with respirator.

REENTRY STATEMENT

Do not apply this product in such a manner as to directly or through drift expose workers or other persons. The area being treated must be vacated by unprotected persons.

Do not enter treated areas for 48 hours unless appropriate protective clothing is worn.

Because certain states may require more restrictive reentry intervals for various crops treated with this product, consult your State Department of Agriculture for further information.

Written or oral warnings must be given to workers who are expected to be in a treated area or in an area about to be treated with this product.

ORAL WARNINGS shall inform workers of areas or fields that may not be entered, without protective equipment and clothing specified on this label, during treatment or for 48 hours following treatment. Follow First Aid Treatment instructions shown on this label in case of accidental exposure.

When oral warnings are given, warnings shall be given in a language customarily understood by workers. Oral warnings must be given if there is reason to believe that written warnings cannot be understood by workers.

WRITTEN WARNINGS must include the following information.

DANGER—Area treated with Methyl parathion on (date). Do not enter without appropriate clothing for 48 hours. Follow First Aid Treatment instructions shown on this product label in case of accidental exposure.

ATTENTION

To avoid excessive residues of methyl parathion on food or forage crops always observe the statements found under "Directions for Use," limiting the time before harvest when methyl parathion may be applied.

Because this material is poisonous by skin contact, inhalation or swallowing it should not be used in such a manner or under weather conditions as will permit drift of the spray onto areas not intended to be sprayed.

Do not apply or allow drift to areas occupied by unprotected humans or beneficial animals.

ENVIRONMENTAL HAZARDS

This product is toxic to fish, shrimp, crab, birds, and other wildlife. Birds and other wildlife in treated areas may be killed. Keep out of lakes, streams, ponds, tidal marshes and estuaries. Shrimp and crab may be killed at application rates recommended on this label. Do not apply where these are important resources. Do not apply where weather conditions favor drift from areas treated. Do not contaminate water by cleaning of equipment, or disposal of wastes. Apply this product only as specified on this label.

This product is highly toxic to bees exposed to direct treatment or residues on crops. Protective information may be obtained from your Cooperative Agricultural Extension Service.

PHYSICAL OR CHEMICAL HAZARDS

Combustible: Do not use, pour, spill or store near heat or flame.

In case of:

FIRE, use water spray, foam, dry chemical or CO_2.

SPILL or LEAK, keep all unprotected persons away. Cover with absorbent such as soda ash, lime, clay or saw dust. Sweep up and place in a container for proper disposal. Wash area thoroughly with strong lye solution.

STORAGE AND DISPOSAL

Do not contaminate water, food or feed by storage or disposal.

STORAGE AND HANDLING

1. Handle container carefully so as to avoid damage and prevent leakage.
2. Do not use or store in or around the home.
3. Store container in a well ventilated place.
4. Empty container retains vapor and product residue. Observe all

labeled safeguards until container is cleaned, reconditioned or destroyed.

5. Keep out of reach of children and domestic animals.

DISPOSAL

Wastes of this pesticide are acutely hazardous. Improper disposal of excess pesticide, spray mixtures, or rinsate is a violation of federal law. If these wastes cannot be disposed of according to label instructions, contact your state pesticide or Environmental Control Agency, or the Hazardous Waste Representative at the nearest EPA Regional Office for guidance.

Triple rinse empty container with 5% caustic soda solution. Flush with water. Then offer for recycling or reconditioning, or puncture to prevent other use and dispose of in an approved landfill, or by other procedures approved by state and local authorities.

DIRECTIONS FOR USE

It is a violation of Federal Law to use this product in any manner inconsistent with its labeling.

Be sure to read the precautionary statements before using!

This product is designed for application after dilution with water and for use by trained operators using airplane or power ground equipment. The hazards and precautions for handling the product in this container are equally applicable to it after dilution with water for spray application. Add the concentrate to the spray tank while filling with water, and mix thoroughly either by means of a tank agitator or pump by-pass. For best results, thoroughly cover all surfaces to be treated with spray. Rates of application given below should not be exceeded. Never apply later than indicated to assure residue levels at harvest are below tolerances established by the Environmental Protection Agency.

Consult the State Agricultural Extension Service or Experimental Station for specific recommendations regarding application, dosage and timing of sprays. For application by ground equipment, add the desired amount of concentrate to sufficient water to apply at least 3 gallons of water per acre. For application by aircraft, add the amount of concentrate desired per acre to ½ to 3 gallons of water consistent with crop growth and good coverage. Greater quantities of water may be required to give sufficient coverage of orchard trees.

CEREAL

BARLEY, OATS, RYE AND WHEAT—For control of barley thrips, use ½ to ¾ pint per acre. For aphids (including greenbugs), armyworms, leafhoppers and winter grain mites, use ½ to 1½ pints per acre. For climbing cutworms and grasshoppers use ¾ to 1 pint per acre. For chinch bugs, false chinch bugs and Say's plant bugs, use 1½ pints per acre. For western or brown wheat mites and grasshoppers, use 1 pint per acre. Do not use more than ½ pint per acre within 15 days of harvest.

CORN—For control of armyworms, climbing cutworms, corn leaf aphids, corn rootworm adults, fall armyworms and stink bugs, use ½ pint per acre. For grasshoppers, use 1 pint per acre. Do not apply within 12 days of harvest.

RICE—For rice stink bugs, use ½ to 1 pint per acre. For grasshoppers, use 1 pint per acre. Do not apply within 15 days of harvest or within 5 days after propanil has been applied. Shrimp, crabs and crayfish may be killed. Do not apply where these are important resources.

FIELD AND FORAGE CROPS

ALFALFA AND CLOVER—For control of alfalfa weevil larvae, aphids (including spotted alfalfa aphids), climbing cutworms and grasshoppers use ½ to 1 pint per acre. For alfalfa caterpillars, adult alfalfa weevils, armyworms, clover leaf weevils and webworms, use 1 pint per acre. For Egyptian alfalfa weevils, leafhoppers, Lygus bugs and spider mites, use 1 to 2 pints per acre. Do not treat closer than 15 days before harvest, cutting, or grazing. In California do not use more than ¾ pint per acre.

COTTON—For control of thrips, use ¼ to ½ pint per acre. For cotton leafworms, use ¼ to ¾ pint per acre. For grasshoppers and fall armyworms use ½ pint per acre. For spider mites (does not control all species), use ½ to ¾ pint per acre. For control of fleahoppers, Lygus and other mirids, use ¼ pint to 1 quart per acre. For boll weevils, aphids and garden webworms, use ½ to 1 pint per acre. For false chinch bugs, use 1 pint per acre. For cutworms, use 1 to 2 pints per acre. For bollworms, salt-marsh caterpillars, armyworms, cotton leaf perforators and stink bugs, use 1 quart per acre. Applications should be made at 4 to 5 day intervals until control is obtained. At above dosages, application may be made up to the day before harvest, if harvest will be with mechanical pickers.

Also see "Tank Mixtures On Cotton" section.

CAUTION—When necessary to enter treated cotton fields within 24 hours after application, protective clothing should be worn. Do not apply within 5 days of handpicking cotton.

GRASS (Forage)—For range caterpillar, use 1 pint per acre. For control of armyworms, crested wheat bugs, false chinch bugs, grasshoppers and leafhoppers, use 1½ pints per acre. Do not apply within 15 days of harvest or grazing.

HOPS—For control of aphids and spider mites, use 1 quart per acre. Do not treat closer than 15 days before harvest.

SOYBEANS—For control of climbing cutworms and three-cornered alfalfa hoppers, use ½ pint per acre. For control of garden webworms and grasshoppers use ½ to 1 pint per acre. For control of aphids, blister beetles, Mexican bean beetles, stink bugs, two-spotted mites and velvet bean caterpillars, use 1 pint per acre. For bollworms, fall armyworms and green cloverworms, use 1 quart per acre. Do not apply closer than 20 days before harvest or grazing.

SUGAR BEETS—For aphids, armyworms, flea beetles, grasshoppers, leafhoppers, Lygus bugs, stink bugs and webworms, use ¾ pint per acre. Do not treat closer than 20 days before harvest, 60 days if tops are to be fed to livestock.

TOBACCO—For control of green peach aphids, use ½ pint per acre. For control of surface feeding or climbing cutworms, use ½ to 1 pint per acre.

CAUTION—when necessary to enter treated tobacco fields within 24 hours after application, protective clothing should be worn. Do not apply within 5 days of priming tobacco or within 15 days of cutting tobacco. Avoid contact with plant juices when priming or cutting tobacco.

VETCH—For control of aphids, armyworms, climbing cutworms, leafhoppers, Lygus bugs, spider mites and vetch bruchids, use 1 to 2 pints per acre. Do not treat closer than 15 days before harvest or grazing.

FRUIT

APPLY AT THE RATES INDICATED FOR THE FOLLOWING CROPS. MAKE NO APPLICATION CLOSER THAN 14 DAYS BEFORE HARVEST. APPLY AS FREQUENTLY AS NEEDED TO CONTROL INSECTS.

APPLES—For control of aphids, codling moths, plum curculio, scales and red-banded leaf rollers, use ½ pint to 1 quart per 100 gallons of water. For codling moths, plum curculio and red-banded leaf rollers, apply at petal fall and 3 to 4 applications, 8 to 14 days thereafter, to maintain control. Do not use more than 6 quarts of this product per acre.

GRAPES—For control of aphids, grape leaf folders and grape leafhoppers, use ½ pint to 1 quart per 100 gallons of water. Do not use more than 1½ pints of this product per acre.

PEACHES, PLUMS AND PRUNES—For control of aphids, peach tree borers, plum curculio and Oriental fruit moths, use ½ pint to 1 quart per 100 gallons of water. Spray tree trunks at time of moth flight to control peach tree borers. For plum curculio and Oriental fruit moths, apply at petal fall or shuck split and 3 to 4 applications, 8 to 14 days thereafter, to maintain control. Do not use more than 1 gallon of this product per acre.

STRAWBERRIES—For control of aphids, use 1 to 1½ pints per acre.

OIL SEED CROPS

SAFFLOWER—For aphids and Lygus bugs, use 1 pint per acre before flowering. Do not apply after flowering.

SUNFLOWER—For control of sunflower head moth, sunflower seed weevil and sunflower maggot, use 1 quart per acre. Make no more than 3 applications at 5-day intervals. Do not apply within 30 days of harvest.

VEGETABLES

APPLY AT THE RATES INDICATED FOR THE FOLLOWING CROPS. MAKE NO APPLICATIONS CLOSER THAN 5 DAYS BEFORE HARVEST.

POTATOES—For control of aphids, armyworms, false chinch bugs, flea beetles, leafhoppers, mites and shield bugs, use 1½ quarts per acre.

SWEET POTATOES—For control of aphids, armyworms, surface feeding and climbing cutworms, flea beetles, leafhoppers and mites, use ¾ quart per acre.

APPLY AT THE RATES INDICATED FOR THE FOLLOWING CROPS. MAKE NO APPLICATIONS CLOSER THAN 7 DAYS BEFORE HARVEST.

ARTICHOKE—For control of aphids, armyworms, surface feeding and climbing cutworms, flea beetles, leafhoppers, mites and artichoke plume moths, use 1 quart per acre.

BROCCOLI—For control of aphids, armyworms, flea beetles, leafhoppers and mites, use 1 to 3 pints per acre. For stink bugs, use 2 to 3 pints per acre. Rates above 1 pint should not be applied closer than 21 days before harvest.

BRUSSELS SPROUTS—For control of aphids, armyworms, flea beetles, leafhoppers and mites, use 1 to 3 pints per acre. For stink bugs, use 2 to 3 pints per acre. Rates above 1 pint should not be applied closer than 21 days before harvest.

CAULIFLOWER—For control of aphids, armyworms, flea beetles, leafhoppers and mites, use 1 to 3 pints per acre. For stink bugs, use 2 to 3 pints per acre. Rates above 1 pint should not be applied closer than 21 days before harvest.

KOHLRABI—For control of aphids, armyworms, flea beetles, leafhoppers and mites, use 1 to 3 pints per acre. For stink bugs, use 2 to 3 pints per acre. Rates above 1 pint should not be applied closer than 21 days before harvest.

RUTABAGAS—For control of aphids, armyworms, flea beetles, leafhoppers and mites, use 1 to 3 pints per acre. Rates above 1 pint should not be applied closer than 21 days before harvest.

APPLY AT THE RATES INDICATED FOR THE FOLLOWING CROPS. MAKE NO APPLICATIONS CLOSER THAN 10 DAYS BEFORE HARVEST.

CABBAGE—For control of aphids, armyworms, flea beetles, leafhoppers and mites, use 1 to 3 pints per acre. For stink bugs, use 2 to 3 pints per acre. Rates above 1 pint should not be applied closer than 21 days before harvest.

COLLARDS, KALE AND MUSTARD—For control of aphids, armyworms, flea beetles, leafhoppers and mites, use 1 to 3 pints per acre. For stink bugs, use 2 to 3 pints per acre. Rates above 1 pint should not be applied closer than 21 days before harvest.

PEAS—For control of aphids, armyworms, flea beetles, leafhoppers, Lygus bugs and mites, use 1 to 2 pints per acre. For cowpea curculio, surface feeding and climbing cutworms and stink bugs, use 2 pints per acre. Rates above 1 pint should not be applied closer than 15 days before harvest.

TOMATOES—For control of aphids, armyworms, flea beetles, leafhoppers, mites and psyllids, use 1 to 3 pints per acre. Rates above 1 pint should not be applied closer than 15 days before harvest.

APPLY AT THE RATES INDICATED FOR THE FOLLOWING CROPS. MAKE NO APPLICATIONS CLOSER THAN 15 DAYS BEFORE HARVEST.

BEANS (Dry and Green)—For control of aphids, armyworms, flea beetles, leafhoppers, Lygus bugs and mites, use 1 to 3 pints per acre. For cowpea curculio, surface feeding and climbing cutworms and stink bugs, use 2 to 3 pints per acre. Rates above 1 pint should not be applied closer than 21 days before harvest.

BEETS (Red)—For control of aphids, armyworms, flea beetles, leafhoppers, Lygus bugs, mites and stink bugs, use 1 quart per acre. If tops are to be used for food or feed do not apply closer than 21 days before harvest.

CARROTS—For control of aphids, armyworms, surface feeding and climbing cutworms, flea beetles, leafhoppers and mites, use 1 quart per acre. Do not feed tops.

CELERY—For control of aphids, armyworms, flea beetles, leafhoppers, Lygus bugs, mites and stink bugs, use 1 quart per acre.

CUCUMBERS—For control of aphids and two-spotted mites, use ½ pint per acre. Do not apply before vining.

ONIONS—For control of thrips, use ½ pint per acre.

PEPPERS—For control of aphids, armyworms, surface feeding and climbing cutworms, flea beetles, leafhoppers and mites, use 1 quart per acre.

SPINACH—For control of aphids, armyworms, flea beetles, leafhoppers, mites, seed corn maggots in crown and crown mites, use 1 to 2 pints per acre. Rates above 1 pint should not be applied closer than 21 days before harvest.

TURNIPS—For control of aphids, armyworms, flea beetles, leafhoppers, leaf miners and mites, use 1 to 1½ pints per acre. If tops are to be used for food or feed do not apply closer than 21 days before harvest.

APPLY AT THE RATES INDICATED FOR THE FOLLOWING CROPS. MAKE NO APPLICATIONS CLOSER THEN 21 DAYS BEFORE HARVEST.

LETTUCE—For control of aphids, armyworms, flea beetles, leafhoppers and mites, use 1 quart per acre.

TANK MIXTURES

This product may be tank-mixed with other insecticides to provide control of additional insect species.

Read and carefully observe the cautionary statements and all other information appearing on the labels of all the insecticides used.

When wettable powder, water soluble or flowable formulations are to be tank mixed with emulsifiable formulations:

1. Place a 20 to 35 mesh screen or wetting basket over filling port.
2. Through the screen, fill the sprayer tank one-half full with water and start agitation.
3. If a wettable powder is used, make a slurry with the water carrier, and add it SLOWLY through the screen into the tank. Continue agitation.
4. If a flowable formulation is used, pre-mix one part flowable with one part water. Add diluted mixture SLOWLY through the screen into the tank. Continue agitation.
5. Continue filling the sprayer tank with water and add any remaining amounts of liquid insecticides near the end of filling process. Maintain good agitation at all times until the contents of the tank are sprayed.
6. MAKE NO APPLICATIONS OF THESE TANK MIXTURES CLOSER TO HARVEST THAN THE LONGEST PRE-HARVEST INTERVAL SHOWN FOR EITHER PRODUCT.

For additional species of insects controlled on cotton by NIRAN® M4 see "COTTON" section.

Tank Mixture	Insect	Rate/Acre
NIRAN® M-4 plus EPN 4E	Thrips and plant bugs	⅛ to 1¼ pint* plus ⅛ to ¼ pint*
	Boll weevil	¼ to ½ pint** plus ¼ to ½ pint**

Tank Mixture	Insect	Rate/Acre
	Cotton bollworm and tobacco budworm	1 to 1½ pints*** plus 1 to 1½ pints***

* 0.125 to 0.25 lb. combined total a.i./acre
** 0.25 to 0.5 lb. combined total a.i./acre
*** 1.0 to 1.5 lb. combined total a.i./acre

NIRAN® M-4 plus toxaphene 6E	Armyworms, boll weevil, cotton bollworm, tobacco budworm, cotton leaf perforator, cotton leafworm, cutworms, flea beetles, grasshoppers, plant bugs and stink bugs	1 pint to 1 quart* plus 1⅓ to 2¾ pints*

* 1.5 to 3 lb. combined total a.i./acre

NIRAN® M-4 plus Fundal™ 4EC or Galecron™ 4E	Cotton bollworm, tobacco budworm, and boll weevil	1 pint to 1 quart* plus 4 fluid ounces (⅛ lb. a.i.)*

* 0.625 to 1.125 lb. combined total a.i./acre

NIRAN® M-4 plus Orthene™ 75S	Cotton bollworm, tobacco budworm (in AL, AR, FL, GA, LA, MS and TX)	1 quart* plus ⅔ lb. (½ lb. a.i.)*

* 1.5 lb. combined total a.i./acre

Tank Mixture	Insect	Rate/Acre
NIRAN® M-4 plus Lorsban™ 4E	Cotton bollworm, tobacco budworm and boll weevil (for use in all states except AZ and CA)	1 quart* plus 1 pint*

* 1.5 lb. combined total a.i./acre

NIRAN® M-4 plus Lannate™ L (1.8 lb. a.i./gal.)	Cotton bollworm, tobacco budworm, boll weevil, beet armyworm, cotton leaf perforator (This combination should be used as an occasional spray during early and mid-season, not more often than every 10 days. After boll load is set, up to three applications at 3-to-5-day intervals may be made).	1 to 2 pints* plus 1 to 1½ quarts*

NOTE: Registered for use east of the Rocky Mountains only.
* 0.95 to 1.675 lb. combined total a.i./acre

Tank Mixture	Insect	Rate/Acre
NIRAN® M-4 plus permethrin	Cotton bollworm and tobacco budworm	1 pint to 1 quart* plus 0.05 to 0.1 lb. a.i.*

* .55 to 1.1 lb. combined total a.i./acre

NIRAN® M-4 plus EPN 4E plus Lorsban™ 4E	Cotton bollworm and tobacco budworm	1 pint* plus 1 pint* plus ½ to 1 pint*

* 1.25 to 1.5 lb. combined total a.i./acre

NIRAN® M-4 plus toxaphene 6E plus Lorsban™ 4E	Cotton bollworm and tobacco budworm larvae	1 quart* plus 2¾ pints* plus 1 pint*

* 3.56 lb. combined total a.i./acre

NIRAN® M-4 plus EPN 4E plus Lannate™ L (1.8 lb./gal.)	Cotton bollworm and tobacco budworm larvae	1 pint* plus 1 pint* plus 1 to 1½ quarts*

* 1.45 to 1.675 lb. combined total a.i./acre

NIRAN® M-4 plus toxaphene 6E plus Lannate™ L (1.8 lb./gal.)	Cotton bollworm and tobacco budworm larvae and eggs	1 quart* plus 2¾ pints* plus ½ to 1 pint*

* 3.112 to 3.225 lb. combined total a.i./acre

Tank Mixture	Insect	Rate/Acre
NIRAN® M-4 plus EPN 4E plus Fundal™ 4E or Galecron™ 4E	Cotton bollworm, tobacco budworm and boll weevil	1 to 1½ pints* plus 1 to 1½ pints* plus 4 ounces (⅛ lb. a.i.)*

* 1.125 to 1.625 lb. combined total a.i./acre.

NIRAN® M-4 plus toxaphene 6E plus Fundal™ or Galecron™ 4E	Cotton bollworm, tobacco budworm and boll weevil	1 quart* plus 2¾ pints* plus 4 fluid ounces (⅛ lb. a.i.)*

* 3.19 lb. combined total a.i./acre
™ Fundal is a registered trademark of Nor-Am Agricultural Products Co.
™ Galecron is a registered trademark of Ciba-Geigy Corporation.
™ Lannate is a registered trademark of E.I. DuPont DeNemours & Co.
™ Lorsban is a registered trademark of Dow Chemical Co.
™ Orthene is a registered trademark of Chevron Chemical Co.

RAMROD®

FLOWABLE HERBICIDE

Selective Herbicide for Pre-emergence Weed Control in Field Corn, Hybrid Seed Corn, Silage Corn, Sweet Corn, Grain Sorghum (milo), Processing Green Peas, Pumpkins and Flax.

ACTIVE INGREDIENTS:
*Propachlor, (2-chloro-N-isopropylacetanilide) 42.0%
INERT INGREDIENTS 58.0%
TOTAL 100.0%

*Contains 480 grams/liter or 4 pounds/gallon or propachlor.

EPA Reg. No. 524-331-AA ©Monsanto Company 1983

WARNING

Keep Out of Reach of Children

Complete Directions for Use.

SHAKE WELL BEFORE USING

Read each of these sections of this label for essential product performance information.

Read the entire label.
Use only according to label instructions.
Not for reformulation or repackaging.

Read "LIMIT OF WARRANTY AND LIABILITY" before buying or using. If terms are not acceptable, return at once unopenend.

LIMIT OF WARRANTY AND LIABILITY

This company warrants that this product conforms to the chemical description on the label and is reasonably fit for the purposes set forth in the Complete Directions for Use label booklet ("Directions") when used in accordance with those Directions under the conditions described therein. NO OTHER EXPRESS OR IMPLIED WARRANTY OF FITNESS FOR PARTICULAR PURPOSE OR MERCHANTABILITY IS MADE. This warranty is also subject to the conditions and limitations stated herein.

Buyer and all users shall promptly notify this company of any claims whether based in contract, negligence, strict liability, other tort or otherwise.

Buyer and all users are responsible for all loss or damage from use or handling which results from conditions beyond the control of this company, including, but not limited to, incompatibility with products other than those set forth in the Directions, unusual weather (i. weather conditions which are outside the range considered normal at the application site and for the time period when the product is applied with the normal range being determined on the basis of the average range for the prior 40 years computed from the best available information, and ii. weather perils, including but not limited to hurricanes, tornadoes and floods) as well as weather considerations set forth in the Directions, application in any manner not explicitly set forth in the Directions, moisture conditions outside the moisture range specified in the Directions, or the presence of products other than those set forth in the Directions in or on the soil or crop.

THE EXCLUSIVE REMEDY OF THE USER OR BUYER, AND THE LIMIT OF THE LIABILITY OF THIS COMPANY OR ANY OTHER SELLER FOR ANY AND ALL LOSSES, INJURIES OR DAMAGES RESULTING FROM THE USE OR HANDLING OF THIS PRODUCT (INCLUDING CLAIMS BASED IN CONTRACT, NEGLIGENCE, STRICT LIABILITY, OTHER TORT OR OTHERWISE) SHALL BE THE PURCHASE PRICE PAID BY THE USER OR BUYER FOR THE QUANTITY OF THIS PRODUCT INVOLVED, OR, AT THE ELECTION OF THIS COMPANY OR ANY OTHER SELLER, THE REPLACEMENT OF SUCH QUANTITY, OR, IF NOT ACQUIRED BY PURCHASE, REPLACEMENT OF SUCH QUANTITY. IN NO EVENT SHALL THIS COMPANY OR ANY OTHER SELLER BE LIABLE FOR ANY INCIDENTAL OR CONSEQUENTIAL DAMAGES.

The buyer and all users are deemed to have accepted the terms of this LIMIT OF WARRANTY AND LIABILITY which may not be varied by any verbal or written agreement.

PRECAUTIONARY STATEMENTS

HAZARD TO HUMANS AND DOMESTIC ANIMALS

Keep out of reach of children.

WARNING
HARMFUL IF SWALLOWED OR ABSORBED THROUGH SKIN.
CAUSES EYE AND SKIN IRRITATION.
MAY CAUSE ALLERGIC SKIN REACTION.

Do not get in eyes, on skin, or on clothing.

Keep container closed.

While handling, wear rubber gloves.

Wash thoroughly after handling.

FIRST AID: IN CASE OF CONTACT, immediately flush eyes or skin with plenty of water for at least 15 minutes while removing contaminated clothing. For eyes, call a physician. Wash clothing before reuse.

IF SWALLOWED, drink 1 or 2 glasses of water and induce vomiting by touching back of throat with finger or blunt object. Do not induce vomiting or give anything by mouth to an unconscious person. Get medical attention.

ENVIRONMENTAL HAZARD

This product is toxic to fish and wildlife. Keep out of lakes, streams, or ponds. Do not apply where runoff is likely to occur. Do not apply when weather conditions favor drift from areas treated.

Apply this product only as specified on this label.

STORAGE AND DISPOSAL

STORE ABOVE 20°F. AND BELOW 100°F. TO KEEP PRODUCT IN SUSPENSION.

Below 20°F. the product may freeze. If frozen mix well to resuspend when temperature is above 40°F.

Above 100°F. the product may become thick and unsprayable.

Do not contaminate water, foodstuffs, seed or feed by storage or disposal.

This product, spray mixture or rinsate that cannot be used or chemically reprocessed should be disposed of according to applicable Federal, state or local procedures.

Triple rinse container. Then dispose of in a sanitary landfill, or by incineration if allowed by state and local authorities. Do not reuse container.

Consult federal, state or local disposal authorities for approved alternative procedures.

GENERAL INFORMATION

This product is recommended for control of many annual grasses and certain broadleaf weeds. For best results it should be applied to the soil surface before crop or weeds emerge. The seedbed should be fine, firm and free of clods and trash; application should be made within 5 days after the last tillage for weed control. Do not apply when conditions favor drift. Moisture, either as rain or irrigation, is required after application to activate this product. On coarse soils one-third inch of rainfall, and on medium and fine textured and/or high organic soils one-third to three-fourths inch of rainfall is required depending upon original soil moisture. Better results are obtained when moisture occurs within 10 days after application. Generally lower amounts of rainfall are required when soil moisture is high at time of planting.

NOTE: Use of this product in any manner not consistent with this label may result in injury to persons, animals or crops, or other unintended consequences. Keep container closed to prevent spills and contamination.

Flush sprayer with clean water after use.

Do not use this product or any of its labeled tank mixtures in fields where flood irrigation is to be practiced.

DO NOT apply in sprayable fluid fertilizer after crop emergence since crop injury may result.

Do not graze or feed sorghum (milo) forage, silage or fodder (stubble) from treated fields to dairy animals.

Read and carefully observe the cautionary statements and all other information appearing on labeling of all products used in mixtures.

MIXING AND SPRAYING INSTRUCTIONS

Always predetermine the compatibility of this product or its labeled tank mix with water carrier or sprayable fluid fertilizer carrier by mixing small proportional quantities in advance. See the Standard Sprayable Fluid Fertilizer Compatibility Test section on this label.

Mix this product or its labeled tank mix with the appropriate carrier as follows:

1. Place a 20 to 35 mesh sceeen or wetting basket over the filling port.
2. Through the screen, fill the sprayer tank one-half full with the appropriate carrier.
3. If a compability agent is necessary to improve mixing or to prevent the formation of undesirable and unsprayable gels or precipitates, while agitating add it to the carrier already in the tank. Use only compatibility agents cleared by Federal Food and Drug Administration for this use. Read and follow all directions for use, cautionary statements and all other information appearing on the selected compatibility agent. Check for adequate agitation.
4. Add the required amount of this product. Continue agitation.
5. If a labeled tank mix is being prepared, premix the other product in water. If a wettable powder is used, make a slurry. If flowable atrazine, propazine or Modown are used, premix one part flowable with one part water. Add to tank and continue agitation.
6. Complete filling the sprayer tank with carrier. Remove hose from tank immediately after filling to avoid siphoning back into carrier source. NOTE: Maintain good agitation at all times until the contents of the tank are sprayed, otherwise, settling which can lead to clogging may occur. Close tank valves and drain all lines when spraying is delayed.

Keep by-pass line on or near bottom of tank to minimize foaming. Screen size in nozzle or line strainers should be no finer than 50 mesh. Carefully select proper nozzle to avoid spraying a fine mist. For best results, use flat fan or flood jet nozzles. Check for even distribution of spray droplets. To reduce loss of the chemical due to drift of a fine mist, apply at pressures less than 40 psi.

STANDARD SPRAYABLE FLUID FERTILIZER COMPATIBILITY TEST

Herbicides may not always mix evenly throughout a sprayable fluid fertilizer or the components may separate too quickly to make their combined use of practical value. This may be due to certain characteristics of the different fluid fertilizers. A simple test using small quantities of the components is suggested to provide compatibility potential.

The test follows:

A. Materials Required For A Compatibility Test

1. Two one-quart jars with lid or stopper (marked "with" and "without").
2. Teaspoons (for a more exacting test, a five to ten milliliter (ml.) pipette or graduated cylinder is desirable).
3. A minimum of one quart of the sprayable fluid fertilizer to be tested.
4. The herbicide(s) to be mixed.
5. A compatability agent (the purpose of the adjuvant is to help keep the fertilizer and herbicide(s) in suspension, if this assistance is needed).

B. Procedure

1. Add one pint of the sprayable fluid fertilizer that will be used or other herbicide carrier to each jar marked "with" and "without".

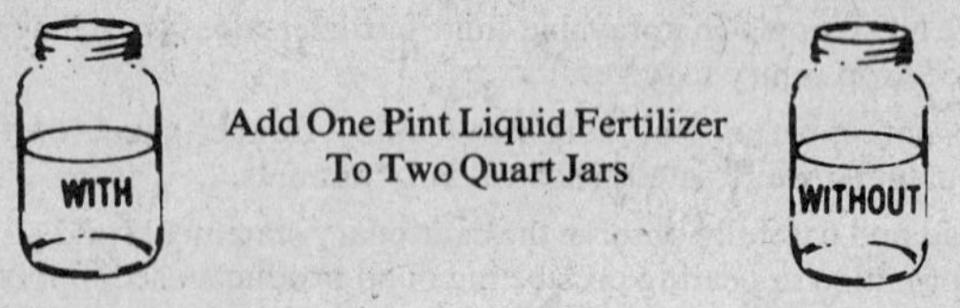

2. To the jar marked "with", add $\frac{1}{4}$ TEAspoon or 1.2 milliliters of a suitable compatibility agent; shake gently for five to ten seconds to mix. ($\frac{1}{4}$ TEAspoon is the equivalent of two pints per 100 gallons of liquid fertilizer.)

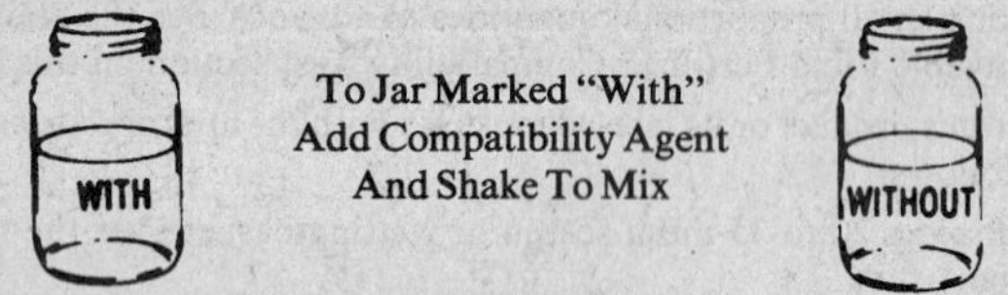

3. To each jar add the appropriate amount of herbicide(s). If more than one is used, add them separately with the wettable powders added first, flowables second and liquid last. Shake gently five to ten seconds after each addition.

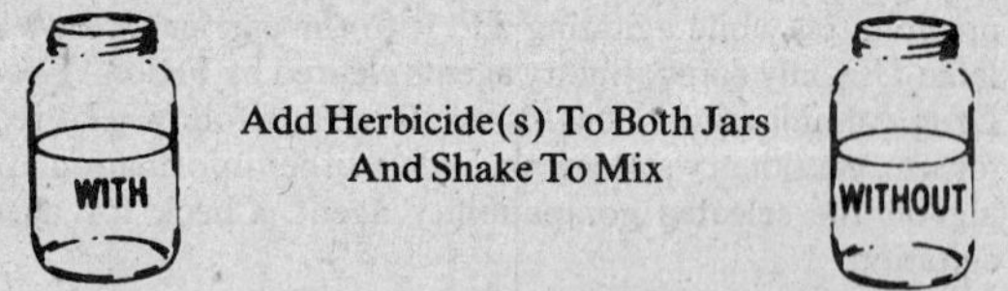

Amount To Be Added Per Pint Of Sprayable Fluid Fertilizer (Assuming Volume is 25 gal/A)

Herbicide	Rate/A		Level TEAspoons		Milliliters
This Product	2.5 qts	=	2.4	or	11.8
	3.0 qts	=	2.9	or	14.2
	3.5 qts	=	3.4	or	16.6
	4.0 qts	=	3.8	or	19.0
	4.5 qts	=	4.3	or	21.4
	5.0 qts	=	4.8	or	23.8
	5.5 qts	=	5.3	or	26.2
	6.0 qts	=	5.8	or	29.6

This compatibility test is designed for 25 gallons of spray per acre with the maximum labeled rate of herbicide. For changes in spray volume or herbicide concentration, make appropriate changes in the ingredients of the test. Regardless of spray volume, the amount of compatibility agent should be equal to the two or three pints (two pints = $\frac{1}{4}$ TEAspoon or 1.2 milliliters, three pints = $\frac{3}{8}$ TEAspoons or 1.8 milliliters per pint of sprayable fluid fertilizer) per 100 gallons of liquid fertilizer.

C. Observations And Decisions

1. If the herbicide(s) and the sprayable fluid fertilizer are compatible.
2. If a compatibility agent is necessary.

Five minutes after the final addition and mixing, observe both jars for the formation of large flakes, sludge, gels or other precipitates. Observe if the herbicide(s) cannot be physically mixed with the liquid fertilizer but remains as small oily particles in the solution.

If incompatibility in any form described above occurs in the jar "with" the compatibility agent added, the liquid fertilizer and the herbicide(s) should not be used together in the same spray tank.

If incompatibility as described above occurs in the jar "without" the adjuvant but not in the jar "with" adjuvant, the use of a compatibility adjuvant is recommended.

Both jars should be allowed to stand and be observed periodically for one-half hour. If the separate layers of liquid fertilizer and additives can be resuspended by shaking, commercial application is possible. Wettable powders will either settle to the bottom of the tank or jar, or float to the top, depending upon the density of the fertilizers.

If the herbicide(s) is compatible with fluid fertilizer in the foregoing test without having to use a compatibility agent, fluid fertilizer may be used for the premixing. If it is not compatible, the herbicide(s) should be premixed with water before adding to the spray tank.

APPLICATION INFORMATION

GROUND APPLICATION

Broadcast Treatment—Apply in 15 or more gallons of water or sprayable fluid fertilizer per acre according to this label. Do not apply during periods of gusty winds or winds in excess of 15 mph.

Band Treatment—Apply a broadcast equivalent rate and volume per acre. To determine these:

$$\frac{\text{Band width in inches}}{\text{Row width in inches}} \times \text{Broadcast RATE per acre} = \text{Band RATE per acre}$$

$$\frac{\text{Band width in inches}}{\text{Row width in inches}} \times \text{Broadcast VOLUME per acre} = \text{Band VOLUME per acre}$$

Mount spray equipment on the planter with nozzles behind press wheel and apply this product at planting time.

NOTE: For band application nozzle placement, refer to the nozzle manufacturer's specification as a guide for determining proper nozzle height.

AERIAL APPLICATION

Only RAMROD FLOWABLE or the labeled tank mixtures of RAMROD FLOWABLE plus atrazine and RAMROD FLOWABLE plus propazine may be applied by aerial spray equipment.

AVOID DRIFT—Do not apply when winds are gusty or in excess of 5 miles per hour or when other conditions, including lesser wind velocities, will allow drift to occur. Any resulting drift can damage nearby crops, vegetation or other areas on which treatment is not intended, and can cause application to be streaked or uneven, reducing the effectiveness of this product. In order to ensure uniform aerial application and to avoid overlapping and possible crop injury, it is recommended that 2 flagmen be used, one at each end of the field. Apply this product in 5 or more gallons of water per acre.

Coarse sprays are less likely to drift; therefore, do not use nozzles or nozzle configurations which dispense spray as fine spray droplets. Do not angle nozzles forward into the airstream and do not increase spray volume by increasing nozzle pressure.

CULTIVATION INFORMATION

As long as weeds are controlled, delay cultivation as long as possible.

If weed seedlings emerge, shallow cultivate or rotary hoe immediately to improve weed control.

If cultivation or rotary hoeing is necessary because of soil crusting or compaction, tillage should be shallow to minimize dilution and displacement of this herbicide or tank mixture.

WEED CONTROL WITH RAMROD FLOWABLE

When applied as directed, under the conditions described, this product CONTROLS the following weeds:

Barnyardgrass	*Echinochloa crus-galli*
Carpetweed	*Mollugo verticillata*
Crabgrass	*Digitaria* spp.
Foxtail (giant)	*Setaria faberi*
Foxtail (green)	*Setaria viridis*
Foxtail (yellow)	*Setaria lutescens*
Goosegrass	*Eleusine indica*
Panicum (fall)	*Panicum dichotomiflorum*
Pigweed	*Amaranthus* spp.
Purslane	*Portulaca oleracea*
Pusley (Florida)	*Richardia scabra*
Ryegrass (annual)	*Lolium* spp.

When applied immediately after planting, under the conditions described, this product at a rate of 5 to 6 quarts per acre on a broadcast basis REDUCES COMPETITION from the following hard-to-control weeds:

Buckwheat (wild)	*Polygonum convolvulus*
Groundsel	*Senecio vulgaris*
Lambsquarters	*Chenopodium album*
Ragweed (common)	*Ambrosia artemisiifolia*
Signalgrass (broadleaf)	*Brachiaria platyphylla*
Smartweed, Ladysthumb	*Polygonum persicaria*

Read each of these sections of this label for essential product performance information.

CROPS—DIRECTIONS FOR USE

It is a violation of Federal Law to use this product in any manner inconsistent with its labeling.

SHAKE WELL BEFORE USING

RAMROD FLOWABLE

Field Corn, Hybrid Seed Corn, Silage Corn and Sweet Corn

Pre-emergence Surface Treatment—For control of listed annual grasses and broadleaf weeds, apply this product in 15 or more gallons of water or sprayable fluid fertilizer per acre on the soil surface any time from immediately after planting and before the crop or weeds emerge. See the following table for recommended rates. In areas of heavy grass or broadleaf weed infestations or when listed hard-to-control weeds are present, use the higher rate in the recommended ranges.

Soil Texture	Broadcast Rate in Quarts per Acre: Less than 3% organic matter	Broadcast Rate in Quarts per Acre: 3% or more organic matter
COARSE Sand thru Sandy Loam	4 to 4.5	4.5 to 5
MEDIUM Loam thru Sandy Clay Loam	4.5 to 5	5.25 to 5.5
FINE Silty Clay Loam thru Clay	5.25 to 5.5	5.5 to 6

For Lister or Deep Furrow Planted Corn—Apply this product at the broadcast rate of 5.25 to 6 quarts per acre in 15 or more gallons of water. If heavy rains follow application, the treatment on the top and sides of the furrow may not provide as good weed control as that obtained in the bottom of the furrow.

Grain Sorghum (Milo)

Pre-emergence Surface Treatment—For control of listed annual grasses and broadleaf weeds apply 4 to 5 quarts of this product in 15 or more gallons of water or sprayable fluid fertilizer solutions per acre any time from immediately after planting and before the crop or weeds emerge. The higher rate should be used on high organic matter soils (greater than 3%) or in areas of heavy grass and broadleaf weed infestations or when listed hard-to-control weeds are present.

For Lister or Deep Furrow Planted Milo—Apply this product at the broadcast rate of 5 quarts per acre in 15 or more gallons of water. If heavy rains follow application, the treatment on the top and sides of the furrow may not provide as good weed and grass control as that obtained in the bottom of the furrow.

Peas, Green (for Processing)

Pre-emergence Surface Treatment—Apply 4 to 5 quarts in 15 or more gallons of water per acre immediately after planting or prior to emergence of crop or weeds.

Pumpkins

Pre-emergence Surface Treatment—For control of listed annual grasses and broadleaf weeds, apply this product in 15 or more gallons of water per acre on the soil surface any time from immediately after planting and before the crop or weeds emerge. See the following table for recommended rates. In areas of heavy grass or broadleaf weed infestations or when listed hard-to-control weeds are present, use the higher rate in the recommended ranges.

Soil Texture*	Broadcast Rate in Quarts Per Acre: 1 to 3% organic matter	Broadcast Rate in Quarts Per Acre: More than 3% organic matter
COARSE Sand thru Sandy Loam	4 to 4.5	4.5 to 5
MEDIUM Loam thru Sandy Clay Loam	4.5 to 5	5.25 to 5.5
FINE Silty Clay Loam thru Clay	5.25 to 5.5	5.5 to 6

* Do not apply to soil with less than 1% organic matter.

NOTE: Apply this product only once per growing season. Do not apply within 90 days of harvest.

FLAX

Pre-emergence Surface Treatment—For control of listed annual grasses and broadleaf weeds, apply 4 quarts of this product in 15 or more gallons of water or liquid fertilizer per acre. This product can be applied on the soil surface any time from immediately after planting and before the crop or weeds emerge.

RAMROD FLOWABLE PLUS ATRAZINE (Tank Mixture)

Field Corn, Hybrid Seed Corn, Silage Corn, Sweet Corn and Grain Sorghum

NOTE FOR GRAIN SORGHUM:

IN TEXAS USE ONLY ON THE FINE TEXTURE SOILS OF THE GULF COAST.

When applied as directed under the conditions described, this tank mixture CONTROLS these additional annual broadleaf weeds.

Kochia	*Kochia scoparia*
Mustard	*Brassica* spp.
Nightshade (black)	*Solanum nigrum*
Nightshade (hairy)	*Solanum sarachoides*
Smartweed (Pennsylvania)	*Polygonum pensylvanicum*

When applied as directed under the conditions described, this tank mixture REDUCES COMPETITION from these additional hard-to-control weeds.

Cocklebur	*Xanthium pensylvanicum*
Jimsonweed	*Datura stramonium*
Morningglory (annual)	*Ipomoea purpurea*
Signalgrass (broadleaf)	*Brachiaria platyphylla*
Velvetleaf, Buttonweed	*Abutilon theophrasti*

Pre-emergence Surface Treatment—For control of listed annual grasses and broadleaf weeds, apply this tank mixture in 15 or more gallons of water or sprayable fluid fertilizer per acre on the soil surface any time from immediately after planting and before either the crop or weeds emerge. See the following table for recommended rates. In areas of heavy grass or broadleaf weed infestations or when listed hard-to-control weeds are present, use the higher rate in recommended range.

	Broadcast Rate In Pounds Per Acre			
	Less than 3% organic matter		3% or more organic matter	
Soil Texture	RAMROD Flowable + (Quarts)	ATRAZINE* 4L (Quarts)	RAMROD Flowable + (Quarts)	ATRAZINE* 4L (Quarts)
COARSE** Sand thru Sandy Loam	2.5	1.0	3	1.2
MEDIUM Loam thru Sandy Clay Loam	3	1.2	3.4	1.4
FINE Silty Clay Loam thru Clay	3	1.2 to 1.4	3.4 to 4	1.4 to 1.6

* Use rates listed in this label when using atrazine 4L. Use equivalent rates when using atrazine 80W or 90% dry flowable formulations. One quart of atrazine 4L equals 1.25 pounds of atrazine 80W or 1.1 pounds of atrazine 90% dry flowable.

** Do not use on grain sorghum on coarse textured soils with less than 3% organic matter.

PLANT ONLY CORN, SORGHUM OR SOYBEANS THE YEAR FOLLOWING THE USE OF THIS TANK MIXTURE. IF SOYBEANS ARE TO BE PLANTED, THERE IS A POSSIBILITY OF CROP INJURY DUE TO CARRYOVER OF ATRAZINE. Do not plant soybeans in areas where furrow irrigation is practiced. Applications made on alkali soils or where cuts, fills or erosion have exposed calcareous or alkali subsoils may result in crop injury to milo.

Emerged Weed Control—For Corn Only—This tank mixture can be applied until weeds reach the 2-leaf stage, but prior to corn emergence. Applications beyond the 2-leaf stage will not provide satisfactory results.

Better results will be obtained if rainfall occurs within 5 to 7 days after application. If rainfall does not occur, use shallow cultivation to remove emerged weeds. Avoid cultivation methods which will displace the treated soil layer.

See the rate table in the Pre-emergence Treatment section of this label for the recommended rate for the soil texture and organic matter content to be treated.

For Lister or Deep Furrow Planted Corn—Apply this tank mixture at the higher broadcast rate listed in the rate table in the pre-emergence section of this label for the soil texture and organic matter content to be treated.

If heavy rains follow application, the treatment on the top and sides of the furrow may not provide as good weed control as that obtained in the bottom of the furrow.

RAMROD FLOWABLE PLUS PROPAZINE (Tank Mixture)

Grain Sorghum (Milo)

When applied as directed under the conditions described, this tank mixture CONTROLS these additional annual broadleaf weeds:

Kochia	*Kochia scoparia*
Lambsquarters	*Chenopodium album*

When applied as directed under conditions described, this tank mixture REDUCES COMPETITION from this additional annual broadleaf weed:

Morningglory (annual)	*Ipomoea purpurea*

Pre-emergence Surface Treatment—For control of listed annual grasses and broadleaf weeds, apply this tank mixture in 15 or more gallons of water per acre anytime after planting and before either crop or weeds emerge. Application must be made within 5 days of last tillage operation for best results. See the following table for recommended rates:

	Broadcast Rate Per Acre Quarts	
Soil Texture	RAMROD Flowable +	PROPAZINE 4L More than 1% organic matter**
COARSE Sand thru Sandy loam	2.5 to 3	.8 to 1*
MEDIUM THRU FINE Loam thru Clay	3	1

* Use the higher rates within the recommended range for this tank mixture when heavy weed infestations exist, harder-to-control species are present, or soil organic matter is greater than 3 percent.

** In coarse soil types with 1% or less organic matter, use the following rates of propazine 80W in the tank mixture:

Coarse Soils Percent Organic Matter	Quarts PROPAZINE 4L per acre***
Less than .75%	.4
.75% to 1%	.6

*** Use rates listed in this label when using propazine 4L. Use equivalent rates when using atrazine 80W or 90% dry flowable formulations. One quart 4L equals 1.25 pounds of 80W or 1.1 pounds of 90% dry flowable.

NOTE: Applications of this tank mixture to alkali soils where cuts, fills, or erosions have exposed calcareous or alkali subsoils may result in crop injury. Do not plant crops other than corn or sorghum for 18 months following application of this tank mixture due to possibility of crop injury from propazine carryover.

Whenever using reduced rates of propazine recommended for coarse soils with less than 1% organic matter, duration of weed control may be reduced. Weed control may be reduced if either an irrigation or a rainfall in excess of one inch occurs after application.

All applicable directions, restrictions and precautions on the container labels for RAMROD FLOWABLE herbicide and propazine are to be followed.

RAMROD FLOWABLE PLUS MODOWN™ 4F FLOWABLE (Tank Mixture)

Grain Sorghum (Milo)

When applied as directed under the conditions described, this tank mixture CONTROLS these additional annual broadleaf weeds:

Kochia	*Kochia scoparia*
Lambsquarters	*Chenopodium album*

When applied as directed under the conditions described, this tank mixture REDUCES COMPETITION from these additional annual broadleaf weeds:

Jimsonweed	*Datura stramonium*
Morningglory (annual)	*Ipomoea purpurea*
Sida (prickly), Teaweed	*Sida spinosa*
Velvetleaf, Buttonweed	*Abutilon theophrasti*

Pre-emergence Surface Treatment—For control of listed annual grasses and broadleaf weeds, apply this tank mixture in 15 or more gallons of water per acre any time after planting and before either crop or weeds emerge. Application must be made within 5 days of last tillage operation for best results. See the following table for recommended rates:

Soil Texture	Broadcast Rate In Quarts Per Acre* RAMROD Flowable	+	MODOWN 4F Flowable
COARSE Sand thru Sandy Loam	3		1.25
MEDIUM AND FINE Loam thru Clay	3 to 4		1.25 to 1.5

* Use the higher rate within the recommended range for this tank mixture when heavy weed infestations exist, harder-to-control species are present, or soil organic matter is greater than 3 percent.

™Trademark of Mobil Chemical Company

NOTE: Do not apply this tank mixture after milo has begun to crack the soil surface just prior to emergence. Rainfall which splashes treated soil onto the crop at time of emergence may result in crop injury from Modown. Do not rotate to crops other than corn, soybeans, grain sorghums, small grains or rice prior to one year following treatment with this tank mixture.

All applicable directions, restrictions and precautions on the container labels for RAMROD FLOWABLE herbicide and Modown herbicide are to be followed.

In case of an emergency involving this product, Call Collect, day or night, (314) 694-4000

RAMROD®AND ATRAZINE FLOWABLE

HERBICIDE MIXTURE

Selective Herbicide for Pre-emergence Weed Control in Grain Sorghum (milo), Field Corn, Hybrid Seed Corn, Silage Corn and Sweet Corn.

ACTIVE INGREDIENTS:	
Propachlor, (2-chloro-N-isopropylacetanilide	31.5%
Atrazine, [2-Chloro-4-(ethylamino)-6-(isopropylamino)-s-triazine]	10.0%
Related atrazine compounds	0.5%
INERT INGREDIENTS	58.0%
TOTAL	100.0%

* Contains 360 grams/litre or 3 pounds/gallon of propachlor and 120 grams/litre or 1 pound/gallon of atrazine and related atrazine compounds.

EPA Reg. No. 524-328-AA

WARNING

Keep Out of Reach of Children

Complete Directions for Use

SHAKE WELL BEFORE USING

Read each of these sections of this label for essential product performance information.

Read the entire label.

Use only according to label instructions.

Not for reformulation or repackaging.

Read "LIMIT OF WARRANTY AND LIABILITY" before buying or using. If terms are not acceptable, return at once unopened.

LIMIT OF WARRANTY AND LIABILITY

This company warrants that this product conforms to the chemical description on the label and is reasonably fit for the purposes set forth in the Complete Directions for Use label booklet ("Directions") when used in accordance with those Directions under the conditions described therein. NO OTHER EXPRESS OR IMPLIED WARRANTY OF FITNESS FOR PARTICULAR PURPOSE OR MERCHANTABILITY IS MADE. This warranty is also subject to the conditions and limitations stated herein.

Buyer and all users shall promptly notify this company of any claims whether based in contract, negligence, strict liability, other tort or otherwise.

Buyer and all users are responsible for all loss or damage from use or handling which results from conditions beyond the control of this company, including, but not limited to, incompatibility with products other than those set forth in the Directions, unusual weather (i. weather conditions which are outside the range considered normal at the application site and for the time period when the product is applied with the normal range being determined on the basis of the average range for the prior 40 years computed from the best available information, and ii. weather perils, including but not limited to hurricanes, tornadoes and floods) as well as weather considerations set forth in the Directions, application in any manner not explicitly set forth in the Directions, moisture conditions outside the moisture range specified in the Directions, or the presence of products other than those set forth in the Directions in or on the soil or crop.

THE EXCLUSIVE REMEDY OF THE USER OR BUYER, AND THE LIMIT OF THE LIABILITY OF THIS COMPANY OR ANY OTHER SELLER FOR ANY AND ALL LOSSES, INJURIES OR DAMAGES RESULTING FROM THE USE OR HANDLING OF THIS PRODUCT (INCLUDING CLAIMS BASED IN CONTRACT, NEGLIGENCE, STRICT LIABILITY, OTHER TORT OR OTHERWISE) SHALL BE THE PURCHASE PRICE PAID BY THE USER OR BUYER FOR THE QUANTITY OF THIS PRODUCT INVOLVED, OR, AT THE ELECTION OF THIS COMPANY OR ANY OTHER SELLER, THE REPLACEMENT OF SUCH QUANITY, OR, IF NOT ACQUIRED BY PURCHASE, REPLACEMENT OF SUCH QUANTITY. IN NO EVENT SHALL THIS COMPANY OR ANY OTHER SELLER BE LIABLE FOR ANY INCIDENTAL OR CONSEQUENTIAL DAMAGES.

The buyer and all users deemed to have accepted the terms of this LIMIT OF WARRANTY AND LIABLITY which may not be varied by any verbal or written agreement.

PRECAUTIONARY STATEMENTS

HAZARDS TO HUMANS AND DOMESTIC ANIMALS

KEEP OUT OF REACH OF CHILDREN.

WARNING!

HARMFUL IF SWALLOWED OR ABSORBED THROUGH SKIN.

CAUSES EYE AND SKIN IRRITATION.

MAY CAUSE ALLERGIC SKIN REACTION.

Do not get in eyes, on skin, or on clothing.

Keep container closed.

While handling, wear rubber gloves.

Wash thoroughly after handling.

FIRST AID: IN CASE OF CONTACT, immediately flush eyes or skin with plenty of water for at least 15 minutes while removing contaminated clothing. For eyes, call a physician. Wash clothing before reuse.

IF SWALLOWED, drink 1 or 2 glasses of water and induce vomiting by touching back of throat with finger or blunt object. Do not induce vomiting or give anything by mouth to an unconscious person. Get medical attention.

ENVIRONMENTAL HAZARD

This product is toxic to fish and wildlife. Keep out of lakes, streams, or ponds. Do not apply where runoff is likely to occur. Do not apply when weather conditions favor drift from areas treated. Apply this product only as specified on this label.

STORAGE AND DISPOSAL

STORE ABOVE 20°F. AND BELOW 100°F. TO KEEP PRODUCT IN SUSPENSION.

Below 20°F. the product may freeze. If frozen, mix well to resuspend when temperature is above 40°F.

Above 100°F. the product may become thick and unsprayable.

Do not contaminate water, foodstuffs, seed or feed by storage or disposal.

This product, spray mixture or rinsate that cannot be used or chemically reprocessed should be disposed of according to applicable Federal, state or local procedures.

Triple rinse container. Then dispose of in a sanitary landfill, or by incineration if allowed by state and local authorities. Do not reuse container.

Consult federal, state or local disposal authorities for approved alternative procedures.

GENERAL INFORMATION

This herbicide mixture is recommended for control of many annual grasses and certain broadleaf weeds. For best results it should be applied to the soil surface before crop or weeds emerge. The seedbed should be fine, firm and free of clods and trash; application should be made within 5 days after the last tillage for weed control. Do not apply when conditions favor drift. Some moisture, either as rain or irrigation, is required soon after application to activate this herbicide mixture. On coarse texture soil one-third inch of rainfall, and on medium and fine textured and/or high organic soils one-third to three-fourths inch of rainfall is required depending upon original soil moisture. Better results are obtained when moisture occurs within 10 days after application. Generally lower amounts of rainfall are required where soil moisture is high at time of planting.

NOTE: Use of this product in any manner not consistent with this label may result in injury to persons, animals or crops, or other unintended consequences. Keep container closed to prevent spills and contamination.

Flush sprayers with clean water after use.

Do not use this product in fields where flood irrigation is to be practiced.

DO NOT apply in sprayable fluid fertilizer after crop emergence since crop injury may result.

Do not gaze or feed sorghum (milo) forage, silage or fodder (stubble) from treated fields to dairy animals.

PLANT ONLY CORN, SORGHUM OR SOYBEANS THE YEAR FOLLOWING THE USE OF THIS TANK MIXTURE. IF SOYBEANS ARE TO BE PLANTED, THERE IS A POSSIBILITY OF CROP INJURY DUE TO CARRYOVER OF ATRAZINE. Do not plant soybeans in areas where furrow irrigation is practiced.

In Texas, on grain sorghum, use only on the fine textured soils of the Gulf Coast.

Applications made to sorghum growing on alkali soils or where cuts, fills or erosions have exposed calcareous or alkali subsoils may result in crop injury.

MIXING AND SPRAYING INSTRUCTIONS

Always predetermine the compatibility of this product with water carrier or sprayable fluid fertilizer carrier by mixing small proportional quantities in advance. See the Standard Sprayable Fluid Fertilizer Compatibility. Test section on this label.

Mix this product with the appropriate carrier as follows:

1. Place a 20 to 34 mesh screen or wetting basket over the filling port.
2. Through the screen, fill the sprayer tank one-half full with the appropriate carrier.
3. If a compatibility agent is necessary to improve mixing or to prevent the formation of undesirable and unsprayable gels or precipitates, while agitating add it to the carrier already in the tank. Use only compatibility agents cleared by Federal Food and Drug Administration for this use. Read and follow all directions for use, cautionary statements and all other information appearing on the selected compatibility agent. Check for adequate agitation.
4. Add the required amount of this product. Continue agitation.
5. Complete filling the sprayer tank with carrier. Remove hose from tank immediately after filling to avoid siphoning back into carrier source. Maintain good agitation at all times until the contents of the tank are sprayed, otherwise, settling which can lead to clogging may occur. Close tank valves and drain all lines when spraying is delayed.

Keep by-pass line on or near bottom of tank to minimize foaming. Screen size in nozzle or line strainers should be no finer than 50 mesh. Carefully select proper nozzle to avoid spraying a fine mist. For best results, use flat fan or flood jet nozzles. Check for even distribution of spray droplets. To reduce loss of the chemical due to drift of a fine mist, apply at pressures less than 40 psi.

STANDARD SPRAYABLE FLUID FERTILIZER COMPATIBILITY TEST

Herbicides may not always mix evenly throughout a sprayable fluid fertilizer or the components may separate too quickly to make their combined use of practical value. This may be due to certain characteristics of the different fluid fertilizers. A simple test using small quantities of the components is suggested to provide compatibility potential.

The test follows:

A. Materials Required For A Compatibility Test

1. Two one-quart jars with lid or stopper (marked "with" and "without").
2. Teaspoons (for a more exacting test, a five to ten milliliter (ml.) pipette or graduated cylinder is desirable).
3. A minimum of one quart of the sprayable fluid fertilizer to be tested.
4. The herbicides(s) to be mixed.
5. A compatability agent (the purpose of the adjuvant is to help keep the fertilizer and herbicide(s) in suspension, if this assistance is needed).

B. Procedure

1. Add one pint of the sprayable fluid fertilizer that will be used or other herbicide carrier to each jar marked "with" and "without".

   Add One Pint Liquid Fertilizer
To Two Quart Jars

2. To the Jar marked "with", add ¼ TEAspoon or 1.2 milliliters of a suitable compatibility agent; shake gently for five to ten seconds to mix. (¼ TEAspoon is the equivalent of two pints per 100 gallons of liquid fertilizer.)

   To Jar marked "With"
Add Compatibility Agent
And Shake To Mix

3. To each jar add the appropriate amount of this product. Shake gently for five to ten seconds.

   Add Herbicide(s) To Both Jars
And Shake To Mix

Amount To Be Added Per Pint Of Sprayable Fluid Fertilizer (Assuming Volume is 25 gal/A)

Herbicide	Rate/A		Level TEAspoons		Milliliters
This Product	3.5 qts.	=	3.4	or	16.6
	4.0 qts.	=	3.8	or	19.0
	4.5 qts.	=	4.3	or	21.4
	5.0 qts.	=	4.8	or	23.8
	5.5 qts.	=	5.3	or	26.2

This compatibility test is designed for 25 gallons of spray per acre with the maximum labeled rate of herbicide. For changes in spray volume or herbicide concentration, make appropriate changes in the ingredients of the test. Regardless of spray volume, the amount of compatibility agent should be equal to two or three pints (two pints = 1/4 TEAspoon or 1.2 milliliters, three pints = 3/8 TEAspoon or 1.8 milliliters per pint of sprayable fluid fertilizer) per 100 gallons of liquid fertilizer.

C. Observations And Decisions

1. If the herbicide(s) and the sprayable fluid fertilizer are compatible.
2. If a compatibility agent is necessary.

Five minutes after the final addition and mixing observe both jars for the formation of large flakes, sludge, gels or other precipitates. Observe if this product cannot be physically mixed with the liquid fertilizer but remains as small oily particles in the solution.

If incompatibility in any form described above occurs in the jar "with" the compatibility agent added, the liquid fertilizer and this product should not be used together in the same spray tank.

If incompatibility as described above occurs in the jar "without" the adjuvant but not in the jar "with" adjuvant, the use of a compatibility adjuvant is recommended.

Both jars should be allowed to stand and be observed periodically for one-half hour. If the separate layers of liquid fertilizer and additives can be resuspended by shaking, commercial application is possible. Wettable powders will either settle to the bottom of the tank or jar, or float to the top, depending upon the density of the fertilizers.

If this product is compatible with fluid fertilizer in the foregoing test without having to use a compatibility agent, fluid fertilizer may be used for the premixing. If it is not compatible, this product should be premixed with water before adding to the spray tank.

APPLICATION INFORMATION

GROUND APPLICATION

Broadcast Treatment—Apply in 15 or more gallons of water or sprayable fluid fertilizer per acre according to this label. Do not apply during periods of gusty winds or winds in excess of 15 mph.

$$\frac{\text{Band width in inches}}{\text{Row width in inches}} \times \frac{\text{Broadcast RATE}}{\text{per acre}} = \frac{\text{Band RATE}}{\text{per acre}}$$

$$\frac{\text{Band width in inches}}{\text{Row width in inches}} \times \frac{\text{Broadcast VOLUME}}{\text{per acre}} = \frac{\text{Band VOLUME}}{\text{per acre}}$$

Mount spray equipment on the planter with nozzles behind press wheel and apply this product at planting time.

NOTE: For band application nozzle placement, refer to the nozzle manufacturer's specification as a guide for determinig proper nozzle height.

AERIAL APPLICATION

AVOID DRIFT—Do not apply when winds are gusty or in excess of 5 miles per hour or when other conditions, including lesser wind velocities, will allow drift to occur. Any resulting drift can damage nearby crops, vegetation or other areas on which treatment is not intended, and can cause application to be streaked or uneven, reducing the effectiveness of this product. In order to ensure uniform aerial application and to avoid overlapping and possible crop injury, it is recommended that 2 flagmen be used, one at each end of the field. Apply this product in 5 or more gallons of water per acre.

Coarse sprays are less likely to drift; therefore, do not use nozzles or nozzle configurations which dispense spray as fine spray droplets. Do not angle nozzles forward into the airstream and do not increase spray volume by increasing nozzle pressure.

CULTIVATION INFORMATION

As long as weeds are controlled, delay cultivation as long as possible.

If weed seedlings emerge, shallow cultivate or rotary hoe to improve weed control.

If cultivation or rotary hoeing is necessary because of soil crusting or compaction, tillage should be shallow to minimize dilution and displacement of this product.

WEEDS CONTROLLED

When applied as directed, under the conditions described, this product CONTROLS the following weeds:

Barnyardgrass, Watergrass	*Echinochloa crus-galli*
Buckwheat (wild)	*Polygonum convolvulus*
Carpetweed	*Mollugo verticillata*
Crabgrass	*Digitaria* spp.
Foxtail (giant)	*Setaria faberi*
Foxtail (green)	*Setaria viridis*
Foxtail (yellow)	*Setaria lutescens*
Goosegrass	*Eleusine indica*
Groundsel	*Senecio vulgaris*
Kochia	*Kochia scoparia*
Lambsquarters	*Chenopodium album*
Mustard	*Brassica* spp.
Nightshade (black)	*Solanum nigrum*
Nightshade (hairy)	*Solanum sarachoides*
Panicum (fall)	*Panicum dichotomiflorum*
Pigweed	*Amaranthus* spp.
Purslane	*Portulaca oleracea*
Pusley (Florida)	*Richardia scabra*
Ragweed	*Ambrosia artemisiifolia*
Ryegrass (annual)	*Lolium* spp.
Smartweed	*Polygonum pensylvanicum*
	Polygonum persicaria

When applied immediately after planting, under the conditions described, this product at a rate of 4 to 5.5 quarters per acre REDUCES COMPETITION from the following hard to control weeds:

Buttonweed, Velvetleaf	*Abutilon theophrasti*
Cocklebur	*Xanthium pensylvanicum*
Jimsonweed	*Datura stramonium*
Morningglory (annual)	*Ipomea purpurea*
Signalgrass (broadleaf)	*Brachiaria platyphylla*

DIRECTIONS FOR USE

It is a violation of Federal Law to use this product in any manner inconsistent with its labeling.

SHAKE WELL BEFORE USING

FOR WEED CONTROL IN GRAIN SORGHUM (milo), FIELD CORN, HYBRID SEED CORN, SILAGE CORN AND SWEET CORN.

Pre-emergence Surface Treatment—For control of listed annual grasses and broadleaf weeds, apply this product in 15 or more gallons of water or sprayable fluid fertilizer per acre on the soil surface any time from immediately after planting and before the crop or weeds emerge. See the following table for recommended rates. In areas of heavy grass or broadleaf weed infestations or when listed hard to control weeds are present use the higher rate in the recommended ranges

Soil Texture	Broadcast Rate In Quarts Per Acre: Less than 3% organic matter	3% or more organic matter
COARSE* Sand thru Sandy Loam	3.5	4
MEDIUM Loam thru Sandy Clay Loam	4	4.5
FINE Silty Clay Loam thru Clay	4 to 4.5	4.5 to 5.5

* Do not use on grain sorghum on coarse texture soils with less than 3% organic matter.

Note for Corn only—This product can be applied until weeds reach the 2-leaf stage but prior to corn emergence. Applications beyond the 2-leaf stage will not provide satisfactory results. Better results will be obtained if rainfall occurs within 5 to 7 days after application. If rainfall does not occur, use shallow cultivation with a rolling cultivator or similar implement to remove emerged weeds. Avoid cultivation methods which will displace the treated soil layer. See the rate table in the pre-emergence treatment secton of this label for the recommended rate for the soil texture and organic matter content to be treated.

For Lister or Deep Furrow Planted Corn—Apply this product at the higher broadcast rate listed in the rate table in the pre-emergence section of this label for the soil texture and organic matter content to be treated.

If heavy rains follow application, the treatment on the top and sides of the furrow may not provide as good weed control as that obtained in the bottom of the furrow.

In case of an emergency involving this product, Call Collect, day or night, (314) 694-4000

RAMROD® 20 Granular

SELECTIVE HERBICIDE

For Pre-emergence Weed Control in Field Corn, Hybrid Seed Corn, Silage Corn, Sweet Corn, and Grain Sorghum (milo); and Soybeans (for seed only).

ACTIVE INGREDIENT:

2-chloro-N-isopropylacetanilide	20%
INERT INGREDIENTS:	80%
TOTAL	100%

EPA Reg. No. 524-152-AA EPA Est. 524-1A-1

WARNING

Keep Out of Reach of Children

READ "LIMIT OF WARRANTY AND LIABILITY" BEFORE BUYING OR USING. IF TERMS ARE NOT ACCEPTABLE, RETURN AT ONCE UNOPENED.

Use only according to label instructions.
Not for reformulation or repackaging.

KEEP OUT OF REACH OF CHILDREN

WARNING!

IRRITATING TO SKIN, EYES, NOSE AND THROAT. MAY CAUSE ALLERGIC SKIN REACTION.
Avoid contact with eyes and skin.
While handling, wear freshly laundered cotton or clean rubber gloves.
Avoid breathing dust.
Do not take internally.
In case of contact, flush skin or eyes with plenty of water; for eyes, get medical attention.
Wash thoroughly after handling.
Launder clothing before re-use.
Avoid contamination of seed, feed, and foodstuffs.

LIMIT OF WARRANTY AND LIABILITY

This company warrants that this product conforms to the chemical description on the label and is reasonably fit for the purposes set forth in the Complete Directions for Use ("Directions"), when used in accordance with those Directions under the conditions described therein. NO OTHER EXPRESS OR IMPLIED WARRANTY OF FITNESS FOR PARTICULAR PURPOSE OR MERCHANTABILITY IS MADE. This warranty is also subject to the conditions and limitations stated herein.

Buyer and all users shall promptly notify this company of any claims whether based in contract, negligence, strict liability, other tort or otherwise.

Buyer and all users are responsible for all loss or damage from use or handling which results from conditions beyond the control of this company, including but not limited to incompatibility with products other than those set forth in the Directions, unusual weather (i. weather conditions which are outside the range considered normal at the application site and for the time period when the product is applied with the normal range being determined on the basis of the average range for the prior 40 years computed from the best available information, and ii. weather perils, including but not limited to hurricanes, tornadoes and floods) as well as weather considerations set forth in the Directions, application in any manner not explicitly set forth in the Directions, moisture conditions outside the moisture range specified in the Directions, or the presence of products other than those set forth in the Directions in or on the soil or crop.

THE EXCLUSIVE REMEDY OF THE USER OR BUYER, AND THE LIMIT OF THE LIABILITY OF THIS COMPANY OR ANY OTHER SELLER FOR ANY AND ALL LOSSES, INJURIES OR DAMAGES RESULTING FROM THE USE OR HANDLING OF THIS PRODUCT (INCLUDING CLAIMS BASED IN CONTRACT, NEGLIGENCE, STRICT LIABILITY, OTHER TORT OR OTHERWISE) SHALL BE THE PURCHASE PRICE PAID BY THE USER OR BUYER FOR THE QUANTITY OF THIS PRODUCT INVOLVED, OR, AT THE ELECTION OF THIS COMPANY OR ANY OTHER SELLER, THE REPLACEMENT OF SUCH QUANTITY, OR, IF NOT ACQUIRED BY PURCHASE, REPLACEMENT OF SUCH QUANTITY, IN NO EVENT SHALL THIS COMPANY OR ANY OTHER SELLER BE LIABLE FOR ANY INCIDENTAL OR CONSEQUENTIAL DAMAGES.

The buyer and all users are deemed to have accepted the terms of this LIMIT OF WARRANTY AND LIABILITY which may not be varied by any verbal or written agreement.

GENERAL INFORMATION

RAMROD 20 is recommended for use as a pre-emergence herbicide to control most annual grasses and certain broadleaf weeds. It should be applied to the soil suface prior to weed or crop emergence. The soil surface should be freshly worked and free of weed top growth. The seed bed should be fine, firm and free of clods and trash.

Some moisture either as rain or irrigation is required soon after application to activate RAMROD 20. On light sandy loam soil one-third inch of rainfall and on heavy clay and or high organic soils one-third to three-fourths inch of rainfall is required depending upon original soil moisture. Best results are obtained when moisture occurs within 10 days after application. Generally lower amounts of rainfall are required where soil moisture is high at time of planting.

RAMROD 20 with lesser amounts of rainfall in some instances will allow the grass seedlings to emerge, but they will be severely stunted and with a subsequent cultivation they can be covered up.

If soil crusting becomes a problem and hoeing is necessary, hoe in the same direction the rows are planted.

Where deep lister planting is practiced only partial weed control can be expected on steep furrow sides. It is suggested that a 7 inch band be applied at the bottom of the furrow. Best results can be expected where the granules lie on an even flat surface not subject to shifting by wind or rain.

With broadcast treatment, as long as RAMROD 20 is controlling the grasses and weeds, there is no need for cultivation. For band treated fields, when first cultivation is made, set cultivator to throw soil into the row covering the band.

NOTE: Rates of application given on this label are general. Consult your Agricultural Experiment Station or Extension Service Weed Specialists concerning specific usage, dosages and methods of application.

Use of this product in any manner not consistent with this label may result in injury to persons, animals or crops, or other unintended consequences. Keep container closed to prevent spills and contamination. Do not contaminate fertilizers, insecticides, fungicides and seed with RAMROD.

This product is toxic to fish. Keep out of lakes, ponds, and streams. Do not contaminate water by cleaning of equipment or disposal of wastes. Apply this product only as specified on this label. Do not reuse container. Destroy when empty.

WEEDS CONTROLLED

Annual Ryegrass	*Lolium* spp.
Barnyardgrass (Watergrass)	*Echinochloa crus-galli*
Carpetweed	*Mollugo verticillata*
Crabgrass	*Digitaria* spp.
Fall Panicum	*Panicum dichotomiflorum*
Florida Pusley	*Richardia scabra*
Giant Foxtail (Wild Millet)	*Setaria faberi*
Green Foxtail	*Setaria viridis*
Yellow Foxtail	*Setaria glauca*
Goosegrass	*Eleusine indica*
Groundsel	*Senecia vulgaris*
Lambsquarters	*Chemopodium album*
Pigweed	*Amaranthus* spp.
Purslane	*Portulaca oleracea*
Ragweed	*Ambrosia artemisiifolia*
Sandbur	*Cenchrus* spp.
Shepherdspurse*	*Capsella bursa pastoris*
Smartweed*	*Polygonum persicaria* only
Wild Buckwheat*	*Polygonum convolvulus*

* At 25 to 30 pounds per acre granular RAMROD 20.

Granular RAMROD 20 WILL NOT CONTROL: Johnsongrass, Canada thistle, field bindweed, quackgrass, morningglory, cocklebur and bull nettle.

CALIBRATION AND APPLICATION INSTRUCTIONS

It is important that that applicator be calibrated properly to deliver the desired amount of RAMROD 20 to avoid applying too little or too much material. To calibrate, attach a cloth bag or strong plastic bag over the spreader plate or remove the delivery tube and attach a plastic nursing bottle over the spout. Operating over normal terrain to be planted, at planting speed, collect the granules for five hundred (500) linear feet and measure the VOLUME ounces collected. Use the following table to determine the correct volume of granules required for a 14-inch band at a given broadcast application rate.

Lbs. Per Acre To Be Applied (Broadcast Rate)	Volume To Be Collected (In 500 Feet)
20	$6\frac{1}{4}$ oz.
25	$7\frac{3}{4}$ oz.
30	$9\frac{1}{2}$ oz.

Tap plastic nursing bottle or other measuring container gently to settle the granules when taking a reading. Adjust the application equipment so that each spout dispenses the proper amount.

For broadcast (over-all treatment)—use broadcast applicators or fertilizer spreaders that can apply small amounts of granules evenly.

Fox band treatment—use applicators designed for this purpose. Calculate the amount of granules per acre needed for band treatment as follows:

$$\frac{\text{Band Width in Inches}}{\text{Inches Between Crop Rows}} \times \text{Recommended Over-all Rate} = \text{Pounds of RAMROD 20 needed per acre for band treatment}$$

For example: If 25 pounds per acre are recommended for over-all treatment and you wish to treat a band of 14 inches wide in crop rows spaced 38 inches apart:

$$\frac{14 \times 25}{38} = 9.2 \text{ pounds of granules needed per acre of crop for the band treatment.}$$

For application of granular RAMROD 20 do not rely on settings used for other granular herbicides.

CROP USE DIRECTIONS
FIELD CORN, HYBRID SEED CORN, SILAGE CORN AND SWEET CORN.

Broadcast (over-all treatment)—Apply 20 to 30 pounds of granular RAMROD 20 per acre on soil surface after planting and prior to emergence of either crop or weeds. See table below for recommended rates on various soil textures and levels of organic matter. In areas of heavy grass or broadleaf weed infestations use the higher rate shown for your soil texture.

Soil Texture	Broadcast Rates per Acre: Less than 3% organic matter	More than 3% organic matter
Sand to Sandy Loam	20 lbs.	25 lbs.
Silt Loam	22 lbs.	27 lbs.
Clay Loam to Clay	25 lbs.	30 lbs.

Band Treatment (Behind Press Wheel)—Apply granular RAMROD 20 on soil surface after planting and prior to emergence of either crop or weeds. Use the higher broadcast rate for higher organic soils, or in areas of heavy grass or broadleaf weed infestations. With the broadcast rate for your soil texture, the following table gives the pounds per acre needed for your row spacing.

Broadcast Rate/Acre	Band Treatments 20 lbs.	22 lbs.	25 lbs.	27 lbs.	30 lbs.
	14" Band on Various Row Spacings				
Row Spacings	lb./acre	lb./acre	lb./acre	lb./acre	lb./acre
40	7.0	7.7	8.8	9.5	10.5
38	7.4	8.1	9.2	9.9	11.0
36	7.8	8.6	9.7	10.5	11.7
30	9.3	10.3	11.7	12.6	14.0

FOR LISTER OR DEEP FURROW PLANTED CORN.

Broadcast (Over-all-treatment)—Apply granular RAMROD 20 at the broadcast rate of 25 to 30 pounds per acre.

Band Treatment (Behind Press Wheel)—For a 7 inch band treatment over the row (38 inch spacing), apply 4.6 to 5.5 pounds of granular RAMROD 20 per acre.

If heavy rains follow application, the treatment on the top and sides of the furrow may not provide as good weed and grass control as that obtained in the bottom of the furrow.

FOR FURROW IRRIGATED CORN.

Granular RAMROD 20 must be incorporated into the soil by harrow, finger weeder or rotary hoe for weed control on furrow irrigated land. Shallow incorporation of at least one-half inch but not more than 2 inches is most effective for weed control. Incorporation deeper than 2 inches will result in unsatisfactory weed control. If applied on very dry soil, irrigation is necessary to activate granular RAMROD 20. See the table above for recommended rates on various soil textures and levels of organic matter.

GRAIN SORGHUM (milo)
SOYBEANS (to be grown for seed only)

Broadcast (over-all treatment)—Apply 20 to 25 pounds of granular RAMROD 20 per acre on soil surface after planting and prior to emergence of either crop or weeds. In areas of heavy grass or broadleaf weed infestations and on high organic matter soils, use 25 pounds per acre.

Band Treatment (Behind Press Wheel)—Apply granular RAMROD 20 on the soil surface after planting and prior to emergence of either crop or weeds. Use the higher rate for high organic soils or in areas of heavy grass or broadleaf weed infestations.

Broadcast Rate/Acre	Band Treatments 20 lbs.	22 lbs.	25 lbs.
	14" Band on Various Row Spacings		
Row Spacings	lb./acre	lb./acre	lb./acre
40	7.0	7.7	8.8
38	7.4	8.1	9.2
36	7.8	8.6	9.7
30	9.3	10.3	11.7

NOTE: Do not graze or feed sorghum (milo) forage or silage from treated fields to dairy animals.

Do not use soybean seed for food, feed or oil purposes.
Do not graze or feed soybean forage from treated area.

RODEO®

HERBICIDE

For Broad-Spectrum Control of Emerged Weeds.
Complete Directions for Use in Aquatic Sites.
AVOID CONTACT WITH FOLIAGE, GREEN STEMS, OR FRUIT OF CROPS, DESIRABLE PLANTS AND TREES, SINCE SEVERE INJURY OR DESTRUCTION MAY RESULT.

ACTIVE INGREDIENT:	
* Isopropylamine salt of glyphosate	53.5%
INERT INGREDIENTS:	46.5%
TOTAL	100.0%

* Contains 648 grams per litre or 5.4 pounds of the active ingredient isopropylamine salt of N-(phosphonomethyl) glycine per U.S. gallon. Equivalent to 480 grams per litre or 4 pounds per U.S. gallon of the acid glyphosate.

EPA Reg. No. 524-343
®RODEO is a registered trademark of Monsanto Company
1985-1 892.38-000.06/53
Rodeo® herbicide is protected by U.S. Pat. No. 3,799,758 and U.S. Pat. No. 4,405,536.

In case of an emergency involving this product, Call Collect, day or night, (314) 694-4000.

CAUTION

Keep Out of Reach of Children

Read the entire label before using this product.
Use only acccording to label instructions.
NOT FOR REFORMULATION OR REPACKAGING.
Read "LIMIT OF WARRANTY AND LIABILITY" before buying or using.
If terms are not acceptable, return at once unopened.

LIMIT OF WARRANTY AND LIABILITY

This company warrants that this product conforms to the chemical description on the label and is reasonably fit for the purposes set forth in the Complete Directions for Use label booklet ("Directions") when used in accordance with those Directions under the conditions described therein. NO OTHER EXPRESS OR IMPLIED WARRANTY OF FITNESS FOR PARTICULAR PURPOSE OR MECHANTABILITY IS MADE. This warranty is also subject to the conditions and limitations stated herein.

Buyer and all users shall promptly notify this company of any claims whether based in contract, negligence, strict liability, other tort or otherwise.

Buyer and all users are responsible for all loss or damage from use or handling which results from conditions beyond the control of this company, including but not limited to incompatibility with products other than those set forth in the Directions, application to or contact with desirable vegetation, unusual weather (i. weather conditions which are outside the range considered normal at the application site and for the time period when the product is applied with the normal range being determined on the basis of the average range for the prior 40 years computed from the best available information, and ii, weather perils, including but not limited to hurricanes, tornadoes and floods) as well as weather considerations set forth in the Directions, application in any manner not explicitly set forth in the Directions, moisture conditions outside the moisture range specified in the Directions, or the presence of products other than those set forth in the Directions in or on the soil, crop or treated vegetation.

THE EXCLUSIVE REMEDY OF THE USER OR BUYER, AND THE LIMIT OF THE LIABILITY OF THIS COMPANY OR ANY OTHER SELLER FOR ANY AND ALL LOSSES, INJURIES OR DAMAGES RESULTING FROM THE USE OR HANDLING OF THIS PRODUCT (INCLUDING CLAIMS BASED IN CONTRACT, NEGLIGENCE, STRICT LIABILITY, OTHER TORT OR OTHERWISE) SHALL BE THE PURCHASE PRICE PAID BY THE USER OR BUYER FOR THE QUANTITY OF THIS PRODUCT INVOLVED, OR, AT THE ELECTION OF THIS COMPANY OR ANY OTHER SELLER, THE REPLACEMENT OF SUCH QUANTITY OR, IF NOT ACQUIRED BY PURCHASE, REPLACEMENT OF SUCH QUANTITY. IN NO EVENT SHALL THIS COMPANY OR ANY OTHER SELLER BE LIABLE FOR ANY INCIDENTAL OR CONSEQUENTIAL DAMAGES.

The buyer and all users are deemed to have accepted the terms of this LIMIT OF WARRANTY AND LIABILITY which may not be varied by any verbal or written agreement.

PRECAUTIONARY STATEMENTS

HAZARDS TO HUMANS AND DOMESTIC ANIMALS
Keep out of reach of children.

CAUTION!

MAY CAUSE EYE IRRITATION.

Avoid contact with eyes, skin or clothing.

FIRST AID: IF IN EYES, flush with plenty of water for at least 15 minutes. Call a physician.

IF ON SKIN, flush with water. Wash clothing before reuse.

ENVIRONMENTAL HAZARDS

Do not contaminate water by disposal of waste or cleaning of equipment.

In case of:

SPILL or LEAK, soak up and remove to a landfill.

PHYSICAL OR CHEMICAL HAZARDS

Spray solutions of this product should be mixed, stored and applied only in stainless steel, aluminum, fiberglass, plastic aand plastic-lined steel containers.

DO NOT MIX, STORE OR APPLY THIS PRODUCT OR SPRAY SOLUTIONS OF THIS PRODUCT IN GALVANIZED STEEL OR UNLINED STEEL (EXCEPT STAINLESS STEEL) CONTAINERS OR SPRAY TANKS. This product or spray solutions of this product react with such containers and tanks to produce hydrogen gas which may form a highly combustible gas mixture. This gas mixture could flash or explode causing serious personal injury if ignited by open flame, spark, welder's torch, lighted cigarette or other ignition source.

STORAGE AND DISPOSAL

Do not contaminate water, foodstuffs, seed or feed by storage and disposal.

STORAGE

STORE ABOVE 10°F. (−12°C.) TO KEEP FROM CRYSTALIZING.

Crystals will settle to the bottom. If allowed to crystalize, place in a warm room 68°F. (20°C.) for several days to dissolve and mix well before using.

DISPOSAL

Wastes resulting from the use of this product that cannot be used or chemically reprocessed should be disposed of in a landfill approved for pesticide disposal or buried on site in a safe place away from water supplies. All disposal should be in accordance with applicable Federal, state or local procedures.

Emptied container retains vapor and product residue. Observe all labeled safeguards until container is destroyed. Do not reuse container, destroy when empty.

Triple rinse container, then puncture and dispose of in a sanitary landfill, or by incineration, or, if allowed by state and local authorities, by burning. If burned, stay out of smoke.

GENERAL INFORMATION

This product, a water soluble liquid, mixes readily with water and nonionic surfactant to be applied as a foliage spray for the control or destruction of most herbaceous plants. It may be applied through most standard industrial or field type sprayers after dilution and thorough mixing with water and surfactant in accordance with label instructions.

This product moves through the plant from the point of foliage contact to and into the root system. Visible effects on most annual weeds occur within 2 to 4 days but on most perennial weeds may not occur for 7 days or more. Extremely cool or cloudy weather following treatment may slow the activity of this product and delay visual effects of control. Visible effects are a gradual wilting and yellowing of the plant which advances to complete browning of above ground growth and deterioration of underground plant parts.

Unless otherwise directed on this label, delay application until vegetation has emerged and reached the stages described for control of such vegetation under the "Weeds Controlled" section of this label.

Unemerged plants arising from unattached underground rhizomes or root stocks of perennials will not be affected by the spray and will continue to grow. For this reason best control of most perennial weeds is obtained when treatment is made at late growth stages approaching maturity.

Always use the higher rate of this product per acre within the recommended range when weed growth is heavy or dense.

Do not treat weeds under poor growing conditions such as drought stress, disease or insect damage, as reduced weed control may result. Reduced results may also occur when treating weeds heavily covered with dust.

Reduced control may result when applications are made to any weed or brush species that have been mowed, grazed, or cut, and have not been allowed to regrow to the recommended stage for treatment.

Rainfall or irrigation occurring within 6 hours after application may reduce effectiveness. Heavy rainfall or irrigation within 2 hours after application may wash the product off the foliage and a repeat treatment may be required.

This product does not provide residual weed control. For subsequent residual weed control, follow a label-approved herbicide program. Read and carefully observe the cautionary statements and all other information appearing on the labels of all herbicides used.

Buyer and all users are responsible for all loss or damage in connection with the use or handling of mixtures of this product or other materials that are not expressly recommended in this label. Mixing this product with herbicides or other materials not recommended on this label may result in reduced performance.

ATTENTION

AVOID DRIFT. EXTREME CARE MUST BE USED WHEN APPLYING THIS PRODUCT TO PREVENT INJURY TO DESIRABLE PLANTS AND CROPS.

Do not allow the herbicide solution to mist, drip, drift, or splash onto desirable vegetation since minute quantities of this product can cause severe damage or destruction to the crop, plants, or other areas on which treatment was not intended. The likelihood of plant or crop injury occurring from the use of this product is greatest when winds are gusty or in excess of 5 miles per hour or when other conditions, including lesser wind velocities, will allow spray drift to occur. When spraying avoid combinations of pressure and nozzle type that will result in splatter or fine particles (mist) which are likely to drift. AVOID APPLYING AT EXCESSIVE SPEED OR PRESSURE.

NOTE: Use of this product in any manner not consistent with this label may result in injury to persons, animals or crops, or other unintended consequences. When not in use, keep container closed to prevent spills and contamination.

Clean sprayer and parts immediately after using this product by thoroughly flushing with water.

MIXING AND APPLICATION INSTRUCTIONS

APPLY THESE SPRAY SOLUTIONS IN PROPERLY MAINTAINED AND CALIBRATED EQUIPMENT CAPABLE OF DELIVERING DESIRED VOLUMES. HAND GUN APPLICATIONS SHOULD BE PROPERLY DIRECTED TO AVOID SPRAYING DESIRABLE PLANTS. NOTE: REDUCED RESULTS MAY OCCUR IF WATER CONTAINING SOIL IS USED such as WATER FROM PONDS AND UNLINED DITCHES.

MIXING

This product mixes readily with water. Mix spray solutions of this product as follows: fill the mixing or spray tank with the required amount of water while adding the required amount of this product (see "Directions for Use" and "Weeds Controlled" sections of this label). Near the end of the filling process, add the required surfactant and mix well. Remove hose from tank immediately after filling to avoid siphoning back into the water source. During mixing and application, foaming of the spray solution may occur. To prevent or minimize foam, avoid the use of mechanical agitators, place the filling hose below the surface of the spray solution, terminate by-pass and return lines at the bottom of the tank and if needed use an approved anti-foam or defoaming agent.

Keep by-pass line on or near bottom of tank to minimize foaming. Screen size in nozzle or line strainers should be no finer than 50 mesh. Carefully select correct nozzle to avoid spraying a fine mist. For best results with conventional ground application equipment, use flat fan nozzles. Check for even distribution of spray droplets.

THE USE OF A NONIONIC SURFACTANT APPROVED FOR THE SITE OF THE DESIRED APPLICATION IS REQUIRED FOR USE WITH THIS PRODUCT. Use ¼ to ½ percent surfactant by total spray volume. Carefully observe all cautionary statements and other information appearing on the surfactant label.

APPLICATION EQUIPMENT AND TECHNIQUES

AERIAL EQUIPMENT

DO NOT APPLY THIS PRODUCT BY AIR IN CALIFORNIA.

Use the recommended rates of this product and surfactant in 3 to 20 gallons of water per acre as a broadcast spray. See "WEEDS CONTROLLED" section of this label for specific rates. Aerial applications of this product may only be made as specifically recommended on this label.

AVOID DRIFT—DO NOT APPLY DURING INVERSION CONDITIONS, WHEN WINDS ARE GUSTY, OR UNDER ANY OTHER CONDITION WHICH WILL ALLOW DRIFT. DRIFT MAY CAUSE DAMAGE TO ANY VEGETATION CONTACTED TO WHICH TREATMENT IS NOT INTENDED. TO PREVENT INJURY TO ADJACENT DESIRABLE VEGETATION, APPROPRIATE BUFFER ZONES MUST BE MAINTAINED.

Coarse sprays are less likely to drift, therefore, do not use nozzles or nozzle configurations which dispense spray as fine spray droplets. Do not angle nozzles forward into the airstream and do not increase spray volume by increasing nozzle pressure.

Drift control additives may be used. When a drift control additive is used, read and carefully observe the cautionary statements and all other information appearing on the additive label.

Ensure uniform application—To avoid streaked, uneven or overlapped application, use appropriate marking devices.

Thoroughly wash aircraft, especially landing gear, after each day of spraying to remove residues of this product accumulated during spraying or from spills. PROLONGED EXPOSURE OF THIS PRODUCT TO UNCOATED STEEL SURFACES MAY RESULT IN CORROSION AND POSSIBLE FAILURE OF THE PART. LANDING GEAR ARE MOST SUSCEPTIBLE. The maintenance of an organic coating (paint) which meets aerospace specification MIL-C-38413 may prevent corrosion.

BOOM EQUIPMENT

For control of weed or brush species listed on this label using conventional boom equipment—Use the recommended rates of this product and surfactant in 3 to 20 gallons of water per acre as a broadcast spray. See the "Weeds Controlled" section of this label for specific rates. As density of weeds increases, spray volume should be increased within the recommended range to insure complete coverage. Carefully select correct nozzle to avoid spraying a fine mist. For best results with ground application equipment, use flat fan nozzles. Check for even distribution of spray droplets.

HAND-HELD and HIGH-VOLUME EQUIPMENT

Use coarse sprays only

For control of weeds listed on this label using knapsack sprayers or high volume spraying equipment utilizing handguns or other suitable nozzle arrangements—Prepare a ¾ to 1½ percent solution of this product in water, add surfactant and apply to foliage of vegetation to be controlled. For specific rates of application and instructions for control of various annual and perennial weeds, see the "Weeds Controlled" section of this label.

Applications should be made on a spray-to-wet basis. Spray coverage should be uniform and complete. Do not spray to point of runoff.

Prepare the desired volume of spray solution by mixing the amount of this product in water, shown in the following table.

Spray solution

Desired	Amount of RODEO®			
Volume	¾%	1%	1¼%	1½%
1 gallon	1 oz.	1⅓ oz.	1⅔ oz.	2 oz.
25 gallons	1½ pt.	1 qt.	1¼ qt.	1½ qt.
100 gallons	3 qt.	1 gal.	1¼ gal.	1½ gal.

2 tablespoons = 1 ounce

For use in knapsack sprayers, it is suggested that the recommended amount of this product be mixed with water in a larger container. Fill sprayer with the mixed solution and add the correct amount of surfactant.

WEEDS CONTROLLED

ANNUAL WEEDS

Apply to actively growing annual grasses and broadleaf weeds.

Allow at least 3 days after application before disturbing treated vegetation. After this period the weeds may be mowed, tilled or burned. See "Directions for Use," "General Information," and "Mixing and Application Instructions" for labeled uses and specific application instructions.

Broadcast Application—Use 1½ pints of this product per acre plus ¼ to ½ percent surfactant by total spray volume if weeds are less than 6 inches tall. If weeds are greater than 6 inches tall, use 2½ pints of this product per acre plus ¼ to ½ percent surfactant by total spray volume.

Hand-Held High-Volume Application—Use a ¾ to 1½ percent solution of this product in water plus ¼ to ½ percent surfactant by total spray volume and apply to foliage of vegetation to be controlled.

When applied as directed under the conditions described in this label, this product plus nonionic surfactant WILL CONTROL the following ANNUAL WEEDS:

Barley	*Hordeum vulgare*
Bluegrass (annual)	*Poa annua*
Brome (downy)	*Bromus tectorum*
Cocklebur	*Xanthium pensylvanicum*
Corn (volunteer)	*Zea mays*
Crabgrass	*Digitaria* spp.
Falseflax (smallseed)	*Camelina microcarpa*
Fiddleneck	*Amsinckia* spp.
Fleabane	*Erigeron* spp.
Foxtail	*Setaria* spp.
Kochia	*Kochia scoparia*
Lambsquarters (common)	*Chenopodium album*
Lettuce (prickly)	*Lactuca serriola*
Mustard (tansy)	*Descurainia pinnata*
Oats (wild)	*Avena fatua*
Panicum	*Panicum* spp.
Pennycress (field)	*Thlaspi arvense*
Pigweed, Redroot	*Amaranthus retroflexus*
Pigweed (smooth)	*Amaranthus hybridus*
Ragweed (common)	*Ambrosia artemisiifolia*
Ragweed (giant)	*Ambrosia trifida*
Rye	*Secale cereale*
Ryegrass (Italian)*	*Lolium multiflorum*
Sandbur (field)	*Cenchrus* spp.
Shattercane	*Sorghum bicolor*
Smartweed (Pennsylvania)	*Polygonum pensylvanicum*
Spanishneedles*	*Bidens bipinnata*
Sunflower	*Helianthus annuus*
Thistle (Russian)	*Salsola kali*
Velvetleaf	*Abutilon theophrasti*

* Apply 3 pints of this product per acre.

Annual weeds will generally continue to germinate from seed throughout the growing season. Repeat treatments will be necessary to control later germinating weeds.

PERENNIAL WEEDS

Apply this product as follows to control or destroy most vigorously growing perennial weeds. Unless otherwise directed, allow at least 7 days after application before disturbing vegetation.

Add ¼ to ½ percent nonionic surfactant by total spray volume to the rates of this product given in this list. See the "General Information," "Directions for Use," "Mixing and Application" sections of this label for specific uses and application instructions.

NOTE: If weeds have been mowed or tilled, do not treat until regrowth has reached the recommended stages. Fall treatments must be applied before a killing frost.

Repeat treatments may be necessary to control weeds regenerating from underground parts or seed.

When applied as recommended under the conditions described, this product plus surfactant WILL CONTROL the following PERENNIAL WEEDS:

Alfalfa	*Medicago sativa*
Alligatorweed*	*Alternanthera philoxeroids*
American Lotus	*Nelumbo lutea*
(FOR USE IN TEXAS ONLY)	
Artichoke (Jerusalem)	*Helianthus tuberosus*
Bahiagrass	*Paspalum notatum*
Bermudagrass	*Cynodon dactylon*
Bindweed (field)	*Convolvulus arvensis*
Bluegrass (Kentucky)	*Poa* spp.
Brackenfern	*Pteridium aquilinum*
Bromegrass (smooth)	*Bromus inermis*
Cattail	*Typha* spp.
Clover (red)	*Trifolium pratense*
Clover (white)	*Trifolium repens*
Cutgrass (giant)*	*Zizaniopsis miliacea*
Dallisgrass	*Paspalum dilatatum*
Dandelion	*Taraxacum officinale*
Dock (curly)	*Rumex crispus*
Dogbane (hemp)	*Apocynum cannabinun*
Fescues	*Festuca* spp.
Guineagrass	*Panicum maximum*
Horsenettle	*Solanum carolinense*
Horseradish	*Armoracia rusticana*
Johnsongrass	*Sorghum halepense*
Kikuyugrass	*Pennisetum clandestinum*
Knapweed	*Centaurea repens*
Lantana	*Lantana camara*
Maidencane	*Panicum hematomon*
Milkweed	*Asclepias* spp.
Muhly (wirestem)	*Muhlenbergia frondosa*
Mullein (common)	*Verbascum thapsus*
Napiergrass	*Pennisetum purpureum*

Nightshade (silverleaf)	*Solanum elaeagnifolium*
Nutsedge (purple, yellow)	*Cyperus rotundus*
	Cyperus esculentus
Orchardgrass	*Dactylis glomerata*
Paragrass	*Brachiaria mutica*
Phragmites**	*Phragmites* spp.
Quackgrass	*Agropyron repens*
Reed canarygrass	*Phalaris arundinacea*
Ryegrass (perennial)	*Lolium perenne*
Smartweed (swamp)	*Polygonum coccineum*
Spatterdock	*Nuphar luteum*
Texas Blueweed	*Helianthus ciliaris*
Thistle (Canada)	*Cirsium arvense*
Timothy	*Phleum pratense*
Torpedograss*	*Panicum repens*
Vaseygrass	*Paspalum urvillei*
Waterhyacinth	*Eichornia crassipes (B.F.)*
Waterlettuce	*Pistia stratiotes*
Waterprimrose	*Jussiaea* spp.
Wheatgrass (western)	*Agropyron smithii*

*Partial control.

**Partial control in Southeastern states. See specific recommendation below.

ALLIGATORWEED—Apply 6 pints of this product per acre as a broadcast spray or as the 1¼ percent solution with hand-held equipment, to provide partial control of alligatorweed. Apply when most of the target plants are in bloom. Repeat applications will be required to maintain such control.

AMERICAN LOTUS—(FOR USE IN TEXAS ONLY) Apply 4 pints of this product per acre as a broadcast spray or as a 1 percent solution with hand-held equipment. Treat when plants are actively growing at or beyond the bloom stage of growth. Best results are achieved when application is made during summer or fall months. Fall treatments must be applied before a killing frost. Repeat treatment may be necessary to control regrowth from underground parts and seeds.

BERMUDAGRASS—Apply 7½ pints of this product per acre as a broadcast spray or as a 1½ percent solution with hand-held equipment. Apply when target plants are actively growing and when seed heads appear.

BRACKENFERN—Apply 4½ to 6 pints of this product per acre as a broadcast spray or as a ¾ to 1 percent solution with hand-held equipment. Apply to fully expanded fronds which are at least 18 inches long.

CANADA THISTLE—Apply 3 to 4½ pints of this product per acre as a broadcast spray or as a 1½ percent solution with hand-held equipment. Apply when target plants are actively growing and are at or beyond the bud stage of growth. Allow 3 or more days after application before tillage.

CATTAIL—Apply 4½ to 6 pints of this product per acre as a broadcast spray or as a ¾ percent solution with hand-held equipment. Apply when target plants are actively growing and are at or beyond the early-to-full bloom stage of growth. Best results are achieved when application is made during the summer or fall months.

CUTGRASS (giant)—Apply 6 pints of this product per acre as a broadcast spray or as 1 percent solution with hand-held equipment to provide partial control of giant cutgrass. Repeat applications will be required to maintain such control, especially where vegetation is partially submerged in water. Allow for substantial regrowth to the seven to ten leaf stage prior to retreatment.

FIELD BINDWEED/SILVERLEAF NIGHTSHADE/TEXAS BLUEWEED—Apply 6 to 7½ pints of this product per acre as a broadcast spray west of the Mississippi River and 4½ to 6 pints of this product per acre east of the Mississippi River. With hand-held equipment, use a 1½ percent solution. Apply when target plants are actively growing and are at or beyond full bloom. For silverleaf nightshade, best results can be obtained when application is made after berries are formed. Do not treat when weeds are under drought stress. New leaf development indicates active growth. For best results, apply in late summer or fall.

GUINEAGRASS—Apply 4½ pints of this product per acre as a broadcast spray or when target plants are actively growing and a ¾ percent solution with hand-held equipment. Apply when most have reached at least the 7-leaf stage of growth.

HEMP DOGBANE/KNAPWEED/HORSERADISH—Apply 6 pints of this product per acre as a broadcast spray or as a 1½ percent solution with hand-held equipment. Apply when target plants are growing and most weeds have reached the late bud to flower stage of growth. For best results, apply in late summer or fall.

JOHNSONGRASS/BROMEGRASS (Smooth)/REED CANARYGRASS/RYEGRASS (Perennial)/TIMOTHY/ WHEATGRASS (Western)—Apply 3 to 4½ pints of this product per acre as a broadcast spray or as a ¾ percent solution with hand-held equipment. Apply when target plants are actively growing and most have reached the boot to head stage of growth. When applied prior to the boot stage, less desirable control may be obtained. Allow Johnsongrass to reach at least 18 inches average height. In the fall, apply before plants have turned brown.

LANTANA—Apply this product as a ¾ to 1 percent solution with hand-held equipment. Apply to actively growing lantana at or beyond the bloom stage of growth. Use the higher application rate for plants that have reached the woody state of growth.

MAIDENCANE/PARAGRASS—Apply 6 pints of this product per acre as a broadcast spray or as a ¾ percent solution with hand-held equipment. Repeat treatments will be required especially to vegetation partially submerged in water. Under these conditions, allow for regrowth to the seven to ten leaf stage prior to retreatment.

MILKWEED (Common)—Apply 4½ pints of this product per acre as a broadcast spray or as a 1½ percent solution with hand-held equipment. Apply when target plants are actively growing and most have reached the late bud to flower stage of growth.

NUTSEDGE (Purple, Yellow)—Apply 4½ pints of this product per acre as a broadcast spray, or as a ¾ percent solution with hand-held equipment to control existing nutsedge plants and immature nutlets attached to treated plants. Apply when target plants are in flower or when new nutlets can be found at rhizome tips. Nutlets which have not germinated will not be controlled and may germinate following treatment. Repeat treatments will be required for long-term control.

PHRAGMITES—For partial control of phragmites in Florida and the counties of other states bordering the Gulf of Mexico, apply 7.5 pints per acre as a broadcast spray or apply a 1½ percent solution with hand-held equipment. In other areas of the U.S., apply 4 pints per acre as a broadcast spray or apply a ¾ percent solution with hand-held equipment for partial control. For best results, treat during late summer or fall months when plants are actively growing and in full bloom. Due to the dense nature of the vegetation, which may prevent good spray coverage and uneven stages of growth, repeat treatments may be necessary to maintain control. Visual control symptoms will be slow to develop.

QUACKGRASS/WIRESTEM MUHLY/KIKUYUGRASS — Apply 3 to 4½ pints of this product per acre as a broadcast spray or as a ¾ percent solution with hand-held equipment when most quackgrass or wirestem muhly is at least 8 inches in height (3 or 4 leaf stage of growth), and actively growing. Allow 3 or more days after application before tillage.

SPATTERDOCK—Apply 6 pints of this product per acre as a broadcast spray or apply a ¾ percent solution with hand-held equipment. Apply when most plants are in full bloom. For best results, apply during the summer or fall months.

TORPEDOGRASS—Apply 6 to 7½ pints of this product per acre as a broadcast spray or as a ¾ to 1½ percent solution with hand-held equipment to provide partial control of torpedograss. Use the lower rates under terrestial conditions, and the higher rates under partially submerged or a floating mat condition. Repeat treatments will be required to maintain such control.

WATER HYACINTH—Apply 5 to 6 pints of this product per acre as a broadcast spray or as a 1 percent solution with hand-held equipment. Apply when target plants are actively growing and at or beyond the early bloom stage of growth. After application, visual symptoms may require 3 or more weeks to appear with complete necrosis and decomposition usually occuring within 60 to 90 days.

WATERLETTUCE—For control, apply ¾ to 1 percent solution of this product with hand-held equipment to actively growing plants. Use higher rates where infestations are heavy. Best results are obtained from mid-summer through winter applications. Spring applications may require retreatment.

WATERPRIMROSE—Apply this product as a 1 percent solution using hand-held equipment. Apply to plants that are actively growing at or beyond the bloom stage of growth, but before fall color changes occur. Thorough coverage is necessary for best control.

OTHER PERENNIALS LISTED ON THIS LABEL—Apply 4½ to 7½ pints of this product per acre as a broadcast spray or as a ¾ to 1½ percent solution with hand-held equipment. Apply when target plants are actively growing and most have reached early head or early bud stage of growth.

WOODY BRUSH AND TREES

Apply this product as follows to control actively growing woody brush and trees listed below. Unless otherwise specified, allow at least 7 days after application before disturbing treated vegatation.

Add ¼ to ½ percent surfactant by total spray volume to the rates of this product given in this list. See the "General Information", "Directions for Use" and "Mixing and Application" sections of this label for specific uses and application instructions.

When applied as recommended under the conditions described, this product plus surfactant CONTROLS the following woody brush plants and trees:

Alder	*Alnus* spp.
Berries*	*Rhubus* spp.
Elderberry	*Sambucus* spp.
Florida Holly (Brazil peppertree)†	*Schinus terebinthifolius*
Honeysuckle	*Lonicera* spp.
Kudzu	*Pueraria lobata*
Maple**	*Acer* spp.
Multiflora rose	*Rosa multiflora*
Oak***	*Quercus* spp.
Poison Ivy	*Rhus radicans*
Poison Oak	*Rhus toxicodendron*
Saltbush	*Braccaharis spp.*
Tallowtree (Chinese)	*Sapium sebiferum*
Trumpet creeper	*Campsis radicans*
Waxmyrtle (Southern)†	*Myrica cerifera*
Willow	*Salix* spp.

* Includes blackberry, dewberry and raspberry.
** Includes sugar maple and red maple.
*** Includes red oak, white oak and Northern pin oak.
† Partial control.

Apply this product when plants are vigorously growing, and unless otherwise directed, after full leaf expansion. Use the higher rate for larger plants and/or dense areas of growth. On vines, use the higher rate for plants that have reached the woody stage of growth. Best results are obtained when application is made in late summer or fall after fruit formation. Ensure thorough coverage when using hand-held equipment. Symptoms may not appear prior to frost or senescence with fall treatments.

Repeat treatments may be necessary to control plants regenerating from underground parts or seed. Some autumn colors on undesirable deciduous species are acceptable provided no major leaf fall has occurred.

Apply this product as follows to control the following listed plants and trees.

ALDER / BERRIES / ELDERBERRY / HONEYSUCKLE — Apply 4½ to 6 pints of this product per acre as a broadcast spray or as a ¾ to 1 percent solution with hand-held equipment.

FLORIDA HOLLY/WAXMYRTLE—Apply this product as a 1½ percent solution with hand-held equipment for partial control.

KUDZU—Apply 6 pints of this product per acre as a broadcast spray or as a 1½ percent solution with hand-held equipment.

MAPLES/OAKS—Apply this product as a ¾ to 1 percent solution with hand-held equipment. Apply when at least 50 percent of the new leaves are fully developed. Use the higher rate for large mature trees.

MULTIFLORA ROSE—Apply 3 pints of this product per acre as a broadcast spray or as a ¾ percent solution with hand-held equipment. Treatments should be made prior to leaf deterioration by leaf-feeding insects.

POISON IVY/POISON OAK—Apply 6 to 7½ pints of this product per acre as a broadcast spray or as a 1½ percent solution with hand-held equipment. Treatments must be applied before leaves lose green color. Repeat applications may be required to maintain control.

SALTBUSH—Apply this product as a 1 percent solution with hand-held equipment.

TRUMPET CREEPER—Apply 3 to 4½ pints of this product per acre as a broadcast spray or as a ¾ to 1 percent solution with hand-held equipment.

WILLOW/TALLOWTREE (Chinese)—Apply this product as a ¾ percent solution with hand-held equipment. Thorough coverage of foliage is necessary for best results.

DIRECTIONS FOR USE

It is a violation of Federal law to use this product in any manner inconsistent with its labeling.

AQUATIC SITES

When applied as directed under conditions described, this product plus nonionic surfactant will control emerged annual weeds and control or partially control emerged perennial weeds, woody brush and trees listed in this label. This product does not control plants which are either completely submerged or have a majority of the foliage under water. See the "Weeds Controlled" section of this label for rates and degree of control provided.

This product may be used in and around aquatic sites, including all bodies of fresh and brackish water, which may be flowing, non-flowing, or transient. This includes lakes, rivers, streams, ponds, seeps, irrigation and drainage ditches, canals, reservoirs, and similar sites. There is no restriction on the use of water for irrigation, recreation, or domestic purposes.

Consult local state fish and game agency and water control authorities before applying this product to public water. Permits may be required to treat such water.

NOTE: Do not apply this product within ½ mile upstream of potable water intakes.

Do not apply this product on rice levees when flood water is present.

Do not apply in estuaries.

For treatments after drawdown of water or in dry ditches, allow 7 or more days after treatment before reintroduction of water. Apply this product within one day after drawdown to ensure application to actively growing weeds.

When using this product in aquatic sites where water is present, add 1 to 2 quarts of an agriculturally approved surfactant per 100 gallons of spray solution (¼ to ½% surfactant by total spray volume). The following surfactants are approved for use with this product in aquatic sites: Agri-Dex™, Induce™, Liqua-wet™, Ortho™ X-77, Passage™, R-11™, Spreader Sticker™, Super Spread™ 200, and Widespread™.

When using this product in sites where water is not present (dry ditches, ditchbanks, dry canals), use 1 to 2 quarts of nonionic surfactant per 100 gallons of spray solution (¼ to ½% surfactant by total spray volume).

Floating mats of vegetation may require retreatment. Avoid washoff of sprayed foliage by spray boat or recreational boat backwash or by rainfall within six hours of application. Do not retreat within 24 hours following the initial treatment.

Applications made to moving bodies of water must be made while traveling upstream to prevent concentration of this herbicide in water. When making any bankside applications, do not overlap more than 1 foot into open water. Do not spray across open moving bodies of water, or where weeds do not exist. The maximum application rate of 7½ pints per acre must not be exceeded in any single application.

When emerged infestations require treatment of the total surface area of impounded water, treating the area in strips may avoid oxygen depletion due to decaying vegetation. Oxygen depletion may result in fish kill.

TM Agri-Dex and Induce are trademarks of the Helena Chemical Company.

TM Liqua-wet is a trademark of the Woodbury Chemical Company.

TM Ortho X-77 is a trademark of Chevron Chemical Company.

TM Passage is a trademark of Asgrow Florida Company.

TM R-11 and Super Spread 200 are trademarks of the Wilber-Ellis Company.

TM Spreader-Sticker is a trademark of the Southern Mill Creek Products Company.

TM Widespread is a trademark of the FMC Corporation.

ROUNDUP®
HERBICIDE

ACTIVE INGREDIENT

*Isopropylamine salt of glyphosate	41.0%
INERT INGREDIENTS	59.0%
TOTAL	100.0%

*Contains 480 grams per litre or 4 pounds of the active ingredient isopropylamine salt of N-(phosphonomethyl) glycine per U.S. gallon. Equivalent to 356 grams per litre or 3 pounds per U.S. gallon of the acid, glyphosate.

EPA Reg. No. 524-308-AA

Roundup Herbicide is protected by U.S. Pat. No. 3,799,758 and U.S. Pat. No. 4,405,531

Other patents are pending.

®Registered Trademark of Monsanto Company

1985-2 897.10-002.83/53

WARNING
Keep Out of Reach of Children

Complete Directions for Use

This product has been approved for use in California except as stated otherwise on page 103.

AVOID CONTACT WITH FOLIAGE, GREEN STEMS, OR FRUIT OF CROPS, DESIRABLE PLANTS AND TREES, SINCE SEVERE INJURY OR DESTRUCTION MAY RESULT.

Read each of these sections of this label for essential product performance information.

Read the entire label before using this product.

Use only according to label instructions.

Read "LIMIT OF WARRANTY AND LIABILITY" before buying or using. If terms are not acceptable, return at once unopened.

LIMIT OF WARRANTY AND LIABILITY

(Not applicable to consumer applications applied by the homeowner for noncommercial purposes as permitted by the supplemental labeling for one-quart containers.)

This company warrants that this product conforms to the chemical description on the label and is reasonably fit for the purposes set forth in the complete Directions for Use label booklet ("Directions") when used in accordance with those Directions under the conditions described therein. NO OTHER EXPRESS OR IMPLIED WARRANTY OF FITNESS FOR PARTICULAR PURPOSE OR MERCHANTABILITY IS MADE. This warranty is also subject to the conditions and limitations stated herein.

Buyer and all users shall promptly notify this company of any claims whether based in contract, negligence, strict liability, other tort or otherwise.

Buyer and all users are responsible for all loss or damage from use or handling which results from conditions beyond the control of this company, including but not limited to incompatibility with products other than those set forth in the Directions, application to or contact with desirable vegetation, unusual weather, weather conditions which are outside the range considered normal at the application site and for the time period when the product is applied as well as weather conditions which are outside the application ranges set forth in the Directions, application in any manner not explicitly set forth in the Directions, moisture conditions outside the miosture range specified in the Directions, or the presence of products other than those set forth in the Directions in or on the soil or crop.

THE EXCLUSIVE REMEDY OF THE USER OR BUYER, AND THE LIMIT OF THE LIABILITY OF THIS COMPANY OR ANY OTHER SELLER FOR ANY AND ALL LOSSES, INJURIES OR DAMAGES RESULTING FROM THE USE OR HANDLING OF THIS PRODUCT (INCLUDING CLAIMS BASED IN CONTRACT, NEGLIGENCE, STRICT LIABILITY, OTHER TORT OR OTHERWISE) SHALL BE THE PURCHASE PRICE PAID BY THE USER OR BUYER FOR THE QUANTITY OF THIS PRODUCT INVOLVED, OR, AT THE ELECTION OF THIS COMPANY OR ANY OTHER SELLER, THE REPLACEMENT OF SUCH QUANTITY OR, IF NOT ACQUIRED BY PURCHASE, REPLACEMENT OF SUCH QUANTITY. IN NO EVENT SHALL THIS COMPANY OR ANY OTHER SELLER BE LIABLE FOR ANY INCIDENTAL OR CONSEQUENTIAL DAMAGES.

The buyer and all users are deemed to have accepted the terms of this LIMIT OF WARRANTY AND LIABILITY which may not be varied by any verbal or written agreement.

PRECAUTIONARY STATEMENTS

HAZARDS TO HUMANS AND DOMESTIC ANIMALS

Keep Out of Reach of Children
WARNING!

CAUSES EYE IRRITATION.
HARMFUL IF SWALLOWED.
MAY CAUSE SKIN IRRITATION

Do not get in eyes, on skin or on clothing.
Wash thoroughly after handling.

FIRST AID: IF IN EYES, immediately flush with plenty of water for at least 15 minutes. Call a physician.

IF ON SKIN, immediately flush with plenty of water. Remove contaminated clothing. Wash clothing before reuse.

IF SWALLOWED, this product will cause gastrointestinal tract irritation. Immediately dilute by swallowing water or milk. Call a physician.

In case of an emergency involving this product, Call Collect, day or night, (314) 694-4000.

ENVIRONMENTAL HAZARDS

Avoid direct applications to any body of water. Do not contaminate water by disposal of waste or cleaning of equipment.

PHYSICAL OR CHEMICAL HAZARDS

Spray solutions of this product should be mixed, stored and applied only in stainless steel, aluminum, fiberglass, plastic and plastic-lined steel containers.

DO NOT MIX, STORE OR APPLY THIS PRODUCT OR SPRAY SOLUTIONS OF THIS PRODUCT IN GALVANIZED STEEL OR UNLINED STEEL (EXCEPT STAINLESS STEEL) CONTAINERS OR SPRAY TANKS. This product or spray solutions of this product react with such containers to produce hydrogen gas which may form a highly combustible gas mixture. This gas mixture could flash or explode, causing serious personal injury, if ignited by open flame, spark, welder's torch, lighted cigarette or other ignition source.

STORAGE AND DISPOSAL

Do not contaminate water, foodstuffs, seed or feed by storage and disposal.

See container label for STORAGE AND DISPOSAL instructions.

GENERAL INFORMATION

DO NOT APPLY THIS PRODUCT USING AERIAL SPRAY EQUIPMENT EXCEPT UNDER CONDITIONS AS SPECIFIED WITHIN THIS LABEL.

ROUNDUP® herbicide, a water soluble liquid, mixes readily with water to be applied as a foliage spray for the control or destruction of most herbaceous plants. It may be applied through most standard industrial or field type sprayers after dilution and thorough mixing with water in accordance with label instructions.

This product moves through the plant from the point of foliage contact to and into the root system. Visible effects on most annual weeds occur within 2 to 4 days but on most perennial weeds may not occur for 7 days or more. Extremely cool or cloudy weather following treatment may slow down activity of this product and delay visual effects of control. Visible effects are a gradual wilting and yellowing of the plant which advances to complete browning of above ground growth and deterioration of underground plant parts.

Unless otherwise specified on this label delay application until vegetation has emerged and reached the stages described for control of such vegetation under the "Weeds Controlled" section of this label. Unemerged plants arising from unattached underground rhizomes or root stocks of perennials will not be affected by the spray and will continue to grow. For this reason best control of most perennial weeds is obtained when treatment is made at late growth stages approaching maturity.

Always use the higher rate of this product per acre within the recommended range when (1) weed growth is heavy or dense, or (2) weeds are growing in an undisturbed (non-cultivated) area.

Do not treat weeds under poor growing conditions such as drought stress, disease or insect damage, as reduced weed control may result. Reduced results may also occur when treating weeds heavily covered with dust.

Reduced control may result when applications are made to annual or perennial weeds that have been mowed, grazed, or cut, and have not been allowed to regrow for the recommended stage for treatment.

Rainfall or irrigation occurring within 6 hours after application may reduce effectiveness. Heavy rainfall or irrigation within 2 hours after application may wash the chemical off the foliage and repeat treatment may be required.

ROUNDUP herbicide does not provide residual weed control. For subsequent residual weed control, follow a label-approved herbicide program. Read and carefully observe tthe cautionary statements and all other information appearing on the labels of all herbicides used.

Buyer and all users are responsible for all loss or damage in connection with the use or handling of mixtures of ROUNDUP with herbicides or other materials that are not expressly recommended in this labeling. Mixing this product with herbicides or other materials not recommended on this label, may result in reduced ROUNDUP performance.

For best results, spray coverage should be uniform and complete. Do not spray weed foliage to the point of runoff.

Keep people and pets off treated areas until spray solution has dried to prevent transfer of this product onto desirable vegetation.

ATTENTION

AVOID DRIFT. EXTREME CARE MUST BE USED WHEN APPLYING THIS PRODUCT TO PREVENT INJURY TO DESIRABLE PLANTS AND CROPS.

Do not allow the herbicide solution to mist, drip, drift, or splash onto desirable vegetation since minute quantities of this herbicide can cause severe damage or destruction to the crop, plants, or other areas on which treatment was not intended. The likelihood of injury occurring from the use of this product is greatest when winds are gusty or in excess of 5 miles per hour or when other conditions, including lesser wind velocities, will allow spray drift to occur. When spraying avoid combinations of pressure and nozzle type that will result in splatter or fine particles (mist) which are likely to drift. AVOID APPLYING AT EXCESSIVE SPEED OR PRESSURE.

NOTE: Use of this product in any manner not consistent with this label may result in injury to persons, animals or crops, or other unintended consequences. Keep container closed to prevent spills and contamination.

MIXING, ADDITIVES AND APPLICATION INSTRUCTIONS

APPLY THESE SPRAY SOLUTIONS IN PROPERLY MAINTAINED AND CALIBRATED EQUIPMENT CAPABLE OF DELIVERING DESIRED VOLUMES. DO NOT APPLY UNDER WIND OR OTHER CONDITIONS WHICH ALLOW DRIFT TO OCCUR. HAND GUN APPLICATIONS SHOULD BE PROPERLY DIRECTED TO AVOID SPRAYING DESIRABLE PLANTS. NOTE: REDUCED RESULTS MAY OCCUR IF WATER CONTAINING SOIL IS USED, such as WATER FROM PONDS AND UNLINED DITCHES.

MIXING

This product mixes readily with water. Mix spray solutions of this product as follows: Fill the mixing or spray tank with the required amount of water. Add the proper amount of this product (see "Directions for Use" and "Weeds Controlled" sections of this label) near the end of the filling process and mix well. Remove hose from tank immediately after filling to avoid siphoning back into the carrier source. During mixing and application foaming of the spray solution may occur. To prevent or minimize foam, avoid the use of mechanical agitators, place the filling hose below the surface of the spray solution, terminate by-pass and return lines at the bottom of the tank and if needed use an approved anti-foam or defoaming agent.

TANK MIXTURES

Always predetermine the compatibility of labeled tank mixtures of this herbicide with water carrier by mixing small proportional quantities in advance.

Mix labeled tank mixtures of ROUNDUP herbicide with water as follows:

1. Place a 20 to 35 mesh screen or wetting basket over filling port.
2. Through the screen, fill the sprayer tank one-half full with water and start agitation.
3. If a wettable powder is used, make a slurry with the water carrier, and add it SLOWLY through the screen into the tank. Continue agitation.
4. If a flowable formulation is used, pre-mix one part flowable with one part water. Add diluted mixture SLOWLY through the screen into the tank. Continue agitation.
5. If an emulsifiable concentrate formulation is used, pre-mix one part emulsifiable concentrate with two parts water. Add diluted mixture slowly through the screen into the tank. Continue agitation.
6. Continue filling the sprayer tank with water and add the required amount of ROUNDUP herbicide near the end of the filling.
7. Where nonionic surfactant is recommended, add this to the spray before completing the filling process.
8. Add individual formulations to the spray tank as follows: wettable powder, flowable, emulsifiable concentrate, drift control additive, water soluble liquid followed by surfactant.

Maintain good agitation at all times until contents of the tank are sprayed. If the spray mixture is allowed to settle, thorough agitation is required to resuspend the mixture before spraying is resumed.

Keep by-pass line on or near bottom of tank to minimize foaming. Screen size in nozzle or line strainers should be no finer than 50 mesh. Carefully select proper nozzle to avoid spraying a fine mist. For best results with conventional ground applications equipment, use flat fan nozzles.

Clean sprayer and parts immediately after using this product by thoroughly flushing with water.

ADDITIVES

SURFACTANTS

Nonionic surfactants which are labeled for use with herbicides may be used. Do not reduce rates of this herbicide when adding surfactant. Use 0.5 percent surfactant concentration (2 quarts per 100 gallons of spray solution) when using surfactants which contain at least 50 percent active ingredient or a 1 percent surfactant concentration (4 quarts per 100 gallons of spray solution) for those surfactants containing less than 50 percent active ingredient. Read and carefully observe surfactant cautionary statements and other information appearing on the surfactant label.

AMMONIUM SULFATE

The addition of 2 percent dry ammonium sulfate by weight or 17 pounds per 100 gallons of water may increase the performance of Roundup and Roundup plus 2, 4-D amine or Banvel tank mixtures on annual weeds. The improvement in performance may be apparent where environmental stress is a concern. Low quality ammonium sulfate may contain material that will not readily dissolve which could result in nozzle tip plugging. To determine quality, perform a jar test by adding ⅓ cup of ammonium sulfate to 1 gallon of water and agitate for 1 minute. If undissolved sediment is observed, predissolve the ammonium sulfate in water and filter prior to addition to the spray tank. If ammonium sulfate is added directly to the spray tank, add slowly with agitation. Adding too quickly may clog outlet lines. Ensure that ammonium sulfate is completely dissolved in the tank before adding herbicides or surfactant.

Thoroughly rinse the spray system with clean water after use to reduce corrosion.

NOTE: The use of ammonium sulfate as an additive does not preclude the need for additional surfactant. Do not use herbicide rates lower than recommended in this label.

APPLICATION EQUIPMENT AND TECHNIQUES

This product may be applied with the following application equipment:

Aerial—fixed wing and helicopter

Boom—conventional broadcast spray

Controlled Droplet Applicator (CDA)—hand-held or boom mounted applicators which produce a spray consisting of a narrow range of droplet sizes.

Hand-held and high-volume spray equipment—knapsack and backpack sprayers, pump-up pressure sprayers, handguns, handwands, lances, and other hand-held spray equipment used to direct the spray onto weed foliage and vehicle mounted high-volume spray equipment for spray-to-wet applications.

Selective equipment—recirculating sprayers, shielded sprayers and wiper applicators.

See the appropriate part of this section for specific rates of application and instructions.

AERIAL EQUIPMENT

Use the recommended rates of this herbicide in 3 to 15 gallons of water per acre unless otherwise specified on this label. See "WEEDS CONTROLLED" section of this label for specific rates. **Aerial applications of this product may only be made in fallow and reduced tillage systems, silvicultrual sites, and rights-of-way. Refer to the individual use area sections of this label for recommended volumes and application rates.**

Avoid direct application to any body of water.

AVOID DRIFT—DO NOT APPLY DURING INVERSION CONDITIONS, WHEN WINDS ARE GUSTY, OR UNDER ANY OTHER CONDITION WHICH WILL ALLOW DRIFT. DRIFT MAY CAUSE DAMAGE TO ANY VEGETATION CONTACTED TO WHICH TREATMENT IS NOT INTENDED. TO PREVENT INJURY TO ADJACENT DESIRABLE VEGETATION, APPROPRIATE BUFFER ZONES MUST BE MAINTAINED.

Coarse sprays are less likely to drift, therefore, do not use nozzles or nozzle configurations which dispense spray as fine spray droplets. Do not angle nozzles forward into the airstream and do not increase spray volume by increasing nozzle pressure.

Drift control additives may be used. When a drift control additive is used, read and carefully observe the cautionary statements and all other information appearing on the additive label.

Ensure uniform application—To avoid streaked, uneven or overlapped application, use appropriate marking devices.

Thoroughly wash aircraft, especially landing gear, after each day of spraying to remove residues of this product accumulated during spraying from spills.

PROLONGED EXPOSURE OF THIS PRODUCT TO UNCOATED SURFACES MAY RESULT IN CORROSION AND POSSIBLE FAILURE OF THE PART. LANDING GEAR ARE MOST SUSCEPTIBLE. The maintenance of an organic coating (Paint) which meets aerospace specification MIL-C-38413 may prevent corrosion.

BOOM EQUIPMENT

For control of annual or perennial weeds listed on this label using conventional boom equipment—Use the recommended rates of this product in 10 to 40 gallons of water per acre as a broadcast spray unless otherwise specified on this label. See "Weeds Controlled" section of this label for specific rates. As density of weeds increases, spray gallonage should be increased within the recommended range to insure complete coverage. Carefully select proper nozzle to avoid spraying a fine mist. For best results with ground application equipment, use flat fan nozzles. Check for even distribution of spray droplets.

CONTROLLED DROPLET APPLICATION (CDA)

The rate of this product applied per acre for vehicle mounted CDA equipment must not be less than the amount recommended in this label applied by conventional broadcast equipment. For vehicle mounted equipment apply 3 to 15 gallons of water per acre.

For the control of labeled annual weeds with hand-held CDA units, apply a 20 percent solution of this product at a flow rate of 2 fluid ounces per minute and a walking speed of 1.5 MPH (1 quart per acre). For the control of labeled perennial weeds, apply a 20 to 40 percent solution of this product at a flow rate of 2 fluid ounces per minute and a speed of 0.75 MPH (2 to 4 quarts per acre).

Controlled droplet application equipment produces a spray pattern which is not easily visible. Extreme care must be exercised to avoid spray or drift contacting the foliage or any other green tissue of desirable vegetation as damage or destruction may result.

HAND-HELD and HIGH-VOLUME EQUIPMENT

Use coarse sprays only

Mix this product in water and apply to foliage of vegetation to be controlled. For applications made on a spray-to-wet basis, spray coverage should be uniform and complete. Do not spray to the point of runoff.

Use a 1 percent solution, unless otherwise specified. For best results, use a 2 percent solution on harder-to-control perennials, such as bermudagrass, dock, field bindweed, hemp dogbane, milkweed and Canada thistle.

Less than complete coverage of weeds may result from the use of spray equipment designed for motorized spot treatments. Where less than complete coverage of annual weeds occurs, use a 5 percent solution. Do not reduce recommended concentrations of ROUNDUP when adding surfactant.

An agriculturally approved marking dye may be added to recommended rates of ROUNDUP when making applications with hand-held equipment. Reduced control of perennial weeds may result from applications made with a dye. Use dyes according to manufacturer's recommendations.

Prepare the desired volume of spray solution by mixing the amount of this product in water, shown in the following table:

Spray solution

Desired Volume	Amount of ROUNDUP® 1%	1½%	2%	5%
1 gallon	1⅓ oz	2 oz	2⅔ oz	6½ oz
25 gallons	1 qt	1½oz	2 qt	5 qt
100 gallons	1 gal.	1½ gal.	2 gal.	5 gal.

2 tablespoons = 1 ounce

For use in knapsack sprayers, it is suggested that the proper amount of this product be mixed with water in a larger container. Fill sprayer with the mixed solution.

For hand-held WIPER APPLICATORS see the "Selective Equipment" section and for hand-held CONTROLLED DROPLET APPLICATORS see the "Controlled Droplet Applicator (CDA)" section of this label.

SELECTIVE EQUIPMENT

This product may be applied through a recirculating spray system, a shielded applicator, or a wiper applicator after dilution and thorough mixing with water to listed weeds growing in any non-crop site specified on this label and only when specifically recommended in cropping systems.

A recirculating spray system directs the spray solution onto weeds growing above desirable vegetation, while spray solution not intercepted by weeds is collected and returned to the spray tank for reuse.

A shielded applicator directs the herbicide solution onto weeds while shielding desirable vegetation from the herbicide.

A wiper applicator applies the herbicide solution onto weeds by rubbing the weed with an absorbent material containing the herbicide solution.

AVOID CONTACT WITH DESIRABLE VEGETATION.

Contact of the herbicide solution with the desirable vegetation may result in damage or destruction. Applicators used above desired vegetation should be adjusted so that the lowest spray stream or wiper contact point is at least 2 inches above the desirable vegetation. Droplets, mist, foam, or splatter of the herbicide solution settling on desirable vegetation may result in discoloration, stunting, or destruction.

Applications made above the crops should be made when the weeds are a minimum of 6 inches above the desirable vegetation. Better results may be obtained when more of the weed is exposed to the herbicide solution. Weeds not contacted by the herbicide solution will not be affected. This may occur in dense clumps, severe infestations, or when the height of the weeds varies so that not all weeds are contacted. In these instances, repeat treatment may be necessary.

See the "Weeds Controlled" section of this label for recommended stage of growth for specific weeds.

NOTE

- Maintain equipment in good operating condition. Avoid leakage or dripping onto desirable vegetation.
- Adjust height of applicator to insure proper contact with weeds.
- Keep nozzle tips and wiping surfaces clean.
- Keep spray patterns aligned into recovery chamber of the recirculating sprayer.
- Keep shields on shielded applicators adjusted to protect desirable vegetation.
- Maintain recommended roller RPM on roller applicators while in use.
- Keep wiper material at proper degree of saturation with herbicide solution.
- DO NOT use wiper equipment when weeds are wet.
- DO NOT operate equipment at ground speeds greater than 5 mph. Weed control may be affected by speed of application equipment. As weed density increases, reduce equipment ground speed to insure good coverage of weeds.
- Be aware that on sloping ground the herbicide solution may migrate, causing dripping on the lower end and drying on the upper end of a wiper applicator.
- Variation in equipment design may affect weed control. With wiper applicators, the wiping material and its orientation must allow delivery of sufficient quantities of the recommended herbicide solution directly to the weed.
- Care must be taken with all types of wipers to insure that the absorbent material does not become oversaturated, causing the herbicide to drip on desirable vegetation.
- Mix only the amount of solution to be used during a one day period, as reduced activity may result from use of leftover solutions. With all equipment, drain and clean wiper parts immediately after using this product by thoroughly flushing with water.

RECIRCULATING SPRAYERS

Recirculating sprayer calibration is made on the basis of ground speed and delivery volume. Two procedures can be used to calibrate: (1) determine the discharge being delivered per minute, then operate the designated ground speed, or (2) select the desired ground speed and then adjust the sprayer to deliver the proper volume per minute (this may require nozzle changes). Use the appropriate table below.

Do not operate at nozzle pressure above 20 PSI.

Table 1. Use this table when calibrating Box or Row type recirculating sprayers. Box or Row type sprayer calibration is based on the total discharge collected per row. Use only straight stream or 15° fan type nozzles.

*Volume Per Minute Per Row MPH	Ounces
2	26 to 35
3	38 to 51
4	51 to 68
5	65 to 86

* NOTE: Be certain the amount collected is for all spray streams treating one row.

Table 2. Use this table when calibrating Broadcast type recirculating sprayers. Broadcast recirculating sprayer calibration is based on the discharge collected per minute from one nozzle on a 20 inch spacing.

Volume Per Minute Per Nozzle MPH	Ounces
2	7 to 9
3	10 to 13
4	13 to 18
5	16 to 22

When applied as recommended under the conditions described for recirculating sprayers, this product will control the following weeds growing a minimum of 6 inches above desirable vegetation.

Perennial Broadleaf Weeds—To SUPPRESS the following weeds, mix in a ratio of 4 quarts of this product to 20 gallons of water and apply as directed.

Dogbane (hemp)	*Apocynum cannabinum*
Milkweed	*Asclepias syriaca*

Perennial Grasses and Annual Broadleaf Weeds—To control the following weeds, mix in a ratio of 3 quarts of this product in 20 gallons of water and apply as directed:

Cocklebur	*Xanthium pensylvanicum*
Johnsongrass	*Sorghum halepense*
Pigweed, Redroot	*Amaranthus retroflexus*
Sunflower	*Helianthus annuus*

Annual Grasses—To control the following weeds, mix in a ratio of 2 quarts of this product in 20 gallons of water and apply as directed:

Corn (volunteer)	*Zea mays*
Shattercane	*Sorghum bicolor*

SHIELDED APPLICATORS

When applied as directed under conditions described for shielded applicators, this product will control those weeds listed in the "Weeds Controlled" section of this label.

Shielded applicators which apply the herbicide solution as a spray band should be calibrated on a broadcast equivalent rate and volume basis. To determine these:

$$\frac{\text{Band width in inches}}{\text{Row width in inches}} \times \text{Herbicide Broadcast RATE per acre} = \text{Herbicide Band RATE per acre}$$

$$\frac{\text{Band width in inches}}{\text{Row width in inches}} \times \text{Broadcast VOLUME of solution per acre} = \text{Band VOLUME of solution per acre}$$

Use nozzles that provide uniform coverage within the treated area. EXTREME CARE MUST BE EXERCISED TO AVOID CONTACT WITH DESIRABLE VEGETATION.

For specific rates of application and instructions for control of various annual and perennial weeds, see the "Weeds Controlled" section of this label.

WIPER APPLICATORS

Wiper applicators include either roller or wick devices which physically wipe appropriate concentrations or amounts of this product directly onto the weed. Equipment must be designed, maintained, and operated to prevent the herbicide solution from contacting desirable vegetation. Operate this equipment at ground speeds no greater than 5 mph. Performance may be improved by reducing speed in areas of heavy weed infestations to insure adequate wiper saturation. Better results may be obtained if 2 applications are made in opposite directions.

Do not add surfactant to the herbicide solution.

For Roller Applicators—Mix 1 gallon of this product in enough water to prepare 10 gallons of herbicide solution (10% solution). Apply this solution to perennial weeds or annual broadleaf weeds listed in this "Wiper Applicators" section.

Mix 1 gallon of this product in enough water to provide 20 gallons of herbicide solution (5% solution). Apply this solution to annual grasses listed in this "Wiper Applicators" section.

Roller speed should be maintained at 40 to 60 RPM.

For Wick or other Wiper Applicators—Mix 1 gallon of this product in 2 gallons of water to prepare a 33% solution. Apply this solution to weeds listed in this "Wiper Applicators" section.

In severe infestations, reduce equipment ground speed to insure that adequate amounts of this product are wiped on the weeds. A second treatment in the opposite direction may be beneficial.

Do not permit herbicide solution to contact desirable vegetation.

When applied as recommended under the conditions described for "Wiper Applicators", this product CONTROLS the following weeds:

Annual Grasses

Corn (volunteer)	*Zea mays*
Rye (common)	*Secale cereale*
Shattercane	*Sorghum bicolor*
Texas Panicum	*Panicum texanum*

Annual Broadleaves

Bristly Starbar	*Acanthospermum hispidum*
Sicklepod	*Cassia obtusifolia*
Spanishneedles	*Bidens bipinnata*

When applied as recommended under the conditions described for "Wiper Applicators", this product SUPPRESSES the following weeds:

Annual Broadleaves

Dog fennel	*Eupatorium capillifolium*
Florida beggarweed	*Desmodium tortuosum*
Pigweed, Redroot	*Amaranthus retroflexus*
Ragweed, (common)	*Ambrosia artemisifolia*
Ragweed (giant)	*Ambrosia trifida*
Sunflower	*Helianthus annuus*
Thistle (musk)	*Carduus nutans*
Velvetleaf	*Abutilon theophrasti*

Perennial Grasses

Bermudagrass	*Cynodon dactylon*
Guineagrass	*Panicum maximum*
Johnsongrass	*Sorghum halepense*
Smutgrass	*Sporobolus poiretii*
Vaseygrass	*Paspalum urvillei*

Perennial Broadleaves

Dogbane (hemp)	*Apocynum cannabinum*
Milkweed	*Ascelepias syriaca*
Nightshade (silverleaf)	*Solanum elaeagnifolium*
Thistle (Canada)	*Cirsium arvense*

WEEDS CONTROLLED

This herbicide controls many emerged annual and perennial grasses and broadleaf weeds.

ANNUAL WEEDS

- Apply to actively growing grass and broadleaf weeds.
- Allow at least 3 days after treatment before tillage.
- For maximum agronomic benefit, apply when weeds are 6 inches or less in height.
- To prevent seed production, applications should be made prior to seedhead formation.
- This product does not provide residual control; therefore, delay application until maximum weed emergence. Repeat treatments may be necessary to control later germinating weeds.

LOW-VOLUME BROADCAST APPLICATION (LOW-RATE TECHNOLOGY)

When applied as directed under the conditions described, this product will control the weeds listed below when:

1. Water carrier volumes are 3 to 10 gallons per acre for ground applications and 3 to 5 gallons per acre for aerial applications. (See the "Aerial Application" section of this label for approved sites.)
2. A nonionic surfactant is added at 0.5 to 1 percent by total spray volume. Use 0.5 percent surfactant concentration when using surfactants which contain at least 50 percent active ingredient or a 1 percent surfactant concentration for those surfactants containing less than 50 percent active ingredient.

NOTE

- The addition of 2 percent dry ammonium sulfate by weight or 17 pounds per 100 gallons of water may increase the performance of this herbicide on annual weeds. The improvement in performance may be apparent where environmental stress is a concern. Refer to the "Mixing, Additives and Application Instruction" section of this label.
- Do not tank-mix with soil residual herbicides when using these rates unless otherwise specified.
- Do not use these rates for weeds that have been mowed, grazed, or cut. (Refer to the "High-Volume Application" portion of this section for recommended rates.)
- Refer to the "Tank Mix" portion of this section for control of additional broadleaf weeds.

Weed Species	Maximum Height	Rate Per Acre**
Foxtails *Setaria* spp.	12"	8 oz.
Barnyardgrass *Echinochloa crus-galli*	6"	12 oz.
Bluegrass (annual) *Poa annua*		
Brome (downy)* *Bromus tectorum*		
Chickweed (jagged) *Holosteum umbellatum*		
Mustard (blue) *Chorispora tenella*		
Mustard (tansy) *Descurainia pinnata*		
Mustard (tumble) *Sisymbrium altissimum*		
Mustard (wild) *Brassica kaber*		
Barley *Hordeum vulgare*	12"	
Rye *Secale cereale*		
Sandbur (field) *Cenchrus spp.*		
Shattercane *Sorghum bicolor*		
Stinkgrass *Eragrostis ciliianensis*		
Wheat *Triticum aestivum*	18"	
Morningglory *Ipomea* spp.	3"	16 oz.
Bluegrass (bulbous) *Poa bulbosa*	6"	
Cheat *Bromus secalinus*		
Chickweed (mouseear) *Cerastium vulgatum*		
Corn (volunteer) *Zea mays*		
Groundsel (common) *Senecio vulgaris*		
Horseweed/Marestail *Conyza canadensis*		
Pennycress (field) Fanweed *Thlaspi arvense*		
Rocket (London) *Sisymbrium irio*		
Ryegrass (Italian) *Lolium multiflorum*		
Shepherdspurse *Capsella bursa-pastoris*		
Buttercup *Ranunculus* spp.	12"	
Cocklebur *Xanthium pensylvanicum*		
Crabgrass *Digitaria* spp.		
Dwarfdandelion *Krigia cespitosa*		
Falseflax (smallseed) *Camelina microcarpa*	12"	16 oz.
Foxtail (Carolina) *Alopecurus carolinianus*		
Johnsongrass (seedling) *Sorghum halepense*		
Lambsquarters (common) *Chenopodium album*		
Oats (wild) *Avena fatua*		
Panicum (fall) *Panicum dichotomiflorum*		
Panicum (Texas) *Panicum texanum*		
Pigweed (redroot) *Amaranthus retroflexus*		
Pigweed (smooth) *Amaranthus hybridus*		
Witchgrass *Panicum capillare*		
Signalgrass (broadleaf) *Brachiaria platyphylla*	6"	24 oz.
Teaweed *Sida spinosa*	3"	32 oz.
Geranium (Carolina) *Geranium carolinianum*	12"	
Goosegrass *Eleusine indica*		
Primrose (cutleaf evening) *Oenothera laciniate*		
Pusley (Florida) *Richardia scabra*		
Sicklepod *Cassia obtusifolia*		
Spanishneedles *Bidens bipinnata*		
Filaree *Erodium* spp.	12"	48 oz.

* For control in no-till systems, use 16 fluid ounces per acre.

** For those rates less than 32 ounces per acre, Roundup at rates up to 32 ounces per acre may be used where heavy weed densities exist.

TANK MIXTURES

ROUNDUP® plus BANVEL plus NONIONIC SURFACTANT

ROUNDUP plus 2,4-D AMINE plus NONIONIC SURFACTANT

DO NOT APPLY BANVEL OR 2,4-D AMINE TANK MIXTURES BY AIR IN CALIFORNIA.

These tank mixtures are recommended for use in fallow and reduced tillage areas only. Follow use directions as given in the "Low-Volume Application" section.

Roundup plus Banvel or 2,4-D amine will control the annual grasses and broadleaf weeds previously listed for Roundup alone at the indicated heights (except 8 ounces per acre applications), plus the following broadleaf weeds. For those weeds previously listed at 8 ounces of Roundup herbicide alone per acre, use the 12-ounce rate in these tank mixtures.

NOTE: Refer to the specific product labels for crop rotation restrictions and cautionary statements of all products used in tank mixtures. Some crop injury may occur if Banvel is applied within 45 days of planting. The addition of Banvel in a mixture with this product may provide short-term residual control of selected weed species.

Apply 12 to 16 ounces of this product plus 0.25 lb. a.i. of Banvel or 0.5 pound a.e. of 2,4-D amine, plus 0.5 to 1 percent nonionic surfactant by total spray volume per acre to control dense populations of the following annual broadleaf weeds when less than 6 inches in height:

Kochia*	*Kochia scoparia*
Lambsquarters	*Chenopodium album*
Lettuce (prickly)	*Lactuca serriola*
Pigweed (redroot)	*Amaranthus retroflexus*
Thistle (Russian)	*Salsola kali*

* Controlled with Banvel tank mixture only.

HIGH-VOLUME BROADCAST APPLICATIONS

When applied as directed under the conditions described, this product will control the weeds listed below when water carrier volumes are 10 to 40 gallons per acre for ground applications.

Use 1 quart of this product per acre if weeds are less than 6 inches tall. If weeds are over 6 inches tall, use 1.5 quarts of this product per acre. If weeds have been mowed, grazed, or cut allow adequate time for new growth to recommended stages prior to treatment. These rates will also provide control of weeds listed in the "Low-Volume Application" section.

WEED SPECIES

Balsamapple*	*Momordica charantia*
Bassia (fivehook)	*Bassia hyssopifolia*

Brome	*Bromus* spp.
Fiddleneck	*Amsinckia* spp.
Flaxleaf Fleabane	*Conyza banariansis*
Fleabane	*Erigeron* spp.
Kochia	*Kochia scoparia*
Lettuce (prickly)	*Lactuca serriola*
Panicum	*Panicum* spp.
Ragweed (common)	*Ambrosia artemisiifolia*
Ragweed (giant)	*Ambrosia trifida*
Smartweed (Pennsylvania)	*Polygonum pensylvanicum*
Sowthistle (annual)	*Sonchus oleraceus*
Sunflower	*Helianthus annuus*
Thistle (Russian)	*Salsola kali*
Velvetleaf	*Abutilon theophrasti*

* Apply with hand-held equipment only.

PERENNIAL WEEDS

Apply this product as follows to control or destroy most perennial weeds:

NOTE: If weeds have been mowed or tilled, do not treat until plants have resumed active growth and are at the recommended stages.

Repeat treatments may be necessary to control weeds regenerating from underground parts or seed. Repeat treatments must be made prior to crop emergence.

When applied as recommended under the conditions described, this product WILL CONTROL the following PERENNIAL WEEDS:

Alfalfa	*Medicago sativa*
Alligatorweed*	*Alternanthera philoxeroides*
Artichoke (Jersusalem)	*Helianthus tuberosus*
Bahiagrass	*Paspalum notatum*
Bermudagrass	*Cynodon dactylon*
Bindweed (field)	*Convolvulus arvensis*
Bluegrass (Kentucky)	*Poa* spp.
Blueweed (Texas)	*Helianthus ciliaris*
Brackenfern	*Pteridium aquilinum*
Bromegrass (smooth)	*Bromus inermis*
Canarygrass (reed)	*Dactylis glomerata*
Cattail	*Typha* spp.
Clover (red)	*Trifolium pratense*
Clover (white)	*Trifolium repens*
Dallisgrass	*Paspalum dilatatum*
Dandelion	*Taraxacum officinale*
Dock (curly)	*Rumex crispus*
Dogbane (hemp)	*Apocynum cannabinum*
Fescues	*Festuca* spp.
Fescue (tall)	*Festuca arundinaceae*
Guineagrass	*Panicum maximum*
Horsenettle	*Solanum carolinense*
Horseradish	*Armoracia rusticana*
Johnsongrass	*Sorghum halepense*
Kikuyugrass	*Pennisetum clandestinum*
Knapweed	*Centaurea repens*
Lantana	*Lantana camàra*
Milkweed	*Asclepias* spp.
Muhly (wirestem)	*Muhlenbergia frondosa*
Mullein (common)	*Verbascum thapsus*
Napiergrass	*Pennisetum purpureum*
Nightshade (silverleaf)	*Solanum elaeagnifolium*
Nutsedge (purple)	*Cyperus rotundus*
(yellow)	*Cyperus esculentus*
Orchardgrass	*Dactylis glomerata*
Paragrass	*Brachiaria mutica*
Quackgrass	*Agropyron repens*
Ryegrass (perennial)	*Lolium perenne*
Smartweed (swamp)	*Polygonum coccineum*
Sweet potato (wild)*	*Ipomoea pandurata*
Thistle (Canada)	*Cirsium arvense*
Timothy	*Phleum pratense*
Torpedograss*	*Panicum repens*
Vaseygrass	*Paspalum urvillai*
Waterprimrose	*Jussiaea* spp.
Wheatgrass (western)	*Agropyron smithii*
Woollyleaf bursage	*Franseria tomentosa*

* Partial control.

Alligatorweed—Apply 4 quarts of this product per acre or apply a 1½ percent solution with hand-held equipment to provide partial control. Apply when most of the plants are in bloom. Repeat applications will be required to maintain such control.

Bermudagrass—For control, apply 5 quarts of this product per acre and for partial control, apply 3 quarts per acre. Treat when bermudagrass is actively growing and seedheads are present. Retreatment may be necessary to maintain control. Allow 7 or more days after application before tillage. See "Directions for Use" and "Mixing and Application" sections of this label for labeled uses and specific application instructions.

Bindweed (field)/Nightshade (silverleaf)/Blueweed (Texas)—Apply 4 to 5 quarts of this product per acre west of the Mississippi River and 3 to 4 quarts per acre east of the Mississippi River. Apply when weed is actively growing and is at or beyond full bloom. For silverleaf nightshade, best results can be achieved when application is made after berries are formed. Do not treat when weed is under drought stress as good soil moisture is necessary for active growth. New leaf development indicates active growth. For best results, apply in late summer or fall. Fall treatments must be applied before a killing frost. Allow 7 or more days after application before tillage. See "Directions for Use" and "Mixing and Application" sections of this label for labeled uses and specific application instructions.

Bluegrass (Kentucky)/Bromegrass (Smooth)/ Orchardgrass—Apply 2 quarts of this product in 10 to 40 gallons of water per acre when the grasses are actively growing and most plants have reached boot-to-early seedhead stage of development. For partial control in pasture or hay crop renovation apply 1 to 1½ quarts of this product plus 0.5 to 1 percent nonionic surfactant by total spray volume in 5 to 10 gallons of water per acre. Use 0.5 percent surfactant concentration when using surfactants which contain at least 50 percent active ingredient or a 1 percent surfactant concentration for those surfactants containing less than 50 percent active ingredient. Apply to actively growing plants when most have reached 4 to 12 inches in height. Allow 7 or more days after application before tillage. See "Directions for Use" and "Mixing and Application" sections and "PASTURES" in the "Cropping Systems" section of this label for labeled uses and specific application instructions.

Brackenfern—Apply 3 to 4 quarts of this product per acre as a broadcast spray or as a 1 to 1½ percent solution with hand-held equipment. Apply to brackenfern after fronds are at least 18 inches long.

Bursage (woollyleaf)—For control apply 2 quarts of this product plus 1 pint of Banvel™ per acre. For partial control apply 1 quart of this product plus 1 pint of Banvel per acre. Add 0.5 to 1 percent nonionic surfactant by total spray volume and apply 3 to 20 gallons of water per acre. Use 0.5 percent surfactant concentration when using surfactants which contain at least 50 percent active ingredient or a 1 percent surfactant concentration for those surfactants containing less than 50 percent active ingredient. Apply when plants are producing new active growth which has been initiated by moisture for at least 2 weeks and when plants are at or beyond flowering. See "Directions for Use" and "Mixing and Application" sections of this label for labeled uses and specific application instructions.

Canarygrass (reed)/Ryegrass (perennial)/Timothy/Wheatgrass (western)—Apply 2 to 3 quarts of this product per acre. For best results, apply to actively growing plants when most have reached the boot to head stage of growth. Allow 7 or more days after application before tillage. See "Directions for Use" and "Mixing and Application" sections of this label for labeled uses and specific application instructions.

Dogbane (hemp)/Knapweed/Horseradish—Apply 4 quarts of this product per acre. Apply when actively growing and most weeds have reached the late bud to flower stage of growth. Following crop harvest or mowing, allow weeds to regrow to a mature stage prior to treatment. For best results, apply in late summer or fall. Allow 7 or more days after application before tillage. See "Directions for Use" and "Mixing and Application" sections of this label for labeled uses and specific application instructions.

Fescue (tall)—Apply 3 quarts of this product in 10 to 40 gallons of water per acre to actively growing plants when most have reached boot-to-early seedhead stage of development. For partial control in pasture or hay renovation apply 2 quarts of this product plus 0.5 to 1 percent by total spray volume of nonionic surfactant in 5 to 10 gallons of water per acre. Use 0.5 percent surfactant concentration when using surfactants which contain at least 50 percent active ingredient or a 1 percent surfactant concentration for those surfactants containing less than 50 percent active ingredient. Apply to actively growing plants when most have reached 4 to 12 inches in height. Allow 7 or more days after application before tillage. See "Directions for Use" and "Mixing and Application" sections and "PASTURES" in the "Cropping Systems" section of this label for labeled uses and specific application instruction.

Guineagrass—Apply 3 quarts of this product per acre or use a 1 percent solution with hand-held equipment. Apply to actively growing guineagrass when most has reached at least the 7-leaf stage of growth. Ensure thorough coverage when using hand-held equipment. Allow 7 or more days after application before tillage. See "Directions for Use" and "Mixing and Application" sections of this label for labeled uses and specific application instructions.

Johnsongrass—Apply 1 to 3 quarts of this product per acre. In annual cropping systems apply 1 to 2 quarts of this product per acre. Apply 1 quart of this product plus 0.5 to 1 percent nonionic surfactant by total spray volume in 5 to 10 gallons of water per acre. Use 0.5 percent surfactant concentration when using surfactant which contain at least 50 percent active ingredient or a 1 percent surfactant concentration for those surfactants containing less than 50 percent active ingredient. Use 2 quarts of this product when applying 10 to 40 gallons of water per acre. In non-crop or areas where annual tillage (no-till), is not performed, apply 2 to 3 quarts of this product in 10 to 40 gallons of water per acre. For best results, apply to actively growing plants when most are at least 18 inches in height and have reached the boot-to-head stage of growth. Allow 7 or more days after application before tillage. Do not tank mix with residual herbicides when using the 1 quart per acre rate. See "Directions for Use" and "Mixing and Application" sections of this label for labeled uses and specific application instructions.

Kikuyugrass—Apply 2 to 3 quarts of this product per acre. Spray when most kikuyugrass is at least 8 inches in height (3 or 4-leaf stage of growth) and actively growing. Allow 3 or more days after application before tillage. See "Directions for Use" and "Mixing and Application" sections of this label for labeled uses and specific application instructions.

Lantana—Apply this product as a 1 to $1\frac{1}{4}$ percent solution using hand-held equipment only. Apply to actively growing lantana at or beyond the bloom stage of growth. Use the higher application rate for plants that have reached the woody stage of growth. Allow 7 or more days after application before tillage.

Milkweed (common)—Apply 3 quarts of this product per acre. Apply when actively growing and most of the milkweed has reached the late bud to flower stage of growth. Following small grain harvest or mowing, allow milkweed to regrow to a mature stage prior to treatment. Allow 7 or more days after application before tillage. See "Directions for Use" and "Mixing and Application" sections of this label for labelled uses and specific application instructions.

Muhly (wirestem)—Apply 1 to 2 quarts of this product per acre. Use 1 quart of this product plus 0.5 to 1 percent nonionic surfactant by total spray volume in 5 to 10 gallons of water per acre. Use 0.5 percent surfactant concentration when using surfactants which contain at least 50 percent active ingredient or a 1 percent surfactant concentration for those surfactants containing less than 50 percent active ingredient. Use 2 quarts of this product when applying 10 to 40 gallons of water per acre or in pasture, sod, or noncrop areas. Spray when the wirestem muhly is 8 inches or more in height and actively growing. Do not till between harvest and fall applications or in the fall or spring prior to spring applications. Allow 3 or more days after application before tillage. This product will not provide residual control of wirestem muhly from seeds which germinate after a Roundup herbicide application. Do not tank mix with residual herbicides when using the 1 quart per acre rate. See "Directions for Use" and "Mixing and Application" sections of this label for labeled uses and specific application instructions.

Nutsedge (purple, yellow)—Apply 3 quarts of this product per acre as a broadcast spray, or apply a 1 percent solution from hand-held equipment to control existing nutsedge plants and immature nutlets attached to treated plants. Treat when plants are in flower or when new nutlets can be found at rhizome tips. Nutlets which have not germinated will not be controlled and may germinate following treatment. Repeat treatments will be required for long-term control. Wait 7 days after treatment before tillage. Tillage will stimulate nutlet germination.

Apply 1 quart of this product per acre, plus 0.5 to 1 percent nonionic surfactant in 3 to 40 gallons of water per acre to provide partial control of existing nutsedge plants. Use 0.5 percent surfactant concentration when using surfactants which contain at least 50 percent active ingredient or a 1 percent surfactant concentration for those surfactants containing less than 50 percent active ingredient. Treat when plants have 3 to 5 leaves (less than 6 inches). Repeat treatments at this rate may be applied as necessary to control subsequent emerging plants. Wait 7 days after treatment before tillage or mowing.

Phragmites—For partial control of phragmites in Florida and the counties of other states bordering the Gulf of Mexico, apply 5 quarts per acre as a broadcast spray or apply a 2 percent solution from hand-held equipment. In other areas of the U.S., apply 3 quarts per acre as a broadcast spray or apply a 1 percent solution from hand-held equipment for partial control. For best results, treat during late summer or fall months.

Quackgrass—in annual cropping systems, or in pastures and sods followed by deep tillage: Apply 1 to 2 quarts of this product per acre. For the one quart rate, apply 0.5 to 1 percent nonionic surfactant by total spray volume in 5 to 10 gallons of water per acre. Use the 0.5 percent surfactant concentration when using surfactants which contain at least 50 percent active ingredient or a 1 percent surfactant concentration for those surfactants which contain at least 50 percent active ingredient or a 1 percent surfactant concentration for those surfactants containing less than 50 percent active ingredient. For the 2 quart rate, apply in 10 to 40 gallons of water per acre. Do not tank mix with residual herbicides when using the 1 quart rate. Spray when quackgrass is 8 to 12 inches in height and actively growing. Do not till between harvest and fall applications or in the fall or spring prior to spring application. Allow 3 or more days after application before tillage. In pastures or sods, for best results use a moldboard plow.

Quackgrass—pasture or sod or other noncrop areas where deep tillage is not planned following application: Apply 2 to 3 quarts in 10 to 40 gallons of water per acre. Spray when the quackgrass is greater than 8 inches tall and actively growing. Do not till between harvest and fall application or in fall or spring prior to spring application. Allow 3 or more days after application before tillage.

See Directions for use and "Mixing and Application" section of this label for labeled uses and specific application instructions.

Sweet potato, wild—Apply this product as a 2% solution using hand-held equipment. Apply to actively growing weeds that are at or beyond the bloom stage of growth. Repeat applications will be required. Allow the plant to reach the recommended stage of growth before retreatment. Allow 7 or more days before tillage.

Thistle (Canada)—Apply 2 to 3 quarts of this product per acre. Apply to actively growing thistles when most are at or beyond the bud stage of growth. After harvest, mowing or tillage in the late summer or fall, allow at least 4 weeks for initiation of active growth and rosette development prior to the application of this product. Fall treatments must be applied before a killing frost. Allow 3 or more days after application before tillage. See "Directions for Use" and "Mixing and Application" sections of this label for labeled uses and specific application instructions.

Torpedograss—Apply 4 to 5 quarts of this product per acre to provide partial control of torpedograss. Apply to actively growing torpedograss when most plants are at or beyond the seedhead stage of growth. Repeat applications will be required to maintain control. Fall treatments must be applied before frost. Allow 7 or more days after application before tillage. See "Directions for Use" and "Mixing and Application" sections of this label for labeled uses and specific application instructions.

Waterprimrose—Apply this product as a 1 percent solution using hand-held equipment. Apply to plants that are actively growing at or beyond the bloom stage of growth, but before fall color changes. Thorough coverage is necessary for best control.

Other perennials listed on this label—Apply 3 to 5 quarts of this product per acre. Apply when actively growing and most have reached early head or early bud stage of growth. Allow 7 or more days after application before tillage. See "Directions for Use" and "Mixing and Application" sections of this label for labeled uses and specific application instructions.

™Banvel is a trade mark of the Velsicol Chemical Company.

TANK MIXTURES

When applied as directed, under the conditions described, Roundup plus Banvel or 2,4-D amine will suppress or control the following perennial broadleaf weed:

Bindweed (field)

Convolvulus arvensis

For suppression, apply 16 ounces of this product plus 0.5 lb a.e. of 2,4-D amine or 0.25 lb a.i. of Banvel, plus 0.5 to 1 percent nonionic surfactant by total spray volume in 5 to 10 gallons of water per acre for ground applications and 3 to 5 gallons of water per acre for aerial applications. Applications should be delayed until maximum emergence has occurred and when vines are between 6 to 18 inches in length. These tank mixtures may be applied using aerial application equipment (except in California) in fallow and reduced tillage systems only.

For control, apply 2 quarts of this product plus 0.5 pound a.i. of Banvel in 10 to 20 gallons of water per acre. At these rates, apply using ground application equipment only.

Refer to the specific product labels for crop rotation restrictions and cautionary statements of all products in tank mixture.

WOODY BRUSH AND TREES

When applied as recommended under the conditions described this product CONTROLS or PARTIALLY CONTROLS the following woody brush plants and trees:

Alder	*Alnus* spp.
Ash*	*Fraxinus* spp.
Aspen (quaking)	*Populus tremuloides*
Birch	*Betual* spp.
Blackberry	*Rubus* spp.
Broom:	
French	*Cytisus monspessulanus*
Scotch	*Cytisus scoparius*
Buckwheat (California)*	*Eriogonum fasciculatum*
Cascara*	*Rhamnus purhiana*
Catsclaw*	*Acacia greggi*
Ceanothus:	
Deerbrush*	*Ceanothus integerrimus*
Redstem*	*Ceanothus sanguineus*
Chamire	*Adenostoma fasciculatum*
Cherry:	
Bitter	*Prunus emarginata*
Black	*Prunus serotina*
Pin	*Prunus Pensylvanica*
Coyotebrush	*Baccharis consanguinea*
Dewberry	*Rubus trivialis*
Elderberry	*Sambucus* spp.
Elm*	*Ulmus* sp.
Eucalyptus (bluegum)	*Eucalyptus glotulus*
Hasardia*	*Haplopappus squamosus*
Hawthorn	*Crataegus* sp.
Hazel	*Corylus cornuta californica*
Honeysuckle	*Lonicera* spp.
Kudzu	*Pueraria lobata*
Locust* (black)	*Robinia pseudoacacia*
Maple:	
Red**	*Acer rubrum*
Sugar	*Acer saccharum*
Vine*	*Acer circinatum*
Monkey Flower*	*Mimulus guttatus*
Oak:	
Black*	*Quercus velutina*
Northern Pin	*Quercus palustris*
Post	*Quercus stellata*
Red	*Quercus rubra*
Southern Red	*Quercus falcata*
White*	*Quercus alba*
Persimmon*	*Diospyros* spp.
Poison Ivy	*Rhus radicans*
Poison Oak	*Rhus toxicodendron*
Poplar* (yellow)	*Liriodendron tulipifera*
Raspberry	*Rubus* spp.
Rose (mutiflora)	*Rosa mutiflora*
Sage (black)	*Salvia mellifera*
Sagebrush (California)	*Artemisia Californica*
Salmonberry	*Rubus spectabilis*
Sassafras	*Sassafras aibidrum*
Sourwood*	*Oxydendrum arboreum*
Sumac:	
Poison*	*Rhus vernix*
Smooth*	*Rhus glabra*
Winged*	*Rhus copallina*
Sweetgum	*Liquidambar styraciflua*
Swordfern*	*Polystichum minitum*
Tallowtree (Chinese)	*Sapium sebiferum*
Tobacco (tree)*	*Nicotiana glauca*
Thimbleberry	*Rubus parviflorus*
Trumpet Creeper	*Campsis radicans*
Virginia Creeper*	*Parthenocissus quinquefolia*
Willow	*Salix* spp.

* Partial control.

** See below for Control or Partial Control instructions.

**See below for Control or Partial Control instructions.

NOTE: If brush has been mowed or tilled or trees have been cut, do not treat until regrowth has reached the recommended stages of growth.

Apply this product when plants are actively growing, and unless otherwise directed, after full leaf expansion. Use the higher rate for larger plants and/or dense areas of growth. On vines, use the higher rate for plants that have reached the woody stage of growth. Best results are obtained when application is made in late summer or fall after fruit formation.

In arid areas, best results are obtained when application is made in the spring to early summer when brush species are at high moisture content and are flowering.

Ensure thorough coverage when using hand-held equipment. Symptoms may not appear prior to frost or senescence with fall treatments.

Allow 7 or more days after application before tillage, mowing or removal. Repeat treatments may be necessary to control plants regenerating from underground parts or seed. Some autumn colors on undesirable deciduous species are acceptable provided no major leaf drop has occurred. Reduced performance may result if fall treatments are made following a frost.

See "Directions for Use", and "Mixing Application" sections of this label for labeled uses and specific application instructions.

Apply this product as follows to control or partially control the following woody brush and trees.

Alder/Dewberry/Honeysuckle/Post Oak/Raspberry —For control, apply 3 to 4 quarts per acre of this product as a broadcast spray or as a 1 to 1½ percent solution with hand-held equipment.

Aspen (quaking)/Bitter Cherry/Black Cherry/Hawthorn/Pin Cherry/Southern Red Oak/Sweetgum/Trumpet Creeper—For control, apply 2 to 3 quarts of this product per acre as a broadcast spray or as a 1 to 1½ percent solution with hand-held equipment.

Birch/Elderberry/Hazel/Salmonberry/Thimbleberry—For control, apply 2 quarts per acre of this product as a broadcast spray or as a 1 percent solution with hand-held equipment.

Blackberry—For control, apply 3 to 4 quarts per acre of this product as a broadcast spray or 0.75 percent solution of this product plus a nonionic surfactant with hand-held equipment. Add 0.5 percent surfactant concentration when using surfactant which contains at least 50 percent active ingredient or a 1 percent surfactant concentration for those surfactants containing less than 50 percent active ingredient. When using percent solution, thorough coverage of plant is necessary. Apply in the fall to actively growing plants when most foliage is still green. For best results, apply to blackberries after fruits set or drop. Late applications can be made as long as most of the foliage is still green and actively growing.

Broom (French, Scotch)—For control, apply a 1½ to 2 percent solution with hand-held equipment.

Buckwheat (California)/Hasardia/Monkey Flower/Tobacco (tree)—For partial control of these species, apply a 1 to 2 percent solution of this product as a foliar spray with hand-held equipment. Thorough coverage of foliage is necessary for best results.

Catsclaw—For partial control, apply as a 1 to 1½ percent solution with hand-held equipment.

Coyote Brush—For control, apply a 1½ to 2 percent solution with hand-held equipment when at least 50 percent of the new leaves are fully developed.

Eucalyptus (bluegum)—For control of Eucalyptus resprouts, apply a 2 percent solution of this product with hand-held equipment when resprouts are 6 to 12 feet tall. For control of roots and resprouts of cut stumps, apply a 50 percent solution of this product to freshly cut surface immediately after cutting. Delay in applying this product may result in poor performance.

Kudzu—For control, apply 4 quarts of this product per acre as a broadcast spray or as a 2 percent solution with hand-held equipment. Repeat applications will be required to maintain control.

****Maple (red)**—For control, apply as a 1 to 1½ percent solution with hand-held equipment when at least 50 percent of the new leaves are fully developed.

For partial control, apply 2 to 4 quarts of this product per acre as a broadcast spray when plants are at or beyond full flowering.

Maple (Sugar)/Oak (Northern pin)/Oak (red)—For control, apply as a 1 to 1½ percent solution with hand-held equipment when at least 50 percent of the new leaves are fully developed.

Poison Ivy/Poison Oak—For control, apply 4 to 5 quarts of this product per acre as a broadcast spray or as a 2 percent solution with hand-held equipment. Repeat applications may be required to maintain control. Fall treatments must be applied before leaves lose green color.

Rose (mutiflora)—For control, apply 2 quarts of this product per acre as a broadcast spray or as a 1 percent solution with hand-held equipment. Treatments should be made prior to leaf deterioration by leaf-feeding insects.

Sage (black)/Sagebrush (California)/Chamise/Tallowtree (Chinese)—For control of these species, apply a 1 percent solution of this product as a foliar spray with hand-held equipment. Thorough coverage of foliage is necessary for best results.

Willow—For control, apply 3 quarts of this product per acre as a broadcast spray or as a 1 percent solution with hand-held equipment.

***Other Woody Brush and Trees listed on this label**—For partial control, apply 2 to 4 quarts of this product per acre as a broadcast spray or as a 1 to 1½ percent solution with hand-held equipment.

DIRECTIONS FOR USE

It is a violation of Federal law to use this product in any manner inconsistent with its labeling.

NON-CROP USES

See "General Information" and "Mixing and Application Instructions" sections of this label for essential product performance information and the following NON-CROP SECTIONS for specific recommended uses.

EXTREME CARE MUST BE EXERCISED TO AVOID CONTACT OF SPRAY WITH FOLIAGE OF DESIRABLE TURFGRASSES, TREES, SHRUBS, OR OTHER DESIRABLE VEGETATION SINCE SEVERE DAMAGE OR DESTRUCTION MAY RESULT.

NOTE: If spraying areas adjacent to desirable plants, use a shield made of cardboard, sheet metal or plyboard while spraying to help prevent spray from contacting foliage of desirable plants.

Repeat treatments may be necessary to control weeds regenerating from underground parts or seeds.

ROUNDUP herbicide does not provide residual weed control. For subsequent weed control, follow a label approved herbicide program.

Read and carefully observe the cautionary statements and all other information appearing on the labels of all herbicides used.

INDUSTRIAL, RECREATIONAL AND PUBLIC AREAS

When applied as directed for "Non-Crop Uses", under conditions described, this product controls annual and perennial weeds listed on this label growing in areas such as airports, ditch banks, dry ditches, dry canals, fencerows, golf courses, highways, industrial plant sites, lumberyards, parking areas, parks, petroleum tank farms and pumping installations, pipelines, power and telephone rights-of-way, railroads, roadsides, schools, storage areas, other public areas and similar industrial or non-crop areas.

For specific rates of application and instructions for control of various annual and perennial weeds and woody brush and trees, see the "Weeds Controlled" section of this label.

This product may be applied with recirculating sprayers, shielded applicators, or wiper applicators in any non-crop site specified on this label. See the "Selective Equipment" part of "APPLICATION EQUIPMENT AND TECHNIQUES" section of this label for information on proper use and calibration of this equipment.

TANK MIXTURES FOR INDUSTRIAL SITES

ROUNDUP® plus OUST™

Use on industrial sites including airports, industrial plants, lumberyards, petroleum tank farms, pumping stations, pipelines, railroads, roadsides, storage areas or other similar sites where bare ground is desried.

NOTE: Do not apply with aerial equipment.

When applied as directed for "Noncrop Uses" under the conditions described, this product plus Oust provides control of annual weeds listed in the "Weeds Controlled" section of this and the Oust label and control or partial control of the following perennial weeds.

Apply 1 to 2 quarts of this product with 2 to 4 ounces of Oust in 10 to 40 gallons of spray solution per acre as a broadcast spray to actively growing weeds.

For control of annual weeds, use the lower rate of these products.

For control of the listed perennial weeds use the higher rates of both products. For partial control, use the lower rates.

Bahiagrass	*Paspalum notatum*
Bermudagrass*	*Cynodon dactylon*
Broomsedge	*Andropogon virginicus*
Dock (curly)	*Rumex crispus*
Dogfennel	*Eupatorium capilliforium*
Fescue (tall)	*Festuca arundinacea*
Johnsongrass**	*Sorghum halepense*

Poorjoe**	*Diodia teres*
Quackgrass	*Agropyron repens*
Trumpetcreeper*	*Campsis radicans*
Vaseygrass	*Paspalum urvillei*
Vervain (blue)	*Verbena hastata*

* Suppression at the higher rates only.
** Control at the lower rates.
Read and carefully observe the cautionary statements and all other information appearing on the labels of all herbicides used.
™ Oust is a trademark of E.I. duPont deNemours and Company.

TANK MIXTURES—NON-CROP SITES

When applied as a tank mixture, this product provides control of the emerged annual weeds and partial control of the emerged perennial weeds listed in this label. When applied as a tank mixture, the following residual herbicides will provide pre-emergence control of the weeds listed in the individual product labels.

ROUNDUP® plus KROVAR™ I
ROUNDUP plus KROVAR II
ROUNDUP plus SIMAZINE CALIBER™ 90
ROUNDUP plus SIMAZINE 4L
ROUNDUP plus SIMAZINE 80W
ROUNDUP plus SURFLAN 75W
ROUNDUP plus SURFLAN AS

When tank mixing with residual herbicides add an agriculturally approved nonionic surfactant at 0.5 to 1 percent by volume of spray solution. Use 0.5 percent surfactant concentration when using surfactants which contain at least 50 percent active ingredient or a 1 percent concentration for those surfactants containing less than 50 percent active ingredients. See the "Mixing and Application Instructions" section of this label before preparing these tank mixtures.

Read and carefully observe the label claims, cautionary statements, recommended use rates and all other information on the labels of all products used in these tank mixtures. Use according to the most restrictive label directions for each product in the mixture.

CONTROL OF EMERGED WEEDS

Annual Weeds—Apply 1 quart per acre of this product in these tank mixtures when weeds are less than 6 inches tall and 1.5 quarts per acre when weeds are more than 6 inches tall.

Perennial Weeds—For partial control of perennial weeds using these tank mixtures apply 2 to 5 quarts per acre of this product. Follow the recommendations in the "Weeds Controlled" section of this label for stage of growth and rate of application for specific perennial weeds.

PREEMERGENCE WEED CONTROL

The following are the labeled rates for the recommended residual herbicides for non-crop uses:

Product	Rate/Acre
Krovar I	4 to 6 lbs.
Krovar II	2 to 5 lbs.
Simazine, Caliber 90	5.3 to 11.1 lbs.
Simazine 80W	6 to 12½ lbs.
Simazine 4L	4.8 to 10.0 qts.
Surflan 75W	2⅔ to 5⅓ lbs.
Surflan AS	2 to 4 qts.

Refer to the individual product labels for specific noncrop sites, rates, carrier volumes and precautionary statements.

Mix only the quantity of spray solution which can be used during the same day. Do not allow these tank mixtures to stand overnight as this may result in reduced weed control.

APPLY THESE TANK MIXTURES THROUGH CONVENTIONAL BROADCAST EQUIPMENT ONLY.

TM Krovar is a trademark of E.I. du Pont de Nemours and Company.
TM Caliber is a trademark of Ciba-Geigy Corporation.
TM Surflan is a trademark of Elanco Products Company.

FARMSTEAD WEED CONTROL

When applied as directed for "Non-Crop Uses," under conditions described, this product controls undesirable vegetation listed on this label around farmstead building foundations, along and in fences, shelterbelts, and for general non-selective farmstead weed control.

For specific rates of application and instructions for control of various annual and perennial weeds, see the "Weeds Controlled" section of this label.

ORNAMENTALS

THIS PRODUCT IS NOT RECOMMENDED FOR USE AS AN OVER-THE-TOP BROADCAST SPRAY IN ORNAMENTALS.

When applied as directed for the conditions described for "Non-Crop Uses", this product controls undesirable vegetation listed on this label prior to planting ornamentals, in established ornamentals, and within and around greenhouses and shadehouses.

For specific rates of application and instructions for control of various annual and perennial weeds, see the "Weeds Controlled" section of this label.

Where repeat applications are necessary, do not exceed 10.6 quarts of this product per acre per year.

Site Preparation—Following preplant applications of this product, any ornamental species may be planted. Precautions should be taken to protect non-target plants during site preparation applications.

Greenhouse/Shadehouse Use—This product may be used to control weeds listed on this label which are growing in greenhouses. Desirable vegetation must not be present during application and air circulation fans must be turned off.

Post Directed Spray—Use as a directed spray toward the base of established woody ornamentals species listed below.

Arborvitae	*Thuja* spp.
Azalea	*Rhododendron* spp.
Boxwood	*Buxus* spp.
Crabapple	*Malus* spp.
Euonymus	*Euonymus* spp.
Fir	*Abies* spp.
	Pseudotsuga spp.
Hollies	*Ilex* spp.
Lilac	*Syringa* spp.
Magnolia	*Magnolia* spp.
Maple	*Acer* spp.
Oak	*Quercus* spp.
Privet	*Ligustrum* spp.
Pine	*Pinus* spp.
Spruce	*Picea* spp.
Yew	*Taxus* spp.

SILVICULTURAL SITES, RIGHTS-OF-WAY and CHRISTMAS TREES

NOTE: NOT RECOMMENDED FOR USE AS AN OVER-THE-TOP BROADCAST SPRAY IN SILVICULTURAL NURSERIES, OR CHRISTMAS TREE PLANTATIONS.

When applied as directed for "Non-Crop Uses" under conditions described, this product controls undesirable vegetation listed on this label. This product also suppresses or controls undesirable vegetation listed on this label when applied at recommended rates for release of established coniferous species listed on this label.

For specific rates of application and instructions for control of various brush, annual and perennial weeds, see the "Weeds Controlled" section of this label. For specific rates of application for release of listed coniferous species, see the "Conifer Release" part of this section of the label.

Where repeat applications are necessary do not exceed 10.6 quarts of this product per acre per year.

Aerial Application—This product may be applied using aerial spray equipment for silvicultural site preparation, conifer release and rights-of-way treatments. See the "Application Equipment and

Techniques" part of the "Mixing and Application" section of this label for information on how to properly spray this product by air. DO NOT APPLY THIS PRODUCT BY AIR TO RIGHTS-OF-WAY SITES IN CALIFORNIA.

In order to reduce the aerial application drift hazard to aquatic sites*, non-targeted sites or any site containing desirable vegetation always maintain appropriate buffer zones. A buffer zone of the following minimum distances should be maintained:

- Helicopter using a Microfoil™ boom should maintain at least a 50-foot buffer zone.
- When using other aerial equipment:
  1. Maintain at least a 75-foot buffer zone for applications using 2 quarts or less per acre of this product.
  2. Maintain at least a 125-foot buffer zone for application using more than 2 quarts per acre of this product.
  3. Maintain at least a 400-foot buffer zone for applications on rights-of-way when applied from 75 feed or more above ground level.

These distances should be increased if conditions favoring drift exist.

*Aquatic sites include all lakes, ponds, and streams used for significant domestic purposes or angling.

™Microfoil is a trademark of Union Carbide Agricultural Products, Inc.

SITE PREPARATION

Following preplant applications of this product, any silvicultural species may be planted.

POST DIRECTED SPRAY

In established silvicultural sites, use as a spray on the foliage of undesirable vegetation. Care must be exercised to avoid contact of spray, drift or mist with foliage of green bark of desirable species.

CONIFER RELEASE

For release, apply only where conifers have been established for more than one year. Vegetation should not be disturbed prior to treatment or until visual symptoms appear after treatment. Symptoms of treatment are slow to appear, especially in woody species treated in late fall. Injury may occur to conifers treated for release, especially where spray patterns overlap or the higher rates are applied or when applications are made during periods of active conifer growth.

Applications must be made after formulation of final conifer resting buds in the fall or prior to initial bud swelling in spring. Some autumn colors on undesirable deciduous species are acceptable provided no major leaf drop has occurred. Use of the following rates for conifer release to control or partially control the weeds listed in the "Weeds Controlled" section of the label.

For release of the following conifer species:

Douglas fir	*Pseudotsuga menziesii*
Fir	*Abies* spp.
Hemlock	*Tsuga* spp.
Pines*	*Pinus* spp.
Spruce	*Picea* spp.

*Includes all species except Eastern White Pine, Loblolly Pine or Slash Pine.

Apply 1½ to 2 quarts of this product per acre except in Washington and Oregon, west of the crest of the Cascade Mountains. For spring treatments west of the crest of the Cascade Mountains apply 1 quart of this product per acre before conifer bud swell for control of annual weeds. For fall treatments in Washington and Oregon west of the crest of the Cascade Mountains, apply 1 to 1½ quarts of this product per acre before any major leaf drop of deciduous species.

For release of western hemlock apply 1 quart of this product per acre.

For release of the following conifer species:

Loblolly Pine	*Pinus taeda*
Eastern White Pine	*Pinus strobus*
Slash Pine	*Pinus elliotti*

Late Season Application—Apply 1½ to 2 quarts of this product in a minimum of 5 gallons of spray solution per acre during early autumn. Applications made prior to September 1 or when conditions are conducive to rapid growth of conifers will result in potential for increased injury in the form of tip and/or needle burn. Injury may decrease with later applications. Some autumn colors are acceptable at time of application. Apply prior to frost or leaf drop of undesirable plants. Applications made according to label direction will release Loblolly Pine, Eastern White Pine and Slash Pine by reducing competition from the following species:

Ash	*Fraxinus* spp.
Cherry:	
Black	*Prunus serotina*
Pin	*Prunus Pensylvanica*
Elm	*Ulmus* spp.
Hawthorn	*Crataegus* spp.
Locust (Black)	*Robinia pseudoacacia*
Maple (red)	*Acer rubra*
Oak:	
Black	*Quercus velutina*
Post	*Quercus stellata*
Southern Red	*Quercus falcata*
White	*Quercus alba*
Persimmon	*Diospyros* spp.
Poplar (yellow)	*Liriodendron tulipifera*
Sassafras	*Sassafras albidum*
Sourwood	*Oxydendrum arboreum*
Sumac:	
Poison	*Rhus vernix*
Smooth	*Rhus glabra*
Winged	*Rhus copallina*
Sweetgum	*Liquidambar styraciflua*

Apply only to those sites where woody brush and trees listed in this label constitute the majority of the undesirable species.

Roundup plus Oust Tank Mixtures for Conifer Release from Herbaceous Weeds

To release **loblolly pines** from herbaceous weeds, tank mixtures of this product with Oust will provide control of annual weeds listed in the "Weeds Controlled" section of this and the Oust label, and partial control of the following perennial weeds listed below.

Apply 16 to 24 fluid ounces of Roundup with 2 to 4 ounces of Oust in 10 to 30 gallons of spray solution per acre. Make application to actively growing weeds as a broadcast spray over the top of the young loblolly pines.

For control of annual weeds below 12 inches in height (or runner length on annual vines) use the low rates of both products. Use the higher rates of both products when annual weeds are in more advanced stages of growth and approaching flower or seed formation.

Use the higher rates of both products for partial control of the following perennial weeds. Use the lower rates for suppression of growth.

Bahiagrass	*Paspalum notatum*
Broomsedge	*Andropogon virginicus*
Dock (curly)	*Rumex crispus*
Dogfennel	*Eupatorium capilliforium*
Fescue (tall)	*Festuca arundinacea*
Johnsongrass**	*Sorghum halepense*
Poorjoe**	*Diodia teres*
Trumpetcreeper*	*Campsis radicans*
Vaseygrass	*Paspalum urvillei*
Vervain (blue)	*Verbena hastata*

* Suppression at the higher rates only.
** Control at the higher rates.

Pine damage may occur or can be accentuated if treatment takes place when young trees are under strees from drought, flood water, insects or diseases.

Read and observe the cautionary statements and all other information appearing on the label of all herbicides used.

NOTE TO USER

This product must not be used in areas where adverse impact on Federally designated endangered/threatened plant or aquatic species is likely.

Prior to making applications the user of this product must determine that no such species are located in or immediately adjacent to the area to be treated.

INJECTION AND FRILL APPLICATIONS

Woody vegetation may be controlled by injection or frill application of this product. Apply this product using suitable equipment which must penetrate into living tissue. Use this product without dilution and apply at least 1 ml (1/29 oz.) of herbicide solution for each 2 to 3 inches of trunk diameter breast height (DBH). Space applications evenly around the circumference of the trunk. Application should be made during periods of active growth and full leaf expansion.

This treatment WILL CONTROL the following woody species:

Oak	*Quercus* spp.
Poplar	*Populus* spp.
Sweetgum	*Liquidambar styraciflua*
Sycamore	*Platanus occidentalis*

This treatment WILL SUPPRESS the following woody species:

Black gum	*Nyssa sylvatica*
Dogwood	*Cornus* spp.
Hickory	*Carya* spp.
Maple (red)	*Acer rubrum*

TURFGRASSES AND GRASSES FOR SEED PRODUCTION

PREPLANT AND RENOVATION

When applied as directed for "Non-Crop Uses," under conditions described, this product controls most existing vegetation prior to the planting or renovation of either turfgrasses or grass seed production areas.

For specific rates of application and instructions for control of various annual and perennial weeds, and woody brush and trees, see the "Weeds Controlled" section of this label.

For maximum control of existing vegetation, delay planting to determine if any regrowth from escaped underground plant parts occurs. Where repeat treatments are necessary, sufficient regrowth must be attained prior to application. For warm-season grasses, such as bermudagrass, summer or fall application provide best control.

DO NOT DISTURB SOIL OR UNDERGROUND PLANT PARTS BEFORE TREATMENT. Tillage or renovation techniques such as vertical mowing, coring or slicing should be delayed for 7 days after application to allow proper translocation into underground plant parts.

TURFGRASSES

Where existing vegetation is growing in a field or unmowed situation, apply this product to actively growing weeds at stages of growth given in the "Weeds Controlled" section of this label.

Where existing vegetation is growing under mowed turfgrass management, apply this product after omitting at least one regular mowing to allow sufficient growth for good interception of the spray.

Desirable turfgrasses may be planted following the above procedures.

GRASSES FOR SEED PRODUCTION

Apply this product to actively growing weeds at the stages of growth given in the "Weeds Controlled" section of this label prior to planting or renovation of turf or forage grass areas grown for seed production.

DO NOT feed or graze treated areas within 8 weeks after application.

ANNUAL WEED CONTROL IN DORMANT BERMUDAGRASS AND BAHIAGRASS TURF

When applied as directed for "NONCROP USES", this product will provide control or suppression of many winter annual weeds and tall fescue for effective release of dormant bermudagrass and bahiagrass turf. Refer to the rate table for Roundup alone under the "RELEASE OF DORMANT BERMUDAGRASS AND BAHIAGRASS" section of this label for recommended rates and volumes on the species to be suppressed or controlled. Treat only when turf is dormant and prior to spring greenup. Do not exceed 16 ounces per acre of Roundup herbicide for this use. Do not tank mix with Oust as turf injury or delayed greenup may result.

RELEASE OF BERMUDAGRASS OR BAHIAGRASS

NOTE: Use only in areas where bermudagrass or bahiagrass are desirable ground covers and where some temporary injury or discoloration can be tolerated. Use Roundup plus Oust tank mixtures only on railroads, highways, utility plant sites, or other right-of-way areas.

When applied as directed for "Noncrop Uses", this product will provide control or suppression of many winter annual weeds and tall fescue for effective release of dormant bermudagrass or bahiagrass. This product may be tank-mixed with Oust as recommended for residual control. Make applications to dormant bermudagrass or bahiagrass. Roundup plus Oust tank mixture may delay greenup. **To avoid delays in greenup and minimize injury, do not add more than 1 ounce per acre of Oust on bermudagrass or more than ½ ounce per acre on bahiagrass, or treat when these grasses are in a semidormant condition.**

For best results on winter annuals, treat when plants are in an early growth stage (below 6 inches in height) after most have germinated. For best results on tall fescue, treat when fescue is in or beyond the 4 to 6-leaf stage.

WEEDS CONTROLLED

Rate recommendations for control or suppression of winter annuals and tall fescue are listed below.

Apply the recommended rates of this product alone or as a tank mixture in 10 to 25 gallons of water, plus 0.5 to 1 percent nonionic surfactant by total spray volume per acre. Use 0.5 percent surfactant concentration when using surfactants which contain at least 50 percent active ingredient or a 1 percent surfactant concentrate for those surfactants containing less than 50 percent active ingredient.

For the best recommendation for the mixture of weeds within your geographic areas, contact your Monsanto sales representative.

WEEDS CONTROLLED OR SUPPRESSED WITH ROUNDUP® ALONE*

NOTE: C=Control

S=Suppression

	ROUNDUP® Oz/Acre					
Weed Species	**8**	**12**	**16**	**24**	**32**	**64**
Barley (little) *Hordeum Pusilium*	S	C	C	C	C	C
Bedstraw (catchweed) *Galium aparine*	S	C	C	C	C	C
Bluegrass (annual) *Poa annua*	S	C	C	C	C	C
Chervil *Chaerophyllum tainturieri*	S	C	C	C	C	C
Chickweed (common) *Stellaria media*	S	C	C	C	C	C
Clover (crimson) *Trifolium incarnatum*	•	S	S	C	C	C
Clover (large hop) *Trifolium campestre*	•	S	S	C	C	C
Fescue (tall) *Festuca arundinaceae*	•	•	•	•	S	S
Geranium (Carolina) *Geranium carolinianum*	•	•	S	S	C	C
Henbit *Lamium amplexicaule*	•	S	C	C	C	C
Ryegrass (Italian) *Lolium multiflorum*	•	•	S	C	C	C
Speedwell (corn) *Veronica arvensis*	S	C	C	C	C	C
Vetch (common) *Vicia sativa*	•	•	S	C	C	C

* These rates apply only to sites where an established competitive turf is present.

WEEDS CONTROLLED OR SUPPRESSED WITH ROUNDUP® PLUS OUST*

NOTE: C=Control
S=Suppression ROUNDUP® + OUST (OZ/ACRE)

Weed Species	Roundup + Oust	8 + ¼	12 + ¼	12 + ½	16 + ¼	16 + ½	12 + 1	16 + 1
Barley (little) *Hordeum pusilium*		C	C	C	C	C	C	C
Bedstraw (catchweed) *Galium aparine*		C	C	C	C	C	C	C
Bluegrass (annual) *Poa annua*		S	C	C	C	C	C	C
Chervil *Chaerophyllum tainturieri*		C	C	C	C	C	C	C
Chickweed (common) *Stellaria media*		S	C	C	C	C	C	C
Clover (crimson) *Trifolium incarnatum*		S	S	S	S	C	C	C
Clover (large hop) *Trifolium campestre*		•	•	S	S	S	C	C
Fescue (tall) *Festuca arundinaceae*		•	•	•	•	•	S	S
Geranium (Carolina) *Geranium carolinianum*		•	S	S	C	C	C	C
Henbit *Lamium amplexicaule*		•	S	C	C	C	C	C
Ryegrass (Italian) *Lolium multiflorum*		•	S	S	C	C	C	C
Speedwell (corn) *Veronica arvensis*		S	C	C	C	C	C	C
Vetch (common) *Vicia sativa*		C	C	C	C	C	C	C

* These rates or mixtures of rates apply only to sites where an established competitive turf is present.

RELEASE OF ACTIVELY GROWING BERMUDAGRASS

When applied as directed, this product will aid in the release of bermudagrass by providing control of annual species listed in the "Weeds Controlled" section of this and the Oust label, and suppression or partial control of certain perennial weeds.

For control or supression of those annual species listed on the Roundup label, use 1 to 3 pints of this product as a broadcast spray in 10 to 25 gallons of spray solution per acre. Use the lower rate when treating annual weeds below 6 inches in height (or length of runner in annual vines). Use higher rate as size of plants increases or as they approach flower or seedhead formation.

Use the higher rate of this product for partial control of the following perennial species. Use the lower rates for suppression of growth. For best results, see the "Weeds Controlled" section for proper stage of growth.

Bahiagrass	*Paspalum notatum*
Fescue (tall)	*Festuca arundinacea*
Johnsongrass**	*Sorghum halepense*
Trumpetcreeper*	*Campsis radicans*
Vaseygrass	*Paspalum urvillei*

* Suppression at higher rates only.
** Control at the higher rates.

This product may be tank-mixed with Oust. If tank-mixed, use no more than 1 to 2 pints per acre of Roundup herbicide with 1 to 2 ounces of Oust per acre.

Use the lower rates of both mixtures to control annual weeds below 6 inches in height (or runner length in annual vines) that are listed in the "Weed Controlled" section of this booklet and the Oust label. Use the higher rates as annual weeds increase in size and approach the flower or seedhead stages.

Use the higher rates of this product to provide partial control of the following perennial weeds. Use the lower rates for suppression of growth.

Bahiagrass	*Paspalum notatum*
Broomsedge	*Andropogun virginicus*
Dock (curly)	*Rumex crispus*
Dogfennel	*Eupatorium capilliforium*
Fescue (tall)	*Festuca arundinacea*
Johnsongrass**	*Sorghum halepense*
Poorjoe**	*Diodia teres*
Trumpetcreeper*	*Campsis radicans*
Vaseygrass	*Paspalum urvillei*
Vervain (blue)	*Verbena hastata*

* Suppression at higher rates only.
** Control at the higher rates.

Use only on well-established bermudagrass. Bermudagrass injury may result from the treatment but regrowth will occur under moist conditions. Repeat applications in the same season are not recommended, since severe injury may result.

Read and carefully observe the cautionary statements and all other information appearing on the labels of all herbicides used.

BAHIAGRASS SEEDHEAD AND VEGETATIVE SUPPRESSION

When applied as directed in the indicated non-crop areas (roadsides, airports, golf course roughs, and plant sites) and "TREE AND VINE CROPS", this product will provide significant inhibition of seedhead emergence and will suppress vegetative growth for a period of approximately 45 days with single applications and approximately 120 days with sequential applications.

Apply Roundup herbicide 1 to 2 weeks after full greenup of bahiagrass or after the bahiagrass has been mowed to a uniform height of 3 to 4 inches. Applications must be made prior to seedhead emergence. Apply 6 fluid ounces per acre of this product plus 0.5 to 1 percent nonionic surfactant by total spray volume in 10 to 25 gallons of water per acre. Use the 0.5 percent surfactant concentration when using surfactants which contain at least 50 percent active ingredient or a 1 percent surfactant concentration for those surfactants containing less than 50 percent active ingredients.

Sequential applications of this product plus 0.5 to 1 percent nonionic surfactant by total spray volume may be made at approximately 45 day intervals to extend the period of seedhead and vegetative growth suppression. For continued seedhead suppression, sequential applications must be made prior to seedhead emergence. Apply no more than 2 sequential applications per year. As a first sequential application, apply 4 fluid ounces of this product per acre plus nonionic surfactant. A second sequential application of 2 to 4 fluid ounces per acre plus nonionic surfactant may be made approximately 45 days after the last application.

A tank mixture of Roundup plus Oust **may be applied only on roadsides** for seedhead inhibition and vegetative suppression. Apply 6 fluid ounces per acre of this product plus 0.25 ounces per acre of Oust, plus 0.5 to 1 percent nonionic surfactant by total spray volume 1 to 2 weeks following an initial spring mowing. When using Roundup plus Oust for suppression of bahiagrass, make only one application per year.

CROPPING SYSTEMS

See "General Information" and "Mixing and Application Instructions" sections of this label for essential product performance information.

See the following CROPPING SYSTEMS SECTIONS for specific uses.

EXTREME CARE MUST BE EXERCISED TO AVOID CONTACT OF SPRAY WITH FOLIAGE. GREEN STEMS OR FRUIT OF DESIRABLE CROPS, PLANTS, TREES OR OTHER DESIRABLE VEGETATION SINCE SEVERE DAMAGE OR DESTRUCTION MAY RESULT.

Repeat treatments may be necessary to control weeds regenerating from under ground parts or seed. Except as otherwise specified on this label, repeat treatments must be made before the crop emerges in accordance with the instructions on this label.

Except as otherwise specified in a Crop section of this label the combined total of all treatments must not exceed 8 quarts per acre of this product per year.

Do not plant subsequent crops other than those on the label for one year following application.

Do not graze treated cotton fields or feed treated cotton forage to livestock. For other cropping systems, except following spot treatment in forage grasses and legumes, do not harvest or feed treated crops for 8 weeks after application. Allow 14 days following spot treatment or selective equipment use before grazing domestic livestock or harvesting forage grasses and legumes.

When applied as directed for "Cropping Systems", under the conditions described, this product controls annual and perennial weeds listed on this label, prior to the emergence of these crops.

Alfalfa*	Lentils
Artichoke (Jerusalem)	Lettuce
Barley*	Mustard Greens
Beans (all)	Oats*
Beet Greens	Okra
Beets (Red, Sugar)	Onions
Broccoli	Parsnips
Cabbage	Peanuts*
Carrot	Peas (all)
Cauliflower	Potato (Irish, Sweet)
Celery	Radish
Chicory	Rice**
Corn (all)*	Rutabaga
Cotton*	Sorghum (Milo)*
Forage Grasses*	Soybeans*
Forage Legumes*	Spinach
Horseradish	Watercress***
Kale	Wheat*

* Spot treatments may be applied in these crops.

** Do not treat rice fields or levees when fields contain flood water.

*** Apply only prior to planting, allow at least 3 days between application and planting.

Spot treatment (Only those crops with a "*" can be spot treated) — Applications in growing crops must be made prior to heading of small grains and milo, initial pod set in soybeans, silking of corn, boll opening on cotton and pegging of peanuts.

For forage grasses and forage legumes see Spot Treatment in the "Pastures" section of "CROPPING SYSTEMS" in this label.

For dilution and rates of application using Boom or Hand-Held Equipment, see "Mixing and Application" and "Weeds Controlled" sections of this label.

NOTE: FOR FORAGE GRASSES AND FORAGE LEGUMES, NO MORE THAN ONE-TENTH OF ANY ACRE SHOULD BE TREATED AT ONE TIME. FOR ALL OTHER CROPS, DO NOT TREAT MORE THAN 10% OF THE TOTAL FIELD AREA TO BE HARVESTED.
THE CROP RECEIVING SPRAY IN TREATED AREA WILL BE KILLED. TAKE CARE TO AVOID DRIFT OR SPRAY OUTSIDE TARGET AREA FOR SAME REASON.

Selective Equipment—This product may be applied through recirculating sprayers, shielded applicators, or wiper applicators in cotton and soybeans. Shielded and wiper applicators may also be used in tree crops and grapes. Wiper applicators may be used in parsnips, rutabagas, forage grasses, and forage legumes including pasture sites.

See the "Selective Equipment" part of the "APPLICATION EQUIPMENT AND TECHNIQUES" section of this label for information on proper use and calibration of this equipment.

Allow at least the following time intervals between application and harvest:

Cotton, soybeans	7 days
Apples, Avocado, Cherry, Citrus, Grapes, Parsnips, Pear, Rutabagas, Stone Fruit	14 days
Nut Crop	21 days

ASPARAGUS

When applied as directed for "Cropping Systems" under the conditions described, this product controls weeds listed on this label in asparagus.

For specific rates of applications and instructions for control of various annual and perennial weeds, see the "Weeds Controlled" section on this label.

Prior to Crop Emergence—Apply this product prior to crop emergence for the control of emergence for the control of emerged labeled annual and perennial weeds. DO NOT APPLY WITHIN A WEEK BEFORE THE FIRST SPEARS EMERGE.

Post Harvest—Apply this product after the last harvest and all spears have been removed. If spears are allowed to regrow, delay application until ferns have developed. Delayed treatments should be applied as a directed or shielded spray in order to avoid contact of the spray with ferns, stems or spears. Direct contact of the spray with the asparagus may result in serious crop injury.

NOTE: Select and use recommended types of spray equipment for post-emergence post harvest applications. A directed spray is any application where the spray pattern is aligned in such a way as to avoid direct contact of the spray with the crop. A shielded spray is any application where a physical barrier is positioned and maintained between the spray and the crop to prevent contact of spray with the crop.

CRANBERRIES

Wiper applicators may be used in cranberries in accordance with instructions in this section.

See "General Information" and "Mixing and Application Instructions" sections of this label for essential product performance information.

See the "Selective Equipment" part of the "APPLICATION EQUIPMENT AND TECHNIQUES" section of this label for information on proper use and calibration of this equipment.

For Wick or other Wiper Applications—Mix 1 gallon of this product in 4 gallons of water to prepare a 20 percent solution. Apply the solution to emerged weeds. Apply after cranberry fruit set and no later than 30 days before harvest.

In severe infestations, reduce equipment ground speed to insure that adequate amounts of this product are wiped on the weeds. A second treatment in the opposite direction may be beneficial.

Do not permit herbicide solution to contact desirable vegetation.

FALLOW AND REDUCED TILLAGE SYSTEMS

FOR AERIAL APPLICATION IN CALIFORNIA, REFER TO SUPPLEMENTAL LABEL.

Use this product in fallow and reduced tillage systems for control of annual weeds prior to emergence of crops listed in this label. Refer to the "Weeds Controlled" section of this label for specific rates and instructions. This product may be applied using ground or aerial spray equipment. See the "APPLICATION EQUIPMENT AND TECHNIQUES" section of this label for instructions.

TANK MIXTURES

ROUNDUP® plus BANVEL

plus NONIONIC SURFACTANT

ROUNDUP plus 2,4-D AMINE

plus NONIONIC SURFACTANT

ROUNDUP plus GLEAN™

plus NONIONIC SURFACTANT

DO NOT APPLY BANVEL, GLEAN OR 2,4-D AMINE TANK MIXTURES BY AIR IN CALIFORNIA.

The addition of Banvel in a mixture with this product may provide short-term residual control of selected weed species. Some crop injury may occur if Banvel is applied within 45 days of planting. Refer to the Banvel and 2,4-D labels for cropping restrictions and other use instructions.

The Roundup plus Glean plus nonionic surfactant tank mixture provides postemergence control and soil residual control of labeled annual weeds in fallow and reduced tillage areas to be planted only to wheat. Use ⅓ to ½ ounces of Glean with the recommended rates of Roundup and nonionic surfactant as instructed in the "Low-Volume Application" section of this label booklet. Refer to the Glean label for planting, cropping and other restrictions. Follow all precautionary statements on this and the Glean label.

These recommended tank mixtures may be applied using ground or aerial spray equipment. Refer to the "Weeds Controlled" section of this label for specific rates and instructions.

NOTE: Do not exceed the 16 fluid ounces per acre rate of Roundup when applying this product alone or in a tank mix to these systems.

™Glean is a trademark of E.I. dePont deNemours and Company.

PASTURES

Apply this product prior to planting forage grasses and legumes.

Pasture or Hay Crop renovation— When applied as a broadcast spray this product controls the annual and perennial weeds listed in this label prior to planting forage grasses or legumes. Remove domestic livestock before application and wait 8 weeks after application before grazing or harvesting.

Spot Treatment—When applied as a spot treatment as recommended, this product controls annual and perennial weeds listed in this label which are growing in pastures, forage grasses and forage legumes composed of bahiagrass, bermudagrass, bluegrass, brome, fescue, orchardgrass, ryegrass, timothy, wheatgrass, alfalfa or clover.

Wiper Application—When applied as directed this product controls or suppresses the weeds listed under "Wiper Applicators" in the "Selective Equipment" section of this label.

For spot treatment and wiper application, apply in areas where the movement of domestic livestock can be controlled. No more than one-tenth of any acre should be treated at one time. Further applications may be made in the same area at 30 day intervals. Remove domestic livestock before application and wait 14 days after application before grazing livestock or harvesting.

See "General Information", "Mixing and Application Instructions" and "Weeds Controlled" sections of this label.

SUGARCANE

When applied as directed for "Cropping Systems," under the condition described, this product controls those emerged annual and perennial weeds listed on this label growing in or around sugarcane or in fields to be planted to sugarcane. This product will also control undesirable sugarcane.

NOTE: Where repeat treatments are necessary, do not exceed a total of 10.6 quarts of this product per acre per year. Do not apply to vegetation in or around ditches, canals or ponds containing water to be used for irrigation.

Broadcast Treatment—Apply this product in 10 to 40 gallons of water per acre on emerged weeds growing in fields to be planted to sugarcane.

For specific rates of application and instructions for control of various annual and perennial weeds see the "Weeds Controlled" section of this label.

For removal of last stubble or ratoon cane, apply 4 to 5 quarts of this product in 10 to 40 gallons of water per acre to new growth having at least 7 or more new leaves. Allow 7 or more days after application before tillage.

Spot Treatment in or Around Sugarcane Fields—For dilution and rates of application using Hand-Held Equipment, see "Mixing and Application" and "Weeds Controlled" sections of this label.

For control of volunteer or diseased sugarcane, make a 1 percent solution of this product in water and spray to wet the foliage of vegetation to be controlled.

NOTE: When spraying volunteer or diseased sugarcane, the plants should have at least 7 new leaves.

Avoid spray contact with healthy cane plants since severe damage or destruction may result.

MINIMUM TILLAGE SYSTEMS
Tank Mixtures
CORN AND SOYBEANS

When applied as recommended under the conditions described, these tank mixtures control many emerged weeds and give preemergence control of many annual weeds when corn or soybeans will be planted directly into a cover crop, established sod, or in previous crop residues.

Refer to specific product labels for crop rotation restrictions and cautionary statements of all products used in these tank mixtures. Lasso® EC herbicide may be substituted for Lasso herbicide in these tank mixtures. For mixing instructions, see the "MIXING, ADDITIVES AND APPLICATION INSTRUCTIONS" section of this label.

CORN

ROUNDUP® plus LASSO®

ROUNDUP plus LASSO plus ATRIZINE

ROUNDUP plus LASSO plus BLADEX™

ROUNDUP plus LASSO plus SIMAZINE

ROUNDUP plus ATRAZINE plus SIMAZINE

SOYBEANS

ROUNDUP® plus LASSO®

ROUNDUP plus LASSO plus LINURON

ROUNDUP plus LASSO plus LEXONE™

ROUNDUP plus LASSO plus SENCOR™

Apply these tank mixtures in 10 to 40 gallons of water per acre after planting or during planting in such manner that the planter does not disturb the treated soil. Do not apply these mixtures after crop emergence.

REDUCED CONTROL MAY RESULT IF THIS PRODUCT IS USED IN TANK MIXTURES CONTAINING FLUID FERTILIZERS.

CONTROL OF EMERGED WEEDS

Annual Weeds—Apply to actively growing grasses and broadleaf weeds. Use 1 quart of Roundup herbicide per acre in these tank mixtures if weeds are less than 6 inches tall. If weeds are over 6 inches tall, apply 1.5 quarts of this product per acre. For emerged annual weeds controlled, see the "Weeds Controlled" section of this label.

Perennial Weeds—At normal application dates in minimum tillage systems, perennial weeds may not be at the proper stage of growth for control. See the "General Information" section of this label for the proper stage of growth for perennial weeds.

Use 2 to 4 quarts of Roundup herbicide per acre in these tank mixtures, under these conditions provides top kill and reduces competition from many emerged perennial grass and broadleaf weeds. For emerged perennial weeds controlled, see the "Weeds Controlled" section of this label.

To obtain the desired stage of growth, it may be necessary to apply Roundup herbicide alone in the late summer or fall and then follow with a label-approved, seedling weed-control program at planting.

NOTE: When using these tank mixtures, do not exceed 4 quarts of Roundup herbicide per acre.

USE OF THESE TANK MIXTURES FOR BERMUDAGRASS OR JOHNSONGRASS CONTROL IN MINIMUM TILLAGE SYSTSMS IS NOT RECOMMENDED. For bermudagrass control, follow the instructions under "Control of Perennial Weeds" section of this label and then use a label-approved, seedling weed-control program in a minimum tillage or conventional tillage system. For Johnsongrass control, follow the instructions under the "Control of Perennial Weeds" section of the label, and then use a label-approved seedling weed-control program with conventional tillage.

The following are labeled rate ranges for the recommended preemergence herbicides:

Product	Rates/Acre (Quarts)	(Pounds)
Lasso 4 EC	2 to 4	•
Atrazine 4L	1 to 2	•
Atrazine 80WP	•	1.25 to 2.3
Atrizine 90DF	•	1.1 to 2
Bladex 4L	1 to 2.2	•
Bladex 80WP	•	1.25 to 2.75
Simazine 4L	1 to 2.2	•
Simazine 80WP	•	1.25 to 2.75
Linuron 4L	0.5 to 1.5	•
Linuron 50WP	•	1 to 3
Lexone 4L or Sencor 4F	0.25 to 0.75	•
Lexone or Sencor 50WP	•	0.5 to 1.55
Lexone or Sencor 75DF	•	0.33 to 1

Read and carefully observe the label claims, cautionary statements, rates and all other information on the labels of all products. Use according to the most restrictive label directions for each product in the mixture.

® Lasso is a registered trademark of Monsanto Company.

TM Bladex is a trademark of Shell Chemical Company.

TM Lexone is a trademark of E.I. duPont de Nemours and Company.

TM Sencor is a trademark of the parent company of Farbenfabriken Bayer GmbH, Leverkusen.

PREHARVEST APPLICATIONS

When applied as directed under the conditions described, this product controls annual and perennial weeds listed on this label prior to the harvest of COTTON.

Weed Control—For specific rates of application and instructions for control of various annual and perennial weeds, see the "Weeds Controlled" section of this label.

To control Johnsongrass using multiple nozzle-directed spray equipment, apply 1 quart per acre plus 0.5 to 1 percent nonionic surfactant by total spray volume in 10 to 20 gallons of water per acre. Insure that complete coverage is obtained. Use 0.5 percent surfactant concentration when using surfactants which contain at least 50 percent active ingredient or a 1 percent surfactant concentration for those surfactants containing less than 50 percent active ingredient. Refer to the "Weeds Controlled" section of this label for instructions when other types of equipment are used.

Broadcast Applications—This product may be applied using either aerial or ground spray equipment. For ground applications apply this product in 10 to 40 gallons of water per acre. For aerial applications, apply this product in 3 to 15 gallons of water per acre. Do not apply by air in California. Do not exceed 1 quart per acre of Roundup when applying this product by air.

Timing of Application—Apply this product for preharvest weed control after 60 percent of the cotton bolls have opened.

NOTE: DO NOT APPLY TO CROPS GROWN FOR SEED. Allow a minimum of 7 days between application and harvest.

Do not feed or graze treated areas within 8 weeks after application.

TREE AND VINE CROPS

This product is recommended for weed control in established groves, vineyards, or orchards, or for site preparation prior to transplanting crops listed in this section. Applications may be made with boom equipment, CDA, shielded sprayers, hand-held and highvolume wands, lances, or orchard guns, or with wiper applicator equipment, except as directed in this section. See the "Application Equipment and Techniques" section of this label for specific information on use of equipment.

When applying Roundup herbicide alone, refer to the "WEEDS CONTROLLED" section for recommended rates to be used.

NOTE

Repeat treatments may be necessary to control weeds originating from underground parts of untreated weeds or from seeds. This product does not provide residual weed control. For subsequent weed control, follow a program using residual herbicides or use repreated applications of this product. Do not apply more than 10.6 quarts of this product per acre per year.

EXTREME CARE MUST BE EXERCISED TO AVOID CONTACT OF HERBICIDE SOLUTION, SPRAY, DRIFT, OR MIST WITH FOLIAGE OR GREEN BARK OF TRUNK, BRANCHES, SUCKERS, FRUIT, OR OTHER PARTS OF TREES OR VINES. CONTACT OF THIS PRODUCT WITH OTHER THAN MATURED BROWN BARK CAN RESULT IN SERIOUS CROP DAMAGE.

Reduced control may result when applications are made to annual or perennial weeds that have been mowed, grazed or cut and have not been allowed to regrow to the recommended stage for treatment.

For specific rates of applications and instructions, see the "Weeds Controlled" section of this label.

TANK MIXTURES TREE AND VINE CROPS

When applied as a tank mixture, this product provides control of the emerged annual weeds and partial control of the emerged perennial weeds listed in this label. The following residual herbicides will provide preemergence control of the weeds listed in the individual product labels.

ROUNDUP® plus GOAL™ 1.6E

ROUNDUP plus KARMEX™ WP

ROUNDUP plus KROVAR™ I

ROUNDUP plus KROVAR™ II

ROUNDUP plus SIMAZINE™ CALIBER™ 90

ROUNDUP plus SIMAZINE 4L

ROUNDUP plus SIMAZINE 80W

ROUNDUP plus SOLICAM™ 80WP

ROUNDUP plus SURFLAN™ AS

ROUNDUP plus SURFLAN 75W

ROUNDUP plus SIMAZINE
(80W, OR 4L, OR CALIBER 90)
plus SURFLAN (AS OR 75W)

ROUNDUP plus GOAL (1.6E)
plus SURFLAN (AS or 75W)

ROUNDUP plus GOAL (1.6E)
plus SIMAZINE (80W, or 4L, or CALIBER 90)*

*See the ROUNDUP plus Goal plus Simazine tank mixtures section below for specific use instructions.

When tank mixing with residual herbicides, add an agriculturally approved nonionic surfactant. Use 0.5 percent surfactant concentration when using surfactants which contain at least 50 percent active ingredient or a 1 percent surfactant concentration for those surfactants containing less than 50 percent active ingredient. See the "Mixing and Application Instructions" section of this label.

Do not apply these tank mixtures in Puerto Rico.

CONTROL OF EMERGED WEEDS

Annual Weeds—Apply 1 quart per acre of this product in these tank mixtures when weeds are less than 12 inches tall. For weeds greater than 12 inches tall, apply 1.5 quarts per acre.

Perennial Weeds—For partial control of perennial weeds using these tank mixtures apply 2 to 5 quarts per acre of this product. Follow the recommendations in the "Weeds Controlled" section of this label for stage growth and rate of application for specific perennial weeds.

PREEMERGENCE WEED CONTROL

The following are the labeled rates for the recommended residual herbicides:

Product	Rate/Acre
Goal 1.6E	1.25 to 5 qts.
Karmex WP	2 to 8 lbs.
Krovar I	2 to 8 lbs.
Krovar II	2 to 8 lbs.
Simazine, Caliber 90	1.1 to 10.6 lbs.
Simazine 80W	1¼ to 12 lbs.
Simazine 4L	2 to 9.6 qts.
Solicam 80WP	2½ to 5 lbs.
Surflan AS	2 to 4 qts.
Surflan 75W	2⅔ to 5⅓ lbs.

NOTE: These residual herbicides may provide postemergence activity on certain annual weed species.

Refer to the individual product labels for specific crops, rates, geographical restrictions and precautionary statements.

Read and carefully observe the label claims, cautionary statements, rates and all other information on the labels of all products. Use according to the most restrictive label directions for each product in the mixture.

For specific rates of applications and instructions see the "Weeds Controlled" section of this label.

ROUNDUP® plus GOAL plus SIMAZINE
(California only)

ROUNDUP may be tank mixed with Goal plus Simazine for broad spetrum postemergence and preemergence control. Refer to the following table for approved crops and rate ranges for each product in this tank mix. Refer to the individual product labels for weeds controlled, geographical restrictions, precautionary statements and specific use rates.

	Rate Ranges*				
Crop	ROUNDUP® (quarts)	+	GOAL 1.6E* (quarts)	+	SIMAZINE 80W** (pounds)
Grapes	1 to 1.5		1 to 4		2.5 to 5
Almonds	1 to 1.5		1 to 4		1.25 to 2.5

Use equivalent rates of Caliber 90 or Simazine 4L.

* Rate dependent on weeds present.

** Rate dependent on soil type and weeds present.

TM Caliber is a trademark of Ciba-Geigy Corporation.
TM Goal is a trademark of Rohm and Haas Company.
TM Karmex and TM Krovar are trademarks of E.I. du Pont de Nemours and Company.
TM Solicam is a trademark of Sandoz, Inc.
TM Surflan is a trademark of Elanco Products Company.

TREE CROPS

CITRUS*: citron, grapefruit, kumquat, lemon, lime, orange, tangelo, tangerine.

NUTS:** almond, filbert, macadamia, pecan, pistachio, walnut.

POME FRUIT*: apple, pear.

STONE FRUIT*: apricots, cherries, nectarines, peaches, plums/prunes.

For cherries, any application equipment listed in this section may be used in all states.

For citron, apply as a directed spray only.

Any application equipment listed in this section may be used in apricots, nectarines, peaches, and plums/prunes growing in Arizona, California, Colorado, Idaho, Kansas, Kentucky, New Jersey, North Dakota, Oklahoma, Oregon, Texas, Utah, and Washington, except for peaches grown in the states specified in the following paragraph. In all other states use wiper equipment only.

For PEACHES grown in Alabama, Arkansas, Florida, Georgia, Louisiana, Mississippi, North Carolina, South Carolina, and Tennessee only, apply with a shielded boom sprayer or shielded wiper applicator which prevents any contact of this product with the foliage or bark of trees. Apply no later than 90 days after first bloom. Applications made after this time may result in severe damage. Remove suckers and low hanging limbs at least 10 days prior to application. Avoid applications near trees with recent pruning wounds or other mechanical injury. Apply only near trees which have been planted in the orchard for 2 or more years. EXTREME CARE MUST BE TAKEN TO ENSURE NO PART OF THE PEACH TREE IS CONTACTED.

Tropical Fruit: avocado*, banana, coffee*, guava, mango*, papaya, tea*. Allow a minimum of 1 day between last application and harvest of guava and papaya. In coffee and banana delay applications 3 months after transplanting to allow the new coffee or banana plant to become established.

NOTE:

* Allow a minimum of 14 days between last application and harvest.

**Allow a minimum of 21 days between last application and harvest of these crops.

VINE CROPS

Grapes*: any variety of table, wine, or raisin grape may be treated with any equipment listed in this section. Applications should not be made when green shoots, canes, or foliage are in the spray zone.

In the northeast and Great Lakes regions, applications must be made prior to the end of bloom stage of grapes to avoid injury.

NOTE:

*Allow a minimum of 14 days between last application and harvest.

CALIFORNIA

ROUNDUP® herbicide has been approved by the U.S. Environmental Protection Agency for the uses, crops and sites listed in this label and by California under label designation 1985-1. Approval of the items listed below is pending under the State of California registration requirements. With the exception of these items, this booklet contains the material approved by California in label 1985-1.

These use conditions, crops and sites may not be treated with this product in California until approval is received:

- Use of ammonium sulfate with this product alone, and with tank mixes of Roundup herbicide with 2,4-D or Banvel.
- Preharvest cotton applications at 60 percent boll opening.
- Partial control of netsedge using 1 quart per acre.
- Roundup plus Oust for use in industrial sites.
- Roundup plus Oust for conifer release from herbaceous weeds.
- Roundup plus Oust for release of actively growing bermudagrass.
- Roundup plus Sencor or Lexone for use in minimum tillage soybeans.
- Control of Johnsongrass at 1 quart per acre using multiple nozzle-directed equipment in 10 to 20 gallons of water per acre.
- Roundup plus Glean plus nonionic surfactant in fallow and reduced tillage systems.

MSD AGVET

Division of Merck & Co., Inc.
P.O. Box 2000
Rahway, NJ 07065-0912
(201) 750-4900

MERTECT® 340-F

FUNGICIDE

Flowable (Water Dispersible Suspension)

ACTIVE INGREDIENT—3.8 lb./gal.	
2-(4-thiazolyl) benzimidazole	42.28%
INERT INGREDIENTS	57.72%
TOTAL	100.00%

EPA Reg. No. 618-75-AA EPA Est. No. 618-NJ-1
U.S. Pat. 3,370,957 Product 48015

PRECAUTIONARY STATEMENTS

HAZARDS TO HUMANS AND DOMESTIC ANIMALS

Keep Out of Reach of Children
CAUTION

Harmful if swallowed

Do not get in eyes or on skin

FIRST AID

If in eyes or on skin: Flush with water; for eyes, get medical attention.

ENVIRONMENTAL HAZARDS

This pesticide is toxic to fish. Do not apply when weather conditions favor runoff or drift from the target area. Do not apply directly to lakes, streams or ponds. Do not contaminate water by cleaning of equipment or disposal of wastes.

DIRECTIONS FOR USE

It is a violation of Federal (U.S.A.) law to use this product in a manner inconsistent with its labeling.

GENERAL—Shake or stir well before use.

MERTECT 340-F is a systemic fungicide recommended for the control of many important plant diseases. If treatment is not effective following use of MERTECT 340-F as recommended, a tolerant strain of fungus may be present. Consideration should be given to prompt use of suitable fungicides.

Clean equipment before using MERTECT 340-F. **When MERTECT 340-F is dispersed in water, the resulting suspension must be constantly agitated.** Avoid foaming.

Compatibility information for MERTECT 340-F with other chemicals is limited. Consult your supplier or agricultural equipment station for specific recommendations.

SOYBEANS—Reduce the Severity of Pod and Stem Blight, Anthracnose, Brown Spot, Frog Eye Leaf Spot and Purple Seed Stain (Ground and Aerial Applications): Apply MERTECT 340-F at 6.0–10.0 fl. oz. per acre per application in sufficient water for coverage. Make two applications per season, the first application at late flowering to early pod set and the second application two weeks later. Make aerial applications in a minimum of 5 gallons of water per acre. Do not apply MERTECT 340-F within 21 days of harvest. Do not graze or feed treated soybean vines or hay to livestock. Do not tank mix with copper fungicides.

SOYBEANS GROWN FOR SEED PURPOSES—To Improve Seed Quality by Reducing the Severity of Pod and Stem Blight, Anthracnose, Brown Spot, Frog Eye Leaf Spot and Purple Seed Stain: Make a single application of MERTECT 340-F at 6 to 32 fl oz per acre to soybeans at growth stage R6 (pod fill) in sufficient water for coverage. Make aerial application in a minimum of 5 gallons of water per acre. Use higher rates when disease potential is high as a result of greater than normal rainfall in August. Do not apply MERTECT 340-F within 21 days of harvest. Do not graze or feed soybean vines or hay to livestock.

POME FRUIT (Apples and Pears)—Control Blue Mold Rot, Bull's Eye Rot, and Gray Mold (Nest Rot, Cluster Rot), Stem End and Neck Rot: Dip, flood, or spray harvested fruit with a suspension of 16 fl. oz. MERTECT 340-F in 100 gal. water. Do not treat over 3 minutes. Treat apples only before and after storage for maximum decay control. Treat pears once.

SUGAR BEETS (Post Harvest Treatment)—**Reduce rot caused by *Penicillium, Botrytis,* and *Fusarium* species:** Mist sugar beet roots as they enter storage with 0.42 fl. oz. of MERTECT 340-F to each 2,000 lb. of sugar beet roots in sufficient water for complete coverage. (0.42 fl. oz. of MERTECT 340-F is equivalent to 0.2 oz. of 2-(4-thiazolyl) benzimidazole.) Roots should be treated within 72 hours after lifting.

ORNAMENTAL BULBS AND CORMS—Control Fusarium Basal Rot and Penicillium Blue Mold: Clean and treat bulbs or corms within 24–48 hours of digging. Prepare suspension of 30 fl. oz. MERTECT 340-F per 100 gal water. Submerge stock completely in dipping suspension. Suspension temperature should be 55–75°F (13–24°C). Discard suspension (1) when it becomes dirty, (2) after using five times, or (3) after 24 hours, whichever occurs first.

Fusarium (Bulbs and Corms): Dip bulbs 15–30 minutes, corms 15 minutes.

Blue Mold (Bulbs): Dip 10–15 minutes.

After treatment, dry bulbs or corms in a shaded, well ventilated area. Curing or retarding may precede or follow treatment.

SWEET POTATO "SEED ROOTS"—Control Black Rot, Scurf, and Foot Rot: Dip the seed roots in a suspension containing 8 fl. oz. MERTECT 340-F per 7.5 gal. water (107 fl. oz. per 100 gal). Treat the seed roots for 1–2 minutes and plant immediately. Discard the suspension when the volume is too low or when it becomes dirty. Do not use the treated roots for food or feed.

POTATO—Control Fusarium Tuber Rot: Mist unwashed tubers on a conveyor line, with tumbling action, entering storage with 0.42 fl. oz. of MERTECT 340-F to each 2000 lb. of tubers in sufficient water for complete coverage. If an additional treatment is necessary before shipping, mist the tubers at the same rate or dip the tubers for 20 seconds in a solution containing 0.42 fl oz of MERTECT 340-F per gallon of water. Seed potatoes should be treated before cutting.

WHEAT—Control of *Cerosporella* Foot Rot—Apply MERTECT 340-F after the development of 2–3 tillers in the Fall but prior to the time the first node is visible in the Spring. AERIAL Application: Apply 16–24 fluid ounces of MERTECT 340-F in a minimum of 5 gallons water per acre. Use the high rate in fields with a previous history of foot rot. GROUND Application: Apply 16–24 fluid ounces of MERTECT 340-F per acre in sufficient water for coverage. Use the high rate in fields with a previous history of foot rot. Do not allow livestock to graze or feed on treated green wheat forage.

RICE—Reduce Severity of Rice Blast, Stem Rot, Sheath Blight, and Narrow Brown Leaf Spot: Apply MERTECT 340-F at 6.0–12.0 fl. oz. per acre per application in sufficient water for coverage (minimum of 5 gallons of water per acre for aerial application). Make two applications per season, the first application at boot initiation followed by a second application 14–21 days later. Use the higher rate when disease infestations are moderate to severe. Do not use in California.

WHEAT—To reduce the severity of Glume Blotch—Apply MERTECT 340-F at 8 to 12 fl oz per acre per application. For aerial application use a minimum of 5 gallons of water per acre. Make the first application when flag leaf first emerges, repeat application 10 to 14 days later. Use the higher rate under conditions of heavier disease pressure. Do not make more than a total of two foliar applications of MERTECT 340-F in any one season.

CARROTS—Control of Botrytis cineria gray mold and Sclerotinia sclerotiorum (sclerotinia rot)—Dip carrots, before storage, for

5–10 seconds in a suspension containing 42 fl. oz. of MERTECT 340-F per 100 gal. of water. Replenish the suspension when the volume is too low or when it becomes dirty.

The run-off and wastes from the dipping operation should not be discarded in a drainage which could contaminate public water systems.

BANANAS AND PLANTAINS—Control of Crown Rot—Use 45 ml. (1.5 fl. oz.) to 100 ml. (3.4 fl. oz.) per 227 liters (60 gallons) of water. Dip fruit for 2 to 4 minutes after it has been dehanded from the stem and passed through the delatexing operation. Use the low rate where rot incidence is low. Stir the mixture frequently to maintain the suspension.

DRY BEANS—Aids in the controls of White Mold (Sclerotinia Sclerotiorum): Apply 16-32 fl oz of MERTECT 340-F in a minimum of 5 gallons of water per acre. Make the first application when beans are at 10% bloom. Repeat application at peak bloom. Do not make more than 2 applications per season. Do not treat within 28 days of harvest. Do not graze or feed foliage from treated areas to livestock. Do not tank mix with copper fungicides.

MUSHROOMS (Agarticus sp.)—To aid in the control of Dry Bubble (Verticillium fungicola), Wet Bubble (Mycogone perniciosa), Cobweb (Dactylium dendroides) and Green Mold (Trichoderma sp.): Apply in water during normal watering or in hand sprayer. MERTECT 340-F may be applied at casing, fuzzing, pinning and between breaks. A maximum of 16 fl oz of MERTECT 340-F per 1000 ft may be applied to each crop. A single dose should not exceed 4 fl oz per 1000 ft. Do not apply within one day prior to harvesting.

STORAGE AND DISPOSAL

DO NOT STORE BELOW 32°F (0°C).

DISPOSAL

Prohibitions—Do not contaminate water, food, or feed by storage or disposal. Open dumping is prohibited. Do not reuse empty container.

Pesticide Disposal—Pesticide, spray mixture, or rinsate that cannot be used or chemically reprocessed should be disposed of in a landfill approved for pesticides or buried in a safe place away from water supplies.

Container Disposal—Triple rinse (or equivalent) and dispose in an incinerator or landfill approved for pesticide containers, or bury in a safe place.

General—Consult Federal, state, or local disposal authorities for approved alternative procedures such as limited open burning.

NOTICE OF WARRANTY

The **Directions for Use** of this product reflect the opinion of experts based on field use and tests. The directions are believed to be reliable and should be followed carefully. However, it is impossible to eliminate all risks inherently associated with use of this product. Crop injury, ineffectiveness, or other unintended consequences may result because of such factors as weather conditions, presence of other materials, or the manner of use or application, all of which are beyond the control of MSD AGVET or the Seller. All such risks shall be assumed by the Buyer.

MSD AGVET warrants that this product conforms to the chemical description on the label and is reasonably fit for the purposes referred to in the **Directions for Use,** subject to the inherent risks referred to above. Buyer assumes all risk of use and handling which are at variance in any way with the directions hereon. **MSD AGVET makes no other express or implied warranty of Fitness or Merchantability or any other express or implied warranty. In no case shall MSD AGVET or the Seller be liable for consequential, special, or indirect damages resulting from the use or handling of this product.** MSD AGVET and the Seller offer this product and the Buyer and user accept it, subject to the foregoing **Notice of Warranty** which may be varied only by agreement in writing signed by a duly authorized representative of MSD AGVET.

NOR-AM Chemical Company

3509 Silverside Road
P.O. Box 7495
Wilmington, DE 19803
(302) 575-2000

ANTOR® 4ES

EMULSIFIABLE SOLUTION
HERBICIDE

FOR AGRICULTURAL USE ONLY

For Selective Weed Control in Sugarbeets, Red Beets, and Spinach (only intended for processing)

ACTIVE INGREDIENTS	
Diethatyl ethyl: [N (chloroacctyl)-N-(2,6-diethylphenyl)glycine, ethyl ester]	43.3%
INERT INGREDIENTS	56.7%
TOTAL	100.0%

This product contains 4 lbs. active ingredient per gallon.

EPA Reg. No. 45639-54 EPA Est. No. 407-MN-1

WARNING — AVISO

Keep Out of Reach of Children

STATEMENT OF PRACTICAL TREATMENT

IF SWALLOWED: Do not induce vomiting. Call a physician immediately. See "Note to Physician."

IF IN EYES: Flush with plenty of water. Get medical attention.

IF ON THE SKIN: Wash skin with soap and plenty of water.

NOTE TO PHYSICIAN—Contains monochlorotoluene solvent.

PRECAUCION AL USUARIO: Si usted no lee ingles, no use este producto hasta que el etiqueta haya sido explicado ampliamente.

PRECAUTIONARY STATEMENTS

HAZARDS TO HUMANS AND DOMESTIC ANIMALS

WARNING

The active ingredient in ANTOR may cause skin sensitization reactions in certain individuals. Wear coveralls and gloves while handling or using this product. Causes substantial but temporary eye injury.

Harmful if swallowed. Do not get in eyes or on clothing. Wear goggles, face shield, or safety glasses when handling. Wash thoroughly with soap and water after handling. Avoid breathing of spray mist. Avoid contamination of food.

ENVIRONMENTAL HAZARDS

Do not apply directly to any body of water.

Do not apply when weather conditions favor drift from areas treated.

Do not contaminate water by cleaning of equipment or disposal of wastes.

PHYSICAL OR CHEMICAL HAZARDS

Do not use, pour, or store near heat or open flame.

GENERAL USE PRECAUTIONS

Unusually dry, windy weather following application of ANTOR 4ES or tank-mixes, which dries the upper soil layer, may reduce effectiveness.

If crop is lost due to climatic or soil conditions following application of ANTOR 4ES or tank-mixes, do not retreat field with ANTOR 4ES or tank-mixes within the same season.

Do not allow spray mixture to stand in tank overnight. Flush and drain equipment after each day's use.

Where ANTOR 4ES is used in tank-mix with TCA, do not use treated sugarbeet tops for feed or forage.

Do not mix ANTOR 4ES or tank mixes with liquid fertilizers.

ROTATIONAL CROPS:
Crops, except wheat, may be rotated at 3 months post-application. Wheat may be planted 9 months after application.

DIRECTIONS FOR USE

It is a violation of federal law to use this product in a manner inconsistent with its labeling.
Read entire Directions for Use and Disclaimer of Warranties before using this product.

REENTRY STATEMENT
Do not apply this product in such a manner as to directly or through drift expose workers or other persons. The area being treated must be vacated by unprotected persons.

When ANTOR 4ES is applied preemergence, the following statements apply:

Reentry Statement
Do not enter treated areas without protective clothing until sprays have dried.

Because certain states may require more restrictive reentry intervals for various crops treated with this product, consult your State Department of Agriculture for further information.

Written or oral warnings must be given to workers who are expected to be in a treated area or in an area about to be treated with this product. (Indicate specific oral warnings which inform workers of areas or fields that may not be entered without specific protective clothing, period of time field must be vacated, and appropriate actions to take in case of accidental exposure.) When oral warnings are given, warnings shall be given in a language customarily understood by workers. Oral warnings must be given if there is a reason to believe that written warnings cannot be understood by workers. Written warnings must include the following information:

WARNING: Area treated with ANTOR 4ES herbicide on (date of application). Do not enter without appropriate clothing until spray has dried. In case of accidental exposure, see Statement of Practical Treatment on the label.

GENERAL INFORMATION
ANTOR 4ES is a selective soil-applied herbicide for use in sugarbeets, red beets and spinach. It can be applied either preplant incorporated or preemergence. The seedbed should be firm, fine and free of clods and trash. When applied preemergence, ANTOR 4ES requires rainfall or irrigation prior to weed germination for optimum weed control. Similarly, when applied preplant incorporated, moisture in the form of rainfall, irrigation or existing soil moisture must be available to activate the product.

When dry weather conditions prevail or furrow irrigation is used, preplant incorporation of ANTOR 4ES is recommended for best results.

SPRAY EQUIPMENT
Apply ANTOR 4ES alone or in tank mixes using standard low pressure (20–50 psi) spray equipment. The equipment should be carefully calibrated before use and checked frequently during application to ensure that it is functioning properly. Uniformily apply the recommended rates. Use a spray volume of 10 to 50 gallons per acre broadcast basis. Use proportional amounts for band spraying. See section on Band Applications for calculations.

MIXING THE SPRAY
For ANTOR 4ES alone, add $\frac{1}{4}$ to $\frac{1}{2}$ the total amount of water to the tank. Then add the required amount of ANTOR 4ES and the remaining quantity of water with the bypass agitator running. When tank-mixing ANTOR 4ES with wettable powders, mix the proper amount of wettable powder formulation into half the required amount of water in the spray tank with the agitator running. Add the required amount of ANTOR 4ES and then add the remaining

required amount of water. Before tank-mixing with other herbicides, refer to the specific product labels for additional mixing directions.

BAND APPLICATION

To determine the amount of ANTOR 4ES or tank-mixed emulsifiable concentrates needed per acre by band application, divide band width in inches by row width in inches and multiply the answer by the recommended broadcast rate in quarts. To determine the pounds of wettable powder needed per acre when tank-mixed with ANTOR 4ES, divide band width in inches by row width in inches and multiply the answer by the recommended broadcast rate in pounds.

WEEDS SPECIES CONTROLLED: SUGARBEETS

Annual Grass Weeds	
Barnyardgrass (watergrass)	*Echinochloa crus-galli*
Yellow foxtail	*Setaria glauca*
Green foxtail	*Setaria viridis*
Giant foxtail	*Setaria faberi*
Foxtail millet	*Setaria italica*
Wild oats	*Avena fatua*
Canarygrass	*Phalaris canariensis*
Johnsongrass (seedling)	*Sorghum halepense*
Large crabgrass	*Digitaria sanguinalis*
Italian ryegrass (annual)	*Lolium multiflorum*
Broadleaf signalgrass	*Brachiaria platyphylla*
Witchgrass	*Panicum capillare*
Junglerice	*Echinochloa colonum*
Fall panicum	*Panicum dichotomiflorum*
Annual Broadleaf Weeds	
Redroot pigweed	*Amaranthus retroflexus*
Prostrate pigweed	*Amaranthus blitoides*
Black nightshade	*Solanum nigrum*
Annual sowthistle	*Sonchus oleracaus*
Groundcherry	*Physalis longifolia*
Sheperdspurse	*Capsella bursa-pastoris*

ANTOR 4ES USED ALONE—ALL AREAS—SUGARBEETS

PREPLANT INCORPORATED

In arid regions where furrow irrigation is practiced or where the rainfall pattern is erratic, soil incorporation of this product is required to provide optimum weed control.

Apply ANTOR 4ES to the soil surface broadcast or as a band treatment preplant or at the time of planting and incorporate. Use the recommended rates in Dosage Table 1.

Incorporation immediately after application is not required but must be done prior to weed germination. Use a hooded-power or ground-driven rotary tiller, rolling cultivator or similar equipment properly adjusted to uniformly incorporate ANTOR 4ES to a depth of 1 to 2 inches. Deeper incorporation may reduce effectiveness.

Where sugarbeets are grown in beds, apply ANTOR 4ES after bedding and incorporate. Irrigate until tops of beds are thoroughly wetted. Incorporation is recommended in the Red River Valley of Minnesota and North Dakota where rainfall is erratic and marginal.

DOSAGE TABLE 1—SUGARBEETS
ANTOR 4ES Alone—All Areas

Soil Texture	Broadcast Rate of ANTOR 4ES Per Acre(1)
Coarse:	
Loamy sands and sandy loams	3 qts.
Medium/Fine:	
Loams, silt loams, clay loams and clays containing less than 5% organic matter	4 to 6 qts.(2)
Fine:	
Clay loams and clays containing more than 5% organic matter	6 qts.

(1) See section on Band Application for calculation of proportional amounts.

(2) Use higher rates on heavy clay soils or where black nightshade or groundcherry are expected to be a problem.

PREEMERGENCE

Preemergence application of ANTOR 4ES is recommended where sprinkler irrigation is used for crop germination and where rainfall can be depended on to activate the product.

Apply ANTOR 4ES to the soil surface, either broadcast or as a band treatment at time of planting or shortly after, but prior to weed germination. One-half inch of rainfall or irrigation is usually adequate for activation. It is important that the top 2 inches of soil be thoroughly wetted. Refer to Dosage Table 1 for recommended rates.

ANTOR 4ES + PYRAMIN®[1] FL TANK-MIX—SUGARBEETS

A tank-mix of ANTOR 4ES with Pyramin FL may be used in specific geographic areas for a broader spectrum of weed control. In addition to the weeds controlled by ANTOR 4ES alone, this mixture controls:

Common ragweed	*Ambrosia artemisiifolia*
Wild mustard	*Sinapis arvensis*
Pennsylvania smartweed	*Polygonum pensylvanicum*
Common lambsquarters	*Chenopodium album*
Common purslane	*Portulaca oleracea*
Nettleleaf goosefoot	*Chenopodium murale*
Henbit	*Lamium amplexicaule*
Fanweed (stinkweed)	*Thlaspi arvense*

Do not use this mixture under conditions where Pyramin FL alone is not recommended. Before use, read the Pyramin FL label for additional information and precautions.

[1] Pyramin® is a registered trademark of BASF Atkiengesellschaft.

CALIFORNIA

For preplant incorporation with furrow irrigation, apply the tank-mix to the soil surface prior to, or at time of, planting and incorporate to a depth of 1 to 2 inches using a rotary type tiller. Use the recommended rates in Dosage Table 2. Irrigate until tops of beds are thoroughly wetted.

For preemergence application with sprinkler irrigation, apply the tank-mix to the soil surface at time of planting or shortly after, but prior to weed germination, using the rates recommended in Dosage Table 2. Irrigate prior to crop and weed germination. Follow directions and precautions on the Pyramin FL label for use preemergence with sprinkler irrigation. Do not exceed 3/4 inch of sprinkler irrigation per set until the sugarbeets have two true leaves.

NORTHWEST, INTERMOUNTAIN STATES

For use preemergence under sprinkler irrigation only. Apply tank-mix to the soil surface at the time of planting or shortly after, but prior to weed germination, using the rates recommended in Dosage Table 2. Irrigate prior to crop and weed germination. Follow directions and precautions on the Pyramin FL label for use preemergence with sprinkler irrigation. Do not exceed 3/4 inch of sprinkler irrigation per set until the sugarbeets have two true leaves.

MICHIGAN AND OHIO

For preemergence without irrigation, apply the tank-mix to the soil surface at the time of planting or shortly after, but prior to weed germination using rates recommended in Dosage Table 2.

DOSAGE TABLE 2—SUGARBEETS
ANTOR 4ES + Pyramin FL Tank-Mix

Area	Soil Texture	Broadcast Rate Per Acre(1) ANTOR 4ES (qts.)	Pyramin FL (qts.)
California, Northwest, Intermountain states, Michigan and Ohio	**Coarse:** Sandy loams only	2	2.5
	Medium: Loams, silt loams and clay loams containing less than 3% organic matter	2 to 3(2)	3
	Fine: Clay loams and clays containing more than 3% organic matter	3	3

(1) See section on Band Applications for calculations of proportional amounts.
(2) Use higher rate on clay loam soils.

ANTOR 4ES + TCA TANK-MIX (MN and ND Only)—SUGARBEETS

Tank-mixes of ANTOR 4ES + TCA may be used in the Red River Valley of North Dakota and Minnesota for improved control of foxtails (*Setaria* spp.).

This tank-mix may be applied either preemergence or preplant incorporated; however, shallow (up to 2") incorporation will ensure more consistent weed control where rainfall is erratic or marginal.

TCA is not a registered trademark

Ref: Farm Chemicals Handbook 1984

See Dosage Table 3 for recommended rates.

DOSAGE TABLE 3—SUGARBEETS
ANTOR 4ES + TCA Tank-Mix

Soil Texture	Broadcast Rates Per Acre(1) ANTOR 4ES (qts.)	TCA 4.76 lb. ai/gal(2)	TCA 79.3% WSP(3)
Coarse: Loamy sands, sandy loams	NOT RECOMMENDED		
Medium/Fine: Loams, silt loams, clay loams and clays containing less than 5% organic matter	4 to 6(4)	10 pts.	or 7.5 lbs.
Fine: Clay loams and clays containing more than 5% organic matter	6	10 pts.	or 7.5 lbs.

(1) See section on Band Applications for calculations or proportional amounts.
(2) Solution.
(3) Water Soluble Powder.
(4) Use higher rates on heavy clay soils.

ANTOR 4ES + EPTAM®[1] TANK-MIX OR SEQUENCE (MN and ND Only)—SUGARBEETS

The tank-mix or sequence of ANTOR 4ES and Eptam 7E may be used in the Red River Valley of Minnesota and North Dakota for improved control of:

Green foxtail	*Setaria viridis*
Yellow foxtail	*Setaria glauca*
Wild oats	*Avena fatua*

In addition, this tank mix will reduce competition from the following weeds:

Kochia	*Kochia scoparia*
Common lambsquarters	*Chenopodium album*
Wild buckwheat	*Polygonum convolvulus*

Tank-mixes of ANTOR 4ES + Eptam 7E may be used in the spring. Apply preplant incorporated with incorporation of 2 to 4 inches.

See Dosage Table 4 for recommended rates.

[1] Eptam® is a registered trademark of Stauffer Chemical Co.

DOSAGE TABLE 4—SUGARBEETS
ANTOR 4ES + Eptam 7E Tank-Mix (Spring Applications Only)

Soil Texture	Broadcast Rate Per Acre(1) ANTOR 4ES (qts.)	Eptam 7E (pts.)
Coarse: Loamy sands, sandy loams, or soils, containing less than 4% organic matter	NOT RECOMMENDED	
Medium/Fine: Loams, silt loams, clay loams and clays containing more than 4% organic matter	4	2.25

(1) See section on Band Applications for calculations of proportioned amounts.

Eptam 7E or 10G (Fall Applied) + ANTOR 4ES (Spring Applied).

Fall applications of Eptam 7E or 10G at labeled rates may be followed by spring applications of ANTOR 4ES at 4 qts. per acre or a broadcast basis.

Note: ANTOR/Eptam tank-mixes or sequences are not recommended for use on soils containing less than 4% organic matter.

Do not use this mixture under conditions where Eptam alone is not recommended. Before use, read the Eptam label for additional information and precautions.

SEQUENTIAL TREATMENT WITH BETANEX ® OR BETAMIX®—ALL AREAS—SUGARBEETS

A sequential treatment of ANTOR 4ES applied preplant incorporated or preemergence at recommended rates followed by a postemergence treatment of BETANEX or BETAMIX will control the weeds listed under the Weed Species Controlled section, and the additional weeds:

Kochia	*Kochia scoparia*
Common lambsquarters	*Chenopodium album*
Common ragweed	*Ambrosia artemisiifolia*
Wild buckwheat	*Polygonum convolvulus*
Common chickweed	*Stellaria media*
Common purslane	*Portulaca oleracea*
Londonrocket	*Sisymbrium irio*
Nettleleaf goosefoot	*Chenopodium murale*

The postemergence treatment should be applied according to the Directions for Use on the BETANEX and BETAMIX labels.

Read and follow the precautionary statements and all other information appearing on the BETANEX and BETAMIX labels.

APPLICATION OF ANTOR 4ES VIA SPRINKLER SYSTEM INJECTION—SUGARBEETS

In arid regions where overhead sprinkler irrigation is used to germinate the sugarbeet crop, ANTOR 4ES may be injected through sprinkler irrigation systems. Diluted ANTOR should be metered into the central pivot irrigation system at the rates recommended in Dosage Table 5. Applications should be made after seeding and before emergence of sugarbeets or weeds. Apply ANTOR 4ES in ⅓ to ¾ inch of water per acre.

Agitation of the diluted ANTOR 4ES in the chemical source (or slurry) tank should be maintained during the entire application period.

Apply ANTOR 4ES only through systems with anti-siphon and check valves which will prevent water source contamination and overflow of the chemical source (or slurry) tank, and systems having interlocking controls between the metering device and the water pump to ensure simultaneous shut-off. Inject ANTOR 4ES with a positive displacement pump into the main water line ahead of a right-angle turn to insure adequate mixing. Do not apply ANTOR 4ES in greater than the recommended quantities of irrigation water per acre or decreased activity will result.

DOSAGE TABLE 5—SUGARBEETS
ANTOR 4ES Applied Via Sprinkler Systems

Soil Texture	Broadcast Rate Per Acre ANTOR 4ES (qts.)
Coarse:	
Loamy sands and sandy loams	3
Medium/Fine:	
Loams, silt loams, clay loams and clay	4

Do not apply ANTOR 4ES when wind speed favors drift. Do not use ANTOR 4ES in irrigation systems where connections or fittings leak, or where nozzles do not give uniform distribution. Flush the irrigation system completely with water upon completion of the application to remove ANTOR 4ES from the line before dismantling or draining. When sprinkler distribution patterns do not overlap, unacceptable weed control may result. Where overlap is excessive, crop injury may result.

ANTOR 4ES + NORTRON®EC TANK-MIX (CA only)—SUGARBEETS

Tank-mixes of ANTOR 4ES with Nortron EC can be used in specified geographic areas for control of the additional weeds:

Kochia	*Kochia scoparia*
Common lambsquarters	*Chenopodium album*
Russian thistle	*Salsola iberica*
Volunteer barley	*Hordeum* sp.
Volunteer wheat	*Triticum* sp.
Common purslane	*Portulaca oleracea*
Common chickweed	*Stellaria media*
Ladysthumb	*Polygonum persicaria*
Pennsylvania smartweed	*Polygonum pensylvanicum*

ANTOR 4ES + Nortron tank-mixes may be applied preplant incorporated where furrow irrigation is practiced using shallow incorporation implements such as rolling cultivators or harrows, or using power tillers set to a depth of 1–2 inches.

Preemergence applications of ANTOR 4ES + Nortron tank-mixes may be used when application is followed by sprinkler irrigation prior to weed germination and crop emergence, or where rainfall can be expected to occur within 7–10 days after application.

ANTOR 4ES may also be utilized in sequential applications with Nortron where a broadcast preemergence treatment of ANTOR 4ES is followed by a band overlay of Nortron preemergence. The same dosage rates should be used as in the tank-mix.

Refer to Dosage Table 6 for recommended rates.

Before use, read the Nortron®EC label for additional information and precautions.

DOSAGE TABLE 6—SUGARBEETS
ANTOR 4ES + NORTRON® EC Tank—California

Soil Texture	Broadcast Rates Per Acre(1) ANTOR 4ES (qts.)	Nortron 1.5EC (qts.)
Coarse:		
Loamy sands and sandy loams	1.5	4
Medium:		
Loams and silt loams	2	5.3
Fine:		
Clay loams and clays	3 to 4(2)	6.6

(1) See section on Band Applications for calculations or proportional amounts.
(2) Use higher rates on heavy clay soils.

ANTOR 4ES + NORTRON + PYRAMIN TANK-MIX (MI)—SUGARBEETS

A tank-mix of ANTOR 4ES + Nortron EC + Pyramin FL may be used in Michigan for control of the additional weeds:

Common ragweed	*Ambrosia artemisiifolia*
Wild mustard	*Sinapis arvensis*
Pennsylvania smartweed	*Polygonum pensylvanicum*
Common lambsquarter	*Chenopodium album*
Common purslane	*Portulaca oleracea*
Nettleleaf goosefoot	*Chenopodium murale*
Common chickweed	*Stellaria media*
Henbit	*Lamium amplexicaule*
Fanweed (stinkweed)	*Thlaspi arvense*
Kochia	*Kochia scoparia*
Russian thistle	*Salsola iberica*
Volunteer barley	*Hordeum* sp.
Volunteer wheat	*Tritcum* sp.
Ladysthumb	*Polygonum persicaria*

For preemergence use without irrigation, apply the tank-mix to the soil surface at the time of planting or shortly after, but prior to weed germination. Use rates recommended in Dosage Table 7.

Do not use this mixture under conditions where Pyramin alone is recommended. Before use, read the Pyramin FL and Nortron EC labels for additional information and precautions.

DOSAGE TABLE 7—SUGARBEETS
ANTOR 4ES + Nortron + Pyramin—Michigan

Soil Texture	Broadcast Rates Per Acre ANTOR 4ES (qts.)	Nortron EC (qts.)	Pyramin FL (qts.)
Coarse:			
Sandy loams only	1.5	4	2.25
Medium:			
Loams and silt loams and clay loams containing less than 3% organic matter	2	5.3	2.25
Fine:			
Clay loams and clays containing more than 3% organic matter	3	5.3	2.25

STORAGE AND DISPOSAL

STORAGE

Protect ANTOR 4ES from frost.
Open dumping is prohibited.

PESTICIDE DISPOSAL

(Do not contaminate water, food, or feed by storage or disposal.) Pesticide wastes are toxic. Improper disposal of excess pesticides, spray mixture, or rinsate is a violation of Federal law. If these wastes cannot be disposed of by use according to label instructions, contact your State Pesticide or Environmental Control Agency, or the Hazardous Waste representative at the nearest EPA Regional Office for guidance.

CONTAINER DISPOSAL

Triple rinse (or equivalent). Then offer for recycling or reconditioning, or puncture and dispose of in a sanitary landfill, or by other procedures approved by state and local authorities.

IMPORTANT: READ BEFORE USE

By using this product, user accepts the following conditions, warranty, disclaimer of warranties and limitations of liability.

CONDITIONS:

The directions for use of this product are believed to be adequate and should be followed carefully. However, because of extreme weather and soil conditions, manner of use and other factors beyond NOR-AM Chemical Company's control, it is impossible for NOR-AM to eliminate all risks associated with the use of this product. As a result, crop injury or ineffectiveness is always possible.

DISCLAIMER OF WARRANTIES:

THERE ARE NO WARRANTIES, EXPRESS OR IMPLIED, OF MERCHANTABILITY OR OF FITNESS FOR A PARTICULAR PURPOSE OR OTHERWISE, WHICH EXTEND BEYOND THE STATEMENTS MADE ON THIS LABEL. No agent of NOR-AM Chemical Company is authorized to make any warranties beyond those contained herein or to modify the warranties contained herein. UNDER NO CIRCUMSTANCES SHALL NOR-AM CHEMICAL COMPANY BE LIABLE FOR ANY INCIDENTAL OR CONSEQUENTIAL DAMAGES RESULTING FROM FAILURE TO USE THIS PRODUCT IN ACCORDANCE WITH LABEL DIRECTIONS.

LIMITATIONS OF LIABILITY:

In no event shall NOR-AM Chemical Company's liability for damages arising out of the use of this product exceed the purchase price of the product used.

SUPPLEMENTAL LABEL

ANTOR 4ES

EPA Reg. No. 45639-54

This labeling must be in possession of the user at the time of pesticide application. All applicable directions, restrictions, and precautions on the EPA registered label are to be followed.

WEED CONTROL IN RED BEETS AND SPINACH (ONLY FOR BEETS INTENDED FOR PROCESSING)

DIRECTIONS FOR USE

It is a violation of federal law to use this product in a manner inconsistent with its labeling. Read entire Directions for Use and Disclaimer of Warranties before using this product.

GENERAL INFORMATION

Antor® 4ES is a selective soil-applied herbicide for use in red beets and spinach. It can be applied either preplant incorporated or preemergence. The seedbed should be firm, fine and free of clods and trash. When applied preemergence, Antor® 4ES requires rainfall or irrigation prior to weed germination for optimum weed control. Similarly, when applied preplant incorporated, moisture in the form of rainfall, irrigation or existing soil moisture must be available to activate the product. When dry weather conditions prevail or furrow irrigation is used, preplant incorporation of Antor® 4ES is recommended for best results.

WEED SPECIES CONTROLLED:

Annual Grass Weeds	
Barnyardgrass (watergrass)	*Echinochloa crus-galli*
Yellow foxtail	*Setaria glauca*
Green foxtail	*Setaria viridis*
Giant foxtail	*Setaria faberi*
Foxtail millet	*Setaria italica*
Wild oats	*Avena fatua*
Canarygrass	*Phalaris canariensis*
Johnsongrass (seedling)	*Sorghum halepense*
Large crabgrass	*Digitaria sanguinalis*
Italian ryegrass (annual)	*Lolium multiflorum*
Broadleaf signalgrass	*Brachiaria platyphylla*
Witchgrass	*Panicum capillare*
Junglerice	*Echinochloa colonum*
Fall panicum	*Panicum dichotomiflorum*
Annual Broadleaf Weeds	
Redroot pigweed	*Amaranthus retroflexus*
Prostrate pigweed	*Amaranthus blitoides*
Spiny pigweed	*Amaranthus spinosus*
Annual sowthistle	*Sonchus oleraceus*
Sibara	*Sibara virginica*
Henbit	*Lamium amplexicaule*
Smallflower galinsoga	*Galinsoga parviflora*
Hairy galinsoga	*Galinsoga ciliata*
Groundcherry	*Physalis longifolia*
Shepherdspurse	*Capsella bursa-pastoris*
Pineappleweed	*Matricaria matricarioides*

SPRAY EQUIPMENT

Apply Antor® 4ES alone or in tank mixes using standard low pressure (20-50 psi) spray equipment. The equipment should be carefully calibrated before use and checked frequently during application to ensure that it is functioning properly. Uniformly apply the recommended rates. Use a spray volume of 20 to 40 gallons per acre broadcast basis. Use proportional amounts for band spraying. See section on Band Applications on the container label for calculations.

MIXING THE SPRAY

For Antor® 4ES alone, add 1/4 to 1/2 the total amount of water to the tank. Then add the required amount of Antor® 4ES and the remaining quantity of water with the bypass agitator running. When tank-mixing Antor® 4ES with flowable formulations, mix the proper amount of flowable formulation into half the required amount of water in the spray tank with the agitator running. Add the required amount of Antor® 4ES and then add the remaining required amount of water. Before tank-mixing with other herbicides, refer to the specific product labels for additional mixing directions.

ROTATIONAL CROPS

Crops, except wheat, may be rotated at 3 months post application. Wheat may be planted 9 months after application.

METHOD OF APPLICATION

Preplant Incorporated

In arid regions where furrow irrigation is practiced or where the rainfall pattern is erratic, soil incorporation of Antor® 4ES is required to provide optimum weed control.

Incorporation immediately after treatment is not required but must be done prior to weed germination. Use a hooded-power or ground-driven rotary tiller, rolling cultivator, or similar equipment properly adjusted to uniformly incorporate Antor® 4ES to a depth of 1 to 2 inches. Deeper incorporation may reduce effectiveness.

Preemergence

Preemergence application of Antor® 4ES is recommended where sprinkler irrigation is used for crop germination and where rainfall can be depended on to activate the product.

Apply Antor® 4ES to the soil surface at time of planting or shortly after, but prior to weed germination. One-half inch of rainfall or irrigation is usually adequate for activation. It is important that the top 2 inches of soil be thoroughly wetted.

RED BEET

Do not use Antor® 4ES or Antor® 4ES tank mixes on red beets within 75 days of harvest.

Apply Antor® 4ES preplant incorporated or preemergence using rates recommended in Dosage Table 1.

DOSAGE TABLE 1
(RED BEETS) ANTOR® 4ES ALONE

Soil Texture	Broadcast Rate of Antor® 4ES Per Acre(1)
Coarse:	
Loamy sands and sandy loams	2-3 qts.
Medium/Fine:	
Loams, silt loams, clay loams and clays containing less than 5% organic matter	3-4 qts.
Fine:	
Clay loams and clays containing more than 5% organic matter	4-6 qts.

(1) Within each soil texture category use the higher rate on the finer textured soils and/or where shepherdspurse or groundcherry is expected to be a problem.

ANTOR® 4ES + PYRAMIN®(1) TANK-MIX

A tank mix of Antor® 4ES with Pyramin® FL may be used in red beets to control a broader spectrum of weeds. In addition to the weeds controlled by Antor® 4ES alone, this mixture controls:

Common ragweed	*Ambrosia artemisiifolia*
Wild mustard	*Sinapis arvensis*
Pennsylvania smartweed	*Polygonum pensylvanicum*
Fanweed (stinkweed)	*Thlaspi arvense*
Common lambsquarters	*Chemopodium album*
Common purslane	*Portulaca oleracea*
Nettleleaf goosefoot	*Chenopodium murale*

Apply this tank-mix preemergence using rates recommended in Dosage Table 2. Do not use this mixture under conditions where Pyramin FL alone is not recommended. Before use, read the Pyramin FL label for additional information and precautions.

(1) Pyramin® is a registered trademark of BASF Aktiengesellschaft.

IMPORTANT: READ BEFORE USE

By using this product, user accepts the following conditions, warranty, disclaimer of warranties and limitations of liability.

CONDITIONS: The directions for use of this product are believed to be adequate and should be followed carefully. However, because of extreme weather and soil conditions, manner of use and other factors beyond NOR-AM Chemical Company's control, it is impossible for NOR-AM to eliminate all risks associated with the use of this product. As a result, crop injury or ineffectiveness is always possible.

DOSAGE TABLE 2
(RED BEETS) ANTOR® 4ES + PYRAMIN® FL TANK-MIX

Soil Texture	Broadcast Rate Per Acre Antor® 4ES (qts.)	Pyramin® FL (qts.)
Coarse:		
Sandy loams only	2	3
Medium:		
Loams, silt loams and clay loams containing less than 3% organic matter	2 to 3(1)	3
Fine:		
Clay loams and clays containing more than 3% organic matter	3	3

(1) Use the higher rate on clay loam soils.

SPINACH

Apply Antor® 4ES preplant incorporated or preemergence using rates recommended in dosage Table 3.

DOSAGE TABLE 3
(SPINACH)

Soil Texture	Broadcast Rate of Antor® 4ES Per Acre(1) (2)
Coarse:	
Loamy sands and sandy loams	2-3 qts.
Medium/Fine:	
Loams, slit loams, clay loams and clays containing less than 5% organic matter	3-4 qts.

(1) Within each soil texture category use the higher rate on the finer-textured soils and/or where shepherdspurse or groundcherry is expected to be a problem.

(2) The use of Antor® 4ES on clay loams or clays containing more than 5% organic matter has not been fully investigated. Do not use Antor® 4ES on these soils.

DISCLAIMER OF WARRANTIES: THERE ARE NO WARRANTIES, EXPRESS OR IMPLIED, OF MERCHANTABILITY OR OF FITNESS FOR A PARTICULAR PURPOSE OR OTHERWISE, WHICH EXTEND BEYOND THE STATEMENTS MADE ON THIS LABEL. No agent of NOR-AM Chemical Company is authorized to make any warranties beyond those contained herein or to modify the warranties contained herein. UNDER NO CIRCUMSTANCES SHALL NOR-AM CHEMICAL COMPANY BE LIABLE FOR ANY INCIDENTAL OR CONSEQUENTIAL DAMAGES RESULTING FROM FAILURE TO USE THIS PRODUCT IN ACCORDANCE WITH LABEL DIRECTIONS.

LIMITATIONS OF LIABILITY: In no event shall NOR-AM Chemical Company's liability for damages arising out of the use of this product exceed the purchase price of the product used.

SUPPLEMENTAL LABEL

ANTOR® 4ES

EPA Reg. No. 45639-54

This labeling must be in possession of the user at the time of pesticide application. All applicable directions, restrictions, and precautions on the EPA registered label are to be followed.

HERBICIDE FOR SELECTIVE WEED CONTROL IN BERMUDAGRASS GROWN FOR SEED

DIRECTIONS FOR USE

It is a violation of federal law to use this product in a manner inconsistent with its labeling. Read entire Directions for Use and Disclaimer of Warranties before using this product.

GENERAL INFORMATION

Antor® 4ES is an effective preemergence herbicide for use in established bermudagrass grown for seed (*Cynodon dactylon*). Antor® 4ES controls germinating weeds only after shoot and root absorption. It should be applied in the irrigation water (water run) prior to weed germination.

WEED SPECIES CONTROLLED

Annual Grass Weeds	
Sprangletop	*Leptochloa spp.*
Watergrass	*Echinochloa oryzoides*
Canarygrass	*Phalaris spp.*
Annual Broadleaf Weeds	
Redroot pigweed	*Amaranthus retroflexus*

APPLICATION RATE

Spring Application:

For early season weed control meter 2.0-4.0 quarts per acre of Antor® 4ES into the irrigation water of established bermudagrass prior to weed emergence. Use the higher rate for extended residual control, heavy weed pressure, or on heavy soil types. Refer to the flow chart and sample calculation for additional information.

Spring and Summer Application:

For control of weeds in both the spring and early summer months meter Antor® 4ES at a rate of 2.0 quarts per acre in the first spring irrigation followed by a second application in 6-8 weeks of 2.0 quarts per acre applied into the irrigation water. Refer to the flow chart and sample calculation for additional information.

Fall Application:

For control of cool season winter weeds, meter 2.0-4.0 quarts per acre of Antor 4ES into the irrigation water prior to weed emergence. Application timing may coincide with the last irrigation of the year. Use the higher rate for extended residual control, heavy weed pressure or on heavy soil types. Refer to the flow chart and sample calculation for additional information.

Application made after weed emergence may result in unacceptable weed control.

APPLICATION METHOD

Type of Device:

The constant flow device is recommended for metering Antor 4ES into the irrigation water for water-run applications of border type flood irrigation. This device is not recommended for other types of irrigation systems.

The necessary hardware is a length of pipe to reach from the chemical drum to just above the water level, a petcock valve, screen to prevent particles from clogging the flow orifice and a coupling to hold the orifice disc in place. Once the flow has begun by opening the petcock, open the venting petcock above the T coupler. This forms a constant head siphon, the flow rate governed by the orifice aperture.

Set Up and Calibration:

Set the Antor 4ES drum on planks over the ditch with the constant flow device pointed downward. Place the drum behind a gate or near some structure where there is turbulence for mixing of the herbicide. Start the herbicide flow to commence with the start of the irrigation cycle calibrated to finish with the end of the irrigation period.

To calibrate, estimate the length of time of the irrigation and number of acres to be treated. Calculate the flow rate for the amount of Antor 4ES to be applied and check the flow chart for the appropriate size orifice. (Refer to flow rate chart and sample calculation.) Select the appropriate orifice and measure the flow of herbicide for 1 minute in the measuring cup. Compare this to your calculations and adjust the orifice accordingly so that the correct amount is metered during the irrigation cycle. Tail water (run-off) from flood irrigation should be recirculated or controlled so that it does not leave the field.

Flow Rates For Antor® 4ES Using Various Tee Jet Orifices*

Tee Jet Orifice	CC Per Minute	Ounces Per Minute	Quarts Per Hour	Tee Jet Orifice	CC Per Minute	Ounces Per Minute	Quarts Per Hour
.015	10	.34	.64	.048	103	3.55	6.53
.016	11	.37	.69	.049	113	3.92	7.16
.018	13	.44	.83	.051	122	4.13	7.74
.020	17	.57	1.07	.052	126	4.33	7.99
.022	22	.74	1.39	.054	136	4.70	8.63
.024	24	.81	1.52	.055	144	4.97	9.17
.026	30	1.01	1.89	.057	153	5.27	9.69
.028	37	1.25	2.34	.059	168	5.71	10.65
.030	42	1.42	2.66	.061	173	5.98	10.97
.032	47	1.59	2.98	.063	184	6.36	11.66
.034	54	1.83	3.43	.065	190	6.63	12.04
.035	59	2.00	3.75	.067	207	7.13	13.13
.037	61	2.06	3.86	.070	238	8.01	15.09
.040	73	2.47	4.63	.075	270	9.23	17.12
.041	78	2.64	4.95	.080	296	10.21	18.77
.043	84	2.84	5.33	.086	339	11.46	21.49
.045	95	3.21	6.02	.093	384	12.98	24.36
.046	98	3.31	6.21	.103	470	16.13	29.79

* Figures were taken at 60°F and are approximate.

Occasionally check to measure flow in the field to make sure you have the correct orifice and because rates vary with temperatures

Sample Calculation

The following sample calculation is appropriate for Antor® 4ES applied as water-run with a constant flow device in bermudagrass grown for seed.

1. Size of block to be watered in one irrigation setting.
   eg. **10 acres**
2. Antor 4ES rate per acre.
   eg. **2 quarts**
3. Total amount of Antor 4ES to be applied to 10 acre block.
   eg. 2 qts/ Antor 4ES 10 acres = **20 quarts**
4. Amount of time to irrigate the 10 acre block.
   eg. **3 hours**
5. Amount of Antor 4ES applied per hour.
   eg. 20 qts ÷ 3 hours = **6.67 quarts/hr.**
6. Ounces of Antor 4ES applied per hour.
   eg. 6.67 qts/hr 32 oz/qt = **213.4 oz./hr.**
7. Ounces of Antor 4ES applied per minute.
   eg. 213.4 oz/hr ÷ 60 min/hr = **3.56 oz./min.**
8. Cubic centimeters (CC) of Antor 4ES per minute.
   eg. 3.56 oz/min 29.6 ml/oz = **105.37 ml/min.**
9. Check flow chart for flow rate.
   eg. 105.37 ml/min. ≈ **orifice No. 0.48**
10. Always calibrate the orifice before beginning because of differences in orifice flow rate.

GENERAL USE PRECAUTIONS

Do not apply Antor 4ES more than three times per season.

Apply Antor 4ES to bermudagrass which has been established for a minimum of 4 months. If applied earlier, crop injury may occur.

Take steps to insure that tail water does not reach other crops.

RESTRICTION

Do not graze or feed treated crop, forage, or hay to livestock.

The treated crop is to be used only for seed production.

ENVIRONMENTAL HAZARDS

Do not apply directly to any body of water except as specified on this label.

POSTING REQUIREMENTS

Areas being treated with pesticides applied through irrigation systems must be posted at the usual entrances to the area, at bulletin boards where workers assemble and where the treated area lies adjacent to roads, residential areas, or any other areas such as schools, parks or other public facilities. Posted signs must be legible from 25 feet away and be printed in English and any other appropriate language such as Spanish. Signs must be posted prior to application, during application and for at least one week after application, but may remain in place indefinitely. The sign must indicate the following:

NOTICE

Pesticides are applied in irrigation water in this field.

Do not enter this field when the irrigation system is operating.

Do not drink, bathe in or play in water or in furrows, puddles, ponds, canals or ditches associated with this irrigation system.

IMPORTANT: READ BEFORE USE

By using this product, user accepts the following conditions, warranty, disclaimer of warranties and limitations of liability.

CONDITIONS: The directions for use of this product are believed to be adequate and should be followed carefully. However, because of extreme weather and soil conditions, manner of use and other factors beyond NOR-AM Chemical Company's control, it is impossible for NOR-AM to eliminate all risks associated with the use of this product. As a result, crop injury or ineffectiveness is always possible.

DISCLAIMER OF WARRANTIES: THERE ARE NO WARRANTIES, EXPRESS OR IMPLIED, OF MERCHANTABILITY OR OF FITNESS FOR A PARTICULAR PURPOSE OR OTHERWISE, WHICH EXTEND BEYOND THE STATEMENTS MADE ON THIS LABEL. No agent of NOR-AM Chemical Company is authorized to make any warranties beyond those contained herein or to modify the warranties contained herein. UNDER NO CIRCUMSTANCES SHALL NOR-AM CHEMICAL COMPANY BE LIABLE FOR ANY INCIDENTAL OR CONSEQUENTIAL DAMAGES RESULTING FROM FAILURE TO USE THIS PRODUCT IN ACCORDANCE WITH LABEL DIRECTIONS.

LIMITATIONS OF LIABILITY: In no event shall NOR-AM Chemical Company's liability for damages arising out of the use of this product exceed the purchase price of the product used.

BETAMIX®

SUGAR BEET HERBICIDE

FOR AGRICULTURAL USE ONLY

Postemergence Herbicide for Control of Weeds in Sugar Beets

ACTIVE INGREDIENTS:

Phenmedipham*	8.0%
Desmedipham**	8.0%
INERT INGREDIENTS	84.0%
TOTAL	100.0%

Contains 1.3 lbs. active ingredient per gallon.

*Methyl *m* -hydroxycarbanilate *m* -methylcarbanilate (ester)

**Ethyl *m* -hydroxycarbanilate carbanilate (ester)

EPA Reg. No. 45639-87 EPA Est. No. 407-MN-1
U.S. Patent No. 3,404,975 U.S. Patent No. 3,692,820

Keep Out of Reach of Children

WARNING — AVISO

STATEMENT OF PRACTICAL TREATMENT

IF SWALLOWED: Call a physician or Poison Control Center. Drink 1 or 2 glasses of water and induce vomiting by touching back of throat with finger. Do not induce vomiting or give anything by mouth to an unconscious person.

IF IN EYES: Flush with plenty of water. Get medical attention.

IF ON SKIN: Wash immediately with plenty of soap and water.

PRECAUCION AL USUARIO: Si usted no lee ingles, no use este producto hasta que la etiqueta haya.

PRECAUTIONARY STATEMENTS

HAZARDS TO HUMANS

Harmful if swallowed. May produce severe irritation of eyes and irritation of skin. Avoid breathing spray mist. Avoid contact with skin and eyes.

In case of eye contact, flush with plenty of water for 15 minutes and get medical attention; for skin, wash immediately with soap and water.

Avoid contamination of feed and foodstuffs.

ENVIRONMENTAL HAZARDS

This product is toxic to fish. Keep out of lakes, streams or ponds. Do not apply where runoff is likely to occur. Do not apply when weather conditions favor drift from areas treated.

PHYSICAL OR CHEMICAL HAZARDS

Do not use or store near heat or open flame.

DO NOT REUSE THIS CONTAINER.

DESTROY WHEN EMPTY.

GENERAL INFORMATION

BETAMIX is a selective postemergence herbicide for use in sugar beets (including beets grown for seeds):

Annual sowthistle	*Sonchus oleraceus*
Black nightshade	*Solanum nigrum*
Coast fiddleneck	*Amsinckia intermedia*
Common chickweed	*Stellaria media*
Common lambsquarters	*Chenopodium album*
Common ragweed	*Ambrosia artemisliifolia*
Green foxtail	*Setaria viridis*
Groundcherry	*Physalis lanceifolia*
Kochia*	*Kochia scoparia*
London rocket	*Sisymbrium irio*
Nettleleaf goosefoot	*Chenopodium murale*
Green foxtail	*Setaria viridis*
Pigeon grass (yellow foxtail)	*Setaria glauca*
Prostrate pigweed	*Amaranthus gracizans*
Purslane	*Portulaca oleraces*
Redroot pigweed	*Amaranthus retroflexus*
Shepherdspurse	*Capsella bursa-postoris*
Wild buckwheat	*Polygonum convolvulus*
Wild mustard	*Sinapis arvensis*

* Spray Kochia while in the rosette stage, less than 1" in diameter. For best results, spray when the weeds are at the two-leaf stage.

USE PRECAUTIONS

DO NOT APPLY BETAMIX TO SUGAR BEETS LATER THAN 90 DAYS PRIOR TO HARVEST.

DO NOT EXCEED A TOTAL OF 12 PINTS BETAMIX PER ACRE PER SEASON.

BETAMIX MAY CAUSE BEET INJURY IF THE CROP IS UNDER STRESS FROM ONE OR MORE OF THE FOLLOWING CONDITIONS:

- Rapid climatic changes from cool, overcast days, to hot (85° or over), bright days.
- Windy conditions or drought.
- Use of a preplant or postemergence herbicide or other chemicals.
- Insect or disease injury.
- Close cultivation.

If stress conditions are present, delay application in order to give plants a chance to recover.

If extreme weather conditions are of short duration, delay spraying until the end of such a period.

If BETAMIX applications must be made on days with extreme temperature and/or brightness, delay spraying until evening.

DO NOT OVERTREAT: The use of higher than recommended rates may cause beet injury.

Do not spray while dew is present.

Rainfall within 6 hours of spraying may reduce weed kill. Do not allow spray drift to contact adjacent crops which may be injured by spray drift.

IMPORTANT

BETAMIX may cause temporary growth retardation and/or chlorosis or tip-burn on sugar beets. Beets usually resume normal growth within 10 days.

DIRECTIONS FOR USE

It is a violation of Federal Law to use this product in a manner inconsistent with its labeling. Read the entire Directions for Use before using this product.

REENTRY STATEMENT

Reentry into treated areas is prohibited until leaves are completely dry or until 24 hours after application.

Do not enter treated areas without protective clothing.

Because certain states may require more restrictive reentry intervals for various crops treated with this product, consult your State Department of Agriculture for further information.

Written or oral warnings must be given to workers who are expected to be in a treated area or in an area about to be treated with this product. (Indicate specific oral warnings which inform workers of areas or fields that may not be entered without specific protective clothing, period of time field must be vacated and appropriate actions to take in case of accidental exposure.) When oral warnings are given, warnings shall be given in a language customarily understood by workers. Oral warnings must be given if there is reason to believe that written warnings cannot be understood by workers. Written warnings must include the following information:

WARNING. Area treated with BETAMIX on (date of application). Do not enter without appropriate protective clothing until leaves are completely dry. In case of accidental exposure see Statement of Practical Treatment on the label.

When used as directed, BETAMIX is selective in beets past the two-true leaf stage. Beets may be severely injured if treated before the two-true leaf stage. The stage of weed growth is very important for satisfactory control. For best results, spray when weeds are at the two-true leaf stage. Best results are obtained when the weeds are actively growing and are not under water or heat stress.

Mixing the Spray: Make sure the Sprayer is CLEAN.

BETAMIX emulsifiable concentrate formulation contains sufficient wetting agents for optimum coverage. Do not add additional wetting agents or other spray adjuvants. Add sufficient water to fill the lines. Then add the desired amount of BETAMIX and the remaining quantity of water with the bypass agitator running. Bypass agitation is sufficient. Mechanical agitation is not necessary. Only use freshly prepared spray emulsions.

Always spray immediately after preparing the spray solution. Prepare only enough spray solution to last less than four hours.

RATE OF APPLICATION

By Ground: Apply BETAMIX at the rate of 4½ to 7½ pints in 20 to 50 gallons of water broadcast basis. For band application see dosage chart.

By Air: Apply BETAMIX at the rate of 4½ to 7½ pints per acre using 5 to 15 gallons of spray per acre.

Use the 7½-pint rate only on well-established beets which are not under stress.

In order to avoid phytotoxic spray drift to non-target crops during application of BETAMIX, the following buffer zones should be observed:

Cotton—Potatoes—Sunflowers—Sorghum—Wheat	50 feet
Blackeye beans—Cabbage—Flax	100 feet
Lettuce—Rape—Tomatoes	300 feet

DO NOT APPLY WHEN WIND SPEED IS OVER 10 MILES PER HOUR. AVOID APPLICATIONS WHEN CONDITIONS FAVOR DRIFT.

To avoid excessive phytotoxicity to fall-planted sugar beets in Arizona and south of the Tehachapi Mountains in California when temperatures are above 85°F, apply BETAMIX at the rate of 4 pints per acre (broadcast equivalent). Evening applications are recommended.

For further information, contact your County Agricultural Agent, Farm Advisor or NOR-AM Chemical Company.

REPEAT APPLICATION OF BETAMIX

For control of later germinating weeds, make a second application of BETAMIX. Use 4½ to 6 pints of BETAMIX. Allow at least 7 days between first and second applications. Apply when beets have at least 4 leaves. For best results, use the higher rate and spray when weeds are at the two-true leaf stage. Apply lower rates when the beets are under stress as explained on the left panel.

SPLIT (LOW RATE) APPLICATIONS:
(ALL SUGAR BEET AREAS EXCEPT CALIFORNIA)

Split (low rate) applications of BETAMIX at 2 to 3 pints per acre broadcast equivalent in 5 to 10 gallons of water may be applied to sugar beets at any stage of growth when necessary to control early germinating, difficult to control weeds. The spray must be applied when the earliest emerging weeds have reached cotyledon size.

Any weeds which are not completely controlled by the first treatment will usually be checked and controlled by repeat applications. The following application should be made 5 to 7 days after the preceding application or when another flush of weeds germinates. If the second application is delayed, conventional treatment as described above will be necessary.

Apply the 4½ to 7½ pint rates only to sugar beets past the two-true leaf stage.

Dosage Chart For Band Application

Broadcast Equivalent	Band Width	Band Rate — Row Spacing 22"	38"	32"
2 pints/acre	5"	7.3 fl. oz.	5.7 fl. oz.	5.3 fl. oz.
	7"	10.1 fl. oz.	8.0 fl. oz.	7.5 fl. oz.
3 pints/acre	5"	10.9 fl. oz.	8.5 fl. oz.	8.0 fl. oz.
	7"	15.2 fl. oz.	12.0 fl. oz.	11.2 fl. oz.
4½ pints/acre	5"	16.4 fl. oz.	12.9 fl. oz.	12.1 fl. oz.
	7"	22.9 fl. oz.	18.0 fl. oz.	16.9 fl. oz.
6 pints/acre	5"	21.8 fl. oz.	17.1 fl. oz.	16.0 fl. oz.
	7"	30.5 fl. oz.	24.0 fl. oz.	22.5 fl. oz.
7½ pints/acre	5"	27.3 fl. oz.	21.4 fl. oz.	20.1 fl. oz.
	7"	38.2 fl. oz.	30.0 fl. oz.	28.1 fl. oz.

STORAGE AND DISPOSAL

STORAGE

Store in original container and keep closed. Store in cool, dry place. Do not use or store near heat or open flame.

PESTICIDE DISPOSAL

Pesticide wastes are toxic. Improper disposal of excess pesticide, spray mixture, or rinsate is a violation of Federal Law. If these wastes cannot be disposed of by use according to label instructions, contact your State Pesticide or Environmental Control Agency, or the Hazardous Waste representative at the nearest EPA Regional Office for guidance.

CONTAINER DISPOSAL

Triple rinse (or equivalent). Then offer for recycling or reconditioning, or puncture and dispose of in a sanitary landfill, or by other procedures approved by state and local authorities.

IMPORTANT: READ BEFORE USE

By using this product, user accepts the following conditions, warranty, disclaimer of warranties and limitations of liability.

CONDITIONS:

The directions for use of this product are believed to be adequate and should be followed carefully. However, because of extreme weather and soil conditions, manner of use and other factors beyond NOR-AM Chemical Company's control, it is impossible for NOR-AM to eliminate all risks associated with the use of this product. As a result, crop injury or ineffectiveness is always possible.

DISCLAIMER OF WARRANTIES:

THERE ARE NO WARRANTIES, EXPRESS OR IMPLIED, OF MERCHANTABILITY OR OF FITNESS FOR A PARTICULAR PURPOSE OR OTHERWISE, WHICH EXTEND BEYOND THE STATEMENTS MADE ON THIS LABEL. No agent of NOR-AM Chemical Company is authorized to make any warranties beyond those contained herein or to modify the warranties contained herein. UNDER NO CIRCUMSTANCES SHALL NOR-AM CHEMICAL COMPANY BE LIABLE FOR ANY INCIDENTAL OR CONSEQUENTIAL DAMAGES RESULTING FROM FAILURE TO USE THIS PRODUCT IN ACCORDANCE WITH LABEL DIRECTIONS.

LIMITATIONS OF LIABILITY:

In no event shall NOR-AM Chemical Company's liability for damages arising out of the use of this product exceed the purchase price of the product used.

BETANEX®

SUGAR BEET HERBICIDE

FOR AGRICULTURAL USE ONLY

Postemergence Herbicide for Control of Redroot Pigweed And Other Weeds in Sugar Beets

ACTIVE INGREDIENT:
Desmedipham* 16.0%
INERT INGREDIENTS: 84.0%
TOTAL 100.0%

Contains 1.3 lbs. active ingredient per gallon.

*Ethyl *m* -hydroxycarbanilate carbanilate (ester)

EPA Reg. No. 45639-86 EPA Est. No. 407-MN-1

Keep Out of Reach of Children
WARNING — AVISO

STATEMENT OF PRACTICAL TREATMENT

IF SWALLOWED: Call a physician or Poison Control Center. Drink 1 or 2 glasses of water and induce vomiting by touching back of throat with finger. Do not induce vomiting or give anything by mouth to an unconscious person.

IF IN EYES: Flush with plenty of water. Get medical attention.

IF ON SKIN: Wash immediately with plenty of soap and water.

PRECAUCION AL USUARIO: Si usted no lee ingles, no use este producto hasta que le etiqueta haya le sido explicado ampliamente.

PRECAUTIONARY STATEMENTS

HAZARDS TO HUMANS

Harmful if swallowed, may produce severe irritation of eyes and irritation of skin. Avoid breathing spray mist. Avoid contact with skin and eyes.

In case of eye contact: flush with plenty of water for 5 minutes and get medical attention; for skin, wash immediately with soap and water.

Avoid contamination of feed and foodstuffs.

ENVIRONMENTAL HAZARDS

This product is toxic to fish. Keep out of lakes, streams or ponds. Do not apply where runoff is likely to occur. Do not apply when weather conditions favor drift from areas treated.

PHYSICAL OR CHEMICAL HAZARDS

Do not use or store near heat or open flame.

DO NOT REUSE THIS CONTAINER.
DESTROY WHEN EMPTY.

GENERAL INFORMATION

BETANEX is a selective postemergence herbicide for use in sugar beets (including beets grown for seeds):

Annual sowthistle	*Sonchus oleraceus*
Black nightshade	*Solanum nigrum*
Coast fiddleneck	*Amsinckia intermedia*
Common chickweed	*Stellaria media*
Common lambsquarters	*Chenopodium album*
Common ragweed	*Ambrosia artemisliifolia*
Groundcherry	*Physalis lanceifolia*
Kochia*	*Kochia scoparia*
London rocket	*Sisymbrium irio*
Nettleleaf goosefoot	*Chenopodium murale*
Prostrate pigweed	*Amaranthus gracizans*
Purslane	*Portulaca oleraces*
Redroot pigweed	*Amaranthus retroflexus*
Shepherdspurse	*Capsella bursa-postoris*
Wild buckwheat	*Polygonum convolvulus*
Wild mustard	*Sinapis arvensis*

* Spray Kochia while in the rosette stage, less than 1" in diameter. For best results, spray when the weeds are at the two-leaf stage.

USE PRECAUTIONS

DO NOT APPLY BETANEX TO SUGAR BEETS LATER THAN 90 DAYS PRIOR TO HARVEST.

DO NOT EXCEED A TOTAL OF 12 PINTS BETANEX PER ACRE PER SEASON.

BETANEX MAY CAUSE BEET INJURY IF THE CROP IS UNDER STRESS FROM ONE OR MORE OF THE FOLLOWING CONDITIONS:

- Rapid climatic changes from cool, overcast days, to hot (85° or over), bright days.
- Windy conditions or drought.
- Use of a preplant or postemergence herbicide or other chemicals.
- Insect or disease injury.
- Close cultivation.

If stress conditions are present, delay application in order to give plants a chance to recover.

If extreme weather conditions are of short duration, delay spraying until the end of such a period.

If BETANEX applications must be made on days with extreme temperature and/or brightness, delay spraying until evening.

DO NOT OVERTREAT: The use of higher than recommended rates may cause beet injury.

Do not spray while dew is present.

Rainfall within 6 hours of spraying may reduce weed kill. Do not allow spray drift to contact adjacent crops which may be injured by spray drift.

IMPORTANT

BETANEX may cause temporary growth retardation and/or chlorosis or tip-burn on sugar beets. Beets usually resume normal growth within 10 days.

DIRECTIONS FOR USE

It is a violation of Federal Law to use this product in a manner inconsistent with its labeling. Read the entire Directions for Use before using this product.

REENTRY STATEMENT

"Do not apply this product in such a manner as to directly or through drift expose workers or other persons. The area being treated must be vacated by unprotected persons."

Reentry into treated areas is prohibited until leaves are completely dry.

Do not enter treated areas without protective clothing.

Because certain states may require restrictive reentry intervals for various crops treated with this product, consult your State Department of Agriculture for further information.

Written or oral warnings must be given to workers who are expected to be in a treated area or in an area about to be treated with this product. (Indicate specific oral warnings which inform workers of areas or fields that may not be entered without specific protective clothing, period of time field must be vacated and appropriate actions to take in case of accidental exposure.) When oral warnings are given, warnings shall be given in a language customarily understood by workers. Oral warnings must be given if there is reason to believe that written warnings cannot be understood by workers. Written warnings must include the following information:

WARNING. Area treated with BETANEX on (date of application). Do not enter without appropriate protective clothing until leaves are completely dry. In case of accidental exposure see Statement of Practical Treatment on the label.

When used as directed, BETANEX is selective in beets past the two-true leaf stage. Beets may be severely injured if treated before the two-true leaf stage. The stage of growth of the weeds is very important for satisfactory control. For best results, spray when the weeds are at the two-true leaf stage. Best results are obtained when the weeds are actively growing and are not under water or heat stress.

Mixing the Spray: Make sure the Sprayer is CLEAN.

BETANEX emulsifiable concentrate formulation contains sufficient wetting agents for optimum coverage. Do not add additional wetting agents or other spray adjuvants. Add sufficient water to fill the lines. Then add the desired amount of BETANEX and the remaining quantity of water with the bypass agitator running. Bypass agitation is sufficient. Mechanical agitation is not necessary. Only use freshly prepared spray emulsions.

Always spray immediately after preparing the spray solution. Prepare only enough spray solution to last less than four hours.

RATE OF APPLICATION

By Ground: Apply BETANEX at the rate of 4½ to 7½ pints in 20 to 50 gallons of water broadcast basis. For band application, see dosage chart.

By Air: Apply BETANEX at the rate of 4½ to 7½ pints per acre using 5 to 15 gallons of spray per acre.

Use the 7½-pint rate only on well-established beets which are not under stress.

In order to avoid phytotoxic spray drift to non-target crops during application of BETANEX the following buffer zones should be observed:

Cotton—Potatoes—Sunflowers—Sorghum—Wheat	50 feet
Blackeye beans—Cabbage—Flax	100 feet
Lettuce—Rape—Tomatoes	300 feet

DO NOT APPLY WHEN WIND SPEED IS OVER 10 MILES PER HOUR. AVOID APPLICATIONS WHEN CONDITIONS FAVOR DRIFT.

To avoid excessive phytotoxicity to fall-planted sugar beets in Arizona and south of the Tehachapi Mountains in California when temperatures are above 85°F, apply BETANEX at the rate of 4 pints per acre (broadcast equivalent). Evening applications are recommended.

For further information, contact your County Agricultural Agent, Farm Advisor or NOR-AM Chemical Company.

REPEAT APPLICATION OF BETANEX

For control of later germinating weeds, make a second application of BETANEX. Use 4½ to 6 pints of BETANEX. Allow at least 7 days between first and second applications. Apply when beets have at least 4 leaves. For best results, use the higher rate and spray when weeds are at the two-true leaf stage. Apply lower rates when the beets are under stress as explained on the left panel.

SPLIT (LOW RATE) APPLICATIONS:
(ALL SUGAR BEET AREAS EXCEPT CALIFORNIA)
Split (low rate) applications of BETANEX at 2 to 3 pints per acre broadcast equivalent in 5 to 10 gallons of water may be applied to sugar beets at any stage of growth when necessary to control early germinating, difficult to control weeds. The spray must be applied when the earliest emerging weeds have reached cotyledon size.

Any weeds which are not completely controlled by the first treatment will usually be checked and controlled by repeat applications. The following application should be made 5 to 7 days after the preceding application or when another flush of weeds germinates. If the second application is delayed, conventional treatment as described above will be necessary.

Apply the $4\frac{1}{2}$ to $7\frac{1}{2}$ pint rates only to sugar beets past the two-true leaf stage.

Dosage Chart For Band Application

Broadcast Equivalent	Band Width	Band Rate: Row Spacing 22″	38″	32″
2 pints/acre	5″	7.3 fl. oz.	5.7 fl. oz.	5.3 fl. oz.
	7″	10.1 fl. oz.	8.0 fl. oz.	7.5 fl. oz.
3 pints/acre	5″	10.9 fl. oz.	8.5 fl. oz.	8.0 fl. oz.
	7″	15.2 fl. oz.	12.0 fl. oz.	11.2 fl. oz.
$4\frac{1}{2}$ pints/acre	5″	16.4 fl. oz.	12.9 fl. oz.	12.1 fl. oz.
	7″	22.9 fl. oz.	18.0 fl. oz.	16.9 fl. oz.
6 pints/acre	5″	21.8 fl. oz.	17.1 fl. oz.	16.0 fl. oz.
	7″	30.5 fl. oz.	24.0 fl. oz.	22.5 fl. oz.
$7\frac{1}{2}$ pints/acre	5″	27.3 fl. oz.	21.4 fl. oz.	20.1 fl. oz.
	7″	38.2 fl. oz.	30.0 fl. oz.	28.1 fl. oz.

STORAGE AND DISPOSAL

STORAGE
Store in original container and keep closed. Store in a cool, dry place. Do not use or store near heat or open flame.

PESTICIDE DISPOSAL
Pesticide wastes are toxic. Improper disposal of excess pesticide, spray mixture, or rinsate is a violation of Federal Law. If these wastes cannot be disposed of by use according to label instructions, contact your State Pesticide or Environmental Control Agency, or the Hazardous Waste representative at the nearest EPA Regional Office for guidance.

CONTAINER DISPOSAL
Triple rinse (or equivalent). Then offer for recycling or reconditioning, or puncture and dispose of in a sanitary landfill, or by other procedures approved by state and local authorities.

IMPORTANT: READ BEFORE USE

By using this product, user accepts the following conditions, warranty, disclaimer of warranties and limitations of liability.

CONDITIONS:
The directions for use of this product are believed to be adequate and should be followed carefully. However, because of extreme weather and soil conditions, manner of use and other factors beyond NOR-AM Chemical Company's control, it is impossible for NOR-AM to eliminate all risks associated with the use of this product. As a result, crop injury or ineffectiveness is always possible.

DISCLAIMER OF WARRANTIES:
THERE ARE NO WARRANTIES, EXPRESS OR IMPLIED, OF MERCHANTABILITY OR OF FITNESS FOR A PARTICULAR PURPOSE OR OTHERWISE, WHICH EXTEND BEYOND THE STATEMENTS MADE ON THIS LABEL. No agent of NOR-AM Chemical Company is authorized to make any warranties beyond those contained herein or to modify the warranties contained herein. UNDER NO CIRCUMSTANCES SHALL NOR-AM CHEMICAL COMPANY BE LIABLE FOR ANY INCIDENTAL OR CONSEQUENTIAL DAMAGES RESULTING FROM FAILURE TO USE THIS PRODUCT IN ACCORDANCE WITH LABEL DIRECTIONS.

LIMITATIONS OF LIABILITY:
In no event shall NOR-AM Chemical Company's liability for damages arising out of the use of this product exceed the purchase price of the product used.

BOTRAN® 75W

FUNGICIDE

For: Apricot, Blackberry, Boysenberry, Carrot, Celery, Cotton, Cucumber, Endive, Garlic, Grape, Lettuce, Nectarine, Onion, Peach, Plum and Prune, Potato, Red Raspberry, Rhubarb, Snap Bean, Sweet Cherry, Sweet Potato, Tomato and Ornamentals.

See folder in bottom flap of bag for usage directions.

ACTIVE INGREDIENT
2,6-Dichloro-4-nitroaniline .... 75% w/w
INERT INGREDIENTS .... 25% w/w
TOTAL .... 100% w/w

EPA Reg. No. 45639-110

Keep Out of Reach of Children
CAUTION—PRECAUCION

PRECAUCION AL USUARIO: Si usted no lee ingles, no use este producto hasta que el etiqueta haya sido explicado ampliamente.

STATEMENT OF PRACTICAL TREATMENT

IF ON SKIN: REMOVE CONTAMINATED CLOTHING AND WASH WITH SOAP AND WATER.

IF IN EYES: FLUSH WITH WATER. IF EYE IRRITATION PERSISTS, CALL A PHYSICIAN.

IF SWALLOWED: DRINK WARM WATER AND INDUCE VOMITING BY TOUCHING BACK OF THROAT WITH FINGER AND/OR BY ADMINISTERING SYRUP OF IPECAC. FOLLOW WITH MILK AND A MILD CATHARTIC (MILK OF MAGNESIA). CALL A PHYSICIAN OR POISON CONTROL CENTER IMMEDIATELY. NEVER GIVE FLUIDS OR INDUCE VOMITING IF PATIENT IS UNCONSCIOUS OR HAVING CONVULSIONS.

PRECAUTIONARY STATEMENTS

HAZARDS TO HUMANS AND DOMESTIC ANIMALS

CAUTION
Harmful if swallowed. Avoid inhalation of dust or spray mist. Avoid contact with skin and eyes.

ENVIRONMENTAL HAZARDS
Toxic to fish. Do not apply directly to water or wetlands. Do not apply where runoff is likely to occur. Do not apply when weather conditions favor drift from areas treated. Do not contaminate water by cleaning of equipment or disposal of wastes.

STORAGE AND DISPOSAL

Do not contaminate water, food or feed by storage or disposal.

STORAGE
Store in original container and keep closed. Store in a cool, dry place.

PESTICIDE DISPOSAL
Wastes resulting from the use of this product may be disposed of on site or at an approved waste disposal facility.

CONTAINER DISPOSAL
Completely empty bag into application equipment. Then dispose of empty bag in a sanitary landfill or by incineration, or, if allowed by State and local authorities, by burning. If burned, stay out of smoke.

DIRECTIONS FOR USE

It is a violation of Federal law to use this product in a manner inconsistent with its labeling.

Do not apply this product in such a manner as to directly or through drift expose workers or other persons. The area being treated must be vacated by unprotected persons.

EQUIVALENTS

One and one-third pounds as found in Directions for Use equals one 4 pound bag in 300 gallons. One (1) level tablespoonful per gallon equals one (1) pound per 100 gallons.

COMPATIBILITY

Botran 75W is compatible with wettable powder formulations of captan, zineb, Guthion, parathion, carbaryl, and endosulfan. Normally Botran 75W can be mixed with other fungicides, insecticides, or defoliants formulated as wettable powders. However, combination sprays of Botran 75W and miscible oil formulations of insecticides, particularly organic phosphorous compounds may cause plant injury. It is suggested that such combinations be used on a small number of plants before treating large areas.

REENTRY STATEMENT

Do not enter treated areas without protective clothing until sprays have dried. Because certain states may require more restrictive reentry intervals for various crops treated with this product, consult your State Department of Agriculture for further information.

Written or oral warnings must be given to workers who are expected to be in a treated area or in an area about to be treated with this product. When oral warnings are given, warnings shall be given in a language customarily understood by workers. Oral warnings must be given if there is reason to believe that written warnings cannot be understood by workers.

IMPORTANT: READ BEFORE USE

By using this product, user accepts the following conditions, warranty, disclaimer of warranties and limitations of liability.

CONDITIONS

The directions for use of this product are believed to be adequate and should be followed carefully. However, because of extreme weather and soil conditions, manner of use and other factors beyond NOR-AM Chemical Company's control, it is impossible for NOR-AM to eliminate all risks associated with the use of this product. As a result, crop injury or ineffectiveness is always possible.

DISCLAIMER OF WARRANTIES

THERE ARE NO WARRANTIES, EXPRESS OR IMPLIED, OF MERCHANTABILITY OR OF FITNESS FOR A PARTICULAR PURPOSE OR OTHERWISE, WHICH EXTEND BEYOND THE STATEMENTS MADE ON THIS LABEL. No agent of NOR-AM Chemical Company is authorized to make any warranties beyond those contained herein or to modify the warranties contained herein. UNDER NO CIRCUMSTANCES SHALL NOR-AM CHEMICAL COMPANY BE LIABLE FOR ANY INCIDENTAL OR CONSEQUENTIAL DAMAGES RESULTING FROM FAILURE TO USE THIS PRODUCT IN ACCORDANCE WITH LABEL DIRECTIONS.

LIMITATIONS OF LIABILITY

In no event shall NOR-AM Chemical Company's liability for damages arising out of the use of this product exceed the purchase price of the product used.

DIRECTIONS FOR USE ON EDIBLE CROPS

Crop	Disease	Dosage[1] Lb Per 100 Gal. of Water	Use
Apricot	Fruit decay *Monilinia* Brown Rot *Rhizopus* rot	1⅓	Apply 10 days and 1 day prior to harvest. Allow 1 day between application and harvest.
	Postharvest decay *Rhizopus stolonifer* *Monilinia* rot (Brown rot)	1	Apply as combination of fruit wax plus Botran 75W plus Benlate 50W (0.5 lb. per 100 gal. of water) during packing line operations. Mix the combination in a tank equipped with mechanical agitation. Mechanically agitate combination during use. Apply the Botran-Benlate-wax combination to the fruit at the rate of 35–45 gal/hr. Flat-fan or flood jet nozzles should be used. Collect spray runoff in a drain tray and discard. Care should be taken to prevent plugging of nozzles.
Note: Do not treat fruit packed in containers which do not allow adequate drainage. Botran does not control *R. arrhizus.*			
Blackberry Boysenberry Raspberry (red)	Fruit decay *Botrytis* rot	1⅓	Use a maximum of 4 applications. Begin applications (not to exceed 250 gallons per acre) just prior to bloom and continue at 10-day intervals. Allow 1 day between last treatment and harvest. LIMIT USE TO PACIFIC NORTHWEST ONLY.
Carrot (Storage only)	Post-harvest decay *Sclerotinia sclerotiorum* (dip)	1	Dip carrots for 10 seconds in well agitated suspension. Do not rinse after treatment. Add ½ lb. Botran 75W to 100 gallons of dip suspension after 500 bushels are treated. After each 1000 bushels treated, drain and flush tank. Refill with fresh dip suspension.
Cucumber (greenhouse)	White mold *Sclerotinia sclerotiorum*	1⅓	Apply to diseased areas of plants. Additional applications may be required after 14 days.
Grape	Bunch rot *Botrytis* rot Storage rot	1⅓	Botran 75W may be applied at onset of bloom to soon after shatter. Use 2⅔ lb./acre of Botran 75W in 50 to 400 gallons of water per acre for control of Botrytis bunch rot *(Botrytis cinerea).* Up to 5⅓ pounds of Botran 75W may be applied per season. Use Botran dust after cluster tightening (see Botran dust labels for complete directions for use). When combining Botran 75W with gibberellin, use with last gibberellin application.
Peach and Nectarine	Blossom blight *Monilinia*	1⅓	Apply at pink bud and full bloom.

Crop	Disease	Dosage[1] Lb Per 100 Gal. of Water	Use
	Fruit decay *Monilinia* Brown rot *Rhizopus* rot	$1\frac{1}{3}$	Apply 18 days, 10 days and 1 day prior to harvest. Allow 1 day between application and harvest.
Note: When combining with captan, use 1 to $1\frac{1}{3}$ lb. Botran 75W to 1–2 lbs. captan 50W or equivalent per 100 gallons of water as recommended.			
Peach (only)	Post-harvest decay *Rhizopus* rot *Botrytis* rot	1	**Spray or dip**—Immediately after picking, spray or dip peaches with suspension of Botran 75W. Do not recirculate used spray suspension. **Processing line operations**—Botran 75W is most effective when applied during or after the defuzzing process, either by brushes, spray, or in the hydrocooler. **Brushes**—Apply Botran 75W to the brush with traversing feeder nozzle capable of metering approximately 10 gallons of Botran 75W suspension per hour with the valve completely open. Agitate the spray suspension. Unused suspension that drips from the brush should be caught in a drain tray and discarded. **Spray**—Direct a mist spray down upon the rotating peaches. Use pressures between 20–25 psi with delivery system of approximately 45 gallons per hour. TJ80 nozzles are suggested. Agitate the spray suspension. Collect spray runoff in a drain tray and discard. **Hydrocooler**—As volume decreases, make up to original volume by adding Botran 75W suspension of 1 Lb per 100 gal. Drain and flush daily.
Note: Do not treat peaches packed in containers which do not allow adequate drainage.			
	Post-harvest decay *Rhizopus* rot *Botrytis* rot *Monilinia* rot (brown rot)	$2\frac{2}{3}$	For freezing and canning only. Immediately after picking dip peaches in suspension of Botran 75W. Continuous agitation is necessary. As volume decreases, make up to original volume by adding Botran 75W suspension of $2\frac{2}{3}$ lb. per 100 gal. of water. Drain and flush system daily.
WARNING: Peaches treated with this rate of Botran will have residues exceeding the established tolerance of Botran on peaches. Do not use this fruit for the fresh fruit market. Processing for freezing and canning removes all excess residue.			
Nectarine (only)	Post-harvest decay *Rhizopus* rot *Monilinia* rot (brown rot)	$2\frac{2}{3}$	Apply only with fruit wax. Mix the Botran with wax in a tank equipped with mechanical agitation. The Botran-wax combination should be mechanically agitated during use; by-pass agitation will not keep the Botran in suspension. Fruit should be sprayed during packing line operations. Apply the Botran-wax combination to the fruit at the rate of 30–35 gal./hour at 18–30 psi. Flat-fan or flood jet nozzles should be used. Collect spray run-off in a drain tray and discard. Care should be taken to prevent plugging of nozzles.
Note: If wax containing OPP is used, the concentration of OPP in the final mixture should not exceed 0.25%.			
Plum and Prune	Blossom blight *Monilinia* Brown rot	$1\frac{1}{3}$	Apply at popcorn and full bloom.
Plum	Post-harvest decay *Rhizopus* rot *Monilinia* rot (Brown rot)		Apply only with fruit wax. Mix Botran 75W with wax in a tank equipped with mechanical agitation. The Botran-wax combination should be mechanically agitated during use; by-pass agitation will not keep the Botran in suspension. Use flat-fan or flood jet nozzles to uniformly spray fruit during packing line operations. Care should be taken to prevent plugging of nozzles. Collect spray run-off in a drain tray and discard.
		$2\frac{2}{3}$	For use in conventional applicators (30–50 gal/hr), apply 100 gallons of the diluted Botran-wax combination to approximately 50,000 pounds of fruit.
		10–12	For use in low volume applicators (5–8 gal/hr), apply 100 gallons of the Botran-wax combination to approximately 500,000 pounds of fruit.
Rhubarb (hot house)	*Botrytis* leaf rot	$1\frac{1}{3}$	Begin applications when first buds emerge from crowns. Repeat at weekly intervals. Do not use later than 3 days before harvest.
Sweet Cherry	Blossom blight *Monilinia*	$1\frac{1}{3}$	Apply at popcorn, bloom, full bloom and petal fall.

Crop	Disease	Dosage[1] Lb Per 100 Gal. of Water	Use
	Fruit decay *Monilinia* Brown rot *Rhizopus* rot	1⅓	Apply 10 days and 1 day prior to harvest.
	Post-harvest decay *Rhizopus* rot	1⅓	Spray cherries, while sorting, with a suspension of Botran 75W. Do not recirculate used spray suspension.
Note: When combining with captan, use 1 to 1⅓ lb Botran 75W to 1–2 lbs captan 50W or equivalent per 100 gallons of water as recommended.			
Tomato	Stem canker *Botrytis* (greenhouse)	1	Spray stem of plant from ground level up to a height of 18 to 24 inches. Repeat weekly or as necessary. Seedlings or newly set transplants may be injured by drenching.
Sweet potato Post-harvest	Post-harvest decay *Rhizopus* rot (Spray)	1	Spray, using 8001 fan-shaped nozzle at 50 to 70 psi immediately after washing while the potatoes are conveyed on chain rollers or belts. Do not rinse after treatment. In recirculating systems, as volume decreases, make up to original volume with suspension equivalent to 1 lb Botran 75W per 100 gal. of water. Where black rot is a problem, do not recirculate used spray suspension.
	(Dip)	1	Dip roots for 10 to 15 seconds in well agitated suspension of Botran 75W. Do not rinse after treatment. Add ½ lb Botran 75W to 100 gallons of treating suspension after 500 bushels are treated. After each 1000 bushels treated, drain and flush tank. Refill with fresh dip suspension.
Plantbed	Sclerotial blight *Sclerotium rolfsii*		*Seed Dip. Dip seed sweet potatoes 10 to 15 seconds in well agitated suspension containing 1 lb Botran 75W in 7.5 gal of water. Drain sweet potatoes and bed promptly. Prepare fresh Botran 75W suspension daily. **OR**
Seed Dip	Scurf *Monilochaetes infuscans*	*	**Bed Spray. Use 3 to 3¾ lb of Botran 75W in 14 gal of water per 1,000 square feet of plant-bed. Spray or sprinkle Botran 75W suspension over bedded sweet potatoes before covering them with soil.
Bed Spray		**	**Caution:** Do not plant tomatoes as follow-up crop.
Celery	Pink rot *Sclerotinia sclerotiorum*	2	Begin application when disease is anticipated or approximately 10 weeks before harvest. Repeat at 7-day intervals in summer or 14-day intervals in fall and winter. Do not apply within 7 days of harvest. **OR**
		5⅓	If early disease is not likely, make a single application of 5⅓ lbs of Botran 75W per acre 4 to 8 weeks before harvest. Use sprayer with drop nozzle boom to direct spray (100 gal per acre) to base of plants and adjacent soil.
Cotton	Boll rot *Rhizopus*	2–2⅓	Use sufficient water to cover. Spray prior to or as first bolls begin to open; repeat application 14 days later. **Caution:** Treated cotton and gin trash should not be grazed by or fed to livestock.
Potato	*Botrytis* blight White mold *Sclerotinia sclerotiorum*	2	Apply 100 gallons per acre beginning at layby. Continue at 10- to 14-day intervals. Do not apply within 14 days of harvest. Do not feed treated potatoes to livestock. LIMIT USE TO NORTH CENTRAL STATES ONLY*
Lettuce (leaf only) (greenhouse) (field)	*Botrytis* gray mold rot *Botrytis cinerea*	2⅔	Apply 100 gallons of spray mixture per acre 7 days following transplanting. Repeat application when plants are half mature. Do not apply to wilted plants, seedlings, or later than 14 days before harvest.
	Drop or watery soft rot *Sclerotinia sclerotiorum*	5⅓	Apply immediately after thinning. Do not apply this product again after this single application. **OR**
	Botrytis gray mold rot *Botrytis cinerea*	2⅔	Apply Botran 75W spray at thinning and repeat application 7 days later. Do not apply within 14 days of harvest. Use sprayer with drop nozzle boom to direct spray (100 gal per acre) to lower portion of plants and surrounding soil.
Note: Some leaf bronzing may be noted soon after application, but the effect is of a temporary nature.			
Endive (escarole) (field)	*Botrytis* gray mold rot *Botrytis cinerea*	2⅔	Apply 100 gallons of spray mixture per acre 7 days following transplanting. Repeat application when plants are half mature. Do not apply to wilted plants, seedlings, or later than 14 days before harvest. **OR**
		2⅔	Apply Botran 75W spray at thinning and repeat application 7 days later. Do not apply within 14 days of harvest.

Crop	Disease	Dosage[1] Lb Per 100 Gal. of Water	Use
			Use sprayer with drop nozzle boom to direct spray (100 gal per acre) to lower portion of plants and surrounding soil.

Note: Some leaf bronzing may be noted soon after application, but the effect is of a temporary nature. Do not apply more than two applications of Botran 75W per season.

Crop	Disease	Dosage[1] Lb Per 100 Gal. of Water	Use
Onion and Garlic	White rot *Sclerotium*	32–40*** (broadcast, or 43,560 linear feet of furrow)	**Direct seeding**—mix (broadcast-incorporate) Botran 75W in top 1½ inches of soil 1–2 weeks before onions are seeded. Use restricted to Eastern Oregon, Eastern Washington and Western Idaho. **Planting sets or buds—In Furrow:** Use low pressure (30 to 40 psi) spray equipment. Use 50-mesh or coarser screens. Center the front nozzle on the furrow so that spray strikes the soil around the sets or buds and the rear nozzle sprays the covering soil as it fills the furrow. **Broadcast:** Use 100 gallons per acre for broadcast treatment. Do not exceed 40 lbs per acre broadcast.

Note: Plow to 8 inches and cross-disc Botran 75W treated area prior to seeding sensitive follow-up crops to prevent injury (lettuce). Do not plant spinach as follow-up crop in treated soil. ***Use 6–10 lb per acre in North Central States.

Crop	Disease	Dosage[1] Lb Per 100 Gal. of Water	Use
Snap Bean	White mold *Sclerotinia sclerotiorum (Sclerotiniose)*	4 (pole varieties) 3 (bush varieties)	Begin applications when disease is anticipated and continue at 7-day intervals during periods favorable to development of disease. Use 100 gallons (pole varieties) and 75 gallons (bush varieties) per acre. Do not apply within 2 days of any harvest. Do not feed treated foliage to livestock.

DIRECTIONS FOR USE ON ORNAMENTAL CROPS

Crop	Disease	Dosage[1] Lb Per 100 Gal. of Water	Use
Rose, Hydrangea	Blight *Botrytis*	1⅓	**Field**—Apply Botran 75W prior to shipment or storage. **Stored**—Spray storage area thoroughly; spray or dip plants; repeat application during long storage periods.
Chrysanthemum, Geranium, Rose	Blight *Botrytis*	½ to ¾	**Stock cuttings or greenhouse plants**—Begin when disease is anticipated or first appears. Spray foliage and flowers at 7- to 14-day intervals. During prolonged damp, cool periods, apply every 5 to 7 days.
Gladiolus	Dry rot *Stromatinia* (soil)	20	Spray or dust into furrow at planting time, so that treated soil surrounds corms. 20 lb will treat one acre in furrows 1 ft. wide spaced 3 ft apart (14,520 linear ft. of furrow).
	Blight *Botrytis*	¾	Apply spray mixture with enough pressure to force fungicide between spikes. Apply to bunched cut flower spikes after harvest and before shipment or storage.

[1] Use suggested amount of Botran 75W per 100 gallons water (unless otherwise specified) in full coverage sprays in conventional sprayers (not to exceed a maximum of 500 gallons per acre) or equivalent amount in concentrate or aerial equipment.

CARZOL® SP

MITICIDE/INSECTICIDE

FOR AGRICULTURAL USE ONLY

ACTIVE INGREDIENT:

Formetanate hydrochloride*	92%
INERT INGREDIENTS	8%
TOTAL	100%

*N,N-dimethyl-N[3-[[(methylamino)carbonyl]oxy]phenyl] methanimidamide monohydrochloride

EPA Reg. No. 45639-74 EPA Reg. No. 33689-WG-1

U.S. Pat. Nos. 3,336,186; RE 27,347; 3,542,853

Keep Out of Reach of Children

DANGER–POISON

PELIGRO — VENENO

STATEMENT OF PRACTICAL TREATMENT

IF SWALLOWED—drink 1 or 2 glasses of water and induce vomiting by touching back of throat with finger. Do not induce vomiting or give anything by mouth to an unconscious person. Get medical attention.

IF ON SKIN—wash immediately with soap and water.

IF IN EYES—flush with water for at least 15 minutes and get medical attention.

NOTICE TO PHYSICIAN—CARZOL SP is a carbamate which acts as a cholinesterase inhibitor. Specific antidote—Atropine.
Do not use 2-PAM.

Do Not Reuse This Box.

PRECAUCION AL USUARIO: Si usted lee ingles, no use este producto hasta que la etiqueta haya sido explicado ampliamente.

PRECAUTIONARY STATEMENTS

HAZARDS TO HUMANS

DANGER

This product is poisonous if swallowed. Do not get in eyes, on skin or on clothing. Do not breathe dust or spray mist. Wear protective clothing, goggles and rubber (natural or synthetic) gloves when handling this material. Wash before eating or smoking. Wash contaminated clothing before reuse. If exposure to dust or spray mist cannot be avoided, wear a self-contained breathing apparatus.

Do not apply this product in such a manner as to directly or through drift expose workers or other persons. The area being treated must be vacated by unprotected persons.

ENVIRONMENTAL HAZARDS

This product is toxic to birds and other wildlife. Do not apply directly to water or wetlands. Apply only as specified on this label. CARZOL SP is moderately toxic to honey bees. Avoid using when bees are active. Do not allow animals to graze in orchard areas treated with CARZOL SP.

USE PRECAUTIONS

To avoid excessive residues, observe the following limitations: CITRUS: Do not apply more than 5 lbs. CARZOL SP per acre in a 12-month period. Allow 30 days between applications.

NOTE: If brown soft scale is present in the grove, a suitable scalicide should be applied after the CARZOL SP application.

APPLES, PEARS: Do not apply more than 4 lbs. CARZOL SP per acre in one crop season. PEACHES: Do not apply more than 2½ lbs. CARZOL SP per acre in one crop season. NECTARINES: Do not apply more than 4 lbs. CARZOL SP per acre in one crop season. PLUMS, PRUNES: Do not apply more than 3 lbs. CARZOL SP per acre in one crop season.

ALFALFA (seed crop only): Use this product only on pure stands of alfalfa to be harvested for seed. Do not allow spray to drift onto adjacent crops. Do not spray while bees are foraging. Do not allow livestock to graze treated fields. Do not use treated crops for livestock feed. Crop residues are not to be fed to livestock or baled for such use. Fields treated with CARZOL SP must be cleared after harvest by burning or by disking into the soil. CARZOL SP does not control aphids. If aphids appear, add an aphicide to the spray.

DIRECTIONS FOR USE

It is a violation of Federal law to use this product in a manner inconsistent with its labeling.

REENTRY STATEMENT

Reentry into treated areas is prohibited until leaves are completely dry or until 24 hours after application.

Do not enter treated areas without protective clothing.

Because certain states may require more restrictive reentry intervals for various crops treated with this product, consult your State Department of Agriculture for further information.

Written or oral warnings must be given to workers who are expected to be in a treated area or in an area about to be treated with this product. Indicate specific oral warnings which inform workers of areas or fields that may not be entered without specific protective clothing, period of time field must be vacated and appropriate actions to take in case of accidental exposure. When oral warnings are given, warnings shall be given in a language customarily understood by workers. Oral warnings must be given if there is reason to believe that written warnings cannot be understood by workers. Written warnings must include the following information:

DANGER: Area treated with CARZOL SP on (date of application). Do not enter without appropriate protective clothing for 24 hours or until leaves are completely dry. In case of accidental exposure see Statement of Practical Treatment on the label.

CARZOL SP is completely soluble in water. Preparation of a premix or slurry before adding to the spray tank is not necessary. Add the desired amount of CARZOL SP directly to the spray tank ½ full of water with the agitator running. Then add the remainder of the water.

NOTE: CARZOL SP is not stable in alkaline water. Do not prepare the spray mixture until just before application. Do not prepare more spray mixture than will be used within 4 hours.

TO AVOID EXCESSIVE RESIDUES OBSERVE THE PRE-HARVEST INTERVALS SHOWN IN THE TABLE. FOR OTHER USE LIMITATIONS SEE USE PRECAUTIONS.

CROP	PEST	APPLICATION
Oranges, Lemons, Tangerines, Tangelos and Grapefruit (California & Arizona only)	Thrips	Use 8 oz. CARZOL SP per 100 gallons of water when thrips appear. For outside coverage, apply 100 to 300 gallons of water per acre depending on the size of the trees. For aerial application, apply ½–1½ lbs. CARZOL SP per acre in a minimum of 10 gallons of water. Repeat application as needed. Allow 7 days before harvest.
Oranges, Lemons, and Tangerines (California & Arizona only)	Red Mite	Use 8 oz. CARZOL SP per 100 gallons of water. For full cover dilute spray, apply 700 to 1000 gallons of water per acre depending on the size of the trees. Repeat application as needed. Allow 7 days before harvest.
Oranges, Lemons, Tangerines and Grapefruit (Florida)	Rust Mite, Apopka Weevil	Use 1–2 oz. CARZOL SP per 100 gallons of water. For full cover dilute spray, apply 500 to 1500 gallons of water per acre depending on the size of the trees. For concentrate spray, apply 1 lb. CARZOL SP per acre. Repeat application as needed. Allow 7 days before harvest.
Oranges, Lemons, Tangerines and Grapefruit (Texas)	Rust Mite	Use 2–4 oz. CARZOL SP per 100 gallons of water. For full cover dilute spray, apply 500 gallons of water per acre. Repeat application as needed. Allow 7 days before harvest.
Apples	Tentiform Leaf Miner	Use 4–8 oz. CARZOL SP per 100 gallons of water at pink and petal fall. For full cover dilute spray, apply 400 to 800 gallons of water per acre depending on the size of the trees. For concentrate spray, apply 1–4 lbs. CARZOL SP per acre in at least 50 gallons of water. Allow 7 days before harvest.
Apples	White Apple Leafhopper	Use 2–4 oz. CARZOL SP per 100 gallons when leafhoppers appear. For full cover dilute spray, apply 400 gallons of water per acre. For concentrate spray, apply ½–1 lb. CARZOL SP per acre in at least 50 gallons of water. Repeat application as needed. Allow 7 days before harvest.
Apples and Pears	European Red Mite, Two-Spotted Mite	Use 4–8 oz. CARZOL SP per 100 gallons of water. For full cover dilute spray, apply 400 to 800 gallons of water per acre depending on the size of the trees and severity of the mite infestation. Repeat application as needed. Allow 7 days before harvest.
Pears	Pear Rust Mite	Use 4 oz. CARZOL SP per 100 gallons of water. Apply post bloom to harvest at first indication of rust mite infestation. For full cover dilute spray, apply 400 to 800 gallons of water per acre depending on the size of the trees. Repeat application as needed. Allow 7 days before harvest.

TO AVOID EXCESSIVE RESIDUES OBSERVE THE PRE-HARVEST INTERVALS SHOWN IN THE TABLE. FOR OTHER USE LIMITATIONS SEE USE PRECAUTIONS.

CROP	PEST	APPLICATION
Peaches and Nectarines	European Red Mite, McDaniel Mite, Two-Spotted Mite	Use 4 oz. CARZOL SP per 100 gallons of water when mites appear. For full cover dilute spray, apply 400 to 500 gallons of water per acre depending on the size of the trees. For concentrate spray, apply 1–1¼ lbs. CARZOL SP per acre in at least 50 gallons of water. Repeat application as needed. Allow 21 days before harvest for peaches, 14 days for nectarines.
Peaches	Lygus Bugs, Stink Bugs	Use 4 oz. CARZOL SP per 100 gallons of water. For full cover dilute spray, apply 400 to 500 gallons of water per acre. For concentrate spray 1–1¼ lbs. CARZOL SP per acre. Repeat application as needed. Allow 21 days before harvest.
Nectarines	Thrips	Use 4–6 oz. CARZOL SP per 100 gallons of water when thrips infest blossoms through petal fall. For full cover dilute spray, apply 400 to 500 gallons of water per acre depending on the size of the trees. For concentrate or aerial spray, apply 1–2 lbs. CARZOL SP per acre. Apply at night or while bees are not working in blooms. Repeat application as needed. Allow 14 days before harvest.
Plums and Prunes	European Red Mite, McDaniel Mite, Two-Spotted Mite	Use 4 oz. CARZOL SP per 100 gallons of water when mites appear. For full cover dilute spray, apply 400 to 600 gallons of water per acre depending on the size of the trees. For concentrate spray, apply 1–1½ lbs. CARZOL SP per acre in at least 50 gallons of water. Repeat application as needed. Allow 7 days before harvest.
Alfalfa (seed crop only in California, Oregon, Washington, Idaho and Nevada. Use pre-bloom only on alfalfa in all areas except California.)	Lygus Bugs	Use ½–1 lb. CARZOL SP per acre. Apply when lygus bugs reach damaging numbers. Repeat application as needed. Consult your State Agricultural Experiment Station or State Extension Service Specialist for proper timing.
	Consperse Stink Bugs (California only)	Use 1 lb. CARZOL SP per acre for knockdown of a heavy nymphal population. A lygus control program will control stink bugs. When applied by air use a minimum of 5 gallons per acre. When applied by ground use a minimum of 20 gallons per acre.
	Two-Spotted Mite	Use ½–1 lb. CARZOL SP per acre. Apply when mites appear. When applied by air use a minimum of 10 gallons of spray per acre. When applied by ground use a minimum of 20 gallons per acre. Repeat application as needed.

STORAGE AND DISPOSAL

Do not contaminate water, food, or feed by storage or disposal. Open dumping is prohibited.

Pesticide wastes are acutely hazardous. Improper disposal of excess pesticide, spray mixture, or rinsate is a violation of Federal law. If these wastes cannot be disposed of by use according to label instructions, contact your State Pesticide or Environmental Control Agency, or the Hazardous Waste representative at the nearest EPA Regional Office for guidance.

Completely empty box into application equipment. Then dispose of empty box in a sanitary landfill or by incineration, or, if allowed by State and local authorities, by burning. If burned, stay out of smoke.

IMPORTANT: READ BEFORE USE

By using this product, user accepts the following conditions, warranty, disclaimer of warranties and limitations of liability.

CONDITIONS:

The directions for use of this product are believed to be adequate and should be followed carefully. However, because of extreme weather and soil conditions, manner of use and other factors beyond NOR-AM Chemical Company's control, it is impossible for NOR-AM to eliminate all risks associated with the use of this product. As a result, crop injury or ineffectiveness is always possible.

DISCLAIMER OF WARRANTIES:

THERE ARE NO WARRANTIES, EXPRESS OR IMPLIED, OF MERCHANTABILITY OR OF FITNESS FOR A PARTICULAR PURPOSE OR OTHERWISE, WHICH EXTEND BEYOND THE STATEMENTS MADE ON THIS LABEL. No agent of NOR-AM Chemical Company is authorized to make any warranties beyond those contained herein or to modify the warranties contained herein. UNDER NO CIRCUMSTANCES SHALL NOR-AM CHEMICAL COMPANY BE LIABLE FOR ANY INCIDENTAL OR CONSEQUENTIAL DAMAGES RESULTING FROM FAILURE TO USE THIS PRODUCT IN ACCORDANCE WITH LABEL DIRECTIONS.

LIMITATIONS OF LIABILITY:

In no event shall NOR-AM Chemical Company's liability for damages arising out of the use of this product exceed the purchase price of the product used.

COLLEGO®

SELECTIVE POSTEMERGENT HERBICIDE BIOLOGICAL WEED CONTROL AGENT

For Control of Northern Jointvetch (Curly Indigo) in Rice and Soybean

COLLEGO is a two component product.
COMPONENT A:
A fungal spore rehydrating agent.
COMPONENT B:
ACTIVE INGREDIENT

Colletotrichum gloeosporioides f. sp. aeschynomene ATCC20358	15%*
INERT INGREDIENTS	85%
TOTAL	100%

* Contains at least 75.7 × 10^{10} viable fungal spores.

EPA Reg. No: 45639–134.

CAUTION—PRECAUCION

Keep Out of Reach of Children

PRECAUCION AL USUARIO: Si usted no lee ingles, no use este producto hasta que el etiqueta haya sido explicado ampliamente.

STATEMENT OF PRACTICAL TREATMENT

IF IN EYES: Flush with plenty of water. Get medical attention if irritation persists. Wash thoroughly with soap and water after handling.

See side panels for additional precautionary statements

PRECAUTIONARY STATEMENTS

HAZARD TO HUMANS AND DOMESTIC ANIMALS

CAUTION

Causes slight eye irritation. Avoid contact with eyes or clothing.

ENVIRONMENTAL HAZARDS

Do not apply directly to water except as indicated in the directions for use. Do not contaminate water by cleaning of equipment or disposal of wastes.

SPECIFIC PRECAUTIONS

After 7 days following application, the grower should frequently examine the northern jointvetch (curly indigo) plants to determine if disease lesions are developing.

As with other pesticides, effectiveness of Collego may be reduced by mistakes in application, fertilization, cultivation and management practices. Results will be affected by extremes in weather, soil moisture, and temperature.

If the disease lesions caused by Collego do not reach one-half (½) inch in diameter and do not encircle the stems of the northern jointvetch (curly indigo) plants within 14 days after treatment, a second application of Collego should be made. See Application Rate and Recommendations Sections for Collego amounts and timing.

STORAGE AND DISPOSAL

Do not contaminate water, feed, or food by storage or disposal.

STORAGE

Store in original container and keep closed. Store Collego at temperatures of 40°F to 80°F. Collego contains viable fungal spores. Germination of these spores will be reduced by temperatures below 32°F or when Collego is held for 12 hours or more at temperatures above 105°F. Store in original container and keep closed.

PESTICIDE DISPOSAL

Wastes resulting from the use of this product may be disposed of on site or at an approved waste disposl facility.

CONTAINER DISPOSAL

Triple rinse (or equivalent). Then offer for recycling or reconditioning, or puncture and dispose of in a sanitary landfill, or by incineration, or, if allowed by State and local authorities, by burning. If burned, stay out of smoke.

COMPONENT B BAG

Completely empty bag into application equipment. Then dispose of empty bag in a sanitary landfill or by incineration, or, if allowed by State and local authorities, by burning. If burned, stayout of smoke.

DIRECTIONS FOR USE

It is a violation of Federal law to use this product in a manner inconsistent with its labeling.

GENERAL INFORMATION

Collego is a selective postemergent mycoherbicide which is a specific biological weed control agent. Postemergent applications of Collego will selectively control NORTHERN JOINTVETCH (curly indigo) Aeschynomene virginica (L.) B.S.P. in rice (Oryza sativa L.) and soybean [Glycine max (L.) Merr.] To prevent seed production, Collego should be applied when northern jointvetch plants are 12 OR MORE INCHES TALL AND HAVE NOT REACHED THE BLOOM STAGE. Collego will cause disease lesions that will completely encircle the stems of the northern jointvetch plants. The fungus primarily infects the stems of the weed but it also infects the petioles and leaflets.

Disease plants become limp; they may collapse. Plants not killed by Collego are stunted, unthrifty, unable to compete with the rice or soybean and will not be able to seed. Death of northern jointvetch plants may not occur for five (5) weeks after application.

Collego is a two component product. Collego Component A consists of three 1-quart bottles containing a water soluble spare rehydrating agent that allows the spores to take up water prior to germination. Collego Component B consists of three bags that contain a water suspendible dried fungal spore formulation of Colletotrichum gloeosporioides f.sp. aeschynomene ATCC 20358. Both components are packaged in a 5 gallon plastic mixing container with a lid and stirring paddle.

APPLICATION RATE

Collego Component A (1 quart) and Component B (1 bag) will treat 10 acres.

Acreage to be Treated	Amount Required Component A	Component B
10	1 quart	1 bag
20	2 quarts	2 bags
30	3 quarts	3 bags

RECOMMENDATIONS FOR CONTROL OF NORTHERN JOINTVETCH (CURLY INDIGO)

Crop*	When to Apply Collego	Remarks
RICE	At any vegetative stage of rice growth if the northern jointvetch is at least 12 inches tall. Rice fields should be flooded before application	To prevent northern jointvetch from setting seed, Collego must be applied prior to bloom. Application of Collego to northern jointvetch in bloom or postbloom will not completely prevent seed development. Northern jointvetch usually begins to bloom at about the same time as rice heads are emerging from the boot or pods are forming on the lower nodes of soybean plants.
SOYBEAN	After the soybean plants begin to flower if the northern jointvetch is at least 12 inches tall. Soybean fields should be irrigated just prior to application.	

* Collego is exempt from the requirement of a tolerance for residues in or on rice and soybean when used according to these directions for use.

TIMING

Apply Collego when free moisture is on northern jointvetch leaves or relative humidities are above 80% and air temperatures of approximately 80%F are expected for at least 12 hours immediately following application. These conditions, necessary for infection to develop, usually prevail in rice and soybean fields during the evening. Soybean fields should be irrigated just prior to application; rice fields should be flooded.

CAUTION: DO NOT APPLY COLLEGO DURING PERIODS WHEN RICE AND SOYBEANS ARE UNDER STRESS FOR MOISTURE OR WHEN DRYING CONDITIONS ARE LIKELY TO OCCUR. DO NOT APPLY COLLEGO TO NORTHERN JOINTVETCH PLANTS PREVIOUSLY TREATED WITH PHENOXY HERBICIDES. DO NOT APPLY FUNGICIDES FOR AT LEAST THREE WEEKS FOLLOWING COLLEGO APPLICATION.

APPLICATION EQUIPMENT: Chemical pesticide residues that remain in the spray tank and boom may kill the live Collego spores. Therefore, prior to use of Collego, THOROUGHLY CLEAN THE SPRAY TANK.

Be sure the sprayer has been calibrated to deliver proper spray gallonage with a uniform spray pattern as thorough coverage of the northern jointvetch leaves and stems with Collego is essential. Check frequently during application for output of desired gallonage. Use proper nozzle discs and nozzle arrangements on the spray boom. Use 50-mesh or coarser screens in strainers, nozzles and suction units. Clean nozzles and screens frequently. Wind may cause uneven coverage. Do not apply when wind velocity is greater than 10 miles per hour.

FILLING THE SPRAY TANK: Fill the spray tank one-half full with water; add the Collego spore suspension (see Spore Preparation below) to the tank and rinse the mixing container; finish filling the tank to the desired spray volume. Agitate the spores and water to maintain the spore suspension. For best results, maintain continuous agitation during spraying.

SPORE PREPARATION: Collego Component B, dried fungal spores, must be rehydrated just prior to use. The sequence of the steps that must be used in rehydrating the spores for 10 acres:

STEP 1. Pour the contents from the 1 quart container of Collego Component A into the 5 gallon mixing container.

STEP 2. Completely fill the Collego Component A container with water and empty into mixing container.

STEP 3. Repeat Step 2. (Final concentration is 1 part rehydrating solution and 2 parts water.)

STEP 4. Vigorously stir until Collego Component A is completely dissolved.

STEP 5. Add one bag Component B and stir with mixing paddle until the spores are completely wet and suspended.

STEP 6. The suspension of spores may then be added to the spray tank. Rinse the mixing container to remove all of the spore suspension.

STEP 7. Dilute the spore suspension in the spray tank to the desired volume.

SPRAY APPLICATION

For best results, Collego should be applied by aerial application with fixed-wing or helicopter aircraft. Use a spray volume of at least 10 gallons of water per acre.

Ground applications are not prohibited; however they may be impractical in most rice and soybean situations where northern jointvetch is found with the exception of special areas such as levees where ground applications are practical. See Application Equipment section for instructions for preparing the ground spray rig prior to use of Collego. Use a spray volume of at least 15 gallons of water per acre.

CAUTION: DO NOT ALLOW SPRAY SUSPENSION TO REMAIN IN THE SPRAY TANK FOR MORE THAN 12 HOURS OR ALLOW THE SUSPENSION TO HEAT UP.

COMPATIBILITY: Collego contains live fungal spores; care must be exercised in the handling of the spores. Collego is not compatible with liquid nitrogen fertilizers, insecticides, fungicides, and herbicides such as MCP, 2,4-D, 2, 4-DB, 2,45-T and molinate (Ordram®). Germination of the spores will be reduced if combined with these materials.

Collego is compatible with aciflourfen (Blazer®).

CROP ROTATION Food, feed and forage crops may be sown in Collego-treated fields immediately after harvest of rice or soybeans.

Blazer®—Trademark of Rohm and Haas Co.

Ordram®—Trademark of Stauffer Chemical Co.

IMPORTANT: READ BEFORE USE

By using this product, user accepts the following conditions, warranty, disclaimer of warranties and limitations of liability.

CONDITIONS: The directions for use of this product are believed to be adequate and should be followed carefully. However, because of extreme weather and soil conditions, manner of use and other factors beyond NOR-AM Chemical Company's control, it is impossible for NOR-AM to eliminate all risks associated with the use of this product. As a result, crop injury or ineffectiveness is always possible.

DISCLAIMER OF WARRANTIES: THERE ARE NO WARRANTIES, EXPRESS OR IMPLIED, OF MERCHANTABILITY OR OF FITNESS FOR A PARTICULAR PURPOSE OR OTHERWISE, WHICH EXTEND BEYOND THE STATEMENTS MADE ON THIS LABEL. No agent of NOR-AM Chemical Company is authorized to make any warranties beyond those contained herein or to modify the warranties contained herein. UNDER NO CIRCUMSTANCES SHALL NOR-AM CHEMICAL COMPANY BE LIABLE FOR ANY INCIDENTAL OR CONSEQUENTIAL DAMAGES RESULTING FROM FAILURE TO USE THIS PRODUCT IN ACCORDANCE WITH LABEL DIRECTIONS.

LIMITATIONS OF LIABILITY: In no event shall NOR-AM Chemical Company's liability for damages arising out of the use of this product exceed the purchase price of the product used.

DROPP® 50 WP

COTTON DEFOLIANT

FOR AGRICULTURAL USE ONLY

PACKED IN WATER-SOLUBLE BAGS

ACTIVE INGREDIENT:

Thidiazuron*	50%
INERT INGREDIENTS	50%
TOTAL	100%

*N–phenyl–N'–1,2,3–thiadiazol–5–ylurea

EPA Reg. No. 45639-89 EPA Est. No. 33689-WG-1

CAUTION—CAUCION

Keep Out of Reach of Children

STATEMENT OF PRACTICAL TREATMENT

IF SWALLOWED: CALL A PHYSICIAN OR POISON CONTROL CENTER. DRINK 1 OR 2 GLASSES OF WATER AND INDUCE VOMITING BY TOUCHING BACK OF THROAT WITH FINGER. DO NOT INDUCE VOMITING OR GIVE ANYTHING BY MOUTH TO AN UNCONSCIOUS PERSON.

IF INHALED: REMOVE PERSON TO FRESH AIR.

IF IN EYES: FLUSH WITH PLENTY OF WATER FOR FIFTEEN MINUTES.

IF ON SKIN: WASH WITH PLENTY OF SOAP AND WATER.

THIS BOX CONTAINS DROPP 50 WP IN WATER SOLUBLE BAGS. INNER BAGS DISSOLVE IN WATER AND CONTENTS DISPERSE. AFTER OPENING OUTER BOX, OPEN FOIL LINER, IMMEDIATELY DUMP ENTIRE UNOPENED INNER BAGS INTO THE MIX TANK. DO NOT HANDLE SOLUBLE BAGS OR EXPOSE THEM TO MOISTURE, SINCE THIS MAY CAUSE BREAKAGE.

PRECAUCION AL USUARIO: Si usted no lee ingles, no use este producto hasta que el etiqueta haya sido explicado ampliamente.

PRECAUTIONARY STATEMENTS

HAZARDS TO HUMANS AND DOMESTIC ANIMALS

Harmful if swallowed, inhaled, or absorbed through the skin. Avoid breathing spray mist. Avoid contact with skin, eyes, and clothing. Do not contaminate food or feedstuffs.

Do not apply this product in such a manner as to directly or through drift expose workers or other persons. The area being treated must be vacated by unprotected persons.

ENVIRONMENTAL HAZARDS

Do not apply directly to water or wetlands. Do not contaminate water by cleaning of equipment or disposal of wastes. Do not apply when weather conditions favor drift from the target area.

GENERAL INFORMATION

DROPP® 50WP cotton defoliant is used for removal of leaves prior to harvesting. DROPP disperses easily in water and can be used in any ground and aircraft equipment designed for application of wettable powders. It may take several days before the effect of DROPP becomes noticeable. Adverse conditions, such as low temperatures, may require higher dosages and/or longer times for more complete defoliation. DROPP inhibits regrowth after defoliation and thus reduces staining of lint during harvesting and ginning.

USE PRECAUTIONS

Do not allow spray drift to contact trees or crops other than cotton as this product may injure or defoliate other crops. Rainfall within 24 hours after application will reduce the effectiveness of DROPP. Use only freshly prepared sprays. Do not store spray mixture overnight. Do not feed foliage from treated cotton plants or gin trash to livestock. Do not plant small grain, sorghum, corn or root crops earlier than two weeks after application of DROPP. Do not plant legumes, alfalfa or leafy vegetables earlier than two months after application of DROPP. Do not use immature crops for food or feed.

DIRECTIONS FOR USE

It is a violation of Federal law to use this product in a manner inconsistent with its labeling.

REENTRY STATEMENT

Reentry into treated areas is prohibited until leaves are completely dry or until 24 hours after application.

Do not enter treated areas without protective clothing.

Because certain states may require more restrictive reentry intervals for various crops treated with this product, consult your State Department of Agriculture for further information.

Written or oral warnings must be given to workers who are expected to be in a treated area or in area about to be treated with this product. (Indicate specific oral warnings which inform workers of areas or fields that may not be entered without specific protective clothing, period of time field must be vacated and appropriate actions to take in case of accidental exposure.) When oral warnings are given, warnings shall be given in a language customarily understood by workers. Oral warnings must be given if there is reason to believe that written warnings cannot be understood by workers. Written warnings must include the following information:

CAUTION. Area treated with DROPP® 50WP on (date of application). Do not enter without appropriate protective clothing for 24 hours or until leaves are completely dry. In case of accidental exposure, see Statement of Practical Treatment on the label.

MIXING INSTRUCTIONS

Fill the spray tank with one-half of the total amount of water to be used. After opening each box to be used, open its foil liner. Then, with the agitator running, immediately dump the entire contents of the required number of boxes of soluble bags into the mix tank and fill the tank with the additional quantity of water required. Use entire contents of each box.

Prior to reuse of spray equipment following DROPP®50WP applications, washing of all equipment with an appropriate detergent is recommended.

Insure sufficient flushing and rinsing of booms, mix tanks, sump, nozzles, pumps, transfer lines, and related equipment.

APPLICATION

Use the specified dosage of DROPP® 50WP cotton defoliant in sufficient amount of water to give complete coverage of the foliage with uniform wetting of leaf surfaces. Agitate the spray mix during application.

DOSAGE

One box (5×1 lb. bags) of DROPP treats 12.5–25 acres (0.2–0.4 lb. DROPP per acre). Apply in 10–25 gallons of spray per acre by ground equipment and 2–10 gallons per acre by air. Use the higher rates during periods of low temperatures.

Two applications not exceeding a total of 1 box per 8.3 acres (0.6 lb./acre may be necessary to defoliate rank cotton.

TIME OF APPLICATION

Apply DROPP cotton defoliant only to mature cotton plants when 70% or more of the bolls are open. Apply at least 5 days prior to anticipated picking.

NOTE: Do not use when the average 24-hour temperature is expected to be below 70°F for 2–3 days after application because defoliation may be substantially reduced.

BOTTOM DEFOLIATION FOR USE IN ARIZONA ONLY

Dropp 50WP may be used as an aid to improve air circulation and help reduce boll rot caused by moisture.

Apply Dropp 50WP at .1 lbs. formulated product per acre in 10-25 gallons of water per acre.

Application to **bottom** ⅓ of plant **only** recommended.

Timing of application should be based upon the latest date when ground application equipment can enter the field without crop damage.

PRE-CONDITIONING FOR USE IN ARIZONA ONLY

Dropp 50WP may be used as a pre-conditioner to enhance the activity of a defoliant application.

Apply Dropp 50WP at .1 lbs. formulated product per acre. Apply in 10-25 gallons of water per acre by ground application or 2-10 gallons of water per acre by aerial application.

Timing of application is recommended to be 7-14 days prior to the use of another defoliant. Refer to the second product label prior to use for complete recommendations.

TANK MIX OF DROPP® PLUS DEF® 6[1] FOR USE IN LA., MS., ARK., ALA., GA., NC., SC., TN. ONLY

The tank mix of Dropp 50WP and DEF 6 is recommended to improve defoliation and inhibit regrowth under less than ideal conditions. Less than ideal conditions are those where the average 24 hour temperature is expected to be between 60-70°F the day of application and for 2-3 days after application.

Best activity will be obtained where the tank mix is applied to mature cotton plants when 60% or more of the bolls are open.

Apply Dropp 50WP at rates of .1-.2 lbs. formulated product per acre + DEF at .5-1 pint per acre.

Use lower rate of DEF under ideal temperature conditions; higher rate during less than ideal or adverse conditions.

Higher rate of Dropp 50WP should be used where excessive regrowth is anticipated.

Do not apply with adjuvants or use higher than recommended dosages of either product in tank-mix combination, since dessication may occur.

Apply as a dilute spray in 5 to 10 gallons of water per acre by air or 10 to 25 gallons of water per acre with ground equipment. Do not apply in oil.

When applying DEF 6 defoliant and Dropp 50WP defoliant as a tank-mix combination, the following order of mixing is recommended: (1) water, (2) Dropp 50WP (according to directions on Dropp label), and (3) DEF 6 after Dropp has completely dispersed.

A second application of labeled rates of the tank-mix combination, or of Dropp alone, may be made where necessary.

DEF 6 is a registered trademark of Mobay Chemical Corp., Agricultural Chemical Division.

STORAGE AND DISPOSAL

STORAGE

Do not store box under wet conditions. Handle box carefully when stored at temperatures of less than 50°F to avoid breakage of soluble bags.

PESTICIDE DISPOSAL

Do not contaminate water, food or feed by storage or disposal. Wastes resulting from the use of this product may be disposed of on site or at an approved waste disposal facility.

CONTAINER DISPOSAL

Completely empty box into application equipment. Then dispose of empty box in a sanitary landfill or by incineration, or, if allowed by State and local authorities, by burning. If burned, stay out of smoke.

IMPORTANT: READ BEFORE USE

By using this product, user accepts the following conditions, warranty, disclaimer of warranties and limitations of liability.

CONDITIONS:

The directions for use of this product are believed to be adequate and should be followed carefully. However, because of extreme weather and soil conditions, manner of use and other factors beyond NOR-AM Chemical Company's control, it is impossible for NOR-AM to eliminate all risks associated with the use of this product. As a result, crop injury or ineffectiveness is always possible.

DISCLAIMER OF WARRANTIES:
THERE ARE NO WARRANTIES, EXPRESS OR IMPLIED, OF MERCHANTABILITY OR OF FITNESS FOR A PARTICULAR PURPOSE OR OTHERWISE, WHICH EXTEND BEYOND THE STATEMENTS MADE ON THIS LABEL. No agent of NOR-AM Chemical Company is authorized to make any warranties beyond those contained herein or to modify the warranties contained herein. UNDER NO CIRCUMSTANCES SHALL NOR-AM CHEMICAL COMPANY BE LIABLE FOR ANY INCIDENTAL OR CONSEQUENTIAL DAMAGES RESULTING FROM FAILURE TO USE THIS PRODUCT IN ACCORDANCE WITH LABEL DIRECTIONS.

LIMITATIONS OF LIABILITY:
In no event shall NOR-AM Chemical Company's liability for damages arising out of the use of this product exceed the purchase price of the product used.

ENIDE® 90W

SELECTIVE PREEMERGENCE HERBICIDE

FOR: Tomato, Pepper, Sweet Potato, Potato, Okra, Strawberry, Blackberry, Raspberry, Tobacco, Soybean, Peanut, Cotton, Fruit Trees, Bermudagrass, Dichondra, Ice Plant, Ornamentals, Conifers, Evergreens, and Hardwood Seedbeds.

ACTIVE INGREDIENT
Diphenamid
(N,N-Dimethyl-2,2-diphenylacetamide) .... 90% w/w
INERT INGREDIENTS .... 10% w/w

TOTAL .... 100% w/w

EPA Reg. No. 45639-131

Keep Out of Reach of Children
CAUTION—PRECAUCION

PRECAUCION AL USUARIO: Si usted no lee ingles, no use este producto hasta que el etiqueta haya sido explicado ampliamente.

STATEMENT OF PRACTICAL TREATMENT

IF ON SKIN: REMOVE CONTAMINATED CLOTHING AND WASH WITH SOAP AND WATER.

IF IN EYES: FLUSH WITH WATER. IF EYE IRRITATION PERSISTS, CALL A PHYSICIAN.

IF SWALLOWED: CALL A PHYSICIAN OR POISON CONTROL CENTER. DRINK 1 OR 2 GLASSES OF WATER AND INDUCE VOMITING BY TOUCHING BACK OF THROAT WITH FINGER. DO NOT INDUCE VOMITING OF GIVE ANYTHING BY MOUTH TO AN UNCONSCIOUS PERSON.

PRECAUTIONARY STATEMENTS

HAZARDS TO HUMANS AND DOMESTIC ANIMALS

CAUTION

Harmful if swallowed. Avoid inhalation of dust or spray mist. Avoid prolonged or repeated contact with skin and eyes. Do not apply this product in such a manner as to directly or through drift expose workers or other persons. The area being treated must be vacated by unprotected persons.

ENVIRONMENTAL HAZARDS

Toxic to fish. Do not apply directly to water or wetlands. Do not apply where runoff is likely to occur. Do not apply when weather conditions favor drift from areas treated. Do not contaminate water by cleaning of equipment or disposal of wastes.

GENERAL USE PRECAUTIONS

Apply this product only as specified on this label. Edible crops other than those registered for Enide 90W use should not be planted in Enide 90W treated soil for 6 months after date of last application. Small grain cover crops sown within 6 months after date of Enide 90W treatment may be injured. Use band treatments of Enide 90W to reduce the possibility of injury to cover crops or use tolerant cover crops such as vetch, birdsfoot trefoil, Korean lespedeza, alfalfa or rape. Do not use cover crops for food or feed within 6 months after application. Avoid contamination of feed or foodstuffs.

As with all herbicides, effectiveness of Enide 90W may be reduced by mistakes in application, fertilization, cultivation and other crop management practices. Results may be affected by extremes in soil and weather conditions. The grower should watch to note that the treatment is working. In case of poor control, weeds should be removed by mechanical or other chemical methods.

STORAGE AND DISPOSAL

Do not contaminate water, food or feed by storage or disposal.

STORAGE

Store in original container and keep closed. Store in a cool, dry place.

PESTICIDE DISPOSAL

Wastes resulting from the use of this product may be disposed of on site or at an approved waste disposal facility.

CONTAINER DISPOSAL

Completely empty bag into application equipment. Then dispose of empty bag in a sanitary landfill or by incineration, or, if allowed by State and local authorities, by buring. If burned, stay out of smoke.

DIRECTIONS FOR USE

It is a violation of Federal law to use this product in a manner inconsistent with its labeling.

Written or oral warnings must be given to workers who are expected to be in a treated area or in an area about to be treated with this product. When oral warnings are given, warnings shall be given in a language customarily understood by workers. Oral warnings must be given if there is reason to believe that written warnings cannot be understood by workers.

Do not enter treated areas without protective clothing until sprays have dried.

Because certain states may require more restrictive reentry intervals for various crops treated with this product, consult your State Department of Agriculture for further information.

GENERAL INFORMATION

Read all label directions, precautions and conditions of sale before using Enide® 90W. Enide 90W is a 90% wettable powder formulation of diphenamid. It is a preemergence weed control chemical that is selective in activity. It controls may annual grass and broadleaf weeds as they germinate.

To control the following weeds:

Grass Weeds

Annual bluegrass	Goosegrass (Crowfoot grass)
Annual sedge	Johnsongrass (from seed)
Barnyardgrass (Watergrass)	Ryegrass
Cheat	Sandbur (Sandspur)
Crabgrass	Stinkgrass
Fall panicum (Panic grass)	Wild Oat
Foxtail	Witchgrass (Ticklegrass)

Broadleaf Weeds

Carpetweed	Mouseear chickweed
Common chickweed	Pigweed (Careless weed)
Corn spurry	Purslane
Eveningprimrose	Red sorrel
Field pepperweed	Shepherdspurse
Florida pusley	Smartweed
Groundsel	Spiny amaranth (Stickerweed)
Knotgrass (German moss)	Thymeleaf sandwort
Knotweed	White clover (in tobacco seedbeds)
Lambsquarters	

Soil Preparation: Prepare the soil for seeding, planting, or transplanting according to good agricultural practices. All weed growth and crop debris should be worked into the soil before treatment. The soil should be moist and loose with all clods broken down.

Incorporation and Irrigation: To be effective, Enide 90W must be moved into the top few inches of soil and contact germinating weed seeds or very young seedlings before they emerge. This may be accomplished by rainfall, overhead irrigation, or mechanicai incorporation. If rainfall is not expected shortly after application, apply 1 to 2 inches of water by overhead irrigation, or mechanically incorporate to a depth of ½ to 2 inches. Deeper incorporation will dilute the Enide 90W in the soil and may result in poor weed control. Equipment used for incorporation must mix the Enide 90W thoroughly and evenly with the soil.

Cultivation: Weeds which emerge after treatment should be destroyed by shallow cultivation (½ to 2 inches). Deeper cultivation that brings untreated soil to the surface may result in poor weed control.

Spray Equipment: Before use, calibrate sprayer to deliver proper spray gallonage and uniform spray pattern; check frequently during application. Use standard low pressure (25 to 50 psi) boom sprayer with fan-type nozzles for broadcast application or fan-type even-spray nozzles for band application. Use 50 mesh or coarser screens in strainers, nozzles, and suction units. Clean nozzles and screens frequently, replace worn nozzles to insure uniform application. Maintain continuous agitation of spray mixture. Wind can cause loss of spray or uneven coverage. Do not apply in strong wind.

Broadcast (Overall) Application: See recommendations. Evenly apply the recommended rate of Enide 90W in at least 30 gallons of water per acre. Avoid spray overlap that will increase the applied amount of Enide 90W above the recommended rates.

The Number of 2.2 Pound Bags Needed Per Treated Acre For Enide 90W Broadcast (Overall) Applications

Broadcast Rate	2.2 lb/A	3.3 lb/A	4.4 lb/A	5.5 lb/A	6.6 lb/A	8.8 lb/A	11.0 lb/A
Number Bags Per Acre	1	1½	2	2½	3	4	5

The Number of Acres That Can Be Treated With an Enide 90W 52.8 Pound Bulk Package When Broadcast (Overall) Applications Are Made

Broadcast Rate	2.2 lb/A	3.3 lb/A	4.4 lb/A	5.5 lb/A	6.6 lb/A	8.8 lb/A	11.0 lb/A
Number Acres That Can be Treated	24	16	12	9.6	8	6	4.8

Band (Row) Application: The amount of Enide 90W required per crop acre is reduced with band application. Use this formula:

$$\frac{\text{Band width (inches)}}{\text{Row spacing (inches)}} \times \frac{\text{Broadcast rate in}}{\text{pounds per acre}} = \frac{\text{Pounds Enide 90W per}}{\text{acre for band treatment}}$$

Example: If 6.6 lb per acre are recommended for broadcast, and you want to apply a 12-inch band on rows spaced 36 inches apart:

$$\frac{12\text{ inches}}{36\text{ inches}} \times 6.6\text{ lb.} = 2.2\text{ lb per acre of crop}$$

Compatibility: Tank mixes of Enide 90W with other pesticides (except as identified in these directions for use), fertilizers, oils and spray additives may not be compatible. The uniformity of dispersion and the sprayability of these mixtures may be influenced by certain chemical combinations. Therefore, check compatibility of these tank mixtures before spraying. A simple fruit jar test may be used to determine if combinations may be mixed and are compatible.

Tank Mixes: Enide 90W can be tank mixed for the following uses: In *Tobacco* with isopropalin (Paarlan® 6E, Trademark of Eli Lilly & Co.), pebulate (Tillam® 6E, Trademark of Stauffer Chemical Co.), pendamethalin, 4 lbs/gallon (Prowl® Trademark of American Cyanamid), trichlorfon (Proxol® 80SP), metalaxyl (Ridomil,® Trademark of Ciba Geigy). In *Tomato* with napropamide (Devrinol® 50WP, Trademark of Stauffer Chemical Co.), pebulate (Tillam® 6E, Trademark of Stauffer Chemical Co.), trifluralin (Treflan® E.C., Trademark of Eli Lilly & Co.); In *Sweet potato* with chloramben, 2 lbs/gallon (Amiben® Trademark of Union Carbide Agricultural Co.) and oryzalin (Surflan® 75W or 4AS, Trademark of Eli Lilly & Co.). Observe all cautions and limitations on labeling of all products used in mixtures.

RECOMMENDATIONS: FOR CONTROL OF CERTAIN ANNUAL GRASS AND BROADLEAF WEEDS IN CROPS LISTED BELOW.

CROP	WHEN TO APPLY	LB ENIDE 90W PER ACRE (SEE APPLICATION INSTRUCTIONS) COARSE SOIL	MEDIUM/ FINE SOIL	REMARKS
Tomato	From 1 week before to 1 month after direct seeding or transplanting as a broadcast or band application.	4.4	6.6	Incorporate Enide 90W into soil when applied before seeding or transplanting. May be applied over top of plants. The following tank mix combinations may be applied over top of plants. The following tank mix combinations may be applied at the rate(s) specified.
				Tank Mix: Enide 90W, 4.4 lb and Treflan E.C., 8 fl oz/acre. Enide 90W 2.2 to 4.4 lb/acre and Tillam, 1½ to 2½ qts/acre. Enide 90W, 2.2 to 4.4 lb/acre and Devrinol 50WP, 1 to 2 lb/acre. Immediately after application, thoroughly incorporate the material to a depth of 2 to 3 inches. Set disk to cut from 4 to 6 inches. **DO NOT** use Enide 90W in tank mix combination with Treflan or Tillam or direct seeded tomatoes **except in California.**
	At planting or immediately after without incorporation.	4.4	6.6	This method of application is recommended for use in California if sprinkler irrigation is available. Enide 90W should be uniformly applied at the recommended rate to the prepared beds, rows, or entire fields at or immediately after seeding. Delay in seeding and application of Enide 90W following the seedbed preparation may permit germination of weed seeds and result in lack of effectiveness. Immediately after treatment, 1 to 2 inches of water should be applied by sprinkler irrigation. After this first irrigation, follow normal irrigation procedures. *****See Irrigation**
	On prepared beds with power driven tiller incorporation and furrow irrigation.	4.4	6.6	For use in California in fields to be planted in direct-seeded tomatoes. Enide 90W should be applied at the recommended rate in front of the tiller. Seeding may also be included in this operation. Enide 90W should not be applied earlier than 14 days before seeding. It is recommended that Enide 90W be mixed evenly in the soil to a depth of 1 to 2 inches. Immediately following incorporation and seeding, the

CROP	WHEN TO APPLY	LB ENIDE 90W PER ACRE (SEE APPLICATION INSTRUCTIONS) COARSE SOIL	MEDIUM/ FINE SOIL	REMARKS
Tomato (continued)				beds should be thoroughly wet by furrow irrigation. The top of the bed should be wet. *****See Irrigation**
	Broadcast preplant application with disk incorporation and furrow irrigation	4.4	6.6	For use in California in fields to be planted in direct-seeded tomatoes. This method of application requires special care because: (1) in shaping the beds it is extremely difficult to avoid bringing up untreated soil from which weeds may emerge; (2) however, deep incorporation of Enide 90W reduces its concentration in the soil and reduces effectiveness.
				Enide 90W should be broadcast applied to the field not earlier than 14 days before planting. Use the recommended rate of Enide 90W in at least 30 gallons of water per acre. Following application, the field should be double-disked with the second disking at right angles to the first to uniformly incorporate Enide 90W to a depth of 3 or 4 inches. The depth at which the disk is incorporating should be checked carefully before the whole field is worked. The disking should thoroughly mix Enide 90W in the soil and not just turn the soil over. In shaping the bed it is important not to use bed shapers which bring up soil from below the treated soil depth. Immediately following bedding and seeding, the beds should be thoroughly wet using furrow irrigation. The top of the bed should be wet. *****See Irrigation**
	For the control of winter weeds (most grasses and many broadleaf weeds) where fields are prepared in the fall for early spring planting of direct-seeded tomatoes.		6.6	After soil preparation and prior to start of the winter rainy season in California (October or November) apply Enide 90W. **Do not** mechanically incorporate. Allow winter rains to incorporate Enide 90W. Tomatoes may be planted through March 1 in fields treated. Do not use a fall application on fields scheduled to be planted after March 1. Do not cultivate bed tops prior to or just after planting. Weeds that emerge on bed tops prior to tomato germination should be treated with an approved contact herbicide. To insure late season weed control, treat at layby with a herbicide recommended for application at this time. Do not use a second application of diphenamid. *****See Irrigation**
	Enide 90W may be tank mixed with trifluralin, for preemergence control of crabgrass, barnyard grass (water-grass) and pigweed in direct-seeded tomatoes grown for processing in California.		4.4	For broadcast application, mix and apply preplant at the rate of 4.4 lb Enide 90W and 0.25 lb trifluralin per acre in 30 to 50 gallons of water. Immediately after application, thoroughly incorporate the material to a depth of 2–3 inches. This may be accomplished with PTO-driven rotary equipment or by double-disking. Rototiller-type equipment should be set to cut 2–3 inches deep. Disks should be set to cut 4–6 inches deep and the field cross-disked, operating at 4–6 mph. **CAUTION:** Do not use this combination on coarse soils, soils containing more than 5% organic matter, or soils with a salt problem. Use of this combination may result in injury to the tomato seedings. The possibility of damage is increased by seeding disease, cold temperature, excessive moisture, drought, and other adverse conditions.
				If a follow-up treatment with trifluralin is desired, do not use more than 0.5 lb/acre of trifluralin applied after thinning and blocking. Read the complete label on each product before using this tank mixture. *****See Irrigation**
	Enide 90W may be tank mixed with pebulate for use in direct-seeded tomatoes grown for processing in California.		6.6	For broadcast application, mix and apply preplant at the rate of 6.6 lb Enide 90W and 4 lb pebulate per acre in 30 to 50 gallons of water. Immediately after application thoroughly incorporate the material to a depth of 2–3 inches. This may be accomplished with PTO-driven rotary equipment or double-disking. Rototiller-type equipment should be set to cut 2–3 inches deep. Disks should be set to cut 4–6 inches deep and the field cross-disked, operating at 4–6 mph. **CAUTION:** Do not use this combination on coarse soils, or on soils containing more than 20% organic matter. Use of this combination may result in injury to tomato seedings. The possibility of damage is increased by seeding disease, cold temperature, excessive moisture, drought, and other adverse conditions. Read the complete label on each product before using this tank mixture. *****See Irrigation**
Pepper	At direct seeding, transplanting, or within 1 month after transplanting.	3.3	5.5	May be applied over top of transplants.
Sweet potato Transplants	At transplanting	4.4	6.6	Enide 90W may be applied over top of transplant Enide 90W at 2.2 to 4.4 lb/acre may be tank mixed with: Amiben (1 to 2 gal/acre), Surflan 75W (0.67 to 1.33 lb/acre) or Surflan 4AS (1 to 2 pt/acre). These tank mixtures should be applied prior to weed emergence.

CROP	WHEN TO APPLY	LB ENIDE 90W PER ACRE (SEE APPLICATION INSTRUCTIONS) COARSE SOIL	MEDIUM/ FINE SOIL	REMARKS
Plantbed	At bedding	4.4	6.6	Apply to soil after covering potatoes.
Potatoes	Between planting and emergence, or at layby.	4.4	6.6	Shallow cultivation (½ to 2 inches) after application will not destroy effectiveness of Enide 90W. However, any tillage operation such as dragging-off, hilling, etc., that brings untreated soil to the surface will result in poor weed control. At layby, apply as a directed spray to base of plants. Do not apply within 50 days before harvest.
Okra	At planting	3.3	5.5	
Tobacco Seedbed	At seeding	**	**	**Use Enide 90W on coarse soil at 1.7 ounces and on medium/fine soils at 2.5 ounces per 1,000 sq. ft. Applied at seeding, Enide 90W controls white clover. Enide 90W is compatible with Ridomil 2E (EPA Reg. No. 100-607). A tank mixture with Enide 90W and Ridomil 2E can be made at seeding. Observe all restrictions, precautions, and directions for use of Ridomil 2E labels and labeling.
Field Before Transplanting	From 0 to 7 days before transplanting	4.4	6.6	Apply Enide 90W evenly to the soil surface and thoroughly incorporate into the top 2 inches of soil. Plant crop in usual manner. Enide 90W may also be applied preemergence in the following combinations. Enide 90W at 2.2 to 4.4 lb/acre and Prowl at 1 to 2 pt/acre. Enide 90W at 2.2 to 4.4 lb/acre and Paarlan 6E at 1 to 2 pt/acre. Enide 90W at 2.2 to 4.4. lb/acre and Tillam at 1½ to 2⅔ qt/acre.
After Transplanting	From transplanting to 7 days after transplanting, **OR**	4.4	6.6	May be applied over top of transplants. When Enide 90W is applied more than 2 days after transplanting, tillage is required immediately before or at time of application to destroy germinating weed seeds. Enide 90W is compatible with Proxol 80SP when used in combination with a spreader (such as Ortho X-77 spreader). A tank mixture with Enide 90W and 20 ounces per acre of Proxol 80SP can be made within 7 days of transplanting.
Layby	Following last cultivation, usually 4 to 6 weeks after transplanting.	4.4	6.6	If Enide 90W was used before or after transplanting (see above), apply 2.2 to 4.4 pounds of Enide 90W per acre after the last cultivation. Do not use more than a total of 8.8 pounds of Enide 90W per acre in both applications. Enide 90W also may be applied in combination with Tillam at layby. Use a band treatment of Enide 90W alone or in combination. Enide 90W/Surflan tank mix may be applied at layby for ragweed suppression using band application to row middles immediately following last (layby) cultivation. Do not spray tobacco leaves as injury may occur. Combine Enide 90W 2.2 lbs to 4.4 lbs with either 0.67 lb to 1.33 lbs Surflan 75W or 1 pint to 2 pints Surflan 4AS. This labeling must be in the possession of the user at the time of pesticide application. Enide 90W is compatible with Ridomil 2E (EPA Reg. No. 100-607). A tank mixture with Enide 90W and Ridomil 2E can be made at layby. Do not use Ridomil 2E in a foliar application on field grown tobacco as injury may occur. Observe all restrictions, precautions, and directions for use of Ridomil 2E labels and labeling.
Cotton	At planting	3.3	4.4	
Peanut	At planting, **OR**	4.4	6.6	
	At cracking or within 7 days after cracking, **OR**	2.2	3.3	Combine Enide 90W with 1½ pounds of DNBP amine per acre.
	At layby—following last cultivation	2.2	2.2	Will not control established weeds. Do not use if more than 2.2 pounds per acre of Enide 90W were applied at planting or cracking.
Soybean	At planting	3.3	5.5	
	At planting	2.2	3.3	Combine Enide 90W with ½ pound of linuron or 3 pounds of CIPC per acre. **CAUTION:** Do not graze treated areas or feed forage from treated areas to livestock. Do not plant food crops other than soybean within 6 months after treatment.
Strawberry Newly planted	2–6 weeks after planting but not before new foliage has appeared.	4.4	6.6	A temporary delay in rooting of daughter plants may occur but will not affect yield. Do not use on the herbicide sensitive variety Shasta. Do not apply within 60 days before harvest.
Established	While strawberries are dormant or semidormant or after renovation. Additional applications may be made at intervals of 6 months or more.	4.4	6.6	
Blackberry, Erect and Trailing (Non-bearing)	After planting and before weed seeds germinate.	4.4	6.6	Work all weed growth into soil before application. Apply as a directed spray to the soil. **USE ONLY IN SEASON OF PLANTING.** Do not apply within 12 months of harvest.
Raspberry, Red and Black	After planting and before weed seeds germinate.	4.4	6.6	New Plantings: Apply planting before weed seeds germinate. Established: Apply prior to weed emergence; or apply immediately following a cultivation after all weed growth has been worked into the

CROP	WHEN TO APPLY	LB ENIDE 90W PER ACRE (SEE APPLICATION INSTRUCTIONS) COARSE SOIL	MEDIUM/ FINE SOIL	REMARKS
Raspberry (continued)				soil. Apply as a directed spray to the soil. Do Not Apply Within 60 Days of Harvest. Do Not Use more than 6.6 pounds of Enide 90W during the growing season.
Fruit Trees Apple, Peach	Any time of the year **except** when fruit is on tree or within 90 days before harvest.	4.4	6.6	Following clean cultivation, apply to soil around established trees. **CAUTION:** On peaches do not use on coarse soil. Do not allow livestock to graze in treated orchards.
Non-bearing, Cherry, Lime, Orange	Do not apply within 12 months before harvest.	4.4	6.6	
Evergreens, Shrubs and Trees	In fall or spring before weeds emerge.	6.6	8.8	Enide 90W may be used on new plantings and container stock after plants are established. Deciduous and evergreen trees and shrubs tolerant to Enide 90W include: almond, apple†, arborvitae, ash, azalea, bald cypress, barberry, beauty-bush, beech, birch, blackberry, black walnut†, black locust, blueberry†, camellia, ceanothus, cherry laurel, cherry†, chestnut, cotoneaster, cottonweed, crab apple, cream carpet, currant†, dogwood, elaeagnus, eucalyptus, euonymus, firethorn, forsythia, heath, heather, heliotrope, hemlock, holly, honeysuckle, hydrangea, juniper, laurel, lilac, lime, mahonia, maple, mock orange, oak, oleander (marginal leaf chlorosis may develop), orange†, peach, pear†, pecan†, pittosporum, plum†, privet, raspberry (black), raspberry (red), red bud, red cedar, rhododendron, rose, rose-of-sharon, spiraea, St. Johnswort, star of confederate jasmine, sugar maple, sweet gum, sycamore, taxus (yew), tulip tree, viburnum, weigela, white ash, white cedar and willow. **†Non-bearing only**
Annual and Perennial Flowers	At transplanting, in fall or spring before weeds emerge.	6.6	8.8	Do not apply Enide 90W until plants are well established. Ornamental plants tolerant to Enide 90W include: alyssum, aster, baby's breath, begonia, chrysanthemum, dahlia, delphinium, feverfew, foxglove (yellow), geranium, lobelia, loosestrife, marigold, peony, philodendron, phlox, salvia, shasta daisy, snapdragon, stock, sweet william, verbena, and zinnia.
Dichondra New Seeding	At seeding		11.0	Apply Enide 90W to the prepared seedbed and rake in ½ inch deep before seeding.
Established	Before weeds emerge.		11.0	Repeat application in 6 months. Existing bermuda grass will be greatly weekened.
Ground Cover New planting	After cuttings have rooted (usually 1 month after planting		8.8	Ground cover plants tolerant to Enide 90W include: ice plant, English ivy, ornamental strawberry, periwinkle and myrtle. Avoid overwatering. Repeat treatment in 6 months.
Established plantings	Before weeds emerge.			
Bermudagrass Lawns	In fall or early spring, or both fall and early spring, to mature established Bermuda turf.	4.4		Fall applications control annual bluegrass while spring applications are required for good control of crabgrass. Make early spring applications while bermudagrass is still dormant. **CAUTION:** Do not graze treated areas or feed forage from treated areas to livestock.
Seedbeds Conifers Fir, Hemlock Larch and Pine Cypress and Spruce	One day prior to seeding to within one month after seeding and mulching, if mulching is used. Apply after seedings are one month old.	4.4	4.4	Enide 90W may flocculate on contact with petroleum solvents such as mineral spirits. If sprayer has been used for mineral spirits, wash it thoroughly with detergent and water before applying Enide 90W. Work all weed growth into soil before application. After broadcast application of Enide 90W to moist seedbeds, apply ½ to ¾ inch of water by overhead sprinkler irrigation. If Enide 90W is used one day before or within two days after seeding, a second 4.4 lb of Enide 90W may be applied 6 weeks after seeding. *Use only postplanting applications on Austrian pine.* Conifers tolerant to Enide 90W include: bald cypress, douglas fir, fraser fir, grand fir, noble fir, red fir, rocky mountain douglas fir, white fir, eastern hemlock, western hemlock, western larch, Austrian pine, loblolly pine, lodgepole pine, longleaf pine, monterey pine, ponderosa pine, scotch pine, shortleaf pine, slash pine, sugar pine, white pine, englemann spruce, Norway spruce and white spruce.
Evergreens and Hardwood	Apply within two days after seeding and mulching, if mulching is used.	4.4	4.4	Destroy all weed growth before treatment. After broadcast application of Enide 90W to moist seedbeds, apply ½ to ¾ inch of water by overhead sprinkler irrigation. Evergreens and hardwoods tolerant to Enide 90W include: arborvitae, euonymus, multiflora rose and black walnut.

*** Irrigation Following Planting: Following initial watering, the beds should be kept at the optimum moisture level for seed germination and emergence. Allowing the seedbed to dry out reduces the level of herbicide effectiveness as well as the stand and vigor of tomato plants.

IMPORTANT: READ BEFORE USE

By using this product, user accepts the following conditions, warranty, disclaimer of warranties and limitations of liability.

CONDITIONS

The directions for use of this product are believed to be adequate and should be followed carefully. However, because of weather and soil conditions, manner of use and other factors beyond NOR-AM

Chemical Company's control, it is impossible to eliminate all risks associated with the use of this product. As a result, crop injury or ineffectiveness is always possible.

DISCLAIMER OF WARRANTIES

THERE ARE NO WARRANTIES, EXPRESS OR IMPLIED, OF MERCHANTABILITY OR OF FITNESS FOR A PARTICULAR PURPOSE OR OTHERWISE, WHICH EXTEND BEYOND THE STATEMENTS MADE ON THIS LABEL. No agent of NOR-AM Chemical Company is authorized to make any warranties beyond those contained herein or to modify the warranties contained herein. UNDER NO CIRCUMSTANCES SHALL NOR-AM CHEMICAL COMPANY BE LIABLE FOR ANY INCIDENTIAL OR CONSEQUENTIAL DAMAGES RESULTING FROM FAILURE TO USE THIS PRODUCT IN ACCORDANCE WITH LABEL DIRECTIONS.

LIMITATIONS OF LIABILITY

In no event shall NOR-AM Chemical Company's liability for damages arising out of the use of this product exceed the purchase price of the product used.

SUPPLEMENTAL LABEL

ENIDE® 90W

COVER CROPS IN ROTATION WITH TOBACCO

EPA SLN NUMBERS: NC-800006, SC-800002, AND VA-800011

For Use in: North Carolina, South Carolina and Virginia

GENERAL USE PRECAUTIONS

NORTH CAROLINA AND SOUTH CAROLINA—

TOBACCO: Small grains (such as wheat, barley, rye and oats) intended for use as cover crops, when grown in rotation with tobacco, can be sown within less than 6 months after application of Enide® 90W to reduce the possibility of injury to cover crops. **Do Not Use Cover Crops for Food or Feed Within 6 Months After Application of Enide 90W.**

VIRGINIA—

TOBACCO: Small grain (i.e., wheat, barley, rye and oats) cover crops may be sown after tobacco harvest provided Enide 90W was applied as a band treatment reducing the possibility of cover crop injury. **Do Not Use Cover Crops for Food or Feed purposes within 6 months after Enide 90W application.**

24(c) Registrant: NOR-AM Chemical Company, 3509 Silverside Road, P.O. Box 7495, Wilmington, DE 19803

TOBACCO

For Suppression of Common Ragweed *(Ambrosia artemisiifolia L.)*

EPA SLN NUMBERS: NC-800005, SC-800001, and VA-800009

For Use in: North Carolina, South Carolina and Virginia

TOBACCO: For Suppression of Common Ragweed *(Ambrosia artemisiifolia L.)* use:

Recommended Broadcast Rate

Soil Texture	Enide 90W per Acre
Light (Coarse) Heavy	4.4 lb.
(Medium/Fine)	6.6 lb.

Applications may be made before transplanting, after transplanting or at layby. **Do Not Use More Than 8.8 lbs. of Enide 90W if used before or after transplanting and again after the last cultivation layby.**

24(c) Registrant: NOR-AM Chemical Company, 3509 Silverside Road, P.O. Box 7495, Wilmington, DE 19803

TOMATO, POTATO, AND PEPPER

EPA SLN NUMBER: CA-780050

For Use in: California

Target Pest/Problem: Weeds listed on registered labeling.

Dosage: Do not use more than 6.6 pounds of Enide 90W on tomatoes and potatoes. Do not use more than 5.5 pounds of Enide 90W on peppers.

Dilutation Rate: Do not use in less than five gallons of water per acre.

Method of Application: Aircraft.

Frequency/Timing of Application: One time per season.

Field Reentry After Application: Preplant only.

Preharvest Interval: None

Other Requirements: Preplant use only.

Valid until withdrawn, suspended or cancelled by the EPA, NOR-AM Chemical Company or the Department of Food and Agriculture.

Authorization must be obtained from the county agricultural commissioner prior to this use. This does not constitute a recommendation of the Department of Food and Agriculture and will not prevent quarantine action if illegal residues are found on or in the crop. Neither the department nor the county agricultural commissioner, the manufacturer or formulator makes any warranty of merchantability, fitness or purpose, or otherwise, expressed or implied.

24(c) Registrant: Contra Costa County Department of Agriculture

ENIDE® 90W/DEVRINOL®

TANK MIXTURE FOR TOMATOES

EPA SLN NUMBER: CA-790180

For Use in: California

ENIDE® 90W (EPA Reg. No. 45639-131)
DEVRINOL® 50WP (EPA Reg. No. 476-2108-AA)
DEVRINOL® 2E (EPA Reg. No. 476-2150-AA)

TOMATOES: The ENIDE/DEVRINOL combinations applied as preplant incorporated tank mixtures provide control of barnyardgrass, crabgrass, lambsquarters and pigweed.

Not recommended for use on soils with over 10% organic matter or soils with a salt problem.

Recommended Broadcast Rates per Acre for Enide/Devrinol Tank Mixture

Soil Texture	Enide 90W	Devrinol 50WP	Devrinol 2E
Light (Coarse)	4.4 lbs.	2 lbs.	2 qts.
Heavy (Medium/Fine)	6.6 lbs.	4 lbs.	4 qts.

DEVRINOL®—Trademark of Stauffer Chemical Company

MIXING: Combine the recommended application rate of Enide 90W with the recommended application rate of Devrinol 50WP or Devrinol 2E. Add a surfactant to the spray solution. The surfactant rate should be 4 ounces or less per 30 gallons of spray solution.

Use at least 30 gallons of water per acre using a conventional field sprayer with an inline strainer and screens of 50-mesh or coarser. Add the required amount of water to the spray tank. Begin agitation and add, in order, the recommended amount of Enide and Devrinol. Maintain continuous agitation during application with good bypass or paddle agitation.

APPLICATION: Mix and apply Enide/Devrinol tank mixtures before planting. Incorporate immediately after application. Crops should be planted or seeded immediately after incorporation.

GENERAL USE PRECAUTIONS: Do not seed to alfalfa, lettuce, small grains, sorghum, corn, or sugar beets for 12 months after the application of Enide plus Devrinol tank mixtures.

24(c) Registrant: NOR-AM Chemical Company, 3509 Silverside Road, P.O. Box 7495, Wilmington, DE 19803

515–SL*–EE–24C–5/85

FUNDAL® 4EC

INSECTICIDE/OVICIDE
FOR USE ON COTTON

RESTRICTED USE PESTICIDE

For retail sale to and use only by Certified Applicators or persons under their direct supervision and only for those uses covered by the Certified Applicators Certification.

FOR AGRICULTURAL USE ONLY.

ACTIVE INGREDIENTS:

Chlordimeform*	48.5%
Aromatic Petroleum Distillates	43.4%
INERT INGREDIENTS	8.1%
TOTAL	100.0%

*N–(4–chloro–O–tolyl)–N,N–dimethylformamidine

Contains 4 lbs. chlordimeform per gallon.

EPA Reg. No. 45639-75 EPA Est. No. 407-MN-1

WARNING — AVISO

Keep Out of Reach of Children

STATEMENT OF PRACTICAL TREATMENT

If swallowed: Call a physician or Poison Control Center immediately. DO NOT INDUCE VOMITING UNLESS NO OTHER TREATMENT IS AVAILABLE. Vomiting may cause aspiration pneumonia. If it is necessary to induce vomiting, do so under medical supervision. Do not induce vomiting or give anything by mouth to an unconscious person.

If in eyes: Immediately flush with plenty of water. Get medical attention.

If on skin: Immediately remove comtaminated clothing, wiping away excess material from skin. Wash skin with plenty of soap and water. Get medical attention.

If inhaled: Move patient to fresh air.

NOTE TO PHYSICIAN: There is no specific antidote for FUNDAL ingestion. Do not induce emesis. Lavage stomach, give a saline laxative and supportive therapy. Sympathomimetic drugs should be avoided as these may have potentiated effects (e.g., hypertensive crisis).

PRECAUTIONARY STATEMENTS

HAZARDS TO HUMANS

May be harmful or fatal if swallowed or absorbed through the skin. Do not get in eyes, on skin or on clothing. Wear goggles when handling. Avoid inhaling vapors or spray mist. Wash with soap and water after handling and before eating, drinking or smoking. Wear clean clothes and wash before reusing. Do not contaminate food or feedstuffs.

ENVIRONMENTAL HAZARDS

This product is toxic to fish and wildlife. Do not apply directly to water or wetlands. Do not apply where runoff is likely to occur. Do not apply when weather conditions favor drift from areas treated. Do not contaminate water by cleaning of equipment or disposal of wastes. FUNDAL 4EC, when used alone, has little or no effect on predators, beneficial insects, and bees.

NOTE: This product has been shown to cause cancer in laboratory animals. The user must read and follow all precautionary statements and instructions for use on label of the container.

PRECAUCIÓN AL USUARIO: Si usted no lee ingles, no use este producto hasta que el etiqueta haya sido explicado ampliamente.

Indicación: El Producto presente mostro un efecto cancerigeno en animales de laboratorio.

El usuario tiene que leer y seguir estrictamente todas las precauciones e instrucciones que se indican en la etiqueta de envase.

PHYSICAL AND CHEMICAL HAZARDS

Do not use or store near heat or open flame.

DO NOT REUSE THIS CONTAINER.

SPECIAL PRECAUTIONS TO PREVENT EXPOSURE

Wear the standard and special protective clothing as indicated below to ensure against possible exposure during all operations associated with handling and application. Wear clean clothes each day and launder before reusing. Place contaminated clothing in a special designated container until laundering. Keep a complete clean change of clothes at the worksite for use if needed. Replace boots and gloves if contaminated.

STANDARD CLOTHING: Heavy-duty, long-sleeve, one-piece worksuit; cloth cap; underwear including T-type undershirt; rubber or neoprene gloves; socks; heavy-duty, waterproofed workboots.

SPECIAL PROTECTIVE CLOTHING AND EQUIPMENT

When washing aircraft cleaning equipment, in case of spills or accidents, and for other special use when maximum protection is needed, wear a complete waterproof, rubberized or plastic rainsuit with head covering, rubber boots (slip-on, knee-length, waterproof) heavy-duty rubber or neoprene gloves and an approved pesticide respirator such as Willson® 1200 facepiece equipped with Willson R-21 cartridges and R-15 filters. Cartridges and filters are to be changed daily not to exceed 8 hours of use or if breathing becomes difficult. Clean all protective clothing and equipment after each use.

PERSONAL PRECAUTIONS

If FUNDAL 4EC concentrate or spray gets on clothing, immediately remove contaminated clothing and wash affected parts of body with soap and water. If there is any question as to the extent of contamination, bathe entire body thoroughly. Change to clean clothing.

Wash hands with soap and water before eating, drinking or smoking.

At the end of the work day, bathe entire body with soap and plenty of water.

SPECIAL EQUIPMENT

FUNDAL 4EC must be transferred and mixed using closed-system equipment. A closed-system is defined as appropriate connections, meters, pumps and plumbing designed to eliminate human contact. Do not use open mixing vats or tanks, or open pouring.

All plumbing connections must be sufficiently tight to prevent leakage.

Use hoses constructed of chemically resistant, heavy-duty, reinforced materials. Avoid excessive pressures to minimize the possibility of rupture. If any sign of weakness appears, flush the hose thoroughly with clean water and replace immediately.

Transport and storage vessels and transfer equipment must be equipped with drybreak valves.

Any equipment which has been contaminated with FUNDAL 4EC must be thoroughly cleaned and flushed with clean water prior to maintenance or use for other purposes. Wear appropriate protective clothing when performing equipment maintenance operations.

Aircraft must be equipped with low-drift nozzles.

ACCIDENTS

For minor spills, leaks, etc. follow all previously mentioned precautions and clean-up immediately. Take special care to avoid contamination of equipment and facilities during clean-up procedures and disposal of wastes.

In the event of a major spill or accident, phone (800) 424-9300 for assistance.

GENERAL INFORMATION

FUNDAL 4EC is a water emulsifiable insecticide ovicide for control of bollworm and tobacco budworm on cotton. Best results can be expected in fields under a regular scouting program.

USE PRECAUTIONS

Do not spray shortly before rainfall. Do not allow prepared solutions to stand overnight. Flush spray tank thoroughly with water daily after use. Do not allow animals to graze in areas treated with FUNDAL 4EC. Do not feed crop residues. Do not apply near persons or dwellings. Do not enter treated fields for at least 24 hours after application.

NOTE: To avoid excessive residues, do not apply less than 21 days before harvest. Do not feed foliage from treated cotton plants or gin trash to livestock.

DIRECTIONS FOR USE

It is a violation of Federal law to use this product in a manner inconsistent with its labeling.

Apply by air only, using low drift nozzles. Use not less than 1 gallon spray per acre. Read the entire Directions for Use before using this product.

REENTRY STATEMENT

Reentry into treated areas is prohibited until leaves are completely dry or until 24 hours after application.

Do not enter treated areas without protective clothing.

Because certain states may require more restrictive reentry intervals for various crops treated with this product, consult your State Department of Agriculture for further information.

Written or oral warnings must be given to workers who are expected to be in a treated area or in an area about to be treated with this product. (Indicate specific oral warnings which inform workers of areas or fields that may not be entered without specific protective clothing, period of time field must be vacated and appropriate actions to take in case of accidental exposure.) When oral warnings are given, warnings shall be given in a language customarily understood by workers. Oral warnings must be given if there is reason to believe that written warnings cannot be understood by workers. Written warnings must include the following information:

WARNING. Area treated with FUNDAL® 4EC on (date of application). Do not enter without appropriate protective clothing for 24 hours or until leaves are completely dry. In case of accidental exposure see Statement of Practical Treatment on the label.

FUNDAL 4EC ALONE

Apply on a 3–5 day schedule when moth flights begin and/or when eggs first appear and continue application as long as eggs are present. Apply the lower rate when egg deposition is just starting.

Bollworm eggs ¼–½ pint/acre
Tobacco Budworm eggs ½ pint/acre

YIELD ENHANCEMENT: For a yield enhancement effect above that expected from insect control, apply FUNDAL 4EC at ¼–½ pint/acre 4–6 times at 5-day intervals beginning at first square formation.

MIXING DIRECTIONS: Thoroughly clean spray equipment before using FUNDAL 4EC. Prepare no more spray mixture than is needed for the immediate operation. Use a closed system for introduction of FUNDAL 4EC into the spray tank. Add the desired amount of FUNDAL 4EC to the mix tank ¾ full of water with the agitator running. Then add the remainder to the water. By-pass agitation will keep FUNDAL 4EC emulsified.

TANK-MIXES

If egg hatch occurs because of delayed application of FUNDAL 4EC or other causes or if it becomes necessary to begin applications in a field where bollworm and tobacco budworm larvae or other insect pests have reached economic levels, it may be advantageous to apply FUNDAL 4EC in a tank-mix with other insecticides listed on this label.

Use other insecticides listed on this label at rates recommended on the labels for these products and observe all limitations and precautions which appear on these labels.

At low to moderate bollworm and budworm infestation levels and in closely scouted and supervised fields, pyrethroid products may be used in the tank-mix at reduced rates.

FUNDAL 4EC Tank-Mix Rates

Pest	Time of Application	Rate FUNDAL 4EC	Other Insecticide
Bollworm and/or tobacco budworm larvae	Apply on a 3–5 day schedule when larvae reach economic levels. Continue applications as long as egg and larvae are present.	¼–½ pint/acre	Refer to the other insecticide label for recommended rates

FUNDAL 4EC has been found to be effective and compatible when tank-mixed with the following products:

Ambush® Trademark of ICI Americas, Inc
Ammo® Trademark of FMC Corp
Azodrin® Trademark of Shell Chemical Co
Bacillus thuringiensis
Bolstar® Trademark of Mobay Chemical Corp
Bidrin® Trademark of Shell Chemical Co
Curacron® Trademark of Ciba-Geigy Corp
Cymbush® Trademark of ICI Americas, Inc
Dimethoate
Dimilin® Trademark of Thompson-Hayward Chemical Co
Elcar® Trademark of Sandoz Inc
EPN & Methyl parathion
Guthion® Trademark of Mobay Chemical Corp
Lorsban® Trademark of Dow Chemical Co
Methomyl
Methyl parathion
Pay-Off® Trademark of American Cyanamid Co
Pounce® Trademark of FMC Corp
Pydrin® Trademark of Shell Chemical Co
Supracide® Trademark of Ciba Geigy Corp

TANK-MIX DIRECTIONS

Add ¾ of the required amount of water to the mix tank. With the agitator running add the desired amount of the other insecticide. Continue agitating while adding the remainder of the water followed by the desired amount of FUNDAL 4EC. By-pass agitation will keep the tank mix emulsified. Always prepare fresh mixtures just prior to use. Do not allow tank-mixes to stand overnight.

STORAGE AND DISPOSAL

STORAGE

Do not store at temperatures below 32°F. Crystals may form if stored under very cold conditions. Do not contaminate water, food or feed by storage or disposal. Open dumping is prohibited. DO NOT REUSE THIS CONTAINER.

PESTICIDE DISPOSAL

Pesticide wastes are toxic. Improper disposal of excess pesticide, spray mixture, or rinsate is a violation of federal law. If these wastes cannot be disposed of by use according to label instructions, contact your State Pesticide or Environmental Control Agency, or the Hazardous Waste representative at the nearest EPA Regional Office for guidance.

CONTAINER DISPOSAL

Triple rinse (or equivalent). Then offer for recycling or reconditioning, or puncture and dispose of in a sanitary landfill, or by other procedures approved by state and local authorities.

Dispose of pesticide rinsate by application to an already treated area.

IMPORTANT: READ BEFORE USE

By using this product, user accepts the following conditions, warranty, disclaimer of warranties and limitations of liability.

CONDITIONS: The directions for use of this product are believed to be adequate and should be followed carefully. However, because of extreme weather and soil conditions, manner of use and other factors beyond NOR-AM Chemical Company's control, it is impossible for NOR-AM to eliminate all risks associated with the use of this product. As a result, crop injury or ineffectiveness is always possible.

DISCLAIMER OF WARRANTIES: THERE ARE NO WARRANTIES, EXPRESS OR IMPLIED, OF MERCHANTABILITY OR OF FITNESS FOR A PARTICULAR PURPOSE OR OTHERWISE, WHICH EXTEND BEYOND THE STATEMENTS MADE ON THIS LABEL. No agent of NOR-AM Chemical Company is authorized to make any warranties beyond those contained herein or to modify the warranties contained herein. UNDER NO CIRCUMSTANCES SHALL NOR-AM CHEMICAL COMPANY BE LIABLE FOR ANY INCIDENTAL OR CONSEQUENTIAL DAMAGES RESULTING FROM FAILURE TO USE THIS PRODUCT IN ACCORDANCE WITH LABEL DIRECTIONS.

LIMITATIONS OF LIABILITY: In no event shall NOR-AM Chemical Company's liability for damages arising out of the use of this product exceed the purchase price of the product used.

Also available as Fundal® SP Insecticide/ovicide

MITAC® WP

50% WETTABLE POWDER INSECTICIDE FOR PEARS

FOR AGRICULTURAL USE ONLY

RESTRICTED USE PESTICIDE

For retail sale to and use only by Certified Applicators or persons under their direct supervision, and only for those uses covered by the Certified Applicator's certification.

ACTIVE INGREDIENT:

Amitraz (N'–(2,4–dimethylphenyl)–N–[[(2,4–dimethylphenyl) imino]methyl]–N–methylmethanimidamide)	50%
INERT INGREDIENTS:	50%
TOTAL	100%

EPA Reg. No. 45639-61 EPA Est. No. 477-MD-1

Keep Out of Reach of Children

WARNING — AVISO

STATEMENT OF PRACTICAL TREATMENT

IF IN EYES—Flush with clean water. Get medical attention if irritation persists.

IF SWALLOWED—Drink promptly a large quantity of milk, egg whites, gelatin solution, or, if these are not available, drink large quantities of water. Avoid alcohol.

IF ON SKIN—Wash hands with soap and water each time before eating, drinking, or smoking. At the end of the working day, bathe entire body with soap and plenty of water.

ANTIDOTE—None known.

PRECAUCION AL USARIO: Si usted no lee ingles, no use este producto hasta que el etiqueta haya sido explicado ampliamente.

PRECAUTIONARY STATEMENTS

HAZARDS TO HUMANS

WARNING

CAUSES EYE IRRITATION. HARMFUL IF SWALLOWED. DO NOT GET IN EYES. WEAR GOGGLES OR FACE SHIELD. WASH THOROUGHLY AFTER HANDLING. AVOID BREATHING DUST OR SPRAY MIST.

REQUIRED CLOTHING AND EQUIPMENT FOR MIXING, LOADING, GROUND SPRAY APPLICATION AND CLEAN-UP PROCEDURES:

1) Long-Sleeve Shirt (Fine Weave) 3) Rubber Gloves

2) Long Pants (Fine Weave) 4) Boots

Wear clean clothing each day and launder before reusing.

IF ON CLOTHING—Remove contaminated clothing and wash affected parts of the body with soap and water. If the extent of contamination is unknown, bathe entire body thoroughly, change clothing.

ENVIRONMENTAL HAZARDS

This product is toxic to fish. Do not apply directly to water. Do not apply when weather conditions favor drift from treated areas. Do not contaminate water by cleaning of equipment or disposal of wastes. Apply this pesticide only as specified on this label.

DIRECTIONS FOR USE

It is a violation of federal law to use this product in a manner inconsistent with its labeling.

Do not apply this product in such a manner as to directly or through drift expose workers or other persons. The area being treated must be vacated by unprotected persons.

REENTRY STATEMENT

Reentry into treated areas is prohibited until leaves are completely dry or until 24 hours after application.

Do not enter treated areas without protective clothing.

Because certain states may require more restrictive reentry intervals for various crops treated with this product, consult your State Department of Agriculture for further information.

Written or oral warnings must be given to workers who are expected to be in a treated area or in an area about to be treated with this product. (Indicate specific oral warnings which inform workers of areas or fields that may not be entered without specific protective clothing, period of time field must be vacated and appropriate actions to take in case of accidental exposure.) When oral warnings are given, warnings shall be given in a language customarily understood by workers. Oral warnings must be given if there is reason to believe that written warnings cannot be understood by workers. Written warnings must include the following information:

WARNING. Area treated with MITAC W.P. on (date of application). Do not enter without appropriate protective clothing for 24 hours or until leaves are completely dry. In case of accidental exposure see Statement of Practical Treatment on label.

GENERAL INFORMATION

MITAC WP is an insecticide for use on pears. When applied as directed, MITAC WP will control pear psylla.

MITAC WP can be applied by air or ground equipment, as dilute or concentrate sprays. The rate of MITAC WP per acre is based on thorough coverage with a dilute spray of 400 gallons from a conventional sprayer. When using concentrate sprayers, the amount of water is reduced but the amount of MITAC WP per acre should remain the same as when using dilute sprays.

COMPATIBILITY

MITAC WP is compatible with most commonly used pear insecticides, miticides and fungicides; however, do not combine MITAC WP with detergents, sulfur, bordeau mixture, NAA or other highly alkaline materials as a tank mix.

DO NOT GRAZE TREATED AREAS.

PEAR PSYLLA

Apply $1\frac{1}{2}$ to 3 pounds MITAC WP per acre. Where full dilute coverage requires less than 400 gallons per acre, use 6 to 12 ounces per 100 gallons.

For prebloom control of overwintering adults, apply MITAC WP after eggs have moved into the female psylla oviduct, but prior to heavy egg laying. For best results, MITAC WP should be applied when daily maximum temperatures exceed 50°F. Consult your local University Extension Service for further information concerning best timing for control of overwintering adult psylla.

If weather or orchard conditions do not permit spraying before heavy egg laying, spray after the majority of the first generation eggs have hatched.

For postbloom control, apply MITAC WP as needed. Best control will be obtained when the majority of pear psylla are in the adult or young nymphal stages of development.

Rate should not exceed 3 pounds of MITAC WP per acre per application. Do not apply within 7 days of harvest. Do not apply more than 21 pounds per acre of MITAC WP during the use season.

STORAGE AND DISPOSAL

STORAGE

Store in original container and keep closed. Store in a cool, dry place.

PESTICIDE DISPOSAL

Do not contaminate water, food, or feed by storage or disposal. Wastes resulting from the use of this product may be disposed of on site or at an approved waste disposal facility.

CONTAINER DISPOSAL

Completely empty bag into application equipment. Then dispose of empty bag in a sanitary landfill or by incineration, or, if allowed by State and local authorities, by burning. If burned, stay out of smoke.

IMPORTANT: READ BEFORE USE

By using this product, user accepts the following conditions, warranty, disclaimer of warranties and limitations of liability.

CONDITIONS:
The directions for use of this product are believed to be adequate and should be followed carefully. However, because of extreme weather and soil conditions, manner of use and other factors beyond NOR-AM Chemical Company's control, it is impossible for NOR-AM to eliminate all risks associated with the use of this product. As a result, crop injury or ineffectiveness is always possible.

DISCLAIMER OF WARRANTIES:
THERE ARE NO WARRANTIES, EXPRESS OR IMPLIED, OF MERCHANTABILITY OR OF FITNESS FOR A PARTICULAR PURPOSE OR OTHERWISE, WHICH EXTEND BEYOND THE STATEMENTS MADE ON THIS LABEL. No agent of NOR-AM Chemical Company is authorized to make any warranties beyond those contained herein or to modify the warranties contained herein. UNDER NO CIRCUMSTANCES SHALL NOR-AM CHEMICAL COMPANY BE LIABLE FOR ANY INCIDENTAL OR CONSEQUENTIAL DAMAGES RESULTING FROM FAILURE TO USE THIS PRODUCT IN ACCORDANCE WITH LABEL DIRECTIONS.

LIMITATIONS OF LIABILITY:
In no event shall NOR-AM Chemical Company's liability for damages arising out of the use of this product exceed the purchase price of the product used.

Also available as Mitac EC. Insecticide/Miticide

NORTRON® EC

SUGARBEET HERBICIDE

FOR AGRICULTURAL USE ONLY

For selective control of weeds in sugarbeets

ACTIVE INGREDIENT
Ethofumesate:
(2–ethoxy–2,3–dihydro–3,3–dimethyl–5–benzofuranyl methanesulfonate*) 19%
INERT INGREDIENTS: 81%
TOTAL 100%

* Protected by U.S. Patent No. 3,689,507

This product contains 1.5 lb. active ingredient per gallon.

EPA Reg. No. 45639-5 EPA Est. No. 407-MN-1

Keep Out of Reach of Children

DANGER — PELIGRO

STATEMENT OF PRACTICAL TREATMENT

CORROSIVE, CAUSES EYE DAMAGE.

WEAR GOGGLES OR FACE SHIELD WHEN HANDLING.

IF SWALLOWED—Drink promptly a large quantity of milk, egg whites, geletin solution or if these are not available, drink large quantities of water. Avoid alcohol. Call a physician.

IF ON SKIN—Flush with plenty of water for at least 15 minutes.

IF IN EYES—Immediately flush with plenty of water for at least 15 minutes. Call a physician.

PRECAUCION AL USUARIO: Si usted no lee ingles, no use este producto hasta que la etiqueta le haya sido explicade ampliamente.

PRECAUTIONARY STATEMENTS

HAZARDS TO HUMANS AND DOMESTIC ANIMALS
DANGER
Corrosive. Causes eye damage. Do not get in eyes. Wear goggles or face shield when handling. Harmful if absorbed through skin. Avoid contact with skin.

FIRST AID
In case of contact, immediately flush eyes or skin with plenty of water for at least 15 minutes. For eyes, call a physician. Get medical attention if skin irritation persists. Remove and wash contaminated clothing before reuse.

ENVIRONMENTAL HAZARDS
Do not apply NORTRON EC directly to water. Do not apply NORTRON EC when weather conditions favor drift from treated area. Do not contaminate water by cleaning of equipment or disposal of wastes.

PHYSICAL OR CHEMICAL HAZARDS
Do not use, pour, spill or store NORTRON EC near heat or open flame.

GENERAL INFORMATION

NORTRON EC is a selective herbicide for use in sugar beets for the control of the weed species listed below. It provides effective control of these weeds for up to 10 weeks following application.

WEED SPECIES CONTROLLED

Annual Broadleaf Weeds	
Black nightshade	*Solanum nigrum*
Common chickweed	*Stellaria media*
Common lambsquarters	*Chenopodium album*
Common purslane	*Portulaca oleracea*
Kochia	*Kochia scoparia*
Ladysthumb	*Polygonum persicaria*
Pennsylvania smartweed	*Polygonum pensylvanicum*
Redroot pigweed	*Amaranthus retroflexus*
Russian thistle	*Salsola iberica*
Wild buckwheat	*Polygonum convolvulus*
Annual Grass Weeds	
Annual bluegrass	*Poa annua*
Barnyardgrass*	*Echinochloa crus-galli*
Canarygrass	*Phalaris canariensis*
Green foxtail	*Setaria viridis*
Large crabgrass	*Digitaria sanguinalis*
Volunteer barley	*Hordeum sp.*
Volunteer wheat	*Triticum sp.*
Wild oats**	*Avena fatua*
Yellow foxtail	*Setaria glauca*

* Control of barnyardgrass may be reduced with the NORTRON EC + Pyramin because of the lower rate of NORTRON EC recommended.
** Control of wild oats has been inconsistent in Minnesota and North Dakota.

NORTRON EC alone will also reduce competition from these HARD-TO-CONTROL weeds:

Annual sowthistle	*Sonchus oleraceus*
Puncturevine	*Tribulus terrestris*
Shepherdspurse	*Capsella bursa-pastoris*
Purple nutsedge	*Cyperus rotundus*
Yellow nutsedge	*Cyperus esculentus*

NORTRON EC ALONE
DOSAGE TABLE 1 (ALL REGIONS EXCEPT NORTH DAKOTA AND MINNESOTA):

SOIL TEXTURE	RATE OF NORTRON EC PER ACRE[1] BROADCAST	7-INCH BAND WIDTH[2] 22" ROW	28" ROW	30" ROW
Coarse Textured Soils: Sands, lomy sands and sandy loams	3/4 to 1 1/4 Gallons	2 to 3 1/4 Pints	1 1/2 to 2 1/2 Pints	1 1/4 to 2 1/4 Pints
Medium Textured Soils: Silt loams, clay loams which contain less than 3% organic matter	1 1/4 to 2 Gallons	3 1/4 to 5 Pints	2 1/2 to 4 Pints	2 1/4 to 3 3/4 Pints

NORTRON EC ALONE
DOSAGE TABLE 1 (ALL REGIONS EXCEPT NORTH DAKOTA AND MINNESOTA):

SOIL TEXTURE	RATE OF NORTRON EC PER ACRE[1]			
		7-INCH BAND WIDTH[2]		
	BROADCAST	22″ ROW	28″ ROW	30″ ROW
Fine Textured Soils: Clay loams which contain more than 3% organic matter and clays	2 to 2½ Gallons	5 to 6¼ Pints	4 to 5 Pints	3¾ to 4¾ Pints

[1] Use the higher rate within each soil texture category on the finer texture soils and/or where kochia, barnyardgrass or black nightshade are expected to be a problem.

[2] For other band or row widths, adjust the rate in proportion to the area actually treated.

GENERAL USE PRECAUTIONS

NORTRON EC or tank-mixes should be used for recommended purposes and at recommended rates only. (DO NOT OVERTREAT.)

Apply tank-mixes only in specified regions or states and in accordance with directions on label.

Do not rotate with any other crops other than sugarbeets for 12 months after a preplant or preemergence application. Thorough tillage, including moldboard plowing, should precede the planting of crops other than sugarbeets.

If crop is lost due to climatic or soil conditions following application of NORTRON EC or tank-mixes, do not plant crops other than sugarbeets in treated land during the same season. Do not retreat field with NORTRON EC. If fields are replanted to sugarbeets, reseed into treated band.

Do not mix NORTRON EC or tank-mixes with liquid fertilizers.

Do not allow spray mixture to stand in tank overnight. Flush and drain spray equipment after each day's use. Store unused spray mixture in tightly-sealed containers and protect from frost.

Do not use NORTRON EC in water having temperature below 40°F., as crystallization of spray mixture may occur in the nozzles and strainers.

DIRECTIONS FOR USE

It is a violation of federal law to use this product in a manner inconsistent with its labeling. Read entire Directions for Use and Disclaimer of Warranties before using this product.

REENTRY STATEMENT

Do not apply this product in such a manner as to directly or through drift expose workers or other persons. The area being treated must be vacated by unprotected persons.

When NORTRON EC is applied preemergence or postemergence, the following statements apply:

Reentry Statement

Do not enter treated areas without protective clothing until sprays have dried.

Because certain states may require more restrictive reentry intervals for various crops treated with this product, consult your State Department of Agriculture fo further information.

Written or oral warnings must be given to workers who are expected to be in a treated area or in an area about to be treated with this product. (Indicate specific oral warnings which inform workers of areas or fields that may not be entered without specific protective clothing, period of time field must be vacated, and appropriate actions to take in case of accidental exposure.) When oral warnings are given, warnings shall be given in a language customarily understood by workers. Oral warnings must be given if there is reason to believe that written warnings cannot be understood by workers. Written warnings must include the following information:

DANGER: Area treated with NORTRON EC herbicide on (date of application). Do not enter without appropriate protective clothing until spray has dried. In case of accidental exposure, see Statement of Practical Treatment on the label.

PREPLANT INCORPORATED AND PREEMERGENCE APPLICATIONS

SOIL PREPARATION: The soil should be prepared according to good agricultural practices. Large clods can reduce the effectiveness of NORTRON EC and tank-mixes. All existing vegetative growth should be thoroughly worked into the soil before treatment.

SPRAY EQUIPMENT: Apply NORTRON EC alone or in tank-mixes to the soil using standard low pressure (20–50 psi) spray equipment. Spray equipment should be carefully calibrated before use and checked frequently during application to see that it is functioning properly. Do not use smaller than 50-mesh strainer. Uniformly apply the recommended rates of NORTRON EC or NORTRON EC tank-mixes in 10–60 gallons of water per acre on a broadcast basis. Avoid overlaps since crop injury may result. When applying NORTRON EC or tank mixes in a band, check to make certain that the band width is accurate for the dosage rate being applied.

The spray tank and lines should be thoroughly cleaned and rinsed prior to using NORTRON EC.

INCORPORATION EQUIPMENT: Where soil incorporation is recommended use a hooded-power or ground-driven rotary tiller, rolling cultivator or similar equipment properly adjusted to uniformly incorporate NORTRON EC or tank-mixes to a depth of 1 to 2 inches. Deeper incorporation may reduce effectiveness. Do not apply NORTRON EC or tank-mixes through soil injector shanks. Incorporation should be accomplished prior to planting; or if done after planting, proper precautions should be taken to avoid damaging or moving the crop seed. See below for Layering Application.

LAYERING APPLICATION

Spring: Form beds with appropriate bedding equipment. Pre-irrigate field if necessary. Remove bed top with suitable de-ridging machinery to provide a minimum width of 10″ across the top of the bed. Apply NORTRON EC in a band at the recommended rate indicated in the appropriate regional dosage table and cover the treated band with 1 inch of soil using ditchers or discs equipment. Shape the bed with roller shaper and irrigate until the top of the beds are thoroughly wetted. Irrigate from furrows on both sides of the row.

Fall: This method of application can be used when spring moisture is marginal or where irrigation water is not available at planting time. Fall bedding utilizes the winter-accumulated moisture to enhance activation of the herbicide and to aid in germination of the sugarbeet crop.

Prepare the field (as for planting; plow, pack, and float, etc.), in the fall, usually late September or October. Apply NORTRON EC in a band to the soil surface at the recommended rate indicated in the appropriate regional dosage table. Be sure that the soil surface to be treated is free from trash and vegetation.

Cover the treated bands with soil and form beds or ridges using ditchers or discs. In the spring when the soil is sufficiently dry to be worked, de-ridge the beds down to within ½″ to 1″ of the treated layer using suitable equipment such as the Kirchner bedder or Oregon Northslope harrow. When de-ridging, maintain the original bedding guidance system by using a bull tongue chisel, slide guides or similar equipment. This will ensure that the planter will follow in the treated band. Plant sugar beets in the de-ridged area when the soil conditions allow.

MIXING DIRECTIONS: When mixing NORTRON EC in the spray tank with Pyramin or TCA, add NORTRON EC first. Agitate spray solution thoroughly and continuously. See Pyramin label for additional mixing directions.

GENERAL APPLICATION

Sugarbeets grown under rainfall: Apply NORTRON EC alone or in a tank-mix preemergence at time of planting or shortly after, but prior to weed germination. NORTRON EC or tank-mix does not require mechanical soil incorporation provided that sufficient rainfall occurs shortly following application to activate the chemical. One-half inch of rainfall is usually adequate for activation. In areas where rainfall can be marginal for activation, such as the Red River Valley (Minnesota and North Dakota), it is recommended that NORTRON EC or the tank-mix be applied before or at the time of planting and incorporated into the soil.

Sugarbeets grown under furrow irrigation: Apply NORTRON EC alone or in a tank-mix to the soil surface preplant or at time of planting, but prior to weed germination, and incorporate into the soil. Where sugarbeets are grown in beds, apply NORTRON EC or tank-mix after bedding and incorporate. Since NORTRON EC or tank-mix must have moisture to control weeds effectively, irrigate until tops of beds are thoroughly wetted.

Sugarbeets grown under sprinkler irrigation: Apply NORTRON EC alone or in tank-mix preemergence at time of planting or shortly after, and irrigate prior to crop and weed germination. Repeat irrigation as necessary to maintain good moisture in upper soil layer. Apply at least one-half inch of water during first irrigation. Do not mechanically incorporate NORTRON EC or tank-mix into the soil under sprinkler irrigation.

CULTURAL PRACTICES FOLLOWING APPLICATION: When properly applied, NORTRON EC alone or in a tank-mix will provide up to 10 weeks control of susceptible weed species. When cultivating fields in which NORTRON EC or tank-mixes have been banded, care should be exercised to minimize the movement of untreated soil into the treated band. Where a broadcast application has been made, do not cultivate deeper than two inches or this reduces the effectiveness of NORTRON EC or tank-mixes.

NORTRON EC Alone
Dosage Table 2 (North Dakota and Minnesota only):

		NORTRON EC/Acre
		7-inch Band Width[1]
Soil Texture	Broadcast	22" Row
Coarse Textured Soils: Sandy loams only	2 Gallons	5 Pints
Medium Textured Soils: Silt loams and clay loams	2 Gallons	5 Pints
Fine Textured Soils: Heavy clays	$2\frac{1}{2}$ Gallons	$6\frac{1}{4}$ pints

[1] For other band or row widths, adjust the rate in proportion to the area actually treated.

NORTRON EC + ANTOR® 4 ES (tank-mix)

CALIFORNIA ONLY

This tank-mix controls these additional weeds:

Italian ryegrass (annual)	*Lolium multiflorum*
Annual sowthistle	*Sonchus oleraceus*
Fall panicum	*Panicum dichotomoflorum*
Foxtail millet	*Setaria italica*
Groundcherry	*Physalis longifolia*
Junglerice	*Echinochloa colonum*
Prostrate pigweed	*Amaranthus blitoides*
Shepherdspurse	*Capsella bursa-pastoris*
Witchgrass	*Panicum capillare*

NORTRON EC + ANTOR 4 ES tank-mix may be applied preplant incorporated where furrow irrigation is practiced using shallow incorporation implements such as rolling cultivators or harrows, or using power tillers set to a depth of 1–2 inches.

Preemergence applications of NORTRON EC + ANTOR ES tank-mix may be used when application is followed by sprinkler irrigation prior to weed germination and crop emergence, or where rainfall can be expected to occur within 7–10 days after application.

ANTOR 4 ES may also be utilized in sequential applications with NORTRON EC where a broadcast preemergence treatment of ANTOR 4 ES is followed by a band overlay of NORTRON EC preemergence. The same dosage rates should be used as in the tank-mix.

Refer to the Dosage Table 3 for recommended rates.

Before use, read the ANTOR 4 ES label for additional information and precautions.

NORTRON EC + ANTOR 4ES (tank-mix)
Dosage Table 3 (California only):

Soil Texture	ANTOR 4ES Per Acre[1]	NORTRON EC Per Acre[1]
Coarse: Loamy sands and sandy loams	1.5 Quarts	4 Quarts
Medium: Loams and silt loams	2 Quarts	5.3 Quarts
Fine: Clay loams and clays	3 to 4 Quarts[2]	6.6 Quarts

[1] See section on Band Applications for calculations of proportional amounts.
[2] Use higher rate on heavy clay soils.

CALIFORNIA (winter-grown sugarbeets only)

Under sprinkler irrigation or where natural rainfall is adequate, apply this tank-mix preemergence. See Pyramin label for precautions regarding application of sprinkler irrigation. Where furrow irrigation is to be used, apply preplant incorporated. Prepare seedbed or form beds for planting. Then use rotary tiller type of incorporation and incorporate not more than 2 inches deep. Plant sugarbeets and irrigate. Sub-irrigate until tops of beds are thoroughly wetted.

NORTRON EC + PYRAMIN®* (tank-mix)

This tank-mix controls these additional broadleaf weeds:

Annual sowthistle	*Sonchus oleraceus*
Black mustard	*Brassica nigra*
Buffalobur	*Solanum rostratum*
Coast fiddleneck	*Amsinckia intermedia*
Common groundsel	*Senecio vulgaris*
Common ragweed	*Ambrosia artemisiifolia*
Cutleaf nightshade	*Solanum triflorum*
Groundcherry	*Physalis* spp.
Henbit	*Lamium amplexicaule*
Nettleleaf goosefoot	*Chenopodium murale*
Prickly lettuce	*Lactuca serriola*
Prostrate knotweed	*Polygonum aviculare*
Shepherdspurse	*Capsella bursa-pastoris*

* PYRAMIN® is a registered trademark of BASF Atkiengesellschaft.

PRECAUTIONS: Do not exceed $\frac{3}{4}$ inch of sprinkler irrigation per set until the beets have two true leaves. Do not use the tank-mix under conditions where Pyramin alone is not recommended. Before use, read the Pyramin label for additional information and precautions.

CENTRAL AND EASTERN STATES, INCLUDING MICHIGAN AND OHIO

This tank-mix controls the additional weed, common ragweed, and other broadleaf weeds specified in the weed table. Apply preemergence at the time of planting or shortly after, but before weed germination using recommended rates listed in Dosage Table 5. Do not mechanically incorporate the herbicides into the soil as crop injury may result. Do not use this tank-mix where Pyramin alone is not recommended. Before use, read Pyramin label for additional information and precautions.

NORTRON EC & PYRAMIN (TANK-MIX)
DOSAGE TABLE 4 (CALIFORNIA, WINTER-GROWN SUGARBEETS ONLY):

SOIL TEXTURE	NORTRON EC/ACRE BROADCAST (FOR CALIBRATION PURPOSES ONLY)	NORTRON EC/ACRE 10-INCH BAND WIDTH[1,2] 30" ROW	PYRAMIN FL/ACRE BROADCAST (FOR CALIBRATION PURPOSES ONLY)	PYRAMIN FL/ACRE 10-INCH BAND WIDTH[1] 30" ROW
Coarse Textured Soils: Sands, loamy sands and sandy loams	NOT RECOMMENDED			
Medium Textured Soils: Silt loams, clay loams which contain less than 3% organic matter	1 to 1¼ Gallons	2⅔ to 3⅓ Pints	3 Quarts	1 Quart
Fine Textured Soils: Clay loams which contain more than 3% organic matter and clays	1⅓ to 1¾ Gallons	3½ to 4½ Pints	3 Quarts	1 Quart

[1] For other band or row widths, adjust rates in proportion to the area actually treated. Do not apply this mixture broadcast.
[2] Use the higher rate of NORTRON EC within each soil texture category on the finer-textured soils and/or where volunteer barley or wheat are expected to be a problem.

NORTRON EC & PYRAMIN (TANK-MIX)
DOSAGE TABLE 5 (CENTRAL AND EASTERN STATES ONLY):

SOIL TEXTURE	NORTRON EC/ACRE BROAD-CAST	NORTRON EC/ACRE 7-INCH BAND WIDTH[1] 22" ROW	NORTRON EC/ACRE 7-INCH BAND WIDTH[1] 28" ROW	PYRAMIN FL/ACRE BROAD-CAST	PYRAMIN FL/ACRE 7-INCH BAND WIDTH[1] 22" ROW	PYRAMIN FL/ACRE 7-INCH BAND WIDTH[1] 28" ROW
Coarse Textured Soils: Sandy loams only	1 Gallon	2½ Pints	2 Pints	2.25 Quarts	.75 Quart	.5 Quart
Medium Textured Soils: Silt and clay loams which contain less than 3% organic matter	1⅓ Gallons	3⅓ Pints	2⅔ Pints	3 Quarts	1 Quart	.75 Quart
Fine Textured Soils: Clay loams which contain more than 3% organic matter and clays	1⅔ Gallons	4¼ Pints	3⅓ Pints	3 Quarts	1 Quart	.75 Quart

[1] For other band or row widths, adjust the rate in proportion to the area actually treated.

For preemergence use without irrigation, apply the tank-mix to the soil surface at the time of planting or shortly after, but prior to weed germination. Use rates recommended in Dosage Table 6.

Do not use this mixture under conditions where Pyramin alone is not recommended. Before use, read the Pyramin and ANTOR 4ES labels for additional information and precautions.

NORTRON EC + ANTOR 4ES + PYRAMIN (tank-mix)

A tank-mix of NORTRON EC + ANTOR 4ES + Pyramin may be used in Michigan for control of the additional weeds:

Common ragweed	*Ambrosia artemisiifolia*
Wild mustard	*Sinapis arvensis*
Pennsylvania smartweed	*Polygonum pensylvanicum*
Common lambsquarters	*Chenopodium album*
Common purslane	*Portulaca oleracea*
Nettleleaf goosefoot	*Chenopodium murale*
Common chickweed	*Stellaria media*
Henbit	*Lamium amplexicaule*
Fanweed (stinkweed)	*Thlaspi arvense*
Kochia	*Kochia scoparia*
Russian thistle	*Salsola iberica*
Volunteer barley	*Hordeum* sp.
Volunteer wheat	*Triticum* sp.
Ladysthumb	*Polygonum persicaria*

NORTRON EC + TCA (tank-mix)

CENTRAL AND EASTERN STATES, INCLUDING MICHIGAN AND OHIO

The control of certain grass weeds, particularly yellow foxtail and barnyardgrass, can be enhanced by the use of this tank-mix. Apply premergence at time of planting or shortly after but prior to weed germination. Apply at rates recommended in Dosage Table 7. Before use, read TCA label for additional information and precautions.

NORTRON EC + ANTOR 4ES + Pyramin (tank-mix)
Dosage Table 6 (Michigan only):

Soil Texture	Broadcast Rates Per Acre: NORTRON EC (qts)	ANTOR 4ES (qts)	Pyramin FL (qts)
Coarse Textured Soils: Sandy loams only	4	1.5	2.25
Medium Textured Soils: Loams, silt loams and clay loams containing less than 3% organic matter	5.3	2	2.25
Fine Textured Soils: Clay loams and clays containing more than 3% organic matter	5.3	3	2.25

NORTRON EC + TCA (TANK-MIX)
DOSAGE TABLE 7
(CENTRAL AND EASTERN STATES, INCLUDING MICHIGAN AND OHIO):

SOIL TEXTURE	NORTRON EC/ACRE BROADCAST	NORTRON EC/ACRE 7-INCH BAND[2] 28" ROW	TCA/ACRE[1] BROADCAST	TCA/ACRE[1] 7-INCH BAND[2] 28" ROW
Coarse Textured Soils: Sandy loams only	1 Gallon	2 Pints	3 Quarts	1½ Pints
Medium Textured Soils: Silt and clay loams which contain less than 3% organic matter	1⅓ Gallons	2⅔ Pints	1 Gallon	1 Quart
Fine Textured Soils: Clay loams which contain more than 3% organic matter and clays	2 Gallons	4 Pints	1 Gallon	1 Quart

[1] Based on commercial liquid formulation containing the equivalent of 6 pounds of dry 90% sodium TCA per gallon.
[2] For other band or row widths, adjust the rate in proportion to the area actually treated.

NORTH DAKOTA AND MINNESOTA ONLY

This tank-mix enhances the control of yellow foxtail (pigeongrass) and barnyardgrass, especially on fine-textured soils. Apply to the soil surface preplant or at time of planting and incorporate into the soil. Apply at recommended rates listed in Dosage Table 8. Before use, read TCA label for additional information and precautions.

NORTH DAKOTA AND MINNESOTA ONLY (sequential treatment)

The control of certain weeds, particularly yellow foxtail and redroot pigweed, can be improved by the sequential treatment of Eptam 7E applied in the fall followed by a spring application of NORTRON EC. Furthermore, the addition of NORTRON EC provides up to 10 weeks' control of susceptible weeds.

Apply Eptam 7E in the fall in accordance with directions on the label.

Apply NORTRON EC in the spring to the soil surface preplant or at the time of planting and incorporate into the soil. Apply at recommended rates listed in Dosage Table 9. This sequential treatment is not recommended for use on coarse-textured soils.

Do not use in areas or under conditions where Eptam 7E is not recommended. Before use, read Eptam 7E label for additional information and precautions.

** Eptam® is a registered trademark of Stauffer Chemical Company.

NORTRON EC + TCA (TANK-MIX)
DOSAGE TABLE 8 (NORTH DAKOTA AND MINNESOTA ONLY):

	RATE OF NORTRON EC/ACRE		RATE OF TCA/ACRE[1]	
		7-INCH BAND[2]		7-INCH BAND[2]
SOIL TEXTURE	BROADCAST	22" ROW	BROADCAST	22" ROW
Coarse Textured Soils: Sandy loams	1½ Gallons	3¾ Pints	1¼ Gallons	3¼ Pints
Medium Textured Soils: Clay loams and silt loams	2 Gallons	5 Pints	1¼ Gallons	3¼ Pints
Fine Textured Soils: Heavy clays	2½ Gallons	6¼ Pints	1¼ Gallons	3¼ Pints

[1] Based on commercial liquid formulation containing the equivalent of 6 pounds of dry 90% sodium TCA per gallon.
[2] For other band or row widths, adjust the rate in proportion to the area actually treated.

POSTEMERGENCE APPLICATION

(For use in California, Michigan, Ohio, Minnesota, and North Dakota only)

GENERAL INFORMATION

The tank-mixes of NORTRON EC plus BETANEX and NORTRON EC plus BETAMIX applied postemergence broaden and enhance the control of weeds. Furthermore, control of susceptible weeds which may germinate after treatment can be obtained provided overhead moisture is applied to activate NORTRON EC. The choice of tank-mixes is dependent upon weed species present.

NORTRON EC alone is not recommended for postemergence use.

Timing of Application:

Apply the tank-mix NORTRON EC + BETANEX or NORTRON E C + BETAMIX when sugarbeets have at least 4 fully expanded true leaves. Apply at rates recommended in the Dosage Tables. Use the higher rate of NORTRON EC where increased residual weed control is desired. Where Eptam has been applied preplant (fall or spring-applied), do not apply NORTRON EC + BETANEX or BETAMIX tank-mix before the sugarbeets have 6 expanded true leaves. Sugarbeets may be killed or severely injured if treated at earlier stages of crop growth. See Postemergence Use Precautions for additional information on proper use.

Mixing the Spray:

NORTRON EC, BETANEX and BETAMIX are formulated as emulsifiable concentrates and are easily mixed. Add NORTRON EC to the water in the spray tank followed by BETANEX or BETAMIX while agitating the spray solution thoroughly. Refer to the BETANEX or BETAMIX labels for additional precautions and information on mixing.

Spray Equipment:

Apply the mixture using standard low pressure (20–60 psi) spray equipment. Ensure that the sprayer is thoroughly clean. Spray equipment should be carefully calibrated before use and checked frequently during application to see that it is functioning properly. Uniformly apply the recommended rate in 20–60 gallons of water per acre on a broadcast basis. Avoid overlaps, since crop injury may result. When applying in a band, check to make certain that the band width is accurate for the dosage rate being applied. Do not use strainer smaller than 50-mesh.

Moisture Following Application/Residual Weed Control:

Rainfall or sprinkler irrigation within 6 hours of spraying may reduce weed control; however, moisture after this period of time is advantageous for moving NORTRON EC into the 1–2" layer of soil where it can be absorbed by the roots of sprayed and germinating weeds to provide optimum control. One-half inch or more of sprinkler irrigation is required to activate NORTRON EC on most soil types.

Residual control of weeds is dependent upon soil moisture conditions, rate of herbicide used, and soil texture. The activity of NORTRON EC in the soil is reduced as the soil texture becomes finer and organic matter increases.

CAUTION:

MINNESOTA AND NORTH DAKOTA—Do not apply to crop which has been previously treated with NORTRON EC.

CALIFORNIA, MICHIGAN AND OHIO—Following a preemergence treatment of NORTRON EC, do not apply NORTRON EC postemergence where more than 2 gallons were applied preemergence. No more than a total of 2¾ gallons of NORTRON EC should be applied in a single growing season. See Postemergence Use Precautions for additional information on proper use.

NORTRON EC+ BETANEX® (tank-mix)

MICHIGAN, OHIO, MINNESOTA, AND NORTH DAKOTA ONLY

NORTRON EC applied postemergence in a tank-mix with BETANEX broadens and enhances the control of troublesome weeds. Furthermore, preemergence control of susceptible weeds, which may germinate following treatment, can be obtained provided overhead moisture is sufficient to activate NORTRON EC.

EPTAM® (fall-applied) NORTRON EC (spring-applied)
Dosage Table 9 (North Dakota and Minnesota only)

	NORTRON EC Per Acre	
		7-inch Band Width[1]
Soil Texture	Broadcast	22" Row
Coarse Textured Soils: Sands, loamy sands and sandy loams	NOT RECOMMENDED	
Medium Textured Soils: Silt loams and clay loams	2 Gallons	5 Pints
Fine Textured Soils: Heavy clays	2 Gallons	5 Pints

[1] For other band or row widths, adjust the rate in proportion to the area actually treated.

Weed Species Controlled:

Group I: Weeds controlled up to the 6-leaf stage.

Redroot pigweed	*Amaranthus retroflexus*
Wild mustard	*Sinapis arvensis*
Nettleleaf goosefoot	*Chenopodium murale*
London rocket	*Sisymbrium irio*
Common lambsquarters	*Chenopodium album*

Group II: Weeds controlled up to the 4-leaf stage.

Common chickweed	*Stellaria media*
Black nightshade	*Solanum nigrum*
Common ragweed	*Ambrosia artemisiifolia*
Shepherdspurse	*Capsella bursa-pastoris*
Groundcherry	*Physalis lanceifolia*
Pennsylvania smartweed	*Polygonum pensylvanicum*
Ladysthumb	*Polygonum persicaria*

Group III: Weeds controlled up to the 2-leaf stage.

Annual sowthistle	*Sonchus oleraceus*
Kochia*	*Kochia scoparia*
Common purslane	*Portulaca oleracea*
Wild buckwheat	*Polygonum convolvulus*

* Spray kochia while in the rosette stage, less than 1″ in diameter.

NORTRON EC + BETAMIX® (tank-mix)

CALIFORNIA, MICHIGAN, OHIO, MINNESOTA AND NORTH DAKOTA ONLY

NORTRON EC applied postemergence in a tank-mix with BETAMIX broadens and enhances the control of troublesome weeds. Furthermore, preemergence control of susceptible weeds, which may germinate following treatment, can be obtained provided overhead moisture is sufficient to activate NORTRON EC.

NORTRON EC + BETANEX® (TANK-MIX)
DOSAGE TABLE 10 (MICHIGAN, OHIO, MINNESOTA AND NORTH DAKOTA ONLY):

RATE OF NORTRON EC PER ACRE[1]					RATE OF BETANEX PER ACRE[1]				
BROADCAST RATE	BAND[2] WIDTH (IN)	ROW SPACING			BROADCAST RATE	BAND[2] WIDTH (IN)	ROW SPACING		
		22″	28″	30″			22″	28″	30″
3/4–1 Gallons	7	2–2 2/3 Pints	1 1/2–2 Pints	1 1/3–1 3/4 Pints	4 1/2–6 Pints	7	1 1/2–2 Pints	1 1/8–1 1/2 Pints	1–1 1/3 Pints

[1] Use the higher rate on larger weeds and sugar beets.
[2] For other band or row widths, adjust rates in proportion to the area actually treated.

Weed Species Controlled:

Group I: Weeds controlled up to the 6-leaf stage.

Redroot pigweed	*Amaranthus retroflexus*
Wild mustard	*Sinapsis arvensis*
Nettleleaf goosefoot	*Chenopodium murale*
London rocket	*Sisymbrium irio*
Common lambsquarters	*Chenopodium album*

Group II: Weeds controlled up to the 4-leaf stage.

Common chickweed	*Stellaria media*
Black nightshade	*Solanum nigrum*
Common ragweed	*Ambrosia artemisiifolia*
Shepherdspurse	*Capsella bursa-pastoris*
Groundcherry	*Physalis lanceifolia*
Pennsylvania smartweed	*Polygonum pensylvanicum*
Ladysthumb	*Polygonum persicaria*
Prostrate pigweed	*Amaranthus blitoides*
Coast fiddleneck	*Amsinckia intermedia*

Group III: Weeds controlled up to the 2-leaf stage.

Annual sowthistle	*Sonchus oleraceus*
Kochia*	*Kochia scoparia*
Common purslane**	*Portulaca oleracea*
Prostrate knotweed	*Polygonum aviculare*
Wild buckwheat	*Polygonum convolvulus*
Green foxtail***	*Setaria viridis*
Yellow foxtail (pigeongrass)***	*Setaria glauca*
Annual bluegrass***	*Poa annua*
Canarygrass***	*Phalaris canariensis*

* Spray kochia while in the rosette stage, less than 1″ in diameter.
** Group II weed in California.
*** For best control, overhead moisture required.

NORTRON EC + BETAMIX® (TANK-MIX)
DOSAGE TABLE 11 (CALIFORNIA, MICHIGAN, OHIO, MINNESOTA AND NORTH DAKOTA ONLY):

RATE OF NORTRON EC PER ACRE[1]					RATE OF BETAMIX PER ACRE[1]				
BROADCAST RATE	BAND[2] WIDTH (IN)	ROW SPACING			BROADCAST RATE	BAND[2] WIDTH (IN)	ROW SPACING		
		22″	28″	30″			22″	28″	30″
3/4–1 Gallon	7	2–2 2/3 Pints	1 1/2–2 Pints	1 1/3–1 3/4 Pints	4 1/2–6 Pints	7	1 1/2–2 Pints	1–1 1/2 Pints	1–1 1/3 Pints

[1] Use the higher rate on larger weeds and sugarbeets.
[2] For other band or row widths, adjust rates in proportion to the area actually treated.

PREPLANT AND PREEMERGENCE USE PRECAUTIONS:

NORTRON EC applied alone or in tank-mixes according to label directions and under normal growing conditions may cause temporary leaf fusion, distortion and stunting. Crop injury may occur during early growth when crop is stressed due to herbicide residue carryover, highly saline or alkaline soils, unusually cold and wet weather or improperly placed fertilizers or soil insecticides.

Unusually dry, windy weather following application of NORTRON EC, which dries the upper soil layer, may reduce effectiveness.

Where NORTRON EC is used in a tank-mix with TCA, do not use treated sugarbeet tops for feed or forage.

POSTEMERGENCE USE PRECAUTIONS:

Make only one application of NORTRON® E C + BETANEX® or BETAMIX® tank-mix during each growing season.

Do not apply NORTRON EC + BETANEX or BETAMIX tank-mix to sugarbeets later than 90 days prior to harvest.

Crop Planting Precautions: If crop is lost due to unfavorable growth conditions following treatment, do not replant with crops other than sugarbeet in treated land during the same season. If fields are replanted to sugarbeets, reseed into treated band. Do not retreat field with NORTRON EC in the same season.

NORTRON EC + BETANEX OR BETAMIX MAY CAUSE SUGARBEET INJURY OR STAND LOSS IF THE CROP IS UNDER STRESS FROM ONE OR MORE OF THE FOLLOWING CONDITIONS:

Rapid climactic change from cool, overcast days, to hot (80°F or over), bright days. When the air temperature is, or is likely to be, above 80°F on the day of spraying, application should be made in the evening when the temperature is lower.

Frost within seven days following treatment;

Windy or drought conditions;

Use of preplant or preemergence herbicides or other chemicals;

Insect or disease injury;

Close cultivation;

If stress conditions are present, delay application until crop has recovered.

DO NOT OVERTREAT. The use of higher-than-recommended rates may cause crop injury.

Do not spray while dew is present.

Rainfall or spinkler irrigation within 6 hours of application may reduce weed kill.

Do not allow spray drift to contact adjacent crops which may be injured by spray drift.

IMPORTANT

This tank-mix may cause temporary growth retardation and/or chlorosis or tip-burn of sugarbeets. Sugarbeets usually resume normal growth within 10 days.

STORAGE AND DISPOSAL

STORAGE

Store NORTRON EC in a cool place above 34°F.

PESTICIDE DISPOSAL

Do not contaminate water, food or feed by storage or disposal. Pesticide wastes are acutely hazardous. Improper disposal of excess pesticide, spray mixture, or rinsate is a violation of Federal Law. If these wastes cannot be disposed of by use according to label instructions, contact your State Pesticide or Environmental Control Agency, or the Hazardous Waste representative at the nearest EPA Regional Office for guidance.

CONTAINER DISPOSAL

Triple rinse (or equivalent). Then offer for recycling or reconditioning, or puncture and dispose of in sanitary landfill, or by other procedures approved by state and local authorities.

IMPORTANT: READ BEFORE USE

By using this product, user accepts the following conditions, warranty, disclaimer of warranties and limitations of liability.

CONDITIONS: The directions for use of this product are believed to be adequate and should be followed carefully. However, because of extreme weather and soil conditions, manner of use and other factors beyond NOR-AM Chemical Company's control, it is impossible for NOR-AM to eliminate all risks associated with the use of this product. As a result, crop injury or ineffectiveness is always possible.

DISCLAIMER OF WARRANTIES: THERE ARE NO WARRANTIES, EXPRESS OR IMPLIED, OF MERCHANTABILITY OR OF FITNESS FOR A PARTICULAR PURPOSE OR OTHERWISE, WHICH EXTEND BEYOND THE STATEMENTS MADE ON THIS LABEL. No agent of NOR-AM Chemical Company is authorized to make any warranties beyond those contained herein or to modify the warranties contained herein. UNDER NO CIRCUMSTANCES SHALL NOR-AM CHEMICAL COMPANY BE LIABLE FOR ANY INCIDENTAL OR CONSEQUENTIAL DAMAGES RESULTING FROM FAILURE TO USE THIS PRODUCT IN ACCORDANCE WITH LABEL DIRECTIONS.

LIMITATIONS OF LIABILITY: In no event shall NOR-AM Chemical Company's liability for damages arising out of the use of this product exceed the purchase price of the product used.

Also available as Nortron® GS Flowable Herbicide for use on certain grass seed crops in Western Oregon and Western Washington.

PROXOL 80SP

INSECTICIDE

For Farm Crops and Fine Turf Areas

Not for Sale or Use in the State of Maine

ACTIVE INGREDIENT

Dimethyl (2,2,2–trichloro–1–hydroxyethyl) phosphonate	80%
INERT INGREDIENTS	20%
Total	100%

EPA Reg. No. 45639-123

Keep Out of Reach of Children

WARNING—AVISO

PRECAUCION AL USUARIO: Si usted no lee ingles, no use este producto hasta que el etiqueta haya sido explicado ampliamente.

STATEMENT OF PRACTICAL TREATMENT

PROLONGED EXPOSURE WILL RESULT IN CHOLINESTERASE DEPRESSION.

IF SWALLOWED: INDUCE VOMITING AND GET MEDICAL ATTENTION IMMEDIATELY.

IF IN EYES: FLUSH WITH WATER FOR AT LEAST 15 MINUTES. GET MEDICAL ATTENTION IMMEDIATELY.

IF ON SKIN: REMOVE CONTAMINATED CLOTHING. WASH SKIN WITH SOAP AND WATER. GET MEDICAL ATTENTION IMMEDIATELY.

NOTE TO PHYSICIAN: ATROPINE SULFATE IS ANTIDOTAL. 2–PAM IS ALSO ANTIDOTAL AND MAY BE ADMINISTERED IN CONJUNCTION WITH ATROPINE.

PRECAUTIONARY STATEMENTS

HAZARDS TO HUMANS AND DOMESTIC ANIMALS

WARNING

May be fatal if swallowed. Do not get in eyes, on skin, or on clothing. Avoid breathing mist or dust. Wash thoroughly with soap and water after handling. Wash contaminated clothing with soap and hot water before reuse.

PRECAUTIONS

Use with adequate ventilation. Keep children and pets off treated areas until spray has dried. Do not use on other crops used for food or forage. Use only according to label directions. Application at rates above those shown may result in illegal crop residues. Do not treat food crops grown in the greenhouse. Do not apply this product in such a manner as to directly or through drift expose workers or other persons. The area being treated must be vacated by unprotected persons.

REENTRY STATEMENT: Do not enter treated areas without protective clothing until sprays have dried. Because certain states may require more restrictive reentry intervals for various crops treated with this product, consult your State Department of Agriculture for further information. Written or oral warnings must be given to workers who are expected to be in a treated area or in an area about to be treated with this product. When oral warnings are given, warnings shall be given in a language customarily understood by workers. Oral warnings must be given if there is reason to believe that written warnings cannot be understood by workers.

ENVIRONMENTAL HAZARDS: This product is toxic to birds and other wildlife. Keep out of lakes, streams, ponds, tidal marshes and estuaries. Do not apply where runoff is likely to occur. This product is toxic to bees exposed to direct treatment or residues on blooming crops or weeds and should not be applied when weather conditions favor drift from areas treated. Do not contaminate water by cleaning of equipment or disposal of wastes.

STORAGE AND DISPOSAL

Do not contaminate water, food or feed by storage or disposal.

STORAGE
Store in original container and keep closed. Store in a cool, dry place.

PESTICIDE DISPOSAL
Wastes resulting from the use of this product may be disposed of on site or at an approved waste disposal facility.

CONTAINER DISPOSAL
Completely empty bag into application equipment. Then dispose of empty bag in a sanitary landfill or by incineration, or, if allowed by State and local authorities, by burning. If burned, stay out of smoke.

DIRECTIONS FOR USE

It is a violation of Federal law to use this product in a manner inconsistent with its labeling.

IMPORTANT
Read these entire Directions and Conditions of Sale, including the Warranties, and Limitation of Liability provisions, before using Proxol 80 SP. For agricultural use and for sale to service persons. Do not store in or around the house. Proxol insecticide offers the unique advantage of not significantly affecting beneficial insects (parasites, predators, and pollinators) especially when applied at the minimum recommended rates per acre. Thus, Proxol supplements the natural control factors, instead of destroying them. This advantage is lost when Proxol is used in conjunction with parathion or other non-selective insecticides. Accordingly, for most effective and most economical control, Proxol should not be used in conjunction with or alternated in a spray schedule with pesticides destroying parasites, predators, and other beneficial insects. This applies to alfalfa and cotton in particular. While Proxol is not recommended for the control of aphids, mites, and cabbage looper, some suppression of these pests may result from recommended dosages.

LAWN CARE INDUSTRY
For the best results with Proxol 80 SP against white grubs (May beetle, European chafer, Asiatic Garden beetle, Japanese beetle, Northern Masked chafer, Oriental beetle), it is recommended that post treatment watering-in take place immediately after treatment. Post treatment watering-in may also be utilized up to 24 hours after application. Consideration must be given to species of grubs, and the level of infestation before irrigation treatment is performed.

COMPATIBILITY
Proxol 80 SP may be applied with most commonly used insecticides. Do not mix with alkaline materials such as lime and lime sulfur.

MIXING
Proxol 80 SP dissolves readily in water and is suitable for use in all power-operated ground sprayers and aircraft sprayers. To dissolve in water, pour the required amount of Proxol 80 SP into full amount of water and then agitate. Use immediately after mixing. It is recommended that Proxol 80 SP be used in sprayers equipped with 50 mesh screens. If 100 mesh screens are used, some screen clogging may occur.

RECOMMMENDED APPLICATIONS

PREMISE SPRAY FOR FLY CONTROL
For Fly Control in and Around Farm Buildings (Except Dairy Barns and Poultry Houses): Mix 5 pounds of Proxol 80 SP in 40 gallons of water (equivalent to 5 level tablespoonfuls per gallon). Thoroughly wet all areas to be treated by spraying 1 gallon per 500 square feet of surface area or to the point of run-off. Apply to floors, walls, ceiling, around windows, doors, and other areas where flies congregate. Also spray stock pens (after removal of animals), garbage areas, and other locations where refuse has collected. Repeat applications every 1 to 2 weeks or as fly populations increase. For longer residual control, add 1 pound of sugar to each gallon of spray solution.

Restrictions (Premise Spray)
Remove animals before spraying barn. Do not treat portions of buildings where poultry feed. Direct applications to walls, floors, or other surfaces previously treated with lime, whitewash, or other alkaline materials may be ineffective. Do not contaminate feed, water, or milk-handling equipment.

RECOMMENDED APPLICATIONS—TEES, GREENS, FAIRWAYS, LAWNS AND OTHER FINE TURF

INSECT	RATE: AREA	RATE: AMOUNT	REMARKS
Cutworms Sod webworms (lawn moths) Armyworms	1,000 sq. ft. 5,000 sq. ft. 10,000 sq. ft. 20,000 sq. ft. 30,000 sq. ft. 40,000 sq. ft.	$1\frac{1}{2}$ to $3\frac{3}{4}$ oz. $\frac{1}{2}$ lb 1 lb 2 lb 3 lb 4 lb	Apply by means of any commercial power ground sprayer. Do not use Proxol 80 SP in hose proportioners. **For greens or other small areas**—Mix specified amount of Proxol 80 SP in 5 to 10 gallons of water and apply uniformly to 1,000 sq. ft. **For fairways and larger areas**—Apply specified amount using sufficient water for complete coverage but not less than 10 gallons per acre. For best results, mow the turf and rake dead grass from damaged spots: water the turf and allow to dry before spraying. Do not water again until necessary. Apply Proxol 80 SP when insect populations reach levels sufficient to warrant control measures.
White grubs	1,000 sq. ft.	$3\frac{3}{4}$ oz.	Mix specified amount of Proxol 80 SP in 15 to 30 gallons of water and apply uniformly to 1,000 sq. ft. of turf. Thoroughly irrigate turf immediately after application. **LAWN CARE INDUSTRY:** For the best results with Proxol 80 SP against White grubs (May beetle, European chafer, Asiatic Garden beetle, Japanese beetle, Northern Masked chafer, Oriental beetle), it is recommended that post watering-in take place immediately after treatment. Post treatment watering-in may be utilized up to 24 hours after application. Consideration must be given to species of grubs, and the level of infestation, before irrigation treatment is performed. After the grubs have been identified, count the number of grubs per square foot to ascertain timing of the irrigation treatment. The watering-in schedule is shown below:

Type of White Grub	White Grub Counts/Sq. Ft.	Watering-in Treatment
European chafer, Asiatic Garden beetle,	No more than 59	Immediate

RECOMMENDED APPLICATIONS—TEES, GREENS, FAIRWAYS, LAWNS AND OTHER FINE TURF

INSECT	RATE: AREA	RATE: AMOUNT	REMARKS: Type of White Grub*	White Grub Counts/Sq. Ft.	Watering-in Treatment
			Japanese beetle, Northern Masked chafer, Oriental beetle (one-year cycle)	No more than 23	
			May beetle (three-year cycle)	No more than 29	Immediate
				No more than 11	Within 24 hours
			Higher grub counts than listed of either species, may require repeat treatments (late April to early May or early August to September) (late April to early May . . . late April to early May) to reduce grub populations below threshold levels. Threshold levels of grubs, dictating control, are 6–8 or more one-year-cycle larvae (Japanese beetle, Masked chafer, etc. . . .) per square foot, or 3 or more, three-year-cycle grubs (May beetle) per square foot.		
European Crane Fly larvae	1,000 sq. ft.	$3\frac{3}{4}$ oz.	Applications should be made when damage is first noticed in the lawn or other fine turf area. For best results, apply specified amount of Proxol 80 SP in 15 to 30 gallons of water and apply uniformly to 1,000 sq. ft. of turf. Crane fly larvae are usually located in the top one inch of soil and thatch. If necessary, irrigate turf or lawn after application to move Proxol 80 SP into the zone of crane fly larvae infestation.		

FOR FARM CROPS

CROP	INSECT	OUNCES PROXOL 80 SP	REMARKS
FIELD CROPS			
Alfalfa	Alfalfa caterpillar	$7\frac{1}{2}$ to 10	Apply specified dosage per acre using sufficient water for complete coverage but not less than 1 gallon per acre. Three applications may be made per cutting with the last application up to the day of cutting.
Clover	Alfalfa webworm	5 to 20	
(including mixed stands with grasses)	Western yellow-striped armyworm	10	
	Beet armyworm		
	Leafhoppers		
	Variegated cutworm	10 to 20	
	Armyworm		
	Alfalfa plant bug		
	Lygus bugs, Stink bugs		
	Tarnished plant bug	20	
Barley	Armyworms		Apply specified dosage per acre using sufficient water for complete coverage but not less than 1 gallon per acre. Three applications may be made per season but not within 21 days of harvest.
Flax	Beet webworm		
Oats	Variegated cutworm	10 to 20	
Wheat	Bertha armyworm		
	Diamondback moth (NOTE: Applications may be made without removal of grazing livestock.)	20	
Corn Field, Sweet, and Popcorn	Armyworms		For early application when plants are 3 to 12 inches high, direct a band spray 10 to 12 inches wide, toward the soil, over the tops of the plants. Apply specified dosages per acre in sufficient water for complete coverage but not less than 1 gallon per acre. Three applications may be made per season. Later applications may be made as full coverage spray up to day of harvest.
	Cutworms	10 to 20	
FIELD CROPS			
Cotton	Cotton fleahopper	5 to 20	Apply specified dosage per acre using sufficient water for complete coverage but not less than 1 gallon per acre. Do not pasture treated fields within 14 days of application. Do not apply within 7 days of picking. Proxol application may occasionally cause marginal leaf burn. NOTE: For control of light to moderate infestation of cotton fleahoppers, use 5 ounces of Proxol 80 SP per acre. Higher rates may be required if heavy infestations or migrations occur.
	Cotton leafworm		
	Darkling ground beetle		
	Western yellow-striped armyworm	10 to 20	
	Beet armyworm		
	Southern garden leaf-hopper	20	
	Black fleahopper complex		
	Cotton leaf perforator		
	Leaf roller, Lygus bugs		
	Stink bugs	20 to 30	
	Salt-marsh caterpillar	30	

CROP	INSECT	OUNCES PROXOL 80 SP	REMARKS
Grasses (including rangeland and pastures)	Leafhoppers	10 to 20	Apply specified dosage per acre using sufficient water for complete coverage but not less than 1 gallon per acre. Three applications may be made per cutting with the last application up to the day of cutting. Do not apply more than 3 times per season if crop is not cut for hay.
	Armyworm		
	Fall armyworm		
	Plant bugs	20	
	NOTE: Applications may be made without removal of grazing livestock.)		
Safflower	Armyworms		Apply specified dosage per acre in each of two applications, using sufficient water for complete coverage but not less than 1 gallon per acre. Do not apply more than twice per year or later than bloom stage. Make first application approximately 2 weeks prior to bloom—second application at onset of bloom.
	Lygus bugs		
	Thrips		
	Variegated cutworm	20 to 30	
Sugar beets	Beet webworm		Apply specified dosage per acre using sufficient water for complete coverage but not less than 1 gallon per acre. Repeat as necessary. Do not apply within 14 days of harvest.
	Variegated cutworm	10 to 20	
	Dipterous leaf miners	20	
	Alfalfa webworm		
	Beet armyworm	20 to 30	
	Salt-marsh caterpillar	30	
Tobacco	Budworm		Apply specified dosage per acre using sufficient water for complete coverage but not less than 1 gallon per acre. For budworm control make application directly into bud. Repeat as necessary. Do not treat within 3 days of harvest.
	Hornworm	20	
	Green June beetle larvae	10	Apply specified dosage in 100 gallons of water per 100 square yards to uprooted areas of plant beds.
	Black cutworm	20	Apply specified dosage per acre. Application may be made at transplanting or when cutworm damage is observed in the field. For best results, use 25 to 50 gallons of water per acre on a 12 to 14 inch band and spray the entire plant and soil around the plant. Treat in early evening for best results, since cutworms are most active at night. Enide® 90W at 4.4 to 6.6 pounds per acre may be applied over the top after transplanting in combination with Proxol. Enide may be applied from transplanting to 7 days after transplanting. When Enide 90W is applied more than 2 days after transplanting, tillage is required immediately before or at time of application to destroy germinating weed seeds.
SEED FIELD CROPS			
Alfalfa			
Clover	Armyworms		Apply specified dosage per acre using sufficient water for complete coverage but not less than 1 gallon per acre. Repeat as necessary up to 7 days before harvest. Chaff from alfalfa and clover seed crop may be used for feed or forage, however, do not cut green crop for these purposes. Do not pasture or use treated soybeans for feed, food, forage, or oil purposes.
	Lygus bugs		
	Stink bugs		
	Variegated cutworm	20 to 30	
Soybeans	Armyworms		
	Dipterous Leaf Miners		
	Lygus bugs		
	Stink bugs		
	Variegated cutworm		
FRUIT			
Blueberries	Spanworm	16	Apply specified dosage per acre using sufficient water for complete coverage, but no less than 1 gallon per acre. Make a prebloom application as soon as insects appear. A second application may be made during bloom but not less than 7 days after first application. Do not apply after bloom stage.
	Gypsy Moth	20	
Citrus (California only)	Fruit tree leafroller		Apply specified dosage per acre in sufficient water for complete coverage, but not less than 1 gallon per acre. One application may be made by aircraft up to 7 days before harvest. Two applications with ground equipment may be made in addition to one aerial application. However, if more than one application is made do not apply last application within 21 days of harvest.
	Omniverous leafroller		
	Black Swallow tail	40	
	Avocado leafroller		
	Orange Tortrix		
	Citrus cutworm		
	Citrus looper	80	
VEGETABLES			
Beans (dry and succulent)*	Western bean cutworm	10 to 20	Apply specified dosage per acre using sufficient water for complete coverage but not less than 1 gallon per acre. Repeat as necessary. Do not apply within 14 days of harvest. *Identified as *Cicer* sp. (including chickpeas and garbanzo beans); *Phaseolus* sp. (including kidney beans, lima beans, mung beans, navy beans, pinto beans, snap beans, and wax beans); *Vicia* sp. (including broad beans, fava beans); *Vigna* sp. (including asparagus beans).
	Armyworms		
	Dipterous leaf miners		
	Lygus bugs		
	Mexican bean beetle		
	Stink bugs		
	Variegated cutworm	20 to 30	

CROP	INSECT	OUNCES PROXOL 80 SP	REMARKS
Lima beans	Western beat cutworm Armyworms Dipterous leaf miners Lygus bugs Mexican bean beetle Stink bugs Variegated cutworm	10 to 20 20 to 30	Apply specified dosage per acre using sufficient water for complete coverage but not less than 1 gallon per acre. Up to five applications may be made per season. Do not apply within 3 days of harvest.
Cowpeas (Southern peas, Blackeyed peas, Crowder peas)	Armyworms Dipterous leaf miners Imported cabbageworm Lygus bugs, Mexican bean beetle, Stink bugs, Variegated cutworm	20 to 30	Apply specified dosage per acre using suffiicent water for complete coverage but not less than 1 gallon per acre. Do not apply more than twice per season nor within 14 days of harvest.
Brussels sprouts Cabbage Cauliflower	Western yellow-striped armyworm Imported cabbageworm Variegated cutworm Diamondback moth	10 10 to 20 20	Apply specified dosage per acre using sufficient water for complete coverage but not less than 1 gallon per acre. Repeat as necessary. Do not apply within 21 days of harvest.
Carrots	Western yellow-striped armyworm Imported cabbageworm Variegated cutworm Diamondback moth Dipterous leaf miners Beet armyworm Lygus bugs Salt-marsh caterpillar	10 10 to 20 20 20 to 30 30	Apply specified dosage per acre using sufficient water for complete coverage but not less than 1 gallon per acre. Repeat as necessary. Do not apply within 28 days of harvest. Do not use tops for food or feed.
Collards Lettuce	Thrips Variegated cutworm Armyworms Beet Webworm Diamondback moth Dipterous leaf miners Salt-marsh caterpillar	10 to 20 20	Apply specified dosage per acre using sufficient water for complete coverage but not less than 1 gallon per acre. Repeat as necessary until heads begin to form. Do not apply within 28 days of harvest.
Peppers	Dipterous leaf miners Pepper maggot Serpentine leaf miners	20	Apply specified dosage per acre using sufficient water for complete coverage but not less than 1 gallon per acre. Repeat as necessary. Do not apply within 21 days of harvest.
Pumpkin	Variegated cutworm Squash bug	10 to 20 20	Apply specified dosage per acre using sufficient water for complete coverage but not less than 1 gallon per acre. Do not apply more than three times per season nor within 3 days of harvest.
Table beets	Variegated cutworm Alfalfa webworm Beet webworm Dipterous leaf miners Salt-marsh caterpillar Beet armyworms Lygus bugs	10 to 20 20 20 to 30	Apply specified dosage per acre using sufficient water for complete coverage but not less than 1 gallon per acre. Repeat as necessary. Do not apply within 28 days of harvest. Do not eat tops.
Tomatoes	Serpentine leaf miners Tomato hornworms Dipterous leaf miners	20	Apply specified dosage per acre using sufficient water for complete coverage but not less than 1 gallon per acre. Repeat as necessary. Do not apply within 21 days of harvest.
ORNAMENTALS Flowers Shrubs and Trees	Nantucket pine tip moth Zimmerman pine moth Armyworms Bagworms Climbing cutworms Dipterous leaf miners Lygus bugs Stink bugs Tarnished plant bug Tobacco budworm Webworms	2 2 to 3	Mix specified dosage in 10 gallons of water and spray plants until they are thoroughly wet. Repeat as necessary. CAUTION: Phytotoxicity has occurred on certain varieties of carnations and zinnias. When large plantings of these flowers are to be sprayed for the first time, treat only a few plants and observe for 4 to 5 days before spraying the entire planting. Injury from spray droplets may occur on leaves of certain trees.
Narcissus	Narcissus bulb fly	2	Apply specified dosage in 10 gallons of water as a drench per 100 feet of row. Direct steam to base of plants at beginning of adult fly activity (early May to June). Repeat treatment annually.

Rate/Acre Ounces Proxol 80 SP	Pounds of Proxol 80 SP to Treat:		
	5 Acres	10 Acres	15 Acres
5	1½	3⅛	4⅔
7½	2⅓	4⅔	7
8	2½	5	7½
10	3⅛	6¼	9⅜
16	5	10	15
20	6¼	12½	18¾
30	9⅜	18¾	28⅛

NOTE: When using a more convenient whole number or fraction of pounds of formulated material do not exceed maximum registered rate.

DOSAGE

Use specified dosage of Proxol 80 SP in the amount of water necessary to give complete coverage of foliage. The type of equipment used will determine the concentration required.

SPRAYING

Complete coverage and thorough application are essential for most effective results. Schedule applications in accordance with local conditons. Consult your State Agricultural Extension Service or Experiment Station for specific timing of applications in your area. NOTE: DO NOT ALLOW PROXOL SPRAY TO DRIFT ONTO SORGHUM. TO DO SO MAY CAUSE BURNING OF THE SORGHUM OR MILO. Spray droplets from applications of Proxol 80SP may cause spotting of paint finishes on aluminum and steel siding, automobiles, trucks and tractors. Do not allow spray droplets to contact metal painted surfaces. If accidental exposure does occur, the painted surface should be thoroughly washed immediately after exposure.

IMPORTANT: READ BEFORE USE

By using this product, user accepts the following conditions, warranty, disclaimer of warranties and limitations of liability.

CONDITIONS:

The directions for use of this product are believed to be adequate and should be followed carefully. However, because of extreme weather and soil conditions, manner of use and other factors beyond NOR-AM Chemical Company's control, it is impossible for NOR-AM to eliminate all risks associated with the use of this product. As a result, crop injury or ineffectiveness is always possible.

DISCLAIMER OF WARRANTIES:

THERE ARE NO WARRANTIES, EXPRESS OR IMPLIED, OF MERCHANT ABILITY OR OF FITNESS FOR A PARTICULAR PURPOSE OR OTHERWISE, WHICH EXTEND BEYOND THE STATEMENTS MADE ON THIS LABEL. No agent of NOR-AM Chemical Company is authorized to make any warranties beyond those contained herein or to modify the warranties contained herein. UNDER NO CIRCUMSTANCES SHALL NOR-AM CHEMICAL COMPANY BE LIABLE FOR ANY INCIDENTAL OR CONSEQUENTIAL DAMAGES RESULTING FROM FAILURE TO USE THIS PRODUCT IN ACCORDANCE WITH LABEL DIRECTIONS.

LIMITATIONS OF LIABILITY:

In no event shall NOR-AM Chemical Company's liability for damages arising out of the use of this product exceed the purchase price of the product used.

SPIN-AID®

HERBICIDE

FOR AGRICULTURAL USE ONLY

Postemergence herbicide for control of weeds in spinich (grown for processing and seed only) and red beets

ACTIVE INGREDIENT:

Phenmedipham*	15.9%
INERT INGREDIENTS	84.1%
TOTAL	100.0%

* Methyl *m* -hydroxycarbanilate *m* -methylcarbanilate (ester)

Contains 1.3 lbs. active ingredient per gallon.

EPA Reg. No. 45639-76 EPA Est. No. 407-MN-1

U.S. Patent No. 3,404,975 U.S. Patent No. 3,692,820

Keep Out of Reach of Children

WARNING—AVISO

STATEMENT OF PRACTICAL TREATMENT

IF SWALLOWED: Call a physician or Poison Control Center. Drink 1 or 2 glasses of water and induce vomiting by touching back of throat with finger. Do not induce vomiting or give anything by mouth to an unconscious person.

IF IN EYES: Flush with plenty of water. Get medical attention.

IF ON SKIN: Wash immediately with plenty of soap and water.

PRECAUCION AL USUARIO: Si usted no lee ingles, no use este producto hasta que le etiqueta haya le sico explicada ampliamente.

PRECAUTIONARY STATEMENTS

HAZARDS TO HUMANS

Harmful if swallowed. May produce severe irritation of eyes and irritation of skin. Avoid breathing spray mist. Avoid contact with skin and eyes.

In case of eye contact, flush with plenty of water for 15 minutes and get medical attention.

For skin, wash immediately with soap and water.

Avoid contamination of feed and foodstuffs.

ENVIRONMENTAL HAZARDS

This product is toxic to fish. Keep out of lakes, streams or ponds. Do not apply where runoff is likely to occur. Do not apply when weather conditions favor drift from areas treated.

DO NOT REUSE THIS CONTAINER

GENERAL INFORMATION

SPIN-AID + up,TM is a selective postemergence herbicide for use in spinach (grown for processing and seed) and red (table) beets. Note: Use SPIN-AID on spinach only during the months September–May when temperatures are below 75°F. SPIN-AID is effective for control of the following weeds:

Wild mustard	*Sinapis arvensis*
Common lambsquarters	*Chenopodium album*
Shepherdspurse	*Capsella bursa-pastoris*
London rocket	*Sisymbrium irio*
Nettleleaf goosefoot	*Chenopodium murale*
Groundcherry	*Physalis lanceifolia*
Coast fiddleneck	*Amsinckia intermedia*
Common Chickweed	*Stellaria media*
Purslane	*Portulaca oleracea*
Common ragweed	*Ambrosia artemisiifolia*
Annual sowthistle	*Sonchus oleraceus*

For best results spray when the weeds are at the two-leaf stage.

USE PRECAUTIONS

DO NOT APPLY **SPIN-AID** TO SPINACH LATER THAN 40 DAYS PRIOR TO HARVEST. DO NOT APPLY **SPIN-AID** TO RED BEETS LATER THAN 60 DAYS PRIOR TO HARVEST.

SPIN-AID MAY CAUSE INJURY IF THE CROP IS UNDER STRESS FROM ONE OR MORE OF THE FOLLOWING CONDITIONS:

Rapid climatic changes from cool, overcast days to hot (75°F or over), bright days. Windy conditions or drought.

Use of a preplant or preemergence herbicide or other chemicals.

Insect or disease injury.

Close cultivation.

If stress conditions are present, delay application in order to give plants a chance to recover.

If extreme weather conditions are of short duration, delay spraying until the end of such a period.

If SPIN-AID application must be made on days with extreme temperature and/or brightness, delay spraying until evening.

DO NOT OVERTREAT: The use of higher than recommended rates may cause injury.

Do not spray while dew is present.

Rainfall within 6 hours of spraying may reduce weed kill. Do not allow spray drift to contact adjacent crops which may be injured by spray drift.

IMPORTANT

SPIN-AID may cause temporary growth retardation and/or chlorosis or tip-burn. Crops usually resume normal growth within 10 days.

DIRECTIONS FOR USE

It is a violation of Federal Law to use this product in a manner inconsistent with its labeling. Read the entire Directions for Use before using this product.

REENTRY STATEMENT

Reentry into treated areas is prohibited until leaves are completely dry or until 24 hours after application.

Do not enter treated areas without protective clothing.

Because certain states may require more restrictive entry intervals for various crops treated with this product, consult your State Department of Agriculture for further information.

Written or oral warnings must be given to workers who are expected to be in a treated area or in an area about to be treatedwith this product. (Indicate specific oral warnings which inform workers of areas or fields that may not be entered without specific protective clothing, period of time field must be evacuated and appropriate actions to take in case of accidental exposure.) When oral warnings are given, warnings shall be given in a language customarily understood by workers. Oral warnings must be given if there is reason to believe that written warnings cannot be understood by workers. Written warnings must include this information:

WARNING: Areas treated with SPIN-AID on (date of application). Do not enter without appropriate protective clothing for 24 hours or until leaves are completely dry. In case of accidental exposure see Statement of Practical Treatment on the label.

When used as directed SPIN-AID is selective in spinach and red beets past the 4–6 true leaf stage. Crops may be severely injured if treated before the 4–6 true leaf stage. The stage of growth of the weeds is very important for satisfactory control. For best results, spray when the weeds are at the two-true leaf stage. Best results are obtained when the weeds are actively growing and are not under water or heat stress.

MIXING THE SPRAY: MAKE SURE THE SPRAYER IS CLEAN

SPIN-AID is an emulsifiable concentrate. The formulation contains sufficient wetting agents for optimum coverage. Do not add additional wetting agents or other spray adjuvants. Add sufficient water to fill the lines. Then add the desired amount of SPIN-AID and the remaining quantity of water with the bypass agitator running. By-pass agitation is sufficient. Mechanical agitation is not necessary. Only use freshly prepared spray emulsions.

Always spray immediately after preparing the spray solution. Prepare only enough spray solution to last less than four hours.

RATE OF APPLICATION:

By Ground: Apply SPIN-AID at the rate of 3 to 6 pints per acre in 11 to 22 gallons of water broadcast basis. For band application see dosage chart. TOO MUCH WATER MAY CAUSE PRECIPITATION.

By Air: Apply SPIN-AID at the rate of 3 to 6 pints per acre using 5 to 20 gallons of spray per acre.

Use the 6 pint rate only on well established crops which are not under stress.

Avoid phytotoxic spray drift to non-target crops during application of SPIN-AID.

DO NOT APPLY WHEN WIND SPEED IS OVER 10 MILES PER HOUR. AVOID APPLICATIONS WHEN CONDITIONS FAVOR DRIFT.

For further information, contact your County Agricultural Agent, Farm Advisor or NOR-AM Chemical Co.

DOSAGE CHART FOR BAND APPLICATIONS

Broadcast Equivalent	Band Rate: Band Width	Band Rate: Row Spacing 26″
3 pints/acre	5″	9.2 fl. oz.
	7″	12.9 fl. oz.
6 pints/acre	5″	18.5 fl. oz.
	7″	25.8 fl. oz.

STORAGE AND DISPOSAL

STORAGE

Store in original container and keep closed. Store in a cool, dry place. Do not use or store near heat or open flame.

PESTICIDE DISPOSAL

Do not contaminate water, food or feed by storage or disposal. Wastes resulting from the use of this product may be disposed of on site or at an approved waste disposal facility.

CONTAINER DISPOSAL

Triple rinse (or equivalent). Then offer for recycling or reconditioning, or puncture and dispose of in a sanitary landfill, or by other procedures approved by state and local autorities.

IMPORTANT: READ BEFORE USE

By using this product, user accepts the following conditions, warranty, disclaimer of warranties and limitations of liability.

CONDITIONS: The directions for use of this product are believed to be adequate and should be followed carefully. However, because of extreme weather and soil conditions, manner of use and other factors beyond NOR-AM Chemical Company's control, it is impossible for NOR-AM to eliminate all risks associated with the use of this product. As a result, crop injury or ineffectiveness is always possible.

DISCLAIMER OF WARRANTIES:

THERE ARE NO WARRANTIES, EXPRESS OR IMPLIED, OF MERCHANTABILITY OR OF FITNESS FOR A PARTICULAR PURPOSE OR OTHERWISE, WHICH EXTEND BEYOND THE STATEMENTS MADE ON THIS LABEL. No agent of NOR-AM Chemical Company is authorized to make any warranties beyond those contained herein or to modify the warranties contained herein. UNDER NO CIRCUMSTANCES SHALL NOR-AM CHEMICAL COMPANY BE LIABLE FOR ANY INCIDENTAL OR CONSEQUENTIAL DAMAGES RESULTING FROM FAILURE TO USE THIS PRODUCT IN ACCORDANCE WITH LABEL DIRECTIONS.

LIMITATIONS OF LIABILITY:

In no event shall NOR-AM Chemical Company's liability for damages arising out of the use of this product exceed the purchase price of the product used.

VORLEX®

SOIL FUMIGANT

FOR AGRICULTURAL USE ONLY

ACTIVE INGREDIENTS:	
1,3-dichloropropene and related chlorinated	
C_3 hydrocarbons	40.0%
Methyl isothiocyanate	20.0%
INERT INGREDIENTS:	40.0%
TOTAL	100.0%

E.P.A. Reg. No. 45639-90 E.P.A. Est. No. 08800-IL-1

U.S. Patent Nos. 3,205,129 3,113,908 3,240,666

Keep Out of Reach of Children

DANGER–POISON

PELIGRO — VENENO

STATEMENT OF PRACTICAL TREATMENT

IF INHALED, remove to fresh air. If not breathing, give artificial respiration, preferably mouth-to-mouth. If breathing is difficult, give oxygen. Call a physician.

IF ON SKIN, immediately flush skin with plenty of water for at least 15 minutes while removing contaminated clothing and shoes. Call a physician. If water is not immediately available, remove excess chemical from skin with sorbent material such as a towel or dry soil, then proceed at once to location where water is available and thoroughly wash contaminated skin with plenty of water. Call a physician.

IF IN EYES, immediately flush eyes with plenty of water or a 5% solution of sodium bicarbonate for at least 15 minutes: call a physician. Continue treatment with ophthalmic hydro-cortisone ointment.

IF SWALLOWED, immediately call a physician or poison control center. Do not induce vomiting. Note to physician: empty stomach contents by gastric lavage. Never give anything by mouth to an unconscious person.

PRECAUCION AL USARIO: Si usted no lee inglés, no use este producto hasta que la etiqueta haya sido explicada ampliamente.

PRECAUTIONARY STATEMENTS

HAZARDS TO HUMANS AND DOMESTIC ANIMALS

HAZARDOUS VAPOR AND LIQUID. MAY BE FATAL IF INHALED, SWALLOWED, OR ABSORBED THROUGH SKIN. CAUSES BURNS OF SKIN AND EYES. MAY PRODUCE KIDNEY AND LIVER DAMAGE UPON PROLONGED EXPOSURE.

Do Not Breathe Vapor. Do Not Get In Eyes, On Skin, Or On Clothing. Do Not Take Internally. Use Only With Adequate Ventilation. Wear Eye And Skin Protection Necessary To Prevent Contact When Handling. Wash Thoroughly After Handling And Before Eating Or Smoking.

If protective gear, such as boots or gloves, becomes contaminated, immediately wash with soap and water. Never wear protective gear having the odor of VORLEX. Aerate and wash all protective gear thoroughly after use until all odor is gone. Render unusable and dispose of contaminated leather goods, including shoes.

GENERAL INFORMATION

VORLEX SOIL FUMIGANT is used as a preplanting treatment in both field and greenhouse soils.

In the soil, when used as directed, VORLEX vaporizes and diffuses, helping to control plant parasitic nematodes and harmful fungi causing damping-off, root-rot and wilt. When used as directed, VORLEX also helps to control wireworms and symphylans. When used as directed, VORLEX helps control weeds such as pigweed, dandelion, lambsquarters, bermudagrass, purslane, crabgrass, quackgrass, spurge, fleabane, sedge, chickweed, ragweed, mustard, Poa spp., smartweed, wild radish, dodder, velvetleaf and shepherdspurse.

PHYSICAL OR CHEMICAL HAZARDS

FLAMMABLE

DO NOT USE, POUR, SPILL, OR STORE NEAR HEAT OR OPEN FLAME. DO NOT CUT OR WELD CONTAINER. FIRE OR EXPLOSION MAY RESULT EVEN IF CONTAINER IS EMPTY.

ENVIRONMENTAL HAZARDS

This product is toxic to fish and wildlife. Do not spill or empty into streams, ponds or any other body of water. Do not contaminate water or wetlands by cleaning of equipment, disposal of wastes, or direct application.

HANDLING PRECAUTIONS

DO NOT INHALE VAPORS—NIOSH- or MESA-approved respiratory protection should be worn when VORLEX liquid soil fumigant is exposed to the atmosphere or when conducting operations which vent to the atmosphere. A NIOSH- or MESA-approved half-faced respirator with chemical worker's goggles or full-face respirators shall be used during small spills, repairs, calibrations, transfers, sampling and when working in poorly ventilated areas. When in use canisters or cartridges shall be replaced daily or sooner if specified by manufacturer or at first sign of odor breakthrough, whichever comes first. NIOSH-approved cartridges, such as a WILLSON-R21 will be adequate for short-term situations such as listed above.

Where very high concentrations of vapors might be expected (such as large spills in poorly ventilated areas) a self-contained or air-supplied respirator should be used.

DO NOT GET ON SKIN—When handling or working with VORLEX, wear clean body covering including gloves and heavy footwear. Immediately remove contaminated coverings. Aerate and wash all protective clothing and gear thoroughly after use.

There are no protective clothing materials that are completely impervious. Thin layers of polyethylene (minimum 1 mil.), give short-term protection and shall be immediately discarded upon contamination. Heavy (3 + mil) polyethylene, provides longer term protection. Leather gives no protection. If gear becomes contaminated, immediately wash with soap and water. Never wear protective gear having odor of VORLEX. Wash and aerate all protective gear thoroughly after use until all odor is gone.

DO NOT GET IN EYES—Wear eye protection such as chemical worker's goggles or full-face respirator when handling VORLEX.

DO NOT SWALLOW—Do not use the mouth to siphon VORLEX from containers or to blow out clogged lines, nozzles, etc.

For Chemical Emergency—Spill, Leak, Fire, Exposure Or Accident Call 800/424-9300 Day Or Night

DIRECTIONS FOR USE

It is a violation of Federal Law to use this product in a manner inconsistent with its labeling.

USE PRECAUTIONS: As with most soil fumigants, application of VORLEX may temporarily raise the level of ammonia nitrogen and soluble salts in the soil which may result in plant injury. This is most likely to occur when heavy rates of fertilizer or fumigant are applied to soils that are either cold, wet, acid or high in organic matter.

Do not add fertilizers containing ammonia or ammonium salts or undecomposed manure to the soil at or near the time of fumigation. Use only fertilizers containing nitrates until the soil temperature is above 65°F and the crop is well established.

Do not allow liquid or vapors to come in contact with lawn and desirable plants; do not apply on areas beneath which roots of desirable plants extend.

Do not use in greenhouses which contain living plants.

Do not fumigate if the soil temperature at 3 inch depths is over 90°F.

Avoid reinfestation of treated soil. Do not contaminate treated soil with untreated soil. Do not use transplants or tools which could carry soil-borne pests from infested land. Clean equipment carefully before using.

VORLEX may be corrosive under certain conditions. Flush all applicators with fuel oil or kerosene immediately after use. DO NOT USE WATER. Do not use containers or equipment of aluminum, magnesium or their alloys. Close containers tightly.

TIME OF APPLICATION: Treatment can be made at any time the soil is workable; fall or winter application is desirable for crops which are planted early in the spring.

Soil temperature: VORLEX is active in soils as cold as 35°F except against soil insects.

PREPARATION OF SOIL: FOR OPTIMUM CONTROL WITH VORLEX FOLLOW THESE STEPS CAREFULLY. (1) Work crop remains into the soil so that they are well decomposed before treatment. (2) In shrub and tree replant sites remove as much of the root system as possible. (3) Break up all clods and loosen soil thoroughly by cultivation one week before treatment. Deep tillage is recommended, particularly for deep-rooted crops. Always work soil deeper than intended application depth. (4) Keep soil moist from the surface down through the root zone and maintain in this condition for one week before fumigation. DO NOT FUMIGATE DRY SOILS.

APPLICATION: Use injection type applicators.

Broadcast treatment. Set shanks 6" to 8" apart; for band treatment, set shanks 8" apart. Inject at a depth of 6" to 8". (For best weed control, use the higher dosages indicated, inject at 4-inch depth and cover with a tarp for 4 days).

Row treatment. Adjust injection chisels to same spacing as planter. Seal the treated strips by bedding or listing. Always inject the fumigant at least 6-8" below the final soil (planting) surface. Before planting, remove the tops of the beds, being careful that none of this soil falls over on the fumigated band.

DOSAGE:

PEST	SOIL TYPE	OVERALL (BROADCAST) APPLICATION		ROW OR BAND APPLICATION	
		PER ACRE	PER 100 SQ. FT.	PER CHISEL	PER 1000 LINEAR FT.
Nematodes	Mineral	7–15 gal.	2 1/8–4 1/2 fl. oz.	40–87 ml.[6]	1.4–3.0 fl. oz.[6]
Nematodes	Organic[2] Muck	10–25 gal.	3–7 1/2 fl. oz.	58–146 ml.[6]	2–5 fl. oz.[6]
Symphylans[3]	—	10–15 gal.	3–4 1/2 fl. oz.	—	—
Weeds[1], fungi, nematodes and soil insects (wireworms)	Mineral	25–50 gal.[4]	7 1/2–15 fl. oz.[4]	145–290 ml.[4 5]	5–10 fl. oz.[4 5]
Weeds[1], fungi, nematodes and soil insects (wireworms)	Organic[2] Muck	40–60 gal.	3/4–1 1/8 pint	232–348 ml.[5]	8–12 fl. oz.[5]

1. For best weed control, use the higher dosages indicated. Inject at 4-inch depth and cover with a tarp for 4 days.
2. If VORLEX is used in heavy clay soils, an extended aeration period may be required.
3. To obtain best symphylan control remove all residues prior to fumigation.
4. For citrus and other orchard trees where deep penetration is desirable, use 60–70 gallons of VORLEX per acre (1 1/8–1 3/8 pints per 100 sq. ft.)
5. Row or band treatment does not give effective soil insect control.
6. Row fumigation may not effectively control nematodes on root crops where infestation is severe.

IMMEDIATELY after application, drag the soil to remove chisel marks, then firm with a smooth roller. If additional water is required, apply only the amount necessary to wet the soil to a depth of one-half inch. If a tarp seal is used, lower dosages may be effective.

In soils 70°F or higher at 6" depth, special attention should be given to sealing the surface of the soil. Tarping gives the best seal.

FUMIGATION PERIOD: After application at soil temperatures of 60°F or above, keep soil moist and undisturbed for a minimum of 4 days; 7 days is preferable. Use a shorter fumigation period only if it has been shown effective.

Colder soils require a longer fumigation period. If these conditions occur, allow at least one to three weeks for the fumigation period depending upon the temperature.

In fall or winter application the fumigation period should extend until spring. The soil should be undisturbed during this period.

AERATION PERIOD: For soil temperatures above 60°F (at 6" depth), aerate the soil at the end of the fumigation period by cultivation, e.g. discing. Repeated cultivation will aid release of the fumigant and is especially desirable if heavy rains occur during the fumigation period. If soil crusting occurs, it retards aeration and must be broken up. Repeated cultivation may be necessary to prevent soil crusting.

Aerate until the fumigant has completely escaped from the soil. This is usually accomplished by allowing 1 week between cultivation (end of fumigation period) and planting for each 12 gallons per acre of VORLEX.

For soil temperatures below 60°F (at 6" depth), longer aeration periods than those mentioned above will be required. Repeated deep cultivation is especially important when heavy rains or very wet soil accompany low soil temperatures, since these conditions retard aeration.

DO NOT PLANT WHEN THE ODOR OF THE FUMIGANT IS STILL DETECTABLE.

TREATMENT OF ORCHARD TREE REPLACEMENT SITES: Use VORLEX for control of fungi and nematodes in orchard tree replacement sites. For instructions regarding this use, contact NOR-AM for bulletin entitled "Fumigation of Orchard Tree Replacement Sites".

DO NOT REUSE ORIGINAL CONTAINERS UNLESS RECONDITIONED

STORAGE, SHIPMENT AND DISPOSAL

STORAGE
Store in tightly-closed original container in a cool place away from dwellings. Do not allow contamination of seeds, plants, fertilizers, or other pesticide chemicals. Do not contaminate food, feedstuffs, drugs, or domestic water supplies. In outside storage, store drums (make sure that bungs are tightly closed) on sides to avoid accumulation of rain water in top or bottom of recessed areas.

SHIPMENT
Do not ship or store with food, feeds, drugs, or clothing.

PESTICIDE DISPOSAL
Do not contaminate water, food or feed by storage or disposal. Pesticide wastes are acutely hazardous. Improper disposal of excess pesticide, spray mixture, or rinsate is a violation of Federal Law. If these wastes cannot be disposed of by use according to label instructions, contact your State Pesticide or Environmental Control Agency, or the Hazardous Waste Representative at the nearest EPA Regional Office for guidance.

CONTAINER DISPOSAL
Triple rinse. Then offer for recycling or reconditioning, or puncture and dispose of in a sanitary landfill, or by other approved state and local procedures.

DO NOT REUSE ORIGINAL CONTAINERS UNLESS RECONDITIONED

IMPORTANT: READ BEFORE USE

By using this product, user accepts the following conditions, warranty, disclaimer of warranties and limitations of liability.

CONDITIONS:
The directions for use of this product are believed to be adequate and should be followed carefully. However, because of extreme weather and soil conditions, manner of use and other factors beyond NOR-AM Chemical Company's control, it is impossible for NOR-AM to eliminate all risks associated with the use of this product. As a result, crop injury or ineffectiveness is always possible.

DISCLAIMER OF WARRANTIES:
THERE ARE NO WARRANTIES, EXPRESS OR IMPLIED, OF MERCHANTABILITY OR OF FITNESS FOR A PARTICULAR PURPOSE OR OTHERWISE, WHICH EXTEND BEYOND THE STATEMENTS MADE ON THIS LABEL. No agent of NOR-AM Chemical Company is authorized to make any warranties beyond those contained herein or to modify the warranties contained herein. UNDER NO CIRCUMSTANCES SHALL NOR-AM CHEMICAL COMPANY BE LIABLE FOR ANY INCIDENTAL OR CONSEQUENTIAL DAMAGES RESULTING FROM FAILURE TO USE THIS PRODUCT IN ACCORDANCE WITH LABEL DIRECTIONS.

LIMITATIONS OF LIABILITY:
In no event shall NOR-AM Chemical Company's liability for damages arising out of the use of this product exceed the purchase price of the product used.

VORLEX® 201

SOIL FUMIGANT

RESTRICTED USE PESTICIDE

For retail sale to, and use only by, Certified Applicators or persons under their direct supervision and only for those uses covered by the Certified Applicator's certification.

FOR AGRICULTURAL USE ONLY

ACTIVE INGREDIENTS:

1,3-dichloropropene and other related chlorinated C, hydrocarbons	34.0%
Methyl isothiocyanate	17.0%
Chloropicrin	15.0%
INERT INGREDIENTS	34.0%
TOTAL	100.00%

EPA Reg. No. 45639-91 EPA Est. No. 08800 IL-1
PROTECTED BY U.S. PAT. NOs 3,205,129-3,113,908-3,240,666-3,265,561

Keep Out of Reach of Children

DANGER–POISON

PELIGRO — VENENO

PRECAUCION AL USUARIO: Si usted no lee ingles, no use este producto hasta que el etiqueta haya sido explicado ampliamente.

STATEMENT OF PRACTICAL TREATMENT

IF INHALED: Remove to fresh air. If not breathing, give artificial respiration, preferably mouth-to-mouth. If breathing is difficult, give oxygen. Call a physician.

IF ON SKIN: Immediately flush skin with plenty of water for at least 15 minutes while removing contaminated clothing and shoes. Call a physician. If water is not immediately available, remove excess chemical from skin with absorbent material such as towel or dry soil, then proceed at once to location where water is available and thoroughly wash contaminated skin with plenty of water. Call a physician.

IF IN EYES: Immediately flush eyes with plenty of water or a 5% solution of sodium bicarbonate for at least 15 minutes; call a physician. Continue treatment with ophthalmic hydrocortisone ointment.

IF SWALLOWED: Immediately call a physician or poison control center. Do not induce vomiting. Note to physician: Empty stomach contents by gastric lavage.

DO NOT REUSE ORIGINAL CONTAINER UNLESS RECONDITIONED

PRECAUTIONARY STATEMENTS

HAZARDS TO HUMANS AND DOMESTIC ANIMALS
HAZARDOUS VAPOR AND LIQUID. CORROSIVE CAUSES IRREVERSIBLE EYE DAMAGE AND BURNS OF SKIN. HARMFUL OR FATAL IF INHALED, SWALLOWED OR ABSORBED THROUGH SKIN. MAY PRODUCE KIDNEY AND LIVER DAMAGE UPON PROLONGED EXPOSURE.

Do Not Breathe Vapor. Do Not Get in Eyes, On Skin, Or On Clothing. Do Not Take Internally. Use Only With Adequate Ventilation. Wear Eye And Skin Protection Necessary To Prevent Contact When Handling. Wash Thoroughly After Handling And Before Eating Or Smoking.

If protective gear, such as boots or gloves, becomes contaminated, immediately wash with soap and water. Never wear protective gear having the odor of Vorlex® 201. Remove contaminated clothing and wash thoroughly after use until all odor is gone. Render unusable and dispose of contaminated leather goods, including shoes.

STORAGE, SHIPMENT AND DISPOSAL

STORAGE Store in tightly-closed container in a cool place away from dwellings. Do not allow contamination of seeds, plants, fertilizers, or other pesticide chemicals. Do not contaminate food, feedstuffs, drugs or domestic water supplies. In outside storage, store drums (make sure that bungs are tightly closed) on sides to avoid accumulation of rain water in top or bottom of recessed areas.

SHIPMENT Do not ship or store with food, feeds, drugs or clothing.

PESTICIDE DISPOSAL Do not contaminate water, food or feed by storage or disposal. Pesticide wastes are acutely hazardous. Improper disposal of excess pesticide, spray mixture, or rinsate is a violation of Federal Law. If these wastes cannot be disposed of by use according to label instructions, contact your State Pesticide or Environmental Control Agency, or the Hazardous Waste Representative at the nearest EPA Regional Office for guidance.

CONTAINER DISPOSAL Triple rinse. Then offer for recycling or reconditioning or puncture and dispose of in a sanitary landfill or by other procedures approved by state and local authorities.

DO NOT REUSE ORIGINAL CONTAINERS UNLESS RECONDITIONED

GENERAL INFORMATION

Vorlex 201 SOIL FUMIGANT is used as a preplant treatment in both field and greenhouse soils to be planted to ornamentals or flowers such as mums, lilies, gladiola, iris, carnations, evergreens; vegetables such as celery, lettuce, garden beets, tomatoes, eggplant, beans, melons, squash; field crops such as tobacco, cotton, potatoes, onions, mint; small fruits such as strawberries, raspberries; fruit trees such as cherry, peach; citrus trees such as orange, lemon.

In the soil Vorlex 201 vaporizes and diffuses helping to control plant parasitic nematodes and harmful fungi (including Fusarium oxysporum f.lycopersici) causing damping-off, root-rot and wilt. Vorlex 201 also helps control soil insects such as wireworms, symphilids and others. Vorlex 201 helps control weeds such as pigweed, dandelion, lambsquarters, bermudagrass, purslane, crabgrass, quackgrass, spurge, fleabane, sedge, chickweed, ragweed, mustard, Poa spp., smartweed, wild radish, dodder, velvetleaf and shepherds-purse.

PHYSICAL OR CHEMICAL HAZARDS

FLAMMABLE

DO NOT USE, POUR, SPILL, OR STORE NEAR HEAT OR OPEN FLAME. DO NOT CUT OR WELD CONTAINER. FIRE OR EXPLOSION MAY RESULT EVEN IF CONTAINER IS EMPTY.

ENVIRONMENTAL HAZARDS

This product is toxic to fish and wildlife. Do not apply directly to water or wetlands. Do not contaminate water or wetlands by cleaning of equipment or disposal of wastes.

HANDLING PRECAUTIONS

DO NOT INHALE VAPORS—NIOSH- or MSHA-approved respiratory protection should be worn when Vorlex 201 liquid soil fumigant is exposed to the atmosphere or when conducting operations which vent to the atmosphere. A NIOSH- or MSHA-approved half-face respirator with chemical worker's goggles or full-face respirator shall be used during small spills, repairs, calibrations, transfers, sampling and when working in poorly ventilated areas. When in use canisters or cartridges shall be replaced daily or sooner if specified by manufacturer or at first sign of odor breakthrough, whichever comes first. NIOSH-approved cartridges, such as WILLSON R21 will be adequate for short term situations such as listed above.

Where very high concentrations of vapors might be expected (such as large spills in poorly ventilated areas), a self-contained or air-supplied respirator should be used.

DO NOT GET ON SKIN—When handling or working with Vorlex 201, wear clean body covering including gloves and heavy footwear. Immediately remove contaminated coverings. Aerate and wash all protective clothing and gear thoroughly after use.

There are no protective clothing materials that are completely impervious. Thin layers of polyethylene (minimum 1 mil), give short-term protection and shall be immediately discarded upon contamination. Heavy (3 + Mil) polyethylene provides longer term protection. Leather gives no protection. If gear becomes contaminated, immediately wash with soap and water. Never wear protective gear having odor of Vorlex 201. Wash and aerate all protective gear thoroughly after use until all odor is gone.

DO NOT GET IN EYES—Wear eye protection such as chemical worker's goggles or full-face respirator when handling Vorlex 201.

DO NOT SWALLOW—Do not use the mouth to siphon Vorlex 201 from containers or to blow out clogged lines, nozzles, etc.

FOR CHEMICAL EMERGENCY—Spill, Leak, Fire, Exposure or Accident—Call 800/424-9300 DAY OR NIGHT

DIRECTIONS FOR USE

It is a violation of Federal law to use this product in a manner inconsistent with its labeling.

USE PRECAUTIONS

As with most soil fumigants, application of Vorlex 201 may temporarily raise the level of ammonia nitrogen and soluble salts in the soil which may result in plant injury. This is most likely to occur when heavy rates of fertilizer or fumigant are applied to soils that are either cold, wet, acid or high in organic matter.

Do not add fertilizers containing nitrates until the soil temperature is above 65 °F and the crop is well established.

Do not allow liquid or vapors to come in contact with lawn and desirable plants; do not apply on areas beneath which roots of desirable plants extend.

Do not use in greenhouses which contain living plants.

Do not fumigate if the soil temperature at 3 inch depths is over 90 °F

Avoid reinfestation of treated soil. Do not contaminate treated soil with untreated soil. Do not use transplants or tools which could carry soil-borne pests from infested land. Clean equipment carefully before using.

Vorlex® 201 may be corrosive under certain conditions. Flush all applicators with fuel oil or kerosene immediately after use. DO NOT USE WATER. Do not use containers or equipment of aluminum, magnesium or their alloys. Close containers tightly.

TIME OF APPLICATION

Treatment can be made at any time the soil is workable. Fall or winter application is desirable for crops which are planted early in the spring.

PREPARATION OF SOIL

FOR OPTIMUM CONTROL WITH VORLEX 201 FOLLOW THESE STEPS CAREFULLY.

(1) Work crop remains into the soil so that they are well decomposed before treatment. (2) In shrub and tree replant sites remove as much of the root system as possible. (3) Break up all clods and loosen soil thoroughly by cultivation one week before treatment. Deep tillage is recommended, particularly for deep-rooted crops. Always work soil deeper than intended application depth. (4) Keep soil moist from the surface to a depth of six inches and maintain in this condition for one week before fumigation. DO NOT FUMIGATE DRY SOILS.

APPLICATION AND DOSAGE

Use injection type applicators. For broadcast application, set shanks 6″ to 8″ apart; for a band treatment, set shanks 8″ apart. Inject at a depth of 6″ to 8″.

DOSAGE

Pest	Soil Type	Overall (Broadcast) Application		Row or Band (8″ centers) (8″ centers Application)
		per acre	per 100 sq. ft.	per chisel per 100 linear ft.
Nematodes	Mineral	7-15 gal.	2-4⅓ fl. oz.	40-87 ml.
Nematodes	Organic[2] (Muck)	10-25 gal.	3-7⅓ fl. oz.	58-145 ml.
Weeds[1], fungi, nematodes and soil insects	Mineral	25-50 gal.[3]	7⅓-14¾ fl. oz.	145-290 ml.[3]
Weeds[1], fungi, nematodes and soil insects	Organic[2] (Muck)	40-60 gal.	11¾-17½ fl. oz.	232-348 ml.

(The above recommendations may vary because of local conditions. Consult your local agricultural authority for regional recommendations.)

[1] For best weed control, use the higher dosages indicated. Inject at 4-inch depth and cover with tarp for 4 days.

[2] If Vorlex 201 is used in heavy clay soils, an extended aeration period may be required.

[3] For citrus and other orchard trees where deep penetration is desirable, use 60-70 gallons per acre of Vorlex 201 (17½-20½ fl. oz. per 100 sq. ft.)

IMMEDIATELY after application, drag the soil to remove chisel marks, then firm it with a smooth roller. If additional water is required, apply only the amount necessary to wet the soil to a depth of one-half inch. If a tarp seal is used, lower dosages may be effective.

In soils 70 °F or higher at 6″ depth, special attention should be given to sealing the surface of the soil. Tarping gives the best seal.

FUMIGATION PERIOD: After application at soil temperature at 60 °F or above, keep soil moist and undisturbed for a minimum of 4 days; 7 days is preferable. Use a shorter fumigation period if it has been shown effective.

Colder soils require a longer fumigation period. If these conditions occur, allow at least one to three weeks for the fumigation period depending upon the temperature.

In fall or winter application the fumigation period should extend until spring. The soil should be undisturbed during this period.

AERATION PERIOD: FOR SOIL TEMPERATURES ABOVE 60 °F (AT 6" DEPTH), aerate soil at the end of the fumigation period by cultivation e.g., discing. Repeated cultivation will aid release of the fumigant and is especially desirable if heavy rains occur during the fumigation period. If soil crusting occurs, it retards aeration and must be broken up. Repeated cultivation may be necessary to prevent soil crusting.

Aerate until the fumigant has completely escaped from the soil. This is usually accomplished by allowing one week between cultivation (end of fumigation period) and planting for each 12 gallons per acre of Vorlex 201.

FOR SOIL TEMPERATURES BELOW 60 °F (AT 6" DEPTH), longer aeration periods than those mentioned above will be required. Repeated deep cultivation is especially important when heavy rains or very wet soil accompany low soil temperatures, since these conditions retard aeration.

Do Not Plant When The Odor Of The Fumigant Is Still Detectable.

IMPORTANT: READ BEFORE USE

By using this product, user accepts the following conditions, warranty, disclaimer of warranties and limitations of liability.

CONDITIONS

The directions for use of this product are believed to be adequate and should be followed carefully. However, because of extreme weather and soil conditions, manner of use and other factors beyond NOR-AM Chemical Company's control, it is impossible for NOR-AM to eliminate all risks associated with the use of this product. As a result, crop injury or ineffectiveness is always possible.

DISCLAIMER OF WARRANTIES

THERE ARE NO WARRANTIES, EXPRESS OR IMPLIED, OF MERCHANTABILITY OR OF FITNESS FOR A PARTICULAR PURPOSE OR OTHERWISE, WHICH EXTEND BEYOND THE STATEMENTS MADE ON THIS LABEL. No agent of NOR-AM Chemical Company is authorized to make any warranties beyond those contained herein or to modify the warranties contained herein. UNDER NO CIRCUMSTANCES SHALL NOR-AM CHEMICAL COMPANY BE LIABLE FOR ANY INCIDENTAL OR CONSEQUENTIAL DAMAGES RESULTING FROM FAILURE TO USE THIS PRODUCT IN ACCORDANCE WITH LABEL DIRECTIONS.

LIMITATIONS OF LIABILITY

In no event shall NOR-AM Chemical Company's liability for damages arising out of the use of this product exceed the purchase price of the product used.

Pennwalt Corporation

Agchem Division
Three Parkway
Philadelphia, PA 19102
(215) 587-7147

ACCELERATE®

A HARVEST AID FOR COTTON

Contains 0.52 lbs. Endothall* per gallon

ACTIVE INGREDIENT:	
Mono (N,N-dimethylalkylamine)**salt of Endothall*	15.9%
INERT INGREDIENTS:	84.1%
TOTAL	100.0%

* 7-Oxabicyclo (2.2.1)heptane-2,3-dicarboxylic acid equivalent 5.5%

** Alkyl groups as in fatty acids of coconut oil

EPA Reg. No. 4581-284 EPA Est. No. 4581-TX-1

U.S. Patents; 3,246,015; other patents pending

Keep Out of Reach of Children
DANGER—PELIGRO

PRECAUCION AL USUARIO: Si usted no lee ingles, no use este producto hasta que la etiqueta haya sido explicado ampliamente.

STATEMENT OF PRACTICAL TREATMENT

IF IN EYES, immediately flush with plenty of water for at least 15 minutes. Call a physician.

IF ON SKIN, immediately flush with plenty of water for at least 15 minutes. Remove and wash contaminated clothing before reuse.

IF SWALLOWED, drink promptly a large quantity of milk, egg whites, gelatin solution or, if these are not available, drink large quantities of water. Avoid alcohol. Call a physician immediately.

NOTE TO PHYSICIAN: Probable mucosal damage may contraindicate the use of gastric lavage. Measures against circulatory shock, respiratory depression and convulsion may be needed.

PRECAUTIONARY STATEMENTS

HAZARDS TO HUMANS (AND DOMESTIC ANIMALS)

DANGER

CORROSIVE. CAUSES IRREVERSIBLE EYE DAMAGE. CAUSES SKIN IRRITATION. MAY BE FATAL IF ABSORBED THROUGH SKIN. HARMFUL IF SWALLOWED OR INHALED. DO NOT GET IN EYES, ON SKIN OR ON CLOTHING. WEAR PROTECTIVE CLOTHING, RUBBER GLOVES AND GOGGLES OR FACE SHIELD WHEN HANDLING. Wash thoroughly with soap and water after handling and before eating or smoking. Avoid breathing spray mist. Remove contaminated clothing and wash before reuse.

ENVIRONMENTAL HAZARDS

Avoid contact with or drift to other crops, trees, or plants as injury may occur.

This product is toxic to fish and wildlife. Keep out of lakes, streams or ponds. Do not contaminate water by cleaning of equipment, or disposal of wastes. Apply this product only as specified on this label.

STORAGE AND DISPOSAL

Do not contaminate water, food, or feed by storage or disposal.

STORAGE INSTRUCTIONS

Store in the original container. Do not store in a manner where cross-contamination with other pesticides, fertilizers, food or feed could occur. In the event of spillage during handling or storage, absorb with sand or other inert material and dispose of absorbent in accordance with the Pesticide Disposal Instructions listed below.

PESTICIDE DISPOSAL INSTRUCTIONS

Pesticide wastes are acutely hazardous. Improper disposal of excess pesticide, spray mixture, or rinsate is a violation of Federal law. If these wastes cannot be disposed of by use according to label instructions, contact your State Pesticide or Environmental Control Agency, or the Hazardous Waste representative at the nearest EPA Regional Office for guidance.

CONTAINER DISPOSAL INSTRUCTIONS

Triple rinse (or equivalent). Then offer for recycling or reconditioning, or puncture and dispose of in a sanitary landfill, or by incineration, or, if allowed by state and local authorities, by burning. If burned, stay out of smoke.

GENERAL INFORMATION

ACCELERATE is a liquid concentrate that when tank mixed with sodium chlorate type (containing suitable fire suppresants) or organic phosphate cotton defoliants causes a faster and more dependable leaf drop prior to hand picking or machine harvesting. The combination of ACCELERATE and sodium chlorate type or organic phosphate defoliants usually drops leaves in a green condition in approximately one-half the time of sodium chlorate type or organic phosphate defoliants without ACCELERATE. Under favorable defoliation conditions, satisfactory leaf drop will require 3–4 days; whereas, under adverse conditions, such as low temperatures, toughened plants or actively growing plants, it may be extended to 6–7 days. ACCELERATE is non-corrosive and will not damage metal parts normally used in applying spray chemicals. Recommended rates of ACCELERATE used alone without sodium chlorate type or organic Phosphate defoliants will not give satisfactory defoliation. Wash out sprayer immediately after use.

DIRECTIONS FOR USE

It is a violation of Federal law to use this product in a manner inconsistent with its labeling.

RE-ENTRY STATEMENT

Do not apply this product in such a manner as to directly or through drift expose workers or other persons. The area being treated must be vacated by unprotected persons.

Do not enter treated areas without protective clothing until sprays have dried.

Because certain states may require more restrictive reentry intervals for various crops treated with this product, consult your State Department of Agriculture for further information.

Written or oral warnings must be given to workers who are expected to be in an area about to be treated with this product. In case of accidental exposure, workers should wash thoroughly with soap and water and contaminated clothing should be removed and washed before reuse. If symptoms of skin or eye irritation occur call a physician. When oral warnings are given, warnings shall be given in a language customarily understood by workers. Oral warnings shall be given if there is reason to believe that written warnings cannot be understood by workers. Written warnings must include the following information: "DANGER. Area treated with ACCELERATE Cotton Harvest Aid on (date of application). Do not enter without protective clothing until spray has dried. In case of accidental exposure, workers should wash thoroughly with soap and water and contaminated clothing should be removed and washed before reuse. If symptoms of skin or eye irritation occur call a physician."

Under normal conditions, apply the combination of ACCELERATE and Sodium Chlorate type or organic Phosphate defoliants when fiber quality of the top bolls will not be damaged by loss of the top leaves which feed these bolls (top bolls should be firm to thumb pressure or difficult to cut with a sharp knife). The combination of

ACCELERATE and Sodium Chlorate type or organic Phosphate defoliants will not damage the bolls directly or cause them to open prematurely. Ground machines should have at least 3 nozzles per row including one overhead nozzle and have pressure and speed adjusted to give fine droplets yet maintain the correct spray volume of 15 to 25 gallons per acre. For aerial application rates of 5–12 gallons of total spray should be used. Aerial applicators should fly as closely as possible to the plant tops and avoid application during the excessive temperatures of mid-day and when winds may cause drift to adjacent fields. If spraying is interrupted, be sure to start agitation before spraying is resumed.

Apply specified spray dosage per acre in recommended gallons of water to give thorough coverage to mature cotton when 50% or more of bolls are open. Applications can be made until frost.

APPLICATION RATES

Observe all cautions and limitations on labeling of all products used in mixtures.

ACCELERATE WITH ORGANIC PHOSPHATE DEFOLIANTS:

Under ideal defoliation conditions tank mix with water 1 pint of organic Phosphate defoliant with 1½ pints of ACCELERATE per acre. Under adverse defoliation conditions, use 1½ pints of ACCELERATE with an increased rate, up to the maximum label rate, of organic Phosphate defoliant. Always add ACCELERATE to organic Phosphate defoliants previously tank mixed with water. Do not mix together without water. Be sure mixing tank is clean because residues of insecticides may cause curdling.

ACCELERATE WITH SODIUM CHLORATE TYPE DEFOLIANTS:

Under ideal defoliation conditions, tank mix 1 pint of ACCELERATE with label recommended rate of Sodium Chlorate type defoliants per acre. Under adverse defoliation conditions increase rate of ACCELERATE to 1½ pints per acre.

WARRANTY AND DISCLAIMER

Pennwalt Corporation warrants that this material conforms to the chemical description on the label and is reasonably fit for the purposes referred to in the Directions for Use, subject to the risks referred to therein. PENNWALT MAKES NO OTHER EXPRESS OR IMPLIED WARRANTY OF FITNESS OR MERCHANTABILITY OR ANY OTHER EXPRESS OR IMPLIED WARRANTY. IN NO CASE SHALL PENNWALT OR SELLER BE LIABLE FOR CONSEQUENTIAL, SPECIAL OR INDIRECT DAMAGES RESULTING FROM THE USE OR HANDLING OF THIS PRODUCT INCLUDING, BUT NOT LIMITED TO, LOSS OF PROFITS, BUSINESS REPUTATION, OR CUSTOMERS; LABOR COST; OR OTHER EXPENSES INCURRED IN PLANTING OR HARVESTING.

Pennwalt and seller offer this product and the buyer and user accept it subject to the foregoing conditions of sale and warranty which may be varied only by agreement in writing signed by a duly authorized representative of Pennwalt.

AQUATHOL® K

AQUATIC HERBICIDE

ACTIVE INGREDIENT:

Dipotassium salt of endothall*	40.3%
INERT INGREDIENTS:	59.7%
TOTAL	100.0%

* 7-oxabicyclo [2.2.1]heptane-2,3-dicarboxylic acid equivalent 28.6%.

Contains per gallon 4.23 lbs. dipotassium endothall (equivalent to 3.0 lbs. endothal acid)

EPA Reg. No. 4581-204 EPA Est. No. 4581-TX-1

DANGER–POISON

Keep Out of Reach of Children

STATEMENT OF PRACTICAL TREATMENT

IF SWALLOWED, drink promptly a large quantity of milk, egg whites, gelatin solution or if these are not available, drink large quantities of water. Avoid alcohol. Call a physician immediately.

IF ON SKIN, immediately flush with plenty of water for at least 15 minutes. Remove and wash contaminated clothing before reuse.

IF IN EYES, immediately flush with plenty of water for at least 15 minutes. Call a physician.

NOTE TO PHYSICIAN: Probable mucosal damage may contraindicate the use of gastric lavage. Measures against circulatory shock, respiratory depression and convulsion may be needed.

See Side Panel for Additional Precautionary Statements

NOTE: For GENERAL INFORMATION and DIRECTIONS FOR USE refer to accompanying brochure.

PRECAUTIONARY STATEMENTS

HAZARDS TO HUMANS (AND DOMESTIC ANIMALS)

DANGER

FATAL IF ABSORBED THROUGH SKIN. MAY BE FATAL IF SWALLOWED. HARMFUL IF INHALED. CORROSIVE, CAUSES IRREVERSIBLE EYE DAMAGE. DO NOT GET IN EYES, ON SKIN OR ON CLOTHING. WEAR PROTECTIVE CLOTHING, RUBBER GLOVES, AND GOGGLES OR FACE SHIELD WHEN HANDLING. Wash thoroughly with soap and water after handling and before eating or smoking. Remove contaminated clothing and wash before reuse. Avoid breathing spray mist.

ENVIRONMENTAL HAZARDS

Avoid contact with or drift to other crops or plants as injury may result. Wash out spray equipment with water after each operation.

Do not use fish from treated water for food or feed within 3 days of treatment. Do not use treated water for watering livestock, for preparing agricultural sprays for food crops, for irrigation, or for domestic purposes within the following periods:

Up to 0.5 ppm dipotassium salt
(0.35 ppm acid equivalent)—7 days after application

Up to 4.25 ppm dipotassium salt
(3.0 ppm acid equivalent)—14 days after application

Up to 5.0 ppm dipotassium salt
(3.5 ppm acid equivalent)—25 days after application

NOTE: Areas treated with AQUATHOL K may be used for swimming twenty-four hours after treatment.

Treated water can be used for sprinkling bent grass immediately.

GENERAL INFORMATION

AQUATHOL K is a liquid concentrate soluble in water which is effective against a broad range of aquatic plants with a margin of safety to fish.

Dosage rates indicated for the application of AQUATHOL K are measured in "Parts Per Million" (ppm) of dipotassium endothall. Only 0.5 to 5.0 ppm are generally required for aquatic weed control, whereas some fish species are tolerant to approximately 100 ppm or over.

DIRECTIONS FOR USE

It is a violation of Federal law to use this product in a manner inconsistent with its labeling.

AQUATIC WEEDS CONTROLLED AND DOSAGE RATE CHARTS

AQUATHOL K is recommended for the control of the following aquatic weeds in irrigation and drainage canals, ponds and lakes at the rates indicated. Since the active ingredient is water soluble and tends to diffuse from the area treated, select the dosage rate applicable to the area to be treated. Use the lower rate in each range of rates where the growth is young and growing and/or where the weed stand is not heavy. Marginal treatments of large bodies of water require higher rates as indicated.

HOW TO APPLY

AQUATHOL K is a contact killer; consequently, do not apply before weeds are present. Application as early as possible after weeds are present is recommended to permit use of lower application rates. However, for best results water temperature should be at 65°F or above. If an entire pond is treated at one time, or if the dissolved oxygen level is low at time of application, decay of weeds may remove enough oxygen from the water, causing fish to suffocate. Water containing very heavy vegetation should be treated in sections to prevent suffocation of fish. Sections should be treated 5-7 days apart. Carefully measure size and depth of area to be treated and determine amount of AQUATHOL K to apply from chart. For best results apply on a calm day where there is little wave action.

AQUATHOL K should be sprayed on the water or injected below the water surface and should be distributed as evenly as possible. It may be applied as it comes from the container or diluted with water depending on the equipment. Some dilution will give better distribution.

In instances where the nuisance to be controlled is an exposed surface problem (i.e., some of the broad-leaved pond weeds) it is important to get good contact coverage utilizing the highest concentration (least water dilution) compatible with the type of equipment used so that even distribution is achieved.

Necessary approval and/or permits should be obtained in states where required.

Common Name	Latin Name	Entire Pond Or Large Area Treatment	Spot Or Lake Margin Treatment
Bass Weed	Potamogeton amplifolius	2.0–3.0 ppm	3.0–4.0 ppm
Bur Reed	Sparganium spp.	3.0–4.0 ppm	4.0–5.0 ppm
Coontail	Ceratophyllum spp.	1.0–2.0 ppm	2.0–3.0 ppm
Hydrilla	Hydrilla verticillata	2.0–3.0 ppm	3.0–4.0 ppm
Milfoil	Myriophyllum spp.	2.0–3.0 ppm	3.0–4.0 ppm
Pondweed			
Bushy	Najas spp.	0.5–1.5 ppm	2.0–3.0 ppm
Curly-Leaf	Potamogeton crispus	0.5–1.5 ppm	2.0–3.0 ppm
Flat-Stem	Potamogeton zosteriformis	2.0–3.0 ppm	3.0–4.0 ppm
Floating-Leaf	Potamogeton natans	1.0–2.0 ppm	2.0–3.0 ppm
Horned	Zannichellia spp.	1.0–2.0 ppm	2.0–3.0 ppm
Sago	Potamogeton pectinatus	1.0–2.0 ppm	2.0–3.0 ppm
________	Potamogeton americanus	2.0–3.0 ppm	3.0–4.0 ppm
________	Potamogeton diversifolius	1.0–2.0 ppm	2.0–3.0 ppm
________	Potamogeton filiformis	2.0–3.0 ppm	3.0–4.0 ppm
________	Potamogeton pusillus	1.0–2.0 ppm	2.0–3.0 ppm
Water Star Grass	Heteranthera spp.	2.0–3.0 ppm	3.0–4.0 ppm

RATE OF APPLICATION—LAKES AND PONDS

The following chart indicates the total quantity of material to be applied.

APPROXIMATE GALLONS OF AQUATHOL K FOR ONE ACRE (208′ × 208′) TREATMENT

	DOSAGE IN GALLONS FOR VARIOUS CONCENTRATIONS IN PPM						
DEPTH	0.5 ppm	1.0 ppm	1.5 ppm	2.0 ppm	3.0 ppm	4.0 ppm	5.0 ppm
1 ft.	0.3	0.6	1.0	1.3	1.9	2.6	3.2
2 ft.	0.6	1.3	1.9	2.6	3.8	5.1	6.4
4 ft.	1.3	2.6	3.8	5.1	7.7	10.2	12.8
6 ft.	1.9	3.8	5.8	7.6	11.5	15.3	19.2

RATE OF APPLICATION—IRRIGATION AND DRAINAGE CANALS**

The following indicates the total quantity of material to be applied.

GALLONS OF AQUATHOL K REQUIRED TO TREAT 1 MILE OF CANAL 1 FOOT DEEP*

	WIDTH OF CANAL IN FEET			
PPM	5	10	15	20
1.0 ppm	0.4	0.75	1.2	1.5
2.0 ppm	0.75	1.5	2.3	3.0
3.0 ppm	1.2	2.3	3.5	4.5
4.0 ppm	1.5	3.0	4.5	6.0
5.0 ppm	2.0	3.8	5.7	7.5

The minimum contact time with weeds for optimum results should be 2 hours.

* For deeper water, adjust rate accordingly.

** Not for this use in California.

STORAGE AND DISPOSAL

Do not contaminate water, food, or feed by storage or disposal.

STORAGE INSTRUCTIONS

Store in the original container. Do not store in a manner where cross-contamination with other pesticides, fertilizers, food or feed could occur. Storage at temperatures below 32°F may result in the product freezing or crystallizing. Should this occur the product must be warmed to 50°F or higher and thoroughly agitated. In the event of a spillage during handling or storage, absorb with sand or other inert material and dispose of absorbant in accordance with the Pesticide Disposal Instructions listed below.

PESTICIDE DISPOSAL INSTRUCTIONS

Pesticide wastes are acutely hazardous. Improper disposal of excess pesticide, spray mixture, or rinsate is a violation of Federal law. If these wastes cannot be disposed of by use according to label instructions, contact your State Pesticide or Environmental Control Agency, or the Hazardous Waste representative at the nearest EPA Regional Office for guidance.

CONTAINER DISPOSAL INSTRUCTIONS

Triple rinse (or equivalent). Then offer for recycling or reconditioning, or puncture and dispose of in a sanitary landfill, or incineration, or, if allowed by state and local authorities, by burning. If burned, stay out of smoke.

WARRANTY AND DISCLAIMER

Pennwalt Corporation warrants that this material conforms to the chemical description on the label and is reasonably fit for the purposes referred to in the Directions for Use, subject to the risks referred to therein. PENNWALT MAKES NO OTHER EXPRESS OR IMPLIED WARRANTY OF FITNESS OR MERCHANTABILITY OR ANY OTHER EXPRESS OR IMPLIED WARRANTY. IN NO CASE SHALL PENNWALT OR SELLER BE LIABLE FOR CONSEQUENTIAL, SPECIAL OR INDIRECT DAMAGES RESULTING FROM THE USE OR HANDLING OF THIS PRODUCT INCLUDING, BUT NOT LIMITED TO,

LOSS OF PROFITS, BUSINESS REPUTATION, OR CUSTOMERS; LABOR COST; OR OTHER EXPENSES INCURRED IN PLANTING OR HARVESTING.

Pennwalt and seller offer this product and the buyer and user accept it subject to the foregoing conditions of sale and warranty which may be varied only by agreement in writing signed by a duly authorized representative of Pennwalt.

DESICCANT L-10®

HARVEST AID FOR COTTON

ACTIVE INGREDIENT:

Arsenic Acid (H_3AsO_4)	75%
INERT INGREDIENTS:	25%
TOTAL	100%
Total arsenic, all in water soluble form, expressed as metallic	39.6%
Weight of solution per gallon at 68°F	15.7 lb.

EPA Reg. No. 4581-231 EPA Est. No. 4581-TX-1

DANGER–POISON

Keep Out of Reach of Children

PRECAUCION AL USUARIO: Si usted no lee ingles, no use este producto hasta que la etiqueta haya sido explicado ampliamente.

PRECAUTIONARY STATEMENTS

HAZARDS TO HUMANS (AND ANIMAL SAFETY)

DANGER

CONTAINS INORGANIC ARSENIC
CANCER HAZARD
HARMFUL IF INHALED OR SWALLOWED

USE ONLY WITH ADEQUATE VENTILATION OR RESPIRATORY PROTECTION FATAL IF SWALLOWED, ABSORBED THROUGH SKIN, OR INHALED. CORROSIVE. CAUSES IRREVERSIBLE EYE DAMAGE AND SKIN BURNS. Do not get in eyes, on skin, or on clothing. Do not breathe spray mist. Wear protective clothing and rubber gloves. Wear a mask or pesticide air purifying respirator equipped with a high efficiency filter jointly approved by the Mining Enforcement and Safety Administration, and the National Institute for Occupational Safety and Health. Wear goggles or face shield when handling. Wash thoroughly with soap and water after handling and before eating or smoking. Remove contaminated clothing and wash before reuse.

ENVIRONMENTAL HAZARDS

Keep children, pets, livestock, and poultry away from treated areas. Avoid contamination of food and water to be consumed by humans and animals. Do not graze areas treated with DESICCANT L-10 or use treated plants or gin trash for feed or forage. Since DESICCANT L-10 is injurious to all plant foliage, avoid contact with or drift to desirable plants or crops. Special precautions should be taken to avoid air drift to areas and buildings occupied by humans, animals or poultry.

This product is toxic to wildlife. Do not contaminate streams, lakes, ponds, woodlands or similar non-crop areas. This product is highly toxic to bees exposed to direct treatment or residues on crops. Birds feeding on treated areas may be killed. Do not apply where runoff is likely to occur. Do not contaminate water by cleaning of equipment or disposal of wastes. Protective information may be obtained from your Cooperative Agricultural Extension Service.

STORAGE AND DISPOSAL

Do not contaminate water, food, or feed by storage or disposal.

STORAGE INSTRUCTIONS

Store in the original container.

DESICCANT—L-10 has acidic properties and reacts with fabric, galvanized metals, black iron, and certain other metals. Do not use galvanized metal containers for storage as highly toxic arsine gas may be formed.

Do not store in a manner where cross-contamination with other pesticides, fertilizers, food or feed could occur. In the event of spillage during handling or storage, absorb with sand or other inert material and dispose of absorbent in accordance with the Pesticide Disposal Instructions listed below.

PESTICIDE DISPOSAL INSTRUCTIONS

Pesticide wastes are acutely hazardous. Improper disposal of excess pesticide, spray mixture, or rinsate is a violation of Federal law. If these wastes cannot be disposed of by use according to label instructions, contact your State Pesticide or Environmental Control agency, or the Hazardous Waste representative at the nearest EPA Regional Office for guidance.

CONTAINER DISPOSAL INSTRUCTIONS

Triple rinse (or equivalent). Then offer for recycling or reconditioning, or puncture and dispose of in a sanitary landfill, or by incineration, or, if allowed by state and local authorities, by burning. If burned, stay out of smoke.

GENERAL INFORMATION

DESICCANT L-10 is a liquid concentrate containing arsenic acid and is especially effective as a cotton desiccant. DESICCANT L-10 is recommended particularly for use on cotton which is to be stripped. Cotton which has been properly desiccated with DESICCANT L-10 is suitable for machine picking or stripping as normally there will be an absence of green leaves with little or no regrowth.

DESICCANT L-10 has acidic properties and reacts with fabric, galvanized metals, black iron, and certain other metals. Any metallic equipment or nozzles used in application of DESICCANT L-10 should be made of stainless steel as there is little or no reaction with this metal at field strength dilution. Do not use galvanized metal containers or spraying equipment as highly toxic arsine gas may be formed. To reduce possible corrosion to metals and deterioration of fabric and other parts, all equipment including airplanes which have been in contact with or exposed to DESICCANT L-10, or its solution, should be thoroughly washed with water immediately after use.

DIRECTIONS FOR USE

It is a violation of Federal Law to use this product in a manner inconsistent with its labeling.

RE-ENTRY STATEMENT

Do not apply this product in such a manner as to directly or through drift expose workers or other persons. The area being treated must be vacated by unprotected persons.

Do not enter treated areas without protective clothing until sprays have dried.

Because certain states may require more restrictive reentry intervals for various crops treated with this product, consult your State Department of Agriculture for further information.

Written or oral warnings must be given to workers who are expected to be in an area about to be treated with this product.

In case of accidental exposure, workers should wash thoroughly with soap and water and contaminated clothing should be removed and washed before reuse. If symptoms of skin or eye irritation occur call a physician. When oral warnings are given, warnings shall be given in a language customarily understood by workers. Oral warnings shall be given if there is reason to believe that written warnings cannot be understood by workers. Written warnings must include the following information: "DANGER. Area treated with DESICCANT L-10 Harvest Aid for Cotton on (date of application). Do not enter without protective clothing until spray has dried. In case of accidental exposure, workers should wash thoroughly with soap and water and contaminated clothing should be removed and washed before reuse. If symptoms of skin or eye irritation occur call a physician."

Apply DESICCANT L-10 at the rate of $1\frac{1}{2}$ quarts per acre using sufficient water to make 4 to 10 gallons of spray. For best results, thorough uniform coverage is necessary. Application should be made when air is calm and weather is clear and hot. Apply after the crop has matured, as earlier application may reduce yield. Do not apply within 4 days of harvest. Do not apply more than $1\frac{1}{2}$ quarts per acre to avoid excessive residues of arsenic acid on raw cottonseed.

WARRANTY AND DISCLAIMER

Pennwalt Corporation warrants that this material conforms to the chemical description on the label and is reasonably fit for the purposes referred to in the Directions for Use, subject to the risks referred to therein. PENNWALT MAKES NO OTHER EXPRESS OR IMPLIED WARRANTY OF FITNESS OR MERCHANTABILITY OR ANY OTHER EXPRESS OR IMPLIED WARRANTY. IN NO CASE SHALL PENNWALT OR SELLER BE LIABLE FOR CONSEQUENTIAL, SPECIAL OR INDIRECT DAMAGES RESULTING FROM THE USE OR HANDLING OF THIS PRODUCT INCLUDING, BUT NOT LIMITED TO, LOSS OF PROFITS, BUSINESS REPUTATION, OR CUSTOMERS; LABOR COST; OR OTHER EXPENSES INCURRED IN PLANTING OR HARVESTING.

Pennwalt and seller offer this product and the buyer and user accept it subject to the foregoing conditions of sale and warranty which may be varied only by agreement in writing signed by a duly authorized representative of Pennwalt.

HERBICIDE 273

SUGAR BEET HERBICIDE

ACTIVE INGREDIENT:

Dipotassium salt of endothall*	40.3%
INERT INGREDIENTS:	59.7%
TOTAL	100.0%

* 7-oxabicyclo[2.2.1]heptane-2,3-dicarboxylic acid equivalent 28.6%

This product contains 4.23 lbs. dipotassium endothall per gallon (equivalent to 3 lbs. endothall)

EPA Reg. No. 4581-223 EPA Est. No. 228-IL-1

Keep Out of Reach of Children

DANGER–POISON

STATEMENT OF PRACTICAL TREATMENT

If swallowed, drink promptly a large quantity of milk, egg whites, gelatin solution or, if these are not available, drink large quantities of water. Avoid alcohol. Call a physician immediately.

If on skin, immediately flush with plenty of water for at least 15 minutes while removing contaminated clothing and shoes.

If in eyes, immediately flush with plenty of water for at least 15 minutes. Call a physician.

NOTE TO PHYSICIAN: Probable mucosal damage may contraindicate the use of gastric lavage. Measures against circulatory shock, respiratory depression, and convulsion may be needed.

See Back Panel for Precautionary Statements.

PRECAUTIONARY STATEMENTS

HAZARD TO HUMANS (AND DOMESTIC ANIMALS)
DANGER

FATAL IF ABSORBED THROUGH SKIN. MAY BE FATAL IF SWALLOWED. HARMFUL IF INHALED. CORROSIVE, CAUSES IRREVERSIBLE EYE DAMAGE. DO NOT GET IN EYES, ON SKIN OR ON CLOTHING. WEAR PROTECTIVE CLOTHING, RUBBER GLOVES, AND GOGGLES OR FACE SHIELD WHEN HANDLING. Wash thoroughly with soap and water after handling and before eating or smoking. Remove contaminated clothing and wash before reuse. Avoid breathing spray mist.

ENVIRONMENTAL HAZARDS

This product is toxic to wildlife. Do not apply directly to water. Do not contaminate water by cleaning of equipment or disposal of wastes. Avoid contact with or drift to other crops or plants as injury may result.

STORAGE AND DISPOSAL

Do not contaminate water, food or feed by storage or disposal.

STORAGE INSTRUCTIONS

Store in the original container. Do not store in a manner where cross-contamination with other pesticides, fertilizers, food or feed could occur. Storage at temperatures below 32°F may result in the product freezing or crystalizing. Should this occur the product must be warmed to 50°F or higher and thoroughly agitated. In the event of a spillage during handling or storage, absorb with sand or other inert material and dispose of absorbant in accordance with the Pesticide Disposal Instructions listed below.

PESTICIDE DISPOSAL INSTRUCTIONS

Pesticide wastes are acutely hazardous. Improper disposal of excess pesticide, spray mixture, or rinsate is a violation of Federal law. If these wastes cannot be disposed of by use according to label instructions, contact your State Pesticide or Environmental Control Agency, or the Hazardous Waste representative at the nearest EPA Regional Office for guidance.

CONTAINER DISPOSAL INSTRUCTIONS

Triple rinse (or equivalent). Then offer for recycling or reconditioning, or puncture and dispose of in a sanitary landfill, or by incineration, or, if allowed by state and local authorities, by burning. If burned, stay out of smoke.

GENERAL INFORMATION

HERBICIDE 273 is a liquid formulation which mixes readily with water for use in the control of cheatgrass, redroot or pigweed, Kochia, foxtails, burclover, blueweed (henbit), Texas blueweed, volunteer barley, wild buckwheat, barnyardgrass, bullgrass, annual bluegrass, ragweed, purslane, smartweed, wild carrot, volunteer sunflowers, and shepherdspurse in sugar beets.

Mix the recommended amounts of HERBICIDE 273 listed in the tables below in 2 to 15 gallons of water per acre for aerial application and in 10 to 40 gallons of water per acre for ground equipment.

This material is known to be phytotoxic to a wide variety of plants, especially strawberries. For aerial application a 50-foot buffer zone downwind from the wing edge is recommended to avoid injury to desirable plants. Apply when there is a positive wind of less than 10 mph. A height of 6 to 10 feet over the crop is recommended. Keep nozzle pressure below 40 psi.

DIRECTIONS FOR USE

It is a violation of Federal law to use this product in a manner inconsistent with its labeling.

RE-ENTRY STATEMENT

Do not apply this product in such a manner as to directly or through drift expose workers or other persons. The area being treated must be vacated by unprotected persons.

Do not enter treated areas without protective clothing until sprays have dried.

Because certain states may require more restrictive reentry intervals for various crops treated with this product, consult your State Department of Agriculture for further information.

Written or oral warnings must be given to workers who are expected to be in an area about to be treated with this product. In case of accidental exposure, workers should wash thoroughly with soap and water and contaminated clothing should be removed and washed before reuse. If symptoms of skin or eye irritation occur call a physician. When oral warnings are given, warnings shall be given in a language customarily understood by workers. Oral warnings shall be given if there is reason to believe that written warnings cannot be understood by workers. Written warnings must include the following information: "DANGER. Area treated with HERBICIDE 273 Sugar Beet Herbicide on (date of application). Do not enter without protective clothing until spray has dried. In case of accidental exposure, workers should wash thoroughly with soap and water and contaminated clothing should be removed and washed before reuse. If symptoms of skin or eye irritation occur call a physician."

PRE-EMERGENCE APPLICATIONS

Field tests have shown that best pre-emergence results are obtained when HERBICIDE 273 is sprayed onto the seed bed during planting in a manner that spray would be incorporated into the soil not exceeding 1½ inches deep. High moisture content of soil is essential for best weed control. Apply before sugar beets emerge from the soil.

Herbicide 273 Per Acre*

Broadcast Rates	Band Rates—Pints			
	Band Width	Row Spacing 22"	28"	32"
1.0–2.2 Gallons/Acre	5"	2.0–4.0	1.5–3.0	1.25–3.0
	7"	2.5–5.5	2.0–4.5	1.75–4.0
	10"	3.5–8.0	3.0–6.5	2.5 –5.5

* NOTE: For fine soils, use the lower rate. For use on the soils of Eastern Colorado and Western Nebraska, apply 1.5 to 3 pints HERBICIDE 273 for 7 inch band application on rows 22 inches wide. Adjust rate proportionately for other band widths and row spacings.

POST-EMERGENCE APPLICATIONS

Make application when the sugar beets have 2 to 4 true leaves (do not count the cotyledonary leaves, which are the first 2 leaves to appear). Under certain weather conditions, some temporary marginal leaf burn may occur to the sugar beets, especially at the higher rates, but recovery is normally rapid. For best results, apply when the temperature is between 60°F and 80°F. Do not apply later than 60 days after sugar beet emergence.

Herbicide 273 Per Acre*

Broadcast Rates	Band Rates—Fluid Ounces			
	Band Width	Row Spacing 22"	28"	32"
1–2 Quarts/Acre	5"	7–15	6–11	5–10
	7"	10–20	8–16	7–14
	10"	15–29	11–23	10–20

* NOTE: Use the lower rates of HERBICIDE 273 when the sugar beets have only 2 true leaves, or when spraying very susceptible weeds such as smartweed or wild buckwheat at early growth stage (2 to 4 true leaves).

TANK MIXES OF HERBICIDE 273 AND SELECTED HERBICIDES

FOR POST-EMERGENCE USE ON SUGAR BEETS ONLY

Follow all use directions and precautions that appear on both product labels.

In addition to the weeds mentioned above, tank mix combinations of HERBICIDE 273 plus one of the herbicides listed below will control a broader spectrum of annual broadleaf weeds and seedling grasses than either herbicide used alone. For best results, apply when sugar beets are at the 4 to 6 true leaf stage, and when the weeds are actively growing and are not under water or heat stress. Since leaf burn may occur, do not apply tank mixes before sugar beets have 4 expanded true leaves.

Mixing Directions—Fill sprayer tank half full with water. Start agitation with a mechanical or good hydraulic agitator. While agitating, add the required amount of HERBICIDE 273 followed by required amount of one of the other herbicides listed below. Add the remaining quantity of water with the bypass agitator running. Mix only the amount of spray that will be used immediately.

HERBICIDE 273 PLUS BETANAL®—Controls seedlings of annual broadleaf weeds, such as knotweed, lambsquarters, nightshade, mustard, London rocket, goosefoot, ground cherry, fiddleneck, chickweed, and green and yellow foxtail.

Do not add any wetting agents or other spray adjuvants. Do not apply later than 90 days prior to harvest.

HERBICIDE 273 PLUS BETANEX™—Controls seedlings of annual broadleaf weeds such as lambsquarters, mustard, nightshade, goosefoot, and London rocket.

Do not add any wetting agents or other spray adjuvants. Do not apply later than 90 days prior to harvest.

HERBICIDE 273 PLUS DOWPOM™—Controls seedlings of annual broadleaf weeds and grasses, such as foxtails, crabgrass, panicums, wild oats, bermudagrass, bluegrass, Johnsongrass, maidencane, and quackgrass.

HERBICIDE 273 PLUS PYRAMIN W®—Controls seedlings of annual broadleaf weeds, such as lambsquarters, mustard, fanweed, nightshade, neetleleaf, and goosefoot. Use 50 mesh screens in the line and Tee-jet #8004E nozzles or larger, or their equivalent.

Rates of Application—See Example Below

Post-Emergence Herbicide	Broadcast Rates Per Acre	Band Rates per Acre—Pints/Pounds			
		Band Width	Row Spacing 22"	28"	32"
HERBICIDE 273	2–3 pts	5"	0.5–0.7	0.4–0.6	0.3–0.4
		7"	0.7–1.0	0.5–0.8	0.5–0.7
		10"	0.9–1.3	0.7–1.1	0.6–0.9
BETANAL	6–7.5 pts	5"	1.4–1.7	1.0–1.2	0.9–1.1
		7"	2.0–2.5	1.5–1.9	1.3–1.6
		10"	2.7–3.4	2.1–2.6	1.8–2.2
BETANEX	6–7.5 pts	5"	1.4–1.7	1.0–1.2	0.9–1.1
		7"	2.0–2.5	1.5–1.9	1.3–1.6
		10"	2.7–3.4	2.1–2.6	1.8–2.2
DOWPON M (Dalapon)	2–3 lbs	5"	0.5–0.7	0.3–0.5	0.3–0.5
		7"	0.7–1.0	0.5–0.8	0.5–0.7
		10"	0.9–1.4	0.7–1.1	0.6–0.9
PYRAMIN W	4–5 lbs	5"	0.9–1.1	0.7–0.9	0.6–0.7
		7"	1.3–1.6	1.0–1.3	0.8–1.0
		10"	1.8–2.3	1.4–1.9	1.2–1.5

NOTE: Use the low rates for fine soils or when the temperature exceeds 80° F.

Example:

BROADCAST TREATMENT—Tank mix 2 to 3 pints of HERBICIDE 273 and 2 to 3 pounds of DOWPON M per acre.

BAND TREATMENT—For a 10" band and 32" row spacing, tank mix 0.6 to 0.9 pint of HERBICIDE 273 and 0.6 to 0.9 pound of DOWPON M per acre.

® BETANAL is a registered trademark of Schering A.G. Berlin/Bergkamen

™BETANEX is a trademark of Schering A.G. Berling/Bergkamen

™DOWPON M is a trademark of the Dow Chemical Company

® PYRAMIN W is a registered trademark of BASF Wyandotte Corporation

WARRANTY AND DISCLAIMER

Pennwalt Corporation warrants that this material conforms to the chemical description on the label and is reasonably fit for the purposes referred to in the Directions for Use, subject to the risks referred to therein. PENNWALT MAKES NO OTHER EXPRESS OR IMPLIED WARRANTY OF FITNESS OR MERCHANTABILITY OR ANY OTHER EXPRESS OR IMPLIED WARRANTY. IN NO CASE SHALL PENNWALT OR SELLER BE LIABLE FOR CONSEQUENTIAL, SPECIAL OR INDIRECT DAMAGES RESULTING FROM THE USE OR HANDLING OF THIS PRODUCT INCLUDING, BUT NOT LIMITED TO, LOSS OF PROFITS, BUSINESS REPUTATION, OR CUSTOMERS; LABOR COST; OR OTHER EXPENSES INCURRED IN PLANTING OR HARVESTING.

Pennwalt and seller offer this product and the buyer and user accept it subject to the foregoing conditions of sale and warranty which may be varied only be agreement in writing signed by a duly authorized representative of Pennwalt.

HYDROTHOL® 191

AQUATIC ALGICIDE AND HERBICIDE

Contains 2 lb. endothall* per gallon

ACTIVE INGREDIENT:

Mono(N,N-dimethylalkylamine**) salt of endothall*	53.0%
INERT INGREDIENTS:	47.0%
TOTAL	100.0%

* 7-oxabicyclo [2.2.1] heptane-2,3-dicarboxylic acid equivalent 23.36%

** Alkyl groups as derived from Coconut Oil

EPA Reg. No. 4581-174 EPA Est. No. 4581-TX-1

Keep Out of Reach of Children

DANGER–POISON

STATEMENT OF PRACTICAL TREATMENT

IF SWALLOWED, drink promptly a large quantity of milk, egg whites, gelatin solution or if these are not available, drink large quantities of water. Avoid alcohol. Call a physician immediately.

IF ON SKIN, immediately flush with plenty of water for at least 15 minutes. Remove and wash contaminated clothing before reuse.

IF IN EYES, immediately flush with plenty of water for at least 15 minutes. Call a physician.

NOTE TO PHYSICIAN: Probable mucosal damage may contraindicate the use of gastric lavage. Measures against circulatory shock, respiratory depression and convulsion may be needed.

See Side Panel for Additional Precautionary Statements

NOTE: For GENERAL INFORMATION and DIRECTIONS FOR USE refer to accompanying brochure.

PRECAUTIONARY STATEMENTS

HAZARDS TO HUMANS (AND DOMESTIC ANIMALS) DANGER

FATAL IF ABSORBED THROUGH SKIN. MAY BE FATAL IF SWALLOWED. CORROSIVE, CAUSES IRREVERSIBLE EYE DAMAGE AND SKIN BURNS. HARMFUL IF INHALED. DO NOT GET IN EYES, ON SKIN OR ON CLOTHING. WEAR PROTECTIVE CLOTHING, RUBBER GLOVES, AND GOGGLES OR FACE SHIELD WHEN HANDLING. Wash thoroughly with soap and water after handling and before eating or smoking. Remove contaminated clothing and wash before reuse. Avoid breathing spray mist.

ENVIRONMENTAL HAZARDS

Fish will be killed by dosages in excess of 0.3 ppm. Do not use where fish are important resources.

Avoid contact with or drift to desirable plants or crops as injury may result.

Clean out application equipment after each operation.

Do not use fish from treated water for food or feed within three days after treatment.

Do not use treated water for watering livestock or domestic purposes within the following periods:

Up to 0.3 ppm— 7 days after application
Up to 3.0 ppm—14 days after application
Up to 5.0 ppm—25 days after application

GENERAL INFORMATION

HYDROTHOL 191 is a liquid concentrate soluble in water and is a highly effective aquatic herbicide and algicide for use in irrigation and drainage canals, lakes and ponds to control the following weeds and algae: Najas, Elodea, Coontail, Potamogeton, Milfoil sp., Zannichellia, Vallisneria, Cladophora, Pithophora, Spirogyra, and Chara, when weeds are actively growing. Dosage rates indicated are measured in "Parts per Million" (ppm). 1.0 ppm as a dosage rate means that there would be 1 part endothall (acid) in 1,000,000 parts of water.

Necessary approval and/or permits should be obtained in States where required. Consult state water or conservation authorities before applying to public waters or to ponds, canals or streams which flow into public waters.

DIRECTIONS FOR USE

It is a violation of Federal law to use this product in a manner inconsistent with its labeling.

IRRIGATION AND DRAINAGE CANALS

Apply HYDROTHOL 191 as a surface spray with uniform coverage or inject under the surface of the water. The minimum contact time with weeds for optimum results should be two hours.

HEAVY INFESTATION—Apply at rate of 4.0 to 6.75 gal. per acre foot (equivalent to 3.0–5.0 ppm). See table below.

MODERATE OR LIGHT INFESTATIONS—Apply at rate of 1.4 to 2.7 gal. per acre foot (equivalent to 1.0–2.0 ppm). See table below.

GALLONS OF HYDROTHOL 191 REQUIRED TO TREAT 1 MILE OF DITCH, 1 FOOT DEEP

	Width of Ditch or Canal in Feet									
ppm	**4**	**6**	**8**	**10**	**12**	**14**	**16**	**18**	**20**	**22**
1.0 ppm	0.7	1.0	1.3	1.7	1.9	2.3	2.6	2.9	3.3	3.6
2.0 ppm	1.3	2.0	2.6	3.3	3.9	4.6	5.2	5.9	6.5	7.2
3.0 ppm	2.0	2.8	3.9	4.9	5.9	6.8	7.8	8.8	9.8	10.8
4.0 ppm	2.6	3.9	5.2	6.6	7.8	9.1	10.4	11.8	13.1	14.4
5.0 ppm	3.2	4.9	6.5	8.2	9.7	11.5	13.1	14.7	16.4	18.0

Note: 1.1 pt. HYDROTHOL 191 per acre foot equals 0.1 ppm of endothall (acid).

Do not use treated water for irrigation of crops.

LAKES AND PONDS

ALGAE CONTROL: Cladophora, Pithophora, Spirogyra, and Chara can be controlled with applications of 0.6 to 2.2 pints per acre foot (0.05 to 0.2 ppm) applied as a uniform surface spray of injected under water surface. Pithophora may require 0.2 ppm or higher. Dosages may be increased to 0.4 to 1.1 gals (0.3 to 0.8 ppm) where greater longevity of control is desired and for marginal treatments. Repeat treatments when algal growth reappears.

SUBMERGED AQUATICS: Due to the toxicity to fish, the use of HYDROTHOL 191 for submerged aquatic weeds is suggested only by commercial applicators on a marginal or sectional rather that overall type treatment. Use 0.7 to 3.4 gals. per acre foot (0.5 to 2.5 ppm) on a marginal treatment either sprayed on or injected below the surface. Use dosages over 1.0 ppm on very narrow margins or in areas where some fish kill is not objectionable.

Do not treat more than $1/10$ of the lake or pond at one time with doses in excess of 1.0 ppm.

APPROXIMATE QUANTITY OF HYDROTHOL 191 FOR ONE ACRE* TREATMENT

	DOSAGES FOR VARIOUS CONCENTRATIONS IN PPM								
Depth	**0.05**	**0.1**	**0.2**	**0.3**	**0.5**	**0.8**	**1.0**	**2.0**	**2.5**
1 ft.	0.6 pts	1.1 pts	2.2 pts	.4 gal	.7 gal	1.1 gal	1.4 gal	2.7 gal	3.4 gal
2 ft.	1.1 pts	2.2 pts	4.3 pts	.8 gal	1.4 gal	2.2 gal	2.8 gal	5.4 gal	6.8 gal
4 ft.	2.2 pts	4.3 pts	8.6 pts	1.6 gal	2.7 gal	4.4 gal	5.4 gal	10.9 gal	13.6 gal
6 ft.	3.3 pts	6.5 pts	12.9 pts	2.5 gal	4.1 gal	6.5 gal	8.2 gal	16.3 gal	20.4 gal

* One acre equals approximately 208′ X 208′

STORAGE AND DISPOSAL

Do not contaminate water, food, or feed by storage or disposal.

STORAGE INSTRUCTIONS
Store in the original container. Do not store in a manner where cross-contamination with other pesticides, fertilizers, food or feed could occur. In the event of spillage during handling or storage, absorb with sand or other inert material and dispose of absorbant in accordance with the Pesticide Disposal Instructions listed below.

PESTICIDE DISPOSAL INSTRUCTIONS
Pesticide wastes are acutely hazardous. Improper disposal of excess pesticide, spray mixture, or rinsate is a violation of Federal law. If these wastes cannot be disposed of by use according to label instructions, contact your State Pesticide or Environmental Control Agency, or the Hazardous Waste representative at the nearest EPA Regional Office for guidance.

CONTAINER DISPOSAL INSTRUCTIONS
Triple rinse (or equivalent). Then offer for recycling or reconditioning, or puncture and dispose of in a sanitary landfill, or by incineration, or, if allowed by state and local authorities, by burning. If burned, stay out of smoke.

WARRANTY AND DISCLAIMER

Pennwalt Corporation warrants that this material conforms to the chemical description on the label and is reasonably fit for the purposes referred to in the Directions for Use, subject to the risks referred to therein. PENNWALT MAKES NO OTHER EXPRESS OR IMPLIED WARRANTY OF FITNESS OR MERCHANTABILITY OR ANY OTHER EXPRESS OR IMPLIED WARRANTY. IN NO CASE SHALL PENNWALT OR SELLER BE LIABLE FOR CONSEQUENTIAL, SPECIAL OR INDIRECT DAMAGES RESULTING FROM THE USE OR HANDLING OF THIS PRODUCT INCLUDING, BUT NOT LIMITED TO, LOSS OF PROFITS, BUSINESS REPUTATION, OR CUSTOMERS; LABOR COST; OR OTHER EXPENSES INCURRED IN PLANTING OR HARVESTING.

Pennwalt and seller offer this product and the buyer and user accept it subject to the foregoing conditions of sale and warranty which may be varied only by agreement in writing signed by a duly authorized representative of Pennwalt.

KNOX OUT® 2FM

FLOWABLE MICROENCAPSULATED INSECTICIDE

SHAKE WELL BEFORE USING

ACTIVE INGREDIENTS:
Diazinon® (O, O–Diethyl O–[6 methyl–2–(1–methylethyl)–4–pyrimidinyl] phosphorothioate) 23%
INERT INGREDIENTS: 77%
TOTAL 100%

Contains 2 pounds Diazinon per gallon

EPA Reg. No. 4581-335 EPA Est. No. 4581-TX-1

U.S. Patents 3,429,827 and 3,577,515

Diazinon is a registered trademark of Ciba-Geigy Corporation

CAUTION

Keep Out of Reach of Children

HARMFUL IF SWALLOWED, INHALED, OR ABSORBED THROUGH SKIN

For additional Precautionary Statements, Statement of Practical Treatment and Storage and Disposal instructions refer to Side Panel. For Directions for Use refer to accompanying Brochure.

PRECAUTIONARY STATEMENTS

HAZARD TO HUMANS (AND DOMESTIC ANIMALS)

CAUTION

STATEMENT OF PRACTICAL TREATMENT

IF SWALLOWED—Drink 1 or 2 glasses of water and induce vomiting by touching back of throat with finger or blunt object. Do not induce vomiting or give anything by mouth to an unconscious person.

IF ON SKIN—Remove contaminated clothing immediately and wash skin with soap and water.

IF IN EYES—Wash thoroughly with water for at least 15 minutes. Call a physician if symptoms occur.

NOTE TO PHYSICIAN: The active ingredient of KNOX OUT 2FM is an organophosphate cholinesterase inhibitor. Atropine is antidotal only if symptoms of cholinesterase inhibition are present. Morphine is contraindicated.

Harmful if swallowed, inhaled or absorbed through skin. Avoid contact with skin, eyes or clothing. Avoid breathing spray mist. Wash thoroughly with soap and water after handling. Remove contaminated clothing and wash before reuse. Keep out of reach of children.

Do not apply to humans, their clothing or bedding. Do not allow children or pets to contact treated surfaces until spray has dried. Do not contaminate food or use on household pets. Cover fish tanks prior to application.

Avoid contamination of food, feedstuffs, or water supply. Do not contaminate food preparation surfaces, kitchen utensils, dishes, or food storage containers.

Do not spray where electrical short circuits might result, such as wall outlets, conduits, etc.

ENVIRONMENTAL HAZARDS
This product is toxic to fish, birds and other wildlife. Do not apply directly to water and wetlands. Do not contaminate water by cleaning of equipment or disposal of wastes.

This product is highly toxic to bees.

GENERAL INFORMATION

KNOX OUT 2FM is a flowable microencapsulated formulation containing 2 pounds Diazinon per gallon.

It is to be used for residual pest control in and around buildings including, but not limited to, homes, apartments, stores, hospitals, nursing homes, manufacturing plants, warehouses, schools, restaurants, hotels, motels, food processing plants, and on vessels, aircraft, boats, ships, rail cars, buses, trucks, and trailers.

KNOX OUT 2FM can be sprayed on any surface which will not be damaged or stained by water. Under conditions of proper application, KNOX OUT 2FM will not leave a visible deposit. However, if unevenly applied or sprayed to run-off, a visible deposit may occur on some dark surfaces and may be wiped off with a wet cloth or sponge.

Due to its controlled-release nature, KNOX OUT 2FM gives excellent residual kill of the pests listed below; however, effective knockdown may take up to 24 hours after application.

DIRECTIONS FOR USE
GENERAL CLASSIFICATION

It is a violation of Federal law to use this product in a manner inconsistent with its labeling.

Thoroughly wash out spray equipment before using KNOX OUT 2FM. Shake concentrate well before diluting. When diluting, first add approximately half of the water to the spray tank, then add the required quantity of KNOX OUT 2FM, and finally add the rest of the water. Occasionally agitate sprayer during use to assure even coverage. Shake or re-agitate sprayer before use if application is interrupted.

Do not use this formulation with oil.

IF SPRAY SCREENS ARE USED, THEY SHOULD BE 50 MESH OR LARGER.

Unless otherwise directed, dilute KNOX OUT 2FM with water to make a 0.5% to 1% spray** as indicated below:

Amount of Finished Spray	Amount* of KNOX OUT 2FM to use for: 0.5% Spray**	1% Spray**
1 Pint	0.3 fl. oz.	0.6 fl. oz.
2 Pints	0.6 fl. oz.	1.2 fl. oz.
4 Pints	1.2 fl. oz.	2.5 fl. oz.
1 Gallon	2.5 fl. oz.	5.0 fl. oz.
2 Gallons	5.0 fl. oz.	10.0 fl. oz.
3 Gallons	7.5 fl. oz.	15.0 fl. oz.
10 Gallons	25.0 fl. oz. (1.6 pints)	50.0 fl. oz. (1.6 quarts)

* 1 Tablespoon = 0.5 fl. oz.
** Active ingredient basis

Cockroaches and Silverfish—Make spot applications of KNOX OUT 2FM as a 0.5% to 1% fine spray or by paint brush to drawers, shelves, around water pipes, storage areas, closets, behind and under cabinets, refrigerators, sinks, stoves and in similar areas where these insects hide.

Repeat treatments of KNOX OUT 2FM should be made as needed, but generally will not be necessary more than once every 60 days. Use the low rate for light infestations or monthly maintenance sprays, and the higher concentration for heavy infestations, initial clean out, or where insecticide resistance is suspected.

Ants—Make spot applications of 0.5% to 1% KNOX OUT 2FM as a spray to baseboards, window frames, wall voids, crawl spaces, cracks and crevices, and similar hiding or habitation areas. Apply to trails, around doors and windows, and other areas where these pests may crawl or enter the premises. Repeat application as necessary.

Carpenter Ants—Make spot applications of 0.5% to 1% KNOX OUT 2FM to any trails, around doors and windows, wall voids, crawl spaces, and other places where ants enter premises. Where possible, treat ant nest or infested wood directly. Repeat application as necessary.

Carpet Beetles—Make applications of 0.5% to 1% KNOX OUT 2FM as a spray on and under edges of floor coverings, under carpets and furniture, and in closets or other localities where these insects are found. Entire carpets may be treated. When treating entire carpets see directions under **Fleas and Ticks.** Repeat application as necessary.

Crickets—Make spot applications of 0.5% to 1% of KNOX OUT 2FM as a spray to baseboards, behind and under cabinets and appliances, and other hiding places. Areas around doors and water pipes should be treated to help prevent crickets from entering the premises. Repeat application as necessary.

Flies—Make spot applications of 1% KNOX OUT 2FM as a spray to window frames, screens, dumpsters, trash cans, and similar areas where these pests may congregate. Repeat application as necessary.

Fleas and Ticks—Thoroughly apply KNOX OUT 2FM as a 1% fine spray to pet beds and resting quarters, adjacent cracks and crevices, baseboards, carpets and floor coverings, between and under cushions of upholstered furniture, and other areas where these pests may be present. Entire carpets may be treated. Applications to most surfaces should be as light as possible, approximately 2 quarts finished spray per 800 square feet. It may be necessary to use a higher volume of spray, approximately 3 to 4 quarts per 800 square feet, on long, dense shag carpets. Old bedding of pets should be replaced with clean, fresh bedding after treatment. Animals should not be sprayed with this product.

Perimeter Treatment—Residual spray for control of Ants, Carpenter Ants, Cockroaches, Crickets, Fleas, Silverfish, and Ticks:

Apply a 0.5% to 1% KNOX OUT 2FM spray to exterior surfaces of buildings, porches, screens, window frames, eaves, patios, garages, and other areas where these pests congregate or enter premises.

To help prevent infestations of buildings by the above pests, treat a band of soil 6 to 10 feet wide around and adjacent to the building. Also, treat the building foundation to a height of 2 to 3 feet where pests are active and may find entrance. KNOX OUT 2FM may be applied in compressed air equipment or power spray equipment. Apply as a coarse spray to thoroughly and uniformly wet the band area. A rate o 4 to 6 ounces of KNOX OUT 2FM per 1,000 square feet is desirable.

Food Handling Establishments—Places other than private residences in which food is held, processed, prepared, packaged, or served. Applications are limited to spot and/or crack and crevice treatments only.

Make spot and/or crack and crevice applications of KNOX OUT 2FM using a 0.5% to 1% spray in and around both food and nonfood locations, treating localized areas where cockroaches, silverfish, ants, and flies have been seen or are suspected of hiding or entering. Do not use as a space spray or allow spray to contact food, food processing surfaces, or utensils. Make applications using low pressure.

Crack and crevice applications should be made using equipment capable of delivering a pin stream or of applying small amounts of insecticide directly into or between equipment and floors, openings leading to voids in walls, floors, and ceilings, and similar locations where these pests may hide or enter.

Spot treatments should not cover areas of more than 2 square feet and not more than 20% of the floor surface should be treated.

This product is authorized for use in inedible product areas of Federally Inspected Meat and Poultry Plants.

Stored Product Pests—For control of exposed stages of Saw-toothed Grain Beetle, Confused Flour Beetle, Red Flour Beetle, Rice Weevil, Grain Weevil, Warehouse Beetle:

Make spot and/or crack and crevice treatment of 0.5% to 1% KNOX OUT 2FM to localized areas including shelves, cupboards, walls and floors, where food products suspected of being infested are stored. Food should be removed or covered during treatment. This treatment is supplemental to good housekeeping, since it will have no effect on insects breeding within packaging. Infested food packaging should be destroyed.

Use in Greenhouses—For control of Aphids, Fungus Gnat Larvae, Leafminers, Mealybugs, (Citrus, Stem and Root), Mites, Scafe Insects and Thrips:

Apply KNOX OUT 2FM to the ornamentals listed below at a rate of 0.5 to 1.0 fl. oz. per gallon of water (3-6 pints/100 gallons).

Treat when insects appear. Use higher rate for heavy insect infestations. Repeat applicatons as necessary.

ORNAMENTALS

African Violets	Japanese Arelia
Agapanthus	Juniper
Ageratum	Leucothoe
Aluminum Plant	Lilac
Alyssum	Malus
Arrowhead	Maranta
Azalea	Marigold
Begonia	Nandia
Bird's Nest Fern	Oleander
Boston Fern	Orchid

Bottlebrush
Cactus
Calendula
Camelia
Celosia Jewel Box
Chrysanthemum
Cockscomb
Coleus
Cotoneaster
Crape Myrtle
Croton
Daisy
Dianthus
Dogwood
Escalonia
Euonymus
Flat Topped Sedum
Floss Flower
Friendship Plant
Gardenia
Gazinia
Geranium
Golddust
Honeysuckle
Hydrangea
Ilex
Impatiens
Ivy
Jade Plant (Crassula)
Osmanthus
Pachysandra
Periwinkle
Petunia
Photinia
Pineapple Guava
Pittosporum
Podocarpus
Primrose
Privit
Pyracantha
Raphiolepsis
Rhododendron
Rose
Salvia
Sansevieria
Scarlet Sage
Snake Plant
Snapdragon
Spider Plant
Star Jasmine
Strawflower
Tahitian Bridal Veil
Velvet Plant (Gynura)
Verbena
Virvia
Wandering Jew
Wax Plant
Zinnia

Do not use on Bella Palm, Dracaena, Hoya, Maidenhand Fern, Neath, Poinsettia, and Prayer Plant.

KNOX OUT 2FM has demonstrated excellent plant safety; however, all varieties of the plants listed above have not been tested. When treating large numbers of plants of a single variety, spray a few plants and observe for phytotoxicity prior to full scale application. KNOX OUT 2FM may not control some organophosphate resistant ornamental plant pests.

Use In and Around Livestock Housing—For control of Flies in Barns and Animal Sleeping Quarters (except Dairy Barns, Milk Rooms and Poultry Houses):

Apply a 1% KNOX OUT 2FM spray to ceilings and walls of livestock sheds, calf barns, hog barns, loading sheds, and other farm buildings except dairy barns, milkrooms and poultry houses. Areas where flies congregate, such as garbage dumps and corrals, should be sprayed thoroughly. Repeat applications as fly populations warrant.

Remove animals from buildings prior to treatment and keep animals out until spray has dried.

Do not contaminate milk, feed, drinking water, or apply to animals or poultry.

Yellowjackets—Use in Baits for Control of Scavenger Yellowjackets (*Vespula pensylvanica, V. vulgaris*) in the 11 Contiguous Western States Only:

Consult the Supplemental Label available in these states.

STORAGE AND DISPOSAL

Do not contaminate water, food or feed by storage or disposal.

STORAGE

Store in original container only, preferably in a locked storage area. Do not store in a manner where cross contamination with other pesticides, fertilizer, food or feed could occur. If spilled during storage or handling, soak up spillage with absorbant material and dispose of in accordance with Pesticide Disposal instructions listed below.

PESTICIDE DISPOSAL

Wastes resulting from the use of this product may be disposed of on site or at an approved waste disposal facility.

CONTAINER DISPOSAL

Triple rinse container, then offer for recycling or reconditioning, or puncture and dispose of in a sanitary landfill, or by incineration, or, if allowed by state and local authorities, by burning. If burned, stay out of smoke.

WARRANTY AND DISCLAIMER

Pennwalt Corporation warrants that this material conforms to the chemical description on the label and is reasonably fit for the purposes referred to in the Directions for Use, subject to the risks referred to therein. PENNWALT MAKES NO OTHER EXPRESS OR IMPLIED WARRANTY OF FITNESS OR MERCHANTABILITY OR ANY OTHER EXPRESS OR IMPLIED WARRANTY. IN NO CASE SHALL PENNWALT OR SELLER BE LIABLE FOR CONSEQUENTIAL, SPECIAL OR INDIRECT DAMAGES RESULTING FROM THE USE OR HANDLING OF THIS PRODUCT INCLUDING, BUT NOT LIMITED TO, LOSS OF PROFITS, BUSINESS REPUTATION, OR CUSTOMERS; LABOR COST; OR OTHER EXPENSES INCURRED IN PLANTING OR HARVESTING.

Pennwalt and seller offer this product and the buyer and user accept it subject to the foregoing conditions of sale and warranty which may be varied only by agreement in writing signed by a duly authorized representative of Pennwalt.

REGISTRATION INFORMATION

KNOX OUT® 2FM Flowable Microencapsulated Insecticide

EPA Registration No. 4581-335 EPA SLN No. TX-820015

FOR DISTRIBUTION AND USE ONLY WITHIN THE STATE OF TEXAS

DIRECTIONS FOR USE

It is a violation of Federal law to use this product in a manner inconsistent with its labeling.

FIRE ANTS—To aid in the control of fire ants in lawns and other recreation areas, apply 1 fluid ounce (2 tablespoons) of KNOX OUT 2FM in 1 gallon of water as a drench. Gently sprinkle each mound with 1 gallon total of the diluted product. Make application over and 6 inches around each mound. Avoid using high pressure sprays as they may cause the fire ants to migrate to other locations.

For best results, make applications in the early morning or late evening hours, or when daytime temperatures are moderate (65° to 80°F). Make repeat applications as necessary. Treat new mounds promptly as they appear.

For treating more than one mound, larger quantities of the KNOX OUT 2FM solution may be prepared.

KEEP OUT OF REACH OF CHILDREN

CAUTION

HARMFUL IF SWALLOWED, INHALED, OR ABSORBED THROUGH SKIN

For First Aid Statement and Other Precautionary Statements, see label attached to the container.

All applicable directions, restrictions, and precautions on the EPA-registered label are to be followed.

NOTE: This labeling must be in the possession of the user at the time of pesticide application.

REGISTRATION INFORMATION

KNOX OUT® 2FM Flowable Microencapsulated Insecticide

EPA Registration No. 4581-335 EPA SLN No. AR-820024

FOR DISTRIBUTION AND USE ONLY WITHIN THE STATE OF ARKANSAS

DIRECTIONS FOR USE

It is a violation of Federal law to use this product in a manner inconsistent with its labeling.

FIRE ANTS—To aid in the control of fire ants in lawns and other recreation areas, apply 1 fluid ounce (2 tablespoons) of KNOX OUT 2FM in 1 gallon of water as a drench. Gently sprinkle each mound with 1 gallon total of the diluted product. Make application over and 6 inches around each mound. Avoid using high pressure sprays as they may cause the fire ants to migrate to other locations.

For best results, make applications in the early morning or late evening hours, or when daytime temperatures are moderate (65° to 80°F). Make repeat applications as necessary. Treat new mounds promptly as they appear.

For treating more than one mound, larger quantities of the KNOX OUT 2FM solution may be prepared.

KEEP OUT OF REACH OF CHILDREN

CAUTION

HARMFUL IF SWALLOWED, INHALED, OR ABSORBED THROUGH SKIN

For First Aid Statement and Other Precautionary Statements, see label attached to the container.

All applicable directions, restrictions, and precautions on the EPA-registered label are to be followed.

NOTE: This labeling must be in the possession of the user at the time of pesticide application.

KRYOCIDE®

INSECTICIDE

ACTIVE INGREDIENT:

Sodium Fluoaluminate (Fluorine—not less than 50%) ... 96.0%

INERT INGREDIENTS: 4.0%

TOTAL 100.0%

EPA Reg. No. 4581-116 EPA Est. No. 33770-JP-01

CAUTION

Keep Out of Reach of Children

See Specific Caution Statements On Panel Below

PRECAUTIONARY STATEMENTS

HAZARDS TO HUMANS (AND DOMESTIC ANIMALS)

CAUTION

HARMFUL IF SWALLOWED, ABSORBED THROUGH SKIN, OR INHALED, CAUSES MODERATE EYE IRRITATION. Avoid contact with skin, eyes, or clothing. Avoid breathing spray mist or dust. Wash thoroughly with soap and water after handling. Remove contaminated clothing and wash before reuse.

Do not apply this product in such a manner as to directly or through drift expose workers or other persons. The area being treated must be vacated by unprotected persons.

ENVIRONMENTAL HAZARDS

This product is toxic to fish. Keep out of any body of water. Do not apply where runoff is likely to occur. Do not contaminate water by cleaning of equipment or disposal of wastes.

GENERAL INFORMATION

KRYOCIDE is a fine powder and suitable for spray or dust application by ground equipment or air. A thorough application should be made, covering the underside of the leaves as well as the top if possible.

DIRECTIONS FOR USE

It is a violation of Federal law to use this product in a manner inconsistent with its labeling.

Do not enter treated areas without protective clothing until spray has dried or dust has settled. Because certain states may require more restrictive reentry intervals for various crops treated with this product, consult your State Department of Agriculture for further information.

Written or oral warnings must be given to workers who are expected to be in a treated area or in an area about to be treated with this product. When oral warnings are given, such warnings shall be given in a language customarily understood by the workers. Oral warnings must be given if there is reason to believe that written warnings cannot be understood by the workers. Written warnings must include the following information: " 'CAUTION. Area treated with KRYOCIDE® Insecticide on (*date of application*) "

KRYOCIDE® is recommended for control of the insects on the crops listed below.

	Pounds Per Acre		
Crop	Spray	Dust	Insects
Apples Ornamentals	25–50	25–50	Codling moth Flea weevil
Peaches Pears Shade Trees	25–50	25–50	Fuller rose beetle Gypsy moth caterpillar Katydids Plum curculio
Beans	25–50	25–50	Bean leaf beetle Bean leafroller Cucumber beetles Corn earworm Flea beetles Mexican bean beetle
Broccoli Cabbage Cauliflower Collards Mustard Radish Turnips	25–50	25–50	Cabbage looper Climbing cutworms Corn earworm Cucumber beetles Diamondback moth caterpillar Flea beetles Imported cabbage-worm Yellowstriped army-worm
Cantaloupe Cucumber Squash Watermelon	15–25	25–50	Cabbage looper Cucumber beetles Flea beetles Melonworm Pickleworm
Citrus	8–10	25–50	Citrus cutworm
	25–50	25–50	Fruittree leafroller Fuller rose beetle Garden tortrix Halcocera Katydids Orangedog Orange tortrix Variegated cutworm
Cranberries	25–50	25–50	Cranberry weevil False armyworm Gypsy moth caterpillar
Grapes	6–8(1) —by ground 8(2) —by air (helicopter)	10–20 (3)	Cutworms Grape leaffolder Grape leafroller Omnivorous leafroller Orange tortrix Western grapeleaf skeletonizer
Lettuce	15–30	15–30	Cabbage looper Corn earworm Tobacco budworm
Peppers	25–50	25–50	Hornworms Pepper weevil
Strawberries	8–10	25–50	Omnivorous leafroller

Crop	Pounds Per Acre Spray	Dust	Insects
			Strawberry leafroller
Tomatoes	15–30	15–30	Blister beetles Cabbage looper Flea beetles Fruitworms Hornworms Tomato pinworm

(1) in 25 to 200 gallons of water.
(2) In 20 gallons of water.
(3) Do not use as a dust after fruit formation.
Do not make more than one application as a dust to grapes per year.

NOTE: Do not use KRYOCIDE in combination with lime or compounds containing free lime. Do not apply KRYOCIDE immediately before a harvest.

Do not apply to bush and vine fruits, except grapes, after fruit starts to form. Do not apply to cabbage or broccoli after heads or edible parts start to form.

Do not apply to collards, mustard, cauliflower or lettuce after the seedling stage or to turnips after seedling stage if tops are to be used for food.

Remove excess residues on edible portions of fruit and vegetables by washing, brushing or other effective means.

For grapes, apply just as fruit starts to form. When needed to control later broods, make a second application after the fruit is 1/4" or larger in diameter. Do not use as a dust on grapes after fruit starts to form. Do not make more than 2 applications per year. Do not apply later than 30 days before harvest. For application to grapes grown for table use after the fruit is 1/4" or larger in diameter, use of concentrate sprays will reduce visible residue spots which may result from the use of dilute applications at that stage.

STORAGE AND DISPOSAL

Do not contaminate water, food, or feed by storage or disposal.

STORAGE INSTRUCTIONS
Store in the original container in a dry area. Do not store in a manner where cross-contamination with other pesticides, fertilizers, food or feed could occur. If spilled during storage or handling sweep up spillage and dispose of in accordance with the Pesticide Disposal Instructions listed below.

PESTICIDE DISPOSAL INSTRUCTIONS
Wastes resulting from the use of this product may be disposed of on site or at an approved waste disposal facility.

CONTAINER DISPOSAL INSTRUCTIONS
Completely empty bag into application equipment. Then dispose of empty bag in a sanitary landfill or by incineration, or, if allowed by State and local authorities, by burning. If burned, stay out of smoke.

WARRANTY AND DISCLAIMER

Pennwalt Corporation warrants that this material conforms to the chemical description on the label and is reasonably fit for the purposes referred to in the Directions for Use, subject to the risks referred to therein. PENNWALT MAKES NO OTHER EXPRESS OR IMPLIED WARRANTY OF FITNESS OR MERCHANTABILITY OR ANY OTHER EXPRESS OR IMPLIED WARRANTY. IN NO CASE SHALL PENNWALT OR SELLER BE LIABLE FOR CONSEQUENTIAL, SPECIAL OR INDIRECT DAMAGES RESULTING FROM THE USE OR HANDLING OF THIS PRODUCT INCLUDING, BUT NOT LIMITED TO, LOSS OF PROFITS, BUSINESS REPUTATION, OR CUSTOMERS; LABOR COST; OR OTHER EXPENSES INCURRED IN PLANTING OR HARVESTING.

Pennwalt and seller offer this product and the buyer and user accept it subject to the foregoing conditions of sale and warranty which may be varied only by agreement in writing signed by a duly authorized representative of Pennwalt.

PENNCAP-M®
MICROENCAPSULATED INSECTICIDE

RESTRICTED USE PESTICIDE
For retail sale to and use only by Certified Applicators or persons under their direct supervision and only for those uses covered by the Certified Applicator's certification.

ACTIVE INGREDIENTS:

0,0–Dimethyl O–p–nitrophenyl phosphorothioate*	20.9%
Related Isomers	1.1%
Xylene Base Aromatec Solvent	4.9%
INERT INGREDIENTS	73.1%
TOTAL	100.0%

*Methyl Parathion

Contains 2 pounds Methyl Parathion and Related Isomers per Gallon

EPA Reg. No. 4581-292 EPA Est. No. 4581-TX-1
U.S. Patent Nos. 3,429,827; 3,577,515; 3,959,464

WARNING
Keep Out of Reach of Children
MAY BE FATAL IF SWALLOWED, INHALED, OR ABSORBED THROUGH SKIN

Refer to back panel for first aid statement and note to physician

FOR GENERAL INFORMATION AND DIRECTIONS FOR USE REFER TO ACCOMPANYING BROCHURE

PRECAUTIONARY STATEMENTS

HAZARD TO HUMANS AND DOMESTIC ANIMALS
WARNING

FIRST AID: CALL A PHYSICIAN IN ALL CASES OF SUSPECTED POISONING. IT IS SUGGESTED THAT A SUITABLY EQUIPPED LOCAL MEDICAL FACILITY BE IDENTIFIED PRIOR TO USING THIS PRODUCT.

IF SWALLOWED—Drink 1 or 2 glasses of water and induce vomiting by touching back of throat with finger or blunt object. Do not induce vomiting or give anything by mouth to an unconscious person.

IF ON SKIN—Remove contaminated clothing immediately and wash skin with soap and water.

IF IN EYES—Wash thoroughly with water for at least 15 minutes.

IF INHALED—If warning symptoms appear, see below.

Warning symptoms of organophosphate poisoning include weakness, headache, tightness of chest, blurred vision, non-reactive pinpoint pupils, salivation, sweating, nausea, vomiting, diarrhea and abdominal cramps. If these symptoms appear transport the patient immediately to the nearest hospital. Keep patient prone and quiet. Start artificial respiration immediately if patient is not breathing.

Note to Physician: The active ingredient of PENNCAP-M is an organophosphate cholinesterase inhibitor. Atropine is antidotal. 2-PAM is also antidotal and may be administered in conjunction with atropine. Morphine is contraindicated.

Use in ventilated areas since accumulation of xylene vapors may be hazardous. Use with adequate ventilation. Avoid prolonged breathing of vapor.

Keep out of reach of domestic animals. Do not contaminate food, feed, or use on household pets or livestock.

ENVIRONMENTAL HAZARDS
This pesticide is extremely toxic to wildlife and fish. Use with caution when applying in areas frequented by wildlife or adjacent to any body of water. Birds and other wildlife feeding in treated areas

may be killed. Keep out of lakes, ponds, and streams. Do not apply when weather conditions favor runoff and drift from target areas.

THIS PRODUCT IS HIGHLY TOXIC TO BEES exposed in direct treatment or residues on crops or blooming weeds. Do not apply this product or allow it to drift to blooming crops or weeds if bees are visiting the areas to be treated. Refer to and follow the more specific and applicable precautions in the directions for certain crops. Protective information may be obtained from your Cooperative Agricultural Extension Service.

REENTRY STATEMENT

Do not enter treated fields without protective clothing before spray dries. Consult your State Agricultural Experiment Station regarding special local restrictions.

STORAGE AND DISPOSAL

Do not contaminate water, food or feed by storage or disposal. Open dumping is prohibited. Pesticide, spray mixture, or rinsate that cannot be used or chemically reprocessed should be disposed of in a landfill approved for pesticides or buried in a safe place away from water supplies. Do not contaminate water by cleaning of equipment or disposal of wastes.

Puncture container for disposal in approved landfill or bury in a safe place.

Consult Federal, State or local disposal authorities for approved alternative procedures.

GENERAL INFORMATION

PRECAUCION AL USARIO Resultarid peligroso al usar este producto al no ser de acuerdo con las direcciones. Si no lee ingles, no use este producto hasta que las instrucciones le hayan sido explicadas ampliamente.

PENNCAP-M is a flowable formulation consisting of a water suspension of polymeric-type microcapsules which contain methyl parathion. It is applied by air or ground equipment by diluting it with an amount of water suitable for the specific crop and type of spraying equipment. For air application, use at least 2 gallons total spray per acre.

First add approximately half of the water to the spray tank, then add the required quantity of PENNCAP-M with good agitation, and finally add the rest of the water. This procedure should be followed to insure proper mixing of PENNCAP-M

Mixtures of PENNCAP-M with emulsifiable concentrates, organic solvents, or some surfactants may cause increased toxicity and shorter residual effectiveness. This effect is more pronounced with concentrated sprays and extended periods between mixing and application.

DIRECTIONS FOR USE

It is a violation of Federal law to use this product in a manner inconsistent with its labeling.

DO NOT USE SCREEN OR NOZZLES FINER THAN 50 MESH. STIR OR SHAKE BEFORE USING.

PENNCAP-M will not control organophosphate resistant insects.

FIELD CROPS

Crop	Insects	Pints/Acre	Remarks
Alfalfa	Blue Alfalfa Aphid	1½–2	Do not spray alfalfa during bloom to avoid injury to bees. Do not apply to alfalfa grown for seed within 7 days prior to bloom or while bloom is present in the treatment area. Do not apply within 15 days of harvest or grazing.
	Grasshoppers*	1–3	
	Alfalfa Blotch Leafminer Alfalfa Plant Bug Alfalfa Weevil (larvae and adults) Green Cloverworm Potato Leafhopper Pea Aphid Spittlebugs Tarnished Plant Bug	2–3	
	Cutworms Egyptian Alfalfa Weevil	2–4	
Corn (Field)	Corn Rootworm Adults (for protection of silks)	1–2	First Generation European Corn Borer: Whorl Application—2 pints per acre by ground Broadcast Application—4 pints per acre Second Generation European Corn Borer: Broadcast Application—2 to 4 pints per acre. Use the low rate only for light infestations. When environmental conditions favor extended moth flights of the second generation of European Corn Borer two correctly timed applications of PENNCAP-M at the lowest recommended rate may provide improved control over a single application at a higher rate.** Apply PENNCAP-M at 2 pints per acre at the time of the initial hatch of the second generation. Repeat 10-14 days later. For low volume aerial applications to control Corn Rootworm Adults, Aphids, European Corn Borer and Western Bean Cutworm use the indicated rates of PENNCAP-M at a minimum of ½ gallon total spray per acre.** Application of undiluted PENNCAP-M is not recommended. Equipment that has been specially adapted and calibrated for low volume should be used. Best control will be achieved using spray devices that will produce droplets within the size range of 100 to 200 microns. The following conditions must be met 1 Equipment must be adjusted to distribute spray uniformly over the spray swath; 2 Wind conditions and other factors such as temperature and humidity are such that the spray is deposited on the target; 3 Local regulations do not prohibit low volume aerial sprays. For most efficient European Corn Borer and Corn Earworm control, use a spray schedule suitably adapted to the area. For best Black Cutworm control, apply PENNCAP-M under moist soil conditions. To avoid injury to bees, do not apply during pollen shed if bees are visiting the areas to be treated during their normal foraging hours. Do not harvest, cut for
	Grasshoppers*	1–3	
	Aphids Flea Beetles True Armyworm**	2–3	
	Corn Earworm European Corn Borer Southwestern Corn Borer Western Bean Cutworm	2–4	
	Black Cutworm	4	

FIELD CROPS

Crop	Insects	Pints/Acre	Remarks
			forage, or graze within 12 days of application.
Corn (Sweet)	Corn Rootworm Adults (for protection of silks)	1–2	First Generation European Corn Borer: Whorl Application—2 pints per acre by ground Broadcast Application—4 pints per acre Second Generation European Corn Borer: Broadcast Application—2 to 4 pints per acre. Repeat applications as necessary to maintain control. For most efficient European Corn Borer and Corn Earworm control, use a spray schedule suitably adapted to the area. For the best Black Cutworm control, apply PENNCAP-M under moist soil conditions. To avoid injury to bees, do not apply during pollen shed if bees are visiting the areas to be treated during their normal foraging hours. Do not apply later than 3 days before harvest. Do not cut for forage or graze within 12 days of application.
	Grasshoppers*	1–3	
	Aphids Flea Beetles True Armyworm**	2–3	
	Corn Earworm European Corn Borer Sap Beetles	2–4	
	Black Cutworm		
Cotton(1)	Fleahoppers Plant Bugs (including Lygus spp.) Thrips**	½–1	Do not apply within 7 days of hand picking or harvest.
	Boll Weevil	1–3	
	Bollworm Tobacco Budworm	4–6	
Rangeland, Pasture, Grass grown for seed production, and Roadside Areas	Grasshoppers*	1–3	Do not apply within 15 days of harvest or grazing.
	Thrips**	2	
	True Armyworms**	2–3	
Rice	Stink Bugs	1–2	Do not apply within 14 days of application of propanil. Do not apply within 15 days of harvest.
	Grasshoppers*	1–3	
Soybeans	Grasshoppers*	1–3	Do not make more than 2 applications per season.
	Bean Leaf Beetle Green Cloverworm Mexican Bean Beetle Potato Leafhopper Silverspotted Skipper Stink Bugs Threecornered Alfalfa Hopper Thrips** Velvetbean Caterpillar	2–3	Do not apply within 20 days of harvest or grazing.
	Corn Earworm (Podworm) Japanese Beetle**	3–4	
Tobacco	Grasshoppers*	1–3	Do not apply within 5 days of priming. Avoid contact with plant juices when priming or cutting tobacco.
	Aphids	2–4	
	Tobacco Hornworm Vegetable Weevil		
	Flea Beetle Tobacco Budworm	4–6	
Wheat, Oats, and Barley	Aphids (including Greenbug)	1–2	Do not apply within 15 days of harvest or grazing.
	Grasshoppers*	1–3	
	True Armyworm**	2–3	
Apples and Pears	Apple Maggot Apple and Thorn Skeletonizer Codling Moth Fruit Tree Leafroller** Green Fruitworm Lesser Appleworm Oblique-banded Leafroller** Oriental Fruit Moth Oyster Shell Scale** Pandemis Leafroller** Pear Plant Bug (pears only) Plum Curculio Red-banded Leafroller** San Jose Scale Tarnished Plant Bug Tufted Apple Budmoth Variegated Leafroller** White-marked Tussock Moth Woolly Apple Aphid**	4–8 OR 1–2 pts./100 gals. of dilute spray	Under heavy insect pressure, use the higher dosage. Consult your local agricultural experiment station for proper timing of sprays. Do not apply more than 8 pints per acre per treatment. Do not tank-mix PENNCAP-M with streptomycin since injury to fruit and foliage may result. Do not apply to D'Anjou pears as a concentrate or semi-concentrate spray since fruit injury is likely to occur. Do not make initial application before total petal fall. Do not apply when a substantial number of weeds or cover crops in orchards are in bloom and bees are actively visiting the area to be treated. Do not apply within 14 days of havest.
Cherries	Black Cherry Aphid** Fruit Flies (including Cherry Fruit Fly) Green Fruitworms Leafrollers** Plum Curculio	3–6 OR 1–2 pts./100 gals. of dilute spray	For additional control, during post-harvest period, treat interplanted trees, all unpicked fruit and pollinators. Do not apply more than 6 pints per acre per treatment. Do not make initial application before total petal fall. Do not apply when a substantial number of weeds or cover crops in orchards are in bloom and bees are actively visiting the area to be treated. Do not graze or feed cover crops from treated orchard. Do not apply within 14 days of harvest.
Grapes	Grape Aphids Grape Blossom Midge Grape Leafhopper (East of the Rocky Mountains) Potato Leafhopper Red-banded Leafroller**	4 OR 2 pts./100 gals. of dilute spray	Do not apply more than 12 pints per acre per treatment. Do not apply within 14 days of harvest.
	Grape Berry Moth Japanese Beetle**	4–8 OR 2–4 pts./100 gals. of dilute spray	

	Rose Chafer	6–8 OR 3–4 pts./100 gals. of dilute spray	
	Grape Leafhopper (West of the Rocky Mountains)	8–12 OR 4–6 pts./100 gals. of dilute spray	
Nectarines, Peaches, Plums, and Prunes	Lesser Peachtree Borer Peachtree Borer	4½–6 OR 1½–2 pints/100 gals. of dilute spray	Do not apply more than 8 pints per acre per treatment. Do not apply more than 2½ gallons per acre per year to nectarines. Do not make initial application before total petal fall.
	Catfacing Insects Oriental Fruit Moth Plum Curculio	3–6 OR 1–2 pints/100 gals. of dilute spray	Do not apply when a substantial number of weeds or cover crops in orchards are in bloom and bees are actively visiting the area to be treated. Do not graze or feed cover crops from treated orchards. Do not apply later than 14 days before harvest.
Nectarines, Peaches, Plums, and Prunes	Peach Twig Borer	8	Do not apply in less than 200 gallons per acre. Make dormant or delay dormant applications only. May be mixed with suitable spray oil. Do not allow tank mixes of PENNCAP-M and oil to stand, but use immediately. For precautions relating to the protection of pollinators and the grazing or feeding of cover crops, see above.
Nectarines and Peaches (Arizona and California only)	Insects as above	As above	Do not use more than 2½ gallons per acre on peaches from January 1 to harvest. Do not use more than 8 pints per acre per treatment when using dilute sprays on nectarines. Do not apply more than 2½ gallons per acre per year to nectarines. For precautions relating to the protection of pollinators and the grazing or feeding of cover crops, see above. Do not apply within 21 days of harvest.

VEGETABLE CROPS

Crop	Insects	Pints/Acre	Remarks
Artichokes	Artichoke Plume Moth	4	Do not apply within 7 days of harvest.
Beans (Dry & Succulent, including Southern Peas)	Aphids** Cucumber Beetles, (**Diabrotica** spp.) Green Cloverworm Leafhoppers Lygus Bugs Mexican Bean Beetle	2	Beans—To avoid injury to bees, do not apply during the period from 7 days prior to first bloom through peak bloom. After this time, PENNCAP-M may be applied to beans as long as bees are not actively visiting the area. Southern Peas—Start application when blooms are first observed and repeat at 3 to 5 day intervals. Do not apply within 15 days of harvest.
	Cowpea Curculio European Corn Borer	2–4	
	Corn Earworm	4	
Cabbage	Diamondback Moth** Flea Beetles	2–4	Do not apply within 21 days of harvest
	Imported Cabbage-worms	4–6	
Onions	Thrips	2	Do not apply when onions are blooming to avoid injury to bees. Do not apply within 15 days of harvest.
Peas (White Blooming)	Pea Aphid Pea Leaf Weevil Pea Weevil	2	Do not apply more than 4 pints per acre per year. Do not apply within 10 days of harvest for pods. Do not harvest for forage or graze within 15 days of applications.
Peas (Austrian Winter)	Pea Aphid Pea Leaf Weevil Pea Weevil	2	See Peas (White Blooming) Do not apply within 7 days of bloom or if bloom is evident in the area to be treated.
Potatoes	Grasshoppers*	1–3	Do not apply within 5 days of harvest.
	Cutworms European Corn Borer Flea Beetles Potato Leafhopper Tarnished Plant Bug	2–4	
	Colorado Potato Beetle	2–6	
Tomatoes	Flea Beetles Leafminers	2–4	Do not apply within 15 days of harvest.
	Colorado Potato Beetle Hornworms Tomato Fruitworm Tomato Pinworm	4	

* For grasshoppers, use higher rates if the majority of insects are large or weather is cool.
** Not for this use in California.

APPLICATION THROUGH SPRINKLER IRRIGATION SYSTEMS (Except California)

Apply PENNCAP-M Insecticide only through systems containing anti-syphon and check valves designed to prevent water source contamination or overflow of the mix tank and containing interlocking controls between the metering device and the water pump to insure simultaneous shut-off.

Maintain a gentle continuous agitation in mix tank during mixing and application to assure a uniform suspension.

Greater accuracy in calibration and distribution will be achieved by injecting a larger volume of a more dilute suspension per unit time.

Application of more than recommended quantities of irrigation water per acre may result in decreased product performance.

Do not apply when wind speed favors drift, when system connections or fittings leak, when nozzles do not provide uniform distribution or when lines containing the product cannot be flushed and must be dismantled and drained. In a center pivot system, block the nozzle set nearest the well pivot injection unit to prevent spray being applied to this area. All screens should be 50 mesh or larger.

Where sprinkler distribution patterns do not overlap sufficiently, unacceptable insect control may result.

Allow sufficient time for pesticide to be flushed through all lines and all nozzles before turning off irrigation water.

PENNCAP-M may be applied in conjunction with chemically neutral liquid fertilizers. Application in conjunction with highly alkaline fertilizers, such as aqueous ammonia, may cause a degradation of the pesticide, resulting in reduced performance and should be avoided.

Check local restrictions and requirements regarding sprinkler irrigation applications, as they may vary from state to state.

SPRAY PREPARATION

Remove scale, pesticide residues, and other foreign matter from the chemical tank and entire injector system. Flush with clean water.

Prepare a suspension of PENNCAP-M Insecticide in a mix tank. Fill the tank with ½ or ¾ the desired amount of water. Start mechanical or hydraulic agitation. Slowly add the required amount of PENNCAP-M and then the remaining volume of water.

The addition of once-refined vegetable oil in the mix tank (1 part oil and 1 part PENNCAP-M) may improve efficacy. The PENNCAP-M should be added to the mix tank first.

APPLICATION INSTRUCTIONS

Set sprinkler system to deliver 0.1 to 1.25 inches of water per acre. Volumes of water higher than this may reduce efficacy. Start sprinkler and then uniformly inject the suspension of PENNCAP-M into the irrigation water line so as to deliver the desired rate per acre. The suspension of PENNCAP-M should be injected with a positive displacement pump into the main line ahead of a right angle turn to insure adequate mixing.

NOTE: When treatment with PENNCAP-M has been completed, do not irrigate the treated area for 24 to 28 hours to prevent washing the chemical off the crop.

Crop	Insect	Rate Pints/Acre	Remarks
Alfalfa	Egyptian Alfalfa Weevil	2–4	Do not apply to alfalfa during bloom to avoid injury to bees.
	Alfalfa Weevil Pea Aphid Potato Leafhopper	2–3	Do not apply to alfalfa grown for seed within 7 days prior to bloom or while bloom is present in treatment area.
	Grasshoppers	1–3	Do not apply within 15 days of harvest or grazing.
Beans	Cucumber Beetle (Diabrotica sp.)	2	To avoid injury to bees, do not apply during the period from 7 days prior to first bloom to peak bloom. After this time, PENNCAP-M may be applied as long as bees are not actively foraging. Do not apply within 15 days of harvest.
Corn Sweet & Field	Corn Rootworm Adults	1–2	To avoid injury to bees, do not apply during pollen shed if bees are visiting the areas to be treated during their normal foraging hours. First Generation European Corn Borer 4 pints per acre. Second Generation European Corn Borer 2-4 pints per acre. Use the low rate only for light infestations. On sweet corn repeat applications as necessary to maintain control. For the most efficient European Corn Borer and Corn Earworm control, use a spray schedule suitably adapted to the area. Do not cut field or sweet corn for forage or graze within 12 days of application. Do not harvest field corn within 12 days of application. Do not apply to sweet corn later than 3 days before harvest.
	Aphids Flea Beetle	2–3	
	Corn Earworm European Corn Borer Sap Beetle (Sweet only) Western Bean Cutworm (Field only) Southwestern Corn Borer (Field only)	2–4	
Cabbage	Diamondback Moth	2–4	Do not apply within 21 days of harvest.
	Imported Cabbageworm	4–6	
Southern Peas	Cowpea Curculo	2–4	Start application when blooms are first observed and repeat at 3-5 day intervals.
	Cucumber Beetle (Diabrotica sp.)	2	Do not apply within 15 days of harvest.

Multiple applications may be necessary if insect pressure is heavy. Under heavy insect pressure the higher rates are recommended.

SUPPLEMENTAL LABELING FOR TANK MIXES WITH PENNCAP-M

TANK MIX FOR USE ON APPLES AND PEARS
(All Apple and Pear Growing Areas Except California)

Insects	Rates	Remarks
Apple Maggot Apple and Thorn Skeletonizer Codling Moth Fruit Tree Leafroller Green Fruitworm Lesser Appleworm Oblique-banded Leafroller Oriental Fruit Moth Oyster Shell Scale Pandemis Leafroller Pear Plant Bug (pears only) Plum Curculio Red-banded Leafroller San Jose Scale Tarnished Plant Bug Tufted Apple Budmoth Variegated Leafroller White-marked Tussock Moth Woolly Apple Aphid	PENNCAP-M 2–4 pints/acre Plus azinphosmethyl 50W 1 lb./acre OR PENNCAP-M ½–1 pint Plus azinphos-methyl 50W ¼ lb./100 gallons for dilute spray	Do not apply more than 4 pints of PENNCAP-M plus 1 pound of azinphos-methyl 50 W per acre per treatment. Do not apply within 7 days of harvest.

Observe all limitations, precautions, and rate recommendations which appear on the labels for these products.

TANK MIX FOR USE ON APPLES AND PEARS
(All Apple and Pear Growing Areas Except California)

Insects	Rates	Remarks
Apple Maggot Apple and Thorn Skeletonizer Codling Moth Fruit Tree Leafroller Green Apple Aphid Green Fruitworm Lesser Appleworm Oblique-banded Leafroller Oriental Fruit Moth Oyster Shell Scale Pandemis Leafroller Pear Plant Bug (pears only) Plum Curculio Red-banded Leafroller	PENNCAP-M 2–4 pints/acre Plus one of the following ZOLONE®3EC* 2–6 pints/acre ZOLONE 25W 4–8 lbs./acre OR PENNCAP-M ½–1 pint Plus one of the following	Do not apply more than 4 pints PENNCAP-M plus 6 pints of ZOLONE 3EC or 8 pounds of ZOLONE 25W per acre per treatment. Do not apply within 14 days of harvest.

San Jose Scale Tarnished Plant Bug Tentiform Leafminer Tufted Apple Budmoth Variegated Leafroller White Apple Leafhopper White-marked Tussock Moth Woolly Apple Aphid	ZOLONE 3EC ½–1½ pints ZOLONE 25W 1–2 lbs./100gallons for dilute spray	

Observe all limitations, precautions, and rate recommendations which appear on the labels for these products.

* ZOLONE is a registered trademark of Rhone-Poulenc Chemical Company.

TANK MIX FOR USE ON APPLES
(All Apple Growing Areas Except California)

Insects	Rates	Remarks
Apple Maggot Apple and Thorn Skeletonizer Codling Moth Fruit Tree Leafroller Green Apple Aphid Green Fruitworm Lesser Appleworm Oblique-banded Leafroller Oriental Fruit Moth Oyster Shell Scale Pandemis Leafroller Pear Plant Bug (pears only) Plum Curculio Red-banded Leafroller San Jose Scale Tarnished Plant Bug Tentiform Leafminer Tufted Apple Budmoth Variegated Leafroller White Apple Leafhopper White-marked Tussock Moth Woolly Apple Aphid	PENNCAP-M 2–4 pints/acre Plus one of the following methomyl 1.8L 2–4 pints/acre methomyl 90SP ½–1 lb./acre OR PENNCAP-M ½–1 pint Plus one of the following methomyl 1.8L ½–1 pint methomyl 90SP 2–4 ounces/100 gallons for dilute spray	Do not apply more than 4 pints of PENNCAP-M plus 4 pints of methomyl 1.8L or 1 pound methomyl 90SP per acre per treatment. Do not apply within 8 days of harvest.

Observe all limitations, precautions, and rate recommendations which appear on the labels for these products.

TANK MIX FOR USE ON COTTON
(All Cotton Growing Areas Except California)

Insects	Rates/Acres	Remarks
Boll Weevil Bollworm Tobacco Budworm	PENNCAP-M 2-3 pints Plus one of the following GALECRON® SP* FUNDAL® SP** ⅛-¼ pound GALECRON 4E FUNDAL 4E ⅛-½ pint	Always prepare fresh mixtures just prior to use. Do not apply within 21 days of harvest. Do not feed foliage from treated cotton plant or gin trash to livestock.

Observe all limitations, precautions, and rate recommendations which appear on the labels for these products.

* GALECRON is a registered trademark of Ciba-Geigy.

** FUNDAL is a registered trademark of Nor-AM®

TANK MIX FOR USE ON COTTON
(All Cotton Growing Areas Except California)

Insects	Rates/Acre	Remarks
Armyworms Boll Weevil Bollworm Cabbage Looper Plant Bugs Tobacco Budworm Whiteflies	PENNCAP-M 1-3 pints Plus ORTHENE® Insecticide* ½-1 pound	Do not apply within 21 days of harvest.

Observe all limitations, precautions, and rate recommendations which appear on the labels for these products.

* ORTHENE is a registered trademark of Chevron Chemical Company.

TANK MIX FOR USE ON COTTON
(All Cotton Growing Areas Except California)

Insects	Rates/Acre	Remarks
Boll Weevil Bollworm Tobacco Budworm	PENNCAP-M 1–2 pints Plus AMBUSH® Insecticide* ⅕–⅘ pint	Low rates indicated should be used under low insect pressure or in a season long program under close scouting supervision. Do not apply within 14 days of harvest.

Observe all limitations, precautions, and rate recommendations which appear on the labels for these products.

* AMBUSH is a registered trademark of ICI Americas, Inc.

FOR USE ON COTTON ONLY
USE ONLY WATER AS A CARRIER
DO NOT USE VEGETABLE OIL

Use PENNCAP-M at a rate of ½ pint to 6 pints, plus .05 lbs. ai to .2 lbs. ai of Pydrin 2.4 EC.

LOW VOLUME AERIAL APPLICATIONS ON FIELD CORN

(Not for this use in California)

PENNCAP-M Insecticide is recommended for use in low volume aerial applications at a minimum of ½ gallon total spray per acre. Application of undiluted PENNCAP-M is not recommended. Equipment that has been specially adapted and calibrated for low volume application should be used. Best control will be achieved using spray devices that will produce droplets within the size range of 100 to 200 microns. The following conditions must be met:

1. Equipment must be adjusted to distribute spray uniformly over the spray swath;
2. Wind conditions and other factors such as temperature and humidity are such that the spray is deposited on the target:
3. Local regulations do not prohibit low volume aerial sprays.

Crop	Insect	Rate/A	Remarks
Field Corn	Western Bean Cutworm European Corn Borer	2–4 pts	Do not harvest, cut for forage, or graze within 12 days of application.
	Corn Rootworm Adults	1–2 pts	
	Aphids	2–3 pts	

Follow all applicable restrictions and precautions on the PENNCAP-M label.

SPLIT APPLICATION FOR THE CONTROL OF SECOND GENERATION EUROPEAN CORN BORER ON FIELD CORN

(Not for this use in California)

When environmental conditions favor extended moth flights of the second generation of European Corn Borer, two correctly timed applications of PENNCAP-M at the lowest recommended rate may provide improved control over a single application at a higher rate.

Crop	Insect	Rate/A	Remarks
Field Corn	European Corn Borer	2	Apply PENNCAP-M at the rate indicated at the time of the initial hatch of the second generation. Repeat 10-14 days later. Do not harvest, cut for forage, or graze within 12 days of application.

Follow all applicable precautions and limitations on the PENNCAP-M label.

When making applications in low volumes (½-2 gallons per acre) follow additional precautions on the PENNCAP-M supplemental label for low volume aerial applications to field corn.

APPLICATION THROUGH SPRINKLER IRRIGATION SYSTEMS

(Not for this use in California)

Apply PENNCAP-M Insecticide only through systems containing anti-syphon and check valves designed to prevent water source contamination or overflow of the mix tank and containing interlocking controls between the metering device and the water pump to insure simultaneous shut-off.

Maintain a gentle continuous agitation in mix tank during mixing and application to assure a uniform suspension.

Greater accuracy in calibration and distribution will be achieved by injecting a larger volume of a more dilute suspension per unit time.

Application of more than recommended quantities of irrigation water per acre may result in decreased product performance.

Do not apply when wind speed favors drift, when system connections or fittings leak, when nozzles do not provide uniform distribution or when lines containing the product cannot be flushed and must be dismantled and drained. In a center pivot system, block the nozzle set nearest the well/pivot/injection unit to prevent spray being applied to this area. All screens should be 50 mesh or larger.

Where sprinkler distribution patterns do not overlap sufficiently, unacceptable insect control may result.

Allow sufficient time for pesticide to be flushed through all lines and all nozzles before turning off irrigation water.

PENNCAP-M may be applied in conjunction with chemically neutral liquid fertilizers. Application in conjunction with highly alkaline fertilizers, such as aqueous ammonia, may cause a degradation of the pesticide, resulting in reduced performance and should be avoided.

Check local restrictions and requirements regarding sprinkler irrigation applications, as they may vary from state to state.

SPRAY PREPARATION

Remove scale, pesticide residues, and other foreign matter from the chemical tank and entire injector system. Flush with clean water.

Prepare a suspension of PENNCAP-M Insecticide in a mix tank. Fill the tank with ½ or ¾ the desired amount of water. Start mechanical or hydraulic agitation. Slowly add the required amount of PENNCAP-M and then the remaining volume of water.

The addition of once-refined vegetable oil in the mix tank (1 part oil and 1 part PENNCAP-M) may improve efficacy. The PENNCAP-M should be added to the mix tank first.

APPLICATION INSTRUCTIONS

Set sprinkler system to deliver 0.1 to 1.25 inches of water per acre. Volumes of water higher than this may reduce efficacy. Start sprinkler and then uniformly inject the suspension of PENNCAP-M into the irrigation water line so as to deliver the desired rate per acre. The suspension of PENNCAP-M should be injected with a positive displacement pump into the main line ahead of a right angle turn to insure adequate mixing.

NOTE: When treatment with PENNCAP-M has been completed, do not irrigate the treated area for 24 to 48 hours to prevent washing the chemical off the crop.

CROP	INSECT	RATE (PTS/ACRE)	REMARKS
Alfalfa	Egyptian Alfalfa Weevil	2–4	Do not apply to alfalfa during bloom to avoid injury to bees. Do not apply to alfalfa grown for seed within 7 days prior to bloom or while bloom is present in treatment area. Do not apply within 15 days of harvest or grazing.
	Alfalfa Weevil Pea Aphid Potato Leafhopper	2–3	
	Grasshoppers	1-3	
Beans	Cucumber Beetle (*Diabrotica* sp.)	2	To avoid injury to bees, do not apply during the period from 7 days prior to first bloom to peak bloom. After this time, PENNCAP-M may be applied as long as bees are not actively foraging. Do not apply within 15 days of harvest.
Corn, Sweet & Field	Corn Rootworm Adults	1–2	To avoid injury to bees, do not apply during pollen shed if bees are visiting the areas to be treated during their normal foraging hours. First generation European Corn Borer: 4 pints per acre. Second generation European Corn Borer: 2-4 pints per acre. Use the low rate only for light infestations. On sweet corn repeat applications as necessary to maintain control. For the most efficient European Corn Borer and Corn Earworm control, use a spray schedule suitable adapted to the area. Do not cut field or sweet corn for forage or graze within 12 days of application. Do not harvest field corn within 12 days of application. Do not apply to sweet corn later than 3 days before harvest.
	Aphids Flea Beetle	2–3	
	Corn Earworm European Corn Borer Sap Beetle (Sweet only) Western Bean Cutworm (Field only) Southwestern Corn Borer (Field only)	2–4	
Cabbage	Diamondback Moth	2–4	Do not apply within 21 days of harvest.
	Imported Cabbage Worm	4–6	
Southern Peas	Cowpea Curculio	2–4	Start application when blooms are first observed and repeat at 3-5 intervals. Do not apply within 15 days of harvest.
	Cucumber Bettle (*Diabrotica* sp.)	2	

Multiple applications may be necessary if insect pressure in heavy. Under heavy insect pressure the higher rates are recommended.

Do not enter treated fields without protective clothing before spray dies.

Observe all limitations, precautions and rate recommendations which appear on the labels for these products.

TANK MIX FOR USE ON COTTON
(Not for this use in California)

Insects	Rates/Acre	Remarks
Boll Weevil Bollworm Tobacco Budworm	PENNCAP-M 1–2 pints Plus PYDRIN® ⅙–⅔ pint	Low rates indicated should be used under low insect pressure or in a season long program under close scouting supervision. Do not apply within 21 days of harvest.

PYDRIN is a registered trademark of Shell Chemical Company

SPECIAL LOCAL NEED LABEL

PENNCAP-M® INSECTICIDE

FOR DISTRIBUTION AND USE ONLY WITHIN THE STATE OF NEW MEXICO

EPA Reg No. 4581-292 EPA SLN No. NM-820004

DIRECTIONS FOR USE

It is a violation of Federal law to use this product in a manner inconsistent with its labeling.

WHEAT—For the control of armyworm, greenbug, grasshoppers, leafhoppers, and aphids, apply PENNCAP-M at the rate of 1 to 2 pints per acre. Start application when insect damage threatens. Do not apply after heads form. Do not apply within 15 days of grazing.

KEEP OUT OF REACH OF CHILDREN
WARNING
MAY BE FATAL IF SWALLOWED, INHALED, OR ABSORBED THROUGH SKIN

For First Aid Statement and Other Precautionary Statements, see label attached to container.

All applicable directions, restrictions, and precautions on the EPA-registered label are to be followed.

NOTE: This labeling must be in the possession of the user at the time of pesticide application.

SPECIAL LOCAL NEED LABEL

PENNCAP-M® INSECTICIDE

FOR DISTRIBUTION AND USE ONLY WITHIN THE STATE OF DELAWARE

EPA Reg. No. 4581-292 EPA SLN No. DE-800006

DIRECTIONS FOR USE

It is a violation of Federal law to use this product in a manner inconsistent with its labeling.

SOYBEANS—For the control of silverspotted skipper, potato leafhopper, thrips, and green cloverworm, apply PENNCAP-M at the rate of 2 to 3 pints per acre by conventional air or ground equipment. Do not use less than 3 gallons total spray per acre for aerial applications.

For silverspotted skipper and green cloverworm, adhere to the Delmarva insect population and defoliation thresholds for timing of sprays. For potato leafhoppers and thrips, apply when young plants begin to show injury.

Do not apply later than 20 days before harvest or grazing. Do not make more than 2 applications per season.

KEEP OUT OF REACH OF CHILDREN
WARNING
MAY BE FATAL IF SWALLOWED, INHALED, OR ABSORBED THROUGH SKIN.

For First Aid Statement and Other Precautionary Statements, see label attached to the container.

All applicable directions, restrictions, and precautions on the EPA-registered label are to be followed.

NOTE: This labeling must be in the possession of the user at the time of pesticide application.

SPECIAL LOCAL NEED LABEL

PENNCAP-M® INSECTICIDE

FOR DISTRIBUTION AND USE ONLY WITHIN THE STATE OF MISSISSIPPI

EPA Reg. No. 4581-292 EPA SLN No. MS-830004

DIRECTIONS FOR USE

It is a violation of Federal law to use this product in a manner inconsistent with its labeling.

CHRYSANTHEMUMS—For the control of leafminers on chrysanthemums grown in commercial greenhouses, apply PENNCAP-M at the rate of 2 pints per 100 gallons water as a full coverage spray on foliage. If area to be treated is less than 1 acre, reduce finished spray volume accordingly. Begin application when insect damage threatens and continue at 7-day intervals. Use in well ventilated area or use approved respiratory device during application if greenhouse is closed and air circulation is not possible. Apply PENNCAP-M to a few blooms of each cultivator and observe for phytotoxicity for a period of 7 days before applying this product to larger areas of plants in bloom. Do not market plants within one (1) day of last application.

KEEP OUT OF REACH OF CHILDREN
WARNING
MAY BE FATAL IF SWALLOWED, INHALED, OR ABSORBED THROUGH SKIN.

Workers entering treated greenhouse areas within 24 hours after application should wear protective clothing including clean coveralls, gloves, caps, etc.

For First Aid Statement and Other Precautionary Statements, see label attached to the container.

All applicable directions, restrictions, and precautions on the EPA-registered label are to be followed.

NOTE: This labeling must be in the possession of the user at the time of pesticide application.

SPECIAL LOCAL NEED LABEL

PENNCAP-M® INSECTICIDE

FOR DISTRIBUTION AND USE ONLY WITHIN THE STATE OF OHIO

EPA Reg. No. 4581-292 EPA SLN No. OH-810020

DIRECTIONS FOR USE

It is a violation of Federal law to use this product in a manner inconsistent with its labeling.

CHRYSANTHEMUMS—For the control of leafminers on chrysanthemums grown in commercial greenhouses, apply PENNCAP-M at the rate of 2 pints per 100 gallons water as a full coverage spray on foliage. If area to be treated is less than 1 acre, reduce finished spray volume accordingly. Begin application when insect damage threatens and repeat as necessary. Apply PENNCAP-M to a few blooms of each cultivar and observe for phytotoxicity before applying this product to larger areas of plants in bloom.

Use in well ventilated area or use approved respiratory device during application if greenhouse is closed and air circulation is not possible. Do not enter treated area without protective clothing until spray dries. Plants treated are not to be sold within 24 hours of treatment.

KEEP OUT OF REACH OF CHILDREN
WARNING
MAY BE FATAL IF SWALLOWED, INHALED, OR ABSORBED THROUGH SKIN.

For First Aid Statement and Other Precautionary Statements, see label attached to the container.

All applicable directions, restrictions, and precautions on the EPA-registered label are to be followed.

NOTE: This labeling must be in the possession of the user at the time of pesticide application.

SPECIAL LOCAL NEED LABEL

PENNCAP-M® INSECTICIDE

FOR DISTRIBUTION AND USE ONLY WITHIN THE STATE OF VIRGINIA

EPA Reg. No. 4581-292 EPA SLN No. VA-810013

DIRECTIONS FOR USE

It is a violation of Federal law to use this product in a manner inconsistent with its labeling.

CHRYSANTHEMUMS—For the control of leafminers on chrysanthemums grown in commercial greenhouses, apply PENNCAP-

M at the rate of 2 pints per 100 gallons water as a full coverage spray on foliage. If area to be treated is less than 1 acre, reduce finished spray volume accordingly. Begin application when insect damage threatens and continue at 7-day intervals. Use in well ventilated area or use approved respiratory device such as NIOSH + MSHA Approval Nos. TC23(c)-40 or TC23(c)-47 during application if greenhouse is closed and air circulation is not possible. Apply PENNCAP-M to a few blooms of each cultivar and observe for phytotoxicity for a period of 7 days before applying this product to larger areas of plants in bloom. Do not market plants within one (1) day of last application.

KEEP OUT OF REACH OF CHILDREN
WARNING

MAY BE FATAL IF SWALLOWED, INHALED, OR ABSORBED THROUGH SKIN.

Workers entering treated greenhouse areas within 24 hours after application should wear protective clothing including clean coveralls, gloves, caps, etc.

For First Aid Statement and Other Precautionary Statements, see label attached to the container.

All applicable directions, restrictions, and precautions on the EPA-registered label are to be followed.

NOTE: This labeling must be in the possession of the user at the time of pesticide application.

SPECIAL LOCAL NEED LABEL
PENNCAP-M® INSECTICIDE

FOR DISTRIBUTION AND USE ONLY WITHIN THE STATE OF TEXAS

EPA Reg. No. 4581-292 EPA SLN No. TX-820032

DIRECTIONS FOR USE

It is a violation of Federal law to use this product in a manner inconsistent with its labeling.

CHRYSANTHEMUMS—For the control of leafminers on chrysanthemums grown in commercial greenhouses, apply PENNCAP-M at the rate of 1 to 2 quarts per 100 gallons water as a full coverage spray on foliage. If area to be treated is less than 1 acre, reduce finished spray volume accordingly. Begin application when insect damage threatens and continue at 7-day intervals. Apply PENNCAP-M to a few blooms of each cultivar and observe for phytotoxicity before applying this product to larger areas of plants in bloom. Do not pinch or sell treated mums for 24 hours following treatment.

Use in well ventilated area or use approved respiratory device during application if greenhouse is closed and air circulation is not possible.

KEEP OUT OF REACH OF CHILDREN
WARNING

MAY BE FATAL IF SWALLOWED, INHALED, OR ABSORBED THROUGH SKIN.

For First Aid Statement and Other Precautionary Statements, see label attached to the container.

All applicable directions, restrictions, and precautions on the EPA-registered label are to be followed.

NOTE: This labeling must be in the possession of the user at the time of pesticide application.

SPECIAL LOCAL NEED LABEL
PENNCAP-M® INSECTICIDE

FOR DISTRIBUTION AND USE ONLY WITHIN THE STATE OF SOUTH CAROLINA

EPA Reg. No. 4581-292 EPA SLN No. SC-800028

DIRECTIONS FOR USE

It is a violation of Federal law to use this product in a manner inconsistent with its labeling.

CHRYSANTHEMUMS—For the control of leafminers on chrysanthemums grown in commercial greenhouses, apply PENNCAP-M at the rate of 2 pints per 100 gallons water as a full coverage spray on foliage. If area to be treated is less than 1 acre, reduce finished spray volume accordingly. Begin application when insect damage threatens and continue at 7-day intervals. Use in well ventilated area or use approved respiratory device during application if greenhouse is closed and air circulation is not possible. Apply PENNCAP-M to a few blooms of each cultivar and observe for phytotoxicity before applying this product to larger areas of plants in bloom.

KEEP OUT OF REACH OF CHILDREN
WARNING

MAY BE FATAL IF SWALLOWED, INHALED, OR ABSORBED THROUGH SKIN.

For First Aid Statement and Other Precautionary Statements, see label attached to the container.

All applicable directions, restrictions, and precautions on the EPA-registered label are to be followed.

NOTE: This labeling must be in the possession of the user at the time of pesticide application.

SPECIAL LOCAL NEED LABEL
PENNCAP-M® INSECTICIDE

FOR DISTRIBUTION AND USE ONLY WITHIN THE STATE OF PENNSYLVANIA

EPA Reg No. 4581-292 EPA SLN No. PA-820021

DIRECTIONS FOR USE

It is a violation of Federal law to use this product in a manner inconsistent with its labeling.

CHRYSANTHEMUMS—For the control of leafminers in commercial fields and in greenhouses, apply PENNCAP-M at the rate of 2 to 4 pints per acre in 100 gallons of water as a full coverage spray on foliage. If area to be treated is less than 1 acre, reduce finished spray volume accordingly. Begin application when insect damage threatens and repeat as necessary. Apply PENNCAP-M to a few blooms of each cultivar and observe for phytotoxicity before applying this product to larger areas of plants in bloom.

Use in well ventilated area or use approved respiratory device during application if greenhouse is closed and air circulation is not possible. Do not enter treated area without protective clothing until spray dries. Plants treated are not to be sold within 24 hours of treatment.

KEEP OUT OF REACH OF CHILDREN
WARNING

MAY BE FATAL IF SWALLOWED, INHALED, OR ABSORBED THROUGH SKIN

For First Aid Statement and Other Precautionary Statements, see label attached to the container.

All applicable directions, restrictions, and precautions on the EPA-registered label are to be followed.

NOTE: This labeling must be in the possession of the user at the time of pesticide application.

SPECIAL LOCAL NEED LABEL
PENNCAP-M® INSECTICIDE

FOR DISTRIBUTION AND USE ONLY WITHIN THE STATE OF NEW JERSEY

EPA Reg No. 4581-292 EPA SLN No. NJ-820012

DIRECTIONS FOR USE

It is a violation of Federal law to use this product in a manner inconsistent with its labeling.

CHRYSANTHEMUMS—For the control of leafminers in commercial fields and in greenhouses, apply PENNCAP-M at the rate of 2 to 4 pints per acre in 100 gallons of water as a full coverage spray on foliage. If area to be treated is less than 1 acre, reduce finished spray volume accordingly. Begin application when insect damage threatens and repeat as necessary. Apply PENNCAP-M to a few blooms of each cultivar and observe for phytotoxicity before applying this product to larger areas of plants in bloom.

Use in well ventilated area or use approved respiratory device during application if greenhouse is closed and air circulation is not possible. Do not enter treated area without protective clothing until spray dries. Plants treated are not to be sold within 24 hours of treatment.

KEEP OUT OF REACH OF CHILDREN
WARNING

MAY BE FATAL IF SWALLOWED, INHALED, OR ABSORBED THROUGH SKIN

For First Aid Statement and Other Precautionary Statements, see label attached to the container.

All applicable directions, restrictions, and precautions on the EPA-registered label are to be followed.

NOTE: This labeling must be in the possession of the user at the time of pesticide application.

SPECIAL LOCAL NEED LABEL

PENNCAP-M® INSECTICIDE

FOR DISTRIBUTION AND USE ONLY WITHIN THE STATE OF MISSOURI

EPA Reg No. 4581-292 EPA SLN No. MO-81-0008

DIRECTIONS FOR USE

It is a violation of Federal law to use this product in a manner inconsistent with its labeling.

CHRYSANTHEMUMS—For the control of leafminers on chrysanthemums grown in commercial greenhouses, apply PENNCAP-M at the rate of 2 pints per 100 gallons water as a full coverage spray on foliage. If area to be treated is less than 1 acre, reduce finished spray volume accordingly. Begin application when insect damage threatens and continue at 7-day intervals. Use approved respiratory device during application if greenhouse is closed and air circulation is not possible. Apply PENNCAP-M to a few blooms of each cultivar and observe for phytotoxicity before applying this product to larger areas of plants in bloom.

Do not enter treated areas without protective clothing before spray dries.

KEEP OUT OF REACH OF CHILDREN
WARNING

MAY BE FATAL IF SWALLOWED, INHALED, OR ABSORBED THROUGH SKIN

For First Aid Statement and Other Precautionary Statements, see label attached to the container.

All applicable directions, restrictions, and precautions on the EPA-registered label are to be followed.

NOTE: This labeling must be in the possession of the user at the time of pesticide application.

SPECIAL LOCAL NEED LABEL

PENNCAP-M® INSECTICIDE

FOR DISTRIBUTION AND USE ONLY WITHIN THE STATE OF MASSACHUSETTS

EPA Reg No. 4581-292 EPA SLN No. MA820006

DIRECTIONS FOR USE

It is a violation of Federal law to use this product in a manner inconsistent with its labeling.

CHRYSANTHEMUMS—For the control of leafminers on chrysanthemums grown in commercial greenhouses, apply PENNCAP-M at the rate of 2 pints per 100 gallons water as a full coverage spray on foliage. If area to be treated is less than 1 acre, reduce finished spray volume accordingly. Begin application when insect damage threatens and continue at 7-day intervals. Use in well ventilated area or use approved respiratory device during application if greenhouse is closed and air circulation is not possible. Apply PENNCAP-M to a few blooms of each cultivar and observe for phytotoxicity for a period of 7 days before applying this product to larger areas of plants in bloom. Do not market plants within one (1) day of last application.

KEEP OUT OF REACH OF CHILDREN
WARNING

MAY BE FATAL IF SWALLOWED, INHALED, OR ABSORBED THROUGH SKIN

Workers entering treated greenhouse areas within 24 hours after application should wear protective clothing including clean coveralls, gloves, caps, etc.

For First Aid Statement and Other Precautionary Statements, see label attached to the container.

All applicable directions, restrictions, and precautions on the EPA-registered label are to be followed.

NOTE: This labeling must be in the possession of the user at the time of pesticide application.

SPECIAL LOCAL NEED LABEL

PENNCAP-M® INSECTICIDE

FOR DISTRIBUTION AND USE ONLY WITHIN THE STATE OF HAWAII

EPA Reg No. 4581-292 EPA SLN No. HI-810012

DIRECTIONS FOR USE

It is a violation of Federal and State law to use this product in a manner inconsistent with its labeling.

CHRYSANTHEMUMS—For controlling leafminers on chrysanthemums grown in commercial fields and greenhouses, apply PENNCAP-M at the rate of 2 to 4 pints per acre in 100 gallons water as a full coverage spray on foliage. Begin applications when insect damage threatens and continue at 5 to 7 day intervals until petals of flower buds show color. PENNCAP-M may cause bleaching of flower petals. Limit PENNCAP-M use to pre-blooming stages only.

Use the approved respiratory device during application.

THIS PRODUCT IS HIGHLY TOXIC TO BEES EXPOSED TO DIRECT TREATMENT OR RESIDUES ON CROPS. DO NOT APPLY TO FLOWERS IN BLOOM. DO NOT APPLY IF HONEYBEES ARE OBSERVED IN AREA TO BE SPRAYED. DO NOT SPRAY OR ALLOW TO DRIFT ON WEEDS AND OTHER PLANTS IN BLOOM WHICH ARE IN BORDERING AREAS ADJACENT TO CHRYSANTHEMUM FIELDS AND GREENHOUSES.

Workers entering treated areas within 24 hours after application should wear protective clothing including clean coveralls, gloves, caps, etc.

KEEP OUT OF REACH OF CHILDREN
WARNING

MAY BE FATAL IF SWALLOWED, INHALED, OR ABSORBED THROUGH SKIN

For First Aid Statement and Other Precautionary Statements, see label attached to the container.

All applicable directions, restrictions, and precautions on the EPA-registered label are to be followed.

NOTE: This labeling must be in the possession of the user at the time of pesticide application.

SPECIAL LOCAL NEED LABEL

PENNCAP-M® INSECTICIDE

FOR DISTRIBUTION AND USE ONLY WITHIN THE STATE OF GEORGIA

EPA Reg No. 4581-292 EPA SLN GA. 810002

DIRECTIONS FOR USE

It is a violation of both State and Federal law to use this product in a manner inconsistent with its labeling.

CHRYSANTHEMUMS—For the control of leafminers on chrysanthemums grown in commercial greenhouses, apply PENNCAP-M at the rate of 2 pints per 100 gallons water as a full coverage spray on foliage. If area to be treated is less than 1 acre, reduce finished spray volume accordingly. Begin application when insect damage threatens and continue at 7-day intervals. Use in well ventilated area or use approved respiratory device during application if greenhouse is closed and air circulation is not possible. Apply PENNCAP-M to a few blooms of each cultivar and observe for phytotoxicity before applying this product to larger areas of plants in bloom.

KEEP OUT OF REACH OF CHILDREN
WARNING

MAY BE FATAL IF SWALLOWED, INHALED, OR ABSORBED THROUGH SKIN

For First Aid Statement and Other Precautionary Statements, see label attached to the container.

All applicable directions, restrictions, and precautions on the EPA-registered label are to be followed.

NOTE: This labeling must be in the possession of the user at the time of pesticide application.

SPECIAL LOCAL NEED LABEL

PENNCAP-M® INSECTICIDE

FOR DISTRIBUTION AND USE ONLY WITHIN THE STATE OF FLORIDA

EPA Reg No. 4581-292 EPA SLN No. FL-800043

ACTIVE INGREDIENTS:

O,O-Dimethyl O-p-nitrophenyl phosphorothioate*	20.9%
Related Isomers	1.1%
Xylene Base Aromatic Solvent	4.9%
INERT INGREDIENTS:	73.1%
TOTAL	100.0%

*Methyl Parathion
(Contains 2 pounds Methyl Parathion and Related Isomers per gallon)

DIRECTIONS FOR USE

It is a violation of Federal law to use this product in a manner inconsistent with its labeling.

CHRYSANTHEMUMS—For the control of leafminers in the field and in shade houses, apply PENNCAP-M at the rate of 2 pints per 100 gallons water per acre. If area to be treated is less than one acre, reduce finished spray volume accordingly. Begin application when insect damage threatens and continue at 7-day intervals. Apply PENNCAP-M to a few blooms of each cultivar and observe for phytotoxicity before applying this product to larger areas of plants in bloom.

KEEP OUT OF REACH OF CHILDREN
WARNING

MAY BE FATAL IF SWALLOWED, INHALED, OR ABSORBED THROUGH SKIN

For First Aid Statement and Other Precautionary Statements, see label attached to the container.

All applicable directions, restrictions, and precautions on the EPA-registered label are to be followed.

NOTE: This labeling must be in the possession of the user at the time of pesticide application.

SPECIAL LOCAL NEED LABEL

PENNCAP-M® INSECTICIDE

FOR DISTRIBUTION AND USE ONLY WITHIN THE STATE OF WASHINGTON

EPA Reg No. 4581-292 EPA SLN No. WA-820077

DIRECTIONS FOR USE

It is a violation of Federal law to use this product in a manner inconsistent with its labeling.

CHRYSANTHEMUMS—For the control of leafminers on chrysanthemums grown in commercial greenhouses, apply PENNCAP-M at the rate of 2 pints per 100 gallons water as a full coverage spray on foliage. If area to be treated is less than 1 acre, reduce finished spray volume accordingly. Begin application when insect damage threatens and continue at 7-day intervals. Use an approved respiratory device during application. Applicators and workers entering treated areas within 24 hours after application should wear protective clothing, including coveralls, gloves, and caps. Apply PENNCAP-M to a few blooms of each cultivar and observe for phytotoxicity before applying this product to larger areas of plants in bloom.

KEEP OUT OF REACH OF CHILDREN

WARNING

MAY BE FATAL IF SWALLOWED, INHALED, OR ABSORBED THROUGH SKIN

For First Aid Statement and Other Precautionary Statements, see label attached to the container.

All applicable directions, restrictions, and precautions on the EPA-registered label are to be followed.

NOTE: This labeling must be in the possession of the user at the time of pesticide application.

SPECIAL LOCAL NEED LABEL

PENNCAP-M® INSECTICIDE

FOR DISTRIBUTION AND USE ONLY WITHIN THE STATE OF TEXAS

EPA Reg No. 4581-292 EPA SLN No. TX-820024

DIRECTIONS FOR USE

It is a violation of Federal law to use this product in a manner inconsistent with its labeling.

COTTON—For the control of boll weevils, apply 1 to 2 pints PENNCAP-M in an equal amount of once refined vegetable oil (such as soybean or cottonseed) to provide 2 to 4 pints total finished spray per acre.

Where bollworms and budworms (*Heliothis* spp.) are present also, add a pyrethroid to the above tank mix. Use the rate recommended on the label for the specific pyrethroid used. When tank mixing PENNCAP-M/oil/pyrethroids, the use of a suitable adjuvant may be necessary. ORTHO X-77® Spreader at the rate of 1 ounce per acre is suggested.

Repeat application as necessary to maintain insect control.

Aircraft should be outfitted with a delivery system designed to apply droplets with a VMD of 150 to 220 microns. Swath width should not exceed wing span plus 10%. When using conventional hydraulic nozzle systems, the nozzles should be oriented at 90° to the laminar air flow.

Workers entering fields within 24 hours should wear suitable protective clothing, such as cap, long-sleeved cotton shirt, and long-legged trousers. Protective clothing should be laundered daily.

KEEP OUT OF REACH OF CHILDREN

WARNING

MAY BE FATAL IF SWALLOWED, INHALED, OR ABSORBED THROUGH SKIN

For First Aid Statement and Other Precautionary Statements, see labels attached to the containers.

All applicable directions, restrictions, and precautions on the EPA-registered labels are to be followed.

NOTE: This labeling must be in the possession of the user at the time of pesticide application.

RECOMMENDATION*

PENNCAP-M® INSECTICIDE

USE ON COTTON FOR EARLY SEASON CONTROL OF BOLL WEEVIL

For control of emerging populations of overwintering boll weevil apply Penncap-M at a rate of ½–1 pt./A (.125–.25 lb./A a.i.) starting at pin head square. Use the higher rate for aerial application. Repeat as necessary to maintain control. The rate of Penncap-M should be increased to labeled rates when field scouting indicates the presence of first field generation adult boll weevils.

Penncap-M may be tank mixed with chlordimeform when used in early season applications. Refer to product label for rates of chlordimeform.

READ AND FOLLOW ALL PRECAUTIONARY STATEMENTS ON THE PRODUCT LABEL

When Penncap-M is used in mixtures observe the more restrictive limitations of the products used in the mixture.

* Recommendation made under the provisions of Section 2(ee) of FIFRA

SPECIAL LOCAL NEED LABEL

PENNCAP-M® INSECTICIDE

FOR DISTRIBUTION AND USE ONLY WITHIN THE STATE OF IDAHO
FOR CONTROL OF APHIDS AND LYGUS ON LENTILS

EPA Registration No. 4581-292 EPA SLN No. ID-840010

DIRECTIONS FOR USE

It is a violation of Federal law to use this product in a manner inconsistent with its labeling.

LENTILS—For the control of aphids and lygus, apply PENNCAP-M at the rate of 2 pints per acre. Apply by air in a minimum of 5 gallons of water. Start application when insect damage threatens.

Do not apply within 15 days of harvest.

Do not harvest for forage or graze treated area.

THIS PRODUCT IS HIGHLY TOXIC TO BEES exposed to direct treatment or residues on crops or blooming weeds. Do not apply this product or allow it to drift to blooming crops or weeds if bees are visiting the area to be treated.

Protective information may be obtained from your Cooperative Extension Service.

KEEP OUT OF REACH OF CHILDREN
WARNING

MAY BE FATAL IF SWALLOWED, INHALED, OR ABSORBED THROUGH SKIN

For First Aid Statement and Other Precautionary Statements, see label attached to the container.

All applicable directions, restrictions, and precautions on the EPA registered label are to be followed.

NOTE: This labeling must be in the possession of the user at the time of pesticide application.

SPECIAL LOCAL NEED LABEL

PENNCAP-M® INSECTICIDE

FOR DISTRIBUTION AND USE ONLY WITHIN THE STATE OF WASHINGTON
FOR CONTROL OF APHIDS AND LYGUS ON LENTILS

EPA Registration No. 4581-292 EPA SLN No. WA-820054

DIRECTIONS FOR USE

It is a violation of Federal law to use this product in a manner inconsistent with its labeling.

LENTILS—For the control of aphids and lygus, apply PENNCAP-M at the rate of 2 pints per acre. Apply by air in a minimum of 5 gallons of water. Start application when insect damage threatens.

Do not apply within 15 days of harvest.

Do not harvest for forage or graze treated area.

THIS PRODUCT IS HIGHLY TOXIC TO BEES exposed to direct treatment or residues on crops or blooming weeds. Do not apply this product or allow it to drift to blooming crops or weeds if bees are visiting the area to be treated.

Protective information may be obtained from your Cooperative Extension Service.

KEEP OUT OF REACH OF CHILDREN
WARNING

MAY BE FATAL IF SWALLOWED, INHALED, OR ABSORBED THROUGH SKIN

For First Aid Statement and Other Precautionary Statements, see label attached to the container.

All applicable directions, restrictions, and precautions on the EPA registered label are to be followed.

NOTE: This labeling must be in the possession of the user at the time of pesticide application.

SPECIAL LOCAL NEED LABEL

PENNCAP-M® INSECTICIDE

FOR DISTRIBUTION AND USE ONLY WITHIN THE STATE OF GEORGIA

ULV AERIAL APPLICATION
APPLIED UNDILUTED TO COTTON OR IN VEGETABLE OIL TO COTTON AND SOYBEANS

EPA Reg No. 4581-292 EPA SLN No. GA-840003

DIRECTIONS FOR USE

It is a violation of Federal law to use this product in a manner inconsistent with its labeling.

PENNCAP-M UNDILUTED

When applying undiluted, use spray equipment that has been specifically adapted and calibrated for ultra-low-volume spraying of PENNCAP-M. The following conditions are to be met:

1. Equipment must be adjusted to distribute the spray uniformly over the spray swath.
2. Wind conditions and other factors such as temperature, humidity, and aircraft altitude are such that the spray is deposited on the target.

Spray equipment should be cleaned as soon as possible after completion of spraying. Do not allow PENNCAP-M to dry in or on application equipment before cleaning.

PENNCAP-M IN VEGETABLE OIL

Before combining vegetable oil with PENNCAP-M, thoroughly mix the oil with a non-ionic surfactant that is miscible with vegetable oil to promote thorough mixing. Armul 535 surfactant at a rate of 4 oz per gallon of oil (3% by weight of the oil) is recommended.

It is recommended that the oil be filtered prior to mixing with PENNCAP-M to remove any particles which might plug nozzles.

Aircraft should be outfitted with a delivery system designed to apply droplets with a VMD of 150 to 220 microns. Swath width should not exceed wing span plus 10%. When using conventional hydraulic nozzle systems, the nozzles should be oriented at 90° to the laminar air flow.

Crop	Insects	Rates/Acre	Remarks
Cotton	Boll Weevil	PENNCAP-M 1 pint (undiluted) OR	Begin applications when damaging levels of boll weevils appear. Repeat applications as necessary to maintain insect control.*
		PENNCAP-M 1–2 pints PLUS AN EQUAL AMOUNT OF VEGETABLE OIL (1–2 pints)	Begin applications when damaging levels of boll weevils appear. Repeat applications as necessary to maintain insect control.* Add a material for

Crop	Insects	Rates/Acre	Remarks
			Heliothis control to the tank mix when bollworms and/or budworms (*Heliothis* spp.) are present. Use the rate recommended on the label for the specific material used.
Soybeans	Stinkbugs	PENNCAP-M 2–3 pints PLUS AN EQUAL AMOUNT OF VEGETABLE OIL (2–3 pints)	Apply when damaging levels of stinkbugs appear. Do not make more than two applications per season.**

* Do not apply within 7 days of hand picking or harvest.
** Do not apply within 20 days of harvest or grazing.

Workers entering fields within 48 hours after application should wear suitable protective clothing, such as cap, long-sleeved cotton shirt, and long-legged trousers. Protective clothing should be laundered daily.

KEEP OUT OF REACH OF CHILDREN
WARNING

MAY BE FATAL IF SWALLOWED, INHALED, OR ABSORBED THROUGH SKIN

For First Aid Statement and other Precautionary Statements, see labels attached to containers.

All applicable directions, restrictions, and precautions on the EPA-registered labels are to be followed. When PENNCAP-M is used in a tank mix observe the more restrictive precautionary statements of the products used in the tank mix.

NOTE: This label must be in the possession of the user at the time of pesticide application.

SPECIAL LOCAL NEED LABEL

PENNCAP-M® INSECTICIDE

FOR DISTRIBUTION AND USE ONLY WITHIN THE STATE OF ALABAMA

ULV AERIAL APPLICATION TO COTTON

EPA Reg. No. 4581-292 EPA SLN No. AL-840001

DIRECTIONS FOR USE

It is a violation of Federal law to use this product in a manner inconsistent with its labeling.

COTTON—For control of boll weevils, apply PENNCAP-M by air at the rate of 1 pint per acre mixed with 1 to 2 pints of vegetable oil blended as instructed below to provide 2 to 3 pints total spray per acre.

Before combining with PENNCAP-M, thoroughly mix the vegetable oil with the surfactant Armul 535 at a rate of 4 oz. Armul 535 per gallon of oil.

It is recommended that the oil be filtered prior to mixing with PENNCAP-M to remove any particles which might plug nozzles.

When bollworms and budworms (*Heliothis* spp.) are present, also add a material for Heliothis control to the above tank mix. Use the rate recommended on the label for the specific material used.

Begin applications when damaging levels of boll weevils appear. Repeat applications as necessary to maintain insect control.

Aircraft should be outfitted with a delivery system designed to apply droplets with a VMD of 150 to 220 microns. Swath width should not exceed wing span plus 10%. When using conventional hydraulic nozzle systems, the nozzles should be oriented at 90° to the laminar air flow.

Workers entering fields within 48 hours should wear suitable protective clothing, such as cap, long-sleeved cotton shirt, and long-legged trousers. Protective clothing should be laundered daily.

KEEP OUT OF REACH OF CHILDREN

WARNING

MAY BE FATAL IF SWALLOWED, INHALED, OR ABSORBED THROUGH SKIN

For First Aid Statement and other Precautionary Statements, see labels attached to containers.

All applicable directions, restrictions, and precautions on the EPA-registered labels are to be followed. When PENNCAP-M is used in a tank mix observe the more restrictive precautionary statements of the products used in the tank mix.

NOTE: This label must be in the possession of the user at the time of pesticide application.

TOPSIN® M

THIOPHANTE–METHYL FUNGICIDE

70% Wettable Powder

ACTIVE INGREDIENT:

Thiophante-methyl (dimethyl[(1,2-phenylene)-bis(iminocarbonothioyl)]bis[carbamate])*	70%
INERT INGREDIENTS:	30%
TOTAL:	100%

*Also known as dimethyl 4, 4′-o-phenylenebis[3-thioallophante]

®TOPSIN is a registered trademark of Nippon Soda Company, Ltd. and is licensed to Pennwalt Corporation, and is covered by one or more of the following U.S. Patents: 3,769,308;3,856,847, 4,020,095; and 4,029,813.

EPA Reg. No. 4581-322 EPA Est. No. 00000-00-0

CAUTION

Keep Out of Reach of Children

HAZARDS TO HUMANS (AND DOMESTIC ANIMALS)

HARMFUL IF SWALLOWED, ABSORBED THROUGH SKIN, OR INHALED.
CAUSES MODERATE EYE IRRITATION.

IF SWALLOWED, induce vomiting immediately by giving two glasses of water and sticking finger down throat. Call a physician. Never give anything by mouth to an unconscious person.

IF IN EYES, immediately flush eyes with plenty of water for at least 15 minutes. Get medical attention if irritation persists.

IF ON SKIN, immediately flush skin with plenty of water. Get medical attention if irritation perists.

Avoid contact with skin, eyes, or clothing. Avoid breathing dust or spray mist. Wash thoroughly with soap and water after handling. Remove contaminated clothing and wash before use.

Do not apply this product in such a manner as to directly or through drift expose workers or other persons. The area being treated must be vacated by unprotected persons.

ENVIRONMENTAL HAZARD

This pesticide is toxic to fish. Do not apply directly to water or wetlands. Do not apply where runoff is likely to occur. Do not contaminate water by cleaning of equipment or disposal of wastes.

NOTE: For general information and directions for use refer to accompanying brochure.

STORAGE AND DISPOSAL

Do not contaminate water, food, or feed by storage or disposal.

STORAGE INSTRUCTIONS

Store in the original container in a dry area. Do not store in a manner where cross-contamination with other pesticides, fertilizers, food or feed could occur. If spilled during storage or handling, sweep up spillage and dispose of in accordance with the Pesticide Disposal Instructions listed below.

PESTICIDE DISPOSAL INSTRUCTIONS

Wastes resulting from the use of this product may be disposed of on site or at an approved waste disposal facility.

CONTAINER DISPOSAL INSTRUCTIONS

After bag has been emptied, dispose of empty bag in a sanitary landfill or by incineration, or if allowed by State and local authorities, by burning. If burned, stay out of smoke.

GENERAL INFORMATION

Apply TOPSIN M with ground or aerial equipment, using sufficient volume of spray to provide thorough coverage. Add required amount of TOPSIN M to partially filled tank agitated by mechanical or hydraulic means, and then add remaining required amount of water. Continuous agitation is required to keep the material in suspension. Do not tank mix with highly alkaline pesticides, such as Bordeaux mixture or lime sulfur. No claim of compatability with other pesticides is implied. Use the higher rate under conditions of severe disease pressure. Also, see local State Extension Service recommendations for application schedules.

IMPORTANT: If, after using TOPSIN M as recommended, treatment is not effective, a tolerant strain of fungi may be present. Consult your State Agricultural Experiment Station or your State Agricultural Extension Service for advice on the prompt use of some other suitable fungicide.

DIRECTIONS FOR USE

It is a violation of Federal law to use this product in a manner inconsistent with its labeling.

Do not enter treated areas without protective clothing until spray has dried. Because certain states may require more restrictive reentry intervals for various crops treated with this product, consult your State Department of Agriculture for further information.

Written or oral warnings must be given to workers who are expected to be in a treated area or in an area about to be treated with this product. When oral warnings are given, such warnings shall be given in a language customarily understood by the workers. Oral warnings must be given if there is reason to believe that written warnings cannot be understood by the workers. Written warnings must include the following information:

"'CAUTION. Area treated with TOPSIN M Thiophanate-methyl Fungicide on (*date of application*).'"

CROP	DISEASES	LBS/ACRE	LBS/100 GAL	REMARKS
Almonds	Brown Rot Blossom Blight (*Monilinia* sp.)	1½–2		Apply at early pink bud and early petal fall.
Apples (pre-harvest)	Apple Scab (*Venturia* sp.) Bitter Rot (*Glomerella* sp.) Flyspeck (*Schizothyrium* sp.)	1–1½	¼–⅜	Apply at 5– to 10–day intervals from green tip through petal fall: continue at 7– to 14–day intervals in cover sprays. NOTE: For control of apple diseases in the Southeastern States of Florida, Georgia, South Carolina,
Apples (pre-harvest)	Powdery Mildrew (*Podosphaera* sp.) Sooty Blotch (*Gloeodes* sp.) Black Rot (*Physalospora* sp.)			North Carolina, and Alabama, use 3 to 4 pounds TOPSIN M per acre.
Apples (post harvest)	Postharvest Fruit Rots (*Botrytis* sp.) (*Gloeosporium* sp.) (*Penicillium* spp)		½–1 (420–840 ppm a.i.)	Dip or spray fruit thoroughly as soon as possible after harvest. Do not dip for more than 2 minutes. Double the concentration of TOPSIN M when incorporated in wax and applied in flow-through sprays to the fruit.
Beans	Gray Mold (*Botrytis* sp.) White Mold (*Sclerotinia* sp.)	1½–2 OR 1–1½	1–1½ OR ⅔–1	Apply once at 50% to 70% of full bloom. OR Apply twice with the first application at 10% to 30% of full bloom and a second application at 4 to 7 days later or at peak bloom. Do not apply to snap or dry beans within 14 days of harvest or to lima beans within 28 days of harvest.
Celery	Early Blight (*Cercospora* sp.) Late Blight (*Septoria* sp.)	½		Apply to the base of the plants on a 14– to 21–day schedule beginning about 2 weeks after transplanting. OR Apply as a foliar spray every 7 to 14 days beginning about 2 weeks after transplanting. Do not apply within 7 days of harvest. Do not make more than 4 applications per season.
Peanuts	Leaf Spot (*Cercospora* spp.)	½		Begin applications 35 days after planting or when disease first appears and repeat at 14– to 21–day intervals as needed. Use the 14–day interval under severe disease pressure. Do not apply within 14 days of harvest.
Pecans	Brown Spot (*Cercospora* sp.) Downy Spot (*Mycosphaerella* sp.) Powdery Mildew (*Micosphaera* sp.) Scab (Fusicladium sp.)	½–1		Use the higher rate for trees over 30 feet tall and for aerial applications in Arkansas, Louisiana, Mississippi, Oklahoma, and Texas. Begin applications when first leaves are showing and repeat at 3– to 4–weeks intervals until shuck split. Do not apply after shuck split.
Soybeans	Anthracnose (*Colletotrichum* sp.) Brown Spot (*Septoria* sp.)			Apply from full bloom to when pods are ⅛" to ¼" in length. Make a second application 14 to 21 days later. Do not make the second application later than 14 days after pods average ¼" in length or when

CROP	DISEASES	LBS/ACRE	LBS/100 GAL	REMARKS
	Frogeye Leaf Spot (*Cercospora* sp.) Pod and Stem Blight (*Diaporthe* sp. and the imperfect stage. *Phomopsis* sp.) Purple Seed Stain (*Cercospora* sp.)	$\frac{1}{2}$–1		beans form in the pod. Use the high rate under severe disease pressure. FOR SEED BEANS ONLY—For seed quality make a single application at the high rate when beans form in the pod. Do not make more than 2 applications per year. Do not graze or feed treated vines or hay to livestock.
Stone Fruits (pre harvest) —Apricots	Brown Rot Blossom Blight (*Monilina* sp.) Fruit Brown Rot (*Monilinia* sp.)	$1\frac{1}{2}$	$\frac{1}{2}$	Apply at early bloom (red bud). Make a second application at full bloom. In addition, for Fruit Brown Rot, apply 1 or 2 sprays starting 3 weeks before harvest. If needed under severe disease pressure, apply additional sprays at 10– to 14–day intervals between full bloom and final pre-harvest sprays. Do not apply within 1 day of harvest.
—Cherries	Brown Rot Blossom Blight (*Monilinia* sp.) Fruit Brown Rot (*Monilinia* sp.)	$1\frac{1}{2}$	$\frac{1}{2}$	Apply at early bloom (early popcorn). Make a second application at full bloom In addition for Fruit Brown Rot, apply 1 or 2 sprays starting 3 weeks before harvest. If needed under severe disease pressure apply additional sprays at 10– to 14–day intervals between full bloom and final preharvest sprays. Do not apply within 1 day of harvest.
	Cherry Leaf Spot (*Coccomyces* sp.)	$1\frac{1}{8}$–$\frac{1}{2}$	$\frac{3}{8}$–$\frac{1}{2}$	Apply at petal fall or before, when leaves first unfold, and at first, second and third cover at 10– to 14–day intervals and one spray 14 to 21 days after harvest. Do not apply within 1 day of harvest.
	Powdery Mildew (*Podosphaera* sp.) and *Sphaerotheca* sp.)	$1\frac{1}{2}$ PLUS $1\frac{1}{8}$–$1\frac{1}{2}$	$\frac{1}{2}$ PLUS $\frac{3}{8}$–$\frac{1}{2}$	Apply at early bloom (early popcorn). Make a second application at full bloom. PLUS Apply at shuck fall and first cover. Do not apply within 1 day of harvest.
—Nectarines	Brown Rot Blossom Blight (*Monilinia* sp.) Fruit Brown Rot (*Monilinia* sp.)	$1\frac{1}{2}$	$\frac{1}{2}$	Apply at early bloom (pink bud). Make a second application at full bloom if conditions favor disease development. In addition, for Fruit Brown Rot, apply 1 or 2 sprays starting 3 weeks before harvest. If needed under severe disease pressure, apply additional sprays at 10– to 14–day intervals between full bloom and final preharvest sprays. Do not apply within 1 day of harvest.
—Peaches	Brown Rot Blossom Blight (*Monilinia* sp.) Fruit Brown Rot (*Monilinia* sp.)	$1\frac{1}{2}$–$2\frac{1}{4}$	$\frac{1}{2}$–$\frac{3}{4}$	Apply at early bloom (pink bud). Make a second application at full bloom if conditions favor disease development. In addition, for Fruit Brown Rot, apply 1 or 2 sprays starting 3 weeks before harvest. If needed under severe diease pressure, apply additional sprays at 10– to 14–day intervals between full bloom and final preharvest sprays. Do not apply within 1 day of harvest.
	Peach Scab (*Cladosporium* sp.)	$1\frac{1}{2}$–$2\frac{1}{4}$ PLUS	$\frac{1}{2}$–$\frac{3}{4}$ PLUS $\frac{3}{8}$–$\frac{1}{2}$	Apply at early bloom (pink bud). Make a second application at full bloom if conditions favor disease development. PLUS Apply at shuck split and at first cover sprays. Do not apply within 1 day of harvest.
—Plums and Prunes	Brown Rot Blossom Blight (*Monilinia* sp.) Fruit Brown Rot (*Monilinia* sp.)	$1\frac{1}{2}$	$\frac{1}{2}$	Apply at early bloom (green tip). Make a second application at full bloom. In addition, for Fruit Brown Rot, apply 1 or 2 sprays starting 3 weeks before harvest. If needed under severe disease pressure, apply additional sprays at 10– to 14–day intervals between full bloom and final preharvest sprays. Do not apply within 1 day of harvest.
	Black Knot (*Dibotryon* sp.)	$1\frac{1}{2}$	$\frac{1}{2}$	Apply at pre-bloom, petal fall, and at first, second, and third cover sprays at 10– to 14–day intervals. Do not apply within 1 day of harvest.
	Leaf Spot (*Coccomyces* sp.)	$1\frac{1}{2}$	$\frac{1}{2}$	Apply at petal fall, shuck, and at first, second, and third cover sprays at 10– to 14–day intervals and 1 spray 14 to 21 days after harvest. Do not apply within 1 day of harvest.
Stone Fruits (post harvest) —Apricots —Cherries	Postharvest Fruit Rots Fruit Brown Rot (*Monilinia* sp.)		$\frac{3}{4}$-1 (630-840 ppm a.i.)	Dip or spray fruit thoroughly as soon as possible after harvest. Maximum postharvest treatment time is 1 min.

CROP	DISEASES	LBS/ACRE	LBS/100 GAL	REMARKS
—Nectarines —Peaches —Plums —Prunes				NOTE: TOPSIN M will not control fruit rots caused by *Rhizopus* spp. and *Alternaria* spp.
Strawberries	Fruit Rot (*Botrytis* sp.) Leaf Blight (*Dendrophoma* sp.) Leaf Scorch (*Diplocarpon* sp.)	¾–1		Begin applications at early bloom and continue at 7– to 10–day intervals. Use the higher rate East of the Rocky Mountains and under conditions of severe disease pressure. Do not apply within 1 day of harvest.
Sugar Beets	Leaf Spot (*Cercospora* sp.)	⅛–½		Begin applications when disease first appears and repeat at 14– to 21–day intervals as needed. Do not apply within 21 days of harvest.
Sugar Cane (Hawaii only)	Pineapple Disease (*Thielaviopsis* sp. or *Ceratocystis* sp.)		½ (1:1,600)	Apply to cut seed pieces as a cold or hot dip. Cold Dip—Immerse seed pieces to provide thorough coverage; remove and allow to drain
			¼ (1:3,200)	Hot Dip—Maintain temperature of dip at 52°C and soak seed pieces for 20 minutes; remove and allow to drain. Do not use treated seed pieces for food, feed or fodder. Do not graze or feed treated forage to livestock.

TANK MIX FOR USE ON APPLES
(Not for use in California)

DISEASES	LBS/ACRE	LBS/100 GAL	REMARKS
Scab (*Venturia* sp.) Bitter Rot (*Glomerella* sp.) Black Rot (*Physalospora* sp.) Flyspeck (*Schizothyrium* sp.) Powdery Mildew (*Podoshaera* sp.) Sooty Blotch (*Gloeodes* sp.)	TOPSIN M ½–¾ + Captan 50WP 3–4 OR Captan 80WP 2–2½	TOPSIN M ⅛–3/16 + Captan 50WP ¾–1 OR Captan 80WP ½–⅝	Apply at ½" green tip and repeat at 7– to 14–day intervals or as needed through the cover sprays. Use the high rate of TOPSIN M for varieties more susceptible to powdery mildew. If an application is missed during an infection period, apply the higher rates as soon as possible after the infection period in order to deactivate scab and to prevent further infection. NOTE: Spray injury may result if Captan is used with, immediately before, or closely following an oil spray.

Observe the more restrictive cautions and limitations on labeling of products used in mixtures.

TANK MIX FOR USE ON APPLES
(Not for use in California)

DISEASES	LBS/ACRE	LBS/100 GAL	REMARKS
Scab (*Venturia* sp.) Bitter Rot (*Glomerella* sp.) Black Rot (*Physalospora* sp.) Flyspeck (*Schizothyrium* sp.) Powdery Mildew (*Podosphaera* sp.) Sooty Blotch (*Gloeodes* sp.) Brown Rot (*Monilinia* sp.) Cedar Apple Rust (*Gymnosporangium* sp.) Quince Rust (*Gymnosporanigum* sp.)	TOPSIN M ½–¾ + Mancozeb 80WP 3–4	TOPSIN M ⅛–3/16 + Mancozeb 80WP ¾–1	Apply at ½" green tip and repeat at 7 to 14-day intervals or as needed through the cover sprays. One Quart spray oil may be added per 100 gallons. Add oil to the spray tank last. Use the higher rate of TOPSIN M and add spray oil to the spray mixture for varieties more susceptible to powdery mildew, and for scab if an application is missed during an infection period (apply as soon as possible after infection period in order to deactivate scab and to prevent further infection). Do not apply within 30 days of harvest. Do not graze treated areas.

Observe the more restrictive cautions and limitations on labeling of products used in mixtures.

TANK MIX FOR USE ON APPLES
(Not for use in California)

DISEASES	LBS/ACRE	LBS/100 GAL	REMARKS
Scab (*Venturia* sp.) Bitter Rot (*Glomerella* sp.) Black Rot (*Physalospora* sp.) Flyspeck (*Schizothyrium* sp.) Powdery Mildew (*Podosphaera* sp.) Sooty Blotch (*Gloeodes* sp.)	TOPSIN M ½–¾ + Maneb 80WP 6–8	TOPSIN M ⅛–3/16 + Maneb 80WP 1½–2	Begin applications at green tip and repeat at pink and at 7- to 14-day intervals or as needed through the cover sprays. Use the higher rates and closer interval under severe disease conditions. With Golden Delicious apples, adverse effects on fruit finish and color may result. Use the lower rates of TOPSIN M when finish and color are of primary importance. Do not apply within 30 days of harvest. Do not graze livestock in treated area.

DISEASES	LBS/ACRE	LBS/100 GAL	REMARKS
Cedar Apple Rust (*Gymnosporangium* sp.) Bull's Eye Rot (*Gloeosoporium* sp.)			

Observe the more restrictive cautions and limitations on labeling of products used in mixtures.

TANK MIX FOR USE ON APPLES
(Not for use in California)

DISEASES	LBS/ACRE	LBS/100 GAL	REMARKS
Scab (*Venturia* sp.) Bitter Rot (*Glomerella* sp.) Black Rot (*Physalospora* sp.) Flyspeck (*Schizothyrium* sp.) Powdery Mildew (*Podosphaera* sp.) Sooty Blotch (*Gloeodes* sp.)	TOPSIN M ½–¼ + Polyram 80WP 3–4	TOPSIN M ⅛–3/16 + Polyram 80WP ¾–1	Begin application at green tip and repeat applications as needed through the cover sprays. Do not apply within 30 days of harvest. Do not graze livestock in treated areas.

Observe the more restrictive cautions and limitations on labeling of products used in mixtures.

TANK MIX FOR USE ON APPLES
(Not for use in California)

DISEASES	LBS/ACRE	LBS/100 GAL	REMARKS
Scab (*Venturia* sp.) Bitter Rot (*Glomerella* sp.) Black Rot (*Physalospora* sp.) Flyspeck (*Schizothyrium* sp.) Powdery Mildew (*Podosphaera* sp.) Sooty Blotch (*Gloeodes* sp.)	TOPSIN M ½–¾ + Sulfur 12–30 lbs. a.i.	TOPSIN ⅛–3/16 + Sulfur 3–7½ lbs. a.i.	Begin applications at green tip and repeat at pink and at 7- to 14-day intervals or as needed through the cover sprays. Use the higher rates and closer interval under severe disease conditions.

Observe the more restrictive cautions and limitations on labeling of products used in mixtures.

TANK MIX FOR USE ON STONE FRUITS
(Apricots, Cherries, Nectarines, Peaches, Plums and Prunes)
(Not for use in California)

DISEASES	LBS/ACRE	LBS/100 GAL	REMARKS
Brown Rot Blossom Blight (*Monilinia* sp.) Fruit Brown Rot (*Monilinia* spp.) Peach Scab (*Cladosporium* sp.) Powdery mildew (*Podosphaera* sp. and *Sphaerotheca* sp.) Cherry Leaf Spot (*Coccomyces* sp.) Russet Scab Black Knot (*Dibotryon* sp.)	TOPSIN M ¾–1 + Captan 50WP 3–6 OR Captan 80WP 2–3¾	TOPSIN M ¼–⅓ + Captan 50WP 1–2 OR Captan 80WP ⅜–1¼	BROWN ROT BLOSSOM BLIGHT—Apply at early bloom stages before infection occurs (apricots—red bud: cherries—early popcorn; nectarines & peaches—pink bud; plums & prunes—green tip). If blossoming is prolonged or conditions favorable for disease persist, make a second application 14 days later. FRUIT BROWN ROT—After blossom blight sprays, make a preharvest application any time from 3 weeks before harvest to 1 day before harvest. Make a second application if conditions favorable for disease persist or harvest is prolonged. PEACH SCAB, POWDERY MILDEW—Use same schedule as for Brown Rot Blossom Blight and continue at 10- to 14-day intervals to 1 day before harvest. Make an additional application 2 to 3 weeks after harvest. CHERRY LEAF SPOT—Use same schedule as for Brown Rot Blossom Blight and continue at 10- to 14-day intervals to 1 day before harvest. Make an additional application 2 to 3 weeks after harvest. RUSSET SCAB (PLUMS & PRUNES)—Use same schedules as for Brown Rot Blossom Blight; include full coverage spray at full bloom. BLACK KNOT—Apply at first, second, and third cover sprays at 10- to 14-day intervals. Do not apply within 1 day of harvest. NOTE: This tank mixture will not control bacterial blast, but will supplement a dormant program for peach leaf curl and shothole (*Coryneum* blight): refer to other fungicide labels for dormant period control information. Do not use this tank mixture with spray oils.

Observe the more restrictive cautions and limitations on labeling of products used in mixtures.

TANK MIX FOR USE ON STONE FRUITS
(Apricots, Cherries, Nectarines, Peaches, Plums and Prunes)
(Not for use in California)

DISEASES	LBS/ACRE	LBS/100 GAL	REMARKS
Brown Rot Blossom Blight (*Monilinia* sp.) Fruit Brown Rot (*Monilinia* sp.) Powdery mildew (*Podosphaera* sp. and *Sphaerotheca* sp.) Leaf spot (*Coccomyces* sp.) Shot-hole (*Coryneum* sp.) Rust (*Tranzchelia* sp.) Peach Canker Peach Scab (*Cladosporium* sp.) Black Knot (*Dibotryon* sp.)	TOPSIN M $\frac{1}{3}$–$1\frac{1}{4}$ + Sulfur 12–18 lbs a.i.	TOPSIN M $\frac{1}{4}$–$\frac{1}{2}$ + Sulfur 4–6 lbs a.i.	BROWN ROT BLOSSOM BLIGHT—Make application at early bloom & again at full bloom. FRUIT BROWN ROT—In addition to bloom sprays, apply 1 or 2 sprays starting 3 weeks before harvest. If needed under severe disease pressure, apply additional sprays at 10- to 14-day intervals between full bloom and final preharvest sprays. POWDERY MILDEW—In addition to bloom sprays, make applications at shuck fall and first cover. LEAF SPOT—Make application when leaves first unfold and first, second, and third cover at 10- to 14-day intervals and 1 spray 14 to 21 days after harvest. SHOT-HOLE, RUST, PEACH CANKER, & PEACH SCAB—Apply at pink bloom, petal fall, shock split, shuck fall, and at 10- to 14-day intervals as long as necessary. BLACK KNOT—Apply at first, second, and third cover sprays at 10- to 14-day intervals. Do not apply within 1 day of harvest.

Observe the more restrictive cautions and limitations on labeling of products used in mixtures.

TANK MIX FOR USE ON STONE FRUITS
(Apricots, Cherries, Nectarines and Peaches)
(Not for use in California)

DISEASES	CROP	LBS/ACRE	LBS/100 GAL	REMARKS
Brown Rot Blossom Blight (*Monilinia* sp.) Fruit Brown Rot (*Monilinia* sp.) Powdery Mildew (*Podosphaera* sp. and (*Sphaerotheca* sp.) Cherry Leaf Spot (*Coccomyces* sp.) Shot-hole (*Coryneum* sp.) Rust (*Tranzchelia* sp.) Peach Canker Peach Scab (*Cladosporium* sp.)	Apricots and Nectarines	TOPSIN M $\frac{3}{4}$–$1\frac{1}{2}$ + Ziram 3–$4\frac{1}{2}$ lbs a.i.	TOPSIN M $\frac{1}{4}$–$\frac{1}{2}$ + Ziram 1–$1\frac{1}{2}$ lbs a.i.	BROWN ROT BLOSSOM BLIGHT—Make application at early bloom and again at full bloom. FRUIT BROWN ROT—In addition to bloom sprays, apply 1 or 2 sprays starting 3 weeks before harvest. If needed under severe disease pressure, apply additional sprays at 10- to 14-day intervals between full bloom and final pre-harvest sprays. POWDERY MILDEW—In addition to bloom sprays, make applications at shuck fall and first cover. CHERRY LEAF SPOT—Make application when leaves first unfold & first, second, and third cover at 10- to 14-day intervals and 1 spray 14-21 days after harvest. SHOT-HOLE, RUST, PEACH CANKER & PEACH SCAB—Apply at pink, bloom, petal fall, shuck split, shuck fall, and at 10-to 14-day intervals as long as necessary. For apricots, do not apply after early cover periods. For cherries, do not apply within 7 days of harvest. For nectarines & peaches, last application should be made no later than at petal fall.
	Cherries	TOPSIN M $\frac{3}{4}$–$1\frac{1}{2}$ + Ziram $3\frac{3}{8}$ lbs a.i.	TOPSIN M $\frac{1}{4}$–$\frac{1}{2}$ + Ziram $1\frac{1}{8}$ lbs a.i.	
	Peaches	TOPSIN M $\frac{3}{4}$–$1\frac{1}{2}$ + Ziram 3–$4\frac{1}{2}$ lbs a.i.	TOPSIN M $\frac{1}{4}$–$\frac{1}{2}$ + Ziram 1–$1\frac{1}{2}$ lbs a.i.	

Observe the more restrictive cautions and limitations on labeling of products used in mixtures.

TANK MIX FOR USE ON STONE FRUITS
(Apricots, Nectarines and Peaches)
(Not for use in California

DISEASES	LBS/ACRE	LBS/100 GAL	REMARKS
Brown Rot Blossom Blight (*Monilinia* sp.) Fruit Brown Rot (*Monilinia* sp.) Peach Scab (*Cladosporium* sp.) Powdery Mildew (*Podosphaera* sp. and *Sphaerotheca* sp.)	TOPSIN M $1\frac{1}{2}$ + Maneb 80WP 6–8	TOPSIN M $\frac{1}{2}$ + Maneb 80WP 2–$2\frac{2}{3}$	BROWN ROT BLOSSOM BLIGHT—Apply at early bloom stages before infection occurs (apricots—red bud; nectarines & peaches—pink bud). For this application only. TOPSIN M plus Maneb 80WP may be used in combination with superior-type spray oil. If blossoming is prolonged or conditions favorable for disease persist, make a second application 14 days later. FRUIT BROWN ROT—After blossom blight sprays, make an additional application 3 weeks before harvest. PEACH SCAB, POWDERY MILDEW—Use same schedule as for Brown Rot Blossom Blight plus applications at shuck split, shuck fall, and 14 days later. Do not apply within 14 days of harvest. NOTE: This tank mixture will not control bacterial blast, but will supplement a dormant program for peach leaf curl and shothole (*Coryneum* blight); refer to other fungicide labels for dormant period control information.

Observe the more restrictive cautions and limitations on labeling of products used in mixtures.

TANK MIX FOR USE ON PEANUTS
(Not for use in California)

DISEASES	LBS/ACRE	REMARKS
Leaf Spot (*Cercospora* sp.) Rust (*Puccinia* sp.) Web Blotch (*Assochyta* sp.)	TOPSIN M ¼ + Mancozeb 80WP 1–2	Begin application 35 to 40 days after planting or when disease first appears. Repeat at the following intervals: for *Cercospora* leaf spot at 10 to 14 days; for Rust at 7 to 10 days; and for *Ascochyta* web blotch at 7 to 14 days. Spray oil may be added at the rate of 1 pint to 1 quart per acre. Do not graze or feed treated vines, hay, or hulls to livestock. Do not apply within 14 days of harvest

Observe the more restrictive cautions and limitations on labeling of products used in mixtures.

Observe the more restrictive cautions and limitations on labeling of products used in mixtures.

WARRANTY AND DISCLAIMER

Penwalt Corporation warrants that this material conforms to the chemical description on the label and is reasonably fit for the purposes referred to in the Directions for Use, subject to the risks referred to therein. PENNWALT MAKES NO OTHER EXPRESS OR IMPLIED WARRANTY OF FITNESS OR MERCHANTABILITY OR ANY OTHER EXPRESS OR IMPLIED WARRANTY. IN NO CASE SHALL PENNWALT OR SELLER BE LIABLE FOR CONSEQUENTIAL, SPECIAL OR INDIRECT DAMAGES RESULTING FROM THE USE OR HANDLING OF THIS PRODUCT INCLUDING, BUT NOT LIMITED TO, LOSS OF PROFITS, BUSINESS REPUTATION, OR CUSTOMERS; LABOR COST; OR OTHER EXPENSES INCURRED IN PLANTING OR HARVESTING.

Pennwalt and seller offer this product and the buyer and user accept it subject to the foregoing conditions of sale and warranty which may be varied only by agreement in writing signed by a duly authorized representative of Pennwalt.

RECOMMENDATION*

TOPSIN M 70W FUNGICIDE

APPLICATION THROUGH SPRINKLER IRRIGATION SYSTEMS
USE RESTRICTED TO THE STATES OF IDAHO AND WASHINGTON

EPA Reg. No. 4581-322

GENERAL INFORMATION

Apply TOPSIN M Fungicide only through systems containing anti-syphon and check valves designed to prevent water source contamination or overflow of the mix tank and containing interlocking controls between the metering device and the water pump to insure simultaneous shut-off. Allow sufficient time for flushing of TOPSIN M from the irrigation system prior to shutdown.

DIRECTIONS FOR USE

It is a violation of Federal law to use this product in a manner inconsistent with its labeling.

Add required amount of TOPSIN M to a partially filled tank agitated hydraulically or mechanically, then add remaining required amount of water. Constant agitation is required to keep material in suspension. If other pesticides are to be used in combination, TOPSIN M should be added to the mix tank prior to the addition of any emulsifiable concentrates. No claim of compatibility with other pesticides is implied.

CROP	DISEASE	Rate 70% SP (lb/A)	REMARKS
Dry Beans	White Mold (*Sclerotinia* sp.) Gray Mold (*Botrytis* sp.)	1.5–2.0	Apply once when all plants have first opened blossom (100% bloom), or at first petal fall. Do not apply to dry beans within 30 days of harvest.

If conditions are favorable for maximal disease incidence, use the higher rate.

Set center-pivot or linear system to deliver a minimum of 0.1 inch to a maximum of 0.25 inches of water per acre. Operate at 100% or maximum speed. Higher water volume may reduce efficacy.

Inject suspension of TOPSIN M uniformly throughout the entire irrigation period when using center-pivot and linear systems. In solid set, hand-moved, or rolling-type systems, inject TOPSIN M suspension during the last 15–30 minutes.

Allow adequate time after injection for flushing of system. Inject TOPSIN M suspension into system at a right angle to enhance adequate mixing.

Note: Poor or decreased control may occur when 1) white mold infections occur prior to application, or favorable disease conditions persist for an extended time period, 2) irrigation systems do not apply water uniformly, 3) lower rates are used when conditions are favorable for disease.

READ AND FOLLOW THE TOPSIN M LABELS FOR ADDITIONAL PRECAUTIONS, INSTRUCTIONS AND RESTRICTIONS.

*Recommendations made under the provisions of Section 2(ee) of FIFRA.

RECOMMENDATION*

TOPSIN® M 70% W.P.

EPA Registration No. 4581-322

USE ON PECANS

DISEASE CONTROLLED	RATE OF TOPSIN M PER ACRE	REMARKS
Zonate Leaf Spot (*Gloeocercospora* sp.)	½–1 lb.	Apply in sufficient water for thorough coverage. Use the higher rate for trees over 30 ft tall and for aerial applications in Arkansas, Louisiana, Mississippi, Oklahoma and Texas. Begin applications when first leaves are showing and repeat at 3 to 4 weeks intervals until shuck split. Do not apply after shuck split.

FOLLOW ALL CAUTIONS AND LIMITATIONS ON THE PRODUCT LABEL.

* Recommendation made under the provisions of Section 2(ee) of FIFRA.

RECOMMENDATION*

TOPSIN M 70% W

FOR CONTROL OF BROWN ROT BLOSSOM BLIGHT (*Monilinia* sp.) AND FRUIT BROWN ROT (*Monilinia* sp.) ON PEACHES IN THE STATES OF ALABAMA, FLORIDA, GEORGIA, MISSISSIPPI, NORTH CAROLINA AND SOUTH CAROLINA

EPA Reg. No. 4581-322

DIRECTIONS FOR USE

It is a violation of Federal law to use this product in a manner inconsistent with its labeling.

CROP	DISEASES	LBS/ACRE	LBS/100 GAL	REMARKS
Peaches	Brown Rot Blossom Blight (*Monilinia* sp.) Fruit Brown Rot (*Monilinia* sp.)	$1\frac{1}{2}$ to $2\frac{1}{4}$	$\frac{1}{2}$ to $\frac{3}{4}$	Apply at early bloom (pink bud). Make a second application at full bloom if conditions favor disease development. In addition, for Fruit Brown Rot, apply 1 or 2 sprays starting 3 weeks before harvest. If needed under severe disease pressure, apply additional sprays at 10- to 14-day intervals between full bloom and final preharvest sprays. NOTE: When tank mix is considered appropriate, TOP-SIN M 70W is registered for use in combination with Captan, Sulfur, Ziram and Maneb. See supplemental labels for tank mixes on stone fruits.

KEEP OUT OF REACH OF CHILDREN
CAUTION

ALL APPLICABLE DIRECTIONS, RESTRICTIONS, AND PRECAUTIONS ON THE EPA-REGISTERED LABELS ARE TO BE FOLLOWED.

* Recommendation made under the provisions of Section 2(ee) of FIFRA

REGISTRATION INFORMATION
TOPSIN® M

FOR DISTRIBUTION AND USE ONLY WITHIN THE STATE OF SOUTH CAROLINA

EPA Reg. No. 4581-322 EPA SLN No. SC-790033

Tank Mixture With Captan For Use On Stone Fruits

DIRECTIONS FOR USE

It is a violation of Federal law to use this product in a manner inconsistent with its labeling.

APPLICATION INSTRUCTIONS

The use of TOPSIN M is recommended as a tank mixture with captan for the control of Monilinia Brown Rot Blossom Blight and Fruit Brown Rot in South Carolina. Apply with ground or aerial equipment, using sufficient volume of spray to provide thorough coverage.

It is suggested that TOPSIN M not be used alone for Monilinia Brown Rot control in South Carolina.

Mixing Instructions—Add required amount of TOPSIN M to partially filled tank agitated by mechanical or hydraulic means, add captan 50% wettable, and then add remaining required amount of water; continuous agitation is required to keep the material in suspension.

Do not tank mix with copper-containing materials or with highly alkaline pesticides, such as Bordeaux mixture or lime sulfur. No claim of compatibility with any other pesticide is implied.

STONE FRUIT (PEACHES, CHERRIES, APRICOTS, NECTARINES, PRUNES, AND PLUMS)—For the control of Monilinia Brown Rot Blossom Blight, apply 0.75 to 1.5 pounds TOPSIN M in mixture with 4 pounds captan 50% wettable per acre. Make applications at early bloom and again at full bloom.

Monilinia Fruit Brown Rot:

In addition to bloom sprays, for the control of Fruit Brown Rot, apply 1 or 2 pre-harvest sprays at the same rates as above starting 3 weeks before harvest. If needed under severe disease pressure, apply additional sprays at 10 to 14 day intervals between full bloom and final pre-harvest sprays. Do not apply within 1 day of harvest.

Do not graze livestock in treated areas.

All applicable directions, restrictions, and precautions on the EPA-registered label are to be followed.

NOTE: This labeling must be in the possession of the user at the time of pesticide application.

KEEP OUT OF REACH OF CHILDREN
CAUTION

If swallowed, induce vomiting with a mixture of soap or mustard in warm water. Call a physician immediately.

If on skin, wash immediately with soap and water.

If in eyes, flush immediately with plenty of water.

For First Aid Statement and Other Precautionary Statements, see label attached to the container.

SPECIAL LOCAL NEED LABEL
TOPSIN® M

FOR DISTRIBUTION AND USE ONLY IN THE STATE OF LOUISIANA

EPA Reg. No. 4581-322 EPA SLN No. LA-840014 (Rev. 1)

For suppression of aerial blight (*Rhizoctonia solani*) on soybeans

DIRECTIONS FOR USE

It is a violation of Federal law to use this product in a manner inconsistent with its labeling.

CROP	POUNDS/ACRE	REMARKS
Soybean	1	Apply from full bloom to when pods are $\frac{1}{8}''$ to $\frac{1}{4}''$ in length. Make a second application 14 to 21 days later. Do not make the second application later than 14 days after pods average $\frac{1}{4}''$ in length or when beans form in the pod. Use the high rate under severe disease pressure. Do not make more than 2 applications per year. Do not graze or feed treated vines or hay to livestock.

KEEP OUT OF REACH OF CHILDREN
CAUTION

For Statement of Practical Treatment and Other Precautionary Statements see Label attached to container.

All applicable directions, restrictions, and precautions on the EPA-registered labels are to be followed.

NOTE: This label must be in the possession of the user at the time of pesticide application.

SPECIAL LOCAL NEED LABEL
TOPSIN® M

FOR DISTRIBUTION AND USE ONLY IN THE STATE OF MISSISSIPPI

EPA Reg. No. 4581-322 EPA SLN No. MS-840014

For suppression of aerial blight (*Rhizoctonia solani*) on soybeans

DIRECTIONS FOR USE

It is a violation of Federal law to use this product in a manner inconsistent with its labeling.

CROP	POUNDS/ACRE	REMARKS
Soybean	$\frac{1}{2}$–1	Apply from full bloom to when pods are $\frac{1}{8}''$ to $\frac{1}{4}''$ in length. Make a second application 14 to 21 days later. Do not make the second application later than 14 days after pods average $\frac{1}{4}''$ in length or

CROP	POUNDS/ACRE	REMARKS
		when beans form in the pod. Use the high rate under severe disease pressure. Do not make more than 2 applications per year. Do not graze or feed treated vines or hay to livestock.

KEEP OUT OF REACH OF CHILDREN

CAUTION

For Statement of Practical Treatment and Other Precautionary Statements see Label attached to container.

All applicable directions, restrictions, and precautions on the EPA-registered labels are to be followed.

NOTE: This label must be in the possession of the user at the time of pesticide application.

TOPSIN® M 4.5 F

THIOPHANATE–METHYL FUNGICIDE

4.5 Pounds per Gallon Flowable

SHAKE WELL BEFORE USING

ACTIVE INGREDIENT:

Thiophanate-methyl (dimethyl[(1,2-phenylene)—bis(iminocarbonothioyl)]bis[carbamate])* 46.2%

INERT INGREDIENTS: 53.8%

TOTAL: 100.0%

*Also known as dimethyl 4,4′-o-phenylenebis[3-thioallophanate]

EPA Reg. No. 4581-352 EPA Est. No. 35982-TX-1

®TOPSIN is a registered trademark of Nippon Soda Company, Ltd., and is licensed to Pennwalt Corporation. Covered by one or more of the following U.S. patents: 3,769,308; 3,856,847, 4,020,095; and 4,029,813

CAUTION

Keep Out of Reach of Children

May be Harmful if Swallowed, Inhaled, or Absorbed Through Skin. Refer to back panel for additional precautionary statements and statement of practical treatment. For general information and directions for use refer to accompanying brochure.

CAUTION

PRECAUTIONARY STATEMENTS

HAZARD TO HUMANS AND DOMESTIC ANIMALS

May be Harmful if Swallowed, Inhaled, or Absorbed Through Skin. Avoid contact with eyes, skin, or clothing. Avoid breathing of spray mist. Wash thoroughly with soap and water after handling. Wash contaminated clothing before reuse. Do not contaminate feed or food. Keep children and pets out of treated areas until material is completely dry.

STATEMENT OF PRACTICAL TREATMENT

If swallowed, induce vomiting with a mixture of soap or mustard in warm water. Call a physician immediately.

If on skin, wash immediately with soap and water.

If in eyes, flush immediately with plenty of water.

ENVIRONMENTAL HAZARD

This pesticide is toxic to fish. Do not apply directly to water and wetlands. Do not apply where run-off is likely to occur. Do not apply when weather conditions favor drift from treated areas. Do not contaminate water by cleaning of equipment or disposal of wastes.

STORAGE AND DISPOSAL

Do not contaminate water, food, or feed by storage or disposal.

PESTICIDE DISPOSAL

Pesticide, spray mixture, or rinse water that cannot be used according to label instructions must be disposed of according to applicable Federal, State, or local procedures.

CONTAINER DISPOSAL

Triple rinse container, then offer for recycling or reconditioning, or dispose of in a sanitary landfill, or by incineration if allowed by State and local authorities.

GENERAL INFORMATION

Apply TOPSIN® M with ground or aerial equipment, using sufficient volume of spray to provide thorough coverage. Add required amount of TOPSIN® M to partially filled tank agitated by mechanical or hydraulic means, and then add remaining required amount of water. Continuous agitation is required to keep the material in suspension. Do not tank mix with copper-containing materials or with highly alkaline pesticides, such as Bordeaux mixture or lime sulfur. No claim of compatability with other pesticides is implied. Use the higher rate under conditions of severe disease pressure. Also, see local State Extension Service recommendations for application schedules.

IMPORTANT: If, after using TOPSIN® M as recommended, treatment is not effective, a tolerant strain of fungi may be present. Consult your State Agricultural Experiment Station or your State Agricultural Extension Service for advice on the prompt use of some other suitable fungicide.

DIRECTIONS FOR USE

It is a violation of Federal law to use this product in a manner inconsistent with its labeling.

SHAKE WELL BEFORE USING

CROP	DISEASES	FL OZ/A	FL OZ/100 GAL	REMARKS
Almonds	Brown Rot Blossom Blight (*Monilinia* sp.)	30–40		Apply at early pink bud and early petal fall.
Apples (pre-harvest)	Apple Scat (*Venturia* sp.) Bitter Rot (*Glomerella* sp.) Flyspeck (*Schizothyrium* sp.) Powdery Mildew (*Podosphaera* sp.) Sooty Blotch (*Gloeodes* sp.) Black Rot (*Physalospora* sp.)	20–30	5–7½	Apply at 5- to 10-day intervals from green tip through petal fall; continue at 7- to 14-day intervals in cover sprays. NOTE: For control of apple diseases in the Southeastern States of Florida, Georgia, South Carolina, North Carolina and Alabama, use 60–80 fl. oz. TOPSIN M per acre.
Beans	Gray Mold (*Botrytis* sp.)	30–40	20–27	Apply once at 50% to 70% of full bloom.

CROP	DISEASES	FL OZ/A	FL OZ/100 GAL	REMARKS
	White Mold (*Sclerotinia* sp.)	OR 20–30	OR 13–20	OR Apply twice with the first application at 10% to 30% of full bloom and a second application at 4 to 7 days later or at peak bloom. Do not apply to snap or dry beans within 14 days of harvest or to lima beans within 28 days of harvest.
Celery	Early Blight (*Cercospora* sp.) Late Blight (*Septoria* sp.)	10		Apply to the base of the plants on a 14- to 21-day schedule beginning about 2 weeks after transplanting. OR Apply as a foliar spray every 7 to 14 days beginning about 2 weeks after transplanting. Do not apply within 7 days of harvest. Do not make more than 4 applications per season.
Peanuts	Leaf Spot (*Cercospora* sp.)	10		Begin applications 35 days after planting or when disease first appears and repeat at 14- to 21-day intervals as needed. Use the 14-day interval under severe disease pressure. Do not apply within 14 days of harvest.
Pecans	Brown Spot (*Cercospora* sp.) Downy Spot (*Mycosphaerella* sp.) Powdery Mildew (*Micosphaero* sp.) Scab (*Fusicladium* sp.)	10–20		Use the higher rate for trees over 30 feet tall and for aerial applications in Arkansas, Louisiana, Mississippi, Oklahoma, and Texas. Begin applications when first leaves are showing and repeat at 3- to 4-week intervals until shuck split. Do not apply after shuck split.
Soybeans	Anthracnose (*Colletotrichum* sp.) Brown Spot (*Septoria* sp.) Frogeye Leaf Spot (*Cercospora* sp.) Pod and Stem Blight (*Diaporthe* sp. and the imperfect stage. *Phomopsis* sp.) Purple Seed Stain (*Cercospora* sp.)	10–20		Apply from full bloom to when pods are ⅛" to ¼" in length. Make a second application 14 to 21 days later. Do not make the second application later than 14 days after pods average ¼" in length or when beans form in the pod. Use the high rate under severe disease pressure. FOR SEED BEANS ONLY—For seed quality make a single application at the high rate when beans form in the pod. Do not make more than 2 applications per year. Do not graze or feed treated vines or hay to livestock.
Stone Fruits (pre-harvest) —Apricots	Brown Rot Blossom Blight (*Monilinia* sp.) Fruit Brown Rot (*Monilinia* sp.)	30	10	Apply at early bloom (red bud). Make a second application at full bloom. In addition, for Fruit Brown Rot, apply 1 or 2 sprays starting 3 weeks before harvest. If needed under severe disease pressure apply additional sprays at 10- to 14-day intervals between full bloom and final pre-harvest sprays. Do not apply within 1 day of harvest.
—Cherries	Brown Rot Blossom Blight (*Monilinia* sp.) Fruit Brown Rot (*Monilinia* sp.)	30	10	Apply at early bloom (early popcorn). Make a second application at full bloom. In addition, for Fruit Brown Rot, apply 1 or 2 sprays starting 3 weeks before harvest. If needed under severe disease pressure, apply additional sprays at 10- to 14-day intervals between full bloom and final preharvest sprays. Do not apply within 1 day of harvest.
	Cherry Leaf Spot (*Coccomyces* sp.)	24–30	8–10	Apply at petal fall or before, when leaves first unfold, and at first, second and third cover at 10- to 14-day intervals and one spray 14 to 21 days after harvest. Do not apply within 1 day of harvest.
	Powdery Mildew (*Podosphaera* sp.) and (*Sphaerotheca* sp.)	30 PLUS 24–30	10 PLUS 8–10	Apply at early bloom (early popcorn). Make a second application at full bloom. PLUS Apply at shuck fall and first cover. Do not apply within 1 day of harvest.
—Nectarines	Brown Rot Blossom Blight (*Monilinia* sp.) Fruit Brown Rot (*Monilinia* sp.)	30	10	Apply at early bloom (pink bud). Make a second application at full bloom if conditions favor disease development. In addition, for Fruit Brown Rot, apply 1 or 2 sprays starting 3 weeks before harvest. If needed under severe disease pressure, apply additional sprays at 10- to 14-day intervals between full bloom and final preharvest sprays. Do not apply within 1 day of harvest.
—Peaches	Brown Rot Blossom Blight (*Monilinia* sp.) Fruit Brown Rot (*Monilinia* sp.)	30–45	10–15	Apply at early bloom (pind bud). Make a second application at full bloom if conditions favor disease development. In addition, for Fruit Brown Rot, apply 1 or 2 sprays starting 3 weeks before harvest. If needed under severe disease pressure,

CROP	DISEASES	FL OZ/A	FL OZ/100 GAL	REMARKS
				apply additional sprays at 10- to 14-day intervals between full bloom and final preharvest sprays. Do not apply within 1 day of harvest. NOTE: For Florida, Georgia, South Carolina, North Carolina, Mississippi and Alabama use the higher rate.
	Peach Scab (*Cladosporium* sp.)	30–45 **PLUS** 24–30	10–15 **PLUS** 8–10	Apply at early bloom (pink bud). Made a second application at full bloom if conditions favor disease development. **PLUS** Apply at shuck split and at first cover sprays. Do not apply within 1 day of harvest.
—Plums and Prunes	Brown Rot Blossom Blight (*Monilinia* sp.) Fruit Brown Rot (*Monilinia* sp.)	30	10	Apply at early bloom (green tip). Make a second application at full bloom. In addition, for Fruit Brown Rot, apply 1 or 2 sprays starting 3 weeks before harvest. If needed under severe disease pressure, apply additional sprays at 10- to 14-day intervals. Do not apply within 1 day of harvest.
	Black Knot (*Dibotryon* sp.)	30	10	Apply at pre-bloom, petal fall, and at first, second, and third cover sprays at 10- to 14-day intervals. Do not apply within 1 day of harvest.
	Leaf Spot (*Coccomyces* sp.)	30	10	Apply at petal fall, shuck, and at first, second, and third cover sprays at 10- to 14-day intervals and 1 spray 14 to 21 days after harvest. Do not apply within 1 day of harvest.
Strawberries	Fruit Rot (*Botrytis* sp.) Leaf Blight (*Dendrophoma* sp.) Leaf Scorch (*Diplocarpon* sp.)	15–20		Begin applications at early bloom and continue at 7- to 10-day intervals. Use the higher rate East of the Rocky Mountains and under conditions of severe disease pressure. Do not apply within 1 day of harvest.
Sugar Beets	Leaf Spot (*Cercospora* sp.)	8–10		Begin applications when disease first appears and repeat at 14- to 21-day intervals as needed. Do not apply within 21 days of harvest.
Sugar Cane (Hawaii only)	Pineapple Disease (*Thielaviopsis* sp. or *Ceratocystis* sp.)		10 (1:1,600)	Apply to cut seed pieces as a cold or hot dip. Cold Dip—Immerse seed pieces to provide thorough coverage, remove and allow to drain.
			5 (1:3,200)	Hot Dip—Maintain temperature of dip at 52°C and soak seed pieces for 20 minutes, remove and allow to drain. Do not use treated seed pieces for food, feed or fodder. Do not graze or feed treated forage to livestock.

TANK MIX FOR USE ON APPLES
(NOT FOR USE IN CALIFORNIA)

DISEASES	RATE/ACRE	RATE/100 GAL	REMARKS
Scab (*Ventura* sp.) Bitter Rot (*Glomerella* sp.) Black Rot (*Physalospora* sp.) Flyspeck (*Schizothyrium* sp.) Powdery Mildew (*Podosphaera* sp.) Sooty Blotch (*Gloeodes* sp.) Brown Rot (*Monilinia* sp.) Cedar Apple Rust (*Gymnosporangium* sp.) Quince Rust (*Gymnosporangium* sp.)	TOPSIN M 10–16 fl oz + Mancozeb 80WP 3–4 lbs.	TOPSIN M 2¼–4 fl oz + Mancozeb 80WP ¾–1 lb	Apply at ½" green tip and repeat at 7-to 14-day intervals or as needed through the cover sprays One quart spray oil may be added per 100 gallons Add oil to the spray tank last. Use the high rate of TOPSIN M and add spray oil to the spray mixture for varieties more susceptible to powdery mildew, and for scab if an application is missed during an infection period (apply as soon as possible after infection period in order to deactivate scab and to prevent further infection). Do not apply within 30 days of harvest. Do not graze treated areas.

Observe the more restrictive cautions and limitations on labeling of products used in mixtures.

TANK MIX FOR USE ON APPLES
(NOT FOR USE IN CALIFORNIA)

DISEASES	RATE/ACRE	RATE/100 GAL	REMARKS
Scab (*Venturia* sp.) Bitter Rot (*Glomerella* sp.) Black Rot (*Physalospora* sp.) Flyspeck (*Schizothyrium* sp.)	TOPSIN M 10–16 fl oz + Captan 50 WP 3–4 lbs OR	TOPSIN M 2½–4 fl oz + Captan 50 WP ¾–1 lb OR	Apply at ½" green tip and repeat at 7– to 14–day intervals or as needed through the cover sprays. Use the high rate of TOPSIN M for varieties more susceptible to powdery mildew. If an application is missed during an infection period, apply the higher rates as soon as possible after the infection period in order to deactivate scab and to prevent further infection.

DISEASES	RATE/ACRE	RATE/100 GAL	REMARKS
Powdery Mildew (*Podosphaera* sp.) Sooty Blotch (*Gloeodes* sp.)	Captan 80WP 2–2½ lbs	Captan 80WP ½–⅝ lb	Note: Spray injury may result if Captan is used with, immediately before, or closely following an oil spray.

Observe the more restrictive cautions and limitations on labeling of products used in mixtures.

TANK MIX FOR USE ON APPLES
(NOT FOR USE IN CALIFORNIA)

DISEASES	RATE/ACRE	RATE/100 GAL	REMARKS
Scab (*Venturia* sp.) Bitter Rot (*Glomerella* sp.) Black Rot (*Physalospora* sp.) Flyspeck (*Schizothyrium* sp.)	TOPSIN M 10–16 fl oz + Maneb 80WP 6–8 lbs	TOPSIN M 2½–4 fl oz + Maneb 80WP 1½–2 lbs	Begin applications at green tip and repeat at pink and at 7– to 14–day intervals or as needed through the cover sprays. Use the higher rates and closer interval under severe disease conditions. With Golden Delicious apples, adverse effects on fruit finish and color may result. Use the lower rates of TOPSIN M when finish and color are of primary importance.
Powdery Mildew (*Posdosphaera* sp.) Sooty Blotch (*Gloeodes* sp.) Cedar Apple Rust (*Gymnosporangium* sp.) Bull's Eye Rot (*Gloeosporium* sp.)			Do not apply within 30 days of harvest. Do not graze livestock in treated areas.

Observe the more restrictive cautions and limitations on labeling of products used in mixtures.

TANK MIX FOR USE ON APPLES
(NOT FOR USE IN CALIFORNIA)

DISEASES	RATE/ACRE	RATE/100 GAL	REMARKS
Scab (*Venturia* sp.) Bitter Rot (*Glomerella* sp.) Black Rot (*Physalospora* sp.) Flyspeck (*Schizothyrium* sp.) Powdery Mildew (*Podosphaera* sp.) Sooty Blotch (*Gloeodes* sp.)	TOPSIN M 10–16 fl oz + Polyram 80WP 3–4 lbs	TOPSIN M 2½–4 fl oz + Polyram 80WP ¾–1 lb	Begin applications at green tip and repeat applications as needed through the cover sprays. Do not apply within 30 days of harvest. Do not graze livestock in treated areas.

Observe the more restrictive cautions and limitations on labeling of products used in mixtures.

TANK MIX FOR USE ON APPLES
(NOT FOR USE IN CALIFORNIA)

DISEASES	RATE/ACRE	RATE/100 GAL	REMARKS
Scab (*Ventura* sp.) Bitter Rot (*Glomerella* sp.) Black Rot (*Physalospora* sp.) Flyspeck (*Schizothyrium* sp.) Powdery Mildew (*Podosphaera* sp.) Sooty Blotch (*Gloeodes* sp.)	TOPSIN M 10–16 fl oz + Sulfur 12–30 lbs a.i.	TOPSIN M 2½–4 fl oz + Sulfur 3–7½ lbs a.i.	Begin applications at green tip and repeat at pink and at 7– to 14–day intervals or as needed through the cover sprays. Use the higher rates and closer interval under severe disease conditions.

Observe the more restrictive cautions and limitations on labeling of products used in mixtures.

TANK MIX FOR USE ON STONE FRUITS
(APRICOTS, CHERRIES, NECTARINES, PEACHES, PLUMS AND PRUNES)
(NOT FOR USE IN CALIFORNIA)

DISEASES	RATE/ACRE	RATE/100 GAL	REMARKS
Brown Rot Blossom Blight (*Monilinia* spp.) Fruit Brown Rot (*Monilinia* spp.) Peach Scab (*Cladosporium* sp.) Powdery Mildew (*Podosphaera* sp. and *Sphaerotheca* sp.)	TOPSIN M 15–20 fl oz + Captan 50WP 3–6 lbs **OR** Captan 80WP 2–3¾ lbs	TOPSIN M 5–6½ fl oz + Captan 50WP 1–2 lbs **OR** Captan 80WP ⅝–1¼ lbs	BROWN ROT BLOSSOM BLIGHT—Apply at early bloom stages before infection occurs (apricots—red bud; cherries—early popcorn; nectarines & peaches—pink bud; plums & prunes—green tip). If blossoming is prolonged or conditions favorable for disease persist, make a second application 14 days later. FRUIT BROWN ROT—After blossom blight sprays, make a preharvest application any time from 3 weeks before harvest to 1 day before harvest.

DISEASES	RATE/ACRE	RATE/100 GAL	REMARKS
Cherry Leaf Spot (*Coccomyces* sp.) Russet Scab Black Knot (*Dibotryon* sp.)			Make a second application if conditions favorable for disease persist or harvest is prolonged. PEACH SCAB, POWDERY MILDEW—Use same schedule as for Brown Rot Blossom Blight and continue at 10- to 14-day intervals to 1 day before harvest. Make an additional application 2 to 3 weeks after harvest. CHERRY LEAF SPOT—Use same schedule as for Brown Rot Blossom Blight and continue at 10- to 14-day intervals to 1 day before harvest. Make an additional application 2 to 3 weeks after harvest. RUSSET SCAB (PLUMS & PRUNES)—Use same schedule as for Brown Rot Blossom Blight; include full coverage spray at full bloom. BLACK KNOT—Apply at first, second, and third cover sprays at 10- to 14- day intervals. Do not apply within 1 day of harvest. NOTE: This tank mixture will not control bacterial blast, but will supplement a dormant program for peach leaf curl and shothole (*Coryneum* blight) refer to other fungicide labels for dormant period control information. Do not use this tank mixture with spray oils

Observe the more restrictive cautions and limitations on labeling of products used in mixtures.

TANK MIX FOR USE ON STONE FRUITS (APRICOTS, CHERRIES, NECTARINES, PEACHES, PLUMS AND PRUNES) (NOT FOR USE IN CALIFORNIA)

DISEASES	RATE/ACRE	RATE/100 GAL	REMARKS
Brown Rot Blossom Blight (*Monilinia* sp.) Fruit Brown Rot (*Monilinia* sp.) Powdery Mildew (*Podosphaera* sp. and *Sphaerotheca* sp.) Leaf Spot (*Coccomyces* sp.) Shot-hole (*Coryneum* sp.) Rust (*Tranzschelia* sp.) Peach Canker Peach Scab (*Cladosporium* sp.) Black Knot (*Dibatryon* sp.)	TOPSIN M 15–30 fl oz + Sulfur 12–18 lbs a.i.	TOPSIN M 5–10 fl oz + Sulfur 4–6 lbs i.a.	BROWN ROT BLOSSOM BLIGHT—Make application at early bloom & again at full bloom. FRUIT BROWN ROT—In addition to bloom sprays, apply 1 or 2 sprays starting 3 weeks before harvest. If needed under severe disease pressure, apply additional sprays at 10- to 14-day intervals between full bloom and final preharvest sprays. POWDERY MILDEW—In addition to bloom sprays, make applications at shuck fall and first cover. LEAF SPOT—Make application when leaves first unfold and first, second, and third cover at 10- to 14- day intervals and 1 spray 14 to 21 days after harvest. SHOT-HOLE, RUST, PEACH CANKER & PEACH SCAB—Apply at pink bloom, petal fall, shuck split, shuck fall, and at 10- to 14-day intervals as long as necessary. BLACK KNOT—Apply at first, second, and third cover sprays at 10- to 14- day intervals. Do not apply within 1 day of harvest.

Observe the more restrictive cautions and limitations on labeling of products used in mixtures.

TANK MIX FOR USE ON STONE FRUITS (APRICOTS, CHERRIES, NECTARINES AND PEACHES) (NOT FOR USE IN CALIFORNIA)

DISEASES	CROP	RATE/ACRE	RATE/100 GAL	REMARKS
Brown Rot Blossom Blight (*Monilinia* sp.) Fruit Brown Rot (*Monilinia* sp.) Powdery Mildew (*Podosphaera* sp. and *Sphaerotheca* sp.)	Apricots and Nectarines	TOPSIN M 15–30 fl oz + Ziram 3–4½ lbs a.i.	TOPSIN M 5–10 fl oz + Ziram 1–1½ lbs a.i.	BROWN ROT BLOSSOM BLIGHT—Make application at early bloom and again at full bloom. FRUIT BROWN ROT—In addition to bloom sprays, apply 1 or 2 sprays starting 3 weeks before harvest. If needed under severe disease pressure, apply additional sprays at 10- to 14-day intervals between full bloom and final pre-harvest sprays.
Cherry Leaf Spot (*Coccomyces* sp.) Shot-hole (*Coryneum* sp.) Rust (*Tranzschelia* sp.)	Cherries	TOPSIN M 15–30 fl oz + Ziram 3⅜ lbs a.i.	TOPSIN M 5–10 fl oz + Ziram 1⅛ lbs a.i.	POWDERY MILDEW—In addition to bloom sprays, make applications at shuck fall and first cover. CHERRY LEAF SPOT—Make application when leaves first unfold & at first, second, and third cover at 10- to 14-day intervals and 1 spray 14-21 days after harvest.
Peach Canker Peach Scab (*Cladosporium* sp.)	Peaches	TOPSIN M 15–30 fl oz + Ziram 3–4½ lbs. a.i.	TOPSIN M 5–10 fl oz + Ziram 1–1½ lbs. a.i.	SHOT-HOLE, RUST, PEACH CANKER, & PEACH SCAB—Apply at pink, bloom, petal fall, shuck split, shuck fall, and at 10- to 14-day intervals as long as necessary. For apricots, do not apply after early cover periods. For cherries, do not apply within 7 days of harvest. For nectarines & peaches, last application should be made no later than at petal fall.

Observe the more restrictive cautions and limitations on labeling of products used in mixtures.

TANK MIX FOR USE ON STONE FRUITS (APRICOTS, NECTARINES, AND PEACHES) (NOT FOR USE IN CALIFORNIA)

DISEASES	RATE/ACRE	RATE/100 Gal	REMARKS
Brown Rot Blossom Blight (*Monilinia* sp.) Fruit Brown Rot (*Monilinia* sp.) Peach Scab (*Cladosporium* sp.) Powdery Mildew (*Podosphaera* sp. and *Sphaerotheca* sp.)	TOPSIN M 30 fl oz + Maneb 80WP 6–8 lbs	TOPSIN M 10 fl oz + Maneb 80WP 2–2⅔ lbs	BROWN ROT BLOSSOM BLIGHT—Apply at early bloom stages before infection occurs (apricots—red bud; nectarines & peaches—pink bud). For this application only, TOPSIN M plus Maneb 80WP may be used in combination with superior-type spray oil. If blossoming is prolonged or conditions favorable for disease persist, make a second application 14 days later. FRUIT BROWN ROT—After blossom blight sprays, make an application 3 weeks before harvest. PEACH SCAB, POWDERY MILDEW—Use same schedule as for Brown Rot Blossom Blight plus applications at shuck split, shuck fall, and 14 days later. Do not apply within 14 days of harvest. NOTE: This tank mixture will not control bacterial blast, but will supplement a dormant program for each leaf curl and shot-hole (*Coryneum blight*); refer to other fungicide labels for dormant period control information.

Observe the more restrictive cautions and limitations on labeling of products used in mixtures.

TANK MIX FOR USE ON PEANUTS (NOT FOR USE IN CALIFORNIA)

DISEASES	RATE/ACRE	REMARKS
Leaf Spot (*Cerospora* sp.) Rust (*Puccinia* sp.) Web Blotch (*Ascochyta* sp.)	TOPSIN M 5 fl oz + Mancozeb 80WP 1–2 lbs	Begin application 35 to 40 days after planting or when disease first appears. Repeat at the following intervals for *Cercospora* leaf spot at 10 to 14 days; for Rust at 7 to 10 days; and for *Ascochyta* web blotch at 7 to 14 days. Spray oil may be added at the rate of 1 pint to 1 quart per acre. Do not graze or feed treated vines, hay, or hulls to livestock. Do not apply within 14 days of harvest.

Observe the more restrictive cautions and limitations on labeling of products used in mixtures.

WARRANTY AND DISCLAIMER

Pennwalt Corporation warrants that this material confirms to the chemical description on the label and is reasonably fit for the purposes referred to in the Directions for Use, subject to the risks referred to therein PENNWALT MAKES NO OTHER EXPRESS OR IMPLIED WARRANTY OF FITNESS OR MERCHANTABILITY OR ANY OTHER EXPRESS OR IMPLIED WARRANTY. IN NO CASE SHALL PENNWALT OR SELLER BE LIABLE FOR CONSEQUENTIAL, SPECIAL OR INDIRECT DAMAGES RESULTING FROM THE USE OR HANDLING OF THIS PRODUCT INCLUDING, BUT NOT LIMITED TO, LOSS OF PROFITS, BUSINESS REPUTATION, OR CUSTOMERS LABOR COST, OR OTHER EXPENSES INCURRED IN PLANTING OR HARVESTING.

Pennwalt and seller offer this product and the buyer and user accept it subject to the foregoing conditions of sale and warranty which may be varied only by agreement in writing signed by a duly authorized representative of Pennwalt.

RECOMMENDATION*

TOPSIN M 4.5F THIOPHANATE-METHYL FUNGICIDE

EPA Registration No. 4581-352

FOR CONTROL OF BROWN ROT BLOSSOM BLIGHT (*Monilinia* sp.) AND FRUIT BROWN ROT (*Monilinia* sp.) ON PEACHES
IN THE STATES OF ALABAMA, FLORIDA, GEORGIA, MISSISSIPPI, NORTH CAROLINA AND SOUTH CAROLINA

DIRECTIONS FOR USE

It is a violation of Federal law to use this product in a manner inconsistet with its labeling.

KEEP OUT OF REACH OF CHILDREN

CAUTION

ALL APPLICABLE DIRECTIONS, RESTRICTIONS, AND PRECAUTIONS ON THE EPA-REGISTERED LABELS ARE TO BE FOLLOWED.

Crop	Diseases	Fl Oz/A	Fl Oz/ 100 Gal	Remarks
Peaches	Brown Rot Blossom Blight (*Monilinia* sp.) Fruit Brown Rot (*Monilinia* sp.)	30–45	10–15	Apply at early bloom (pink bud. Make a second application at full bloom if conditions favor disease development. In addition, for Fruit Brown Rot, apply 1 or 2 sprays starting 3 weeks before harvest. If needed under severe disease pressure, apply additional sprays at 10- to 14-day intervals between full bloom and final preharvest sprays. NOTE: When tank mix is considered appropriate, TOPSIN M 4.5F is registered for use in combination with Captan, Sulfur, Ziram and Maneb. See supplemental labels for tank mixes on stone fruits.

* Recommendation made under the provisions of Section 2(ee) of FIFRA

RECOMMENDATION*

TOPSIN® M 4.5F FUNGICIDE

EPA Registration No. 4581-352

APPLICATION THROUGH SPRINKLER IRRIGATION SYSTEMS
USE RESTRICTED TO THE STATES OF IDAHO AND WASHINGTON

GENERAL INFORMATION

Apply TOPSIN M Fungicide only through systems containing anti-syphon and check valves designed to prevent water source contamination or overflow of the mix tank and containing interlocking controls between the metering device and the water pump to insure simultaneous shut-off. Allow sufficient time for flushing of TOPSIN M from the irrigation system prior to shutdown.

DIRECTIONS FOR USE

It is a violation of Federal law to use this product in a manner inconsistent with its labeling.

Add required amount of TOPSIN M to a partially filled tank agitated hydraulically or mechanically, then add remaining required amount of water. Constant agitation is required to keep material in suspension. If other pesticides are to be used in combination, TOPSIN M should be added to the mix tank prior to the addition of any emulsifiable concentrates. No claim of compatibility with other pesticides is implied.

Crop	Disease	Rate 4.5 F Flowable (oz/A)	Remarks
Dry Beans	White Mold (*Sclerotinia* sp.) Grey Mold (*Botrytis* sp.)	30–40	Apply once when all plants have first opened blossom (100% bloom), or at first petal fall. Do not apply to dry beans within 30 days of harvest.

If conditions are favorable for maximal disease incidence, use the higher rate.

Set center-pivot or linear system to deliver a minimum of 0.1 inch to a maximum of 0.25 inches of water per acre. Operate at 100% or maximum speed. Higher water volume may reduce efficacy.

Inject suspension of TOPSIN M uniformly throughout the entire irrigation period when using center-pivot and linear systems. In solid set, handmoved, or rolling-type systems, inject TOPSIN M suspension during the last 15-30 minutes.

Allow adequate time after injection for flushing of system. Inject TOPSIN M suspension into system at a right angle to enhance adequate mixing.

Note: Poor or decreased control may occur when 1) white mold infections occur prior to application, or favorable disease conditions persist for an extended time period, 2) irrigation systems do not apply water uniformly, 3) lower rates are used when conditions are favorable for disease.

READ AND FOLLOW THE TOPSIN M LABELS FOR ADDITIONAL PRECAUTIONS, INSTRUCTIONS AND RESTRICTIONS.

*Recommendation made under the provisions of Section 2(ee) of FIFRA.

SPECIAL LOCAL NEED LABEL

TOPSIN® M 4.5F THIOPHANATE-METHYL FUNGICIDE

EPA Reg. No. 4581-352 EPA SLN No. LA-840013 (Rev. 1)

FOR DISTRIBUTION AND USE ONLY IN THE STATE OF LOUISIANA

FOR SUPPRESSION OF AERIAL BLIGHT *(Rhizoctonia solani)* ON SOYBEANS

DIRECTIONS FOR USE

It is a violation of Federal law to use this product in a manner inconsistent with its labeling.

Crop	Fl. Oz/Acre	Remarks
Soybean	16–20	Apply from full bloom to when pods are 1/8" to 1/4" in length. Make a second application 14 to 21 days later. Do not make the second application later than 14 days after pods average 1/4" in length or when beans form in the pod. Use the high rate under severe disease pressure. Do not make more than 2 applications per year. Do not gaze or feed treated vines or hay to livestock.

KEEP OUT OF REACH OF CHILDREN

CAUTION

For Statement of Practical Treatment and Other Precautionary Statements see Label attached to container.

All applicable directions, restrictions, and precautions on the EPA-registered labels are to be followed.

NOTE: This label must be in the possession of the user at the time of pesticide application.

SPECIAL LOCAL NEED LABEL

TOPSIN® M 4.5F THIOPHANATE-METHYL FUNGICIDE

EPA Reg. No. 4581-352 EPA SLN No. MS-840013

FOR DISTRIBUTION AND USE ONLY IN THE STATE OF MISSISSIPPI

FOR SUPPRESSION OF AERIAL BLIGHT *(Rhizoctonia solani)* ON SOYBEANS

DIRECTIONS FOR USE

It is a violation of Federal law to use this product in a manner inconsistent with its labeling.

Crop	Fl. Oz/Acre	Remarks
Soybean	10–20	Apply from full bloom to when pods are 1/8" to 1/4" in length. Make a second application 14 to 21 days later. Do not make the second application later than 14 days after pods average 1/4" in length or when beans form in the pod. Use the high rate under severe disease pressure. Do not make more than 2 applications per year. Do not graze or feed treated vines or hay to livestock.

KEEP OUT OF REACH OF CHILDEN

CAUTION

For Statement of Practical Treatment and Other Precautionary Statements see Label attached to container.

All applicable directions, restrictions, and precautions on the EPA-registered labels are to be followed.

NOTE: This label must be in the possession of the user at the time of pesticide application.

PPG Industries, Inc.

Chemicals Group
One PPG Place
Pittsburgh, PA 15272
(800) 245-6611

BUD NIP®

SYSTEMIC SUCKER CONTROL AGENT PLANT GROWTH REGULATOR

ACTIVE INGREDIENTS:

Ispropyl *m* -chlorocarbanilate	11.9%
INERT INGREDIENTS:	88.1%
TOTAL	100.0%

Contains 1 Pound Chlorpropham Per Gallon

EPA Reg No. 748-233 EPA Est. No. 748-OH-01

DANGER

Keep Out of Reach of Children

Emergency Telephone Number
PPG Industries, Inc., Natrium, WV, (304) 843-1300

HAZARDOUS TO HUMANS AND DOMESTIC ANIMALS

Corrosive. May cause eye damage and skin irritation. Harmful if swallowed. Avoid breathing vapor or mist. Wear goggles or face shield. Wash thoroughly with soap and water after handling. Remove contaminated clothing and wash before reuse.

PRACTICAL TREATMENT

If in eyes flush with plenty of water and get medical attention. If on skin flush with plenty of water for 15 minutes and get medical attention if irritation persists. If swallowed drink 1 or 2 glasses of water and induce vomiting by touching back of throat with finger and get medical attention. Never give anything by mouth to an unconscious person.

ENVIRONMENTAL HAZARDS

Do not apply directly to water.
Do not contaminate water by cleaning of equipment or disposal of wastes.

PHYSICAL OR CHEMICAL HAZARDS

Do not use or store near heat or open flame.

STORAGE AND DISPOSAL

Do not contaminate water, food, or feed by storage or disposal. Pesticide wastes are acutely hazardous. Improper disposal of excess pesticide, spray mixture, or rinsate is a violation of Federal law. If these wastes cannot be disposed of by use according to label instructions, contact your State Pesticide or Environmental Control Agency, or the Hazardous Waste representative at the nearest EPA Regional Office for guidance.

Do not reuse container. Triple rinse, then offer for recycling or reconditioning, or puncture and dispose of in a sanitary landfill, or by incineration, or if allowed by state and local authorities, by burning. If burned, stay out of smoke.

NOTES AND PRECAUTIONS

1. BUD NIP + up,TM plant growth regulator should be used as the final step in a sequential sucker control program, following the application of contact agents and systemic agents used in their normal sequence. These traditional sucker control chemicals should be used as recommended by state and local officials and as labeled. The use of BUD NIP in no way reduces your need to be accurate and conscientious in the appropriate use of these traditional products nor in agronomic practices used for maximum production efficiency. **Read SPECIFIC DIRECTIONS FOR USE section before starting your applications.**
2. Follow dosage recommendations and timing requirements carefully for BUD NIP and for all products used in the sequential sucker control program.
3. Rains occurring within four (4) hours after spraying may reduce effectiveness of BUD NIP.
4. Do not use BUD NIP on tobacco grown for seed production.
5. Do not tank mix BUD NIP with other tobacco sucker control agents.
6. Application of BUD NIP to immature plants may result in leaf distortion on the upper portion of the plant. Follow **WHEN TO SPRAY** section carefully.
7. Avoid application of BUD NIP to severely wilted, drought stricken, or waterlogged tobacco. Best results are obtained with morning and cloudy day applications when the tobacco is actively growing and soil moisture is ample.
8. Do not apply BUD NIP under conditions which favor drift onto nearby crops or nontarget areas.

SPECIFIC DIRECTIONS FOR USE

BUD NIP is a new formulation of chlorpropham used to prevent sucker development in flue-cured tobacco throughout the harvest period, when applied as a sequential treatment, **following** the use of one or more contact (fatty alcohol) applications plus a systemic (such as a potassium salt of maleic hydrazide) application.

Best results are obtained when BUD NIP is applied to a clean field in which the state and local recommendations have been followed closely for fertilization, time of topping, and application rates and timing intervals for contact and other systemic products so that sucker growth can be controlled up to the time BUD NIP is applied. Cleaner stalks at application time will result in cleaner stalks thru harvest.

WHEN TO SPRAY

Match local recommendations to your crop situation with respect to uniformity of flowering and follow this sequence.

1. Starting at button to early flower stage, apply one or more applications of a contact agent. Non-uniform growth will usually require two contact applications 3-5 days apart for best results.
2. Top all plants ready for topping immediately after each contact application.
3. Hand-sucker, to remove any escapes if needed, and top any late maturing plants 5-7 days after the last contact spray is made, and apply a systemic agent (such as a potassium formulation of MH) within 24 hours. Tip leaves to be harvested should be at least 6-8 inches long when systemic products are applied.
   NOTE: Consult the label on your MH-containing product for rate, optimum crop conditions, harvest interval restrictions, and other important information to insure best results.
4. Apply BUD NIP uniformly over the field using 20–40 pounds pressure and 25–40 gallons of water per acre 5–7 days after the systemic agent is applied.

HOW TO SPRAY

1. Use a conventional ground sprayer with nozzles oriented to give uniform, overall coverage of a fine particle spray pattern, similar to insecticide sprayers or a boom used with traditional systemic sucker control agents. Spray tips should be selected from the hollow cone or solid cone series and matched with your pressure range capacity to deliver a fine mist spray which uniformly wets the entire stalk.
2. The recommended rate of BUD NIP should be applied in 25–40 gallons clean water per acre. **Do not use more than 40 gallons of water per acre.** See **WHEN TO SPRAY** section and follow all steps carefully. The formulated product should be thoroughly

mixed in the tank and agitation maintained throughout the application.

AMOUNT TO SPRAY

For fields having plant population ranging from 5000–7000, apply 1 to 1½ gallons BUD NIP plant growth regulator per acre.

The statements and methods presented about the products mentioned are based upon the best available information and practices known to PPG Industries at the present time, but are not representations or warranties of performance, result or comprehensiveness of such data. PPG Industries warrants that the products mentioned herein conform to their chemical description and are reasonably fit for the purposes stated on the label when used in accordance with the label directions under normal conditions of use, or conditions of use reasonably foreseeable to the seller.

The products mentioned herein if not used in accordance with directions can be hazardous. PPG Industries recommends that anyone using and/or handling the products mentioned herein before using them thoroughly read and understand the specific directions for use and the precautions appearing in the product label.

The products mentioned herein, as all potentially hazardous materials, must be kept out of the reach of children.

CHEM HOE® FL4 Flowable

SELECTIVE HERBICIDE

Contains 4 Pounds Propham Per Gallon

ACTIVE INGREDIENT:	
Isopropyl carbanilate	43.0%
INERT INGREDIENTS:	57.0%
TOTAL	100.0%

EPA Reg. No. 748–207

CAUTION

Keep Out of Reach of Children

HARMFUL IF SWALLOWED

PRECAUTIONARY STATEMENTS

HAZARD TO HUMANS AND DOMESTIC ANIMALS

Harmful if swallowed. Avoid contact with skin, eyes, or clothing. Wash thoroughly with soap and water after handling. Avoid breathing of vapor or spray mist. Do not apply in such a manner as to directly or through drift expose workers or other persons. The area being treated must be vacated by unprotected persons.

PRACTICAL TREATMENT

If swallowed, induce vomiting. GET IMMEDIATE MEDICAL ATTENTION. In case of contact, flush skin or eyes with plenty of water.

ENVIRONMENTAL HAZARDS

Keep out of lakes, ponds or streams. Do not contaminate water by cleaning of equipment or disposal of wastes.

STORAGE AND DISPOSAL

Do not freeze. Store at temperature above 32° F. Do not reuse container. Destroy by crushing or perforating and burying in a safe place. Avoid contamination of feeds, fertilizers, seeds, plants and pesticides.

DIRECTIONS FOR USE

It is a violation of Federal law to use this product in a manner inconsistent with the label.

CAUTION! KEEP OUT OF REACH OF CHILDREN.

In case of emergency contact PPG Industries, Inc., Natrium, W. Va. (304) 843-1300.

GENERAL DIRECTIONS FOR USE

NOTICE: Use of this product in a manner or at a time or in quantities or on crops other than in accordance with the directions may result in crop injury.

These directions pertain to application under average conditions, ineffective weed control or crop damage may result from differing conditions of soil, temperature, moisture and crop variety. Therefore, before large-scale use, this product should be applied on a trial basis under conditions actually encountered. Consult state agricultural experimental station or extension service weed specialist for detailed recommendations of application under varying local conditions.

WEEDS CONTROLLED

- Annual bluegrass (Poa annua)
- Annual ryegrass (Lolium multiflorum)
- Bulbous bluegrass (Poa bulbosa)
- Burning nettle (Urtica urens)
- Canarygrass (Phalaris sp.)
- Common chickweed (Stellaria media)
- Downy brome (Bromus tectorum)
- Little barley (Hordeum pusillum)
- Medusahead (Elymus caput-medusae)
- Rabbitfootgrass (Polypogon monspeliensis)
- Rattail fescue (Festuca myuros)
- Red brome (Bromus rubens)
- Stinging nettle (Urtica dioica)
- Volunteer grains (Barley, oats, rye, wheat)
- Mouse barley (Hordeum murinum)
- Wild barley (Hordeum leporium)
- Wild oats (Avena fatua)

NOTE: Established perennial weeds will not be controlled. Soil moisture and temperature are very important for effective use of CHEM HOE FL4. Under dryland conditions, most effective weed control will be obtained when the soil temperature is below 55° F. Under irrigated conditions, application can be made when soil temperature is above 55° F provided irrigation immediately follows the application. Apply ½-inch of water on coarse textured soils and 1-inch on fine textured soils.

MIXING AND APPLICATION

CAUTION: Thoroughly clean tanks, lines, and pumps of spray equipment before use in applying CHEM HOE IPC FL4. Oily residues of previously handled materials can seriously affect the suspension qualities of CHEM HOE IPC FL4. Screens in pump lines and nozzles should be no finer than 50 mesh. Thoroughly mix the recommended amount of CHEM HOE IPC FL4 with clean water and maintain agitation of the mixture during application. Uniform application is essential. Equipment should be properly calibrated and adequate carrier used to assure even distribution. Avoid skips or overlapping which may result in poor control or crop injury. Avoid use of equipment or application techniques or application under weather conditions which favor drift.

CHEM HOE APPLICATION METHODS—Unless otherwise noted, application of CHEM HOE FL4 can be made by:

Ground Equipment—Use 20 or more gallons of total spray solution per acre. All post-emergence directed spray applications should be made with ground equipment.

Aircraft—Use 5 or more gallons of total spray solution per acre.

SPECIFIC DIRECTIONS FOR USE

Rates given are for overall coverage except as otherwise noted. For strip or band treatment reduce the amount of CHEM HOE IPC FL4 and water proportionately.

ALFALFA AND CLOVER—Preplant—Within 2 days before planting, uniformly apply 3 to 4 quarts of CHEM HOE IPC FL4 per acre. Incorporate into the top 2 inches of soil with a power driven rotary tiller or by cross discing. Use the lower rate on coarse and medium textured soils and the high rate on fine textured soils. Rainfall or irrigation is necessary after application.

ALFALFA AND CLOVER—Preemergence—Uniformly apply 3 to 4 quarts of CHEM HOE IPC FL4 per acre within 2 days after planting. Rainfall or overhead irrigation is necessary soon after application. Use the lower rate on coarse and medium textured soils and the high rate on fine textured soils.

ALFALFA AND CLOVER—Postemergence—Uniformly apply 3 to 4 quarts of CHEM HOE IPC FL4 per acre. Make application in fall to early spring. Apply when crop has at least 3 true leaves. Use the lower rate on coarse and medium textured soils and high rate on fine textured soils. Rainfall or irrigation is necessary soon after application (See SLN level in CA, NV, UT, OR, WA.)

LETTUCE—preplant—TANK MIXTURE OF CHEM HOE® PLUS BALAN™ for weed control in irrigated lettuce in Arizona, California, Colorado, and New Mexico.

TANK MIXING DIRECTIONS

Sprayer tank, boom and nozzles should be thoroughly cleaned and flushed with clean water before filling. Mechanical or bypass agitation should be maintaincd during filling and application. Fill the spray tank one-half full of clean water and with the agitator running add the recommended amount of CHEM HOE FL4 first and after thoroughly mixing add the recommended amount of Balan™ EC. If a compatibility agent is necessary (see Test for Compatibility section) it should be added *before* adding the herbicides. Fill the tank to the desired level. Do not let the mixture stand without agitation.

For more information on mixing Balan formulations contact PPG.

COMPATIBILITY TEST

Place one pint of carrier water in a quart jar. Add 1/2 teaspoon of compatibility agent such as E-Z Mix (Loveland Industries, Loveland, CO) or Co-Mix (Rinchem Company Inc., Phoenix, AZ). Add one tablespoon of each herbicide one at a time, and shake well after each addition. Let the mixture stand for one hour. If sludge or lumps develop, the mixture is incompatible and should not be used. A parallel test should be done without the compatibility agent to determine if it is necessary. When required, the compatibility agent should be used in accordance with the manufacturer's recommendations.

APPLICATION

Apply the tank mix uniformly to the soil. Immediately after or simultaneously with application, thoroughly incorporate the herbicides into the top two inches of soil. Several types of equipment do a good job of incorporating.

1. P.T.O.-driven equipment; use power driven rotary equipment (tillers, cultivators, hoes) with knives set to cut 2–3 inches deep and spaced to provide a clean sweep of the soil.

2. Double disc or double disc with harrow in tandem. Set the disc to cut 3–6 inches deep and work the soil in two directions (cross disc) at a speed of 4–6 miles per hour.

CHEM HOE® FL4: Balan™ EC Mixtures
Broadcast Rates Per Acre*
Mix in 40 to 60 Gallons of Clean Water Per Acre

	CHEM HOE® FL4 (4 Lbs./Gals.)	Balan™ EC (1.5 Lb./Gal.)
Coarse (sand, sandy loam) or Medium (silt, silt loam, loam)	3 quarts	3 quarts
Fine (clay, clay loam, silty clay)	4 quarts	4 quarts

* Reduce proportionately for band applications.

LETTUCE—Preplant—Within 2 days before planting, uniformly apply 4 to 6 quarts of CHEM HOE IPC FL4 per acre broadcast. Incorporate with a power driven rotary tiller to mix the herbicide into the top 2 inches of soil. Irrigate within a few days after application. Use the lower rate on coarse textured sandy soils and the higher rate on fine textured clay soils.

LETTUCE—Preemergence—Uniformly apply 4 to 6 quarts of CHEM HOE IPC FL4 per acre broadcast within 2 days after planting. At least 1/2 inch of rainfall or sprinkler irrigation is required within a few days after application for best results. Use the lower rate on coarse textured sandy soils and the higher rate on fine textured clay soils.

LETTUCE—Postemergence—Apply when the lettuce has at least 4 or more leaves. Uniformly apply 4½ to 6 quarts of CHEM HOE IPC FL4 in 20 gallons or more of water per acre. Irrigate following application to move the chemical into the soil and root zone of the weed plants. Use the lower rate on coarse textured sandy soils and the higher rate on fine textured clay soils.

PEAS (Green or Dry) and LENTILS—Preplant—Apply for control of Wild Oats and volunteer grain. Before planting apply 4 quarts of CHEM HOE IPC FL4 per acre broadcast. Best results are obtained by lightly working and leveling the soil to smooth the surface just before application. The CHEM HOE IPC FL4 water mixture is then sprayed directly onto the soil surface and immediately cross disked to a depth of at least 4 inches. Seed may be planted the same day following the CHEM HOE IPC FL4 treatment, but no later than 1 or 2 days after treatment for best results. The application should be uniform, avoiding overlaps and untreated areas.

SAFFLOWER—Preplant—Within 2 days before planting, uniformly apply 3 to 4 quarts of CHEM HOE IPC FL4 per acre. Shallow incorporate 1 to 2 inches and follow with irrigation. Use the lower rate on coarse and medium texture soils and the high rate on fine textured soils.

SAFFLOWER—Preemergence—Uniformly apply 3 to 4 quarts of CHEM HOE IPC FL4 per acre within 2 days after planting. Rainfall or sprinkle irrigation is essential after treatment for weed control. Use the lower rate on coarse and medium texture soils and the high rate on fine textured soils.

SPINACH—Postemergence—Uniformly apply 4½ quarts of CHEM HOE IPC FL4 per acre. Apply when spinach is in 2–4 leaf stage to control germinating annual weeds. At least 1 inch of rainfall or sprinkler irrigation should follow application within a few days.

SUGAR BEETS—Preplant—For control of annual grasses, wild oats and volunteer grains, apply 4 quarts of CHEM HOE FL4 per acre broadcast to lightly worked and leveled soil surface. Immediately after application of the spray mixture, cross disc the soil to a depth of at least 4 inches. Seed can be planted the same day but no later than 2 days after application.

Where beets are to be planted on pre-shaped beds, two methods of application can be used.

1. Prior to listing the soil into beds, apply 4 quarts of CHEM HOE FL4 per acre broadcast. Immediately after application, incorporate 2 to 3 inches into the soil be discing then list into beds. Plant the sugar beets within 2 days of application and furrow irrigate within 2 days of planting.

2. After listing the soil into beds, apply CHEM HOE FL4 at the rate of 4 quarts per acre (broadcast basis) to the bed top and incorporate 2 inches into the soil. Plant the sugar beets within 2 days of application and furrow or sprinkler irrigate within 2 days of planting.

SUGAR BEETS—Postemergence—When sugar beets have 4 to 8 true leaves, apply 4½ to 6 quarts of CHEM HOE IPC FL4 mixed with 20 gallons or more of water per acre. Application should be made as a directed shielded spray. Irrigation should follow application to move the chemical into the soil and root zone of the weed plants. Use the lower rate on coarse textured sandy soils and the higher rate on fine textured clay soils.

SUGAR BEETS—Idaho, Oregon, and Washington—For control of grains (wheat, barley, oats or rye) planted for soil stabilization on sandy soils. Grain may be seeded or volunteer grain may be allowed to grow in the fall to protect the soil from serious wind erosion during the winter months and during the establishment period of the sugar beet crop. To control the grain during the establishment period. CHEM HOE IPC FL4 can be applied: (a) preplant to the sugar beets, (b) preemergence to the sugar beets at planting, or (c) postemergence to the sugar beets when they have 4 to 8 true leaves as a directed shielded spray. Apply CHEM HOE IPC FL4 at 1 gallon per acre. (4 lbs. a.i. A) by:

1. Ground equipment by any of the above three methods using 20 or more gallons of spray solution per acre broadcast. For band applications use proportionately less. Immediately after application, apply ½ to 1 inch of water by sprinkler irrigation to move the CHEM HOE IPC FL4 into the root zone of the grain.
2. Aircraft application using 5 to 10 gallons of spray solution per acre broadcast either preplant or preemergence only. Immediately after application, apply ½ to 1 inch of water by sprinkler irrigation to move the CHEM HOE IPC FL4 into the root zone of the grain.
3. Center Pivot Sprinkler Irrigation Systems—Broadcast either preplant or preemergence only. Thoroughly mix the recommended amount of CHEM HOE IPC FL4 with an equal amount or more of water in the nurse tank to supply the system for one complete revolution. Meter the solution into the irrigation water on the discharge side of the irrigation pump during the entire application cycle using a positive placement metering pump. Apply ½ to 1 inch of water per acre.

CAUTION: Maintain good agitation of the solution in the nurse tank throughout the application period.

Apply the product only through center pivot irrigation systems containing anti-siphon and check valves which will prevent water source contamination and overflow of the slurry tank and containing interlocking controls between the metering device and the water pump to ensure simultaneous shutoff.

Inject the product with a positive displacement pump into the main line ahead of a right angle turn to ensure adequate mixing.

Application of more than label recommended quantities of irrigation water per acre may result in decreased product performance by removing the chemical from the root zone of effectiveness.

Do not apply when wind speed favors drift, when system connections of fittings leak, when nozzles do not provide uniform distribution or when lines containing the product must be dismantled and drained. Greater accuracy in calibration and distribution will be achieved by injecting a larger volume of a more dilute slurry per hour.

Where sprinkler distribution patterns do not overlap sufficiently, unacceptable weed control may result. Where sprinkler patterns overlap excessive crop injury may result.

ESTABLISHED GRASSES GROWN FOR SEED—(For Application in Washington)—For control of annual grasses, apply 3 to 4 quarts of CHEM HOE IPC FL4 per acre. Apply only between September 15 and October 15. Straw and plant residues should be removed before application is made. Best results are obtained when annual grasses are in the one-leaf stage and rainfall or irrigation follows application. Use the lower rate on coarse textured sandy soils and the higher rate on fine textured clay soils.

ESTABLISHED PERENNIAL GRASS—Limited to Oregon-West of Cascade Mountains—For the control of annual weeds and grasses. Straw and plant residues should be removed before application is made. Reduce application rate in low spots where surface water may cause accumulation of the chemical with resulting crop injury. Where possible, avoid application of CHEM HOE IPC FL4 early on warm, windy days. Evaporation losses will be reduced by making treatments during quiet, cool days and as late in the day as practical. Use in other areas may result in crop injury due to varying weather conditions.

Crop	Rate of CHEM HOE IPC FL4 Per Acre	Time of Application
Perennial Ryegrass	3 Quarts*	October after fall rains have started. Do not spray after October 31.
Kentucky Bluegrass Tall Fescue	3 Quarts	October after fall rains have started. Do not spray after October 31.
Orchardgrass		
Creeping Fescue Chewing Fescue	3 Quarts*	Spray only during period from October 1 to October 20 after fall rains have started.
Highland Bentgrass Astoria Bentgrass	4 Quarts*	Spray in October after fall rains have started. Do not spray after October 31.

* Mix in 20 gallons or more of water.

CAUTION: Use only on established fields which have been harvested at least once.

FOR DISTRIBUTION AND USE ONLY WITHIN THE STATE OF CALIFORNIA

EPA Reg. No. 748-207
SLN No. CA-780211 SLN No. CA-780196

CHEM HOE FL4 HERBICIDE

DIRECTIONS FOR USE

It is a violation of Federal law to use this product in a manner inconsistent with the label.

CHEM HOE FL4 WATER-RUN IN ALFALFA AND SUGAR BEETS

CHEM HOE FL4 herbicide metered into irrigation water to control fall and winter volunteer grains (barley, oats, wheat) wild oats, canarygrass, rabbitfootgrass, and annual ryegrass. Refer to the EPA registered label for additional weeds controlled.

ALFALFA

Application can be made to dormant or nondormant alfalfa. New seeding alfalfa should have at least three true leaves before treatment.

When the winter annual grasses are from germination to 3 inches high, meter 4 quarts (4 lbs. a.i.) of CHEM HOE FL4 herbicide per acre into the irrigation water.

NOTE: Do not graze or harvest for feed within 8 days of application.

SUGAR BEETS

When the sugar beets have at least six true leaves and the winter annual grasses are from germinating to 3 inches high, apply 4 to 6 quarts (4 to 6 lbs. a.i.) of CHEM HOE FL4 herbicide per acre in the irrigation water. Use the 4 quart rate on coarse texture (sandy) soils and the 6 quart rate on medium texture (loams) or fine texture (clay) soils. When the weeds are 3 to 6 inches high, or if soil temperatures above 55°F are anticipated within five days after application, use the 6 quart rate on all soil textures.

To prepare a suspension of CHEM HOE FL4 herbicide, add 1 part CHEM HOE FL4 to at least 4 parts water in a tank with vigorous mechanical agitation or high pressure jet bypass agitation. Meter the mixture into the irrigation water during the entire irrigation period maintaining agitation in the tank. More uniform application will be obtained when the mixture is applied under constant pressure through an orifice. See the following table for application rate of the mixture.

NOTE: Calibration should be checked periodically to assure uniform application. Irrigation water must completely wet the field to be treated to assure proper distribution of the CHEM HOE FL4 herbicide. Irrigate to eliminate water runoff (tailwater) from the treated field. Do not use any runoff water to irrigate crops not on the container label or for domestic use. All applicable directions, restrictions, and precautions on the EPA registered label are to be followed. This label must be in the possession of the user at the time of application.

EXAMPLE: Assuming 30 acres irrigated in 24 hours. To treat 30 acres of crop at 4 quarts per acre, mix 30 gallons of CHEM HOE FL4 herbicide into 120 gallons of water under agitation in the spray tank. Calibrate the discharge rate of the suspension into the irrigation water to equal 6.02 gallons per hour or 395 milliliters per minute. Adjust the flow rate if the time period of irrigation varies.

CHEM HOE FL4 HERBICIDE MIX* FLOW RATE FOR 24 HOURS IRRIGATION PERIOD

ACRES IRRIGATED IN 24 HOURS	4 QUART RATE			6 QUART RATE		
	GAL/HOUR	FL. OZ./MIN	MI/MIN	GAL/HOUR	FL. OZ./MIN	MI/MIN
10	2.01	4.3	132	3.13	6.7	197
15	3.01	6.5	198	4.69	10.0	296
20	4.02	8.6	263	6.25	13.4	395
25	5.02	10.8	319	7.82	16.8	494
30	6.02	12.9	395	9.38	20.1	592
35	7.03	15.1	460	10.94	23.5	691
40	8.03	17.2	526	12.50	26.8	789
45	9.04	19.4	642	14.07	30.2	888
50	10.04	21.5	658	15.63	33.5	987
55	11.05	23.7	724	17.19	36.9	1096
60	12.05	25.8	789	18.75	40.2	1184
65	13.06	27.9	855	20.32	43.6	1283
70	14.06	30.1	921	21.88	46.9	1381
75	15.06	32.3	987	23.44	50.3	1480
80	16.06	34.4	1052	25.00	53.6	1578
85	17.07	36.6	1118	26.57	57.0	1677
90	18.07	38.7	1184	28.13	60.3	1776
95	19.08	40.9	1250	29.69	63.7	1875
100	20.08	43.0	1315	31.25	67.0	1973

* Mixed at the rate of 4 gallons of water to 1 gallon of CHEM HOE FL4 herbicide.

In lieu of the above table, the following formula can be used to determine application rate of the suspension under any combination of circumstances:

A = Acres to be treated
R = Rate per acre of CHEM HOE in quarts
D = Dilution ratio of water in tank
T = Hours to irrigate treated acreage
F = Required flow rate gallons per hour

$$\frac{(A \times R)+(A \times R)}{4T} = F$$

$$\frac{F}{60} \times 128 = \text{fl. oz. per minute}$$

$$\frac{F}{60} \times 3785 = \text{ml per minute}$$

Adjust the flow rate if the time of irrigation varies.

EXAMPLE: 76 acres of crop to be treated at 4 quarts per acre. Dilution ratio of 1 part CHEM HOE to 5 parts water. Total time to irrigate the 76 acres 48 hours.

$$\frac{(76 \times 4) + 5(76 \times 4)}{4 \times 48} = \frac{1824}{192} = 9.5 \text{ gallons per hour}$$

$$\frac{9.5}{60} = .158 \times 128 = 20.3 \text{ fluid ounces per minute}$$

OR

$$\frac{9.5}{60} = .158 \times 3785 = 599.3 \text{ milliliters per minute}$$

FOR DISTRIBUTION AND USE ONLY WITHIN THE STATE OF CALIFORNIA

EPA Reg. No. 748-207
SLN No. CA 780058

CHEM HOE FL4 HERBICIDE

DIRECTIONS FOR USE

WEED CONTROL IN ALFALFA AND CLOVER CROPS

For control of winter annual grasses, volunteer grains, wld oats, and chickweed. Refer to the EPA registered label for additional weeds controlled.

It is a violation of Federal law to use this product in a manner inconsistent with the label.

NEW SEEDINGS

Postemergence application in late fall, winter or early spring when the weeds are small and the crop has at least 3 true leaves. Uniformly apply 4 to 6 quarts per acre of CHEM HOE FL4 herbicide in 20 to 60 gallons of water by ground equipment or 5 to 10 gallons of water by aircraft. Use the lower rate on coarse textured (sandy) soils and the higher rate on fine textured (clay) soils.

ESTABLISHED STANDS

Postemergence application in late fall or winter when the weeds are small. Uniformly apply 5 or 6 quarts per acre of CHEM HOE FL4 herbicide in 20 to 60 gallons of water by ground equipment or 5 to 10 gallons of water by aircraft. Use the higher rate on fine textured (clay) soils or under low rainfall conditions if not irrigated.

NOTE: Rainfall or irrigation should occur within 5 days of application to leach the chemical into the soil. Thoroughly mix the CHEM HOE FL4 concentrate to a uniform suspension before adding to the spray tank. Maintain constant agitation during the mixing and spraying operation. All applicable directions, restrictions, and precautions on the EPA registered label are to be followed. THIS LABELING MUST BE IN THE POSSESSION OF THE USER AT THE TIME OF APPLICATION.

FOR USE AND DISTRIBUTION ONLY WITHIN THE STATES OF OREGON AND WASHINGTON

EPA Reg. No. 748-207
SLN No. OR-770015 SLN No. WA-770003

CHEM HOE FL4

APPLIED THROUGH CENTER PIVOT SPRINKLER IRRIGATION SYSTEMS

For control of winter annual grasses and volunteer grains in alfalfa fields of eastern Oregon.

It is a violation of Federal law to use this product in a manner inconsistent with the label.

DIRECTIONS FOR USE

Center pivot sprinkler irrigation systems—broadcast application only. Thoroughly mix 3 to 4 quarts per acre of CHEM HOE FL4 with an equal amount or more of water and add to the nurse tank the required amount to supply the system for one complete revolution. Meter the solution into the irrigation water on the discharge side of the irrigation pump using a positive displacement metering pump. Apply in ½ to 1 inch of water per acre.

CAUTION: Maintain good agitation of the solution in the nurse tank throughout the application period.

Observe all handling and use precautions on the container label.

Apply the product only through center pivot irrigation systems containing anti-siphon and check valves which will prevent water source contamination and overflow of the slurry tank and containing interlocking controls between the metering device and the water pump to insure simultaneous shut-off.

Inject the product with a positive displacement pump into the main line ahead of a right angle turn to insure adequate mixing.

Application of more than label recommended quantities of irrigation water per acre may result in decreased product performance by removing the chemical from the zone of effectiveness.

Do not apply when wind speed favors drift, when system connections or fittings leak, when nozzles do not provide uniform distribution or when lines containing the product must be dismantled and drained.

Greater accuracy in calibration and distribution will be achieved in injecting a larger volume of a more dilute slurry per hour.

Where sprinkler distribution patterns do not overlap sufficiently unacceptable weed control may result. Where sprinkler patterns overlap excessively, crop injury may result.

FOR DISTRIBUTION AND USE ONLY WITHIN THE STATES OF OREGON AND IDAHO

EPA Reg. No. 748-207

SLN No. OR-830005 SLN No. ID-830006

CHEM HOE FL4 HERBICIDE

DIRECTIONS FOR USE

WEED CONTROL IN ALFALFA

For control of winter annual grasses, volunteer grins, and wild oats.* Refer to the EPA registered label for additional weeds controlled.

*Deep germinating wild oats may not be controlled.

It is in violation of Federal law to use this product in a manner inconsistent with the label.

NEW SEEDINGS AND ESTABLISHED STANDS

APPLY BEFORE APRIL 15th. Postemergence application in late fall, winter or early spring when the newly seeded crop has at least 3 true leaves.

For best control use rates recommended in the quick-reference chart which are based on weed size and growth stage. CHEM HOE FL4 can be applied in 20 or more gal/A water by ground equipment, 5 or more gal/A water by air or through center pivot irrigation system. For more mixing and application details see the EPA registered CHEM HOE FL4 label.

Quick-Reference Chart For CHEM HOE FL4 Application Rates

Timing	Crop Stage			Rate of CHEM HOE FL4
	Established	Seedling	Weed Stage	
Late fall (soil temp. below 55 °F)**	Need not be dormant	3 or more true leaves	Germinating & Seedlings	3–4 quarts***
Early winter (soil temp. below 55 °F)**	Need not be dormant	3 or more true leaves	Germinating, Seedlings & Established	4–6 quarts***
Early spring (soil temp. below 55 °F)**	Need not be dormant	3 or more true leaves	Germinating, Seedlings & Established	5–6 quarts**

** In irrigated alfalfa, if soil temperature is 55°F or above, you can apply CHEM HOE but irrigation (½ inch to 2 inches of water) must follow immediately.

*** Use lower rate on coarse textured (sandy) and the higher rate on medium or fine textured (loam and clay) soils.

FOR DISTRIBUTION AND USE ONLY WITHIN THE STATE OF WASHINGTON

EPA Reg. No. 748-207

SLN No. WA-830002

CHEM HOE FL4 HERBICIDE

DIRECTIONS FOR USE

WEED CONTROL IN ALFALFA

For control of winter annual grasses, volunteer grains, and wild oats.* Refer to the EPA registered label for additional weeds controlled.

* Deep germinating wild oats may not be controlled.

It is in violation of Federal law to use this product in a manner inconsistent with the label.

NEW SEEDINGS AND ESTABLISHED STANDS

APPLY BEFORE APRIL 15th. Postemergence application in late fall, winter or early spring when the newly seeded crop has at least 3 true leaves.

For best control use rates recommended in the quick-reference chart which are based on weed size and growth stage. CHEM HOE FL4 herbicide can be applied in 20 or more gal/A water by ground equipment, 5 or more gal/A water by air or through center pivot irrigation system. See Biochemical Product Bulletin—CHEM HOE Herbicide for Control of Grasses and Grains in Alfalfa Applied Through Center Pivot Sprinkler Irrigation Systems A-1169-60—for restrictions and instructions. For more mixing and application details see the EPA registered CHEM HOE FL4 herbicide label.

NOTE: Do not graze or harvest for feed within 50 days of application. Rainfall or irrigation should occur within 5 days of application to leach the chemical into the soil. Thoroughly mix the CHEM HOE FL4 concentrate to a uniform suspension before adding to the spray tank. Maintain constant agitation during the mixing and spraying operation. All applicable directions, restrictions, and precautions on the EPA registered label are to be followed. THIS LABELING MUST BE IN THE POSSESSION OF THE USER AT THE TIME OF APPLICATION.

FOR DISTRIBUTION AND USE ONLY WITHIN THE STATES OF UTAH & NEVADA

EPA Reg. No. 748-207

SLN No. UT 78-0001 SLN No. NV 78-0002

CHEM HOE FL4 HERBICIDE

DIRECTIONS FOR USE

WEED CONTROL IN ALFALFA

For control of winter annual grasses, volunteer grains, and wild oats. Refer to the EPA registered label for additional weeds controlled.

It is a violation of Federal law to use this product in a manner inconsistent with the label.

NEW SEEDINGS

Postemergence application in late fall, winter or early spring when the weeds are small and the crop has at least 3 true leaves. Uniformly apply 4 to 6 quarts per acre of CHEM HOE FL4 herbicide in 20 to 60 gallons of water by ground equipment or 5 to 10 gallons of water by aircraft. Use the lower rate on coarse textured (sandy) soils and the higher rate on fine textured (clay) soils.

ESTABLISHED STANDS

Postemergence application in late fall or winter when the weeds are small. Uniformly apply 5 or 6 quarts per acre of CHEM HOE FL4 herbicide in 20 to 60 gallons of water by ground equipment or 5 to 10 gallons of water by aircraft. Use the higher rate on fine textured (clay) soils or under low rainfall conditions if not irrigated.

NOTE: Rainfall or irrigation should occur within 5 days of application to leach the chemical into the soil. Thoroughly mix the CHEM HOE FL4 concentrate to a uniform suspension before adding to the spray tank. Maintain constant agitation during the mixing and spraying operation. All applicable directions, restrictions, and precautions on the EPA registered label are to be followed. THIS LABELING MUST BE IN THE POSSESSION OF THE USER AT THE TIME OF APPLICATION.

FOR DISTRIBUTION AND USE ONLY WITHIN THE STATE OF OREGON

EPA Reg. No. 748-207 EPA Accession No. 06009

CHEM HOE FL4

FOR CHEMICAL PREPARATION OF SEEDBEDS FOR GRASS SEED CROPS IN WESTERN OREGON PRIOR TO PLANTING GRASSES

It is a violation of Federal law to use this product in a manner inconsistent with the label.

DIRECTIONS FOR USE

Apply four to six quarts of CHEM HOE FL4 in 10 to 20 gallons of water per acre, with one pound of 2,4-D acid equivalent on fall prepared seedbed in late November or early December.

Plant grasses as soon as the weather permits—between February 15 and April 1. Drill the grasses, but do not disturb the seedbed in any way. Best results are obtained if the seeds are planted shallow and, if possible, covered to protect them from birds.

Chemical seedbed preparation will control only those weeds growing from seeds. It will not control perennial weeds. Under certain conditions more weeds may germinate between the time of CHEM HOE FL4 dissipation and crop planting. Should this occur, application of a contact herbicide is recommended. Consult local agricultural authorities for information on materials for rates of application.

FOR DISTRIBUTION AND USE ONLY WITHIN THE STATE OF WASHINGTON

EPA Reg. No. 748-207 EPA Accession No. 06004

WASHINGTON STATE PESTICIDE REGISTRATION

FOR USE OF CHEM HOE FL4 ON CARROTS GROWN FOR SEED

DIRECTIONS FOR USE

CARROTS GROWN FOR SEED ONLY

Apply 3 to 5 quarts (3–5 lbs.) of CHEM HOE FL4 for the control of winter annual grasses and volunteer grain from November 15 to January 15 in 20 to 40 gallons of water per acre. Fields are not to be grazed by livestock, and crop cuttings or residues are not to be fed. Fields must be cleared after harvest by discing into soil.

CONDITIONS OF SALE

The statements and methods presented about the products mentioned are based upon the best available information and practices known to PPG Industries at the present time, but are not representations or warranties of performance, result or comprehensiveness of such data. PPC Industries warrants that the products mentioned herein conform to their chemical description and are reasonably fit for the purposes stated on the label when used in accordance with the label directions under normal conditions of use, or conditions of use reasonably foreseeable to the seller.

The products mentioned herein if not used in accordance with directions can be hazardous. PPG Industries recommends that anyone using and/or handling the products mentioned herein before using them, thoroughly read and understand the specific directions for use and the precautions appearing on the product label.

The products mentioned herein, as all potentially hazardous materials, must be kept out of the reach of children.

FURLOE® CHLORO IPC 4EC

EMULSIFIABLE CONCENTRATE SELECTIVE HERBICIDE

ACTIVE INGREDIENT:

Isopropyl m-chlorocarbanilate	47%
INERT INGREDIENTS	53%
TOTAL	100%

Contains 4 pounds chlorpropham per gallon

EPA Reg. No. 748-161 EPA Est. No. 32761-MO-3

CAUTION

Keep Out of Reach of Children

HAZARD TO HUMANS AND DOMESTIC ANIMALS

Harmful if swallowed. Avoid contact with skin, eyes, or clothing. Avoid breathing of vapor or spray mist.

PRACTICAL TREATMENT

If in eyes or on skin, flush with plenty of water. If swallowed, do not induce vomiting. GET IMMEDIATE MEDICAL ATTENTION. Never give anything by mouth to an unconscious person.

ENVIRONMENTAL HAZARD

This product is toxic to fish. Do not apply directly to water. Do not contaminate water by cleaning of equipment or disposal of wastes.

PHYSICAL AND CHEMICAL HAZARD

Do not use or store near heat, sparks, or open flame.

STORAGE AND DISPOSAL

Do not contaminate water, food, or feed by storage or disposal.

Do not freeze. Store at temperature above 40°F. Prolonged storage below 40° may result in crystal formation. Warming the contents and agitation will restore material to usable condition.

Pesticide, spray mixture or rinsate that cannot be used or chemically processed should be disposed of in an approved hazardous waste management facility.

Pesticide wastes are toxic. Improper disposal of excess pesticide, spray mixture, or rinsate is a violation of Federal Law. If these wastes cannot be disposed of according to label instructions, contact your State Pesticide or Environmental Control Agency or the Hazardous Waste representative at the nearest EPA Regional Office for guidance. Do not reuse container. Triple rinse and destroy by crushing or perforating, and dispose of according to any applicable federal, state, or local regulations.

In case of emergency, contact PPG Industries, Inc., Natrium, W. Va. (304) 843-1300.

CONDITIONS OF SALE

The statements and methods presented about the products mentioned are based upon the best available information and practices known to PPG Industries at the present time, but are not representations or warranties of performance, result or comprehensiveness of such data. PPG Industries warrants that the products mentioned herein conform to their chemical description and are reasonably fit for the purposes stated on the label when used in accordance with the label directions under normal conditions of use, or conditions of use reasonably foreseeable to the seller.

The products mentioned herein if not used in accordance with directions can be hazardous. PPG Industries recommends that anyone using and/or handling the products mentioned herein before using them thoroughly read and understand the specific directions for use and the precautions appearing on the product label.

The products mentioned herein, as all potentially hazardous materials, must be kept out of reach of children.

It is a violation of Federal law to use this product in a manner inconsistent with the label. Do not treat areas while unprotected humans or domestic animals are present in the treatment area. Do not allow entry into treated areas without protective clothing until spray has dried.

GENERAL DIRECTIONS FOR USE

These directions pertain to application under average conditions. Ineffective weed control or crop damage may result from differing conditions of soil, temperature, moisture, and crop variety. Therefore, before large scale use, this product should be applied on a trial basis under conditions actually encountered. Consult State Agricultural Experiment Station or Extension Service weed specialist for recommendations under varying local conditions. In general, coarse textured soils require lower application rates than fine textured or muck soils. Lower soil temperatures also reduce the required rate. Adequate soil moisture (rainfall or irrigation) is necessary for effective use of FURLOE® herbicide. FURLOE controls weed seeds as they germinate; therefore, existing weeds should be removed before application.

FURLOE® 4EC used according to directions will normally give effective control of the following:

GRASSES

- Annual bluegrass *(Poa annua)*
- Winter annual or annual bromegrass *(Bromus sp.)*
- Annual ryegrass *(Lolium multiflorum)*
- Canarygrass *(Phalaris sp.)*
- Little barley *(Hordeum pusillum)*
- Rabbitfootgrass *(Polypogon monspeliensis)*
- Rattail fescue *(Festuca myuros)*
- Stinkgrass *(Eragrostis cilianensis)*
- Witchgrass *(Panicum capillare)*
- Wild oats *(Avena fatua)*
- * Volunteer grain *(Barley, oats, rye, wheat)*

BROADLEAF WEEDS

- Black nightshade *(Solanum nigrum)*
- Carpetweed *(Mollugo verticillata)*
- Chickweed *(Stellaria media)*
- Common mallow *(Malva neglecta)*
- * Curly dock seedings *(Rumex crispus)*
- Dodder *(Cuscuta sp.)*
- Falseflax *(Camelina sp.)*
- Knotweed *(Polygonum sp.)*
- Ladysthumb *(Polygonum persicaria)*
- * London rocket *(Sisymbrium irio)*
- * Miners lettuce *(Montia perfoliata)*
- * Nettleleaf goosefoot *(Chenopodium murale)*
- Purslane *(Portulaca oleracea)*
- Shepherdspurse *(Capsella bursa-pastoris)*
- Pennsylvania smartweed *(Polygonum pensylvanicum)*
- Stinging nettle *(Urtica dioica)*
- * Tansey mustard *(Descurainia pinnata)*
- Wild buckwheat *(Polygonum convolvulus)*
- * Wild mustard *(Brassica kaber)*

* Controlled only when treatment is made in conjunction with irrigation. No control of perennial weeds should be expected.

SPECIFIC DIRECTIONS FOR USE

Rates given are for overall coverage except as otherwise indicated. For strip or band treatment reduce amounts of FURLOE® Chloro IPC 4EC and water proportionately. For use, it is mixed with water to form an emulsion which is sprayed uniformly on the soil surface. During spraying, emulsion should be agitated.

APPLICATION METHODS: Unless otherwise specified, FURLOE® Chloro IPC 4EC can be applied by:

Ground Equipment: Add the recommended amount of FURLOE® Chloro IPC 4EC to clean water in the spray tank. Apply in 20 or more gallons of water per acre broadcast using a properly calibrated low pressure sprayer that will provide even distribution.

Aircraft: Apply the recommended amount of FURLOE® Chloro IPC 4EC in 5 to 10 gallons of water per acre. Even distribution is essential, and use of flagmen recommended. Do not apply by air if conditions favor drift from the target area.

NOTE: When irrigation immediately follows application, minimize runoff water from the field. Do not use runoff water to irrigate other crops or for domestic purposes.

ALFALFA—East of the Rocky Mountains—Postemergence—Late Summer Seeded or Established—For control of Chickweed and Downy Brome. Apply 1 to 2 quarts of FURLOE® Chloro IPC 4EC per acre from October through January. After February 1, apply 2 to 3 quarts per acre. FURLOE® Chloro IPC 4EC may be applied to actively growing, semi-dormant or dormant alfalfa. Use the lower rate on coarse textured sandy soils.

West of the Rocky Mountains—Late Summer Seeded or Established Plantings—For control of winter annual weeds. Apply 2 to 4 quarts per acre. Use the lower rate on coarse textured sandy soils.

NOTE: Do not graze or feed forage from treated area within 40 days after application.

Do not apply to grass-alfalfa mixed stands since grass may be injured.

Do not apply to new seedings before plants have four true leaves.

NOTE: This application will not control attached dodder.

The addition of Sevin® (carbaryl) insecticide to FURLOE® 4EC herbicide as a tank-mix will provide extended dodder control. The application should be made just prior to dodder germination. A shallow cultivation before application is recommended to eliminate weeds and prepare a smooth soil surface. A broadcast application of the tank-mix can be made at the following times:

In first crop seed—in early spring just prior to alfalfa breaking dormancy or immediately after final tillage operation layby.

In second crop seed—in early spring just prior to alfalfa breaking dormancy and/or as soon as possible after removal of a first forage crop.

Uniformly apply 4 to 6 quarts of FURLOE 4EC per acre as a tank-mix with 1 to 1½ pounds of Sevin 50W or .6 to .9 lb. of Sevin Sprayable or 1 to 1½ pints of Sevin XLR. The ratio should be 8 parts FURLOE active ingredient to 1 part Sevin active ingredient. Apply in 20 or more gallons of water per acre by ground sprayer or in 5 or more gallons of water per acre by air. Add the required amount of Sevin product to the partially filled (½ or more) spray tank with the agitator running. After thorough mixing, add the FURLOE and the balance of water required. Maintain agitation during application.

Use the higher rate when only one application is to be made. Use the lower rate for the second crop seed if a second application is to be made after removal of a forage crop. If regrowth is present, use FURLOE® 20-G herbicide for the second application. Refer to the FURLOE 20-G label.

Irrigation is necessary if rainfall does not occur soon after application.

PRECAUTIONS:

This mixture is toxic to pollinator bees. See the Sevin product label for BEE CAUTION.

Use only on alfalfa planted for spring or fall of the year preceding application.

If alfalfa foliage is present at application, growth may be temporarily damaged and/or delayed.

Do not graze or feed within 40 days of application.

Observe all cautions and limitations on labeling of all products used in mixtures.

This label must be in possession of the user at the time of application.

Sevin®—Registered trademark of Union Carbide.

BEANS, LIMA AND SNAP—Preemergence—Apply 4 quarts per acre after planting and prior to emergence of beans. Seed should be planted at least 1 inch deep.

BLACKBERRIES, RASPBERRIES—Dormant—Apply 6 quarts in 40 gallons of water per acre during late winter or early spring. Avoid wetting canes; direct spray to soil.

BLUEBERRIES—Dormant—Apply 8 to 12 quarts in 40 gallons of water per acre during late fall to weed-free soil. Direct spray to base of plants. Application may be repeated in early spring before growth begins.

CARROTS—Preemergence—Apply 4 to 6 quarts in 20 to 40 gallons of water per acre. Use lower rate on medium textured loam soils and higher rate on fine textured, dark mineral or muck soils. THIS APPLICATION NOT RECOMMENDED IN CALIFORNIA.

CLOVERS, RED, WHITE, LADINO, AND TREFOIL—Late Summer Seeded or Established Plantings—Apply 2 quarts per acre for annual winter weed control. Use lower rate on coarse textured sandy soils.

NOTE: Do not graze or feed forage from treated area within 40 days after application.

Do not apply to grass-clover mixed stands as grass may be injured.

Do not apply to new seedings before plants have four true leaves.

GARLIC, CALIFORNIA—For control of annual bluegrass, annual ryegrass, wild oats, wild* and tansey mustard*, London rocket*, shepherdspurse*, chickweed, and knotweed.

* Controlled only when treatment is made in conjunction with irrigation.

Preemergence—Apply 2 to 4 quarts of FURLOE® Chloro IPC 4EC in 20 to 70 gallons of water per acre on fall planted sets immediately after planting. Apply 2 to 3 quarts on coarse (sandy loam, loam, and silty loam); 3 to 4 quarts on medium (clay loam, silty clay loam); or 4 quarts on fine textured soils (clay and silty clay). Use the lower rates if the soil temperature is below 60° F, and the higher rates when the soil temperature is above 60° F. DO NOT USE ON SOILS CONTAINING LESS THAN 1% ORGANIC MATTER. Sprinkle irrigate if rainfall does not occur within 5 days after application.

Postemergence—Apply to a weed-free soil after garlic is 6 inches tall, and has 3 or more true leaves. Apply 4 quarts of FURLOE® Chloro IPC 4EC in 20 to 70 gallons of water per acre. Direct the spray to wet only the lower 1 to 2 inches of garlic, avoiding the tops. Irrigate soon after application. DO NOT USE ON SOILS CONTAINING LESS THAN 1% ORGANIC MATTER. DO NOT APPLY WITHIN 30 DAYS OF HARVEST.

GRASS SEED CROPS, PERENNIAL—Oregon—West of the Cascade Mountains—Uniformly apply in fall after rains have started, but prior to November 1. Application should be made prior to October 20, on fine fescues. Use only on established stands which have been harvested for seed at least once. DO NOT APPLY WITHIN 30 DAYS OF HARVEST.

Perennial ryegrass, fine leaved fescues such as Chewings, Creeping, Pennlawn, Ilahee, Ranier, use 2 quarts in 10 or more gallons of water.

Astoria bentgrass, Highlands bentgrass, varieties of Kentucky bluegrass, Tall fescue, Orchardgrass, use 3 quarts in 10 or more gallons of water.

NOTE: **Do not use on new varieties or selections without small plot testing to determine crop tolerance.** Only true perennial varieties are tolerant. Crosses with annual varieties may be susceptible to crop injury.

ONIONS—SEEDED OR SETS. THIS APPLICATION NOT RECOMMENDED IN CALIFORNIA. Preemergence—and up to early loop stage for seeded onions.

Postemergence—after onions are 6 inches high and have 3 or more true leaves, direct spray to wet only lower 1 to 2 inches of onions.

Fine textured, dark mineral soils—Apply 4 quarts per acre in 40 gallons of water to weed-free soil.

Muck soil—Apply 6 quarts per acre in 40 gallons of water during cool weather. Increase rate to 8 quarts per acre when soil temperature is above 60° F.

NOTE: Do not apply during the "flag" stage as injury may result.

Do not use on coarse textured soils or soils containing less than 4% organic matter.

SAFFLOWER—Preplant Incorporated—Within one to three days before planting, uniformly apply 3 to 6 quarts per acre (3 quarts on coarse and medium, 4 quarts on fine and 6 quarts on peat soils).

Shallow incorporate 1 to 2 inches. Rainfall or irrigation is essential following treatment for best weed control.

SOYBEANS

FURLOE®/Tolban™ Tank Mix—Preplant Incorporated—To control the annual grasses and weeds listed on the Tolban™ 4E label plus smartweed† and similar weeds in soybeans in the north central area of the U.S., uniformly apply and incorporate before planting a tank mix of FURLOE.® Chloro IPC 4EC and Tolban™ 4E at the following rates per acre broadcast in 10 to 40 gallons of water.

†Pennsylvania smartweed and ladysthumb.

Soil Texture	FURLOE 4EC Rate/Acre	Tolban™4E Rate/Acre
Coarse (Sandy, Loamy Sand, Sandy Loam)	2 quarts	1–1½ pints††
Medium (Loam, Silty Loam, Silt)	2½ quarts	1½–2 pints††
Fine (Sandy Clay Loam, Silty Clay Loam, Clay Loam, Silty Clay, Clay)	3 quarts	2–3 pints††

†† Use the higher rate of Tolban™ on soils with over 2% organic matter.

Soil Incorporation—See the Tolban™ 4E label for full instructions. Application and incorporation should be done in the same operation if possible. If not, the tank mix should be applied to dry loose soil surface and incorporated within four hours. Use a disk harrow set to cut 4 to 6 inches deep and operated in two different directions at 4 to 6 mph or other incorporation equipment listed on the Tolban™ 4E label.

IMPORTANT:

See General Directions section on this label and the Application Directions and Precaution statements on Tolban™ 4E label for further information. In case of crop failure follow all precautions and limitations under Rotational Crops on the Tolban™ 4E label.

FURLOE® Overlay on Tolban™

FURLOE® Chloro IPC 4EC can also be applied preemergence following preplant incorporated applications of Tolban™ alone. Follow the above FURLOE® application rates and apply to the soil surface at planting or right after planting.

FURLOE®/Treflan® Tank Mix—Preplant Incorporated—A tank mix of FURLOE Chloro IPC 4EC and Treflan EC will give control of black nightshade**, smartweed† and the grasses and broadleaf weeds listed on the FURLOE 4EC and Treflan EC labels in soybeans in the north central area of the U.S. Uniformly apply a tank mix of FURLOE Chloro IPC 4EC and Treflan EC at the following rates per acre broadcast in 10 to 40 gallons of water.

Soil Texture	Treflan EC Rate/Acre	FURLOE 4EC Rate/Acre
Coarse (Sandy and Sandy Loam)	1 pint	2 quarts
Medium (Loams)	1–2½ pints	2½ quarts
Fine (Clay, Silt and Soil over 5% Organic Matter)	2 pints	3 quarts

** Use three quarts of FURLOE 4EC to control black nightshade.

Soil Incorporation—See Treflan label for detailed directions. Incorporate the chemicals thoroughly into the soil within four hours of application. Good incorporation can be achieved with different equipment including the following:

P.T.O. driven equipment (tillers, cultivators, hoes) set to cut 2 to 3 inches deep with rotors spaced to give a clean sweep of the soil.

Double disk (or double disk with spike tooth harrow in tandem) set to cut 3 to 4 inches deep and operated in two different directions (cross disked) at 4 to 6 m.p.h.

IMPORTANT:
See General Directions section and the Application and Incorporation Directions and Special Precautions on the Treflan EC label for further information.

FURLOE® Overlay on Treflan®—FURLOE Chloro CIPC 4EC can also be applied preemergence following preplant incorporated applications of Treflan. Follow the above application rates and apply the FURLOE to the soil surface at planting or right after planting.

FURLOE®/Vernam® Tank Mix—Preplant Incorporated—A tank mix of FURLOE Chloro IPC 4EC and Vernam 7-E will give control of the annual grasses and broadleaf weeds listed on the Vernam label plus smartweed† and similar weeds in soybeans in the north central area of the U.S. Uniformly apply the following rates per acre broadcast in 10 to 40 gallons of water.

Soil Texture	Vernam 7-E Rate/Acre	FURLOE 4EC Rate/Acre
Coarse (Sandy)	2⅔ pints	2 quarts
Medium (Loams)	3 pints	2½ quarts
Fine (Clay)	3½ pints	3 quarts

Soil Incorporation—See the Vernam label for full instructions. The chemicals must be incorporated into the soil immediately (within minutes) after application using power driven cultivation equipment set to cut a depth of 2 to 3 inches or a tandem disk set to cut 4 to 6 inches deep and operated 4 to 6 m.p.h. For thorough mixing disk in two different directions (cross disk).

FURLOE® Overlay on Vernam®—FURLOE Chloro CIPC 4EC can also be applied preemergence following preplant incorporated applications of Vernam. Follow the above application rates and apply the FURLOE to the soil surface at planting or right after planting.

IMPORTANT:
See General Directions section of this label and the General Use Precautions and Directions of Use sections on the Vernam 7-E label for further information.

FURLOE®/Lasso® Tank Mix—Preemergence
A tank mix of FURLOE Chloro IPC 4EC and Lasso 4EC will give control of black nightshade**, smartweed† and grasses and broadleaf weeds listed on the FURLOE 4EC and Lasso EC labels in soybeans in the north central area of the U.S. At planting or right after planting, apply the mixture in 5 to 40 gallons of water per acre to provide uniform distribution over the soil surface. Apply the following broadcast rates on the indicated soil textures:

Soil Texture	FURLOE 4EC Less than 5% O.M. Rate/Acre	FURLOE 4EC 5% O.M. Or More Rate/Acre	Lasso All O.M. Levels Rate/Acre
Coarse (Sandy)	2 quarts	2 quarts	2 quarts
Medium (Loams)	2–3 quarts	2½–3 quarts	2–2½ quarts
Fine (Clays)	2–3 quarts	3 quarts	2½–3 quarts

† Pennsylvania smartweed and ladysthumb.
** Use three quarts of FURLOE 4EC to control black nightshade.

Band treatment behind press wheel—Apply the tank mixture in 5 or more gallons of water per acre on the soil surface at planting. To calculate rates, multiply the band width in inches by recommended broadcast rate and divide by the inches between rows.

IMPORTANT:
See General Directions section of this label and the General Information and Mixing and Spraying Instructions sections on the Lasso EC label for further information.

NOTE FOR ALL SOYBEAN USES:

Plant the Soybeans between 1 and 2 inches in depth.

Do not use these mixtures in crops other than soybeans. Do not feed forage or pods to livestock.

Observe all cautions and limitations on labeling of all products used in mixtures.

SUGAR BEETS—Postemergence—Imperial Valley of California. For control of winter annual grasses and volunteer grain apply, layby, 4 quarts per acre in 10 to 40 gallons of water when the sugar beets have 8 true leaves. For application in water-run, see state label.

SPINACH—Preemergence—For warm weather application apply 2 quarts per acre in 40 gallons of water immediately after planting. Reduce application to 1 quart per acre during cool weather when temperature is below 60°F. THIS APPLICATION NOT RECOMMENDED IN CALIFORNIA.

TOMATOES—Postemergence—Apply 4 quarts per acre in 20 to 40 gallons of water to weed-free soil when plants are well established. Direct spray to base of tomato plant; avoid wetting the foliage.

WOODY NURSERY STOCK—Dormant—Apply 4 to 8 quarts in 40 gallons of water per acre as a broadcast spray to clean weed-free soil. Direct spray to avoid wetting plants. Do not apply to nursery stock that will be transplanted sooner than one month after treatment.

SUPPLEMENTAL LABELING

ARIZONA

FURLOE® 4EC HERBICIDE
EPA Reg. No. 748-161
EPA SLN No. AZ-800030

It is a violation of Federal law to use this product in a manner inconsistent with the label.

All applicable directions, restrictions, and precautions on the FURLOE® 4EC label are to be followed.

FURLOE® 4EC Herbicide Water-Run on Alfalfa in Yuma County

Metering of FURLOE® 4EC herbicide into irrigation water to control fall and winter annual weeds at germination:
- Volunteer grain (barley, oats, rye, wheat)
- Canarygrass (Phalaris sp.)
- Rabbitfootgrass (Polypogon monspeliensis)
- London rocket (sisymbrium irio)
- Shepherdspurse (Capsella bursa-pastoris)

Refer to FURLOE® 4EC registered label for additional weed information.

FURLOE® 4EC should be applied with a constant flow metering device with the appropriate orifice to give the required flow rate throughout the irrigation period. FURLOE® 4EC should be introduced into the irrigation water where there is sufficient turbulence to insure mixing of the product before entering the field.

Late Summer or Fall Planted Crop
Meter 2–3 quarts per acre of FURLOE® 4EC into the irrigation water in the late fall or winter. Do not apply until the alfalfa has at least 6 true leaves (4–6 inches tall).

Established Crop
Meter 2–4 quarts per acre of FURLOE® 4EC into the irrigation water during the late fall or winter months. Apply 2–3 quarts prior to weed emergence or 3–4 quarts after weeds have germinated. A second application may be required for extended winter annual weed control. In areas of winter rainfall where fields are not irrigated during the winter, apply 4 quarts per acre in the water during the last irrigation in the fall.

NOTE:
Refer to the calibration chart on page 8 for proper orifice selection. Do not apply within 40 days of feeding, grazing, or harvest. Do not apply to mixed alfalfa/grass stands since the grass may be injured. Do not apply to fields saturated from recent rainfall or to alfalfa under stress from insect, disease, or other environmental factors. Irrigate to eliminate water runnoff (tailwater) from the treated field. Do not use any runoff water to irrigate crops not on the container label or for domestic

uses. Use the following formula to determine the required flow rate and select the proper orifice from the calibration chart:

$$\frac{\text{Acres To Be Treated} \times \text{Pounds Of Product Per Acre}}{\text{Hours To Irrigate Acres}} = \text{Flow Rate In Pounds Per Hour}$$

EXAMPLE: 50 acres at 4 lbs. per hours over a 36-hour period:

$$\frac{50 \times 4}{36} = \frac{200}{36} = 5.5 \text{ lbs. per hour}$$

This would require the use of orifice .045 on the chart.

CALIFORNIA

FURLOE® 4EC HERBICIDE
EPA Reg. No. 748-161
EPA Accession No. 06008

WATER INJECTION APPLICATION FOR WEED CONTROL IN ALFALFA AND CLOVER

It is a violation of Federal law to use this product in a manner inconsistent with the label.

Late Summer or Fall Planted Crop
For control of most winter annual weeds, meter 2–3 quarts per acre of FURLOE® 4EC into the irrigation water in the late fall or winter. Do not apply until the crop has at least 6 true leaves (4–6 inches tall).

Established Crop
For control of most winter annual weeds, meter 2–4 quarts per acre of FURLOE® 4EC into the irrigation water during the late fall or winter months. (Apply 2–3 quarts prior to weed emergence or 3–4 quarts after weeds have germinated.) A second application may be required for extended winter annual weed control. In areas of winter rainfall where fields are not irrigated during the winter, apply 4 quarts per acre in the water during the last irrigation in the fall.

FURLOE® 4EC applied into the irrigation water will control most winter annual grasses at germination to 6 inches in height and many winter annual broadleaf weeds from germination to the seeding stage. A constant-flow metering device should be used to accurrately meter the chemical in the water throughout the irrigation period. Refer to the calibration chart on page 8 for proper orifice selection.

NOTE: Do not apply within 40 days of feeding or grazing. Do not apply to mixed legume/grass stands since the grass may be injured. Minimize irrigation water run-off from the field. Do not use run-off water to irrigate susceptible crops or for domestic purposes.

For additional information, consult your local dealer or technical representative of PPG Industries, Inc.

CALIFORNIA

FURLOE® 4EC HERBICIDE
EPA Reg. No. 748-161
EPA Accession No. 10560

FOR WEED CONTROL IN ESTABLISHED ALFALFA AND CLOVER (DORMANT SEASON)

It is a violation of Federal law to use this product in a manner inconsistent with the label.

For the residual control of most Winter annual weeds and burn back of top growth of perennial weeds. Apply 3 to 4 quarts of FURLOE 4EC plus Dinitro General Weed Killer and 10 to 30 gallons of weed oil in sufficient water to make 50 to 100 gallons of spray per acre. Refer to the specific Dinitro General Weed Killer product label for the labeled dosage rate.

The most effective weed control can be expected when application is made to small weeds. Use the higher rates when heavier or more mature weed growth is present. Rainfall or irrigation is necessary for effective residual weed control.

Maintain agitation during the mixing and spraying operation. Spray to thoroughly wet all foliage. Do not graze or feed forage from treated area within 6 weeks after application.

READ AND FOLLOW MIXING AND USE INSTRUCTIONS AND PRECAUTIONS ON THE PRODUCT LABELS.

For additional information, consult your local dealer or technical representative of PPG Industries, Inc.

This label must be in the possession of the user at the time of application.

IDAHO

FURLOE® 4EC HERBICIDE
EPA Reg. No. 748-161
EPA Accession No. 06005

FOR CONTROL OF DODDER IN ALFALFA SEED FIELDS

It is a violation of Federal law to use this product in a manner inconsistent with the label.

FURLOE® 20-G (20% granular chlorpropham) is generally best for controlling dodder in alfalfa and seed fields. However, if conditions exist whereby a field cannot be irrigated so as to wet the entire field, FURLOE® 4EC (an emulsifiable concentrate) will give better dodder control. Apply 4 to 6 quarts FURLOE® 4EC in 20–40 gallons of water per acre just prior to or during actively growing, semi-dormant, or dormant alfalfa. However, results will be most effective if alfalfa growth does not exceed a height of 4 inches. For taller alfalfa use FURLOE® 20-G.

This label must be in the possession of the user at the time of application.

KENTUCKY

FURLOE® 4EC HERBICIDE
EPA Reg. No. 748-161
EPA SLN No. KY 780006

It is a violation of Federal law to use this product in a manner inconsistent with its labeling.

For weed control in alfalfa/forage grass mixed stands and for control of chickweed in mixed stands of alfalfa with the following forage grasses: orchardgrass, tall fescue or smooth bromegrass.

DIRECTIONS FOR USE
Apply 2 quarts (2 lbs. active ingredient) of FURLOE® 4EC herbicide per acre in 20 to 40 gallons of water before chickweed emerges or when chickweed is up to 3 inches tall. When the chickweed is over 3 inches high and/or after February 1, apply 3 quarts (3 lbs. active) per acre. Apply only to established stands. See FURLOE® 4EC registered product label for additional weeds controlled.

NOTE: Observe all handling and use precautions on the container label. Do not graze or harvest treated crop for forage within 40 days of treatment. Treatment may cause slight delay in maturity of forage grass.

This label must be in possession of the user at the time of application.

WASHINGTON

FURLOE® 4EC HERBICIDE
EPA Reg. No. 748-161
EPA Accession No. 06007

FOR DODDER CONTROL IN ALFALFA

It is a violation of Federal law to use this product in a manner inconsistent with the label.

DIRECTIONS FOR USE
Apply six (6) quarts of FURLOE® 4EC in 20–40 gallons of water per acre before alfalfa has more than four (4) inches growth. This treatment will control germinating dodder. It will not control attached dodder.

This label must be in the possession of the user at the time of application.

CALIFORNIA

FURLOE® 4EC HERBICIDE

EPA Reg. No. 748-161

EPA Accession No. 06008

FOR CONTROL OF VOLUNTEER SMALL GRAINS IN DRY AND SUCCULENT BEANS.

It is a violation of Federal law to use this product in a manner inconsistent with the label.

DIRECTIONS FOR USE

Apply broadcast 3 to 4 quarts of FURLOE® 4EC in 20 to 60 gallons of water per acre to a well-prepared soil. Use the higher rate on heavy textured soils. Immediately incorporate with a disc set to cut 4–6 inches and operate in different directions. Or apply to the beds and incorporate with a power tiller to a depth of 2 to 3 inches.

Pre-irrigate on the flat or list into beds and irrigate within 7 days after application.

Do not use on coarse-textured soils (sand, loamy sand, and sandy loams).

Refer to the FURLOE® 4EC label for additional information on general directions-for-use, additional weed control, use precautions, container storage and handling, and conditions of sale.

This label must be in the possession of the user at the time of application.

WASHINGTON

FURLOE® 4EC HERBICIDE

EPA Reg. No. 748-161

EPA SLN No. WA-790012

FOR WEED CONTROL IN ORNAMENTAL BULB CROPS TULIPS, IRIS, NARCISSI

It is a violation of Federal law to use this product in a manner inconsistent with its labeling.

DIRECTIONS FOR USE

Winter annual weed control of wild mustard, chickweed, annual bluegrass, groundsel, shepherdspurse, annual ryegrass, and volunteer grain. Refer to the FURLOE® 4EC registered product labels for additional weeds controlled. Apply a tank mix combination of 4 quarts (4 lbs. a.i.) of FURLOE® 4EC plus 6 quarts (6.5 lbs. a.i.) dinoseb amine in 30 to 60 gallons of water per acre. Make the application anytime after the planting and final hilling operation, but no later than two weeks before emergence of the bulbs. Application is normally in late September through October when surface soil moisture is present, but before soils become waterlogged. Use ground equipment only.

Summer annual weed control of smartweeds, knotweed, redroot pigweed, mustard species, and annual grasses. Refer to the FURLOE® 4EC registered label for additional weeds controlled. Apply 2 quarts (2 lbs. a.i.) of FURLOE® 4EC in 30 to 60 gallons of water per acre post-emergence to the bulbs in early spring. Application should be made just before weed seed germination or to seeding weeds. Do not apply later than April 15. Use ground equipment only.

NOTE: All applicable directions, restrictions, and precautions on the FURLOE® 4EC registered labels are to be followed.

Observe all cautions and limitations on labeling of products used in combinations.

This label must be in the possession of the user at the time of application.

CALIFORNIA

FURLOE® 4EC HERBICIDE

EPA Reg. No. 748-161

EPA SLN No. CA-780028

TANK MIX OF FURLOE ® 4EC AND TREFLAN® IN SAFFLOWER

It is a violation of Federal law to use this product in a manner inconsistent with the label.

DIRECTIONS FOR USE

Preplant incorporated application in drill or row planted safflower to control the following weeds: volunteer grain, annual bluegrass, foxtail, ryegrass, wild oats, mustard, London rocket, shepherdspurse, chickweed, knotweed, and smartweed. Refer to the FURLOE® 4EC and Treflan® labels for additional weeds controlled. Uniformly apply 2 to 3 quarts of FURLOE® 4EC and 1 to 1½ pints of Treflan M.T.F. or 0.8 to 1.2 pints of Treflan Pro 5 as a tank-mix in 35 to 50 gallons of water per acre broadcast. Use the lower rate on coarse and higher rate on medium and fine textured soils. Make the application just prior to planting and immediately incorporate into the top 2 to 3 inches of the final seed bed in the same operation if possible. Follow the incorporation directions on the Treflan® label.

NOTE: Safflower seeds should be planted below the depth of incorporation or crop injury may result. Rainfall or irrigation is necessary soon after the application for best results. All applicable directions, restrictions, and precautions on the FURLOE® 4EC and Treflan® registered labels must be followed.

This label must be in the possession of the user at the time of application.

Treflan® is a registered trademark of Elanco Products Company, Indianapolis, IN., a division of Eli Lilly and Co.

MARYLAND

FURLOE® 4EC HERBICIDE

EPA Reg. No. 748-161

EPA SLN No. MO-780017

FOR WEED CONTROL IN FALL SEEDED SPINACH AND FOR CONTROL OF CHICKWEED.

It is a violation of Federal law to use this product in a manner inconsistent with the label.

Refer to the FURLOE® 4EC registered label for additional weeds controlled.

DIRECTIONS FOR USE

Between November 1 and February 1, when the spinach has 6 to 8 leaves, apply 1 to 2 quarts of FURLOE® 4EC in 20 to 40 gallons of water per acre. Use the lower rate on coarse texture sandy soils and the higher rate on all other soil textures. Do not make the application when air temperatures are above 75°F.

NOTE: All applicable directions, restrictions, and precautions on the FURLOE® 4EC registered label are to be followed.

Prolonged periods of low temperatures of 32°F or below may cause temporary stunting of the treated crop.

Do not apply within 30 days of harvest.

This label must be in the possession of the user at the time of application.

OKLAHOMA, ARKANSAS, TENNESSEE

FURLOE® 4EC HERBICIDE

EPA Reg. No. 748-161

EPA SLN No. OK-780024

EPA SLN No. AR-780025

EPA SLN No. TN-780019

FOR WEED CONTROL IN FALL SEEDED SPINACH AND FOR CONTROL OF HENBIT, CHICKWEED, BITTERCRESS, AND SIBARIA.

It is a violation of Federa law to use this product in a manner inconsistent with its labeling.
Refer to the FURLOE® 4EC registered label for additional weeds controlled.

DIRECTIONS FOR USE
Between November 1 and February 15, when the spinach has 6 to 8 leaves, apply 2 quarts of FURLOE® 4EC in 20 to 40 gallons of water per acre. Do not make the applications when air temperatures are above 75°F.

NOTE: All applicable directions, restrictions, and precautions on the FURLOE® 4EC registered label are to be followed.

Prolonged periods of low temperatures of 32°F or below may cause temporary stunting of the treated crop.

Do not apply within 30 days of harvest.

This label must be in the possession of the user at the time of application.

VIRGINIA

FURLOE® 4EC HERBICIDE
EPA Reg. No. 748-161
EPA SLN No. VA-780027

FOR CHICKWEED CONTROL IN FALL SEEDED SPINACH.

It is a violation of Federal law to use this product in a manner inconsistent with its labeling.
Refer to the FURLOE® 4EC registered label for additional weeds controlled.

DIRECTIONS FOR USE
Between November 1 and February 1, when the spinach has 6 to 8 leaves, apply 1 to 2 quarts of FURLOE® 4EC in 20 to 40 gallons of water per acre. Use the lower rate on coarse texture sandy soils and the higher rate on all other soil textures. Do not make the application when air temperatures are above 75°F. Do not apply within 45 days of harvest.

NOTE: All applicable directions, restrictions, and precautions on the FURLOE® 4EC registered label are to be followed.

Prolonged periods of low temperatures of 32°F or below may cause temporary stunting of the treated crop.

This label must be in the possession of the user at the of application.

CALIFORNIA

FURLOE® 4EC HERBICIDE
EPA Reg. No. 748-161
EPA SLN No. CA-780197

FURLOE ® 4EC HERBICIDE WATER-RUN ON SUGAR BEETS IN THE IMPERIAL VALLEY

It is a violation of Federal law to use this product in a manner inconsistent with the label.

DIRECTIONS FOR USE
Metering of FURLOE® 4EC herbicide into the irrigation water to control fall and winter germinating volunteer grain (barley, oats, wheat) annual ryegrass, London rocket, shepherdspurse, knotweed, lambsquarter, canarygrass, rabbitfootgrass, purslane, nettleleaf goosefoot, and fivehook bassia. Refer to the FURLOE® 4EC registered label for additional weeds controlled.

Sugar beets should have at least eight true leaves at the time of application. Grassy weeds should be no more than 4 inches high and broadleaf weeds no larger than seedlings at the time of application.

Meter 3 to 4 quarts (3 to 4 lbs. a.e.) of FURLOE® 4EC per acre into the irrigation water during the entire irrigation period. Use a constant flow metering device with the appropriate orifice to give the required flow rate. Th product should be introduced at a point where there is water turbulence to enhance mixing. Use the lower rate on coarse textured (sandy) soils and the higher rate on fine texture (clay) soils.

NOTE: Irrigate to eliminate water run-off (tailwater) from the treated field. Do not use any run-off water to irrigate crops not on the container label or for domestic uses. All applicable directions, restrictions, and precautions on the FURLOE® 4EC registered label are to be followed. Use the following formula to determine the required flow rate and select the proper orifice from the calibration chart on page 8:

$$\frac{\text{Acres To Be Treated X Pounds Of Product Per Acre}}{\text{Hours To Irrigate Acres}} = \text{Flow Rate In Pounds Per Hour}$$

EXAMPLE: 50 acres at 4 lbs. per hours over a 36-hour period:

$$\frac{50 \times 4}{36} = \frac{200}{36} = 5.5 \text{ lbs. per hour}$$

This would require the use of orifice .045 on the chart.

CALIFORNIA

FURLOE® 4EC HERBICIDE
EPA Reg. No. 748-161
EPA Accession No. 06008

LAYBY APPLICATION FOR WEED CONTROL IN SUGARBEETS

It is a violation of Federal law to use this product in a manner inconsistent with the label.

DIRECTIONS FOR USE
The application should be made to weed-free soil to control most winter annual grasses and broadleaf weeds. Sugarbeets should have 8 or more true leaves.

Apply 4 quarts of FURLOE® 4EC in 20–40 gallons of water per acre. Direct the spray into each furrow up to the base of the beets without spraying the crown. Use leaf lifters if the leaves are folded into the furrow. Every row should be sprayed. Irrigate soon after application to thoroughly wet the bed.

For additional information, consult your local dealer or technical representative of PPG Industries, Inc.

Refer to the FURLOE® 4EC registered label for additional information on general directions-for-use, weeds, controlled, use precautions, container storage, and handling and conditions of sale.

This label must be in the possession of the user at the time of application.

CALIFORNIA

FURLOE® 4EC HERBICIDE
EPA Reg. No. 748-161
EPA SLN No. CA-790007

FOR WEED CONTROL ON WINTER FALLOW BEDDED SOIL FOR SPRING PLANTING OF SUGAR BEETS OR TOMATOES

It is a violation of Federal law to use this product in a manner inconsistent with the label.

DIRECTIONS FOR USE
Apply in late fall after bed-up of soil prior to weed emergence to control annual bluegrass, ryegrass, wild oats, London rocket, shepherdspurse, mustard, knotweed, smartweed, malva and chickweed. Refer to the FURLOE® 4EC registered label for additional weeds controlled.

Apply 4 quarts (4 lbs. a.i.) of FURLOE® 4EC in 20 to 40 gallons of water per acre by ground equipment or in 10 gallons of water per acre by aircraft. Rainfall or sprinkler irrigation is necessary within one week of application. Shallow incorporation at application will extend the time to two weeks.

NOTE: Do not plant crops listed within 80 days of application. Do not plant crops other than those on the label. All applicable directions, restrictions, and precautions on the FURLOE® 4EC registered label are to be followed.

This label must be in the possession of the user at the time of application.

CALIFORNIA

FURLOE® 4EC HERBICIDE
EPA Reg. No. 748-161
EPA SLN No. CA-790054

FOR WEED CONTROL IN PLUG-PLANTED TOMATOES

DIRECTIONS OF USE
For control of the following weeds in tomatoes seeded in plugs of an artificial growing mix containing an activated charcoal safener: black nightshade, hairy nightshade, and ground cherry. Refer to the FURLOE® 4EC registered label for additional weeds controlled.

PREPLANT: Just prior to planting apply 2 to 4 quarts of FURLOE® 4EC in 20 to 60 gallons of water per acre to the bed top. (Compute rate on a banded basis.) Sprinkler irrigation is necessary within 5 days after application.

APPLICATION GUIDE:

Soil Texture	Rate of Application (Qts./A)
Sand, Loamy Sand	Do Not Use
Sandy Loam, Loam, Silt Loam, Silt	2
Sandy Clay Loam, Clay Loam, Silty Clay Loam, Sandy Clay	3
Silty Clay, Clay	4

NOTE: Use on soils having more than 4% organic matter may result in reduced weed control.

PRECAUTIONS: These recommendations are based on the use of a plug containing 60 ml. (on fine texture or high OM soils) to 120 ml. (on coarse textured soils) of a 50:50 mixture of peat and vermiculite containing at least 5% by volume Gro-Safe® activated charcoal. Deviation from this plug composition or method of herbicide application may result in herbicide injury to seedlings.

Plug planting of tomatoes is a new concept. Contact your county Farm Advisor for further precautions and/or recommendations for your area.

All applicable directions, restrictions, and precautions on the FURLOE® 4EC registered label are to be followed.

This label must be in the possession of the user at the time of application.

Gro-Safe® is a registered trademark of ICI United States, Inc.

FURLOE® HERBICIDE FLOW-RATES FOR VARIOUS TEEJET® ORIFICES (4916)

TEEJET® ORIFICE	MILLILITERS PER MINUTE	OUNCES PER MINUTE	POUNDS PER HOUR	APPROXIMATE DELIVERY TIME (HOURS) 1 GALLON	5 GALLONS
.010	2.84	.10	.18	22	110
.012	5.36	.18	.34	12	60
.014	6.75	.23	.43	9.4	47
.016	10.30	.35	.65	6.1	24
.018	13.2	.45	.84	4.8	24
.020	16.8	.57	1.1	3.8	19
.022	20.0	.68	1.3	3.2	16
.024	24.8	.84	1.6	2.5	13
.026	29.1	.98	1.8	2.2	11
.028	34.2	1.16	2.2	1.8	9.0
.030	38.3	1.29	2.4	1.6	8.0
.032	44.2	1.50	2.8	1.4	7.0
.034	50.5	1.71	3.2	1.3	6.5
.035	53.4	1.80	3.4	1.2	6.0
.037	60.7	2.05	3.8	1.0	5.0
.039	68.2	2.31	4.3	.93	4.65
.041	76.5	2.59	4.8	.83	4.15
.043	83.3	2.82	5.3	.76	3.80
.045	89.0	3.01	5.6	.71	3.55
.047	99.6	3.37	6.3	.63	3.15
.049	107	3.62	6.8	.59	2.95
.051	116	3.92	7.4	.54	2.70
.053	127	4.29	8.1	.50	2.50
.055	136	4.60	8.6	.46	2.30
.057	147	4.97	9.3	.43	2.15
.059	158	5.34	10.0	.40	2.00

Figures were taken at 62°F and are approximate. Be sure to occasionally measure the flow in the field for the correct orifice as the flow varies with temperature. (Flow on an .024 orifice increases from 24.8 ml at 62°F to 26.0 ml at 84°F.) Always use a screen to minimize plugging of the orifice (300 mesh screen on sizes below .016).
TeeJet®—Registered trademark of Spraying Systems Company.
FURLOE®—Registered trademark of PPG Industries, Inc.

TOLBAN™ —Trademark of CIBA-GEIGY for profluralin.

Treflan®—Registered Trademark of Elanco Products Company.

Vernam®—Registered Trademark of Stauffer Chemical Company.

Lasso®—Registered Trademark of Monsanto Company.

DO NOT FREEZE

In case of emergency, contact PPG Industries, Inc., Natrium, W. Va. (304) 843-1300.

FURLOE® CHLORO IPC 20G

20% GRANULAR CHLORPROPHAM SELECTIVE HERBICIDE

Granular base used 16/30 mesh clay

ACTIVE INGREDIENT:
Isopropyl M-chlorocarbanilate 20%
INERT INGREDIENTS: 80%

TOTAL 100%

EPA Reg. No. 748–163 EPA Est. No. 279 CA-1
EPA Est. No. 32761 MO-3

CAUTION

Keep Out of Reach of Children

HAZARD TO HUMANS AND DOMESTIC ANIMALS

Harmful if swallowed. Avoid contact with skin, eyes, or clothing.

PRACTICAL TREATMENT

If in eyes or on skin, flush with plenty of water.

If swallowed, induce vomiting by touching back of throat with finger. GET IMMEDIATE MEDICAL ATTENTION. Never give anything by mouth to an unconscious person.

ENVIRONMENTAL HAZARDS:

This product is toxic to fish. Do not apply directly to water. Do not contaminate water by cleaning of equipment or disposal of wastes.

Hazard to wildlife may be reduced by covering or incorporating granules which are spilled during loading or are visible on soil surface in turn areas.

It is a violation of Federal law to use this product in a manner inconsistent with the label. Do not treat areas while unprotected humans or domestic animals are present in the treatment area. Do not allow entry into treated areas without protective clothing until dusts have settled.

Emergency telephone number, PPG Industries, Inc., Natrium, WV (304) 843-1300.

GENERAL DIRECTIONS

While these directions pertain to application under average conditions, ineffective weed control or crop damage may result from differing conditions of soil, temperature, moisture, and crop variety. Therefore, before large scale use, this product should be applied on a trial basis under conditions actually encountered. Consult State Agricultural Experiment Station or Extension Service weed specialist for recommendations under varying local conditions. In general, coarse textured soils require lower application rates than fine textured or muck soils. Lower soil temperatures also reduce the required rate. Adequate soil moisture (rainfall or irrigation) is necessary for effective use of FURLOE® 20-G herbicide. FURLOE 20-G controls weed seeds as they germinate; therefore, existing weeds should be removed before application.

APPLICATION METHODS

FURLOE 20-G is used as it comes from the container. It may be applied with properly designed and calibrated ground or airborne equipment. Small areas can be treated with calibrated hand equipment. Uniform distribution at the specified rate over the prescribed area is essential. Thoroughly clean the application equipment before and after use. Band applicators must be calibrated to deliver the recommended rate of FURLOE 20-G. To check the calibration, set the applicator in accordance with the manufacturer's instructions and attach a bag over each spreader plate. Operate the unit at a constant planting speed over the field to be planted. Collect the granules for a distance of 500 linear feet and weigh the amount collected for each unit. Refer to the following table for the amount of FURLOE 20-G which should be collected.

Broadcast Rate/Acre Lbs. FURLOE 20-G	Amount Collected Per 500 Feet Of Row	
	14 Band	7 Inch Band
10.0	2.1 oz. (61 grams)	1.1 oz. (31 grams)
12.5	2.7 oz. (75 grams)	1.4 oz. (38 grams)
15.0	3.2 oz. (91 grams)	1.6 oz. (46 grams)

For other band widths, calculate the calibration distance as follows:

$$\frac{14}{\text{Band Width Inches}} \times 500 = \text{Distance in Linear Feet to Collect the Granules}$$

The amount collected in the calculated distance should be the same as the amount shown under 14" band in the table. When necessary, adjust the applicator so that each unit delivers the proper amount.

WEEDS CONTROLLED

FURLOE 20-G used according to directions will normally give effective control of the following weeds:

Grasses	
Annual bluegrass	*Poa annua*
Annual or winter annual bromegrass	*Bromus* sp.
Annual ryegrass	*Lolium multiflorum*
Canarygrass	*Phalaris* sp.
Little barley	*Hordeum pusillum*
Rabbitfootgrass	*Polypogon monspeliensis*
Rattail fescue	*Festuca myuros*
Stinkgrass	*Eragrostis cilianensis*
Witchgrass	*Panicum capillare*
Wild oats	*Avena fatua*
Broadleaf	
Black nightshade	*Solanum nigrum*
Carpetweed	*Mollugo verticillata*
Chickweed	*Stellaria media*
Dodder	*Cuscuta* sp.
Falseflax	*Camelina* sp.
Knotweed	*Polygonum* sp.
Ladysthumb	*Polygonum persicaria*
*London rocket	*Sisymbrium irio*
Pennsylvania smartweed	*Polygonum pensylvanicum*
Purslane	*Portulaca oleracea*
*Shepherdspurse	*Capsella bursa-pastoris*
Wild buckwheat	*Polygonum convolvulus*
*Wild mustard	*Brassica kaber*

* Controlled only when treatment is made in conjunction with irrigation.

STORAGE AND DISPOSAL

Do not contaminate water, food, or feed by storage or disposal.

STORAGE

Store in a dry location. Avoid water or oilsoaked floors. Do not store near seeds or fertilizers.

PESTICIDE DISPOSAL

Pesticide, spray mixture, or rinsate that cannot be used according to label instructions must be disposed of according to Federal, State, or local procedures under the Resource Conservation and Recovery Act.

CONTAINER DISPOSAL

Completely empty bag by shaking and tapping sides and bottom to loosen clinging particles. Empty residue into application equipment. Then dispose of bags in a hazardous waste management facility or by incineration if allowed by State and local authorities.

CROP RECOMMENDATIONS

Rates given are for overall coverage except as otherwise noted. For strip or band applications, the rate should be reduced proportionately.

ALFALFA OR CLOVER

Late Summer Seeded or Established Plantings: Uniformly apply 20 pounds per acre before weeds emerge. For best results apply to moist soil.

NOTE: Seedling crops should have at least 4 true leaves at the time of treatment.

Do not use on grass-legume mixed stands, as grass may be injured.

Do not graze or feed within 40 days of application.

ALFALFA DODDER CONTROL

THIS APPLICATION WILL NOT CONTROL ATTACHED DODDER

To control dodder in forage or seed crops, apply the recommended rate of FURLOE 20-G in the spring just prior to dodder germination. Use the higher rate for single applications. Use the lower rate for repeat application in hay crops and second crop seed. Make the repeat application immediately after removal of the forage crop. The best dodder control is obtained with air application to thoroughly wet soil.

State Recommendations	Lbs. FURLOE 20-G Per Acre
CA, ID, NV, OR, UT, WA	20 to 30

NOTE: Use only on alfalfa planted no later than the spring or fall of the year preceding application.

Do not apply to alfalfa foliage that is wet or damp.

Do not graze or feed within 40 days of application.

See California Special Local Need 24(c) label for 20-day harvest interval.

BLACKBERRIES, RASPBERRIES

Dormant—Apply 30 pounds per acre to weed-free soil during late winter or early spring.

BLUEBERRIES

Dormant—Uniformly apply 40 to 60 pounds per acre during late fall to weed-free soil. Where necessary, applications may be repeated in early spring before growth begins.

CARROTS (Do not use in California)

Preemergence—Uniformly distribute 20 to 30 pounds per acre. Use lower rate on clay loam soils and higher rate on heavy dark mineral or muck soils. A band application may be used to reduce cost of application.

CRANBERRIES

Dormant established plantings—For the control of dodder, horsetail, loosestrife, tearthumb, rushes (Juncus), annual bluegrass, turkeyfootgrass, bentgrass, velvetgrass, bluejointgrass, and sicklegrass. Uniformly distribute 50 to 100 pounds per acre in early winter after harvest or in early spring before cranberry growth begins. Apply only when plants are dry to avoid contact and sticking with possible injury to fruit buds and loss of yield. The best time for application varies according to the area and with the weeds to be controlled. Consult your local authorities as to your best time and rate of application.

LIMA OR SNAP BEANS

Preemergence—Apply 20 pounds per acre after planting and before beans emerge. Seed should be planted at least one inch deep.

ONIONS (Do not use in California)

(Seeded or Sets)—Just prior to emergence or during the early loop stage of seeded onions.

Do not apply during the "flag" stage, as injury may result.

Heavy Dark Mineral Soil—Uniformly distribute 20 pounds per acre.

Muck Soil—Apply 30 pounds per acre in cool weather. Increase rate to 40 pounds per acre when soil temperature is above 60°F.

Postemergence (Seeded, Sets or Transplants) After onions are 6 inches high and have 3 or more true leaves.

Heavy Dark Mineral Soil—Uniformly distribute 20 pounds per acre to weed-free soil.

Muck Soil—Apply 30 pounds per acre in cool weather. Increase rate to 40 pounds per acre when soil temperature is above 60°F. Postemergence applications may be repeated once, but not later than 30 days before harvest.

CAUTION—Do not use on coarse textured soils or soils containing less than 4% organic matter.

SOUTHERN PEAS

(Edible Field Peas)—Preemergence—Uniformly distribute 20 or 30 pounds per acre at planting. Seed should be planted at least 1 inch deep. Use lower rate on coarse textured soil, and higher rate on fine textured soils.

SOYBEANS (Do not use in California)

Preemergence—Uniformly apply FURLOE 20-G following a preplant incorporated application of Lasso® 4 EC herbicide, Tolbam,TM 4E herbicide, or Treflan® EC herbicide to control smartweeds (Pennsylvania smartweed and ladysthumb) and black nightshade in addition to the weeds listed on the Lasso, Tolban, or Treflan labels. After applying the preplant incorporated herbicide according to label directions, apply FURLOE 20-G at the rate recommended below.

NOTE: Plant the soybeans between one and two inches in depth. Do not feed forage or pods to livestock.

Soil Texture	Pounds/Acre
Coarse (Sand, Loamy Sand, Sandy Loam)	10.0
Medium (Loam, Silt Loam, Silt)	12.5
Fine (Sandy Clay Loam, Silty Clay Loam, Clay Loam, Silty Clay, Clay)	15.0

For band application the amount of FURLOE® 20-G may be calculated by the formula:

$$\frac{\text{Band Width In Inches}}{\text{Row Width In Inches}} \times \text{Broadcast Rate per Acre} = \text{Amount Needed}$$

See application section for band applicator calibration.

ORNAMENTALS

Bulbs, Corms, and Tubers—Daffodils, Dahlia, Dutch Iris, Gladioli, Lily, Narcissus, Tulips.

Uniformly apply 20 to 30 pounds per acre on clean cultivated soil at planting or prior to emergence of the crop and weeds. Use the higher rate on medium and fine textured soils.

Field Transplanted—Ageratum, Alyssum*, Begonia, Chrysanthemum, Coleus, Dianthus, Geranium, Impatiens, Marigold, Pansy*, Petunia*, Salvia*, Snapdragon, Zinnia.

* Do not use over 20 pounds per acre on these species.

Uniformly apply 20 to 30 pounds per acre on clean cultivated soil after the transplants have become established. Use the higher rate on medium and fine textured soils.

Ground Covers and Herbaceous Perennials, Established dormant only, Myrtle, English Ivy, Pachysandra, Winter Creeper, Creeping Juniper and Aster, Astilbe, Bluebell, Coralbell, Columbine, Carnation, Delphinium, Forget-me-not, Peony,

Uniformly apply 20 to 30 pounds per acre on clean soil in the spring before new growth starts. Use the higher rate on medium and fine textured soils.

Woody Trees and Shrubs, Established, dormant only. Trees—Balsam, Cedar, Fir, Hemlock, Holly, Juniper, Pine, Spruce, Birch, Cherry, Crabapple, Dogwood, Locust, Magnolia, Maple, Plum, Poplar, Sycamore.

Shrubs—Arborvitae, Euonymus, Heather, Laucothoe, Pieris, Taxus, Azalea, Boxwood, Camellia, Laurel, Mahonia, Rhododendron, Barberry, Forsythia, Fuchsia, Honeysuckle, Hydrangea, Lilac, Rose, Mock Orange, Privet, Spirea, Viburnum, Cotoneaster, Enkianthus.

Uniformly apply 30 to 40 pounds per acre on clean weed-free soil from late fall through early spring before new growth starts. Use the higher rate on medium and fine textured soils. Do not apply to nursery stock that is to be transplanted within 30 days of treatment.

SUPPLEMENTAL LABELING

FURLOE 20-G

(CA, UT)

EPA Reg. No. 748-163
EPA SLN No. CA-810030
EPA SLN No. CA-810064

FOR DODDER CONTROL IN ALFALFA SEED AND FORAGE CROPS

It is a violation of Federal law to use this product in a manner inconsistent with its labeling.

NOTE: This application will not control attached dodder. For effective and extended dodder control with FURLOE 20-G herbicide, the following recommended program should be closely followed.

Shallow cultivation in later winter and early spring prior to application will eliminate weed growth, break up clods and crop debris and leave a clean, smooth soil surface for uniform distribution of the FURLOE 20-G herbicide granules.

An early spring irrigation or rainfall, after soil temperature has warmed, will usually start dodder germination. FURLOE 20-G herbicide should be applied just prior to the dodder germination.

Uniformly apply 20 to 30 pounds of FURLOE 20-G herbicide per acre. Apply the higher rates to fields with heavy dodder infestations. Make the application to thoroughly wet soil surface or apply just prior to rainfall or sprinkler irrigation. Repeat applications within 4 to 6 weeks as required for extended dodder control. Application can be made by aircraft or ground equipment. Properly calibrated equipment is essential for even distribution of the granules.

PRECAUTIONS: Use only on alfalfa planted no later than spring or fall of the year preceding application.

Do not apply to alfalfa foliage when wet or damp, as crop injury may occur and the amount of material reaching the soil may be reduced.

Do not graze, feed or harvest for forage within 20 days of application.

NOTE: All applicable directions, restrictions, and precautions on the EPA registered label are to be followed.

This label must be in the possession of the user at the time of application.

CONDITIONS OF SALE

The statements and methods presented about the products mentioned are based upon the best available information and practices known to PPG Industries at the present time, but are not representations or warranties of performance, result or comprehensiveness of such data. PPG Industries warrants that the products mentioned herein conform to their chemical description and are reasonably fit for the purposes stated on the label when used in accordance with the label directions under normal conditions of use, or conditions of use reasonably foreseeable to the seller.

The products mentioned herein if not used in accordance with directions can be hazardous. PPG Industries recommends that anyone using and/or handling the products mentioned herein before using them thoroughly read and understand the specific directions for use and the precautions appearing in the product label.

The products mentioned herein, as all potentially hazardous materials, must be kept out of the reach of children.

FURLOE® is a registered trademark of PPG Industries, Inc.

Lasso® is a registered trademark of Monsanto Company.

Tolban™ is a registered trademark of Ciba-Geigy Company.

Treflan® is a registered trademark of Elanco Products Company, a Division of Eli Lilly and Company.

GENATE PLUS™ 6.7EC

SELECTIVE HERBICIDE EMULSIFIABLE CONCENTRATE

HERBICIDE GRADE

This product contains 6.7 lbs. active ingredient per gallon.

ACTIVE INGREDIENT:	
S-Ethyl Diisobutylthiocarbamate	85.1%
INERT INGREDIENTS	14.9%
TOTAL	100.0%

EPA Reg. No. 748-236 EPA Est. No. 32761 MO-3

CAUTION

Keep Out of Reach of Children

HAZARD TO HUMANS AND DOMESTIC ANIMALS

Harmful if swallowed. Avoid contact with skin, eyes, or clothing. Avoid breathing spray mist. Wear rubber gloves and fresh clothing. Wash thoroughly with soap and water immediately after contact or use.

PRACTICAL TREATMENT

If in eyes or on skin, flush with plenty of water. If swallowed, call a physician or Poison Control Center. Drink 1 or 2 glasses of water. Do not induce vomiting. Do not induce vomiting or give anything by mouth to an unconscious person.

ENVIRONMENTAL HAZARDS

Do not apply directly to water. Do not apply where runoff is likely to occur. Do not apply when weather conditions favor drift from treatment area. Do not contaminate water by cleaning of equipment or disposal of wastes.

In case of emergency contact PPG Industries, Inc., Natrium, W. Va. (304) 843-1300.

DIRECTIONS FOR USE

GENATE PLUS is a selective herbicide for control of annual grasses in field, sweet, and silage corn. See the Directions For Use pamphlet attached to this container for general use precautions, specific crops, and dosage rates.

It is a violation of federal law to use this product in a manner inconsistent with its labeling. Do not treat areas while unprotected humans or domestic animals are present in the treatment area. Do not allow entry into treated areas without protective clothing until sprays have dried.

STORAGE AND DISPOSAL

Keep container tightly closed when not in use. Do not store near seeds or fertilizers. Store out of reach of children, pets, and domestic animals. Pesticide, spray mixture, or rinse water that cannot be used according to label instructions must be disposed of according to applicable Federal, State or local procedures.

Triple rinse container. Puncture and dispose of in a hazardous waste management facility or by other approved State and local procedures.

GENERAL USE PRECAUTIONS

Real all label directions before using.

GENATE PLUS™ herbicide should be used only for recommended purposes and at recommended rates. DO NOT OVERDOSE!

GENATE PLUS™ herbicide is recommended for use on mineral soils only (soils containing less than 10% organic matter).

Do not use on milo or sorghum. Do not use on corn seed stock.

Do not contaminate irrigation water (used for crops other than corn) or water used for domestic purposes.

Do not contaminate food or feed.

When properly applied and weather conditions exist for normal plant growth through the season, GENATE PLUS™ will not harm the treated crop nor should harmful soil residues remain beyond harvest. However, during germination and early growth, extended periods of unusually cold and wet or hot and dry weather, insect, nematode, or plant disease attack, carry-over soil residues of certain persistent herbicides, the use of certain soil applied systemic insecticides, improperly placed fertilizers or soil insecticides may create abnormal conditions that weaken crop seedlings. GENATE PLUS™ used under these abnormal conditions could result in crop injury.

Do not use in Arizona or in the ten southern-most California counties (except Kern County where field and silage corn may be treated).

WEEDS CONTROLLED

GENATE PLUS™ will not control established weeds.

Annual Grasses:

Barnyardgrass, Watergrass	*Echinochloa crusgalli*
Crabgrass	*Digitaria* spp.
Fall Panicum	*Panicum dichotomiflorum*
Field Sandbur*	*Cenchrus incertus*
Giant Foxtail	*Setaria faberii*
Green Foxtail	*Setaria viridis*
Yellow Foxtail	*Setaria glauca*
Goosegrass	*Eleusine indica*
Rhizome Johnsongrass**	*Sorghum halepense*
Seedling Johnsongrass*	*Sorghum halepense*
Seedling Bermudagrass	*Cynodon dactylon*
Texas Panicum;* †Buffalograss, Coloradograss	*P. texanum*
Volunteer Sorghum*	*Sorghum* spp.
Wild Cane* Shattercane Black Amber Cane Chicken Corn Scatter Cane	*Sorghum bicolor*

* Cultivation is suggested in addition to the GENATE PLUS™ treatment in fields with moderate to heavy infestations of wild cane, sandbur, johnsongrass, buffalograss (Texas panicum) or volunteer sorghum.
** Partial control (suppression) only.
† Southeast Region only.

Perennial Weeds:

Purple Nutgrass (Nutsedge)	*Cyperus rotundus*
Yellow Nutgrass (Nutsedge)	*Cyperus esculentus*

Existing stands of nutgrass must be turned under and chopped up thoroughly before treatment. Cultivation is suggested in addition to GENATE PLUS™ treatment in fields with moderate to heavy nutgrass infestations.

DIRECTIONS FOR USE OF GENATE PLUS™ HERBICIDE

GENATE PLUS™ is a selective herbicide which is mixed (incorporated) or injected into the soil for control of weeds listed on this label.

GENATE PLUS™ controls annual grasses as their seeds germinate by interfering with normal germination and seedling development. It does not control established weeds. All weed growth and crop stubble should be thoroughly worked into the soil before treatment.

COMBINATION WITH FLUID FERTILIZERS

GENATE PLUS™ may be combined with fluid (solution, slurry, or suspension) fertilizers. HOWEVER, PHYSICAL COMPATIBILITY WITH THESE FLUID FERTILIZERS MUST BE DETERMINED BEFORE COMBINING IN THE SPRAY TANK. See Appendix II for directions regarding these combinations. Even though found to be compatible, constant agitation is necessary to keep the GENATE PLUS™ evenly mixed with the fluid fertilizer during application.

IMPREGNATION ON DRY BULK FERTILIZERS

Dry bulk fertilizers (except nitrate fertilizers) may be impregnated or coated with GENATE PLUS™ . However, uniform distribution of GENATE PLUS™ on fertilizer particles and uniform application are necessary to assure good results. See Appendix IV for information and directions regarding impregnation and use for these combinations.

APPLICATION AND SOIL INCORPORATION DIRECTIONS

Spray Application and Mechanical Incorporation

Pour the recommended amount of GENATE PLUS™ into the spray tank during the filling operations, Apply in 10 to 50 gallons of water per acre using a properly calibrated, low-pressure (20–40 psi) boom type sprayer having good agitation (use the higher gallonage in arid areas). The soil should be well-worked and dry enough to permit good soil mixing (incorporation).

GENATE PLUS™ must be incorporated into the soil immediately to prevent loss of the herbicide. Whenever possible, application and incorporation should be done in the same operation. See Appendix 1 for Soil Mixing (Incorporation) before Planting Directions.

IMPORTANT: Thorough mixing of GENATE PLUS™ in the soil is necessary. The soil must be in good tilth and cross discing is recommended. Improper discing or poor tilth (large clods) may result in unsatisfactory control.

Center Pivot Sprinkler Application (Limit use to Western Plains states only) Apply GENATE PLUS™ with irrigation water before planting (see PLANTING), or immediately after planting before either corn or weeds emerge. It is important that application be done immediately after the last tillage operation since poor results will occur if the weeds have germinated before application of the herbicide. Meter the GENATE PLUS™ into the irrigation water during the entire period. Apply in $\frac{1}{2}$ to $\frac{3}{4}$ inch of water.

Subsurface Application Before or at Planting (Limit use to the Southeastern U.S. only)

Special equipment designed for subsurface application MUST be used. Injector units must be rigidly mounted on, or IMMEDIATELY ahead of, or IMMEDIATELY behind the planter unit.

Soil Injection: Coulters and injector shanks must be spaced $3\frac{1}{2}$ inches apart. The two shanks adjacent to the drill row must be $1\frac{3}{4}$ to 2 inches on either side of it. Mount injectors in staggered positions to avoid trash buildup. Set injectors to inject GENATE PLUS™ 3 inches below the soil surface. The width of the band in which weed control is desired will determine the number of injectors required per row. (Example: Four injectors spaced $3\frac{1}{2}$ inches apart give a 14 inch band). A broadcast application can be made by increasing the number of shanks. Seal injector openings by using a roller or chain drag behind the injectors. Apply the recommended rate of GENATE PLUS™ in 20 to 50 gallons of water per acre.

For band applications, calculate the amount to be applied per acre as follows:

$$\frac{\text{Band width in inches}}{\text{Distance between rows in inches}} \times \text{Rate/A for broadcast treatment} = \text{Amount needed for band treatment}$$

PLANTING

Plant seed to a maximum depth of 2 inches. Plant within 2 weeks after preplant application. To avoid removing GENATE PLUS™ from the row (with loss of weed control in the row), do not move or shape soil after treatment, such as by planting in deep furrows.

CULTURAL PRACTICES FOLLOWING APPLICATION

Should weeds develop, a shallow cultivation or rotary hoeing will generally result in better weed control. When cultivating for any reason, it should be shallow, i.e., no more than $\frac{1}{2}$ the depth the herbicide was incorporated or injected. A sequential application of

2,4-D may be necessary to control weeds resistant to GENATE PLUS™.

Under circumstances when a poor stand of corn is obtained after GENATE PLUS™ treatment (caused by flooding, disease, etc.) corn can be planted again without working the field.

RECOMMENDATIONS
For all uses, cultivation is recommended in addition to GENATE PLUS™ treatment in fields with moderate to heavy infestations of wild cane, nutgrass, sandbur, johnsongrass seedlings, buffalograss (Texas panicum) or Volunteer Sorghum.

BROADCAST RATES PER ACRE (Conventional sprays and center pivot sprinklers).

For control of most annual grasses:
Use 3¾ to 4¾ pints of GENATE PLUS™. Use the higher rate on medium (medium textured-silt loam, clay loam, etc.) to fine (fine textured-clay) soils and for sandbur and johnsongrass seedling control.

For control of nutgrass (nutsedge) and Texas panicum (buffalograss) (in addition to the other annual grasses listed):
Use 4¾ to 7⅓ pints of GENATE PLUS™. Use the higher rates on medium (medium textured-silt loam, clay loam, etc.) soils, fine (fine textured-clay) soils and for heavy infestations.

Wild Cane (Shattercane)
Immediately before planting, apply and thoroughly incorporate 7⅓ pints of GENATE PLUS™ per acre, broadcast. To improve control of light infestations or improve suppression of heavy infestations, all of the following practices should be followed:

1. Moldboard plow in the fall or spring.
2. Work soil thoroughly before application to bury trash and crop residue. Apply to soil that is dry enough to insure good mixing action when GENATE PLUS™ is incorporated. Discing is recommended for this operation.
3. Delay application until soil warms to 55°F or greater.
4. Incorporate immediately after application using either 1.) a tandem disk run twice in two different directions (cross disc) or 2.) a two pass system using a tandem disc in one pass and a properly set field cultivator in the other pass. **The use of a field cultivator alone as an incorporation tool should not be used for GENATE PLUS™ applications for wild cane control.**
5. Shallow cultivate (less than incorporation depth) if wild cane emerges.

Johnsongrass:
Using a program of plowing and discing followed by GENATE PLUS™ will give good control of seedlings and greatly reduce competition from plants coming from rhizomes.

In some areas discing 2 to 3 times spaced at 4 to 6 week intervals during fall and winter can be done. If this cannot be done, start with fall plowing followed by several discings in the spring before planting. The discing cuts rhizomes into small pieces and brings them to the surface, where both the weather and GENATE PLUS™ will kill or weaken them.

Next, apply and disc-incorporate GENATE PLUS™ just before planting at the rate of 4¾ to 7⅓ pints per acre. Use the higher rates of GENATE PLUS™ on medium (medium textured-silt loam, clay loam, etc.) soils, fine (fine textured-clay) soils, for rhizome johnsongrass suppression, and for heavy infestations. Disc a second time at right angles to the first. Cultivate at least once, preferably twice.

SUBSURFACE INJECTION APPLICATION (Limit use to the Southeastern U.S. only)
Use 3¾ pints GENATE PLUS™ per broadcast acre, reducing the dosage proportionately depending upon the row spacing and band width to be treated.

GENATE PLUS ™ /ATRAZINE

TANK MIX DIRECTIONS
A tank mix combination of GENATE PLUS™ and ATRAZINE can be applied and mixed into the soil before planting for weed control in corn. This combination will control a broader spectrum of weeds than will either material used alone. In addition, potential hazard of ATRAZINE residues affecting certain succeeding crops is reduced because a lower rate of ATRAZINE can be used when it is combined with GENATE PLUS™ than when it is used alone.

* Follow all use precautions and warnings that appear on the ATRAZINE label, supplementary ATRAZINE literature and elsewhere on this label.
* Make only one application per crop. After treatment with GENATE PLUS™ /ATRAZINE combination, do not plant any crop except corn until the following year or injury may occur.

WEEDS CONTROLLED BY GENATE PLUS™ /ATRAZINE

Annual Grasses:	
Barnyardgrass Watergrass	*Echinochloa crusgalli*
Crabgrass	*Digitaria* spp.
Fall Panicum	*Panicum dichotomiflorum*
Field Sandbur*	*Cenchrus incertus*
Giant Foxtail	*Setaria faberii*
Green Foxtail	*Setaria viridis*
Yellow Foxtail	*Setaria glauca*
Goosegrass	*Eleusine indica*
Seedling Bermudagrass	*Cynodon dactylon*
Rhizome Johnsongrass**	*Sorghum halepense*
Seedling Johnsongrass*	*Sorghum halepense*
Texas Panicum;* †Buffalograss Coloradograss	*P. texanum*
Volunteer Sorghum*	*Sorghum* spp.
Wild Cane*	*Sorghum bicolor*
Wild Oats	*Avena fatua*
Witchgrass	*Panicum capillare*

* Cultivation is suggested in addition to the GENATE PLUS™/ATRAZINE treatment in fields with moderate to heavy infestations of wild cane, sandbur, johnsongrass, (*Texas panicum*) buffalograss or volunteer sorghum. For control of light infestations or suppression of high infestations of wild cane, see directions for rates and cultural practices.

** Partial control (*suppression*) only.

† Southeast Region only.

Annual Broadleaf Weeds:	
Annual Morningglory	*Ipomoea purpurea*
Carpetweed	*Mollugo verticillata*
Cocklebur Common	*Xanthium pensylanicum*
Jimsonweed	*Datura stramonium*
Lambsquarters Common	*Chenopodium album*
Mustards	*Brassica* spp.
Nightshade	*Solanum* spp.
Common Purslane	*Portulaca oleracea*
Ragweed	*Ambrosia* spp.
Redroot Pigweed	*Amaranthus retroflexus*
Prostrate Pigweed	*Amaranthus graecizans*
Smooth Pigweed	*Amaranthus hybridus*
Smartweed	*Polygonum* spp.
Velvetleaf	*Abutilon theophrasti*
Florida Pusley	*Richardia scabra*
Perennial Weeds:	
Purple Nutgrass Nutsedge	*Cyperus rotundus*
Yellow Nutgrass Nutsedge	*Cyperus esculentus*

Existing stands of nutgrass must be turned under and chopped up thoroughly before treatment. Cultivation is suggested in addition to GENATE PLUS™ /ATRAZINE treatment in fields with moderate or heavy nutgrass infestations.

GENATE PLUS™ /ATRAZINE (80W OR 4L) DIRECTIONS FOR USE

The GENATE PLUS™ /ATRAZINE (80W or 4L) combination can be mixed with water or combined with fluid (solution, slurry, or suspension) fertilizers and applied as a spray, or be impregnated on dry bulk fertilizers.

GENATE PLUS™ /ATRAZINE (80W OR 4L) TANK MIX

See Appendix III for Tank Mix Directions

GENATE PLUS™ /ATRAZINE (80W OR 4L)

FLUID FERTILIZER TANK MIX

The GENATE PLUS™ /ATRAZINE (80W or 4L) combination can be combined with fluid fertilizers for applications made before planting. However, the physical compatibility with these fluid fertilizers must be determined before combining in the spray tank, and even though found to be compatible, constant agitation is necessary to keep the GENATE PLUS™ /ATRAZINE evenly mixed with the fluid fertilizer during application.

Refer to Appendix II regarding the procedure for testing the compatibility of GENATE PLUS™ /ATRAZINE combination with fluid fertilizers.

IMPREGNATION ON DRY BULK FERTILIZERS

Dry bulk fertilizers (except single and treble super-phosphate fertilizers) may be impregnated or coated with GENATE PLUS™ / ATRAZINE (80W or 4L). However, uniform distribution of GENATE PLUS™ /ATRAZINE (80W or 4L) on fertilizer particles and uniform field application are necessary to assure good results. See Appendix III for information and directions for impregnation and use for these combinations.

APPLICATION AND SOIL INCORPORATION DIRECTIONS

Spray Application and Mechanical Incorporation

Apply the recommended dosage in 10 to 50 gallons of water per acre (use the higher gallonage in arid areas) using a properly calibrated, low pressure 20–40 psi), boom-type sprayer having good, but not excessive, agitation. Check frequently during application to be sure nozzles are free from clogging and are delivering a uniform spray pattern. Read directions on ATRAZINE label concerning nozzle and screen sizes.

The herbicides must be incorporated into the soil immediately upon application to prevent loss of the GENATE PLUS™ . Whenever possible, application and incorporation should be done in the same operation. Apply the herbicides to well-worked soil, dry enough to permit good soil mixing.

See Appendix I for Soil Mixing (Incorporation) before Planting Directions.

Center Pivot Sprinkler Application (Limit Use to Western Plains States Only) Apply GENATE PLUS™ /ATRAZINE with irrigation water before planting (see PLANTING), or immediately after planting before either corn or weeds emerge. It is important that application be done immediately after the last tillage operation since poor results will occur if the weeds have germinated before application of the herbicide. Meter the pre-mixed GENATE PLUS™ / ATRAZINE into the irrigation water during the entire period. Apply in 1/2 to 3/4 inch of water. Good agitation of the pre-mixed GENATE PLUS™ /ATRAZINE should be maintained during the entire application period. See Appendix III for tank mixing directions for GENATE PLUS™ with ATRAZINE.

PLANTING

Do not plant seed deeper than 2 inches. For a GENATE PLUS™ / ATRAZINE application made before planting, seeding can be done immediately, or it may be delayed up to 2 weeks. To avoid removing GENATE PLUS™ /ATRAZINE from the row (with loss of weed control in the row), do not move or shape soil after treatment, such as by planting in deep furrows. If it is necessary to work the soil following application and incorporation, but before planting, do not work the soil deeper than the depth the herbicides were incorporated.

Do not use on milo or sorghum.

Do not use on corn seed stock (Breeders, Foundation and Increase).

NOTE: The GENATE PLUS™ /ATRAZINE TANK MIX combination is not recommended on corn grown in Arizona or in the ten southernmost California counties.

For all uses, cultivation is recommended in addition to the GENATE PLUS™ /ATRAZINE treatment in fields with moderate to heavy infestations of wild cane, nutgrass, sandbur, johnsongrass seedlings, or Texas panicum (buffalograss).

BROADCAST RATES PER ACRE (Conventional sprays and center pivot sprinklers)

For control of annual grasses and broadleaf weeds: Use 3³/₄ to 4³/₄ pints GENATE PLUS™ and 1¹/₄ to 2* pounds of ATRAZINE 80W or 1 to 1¹/₂* quarts ATRAZINE 4L per acre. Use the higher rate of GENATE PLUS™ on medium (medium textured-silt loam, clay loam, etc.) to fine (fine textured-clay) soils and for sandbur and johnsongrass seedling control.

For control of wild cane, nutgrass (nutsedge) and Texas panicum (buffalograss) (in addition to the broadleaf weeds and other annual grasses listed): Use 4³/₄ to 7¹/₃ pints of GENATE PLUS™ and 1¹/₄ to 2* pounds of ATRAZINE 80W or 1 to 1¹/₂*quarts ATRAZINE 4L per acre. Use the higher rates of GENATE PLUS™ on medium (medium textured-silt loam, clay loam, etc.) soils, fine (fine textured-clay) soils and for heavy infestations.

Johnsongrass:

Using a program of plowing and discing followed by GENATE PLUS™ and ATRAZINE will give good control of seedlings and greatly reduced competition from plants coming from rhizomes.

In some areas discing 2 to 3 times spaced at 4 to 6 week intervals during fall and winter can be done. If this cannot be done, start with fall plowing followed by several discings in the spring before planting. The discing cuts rhizomes into small pieces and brings them to the surface, where both the weather and GENATE PLUS™ will kill or weaken them.

Next, apply and disc-incorporate GENATE PLUS™ just before planting at the rate of 4³/₄ to 7¹/₃ pints plus 1¹/₄ to 2* pounds of ATRAZINE 80W or 1 to 1¹/₂* quarts of ATRAZINE 4L per acre. Use the higher rates of GENATE PLUS™ on medium (medium textured-silt loam, clay loam, etc.) soils, fine (fine textured-clay) soils, and rhizome johnsongrass suppression and for heavy infestations. Disc a second time at right angles to the first. Cultivate at least once, preferably twice.

* Use the higher rate of ATRAZINE on soils with over 5% organic matter.

CULTURAL PRACTICES FOLLOWING APPLICATION

Should weeds develop, a shallow cultivation or rotary hoeing will generally result in better weed control. When cultivating for any reason, it should be shallow, i.e., no more than 1/2 the depth the herbicide was incorporated.

GENATE PLUS™ /BLADEX* COMBINATION

* BLADEX is the registered trademark of Shell Chemical Company.

GENERAL INFORMATION

GENATE PLUS™ may be used in combination with BLADEX (80W or 4L) for weed control in corn. These combinations will control a broader spectrum of weeds than will either of the products used alone.

These combinations may be applied as a tank mix, preplant and soil-incorporated, or the BLADEX (80W or 4L) may be applied preemergence following a preplant incorporated application of GENATE PLUS™ .

IMPORTANT: Observe all cautions and limitations on the BLADEX (80W or 4L) and GENATE PLUS™ labels.

Make only one application per crop.

Do not use on milo or sorghum. Do not use on corn seed stock (Breeders, Foundation and Increase).

Do not contaminate water by cleaning of equipment or disposal of wastes.

WEEDS CONTROLLED BY GENATE PLUS™ /BLADEX

Annual Grasses:

Annual Bluegrass	*Poa annua*
Annual Fescues	*Festuca* spp.
Annual (Italian) Ryegrass	*Lolium multiflorum*
Barnyardgrass Watergrass	*Echinochloa* spp.
Bermudagrass (Seedlings)	*Cynodon dactylon*
Bullgrass	*Pastpalum boscianum*
Crabgrass	*Digitaria* spp.
Fall Panicum	*Panicum dichotomiflorum*
Field Sandbur*	*Cenchrus incertus*
Giant Foxtail	*Setaria faberii*
Goosegrass	*Eleusine indica*
Green Foxtail	*Setaria viridis*
Indian Lovegrass	*Eragrostis pilosa*
Johnsongrass (Seedlings)*	*Sorghum halepense*
Junglerice	*Echinochloa* spp.
Stinkgrass	*Eragrostis cilianensis*
Texas Panicum (Buffalograss)*	*Panicum texanum*
Volunteer Sorghum*	*Sorghum* spp.
Wild Cane (Shattercane)*	*Sorghum bicolor*
Wild Oat	*Avena fatua*
Witchgrass	*Panicum capillare*
Yellow Foxtail	*Setaria glauca*

* Cultivation is suggested in addition to the GENATE PLUS™/BLADEX treatment in the fields with moderate to heavy infestations of wild cane, sandbur, Texas panicum, (buffalograss), johnsongrass seedlings or volunteer sorghum. For control of light infestations or suppression of high infestations of wild cane, see directions for rates and cultural practices.

Annual Broadleaf Weeds:

Annual Groundcherry	*Physalis lanceifolia*
Annual Morningglory	*lpomoea purpurea*
Annual Sedge	*Carex* spp.
Black Mustard	*Brassica nigra*
Buffalobur	*Solanum rostratum*
Buttercup (Annual)	*Ranuncelus* spp.
Carpetweed	*Mollugo verticillata*
Cocklebur	*Xanthium pensylvanicum*
Common Chickweed	*Stellaria media*
Common Groundsel	*Senecio vulgaris*
Common Mallow	*Malva neglecta*
Common Purslane	*Portulaca oleracea*
Corn Spurry	*Spergula arvensis*
Curly Dock (Seedlings)	*Rumex crispus*
Floras Paintbrush	*Emilia sagittata*
Florida Pusley	*Richardia scabra*
Hedge Mustard	*Sisymbrium officinale*
Jimsonweed	*Datura stramonium*
Kochia	*Kochia scoparia*
Lambsquarters (Common)	*Chenopodium album*
Mayweed	*Anthemis cotula*
Pigweed	*Amaranthus* spp.
Pineappleweed	*Matricaria matricarioides*
Plantain	*Plantago* spp.
Poorjoe	*Diodia teres*
Prostrate Knotweed	*Polygonum aviculare*
Prostrate Spurge	*Euphorbia supina*
Ragweed	*Ambrosia* spp.
Shepherdspurse	*Capsella bursapastoris*
Smallflower (Galinsoga)	*Galinsoga parviflora*
Smartweed (Pennsylvania)	*Polygonum pensylvanicum*
Spinysida	*Sida spinosa*
Tarweed	*Hemizonia congesta*
Velvetleaf	*Abutilon theophrasti*
Wild Buckwheat	*Polygonum convolvulus*
Wild Mustard	*Brassica kaber*
Wild Radish	*Raphanus raphanistrum*
Wild Turnip	*Brassica campestris*

Perennial Weeds:

Purple Nutgrass (Nutsedge)	*Cyperis rotundus*
Yellow Nutgrass (Nutsedge)	*Cyperus esculentus*

Existing stands of nutgrass must be turned under and chopped up thoroughly before treatment. Cultivation is suggested in addition to GENATE PLUS™ /BLADEX treatment in fields with moderate to heavy nutgrass infestations.

GENATE PLUS™ /BLADEX (80W or 4L) DIRECTIONS FOR USE

The GENATE PLUS™ /BLADEX (80W or 4L) combination can be mixed with water and applied as a spray.

TANK MIX DIRECTIONS: See Appendix III

Application and Incorporation Directions:

For a tank mix application, apply the recommended dosage (see recommendations) in 10 to 50 gallons of water per acre, (use the higher gallonage in arid areas) using a properly calibrated low-pressure (20-40 psi), boom-type sprayer having good agitation. Check frequently during application to be sure the nozzles are free from clogging and are delivering a uniform spray pattern. Read directions on the BLADEX (80W or 4L) labels concerning nozzle and screen sizes.

The tank mix must be incorporated into the soil immediately after application to prevent loss of the GENATE PLUS™ . Whenever possible, application and incorporation should be done in the same operation. Apply the herbicides to worked soil, dry enough to permit good soil mixing.

See Appendix I for Soil Mixing (Incorporation) before Planting Directions:

For sequential application of GENATE PLUS™ preplant incorporated followed by BLADEX (80W or 4L) preemergence, consult the appropriate application directions on the respective product labels. Follow the dosage recommendations given below.

RECOMMENDATIONS

For all uses, cultivation is recommended in addition to the GENATE PLUS™ /BLADEX treatment in fields with moderate to heavy infestations of wild cane, nutgrass, sandbur, johnsongrass seedlings, Texas panicum (buffalograss), or volunteer sorghum.

BROADCAST RATES PER ACRE

For control of annual grasses and broadleaf weeds: Use 3¾ to 4¾ pints of GENATE PLUS™ and 2 to 2½ pounds of BLADEX 80W or 3 to 4 pints BLADEX 4L per acre. Use the higher rate of GENATE PLUS™ on medium (medium textured-silt loam, clay loam, etc.) to fine (fine textured-clay) soils and for field sandbur control.

For control of wild cane, nutgrass (nutsedge) and buffalograss (in addition to the broadleaf weeds and other annual grasses listed): Use 4¾ pints of GENATE PLUS™ and 2 to 2½ pounds of BLADEX 80W or 3 to 4 pints BLADEX 4L per acre.

NOTE: BLADEX in combination with GENATE PLUS™ is not recommended for use on coarse or medium textured soils containing less than 2% organic matter. Use the lower rate of BLADEX on coarse to medium textured soils containing 2% organic matter and above. Use the higher rate of BLADEX on fine (fine textured-clay) soils with 3% organic matter and above.

The GENATE PLUS™ /BLADEX can be applied either as a tank mix combination and incorporated into the soil or apply GENATE PLUS™ preplant incorporated followed by BLADEX preemergence. If BLADEX is surface-applied on top of a preplant application of GENATE PLUS™ rotary hoeing or shallow cultivation is recommended if rainfall or sprinkler irrigation has not occurred within about 10 days after the application of BLADEX.

PLANTING

Do not plant seed deeper than 2 inches for a GENATE PLUS™ / BLADEX application made before planting. Seeding can be done immediately or it may be delayed up to 2 weeks. To avoid removing GENATE PLUS™ /BLADEX from the row (with loss of weed control in the row), do not move or shape soil after treatment, such as by planting in deep furrows. If it is necessary to work the soil following application and incorporation, but before planting, it should be shallow, i.e., no more than $1/2$ the depth the herbicide was incorporated.

CULTURAL PRACTICES FOLLOWING APPLICATION

Should weeds develop, a shallow cultivation or rotary hoeing will generally result in better weed control. When cultivating for any reason, it should be shallow, i.e., no more than $1/2$ the depth the herbicide was incorporated.

APPENDIX I

SOIL MIXING (INCORPORATION) BEFORE PLANTING DIRECTIONS

The following equipment commonly is used for soil mixing (incorporation) before planting.

Power-Driven Cultivation Equipment (Recommended on all soil types) set to cut to a depth of 2 to 3 inches.

Tandem Discs (Recommended on all soil types) set to cut to a depth of 4 to 6 inches, operated at 4 to 6 mph followed by a spiked-tooth harrow or some other leveling device which extends beyond the ends of the discs. For more thorough mixing (or perennial grasses and in fine textured soils), disc in two different directions (cross disc). When discing the second time, the disc should be reset to cut no deeper than 4 to 6 inches.

Field Cultivator (Recommended on lighter soils in good tilth only). Use 3 to 4 rows of sweeps spaced at 7 inch or less intervals and staggered so that no soil is left unturned, followed by a spiked-tooth harrow pulled behind the cultivator. Do not use chisels or points. Set the cultivator to cut 4 to 6 inches deep, operated at 5 mph or more. Run the equipment over the field twice, the second run at an angle to the first. Care should be taken on the second pass not to incorporate deeper than the first. The field cultivator should not be used alone as an incorporation tool where control of deep germinating weeds such as nutgrass, johnsongrass seedlings or wild cane is desired. For the control of these weeds, use a disc alone for at least one of the passes over the field.

APPENDIX II

PROCEDURE FOR TESTING THE COMPATIBILITY OF GENATE PLUS™ AND GENATE PLUS™ /ATRAZINE [80W OR 4L] WITH FLUID FERTILIZERS

The following procedure is suggested for determining whether GENATE PLUS™ or GENATE PLUS™ /ATRAZINE (80W or 4L) may be combined with a specific fluid fertilizer for spray tank application.

Materials Required:

1. a. GENATE PLUS™
   b. or GENATE PLUS™ /ATRAZINE (80W or 4L) combination
2. Fluid fertilizer to be used.
3. Adjuvant for fertilizer tank mix: Compex*, Sponto 168-D*, Unimix*, or equivalent. The adjuvant which provides the best emulsification depends on the specific fertilizer under consideration.
4. Two one-quart, wide-mouth glass jars with lid or stopper.
5. Measuring spoons (a 25 ml pipette or graduated cylinder provides more accurate measurement).
6. Measuring cup, 8 oz. (237 ml).

* Compex, Kalo Laboratories, Inc., Petuluma, Calif. Sponto 168-D, Witco Chemical Company, Houston, Texas, Uni-mix, Thompson-Hayward Chemical Company, Kansas City, Kansas.

Procedure

1. Pour a pint (about 473 ml) of the fluid fertilizer into each of the quart jars.
2. Add adjuvant to one of the jars and mix (see next rate table).
3. Add the GENATE PLUS™ or GENATE PLUS™ / ATRAZINE combination to both jars (see next rate table).
4. Close both jars with lid or stopper and mix the contents by turning the jars upside down ten times.
5. Inspect the surface and body of the mixtures—
   (A) Five minutes after completing the jar inversions.
   (B) Two hours later after repeating the jar inversions.

A mixture may be incompatible if either of the following conditions are observed at either inspection period:

(1) An oil layer or large oil globules are seen at the surface of the mixture.

(2) Clumps or aggregates are present.

If the mixture has separated at the five minute inspection period but mixes readily with agitation, the combination may be used PROVIDED good agitation is maintained throughout the mixing and application operations. If the oil layer cannot be redispersed with agitation or clumps persist, the mixture is incompatible and should not be used.

If the GENATE PLUS™ or GENATE PLUS™ /ATRAZINE combination is incompatible with the fertilizer, carefully inspect the mixture containing the adjuvant. If this mixture is compatible, then the GENATE PLUS™ or GENATE PLUS™ /ATRAZINE may be used with the fluid fertilizer provided that adjuvant is added to the fluid fertilizer and thoroughly mixed before adding the GENATE PLUS™ or GENATE PLUS™ /ATRAZINE combination. Add the adjuvant at a rate of 3 pints per 100 gallons of fluid fertilizer. Minimize foaming by using moderate agitation.

If the mixture with the adjuvant is also compatible, the GENATE PLUS™ should not be used in the same tank with the fertilizer.

Rate Table for $1/4$ Cup of GENATE PLUS™/ATRAZINE Premix* And Adjuvant With Fluid Fertilizer**

Gallons of Fluid Fertilizer To Be Applied Per Acre	10	15	20	25	30	40
Tsp. of GENATE PLUS™ for Premix with ATRAZINE	5	$3\frac{1}{2}$	$2\frac{1}{2}$	2	$1\frac{3}{4}$	$1\frac{1}{4}$
Tsp. of ATRAZINE 80-WP for Premix with GENATE PLUS™	2	$1\frac{1}{3}$	1	$\frac{3}{4}$	$\frac{2}{3}$	$\frac{1}{2}$
Tsp. of ATRAZINE 4-L premix with GENATE PLUS™	$2\frac{1}{2}$	$1\frac{1}{2}$	$1\frac{1}{4}$	1	$\frac{7}{8}$	$\frac{2}{3}$

* Prepare two $1/4$-cup GENATE PLUS™/ATRAZINE premixes by first adding ATRAZINE to water to make a smooth slurry; then add the GENATE PLUS™ and stir thoroughly. Add one premix to each of the jars.

** Two (2) milliliters or one-half ($1/2$) teaspoon of adjuvant should be added to 1 pint of fluid fertilizer in order to equal the rate of 3 pints of adjuvant per 100 gallons of fluid fertilizers.

Rate Table for GENATE PLUS™ and Adjuvant With The Fluid Fertilizer**

Gallons of Fluid Fertilizer to be Applied Per Acre	ml or Tsp. of GENATE PLUS™ to be Added to 1 Pint of Fertilizer: ml	Tsp.
10	7	$1\frac{1}{3}$
15	5	1
20	4	$\frac{3}{4}$
25	3	$\frac{2}{3}$
30	2	$\frac{1}{2}$
40	2	$\frac{1}{2}$

** Two (2) milliliters or one-half ($1/2$) teaspoon of adjuvant should be added to 1 pint of fluid fertilizer in order to equal the rate of 3 pints of adjuvant per 100 gallons of fluid fertilizers.

** Based on field rate of 1 pound active ingredient per acre in the fertilizer volumes indicated. Increase volume proportionately to correspond with intended field rate in terms of pounds active ingredient per acre (e.g. for field rate of 4 pounds actual GENATE PLUS™ in 40 gallons fertilizer per acre, add 8 ml or 2 tsp. GENATE PLUS™ to each jar for compatibility testing purposes).

APPENDIX III

TANK MIX DIRECTIONS

1. Fill the spray tank at least ⅔ full of appropriate carrier. Be sure tank is clean and not contaminated with 2,4-D, oil residue or other materials.
2. Start agitation and continue moderate agitation throughout mixing.
3. Premix the BLADEX or ATRAZINE, making sure it is thoroughly wetted and dispersed in the tank prior to adding GENATE PLUS™.
4. Add the GENATE PLUS™ to the tank ⅔ full of carrier already mixed with BLADEX or ATRAZINE.
5. Complete filling the remaining portion of the tank with carrier; mix thoroughly, maintain moderate agitation to provide a uniform mix during application.
6. The tank mix combination should not be left in the spray tank for prolonged periods; batches should be mixed and applied the same day.

Do not use excessive agitation as it may cause the BLADEX or ATRAZINE to settle out and set up in the spray tank. All return lines to the tank must discharge below liquid level.

APPENDIX IV

GENATE PLUS™ IMPREGNATION ON DRY BULK FERTILIZERS

GENATE PLUS™ and ATRAZINE (80W or 4L) Impregnation on Dry Bulk Fertilizers

GENATE PLUS™ alone and in combination with ATRAZINE (80W or 4L) may be impregnated on many dry bulk fertilizers and applied and incorporated into the soil before planting for the control of certain grassy weeds by GENATE PLUS™ and many grassy and broadleaf weeds by the GENATE PLUS™/ATRAZINE combination, in corn.

All GENATE PLUS™ and ATRAZINE (80W or 4L) label and supplementary literature instructions and precautions regarding rates per acre, soil type and soil incorporation, application and other directions must be followed.

Test results have shown GENATE PLUS™ and GENATE PLUS™ and ATRAZINE (80W or 4L) on many dry bulk fertilizers give weed control equal to GENATE PLUS™ and a tank-mix of GENATE PLUS™ and ATRAZINE (80W and 4L) respectively applied as a spray in water or liquid fertilizer. However, uniform impregnation of the pesticides on the dry fertilizer particles and uniform application in the field are necessary to assure good results.

A minimum of 200 pounds and a maximum of 450 pounds of approved fertilizer ingredients impregnated with the appropriate amount of GENATE PLUS™ and ATRAZINE (80W or 4L) must be applied per acre.

For impregnating the pesticides on dry fertilizers, use a closed rotary-drum type mixer equipped with suitable spraying equipment. The spray nozzles should be positioned inside of the mixer to provide uniform spray coverage onto the tumbling fertilizer. The GENATE PLUS™ should be sprayed uniformly onto the fertilizer using a fine spray pattern. For the GENATE PLUS™ / ATRAZINE combination product, the ATRAZINE (80W or 4L) can be added and mixed uniformly with the fertilizer prior to the GENATE PLUS™ addition, or alternatively, the GENATE PLUS™ and ATRAZINE (80W or 4L) can be mixed in the proper use ratio to prepare a uniform slurry for joint spray impregnation.

The physical properties of fertilizers vary, particularly in liquid absorptive capacity. When absorptivity is sufficient, simple spray impregnation of the fertilizer with GENATE PLUS™ provides a satisfactory dry mixture.

If the absorptivity is not adequate, use of a highly absorptive powder is required to provide a dry, free-flowing mixture. Microcel E (Johns-Manville Products Corporation) is the recommended absorbent powder. It should be added separately and uniformly to the prepared pesticide-fertilizer mixture, in a quantity that is sufficient to provide a suitably free-flowing mixture. Generally, less than 2% by weight of Microcel E is required.

The amount of GENATE PLUS™ and ATRAZINE (80W or 4L) actually required in the manufacture of individual fertilizer mixtures should be determined carefully for each production operation. This is necessary to insure that the amounts of GENATE PLUS™ and ATRAZINE actually contained in the mixture applied to the soil represent the correct rates of use.

All individual State regulations relating to dry bulk fertilizer blending registration, labeling and application are the responsibility of the individual and/or company selling the GENATE PLUS™ or GENATE PLUS™ and ATRAZINE fertilizer mixtures.

Do not impregante GENATE PLUS™ or GENATE PLUS™ and ATRAZINE (80W or 4L) on ammonium nitrate. Do not use fertilizers containing ammonium, postassium, or sodium nitrate.

Do not combine GENATE PLUS™ and ATRAZINE (80W or 4L) with single superphosphate (0-20-0) or treble super-phosphate (0-46-0).

Bulk fertilizers impregnated with GENATE PLUS™ or GENATE PLUS™ and ATRAZINE (80W or 4L) should be applied immediately, NOT STORED. All bulk containers should be tightly covered while the products are being transported and applied to reduce chances of GENATE PLUS™ loss via volatilization.

GENATE PLUS™ physical data:

Specific gravity, 25/25°C: 0.953	Flashpoint: 107°F (Seta flash closed cup)
Pounds/gallon, 25/25°C: 7.94	Viscosity: Sprayable down to minus 20°F

RATE CHART FOR THE IMPREGNATION OF DRY BULK FERTILIZERS WITH GENATE PLUS™ OR GENATE PLUS™ AND ATRAZINE 80W OR GENATE PLUS™ AND ATRAZINE 4L

FERTILIZER RATE PER ACRE	GENATE PLUS™ RATE/ACRE			ATRAZINE 80W RATE/ACRE		ATRAZINE 4L RATE/ACRE	
	3¾ PTS. PER ACRE	4¾ PTS. PER ACRE	7⅓ PTS. PER ACRE	1¼ LBS. PER ACRE	2 LBS. PER ACRE	1 QT. PER ACRE	1½ QTS. PER ACRE
200 lbs.	18¾ qts./ton	23¾ qts./ton	36⅔ qts./ton	12½ lbs./ton	20 lbs./ton	10 qts./ton	15 qts./ton
250 lbs.	15 qts./ton	19 qts./ton	29⅓ qts./ton	10 lbs./ton	16 lbs./ton	8 qts./ton	12 qts./ton
300 lbs.	12½ qts./ton	15⅘ qts./ton	24⅖ qts./ton	8⅓ lbs./ton	13⅓ lbs./ton	6⅗ qts./ton	10 qts./ton
350 lbs.	10¾ qts./ton	13⅗ qts./ton	21 qts./ton	7⅛ lbs./ton	11⅖ lbs./ton	5⅘ qts./ton	8⅗ qts./ton
400 lbs.	9⅖ qts./ton	11⅞ qts./ton	18⅓ qts./ton	6¼ lbs./ton	10 lbs./ton	5 qts./ton	7½ qts./ton
450 lbs.	8⅓ qts./ton	10½ qts./ton	16⅓ qts./ton	5½ lbs./ton	8⅞ lbs./ton	4⅖ qts./ton	6⅗ qts./ton

Approved Dry Fertilizer Ingredients For Use With GENATE PLUS™

	N	P	K
Ammonium sulfate	21	0	0
Ammonium phosphate-sulfate	16	20	0
Diammonium phosphate	18	46	0
Potassium chloride	0	0	60
Potassium sulfate	0	0	52
Single super-phosphate	0	20	0
Treble super-phosphate	0	46	—
Urea*	45	0	0

* Some ureas may be phytotoxic when high rates are applied to corn. Use only urea rates known to be safe for corn application.

NOTE: Sul-po-mag, K-mag and 11–48–0 have been shown to be compatible with GENATE PLUS™ and are approved for use.

Approved Dry Fertilizer Ingredients For Use With GENATE PLUS™ and ATRAZINE (80W or 4L)

	N	P	K
Ammonium sulfate	21	0	0
Ammonium phosphate-sulfate	16	20	0
Diammonium phosphate	18	46	0
Potassium chloride	0	0	60
Potassium sulfate	0	0	52
Urea*	45	0	0

* Some ureas may be phytotoxic when high rates are applied to corn. Use only urea rates known to be safe for corn application.

NOTE: Sul-po-mag, K-mag and 11–48–0 have been shown to be compatible with GENATE PLUS™ and ATRAZINE 80W and are approved for use.

CONDITIONS OF SALE

The statements and methods presented about the products mentioned are based upon the best available information and practices known to PPG Industries at the present time, but are not representations or warranties of performance, the result or comprehensivness of such data. PPG Industries warrants that the products mentioned herein conform to their chemical description and are reasonably fit for the purposes stated on the label when used in accordance with the label directions under normal conditions of use, or conditions of use reasonably foreseeable to the seller.

The products mentioned herein if not used in accordance with directions can be hazardous. PPG Industries recommends that anyone using and/or handling the products mentioned herein before using them thoroughly read and understand the specific directions for use and the precautions appearing on the product label.

The products mentioned herein, as all potentially hazardous materials, must be kept out of reach of children.

GENEP™ EPTC 7EC

SELECTIVE HERBICIDE EMULSIFIABLE CONCENTRATE

ACTIVE INGREDIENT:

S-Ethyl Dipropylthiocarbamate 87.8%
INERT INGREDIENTS: 12.2%

TOTAL 100.0%

This product contains 7 lbs. active ingredient per gallon.

EPA Reg. No. 748-222 EPA Est. No. 32761-MO-3

CAUTION

Keep Out of Reach of Children

PRECAUTIONARY STATEMENTS

HAZARD TO HUMAN AND DOMESTIC ANIMALS

Harmful if swallowed. Avoid contact with skin, eyes, or clothing. Avoid breathing spray mist. Wash thoroughly after use.

PRACTICAL TREATMENT

If in eyes or on skin, flush with plenty of water. If swallowed, call a physician or Poison Control Center. Drink 1 or 2 glasses of water and do not induce vomiting. Do not induce vomiting or give anything by mouth to an unconscious person.

ENVIRONMENTAL HAZARDS

This product is toxic to shrimp. Keep out of tidal marshes and estuaries. Do not apply directly to water. Do not apply where runoff is likely to occur. Do not apply when weather conditions favor drift from treatment area. Do not contaminate water by cleaning of equipment or disposal of wastes.

DIRECTIONS FOR USE

It is a violation of Federal law to use this product in a manner inconsistent with its labeling.

Do not apply this product in such a manner as to directly or through drift expose workers or other persons. The area treated must be vacated by unprotected persons. Do not enter areas without protective clothing until sprays have dried.

STORAGE AND DISPOSAL

Do not contaminate water, food, or feed by storage or disposal.

Keep container tightly closed when not in use. Do not store near seeds or fertilizers. Store out of reach of children, pets, and domestic animals. Pesticide, spray mixture, and rinse water that cannot be used according to label instructions must be disposed of according to applicable Federal, State or local procedures.

Triple rinse container. Then offer for recycling or reconditioning, or puncture and dispose of in a sanitary landfill, or incinerate if allowed by state and local authorities. If burned, stay out of smoke.

GENERAL USE PRECAUTIONS

Read all label directions before using.

GENEP™ EPTC should be used only for recommended purposes and at recommended rates. DO NOT OVERUSE.

GENEP is recommended for use on mineral soils only (soils containing less than 10% organic matter).

SPECIAL PRECAUTIONS FOR CROP USES

For incorporated applications, use equipment which has been proven to incorporate thoroughly to the recommended depth.

In irrigated areas, do not apply GENEP prior to pre-irrigation.

Do not tank mix this product with fungicides or with insecticides.

When properly applied and when weather conditions exist for normal plant growth through the season, GENEP will not harm the treated crop nor will harmful soil residues remain beyond harvest. However, during germination and early growth, extended periods of unusually cold and wet or hot and dry weather; insect, nematode, or plant disease attack; carry-over soil residues of certain persistent herbicides, the use of certain soil-applied systemic insecticides; highly saline or alkaline soil conditions, and/or improperly placed fertilizers or soil insecticides may create abnormal conditions that weaken crop seedlings. Also, some of these abnormal conditions may weaken established crops: such as alfalfa and almonds. GENEP used under these abnormal conditions could result in crop injury.

WEEDS CONTROLLED

GENEP controls weeds by interfering with normal germination and seedling development. It does not control established weeds. All weed growth and crop stubble should be thoroughly worked into the soil before treatment.

Annual Grasses:

Barnyardgrass (Watergrass, Junglerice)	*Echinochloa species*
Bermudagrass seedlings	*Cynodon dactylon*
Bluegrass, Annual	*Poa annua*
Crabgrass	*Digitaria species*
Foxtail, Giant	*Setaria faberii*
Foxtail, Green	*Setaria viridis*
Foxtail, Yellow	*Setaria glauca*
Goosegrass	*Eleusine indica*
Johnsongrass seedlings	*Sorghum halepense*
Lovegrass (Stinkgrass)	*Eragrostis cilianenis*
Panicum, Fall	*Panicum dichotomiflorum*
*Panicum, Texas	*Panicum texanum*
Rescuegrass	*Bromus willdenowii*
Ryegrass, Annual (Italian ryegrass)	*Lolium multiflorum*
Sandbur, Field	*Cenchrus pauciflorus*
*Shattercane (Wild Cane)	*Sorghum bicolor*
Signalgrass	*Brachiaria species*
Volunteer grains (Barley, Oats, Wheat)	
Wild Oats	*Avena fatua*
*Witchgrass	*Panicum capillare*

* May not be controlled at less than $3\frac{1}{2}$ pints of GENEP per acre.

Annual Broadleaf Weeds:

Carpetwood	*Mollugo verticillata*
Chickweed, Common	*Stellaria media*
Corn Spurry	*Spergula arvensis*
Fiddleneck	*Amsinckia species*
Henbit (Deadnettle)	*Lamium amplexicaule*
Lambsquarter, Common	*Chenopodium album*
Morningglory, Tall	*Ipomoea purpurea*
Nettleleaf Goosefoot	*Chenopodium murale*
Nightshade, Black	*Solanum nigrum*
Nightshade, Hairy	*Solanum sarrachoides*
Pigweed, Prostrate	*Amaranthus blitoides*
Pigweed, Redroot	*Amaranthus retroflexus*
Pigweed, Tumble	*Amaranthus albus*
Purslane, Common	*Portulaca oleracea*
Pusley, Florida	*Richardia scabra*
Shepherdspurse	*Capsella bursa-pastoris*
Sicklepod	*Cassia obtusifolia*
Sida, Prickly (Teaweed)	*Sida spinosa*

The annual broadleaf weeds listed above will be controlled only if treatment is made when conditions are favorable for weed germination and growth. Broadleaf weeds may only be suppressed at less than $3\frac{1}{2}$ pints GENEP per acre in heavier soils, or under very cold soil conditions.

Perennial Weeds:

Bermudagrass	*Cynodon dactylon*
Mugwort (Chrysanthemum weed)	*Artemisia vulgaris*
Nutsedge, Purple (Nutgrass)	*Cyperus rotundus*
Nutsedge, Yellow (Nutgrass)	*Cyperus esculentus*
Quackgrass (Couchgrass, Quitchgrass)	*Agropyron repens*

Perennial weeds must be turned under and chopped up thoroughly prior to treatment. The underground rhizomes of quackgrass and the rhizomes, and stolons of bermudagrass, must be cut up thoroughly so that four or less nodes remain on a strand. For the suppression or control of quackgrass and bermudagrass, the disc must be set to cut 6 inches deep. Use $4\frac{1}{2}$ to 7 pints GENEP per acre for quackgrass, and $3\frac{1}{2}$ to 7 pints for bermudagrass. The GENEP should be incorporated by discing, or run in the irrigation water after the rhizomes and stolons have been cut up. CONSULT RECOMMENDATIONS FOR CROPS ON WHICH THESE HIGHER RATES MAY BE USED. Nutsedge may not be controlled by water-run applications in heavier soils.

DIRECTIONS FOR USE OF GENEP HERBICIDE ON CROPS

APPLICATION/INCORPORATION

Application Directions:

Pour the recommended amount of GENEP into the spray tank during the filling operation. Apply in 10 to 50 gallons of water per acre; using a properly calibrated, low-pressure sprayer having good agitation. The soil should be well-worked and dry enough to permit good soil mixing (incorporation).

GENEP may be combined with solution, slurry or suspension fertilizers. However, physical compatibility with these fluid fertilizers must be determined before combining in the spray tank. See Appendix I on page 10 for special directions regarding these combinations. Even though found to be compatible, constant agitation is necessary to keep the GENEP uniformly mixed with the fluid fertilizer.

For all band applications, reduce dosage proportionately depending upon the row spacing and band width to be treated.

IMPREGNATION ON DRY BULK FERTILIZERS

Dry bulk fertilizers (except single- and triple- super phosphate fertilizers) may be impregnated or coated with GENEP. However, uniform distribution of GENEP on fertilizer particles and uniform application are necessary to insure good results. See Appendix II on page 11 for information and directions regarding impregnation and use for these combinations.

Incorporation directions:

GENEP must be incorporated into the soil immediately to prevent loss of herbicide.

Whenever possible, application and incorporation should be done in the same operation.

Incorporation before planting:

The following equipment typically is used for soil mixing (incorporation) before planting:

Power Driven Cultivation Equipment—(recommended on all soil types) set to cut to a depth of 2 to 3 inches.

Tandem Discs—(recommended on all soil types) set to cut to a depth of 4 to 6 inches, operated at 4 to 6 MPH followed by a spiked-tooth harrow or some other leveling device which extends beyond the ends of the discs. For more thorough mixing (for perennial grasses and in heavier soils) disc in two different directions (cross disc).

Field Cultivators—(recommended for spring application on coarse textured soils, and for fall application on all soils.) Use only on soils in good tilth. Use 3 to 4 rows of sweeps, spaced at 7 inch or less intervals and staggered so that no soil is left unturned, followed by a spiked-tooth harrow pulled behind the cultivator. Do not use chisels or points. Set the cultivator to cut 4 inches deep, operated at 5 MPH or more. Run the equipment over the field twice, with the second run at an angle to the first.

Rotary Ground Driven or Spring-Tooth Cultivators—(recommended on coarse textured soils in good tilth only). Set to penetrate to a depth of 4 to 6 inches and operated at 5 to 8 MPH in two different directions.

Incorporation at or after planting:

Use power-driven cultivation equipment set to cut to a depth of 2 to 3 inches OR, on coarse textured soils only, ground-driven tillers (rolling cultivators, rotary hoe, row wheels, etc.) set to cut to a depth of $1\frac{1}{2}$ inches and operated at 6 to 8 MPH. When incorporating after planting, care must be taken not to disturb the crop seed or seedling.

SUBSURFACE APPLICATION—AT PLANTING OR POSTEMERGENCE

Apply GENEP in 10 or more gallons of water per acre.

Special equipment designed for subsurface application MUST be used. Injector and sweep units must be rigidly mounted ahead of the planters.

Soil Injection—Injector shanks must be spaced $2\frac{1}{2}$ to 3 inches apart and mounted in staggered positions to avoid trash buildup. Set shanks to inject GENEP to 2 to 3 inches below the soil surface. The width of the band in which weed control is desired will determine the number and spacing of injector shanks required per row. (Example: Four injector shanks spaced 3 inches give a 12-inch band). A broadcast application can be made by increasing the number of shanks. The two shanks adjacent to the drill row must be $1\frac{1}{4}$ to $1\frac{1}{2}$ inches on either side of it. EXCEPT IN COTTON, WHERE THE DISTANCE MUST BE 4 INCHES ON EITHER SIDE OF THE DRILL ROW, AND SUGAR BEETS, WHERE THE DISTANCE MUST BE $2\frac{3}{4}$ INCHES ON EITHER SIDE OF THE DRILL ROW.

Covered Sweeps—Set the sweeps to run below the soil surface deep enough to cover the GENEP with 2 to 3 inches of soil. Calibrate by measuring the spray band width at the back of the sweep, not the sweep width. For broadcast applications, stagger sweeps on double tool bar so they overlap sufficiently to allow spray bands to meet.

NOTE: When using either injectors or sweeps, GENEP must be applied deep enough to allow 2 to 3 inches of soil to remain over the treatment after the planting operations.

PLANTING DIRECTIONS

For preplant applications, seeding should be done as soon as possible after treatment to obtain a maximum period of weed control.

IRRIGATION APPLICATION—

POSTPLANTING AND ESTABLISHED CROPS

Meter GENEP into the irrigation water using a metering device that will introduce a constant flow into the water. For flood, furrow, or sprinkler irrigation, meter the GENEP into the water during the entire period OR, for sprinkler irrigation, GENEP may be metered into sufficient water to penetrate to a depth of 3 to 4 inches. Time this GENEP application to insure that proper penetration of the herbicide corresponds with the end of the irrigation period. Flush the lines and then return the water off promptly. Consult "RECOMMENDATIONS" on this label for proper timing or application for each crop for which irrigation is recommended. A flow rate chart for water run applications will be found in Appendix III on back page of this booklet. Tailwater (runoff water) from flood or furrow irrigation should be recirculated or used only on other crops which are registered for this type of application. Do not contaminate water to be used on susceptible crops and ornamentals, or to be used for domestic purposes.

CULTURAL PRACTICES FOLLOWING APPLICATION

Should weeds develop, a shallow cultivation or rotary hoeing will generally result in better weed control. When cultivating for any reason, it should be shallow, (i.e., no more than $\frac{1}{2}$ the depth the herbicide was incorporated or injected). Preemergence or postemergence herbicides may be necessary to control weeds resistant to GENEP.

CROP RECOMMENDATIONS

All crop recommendations are given on a regional basis. There are five regions, as delineated on the U.S. map. USE THE RECOMMENDATIONS IN YOUR REGION ONLY.

REGIONAL USE MAP:

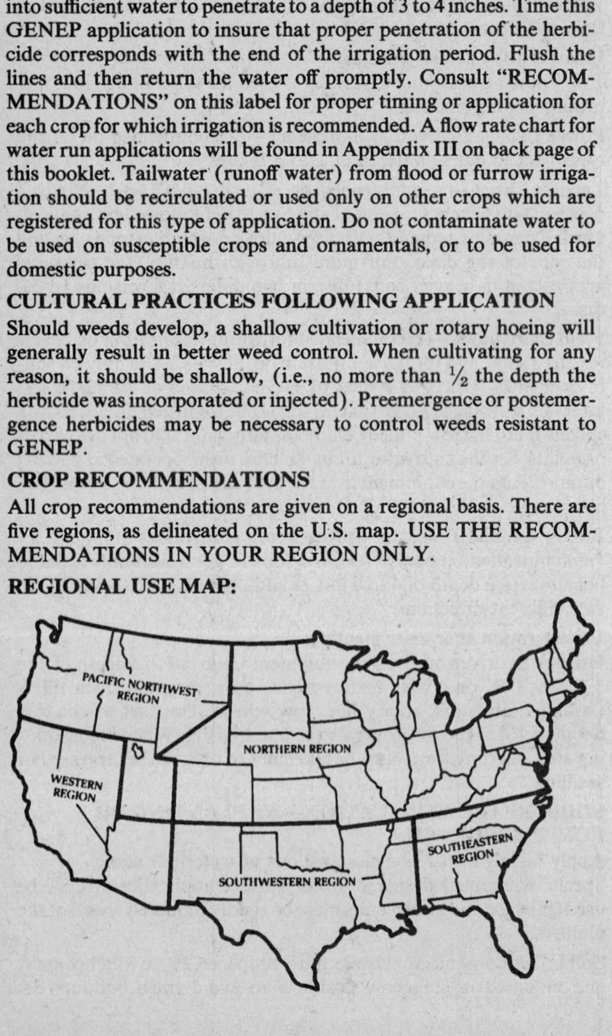

RATE CONVERSION TABLE

Dosage rates in this booklet are expressed as pints GENEP per acre. The following table shows pints GENEP per acre in the left column and the equivalent amount of active ingredient per acre in the center column.

Pints GENEP 7E/Acre	Approximate lb. Active Ingredient/ Acre	Approximate Acres Treated By One Gallon GENEP 7E
$1\frac{1}{4}$	1	$6\frac{1}{2}$
$1\frac{3}{4}$	$1\frac{1}{2}$	$4\frac{2}{3}$
$2\frac{1}{4}$	2	$3\frac{1}{2}$
$3\frac{1}{2}$	3	$2\frac{1}{3}$
$4\frac{1}{2}$	4	$1\frac{3}{4}$
$5\frac{1}{4}$	$4\frac{1}{2}$	$1\frac{1}{2}$
$5\frac{3}{4}$	5	$1\frac{2}{5}$
7	6	$1\frac{1}{6}$
$8\frac{1}{2}$	$7\frac{1}{2}$	1
17	15	$\frac{1}{2}$

RECOMMENDATIONS—FIELD CROPS

Recommendations are given as the broadcast (overall) rates of GENEP per acre. For band treatment, use proportionately less material per acre depending upon the width of the band to be treated and the crop row spacing. Do not use band application on rocky ground because thorough incorporation is not possible.

ALFALFA*, BIRDSFOOT TREFOIL, CLOVERS AND LESPEDEZA

Do not use GENEP if a grass or grain nurse crop is to be planted with the legume. Do not use on white dutch clover. Temporary crop stunting and sealing of the first leaves will occur if conditions for germination and growth are not optimum, or relieved by irrigation or adequate rainfall. See specific recommendation for appropriate region.

* Do not use GENEP on alfalfa if more than 1.2 pounds of actual atrazine was applied within the previous 12 months.

ALFALFA AND LADINO CLOVER—

ESTABLISHED STANDS

All regions except Southeast. Meter $2\frac{1}{4}$ to $3\frac{1}{2}$ pints of GENEP per acre into the irrigation water applied to established stands prior to weed emergence. Use the lower rate on very coarse textured soils.

Do not apply to alfalfa within 14 days of harvest or grazing.

Do not apply to Ladino clover within 45 days of harvest or grazing.

NORTHERN REGION

Alfalfa—Apply and incorporate $2\frac{1}{4}$ to $3\frac{1}{2}$ pints of GENEP per acre before planting. Use the lower rate for annual grass control only.

Alfalfa, Birdsfoot Trefoil, Clovers, Lespedeza—Apply and incorporate $3\frac{1}{2}$ to $4\frac{1}{2}$ pints of GENEP per acre before planting. Use the lower rate on coarse textured soils.

PACIFIC NORTHWEST REGION

Alfalfa, Birdsfoot Trefoil, Clovers, Lespedeza—Apply and incorporate $2\frac{1}{4}$ to $4\frac{1}{2}$ pints of GENEP per acre before planting. Use the lower rate on very coarse textured soils.

Alfalfa only—Meter $2\frac{1}{4}$ to $3\frac{1}{2}$ pints of GENEP per acre into the irrigation water that is applied immediately after planting. Use the lower rate on coarse textured soils. SEE DIRECTIONS FOR USE.

SOUTHEASTERN REGION

Alfalfa, Birdsfoot Trefoil, Clovers, Lespedeza—Apply and incorporate $3\frac{1}{2}$ pints of GENEP per acre before planting.

Alfalfa (South Carolina Only)—Apply and incorporate $1\frac{3}{4}$ pints of GENEP per acre before planting.

SOUTHWESTERN REGION

Alfalfa, Birdsfoot Trefoil, Clovers, Lespedeza—Apply and incorporate $3\frac{1}{2}$ pints of GENEP per acre before planting.

WESTERN REGION

Alfalfa, Birdsfoot Trefoil, Clovers, Lespedeza—Apply and incorporate $2^1/_4$ to $4^1/_2$ pints of GENEP per acre before planting. Use the lower rate on very coarse textured soils.

Alfalfa Irrigated—Meter $2^1/_4$ to $3^1/_2$ pints of GENEP per acre into irrigation water that is applied immediately after planting. Use the lower rate on coarse-textured soils. SEE DIRECTIONS FOR USE.

Alfalfa—Limit use to one application per cutting and 7 pints of GENEP per acre per year. If applied by flood irrigation, up to 14 pints of GENEP per acre per year can be used.

BEANS, GREEN OR DRY

Do not use GENEP on Adzuki beans, cowpeas (blackeye peas, blackeye beans), soybeans, lima beans, or other flat-podded beans except Romano. Under abnormal weather conditions, stunting may occur on Gratiot, Michilite, Sanilac, Seafarer, and Seaway varieties. See specific recommendation for appropriate regions.

Apply GENEP by the method and at the rate shown below.

NORTHERN REGION

Fall Applicaton—(Dry beans Minnesota and North Dakota only) Apply and incorporate in the late fall before soil freezes. Use $4^1/_2$ pints of GENEP per acre on coarse textured soils and $5^1/_4$ pints on medium- and fine-textured soils.

At Planting—Apply and incorporate just before or immediately after planting or meter into the irrigation water before or immediately after planting. Use $3^1/_2$ to $4^1/_2$ pints of GENEP per acre. Use the lower rate on coarse textured soils.

Lay-By Application—At last cultivation apply and incorporate $3^1/_2$ to $4^1/_2$ pints of GENEP per acre. Apply as directed spray to the base of the bean plants before the pods start to form. Use the lower rate on coarse textured soils. Do not feed or graze treated vines within 45 days of treatment.

PACIFIC NORTHWEST REGION

At Planting—Apply and incorporate $3^1/_2$ to $4^1/_2$ pints of GENEP per acre just before planting. Use the lower rate on coarse-textured soils.

SOUTHEASTERN REGION

At Planting—Apply and incorporate $3^1/_2$ pints of GENEP per acre just before planting.

Subsurface Application—Apply $2^1/_4$ pints of GENEP per acre preplant or at planting. SEE DIRECTIONS FOR USE.

Bed Treatments—Pre-Planting

Apply $3^1/_2$ pints of GENEP per acre broadcast and disc in 6 inches deep before forming beds and planting.

Apply $1^3/_4$ pints of GENEP per acre broadcast (do not disc) immediately ahead of bedding disc. Plant 7 days after treatment.

Apply as a band treatment immediately ahead of bedding discs or to partially formed beds immediately in front of rebedding operation. Use a band rate equivalent to $2^1/_4$ pints of GENEP per acre broadcast. Treated band should be covered with 3 to 4 inches of untreated soil. Plant 7 days after treatment.

Lay-By Application—At last cultivation, apply and incorporate $3^1/_2$ pints of GENEP per acre. Apply at a directed spray to the base of the bean plant before the pods start to form. Do not feed or graze vines within 45 days of treatment.

SOUTHWESTERN REGION

At Planting—Apply and incorporate $3^1/_2$ pints of GENEP per acre just before planting.

Subsurface Application—Apply $3^1/_2$ pints of GENEP per acre preplant or at planting. SEE DIRECTIONS FOR USE.

Lay-By Incorporated—At the last cultivation apply and incorporate $3^1/_2$ pints of GENEP per acre. Apply as a directed spray to the base of the bean plant before pods start to form. Do not feed or graze vines within 45 days of treatment.

Subsurface Application—Apply $3^1/_2$ pints of GENEP per acre preplant or at planting. SEE DIRECTIONS FOR USE.

Lay-By Incorporated—At the last cultivation apply and incorporate $3^1/_2$ to $4^1/_2$ pints of GENEP per acre. Use the lower rate on coarse textured soils. Apply as a directed spray to the base of the bean plant before pods have started to form. Do not feed or graze vines within 45 days of treatment.

Lay-By Surface—At last cultivation apply $3^1/_2$ pints of GENEP per acre to clean weed free soil. SEE DIRECTIONS FOR USE. Do not feed or graze vines within 45 days of application.

WESTERN REGION

At Planting—Apply and incorporate $3^1/_2$ pints of GENEP per acre preplant or at planting. SEE DIRECTIONS FOR USE.

Subsurface Application—Apply $3^1/_2$ pints of GENEP per acre preplant or at planting. SEE DIRECTIONS FOR USE.

Lay-By Incorporated.—At the last cultivation apply and incorporate $3^1/_2$ to $4^1/_2$ pints of GENEP per acre. Use the lower rate on coarse-textured soils. Apply as directed spray to the base of the bean before the pods have started to form. Do not feed or graze vines within 45 days of treatment.

Lay-By Subsurface—At last cultivation apply $3^1/_2$ pints of GENEP per acre to clean weed free soil. SEE DIRECTIONS FOR USE. Do not feed or graze vines within 45 days of treatment.

BEANS, DRY

GENEP AND TREFLAN* EC HERBICIDE TANK-MIX

A tank mix combination of GENEP plus Treflan EC will give a broader spectrum of weed control than either product used separately.

CAUTION: The combination of GENEP and Treflan EC should not be used on soybeans, blackeyed peas (beans), lima beans, and other flat-podded beans, except Romano.

In the lighter soils under sprinkler irrigation, when it is necessary to irrigate beans after planting and before emergence, sufficient water should be applied to wet the soil well below the depth of planted seed.

Do not graze or feed forage from treated fields to livestock.

Read both the GENEP and Treflan EC labels carefully before using. Observe all cautions and limitations on labeling of both products.

Do not contaminate water by the cleaning of equipment or disposal of wastes.

ADDITIONAL WEEDS CONTROLLED BY THE COMBINATION

Annual Grasses:	
Bromegrass	*Bromus species*
Cheat	*Bromus secalinus*
Junglerice	*Echinochloa colonum*
Sprangletop	*Leptochloa filiformis*
Annual Broadleaves:	
Henbit	*Lamium amplexicaule*
Knotweed	*Polygonum aviculare*
Kochia	*Kochia scoparia*
Nettle, Stinging	*Urtica divica*
Puncturevine	*Tribulus terrestris*
Pusley, Florida	*Richardia scabra*
Thistle, Russian	*Salola Kali*

DIRECTIONS FOR USE

Mixing

Add the recommended rates of both GENEP and Treflan EC to the spray tank during filling, and mix thoroughly. Apply in 10 to 40 gallons of water per acre.

Spray Equipment

Use any properly calibrated low-pressure boom-type herbicide sprayer which will apply the material uniformly. Check calibration frequently during application and observe the nozzles to be sure they are delivering a uniform spray pattern.

Soil Incorporation

The GENEP and Treflan EC combination must be incorporated (mixed) thoroughly into the top 2 to 3 inches of soil immediately after spraying. Spraying and incorporation should be accomplished in the same operation, if possible. This can be done by mounting the tank and boom right on the incorporation rig.

Thorough incorporation can be achieved with any of the following equipment:

Power-driven rotary cultivators—set to cut 2 to 3 inches deep.

Double disc (or double disc with spike-tooth harrow in tanden)—set to cut 3 to 6 inches deep and operate in two directions (cross disc) at 4 to 6 mph.

Shallow incorporation with implements set to cut less than 2 inches may result in erratic weed control.

Planting

Plant dry beans within 48 hours after incorporation.

RECOMMENDATIONS

Broadcast the combination of GENEP and Treflan EC according to the following rates:

GENEP

Apply 2½ pints of GENEP in combination with Treflan EC for control of annual grasses. Apply 3½ pints of GENEP in combination with Treflan EC for control of nutsedge and labeled broadleaf weeds.

Treflan EC

Soil Type	Organic Matter Content	Rate
Coarse (sand)	0–2%	1 pt.
Coarse (sand)	2-5%	1½–2 pts.
Medium (loam)	0–5%	1½ pts.
Fine (Clay)	0–5%	2 pts.
All Soil Types	5.1–10%	2 pts.

* Treflan-trifluralin is a registered trademark of Elanco Product Co.

BEANS, DRY

GENEP AND

SONALAN* TANK-MIX

For expanded weed control, especially nightshade control in dry beans, apply the tank mix combination before crop planting. Observe application requirements, cautions and limitations for both products and follow label recommendations. Prepare the soil as described in GENEP™ label, apply the tank mix combinations, and incorporate immediately.

Soil Texture	Recommended Rates (Pints) GENEP 7	Sonalan
Coarse (sandy)	3.5	1.25-2.0
Medium (sandy loam)	3.5	1.75-2.5
Fine (clay)	3.5	2.25-3.0

Precaution: Observe all Sonalan label precautions and limitations.

* Sonalan—ethalfluralin is a registered trademark of Elanco Products Co.

CASTOR BEANS

NORTHERN REGION ONLY

Apply and incorporate 2¼ pints of GENEP per acre immediately after planting. Use a rotary hoe for incorporation. Early cultivation after GENEP application enhances weed control.

FLAX

NORTHERN REGION

Fall Application—(Minnesota and North Dakota Only)—Apply and incorporate in late fall before soil freezes. Use 4½ pints of GENEP per acre on coarse-textured soils and 5¼ pints per acre on medium- and fine-textured soils.

PACIFIC NORTHWEST REGION

Preplanted Incorporated—Apply and incorporate 3½ pints of GENEP per acre broadcast just before planting.

PEAS, GREEN PROCESSING

WESTERN WASHINGTON ONLY

Apply and incorporate 2¼ pints of GENEP per acre broadcast just before planting. Early stunting of crop may occur.

POTATOES (IRISH)

ALL REGIONS EXCEPT SOUTHEASTERN

For tank mix combination of GENEP with Metribuzin in Oregon and Washington only, see state label following this section.

Do not exceed 7 pints of GENEP per acre per crop. The Superior variety potato is sensitive to GENEP and under stress conditions, early season stunting may occur.

Preplant Incorporated—Apply and incorporate 3½ pints of GENEP per acre just before planting. For quackgrass and nutgrass control in the Northern and Southwestern regions, use 7 pints per acre.

Fall application—(Minnesota and North Dakota Only)—Apply and incorporate in late fall before soil freezes 5¼ pints of GENEP on coarse-textured soils and 7 pints on medium- or fine-textured soils.

Drag-off Incorporation—Apply and incorporate 3½ pints of GENEP per acre. For nutgrass control in Northern and Southern regions, use 7 pints per acre. The field should be "dragged-off" before application and incorporation. Use spike-toothed harrows or cultivation equipment for incorporation.

POTATOES (IRISH)

SOUTHEASTERN REGION ONLY

Do not exceed 3½ pints of GENEP per acre per crop.

CAUTION: In Florida on winter and early spring potatoes, apply only after potatoes have emerged and formed true leaves.

Preplant Incorporation—Apply and incorporate 3½ pints of GENEP per acre broadcast just before planting. For incorporated applications to beds, apply as a band and incorporate with ground- or power-driven tillers. SEE DIRECTIONS FOR USE: Incorporation.

After Planting and Before Bedding—Apply 1¾ pints of GENEP per acre broadcast over planted crop and bed up immediately with bedding discs set to cover with 3–4 inches of soil. The same application can be made after bed formation by rebedding immediately after application of the GENEP.

Drag-Off Incorporation—Apply and incorporate 3½ pints of GENEP per acre. The field should be "dragged off" before application. Use spike-toothed harrows or cultivation equipment for incorporation.

Lay-By Incorporation—Apply and incorporate 3½ pints of GENEP per acre to clean cultivated soil after potatoes have emerged. Apply as directed spray to the soil. Do not apply within 45 days of harvest.

Lay-By Irrigation—Meter 3½ pints of GENEP per acre into the irrigation water following clean cultivation. Do not apply within 45 days of harvest.

POTATOES (IRISH)

GENEP™ EPTC 7E SELECTIVE HERBICIDE

SUPPLEMENTAL LABELING GENEP™

AND METRIBUZIN TANK MIX

FOR USE ONLY IN OREGON AND WASHINGTON

For Control of Weeds in Irish Potatoes

OREGON SLN No. 830024

WASHINGTON SLN No. 830022

A tank mix combination of GENEP™ EPTC 7E herbicide and metribuzin (Sencor** 4, Sencor 50W, Sencor DF, Lexone***4, Lexone 50W or Lexone Dry Flowable 75%) can be applied to Irish potatoes in the Pacific Northwest (Oregon, Washington, and Idaho) of the United States to give a broader spectrum of weed control than either product used alone.

CAUTION: Before using GENEP and metribuzin (Sencor, Lexone) as a tank mix, read both the GENEP and metribuzin labels carefully. Observe all cautions and limitations on labeling of both products.

Do not contaminate water by the cleaning of equipment or disposal of wastes.

WEEDS CONTROLLED

See both the GENEP and metribuzin (Sencor or Lexone) labels for the list of weeds controlled by the combination.

DIRECTIONS FOR USE

It is a violation of the Federal law to use this product in a manner inconsistent with its labeling.

MIXING: Add the recommended rates of GENEP and metribuzin (Sencor, Lexone) to the spray tank during the filling operation. Add GENEP last. Mix thoroughly, apply in 20 to 30 gallons of water per acre.

SPRAY EQUIPMENT: Use any properly calibrated low pressure boom type herbicide sprayer which will apply the material uniformly. Check calibration frequently and observe to be sure that the nozzles are delivering a uniform spray pattern during application.

SOIL INCORPORATION: The GENEP and metribuzin combination must be incorporated (mixed) thoroughly into the top 2 to 3 inches of soil after spraying. Spraying and incorporation should be accomplished in the same operation if possible. See the GENEP label for complete incorporation directions.

SPRINKLER APPLICATION (Pacific Northwest only): Apply through irrigation sprinkler system after planting as a preemergence application according to the SENCOR or LEXONE labels for preemergence application and/or as early postemergence application until potatoes are 4 to 6 inches tall. Regulate postemergence dosage rates according to soil type on rate chart.

Pre-mix metribuzin first with water in the holding tank at desired rate in 4 to 5 parts water to 1 part chemical. Add GENEP last. Meter chemical-water mixture into sprinkling system at rate proportionate to acreage covered. For center pivot systems, apply 1/2 to 3/4 inch of water per acre. For solid set, wheel lines or hand lines, moisten soil surface lightly first, then apply chemical during first hour of irrigation set.

All irrigation systems for this purpose must have check valves and antisiphon devices to prevent back-flow into water source or supply tank and must be equipped with a power interlock to insure simultaneous shut-off of water pumps and injector.

RECOMMENDATIONS

Broadcast the combination of GENEP and metribuzin (Sencor, Lexone) according to the following rates:

	Pints GENEP 7E	Pints Sencor 4 or Lexone 4 or Pounds Sencor 50 or Lexone 50	Pounds Sencor DF or Lexone Dry Flowable 75%
COARSE SOILS Sand Sandy loam Loamy sand	3 1/2	1/2	1/3
FINE SOILS Loam Silt loam Sandy clay Clay loam	4 1/2	1/2-1	1/3-2/3

GENERAL USE PRECAUTIONS

Before using GENEP plus metribuzin as a tank mix, read and observe the cautionary statements and all other information appearing on both product labels.

Always perform a jar test to insure compatibility of the desired GENEP and metribuzin tank mix.

** SENCOR is a registered trademark of the Parent Company of Farbenfabriken Bayer GmbH, Leverkusen.

*** LEXONE is a registered trademark of E. I. DuPont De Nemours & Co., Inc.

SAFFLOWER

NORTHERN, PACIFIC NORTHWEST, AND WESTERN REGIONS

Apply and incorporate 3 1/2 pints of GENEP per acre broadcast just before planting.

SUGAR BEETS

NORTHERN REGION

Postemergence Irrigation Water—Meter 2 1/4 to 3 1/2 pints of GENEP per acre into the first irrigation application after the last cultivation of the season. Use the lower rate on coarse textured sandy soils.

Postemergence Incorporation—Apply and incorporate 3 1/2 pints of GENEP per acre to a depth of 2 to 3 inches after thinning and clean cultivation.

Postemergence Subsurface Injection—Apply 3 1/2 pints of GENEP per acre following a clean cultivation to destroy all weed growth. For a band application on 22 inch rows, use two injectors spaced 5 1/2 inches apart and centered on the drill row. SEE DIRECTIONS FOR USE.

Fall Application—(Minnesota, North Dakota Only) Apply and incorporate 4 1/2 to 5 1/4 pints of GENEP per acre in the late fall before the soil freezes. Use the lower rate on coarse textured sandy soils.

Preplant—(Iowa, E. Nebraska, Minnesota, Michigan, N. Dakota, S. Dakota only) Apply and incorporate 2 1/4 to 3 1/2 pints of GENEP per acre just before planting. Injury will occur if conditions for germination and growth are not optimum. Use the lower rate on coarse textured sandy soils.

PACIFIC NORTHWEST REGION

Postemergence Irrigation Water—Meter 2 1/4 to 3 1/2 pints of GENEP per acre into the first irrigation applied after the last cultivation of the season. Use the lower rate on coarse-textured sandy soils. Do not apply within 49 days of harvest.

Postemergence Incorporation—Apply and incorporate 3 1/2 pints of GENEP per acre to a depth of 2 to 3 inches after thinning and clean cultivation. Do not apply within 49 days of harvest.

Postemergence Subsurface Injection—Apply 3 1/2 pints of GENEP per acre following a clean cultivation to destroy all weed growth. For a band application on 22 inch rows, use two injectors spaced 5 1/2 inches apart and centered on the drill row. SEE DIRECTIONS FOR USE. Do not apply within 49 days of harvest.

SOUTHWESTERN REGION

Postemergence Irrigation Water—Meter 2 1/4 to 3 1/2 pints of GENEP per acre into the first irrigation applied after the last cultivation of the season. Use the lower rate on coarse-textured sandy soils.

Postemergence Incorporation—Apply and incorporate 2 1/4 pints of GENEP per acre to a depth of 2 to 3 inches after thinning and clean cultivation.

WESTERN REGION

Postemergence Irrigation Water—Meter 2 1/4 to 3 1/2 pints of GENEP per acre into the first irrigation applied after the last cultivation of the season. Use the lower rate on coarse textured sandy soils.

Postemergence Incorporation—Apply and incorporate 3 1/2 pints of GENEP per acre to a depth of 2 to 3 inches after thinning and clean cultivation.

Postemergence Subsurface Injection—Apply 3 1/2 pints of GENEP per acre following a clean cultivation to destroy all weed growth. For a band application on 22 inch rows, use two injectors spaced 5 1/2 inches apart and centered on the drill row. SEE DIRECTIONS FOR USE.

SUNFLOWER

NORTHERN REGION ONLY

Preplant incorporated (Minnesota, North Dakota, and South Dakota)—Apply and incorporate $3\frac{1}{2}$ pints of GENEP per acre just before planting.

Fall application (Minnesota, North Dakota)—Apply and incorporate GENEP in the late fall before soil freezes. Use $4\frac{1}{2}$ pints per acre on coarse-textured soils and $5\frac{1}{4}$ pints per acre on medium- and fine-textured soils.

SWEET POTATOES

SOUTHWESTERN REGION ONLY

Preplant—Apply and incorporate GENEP to a maximum depth of 3 inches just before planting. Use $1\frac{3}{4}$ pints on coarse textured soils and $2\frac{1}{4}$ pints on medium- and fine-textured soils. Immediately after application, cover the treated bed with 2 to 4 inches of untreated soil from the area adjacent to the band using bed shaping equipment.

Preplant Bed-up—After pre-shaped beds have been dragged down, apply GENEP broadcast to the soil. Use $1\frac{3}{4}$ pints per acre on coarse-textured soils or $2\frac{1}{4}$ pints per acre on medium- and fine textured soils. Immediately after application, reform the bed with bed-shaping equipment to leave a band of GENEP 2 to 4 inches below the bed surface.

Postplanting—Apply $8\frac{1}{2}$ pints of GENEP per acre broadcast immediately after or within 2 days after planting slips or vine cutting. Apply to dry soil surface. Do not mix into the soil. If sweet potatoes are to be irrigated, apply GENEP before irrigation.

TABLE BEETS

PACIFIC NORTHWEST REGION ONLY

Apply and incorporate $2\frac{1}{4}$ pints of GENEP per acre just before planting. Note: Under normal use, table beets are susceptible to GENEP injury and the seeding rate should be increased 10 percent.

TOMATOES

WESTERN REGION ONLY

Preplant—Apply $3\frac{1}{2}$ pints of GENEP per treated acre as a spray the soil surface. Incorporate immediately. For band applications, reduce rate proportionately. DO NOT APPLY WITHIN 2 INCHES OF THE CROP ROW. Do not apply where grain will be planted within 90 days. Do not irrigate within 5 days of application. Do not apply within 21 days of harvest.

Lay-by Application—Only in Northern California Counties of Butte, Colusa, Contra Costa, Glenn, Merced (North of Highway 152), Sacramento, San Joaquin, Solano, Stanislaus, Sutter, Yolo, and Yuba.

Use on tomatoes at least 3–4 inches tall grown on clay and clay loam soils only. DO NOT USE ON SANDY SOILS.

TREE CROPS

ALMONDS

WESTERN REGION ONLY

After making the last cultivation of the season, meter $3\frac{1}{2}$ pints of GENEP per acre into the irrigation water. Do not apply within 14 days of harvest.

CITRUS TREES

SOUTHEASTERN, SOUTHWESTERN, AND WESTERN REGIONS

Citrus Nursery Stock and Young Field Plantings (Non-bearing Orange, Grapefruit, and Lemon Groves)—After lining out, apply $3\frac{1}{2}$ to 7 pints of GENEP per acre as a directed spray to the soil. Incorporate with cultivation equipment such as tree hoes or rotary hoes. Use the lower rate only on very coarse textured soils.

Citrus Bearing (Oranges, Tangerines, Grapefruit, Lemons)—After clean cultivation, or before weed emergence in bearing citrus, apply $3\frac{1}{2}$ pints of GENEP per acre by flood or furrow irrigation. Do not apply within 15 days of harvest.

PINE SEEDLING NURSERIES

SOUTHEASTERN AND SOUTHWESTERN REGIONS

Loblolly, Slash, Long Leaf, Short Leaf—Apply and incorporate 7 pints of GENEP per acre 14 days before seeding.

WALNUTS

PACIFIC NORTHWEST AND WESTERN REGIONS

After clean cultivation or before weed emergence on well established trees, meter $3\frac{1}{2}$ pints per acre into the irrigation water during the entire irrigation period.

ORNAMENTALS

SPECIAL PRECAUTIONS FOR ORNAMENTAL USES

GENEP must be thoroughly mixed into the soil for all ornamental uses.

GENEP may cause injury to ornamentals under certain soil and climatic conditions or if directions are not followed.

Soil Preparation: The soil to be treated should be loose and free of clods. All weed growth should be removed or thoroughly worked into the soil before application.

Application: The recommended rate of GENEP should be applied as uniformly as possible. Apply to well-worked soil that is dry enough to permit thorough mixing with incorporation equipment. When treating around established plants, direct spray to soil surface for maximum coverage. Use one of the following appropriate means of application.

Low Pressure Herbicide Sprayer: For broadcast application, use 10 to 50 gallons of water per acre. For band application (in front of power tiller), use less water depending upon row spacing and width of band desired. Check pressure and nozzles frequently to assure uniform application.

Hose Proportioner: Make sure proportioner is working properly. A more uniform application can be made by applying half the required amount of GENEP over the area to be treated, then apply the remainder at right angles or crosswise.

Knapsack Sprayer: Apply as suggested for the hose proportioner.

Incorporation: Immediately after application, thoroughly mix GENEP into the soil to a depth of 2 to 3 inches. Mix to a depth of 6 inches for Nutgrass, Quackgrass, Bermudagrass, and Chrysanthemumweed (Mugwort) control. Thorough soil mixing is necessary for good weed control. Use the following equipment or other equipment which has proven satisfactory under local conditions.

Commercial Nursery: Use nursery cultivator or rototillers for preplant broadcast (overall) applications, preplant band applications, and postplant applications.

Home Garden—Preplant Application—Rototiller—Postplant Application: For annual weeds, use hand rake, or hoe or water-in immediately after application to a depth of 2 to 3 inches. Where incorporation by hand raking is done, light watering after raking is recommended.

For perennial weeds, incorporate to a depth of 6 inches with a rototiller.

GENEP CAN BE USED ON THESE ORNAMENTALS

Herbaceous Plants

Ageratum	Dahlia
Alyssum	Daylilies
Amaranthus	Dianthus
Asters	Marigold
Balsam	Nasturtium
Begonia	Pansy

Chrysanthemum	Petunia
	Zinnia
Ground Covers	
Ajuga	Ivy
Gazania	Pachysandra
Hypericum	Periwinkle (vinca minor)
Ice Plant	Sedum
	Strawberry (ornamental)
Evergreen and Deciduous Trees and Shrubs	
Azalea	Leucothoe
Berberis	Lilac
Boxwood	Linden
Camellia	Magnolia
Chamaecyparis	Maple
Citrus (non-bearing)	Oak
Dogwood	Pieris
Euonymus	Pine
Fir	Podocarpus
Hemlock	Rhododendron
Holly (American and Japanese)	Spruce
Juniper	Viburnum
	Yew (taxus)

NOTE: All flowering bulbs, salvia, phlox, snapdragon, and ornamental pepper are susceptible to injury from an application of GENEP.

For Annual Weed Control—Use GENEP at the rate of 5¾ pints in 10 to 50 gallons of water per acre (2 fl. oz. per 1,000 square feet).

For Quackgrass, Nutgrass, and Bermudagrass Control in Trees and Shrubs Only—Existing stands of these perennial grasses must be turned under and chopped up thoroughly before treatment. Use GENEP at the rate of 7 pints in 10 to 50 gallons of water per acre (2.5 fl. oz. per 1,000 square feet).

For Mugwort (Chrysanthemumweed) Control in the Following Plants: Juniper, Japanese Holly, Ivy, Pachysandra, Petunias—Use 17 pints of GENEP in 10 to 50 gallons of water per acre (6 fl. oz. per 1,000 square feet). Mix thoroughly into the top 6 inches of soil. Apply 4 weeks before desired planting date.

When to Use GENEP

Herbaceous Plants and Ground Covers—Apply 2 weeks after transplanting or after growth starts in the spring.

Trees and Shrubs—Apply 2 weeks before transplanting balled and canned stock (only) and anytime after transplanting. Around established plants, apply after growth starts in the spring.

APPENDIX I

PROCEDURE FOR TESTING THE COMPATIBILITY OF GENEP WITH FLUID FERTILIZERS

The following procedure is suggested for determining whether GENEP may be combined with a specific fluid fertilizer for spray tank application and whether an adjuvant is required.

Materials Required

1. GENEP 7E
2. Fluid fertilizer to be used.
3. Adjuvant for fertilizer tank mix: Compex,* Sponto 168-D,* Unimix,* or equivalent. The adjuvant which provides the best emulsification depends on the specific fertilizer under consideration.
4. Two one-quart, wide-mouth glass jars with lids or stoppers.
5. Measuring spoons (a 25 ml pipette or graduated cylinder provides more accurate measurement).
6. Measuring cup, 8 oz. (237 ml).

Procedure

1. Pour a pint (about 473 ml) of the fluid fertilizer into each of the quart jars.
2. Add adjuvant to one of the jars and mix (See Rate Table).
3. Add the GENEP to both jars (See the Rate Table).
4. Close both jars with lids or stoppers and mix the contents by turning the jars upside down ten times.
5. Inspect the surface and body of the mixtures:
   A. Five minutes after completing the jar inversions.
   B. Two hours later after repeating the jar inversions.

A mixture may not be compatible if either of the following conditions are observed at either inspection period:

1. An oil layer or large oil globules are seen at the surface of the mixture.
2. Clumps or aggregates are present.

If the mixture has separated at the five-minute inspection period, but mixes readily with agitation, the combination may be used PROVIDED good agitation is maintained throughout the mixing and application operations. If the oil layer cannot be redispersed with agitation, or clumps persist, the mixture is incompatible and should not be used.

If the GENEP is incompatible with the fertilizer, carefully inspect the mixture containing the adjuvant. If this mixture is compatible, then the GENEP may be used with the fluid fertilizer provided that adjuvant is added to the fluid fertilizer and thoroughly mixed before adding the GENEP. Add the adjuvant at the rate of 3 pints per 100 gallons of fluid fertilizer. Minimize foaming by using moderate agitation.

If the mixture with the adjuvant is also incompatible, then GENEP should not be used in the same tank with the fertilizer.

* Compex-Kalo Laboratories, Inc., Petaluma, CA; Sponto 168-D—Witco Chemical Company, Houston, TX: Uni-Mix—Thompson Hayward Chemical Co., Kansas City, KS

Rate Table for GENEP and Adjuvant With the Fluid Fertilizer

Gallons of Fluid Fertilizer to be applied per acre	Ml. or Tsp. of GENEP 7E to be Added to 1 Pint of Fertilizer	
	ML.	TSP.
10	7	1⅓
15	4	¾
20	3	⅔
25	3	⅔
30	2	½
40	2	½

* Based on field rate of 1 pound active ingredient per acre in the fertilizer volumes indicated. Increase volume proportionately to correspond with intended field rate in terms of pounds of active ingredient per acre (e.g. for field rate of 4 pounds actual GENEP in 40 gallons fertilizer per acre, add 8 ml. or 2 tsp. GENEP to each jar for compatibility testing purposes).

** Two (2) milliliters or one-half (½) teaspoon of adjuvant to be added to 1 pint of fluid fertilizer in order to equal the rate of 3 pints of adjuvant per 100 gallons of fluid fertilizer.

APPENDIX II

IMPREGNATION ON DRY BULK FERTILIZERS

CAUTION: GENEP alone or in combination with other herbicides must not be impregnated on ammonium nitrate, sodium nitrate, potassium nitrate, or blended fertilizers containing these nitrates. Nitrate fertilizers represent a potential explosive and fire hazard, particularly in contact with organic substances.

GENEP may be impregnated on many dry bulk fertilizers and applied and incorporated in the soil before planting for the control of grass and broadleaf weeds in corn.

Field results have shown that GENEP on bulk dry fertilizers gives weed control equal to GENEP applied as a spray in water or liquid fertilizer. However, uniform impregnation of GENEP on dry fertilizer particles and uniform application in the field are necessary to assure good results.

For impregnating GENEP on dry fertilizers, use a closed rotary-drum mixer or a similar type of closed blender equipped with suitable spray equipment. The spray nozzle (or nozzles) should be positioned inside of the mixer to provide uniform spray coverage of the tumbling fertilizer. The physical properties of fertilizers vary, particularly in liquid absorptive capacity.

When absorptivity is sufficient, simple spray impregnation of the fertilizer with GENEP provides a satisfactory, dry mixture. If the absorptive capacity is inadequate, the use of a highly absorptive powder is required to provide a dry, flowable mixture. Microcel E (Johns-Manville Products Corp.) is the recommended absorbent powder. It should be added separately and uniformly to the prepared GENEP fertilizer mixture, in a quantity that is sufficient to provide a suitably flowable mixture. Generally less than 2% by weight of Microcel E is required. The amount of GENEP actually required in the percent manufacture of individual fertilizer mixtures should be determined carefully for each production operation. This is necessary to ensure that the amount of GENEP actually contained in the mixture applied to the soil represents the correct rate of use.

All GENEP supplementary literature instruction and label recommendations regarding rates per acre, soil incorporation, application, cautions, general use precautions, and other directions must be followed.

All individual state regulations governing bulk dry fertilizer impregnation and application of impregnated fertilizer must be observed and followed.

Bulk fertilizer impregnated with GENEP should be applied immediately, not stored. It is recommended that all bulk containers be tightly covered while the product is being transported and applied to reduce chances of GENEP loss via volatilization.

Approved Dry Fertilizer Ingredients

	N	P	K
Ammonium Sulfate	21	0	0
Diammonium phosphate	18	46	0
Potassium chloride	0	0	60
Potassium sulfate	0	0	52
Super-phosphate (single)	0	20	0
Triple super-phosphate	0	46	0
Urea	45	0	0
Ammonium phosphate-sulfate	16	20	0
11-48-0	11	48	0

NOTE: K-mag has been shown to be compatible with GENEP and is approved for use.

Rate Chart for the Impregnation of Dry Bulk Fertilizers with GENEP 7E

Fertilizer Rate Per Acre	GENEP 7E RATES PER ACRE 3½ Pts. per acre		4½ Pts. per acre		7 pts. per acre	
200 lbs.	17½	qts./ton	22¼	qts./ton	35	qts./ton
250 lbs.	14	qts./ton	18	qts./ton	28	qts./ton
300 lbs.	11⅔	qts./ton	15	qts./ton	23⅓	qts./ton
350 lbs.	10	qts./ton	12½	qts./ton	20	qts./ton
400 lbs.	8¾	qts./ton	11¼	qts./ton	17⅕	qts./ton
450 lbs.	7¾	qts./ton	10	qts./ton	15⅕	qts./ton
500 lbs.	7	qts./ton	9	qts./ton	14	qts./ton
550 lbs.	6⅓	qts./ton	8⅕	qts./ton	12⅔	qts./ton
600 lbs.	5⅞	qts./ton	7½	qts./ton	11¾	qts./ton
650 lbs.	5⅖	qts./ton	7	qts./ton	10⅘	qts./ton
700 lbs.	5	qts./ton	6⅖	qts./ton	10	qts./ton

APPENDIX III

CONSTANT FLOW DEVICE FOR GENEP IN WATER-RUNS

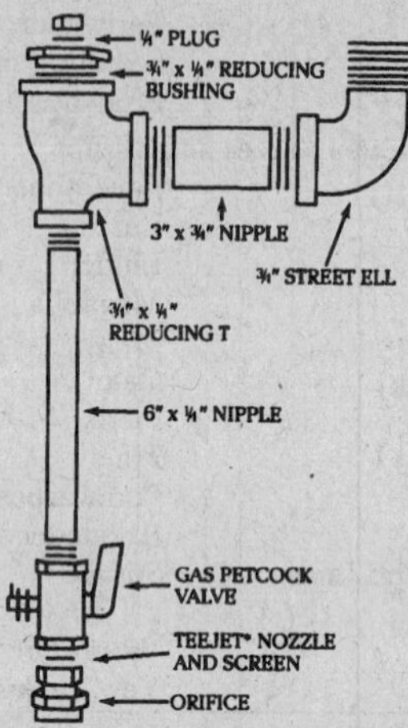

HOW TO FIGURE WHICH ORIFICE TO USE:

Step One—Choose GENEP broadcast rate per acre in pints. Multiply the number of pints times .875 pounds. The results of this calculation will give pounds per acre broadcast. Plug pounds per acre broadcast into equation in Step Two.

Step Two—Calculate pounds per hour using following formula:

$$\frac{\text{Pounds per acre Broadcast (from Step One)} \times \text{Acres to be treated}}{\text{Hours of Irrigation}} = \text{Pounds per hour}$$

Step Three—Refer to the Flow Rate Chart column entitled "Pounds per Hour." Match the number in this column to pounds per hour calculated in Step Two. Read across to the orifice column to determine the proper orifice size. Insert orifice and measure flow rate (column marked cc Per Minute) to assure accuracy.

Flow Rates for GENEP Using Various TeeJet* Orifices**

See Table Following

TeeJet Orifice	Ounces Per Minute	cc Per Minute	Gallons Per Hour	Pounds Per Hour
.012	0.215	6.37	0.101	0.707
.014	0.286	8.45	0.134	0.938
.015	0.324	9.59	0.152	1.064
.016	0.375	11.10	0.176	1.232
.018	0.523	15.46	0.245	1.715
.020	0.610	18.04	0.286	2.002
.022	0.796	23.53	0.373	2.611
.024	0.896	26.50	0.420	2.940
.025	0.996	29.46	0.467	3.269
.026	1.111	32.87	0.521	3.647
.027	1.269	37.54	0.595	4.165
.029	1.284	37.98	0.602	4.214
.030	1.502	44.42	0.704	4.928
.032	1.641	48.52	0.769	5.383
.034	1.871	55.33	0.877	6.139
.035	2.091	61.83	0.980	6.860
.037	2.223	65.74	1.042	7.294
.039	2.539	75.08	1.190	8.330
.040	2.603	76.97	1.220	8.540
.041	2.807	83.03	1.316	9.212
.043	2.882	85.24	1.351	9.457
.045	3.334	98.61	1.563	10.941
.046	3.441	101.77	1.613	11.291
.047	3.678	108.77	1.724	12.068
.048	3.951	116.84	1.852	12.964
.051	4.102	121.32	1.923	13.461
.052	4.437	131.42	2.083	14.581

.054	4.849	143.41	2.273	15.911
.055	5.079	150.22	2.381	16.667
.057	5.333	157.73	2.500	17.500
.059	5.926	175.27	2.778	19.446
.063	6.272	185.49	2.940	20.580
.067	7.110	210.28	3.333	23.331
.070	8.205	242.65	3.846	26.922

* Registered Trademark of Spraying Systems Co.

** Figures were taken at 70°F and are approximate. Be sure to occasionally measure flow in the field to make certain you have the correct orifice, because rates vary with temperature (flow on an .037 orifice increases from 2.2 ozs. at 70°F to 2.4 at 92°F). Use a 300-mess screen on orifice sizes below .014 and a 200 mesh screen on all others.

CONDITIONS OF SALE

The statements and methods presented about the product mentioned are based upon the best available information and practices known to PPG Industries at the present time, but except as expressly set forth in the performance warranty, are not representations of performance, result, or comprehensiveness of such data.

The products mentioned herein, if not used in acocordance with directions, can be hazardous. PPG Industries recommends that anyone using and/or handling the products mentioned herein, before using them, thoroughly read and understand the specific directons for use and the precautions appearing on the product label.

The products mentioned herein, as all potentially dangerous materials, must be kept out of reach of children.

Emergency telephone number: PPG Industries, Inc., Natrium, WV (304) 843-1300.

SPROUT NIP®

AEROSAL GRADE
POTATO SPROUT INHIBITOR

ACTIVE INGREDIENT:

Isopropyl *m* -chlorocarbanilate	46%
INERT INGREDIENTS:	54%
TOTAL	100%

Contains 4 Pounds Active Ingredient Per Gallon

EPA Reg No. 748-178
EPA Est. No. 748-OH-01 EPA Est. No. 32761-MO-3

Manufactured Under U S Pat No. 2,695,225

CAUTION
Keep Out of Reach of Children

HARMFUL IF SWALLOWED
DO NOT TAKE INTERNALLY
ONE GALLON OF SPROUT NIP AEROSAL GRADE WILL TREAT 2,400 CWT OF POTATOES (120 TONS)

HAZARD TO HUMANS AND DOMESTIC ANIMALS

Harmful if swallowed.
Avoid contact with skin, eyes, or clothing.
Avoid breathing of vapor or spray mist.

PRACTICAL TREATMENT

If swallowed: Induce vomiting by placing fingers in back of throat. Get immediate medical attention. Never give anything by mouth to an unconscious person. +

If in eyes: Flush with plenty of water for at least 15 minutes and see a doctor.

If on skin: Wash with plenty of soap and water and see a doctor.

See right panel for additional precautionary statements

DIRECTIONS FOR USE

It is a violation of Federal law to use this product in a manner inconsistent with the label.

CAUTION!

ENVIRONMENTAL HAZARD

Do not apply directly to water. Do not contaminate water by cleaning of equipment or disposal of wastes.

NOTICE

SPROUT NIP Inhibitor is used as an aerosol for treating potatoes for sprout inhibition during storage.

Do *not* apply in the field.

Do not use on seed potatoes. This inhibition of sprouting at recommended rates is usually effective up to a year regardless of removal from storage.

Do not allow vapors to come in contact with, or get near to, storage areas used for seed potatoes.

Let six months elapse before using treated storage area for seed potatoes.

SPROUT NIP will prevent wound periderm formation of potatoes, therefore, it should be used only after bruises and cuts have healed (normally minimum of two weeks.)

FORCED AIR DISTRIBUTION METHOD

1. Assemble unit as shown: Insert aerosol generator intake hose into SPROUT NIP container.
2. Set air ducts for recirculation.
3. Place exhaust end of aerosol generator at center of plenum (air mixing chamber) pointing it in direction of air flow. This will assure the best possible distribution of SPROUT NIP throughout the duct system.

TREATMENT OF STORAGES OR OTHER AREAS THAT DO NOT HAVE RECIRCULATING AIR SYSTEMS

Prior to placing the potatoes in the area to be treated, make the following preparations.

1. On the floor of the area, install an air duct approximately 12 inches by 12 inches running the length or width of the potato pile leaving a false wall space at both ends for air circulation. The ducts should be spaced 10 to 12 feet apart and can be perforated metal pipe, slotted wood construction or if the potatoes are in bags, by bridging a 12 inch space between two rows of bags in the bottom layer with bags placed crosswise the space.
2. At the end of each duct in the false wall space where the fog is to be introduced, place a squirrel cage fan positioned to force air through the duct. The exhaust end of the duct must be blocked to force the air up through the piled potatoes.

When the area is filled and ready to treat, the following steps should be taken:

1. Close off any ventilating systems.
2. Start the squirrel cage fans.
3. Introduce the fog as near as possible to the bottom of the false wall space containing the fans.
4. Operate the fans until the fog has settled.
5. Reactivate the ventilation system 48 hours after treatment.

NOTE: When treating small areas such as trailer trucks or railroad cars, it is recommended that low volume aerosol generators such as a "Swing Fog" be used.

APPLICATION

Application of SPROUT NIP can be made anytime after the curing period and before sprouting of potatoes occurs.

1. Apply at rate of 1 lb. active ingredient per 1,000 bushels (600 cwt.). One gallon treats 4,000 bushels (2,400 cwt.). Treat according to volume of storage (see table below for conversion factors).
2. Use FORCED AIR recirculation through the pile at rate of ½ CFM per bushel of potatoes (.8 CFM per cwt. or potatoes).
3. Keep storage closed during application and of the following 48 hours, and continue to recirculate air in storage until fog has settled. Stop recirculating air until need for normal storage operation requires air flow through the pile.

   1 cwt. = 1.67 bu. = 2.5 cubic feet
   1 bu. = 60 lbs. = 1.5 cubic feet

Treated storages should be monitored in accordance with PPG Industries, Inc. recommended residue sampling procedure.

Do not re-treat the storage until the potatoes have been analyzed for SPROUT NIP residue.

When the SPROUT NIP residue level decreases to 4.0 parts per million, re-treat the storage at the rate of 1 gallon per 2,400 cwt.

NOTE: All applicable directions, restrictions, and precautions on the EPA registered label are to be followed. This label must be in possession of the user at the time of application.

STORAGE AND DISPOSAL

Keep container closed. Do not contaminate water, food, or feed by storage or disposal. This product inhibits germination of seed potatoes.

Avoid storage below freezing temperatures. Prolonged storage may cause crystal formation. Warming contents and agitation will restore material to usuable condition, but do not heat above 250°F.

Any SPROUT NIP or rinse water that cannot be used according to label instructions must be disposed of according to applicable Federal, state or local procedures.

Do not reuse container. Triple rinse, then puncture and dispose of in a sanitary landfill, or by other approved state and local procedures.

CONDITONS OF SALE

PPG Industries, Inc., the manufacturer, warrants that the product conforms to its chemicl description and is reasonably fit for the purpose stated on the label when used in accordance with directions under normal conditions of use, but neither this warranty nor any other warranty of merchantability or fitness for a particular purpose, express or implied, extends to the use of this product contrary to label instructions, or under abnormal conditions or under conditions not reasonably foreseeable to manufacturer, and buyer assumes the risk of any such use.

SUPPLEMENTAL LABELING

SPROUT NIP AEROSOL GRADE

For distribution and use only within the states of Idaho (EPA SLN NO. ID-81005), Oregon (EPA SLN NO. OR-810020), and Washington (EPA SLN NO. WA-800086).

DIRECTIONS FOR USE

POTATO SPROUT INHIBITION IN FORCED AIR OR REFRIGERATED STORAGE

Application rates of SPROUT NIP Aerosol Grade are dependent on storagte conditions and anticipated length of storage. The recommended maximum rates of various conditions are:

Storage for 2 to 3 months at 45°F
 1 gallons per 2,400 cwt.
Storage for 4 to 6 months at 50°F
 2 gallons per 2,400 cwt.
Storage under conditions intermediate to the above
 1 to 2 gallons per 2,400 cwt.

SPROUT NIP®

EMULSIFIABLE CONCENTRATE POTATO SPROUT INHIBITOR

ACTIVE INGREDIENT:	
Isopropyl *m* -chlorocarbanilate	36%
INERT INGREDIENTS:	64%
TOTAL	100%

Contains 3 pounds active ingredient per gallon

EPA Reg. No. 748-182
EPA Est. No. 748-OH-01 EPA Est. No. 32761-MO-3

CAUTION

Keep Out of Reach of Children

HARMFUL IF SWALLOWED

DO NOT TAKE INTERNALLY

ONE GALLON OF SPROUT NIP AEROSOL GRADE WILL TREAT 2,880 CWT OF POTATOES.

HAZARD TO HUMANS AND DOMESTIC ANIMALS

Harmful if swallowed.
Avoid contact with skin, eyes, or clothing.
Avoid breathing of vapor or spray mist.

PRACTICAL TREATMENT

If swallowed: Induce vomiting by placing fingers in back of throat. Get immediate medical attention. Never give anything by mouth to an unconscious person.

If in eyes: Flush with plenty of water for at least 15 minutes and see a doctor.

If on skin: Wash with plenty of soap and water and see a doctor.

See Right Panel For Additional Precautionary Statements

DIRECTIONS FOR USE

It is a violation of Federal Law to use this product in a manner inconsistent with the label.

CAUTION!

ENVIRONMENTAL HAZARD

Do not apply directly to water. Do not contaminate water by cleaning of equipment or disposal of wastes.

NOTICE

SPROUT NIP Emulsifiable Concentrate is used for treating potatoes after storage and washing. SPROUT NIP Emulsifiable Concentrate will prevent wound periderm formation of potatoes, therefore, it should be used only after bruises and cuts have healed, (normally minimum of two weeks).

Do Not apply in field.

Do not use on seed potatoes. This inhibition of sprouting at recommended rates is usually effective up to a year. Wash equipment thoroughly with soap and water before using for seed potatoes.

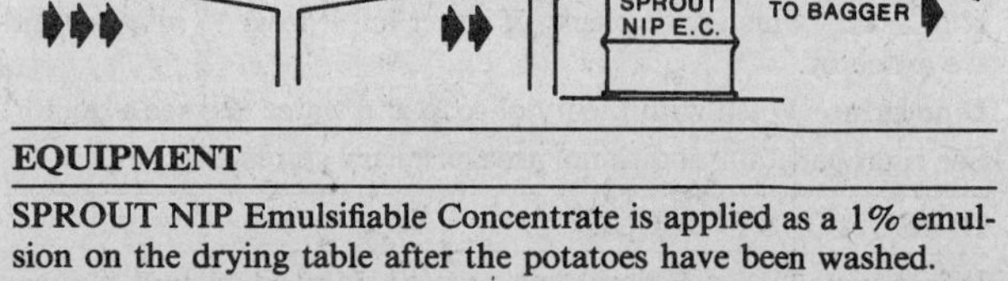

EQUIPMENT

SPROUT NIP Emulsifiable Concentrate is applied as a 1% emulsion on the drying table after the potatoes have been washed.

Spray nozzle should be set to spray evenly across rolls moving the potatoes.

Dryer rolls will distribute SPROUT NIP Emulsifiable Concentrate to assure complete coverage of each potato.

APPLICATION

Dilute SPROUT NIP Emulsifiable Concentrate to a 1% emulsion by adding 1 gallon of SPROUT NIP Emulsifiable Concentrate to 35 gallons of water. Agitate enough to form a uniform emulsion.

Apply at rate of 1 quart of 1% emulsion per 20 bags of potatoes (100 pounds per bag).

STORAGE AND DISPOSAL

Keep container closed. Do not contaminate water, food, or feed by storage or disposal. This product inhibits germination of seed potatoes.

Avoid storage below freezing temperatures. Prolonged storage may cause crystal formation. Warming contents and agitation will restore material to usable condition, but do not heat above 250° F.

Any SPROUT NIP or rinse water that cannot be used according to label instructions must be disposed of according to applicable Federal, state, or local procedures.

Do not reuse container. Triple rinse, then puncture and dispose of in a sanitary landfill, or by other approved state and local procedures.

CONDITIONS OF SALE

PPG Industries, Inc., the seller, warrants that the product conforms to its chemical description and is reasonably fit for the purpose stated on the label when used in accordance with directions under normal conditions of use, but neither this warranty nor any other warranty of merchantability or fitness for a particular purpose, express or implied, extends to the use of this product contrary to label instructions, or under abnormal conditions, or under conditions not reasonably foreseeble to seller, and buyer assumes the risk of any such use.

 L60B-1B-782C

SPROUT NIP® 7A

AEROSOL GRADE
POTATO SPROUT INHIBITOR

ACTIVE INGREDIENT:

Isopropyl m-chlorocarbanilate	78.5%
INERT INGREDIENTS:	21.5%
TOTAL	100.0%

Contains 7 Pounds Active Ingredient Per Gallon

Contains methanol

E.P.A. Reg. No. 748-234

E.P.A. Est. 748-OH-01 E.P.A. Est. 32761MO-3

ONE GALLON OF SPROUT NIP AEROSOL GRADE WILL TREAT 4200 CWT OF POTATOES (210 TONS)

DANGER–POISON

Keep Out of Reach of Children

DANGER!

HARMFUL IF SWALLOWED

DO NOT TAKE INTERNALLY

HAZARDS TO HUMANS AND DOMESTIC ANIMALS

DANGER!

Methanol may cause blindness.

Fatal if swallowed.

Avoid breathing spray mist or vapor.

Avoid contact with skin, eyes, or clothing.

PRACTICAL TREATMENT

IF SWALLOWED: Induce vomiting by placing fingers in back of throat. Get immediate medical attention. Never give anything by mouth to an unconscious person.

IF IN EYES: Flush with plenty of water for at least 15 minutes and see a doctor.

IF ON SKIN: Wash with plenty of soap and water and see a doctor if irritation persists.

See Right Panel For Additional Precautionary Statements

DIRECTIONS FOR USE

It Is A Violation Of Federal Law To Use This Product In A Manner Inconsistent With Its Labeling.

PRECAUTIONARY STATEMENTS

ENVIRONMENTAL HAZARDS

Do not apply directly to water. Do not contaminate water by cleaning of equipment or disposal of wastes.

PHYSICAL AND CHEMICAL HAZARDS

All combustible aerosols present the possibility of explosion hazard in confined areas if the lower limit for combustion is exceeded. Keep the concentration of SPROUT-NIP® aerosol below 3.7 gallons of formulation per 10,000 SCF of air. Consult the PPG SPROUT NIP Aerosol Guide for further information.

Flammable. Do not use or store near heat or open flame.

STORAGE AND DISPOSAL

Keep container closed. Do not contaminate water, food, or feed by storage or disposal. This product inhibits germination of seed potatoes.

Avoid storage below freezing temperatures. Storage below 35°F may cause crystal formation. Warming contents and agitation will restore material to usable condition, but do not heat above 140°F. At 140°F the formulation will lose methanol, the vapors of which are flammable. Notice: Vent container before heating to prevent pressure build-up.

Any SPROUT NIP or rinse water that cannot be used according to label instructions must be disposed of according to approved state procedures under Subtitle C of the Resource Conservation and Recovery Act.

Do not reuse container. Triple rinse, then offer for recycling or reconditioning, or puncture and dispose of in a sanitary landfill, or by other approved state and local procedures.

NOTICE

SPROUT NIP® is used as an aerosol for treating potatoes for sprout inhibition during storage.

Do *not* apply in the field.

Do not use on seed potatoes. This inhibition of sprouting at recommended rates is usually effective up to a year regardless of removal from storage.

Do not allow vapors to come in contact with, or get near to, storage areas used for seed potatoes.

Let six months elapse before using treated storage area for seed potatoes.

SPROUT NIP will prevent wound periderm formation of potatoes, therefore, it should be used only after bruises and cuts have healed (normally minimum of two weeks.)

FORCED AIR DISTRIBUTION METHOD

1. Assemble unit as shown: Insert aerosol generator intake hose to SPROUT NIP container.
2. Set air ducts for recirculation.
3. Place exhaust end of aerosol generator at center of plenum (air mixing chamber) pointing it in direction of air flow. This will assure the best possible distribution of SPROUT NIP throughout the duct system.

TREATMENT OF STORAGES OR OTHER AREAS THAT DO NOT HAVE RECIRCULATING AIR SYSTEMS

Prior to placing the potatoes in the area to be treated, make the following preparations:

1. On the floor of the area, install an air duct approximately 12 inches by 12 inches running the length or width of the potato pile leaving a false wall space at both ends for air circulation. The ducts should be spaced 10 to 12 feet apart and can be perforated metal pipe, slotted wood construction or if the potatoes are in bags, by bridging a 12 inch space between two rows of bags in the bottom layer with bags placed crosswise the space.
2. At the end of each duct in the false wall space where the fog is to be introduced, place a squirrel cage fan positioned to force air through the duct. The exhaust end of the duct must be blocked to force the air up through the piled potatoes.

When the area is filled and ready to treat, the following steps should be taken:

1. Close off any ventilating systems.
2. Start the squirrel cage fans.
3. Introduce the fog as near as possible to the bottom of the false wall space containing the fans.
4. Operate the fans until the fog has settled.
5. Reactivate the ventilation system 48 hours after treatment.

NOTE: When treating small areas such as trailer trucks or railroad cars, it is recommended that low volume aerosol generators such as a "Swing Fog" be used.

APPLICATION

Application of SPROUT NIP can be made anytime *after* the curing period and before sprouting of potatoes occurs.

1. Apply at rate of 1 lb. active ingredient per 1,000 bushels (600 cwt). One gallon treats 7,000 bushels (4,200 cwt). Treat according to volume of storage (see table below for conversion.
2. Use FORCED AIR recirculation through the pile at rate of ½ CFM per bushel of potatoes (.8 CFM per cwt. of potatoes).
3. Keep storage closed during application and for the following 48 hours, and continue to recirculate air in storage until fog has settled. Stop recirculating air until need for normal storage operation requires air flow through the pile.

1 cwt. = 1.67 bu. = 2.5 cubic feet
1 bu. = 60 lbs. = 1.5 cubic feet

CONDITIONS OF SALE

PPG Industries, Inc., the manufacturer, warrants that the product conforms to its chemical description and is reasonably fit for the purpose stated on the label when used in accordance with directions under normal conditions of use, but neither this warranty nor any other warranty of merchantability or fitness for a particular purpose, express or implied, extends to the use of this product contrary to label instructions, or under abnormal conditions or under conditions not reasonably foreseeable to manufacturer, and buyer assumes the risk for any such use.

Rhône-Poulenc Inc.

Agrochemical Division
Monmouth Junction, NJ 08852
(201) 297-0100

ASULOX®

HERBICIDE

For Postemergent Weed Control in Sugarcane, Turf, Ornamentals, Reforestation Areas (Site Preparation Areas and Conifer Release Areas), Christmas Tree Plantings, and Noncropland.

ACTIVE INGREDIENT:
Sodium salt of asulam
(methyl sulfanilylcarbamate)* 36.2%
INERT INGREDIENTS 63.8%
TOTAL 100.0%

* Equivalent to 33.1% asulam or not less than 3.34 lbs. asulam per gallon.

EPA Reg. No. 359-662 EPA Est.

CAUTION

Keep Out of Reach of Children

STATEMENT OF PRACTICAL TREATMENT

IF ON SKIN: Wash with plenty of soap and water. If irritation persists get medical attention.

IF IN EYES: Flush with water for 15 minutes and get medical attention if irritation persists.

See Side Panel for Additional Precautionary Statements.

PRECAUTIONARY STATEMENTS

HAZARDS TO HUMANS AND DOMESTIC ANIMALS

CAUTION

Avoid contact with skin, eyes or clothing. Wash thoroughly after handling.

ENVIRONMENTAL HAZARDS

Do not apply directly to lakes, streams or ponds. Do not apply when weather conditions favor drift from treated area. Do not contaminate water by cleaning of equipment or disposal of wastes.

DIRECTIONS FOR USE

It is a violation of Federal Law to use this product in a manner inconsistent with its labeling.

STORAGE AND DISPOSAL

Do not contaminate water, food, or feed by storage or disposal. Open dumping is prohibited. Do not reuse empty container. Store at temperatures above 20°F.

PESTICIDE DISPOSAL

Wastes resulting from the use of this product may be disposed of on site or at an approved waste disposal facility.

CONTAINER DISPOSAL

Triple rinse (or equivalent). Then offer for recycling or reconditioning, or puncture and dispose of in a sanitary landfill, or incineration, or, if allowed by state and local authorities, by burning. If burned, stay out of smoke.

SUGARCANE

ASULOX can be applied to either plant cane or cane grown from stubble. Apply ASULOX as a water-mix spray for ground applications. Use 15 to 100 gallons of water per acre, depending on local practice. For aerial application, ASULOX should be mixed in 3 to 5 gallons of water per acre, except in Hawaii, where 5 to 10 gallons of water per acre should be used. A surfactant may be added to the spray solution at 0.25% by volume. (Use an approved nonionic surfactant for agricultural uses.)

The rates of ASULOX given below are for broadcast applications. For banded application reduce the rate proportionally to the width of the band according to the following formula:

$$\frac{\text{Band width (inches)}}{\text{Row width (inches)}} \times \text{Broadcast rate} = \text{Band rate/acre.}$$

For spot treatments use 6 to 12 pints of ASULOX in 100 gallons of water and apply in an amount not to exceed 50 gallons of total spray solution per acre. (Do not exceed 8 pints per acre per treatment.)

ASULOX should be used when the weeds are actively growing.

Do not plant crops other than sugarcane in treated area. Cover crops may be planted if plowed under and not grazed.

Do not graze or feed sugarcane fodder and forage to livestock.

Sugarcane
For Single Application per Growing Season:

Weed Species	Special Instructions	Rate
Itchgrass or Raoulgrass *(Rottboellia exaltata)*	Apply when the grass is 12 inches tall or less (addition of surfactant is necessary).	8 Pints/ Acre
Johnsongrass *(Sorghum halepense)*	Apply when the grass is between 18 and 24 inches tall.	
Paragrass or Californiagrass *(Brachiaria mutica = Panicum purpurascens)*	Apply when the grass is 6 to 8 inches tall or less.	
Crabgrass *(Digitaria spp.)*	If treatment is made before the grass reaches seed head formation then the lower rate should be used. If the grass is in early seed head formation then the higher rate should be used.	6 to 8 Pints/ Acre
Alexandergrass *(Brachiaria plantaginea)* Foxtail *(Setaria spp.)* Goosegrass *(Eleusine indica)* Broadleaf Panicum *(Panicum adspersum)* Barnyardgrass *(Echinochloa crus-galli)*	If treatment is made when the grass is 6 to 8 inches tall or less then the lower rate should be used. If the grass is greater than 8 inches tall then the higher rate should be used.	

REFORESTATION AREAS (Site Preparation Areas and Conifer Release Areas) AND CHRISTMAS TREE PLANTINGS

ASULOX may be used as a postemergent treatment in Reforestation Areas and Christmas Tree Plantings where Douglas Fir, Grand Fir, Nobel Fir or Scotch Pine are grown. Do not graze or feed foliage from treated areas to livestock.

ASULOX should be applied as a water-mix spray. For ground application use a minimum of 20 gallons of solution per acre, and for aerial application use a minimum of 10 gallons of solution per acre. Do not use a wetting agent with ASULOX.

Weed Species	Special Instructions	Rate
Western Bracken *(Pteridium aquilinum var. pubescens var. pubescens)*	Apply after bud break and hardening or firming of new tree growth. Bracken should be in full frond prior to treatment.	1 Gal./ Acre

Two Applications per Growing Season:
This may be required when initial weed infestations are heavy or reinfestation from seed occurs. Two applications may also be used when treating weed species which germinate at different times during one growing season.

WEED SPECIES	SPECIAL INSTRUCTIONS	FIRST APPLICATION RATE	SECOND APPLICATION RATE
Crabgrass *(Digitaria spp.)*	At each application the grass should be treated before seed head formation.	5 to 8 Pints/Acre	5 to 8 Pints/Acre
Itchgrass or Raoulgrass *(Rottboellia exaltata)*	At each application the grass should be 12 inches tall or less. (Addition of surfactant is necessary.)	8 Pints/Acre	8 Pints/Acre
Johnsongrass *(Sorghum halepense)*	At each application the grass should be between 18 and 24 inches tall.	8 Pints/Acre	4 to 8 Pints/Acre

Do NOT treat sugarcane within 90 days of harvest.

NONCROPLAND

ASULOX may be used as a postemergent treatment to control weeds on noncropland areas such as:

Boundary fences
Ditchbanks
Fence rows
Highway and roadside rights-of-way
Lumberyards
Pipeline rights-of-way
Railroad rights-of-way and yards
Storage areas and industrial plant sites
Utility rights-of-way and yards
Warehouse lots

A surfactant may be added to the spray solution at 0.25% by volume. (Use an approved non-ionic surfactant.)
Apply ASULOX as a water-mix spray for ground applications using 20 to 100 gallons of solution per acre, depending on local practice to control the following weed species:

Weed Species	Special Instructions	Rate
Crabgrass *(Digitaria spp.)*	Apply before the grass reaches seed head formation.	
Johnsongrass *(Sorghum halepense)*	Apply when the grass is 18 inches or taller. Use the higher rate in well established heavy infestations. For spot treatment in Hawaii, use the higher rate in 100 gallons of solution and apply an amount not to exceed 50 gallons of total solution per acre.	**1 to 2 Gal./Acre**
Paragrass or Californiagrass *(Brachiaria mutica = Panicum purpurascens)*	Apply before the grass reaches seed head formation. For spot treatment *in Hawaii, use the same rate* in 100 gallons of solution and apply an amount not to exceed 50 gallons of total solution per acre.	
Western Bracken *(Pteridium aquilinum* var. *pubescens)*	Apply when the fern is in full frond.	**7 to 8 Pints/Acre**

TURF

ASULOX can be applied on St. Augustinegrass and Tifway 419 Bermudagrass turf. Apply one application per season postemergence to the weeds listed below. Use 20 to 50 gallons of water per acre in the spray solution.

Turf Species	Weed Species	Rate
St. Augustinegrass Tifway 419 Bermudagrass	Bullgrass *(Paspalum supinum)* Crabgrass *(Digitaria sp.)* Goosegrass *(Eleusine indica)* Sandbur *(Crenchrus sp.)*	5 Pints/Acre

Do not use a surfactant. Do not apply to turf which is under stress or freshly mowed.

ORNAMENTALS

ASULOX can be applied as a single, postemergent, broadcast application on the following ornamentals:

JUNIPERS	YEWS
Juniperus andorra	*Taxus cuspidata*
Juniperus chinensis	*Taxus media*
Juniperus conferta	*Podocarpus macrophyllus*
Juniperus cuspidata	
Juniperus horizontalis	
Juniperus litoralis	
Juniperus sabina	

Treatment should be made with a minimum of 20 gallons of water per acre. Do not use a surfactant.

Weed Species	Special Instructions	Rate
Barnyardgrass *(Echinochloa crus-galli)* Crabgrass *(Digitaria sp.)* Fall Panicum *(Panicum dichotomiflorum)* Foxtails *(Setaria sp.)* Goosegrass *(Eleusine indica)* Horseweed (Marestail) *(Conyza canadensis)*	Apply when the weeds are between the stages of early seedling and early seed head formation.	1 Gal./Acre

Local conditions may affect the use of this chemical. Consult State Agricultural Extension or Experiment Station weed specialists for specific recommendations for local weed problems and for information on possible lower dosages.

CONDITION OF SALE, WARRANTY, LIMITATION OF LIABILITY

This product conforms to the chemical description on the label thereof and is reasonably fit for the purpose stated on such label only when used in accordance with directions under normal use conditions. Follow directions carefully. Timing and method of application, weather and crop conditions, mixture with other chemicals not specifically recommended and other influencing factors in the use of this product are beyond the control of Rhone-Poulenc or the seller. Buyer assumes all risks of use, storage, or handling of this material, not in strict accordance with directions given herewith. In no case shall Rhone-Poulenc or the seller be liable for consequential, special, or indirect damages such as loss of profits or values resulting from the use or handling of this product.

BRONATE®

HERBICIDE

For the Control of Certain Broadleaf Weeds in Wheat, Barley, Oats and Rye.

ACTIVE INGREDIENT:	
Octanoic acid ester of bromoxynil* (3,5-dibromo-4-hydroxybenzonitrile)	31.7%
Isooctyl ester of 2-methyl-4-chlorophenoxyacetic acid**	34.0%
INERT INGREDIENTS	34.3%
TOTAL	100.0%

* Bromoxynil octanoate. Equivalent to 21.8% of bromoxynil or not less than 2.0 pounds of bromoxynil per gallon.

** Equivalent to 21.8% 2-methyl-4-chlorophenoxyacetic acid or not less than 2.0 lbs. MCPA acid per gallon.

EPA Reg. No. 359-601 EPA Est.

WARNING

Keep Out of Reach of Children

STATEMENT OF PRACTICAL TREATMENT

IF SWALLOWED: Get medical attention. Do not induce vomiting. Do not give anything by mouth to an unconscious person.

IF ON SKIN: Wash with plenty of soap and water. Get medical attention if irritation persists.

IF IN EYES: Flush with water for 15 minutes. Get medical attention.

See Side Panel for Additional Precautionary Statements.

PRECAUTIONARY STATEMENTS

HAZARDS TO HUMANS AND DOMESTIC ANIMALS

WARNING

Harmful if swallowed, absorbed through skin or inhaled. Avoid contact with skin, eyes or clothing. Avoid breathing spray mist. Wash thoroughly after handling. Remove contaminated clothing and wash before reuse.

ENVIRONMENTAL HAZARDS

This pesticide is toxic to wildlife and fish. Use with care when applying to areas frequented by wildlife or adjacent to any body of water. Do not apply directly to water. Do not apply when weather conditions favor drift from target areas. Do not contaminate water by cleaning of equipment or disposal of wastes.

PHYSICAL AND CHEMICAL HAZARDS

Do not use or store near heat or open flame.

DIRECTIONS FOR USE

It is a violation of Federal Law to use this product in a manner inconsistent with its labeling.

STORAGE AND DISPOSAL

Do not contaminate water, food or feed by storage or disposal.

STORAGE

Do not store near fertilizers or seeds. Store at temperatures above 3°F. If allowed to freeze remix before using.

PESTICIDE DISPOSAL

Pesticide wastes are acutely hazardous. Improper disposal of excess pesticide, spray mixture, or rinsate is a violation of Federal Law. If these wastes cannot be disposed of by use according to label instructions, contact your State Pesticide or Environmental Control Agency, or the Hazardous Waste representative at the nearest EPA Regional Office for guidance.

CONTAINER DISPOSAL

Metal: Triple rinse (or equivalent). Then puncture and dispose of in a sanitary landfill, or by other procedures approved by state and local authorities.

Plastic: Triple rinse (or equivalent). Then puncture and dispose of in a sanitary landfill, or incineration, or, if allowed by state and local authorities, by burning. If burned, stay out of smoke.

GENERAL INFORMATION

Bronate is a selective postemergence product for control of broadleaf weeds in small grains.

To broaden the spectrum of control, Bronate may be tankmixed with Glean®, Avenge®, or MCPA. To ensure maximum safety and weed control, follow all cautions and limitations on the label and the labels of products used in the tank mixture with Bronate.

GENERAL WEED LIST

Postemergence application of Bronate will control the following weeds when sprayed in the seedling stage. Maximum weed stage of growth is listed under BRONATE RECOMMENDATIONS.

MOST SUSCEPTIBLE BROADLEAF WEED SPECIES

Annual Sowthistle *(Sonchus oleraceus)*
Black Mustard *(Brassica nigra)*
Black Nightshade *(Solanum nigrum)*
Common Cocklebur *(Xanthium pensylvanicum)*
Common Lambsquarters *(Chenopodium album)*
Common Tarweed *(Hemizonia congesta)*
Cow Cockle *(Saponaria vaccaria)*
Eastern Black Nightshade *(Solanum ptycanthum)*
Coast Fiddleneck *(Amsinckia intermedia)*
Field Pennycress *(Thlaspi arvense)*
Green Smartweed *(Polygonum scabrum)*
Hairy Nightshade *(Solanum sarachoides)*
Jimsonweed *(Datura stramonium)*
Ladysthumb *(Polygonum persicaris)*
London Rocket *(Sisymbrium irio)*
Marshelder *(Iva xanthifolia)*
Pennsylvania Smartweed *(Polygonum pensylvanicum)*
Pepperweed spp. *(Lepidium spp.)*
Redroot Pigweed *(Amaranthus retroflexus)*
Russian Thistle *(Salsola kali)*
Shepherdspurse *(Capsella bursa-pastoris)*
Silverleaf Nightshade *(Solanum elaeagnifolium)*
Smooth Pigweed *(Amaranthus hybridus)*
Spiny Pigweed *(Amaranthus spinosus)*
[1] Sunflower *(Helianthus annuus)*
Tall Waterhemp *(Amaranthus tuberculatus)*
Tartary Buckwheat *(Fagopyrum tataricum)*
Tumble Mustard *(Sisymbrium altissimum)*
Wild Buckwheat *(Polygonum convolvulus)*
Wild Mustard *(Brassica kaber)*
Yellow Rocket *(Barbarea vulgaris)*

[1] For control of sunflower, delay application until first sunflower seedlings emerging are 4 inches in height.

SUSCEPTIBLE BROADLEAF WEED SPECIES

Blue (purple) Mustard *(Chlorispora tenella)*
Common Groundsel *(Senecio vulgaris)*
Common Ragweed *(Ambrosia artemisiifolia)*
Corn Chamomile *(Anthemis arvensis)*
Corn Gromwell *(Lithospermum arvense)*
Fumitory *(Fumaria officinalis)*
Giant Ragweed *(Ambrosia trifida)*
Hemp Sesbania *(Sesbania exaltata)*
[2] Henbit *(Lamium amplexicaule)*
Ivyleaf Morningglory *(Ipomoea hederacea)*
[2] Knawel *(Scleranthus annuus)*
Kochia *(Kochia scoparia)*
[2] Mayweed *(Anthemis cotula)*
Prostrate Knotweed *(Polygonum aviculare)*
Tall Morningglory *(Ipomoea purpurea)*
Tansy Mustard *(Descurainia pinnata)*
Tarweed *(Hemizonia spp.)*
Velvetleaf *(Abutilon theophrasti)*
Wild Radish *(Raphanus raphanistrum)*

[2] Use Bronate at 2 pints/A for control of henbit, knawel and mayweed up

to the 4 leaf stage or 2 inches in height, whichever comes first. Weeds germinating after spraying will not be controlled.

WEED SUPPRESSION

Canada Thistle *(Cirsium arvense)*

Bronate applied at 1½-2 pints per acre provides burn down of top growth. Regrowth may occur. Make applications when Canada Thistle is 8 inches tall to the bud stage.

WHEAT, BARLEY, OATS, AND RYE

Bronate may be applied to wheat and barley from the 3 leaf stage up to the boot stage. Applications should be made to weeds soon after emergence for best weed control and to prevent competition with the crop. Weed control may be reduced when weeds are stressed from lack of moisture or cold temperatures. Avoid spray drift to non-target areas. Do not graze treated field for 30 days following application.

Bronate can be applied to wheat and barley alone or in a tank mixture with Glean®, Avenge®, or MCPA to broaden the spectrum of weed control. Bronate can also be applied to oats and rye alone or in a tank mixture with MCPA. To ensure maximum safety and weed control, follow all cautions and limitations on this label and the labels of products used in tank mixtures with Bronate.

MIXING INSTRUCTIONS

BRONATE ALONE: Fill the spray tank ½ to ¾ full with clean water. Begin agitation and add the recommended amount of Bronate. Add water to the spray tank to the desired level. Maintain sufficient agitation to ensure a uniform spray mixture during application.

TANK MIXTURES: Fill the spray tank ½ to ¾ full with clean water and begin agitation. If tankmixing with Glean, add the recommended amount of Glean first. After Glean is thoroughly mixed with water, add the recommended amount of Bronate and add water to the spray tank to the desired level. If tankmixing with Avenge or MCPA, add the recommended amount of Bronate first. After Bronate is thoroughly mixed with water, add the recommended amount of the other herbicide to the spray tank and add water to the spray tank to the desired level. Maintain sufficient agitation while mixing and during application to ensure a uniform spray mixture.

APPLICATION PROCEDURES

Bronate alone or in a tank mixture with Glean, Avenge, or MCPA can be applied by both ground and aerial equipment.

GROUND APPLICATION

Use a standard herbicide boom sprayer that provides uniform and accurate application. Sprayer should be equipped with flat fan nozzles spaced a maximum of 20 inches apart and screens no finer than 50 mesh in the nozzle tips and in-line strainers.

Applications of Bronate used alone and in tank mixtures should be used in a minimum spray volume of 10 gallons per acre with a minimum spray pressure of 30 psi at the nozzle tips.

AERIAL APPLICATION

Bronate and tank mixtures with Bronate can be applied by aerial equipment. Use a minimum of 5 gallons of spray volume per acre using diaphragm-type nozzles that produce cone or fan spray patterns with a maximum spray pressure of 40 psi at the nozzle tips. Nozzles should be located no farther than ¾ the distance from the center of the aircraft to the end of the wing or rotor. Nozzles should be oriented at an angle between straight down and straight back.

The aircraft should discharge the spray a maximum of 10 feet above the crop. To minimize drift, do not apply during periods of gusty winds or when wind exceeds 5 mph.

SPRAYABLE LIQUID FERTILIZERS

Bronate can be applied to small grains using sprayable liquid fertilizer as the carrier. Predetermine the compatibility with the liquid fertilizer by mixing small proportional quantities in advance. Agitation must be maintained during filling and application operations to ensure that Bronate is evenly mixed with the fertilizer.

RECOMMENDED USES

The recommended rate of Bronate used alone or in a tank mixture with another herbicide is determined by the crop (fall or spring seeded), geographic area, application equipment (ground or aerial application), weed species and stage of growth of weeds at the time of application.

WHEAT AND BARLEY (Ground and Aerial Applications)
BRONATE RECOMMENDATIONS

			TIMING OF APPLICATION	
PRODUCT	CROP/GEOGRAPHICAL AREA	RATE	CROP	WEED
Bronate	Fall seeded wheat and barley throughout the United States and spring seeded wheat and barley in Idaho, Oregon and Washington.	1 Pint/A	Apply to wheat and barley after the 3 leaf stage but before the crop reaches the boot stage.	**Most Susceptible Broadleaf Weeds** Apply to weeds up to the 8 leaf stage or 4 inches in height, whichever comes first. If weed forms rosette, apply before weeds exceed 2 inches in diameter.
		1½ Pints/A		**Susceptible Broadleaf Weeds** Apply to weeds up to the 4 leaf stage or 2 inches in height, whichever comes first. If weed forms rosette, apply before weeds exceed 1 inch in diameter.
		2 Pints/A		Apply to henbit, knawel and mayweed up to the 4 leaf stage or 2 inches in height, whichever comes first.
	Spring seeded wheat and barley except Idaho, Oregon and Washingon.	1 Pint/A		**Most Susceptible and Susceptible Broadleaf Weeds** Apply to weeds that do not exceed 8 leaf stage or 4 inches in height, whichever comes first. If weed forms rosette, apply before weeds exceed 2 inches in diameter. Apply to kochia up to 2 inches in height.
		1½ Pints/A		Apply to control kochia that is 2 to 4 inches in height.
Bronate + Glean (Tankmix)	Fall seeded wheat and barley throughout the United States and spring seeded wheat and barley in Idaho, Oregon and Washington.	¾ Pint/A + ⅙–⅓ ounce/A	Apply to wheat and barley after the 3 leaf stage but before the crop reaches the boot stage.	**Most Susceptible Broadleaf Weeds** Apply to weeds up to the 8 leaf stage or 4 inches in height, whichever comes first. If weed froms rosette, apply before weeds exceed 2 inches in diameter.
		1 Pint/A + ⅙–⅓ ounce/A		**Susceptible Broadleaf Weeds** Apply to weeds up to the 4 leaf stage or 2 inches in height, whichever comes first. If weed forms rosette, apply before weeds exceed 1 inch in diameter.

PRODUCT	CROP/GEOGRAPHICAL AREA	RATE	TIMING OF APPLICATION: CROP	TIMING OF APPLICATION: WEED
	Spring seeded wheat and barley except Idaho, Oregon and Washington.	1 Pint/A + ⅙–⅓ ounce/A		**Most Susceptible and Susceptible Broadleaf Weeds** Apply to weeds that do not exceed 8 leaf stage or 4 inches in height, whichever comes first. If weed forms rosette, apply before weeds exceed 2 inches in diameter. Apply to kochia up to 2 inches in height.
		1½ Pints/A + ⅙–⅓ ounce/A		Apply to control kochia that is 2 to 4 inches in height.

A tankmix application of Bronate and Glean broadens the spectrum of weed control. See Glean label for listing of additional susceptible weed species and crop rotation instructions.
Glean is a registered trademark of E.I. DuPont de Nemours & Co., Inc.

BRONATE and AVENGE

A tank mixture of Bronate + Avenge can be used to control wild oats in addition to the broadleaf weeds controlled by Bronate. The rate and application timing for weed control with Bronate in the tank mixture is the same as when Bronate is used alone. See BRONATE RECOMMENDATIONS.

Apply Bronate + Avenge tank mixtures when the majority of the wild oats are in the 3 to 5 leaf stage of growth. This frequently coincides with barley in the 3 to 7 leaf stage, spring seeded wheat in the 5 to 6 leaf stage, and fall seeded wheat in the 4 leaf to tillered stage of growth. See Avenge label for varietal restrictions and recommended Avenge rates as related to wild oat density.

Avenge is a trademark of American Cyanamid Company.

BRONATE and MCPA

A tank mixture of Bronate at the recommended rate and MCPA at ¼ lb ai/A is recommended when the weed infestations of MCPA susceptible weeds are heavy or the weeds are at the maximum recommended stage of growth for Bronate. The tank mixture will control the weeds listed on the MCPA label in addition to the weeds listed on the Bronate label. The rate of Bronate in the tank mixture is the same as when Bronate is used alone. See BRONATE RECOMMENDATIONS.

A tank mixture of Bronate + MCPA can be applied after the 3 leaf stage but before the crop reaches the boot stage.

OATS AND RYE

(Ground and Aerial Application)
Bronate alone and in a tank mixture with MCPA can be applied to oats and rye. Recommendations for Bronate alone and in a tank mixture with MCPA for use on oats and rye are the same as for wheat and barley.

Local conditions may affect the use of this chemical. Consult State Agricultural Extension or Experiment Station weed specialists for specific recommendations for local weed problems.

CONDITION OF SALE, WARRANTY, LIMITATION OF LIABILITY

This product conforms to the chemical description on the label thereof and is reasonably fit for the purpose stated on such label only when used in accordance with directions under normal use conditions. Follow directions carefully. Timing and method of application, weather and crop conditions, mixture with other chemicals not specifically recommended and other influencing factors in the use of this product are beyond the control of Rhone-Poulenc or the seller. Buyer assumes all risks of use, storage, or handling of this material, not in strict accordance with directions given herewith. In no case shall Rhone-Poulenc or the seller be liable for consequential, special, or indirect damages such as loss of profits or values resulting from the use or handling of this product.

BUCTRIL®
HERBICIDE

For the Control of Certain Broadleaf Weeds in Small Grains, Corn, Sorghum, Flax, Garlic, Turf and Other Noncrop Areas.

ACTIVE INGREDIENT:
Octanoic acid ester of bromoxynil*
(3,5-dibromo-4-hydroxybenzonitrile) 33.4%
INERT INGREDIENTS 66.6%
TOTAL 100.0%

*Bromoxynil octanoate equivalent to 22.9% of bromoxynil or not less than 2.0 pounds of bromoxynil per gallon.

EPA Reg. No. 359-564

Keep Out of Reach of Children
WARNING—AVISO

PRECAUCION AL USUARIO: Si usted no lee inglés, no use este producto hasta que le etiqueta haya sido explicada ampliamente.

STATEMENT OF PRACTICAL TREATMENT
IF SWALLOWED: Get medical attention. Do not induce vomiting. Do no give anything by mouth to an unconscious person.

IF ON SKIN: Wash with plenty of soap and water. Get medical attention if irritation persists.

IF IN EYES: Flush with water for 15 minutes. Get medical attention.

See Side Panel for Additional Precautionary Statements.

PRECAUTIONARY STATEMENTS

HAZARDS TO HUMANS AND DOMESTIC ANIMALS
WARNING
Harmful if swallowed, absorbed through skin or inhaled. Avoid contact with skin, eyes or clothing. Avoid breathing spray mist. Wash thoroughly after handling. Remove contaminated clothing and wash before reuse.
ENVIRONMENTAL HAZARDS
This pesticide is toxic to wildlife and fish. Use with care when applying to areas frequented by wildlife or adjacent to any body of water. Do not apply directly to water. Do not apply when weather conditions favor drift from target areas. Do not contaminate water by cleaning of equipment or disposal of wastes.
PHYSICAL AND CHEMICAL HAZARDS
Do not use or store near heat or open flame.

DIRECTIONS FOR USE

It is a violation of Federal Law to use this product in a manner inconsistent with its labeling.
REENTRY STATEMENT
Do not apply this product in such a manner as to directly or through drift expose workers or other persons. The area being treated must be vacated by unprotected persons.

Do not enter treated areas without protective clothing until sprays have dried.

Because certain states may require more restrictive reentry intervals for various crops treated with this product, consult your State Department of Agriculture for further information.

Written or Oral warnings must be given to workers who are expected to be in a treated area or in an area about to be treated with this product. Oral warnings must be given if there is reason to believe that written warnings cannot be understood by workers and given in a language customarily understood by the workers.

Oral and Written warnings must include the Statement of Practical Treatment and Reentry Statements as they appear on the label.

STORAGE AND DISPOSAL

Do not contaminate water, food or feed by storage or disposal.

STORAGE

Do not store near fertilizers or seeds. Store at temperatures above 3°F. If allowed to freeze remix before using.

PESTICIDE DISPOSAL

Pesticide wastes are acutely hazardous. Improper disposal of excess pesticide, spray mixture, or rinsate is a violation of Federal Law. If these wastes cannot be disposed of by use according to label instructions, contact your State Pesticide or Environmental Control Agency, or the Hazardous Waste representative at the nearest EPA Regional Office for guidance.

CONTAINER DISPOSAL

Metal: Triple rinse (or equivalent). Then puncture and dispose of in a sanitary landfill, or by other procedures approved by state and local authorities.

Plastic: Triple rinse (or equivalent). Then puncture and dispose of in a sanitary landfill, or incineration, or, if allowed by state and local authorities, by burning. If burned, stay out of smoke.

GENERAL INFORMATION

Buctril is a selective postemergence herbicide for control of important broadleaf weeds infesting small grains (wheat, barley, oats, and rye), field corn, popcorn, grain sorghum, flax, garlic, turf and noncrop areas. Optimum weed control is obtained when Buctril is applied to actively growing weed seedlings. Buctril is primarily a contact herbicide, therefore thorough coverage of the weed seedlings is essential for optimum control.

Buctril is formulated as an emulsifiable concentrate of octanoic acid ester of bromoxynil containing the equivalent of 2 pounds of bromoxynil per gallon. To broaden the spectrum of control in wheat and barley, Buctril may be tankmixed with Glean,® Avenge,® MCPA, 2,4-D or Hoelon.® In oats and rye, Buctril may be tankmixed with MCPA or 2,4-D. In field corn and popcorn, Buctril can be tankmixed with atrazine or 2,4-D. Buctril may also be tankmixed with atrazine in grain sorghum. On turf, Buctril can be tankmixed with MCPP, Banvel, MCPP + Banvel, or 2,4-D + MCPP. See individual crop recommendations for proper chemical rates to be used. To ensure maximum crop safety and weed control, follow all cautions and limitations on this label and the labels of products used in the tank mixture with Buctril.

MIXING INSTRUCTIONS

Buctril Alone: Fill the spray tank ½ to ¾ full with clean water. Begin agitation and add the recommended amount of Buctril. Add water to the spray tank to the desired level. Maintain sufficient agitation to ensure a uniform spray mixture during application.

Tank Mixtures: Fill the spray tank ½ to ¾ full with clean water and begin agitation. If tankmixing with either Glean or atrazine, add the recommended amount of Glean or atrazine first. After the herbicide is thoroughly mixed with water add the recommended amount of Buctril and add water to the spray tank to the desired level.

If tankmixing with Avenge, MCPA, 2,4-D, Hoelon, MCPP, or Banvel add the recommended amount of Buctril first. After Buctril is thoroughly mixed with water, add the recommended amount of the other herbicide(s) to the spray tank and add water to the spray tank to the desired level.

Maintain sufficient agitation while mixing and during application to ensure a uniform spray mixture. If spray mixture is allowed to remain without agitation for short periods of time, be sure to agitate until uniformly mixed before application.

APPLICATION PROCEDURES

Buctril alone and in a tank mixture with Glean, Avenge, MCPA, 2,4-D, Hoelon and atrazine can be applied by both ground and aerial equipment. In addition, Buctril alone can be applied by automated sprinkler irrigation systems. The following provides recommended methods of application for each crop.

	Type of Application Equipment		
Crop	Ground	Aerial	Automated Sprinkler Irrigation
Small Grains	X	X	X
Field and Popcorn	X	X	X
Grain Sorghum	X	X	X
Flax	X	X	
Garlic	X	X	
Turf	X		
Noncrop Areas	X		

X indicates recommended application use

GROUND APPLICATION

Use a standard herbicide boom sprayer that provides uniform and accurate application. Sprayer should be equipped with screens no finer than 50 mesh in the nozzle tips and in-line strainers. Specific application instructions are described under each crop heading.

AERIAL APPLICATION

Use a minimum of 5 gallons of spray volume per acre using diaphragm-type nozzles that produce cone or fan spray patterns with a maximum spray pressure of 40 psi at the nozzle tips. Nozzles should be located no farther than ¾ the distance from the center of the aircraft to the end of the wing or rotor. Nozzles should be oriented at an angle between straight down and straight back.

The aircraft should discharge the spray a maximum of 10 feet above the crop. To minimize drift, do not apply during periods of gusty winds or when wind exceeds 5 mph.

AUTOMATED SPRINKLER IRRIGATION SYSTEMS

Buctril may be applied through automated sprinkler irrigation systems to small grains, field and popcorn and grain sorghum. Use a properly engineered chemical injection system which includes anti-siphon and check valves to prevent water source contamination and overflow of the mix tank. The system should also include interlocking controls between the metering device and the water pump to ensure simultaneous shutoff. The sprinkler system should provide uniform coverage and should have no leaks. The chemical injection tank should be cleaned before application to remove all fertilizer, pesticide and other foreign matter. Add Buctril or a Buctril-water mixture to the injector tank. A solution of Buctril diluted with water can be used only where agitation is present in the mix tank to achieve higher volumes needed in some irrigation systems. Carefully calibrate the system before and during application. Allow sufficient time for Buctril to be flushed through the system before turning off the irrigation water.

GENERAL WEED LIST

Postemergence application of Buctril will control the following weeds when sprayed in the seedling stage. Maximum weed stage of growth is listed under RECOMMENDED USES for each crop.

MOST SUSCEPTIBLE BROADLEAF WEED SPECIES:

Annual Sowthistle	*Sonchus oleraceus*
Black Nightshade	*Solanum nigrum*
Blue Mustard	*Chorispora tenella*
Coast Fiddleneck	*Amsinckia intermedia*
Common Cocklebur	*Xanthium pensylvanicum*
Common Lambsquarters	*Chenopodium album*
Common Tarweed	*Hemizonia congesta*
Eastern Black Nightshade	*Solanum ptycanthum*
Field Pennycress	*Thlaspi arvense*
Green Smartweed	*Polygonum scabrum*

Hairy Nightshade	*Solanum sarachoides*
Jimsonweed	*Datura stramonium*
Ladysthumb	*Polygonum persicaria*
Pennsylvania Smartweed	*Polygonum pensylvanicum*
Pepperweed spp. (annual)	*Lepidium* spp.
Shepherdspurse	*Capsella bursa-pastoris*
Silverleaf Nightshade	*Solanum elaeagnifolium*
Tartary Buckwheat	*Fagopyrum tatoricum*
Sunflower	*Helianthus annuus*[1]
Wild Buckwheat	*Polygonum convolvulus*

1 For control of sunflower, delay application until first emerging sunflower seedlings are 4 inches in height.

SUSCEPTIBLE BROADLEAF WEED SPECIES:

Buffalobur	*Solanum nostratum*
Common Groundsel	*Senecio vulgaris*
Common Ragweed	*Ambrosia artemisiifolia*
Corn Chamomile	*Anthemis arvensis*
Corn Gromwell	*Lithospermum arvense*
Cow Cockle	*Saponaria vaccaria*
Giant Ragweed	*Ambrosia trifida*
Hemp Sesbania	*Sesbania exaltata*
Ivyleaf Morningglory	*Ipomoea hederacea*
Knawel	*Scleranthus annuus*
Kochia	*Kochia scoparia*[2]
London Rocket	*Sisymbrium irio*
Mayweed	*Anthemis cotula*
Prostrate Knotweed	*Polygonum aviculare*
Redroot Pigweed	*Amaranthus retroflexus*[2]
Russian Thistle	*Salsola kali*
Spiny Pigweed	*Amaranthus spinosus*[2]
Tall Morningglory	*Ipomoea purpurea*
Tall Waterhemp	*Amaranthus tuberculatus*[2]
Tumble Mustard	*Sisymbrium altissimum*
Velvetleaf	*Abutilon theophrasti*[2]
Wild Mustard	*Brassica kaber*
Wild Radish	*Raphanus raphanistrum*

2 For effective control, these weeds should not exceed the 4 leaf stage or 2 inches in height, whichever comes first.

WEED SUPPRESSION:

Buctril suppresses the growth of Canada Thistle (Cirsium arvense) by burning down of top growth. Regrowth may occur.

WHEAT, BARLEY, OATS, AND RYE

Buctril can be applied to small grains from emergence up to the boot stage. Applications should be made to weeds soon after emergence for best weed control and to prevent competition to the crop. Reduced weed control may occur when weeds are stressed from lack of moisture or cold temperatures. Avoid spray drift to nontarget areas. Do not graze treated fields for 30 days following application.

Buctril can be applied to wheat and barley alone or in a tank mixture with Glean, Avenge, MCPA, 2,4-D, or Hoelon to broaden the spectrum of weed control. Buctril can also be applied to oats and rye alone or in a tank mixture with MCPA or 2,4-D. To ensure maximum crop safety and weed control, follow all cautions and limitations on this label and the labels of products used in tank mixtures with Buctril.

APPLICATION PROCEDURES

GROUND APPLICATION

Buctril used alone and in tank mixtures can be applied by ground equipment using a minimum spray volume of 10 gallons per acre. Use flat fan nozzles spaced a maximum of 20 inches apart with a minimum spray pressure of 30 psi at the nozzle tips. Other nozzle types may not provide adequate coverage of the weeds to ensure optimum weed control.

AERIAL APPLICATION

Buctril and tank mixtures with Buctril can be applied by aerial equipment using a minimum spray volume of 5 gallons per acre with a maximum spray pressure of 40 psi at the nozzle tips. Refer to AERIAL APPLICATION under the GENERAL INFORMATION section for complete use directions.

AUTOMATED SPRINKLER IRRIGATION SYSTEMS

Apply Buctril by automated sprinkler irrigation systems using $\frac{1}{4}$ to $\frac{1}{2}$ acre-inch of water per acre. Refer to GENERAL INFORMATION section for complete use directions and precautions.

SPRAYABLE LIQUID FERTILIZERS

Buctril can be applied to small grains using sprayable liquid fertilizer as the carrier. Predetermine the compatibility with the liquid fertilizer by mixing small proportional quantities in advance. Agitation must be maintained during filling and application operations to ensure that Buctril is evenly mixed with the fertilizer. Leaf burn may occur when Buctril is applied with liquid fertilizer, but new leaves are not adversely affected.

RECOMMENDED USES

The recommended rate of Buctril used alone or in a tank mixture with another herbicide is determined by the crop (winter or spring seeded), geographic area, application equipment (ground, aerial, or automated sprinkler irrigation systems), weed species and stage of growth of weeds at time of application.

Recommendations are as follows:

WHEAT and BARLEY
(Ground and Aerial Applications)
BUCTRIL RECOMMENDATIONS

PRODUCT	CROP/GEOGRAPHICAL AREA	RATE	TIMING OF APPLICATION: CROP	TIMING OF APPLICATION: WEED
Buctril	Fall seeded wheat and barley throughout the United States and spring seeded wheat and barley in Idaho, Oregon and Washington.	$1\frac{1}{2}$ Pints/A	Apply to wheat and barley from emergence to the boot stage.	**Most Susceptible Broadleaf Weeds** Apply to weeds up to the 8 leaf stage or 4 inches in height whichever comes first. If weed forms rosette, apply before weeds exceed 2 inches in diameter.
		2 Pints/A		**Susceptible Broadleaf Weeds** Apply to weeds up to the 4 leaf stage or 2 inches in height whichever comes first. If weed forms rosette, apply before weeds exceed 1 inch in diameter.
	Spring seeded wheat and barley except Idaho, Oregon and Washington.	1 Pint/A	Apply to wheat and barley from emergence to the boot stage.	**Most Susceptible and Susceptible Broadleaf Weeds** Apply Buctril at 1 pint/A to weeds that do not exceed the 4 leaf stage or 2 inches in height whichever comes first. If weed forms rosette, apply before weeds exceed 1 inch in diameter.

PRODUCT	CROP/GEOGRAPHICAL AREA	RATE	TIMING OF APPLICATION: CROP	TIMING OF APPLICATION: WEED
		1½ Pints/A		Use Buctril at 1½ pints/A for control of kochia that is 2-4 inches in height and pigweed (*Amaranthus* spp.) that does not exceed the 4 leaf stage or 2 inches in height, whichever comes first.

BUCTRIL TANK MIXTURE RECOMMENDATIONS

PRODUCT	CROP/GEOGRAPHICAL AREA	RATE	TIMING OF APPLICATION: CROP	TIMING OF APPLICATION: WEED
Buctril + GLEAN (Tankmix)	Fall seeded wheat and barley throughout the United States and spring seeded wheat and barley in Idaho, Oregon and Washington.	¾ Pint/A + ⅙-⅓ ounce/A	Apply to wheat and barley in the fall or spring at any time after the crop is in the 2 to 3 leaf stage, but before the boot stage.	**Most Susceptible Broadleaf Weeds** Apply to weeds up to the 8 leaf stage or 4 inches in height whichever comes first. If weed forms rosette, apply before weeds exceed 2 inches in diameter.
		1 Pint/A + ⅙-⅓ ounce/A		**Susceptible Broadleaf Weeds** Apply to weeds up to the 4 leaf stage or 2 inches in height whichever comes first. If weed forms rosette, apply before weeds exceed 1 inch in diameter.
	Spring seeded wheat and barley except Idaho, Oregon and Washington.	¾ Pint/A + ⅙-⅓ ounce/A		**Most Susceptible and Susceptible Broadleaf Weeds** Apply Buctril + Glean at ¾ pint/A + ⅙-⅓ ounces/A to weeds that do not exceed the 4 leaf stage or 2 inches in height whichever comes first. If weed forms rosette, apply before weeds exceed 1 inch in diameter.
		1½ Pints/A + ⅙-⅓ ounce/A		Use Buctril + Glean at 1½ pints/A + ⅙-⅓ ounces/A for control of kochia that is 2-4 inches in height and pigweed (*Amaranthus* spp.) that does not exceed the 4 leaf stage or 2 inches in height, whichever comes first.

A tankmix application of Buctril and Glean broadens the spectrum of weed control. See Glean label for listing of additional susceptible weed species and crop rotation instructions.

Glean is a registered trademark of E.I. DuPont de Nemours & Co., Inc.

BUCTRIL and AVENGE

A tank mixture of Buctril + Avenge can be used to control wild oats in addition to the broadleaf weeds controlled by Buctril. The rate and application timing for broadleaf weed control with Buctril in the tank mixture is the same as when Buctril is used alone. See BUCTRIL RECOMMENDATIONS.

Apply Buctril + Avenge tank mixtures when the majority of the wild oats are in the 3 to 5 leaf stage of growth. This frequently coincides with barley in the 2 to 7 leaf stage, spring seeded wheat in the 5 to 6 leaf stage, and fall seeded wheat in the 4 leaf to tillered stage of growth.

See Avenge label for varietal restrictions and recommended Avenge rates as related to wild oat density.

Avenge is a trademark of American Cyanamid Company.

BUCTRIL and MCPA or 2,4-D

A tank mixture of Buctril at the recommended rate and MCPA or 2,4-D at ¼ to ½ pound of active ingredient per acre is used to broaden the spectrum of broadleaf weed control. The tank mixture will control the weeds listed on the MCPA or 2,4-D labels in addition to the weeds listed on the Buctril label. The rate of Buctril in the tank mixture is the same as when Buctril is used alone. See BUCTRIL RECOMMENDATIONS.

Buctril + MCPA tank mixtures can be applied after the 3 leaf stage but before the crop reaches the boot stage. Buctril + 2,4-D tank mixtures can be applied after the crop has tillered but before the forming of joints in the stem.

BUCTRIL and HOELON

A tank mixture of Buctril + Hoelon can be used to control annual grasses in addition to the broadleaf weeds controlled by Buctril. The rate and application timing for broadleaf weed control with Buctril in the tank mixture is the same as when Buctril is used alone. See BUCTRIL RECOMMENDATIONS.

Buctril + Hoelon tank mixtures can be applied to fall seeded wheat and spring seeded wheat and barley up to the jointing stage of growth. Annual grasses should be in the 1 to 3 leaf stage of growth; up to the 4 leaf stage in wheat. See Hoelon label to determine recommended Hoelon rate in relation to the annual grassy weed stage of growth.

Hoelon is a trademark of Hoechst AG.

OATS AND RYE

Buctril used alone or in a tank mixture with MCPA or 2,4-D can be applied to oats and rye. Use the same rates and application timing as recommended for wheat and barley. Refer to previous section for WHEAT AND BARLEY.

WHEAT, BARLEY, OATS AND RYE

(Automated Sprinkler Irrigation Application)

Buctril can be applied through automated sprinkler irrigation systems to fall and spring seeded wheat, barley, oats and rye from emergence to the boot stage. Apply Buctril at 2 pints/A in ¼ to ½ acre-inch of water. See GENERAL INFORMATION section for complete use directions and precautions. Refer to GENERAL WEED LIST in GENERAL INFORMATION section for list of MOST SUSCEPTIBLE BROADLEAF WEEDS and SUSCEPTIBLE BROADLEAF WEEDS to postemergence applications of Buctril.

MOST SUSCEPTIBLE BROADLEAF WEEDS

Apply to weeds up to the 8 leaf stage or 4 inches in height whichever comes first. If weed forms rosette apply before weeds exceed 2 inches in diameter.

SUSCEPTIBLE BROADLEAF WEEDS

Apply to weeds up to the 4 leaf stage or 2 inches in height whichever comes first. If weed forms rosette apply before weeds exceed 1 inch in diameter.

WEED SIZE IS CRITICAL WHEN BUCTRIL IS APPLIED THROUGH AUTOMATED SPRINKLER IRRIGATION SYSTEMS. WEEDS SHOULD NOT EXCEED THE MAXIMUM INDICATED STAGE OF GROWTH.

Do not apply Buctril through automated sprinkler irrigation systems in California.

FIELD CORN, POPCORN AND GRAIN SORGHUM

Buctril is applied as a broadcast application in field corn, popcorn or grain sorghum grown for grains, seed or silage under conventional, minimum tillage, ridge tillage or no-till systems. For optimum control spray seedling weeds that are actively growing and not stressed from lack of moisture or low temperatures. Avoid spray drift to non-target areas.

A tank mixture of Buctril + atrazine may be used in field corn, popcorn or grain sorghum to broaden the spectrum of control and provide residual control of germinating seedlings. In field corn, Buctril may also be used in a tank mixture with 2,4-D.

When environmental conditions are quite cool and cloudy or hot and humid, some leaf burn may occur on the corn or grain sorghum foliage. The crop rapidly outgrows this condition and new growth is unaffected. To ensure maximum crop safety and weed control, follow all cautions and limitations on this label and the labels of products used in tank mixtures with Buctril.

APPLICATION PROCEDURES

Buctril used alone and in tank mixtures can be applied by ground equipment using a minimum of 20 gallons of spray volume per acre. Use flat fan nozzles spaced a maximum of 20 inches apart with a minimum spray pressure of 30 psi at the nozzle tips. Other nozzle types may not provide adequate coverage of the weeds to ensure optimum control. When weed infestations are heavy, use of a higher spray volume and spray pressure will be helpful to ensure uniform weed coverage.

For aerial application use a minimum of 5 gallons of spray volume per acre with a maximum spray pressure of 40 psi at the nozzle tips. Refer to AERIAL APPLICATION under GENERAL INFORMATION section for complete use directions.

Buctril can also be applied to field corn, popcorn and grain sorghum through automated sprinkler irrigation systems. Refer to GENERAL INFORMATION section for complete use directions and precautions.

RESTRICTIONS AND LIMITATIONS

Buctril does not control grasses. Therefore, it is recommended that a suitable grass control program be used to provide any required grass control.

Do not add a spray additive, or mix with liquid fertilizers, because excessive crop injury may occur.

Do not apply Buctril if heavy rainfall is expected within 1 hour of application.

Seed corn producers should consult the respective seed corn company regarding tolerance of certain seed production inbred lines to Buctril.

Do not plant rotational crops until the following use season.

Do not cut for feed or graze within 30 days after application.

A second application of Buctril is recommended if a new flush of weeds occurs following the first application. The total cumulative rate should not exceed 3 pints/A per season. Application may be made to corn up to the 8-leaf stage regardless of height. Application may be made to grain sorghum up to the 10-leaf stage not to exceed 12 inches in height.

BUCTRIL RECOMMENDATIONS

Postemergence applications of Buctril at 1 and 1½ pints/A can be applied to field corn, popcorn and grain sorghum using the following use directions. Refer to APPLICATION RATE TABLE for weeds controlled.

APPLICATION RATE TABLE FOR CORN AND GRAIN SORGHUM

WEED SPECIES When determining leaf stage, count all leaves except cotylendonary leaves		1 Pint/A Maximum Leaf Stage	1 Pint/A Maximum Weed Height (Inches)	1½ Pints/A Maximum Leaf Stage	1½ Pints/A Maximum Weed Height (Inches)
Black Nightshade	*Solanum nigrum*	6	6	6	6
Buffalobur	*Solanum rostratum*	4	2	6	4
Common Cocklebur	*Xanthium pensylvanicum*	6	8	8	10
Common Lambsquarters	*Chenopodium album*	—	6	—	8
Common Ragweed	*Ambrosia artemisiifolia*	6	4	8	6
Eastern Black Nightshade	*Solanum ptycanthum*	6	6	6	6
Giant Ragweed	*Ambrosia trifida*	6	4	6	6
Hemp Sesbania	*Sesbania exaltata*	—	—	4	4
Ivyleaf Morningglory	*Ipomoea hederacea*	3	3	4	4
Jimsonweed	*Datura stramonium*	4	4	6	6
Ladysthumb	*Polygonum persicaria*	4	4	6	6
Pennsylvania Smartweed	*Polygonum pensylvanicum*	4	4	6	6
Redroot Pigweed	*Amaranthus retroflexus*	—	—	4	2
Spiny Pigweed	*Amaranthus spinosus*	—	—	4	2
Sunflower	*Helianthus annuus*	4	6	6	8
Tall Morningglory	*Ipomoea purpurea*	3	3	4	4
Tall Waterhemp	*Amaranthus tuberculatus*	—	—	4	2
Velvetleaf	*Abutilon theophrasti*	4	3	6	5
Wild Buckwheat	*Polygonum convolvulus*	4	6	6	8
Wild Mustard	*Brassica kaber*	—	—	4	4
WEEDS SUPPRESSED					
Canada Thistle Buctril suppresses the growth by burning down of top growth. Regrowth may occur.	*Cirsium arvense*	not recommended		8 inch to bud stage	

FIELD CORN and POPCORN
(Ground and Aerial Applications)
BUCTRIL RECOMMENDATIONS

PRODUCT	RATE	TIMING OF APPLICATION: CROP	TIMING OF APPLICATION: WEED[1]
Buctril	1 Pint/A	Apply to corn from the 3-leaf stage to the 8-leaf stage—maximum of two applications per season.	See APPLICATION RATE TABLE for list of weeds and corresponding maximum stage of growth that are controlled by Buctril at 1 pint/A.
	1½ Pints/A	Apply to corn from the 4-leaf stage to the 8-leaf stage—maximum of two applications per season.	See APPLICATION RATE TABLE for list of weeds and corresponding maximum stage of growth that are controlled by Buctril at 1½ pints/A.

FIELD CORN and POPCORN
(Ground and Aerial Applications)
BUCTRIL TANKMIX RECOMMENDATIONS

PRODUCT	RATE	TIMING OF APPLICATION: CROP	TIMING OF APPLICATION: WEED
Buctril +	1 Pint/A + ½-1⅕ lbai/A	Apply to corn from the 3-leaf stage to the 8-leaf stage.	All weeds controlled by Buctril at 1 pint/A plus control of pigweeds (*Amaranthus* spp.) up to the 6-leaf stage or 4 inches in height, whichever comes first. This tank mixture also provides some residual control of germinating seedlings.
Atrazine[1] (Tankmix)	1½ Pints/A + ½-1⅕ lbai/A	Apply to corn from the 4-leaf stage to the 8-leaf stage.	All weeds controlled by Buctril at 1½ pints/A plus control of pigweeds (*Amaranthus* spp.) up to the 6-leaf stage or 4 inches in height, whichever comes first. This tank mixture also provides some residual control of germinating seedlings.

[1] For control of additional weeds not listed in the APPLICATION RATE TABLE for field corn, popcorn and grain sorghum, see GENERAL WEED LIST: Use Buctril at 1 pint/A for control of MOST SUSCEPTIBLE WEEDS and 1½ pints/A for control of SUSCEPTIBLE WEEDS. Apply Buctril to weeds that do not exceed the 4-leaf stage or 2 inches in height, whichever comes first. If weed forms rosette, apply before weed exceeds 1 inch in diameter.

FIELD CORN
(Ground and Aerial Applications)
BUCTRIL TANKMIX RECOMMENDATIONS

PRODUCT	RATE	TIMING OF APPLICATION: CROP	TIMING OF APPLICATION: WEED
Buctril + 2,4-D[2] (Tankmix)	1 Pint/A + ¼ lbai/A	Apply to corn from the 3-leaf stage to the 8-leaf stage. When corn exceeds 8 inches in height, drop nozzles must be used to apply the Buctril + 2,4-D tank mixture as a directed spray.	All weeds controlled by Buctril at 1 pint/A plus control of wild mustard that is larger than the 4-leaf stage or 4 inches in height, whichever comes first.
	1½ Pints/A + ¼ lbai/A	Apply to corn from the 4-leaf stage to the 8-leaf stage. When corn exceeds 8 inches in height, drop nozzles must be used to apply the Buctril + 2,4-D tank mixture as a directed spray.	All weeds controlled by Buctril at 1½ pints/A plus control of wild mustard that is larger than the 4-leaf stage or 4 inches in height, whichever comes first.

ATRAZINE CONVERSION TABLE

Atrazine Formulation	Atrazine Rate Pounds of active Ingredient per acre	Atrazine Formulation Rate per acre
Atrazine 4L	½	1 Pint
	1⅕	2⅖ Pints
Atrazine 80 WP	½	⅝ Pound
	1⅕	1½ Pounds
Aatrex® Nine-0	½	⅗ Pound
	1⅕	1⅓ Pounds

2,4-D CONVERSION TABLE

2,4-D Formulation	2,4-D Rate Pounds of active Ingredient per acre	2,4-D Formulation Rate per acre
2,4-D 4lb/gallon	¼	½ Pint
2,4-D 6lb/gallon	¼	⅓ Pint

SPECIAL PRECAUTIONS FOR BUCTRIL + 2,4-D TANK MIXTURE: Postemergence applications of 2,4-D cause brittleness to corn. Winds or cultivations may cause breakage while corn is brittle. Refer to 2,4-D label for other use directions and cautions.

FIELD CORN and POPCORN
(Automated Sprinkler Irrigation Application)

Buctril can be applied through automated sprinkler irrigation systems to field and popcorn from the 3-leaf to the 8-leaf stage. Apply Buctril at 2 pints/A in ¼ to ½ acre-inch of water. See GENERAL INFORMATION section for complete use directions and precautions. Refer to GENERAL WEED LIST in the GENERAL INFORMATION section for list of MOST SUSCEPTIBLE BROADLEAF WEEDS and SUSCEPTIBLE BROADLEAF WEEDS to postemergence applications of Buctril.

MOST SUSCEPTIBLE BROADLEAF WEEDS

Apply to weeds up to the 8 leaf stage or 4 inches in height whichever comes first. If weed forms rosette apply before weeds exceed 2 inches in diameter.

SUSCEPTIBLE BROADLEAF WEEDS

Apply to weeds up to the 4 leaf stage or 2 inches in height whichever comes first. If weed forms rosette apply before weeds exceed 1 inch in diameter.

WEED SIZE IS CRITICAL WHEN BUCTRIL IS APPLIED THROUGH AUTOMATED SPRINKLER IRRIGATION SYSTEMS. WEEDS SHOULD NOT EXCEED THE MAXIMUM INDICATED STAGE OF GROWTH.

Do not apply Buctril through automated sprinkler irrigation systems in California.

GRAIN SORGHUM
(Ground and Aerial Applications)
BUCTRIL RECOMMENDATIONS

PRODUCT	RATE	TIMING OF APPLICATION: CROP	TIMING OF APPLICATION: WEED[1]
Buctril	1 Pint/A	Apply to grain sorghum from the 3-leaf stage to the 10-leaf stage, not to exceed 12 inches in height.	See APPLICATION RATE TABLE for list of weeds and corresponding maximum stage of growth that are controlled by Buctril at 1 pint/A.
	1½ Pints/A	Apply to grain sorghum from the 4-leaf stage to the 10-leaf stage, not to exceed 12 inches in height.	See APPLICATION RATE TABLE for list of weeds and corresponding maximum stage of growth that are controlled by Buctril at 1½ pints/A.
Buctril + Atrazine (Tankmix)	1 Pint/A + ½-1⅕ lbai/A	Apply to grain sorghum from the 3-leaf stage to the 10-leaf stage, not to exceed 12 inches in height.	All weeds controlled by Buctril at 1 pint/A plus control of pigweeds (*Amaranthus* spp.) up to the 6-leaf stage or 4 inches in height, whichever comes first. This tank mixture also provides some residual control of germinating seedlings.

PRODUCT	RATE	TIMING OF APPLICATION: CROP	TIMING OF APPLICATION: WEED[1]
	1½ Pints/A + ½-1⅕ lbai/A	Apply to grain sorghum from the 4-leaf stage to the 10-leaf stage, not to exceed 12 inches in height.	All weeds controlled by Buctril at 1½ pints/A plus control of pigweeds (*Amaranthus* spp.) up to the 6-leaf stage or 4 inches in height, whichever comes first. This tank mixture also provides some residual control of germinating seedlings.

[1] For control of additional weeds not listed in the APPLICATION RATE TABLE for field corn, popcorn and grain sorghum, see GENERAL WEED LIST: Use Buctril at 1 pint/A for control of MOST SUSCEPTIBLE WEEDS and 1½ pints/A for control of SUSCEPTIBLE WEEDS. Apply Buctril to weeds that do not exceed the 4-leaf stage or 2 inches in height, whichever comes first. If weed forms rosette, apply before weed exceeds 1 inch in diameter.

GRAIN SORGHUM
(Automated Sprinkler Irrigation Application)

Buctril can be applied through automated sprinkler irrigation systems to grain sorghum from the 3-leaf to the 10-leaf stage, not to exceed 12 inches in height. Apply Buctril at 2 pints/A in ¼ to ½ acre-inch of water. See GENERAL INFORMATION section for complete use directions and precautions. Refer to GENERAL WEED LIST in the GENERAL INFORMATION section for list of MOST SUSCEPTIBLE BROADLEAF WEEDS and SUSCEPTIBLE BROADLEAF WEEDS to postemergence applications of Buctril.

MOST SUSCEPTIBLE BROADLEAF WEEDS
Apply to weeds up to the 8 leaf stage or 4 inches in height whichever comes first. If weed forms rosette apply before weeds exceed 2 inches in diameter.

SUSCEPTIBLE BROADLEAF WEEDS
Apply to weeds up to the 4 leaf stage or 2 inches in height whichever comes first. If weed forms rosette apply before weeds exceed 1 inch in diameter.

WEED SIZE IS CRITICAL WHEN BUCTRIL IS APPLIED THROUGH AUTOMATED SPRINKLER IRRIGATION SYSTEMS. WEEDS SHOULD NOT EXCEED THE MAXIMUM INDICATED STAGE OF GROWTH.

Do not apply Buctril through automated sprinkler irrigation systems in California.

FLAX
Apply Buctril when weed seedlings are actively growing. For optimum control apply Buctril in sufficient spray volume to ensure good coverage of the weeds. Do not apply to flax under humid conditions and when air temperature is above 85°F as injury may occur. Do not add spray additive with Buctril.

BUCTRIL RECOMMENDATIONS
Buctril at 1 pint/A should be applied postemergence to flax that is 2 to 8 inches in height. Do not apply Buctril to flax during or after the bud stage.

Buctril at 1 pint/A will control the MOST SUSCEPTIBLE AND SUSCEPTIBLE BROADLEAF WEEDS (see GENERAL WEED LIST) when applied at the recommended weed stage of growth. Weeds should not exceed the 4 leaf stage or 2 inches in height, whichever comes first. If weed forms rosette, apply before weeds exceed 1 inch in diameter.

For ground application, use a minimum of 10 gallons of spray volume per acre. Use flat fan nozzles spaced a maximum of 20 inches apart with a minimum spray pressure of 30 psi at the nozzle tips. Other nozzle types may not provide adequate coverage of weeds to ensure optimum weed control. For aerial application, use a minimum of 5 gallons of spray volume per acre with a maximum spray pressure of 40 psi at the nozzle tips. Refer to AERIAL APPLICATION under GENERAL INFORMATION section for complete use directions.

GARLIC
Apply Buctril when weed seedlings are actively growing. For optimum control apply Buctril in sufficient spray volume to ensure good coverage of the weeds.

BUCTRIL RECOMMENDATIONS
Buctril can be applied after the crop emerges but before garlic is 12 inches in height.

For control of MOST SUSCEPTIBLE BROADLEAF WEEDS (See GENERAL WEED LIST) apply Buctril at 2 pints/A to weeds up to the 8 leaf stage or 4 inches in height, whichever comes first. If weed forms rosette, apply before weeds exceed 2 inches in diameter.

For control of SUSCEPTIBLE WEEDS, apply Buctril at 4 pints/A to weeds up to the 4 leaf stage or 2 inches in height, whichever comes first. If weed forms rosette, apply before weeds exceed 1 inch in diameter.

For ground application, use a minimum of 20 gallons of spray volume per acre. Use flat fan nozzles spaced a maximum of 20 inches apart with a minimum spray pressure of 30 psi at the nozzle tips. Other nozzle types may not provide adequate coverage of the weeds to ensure optimum weed control. For aerial application, use a minimum of 5 gallons of spray volume per acre with a maximum spray pressure of 40 psi at the nozzle tips. Refer to AERIAL APPLICATION under GENERAL INFORMATION section for complete use directions. Do not apply Buctril within 112 days of anticipated harvest.

ESTABLISHED TURFGRASSES AND NEWLY PLANTED GRASSES FOR SOD AND SEED PRODUCTION
Postemergence applications of Buctril controls many seedling broadleaf weeds commonly found in turfgrasses. Established grasses tolerant to Buctril include bentgrasses, Kentucky Bluegrass, Fescues, Ryegrass, Bermudagrass, St. Augustinegrass and Zoysiagrass. Buctril may also be used on seedling grasses such as Merion, Park, Delta, or common Kentucky Bluegrasses, Pennlawn, Chewings, Illahee or Alta Fescues, Orchard grass, Highland, Seaside, or Astoria Bentgrasses, perennial Ryegrasses, Bahiagrass and newly sprigged Zoysiagrass planted for seed or sod production, or forage grasses planted for seed production only.

APPLICATION PROCEDURES
Buctril used alone and in tank mixtures should be applied in 10 to 100 gallons of spray volume per acre. When using a spray volume from 10 to 40 GPA, use flat fan nozzles spaced a maximum of 20 inches apart with a minimum spray pressure of 30 psi at the nozzle tips. When using higher spray volumes, flood jet nozzles may be used with a minimum spray pressure of 30 psi at the nozzle tips.

BUCTRIL[1] RECOMMENDATIONS

PRODUCT	RATE PER ACRE	RATE PER 1000 FT.[2]	WEED SPECIES
SEEDLING AND ESTABLISHED TURF			
Buctril	1½-2 Pints	0.6-0.75 fl. oz.	See GENERAL WEED LIST
ESTABLISHED TURF ONLY; EXCLUDING BENTGRASS PUTTING GREENS			
Buctril	4 Pints	1.5 fl. oz.	As previously listed for Buctril plus the following: Spurweed (*Soliva* spp.) Yellow Woodsorrel (*Oxalis* spp.)[2]
	8 Pints	3.0 fl. oz.	As previously listed for Buctril plus the following: Prostrate Spurge (Euphorbia supina)

BUCTRIL TANK MIXTURE RECOMMENDATIONS

BUCTRIL TANK MIXTURES ESTABLISHED TURF ONLY: EXCLUDING BENTGRASSES

PRODUCT	RATE PER ACRE	RATE PER 1000 FT.[2]	WEED SPECIES
Buctril + MCPP	2–4 Pints + 1.0 LBAI	0.75–1.5 fl. oz. + 0.025 LBAI	As previously listed for Buctril plus the following: Common Chickweed (*Stellaria media*) Mouseear Chickweed (*Cerastium vulgatum*) Red Clover (*Trifolium pratense*) White Clover (*Trifolium repens*) Ground Ivy (*Glechoma hederacea*) Stitchwort (*Stellaria graminea*) Knotweed (*Polygonum aviculare*) Prostrate Spurge (*Euphorbia supina*)
Buctril + dicamba	2 Pints + 0.25–0.5 LBAI	0.75–1.5 fl. oz. + 0.006–0.012 LBAI	As previously listed for Buctril plus the following: Common Chickweed (*Stellaria media*) Mouseear Chickweed (*Cerastium vulgatum*) Red Clover (*Trifolium pratense*) White Clover (*Trifolium repens*) Knotweed (*Polygonum aviculare*) Red Sorrel (*Rumex acetosella*) Pepperweed (*Lepidium* spp.)
	4 Pints + 0.25–0.5 LBAI	1.5 fl. oz. + 0.006–0.012 LBAI	As previously listed for Buctril and Buctril + dicamba *plus the following*: Prostrate Spurge (*Euphorbia supina*) Yellow Woodsorrel (*Oxalis* spp.) Spurweed (*Soliva* spp.)
Buctril + MCPP + dicamba	2 Pints + 0.5–1.0 LBAI + 0.125–0.25 LBAI	0.75 fl. oz. + 0.0125–0.025 LBAI + 0.003–0.006 LBAI	As previously listed for Buctril combinations plus the following Plaintains (*Plantago* spp.) Dandelion (*Taraxacum officinale*)
	4 Pints + 0.5–1.0 LBAI + 0.125–0.25 LBAI	1.5 fl. oz. + 0.0125–0.025 LBAI + 0.003–0.006 LBAI	Prostrate Spurge (*Euphorbia supina*) Spurweed (*Soliva* spp.)
Buctril + 2,4-D + MCPP[4]	2 Pints + 0.5–1.0 LBAI + 0.5–1.0 LBAI	0.75 fl. oz. + 0.0125–0.025 LBAI + 0.0125–0.025 LBAI	As previously listed for Buctril tank mixtures plus the following: Plaintains (*Plantago* spp.) Knotweed (*Polygonum aviculare*) Dandelion (*Taraxacum officinale*) Red Sorrel (*Rumex acetosela*)[3]

[1] Buctril use rates are expressed as pints/A and fl. oz./1000 ft.[2] and the tankmix herbicides (dicamba, MCPP, and 2, 4-D) use rates are given in pounds of active ingredient per acre (LBAI/A) and pounds of active ingredient per 1000 ft.[2] (LBAI/1000 ft.[2]) due to the various formulations comercially available. Make the necessary calculations to liquid measure based on the formulation used.

[2] Except in California.

[3] Use high rate of MCPP or 2,4-D.

[4] Including prepackaged mixtures such as Chipco Turf Kleen.

A second application of Buctril alone 2 weeks after the initial treatment may be needed to get complete control of established Prostrate Spurge or Yellow Woodsorrel.

In order to ensure maximum safety and weed control follow all cautions and limitations on this label and the labels of products used in tank mixtures with Buctril.

INDUSTRIAL SITES AND NONCROP AREAS

FOR BROADCAST TREATMENT OF SMALL WEEDS:

Apply 2 quarts of Buctril with either 2 quarts of surfactant or 2 gallons of diesel oil in 150 gallons of water for each acre sprayed. Thorough coverage is important. Make sure spray boom is high enough to contact tops of all plants. Use adequate spray pressure to contact all leaf surfaces.

FOR SPOT TREATMENT OF SCATTERED INFESTATIONS OR LARGE WEEDS ALONG FENCEROWS IN VACANT LOTS AND ON INDUSTRIAL SITES:

Mix 1 quart of Buctril with either 1 quart of surfactant or 1 gallon of diesel oil per 100 gallons of spray solution. Cover all the weed foliage thoroughly. Use at least 200 gallons of spray solution for each acre sprayed.

When using oil, add Buctril to oil—then add water to this mixture while agitating vigorously.

CONTROLS THESE WEEDS IN THE SEEDLING STAGE:

In the Southwestern United States, Buctril effectively controls Russian thistle (*Salsola kali*), Saltbush (*Atriplex* spp.), Marestail (*Conyza canadensis*), Puncturevine (*Tribulus terrestris*), Prostrate Spurge (*Euphorbia supina*) and Bassia (*Bassia* spp.).

Buctril should be applied in early summer when Russian Thistle is immature and growing vigorously. Best results are obtained when temperatures are 80°F or higher.

Local conditions may affect the use of this chemical. Consult State Agricultural Extension or Experiment Station weed specialists for specific recommendations for local weed problems.

CONDITION OF SALE, WARRANTY, LIMITATION OF LIABILITY

This product conforms to the chemical description on the label thereof and is reasonably fit for the purpose stated on such label only when used in accordance with directions under normal use conditions. Follow directions carefully. Timing and method of application, weather and crop conditions, mixture with other chemicals not specifically recommended and other influencing factors in the use of this product are beyond the control of Rhone-Poulenc or the seller.

Buyer assumes all risks of use, storage, or handling of this material, not in strict accordance with directions given herewith. In no case shall Rhone-Poulenc or the seller be liable for consequential, special, or indirect damages such as loss of profits or values resulting from the use or handling of this product.

LBL 5149

FOLEX®

COTTON DEFOLIANT

Emulsifiable Concentrate Contains 6 lbs. Merphos per Gallon.

ACTIVE INGREDIENT:	
Tributyl Phosphorotrithioite and Related Compounds	73.8%
INERT INGREDIENTS	26.2%
TOTAL	100.0%

EPA Reg. No. 359-691 EPA Est.
U.S. Patent No. 2,955,803

WARNING

Keep Out of Reach of Children

STATEMENT OF PRACTICAL TREATMENT

IF SWALLOWED: Drink promptly a large quantity of milk, egg whites, gelatin solution or if these are not available drink large quantities of water. Avoid alcohol. Get medical attention.

IF ON SKIN: Wash with plenty of soap and water. Get medical attention.

IF IN EYES: Flush with plenty of water for at least 15 minutes. Get medical attention if irritation persists.

PRECAUTIONARY STATEMENTS

HAZARDS TO HUMANS AND DOMESTIC ANIMALS

WARNING

Harmful if swallowed. Causes skin irritation. May be fatal if absorbed through skin. Causes eye irritation. Do not get in eyes, on skin or on clothing. For protection during application wear a hat, a long sleeved shirt and trousers. In addition, mixer/loaders must wear rubber or neoprene gloves. Wash thoroughly with soap and water after handling and before eating or smoking. Remove contaminated clothing and wash before reuse.

DIRECTIONS FOR USE

GENERAL CLASSIFICATION

It is a violation of Federal Law to use this product in a manner inconsistent with its labeling.

STORAGE AND DISPOSAL

STORAGE

Do not contaminate water, food or feed by storage or disposal.

PESTICIDE DISPOSAL

Pesticide wastes are toxic. Improper disposal of excess pesticide, spray mixture, or rinsate is a violation of Federal Law. If these wastes cannot be disposed of by use according to label instructions, contact your State Pesticide or Environmental Control Agency, or the Hazardous Waste representative at the nearest EPA Regional Office for guidance.

CONTAINER DISPOSAL

Triple rinse (or equivalent). Then puncture and dispose of in a sanitary landfill, or by other procedures approved by state and local authorities.

GENERAL INFORMATION

Folex is a cotton defoliant that is used to remove leaves from cotton plants prior to anticipated harvesting. Formulated as an emulsifiable concentrate, Folex contains 6 pounds of merphos per gallon. Folex is usually mixed with water and applied to cotton using either ground or aerial equipment. However, white diesel oil may be substituted for water under adverse weather conditions. Folex is non-corrosive and will not damage metal parts normally used in applying spray chemicals.

Folex usually drops leaves in a green condition and will not appear effective until defoliation actually begins, approximately 3 days after application. By this time the field will have an "off-cast" color. Under favorable conditions, satisfactory leaf drop will require 5 to 7 days. Adverse conditions, such as low temperatures and/or toughened plants (drought stressed, etc.), may extend this to 10 to 14 days.

Folex is effective even when applied to cotton with a heavy dew. Once Folex has dried on the leaf, subsequent rainfall or dew does not adversely affect the activity of Folex. Application of Folex is not recommended when a heavy rainfall is expected within 1 hour after application. Conditions which delay absorption into the leaves are primarily those which cause the cotton leaves to be wilted, toughened, or leathery. When these conditions prevail use white diesel oil instead of water, or make two applications 10 to 14 days apart.

Folex does not suppress second growth especially where rainfall follows defoliation. A second application of Folex will be effective in defoliating this second growth if excessive regrowth occurs between the first application and first and/or second harvest. Folex must be applied in sufficient water or white diesel oil to give thorough coverage of the leaves. Larger droplets more effectively reach the lower leaves.

Folex may be used alone for bottom defoliation of the leaves on the lower portion of the plant, as a preconditioner prior to total defoliation, or as a total defoliation treatment. Folex may also be used in a combination with Accelerate® (tankmix), Dropp® (tankmix), and Prep® Plant Regulator (tankmix or sequential application).

MIXING INSTRUCTIONS

FOLEX ALONE: Fill the spray tank ½ to ¾ full with clean water. Begin agitation and add the recommended amount of Folex. Add water to the spray tank to the desired level.

Maintain sufficient agitation to ensure a uniform spray mixture during application. Follow same mixing instructions when white diesel oil is substituted for water.

TANK MIXTURES: Fill the spray tank ½ to ¾ full with clean water and begin agitation. If tankmixing with Accelerate or Prep Plant Regulator, add the recommended rate of Folex first. After Folex is thoroughly mixed with water, add the recommended amount of Accelerate or Prep Plant Regulator and add water to the spray tank to the desired level.

When tankmixing with Dropp, add the recommended amount of Dropp first. After Dropp is thoroughly mixed with water, add the recommended amount of Folex. Add water to the spray tank to the desired level.

Maintain sufficient agitation while mixing and during application to ensure a uniform spray mixture. If spray mixture is allowed to remain without agitation for short periods of time, be sure to agitate until uniformly mixed before application. Do not use white diesel oil when Folex is used in a tank mixture with other products.

APPLICATION PROCEDURES

Folex alone and in a tank mixture with Accelerate, Dropp or Prep Plant Regulator can be applied by both ground and aerial equipment.

GROUND: Use a standard high clearance sprayer that provides uniform and accurate application. Sprayer should be equipped with screens no finer than 50 mesh in the nozzle tips and in-line strainers. A spray volume of 15 to 25 gallons per acre is recommended when Folex is applied with water and 10 to 15 gallons per acre is recommended when Folex is applied with white diesel oil.

AERIAL: Use aerial equipment with diaphragm-type nozzles that produce cone or fan spray patterns with a maximum spray pressure of 40 psi at the nozzle tips. Nozzles should be located no farther than ¾ the distance from the center of the aircraft to the end of the wing or rotor. The aircraft should discharge the spray a maximum of 10 feet above the crop. To minimize drift, do not apply when winds may cause drift to adjacent fields.

A spray volume of 5 to 12 gallons per acre should be used when Folex and tank mixtures with Folex are applied with water. Folex alone may be applied in 5 to 8 gallons of spray volume per acre when using white diesel oil.

FOLEX RECOMMENDATIONS

BOTTOM DEFOLIATION: Bottom defoliation is used to remove the leaves from the lower portion of the cotton plant when the bottom bolls are mature. This is determined when (1) youngest bolls cannot be dented by pressure between the thumb and forefinger, or (2) youngest bolls cannot be cut through with a sharp knife. Removal of leaves from this portion of the plant allows faster opening of mature bolls, more air movement and sunlight to the lower portion of the plant, resulting in less boll rot. Bottom defoliation has no effect on the maturity development of the less mature bolls on the upper portion of the plant. Recommendations are as follows:

Product	Rate	Directions
Folex	1⅓ Pts/A (broadcast equivalent)	Direct the Folex spray to thoroughly cover the leaves on the desired portion of the plant to be defoliated. The rate of Folex and spray volume should be proportional to the part of the plant to be defoliated.

PRE-CONDITIONING APPLICATION: Apply Folex as a preconditioning application to aerate rank growth and induce maturity of green plants prior to total plant defoliation. Use Folex as a preconditioner only after fiber quality of the youngest bolls which are to be harvested will not be damaged. Pre-conditioning of early maturing cotton varieties is generally not recommended.

Recommendations are as follows:

Product	Rate	Directions
Folex	½ Pint/A	Apply Folex as a broadcast spray approximately 7 to 10 days before making the total defoliation application.

TOTAL DEFOLIATION APPLICATION: Apply Folex when fiber quality of the top bolls will not be damaged by loss of the top leaves which feed these bolls (top bolls firm to thumb pressure). Folex will not damage the bolls directly or cause them to open prematurely. If pre-conditioning application was made, wait 7-10 days before applying total defoliation treatment. Recommendations are as follows:

Product	Rate	Directions
Folex	1½ Pints/A	Mix Folex with water and apply when ideal conditions occur (minimum night temperature is above 65°F).
	1½ to 2 Pints/A	Mix Folex with water and apply when average conditions occur (minimum night temperature is 55° to 65°F).
	2 Pints/A	Mix Folex with white diesel oil and apply when adverse conditions occur (minimum night temperature below 55°F).
	1½ Pints/A	**TOUCH-UP APPLICATIONS:** Pre-conditioning will usually eliminate the necessity for two total defoliation applications. However, when leaves are extremely tough or when cotton is badly lodged, it may be desirable to follow the initial total defoliation application in approximately 5 to 7 days with a touch-up treatment.
	2 to 3 Pints/A	**DEFOLIATION OF "RANK" COTTON:** For defoliation of rank cotton, mix Folex with water and apply when 50% or more of the bolls are open. Two applications can be made two to six weeks apart for complete defoliation. Applications can be made until first frost.

FOLEX COMBINATIONS

ACCELERATE®: The tank mixture of Folex + Accelerate results in a 2 to 3 day reduction in time required for maximum defoliation but results in little or no increase in total defoliation when compared to Folex used alone. Compatibility tests should be conducted before mixing the two products in the spray tank. Recommendations for Folex + Accelerate tank mixtures are:

Product	Rate	Directions
Folex + Accelerate	1 Pint/A + 1½ Pints/A	Mix Folex and Accelerate with water and apply the tank mixture by ground or aerial equipment for total defoliation under *ideal conditions* for defoliation.
	1½-2 Pints/A + 1½ Pints/A	Mix Folex and Accelerate with water and apply the tank mixture by ground or aerial equipment for total defoliation under *adverse conditions* for defoliation.

Refer to Accelerate label for additional use directions and cautions when using tank mixtures of Folex with Accelerate.
Accelerate is a registered trademark of Pennwalt Corporation.

DROPP®: Tank mixtures of Folex + Dropp is recommended for total defoliation when heavy regrowth is expected. Optimum defoliation and regrowth control is obtained when the tank mixture is applied to mature cotton plants when 60% or more bolls are open. A second application of labeled rates of Folex + Dropp, or Folex alone may be made where necessary. Folex + Dropp tank mixtures are recommended for use in Alabama, Arkansas, Georgia, Louisiana, Mississippi, North Carolina, South Carolina and Tennessee.

Recommendations for Folex + Dropp tank mixtures are:

Product	Rate	Directions
Folex + Dropp	1 Pint/A + 0.1 to 0.2 lb/A	Mix Folex and Dropp with water and apply the tank mixture by ground or aerial equipment for total defoliation and regrowth control when minimum night temperature is *above 65°F.*
	1½ Pints/A + 0.1 to 0.2 lb/A	Use when minimum night temperature is *60 to 65°F.*

Use the high rate of Dropp when environmental conditions favor heavy regrowth.
When minimum night temperature is *below 60°F*, use Folex alone.
Do not add an adjuvant to Folex + Dropp tank mixtures or use higher recommended rates of either Folex or Dropp because desiccation may occur.
Refer to Dropp label for additional use directions and cautions when using tank mixtures of Folex with Dropp.
Dropp is a registered trademark of Schering AG West Germany, Agrochemical Division.

PREP® PLANT REGULATOR: The combination of Folex and Prep Plant Regulator is used for total plant defoliation and to accelerate boll opening. Folex and Prep Plant Regulator may be applied as a tank mixture or used in two separate applications where Prep Plant Regulator is applied first and followed by Folex a minimum of 4 days later.

TANK MIXTURES AND SEQUENTIAL APPLICATIONS
Tank mixtures of Folex + Prep Plant Regulator is recommended for use in all cotton growing states except Arizona and California, while applications of Folex sequential to Prep Plant Regulator

(minimum of 4 days between applications) is recommended in all cotton growing states.

Folex + Prep Plant Regulator tank mixtures or sequential applications should be applied to cotton with a minimum of 60% opened bolls and/or when there is sufficient mature unopened bolls to produce the desired yield. If used in 'rank' cotton, a pretreatment with Folex (See FOLEX RECOMMENDATIONS for use rate) may be useful to ensure thorough coverage of the unopened bolls.

Recommendations for Folex + Prep Plant Growth Regulator tank mixtures and sequential applications are:

Product	Rate	Directions
Folex + Prep Plant Regulator (tankmix) or Folex sequential to Prep Plant Regulator	$1\frac{1}{2}$ to 2 Pints/A + 2 to 4 Pints/A; $1\frac{1}{2}$ to 2 Pints/A sequential to 2 to 4 Pints/A	**FOR USE IN THE SOUTHEAST*, SOUTHWEST* AND MIDSOUTH*:** Use Folex and Prep Plant Regulator as a tank mixture or as a sequential application. For use in the SOUTHEAST AND SOUTHWEST, use the low rates of Folex and Prep Plant Regulator when ideal conditions occur (minimum night temperatures above 65°F) and use the high rates when average conditions occur (minimum night temperature between 55° and 65°F). When used in the MIDSOUTH, use only the high rate of Prep Plant Regulator.

* SOUTHEAST AND SOUTHWEST include states of AL, GA, LA, NM, NC, OK, SC, central TN and TX. MIDSOUTH includes states of AR, MS, MO, WEST TN.

Folex + Prep Plant Regulator (tankmix)	$1\frac{1}{2}$ to 2 Pints/A + 1 to 2 Pints/A	**FOR USE IN OK, TX—EXCLUDING THE RIO GRANDE VALLEY:** Use the low rate of Folex and Prep Plant Regulator when ideal conditions occur (minimum night temperature above 65°F) and use the high rates when average conditions occur (minimum night temperature between 55° and 65°F).
Folex sequential to Prep Plant Regulator	$1\frac{1}{2}$ to 2 Pints/A sequential to 1 to 2 Pints/A	**FOR USE IN SOUTHWEST (OK, TX—EXCLUDING THE RIO GRANDE VALLEY), AZ AND CA:** Use the low rate of Folex and Prep Plant Regulator when ideal conditions occur (minimum night temperatures above 65°F) and use the high rates when average conditions occur (minimum night temperature between 55°F and 65°F).

Do not use Folex + Prep Plant Regulator tank mixtures in Arizona or California.

Refer to Prep Plant Regulator label for additional use directions and cautions when using Prep Plant Regulator as a tank mixture or as a sequential application with Folex.

Prep Plant Regulator is a registered trademark of Union Carbide Agricultural Products Company.

RESTRICTIONS AND LIMITATIONS

Application of Folex should not be made within 7 days of harvest. The total cumulative rate of Folex should not exceed $3\frac{2}{3}$ pints/A per use season. To ensure maximum safety and efficacy, follow all cautions and limitations on this label and the labels of products used in tank mixtures with Folex.

CONDITIONS OF SALE, WARRANTY, LIMITATION OF LIABILITY

Rhone-Poulenc warrants that this product conforms to the chemical description on the label and is reasonably fit for purposes described on the label. Rhone-Poulenc makes no other express or implied warranty including any warranty of fitness of merchantability. In no case shall Rhone-Poulenc or the seller be liable for consequential, special, or indirect damages such as loss of profits or values resulting from the use or handling of this product.

MOCAP®
10% Granular
NEMATICIDE–INSECTICIDE

RESTRICTED USE PESTICIDE

Acute dermal toxicity.

For retail sale to and use by Certified Applicators or persons under their direct supervision, and only for those uses covered by the Certified Applicator's certification.

ACTIVE INGREDIENT:

Ethoprop O-Ethyl S,S-Dipropyl Phosphorodithioate 10%

INERT INGREDIENTS 90%

TOTAL 100%

EPA Reg. No. 359-703 EPA Est.

U.S. Patent Nos. 3,112,244; 3,268,393

WARNING
Keep Out of Reach of Children

See antidote and warning information on back panel

STATEMENT OF PRACTICAL TREATMENT

SWALLOWED: Call a physician or Poison Control Center. Drink 1 or 2 glasses of water and induce vomiting by touching back of throat with finger. Do not induce vomiting or give anything to an unconscious person.

INHALED: Remove to fresh air and apply artificial respiration if indicated.

ON SKIN: Remove all contaminated clothing and wash skin with soap and water. Wash clothing before reuse. Discard contaminated shoes.

IN EYES: Flush with plenty of water for at least 15 minutes. Call a physician immediately.

See side or back panel for additional precautionary statements including note to physician.

HAZARDS TO HUMANS AND DOMESTIC ANIMALS

WARNING: May be fatal if swallowed, inhaled or absorbed through skin. Rapidly absorbed through skin. Wear protective clothing. In case of contact, wash immediately with soap and water. Do not breathe fumes or dust. After use, wash hands, arms and face thoroughly with soap and warm water before eating or smoking. Wash all clothing with soap and hot water before reuse. Discard contaminated shoes. Do not contaminate feed or food.

Do not store in or around the home.

SYMPTOMS OF POISONING: Nausea, vomiting, abdominal cramps, diarrhea, excessive salivation, headache, dizziness, weakness, blurring or dimness of vision, excessive tearing, loss of muscular coordination, slurring of speech, twitching of muscles (especially of tongue and eyelids), mental confusion, disorientation, drowsiness, difficulty in breathing (chest tightness), runny nose.

STATEMENT OF PRACTICAL TREATMENT

IF SWALLOWED: Give 1 or 2 glasses of water and induce vomiting by touching back of throat with finger. Do not induce vomiting or give anything by mouth to an unconscious person. Call a physician, Poison Control Center, or hospital emergency room. If not breathing, give mouth to mouth artifical respiration.

IN CASE OF SKIN CONTACT: Immediately remove all contaminated clothing, and wash skin with soap and water. Wash clothing before reuse. Discard contaminated shoes.

FOR EYES: Flush with plenty of water for at least 15 minutes. Call a physician immediately in all cases of suspected poisoning.

Persons providing first aid should prevent contamination of themselves.

NOTE TO PHYSICIAN

This product is a cholinesterase inhibitor. Severe symptoms and signs include diarrhea, pinpoint and non-reactive pupils, respiratory difficulty, pulmonary edema, cyanosis, loss of sphincter control, convulsions, and coma.

Support respiration as needed. Measures should include removal of secretions, maintenance of a patent airway, and if necessary, artificial ventilation.

If cyanosis is absent, give ATROPINE 2-4 mg intravenously (0.05 mg/kg for children). Repeat ATROPINE at 5-10 minute intervals until atropinization occurs (dry, flushed skin, tachycardia, pupillary dilitation), and maintain for 48 hours. If cyanotic, give initial ATROPINE intramuscularly and start measures to improve ventilation.

Start 2-PAM (PROTOPAM, Ayerst) at the same time. Give 1-2 grams PROTOPAM (20-40 mg/kg for children) in 100 cc saline over 15-30 minutes. If pulmonary edema is present, give intravenously slowly as a 5% solution in water over a period of at least 5 minutes.

A second dose may be given after one hour if muscle weakness persists. Additional doses may be given cautiously for persistent muscle weakness. May be given by intramuscular or subcutaneous routes if intravenous administration is not feasible.

In case of skin contact wash patient with soap and water followed by wash with 95% ethyl alcohol. Keep patient under constant observation for 24-36 hours. Symptoms may persist for one month.

The use of theophylline, morphine, barbiturates, phenothiazines, reserpine, and succinyl choline is contraindicated.

ENVIRONMENTAL HAZARDS

This pesticide is toxic to fish and wildlife. Birds feeding in treated areas may be killed. Cover or incorporate granules that are spilled. Do not apply directly to water or wetlands. Drift or runoff from treated areas may be hazardous to aquatic organisms in neighboring areas. Do not contaminate water by cleaning of equipment or disposal of wastes.

DIRECTIONS FOR USE

It is a violation of Federal Law to use this product in a manner inconsistent with its labeling.

Unprotected persons may enter treated fields after chemical has been mixed into the soil.

Granules lying on the soil surface in turn areas at row ends should be incorporated to remove possible hazard to birds and other wildlife.

STORAGE AND DISPOSAL

PROHIBITIONS

Do not contaminate water, food, or feed by storage or disposal. Do not store in or around the home.

PESTICIDE DISPOSAL

Wastes resulting from the use of this product may be disposed of on site or at an approved waste disposal facility.

CONTAINER DISPOSAL

Completely empty bag into application equipment. Then dispose of empty bag in a sanitary landfill or by incineration, or, if allowed by State and local authorities, by burning. If burned, stay out of smoke.

INCORPORATION INFORMATION

MOCAP nematicide-insecticide can be incorporated with a rotary tiller, rotary hoe, spring-tooth harrow, by double-discing or with other equipment for mixing the chemical with the soil to the depth recommended. Follow any specific incorporating directions given under the crop heading.

If moisture is applied immediately after application to carry the treatment into the soil, only shallow incorporation (¼ to ½ inch) with mechanical equipment such as drag chains is required.

NEMATODES CONTROLLED

	Banana & Plantain	Cabbage	Field Corn & Sweet Corn	Cucumbers	Peanuts (planting)	Peanuts (pegging)	Pineapple	White (Irish) Potato	Sweet Potato	Snap Beans & Lima Beans	Soybeans	Sugarcane	Tobacco	Commercial Turf
Burrowing Nematodes														
Radopholus similis	C													
Dagger Nematodes														
Ciphinema spp.			C								S			
Cyst Nematodes														
Heterodera glucines											S			
Heterodera cruciferae		C												
Heterodera schactii		C												
Lesion Nematodes														
Pratylenchus spp.	C	C	C	C	C	C		C		C	C	C	C	C
Root-Knot Nematodes														
Meloidogyne spp.	C										C	C		C
Meloidogyne hapla		C	C	C	C	C		C	C	C			C	
Meloidogyne incognita		C	C	C				C	C	C			C	
Meloidogyne arenaria					C	C							C	
Meloidogyne javanica				C									C	
Lance Nematodes														
Hoplolaimus spp.		S	C	S						S				
Hoplolaimus tylenchiformis				C										
Hoplolaimus columbus											S			

NEMATODES CONTROLLED

	Banana & Plantain	Cabbage	Field Corn & Sweet Corn	Cucumbers	Peanuts (planting)	Peanuts (pegging)	Pineapple	White (Irish) Potato	Sweet Potato	Snap Beans & Lima Beans	Soybeans	Sugarcane	Tobacco	Commercial Turf
Spiral Nematodes														
Helicotylenchus spp.	C		C	C				C	C	C	C			C
Rotylenchus spp.			C											
Sting Nematodes														
Belonolaimus spp.			C		C	C								
Belonolaimus longicaudatus		C		C				C		C	C			C
Ring Nematodes														
Criconemoides spp.				C	C	C		C			C	C		C
Criconemoides ornatus			C							C				
Reniform Nematodes														
Rotylenchulus reniformis							C		C					
Stunt Nematodes														
Tylenchorhynchus spp.			C	C	C	C		C	C		C		C	C
Tylenchorhynchus claytoni										C				
Tylenchorhynchus martini												C		
Stubby Root Nematodes														
Trichodorus spp.			C		C	C		C	C					C
Trichodorus christiei		C		C							C	C		
Pin Nematodes														
Paratylenchus spp.														C

C = Control S = Suppression

INSECTS CONTROLLED

	Banana & Plantain	Field Corn & Sweet Corn	Peanuts (pegging)	White (Irish) Potato	Sweet Potato	Sugarcane	Tobacco	Commercial Turf
Corn Rootworm								
Diabrotica virgifera		C						
Diabrotica longicoris		C						
Diabrotica undecimpuctata		C	C					
Diabrotica howardi		C						
Wireworms								
Conoderus spp.							C	
Conoderus vespertinus		C			C			
Conoderus falli		C						
Melanotus spp.		C			C			
Melanotus communis				C		C		
Limonius dubitan		C						
Limonius canus				C				
Limonius californicus				C				
Aeolus mellius		C						
Ctenicera. spp.		C						
Ctenicera pruinina		C		C				
Eleodes spp.				C				
Black Cutworm								
Agrotis ipsilon		C*						
Sandhill Cutworm								
Euxoa detersa		C*						
White Grubs								
Phyllophaga spp.								
Phyllophaga crinita		S						
Phyllophaga ephilida		S			C			

INSECTS CONTROLLED

	Banana & Plantain	Field Corn & Sweet Corn	Peanuts (pegging)	White (Irish) Potato	Sweet Potato	Sugarcane	Tobacco	Commercial Turf
Banded Cucumber Beetle								
Diabrotica balteata					C			
Fleabeetle larvae								
Chaetocnema confinis					C			
Systena blanda					C			
Lesser Cornstalk Borer								
Elasmopalpus lignosellus			S					
Banana Root Borer								
Cosmopolites sordidus	C							
Japanese Beetle Grub								
Popillia japonica								C
European Chafer								
Rhizotrogus majalis								C
Black Turfgrass Ataenius Grub								
Ataenius Spretulus								C
Chinchbug								
Blissus leucopterus								C
Bluegrass Billbug								
Sphenophorus parvulus								C
Sod Webworm								
Crambus spp.								C
Herpetogramma spp.								C
Mole Crickets								
Scapteriscus vicinus								C
Scapteriscus acletus								C

C—Control S—Suppression

* **NOTE:** Under heavy infestations of cutworms, enough cutworms may survive to cause some crop damage. If this occurs, it may be necessary to apply a rescue treatment with another registered pesticide to control the surviving worms.

BANANA AND PLANTAIN

For nematode and banana root borer *(Cosmopolites sordidue)* control: Apply 60 grams of 10% MOCAP in a radius of 3/4 meter around each producing stem or 2 ounces of 10% MOCAP in a radius of 30 inches. Before application, remove ground litter from area to be treated and apply granules evenly. For best results, make application just before irrigation or a rainy period. If application is made during a dry period, mix granules into the top inch of soil with a rake. Repeat application in 6 months.

BEAN (SNAP AND LIMA)

Row treatment for nematode control: Use 20 to 30 pounds of 10% MOCAP per acre (36-inch row spacing) or 1/7 to 1/5 pound per 100 linear feet of row. Make application in band 12 to 15 inches wide on the row. **Do not** use as a seed furrow treatment or allow granules to contact the seed.

Broadcast application for nematode control: Use 60 to 80 pounds per acre. Apply anytime from 3 days before planting to at-planting time. Mix with the top 4 inches of soil right after application.

NOTE: Beans grown in treated soil may remain green longer than other beans and mature 1 to 2 weeks later.

CABBAGE

Row treatment for nematode control: Use 20 pounds of 10% MOCAP per acre on a band 15 inches wide on the row (36-inch row spacing) **or 1/8 pound per 100 linear feet of row. Do not** use as a seed furrow treatment or allow granules to contact the seed.

Broadcast treatment for nematode control: Use 50 pounds of 10% MOCAP per acre. Apply evenly over the soil area to be treated from 1 week before planting to at-planting time. Mix with the top 3 inches of soil right after application.

CORN (FIELD AND SWEET)

At-planting application for control of corn rootworms, wireworms, cutworms, Symphylans *(Scutigerella maculata)*, nematodes and suppression of white grubs: Use 12 ounces of 10% MOCAP per 1,000 feet of row. Rate per acre depends on row spacing as shown below.

Row spacing:	**40″**	**38″**	**36″**	**30″**	**20″**
Lbs. per acre:	10	10.5	11.1	13.3	20

Apply the granules in a 6 to 7-inch band on the row over a closed seed furrow. Mix the granules with the top 1/2 inch of soil with drag chains, spring-tooth incorporators, or similar equipment.

For commonly used applicators see calibrations below:

Gauge settings for 12 ounces per 1000 feet of row (Do not change with row width.)

BRAND OF APPLICATOR	SPEED IN MILES PER HOUR 2	3	4	5	6	7	8	9	10
Allis Chalmers	At all Speeds: GA9 = 11.5 Ounces . . . GA10 = 12.7 Ounces								
Gandy	19.1	23.5	26.7	29.5	32.1	34.5	36.6	38.7	40.8
International Harvester*	1/8.0	2/2.7	2/8.5	3/4.0	3/8.7	4/5.0	5/1.3	5/6.0	6/1.5
International Harvester (Cyclo)	1/5.7	1/9.3	2/4.7	3/0.5	3/5.1	4/0.6	4/5.8	5/1.5	5/7.0
John Deere**	1/29	2/6	2/13	2/19	2/27	3/1	3/6	3/11	3/15
John Deere (Max. Emerge)	13	19	24	29	32	33	35	36	37
Noble	13	19	24	28	31	34	36	38	39
Noble (New)	9	11	14	17	20	22	25	27	29

* 1/2.0 means Gate 1. Dial 8.0
** 1/29 means Range 1. Notch 29

Cultivation application for corn rootworm control: Apply after planting until layby. Use the rates shown above for at-planting application. On small corn, apply in a 6 to 7-inch band over the row. On taller corn, apply in bands 3 to 4 inches wide on each side of the corn, directing the granules toward the base of the plants. Cover the granules with ½ to 1 inch of soil right after application.

Broadcast treatment for cutworm control: Use 30 pounds of 10% MOCAP per acre or ¾ pounds per 1,000 square feet. Apply treatment from 3 days before planting to at-planting time. Mix treatment with the top 2 inches of soil right after application.

Nematode control: Row treatment: Use 1⅛ to 1½ pounds of 10% MOCAP per 1,000 feet of row. Rate per acre depends on row spacing as shown below. Make application in a band 12 to 15 inches wide on the row.

Row Spacing	40″	38″	36″	30″	20″
Lbs. per acre:	15-20	15.5-21	16.3-22.2	19.6-26.6	30-40

Broadcast treatment: Use 60 pounds of 10% MOCAP per acre or 1½ pounds per 1,000 square feet. Apply treatment from 3 days before planting to at-planting time. Mix treatment with the top 2 inches of soil right after application. The treatments for nematode control will also control rootworms, wireworms, cutworms and Symphylans (*Scutigerella maculata*), and suppress white grubs.

CUCUMBER

Row treatment for nematode control: Use 20 pounds of 10% MOCAP per acre (7-foot row spacing) or $^3/_{10}$ pound per 100 linear feet of row. Make application in a band 12 to 15 inches wide on the row at or just before planting. Prepare bed and knock off top to provide a level area for treatment. Mix with the top 2 inches of soil right after application. **Do not** use as a seed furrow treatment or allow granules to contact the seed.

PEANUT

Row treatment for nematode control planting: Use 20 to 40 pounds of 10% MOCAP per acre (2 to 4 ounces per 100 linear feet of row) in a band 12 inches wide on the row (36 inch row spacing). For control of peanut root-knot nematode (*M. arenaria*) use the higher rate of 10% MOCAP. **Do not** use as a seed furrow treatment or allow granules to contact the seed.

Broadcast treatment at planting: Use 40 to 60 pounds of 10% MOCAP per acre. Make application from 1 week before planting to at-planting time. Mix with the top 4 inches of soil right after application.

Nematode, corn rootworm control and suppression of Lesser cornstalk borer and white mold *(Sclerotium rolfsii)* at pegging: Use 30 pounds of 10% MOCAP per acre (36-inch row spacing) or ⅕ pound per 100 linear feet of row. Make application in a band 15 to 18 inches wide on the row.

Corn rootworm control at pegging: Use 25 pounds of 10% MOCAP per acre (36-inch row spacing) or ⅙ pound per 100 linear feet of row. Make application in a band 18 inches wide on the row.

NOTE: Pegging treatment should be incorporated lightly into the soil.

PINEAPPLE

Broadcast treatment for nematode control: Use 200 to 400 pounds of 10% MOCAP per acre. Use higher rates where longer residual activity is desired or in field where nematodes have been especially severe. Apply as broadcast treatment anytime from 3 weeks before planting to at-planting time. Mix with the top 4 inches of soil right after application.

POTATOES

Wireworm control: Row treatment: Use 30 pounds of 10% MOCAP per acre (40-inch row spacing) or ¼ pound per 100 linear feet of row. Make application in a band 12 inches wide on the row at planting. The following broadcast treatment is recommended for fields where wireworm injury has been (or is suspected to be) especially severe (i.e., potatoes planted following sod, pasture or new ground).

Broadcast treatment: Use 40-60 pounds of 10% MOCAP per acre. Mix treatment with the top 4 inches of soil right after application.

Nematode control: Row treatment: (Control of Lesion, Spiral, Sting, Ring, Stunt, and Stubby Root Nematodes and Suppression of Northern Root Knot and Southern Root Knot Nematodes.) Use 30 pounds of 10% MOCAP per acre (40-inch row spacing) or ¼ pound per 100 linear feet of row. Make application in a band 12 inches wide on the row at planting.

Broadcast treatment: (Control of Northern Root Knot, Southern Root Knot, Lesion, Spiral, Sting, Ring, Stunt and Stubby Root Nematodes.) Use 60-90 pounds 10% MOCAP per acre. Mix treatment with the top 4 inches of soil right after application. The higher rate treatment is recommended for fields in which nematode injury has been especially severe or where soil is sandy.

SOYBEAN

Row treatment for nematode control: Use 15 to 30 pounds of 10% MOCAP per acre or ⅛ to ¼ pound per 100 linear feet of row. Make application in a band 12 to 15 inches wide on the row any time from 1 week before planting to at-planting time. Mix with the top 3 inches of soil right after application. **Do not** use as a seed furrow treatment or allow granules to contact seed.

NOTE: Soybeans treated with organo phosphate chemicals such as MOCAP may be more susceptible to injury from metribuzin and propanil herbicides.

SUGARCANE

Furrow treatment for nematode control: Use 20 to 40 pounds of 10% MOCAP per acre at planting time (6-foot row spacing). Use the higher rates where nematodes have been especially severe. Apply evenly in a band 12 to 15 inches wide, centered over the seed pieces in the opened furrow, immediately before closing. Mix with the soil with a disc hiller or other suitable equipment as the cane is covered.

SWEET POTATO

For nematode, wireworm, white grub, banded cucumber beetle and fleabeetle larvae (pale striped and sweet potato fleabeetle) control:

Row treatment: Use 30 to 40 pounds of 10% MOCAP per acre or ¼ to ⅓ pound per 100 linear feet of row (minimum 42-inch row spacing). Make application in a band 12 to 15 inches wide on the row 2 to 3 weeks before planting. Mix with the top 4 inches of soil right after application.

Good results have been obtained by opening the row, applying 10% MOCAP in a 12- to 15- inch band on the row, bedding over and dragging.

Broadcast treatment: Use 60 to 80 pounds of 10% MOCAP per acre. Apply the treatment evenly on top of the soil 2 to 3 weeks before planting. Mix with the top 4 inches of soil right after application. Use a double gang disc harrow or other equipment which will do a good job mixing the 10% MOCAP with the soil.

TOBACCO

Row treatment for nematode and wireworm control: Use 60 to 120 pounds of 10% MOCAP per acre or ½ to 1 pound per 100 linear feet of row. Apply after fertilizer has been listed or bedded. Make application in a band 18 to 24 inches wide on the row, anytime from 1 week before planting to at-planting time. Mix with the top 4 inches of soil right after application.

The higher rate treatment is recommended for fields in which nematode injury to tobacco has been especially severe, or where soil is sandy.

Broadcast treatment for nematode and wireworm control: Use 60 to 120 pounds of 10% MOCAP per acre. Apply the treatment evenly on the top of the soil anytime from 1 week before planting to at-planting time. Mix with the top 3 inches of soil right after application. Use a double gang disc harrow or other equipment which will do a good job of mixing the 10% MOCAP with the soil. After mixing, apply fertilizer and, for best results, shape a wide bed with a tractor equipped with 4 disc hillers and wide sweeps for running out the middles. Be sure that only the treated soil (top 3 inches) is used to make beds.

The higher rate treatment is recommended for fields in which nematode injury to tobacco has been especially severe, or where soil is sandy.

Broadcast treatment for wireworm control: Use 20 pounds of 10% MOCAP per acre. Apply as a broadcast treatment anytime from 2 weeks before planting to at-planting time. Mix with the top 2 inches of soil right after application. Use a disc harrow or other cultivating equipment that will do a good job of mixing 10% MOCAP with the soil.

Rhone-Poulenc does not recommend the use of MOCAP on tobacco in Florida.

COMMERCIAL TURF
(Such as Golf Courses, Sod Farms, Cemeteries)

FOR USE ONLY BY PROFESSIONAL TURFMEN

For Nematode Control (Bahia, Bermuda, Centipede, Fescue, Kentucky Blue, St. Augustine, and Zoysia grasses): Use 4.6 to 6.9 pounds of MOCAP 10% G per 1,000 square feet (200–300 pounds per acre) of established turf. Apply uniformly with a fertilizer spreader or other suitable equipment. Severe infestations may require re-treatment after 6 months.

For Insect Control (Bent, Bahia, Bermuda, Centipede, Fescue, Kentucky Blue, St. Augustine, and Zoysia grasses): Use 2.3 pounds of MOCAP 10% G per 1,000 square feet (100 pounds per acre) of established turf to control mole crickets. Use 1.15 pounds per 1,000 square feet (50 pounds per acre) of established turf to control chinch bugs and larvae of Japanese beetle, European chafer, black turfgrass Ataenius beetle, bluegrass billbug, and sod webworms. Apply uniformly with a fertilizer spreader or other suitable equipment.

Notes: Immediately after application, apply ½ inch of water to treated grass. Failure to apply water may result in a hazard to animals, poor pest control, injury to grass and longer presence of chemical odor. DO NOT allow unprotected humans or animals to walk on treated areas until grass and soil surface are dry.

Do not apply with equipment carried on the chest. Do not use cyclone or rotary-type applicators which are likely to throw the granules to areas outside the lawn area to be treated. Any granules on non-lawn areas, such as walkways and driveways, must be removed immediately by sweeping or washing the granules onto the lawn.

For use on home lawns only by Professional Turfmen. Do not use on newly seeded areas until plants are well established. Do not apply to wet foliage. Do not cut sod or handle treated turf for 21 days after treatment. Birds and other wildlife in treated areas may be killed. Do not apply where runoff is likely to occur.

CONDITION OF SALE, WARRANTY, LIMITATION OF LIABILITY

This product conforms to the chemical description on the label thereof and is reasonably fit for the purpose stated on such label only when used in accordance with directions under normal use conditions. Follow directions carefully. Timing and method of application, weather and crop conditions, mixture with other chemicals not specifically recommended and other influencing factors in the use of this product are beyond the control of Rhone-Poulenc or the seller. Buyer assumes all risks of use, storage, or handling of this material, not in strict accordance with directions given herewith. In no case shall Rhone-Poulenc or the seller be liable for consequential, special, or indirect damages such as loss of profits or values resulting from the use or handling of this product.

MOCAP®
15% Granular

NEMATICIDE–INSECTICIDE

RESTRICTED USE PESTICIDE

Acute Dermal Toxicity

For retail sale to, and use by Certified Applicators or persons under the direct supervision of a Certified Applicator, and only for those uses covered by the Certified Applicator's certification.

ACTIVE INGREDIENT:

Ethoprop: *O*-ethyl *S,S*-dipropyl phosphorodithioate	15%
INERT INGREDIENTS	85%
TOTAL	100%

EPA Reg. No. 359-695 EPA Est. 359-TN-1

Keep Out of Reach of Children

DANGER–POISON

See Antidote And Warning Information On Back Panel

STATEMENT OF PRACTICAL TREATMENT

SWALLOWED: Call a physician or Poison Control Center. Drink 1 or 2 glasses of water and induce vomiting by touching back of throat with finger. Do not induce vomiting or give anything to an unconscious person.

INHALED: Remove to fresh air and apply artificial respiration if indicated.

ON SKIN: Remove all contaminated clothing and wash skin with soap and water. Wash clothing before reuse. Discard contaminated shoes.

IN EYES: Flush with plenty of water for at least 15 minutes. Call a physician immediately.

See side or back panel for additional precautionary statements including note to physician.

HAZARDS TO HUMANS AND DOMESTIC ANIMALS

DANGER: Poisonous if swallowed, inhaled or absorbed through skin. Rapidly absorbed through skin. Wear protective clothing. In case of contact, wash immediately with soap and water. Do not breathe fumes or dust. After use, wash hands, arms and face thoroughly with soap and warm water before eating or smoking. Wash all clothing with soap and hot water before reuse. Discard contaminated shoes. Do not contaminate feed or food.

Do not store in or around the home.

SYMPTOMS OF POISONING: Nausea, vomiting, abdominal cramps, diarrhea, excessive salivation, headache, dizziness, weakness, blurring or dimness of vision, excessive tearing, loss of muscular coordination, slurring of speech, twitching of muscles (especially of tongue and eyelids), mental confusion, disorientation, drowsiness, difficulty in breathing (chest tightness), runny nose.

STATEMENT OF PRACTICAL TREATMENT

IF SWALLOWED: Give one or 2 glasses of water and induce vomiting by touching back of throat with finger. Do not induce vomiting or give anything by mouth to an unconscious person. Call a physician, Poison Control Center, or hospital emergency room. If not breathing, give mouth to mouth artificial respiration.

IN CASE OF SKIN CONTACT: Immediately remove all contaminated clothing, and wash skin with soap and water. Wash clothing before reuse. Discard contaminated shoes.

FOR EYES: Flush with plenty of water for at least 15 minutes. Call a physician immediately in all cases of suspected poisoning.

Persons providing first aid should prevent contamination of themselves.

NOTE TO PHYSICIAN: This product is a cholinesterase inhibitor. Severe symptoms and signs include diarrhea, pinpoint and nonreactive pupils, respiratory difficulty, pulmonary edema, cyanosis, loss of sphincter control, convulsions, and coma.

Support respiration as needed. Measures should include removal of secretions, maintenance of a patent airway, and if necessary, artificial ventilation.

If cyanosis is absent, give ATROPINE 2-4 mg intravenously (0.05 mg/kg for children). Repeat ATROPINE at 5-10 minute intervals until atropinization occurs (dry, flushed skin, tachycardia, pupillary dilatation), and maintain for 48 hours. If cyanotic, give initial ATROPINE intramuscularly and start measures to improve ventilation.

Start 2-PAM (PROTOPAM, Ayerst) at the same time. Give 1-2 grams PROTOPAM (20-40 mg/kg for children) in 100cc saline over 15-30 minutes. If pulmonary edema is present, give intravenously slowly as a 5% solution in water over a period of at least 5 minutes.

A second dose may be given after one hour if muscle weakness persists. Additional doses may be given cautiously for persistent muscle weakness. May be given by intramuscular or subcutaneous routes if intravenous administration is not feasible.

In case of skin contact wash patient with soap and water followed by wash with 95% ethyl alcohol. Keep patient under constant observation for 24-36 hours. Symptoms may persist for one month.

The use of theophylline, morphine, barbiturates, phenothiazines, reserpine, and succinyl choline is contraindicated.

ENVIRONMENTAL HAZARDS

This pesticide is toxic to fish and wildlife. Birds feeding in treated areas may be killed. Cover or incorporate granules that are spilled. Do not apply directly to water or wetlands. Drift or runoff from treated areas may be hazardous to aquatic organisms in neighboring areas. Do not contaminate water by cleaning of equipment or disposal of wastes.

DIRECTIONS FOR USE

It is a violation of Federal law to use this product in a manner inconsistent with its labeling.

Unprotected persons may enter treated fields after chemical has been mixed into the soil.

Granules lying on the soil surface in turn areas at row ends should be incorporated to remove possible hazard to birds and other wildlife.

STORAGE AND DISPOSAL

PROHIBITIONS

Do not contaminate water, food, or feed by storage or disposal. Do not store in or around the home.

PESTICIDE DISPOSAL

Wastes resulting from the use of this product may be disposed of on site or at an approved waste disposal facility.

CONTAINER DISPOSAL

Completely empty bag into application equipment. Then dispose of empty bag in a sanitary landfill or by incineration, or, if allowed by State and local authorities, by burning. If burned, stay out of smoke.

INCORPORATION INFORMATION

MOCAP nematicide-insecticide can be incorporated with a rotary tiller, rotary hoe, spring-tooth harrow, by double-discing or with other equipment for mixing the chemical with the soil to the depth recommended. Follow any specific incorporating directions given under the crop heading.

If moisture is applied immediately after application to carry the treatment into the soil, only shallow incorporation (¼ to ½ inch) with mechanical equipment such as drag chains is required.

NEMATODES CONTROLLED

	Banana & Plantain	Cabbage	Field Corn & Sweet Corn	Cucumbers	Peanuts (planting)	Peanuts (pegging)	Pineapple	White (Irish) Potato	Sweet Potato	Snap Beans & Lima Beans	Soybeans	Sugarcane	Tobacco	Commercial Turf
Burrowing Nematodes														
Radopholus similis	C													
Dagger Nematodes														
Ciphinema spp.			C								S			
Cyst Nematodes														
Heterodera glucines											S			
Heterodera cruciferae		C												
Heterodera schactii		C												
Leslon Nematodes														
Pratylenchus spp.	C	C	C	C	C	C		C		C	C	C	C	C

NEMATODE CONTROLLED

	Banana & Plantain	Cabbage	Field Corn & Sweet Corn	Cucumbers	Peanuts (planting)	Peanuts (pegging)	Pineapple	White (Irish) Potato	Sweet Potato	Snap Beans & Lima Beans	Soybeans	Sugarcane	Tobacco	Commercial Turf
Root-Knot Nematodes														
Meloidogyne spp.	C										C	C		C
Meloidogyne hapla		C	C	C	C	C		C	C	C			C	
Meloidogyne incognita		C	C	C				C	C	C			C	
Meloidogyne arenaria					C	C							C	
Meloidogyne javanica				C									C	
Lance Nematodes														
Hoplolaimus spp.		S	C	S						S				
Hoplolaimus tylenchiformis				C										
Hoplolaimus columbus											S			
Spiral Nematodes														
Helicotylenchus spp.	C		C	C				C	C	C	C			C
Rotylenchus spp.			C											
Sting Nematodes														
Belonolaimus spp.			C		C	C								
Belonolaimus longicaudatus		C		C				C		C	C			C
Ring Nematodes														
Criconemoides spp.				C	C	C		C			C	C		C
Criconemoides ornatus			C							C				
Reniform Nematodes														
Rotylenchulus reniformis							C		C					
Stunt Nematodes														
Tylenchorhynchus spp.			C	C	C	C		C	C		C		C	C
Tylenchorhynchus claytoni										C				
Tylenchorhynchus martini												C		
Stubby Root Nematodes														
Trichodorus spp.			C		C	C		C	C					C
Trichodorus christiei		C		C							C	C		
Pin Nematodes														
Paratylenchus spp.														C

INSECTS CONTROLLED

	Banana & Plantain	Field Corn & Sweet Corn	Peanuts (pegging)	White Irish Potatoes	Sweet Potato	Sugarcane	Tobacco	Commercial Turf
Corn Rootworm								
Diabrotica virgifera		C						
Diabrotica longicornis		C						
Diabrotica undecimpuctata		C	C					
Diabrotica howardi		C						
Wireworms								
Conoderus spp.							C	
Conoderus vespertinus		C			C			
Conoderus falli		C						
Melanotus spp.		C			C			
Melanotus communis				C		C		
Limonius dubitan		C						
Limonius canus				C				
Limonius californicus				C				
Aeolus mellius		C						
Ctenicera spp.		C						
Ctenicera pruinina		C		C				
Eleodes spp.				C				
Black Cutworm								
Agrotis ipsilon		C*						
Sandhill Cutworm								
Euxoa detersa		C*						

INSECTS CONTROLLED

	Banana & Plantain	Field Corn & Sweet Corn	Peanuts (pegging)	White Irish Potatoes	Sweet Potato	Sugarcane	Tobacco	Commercial Turf
White Grubs								
Phyllophaga spp.								
Phyllophaga crinita		S						
Phyllophaga ephilida		S			C			
Banded Cucumber Beetle								
Diabrotica balteata					C			
Fleabeetle larvae								
Chaetocnema confinis					C			
Systena blanda					C			
Lesser Cornstalk Borer								
Elasmopalpus lignosellus			S					
Banana Root Borer								
Cosmopolites sordidus	C							
Japanese Beetle								
Popillia japonica								C
European Chafer								
Rhizotrogus majalis								C
Black Turfgrass Ataenius								
Ataenius Spretulus								C
Chinchbug								
Blissus leucopterus								C
Bluegrass Billbug								
Sphenophorus parvulus								C
Sod Webworm								
Crambus spp.								C
Herpetogramma spp.								C
Mole Crickets								
Scapteriscus vicinus								C
Scapteriscus acletus								C

* NOTE: Under heavy infestations of cutworms, enough cutworms may survive to cause some crop damage. If this occurs, it may be necessary to apply a rescue treatment with another registered pesticide to control the surviving worms.

C=Control S=Suppression

BANANA AND PLANTAIN

For nematode and banana root borer *(Cosmopolites sordidue)* control: Apply 40 grams of 15% MOCAP in a radius of ¾ meter around each producing stem or 1.3 ounce of 15% MOCAP in a radius of 30 inches around each producing stem. Before application, remove ground litter from area to be treated and apply granules evenly. For best results, make application just before irrigation or a rainy period. If application is made during a dry period, mix granules into the top inch of soil with a rake. Repeat application in 6 months.

BEAN (SNAP AND LIMA)

Row treatment for nematode control: Use 13 to 20 pounds of 15% MOCAP per acre (36-inch row spacing) or 1½-2 ounces per 100 linear feet of row. Make application in band 12 to 15 inches wide on the row. **Do not** use as a seed furrow treatment or allow granules to contact the seed.

Broadcast application for nematode control: Use 40 to 54 pounds per acre. Apply anytime from 3 days before planting to at-planting time. Mix with the top 6 inches of soil right after application.

NOTE: Beans grown in treated soil may remain green longer than other beans and mature 1 to 2 weeks later.

CABBAGE

Row treatment for nematode control: Use 13 pounds of 15% MOCAP per acre on a band 15 inches wide on the row (36-inch row spacing) or 1.3 ounce per 100 linear feet of row. **Do not** use as a seed furrow treatment or allow granules to contact the seed.

Broadcast treatment for nematode control: Use 34 pounds of 15% MOCAP per acre. Apply evenly over the soil area to be treated from 1 week before planting to at-planting time. Mix with the top 3 inches of soil right after application.

CORN (FIELD AND SWEET)

At-planting application for control of corn rootworms, wireworms, cutworms, Symphylans *(Scutigerella maculata)* nematodes and suppression of white grubs: Use 8 ounces of 15% MOCAP per 1,000 feet of row. Rate per acre depends on row spacing as shown below:

Row Spacing:	40″	38″	36″	30″	20″
Lbs. per acre:	6.6	7	7.3	8.8	13.3

Apply the granules in a 6 to 7-inch band on the row over a closed seed furrow. Mix the granules with the top ½ inch of soil with drag chains, spring-tooth incorporators, or similar equipment.

For commonly used applicators, see calibrations following.

Gauge settings for 8 ounces per 1,000 feet of row.
(Do not change row width.)

For Application Rate of 8 oz./1000 feet of Row

	Speed in Miles Per Hour								
Brand of Applicator	2	3	4	5	6	7	8	9	10
Allis Chalmers	Gauge 7	—Rate is uniform for all speeds—							Gauge 7
Gandy snap-on bottom	15.3	18.5	21.0	23.1	25.1	26.0	28.6	30.4	31.9
International Harvester (old)	1/5.3	1/7.7	1/9.7	2/3.0	2/6.3	2/9.5	3/2.7	3/6.0	3/9.2

Gauge settings for 8 ounces per 1,000 feet of row.
(Do not change row width.)

For Application Rate of 8 oz./1000 feet of Row

	Speed in Miles Per Hour								
Brand of Applicator	2	3	4	5	6	7	8	9	10
	Gauge	—Rate is uniform for all speeds—							Gauge
International Harvester (new)	1/7.2	1/9.4	2/2.4	2/5.7	2/9.9	3/2.3	3/5.4	3/9.7	4/2.2
John Deere (metal boxes)	1/22	1/24	1/26	1/29	2/4	2/8	2/13	2/17	2/22
John Deere new (1)	21	22.5	24	25	27	28	29	31	32
John Deere newest (2)	8	11	13	15	17	18	20	21	23
Noble (3)	7	11	14	18	21	24	27	29	31
Noble (new) (4)	5	7	9	12	14	17	19	21	23

(1) Gauge Reads 10-20-30-40
(2) Gauge Reads 5-15-25-35-45
(3) Round hole in bottom of hopper
(4) Rectangular hole in bottom of hopper

Cultivation application for corn rootworm control: Apply after planting until layby. Use the rates shown above for at-planting application. On small corn, apply in a 6 to 7-inch band over the row. On taller corn, apply in bands 3 to 4 inches wide on each side of the corn, directing the granules toward the base of the plants. Cover the granules with soil by making application immediately ahead of cultivation equipment.

Broadcast treatment for Cutworm control: Use 20 pounds of 15% MOCAP per acre or ½ pound per 1,000 square feet. Apply treatment from 3 days before planting to at-planting time. Mix treatment with the top 2 inches of soil right after application.

Nematode control: Row treatment: Use ¾ to 1 pound of 15% MOCAP per 1,000 feet of row. Rate per acre depends on row spacing as shown below. Make application in a band 12 to 15 inches wide on the row. Mix treatment with the top 2 inches of soil right after application.

Row Spacing:	40″	38″	36″	30″	20″
Lbs. per acre:	10-13.3	10.3-14	10.8-14.8	13-17.7	20-26.6

Broadcast treatment: Use 40 pounds of 15% MOCAP per acre or 1 pound per 1,000 square feet. Apply treatment from 3 days before planting to at-planting time. Mix treatment with the top 2 inches of soil right after application. The treatments for nematode control will also control rootworms, wireworms, cutworms, Symphylans (*Scutigerella maculata*), and suppress white grubs.

CUCUMBER

Row treatment for nematode control: Use 13 pounds of 15% MOCAP per acre (7-foot row spacing) or 3 ounces per 100 linear feet of row. Make application in a band 12 to 15 inches wide on the row at or just before planting. Prepare bed and knock off top to provide a level area for treatment. Mix with the top 2 to 3 inches of soil right after application. **Do not** use as a seed furrow treatment or allow granules to contact the seed.

PEANUT

Row treatment for nematode control at planting: Use 13.3 to 26.7 pounds of 15% MOCAP per acre (1.3 to 2.7 ounces per 100 linear feet of row) in a band 12 inches wide on the row (36 inches row spacing). For control of peanut root-knot nematode (*M. arenaria*) use the higher rate of 15% MOCAP. **Do not** use as a seed furrow treatment or allow granules to contact the seed.

Broadcast treatment at planting: Use 26.7 to 40 pounds of 15% MOCAP per acre. Make application from 1 week before planting to at-planting time. Mix with the top 4 inches of soil right after application.

Nematode, corn rootworm control and suppression of lesser cornstalk borer and white mold *(Sclerotium rolfsii)* at pegging: Use 20 pounds of 15% MOCAP per acre (36-inch row spacing) or 2.1 ounces per 100 linear feet of row. Make application in a band 15 to 18 inches wide on the row.

Corn rootworm control at pegging: Use 16.7 pounds of 15% MOCAP per acre (36-inch row spacing) or 1.8 ounces per 100 linear feet of row. Make application in a band 18 inches wide on the row.

NOTE: Pegging treatment should be incorporated lightly into the soil.

POTATOES

WIREWORM CONTROL: Row Treatment: Use 20 pounds of 15% MOCAP per acre (40-inch spacing) or 2.7 ounces per 100 linear feet of row. Make application in a band 12 inches wide on the row at planting. The following broadcast treatment is recommended for fields where wireworm injury has been (or is suspected to be) especially severe (i.e., potatoes planted following sod, pasture or new ground).

Broadcast Treatment: Use 26.7 to 40 pounds of 15% MOCAP per acre. Mix treatment with the top 4 inches of soil right after application.

NEMATODE CONTROL: Row treatment: (Control of Lesion, Spiral, Sting, Ring, Stunt, and Stubby Root Nematodes and Suppression of Northern Root-Knot and Southern Root-Knot Nematodes.) Use 20 pounds of 15% MOCAP per acre (40 inch row spacing) or 2.7 ounces per 100 linear feet of row. Make application in a band 12 inches wide on the row at planting.

Broadcast Treatment: (Control of Northern Root-Knot, Southern Root-Knot, Lesion, Spiral, Sting, Ring, Stunt and Stubby Root Nematodes.) Use 40 to 60 pounds 15% MOCAP per acre. Mix treatment with the top 4 inches of soil right after application. The higher rate treatment is recommended for fields in which nematode injury has been especially severe or where soil is sandy.

SOYBEAN

Row treatment for nematode control: Use 10 to 20 pounds of 15% MOCAP per acre or 1.3 to 2.6 ounces per 100 linear feet of row. Make application in a band 12 to 15 inches wide on the row anytime from 1 week before planting to at-planting time. Mix with the top 3 to 6 inches of soil right after application. **Do not** use as a seed furrow treatment or allow granules to contact seed.

NOTE: Soybeans treated with organophosphate chemicals such as MOCAP may be more susceptible to injury from metribuzin and propanil herbicides.

SUGARCANE

Furrow treatment for nematode and wireworm control: Use 13 to 27 pounds of 15% MOCAP per acre at planting time (6-foot row spacing). Use the higher rates where nematodes have been especially severe. Apply evenly in a band 12 to 15 inches wide, centered over the seed pieces in the opened furrow, immediately before closing. Mix with the soil with a disc hiller or other suitable equipment as the cane is covered.

SWEET POTATO

For nematode, wireworm, white grub, banded cucumber beetle and fleabeetle larvae (pale striped and sweet potato fleabeetle) control:

Row treatment: Use 20 to 27 pounds of 15% MOCAP per acre or 2.6 to 3.6 ounces per 100 linear feet of row (minimum 42-inch row spacing). Make application in a band 12 to 15 inches wide on the row 2 to 3 weeks before planting. Mix with the top 4 inches of soil right after application.

Good results have been obtained by opening the row, applying 15% MOCAP in a 12 to 15-inch band on the row, bedding over and dragging.

Broadcast treatment: Use 40 to 54 pounds of 15% MOCAP per acre. Apply the treatment evenly on top of the soil 2 to 3 weeks before planting. Mix with the top 4 to 8 inches of soil right after application. Use a double gang disc harrow or other equipment which will do a good job of mixing the 15% MOCAP with the soil.

TOBACCO

Row treatment for nematode and wireworm control: Use 40 to 80 pounds of 15% MOCAP per acre or ⅓ to ⅔ pound per 100 linear feet of row. Apply after fertilizer has been listed or bedded. Make application in a band 18 to 24 inches wide on the row, anytime from

1 week before planting to at-planting time. Mix with the top 4 to 6 inches of soil right after application.

The higher rate treatment is recommended for fields in which nematode injury to tobacco has been especially severe, or where soil is sandy.

Broadcast treatment for nematode and wireworm control: Use 40 to 80 pounds of 15% MOCAP per acre. Apply the treatment evenly on the top of the soil anytime from 1 week before planting to at-planting time. Mix with the top 3 to 4 inches of soil right after application. Use a double gang disc harrow or other equipment which will do a good job of mixing the 15% MOCAP with the soil. After mixing, apply fertilizer and, for best results, shape a wide bed with a tractor equipped with 4 disc hillers and wide sweeps for running out the middles. Be sure that only the treated soil (top 3 to 4 inches) is used to make beds.

The higher rate treatment is recommended for fields in which nematode injury to tobacco has been especially severe, or where soil is sandy.

Broadcast treatment for wireworm control: Use 13 pounds of 15% MOCAP per acre. Apply as a broadcast treatment anytime from 2 weeks before planting to at-planting time. Mix with the top 2 to 4 inches of soil right after application. Use a disc harrow or other cultivating equipment that will do a good job of mixing 15% MOCAP with the soil.

Rhone-Poulenc does not recommend the use of MOCAP on tobacco in Florida.

CONDITION OF SALE, WARRANTY, LIMITATION OF LIABILITY

This product conforms to the chemical description on the label thereof and is reasonably fit for the purpose stated on such label only when used in accordance with directions under normal use conditions. Follow directions carefully. Timing and method of application, weather and crop conditions, mixture with other chemicals not specifically recommended and other influencing factors in the use of this product are beyond the control of Rhone-Poulenc or the seller. Buyer assumes all risks of use, storage, or handling of this material, not in strict accordance with directions given herewith. In no case shall Rhone-Poulenc or the seller be liable for consequential, special, or indirect damages such as loss of profits or values resulting from the use or handling of this product.

MOCAP® EC

NEMATICIDE–INSECTICIDE

RESTRICTED USE PESTICIDE

Acute dermal toxicity

For retail sale to and use only by Certified Applicators or persons under their direct supervision, and only for those uses covered by the Certified Applicator's certification.

ACTIVE INGREDIENTS:	
Ethoprop (*O*-Ethyl *S,S*-Dipropyl Phosphorodithioate)	69.6%
Xylene	21.0%
INERT INGREDIENTS	9.4%
TOTAL	100.0%

(Contains 6 pounds active ingredient per gallon)

EPA Reg. No. 359-696 EPA Est.

U.S. Patent Nos. 3,112,244; 3,268,393

Keep Out of Reach of Children

DANGER–POISON

See antidote and warning information on back panel

STATEMENT OF PRACTICAL TREATMENT

IF SWALLOWED: Call a physician or Poison Control Center. Drink 1 or 2 glasses of water and induce vomiting by touching back of throat with finger. Do not induce vomiting or give anything by mouth to an unconscious person.

IF INHALED: Remove to fresh air and apply artificial respiration if indicated.

IF ON SKIN: Remove all contaminated clothing and wash skin with soap and water. Wash clothing before reuse. Discard contaminated shoes.

IF IN EYES: Flush with plenty of water for at least 15 minutes. Call a physician immediately.

See side or back panel for additional precautionary statements including note to physician.

HAZARDS TO HUMANS AND DOMESTIC ANIMALS

DANGER

Poisonous if swallowed, inhaled or absorbed through skin. Rapidly absorbed through skin. Wear waterproof protective clothing, rubber gloves and goggles. In case of contact, wash immediately with soap and water.

Do not breathe fumes or spray mist. Wear an AO R-6058 respirator with an R-58 cartridge, or equivalent, for protection during field handling and field exposures. After use, wash hands, arms and face thoroughly with soap and warm water before eating or smoking. Wash all clothing with soap and hot water before reuse. Discard contaminated shoes. Do not contaminate feed or food.

Do not store in or around the home.

SYMPTOMS OF POISONING: Nausea, vomiting, abdominal cramps, diarrhea, excessive salivation, headache, dizziness, weakness, blurring or dimness of vision, excessive tearing, loss of muscular coordination, slurring of speech, twitching of muscles (especially of tongue and eyelids), mental confusion, disorientation, drowsiness, difficulty in breathing (chest tightness), runny nose.

FIRST AID

If Swallowed—Give 1 or 2 glasses of water and induce vomiting by touching back of throat with finger. Do not induce vomiting or give anything by mouth to an unconscious person. Call a physician, Poison Control Center, or hospital emergency room.

If not breathing, give mouth to mouth artificial respiration.

In Case of Skin Contact—Immediately remove all contaminated clothing, and wash skin with soap and water. Wash clothing before reuse. Discard contaminated shoes.

For Eyes—Flush with plenty of water for at least 15 minutes. Call a physician immediately in all cases of suspected poisoning.

Persons providing first aid should prevent contamination of themselves.

NOTE TO PHYSICIAN

This product is a cholinesterase inhibitor. Severe symptoms and signs include diarrhea, pinpoint and non-reactive pupils, respiratory difficulty, pulmonary edema, cyanosis, loss of sphincter control, convulsions, and coma.

Support respiration as needed. Measures should include removal of secretions, maintenance of a patent airway, and if necessary, artificial ventilation.

If cyanosis is absent, give atropine 2–4 mg intravenously (0.05 mg/kg for children). Repeat Atropine at 5–10 minute intervals until atropinization occurs (dry, flushed skin, tachycardia, pupillary dilation), and maintain for 48 hours.

If cyanotic, give initial ATROPINE intramuscularly and start measures to improve ventilation.

Start 2-PAM (PROTOPAM, Ayerst) at the same time. Give 1–2 grams PROTOPAM (20–40 mg/kg for children) in 100 cc saline over 15–30 minutes. If pulmonary edema is present, give intravenously slowly as a 5% solution in water over a period of at least 5 minutes.

A second dose may be given after one hour if muscle weakness persists. Additional doses may be given cautiously for persistent muscle weakness. May be given by intramuscular or subcutaneous routes if intravenous administration is not feasible.

In case of skin contact wash patient with soap and water followed by wash with 95% ethyl alcohol. Keep patient under constant observation for 24–36 hours. Symptoms may persist for one month.

The use of theophylline, morphine, barbiturates, phenothiazines, reserpine, and succinyl choline is contraindicated.

ENVIRONMENTAL HAZARDS

This product is toxic to fish, birds and other wildlife. Birds feeding in treated areas may be killed. Keep out of lakes, streams or ponds. Do not apply where runoff is likely to occur.

Do not apply within 140 feet of inland freshwater habitats.

Along the Atlantic seaboard, do not apply within 850 feet of brackish water habitats.

Do not apply when weather conditions favor drift from areas treated. Do not contaminate water by cleaning of equipment or disposal of wastes. Apply this product only as specified on this label.

STORAGE AND DISPOSAL

PROHIBITIONS

Do not contaminate water, food, or feed by storage or disposal. Open dumping is prohibited. Do not store in or around the home.

PESTICIDE DISPOSAL

Pesticide wastes are toxic. Improper disposal of excess pesticide, spray mixture, or rinsate is a violation of Federal Law. If these wastes cannot be disposed of by use according to label instructions, contact your State Pesticide or Environmental Contol Agency, or the Hazardous Waste representative at the nearest EPA Regional Office for guidance.

CONTAINER DISPOSAL

Triple rinse (or equivalent). Then offer for recycling or reconditioning, or puncture or dispose of in sanitary landfill, or by other procedures approved by state and local authorities.

PHYSICAL OR CHEMICAL HAZARDS

Do not use or store near heat or open flame.

DIRECTIONS FOR USE

It is a violation of Federal law to use this product in a manner inconsistent with its labeling.

Unprotected persons may enter treated field or area after chemical has been mixed into the soil.

INCORPORATION INFORMATION

MOCAP nematicide-insecticide can be incorporated with a rotary tiller, rotary hoe, spring-tooth harrow, by double-discing or with other equipment for mixing the chemical with the soil to the depth recommended. Follow any specific incorporating directions given under the crop heading.

If moisture is applied immediately after application to carry the treatment into the soil, only shallow incorporation ($\frac{1}{4}$ to $\frac{1}{2}$ inch) with mechanical equipment such as drag chains is required.

EQUIPMENT

MOCAP spray mixtures may affect some plastic materials after prolonged use. Spray tubing, hoses and gaskets should be inspected frequently, particularly black-colored materials. Replace any material which is softened or swollen. Polyethylene, polypropylene, nylon and Teflon materials are recommended. Polyvinyl chloride (PVC) material is not recommended.

Mixing: Add the recommended amount of MOCAP EC to the water in the spray tank and mix well. Use enough water in the sprayer to thoroughly and evenly cover the area to be treated. Agitate frequently during use.

BANANA AND PLANTAIN

For Nematode and Banana Root Borer *(Cosmopolites sordidus)* control: Apply 8 ml of MOCAP EC in a radius of $\frac{3}{4}$ meter around each producing stem. Before application, remove ground litter from area to be treated. For best resuls, make application just before irrigation or a rainy period. If application is made during a dry period, mix MOCAP EC into the top 2 cm of soil with a rake. Repeat application in 6 months.

BEAN (SNAP AND LIMA)

Row treatment for nematode control: Use $1\frac{1}{3}$ to 2 quarts of MOCAP EC per acre (36-inch row spacing). Make application in band 12 to 15 inches wide on the row. **Do not** use as a seed furrow treatment or allow spray to contact the seed.

Broadcast treatment for nematode control: Use 1 to $1\frac{1}{3}$ gallons of MOCAP EC per acre. Apply anytime from 3 days before planting to at-planting time. Mix with the top 4 inches of soil right after application.

Note: Beans grown in treated soil may remain green longer than other beans and mature 1 to 2 weeks later.

CABBAGE

Row treatment for nematode control: Use $1\frac{1}{3}$ quarts of MOCAP EC per acre on a band 15 inches wide on the row (36-inch row spacing). **Do not** use as a seed furrow treatment or allow spray to contact the seed.

Broadcast treatment for nematode control: Use $3\frac{1}{3}$ quarts of MOCAP EC per acre. Apply evenly over the soil area to be treated from 1 week before planting to at-planting time. Mix with the top 3 inches of soil right after application.

NEMATODES CONTROLLED

	Bananas and Plantain	Cabbage	Field Corn & Sweet Corn	Cucumbers	Peanuts (planting)	Pineapple	White (Irish) Potatoes	Sweet Potatoes	Snap Beans & Lima Beans	Soybeans	Sugarcane	Tobacco	Commercial Turf
Burrowing Nematodes													
Radopholus similis	C												
Dagger Nematodes													
Ciphinema spp.			C							S			
Cyst Nematodes													
Heterodera glucines										S			
Heterodera cruciferae		C											
Heterodera schactii		C											
Lesion Nematodes													
Pratylenchus spp.	C	C	C	C	C		C		C	C	C	C	C

NEMATODES CONTROLLED

	Bananas and Plantain	Cabbage	Field Corn & Sweet Corn	Cucumbers	Peanuts (planting)	Pineapple	White (Irish) Potatoes	Sweet Potatoes	Snap Beans & Lima Beans	Soybeans	Sugarcane	Tobacco	Commercial Turf
Root-Knot Nematodes													
Meloidogyne spp.	C									C	C		C
Meloidogyne hapla		C	C	C	C		C	C	C			C	
Meloidogyne incognita		C	C	C			C	C	C			C	
Meloidogyne arenaria					C							C	
Meloidogyne javanica				C								C	
Lance Nematodes													
Hoplolaimus spp.		S	C	S					S				
Hoplolaimus tylenchiformis				C									
Hoplolaimus columbus										S			
Spiral Nematodes													
Helicotylenchus spp.	C		C	C			C	C	C	C			C
Rotylenchus spp.			C										
Sting Nematodes													
Belonolaimus spp.			C		C								
Belonolaimus longicaudatus		C		C			C		C	C			C
Ring Nematodes													
Criconemoides spp.				C	C		C			C	C		C
Criconemoides ornatus			C						C				
Reniform Nematodes													
Rotylenchulus reniformis						C		C					
Stunt Nematodes													
Tylenchorhynchus spp.			C	C	C		C	C		C		C	C
Tylenchorhynchus claytoni									C				
Tylenchorhynchus martini											C		
Stubby Root Nematodes													
Trichodorus spp.			C		C		C	C					C
Trichodorus christiei		C		C						C	C		
Pin Nematodes													
Paratylenchus spp.													C

INSECTS CONTROLLED

	Bananas and Plantain	Field Corn & Sweet Corn	White (Irish) Potatoes	Sweet Potato	Sugarcane	Tobacco	Commercial Turf
Corn Rootworm							
Diabrotica virgifera		C					
Diabrotica longicornis		C					
Diabrotica undecimpuctata		C					
Diabrotica howardi		C					
Wireworms							
Conoderus spp.						C	
Conoderus vespertinus		C		C			
Conoderus falli		C					
Melanotus spp.		C		C			
Melanotus communis			C		C		
Limonius dubitan		C					
Limonius canus			C				
Limonius californicus			C				
Aeolus mellius		C					
Ctenicera spp.		C					
Ctenicera pruinina		C	C				
Eleodes spp.			C				
Black Cutworm							
Agrotis ipsilon		C*					

INSECTS CONTROLLED

	Bananas and Plantain	Field Corn & Sweet Corn	White (Irish) Potatoes	Sweet Potato	Sugarcane	Tobacco	Commercial Turf
Sandhill Cutworm							
Euxoa detersa		C*					
White Grubs							
Phyllophaga spp.							
Phyllophaga crinita		S			C		
Phyllophaga ephilida		S		C			
Banded Cucumber Beetle							
Diabrotica balteata				C			
Fleabeetle Larvae							
Chaetocnema confinis				C			
Systena blanda				C			
Banana Root Borer							
Cosmopolites sordidus	C						
Japanese Beetle Grub							
Popillia japonica							C
European Chafer							
Rhizotrogus majalis							C
Black Turfgrass Ataenius Grub							
Ataenius spretulus							C
Chinchbug							
Blissus Leucopterus							C
Bluegrass Billbug							
Sphenophorus parvulus							C
Sod Webworm							
Crambus spp.							C
Herpetogramma spp.							C
Mole Crickets							
Scapteriscus vicinus							C
Scapteriscus acletus							C

* NOTE: Under heavy infestations of cutworms, enough cutworms may survive to cause some crop damage. If this occurs, it may be necessary to apply a rescue treatment with another registered pesticide to control the surviving worms.
C = Control S = Suppression

CITRUS SEEDLINGS

Bare root dip treatment for nematode control: Mix ½ pint (1 cupful) of MOCAP EC with 50 gallons of water* (or 2 teaspoonfuls with 2 gallons of water). Dip the washed roots of citrus seedlings in the mixture so that only the roots are covered. Allow the roots to soak for 30 minutes then remove for planting. *(*MOCAP EC concentration is 900 parts per million in this mixture.)*

Treated plants should be tagged: "Warning. Wear rubber gloves when handling treated plants."

Pot drench treatment for nematode control: Mix ½ pint (1 cupful) of MOCAP EC with 50 gallons of water (or 5 teaspoonfuls with 5 gallons of water). Pour enough of the mixture into each pot to completely drench the soil (about ½ pint for a 6-inch pot; 1 pint for an 8-inch pot).

CORN (FIELD AND SWEET)

At-planting application: For corn rootworms, wireworms, cutworms, Symphylans (*Scutigerella maculata*), nematode and suppression of white grubs: Use 1⅔ ounces per 1,000 feet of row or ⅔ quarts MOCAP EC per acre (minimum 40-inch row spacing) at planting.

Apply the spray in a 6 to 7 inch band on the row over a closed seed furrow. Mix the spray with the top ½ inch of soil with drag chains, spring-tooth incorporators, or similar equipment.

Broadcast treatment for cutworm control: Use 2 quarts of MOCAP EC per acre. Apply treatment from 3 days before planting to at-planting time. Mix treatment with the top 2 inches of soil right after application.

Nematode Control: Row treatment: Use 1 to 1⅓ quarts of MOCAP EC per acre (minimum 40-inch row spacing). Apply at planting in a band 12 to 15 inches wide on the row.

Broadcast treatment: Use 1 gallon of MOCAP EC per acre. Apply treatment from 3 days before planting to at-planting time. Mix treatment with the top 2 inches of soil right after application.

The treatments for nematode control will also control corn rootworms, wireworms, cutworms and Symphylans (*Scutigerella maculata*) and suppress white grubs.

CUCUMBER

Row treatment for nematode control: Use 1⅓ quarts of MOCAP EC per acre (7-foot row spacing). Make application in a band 12 to 15 inches wide on the row at or just before planting. Prepare bed and knock off top to provide a level area for treatment. Mix with the top 2 inches of soil right after application. **Do not** use as a seed furrow treatment or allow spray to contact the seed.

ORNAMENTAL PLANTS

(See Note)

(Aglaonema, Azalea, Boxwood, Bromeliads, Cacti, Caladium, Camellia, Cape Jasmine, Fern, Gardenia, Holly, Philodendron, Pothos, Sansevieria and Yew.)

Bare root and tuber dip treatment for nematode control: Mix ½ pint (1 cupful) of MOCAP EC with 50 gallons of water* (or 2 teaspoonfuls with 2 gallons of water). Dip the washed roots of plants in the mixture so that only the roots are covered. Allow the roots to soak for 30 minutes then remove for planting. For Leatherleaf Fern, mix 1/4 pint (1/2 cupful) with 100 gallons of water (or 1 teaspoonful with 4 gallons of water), and dip washed roots for 15 minutes. **MOCAP EC concentration is 900 parts per million in this mixture.)*

Treated plants should be tagged: "Warning. Wear rubber gloves when handling treated plants."

Pot drench treatment for nematode control: Mix 1/4 pint (1/2 cupful) of MOCAP EC with 50 gallons of water (or 1 teaspoonful with 2 gallons of water). Pour enough of the mixture into each pot to completely drench the soil (about 1/2 pint for a 6-inch pot; 1 pint for an 8-inch pot). If the mixture contacts foliage wash it off immediately with fresh water to avoid foliage injury. For Leatherleaf Fern, mix 1/8 pint (1/4 cupful) with 50 gallons of water (or 1 teaspoonful with 4 gallons of water). Apply after harvesting and follow directions above.

Bed and bench treatment for nematode control: Mix 1/2 to 1 pint (1 to 2 cupfuls) of MOCAP EC with enough water to completely cover 1,000 sq. feet of bed or bench area (or 2 to 4 tablespoonfuls for 100 sq. feet). Wash the mixture off contacted foliage immediately to avoid foliage injury. After treatment apply enough water to soak the treatment 4 to 6 inches into the soil. For Leatherleaf Fern, use 1/3 pint (2/3 cupful) for 1,000 sq. feet (or 2 teaspoonfuls for 100 sq. feet). Apply after harvesting, and follow with enough water to soak the treatment 4 to 6 inches into the soil.

Field nursery stock treatment for nematode control: Mix 1/2 pint (1 cupful) of MOCAP EC with 100 gallons of water (or 2 1/2 teaspoonfuls with 5 gallons of water). Apply 1 gallon of mixture to the soil in a square yard area around each plant. Wash the mixture off contacted foliage immediately to avoid injury. After treatment apply enough additional water to wet the soil to a depth of 6 to 12 inches, depending upon depth of the roots.

NOTE: If plants other than those mentioned on this label require treatment, only a few should be treated until its effects can be determined. Some plants or varieties may show temporary growth retardation after treatment.

PEANUT

Row treatment for nematode control at planting: Use 1 1/3 to 2 2/3 quarts of MOCAP EC per acre (36-inch row spacing) in a band 12 inches wide on the row. For control of peanut root-knot nematode (*M. arenaria*) use the higher rate of MOCAP EC. **Do not** use as a seed furrow treatment or allow spray to contact the seed.

Broadcast treatment for nematode control at planting: Use 3 to 4 quarts of MOCAP EC per acre. Make application from 1 week before planting to at-planting time. Mix with the top 4 inches of soil right after application.

PINEAPPLE

Broadcast treatment for nematode control: Use 3 1/3 to 6 2/3 gallons of MOCAP EC per acre. Use higher rates where longer residual activity is desired or in fields where nematodes have been especially severe. Apply as broadcast treatment anytime from 3 weeks before planting to at-planting time. Mix with the top 4 inches of soil right after application.

POTATOES

Wireworm control: Row treatment: Use 2 quarts MOCAP EC per acre (40-inch row spacing). Make application in a band 12 inches wide on the row at planting. The following broadcast treatment is recommended for fields where wireworm injury has been (or is suspected to be) especially severe (i.e., potatoes planted following sod, pasture or new ground).

Broadcast treatment: Use 2 2/3 to 4 quarts of MOCAP EC per acre. Mix treatment with the top 4 inches of soil right after application.

Nematode control: Row Treatment: (Control of Lesion, Spiral, Sting, Ring, Stunt, and Stubby Root Nematodes and Suppression of Northern Root-Knot and Southern Root-Knot Nematodes.) Use 2 quarts of MOCAP EC per acre (40-inch row spacing). Make application in a band 12 inches wide on the row at planting.

Broadcast treatment: (Control of Northern, Root-Knot, Southern Root-Knot, Lesion, Spiral, Sting, Ring, Stunt and Stubby Root Nematodes. Use 1 to 1 1/2 gallons of MOCAP EC per acre. Mix treatment with the top 4 inches of soil right after application. The higher rate treatment is recommended for fields in which nematode injury has been especially severe or where soil is sandy.

SOYBEAN

Row treatment for nematode control at planting: Use 1 to 2 quarts of MOCAP EC per acre. Make application in a band 12 to 15 inches wide on the row. **Do not** use as a seed furrow treatment or allow spray to contact the seed.

Broadcast treatment for nematode control at planting: Use 3 to 4 quarts of MOCAP EC per acre. Make application from 1 week before planting to at-planting time. Mix with the top 3 inches of soil right after application.

Tankmix with Treflan® Herbicide: MOCAP EC may be tank-mixed with Treflan (trifluralin) Herbicide for control of the nematodes, grasses, and weeds named on the labels. Follow all directions and precautions on each label.

NOTE: Soybeans treated with organophosphate chemicals such as MOCAP may be more susceptible to injury from metribuzin and propanil herbicides.

SUGARCANE

Furrow treatment for nematode and wireworm control: Use 1 1/3 to 2 2/3 quarts of MOCAP EC per acre at planting time (6-foot row spacing). Use the higher rates where nematodes and wireworms have been especially severe. Apply evenly in a band 12 to 15 inches wide, centered over the seed pieces in the opened furrow immediately before closing. Mix with the soil with a disc hiller or other suitable equipment as the cane is covered.

SWEET POTATO

For Nematode, Wireworm, White Grub, Banded Cucumber Beetle and Fleabeetle Larvae (pale-striped and sweet potato fleabeetle) control:

Row treatment: Use 2 to 2 2/3 quarts of MOCAP EC per acre (minimum 42-inch row spacing). Make application in a band 12 to 15 inches wide on the row 2 to 3 weeks before planting. Mix with the top 4 inches of soil right after application. Good results have been obtained by opening the row, applying MOCAP EC in a 12 to 15 inch band on the row, bedding over and dragging.

Broadcast treatment: Use 1 to 1 1/3 gallons of MOCAP EC per acre. Apply the treatment evenly on the top of the soil 2 to 3 weeks before planting. Mix the MOCAP EC with the top 4 inches of soil right after application. Use a double gang disc harrow or other equipment which will do a good job of mixing the MOCAP with the soil.

TOBACCO

Row treatment for both nematode and wireworm control: Use 1 to 2 gallons of MOCAP EC per acre (or 2 to 4 tablespoonfuls per 100 linear feet of row). Apply after fertilizer has been listed or bedded. Make application in a band 18 to 24 inches wide on the row, any time from one week before planting to at-planting time. Mix with the top 4 inches of soil right after application. The higher rate treatment is recommended for fields in which nematode injury to tobacco has been especially severe or where the soil is sandy.

Broadcast treatment for both nematodes and wireworm control: Use 1 to 2 gallons of MOCAP EC per acre. Apply the treatment evenly on the top of the soil anytime from one week before planting to at-planting time. Mix the MOCAP EC with the top 3 inches of soil right after application. Use a double gang disc harrow or other equipment which will do a good job of mixing the MOCAP EC with the soil. After mixing, apply fertilizer and for best results, shape a wide bed with a tractor equipped with 4 disc hillers and wide sweeps for running out the middles. Be sure that only the treated soil (top 3 inches) is used to make beds. The higher rate treatment is recommended for fields in which nematode injury to tobacco has been especially severe, or where soil is sandy.

Broadcast treatment for wireworm control: Use 1 1/3 quarts of MOCAP EC per acre. Apply as a broadcast treatment anytime from 2 weeks before planting to at-planting time. Mix the MOCAP EC with the top 2 inches of soil right after application. Use a disc harrow or other cultivating equipment that will do a good job of mixing the MOCAP EC with the soil.

For Fleabeetle control: Add $\frac{1}{2}$ gallon (4 pints) per acre of Di-Syston® 8 Liquid Systemic Insecticide to the spray tank containing the MOCAP EC mixture for nematode and wireworm control (see above). Observe all precautions on the label. Do not make more than one field application per crop season regardless of the formulation of Di-Syston used. *(Di-Syston is a registered trademark of Farbenfabriken Bayer A.G.)*

Rhone-Poulenc does not recommend the use of MOCAP on tobacco in Florida.

COMMERCIAL TURF

(Such as Golf Courses, Sod Farms, Cemeteries) FOR USE ONLY BY PROFESSIONAL TURFMEN

On Bahia, Bermuda, Centipede, St. Augustine, Zoysia grasses. MOCAP is an effective contact nematicide for control of nematodes that attack turf such as golf courses, sod farms and cemeteries. Established turf infested with sting and other parasitic nematodes shows increased vigor and improved appearance within 7 to 10 days following treatment. MOCAP treatment at recommended rates will protect turf grasses from further nematode injury for 6 months to an entire growing season.

For nematode control: Use $\frac{2}{3}$ to 1 pint of MOCAP EC per 1,000 square feet of established turf ($3\frac{1}{2}$ to 5 gallons per acre). Apply uniformly, and immediately after application apply 1 inch of water to turf. Failure to apply water may result in poor nematode control, plant injury, and toxicity hazard.

NOTE: Do not use on home lawns. Do not use on newly seeded areas until plants are well established. Do not cut sod or handle treated turf for 21 days after treatment. Birds and other wildlife in treated areas may be killed. Keep unprotected persons and pets off treated turf until water has washed the spray off plants into soil and the area is thoroughly dry. Do not apply where runoff is likely to occur.

Compatibility: MOCAP EC is compatible with Balan®, Paarlan®, Ridomil®, Tillam® and Treflan® emulsifiable or liquid concentrates. Balan®, Paarlan® and Treflan® are registered trademarks of Elanco products. Ridomil® is a registered trademark of Ciba-Geigy. Tillam® is a registered trademark of Stauffer Chemical Company.

CONDITION OF SALE, WARRANTY, LIMITATION OF LIABILITY

This product conforms to the chemical description on the label thereof and is reasonably fit for the purpose stated on such label only when used in accordance with directions under normal use conditions. Follow directions carefully. Timing and method of application, weather and crop conditions, mixture with other chemicals not specifically recommended and other influencing factors in the use of this product are beyond the control of Rhone-Poulenc or the seller. Buyer assumes all risks of use, storage, or handling of this material, not in strict accordance with directions given herewith. In no case shall Rhone-Poulenc or the seller be liable for consequential, special, or indirect damages such as loss of profits or values resulting from the use or handling of this product.

MOCAP® PLUS 4-2 EC

NEMATICIDE–INSECTICIDE

RESTRICTED USE PESTICIDE

Acute human oral, dermal and inhalation toxicity.

For retail sale to and use by Certified Applicators or persons under their direct supervision, and only for those uses covered by the Certified Applicator's certification.

ACTIVE INGREDIENTS:	
Ethoprop (*O*-Ethyl *S,S*-Dipropyl Phosphorodithioate)	46.0%*
O,O-Diethyl *S*-[2-(ethylthio) ethyl] Phosphorodithioate	23.0%**
Xylene range aromatic solvent	20.9%
INERT INGREDIENTS	10.1%
TOTAL	100.0%

* Equivalent to 4 pounds per gallon.

** Equivalent to 2 pounds per gallon.

EPA Reg. No. 359-702 EPA Est.

U.S. Patent Nos. 2,759,010; 3,112,244; 3,268,393

DANGER–POISON

Keep Out of Reach of Children

See Antidote and Warning Information on Back Panel

STATEMENT OF PRACTICAL TREATMENT

IF SWALLOWED: Call a physician or Poison Control Center. Drink 1 or 2 glasses of water and induce vomiting by touching back of throat with finger. Do not induce vomiting or give anything to an unconscious person.

IF INHALED: Remove to fresh air and apply artificial respiration if indicated.

IF ON SKIN: Remove all contaminated clothing and wash skin with soap and water. Wash clothing before reuse. Discard contaminated shoes.

IF IN EYES: Flush with plenty of water for at least 15 minutes. Call a physician immediately.

See side or back panel for additional precautionary statements including note to physician.

PRECAUTIONARY STATEMENTS

HAZARDS TO HUMANS AND DOMESTIC ANIMALS

DANGER: Poisonous if swallowed, inhaled or absorbed through skin. Rapidly absorbed through skin. Wear waterproof protective clothing, rubber gloves and goggles. In case of contact, wash immediately with water.

Do not breathe fumes or spray mist. Wear an AO R-6058 respirator with an R-58 cartridge, or equivalent, for protection during field handling and field exposure. After use, wash hands, arms and face thoroughly with soap and warm water before eating or smoking. Wash all clothing with soap and warm water before reuse. Discard contaminated shoes. Do not contaminate feed or food.

Do not store in or around the home.

SYMPTOMS OF POISONING: Nausea, vomiting, abdominal cramps, diarrhea, excessive salivation, headache, dizziness, weakness, blurring or dimness of vision, excessive tearing, loss of muscular coordination, slurring of speech, twitching of muscles (especially of tongue and eyelids), mental confusion, disorientation, drowsiness, difficulty in breathing (chest tightness), runny nose.

FIRST AID: If Swallowed—give 1 or 2 glasses of water and induce vomiting by touching back of throat with finger. Do not induce vomiting or give anything by mouth to an unconscious person. Call a physician, Poison Control Center, or hospital emergency room.

If not breathing, give mouth to mouth artficial respiration.

In Case of Skin Contact, immediately remove all contaminated clothing, and wash skin with soap and water. Wash clothing before reuse. Discard contaminated shoes.

For Eyes, flush with plenty of water for at least 15 minutes. Call a physician immediately in all cases of suspected poisoning.

Persons providing first aid should prevent contamination of themselves.

NOTE TO PHYSICIAN

This product is a cholinesterase inhibitor. Severe symptoms and signs include diarrhea, pinpoint and non-reactive pupils, respiratory difficulty, pulmonary edema, cyanosis, loss of sphincter control, convulsions, and coma.

Support respiration as needed. Measures should include removal of secretions, maintenance of a patent airway, and if necessary, artificial ventilation.

If cyanosis is absent, give Atropine 2–4 mg intravenously (0.5mg/kg for children). Repeat Atropine at 5–10 minute intervals until atropinization occurs (dry, flushed skin, tachycardia, pupillary dilation), and maintain for 48 hours. If cyanotic, give initial atropine intramuscularly and start measures to improve ventilation.

Start 2-PAM at the same time. Give 1–2 grams 2-PAM (20–40mg/kg for children) in 100cc saline over 15–30 minutes. If pulmonary edema is present, give intravenously slowly as a 5% solution in water over 5 minutes.

A second dose may be given after one hour if muscle weakness persists. Additional doses may be given cautiously for persistent muscle weakness. May be given by intramuscular or subcutaneous routes if intravenous administration is not feasible.

In case of skin contact wash patient with soap and water followed by wash with 95% ethyl alcohol. Keep patient under constant observation for 24–36 hours. Symptoms may persist for one month.

The use of theophylline, morphine, barbiturates, phenothiazines, reserpine, and succinyl choline is contraindicated.

ENVIRONMENTAL HAZARDS

This product is toxic to fish, birds and other wildlife. Birds and other wildlife feeding in treated areas may be killed. Keep out of lakes, streams or ponds.

Do not apply within 140 feet of inland freshwater habitats.

Along the Atlantic seaboard, do not apply within 850 feet of brackish water habitats.

Do not apply when weather conditions favor drift from areas treated. Do not contaminate water by cleaning of equipment or disposal of wastes. Apply this product only as specified on this label.

DIRECTIONS FOR USE

It is a violation of Federal law to use this product in a manner inconsistent with its labeling.

This product may be applied under the direct supervision of a certified applicator who is not physically present at the site of application, but is nonetheless, available to the person applying the pesticide.

Unprotected persons may enter treated fields or area after chemical has been mixed into the soil.

REENTRY STATEMENT

Reentry into treated fields is prohibited within 24 hours of application. State or local use and reentry regulations may be more restrictive than those stated on this label.

Do not apply this product in such a manner as to directly or through drift expose workers or other persons. The area being treated must be vacated by unprotected persons.

Do not enter treated areas for 24 hours unless appropriate protective clothing is worn.

Because certain states may require more restrictive reentry intervals for various crops treated with this product, consult your State Department of Agriculture for further information.

Written or Oral warnings must be given to workers who are expected to be in a treated area or in an areaa about to be treated with this product. Oral warnings must be given if there is reason to believe that written warnings cannot be understood by workers and given in a language customarily understood by the workers.

Oral and Written warnings must include the Statement of Practical Treatment and notification that unprotected persons may enter treated fields or area only after chemical has been mixed in the soil.

STORAGE AND DISPOSAL

Do not contaminate water, food or feed by storage or disposal.

STORAGE

Do not use or store near heat or open flame.

PESTICIDE DISPOSAL

Pesticide wastes are toxic. Improper disposal of excess pesticide, spray mixture, or rinsate is a violation of Federal Law. If these wastes cannot be disposed of by use according to label instructions, contact your State Pesticide or Environmental Control Agency, or the Hazardous Waste representative at the nearest EPA Regional Office for guidance.

CONTAINER DISPOSAL

Triple rinse (or equivalent). Then offer for recycling or reconditioning, or puncture and dispose of in a sanitary landfill, or by other procedures approved by state and local authorities.

EQUIPMENT

MOCAP spray mixtures may affect some plastic materials after prolonged use. Spray tubing, hoses and gaskets should be inspected frequently, particularly black-colored materials. Replace any material which is softened or swollen. Polyethylene, polypropylene, nylon and Teflon materials are recommended. Polyvinyl chloride (PVC) material is not recommended.

MIXING

Add the recommended amount to the water in the spray tank and mix well. Use enough water in the sprayer (usually 20 to 25 gallons) to thoroughly and evenly cover the area to be treated. Agitate frequently during use.

INCORPORATION INFORMATION

MOCAP nematicide-insecticide can be incorporated with a rotary tiller, rotary hoe, springtooth harrow, by double-discing or with other equipment for mixing the chemical with the soil to the depth recommended. Follow any specific incorporating directions given under the crop heading.

If moisture is applied immediately after application to carry the treatment into the soil, only shallow incorporation (1/4 to 1/2 inch) with mechanical equipment such as drag chains is required.

Tobacco Nematodes Controlled	Tobacco Insects Controlled
DAGGER NEMATODE *Xiphinema* spp.	WIREWORM *Conoderus* spp.
LESION NEMATODE *Pratylenchus* spp.	TOBACCO FLEA BEETLE *Epitrix hirtipennis*
ROOT-KNOT NEMATODE *Meloidogyne hapla* *M. Incognita*	GREEN PEACH APHID *Myzus persicae*
STUNT NEMATODE *Tylenchorhynchus* spp.	

TOBACCO

Row treatment for control of nematodes, aphids, fleabeetles and wireworms: Use 1½ gallons of MOCAP PLUS per acre (42-inch row spacing).

Apply after fertilizer has been listed or bedded. Make application in a band 18 inches wide on the row, anytime from 1 week before planting to at-planting time. Mix MOCAP PLUS with the top 4 to 6 inches of soil right after application. Use a rotary tiller, rotary hoe or other cultivating equipment that will do a good job of mixing MOCAP PLUS with the soil.

Broadcast treatment for control of nematodes, aphids, fleabeetles and wireworms: Use 1½ to 2 gallons of MOCAP PLUS per acre. Apply the treatment evenly on top of the soil anytime from 1 week before planting to at-planting time. Mix MOCAP PLUS with the top 3 to 4 inches of soil right after application. Use a double gang disc harrow or other equipment that will do a good job of mixing MOCAP PLUS with the soil. After mixing, apply fertilizer and, for best results, shape a wide bed with a tractor equipped with 4 disc hillers and wide sweeps for running out the middles. Be sure that only the treated soil (top 3 to 4 inches) is used to make beds.

The higher rate treatment is recommended for fields in which nematode injury to tobacco has been especially severe, or where the soil is sandy.

Rhone-Poulenc does not recommend the use of MOCAP on tobacco in Florida.

Broadcast treatment in tank mix with Ridomil® 2EC for control of nematodes, aphids, fleabeetles, wireworms, black shank and blue mold: Use 1½ to 2 gallons of MOCAP PLUS mixed with 2 to 4 quarts of Ridomil 2EC per acre. Mix the treatment with the top 3 to 4 inches of soil right after application, then form beds. Read both product labels for specific use and rate information. Follow all precautions, directions, and restrictions on the labels.

Ridomil® Trademark of Ciba-Geigy for metalaxyl. U.S. Patent No. 4,151,299.

CONDITION OF SALE, WARRANTY, LIMITATION OF LIABILITY

This product conforms to the chemical description on the label thereof and is reasonably fit for the purpose stated on such label only when used in accordance with directions under normal use conditions. Follow directions carefully. Timing and method of application, weather and crop conditions, mixture with other chemicals not specifically recommended and other influencing factors in the use of this product are beyond the control of Rhone-Poulenc or the seller. Buyer assumes all risks of use, storage, or handling of this material, not in strict accordance with directions given herewith. In no case shall Rhone-Poulenc or the seller be liable for consequential, special, or indirect damages such as loss of profits or values resulting from the use or handling of this product.

MODOWN®
4 Flowable
HERBICIDE

ACTIVE INGREDIENT:

Bifenox [Methyl 5-(2′,4′-dichlorophenoxy)-2-nitrobenzoate]	40%
INERT INGREDIENTS	60%
TOTAL	100%

Contains 4 pounds bifenox per gallon

EPA Reg. No. 359-701 EPA Est. No.

DANGER
Keep Out of Reach of Children
PELIGRO

PRECAUCION AL USUARIO:
Si usted no lee ingles, no use este producto hasta que le etiqueta haya sido explicada ampliamente

STATEMENT OF PRACTICAL TREATMENT

IF IN EYES: Flush with water for 15 minutes. Get medical attention.

IF ON SKIN: Wash with plenty of soap and water. Get medical attention if irritation persists.

IF SWALLOWED: Drink 1 or 2 glasses of water and induce vomiting by touching back of throat with finger. Do not give anything by mouth to an unconscious person. Get medical attention.

PRECAUTIONARY STATEMENTS

HAZARDS TO HUMANS AND DOMESTIC ANIMALS

DANGER
Do not get in eyes. Wear goggles or face shield when handling. Harmful if swallowed, inhaled, or absorbed through skin. Avoid breathing vapors or spray mist. Avoid contact with skin or clothing. Wash thoroughly after handling. Remove contaminated clothing and wash before reuse.

ENVIRONMENTAL HAZARDS:
This product is toxic to fish. Use with care when applying to areas adjacent to any body of water. Do not apply directly to water except as specified on this label. Do not apply when weather conditions favor drift from target areas. Do not contaminate water by cleaning of equipment or disposal of wastes.

DIRECTIONS FOR USE

It is a violation of Federal Law to use this product in a manner inconsistent with its labeling.

STORAGE AND DISPOSAL

Do not contaminate water, food or feed by storage or disposal.

STORAGE
If exposed to freezing temperatures, store at temperatures above 55°F (13°C) for 24 hours or until completely thawed. Shake well before using.

PESTICIDE DISPOSAL
Pesticide wastes are acutely hazardous. Improper disposal of excess pesticide, spray mixture, or rinsate is a violation of Federal Law. If these wastes cannot be disposed of by use according to label instructions, contact your State Pesticide or Environmental Control Agency, or the Hazardous Waste representative at the nearest EPA Regional Office for guidance.

CONTAINER DISPOSAL
Triple rinse (or equivalent). Then dispose of in a sanitary landfill, or incineration, or, if allowed by state and local authorities, by burning. If burned, stay out of smoke.

GENERAL INFORMATION

MODOWN® herbicide is formulated as a flowable product containing 4 pounds of bifenox per gallon. MODOWN provides control of many broadleaf and grassy weeds infesting soybeans, corn, grain sorghum, rice, small grains, and forestry nursery seedbeds.

In soybeans, MODOWN can be used alone or with Treflan, Lasso, Dual, Basalin, Prowl, Surflan and/or Lexone or Sencor. These combinations control a broader spectrum of weeds than when either product is used alone.

MODOWN can be applied as a preemergent to both corn and grain sorghum alone or in combination with Lasso or Dual; in addition, the combination of MODOWN with Ramrod can be used in grain sorghum.

Uses in rice include preemergent, early postemergent and postflood applications. The early postemergent application can be made alone or in combination with propanil.

Other registered uses of MODOWN include postemergent applications in small grains and both preemergent and postemergent applications to forest tree nursery seedbeds.

Do not rotate treated areas until the following year except corn, soybeans, grain sorghum, small grains and rice. Cover crops may be planted if plowed under and not grazed.

MIXING INSTRUCTIONS

MODOWN Alone: Fill the spray tank ½ to 3/4 full with clean water. Begin agitation and add the recommended amount of MO-

DOWN. Add the necessary amount of water required to fill tank. Maintain sufficient agitation to insure a uniform spray mix during application. Do not allow mixture to stand without agitation.

Tank Mixtures: Fill the spray tank ½ to ¾ full with clean water. Begin agitation and add the recommended amounts of Lexone, Ramrod, Sencor, or Surflan if any of these products are being used. Allow chemicals to become fully mixed then add the recommended amount of MODOWN. Next add the recommended amount of Dual, Lasso, paraquat, Roundup or propanil if any of these products are being used. Add the necessary amount of water required to fill the tank. Maintain sufficient agitation to insure a uniform spray mixture during application. Do not allow mixture to stand without agitation. If there is any evidence of incompatibility, add a suitable compatibility agent. Predetermine the compability of tank mixtures containing propanil by mixing small proportional quantities in advance.

NOTE: SHAKE MODOWN CONTAINER WELL BEFORE USING. CLEAN SPRAYER PRIOR TO AND AFTER USE.

APPLICATION PROCEDURES

Ground Applications: Use a standard herbicide boom sprayer that provides uniform and accurate application. Sprayer should be equipped with screens no finer than 50 mesh in the nozzle tips and the in-line strainers. Calibrate sprayer to deliver 30 to 40 gallons of spray mixture per acre at 20 to 40 psi. Do not apply sprays under wind conditions which will allow drift to areas outside the area being treated. Wind velocity above 5 mph may cause improper application.

Calculate the amount of herbicide needed for band treatment by using this formula:

$$\frac{\text{band width in inches}}{\text{row width in inches}} \times \text{broadcast rate per acre} = \text{amount needed per acre in band}$$

Banding requires use of even flat fan nozzles to assure uniform rate across the width of the band.

Aerial Application: MODOWN and tank mixtures can be applied by aircraft where specified under individual crop use directions. Use a minimum of 10 gallons of total volume per acre. Do not apply sprays under wind conditions which will allow drift to areas outside the area being treated. Wind velocity above 5 mph may cause improper application.

SOYBEANS, GRAIN SORGHUM AND CORN
Conventional Tillage Systems

GENERAL INFORMATION

MODOWN is a selective herbicide recommended as a surface mix or preemergent treatment. MODOWN can be used either alone or in combination with other herbicides as specified in the following directions. MODOWN controls weeds by shoot uptake. Proper application is necessary to obtain good soil surface coverge and uniform distribution near the soil surface. For best results when MODOWN is applied as a preemergent surface treatment there should be a good, well prepared seedbed. Be sure large soil clods are thoroughly broken up before application. When properly applied, MODOWN forms a chemical zone on or near the surface of the soil. MODOWN controls weeds as their shoots come in contact with the herbicide while growing through the chemical zone. MODOWN works at one rate on all soils regardless of the soil type, organic matter content or pH. The rate of MODOWN used per acre should be based on weed density. USE THE HIGHER RECOMMENDED RATE OF MODOWN WHERE PREVIOUS INFESTATIONS OF THE WEEDS LISTED ON THIS LABEL WERE MODERATE TO HEAVY.

Broadleaf Weeds Controlled:

Carpetweed	*Mollugo verticillata*
Eastern Black Nightshade	*Solanum ptycanthum*
Jimsonweed	*Datura stramonium*
Kochia	*Kochia scoparia*
Lambsquarters	*Chenopodium album*
Ladysthumb	*Polygonum persicaria*
Pennsylvania Smartweed	*Polygonum pensylvanicum*
Prickly sida	*Sida spinosa*
Redroot pigweed	*Amaranthus retroflexus*
Velvetleaf	*Abutilon theophrasti*

Broadleaf Weeds Suppressed:

Tall morningglory	*Ipomoea purpurea*
Ivyleaf morningglory	*Ipomoea hederacea*

SOYBEANS—PREEMERGENCE TREATMENTS

Apply MODOWN or MODOWN tank mixtures after planting but before the crop and weed seedlings emerge.

The tank mixtures will control the weeds listed on the label of the product used in the tank mixture. See the specific product labels for information on directions and precautions for use.

Product	Rate Per Acre: Unit Formulated	Rate Per Acre: Pound(s) Active
MODOWN 4F	3 to 4 pts.	1.5 to 2.0
MODOWN 4F + Lasso 4 EC (tank mixture)	3 to 4 pts. + 2 to 4 qts.	1.5 to 2.0 + 2.0 to 4.0
MODOWN 4F + Dual 8E (tank mixture)	3 to 4 pts. + 1½ to 3 pts.	1.5 to 2.0 + 1.5 to 3.0
MODOWN 4F + Surflan 75 WP or Surflan 4AS (tank mixture)	3 to 4 pts. + 1 to 2 lbs. or ¾ to 1½ qts.	1.5 to 2.0 + + 0.75 to 1.5

SOYBEANS—SURFACE MIX TREATMENTS

Apply MODOWN tank mixtures within 7 days prior to planting. Surface mix (one pass incorporation) the treatment with the upper one inch of soil. Use rotary hoe, rolling cultivator or suitable equipment other than a disk that will insure thorough mixing and distribution of the treatment. Suitable soil conditions are necessary for proper incorporation. Improper incorporation may result in poor or erratic weed control. The tank mixtures will control the weeds listed on the MODOWN label plus those weeds listed on the label of product(s) used in the tank mixture. See the specific product labels for information on directions and precautions for use. When using Sencor 50WP or Lexone 50WP, substitute pounds for pints. For Sencor DF or Lexone DF, multiply pints by 0.67.

Product	Rate Per Acre: Unit Formulated	Rate Per Acre: Pound(s) Active
MODOWN 4F + Lasso 4EC (tank mixture)	3 to 4 pts. + 2½ to 4 qts.	1.5 to 2.0 + 2.5 to 4.0
MODOWN 4F + Dual 8E (tank mixture)	3 to 4 pts. + 1½ to 3 pts.	1.5 to 2.0 + 1.5 to 3.0
MODOWN 4F + Lasso 4EC + Sencor 4 or Lexone 4L (tank mixture)	2½ to 3 pts. + 2½ to 3 qts. + ½ to 1 pts.	1.25 to 1.5 + 2.5 to 3.0 + 0.25 to 0.5
MODOWN 4F + Dual 8E + Sencor 4 or Lexone 4L (tank mixture)	2½ to 3 pts. + 1¼ to 2½ pts. + ½ to 1 pt.	1.25 to 1.5 + 1.25 to 2.5 + 0.25 to 0.5

SOYBEANS—SEQUENTIAL TREATMENTS

Apply MODOWN as a preemergent treatment over the top of a preplant incorporated treatment of either Treflan, Basalin, Prowl or a Treflan and Sencor/Lexone tank mixture. These treatments will control the weeds listed on the MODOWN label plus those weeds listed on the label of the products used in conjunction with MODOWN. See the specific product labels for information on directions and precautions for use. When using Sencor 50WP or Lexone 50WP, substitute pounds for pints. For Sencor DF or Lexone DF, multiply pints by 0.67.

Product	Rate Per Acre	
	Unit Formulated	Pound(s) Active
MODOWN 4F sequential to Treflan 4EC	3 to 4 pts. sequential to 1 to 2 pts.	1.5 to 2 sequential to 0.5 to 1.0
MODOWN 4F sequential to Basalin 4EC	3 to 4 pts. sequential to 1 to 3 pts.	1.5 to 2.0 sequential to 0.5 to 1.5
MODOWN 4F sequential to Prowl 4EC	3 to 4 pts. sequential to 1 to 3 pts.	1.5 to 2.0 sequential to 0.5 to 1.5
MODOWN 4F sequential to tank mixture of Treflan 4EC + Sencor 4 or Lexone 4L	2½ to 3 pts. sequential to tank mixture of 1 to 2 pts. + ½ to 1 pt.	1.25 to 1.5 sequential to tank mixture 0.5 to 1.0 + 0.25 to 0.5

GRAIN SORGHUM—PREEMERGENT TREATMENTS

Apply MODOWN or MODOWN tank mixtures after planting but before the crop and weed seedlings emerge. The tank mixtures will control the weeds listed on the MODOWN label plus those weeds listed on the label of the product used in the tank mixture. See the specific product labels for information on directions and precautions for use.

Do not use on sweet sorghum or sorghum sudans.

Product	Rate Per Acre	
	Unit Formulated	Pound(s) Active
MODOWN 4F	3 to 4 pts.	1.5 to 2.0
MODOWN 4F + Ramrod 4L (tank mixture)	3 to 4 pts. + 3 to 4 qts.	1.5 to 2.0 + 3.0 to 4.0
MODOWN 4F + Lasso 4EC* (tank mixture)	3 to 4 pts. + 2 to 4 qts.	1.5 to 2.0 + 2.0 to 4.0
MODOWN 4F + Dual 8E** (tank mixture)	3 to 4 pts. + 1½ to 2½ pts.	1.5 to 2.0 + 1.5 to 2.5

* Only use the MODOWN plus Lasso tank mixtures when the grain sorghum has been treated with Screen seed protectant.

** Only use the MODOWN plus Dual tank mixtures when the grain sorghum has been treated with Concept seed protectant.

CORN—PREEMERGENT TREATMENTS

Apply MODOWN or MODOWN tank mixtures after planting but before the crop and weed seedlings emerge. The tank mixtures will control the weeds listed on the MODOWN label plus those weeds listed on the label of product used in the tank mixture. See the specific product labels information on directions and precautions for use.

Product	Rate Per Acre	
	Unit Formulated	Pound(s) Active
MODOWN 4F	3 to 4 pts.	1.5 to 2.0
MODOWN 4F + Lasso 4EC (tank mixture)	3 to 4 pts. + 2 to 4 qts.	1.5 to 2.0 + 2.0 to 4.0
MODOWN 4F + Dual 8E (tank mixture)	3 to 4 pts. + 1½ to 3 pts.	1.5 to 2.0 + 1.5 to 3.0

SOYBEANS, GRAIN SORGHUM AND CORN—REDUCE-TILL OR MINIMUM-TILL SYSTEMS

The above mentioned preemergent treatments recommended for conventional tillage systems may also be used when soybeans, grain sorghum or corn are planted in reduce-till or minimum-till systems. A contact herbicide (such as paraquat or Roundup) should be used to control emerged annual weeds and grasses and for top kill and suppression of perennials. It is important that the MODOWN spray mixture reaches the soil surface to create a chemical zone. To assure complete coverage, use the higher recommended spray volumes as the density of cover crop residue or weeds increases. See paraquat or Roundup labels for specific directions and precautions for use.

NOTE

Do not apply MODOWN as a "cracking" or "spiking" treatment.

If rain does not occur within 7 days following application, a shallow cultivation or rotary hoeing is recommended.

Cultivation will help control weeds which escape herbicide treatment.

Rainfall which splashes MODOWN treated soil onto the crop at time of emergence may result in crop injury. Leaves of plants may crinkle and discolor but injury is usually temporary. MODOWN is not root absorbed. With good growing conditions, plants will recover quickly and yields should not be affected. As plants grow taller, their foliage is no longer subject to this type of injury. This temporary injury is less likely to occur on sobyeans where MODOWN is surface mixed.

In order to assure maximum safety and weed control, follow all cautions and limitations on this label and the labels of products used in tank mixtures with MODOWN.

Basalin, registered trademark of BASF AG.

Dual, registered trademark of Ciba-Geigy.

Lasso, Ramrod and Roundup are registered trademarks of Monsanto Co.

Lexone, registered trademark of E.I. du Pont de Nemours & Co., Inc.

Prowl, registered trademark of American Cyanamid Co.

Sencor, registered trademark of the Parent Company of Farbenfabriken Bayer GmbH, Leverkusen.

Surflan and Treflan are registered trademarks of Elanco Products Co.

RICE

GENERAL INFORMATION

MODOWN is a selective herbicide recommended as a preemergent, postemergent or postflood treatment. MODOWN can be applied either by air or ground equipment for postemergent or postflood treatment of rice. MODOWN can be used either alone or in conjunction with propanil as specified in the following directions. Water management is a key to successful use of MODOWN. Maintaining soil moisture is important. Proper application is necessary to obtain good coverage and uniform distribution. For best results when MODOWN is applied as a preemergent surface treatment, there should be a good, well prepared seed bed. Large soil clods should be thoroughly broken up before application. When applied preemergent, MODOWN forms a chemical zone on the surface of the soil controlling weeds as their shoots come in contact with the herbicide while growing through the chemical zone. When applied postemergent MODOWN controls susceptible weeds by contact action and provides a 4 to 6 week residual. This may allow delay of flooding. When applied postflood, MODOWN controls susceptible weeds above the water level by contact action. The rate of MODOWN used per acre should be based on weed density. USE THE HIGHER RECOMMENDED RATE OF MODOWN WHEN INFESTATIONS OF THE WEEDS LISTED ON THIS LABEL ARE MODERATE TO HEAVY.

PREEMERGENT
Weeds Controlled:

Dayflower	*Commelina communis*
Ducksalad	*Heteranthera limosa*
Hemp sesbania	*Sesbania exaltata*
Redstem	*Ammannia auriculata*

Sprangletop	*Leptochloa spp.*
Waterhyssop	*Bacopa rotundifolia*
Water sedge	*Carex aquatilis*

Weeds Suppressed:

Barnyardgrass	*Echinochloa crus-galli*

POSTEMERGENT
Weeds Controlled:

Dayflower	*Commelina communis*
Ducksalad	*Heteranthera limosa*
Hemp sesbania	*Sesbania exaltata*
Redstem	*Ammannia auriculata*
Waterhyssop	*Bacopa rotundifolia*
Waterprimrose	*Jussiaea sp.*

Weeds Suppressed:

Alligator weed	*Alternanthera philoxeroides*
Sprangletop	*Leptochloa spp.*

POSTFLOOD
Weeds Suppressed:

Alligatorweed	*Alternanthera philoxeroides*

DRY-SEEDED RICE—PREEMERGENT TREATMENT

Apply MODOWN anytime after seeding until rice reaches early spike stage and before the weed seedlings have emerged. For best results, apply to wet soil. Flush field if necessary to wet soil but drain field before application. Do not apply MODOWN to fields with standing water. Follow usual cultural practices after application. Fields with dayflower showing chemical symptoms or injury should be flushed to activate control if dayflower reaches 2 to 3 leaf stage. Flushing field during dry periods may improve preemergent performance of MODOWN.

Product	Rate Per Acre	
	Unit Formulated	Pound(s) Active
MODOWN 4F	2 to 3 qts.	2.0 to 3.0

DRY-SEEDED OR WATER-SEEDED RICE—POSTEMERGENT TREATMENTS

Apply MODOWN plus propanil tank mixture when rice has a minimum of 2 to 3 leaves and before the weeds listed above have reached 1 inch in height. (Sprangletop not to exceed ½ inch.) Drain field before treatment. Wait a minimum of 48 hours after treatment before flushing or setting permanent flood. If permanent flood is delayed, flushing field during dry periods may improve residual preemergent performance of MODOWN until permanent flood is set.

Reinfestation of the grassy weeds that propanil controls by contact action can be prevented by setting of permanent flood 48 hours after treatment. This tank mixture will control the weeds listed on the MODOWN label plus the weeds listed on the propanil label. See the propanil label for specific information on directions and precautions for use.

For alligatorweeds suppression MODOWN can be used alone. Drain field before treatment. Apply when rice has a minimum of 2 to 3 leaves. Flood the field 48 hours after treatment. Maintain the flood at 4 to 6 inches deep. Do not completely cover the rice plants with water because serious injury to the rice might occur.

Product	Rate Per Acre	
	Unit Formulated	Pound(s) Active
MODOWN 4F	2 to 3 qts.	2.0 to 3.0
MODOWN 4F + propanil*	2 to 3 qts. + 3 to 5 qts.	2.0 to 3.0 + 3.0 to 5.0

* Predetermine the compatibility of this tank mixture.

DRY-SEEDED OR WATER-SEEDED RICE—POSTFLOOD TREATMENT

Apply MODOWN when rice has 4 to 6 leaves and is 8 to 10 inches tall. The alligatorweed should be exposed 3 to 6 inches above the permanent flood at the time of application. A permanent flood 4 to 6 inches deep should be maintained before and after treatment. If the alligatorweed is totally submerged, lower the flood level to expose approximately ½ the weed height. The flood then can be raised to 4 to 6 inches 3 to 4 days after treatment.

Product	Rate Per Acre	
	Unit Formulated	Pound(s) Active
MODOWN 4F	2 to 3 qts.	2.0 to 3.0

NOTE

Do not apply to rice after early spike stage until rice has 2 true leaves.
Do not apply to a second rice crop in areas where double cropping is an agricultural practice.
Do not use where crayfish farming is practiced.
Do not use water from treated fields to irrigate upland crops.
Do not use in California.
In order to assure maximum safety and weed control, follow all cautions and limitations on this label and the propanil label.

SMALL GRAINS
Wheat, Barley, Oats

GENERAL INFORMATION

MODOWN is a selective herbicide recommended as a postemergent treatment. MODOWN contols weeds by shoot uptake. Proper application is important to obtain good soil surface coverage. When properly applied, MODOWN forms a chemical zone on or near the surface of the soil. MODOWN controls weeds as their shoots come in contact with the herbicide while growing through the chemical zone. MODOWN works at one rate on all soils regardless of the soil type, organic matter content, or pH. The rate of MODOWN used per acre should be based on weed density. USE THE HIGHER RECOMMENDED RATE OF MODOWN WHERE PREVIOUS INFESTATIONS OF THE WEEDS LISTED ON THIS LABEL WERE MODERATE TO HEAVY.

Apply MODOWN when wheat, barley or oats are in the 2 to 4 leaf stage and before the weed seedlings have emerged. Do not allow livestock to graze on green grain forage.

Weeds Controlled:

Foxtail	*Setaria sp.*
Lambsquarters	*Chenopodium album*
Ladysthumb	*Polygonum persicaria*
Pennsylvania smartweed	*Polygonum pensylvanicum*
Redrood pigweed	*Amaranthus retroflexus*
Wild buckwheat	*Polygonum convolvulus*
Wild mustard	*Brassica kaber*

Product	Rate Per Acre	
	Unit Formulated	Pound(s) Active
MODOWN 4F	1½ to 2 pts.	0.75 to 1.0

FOREST TREE NURSERY SEEDBEDS

GENERAL INFORMATION

MODOWN is a selective herbicide recommended as a preemergent or postemergent treatment as specified in the following directions. MODOWN controls weeds by shoot uptake. Proper application is important to obtain good soil surface coverage. When properly applied, MODOWN forms a chemical zone on or near the surface of the soil. MODOWN controls weeds as their shoots come in contact with the herbicide while growing through the chemical zone. MODOWN can be used on: fir (balsam, California red, Douglas, Fraser, grand, noble, and white), hemlock (Eastern and Western), pine (loblolly, long leaf, monterey, ponderosa, short leaf, slash, sugar, and white), redwood, Russian olive, spruce (blue, Engelmann, and white).

Do not apply more than 6 quarts of MODOWN per season.

Do not use on seedbeds with less than 1% organic matter.

Treat only small areas until the effects of soil and other local environmental conditions on herbicidal activity and the reaction of forest tree seedlings have been determined.

Weeds Controlled:

Annual bluegrass	*Poa annua*
Australian burnweed	*Erechtites prenanthoides*
Carpetweed	*Mollugo verticillata*
Common chickweed	*Stellaria media*
Common goundsel	*Senecio arvense*
Common mallow	*Malva neglecta*
Common purslane	*Portulaca oleracea*
Corn spurry	*Spergula arvensis*
Filaree	*Erodium spp.*
Goosefoot	*Chenopodium spp.*
Kochia	*Kochia scoparia*
Lambsquarters	*Chenopodium album*
Pennycress	*Thlaspi arvense*
Pineappleweed	*Matricaria matricarioides*
Prickly lettuce	*Lactuca serriola*
Puncturevine	*Tribulus terrestris*
Red sand spurry	*Spergularia rubra*
Redroot pigweed	*Amaranthus retroflexus*
Sheperdspurse	*Capsella bursa-pastoris*
Tall Morningglory	*Ipomoea purpurea*
Tansymustard	*Descurania pinnata*
Whitetop	*Cardaria pubescens*

PREEMERGENT TREATMENT

Apply MODOWN to dry seedbeds within 48 hours after sowing seed and mulching (if mulch is used) and before the weed seedlings have emerged. Use low pressure (20 to 40 psi) spray with 25 to 50 gallons of water per acre. For smaller areas, use 1⅓ pints of MODOWN per 10,000 square feet in 6 to 12 gallons of water. Sprinkler irrigate with ½ to ¾ inch of water immediately after application.

	Rate Per Acre	
Product	Unit Formulated	Pound(s) Active
MODOWN 4F	3 qts.	3.0

POSTEMERGENT TREATMENT

Apply MODOWN to dry seedbeds when the tree seedlings are at least 4 weeks old and before the weed seedlings have emerged. Use low pressure (20 to 40 psi) spray with 25 to 50 gallons of water per acre. For smaller areas, use 1⅓ pints of MODOWN per 10,000 square feet in 6 to 12 gallons of water. Sprinkler irrigate immediately after application to wash the herbicide off the seedlings.

	Rate Per Acre	
Product	Unit Formulated	Pound(s) Active
MODOWN 4F	3 qts.	3.0

MODOWN® is a registered trademark of Rhone-Poulenc Inc.

CONDITIONS OF SALE, WARRANTY, LIMITATION OF LIABILITY

This product conforms to the chemical description on the label thereof and is reasonably fit for the purposes stated on such label only when used in accordance with directions under normal use conditions. Follow directions carefully. Timing and method of application, weather, and crop conditions, mixture with other chemicals not specifically recommended and other influencing factors in the use of this product are beyond the control of Rhone-Poulenc or the seller. Buyer assumes all risks of use, storage, or handling of this material, not in strict accordance with directions given herewith. In no case shall Rhone-Poulenc or the seller be liable for consequential, special, or indirect damages such as loss of profits or values resulting from the use or handling of this product.

ROVRAL®

FUNGICIDE

ACTIVE INGREDIENT:

Iprodione: 3-(3,5-dichlorophenyl)-N-(1-methylethyl)-2,4-dioxo-1-imidazolidinecarboxamide	50.0%
INERT INGREDIENTS	50.0%
TOTAL	100.0%

EPA Reg. No. 359-685 EPA Est.

CAUTION

Keep Out of Reach of Children

STATEMENT OF PRACTICAL TREATMENT

In case of contact, wash skin with soap and water; for eyes flush with water for 15 minutes and get medical attention.

See Back Panel for Additional Precautionary Statements.

PRECAUTIONARY STATEMENTS

HAZARDS TO HUMANS AND DOMESTIC ANIMALS

CAUTION

Avoid contact with skin, eyes, or clothing. In case of contact immediately flush eyes or skin with plenty of water. Get medical attention if irritation persists.

ENVIRONMENTAL HAZARDS

Do not apply directly to water. Do not contaminate water by cleaning of equipment or disposal of wastes.

DIRECTIONS FOR USE

GENERAL USE

It is a violation of Federal Law to use this product in a manner inconsistent with its labeling.

REENTRY STATEMENT

Do not apply this product in such a manner as to directly or through drift expose workers or other persons. The area being treated must be vacated by unprotected persons.

Do not enter treated areas without protective clothing until sprays have dried.

Because certain states may require more restrictive reentry intervals for various crops treated with this product, consult your State Department of Agriculture for further information.

Written or Oral warnings must be given to workers who are expected to be in a treated area or in an area about to be treated with this product. Oral warnings must be given if there is reason to believe that written warnings cannot be understood by workers and given in a language customarily understood by the workers.

Oral and Written warning must include the Statement of Practical Treatment and Reentry Statements as they appear on the label.

STORAGE AND DISPOSAL

Do not contaminate water, food or feed by storage or disposal.

PESTICIDE DISPOSAL

Wastes resulting from the use of this product may be disposed of on site or at an approved waste disposal facility.

CONTAINER DISPOSAL

Completely empty bag into application equipment. Then dispose of empty bag in a sanitary landfill or by incineration, or, if allowed by State and local authorities, by burning. If burned, stay out of smoke.

GENERAL INFORMATION

ROVRAL is a wettable powder formulation of iprodione for the control of certain diseases on Stone Fruits and Nuts, Grapes, Leafy Vegetables, Vegetables and Certain Field Crops.

HOW TO USE ROVRAL

Partially fill the spray tank with clean water. Measure the required amount of ROVRAL and pre-mix with a small volume of water, add this to the tank. Agitate to ensure thorough mixing while filling tank with remaining water. Maintain agitation during application and apply with properly calibrated application equipment. Do not allow spray mixture to stand overnight or for prolonged periods, as some chemical breakdown may occur, particularly in water with a high pH.

HOW MUCH ROVRAL TO USE

The rate chosen will depend on the severity of disease pressure and weather conditions at the time of application (if conditions are severe use the highest rate). Apply ROVRAL in accordance with the directions in the following tables.

STONE FRUITS and NUTS

CROP	HOW TO USE	DISEASE	DOSAGE RATE LBS. PRODUCT/ 100 GAL.	LBS. PRODUCT/ ACRE	WHEN TO USE
APRICOTS, CHERRIES, NECTARINES, PEACHES, PLUMS, and PRUNES*	Apply Rovral as an overall spray in sufficient water to obtain THOROUGH COVERAGE of bloom, foliage, and/or fruit (20-400 gallons per acre by ground equipment and in a minimum of 15 gallons per acre by air). Aerial applications are only recommended during the BLOOM PERIOD.	Brown Rot Blossom Blight *(Monilinia sp.)* Fruit Brown Rot *(Monilinia sp.)*	0.25 to 0.50	1.0 to 2.0	The rates of Rovral per 100 gallons are based on a standard of 400 gallons per acre dilute spray for mature trees. For less than mature trees, apply the rate per 100 gallons until run-off. If less than 400 gallons of spray solution per acre is applied to mature trees, refer to the rate per acre to insure that the proper amount of material is applied. **Brown Rot Blossom Blight** Apply first at early bloom (approximately 5% bloom). If conditions are favorable for disease development, apply again at full bloom and at petal fall. **Fruit Brown Rot** An application should be made whenever temperatures and moisture conditions favor disease infection in the five week period prior to harvest. If these conditions persist or reoccur, a second application should be made. This second application should be made no sooner than 7 days following the first preharvest application. Preharvest applications may be made up to and including the day of harvest. Do not apply more than 2.0 lbs. of Rovral per acre per application. Do not make more than 5 applications of Rovral per season.
ALMONDS*	Apply Rovral as an overall spray in sufficient water to obtain THOROUGH COVERAGE of bloom and foliage (100-400 gallons per acre by ground equipment and in a MINIMUM of 20 gallons per acre by air).	Brown Rot Blossom Blight *(Monilinia laxa)* Shothole *(Stigmina carpophila)*	0.25	1.0	The rates of Rovral per 100 gallons are based on a standard of 400 gallons per acre dilute spray for mature trees. For less than mature trees, apply the rate per 100 gallons until run-off. If less than 400 gallons of spray solution per acre is applied to mature trees, refer to the rate per acre to insure that the proper amount of material is applied. **Brown Rot Blossom Blight: Shothole** Apply first at pink bud and, if conditions favorable for disease development persist or reoccur, again at full bloom. Do not make more than 2 applications of Rovral per season. If conditions favor the development of shothole during the bloom period when Rovral may be applied, i.e. leaf initiation and appropriate weather conditions, Rovral will control shothole. If conditions persist or reoccur after bloom is over, additional sprays with a fungicide labelled for application at that growth stage may be required.

* **NOTE TO USER:** Do not graze animals in treated orchards. Do not feed cover crops grown in treated orchards to livestock.

GRAPES

CROP	HOW TO USE	DISEASE	DOSAGE RATE LBS. PRODUCT/ 100 GAL.	DOSAGE RATE LBS. PRODUCT/ ACRE	WHEN TO USE
GRAPES	Rovral is a wettable power formulation for the control of Bunch Rot on GRAPES. Apply as a foliar spray in sufficient water to obtain THOROUGH COVERAGE (50-200 gallons per acre). The spray mixture should not be allowed to stand overnight as some breakdown of the product may occur from prolonged delays, particularly in water with a high pH. Maintain agitation during spray operations and apply with properly calibrated application equipment. Do not make more than 4 applications of Rovral per season.	Bunch Rot *(Botrytis sp.)*	—	1.5-2.0	Apply Rovral in accordance with directions provided in the table: **SPRAY SCHEDULE** 1. Early—midbloom 2. Prior to bunch closing. 3. Beginning of fruit ripening (veraison). 4. Final application prior to harvest as needed. The last application may be made up to and including the day of harvest. **Thorough coverage of the bunches is essential.** Under severe disease conditions the higher rate is recommended. Rovral must be used in conjunction with good cultural practice designed to minimize conditions conducive to bunch rot *(Botrytis cinerea)*. It should be applied with equipment calibrated and adjusted to direct the spray at the bunches to ensure thorough bunch coverage.

LEAFY VEGETABLES

CROP	HOW TO USE	DISEASE	DOSAGE RATE LBS. PRODUCT/ 100 GAL.	DOSAGE RATE LBS. PRODUCT/ ACRE	WHEN TO USE
HEAD LETTUCE*	Apply as a foliar spray in sufficient water to obtain thorough coverage (50-100 gallons per acre).	Lettuce Drop *(Sclerotinia sp.)* Bottom Rot *(Rhizoctonia solani)*	—	1.5-2.0	Application should be made with tractor mounted boom sprayers equipped with flat fan or hollow cone nozzles directed to insure thorough coverage of the lower portion of the plants and the surrounding soil surface. Do not drench. The higher rate should be used under severe disease conditions. Apply at the 3 leaf stage of growth and again 10 days later. If conditions still favor disease development a third application should be made 10 days after the second spray. Do not apply within 14 days of harvest. Do not make more than 3 applications of Rovral per season.

***NOTE TO USER:** The following crops may be rotated after harvest: Garlic, Dry Bulb Onions, Leafy Vegetables and Peanuts. The following crops may be rotated the year following treatment: Root Crops, Cereal Grains, Soybeans and Tomatoes.

VEGETABLES and CERTAIN FIELD CROPS

CROP	HOW TO USE	DISEASE	DOSAGE RATE LBS. PRODUCT/ ACRE	DOSAGE RATE GALS. WATER PER ACRE	WHEN TO USE
DRY BULB ONIONS*	A single, flat fan, or cone nozzle should be centered and adjusted to provide complete coverage of each row. Up to 10 applications can be made per season. Do not apply within 7 days of harvest. Apply Rovral using ground equipment in accordance with the directions in the table	Botrytis leaf blight (*Botrytis squamosa*) Purple blotch (*Alternaria porri*)	1.5	50-100	Apply as a foliar spray as soon as conditions become favorable for disease. Continue applications on a 7 day interval as long as conditions favor disease development.
GARLIC*	Apply as an in furrow spray in sufficient water to obtain thorough coverage.	White Rot (*Sclerotium cepivorum*)	4.0**	50-100	Apply Rovral in the furrow at planting. Position nozzles to treat the cloves and the covering soil used to fill the furrow. **Using a 38-40 inch bed spacing.
PEANUTS*	Apply Rovral using a tractor mounted boom sprayer equipped with low pressure nozzles (e.g. 8008 LP; 8010 LP, and TK 7.5) that produce large droplets. A single nozzle should be centered and adjusted to provide complete coverage of the row.	Sclerotinia blight (*Sclerotinia sp.*)	2.0	40	Make an initial application when conditions first become favorable for disease development. Up to 2 subsequent applications should be made at 4 week intervals. Do not apply within 10 days of harvest.

***NOTE TO USER:** The following crops may be rotated after harvest: Garlic, Dry Bulb Onions, Leafy Vegetables and Peanuts. The following crops may be rotated the year following treatment: Root Crops, Cereal Grains, Soybeans and Tomatoes.

CONDITION OF SALE, WARRANTY, LIMITATION OF LIABILITY

This product conforms to the chemical description on the label thereof and is reasonably fit for the purpose stated on such label only when used in accordance with directions under normal use conditions. Follow directions carefully. Timing and method of application, weather and crop conditions, mixture with other chemicals not specifically recommended and other influencing factors in the use of this product are beyond the control of Rhone-Poulenc or the seller. Buyer assumes all risks of use, storage, or handling of this material, not in strict accordance with directions given herewith. In no case shall Rhone-Poulenc or the seller be liable for consequential, special, or indirect damages such as loss of profits or values resulting from the use or handling of this product.

ZOLONE® EC

INSECTICIDE

For Agricultural Or Commercial Use Only

For the Control of Insect Pests and Spider Mites

ACTIVE INGREDIENT:	
Phosalone [0,0-diethyl S-[(6-chloro-2-oxo benzoxazolin-3-yl) methyl] phosphorodithioate]	34.4%
INERT INGREDIENTS	65.6%
TOTAL	100.0%

This product contains 3 pounds of phosalone per gallon.

EPA Reg. No. 359-620 EPA Est.

WARNING
Keep Out of Reach of Children
AVISO

PRECAUCION AL USUARIO: Si usted no lee ingles, no use este producto hasta que le etiqueta haya sido explicado ampliamente.

STATEMENT OF PRACTICAL TREATMENT

IF SWALLOWED: Call a physician or Poison Control Center. Drink 1 or 2 glasses of water and induce vomiting by touching finger to back of throat. Do not give anything by mouth or induce vomiting to an unconscious or convulsing person. Get medical attention.

IF ON SKIN: Wash with plenty of soap and water. Get medical attention if irritation persists.

IF IN EYES: Flush with water for 15 minutes. Get medical attention.

See Side Panel for Additional Precautionary Statements.

PRECAUTIONARY STATEMENTS

HAZARDS TO HUMANS AND DOMESTIC ANIMALS

WARNING

WARNING, may be fatal if swallowed. Harmful if absorbed through skin or inhaled. Causes eye irritation. May cause allergic skin reactions. Do not get in eyes, on skin or on clothing. Wear protective clothing, natural rubber gloves, and goggles. Do not breathe fumes or spray mist. Wear a pesticide respirator jointly approved by the Mining Enforcement and Safety Administration (formerly the U.S. Bureau of Mines) and by the National Institute for Occupational Safety under the provisions of 30 CFR Part II.

NOTE TO PHYSICIAN

Causes cholinesterase inhibition. Atropine by injection is antidotal only if symptoms of cholinesterase inhibition are present. 2-PAM is also antidotal and may be administered in conjunction with atropine. Warning symptoms include weakness, headache, tightness in the chest, blurred vision, nonreactive pinpoint pupils, salivation, sweating, nausea, vomiting, diarrhea, and abdominal cramps.

ENVIRONMENTAL HAZARDS

This product is toxic to fish, shrimp, crabs and other wildlife. Do not apply directly to lakes, streams or ponds. Do not apply where run off is likely to occur. Do not contaminate water by cleaning of equipment or disposal of wastes. Do not apply when weather conditions favor drift from areas treated.

This product is toxic to bees exposed to direct treatment or residues on blooming crops or weeds. Do not apply this product if bees are visiting the treatment area.

DIRECTIONS FOR USE

It is a violation of Federal law to use this product in a manner inconsistent with its labeling.

REENTRY STATEMENT

Reentry into treated fields is prohibited within 24 hours of application. State or local use and reentry regulations may be more restrictive than those stated on this label.

Do not apply this product in such a manner as to directly or through drift expose workers or other persons. The area being treated must be vacated by unprotected persons.

Do not enter treated areas without protective clothing until sprays have dried.

Do not enter treated areas for 24 hours unless appropriate protective clothing is worn.

Because certain states may require more restrictive reentry intervals for various crops treated with this product, consult your State Department of Agriculture for further information.

STORAGE AND DISPOSAL

Do not contaminate water, food, or feed by storage or disposal.

STORAGE

Do not use, pour, spill or store near heat or open flame. Do not store below 32°F.

PESTICIDE DISPOSAL

Wastes resulting from the use of this product may be disposed of on site or at an approved waste disposal facility.

CONTAINER DISPOSAL

Triple rinse (or equivalent). Then puncture and dispose of in a sanitary landfill, or by other procedures approved by state and local authorities.

GENERAL INFORMATION

Special Notes: Do not allow livestock to graze cover crops in orchards or nut groves treated with ZOLONE EC; do not feed cover crops treated with ZOLONE EC to livestock. Apply this product only as specified on this label.

ZOLONE EC is an emulsifiable concentrate intended for application as a water mix spray. This product is suitable for use in all power operated ground sprayers, including concentrate sprayers and by aircraft. Keep spray mix agitated when spraying. For application as a concentrate spray to all crops, mix the amount shown per acre and apply in sufficient water to obtain adequate coverage. The use of surfactants (wetting agents) with ZOLONE EC is not recommended. Crop injury resulting from adverse environmental conditions may be aggravated by the use of this product.

ORCHARD AND VINEYARD CROPS

CROP	PEST	DILUTE PTS./ 100 GAL.	CONC. PTS./ACRE	SPECIAL INSTRUCTIONS
Apples and Pears	Aphids, Codling Moth	$\frac{2}{3}$ to $1\frac{1}{3}$	$2\frac{2}{3}$ to $5\frac{1}{3}$	Time applications to local pest control programs. Use minimum recommended rate for preventive applications. For severe infestation use higher recommended rates. For application as dilute sprays, mix the recommended amount per 100 gallons of water and apply as a complete coverage spray (normally 400 to 600 gallons per acre, depending upon species, size of trees, and row spacing). Do not apply more than $17\frac{1}{2}$ lbs. ($46\frac{2}{3}$ pints) per acre per season. Some russeting of golden delicious and other yellow varieties of apples may follow use of ZOLONE EC in some areas. Do not apply within 14 days of harvest. *Except phosphate resistant strains. **Control when used in a seasonal program.
	Redbanded Leafroller, European Red Mite	1 to $2\frac{2}{3}$	4 to $10\frac{2}{3}$	
	Plum Curculio, Twospotted Mite*	$1\frac{1}{3}$ to $2\frac{2}{3}$	$5\frac{1}{3}$ to $10\frac{2}{3}$	
	Green Fruitworm	1	4	
Apples	Apple Maggot, Apple Rust Mite**	$1\frac{1}{2}$ to $2\frac{2}{3}$	$5\frac{1}{2}$ to $10\frac{2}{3}$	
Pears	Pear Psylla	2	8	
Grapes	Leafhopper	2 to 4	4 to 8	Apply minimum recommended rate for preventive applications. For severe infestations use higher rates. For dilute sprays to grapes, mix the quantity shown per 100 gallons of water and apply as a complete coverage spray (normally 100 to 200 gallons per acre). Repeat applications as required but do not use more than 8 lbs. ($21\frac{1}{3}$ pints) per acre per season. Do not apply within 14 days of harvest. Use only concentrate sprays on grape varieties intended for packing or canning. *Use ground equipment, full coverage spray.
	Mites, Omnivorous Leafroller*	$2\frac{2}{3}$ to $5\frac{1}{3}$	$5\frac{1}{3}$ to $10\frac{2}{3}$	
	Grape Berry Moth	2 to 4	4 to 6	
	Grape Phylloxera	$1\frac{1}{3}$ to 2	$2\frac{2}{3}$ to 4	
Almonds	Pacific Mite* Twospotted Mite*	$1\frac{1}{3}$	$5\frac{1}{3}$ to 8	Mix recommended amount in 100 gallons of water and apply as a complete coverage spray (normally 400 to 600 gallons per acre). Apply a minimum of 2 pounds ($5\frac{1}{3}$ pints) per acre, per application. Do not apply more than one application during the fruiting season. Do not apply more than 6 pounds (16 pints) per acre per season, of which no more than 3 pounds (8 pints) should be applied when fruit is present. Do not apply within 60 days of harvest. *Except phosphate resistant strains.
	Peach Twig Borer, Brown Mite*	$1\frac{1}{3}$ to 2	$5\frac{1}{3}$ to 8	
English Walnuts	Walnut Aphid	$\frac{2}{3}$ to $1\frac{1}{3}$	2 to 4	Mix recommended amount in 100 gallons of water and apply as a complete coverage spray (normally from 300 to 600 gallons per acre). Apply in a seasonal program as required, but do not apply more than 20 lbs. ($53\frac{1}{3}$ pints) per acre per season. Apply a minimum of 1 pound ($2\frac{2}{3}$ pints) per acre when used in air blast sprayers. Do not apply after the husks split. * Except phosphate resistant strains. ** Apply before larval feeding causes the husks to darken or shortly after oviposition.
	Twospotted Mite*	$1\frac{1}{3}$ to 2	$5\frac{1}{3}$ to 8	
	Walnut Husk Fly**	$1\frac{1}{3}$ to 2	6	
	Codling Moth	$1\frac{1}{2}$ to 2	6	
Filberts*	Filbert Worm, Filbert Aphid, Filbert Leafroller	1 to $1\frac{1}{2}$	4 to 8	Mix recommended amount in 100 gals. of water and apply as a complete cover spray. Apply a minimum of 4 pts. per acre per application. Do not apply later than 30 days prior to harvest. * Recommendations limited to Oregon and Washington only.
Pecans	Pecan Aphid *(Monellia sp.)*, Black Pecan Aphid, Hickory Shuckworm, Pecan Nut Casebearer, Pecan Spittlebug, Pecan Weevil, Pecan Phylloxera*	$1\frac{1}{3}$	$2\frac{2}{3}$ to $5\frac{1}{3}$	Apply specified dosage per acre by air or ground equipment in sufficient water for complete coverage (normally 5–15 gallons per acre by air; 200-400 gallons per acre by ground). Apply in a seasonal program as required, but do not apply more than 20 lbs. ($53\frac{1}{3}$ pints) per acre per season. Do not apply after shucks split. NOTE: With aerial application control may be reduced due to poorer coverage. Do not tank mix this product with Disulfoton * Suppression
Plums/ Prunes Apricots Peaches and Nectarines	Aphids	$\frac{2}{3}$ to $1\frac{1}{3}$	$2\frac{2}{3}$ to $5\frac{1}{3}$	Mix recommended amount in 100 gallons of water and apply as a complete cover spray (normally 400 gallons per acre based on mature trees with normal spacing). Apply in a seasonal program as required, but do not apply more than 20 pounds of phosalone ($53\frac{1}{3}$ pints) per acre per season. Do not apply within 7 days of harvest. * Except phosphate resistant strains.
	Peach Twig Borer	$1\frac{1}{3}$	$5\frac{1}{3}$	
	Mites (including European Red Mite, Twospotted Mite,* Brown Almond Mite)	2 to $2\frac{2}{3}$	8 to $10\frac{2}{3}$	
	Leafhoppers, Plum Curculio	$1\frac{1}{3}$ to $2\frac{2}{3}$	$5\frac{1}{3}$ to $10\frac{2}{3}$	
Peaches and Nectarines, Apricots	Oriental Fruit Moth	$1\frac{1}{3}$	$5\frac{1}{3}$	
Cherries	Fruit Tree Leafroller	1 to $2\frac{2}{3}$	8 to $21\frac{1}{3}$	Mix recommended amount in 100 gallons of water and apply as a complete cover spray (normally 600 to 800 gallons per acre based on mature trees with normal spacing). Apply in a seasonal program as required, but do not apply more than 20 pounds of phosalone ($53\frac{1}{3}$ pints) per acre per season. Do not apply within 7 days of harvest.
	Cherry Fruit Fly	$1\frac{1}{3}$	$10\frac{2}{3}$	
Citrus (Oranges, Lemons, and Grapefruit)	Citrus Red Mite*	$\frac{2}{3}$ to 1	$13\frac{1}{3}$ to 24	Mix recommended amount in 100 gallons of water and apply as a complete cover spray (on mature trees apply a minimum of 5 lbs. ($13\frac{1}{3}$ pints) of ZOLONE EC per acre for Citrus Red Mite control). Do not make more than one application every 30 days, nor more than 2 applications per season. Do not apply more than 25 lbs. ($66\frac{2}{3}$ pints) per acre per season. Do not apply within 14 days of harvest. * Recommendations limited to California and Arizona.
	Orange Dog	$\frac{2}{3}$	4 to 8	
	Citrus Thrip	$1\frac{1}{3}$	$5\frac{1}{3}$ to 8	
	White Fly	$1\frac{1}{3}$	8 to 12	

FIELD CROPS

CROP	PEST	DILUTE PTS./ACRE	CONC. LB.AI/A	SPECIAL INSTRUCTIONS
Artichokes	Chrysanthemum Leaf Miner, Aphids	5⅓	—	Apply specified dosage per acre by air or ground equipment in sufficient water for complete coverage (normally 5-10 gallons per acre by air, 50-100 gallons per acre by ground). Apply in a seasonal program as required, but do not apply more than 12 pounds (32 pints) per acre per season. Do not apply within 7 days of harvest.
	Plume Moth	2⅔ to 5⅓	—	
Potatoes	Control: Colorado Potato Beetle	1⅓ to 4	0.5 to 1.5	Apply specified dosage per acre in sufficient water for complete coverage (normally 60 to 150 gallons per acre by ground; 5 gallons per acre by air). Apply in a seasonal program as required, but do not apply more than 20 lbs. (53⅓ pints) per acre per season. Do not forage or feed potato vines to livestock. Do not use on sweet potatoes. Do not tank mix this product with Dioulfoton.
	European Corn Borer	2⅔	1	
	Suppression: Buckthorn Aphid, Green Peach Aphid, Potato Aphid, Potato Leafhopper, Tuberworm	1⅓ to 4	0.5 to 1.5	

ORNAMENTALS: This product is not for use in greenhouses.

CROP	PEST	DILUTE PTS./100 GAL.	CONC. LB. AI/A	SPECIAL INSTRUCTIONS
Arborvitae	Spruce Spider Mite	1⅓ to 2⅔	—	Apply specified dosage in sufficient water for complete coverage (spray to wet). Apply when pest is present and repeat as necessary.
Roses	Rose Aphid	1⅓ to 2⅔	—	Apply specified dosage in sufficient water for complete coverage (spray to wet). Apply when pest is present and repeat as necessary. Slight injury to the leaves may occur after the use of ZOLONE EC.

Local conditions may affect the use of the insecticide. Consult state Agricultural Extension or Experiment station specialists for specific recommendations, consistent with label, for local pest problems.

CONDITION OF SALE, WARRANTY, LIMITATION OF LIABILITY

This product conforms to the chemical description on the label thereof and is reasonably fit for the purpose stated on such label only when used in accordance with directions under normal use conditions. Follow directions carefully. Timing and method of application, weather and crop conditions, mixture with other chemicals not specifically recommended and other influencing factors in the use of this product are beyond the control of Rhone-Poulenc or the seller. Buyer assumes all risks of use, storage, or handling of this material, not in strict accordance with directions given herewith. In no case shall Rhone-Poulenc or the seller be liable for consequential, special, or indirect damages such as loss of profits or values resulting from the use or handling of this product.

Rohm and Haas Company

Independence Mall West
Philadelphia, PA 19105
(215) 592-3000

BLAZER® 2L

POSTEMERGENCE HERBICIDE

For Use on Soybeans, Peanuts and Rice.

ACTIVE INGREDIENT
Sodium salt of acifluorfen
Sodium 5-[2-chloro-4-(trifluoromethyl)

phenoxy]-2-nitrobenzoate	20.1%*
INERT INGREDIENTS	79.9%
TOTAL	100.0%

*Equivalent to 2 lbs active ingredient per gallon

EPA Reg. No. 707-149 EPA Est. No. 707-TX-1
U.S. Pat 31455
8815-R6

DANGER

Keep Out of Reach of Children

NOT FOR HOME USE

NOTICE: Before using this product, read the entire Precautionary Statements, Conditions of Sale and Warranty, Directions for Use, Storage and Disposal Instructions, and Use Restrictions on side and back panels and attached use directions. If the Conditions of Sale and Warranty are not acceptable, return the product unopened within thirty days of purchase to the place of purchase.

PRECAUTIONARY STATEMENTS

HAZARDS TO HUMANS AND DOMESTIC ANIMALS

DANGER

Causes eye damage. Harmful if swallowed, inhaled or absorbed through the skin. Do not get in eyes. Wear goggles or face shield when handling. For protection during application wear a hat, long-sleeved shirt and trousers. Mixer-loaders must wear rubber gloves. Avoid breathing vapor or spray mist and contact with skin or clothing. In case of contact, immediately remove contaminated clothing and shoes. Wash contaminated clothing with soap and hot water before re-use.

STATEMENT OF PRACTICAL TREATMENT

If in eyes: Flush with large amounts of water for at least 15 minutes. Get medical attention.

If on skin: Wash with plenty of soap and water. Consult a physician if irritation persists.

If swallowed: Dilute by giving 2 glasses of water to drink and call a physician. Never give anything by mouth to an unconscious person.

Note to Physician: Emesis is recommended.

ENVIRONMENTAL HAZARDS

Do not apply directly to water. Do not contaminate water by cleaning of equipment or disposal of wastes. Do not apply when weather conditions favor drift from target area.

(SEE ATTACHED BOOKLET FOR COMPLETE USE DIRECTIONS.)

SHAKE WELL BEFORE USING

RE-ENTRY AND WORKERS PROTECTION STATEMENTS

Do not apply this product in such a manner as to directly or through drift, expose workers or other persons. The area being treated must be vacated by unprotected persons. Do not enter treated areas without protective clothing until sprays have dried. Because certain states may require more restrictive re-entry intervals for various crops treated with this product, consult your State Department of Agriculture for further information.

STORAGE AND DISPOSAL

STORAGE

Keep from freezing. Store above 32°F.

PESTICIDE DISPOSAL

Do not contaminate water, food or feed by storage or disposal. Pesticide wastes are acutely hazardous. Improper disposal of excess pesticide spray mixture, or rinsate is a violation of Federal law. If these wastes cannot be disposed of by use according to label instructions, contact your State Pesticide or Environmental Control Agency, or the Hazardous Waste representative at the nearest EPA Regional Office for guidance.

CONTAINER DISPOSAL

Triple rinse (or equivalent). Puncture and dispose of in a sanitary landfill, or by incineration, or if allowed by state and local authorities, by burning. If burned, stay out of smoke.

STEPS TO BE TAKEN IN CASE MATERIAL IS RELEASED OR SPILLED

Dike and contain spill with inert material (sand, earth, etc.) and transfer liquid and solid diking material to separate containers for disposal. Remove contaminated clothing and wash affected skin areas with water. Wash clothing before re-use. Keep spill out of all sewers and open bodies of water.

DIRECTIONS FOR USE

General Classification.

It is a violation of Federal law to use this product in a manner inconsistent with its labeling.

CONDITIONS OF SALE AND WARRANTY

Rohm and Haas warrants that the product conforms to its chemical description and is reasonably fit for the purpose stated on the label only when used in accordance with label directions under normal conditions of use. **ROHM AND HAAS MAKES NO OTHER EXPRESS OR IMPLIED WARRANTIES EITHER OF MERCHANTABILITY OR FITNESS FOR A PARTICULAR USE.** Handling, storage and use of this product by Buyer or User are beyond the control of Rohm and Haas and Seller. Risks such as crop injury, ineffectiveness or other unintended consequences resulting from, but not limited to, weather or soil conditions, presence of other materials, disease, pests, drift to other crops or property or failure to follow label directions will be assumed by the Buyer or User. **IN NO CASE WILL ROHM AND HAAS OR SELLER BE HELD LIABLE FOR CONSEQUENTIAL, SPECIAL OR INDIRECT DAMAGES RESULTING FROM THE HANDLING, STORAGE OR USE OF THIS PRODUCT.**

GENERAL INFORMATION—POSTEMERGENCE SOYBEANS, PEANUTS AND RICE

BLAZER 2L is a selective broad-spectrum herbicide recommended for postemergence applications to soybeans, peanuts and rice to control susceptible weeds. Optimum weed control is achieved when young actively growing weed seedlings are treated. It is important to cover all weed parts thoroughly, as BLAZER 2L herbicide works primarily by contact action. Failure to follow the suggested dosages on maximum leaf stage limits may result in unsatisfactory control. See the weed charts and special use directions for recommended rates and timing of applications.

BLAZER 2L herbicide is formulated as a liquid concentrate containing 2 pounds of active ingredient per gallon of product. In this label text TRITON AG-98™ is recommended as the choice of surfactant.

The use of comparable 80% active nonionic surfactants, cleared for application on growing crops, is also satisfactory. TRITON AG-98 surfactant should be used at a rate of 1 pint per 100 gallons of spray mixture. Some weeds respond to additional surfactant. See special use directions for specific weeds and amounts of TRITON AG-98 surfactant necessary for control.

Crop oil concentrate may be used in place of TRITON AG-98 surfactant where specific use directions are detailed on this label or on other approved Rohm and Haas Company supplemental labeling. The use of crop oil concentrate will result in greater crop response.

Soybeans, peanuts and rice are tolerant to postemergence applications of BLAZER 2L herbicide when applied at the recommended growth stages and suggested dosage rates. Leaves may respond to BLAZER 2L herbicide and exhibit burning, crinkling, and bronzing, particularly on the youngest leaves present at the time of application. Exposed stems may also exhibit external spotting and bronzing. The addition of TRITON AG-98 surfactant may intensify the crop response. Crops outgrow this condition and continue to develop normally.

MIXING

BLAZER 2L herbicide causes eye damage. Do not get in eyes. Wear goggles or face shield when handling. In case of eye contact, flush eyes with plenty of water for at least 15 minutes and call a physician.

Refer to container label for additional precautionary statements.

For the control of most annual broadleaf weed species, a use rate of 2 pints (0.5 lb. active) per acre on a broadcast or overall basis is recommended. See specific use directions for tank mixes. Fill the spray tank about two-thirds full with water. Add the required amount of BLAZER 2L herbicide and TRITON AG-98 surfactant, mix thoroughly and continue filling the spray tank. Maintain sufficient agitation during mixing and spraying to insure a uniform spray mixture.

GROUND APPLICATION

BLAZER 2L herbicide may be applied by ground equipment using standard herbicide sprayers equipped with hollow cone or flat fan nozzles. Spray equipment should be calibrated to deliver a minimum of 20 gallons of spray mixture per acre. A spray pressure of 40 to 60 psi at the nozzle tip is recommended. Adjust the height of the boom above the crop to give complete coverage of all weeds. The high gallonage and high pressure will promote necessary coverage of weeds. For further information on optimum spray pressures for specific nozzles, refer to manufacturers' charts for recommendations.

Cultivation before or during the application is not recommended. Cultivation may put weeds under stress, thus making control more difficult to obtain. Timely cultivation 7 days after application will usually assist in weed control.

When row banding equipment is employed it should be adjusted to provide maximum coverage of weeds in the row.

Thorough coverage of the weeds can be obtained with two nozzles directed from either side of the crop row toward the weeds in the center rows. Recommended minimum band width is 15 inches with a minimum of 15 gallons of water per acre on this band. Application with a single nozzle over the top of weeds in the center row is not recommended.

LOW PRESSURE SPRAY NOZZLE TIPS (flood type) ARE NOT RECOMMENDED.

AERIAL APPLICATION

For the control of weeds on this label, use the rates listed for BLAZER 2L herbicide and TRITON AG-98 surfactant, and do not exceed the maximum leaf stages. Best coverage of weeds occurs when soybeans are not past the first to second trifoliate leaf stage, or before peanuts start to cover the soil. BLAZER 2L herbicide should be applied in a minimum of 5 to 10 gallons of water per acre. BLAZER 2L applied in 5 gallons per acre has been effective for control of hemp sesbania. For control of other weeds, the 10 gallon per acre spray volume is recommended. Nozzles on spray booms should not be placed any closer to wing tips than ¾ of the wing span; this will minimize the formation of spray or wing-tip vortice roll. A height of 6 to 10 feet over the crop is recommended. Swirljet or hollow cone nozzles are suggested. Keep pressure below 40 psi. Nozzles should point to the rear of the aircraft and not be pointed downward more than 20 degrees.

AVOID DRIFT

When spraying labeled crops, care must be exercised to prevent spray drift which could result in damage to other crops. Aerial spraying when other crops are closer than 100 yards downwind or 50 yards upwind from the point of application is not recommended. The use of any cleared drift control agent may reduce this hazard; however, the drift control agent may also decrease the weed control activity.

IMPORTANT

Aerial applicators must be familiar with the EPA registered label and follow the use precautions. Spraying of BLAZER 2L herbicide in a manner other than as recommended is done at the user's risk. Users are responsible for all loss or damage which results from such spraying. In addition, aerial applicators should follow all applicable state and local regulations and ordinances. In interpreting the label and local regulations, the most restrictive situations should apply to avoid drift hazards.

TIMING

For best control of susceptible weeds, apply BLAZER 2L herbicide when seedling weeds are in the 2 to 4 leaf stage and actively growing; this usually occurs 14 to 21 days after planting the crop. In solid seeded narrow row soybean planting, BLAZER 2L herbicide should be applied when soybeans are in the 1 to 2 trifoliate leaf stage in order to insure good spray coverage of weeds. For suggestions after the weeds exceed the maximum leaf stage of development listed in the weed table, see use instructions for tank mixture with Basagran® (soybeans) and tank mixture with 2,4-DB (soybeans and peanuts).

MULTIPLE APPLICATIONS

SOYBEANS

Multiple applications of BLAZER 2L herbicide, not exceeding a combined total of 4 pints (1.0 lb. active) per acre per season, can be used as needed to control new weed flushes. Soybeans (conventionally planted) past the third trifoliate stage of development may interfere with the spray pattern, thus preventing adequate spray coverage of weeds and reducing effectiveness. When using multiple applications for the control of subsequent weed flushes, spray equipment should be further adjusted to direct the spray toward the weeds. Specific use directions for multiple applications are covered under the section Special Use Directions for Additional Weed Problems.

PEANUTS

Multiple applications are not suggested in peanuts since the maximum allowable rate per acre is 2 pints of BLAZER 2L herbicide (0.5 lb. active).

RICE (NOT FOR USE IN CALIFORNIA)

BLAZER 2L herbicide is suggested for aerial application to rice for control of hemp sesbania (*Sesbania exaltata*). BLAZER 2L herbicide should be applied to actively growing sesbania plants but before the sesbania is in the flowering stage. Best coverage is usually obtained when the sesbania is above the rice plants. A treatment may be applied when rice is at the late tillering stage up to the early boot stage, which is normally in June and July. A second application of ½ pint BLAZER 2L herbicide (0.125 lb. active) can be applied to control late germinating sesbania. Do not apply to rice after it reaches the boot stage. Apply BLAZER 2L herbicide by aircraft using nozzling to deliver from 5 to 10 gallons of total spray per acre. Use ½ pint BLAZER 2L herbicide (0.125 lb. active) per acre per application. In addition add 2 pints of TRITON AG-98 surfactant to each 100 gallons of spray mix containing BLAZER 2L herbicide. The use of TRITON AG-98 surfactant is important for effective control of hemp sesbania.

When rice has at least 3 leaves, a tank mixture of STAM® M-4 herbicide plus BLAZER 2L herbicide can be used for all the weeds on the STAM M-4 herbicide label plus suppression of northern jointvetch (*Aeschynomene virginica*) and 4 to 6 leaf annual morningglories (*Ipomoea* species). Apply 1 pint of BLAZER

2L herbicide (0.25 lb. active) per acre plus 3 to 4 quarts (3 to 4 lb. active) per acre STAM M-4 herbicide. Add 2 pints of TRITON AG-98 surfactant to each 100 gallons of spray mix. When using the tank mixture an increase in foliage burn may be noticed.

IMPORTANT

DO NOT ADD CROP OIL CONCENTRATES.

TO AVOID EXCESSIVE RESIDUES AT HARVEST, DO NOT APPLY STAM M-4 IN A TANK MIX WITH BLAZER 2L HERBICIDE AFTER THE END OF TILLERING. ALWAYS READ AND FOLLOW THE MOST RESTRICTIVE LABELING USE RESTRICTIONS WHEN TANK MIXING.

SOYBEANS, PEANUTS AND RICE

POSTEMERGENCE USE RESTRICTIONS

- Do not apply BLAZER 2L herbicide within 50 days of harvest for soybeans or 75 days of harvest for peanuts. Do not apply BLAZER 2L herbicide to rice after it reaches the boot stage, or within 50 days of harvest.
- Do not apply more than 4 pints (1.0 lb. active) per acre of BLAZER 2L herbicide per growing season for soybeans or 2 pints (0.5 lb. active) per growing season for peanuts. Do not apply more than two applications to rice per season nor exceed 1 pint (0.25 lb. active total) per acre per season.
- Do not use treated plants for feed or forage.
- Crop Rotation Restriction: root crops (such as carrots, turnips, sweet potatoes, etc.) must *not* be planted in fields treated with BLAZER 2L herbicide for a period of 18 months following treatment.
- In the case of crop failure, only peanuts or soybeans may be immediately replanted.
- Do not harvest crayfish from treated rice areas for food.
- Do not use water from treated rice fields for irrigation purposes for other than BLAZER 2L herbicide labeled crops.
- Avoid drift to all other crops and non-target areas.
- Thoroughly flush spray equipment (tank, hoses, boom, pump) with water before and after each use.
- Do not mix BLAZER 2L herbicide with liquid fertilizers or pesticides except as specifically directed on this label or on other approved Rohm and Haas Company supplemental labeling.
- BLAZER 2L herbicide requires a six hour rain-free period for best results. Do not apply if rain is threatening.
- Do not use during periods of dry weather when crop and weeds are under stress and not actively growing, as weed control may be reduced.
- Do not apply BLAZER 2L herbicide when crop and weeds are under stress from cold temperature, when maximum day temperature is below 70°F, or when soil temperature is below 60°F, as weeds will not be actively growing and control may be reduced.
- Do not apply BLAZER 2L herbicide when crops are exhibiting injury from previously applied pesticides or are exhibiting stress symptoms from disease, nematodes or insects, as increased crop response may result.
- Do not apply BLAZER 2L herbicide when crop and weeds are under stress from excessive fertilizer or soil salts, wind injury, frost damage or high temperature stress or wilt, as increased crop response and/or decreased weed control may result.

SOYBEANS AND PEANUTS
BLAZER 2L HERBICIDE WEED CONTROL AND USE RATE TABLE

Rate: The rate for **broad spectrum** control is 2 pints of BLAZER 2L herbicide per acre plus 1 pint of TRITON AG-98 surfactant per 100 gallons of spray mix.

WEED SPECIES		1.5 PINTS/ACRE LEAF STAGE (MAXIMUM)*	2.0 PINTS/ACRE LEAF STAGE (MAXIMUM)*	3.0 PINTS/ACRE** LEAF STAGE (MAXIMUM)*
ANNUAL BROADLEAF WEEDS				
Amaranth, palmer	*Amaranthus palmeri*	4	6	—
Amaranth, spiny	*Amaranthus spinosus*	—	6	—
Balloonvine	*Cardiospermum halicacabum*	—	4	—
Buckwheat, wild	*Polygonum convolvulus*	—	4	—
Buffalobur	*Solanum rostratum*	—	3***	—
Burgherkin	*Cucumis anguria*	—	4***	—
Carpetweed	*Mollugo verticillata*	multi 6″ dia	multi 8″ dia	—
Citron (wild watermelon)	*Citrullus vulgaris*	—	4***	—
Cocklebur, common	*Xanthium pensylvanicum*	—	4***	8***
Cocklebur, heartleaf	*Xanthium strumarium*	—	4***	8***
Copperleaf, hophornbeam	*Acalypha ostryaefolia*	2	4	—
Copperleaf, Virginia	*Acalypha virginica*	—	4	—
Cowpea, volunteer	*Vigna sinensis*	—	1***	—
Crotalaria, showy	*Crotalaria spectabilis*		See special use directions for rate	
Croton, tropic	*Croton glandulosus*	—	2	—
Croton, woolly	*Croton capitatus*	—	2	—
Cucumber, wild spiny	*Cucumis dipsaceus*	—	4***	—
Devilsclaw	*Proboscidea lousianica*	—	2	—
Galinsoga, hairy	*Galinsoga ciliata*	—	4	6
Galinsoga, smallflower	*Galinsoga parviflora*	—	4	6
Gourd, Texas	*Cucurbita texana*	—	3***	—
Groundcherry, cutleaf	*Physalis angulata*	2	4	6
Groundcherry, lanceleaf	*Physalis lanceifolia*	—	4	6
Indigo, hairy	*Indigofera hirsuta*	—	3	—
Jimsonweed	*Datura stramonium*	6	8	12
Ladysthumb	*Polygonum persicaria*	—	4	—
Lambsquarters	*Chenopodium album*	—	3***	—
Morningglory				
Cypressvine	*Ipomoea quamoclit*	—	4	6
Entireleaf	*Ipomoea hederacea var. integriuscula*	—	3	4
Ivyleaf	*Ipomoea hederacea var. hederacea*	—	3	4
Purple moonflower	*Ipomoea muricata*	4	4	8
Scarlet	*Ipomoea coccinea*	—	4	6

WEED SPECIES		1.5 PINTS/ACRE LEAF STAGE (MAXIMUM)*	2.0 PINTS/ACRE LEAF STAGE (MAXIMUM)*	3.0 PINTS/ACRE** LEAF STAGE (MAXIMUM)*
Smallflower	*Jacquemontia tamnifolia*	—	4	8
Small white (pitted)	*Ipomoea lacunosa*	4	4	8
Tall (common)	*Ipomoea purpurea*	—	3	4
Willowleaf (palmleaf)	*Ipomoea wrightii*	—	4	6
Mustard, wild	*Brassica kaber*	4	6	10
Nightshade, eastern black	*Solanum ptycanthum*	—	4	8
Nightshade, black	*Solanum nigrum*	—	4	8
Pigweed, prostrate	*Amaranthus blitoides*	—	6	8
Pigweed, redroot	*Amaranthus retroflexus*	4	6	8
Pigweed, smooth	*Amaranthus hybridus*	4	6	8
Poinsettia, wild	*Euphorbia heterophylla*	—	2***	—
Poorjoe	*Diodia teres*	—	4	6
Purslane, common	*Portulaca oleracea*	multi 6" dia	multi 8" dia	—
Pusley, Florida	*Richardia scabra*	—	2	4
Ragweed, common	*Ambrosia artemisiifolia*	4	6	8
Ragweed, giant	*Ambrosia trifida*	—	4	—
Senna, coffee	*Cassia occidentalis*	—	—	2***
Sesbania, hemp	*Sesbania exaltata*		See special use directions for rate	
Smartweed, Pennsylvania	*Polygonum pensylvanicum*	—	4	8
Smellmelon	*Cucumis melo*	—	4***	—
Spurge, prostrate	*Euphorbia supina*	—	multi 0.5" dia	—
Spurge, spotted	*Euphorbia maculata*	—	multi 0.5" dia	—
Starbur, bristly	*Acanthospermum hispidum*	—	—	2***
Waterhemp, tall	*Amaranthus tuberculatos*	—	4	6
ANNUAL GRASSES				
Foxtail, giant	*Setaria faberi*	—	3***	4***
Foxtail, green	*Setaria viridis*	—	3***	4***
Foxtail, yellow	*Setaria lutescens*	—	3***	4***
Johnsongrass, seedling	*Sorghum halepense*	—	3***	4***
Panicum, fall	*Panicum dichotomiflorum*	—	3***	4***
Shattercane	*Sorghum bicolor*	—	3***	4***
Volunteer small grains	*Barley, oats, rye, wheat*	—	3***	4***
PERENNIAL WEEDS				
Bindweed, field	*Convolvulus arvensis*	—	multi***	—
Bindweed, hedge	*Convolvulus sepium*	—	multi***	—
Milkweed, climbing	*Sarcostemma cyanchoides*	—	multi***	—
Milkweed, common	*Asclepias syriaca*	—	multi***	—
Redvine	*Brunnichia cirrhosa*	—	multi***	—
Thistle, Canada	*Cirsium arvense*	—	multi***	—
Trumpetcreeper	*Campsis radicans*	—	multi***	—

* Do not count leaves as pairs ... count each leaf separately. Do not count cotyledon leaves. Spraying weeds in the cotyledon growth stage is not recommended.

** For peanuts do not exceed 2 pints per acre.

*** See Special Use Directions for these weed problems.

SPECIAL USE DIRECTIONS FOR ADDITIONAL WEED PROBLEMS IN SOYBEANS AND PEANUTS

ANNUAL BROADLEAF WEEDS:

Buffalobur

Partial control of buffalobur can usually be obtained when the seedlings have less than 4 true leaves. Use BLAZER 2L herbicide at 2 pints (0.5 lb. active) in 30 gallons of water per acre. The use of 2 pints of TRITON AG-98 surfactant per 100 gallons of spray mix is recommended. If regrowth or new seedlings appear, repeat application. Regrowth usually occurs when day or night temperatures fall below 70°. **NOTE: For PEANUTS do not apply more than one application per growing season.**

Cocklebur, Common

Cocklebur, Heartleaf

The most consistent control of cocklebur is obtained when 2 pints per acre of BLAZER 2L herbicide (0.5 lb. active) is applied to seedlings in the 2 leaf stage which are actively growing under conditions of high soil moisture and high relative humidity. The use of 1 pint of TRITON AG-98 surfactant per 100 gallons of spray mix is recommended. As cocklebur approaches the 8 leaf stage, the use of 3 pints (0.75 lb. active) per acre of BLAZER 2L herbicide is suggested. Control of cocklebur may be partial or inconsistent when sprayed during periods of dry weather or other stress conditions. **NOTE: Use the 3 pint rate for soybeans only.**

Cowpea, Volunteer

Volunteer cowpea germinates over a long period of time thus making control difficult. Partial control can be obtained by applying 2 pints (0.5 lb. active) per acre of BLAZER 2L herbicide just past the cowpea cotyledon growth stage. Do not wait for the first leaves to fully expand. Use 2 pints of TRITON AG-98 surfactant per 100 gallons of spray mix for maximum activity. A repeat application is normally required for new seedling weeds. **NOTE: For PEANUTS do not apply more than one application per growing season.**

Sesbania, Hemp

Crotalaria, Showy

Sesbania and crotalaria are very sensitive to BLAZER 2L herbicide with TRITON AG-98 surfactant. The use of 2 pints of TRITON AG-98 surfactant per 100 gallons of spray mix is required for control. Effective control can be obtained at just about all plant heights. It is important, however, that BLAZER 2L herbicide be applied prior to bloom. Applications after bloom are usually not effective and therefore not recommended. The use of 1 pint (0.25 lb. active) per acre has given excellent control of both sesbania and crotalaria. During or after periods of dry weather, control may be erratic. Best control is then obtained when 1.5 pints (0.38 lb. active) are applied to each acre. Application for control of these weeds should be timed to occur after maximum weed emergence has taken place. Care must be exercised to make certain that crops do not shade this weed from spray deposits. Waiting for the sesbania to break through the crop canopy may be advisable for control of late season infestations.

CUCURBITS:

Burgherkin

Citron (Wild Watermelon)

Cucumber, Wild Spiny

Gourd, Texas

Smellmelon

Members of the cucumber family germinate over an extended period of time. Control is therefore difficult to obtain with a single spray. In order for BLAZER 2L herbicide to be effective, initial application should be made to weeds no later than the 4 leaf growth stage. Use 2 pints of BLAZER 2L herbicide (0.5 lb. active) per acre plus 2 pints of TRITON AG-98 surfactant per 100 gallons of spray mix. If subsequent weed flushes occur in severely infested soybean fields, a second application should be timed much like the first application. Waiting for several weed flushes to appear and then spraying will result in poor control. The most effective control is obtained when weeds have no more than 4 true leaves. **NOTE: For PEANUTS do not apply more than one application per growing season.**

Lambsquarters, Common

BLAZER 2L herbicide, at the 2 pint rate (0.5 lb. active) per acre, will usually cause spotting, stunting or death of many seedlings not exceeding 3 true leaves. The use of 2 pints of TRITON AG-98 surfactant per 100 gallons of spray mix is suggested for maximum kill of seedlings. Cultivation 7 to 10 days after the BLAZER 2L herbicide/surfactant application will usually assist in control. **NOTE: For PEANUTS do not apply more than one application per growing season.**

Poinsettia, Wild

BLAZER 2L herbicide, at the 2 pint rate (0.5 lb. active) per acre plus 2 pints of TRITON AG-98 surfactant per 100 gallons of spray mix, will usually kill or severely stunt wild poinsettia. Application must be made prior to the formation of the third true leaf. In addition, the seedlings must be actively growing. This treatment will usually result in a height differential between soybeans and surviving wild poinsettia, thus allowing post directed applications for additional control. **NOTE: For PEANUTS do not apply more than one application per growing season.**

Starbur, Bristly

Senna, Coffee

BLAZER 2L herbicide, at the 3 pint rate (0.75 lb. active) per acre plus 2 pints of TRITON AG-98 surfactant per 100 gallons of spray mix, will kill or suppress seedlings that are not past the 2 leaf stage. Applications after the 2 leaf stage are usually ineffective. **NOTE: Use the 3 pint rate for soybeans only.**

ANNUAL GRASSES:

Foxtail, Giant

Foxtail, Green

Foxtail, Yellow

Johnsongrass, Seedling

Panicum, Fall

Shattercane

BLAZER 2L herbicide must not be the basic portion of the grass herbicide program. For additional control of escape grasses following a preplant incorporated or preemergence herbicide, use BLAZER 2L herbicide at 2 pints (0.5 lb. active) per acre. The use of 2 to 4 pints of TRITON AG-98 surfactant per 100 gallons of spray mix is required. Grasses should not exceed the 2 to 3 leaf stage for this treatment. When escape grasses are in the 4 leaf growth stage use 3 pints of BLAZER 2L herbicide (0.75 lb. active) per acre plus 2 pints of TRITON AG-98 surfactant per 100 gallons of spray. Activity is dependent upon good soil moisture during the following spray application. **NOTE: Use the 3 pint rate for soybeans only.**

Volunteer Small Grains
(Barley, Oats, Rye, Wheat)

When BLAZER 2L herbicide is applied to emerging volunteer small grains in the 2 to 3 leaf stage, many plants will die or remain stunted. BLAZER 2L herbicide should be applied at 2 pints (0.5 lb. active) for each acre plus 2 to 4 pints of TRITON AG-98 surfactant for each 100 gallons of spray mix. Activity is dependent upon good soil moisture during and following spray applications. When volunteer small grains are in the 4 leaf growth stage use 3 pints of BLAZER 2L herbicide (0.75 lb. active) per acre plus 2 pints of TRITON AG-98 surfactant per 100 gallons of spray. **NOTE: Use the 3 pint rate for soybeans only.**

PERENNIAL WEEDS:

Bindweed, Field

Bindweed, Hedge

Milkweed, Climbing

Milkweed, Common

Redvine

Thistle, Canada

Trumpetcreeper

Growth of perennial weeds resulting from underground rootstocks is very difficult to control. BLAZER 2L herbicide at the 2 pint rate (0.5 lb. active) per acre, applied under favorable environmental conditions, will burn back the above-ground plant parts and retard regrowth. A second application may be applied to soybeans for additional burndown. BLAZER 2L will not kill the underground rootstocks of these weeds. **NOTE: For PEANUTS do not apply more than one application per growing season.**

TANK MIXTURE WITH AMIBEN®

SOYBEANS

When applied to young, actively growing broadleaf weed seedlings, the tank mixture of BLAZER 2L herbicide plus Amiben will control all weeds listed on the BLAZER 2L herbicide label and will also provide preemergence control of those grasses and broadleaf weeds listed in the General Information section of the Amiben label which have not yet germinated at time of treatment.

Rate: Apply a tank mixture of 1½ to 2 pints of BLAZER 2L herbicide (0.38 to 0.5 lb. active) per acre tank plus 5 to 6 quarts of Amiben (2.5 to 3.0 lb. active) per acre or 3.0 to 3.6 pounds Amiben DS (2.5 to 3.0 lb. active) per acre. In order to determine the correct application rate of BLAZER 2L herbicide to use in the tank mixture, see the BLAZER 2L herbicide use rate table.

USE RESTRICTIONS FOR TANK MIXTURE WITH AMIBEN

SOYBEANS

- Follow all use restrictions detailed on the BLAZER 2L herbicide and Amiben labels.
- Do not add crop oil when using the tank mix. TRITON AG-98 surfactant should be added at the rate of 2 to 4 pints per 100 gallons of spray for maximum control of certain susceptible weeds such as escape grasses, sesbania and crotalaria. A complete listing of these weeds is found on the BLAZER 2L herbicide label under the Special Use Directions.
- Do not make applications later than 33 days after soybean planting.

TANK MIXTURE WITH BASAGRAN®

SOYBEANS

BLAZER 2L herbicide may be tank mixed with Basagran for postemergence control of the major broadleaf weed species in soybeans.

Rate: Use BLAZER 2L herbicide at the rate of 1.5 to 2.0 pints (0.38 to 0.5 lb. active ingredient) per acre tank mixed with Basagran at the rate of 1.0 to 1.5 pints (0.5 to 0.75 lb. active) per acre. In order to determine the correct application rate of BLAZER 2L herbicide to use in the tank mixture, see the BLAZER 2L use rate table.

For control of cocklebur up to the 6 leaf stage and velvetleaf up to the 4 leaf stage, add 1 pint (0.5 lb. active) per acre of Basagran to BLAZER 2L herbicide. For prickly sida (teaweed) up to the 6 leaf stage, add 1.5 pints (0.75 lb. active) per acre of Basagran to BLAZER 2L herbicide.

Basagran may also be used in the tank mixture at rates up to 2 pints (1.0 lb. active) per acre.

Surfactant: The addition of TRITON AG-98 surfactant is recommended and should be used at a minimum rate of 1 pint per 100 gallons of spray mixture. For control of certain weeds such as escape grasses and hemp sesbania with the tank mix, the addition of 2 to 4 pints of TRITON AG-98 surfactant per 100 gallons of spray mix is required. See the BLAZER 2L herbicide special use directions for additional weeds and specific amounts of TRITON AG-98 surfactant.

Another recommended substitute for TRITON AG-98 surfactant is diammonium phosphate, commonly sold as a solution (10-34-0). This completely water-soluble material should be used at the rate of 1 quart per acre specifically for improved velvetleaf control. The leaf burn commonly associated with crop oil concentrate does not occur with diammonium phosphate which functions solely as a surfactant. Do not add oil concentrate or additional surfactant when using diammonium phosphate (10-34-0) with BLAZER/Basagran tank mix.

Crop Oil Concentrate: The addition of 1 pint per acre of crop oil concentrate may be used as a substitute for the surfactant for improved control of velvetleaf. See your Basagran supplier for a list of oil concentrates recommended by BASF Wyandotte Corporation. Soybean leaf burn will be increased by the addition of oil concentrate.

TANK MIXTURE WITH BASAGRAN

PEANUTS

BLAZER 2L herbicide may be tank mixed with Basagran for postemergence control of the major broadleaf weed species in peanuts.

Rate: Use BLAZER 2L herbicide at the rate of 1.5 to 2.0 pints (0.38 to 0.5 lb. active) per acre tank mixed with Basagran at the rate of 1.5 pints (0.75 lb. active) per acre. In order to determine the correct application rate of BLAZER 2L herbicide to use in the tank mixture, set the BLAZER 2L herbicide use rate table.

For the control of spurred anoda, beggarticks, common cocklebur, dayflower, redweed and prickly sida up to the 6 leaf stage and bristly starbur up to the 4 leaf stage and 1.5 pints (0.75 lb. active) per acre of Basagran to BLAZER 2L herbicide. Add ½ pint of TRITON AG-98 for each 100 gallons of the tank mix spray solution.

Basagran may also be used in the tank mixture at 2 pints (1.0 lb. active) per acre.

Mixing: Fill the spray tank half full with water and add the suggested amount of BLAZER 2L and Basagran while the agitator is running. Add ½ pint of TRITON AG-98 per 100 gallons of spray mix and finish filling the spray tank.

USE RESTRICTIONS FOR BLAZER 2L/BASAGRAN PEANUTS

- Read and follow all applicable directions and use restrictions on the BLAZER 2L herbicide and Basagran labels.
- Do not add crop oil concentrate when using the tank mix.
- Do not apply the tank mixture within 75 days of harvest for peanuts.
- Do not apply more than one application of the tank mixture to peanuts per growing season.

TANK MIXTURE WITH POAST™

SOYBEANS

BLAZER 2L herbicide may be tank mixed with Poast herbicide for postemergence broadleaf and annual grass weed control.

Rate: Use BLAZER 2L herbicide at the rate of 1.5 to 2.0 pints (0.38 to 0.5 lb. active) per acre may be tank mixed with Poast herbicide for postemergence control of selected annual broadleaf/grass weeds in soybeans. The rate of Poast recommended in the tank mix is 50% greater than the rate of Poast used alone. See the Poast label. In order to determine the correct application rate of BLAZER 2L herbicide to use in the tank mixture, see the BLAZER 2L herbicide use rate table.

Timing: For grass species and timing refer to Poast label. Sequential applications should always be used in place of tank mixtures where perennial or certain grasses are present. Refer to the Poast label for details.

Always add Poast herbicide in the tank mixture with 2 pints per acre of a recommended nonphytotoxic oil concentrate. Use 20 gallons of water per acre with a minimum of 40 psi pressure at the nozzle. Air applications should be made in 10 gallons of water per acre. See your Poast supplier for a list of oil concentrates recommended by BASF Wyandotte Corporation. Oil concentrate must be used with the tank mixture in place of TRITON AG-98 surfactant or any other surfactant.

SEQUENTIAL APPLICATIONS— BLAZER 2L/POAST

SOYBEANS

When using BLAZER 2L herbicide as a separate application be sure to allow a minimum of 7 days between the BLAZER 2L herbicide and the Poast grass herbicide application. This applies when BLAZER 2L herbicide is applied first. When Poast herbicide is applied first, wait 24 hours before applying BLAZER 2L herbicide. It is important that treated grass has resumed active growth before spraying with Poast grass herbicide. This waiting period is important in achieving maximum activity with the Poast grass herbicide. For additional information on rate, timing and use restrictions for Poast herbicide refer to the Poast label.

TANK MIXTURE WITH BASAGRAN/POAST

SOYBEANS

BLAZER 2L herbicide may be tank mixed with Basagran and Poast herbicides for postemergence broadleaf and annual grass weed control.

Rate: BLAZER 2L herbicide at 1.5 to 2.0 pints (0.38 to 0.5 lb. active) per acre may be tank mixed with Basagran herbicide at 1.0 to 1.5 pints (0.5 to 0.75 lb. active) per acre and Poast herbicide for postemergence control of selected annual broadleaf/grass weeds in soybeans. The rate of Poast recommended in the tank mix is 50% greater than the rate of Post used alone. See the Poast label. In order to determine the correct application rate of BLAZER 2L herbicide to use in the tank mixture, see the BLAZER 2L herbicide use rate table.

Timing: For control of cocklebur up to the 6 leaf stage and velvetleaf up to the 4 leaf stage add 1 pint (0.5 lb. active) per acre of Basagran to BLAZER 2L herbicide and Poast herbicide. For prickly sida (teaweed) up to the 6 leaf stage, add 1.5 pints (0.75 lb. active) per acre of Basagran to BLAZER 2L herbicide and Poast herbicide. Basagran may also be used in the tank mixture at rates up to 2 pints (1.0 lb. active) per acre.

For individual grass species and timing refer to Basagran/Poast labels. When tank mixing Poast it is recommended that the Poast rate be increased by 50% over the rate of Poast used alone. Sequential applications should always be used in place of tank mixtures where perennial or certain grasses are present. Refer to the Poast label for details.

Always add Poast herbicide in the tank mixture with 2 pints per acre of a recommended nonphytotoxic oil concentrate. Use 20 gallons of water per acre with a minimum of 40 psi pressure at the nozzle. Air applications should be made in 10 gallons of water per acre. See your Poast/Basagran suppliers for a list of oil concentrates recommended by BASF Wyandotte Corporation. Oil concentrate must be used with the tank mixture in place of TRITON AG-98 surfactant or any other surfactant.

SEQUENTIAL APPLICATIONS—

BLAZER 2L HERBICIDE TANK MIX WITH BASAGRAN/POAST

SOYBEANS

When using BLAZER 2L/Basagran herbicides as a separate application be sure to allow a minimum of 7 days between the BLAZER 2L/Basagran herbicides in the tank mix and the Poast grass herbicide application. This applies when BLAZER 2L/Basagran herbicides are applied first. When Poast herbicide is applied first, wait 24 hours before applying a tank mixture of BLAZER 2L/Basagran herbicides. It is important that treated grass has resumed active growth before spraying with Poast grass herbicide. This waiting period is important in achieving maximum activity with the Poast grass herbicide. For additional information on rate, timing and use restrictions for Poast herbicide refer to Poast label.

USE RESTRICTIONS FOR TANK MIXTURES AND SEQUENTIAL APPLICATIONS—

BLAZER 2L/BASAGRAN/POAST

SOYBEANS

- Always read and follow the restrictions and limitations for all products when used alone or in a tank mix. The most restrictive labeling must apply in tank mixtures.
- Do not apply BLAZER 2L herbicide or Poast herbicide when soybeans are exhibiting injury from previously applied pesticides or are exhibiting stress symptoms from disease, nematodes or insects, as increased crop response may result.
- Do not apply BLAZER 2L herbicide within 50 days of harvest and do not apply Poast within 90 days of harvest.
- Do not graze treated soybean fields and do not feed treaed soybean forage, ensilage or hay to livestock (see BLAZER 2L herbicide and Poast herbicide labels).
- Do not apply BLAZER 2L herbicide or Poast herbicide when soybeans and weeds are under stress from excessive fertilizer or soil salts, wind injury, frost damage or high temperature stress or wilt, as increased crop response and/or decreased weed control may result.

BLAZER 2L AND FUSILADE ®

SOYBEANS

BLAZER 2L herbicide and Fusilade may be used together in a postemergence program for broad spectrum weed control. Use either Fusilade 4E or Fusilade 2000 1E.

BLAZER 2L is a selective postemergence herbicide for the control of certain broadleaf weeds. Fusilade is a selective postemergence herbicide for control of annual and perennial grass weeds in soybeans. BLAZER 2L herbicide and Fusilade may be applied sequentially or in a tank mix. The growth stage of weeds at the time of application will determine which method of application will provide the most satisfactory results. Both BLAZER 2L herbicide and Fusilade should be applied to actively growing weeds. See BLAZER 2L label for defined environmental conditions that promote active growth. Do not apply either herbicide if weeds appear stressed due to unfavorable temperatures, drought and/or low soil fertility.

Do not apply BLAZER 2L or BLAZER 2L/Fusilade tank mix if soybeans show injury from prior herbicide applications. Read and observe all applicable label directions and limitations for both BLAZER 2L herbicide and the appropriate Fusilade formulation before using.

METHOD 1: Sequential Application—Fusilade Followed by BLAZER 2L

(Annual and/or perennial grass weeds at proper growth stage for treatment, prior to broadleaf weed treatment)

Use either Fusilade 4E or Fusilade 2000 1E. Apply Fusilade and an approved adjuvant at the recommended rate and growth stage for the grass weeds being treated, as per the Fusilade label directions. Fusilade is a systemic herbicide, therefore, when treating annual grasses, allow at least 3 days, and when treating perennial grasses allow at least 5 days to transpire prior to a BLAZER 2L herbicide application. This will enable Fusilade to be adequately translocated to the meristematic regions of the plant. After the appropriate time interval has elapsed, apply BLAZER 2L herbicide and TRITON AG-98 surfactant for susceptible broadleaf weeds as outlined on the BLAZER 2L herbicide label.

METHOD 2: Sequential Application—BLAZER 2L Followed by Fusilade

(Broadleaf weeds at proper growth stage for treatment, prior to annual and/or perennial grass weed treatment)

Use either Fusilade 4E or Fusilade 2000 1E. Apply BLAZER 2L herbicide with TRITON AG-98 surfactant at the recommended rate and growth stage for susceptible broadleaf weeds, as per the BLAZER 2L herbicide label directions. Broadleaf herbicides can stress and affect the growth of grass species. Fusilade must be applied to actively growing grasses for best results. A sequential application of Fusilade may be applied following a BLAZER 2L herbicide application when grasses resume active growth with the development of a new leaf. Follow the recommended rates and growth stage for the grass weeds being treated, as per the appropriate Fusilade label.

METHOD 3: Tank Mix Applications of BLAZER 2L herbicide and Fusilade

(Only annual grasses and broadleaf weeds are present and at the proper growth stage for treatment, as per the respective labels)

Use either Fusilade 4E or Fusilade 2000 1E. A tank mix of BLAZER 2L herbicide and Fusilade may be applied at the recommended rates and growth stages to susceptible broadleaf weeds and annual grass species in a manner as described on the respective labels. Use an approved nonionic surfactant, as per the Fusilade label recommendations. Follow water volume and spray pressure recommendations, as per the BLAZER 2L herbicide label. Load the spray tank with half the amount of required water and add the recommended amounts of Fusilade, BLAZER 2L, and nonionic surfactant while the agitator is running. Add the remaining quantity of water. Do not make more than one application of this tank mix per season.

NOTE: Tank mix applications sometimes have resulted in reduced grass weed control and possible increases in crop injury as compared to either product used alone. If grass regrowth occurs following an application of the tank mix or an additional flush of grasses emerge, make a second application of Fusilade to actively growing annual grass weeds, as per the label recommendations. A tank mix application is not recommended if perennial grass weeds are the predominant grasses to be controlled.

USE RESTRICTIONS FOR BLAZER 2L/FUSILADE

SOYBEANS

- Always read and follow the restrictions and limitations for all products whether used alone or in a tank mix. The most restrictive labeling applies in tank mixtures.
- Do not apply a total per season of more than 1 pint of Fusilade 4E per acre or 4 pints of Fusilade 2000 1E per acre per season to soybeans.
- Do not apply more than 4 pints (1.0 lb. active) per acre of BLAZER 2L herbicide per growing season for soybeans.
- Make the last Fusilade application before soybean bloom.
- Do not apply BLAZER 2L herbicide within 50 days of harvest for soybeans.
- Do not use treated plants for feed or forage.
- Do not plant rotational crops other than cotton and soybeans within 60 days after last Fusilade application.
- Crop Rotation Restriction: root crops (such as carrots, turnips,

sweet potatoes, etc.) must not be planted in fields treated with BLAZER 2L herbicide for a period of 18 months following treatment.

- In the case of crop failure, only soybeans may be immediately replanted.
- Avoid drift to all other crops and non-target areas.
- Thoroughly flush spray tanks with water before and after each use.
- BLAZER 2L herbicide requires a six hour rain-free period for best results. Do not apply BLAZER 2L herbicide or BLAZER 2L/Fusilade tank mix if rain is threatening.
- Do not make more than one application of BLAZER 2L/Fusilade herbicide tank mix in a single season.
- Do not use during dry weather conditions when crop and weeds are under stress or wilt and not actively growing, as weed control may be reduced.
- Do not apply BLAZER 2L herbicide or BLAZER 2L/Fusilade herbicide tank mix when crop and weeds are under stress from cold temperature, when maximum day temperature is below 70°F, or when soil temperature is below 60°F, as weeds will not be actively growing and control may be reduced.
- Do not apply BLAZER 2L herbicide or BLAZER 2L/Fusilade tank mix when crops are exhibiting injury from previously applied pesticides or are exhibiting stress symptoms from disease, nematodes or insects, as increased crop response may result.
- Do not apply BLAZER 2L herbicide or BLAZER 2L/Fusilade tank mix when crop and weeds are under stress from excessive fertilizer or soil salt, wind injury, frost damage, or high temperature stress or wilt, as increased crop response may result.
- Do not mix BLAZER 2L herbicide or BLAZER 2L/Fusilade tank mix with other pesticides or liquid fertilizers except as specified on this label or on other approved Rohm and Haas Company supplemental labeling.

TANK MIXTURE WITH 2,4-DB SOYBEANS AND PEANUTS

A tank mixture of BLAZER 2L herbicide plus 2,4-DB is recommended for control of morningglory, cocklebur, common ragweed, redroot pigweed, jimsonweed, burgherkin and citron in soybeans and peanuts when the weed size exceeds that specified on the BLAZER 2L herbicide label. Control with this mixture may decrease with increasing weed size or density of weed or soybean canopy, due to poor spray coverage. For control of other weeds refer to the BLAZER 2L herbicide label directions.

SOYBEANS

Rate: Mix 2 fluid ounces of the product Butyrac® 200, EPA Reg. No. 264-105, or 2 fluid ounces of the product Butoxone®, EPA Reg. No. 39511-122, with 1.5 to 2 pints (0.38 to 0.5 lb. active) of BLAZER 2L herbicide for each acre being treated. Add 1 pint of TRITON AG-98 surfactant per 100 gallons to increase control of weeds. Do not add crop oils to the tank mixture. The addition of TRITON AG-98 will increase the hormonal 2,4-DB crop response. Read and follow all applicable directions and use instructions on BLAZER 2L herbicide and 2,4-DB herbicide labels.

Timing: For optimum control apply the tank mix to actively growing weeds up to 12 inches in height or length. Applications at later stages will result in partial control or suppression.

The use of this tank mix will cause foliage injury and may reduce yields. Applications at the third or greater trifoliate leaf stage will assist in minimizing foliar injury.

PEANUTS

Rate: Mix 1 pint of the product Butyrac 200, EPA Reg. No. 264-105, or 1 pint of the product Butoxone, EPA Reg. No. 39511-122, with 1.5 to 2 pints (0.38 to 0.5 lb. active) of BLAZER 2L herbicide for each acre being treated. Add 1 pint of TRITON AG-98 surfactant per 100 gallons to increase control of weeds. Do not add crop oils to the tank mixture. The addition of TRITON AG-98 will increase the hormonal 2,4-DB crop response. Read and follow all applicable directions and use instructions on BLAZER 2L herbicide and 2,4-DB herbicide labels.

Timing: For optimum control apply BLAZER 2L herbicide plus 2,4-DB tank mix to actively growing weeds up to 12 inches in height or length, usually 2 to 12 weeks after planting. Applications at later weed stages will result in partial control or suppression.

Peanuts should be at least 2 weeks old when using a tank mixture of BLAZER 2L herbicide and 2,4-DB. Do not use after pod-filling stage begins.

SOYBEANS AND PEANUTS

Application: For best results, the tank mixture should be applied with ground equipment. For thorough coverage of weeds apply with flat fan or hollow cone nozzles in a minimum of 20 gallons of water per acre with a spray pressure of 40 psi.

Drift Hazard: Care must be taken when applying the tank mixture to prevent drift to all non-target crops. Tobacco, ornamentals, mustards, sugarbeets, potatoes, vegetables, and cotton are a few of the crops known to be sensitive to this tank mixture. Hormone type injury in non-target crops can result from trace amounts of 2,4-DB drift. The use of any cleared drift control agent may reduce this hazard; however, the drift control agent may also decrease the weed control activity.

USE RESTRICTIONS FOR TANK MIXTURE WITH 2,4-DB HERBICIDE

SOYBEANS AND PEANUTS

- Read and follow all directions and use restrictions on BLAZER 2L herbicide and 2,4-DB labels.
- Do not use rates of BLAZER 2L herbicide or 2,4-DB in excess of that recommended on this label, or excessive injury and possible yield reduction could result.
- Do not apply the tank mixture within 60 days of harvest for soybeans or 75 days of harvest for peanuts.
- Do not mix oils, liquid fertilizers or other pesticides with the tank mixture except as specifically directed on this label.
- Do not apply more than one application of the tank mixture to soybeans or peanuts per growing season.
- Do not apply the tank mixture when soybeans or peanuts are exhibiting injury from previously applied pesticides or are exhibiting stress symptoms from disease, nematodes, insects, excessive fertilizer or soil salts, wind injury, frost damage or high temperature stress or wilt, as increased crop response will result.
- Aerial applicators must be familiar with the EPA registered labels and follow the use precautions. In addition, aerial applicators should follow all applicable state and local regulations. In interpreting the label and local regulations, the most restrictive situations should apply to avoid drift hazards.

BLAZER®, STAM®, TRITON® and AG-98 are trademarks of the Rohm and Haas Company.

AMIBEN® and BUTYRAC® are registered trademarks of Union Carbide Corporation.

BASAGRAN® and POAST® are registered trademarks of BASF Wyandotte Corporation.

BUTOXONE® is a registered trademark of Vertac Chemical Company.

FUSILADE® is a registered trademark of ICI Americas Inc.

8815-B3

SPECIAL LOCAL NEED REGISTRATIONS FOR BLAZER 2L

State	Use	SLN No.
Alabama	Attac 6 or Attac 8 Tank Mix	AL-810024
Arkansas	Attac 6 or Attac 8 Tank Mix	AR-810046
Georgia	Attac 6 or Attac 8 Tank Mix	GA-810020
Mississippi	Attac 6 or Attac 8 Tank Mix	MS-810047
North Carolina	Attac 6 or Attac 8 Tank Mix	NC-830022

DIKAR®

FUNGICIDE-MITICIDE

For Agricultural Use Only Or Only For Sale To, Use And Storage By Service Persons

ACTIVE INGREDIENTS

A coordination product of zinc ion and manganese ethylene bisdithiocarbamate		72.0%
In which the ingredients are:		
Manganese ++	14.4%	
Zinc ++	1.8%	
Ethylene bisidithiocarbamate	55.8%	
2,4-Dinitro-6-octyl* phenyl crotonate		
2,6 Dinitro-4-octyl* phenyl crotonate		4.4%
Nitrooctyl* phenols (principally dinitro)		0.3%
INERT INGREDIENTS		23.3%
TOTAL		100.0%

* A mixture of 1-methylheptyl 1-ethylhexyl and 1-propylpentyl isomers.

EPA Reg. No. 707-93 EPA Est. No. 707-PA-1
U.S. Patent 3,379,610

CAUTION

Keep Out of Reach of Children

NOTICE: Before using this product read the entire Precautionary Statements, Conditions of Sale and Warranty, Directions for Use, Use Precautions, and Storage and Disposal Instructions on side and back panels. If the Conditions of Sale and Warranty are not acceptable, return the product unopened within thirty days of purchase to the place of purchase.

PRECAUTIONARY STATEMENTS

HAZARDS TO HUMANS AND DOMESTIC ANIMALS

CAUTION

Harmful if swallowed, absorbed through the skin or inhaled. Causes moderate eye, skin, nose and throat irritation. Avoid contact with skin, eyes and clothing. Avoid breathing dust or spray mist.

Wash thoroughly with soap and water after handling and before eating or smoking.

When mixing/loading and spraying wear midforearm-to-elbow-length chemical-resistant gloves, boots or overshoes, one-piece overalls that have long sleeves and long pants, and face shield or goggles. Mixer/loaders should, in addition, wear a chemical-resistant apron.

Any article of clothing worn while applying product must be cleaned before re-using. Clothing that has been drenched or heavily contaminated should be disposed of in accordance with State or local regulations.

Instead of clothing and equipment specified above, the applicator can use an enclosed tractor cab with filtered air supply.

STATEMENT OF PRACTICAL TREATMENT

IF SWALLOWED: Dilute by giving 2 glasses of water to drink and call a physician. Never give anything by mouth to an unconscious person.

IF INHALED: Move subject to fresh air.

IF ON SKIN: Wash exposed areas with plenty of soap and water. If irritation persists, call a physician.

IF IN EYES: Flush eyes with plenty of water for at least 15 minutes. If irritation persists, call a physician.

ENVIRONMENTAL HAZARDS

This product is toxic to fish. Do not apply where runoff is likely to occur. Do not apply when weather conditions favor drift from areas treated. Do not contaminate water by cleaning of equipment or disposal of wastes. Apply this product only as specified on this label.

CONDITIONS OF SALE AND WARRANTY

Rohm and Haas warrants that the product conforms to its chemical description and is reasonably fit for the purpose stated on the label only when used in accordance with label directions under normal conditions of use. **ROHM AND HAAS MAKES NO OTHER EXPRESS OR IMPLIED WARRANTIES EITHER OF MERCHANTABILITY OR FITNESS FOR A PARTICULAR USE.** Handling, storage and use of the product by Buyer or User are beyond the control of Rohm and Haas and Seller. Risks such as crop injury, ineffectiveness or other unintended consequences resulting from, but not limited to, weather or soil conditions, presence of other materials, disease, pests, drift to other crops or property or failure to follow label directions will be assumed by the Buyer or User. **IN NO CASE WILL ROHM AND HAAS OR SELLER BE HELD LIABLE FOR CONSEQUENTIAL, SPECIAL OR INDIRECT DAMAGES RESULTING FROM THE HANDLING, STORAGE OR USE OF THIS PRODUCT.**

DIRECTIONS FOR USE

It is a violation of Federal law to use this product in a manner inconsistent with its labeling.

DIKAR is a fungicide-miticide preparation for use on apples and when used as recommended is effective for the control of the following diseases and suppression of the following mites:

Diseases:	Apple Scab	Brown Rot	Mites:	European Red
	Powdery Mildew	Fly Speck		Schoene
	Cedar Apple Rust	Sooty Blotch		Two-spotted
	Black Rot	Bitter Rot		Clover

Use DIKAR fungicide-miticide throughout the season in a regular apple spray program in accordance with local state spray schedule. Supplement with insecticides or other fungicides as needed.

DIKAR is believed to be compatible with the various insecticides recommended for use on apples. If emulsifiable formulations of any pesticide are used with DIKAR, care should be taken to check compatibility. For further details, see recommendations on this label and then consult the current State Agricultural Experiment Station spray schedules or State Extension Service Specialists for information on compatibility, dosage, timing and frequency of spray.

APPLES—Use in a regular spray program beginning early with the green tip or delayed dormant and repeat as sprays are required in accordance with the local or state spray schedule. In natural rainfall areas of high scab incidence, sprays are generally required every 7 to 10 days through petal fall and every 10 to 14 days in the cover sprays.

Use $1\frac{1}{2}$ to 2 pounds per 100 gallons of water in dilute, single strength, full coverage sprays. Use the higher dosage early in the season especially during primary scab and rapid mildew development. The lower dosage may be used in cover sprays.

For concentrate application adjust the mixing rate accordingly. Use approximately 8 to 10 pounds per acre for mature trees of average density. DIKAR fungicide-miticide has performed satisfactorily in high concentrate sprays in both ground sprayers and in aircraft.

NOTE

Do not apply to apples within 21 days of harvest in Arkansas, Delaware, Illinois, Indiana, Kansas, Kentucky, Maryland, Missouri, New Jersey, North Carolina, Ohio, Pennsylvania, South Carolina, Tennessee, Virginia and West Virginia, or within 30 days of harvest in other states not specifically listed above.

Do not graze livestock in the treated areas.

Do not use within 7 days before or after an oil application.

To avoid uneven, blotchy residues from high volume dilute sprays add 2 to 3 ounces TRITON® B-1956® per 100 gallons of spray in the last 1 or 2 applications before harvest. Concentrate applications usually leave a uniform, even deposit.

NOTE: DIKAR fungicide-miticide has about 24 to 30 hours "backup" or eradicative action against apple scab. If weather conditions or other delays interfere with the normal timing of a protective scab schedule, DIKAR may be augmented with dodine in one or more combination or separate sprays.

If the normal spray interval is extended to 2 weeks or more and powdery mildew and mites are especially severe, 1 or 2 sprays of KARATHANE® LC or KARATHANE® WD should be used between the regular DIKAR applications for added mildew or mite control.

RE-ENTRY AND WORKER PROTECTION STATEMENTS

Do not enter treated areas without protective clothing until sprays have dried. Because certain states may require more restrictive re-entry intervals for various crops treated with this product, consult your State Department of Agriculture for further information. Do not apply this product in such a manner as to directly or through drift expose workers or other persons. The area being treated must be vacated by unprotected persons.

Written or oral warnings must be given to workers who are expected to be in a treated area or in an area about to be treated with this product. Oral warnings must be given if there is reason to believe that written warnings cannot be understood by workers. When oral warnings are given, warnings shall be given in a language customarily understood by workers. Written warnings must include the following information.

CAUTION

(Insert area or field description) treated with DIKAR on (insert date of application). Do not enter treated areas without protective clothing until sprays have dried. In case of accidental exposure: Call a doctor (physician), clinic or hospital immediately. Explain that the victim has been exposed to DIKAR and describe his condition. For further information see the "Statement of Practical Treatment." Oral warnings must include the same information that is in the written warnings.

STORAGE AND DISPOSAL

STORAGE

Keep away from fire and sparks. Store in a cool, dry, well-ventilated area. Do not allow to become wet or overheated in storage, decomposition, impaired activity, or fire may result. Keep container closed when not in use.

PESTICIDE DISPOSAL

Do not contaminate water, food or feed by storage or disposal. Pesticide wastes are toxic. Improper disposal of excess pesticide, spray mixture, or rinsate is a violation of Federal law. If these wastes cannot be disposed of by use according to label instructions, contact your State Pesticide or Environmental Control Agency, or the Hazardous Waste representative at the nearest EPA Regional Office for guidance.

CONTAINER DISPOSAL

Completely empty bag into application equipment. Then dispose of empty bag in a sanitary landfill or by incineration, or, if allowed by State and local authorities, by burning. If burned, stay our of smoke.

STEPS TO BE TAKEN IN CASE MATERIAL IS RELEASED OR SPILLED: Sweep or shovel into containers for disposal or reworking. Keep dusting to a minimum. Flush contaminated area with a large amount of water to a chemical or sanitary sewer containing a settling pit.

8118-D10

DITHANE® FZ

FLOWABLE AGRICULTURAL FUNGICIDE WITH ZINC

ACTIVE INGREDIENTS

Maneb (manganese ethylene bisdithiocarbamate)		37%*
Total metallic manganese	7.6%	
INERT INGREDIENTS		63%
TOTAL		100%

* Equivalent to 4 lbs of active ingredient per gallon

EPA Reg. No. 707-170 EPA Est. No. 8590-PA-1

SHAKE WELL—grasp container by handle and mix contents with a twisting motion followed by inverting. Product will usually become homogeneous by repeating this action 5 to 10 times.

9624-R3

CAUTION

Keep Out of Reach of Children

NOTICE: Before using this product, read the entire Precautionary Statements, Conditions of Sale and Warranty, Use Restrictions, and Storage and Disposal Instructions on side panel and attached use directions. If the Conditions of Sale and Warranty are not acceptable, return the product unopened within thirty days of purchase to the place of purchase.

PRECAUTIONARY STATEMENTS

HAZARDS TO HUMANS AND DOMESTIC ANIMALS

May cause irritation of nose, throat, eyes and skin. Avoid contact with skin, eyes or clothing. Avoid breathing spray mist. In case of contact immediately flush eyes or skin with plenty of water. Get medical attention if irritation persists.

Protective clothing consisting of long pants, long-sleeved shirt, gloves, hat and boots must be worn during mixing and loading. Wash thoroughly after using and before eating or smoking. Remove and wash contaminated clothing before re-use.

STATEMENTS OF PRACTICAL TREATMENT

IF SWALLOWED: Dilute by giving 2 glasses of water to drink and call a physician. Never give anything by mouth to an unconscious person.

IF INHALED: Move subject to fresh air.

IF IN EYES: Flush eyes with large amounts of water for at least 15 minutes. Consult a physician if irritation persists.

IF ON SKIN: Wash affected skin areas with soap and water.

ENVIRONMENTAL HAZARD

This product is toxic to fish. Do not apply directly to water or wetlands. Do not apply when weather conditions favor drift from areas treated. Do not contaminate water by cleaning of equipment or disposal of wastes.

PHYSICAL AND CHEMICAL HAZARDS

Keep away from fire and sparks.
Keep from freezing.

DIRECTIONS FOR USE

It is a violation of Federal law to use this product in a manner inconsistent with its labelling.

RE-ENTRY AND WORKER PROTECTION STATEMENTS

Do not enter treated areas without protective clothing until sprays have dried. Because certain states may require more restrictive re-entry intervals for various crops treated with this product, consult your State Department of Agriculture for further information. Do not apply this product in such a manner as to directly or through drift expose workers or other persons. The area being treated must be vacated by unprotected persons.

Written or oral warnings must be given to workers who are expected to be in a treated area or in an area about to be treated with this product. Oral warnings must be given if there is reason to believe that written warnings cannot be understood by workers. When oral warnings are given, warnings shall be given in a language customarily understood by workers. Written warnings must include the following information:

CAUTION

(Insert area or field description) treated with DITHANE FZ fungicide on (insert date of application). Do not enter treated areas without protective clothing until sprays have dried. In case of accidental exposure: Call a doctor (physician), clinic or hospital immediately. Explain that the victim has been exposed to DITHANE FZ fungicide and describe his condition. For further information see the "Statement of Practical Treatment."

Oral warnings must include the same information that is in the written warnings.

GENERAL USE INFORMATION

DITHANE FZ fungicide is a broad-spectrum protectant fungicide recommended for the control of fruit, field, vegetable, and ornamental crop diseases. Optimum disease control is achieved when the fungicide is applied in a regularly scheduled preventative spray program. The addition of an agricultural surfactant will improve fungicide performance by providing a more uniform spray deposit, increased foliar redistribution, and fungicide retention during periods of wet weather.

USE RATE DETERMINATION

Carefully read, understand, and follow label use rates and limitations.

Under low disease conditions, minimum per application label rates can be used while maximum label rates and shortened spray schedules are recommended for severe or threatening disease conditions.

For proper application, determine the number of acres to be treated, the recommended label use rate, and the gallonage to be applied per acre. Prepare only the amount of spray solution required to treat the measured acreage. Careful calibration of spray equipment is recommended prior to use.

If only a portion of the container's contents are to be used, thoroughly shake the container prior to measuring. When small quantities of spray solution are being prepared for use in hand or power sprayers, the following conversion table should be followed:

Recommended Label Use Rate Per Acre or 100 Gallons	Fluid Ounces DITHANE FZ Flowable Agricultural Fungicide with Zinc Required for: 10 Gallons	5 Gallons	2 Gallons	1 Gallon
0.5 qt.	1.6	0.8	0.3	0.2
1.0 qt.	3.2	1.6	0.7	0.3
1.5 qts.	4.8	2.4	1.1	0.4
2.0 qts.	6.4	3.2	1.3	0.6
2.5 qts.	8.0	4.0	1.6	0.8
3.0 qts.	9.6	4.8	2.0	0.9
3.5 qts.	11.2	5.6	2.3	1.1
4.0 qts.	12.8	6.4	2.6	1.3

1 cup = 8 fluid ounces or 237 milliliters
1 fluid ounce = 2 tablespoons or 30 milliliters
1 tablespoon = 3 teaspoons or 15 milliliters

MIXING

Slowly pour into spray tank as it is being filled or thoroughly premix in a nurse tank for concentrate or aircraft sprayers. Add other co-applied fungicides, insecticides, growth regulators, micronutrients, and spray adjuvants after DITHANE FZ fungicide has been placed into suspension.

When preparing spray solutions for use in a hand sprayer, premix as a slurry in a small container, and then add to sprayer containing ⅓ to ½ the desired final water volume.

COMPATIBILITY

DITHANE FZ fungicide is compatible with most commonly used agricultural fungicides, insecticides, and growth regulators. When preparing tank mixes, user should consult spray compatibility charts or State Extension Service Specialists prior to actual use.

SPRAY ADJUVANTS

The addition of agricultural surfactants, to DITHANE FZ fungicide sprays, will improve initial spray deposits, fungicide redistribution, and weatherability. The following spray adjuvants have been especially formulated to optimize the performance of foliar applied agricultural chemicals:

TRITON® CS-7™	A spreader-binder designed specifically for use in concentrate and low volume sprays applied by aircraft or ground equipment.
TRITON® B-1956®	A water-dispersible, resin-based nonionic surfactant which resists re-wetting and removal by rain. Effective with dilute sprays applied by ground equipment.
TRITON AG-98™	A low foam nonionic general purpose spray adjuvant. May be used in dilute or concentrate sprays applied by aircraft or ground equipment.

Place DITHANE FZ fungicide into suspension prior to adding an adjuvant to the spray mixture. Read and carefully observe the precautionary statements and all other information appearing on both product labels prior to spray preparation.

APPLICATION

Ground:

Apply in sufficient volume of water to provide good coverage with available equipment in either dilute spray or concentrate ground sprays.

Thorough coverage foliar sprays generally result in optimum disease control. To achieve good coverage use proper spray pressure, gallonage per acre, nozzles (generally hollow cone), discs (generally D-5 to D-7), nozzle spacing, and tractor speed. Consult spray nozzle and accessory catalogues for specific information on proper equipment calibration.

Hand Sprayers:

Thoroughly spray plant foliage until run-off.

Aerial:

A uniform initial spray deposit over the crop canopy generally results in optimum disease control. Each aircraft should be prechecked for droplet size, uniformity of spray pattern, swath width, and spray volume.

Nozzle selection—Hollow cone brass nozzles with a D series orifice disc and core (whirlplate) are recommended. Nozzles should point straight down or slightly backward.

Swath width—For most field and vegetable crops, swaths just beyond the wingspan of 36 to 40 feet for light aircraft and up to 45 feet for heavier aircraft are suggested. Optimum swath for helicopters is usually 5 to 10 feet beyond normal boom length.

Spray volume—On vegetable and field crops, 2 to 3 gallons of spray per acre are generally optimum; orchards and vineyards can be handled with spray volumes of 5 gallons per acre. Some tall or dense foliage crops, requiring greater penetration to the lower leaf surface, will require higher spray volumes.

Altitude—For most crops, the spray boom should be positioned 5 to 10 feet above the crop canopy.

Flagging—Swaths should be marked at the end of the field with permanent flags or by a flagman. Swaths should be measured accurately with a chain or other device except when rows can be accurately counted.

Sprinkler Irrigation:

DITHANE FZ fungicide must be applied on a regular protectant fungicide schedule, **not an irrigation schedule.** If irrigation cycles are used, ground or aerial applications must supplement chemigation applications to achieve adequate disease control. With the exception of potatoes and tomatoes DITHANE FZ fungicide has not been sufficiently tested to assure consistent product performance for all labeled uses. The following calibration and application techniques are provided for user reference, but do not constitute a warranty of fitness for application through sprinkler irrigation equipment. Users must check with State and local regulatory agencies, for potential use restrictions, prior to applying any agricultural chemical by sprinkler irrigation equipment.

Center-pivot Irrigation Equipment (Use only with electric or oil hydraulic drive systems which provide a uniform water distribution)

- Determine size of area to be treated.
- Determine the time required to add ¼ inch water over the area to be treated when the system and injection equipment are operated at normal pressures recommended by the equipment manufacturer. Run system at 80 to 95% of manufacturer's rated capacity.
- Using only water, determine the injection pump output when operated at normal line pressure.
- Determine the amount of DITHANE FZ fungicide required to treat area.
- Add the required amount of DITHANE FZ fungicide, and sufficient water to meet the injection time requirements, to the solution tank.
- Maintain constant solution tank agitation during the injection period.
- Stop injection equipment after treatment is completed. Continue to operate the system until the DITHANE FZ fungicide solution has cleared the sprinkler head.

Solid-set and Portable (Wheel Move) Irrigation Equipment

- Determine acreage covered by sprinkler.
- Fill injector solution tank with water and adjust flow rate to use contents over a ten to thirty minute interval.
- Determine the amount of DITHANE FZ fungicide required to treat area.
- Add the required amount of DITHANE FZ fungicide into the same quantity of water used to calibrate the injection equipment.
- Maintain constant solution tank agitation during the injection period.
- Operate system at normal pressure recommended by the manufacturer of the injection equipment and used for the time interval established during calibration.
- Inject DITHANE FZ fungicide at the end of the irrigation cycle or as a separate application to maximize foliar fungicide retention.
- Stop injection equipment after treatment is completed. Continue to operate the system until the DITHANE FZ fungicide solution has cleared the last sprinkler head.

PRECAUTIONS

- Irrigation equipment must have antisiphon devices and automatic shutoff valves to prevent contamination of water sources.
- Have injector pump electrical connections coupled with the system's shutdown controls.
- Do not chemigate if wind speed exceeds 10 mph.
- Irrigation water treated with DITHANE FZ fungicide should be contained on the treated area until the water is absorbed by the soil.

DISEASE MONITORING

DITHANE FZ fungicide is a broad-spectrum protectant fungicide. If not applied on a routine protectant spray schedule, crops should be scouted on a weekly basis. Fungicide application should be made, at the recommended label use rate and spray schedule, at the first sign of disease, report of disease in the area, or during environmental conditions favorable for disease development.

RESTRICTIONS

Users should carefully read, understand, and follow all use restrictions prior to using DITHANE FZ fungicide.

VEGETABLES

CROP	DISEASES CONTROLLED	RATE OF DITHANE FZ PER APPLICATION QTS./A	REMARKS (ALSO REFER TO DIRECTIONS FOR USE)	RESTRICTIONS
Asparagus	Rust	2.4	Start applications when rust first appears and repeat at 10 day intervals. Four applications are usually sufficient.	Apply only to asparagus ferns after harvest has been completed.
Bean (green, dry)	Anthracnose Downy mildew Rust	0.8 to 2.4	Apply when disease first appears and repeat at 7 day intervals.	Do not apply within 4 days of harvest. Do not feed treated forage or hay to livestock.
Broccoli	Alternaria leaf spot Downy mildew	1.5 to 2.25	Apply up to 150 gallons spray per acre, or apply equivalent dosage in concentrate sprays. Add TRITON B-1956 agricultural spreader sticker to sprays. In plant beds and direct seeded fields, apply 7 to 10 days after planting or earlier if disease is present. Repeat every 3 days until transplanting or thinning. In field application apply as soon as disease threatens and repeat weekly as needed. Reduce spray intervals to 3 to 5 days under severe conditions.	Do not apply within 3 days of harvest.
Brussel Sprouts	Refer to broccoli			
Cabbage	Alternaria leaf spot Downy mildew	1.2 to 1.8	Refer to broccoli.	Do not apply more than 1.8 qts. per acre per application. Do not apply within 7 days of harvest.
Cantaloupe	Alternaria leaf spot Anthracnose Cercospora leaf spot Downy mildew Gummy stem blight Scab	0.8 to 2.4	Start applications when plants are in the two leaf stage and repeat at 5 to 7 day intervals. Use sufficient water and direct spray to provide thorough coverage of both upper and lower leaf surfaces.	Do not apply within 5 days of harvest.
Carrot	Alternaria leaf spot Cercospora leaf spot	0.8 to 2.4	Apply in 75 to 125 gallons spray per acre. Start applications when disease first threatens and repeat every 7 to 10 days as needed. The addition of TRITON B-1956 or TRITON CS-7 to spray solutions will improve performance.	

CROP	DISEASES CONTROLLED	RATE OF DITHANE FZ PER APPLICATION QTS./A	REMARKS (ALSO REFER TO DIRECTIONS FOR USE)	RESTRICTIONS
Casaba Melon	Refer to Melon			
Cauliflower	Refer to broccoli			
Celery	Early blight Late blight	0.8 to 2.4	Start applications in the plant bed when plants first come through the ground. Make further applications at 3 to 5 day intervals until plants are set in the field. Continue field applications on a 7 day schedule. Remove excess residue remaining at harvest by stripping, trimming, and washing. The addition of a TRITON surfactant to spray solutions will improve performance.	Do not apply within 14 days of harvest.
Collard	Refer to broccoli		Remove residues by washing.	Do not apply within 10 days of harvest.
Corn (sweet)	Helminthosporium leaf blights	1.6 to 2.0	Use sufficient water for thorough coverage. Start applications when disease first appears and repeat at 4 to 7 day intervals. The addition of TRITON B–1956 or TRITON CS–7 to spray solutions will improve performance.	Do not feed treated forage to livestock.
Crenshaw Melon	Refer to melon			
Cucumber	Refer to melon			
Eggplant	Anthracnose Early blight Phomopsis blight or fruit rot	0.8 to 2.4	Begin sprays when fruit starts to form and repeat at 7 to 10 day intervals.	
Endive	Downy mildew	0.8 to 2.4	Begin sprays when disease appears and repeat at 7 to 10 day intervals. Remove residue by washing, trimming or other effective means.	Do not apply within 10 days of harvest.
Honey Ball Melon	Refer to melon			
Honey Dew Melon	Refer to melon			
Kale	Refer to broccoli		Remove residue by washing.	Do not apply within 10 days of harvest.
Kohlrabi	Refer to broccoli			
Lettuce	Refer to endive		Remove residue from head lettuce by stripping and trimming and from leaf lettuce by washing or other effective means.	
Melon	Anthracnose Cercospora leaf spot Downy mildew Gummy stem blight	0.8 to 2.4	Start applications when plants are in the two leaf stage and repeat at 5 to 7 day intervals. Use sufficient water and direct spray, to provide thorough coverage of both upper and lower leaf surfaces.	Do not apply within 5 days of harvest.
Muskmelon	Refer to melon			
Mustard Greens	Refer to broccoli		Remove residue by washing.	Do not apply within 10 days of harvest.
Onion	Botrytis leaf blight Downy mildew Purple blotch	0.8 to 2.4	Follow a protective spray schedule starting when diseases are first reported in the area and repeat at 7 day intervals throughout the season. The addition of a TRITON surfactant to spray solutions will improve performance.	
Pepper	Refer to eggplant			
Persian Melon	Refer to melon			
Potato	Early blight Late blight	0.8 to 1.6	Begin application when plants are 4 to 6 inches high by applying 0.8 to 1 qt. per acre. As the vines increase in size, apply 1.2 to 1.6 qts. per acre. The addition of a TRITON surfactant to spray solutions will improve performance.	
(seedpieces)	Fusarium decay	1 qt./10 gal.	Dip whole or cut tubers into solution. Spread treated seed pieces in a cool place if held before planting.	Do not use treated seed pieces for food or feed.
Pumpkin	Angular leaf spot Downy mildew	2.4	Start applications when disease threatens and repeat at 7 to 10 day intervals as needed. Use sufficient water and direct spray to provide thorough coverage of both upper and lower leaf surfaces.	

CROP	DISEASES CONTROLLED	RATE OF DITHANE FZ PER APPLICATION QTS./A	REMARKS (ALSO REFER TO DIRECTIONS FOR USE)	RESTRICTIONS
Rhubarb (greenhouse, Mich.)	Botrytis leaf rot	1.6 qts./100 gals.	Begin 4 weeks before harvest and make 4 applications at 7 day intervals.	
Spinach	Downy mildew (blue mold) White rust	0.8 to 2.4	Start applications when disease symptoms first appear and repeat at 7 to 10 day intervals. Remove spray residue by washing or other effective means.	Do not apply within 10 days of harvest.
Squash, summer and winter	Refer to melons			Do not apply within 5 days of harvest.
	Pythium fruit rot	1.5 to 3.5	Begin when runners form or at first sign of disease at weekly intervals as needed.	
Tomato	Anthracnose Early blight Gray leaf spot Late blight Leaf mold Septoria leaf spot	0.8 to 2.4	Start applications when seedlings emerge or transplants are set and repeat at 7 to 10 day intervals throughout the season. In greenhouses use 4.5 to 6.1 fluid ozs. per 5,000 sq. ft. and start applications when first cluster fruit is well formed. Repeat applications at 7 to 10 day intervals. The addition of a TRITON surfactant will improve spray performance.	Do not apply within 5 days of harvest.
Turnip	Refer to broccoli		Remove excess residue by washing.	Do not apply within 7 days of harvest.
Watermelon	Refer to melon			

FRUITS AND NUTS

CROP	DISEASES CONTROLLED	RATE OF DITHANE FZ PER APPLICATION QTS./A	QTS./100 GALS.	REMARKS (ALSO REFER TO DIRECTIONS FOR USE)	RESTRICTIONS
Almond (California)	Brown rot (blossom and twig blight) Coryneum blight (shot hole) Leaf blight Scab	4 to 6.4	1.2 to 1.6	Apply at popcorn, full bloom and petal fall, or every 7 to 10 days if bloom is staggered and weather is rainy. For Coryneum blight, begin applications in dormant stage and repeat at 7 to 10 day intervals through petal fall. Use 4 ounces TRITON B-1956 per 100 gallons spray.	Do not apply later than 5 weeks after petal fall. If applied after petal fall, do not feed hulls to dairy animals or animals being finished for slaughter.
Apple	Bitter rot Black rot Brown rot Bullseye rot Cedar apple rust Fly speck Scab Sooty blotch White rot	8.0	0.8 to 1.6	Apply every 7 to 14 days throughout the season starting with a green tip or delayed dormant spray and continue through cover sprays. For low volume air blast sprays adjust the mixing rate accordingly. To avoid uneven, blotchy residues from high volume dilute sprays add 2 to 3 ounces TRITON B-1956 per 100 gallons of spray in the last 1 or 2 applications before harvest. Concentrate applications usually leave a uniform, even deposit.	Do not apply within 15 days of harvest in Arkansas, Delaware, Illinois, Indiana, Kansas, Kentucky, Maryland, Missouri, New Jersey, North Carolina, Ohio, Pennsylvania, South Carolina, Tennessee, Virginia, and West Virginia, or within 30 days of harvest in other states. Do not graze livestock in treated areas.
Apricot (Western States)	Brown rot (blossom, Twig, fruit) Coryneum blight (shot hole) Green or Jacket rot Leaf curl	8.0	1.2 to 1.6	Apply at dormant to popcorn stage, repeat every 7 to 14 days through petal fall according to local timing recommendations and disease severity. For fruit brown rot use again at 4 to 6 weeks before harvest, repeating applications every 7 to 14 days. For Coryneum blight, begin applications in dormant stage and repeat at 7 to 10 day intervals through petal fall. Use 4 ounces TRITON B-1956 per 100 gallons spray.	Do not apply within 14 days of harvest. Do not use in combination with DIAZINON®
Banana	Sigatoka	1.6 to 4.0		Apply when leaves first appear and repeat every 2 to 3 weeks or as required. Use sufficient water to provide adequate coverage. The addition of a TRITON surfactant to spray solutions will improve performance.	
Cranberries	Fruit rot Twig blight	3.0 to 6.0		Start applications at mid-bloom and repeat at 7 to 10 day intervals as required.	Do not apply within 30 days of harvest.

CROP	DISEASES CONTROLLED	RATE OF DITHANE FZ PER APPLICATION QTS./A	QTS./100 GALS.	REMARKS (ALSO REFER TO DIRECTIONS FOR USE)	RESTRICTIONS
Grape	Black rot Bunch rot	1.2 to 4.0		Apply in sufficient water to provide thorough coverage starting when new shoots are 8 to 10 inches long (prebloom) and continuing until 10 days after bloom. When fruit is present, use maximum of 1.5 qts. per acre.	Do not apply within 7 days of harvest.
Nectarine (Western States)	Refer to apricot				
Peach (Western States)	Refer to apricot				Do not apply within 2 days of harvest. If applied within 14 days of harvest, remove residues by brushing or washing.
FIELD CROPS Grass (Seed Production)	Leaf rust Stem rust Stripe rust	1.6 to 2.4		Begin application when disease is first noted or observed in the area and repeat application every 7 to 10 days until harvest or as long as disease threatens.	Do not feed treated grass to livestock.
Sugar Beets	Cercospora leaf spot	1.2 to 2.4		Start application when disease first threatens and repeat every 7 to 10 days as needed.	Up to 1.6 qts. per acre per application—do not apply within 14 days before harvest. Over 1.6 qts. (2.4 qts. max) per acre per application—do not apply within 14 days of harvest. Do not feed treated tops to livestock.
Tobacco (transplant beds)	Anthracnose Blue mold		0.4 to 3.2	Spray plant beds when plants are the size of a dime or when blue mold is first reported in the area, applying 2½ to 5 gallons spray per 100 square yards, depending on plant size. Repeat sprays twice weekly or more frequently during wet weather. Continue treatment until seedlings have been transplanted.	
(field)	Anthracnose Blue mold	1.4 to 3.2		Begin application when disease is first noted or observed in area and repeat application as long as disease threatens.	

SEED TREATMENT

CROP	DISEASES CONTROLLED	RATE OF DITHANE FZ PER APPLICATION FL. OZ./ BU.	FL. OZ./ 100 LB.	REMARKS (ALSO REFER TO DIRECTIONS FOR USE)	RESTRICTIONS
Beans	Damping-off Seed rots Seedling blights	2.0 to 4.0	3.0 to 6.0	**GENERAL USE INSTRUCTIONS:** Seeds to be treated should be cleaned and well cured prior to treatment. DITHANE FZ fungicide may be applied to dry seed with conventional slurry or mist seed treating equipment. The seed must be completely and uniformly covered with fungicide. A nonacetic dye must be added to DITHANE FZ fungicide to impart an unnatural color to commercially treated seed. Grower treated seed, which is immediately planted after treatment, does not have to include a nonacetic dye.	Treated seed should be labeled, "Must not be used for food, feed, or oil purposes."
Corn	Damping-off Seed rots Seedling blights	2.0 to 4.0	3.4 to 7.6		
Cotton (acid delinted)	Damping-off Seedling blights		5.0 to 8.0		
(reginned)	Damping-off Seedling blights		9.6 to 16.0		
Flax	Damping-off Seed rots Seedling blights	1.5 to 4.0	2.7 to 7.0		
Peanut (shelled)	Damping-off Seed rots Seedling blights		2.8 to 4.5		
Pea	Damping-off Seed rots Seedling blights	2.0 to 4.0	2.0 to 4.2		
Soybean	Damping-off Seed rots Seedling blights	1.5 to 4.0	2.5 to 6.7		
Sorghum	Covered kernel smut Damping-off Seed rots Seedling blights	1.5 to 4.0	3.0 to 8.0		

TURF

CROP	DISEASES CONTROLLED	RATE OF DITHANE FZ FLOWABLE FUNGICIDE WITH ZINC PER APPLICATION FL. OZ./1000 FT.^{2}	REMARKS (ALSO REFER TO DIRECTIONS FOR USE)	RESTRICTIONS
Assorted Grasses	Helminthosporium blight and melting-out	4.8 to 6.4	GENERAL USE DIRECTIONS: Start application when grass greens up in spring or when disease first appears and repeat at 7 to 14 day intervals or until disease threat is past. When conditions are especially favorable for disease development, apply maximum fungicide use rate on a 7 day spray schedule. Apply in sufficient water to provide adequate coverage.	Do not graze treated areas. Do not feed clippings to livestock.
	Rhizoctonia brown patch Rust (leaf, stem, stripe)	4.4 to 12.8		
	Dollar spot	9.6 to 12.8		

ORNAMENTALS

CROP	DISEASES CONTROLLED	REMARKS (ALSO REFER TO DIRECTIONS FOR USE)
		GENERAL INSTRUCTIONS: Apply the equivalent of 1.2 qts. per 100 gals. The addition of TRITON B–1956 to spray solutions will improve performance. Begin spraying when plants are well leafed out or at first sign of disease in a full coverage spray at 7 to 10 day intervals throughout season or follow State Extension Service recommendations for disease control on the following ornamental plants:
Buffaloberry (Nursery)	Leaf spot	
Carnation	Leaf spot Rust	
Chrysanthemum	Ascochyta blight Botrytis petal spot Rust	Apply twice weekly during blooming period.
Conifers	Needle cast Brown spot	Begin application in spring or early summer before infection occurs. Repeat after heavy rains or at two-week intervals as long as necessary.
Dahlia	Botrytis blight	
Dogwood, flowering	Anthracnose	Apply when buds begin to open, when bracts have fallen, 4 weeks later and again in late summer after flower buds for next season are formed.
Dracaena	Fusarium leaf spot	
Geranium	Puccinia rust	
Gladiolous	Curvularia leaf spot Botrytis blossom blight Stemphylium leaf blight	Make regular weekly applications starting before diseases appear and increase to 2 or 3 applications per week during periods of heavy disease and during rainy weather. On flower spikes, reduce spray concentration to $^3/_4$ qts. per 100 gals.
Hydrangeas	Botrytis blight	
Iris	Ink spot Leaf spot (fire)	
Lily	Botrytis blight	
Pansy	Anthracnose	
Peony	Botrytis blossom blight Phytophthora blight	Apply in early spring and early fall, drenching soil around plants as well as the foliage. Promptly destroy all infected plant parts.
Skunkbush sumac	Leaf spots	
Rose	Black spot Cercospora leaf spot Rust	
Snapdragon	Rust	
Zinnia	Leaf blight	

FUNGICIDE TANK MIXES

The following tank mix recommendations are made for the improved control of specific diseases. Before mixing DITHANE FZ Flowable fungicide with zinc with any other fungicide, read and carefully observe the precautionary and all other information appearing on both product labels. User must abide by the most restrictive limitations appearing on the product labels.

CROP	CO-APPLIED FUNGICIDE	FOR IMPROVED CONTROL OF:	USE RATE PER APPLICATION	DITHANE FZ FLOWABLE FUNGICIDE WITH ZINC FUNGICIDE USE RATE PER APPLICATION	REMARKS (ALSO REFER TO DIRECTIONS FOR USE)
Apple	BENLATE 50 WP*	Powdery mildew	2 to 3 ozs./100 gals.	0.6 to 0.8 qts./100 gals.	
	TOPSIN-M 70WP*				
	BAYLETON 50 WP		½ to 2 ozs./100 gals.	0.8 to 1.6 qts./100 gals.	
	KARATHANE WD	Powdery mildew Mite suppression	½ to ¾ lbs./100 gals.		
	KARATHANE LC		4 to 6 fl. ozs./100 gals.		
Celery	BENLATE 50 WP*	Early blight Late blight	4 ozs./A	1.2 to 1.6 qts./A	
	TOPSIN-M 70WP*				
	FIXED COPPERS	Bacterial leaf spot	Manufacturers label use rate	1.6 qts./A	
Cucumber	Refer to melons				
Grape	BAYLETON 50 WP	Powdery mildew	2 to 6 ozs./A	1.2 to 3.2 qts./A	
	KARATHANE WD		1 to 2½ lbs./A		Do not use in California
	KARATHANE LC		4 to 8 fl. ozs./100 gals.		
	BENLATE 50 WP		¾ to 1 lb./A		
Melon	BENLATE 50 WP*	Target spot Gummy stem blight Powdery mildew Anthracnose	4 to 8 ozs./A	1.6 to 2.4 qts./A	
	KARATHANE WD	Powdery mildew	8 to 12 ozs./A		Do not use in California.
	KARATHANE LC		3 to 4 fl. ozs./100 gals.		
	DIFOLATAN 80 SPRILLS	Gummy stem blight	1½ to 2 lb./A		DIFOLATAN 80 SPRILLS may not be used on squash or pumpkins.
Pepper	Fixed Coppers	Bacterial leaf spot	Manufacturers label use rate	0.8 to 2.4 qts./A	
Potato	Triphenyl Tin Hydroxide (TPTH)	Early blight	½ to full label use rate	1.2 to 1.6 qts./A	
Pumpkin	Refer to melons				DIFOLATAN 80 SPRILLS may not be used on pumpkins.
Roses, Flowers Ornamentals, and shade trees	BENLATE 50 WP	Powdery mildew Leaf spots	½ to 1 lb./100 gals.	1.2 qts./100 gals.	
	TOPSIN-M 70WP				
	KARATHANE WD	Powdery mildew Mite suppression	6 to 8 oz./100 gals.		
Squash	Refer to melons				DIFOLATAN 80 SPRILLS may not be used on squash.
Sugar Beet	BENLATE 50 WP*	Cercospora leaf spot	4 oz./A	1.2 to 1.6 qts./A	
	TOPSIN-M 70WP*				
	MERTECT 340-F*		6 fl. oz./A		
	Triphenyl Tin Hydroxide (TPTH)		½ to full label rate		
Tomato	Fixed Coppers	Bacterial leaf spot	Manufacturers label use rate	1.2 to 2.4 qts./A	
	DYRENE 50 WP	Gray leaf spot	½ to 1 lb./A		
	BENLATE 50 WP	Botrytis gray mold Sclerotinia white mold			

* Do not use if fungal resistance to this fungicide is present in field to be treated or has been reported in the area.

STORAGE AND DISPOSAL

STORAGE

Keep away from fire and sparks. Store in a cool, well ventilated area, but not below 32°F. Do not allow to become overheated in storage. This may bring on chemical changes which will impair the fungicidal effectiveness of DITHANE FZ fungicide. Keep container closed when not in use.

PESTICIDE DISPOSAL

Do not contaminate water, food or feed by storage or disposal. Pesticide wastes are toxic. Improper disposal of excess pesticide, spray mixture, or rinsate is a violation of Federal law. If these wastes cannot be disposed of by use according to label instructions, contact your State Pesticide or Environmental Control Agency, or the Hazardous Waste representative at the nearest EPA Regional office for guidance.

CONTAINER DISPOSAL

Plastic Containers—Triple rinse (or equivalent), puncture and then dispose of in a sanitary landfill or by incineration or if permitted by State and local authorities by burning. If burned stay out of smoke.

Metal Containers—Triple rinse (or equivalent) then offer for recyling or reconditioning or puncture and dispose of in a sanitary landfill or by other procedures approved by State and local authorities.

Bulk Containers—Drain thoroughly and return to specified destination for cleaning and re-use.

Steps to be Taken in Case Material is Released or Spilled: Dike and contain the spill. Transfer liquid and solid diking material to separate containers for recovery or disposal. Flush contaminated area with a large amount of water to a chemical or sanitary sewer containing a settling pit. Remove contaminated clothing and wash affected skin areas with soap and water. Wash clothing before re-use. Keep the solids out of the municipal sewers and open bodies of water. Refer to Precautionary Statements.

CONDITIONS OF SALE AND WARRANTY

Rohm and Haas warrants that the product conforms to its chemical description and is reasonably fit for the purpose stated on the label only when used in accordance with label directions under normal conditions of use. **ROHM AND HAAS MAKES NO OTHER EXPRESS OR IMPLIED WARRANTIES EITHER OF MERCHANTABILITY OR FITNESS FOR A PARTICULAR USE.** Handling, storage and use of the product by Buyer or User are beyond the control of Rohm and Haas and Seller. Risks such as crop injury, ineffectiveness or other unintended consequences resulting from, but not limited to, weather or soil conditions, presence of other materials, disease, pests, drift to other crops or property or failure to follow label directions will be assumed by the Buyer or User. **IN NO CASE WILL ROHM AND HAAS OR SELLER BE HELD LIABLE FOR CONSEQUENTIAL, SPECIAL OR INDIRECT DAMAGES RESULTING FROM THE HANDLING, STORAGE OR USE OF THIS PRODUCT.**

BAYLETON and DYRENE are registered trademarks of the parent company of Farbenfabriken Bayer GmbH, Leverkusen.

BENLATE is a registered trademark of E.I. duPont de Nemours and Co., Inc.

DIFOLATAN and SPRILLS are registered trademarks of Chevron Chemical Co.

DITHANE, KARATHANE, TRITON and B-1956 are registered trademarks of the Rohm and Haas Company.

MERTECT is a registered trademark of Merck and Co., Inc.

TOPSIN-M is a registered trademark of Nippon Soda Co., Ltd.

9624-B3

SPECIAL LOCAL NEED REGISTRATIONS FOR DITHANE FZ

State	Use	SLN No.
Indiana	Corn, Hybrid Seed Corn, Popcorn	IN-830002
Florida	Corn, Hybrid Seed Corn, Popcorn	FL-820092
Pennsylvania	Corn, Hybrid Seed Corn, Popcorn	PA-830006

DITHANE® M-22 SPECIAL with zinc

AGRICULTURAL FUNGICIDE

ACTIVE INGREDIENT

Maneb (Manganese ethylene bisdithiocarbamate) 80%

INERT INGREDIENTS

(including $ZnSO_4$) 20%

TOTAL 100%

EPA Reg. No. 707-78

NOTICE: Before using this product, read the entire Precautionary Statements, Conditions of Sale and Warranty, Directions for Use, Use Restrictions, and Storage and Disposal Instructions. If the Conditions of Sale and Warranty are not acceptable, return the product unopened within thirty days of purchase to the place of purchase.

CAUTION

Keep Out of Reach of Children

PRECAUTIONARY STATEMENTS

HAZARDS TO HUMANS AND DOMESTIC ANIMALS

May cause irritation of nose, throat, eyes and skin. Do not breathe dust or spray mist. PROTECTIVE CLOTHING consisting of long pants, long-sleeved shirt, gloves, hat and boots must be worn during mixing and loading.

STATEMENTS OF PRACTICAL TREATMENT

IF SWALLOWED: Dilute by giving 2 glasses of water to drink and call a physician. Never give anything by mouth to an unconscious person.

IF INHALED: Move subject to fresh air.

IF IN EYES: Flush eyes with large amounts of water for at least 15 minutes. Consult a physician if irritation persists.

IF ON SKIN: Wash affected skin areas with soap and water.

ENVIRONMENTAL HAZARDS

This product is toxic to fish. Do not apply directly to water or wetlands. Do not apply when weather conditions favor drift from areas treated. Do not contaminate water by cleaning of equipment or disposal of wastes.

DIRECTIONS FOR USE

It is a violation of Federal law to use this product in a manner inconsistent with its labeling.

RE-ENTRY AND WORKER PROTECTION STATEMENTS

Do not enter treated areas without protective clothing until sprays have dried. Because certain states may require more restrictive reentry intervals for various crops treated with this product, consult your State Department of Agriculture for further information. Do not apply this product in such a manner as to directly or through drift expose workers or other persons. The area being treated must be vacated by unprotected persons.

Written or oral warnings must be given to workers who are expected to be in a treated area or in an area about to be treated with this product. Oral warnings must be given if there is reason to believe that written warnings cannot be understood by workers. When oral warnings are given, warnings shall be given in a language customarily understood by workers. Written warnings must include the following information:

CAUTION

(Insert area or field description) treated with DITHANE M-22 SPECIAL on (insert date of application). Do not enter treated areas without protective clothing until sprays have dried. In case of accidental exposure: Call a doctor (physician), clinic or hospital immediately. Explain that the victim has been exposed to DITHANE M-22 SPECIAL and describe his condition. For further information see the "Statements of Practical Treatment".

Oral warnings must include the same information that is in the written warnings.

GENERAL USE INFORMATION

DITHANE M-22 SPECIAL is a broad-spectrum protectant fungicide recommended for the control of fruit, field, vegetable, and ornamental crop disease. Optimum diseases control is achieved when the fungicide is applied in a regularly scheduled preventative spray program. The addition of an agricultural surfactant will improve fungicide performance by providing a more uniform spray deposit, increased foliar redistribution, and improved fungicide retention during periods of wet weather.

USE RATE DETERMINATION

Carefully read, understand, and follow label use rates and limitations.

Under low disease conditions, minimum per application label rates can be used while maximum label rates and shortened spray schedules are recommended for severe or threatening disease conditions.

For proper application, determine the number of acres to be treated, the recommended label use rate, and the gallonage to be applied per acre. Prepare only the amount of spray solution required to treat the measured acreage. Careful calibration of spray equipment is recommended prior to use.

When applied by hand sprayers, 1 pound DITHANE M-22 SPECIAL fungicide per 100 gallons or acre is equivalent to 1 level tablespoon per gallon spray solution.

MIXING

Slowly place into spray tank as it is being filled or thoroughly premix in a nurse tank for concentrate on aircraft sprayers. Add other coapplied fungicides, insecticides, growth regulators, micronutrients, and spray adjuvants after DITHANE M-22 SPECIAL fungicide has been placed into suspension.

When preparing spray solutions for use in a hand sprayer, premix as a slurry in a small container and then add to sprayer containing ⅓ to ½ the desired final water volume.

COMPATIBILITY

DITHANE M-22 SPECIAL fungicide is compatible with most commonly used agricultural fungicides, insecticides, and growth regulators. When preparing tank mixes, user should consult spray compatibility charts or State Extension Service Specialists prior to actual use.

SPRAY ADJUVANTS

The addition of agricultural surfactants, to DITHANE M-22 SPECIAL fungicide sprays, will improve initial spray deposits, fungicide redistribution, and weatherability. The following spray adjuvants have been especially formulated to optimize the performance of foliar applied agricultural chemicals:

TRITON AG-98™	A low foam nonionic general purpose spray adjuvant. May be used in dilute or concentrate sprays applied by aircraft or ground equipment.
TRITON® B-1956®	A water-dispersible, resin-based nonionic surfactant which resists re-wetting and removal by rain. Effective with dilute sprays applied by ground equipment.
TRITON CS-7™	A spreader-binder designed specifically for use in concentrate and low volume sprays applied by aircraft or ground equipment.

Place DITHANE M-22 SPECIAL fungicide into suspension prior to adding an adjuvant to the spray mixture. Read and carefully observe the precautionary statements and all other information appearing on both product labels prior to spray preparation.

APPLICATION

Ground:

Apply in sufficient volume of water to provide good coverage with available equipment in either dilute sprays or concentrate ground sprays.

Thorough coverage foliar sprays generally result in optimum disease control. To achieve good coverage use proper spray pressure, gallonage per acre, nozzles (generally hollow cone), discs (generally D-5 to D-7), nozzle spacing, and tractor speed. Consult spray nozzle and accessory catalogues for specific information on proper equipment calibration.

Hand Sprayers:

Thoroughly spray plant foliage until runoff.

Aerial:

A uniform initial spray deposit over the crop canopy generally results in optimum disease control. Each aircraft should be prechecked for droplet size, uniformity of spray pattern, swath width, and spray volume.

Nozzle selection—Hollow cone brass nozzles with a D series orifice disc and core (whirlplate) are recommended. Nozzles should point straight down or slightly backward.

Swath width—For most field and vegetable crops, swaths just beyond the wingspan of 36 to 40 feet for light aircraft and up to 45 feet for heavier aircraft are suggested. Optimum swath for helicopters is usually 5 to 10 feet beyond normal boom length.

Spray volume—On vegetable and field crops, 2 to 3 gallons of spray per acre are generally optimum; orchards and vineyards can be handled with spray volumes of 5 gallons per acre. Some tall or dense foliage crops, requiring greater penetration to the lower leaves, will require higher spray volumes.

Altitude—For most crops, the spray boom should be positioned 5 to 10 feet above the crop canopy.

Flagging—Swaths should be marked at the end of the field with permanent flags or by a flagman. Swaths should be measured accurately with a chain or other device except when rows can be accurately counted.

DISEASE MONITORING

DITHANE M-22 SPECIAL is a broad-spectrum protectant fungicide. If not applied on a routine protectant spray schedule, crops should be scouted on a weekly basis. Fungicide application should be made, at the recommended label use rate and spray schedule, at the first sign of disease, report of disease in the area, or during environmental conditions favorable for disease development.

RESTRICTIONS

Users should carefully read, understand, and follow all use restrictions prior to using DITHANE M-22 SPECIAL fungicide.

VEGETABLES

CROP	DISEASES CONTROLLED	RATE OF DITHANE M-22 SPECIAL PER APPLICATION LB./A	REMARKS (ALSO REFER TO DIRECTIONS FOR USE)	RESTRICTIONS
Asparagus	Rust	2 to 3	Start applications when rust first appears and repeat at 10 day intervals. Four applications are usually sufficient	Apply only to asparagus ferns after spears have been harvested.
Beans (green, dry)	Anthracnose Downy mildew Rust	1 to 3	Apply when disease first appears and repeat at 7 days intervals.	Do not apply within 4 days of harvest. Do not feet treated forage or hay to livestock.
Broccoli	Alternaria leaf spot Downy mildew	1 to 3	Apply up to 150 gallons spray per acre or apply equivalent dosage in concentrate sprays. Add TRITON B-1956 agricultural spreader sticker to sprays. In plant beds and direct-seeded fields, apply 7 to 10 days after planting or earlier if disease is present. Repeat every 3 days until transplanting or thinning. In field application apply as soon as disease threatens and repeat weekly as needed. Reduce spray intervals to 3 to 5 days under severe conditions.	Do not apply within 7 days of harvest.
Brussels Sprouts	Refer to broccoli			
Cabbage	Refer to broccoli	1½	Refer to broccoli	Do not apply more than 1½ lbs. per acre per application.
Cantaloupe	Alternaria leaf spot Anthracnose Cercospora leaf spot Downy mildew Gummy stem blight Scab	1 to 3	Start applications when plants are in the two leaf stage and repeat at 5 to 7 day intervals. Use sufficient water and direct spray to provide thorough coverage of both upper and lower leaf surfaces.	Do not apply within 5 days of harvest.
Carrots	Alternaria leaf spot Cercospora leaf spot	1½ to 3	Apply in 75 to 125 gallons spray per acre. Start applications when disease first threatens and repeat every 7 to 10 days as needed. The addition of TRITON B-1956 or TRITON CS-7 to spray solutions will improve performance.	
Casaba Melon	Refer to melons			
Cauliflower	Refer to broccoli			
Celery	Early blight Late blight	1 to 3	Start applications in the plant bed when plants first come through the ground. Make further applications at 3 to 5 day intervals until plants are set in the field. Continue field applications on a 7 day schedule. Remove excess residue remaining at harvest by stripping, trimming, and washing. The addition of a TRITON surfactant to spray solutions will improve performance.	Do not apply within 14 days of harvest.
Collards	Refer to broccoli		Remove residues by washing	Do not apply within 10 days of harvest.
Corn (sweet)	Helminthosporium leaf blights	2 to 2½	Use sufficient water for thorough coverage. Start applications disease first appears and repeat at 4 to 7 day intervals. The addition of TRITON B-1956 or TRITON CS-7 to spray solutions will improve performance.	Do not feed treated forage to livestock.
Crenshaw Melon	Refer to melons			
Cucumber	Refer to melons			
Eggplant	Anthracnose Early blight Phomopsis blight or fruit rot	1 to 3	Begin sprays when fruit starts to form and repeat at 7 to 10 day intervals.	
Endive	Downy mildew	1 to 3	Begin sprays when disease appears and repeat at 7 to 10 day intervals. Remove residue by washing, trimming or other effective means.	Do not apply within 10 days of harvest.
Honey Ball Melons	Refer to melons			
Honey Dew Melons	Refer to melons			
Kale	Refer to broccoli		Remove residue by washing.	Do not apply within 10 days of harvest.
Kohlrabi	Refer to broccoli			

CROP	DISEASES CONTROLLED	RATE OF DITHANE M-22 SPECIAL PER APPLICATION LB./A	REMARKS (ALSO REFER TO DIRECTIONS FOR USE)	RESTRICTIONS
Lettuce	Refer to endive		Remove residue from head lettuce by stripping and trimming and from leaf lettuce by washing or other effective means.	
Melons	Anthracnose Cercospora leaf spot Downy mildew Gummy stem blight	1 to 3	Start applications when plants are in the two leaf stage and repeat at 5 to 7 day intervals. Use sufficient water and direct spray, to provide thorough coverage of both upper and lower leaf surfaces.	Do not apply within 5 days of harvest.
Muskmelons	Refer to melons			
Mustard Greens	Refer to broccoli		Remove residue by washing.	Do not apply within 10 days of harvest.
Onions	Botrytis leaf blight Downy mildew Purple blotch	1 to 3	Follow a protective spray schedule starting when diseases are first reported in the area and repeat at 7 day intervals throughout the season. The addition of a TRITON surfactant to spray solutions will improve performance.	
Peppers	Refer to eggplant			
Persian Melons	Refer to melons			
Potatoes	Early blight Late blight	1 to 2	Begin application when plants are 4 to 6 inches high by applying 1 lb. per acre. As the vines increase in size, apply $1\frac{1}{2}$ to 2 lbs. per acre. The addition of a TRITON surfactant to spray solutions will improve performance.	
(seed pieces)	Fusarium decay		Use $1\frac{1}{4}$ lbs. per 10 gals. water. Dip whole or cut tubers and spread treated seed pieces in a cool place if held before planting.	Do not use treated seed pieces for food or feed.
Pumpkins	Angular leaf spot Downy mildew	3	Start applications when disease threatens and repeat at 7 to 10 day intervals as needed. Use sufficient water and direct spray to provide thorough coverage of both upper and lower leaf surfaces.	
Spinach	Downy mildew (blue mold) White rust	1 to 3	Start applications when disease symptoms first appear and repeat at 7 to 10 day intervals. Remove spray residue by washing or other effective means.	Do not apply within 10 days of harvest.
Squash	Refer to melons			Do not apply within 5 days of harvest.
	Pythium fruit rot	2 to 4	Begin when runners form or at first sign of disease at weekly intervals as needed.	
Tomatoes	Anthracnose Early blight Gray leaf spot Late blight Leaf mold Septoria leaf spot	1 to 3	Start applications when seedlings emerge or transplants are set and repeat at 7 to 10 day intervals throughout the season. In greenhouse use $\frac{1}{4}$ to $\frac{1}{3}$ pound per 5,000 sq. ft. and start applications when first cluster fruit is well formed. Repeat applications at 7 to 10 day intervals. The addition of a TRITON surfactant will improve spray performance.	Do not apply within 5 days of harvest.
Turnips	Refer to broccoli		Remove excess residue by washing.	Do not apply within 7 days of harvest.
Watermelons	Refer to melons			

FRUITS AND NUTS

CROP	DISEASES CONTROLLED	RATE OF DITHANE M-22 SPECIAL FUNGICIDE PER APPLICATION LB./A	LB./100 GALS.	REMARKS (ALSO REFER TO DIRECTIONS FOR USE)	RESTRICTIONS
Almonds (California)	Brown rot (blossom and twig blight) Coryneum blight (shot hole) Leaf blight scab	6 to 8	$1\frac{1}{2}$ to 2	Apply at dormant to popcorn stage, full bloom, and petal fall, or every 7 to 10 days if bloom is staggered and weather is rainy. For Coryneum blight, begin application in dormant stage and repeat at 7 to 10 day intervals through petal fall. Use 4 ounces TRITON B-1956 per 100 gallons spray.	Do not apply later than 5 weeks after petal fall. Do not apply after petal fall if hulls are fed to dairy animals or animals being finished for slaughter.

CROP	DISEASES CONTROLLED	RATE OF DITHANE M-22 SPECIAL FUNGICIDE PER APPLICATION LB./A	LB./100 GALS.	REMARKS (ALSO REFER TO DIRECTIONS FOR USE)	RESTRICTIONS
Apples	Bitter rot Black rot Brown rot Bullseye rot Fly speck Rusts Scab Sooty blotch White rot	10	$1\frac{1}{2}$ to 2	Apply every 7 to 14 days throughout the season starting with a green tip or delayed dormant spray and continue through cover sprays. For low volume air blast sprays adjust the mixing rate accordingly. To avoid uneven, blotchy residues from high volume dilute sprays add 2 to 3 ounces TRITON B-1956 per 100 gallons of spray in the last 1 or 2 applications before harvest. Concentrate applications usually leave a uniform, even deposit.	Do not apply within 15 days of harvest in Arkansas, Delaware, Illinois, Indiana, Kansas, Kentucky, Maryland, Missouri, New Jersey, North Carolina, Ohio, Pennsylvania, South Carolina, Tennessee, Virginia and West Virginia, or within 30 days of harvest in other states. Do not graze livestock in treated areas.
Apricots (Western States)	Brown rot (blossom blight) Coryneum blight (shot hole) Green or Jacket rot Leaf curl	6 to 10	$1\frac{1}{2}$ to 2	Apply at dormant to popcorn stage, repeat every 7 to 14 days through petal fall according to local timing recommendations and disease severity. For fruit brown rot use again at 4 to 6 weeks before harvest, repeating applications 7 to 14 days. For Coryneum blight, begin application in dormant stage and repeat at 7 to 10 day intervals through petal fall. Use 4 ounces TRITON B-1956 per 100 gallons spray.	Do not apply within 14 days of harvest. Do not use in combination with diazinon.
Bananas	Sigatoka	2 to 4	—	Apply when leaves first appear and repeat every 2 to 3 weeks or as required. Use sufficient water to provide adequate coverage. The addition of a TRITON surfactant to spray solutions will improve performance	
Cranberries	Fruit rots Twig blight	3 to 6	—	Start applications at mid-bloom and repeat at 7 to 10 day intervals as required.	Do not apply within 30 days of harvest.
Figs (Kadota)	Alternaria and Cladosporium surface molds and rots	—	$\frac{3}{4}$	Apply once, 10 to 20 days before harvest.	Do not apply within 10 days of harvest.
Grapes	Black rot Bunch rot	$1\frac{1}{2}$ to 5	—	Apply in sufficient water to provide thorough coverage starting when new shoots are 8 to 10 inches long (pre-bloom) and continuing until 10 days after bloom. When fruit is present, use maximum of $1\frac{3}{4}$ per acre.	Do not apply within 7 days of harvest.
Nectarines	Refer to peaches				Do not apply within 14 days of harvest.
Peaches	Blossom blight (brown rot) Brown rot Leaf curl (shothole) Scab	10	$1\frac{1}{2}$ to 2	Apply as dormant and popcorn spray. Apply at pink bud and 25% to 75% of full bloom. Apply at pink bud, bloom, petal fall, shuck and in cover sprays at 7 to 14 day intervals. Apply 4 weeks harvest, or earlier if rot is found in green fruit, and repeat at 7 to 14 day intervals as needed.	Do not apply within 2 days of harvest. If applied within 14 days of harvest, remove residues brushing or washing.

FIELD CROPS

CROP	DISEASES CONTROLLED	RATE OF DITHANE M-22 SPECIAL FUNGICIDE PER APPLICATION LB./A	LB./100 GALS.	REMARKS (ALSO REFER TO DIRECTIONS FOR USE)	RESTRICTIONS
Grasses (Seed Production)	Leaf rust Stem rust Stripe rust	2 to 3	—	Begin application when disease is first noted or observed in the area and repeat application every 7 to 10 days until harvest or as long as disease threatens.	Do not feed treated grass to livestock.
Sugar Beets	Cercospora leaf spot	$1\frac{1}{2}$ to 2	—	Start application when disease first threatens and repeat every 7 to 10 days as needed.	Do not feed treated tops to livestock. Do not apply within 14 days of harvest.
Tobacco (transplant beds)	Anthracnose Blue mold	—	$\frac{1}{2}$ to 1	Spray plant beds when plants are the size of a dime or when blue mold is first reported in the area, applying $2\frac{1}{2}$ to 5 gallons per 100 square yards, depending on plant size. Repeat sprays twice weekly or more frequently during wet weather. Continue treatment until seedings have been transplanted.	
(field)	Anthracnose Blue mold	—	$2\frac{1}{4}$ to 4	Begin application when disease is first noted or observed in area and repeat application as long as disease threatens.	

SEED TREATMENT

CROP	DISEASES CONTROLLED	RATE OF DITHANE M-22 SPECIAL FUNGICIDE PER APPLICATION OZ./BU.	OZ./100 LB.	REMARKS (ALSO REFER TO DIRECTIONS FOR USE)	RESTRICTIONS
Barley	Covered smut False loose smut Damping-off Seed rots Seedling blights	0.9 to 2.5	1.9 to 5.2	GENERAL USE INSTRUCTIONS: Seeds to be treated should be cleaned and well cured prior to treatment. DITHANE M-22 SPECIAL fungicide may be applied to dry seed with conventional slurry or mist seed treating equipment. The seed must be completely and uniformly covered with fungicide. For seed treatment a dye must be added to DITHANE M-22 SPECIAL fungicide which will impart an unnatural color to the seed.	Treated seed should be labeled "Must not be food, used for food, feed, or oil purposes."
Beans	Damping-off Seed rots Seedling blights	1.3 to 2.5	2.0 to 4.2		
Corn	Damping-off Seed rots Seedling blights	1.3 to 2.5	2.2 to 4.5		
Cotton (acid delinted)	Damping-off Seedling blights	—	3.0		
(reginned)	Damping-off Seedling blights	—	6.0		
Flax	Damping-off Seed rots Seedling blights	0.9 to 2.5	1.6 to 4.5		
Oats	Damping-off Seed rots Seedling blights Smuts	0.9 to 2.5	2.8 to 7.8		
Peanuts (shelled)	Damping-off Seed rots Seedling blights	—	2.8		
Peas	Damping-off Seed rots Seedling blights	1.25 to 2.5	2.0 to 4.2		
Rye	Bunt Damping-off Seed rots Seedling blights	0.9 to 2.5	2.3 to 3.6		
Soybeans	Damping-off Seed rots Seedling blights	0.9 to 2.5	1.5 to 4.2		
Sorghum	Covered kernel smut Damping-off Seed rots Seedling blights	0.9 to 2.5	1.8 to 4.5		
Sugar Beets	Damping-off Seed rots Seedling blights	—	6.0 to 10.0		
Wheat	Bunt Damping-off Seed rots Seedling blights	0.9 to 2.5	1.5 to 4.2		

TURF

CROP	DISEASES CONTROLLED	RATE OF DITHANE M-22 SPECIAL PER APPLICATION OZ./100 SQ. FT.	REMARKS (ALSO REFER TO DIRECTIONS FOR USE)	RESTRICTIONS
Assorted Grasses	Helminthosporium blight and melting-out Rust (leaf, stem, stripe) Rhizoctonia brown patch	3 to 8	GENERAL USE INSTRUCTIONS: Start application when grass greens-up in spring or when disease first appears and repeat at 7 to 14 day intervals or until disease threat is past.	Do not graze treated areas. Do not feed clippings to livestock.

CROP	DISEASES CONTROLLED	RATE OF DITHANE M-22 SPECIAL PER APPLICATION OZ./100 SQ. FT.	REMARKS (ALSO REFER TO DIRECTIONS FOR USE)	RESTRICTIONS
Assorted Grasses	Dollar spot	6 to 8	When conditions are especially favorable for disease development, apply maximum fungicide use rate on a 7 day spray schedule. Apply in sufficient water to provide adequate coverage.	

ORNAMENTALS

CROP	DISEASES CONTROLLED	REMARKS (ALSO REFER TO DIRECTIONS FOR USE)
GENERAL USE INSTRUCTIONS:		Apply the equivalent of 1½ lbs. DITHANE M-22 SPECIAL fungicide per 100 gals. The addition of TRITON B-1956 to spray solutions will improve performance. Begin spraying when plants are well leafed out or at first sign of disease in a full coverage spray at 7 to 10 day intervals throughout season or follow State Extension Service recommendations for disease control on the following ornamental plants:
Buffaloberry (Nursery)	Leaf spot	
Carnation	Leaf spot Rust	
Chrysanthemum	Ascochyta blight Botrytis petal spot Rust	Apply twice weekly during blooming period.
Conifer	Needle cast Brown spot	Begin application in spring early or summer before infection occurs. Repeat after heavy rain or at two-week intervals as long as necessary.
Dahlia	Botrytis blight	
Dogwood, flowering	Anthracnose	Apply when buds begin to open, when bracts have fallen, 4 weeks later and again in late summer after flower buds for next season are formed.
Dracaena	Fusarium leaf spot	
Geranium	Puccinia rust	
Gladiolus	Curvularia leaf spot Botrytis blossom blight Stemphylium leaf blight	Make regular weekly applications starting before diseases appear and increase to 2 or 3 applications per week during periods of heavy disease and during rainy weather. On flower spikes, reduce spray concentration ¾ lbs. per 100 gals.
Hydrangea	Botrytis blight	
Iris	Ink spot Leaf spot (fire)	
Lily	Botrytis blight	
Pansy	Anthracnose	
Peony	Botrytis blossom blight Phytophthora blight	Apply in early spring and early fall, drenching soil around plants as well as the foliage. Promptly destroy all infected plant parts.
Rose	Black spot Cercospora leaf spot Rust	
Skunkbush sumac	Leaf spots	
Snapdragon	Rust	
Zinnia	Leaf blight	

FUNGICIDE TANK MIXES

The following tank mix recommendations are made for the improved control of specific diseases. Before mixing DITHANE M-22 with any other fungicide, read and carefully observe the precautionary and all other information appearing on both product labels. User must abide by the most restrictive limitations appearing on the product labels.

CROP	CO-APPLIED FUNGICIDE	FOR IMPROVED CONTROL OF:	USE RATE PER APPLICATION	DITHANE M-22 SPECIAL FUNGICIDE USE RATE PER APPLICATION	REMARKS (ALSO REFER TO DIRECTIONS FOR USE)
Apples	BENLATE 50 WP*	Powdery mildew	2 to 3 ozs./100 gals.	¼ to 1 lb./100 gals.	
	TOPSIN-M 70WP*				
	BAYLETON 50 WP		½ to 2 ozs./100 gals.	1 to 2 lbs./100 gals.	
	KARATHANE WD	Powdery mildew Mite suppression	½ to ¾ lbs./100 gals.		
	KARATHANE LC		4 to 6 fl. ozs./100 gals.		
Celery	BENLATE 50 WP*	Early blight Late blight	4 ozs./A	1½ to 2 lbs./A	
	TOPSIN-M 70WP*				

CROP	CO-APPLIED FUNGICIDE	FOR IMPROVED CONTROL OF:	USE RATE PER APPLICATION	DITHANE M-22 SPECIAL FUNGICIDE USE RATE PER APPLICATION	REMARKS (ALSO REFER TO DIRECTIONS FOR USE)
	FIXED COPPERS	Bacterial leaf spot	Manufacturers label use rate	2 lbs./A	
Cucumbers	Refer to melons				
Grapes	BAYLETON 50 WP	Powdery mildew	2 to 6 ozs./A	1½ to 4 lbs./A	
	KARATHANE WD	Powdery mildew	1 to 2½ lbs./A		Do not use in California
	KARATHANE LC		4 to 8 fl. ozs./100 gals.		
	BENLATE 50 WP		¾ to 1 lb./A		
Melons	BENLATE 50 WP*	Target spot Gummy stem blight Powdery mildew Anthracnose	4 to 8 ozs./A	2 to 3 lbs./A	
	KARATHANE WD	Powdery mildew	8 to 12 ozs./A		Do not use in California
	KARATHANE LC		3 to 4 fl ozs./100 gals.		
	DIFOLATAN 80 SPRILLS	Gummy stem blight	2½ pts./A		DIFOLATAN may not be used on squash or pumpkins
Peppers	Fixed Coppers	Bacterial leaf spot	Manufacturers label use rate	1 to 3 lbs./A	
Potatoes	Triphenyl Tin Hydroxide (TPTH)	Early blight	½ to full label use rate	1½ to 2 lbs./A	
Pumpkins	Refer to melons				Do not use DIFOLATAN
Roses, Flowers, Ornamentals and shade	BENLATE 50 WP	Powdery mildew Leaf spots	½ to 1 lb./100 gals.	1½ lbs./100 gals.	
	TOPSIN-M 70 WP				
	KARATHANE WD	Powdery Mildew Mite suppression	6 to 8 ozs./100 gals.		
Squash	Refer to melons				
Sugar Beets	BENLATE 50 WP*	Cercospora leaf spot	4 oz./A	1½ to 2 lbs./A	
	TOPSIN-M 70WP*				
	MERTECT 340-F*		6 fl. ozs./A		
	Triphenyl Tin Hydroxide (TPTH)		½ to full label rate		
Tomatoes	Fixed Coppers	Bacterial leaf spot	Manufacturers label use rate	1½ to 3 lbs./A	
	DYRENE 50 WP	Gray leaf spot	½ to 1 lb./A		
	BENLATE 50 WP	Botrytis gray mold Sclerotinia white mold			

* Do not use if fungal resistance to this fungicide is present in field to be treated or has been reported in the area.

STORAGE AND DISPOSAL

STORAGE

Keep away from fire and sparks. Store in a cool, dry, well-ventilated area. Do not allow to become wet or overheated in storage, decomposition, impaired activity, or fire may result. Keep container closed when not in use.

Pallets of containers should not be stacked more than three high. Provide access aisle for each two rows. Decomposition produces a foul odor. If observed, check for hot containers and immediately remove to open areas for disposal.

PESTICIDE DISPOSAL

Do not contaminate water, food or feed by storage or disposal. Pesticide wastes are toxic. Improper disposal of excess pesticide, spray mixture, or rinsate is a violation of Federal law. If these wastes cannot be disposed of by use according to label instructions, contact your State Pesticide or Environmental Control Agency, or the Hazardous Waste representative at the nearest EPA Regional Office for guidance.

CONTAINER DISPOSAL

Completely empty bag into application equipment. Then dispose of it in a sanitary landfill or by incineration, or if allowed by state and local authorities, by burning. If burned, stay out of smoke.

STEPS TO BE TAKEN IN CASE MATERIAL IS RELEASED OR SPILLED: Sweep or shovel into containers for disposal or reworking. Keep dusting to a minimum. Flush contaminated area with a large amount of water to a chemical or sanitary sewer containing a settling pit. REFER TO PRECAUTIONARY STATEMENTS.

CONDITIONS OF SALE AND WARRANTY

Rohm and Haas warrants that the product conforms to its chemical description and is reasonably fit for the purpose stated on the label only when used in accordance with label directions under normal conditions of use. **ROHM AND HAAS MAKES NO OTHER EXPRESS OR IMPLIED WARRANTIES EITHER OF MERCHANTABILITY OR FITNESS FOR A PARTICULAR USE.** Handling, storage and use of the product by Buyer or User are beyond the control of Rohm and Haas and Seller. Risks such as crop injury, ineffectiveness or other unintended consequences resulting from, but not limited to, weather or soil conditions, presence of other materials, disease, pests, drift to other crops or property or failure to follow label directions will be assumed by the Buyer or User. **IN NO CASE WILL ROHM AND HAAS OR SELLER BE HELD LIABLE FOR CONSEQUENTIAL, SPECIAL OR INDIRECT DAMAGES RESULTING FROM THE HANDLING, STORAGE OR USE OF THIS PRODUCT.**

BAYLETON and DYRENE are registered trademarks of the parent company of Farbenfabriken Bayer GmbH, Leverkusen.

BENLATE is a registered trademark of E.I. DuPont de Nemours and Co., Inc.

DIFOLATAN and SPRILLS are registered trademarks of Chevron Chemical Co.

DITHANE, KARATHANE, TRITON and B-1956 are registered trademarks of the Rohm and Haas Company.

MERTECT is a registered trademark of Merck and Co., Inc.

TOPSIN-M is a registered trademark of Nippon Soda Co., Ltd.

8122-BO

SPECIAL LOCAL NEED REGISTRATIONS FOR DITHANE M-22 SPECIAL

STATE	USE	SLN NO.
California	Chinese Cabbage	CA-800170
Connecticut	Tobacco Blue Mold	CT-800004
Georgia	Tobacco Blue Mold	GA-800007
Kentucky	Tobacco Blue Mold	KY-800014
Maryland	Tobacco Blue Mold	MD-800008
Missouri	Tobacco Blue Mold	MO-820009
North Carolina	Tobacco Blue Mold	NC-800008
Ohio	Tobacco Blue Mold	OH-810009
Pennsylvania	Tobacco Blue Mold	PA-800014
South Carolina	Tobacco Blue Mold	SC-800007
Tennessee	Tobacco Blue Mold	TN-800014
Virginia	Tobacco Blue Mold	VA-800014

DITHANE M-45™

AGRICULTURAL FUNGICIDE

ACTIVE INGREDIENTS

Mancozeb

A coordination product of zinc ion and manganese ethylene bisdithiocarbamate 80%

In which the ingredients are

Manganese ++ 16%

Zinc ++ 2%

Ethylene bisdithiocarbamate ion ($C_4H_6N_2S_4$) ... 62%

INERT INGREDIENTS 20%

TOTAL 100%

EPA Reg. No. 707-78 EPA Est. No. 8590-PA-1

8120-D12

CAUTION

Keep Out of Reach of Children

PRECAUTIONARY STATEMENTS

HAZARDS TO HUMANS AND DOMESTIC ANIMALS

May cause irritation of nose, throat, eyes and skin. Do not breathe dust or spray mist. PROTECTIVE CLOTHING consisting of long pants, long-sleeved shirt, gloves, hat and boots must be worn during mixing and loading.

STATEMENTS OF PRACTICAL TREATMENT

IF SWALLOWED: Dilute by giving 2 glasses of water to drink and call a physician. Never give anything by mouth to an unconscious person.

IF INHALED: Move subject to fresh air.

IF IN EYES: Flush eyes with large amounts of water for at least 15 minutes. Consult a physician if irritation persists.

IF ON SKIN: Wash affected skin areas with soap and water.

ENVIRONMENTAL HAZARDS

This product is toxic to fish. Do not apply directly to water or wetlands. Do not apply when weather conditions favor drift from areas treated. Do not contaminate water by cleaning of equipment or disposal of wastes.

NOTICE: **Before using this product, read the entire Precautionary Statements, Conditions of Sale and Warranty, Directions for Use, Use Restrictions and Storage and Disposal Instructions on back panel and attached use directions. If the Conditions of Sale and Warranty are not acceptable, return the product unopened within thirty days of purchase to the place of purchase.**

DIRECTIONS FOR USE

It is a violation of Federal law to use this product in a manner inconsistent with its labeling.

RE-ENTRY AND WORKER PROTECTION STATEMENTS

Do not enter treated areas without protective clothing until sprays have dried. Because certain states may require more restrictive re-entry intervals for various crops treated with this product, consult your State Department of Agriculture for further information. Do not apply this product in such a manner as to directly or through drift expose workers or other persons. The area being treated must be vacated by unprotected persons.

Written or oral warnings must be given to workers who are expected to be in a treated area or in an area about to be treated with this product. Oral warnings must be given if there is reason to believe that written warnings cannot be understood by workers. When oral warnings are given, warnings shall be given in a language customarily understood by workers. Written warnings must include the following information:

CAUTION

(Insert area or field description) treated with DITHANE M-45 on (insert date of application). Do not enter treated areas without protective clothing until sprays have dried. In case of accidental exposure: Call a doctor (physician), clinic or hospital immediately. Explain that the victim has been exposed to DITHANE M-45 and describe his condition. For further information see the "Statements of Practical Treatment." Oral warnings must include the same information that is in the written warnings.

GENERAL USE INFORMATION

DITHANE M-45 fungicide is a broad-spectrum protectant fungicide recommended for the control of fruit, field, vegetable, and ornamental crop diseases. Optimum disease control is achieved when the fungicide is applied in a regularly scheduled preventative spray program. The addition of an agricultural surfactant will improve fungicide performance by providing a more uniform spray deposit, increased foliar redistribution, and improved fungicide retention during periods of wet weather.

USE RATE DETERMINATION

Carefully read, understand, and follow label use rates and limitations.

Under low disease conditions, minimum label rates per application can be used while maximum label rates and shortened spray schedules are recommended for severe or threatening disease conditions.

For proper application, determine the number of acres to be treated, the recommended label use rate, and the gallonage to be applied per acre. Prepare only the amount of spray solution required to treat the measured acreage. Careful calibration of spray equipment is recommended prior to use.

When applied by hand sprayers, 1 pound DITHANE M-45 fungicide per 100 gallons per acre is equivalent to 1 level tablespoon per gallon spray solution.

MIXING

Slowly place into spray tank as it is being filled or thoroughly premix in a nurse tank for concentrate or aircraft sprayers. Add other co-applied fungicides, insecticides, growth regulators, micronutrients, and spray adjuvants after DITHANE M-45 fungicide has been placed into suspension.

When preparing spray solutions for use in a hand sprayer, premix as a slurry in a small container, and then add to sprayer containing 1/3 to 1/2 the desired final water volume.

COMPATIBILITY

DITHANE M-45 fungicide is compatible with most commonly used agricultural fungicides, insecticides, and growth regulators. When preparing tank mixes, user should consult spray compatibility charts or State Cooperative Extension Service Specialists prior to actual use.

SPRAY ADJUVANTS

The addition of agricultural surfactants to DITHANE M-45 fungicide sprays will improve initial spray deposits, fungicide redistribution, and weatherability. The following spray adjuvants have been especially formulated to optimize the performance of foliar applied agricultural chemicals:

TRITON AG-98™	A low foam, nonionic, general purpose spray adjuvant. May be used in dilute or concentrate sprays applied by aircraft or ground equipment.
TRITON® B-1956®	A water-dispersible, resin-based nonionic surfactant which resists re-wetting and removal by rain. Effective with dilute sprays applied by ground equipment.
TRITON® CS-7	A spreader-binder designed specifically for use in concentrate and low volume sprays applied by aircraft or ground equipment.

Place DITHANE M-45 fungicide into suspension prior to adding an adjuvant to the spray mixture. Read and carefully observe the precautionary statements and all other information appearing on both product labels prior to spray preparation.

APPLICATION

Ground:

Thorough coverage foliar sprays generally result in optimum disease control. To achieve good coverage use proper spray pressure, gallonage per acre, nozzles (generally hollow cone), discs (generally D-5 to D-7), nozzle spacing, and tractor speed. Consult spray nozzle and accessory catalogues for specific information on proper equipment calibration.

Hand Sprayers:

Thoroughly spray plant foliage until run-off.

Aerial:

A uniform initial spray deposit over the crop canopy generally results in optimum disease control. Each aircraft should be prechecked for droplet size, uniformity of spray pattern, swath width, and spray volume.

Nozzle selection—Hollow cone brass nozzles with a D series orifice disc and core (whirlplate) are recommended. Nozzles should point straight down or slightly backward.

Swath width—For most field and vegetable crops, swaths just beyond the wingspan of 36 to 40 feet for light aircraft and up to 45 feet for heavier aircraft are suggested. Optimum swath for helicopters is usually 5 to 10 feet beyond normal boom length.

Spray volume—On vegetable and field crops. 2 to 3 gallons of spray per acre are generally optimum, orchards and vineyards can be handled with spray volumes of 5 gallons per acre. Some tall or dense foliage crops, requiring greater penetration to the lower leaf surface, will require higher spray volumes.

Altitude—For most crops, the spray boom should be positioned 5 to 10 feet above the crop canopy.

Flagging—Swaths should be marked at the end of the field with permanent flags or by a flagman. Swaths should be measured accurately with a chain or other device except when rows can be accurately counted.

Sprinkler Irrigation:

DITHANE M-45 fungicide must be applied on a regular protectant fungicide schedule, **not an irrigation schedule.** If irrigation cycles are used, ground or aerial applications must supplement chemigation applications to achieve adequate disease control.

With the exception of peanuts, potatoes, and tomatoes, DITHANE M-45 fungicide has not been sufficiently tested to assure consistent product performance for all labeled uses. The following calibration and application techniques are provided for user reference, but do not constitute a warranty of fitness for application through sprinkler irrigation equipment. Users must check with State and local regulatory agencies, for potential use restrictions, prior to applying any agricultural chemical by sprinkler irrigation equipment.

Center-pivot Irrigation Equipment: (Use only with electric or oil hydraulic drive systems which provide a uniform water distribution.)

- Determine size of area to be treated.
- Determine the time required to apply 1/4 inch water over the area to be treated when the system and injection equipment is operated at normal pressures recommended by the equipment manufacturer. Run system at 80 to 95% of manufacturer's rated capacity.
- Using only water, determine the injection pump output when operated at normal line pressure.
- Determine the amount of DITHANE M-45 fungicide required to treat area.
- Add the required amount of DITHANE M-45 fungicide and sufficient water to meet the injection time requirements to the solution tank.
- Maintain constant solution tank agitation during the injection period.
- Stop injection equipment after treatment is completed. Continue to operate the system until the DITHANE M-45 fungicide solution has cleared the sprinkler head.

Solid-set and Portable (Wheel Move) Irrigation Equipment

- Determine acreage covered by sprinkler.
- Fill injector solution tank with water and adjust flow rate to use contents over a 10 to 30 minute interval.
- Determine the amount of DITHANE M-45 fungicide required to treat area.
- Add the required amount of DITHANE M-45 fungicide into the same quantity of water used to calibrate the injection equipment.
- Maintain constant solution tank agitation during the injection period.
- Operate system at normal pressures recommended by the manufacturer of the injection equipment and used for the time interval established during calibration.
- Inject DITHANE M-45 fungicide at the end of the irrigation cycle or as a separate application to maximize foliar fungicide retention.
- Stop injection equipment after treatment is completed. Continue to operate the system until the DITHANE M-45 fungicide solution has cleared the last sprinkler head.

PRECAUTIONS

Irrigation equipment must have antisiphon devices and automatic shutoff valves to prevent contamination of water sources.

Have injector pump electrical connections coupled with the system's shutdown controls.

Do not chemigate if wind speed exceeds 10 mph.

Irrigation water treated with DITHANE M-45 fungicide should be contained on the treated area until the water is absorbed by the soil.

DISEASE MONITORING

DITHANE M-45 fungicide is a broad-spectrum protectant fungicide. If not applied on a routine protectant spray schedule, crops should be scouted on a weekly basis. Fungicide application should be made, at the recommended label use rate and spray schedule, at the first sign of disease, report of disease in the area, or during

RESTRICTIONS

Users should carefully read, understand and follow all use restrictions prior to using DITHANE M-45 fungicide.

VEGETABLES

CROP	DISEASES CONTROLLED	RATE OF DITHANE M-45 PER APPLICATION LB./A	REMARKS (ALSO REFER TO DIRECTIONS FOR USE)	RESTRICTIONS
Asparagus	Rust	2	Start applications when rust first appears and repeat at 10 day intervals. Four applications are usually sufficient.	Apply only on asparagus ferns after spears have been harvested.
Cantaloupes	Refer to melons	2 to 3	Some varieties are sensitive to DITHANE M-45 fungicide. Consult State Cooperative Extension Service Specialist prior to use.	
Carrots	Alternaria leaf Spot Cercospora leaf spot	1½ to 2	Apply in 75 to 125 gallons spray per acre. Start applications when disease first threatens and repeat every 7 to 10 days as needed. The addition of TRITON B-1956 or TRITON CS-7 to spray solutions will improve performance.	Do not apply within 7 days of harvest. Do not use treated tops for food or feed.
Celery	Early blight Late blight	2	Start applications in the plant bed when plants first come through the ground. Make further applications at 3 to 5 day intervals until plants are set in the field. Continue field applications on a 7 day schedule. Remove excess residue remaining at harvest by stripping, trimming and washing. The addition of a TRITON surfactant to spray solutions will improve performance.	Do not apply within 14 days of harvest.
Corn (sweet, popcorn)	Common Corn Rust Helminthosporium leaf blight	1½	Use sufficient water for thorough coverage. Start applications when disease first appears and repeat at 4 to 7 day intervals. The addition of TRITON B-1956 or TRITON CS-7 to spray solutions will improve performance.	Do not apply within 7 days of harvest. Do not feed treated forage to livestock.
Cucumbers	Refer to melons	2 to 3		
Fennel	Refer to celery	2		Do not apply within 7 days of harvest.
Melons	Alternaria leaf spot Anthracnose Cercospora leaf spot Downy mildew Gummy stem blight Scab	2 to 3	Start applications when plants are in the two leaf stage and repeat at 5 to 7 day intervals. Use sufficient water and direct spray, to provide thorough coverage of both upper and lower leaf surfaces.	Do not apply within 5 days of harvest.
Muskmelons	Refer to melons	2 to 3		
Onions (dry bulb)	Botrytis leaf blight Downy mildew Purple blotch Neck rot	3	Follow a protective spray schedule starting when diseases are first reported in the area and repeat at 7 day intervals throughout the season. The addition of a TRITON surfactant to spray solutions will improve performance. Do not allow spray or drift to contact bulbs after lifting from soil.	Do not apply within 7 days of harvest.
(furrow drench)	Smut		Apply 3 lbs. per acre as a furrow drench at time of planting onion seeds. Use 75 to 125 gallons water per acre.	Do not use more than 2.4 lbs. actual per acre (29,000 linear feet of furrow).
Potatoes	Early blight Late blight	1 to 2	Begin application when plants are 4 to 6 inches high by applying 1 lb. per acre. As the vines increase in size, apply 1½ to 2 lbs. per acre. The addition of a TRITON surfactant to spray solutions will improve performance.	
Seedpiece (treatment)	Fusarium decay Seedborne common scab		Dip whole or cut potato tubers in 1¼ lbs. DITHANE M-45 fungicide per 50 gallons of water. Place treated tubers in a clean container following treatment and plant as soon as possible. Spread treated seedpieces in a cool place if held before planting.	Do not use treated seed potatoes for food or feed purposes.
Squash, summer	Refer to melons	2 to 3		
Tomatoes	Anthracnose Early blight Gray leaf spot Late blight Leaf mold	1½ to 3	Start applications when seedlings emerge or transplants are set and repeat at 7 day intervals throughout the season. The addition of a TRITON surfactant to spray solutions will improve performance.	Do not apply within 5 days of harvest.
Watermelons	Refer to melons	2 to 3		

FRUITS AND NUTS

CROP	DISEASES CONTROLLED	RATE OF DITHANE M-45 FUNGICIDE PER APPLICATION LB./A	LB./100 GALS.	REMARKS (ALSO REFER TO DIRECTIONS FOR USE)	RESTRICTIONS
Apples	Bitter rot Black rot Brown rot Fly speck Rusts Scab Sooty blotch	8 to 10	1 to 2	Apply every 7 to 14 days throughout the season starting with a green tip or delayed dormant spray and continue through cover sprays. For low volume air blast sprays adjust the mixing rate accordingly. To avoid uneven, blotchy residues from high volume dilute sprays add 2 to 3 ounces TRITON B-1956 per 100 gallons of spray in the last 1 or 2 applications before harvest. Concentrate applications usually leave a uniform, even deposit.	Do not apply within 21 days of harvest in Arkansas, Delaware, Illinois, Indiana, Kansas, Kentucky, Maryland, Missouri, New Jersey, North Carolina, Ohio, Pennsylvania, South Carolina, Tennessee, Virginia, and West Virginia, or within 30 days of harvest in other states not mentioned above. Do not graze livestock in treated areas.
Bananas	Sigatoka	2 to 4		Apply when leaves first appear and repeat every 2 to 3 weeks or as required. Use sufficient water to provide adequate coverage. The addition of a TRITON surfactant to spray solutions will improve performance.	
Crabapples	Refer to apples	8–10	1 to 2		Do not apply within 15 days of harvest. Do not graze livestock in treated areas.
Cranberries	Fruit rot	3 to 6		Start applications at mid-bloom and repeat at 7 to 10 day intervals as required.	Do not apply within 30 days of harvest.
Grapes (California only)	Bunch rot Deadarm Downy mildew	1½ to 4		Apply in sufficient water to provide thorough coverage starting when new shoots are ½ to 1½ inches long. Repeat when shoots are 3 to 5 inches long, when shoots are 8 to 10 inches long, and then at 10 to 14 day intervals until fruit is set or 66 days before harvest.	Do not apply after fruit is set.
(Other areas)	Black rot Deadarm Downy mildew			For late season control of black rot, deadarm and downy mildew the use of other approved and recommended fungicides is suggested.	Do not apply within 66 days of harvest.
Papaya	Anthracnose Phytophthora fruit rot	2 to 2½		Use 20 to 100 gallons water per acre. Start applications at flowering and continue at 14 day intervals and at 7 day intervals during periods of heavy disease. Direct spray to crown and blossom area. Use 6 to 8 ounces TRITON B-1956 spreader-sticker per acre.	
Pears	Refer to apples	8 to 10	1 to 2		Do not apply within 15 days of harvest. Do not graze livestock in treated areas.
	Pear psylla nymphs	8 to 10	1½	Effective control of young nymphs can be obtained by using 2 to 3 consecutive sprays. Applications should start at petal fall and continue at 7 to 10 day intervals. As new growth occurs, it must be protected by spraying. Timing of sprays should coincide with the initial appearance of pear psylla nymphs. Specific timing of sprays should be obtained from State Agricultural Experiment Station workers or State Extension Service specialists.	
(postharvest sprays)			1½ to 2	Use a full coverage spray of 800 to 1000 gallons spray per acre. Timing of spray should be under direction of State Agricultural or Extension Service workers.	Do not graze livestock in treated areas.
Quince	Refer to apples	8 to 10	1 to 2		Do not apply within 15 days of harvest. Do not graze livestock in treated areas

FIELD CROPS

CROP	DISEASES CONTROLLED	RATE OF DITHANE M-45 PER APPLICATION LB./A	REMARKS (ALSO REFER TO DIRECTIONS FOR USE)	RESTRICTIONS
Barley	Refer to wheat	2		
Corn (field, hybrid seedcorn)	Common Corn Rust Helminthosporium leaf blight	1½	Start application when disease symptoms first appear and, depending on severity of infection, continue on a 4 to 14 day schedule. The addition of TRITON B-1956 or TRITON CS-7 to spray solutions will improve performance.	Do not apply within 40 days of harvest.
Cotton (Southwest U.S. only)	Rust	1¼ to 2	Start applications when rust first appears and repeat at 10 to 14 day intervals until disease threat is past.	Do not apply after bolls are open. Do not graze livestock in treated areas or feed gin trash to livestock.
Oats	Refer to wheat	2		
Peanuts	Cercospora leaf spot Rust	1 to 2	Start applications when disease first appears or is reported in area. Repeat sprays at 10 to 14 day intervals. Reduce sprays to a 7 day interval during humid weather.	Do not apply within 14 days of harvest. Do not feed treated vines to livestock.
Rye	Refer to wheat	2		
Sugar Beets	Cercospora leaf spot	1½ to 2	Start application when disease first threatens and repeat every 7 to 10 days as needed. The addition of a TRITON surfactant to spray solutions will improve performance	Do not apply within 14 days of harvest. Do not feed treated tops to livestock.
Wheat	Helminthosporium leaf spots Septoria leaf spots Septoria glume blotch Leaf rust Tan spot	2	Start application at onset of disease or when plants are in the tillering to jointing stage and repeat at 7 to 10 day intervals. The addition of TRITON CS-7 to spray solutions will improve performance.	Do not make more than three applications during the season. Do not apply within 26 days of harvest. Do not graze livestock in treated areas within 26 days after application.

SEED TREATMENT

CROP	DISEASES CONTROLLED	RATE OF DITHANE M-45 PER APPLICATION OZ./BU.	OZ./100 LB.	REMARKS (ALSO REFER TO DIRECTIONS FOR USE)	RESTRICTIONS
Barley	Covered smut False loose smut Damping-off Seed rots Seedling blights	1.3 to 2.0	2.7 to 4.2	Seeds to be treated should be cleaned and well cured prior to treatment. DITHANE M-45 fungicide may be applied to dry seed with conventional slurry or mist seed treating equipment or as a planter-box application. For best results, the seed must be completely and uniformly covered with fungicide. For seed treatment, a dye must be added to DITHANE M-45 fungicide which will impart an unnatural color to the seed.	Treated seed should be labeled: "Must not be used for food, feed, or oil purposes."
Corn (field)	Damping-off Seed rots Seedling blights	1.5 to 3.0	2.7 to 5.4		
Cotton (acid delinted)	Damping-off Seedling blights		3.0		
(reginned)	Damping-off Seedling blights		6.0		
Flax	Damping-off Seed rots Seedling blights	2.0 to 4.0	3.6 to 7.1		
Oats	Damping-off Seed rots Seedling blights Smuts	1.3 to 2.0	4.0 to 6.3		
Peanuts (shelled)	Damping-off Seed rots Seedling blights	2.0 to 4.0	8.0 to 16.0		

CROP	DISEASES CONTROLLED	RATE OF DITHANE M-45 PER APPLICATION OZ./BU.	OZ./100 LB.	REMARKS (ALSO REFER TO DIRECTIONS FOR USE)	RESTRICTIONS
Rice	Damping-off Seed rots Seedling blights		2.0 to 4.0	Apply before, during or after soaking in water.	
Rye	Bunt Damping-off Seed rots Seedling blights	1.3 to 2.0	2.3 to 3.6		
Safflower	Seedborne rust (*Puccinia carthami*)		2.0		
Sorghum	Covered kernal smut Damping-off Seed rots Seedling blights	1.5 to 2.5	2.7 to 4.5		
Tomatoes	Damping-off Seed rots Seedling blights		8.0		
Wheat	Bunt Damping-off Seed rots Seedling blights	1.3 to 2.0	2.2 to 3.3		

MISCELLANEOUS

CROP	DISEASES CONTROLLED	RATE OF DITHANE M-45 FUNGICIDE PER APPLICATION	REMARKS (ALSO REFER TO DIRECTIONS FOR USE)
Asparagus Crowns	Crown rot	1 lb. per 100 gals.	Place loosely packed crowns into a burlap bag and soak, with gentle agitation, in the fungicide solution for 5 minutes. Remove bag, drain well, and plant crowns as soon as possible. A tank large enough to hold a single burlap bag will treat 2 bags of crowns. Clean dipping suspension should then be prepared in a clean tank. Dirty crowns should be pre-washed to remove excess soil.
Caprifig	Endosepsis (Fusarium) Assorted molds	1 lb. per 25 gals.	Prepare mamme figs by making a shallow cut through the eye and then hand dividing to avoid wasp injury. Submerge mamme figs in the fungicide suspension for a minimum of 15 minutes. The fungicide suspension should be stirred frequently to prevent settling out. Fresh dipping solutions should be used after treating 4 or 5 batches of figs. After treatment, figs should be drained prior to placement in trees.
Conifers	Lophodermium needle cast	4 lbs. per 100 gals.	For use in nursery and newly transplanted plantings. Apply at 21 day intervals during periods of sprinkler irrigations and at monthly intervals during the rest of the year when beds are not covered with snow. The addition of TRITON B-1956 to spray solutions is recommended.
Pineapples	Phytophthora heart rot	3.2 lbs. per 10 gals.	Dip planting material in fungicide solution prior to planting. Stir solution frequently to prevent settling out. A new solution should be prepared when at least two-thirds of the volume has been used or sooner if soil from plant material has noticeably discolored the solution. Depending on the cleanliness and size of planting stock, up to 100 gallons of fungicide solution should be used to treat the propagative materials used to plant one acre.

TURF

CROP	DISEASES CONTROLLED	RATE OF DITHANE M-45 PER APPLICATION OZ./1000 SQ. FT.	REMARKS (ALSO REFER TO DIRECTIONS FOR USE)	RESTRICTIONS
GENERAL USE INSTRUCTIONS			Start application when grass greens-up in spring when disease first appears and repeat at 7 to 14 day intervals or until disease threat is past. When conditions are especially favorable for disease development, apply maximum fungicide use rate on a 7 day spray schedule. Apply in sufficient water to provide adequate coverage.	Do not graze treated areas. Do not feed clippings to livestock.
Assorted Grasses	Helminthosporium melting out Rust (leaf stem, stripe)	4		
	Copper spot Fusarium blight Red thread Slime mold	4 to 8		

CROP	DISEASES CONTROLLED	RATE OF DITHANE M-45 PER APPLICATION OZ./1000 SQ. FT.	REMARKS (ALSO REFER TO DIRECTIONS FOR USE)	RESTRICTIONS
Assorted Grasses—cont'd.	Algae	6		
	Dollar spot	6 to 8		
	Rhizoctonia brown patch	4	Apply on a 7 day spray schedule.	
	Pythium blight	8	Apply at 5 day intervals, or more frequently if conditions are especially favorable for disease development.	
	Fusarium snow mold	6 to 8	Apply at 2 to 6 week intervals during winter.	

ORNAMENTALS

CROP	DISEASES CONTROLLED	REMARKS (ALSO REFER TO DIRECTIONS FOR USE)	RESTRICTIONS
GENERAL USE INSTRUCTIONS		Apply the equivalent of 1½ lbs. DITHANE M-45 fungicide per 100 gals. The addition of TRITON B-1956 to spray solutions will improve performance. Begin spraying when plants are well leafed out or at first sign of disease, in a full coverage spray at 7 to 10 day interval, throughout season or follow State Extension Service recommendations for disease control on the following ornamental plants:	
African violet	Botrytis blight		
Anthurium	Anthracnose spadix rot		
Arborvitae	Cercospora blight		
Ash, mountain	Entomosporium leaf spot Guignardia leaf blotch		
Ash, white	Anthracnose Cylindrosporium leaf spot		
Aster, perennial	Puccinia rusts		
Aucuba, japonica	Alternaria leaf spot Anthracnose		
Azalea	Cylindrocladium rot Petal blight Phytophthora twig and bug blight.	Apply in a full coverage spray, 2 to 3 times a week while flowers are opening. Direct spray into flowers and thoroughly spray ground under bushes.	
Begonia	Botrytis blight		
Buffaloberry	Cylindrosporium leaf spot		Do not use fruit for food or feed purposes.
Camelias	Petal blight	Refer to azalea	
Carnation	Rust Septoria leaf spot		
Cedar, red (Juniper)	Cercospora blight Phomopsis blight		
Chrysanthemum	Ascochyta blight Botrytis petal spot Rust	Apply twice weekly during blooming period.	
Conifers	Lophodermium needle cast Pine gall rust Scirrhia brown spot	Begin application in spring or early summer before infection occurs. Repeat after heavy rains and at two week intervals as long as needed.	
Cordyline	Cercospora leaf spot		
Crabapple (Ornamental)	Cedar-apple rust Scab Sphaeropsis leaf spot		
Cypress Arizona (Cupressus sp.)	Cercospora blight Monochaetia canker		
Dahlia	Botrytis blight		
Dieffenbachia	Leptosphaeria brown spot		

CROP	DISEASES CONTROLLED	REMARKS (ALSO REFER TO DIRECTIONS FOR USE)	RESTRICTIONS
Dogwood, flowering	Anthracnose Elsinoe leaf spot Septoria leaf spot	Apply when buds begin to open, when bracts have fallen, 4 weeks later and again in late summer after flower buds for next season have formed.	
Dracaena	Fusarium leaf spot		
Elm	Black leaf spot		
Euonymus	Anthracnose		
Fatsia	Anthracnose		
Fern	Rhizoctonia blight		
Ficus	Cercospora leaf spot		
Firethorn (Pyracantha)	Fusicladium scab		
Fir, Douglas	Swiss needle cast		
Fuschia	Botrytis blight Rust		
Geranium	Rust		
Gladiolus	Botrytis blossom blight Curvularia leaf spot	Make regular weekly applications starting before diseases appear and increase to 2 or 3 applications per week during periods of heavy disease and during rainy weather. On flower spikes, reduce spray concentration to $^3/_4$ lb. per 100 gals.	
Hawthorn	Cedar-apple rust Fabraea leaf spot Frogeye leaf spot Hawthorn rust Scab		
Holly	Purple spot		
Hollycock	Anthracnose Cercospora leaf spot Puccinia rusts		
Honeysuckle	Herpobasidium blight		
Horsechestnut Buckeye	Alternaria leaf spot Guignardia leaf blotch		
Hydrangea	Botrytis blight Cercospora leaf spot		
Iris	Didymellina leaf spot Mystrosporium ink spot		
Juniper	Phomopsis blight		
Laurel, mountain	Cercospora leaf spot Petal blight	Refer to azalea.	
Ligustrum	Cercospora leaf spot		
Lily	Botrytis blight		
Magnolia	Gloeosporium leaf spot		
Maple	Alternaria leaf spot Phyllosticta leaf spot		Do not use on Sugar maples intended for the production of maple syrup.
Marigold	Botrytis blossom blight		
Narcissus	Botrytis blight (fire) Smoulder		
Oak	Actinopelte leaf spot Taphrina leaf blister		
Orchid (Dendrobium)	Botrytis blossom blight		
Pachysandra	Volutella blight	Use a drenching spray of 2 lbs. per 50 gallons of water per 5,000 sq. ft. of bed. Start application at first sign of disease and apply at least 5 applications at 10 to 14 day intervals.	
Pansy	Anthracnose		
Peony	Botrytis blossom blight Phytophthora blight	Apply in early spring and early fall, drenching soil around plants as well as the foliage. Promptly destroy all infected plant parts.	
Peperomia	Cercospora leaf spot		
Philodendron	Dactylaria leaf spot Phytophthora leaf spot		
Photinia	Entomosporium leaf spot		
Pleomele	Fusarium leaf spot		

CROP	DISEASES CONTROLLED	REMARKS (ALSO REFER TO DIRECTIONS FOR USE)	RESTRICTIONS
Poinsettia	Sphaceloma scab		
Rhododendron	Cercospora leaf spot Discosia leaf spot Petal blight	Refer to azalea.	
Rose	Black spot Cercospora leaf spot Rust		
Schefflera	Alternaria blight		
Skunkbush, sumac	Cylindrosporium leaf spot		
Snapdragon	Rust		
Statice	Cercospora frogeye		
Syngonium	Cephalosporium leaf spot		
Tulip	Botrytis blight (fire)		
Venus, flytrap	Anthracnose		
Viburnum	Downy mildew Ramularia leaf spot		
Walnut	Anthracnose		Do not use treated walnuts for food or feed purposes.
Zinnia	Alternaria leaf blight		

FUNGICIDE TANK MIXES

The following tank mix recommendations are made for the improved control of specific diseases. Before mixing DITHANE M-45 with any other fungicide, read and carefully observe the precautionary and all other information appearing on both product labels. User must abide by the most restrictive limitations appearing on the product labels.

CROP	CO-APPLIED FUNGICIDE	FOR IMPROVED CONTROL OF:	USE RATE PER APPLICATION	DITHANE M-45 FUNGICIDE USE RATE PER APPLICATION	REMARKS (ALSO REFER TO DIRECTIONS FOR USE)
Apples	BENLATE 50 WP*	Powdery mildew	2 to 4 ozs./100 gals.	¼ to 1 lb./100 gals.	
	TOPSIN-M 70WP*				
	BAYLETON 50 WP		½ to 2 ozs./100 gals.	1 to 2 lbs./100 gals.	
	KARATHANE WD	Powdery mildew Mite suppression	½ to ¾ lbs./100 gals.		
	KARATHANE LC		4 to 6 fl. ozs./100 gals.		
Barley	BAYLETON 50 WP	Powdery mildew Rust (leaf, stem, stripe)	2 to 4 ozs./A	2 lbs./A	
Celery	BENLATE 50 WP*	Early blight Late blight	4 ozs./A	1½ to 2 lbs./A	
	TOPSIN-M 70WP*				
	FIXED COPPERS	Bacterial leaf spot	Manufacturer's label use rate	2lbs./A	
Cucumbers	Refer to melons				
Grapes	BAYLETON 50 WP	Powdery mildew	2 to 6 ozs./A	1½ to 4 lbs./A	
	KARATHANE WD		1 to 2½ lbs./A		Do not use in California.
	KARATHANE LC		4 to 8 fl. ozs./100 gals.		
	BENLATE 50 WP		¾ to 1 lb./A		
Melons	BENLATE 50 WP*	Target spot Gummy stem blight Powdery mildew Anthracnose	4 to 8 ozs./A	2 to 3 lbs./A	
	KARATHANE WD	Powdery mildew	8 to 12 ozs./A		Do not use in California.
	KARATHANE LC		3 to 4 fl. ozs./100 gals.		
	DIFOLATAN 80 SPRILLS	Gummy stem blight	1½ to 2 lb./A		
Peanuts	BENLATE 50 WP*	Early leaf spot Late leaf spot Ascochyta web blotch	4 ozs./A	1 to 2 lbs./A	
	TOPSIN-M 70WP*	Early leaf spot			
Pears	BAYLETON 50 WP	Powdery mildew	½ to 2 ozs./100 gals.	1 to 2 lbs./100 gals.	
	KARATHANE WD	Powdery mildew Mite suppression	½ to ¾ lbs./100 gals.		Do not use in California.
	KARATHANE LC		4 to 6 fl. ozs./100 gals.		

CROP	CO-APPLIED FUNGICIDE	FOR IMPROVED CONTROL OF:	USE RATE PER APPLICATION	DITHANE M-45 FUNGICIDE USE RATE PER APPLICATION	REMARKS (ALSO REFER TO DIRECTIONS FOR USE)
Potatoes	Triphenyl Tin Hydroxide (TPTH)	Early blight	½ to full label use rate	1½ to 2 lbs./A	
Squash, summer	Refer to melons				DIFOLATAN may not be used on squash.
Sugar Beets	BENLATE 50 WP*	Cercospora leaf spot	4 ozs./A	1½ to 2 lbs./A	
	TOPSIN-M 70WP*				
	MERTECT 340-F*		6 fl. ozs./A		
	Triphenyl Tin Hydroxide (TPTH)		½ to full label rate		
Tomatoes	Fixed Coppers	Bacterial leaf spot	Manufacturer's label use rate	1½ to 3 lbs./A	
	DYRENE 50 WP	Grayleaf spot	½ to 1 lb./A		
	BENLATE 50 WP	Botrytis gray mold Sclerotinia white mold			
Wheat	BAYLETON 50 WP	Powdery mildew Rust (leaf, stem, stripe)	2 to 4 oz./A	2 lbs./A	

* Do not use if fungal resistance to this fungicide is present in field to be treated or has been reported in the area.

STORAGE AND DISPOSAL

STORAGE

Keep away from fire and sparks. Store in a cool, dry, well-ventilated area. Do not allow to become wet or overheated in storage; decomposition, impaired activity, or fire may result. Keep container closed when not in use.

Pallets of containers should not be stacked more than three high. Provide access aisle for each two rows. Decomposition produces a foul odor; if observed, check for hot containers and immediately remove to open areas for disposal.

PESTICIDE DISPOSAL

Do not contaminate water, food or feed by storage or disposal. Pesticide wastes are toxic. Improper disposal of excess pesticide, spray mixture, or rinsate is a violation of Federal law. If these wastes cannot be disposed of by use according to label instructions, contact your State Pesticide or Environmental Control Agency, or the Hazardous Waste representative at the nearest EPA regional office for guidance.

CONTAINER DISPOSAL

Completely empty bag into application equipment. Then dispose of it in a sanitary landfill or by incineration, or, if allowed by state and local authorities, by burning. If burned, stay out of smoke.

STEPS TO BE TAKEN IN CASE MATERIAL IS RELEASED OR SPILLED: Sweep or shovel into containers for disposal or reworking. Keep dusting to a minimum. Flush contaminated area with a large amount of water to a chemical or sanitary sewer containing a settling pit. Refer to Precautionary Statements.

CONDITIONS OF SALE AND WARRANTY

Rohm and Haas warrants that the product conforms to its chemical description and is reasonably fit for the purpose stated on the label only when used in accordance with label directions under normal conditions of use. **ROHM AND HAAS MAKES NO OTHER EXPRESS OR IMPLIED WARRANTIES EITHER OF MERCHANTABILITY OR FITNESS FOR A PARTICULAR USE.** Handling, storage and use of the product by Buyer or User are beyond the control of Rohm and Haas and Seller. Risks such as crop injury, ineffectiveness, or other unintended consequences resulting from, but not limited to, weather or soil conditions, presence of other materials, disease, pests, drift to other crops or property or failure to follow label directions will be assumed by the Buyer or User. **IN NO CASE WILL ROHM AND HAAS OR SELLER BE HELD LIABLE FOR CONSEQUENTIAL, SPECIAL OR INDIRECT DAMAGES RESULTING FROM THE HANDLING, STORAGE OR USE OF THIS PRODUCT.**

BAYLETON and DYRENE are registered trademarks of the parent company of Farbenfabriken Bayer GmbH, Leverkusen.

BENLATE is a registered trademark of E.I. duPont de Nemours and Co., Inc.

DIFOLATAN and SPRILLS are registered trademarks of the Chevron Chemical Co.

DITHANE M-45, KARATHANE, TRITON, AG-98, CS-7 and B-1956 are registered trademarks of the Rohm and Haas Company.

MERTECT is a registered trademark of Merck and Co., Inc.

TOPSIN-M is a registered trademark of Nippon Soda Co., Ltd.

8120-B1

SPECIAL LOCAL NEED REGISTRATIONS FOR DITHANE M-45

STATE	USE	SLN OR FEDERAL ACCESSION NO.
California	Tomato Black Mold	CA-790124
California	Tomato Water Mold	CA-800107
Connecticut	Tobacco Blue Mold	CT-800002
Georgia	Corn	07618
Georgia	Tobacco Blue Mold	GA-800007
Hawaii	Ornamentals	HI-830006
Hawaii	Taro	07617
Idaho	Fungigation, Potatoes	ID-790030
Kentucky	Tobacco Blue Mold	KY-840001
Maryland	Tobacco Blue Mold	MD-800007
Michigan	Fungigation, Botrytis, Potatoes	MI-790007
Michigan	Fungigation, Tomatoes	MI-800004
Minnesota	Wild Rice	07618
Missouri	Tobacco Blue Mold	MO-820010
North Carolina	Tobacco Blue Mold	NC-800009
Ohio	Tobacco Blue Mold	OH-810010
Oklahoma	Web Blotch, Peanuts	OK-770011
Oregon	Storage Rot, Pears	OR-770059
Oregon	Swiss Needle Cast on Douglas Fir	OR-810089
Oregon	Onion Smut	OR-820009
Pennsylvania	Scots Pine	PA-770007
Pennsylvania	Tobacco Blue Mold	PA-800013
South Carolina	Tobacco Blue Mold	SC-800006
Tennessee	Tobacco Blue Mold	TN-800013
Texas	With Benlate, Sprinkler Irrigation, Peanuts	TX-790026

Texas	Web Blotch, Peanuts	TX-770012
Virginia	Tobacco Blue Mold	VA-800015
Washington	Bull's Eye Rot and Storage Rot, Pears	WA-770048
Washington	Bulb Crops	WA-780067

GOAL®
1.6E HERBICIDE

USE DIRECTIONS FOR:

treefruit/nut/grape • conifer • corn • cotton • fallow bed • globe artichoke • mint • nonbearing citrus • onion • soybean

ACTIVE INGREDIENT	
Oxyfluorfen	
2-chloro-1-(3-ethoxy-4-nitrophenoxy)-4-(trifluoromethyl) benzene	19.4%*
INERT INGREDIENTS	80.6%
TOTAL	100.0%

*Equivalent to 1.6 lbs. active ingredient per gallon

SHAKE WELL BEFORE USING.

EPA Reg. No. 707-174
U.S. Pat. No. 3,798,276

Keep Out of Reach of Children
WARNING AVISO

PRECAUCION AL USUARIO: Si usted no lee ingles, no use este producto hasta que le etiqueta haya sido explicado ampliamente.

NOTICE: Before using this product, read the entire Precautionary Statements, Conditions of Sale and Warranty, Directions for Use, Use Restrictions, and Storage and Disposal Instructions. If the Conditions of Sale and Warranty are not acceptable, return the product unopened within thirty days of purchase to the place of purchase.

DIRECTIONS FOR USE

It is a violation of Federal law to use this product in a manner inconsistent with its labeling.

PRECAUTIONARY STATEMENTS

HAZARDS TO HUMANS AND DOMESTIC ANIMALS

Causes substantial but temporary eye injury and skin irritation. Do not get in eyes, on skin, or on clothing. Wear goggles or face shield. This material may be harmful if swallowed. Wash thoroughly with soap and water after handling. Remove contaminated clothing and wash before re-use.

STATEMENT OF PRACTICAL TREATMENT

IF IN EYES: Flush eyes with a large amount of water for at least 15 minutes. Get prompt medical attention.

IF ON SKIN: Wash thoroughly with plenty of soap and water. Get medical attention.

IF SWALLOWED: Do not induce vomiting. Dilute by giving 2 glasses of water to drink. Call a physician. Do not induce vomiting or give anything by mouth to an unconscious person.

IF INHALED: Move victim to fresh air.

NOTE TO PHYSICIAN: Contains petroleum distillates. Aspiration hazard may contraindicate the use of gastric lavage.

ENVIRONMENTAL HAZARDS

Do not apply directly to water or wetlands. Do not contaminate water by cleaning of equipment or disposal of wastes.

This product is highly toxic to aquatic invertebrates, aquatic plants, wildlife and fish. Use with care when applying in areas frequented by wildlife or adjacent to any body of water or wetland area. Do not apply when weather conditions favor drift or erosion from target areas.

PHYSICAL OR CHEMICAL HAZARDS

Combustible.

Do not use or store near heat or open flame.

CONDITIONS OF SALE AND WARRANTY

Rohm and Haas warrants that the product conforms to its chemical description and is reasonably fit for the purpose stated on the label only when used in accordance with label directions under normal conditions of use. **ROHM AND HAAS MAKES NO OTHER EXPRESS OR IMPLIED WARRANTIES EITHER OF MERCHANTABILITY OR FITNESS FOR A PARTICULAR USE.** Handling, storage and use of the product by Buyer or User are beyond the control of Rohm and Haas and Seller. Risks such as crop injury, ineffectiveness or other unintended consequences resulting from, but not limited to, weather or soil conditions, presence of other materials, disease, pests, drift to other crops or failure to follow label directions will be assumed by the Buyer or User. **IN NO CASE WILL ROHM AND HAAS OR SELLER BE HELD LIABLE FOR CONSEQUENTIAL, SPECIAL OR INDIRECT DAMAGES RESULTING FROM THE HANDLING, STORAGE OR USE OF THIS PRODUCT.**

RE-ENTRY AND WORKER PROTECTION STATEMENTS

Do not enter treated areas without protective clothing until sprays have dried. Because certain states may require more restrictive reentry intervals for various crops treated with this product, consult your State Department of Agriculture for further information. Do not apply this product in such a manner as to directly or through drift expose workers or other persons. The area being treated must be vacated by unprotected persons. Written or oral warnings must be given to workers who are expected to be in a treated area or in an area about to be treated with this product. Oral warnings must be given if there is reason to believe that written warnings cannot be understood by workers. When oral warnings are given, warnings shall be given in a language customarily understood by workers. Written warnings must include the following information:

WARNING

(Insert area or field description) treated with GOAL 1.6E herbicide on (insert date of application). Do not enter treated areas without protective clothing until sprays have dried. In case of accidental exposure: Call a doctor (physician), clinic or hospital immediately. Explain that the victim has been exposed to GOAL 1.6E herbicide and describe his condition. For further information see the "Statement of Practical Treatment." Oral warnings must include the same information that is in the written warnings.

STORAGE AND DISPOSAL

Do not contaminate water, food or feed by storage or disposal.

STORAGE: Keep from freezing. Store above 32° F.

PESTICIDE DISPOSAL: Pesticide wastes are toxic. Improper disposal of excess pesticide, spray mixture or rinsate is a violation of Federal law. If these wastes cannot be disposed of by use according to label instructions, contact your State Pesticide or Environmental Control Agency, or the Hazardous Waste representative at the nearest EPA Regional Office for guidance.

CONTAINER DISPOSAL

Plastic Containers: Triple rinse (or equivalent). Puncture and dispose of in a sanitary landfill, or incinerate, or, if allowed by State and local authorities, by burning. If burned, stay out of smoke.

Metal Containers: Triple rinse (or equivalent). Then offer for recycling or reconditioning, or puncture and dispose of in a sanitary landfill, or by other procedures approved by State and local authorities.

STEPS TO BE TAKEN IN CASE MATERIAL IS RELEASED OR SPILLED: Ventilate area. Avoid breathing vapors. Use MSHA/NIOSH self-contained breathing apparatus or airline respirator for large spills in confined areas. Dike the spill with inert material (sand, earth, etc.) and transfer the liquid or solid diking material to separate containers for recovery or disposal. Remove contaminated clothing promptly and wash exposed skin areas with soap and water. Wash clothing before re-use. Keep spill out of all sewers and bodies of water.

GENERAL CLASSIFICATION

GOAL 1.6E herbicide is an emulsifiable concentrate containing 1.6 lb. of active ingredient per gallon. It is useful in preemergence, postemergence or post-directed applications.

DIRECTIONS FOR USE
TREEFRUIT

DORMANT APPLICATION

—ARIZONA AND CALIFORNIA—ALMOND, APRICOT, CHERRY, FIG, NECTARINE, PEACH, PEAR, PISTACHIO, PLUM, PRUNE, WALNUT AND GRAPE

GENERAL INFORMATION

GOAL 1.6E herbicide is effective as a preemergence and/or postemergence herbicide when used alone, or in recommended combinations, for the control of certain annual broadleaf weeds, in all almond, apricot, cherry, fig, nectarine, peach, pear, pistachio, plum, prune, walnut and grape plantings. The most effective postemergence weed control is achieved when GOAL 1.6E herbicide is applied to seedling weeds (less than 4 inches in height). Preemergence control is most effective when spray is applied to clean, weed-free established soil surfaces. Treated berms or soil surfaces should not be disked or disturbed in any manner as the herbicidal effectiveness of GOAL 1.6E may be decreased. Seedling weeds are controlled as they come in contact with the soil-applied herbicide during emergence.

For postemergence control of certain grassy and broadleaf weeds, a tank mixture of either paraquat (Gramoxone® or Ortho® paraquat) or Roundup® with GOAL 1.6E herbicide can be used.

For residual grass control in dormant almond, apricot, cherry, fig, nectarine, peach, pear, pistachio, plum, prune, walnut and grape, a tank mixture of Devrinol®, KERB®, simazine or Surflan® with GOAL 1.6E herbicide or combinations of GOAL 1.6E herbicide plus paraquat or Roundup can be used.

GOAL 1.6E HERBICIDE USED ALONE

DOSAGE

GOAL 1.6E herbicide is recommended for postemergence control at 2.5 to 10 pints (0.5 to 2.0 lb. active) per broadcast acre. The lower rate is recommended for the control of susceptible seedling weeds less than 4 inches in height. The higher rate (2.0 lb. active) should be used for weeds more than 12 inches in height. For preemergence control of susceptible weeds, use 10.0 pints (2.0 lb. active) per broadcast acre.

WEEDS CONTROLLED POSTEMERGENCE

CHEESEWEED (MALVA)	MINERSLETTUCE
FIDDLENECK, COAST	NETTLE, BURNING
FILAREE, BROADLEAF	PIGWEED, REDROOT
FILAREE, REDSTEM	REDMAIDS
GROUNDSEL, COMMON	SHEPHERDSPURSE
HENBIT	SOWTHISTLE, ANNUAL

WEEDS CONTROLLED PREEMERGENCE

BURCLOVER	LAMBSQUARTERS, COMMON
CHEESEWEED (MALVA)	LETTUCE, PRICKLY
FIDDLENECK, COAST	PIGWEED, REDROOT
FILAREE, BROADLEAF	LONDON ROCKET
FILAREE, REDSTEM	PURSLANE, COMMON
GROUNDSEL, COMMON	REDMAIDS
HENBIT	SHEPHERDSPURSE
KNOTWEED, PROSTRATE	SOWTHISTLE, ANNUAL

EARLY POSTEMERGENCE (weeds up to 4 inches high) apply 2.5 pints (0.5 lb. active) of GOAL 1.6E herbicide per broadcast acre*
CHEESEWEED (MALVA); FIDDLENECK, COAST; FILAREE, REDSTEM; HENBIT; REDMAIDS; SHEPHERDSPURSE

POSTEMERGENCE (weeds up to 8 inches high) apply 5 pints (1.0 lb. active) of GOAL 1.6E herbicide per broadcast acre*
CHEESEWEED (MALVA); FIDDLENECK, COAST; FILAREE, REDSTEM; HENBIT; MINERSLETTUCE; NETTLE, BURNING; REDMAIDS; SHEPHERDSPURSE; SOWTHISTLE, ANNUAL

LATE POSTEMERGENCE (weeds up to 12 inches high) apply 10 pints (2.0 lb. active) of GOAL 1.6E herbicide per broadcast acre*
CHEESEWEED (MALVA); FIDDLENECK, COAST; FILAREE, BROADLEAF; FILAREE, REDSTEM; GROUNDSEL, COMMON; HENBIT; MINERSLETTUCE; NETTLE, BURNING; PIGWEED, REDROOT; REDMAIDS; SHEPHERDSPURSE; SOWTHISTLE, ANNUAL

PREEMERGENCE apply 10 pints (2.0 lb. active) of GOAL 1.6E herbicide per broadcast acre*
BURCLOVER; CHEESEWEED (MALVA); FIDDLENECK, COAST; FILAREE, BROADLEAF; FILAREE, REDSTEM; GROUNDSEL, COMMON; HENBIT; KNOTWEED, PROSTRATE; LAMBSQUARTERS, COMMON; LETTUCE, PRICKLY; PIGWEED, REDROOT; PURSLANE, COMMON; REDMAIDS; ROCKET, LONDON; SHEPHERDSPURSE; SOWTHISTLE, ANNUAL

* Dosages listed are for broadcast application. For banded application, the amount of GOAL 1.6E herbicide used per acre should be reduced according to the following formula:

$$\frac{\text{Band Width (in.)}}{\text{Row Width (in.)}} \times \text{Rate per Broadcast Acre} = \text{Amount Needed per Acre for Banded Application}$$

TIMING AND METHOD OF APPLICATION

Treatments should be made OCTOBER 1 through FEBRUARY 15. In Coachella Valley do not apply to grapes after FEBRUARY 1.
As a preemergence treatment, apply in 50 to 200 gallons of water per acre. Use higher volumes to assure adequate coverage in high densities of emerged weeds or heavy trash. Best preemergence results are achieved when spray is applied to a relatively weed-free established berm or soil surface. GOAL 1.6E herbicide should be directed to the soil and the base of dormant trees or vines. Use a low-pressure sprayer equipped with a breakaway boom and flat fan or off-center (OC) nozzles. Spray equipment should be calibrated carefully before each use. See SPECIFIC USE RESTRICTIONS FOR GOAL 1.6E HERBICIDE USED ALONE for applications on dormant tree or vine plantings.

SPRAY VOLUME:

Weed Stage	Gallons of Water Per Acre
Early Postemergence (weeds up to 4 inches high)	50
Postemergence (weeds up to 8 inches high)	100 to 200
Late Postemergence (weeds up to 12 inches high)	100 to 200
Preemergence	50 to 200

MIXING DIRECTIONS: Fill the spray tank at least one-third full of clean water and add the recommended amount of GOAL 1.6E herbicide while the pump and agitator are running. Complete filling of the spray tank with water and then add 1 quart of a nonionic surfactant such as TRITON AG-98™, or comparable 80% active nonionic surfactant cleared for application to growing crops, per each 100 gallons of spray. Maintain agitation until spraying is completed.

CULTURAL CONSIDERATIONS: In order to provide maximum effectiveness of preemergence activity of GOAL 1.6E herbicide, the berm or soil surface should be smooth and free of crop or weed trash (decaying leaves, clippings, dead weeds, etc.). Leaves and trash may be removed by blowing the area to be treated or by thoroughly mixing the trash into the soil through cultivation prior to herbicide application.

For best preemergence activity, at least one-quarter inch ($\frac{1}{4}$ inch) of water should be applied within 3 to 4 weeks after application if no rainfall occurs.

Cultural practices that result in redistribution or disturbance of the soil surface after treatment will decrease the herbicidal effectiveness of GOAL 1.6E. Cutting water furrows or cultivations that mix untreated soil into treated areas will also reduce the effectiveness of the treatment. The best results are from applications to established berms or soil surfaces that are left undisturbed during the time period for which weed control is desired.

TREEFRUIT

SPECIFIC USE RESTRICTIONS

GOAL 1.6E HERBICIDE

USED ALONE

The following use restrictions should be observed when GOAL 1.6E herbicide is used alone or in any spray combination recommended on this label. In addition to the following, also observe **GENERAL USE RESTRICTIONS** listed at the end of this label.

- Dormancy determines the suitability and safety of using GOAL 1.6E herbicide on treefruit/grapes. Should bud swell occur prior to the dates listed, GOAL 1.6E herbicide must not be used.
- **DO NOT APPLY GOAL 1.6E HERBICIDE (EITHER ALONE OR IN ANY COMBINATION RECOMMENDED ON THIS LABEL) TO TREE OR VINE PLANTINGS BEFORE OCTOBER 1 OR AFTER FEBRUARY 15. DO NOT APPLY GOAL 1.6E HERBICIDE (EITHER ALONE OR IN ANY COMBINATION RECOMMENDED ON THIS LABEL) TO GRAPES IN THE COACHELLA VALLEY (CALIFORNIA) BEFORE OCTOBER 1 OR AFTER FEBRUARY 1.**
- Do not apply to grapes established less than 3 years.
- Apply only to dormant fruit or nut trees or grape vines. Do not apply after buds start to swell or when foliage or fruit/nuts are present.
- Do not apply more than 10 pints (2.0 lb. active) per broadcast acre of GOAL 1.6E herbicide in one season.
- Do not apply to grapes that are not staked or trellised, unless grapes are free-standing.
- GOAL 1.6E herbicide or any of the combinations recommended on this label should be applied to only healthy growing trees or vines.
- Direct spray toward the base of tree or vines. Avoid direct plant contact.

TANK MIXES WITH GOAL 1.6E HERBICIDE

For preemergence control of susceptible grassy weeds in certain dormant treefruit, nut crops and grapes, a tank mix of Devrinol, KERB, simazine, or Surflan can be applied.

For postemergence control of certain grassy and broadleaf weeds, a tank mix of either paraquat or Roundup with GOAL 1.6E herbicide or combinations of GOAL 1.6E herbicide plus Devrinol, KERB, simazine or Surflan can be used.

IMPORTANT: Read and observe all label directions before using. When tank mixing, always read all individual manufacturers' labels. In interpreting all labels for the tank mixture, the most restrictive situations must apply.

DOSAGE

Refer to the following table for labeled use rates

RATE OF PRODUCT PER ACRE

CROP	GOAL 1.6E	KERB 50-W	Devrinol 50-WP	simazine 80-W*	Surflan 75-W**	paraquat	Roundup
Almond	2.5 to 10 pts.	Not labeled	8 lbs.	$1\frac{1}{4}$ to $2\frac{1}{2}$ lbs.***	$2\frac{2}{3}$ to $5\frac{1}{3}$ lbs.	1 qt.	1 to 2 qts.
Apricot	2.5 to 10 pts.	Not labeled	8 lbs.	Not labeled	$2\frac{2}{3}$ to $5\frac{1}{3}$ lbs.	1 qt.	1 to 2 qts.
Cherry	2.5 to 10 pts.	2 to 8 lbs.	8 lbs.	2 to 5 lbs.	$2\frac{2}{3}$ to $5\frac{1}{3}$ lbs.	1 qt.	1 to 2 qts.
Fig	2.5 to 10 pts.	Not labeled	8 lbs.	Not labeled	$2\frac{2}{3}$ to $5\frac{1}{3}$ lbs.	1 qt.	1 to 2 qts.
Nectarine	2.5 to 10 pts.	2 to 8 lbs.	8 lbs.	Not labeled	$2\frac{2}{3}$ to $5\frac{1}{3}$ lbs.	1 qt.	1 to 2 qts.
Peach	2.5 to 10 pts.	2 to 8 lbs.	8 lbs.	$1\frac{1}{4}$ to $2\frac{1}{2}$ lbs.***	$2\frac{2}{3}$ to $5\frac{1}{3}$ lbs.	1 qt.	1 to 2 qts.
Pear	2.5 to 10 pts.	2 to 8 lbs.	8 lbs.	$2\frac{1}{2}$ to 5 lbs.	$2\frac{2}{3}$ to $5\frac{1}{3}$ lbs.	1 qt.	1 to 2 qts.
Pistachio	2.5 to 10 pts.	Not labeled	8 lbs.	Not labeled	$2\frac{2}{3}$ to $5\frac{1}{3}$ lbs.	1 qt.	1 to 2 qts.
Plum	2.5 to 10 pts.	2 to 8 lbs.	8 lbs.	2 to 5 lbs.	$2\frac{2}{3}$ to $5\frac{1}{3}$ lbs.	1 qt.	1 to 2 qts.
Prune	2.5 to 10 pts.	2 to 8 lbs.	8 lbs.	Not labeled	$2\frac{2}{3}$ to $5\frac{1}{3}$ lbs.	1 qt.	1 to 2 qts.
Walnut	2.5 to 10 pts.	Not labeled	8 lbs.	$2\frac{1}{2}$ to 5 lbs.	$2\frac{2}{3}$ to $5\frac{1}{3}$ lbs.	1 qt.	1 to 2 qts.
Grape	2.5 to 10 pts.	2 to 8 lbs.	8 lbs.	$2\frac{1}{2}$ to 6 lbs.	$2\frac{2}{3}$ to $5\frac{1}{3}$ lbs.	1 qt.	1 to 2 qts.

* GOAL 1.6E herbicide can also be tank mixed with Princep® 80-W, Princep 4L and Caliber® 90. When tank mixing with these formulations, adjust the amount of product used to conform with labeled recommendations.

** GOAL 1.6E herbicide can also be tank mixed with Surflan A.S. Multiply the rate of Surflan 75-W by 0.75 to obtain the number of quarts of Surflan A.S. required.

*** Per treated acre in a 2 to 4 ft. band on each side of the tree row.

Herbicide rates listed are for broadcast application unless otherwise noted. For banded application, the amount of herbicide used per acre should be reduced as outlined under GOAL 1.6E HERBICIDE USED ALONE.

For preemergence control of susceptible weeds, use 10 pints (2.0 lb. active) of GOAL 1.6E herbicide per broadcast acre.

WEEDS CONTROLLED PREEMERGENCE

In addition to the weeds controlled preemergence by GOAL 1.6E herbicide used alone, control of the following weeds is also obtained when tank mixed with the products listed below:

KERB 50-W

BLUEGRASS, ANNUAL
BLUEGRASS, KENTUCKY
CHICKWEED, COMMON
ORCHARDGRASS
QUACKGRASS
RYEGRASS, PERENNIAL

Devrinol 50-WP

BARNYARDGRASS (WATERGRASS)
BLUEGRASS, ANNUAL
BROME, RIPGUT
CHICKWEED, COMMON
CRABGRASS, LARGE (HAIRY)
PINEAPPLEWEED

simazine

BARNYARDGRASS (WATERGRASS)
BLUEGRASS, ANNUAL
CHICKWEED, COMMON
CRABGRASS, LARGE (HAIRY)
CRABGRASS, SMOOTH
HORSEWEED (MARESTAIL)

Surflan

BARNYARDGRASS (WATERGRASS)
BLUEGRASS, ANNUAL
CHICKWEED, COMMON
CRABGRASS, LARGE (HAIRY)
CRABGRASS, SMOOTH
CUPGRASS, SOUTHWESTERN

WEEDS CONTROLLED POSTEMERGENCE

In addition to the weeds controlled postemergence by GOAL 1.6E herbicide used alone, control of the following weeds is also obtained when tank mixed with the products listed below:

Roundup or paraquat

BARNYARDGRASS (WATERGRASS)
BLUEGRASS, ANNUAL
BROME, RIPGUT
CHICKWEED, COMMON
HORSEWEED (MARESTAIL)
ROCKET, LONDON

APPLICATION
As a preemergence treatment, apply in 50 to 200 gallons of water per acre, depending upon density of emerged weeds. Best preemergence results are achieved when spray is applied to a relatively weed-free established berm of soil surface. Tank mixtures containing GOAL 1.6E herbicide should be directed to the soil and the base of trees or dormant grape vines. Use a low-pressure sprayer equipped with a breakaway boom and flat fan or off-center (OC) nozzles. Spray equipment should be calibrated carefully before each use.

TANK MIXING DIRECTIONS: Fill the spray tank at least one-third full of clean water. With the pump and agitator running, add the recommended amount of Devrinol, KERB, simazine or Surflan to the spray tank. Then add the recommended amounts of GOAL 1.6E herbicide and paraquat of Roundup during the filling operation. Complete filling of the spray tank with water and then add 1 quart of a nonionic surfactant such as TRITON AG-98 or comparable 80% active nonionic surfactant cleared for application to growing crops, per each 100 gallons of spray. Maintain agitation until spraying is completed.

CULTURAL CONSIDERATIONS: In addition to details previously listed under "GOAL 1.6E HERBICIDE USED ALONE," read all individual manufacturers' labels. In interpreting all labels for the tank mixture, the most restrictive cultural situations must apply.

TREEFRUIT
ARIZONA AND CALIFORNIA
SPECIFIC USE RESTRICTIONS
GOAL 1.6E HERBICIDE TANK MIXES

In addition to the following, also observe all use restrictions previously listed under "GOAL 1.6E HERBICIDE USED ALONE" and GENERAL USE RESTRICTIONS listed at the end of this label.

- When tank mixing, always read all individual manufacturers' labels. In interpreting all labels for the tank mixture, the most restrictive situations must apply.

DIRECTIONS FOR USE
TREEFRUIT

DORMANT APPLICATION
ALL STATES (EXCEPT ARIZONA AND CALIFORNIA)
CHERRY, GRAPE, PEACH

GENERAL INFORMATION

GOAL 1.6E herbicide is effective as a preemergence and/or postemergence herbicide when used alone or in recommended tank mix combinations, for the control of certain annual broadleaf weeds, in bearing and nonbearing cherry, peach and grape plantings.

The most effective postemergence weed control is achieved when GOAL 1.6E herbicide is applied to seedling weeds (less than 4 leaf stage). For postemergence control of certain grassy and broadleaf weeds, a tank mixture of GOAL 1.6E herbicide with either paraquat (Gramoxone or Ortho paraquat) or Roundup can be used.

Preemergence control is most effective when spray is applied to clean, weed-free oil surfaces. Treated berms or soil surfaces should not be disked or distributed in any manner as the herbicidal effectiveness of GOAL 1.6E may be decreased. Seedling weeds are controlled as they come in contact with the soil-applied herbicide during emergence. For residual grass control in dormant treefruit or vine plantings, a tank mixture of GOAL 1.6E herbicide with Devrinol, KERB, simazine or Surflan can be used. Contact herbicides such as paraquat or Roundup may also be added to the tank mixture.

Check individual product labels to determine suitability and use rates for various crops.

GOAL 1.6E HERBICIDE USED ALONE
DOSAGE

GOAL 1.6E herbicide is recommended for postemergence control at 2.5 to 10.0 pints* (0.5 to 2.0 lb. active) per broadcast acre. The lower rate is recommended for the control of susceptible seedling weeds up to the 4 leaf stage. The higher rate (2.0 lb. active) should be used for weeds up to the 6 leaf stage. For preemergence control of susceptible weeds, use 10.0 pints (2.0 lb. active) per broadcast acre.

* Dosages listed are for broadcast application. For banded application, the amount of GOAL 1.6E herbicide used per acre should be reduced according to the following formula:

$$\frac{\text{Band Width (in.)}}{\text{Row Width (in.)}} \times \text{Rate per Broadcast Acre} = \text{Amount Needed per Acre for Banded Application}$$

WEEDS CONTROLLED POSTEMERGENCE

BALSAMAPPLE	NIGHTSHADE, AMERICAN BLACK
COCKLEBUR, COMMON	NIGHTSHADE, BLACK
*CUDWEED, NARROWLEAF	PEPPERWEED, VIRGINIA
**EVENINGPRIMROSE, CUTLEAF	PIGWEED, REDROOT
GROUNDCHERRY, CUTLEAF	POINSETTIA, WILD
GROUNDCHERRY, WRIGHT	PURSLANE, COMMON
JIMSONWEED	SESBANIA, HEMP
LAMBSQUARTERS, COMMON	SHEPHERDSPURSE
MORNINGGLORY, ANNUAL	SIDA, PRICKLY (TEA WEED)
	SMARTWEED, PENNSYLVANIA
	SOWTHISTLE, ANNUAL
	VELVETLEAF

WEEDS CONTROLLED PREEMERGENCE

CAMPHORWEED	NIGHTSHADE, BLACK
CUDWEED, NARROW-LEAF	PEPPERWEED, VIRGINIA
**EVENINGPRIMROSE, CUTLEAF	PIGWEED, REDROOT
GROUNDCHERRY, CUTLEAF	POINSETTIA, WILD
JIMSONWEED	SIDA, PRICKLY
LAMBSQUARTERS, COMMON	SMARTWEED, PENNSYLVANIA
NIGHTSHADE AMERICAN BLACK	SOWTHISTLE, ANNUAL
	SPURGE, PROSTRATE
	SPURGE, SPOTTED
	VELVETLEAF

* Maximum 0.5 inch diameter.

** Highest rate and/or multiple applications may be required for acceptable control. Do not apply more than 10 pints (2.0 lb. active) per broadcast acre of GOAL 1.6E herbicide in one season.

TIMING AND METHOD OF APPLICATION

Dormancy determines the suitability and safety of using GOAL 1.6E herbicide on cherry, peach and grape plantings. Do not apply GOAL 1.6E herbicide after buds start to swell or when foliage or fruit are present.

As a preemergence treatment, apply in a minimum of 40 gallons of water per acre. Use higher volumes to assure adequate coverage in high densities of emerged weeds or heavy trash. Best preemergence results are achieved when spray is applied to a relatively weed-free soil surface or berm. GOAL 1.6E herbicide should be directed to the soil and the base of dormant trees or vines. Use a low-pressure sprayer equipped with a breakaway boom and flat fan nozzles. An off-center (OC) nozzle positioned at the end of the boom may be desired. Spray equipment should be calibrated carefully before each use. See SPECIFIC USE RESTRICTIONS for GOAL 1.6E herbicide application on dormant tree or vine plantings.

SPRAY VOLUME: Weed Stage	Gals. of Water per Acre
Early Postemergence (weeds up to 4 leaf stage)	40 or more
Postemergence (weeds up to 6 leaf stage)	100 or more
Preemergence	40 or more

MIXING DIRECTIONS

Fill the spray tank at least one-third full of clean water and add the recommended amount of GOAL 1.6E herbicide while the pump and agitator are running. Complete filling of the spray tank with water and then add 1 quart of a nonionic surfactant such as TRITON AG-

98, or comparable 80% active nonionic surfactant cleared for application to growing crops, per each 100 gallons of spray. Maintain agitation until spraying is completed.

CULTURAL CONSIDERATIONS

In order to provide maximim effectiveness of preemergence activity of GOAL 1.6E herbicide, the berm or soil surface should be smooth and free of crop or weed trash (decaying leaves, clippings, dead weeds, etc.). Leaves and trash may be removed by blowing the area to be treated or by thoroughly mixing the trash into the soil through cultivation prior to herbicide application.

For best preemergence activity, at least one-quarter inch (1/4 inch) of irrigation water should be applied within 3 to 4 weeks after application if no rainfall occurs.

Cultural practices that result in redistribution or disturbance of the soil surface after treatment will decrease the herbicidal effectiveness of GOAL 1.6E. Cutting water furrows or cultivations that mix untreated soil into treated areas will also reduce the effectiveness of the treatment. The best results are from applications to soil surfaces or berms that are left undisturbed during the time period for which weed control is desired.

TREEFRUIT
ALL STATES (EXCEPT ARIZONA AND CALIFORNIA)
SPECIFIC USE RESTRICTIONS
GOAL 1.6E HERBICIDE

The following specific use restrictions should be observed when GOAL 1.6E herbicide is used alone or in any tank mix spray combination recommended on this label. In addition to the following, also observe GENERAL USE RESTRICTIONS listed at the end of this label.

- Read and observe all label directions before using. When tank mixing, always read all individual manufacturers' labels. In interpreting all labels for the tank mixture, the most restrictive situations must apply.
- Do not apply to grapes established less than 3 years unless vines are on the trellis wire a minimum of 3 feet above the soil surface.
- Apply only to dormant fruit or vines. Dormancy determines the suitability and safety of using GOAL 1.6E herbicide on treefruit and vines. Do not apply after buds start to swell or when foliage or fruit/nuts are present.
- Do not apply more than 10 pints (2.0 lb. active) per broadcast acre of GOAL 1.6E herbicide in one season.
- Do not apply to grapes that are not staked or trellised unless grapes are free-standing.
- GOAL 1.6E herbicide or any of the tank mix combinations recommended on this label should be applied to only healthy growing trees or vines.

TANK MIXES WITH GOAL 1.6E HERBICIDE

IMPORTANT: Read and observe all label directions before using. When tank mixing, always read all individual manufacturers' labels. In interpreting all labels for tank mixture, the most restrictive situations must apply.

DOSAGE

For postemergence control of susceptible grassy and broadleaf weeds in certain dormant treefruit and vines, a tank mix of GOAL 1.6E herbicide with either paraquat or Roundup can be used. Apply at the recommended rates and growth stages to susceptible weed species in a manner described on the respective labels.

For preemergence control of susceptible grassy weeds in certain dormant treefruit and vines, a tank mixture of GOAL 1.6E herbicide with Devrinol, KERB, simazine, or Surflan may be applied. Contact herbicides such as paraquat or Roundup may be added to the tank mixture. Apply at the recommended rates and growth stages to susceptible weed species in a manner described on the respective labels.

RATE OF PRODUCT PER ACRE

CROP	GOAL 1.6E	KERB 50-W	Devrinol 50-WP	Simazine 80-W*	Surflan 75-W**	Paraquat	Roundup
Cherry	2.5 to 10 pts	2 to 8 lbs.	8 lbs.	2 to 5 lbs.	2⅔ to 5⅓ lbs.	1 qt.	1 to 2 qts.
Grape	2.5 to 10 pts.	2 to 8 lbs.	8 lbs.	2½ to 6 lbs.	2⅔ to 5⅓ lbs.	1 qt.	1 to 2 qts.
Peach	2.5 to 10 pts.	2 to 8 lbs.	8 lbs.	2 to 5 lbs.	2⅔ to 5⅓ lbs.	1 qt.	1 to 2 qts.

* GOAL 1.6E herbicide can also be tank mixed with Princep 80-W, Princep 4L or Caliber 90. When tank mixing with these formulations, adjust the amount of product used to conform with labeled recommendations.

** GOAL 1.6E herbicide can also be tank mixed with Surflan A.S. Multiply the rate of Surflan 75-W by 0.75 to obtain the number of quarts of Surflan A.S. required.

Herbicide rates listed are for broadcast application. For banded application, the amount of herbicide used per acre should be reduced as outlined under GOAL 1.6E HERBICIDE USED ALONE.

For preemergence control of susceptible weeds, use 10 pints (2.0 lb. active) of GOAL 1.6E herbicide per broadcast acre.

WEEDS CONTROLLED PREEMERGENCE

In addition to the weeds controlled preemergence by GOAL 1.6E herbicide used alone, control of the following weeds is also obtained when tank mixed with the products listed below:

KERB 50-W
BLUEGRASS, ANNUAL
BLUEGRASS, KENTUCKY
CHICKWEED, COMMON
ORCHARDGRASS
QUACKGRASS
RYEGRASS, PERENNIAL

Devrinol 50-WP
BARNYARDGRASS (WATERGRASS)
BLUEGRASS, ANNUAL
BROME, RIPGUT
CHICKWEED, COMMON
CRABGRASS, LARGE (HAIRY)
PINEAPPLEWEED

simazine
BARNYARDGRASS (WATERGRASS)
BLUEGRASS, ANNUAL
CHICKWEED, COMMON
CRABGRASS, LARGE (HAIRY)
CRABGRASS, SMOOTH
HORSEWEED (MARESTAIL)

Surflan
BARNYARDGRASS (WATERGRASS)
BLUEGRASS, ANNUAL
CHICKWEED, COMMON
CRABGRASS, LARGE (HAIRY)
CRABGRASS, SMOOTH
CUPGRASS, SOUTHWESTERN

WEEDS CONTROLLED POSTEMERGENCE

In addition to the weeds controlled postemergence by GOAL 1.6E herbicide used alone, control of the following weeds is also obtained when tank mixed with the products listed below:

Roundup or paraquat
BARNYARDGRASS (WATERGRASS)
BLUEGRASS, ANNUAL
BROME, RIPGUT
CHICKWEED, COMMON
HORSEWEED (MARESTAIL)
ROCKET, LONDON

APPLICATION

As a preemergence and/or postemergence treatment, apply in a minimum of 40 gallons of water per acre depending upon density of emerged weeds. Spray coverage is essential for good postemergence weed results. Best preemergence results are achieved when spray is applied to a relatively weed-free soil surface or berm. Tank mixtures containing GOAL 1.6E herbicide should be directed to the soil and the base of trees or dormant vines. Use a low-pressure sprayer equipped with a breakaway boom and flat fan nozzles. An off-center (OC) nozzle positioned at the end of the boom may be desired. Spray equipment should be calibrated carefully before each use.

TANK MIXING DIRECTIONS

Fill the spray tank at least one-third full of clean water. With the pump and agitator running, add the recommended amount of Devrinol, KERB, simazine or Surflan to the spray tank. Then add the recommended amounts of GOAL 1.6E herbicide and paraquat or Roundup during the filling operation. Complete filling of the spray tank with water and then add 1 quart of a nonionic surfactant such as TRITON AG-98 or comparable 80% active nonionic surfactant cleared for application to growing crops, per each 100 gallons of spray. Maintain agitation until spraying is completed.

CULTURAL CONSIDERATIONS

In addition to details previously listed under GOAL 1.6E HERBICIDE USED ALONE, read all individual manufacturers' labels. In interpreting all labels for the tank mixture, the most restrictive label specifications must apply.

TREEFRUIT
(ALL STATES EXCEPT ARIZONA AND CALIFORNIA)
SPECIFIC USE RESTRICTIONS
GOAL 1.6E HERBICIDE TANK MIXES

In addition to the following, also observe all specific use restrictions previously listed under GOAL 1.6E HERBICIDE USED ALONE and GENERAL USE RESTRICTIONS on the GOAL 1.6E herbicide label.

- When tank mixing, always read all individual manufacturer's labels. In interpreting all labels for the tank mixture, the most restrictive situations must apply.

DIRECTIONS FOR USE
CONIFER SEEDBEDS, TRANSPLANTS
CONTAINER STOCK

GENERAL INFORMATION

GOAL 1.6E is effective as a preemergence and/or postemergence herbicide for the control of certain annual grass and broadleaf weeds in conifer seedbeds, transplant and conifer stock. The most effective postemergence weed control is achieved when GOAL 1.6E herbicide is applied to seedling weeds less than 4 inches in height. Preemergence control is most effective when spray is applied to clean, weed-free soil surfaces. Treated soil surfaces should not be disturbed as the herbicidal effectiveness of GOAL 1.6E may be decreased. Seedling weeds are controlled during emergence as they come in contact with the soil-applied herbicide.

IMPORTANT: Some varieties or cultivars of conifers and ornamental species listed may be susceptible to GOAL 1.6E herbicide. Care should be taken to insure that the particular variety to be sprayed with GOAL 1.6E herbicide is tolerant. It is suggested that unfamiliar species be tested in limited areas prior to application for preemergence and postemergence weed control.

WEEDS CONTROLLED

When GOAL 1.6E herbicide is applied preemergence or postemergence at recommended dosages and weed stages, the following grasses and broadleaf weeds are controlled:

*BARNYARDGRASS
BEDSTRAW, CATCHWEED
BITTERCRESS, LESSER
*BLUEGRASS, ANNUAL
BUCKWHEAT, WILD
BURCLOVER
CARPETWEED
*CLOVER, RED
*CLOVER, WHITE
COCKLEBUR, COMMON
*CRABGRASS, LARGE
*FIDDLENECK, COAST
FILAREE, BROADLEAF
FILAREE, REDSTEM
FIREWEED (FROM SEED)
FLIXWEED
*FOXTAIL, GIANT
*GOOSEGRASS
GROUNDCHERRY, CUTLEAF
GROUNDCHERRY, WRIGHT
MUSTARD, BLUE
MUSTARD,TUMBLE
MUSTARD, WILD
NETTLE, BURNING
NIGHTSHADE, BLACK
NIGHTSHADE, HAIRY
OATS, WILD
ORACH, RED
PEPPERWEED, YELLOWFLOWER
PIGWEED, PROSTRATE
PIGWEED, REDROOT
PIMPERNEL, SCARLET
PURSLANE, COMMON
REDMAIDS
ROCKET, LONDON
SANDSPURRY, RED
*SHEPHERDSPURSE
SIDA, PRICKLY
SMARTWEED, PENNSYLVANIA
SORREL, RED (FROM SEED)
GROUNDSEL, COMMON
HENBIT
JIMSONWEED
KNOTWEED, PROSTRATE
LADYTHUMB
LAMBSQUARTERS, COMMON
LETTUCE, PRICKLY
MALLOW, LITTLE
MAYWEED
MINERSLETTUCE
*MORNINGGLORY, IVYLEAF
*MORNINGGLORY, TALL
SOWTHISTLE, ANNUAL
SPEEDWELL, BIRDSEYE
**SPURGE, PROSTRATE
**SPURGE, SPOTTED
SPURRY, CORN
TANSYMUSTARD
**THISTLE, BULL
THISTLE, RUSSIAN
VELVETLEAF
WITCHGRASS
**WOODSORREL, YELLOW

* Highest rate and /or multiple applications may be required for acceptable control.

** Preemergence control only.

GOAL 1.6E herbicide is most effective when applied preemergence to annual grasses. Postemergence applications should be made to seedling grasses not exceeding the 2-leaf stage. The addition of 0.25% (2 pints/100 gallons of spray solution) of TRITON AG-98 or comparable 80% active nonionic surfactant, cleared for application on growing crops, enhances the GOAL 1.6E herbicide activity on emerged weeds.

CONIFER SEEDBEDS

To assist in the establishment of conifer seedbeds, a preemergence application should be made after seeding but prior to conifer emergence. Postemergence applications should not be made until a minimum of 5 weeks after emergence of the conifer seedlings. Additional care should be taken if cool, cloudy weather occurs during emergence to make certain that seedlings have hardened off prior to spraying.

Conifers are tolerant to preemergence and postemergence applications of GOAL 1.6E herbicide. Applied postemergence, GOAL 1.6E herbicide will provide both postemergence and residual preemergence control of many broadleaf weeds and annual grass species.

CONIFER SPECIES

GOAL 1.6E herbicide may be applied to conifer seedbeds of numerous species including the following:

DOUGLAS-FIR	*Pseudotsuga menziesii*
FIR	
FRASER	*Abies fraseri*
GRAND	*Abies grandis*
NOBLE	*Abies procera*
HEMLOCK	
EASTERN HEMLOCK	*Tsuga canadensis*
PINE	
AUSTRIAN	*Pinus nigra*
EASTERN WHITE	*Pinus stobus*
HIMALAYAN	*Pinus graffithii*
JACK	*Pinus banksiana*
LOBLOLLY	*Pinus taeda*
LODGEPOLE	*Pinus contorta*
LONGLEAF	*Pinus palustris*
MONTEREY	*Pinus radiata*
MUGHO	*Pinus mugo*
PONDEROSA	*Pinus ponderosa*
SCOTCH	*Pinus sylvestris*
SHORTLEAF	*Pinus echinata*
SLASH	*Pinus elliottii*
VIRGINIA	*Pinus virginiana*
SPRUCE	
BLUE	*Picea pungens*
DWARF ALBERTA	*Picea glauca conica*
NORWAY	*Picea abies*
SITKA	*Picea sitchensis*

PREEMERGENCE DOSAGE

Apply 1.25 to 5 pints (0.25 to 1.0 lb. active) of GOAL 1.6E herbicide per broadcast acre as a preemergence application. Where grassy weeds are present, a minimum rate of 2.5 pints (0.5 lb. active) of GOAL 1.6E herbicide per broadcast acre is suggested. In known areas of high weed competition, 5 pints (1.0 lb. active) of GOAL 1.6E herbicide per broadcast acre are recommended.

TIMING AND METHOD OF APPLICATION

Apply recommended dosage after seeding but prior to conifer emergence. GOAL 1.6E herbicide should be thoroughly mixed with clean water at recommended concentration and applied at 20 to 40 psi in a minimum of 20 gallons of water per treated acre. Broadcast to beds and irrigate prior to weed emergence with ½ to ¾ inch of sprinkler irrigation.

POSTEMERGENCE DOSAGE

Apply 1.25 to 2.5 pints (0.25 to 0.5 lb. active) of GOAL 1.6E herbicide per broadcast acre with each postemergence application. Two or three postemergence applications may be necessary for season-long weed control.

TIMING AND METHOD OF APPLICATION

Apply recommended dosage to seedbeds no sooner than 5 weeks after emergence of conifer seedlings. Additional care should be taken if cool, cloudy weather occurs during emergence to make certain that seedlings have hardened off prior to spraying. Application should be made to seedling weeds (less than 4 inches in height). GOAL 1.6E herbicide should be thoroughly mixed with clean water at recommended concentration and applied as a broadcast application of 20 to 40 psi in a minimum of 20 gallons of water per treated acre.

CONIFER TRANSPLANTS AND CONTAINER STOCK (INCLUDES 2-0 SEEDLING AND CHRISTMAS TREE PLANTINGS)

Many container-grown conifers and conifer transplants are tolerant to preemergence and postemergence applications of GOAL 1.6E herbicide. Applied postemergence, GOAL 1.6E herbicide will provide both postemergence and preemergence control of many broadleaf weeds and grasses. Postemergence applications should be applied before bud break or after foliage has had an opportunity to harden off. Conifers may be transplanted from seedbeds and sprayed directly providing bud break has not occurred.

The following conifer species in addition to species listed under the CONIFER SEEDBED section have been shown to be tolerant to GOAL 1.6E herbicide.

ARBORVITAE
Thuja occidentalis
Thuja orientalis

JUNIPER
Juniperus chinesis
Juniperus horizontalis
Juniperus procumbens
Juniperus sabina
Juniperus scopulum

RED CEDAR
Juniperus virginiana

WESTERN HEMLOCK
Tsuga heterophylla

YEW
Taxus species

DOSAGE

For preemergence or postemergence weed control apply 5 to 10 pints (1.0 to 2.0 lb. active) of GOAL 1.6E herbicide per broadcast acre.

TIMING AND METHOD OF APPLICATION

For optimum weed control, preemergence applications should be made immediately after transplanting seedlings or to weed-free container stock. Postemergence applications should be made to weeds less than 4 inches in height. Two applications may be necessary, in fall-transplanted conifer fields, for season-long weed control. The addition of 0.25% (2 pints/100 gals. of spray solution) of TRITON AG-98 or comparable 80% active nonionic surfactant, cleared for application to growing crops, enhances GOAL 1.6E herbicide activity on emerged weeds. GOAL 1.6E herbicide must be applied only to conifer transplants prior to bud break or after foliage has had an opportunity to harden off. Thoroughly mix with clean water at recommended concentration and apply at 20 to 40 psi in a minimum of 20 gallons of water per treated acre. Spray over the top of transplants. Heavy rainfall immediately following application to emerged weeds may reduce effectiveness.

CONIFER SPECIFIC USE RESTRICTIONS

In addition to the following, also observe GENERAL USE RESTRICTIONS listed at the end of this label.

- NOT FOR CONIFER RELEASE IN FOREST MANAGEMENT PROGRAMS OR FOR FOREST REGENERATION APPLICATIONS.
- Do not apply GOAL herbicide in an enclosed greenhouse structure as injury to plant foliage may result.
- Do not store or transport treated container stock in an enclosed structure until completion of 4 irrigations (minimum 21 days) as injury to non-labeled plants may occur.
- Always apply GOAL 1.6E herbicide only to healthy conifer stock. Do not apply GOAL 1.6E herbicide to conifers that are under stress from excessive fertilizer or soil salts, disease, nematodes, frost, drought, flooding, previously applied pesticides, soil insects, or winter injury, as severe injury may result.

DIRECTIONS FOR USE—CORN

For Use Only As A Directed Spray On Field Corn In Conjunction With The USDA Witchweed Eradication Program In North and South Carolina.

GENERAL INFORMATION

GOAL 1.6E is a selective herbicide for the control of witchweed (*Striga asiatica*), and works both preemergence and postemergence against witchweed.

DOSAGE

Use 2.5 to 3.75 pints of GOAL 1.6E herbicide (0.5 to 0.75 lb. active) per acre for the first application. The 2.5 pint rate (0.5 lb. active) per acre should be the standard use rate, with the 3.75 pint rate (0.75 lb. active) per acre for isolated infestations. Repeat treatments should be made at rates of 1.25 to 2.5 pints (0.25 to 0.5 lb. active) per acre. Applications should be made in 10 to 30 gallons of water per acre. Use TRITON AG-98 spreader in the spray mixture at the rate of 0.25% by water volume or 1 quart in 100 gallons of spray mix.

TIMING AND METHOD OF APPLICATION

Fields in the witchweed infested area selected for treatment with GOAL 1.6E herbicide should be examined during the early part of the growing season to determine uniformity of corn stand and grassy weed pressure. Weedy fields should be cultivated prior to the initial application so as to obtain the best possible soil coverage in the first spray application. Apply during May-August to emerged witchweed before bloom or as soon as possible after bloom appears, to avoid seed set. Corn should have a minimum height of 24 inches at the first application. After this application has been made, the fields should be inspected regularly for any breakthrough of the witchweed. If breakthrough occurs, then a second spray should be applied like the first. This application will be made postemergence to the witchweed, preferably before bloom or as soon as possible past the first appearance of witchweed bloom, to avoid seed set.

In all applications direct the GOAL 1.6E herbicide spray at the base of the corn plant and uniformly over the entire row surface. Do not spray over the top of the corn, as this may result in severe corn injury. Spray droplets contacting the lower leaves will cause necrotic spotting or streaking of sprayed tissue. Spray should contact only the lower 3 to 8 inches of the corn stalk and any leaves in this zone.

CORN—SPECIFIC USE RESTRICTIONS

- In addition to the following, also observe GENERAL USE RESTRICTIONS listed at the end of this label.
- Do not apply more than 6.25 pints of GOAL 1.6E herbicide (1.25 lb. active) per acre to a corn crop during the growing season.
- Do not apply any application within 30 days of harvest.
- Do not use corn plants from a treated field for green chop, ensilage, forage or fodder.
- Do not rotate any crops other than corn or soybeans within a 10-month period after treatment.

DIRECTIONS FOR USE—COTTON

POST-DIRECTED SPRAY

GENERAL INFORMATION

GOAL 1.6E is a selective herbicide for use as a post-directed application for broadleaf weed control in cotton. Cotton leaves that are accidently sprayed will exhibit necrotic spotting and may drop from the plant, therefore, care must be exercised to avoid spray contact with the cotton leaves. Applications made when excessive soil moisture is present may result in enhanced initial crop response.

GOAL 1.6E herbicide may be applied to cotton grown in the Southern Cotton States (Alabama, Arkansas, Georgia, Louisiana, Mississippi, Missouri, New Mexico, North Carolina, Oklahoma, South Carolina, Tennessee and Texas) and in the Western Cotton States (Arizona and California). For specific use instructions, see GEOGRAPHIC LOCATION USE DIRECTIONS.

WEEDS CONTROLLED POSTEMERGENCE

When GOAL 1.6E herbicide is applied as a post-directed application at the recommended weed stage and dosage in cotton, the following weeds are controlled:

COCKLEBUR, COMMON	NIGHTSHADE, BLACK
GROUNDCHERRY, CUTLEAF	PIGWEED, REDROOT
GROUNDCHERRY, WRIGHT	*POINSETTIA, WILD
JIMSONWEED	PURSLANE, COMMON
LAMBSQUARTERS, COMMON	SESBANIA, HEMP
MORNINGGLORY, ANNUAL	SIDA, PRICKLY (TEAWEED)
NIGHTSHADE, AMERICAN BLACK	SMARTWEED, PENNSYLVANIA
	VELVETLEAF

* Multiple applications may be required for acceptable control.

GEOGRAPHIC LOCATION USE DIRECTIONS

DOSAGE

SOUTHERN AND WESTERN COTTON

GOAL 1.6E herbicide is recommended as a post-directed application at 1.25 to 2.5 pints (0.25 to 0.5 lb. active) per acre*. Optimum control is achieved when 2.5 pints of GOAL 1.6E herbicide (0.5 lb. active) per acre* are applied to weed seedlings not exceeding 4 true leaves. Effective control of succulent weed seedlings in the 2 to 3 leaf stage can usually be obtained when 1.25 pints of GOAL 1.6E herbicide (0.25 lb. active) per acre* are applied. Weeds should be in the seedling stage, young and actively growing. Do not count cotyledon leaves.

*Dosages listed are for broadcast application. For banded application, the amount of GOAL 1.6E herbicide used per acre should be reduced according to the following formula:

$$\frac{\text{Band Width (in.)}}{\text{Row Width (in.)}} \times \text{Rate per Broadcast Acre} = \text{Amount Needed per Acre for Banded Application}$$

TIMING

SOUTHERN COTTON

(ALABAMA, ARKANSAS, GEORGIA, LOUISIANA, MISSISSIPPI, MISSOURI, NEW MEXICO, NORTH CAROLINA, OKLAHOMA, SOUTH CAROLINA, TENNESSEE AND TEXAS)

Cotton plant height must be a minimum 6 inches or greater. Application to cotton plants less than 6 inches tall may result in severe crop injury and is not recommended. In cotton 6 to 8 inches tall, GOAL 1.6E herbicide must be applied using rigid precision ground sprayer equipment. Spray shields are recommended to avoid spray contact with cotton foliage. Use branch lifters or shields if excessive spray contact on larger cotton plants (8 inches or greater) cannot be avoided by the directed spray.

WESTERN COTTON

(ARIZONA AND CALIFORNIA)

Cotton plant height must be a minimum 8 inches or greater. Use branch lifters or shields if excessive spray contact to the cotton plant cannot be avoided. To obtain the maximum benefit of postemergence activity, encourage weed emergence by irrigating prior to spraying. Irrigate within two weeks following herbicide application to obtain greatest benefit of preemergence activity from GOAL 1.6E herbicide on nightshade and groundcherry species.

METHOD OF APPLICATION

SOUTHERN AND WESTERN COTTON

Accurate, uniform placement of GOAL 1.6E herbicide spray is essential for effective weed control and to minimize cotton injury. As a directed postemergence application, GOAL 1.6E herbicide should be applied at 20 to 25 psi using 20 to 40 gallons of spray on a broadcast acre basis. Do not exceed 25 psi. Spray should be directed towards the base of the cotton plant. Cotton foliage receiving accidental spray or drift may be injured. Weeds should be in the seedling stage, young and actively growing.

Fill the spray tank at least one-third full of clean water and add the recommended amount while the pump and agitator are running. Complete filling of the spray tank with water and then add 2 to 4 pints of TRITON AG-98, or comparable 80% active nonionic surfactant cleared for use on growing crops, per each 100 gallons of spray to assist in contact activity on dusty weeds. Maintain agitation until spraying is completed.

GOAL 1.6E herbicide can be applied using a post-direct spraying rig with 2 flat flan nozzles per row, 1 nozzle on each side of the row. Additional care should be taken to insure proper spray adjustment prior to application. For best coverage, it is suggested to use 4 flat fan nozzles per row, 2 nozzles on each side of the row. The two forward nozzles should point forward and downward while the rear nozzles should point to the rear and downward. With either sprayer system, nozzles should be adjusted to cover the weed foliage with minimum contact to the cotton plant. **Do not use cone nozzles.**

COTTON—SPECIFIC USE RESTRICTIONS

SOUTHERN AND WESTERN COTTON

In addition to the following, also observe GENERAL USE RESTRICTIONS listed at the end of this label.

- **Southern Cotton:** Do not apply more than 2.5 pints (0.5 lb. active) per broadcast acre of GOAL 1.6E herbicide per season as a result of a single application or multiple applications.
- **Western Cotton:** Do not apply more than 2.5 pints (0.5 lb. active) of GOAL 1.6E herbicide per broadcast acre in a single application, or more than a total of 5.0 pints (1.0 lb. active) of GOAL 1.6E herbicide per broadcast acre per season as a result of multiple applications.
- Do not apply within 75 days of harvest.

DIRECTIONS FOR USE
FALLOW BED
AIR APPLICATION

FOR USE ONLY IN THE FOLLOWING CALIFORNIA COUNTIES: FRESNO, KERN, KINGS, MADERA, MERCED, TULARE

TREATMENTS MUST BE MADE OCTOBER 1 THROUGH MARCH 31.

GENERAL INFORMATION

GOAL 1.6E is effective as a preemergence and/or postemergence herbicide when used alone or in a tank mix combination with Roundup for the control of winter annual broadleaf weeds in fallow beds. Do not apply within 14 days of planting. The fallow bed should be worked with a Lilliston, or similar incorporation tool, to a depth of at least 2 inches prior to planting. It is important to thoroughly break the soil surface prior to planting; weed control should not be expected following breaking of the soil surface.

EXERCISE EXTREME CARE TO AVOID HERBICIDE CONTACT WITH ANY DESIRABLE DORMANT OR NON-DOR-

MANT CROP, PLANT, TREE OR VEGETATION, AS SEVERE INJURY MAY RESULT.

Cotton can be planted 14 days or more following a GOAL 1.6E herbicide fallow bed application.

GOAL 1.6E HERBICIDE USED ALONE
DOSAGE

GOAL 1.6E herbicide should be used at 1.25 to 2.5 pints (0.25 to 0.5 lb. active) per broadcast acre.The lower rate (1.25 pints per acre) should provide postemergence control of susceptible weeds (up to 4 leaf stage) and provide up to 4 weeks of preemergence control of susceptible weeds. The higher rate (2.5 pints per acre) should provide postemergence control of susceptible weeds (up to 6 leaf stage) and preemergence control of susceptible weeds for up to 8 weeks. Best preemergence control is achieved when pre-irrigation or rainfall occurs within 3 or 4 weeks following application.

WEEDS CONTROLLED

GOAL 1.6E herbicide will provide preemergence and postemergence* control of the following weeds when used at recommended dosages and weed stage:

CHEESEWEED (MALVA)	MUSTARD, BLACK
FIDDLENECK, COAST	NETTLE, BURNING
FILAREE, BROADLEAF	REDMAIDS
FILAREE, REDSTEM	ROCKET, LONDON
GROUNDSEL, COMMON	SHEPHERDSPURSE
HENBIT	SOWTHISTLE, ANNUAL
MINERSLETTUCE	

* Thorough spray coverage is essential to maximize the postemergence activity of GOAL 1.6E herbicide. For postemergence control, a tank mixture of GOAL 1.6E herbicide with Roundup is recommended.

AERIAL APPLICATION

GOAL 1.6E herbicide should be applied using swirl jet or hollow cone nozzles and a spray pressure below 40 psi to deliver a minimum spray volume of 10 gallons of water per acre (minimum 5 gallons for GOAL 1.6E/Roundup tank mix). Application should be made at a height of 6 to 10 feet above the soil surface. It is suggested that the nozzles on the spray booms should not be placed any closer to the wing or rotor tips than $^3/_4$ of the span; this will minimize the formation of spray or wing tip vortice roll. Nozzles should be spaced and positioned to produce a uniform spray pattern and to minimize or eliminate the formation of droplets 100 microns or less in diameter.

AVOID DRIFT

WHEN APPLYING TO FALLOW BEDS, EXTREME CARE MUST BE EXERCISED TO PREVENT SPRAY DRIFT WHICH COULD RESULT IN DAMAGE TO OTHER CROPS OR DESIRABLE VEGETATION. USE THE FOLLOWING GUIDELINES WHEN AERIAL APPLICATIONS ARE TO BE MADE:

1. Do not apply when the wind direction is not stable, when inversion conditions exist, or when wind velocity exceeds 10 mph.
2. When wind speeds are 5 mph or less, maintain a minimum downwind buffer zone of at least $^1/_2$ mile from all crops and desirable vegetation, except the following:

Maintain a minimum downwind buffer zone of:

- 150 feet from dormant treefruit, dormant vines and overwintering sugarbeets.
- 650 feet from garlic, jojoba, legumes, onions, pastures, small grains, seedling sugarbeets and vegetable fallow beds.

3. When wind speeds are between 5 to 10 mph, downwind buffer zones in excess of those listed above are suggested.
4. For upwind and side borders, maintain a minimum buffer zone of 150 feet from any vegetable fallow bed, crop, or desirable vegetation.

The use of a drift control agent may be required by local regulations. However, the drift control agent may decrease the weed control activity.

IMPORTANT

Aerial applicators must be familiar with the EPA-registered label and follow the use precautions. Spraying GOAL 1.6E herbicide in a manner other than as recommended is done at the user's risk. Users are responsible for all loss or damage that results from such spraying. In addition, aerial applicators should follow all applicable state and local regulations and ordinances. In interpreting the label and local regulations, the most restrictive situations should apply to avoid drift hazards.

MIXING DIRECTIONS

Fill the spray tank at least one-third full of clean water and add the recommended amount of GOAL 1.6E herbicide while the pump and agitator are running. Complete filling of the spray tank with water and then add 1 quart of TRITON AG-98 or comparable 80% active nonionic surfactant, cleared for use on growing crops, per 100 gallons of spray. Maintain agitation until spraying is complete.

CULTURAL CONSIDERATIONS

In order for GOAL 1.6E herbicide to provide maximum preemergence activity, the bed or soil surface should be smooth and free of crop and weed trash (decaying leaves, clippings, dead weeds, etc.)

For best preemergence activity, pre-irrigation or rainfall should occur within 3 or 4 weeks after application. Cultural practices that result in redistribution or disturbance of the soil surface after treatment will decrease the herbicidal effectiveness of GOAL 1.6E. The best results from GOAL 1.6E herbicide are from applications to established beds or soil surfaces that are left undisturbed during the time period for which weed control is desired.

TANK MIXES
GOAL 1.6E HERBICIDE WITH ROUNDUP
DOSAGE

GOAL 1.6E herbicide can be tank mixed with Roundup to obtain postemergence control of annual grassy weeds. Tank mix 1.25 to 2.5 pints (0.25 to 0.5 lb. active) of GOAL 1.6E herbicide with 0.75 to 1.0 pint (0.38 to 0.5 lb. active) of Roundup for each acre treated. Apply in a minimum of 5 gallons of water per acre at the recommended rates and growth stages to susceptible weed species in a manner consistent with the respective labels.

TANK MIXING DIRECTIONS

Fill the spray tank at least one-third full of clean water and add the recommended amounts of GOAL 1.6E herbicide and Roundup while the pump and agitator are running. Complete filling of the spray tank with water. Add 1 quart of TRITON AG-98 or comparable 80% active nonionic surfactant, cleared for use on growing crops, per 100 gallons of spray. Maintain agitation until spraying is complete.

FALLOW BED AIR APPLICATION
SPECIFIC USE RESTRICTIONS

In addition to the following, also observe GENERAL USE RESTRICTIONS listed at the end of this label.

- Read and observe all label directions before using. When tank mixing, always read all individual manufacturers' labels. In interpreting all labels for the tank mixture, the most restrictive situations must apply.
- DO NOT AERIALLY APPLY GOAL 1.6E HERBICIDE (EITHER ALONE OR IN ANY COMBINATION RECOMMENDED ON THIS LABEL) TO FALLOW BEDS BEFORE OCTOBER 1 OR AFTER MARCH 31.
- Do not apply more than 2.5 pints (0.5 lb. active) of GOAL 1.6E herbicide per acre, per fallow season.
- Do not apply GOAL 1.6E herbicide within 14 days of planting.
- Do not rotate to any crops other than cotton or soybeans for 10 months following a GOAL 1.6E herbicide fallow bed application.

DIRECTIONS FOR USE
FALLOW BED GROUND APPLICATION

GENERAL INFORMATION

GOAL 1.6E is effective as a preemergence and/or postemergence herbicide when used alone or in a tank mix combination with

Roundup or paraquat (Gramoxone or Ortho paraquat) for the control of winter annual broadleaf weeds in fallow beds. Do not apply within 14 days of planting. The fallow beds should be worked with a Lilliston, or similar incorporation tool, to a depth of at least 2 inches prior to planting. It is important to thoroughly break the soil surface prior to planting; weed control should not be expected following breaking of the soil surface.

Cotton can be planted 14 days or more following a GOAL 1.6E herbicide fallow bed application.

GOAL 1.6E HERBICIDE USED ALONE

DOSAGE

GOAL 1.6E herbicide should be used at 1.25 to 2.5 pints (0.25 to 0.5 lb. active) per broadcast acre. The lower rate (1.25 pint per acre) should provide postemergence control of susceptible weeds (up to 4 leaf stage) and provide up to 4 weeks of preemergence control of susceptible weeds. The higher rate (2.5 pints per acre) should provide postemergence control of susceptible weeds (up to 6 leaf stage) and preemergence control of susceptible weeds for up to 8 weeks. Best preemergence control is achieved when pre-irrigation or rainfall occurs within 3 or 4 weeks following application.

WEEDS CONTROLLED

GOAL 1.6E herbicide will provide preemergence and postemergence control of the following weeds when used at recommended dosages and weed stage:

CHEESEWEED (MALVA)	MINERSLETTUCE
FIDDLENECK, COAST	MUSTARD, BLACK
FILAREE, BROADLEAF	NETTLE, BURNING
FILAREE, REDSTEM	REDMAIDS
GROUNDSEL, COMMON	ROCKET, LONDON
HENBIT	SOWTHISTLE, ANNUAL

GOAL 1.6E herbicide is a contact herbicide, therefore, coverage is essential for acceptable postemergence control. If dense weed populations, oversized weed seedlings, volunteer grains, annual grasses or unfavorable environmental conditions exist, a tank mixture of GOAL 1.6E herbicide with paraquat or Roundup for postemergence control is recommended.

METHOD OF APPLICATION

GOAL 1.6E herbicide should be applied in a minimum of 20 gallons of water per acre. The volume of water used should be increased as the weeds become taller and more dense. Use a low-pressure sprayer equipped with flat fan nozzles. Spray equipment should be calibrated carefully before each use.

MIXING DIRECTIONS

Fill the spray tank at least one-third full of clean water and add the recommended amount of GOAL 1.6E herbicide while the pump and agitator are running. Complete filling of the spray tank with water and then add 1 quart of TRITON AG-98 or comparable 80% active nonionic surfactant, cleared for use on growing crops, per 100 gallons of spray. Maintain agitation until spraying is complete.

CULTURAL CONSIDERATIONS

In order for GOAL 1.6E herbicide to provide maximum preemergence activity, the bed or soil surface should be smooth and free of crop and weed trash (decaying leaves, clippings, dead weeds, etc.)

For best preemergence activity, pre-irrigation or rainfall should occur within 3 or 4 weeks after application. Cultural practices that result in redistribution or disturbance of the soil surface after treatment will decrease the herbicidal effectiveness of GOAL 1.6E. The best results from GOAL 1.6E herbicide are from applications to established beds or soil surfaces which are left undisturbed during the time period for which weed control is desired.

FALLOW BED GROUND APPLICATION
SPECIFIC USE RESTRICTIONS
GOAL 1.6E HERBICIDE USED ALONE

In addition to the following, also observe GENERAL USE RESTRICTIONS listed at the end of this label.

- Do not apply more than 2.5 pints (0.5 lb. active) of GOAL 1.6E herbicide per acre, per fallow season.
- Do not apply GOAL 1.6E herbicide within 14 days of planting.
- Do not rotate to any crops other than cotton or soybeans for 10 months following a GOAL 1.6E herbicide fallow bed application.

TANK MIXES WITH GOAL 1.6E HERBICIDE

IMPORTANT: Read and observe all label directions before using. When tank mixing, always read all individual manufacturers' labels. In interpreting all labels for the tank mix, the most restrictive situations must apply.

DOSAGE

GOAL 1.6E herbicide can be tank mixed with paraquat or Roundup to obtain postemergence control of annual grassy weeds. Tank mix 1.25 to 2.5 pints (0.25 to 0.5 lb. active) of GOAL 1.6E herbicide with 1 to 2 quarts (0.5 to 1.0 lb. active) of paraquat or 0.75 to 1 pint (0.38 to 0.5 lb. active) of Roundup for each acre treated. Refer to the PREPLANT OR PREEMERGENCE USES section on the Ortho Paraquat CL label or the Gramoxone label or the FALLOW AND REDUCED TILLAGE SYSTEM section on the Roundup label for specific use directions and restrictions. Apply at the recommended rates and growth stages to susceptible weed species in a manner described on the respective labels.

TANK MIXING DIRECTIONS

Fill the spray tank at least one-third full of clean water and add the recommended amounts of GOAL 1.6E herbicide and paraquat or Roundup while the pump and agitator are running. Complete filling of the spray tank with water. Add 1 quart of TRITON AG-98 or comparable 80% active nonionic surfactant, cleared for use on growing crops, per 100 gallons of spray. Maintain agitation until spraying is complete.

CULTURAL CONSIDERATIONS

In addition to details previously listed under GOAL 1.6E HERBICIDE USED ALONE, read all individual manufacturers' labels. In interpreting all labels for the tank mixture, the most restrictive cultural situations must apply.

FALLOW BED GROUND APPLICATION
SPECIFIC USE RESTRICTIONS
GOAL 1.6E HERBICIDE/TANK MIXES

In addition to the following, also observe all specific use restrictions previously listed under GOAL 1.6E HERBICIDE USED ALONE and follow GENERAL USE RESTRICTIONS listed at the end of this label.

- When tank mixing, always read all individual manufacturers' labels. In interpreting all labels for the tank mixture, the most restrictive situations must apply.

DIRECTIONS FOR USE—GLOBE ARTICHOKE

POST-DIRECTED SPRAY
GENERAL INFORMATION

GOAL 1.6E is an effective herbicide for postemergence and preemergence control of certain broadleaf weeds in artichokes. GOAL 1.6E herbicide should be directed towards the winter ditch, levees or flat rows between the artichoke rows. Artichoke fronds receiving accidental spray or drift will be injured. Over the top applications may exhibit severe injury to the foliage and flower bud and are not recommended.

WEEDS CONTROLLED POSTEMERGENCE

CHEESEWEED (MALVA)	OXALIS (BERMUDA BUTTERCUP)
GROUNDSEL, COMMON	SHEPHERDSPURSE
MUSTARD, COMMON YELLOW	SOWTHISTLE, ANNUAL
NETTLE, BURNING	

WEEDS CONTROLLED PREEMERGENCE

CHEESEWEED (MALVA)	*OXALIS (BERMUDA BUTTERCUP)
GROUNDSEL, COMMON	SHEPHERDSPURSE
LAMBSQUARTERS, COMMON	SOWTHISTLE, ANNUAL
MUSTARD, COMMON YELLOW	

* Suppression.

DOSAGE

GOAL 1.6E herbicide is recommended as a post-directed application at 5 to 10 pints (1.0 to 2.0 lb. active) per acre. Optimum control is achieved when two applications of GOAL 1.6E herbicide are applied at 5 pints (1.0 lb. active) per acre. The initial application should be made to susceptible weed seedlings (up to 8 leaf stage). It is recommended that a second application be made 8 to 10 weeks later. Good results may be achieved when a single application of 10 pints (2.0 lb. active) of GOAL 1.6E herbicide is applied to susceptible weed seedlings (up to 8 leaf stage). Do not apply more than 10 pints (2.0 lb. active) of GOAL 1.6E herbicide per treated acre per season as a result of a single application or multiple applications. Do not apply within 5 days of harvest.

TIMING AND METHOD OF APPLICATION

TREATMENTS SHOULD BE MADE OCTOBER 1 THROUGH FEBRUARY 15 (AFTER COMPLETION OF THE DITCHING OPERATION).

GOAL 1.6E herbicide should be applied in a minimum of 40 gallons of water per acre depending upon density of emerged weeds. Spray volume should be increased as weed height and density increase. Use a low-pressure sprayer equipped with flat fan nozzles. Spray equipment should be calibrated carefully before each use. Spray should be directed towards the winter ditch, levees or flat rows between the artichoke rows. ARTICHOKE FRONDS RECEIVING ACCIDENTAL SPRAY OR DRIFT WILL BE INJURED.

MIXING DIRECTIONS

Fill the spray tank at least one-third full of clean water and add the recommended amount of GOAL 1.6E herbicide while the pump and agitator are running. Complete filling of the spray tank with water and then add 1 quart of a nonionic surfactant such as TRITON AG-98, or comparable 80% active nonionic surfactant, per 100 gallons of spray. Maintain agitation until spraying is completed.

CULTURAL CONSIDERATIONS

In order to provide maximum effectiveness of preemergence activity of GOAL 1.6E herbicide, the soil surface should be smooth and free of crop or weed trash.

For best preemergence activity, at least $\frac{1}{4}$ inch of irrigation water is necessary within 3 to 4 weeks after application if no rainfall occurs. Cultural practices that result in redistribution or disturbance of the soil surface after treatment will decrease the herbicidal effectiveness of GOAL 1.6E. Cutting water furrows or cultivations that mix untreated soil into treated areas will also reduce the effectiveness of the treatment. The best results from GOAL 1.6E herbicide are from applications to soil surfaces that are left undisturbed during the time period for which weed control is desired.

GLOBE ARTICHOKE
SPECIFIC USE RESTRICTIONS

The following specific use restrictions should be observed when GOAL 1.6E herbicide is used alone or in any spray combination recommended on this label. In addition to the following, also observe GENERAL USE RESTRICTIONS listed at the end of this label.

- Do not apply more than 10 pints (2.0 lb. active) of GOAL 1.6E herbicide per treated acre per season as a result of a single application or multiple applications.
- Do not apply GOAL 1.6E herbicide within 5 days of harvest.
- DO NOT APPLY GOAL 1.6E HERBICIDE TO GLOBE ARTICHOKE PLANTINGS BEFORE OCTOBER 1 OR AFTER FEBRUARY 15.
- Avoid direct spray or drift contact of GOAL 1.6E herbicide with artichoke flowers or buds, as severe injury may result.
- Do not apply GOAL 1.6E herbicide to artichoke plantings within 60 days after cutting back or transplanting.

DIRECTIONS FOR USE
NONBEARING CITRUS

CALAMONDIN, CHIRONJA, CITRUS CITRON, GRAPEFRUIT, KUMQUAT, LEMON, LIME, MANDARIN, PUMMELO, SATSUMA MANDARIN, SOUR ORANGE, SWEET ORANGE, TANGELO, TANGERINE, TANGOR

FOR USE ONLY IN PERMANENTLY ESTABLISHED GROVES IN AZ, CA, FL, LA AND TX

GENERAL INFORMATION

GOAL 1.6E is effective as a preemergence and/or postemergence herbicide when used alone or in recommended tank mix combinations, for the control of certain annual broadleaf weeds, in nonbearing citrus plantings. GOAL 1.6E herbicide may be applied to newly planted trees or to young trees that will not bear fruit within one year.

The most effective postemergence weed control is achieved when GOAL 1.6E herbicide is applied to seedling weeds at the recommended growth stage. For postemergence control of certain grassy and broadleaf weeds, a tank mix of GOAL 1.6E herbicide with paraquat (Gramoxone or Ortho paraquat) or Roundup can be used.

Preemergence control is most effective when spray is applied to clean, weed-free soil surfaces. Treated soil surfaces or berms should not be disked or disturbed in any manner as the herbicidal effectiveness of GOAL 1.6E may be decreased. Seedling weeds are controlled as they come in contact with the soil-applied herbicide during emergence. For residual grass control in citrus, a tank mixture of GOAL 1.6E herbicide with Devrinol, simazine, Solicam® or Surflan can be used. Contact herbicides such as paraquat or Roundup may also be added to the tank mixture.

Check individual product labels to determine suitability and use rates for various crops.

GOAL 1.6E HERBICIDE USED ALONE

GEOGRAPHIC USE DIRECTIONS
ARIZONA AND CALIFORNIA
DOSAGE

GOAL 1.6E herbicide is recommended for postemergence control at 2.5 to 10 pints (0.5 to 2.0 lb. active) per broadcast acre. The lower rate is recommended for the control of susceptible seedling weeds less than 4 inches in height. See weed stage rate and weeds controlled for susceptible weeds up to 8 inches in height. The higher rate (2.0 lb. active) should be used for weeds less than 12 inches in height. For preemergence control of susceptible weeds, use 10 pints (2.0 lb. active) per broadcast acre.

WEEDS CONTROLLED POSTEMERGENCE

CHEESEWEED (MALVA)	MINERSLETTUCE
FIDDLENECK, COAST	NETTLE, BURNING
FILAREE, BROADLEAF	PIGWEED, REDROOT
FILAREE, REDSTEM	REDMAIDS
FILAREE, WHITESTEM	SHEPHERDSPURSE
GROUNDSEL, COMMON	SOWTHISTLE, ANNUAL
HENBIT	

WEEDS CONTROLLED PREEMERGENCE

BURCLOVER	LETTUCE, PRICKLY
CHEESEWEED (MALVA)	PIGWEED, REDROOT
FIDDLENECK, COAST	PURSLANE, COMMON
FILAREE, BROADLEAF	REDMAIDS
FILAREE, REDSTEM	ROCKET, LONDON
FILAREE, WHITESTEM	SHEPHERDSPURSE
GROUNDSEL, COMMON	SOWTHISTLE, ANNUAL
HENBIT	SPURGE, PROSTRATE
KNOTWEED, PROSTRATE	SPURGE, SPOTTED
LAMBSQUARTERS, COMMON	

EARLY POSTEMERGENCE (weeds up to 4 inches high) apply 2.5 pints (0.5 lb. active) of GOAL 1.6E herbicide per broadcast acre*

CHEESEWEED (MALVA); FIDDLENECK, COAST; FILAREE, REDSTEM; HENBIT; REDMAIDS; SHEPHERDSPURSE

POSTEMERGENCE (weeds up to 8 inches high) apply 5 pints (1.0 lb. active) of GOAL 1.6E herbicide per broadcast acre*

CHEESEWEED (MALVA); FIDDLENECK, COAST; FILAREE, REDSTEM; HENBIT; MINERSLETTUCE; NETTLE, BURNING; REDMAIDS; SHEPHERDSPURSE; SOWTHISTLE, ANNUAL

LATE POSTEMERGENCE (weeds up to 12 inches high) apply 10 pints (2.0 lb active) of GOAL 1.6E herbicide per broadcast acre*

CHEESEWEED (MALVA); FIDDLENECK, COAST; FILAREE, BROADLEAF; FILAREE, REDSTEM; FILAREE, WHITESTEM; GROUNDSEL, COMMON; HENBIT; MINERSLETTUCE; NETTLE, BURNING; PIGWEED, REDROOT; REDMAIDS; SHEPHERDSPURSE; SOWTHISTLE, ANNUAL

PREEMERGENCE apply 10 pints (2.0 lb. active) of GOAL 1.6E herbicide per broadcast acre*

BURCLOVER; CHEESEWEED (MALVA); FIDDLENECK, COAST; FILAREE, BROADLEAF; FILAREE, REDSTEM; FILAREE, WHITESTEM; GROUNDSEL, COMMON; HENBIT; KNOTWEED, PROSTRATE; LAMBSQUARTERS, COMMON; LETTUCE, PRICKLY; PIGWEED, REDROOT; PURSLANE, COMMON; REDMAIDS; ROCKET, LONDON; SHEPHERDSPURSE; SOWTHISTLE, ANNUAL; SPURGE, PROSTRATE; SPURGE, SPOTTED

* Dosages listed are for broadcast application. For banded application, the amount of GOAL 1.6E herbicide used per acre should be reduced according to the following formula:

$$\frac{\text{Band Width (in.)}}{\text{Row Width (in.)}} \times \text{Rate per Broadcast Acre} = \text{Amount Needed per Acre for Banded Application}$$

TIMING AND METHOD OF APPLICATION

As a preemergence treatment, apply in a minimum of 40 gallons of water per acre depending upon density of emerged weeds. Best preemergence results are achieved when spray is applied to a relatively weed-free berm or soil surface. GOAL 1.6E herbicide should be directed to the soil and the base of trees. Avoid direct spray contact on the citrus foliage. Use a low-pressure sprayer equipped with a breakaway boom and flat fan nozzles. An off-center (OC) nozzle positioned at the end of the boom may be desired. Spray equipment should be calibrated carefully before each use.

SPRAY VOLUME: Weed Stage	Gals. of Water per Acre
Early Postemergence (weeds up to 4 in. high)	40 or more
Postemergence (weeds up to 8 in. high)	100 or more
Late Postemergence (weeds up to 12 in. high)	100 or more
Preemergence	40 or more

FLORIDA, LOUISIANA AND TEXAS

DOSAGE

GOAL 1.6E herbicide is recommended for postemergence control at 2.5 to 10.0 pints* (0.5 to 2.0 lb. active) per broadcast acre. The lower rate is recommended for the control of susceptible seedling weeds in the early postemergence stage up to the 4 leaf stage. The higher rate (2.0 lb. active) should be used for weeds up to the 6 leaf stage. For preemergence control of susceptible weeds, GOAL 1.6E herbicide is recommended at 10.0 pints (2.0 lb. active) per broadcast acre.

* Dosages listed are for broadcast application. For banded application, the amount of GOAL 1.6E herbicide used per acre should be reduced according to the following formula:

$$\frac{\text{Band Width (in.)}}{\text{Row Width (in.)}} \times \text{Rate per Broadcast Acre} = \text{Amount Needed per Acre for Banded Application}$$

WEEDS CONTROLLED POSTEMERGENCE

BALSAMAPPLE	NIGHTSHADE, AMERICAN BLACK
*CUDWEED, NARROWLEAF	NIGHTSHADE, BLACK
**EVENINGPRIMROSE, CUTLEAF	PEPPERWEED, VIRGINIA
GROUNDCHERRY, CUTLEAF	PIGWEED, REDROOT
GROUNDCHERRY, WRIGHT	POINSETTIA, WILD
LAMBSQUARTERS, COMMON	PURSLANE, COMMON
MORNINGGLORY, ANNUAL	PUSLEY, FLORIDA
	SIDA, PRICKLY (TEAWEED)
	SMARTWEED, PENNSYLVANIA
	SOWTHISTLE, ANNUAL

WEEDS CONTROLLED PREEMERGENCE

CUDWEED, NARROWLEAF	PEPPERWEED, VIRGINIA
**EVENINGPRIMROSE, CUTLEAF	PIGWEED, REDROOT
GROUNDCHERRY, CUTLEAF	POINSETTIA, WILD
LAMBSQUARTERS, COMMON	PUSLEY, FLORIDA
NIGHTSHADE, AMERICAN BLACK	SIDA, PRICKLY (TEAWEED)
NIGHTSHADE, BLACK	SMARTWEED, PENNSYLVANIA
	SOWTHISTLE, ANNUAL
	SPURGE, PROSTRATE
	SPURGE, SPOTTED

* Maximum 0.5 inch diameter.

** Highest rate and/or multiple applications may be required for acceptable control. Do not apply more than 20 pints (4.0 lb active) per broadcast acre during any 12-month period as a result of multiple applications.

TIMING AND METHOD OF APPLICATION

As a preemergence treatment, apply in a minimum of 40 gallons of water per acre depending upon density of emerged weeds. Best preemergence results are achieved when spray is applied to a relatively weed-free soil surface. GOAL 1.6E herbicide should be directed to the soil and the base of trees. Avoid direct spray contact on the citrus foliage. Use a low-pressure sprayer equipped with a breakaway boom and flat fan nozzles. An off-center (OC) nozzle positioned at the end of the boom may be desired. Spray equipment should be calibrated carefully before each use.

Spray Volume: Weed Stage	Gals. of Water per Acre
Early Postemergence (weeds up to 4 leaf stage)	40 or more
Postemergence (weeds up to 6 leaf stage)	100 or more
Preemergence	40 or more

ALL STATES (AZ, CA, FL, LA, TX)

MIXING DIRECTIONS

Fill the spray tank at least one-third full of clean water and add the recommended amount of GOAL 1.6E herbicide while the pump and agitator are running. Complete filling of the spray tank with water and then add 1 quart of a nonionic surfactant such as TRITON AG-98, or comparable 80% active nonionic surfactant cleared for application to growing crops, per each 100 gallons of spray. Maintain agitation until spraying is completed.

CULTURAL CONSIDERATIONS

In order to provide maximum effectiveness of preemergence activity of GOAL 1.6E herbicide, the berm or soil surface should be smooth and free of crop or weed trash (decaying leaves, clippings, dead weeds, etc.). Leaves and trash may be removed by blowing the area to be treated or by thoroughly mixing the trash into the soil through cultivation prior to herbicide application.

For best preemergence activity, at least one-quarter inch (1/4 inch) of irrigation water should be applied within 3 to 4 weeks after application if no rainfall occurs.

Cultural practices that result in redistribution or disturbance of the soil surface after treatment will decrease the herbicidal effectiveness of GOAL 1.6E herbicide. Cutting water furrows or cultivations that mix untreated soil into treated areas will also reduce the effectiveness of the treatment. The best results are from applications to soil surfaces or berms that are left undisturbed during the time period for which weed control is desired.

NONBEARING CITRUS
SPECIFIC USE RESTRICTIONS
GOAL 1.6E HERBICIDE USED ALONE

The following specific use restrictions should be observed when GOAL 1.6E herbicide is used alone or in any spray combination recommended on this label. In addition to the following, also observe GENERAL USE RESTRICTIONS listed at the end of this label.

- Apply GOAL 1.6E herbicide only to nonbearing citrus trees.
- Do not apply more than 10 pints of GOAL 1.6E herbicide (2.0 lb. active) per broadcast acre in a single application or more than 20 pints (4.0 lb. active) per broadcast acre during any 12-month period as a result of multiple applications.
- GOAL 1.6E herbicide or any of the combinations recommended on this label should be applied only to healthy growing trees.
- Do not apply during periods of new foliage growth. Applications should be made after foliage has fully expanded and hardened-off.
- Direct spray toward the base of trees. Avoid direct spray contact on the citrus foliage.

TANK MIXES WITH GOAL 1.6E HERBICIDE

IMPORTANT: Read and observe all label directions before using. When tank mixing, always read all individual manufacturers' labels. In interpreting all labels for the tank mixture, the most restrictive situations must apply.

DOSAGE

For postemergence control of grassy and broadleaf weeds in certain citrus plantings, a tank mixture of GOAL 1.6E herbicide with either paraquat or Roundup can be used. Apply at the recommended rates and growth stages to susceptible weed species in a manner described on the respective labels.

For preemergence control of certain grassy and broadleaf weeds in certain citrus plantings, a tank mixture of GOAL 1.6E herbicide with Devrinol, simazine, Solicam or Surflan can be applied. Contact herbicides such as paraquat or Roundup may be added to the tank mixture. Apply at the recommended rates and growth stages to susceptible weed species in a manner described on the respective labels.

For preemergence control of susceptible weeds, use 10 pints (2.0 lb active) of GOAL 1.6E herbicide per broadcast acre.

WEEDS CONTROLLED PREEMERGENCE

In addition to the weeds controlled preemergence by GOAL 1.6E herbicide used alone, control of the following weeds is also obtained when tank mixed with the products listed below:

Devrinol 50-WP
BARNYARDGRASS (WATERGRASS)
BLUEGRASS, ANNUAL
BROME, RIPGUT
CHICKWEED, COMMON
CRABGRASS, LARGE (HAIRY)
PINEAPPLEWEED

Simazine
BARNYARDGRASS (WATERGRASS)
BLUEGRASS, ANNUAL
CHICKWEED, COMMON
CRABGRASS, LARGE (HAIRY)
CRABGRASS, SMOOTH
HORSEWEED (MARESTAIL)

Solicam
BARNYARDGRASS (WATERGRASS)
BLUEGRASS, ANNUAL
BROME, RIPGUT
CUPGRASS, SOUTHWESTERN
PINEAPPLEWEED

Surflan
BARNYARDGRASS (WATERGRASS)
BLUEGRASS, ANNUAL
CHICKWEED, COMMON
CRABGRASS, LARGE (HAIRY)
CRABGRASS, SMOOTH
CUPGRASS, SOUTHWESTERN

WEEDS CONTROLLED POSTEMERGENCE

In addition to the weeds controlled postemergence by GOAL 1.6E herbicide used alone, control of the following weeds is also obtained when tank mixed with the products listed below:

Roundup or paraquat
BARNYARDGRASS (WATERGRASS)
BLUEGRASS, ANNUAL
BROME, RIPGUT
CHICKWEED, COMMON
HORSEWEED (MARESTAIL)
ROCKET, LONDON
SPURGE, PROSTRATE

METHOD OF APPLICATION

As a preemergence and/or postemergence treatment, apply in a minimum of 40 gallons of water per acre depending upon density of emerged weeds. Spray coverage is essential for good postemergence results. Best preemergence results are achieved when the tank mix is applied to a relatively weed-free soil surface or berm. Spray should be directed to the soil and the base of trees. Use a low-pressure sprayer equipped with a breakaway boom and flat fan nozzles. An off-center (OC) nozzle positioned at the end of the boom may be desired. Spray equipment should be calibrated carefully before each use.

MIXING DIRECTIONS

Fill the spray tank at least one-third full of clean water. With the pump and agitator running, add the recommended amounts of Devrinol, simazine, Solicam or Surflan to the spray tank, then add the recommended amounts of GOAL 1.6E herbicide and paraquat or Roundup while the pump and agitator are running. Complete filling of the spray tank with water and then add 1 quart of a nonionic surfactant such as TRITON AG-98 or comparable 80% active nonionic surfactant cleared for application to growing crops, per each 100 gallons of spray. Maintain agitation until spraying is completed.

CULTURAL CONSIDERATIONS

In addition to details previously listed under GOAL 1.6E HERBICIDE USED ALONE, read all individual manufacturers' labels. In interpreting all labels for the tank mixture, the most restrictive label situations must apply.

NONBEARING CITRUS
SPECIFIC USE RESTRICTIONS
GOAL 1.6E HERBICIDE TANK MIXES

In addition to the following, also observe all specific use retrictions previously listed under GOAL 1.6E HERBICIDE USED ALONE and GENERAL USE RESTRICTIONS listed at the end of this label.

- When tank mixing, always read all individual manufacturers' labels. In interpreting all labels for the tank mixture, the most restrictive situations must apply.

DIRECTIONS FOR USE
ONIONS

GENERAL INFORMATION

GOAL 1.6E is a selective herbicide for postemergence application to direct-seeded and transplanted onions for early postemergence control of certain annual broadleaf and grass weeds. Initial spray application should be made when the onions have reached the development stage specified in the DOSAGE section and the SPECIFIC USE RESTRICTIONS section of this label. On onion transplants, spray as soon after transplanting as practical. GOAL 1.6E herbicide can cause necrotic lesions, twisting, pigtailing or stunting of the onion plants. Injury will be more severe if applications are made immediately following or during cool, wet weather and/or if applications are made prior to the development stage of the onion plants as specified in the DOSAGE section and the SPECIFIC USE RESTRICTIONS section of this label.

WEEDS CONTROLLED

GOAL 1.6E herbicide will provide postemergence control of the following weeds when applied at the recommended dosage and leaf stage (2 to 4 leaves):

CANARYGRASS (ANNUAL)	*PIGWEED, PROSTRATE
EVENING PRIMROSE, CUTLEAF	*PIGWEED, REDROOT
GROUNDSEL, COMMON	PUNCTUREVINE
MALLOW, LITTLE (MALVA)	PURSLANE, COMMON
NIGHTSHADE, BLACK	ROCKET, LONDON
	SAGE, LANCELEAF
	*SHEPHERDSPURSE
	SOWTHISTLE, ANNUAL

* Specific weeds controlled at rates recommended for use in Northeastern states (see DOSAGE section).

DOSAGE

SEEDED ONIONS

NORTHEASTERN STATES (CONNECTICUT, MAINE, MASSACHUSETTS, NEW HAMPSHIRE, NEW JERSEY, NEW YORK, RHODE ISLAND, VERMONT)

GOAL 1.6E herbicide is recommended for postemergence control at 2.4 to 4.8 fluid ounces (0.03 to 0.06 lb. active) per acre when applied postemergence to seeded onions that have *at least three (3) true leaves.* For transplanted onions apply the same rates within two days after transplanting. Multiple treatments at the aforementioned rates may be applied. Do not apply more than 2.5 pints (0.5 lb. active) per broadcast acre of GOAL 1.6E herbicide as a result of multiple applications in one season.

WESTERN STATES

(ARIZONA, CALIFORNIA, COLORADO, IDAHO, NEVADA, NEW MEXICO, OREGON, TEXAS, UTAH, WASHINGTON)

GOAL 1.6E herbicide is recommended for postemergence control at 0.6 to 1.25 pints (0.12 to 0.25 lb. active) per acre when applied postemergence to onions that have *at least two (2) true leaves.* Multiple treatments at the aforementioned rates may be applied. Do not apply more than 2.5 pints (0.5 lb. active) per broadcast acre of GOAL 1.6E herbicide in one season.

ALL OTHER STATES

GOAL 1.6E herbicide is recommended for postemergence control at 0.6 pints (0.12 lb. active) per acre when applied postemergence to onions that have *at least two (2) true leaves.* Multiple treatments at the aforementioned rate may be applied. Do not apply more than 2.5 pints (0.5 lb. active) per broadcast acre of GOAL 1.6E herbicide in one season.

DOSAGE

TRANSPLANTED ONIONS (ALL STATES)

Transplanted onions are most tolerant of a GOAL 1.6E herbicide application immediately after transplanting. For all states except the Northeastern states listed under the DOSAGE—SEEDED ONIONS section, an application of up to 2.5 pints per acre (0.5 lb. active) within two days after transplanting may be made. If less than 2.5 pints per acre is applied, a second application can be made two weeks or more after transplanting. Do not exceed the maximum use rate at 2.5 pints of GOAL 1.6E herbicide per acre (0.5 lb. active) per season. For transplanted onions in the Northeastern states, see DOSAGE—SEEDED ONIONS section.

TIMING AND METHOD OF APPLICATION

For best postemergence control of susceptible weeds, apply when the weeds are in the 2 to 4 leaf stage. Application of GOAL 1.6E herbicide after the weeds exceed the maximum leaf stage may result in reduced weed control. More than one postemergence application may be necessary to control subsequent weed flushes.

GOAL 1.6E herbicide should be thoroughly mixed with clean water, at recommended concentrations, and applied in a minimum of 40 gallons of water per acre. Use conventional ground spray equipment with flat fan spray nozzles, at 20 to 40 psi. Accurately calibrate spray equipment prior to each use. Avoid drift to all other crops and non-target areas. Thoroughly flush the spray equipment (tank, hose, pump, boom) with water before and after each use. Residual GOAL 1.6E herbicide remaining in spray equipment may damage other crops.

CULTURAL CONSIDERATIONS

On mineral soils, in order to provide maximum preemergence weed control, the soil surface should be smooth and free of excessive trash (clippings, dead weeds, etc.). Rain or irrigation is not necessary immediately following application; however, preemergence weed control is best when application is followed within two weeks by rainfall.

Cultural practices that result in redistribution or disturbance of the soil surface after spraying will destroy the herbicide effectiveness of the treatment. Cultivations that mix untreated soil in treated areas will also reduce the effectiveness of the treatment. The best results from GOAL 1.6E herbicide are from applications on established beds that are left undisturbed during the time period for which weed control is desired.

ONION—SPECIFIC USE RESTRICTIONS

In addition to the following, also observe GENERAL USE RESTRICTIONS listed at the end of this label.

- In all states except Northeastern states, do not start spraying until the onions *have two (2) fully developed true leaves.* In the Northeastern states (Connecticut, Maine, Massachusetts, New Hampshire, New Jersey, New York, Rhode Island, Vermont), do not start spraying until the onions *have three (3) fully developed true leaves.* Applications made prior to the recommended onion development stage may result in serious injury and is not recommended.
- Do not apply more than a total of 2.5 pints (0.5 lb. active) per acre of GOAL 1.6E herbicide during one use season.
- Do not apply within 60 days of harvest.
- Use only on dry bulb onions.
- Do not mix GOAL 1.6E herbicide with oils, surfactants, liquid fertilizers or pesticides.
- Do not apply to onion plants that are under stress due to drought, flooding, excessive fertilizer or soil salts, wind injury, hail, frost damage, injury from previously applied pesticides, or injury due to insects or diseases.

DIRECTIONS FOR USE—SOYBEAN

(NOT FOR USE IN CALIFORNIA)

GENERAL INFORMATION

GOAL 1.6E is effective as a preemergence and postemergence (post-directed) herbicide for the control of broadleaf weeds in soybeans. Applications can be made early preplant in conservation tillage soybeans, preemergence in no-till (double-crop) soybeans, or post-directed in conventional till soybeans. Seedling weeds are controlled as they come in contact with the herbicide either during emergence or through a post-directed application. Follow specific use directions and restrictions for recommended use and timing of applications.

Soybeans are tolerant to preemergence applications of recommended dosages of GOAL 1.6E herbicide; however, under certain conditions, GOAL 1.6E herbicide can cause temporary injury. Heavy splashing rain shortly after crop emergence or cold, wet soil conditions during early growth stages can produce leaf cupping and crinkling. When injury occurs, it is generally limited to the first few leaves that develop shortly after crop plants emerge from the soil. Soybeans recover from this injury and yields are not adversely affected. Soybean leaves that are accidentally sprayed during a post-direct application will exhibit necrotic spotting and injury to the soybean plant. Therefore, care must be exercised to avoid spray contact with the soybean leaves.

WEEDS CONTROLLED PREEMERGENCE

GOAL 1.6E herbicide used alone, at recommended dosages, provides preemergence control of the following broadleaf weeds:

GROUNDCHERRY, CUT LEAF	PIGWEED, REDROOT
	POINSETTIA, WILD

JIMSONWEED
LAMBSQUARTERS, COMMON
NIGHTSHADE, AMERICAN BLACK
NIGHTSHADE, BLACK
SHEPHERDSPURSE
SIDA PRICKLY (TEAWEED)
SMARTWEED, PENNSYLVANIA
SOWTHISTLE, COMMON
VELVETLEAF

WEEDS CONTROLLED POSTEMERGENCE (POST-DIRECTED APPLICATION)

GOAL 1.6E herbicide, when applied as a post-directed application at recommended dosages to seedling weeds (not exceeding the 4 leaf stage), will provide postemergence control of the following broadleaf weeds:

COCKLEBUR, COMMON
GROUNDCHERRY, CUT LEAF
GROUNDCHERRY, WRIGHT
JIMSONWEED
LAMBSQUARTERS, COMMON
MORNINGGLORY, ANNUAL
NIGHTSHADE, AMERICAN BLACK
NIGHTSHADE, BLACK
PIGWEED, REDROOT
*POINSETTIA, WILD
PURSLANE, COMMON
SESBANIA, HEMP
SHEPHERDSPURSE
SIDA, PRICKLY (TEAWEED)
SMARTWEED, PENNSYLVANIA
VELVETLEAF

* Multiple applications may be required for acceptable control.

Two pints of TRITON AG-98 or comparable 80% active nonionic surfactant cleared for application to growing crops, per each 100 gallons of spray solution, are suggested in all tank mixtures containing GOAL 1.6E herbicide when postemergence weed control is desired.

DOSAGE AND TIMING

CONSERVATION TILLAGE SOYBEANS—EARLY PREPLANT

GOAL 1.6E herbicide is effective for preemergence and postemergence control of susceptible broadleaf weeds when surface-applied at 1.9 to 2.5 pints (0.38 to 0.5 lb. active) per broadcast acre to the stale seedbed prior to the planting of conservation tillage soybeans. It is suggested that applications be made approximately 7 to 14 days prior to planting. The higher rate of 2.5 pints (0.5 lb. active) will assist in early season annual grass control. However, GOAL 1.6E herbicide must not be a basic portion of the grass herbicide program. A planned herbicide program for early preplant, preemergence or postemergence grass control is recommended.

The use of ridge planters is recommended. Soil surfaces should not be disturbed prior to plantings as the herbicidal effectiveness of GOAL 1.6E herbicide may be decreased. Seedling weeds are controlled as they come in contact with the soil-applied herbicide during emergence. Timely cultivations will usually assist in weed control.

NO-TILL (DOUBLE CROP) SOYBEANS—PREEMERGENCE

GOAL 1.6E herbicide is effective for preemergence and postemergence control of susceptible broadleaf weeds when applied at 1.9 to 2.5 pints (0.38 to 0.5 lb. active) per broadcast acre.

For postemergence control of certain grassy and broadleaf weeds a tank mix of either paraquat (Gramoxone or Ortho paraquat) or Roundup with GOAL 1.6E herbicide can be used.

For residual grass control in no-till soybeans, a tank mixture of either Bronco, Dual, Lasso, or Surflan with GOAL 1.6E herbicide or combinations of GOAL 1.6E herbicide plus paraquat or Roundup can be used. Follow specific use directions and restrictions for these combination tank mixes.

Application should be made within one day after planting. Later applications may result in severe crop injury and are not recommended.

TANK MIXES WITH GOAL 1.6E HERBICIDE

IMPORTANT: Read and observe all label directions before using. When tank mixing, always read all individual manufacturers' labels. In interpreting all labels for the tank mixture, the most restrictive situations must apply.

DOSAGE

Refer to the following tables for labeled use rates.

		RATE OF PRODUCT PER BROADCAST ACRE (PINTS PER ACRE)					
SOIL TEXTURE	GOAL 1.6E	DUAL 8E*	LASSO 4E*	SURFLAN A.S.**	PARAQUAT	ROUNDUP	BRONCO*
Coarse	1.9	1.5	4.0 to 5.0	1.5	1.0 to 2.0	2.0 to 3.0	6.5 to 10.0
Medium	2.5	2.0	5.0 to 6.0	2.0	1.0 to 2.0	2.0 to 3.0	8.0 to 10.0
Fine	2.5	2.0 to 2.5	5.0 to 6.0	3.0	1.0 to 2.0	2.0 to 3.0	8.0 to 10.0
Muck or Peat	***	***	***	***	***	***	***

* Use the higher rate of Bronco, Dual or Lasso on soils containing more than 3 percent organic matter.
** When using Surflan 75 WP, multiply pints by 0.67 to obtain the amount of Surflan 75 WP product required. Do not use Surflan on soils containing more than 5 percent organic matter.
*** Not labled.

NO-TILL (DOUBLE-CROP) SOYBEANS—PREEMERGENCE

WEEDS CONTROLLED PREEMERGENCE

When GOAL 1.6E herbicide is tank mixed with Bronco, Dual, Lasso or Surflan and applied preemergence, in addition to the weeds controlled preemergence by GOAL 1.6E herbicide alone, control of the following weeds is also obtained:

BARNYARDGRASS
CRABGRASS, LARGE
FOXTAIL, GIANT
FOXTAIL, YELLOW
JOHNSONGRASS, SEEDLING
PANICUM, FALL
RAGWEED, COMMON
SIGNALGRASS, BROADLEAF

WEEDS CONTROLLED POSTEMERGENCE

When GOAL 1.6E herbicide is tank mixed with Bronco, paraquat or Roundup and applied postemergence, in addition to the weeds controlled postemergence by GOAL 1.6E herbicide alone, control of the following weeds is also obtained:

BLUEGRASS, ANNUAL
CRABGRASS, LARGE
FOXTAIL, GIANT
FOXTAIL, GREEN
FOXTAIL, YELLOW
LAMBSQUARTERS, COMMON
RAGWEED, COMMON
SANDBUR, FIELD

TIMING AND METHOD OF APPLICATION

As a preemergence treatment, apply in 20 to 60 gallons of water per acre. If Bronco or Roundup are included in the tank mix, apply in 20 to 40 gallons of water per acre. To insure complete coverage, spray volume should be increased as the density of emerged weeds, crop residue or stubble increases. Use conventional spray equipment with flat fan or flood jet nozzles. Spray equipment should be calibrated carefully before each use.

MIXING DIRECTIONS

Fill the spray tank at least one-third full of clean water. With the pump and agitator running, add the recommended amounts of herbicide to be used. When tank mixing with Surflan, add the flowable or a slurry of the wettable powder formulation to the spray tank first. The emulsifiable concentrates (EC) formulation should be added next. No particular order need be followed when adding the EC formulations. Complete filling of the spray tank with water. When postemergence weed control is desired, add 2 pints of TRITON AG-98 or comparable 80% active nonionic surfactant cleared for application to growing crops, per each 100 gallons of spray solution. Maintain agitation until spraying is completed.

POST-DIRECTED SPRAY—

GOAL 1.6E HERBICIDE USED ALONE

DOSAGE

GOAL 1.6E herbicide is recommended as a post-directed application at 1.25 pints (0.25 lb. active) per acre*. When applied to seedling weeds not exceeding 4 true leaves that are young and actively growing, do not count cotyledon leaves.

* Dosages listed are for broadcast application. For banded application, the amount of GOAL 1.6E herbicide used per acre should be reduced according to the following formula:

$$\frac{\text{Band Width (in.)}}{\text{Row Width (in.)}} \times \text{Rate per Broadcast Acre} = \text{Amount Needed per Acre for Banded Application}$$

TANK MIXES WITH GOAL 1.6E HERBICIDE

For improved broadleaf weed control, a tank mixture of GOAL 1.6E herbicide plus Butoxone or Butyrac 200 is suggested. Use 1.25 pints of GOAL 1.6E herbicide (0.25 lb. active) with 1 pint of Butoxone (0.22 lb. active) or 0.7 to 0.9 pint of Butyrac 200 (0.175 to 0.22 lb. active) per broadcast acre. When tank mixing, always read all individual manufacturers' labels. In interpreting all labels for the tank mixture, the most restrictive situations must apply.

Two pints of TRITON AG-98 or comparable 80% active nonionic surfactant, cleared for use on growing crops, per 100 gallons of spray solution are suggested in all tank mixtures containing GOAL 1.6E herbicide when postemergence weed control is desired.

TIMING

Soybean plant height must be a minimum 8 inches or greater. Use branch lifters or shields if excessive spray contact to the soybean plant cannot be avoided. Irrigation or rainfall should occur within two weeks of application to receive greatest benefit of preemergence activity from GOAL 1.6E herbicide on nightshade, groundcherry and wild poinsettia species.

METHOD OF APPLICATION

Accurate, uniform placement of GOAL 1.6E herbicide spray is essential for effective weed control and to minimize soybean injury. As a directed postemergence application, GOAL 1.6E herbicide should be applied at 20 to 25 psi using 20 to 40 gallons of spray on a broadcast acre basis. Do not exceed 25 psi. Spray should be directed towards the base of the soybean plant. Soybean foliage receiving accidental spray or drift may be injured. Weeds should be in the seedling stage, young and actively growing.

Fill the spray tank at least one-third full of clean water and add the recommended amount of GOAL 1.6E herbicide while the pump and agitator are running. If tank mix is desired, add recommended amount of Butoxone or Butyrac 200. Complete filling of the spray tank with water and then add 2 pints of TRITON AG-98, or comparable 80% active nonionic surfactant cleared for use on growing crops, per each 100 gallons of spray to assist in contact activity on dusty weeds. Maintain agitation until spraying is completed.

GOAL 1.6E herbicide can be applied using a post-direct spray rig with 2 flat fan nozzles per row, 1 nozzle on each side of the row. Additional care should be taken to insure proper sprayer adjustment prior to application. For best coverage, it is suggested to use 4 flat fan nozzles per row, 2 nozzles on each side of the row. The two forward nozzles should point forward and downward while the rear nozzles should point to the rear and downward. With either sprayer system, nozzles should be adjusted to cover the weed foliage with minimum contact to the soybean plant. **Do not use cone nozzles.**

SOYBEANS—SPECIFIC USE RESTRICTIONS

In addition to the following, also observe GENERAL USE RESTRICTIONS listed at the end of this label.

- Read and observe all label directions before using. When tank mixing, always read all individual manufacturers' labels. In interpreting all labels for the tank mixture, the most restrictive situations must apply.
- Do not make more than two applications of GOAL 1.6E herbicide per growing season.
- Do not apply more than 2.5 pints (0.5 lb. active) of GOAL 1.6E herbicide per acre during one growing season as a result of preemergence application in no-till (double-crop) or conventional till soybeans, or post-directed in conventional till soybeans. If early preplant application is made, do not apply more than 2.5 pints (0.5 lb. active) of GOAL 1.6E herbicide per acre during one growing season.
- Do not rotate with crops other than soybeans, cotton, spearmint or peppermint for 10 months following a GOAL 1.6E herbicide application. In the event of crop failure, do not plow field under. Field may be replanted to soybeans without tillage.
- Do not apply a post-directed application of GOAL 1.6E herbicide to soybeans after the initial appearance of blooms.

SOYBEAN—SPECIFIC ENVIRONMENTAL HAZARDS

This product is highly toxic to freshwater clams, oysters, aquatic invertebrates and aquatic plants. Do not apply GOAL 1.6E herbicide where visible erosion to aquatic habitats and/or wetlands occurs. (See container for further information on Environmental Hazards.)

DIRECTIONS FOR USE

SPEARMINT AND PEPPERMINT

(FOR USE ONLY IN IDAHO, OREGON AND WASHINGTON)

GENERAL INFORMATION

GOAL 1.6E is a selective herbicide for the control of certain annual grasses and broadleaf weeds in spearmint and peppermint grown in Idaho, Oregon and Washington. Applications should be made only to spearmint and peppermint which are dormant.

WEEDS CONTROLLED

When GOAL 1.6E herbicide is applied as a dormant application at recommended dosages in spearmint and peppermint, the following annual weeds are controlled:

BROADLEAF WEEDS CONTROLLED:

BEDSTRAW, CATCHWEED	NIGHTSHADE, HAIRY
FLIXWEED	ORACH, RED
GROUNDSEL, COMMON	PEPPERWEED, YELLOWFLOWER
LAMBSQUARTERS, COMMON	PIGWEED, REDROOT
LETTUCE, PRICKLY (CHINA LETTUCE)	SHEPHERDSPURSE
	SOWTHISTLE, ANNUAL
MUSTARD, BLUE (PURPLE MUSTARD)	TANSYMUSTARD
	THISTLE, RUSSIAN
MUSTARD, TUMBLE (JIM HILL MUSTARD)	

ANNUAL GRASSES CONTROLLED:

BLUEGRASS, ANNUAL	RYEGRASS, ITALIAN
OATS, WILD	

Control of annual grasses is best obtained when GOAL 1.6E herbicide is applied prior to emergence. Postemergence control of winter annual grasses is generally unsatisfactory if applications are made after the 1 to 2 leaf stage.

DOSAGE

In Western Oregon (Willamette Valley) apply 2.5 to 3.75 pints (0.5 to 0.75 lb. active) of GOAL 1.6E herbicide per acre. In other areas, apply 5 to 10 pints (1 to 2 lb. active) of GOAL 1.6E herbicide per acre before weeds exceed 2 to 4 inches in diameter or height. Add TRITON AG-98 or any comparable 80% active nonionic surfactant cleared for use on growing crops. Use at a rate of 1 quart per 100 gallons of spray solution. Longer residual preemergence control will be obtained when the highest use rate of GOAL 1.6E herbicide is applied.

TIMING AND METHOD OF APPLICATION

Apply only to dormant spearmint or peppermint. Application must be made prior to new spring growth or severe corp injury may result. GOAL 1.6E herbicide should be thoroughly mixed with clean water at recommended concentration and applied at 20 to 40 psi in 20 to 40 gallons of water per acre.

GEOGRAPHIC LOCATION USE DIRECTIONS

WESTERN OREGON

PEPPERMINT
(WILLAMETTE VALLEY)

Apply 2.5 to 3.75 pints (0.5 to 0.75 lb. active) of GOAL 1.6E herbicide from November through February to dormant peppermint only. Treatments in January or February generally provide better residual preemergence control of annual broadleaf weeds. Full season weed control should not be expected from this treatment.

DO NOT APPLY GOAL 1.6E HERBICIDE IN THE WILLAMETTE VALLEY TO MINT THAT HAS BEEN PLOWED.

OREGON AND WASHINGTON (EAST OF CASCADES) AND WESTERN IDAHO

SPEARMINT AND PEPPERMINT

Apply GOAL 1.6E herbicide from December through March to dormant mint only. Later winter applications will provide maximum activity on summer weeds; however, under severe infestations summer grass control may be unsatisfactory.

For best results, fall plowed fields should be harrowed to provide a smooth surface prior to application. Plowed fields should not be harrowed after GOAL 1.6E herbicide has been applied, as soil disturbance will decrease the herbicidal effectiveness. In furrow irrigated fields, corrugating must be done prior to application. Corrugating after application can cover treated rows with untreated soil, resulting in poor weed control.

SPEARMINT AND PEPPERMINT
SPECIFIC USE RESTRICTIONS

In addition to the following, also observe GENERAL USE RESTRICTIONS listed at the end of this label.

- Apply GOAL 1.6E herbicide only to healthy spearmint and peppermint. Do not apply to spearmint or peppermint that has been weakened by disease, nematodes, soil insects, or winter injury, as severe injury may result.

SPECIFIC USE RESTRICTIONS FOR INDIVIDUAL CROPS ARE FOUND UNDER DIRECTIONS FOR USE IN EACH CROP GROUP SECTION.

USE RESTRICTIONS THAT APPLY TO ALL REGISTERED APPLICATIONS ARE LISTED BELOW:

GENERAL USE RESTRICTIONS

- Read and observe all label directions before using. When tank mixing, always read all individual manufacturers' labels. In interpreting all labels for the tank mixture, the most restrictive situations must apply.
- Do not contaminate irrigation water or water used for domestic purposes.
- Do not use any plants treated with GOAL 1.6E herbicide for feed or forage.
- Do not feed or allow animals to graze on any areas treated with GOAL 1.6E herbicide.
- GOAL 1.6E herbicide should be applied only by ground application equipment except as specifically directed on this label or on other approved Rohm and Haas company supplemental labeling.
- Do not apply when weather conditions favor drift. Avoid drift to all non-target areas. GOAL 1.6E herbicide is phytotoxic to plant foliage.
- Thoroughly flush spray equipment (tank, pump, hoses and boom) with clean water before and after each use. Residual GOAL 1.6E herbicide remaining in spray equipment may damage other crops. To assist removal of GOAL 1.6E herbicide residues in spray equipment, TRITON AG-98 or TRITON CS-7™ may be added at the rate of 1 quart per 100 gallons of water during flushing.
- Do not rotate any crops other than cotton, onions, soybeans, or spearmint/peppermint within a 10-month period after treatment.
- Use GOAL 1.6E herbicide only for recommended purposes and at recommended rates.
- Do not treat ditch banks or waterways with GOAL 1.6E herbicide.

WEED CONTROL LIST

BALSAMAPPLE	*Momordica charantia*
BARNYARDGRASS (WATERGRASS)	*Echinochloa crus-galli*
BEDSTRAW, CATCHWEED	*Galium aparine*
BITTERCRESS, LESSER	*Cardamine oligosperma*
BLUEGRASS, ANNUAL	*Poa annua*
BLUEGRASS, KENTUCKY	*Poa pratensis*
BROME, RIPGUT	*Bromus rigidus*
BUCKWHEAT, WILD	*Polygonum convolvulus*
BURCLOVER	*Medicago hipida*
CAMPHORWEED	*Heterotheca subaxillaris*
CANARYGRASS (ANNUAL)	*Phalaris canariensis*
CARPETWEED	*Mollugo verticillata*
CHEESEWEED (MALVA)	*Malva parviflora*
CHICKWEED, COMMON	*Stellaria media*
CHICKWEED, STICKY	*Cerastium viscosum*
CLOVER, RED	*Trifolium pratense*
CLOVER, WHITE	*Trifolium repens*
COCKLEBUR, COMMON	*Xanthium pensylvanicum*
CRABGRASS, LARGE (HAIRY)	*Digitaria sanguinalis*
CRABGRASS, SMOOTH	*Digitaria ischaemum*
CUDWEED, NARROWLEAF	*Gnaphalium falcatum*
CUPGRASS, SOUTHWESTERN	*Eriochloa gracilis*
EVENINGPRIMROSE, CUTLEAF	*Oenothera laciniata*
FIDDLENECK, COAST	*Amsinckia intermedia*
FILAREE, BROADLEAF	*Erodium botrys*
FILAREE, REDSTEM	*Erodium cicutarium*
FILAREE, WHITESTEM	*Erodium moschatum*
FIREWEED (FROM SEED)	*Epilobium angustifolium*
FLIXWEED	*Descurainia sophia*
FOXTAIL, GIANT	*Setaria faberi*
FOXTAIL, GREEN	*Setaria viridis*
FOXTAIL, YELLOW	*Setaria lutescens*
GOOSEGRASS	*Eleusine indica*
GROUNDCHERRY, CUTLEAF	*Physalis angulata*
GROUNDCHERRY, WRIGHT	*Physalis wrightii*
GROUNDSEL, COMMON	*Senecio vulgaris*
HENBIT	*Lamium amplexicaule*
HORSEWEED (MARESTAIL)	*Conyza canadensis*
JIMSONWEED	*Datura stramonium*
JOHNSONGRASS, SEEDLING	*Sorghum halepense*
KNOTWEED, PROSTRATE	*Polygonum aviculare*
LADYSTHUMB (SMARTWEED)	*Polygonum persicaria*
LAMBSQUARTERS, COMMON	*Chenopodium album*
LETTUCE, PRICKLY (CHINA LETTUCE)	*Lactuca serriola*

MALLOW, LITTLE (MALVA)	*Malva parviflora*
MAYWEED (DOG FENNEL)	*Anthemis cotula*
MINERSLETTUCE	*Montia perfoliata*
MORNINGGLORY SPECIES, ANNUAL	*Ipomoea species*
MORNINGGLORY, TALL	*Ipomoea purpurea*
MUSTARD, BLACK	*Brassica nigra*
MUSTARD, BLUE (PURPLE MUSTARD)	*Chorispora tenella*
MUSTARD, COMMON YELLOW	*Brassica campestris*
MUSTARD, TUMBLE (JIM HILL MUSTARD)	*Sisymbrium altissimum*
MUSTARD, WILD	*Brassica kaber*
NETTLE, BURNING	*Urtica urens*
NIGHTSHADE, AMERICAN BLACK	*Solanum nodiflorum*
NIGHTSHADE, BLACK	*Solanum nigrum*
NIGHTSHADE, HAIRY	*Solanum sarachoides*
OATS, WILD	*Avena fatua*
ORACH, RED	*Atriplex rosea*
ORCHARDGRASS	*Dactylis glomerata*
OXALIS (BERMUDA BUTTERCUP)	*Oxalis pes-caprae*
PANICUM, FALL	*Panicum dichotomiflorum*
PEPPERWEED, VIRGINIA	*Lepidium virginicum*
PEPPERWEED, YELLOWFLOWER	*Lepidium perfoliatum*
PIGWEED, PROSTRATE	*Amaranthus blitoides*
PIGWEED, REDROOT	*Amaranthus retroflexus*
PIMPERNEL, SCARLET	*Anagallis arvensis*
PINEAPPLEWEED	*Matricaria matricarioides*
POINSETTIA, WILD	*Euphorbia heterophylla*
PUNCTUREVINE	*Tribulus terrestris*
PURSLANE, COMMON	*Portulaca oleracea*
PUSLEY, FLORIDA	*Richardia scabra*
QUACKGRASS	*Agropyron repens*
RAGWEED, COMMON	*Ambrosia artemisiifolia*
REDMAIDS	*Calandrinia caulescens*
ROCKET, LONDON	*Sisymbrium irio*
RYEGRASS, ITALIAN	*Lolium multiflorum*
RYEGRASS, PERENNIAL	*Lolium perene*
SAGE, LANCELEAF	*Salvia lanceplata*
SANDBUR, FIELD	*Cenchrus incertus*
SANDSPURRY	*Spergularia rubra*
SESBANIA, HEMP	*Sesbania exaltata*
SHEPHERDSPURSE	*Capsella bursa-pastoris*
SIDA, PRICKLY (TEAWEED)	*Sida spinosa*
SIGNALGRASS, BROADLEAF	*Brachiaria platyphylla*
SMARTWEED, PENNSYLVANIA	*Polygonum pensylvanicum*
SORREL, RED (FROM SEED)	*Rumex acetosella*
SOWTHISTLE, ANNUAL	*Sonchus oleraceus*
SOWTHISTLE, COMMON	*Sonchus oleraceus*
SPEEDWELL, BIRDSEYE	*Veronica persica*
SPURGE, PROSTRATE	*Euphorbia supina*
SPURGE, SPOTTED	*Euphorbia maculata*
SPURRY, CORN	*Spergula arvensis*
TANSYMUSTARD	*Descurainia pinnata*
THISTLE, BULL	*Cirsium vulgare*
THISTLE, RUSSIAN	*Salsola kali*
VELVETLEAF	*Abutilon theophrasti*
WITCHGRASS	*Panicum capillare*
WITCHWEED	*Striga asiatica*
WOODSORREL, COMMON YELLOW	*Oxalis stricta*

Made in U.S.A.

9541-B2

SPECIAL LOCAL NEED REGISTRATIONS FOR GOAL 1.6E

State	Use	SLN No.
Arizona	Jojoba	AZ-830013
California	JoJoba	CA-830060
California	Noncrop Uses (Fence Rows, etc.)	CA-830065
Indiana	Mint	IN-840003
Michigan	Dormant Treefruit, Grape	MI-840003
North Carolina	Clary Sage	NC-830023
Wisconsin	Mint	WI-840002

KARATHANE® Liquid Concentrate

AGRICULTURAL FUNGICIDE AND MITICIDE

For Agricultural Use Only or Only for Sale to, Use and Storage by Servicepersons

ACTIVE INGREDIENTS	
2,4-Dinitro-6-octyl* phenyl crotonate	
2,6-Dinitro-4-octyl* phenyl crotonate	35.0%**
Nitrooctyl* phenols (principally dinitro)	2.4%
INERT INGREDIENTS	62.6%
TOTAL	100.0%

* A mixture of 1-methylheptyl, 1-ethylhexyl, and 1-propylpentyl isomers.

** Equivalent to 48% or 4 lbs. Technical Karathane.

EPA Reg. No. 707-71

 9180-R2

Keep Out of Reach of Children
WARNING—AVISO

PRECAUCION AL USUARIO: Si usted no lee ingles, no use este producto hasta que le etiqueta haya sido explicado ampliamente.

NOTICE: Before using this product read the entire Precautionary Statements, Conditions of Sale and Warranty, Directions for Use. Use Restrictions and Storage and Disposal Instructions. If the Conditions of Sale and Warranty are not acceptable, return the product unopened within thirty days of purchase to the place of purchase.

PRECAUTIONARY STATEMENTS

HAZARDS TO HUMANS AND DOMESTIC ANIMALS

CAUTION

Harmful if swallowed, absorbed through the skin or inhaled. Causes severe eye irritation. May cause allergic skin reaction in susceptible individuals. Avoid breathing vapors or spray mist. Do not get in eyes, or on skin, or on clothing. Wash thoroughly with soap and water after handling and before eating or smoking.

When mixing/loading and spraying wear midforearm-to elbow-length chemical-resistant gloves, boots or overshoes, one-piece overalls that have long sleeves and long pants, and face shield or goggles. Mixer/loaders should, in addition, wear a chemical-resistant apron.

Any article of clothing worn while applying product must be cleaned before re-using. Clothing that has been drenched or heavily contaminated should be disposed of in accordance with State or local regulations.

Instead of clothing and equipment specified above, the applicator can use an enclosed tractor cab with filtered air supply.

STATEMENT OF PRACTICAL TREATMENT

IF SWALLOWED: Dilute by giving 2 glasses of water to drink. Call a physician. Never give anything by mouth to an unconscious person.

NOTE TO PHYSICIAN: Because of the presence of xylene (petroleum distillates) it may not be advisable to induce emesis. Careful gastric lavage may be indicated depending on the extent of involvement.

IF INHALED: Remove the victim to fresh air. Treat the person symptomatically. Call a physician.

IF ON THE SKIN: Wash exposed areas with plenty of soap and water. If irritation persists, call a physician.

IF IN EYES: Flush eyes with plenty of water for at least 15 minutes. If irritation persists, call a physician.

ENVIRONMENTAL HAZARDS

This product is toxic to fish. Do not apply directly to water or wetlands. Do not apply when weather conditions favor drift from areas treated. Do not apply where runoff is likely to occur. Do not contaminate water by cleaning of equipment or disposal of wastes. Apply this product only as specified on this label.

PHYSICAL AND CHEMICAL HAZARDS

FLAMMABLE

Do not use, pour, spill or store near heat or open flame. Do not use in the vicinity of pilot lights.

DIRECTIONS FOR USE

It is a violation of Federal law to use this product in a manner inconsistent with its labeling.

RE-ENTRY AND WORKERS PROTECTION STATEMENTS

Do not enter treated areas without protective clothing until sprays have dried. Because certain states may require more restrictive re-entry intervals for various crops treated with this product, consult your State Department of Agriculture for further information. Do not apply this product in such a manner as to directly or through drift, expose workers or other persons. The area being treated must be vacated by unprotected persons.

Written or oral warnings must be given to workers who are expected to be in a treated area or in an area about to be treated with this product. Oral warnings must be given if there is reason to believe that written warnings cannot be understood by workers. When oral warnings are given, warnings shall be given in a language customarily understood by workers. Written warnings must include the following information:

WARNING

(Insert area or field description) treated with KARATHANE LIQUID CONCENTRATE fungicide-miticide on (insert date of application). Do not enter treated areas without protective clothing until sprays have dried. In case of accidental exposure: Call a doctor (physician), clinic or hospital immediately. Explain that the victim has been exposed to KARATHANE LIQUID CONCENTRATE fungicide-miticide and describe his condition. For further information see the "Statement of Practical Treatment."

Oral warnings must include the same information that is in the written warnings.

GENERAL USE INFORMATION

KARATHANE LIQUID CONCENTRATE fungicide-miticide will protect against and eradicate powdery mildew when used according to the label directions.

KARATHANE LIQUID CONCENTRATE fungicide-miticide also provides effective control of suppression of the following mites, including those that may be resistant to other miticides:

Citrus Red	European Red	Plum Nursery
Clover	Peach Silver	Red Spider (two-spotted)

KARATHANE LIQUID CONCENTRATE fungicide-miticide possesses short residual activity and is effective only against motile mite forms. Although of short residual activity, effective reduction of motile mite populations may provide economic mite control for many weeks. When used alone, application should be repeated every 7 to 10 days to kill mites emerging from eggs unaffected by previous sprays. Alternating sprays or tank-mix combinations with other miticides will delay the development of mite resistance to other miticides and improve initial mite kill.

USE RATE DETERMINATION

Carefully read, understand, and follow label rates and use limitations.

Label use rates are for single strength (dilute) sprays. KARATHANE LIQUID CONCENTRATE fungicide-miticide may be used in concentrate or airblast sprayers but correct use rates must be maintained. *Do not use in concentrate sprays in California.*

When spray volumes of less than 100 gallons are required to provide a thorough full coverage spray, the following table may be used to determine the proper use rate:

Recommended KARATHANE LIQUID CONCENTRATE Label Use Rate Per 100 Gals. Spray	Milliliters KARATHANE LIQUID CONCENTRATE Required in Varied Single-Strength Spray Gallonages*				
(Oz.)	1 gal.	2 gals.	3 gals.	5 gals.	10 gals.
3	0.9	1.8	2.6	4.4	5.9
4	1.2	2.4	3.5	5.9	11.8
6	1.8	3.6	5.4	9.0	18.0
8	2.4	4.8	7.0	11.8	23.6

* 15 milliters = 1 teaspoon

COMPATABILITY

KARATHANE LIQUID CONCENTRATE fungicide-miticide is compatible with DIKAR®, DITHANE®, captan, ferbam, glyodin, and thiram fungicides and with most insecticides and miticides including azinphosmethyl, demeton, and parathion. Exceptions are notes under specific use directions. **Do not use in combination with phosphate insecticides or nutritional sprays in California.** When preparing tank-mixes, user should consult current spray compatibility charts or State Cooperative Extension Service Specialists prior to actual use.

SPRAY ADJUVANTS

The addition of agricultural surfactants to KARATHANE LIQUID CONCENTRATE fungicide-miticide sprays will improve initial spray deposits, redistribution, and weatherability. The following spray adjuvants have been especially formulated to optimize the performance of foliar applied agricultural chemicals:

TRITON AG-98™	A low foam nonionic general purpose spray adjuvant. May be used in dilute or concentrate sprays applied by aircraft or ground equipment.
TRITON B-1956®	A water-dispersible, resin-based nonionic surfactant which resists re-wetting and removal by rain. Effective with dilute sprays applied by ground equipment.
TRITON CS-7™	A spreader-binder designed specifically for use in concentrate and low volume sprays applied by aircraft or ground equipment.

Place KARATHANE LIQUID CONCENTRATE fungicide-miticide into solution prior to adding an adjuvant. Read and carefully observe the precautionary statements and all other information appearing on both product labels prior to spray preparation.

APPLICATION

Thorough full coverage foliar sprays generally result in optimum powdery mildew and mite control. To achieve good foliar coverage use proper spray pressure, gallonage per acre, spray nozzles, discs, nozzle spacing and tractor speed. Consult spray nozzle and accessory catalogues for specific information on proper equipment calibration. When applied with hand pressurized sprayers, thoroughly spray plant foliage until runoff.

PEST MONITORING

KARATHANE LIQUID CONCENTRATE fungicide-miticide should be applied when the first signs of powdery mildew are noted or when adult mites are observed on plant foliage.

RESTRICTIONS

Users should carefully read, understand, and follow all use restrictions prior to using KARATHANE LIQUID CONCENTRATE.

FRUITS

CROP	CONTROL OF:	RATE OF KARATHANE LIQUID CONCENTRATE PER APPLICATION OZ./100 GAL.	OZ./A	TIMING AND REMARKS	RESTRICTIONS
Apples (orchard)	Powdery mildew	4 to 6	20 to 48	For powdery mildew begin application at delayed dormant to pink stage and continue sprays at 7 to 14 day intervals until terminal growth is completed or mildew activity has ceased. Applications should not be more than 14 days apart. Control measures are needed from the time buds break or at first signs of mildew activity, and should be continued through terminal growth and new bud formation to reduce over-wintering infection in the buds. From 4 to 8 applications are usually required per season depending on variety, location and disease severity. Thorough coverage is necessary. When mildew growth is heavy add a spreader-sticker for thorough wetting of mildew spores and mycelium. Certain spreader-stickers will reduce the deposit on maturing fruit. TRITON B-1956 is recommended at 1 to 2 ozs. per 100 gals. of dilute spray.	Do not apply when day temperatures are expected to exceed 90°F. Do not use with oil or oil-based pesticides or apply to foliage within 30 days after an oil spray application. Do not use within 21 days of harvest. Do not graze treated areas to livestock or feed clover crops grown in treated orchards to animals.
	Clover mites European red mites Red spider (two-spotted) mites			For mites, apply at first signs of an infestation and continue on a 7 to 14 day schedule as long as necessary. Do not use more than ½ lb. of 25% wettable parathion with KARATHANE LIQUID CONCENTRATE from pink through second cover or apply in separate applications. Other phosphate insecticides such as azinphosmethyl, demeton, and malathion, appear to be compatible with KARATHANE LIQUID CONCENTRATE.	
Apples (nursery)	Powdery mildew Red spider mites Clover mites	4	—	On non-bearing nursery stock, apply at first sign of mildew powdery mildew and/or mites, at 7 to 10 day intervals as necessary.	Do not graze treated areas to livestock or feed cover crops grown in treated areas to animals.
Apricots	Powdery mildew Clover mites European red mites Peach silver mites Plum rust mites Two-spotted spider mites	4	20 to 32	Begin application at pink bud and petal fall and in subsequent sprays if disease persists, and post harvest.	Do not apply within 45 days of harvest. Do not graze treated areas to livestock or feed cover crops grown in treated orchards to animals. Avoid drift to bearing trees. Do not use in California.
Cherries (nursery)	Refer to Apples (nursery)				Do not apply to trees that will bear fruit within one year of treatment.
Citrus (nursery)	Citrus red mites	4 to 8	—	Apply two sequential sprays; the first when mites appear and the second 10 to 14 days later.	Apply only to nursery trees or to young trees that will not bear fruit within one year. Avoid drift to bearing trees. Do not use in California.
Grapes	Powdery mildew	4 to 8	12 to 24	Begin applications when mildew first appears (usually during early stages of foliage growth) and repeat at 10 to 14 day intervals. From 2 to 6 applications may be required.	Do not use within 21 days of harvest. Do not use in California.
Peaches (orchard)	Refer to Apricots				Do not use in California.

CROP	CONTROL OF:	RATE OF KARATHANE LIQUID CONCENTRATE PER APPLICATION OZ./100 GAL.	OZ./A	TIMING AND REMARKS	RESTRICTIONS
Peaches (nursery)	Refer to Apples (nursery)				Do not use in California.
Pears	Refer to Apples (orchard)				

VEGETABLES

CROP	CONTROL OF:	RATE OF KARATHANE LIQUID CONCENTRATE PER APPLICATION OZ./100 GAL.	OZ./A	TIMING AND REMARKS	RESTRICTIONS
Cantaloupes	Powdery mildew	3 to 4	3 to 6	Apply when powdery mildew signs are first observed and repeat at 7 to 10 day intervals. From 3 to 10 applications may be required.	Do not use within 7 days of harvest. Do not use in California.
Cucumbers	Powdery mildew	3 to 4	3 to 6	Apply in a light, thorough spray. Begin when mildew first appears and repeat at 5 to 7 day intervals or as necessary. Spray a maximum of 3 to 4 times to an individual fruit.	Do not apply when temperatures exceed 90°F. Do not use within 7 days of harvest. Do not use in California.
Melons	Refer to Cantaloupes				Do not use in California.
Pumpkins	Refer to Cantaloupes				Do not use in California.
Squash	Refer to Cantaloupes			On summer squash, spray a maximum of 3 to 4 times to an individual fruit.	Do not use in California

STORAGE AND DISPOSAL

STORAGE
Do not store at temperature above 90°F for periods longer than 48 hours.

PESTICIDE DISPOSAL
Do not contaminate water, food or feed by storage or disposal. Pesticide wastes are toxic. Improper disposal of excess pesticide, spray mixture or rinsate is a violation of Federal law. If these wastes cannot be disposed of by use according to label instructions, contact your State Pesticide or Environmental Control Agency, or the Hazardous Waste representative at the nearest EPA Regional Office for guidance.

CONTAINER DISPOSAL
Triple rinse (or equivalent). Then puncture and dispose of in a sanitary landfill, or by incineration, or if allowed by state and local authorities, by burning. If burned, stay out of smoke.

STEPS TO BE TAKEN IN CASE MATERIAL IS RELEASED OR SPILLED—Eliminate ignition sources. Ventilate area. Avoid breathing vapors. Use approved self-contained breathing apparatus or air respirators for large spills in confined areas. Dike or contain liquid. Scoop or shovel into a suitable container for recovery or disposal. Remove contaminated clothing promptly and wash exposed skin areas with soap and water. Wash clothing before re-use. Keep spill out of all sewers and open bodies of water. REFER TO PRECAUTIONARY STATEMENTS.

CONDITIONS OF SALE AND WARRANTY

Rohm and Haas warrants that the product conforms to its chemical description and is reasonably fit for the purpose stated on the label only when used in accordance with label directions under normal conditions of use. **ROHM AND HAAS MAKES NO OTHER EXPRESS OR IMPLIED WARRANTIES EITHER OF MERCHANTABILITY OR FITNESS FOR A PARTICULAR USE.** Handling, storage and use of the product by Buyer or User are beyond the control of Rohm and Haas and Seller. Risks such as crop injury, ineffectiveness or other unintended consequences resulting from, but not limited to, weather or soil conditions, presence of other materials, disease, pests, drift to other crops or property or failure to follow label directions will be assumed by the Buyer or User. **IN NO CASE WILL ROHM AND HAAS OR SELLER BE HELD LIABLE FOR CONSEQUENTIAL, SPECIAL OR INDIRECT DAMAGES RESULTING FROM THE HANDLING, STORAGE OR USE OF THIS PRODUCT.**

DIKAR, DITHANE, KARATHANE, TRITON, AND B-1956 are registered trademarks of the Rohm and Haas Company.

9180-B2

KARATHANE® WD
AGRICULTURAL FUNGICIDE AND MITICIDE

For Agricultural Use Only or Only For Sale to, Use, and Storage by Servicepersons

ACTIVE INGREDIENTS

2,4-Dinitro-6-octyl* phenylcrotonate 2,6-Dinitro-4-octyl* phenylcrotonate	18.25%
Nitrooctyl* phenol (principally dinitro)	1.25%
INERT INGREDIENTS	80.50%
TOTAL	100.00%

* A mixture of 1-methylheptyl, 1-ethylhexyl, and 1-propylpentyl isomers

EPA Reg. No. 707-53

CAUTION
Keep Out of Reach of Children

NOTICE: Before using this product, read the entire Precautionary Statements, Conditions of Sale and Warranty, Directions for Use, Use Restrictions and Storage and Disposal Instructions. If the Conditions of Sale and Warranty are not acceptable, return the product unopened within thirty days of purchase to the place of purchase.

PRECAUTIONARY STATEMENTS

HAZARDS TO HUMANS AND DOMESTIC ANIMALS
Harmful if swallowed, absorbed through the skin or inhaled. Causes moderate eye, skin, nose and throat irritation. May cause allergic skin reaction in susceptible individuals. Avoid contact with skin, eyes, and clothing. Avoid breathing dust or spray mist. Wash thor-

oughly with soap and water after handling and before eating or smoking.

When mixing/loading and spraying wear midforearm- to-elbow-length chemical-resistant gloves, boots or overshoes, one-piece overalls that have long sleeves and long pants, and face shield or goggles. Mixer/loaders should, in addition, wear a chemical-resistant apron.

Any article of clothing worn while applying product must be cleaned before re-using. Clothing that has been drenched or heavily contaminated should be disposed of in accordance with State or local regulations.

Instead of clothing and equipment specified above, the applicator can use an enclosed tractor cab with filtered air supply.

STATEMENT OF PRACTICAL TREATMENT

IF SWALLOWED: Drink 1 or 2 glasses of water. Call a physician. Do not give anything by mouth to an unconscious person.

NOTE TO PHYSICIAN: Administer sedatives if necessary to control apprehension and excitement. Be prepared to counteract respiratory depression and hypotension, which may occur following the administration of anticonvulsants.

IF INHALED: Remove victim to fresh air. Treat the victim symptomatically. If irritation persists, call a physician.

IF ON SKIN: Wash exposed areas with plenty of soap and water. If irritation persists, call a physician.

IF IN EYES: Flush eyes with plenty of water for at least 15 minutes. If irritation persists, call a physician.

ENVIRONMENTAL HAZARDS

This product is toxic to fish. Do not apply directly to water or wetlands. Do not apply where runoff is likely to occur. Do not apply when weather conditions favor drift from areas treated. Do not contaminate water by cleaning of equipment or disposal of wastes. Apply this product only as specified on this label.

DIRECTIONS FOR USE

It is a violation of Federal law to use this product in a manner inconsistent with its labeling.

RE-ENTRY AND WORKER PROTECTION STATEMENTS

Do not enter treated areas without protective clothing until sprays have dried. Because certain states may require more restrictive reentry intervals for various crops treated with this product, consult your State Department of Agriculture for further information. Do not apply this product in such a manner as to directly or through drift, expose workers or other persons. The area being treated must be vacated by unprotected persons.

Written or oral warnings must be given to workers who are expected to be in a treated area or in a area about to be treated with this product. Oral warnings must be given if there is reason to believe that written warnings cannot be understood by workers. When oral warnings are given, warnings shall be given in a language customarily understood by the workers. Written warnings must include the following information:

CAUTION

(Insert area or field description) treated with KARATHANE WD on (insert date of application). Do not enter treated areas without protective clothing until sprays have dried. In case of accidental exposure: Call a doctor (physician), clinic or hospital immediately. Explain that the victim has been exposed to KARATHANE WD and describe his condition. For further information see the "Statement of Practical Treatment."

Oral warnings must include the same information that is in the written warnings.

GENERAL USE INFORMATION

KARATHANE WD fungicide-miticide will protect against and eradicate powdery mildew when used according to the label directions.

KARATHANE WD fungicide-miticide also provides effective control or suppression of the following mites, including those that may be resistant to other miticides:

Citrus Red	European Red	Plum Nursery
Clover	Peach Silver	Red Spider (two-spotted)

KARATHANE WD fungicide-miticide possesses short residual activity and is effective only against motile mite forms. Although of short residual activity, effective reduction of motile mite populations may provide economic mite control for many weeks. When used alone, application should be repeated every 7 to 10 days to kill mites emerging from eggs and unaffected by previous sprays. Alternating sprays or tank mix combinations with other miticides will delay the development of mite resistance to other miticides and improve initial mite kill.

USE RATE DETERMINATION

Carefully read, understand, and follow label rates and use limitations.

Label use rates are for single strength (dilute) sprays. KARATHANE WD fungicide-miticide may be used in concentrate or airblast sprayers but correct use rates must be maintained. Do not use in concentrate sprays in California.

When spray volumes of less than 100 gallons are required to provide a thorough full coverage spray, the following table may be used to determine the proper use rate:

Recommended KARATHANE WD Fungicide-Miticide Label Use Rate Per 100 Gals. Spray	Teaspoons KARATHANE WD Fungicide-Miticide Required in Varied Single Strength Spray Gallonages*				
(Oz.)	1 gal.	2 gals.	3 gals.	5 gals.	10 gals.
4	1/2	1	1 1/2	2 1/2	5
6	3/4	1 1/2	2 1/4	3 3/4	7 1/2
8	1	2	3	5	10
12	1 1/2	3	4 1/2	7 1/2	15
16	2	4	6	10	20

* 1 tablespoon = 3 teaspoons

MIXING

Slowly place into spray tank as it is being filled or premix as a slurry prior to adding to spray tank. Add other co-applied fungicides, miticides, insecticides, growth regulators, micro-nutrients, and spray adjuvants after KARATHANE WD fungicide-miticide has been placed into suspension.

COMPATIBILITY

KARATHANE WD fungicide-miticide is compatible with KI-KAR®, DITHANE®, captan, ferbam, glyodin, and thiram fungicides and with most insecticides and miticides including azinphos-methyl, demeton, and parathion. Exceptions are noted under specific use directions. When preparing tank mixes, user should consult current spray compatibility charts or State Extension Service Specialists prior to actual use.

SPRAY ADJUVANTS

The addition of agricultural surfactants to KARATHANE WD fungicide-miticide sprays will improve initial spray deposits, redistribution, and weatherability. The following spray adjuvants have been especially formulated to optimize the performance of foliar-applied agricultural chemicals:

TRITON AG-98™	A low foam, nonionic, general purpose spray adjuvant. May be used in dilute or concentrate sprays applied by aircraft or ground equipment.
TRITON B-1956®	A water-dispersible, resin-based nonionic surfactant which resists rewetting and removal by rain. Effective with dilute sprays applied by ground equipment.
TRITON CS-7™	A spreader-binder designed specifically for use in concentrate and low volume sprays applied by aircraft or ground equipment.

Place KARATHANE WD fungicide-miticide into suspension prior to adding an adjuvant to the spray mixture. Read and carefully observe the precautionary statements and all other information appearing on both product labels prior to spray preparation.

APPLICATION

Thorough, full-coverage foliar sprays generally result in optimum powdery mildew and mite control. To achieve good foliar coverage use proper spray pressure, gallonage per acre, spray nozzles, discs, nozzle spacing, and tractor speed. Consult spray nozzle and accessory catalogues for specific information on proper equipment calibration.

When applied with hand pressurized sprayers, thoroughly spray plant foliage until runoff.

PEST MONITORING

KARATHANE WD fungicide-miticide should be applied when the first signs of powdery mildew are noted or when mites are observed on plant foliage.

RESTRICTIONS

Users should carefully read, understand and follow all use restrictions prior to using KARATHANE WD fungicide-miticide.

FRUITS

CROP	CONTROL OF:	RATE OF KARATHANE WD PER APPLICATION OZ./100 GAL.	LB./A	TIMING AND REMARKS	RESTRICTIONS
Apples (orchard)	Powdery mildew	8 to 12	2½ to 3½	For powdery mildew begin application at delayed dormant to pink stage and continue sprays at 7 to 14 day intervals until terminal growth is completed or mildew activity has ceased. Applications should not be more than 14 days apart. Control measures are needed from the time buds break or at first signs of mildew activity, and should be continued through terminal growth and new bud formation to reduce overwintering infection in the buds. From 4 to 8 applications are usually required per season depending on variety, location and disease severity. Thorough coverage is necessary. In the East, ¼ lb. per 100 gals. water may be used on varieties (such as Delicious, McIntosh & Stayman) which are slightly to moderately susceptible to powdery mildew, and on susceptible varieties under light infection conditions. Apply every 5 to 7 days as needed or until terminal growth is completed. When mildew growth is heavy add a spreader-sticker for thorough wetting of mildew spores and mycelium. Certain spreader-stickers will reduce the deposit on maturing fruit. TRITON B-1956 is recommended at 1 to 2 ozs. per 100 gals. of dilute spray	Do not use within 21 days of harvest. Do not graze livestock in treated areas or feed cover crops grown in treated orchards to animals.
	Clover mites European red mites Red spider (two-spotted) mites			For mites, apply at first signs of an infestation and continue on a 7 to 14 day schedule as long as necessary. Do not use more than ½ lb. of 25% wettable parathion with KARATHANE WD fungicide-miticide from pink through second cover or apply in separate applications. Other phosphate insecticides such as azinphosmethyl, demeton, and malathion, are compatible with KARATHANE WD fungicide-miticide.	
Apples (nursery)	Powdery mildew Red spider mites Clover mites	8	—	On non-bearing nursery stock, apply at first sign of powdery mildew and/or mites, at 7 to 10 day intervals or as necessary.	Do not graze livestock in treated areas or feed cover crops grown in treated areas to animals.
Apricots	Powdery mildew Clover mite European red mite Peach silver mite Plum rust mite Two-spotted spider mite	8 to 12	2½ to 3¼	Begin application at pink bud and petal fall and in subsequent spring sprays if disease persists, and post harvest.	Do not apply within 45 days of harvest. Do not graze livestock in treated areas or feed cover crops grown in treated orchards to animals. Do not use in California.
Cherries (nursery)	Refer to apples (nursery)				Do not apply to trees that will bear fruit within one year of treatment.
Citrus (nursery)	Citrus red mite	8	—	First apply when mites are noted and repeat 10 to 14 days later.	Apply only to nursery trees or to young trees which will not bear fruit within one year.
Grapes	Powdery mildew Mites	8	1 to 2½	Begin applications when mildew first appears (usually during early stages of foliage growth) and repeat at 10 to 14 day intervals. From 2 to 6 applications may be required.	Do not use within 21 days of harvest. Do not use in California.
Peaches (orchard)	Refer to Apricots				Do not use in California.

CROP	CONTROL OF:	RATE OF KARATHANE WD PER APPLICATION OZ./100 GAL.	LB./A	TIMING AND REMARKS	RESTRICTIONS
Peaches (nursery)	Refer to Apples (nursery)				Do not use in California.
Pears	Refer to Apples (orchard)				

VEGETABLES

CROP	CONTROL OF:	RATE OF KARATHANE WD PER APPLICATION OZ./100 GAL.	LB./A	TIMING AND REMARKS	RESTRICTIONS
Cantaloupes	Powdery mildew	6 to 8	1/3 to 3/4	Apply when powdery mildew signs are first observed and repeat at 7 to 10 day intervals. From 3 to 10 applications may be required. On cucumbers and summer squash, spray a maximum of 3 to 4 times to an individual set of fruit.	Do not use within 7 days of harvest. Do not use in California.
Cucumbers (field)	Refer to Cantaloupes				Do not use in California.
Cucumbers (greenhouse)	Powdery mildew	4 to 8	—	Apply in a light thorough spray. Begin when mildew first appears and repeat at 5 to 7 day intervals or as necessary.	Do not apply when temperatures exceed 90°F. Do not use within 7 days of harvest. Do not use in California.
Melons	Refer to Cantaloupes				Do not use in California.
Pumpkins	Refer to Cantaloupes				Do not use in California.
Squash	Refer to Cantaloupes				Do not use in California.

FLOWERS

CROP	CONTROL OF	TIMING AND REMARKS
African violets, Asters, Begonias, Calendulas, Chrysanthemums, Cinerarias, Dahlias, Delphiniums, Euonymus, Gardenia, Hydrangeas, Larkspur, Lilacs	Powdery mildew, Red spider mites, Clover mites	GENERAL USE DIRECTIONS **Greenhouse**—Use 4 to 6 ozs./100 gals. water or 2 level tablespoons per 12½ gals. water. Begin applications when first signs of powdery mildew or mites are observed and apply thorough coverage sprays at 7 to 10 intervals. **Outdoors**—Use 6 to 8 ozs./100 gals. water or 2 level teaspoons per 3 gals. water. Begin applications when first signs of mildew or mites are observed and apply thorough coverage sprays at 5 to 7 day intervals or as necessary. Do not apply when greenhouse or outdoor temperatures are expected to reach or exceed 90°F. Do not use concentrate sprays in California.
Roses (greenhouse)		Apply in a light misting spray when signs of powdery mildew or mites are first observed and repeat at 5 to 7 day intervals or as long as necessary. The use of TRITON B-1956 will improve uniformity of spray deposits. Do not apply if temperatures are expected to reach or exceed 90°F.
Roses (outdoors)		Use includes multiflora roses. Apply 6 to 8 ozs. per 100 gals. when first signs of powdery mildew or mites are observed and repeat at 7 to 10 day intervals. Use thorough full coverage sprays. The addition of TRITON B-1956 will improve uniformity of spray deposits.

STORAGE AND DISPOSAL

STORAGE

Do not store at temperatures above 90°F for periods longer than 48 hours.

PESTICIDE DISPOSAL

Do not contaminate water, food or feed by storage or disposal. Pesticide wastes are toxic. Improper disposal of excess pesticide, spray mixture or rinsate is a violation of Federal law. If these wastes cannot be disposed of by use according to label instructions, contact your State Pesticide or Environmental Control Agency, or the Hazardous Waste Representative at the nearest EPA Regional Office for guidance.

CONTAINER DISPOSAL

PAPER BAGS: Completely empty bag into application equipment. Then dispose of in a sanitary landfill or by incineration, or, if allowed by State and local authorities, by burning. If burned stay out of smoke.

FIBER DRUMS: Completely empty liner by shaking and tapping sides and bottom to loosen clinging particles. Empty residue into application equipment. Then dispose of drum and liner in a sanitary landfill, or by incineration if allowed by State and Local authorities.

STEPS TO BE TAKEN IN CASE MATERIAL IS RELEASED OR SPILLED: Scoop or shovel solid material into a suitable container for recovery or disposal. Remove contaminated clothing promptly and wash exposed skin areas with soap and water. Wash clothing before re-use. Keep spill out of all sewers and open bodies of water. Refer to Precautionary Statements.

CONDITIONS OF SALE AND WARRANTY

Rohm and Haas warrants that the product conforms to its chemical description and is reasonably fit for the purpose stated on the label when used in accordance with label directions under normal conditions of use. **ROHM AND HAAS MAKES NO OTHER EXPRESS OR IMPLIED WARRANTIES EITHER OF MERCHANTABILITY OR FITNESS FOR A PARTICULAR USE.** Handling, storage and use of the product by Buyer or User are beyond the control of Rohm and Haas and Seller. Risks such as crop injury, ineffectiveness or other unintended consequences resulting from, but not limited to, weather or soil conditions, presence of other materials, disease, pests, drift to other crops or property or failure to follow label directions will be assumed by Buyer or User. **IN NO CASE WILL ROHM AND HAAS OR SELLER BE HELD LIABLE FOR CONSEQUENTIAL, SPECIAL OR INDIRECT DAMAGES RESULTING FROM THE HANDLING, STORAGE OR USE OF THIS PRODUCT.**

DIKAR, DITHANE, KARATHANE, TRITON and B-1956 are registered trademarks of the Rohm and Haas Company.

8157-B1

SPECIAL LOCAL NEED REGISTRATION FOR KARATHANE WD

STATE	USE	SLN NO.
Oregon	Caneberries and Gooseberries	OR-790027

KELTHANE® EC

AGRICULTURAL MITICIDE EMULSIFIABLE LIQUID

For Agricultural Use Only or Only for Sale To, Use, and Storage By Servicepersons

ACTIVE INGREDIENTS	
1,1-Bis (chlorophenyl)-2,2,2-trichloroethanol	18.5%*
INERT INGREDIENTS	81.5%
TOTAL	100.0%

* Equivalent to 1.6 lbs. Active Ingredient Per Gallon

** Contains Xylene Range Petroleum Distillates

EPA Reg. No. 707-59 EPA Est. No. 279-CA-1

Kelthane® is a registered trademark of Rohm and Haas Company.

8765-R7

WARNING
Keep Out of Reach of Children
AVISO

PRECAUCION AL USUARIO: Si usted no lee ingles, no use este producto hasta que le etiqueta haya sido explicado ampliamente.

NOTICE: Before using this product, read the entire Precautionary Statements, Conditions of Sale and Warranty, Directions for Use, and Storage and Disposal Instructions on side panel and attached use directions. If the Conditions of Sale and Warranty are not acceptable, return the product unopened within thirty days of purchase to the place of purchase.

PRECAUTIONARY STATEMENTS

HAZARDS TO HUMANS AND DOMESTIC ANIMALS

WARNING

Can cause eye injury and skin irritation. Harmful if swallowed, absorbed through the skin or inhaled. Do not get in eyes, on skin or on clothing. Avoid breathing vapors or spray mist. Wear goggles or face shield when handling. Wash thoroughly with soap and water after handling and before eating or smoking. Remove contaminated clothing and wash before re-use.

STATEMENT OF PRACTICAL TREATMENT

IF SWALLOWED: Drink 1 to 2 glasses of water. Call a physician and have the stomach emptied immediately. Do not induce vomiting or give anything by mouth to an unconscious person.

NOTE TO PHYSICIAN: Because this product contains a chlorinated hydrocarbon insecticide, vomiting is recommended. Probable mucosal damage may contraindicate the use of gastric lavage. Adrenergic amines may cause myocardial irritability in the presence of chlorinated hydrocarbons and convulsions may be a prominent symptom of poisoning. This product contains xylene (petroleum distillate).

IF INHALED: Remove victim to fresh air. Treat the victim symptomatically. Call a physician if problem continues.

IF ON SKIN: Wash exposed areas with plenty of soap and water. If irritation persists, call a physician.

IF IN EYES: Flush eyes with plenty of water for at least 15 minutes. Call a physician.

ENVIRONMENTAL HAZARD

This product is toxic to fish. Do not apply directly to lakes, streams or ponds. Do not apply where runoff is likely to occur. Do not contaminate water by cleaning of equipment or disposal of wastes. Do not apply when weather conditions favor drift from areas treated. Apply this product only as specified on this label.

PHYSICAL AND CHEMICAL HAZARDS

FLAMMABLE

Do not use, pour, spill or store near heat or open flame. Do not use in the vicinity of pilot lights.

CONDITIONS OF SALE AND WARRANTY

Rohm and Haas warrants that the product conforms to its chemical description and is reasonably fit for the purpose stated on the label only when used in accordance with label directions under normal conditions of use. **ROHM AND HAAS MAKES NO OTHER EXPRESS OR IMPLIED WARRANTIES EITHER OF MERCHANTABILITY OR FITNESS FOR A PARTICULAR USE.** Handling, storage and use of the product by Buyer or User are beyond the control of Rohm and Haas and Seller. Risks such as crop injury, ineffectiveness or other unintended consequences resulting from, but not limited to, weather or soil conditions, presence of other materials, disease, pests, drift to other crops or property or failure to follow label directions will be assumed by the Buyer or User. **IN NO CASE WILL ROHM AND HAAS OR SELLER BE HELD LIABLE FOR CONSEQUENTIAL, SPECIAL OR INDIRECT DAMAGES RESULTING FROM THE HANDLING, STORAGE OR USE OF THIS PRODUCT.**

STORAGE AND DISPOSAL

Do not contaminate water, food or feed by storage or disposal.

STORAGE

Keep from freezing. Store at 32°–110°F.

PESTICIDE DISPOSAL

Pesticide wastes are toxic. Improper disposal of excess pesticide spray mixture or rinsate is a violation of Federal law. If these wastes cannot be disposed of by use according to label instructions, contact your State Pesticide or Environmental Control Agency or the Hazardous Waste representative at the nearest EPA Regional Office for guidance.

CONTAINER DISPOSAL

Triple rinse (or equivalent). Puncture and dispose of in a sanitary landfill, or incinerate or if allowed by state and local authorities, by burning. If burned, stay out of smoke.

STEPS TO BE TAKEN IN CASE MATERIAL IS RELEASED OR SPILLED: Eliminate ignition source. Ventilate area. Avoid breathing vapors. Use MSHA/NIOSH self-contained breathing apparatus or air mask for large spills in confined areas. Dike the spill with inert material (sand, earth, etc.) and transfer the liquid or solid diking material to separate containers for recovery or disposal.

Remove contaminated clothing promptly and wash exposed skin areas with soap and water. Wash clothing before re-use. Keep spill out of all sewers and open bodies of water. Refer to Precautionary Statements.

RE-ENTRY AND WORKER PROTECTION STATEMENTS

Do not enter treated areas without protective clothing until sprays have dried. Because certain states may require more restrictive re-entry intervals for various crops treated with this product, consult your State Department of Agriculture for further information. Do not apply this product in such a manner as to directly or through drift, expose workers or other persons. The area being treated must be vacated by unprotected persons.

Written or oral warnings must be given to workers who are expected to be in a treated area or in an area about to be treated with this product. Oral warnings must be given if there is reason to believe that written warnings cannot be understood by workers. When oral warnings are given, warnings shall be given in a language customarily understood by workers. Written warnings must include the following information:

WARNING

(Insert area or field description) treated with KELTHANE EC miticide on (insert date of application). Do not enter treated areas without protective clothing until sprays have dried. In case of accidental exposure: Call a doctor (physician), clinic or hospital immediately. Explain that the victim has been exposed to KELTHANE EC and describe his condition. For further information see the "Statement of Practical Treatment."

Oral warnings must include the same information that is in the written warnings.

DIRECTIONS FOR USE

It is a violation of Federal law to use this product in a manner inconsistent with its labeling.

GENERAL INFORMATION

KELTHANE EC gives high initial kill against most species of agricultural mites including the following:

apple rust	plum nursery
Bermuda	privet
Bryobia (clover, brown, almond)	Schoene
	six-spotted
Carmine	spruce
citrus flat	strawberry (Atlantic)
citrus red (purple)	
citrus rust	Texas citrus
cyclamen	tomato russet
desert	tropical
dryberry	two-spotted (red spider)
European red	
McDaniel	Willamette
Pacific	yellow (carpini)
peach silver	Yuma

KELTHANE EC is a specific miticide and will not kill bees and beneficial insect predators when used as directed.

KELTHANE EC miticide can be applied alone or with other pesticides in recommended spray schedules. It should not be used with lime or with materials not compatible with solvent-based pesticide formulations. For further details see specific recommendations on this label; then consult current State Agricultural Experiment Station spray schedules or State Extension Service Specialists for additional details on compatibility, dosage, timing and uses.

KELTHANE EC miticide is mixed with water and may be applied with conventional dilute or concentrate ground sprayers or with aircraft. Follow the dosage rates per acre listed below regardless of the gallonage of the spray applied. The higher dosages suggested are for larger trees or plants or heavier mite populations. Uniform coverage is necessary but the optimum spray gallonage varies with the type of equipment, size and density of the crop plants. Use in 5 to 10 gallons of spray per acre in aircraft.

To avoid excessive residue at harvest do not exceed the maximum dosage per acre, or the time restriction between last application and harvest for each crop group listed in the table below.

SUMMARY OF DOSAGE RATES AND LIMITATIONS

Crop	Dosage Qts. per Acre Min.	Max.	Qts. Per 100 Gals. Dilute Sprays*	Cutoff Days Before Harvest
Alfalfa and Clover (both grown for seed only—use restricted to California, Idaho, Nevada, Oregon, and Washington)	1½	3¾	1½	—
Chestnuts, Filberts, Hickory Nuts Pecans, Walnuts	4	10	1	14
Apples, Crabapples, Pears, Quinces	4	10	1	7
Beans (dry)	1½	2	—	45
Beans (dry, green, lima)	1	1½	—	7
Beans (dry, green, lima) (California only)	2	2½	—	21
Cantaloupes, Cucumbers, Melons, Pumpkins, Squash, Watermelons	¾	1	—	2
Cherries	4	6	1	7
Citrus (all)	4	10	1	7
Cotton	2	4	—	14
Grapes, Hops	1½	3	1	7
Mint	2	3	—	30
Peaches, Apricots, Nectarines	4	6	1	14
Plums, Prunes, Figs	4	5	1	7
Raspberries, Caneberries	1½	3	1	2
Strawberries—Two-spotted and similar mites	1	2	2	2
Strawberries—Cyclamen mite	—	6	1½	2
Tomatoes, Peppers	1	2	—	2
Clover Mites on buildings	Full Coverage		¾–1	—
Lawn and Turf Grasses	Full Coverage		¾–1	—

* Gallonage of spray will vary with size and number of trees or plants per acre and the density of the foliage. In any case, thorough coverage of all parts of the plant is essential for good mite control

Do not apply to any crop within 2 weeks following sulfur dust or spray.

In general, KELTHANE EC miticide applications should be made as the various mite species begin to appear in threatening numbers and repeated as necessary for adequate seasonal control, except as noted or recommended locally.

VEGETABLE AND FIELD CROPS

ALFALFA AND CLOVER (BOTH GROWN FOR SEED ONLY—USE RESTRICTED TO CALIFORNIA, IDAHO, NEVADA, OREGON AND WASHINGTON)—For carmine, Pacific, strawberry (Atlantic), and two-spotted mites. Use 1½ to 3¾ quarts of KELTHANE EC miticide per acre in sufficient water for thorough coverage. Apply when mites become active and repeat if necessary. Treated fields are not to be grazed, and crop residues and cuttings are not to be fed to livestock or baled for such use. The fields must be cleared by burning crop residues or by discing into the soil.

BEANS, CANTALOUPES, CUCUMBERS, HOPS, MELONS, PEPPERS, PUMPKINS, SQUASH (WINTER AND SUMMER), TOMATOES, WATERMELONS—Pacific, tomato russet, tropical and two-spotted mites. Apply at first signs of mite buildup and repeat as necessary. Do not feed crop residues to dairy or meat animals.

COTTON—Desert, Pacific and two-spotted mites. Use 2 to 4 quarts KELTHANE EC miticide per acre, depending on size of the plant and density of the foliage. Apply with ground sprayers in sufficient water to provide thorough coverage. Begin application when mites are first noticed and repeat if necessary. Do not feed cotton trash to meat or dairy animals.

MINT—Peppermint and spearmint for commercial oil production. For control of two-spotted mites use KELTHANE EC miticide at 2 to 3 quarts per acre. Make application when mites become active. Do not apply within 30 days of harvest and do not make more than one application per crop season. Do not feed treated fresh or spent hay to livestock.

FRUITS AND NUTS

APPLES, CRABAPPLES, PEARS, QUINCES—To clean up clover and European red mites and to control apple rust, McDaniel, Pacific, Schoene, two-spotted, Willamette and yellow (carpini) mites. Make application in second to later cover sprays or whenever mite population appears. In areas where the preventive schedule is accepted practice, use two successive applications at ½ quart per 100 gallons dilute spray applied at 10 to 14 day intervals. Make thorough full coverage application. Note: KELTHANE EC miticide may cause fruit spots on pears under certain conditions and should be used at the lower dosages and with caution on this crop. To avoid possible injury do not use glyodin fungicide with KELTHANE EC miticide in late cover sprays. Do not apply to apples, crabapples, pears, quinces after fruit formation in California, Washington or Oregon. Use KELTHANE 35 miticide instead.

APRICOTS, CHERRIES, FIGS, NECTARINES, PEACHES, PLUMS, PRUNES—Clover, brown, almond, European red, McDaniel, Pacific, peach silver, plum nursery, Schoene, two-spotted, Willamette and yellow mites. Apply at petal fall to first cover or whenever mite population appears. If repeat application is necessry, wait 30 days to avoid excessive residues. Do not apply to stone fruits and figs after fruit formation in California, Washington or Oregon. Use KELTHANE 35 miticide instead.

BLACKBERRIES, BOYSENBERRIES, DEWBERRIES, LOGANBERRIES, OLALLIEBERRIES, RASPBERRIES—Clover, brown, almond, European red, McDaniel, Pacific, two-spotted, Willamette and yellow mites. Apply just after bloom or whenever mites appear and repeat if necessary.

CITRUS (GRAPEFRUIT, KUMQUATS, LEMONS, LIMES, ORANGES, TANGELOS, TANGERINES)—Citrus flat, citrus red (purple), citrus rust, six-spotted, Texas citrus and Yuma mites. Follow local timing recommendations.

GRAPES, CHESTNUTS, FILBERTS, HICKORY NUTS, PECANS, WALNUTS—Clover, brown, almond, European red, McDaniel, Pacific, two-spotted, Willamette and yellow mites. Apply just after bloom or whenever mites appear and repeat if necessary. Do not apply to grapes after fruit formation in California, Washington or Oregon. Use KELTHANE 35 miticide instead.

STRAWBERRIES—Six-spotted, strawberry (Atlantic) and two-spotted mites. Apply when mites appear and repeat as necessary. For cyclamen mite apply 2 or 3 thorough, high-pressure, drenching sprays (400 gallons per acre) at 10 to 20 day intervals when new growth starts in spring, or use as a spot treatment in infested areas during production season. Do not apply to strawberries after fruit formation in California, Washington or Oregon. Use KELTHANE 35 miticide instead.

OTHER USES

BERMUDAGRASS GROWN FOR SEED—For control of Banks grass mite, Bermuda, brown and clover mites, use 0.75 to 1 quart per acre. For ground application mix KELTHANE EC in water and apply in full coverage applications at the rate of 40 to 100 gallons per acre depending on density and height of grass stand. For aerial applications apply in sufficient water to provide thorough coverage. Applications should be made as soon as mites appear and repeat applications made in 10 to 15 days if necessary. Do not feed clippings or allow livestock to graze treated areas.

CLOVER MITES ON BUILDINGS—Use ¾ to 1 quart KELTHANE EC miticide per 100 gallons. Thoroughly spray outside walls, foundations, windowsills, plants and lawns at base of infested buildings.

LAWN AND TURF GRASSES—Bermuda, brown and clover mites. Use ¾ to 1 quart KELTHANE EC miticide per 100 gallons (2 teaspoons per gallon). Apply 2½ gallons dilute spray per 1,000 square feet. Make application when mites appear, covering the entire lawn area with a thorough spray and repeat in 10 to 14 days if necessary.

NOTE: KELTHANE EC miticide may be used without any spreader-sticker. When necessary to improve the spray coverage on foliage difficult to wet, and to smooth out spray deposits on maturing fruit, TRITON® B-1956® is recommended at 1 to 2 ounces per 100 gallons of spray or equivalent for concentrate sprays. Other suitable pesticide spreader-stickers may also be used, except in situations where TRITON B-1956 is recommended as the only acceptable spreader-sticker.

8765-B2

SPECIAL LOCAL NEED REGISTRATIONS FOR KELTHANE EC

State	Use	SLN or Federal Accession No.
Arizona	Bermudagrass	07622
California	Beans	CA-780130
California	Corn	07624
California	Citrus	CA-790114
Texas	False Spider Mites, Citrus	TX-780026

KELTHANE® 4F

FLOWABLE AGRICULTURAL MITICIDE

For Agricultural Use Only or Only for Sale to, Use, and Storage by Servicepersons

ACTIVE INGREDIENT

1,1-Bis(chlorophenyl)-2,2,2-trichloroethanol	41%*
INERT INGREDIENTS	59%
TOTAL	100%

* Equivalent to 4 lbs of active ingredient per gallon.

EPA Reg. No. 707-164 EPA Est. No. 707-PA-1

SHAKE WELL BEFORE USING—Grasp container by handle and mix contents by inverting end-over-end. Product will usually become homogenous by this action 15 to 20 times.

9266-R6

WARNING
Keep Out of Reach of Children
AVISO

PRECAUCION AL USUARIO: Si usted no lee ingles, no use este producto hasta que le etiqueta haya sido explicado ampliamente.

NOTICE: Before using this product, read the entire Precautionary Statements, Conditions of Sale and Warranty, Directions for Use, Use Precautions and Storage and Disposal Instructions on side panels and attached use directions. If the Conditions of Sale and Warranty are not acceptable, return the product unopened within thirty days of purchase to the place of purchase.

DIRECTIONS FOR USE

It is a violation of Federal law to use this product in a manner inconsistent with its labeling.

PRECAUTIONARY STATEMENTS

HAZARDS TO HUMANS AND DOMESTIC ANIMALS

Causes substantial but temporary eye injury and skin irritation. Harmful if swallowed or inhaled. Do not get in eyes, on skin or on clothing. Avoid breathing spray mist. Wear goggles or face shield when applying. Wash thoroughly with soap and water after handling. Remove contaminated clothing and wash before re-use.

STATEMENT OF PRACTICAL TREATMENT

If In Eyes: Flush with plenty of water. Get immediate medical attention.

If Inhaled: Move subject to fresh air.

If On Skin: Wash with plenty of soap and water. Get medical attention if irritation continues.

If Swallowed: Drink 1 or 2 glasses of water and induce vomiting by touching back of throat with finger. Avoid alcohol and fat-based liquids like milk. Get medical attention. Do not induce vomiting or give anything by mouth to an unconscious person.

NOTE TO PHYSICIAN: Adrenergic amines may cause myocardial irritability in the presence of chlorinated hydrocarbons and convulsions may be a prominent symptom of poisoning.

ENVIRONMENTAL HAZARD

This pesticide is toxic to fish. Do not apply directly to water. Do not apply when weather conditions favor drift from areas treated. Do not contaminate water by cleaning of equipment or disposal of wastes.

CONDITIONS OF SALE AND WARRANTY

Rohm and Haas warrants that the product conforms to its chemical description and is reasonably fit for the purpose stated on the label only when used in accordance with label directions under normal conditions of use. **ROHM AND HAAS MAKES NO OTHER EXPRESS OR IMPLIED WARRANTIES EITHER OF MERCHANTABILITY OR FITNESS FOR A PARTICULAR USE.** Handling, storage and use of the product by Buyer or User are beyond the control of Rohm and Haas and Seller. Risks such as crop injury, ineffectiveness or other unintended consequences resulting from, but not limited to, weather or soil conditions, presence of other materials, disease, pests, drift to other crops or property or failure to follow label directions will be assumed by the Buyer or User. **IN NO CASE WILL ROHM AND HAAS OR SELLER BE HELD LIABLE FOR CONSEQUENTIAL, SPECIAL OR INDIRECT DAMAGES RESULTING FROM THE HANDLING, STORAGE OR USE OF THIS PRODUCT.**

RE-ENTRY AND WORKER PROTECTION STATEMENTS

Do not enter treated areas without protective clothing until sprays have dried. Because certain states may require more restrictive re-entry intervals for various crops treated with this product, consult your State Department of Agriculture for further information. Do not apply this product in such a manner as to directly or through drift expose workers or other persons. The area being treated must be vacated by unprotected persons.

Written or oral warnings must be given to workers who are expected to be in a treated area or in an area about to be treated with this product. Oral warnings must be given if there is reason to believe that written warnings cannot be understood by workers. When oral warnings are given, warnings shall be given in a language customarily understood by workers. Written warnings must include the following information:

WARNING

(Insert area or field description) treated with KELTHANE 4F on (insert date of application). Do not enter treated areas without protective clothing until sprays have dried. In case of accidental exposure: Call a doctor (physician), clinic or hospital immediately. Explain that the victim has been exposed to KELTHANE 4F and describe his condition. For further information see the "Statement of Practical Treatment."

Oral warnings must include the same information that is in the written warnings.

GENERAL INFORMATION

KELTHANE 4F miticide is a flowable formulation that gives high initial kill against most species of agricultural mites, including the following:

apple rust	peach silver
Bermuda	plum nursery
brown	privet
Carmine	Schoene
citrus flat	six-spotted
citrus red (purple)	spruce
citrus rust	strawberry (Atlantic)
clover	Texas citrus
cyclamen	tomato russet
desert	tropical
dryberry	two-spotted (red spider)
European red	Willamette
McDaniel	yellow (carpini)
Pacific	Yuma

KELTHANE 4F miticide is a specific miticide and will not kill bees and beneficial insect predators, when used as recommended.

KELTHANE 4F miticide may be applied in dilute or in concentrate ground sprays. For aerial sprays, use 3 to 10 gallons of water per acre and provide thorough coverage of all plant parts. Coverage on the underside of leaves is especially important. For further details, see specific recommendations in these use directions, then consult current State Agricultural Experiment Station spray schedules or State Extension Service Specialists for additional details on compatibility, dosage, timing and uses.

PESTICIDE COMPATIBILITY

KELTHANE 4F miticide is believed to be compatible with most commonly used insecticides and fungicides with the exception of the following: lime and other highly alkaline materials, Cyprex fungicide, and Alar 85 growth regulator. In addition, KELTHANE 4F should not be used with Glyodin fungicide in late cover sprays in apples.

USE PRECAUTIONS

- Read and observe all label directions including use precautions before using.
- Use only for recommended purposes and at recommended rates.

DIRECTIONS FOR USE

ALFALFA (GROWN FOR SEED ONLY—USE RESTRICTED TO CALIFORNIA, IDAHO, NEVADA, OREGON AND WASHINGTON)—For Carmine, Pacific, strawberry (Atlantic) and two-spotted mites. Use 2 to 3 pints KELTHANE 4F miticide per acre in sufficient water for thorough coverage. Apply when mites appear and repeat if necessary. Treated fields are not to be grazed and crop residues and cuttings are not to be fed to livestock or baled for such use. The fields must be cleared by burning crop residues or by discing into the soil.

APPLES—Apple rust, clover, European red, McDaniel, Pacific, Schoene, two-spotted, Willamette and yellow (carpini) mites.

Dosage—For dilute sprays use $^3/_4$ to 1 pint KELTHANE 4F miticide per 100 gallons. Use the 1 pint rate per 100 gallons spray for large trees or heavy mite populations. KELTHANE 4F miticide is recommended at a minimum rate of 3 pints per acre in either dilute or concentrate sprays. The maximum rate for KELTHANE 4F miticide is 8 pints per acre. For concentrate ground applications use 3 to 8 pints of KELTHANE 4F miticide per acre depending on the size and number of trees per acre, foliage density and mite pressure. When heavy mite populations are encountered, two successive sprays applied 7 to 10 days apart are recommended.

Application—Dilute applications should be made in sufficient volume to provide thorough coverage of all parts of the tree. Concentrate applications are recommended at 50 gallons of water per acre or greater with sufficient air volume to provide thorough coverage of all tree parts. The amount of KELTHANE 4F miticide applied per acre should be the same whether dilute or concentrate sprays are used.

Timing of Application—Applications are recommended in the second or later cover sprays at the first sign of mite buildup. Repeat applications should be made as necessary for adequate season control.

Use Restrictions—Apples

- Do not apply KELTHANE 4F miticide to Idared variety before the second cover spray.
- Applications of KELTHANE 4F miticide to Golden Delicious varieties should not exceed 4 pints per acre per spray.
- Do not apply KELTHANE 4F miticide within 7 days of harvest.
- Refer to label section for pesticide compatibility.

BEANS (dry, green, lima)—CALIFORNIA ONLY—Pacific, tomato russet, tropical and two-spotted mites. Use 3 pints per acre. Apply at first signs of mite buildup and repeat as necessary. Do not apply within 21 days of harvest. Do not feed treated vines (or crop residues) to meat or dairy animals.

BEANS (dry, green, lima)—Pacific, tropical and two-spotted mites. Use $^3/_4$ to 1 pint per acre. Apply at first signs of mite buildup and repeat if necessary. Do not apply within 7 days of harvest. Do not feed treated vines (or crop residues) to meat or dairy animals.

CITRUS (grapefruit, kumquats, lemons, limes, oranges, tangelos, tangerines)—Citrus flat, citrus red, citrus rust, six-spotted, Texas citrus and Yuma mites. Apply in full coverage sprays using 1 pint per 100 gallons of dilute spray (2 gallons maximum per acre). If concentrate sprays are used, adjust dosage accordingly. Apply when mites appear in threatening numbers and repeat as required for satisfactory control. Do not apply within 7 days of harvest.

COTTON—Desert, Pacific and two-spotted mites. Use 1 to $1^1/_2$ quarts per acre, depending on size of plant and density of foliage. Apply in sufficient water to provide thorough coverage. Begin application when mites are first noted and repeat if necessary, and do not apply within 14 days of harvest. Do not feed cotton stalks or trash to meat or dairy animals.

CUCURBITS (CANTALOUPES, CUCUMBERS and SQUASH) (Winter and Summer)—For Pacific and two-spotted mites. Use $^2/_3$ to $1^1/_8$ pints KELTHANE 4F miticide per acre in sufficient water to provide thorough coverage of all plant parts. Adjust spray nozzles to provide adequate coverage to underside of leaves. Use the higher maximum use rate of $1^1/_8$ pints when vines begin to cover ground. Make applications at first signs of mite buildup and repeat if necessary. Do not feed treated vines or crop residues to meat or dairy animals. Do not apply KELTHANE 4F miticide within 2 days of harvest.

GRAPES—For European red, McDaniel, Pacific, two-spotted and Willamette mites. Use $1^1/_3$ to $2^1/_3$ pints KELTHANE 4F miticide per acre in sufficient water to provide thorough coverage. Adjust spray nozzles to provide adequate coverage to underside of leaves. Make applications during cover sprays when mites appear and repeat as necessary. Do not apply within 7 days of harvest.

HOPS—For Pacific and two-spotted mites. Use 2 to $2^1/_3$ pints KELTHANE 4F miticide per acre in sufficient water to provide full coverage to all plant parts. Use lower recommended dosage rate when plants are small and increase dosage rate with development of vines. Apply at first sign of mite buildup and repeat when necessary. Do not apply KELTHANE 4F miticide within 7 days of harvest. Do not feed treated vines or crop residues to meat or dairy animals.

MINT (peppermint for oil production)—For two-spotted mites. Use $1^3/_4$ to $2^1/_3$ pints KELTHANE 4F miticide per acre in sufficient water for thorough coverage. Make application when mites become active. Do not apply within 30 days of harvest and do not make more than one application per crop per season. Do not feed treated fresh or spent hay to livestock.

PEACHES—Clover (brown almond), European red, McDaniel, Pacific, peach silver, plum nursery, Schoene, two-spotted, Willamette and yellow (carpini) mites.

Dosage—For dilute sprays use $^3/_4$ to 1 pint KELTHANE 4F miticide per 100 gallons. Use the 1 pint rate per 100 gallons spray for large trees or heavy mite populations. KELTHANE 4F miticide is recommended at a minimum rate of 3 pints per acre in either dilute or concentrate sprays. The maximum rate of KELTHANE 4F miticide in peaches is 6.4 pints per acre. For concentrate ground applications, use 3 to 6.4 pints of KELTHANE 4F miticide per acre depending on the size and number of trees per acre, foliage density and mite pressure.

Application—Application of KELTHANE 4F miticide in peaches is recommended in either dilute or concentrate sprays using ground equipment. For dilute applications, use sufficient volume to provide thorough coverage of all parts of the tree. Concentrate applications are recommended at volume rates of 50 gallons or more of water per acre with sufficient air volume to provide thorough coverage to all tree parts. The amount of KELTHANE 4F miticide applied per acre should be the same whether dilute or concentrate sprays are used.

Application Timing—Application of KELTHANE 4F miticide in peaches is recommended from shuck fall through the cover sprays at the first sign of mite population buildup.

Use Restrictions—Peaches

- Do not apply KELTHANE 4F miticide within 14 days of harvest.

NOTE: KELTHANE 4F miticide may be used without any spreader-sticker. When necessary to improve the spray coverage on foliage difficult to wet and to smooth out spray deposits on maturing fruit, TRITON® B-1956® is recommended. Other suitable pesticide spreader-stickers may also be used except in situations where TRITON B-1956 is recommended as the only acceptable spreader-sticker.

STORAGE AND DISPOSAL

PROHIBITIONS AND RECOMMENDATIONS

Do not contaminate water, food or feed by storage or disposal. Open dumping is prohibited. Do not re-use empty container. Keep container closed when not in use.

STORAGE: Keep from freezing. Store 32°F–110°F.

PESTICIDE DISPOSAL

Pesticide wastes are toxic. Improper disposal of excess pesticide, spray mixture, or rinsate is a violation of Federal law. If these wastes cannot be disposed of by use according to label instructions, contact your State Pesticide or Environmental Control Agency, or the Hazardous Waste representative at the nearest EPA regional office for guidance.

CONTAINER DISPOSAL

Triple rinse (or equivalent). Puncture and dispose of in a sanitary landfill, or incinerate, or, if allowed by state and local authorities, by burning. If burned, stay out of smoke.

STEPS TO BE TAKEN IN CASE MATERIAL IS RELEASED OR SPILLED: Ventilate area. Avoid breathing vapors. Use MSHA/NIOSH self-contained breathing apparatus or airline respirator for large spills in confined areas. Dike the spill with inert material (sand, earth, etc.) and transfer the liquid or solid diking material to separate containers for recovery or disposal. Remove contaminated clothing promptly and wash exposed skin areas with soap and water. Wash clothing before re-use. Keep spill out of all sewers and open bodies of water.

ALAR® is a registered trademark of Uniroyal, Inc.

CYPREX® is a registered trademark of American Cyanamid Company.

KELTHANE® and TRITON® B-1956® are registered trademarks of Rohm and Haas Company.

9266-B7

KELTHANE® MF

AGRICULTURAL MITICIDE EMULSIFIABLE CONCENTRATE

ACTIVE INGREDIENT

1,1-Bis(chlorophenyl)2,2,2-trichloroethanol	42%*
INERT INGREDIENTS	58%
TOTAL	100%

*Equivalent to 4 lbs. active ingredient per gallon

EPA Reg. No. 707-73 EPA Est. No. 279-CA-1

9603-CA-R3

CAUTION

Keep Out of Reach of Children

NOTICE: **Before using this product, read the entire Precautionary Statements, Conditions of Sale and Warranty, Directions for Use, Use Precautions and Storage and Disposal Instructions on side and back panels and attached use directions. If the Conditions of Sale and Warranty are not acceptable, return the product unopened within thirty days of purchase to the place of purchase.**

PRECAUTIONARY STATEMENTS

HAZARDS TO HUMANS AND DOMESTIC ANIMALS

CAUTION

Harmful if swallowed, absorbed through the skin or inhaled. Causes moderate eye, skin, nose and throat irritation. Avoid contact with skin, eyes and clothing. Avoid breathing vapors or spray mist.

Wash thoroughly with soap and water after handling and before eating or smoking.

Remove contaminated clothing and wash before re-use.

STATEMENT OF PRACTICAL TREATMENT

IF SWALLOWED: Drink 1 to 2 glasses of water. Call a physician and have the stomach emptied immediately. Do not induce vomiting or give anything by mouth to an unconscious person.

NOTE TO PHYSICIAN: Because this product contains a chlorinated hydrocarbon insecticide, vomiting or careful gastric lavage is recommended. Epinephrin or other adrenergic amines can cause myocardial irritability.

IF INHALED: Remove victim to fresh air. Treat the victim symptomatically. Call a physician if problem continues.

IF ON SKIN: Wash exposed areas with plenty of soap and water. If irritation persists, call a physician.

IF IN EYES: Flush eyes with plenty of water for at least 15 minutes. Call a physician if irritation persists.

ENVIRONMENTAL HAZARDS

This product is toxic to fish. Do not apply directly to lakes, streams or ponds. Do not apply where runoff is likely to occur. Do not apply when weather conditions favor drift from areas treated. Do not contaminate water by cleaning of equipment or disposal of wastes. Apply this product only as specified on this label.

PHYSICAL AND CHEMICAL HAZARDS

FLAMMABLE

Do not use, pour, spill or store near heat or open flame. Do not use in the vicinity of pilot lights.

CONDITIONS OF SALE AND WARRANTY

Rohm and Haas warrants that the product conforms to its chemical description and is reasonably fit for the purpose stated on the label only when used in accordance with label directions under normal conditions of use. **ROHM AND HAAS MAKES NO OTHER EXPRESS OR IMPLIED WARRANTIES EITHER OF MERCHANTABILITY OR FITNESS FOR A PARTICULAR USE.** Handling storage and use of the product by Buyer or User are beyond the control of Rohm and Haas and Seller. Risks such as crop injury, ineffectiveness or other unintended consequences resulting from, but not limited to, weather or soil conditions, presence of other materials, disease, pests, drift to other crops or property or failure to follow label directions will be assumed by the Buyer or User. **IN NO CASE WILL ROHM AND HAAS OR SELLER BE HELD LIABLE FOR CONSEQUENTIAL, SPECIAL OR INDIRECT DAMAGES RESULTING FROM THE HANDLING, STORAGE OR USE OF THIS PRODUCT.**

DIRECTIONS FOR USE

It is a violation of Federal law to use this product in a manner inconsistent with its labeling.

RE-ENTRY AND WORKER PROTECTION STATEMENTS

Do not enter treated areas without protective clothing until sprays have dried. Because certain states may require more restrictive reentry intervals for various crops treated with this product, consult your State Department of Agriculture for further information. Do not apply this product in such a manner as to directly or through drift expose workers or other persons. The area being treated must be vacated by unprotected persons.

Written or oral warnings must be given to workers who arc expected to be in a treated area or in an area about to be treated with this product. Oral warnings must be given if there is reason to believe that written warnings cannot be understood by workers. When oral warnings are given, warnings shall be given in a language customarily understood by workers. Written warnings must include the following information:

CAUTION

(Insert area or field description) treated with KELTHANE MF on (insert date of application). Do not enter treated areas without protective clothing until sprays have dried. In case of accidental exposure: Call a doctor (physician), clinic or hospital immediately. Explain that the victim has been exposed to KELTHANE MF and describe his condition. For further information see the "Statement of Practical Treatment."

Oral warnings must include the same information that is in the written warnings.

GENERAL

KELTHANE MF miticide is an emulsifiable formulation that gives high initial kill against most species of agricultural mites. KELTHANE MF is a specific miticide and will not kill bees and beneficial insect predators, when used as recommended.

KELTHANE MF miticide is compatible with most other commonly used insecticides and fungicides. It should not be combined with lime. KELTHANE MF miticide may be applied in dilute or in concentrate ground sprays. For aerial sprays use 3 to 10 gallons of water per acre to provide thorough coverage of all plant parts. Coverage on the underside of leaves is especially important. For further details see specific recommendations on this label, then consult current State Agricultural Experiment Station spray schedules or State Extension Service Specialists for additional details on compatibility, dosage, timing and uses.

VEGETABLES AND FIELD CROPS

ALFALFA AND CLOVER (BOTH GROWN FOR SEED ONLY — USE RESTRICTED TO CALIFORNIA, IDAHO, NEVADA, OREGON AND WASHINGTON)—For carmine, Pacific, strawberry (Atlantic), and two-spotted mites. Use 1 to 1½ quarts of KELTHANE MF miticide per acre in sufficient water for thorough coverage. Apply when mites become active and repeat if necessary. Treated fields are not to be grazed, and crop residues and cuttings are not to be fed to livestock or baled for such use. The fields must be cleared by burning crop residues or by discing into the soil.

BEANS (dry, green, lima)—Pacific, tropical and two-spotted mites. Use ¾ to 1 pint KELTHANE MF miticide per acre. Apply at first signs of mite buildup and repeat as necessary. Do not apply within 7 days of harvest. Do not feed treated vines (or crop residues) to meat or dairy animals.

BEANS (dry, green, lima) California only—Pacific, tomato russet, tropical and two-spotted mites. Use 3 pints KELTHANE MF miticide per acre. Apply at first signs of mite buildup and repeat as necessary. Do not apply within 21 days of harvest. Do not feed treated vines (or crop residues) to meat or dairy animals.

COTTON—Desert, Pacific and two-spotted mites. Use 1 to $1\frac{1}{2}$ quarts KELTHANE MF miticide per acre, depending on size of plant and density of foliage. Apply in sufficient water to provide thorough coverage. Begin application when mites are first noticed and repeat if necessary but do not apply within 14 days of harvest. Do not feed cotton stalks or trash to meat or dairy animals.

HOPS—Pacific, tropical and two-spotted mites. Use 2 to $2\frac{1}{3}$ pints KELTHANE MF miticide per acre. Apply at first signs of mite buildup and repeat as necessary. Do not apply within 7 days of harvest. Do not feed treated vines (or crop residues) to meat or dairy animals.

MINT—Peppermint and spearmint for commercial oil production. For control of two-spotted mites use KELTHANE MF miticide at $1\frac{3}{4}$ and $2\frac{1}{2}$ pints per acre in sufficient water for thorough coverage. Make application when mites become active. Do not apply within 30 days of harvest and do not make more than one application per crop per season. Do not feed treated fresh or spent hay to livestock.

FRUITS AND NUTS

APRICOTS, CHERRIES, FIGS, NECTARINES, PEACHES, PLUMS, PRUNES—Clover (brown almond), European red, McDaniel, Pacific, peach silver, plum nursery, Schoene, two-spotted, Willamette and yellow mites. Use $\frac{3}{4}$ to 1 pint KELTHANE MF miticide per 100 gallons dilute spray (6.4 pints maximum/acre for Apricots, Cherries, Nectarines and Peaches and 4.0 pints maximum/acre for Figs, Plums and Prunes). Apply in full coverage spray at petal fall to first cover or whenever mite population appears. KELTHANE MF miticide may cause fruit spotting on figs. When fruit is present use KELTHANE 35 wettable formulation. For repeat application on Cherries, Figs, Plums and Prunes wait 30 days to avoid excessive residues, but do not apply within 7 days of harvest. On Apricots, Nectarines and Peaches do not apply within 14 days of harvest.

CITRUS (Grapefruit, Kumquats, Lemons, Limes, Oranges, Tangelos, Tangerines)—Citrus flat, citrus red (purple), citrus rust, six-spotted, Texas citrus and Yuma mites. Apply KELTHANE MF miticide in full coverage sprays using 1 pint per 100 gallons of dilute spray (2 gallons maximum/acre). If concentrate sprays are used adjust dosage accordingly. Apply when mites appear in threatening numbers and repeat as required for satisfactory control. Do not apply within 7 days of harvest.

GRAPES—For clover (brown almond), European red, McDaniel, Pacific, two-spotted, Willamette and yellow mites. Use $1\frac{1}{3}$ to $2\frac{1}{3}$ pints KELTHANE MF miticide per acre. Apply just after bloom or whenever mites appear and repeat if necessary. Do not apply within 7 days of harvest.

STRAWBERRIES—Six-spotted, strawberry (Atlantic) and two-spotted mites. Use 1 quart KELTHANE MF miticide per acre. Apply when mites appear and repeat as necessary but do not apply within 2 days of harvest, or if temperature exceeds 90°F.

WALNUTS—For clover (brown almond), European red, McDaniel, Pacific, two-spotted, Willamette and yellow mites. Use a full coverage spray of $\frac{3}{4}$ to 1 pint KELTHANE MF miticide per 100 gallons dilute spray (1 gallon maximum/acre). Apply just after bloom or whenever mites appear and repeat if necessary. Do not apply within 14 days of harvest.

NOTE: KELTHANE MF miticide may be used without any spreader-sticker. When necessary to improve the spray coverage on foliage difficult to wet and to smooth out spray deposits on maturing fruit, TRITON® B-1956® is recommended. Other suitable pesticide spreader-stickers may also be used except in situations where TRITON® B-1956® is recommended as the only acceptable spreader-sticker.

STORAGE AND DISPOSAL

STORAGE—Keep from freezing. Store at 32°F–110°F.

PESTICIDE DISPOSAL

Do not contaminate water, food, or feed by storage or disposal. Pesticide wastes are toxic. Improper disposal of excess pesticide, spray mixture, or rinsate is a violation of Federal law. If these wastes cannot be disposed of by use according to label instructions, contact your State Pesticide or Environmental Control agency, or the Hazardous Waste representative at the nearest EPA Regional Office for guidance.

CONTAINER DISPOSAL

Triple rinse (or equivalent). Then puncture and dispose of in a sanitary landfill, or by incineration, or if allowed by state and local authorities, by burning. If burned, stay out of smoke.

STEPS TO BE TAKEN IN CASE MATERIAL IS RELEASED OR SPILLED: Eliminate ignition sources. Ventilate area. Avoid breathing vapors. Wear respirator suitable for vapor concentration encountered (MSHA/NIOSH approved or equivalent). Dike the spill with inert material and transfer the liquid or solid diking materials to separate containers. Wash clothing before re-use. REFER TO PRECAUTIONARY STATEMENTS.

9603-BO

SPECIAL LOCAL NEED REGISTRATIONS FOR KELTHANE MF

State	Use	SLN or Federal Accession No.
Alabama	Peanuts	06610
California	Field Corn	06611
California	Strawberries	CA-770053
Florida	Peanuts	07614
Georgia	Peanuts	06606
South Carolina	Peanuts	06607

KELTHANE® 35

AGRICULTURAL MITICIDE WETTABLE POWDER

ACTIVE INGREDIENT	
1,1-Bis(chlorophenyl) 2,2,2-trichloroethanol	35%
INERT INGREDIENTS	65%
TOTAL	100%

EPA Reg. No. 707–89 EPA Est. No. 477–MD–1

CAUTION

Keep Out of Reach of Children

NOTICE: Before using this product, read the entire Precautionary Statements, Conditions of Sale and Warranty, Directions for Use, Use Restrictions and Storage and Disposal Instructions. If the conditions of Sale and Warranty are not acceptable, return the product unopened within thirty days of purchase to the place of purchase.

PRECAUTIONARY STATEMENTS

HAZARDS TO HUMANS AND DOMESTIC ANIMALS

CAUTION

Harmful if swallowed, absorbed through the skin or inhaled. Causes moderate eye and skin irritation. Avoid contact with skin, eyes and clothing and avoid breathing dust or spray mist.

Wash thoroughly with soap and water after handling and before eating or smoking. Remove contaminated clothing and wash before reuse.

STATEMENT OF PRACTICAL TREATMENT

IF SWALLOWED: Dilute by giving 2 glasses of water to drink. See a physician. Never give anything by mouth to an unconscious person.

NOTE TO PHYSICIAN. Because this product is a chlorinated miticide vomiting is recommended. The administration of milk or other fat based demulcents which might enhance absorption is to be avoided. Epinephrine or other adrenergic amines can cause myocardial irritability in persons poisoned with chlorinated hydrocarbons.

IF INHALED: Remove victim to fresh air. Treat the victim symptomatically. If irritation persists call a physician.

IF ON SKIN: Wash exposed areas with plenty of soap and water. If irritation persists call a physician.

IF IN EYES: Flush eyes with plenty of water for at least 15 minutes. If irritation persists call a physician.

ENVIRONMENTAL HAZARDS

This product is toxic to fish. Do not apply directly to lakes, streams or ponds. Do not contaminate water by cleaning of equipment or disposal of wastes. Do not apply where runoff is likely to occur. Do not apply when weather conditions favor drift from treated areas. Apply this product only as specified on this label.

CONDITIONS OF SALE AND WARRANTY

Rohm and Haas warrants that the product conforms to its chemical description and is reasonably fit for the purpose stated on the label only when used in accordance with label directions under normal conditions of use. **ROHM AND HAAS MAKES NO OTHER EXPRESS OR IMPLIED WARRANTIES EITHER OF MERCHANTABILITY OR FITNESS FOR A PARTICULAR USE.** Handling, storage and use of the product by Buyer or User are beyond the control of Rohm and Haas and Seller. Risks such as crop injury, ineffectiveness or other unintended consequences resulting from, but not limited to, weather or soil conditions, presence of other materials, disease, pests, drift to other crops or property or failure to follow label directions will be assumed by the Buyer or User. **IN NO CASE WILL ROHM AND HAAS OR SELLER BE HELD LIABLE FOR CONSEQUENTIAL, SPECIAL OR INDIRECT DAMAGES RESULTING FROM THE HANDLING, STORAGE OR USE OF THIS PRODUCT.**

RE-ENTRY AND WORKER PROTECTION STATEMENTS

Do not enter treated areas without protective clothing until sprays have dried. Because certain states may require more restrictive reentry intervals for various crops treated with this product consult your State Department of Agriculture for further information. Do not apply this product in such a manner as to directly or through drift, expose workers or other persons. The area being treated must be vacated by unprotected persons.

Written or oral warnings must be given to workers who are expected to be in a treated area or in an area about to be treated with this product. Oral warnings must be given if there is reason to believe that written warnings cannot be understood by workers. When oral warnings are given, warnings shall be given in a language customarily understood by workers. Written warnings must include the following information:

CAUTION

(Insert area or field description) treated with KELTHAN 35 miticide on (insert date of application). Do not enter treated areas without protective clothing until sprays have dried. In case of accidental exposure: Call a doctor (physician), clinic or hospital immediately. Explain that the victim has been exposed to KELTHANE 35 miticide and describe his condition. For further information see the "Statement of Practical Treatment."

Oral warnings must include the same information that is in the written warnings.

STORAGE AND DISPOSAL

PESTICIDE DISPOSAL

Do not contaminate water, food or feed by storage or disposal. Open dumping is prohibited. Pesticide wastes are toxic. Improper disposal of excess pesticide, spray mixture, or rinsate is a violation of Federal Law. If these wastes cannot be disposed of by use according to label instructions, contact your State Pesticide or Environmental Control Agency or the Hazardous Waste representative at the nearest EPA Regional Office for guidance.

CONTAINER DISPOSAL

Completely empty bag into application equipment. Then dispose of empty bag in a sanitary landfill or by incineration or, if allowed by State and local authorities, by burning. If burned, stay out of smoke.

STEPS TO BE TAKEN IN CASE MATERIAL IS RELEASED OR SPILLED: Sweep or shovel solid material into a suitable container for recovery or disposal. Remove contaminated clothing promptly and wash exposed skin areas with soap and water. Wash clothing before re-use. Keep spill out of all sewers and open bodies of water. REFER TO PRECAUTIONARY STATEMENTS.

DIRECTIONS FOR USE

It is a violation of Federal law to use this product in a manner inconsistent with its labeling.

KELTHANE 35 miticide gives high initial kill against most species of agricultural mites including European red, two-spotted (red spider), six-spotted, Pacific, privet, Schoene, yellow (carpini), McDaniel, Willamette, spruce, desert and tropical mites. KELTHANE 35 miticide is a specific miticide and is harmless to predacious and other beneficial insects.

KELTHANE 35 miticide can be applied alone or it can be applied with various other pesticides in recommended spray schedules. For further details, see recommendations below and consult current State Agricultural Experiment Station spray schedules, or State Extension Service Specialists for additional details on compatibility, dosage, timing and uses. KELTHANE 35 miticide is believed to be compatible with other insecticides and fungicides in common use.

For proper application, mix KELTHANE 35 miticide with water and apply as a wet spray with suitable spray equipment. The following chart is offered as a guide in preparing dilute, high gallonage sprays. For concentrate ground sprayers, the amount of water used should be reduced in accordance with instructions from the sprayer manufacturer but the amount of KELTHANE 35 miticide applied per acre should be the same whether dilute or concentrate sprays are used. For serial sprays use 3 to 10 gallons of water per acre to provide thorough coverage of all plant parts.

To avoid excessive residues at harvest follow the use recommendations. Do not exceed the maximum dosage per acre per application, or the time restriction between last application and harvest for each crop listed in the table below.

Summary of KELTHANE 35 miticide Dosage per Acre and Limitations Between Last Application & Harvest

Crop	KELTHANE 35 miticide/lbs. per Acre Each Application: Small Plants	Large Plants	Mixing Rate lbs. per 100 gal. Dilute Spray*	Days Before Harvest
Alfalfa and clover (both grown for seed only—use restricted to California, Idaho, Nevada, Oregon, and Washington)	1½	4	1-1⅓	—
Chestnuts, filberts, hickory nuts, pecans, walnuts	4	11	1-1⅓	14
Apples, crabapples, pears, quince	6	10-12	1-1⅓	7
Beans (dry)	1	2¼	1-1⅓	45
Beans (dry, green, lima)	1	1⅔	1	7

KELTHANE 35 miticide/lbs. per Acre

Crop	Each Application Small Plants	Each Application Large Plants	Mixing Rate lbs. per 100 gal. Dilute Spray*	Days Before Harvest
Beans (dry, green, lima) (California only)	1½	3	1-1⅓	21
Cantaloupe, cucumbers, melons, pumpkins, squash, watermelons	1	1⅔	1	2
Cherries	6	9	1-1½	7
Citrus (all)	7	22	1-1⅓	7
Grapes, hops	1½		1-1⅓	7
Peaches, apricots, nectarines	6	9	1-1⅓	14
Plums, prunes, figs	4	5⅔	1-1⅓	7
Raspberries, cane berries	1½	3½	1⅓	2
Strawberries (see use instructions)	1	6¾	1-1⅓	2
Tomatoes, peppers	1½	2¼	1-1⅓	2
Ornamentals, flowers, nursery stock, shade trees	—	—	1-1⅓	—
Buildings (see use directions)				
Clover mites on buildings	—	—	1-1⅓	—

* Adjust water gallonage if concentrate ground or aerial sprays are used. Amount of spray applied per acre must be adjusted to the size and number of trees or plants per acre and the foliage density.

FRUITS AND NUTS

APPLES, PEARS, CRABAPPLES, QUINCES—To control two-spotted, Schoene, McDaniel, yellow (carpini), Willamette, Pacific and apple rust mite and to clean up late spring European red mite and clover mite, apply at petal fall to second cover or whenever mite population appears. Make thorough full coverage application. Repeat as necessary during the season for adequate control. To avoid possible injury do not use glyodin fungicide with KELTHANE 35 miticide in late cover sprays.

GRAPES, WALNUTS, FILBERTS, PECANS, CHESTNUTS, HICKORY NUTS—For brown almond (clover), two-spotted, European red, Pacific, McDaniel, Willamette and yellow mites, apply just after bloom or whenever mites appear and repeat if necessary.

ORANGES, GRAPEFRUIT, LEMONS, LIMES, TANGELOS, TANGERINES, KUMQUATS—For citrus red (purple), citrus flat, citrus rust, Texas citrus, six-spotted and Yuma mite, follow State Agricultural Experiment Station or State Extension Service Specialist recommendations. Do not apply within seven days of harvest.

PEACHES, APRICOTS, NECTARINES, CHERRIES, PLUMS, PRUNES, FIGS—For European red, brown almond (clover), two-spotted, Schoene, McDaniel, Pacific, Willamette, peach silver, plum nursery and yellow mites, apply at petal to first cover or whenever mite population appears. If needed, a repeat application can be made after 30 days.

RASPBERRIES, DEWBERRIES, LOGANBERRIES, BOYSENBERRIES, BLACKBERRIES, OLALLIE BERRIES—Brown almond (clover), two-spotted, European red, Pacific, McDaniel, Willamette and yellow mites, apply when mites appear and repeat as necessary.

STRAWBERRIES—Two-spotted, strawberry (Atlantic), and six-spotted mite, use 1 to 3 lbs. KELTHANE 35 miticide per acre. Apply when mites appear and repeat as necessary.

For Cyclamen mite use 4 to 6¼ lbs. KELTHANE 35 miticide per acre. Apply 2 or 3 thorough high pressure, drenching sprays (400 gallons per acre) at 10 to 30 day intervals when new growth starts in spring, or use as a spot treatment in infested areas during production season.

VEGETABLES AND FIELD CROPS

ALFALFA AND CLOVER (BOTH GROWN FOR SEED ONLY—USE RESTRICTED TO CALIFORNIA, IDAHO, NEVADA, OREGON, WASHINGTON)—For Carmine, Pacific, strawberry (Atlantic), and two-spotted mites. Use 1½ to 4 lbs. of KELTHANE 35 miticide per acre in sufficient water for thorough coverage. Apply when mites become active and repeat if necessary. Treated fields are not to be grazed, and crop residues and cuttings are not be fed to livestock or baled for such use. The fields must be cleared by burning crop residues or by discing into the soil.

BEANS, HOPS, CUCUMBERS, CANTALOUPES, MELONS, WATERMELONS, PUMPKINS, WINTER AND SUMMER SQUASH, TOMATOES, PEPPERS—For two-spotted, Pacific, tropical and tomato russet mite, apply at first signs of mite buildup and repeat as necessary for adequate control. Do not feed treated vines, forage or crop residues to meat or dairy animals. In California mite control on cantaloupe, cucumbers, melons, pumpkins, squash, and watermelons is inadequate when vine size exceeds 14 inch runner length. Apply with ground rig only in 40 to 100 gallons of water spray mix. Use 2 qts. TRITON® B-1956® per 100 gals. spray mix.

FOR USE ON:

CLOVER MITE ON BUILDINGS—Use 1 to 1½ pounds KELTHANE 35 miticide per 100 gallons of water. Thoroughly spray outside walls, foundations and windowsills, and plants and lawn at base of infested buildings.

LAWN AND TURF GRASSES—Clover, brown and Bermuda grass mites. Use 1 to 1½ pounds KELTHANE 35 miticide per 100 gallons of water (1½ tablespoons per gallon). Apply at 2½ gallons per 1000 sq. ft. Make application when mites appear, covering the entire lawn area with a thorough spray and repeat in 10 to 14 days if necessary.

ORNAMENTALS, FLOWERS, NURSERY STOCK AND SHADE TREES—For control of mites on greenhouse or outdoor plants use 1 to 1½ pounds KELTHANE 35 miticide per 100 gallons (1½ tablespoons per gallon of water). Apply when mites are present and repeat as necessary.

NOTE: KELTHANE 35 miticide may be used without any spreader-sticker. When necessary to improve the spray coverage on foliage difficult to wet and to smooth out spray deposits on maturing fruit. TRITON® B-1956® is recommended. Other suitable pesticide spreader-stickers may also be used, except in situations where TRITON B-1956 is recommended as the only acceptable spreader-sticker.

KELTHANE, TRITON and B-1956 are registered trademarks of Rohm and Haas Company.

8114-D15

KERB® 50-W

A SELECTIVE HERBICIDE IN WATER SOLUBLE POUCHES

NET CONTENTS: 3 × 1 LB WATER SOLUBLE POUCHES
3 LBS (1.36 KG)

ACTIVE INGREDIENT
Pronamide
3,5-dichloro-N-(1, 1-demthyl-2-propynyl)-benzamide ... 50.0%
INERT INGREDIENTS ... 50.0%
TOTAL ... 100.0%

EPA Reg. No. 707-159 EPA Est. No. 33967-NJ-1
U.S. Patents 3,534,098 and 3,640,699 8116-D4

CAUTION
Keep Out of Reach of Children

Rohm and Haas Company has instituted the following use restriction:
For lettuce—Consult local authorities (extension specialists) regarding safety to the crop before using on newly introduced lettuce varieties.

PRECAUTIONARY STATEMENTS

HAZARDS TO HUMANS AND DOMESTIC ANIMALS
Harmful if swallowed. Take special care to avoid contact with eyes, skin or clothing. Do not breathe the dust or spray mist. Wash clothing and gloves after use.

STATEMENT OF PRACTICAL TREATMENT
If Swallowed: Dilute by giving 2 glasses of water to drink and call a physician. Never give anything by mouth to an unconscious person.

Note to Physician: Emesis is recommended.

If Inhaled: Move subject to fresh air.

If on Skin: Wash affected skin areas with soap and water. Consult a physician if irritation persists.

If in Eyes: Flush eyes with large amounts of water for at least 15 minutes. Consult a physician if irritation persists.

Protective Clothing: Use the following items of clothing when mixing or applying this product: long-sleeved shirts and long pants, preferably one-piece (overalls); hat with brim; heavy-duty fabric or rubber work gloves. Use heavy-duty leather or rubber boots when applying with hand spray equipment.

ENVIRONMENTAL HAZARDS
Keep out of lakes, ponds or streams. Do not contaminate water by cleaning of equipment or disposal of wastes.

NOTICE: Before using this product read the entire Precautionary Statements, Conditions of Sale and Warranty, Directions for Use, Use Restrictions and Storage and Disposal Instructions on front and back panel and attached use directions. If the Conditions of Sale and Warranty are not acceptable, return the product unopened within thirty days of purchase to the place of purchase.

CONDITIONS OF SALE AND WARRANTY

Rohm and Haas warrants that the product conforms to its chemical description and is reasonably fit for the purpose stated on the label only when used in accordance with label directions under normal conditions of use. **ROHM AND HAAS MAKES NO OTHER EXPRESS OR IMPLIED WARRANTIES EITHER OF MERCHANTABILITY OR FITNESS FOR A PARTICULAR USE.** Handling, storage and use of the product by Buyer or User are beyond the control of Rohm and Haas and Seller. Risks such as crop injury, ineffectiveness or other unintended consequences resulting from, but not limited to, weather or soil conditions, presence of other materials, disease, pests, drift to other crops or property or failure to follow label directions will be assumed by the Buyer or User. **IN NO CASE WILL ROHM AND HAAS OR SELLER BE HELD LIABLE FOR CONSEQUENTIAL, SPECIAL OR INDIRECT DAMAGES RESULTING FROM THE HANDLING, STORAGE OR USE OF THIS PRODUCT.**

DIRECTIONS FOR USE

It is a violation of Federal law to use this product in a manner inconsistent with its labeling.

RE-ENTRY AND WORKER PROTECTION STATEMENTS

Do not enter treated areas without protective clothing until sprays have dried. Because certain states may require more restrictive re-entry intervals for various crops treated with this product, consult your State Department of Agriculture for further information. Do not apply this product in such a manner as to directly or through drift expose workers or other persons. The area being treated must be vacated by unprotected persons.

Written or oral warnings must be given to workers who are expected to be in a treated area or in an area about to be treated with this product. Oral warnings must be given if there is reason to believe that written warnings cannot be understood by workers. When oral warnings are given, warnings shall be given in a language customarily understood by workers. Written warnings must include the following information:

CAUTION
(Insert area or field description) treated with KERB 50W on (insert date of application). Do not enter treated areas without protective clothing until sprays have dried. In case of accidental exposure: Call a doctor (physician), clinic or hospital immediately. Explain that the victim has been exposed to KERB 50W and describe his condition. For further information see the "Statement of Practical Treatment."

Oral warnings must include the same information that is in the written warnings.

GENERAL INFORMATION

MIXING: The enclosed pouches of KERB 50-W herbicide are water soluble. **Do not** allow pouches to become wet prior to adding to the spray tank. **Do not** handle the pouches with wet hands or gloves. Always reseal overwrap bag to protect remaining unused pouches. **Do not** remove water soluble pouches from overwrap except to add directly to the spray tank.
Add the required number of unopened pouches as determined by the dosage recommendations into the spray tank with agitation. Depending on the water temperature and the degree of agitation, the pouches should dissolve completely within approximately five minutes from the time they are added to the water.

DOSAGE: Refer to the specific crop for dosage recommendations. Determine the total amount of KERB 50-W to be added to the spray tank. For each pound of KERB 50-W, use one one-pound pouch.
Example: If two pounds of KERB 50-W should be added to the spray tank, add two one-pound pouches.

The type of equipment used will determine the concentration required; however, the use of these water soluble pouches is not recommended for making highly concentrated mixtures such as used in aircraft spraying unless pouches are premixed in nurse tanks with adequate agitation.

STORAGE AND DISPOSAL

PROHIBITIONS
Do not contaminate water, food or feed by storage or disposal. Open dumping is prohibited.

STORAGE
Store in a cool, dry place but not below 32°F (0°C). Do not remove package from container except for immediate use.

PESTICIDE DISPOSAL
Pesticide wastes are toxic. Improper disposal of excess pesticide, spray mixture, or rinsate is a violation of Federal law. If these wastes cannot be disposed of by use according to label instructions, contact your State Pesticide or Environmental Control Agency, or the Hazardous Waste representative at the nearest EPA Regional Office for guidance.

CONTAINER DISPOSAL
Completely empty bag into application equipment. Then dispose of in a sanitary landfill or by incineration, or, if allowed by State and local authorities by burning. If burned, stay out of smoke.

STEPS TO BE TAKEN IN CASE MATERIAL IS RELEASED OR SPILLED: Sweep up and place in containers for recovery or disposal. Keep dusting to a minimum. Wear MSHA/NIOSH respirator for dusts encountered. Flush area with water. Keep spill out of all sewers and open bodies of water. **Protective Clothing**—see Precautionary Statements.

USE DIRECTIONS FOR

LETTUCE, ENDIVE AND ESCAROLE
LEGUMES FOR FORAGE AND SEED
ORNAMENTAL BERMUDAGRASS TURF
WOODY ORNAMENTALS, NURSERY STOCK AND

CHRISTMAS TREES

TREEFRUIT (APPLE, CHERRY, NECTARINE, PEACH, PEAR, PLUM, PRUNE) AND GRAPES

BLUEBERRY (OREGON AND WASHINGTON ONLY)

RASPBERRY AND BLACKBERRY (OREGON AND WASHINGTON ONLY)

FALLOW LAND (IDAHO, OREGON AND WASHINGTON ONLY)

GENERAL INFORMATION

KERB 50-W is a selective herbicide formulated as a wettable powder containing 50% active ingredient. It is effective for the control of a wide range of grasses and certain broadleaf weeds. The product is a soil active herbicide with uptake by sensitive weeds occurring through the roots. Before using this herbicide for a specific crop use, study the following general use information which provides important instructions for the safe and effective application of the product.

WEED SPECTRUM

KERB 50-W herbicide is recommended for both preemergence and early postemergence control of winter annual and perennial grasses and chickweed and for preemergence control only of certain other broadleaf weeds and certain other grasses listed below.

WEEDS CONTROLLED BOTH PREEMERGENCE AND EARLY POSTEMERGENCE

Barley, Foxtail	*Hordeum jubatum*
Barley, Volunteer	*Hordeum vulgare*
Bentgrass	*Agrostis species*
Bluegrass, Annual	*Poa annua*
Bluegrass, Bulbous	*Poa bulbosa*
Bluegrass, Kentucky	*Poa pratensis*
Brome, Downy (Cheatgrass)	*Bromus tectorum*
Chickweed, Common	*Stellaria media*
Chickweed, Mouse-Ear	*Cerastium vulgatum*
Fescue, Tall	*Festuca arundinaceae*
Goatgrass, Jointed	*Aegilops cylindrica*
Oat, Volunteer	*Avena sativa*
Oat, Wild	*Avena fatua*
Orchardgrass	*Dactylis glomerata*
Quackgrass	*Agropyron repens*
Rye, Volunteer	*Secale cereale*
Ryegrass, Italian	*Lolium multiflorum*
Ryegrass, Perennial	*Lolium perenne*
Velvetgrass	*Holcus lanatus*
Wheat, Volunteer	*Triticum aestivum*

WEEDS CONTROLLED ONLY PREEMERGENCE

Barnyardgrass	*Echinochloa crus-galli*
Canarygrass	*Phalaris canariensis*
Carpetweed	*Mollugo verticillata*
Crabgrass, Large	*Digitaria sanguinalis*
Dodder, Field	*Cuscuta campestris*
Foxtail, Yellow	*Setaria lutescens*
Goosegrass	*Eleusine indica*
Goosefoot, Nettleleaf	*Chenopodium murale*
Henbit	*Lamium amplexicaule*
Knotweed, Prostrate	*Polygonum aviculare*
Lambsquarters, Common	*Chenopodium album*
Lovegrass	*Eragrostis diffusa*
Rocket, London	*Sisymbrium irio*
Mallow, Little (Cheeseweed)	*Malva parviflora*
Morningglory, Annual	*Ipomoea purpurea*
Mustard, Wild	*Brassica kaber*
Nettle, Burning	*Urtica urens*
Nightshade, Black	*Solanum nigrum*
Nightshade, Hairy	*Solanum sarachoides*
Panicum, Fall	*Panicum dichotomiflorum*
Purslane, Common	*Portulaca oleracea*
Shepherdspurse	*Capsella bursa-pastoris*
Smartweed, Pale	*Polygonum lapathifolium*
Sorrel, Red (from seed)	*Rumex acetosella*
Tomato, Volunteer	*Solanum esculentum*
Radish, Wild	*Raphanus sativus*

NOTE: The weed species controlled by KERB 50-W herbicide are dependent on the rate used, specific crop culture involved, and the associated conditions of temperature, soil type and moisture availability. Refer to specific crop use directions for weed species controlled.

DOSAGE

The rate of KERB 50-W herbicide required will vary depending on the crop culture involved and weed species to be controlled. See specific crop use directions for all dosage recommendations. All dosage recommendations listed in this label are in terms of pounds KERB 50-W herbicide per broadcast acre. For banded application, the amount of KERB 50-W herbicide used per acre should be reduced according to the following formula:

$$\frac{\text{Band width (in.)}}{\text{Row width (in.)}} \times \text{Rate Per Acre Broadcast} = \text{Amount needed per Acre for Band Application}$$

TIMING AND APPLICATION

Unless specific directions are given under the crop to be treated, KERB 50-W herbicide should be applied in the fall or early winter, when temperatures do not exceed 55°F, BUT PRIOR TO SOIL FREEZE UP. Best weed control results when KERB 50-W herbicide is applied preemergence to the weeds and when application is followed by rainfall or irrigation to move the product into the root zone of the germinating weeds.

KERB 50-W herbicide should be thoroughly mixed in clean water at the recommended concentration and applied uniformly as a spray. For ground application, use a conventional low pressure herbicide sprayer equipped with flat fan nozzles spaced and calibrated to uniformly deliver 20 to 50 gallons of spray per acre. For aerial applications apply in a coarse droplet spray at 5 to 10 gallons per acre. Accurately calibrate spray equipment prior to each use.

EFFECT OF SOIL TYPE, MOISTURE AND TEMPERATURE

KERB 50-W herbicide is most active in coarse to medium textured soils of low organic matter and is relatively inactive in peat or muck soils or mineral soils high in organic matter content at rates recommended in this label. Herbicidal activity is best in soils containing less than 4 percent organic matter. Use in soils with higher organic matter may resullt in inconsistent or incomplete weed control.

The herbicidal activity of KERB 50-W herbicide is mainly through root absorption in sensitive weed species. Rain, melting snow or irrigation is ESSENTIAL following treatment to move KERB 50-W herbicide into the root zone of germinating weeds.

Under field conditions, KERB 50-W herbicide will remain stable with little loss of herbicidal activity when soil temperatures are less than 55°F. As soil temperatures increase, degradation of the active ingredient takes place. KERB 50-W herbicide may degrade rather quickly if left exposed on the soil surface in warm weather. If KERB 50-W herbicide is applied when air temperatures exceed 85°F, it should be soil incorporated to a shallow depth (top two to three inches) or watered into the soil as soon as possible.

CULTURAL CONSIDERATIONS

For best results apply KERB 50-W to a trash free soil surface. Clean cultivation before application is preferable, but not necessary. To obtain optimum weed control in areas not clean cultivated, the area to be treated should be free of surface litter (dead or decaying crop and weed debris, mowing clippings, etc.) Trash free areas create ideal conditions for rapid movement of KERB 50-W into the weed root zone following rain or irrigation.

ROTATION CROP PLANTING INFORMATION

Follow the directions given below when rotation crops will be planted to areas previously treated with KERB 50-W herbicide:

1. Whether KERB 50-W herbicide is bed-topped, banded or broadcast, the beds should be knocked down and the field cross-disced before rotation crops other than artichokes, lettuce, endive or escarole are planted.
2. Where the KERB 50-W herbicide treatment is to be followed by a rotation crop within six months of application, bed-topped or banded applications are suggested.

3. Where small grains or grasses will be planted within one year to areas where KERB 50-W herbicide is to be applied at a dosage rate greater than 1.0 pound of product per acre, KERB 50-W herbicide should not be broadcast. Treatments should be limited to band or bed-top applications only.

4. Where rotation crops are to follow within one year of the KERB 50-W herbicide treatment, the following rotational plan applies:

Amount of KERB 50-W Applied /Planted Acre[2]	Waiting Period in Months Before Planting the Crops Indicated[1]			
	Group A	Group B	Group C	Group D
1.0 lb.	0	3	3	6
2.0 lb.	0	3	4	9
3.0 lb.	0	4	6	9
4.0 lb.	0	5	7	12

[1] Crop Group A—Artichokes, lettuce, endive, escarole
Crop Group B—Beans, corn, cotton, sorghum, carrots, celery
Crop Group C—Broccoli, cabbage, cauliflower, cucurbits, spinach, sugarbeets, onion, tomatoes
Crop Group D—Wheat, barley, oats, grasses

NOTE: "Crops not listed in Crop Groups A, B, C or D (above) should not be planted back sooner than 12 months after the last KERB herbicide application regardless of the rate of KERB used."

[2] Planted Acre refers to Acreage planted. KERB at 4.0 pounds of product per treated acre on two 5 inch treated bands (40 inch centers) is the equivalent of one pound KERB 50-W per planted acre. KERB at 4.0 pounds of product as a broadcast treatment (total acreage sprayed) is the equivalent of 4 pounds of KERB 50-W per planted acre.

DIRECTIONS FOR USE
LETTUCE, ENDIVE AND ESCAROLE

GENERAL INFORMATION
KERB 50-W herbicide is a selective herbicide for the control of certain annual grasses and broadleaf weeds in direct seeded or transplanted lettuce, endive and escarole.

WEEDS CONTROLLED
KERB 50-W herbicide is effective at 2.0 to 4.0 pounds of product (1.0 to 2.0 pounds active ingredient) per treated acre for the preemergence control of the following weeds:

GRASSES	BROADLEAF WEEDS
Barley, Foxtail	Carpetweed
Barley, Volunteer	Chickweed, Common
Barnyardgrass	Goosefoot, Nettleleaf
Bluegrass, Annual	Henbit
Brome, Downy (Cheatgrass)	Knotweed
	Lambsquarters, Common
Canarygrass	Morningglory, Annual
Crabgrass	Mustard, Wild
Foxtail, Yellow	Nettle, Burning
Goosegrass	Nightshade, Black
Lovegrass	Nightshade, Hairy
Oats, Volunteer	Purslane, Common
Panicum, Fall	Rocket, London
Ryegrass, Italian	Shepherdspurse
Rye, Volunteer	Smartweed, Pale
Wheat, Volunteer	Tomato, Volunteer

DOSAGE
KERB 50-W herbicide is recommended at the rate of 2.0 to 4.0 pounds of product (1.0 to 2.0 pounds active ingredient) per acre broadcast application. The dosage rate required is dependent on soil texture and method of irrigation. At rates recommended on this label, KERB 50-W is not recommended for weed control on highly organic (peat and muck) soils. For lettuce, endive and escarole follow the dosage recommendations listed in chart below:

Weeds	Pounds KERB 50-W Per Acre Broadcast Application[1]: Dependable Rainfall or Overhead Irrigation	Less Dependable Rainfall or Furrow Irrigation	Soil Texture Group[2]
Susceptible annual grasses and broadleaf weeds	2.0-3.0 (Surface application)	3.0-4.0 (Soil incorporation)	Coarse and medium textured soils
	3.0-4.0 (Surface application)	4.0 (Soil incorporation)	Fine textured soils

[1] Reduce dosage rate accordingly for banded applications.
[2] Soil Texture Group
Coarse: sand, loamy sand, sandy loam
Medium: loam, silt loam, silt, sandy clay loam
Fine: silty clay loam, clay loam, sandy clay, silty clay, clay

CROP TOLERANCE
Most varieties of lettuce are highly tolerant of the recommended rates of KERB 50-W herbicide. Do not use more than 3 pounds KERB 50-W on Val Temp, Grande Verde and Prima Verde varieties of crisp lettuce, or on endive and escarole.

TIMING AND APPLICATION
KERB 50-W herbicide can be applied either pre-plant, post-plant or postemergence to lettuce in banded, bed-topped or broadcast applications. Most applications will be made preemergence to the lettuce just before or after planting and preemergence to the weeds. Applications can be made before or after lettuce thinning but should be made prior to weed emergence. Do not apply KERB 50-W herbicide to lettuce within 55 days of harvest and do not make more than one application to each crop of lettuce.

Mix the recommended amount of KERB 50-W herbicide in clean water and apply uniformly with a ground sprayer in 20 to 50 gallons of water per treated acre. Reduce dosage and volume accordingly for banded treatments. Use a standard low pressure sprayer equipped with flat fan nozzles that provide uniform spray distribution.

APPLICATION-MOISTURE REQUIREMENTS
KERB 50-W herbicide acts mainly through root absorption, therefore it is necessary to move KERB into the root zone of germinating weeds to provide effective control. This can be accomplished by overhead sprinkler irrigation, by rainfall or by shallow mechanical incorporation.

Sprinkler Irrigation
KERB 50-W herbicide can be applied to the soil surface without mechanical incorporation after planting or transplanting if overhead irrigation is used. An initial irrigation of 1 to 2 inches should promptly follow the application of KERB 50-W herbicide, especially in hot weather.

Applications Dependent on Natural Rainfall
In areas of dependable natural rainfall, KERB 50-W herbicide can be applied as a surface treatment preemergence to the weeds. Applications to direct seeded or transplanted lettuce are most successful when followed by ½ to 1 inch of rainfall within two to three days after application.

Furrow Irrigation—Mechanical Incorporation
Where rainfall is not dependable or supplementary overhead irrigation is not used, shallow pre-plant incorporation is recommended. PTO driven incorporators or rolling cultivators that thoroughly mix KERB 50-W herbicide into the top 2 inches of soil are suggested.

Incorporation should be simultaneous or immediately after application of KERB 50-W herbicide, especially in hot weather. Irrigation should be started as soon as possible.

Where furrow irrigation is used, spray application and mechanical incorporation should be made after beds have been formed. KERB 50-W herbicide will not be as effective if disced in prior to bed shaping.

Hoeing, thinning or shallow cultivation of soil treated with KERB 50-W herbicide will not destroy its herbicidal activity.

TEMPERATURE

KERB 50-W herbicide is not highly volatile, but it may degrade rather quickly if left exposed on the soil surface in warm weather. If applied when air temperatures exceed 85°F it should be shallow incorporated or watered into the soil as soon as possible, preferably within 1 to 2 days.

ROTATION CROPS

Where rotation crops are to follow within one year of the KERB 50-W herbicide treatment to lettuce, endive or escarole, follow the directions given in the General Information section of this label under ROTATION CROP PLANTING INFORMATION.

LETTUCE—SPECIFIC USE RESTRICTIONS

- Do not apply KERB 50-W herbicide to lettuce varieties which will be harvested less than 55 days after treatment.
- Do not apply more than one application of KERB 50-W herbicide to each crop of lettuce.

DIRECTIONS FOR USE

LEGUMES GROWN FOR FORAGE AND SEED — ALFALFA, CLOVER, BIRDSFOOT TREFOIL, CROWN VETCH AND SAINFOIN

GENERAL INFORMATION

KERB 50-W is a selective herbicide recommended for fall or winter applications to alfalfa, clover, birdsfoot trefoil, crown vetch and sainfoin for both preemergence and postemergence control of susceptible winter annual and perennial grasses and for preemergence control of certain broadleaf weeds.

DOSAGE

KERB 50-W herbicide is recommended at the rate of 1.0 to 4.0 pounds of product (0.5 to 2.0 pounds active ingredient) per broadcast acre application. The required rate will depend on the weed species present as well as the type of irrigation used or the dependability of rainfall following application. The effective rate will be higher in low rainfall areas or where furrow irrigation is used than in areas of dependable rainfall or where overhead irrigation is practiced. Follow the weed control recommendations given in the chart below for fall or winter applications of KERB 50-W herbicide:

Weeds Controlled		Pounds KERB 50-W Per Acre Broadcast Application: Dependable Rainfall or Overhead Irrigation	Low Rainfall or Furrow Irrigation
Apply Preemergence or Postemergence to These Weeds:			
Barley, Foxtail Bluegrass, Annual Brome, Downy (Cheatgrass) Chickweed	Grain Volunteer Oat, Wild Ryegrass, Italian	1.0–1.5	1.5–2.0
Bluegrass, Kentucky Orchardgrass Ryegrass, Perennial		1.5–2.0	2.0–3.0
Quackgrass		2.0–3.0	3.0–4.0
Apply Preemergence Only to These Weeds:			
Sorrel, Red (from Seed)		1.5–2.0	2.0–3.0
Mustard, Wild Radish, Wild	Rocket, London Shepherdspurse	3.0	4.0

For control of spring germinating dodder refer to specific instructions under Spring Use Directions for Established Alfalfa.

TIMING AND APPLICATION

KERB 50-W herbicide should be applied during the fall or winter months. Optimum herbicidal activity occurs when applications are made under cool temperature conditions (55°F or less) and are followed by rainfall or overhead irrigation. Applications should always be made BEFORE SOIL FREEZE-UP.

Applications are recommended postemergence to established, actively growing or dormant forage legumes or to new plantings after the legume has reached the trifoliate leaf stage. In established forage legume stands, applications should be made after the last cutting when the weather and soil temperatures are cool. In fall-seeded forage legumes, applications should be made after legumes have reached the trifoliate leaf stage. In spring-seeded forage legumes, applications of KERB 50-W herbicide should be made the following fall or early winter to control winter annual and perennial grasses.

Do not use KERB 50-W herbicide as a preplant or preemergence treatment or before the trifoliate leaf stage of the legume has developed in new plantings as injury to the legume stand may result. Remove or disperse trash, crop residues and ashes before treatment.

Mix the recommended amount of KERB 50-W herbicide in clean water and apply uniformly with a ground sprayer at 20 to 50 gallons per acre. Use a conventional herbicide sprayer equipped with flat fan nozzles at 40 to 60 psi.

ROTATION CROPS

Where rotation crops are to follow within one year of the KERB 50-W herbicide treatment to alfalfa, clover, birdsfoot trefoil, crown vetch or sainfoin, follow the directions given in the General Information section of this label under ROTATION CROP PLANTING INFORMATION.

SPECIFIC USE RESTRICTIONS—ALFALFA, CLOVER, BIRDSFOOT TREFOIL, CROWN VETCH AND SAINFOIN

- Do not use more than 4.0 pounds KERB 50-W herbicide per acre per season.
- Do not graze or harvest for forage or dehydration within the following intervals after application:

  Alfalfa—below 3.0 pounds KERB 50-W/A (West of Mississippi River) — 25 days

  Alfalfa—3.0 to 4.0 pounds KERB 50-W/A (West of Mississippi River) — 45 days

  Clover, birdsfoot trefoil, crown vetch, sainfoin (entire U.S.) and alfalfa (East of Mississippi River) up to 4.0 pounds KERB 50-W/A. — 120 days

SPRING USE DIRECTIONS FOR ESTABLISHED ALFALFA—

Dodder Control In Alfalfa, Seed Crops Only In California, Idaho, Nevada, Oregon, Utah And Washington

GENERAL INFORMATION

For effective control KERB 50-W herbicide must be moved into the soil either by rainfall or irrigation before the germination of dodder. Preferably, irrigation should be made within 1 to 3 days following the KERB 50-W application, but can be delayed up to 2 weeks if necessary provided that irrigation preceeds dodder germination. If irrigation of the field treated with KERB 50-W herbicide must be delayed, a light mechanical incorporation (maximum 1 inch depth) should follow the KERB 50-W application and the field irrigated within two weeks.

When using flood type or overhead sprinkler irrigation systems the amount of irrigation following the KERB 50-W application should not exceed one inch of water. Excess irrigation following the KERB 50-W application and prior to germination of dodder may decrease the effectiveness of KERB 50-W herbicide.

DOSAGE AND TIMING

For effective control, KERB 50-W herbicide must be applied before dodder germinates. Follow directions given below depending on method of irrigation used:

Furrow Irrigation—Apply KERB 50-W at the rate of 3.0 to 4.0 pounds of product (1.5 to 2.0 pounds active ingredient) per acre. Incorporate lightly at time of application and irrigate within seven days.

Flood Irrigation—Apply KERB 50-W at the rate of 3.0 pounds of product (1.5 pounds active ingredient) per acre. Flood field with 0.5 to 1.0 inch of water within 1–3 days after application.

Overhead Sprinkler Irrigation—Use same directions as given above for flood irrigation.

Excessive amounts of irrigation water following KERB 50-W herbicide application may adversely affect the herbicidal activity.

NOTE: If the legume crop treated with KERB 50-W herbicide should be winter killed or otherwise destroyed, do not plant the field in the spring to small grain or grass crops since these crops may be injured from residual KERB in the soil. For additional information see Rotation Crop Planting Chart in the General Information section of this label.

DIRECTIONS FOR USE ORNAMENTAL BERMUDAGRASS— GROWN FOR TURF OR SEED

GENERAL INFORMATION

KERB 50-W is a selective herbicide recommended for the preemergence and postemergence control of annual bluegrass (Poa annua) in ornamental bermudagrass (Cynodon dactylon) turf or bermudagrass grown for seed.

WEED CONTROL

KERB 50-W herbicide will control anual bluegrass from pregermination and seedling stages through tillering, heading and seed formation. KERB 50-W herbicide acts slowly on seedling to mature annual bluegrass. Following application of KERB 50-W annual bluegrass may first become dark green and then gradually turn yellow and die over a 3 to 5 week period.

For effective control of annual bluegrass in bermudagrass turf or seed crops moisture is necessary to move KERB 50-W herbicide in the weed root zone. Refer to the Moisture Requirement section of this label for details.

DOSAGE AND TIMING

KERB 50-W herbicide is recommended at the rate of 1.0 to 3.0 pounds of product (0.5 to 1.5 pounds active ingredient) per acre broadcast application. The dosage rate required is dependent on the growth stage of annual bluegrass at time of application. Follow the dosage rate and timing recommendations given below:

Annual Bluegrass Growth Stage	Pounds KERB 50-W Per Acre Broadcast Application
Preemergence or early postemergence	1.0–2.0*
Postemergence-early tillering to heading	1.5–2.0
Postemergence-seed forming stages	2.0–3.0

* Use the higher rate when longer preemergence residual control is desired.

APPLICATION

Mix the recommended amount of KERB 50-W herbicide in clean water and apply uniformly with a low pressure ground sprayer in 20 to 50 gallons of water per acre. The sprayer should be equipped with flat fan nozzles, spaced to provide uniform distribution without skips or excessive overlapping of spray patterns.

NOTE—Avoid spraying on fairways or approaches that may drain onto bentgrass greens or to areas overseeded with sensitive cool season grasses.

MOISTURE REQUIREMENTS

KERB 50-W herbicide acts mainly through root absorption in sensitive weed species. If no rainfall occurs within a day or so of the application, a light overhead irrigation should be made to move the chemical into the weed root zone. Avoid heavy irrigations of more than 1 inch to reduce the possibility of excess washing or leaching of the chemical from the area of application.

ORNAMENTAL BERMUDAGRASS TURF OR SEED CROP—SPECIFIC USE RESTRICTIONS

- Do not apply KERB 50-W herbicide to areas that are to be overseeded with susceptible cool season grasses within 90 days of treatment.
- Do not apply KERB 50-W to dicondra, perennial bluegrass, annual and perennial ryegrasses, fescues and bentgrasses.
- KERB 50-W herbicide is not recommended for use on golf greens.
- Do not graze treated areas and do not feed clippings to livestock.

DIRECTIONS FOR USE WOODY ORNAMENTALS, NURSERY STOCK AND CHRISTMAS TREES

GENERAL INFORMATION

KERB 50-W is a selective herbicide recommended for fall applications to established woody ornamentals, nursery stock and Christmas trees for the control of winter annual and perennial grasses and certain broadleaf weeds.

CROP TOLERANCE

At recommended rates of KERB 50-W herbicide the following trees and shrubs are tolerant to topical applications made in the fall:

Ash	Firethorn	Mountain Ash
Arborvitae	Flowering Cherry	Mountain Laurel
Azalea	Flowering Crabapple	Ohio Buckeye
Barberry	Flowering Quince	Oak
Basswood	Forsythia	Pine
Beech	Ginko	Poplar
Birch	Hawthorn	Privet
Boxwood	Hemlock	Rhododendron
Bradford Pear	Holly	Spirea
Cedar	Honey Locust	Spruce
Cotoneaster	Juniper	Sweetgum
Dogwood	Lilac	Sycamore
Douglas Fir	Linden	Tuliptree
Eastern Redbud	London Plane	Viburnum
Elm	Magnolia	Walnut
Euonymus	Maple	Willow
Fir	Mock Orange	Yew

KERB 50-W herbicide should only be used on established trees and woody ornamentals. KERB 50-W herbicide is not recommended for use on seedling trees or shrubs less than one year old or to fall transplanted stock transplanted less than one year or to spring transplanted stock transplanted less than six months.

WEED CONTROL

KERB 50-W herbicide is effective in fall applications at the rate of 2.0 to 4.0 pounds of product (1.0 to 2.0 pounds active ingredient) per treated acre for the preemergence and postemergence control of susceptible winter annual and perennial grasses and chickweed and for preemergence control only of other broadleaf weeds listed on this label. Refer to chart in dosage rate section below for specific weeds controlled.

DOSAGE AND TIMING

KERB 50-W herbicide is recommended in a single, fall application, either directed or topically applied, to woody ornamentals, nursery stock or Christmas trees at the rate of 2.0 to 4.0 pounds of product (1.0 to 2.0 pounds active ingredient) per treated acre. Application of KERB 50-W herbicide should be made in the fall prior to leaf drop and soil freeze-up. For control of winter annual or perennial grasses or chickweed, applications can be made either preemergence or postemergence to the weeds. For control of other labeled broadleaf weeds, preemergence applications must be used to achieve control.

The dosage rate required will depend on the weed species present in the area to be treated. Follow the weed control recommendations given in the chart below:

Weeds Controlled		Pounds KERB 50–W Per Acre Broadcast Application
Barley, Foxtail Bluegrass, Annual Brome, Downy (Cheatgrass) Chickweed	Grain Volunteer Ryegrass, Italian Sorrel, Red (from seed)	2.0
Mustard, Wild Rocket, London	Shepherdspurse	3.0
Bluegrass, Kentucky Orchardgrass Quackgrass	Ryegrass, Perennial	4.0

APPLICATION

Mix the recommended amount of KERB 50-W herbicide in clean water and apply uniformly in 20 to 50 gallons per acre. Use a low pressure ground sprayer equipped with flat fan nozzles spaced to provide uniform distribution.

Dosage recommendations listed on this label are for surface broadcast application. For banded treatments down the row, the amount of KERB 50-W used per acre should be reduced according to the following formula:

$$\frac{\text{Band width (in.)}}{\text{Row width (in.)}} \times \text{Rate per Acre Broadcast} = \text{Amount Needed per Acre for Band Application}$$

KERB 50-W herbicide should not be soil incorporated.

NOTE: Most ornamental turf grass species and ground covers are sensitive to KERB 50-W. Care should be exercised to avoid contact of KERB 50-W herbicide with these plants from either direct application, spray drift or from applications to areas that may drain onto established ornamental turf and ground cover.

SOIL AND MOISTURE REQUIREMENTS

KERB 50-W herbicide is most active in coarse to medium textured soils of low organic matter and is relatively inactive in peat or muck soils or mineral soils high in organic matter content at rates recommended in this label. Herbicidal activity is best in soils containing less than 4 percent organic matter. Use in soils of higher organic matter content may result in inconsistent or incomplete weed control.

KERB 50-W herbicide acts mainly through root absorption in sensitive weed species. Dependable rainfall or overhead irrigation is essential following application for effective weed control.

Woody Ornamentals, Nursery Stock and Christmas Trees

SPECIFIC USE RESTRICTIONS

- Apply KERB 50-W herbicide in the fall prior to soil freeze-up.
- Do not soil incorporate KERB 50-W herbicide.

DIRECTIONS FOR USE
TREEFRUIT (APPLE, CHERRY, NECTARINE, PEACH, PEAR, PLUM, PRUNE) AND GRAPE PLANTINGS

GENERAL INFORMATION

KERB 50-W is a selective herbicide for use in directed spray applications for the control of winter annual and perennial grasses and certain broadleaf weeds in non-bearing and bearing apples, cherries, nectarines, peaches, pears, plums, prunes and grape plantings.

WEED CONTROL

KERB 50-W herbicide is effective at 2.0 to 8.0 pounds of product (1.0 to 4.0 pounds active ingredient) per treated acre for the preemergence and postemergence control of susceptible winter annual and perennial grasses and chickweed and for preemergence control only of other broadleaf weeds listed on this label. Refer to chart in dosage rate section below for specific weeds controlled.

DOSAGE AND TIMING

KERB 50-W herbicide is recommended in a single, directed application to labeled fruit trees and grape plantings at dosage rates of 2.0 to 8.0 pounds of product (1.0 to 4.0 pounds active ingredient) per treated acre. Application of KERB 50-W should be in the fall, after the fruit is harvested, but prior to leaf drop and soil freeze-up.

The dosage rate required for effective weed control will depend on the weed species present and the soil texture of the area being treated. Follow the specific rate recommendations given in the chart below for the use of KERB 50-W herbicide in labeled fruit trees and grapes:

Weeds Controlled		Pounds of KERB 50–W Per Acre Dependable Rainfall or Overhead Irrigation — Soil Texture Group* Coarse	Medium	Fine
Bluegrass, Annual	Oat, Wild	2.0	3.0	4.0
Brome, Downy (Cheatgrass)	Ryegrass, Italian			
Chickweed	Sorrel, Red (from seed)			
Grain Volunteer				
Bluegrass, Kentucky	Quackgrass	3.0–4.0	4.0–6.0	6.0–8.0
Fescue, Tall	Ryegrass, Perennial			
Orchardgrass				

* Soil Texture Group

Coarse: sand, loamy sand, sandy loam

Medium: loam, silt loam, silt, sandy clay loam

Fine: silty clay loam, clay loam, sandy clay, silty clay, clay

APPLICATION

Mix the recommended amount of KERB 50-W herbicide in clean water and apply uniformly in 40 to 50 gallons of water per acre. Use of a low pressure ground sprayer equipped with a breakaway boom and flat fan or off-center (OC) nozzles is recommended. KERB 50-W should be directed to the soil and the base of trees and vines.

NOTES: Dosage recommendations listed on this label are for surface broadcast application. For banded treatments, the amount of KERB 50-W used per acre should be reduced according to the following formula:

$$\frac{\text{Band width (in.)}}{\text{Row width (in.)}} \times \text{Rate per Acre Broadcast} = \text{Amount Needed per Acre for Band Application}$$

KERB 50-W herbicide should not be soil incorporated.

CROP TOLERANCE

When used as recommended, established non-bearing or bearing fruit trees and grapes listed on this label are very tolerant to KERB 50-W herbicide. KERB 50-W herbicide should not be applied to seedling trees or vines less than 1 year old or to fall transplanted stock transplanted less than 1 year or to spring transplanted stock transplanted less than 6 months.

CULTURAL CONSIDERATIONS

KERB 50-W herbicide acts mainly through root absorption in sensitive weed species. Dependable rainfall or overhead irrigation is essential following the application for effective weed control. Trash-free areas create ideal conditions for rapid movement of KERB 50-W herbicide into the weed root zone following rain or irrigation. Clean cultivation before application is preferable but not necessary.

To obtain optimum weed control in areas not clean cultivated, the area to be treated should be free of surface litter (dead or decaying weeds, leaves, mowing clippings, etc.). If area to be treated is under a mixed grass or weed sod, it should be mowed and the clippings removed.

TREEFRUIT (APPLE, CHERRY, NECTARINE, PEACH, PEAR, PLUM, PRUNE) AND GRAPE PLANTINGS

SPECIFIC USE RESTRICTIONS

- Do not feed or allow livestock to graze areas treated with KERB 50-W herbicide.
- Do not apply KERB 50-W herbicide to labeled fruit trees or grapes at rates higher than 8.0 pounds product (4.0 pounds active ingredient) per acre and make only one application per season.

DIRECTIONS FOR USE
BLUEBERRIES
(OREGON AND WASHINGTON ONLY)

GENERAL INFORMATION

KERB 50-W is a selective herbicide recommended for fall and winter applications to established blueberries for both preemergence and postemergence control of winter annual and perennial grasses and chickweed and preemergence control of certain broadleaf weeds.

DOSAGE RECOMMENDATIONS

KERB 50-W herbicide is recommended at the rate of 2.0 to 4.0 pounds of product (1.0 to 2.0 pounds active ingredient) per acre broadcast application. The rate will depend on the weed species present. Follow the weed control recommendations listed in the chart below:

Weeds Controlled		Pounds KERB 50–W Per Acre[1] Dependable Rainfall or Overhead Irrigation[2]
Bluegrass, Annual Brome, Downy (Cheatgrass) Chickweed	Oat, Wild Sorrel, Red (from seed)	2.0
Bentgrass Bluegrass, Kentucky Fescue, Tall Orchardgrass	Quackgrass Ryegrass, Perennial Velvetgrass	4.0

[1] Dosage rates recommended are in pounds of KERB 50-W per acre broadcast application. Reduce rates accordingly for banded applications.

[2] For effective weed control, rainfall or overhead irrigation is essential following the application of KERB 50-W.

CROP TOLERANCE

Established blueberry plants are tolerant to recommended rates of KERB 50-W herbicide. Do not apply KERB 50-W to newly transplanted blueberries until roots are well established.

TIMING AND APPLICATION

Apply KERB 50-W herbicide only during the fall or winter months. For optimum results, apply KERB 50-W during November or December. Do not make applications when ground is frozen.

Mix the recommended amount of KERB 50-W in clean water and apply uniformly with a low pressure ground sprayer in 20 to 50 gallons of water per acre.

BLUEBERRIES—SPECIFIC USE RESTRICTIONS

- Do not make more than one application of KERB 50-W herbicide per year and do not apply KERB 50-W at rates higher than 4.0 pounds product (2.0 pounds active ingredient) per acre.

DIRECTIONS FOR USE BLACKBERRIES AND RASPBERRIES (OREGON AND WASHINGTON ONLY)

GENERAL INFORMATION

KERB 50-W herbicide is a selective herbicide recommended for fall and winter applications to established blackberries and raspberries for both preemergence and postemergence control of certain winter annual and perennial grasses.

DOSAGE

KERB 50-W herbicide is recommended at the rate of 2.0 to 6.0 pounds of product (1.0 to 3.0 pounds active ingredient) per acre broadcast application. The rate will depend on the weed species present and the soil texture of the site being treated. Follow the weed control recommendations listed in the chart below:

Weeds Controlled	Pounds of KERB 50-W Per Acre[1] Dependable Rainfall or Overhead Irrigation[2]	Comments
Bluegrass, Annual	2.0–4.0	Use low rates on light to medium soils and high rates on heavy soils
Ryegrass, Perennial	4.0–6.0	
Quackgrass	4.0–6.0	

[1] Dosage rates recommended are in pounds of KERB 50-W per acre broadcast application. Reduce rates accordingly for banded applications.

[2] For effective weed control, rainfall or overhead irrigation is essential following the application of KERB 50-W.

CROP TOLERANCE

Established canefruit are tolerant to recommended rates of KERB 50-W herbicide. Newly transplanted blackberries or raspberries should be well rooted and transplanted for at least 3 months prior to the application of KERB 50-W herbicide.

TIMING AND APPLICATION

Apply KERB 50-W herbicide only during the fall or winter months. For optimum results, apply KERB 50-W during November or December. Do not make applications when the ground is frozen. Mix the recommended amount of KERB 50-W in clean water and apply uniformly with a low pressure ground sprayer in 20 to 50 gallons of water per acre.

BLACKBERRIES AND RASPBERRIES—

SPECIFIC USE RESTRICTIONS

- Do not make more than one application of KERB 50-W herbicide per year and do not apply KERB 50-W at rates higher than 6.0 pounds product (3.0 pounds active ingredient) per acre.

DIRECTIONS FOR USE FALLOW LAND (IDAHO, OREGON AND WASHINGTON ONLY)

GENERAL INFORMATION

KERB 50-W herbicide is a selective herbicide for use on fallow land to control certain winter annual grasses and volunteer grains.

Use of the product is restricted to summer fallow land which will be planted back the following year to winter wheat, barley or oats.

Use rates and plant-back restrictions for KERB 50-W herbicide should be strictly followed or severe injury to the following small grain crop may occur.

KERB 50-W is a soil active herbicide that is absorbed through weed roots. In order to achieve good weed control, sufficient soil moisture from rain, melting snow or irrigation is required following treatment to move KERB 50-W herbicide into the soil to the weed root zone.

Do not use any fall tillage prior to or after application of KERB 50-W herbicide to fallow land, since weed control will be reduced. To ensure best weed control results with KERB 50-W, use a straw spreader when combining to eliminate chaff rows.

WEEDS CONTROLLED

KERB 50-W herbicide will provide preemergence and postemergence control of the following weeds when used according to label directions:

Barley, Volunteer	*Hordeum vulgare*
Bluegrass, Bulbous	*Poa bulbosa*
Brome, Downy (Cheatgrass)	*Bromus tectorum*
Goatgrass, Jointed	*Aegilops cylindrica*
Rye, Volunteer	*Secale cereale*
Wheat, Volunteer	*Triticum aestivum*

DOSAGE AND TIMING

Apply KERB 50-W herbicide in a single application at a rate of 0.5 to 1.0 pound (0.25 to 0.50 pound active ingredient) per acre. For optimum weed control, apply KERB 50-W herbicide between mid-October and mid-December, prior to soil freeze-up. Applications made outside these dates could result in poor weed control.

APPLICATION

Mix the recommended amount of KERB 50-W herbicide in clean water and apply uniformly in a broadcast application. For ground application KERB 50-W herbicide should be applied in 10 to 20 gallons of water per acre with a conventional low pressure herbicide sprayer equipped with flat fan nozzles. For aerial applications apply in a coarse droplet spray at 5 to 10 gallons per acre. Spray equipment should be carefully calibrated before each use.

ROTATION CROPS

Winter wheat, barley and oats can be planted into fallow land treated with KERB 50-W herbicide 9 months (270 days) or more after application.

FALLOW LAND—SPECIFIC USE RESTRICTIONS

- Do not plant winter wheat, barley or oats within 270 days of the application of KERB 50-W herbicide.
- Do not use any fall tillage prior to or after the application of KERB 50-W herbicide.

TANK MIXTURE WITH GLEAN® HERBICIDE FOR FALLOW LAND

(IDAHO, OREGON AND WASHINGTON ONLY)

For control or suppression of certain broadleaf weeds not controlled by KERB 50-W herbicide in fallow land areas. Glean herbicide may be tank mixed with KERB 50-W. See Glean label for list of weeds claimed.

DOSAGE

Use KERB 50-W herbicide at the rate of 0.5 to 0.75 pound (0.25 to 0.38 lb. active ingredient) per acre tank mixed with Glean herbicide at $\frac{1}{3}$ ounce of product per acre (when soil pH is 6.6 to 7.5) or $\frac{1}{3}$ to $\frac{1}{2}$ ounce of product per acre (when soil pH is 6.5 or less).

Important: Before applying a tank mixture of KERB 50-W and Glean herbicides on fallow land areas, read and observe all label directions for each product. Use rates and plant back restrictions should be strictly followed or severe injury to the following small grain crops may occur. In interpreting product labels for tank mixtures, the most restrictive situations must apply.

For crop rotation flexibility, do not use Glean herbicide on all your fallow acreage.

8116-B4

KERB is a registered trademark of Rohm and Haas Company.

GLEAN is a registered trademark of E.I. duPont de Nemours & Company, Inc.

SPECIAL LOCAL NEED REGISTRATIONS FOR KERB 50-W

STATE	USE	SLN OR FEDERAL ACCESSION NO.
California	Lettuce, Aerial Application	CA-790002
California	With Dinoseb	04595
Idaho	Fallow Land	ID-830002
Kansas	With Liquid Fertilizers	04596
Kentucky	With Granular Potash Fertilizer, Alfalfa	LTR 12/10/76
Michigan	Blueberries	MI-770008
New Mexico	With Balan	07623
New York	With Granular Potash Fertilizer, Alfalfa	NY-810026
New York	Lettuce in Muck Soil	NY-830003
Oregon	Tank Mix with Herbicide 273 for Weed Control in Sugar Beets	OR-780057
Oregon	Sugar Beets Grown for Seed	06619
Oregon	With Granular Potash Fertilizer, Alfalfa	OR-810027
Oregon	Christmas Tree Plantation	OR-830013
Oregon	Fallow Land	OR-820070
Oregon	Aerial Fallow Land	OR-840047
Pennsylvania	With Granular Potash Fertilizer, Alfalfa	PA-800032
Washington	Aerial Application Christmas Trees at Increased Rate	WA-780025
Washington	Fallow Land	WA-840069
Washington	With Liquid Fertilizers	04594
Wisconsin	With Granular Potash Fertilizer, Alfalfa	LTR 9/17/76

STAM® M-4

HERBICIDE

4 LB. E.C.

For Postemergence Control of Barnyardgrass and Other Weeds in Rice.

ACTIVE INGREDIENT	
Propanil 3′,4′-dichloropropionanilide	43.5%*
INERT INGREDIENTS	56.5%
TOTAL	100.0%

*Equivalent to 4 lbs. active ingredient per gallon

EPA Reg. No. 707-109 EPA Est. No. 4091-TN-2

U.S. Patent 3,816,092 1377-D5

STAM® is a registered trademark of Rohm and Haas Company.

WARNING

Keep Out of Reach of Children

NOTICE: Before using this product, read the entire Precautionary Statements, Conditions of Sale and Warranty, Directions for Use, Use Restrictions and Storage and Disposal Instructions on back panel. If the Conditions of Sale and Warranty are not acceptable, return the product unopened within thirty days of purchase to the place of purchase.

PRECAUTIONARY STATEMENTS

HAZARDS TO HUMANS AND DOMESTIC ANIMALS

WARNING

IRRITATING TO EYES AND SKIN

HARMFUL IF SWALLOWED OR ABSORBED THROUGH SKIN

MAY BE HARMFUL IF INHALED

Do not get in eyes or on skin. Wear eye and skin protection when handling. Do not take internally. Avoid breathing spray mist.

STATEMENT OF PRACTICAL TREATMENT

IF IN EYES: Flush eyes with large amounts of water for at least 15 minutes. Consult a physician if irritation persists.

IF ON SKIN: Wash affected area with soap and water.

IF SWALLOWED: Induce vomiting by immediately giving two glasses of water to drink and touch back of throat with finger. Call a physician. Never give anything by mouth to an unconscious person.

IF INHALED: Remove victim to fresh air. Treat the person symptomatically. If irritation persists, call a physician.

ENVIRONMENTAL HAZARDS

This pesticide is toxic to fish. Keep out of lakes, ponds or streams. Do not contaminate water by cleaning of equipment or disposal of wastes.

CONDITIONS OF SALE AND WARRANTY

Rohm and Haas warrants that the product conforms to its chemical description and is reasonably fit for the purpose stated on the label only when used in accordance with label directions under normal conditions of use. **ROHM AND HAAS MAKES NO OTHER EXPRESS OR IMPLIED WARRANTIES EITHER OF MERCHANTABILITY OR FITNESS FOR A PARTICULAR USE.** Handling, storage and use of the product by Buyer or User are beyond the control of Rohm and Haas and Seller. Risks such as crop injury, ineffectiveness or other unintended consequences resulting from, but not limited to, weather or soil conditions, presence of other materials, disease, pests, drift to other crops or property or failure to follow label directions will be assumed by the Buyer or User. **IN NO CASE WILL ROHM AND HAAS OR SELLER BE HELD LIABLE FOR CONSEQUENTIAL, SPECIAL OR INDIRECT DAMAGES RESULTING FROM THE HANDLING, STORAGE OR USE OF THIS PRODUCT.**

DIRECTIONS FOR USE—(FOR RICE GROWN IN SOUTHERN UNITED STATES ONLY)

It is a violation of Federal law to use this product in a manner inconsistent with its labeling.

RE-ENTRY AND WORKER PROTECTION STATEMENTS

Do not enter treated areas without protective clothing until sprays have dried. Because certain states may require more restrictive re-entry intervals for various crops treated with this product, consult your State Department of Agriculture for further information. Do not apply this product in such a manner as to directly or through drift expose workers or other persons. The area being treated must be vacated by unprotected persons.

GENERAL INFORMATION

STAM M-4 is a selective postemergence herbicide for use in rice only for control of the following weeds:

Barnyardgrass (Watergrass)
Echinochloa crus-galli or *Echinochloa colonum*
Brachiaris
Brachiaria spp.
Crabgrass, large
Digitaria sanguinalis
Croton, woolly
Croton capitatus
Foxtail
Sertaria spp.
Goosegrass
Eleusine indica
Gulf Cockspur
Echinochloa crus-pavonis
Hoorahgrass
Fimbristylis miliaceae
Mexican Weed
Caperonia palustris
Paragrass
Panicum purpurascens
Pigweed, redroot
Amaranthus retroflexus
Redweed
Melochia corchorifolia
Sour Dock
Rumex crispus
Spearhead
Rhynchospora corniculata
Tall Indigo or Coffee Bean
Sesbania exaltata
Texas Millet
Panicum texanum
Wiregrass
Eleocharis spp.

STAM M-4 herbicide is an emulsifiable concentrate containing 4 pounds active ingredient per U.S. gallon. The product is not a hormone-type herbicide, but kills susceptible weeds by direct contact action. For this reason, thorough coverage of emerged weeds is essential for best results. Only weeds which have emerged and are exposed at time of application will be controlled. The product has no preemergence or residual herbicidal activity. Apply STAM M-4 herbicide only to fields which have been drained of flood water. The product is most effective if applied when susceptible grasses and broadleaf weeds are small and growing actively under favorable soil moisture and weather conditions. Early weed control removes competition, saves moisture and generally contributes to increased yields.

USE RESTRICTIONS

Do not apply to any crop other than rice. STAM M-4 herbicide injures most crops except cereal grains and perennial grasses. Avoid drift or accidental application from turning aircraft on cotton, soybeans, corn, safflower, seedling legumes, vegetables, orchards, vineyards, gardens, shrubs, and ornamentals. Once applied, it does not release fumes hazardous to nearby crops.

TIMING AND DOSAGE RECOMMENDATIONS

Treat grassy and weedy fields when a satisfactory stand of rice that will tolerate flooding is established. The amount of STAM M-4 herbicide to apply depends primarily upon the stage and growth condition of the grasses. The growth stage of the rice is also a factor in dosage and timing limitations so as to avoid the possibility of excessive residues.

For best results apply STAM M-4 herbicide at the rate of 3 to 4 quarts (3 to 4 lbs. active) per acre when the grasses are actively growing in the 1 to early 4 leaf stage. This rate will also control many seedling broadleaf and aquatic weeds. Generally this will be 15 to 25 days after planting of the rice. In order to insure satisfactory weed control, do not apply less than 3 quarts of STAM M-4 herbicide per acre in a single spray application.

Apply STAM M-4 herbicide at the rate of 4 to 6 quarts (4 to 6 lbs. active) per acre to actively growing grasses in the 4 to 6 leaf and early tillering stage or when they are in the 2 to 4 leaf stage but stressed under dry soil conditions. Generally this will be 20 to 30 days after planting of the rice.

EMERGENCY TREATMENT: Apply STAM M-4 herbicide at the rate of 5 to 6 quarts (5 to 6 lbs. active) in 15 gallons of spray per acre for emergency control of older tillering grass. Generally this will be 30 to 40 days after planting. If the field is already flooded, the water should be lowered or drained before spraying to expose more of the grass and weeds. Emergency treatment should be considered as a salvage operation only and cannot be relied upon for total control of grass and weeds.

CAUTION: TO AVOID EXCESSIVE RESIDUES AT HARVEST, DO NOT APPLY AFTER THE END OF TILLERING FOR THE RICE VARIETY BEING TREATED. DO NOT APPLY MORE THAN A MAXIMUM OF SIX POUNDS ACTIVE INGREDIENT PER ACRE IN A SINGLE APPLICATION OR EXCEED EIGHT POUNDS ACTIVE INGREDIENT PER ACRE TOTAL DOSAGE PER SEASON.

APPLICATION EQUIPMENT

Aircraft—Fixed wing aircraft or helicopters should have well designed spray systems that produce a uniform pattern of medium fine spray droplets. Apply STAM M-4 herbicide on small grass in no less than 10 gallons of total spray per acre with boom-nozzle sprayers. Increase volume to 12 to 15 gallons per acre for larger or denser stands of grass or during periods of low humidity.

The optimum effective spray swath width depends on operating conditions and type of aircraft being used. For uniform spray coverage with fixed wing aircraft, do not exceed a spray swath width 10 percent greater than the wingspan or the length of the boom in helicopters. Measure the swaths accurately for flagging.

Ground Sprayers—Use standard low pressure herbicide sprayers equipped with boom and flat fan nozzles. Use nozzle sizes that deliver a medium fine droplet in 15 to 20 gallons total spray per acre at 40 to 50 p.s.i. and at ground speeds not in excess of 3 to 4 m.p.h. Avoid raising boom too high. Spray patterns should meet uniformly.

Flush all equipment with clear water after each day's use. Clean all equipment including nurse tanks used for STAM M-4 herbicide with a detergent wash followed by a water rinse, BEFORE AND AFTER spraying other pesticides or other crops. Applicators and flagmen should avoid contact with spray mist from STAM M-4 herbicide or any pesticides and should wear protective clothing and goggles. Wash thoroughly after exposure.

CROP TOLERANCE AND GROWING CONDITIONS

All leading commercial varieties of rice are exceptionally tolerant to STAM M-4 herbicide. A temporary yellowing or tip burn of rice may be noted after treatment, but new growth is normal. Severe leaf burn and partial killing of rice may occur if the product is applied when rice is under stress and in a weakened growth condition due to disease or insect infestations, excessive soil salts, overwatering, or prolonged drought and extremely hot weather. Growers are cautioned not to spray under such conditions and/or when maximum daily temperatures have been or are expected to go above 100°F.

EFFECT OF CLIMATIC CONDITIONS AND CULTURAL PRACTICES ON WEED CONTROL

Field and Seedbed Preparation

Fields should be accurately leveled and contoured and have well prepared seedbeds free of clods. This encourages uniform and rapid emergence of rice, grass and broadleaf weeds and permits better timing and coverage of STAM M-4 herbicide sprays resulting in optimum weed control.

Water Management

Before application of STAM M-4 herbicide, drained or dry planted fields should be flushed as often as needed to prevent drying and crusting. Flushing encourages uniform emergence and vigorous growth of grass, broadleaf weeds and rice which is essential for best results. Flush fields in sufficient time so that weeds and rice are actively growing at time of treatment. Make sure the field is drained prior to treatment so that grasses and broadleaf weeds are fully exposed. Weeds that are partially submerged in standing water at time of application will not be satisfactorily controlled.

After treatment, treated fields should always be flooded before a second infestation of grass has a chance to develop. To prevent more grass from germinating after treatment, fields should be flooded within 24 hours after spraying, or as soon as possible after 24 hours.

Temperature

The temperature a few days before and after applying STAM M-4 herbicide has an important bearing on the weed killing activity. The activity increases as daily maximum temperatures increase above 75°F and decreases as the daily maximum temperatures decline below 75°F. Do not apply STAM M-4 herbicide when maximum temperatures have been or are expected to stay below 65°F or to go above 100°F. Low temperature at time of application is not so important as long as it warms up later during the day.

Relative Humidity and Rain

Grasses and weeds are more responsive to STAM M-4 herbicide during periods of high humidity when the foliage is moist or covered by dew. When the humidity is very low some of the spray may crystalize in the air. When this condition exits, increase spray volume to 12–15 gallons per acre for best results. Do not spray when rain threatens within six hours to avoid loss of the spray deposit before adsorption by the grass.

Wind

Avoid applications when the wind speed exceeds 10 m.p.h. because of drift hazard to sensitive crops and the possiblity of uneven (streaked) application.

COMPATABILITY WITH OTHER CHEMICALS

Tank-mix applications of STAM M-4 herbicide with other herbicides, insecticides, spray adjuvants, or liquid fertilizers may reduce crop tolerance and/or weed control or impair mixing properties. Use of these products in tank-mix applications with STAM M-4 herbicide is done at the user's risk.

Insecticides

Severe injury or kill of rice plants may result from tank-mix combinations or separate sprays of STAM M-4 herbicide and certain insecticides. Do not combine STAM M-4 herbicide with carbamate insecticides such as carbaryl (Sevin, etc.), methomyl (Lannate, Nudrin, etc.) or organophosphorus insecticides such as parathion, methyl parathion, Guthion, malathion, Systox, EPN, Phosphamidon, etc. Do not apply any of the above insecticides to rice fields within 14 days before or after STAM M-4 herbicide. Do not use carbamate or systemic organophosphorus insecticides on rice fields to be treated with STAM M-4 herbicide. **WARNING:** Do not apply STAM M-4 herbicide to rice fields which were planted with rice seed treated with bird repellents containing methiocarb such as Mesurol, Borderland Red, etc. Consult local Extension specialist for current recommendations of approved insecticides on rice.

STORAGE AND DISPOSAL

STORAGE

Do not use, pour, spill or store near heat or open flame. Ground all metal containers when transferring product. Protect from freezing, if stored below 32°F and crystals form, warm to 72°F for 24 hours, periodically rolling drum to reconstitute.

PESTICIDE DISPOSAL

Do not contaminate water, food, or feed by storage or disposal. Pesticide wastes are toxic. Improper disposal of excess pesticide, spray mixture, or rinsate is a violation of Federal law. If these wastes cannot be disposed of by use according to label instructions, contact your State Pesticide or Environmental Control Agency, or the Hazardous Waste representative at the nearest EPA regional office for guidance.

CONTAINER DISPOSAL

Triple rinse (or equivalent). Then offer for recycling or reconditioning, or puncture and dispose of in a sanitary landfill, or by other procedures approved by state and local authorities.

STEPS TO BE TAKEN IN CASE MATERIAL IS RELEASED OR SPILLED: Eliminate ignition sources. Ventilate area. Avoid breathing vapors. Use MSHA/NIOSH self-contained breathing apparatus or air mask for large spills in confined areas. Dike the spill with inert material (sand, earth, fuller's earth, etc.) and if appropriate transfer the liquid and solid diking material to separate containers for recovery or disposal. Remove contaminated clothing promptly and wash affected skin areas with soap and water. Wash clothing before re-use. Keep out of all sewers and open bodies of water. REFER TO PRECAUTIONARY STATEMENTS.

1377-D5

STAMPEDE™ CM
HERBICIDE

ACTIVE INGREDIENTS	48%
propanil (3′,4′-Dichloropropionanilide ... 33%*	
2-Methyl-4-chlorophenoxyacetic acid, isooctyl ester ... 15%**	
INERT INGREDIENTS	52%
TOTAL	100%

* Equivalent to 3 lbs. active ingredient per gallon

** Equivalent to 1.4 lbs. per gallon of 2-Methyl-4-chlorophenoxyacetic acid, isooctyl ester (0.85 lbs. 2-Methyl-4-chlorophenoxyacetic acid equivalent)

EPA Reg. No. 707-182

NOTICE: Before using this product, read the entire Precautionary Statements, Conditions of Sale and Warranty, Directions for Use, Use Restrictions and Storage and Disposal Instructions. If the Conditions of Sale and Warranty are not acceptable, return the product unopened within thirty days of purchase to the place of purchase.

Keep Out of Reach of Children
CAUTION

PRECAUTIONARY STATEMENTS

HAZARDS TO HUMANS AND DOMESTIC ANIMALS

Harmful if swallowed, inhaled, or absorbed through the skin. Avoid breathing vapors or spray mist. Avoid contact with skin, eyes or clothing. Remove and wash clothing before re-use.

STATEMENT OF PRACTICAL TREATMENT

IF IN EYES—Flush eyes with plenty of water for at least 15 minutes. Call a physician if irritation persists.

IF INHALED—Move subject to fresh air.

IF ON SKIN—Wash affected area with soap and water.

IF SWALLOWED—Dilute by giving 2 glasses of water to drink and call a physician. Never give anything by mouth to an unconscious person.

NOTE TO PHYSICIAN—Emesis is recommended.

ENVIRONMENTAL HAZARDS

This pesticide is toxic to fish. Keep out of lakes, ponds or streams. Do not apply when weather conditions favor drift from areas treated. Do not contaminate water by cleaning of equipment or disposal of wastes.

PHYSICAL AND CHEMICAL HAZARDS

COMBUSTIBLE

Do not use, pour, or store near heat or open flame.

DIRECTIONS FOR USE

It is a violation of Federal law to use this product in a manner inconsistent with its labeling.

RE-ENTRY AND WORKER PROTECTION STATEMENTS

Do not enter treated areas without protective clothing until sprays have dried. Because certain states may require more restrictive re-entry intervals for various crops treated with this product, consult your State Department of Agriculture for further information. Do not apply this product in such a manner as to directly or through drift, expose workers or other persons. The area being treated must be vacated by unprotected persons.

GENERAL INFORMATION

STAMPEDE CM is a selective postemergence herbicide for the control of green and yellow foxtail grass (wild millet, pigeongrass) and specific broadleaf weeds in spring barley, durum and hard red spring wheat in North Dakota, South Dakota, Minnesota and Montana. When used according to label directions the product will provide effective control of the following weeds:

PIGEONGRASS	
GREEN FOXTAIL	*Setaria viridis*
YELLOW FOXTAIL	*Setaria lutescens*
WILD BUCKWHEAT	*Polygonum convolvulus*
REDROOT PIGWEED	*Amaranthus retroflexus*
PROSTRATE PIGWEED	*Amaranthus blitoides*
LAMBSQUARTERS	*Chenopodium album*
WILD MUSTARD	*Brassica kaber*
KOCHIA	*Kochia scoparia*

STAMPEDE CM herbicide is an emulsifiable concentrate containing 4.4 pounds active ingredient per U.S. gallon. The product kills susceptible weeds mainly by direct contact action. For this reason thorough coverage of emerged weeds is essential for optimum results. Only weeds that have emerged and are exposed at time of application will be controlled. The product has no preemergence or residual herbicidal activity. STAMPEDE CM is most effective if applied when pigeongrass and susceptible broadleaf weeds are small and growing actively under favorable soil moisture and weather conditions. Early weed control removes competition, saves moisture and generally contributes to increased yields.

DOSAGE RATE RECOMMENDATIONS—SPRING BARLEY, DURUM AND HARD RED SPRING WHEAT

For control of pigeongrass, wild buckwheat and other labeled broadleaf weeds growing under favorable soil moisture conditions, apply STAMPEDE CM herbicide in a single broadcast application at the rate of 2½ pints per acre. Do not apply STAMPEDE CM at rates higher than 2½ pints per acre.

APPLICATION TIMING

Weed Control—For optimum weed control and crop tolerance, the importance of *early* spray timing cannot be overemphasized. STAMPEDE CM herbicide is most effective in controlling pigeongrass if spray applications are made when the majority of the emerged pigeongrass are in the 2 to 3 LEAF STAGE. Effectiveness of STAMPEDE CM in controlling pigeongrass declines rapidly as the 4th leaf begins to emerge. Since pigeongrass at these early stages is often less than one inch tall, close inspection of the field is essential to determine correct application timing. Fields should be checked frequently for emerging pigeongrass starting 2 to 2½ weeks after planting. The best time to apply STAMPEDE CM often will be 10 to 17 days after crop emergence when the grain is in the 3 to 4 LEAF STAGE.

For effective control of labeled weeds, applications of STAMPEDE CM herbicide must be made to actively growing weeds in the seedling stages listed below.

Weeds Controlled	Seedling Stage*
Pigeongrass	1 to 3 leaf
Buckwheat (wild)	1 to 4 leaf
Kochia	1 to 4 leaf
Mustard (wild)	1 to 4 leaf
Lambsquarters	1 to 4 leaf
Pigweed (prostrate)	1 to 4 leaf
Pigweed (redroot)	1 to 4 leaf

* Refers to weed seedling stages for control. Pigeongrass greater than the 3 leaf stage or listed broadleaf weeds beyond the 4 leaf stage will not be controlled.

CROP TOLERANCE

STAMPEDE CM herbicide causes temporary yellowing or tip browning to the grain crop 2 to 5 days after application. The effects are temporary and usually will disappear 10 to 14 days after application. New leaves will have normal green color. Application of STAMPEDE CM under adverse growing conditions may result in greater crop injury and slower recovery. Care should be taken not to spray STAMPEDE CM on grain crops that are under growth stress caused by drought, flooding, excess soil salts, hail damage, or extreme temperatures. Do not apply STAMPEDE CM if frost is expected within 24 hours or when temperatures are above 85°F, especially with drying winds.

Applications of STAMPEDE CM herbicide should be made only to actively growing (healthy) grain crops in the seedling stages recommended below:

Crop	Seedling Stage*
Hard Red Spring Wheat	2 to 5 leaf
Durum Wheat	2 to 4 leaf
Barley	2 to 4 leaf

* Refers to crop seedling stages for best tolerance. To avoid crop injury do not apply STAMPEDE CM herbicide to hard red spring wheat beyond the 5 leaf stage or to spring barley or durum wheat beyond the 4 leaf stage.

MIXING AND EQUIPMENT

STAMPEDE CM herbicide is an emulsifiable concentrate that mixes readily with water. With agitator running, pour the measured quantity of STAMPEDE CM into partially filled spray tank and complete filling. Spray water should be clean and 50°F or warmer when mixed. STAMPEDE CM may be applied with conventional low pressure herbicide sprayers.

Field Sprayers: Spra-Coupe, tractor-drawn or truck-mounted ground sprayers.

Use flat fan nozzles of sufficient size to provide a minimum of 10 to 12 gallons volume per acre at a pressure of not less than 40 p.s.i. Recommended ground speeds are 5 to 8 mph (1st or 2nd gears only) for Spra-Coupe, 4 to 5 mph for tractor-drawn sprayers and 8 to 10 mph for truck-mounted sprayers. Hollow cone or floodjet nozzles are not recommended on these sprayers.

Floaters: Big A, Terragator, etc.

Floaters equipped with floodjet nozzles are satisfactory for STAMPEDE CM herbicide application. A minimum of 15 to 20 gallons spray per acre is recommended because of the coarser droplet size produced by the floodjet nozzles. The nozzles should be no more than 60" apart and of such a size to provide the recommended volume at normal operating speeds of 12 to 15 mph. Pressure should be sufficient to produce proper overlap of spray patterns between nozzles to avoid skips.

If floaters are equipped with flat fan nozzles use a minimum pressure of 40 psi to insure proper overlap of spray patterns between nozzles.

AERIAL APPLICATION

STAMPEDE CM herbicide can be applied either by fixed wing aircraft or helicopter. Make applications at the recommended dosage and timing in a minimum of 5 to 10 gallons of water per acre. Aircraft should be equipped with a well designed spray distribution system that is adjusted and operated to provide a uniform pattern of medium sized droplets (400 micron size) distributed over an optimum width swath. Only hollow cone nozzles with cores are recommended for STAMPEDE CM aerial application. Nozzles should be angled 0° to 20° downward and to the rear, depending on air speed. Nozzles should not be placed within 2 to 3 feet of the wing tips in fixed wing aircraft and extras should be placed on the right side of fuselage. Use a spray pump of at least 1½ to 2 inches in diameter and operate at 20 to 50 p.s.i. Application should be made with boom height 6 to 10 feet above crop. The optimum or effective spray swath width depends on operating conditions and type of aircraft being used. For uniform spray coverage with fixed wing aircraft do not exceed a spray swath width 10% greater than the wingspan or the length of the boom in helicopters. Measure the swaths accurately for flagging.

CAUTION: DO NOT APPLY STAMPEDE CM HERBICIDE IN AERIAL APPLICATIONS WITHIN 600 FEET OF ANY PERMANENT BODY OF WATER OR WHEN WIND SPEED EXCEEDS 10 MPH.

CALIBRATION

Sprayers should be checked carefully for pressure, ground speed, and uniformity of spray pattern, and calibrated accurately for spray volume. Replace any worn nozzles.

CLEANING SPRAYERS

Drain and flush spray tank and lines thoroughly before mixing STAMPEDE CM herbicide. Repeat this procedure at the end of each day's spraying and before changing to or from another pesticide. This cleaning operation of tank and lines is especially important if insecticides, such as SEVIN®, or any of the organophosphorus insecticides are used in the sprayer prior to or after STAMPEDE CM. Interaction of these insecticides and STAMPEDE CM can injure the crop severely. See COMPATIBILITY WITH OTHER CHEMICALS section.

EFFECT OF CLIMATIC CONDITIONS AND CULTURAL PRACTICES ON WEED CONTROL

Field and Seed Bed Preparation

Fields under conventional tillage should have well prepared seed beds, free of large clods, to encourage more uniform and rapid emergence of the crop, pigeongrass, and broadleaf weeds. This permits better timing of sprays. In no-till grain fields, pigeongrass and other weeds may emerge rapidly and very thickly because of the firm seed bed. Monitoring of no-till fields should begin early to assure proper timing of sprays.

Soil Moisture and Rainfall

Weed control with STAMPEDE CM herbicide is strongly influenced by soil moisture. Weed control and crop tolerance is best when adequate moisture is available for active weed and crop growth. The performance of STAMPEDE CM is reduced substantially during periods of prolonged dry weather. STAMPEDE CM IS NOT RECOMMENDED FOR USE WHEN SOIL MOISTURE LEVELS ARE DEEPER THAN 1½ INCHES FROM THE SOIL SURFACE AND ACTIVE PLANT GROWTH STOPS.

STAMPEDE CM herbicide enters the leaf rapidly. Spraying of weeds at the correct stage should not be postponed even if showery weather is expected; however, a heavy rain of 1 inch or more, within 4 hours after application, may reduce control. Fields may be sprayed when the plants are dry or wet with dew or rain.

Temperature

Providing soil moisture is not a limiting factor, the herbicidal activity of STAMPEDE CM herbicide improves as daily maximum temperatures increase. Best weed control occurs when the daily maximum temperatures are 65°F and above. Do not apply STAMPEDE CM when daily maximum temperatures will remain below 50°F or when they are expected to go higher than 85°F.

COMPATIBILITY WITH OTHER CHEMICALS

Foliar Applied Chemicals

If another herbicide is required, for example, a postemergence wild oat herbicide, a 3 day interval should be allowed between the application of STAMPEDE CM herbicide and the other herbicide.

Severe injury or kill of spring barley, durum or hard red spring wheat plants may result from tank mix or separate applications of STAMPEDE CM herbicide and certain insecticides. Grain crops that have been or will be treated with STAMPEDE CM, should not be treated with carbamate insecticides such as carbaryl (Sevin, etc.), methomyl (Lannate®, Nudrin®, etc.) or organophosphorus insecticides such as methyl parathion, Guthion®, etc.

If a foliar insecticide is necessary, malathion or toxaphene can be safely applied 14 days before or after the STAMPEDE CM herbicide treatment.

Soil Applied Chemicals

Do not spray spring barley, durum or hard red spring wheat with STAMPEDE CM herbicide if the field was treated with soil-applied systemic organophosphorus insecticides (Disyston®, Thimet®, Counter®, etc.) the previous year as used for potatoes or sugar beets.

Do not spray STAMPEDE CM herbicide on spring barley, durum or hard red spring wheat that has been treated at planting time with soil-applied systemic insecticides for Hessian Fly control, e.g., Thimet, Disyston, Furadan®, etc.

Grain crops protected with maneb/lindane seed dressings may be treated with STAMPEDE CM HERBICIDE.

USE RESTRICTIONS

- Do not apply this product to any crops other than spring barley, durum or hard red spring wheat.
- Do not mix or apply this product with any other pesticide, spray adjuvant or with fertilizer except as specifically recommended on this label.
- Do not apply this product under windy conditions that will allow spray drift to adjacent susceptible crops such as sunflowers, soybeans, sugarbeets, potatoes, forage legumes, gardens, orchards, and shelter belts.
- DO NOT GRAZE TREATED CROP OR CUT FOR GREEN CHOP FEED.

STORAGE AND DISPOSAL

Do not store at temperatures below 15°F. If stored below 15°F and crystals form, warm to 60°F for 24 hours, periodically shaking container to reconstitute.

PESTICIDE DISPOSAL

Do not contaminate water, food or feed by storage or disposal. Pesticide wastes are toxic. Improper disposal of excess pesticide, spray mixture or rinsate is a violation of Federal law. If these wastes cannot be disposed of by use according to label instructions, contact your State Pesticide or Environmental Control Agency, or the Hazardous Waste representative at the nearest EPA Regional Office for guidance.

CONTAINER DISPOSAL

Triple rinse (or equivalent). Then puncture and dispose of in a sanitary landfill, or by incineration, or if allowed by state and local authorities, by burning. If burned, stay out of smoke.

STEPS TO BE TAKEN IN CASE MATERIAL IS RELEASED OR SPILLED—Eliminate ignition sources. Ventilate area. Avoid breathing vapors. Use MSHA/NIOSH self-contained breathing apparatus or air mask for large spills in confined areas. Dike the spill with inert material (sand, earth, fuller's earth, etc.) and if appropriate, transfer the liquid and solid diking material to separate containers for recovery or disposal. Keep spill out of all sewers and open bodies of water. Refer to Precautionary Statements.

CONDITIONS OF SALE AND WARRANTY

Rohm and Haas warrants that the product conforms to its chemical description and is reasonably fit for the purpose stated on the label only when used in accordance with label directions under normal conditions of use. **ROHM AND HAAS MAKES NO OTHER EXPRESS OR IMPLIED WARRANTIES EITHER OF MERCHANTABILITY OR FITNESS FOR A PARTICULAR USE.** Handling, storage and use of the product by Buyer or User are beyond the control of Rohm and Haas and Seller. Risks such as crop injury, ineffectiveness or other unintended consequences resulting from, but not limited to, weather or soil conditions, presence of other materials, disease, pests, drift to other crops or property or failure to follow label directions will be assumed by the Buyer or User. **IN NO CASE WILL ROHM AND HAAS OR SELLER BE HELD LIABLE FOR CONSEQUENTIAL, SPECIAL OR INDIRECT**

DAMAGES RESULTING FROM THE HANDLING, STORAGE OR USE OF THIS PRODUCT.

STAMPEDE™ is a trademark of Rohm and Haas Company
COUNTER® and THIMET® are registered trademarks of American Cyanamid Co.
DISYSTON® is a registered trademark of Bayer AG
FURADAN® is a registered trademark of FMC Corp.
GUTHION® is a registered trademark of Mobay Chemical Corp.
LANNATE® is a registered trademark of E. I. duPont de Nemours & Co.
NUDRIN® is a registered trademark of Shell Chemical
SEVIN® is a registered trademark of Union Carbide Corp.

10012-B1 MADE IN U.S.A.

SDS Biotech Corporation

(FERMENTA PLANT PROTECTION COMPANY)

Agricultural Chemicals Business
7528 Auburn Road
P.O. Box 348
Painesville, OH 44077
(216) 357-3000

ARSONATE

LIQUID

For Postemergence Weed Control

Contains 6.6 Pounds MSMA Per Gallon

ACTIVE INGREDIENT:	
Monosodium Acid Methanearsonate	51.0%
INERT INGREDIENTS	49.0%
TOTAL	100.0%
Total Arsenic, all in water-soluble form, expressed as elemental	23.61%

EPA Reg. No. 50534-16

CAUTION

Keep Out of Reach of Children.

PRECAUTIONARY STATEMENTS

HAZARDS TO HUMANS AND DOMESTIC ANIMALS
CAUTION

Harmful if swallowed or absorbed through the skin. Avoid contact with eyes, skin or clothing. Avoid breathing vapors or spray mist.

If swallowed, induce vomiting immediately by giving two glasses of water and sticking finger down throat. **Never** give anything by mouth to an unconscious person. Seek medical attention.

In case of contact, flush eyes with plenty of water for at least 15 minutes. Seek medical attention.

Wash exposed skin gently with plenty of soap and water.

Keep children and domestic animals off treated areas until this material has been washed into the soil.

Wear protective clothing when handling or applying this product including long pants, long-sleeve shirt, and impermeable gloves and boots. Mixer-loaders should include an apron and full-face shield when handling or mixing concentrate. Flagmen should be fully protected during spray operations or mechanical flagmen used. Pilots and ground spray rig applicators should wear a mask or respirator approved by the Mining Enforcement and Safety Administration and the National Institute for Occupational Safety and Health.

ENVIRONMENTAL HAZARDS

DO NOT feed treated foliage to livestock or graze treated areas. **DO NOT** spray or allow drift onto vegetables, ornamentals or other desirable plants. **DO NOT** apply when weather conditions favor drift from target area. **DO NOT** apply directly to water. **DO NOT** contaminate water by cleaning of equipment or disposal of wastes.

ARSONATE Liquid
Read entire label before using this product.

GENERAL INFORMATION

ARSONATE Liquid is a postemergence herbicide for use in cotton, turf and non-crop areas. ARSONATE Liquid does not contain a surfactant (wetting agent). A suitable non-ionic surfactant should be added to spray tank. Local conditions and recommendations vary; consult local agricultural experiment station or extension service weed specialists for recommendations in your area.

Mixing Instructions: ARSONATE Liquid herbicide must be thoroughly dissolved. Fill the spray tank with half the required amount of water and, with pump or agitator operating, add the recommended amount of ARSONATE Liquid herbicide and continue filling the tank with balance of water. Add surfactant and mix thoroughly before using.

DO NOT store spray solution in tank overnight. Clean application equipment thoroughly after use by flushing with water in a safe place.

Application Methods: ARSONATE Liquid should be applied with a low-volume, properly calibrated sprayer having satisfactory pumping and bypass action. **Adjust Nozzles to Keep Spray Off Cotton Foliage and in a Manner to Allow Maximum Coverage of Weeds.** Adequate coverage is very important for effective weed control.

Apply during warm weather when weeds are in an active stage of growth.

DIRECTIONS FOR USE

It is a violation of Federal law to use this product in a manner inconsistent with its labeling.

COTTON: ARSONATE Liquid is useful as a directed postemergence spray for controlling weeds in cotton such as barnyardgrass, cocklebur, smooth and large crabgrass, dallisgrass, goosegrass, johnsongrass and nutsedge. Apply as a **Directed** broadcast spray at a rate of 40 gallons per acre of a solution containing 3 pints of ARSONATE Liquid plus 1½ to 2 quarts of suitable surfactant, cleared for application to growing crops, in 50 gallons of water when weeds are small. For band applications, apply 1 gallon of above diluted spray per acre for each 1 inch band width to be treated of cotton grown on 40 inch row spacing. A second or repeat application, if needed, should be timed about 1 to 3 weeks after the first application. Keep spray off cotton foliage.

Apply Only When Cotton is 3 Inches High to First Bloom. DO NOT Apply After First Bloom.

Slight burning and reddish discoloration of cotton foliage may occasionally be seen following recommended treatment; however, cotton plants will develop normally and yield will not be affected.

ORNAMENTAL TURFGRASS: ARSONATE Liquid can be used for selective control of Bahiagrass, barnyardgrass, chickweed, smooth and large crabgrass, dallisgrass, nutsedge, sandbur and wood sorrel with little or no injury to well-established, actively growing turfgrasses. Mow turfgrass to height of 1 to 1½ inches before treatment. Mix 1 fluid ounce (2 tablespoons) of ARSONATE Liquid plus 2 to 3 fluid ounces of suitable surfactant in 5 gallons of water for application to an area of 1,000 square feet. Application should be uniform and thorough to adequately wet all undesirable plants. Two or more repeat treatments at 14 day intervals may be necessary. Make applications during warm weather when temperature is between 80 and 90°F. **DO NOT** water turf for at least 24 hours after application. Turfgrasses may be temporarily discolored. Bermudagrass, bluegrasses and zoysiagrass have shown tolerance to properly applied ARSONATE Liquid. Injury may result if applied to bentgrasses and fescues. **DO NOT** apply to St. Augustinegrass, carpetgrass, centipedegrass or to Dichondra. **DO NOT** reseed until 2 weeks after last application. **DO NOT** apply with hose-end applicators.

NON-CROP: ARSONATE Liquid is effective in controlling barnyardgrass, cocklebur, smooth and large crabgrass, dallisgrass, goosegrass, johnsongrass, nutsedge and many similar weeds on drainage ditchbanks, fence rows, rights-of-way, storage yards and similar non-crop areas. Application should be made when weeds are small and conditions are favorable for good weed growth. Mix at a rate of 3 pints of ARSONATE Liquid plus 1½ to 2 quarts of suitable surfactant in 50 gallons of water. Spray undesirable vegetation thoroughly to point of run-off. Adequate coverage and complete wetting of foliage is important for effective control. Repeat application may be necessary if regrowth occurs. **DO NOT** contaminate waters used for domestic purposes, or by animals, including wildlife and aquatic life, or water used for irrigation purposes.

STORAGE AND DISPOSAL

DO NOT contaminate water, food or feed by storage or disposal. Open dumping is prohibited.

PESTICIDE DISPOSAL

Wastes resulting from the use of this product may be disposed of on site or at an approved waste disposal facility.

CONTAINER DISPOSAL

DO NOT reuse empty container. Triple rinse (or equivalent) and offer for recycling or reconditioning, or puncture and dispose of in a sanitary landfill, or by other procedures approved by state and local authorities.

WARRANTY AND LIMITATION OF DAMAGES

Seller warrants that this material conforms to its chemical description and is reasonably fit for the purposes stated on the label when used in accordance with directions under normal conditions of use and Buyer assumes the risk of any use contrary to such directions. SELLER MAKES NO OTHER EXPRESS OR IMPLIED WARRANTY, INCLUDING ANY OTHER EXPRESS OR IMPLIED WARRANTY OF FITNESS OR OF MERCHANTABILITY, AND NO AGENT OF SELLER IS AUTHORIZED TO DO SO EXCEPT IN WRITING WITH A SPECIFIC REFERENCE TO THIS WARRANTY. In no event shall Seller's liability for any breach of warranty exceed the purchase price of the material as to which a claim is made.

SPECIAL LOCAL NEEDS PURSUANT TO SECTION 24(c) OF AMENDED FIFRA

CROP	REGISTRATIONS	STATES	SLN NO.	DIRECTIONS FOR USE
Cotton	Application on Cotton	Arkansas Louisiana Mississippi	3953 3954 3955	ARSONATE Liquid is useful for the control of weeds listed on the product label and many similar weeds. Application can be made as follows: 1. Preplant or postplant up to cracking of soil before cotton emergence using ground or aircraft equipment. 2. Postemergent, over-the-top, when cotton is 3 to 6 inches high or up to early first square stage, whichever occurs first, using ground or aircraft equipment. 3. Postemergent as a directed spray with ground equipment when cotton is 3 inches high to first bloom. PREPLANT OR POSTPLANT UP TO CRACKING APPLICATION ON COTTON: A single ground or aircraft application of ARSONATE Liquid can be made to prepared cotton seedbeds when planting has been delayed and weeds have emerged, or as a postplant treatment BUT NO LATER THAN INITIAL CRACKING of soil in field before emergence of cotton. Planting of cotton may immediately follow the preplant application. Mix at the rate of 2½ pints of ARSONATE Liquid plus a suitable surfactant in 40 gallons of water for broadcast ground application, or in 5 to 10 gallons of water for aircraft application and apply using a properly calibrated sprayer to one acre. The surfactant should be added to the total spray mix at manufacturer's recommended rate for type of application. See SPECIAL PRECAUTIONS below. POSTEMERGENT, OVER-THE-TOP, APPLICATION ON COTTON: Application may be made using ground or aircraft equipment as an over-the-top broadcast spray to actively growing cotton as a salvage operation when cotton is 3 to 6 inches high or up to early first square stage, whichever occurs first. Mix at the rate of 1 to 1¼ pints of ARSONATE Liquid plus manufacturer's recommended rate of a suitable surfactant in 40 gallons of water for ground application, or in 5 to 10 gallons of water for aircraft application and apply using a properly calibrated sprayer to one acre. The surfactant should be added to the total spray mix at manufacturer's recommended rate for type of application. A second or repeat application, if needed, should be made 1 to 3 weeks after the first treatment. APPLY AS AN OVER-THE-TOP BROADCAST SPRAY ONLY WHEN COTTON IS 3 TO 6 INCHES HIGH OR UP TO EARLY FIRST SQUARE STAGE, WHICHEVER OCCURS FIRST. **DO NOT** APPLY AS AN OVER-THE-TOP BROADCAST SPRAY AFTER COTTON IS MORE THAN 6 INCHES HIGH OR AFTER EARLY FIRST SQUARE STAGE. See SPECIAL PRECAUTIONS below. POSTEMERGENT, DIRECTED, APPLICATION ON COTTON: Mix at the rate of 2½ pints of ARSONATE Liquid plus manufacturer's recommended rate of a suitable surfactant in 40 gallons of water for application as a direct broadcast spray to one acre when weeds are small using ground equipment. For band applications, apply one gallon of above diluted spray per acre for each 1-inch band width to be treated of cotton grown on 40-inch row spacing. Keep spray off cotton foliage. APPLY ONLY WHEN COTTON IS 3 INCHES HIGH TO FIRST BLOOM. DO NOT APPLY AFTER FIRST BLOOM. Slight burning and reddish discoloration of cotton foliage may occasionally be seen following recommended treatment; however, cotton plants will develop normally. SPECIAL PRECAUTIONS—**DO NOT** allow spray or spray drift to contact adjacent crops or injury will result. Apply only on still days when weather conditions **DO NOT** favor drift from areas being treated. Aircraft applications of ARSONATE Liquid should only be made by applicators experienced in use of herbicides, and applications should be made in accordance with state and Federal regulations.

BRAVO® 500

AGRICULTURAL FUNGICIDE

ACTIVE INGREDIENT:

Chlorothalonil (tetrachloroisophthalonitrile)*	40.4%
INERT INGREDIENTS	59.6%
TOTAL	100.0%

* Covered under U.S. Patent No. 3,290,353, No. 3,331,735 and No. 3,948,636

Contains 4.17 Pounds Chlorothalonil Per Gallon (500 grams per liter)

EPA Reg. No. 50534-8

WARNING — AVISO

Keep Out of Reach of Children

PRECAUCION AL USUARIO: Si usted no lee ingles, no use este producto hasta que la etiqueta haya sido explicado ampliamente.

"See precautionary statements."

PRECAUTIONARY STATEMENTS

HAZARDS TO HUMANS AND DOMESTIC ANIMALS

WARNING

Causes eye irritation. May be a potential skin sensitizer.

Do not get in eyes. Wear goggles or eye shield when handling this product. In case of contact with eyes, flush with plenty of water immediately for 15 minutes. Seek medical attention for eyes immediately.

Avoid contact with skin or clothing. Wash exposed areas of skin with soap and warm water after handling or using.

Do not take internally.

Avoid breathing spray mist.

Do not apply this product in such a manner as to directly, or through drift, expose workers or other persons. The area being treated must be vacated by unprotected persons.

Note to User: This product may produce temporary allergic side effects characterized by redness of the eyes, mild bronchial irritation and redness or rash on exposed skin areas. Persons having allergic reaction should contact a physician.

First Aid:

Note to physician: Persons having an allergic reaction responds to treatment with antihistamines or steroid creams and/or systemic steroids.

ENVIRONMENTAL HAZARDS

This product is toxic to fish, aquatic invertebrates, and marine/estuarine organisms. Runoff from treated areas may be hazardous to aquatic organisms in neighboring areas. **Do not** apply directly to water. **Do not** apply when weather conditions favor drift from treated areas.

Apply only to areas specified on label.

DIRECTIONS FOR USE

It is a violation of Federal law to use this product in a manner inconsistent with its labeling.

Note to User: This product may produce temporary allergic side effects characterized by redness of the eyes, mild bronchial irritation and redness or rash on exposed skin areas. Persons having allergic reaction should contact a physician.

Note to User: Wear long sleeve shirt, long pants, and gloves while mixing, loading and applying this product.

Dosage rates on this label indicate pints of BRAVO 500 per acre unless otherwise stated. Under conditions favoring disease development the high rate specified and shortest application interval should be used. Applications should be made in sufficient water to obtain adequate coverage of foliage. Gallonage to be used will vary with crop and amount of plant growth. Spray volume usually will range from 20 to 150 gallons (approximately 80 to 600 liters) per acre for dilute sprays and 5 to 10 gallons (approximately 20 to 40 liters) per acre for concentrate ground sprays and aircraft applications. Both ground and aircraft methods of application are recommended unless specific directions for ground application only are given for a crop. Application through sprinkler irrigation systems is not recommended unless specific directions are given for a crop. See application and calibration instructions below.

Note to User: Do not rotate to crops other than those listed on labels within 12 months of the last treatment. After 12 months from the last application, leafy vegetables (i.e. spinach, lettuce, kale, etc.) may also be rotated.

Reentry: Note to User: **Do not** enter treated area to perform hand labor within 24 hours of application unless protective clothing is worn.

Because certain states may require more restrictive reentry intervals for various crops treated with this product, consult your State Department of Agriculture for further information.

Written or oral warnings must be given to workers who are expected to be in a treated area or in an area about to be treated with this product. When oral warnings are given, warnings shall be given in a language customarily understood by workers. Oral warnings must be given if there is reason to believe that written warnings cannot be understood by workers. Oral and written warnings must include the following information: "**WARNING.** Area treated with BRAVO 500 on (date of application). **Do not** enter without appropriate protective clothing until the sprays have dried. In case of accidental exposure, wash exposed area with plenty of water and get medical attention. For further information see 'Precautionary Statements on the label."

APPLICATION AND CALIBRATION TECHNIQUES FOR SPRINKLER IRRIGATION

A. Center Pivot and Traveling Gun Irrigation Equipment: Operate system and injection equipment at normal pressures recommended by the manufacturer of injection equipment used. Fill tank of injection equipment with water. Operate system for one complete circle for center pivot or one complete run for traveling gun equipment, measuring time required, amount of water injected, and acreage contained in circle or run. Mix recommended amount of BRAVO 500 for acreage to be covered into same amount of water used during calibration and inject into system continuously for one revolution or run. Shut off injection equipment after one revolution or run, but continue to operate irrigation system until BRAVO 500 has been cleared from last sprinkler head.

B. Solid Set and Portable (Wheel Move) Irrigation Equipment: Determine acreage covered by sprinkler. Fill tank of injection equipment with water and adjust flow to use contents over a thirty to forty-five minute period. Mix desired amount of BRAVO 500 for acreage to be covered into quantity of water used during calibration and operate entire system at normal pressures recommended by the manufacturer of injection equipment used for amount of time established during calibration. BRAVO 500 can be injected at the beginning or end of the irrigation cycle or as a separate application. Stop injection equipment after treatment is completed and continue to operate irrigation system until BRAVO 500 has been cleared from last sprinkler head.

DO NOT operate irrigation system without safety valves or other devices to prevent back-siphoning of BRAVO 500 into water source. Irrigation water treated with BRAVO 500 should be maintained on the treated area until the water is absorbed by the soil. Tank containing BRAVO 500 should be connected to the discharge side of irrigation pump or other pressurized equipment attached to irrigation line.

Metric Conversion Guide

2⅛ pints equals 1.0 liter 4¼ pints equals 2.0 liters

STORAGE AND DISPOSAL

DO NOT contaminate water, food or feed by storage or disposal. Open dumping is prohibited.

STORAGE

Store in a cool place. Protect from excessive heat.

PESTICIDE DISPOSAL

Pesticide wastes are toxic. Improper disposal of excess pesticide, pesticide spray or rinsate is a violation of Federal Law. If these wastes cannot be disposed of by use according to label instructions, contact your State Pesticide or Environmental Control Agency, or the Hazardous Waste representative at the nearest EPA Regional Office for guidance.

CONTAINER DISPOSAL

DO NOT reuse empty container. Triple rinse (or equivalent). Then offer for recycling or reconditioning, or puncture and dispose of in a sanitary landfill, or incineration, or, if allowed by state and local authorities, by burning. If burned, stay out of smoke.

WARRANTY AND LIMITATION OF DAMAGES

Seller warrants that this material conforms to its chemical description and is reasonably fit for the purposes stated on the label when used in accordance with directions under normal conditions of use and Buyer assumes the risk of any use contrary to such directions. SELLER MAKES NO OTHER EXPRESS OR IMPLIED WARRANTY, INCLUDING ANY OTHER EXPRESS OR IMPLIED WARRANTY OF FITNESS OR OF MERCHANTABILITY, AND NO AGENT OF SELLER IS AUTHORIZED TO DO SO EXCEPT IN WRITING WITH A SPECIFIC REFERENCE TO THIS WARRANTY. In no event shall Seller's liability for any breach of warranty exceed the purchase price of the material as to which a claim is made.

CROP	DISEASES	RATE PER ACRE	APPLICATION DIRECTIONS
Bean (Snap)	Rust	2 to 4¼ pts.	Use in sufficient water to obtain adequate coverage. Begin applications during early bloom stage or when disease first threatens and repeat at weekly intervals or as necessary to maintain control. **DO NOT** apply within 7 days of harvest. **DO NOT** graze treated areas or feed treated plant parts to livestock.
	Botrytis Blight (gray mold)	4¼ pts.	
Beans (Dry) Navy, Pinto, Kidney, Lima Blackeye	Rust, Anthracnose, Downy mildew, Cercospora leaf spot (blackeye only)	2 to 3 pts.	Use in sufficient water to obtain adequate coverage. Begin applications during early bloom stage and repeat at 7 to 10 day intervals. For use only on beans harvested dry with pods removed. **DO NOT** apply within 6 weeks before harvest. **DO NOT** allow livestock to graze in treated areas or feed treated plant parts to livestock.
Cabbage, Cauliflower, Broccoli, Brussels Sprouts	Alternaria leaf spot, Downy mildew	2¼ pts.	Use in sufficient water to obtain adequate coverage. Begin applications after transplants are set in field, or shortly after emergence of field-seeded crop, or when conditions favor disease development. Repeat at 7 to 10 day intervals or as necessary to maintain control. For field seeded brussels sprouts, begin applications at time of early sprout development or when conditions favor disease development. Repeat at 7 to 10 day intervals or as necessary to maintain control.
	Ring spot (California only)	2¾ pts.	
Carrot	Cercospora (Early) blight, Alternaria (Late) blight	2¼ to 2¾ pts.	Use in sufficient water to obtain adequate coverage. Start applications when disease threatens and repeat at 7 to 10 day intervals or as necessary to maintain control. BRAVO 500 may be applied through sprinkler irrigation equipment (solid set, portable wheel move or center pivot systems only). See calibration directions preceding this section.
Celery	Cercospora (Early) blight, Septoria (Late) blight	1½ to 2⅛ pts.	Use 1½ to 2⅛ pts. per acre on a 3 to 5 day spray schedule or 3 to 4¼ pts. per acre on a 7 day schedule. Start applications when transplants are set in the field. Apply in sufficient water to obtain adequate coverage. BRAVO 500 may be applied through sprinkler irrigation equipment (solid set, portable wheel move or center pivot systems only). See calibration directions preceding this section. **DO NOT** apply within 7 days of harvest.
	Basal stalk rot (Rhizoctonia solani)	3 to 4¼ pts.	
	Pink rot (Suppression)	4¼ pts.	
	Early blight Late blight	2¼ to 2¾ pts. 100 gal.	For celery seedbeds, apply 125 gallons per acre twice weekly or as needed to maintain control. Start applications shortly after crop emergence. Use the higher rate under severe disease conditions.
Corn (Sweet), Corn grown for seed	Helminthosporium leaf blights, Rust	1⅛ to 2¾ pts.	Use in sufficient water to obtain adequate coverage. Begin applications when conditions favor disease development and repeat at 4 to 7 day intervals or as required to maintain control. Under severe disease conditions, use 2¼ to 2¾ pts. per acre. **DO NOT** apply within 14 days of harvest. **DO NOT** apply to sweet corn to be processed. **DO NOT** allow livestock to graze in treated fields. **DO NOT** ensile treated corn or use as livestock forage.
Cranberry	Fruit rots, Lophodermium leaf/twig blight	6 to 10 pts	Apply at late bloom and repeat at 10 to 14 day intervals. Under severe disease conditions use the 10 pts/acre rate on a 10 day schedule. **Do not** apply **Bravo 500** more than 3 times per season, or within 50 days before harvest. **Do not** apply to bogs when flooded or allow release of irrigation water from bogs for at least 3 days following application. **Bravo 500** may be applied through sprinkler irrigation equipment. Use 300 gallons of water per acre through solid set systems only. See calibration directions following this section.
Cucumber	Target spot, Anthracnose, Downy mildew	2¼ to 2¾ pts.	Use in sufficient water to obtain adequate coverage. Begin applications when plants are in first true leaf stage or when conditions are favorable for disease development. Repeat applications at 7 day intervals. Under severe disease conditions, shorten spray interval. BRAVO 500 may be applied through sprinkler irrigation equipment (solid set, portable wheel move or center pivot systems only). See calibration directions preceding this section.
	Powdery mildew (except south-western states), Gummy stem blight, Leaf blight, Scab	2¾ to 4¼ pts.	
	Fruit belly rot (Rhizoctonia solani)	12 pts.	Use BRAVO 500 in sufficient water to obtain runoff to soil surface. Make a single application when vines begin to form. BRAVO 500 may be applied through sprinkler irrigation equipment as directed above.

CROP	DISEASES	RATE PER ACRE	APPLICATION DIRECTIONS
Cantaloupe, Muskmelon, Honeydew melon, Watermelon, Squash, Pumpkin	Anthracnose, Downy mildew	2¼ to 2¾ pts.	Use in sufficient water to obtain adequate coverage. Begin applications when plants are in first true leaf stage or when conditions are favorable for disease development. Repeat applications at 7 day intervals. Under severe disease conditions, shorten spray interval. BRAVO 500 may be applied through sprinkler irrigation equipment (solid set, portable wheel move and center pivot systems only). See calibration directions preceding this section.
	Cercospora leaf spot, Gummy stem blight (black rot), Leaf blight, Scab, Powdery mildew (except south-western states)	2¾ to 4¼ pts.	
Grasses Grown for Seed	Stem rust, Leaf rust, Stripe rust,	1½ to 2 pts.	Use in sufficient water to obtain adequate coverage. Begin applications during stem elongation when conditions favor disease development. Re-apply at flag (top) leaf emergence and at head emergence. Under severe disease conditions use the higher rates and repeat applications at 14 day intervals. **DO NOT** apply within 14 days of harvest. **DO NOT** allow livestock to graze in treated areas or feed treated plant parts to livestock.
	Selenophoma (eye spot)	1½ to 3 pts.	
Mint	Rust, Septoria leaf spot	2 pts.	Use in sufficient water to obtain adequate coverage, normally 20 to 150 gallons per acre for dilute sprays and 5 to 10 gallons per acre for concentrate ground and aircraft applications. Begin applications when emerging plants are 4–8 inches high. Repeat applications at 7 to 10 day intervals or as necessary to maintain control. **DO NOT** apply more than 3 times per season. **DO NOT** apply within 80 days of harvest. **DO NOT** feed fresh or extracted mint hay from treated fields to livestock. Based on available residue data, use of BRAVO 500 on mint is restricted to Indiana, Michigan and Wisconsin.
Onion (dry bulb)	Botrytis leaf blight (blast), Purple blotch	2 to 3 pts.	Apply in sufficient water to obtain adequate coverage. Make the first application at first sign of disease or when dew or rain occurs. Repeat at 7 to 10 day intervals for as long as conditions favor disease. Use the high rate and a 7 day schedule of applications when heavy dew or rain persists. **DO NOT** apply within 7 days before harvest of dry bulb onions or garlic. **DO NOT** apply more than 3 times per season or within 14 days of harvest on green bunching onions, leeks or shallots. If additional disease control is needed before harvest, use another registered fungicide.
Onion (green bunching), Garlic, Leek Shallot, Onion grown for seed	Botrytis leaf blight (blast), Purple blotch, Downy mildew (suppression)	2 to 4¼ pts.	
Papaya	Alternaria fruit spot, Anthracnose, Stem end rot	3 to 5¾ pts.	Apply with ground equipment only, in sufficient water to obtain adequate coverage of fruit and leaves. Begin treatment when conditions favor development of disease and continue treatments at 14 day intervals until weather conditions no longer favor disease development. **DO NOT** graze livestock in treated area or feed processing by-products to livestock.
Parsnip	Alternaria leafspot, Downy mildew, Anthracnose, Botrytis blight (gray mold), Bottom rot (Rhizoctonia)	2 to 3 pts.	Apply in sufficient water to obtain adequate coverage. Make the first application at the first sign of disease or when conditions are favorable for infection. Continue applications on a 7 to 10 day schedule. **DO NOT** apply more than 4 times per season or within 10 days of harvest. **DO NOT** feed treated plant parts to livestock.
Passion Fruit (Hawaii only)	Alternaria fruit and leaf spot (passion fruit brown spot)	2¾ pts.	Apply with ground equipment in sufficient water to obtain adequate coverage of leaves. Begin treatment when fruit spots appear (April to July) and continue treatments at 14 day intervals until weather conditions no longer favor disease development. **DO NOT** graze in treated area or feed vines or processing by-product to livestock used for food.
Peanut	Cercospera (early) leafspot Cercosporidium (late) leafspot	1½ to 2⅛ pts.	Apply in sufficient water for coverage when leaf wetness first occurs or 30 to 40 days after planting; Repeat at 10 to 14 day intervals. When conditions favor late leafspot or when rust or web blotch occur, apply 2⅛ pts. per acre at 10 day intervals for the remainder of the season. **DO NOT** apply within 14 days of harvest. **DO NOT** allow livestock to graze in treated areas. **DO NOT** feed hay or threshings from treated fields to livestock. BRAVO 500 may be applied through sprinkler irrigation equipment. Use 2⅛ pints per acre in solid set, portable wheel move, center pivot or traveling gun sprinkler irrigation equipment. See calibration directions preceding this section.
	Rust, Web blotch	2⅛ pts.	
Potato	Early blight, Late blight, Botrytis vine rot (Botrytis spp.)	1½ to 2⅛ pts.	Use in sufficient water to obtain adequate coverage. Begin applications when plants are 6 to 8 inches high or when disease threatens and continue at 7 to 10 day intervals or as needed to maintain disease control. Under severe disease conditions, use 2⅛ pts. per acre on a 7 day schedule. BRAVO 500 may be applied through sprinkler irrigation equipment (solid set, portable wheel move or center pivot systems only). **DO NOT** exceed a 10 day interval between applications when using this technique. See calibration directions preceding this section.
	Dryland culture only: Early blight, Late blight	1⅛ to 2⅛ pts.	
Soybean Determinate (Southern) Varieties	Anthracnose, Diaporthe pod and stem blight, Frogeye leafspot *(Cercospora sojina)*, Purple seed stain, Cercospora leaf blight *(Cercospora kikuchii)*, Septoria brown spot		Apply in sufficient water to obtain complete coverage, using at least five gallons of water per acre for aerial application. Use the three application program in areas having a history of moderate to severe disease intensity. **Bravo 500** may be applied through sprinkler irrigation equipment. Follow application and calibration directions preceding this section.
		2.0 to 3½ pts	**Two application program** — Make the first application at early pod set (R3 stage, when majority of pods are ⅛ to ⅜ inch in length) and the second at beginning of seed formation (R5) which occurs about 14 days later.
		1½ to 2¾ pts	**Three application program** — Make the first application at the beginning of flowering (R1), the second at early pod set (R3) and the third at beginning of seed formation (R5).

NOTE: Do not exceed total of 3 applications per season. **Do not** apply within 6 weeks of harvest. **Do not** feed treated parts to livestock or allow grazing in treated fields.

CROP	DISEASES	RATE PER ACRE	APPLICATION DIRECTIONS
Soybean Indeterminate (Northern) Varieties	Anthracnose, Diaporthe pod and stem blight, Frogeye leafspot		Apply in sufficient water to obtain complete coverage, using at least five gallons of water per acre for aerial application. Use the three application program in areas having a history of moderate to severe disease intensity. **Bravo 500** may be applied through sprinkler irrigation equipment. Follow application and calibration directions preceding this section.

CROP	DISEASES	RATE PER ACRE	APPLICATION DIRECTIONS
	(Cercospora sojina), Purple seed stain, Cercospora leaf blight *(Cercospora kikuchii),* Septoria brown spot	2.0 to $3\frac{1}{2}$ pts	**Two application program** — Make the first application when largest pods are 1 to $1\frac{1}{2}$ inches in length and make the second application 14 days later. **Bravo 500** may be co-applied with **Benlate® 50WP** as a tank mix for disease control in indeterminate (northern) soybeans. Use $1\frac{1}{2}$ pts of **Bravo 500** plus 8 ounces of **Benlate 50WP** per acre. Make the first application when pods near the top of plants are $\frac{1}{2}$ to 1 inch in length and a second application 14 days later.
		$1\frac{1}{2}$ to $2\frac{3}{4}$ pts	**Three application program** — Make the first application one week after first flowering and continue applications at 14 day intervals. ®Benlate is a registered trademark of E.I. DuPont de Nemours and Co., Inc.

NOTE: **Do not** exceed total of 3 applications per season. **Do not** apply within 6 weeks of harvest. **Do not** feed soybean hay or threshings from treated fields to livestock.

Tomato	FOLIAGE (apply every 7–10 days): Early blight, Late blight, Gray leafspot, Gray leaf mold, Septoria leafspot	2 to 3 pts.	Apply in sufficient water to obtain adequate coverage. Begin applications when dew or rain occur and disease threatens. Use the highest rate and shortest interval specified when disease conditions are severe. BRAVO 500 may be combined in the spray tank with EPA-registered pesticide products that claim copper as the active ingredient and are labeled for control of bacterial diseases of tomatoes. Check the copper manufacturer's label for specific instructions, precautions and limitations prior to mixing with BRAVO 500. **DO NOT** use with Copper-Count*N in concentrated spray suspensions. BRAVO 500 may be applied through sprinkler irrigation equipment (solid set or portable wheel move systems only). See calibration directions preceding this section. * *Copper-Count is a registered trademark of Mineral Research & Development Corp.*
	FRUIT (apply every 7–14 days beginning at fruit set): Anthracnose, Alternaria fruit rot (black mold), Rhizoctonia fruit rot, Botrytis gray mold, Late blight fruit rot	3 to 4 pts.	

TREE AND ORCHARD CROPS

Apply BRAVO 500 in sufficient water and with proper calibration to obtain uniform coverage of tree canopy. Application with ground equipment is preferable to aerial application because ground applications generally give better coverage of the tree canopy. If application with ground equipment is not feasible, BRAVO 500 may be applied with aircraft using at least 20 gallons per acre. When concentrate sprays are used or when treating non-bearing or immature trees, the lower rate of BRAVO 500 listed may be used. **DO NOT** allow livestock to graze in treated areas. The following spray volumes are recommended as gallons of spray per acre:

	Spray Volume (Gallons per Acre)	
Crop	**Dilute**	**Concentrate**
Peach, Nectarine, Apricot, Tart Cherry, Plum, Prune	300	20 to 150
Sweet Cherry	400	20 to 200
Conifers		
Forest stands	Not Used	10 to 20 (aircraft)
Christmas trees	100	10 to 50 (aircraft or ground equipment)
Nursery beds	100	5 to 10 (ground equipment only)

CROP	DISEASES	BRAVO 500 RATE PER ACRE	BRAVO 500 RATE PER 100 GAL*	APPLICATION DIRECTIONS
Peach, Nectarine, Apricot, Cherry, Plum, Prune	Leaf curl, Coryneum blight (shothole)	$4\frac{1}{2}$ to 6 pts.	$1\frac{1}{2}$ to 2 pts.	For best control of both diseases apply at leaf fall in late autumn, using sufficient water and proper sprayer calibration to obtain uniform coverage. When conditions favor high disease levels use the high rate of application and apply once or twice more in mid to late winter before budswell. If the leaf fall application is not practical, application of BRAVO 500 for control of leaf curl may be made at any time prior to budswell the following spring. Where Coryneum blight (shot-hole) occurs, also apply at budbreak to protect newly emerging leaves and at shuck split to prevent fruit infections.
	Brown rot blossom blight	$4\frac{1}{2}$ to 8 pts.	$1\frac{1}{2}$ to 2 pts.	Use 6 to 8 pints per acre on trees taller than 20 ft. and $4\frac{1}{2}$ to 6 pints per acre on smaller trees. Make one application at popcorn (pink, red or early white bud) and a second application at full bloom. If weather conditions favor disease development, make an additional application at petal fall.
	Cherry leafspot; peach, nectarine, apricot scab	$4\frac{1}{2}$ to 6 pts.	$1\frac{1}{2}$ to 2 pts.	In addition to the bloom applications listed above, make one application at shuck-split. **DO NOT** apply BRAVO 500 after shuck-split and before harvest. If additional disease control is needed before harvest, use another registered fungicide. For control of cherry leafspot after harvest, make one application to foliage within 7 days after fruit is removed. In orchards with a history of high leafspot incidence, make a second application 10–14 days later.
Conifers	Swiss needlecast	4 to 8 pts.	4 to 8 pts.	Single application technique: In Christmas tree plantations or forest stands make one application in the spring when new shoot growth is $\frac{1}{2}$ to 2 inches in length.
	Scleroderris cankers (pines) Swiss needlcast and Rhabdocline needlecast (Douglas-Fir)	2 to 4 pts.	2 to 4 pts.	Make the first application in spring when shoot growth is $\frac{1}{2}$ to 2 inches in length. Make additional applications at 3 to 4 week intervals until conditions no longer favor disease development. For use in nursey beds, apply the highest rate specified on a 3 week schedule.
	Sirococcus tip blight	3 to 5 pts.	3 to 5 pts.	

CROP	DISEASES	BRAVO 500 RATE PER ACRE	100 GAL*	APPLICATION DIRECTIONS
	Rhizosphaera needlecast (spruces) Scirrhia brown spot (pines)	8 pts.	8 pts.	
	Lophodermium needlecast	2 to 4 pts.	2 to 4 pts.	North Central and Northeastern States: Begin applications in mid-July to early August before infection occurs. Make additional applications at 3 to 4 week intervals until conditions no longer favor disease development. For use in nursery beds, apply 4 pts. per acre on a 3 week schedule.
	Botrytis seedling blight Phoma twig blight	2 to 4 pts.	2 to 4 pts.	Begin applications in nursery beds when seedlings are 4 inches tall and when cool, moist conditions favor disease development. Make additional applications at 7 to 14 day intervals as long as disease favorable conditions persist.

* Volumetric rates to be used only with full dilute spray volume specified on this label for tree and orchard crops.

SPECIAL LOCAL NEEDS PURSUANT TO SECTION 24(c) OF AMENDED FIFRA AND STATE REGISTRATIONS

CROP	REGISTRATION	STATES	SLN OR REG. NO.	DIRECTIONS FOR USE
Forest Tree Nursery and Greenhouse Management	Control of Boytrytis spp.	Oregon	OR-770060	Apply 2¾ pts. per 100 gallons of water for control of Botrytis spp. on conifer seedlings such as Douglas fir, noble fir, western hemlock, lodgepole pine and ponderosa pine. Begin application when seedlings are 4 inches tall or when disease threatens. Repeat applications at 7 to 14 day intervals as required to maintain disease control. **CAUTIONS:** Enter treated greenhouse only after spray has settled. Wear protective clothing to avoid skin contact in treated greenhouses until spray has dried.
Sugarbeet grown for seed	Control of Downy mildew	Oregon	OR-810032	Use 2½ pts. per acre in sufficient water to obtain adequate coverage, generally 25 to 50 gallons per acre. Make the first application in mid-October when disease conditions become favorable, and make additional applications as favorable disease conditions warrant, through April of the following year. **DO NOT** allow livestock to graze in treated areas. **DO NOT** feed treated crop residues to livestock. **DO NOT** apply Bravo 500 to sugarbeet plants after bolting and seedhead emergence has occurred.
Tomatoes	Weekly and Biweekly applications	Indiana Ohio	3966 3965	WEEKLY: Application with ground or aircraft equipment—Early blight, late blight, gray leaf spot and gray leaf mold—Use 2¼ to 2¾ pts. per acre. Anthracnose—Use 2¾ to 4¼ pts. per acre. Rhizoctonia fruit rot and Botrytis gray mold—Use 4¼ pts. per acre. Apply BRAVO 500 in sufficient water to obtain adequate coverage. Begin applications when disease threatens and repeat at 7 to 10 day intervals. Under severe disease conditions, shorten spray interval. BIWEEKLY: Application with ground equipment only—Early blight, Botrytis gray mold, gray leaf spot, Septoria leaf spot, gray leaf mold, Rhizoctonia fruit rot and anthracnose—Use 4¼ to 7 pts. per acre. Begin applications when disease threatens and repeat at 14 day intervals. Under severe disease conditions, resume weekly schedule at 2¾ to 4¼ pts. per acre.
Tomatoes	Control of Anthracnose (DAT)	Delaware Maryland New Jersey Virginia	DE-840005 MD-840005 NJ-840013 VA-840006	Use 14 pts. per acre in 50 to 100 gallons of water. Apply when crown fruit are ¼ to ½ full size. Three weeks later make a second application to provide an additional 3 weeks of protection. For control of foliage and fruit diseases on alternative schedules, refer to the BRAVO 500 product label. Apply only as directed.

For directions for use through sprinkler irrigation systems, refer to section of label entitled **Application and Calibration Techniques for Sprinkler Irrigation Systems.**

BRAVO® W-75

AGRICULTURAL FUNGICIDE

75% Wettable Powder

Only for Sale To, Use and Storage by Agricultural Personnel or Commercial Applicators

ACTIVE INGREDIENT:

Chlorothalonil (tetrachloroisophthalonitrile)*	75.0%
INERT INGREDIENTS	25.0%
TOTAL	100.0%

* Covered under U.S. Patent No. 3,290,353 and No. 3,331,735

EPA Reg. No. 50534-23

DANGER—PELIGRO

Keep Out of Reach of Children

PRECAUCION AL USUARIO: Si usted no lee ingles, no use este producto hasta que la etiqueta haya sido explicado ampliamente.

See precautionary statements.

PRECAUTIONARY STATEMENTS

HAZARDS TO HUMANS AND DOMESTIC ANIMALS

DANGER

Corrosive, causes severe eye damage. May be a potential skin sensitizer.

Do not get in eyes. Wear goggles or eye shield when handling this product. In case of contact with eyes, flush with plenty of water immediately for 15 minutes. Seek medical attention for eyes immediately.

Avoid contact with skin or clothing. Wash exposed areas of skin with soap and warm water after handling or using.

Do not take internally.

Avoid breathing dust or spray mist.

Do not apply this product in such a manner as to directly, or through drift, expose workers or other persons. The area being treated must be vacated by unprotected persons.

Note to User: This product may produce temporary allergic side effects characterized by redness of the eyes, mild bronchial irritation and redness or rash on exposed skin areas. Persons having allergic reaction should contact a physician.

First Aid:

Note to physician: Persons having an allergic reaction respond to treatment with antihistamines or steroid creams and/or systemic steroids.

ENVIRONMENTAL HAZARDS

This product is toxic to fish, aquatic invertebrates, and marine/estuarine organisms. Runoff from treated areas may be hazardous to aquatic organisms in neighboring areas. **Do not** apply directly to water. **Do not** apply when weather conditions favor drift from treated areas.

Apply only to areas specified on label.

BRAVO W-75

A Broad-Spectrum Agricultural Fungicide.

Read entire label carefully and use only as directed.

GENERAL INFORMATION

BRAVO W-75 is an excellent fungicide when used according to label directions for control of a broad spectrum of plant diseases. BRAVO W-75 can be used effectively in dilute or concentrate sprays. Thorough, uniform coverage is essential for disease control. **DO NOT** combine BRAVO W-75 in the spray tank with pesticides, surfactants or fertilizers, unless your prior use has shown the combination physically compatible and noninjurious under your conditions of use. The required amount of BRAVO W-75 should be added slowly into the spray tank during filling. With concentrate sprays, premix the required amount of BRAVO W-75 in a clean container and add to the spray tank as it is being filled. Keep agitator running when filling spray tank and during spray operations. **DO NOT** use on greenhouse-grown crops.

DIRECTIONS FOR USE

Only for sale to, use, and storage by agricultural personnel or commercial applicators. It is a violation of Federal law to use this product in a manner inconsistent with its labeling.

Note to User: This product may produce temporary allergic side effects characterized by redness of the eyes, mild bronchial irritation and redness or rash on exposed skin areas. Persons having allergic reaction should contact a physician.

Note to User: Wear long sleeve shirt, long pants, and gloves while mixing, loading and applying this product.

Dosage rates on this label indicate pounds of BRAVO W-75 per acre, unless otherwise stated. Applications should be made in sufficient water to obtain adequate coverage of foliage. Gallonage to be used will vary with crop and amount of plant growth. Spray volume usually will range from 20 to 150 gallons per acre for dilute sprays and 5 to 10 gallons per acre for concentrate ground sprays and aircraft applications.

Note to User: Do not rotate to crops other than those listed on labels within 12 months of the last treatment. After 12 months from the last application, leafy vegetables (i.e. spinach, lettuce, kale, etc.) may also be rotated.

Reentry: Note to User: **Do not** enter treated area to perform hand labor within 24 hours of application unless protective clothing is worn.

Because certain states may require more restrictive reentry intervals for various crops treated with this product, consult your State Department of Agriculture for further information.

Written or oral warnings must be given to workers who are expected to be in a treated area or in an area about to be treated with this product. When oral warnings are given, warnings shall be given in a language customarily understood by workers. Oral warnings must be given if there is reason to believe that written warnings cannot be understood by workers. Oral and written warnings must include the following information: **"DANGER.** Area treated with BRAVO W-75 on (date of application). **DO NOT** enter without appropriate protective clothing until the sprays have dried. In case of accidental exposure, wash exposed area with plenty of water and get medical attention. For further information see 'Precautionary Statements' on the label."

BEAN (Snap): Rust—Use 3 lbs. per acre in sufficient water to obtain adequate coverage. Begin applications during early bloom stage or when disease first threatens and repeat at weekly intervals or as necessary to maintain control. **DO NOT** apply within 7 days of harvest. **DO NOT** graze treated areas or feed treated plant parts to livestock.

CABBAGE, CAULIFLOWER, BROCCOLI, BRUSSELS SPROUTS: Alternaria leaf spot (Southeastern States only) and downy mildew—Use 1½ lbs. per acre in sufficient water to obtain adequate coverage. Begin applications after transplants are set in field, or shortly after emergence of field-seeded crop or when conditions favor disease development. Repeat at 7 to 10 day intervals or as necessary to maintain control. Ring spot in field-seeded brussels sprouts (California only)—Use 2 lbs. per acre. Begin applications at time of early sprout development or when conditions favor disease development. Repeat at 7 to 10 day intervals or as necessary to maintain control.

CARROT: Early blight and late blight—Use 1½ to 2 lbs. per acre in sufficient water to obtain adequate coverage. Start applications when disease threatens and repeat at 7 to 10 day intervals or as necessary to maintain control.

CELERY: Early blight, late blight, basal stalk rot *(Rhizoctonia solani)* —Use 1 to 1½ lbs. per acre on a 3 to 5 day spray schedule, or use 2 to 3 lbs. per acre on a 7 day schedule. Suppression of pink rot — Use 3 lbs. per acre on a 7 day schedule. Start applications when transplants are set in the field. Apply BRAVO W-75 in sufficient water to obtain adequate coverage. **DO NOT** apply within 7 days of harvest.

For control of early blight and late blight in celery seedbeds, use 1½ to 2 lbs. BRAVO W-75 per 100 gallons of water and apply 125 gallons per acre twice weekly or as needed to maintain control. Start applications shortly after crop emergence. Use the higher rate under severe disease conditions.

CORN (Sweet): Helminthosporium leaf blight—Use 1½ to 2 lbs. per acre in sufficient water to obtain adequate coverage. Begin applications when conditions favor disease development and repeat at 4 to 7 day intervals or as required to maintain control. **DO NOT** apply within 14 days of harvest. For use on fresh market crop only. **DO NOT** apply to sweet corn to be processed. **DO NOT** allow livestock to graze in treated fields. **DO NOT** ensile treated corn or use as livestock forage.

CUCUMBER: Target spot (Florida only), anthracnose and downy mildew—Use 1½ to 2 lbs. per acre. Powdery mildew (except Southwestern States), gummy stem blight, leaf blight and scab—Use 2 to 3 lbs. per acre. For suppression of fruit rot *(Rhizoctonia solani)* —Use 3 lbs. per acre. Apply BRAVO W-75 in sufficient water to obtain adequate coverage. Begin applications when plants are in first true leaf stage or when conditions are favorable for disease development. Repeat applications at 7 day intervals. Under severe disease conditions, shorten spray interval.

CANTALOUPE, MUSKMELON, HONEYDEW: Anthracnose and downy mildew—Use 1½ to 2 lbs. per acre. Powdery mildew (except in Southwestern States), Cercospora leaf spot, gummy stem blight, leaf blight and scab—Use 2 to 3 lbs. per acre. Apply BRAVO W-75 in sufficient water to obtain adequate coverage. Begin applications when plants are in first true leaf stage or when conditions are favorable for disease development. Repeat applications at 7 day intervals. Under severe disease conditions, shorten spray interval.

WATERMELON, SQUASH, PUMPKIN: Anthracnose and downy mildew —Use 1½ to 2 lbs. per acre. Cercospora leaf spot, gummy stem blight, leaf blight and scab—Use 2 to 3 lbs. per acre. Apply BRAVO W-75 in sufficient water to obtain adequate coverage. Begin applications when plants are in first true leaf stage or when conditions are favorable for disease development. Repeat applications at 7 day intervals. Under severe disease conditions, shorten spray interval.

ONION (Green and dry bulb): Botrytis leaf blight (blast), Downy mildew (Suppression) and Purple blotch—Use 1½ to 3 lbs. per acre. Apply before disease appears and repeat at 7 to 10 day intervals. Under severe disease conditions, use the higher application rate and shorten the interval between applications. **DO NOT** apply on dry bulb onions within 7 days of harvest or on green onions within 14 days of harvest. **DO NOT** apply more than 3 times per season on green onions. If more than 14 days remain before harvest of green onions after 3 applications of BRAVO W-75 have been made, it is recommended that another labeled fungicide be used to complete the application season. **DO NOT** apply to sweet Spanish onions.

PEANUT: Early leaf spot *(Cercospora)*, late leaf spot *(Cercosporidium)*—Use 1 to 1½ lbs. per acre. Rust—Use 1½ lbs. per acre. Apply BRAVO W-75 in sufficient water to obtain adequate coverage. Start applications when disease first appears and repeat at 10 to 14 day intervals or as necessary to maintain control. Under severe disease conditions, use the 1½ lbs. per acre rate. **DO NOT** apply within 14 days of harvest. **DO NOT** allow livestock to graze treated areas. **DO NOT** feed hay or threshings from treated fields to livestock.

POTATO: Early blight, late blight, Botrytis vine rot *(Botrytis spp.)*—Use 1 to 1½ lbs. per acre in sufficient water to obtain adequate coverage. Begin applications when plants are 6 inches high or when disease threatens and continue at 7 to 10 day intervals or as needed to maintain disease control. Under severe disease conditions, use the 1½ lbs. per acre rate and shorten spray interval.

TOMATO: Early blight, late blight, gray leaf spot, gray leaf mold —Use 1½ to 2 lbs. per acre. Anthracnose—Use 2 to 3 lbs. per acre. Rhizoctonia fruit rot and Botrytis gray mold—Use 3 lbs. per acre. Apply BRAVO W-75 in sufficient water to obtain adequate coverage. Begin applications when disease threatens and repeat at 7 to 10 day intervals. Under severe disease conditions, shorten spray interval. Alternaria fruit rot or blackmold (California only)—Make three applications at the rate of 3 lbs. per acre on a 10 to 14 day schedule beginning 5 to 6 weeks before anticipated harvest.

BLUEGRASS, PERENNIAL RYEGRASS GROWN FOR SEED: Stem rust, leaf rust, and stripe rust—Use 1 to 1½ lbs. per acre in sufficient water to obtain adequate coverage. Begin applications at the time of seed head elongation (early to mid-April) or when conditions favor disease development. Repeat at 7 to 10 day intervals or as necessary to maintain control. Use the 1 lb. per acre rate under a light incidence of rust. Under severe disease conditions, use the 1.5 lb. per acre rate at a 7 day spray interval. **DO NOT** apply within 14 days of harvest. **DO NOT** allow livestock to graze in treated areas. **DO NOT** feed hay or threshings from treated fields to livestock.

CONIFER TREE AND ORCHARD CROPS

Apply BRAVO W-75 in sufficient water and with proper calibration to obtain uniform coverage of tree canopy. Application through either ground equipment or aircraft is recommended except where otherwise indicated. When concentrate sprays are used the lower rate per acre of BRAVO W-75 listed may be used. The following spray volumes are recommended as gallons of spray per acre:

	Spray Volume (Gallons per Acre)	
Crop	**Dilute**	**Concentrate**
CONIFERS:		
Forest stands	Not used	10 to 20 (aircraft)
Christmas trees	100	10 to 50 (aircraft or ground equipment)
Nursery beds	100	5 to 10 (ground equipment only)

STORAGE AND DISPOSAL

DO NOT contaminate water, food or feed by storage or disposal. Open dumping is prohibited.

STORAGE

Store in a dry place.

PESTICIDE DISPOSAL

Pesticide wastes are toxic. Improper disposal of excess pesticide, spray mixture, or rinsate is a violation of Federal Law. If these wastes cannot be disposed of by use according to label instructions, contact your State Pesticide or Environmental Control Agency, or the Hazardous Waste representative at the nearest EPA Regional Office for guidance.

CONTAINER DISPOSAL

Completely empty bag into application equipment. Then dispose of empty bag in a sanitary landfill or by incineration, or, if allowed by State and local authorities, by burning. If burned, stay out of smoke.

WARRANTY AND LIMITATION OF DAMAGES

Seller warrants that this material conforms to its chemical description and is reasonably fit for the purposes stated on the label when used in accordance with directions under normal conditions of use and Buyer assumes the risk of any use contrary to such directions. SELLER MAKES NO OTHER EXPRESS OR IMPLIED WARRANTY, INCLUDING ANY OTHER EXPRESS OR IMPLIED WARRANTY OF FITNESS OR OF MERCHANTABILITY, AND NO AGENT OF SELLER IS AUTHORIZED TO DO SO EXCEPT IN WRITING WITH A SPECIFIC REFERENCE TO THIS WARRANTY. In no event shall Seller's liability for any breach of warranty exceed the purchase price of the material as to which a claim is made.

CROP	DISEASES	BRAVO W-75 RATE PER ACRE	BRAVO W-75 RATE PER 100 GAL*	APPLICATION DIRECTIONS
Conifers	Swiss needlecast	2¾ to 5½ lbs.	2¾ to 5½ lbs.	Single application technique: In Christmas tree plantations or forest stands make one application in the spring when new shoot growth is ½ to 2 inches in length.
	Scleroderris canker (pines) Swiss needlecast and Rhabdocline needlecast (Douglas-fir)	1½ to 2¾ lbs.	1½ to 2¾ lbs.	Make the first application in spring when new shoot growth is ½ to 2 inches in length. Make additional applications at 3 to 4 week intervals until conditions no longer favor disease development. For use in nursery beds, apply the highest rate specified on a three week schedule.
	Sirococcus tip blight	2 to 3½ lbs.	2 to 3½ lbs.	
	Rhizosphaera neddlecast (spruces) Scirrhia brown spot (pines)	5½ lbs.	5½ lbs.	
	Lophodermium neddlecast	1½ to 2¾ lbs.	1½ to 2¾ lbs.	North Central and Northeastern States: Begin applications in mid-July to early August before infection occurs. Make additional applications at 3 to 4 week intervals until conditions no longer favor disease development. For use in nursery beds, apply 2¾ lbs. per acre on a 3 week schedule.

CROP	DISEASES	BRAVO W-75 RATE PER ACRE	BRAVO W-75 RATE PER 100 GAL*	APPLICATION DIRECTIONS
	Botrytis seedling blight Phoma twig blight	$1\frac{1}{2}$ to $2\frac{3}{4}$ lbs.	$1\frac{1}{2}$ to $2\frac{3}{4}$ lbs.	Begin applications in nursery beds when seedlings are 4 inches tall and when cool, moist conditions favor disease development. Make additional applications at 7 to 14 day intervals as long as disease favorable conditions persist.

* Volumetric rates to be used only with full dilute spray volume specified on this label for tree and orchard crops.

SPECIAL LOCAL NEEDS PURSUANT TO SECTION 24(c) OF AMENDED FIFRA

CROP	REGISTRATION	STATES	SLN NO.	DIRECTIONS FOR USE
Forest Tree Nursery	Control of Botrytis	California	CA-780209	Apply 1 lb. per 100 gallons of water for control of Botrytis spp. on conifer seedling such as redwood, giant sequoia, and Douglas Fir. Begin application when disease threatens. Repeat applications at 7 to 14 day intervals as required to maintain disease control.

BUENO® 6

HERBICIDE

MSMA Plus Surfactant For Postemergence Weed Control
Contains 6.0 Pounds MSMA Per Gallon

ACTIVE INGREDIENT:

Monosodium Acid Methanearsonate	48.3%
INERT INGREDIENTS:	51.7%
TOTAL	100.0%

Total Arsenic, all in water-soluble form, expressed as elemental 22.3%

EPA Reg. No. 50534-6

CAUTION

Keep Out of Reach of Children

See precautionary statements.

PRECAUTIONARY STATEMENTS

HAZARDS TO HUMANS AND DOMESTIC ANIMALS

CAUTION

Harmful if swallowed or absorbed through the skin. Avoid contact with eyes, skin or clothing. Avoid breathing vapors or spray mist.

If swallowed, induce vomiting immediately by giving two glasses of water and sticking finger down throat. **Never** give anything by mouth to an unconscious person. Seek medical attention.

In case of contact, flush eyes with plenty of water for at least 15 minutes. Seek medical attention.

Wash exposed skin gently with plenty of soap and water.

Keep children and domestic animals off treated areas until this material has been washed into the soil.

Wear protective clothing when handling or applying this product including long pants, long-sleeve shirt, and impermeable gloves and boots. Mixer-loaders should include an apron and full-face shield when handling or mixing concentrate. Flagmen should be fully protected during spray operations or mechanical flagmen used. Pilots and ground spray rig applicators should wear a mask or respirator approved by the Mining Enforcement and Safety Administration and the National Institute for Occupational Safety and Health.

ENVIRONMENTAL HAZARDS

DO NOT feed treated foliage to livestock or graze treated areas. **DO NOT** spray or allow drift onto vegetables, ornamentals or other desirable plants. **DO NOT** apply when weather conditions favor drift from target area. **DO NOT** apply directly to water. **DO NOT** contaminate water by cleaning of equipment or disposal wastes.

BUENO 6

Read entire label before using this product.

GENERAL INFORMATION

BUENO 6 is a postemergence herbicide for use in cotton, turf, fruit, nuts and non-crop areas. BUENO 6 contains a surfactant (wetting agent) in proper ratio for best results. It is not necessary to add additional surfactant to the spray tank. Local conditions and recommendations vary; consult local agricultural experiment station or extension service weed specialists for recommendations in your area.

A partial list of weeds, controlled with BUENO 6 includes:

Barnyardgrass	Crabgrass	Nutsedge
Brachiaria spp.	Dallisgrass	Pigweed
Chickweed	Goosegrass	Puncturevine
Cocklebur	Johnsongrass	Sandbur

Mixing Instructions: Fill the spray tank with half the required amount of water and, with pump or agitator operating, add the recommended amount of BUENO 6 herbicide and continue filling the tank with the balance of water and mix thoroughly.

DO NOT store spray solution in tank overnight. Clean application equipment thoroughly after use by flushing with water in a safe place.

Application Methods: BUENO 6 should be applied with a low-pressure, properly calibrated sprayer having satisfactory pumping and bypass action. **Adjust Nozzles to Keep Spray Off Cotton Foliage and in a Manner to Allow Maximum Coverage of Weeds.** Adequate coverage is very important for effective weed control.

Apply during warm weather when weeds are in an active stage of growth.

DIRECTIONS FOR USE

It is a violation of Federal law to use this product in a manner inconsistent with its labeling.

COTTON: BUENO 6 is useful as a directed postemergence spray for control of the weeds listed above. Mix BUENO 6 at the rate of $2\frac{2}{3}$ pints in 40 gallons of water for application as a **Directed** broadcast spray to one acre when weeds are small. For band applications, apply 1 gallon of above diluted spray per acre for each 1 inch band width to be treated of cotton grown on 40 inch row spacing. A second or repeat application, if needed, should be timed about 1 to 3 weeks after the first application. Keep spray off cotton foliage.

Apply Only When Cotton is 3 Inches High to First Bloom. DO NOT Apply after First Bloom.

Slight burning and reddish discoloration of cotton foliage may occasionally be seen following recommended treatment; however, cotton plants will develop normally and yield will not be affected.

Emergency Preplant Application: When planting of cotton to prepared seedbeds has been delayed and weeds have emerged, one application of BUENO 6 at the above rate can be made prior to seeding.

NOTE: Applications to cotton in Florida should be confined to band treatments.

ORNAMENTAL TURFGRASS: BUENO 6 can be used for selective control of Bahiagrass, barnyardgrass, chickweed, smooth and large crabgrass, dallisgrass, nutsedge, sandbur and wood sorrel with little or no injury to well-established, actively growing turfgrasses. Mow turfgrass to height of 1 to $1\frac{1}{2}$ inches before treatment. Mix 1 fluid ounce (2 tablespoons) of BUENO 6 in 5 gallons of water for application to an area of 1,000 square feet. **Established bermu-**

dagrass and zoysiagrass: Mix 1 to 2 fluid ounces (2 to 4 tablespoons) of BUENO 6 in 5 gallons of water for application to an area of 1,000 square feet. Application should be uniform and thorough to adequately wet all undesirable plants. Two or more repeat treatments at 14 day intervals may be necessary. Make applications during warm weather when temperature is between 80 and 90°F. **DO NOT** water turf for at least 24 hours after application. Turfgrasses may be temporarily discolored. Bermudagrass, bluegrasses and zoysiagrass have shown tolerance to properly applied BUENO 6. Injury may result if applied to bentgrasses and fescues. **DO NOT** apply to St. Augustinegrass, carpetgrass, centipedegrass or to Dichondra. **DO NOT** reseed until 2 weeks after last application. **DO NOT** apply with hose-end applicators.

NON-BEARING FRUIT AND NUTS: BUENO 6 is effective as a directed postemergence spray for control of the above listed weeds in non-bearing almond, walnut, grapefruit, lemon, lime, orange and tangerine orchards and plantings of non-bearing apples, cherries, peaches, pears, plums and prunes. Mix at a rate of 2⅔ pints BUENO 6 in 100 gallons of water and apply to an area of one acre. For spot treatment of weeds, mix 2⅔ pints of BUENO 6 in 50 gallons of water and apply to point of run-off. Application should be made when weeds are small, during warm weather and when conditions are favorable for good weed growth. If regrowth occurs, repeat applications should be made, but not more than three applications per year. **DO NOT** allow spray solution to contact foliage, stems or bark of trees or vines. **DO NOT** use around trees or vines from which crops will be harvested within one year. **DO NOT** graze treated areas. In Florida, use only as a spot treatment.

NON-CROP: BUENO 6 is effective in controlling the above listed weeds and many similar weeds on drainage ditchbanks, fence rows, rights-of-way, storage yards and similar non-crop areas. Application should be made when weeds are small and conditions are favorable for good weed growth. Mix at a rate of 2½ to 2⅔ pints of BUENO 6 in 40 gallons of water. Spray undesirable vegetation thoroughly to point of run-off. Adequate coverage and complete wetting of foliage is important for effective control. Repeat applications may be necessary if regrowth occurs. **DO NOT** contaminate waters used for domestic purposes, or by animals including wildlife and aquatic life, or water used for irrigation purposes.

STORAGE AND DISPOSAL

DO NOT contaminate water, food or feed by storage or disposal. Open dumping is prohibited.

PESTICIDE DISPOSAL

Wastes resulting from the use of this product may be disposed of on site or at an approved waste disposal facility.

CONTAINER DISPOSAL

DO NOT reuse empty container. Triple rinse (or equivalent) and offer for recycling or reconditioning, or puncture and dispose of in a sanitary landfill, or incineration, or, if allowed by state and local authorities, by burning. If burned. stay out of smoke.

WARRANTY AND LIMITATION OF DAMAGES

Seller warrants that this material conforms to its chemical description and is reasonably fit for the purposes stated on the label when used in accordance with directions under normal conditions of use and Buyer assumes the risk of any use contrary to such directions. SELLER MAKES NO OTHER EXPRESS OR IMPLIED WARRANTY, INCLUDING ANY OTHER EXPRESS OR IMPLIED WARRANTY OF FITNESS OR OF MERCHANTABILITY, AND NO AGENT OF SELLER IS AUTHORIZED TO DO SO EXCEPT IN WRITING WITH A SPECIFIC REFERENCE TO THIS WARRANTY. In no event shall Seller's liability for any breach of warranty exceed the purchase price of the material as to which a claim is made.

SPECIAL LOCAL NEEDS PURSUANT TO SECTION 24(c) OF AMENDED FIFRA DIRECTIONS FOR USE

CROP	REGISTRATIONS	STATES	SLN NO.	DIRECTIONS FOR USE
Bluegrass, Fescue, Ryegrass	Control of wild oats and certain other broadleaf and grassy weeds in bluegrass, fescue and ryegrass grown for seed only.	Idaho Oregon Washington	ID-770019 OR-770025 WA-770025	Apply 3 to 4 quarts BUENO 6 per acre in sufficient water for good coverage. Application can be made anytime after weeds emerge and before grass has reached boot stage. **DO NOT USE MORE THAN ONE APPLICATION PER YEAR. DO NOT APPLY AFTER BOOT STAGE. DO NOT FEED TREATED FOLIAGE OR SEED PRODUCED FROM TREATED AREAS.**
Citrus	Citrus—Bearing	California	CA-790189	BUENO 6 should be applied as a directed application in citrus orchards, such as orange, grapefruit, tangerine, lemon and lime orchards. It should be applied at the rate of 2⅔ to 5⅓ pints per acre. Mix BUENO 6 at the rate of 2⅔ pints in 75 gallons of water. Apply as a directed spray in interspaces and around base of trees. Spray unwanted vegetation to just short of run-off. If regrowth occurs, reapply as required; however, **DO NOT** exceed 3 applications per year. **DO NOT** allow spray solution to contact fruit leaves, stem or bark. Use a shield, if necessary, for nursery plantings or young trees. SPECIAL PRECAUTIONS—**DO NOT** allow spray or spray drift to contact adjacent crops or injury will occur. Apply only on still days when weather conditions do not favor drift from areas being treated. Applications should be made in accordance with state and Federal regulations.
Cotton	Application on Cotton	South Carolina	SC-770015	POSTEMERGENT APPLICATION ON COTTON: A single ground or aircraft application may be made as an over-the-top broadcast spray to actively growing cotton as a salvage operation when cotton is 3 to 6 inches high or up to early first square stage, whichever occurs first. Mix BUENO 6 at a rate of 1 to 1⅓ pints in 40 gallons of water for ground application, or in 5 to 10 gallons of water for aircraft application and apply using a properly calibrated sprayer to one acre. APPLY AS AN OVER-THE-TOP BROADCAST SPRAY ONLY WHEN COTTON IS 3 to 6 INCHES HIGH OR UP TO EARLY FIRST SQUARE STAGE, WHICHEVER OCCURS FIRST. **DO NOT APPLY AS AN OVER-THE-TOP BROADCAST SPRAY AFTER COTTON IS MORE THAN 6 INCHES HIGH OR AFTER EARLY FIRST SQUARE STAGE. DO NOT** make more than two applications total of either DSMA or MSMA (or a combination) per season. See SPECIAL PRECAUTIONS below. Slight burning and reddish discoloration of cotton foliage may occasionally be seen following recommended treatment; however, cotton plants will develop normally.

CROP	REGISTRATIONS	STATES	SLN NO.	DIRECTIONS FOR USE
				SPECIAL PRECAUTIONS—**DO NOT** feed treated forage to livestock or graze treated areas. **DO NOT** allow spray or spray drift to contact adjacent crops or injury and residues will result. Apply only on still days when weather conditions **DO NOT** favor drift from areas being treated. Follow other precautions listed on this label. Aircraft applications of BUENO 6 should only be made by applicators experienced in the use of herbicides, and applications should be made in accordance with state and Federal regulations.
Cotton	Application on Cotton	Arkansas Georgia Louisiana Mississippi Missouri New Mexico North Carolina Tennessee Texas	3972 GA-770008 3973 3974 MO-770005 NM-780004 NC-770018 TN-830017 TX-770026	BUENO 6 is useful for the control of weeds listed on the product label and many similar weeds. Application can be made as follows: 1. Preplant or postplant up to cracking of soil before cotton emergence using ground or aircraft equipment. 2. Postemergent, over-the-top, when cotton is 3 to 6 inches high or up to early first square stage, whichever occurs first, using ground or aircraft equipment. 3. Postemergence as a directed spray with ground equipment when cotton is 3 inches high to first bloom. PREPLANT OR POSTPLANT UP TO CRACKING APPLICATION ON COTTON: A single ground or aircraft application of BUENO 6 can be made to prepared cotton seedbeds when planting has been delayed and weeds have emerged, or as a postplant treatment, BUT NO LATER THAN INITIAL CRACKING of soil in field before emergence of cotton. Planting of cotton may immediately follow the preplant application. Mix at the rate of $2\frac{2}{3}$ pints of BUENO 6 in 40 gallons of water for broadcast ground application, or in 5 to 10 gallons of water for aircraft application and apply using a properly calibrated sprayer to one acre. See SPECIAL PRECAUTIONS below. POSTEMERGENT, OVER-THE-TOP, APPLICATION ON COTTON: Application may be made using ground or aircraft equipment as an over-the-top broadcast spray to actively growing cotton as a salvage operation when cotton is 3 to 6 inches high or up to early first square stage, whichever occurs first. Mix BUENO 6 at rate of 1 to $1\frac{1}{3}$ pints to 40 gallons of water for ground application, or in 5 to 10 gallons of water for aircraft application and apply using a properly calibrated sprayer to one acre. A second application, if needed, should be made 1 to 3 weeks after the first treatment. APPLY AS AN OVER-THE-TOP BROADCAST SPRAY ONLY WHEN COTTON IS 3 to 6 INCHES HIGH OR UP TO EARLY FIRST SQUARE STAGE, WHICHEVER OCCURS FIRST. DO NOT APPLY AS AN OVER-THE-TOP BROADCAST SPRAY AFTER COTTON IS MORE THAN 6 INCHES HIGH OR AFTER EARLY FIRST SQUARE STAGE WHICHEVER OCCURS FIRST. See SPECIAL PRECAUTIONS below. POSTEMERGENT, DIRECTED, APPLICATION ON COTTON: Mix BUENO 6 at the rate of $2\frac{2}{3}$ pints in 40 gallons of water for application as a DIRECTED broadcast spray to one acre when weeds are small. For band applications, apply 1 gallon of above diluted spray per acre for each 1-inch band width to be treated of cotton grown on 40-inch row spacing. Keep spray off cotton foliage. APPLY ONLY WHEN COTTON IS 3 INCHES HIGH TO FIRST BLOOM. DO NOT APPLY AFTER FIRST BLOOM. Slight burning and reddish discoloration of cotton foliage may occasionally be seen following recommended treatment; however, cotton plants will develop normally. SPECIAL PRECAUTIONS—**DO NOT** allow spray or spray drift to contact adjacent crops or injury will result. Apply only on still days when weather conditions **DO NOT** favor drift from areas being treated. Follow other precautions listed on this label. Aircraft applications of BUENO 6 should only be made by applicators experienced in use of herbicides, and applications should be made in accordance with state and Federal regulations.
Cotton	Application on Cotton	California	CA-770095	BUENO 6 is useful for the control of weeds listed on the product label and many similar weeds. Application can be made as: 1. Preplant or postplant up to cracking of soil before cotton emergence using ground or aircraft equipment. 2. Postemergent, over-the-top, when cotton is 3 to 6 inches high or up to early first square stage, whichever occurs first, using ground or aircraft equipment. 3. Postemergence as a directed spray with ground equipment when cotton is 3 inches high to first bloom. PREPLANT OR POSTPLANT UP TO CRACKING APPLICATION ON COTTON: A single ground or aircraft application of BUENO 6 can be made to prepared cotton seedbeds when planting has been delayed and weeds have emerged, or as a postplant treatment, BUT NO LATER THAN INITIAL CRACKING of soil in field before emergence of cotton. Planting of cotton may immediately follow the preplant application. Mix at the rate of $2\frac{2}{3}$ pints of BUENO 6 in 40 gallons of water for broadcast ground application, or in 5 to 10 gallons f water for aircraft application,

CROP	REGISTRATIONS	STATES	SLN NO.	DIRECTIONS FOR USE
				and apply using a properly calibrated sprayer to one acre. See SPECIAL PRECAUTIONS below. **POSTEMERGENT, OVER-THE-TOP, APPLICATION ON COTTON:** One or two over-the-top applications may be applied by ground or aircraft equipment to actively growing cotton as a salvage operation when crop is 3 to 6 inches high or to early first square, whichever occurs first. **ONE APPLICATION:** Use 2⅔ pints BUENO 6 in 40 gallons of water per acre for ground application, or in 5 to 10 gallons of water for aircraft application. **DO NOT REPEAT THIS TREATMENT. TWO APPLICATIONS:** Use 1⅓ pints BUENO 6 in 40 gallons of water per acre for ground application, or in 5 to 10 gallons of water for aircraft application. Apply second application, if needed, using the 1⅓ pint rate, 1 to 3 weeks following the first treatment. Use a properly calibrated sprayer. DO NOT APPLY OVER-THE-TOP SPRAYS AFTER COTTON IS MORE THAN 6 INCHES HIGH OR AFTER EARLY FIRST SQUARE STAGE. WHICHEVER OCCURS FIRST. **POSTEMERGENT, DIRECTED, APPLICATION ON COTTON:** Mix BUENO 6 at the rate of 2⅔ pints in 40 gallons of water for application as a DIRECTED broadcast spray to one acre when weeds are small. For band applications, apply 1 gallon of above diluted spray per acre for each 1-inch band width to be treated of cotton grown on 40-inch row spacing. Keep spray off cotton foliage. APPLY ONLY WHEN COTTON IS 3 INCHES HIGH TO FIRST BLOOM. **DO NOT APPLY AFTER** FIRST BLOOM. Slight burning and reddish discoloration of cotton foliage may occasionally be seen following recommended treatment; however, cotton plants will develop normally. SPACIAL PRECAUTIONS—**DO NOT** allow spray or spray drift to contact adjacent crops or injury will result. Apply only on still days when weather conditions. **DO NOT** favor drift from areas being treated. Follow other precautions listed on this label. Aircraft applications of BUENO 6 should only be made by applicators experienced in use of herbicides, and applications should be made in accordance with state and Federal regulations.

DACAMINE® 4D

HERBICIDE

Only for Sale To, Use and Storage by Agricultural Personnel or Commercial Applicators.

ACTIVE INGREDIENT:
N-Oleyl-1,3-Propylenediamine Salt of

2,4-Dichlorophenoxyacetic Acid*	31.7%
2,4-Dichlorophenoxyacetic Acid*	21.6%
INERT INGREDIENTS:	46.7%
TOTAL	100.0%

* 2,4-Dichlorophenoxyacetic Acid Equivalent: 39.6%, 3.6 lb./gal.
* Isomer specific by AOAC Method

EPA Reg. No. 50534-13

CAUTION

Keep Out of Reach of Children

See precautionary statements.

PRECAUTIONARY STATEMENTS

HAZARDS TO HUMANS AND DOMESTIC ANIMALS

CAUTION

Harmful if swallowed or absorbed through the skin.

If swallowed, DO NOT induce vomiting. If vomiting occurs spontaneously, keep airway clear. Never give anything by mouth to an unconscious person. Seek medical attention. **Causes eye and skin irritation. DO NOT** get in eyes.

In case of contact, flush eyes with plenty of water for at least 15 minutes and seek medical attention. **DO NOT** allow contact with skin or clothing. Wash exposed skin gently with plenty of soap and water. Avoid breathing spray mist.

ENVIRONMENTAL HAZARDS

DO NOT apply directly to water. **DO NOT** contaminate irrigation ditches or water used for domestic purposes. Use care to avoid spray contact or drift to 2,4-D susceptible plants such as cotton, tomatoes, flowers, grapes, fruit trees and ornamentals. **DO NOT** permit spray mist containing DACAMINE 4D to drift onto them, since even very small quantities of the spray, which may not be visible, can cause severe injury during both growing and dormant periods. **DO NOT** spray when the wind is blowing towards susceptible crops or ornamental plants. Use coarse sprays to minimize drift. With ground equipment, spray drift can be lessened by keeping the spray boom as low as possible; by applying 20 gallons or more of spray per acre; by using no more than 20 pounds spraying pressure with flat fan or flooding flat fan nozzle tips; by spraying when wind velocity is low; and by stopping all spraying when wind exceeds 8 miles per hour. **DO NOT** apply with hollow cone-type insecticide or other nozzles that produce a fine droplet spray. With aircraft application, drift can be lessened by applying not less than 5 gallons of spray per acre; by using no more than 20 pounds spray pressure at the nozzles; by using nozzles which produce a coarse spray pattern; and by spraying only when the wind velocity is less than 5 miles per hour. Flush sprayer out on suitable non-crop area after use. **DO NOT** use the same spray equipment for applying other materials to 2,4-D susceptible crops as injury may result.

DACAMINE 4D

Read entire label before using this product.

GENERAL INFORMATION

DACAMINE 4D is an oil-soluble diamine form of 2,4-D which is formulated to be used only with water. It combines the advantages of both the ester and amine forms of 2,4-D. DACAMINE 4D contains both the non-volatile features of amines and the effectiveness of esters. It is designed to control weeds in croplands, pear and apple orchards and lawn areas. The following is a partial list of weeds controlled by DACAMINE 4D:

Artemisia (common mugwort)	Leafy Spurge
Canada Thistle	Milkweed
Field Bindweed (creeping jenny)	Russian Knapweed
Horse Nettle	Yarrow

Annual chickweed	Henbit	Ragweed (common, giant, western)
Beggartick	Ironweed	Shepherdspurse
Bitterweed	Jimsonweed	Slender aster
Blackeyed Susan	Knotweed	Snow-on-the-mountain
Broomweed	Kochia	Sour dock
Buckhorn plantain	Lambsquarters	Spike rush
Bull thistle	Mallows	Spurge
Burdock	Marestail	Star thistle
Carpetweed	Marshelder	Sumac
Chicory	Mexicanweed	Sunflower
Cocklebur	Morningglory	Tarweed
Coffeebean	Mustards	Velvetleaf
Croton	Nettle, stinging	Wild carrot
Curly dock	Pennycress	Wild cucumber
Curly indigo	Pennywort	Wild lettuce
Dandelion	Peppergrass	Wild parsnip
Dayflower	Pigweed	Wild radish
Fanweed	Plantain	Willow
Fiddleneck	Poison ivy	Wintercress
Flower-of-an-hour	Poorjoe	Yankeeweed
Gooseweed	Puncturevine	
Hemp		

NOTE: Local conditions, crop varieties and application regulations vary and may affect use of this herbicide. Consult local agricultural experiment station or extension service weed specialists and state regulatory agencies for recommendations in your area.

Aerial application may be of use for control of weeds on certain crops where there would be no danger of drift to susceptible crops. Applications should only be made by applicators experienced in the use of 2,4-D formulations. Regulations governing aerial application of herbicides are in effect in many states.

Consult local regulatory agencies concerning requirements before making applications.

Treating Small Areas: One tablespoonful of DACAMINE 4D in 1½ gallons of water is almost equal to 1 quart in 100 gallons.

NOTE: When stored at temperatures below freezing, it may be necessary to warm contents to 45°F and mix thoroughly before using.

DIRECTIONS FOR USE

It is a violation of Federal law to use this product in a manner inconsistent with its labeling.

Preparation of the Spray: Fill the spray tank with half the required amount of water, then add the recommended amount of DACAMINE 4D and continue filling spray tank with balance of water. Keep agitator running when filling spray tank and during spray operations. Use enough spray volume for uniform coverage by ground or air application. For ground application, when weeds are 1 to 3 inches tall, use 15 to 25 gallons of water per acre. When weeds are taller than 3 inches, use 25 to 50 gallons of water per acre. For aerial application, when weeds are 1 to 3 inches tall, use 5 to 10 gallons of water per acre. When weeds are taller than 3 inches, use 10 to 15 gallons of water per acre. In any case, use the same amount of DACAMINE 4D recommended per acre. For crop uses, **DO NOT** mix with oil, surfactant or other adjuvants unless specifically recommended on this label as this may reduce selectivity to crops resulting in crop damage.

Time of Application: Best results are obtained when DACAMINE 4D is used on young weeds that are actively growing. Applications of lower listed rates to susceptible annual weeds usually will be satisfactory, but for perennial weeds and other conditions, such as in very dry areas where kill is difficult, use the higher recommended rates. When used as a selective spray on crops, the stage of growth of the crop must be considered. Some woody plants and weeds are hard to kill and repeat applications may be necessary.

SMALL GRAINS (Wheat, Barley, Rye): Apply DACAMINE 4D in sufficient water for uniform coverage on small grains when fully tillered or stooled (4 to 8 inches tall), but before the "boot" stage. Crop injury may result if applied earlier than "tiller" or from early "boot" through the "milk" stage. **DO NOT** use on grains undersown with legumes, such as alfalfa or clovers, except where some legume injury can be tolerated. **DO NOT** graze or feed forage from treated fields within 2 weeks after treatment. See chart for recommended rates. For late season use to control large succulent weeds, apply 1 to 2 pints per acre in sufficient water for uniform coverage on small grain from dough stage to harvest—use only when weeds threaten to interfere with harvest operations. **DO NOT** use treated straw for livestock feed.

RICE: Apply DACAMINE 4D when the rice is in the late tillering stage, but before it has started to joint. This should be when the rice seedlings are about 6 to 10 weeks old. Damage to rice may occur if DACAMINE 4D is applied after rice is 12 weeks old. Treatment after flooding is usually safer than treatment before flooding. See chart for recommended rates.

CORN: **Postemergence**—Apply DACAMINE 4D from emergence to tasseling. When spraying corn above 10 inches in height, use nozzle extensions ("corn drops"), directing the spray at base of the corn plant to keep the spray off the leaves as much as possible. **DO NOT** apply from tassel emergence to dough stage, Crop injury is more likely to occur if corn is growing rapidly under high temperature and high soil moisture conditions. Under such conditions, use the lowest recommended rate. Delay cultivation for 8 to 10 days after application to reduce stalk breakage resulting from temporary brittleness caused by 2,4-D. Hybrids vary in tolerance to 2,4-D. Consult local agricultural experiment station or extension service weed specialists regarding the use of 2,4-D on your specific hybrid. See chart for recommended rates.

SORGHUM: Apply DACAMINE 4D when sorghum plants are 6 to 12 inches tall and secondary roots are well established. Use drop nozzles when crop is over 10 inches tall. **DO NOT** apply from flowering to dough stage. See chart for recommended rates.

SUGARCANE: Apply as a postemergence application in spring after cane emerges and through layby. See chart for recommended rates.

Amount of DACAMINE 4D per Acre

Crop (See Detailed Directions Above)	For Average Conditions	For Dry Conditions as in Western States*
Wheat, Barley, Rye		
Annual Weeds	½ to 1 pint	1 to 2 pints
Perennial Weeds	1 pint	1¼ to 2 pints
Rice	1 to 2 pints	
Corn**		
Postemergence	½ pint	½ to ¾ pint
Sorghum**		
Postemergence	½ pint	½ to ¾ pint
Sugarcane	1 to 2 quarts	

* Arizona, California, Idaho, Montana, Nevada, Oregon, Utah, Washington and Wyoming

** If only rows or bands are treated, leaving middles unsprayed, reduce dosage rate per crop acre proportionate to the ground area actually sprayed.

FORAGE CROPS (Pastures and Rangelands): To control many broadleaved weeds in pastures, meadows and rangelands, use 1 quart per acre of DACAMINE 4D in sufficient water to provide for uniform application. Treat pastures when weeds are growing actively. **DO NOT** apply when grass is in "boot" to "milk" stage if grass seed production is desired. **DO NOT** apply to newly-seeded pastures until grass is well established or after heading begins. Most legumes are usually injured or killed at the rates recommended. For control of wild garlic and wild onion in pastures, apply 2 to 3 quarts per acre in very early spring and repeat annually until weeds are eradicated. **DO NOT** graze dairy animals on treated areas within 7 days after application.

LAWNS, GOLF COURSES AND SIMILAR TURF: Apply 1½ to 2 pints of DACAMINE 4D per acre in 100 gallons of water or sufficient water to provide adequate uniform coverage. For treating small areas, use 1½ tablespoonsful of DACAMINE 4D in 2 gallons of water per 1,000 sq. ft. **DO NOT** apply during periods of drought.

DO NOT apply DACAMINE 4D to newly-seeded turf until grasses are well estalished and after making at last two mowings. Bare areas can be reseeded in 3 to 6 weeks. Bentgrass varieties vary in tolerance to DACAMINE 4D under different conditions of growth. Use rates no higher than 2 pints per acre on bentgrass and apply only where experience has shown application to be safe. Injury may result if applied to St. Augustinegrass, centipedegrass, carpetgrass, clovers and Dichondra.

PEAR AND APPLE ORCHARDS: Apply at rate of 1 to 2 quarts DACAMINE 4D per acre in sufficient water to give good coverage and thorough wetting of weeds. Use the higher rate for control of perennial weeds. Repeat applications may be necessary to control some woody plants and perennial weeds. **DO NOT** apply to bare ground areas as injury may result. When bare areas do occur, spot treat by directly spraying only at undesirable plants. **DO NOT** apply immediately before irrigation. Withhold irrigation for three (3) days following application. Best results are obtained when applied within two (2) days following irrigation and when weeds are growing actively. **DO NOT** allow spray drift to contact foliage, fruit, stems or trunks of fruit trees. Use low pressure and coarse nozzles, and spray on still days to minimize drift.

WOODY PLANT CONTROL: To control 2,4-D susceptible plants such as alder, buckbrush, elderberry, sumac and willow in waste areas or non-crop land, such as powerlines and highway rights-of-way, vacant lots and fence rows, use 4 to 6 quarts in 100 gallons of water in spring and early summer. For treating small areas use 1½ to 2 ounces of DACAMINE 4D in one gallon of water. Wet thoroughly all parts of the plants, including foliage and stems to point of runoff. **DO NOT** treat during periods of severe drought or in early fall when leaves have lost their healthy green color. Hard-to-kill species may need retreatment the following season.

GENERAL WEED CONTROL: On vacant lots, roadsides, along fence rows and around buildings, use 1 to 2 quarts of DACAMINE 4D in 100 gallons of water per acre. Thoroughly wet all foliage to run-off. Application should be made to weeds in an early and active stage of growth.

STORAGE AND DISPOSAL

DO NOT contaminate water, food or feed by storage or disposal. Open dumping is prohibited.

PESTICIDE DISPOSAL

Pesticide wastes are toxic. Improper disposal of excess pesticide, spray mixture or rinsate is a violation of Federal Law. If these wastes cannot be disposed of by use according to label instructions, contact your State Pesticide or Environmental Control Agency or the Hazardous Waste Representative at the nearest EPA Regional Office for guidance.

CONTAINER DISPOSAL

DO NOT reuse empty container. Triple rinse (or equivalent) and offer for recycling or reconditioning or puncture and dispose of in a sanitary landfill, or by other procedures approved by state and local authorities.

WARRANTY AND LIMITATION OF DAMAGES

Seller warrants that this material conforms to its chemical description and is reasonably fit for the purposes stated on the label when used in accordance with directions under normal conditions of use and Buyer assumes the risk of any use contrary to such directions. SELLER MAKES NO OTHER EXPRESS OR IMPLIED WARRANTY, INCLUDING ANY OTHER EXPRESS OR IMPLIED WARRANTY OF FITNESS OR OF MERCHANTABILITY, AND NO AGENT OF SELLER IS AUTHORIZED TO DO SO EXCEPT IN WRITING WITH A SPECIFIC REFERENCE TO THIS WARRANTY. In no event shall Seller's liability for any breach of warranty exceed the purchase price of the material as to which a claim is made.

SPECIAL LOCAL NEEDS PURSUANT TO SECTION 24(c) OF AMENDED FIFRA

CROP	REGISTRATIONS	STATES	SLN NO.	DIRECTIONS FOR USE
Rice Stubble Crop	Control of Certain broadleaf weeds	Louisiana	LA-830022	For control of certain broadleaf weeds such as red stem, dayflower, ducksalad, northern jointvetch (curly indigo), and hemp sesbania (coffeebean, coffeeweed, tall indigo), and suppression of alligator weed. Apply DACAMINE 4D, 1 to 2 pints per acre to vigorously growing weeds within two weeks following harvest to the first crop. To insure adequate spray coverage, apply when weeds have fully emerged through crop residue left from first harvest. **DO NOT** apply DACAMINE 4D after rice has reached the panicle—differentiation stage of growth, or crop injury may result. Weeds that are fully grown at time of first harvest or under drought stress may not be controlled by this application.

DACONATE® 6

HERBICIDE

MSMA Plus Surfactant For Postemergence Weed Control
Contains 6.0 Pounds MSMA Per Gallon

ACTIVE INGREDIENT:
Monosodium Acid Methanearsonate 48.3%
INERT INGREDIENTS: 51.7%
TOTAL 100.0%

Total Arsenic, all in water-soluble form, expressed as elemental 22.3%

EPA Reg. No. 50534-5

CAUTION

Keep Out of Reach of Children

See precautionary statements.

PRECAUTIONARY STATEMENTS

HAZARDS TO HUMANS AND DOMESTIC ANIMALS
CAUTION

Harmful if swallowed or absorbed through the skin. Avoid contact with eyes, skin or clothing. Avoid breathing vapors or spray mist.

If swallowed, induce vomiting immediately by giving two glasses of water and sticking finger down throat. **Never** give anything by mouth to an unconscious person. Seek medical attention.

In case of contact, flush eyes with plenty of water for at least 15 minutes. Seek medical attention.

Wash exposed skin gently with plenty of soap and water.

Keep children and domestic animals off treated areas until this material has been washed into the soil.

Wear protective clothing when handling or applying this product including long pants, long-sleeve shirt, and impermeable gloves and boots. Mixer-loaders should include an apron and full-face shield when handling or mixing concentrate. Flagmen should be fully protected during spray operations or mechanical flagmen used. Pilots and ground spray rig applicators should wear a mask or respirator approved by the Mining Enforcement and Safety Administration and the National Institute for Occupational Safety and Health.

ENVIRONMENTAL HAZARDS

DO NOT feed treated foliage to livestock or graze treated areas. **DO NOT** spray or allow drift onto vegetables, ornamentals or other desirable plants. **DO NOT** apply when weather conditions favor drift from target area. **DO NOT** apply directly to water. **DO NOT** contaminate water by cleaning of equipment or disposal wastes.

DACONATE 6

Read entire label before using this product.

GENERAL INFORMATION

DACONATE 6 is a postemergence herbicide for use in cotton, turf and non-crop areas. DACONATE 6 contains a surfactant (wetting agent) in proper ratio for best results. It is not necessary to add additional surfactant to the spray tank. Local conditions and recommendations vary; consult local agricultural experiment station or extension service weed specialists for recommendations in your area.

Mixing Instructions: DACONATE 6 herbicide must be thoroughly dissolved. Fill the spray tank with half the required amount of water and, with pump or agitator operating, add the recommended amount of DACONATE 6 herbicide and continue filling the tank with balance of water.

DO NOT store spray solution in tank overnight. Clean application equipment thoroughly after use by flushing with water in a safe place.

Application Methods: DACONATE 6 should be applied with a low-volume, properly calibrated sprayer having satisfactory pumping and bypass action. **Adjust Nozzles to Keep Spray Off Cotton Foliage and in a Manner to Allow Maximum Coverage of Weeds.** Adequate coverage is very important for effective weed control.

Apply during warm weather when weeds are in an active stage of growth.

DIRECTIONS FOR USE

It is a violation of Federal law to use this product in a manner inconsistent with its labeling.

COTTON: DACONATE 6 is useful as a directed postemergence spray for controlling weeds in cotton such as barnyardgrass, cocklebur, smooth and large crabgrass, dallisgrass, goosegrass, johnsongrass and nutsedge. Apply as a **Directed** broadcast spray at rate of 40 gallons per acre of a solution containing 2⅔ pints of DACONATE 6 in 40 gallons of water when weeds are small. For band applications, apply 1 gallon of above diluted spray per acre for each 1 inch band width to be treated of cotton grown on 40 inch row spacing. A second or repeat application, if needed, should be timed about 1 to 3 weeks after the first application. Keep spray off cotton foliage.

Apply Only When Cotton is 3 Inches High to First Bloom. DO NOT Apply After First Bloom.

Slight burning and reddish discoloration of cotton foliage may occasionally be seen following recommended treatment; however, cotton plants will develop normally and yield will not be affected.

ORNAMENTAL TURFGRASS: DACONATE 6 can be used for selective control of Bahiagrass, barnyardgrass, chickweed, smooth and large crabgrass, dallisgrass, nutsedge, sandbur and wood sorrel with little or no injury to well-established, actively growing turfgrasses. Mow turfgrass to height of 1 to 1½ inches before treatment. Mix 1 fluid ounce (2 tablespoons) of DACONATE 6 in 5 gallons of water for application to an area of 1,000 square feet. **Established bermudagrass and zoysiagrass.** Mix 1 to 2 fluid ounces (2 to 4 tablespoons) of DACONATE 6 in 5 gallons of water for application to an area of 1,000 square feet. Application should be uniform and thorough to adequately wet all undesirable plants. Two or more repeat treatments of 14 day intervals may be necessary. Make applications during warm weather when temperature is between 80 and 90°F. **DO NOT** water turf for at least 24 hours after application. Turfgrass may be temporarily discolored. Bermudagrass, bluegrass and zoysiagrass have shown tolerance to properly applied DACONATE 6. Injury may result if applied to bentgrasses and fescues. **DO NOT** apply to St. Augustinegrass, carpetgrass, centipedegrass or to Dichondra. **DO NOT** reseed until 2 weeks after last application. **DO NOT** apply with hose-end applicators.

NON-CROP: DACONATE 6 is effective in controlling barnyardgrass, cocklebur, smooth and large crabgrass, dallisgrass, goosegrass, johnsongrass, nutsedge and many similar weeds on drainage ditchbanks, fence rows, rights-of-way, storage yards and similar non-crop areas. Application should be made when weeds are small and conditions are favorable for good weed growth. Mix at a rate of 2½ to 2⅔ pints of DACONATE 6 in 40 gallons of water. Spray undesirable vegetation thoroughly to point of run-off. Adequate coverage and complete wetting of foliage is important for effective control. Repeat applications may be necessary if regrowth occurs. **DO NOT** contaminate waters used for domestic purposes, or by animals including wildlife and aquatic life, or water used for irrigation purposes.

STORAGE AND DISPOSAL

DO NOT contaminate water, food or feed by storage or disposal. Open dumping is prohibited.

PESTICIDE DISPOSAL

Wastes resulting from the use of this product may be disposed of on site or at an approved waste disposal facility.

CONTAINER DISPOSAL

DO NOT reuse empty container. Triple rinse (or equivalent) and offer for recycling or reconditioning, or puncture and dispose of in a sanitary landfill, or by other procedures approved by state and local authorities.

WARRANTY AND LIMITATION OF DAMAGES

Seller warrants that this material conforms to its chemical description and is reasonably fit for the purposes stated on the label when used in accordance with directions under normal conditions of use and Buyer assumes the risk of any use contrary to such directions. SELLER MAKES NO OTHER EXPRESS OR IMPLIED WARRANTY, INCLUDING ANY OTHER EXPRESS OR IMPLIED WARRANTY OF FITNESS OR OF MERCHANTABILITY, AND NO AGENT OF SELLER IS AUTHORIZED TO DO SO EXCEPT IN WRITING WITH A SPECIFIC REFERENCE TO THIS WARRANTY. In no event shall Seller's liability for any breach of warranty exceed the purchase price of the material as to which a claim is made.

DACTHAL® W-75

HERBICIDE

75% Wettable Powder

ACTIVE INGREDIENT:

Dimethyl Tetrachloroterephthalate	75.0%
INERT INGREDIENTS:	25.0%
TOTAL	100.0%

EPA Reg. No. 50534-1

CAUTION

Keep Out of Reach of Children

See precautionary statements.

PRECAUTIONARY STATEMENTS

HAZARDS TO HUMANS AND DOMESTIC ANIMALS
CAUTION

Harmful if swallowed. Avoid breathing dust. Avoid contact with eyes. In case of contact flush eyes with plenty of water for at least 15 minutes and seek medical attention. Avoid contact with skin or clothing. Wash exposed skin gently with plenty of soap and water.

ENVIRONMENTAL HAZARDS

DO NOT apply directly to water. **DO NOT** contaminate water by cleaning of equipment or disposal of wastes.

DACTHAL W-75 Herbicide

A selective preemergence herbicide for control of annual grasses and certain broadleaved weeds in vegetable crops and ornamental turf.

DACTHAL W-75

Read entire label carefully and use only as directed.

seed germination usually provides seasonal control. Crabgrass seed germination usually coincides with the time when Forsythia blossoms start to fall. In areas which commonly experience stands of late-germinating crabgrass and other annual weedy grasses, a second application at half the regular rate can be safely made two months after the first application. In order to control annual bluegrass *(Poa annua)* and other erratic or late-germinating grasses and weeds in northern states, a late summer or early fall application before weed seed germination of 20 pounds of DACTHAL W-75 in 40 to 100 gallons of water per acre (½ lb./1 to 2 gals./1000 sq. ft.) should supplement the early spring application. Earlier applications are required in southern United States. Consult agricultural experiment station or extension service weed specialist for date *Poa annua* seed germinates in your area and make application at least two weeks prior to this date.

Early spring application of DACTHAL may be made to new turfgrass seedlings after the grasses have exhibited a uniform greening of the newly sprouted grass, preferably when 1 to 2 inches in height. Such DACTHAL applications give control of crabgrass without injury to newly-emerged grass. Fall sowing of permanent grasses can follow early spring application of DACTHAL.

Where some spring seeding is necessary, there should be a delay of approximately 60 days after the application of most preemergence herbicides, including DACTHAL. This delay increases the survival of desirable grasses.

DACTHAL is not recommended for use on putting-greens. DACTHAL is not recommended for use on bentgrasses mowed at putting-green height. For control of creeping speedwell *(Veronica filiformis)*, apply 16 pounds of DACTHAL W-75 in 40 to 100 gallons of water per acre as a postemergence spray when creeping speedwell is growing vigorously. Spring or early fall applications are usually suitable, provided temperatures are between 65° and 90°F, and there is ample soil moisture. Thorough coverage is essential. Delayed control of creeping speedwell is to be expected and will be manifested first by a gradual loss of color and vigor.

Nursery Stock: DACTHAL W-75 can be applied to a wide range of nursery stock at the rate of 14 to16 pounds in 50 to 100 gallons of water per acre of area treated (1 pound in 3–7 gallons of water/3000 square feet).

Applications should be made to soil recently cultivated to a uniform texture. These can be made immediately following lining-out of stock. With established plantings, application should be made following proper cultivation to remove existing weeds. Where possible, this should be done early in the spring. Late summer application may prove beneficial for control of fall germinating weeds, if made following cultivation. Weed control up to three months or more may be expected following proper application.

FOR USE ON TURF:	TO CONTROL	WHEN TO APPLY	RATE OF DACTHAL W–75	REMARKS
Established turf	Crabgrass	Early Spring, before crabgrass and other Spring annual weed seeds germinate.	14 lbs./Acre or 1 lb./3 to 7 gals./3000 sq. ft	Apply before Forsythia blossoms start to fall. A second application at half the recommended rate may be made 2 months after the first application if necessary.
	Annual bluegrass (*Poa annua*)	Late summer or early fall in Northern states before weed seed germination. Consult agricultural experiment station or extension weed specialist for date **Poa annua** germinates in your area and make application at least two weeks prior to this date.	20 lbs./Acre or ½ lb./1 to 2 gals./1000 sq. ft.	Should be used following an early spring application for crabgrass control.
	Creeping Speedwell (*Veronica filiformis*)	As a postemergence spray in spring or early fall when creeping speedwell is growing vigorously, there is ample soil moisture and temperature is between 70° and 90°F.	16 lbs./Acre or ⅓ lb./1 to 2 gals./1000 sq. ft.	Apply spray to obtain thorough coverage of foliage. Delayed control of creeping speedwell, approximately 30 days, is to be expected and will be noticed first by a gradual loss of color and vigor.
	Spotted Spurge or Prostrate Spurge (*Euphorbia maculata, E. Supina*)	Mid-April or before seed germination of annual weeds. Repeat application in 6 to 8 weeks.	14 to 16 lbs./Acre or 5 to 6 ounces/1 to 2 gals./1000 sq. ft.	Apply first application before germination of annual weed seeds, such as crabgrass. Repeat application at the same rate 6 to 8 weeks later. In addition to control of spotted spurge or prostrate spurge, crabgrass control is also obtained.
Newly seeded turf	Crabgrass	After new turfgrass seedlings have exhibited a greening of the newly-sprouted grass (about 1 to 2 inches high)	14 lbs./Acre or 1 lb./3 to 7 gals./3000 sq. ft.	Allows crabgrass control without injury to the new turf.

Caution: After **DACTHAL** treatment, wait approximately 60 days before any new seeding is undertaken.
DO NOT use on Dichondra
See recommendations on bentgrass before using on this species.

DACTHAL W-75 is specifically recommended for weed control with the following:

Abelia
Ageratum
Alyssum
Andromeda, Jap.
Arborvitae
Ash
Aster
Azalea
Babysbreath
Feverfew
Fir
Forget-me-not
Forsythia
Four-O'Clock
Foxglove
Gaillardia
Geranium
Gladiolus
Nasturtium
Oak
Orpine
Pachistima
Pachysandra
Peony
Petunia
Pine
Pittosporum
Barberry
Bellflower
Birch
Bleedingheart
Bloodleaf
Boxwood
Bugloss
Camellia
Candytufts
Chestnut
Chrysanthemum
Golddust
Goldentuft
Gum
Hawthorn
Heath, pink
Holly
Hydrangea
Honeysuckle
Iris
Ivy, Baltic
Ivy, Boston
Podocarpus
Poker plant
Poplar
Privet
Privet, variegated
Redbud
Rhododendron
Rose
Russian olive
Sage, scarlet
Snapdragon

Cinquefoil
Coleus
Columbine
Coneflower, purple
Coreopsis
Corabells
Cosmos
Cotoneaster
Cottonwood
Crabapple
Cuphea
Cypress
Dahlia
Delphinium
Deutzia
Dogwood
Elaeagnus
Elm
Euonymus
Evening primrose
Ivy, English
Ivy, Wilson
Juniper
Lantana
Larkspur, candle
Lavendercotton
Lilac
Lily
Locust
Lupine
Magnolia
Maple
Marguerite, golden
Marigold
Mockorange
Morningglory
Moss, rose
Mother-of-Thyme
Mountain Laurel
Mourning Bride
Spiderwort, Va.
Spirea
Spruce
Stonecrop
Strawflower
Sundrops
Sunflower
Sweet pea
Sycamore
Tree peony
Tulip tree
Viburnum
Violet, African
Walnut
Weigela
Willow
Wormwood
Yarrow, fernleaf
Yew
Zinnia

Do not use on the following plants:

Bugleweed
Button pink
Carnation
Geum
Germander
Mesembryanthemum
Pansy
Phlox
Sweet William
Telanthera

Amount of DACTHAL W-75 to Use in Treating Small Areas

Area to be treated	Recommended rate of DACTHAL W-75 lb./A	Amount of DACTHAL W-75 to Apply	Struck-off (level) Tablespoons
1000 sq. ft. (From	14 (heavy soil)	5 oz.	22
1 to 2 gallons of	12	4½ oz.	19
water should be	10 (medium soil)	3½ oz.	16
used to spray	8	3 oz.	13
1000 sq. ft.)	6 (sandy soil)	2 oz.	10

Amount of DACTHAL W-75 to Use in Treating Small Areas

Area to be treated	Recommended rate of DACTHAL W-75 lb./A	Amount of DACTHAL W-75 to Apply	Struck-off (level) Tablespoons
250 sq. ft. (From	14 (heavy soil)	1¼ oz.	5½
¼ to ½ gallon of	12	1 oz.	4½
water should be	10 (medium soil)	¾-1 oz.	4
used to spray	8	¾ oz.	3
250 sq. ft.)	6 (sandy soil)	½ oz.	2½

Note: In treating small areas it is important to use care in measuring out required quantity for area to be treated, properly mixing it in water, and keeping it adequately agitated during application.

* Other size areas to be treated can be calculated from the above table.

For example:

To treat 5000 sq. ft. at the 10 lbs./acre rate, use 80 tbsp. (5 × 16) in 5 to 10 gallons of water (5 × 1 × 2).

To treat 600 sq. ft. at the 12 lbs./acre rate, use 11.4 tbsp. (0.6 × 19) in 0.6 to 1.2 gallons of water (0.6 × 1 to 2).

Rates of application can be reduced by banding.

See conversion chart.

Conversion Chart for DACTHAL W-75
For a 12-inch band of spray, use this amount of **DACTHAL W-75** per acre.

Row Width	If overall rate is: 6 lbs. per acre	8 lbs. per acre	10 lbs. per acre	12 lbs. per acre
24 inches	3 pounds	4 pounds	5 pounds	6 pounds
32 inches	2¼ pounds	3 pounds	3¾ pounds	4½ pounds
36 inches	2 pounds	2⅔ pounds	3⅓ pounds	4 pounds
40 inches	1¾ pounds	2½ pounds	3 pounds	3⅔ pounds

Special Precautions: DO NOT feed treated foliage to livestock or graze treated areas. Applied according to directions and under conditions favorable to good plant growth, **DACTHAL W-75** will not harm crops for which its use is recommended. However, conditions such as high salt concentration, seedling disease, cold weather, deep planting, excessive moisture or drought may injure or weaken crops normally tolerant to **DACTHAL W-75,** thereby increasing the possibility of herbicide damage. Under any of these conditions, one or more of the following may result: delayed crop development, reduced yields, or reduced quality.

STORAGE AND DISPOSAL

DO NOT contaminate water, food or feed by storage or disposal. Open dumping is prohibited.

STORAGE

Store in a dry place.

PESTICIDE DISPOSAL

Wastes resulting from the use of this product may be disposed of on site or at an approved waste disposal facility.

CONTAINER DISPOSAL

Completely empty bag into application equipment and dispose of empty bag in a sanitary landfill or by incineration, or, if allowed by state and local authorities, by burning. If burned, stay out of smoke.

WARRANTY AND LIMITATION OF DAMAGES

Seller warrants that this material conforms to its chemical description and is reasonably fit for the purposes stated on the label when used in accordance with directions under normal conditions of use and Buyer assumes the risk of any use contrary to such directions. SELLER MAKES NO OTHER EXPRESS OR IMPLIED WARRANTY, INCLUDING ANY OTHER EXPRESS OR IMPLIED WARRANTY OF FITNESS OR OF MERCHANTABILITY, AND NO AGENT OF SELLER IS AUTHORIZED TO DO SO EXCEPT IN WRITING WITH A SPECIFIC REFERENCE TO THIS WARRANTY. In no event shall Seller's liability for any breach of warranty exceed the purchase price of the material as to which a claim is made.

SPECIAL LOCAL NEEDS PURSUANT TO SECTION 24(c) OF AMENDED FIFRA AND STATE REGISTRATIONS

CROP	REGISTRATIONS	STATES	SLN OR REG. NO.	DIRECTIONS FOR USE
Alfalfa	Control of dodder on Alfalfa grown only for seed purposes	Colorado Idaho Nevada New Mexico Oklahoma Oregon Utah Washington	3977 3978 3979 3950 3981 3982 3983 3984	Apply **DACTHAL W-75** Herbicide uniformly to the soil at the rate of 10 to 14 lbs. per acre in the early spring before weed seeds germinate. It is essential that good coverage of soil surface be obtained. If a first cutting of hay is to be made, **DO NOT** apply **DACTHAL** until immediately after cutting, but application should be made as soon as the hay has been removed and before regrowth occurs (not longer than 10 days after cutting). **DACTHAL** will not control germinated weeds. After **DACTHAL** has been applied, the crop must not be cut for hay. **DO NOT** graze or feed hay, plant refuse, threshings or seed screenings to livestock from alfalfa fields treated with **DACTHAL**.

CROP	REGISTRATIONS	STATES	SLN OR REG. NO.	DIRECTIONS FOR USE
Corn	Control of witchweed in corn interplanted to field beans, snap beans, southern peas or soybeans	North Carolina	NC–780025	**DACTHAL** should be applied uniformly to the soil at a rate of 8 to 14 lbs. in 20 to 30 gallons of water per acre. It should be applied directed postemergence to field corn 18 inches or greater in height before emergence of the witchweed. Use drop nozzles between the rows equipped with whirl jet nozzles (B-type), Flood Jet Nozzles (K-type) or tee jet nozzles equipped with Flood Jet tips (TK-type). If weeds have emerged, soil should be clean cultivated or weeded prior to application to remove weeds which might intercept spray material. **DO NOT** graze treated areas or feed plant refuse to livestock.
Corn	Control of witchweed in corn interplanted to southern peas or soybeans	South Carolina	SC–780015	**DACTHAL** should be applied uniformly to the soil at a rate of 8 to 14 lbs in 20 to 30 gallons of water per acre. It should be applied directed postemergence to corn 18 inches or greater in height before emergence of the witchweed. Use drop nozzles between the rows equipped with whirl jet nozzles (B-type), Flood Jet Nozzles (K-type) or tee jet nozzles equipped with Flood Jet tips (TK-type). If weeds have emerged, soil should be clean cultivated or weeded prior to application to remove weeds which might intercept spray materials. **DO NOT** graze treated areas or feed plant refuse to livestock.
Radish	Application on radish grown for seed	Washington	WA–820004	Apply uniformly to the soil as a spray. Incorporate **DACTHAL** thoroughly into the soil to a depth of 3 to 4 inches using a rotary tiller or double disk. **DO NOT** allow seed screenings to be used for feed. **DO NOT** graze on treated areas or feed plant refuse to livestock.
Upland Cress	Control of crabgrass and other annual grasses & certain broadleaved weeds in upland cress	North Carolina Tennessee	NC–810030 TN–810022	Apply uniformly to the soil as a spray at time of seeding. Can be preplant incorporated.

DALAPON 85

HERBICIDE

ACTIVE INGREDIENTS:
2,2 Dichloropropionic Acid,
Sodium Salt of Dalapon* 85%
INERT INGREDIENTS: 15%
TOTAL 100%

*(2,2 Dichloropropionic Acid equivalent 74%)

EPA Reg. No. 50534-55

WARNING — AVISO

Keep Out of Reach of Children

PRECAUCION AL USUARIO: Si usted no lee ingles, no use este producto hasta que la etiqueta haya sido explicado ampliamente.

See precautionary statements.

PRECAUTIONARY STATEMENTS

HAZARDS TO HUMANS AND DOMESTIC ANIMALS
CAUTION

Avoid contact with eyes. Causes eye irritation. In case of contact, flush eyes with plenty of water and get medical attention. Avoid contact with skin or clothing. May cause skin irritation. In case of contact, remove contaminated clothing and wash skin with plenty of water. Wash clothing before reuse. **DO NOT** apply this product in such a manner as to directly, or through drift, expose workers or other persons. The area being treated must be vacated by unprotected persons.

ENVIRONMENTAL HAZARDS

DO NOT feed treated foliage to livestock or graze treated areas. **DO NOT** spray or allow drift onto vegetables, ornamentals or other desirable plants. **DO NOT** apply when weather conditions favor drift from target area. Keep out of lakes, streams or ponds. **DO NOT** contaminate water by cleaning of equipment or disposal of wastes.

GENERAL INFORMATION

Use DALAPON 85 for control of cattails and annual and perennial grasses such as:

Alexandergrass	Foxtails	Paragrass
Barnyardgrass	Guineagrass	Phragmites
Bluegrass	Johnsongrass	Quackgrass
Bermudagrass	Maidencane	Wild Oats
Cattails	Pangolagrass	Witchgrass
Crabgrass		

A pre-plant application of 8 to 10 lbs. of DALAPON 85 per acre followed by plowing or deep disking has generally controlled established perennial grasses. Specific directions are given below under Pre-Plant Applications. Follow directions carefully especially concerning time interval between plowing or deep disking treated grasses and crop seeding. Spot treatments of perennial grasses using repeated applications at lower rates are also effective.

Postemergence application, broadcast or directed, has generally controlled seedling grasses in crops. Specific directions for use are cited under each crop noted. Follow directions carefully.

Mode of Action: DALAPON 85 is absorbed primarily through the leaves and, therefore, is most effective when applied as a foliar spray. Best results are experienced when grasses are in a vigorous stage of growth, but prior to seed head formation. Several weeks may be required to observe maximum effect. Repeat application may be necessary to maintain control of perennial grasses. Control may be poor if grasses are not growing well at time of treatment due to drought, cold weather and over-maturity. DALAPON 85 disappears in the soil in 4 to 6 weeks due to soil degradation by microorganisms and leaching by rainfall, permitting new seedlings to develop in the treated area. Breakdown of DALAPON 85 requires a longer period of time in cold or dry soil. Under these conditions more time will be required after application before planting sensitive crops. Therefore, DALAPON 85 applications should be repeated as necessary but in accordance with **Directions for Use** described below.

DIRECTIONS FOR USE

It is a violation of Federal law to use this product in a manner inconsistent with its labeling.

Mixing: Fill spray tank at least half full before adding DALAPON 85. Add DALAPON 85 slowly while mixing to dissolve completely. Use sufficient water (10 to 40 gallons per acre) to allow good spray coverage. DALAPON 85 contains a wetting agent which is adequate for most uses. However, for certain perennial grasses that are difficult to wet or when DALAPON 85 is applied at low rates, especially in highly dilute sprays, the addition of a non-ionic wetting agent in the proportion of 2 to 3 pints per 100 gallons of spray may be helpful in obtaining maximum control. **Combination Sprays:** Observe cautions and directions for use on labels of all products used in combination with DALAPON 85. When used in combination with other herbicides, including 2,4-D type materials, completely dissolve DALAPON 85 in the spray tank before the addition of these products. **DO NOT** use more than 8 lbs. of DALAPON 85 in 10 gallons of water when mixing with 2,4-D.

Aerial Applications: Use the suggested rates of DALAPON 85 in 5 to 10 gallons of water per acre for pre-plant treatment of croplands.

CROP	APPLICATION	DALAPON 85 LBS./A	TREATMENT	FOOT-NOTES
Apples and Pears	Directed under trees: broadcast, banded or spot	5–10	For control of most annual grasses and certain established perennial grasses in orchards under 4 years old, use 5 lbs. per acre broadcast and 10 lbs. per acre broadcast in orchards at least 4 years old. Use sufficient water for thorough, uniform coverage. As a directed spray, apply in a 4 to 8 ft. band on each side of the tree row. Spot treatment may be used. For band or spot treatment, reduce amount of DALAPON 85 and water in proportion to the area treated. Apply to grass 6 to 10 inches tall and growing well. Repeat as needed. Two applications 2 to 4 weeks apart usually provide adequate control of most grasses. Avoid spray contacting crop foliage and fruit. **DO NOT** apply more than 2 applications per season. **DO NOT** spray within 30 days of harvest.	2
Apricots	Directed under trees	6–7	WESTERN AREAS—Apply as a spot, band, or broadcast spray.	1,2
Asparagus	Before and after cutting	10–20	QUACKGRASS: Use 10 lbs. per acre before cutting and 3 to 4 weeks later as required. If a single application is preferred, use 20 lbs. per acre during or after the cutting season.	
		10–20	BERMUDAGRASS: Use 20 lbs. during or after the cutting season and a second treatment of 10 lbs. as needed. NOTE: During the cutting season, apply immediately after cutting. Later, sprays should be directed under the ferns. Use enough water to enable uniform wetting of the grass. For spot treatment of patches of grass, use 1 lb. in 10 gallons of water.	
Beans, Field kidney, lima, snap	Spring preplant	8	QUACKGRASS: Apply in spring when grass is 4–6 inches high. Wait 4 days and then plow or deep disc. Wait 4 to 5 weeks following plowing or deep discing before planting.	2,3,4
	Fall preplant	15	QUACKGRASS: Apply in fall (usually September or October) on growing grass. Plow or deep disc 10 days after application or the following spring before planting.	2,3
Banana	Directed under crop plants	10–15	GUINEA, AMARGO AND GAMELOTS: Apply when grasses are growing well but before heading. Repeat treatments, usually spot applications, may be made 4–6 weeks or later as warranted by regrowth. **DO NOT** exceed 30 lbs. of DALAPON 85 per acre. Often rates of 6–8 lbs. of DALAPON 85 per acre has given satisfactory results.	10
Birdsfoot Trefoil (grown for seed production only)	Spring or Fall	3–5	VERMONT AND NEW YORK ONLY: For control or suppression of quackgrass and other perennial grasses, annual grasses and clovers. Apply 20–30 gallons of water per acre when weeds are 4–6 inches long. May cause temporary crop injury but seed yield and quality are usually increased.	2,10
Citrus Grapefruit, lime, orange, tangerine	Directed under trees at least 4 years old	2–4	FLORIDA AND TEXAS: For control of such grasses as bermudagrass, guineagrass, johnsongrass, maidencane, pangolagrass or paragrass, use 2 to 4 lbs. per acre in 50 to 75 gallons of water as a broadcast spray directed to grasses growing on orchard floor or apply a spot treatment not to exceed 4 lbs. per acre using 1 lb. per 20 gallons of water. Spray to wet grass without runoff. Apply when grasses are growing well, but prior to seed head formation. Repeat applications at 2 to 10 day intervals for a maximum of 3 treatments per season. NOTE: **DO NOT** use on newly transplanted citrus trees or trees less than 4 years old. **DO NOT** apply to bare soil under citrus trees. Keep spray off the trees and **DO NOT** allow grazing on treated areas. **DO NOT** apply in groves affected by disease, drought, cold damage or groves in decline. Use only the lower rates when spraying grasses growing in very light soil.	
	Grove ditches and non-grove areas	5–10	Apply 5 to 10 lbs. per acre using about 5 lbs. per 100 gallons of water and spray to wet foliage uniformly. Make two applications two weeks apart for control of guineagrass, phragmite, paragrass or cattails. For overall control of grasses, cattails, weeds and woody plants. DALAPON 85 can be combined with 2,4–D. (See DIRECTIONS FOR USE concerning use of DALAPON 85 in combination with 2,4–D.) **DO NOT** allow grazing on treated areas.	
Coffee	Directed under crop plant	10	HAWAII ONLY: For control of crabgrass, bermudagrass, kikuyagrass, para (California) grass and other grasses in established coffee plantations. Apply at rate of 1 lb. DALAPON 85 in 10 gallons of water and when grasses are growing well and before seed head formation. **DO NOT** exceed 100 gallons of spray per acre. Respray 4–6 weeks later if regrowth warrants. **DO NOT** make more than 2 sprays in any one crop year and **DO NOT** apply within 8 months of harvest.	10
Cotton	Spot treatment	—	SPOT TREATMENT TO CONTROL PERENNIAL GRASSES: Use 1 lb. in 10 gallons of water and spray to wet grass without runoff. Apply when grasses are growing well. Avoid direct contact on cotton foliage as crop injury may result.	
	Layby directed spray	5	LAYBY DIRECTED SPRAY TO CONTROL ANNUAL GRASSES: Use 5 lbs. per acre in 50 gallons of water. Cover entire soil area with a directed spray to wet the grass. Spray after the last cultivation when grass seedlings have emerged but are still small.	
	Preplant application	8	See SPRING TREATMENT TO CONTROL JOHNSONGRASS.	

CROP	APPLICATION	DALAPON 85 LBS./A	TREATMENT	FOOT-NOTES
Corn	Postemergence Directed Spray	2	For control of seedlings of yellow, green and giant foxtail, barnyardgrass, witchgrass, wild oats, or large (hairy) crabgrass and annual broadleaved weeds such as pigweed and ragweed. Use 2 lbs. of DALAPON 85 and 1/2 lb. of 2,4–D per acre in a 14–inch band when crop is 8–20 inches tall (soil to whorl). (See DIRECTIONS FOR USE concerning use of DALAPON 85 in combination with 2,4–D.) Avoid spray contact on corn foliage by using a leaf lifter and directing the spray. Spray only the lower part of the corn stalks below the lifted leaves and in no case higher than 7 inches above the ground. Apply carefully with emphasis on sprayer calibration, spray pattern width, height of spray on corn plant and rate of DALAPON 85. Best results are obtained by spraying when grasses and weeds are not as tall as the corn. Use of a rotary hoe soon after corn emergence will delay weed and grass growth so that the corn will be taller than grasses and weeds. Treated grasses and weeds will be stunted and die back reducing or eliminating water and nutrient competition with the corn plant. Quick kill should not be expected. **DO NOT** use on corn grown for seed.	
	Preplant Application	8	See SPRING TREATMENT TO CONTROL JOHNSONGRASS.	
Cranberries	Early June	20	DITCH AND SHORE: For control of cattails, cutgrass, povertygrass, burreed, common rushes and sedges in irrigation and drainage ditches and along shore areas of cranberry bogs. Apply DALAPON 85 in 200–300 gallons of water per acre in early June when weeds are growing well and before seed head formation. Apply second application if needed but before July 31. Avoid spraying cranberry foliage.	
Cranberries	After harvest in October and early November	10	VINE TREATMENT: Apply DALAPON 85 in about 300 gallons of water per acre on cranberry bogs for control of summergrass, cutgrass, povertygrass, wamsy, panicgrass, bunchgrass, wide leaf grasses, round rush and sedges. Use no more than 1/2 pt. surfactant per 100 gallons of spray.	11
Flax	When flax is 1–6 inches tall	1	To control pigeongrass or yellow, green and giant foxtails, use 1 lb. per acre in 5 to 10 gallons of water when flax is 1 to 6 inches tall. For best results, apply when young grass seedlings are less than 2 inches tall and actively growing. **DO NOT** expect rapid grass kill. MCP Amine can be used with DALAPON 85 where broadleaf weed control is desired. (See DIRECTIONS FOR USE concerning use of DALAPON 85 with other herbicides).	
Grapes	Directed spray	10	CALIFORNIA AND ARIZONA: For control of many annual grasses and certain established perennial grasses, use 10 lbs. per acre broadcast. Use 100 gallons of spray per acre treated. Apply when grasses are growing vigorously and before heading. As a directed spray, apply in a band 30 inches wide centered on the row. Reduce amount of DALAPON 85 and volume of water proportional to reduction in area sprayed. Repeat application if needed, but **DO NOT** apply more than twice in a single growing season. **DO NOT** apply within 30 days of harvest. Keep spray out of irrigation furrows. **DO NOT** graze livestock on treated areas. PRECAUTIONS TO AVOID GRAPE INJURY: Avoid spray contacting grape foliage and fruit. **DO NOT** use on young non-bearing orchards. *DO NOT* exceed dosages given. French-plowing (Kirby plow) before spraying is not advisable as grass emergence is delayed and second application cannot be made wihout spraying vine foliage.	
Macadamia nuts	Directed spray under trees	10	HAWAII: To control crabgrass, bermudagrass, kikuyugrass, para (California) grass, dallisgrass. Apply at rates up to 10 lbs. per acre in sufficient water to wet the grass without runoff. **DO NOT** exceed 100 gallons of spray per acre. Make first application before harvest and respray at intervals needed to control regrowth. **DO NOT** exceed 24 pounds per acre. CAUTION: During harvest, spray only immediately after gleaning nuts from the ground. **DO NOT SPRAY DIRECTLY ON FALLEN NUTS.**	10
Pasture	Broadcast or spot application	4–6	For control of established perennial grasses such as pangola-, bermuda-, bahia-, dallis-, and fescue grasses. Apply DALAPON 85 in 100 gallons of water per acre when grasses are growing well.	3, 4, 8, 9, 13, 14, 15, 16
			GULF COAST STATES, INCLUDING GEORGIA: For smutgrass control, apply 4–6 pounds DALAPON 85 per acre broadcast. As a spot treatment, use 1 lb. DALAPON 85 per 16 gallons of water, Wet smutgrass foliage thoroughly.	4, 8, 9, 12, 13, 15, 16
Peaches	Directed under trees	6–7	WESTERN AREAS: Apply as a spot, band, or broadcast spray.	1, 2
		2 to 3½	OTHER AREAS: Apply 1 lb. DALAPON 85 per acre in 20 gallons of water **DO NOT** exceed 70 gallons of spray per acre.	1, 2
Peas	When peas are 3–6 inches tall	1	For control of seedling grasses, use 1 lb. per acre. Apply when seedlings are small and peas are 3–6 inches tall and have 4 to 6 nodes. **DO NOT** apply within 25 days of harvest. **DO NOT** feed pea vine hay to livestock.	
Pineapples	Spot treatment	10	To control annual and perennial grasses, use 10 lbs. per acre as a spot treatment. Avoid spraying pineapple foliage. Make application at least one month before harvest.	
Plums (Prunes)	Directed under trees	6–7	WESTERN AREAS: Apply as a spot, band, or broadcast spray.	1, 2
Potatoes	Preemergence	3–10	To control seedling grasses, use 3 lbs. per acre in adequate water for good coverage. Apply before emergence of potatoes when grasses are small. Where quackgrass comes up after planting, but before potatoes emerge, use 10 lbs. per acre and wet grasses well. NOTE: **DO NOT** apply after potato plants emerge. **DO NOT** use on land planted to red skinned varieties or White Rose.	6

CROP	APPLICATION	DALAPON 85 LBS./A	TREATMENT	FOOT-NOTES
	Spring Preplant	8	For quackgrass control, use 8 lbs. per acre. After 4 days, land can be plowed and potatoes planted immediately, if normal soil moisture is present.	6
	Late Season Grass Control	5	Apply 5 lbs per acre in 15 to 25 gallons of water after lay-by when grass seeds have germinated but before grass seedlings are 3 inches tall. Use a vine lifter and direct the spray to cover row tops and middles and minimize spray contact on potato foliage.	6
Sorghum	Spring preplant	5–7	For johnsongrass control, spray when grass is 8–12 inches tall. Wait 3 days and plow or deep disc. Wait at least 2½ weeks before planting.	2, 3, 4, 5
Soybeans	Spring preplant	5–7	For johnsongrass control, spray when grass is 8–12 inches tall. Wait 3 days and plow or deep disc. Wait 5 days after plowing or discing before planting.	2, 3, 4, 5
Sugar Beets	Emergence to 6-leaf stage	3–8	To control small grass seedlings, use 3 to 4 lbs. per acre broadcast. For wild oats and quackgrass control apply 6 to 8 lbs. per acre broadcast. These high rates may cause crop stunting and reduce yields. To minimize this effect, rate applied per acre may be reduced by using directed sprays.	
	Beets 6–leaf state to 14–inches tall	3–4	To control annual grasses, apply as a directed spray at 3 to 4 lbs. per broadcast acre. Apply with leaf lifting equipment to remove leaves from the spray pattern and minimize damage to crop foliage. NOTE: Application may be made as a broadcast or directed spray. In directed treatments, the amount of DALAPON 85 and volume of water should be reduced in proportion to reduction in area sprayed. **DO NOT** use more than 8 lbs. of DALAPON 85 postemergence per acre in a single growing season.	
Sugar Cane	Preemergence or directed postemergence	4–10	To control such grasses as alexandergrass, bermudagrass and johnson grass, use 4 to 10 lbs. per acre broadcast as a preemergence or directed postemergence application. Reduce amount of DALAPON 85 per acre and volume of water, proportional to reduction of sprayed area in band or drill application. Directed sprays designed to wet the grass thoroughly and uniformly and the cane as little as possible are suggested.	
Fallow or stubble land	Preplant	5–7	Use in grain stubble or on fallow land to control quackgrass, bermudagrass, and johnsongrass during the growing season prior to planting beans, corn, cotton, potatoes, sorghum, soybeans, and sugarbeets.	2, 3, 5, 6, 7

These footnotes are an Important Part of the Use Directions

1. Apply when grasses have begun growing well in spring, but before heading. Spray to wet the grass without runoff. Apply when soil moisture is high such as after a rain or irrigation. Use one or two applications per season. For band or spot treatment, the amount of spray and DALAPON 85 should be reduced in proportion to the area actually treated. **DO NOT** spray bare soil under fruit trees or vines. Keep spray off crop foilage and fruit. **DO NOT** apply within 30 days of harvest.
2. **DO NOT** graze livestock on treated areas during application season or feed forage, hay, or grass.
3. Apply by air or ground equipment.
4. If soil remains unusually dry after preplant treatment, a longer waiting period may be necessary to avoid residue injury to the crop. A rain of half an inch or more will help to move the herbicide into the soil so that microbial degradation (breakdown) will take place. Wait 5 days after the rain before planting susceptible crops.
5. This treatment will not control unemerged seedling grasses.
6. **DO NOT** use on land planted or to be planted to red-skinned varieties or White Rose. **DO NOT** spray potato plants.
7. To control established bermudagrass or johnsongrass on grain stubble or fallow land, use 5–7 pounds per acre in 2 or 3 applications 2–3 weeks apart, or as needed. Use tillage prior to planting the following spring on such crops as beans, corn, cotton, potatoes, sorghum, soybeans and sugarbeets.
8. Best results obtained if mowed 2–4 weeks after treatment and fertilized to hasten pasture recovery.
9. Pastures recover most rapidly if mowed, adequately fertilized, ample moisture is made available and other good management practices are followed. Desirable perennial grasses and forbs (such as white clover) may be temporarily reduced in stand or vigor but spring treated pastures usually recover within 2–4 months under favorable conditions. Fall treated pastures usually show excellent recovery the following spring under good management and favorable conditions. Any annual grasses present may be seriously injured or killed.
10. Avoid spraying crop. Use sufficient spray to wet grass foliage thoroughly but avoid runoff.
11. This treatment may reduce following season's crop: e.g., yield from variety. Howes may be reduced only slightly, whereas on Early Blacks and certain other varieties, a reduction of 75–90% may occur. Follow recommendations of agricultural experiment station or weed extension specialists.
12. When less than 10% of the pasture area is treated, grazing need not be discontinued.
13. **Grazing Restrictions**

    Pastures Unmowed: **DO NOT** graze meat animals on treated pastures until 8 weeks after treatment and withhold from such grazing for 1 week before slaughter. Fourteen weeks after treatment no withholding period is necessary. **DO NOT** graze lactating dairy animals on treated pastures until 14 weeks after treatment.

    Pastures Mowed: Mow 2–4 weeks after treatment, if desired. **DO NOT** graze lactating or meat animals on treated pastures until 4 weeks after mowing. **DO NOT** havest hay from treated pastures during the season of application.
14. Use only the lower rate of DALAPON 85 on bahiagrass pastures since it is more sensitive than other desirable pasture grasses and may suffer considerable reduction under unfavorable conditions.
15. **DO NOT** sow grass seed until 2–4 weeks after treatment. Species such as annual ryegrass can be fall seeded to provide winter pasture. Defer grazing for 10 weeks or longer after treatment allowing adequate time for sufficient growth to sustain grazing.
16. Apply only once per year.

PRE-PLANTING APPLICATIONS

Spring Treatment to Control Johnsongrass: Use 8 lbs. of DALAPON 85 per acre on land to be planted to corn, cotton, sorghum or soybeans. Often rates of 5 to 7 lbs. of DALAPON 85 will provide effective contol. Spray when grass is 8 to 12 inches high and growing vigorously. The treated area can be plowed or deep disked 3 days following application. One or more diskings is recommended prior to planting. Cotton may be planted 10 to 14 days after spraying. Wait at least 17 days after plow down before planting corn, sorghum and soybeans. If soil remains unusually dry after plowing or deep disking, a longer period will be necessary to avoid crop injury. Under such conditions delay planting these crops at least 5 days after a minimum of 0.5 inch rainfall.

Spring Treatment to Control Quackgrass: Use 8 lbs. of DALAPON 85 per acre as an early spring application when quackgrass is 4 to 6 inches tall and growing vigorously. Wait at least 4 days after spraying before plowing down. Beans, beets, corn and potatoes can be planted 4 to 5 weeks after plow down.

Fall Treatment to Control Quackgrass: Use 15 lbs. of DALAPON 85 per acre on land to be planted in beans, beets, corn and potatoes. Application can be made in September, October or possibly later in some areas. Quackgrass should be growing vigorously when treated. Treated areas can be plowed in 10 to 20 days or delayed until spring. With established sods more effective control will be obtained if the sod is first plowed and quackgrass regrowth sprayed with DALAPON 85.

NON-CROP USES

Grass Control in Drainage Ditches and Non-Agricultural Lands (Industrial Areas, Railroads, Roadsides, Fence Rows, Sidewalks, etc.): When using a spray boom, apply 15 lbs. of DALAPON 85 per acre in a minimum of 40 gallons of water. Repeat application in 2 to 4 weeks and follow up with spot treatments as needed. For application using an orchard-type spray gun on vegetation such as johnsongrass, cattails and phragmites, use 10 lbs. of DALAPON 85 in 100 gallons of water and wet all foliage thoroughly. Spray grasses prior to heading. Cattails and phragmites are more effectively controlled by spraying when new sprouts are 2 to 3 feet tall. In western areas, fall spraying of cattails during catkin development is effective. Spray grass regrowth using 10 to 15 lbs. of DALAPON 85 per acre in 100 gallons of water.

Spot Treatment: Use 1 lb of DALAPON 85 in 10 to 20 gallons of water. Spray foliage thoroughly. Repeat as required.

Repeat Treatments with Low Rates for Johnsongrass and Bermudagrass: Use 5 to 7 lbs. of DALAPON 85 in 40 to 50 gallons of water. Make 2 to 3 applications at 10 to 20 day intervals when grass is 8 to 16 inches tall and growing vigorously. If land is to be planted to a late fall or spring crop, delay tillage 2 weeks after the last application.

Seed Grass Control: Use 5 lbs. of DALAPON 85 per acre in 50 gallons of water. Apply prior to heading and wet foliage thoroughly.

RE-ENTRY

DO NOT enter treated areas without protective clothing until sprays have dried. Because certain states may require more restrictive reentry intervals for various crops treated with this product, consult your State Department of Agriculture for further information.

Written or oral warnings must be given to workers who are expected to be in a treated area or in an area about to be treated with this product. When oral warnings are given, warning shall be given in a language customarily understood by workers. Oral warnings must be given if there is reason to believe that written warnings cannot be understood by workers. Oral and written warnings must include the following information: WARNING, Area treated with **DALAPON 85** on (date of application). Do not enter without appropriate protective clothing until sprays have dried. In case of accidental exposure, wash area with plenty of water and get medical attention. For further information see "Precautionary Statements" on the label.

STORAGE AND DISPOSAL

DO NOT contaminate water, food or feed by storage or disposal.

PESTICIDE DISPOSAL

Wastes resulting from the use of this product may be disposed of on site or at an approved waste disposal facility.

CONTAINER DISPOSAL

Completely empty bag into application equipment. Then dispose of empty bag in a sanitary landfill or by incineration, or, if allowed by state and local authorities, by burning. If burned, stay out of smoke.

WARRANTY AND LIMITATION OF DAMAGES

Seller warrants that this material conforms to its chemical description and is reasonably fit for the purposes stated on the label when used in accordance with directions under normal conditions of use and Buyer assumes the risk of any use contrary to such directions. SELLER MAKES NO OTHER EXPRESS OR IMPLIED WARRANTY, INCLUDING ANY OTHER EXPRESS OR IMPLIED WARRANTY OF FITNESS OR OF MERCHANTABILITY, AND NO AGENT OF SELLER IS AUTHORIZED TO DO SO EXCEPT IN WRITING WITH A SPECIFIC REFERENCE TO THIS WARRANTY. In no event shall Seller's liability for any breach of warranty exceed the purchase price of the material as to which a claim is made.

DSMA

LIQUID HERBICIDE

DSMA Plus Surfactant For Postemergence Weed Control
Contains 3.6 Pounds Disodium Methanearsonate Hexahydrate Equivalent Per Gallon

ACTIVE INGREDIENT:

Disodium Methanearsonate	21.8%
INERT INGREDIENTS:	78.2%
TOTAL	100.0%

Total Arsenic, all in water-soluble form, expressed as elemental 8.9%

EPA Reg. No. 50534-27

CAUTION

Keep Out of Reach of Children

See precautionary statements.

PRECAUTIONARY STATEMENTS

HAZARDS TO HUMANS AND DOMESTIC ANIMALS
CAUTION

Harmful if swallowed or absorbed through the skin. Avoid contact with eyes, skin or clothing. Avoid breathing vapors or spray mist.

If swallowed, induce vomiting immediately by giving two glasses of water and sticking finger down throat. **Never** give anything by mouth to an unconscious person. Seek medical attention.

In case of contact, flush eyes with plenty of water for at least 15 minutes. Seek medical attention.

Wash exposed skin gently with plenty of soap and water.

Keep children and domestic animals off treated areas until this material has been washed into the soil.

Wear protective clothing when handling or applying this product including long pants, long-sleeve shirt, and impermeable gloves and boots. Mixer-loaders should include an apron and full-face shield when handling or mixing concentrate. Flagmen should be fully protected during spray operations or mechanical flagmen used. Pilots and ground spray rig applicators should wear a mask or respirator approved by the Mining Enforcement and Safety Administration and the National Institute for Occupational Safety and Health.

ENVIRONMENTAL HAZARDS

DO NOT feed treated foliage to livestock or graze treated areas. **DO NOT** spray or allow drift onto vegetables, ornamentals or other desirable plants. **DO NOT** apply when weather conditions favor drift from target area. **DO NOT** apply directly to water. **DO NOT** contaminate water by cleaning of equipment or disposal of wastes.

DSMA Liquid

Read entire label before using this product.

GENERAL INFORMATION

DSMA Liquid is a postemergence herbicide for use in cotton, turf and non-crop areas. DSMA Liquid contains a surfactant (wetting agent) in proper ratio for best resuts. It is not necessary to add additional surfactant to the spray tank. Local conditions and recommendations vary; consult local agricultural experiment station or extension service weed specialists for recommendations in your area.

Mixing Instructions: DSMA Liquid herbicide must be thoroughly dissolved. Fill the spray tank with half the required amount of water and, with pump or agitator operating, add the recommended amount of DSMA Liquid herbicide and continue filling the tank with balance of water.

DO NOT store spray solution in tank overnight. Clean application equipment thoroughly after use by flushing with water in a safe place.

Application Methods: DSMA Liquid should be applied with a low-volume, properly calibrated sprayer having satisfactory pumping and bypass action. **Adjust Nozzles to Keep Spray Off Cotton Foliage and in a Manner to Allow Maximum Coverage of Weeds.** Adequate coverage is very important for effective weed control.

Apply during warm weather when weeds are in an active stage of growth.

DIRECTIONS FOR USE

It is a violation of Federal law to use this product in a manner inconsistent with its labeling.

COTTON: DSMA Liquid is useful as a directed postemergence spray for controlling weeds in cotton such as barnyardgrass, cocklebur, smooth and large crabgrass, dallisgrass, goosegrass, johnsongrass and nutsedge. Apply as a **Directed** broadcast spray at rate of 40 gallons per acre of a solution containing 1 gallon of DSMA Liquid in 40 gallons of water when weeds are small. For band applications, apply 1 gallon of above diluted spray per acre for each 1 inch band width to be treated of cotton grown on 40 inch row spacing. A second or repeat application, if needed, should be timed about 1 to 3 weeks after the first application. Keep spray off cotton foliage.

Apply Only When Cotton is 3 Inches High to First Bloom. DO NOT Apply After First Bloom. Slight burning and reddish discoloration of cotton foliage may occasionally be seen following recommended treatment; however, cotton plants will develop normally and yield will not be affected.

ORNAMENTAL TURFGRASS: DSMA Liquid can be used for selective control of bahiagrass, barnyardgrass, chickweed, smooth and large crabgrass, dallisgrass, nutsedge, sandbur and wood sorrel with little or no injury to well-established, actively growing turfgrasses. Mow turfgrass to height of 1 to $1\frac{1}{2}$ inches before treatment. Mix 3 fluid ounces (6 tablespoons) of DMSA Liquid in 5 gallons of water for application to an area of 1,000 square feet. Application should be uniform and thorough to adequately wet all undesirable plants. Two or more repeat treatments at 14 day intervals may be necessary. Make applications during warm weather when temperature is between 80 and 90°F. **DO NOT** water turf for at least 24 hours after application. Turfgrasses may be temporarily discolored. Bermudagrass, bluegrasses and zoysiagrass have shown tolerance to properly applied DSMA Liquid. Injury may result if applied to bentgrasses and fescues. **DO NOT** apply to St. Augustinegrass, carpetgrass, centipedegrass or to Dichondra. **DO NOT** reseed until 2 weeks after last application. **DO NOT** apply with hose-end applicators.

NON-CROP: DSMA Liquid is effective in controlling barnyardgrass, cocklebur, smooth and large crabgrass, dallisgrass, goosegrass, johnsongrass, and nutsedge and many similar weeds on drainage ditchbanks, fence rows, rights-of-way, storage yards and similar non-crop areas. Application should be made when weeds are small and conditions are favorable for good weed growth. Mix at a rate of 1 gallon of DSMA Liquid in 40 gallons of water. Spray undesirable vegetation thoroughly to point of run-off. Adequate coverage and complete wetting of foliage is important for effective control. Repeat applications may be necessary if regrowth occurs. **DO NOT** contaminate waters used for domestic purposes, or by animals including wildlife and aquatic life, or water used for irrigation purposes.

STORAGE AND DISPOSAL

DO NOT contaminate water, food or feed by storage or disposal. Open dumping is prohibited.

PESTICIDE DISPOSAL

Wastes resulting from the use of this product may be disposed of on site or at an approved waste disposal facility.

CONTAINER DISPOSAL

DO NOT reuse empty container. Triple rinse (or equivalent) and offer for recycling or reconditioning or puncture and dispose of in a sanitary landfill, or by other procedures approved by state and local authorities.

WARRANTY AND LIMITATION OF DAMAGES

Seller warrants that this material conforms to its chemical description and is reasonably fit for the purposes stated on the label when used in accordance with directions under normal conditions of use and Buyer assumes the risk of any use contrary to such directions. SELLER MAKES NO OTHER EXPRESS OR IMPLIED WARRANTY, INCLUDING ANY OTHER EXPRESS OR IMPLIED WARRANTY OF FITNESS OR OF MERCHANTABILITY, AND NO AGENT OF SELLER IS AUTHORIZED TO DO SO EXCEPT IN WRITING WITH A SPECIFIC REFERENCE TO THIS WARRANTY. In no event shall Seller's liability for any breach of warranty exceed the purchase price of the material as to which a claim is made.

SPECIAL LOCAL NEEDS PURSUANT TO SECTION 24(c) OF AMENDED FIFRA AND STATE REGISTRATIONS

CROP	REGISTRATIONS	STATES	SLN NO.	DIRECTIONS FOR USE
Cotton	Application on Cotton	Arkansas Louisiana Mississippi Tennessee	3969 3971 3970 TN–830016	DSMA Liquid is useful for the control of weeds listed on the product label and many similar weeds. Application can be made as follows: 1. Preplant or postplant up to cracking of soil before cotton emergence using ground or aircraft equipment. 2. Postemergent, over-the-top, when cotton is 3 to 6 inches high or up to early first square stage, whichever occurs first, using ground or aircraft equipment. 3. Postemergence as a directed spray with ground equipment when cotton is 3 inches high to first bloom. PREPLANT OR POSTPLANT UP TO CRACKING APPLICATION ON COTTON: A single ground or aircraft application of DSMA Liquid can be made to prepared cotton seedbeds when planting has been delayed and weeds have emerged, or as a postplant treatment, BUT NO LATER THAN INITIAL CRACKING of soil in field before emergence of cotton. Planting of cotton may immediately follow the preplant application. Mix at the rate of 1 gallon of DSMA Liquid in 40 gallons of water for broadcast ground application, or in 5 to 10 gallons of water for aircraft application and apply using a properly calibrated sprayer to one acre. See SPECIAL PRECAUTIONS below.

CROP	REGISTRATIONS	STATES	SLN NO.	DIRECTIONS FOR USE
				POSTEMERGENT, OVER-THE-TOP, APPLICATION ON COTTON: Application may be made using ground or aircraft equipment as an over-the-top broadcast spray to actively growing cotton as a salvage operation when cotton is 3 to 6 inches high or up to early first square stage, whichever occurs first. Mix DSMA Liquid at rate of 3 to 4 pints in 40 gallons of water for ground application, or in 5 to 10 gallons of water for aircraft application and apply using a properly calibrated sprayer to one acre. A second or repeat application, if needed, should be made 1 to 3 weeks after the first treatment. APPLY AS AN OVER-THE-TOP BROADCAST SPRAY ONLY WHEN COTTON IS 3 to 6 INCHES HIGH OR UP TO EARLY FIRST SQUARE STAGE, WHICHEVER OCCURS FIRST. **DO NOT** APPLY AS AN OVER-THE-TOP BROADCAST SPRAY AFTER COTTON IS MORE THAN 6 INCHES HIGH OR AFTER EARLY FIRST SQUARE STAGE. See SPECIAL PRECAUTIONS below. POSTEMERGENT, DIRECTED, APPLICATION ON COTTON: Apply as DIRECTED broadcast spray at a rate of 40 gallons per acre of a solution containing 1 gallon of DSMA Liquid in 40 gallons of water when weeds are small. For band application, apply 1 gallon of above diluted spray per acre for each 1-inch band width to be treated of cotton grown on 40-inch row spacing. Keep spray off cotton foliage. APPLY ONLY WHEN COTTON IS 3 INCHES HIGH TO FIRST BLOOM. **DO NOT** APPLY AFTER FIRST BLOOM. Slight burning and reddish discoloration of cotton foliage may occasionally be seen following recommended treatment; however, cotton plants will develop normally. SPECIAL PRECAUTIONS—**DO NOT** allow spray or spray drift to contact adjacent crops or injury will result. Apply only on still days when weather conditions **DO NOT** favor drift from areas being treated. Follow other precautions listed on this label. Aircraft applications of DSMA Liquid should only be made by applicators experienced in use of herbicides, and applications should be made in accordance with state and Federal regulations.
Cotton	Application on Cotton	Georgia Missouri New Mexico North Carolina Texas	GA-770007 MO-770006 NM-780005 NC-770017 TX–770025	DSMA Liquid is useful for the control of weeds listed on the product label and many similar weeds. Application can be made as follows: 1. Preplant or postplant up to cracking of soil before cotton emergence using ground or aircraft equipment. 2. Postemergent, over-the-top, when cotton is 3 to 6 inches high or up to early first square state, whichever occurs first, using ground or aircraft equipment. 3. Postemergence as a directed spray with ground equipment when cotton is 3 inches high to first bloom. PREPLANT OR POSTPLANT UP TO CRACKING APPLICATION ON COTTON: A single ground or aircraft application of DSMA Liquid can be made to prepared cotton seedbeds when planting has been delayed and weeds have emerged, or as a postplant treatment, BUT NO LATER THAN INITIAL CRACKING of soil in field before emergence of cotton. Planting of cotton may immediately follow the preplant application. Mix at the rate of 1 gallon of DSMA Liquid in 40 gallons of water for broadcast ground application, or in 5 to 10 gallons of water for aircraft application and apply using a properly calibrated sprayer to one acre. See SPECIAL PRECAUTIONS below. POSTEMERGENT, OVER-THE-TOP, APPLICATION ON COTTON: Application may be made using ground or aircraft equipment as an over-the-top broadcast spray to actively growing cotton as a salvage operation when cotton is 3 to 6 inches high or up to early first square stage, whichever occurs first. Mix DSMA Liquid at rate of 2 to 3 quarts in 40 gallons of water for ground application and apply using a properly calibrated sprayer to one acre. A second or repeat application, if needed, should be made 1 to 3 weeks after the first treatment. APPLY AS AN OVER-THE-TOP BROADCAST SPRAY ONLY WHEN COTTON IS 3 TO 6 INCHES HIGH OR UP TO EARLY FIRST SQUARE STAGE, WHICHEVER OCCURS FIRST. **DO NOT** APPLY AS AN OVER-THE-TOP BROADCAST SPRAY AFTER COTTON IS MORE THAN 6 INCHES HIGH OR AFTER EARLY FIRST SQUARE STAGE. See SPECIAL PRECAUTIONS below. POSTEMERGENT, DIRECTED, APPLICATION ON COTTON: Apply as DIRECTED broadcast spray at a rate of 40 gallons per acre of a solution containing 1 gallon of DSMA Liquid in 40 gallons when weeds are small. For band application, apply 1 gallon of above diluted spray per acre for each 1-inch band width to be treated of cotton grown on 40-inch row spacing. Keep spray off cotton foliage. APPLY ONLY WHEN COTTON IS 3 INCHES HIGH TO FIRST BLOOM. **DO NOT** APPLY AFTER FIRST BLOOM. Slight burning and reddish discoloration of cotton foliage may occasionally be seen following recommended treatment; however, cotton plants will develop normally.

CROP	REGISTRATIONS	STATES	SLN NO.	DIRECTIONS FOR USE
				SPECIAL PRECAUTIONS—**DO NOT** allow spray or spray drift to contact adjacent crops or injury will result. Apply only on still days when weather conditions **DO NOT** favor drift from areas being treated. Follow other precautions listed on this label. Aircraft applications of DSMA Liquid should only be made by applicators experienced in use of herbicides, and applications should be made in accordance with state and Federal regulations.
Cotton	Application on Cotton	Alabama	AL-770010	DSMA Liquid is useful for the control of weeds listed on the product label and many similar weeds. Application can be made as: 1. Preplant or postplant up to cracking of soil before cotton emergence using ground or aircraft equipment. 2. Postemergent, over-the-top, when cotton is 3 to 6 inches high or up to early first square stage, whichever occurs first, using ground or aircraft equipment. 3. Postemergence as a directed spray with ground equipment when cotton is 3 inches high to first bloom. PREPLANT OR POSTPLANT UP TO CRACKING APPLICATION ON COTTON: A single ground or aircraft application of DSMA Liquid can be made to prepared cotton seedbeds when planting has been delayed and weeds have emerged, or as a postplant treatment. BUT NO LATER THAN INITIAL CRACKING of soil in field before emergence of cotton. Planting of cotton may immediately follow the preplant application. Mix at the rate of 1 gallon of DSMA Liquid in 40 gallons of water for broadcast ground application, or in 5 to 10 gallons of water for aircraft application and apply using a properly calibrated sprayer to one acre. See SPECIAL PRECAUTIONS below. POSTEMERGENT, OVER-THE-TOP, APPLICATION ON COTTON: A single ground or aircraft application may be made as an over-the-top broadcast spray to actively growing cotton as a SALVAGE operation when cotton is 3 to 6 inches high or up to early first square stage, whichever occurs first. Mix DSMA Liquid at rate of 3 to 4 pints in 40 gallons of water for ground application, or in 5 to 10 gallons of water for aircraft application and apply using a properly calibrated sprayer to one acre. APPLY AS AN OVER-THE-TOP BROADCAST SPRAY ONLY WHEN COTTON IS 3 TO 6 INCHES HIGH OR UP TO EARLY FIRST SQUARE STAGE, WHICHEVER OCCURS FIRST, **DO NOT** APPLY AS AN OVER-THE-TOP BROADCAST SPRAY AFTER COTTON IS MORE THAN 6 INCHES HIGH OR AFTER EARLY FIRST SQUARE STAGE. See SPECIAL PRECAUTIONS below. POSTEMERGENT, DIRECTED, APPLICATION ON COTTON: Apply as a DIRECTED broadcast spray at a rate of 40 gallons per acre of a solution containing 1 gallon of DSMA Liquid in 40 gallons of water when weeds are small. For band applications, apply 1 gallon of above diluted spray per acre for each 1-inch band width to be treated of cotton grown on 40-inch row spacing. Keep spray off cotton foliage. APPLY ONLY WHEN COTTON IS 3 INCHES HIGH TO FIRST BLOOM. **DO NOT** APPLY AFTER FIRST BLOOM. Slight burning and reddish discoloration of cotton foliage may occasionally be seen following recommended treatment; however, cotton plants will develop normally. SPECIAL PRECAUTIONS—**DO NOT** allow spray or spray drift to contact adjacent crops or injury will result. Apply on still days when weather conditions **DO NOT** favor drift from areas being treated. Follow other precautions listed on this label. Aircraft applications of DSMA Liquid should only be made by applicators experienced in use of herbicides, and applications should be made in accordance with state and Federal regulations.
Cotton	Application on Cotton	Arizona	AZ-770024	DSMA Liquid is useful for the control of weeds listed on the product label and many similar weeds. Application can be made as: 1. Preplant or postplant up to cracking of soil before cotton emergence using ground or aircraft equipment. 2. Postemergent, over-the-top, when cotton is 3 to 6 inches high or up to early first square stage, whichever occurs first, using ground or aircraft equipment. 3. Postemergence as a directed spray with ground equipment when cotton is 3 inches high to first bloom. PREPLANT OR POSTPLANT UP TO CRACKING APPLICATION ON COTTON: A single ground or aircraft application of DSMA Liquid can be made to prepared cotton seedbeds when planting has been delayed and weeds have emerged, or as a postplant treatment, BUT NO LATER THAN INITIAL CRACKING of soil in field before emergence of cotton. Planting of cotton may immediately follow the preplant application. Mix at the rate of 1 gallon of DSMA Liquid in 40 gallons of water for broadcast ground application, or in 5 to 10 gallons of water for aircraft application and apply using a properly calibrated sprayer to one acre. See SPECIAL PRECAUTIONS below.

CROP	REGISTRATIONS	STATES	SLN NO.	DIRECTIONS FOR USE
				POSTEMERGENT, OVER-THE-TOP, APPLICATION ON COTTON: A single ground or aircraft application may be made as an over-the-top broadcast spray to actively growing cotton as a salvage operation when cotton is 3 to 6 inches high or up to early first square stage, whichever occurs first. Mix DSMA Liquid at rate of 2 to 3 quarts in 40 gallons of water for ground application, or in 5 to 10 gallons of water for aircraft application and apply using a properly calibrated sprayer to one acre. APPLY AS AN OVER-THE-TOP BROADCAST SPRAY ONLY WHEN COTTON IS 3 TO 6 INCHES HIGH OR UP TO EARLY FIRST SQUARE STAGE, WHICHEVER OCCURS FIRST. **DO NOT** APPLY AS AN OVER-THE-TOP BROADCAST SPRAY AFTER COTTON IS MORE THAN 6 INCHES HIGH OR AFTER EARLY FIRST SQUARE STAGE. See SPECIAL PRECAUTIONS below. POSTEMERGENT, DIRECTED, APPLICATION ON COTTON: Apply as a DIRECTED broadcast spray at a rate of 40 gallons per acre of a solution containing 1 gallon of DSMA Liquid in 40 gallons of water when weeds are small. For band applications, apply 1 gallon of above diluted spray per acre for each 1-inch band width to be treated of cotton grown on 40-inch row spacing. Keep spray off cotton foliage. APPLY ONLY WHEN COTTON IS 3 INCHES HIGH TO FIRST BLOOM. **DO NOT** APPLY AFTER FIRST BLOOM. Slight burning and reddish discoloration of cotton foliage may occasionally be seen following recommended treatment; however, cotton plants will develop normally. SPECIAL PRECAUTIONS—**DO NOT** allow spray or spray drift to contact adjacent crops or injury will result. Apply only on still days when weather conditions **DO NOT** favor drift from areas being treated. Follow other precautions listed on this label. Aircraft applications of DSMA Liquid should only be made by applicators experienced in use of herbicides, and applications should be made in accordance with state and Federal regulations.
Cotton	Application on Cotton	California	CA-770096	DSMA Liquid is useful for the control of weeds listed on the product label and many similar weeds. Application can be made as: 1. Preplant or postplant up to cracking of soil before cotton emergence using ground or aircraft equipment. 2. Postemergent, over-the-top, when cotton is 3 to 6 inches high or up to early first square stage, whichever occurs first, using ground or aircraft equipment. 3. Postemergence as a directed spray with ground equipment when cotton is 3 inches high to first bloom. PREPLANT OR POSTPLANT UP TO CRACKING APPLICATION ON COTTON: A single ground or aircraft application of DSMA Liquid can be made to prepared cotton seedbeds when planting has been delayed and weeds have emerged, or as a postplant treatment, BUT NO LATER THAN INITIAL CRACKING of soil in field before emergence of cotton. Planting of cotton may immediately follow the preplant application. Mix at the rate of 1 gallon of DSMA Liquid in 40 gallons of water for broadcast ground application, or in 5 to 10 gallons of water for aircraft application and apply using a properly calibrated sprayer to one acre. See SPECIAL PRECAUTIONS below. POSTEMERGENT, OVER-THE-TOP, APPLICATION ON COTTON: One or two over-the-top applications may be applied by ground or aircraft equipment to actively growing cotton as a salvage operation when crop is 3 to 6 inches high or to early first square stage, whichever occurs first. ONE APPLICATION: Use 1 gallon DSMA Liquid in 40 gallons of water per acre for ground application, or in 5 to 10 gallons of water for aircraft application. **DO NOT** REPEAT THIS TREATMENT. TWO APPLICATIONS: Use 2 to 3 quarts DSMA Liquid in 40 gallons of water per acre for ground application, or in 5 to 10 gallons of water for aircraft application. Apply second applications, if needed, using the 2 to 3 quart rate, 1 to 3 weeks following the first treatment. Use a properly calibrated sprayer. **DO NOT** APPLY OVER-THE-TOP SPRAYS AFTER COTTON IS MORE THAN 6 INCHES HIGH OR AFTER EARLY FIRST SQUARE STAGE, WHICHEVER OCCURS FIRST. POSTEMERGENT, DIRECTED, APPLICATION ON COTTON: Apply as a DIRECTED broadcast spray at a rate of 40 gallons per acre of a solution containing 1 gallon of DSMA Liquid in 40 gallons of water when weeds are small. For band applications, apply 1 gallon of above diluted spray per acre for each 1-inch band width to be treated of cotton grown on 40-inch row spacing. Keep spray off cotton foliage. APPLY ONLY WHEN COTTON IS 3 INCHES HIGH TO FIRST BLOOM. Slight burning and reddish discoloration of cotton foliage may occasionally be seen following recommended treatment; however, cotton plants will develop normally.

CROP	REGISTRATIONS	STATES	SLN NO.	DIRECTIONS FOR USE
				SPECIAL PRECAUTIONS—**DO NOT** allow spray or spray drift to contact adjacent crops or injury will result. Apply only on still days when weather conditions **DO NOT** favor drift from areas being treated. Follow other precautions listed on this label. Aircraft applications of DSMA Liquid should only be made by applicators experienced in use of herbicides, and applications should be made in accordance with state and Federal regulations.
Cotton	Application on Cotton	Oklahoma	OK-780002	DSMA Liquid is useful for the control of weeds listed on the product label and many similar weeds. POSTEMERGENT, OVER-THE-TOP, APPLICATION ON COTTON: Application may be made using ground or aircraft equipment as an over-the-top broadcast spray to actively growing cotton as a salvage operation when cotton is 3 to 6 inches high or up to early first square stage, whichever occurs first. Mix DSMA Liquid at rate of 2 to 3 quarts in 40 gallons of water for ground application, or in 5 to 10 gallons of water for aircraft application and apply using a properly calibrated sprayer to one acre. A second or repeat application, if needed, should be made 1 to 3 weeks after the first treatment. APPLY AS AN OVER-THE-TOP BROADCAST SPRAY ONLY WHEN COTTON IS 3 TO 6 INCHES HIGH OR UP TO EARLY FIRST SQUARE STAGE, WHICHEVER OCCURS FIRST. **DO NOT** APPLY AS AN OVER-THE-TOP BROADCAST SPRAY AFTER COTTON IS MORE THAN 6 INCHES HIGH OR AFTER EARLY FIRST SQUARE STAGE. See SPECIAL PRECAUTIONS below. Slight burning and reddish discoloration of cotton foliage may occasionally be seen following recommended treatment; however, cotton plants will develop normally. SPECIAL PRECAUTIONS—**DO NOT** allow spray or spray drift to contact adjacent crops or injury will result. Apply only on still days when weather conditions **DO NOT** favor drift from areas being treated. **DO NOT** feed treated forage or gin trash to ruminant animals or graze treated areas. Follow other precautions listed on this label. Aircraft applications of DSMA Liquid should only be made by applicators experienced in use of herbicides and applications should be made in accordance with state and Federal regulations.
Cotton	Application on Cotton	South Carolina	SC-770014	DSMA Liquid is useful for the control of weeds listed on the product label. POSTEMERGENT APPLICATION ON COTTON: A single ground or aircraft application may be made as an over-the-top broadcast spray to actively growing cotton as a salvage operation when cotton is 3 to 6 inches high or up to early first square stage, whichever occurs first. Mix DSMA Liquid at rate of 3 to 4 pints in 40 gallons of water for ground application, or in 5 to 10 gallons of water for aircraft application and apply using a properly calibrated sprayer to one acre. APPLY AS AN OVER-THE-TOP BROADCAST SPRAY ONLY WHEN COTTON IS 3 to 6 INCHES HIGH OR UP TO EARLY FIRST SQUARE STAGE, WHICHEVER OCCURS FIRST. **DO NOT** APPLY AS AN OVER-THE-TOP BROADCAST SPRAY AFTER COTTON IS MORE THAN 6 INCHES HIGH OR AFTER EARLY FIRST SQUARE STAGE. **DO NOT** make more than two applications total of either DSMA or MSMA (or a combination) per season. See SPECIAL PRECAUTIONS below. Slight burning and reddish discoloration of cotton foliage may occasionally be seen following recommended treatment; however, cotton plants will develop normally. SPECIAL PRECAUTIONS—**DO NOT** feed treated forage to livestock or graze treated areas. **DO NOT** allow spray or spray drift to contact adjacent crops or injury and residues will result. Apply only on still days when weather conditions **DO NOT** favor drift from areas being treated. Follow other precautions listed on this label. Aircraft applications of DSMA Liquid should only be made by applicators experienced in use of herbicides, and applications should be made in accordance with state and Federal regulations.

Shell Chemical Company

A Division of Shell Oil Company
One Shell Plaza
P.O. Box 3871
Houston, TX 77253
(713) 241-6161

ATRAZINE 4L

HERBICIDE

For season-long weed control in corn, sorghum, and certain other crops.

ACTIVE INGREDIENTS	BY WEIGHT
Atrazine (2-chloro-4-ethylamino-6-isopropylamino-s-triazine)	40.8%
Related compounds	2.2%
INERT INGREDIENTS	57.0%
TOTAL	100.0%

Contains 4 lbs. active ingredients per gallon.

EPA Reg. No. 201-409

Keep Out of Reach of Children

CAUTION

NOTE CONDITIONS OF SALE AND WARRANTY BEFORE OPENING THE CONTAINER. IF THEY ARE NOT ACCEPTABLE, RETURN UNOPENED PRODUCT.

STATEMENT OF PRACTICAL TREATMENT

IF SWALLOWED, drink one or two glasses of water immediately and induce vomiting by sticking finger down back of throat.

IN CASE OF CONTACT, immediately flush eyes or skin with plenty of water. Get medical attention if irritation persists.

SEE ADDITIONAL PRECAUTIONS

PRECAUTIONARY STATEMENTS

HAZARD TO HUMANS AND DOMESTIC ANIMALS
CAUTION

May be fatal if swallowed. Harmful if absorbed through the skin. Avoid breathing vapors or spray mist. Avoid contact with eyes, skin or clothing. Wash thoroughly after handling. Remove and wash contaminated clothing before reuse. For 24 hour emergency medical assistance, call (713) 473-9461.

Do not apply this product in such a manner as to directly or through drift expose workers or other persons. The area being treated must be vacated by unprotected persons.

ENVIRONMENTAL HAZARDS

Atrazine can travel (seep or leach) through soil and can enter groundwater which may be used as drinking water. Atrazine has been found in groundwater. Users are advised not to apply Atrazine to sand and loamy sand soils where the water table (groundwater) is close to the surface and where the soils are very permeable. Your local agriculture agencies can provide further information on the type of soil in your area and the location of groundwater.

Atrazine is toxic to aquatic invertebrates. Do not apply directly to water or wetlands. Do not apply where runoff is likely to occur. Do not apply when weather conditions favor drift from treated areas. Runoff and drift from treated areas may be hazardous to aquatic organisms in neighboring areas. Do not contaminate water by cleaning of equipment or disposal of wastes. In case of significant spill, call CHEMTREC (800) 424-9300 or (713) 473-9461.

DIRECTIONS FOR USE

It is a violation of Federal law to use this product in a manner inconsistent with its labeling.

FOR END USE ONLY. DO NOT REPACKAGE OR REFORMULATE WITHOUT MANUFACTURER'S WRITTEN APPROVAL.

REENTRY STATEMENTS

Do not enter treated areas without protective clothing until sprays have dried.

Because certain states may require more restrictive reentry intervals for various crops treated in this product, consult your State Department of Agriculture for further information.

Written or oral warnings must be given to workers who are expected to be in a treated area or in an area about to be treated with this product. (Indicate specific oral warnings which inform workers of areas or fields that may not be entered without specific protective clothing, period of time field must be vacated and appropriate actions to take in case of accidental exposure.) When oral warnings are given warnings shall be given in a language customarily understood by workers. Oral warnings must be given if there is reason to believe that written warnings cannot be understood by workers. Written warnings must include the following information:

"CAUTION! Area treated with Atrazine Herbicide on (date of application). Do not enter without appropriate protective clothing until sprays have dried (Insert here actions to take in case of accidental exposure. See STATEMENT OF PRACTICAL TREATMENT for details.)"

STORAGE AND DISPOSAL

STORAGE

Do not contaminate water, food, or feed by storage or disposal. Do not use or store around the home environment. Avoid contact with water in case of spill or leak, soak up with sand, earth or synthetic absorbent (do not use alkaline absorbents) and dispose of wastes in compliance with local, state and Federal regulations.

PESTICIDE DISPOSAL

Wastes resulting from the use of this product may be disposed of on site or at an approved waste disposal facility.

CONTAINER DISPOSAL

Triple rinse (or equivalent). Then offer for recycling or reconditioning, or puncture and dispose of in a sanitary landfill, or by incineration, or, if allowed by state and local authorities by burning. If burned, stay out of smoke.

GENERAL INFORMATION

ATRAZINE 4L Herbicide controls many annual broadleaf and grass weeds in corn and sorghum including barnyardgrass, witchgrass, (Panicum capillare), yellow foxtail, green foxtail, wild oats, large (hairy) crabgrass, giant foxtail, velvetleaf, morningglory, lambsquarters, pigweed, ragweed, nightshade, purslane, and mustard. ATRAZINE 4L may be applied before or after weeds emerge.

Apply at the rates given under each crop before or after weeds emerge. Use lower rates on light soils and low organic matter soils. Use higher rates on heavy soils and high organic matter soils.

ATRAZINE 4L acts mainly through root absorption so rainfall or irrigation is needed to move it into the root zone. Should weeds develop, a shallow cultivation or rotary hoeing will generally result in better weed control.

Avoid using ATRAZINE 4L near adjacent desirable trees, shrubs, or plants or injury may occur. ATRAZINE 4L is non-flammable.

APPLICATION PROCEDURES

ATRAZINE 4L, a liquid formulation, should be mixed with water or liquid fertilizer and applied as a spray. Pour into the tank during or after filling. Hydraulic (jet) or mechanical agitation is recommended during mixing and application to keep the material in suspension. All return lines to the tank must discharge below liquid level and agitation should not be so violent as to cause air bubbles to form in the liquid. Wash sprayer thoroughly after use.

GROUND APPLICATION

Apply ATRAZINE 4L in 5–40 gallons of water per acre to assure even coverage unless otherwise specified. For uniform distribution, use 80° flat fan-type nozzles. Screens in nozzles, as well as those in suction and in-line strainers, should be no finer than 50-mesh. Use a minimum of 10 gal. of water per acre for all postemergence applications combined with oil or surfactant.

Use a suitable pump with capacity to deliver (1) the necessary volume to the nozzles at 35–40 p.s.i. and (2) an additional volume to maintain the mixture in suspension in the spray tank.

For band applications, calculate amount to be applied per acre as follows:

$$\frac{\text{Band Width In Inches}}{\text{Row Width in inches}} \times \text{Broadcast rate per acre} = \text{Amount Needed Per acre of field}$$

AERIAL APPLICATION

Use only where broadcast applications are specified. Apply a minimum of 1 gallon of water for each 2 pints of ATRAZINE 4L to be applied per acre. For postemergence treatments, apply recommended rate in a minimum of 2 gal. water per acre. Do not apply where uniform coverage cannot be obtained or where excessive spray drift may occur. Avoid application directly to animals or humans. Flagmen or loaders should avoid inhalation or spray mist or contact with skin and should wash thoroughly before eating and at the end of each day's operation.

APPLICATION IN LIQUID FERTILIZER

Nitrogen solutions or complete liquid fertilizers may replace all or part of the water as a carrier for preemergence and preplant applications. Mix as described under Application Procedures. Do not apply after corn or sorghum emerges as there is danger of liquid fertilizers causing crop injury.

Check to ensure the compatibility of ATRAZINE 4L with liquid fertilizer prior to mixing in the spray tank. A convenient method for testing the compatibility of ATRAZINE 4L with liquid fertilizer follows:

A. Add one pint of liquid fertilizer or other carrier to be used to two quart jars. Mark one quart jar with and the other jar without.

B. To the jar marked "with" add 1/4 teaspoon of a suitable compatibility agent. Shake gently for 5 to 10 seconds to mix. (1/4 teaspoon/pint = 2 pints/100 gallons.)

C. To each jar mix the number of teaspoons of herbicide shown on the following chart times the number of qts. per acre of ATRAZINE 4L to be applied corresponding to the gallons per acre of liquid fertilizer to be used. (For example, if 4 qts. per acre of ATRAZINE 4L are to be applied in 20 gallons of liquid fertilizer per acre, then 4 × 1 1/4 teaspoons, or 5 teaspoons should be mixed with one pint of liquid fertilizer.) If more than one herbicide is used, add them separately; wettable powders first, flowables second and emulsifiable concentrates last. Shake gently for 5 to 10 seconds after each addition.

D. Let stand one-half hour and make observations. If any separation, agglomeration or precipitation has occurred, again shake the jar 10 to 15 seconds.
   1. If mixtures appear to be compatible in both jars, the herbicide(s) can probably be used without a compatibility agent.
   2. If the mixture in the jar marked "with" is compatible, but the one in the jar marked "without" is incompatible, a compatibility agent should be used.
   3. If both mixtures are incompatible, the liquid fertilizer and the herbicide(s) should not be used in the same spray tank.

If a test such as outlined indicates that components of a proposed mix are compatible, materials to be combined in the spray tank should still be added in the sequence according to directions prescribed on the label of the herbicides or pesticides involved.

JAR TEST FOR ATRAZINE 4L COMPATIBILITY WITH LIQUID FERTILIZER

Gal. Liquid Fert/Acre	10	15	20	25	30	40
Teaspoons Atrazine 4L Pint of Fertilizer	2 1/2	1 1/2	1 1/4	1	3/4	1/2

APPLICATION IN WATER PLUS EMULSIFIABLE OIL OR OIL CONCENTRATE

The addition of oil to ATRAZINE 4L in water sprays may speed the activity of ATRAZINE 4L and provide quicker kill of weeds. However, under certain conditions, the use of oil may seriously damage corn or sorghum. To minimize the possibility of damage, follow the directions, procedures and precautions given below.

Use a crop oil containing at least 1% suitable emsulifier or a crop oil concentrate containing not more then 20% emulsifier or surfactant blend designated for use with Atrazine. Emulsifiable oil or oil concentrate contaminated with water or other materials can cause compatibility problems and/or crop injury.

MIXING PROCEDURES—ALL USES

(1) Thoroughly clean sprayer prior to use. Do not use sprayer contaminated with 2,4-D or other materials as crop damage or sprayer clogging may result. (2) Fill spray tank with water, allowing for the addition of the ATRAZINE 4L. (3) With the agitation system in operation, add the ATRAZINE 4L. (4) Add the emulsifiable oil or oil concentrate last. (5) Empty the tank as completely as possible before refilling to prevent a buildup of oil in the tank. Keep the agitation system in operation to avoid a separation of the oil from the water in the spray mixture still remaining in the tank. (6) If an oil film starts to build up in the tank, drain the tank and clean out with a strong detergent solution or solvent. (7) Clean the sprayer thoroughly immediately after use by flushing the system with water containing a detergent. (8) All return lines to the tank must discharge below liquid level.

PRECAUTIONS FOR APPLICATIONS WITH EMULSIFIABLE OIL OR OIL CONCENTRATE IN WATER

(1) Do not use oil in ATRAZINE 4L sprays when corn or sorghum is under stress from prolonged cold, wet weather, poor fertility, or other factors or when corn or sorghum is wet and succulent from recent rainfall as crop injury may occur. (2) Do not use oils in sprays when treating inbred lines or any breeding stock as injury may occur. (3) Adding other insecticides, herbicides, liquid fertilizers, or other materials is not recommended because they may cause compatibility problems or crop injury. (4) Store and handle emsulsifiable oil or oil concentrate carefully. Oil contaminated with even a small amount of water may not emulsify properly when added to the tank. (5) Do not make more than one application per season except as recommended for control of yellow nutsedge and Canada thistle on this label.

CORN

ATRAZINE 4L may be applied either before planting, at planting or after planting at the rate indicated below.

Preplant

Broadcast in the spring after plowing at the rate indicated in Table 1. Apply before, during or after final seedbed preparation. If soil is tilled or worked after application, avoid deep incorporation of ATRAZINE 4L. Best results have been obtained when ATRAZINE 4L is applied within two weeks prior to planting.

Preemergence

Apply during or shortly after planting prior to weed emergence at the rate indicated in Table 1.

Postemergence

Apply before weeds exceed 1.5 inches in height at the rate indicated in Table 1.

Lay-by Treatment (Broadleaf and grass control)

Broadcast 2–4 pt. per acre in a minimum of 5 gals. of water or nitrogen solution, before weeds are 1.5 inches high and corn is 20–30 inches high. When using nitrogen solution, direct the spray to the lower 3–4 inches of cornstalks to avoid corn foliage injury. Maintain agitation in spray tank during application.

Postemergence with Emulsifiable Oil or Oil Concentrate in Water Broadleaf and Grass Control

Ground Application

Broadcast 4 pts. of ATRAZINE 4L and 1 gal. of emulsifiable oil or 1 qt. of oil concentrate in 20–40 gals. of water per acre after emergence but before weeds reach 1.5 inches in height.

Aerial Application

Broadcast 4 pts. of ATRAZINE 4L and 0.5 gal of emulsifiable oil in no less than 5 gals. of water per acre after weed emergence but before weeds reach 1.5 inches in height.

Table 1
Broadleaf and Grass Weed Control in Corn*

Soil Texture	Broadcast Rate Per Acre of ATRAZINE 4L
COARSE soils: Sands, loamy sands, and sandy loams	4 pints
MEDIUM soils: Silt and clay loams that are low in organic matter	4.75 pints
FINE soils: Silt and clay loams with medium to high organic matter and clays	6 pints
Peats, mucks, and high organic clays (Apply postemergence only)	6 pints

* Broadleaf weed control (eastern CO, western KS, western NE, NM, OK Panhandle, west TX, and eastern WY). On sand, loamy sand, sandy loam, mild to strongly alkaline soil and all recently leveled soil apply 2.4 pt. per acre, either preplant or preemergence. Controls many broadleaf weeds including pigweed, lambsquarters, nightshade, purslane, and kochia. On other soil types in these areas, apply rate in Table 1 for broadleaf and grass control.

Broadleaf Control

Broadcast 2.4 pt. per acre for control of broadleaf weeds, such as annual morningglory, cocklebur, lambsquarters, mustards, pigweed, ragweed, smartweed, and wild buckwheat. Add emulsifiable oil at the rate of 1 gal. per acre for ground application and 0.5 gal. per acre for aerial applications. Apply before pigweed and lambsquarters reach 6 inches in height and before all other weeds reach 4 inches in height. A cultivation may be necessary if all weeds are not controlled or if weeds regrow.

Center Pivot Sprinkler Application—Preemergence of Postemergence (CO, KS, NE, SD, and WY)

Apply with irrigation water either after planting before corn and weeds emerge, or after corn emergence, but before lay-by (20–30 inches) and before weeds exceed 1.5 inches in height, at rates in Table 1. Prepare mixture with a minimum ratio of 1 part product to 1 part water. Injecting a larger volume of a more dilute slurry per hour will assure more accurate calibration of metering equipment. Maintain sufficient agitation to keep herbicide in suspension. Meter slurry into irrigation water during entire period. Apply in 0.5–1 inch of water. Use the lower volume on coarser textured soils, the higher volume on finer textured soils. More than 1 inch of water may reduce weed control by moving herbicide below the effective zone in the soil. Inject dilute slurry into system through a positive displacement pump.

Precautions

(1) Apply only through irrigation systems containing anti-siphon and check valves to prevent contamination of well during shutdown and overflow of solution tank. (2) Inject ahead of any right angle turn in the main line to insure adequate mixing. (3) Chemical injection pumps and water pumps must have interlocking controls to insure simultaneous shutoff. (4) Application when drift may occur from windly conditions, when system joints and connections are leaking, or when nozzles are not providing uniform distribution may cause crop injury. (5) Where sprinkler distribution patterns do not overlap sufficiently, unacceptable weed control may result. Where sprinkler patterns overlap excessively, crop injury may result.

Problem Weeds

ATRAZINE 4L will control yellow nutsedge (Cyperus esculentus) and Canada thistle (Cirsium arvense) when applied according to directions. For best results, apply each year until yellow nutsedge or Canada thistle is eliminated or reaches a level of infestation where neither weed is a problem. If yellow nutsedge or Canada thistle regrow following the last application, cultivate once. When ATRAZINE 4L is applied postemergence to the weeds, add 1 gal. of emulsifiable crop oil or 1 qt. of oil concentrate per acre.

Six alternative methods of use for control of yellow nutsedge and Canada thistle are listed below in order of preference. If other weed species, such as annual grasses, are also expected, use alternative 2, 3, 5, or 6.

1. Broadcast 4 pt. plus 1 gal. of emsulsifiable oil or 1 qt. of oil concentrate per acre after crop and yellow nutsedge or Canada thistle emerge, but before yellow nutsedge reaches a height of 3 inches or Canada thistle reaches a height of 6 inches. Repeat application before lay-by (20–30 inches), 10–20 days after the first application has been made.
2. Broadcast 4 pt. per acre preplant. Follow with an application of 4 pt. plus 1 gal. of emulsifiable oil or 1 qt. of oil concentrate per acre, after corn and weeds emerge, but before yellow nutsedge reaches a height of 3 inches (yellow nutsedge control only).
3. Broadcast 4 pt. per acre during or shortly after planting, but prior to crop or weed emergence. Follow with an application of 4 pt. plus 1 gal. of emulsifiable oil or 1 qt. of oil concentrate per acre, after corn and weeds emerge, but before yellow nutsedge reaches a height of 3 inches or Canada thistle reaches a height of 6 inches.
4. Broadcast 8 pt. plus 1 gal. of emulsifiable oil or 1 qt. of oil concentrate per acre after crop emerges, but prior to lay-by (20–30 inches) and after yellow nutsedge and Canada thistle emerge, but before yellow nutsedge is 3 inches tall or Canada thistle is 6 inches tall.
5. Broadcast 8 pt. per acre preplant (yellow nutsedge control only).
6. Broadcast 8 pt. per acre during or shortly after planting, but prior to crop or weed emergence (yellow nutsedge control only).

NOTE: Do not use emulsifiable oil or oil concentrate when corn is wet or under stress, especially when using 8 pt. of this product per acre. See "Precautions for Applications with Emulsifiable Oil or Oil Concentrate in Water" for additional directions.

Quackgrass Control on Land Going Into Corn Production

Split application: Broadcast 4 pt. per acre in the fall or spring. Plow 1–3 weeks later. Broadcast a second application at 4 pt. per acre in the spring before, during or after planting but before weeds are 1.5 inches high. This split application will control both quackgrass and most annual broadleaf and grass weeds.

Single application: Broadcast 6–8 pt. per acre in the fall or spring. Plow 1–3 weeks later.

TANK MIXTURES ON CORN

DUAL 8E

Use as tank mixture for control of certain broadleaf and grass weeds in corn. Refer to Dual 8E label for all directions, weeds controlled, precautions and limitations.

LASSO or LASSO EC (alachlor)

Use as tank mixture for control of many annual broadleaf and grass weeds in corn (field, hybrid seed, silage, sweet, popcorn) including barnyardgrass, carpetweed, crabgrass, fall panicum, Florida pusley, foxtails (giant, green, yellow), goosegrass, jimsonweed, kochia, lambsquarters, mustard, nightshade, pigweed, purslane, ragweed, signalgrass (Brachiaria), smartweed, and witchgrass. This tank mix will reduce competition from the hard to control annual weeds—annual morningglory, cocklebur, and velvetleaf (buttonweed).

For preemergence or early postemergence broadcast application, use the rates in Table 2. Apply tank mixture any time from immediately after planting until weeds reach 2-leaf state and corn is no more than 5 inches tall. On sweet corn, apply before crop and weeds emerge. Application after the 2-leaf stage will not give satisfactory control. With postemergence application, occasional corn leaf burn should not affect growth or yield. If applied with liquid fertilizers, spray before corp emerges.

For preplant incorporated applications, use 0.4 pt. more of ATRAZINE 4L plus 0.5 qt. of Lasso per acre than indicated in Table 2. Use no more than 3.2 pts. of ATRAZINE 4L plus 2.5 qt. of Lasso per acre. Broadcast within 7 days before planting and incorporate 2 inches deep.

Apply tank mixture in a minimum of 20 gal. of water per acre. Liquid fertilizer may replace all or part of the water. Add ATRAZINE 4L to spray tank, thoroughly mix with water, then add Lasso. Follow mixing procedures in the section "Application Procedures" of this label.

Refer to Lasso label for information concerning center pivot irrigation applications and other directions, limitations, and cautions.

Table 2
Tank Mixtures With Lasso On Corn

Soil Texture	Broadcast Rate/Acre			
	Less than 3% Organic Matter		3% or more Organic Matter	
	ATRAZINE 4L	Lasso	ATRAZINE 4L	Lasso
COARSE* Sand, loamy sand, sandy loam	2 pt.	1.5 qt.	2 pt.	1.5 qt.
MEDIUM Loam, silt loam, silt	2–2.4 pt.	1.75 qt.	2.4–2.8 pt.	2 qt.
FINE Silty clay loam, sandy clay loam, silty clay, sandy clay, clay loam, clay	2.4–3.2 pt.	2.25 qt.	2.4–3.2 pt.	2.5 qt.

* When applied via center pivot irrigation on coarse soils, apply 2 pt. of ATRAZINE 4L and 2 qt. of LASSO per acre.

PRINCEP 80W or PRINCEP 4L (Simazine)

Use as a tank mixture for control of many annual weeds including carpetweed, crabgrass, fall panicum, foxtail, lambsquarters, morningglory, pigweed, ragweed, and velvetleaf. Apply before planting, at planting, or after planting, but before crop and weeds emerge, at rates in Table 3. Use 1:1 ratio for control of above weeds. Use 1:2 ratio for expected heavy infestations of crabgrass and fall panicum. Apply in a minimum of 20 gal. water per acre. Cultivate shallowly if weeds develop.

Preplant

Apply in spring during or after final seedbed preparation. If soil is tilled or worked after application, avoid deep incorporation. For best results, apply within two weeks before planting.

Preemergence

Apply during or shortly after planting, but before crop and weeds emerge.

Refer to Corn sections of this label and to Princep 80W or Princep 4L label for further directions, limitations, and cautions.

Table 3
Tank Mixtures With Princep On Corn

Soil Texture	Broadcast Rate/Acre			
	1:1 Ratio*		1:2 Ratio**	
	ATRAZINE 4L	Princep 80W or Princep 4L	ATRAZINE 4L	Princep 80W or Princep 4L
COARSE: Sand, loamy sand, sandy loam	2 pt.	1.25 lb. or 2 pt.	1.32 pt.	1.67 lb. or 2.67 pt.
MEDIUM: Loam, silt loam, silt, clay loam, sandy clay loam, silty clay loam, sandy clay, or silty clay with low organic matter	2.4 pt.	1.5 lb. or 2.4 pt.	1.6 pt.	2 lb. or 3.2 pt.
FINE: Loam, silt loam, silt, clay loam, sandy clay loam, silty clay loam, sandy clay or silty clay with medium to high organic matter and clay (including dark prairie soils of the Corn Belt)	3 pt.	1.8 lb. or 3 pt.	1.92 pt.	2.4 lb. or 3.84 pt.

* For control of above weeds.
** For control of expected heavy infestations of crabgrass and fall panicum.

PROPACHLOR 65WP

Use a tank mixture for control of many annual broadleaf and grass weeds in corn (field, hybrid seed, silage and sweet corn only), including annual morningglory, annual ryegrass, barnyardgrass (watergrass), buttonweed (velvetleaf), carpetweed, cocklebur, crabgrass, fall panicum, Florida pusley, giant foxtail, green foxtail, yellow foxtail, goosegrass, groundsel, jimsonweed, lambsquarters, mustard, nightshade, pigweed, purslane, ragweed, smartweed, and sunflower. Broadcast 2.1–3.2 pt. of ATRAZINE 4L plus 3.8–6 lb. of propachlor 65WP per acre on soil surface of any time after planting until broadleaf and grass weeds reach the 2-leaf stage. Use lower rates on coarse-textured soil low in organic matter. Use higher rates on fine-textured soil high in organic matter.

Use a minimum of 2.8 pt. per acre of ATRAZINE 4L in tank mixture for better control of annual morningglory, buttonweed (velvetleaf), cocklebur, and sunflower.

Apply in a minimum of 20 gal. per acre. Non-pressure fluid fertilizer may replace all or part of the water used as a carrier for applications before crop and weeds emerge. Add ATRAZINE 4L to spray tank and thoroughly mix with water. Then, cut the propachlor 65WP bag and fill with water to fill line. Grasp neck of bag firmly, shake vigorously, and add contents to spray tank.

Refer to propachlor 65WP label for further directions, limitations, and cautions.

PARAQUAT CL

For control of existing vegetation and residual control where corn will be planted directly into cover crop, established sod, or in previous crop residues, broadcast 4–6 pt. of ATRAZINE 4L and 1–2 pt. Paraquat CL in 20–60 gal. of water per acre. Add ATRAZINE 4L to water in spray tank, agitate until thoroughly mixed with water, and then add Paraquat CL and surfactant.

PRINCEP 80W plus PARAQUAT CL

Use as a tank mixture with Princep 80W and Paraquat CL for kill of existing vegetation and residual weed control where corn will be planted directly into a cover crop, established sod, or in previous crop residues. Add ATRAZINE 4L and Princep 80W to water in spray tank, agitating until thoroughly mixed. Then add Paraquat CL and a nonionic surfactant such as X-77. Continue agitation during application. Apply 2–4 pt. of ATRAZINE 4L plus 1.25–2.5 lb. of Princep 80W plus 1–2 pt. Paraquat CL in 20–60 gal. of water per sprayed acre with 3 weeks before, during, or after planting, but before corn crop emerges. Add 8 fl. oz. of a nonionic surfactant, such as X-77, per 100 gal. diluted spray. Use lower rates of ATRAZINE 4L and of Princep 80W on coarse soil, and higher rates on fine-textured soil. Use the 2 pt. rate of Paraquat CL if existing vegetation is 4–6 inches tall. This mixture will not control weeds taller than 6 inches.

Refer to further limitations and precautions on labels for this product, Princep 80W, and Paraquat CL.

PRECAUTIONS FOR ALL APPLICATIONS TO CORN

(1) Do not apply more than 8 pts. ATRAZINE 4L to corn in any one year. (2) Land treated with ATRAZINE 4L should not be planted to any crop except corn or sorghum until the following year or injury may occur. (3) Injury may occur in any crop other than

corn or sorghum planted the year following treatment. (4) Following harvest of a treated crop, plow (moldboard or disk-plow) and thoroughly till the soil in the fall or spring to minimize possible injury to rotational spring-seeded crops, regardless of the rate used. (5) To avoid the possibility of injury to following fall or spring seeded crops, the use of BLADEX® Herbicide in place of ATRAZINE 4L is recommended. Follow directions on the BLADEX label. (6) The use of ATRAZINE 4L on land that is to be rotated the following year to soybeans that will be treated with metribuzin (SENCOR or LEXONE) may increase the chance of injury. (7) If ATRAZINE 4L is applied after June 10, do not rotate with crops other than corn or sorghum the next year or injury may occur. (8) In the High Plains and Intermountain areas of the West where rainfall is sparse and erratic or where irrigation is required, use ATRAZINE 4L to control weeds in corn or sorghum only where corn or sorghum only where corn or sorghum is to follow treated corn or sorghum is to precede other rotational crops. (9) Do not graze treated area or feed treated forage to livestock for 21 days following application.

SORGHUM AND SORGHUM-SUDAN HYBRIDS (Grain and Forage Types)

Apply before planting, or after planting. See Corn section for weeds controlled.

Preplant (Broadleaf and Grass Control)
Broadcast in spring after plowing at rate in Table 4. Apply before, during, or after final seedbed preparation. If soil is tilled or worked after application, avoid deep incorporation. For best results, apply within two weeks before planting.

Premergence (Broadleaf and Grass Control)
Apply during or shortly after planting before weed or crop emergence, at rate in Table 4.

Table 4
Preplant and Preemergence Broadleaf and Grass Weed Control in Sorghum*

Soil Texture	Organic Matter	Broadcast Rate/Acre
COARSE: Sand, loamy sand, sandy loam, sandy clay loam	any level	Do not use (except for pre-emergent use on bedded sorghum in AZ and CA as indicated below)
MEDIUM and FINE: Silt loam clay loam clay	less than 1%	
	1–1.5%	3.2–4 pt.
	more than 1.5%	4–4.75 pt.

* Do not apply preplant in AL, AR, FL, GA, LA, MS, NC, NM, SC, TN, or TX.
Do not apply preemergence in NM, OK, or TX, except in northeast OK and the TX Gulf Coast.

In case of planting failure, sorghum may be replanted. Do not make a second broadcast application, as injury may occur. If originally applied in a band and sorghum is replanted in untreated row middles, ATRAZINE 4L may be applied in a band to the second planting.

Preemergence Broadleaf Control in Furrow Irrigated Bedded Sorghum (AZ and CA only):
For preemergence control of many broadleaf weeds, including groundcherry, lambsquarters, morningglory, mustard, pigweed, and purslane, broadcast 1.6–2.4 pt. per acre. Apply after bed preparation, during or after planning, but before sorghum and weeds emerge and before the first furrow irrigation. Follow with several regular irrigations to thoroughly wet all soil. Use lower rate on coarse-textured soil and soil low in organic matter; use higher rate on fine-textured soil and soil high in organic matter.

Precautions for Preemergence Application to Furrow Irrigated Bedded Sorghum Grown in AZ and CA.
To avoid possible sorghum injury, do not use on sand or loamy sand soil or on sorghum planted in the furrow. Applications to sorghum growing on alkali soils or where cuts, fills, or erosions have exposed calcareous or alkali subsoils may cause crop injury. Do not replant sorghum for 8 months following application. Corn may be planted immediately.

Postemergence (Broadleaf and grass control)
Apply at rate in Table 5 before weeds, exceed 1.5 inches in height. Apply up to "close-in."

Table 5
Postemergence Broadleaf And Grass Weed Control In Sorghum

Soil Texture	Minimum Sorghum Height at Treatment	Broadcast Rate/Acre
Sand or loamy sand	Do not use	
Sandy loam	See directions for postemergence broadleaf weed control below	
Silt loam to sandy clay loam	Completely Emerged	4–4.75 pt.
Olton and Pullman clay soil	At least 6 inches	4–4.75 pt.
Silty clay loam and finer textured soil	Completely Emerged	6 pt.

Postemergence Broadleaf Control with Emulsifiable Oil or Oil Concentrate in Water
Broadcast 2.4 pt. per acre for control of many broadleaf weeds, inluding annual morningglory, cocklebur, lambsquarters, mustard, pigweed, ragweed, smartweed, and wild buckwheat. Apply before pigweed and lambsquarters reach 6 inches in height and before all other weeds reach 4 inches in height. In CO, western KS, NM, OK, TX and desert regions of AZ and CA, apply when sorghum is 6–10 inches in height, but before it reaches boot stage. In all other areas, apply after sorghum reaches the 3-leaf stage. Add emulsifiable oil at 1 gal. per acre for ground application and 0.5 gal. per acre for aerial applications. A cultivation may be necessary if all weeds are not controlled or if weeds regrow.

Precautions for Applications with Emulsifiable Oil or Oil Concentrate in Water

See "Precautions for Applications with Emulsifiable Oil or Oil Concentrate in Water" in Application Procedures section of this label.

Postemergence Broadleaf Control with Surfactant (CO, Western KS, NM, OK, TX and Desert Regions of AZ and CA only)
Broadcast 2.4 pt. plus 0.75–1.5 pt. of surfactant per acre after sorghum reaches 6 inches in height, but before weeds reach 1.5 inches in height. Apply only on sandy loam and finer textured oil.

Precautions for All Applications to Sorghum
(1) Heavy rain immediately following application tends to cause excessive concentrations of herbicide in seed furrow, resulting in possible crop injury. Do not apply to furrow-planted sorghum until furrows are leveled (plowed-in). Level deep planter marks or seed furrows before application. (2) Application to sorghum growing under stress caused by minor element deficiency or to sorghum growing on highly calcareous soil may result in crop injury. (3) Injury may occur in any crop other than corn or sorghum planted the year following treatment. (4) Following harvest, plow (moldboard or disk-plow) and thoroughly till soil in fall or spring to minimize possible injury to rotational spring-seeded crops, regardless of rate used. (5) Injury may occur if both this herbicide, preplant or preemergence, and an at-planting systemic insecticide are used.

NOTE: Do not graze or feed forage from treated areas for 21 days following application

Additional Precautions
See "Precautions for All Applications to Corn" in Corn section of this label.

CHEMICAL FALLOW

Wheat-Sorghum-Fallow

This treatment controls annual broadleaf and grass weeds following wheat harvest and in the following sorghum crop when grown under minimum tillage.

Apply 6 pt. to wheat stubble immediately following wheat harvest. If weeds are present, remove them with a sweep plow or other suitable implement after application. Plant sorghum into wheat stubble the following spring with minimum disturbance of the soil. Use a surface planter or a planter leaving a shallow furrow. If weeds are present at planting, remove them with a sweep plow or other suitable implement before planting.

Precautions

(1) Use only on a silt loam or finer textured soil. (2) Wheat-sorghum-fallow cropping sequence must be followed. (3) Do not apply following sorghum harvest.

NOTE: Do not graze or feed forage from treated area to livestock. Do not plant any crop other than those on this label within 18 months following treatment.

Wheat-Corn-Fallow (KS, NE)

ATRAZINE 4L controls cheatgrass (downy brome, chess), kochia, mustards, pigweed, Russian thistle, wild lettuce, wild sunflower, and volunteer wheat during period after wheat harvest. Weed control may extend into following corn crop grown under minimum tillage.

Apply with ground equipment only.

Follow directions for use, notes, and precautions in the "Wheat-Sorghum-Fallow" section above, substituting corn for references to sorghum.

Wheat-Fallow-Wheat (CO, KS, MT, NE, ND, SD, and WY)

For preemergence control of cheatgrass (downy brome, chess) common lambsquarters, field pennycress, kochia, mustard, Russian thistle, wild lettuce, and suppression of volunteer wheat during fallow period of a wheat-fallow-wheat rotation, apply 1–2 pts. in 10–40 gal. of water per acre for ground application, or in a minimum of 5 gal. of water per acre for aerial application, on all soils except those listed under "Precautions." For control of pigweed and wild sunflower, use the higher rate. Apply to stubble ground. Treat only once during same fallow period.

If weeds are present at application, a tank mix with paraquat CL may be used. Broadcast 1–2 pts. of ATRAZINE 4L plus 1–2 pts. of a nonionic surfactant, such as X-77, per 100 gal. of spray mixture. Add ATRAZINE 4L to spray tank first and thoroughly mix with water. Then add paraquat CL, followed by surfactant. Use the 2 pt. rate of paraquat CL if weeds are 4–6 inches tall. This mixture will not control weeds taller than 6 inches. Apply to stubble ground. Treat only once during same fallow period. Refer to paraquat CL label for further directions, precautions, and limitations.

If weeds are present at application and ATRAZINE 4L is used alone, use either an approved contact herbicide before or after treatment, or tillage after treatment.

Use tillage to control weeds which escape during fallow period. Till before planting. For ATRAZINE 4L applied alone or in tank mixture with paraquat CL, plant at least 2 inches deep and 12 months or more after application.

Precautions

(1) Do not use on sand soil. (2) Do not treat eroded hillsides, caliche and rocky outcroppings, or exposed calcareous subsoil. (3) Do not treat soils of the Rosebud and Canyon Series in Western NE and adjoining counties in CO and WY. (4) Do not treat soils with calcareous surface layers. (5) Avoid spray overlap.

NOTE: Do not graze treated areas within 6 months after application.

RANGELAND

ATRAZINE 4L controls certain weeds, including cheatgrass (downy brome, chess), tumble mustard, sagewort, annual or common broomweed. Use lower rate on areas where cheatgrass is the major problem weed; use higher rate for other weeds.

For ground applications, use a minimum of 10 gals. of water per acre.

For aerial applications, use a minimum of 5 gals. of water per acre. See "Application Procedures" for additional information.

Establishment of Perennial Range Grasses (CA, ID, NV, OR, UT, WA)

Apply 1.6–2 pts. per acre in fall after rains begin, but before ground freezes. Perennial range grass may be seeded when rain begins the next fall following treatment. Plant in deep furrows such as those made by a deep furrow rangeland drill. Treated areas may be grazed 7 months after seeding. However, to provide sufficient time for establishment of desirable range grasses, treated areas should not be grazed until the fall following seeding.

Renovation of Existing Stands of Perennial Range Grasses

Central Great Plains (CO, KS, MT, ND, NE, SD, WY): Apply 1–2 pts. per acre in fall before ground freezes.

Southern Great Plains (OK, TX): Apply 2 pt. per acre in spring before April 30.

West (CA, ID, NV, OR, UT, WA): Do not graze area to be treated for six months prior to application. Apply 1–1.5 pt. per acre by ground equipment in fall before the ground freezes.

Bluegrass and intermediate wheatgrass may be injured by this product at recommended rates.

Do not apply to crested and intermediate wheatgrasses. Do not cut or feed range grass hay.

Do not cut or feed areas within seven months following fall application or three months following spring application. Apply only once per year.

Warm Season Perennial Range Grasses—Midwest Switchgrass

For control of many annual broadleaf and grass weeds including cheatgrass, foxtails (giant, green, and yellow), lambsquarters, pigweed, ragweed, smartweed, and velvetleaf.

For ground application use a minimum of 15 gal. of water per acre.

Use only on loam, silt loam, silty clay loam, clay loam and silty clay soils with at least 1 percent organic matter.

New Seedling

Apply 4 pt. per acre after planting but before weeds emerge. Prepare a good, firm seedbed. Plant 0.5 inch deep with a grassland drill (preferred method) or a conventional drill. If a conventional drill is used on prepared seedbeds, remove all tension from the disk openers. For best results, cultipack or roll after planting. Clip weeds that escape in July or August. Avoid clipping switchgrass seedlings.

Established Stands

Apply 4 pt. per acre in April or early May, before weed emergence.

Precautions

See "Precautions for All-Applications To Corn" in Corn section of this label.

NOTE: Do not cut for hay. Do not graze treated areas within 4 months following application at seeding or 3 months following application to establish switchgrass. Treat only once per year.

SUGARCANE

For control of many broadleaf and grass weeds, including amaranths, crabgrass, fireweed, Flora's paintbrush, foxtails, junglerice, and wiregrass: Broadcast 4–8 pts. in 20–50 gals. of water for adequate coverage of the soil surface at time of planting or ratooning, but before the cane emerges. One additional application may be made over the cane as it emerges, and two additionals may be made interline after emergence as directed sprays.

For control of emerged pellitory weed (Florida only): Apply 0.8–1.2 pt. per acre in at least 40 gals. of water per acre as a directed spray. Add 2 qts. of surfactant for each 50 gals. of spray. Thoroughly cover weed foliage.

Precautions:

(1) Do not apply after "close-in". (2) Do not apply more than 20 pt. per acre to any one crop of cane.

For control of alexandergrass, large crabgrass, pellitory (artillery) weed, and spiny amaranth, use one of the following methods at planting or ratooning (Florida only).

1. Apply 8 pt. in 20–50 gal. of water per acre, preemergence, broadcast or banded. Follow with one or two broadcast or banded over-the-top applications as needed, postemergence to sugarcane and weeds, at 4 pt. in 20–50 gal. of water per acre. Treat before weeds exceed 1.5 inches in height.
2. Apply 1–3 times as needed, at 4 pt. in 20–50 gal. of water per acre either broadcast or banded over-the-top, postemergence to sugarcane and weeds. Treat before weeds exceed 1.5 inches in height.

FORESTS AND CHRISTMAS TREE PLANTATIONS

Douglas Fir, Grand Fir, Noble Fir, White Fir, Lodgepole Pine, Ponderosa Pine and Scotch Pine

Annual broadleaf and grass weed control

Broadcast 4–8 pt. in 20–40 gals. of water per acre to assure thorough coverage. Apply between fall and early spring while trees are dormant or soon after transplanting and before weeds are 1.5 inches high.

Quackgrass Control

Broadcast 8 pt. per acre in 20–40 gals. of water for adequate coverage. Apply in fall or early spring while trees are dormant and before weed seedlings are more than 1.5 inches high.

Precautions

(1) Do not graze treated areas. (2) Do not apply to seedbeds. (3) Do not make more than one application per year.

CONDITIONS OF SALE AND WARRANTY

Shell and the seller offer this product, and the buyer and user accept this product, only under the following agreed conditions of sale and warranty.

The directions for use of this product are believed to be reliable and should be followed carefully. However, it is impossible to take into account all variables and to eliminate all risks associated with its use. Injury or damage may result because of conditions which are beyond the control of Shell or the seller, including soil texture, organic matter content of soil, weather, presence of other materials, and the manner of use or application. Shell warrants only that this product conforms to the chemical description on the label and is believed to be reasonably fit for the purposes referred to in the Directions for Use when used as directed under normal conditions. SHELL MAKES NO OTHER EXPRESS OR IMPLIED WARRANTY OF FITNESS OR MERCHANTABILITY OR ANY OTHER EXPRESS OR IMPLIED WARRANTY. In no case shall Shell or the seller be liable for consequential, special, or indirect damages, such as loss of crops, resulting from the use or handling of this product. Any variation or exception from this warranty must be in writing and signed by an authorized Shell representative.

BLADEX® Herbicide is a registered trademark of Shell Chemical Company.
LASSO is a registered trademark of Monsanto Company.
PRINCEP is a registered trademark of Ciba-Geigy.
X-77 is a registered trademark of Kalo Laboratories, Inc.
SENCOR is a registered trademark of Mobay Chemical Company.
LEXONE is a registered trademark of E.I. DuPont de Nemours and Co.

ATRAZINE 80W

HERBICIDE

For Season-Long Weed Control in Corn and Sorghum. For Weed Control in Certain Other Crops; In Noncrop Areas; And Industrial Sites.

ACTIVE INGREDIENTS	BY WEIGHT
Atrazine: 2-chloro-4-ethylamino-6-isopropylamino-s-triazine	76%
Related compounds	4%
INERT INGREDIENTS	20%
TOTAL	100%

EPA Reg. No. 201-410

Keep Out of Reach of Children
CAUTION

NOTE CONDITIONS OF SALE AND WARRANTY BEFORE OPENING THIS CONTAINER. IF THEY ARE NOT ACCEPTABLE RETURN UNOPENED PRODUCT.

SEE ADDITIONAL PRECAUTIONS

PRECAUTIONARY STATEMENTS

HAZARD TO HUMANS AND DOMESTIC ANIMALS
CAUTION

Causes eye and skin irritation. Do not get in eyes, on skin, or on clothing. In case of contact with eyes, immediately flush eyes with plenty of water for at least 15 minutes. Call a physician. Harmful if swallowed. Wash thoroughly with soap and water after handling and before eating or smoking. Avoid inhalation of dust and contamination of food and feed. For 24 hour emergency medical assistance, call (713) 473-9461.

ENVIRONMENTAL HAZARDS

Atrazine can travel (seep or leach) through soil and can enter groundwater which may be used as drinking water. Atrazine has been found in groundwater. Users are advised not to apply Atrazine to sand and loamy sand soils where the water table (groundwater) is close to the surface and where the soils are very permeable. Your local agriculture agencies can provide further information on the type of soil in your area and the location of groundwater.

Atrazine is toxic to acquatic invertebrates. Do not apply directly to water or wetlands. Do not apply where runoff is likely to occur. Do not apply when weather conditions favor drift from treated areas. Runoff and drift from treated areas may be hazardous to aquatic organisms in neighboring areas. Do not contaminate water by cleaning of equipment or disposal of wastes. In case of significant spill, call CHEMTREC (800) 424-9300 or (713) 473-9461.

DIRECTIONS FOR USE

It is a violation of Federal law to use this product in a manner inconsistent with its labeling.

REENTRY STATEMENTS

Do not apply this product in such a manner as to directly or through drift expose workers or other persons. The area being treated must be vacated by unprotected persons.

Do not enter treated areas without protective clothing until sprays have dried.

Because certain states may require more restrictive reentry intervals for various crops treated in this product, consult your State Department of Agriculture for further information.

Written or oral warnings must be given to workers who are expected to be in a treated area or in an area about to be treated with this product. (Indicate specific oral warnings which inform workers of areas or fields that may not be entered without specific protective clothing, period of time field must be vacated and appropriate actions to take in case of accidental exposure.) When oral warnings are given warnings shall be given in a language customarily understood by workers. Oral warnings must be given if there is reason to believe that written warnings cannot be understood by workers. Written warnings must include the following information:

"CAUTION! Area treated with Atrazine Herbicide on (date of application). Do not enter without appropriate protective clothing until sprays have dried. (Insert here actions to take in case of accidental exposure. See STATEMENT OF PRACTICAL TREATMENT for details.)"

STORAGE AND DISPOSAL

STORAGE

Do not contaminate water, fôod, or feed by storage or disposal. Do not use or store around the home environment. Avoid contact with water. In case of spill or leak, avoid breathing dust or vapors. Clean up and dispose of wastes in compliance with local, state and Federal regulations.

PESTICIDE DISPOSAL

Wastes resulting from the use of this product may be disposed of on site or at an approved waste disposal facility.

CONTAINER DISPOSAL

Completely empty bag into application equipment. Then dispose of empty bag in a sanitary landfill or by incineration, or, if allowed by state and local authorities, by burning. If burned, stay out of smoke.

GENERAL INFORMATION

ATRAZINE 80W will control many annual broadleaf and grass weeds in corn, sorghum, sugarcane, pineapple, and certain other crops. It is also effective for control of many annual and many perennial broadleaf and grass weeds in non-crop areas and industrial sites. ATRAZINE 80W may be applied before or after weeds emerge.

In each case where a range of rates is given, the lower rate should be used on light soils and soils low in organic matter; the higher rate should be used on heavy soils and soils high in organic matter.

Since ATRAZINE 80W acts mainly through root absorption, its effectiveness depends on rainfall or irrigation to move it into the root zone. Should weeds develop, a shallow cultivation or rotary hoeing will generally result in better weed control.

Wash sprayer thoroughly after use.

ATRAZINE 80W is noncorrosive to equipment and metal surfaces, nonflammable, and has low electrical conductivity.

Care should be taken to avoid using ATRAZINE 80W where adjacent desirable trees, shrubs, or plants might be injured.

APPLICATION PROCEDURES

ATRAZINE 80W is a wettable powder that should be mixed with water and applied as a spray as follows: (1) Fill spray tank ½ to ⅔ full of clean water. Be sure tank is clean and not contaminated with 2,4-D, residual oil or other materials. (2) Start agitation. (3) Pour ATRAZINE 80W directly from the bag into the tank. Then let it wet and settle into the water. (4) Fill the tank the rest of the way with water. Continue agitation during mixture and application.

GROUND APPLICATION

Use 10–40 gals. of water per acre to assure even coverage unless otherwise specified. For uniform distribution, use 80° flat fan-type nozzles. Screens in nozzles, as well as those in suction and in-line strainers, should be no finer than 50-mesh.

Use a suitable pump with capacity to deliver (1) the necessary volume to the nozzles at 35–40 psi and (2) an additional volume to maintain the mixture in suspension in the spray tank.

For band applications, calculate amount to be applied per acre as follows:

$$\frac{\text{Band width in inches}}{\text{Row width in inches}} \times \text{Broadcast rate per acre} = \text{Amount needed per acre of field}$$

AERIAL APPLICATION

Use only where broadcast applications are specified. Apply minimum of 1 gallon of water for each 1 lb. to 1.5 lbs. of ATRAZINE 80W to be applied per acre.

For postemergence treatments on corn and sorghum apply recommended rate in a minimum of 2 gallons of water/acre.

Avoid applications under conditions where uniform coverage cannot be obtained or where excessive spray drift may occur.

APPLICATION IN LIQUID FERTILIZER

For preplant and preemergence applications on corn or sorghum, nitrogen solutions or complete liquid fertilizers may replace all or part of the water as a carrier for ATRAZINE 80W. Do not apply after corn or sorghum has emerged as there is danger of liquid fertilizers causing crop injury.

Check to ensure the compatibility of ATRAZINE 80W with liquid fertilizer prior to mixing in the spray tank. A convenient method for testing the compatibility of ATRAZINE 80W with liquid fertilizer follows:

A. Add one pint of liquid fertilizer or other carrier to be used to two quart jars. Mark one quart jar "with" and the other jar "without."

B. To the jar marked "with" add ¼ teaspoon of a suitable compatibility agent. Shake gently for 5 to 10 seconds to mix. (¼ teaspoon/pint = 2 pints/100 gallons.)

C. To each jar mix the number of teaspoons of herbicide shown on the following chart times the number of lbs. per acre of ATRAZINE 80W to be applied corresponding to the gallons per acre of liquid fertilizer to be used. (For example, if 3 lbs. per acre of ATRAZINE 80W are to be applied in 20 gallons of liquid fertilizer per acre, then 3 × 1½ teaspoons, or 4½ teaspoons should be mixed with one pint of liquid fertilizer.) If more than one herbicide is used, add them separately; wettable powders first, flowables second and emulsifiable concentrates last. Shake gently for 5 to 10 seconds after each addition.

D. Let stand one-half hour and make observations. If any separation, agglomeration or precipitation has occurred, again shake the jar 10 to 15 seconds.

1. If mixtures appear to be compatible in both jars, the herbicide(s) can probably be used without a compatibility agent.
2. If the mixture in the jar marked "with" is compatible, but the one in the jar marked "without" is incompatible, a compatibility agent should be used.
3. If both mixtures are incompatible, the liquid fertilizer and the herbicide(s) should not be used in the same spray tank.

If a test such as outlined indicates that components of a proposed mix are comptatible, materials to be combined in the spray tank should still be added in the sequence according to directions prescribed on the label of the herbicides or pesticides involved.

JAR TEST FOR ATRAZINE 80W COMPATIBILITY WITH LIQUID FERTILIZER

Gal. Liquid Fert/Acre	10	15	20	25	30	40
Teaspoons ATRAZINE 80W/ Pint of Fertilizer	3½	2½	1¾	1½	1¼	1

APPLICATION IN WATER PLUS EMULSIFIABLE OIL OR OIL CONCENTRATE

The addition of oil to ATRAZINE 80W in water sprays may speed the activity of ATRAZINE 80W and provide quicker kill of weeds. However, under certain conditions the use of oil may seriously damage corn. To minimize the possibility of damage to corn, follow the directions, procedures and precautions given below.

Use a crop oil containing at least 1 percent suitable emulsifier or a crop oil concentrate containing not more than 20% emulsifier or surfactant blend designated for use with atrazine. Several oils of this type are on the market. Emsulsifiable oil and oil concentrate contaminated with water or other materials can cause compatibility problems and/or crop injury.

MIXING PROCEDURES

(1) Thoroughly clean sprayer prior to use. Do not use sprayer contaminated with 2,4-D or other materials as crop damage or sprayer clogging may result. (2) Fill spray tank ½–⅔ full of clean

water. (3) Start agitation. (4) Pour ATRAZINE 80W directly from bag into tank. Let the ATRAZINE 80W wet and settle into the water. (5) Add emusifiable oil or oil concentrate after the ATRAZINE 80W and water are mixed thoroughly. (6) Fill the tank the rest of the way with water. (7) Empty the tank as completely as possible before refilling to prevent a buildup of emulsifiable oil in the tank. Keep the agitation system in operation to avoid a separation of the oil from the water in the spray mixture still remaining in the tank. (8) If an oil film starts to build up in the tank, drain the tank and clean out with a strong detergent solution or solvent. (9) Clean the sprayer thoroughly immediately after use by flushing the system with water containing a detergent.

PRECAUTIONS FOR APPLICATIONS WITH EMULSIFIABLE OIL OR OIL CONCENTRATE IN WATER

(1) Do not use when corn is under stress from prolonged cold, wet weather, poor fertility, or other factors or when corn is wet and succulent from recent rainfall as crop injury may occcur. (2) Do not use when treating inbred lines or any breeding stock as injury may occur. (3) Adding other insecticides, herbicides, liquid fertilizers, or other materials is not recommended because they may cause compatibility problems or crop injury. (4) Store and handle emulsifiable oil and oil concentrate carefully. Oil contaminated with even a small amount of water may not emulsify properly when added to the tank. (5) Do not make more than one application per season except as recommended for control of yellow nutsedge and Canada thistle on this label.

INSTRUCTIONS FOR SAFE HANDLING

Avoid application directly to humans or animals. Although it is unnecessary for flagmen or loaders to wear special protective clothing or equipment, care should be taken to avoid inhalation of dust or spray mist, or prolonged contact with skin. Flagmen and loaders should wash thoroughly before eating and at the end of each day's operation.

CORN

One application will control many annual broadleaf and grass weeds such as barnyardgrass, witchgrass (Panicum capillare), yellow foxtail, green foxtail, wild oats, large (hairy) crabgrass, giant foxtail, velvetleaf, morningglory, lambsquarters, pigweed, ragweed, nightshade, purslane, and mustard. ATRAZINE 80W may be applied either before planting, at planting, or after planting using 1.25–5 lbs. per acre as indicated below.

PREPLANT

Broadcast in the spring after plowing at the rate indicated in Table 1. Application may be made before, during or after final seedbed preparation. If soil is tilled or worked after application, avoid deep incorporation of ATRAZINE 80W. Best results have been obtained when ATRAZINE 80W is applied within two weeks prior to planting.

PREEMERGENCE

Apply during or shortly after planting to weed emergence at the rate indicated in Table 1.

POSTEMERGENCE

Apply before weeds exceed 1.5 inches in height, at the rate indicated in Table 1.

Table 1
Broadleaf and Grass Weed Control in Corn*

Soil Texture	Broadcast Rate Per Acre of ATRAZINE 80 W
COARSE SOILS:	
Sandy, loamy sands and sandy loams	2.5 lbs.
MEDIUM SOILS:	
Silt and clay loams that are low in organic matter	3 lbs.
FINE SOILS:	
Silt and clay loams with medium to high organic matter and clays (including the dark prairie soils of the Corn Belt)	3.75 lbs.
Peats, mucks, and high organic clays, (Apply postemergence only)	3.75 lbs.

* Broadleaf control (Eastern CO, Western KS, Western NE, NM, OK Panhandle, West TX, and Eastern WY). On sand, loamy sand, sandy loam, mild to strongly alkaline soil and all recently leveled soil apply no more than 1.5 lbs./A. either preplant or preemergence. Controls many broadleaf weeds including pigweed, lambsquarters, night-shades, purslane, and kochia. On other soils in these areas apply rate in Table 1 for broadleaf and grass control.

LAYBY TREATMENT

Broadcast 1.25–2.5 lbs. per acre in 20–40 gals. of water or sufficient nitrogen solution before weeds are 1.5 inches high. Apply when corn is 20–30 inches high. When nitrogen solutions are used, direct the spray to the lower 3–4 inches of cornstalks to avoid corn foliage injury. Agitation in the spray tank during application is essential.

POSTEMERGENCE WITH EMULSIFIABLE OIL OR OIL CONCENTRATE IN WATER

Ground Application:
Broadcast 2.5 lbs. of ATRAZINE 80W and 1 gal. of emulsifiable oil or 1 qt. of oil concentrate/acre after emergence but before weeds reach 1.5 inches in height.

Aerial Application:
Broadcast 2.5 lbs. of ATRAZINE 80W and 0.5 gal. of emulsifiable oil per acre after weed emergence but before weeds reach 1.5 inches in height.

Broadleaf Control:
Broadcast 1.5 lbs. per acre for control of many broadleaf weeds, including morningglory, cocklebur, lambsquarters, mustards, pigweed, ragweed, smartweed, and wild buckwheat. Add 1 gal. of emulsifiable oil per acre for ground application and 0.5 gal. per acre for aerial application. Add 1 qt. of oil concentrate per acre for ground application. Apply before pigweed and lambsquarters reach 6 inches in height and before all other weeds reach 4 inches in height. A cultivation may be necessary if all weeds are not controlled or if weeds regrow.

CENTER PIVOT SPRINKLER APPLICATION

Preemergence or Postemergence (CO, KS, NE, SD, and WY): Apply with irrigation water either after planting before corn and weeds emerge, or after corn emergence, but before layby (20–30 inches) and before weeds exceed 1.5 inches in height, at rates in Table 1. Prepare mixture with minimum ratio of 1 part product to 1 part water. Injecting a larger volume of a more dilute slurry per hour will assure more accurate calibration of metering equipment. Maintain sufficient agitation to keep herbicide in suspension. Meter slurry into irrigation water during entire period. Apply in 0.5–1 inch of water. Use the lower volume on coarser textured soils, the higher volume on finer textured soils. More than 1 inch of water may reduce weed control by moving herbicide below the effective zone in the soil. Inject dilute slurry into system through a positive displacement pump.

Precautions: (1) Apply only through irrigation systems containing antisiphon and check valves to prevent contamination of well during shutdown and overflow of solution tank. (2) Inject ahead of any right angle turn in the main line to insure adequate mixing. (3) Chemical injection pumps and water pumps must have interlocking controls to insure simultaneous shutoff. (4) Application when drift may occur from windy conditions, when system joints and connections are leaking, or when nozzles are not providing uniform distribution may cause crop injury. (5) Where sprinkler distribution patterns do not overlap sufficiently, unacceptable weed control may result. Where sprinkler distribution patterns overlay excessively, crop injury may result.

PROBLEM WEEDS

Yellow Nutsedge and Canada thistle: ATRAZINE 80W will control yellow nutsedge (Cyperus esculentus) and Canada thistle (Cirsium arvense) when applied according to directions. For best results apply each year until yellow nutsedge or Canada thistle is eliminated or reaches a level of infestation where neither weed is a problem. If yellow nutsedge or Canada thistle regrow following last application, cultivate once.

Six alternative methods of use for control of yellow nutsedge and Canada thistle are listed below in order of preference. If other weed species including annual grasses are also expected use alternative 2, 3, 5 or 6.

(1) Broadcast 2.5 lb. plus 1 gal. of emulsifiable oil or 1 qt. of oil concentrate/acre after crop and yellow nutsedge or Canada thistle emerge, but before yellow nutsedge reaches a height of 3 inches or Canada thistle reaches a height of 6 inches. Repeat application before lay-by (20–30 inches) 10–20 days after the first application.

(2) Broadcast 2.5 lb/acre preplant. Follow with an application of 2.5 lb. plus 1 gal. of emulsifiable oil or 1 qt. of oil concentrate/acre after corn and weeds emerge, but before yellow nutsedge reaches a height of 3 inches (yellow nutsedge control only).

(3) Broadcast 2.5 lb. during or shortly after planting, but before crop or weed emergence. Follow with an application of 2.5 lb. plus 1 gal. of emulsifiable oil or 1 qt. of oil concentrate/acre after corn and weeds emerge, but before yellow nutsedge reaches a height of 3 inches or Canada thistle reaches a height of 6 inches.

(4) Broadcast 5 lb. plus 1 gal. of emulsifiable oil or 1 qt. of oil concentrate/acre after crop emerges, but prior to lay-by (20–30 inches) and after yellow nutsedge and Canada thistle emerge, but before yellow nutsedge is 3 inches tall or Canada thistle is 6 inches tall.

(5) Broadcast 5 lb./acre preplant (yellow nutsedge control only).

(6) Broadcast 5 lb./acre during or shortly after planting, but before crop or weed emergence (yellow nutsedge control only).

NOTE: Do not use emulsifiable oil or oil concentrate when corn is wet or under stress, especially when using 5 lb. of this product/acre. See "Precautions for Applications with Emsulsifiable Oil or Oil Concentrate in Water" for additional directions.

QUACKGRASS CONTROL ON LAND GOING INTO CORN PRODUCTION

Split application: Broadcast 2.5 lbs. per acre in the fall or spring and plow 1–3 weeks later. Broadcast a second application at the rate of 2.5 lbs. per acre in the spring before, during or after planting but before weeds are 1.5 inches high. This split application will control both quackgrass and most annual broadleaf and grass weeds.

Single application: Broadcast 3.75 lbs. to 5 lbs. per acre in the fall or spring. Plow 1–3 weeks after application.

TANK MIXTURES ON CORN

DUAL 8E: Use as tank mixture for control of certain broadleaf and grass weeds in corn. Refer to Dual 8E label for all directions, weeds controlled, precautions and limitations.

LASSO or LASSO EC (alachlor): Use as tank mixture with Lasso for control of many annual broadleaf and grass weeds in corn (field, hybrid seed, silage, sweet, popcorn), including barnyardgrass, carpetweed, crabgrass, fall panicum, Florida pusley, foxtails (giant, green, yellow), goosegrass, jimsonweed, kochia, lambsquarters, mustard, nightshade, pigweed, purslane, ragweed, signalgrass (Brachiaria) smartweed, and witchgrass. This tank mix will reduce competition from the hard to control annual weeds—annual morningglory, cocklebur, and velvetleaf (buttonweed).

For preemergence or early postemergence broadcast application, use rates in Table 2. Apply tank mixture any time from immediately after planting until weeds reach 2-leaf stage and corn is no more than 5 inches tall. On sweet corn, apply before crop and weeds emerge. Application after the 2-leaf weed stage will not give satisfactory control. With postemergence application, occasional corn leaf burn should not affect growth or yield. If applied with liquid fertilizers, spray before crop emerges.

Table 2
Tank Mixtures With LASSO On Corn

Soil Texture	Broadcast Rate/Acre: Less Than 3% Organic Matter — ATRAZINE 80W	Less Than 3% Organic Matter — LASSO	3% or More Organic Matter — ATRAZINE 80W	3% or More Organic Matter — LASSO
COARSE* Sand, loamy sand, sandy loam	1.25 lb.	1.5 qt.	1.25 lb.	1.5 qt.
MEDIUM Loam, silt loam, silt	1.25–1.5 lb.	1.75 qt.	1.5–1.75 lb.	2 qt.
FINE Silty clay loam, sandy clay loam, silty clay, sandy clay, clay loam, clay	2–2.5 lb.	2.25 qt.	1.5–2 lb.	2.5 qt.

* When applied via center pivot irrigation on coarse soils, apply 1.25 lb. of this product and 2 qt. of Lasso/acre.

For preplant incorporated applications, use 0.25 lb. more of ATRAZINE 80W plus 0.5 qt. of Lasso/acre than indicated in Table 2. Use no more than 1.75 lb. of ATRAZINE 80W plus 2.5 qt. of Lasso/acre. Broadcast within 7 days before planting and incorporate 2 inches deep.

Apply tank mixture in a minimum of 20 gal. of water/acre. Liquid fertilizer may replace all or part of the water. Add ATRAZINE 80W to spray tank, thoroughly mix with water, then add Lasso. Follow mixing procedures in the section "Application Procedures" of this label.

Refer to Lasso label for information concerning center pivot irrigation applications and other directions, limitations, and cautions.

Propachlor 65WP: Use as tank mixture for control of many annual broadleaf and grass weeds in corn (field, hybrid seed, silage, and sweet corn only), including annual morningglory, annual ryegrass, barnyardgrass (watergrass), velvetleaf, carpetweed, cocklebur, crabgrass, fall panicum, Florida pusley, giant foxtail, green foxtail, yellow foxtail, goosegrass, groundsel, jimsonweed, lambsquarters, mustard, nightshade, pigweed, purslane, ragweed, smartweed, and sunflower. Broadcast 1.3–2 lb. of ATRAZINE 80W plus 3.8–6 lb. of propachlor 65WP/acre on soil surface any time from immediately after planting until after broadleaf and grass weeds reach the 2-leaf stage. Use lower rates on coarse-textured soil low in organic matter. Use higher rates on fine-textured soil high in organic matter.

A minimum of 1.75 lb./acre of ATRAZINE 80W in the tank mixture will give better control of annual morningglory, velvetleaf, cocklebur, and sunflower.

Apply in a minimum of 20 gal. of water/acre. Liquid fertilizer may replace all or part of the water as a carrier for applications before crop and weeds emerge. Add ATRAZINE 80W to spray tank and thoroughly mix with water. Then, cut the propachlor 65WP bag and fill with water to fill line. Grasp neck of bag firmly, shake vigorously, and add contents to spray tank.

Refer to propachlor 65WP label for further directions, limitations, and cautions.

PRINCEP 80W or PRINCEP 4L (Simazine): Use as tank mixture for control of many annual weeds, including carpetweed, crabgrass, fall panicum, foxtail, lambsquarters, morningglory, pigweed, ragweed, and velvetleaf. Apply before planting, at planting or after planting, but before crop and weeds emerge, at rates in Table 3. Use 1:1 ratio for control of above weeds. Use 1:2 ratio for expected heavy infestations of crabgrass and fall panicum. Apply in a minimum of 20 gal. of water per acre. Cultivate shallowly if weeds develop.

Preplant: Apply in spring during or after final seedbed preparation. If soil is tilled or worked after application, avoid deep incorporation. For best results, apply within two weeks before planting.

Preemergence: Apply during or shortly after planting, but before crop and weeds emerge.

Refer to Corn sections of this label and to Princep 80W or Princep 4L label for further directions, limitations, and cautions.

If weeds develop a shallow cultivation will generally result in better weed control.

Table 3
Tank Mixtures With Princep On Corn

Soil Texture	Broadcast Rate/Acre 1:1 Ratio* ATRAZINE 80W	1:1 Ratio* Princep 80W or Princep 4L	1:2 Ratio** ATRAZINE 80W	1:2 Ratio** Princep 80W or Princep 4L
Sand, loamy sand, sandy loam	1.25 lb.	1.25 lb. or 2 pt.	0.83 lb.	1.67 lb. or 2.67 pt.
Loam, silt loam, silt, clay loam, sandy clay loam, silty clay loam, sandy clay, or silty clay with low organic matter	1.5 lb.	1.5 lb. or 2.4 pt.	1 lb.	2 lb. or 3.2 pt.
Loam, silt loam, silt, clay loam, sandy clay loam, silty clay loam, sandy clay or silty clay with medium to high organic matter, and clay (including dark prairie soils of the Corn Belt)	1.8 lb.	1.8 lb. or 3 pt.	1.2 lb.	2.4 lb. or 3.84 pt.

* For control of above weeds.
** For control of expected heavy infestations of crabgrass and fall panicum.

Paraquat CL: For control of existing vegetation and residual control where corn will be planted directly into a cover crop, established sod, or in previous crop residues, apply 2.5–3.75 lb. of ATRAZINE 80W and 1–2 pt. Paraquat CL in 20–30 gal. of water per acre. Add 8 fl. oz. of a nonionic surfactant, such as X-77, per 100 gal. of spray mixture. Add ATRAZINE 80W to spray tank, thoroughly mix with water, then add the Paraquat CL and surfactant.

Refer to the Paraquat CL label for further directions, limitations, and cautions.

PRINCEP 80W and PARAQUAT CL: Use as tank mixture with Princep 80W and Paraquat CL for kill of existing vegetation and residual weed control where corn will be planted directly into a cover crop, established sod, or in previous crop residues. Add this product and Princep 80W to water in spray tank, agitating until thoroughly mixed. Then add Paraquat CL and a nonionic surfactant such as X-77. Continue agitation during application. Apply 1.25–2.5 lb. of this product plus 1.25–2.5 lb. of Princep 80W plus 1–2 pt. Paraquat CL in 20–60 gal. of water per sprayed acre within 3 weeks before, during, or after planting, but before corn crop emerges. Add 8 fl. oz. of a nonionic surfactant, such as X-77, per 100 gal. diluted spray. Use lower rates of this product and of Princep 80W on coarse soil, and higher rates on fine-textured soil. Use the 2 pt. rate of Paraquat CL if existing vegetation is 4–6 inches tall. This mixture will not control weeds taller than 6 inches.

Refer to further limitations and precautions on labels for ATRAZINE 80W, Princep 80W, and Paraquat CL.

PRECAUTIONS FOR ALL APPLICATIONS TO CORN

(1) Do not apply more than 5 lbs. ATRAZINE 80W to corn in any one year. (2) Land treated with ATRAZINE 80W should not be planted to any crop except corn or sorghum until the following year or injury may occur. (3) Injury may occur in any crop other than corn or sorghum planted the year following treatment. (4) Following harvest of a treated crop, plow (moldboard or disk-plow) and thoroughly till the soil in the fall or spring to minimize possible injury to rotational spring-seeded crops, regardless of the rate used. (5) To avoid the possibility of injury to following fall or spring-seeded crops, the use of BLADEX® Herbicide in place of ATRAZINE 80W is recommended. Follow directions on the BLADEX label. (6) The use of ATRAZINE 80W on land that is to be rotated the following year to soybeans that will be treated with metribuzin (SENCOR or LEXONE) may increase the chance of injury. (7) If ATRAZINE 80W is applied after June 10, do not rotate with crops other than corn or sorghum the next year or injury may occur. (8) In the High Plains and Intermountain areas of the West where rainfall is sparse and erratic or where irrigation is required, use ATRAZINE 80W to control weeds in corn or sorghum only when corn or sorghum is to follow corn or a crop of untreated corn or sorghum is to precede other rotational crops. (9) Do not graze treated areas or feed treated forage to livestock for 21 days following application.

SORGHUM AND SORGHUM-SUDAN HYBRIDS (Grain and Forage Types)

Apply before planting, at planting, or after planting as indicated in the directions below. See Corn section for weeds controlled.

Preplant (Broadleaf and grass control): Broadcast in spring after plowing at rate in Table 4. Apply before, during, or after final seedbed preparation. If soil is tilled or worked after application, avoid deep incorporation. For best results, apply within two weeks before planting.

Preemergence (Broadleaf and grass control): Apply during or shortly after planting before weed or crop emergence, at rate in Table 4.

Table 4
Preplant and Preemergence Broadleaf and Grass Weed Control in Sorghum*

Soil Texture	Organic Matter	Broadcast Rate/Acre
COARSE: Sand, loamy sand, sandy loam, sandy clay loam	Any level	Do not Use (except for pre-emergent use on bedded sorghum in AZ and CA)
MEDIUM AND FINE: Silt loam, loam, clay loam, clay	Less than 1%	Do not Use (except for pre-emergent use on bedded sorghum in AZ and CA)
	1–1.5%	2–2.5 lb.
	More than 1.5%	2.5–3 lb.

* Do not apply preplant in AL, AR, FL, GA, LA, MS, NC, NM, OK, SC, TN, or TX. Do not apply preemergence in NM, OK, or TX, except in northeast OK and the TX Gulf Coast.

In case of planting failure, sorghum may be replanted. Do not make a second broadcast application, as injury may occur. If originally applied in a band and sorghum is replanted in untreated row middles, ATRAZINE 80W may be applied in a band to the second planting.

Preemergence Broadleaf Weed Control in Furrow Irrigated Bedded Sorghum (AZ and CA only): For preemergence control of many broadleaf weeds, including groundcherry, lambsquarters, morningglory, mustard, pigweed, and purslane, broadcast 1–1.5 lb./acre after bed preparation, during or after planting, but before sorghum and weeds emerge and before the first furrow irrigation. Follow with several regular irrigations to thoroughly wet all soil. Use lower rate on coarse-textured soil and soil low in organic matter; use higher rate on fine-textured soil and soil high in organic matter.

Precautions for Preemergence Application to Furrow Irrigated Bedded Sorghum Grown in AZ and CA: To avoid possible sorghum injury, do not use on sand or loamy sand soils or on sorghum planted in furrow. Applications to sorghum growing on alkali soils or where cuts, fills, or erosions have exposed calcareous or alkali subsoils may cause crop injury. In case of crop failure, do not replant sorghum for 8 months following application. Corn may be planted immediately.

Postemergence (Broadleaf and grass control): Apply at rate in Table 5 before weeds exceed 1.5 inches in height. Apply up to "close-in".

Table 5
Postemergence Broadleaf and Grass Weed Control in Sorghum

Soil Texture	Minimum Sorghum Height at Treatment	Broadcast Rate/Acre
Sand or loamy sand	Do Not Use	
Sandy loam	See directions for postemergence broadleaf weed control	
Silt loam to sandy clay loam	Completely Emerged	2.5–3 lb.
Olton and Pullman clay soil	6 inches	2.5–3 lb.
Silty clay loam and finer textured soil	Completely Emerged	3.75 lb.

Postemergence Broadleaf Weed Control with Emulsifiable Oil or Oil Concentrate in Water: Broadcast 1.5 lb./acre for control of many broadleaf weeds, including annual morningglory, cocklebur, lambsquarters, mustard, pigweed, ragweed, smartweed, and wild buckwheat. Apply before pigweed and lambsquarters reach 6 inches in height and before all other weeds reach 4 inches in height. In CO, Western KS, NM, OK, TX and desert regions of AZ and CA, apply when sorghum is 6–10 inches in height but before it reaches boot stage; in all other areas apply after sorghum reaches the 3-leaf stage. Add 1 gal. of emulsifiable oil/acre for ground application and 0.5 gal./acre for aerial application. Add 1 qt. of oil concentrate/acre for ground application. A cultivation may be necessary if all weeds are not controlled or if weeds regrow.

Precautions for Applications with Emulsifiable Oil or Oil Concentrate in Water: See "Precautions for Applications with Emulsifiable Oil or Oil Concentrate in Water" in the Application Procedures Section of this label.

Postemergence Broadleaf Weed Control with Surfactant (CO, Western KS, NM, OK, TX, and Desert Regions of AZ and CA Only): Broadcast 1.5 lb. plus 0.75–1.5 pt. of surfactant/acre after sorghum reaches 6 inches in height, but before weeds exceed 1.5 inches in height. Apply only on sandy loam and finer textured soil.

Precautions for All Applications to Sorghum: (1) Heavy rain immediately following application tends to cause excessive concentrations of herbicide in seed furrow, resulting in possible crop injury. Do not apply to furrow-planted sorghum until furrows are leveled (plowed-in). Level deep planter marks or seed furrows before application. (2) Application to sorghum growing under stress caused by minor element deficiency or to sorghum growing on highly calcareous soil may result in crop injury. (3) Injury may occur in any crop other than corn or sorghum planted the year following treatment. (4) Following harvest, plow (mold-board or disk-plow) and thoroughly till soil in fall or spring to minimize possible injury to rotational spring-seeded crops, regardless of rate used. (5) Injury may occur if both this herbicide, preplant or preemergence, and an at-planting systemic insecticide are used.

NOTE: Do not graze or feed forage from treated areas for 21 days following application.

Additional Precautions: See "Precautions for all applications to corn" in Corn section of this label.

WHEAT-SORGHUM FALLOW

To control annual broadleaf and grass weeds following wheat harvest and continuing into the following sorghum crop when grown under minimum tillage:

Broadcast 3.75 lbs. per acre to the wheat stubble immediately following wheat harvest. Use 10–40 gals. of water per acre to assure thorough coverage. If weeds are present, remove them with a sweep plow or other suitable implement after application.

NOTE: (1) Sorghum should be planted into the wheat stubble the spring following treatment with minimum disturbance of the soil. Use a surface planter or a planter leaving a shallow furrow. If weeds are present at planting, remove them with a sweep plow or other suitable implement before planting. Do not apply following sorghum harvest.

PRECAUTIONS

(1) Use only on a silt loam or heavier soil. (2) Wheat-sorghum-fallow cropping sequence must be followed. (3) Do not graze treated area or feed treated forage to livestock. (4) Do not plant treated area to any crop other than those on this label within 18 months following treatment.

WHEAT-CORN-FALLOW (KS, NE)

ATRAZINE 80W controls cheatgrass (downy brome, chess), kochia, mustards, pigweed, Russian thistle, wild lettuce, wild sunflower, and volunteer wheat during period after wheat harvest. Weed control may extend into following corn crop grown under minimum tillage.

Apply with ground equipment only.

Follow directions for use, notes, and precautions in the "Wheat-Sorghum-Fallow" section above, substituting corn for references to sorghum.

WHEAT-FALLOW-WHEAT (CO, KS, MT, NE, ND, SD, AND WY)

For preemergence control of cheatgrass (downy brome, chess), common lambsquarters, field pennycress, kochia, mustard, Russian thistle, wild lettuce, and suppression of volunteer wheat during fallow period of a wheat-fallow-wheat rotation, apply 0.6–1.25 lbs. in 10–40 gal. of water per acre for ground application, or in a minimum of 5 gal. of water per acre for aerial application, on all soils except those listed under "Precautions." For control of pigweed and wild sunflower, use the higher rate. Apply to stubble ground. Treat only once during same fallow period.

If weeds are present at application, a tank mix with Paraquat CL may be used. Broadcast 0.6–1.25 lbs. of ATRAZINE 80W plus 1–2 pt. of Paraquat CL in 20–60 gal. of water per acre by ground equipment. Add 0.5–1 pt. of a nonionic surfactant, such as X-77, per 100 gal. of spray mixture. Add ATRAZINE 80W to spray tank first and thoroughly mix with water. Then add Paraquat CL, followed by surfactant. Use the 2 pt. rate of Paraquat CL if weeds are 4–6 inches tall. This mixture will not control weeds taller than 6 inches. Apply to stubble ground. Treat only once during same fallow period. Refer to Paraquat CL label for further directions, precautions, and limitations.

If weeds are present at application and this product is used alone, use either an approved contact herbicide before or after treatment, or tillage after treatment.

Use tillage to control weeds which escape during fallow period. Till before planting. For ATRAZINE 80W applied alone or in tank mixture with Paraquat CL, plant at least 2 inches deep and 12 months or more after application.

Precautions: (1) Do not use on sand soil. (2) Do not treat eroded hillsides, caliche and rocky outcroppings, or exposed calcareous subsoil. (3) Do not treat soils of the Rosebud and Canyon Series in Western NE and adjoining counties in CO and WY. (4) Do not treat soils with calcareous surface layers. (5) Avoid spray overlap.

NOTE: Do not graze treated areas within 6 months after application.

RANGELAND

To control certain weeds, including cheatgrass (downy brome, chess), tumble mustard, sagewort, and annual or common broomweed, use lower rate on areas where cheatgrass is the major problem weed; use higher rate to control other weeds.

For ground applications, use a minimum of 10 gals. of water per acre.

For aerial applications, use a minimum of 5 gals. of water per acre.

See "Application Procedures" for additional information.

Establishment of Perennial Range Grasses (CA, ID, NV, OR, UT, WA): Apply 1–1.25 lb. per acre in fall after rains begin, but before ground freezes. Perennial range grass may be seeded when rain begins the next fall following treatment. Plant in deep furrows such as those made by a deep furrow rangeland drill. Treated areas may be grazed 7 months after seeding. However, to provide sufficient time for establishment of desirable range grasses, treated areas should not be grazed until the fall following seeding.

Renovation of Existing Stands of Perennial Range Grasses: Central Great Plains (CO, KS, MT, ND, NE, SD, WY). Apply 1–1.25 lbs. per acre in fall before ground freezes.

Southern Great Plains (OK, TX): Apply 1.25 lbs. per acre in spring before April 30.

West (CA, ID, NV, OR, UT, WA): Do not graze areas to be treated for six months prior to application. Apply 0.6–0.9 lbs. per acre by ground equipment in fall before the ground freezes.

Bluegrass and intermediate wheatgrass may be injured by this product at recommended rates.

Do not apply to crested and intermediate wheatgrasses. Do not cut or feed range grass hay. Do not graze treated areas within seven months following fall application or three months following spring application. Apply only once per year.

WARM SEASON PERENNIAL RANGE GRASSES (MIDWEST)

Switchgrass: For control of many annual broadleaf and grass weeds including cheatgrass, foxtails, (giant, green, and yellow), lambsquarters, pigweed, ragweed, smartweed, and velvetleaf.

For ground application use a minimum of 15 gal. of water per acre.

Use only on loam, silt loam, silty clay loam, clay loam, and silty clay soils with at least 1 percent organic matter.

New Seedling: Apply 2.5 lb./acre after planting but before weeds emerge. Prepare a good, firm seedbed. Plant 0.5 inch deep with a grassland drill (preferred method) or a conventional drill. If a conventional drill is used on prepared seedbeds, remove all tension from the disk openers. For best results, cultipack or roll after planting. Clip weeds that escape in July or August. Avoid clipping switchgrass seedlings.

Established Stands: Apply 2.5 lb./acre in April or early May before weed emergence.

Precautions: See "Precautions" in Corn Section.

NOTE: Do not cut for hay. Do not graze treated areas within four months following application at seeding or three months following application to established switchgrass. Treat only once per year.

Perennial Ryegrass for Seed (Pacific Northwest Only): For the control of annual broadleaf and grass weeds such as annual ryegrass, rattail fescue, and similar species: Broadcast 1.5 lbs. in 15 gals. or more water per acre, after first fall rains.

PRECAUTIONS

(1) Apply only to perennial ryegrass stands from which a seed crop has been harvested. (2) Do not apply ATRAZINE 80W to ryegrass more than 2 years in succession. (3) Do not graze treated land within 10 days of application. (4) Do not plant treated land to other crops for at least 16 months.

SUGARCANE

For control of any nonrhizomatous weeds, such as crabgrass, junglerice, wiregrass, foxtail, amaranths, Flora's paintbrush, fireweed and similar plants: Broadcast 2.5–5 lbs. per acre in 20–50 gals. of water for adequate coverage of the soil surface at time of planting or ratooning, but before the cane emerges. One additional application may be made over the cane as it emerges, and two additional applications may be made interline after emergence, as directed sprays.

For control of emerged Pellitory Weed (artillery weed)—Florida only: Apply 0.5–0.75 lb. per acre in at least 40 gals. of water per acre as a directed spray. Add 2 qts. of surfactant for each 50 gals. of spray and be sure weed foliage is thoroughly covered.

PRECAUTIONS

(1) Do not apply ATRAZINE 80W after "close in". (2) Do not apply more than 12.5 lbs. of ATRAZINE 80W to any one crop of cane. For control of alexandergrass, large crabgrass, pellitory (artillery) weed, and spiny amaranth, use one of the following methods at planting or ratooning (Florida only).

(1) Apply 5 lb. in 20–50 gal. of water per acre, preemergence, broadcast or banded. Follow with one or two broadcast or banded over-the-top applications as needed, postemergence to sugarcane and weeds, at 2.5 lb. in 20–50 gal. of water per acre. Treat before weeds exceed 1.5 inches in height.

(2) Apply 1–3 times as needed, at 2.5 lb. in 20–50 gal. of water per acre either broadcast or banded over-the-top, postemergence to sugarcane and weeds. Treat before weeds exceed 1.5 inches in height.

TURF GRASSES FOR SOD PRODUCTION (St. Augustine, Centipede, and Zoysia Grass)

Broadcast 2.5–5 lbs. per acre according to soil type as indicated below. Apply in 20–40 gallons of water per acre for thorough coverage.

Muck or Peat	5 lbs.	Old Beds: Within 2 days after lifting of sod.
		New Beds: 3–4 days after sprigging or plugging.
Sandy Soils	2.5 lbs.	Old Beds: Within 2 days after lifting of sod.
		New Beds: 7–10 days after sprigging or plugging

Apply additional 2.5 lbs. on muck or peat, or 1.25 lbs. on sandy soils, if weed growth recurs.

PRECAUTIONS

(1) Do not apply within 30 days of cutting or lifting. (2) Do not apply ATRAZINE 80W in combination with surfactants or other spray additives, as these combinations may cause injury. (3) Do not use north of North Carolina.

FRUIT AND NUT CROPS

MACADAMIA NUTS

For preemergence control of many broadleaf and grass weeds such as crabgrass, foxtail, wiregrass, Flora's paintbrush, Spanish needles, and fireweed.

Broadcast 2.5 lbs. in 50 gals. of water per acre before harvest and just prior to weed emergence. Repeat applications as necessary. Do not spray when nuts are on the ground during the harvest period.

Do not make aerial applications.

PINEAPPLE

For the control of purslane, Spanish needles, annual grasses, annual bindweed, ageratum, amaranths, rattlepod, Flora's paintbrush, fireweed, spurge, indigo, papalo, and other similar weed species:

Make an initial application of up to 8 lbs. per acre as a blanket spray immediately after planting or following plant crop harvest. Make additional blanket or interspace applications of up to 2 lbs. per acre at one or two month intervals as needed, prior to differentiation. Applications should be made in 20–40 gals. of water per acre to assure thorough coverage.

PRECAUTIONS

(1) Do not apply more than 37.5 lbs. ATRAZINE 80W per acre per cycle. (2) Do not apply within 45 days of fruit harvest, or forage harvest if forage is to be fed to livestock. (3) Repeated monthly applications to plant foliage may slow plant growth and delay fruiting.

FORESTS AND CHRISTMAS TREE PLANTATIONS

Douglas Fir, Grand Fir, Noble Fir, White Fir, Lodgepole Pine, Ponderosa Pine and Scotch Pine

Annual broadleaf and grass weed control: Broadcast 2.5–5 lbs. in 20–40 gals. of water per acre to assure thorough coverage. Apply between fall and early spring while trees are dormant or soon after transplanting and before weeds are 1.5 inches high.

Quackgrass control: Broadcast 5 lbs. per acre in 20–40 gals. of water for adequate coverage. Apply in fall or early spring while trees are dormant and before weed seedlings are more than 1.5 inches high.

PRECAUTIONS
(1) Do not graze treated areas. (2) Do not apply to seedbeds. (3) Do not make more than one application per year.

NONSELECTIVE WEED CONTROL ON NONCROP LAND

Apply ATRAZINE 80W at rates recommended below, before or soon after weeds begin growth. Postemergence applications should be made when weeds are young and actively growing. Use the higher rates on heavy clay and muck soils.

ATRAZINE 80W can be used to provide long-term weed control on industrial sites, highway medians and shoulders, railroad rights-of-way, lumberyards, petroleum tank farms, and in non-crop areas on farms, such as around buildings, equipment and fuel storage areas, along fences, and lanes.

Use sufficient water to assure thorough coverage. Use at least 1 gal. of water for each pound of ATRAZINE 80W, more if practical.

To control many annual broadleaf and grass weeds (such as barn yardgrass, cheatgrass, crabgrass, lambsquarters, foxtail, ragweed, puncturevine, and turkey mullein): Broadcast 6–12.5 lbs. per acre.

To control hard-to-kill annual and many perennial broadleaf and grass weeds (such as bluegrass, burdock, Canada thistle, dogfennel, orchardgrass, plantain, quackgrass, purple top, redtop and smooth brome): Broadcast 12.5–25 lbs. per acre.

To control hard-to-kill biennial and perennial weeds (such as bull thistle and sowthistle): Broadcast 25–50 lbs. per acre.

For longer residual control in regions of high rainfall and a long growing season: Broadcast 25–50 lbs. per acre.

For small areas, 4.5 oz. per 1,000 sq. ft. is equivalent to 12.5 lbs. per acre.

CONDITIONS OF SALE AND WARRANTY

Shell and the seller offer this product, and the buyer and user accept this product, only under the following agreed conditions of sale and warranty.

The directions for use of this product are believed to be reliable and should be followed carefully. However, it is impossible to take into account all variables and to eliminate all risks associated with its use. Injury or damage may result because of conditions which are beyond the control of Shell or the seller, including soil texture, organic matter content of soil, weather, presence of other materials, and the manner of use or application. Shell warrants only that this product conforms to the chemical description on the label and is believed to be reasonably fit for the purposes referred to in the Directions for Use when used as directed under normal conditions. SHELL MAKES NO OTHER EXPRESS OR IMPLIED WARRANTY OF FITNESS OR MERCHANTABILITY OR ANY OTHER EXPRESS OR IMPLIED WARRANTY. In no case shall Shell or the Seller be liable for consequential, special, or indirect damages, such as loss of crops, resulting from the use or handling of this product. Any variation or exception from this warranty must be in writing and signed by an authorized Shell representative.

BLADEX® Herbicide is a registered trademark of Shell Chemical Company.
DUAL and PRINCEP are registered trademarks of Ciba-Geigy.
LASSO is a registered trademark of Monsanto Company.
X-77 is a registered trademark of Kalo Laboratories, Inc.
SENCOR is a registered trademark of Mobay Chemical Company.
LEXONE is a registered trademark of E. I. DuPont de Nemours and Co.

AZODRIN® 5

WATER MISCIBLE INSECTICIDE

RESTRICTED USE PESTICIDE

For retail sale to and use only by Certified Applicators or persons under their direct supervision and only for those uses covered by the Certified Applicator's certification.

ORGANIC PHOSPHATE COMPOUND MIXTURE, LIQUID—POISON B (FLAMMABLE LIQUID), NA-2783

ACTIVE INGREDIENT	BY WEIGHT
Dimethyl phosphate of 3-hydroxy-*N*-methyl-*cis*-crotonamide	55%
INERT INGREDIENTS	45%
TOTAL	100%

Contains 5 lb. active ingredient per gallon

EPA Reg. No. 201-157 U.S. Patent Number 3,258,394

DANGER–POISON

Keep Out of Reach of Children

PELIGRO

PRECAUCION AL USUARIO: Si usted no lee ingles, no use este producto hasta que le etiqueta haya sido explicado ampliamente.

STATEMENT OF PRACTICAL TREATMENT

ATROPINE IS THE EMERGENCY ANTIDOTE FOR AZODRIN INSECTICIDE POISONING. 2-PAM is also antidotal and may be used in conjunction with atropine. CALL A PHYSICIAN IN ALL CASES OF SUSPECTED POISONING.

IF SWALLOWED, induce vomiting immediately by giving two glasses of water and sticking a finger down throat. Call a physician. NEVER GIVE ANYTHING BY MOUTH TO AN UNCONSCIOUS PERSON.

IF ON SKIN, immediately remove patient from the vicinity of the insecticide, remove all contaminated clothing, and wash skin with soap and running water.

IF IN EYES, wash immediately with plenty of water.

IF WARNING SYMPTOMS APPEAR: (See WARNING SYMPTOMS below.) Keep patient prone and quiet. Start artificial respiration immediately if patient is not breathing. Transport the patient immediately to the nearest hospital.

PRECAUTIONARY STATEMENTS

HAZARDS TO HUMANS AND DOMESTIC ANIMALS

DANGER! Poisonous if swallowed, inhaled or absorbed through skin. Rapidly absorbed through skin. Moderately irritating to eyes. Repeated inhalation or skin contact may, without symptoms, progressively increase susceptibility to AZODRIN Insecticide poisoning.

Do not swallow or get in eyes, on skin, or on clothing. Do not breathe vapors.

Keep out of reach of children. Do not contaminate food or feed products.

PRECAUTIONS IN USING

Wear clean rubber gloves, water-proof protective clothing, and goggles. Replace gloves frequently and destroy used gloves. During commercial or prolonged exposures in spray mixing and loading operations, wear a pesticide respirator jointly approved by the Mining Enforcement and Safety Administration (formerly U.S. Bureau of Mines) and by the National Institute for Occupational Safety and Health under the provision of 30 CFR Part II. Wash thoroughly with soap and water after handling and before eating or smoking. Decontaminate area and equipment by washing with dilute of alkali (less than 5%) and detergent and rinse with water.

NOTE TO PHYSICIAN

Warning Symptoms: Symptoms include weakness, headache, tightness in chest, blurred vision, non-reactive pin-point pupils, salivation, sweating, nausea, vomiting, diarrhea, and abdominal cramps.

Treatment: Atropine is the specific therapeutic antagonist of choice against parasympathetic nervous stimulation. If there are signs of parasympathetic stimulation, atropine sulfate should be injected at ten-minute intervals, in doses of 1 to 2 milligrams, until complete atropinization has occurred.

Pralidoxime chloride (2-PAM chloride) may also be used as an effective antidote in addition to and while maintaining full atropinization. In adults, an initial dose of 1 gram of 2-PAM should be injected, preferably as an infusion in 250 cc of saline over a 15- to 30-minute period. If this is not practical, 2-PAM may be administered slowly by intravenous injection as a 5 percent solution in water over not less than two minutes. After about an hour, a second dose of 1 gram of 2-PAM will be indicated if muscle weakness has not been relieved. For infants and children, the dose of 2-PAM is 0.25 grams. **Morphine is an improper treatment.**

Clear chest by postural drainage. Oxygen administration may be necessary. Observe patient continuously for 48 hours. Repeated exposure to cholinesterase inhibitors may, without warning, cause prolonged susceptibility to very small doses of any cholinesterase inhibitor. Allow no further exposure until cholinesterase regeneration has been attained as determined by blood test.

For 24-hour emergency medical assistance, call (713) 473-9461.

ENVIRONMENTAL HAZARDS

Wildlife Precautions

This product will kill birds and other wildlife. This product is toxic to shrimp and crab. Keep out of lakes, streams, ponds, tidal marshes and estuaries. Use only as directed on this label. Note use restrictions for the San Joaquin Valley of California. Do **NOT** spray border areas. Keep border areas within 25 feet of field free of bird feed and shelter. Birds feeding on treated area may be killed. Do **NOT** use whenever large flocks of birds are in the vicinity or when they are noted feeding in or around fields. Avoid spray drift. Spray under low pressure. To minimize spray drift, do **NOT** spray when windy. Keep spray nozzles close to foliage. This product is highly toxic to bees exposed to direct treatment or residues on crops. Protective information may be obtained from your Cooperative Agricultural Extension Service.

WILDLIFE PROTECTION IS EVERYONE'S RESPONSIBILITY

PHYSICAL OR CHEMICAL HAZARDS

DANGER—EXTREMELY FLAMMABLE. KEEP AWAY FROM FIRE, SPARKS AND HEATED SURFACES.

DIRECTIONS FOR USE

It is a violation of Federal law to use this product in a manner inconsistent with the labeling.

FOR END USE ONLY. DO NOT REPACKAGE OR REFORMULATE WITHOUT MANUFACTURER'S WRITTEN APPROVAL.

REENTRY STATEMENTS

Do not apply this product in such a manner as to directly or through drift expose workers or other persons. The area being treated must be vacated by unprotected persons.

Do not enter treated areas for forty-eight (48) hours unless protective clothing is worn.

Because certain States may require more restrictive reentry intervals for various crops treated with this product, consult your State Department of Agriculture for further information.

Written or oral warnings must be given to workers who are expected to be in a treated area or in an area about to be treated with this product. (*Indicate specific oral warnings which inform workers of areas or fields that may not be entered without specific protective clothing, period of time field must be vacated and appropriate actions to take in case of accidental exposure.*)

When oral warnings are given, warnings shall be given in a language customarily understood by workers. Oral warnings must be given if there is reason to believe that written warnings cannot be understood by workers.

Written warnings must include the following information: "DANGER! Area treated with AZODRIN® Insecticide on (*date of application*). Do not enter without appropriate protective clothing within 48 hours after treatment. (*Insert here actions to take in case of accidental exposure. See* STATEMENT OF PRACTICAL TREATMENT *for details.*)"

APPLICATION INSTRUCTIONS

For foliage applications, mix the required amounts of AZODRIN 5 Water Miscible Insecticide in sufficient water to provide uniform coverage. If the contents of the product container have been emptied, triple rinse the container and add the rinsate to the spray tank. AZODRIN 5 Water Miscible Insecticide may be applied to tobacco, cotton, potatoes and peanuts by ground or air application and to sugarcane by air only. **Please refer to the Wildlife Precaution Statements.** Consult your State Agricultural Extension Service or Experiment Station for specific use information.

STORAGE AND DISPOSAL

Do not contaminate water, food, or feed by storage or disposal. Do not reuse empty container. Open dumping is prohibited.

STORAGE

Do not store in or around the home. Store in a secure, dry and temperate area. Store in original container. Keep away from fire, sparks and heated surfaces. Do not store below 70 ° F or above 80 ° F for prolonged periods. If spilled avoid exposure. If significant spill, do not attempt to clean up, call CHEMTREC (800) 424-9300 or (713) 473-9461. Spill can be contained by covering with a sufficient amount of absorbent (sand, soil, sawdust, etc.) to collect all liquid. Collected waste must be disposed of according to pesticide disposal below. CAUTION: Wear a pesticide respirator and protective clothing.

PESTICIDE DISPOSAL

Pesticide wastes are acutely hazardous. Improper disposal of excess pesticide, spray mixture or rinsate is a violation of Federal Law. If these wastes cannot be disposed of by use according to label instructions, contact your State Pesticide or Environmental Control Agency, or the Hazardous Waste representative at the nearest EPA Regional Office for guidance.

CONTAINER DISPOSAL

Triple rinse (or equivalent). Then dispose of in a sanitary landfill, or by other approved State and local procedures.

NOTICE OF WARRANTY

SHELL CHEMICAL COMPANY MAKES NO WARRANTY OF MERCHANTABILITY, FITNESS FOR ANY PURPOSE, OR OTHERWISE, EXPRESS OR IMPLIED, concerning this product or its uses which extend beyond the use of the product under normal conditions in accord with the statements made in this label.

CROP	PESTS CONTROLLED	RATE (PINTS/ACRE)	USE INSTRUCTIONS	USE RESTRICTIONS
Tobacco	Tobacco Hornworm, Vegetable Weevil	7/8	Apply with sufficient water to give uniform coverage. Where specified, use lower dosage for light infestations and the higher dosage for heavy infestations. Repeat as necessary to maintain control.	DO NOT USE ON SHADE-GROWN TOBACCO. Do not apply within 5 days of harvest.
	Green Peach Aphid	7/8		
	Tobacco Budworm	7/8–15/8		

CROP	PESTS CONTROLLED	RATE (PINTS/ACRE)	USE INSTRUCTIONS	USE RESTRICTIONS
Cotton	Plant Bugs (Including Lygus and Leafhoppers)‡	$\frac{7}{8}$–$1\frac{5}{8}$	For early season control use $\frac{7}{8}$ pts. per acre. For high populations use 1 to $1\frac{5}{8}$ pts. per acre.	Do not apply within 21 days of harvest. Do not graze livestock on treated fields or feed treated trash
	Bollworms, Pink Bollworms Loopers, Mites, Bool Weevils and Aphids‡	$\frac{7}{8}$–$1\frac{5}{8}$	To control light infestation use $\frac{7}{8}$ pts. per acre. For higher populations use 1 to $1\frac{5}{8}$ pts. per acre. One gallon will treat 5 acres at the $1\frac{5}{8}$ pt./acre dosage, 9.14 acres at the $\frac{7}{8}$ pt./acre dosage. Apply to give uniform coverage. Repeat as necessary to maintain control (but not more frequently than every 5-7 days).	
	Thrips	3–5ozs.		NOTE: California-San Joaquin Valley Only: Do not apply later than 2 weeks after initiation of bloom and definitely not after July 15 in order to avoid injury to wildlife.
	Beet Armyworm	$\frac{5}{8}$–$1\frac{1}{4}$		
	White Flies	$\frac{3}{8}$–$1\frac{5}{8}$		
Sugarcane	Sugarcane Borers Florida only: Aphids Mites	$1\frac{1}{5}$	For aerial applications use 19.2 ozs. in minimum of 2 gals. of water per acre. To insure uniform coverage aircraft applications should be made as close as possible to cane-top level. Repeat applications as necessary to maintain control.	Do not apply within 30 days of harvest. Applications are restricted to the Continental United States. Use for control of aphids and mites is restricted to the State of Florida only.
Peanuts	Lesser Corn Stalk Borer	$1\frac{5}{8}$	Apply the spray in a 10 to 12 inch band which covers only the soil surface and lowest stems and foliage.	Repeat applications as necessary to maintain control, applying no more than 2 applications per season. Do not feed treated vines or hay to livestock. Do not graze livestock on treated fields.
	Mites (Southwestern United States)	$1\frac{1}{4}$–$1\frac{5}{8}$	Use the lower dosage for light infestations and the higher dosage for heavy infestations.	
	Mites (Southeastern United States)	$\frac{7}{8}$–$1\frac{5}{8}$		
	Granulate Cutworms	$1\frac{5}{8}$		
Potatoes (White Irish)	Colorado Potato Beetle, Leafhoppers	$\frac{2}{5}$–$\frac{7}{8}$	Apply in not more than 100 gallons of water to give uniform coverage. Where specified, use the lower dosage for light infestations and the higher dosage for heavy infestations. Repeat application as necessary to maintain control.	Do not apply within 7 days of harvest.
	Flea Beetle	$\frac{7}{8}$		
	Aphids	$\frac{7}{8}$–$1\frac{5}{8}$		
	Mites (Northwestern United States)	$\frac{7}{8}$–$1\frac{5}{8}$		
	Potato Tuberworm	$1\frac{5}{8}$		
Tomatoes Florida Only	Aphids	$\frac{7}{8}$	Apply with ground equipment only. Apply with sufficient water to give uniform coverage. Applications should begin at first appearance of insect populations. Repeat as necessary to maintain control.	Do not apply within 21 days of harvest.
	Leafminers, Cabbage Loopers, Tomato Fruitworm	$1\frac{5}{8}$		

SUPPLEMENTAL LABELING

AZODRIN 5

FOR USE BY COMMERCIAL APPLICATORS ONLY IN THE PRODUCTION OF NURSERY GROWN ORNAMENTAL PLANTS

IMPORTANT

Before using AZODRIN 5 Water Miscible Insecticide, read and carefully observe the cautionary statements and all other information on the product label.

This bulletin contains new or supplemental instructions which may not appear on the container label. Follow the container label carefully.

PRECAUTIONARY STATEMENTS

Post AZODRIN-treated areas with signs stating date of treatment and warning against re-entry for at least two (2) days post-treatment. If entry is made prior to end of interval, protective clothing is required. Wear clean rubber gloves, water-proof protective clothing, goggles and a face shield. Replace gloves frequently and destroy used gloves. During commercial or prolonged exposures in spray mixing, and loading operations, wear a pesticide respirator jointly approved by the Mining Enforcement and Safety Administration (formerly U.S. Bureau of Mines) and by the National Institute for Occupational Safety and Health under the provisions of 30 CFR Part II. Wash thoroughly with soap and water after handling and before eating or smoking. If the material gets into the eyes, immediately flush with water for at least fifteen minutes and get medical attention. In case of contact, remove clothing and wash skin with soap and water; wash clothing before reuse. In case of spillage, wash down with large amounts of water. Decontaminate areas and equipment by washing with a dilute solution of alkali (less than 5%) and detergent and rinse with water.

GENERAL DIRECTIONS

It is a violation of federal law to use this product in a manner inconsistent with its labeling.

AZODRIN 5 is a water-miscible product which contains 5 pounds of active ingredient per gallon. For foliage applications, mix the required amount of AZODRIN 5 in no more than 100 gallons of water per acre to provide uniform and thorough coverage.

It is suitable for use in conventional hydraulic sprayers and low-gallonage ground equipment.

USE AND DIRECTIONS FOR APPLICATION

Outdoor Ornamental Insect Control

AZODRIN 5 is effective against aphids, mites, leafhoppers, thrips, leafminers, cabbage loopers, tobacco budworm, plant bugs and sawflies. At recommended concentrations it can be used safely on a wide range of shrubs, flowers and shade trees (including rose, chrysanthemum, carnations, azalea, fuchsia, philodendron, juniper, pines, gloxinia, daisies, gardenias, orchid, camellias, nasturtium, arborvitae, maple, Japanese boxwood, Euonymus, privet, magnolia, viburnum, and others). Use $\frac{1}{2}$ to $1\frac{1}{2}$ pints of AZODRIN 5 in up to but no more than 100 gallons of spray per acre when insects or their damage appear. Repeat every two weeks or as needed. Do not transplant, pick or otherwise handle the treated ornamentals until 2 days after applications.

ORNAMENTALS: FOR USE OUTDOORS BY PROFESSIONAL APPLICATORS ONLY IN THE PRODUCTION OF NURSERY GROWN ORNAMENTAL PLANTS

INSECTS	FLOWERS AND HERBACEOUS PLANTS, SHRUBS, TREES AND WOODY PLANTS	AMOUNT OF AZODRIN 5 IN UP TO BUT NOT MORE THAN 100 GALS. WATER PER ACRE. SPRAY PLANTS UNTIL POINT OF DRIP.
Aphids, including Green Peach Aphid	Spruce, Douglas Fir, Pine, Numerous Deciduous Trees, and Flowers	1 pint When young observed
Plant Bugs	Flowers and Shrubs	3/4 pint
Cabbage Looper	Flowers	1 pint When young begin feeding. Second spray two weeks later may be needed
Tobacco Budworm	Geraniums	1 1/2 pint May require up to three 1 1/2 pint applications at 30-day intervals.
Leafminers	Azaleas Birch Roses	1/2 pint When leaves are expanded and small mines begin to appear.
Mites	Conifers, Deciduous Trees and Flowers	1 pint Second spray two weeks later may be needed.
Leafhoppers	Deciduous Trees and Flowers	1 pint Second spray two weeks later may be needed.
Sawflies	Deciduous Trees and Conifers	1/2 pint
Thrips, including Cuban Laurel Thrips	Flowers and Shrubs	3/4 pint Repeat every two weeks, or as necessary.

NOTICE OF WARRANTY: SHELL CHEMICAL COMPANY MAKES NO WARRANTY OF MERCHANTABILITY, FITNESS FOR ANY PURPOSE, OR OTHERWISE, EXPRESS OR IMPLIED, concerning this product or its uses which extend beyond the use of the product under normal conditions in accord with the statements made on this label.

BIDRIN® 8

WATER MISCIBLE INSECTICIDE

RESTRICTED USE PESTICIDE

For retail sale to and use only by Certified Applicators or persons under their direct supervision and only for those uses covered by the Certified Applicator's certification.

ORGANIC PHOSPHATE COMPOUND MIXTURE, LIQUID, POISON B, NA-2783

ACTIVE INGREDIENT	BY WEIGHT
Dimethylphosphate of 3-hydroxy-N, N-dimethyl-*cis*-crotonamide	82%
INERT INGREDIENTS	18%
TOTAL	100%

Contains 2 lb. active ingredient per qt.

EPA Reg. No. 201-274
U.S. Patent No. 2,802,855

DANGER–POISON

Keep Out of Reach of Children

PELIGRO

See other precautions.

PRECAUCION AL USUARIO: Si usted no lee ingles, no use este producto hasta que le etiqueta haya sido explicado ampliamente.

STATEMENT OF PRACTICAL TREATMENT

ATROPINE IS THE EMERGENCY ANTIDOTE FOR BIDRIN INSECTICIDE POISONING. 2-PAM is also antidotal and may be used in conjunction with atropine. CALL A PHYSICIAN IN ALL CASES OF SUSPECTED POISONING.

IF SWALLOWED, induce vomiting immediately by giving two glasses of water and sticking a finger down throat. Call a physician. NEVER INDUCE VOMITING OR GIVE ANYTHING BY MOUTH TO AN UNCONSCIOUS PERSON.

IF ON SKIN, immediately remove patient from the vicinity of the insecticide, remove all contaminated clothing, and wash skin with soap and running water.

IF IN EYES, wash immediately with running water for fifteen minutes.

IF WARNING SYMPTOMS APPEAR, (See "Warning Symptoms") keep patient prone and quiet. Start artificial respiration immediately if patient is not breathing. Transport the patient immediately to the nearest hospital.

PRECAUTIONARY STATEMENTS

HAZARDS TO HUMANS AND DOMESTIC ANIMALS

DANGER! Poisonous if swallowed, inhaled or absorbed through skin. Rapidly absorbed through skin. Inhalation or skin contact may, without symptoms, progressively increase susceptibility to BIDRIN Insecticide poisoning. Do not swallow or get in eyes, on skin, or on clothing. Do not breathe vapors. Keep out of reach of children. Do not contaminate food or feed products.

PRECAUTIONS IN USING

Wear clean natural rubber gloves, clean protective clothing and goggles. Replace gloves frequently and destroy used gloves. During commercial or prolonged exposure in spray mixing and loading operations, wear a pesticide respirator approved by the National Institute for Occupational Safety and Health. Wash thoroughly with soap and water after handling and before eating or smoking. Decontaminate areas and equipment by washing with a dilute solution of alkali (less than 5% in water) and detergent. Absorb waste solution. Dispose of collected waste materials according to pesticide disposal instructions.

NOTE TO PHYSICIAN

Warning Symptoms: Symptoms include weakness, headache, tightness in chest, blurred vision, non-reactive pin-point pupils, salivation, sweating, nausea, vomiting, diarrhea, and abdominal cramps.

Treatment: Atropine is the specific therapeutic antagonist of choice against parasympathetic nervous stimulation. If there are signs of parasympathetic stimulation, atropine sulfate should be injected at 10-minute intervals, in doses of 1 to 2 milligrams, until complete atropinization has occured.

Pralidoxime chloride (2-PAM chloride) may also be used as an effective antidote in addition to and while maintaining full atropinization. In adults, an initial dose of 1 gram of 2-PAM should be injected, preferably as an infusion in 250 cc of saline over a 15-to 30-minute period. If this is not practical, 2-PAM may be administered slowly by intravenous injection as a 5 percent solution in water over not less than two minutes. After about an hour, a second dose of 1 gram of 2-PAM will be indicated if muscle weakness has not been relieved. For infants and children, the dose of 2-PAM is 0.25 grams.

Morphine is an improper treatment.

Clear chest by postural drainage. Oxygen administration may be necessary. Observe patient continuously for 48 hours. Repeated exposure to cholinesterase inhibitors may, without warning, cause prolonged susceptibility to very small doses of any cholinesterase inhibitor. Allow no further exposure until cholinesterase regeneration has been attained as determined by blood test.

For 24-hour emergency medical assistance, call (713) 473-9461.

ENVIRONMENTAL HAZARDS

WILDLIFE PRECAUTIONS

This pesticide is extremely toxic to fish, birds, and other wildlife. DO NOT SPRAY BORDER AREAS. Keep border areas within 25 feet of field free of bird feed and shelter. Birds feeding on treated area may be killed. Do NOT use whenever large flocks of birds are in the vicinity or when they are noted feeding in or around fields. Avoid spray drift. To minimize spray drift, do NOT spray when windy. Do not apply directly to lakes, streams, ponds, or wetlands. Runoff from treated areas may be hazardous to aquatic organisms in neighboring areas. Do not contaminate water by cleaning of equipment or disposal of wastes.

This product is highly toxic to bees exposed to direct treatment or residues on blooming crops or weeds. Do not apply this product or allow it to drift to blooming crops or weeds if bees are visiting the treatment area. Protective information may be obtained from your Cooperative Agricultural Extension Service.

WILDLIFE PROTECTION IS EVERYONE'S RESPONSIBILITY

PHYSICAL OR CHEMICAL HAZARDS

Keep away from heat and open flame.

DIRECTIONS FOR USE

It is a violation of Federal law to use this product in a manner inconsistent with its labeling.

FOR END-USE ONLY DO NOT REPACKAGE OR REFORMULATE WITHOUT MANUFACTURER'S WRITTEN APPROVAL.

REENTRY STATEMENTS

Do not apply this product in such a manner as to directly or through drift expose workers or other persons. The area being treated must be vacated by unprotected persons.

Do not enter treated areas for forty-eight(48) hours unless appropriate protective clothing is worn.

Because certain States may require more restrictive reentry intervals for various crops treated with this product, consult your State Department of Agriculture for further information.

Written or oral warnings must be given to workers who are expected to be in a treated area or in an area about to be treated with this product. *(Indicate specific oral warnings which inform workers of areas or fields that may not be entered without specific protective clothing, period of time field must be vacated and appropriate actions to take in case of accidental exposure.)*

When oral warnings are given, warnings shall be given in a language customarily understood by workers. Oral warnings must be given if there is reason to believe that written warnings cannot be understood by workers.

Written warnings must include the following information:

"DANGER! Area treated with BIDRIN® Insecticide on (*date of application*). Do not enter without appropriate protective clothing until sprays have dried. (*Insert here actions to take in case of accidental exposure. See STATEMENT OF PRACTICAL TREATMENT for details.*)"

STORAGE AND DISPOSAL

Do not contaminate water, food, or feed by storage or disposal. Do not reuse empty container. Open dumping is prohibited.

STORAGE

Do not store in or around the home. Store in a secure, dry, and temperate area. Store in original container. Keep away from heat and open flame. If spilled, avoid exposure. If significant spill, do not attempt to cleanup; call CHEMTREC (800) 424-9300 or (713) 473-9461. Spill can be contained by covering with a sufficient amount of absorbent (sand, soil, sawdust, etc.) to collect all liquid. Collected waste must be disposed of according to pesticide disposal below. CAUTION: Wear a pesticide respirator and protective clothing.

PESTICIDE DISPOSAL

Pesticide wastes are acutely hazardous. Improper disposal of excess pesticide, spray mixture, or rinsate is a violation of Federal law. If these wastes cannot be disposed of according to label instructions, contact your State Pesticide or Environmental Control Agency or the Hazardous Waste Representative at the nearest EPA Regional Office for guidance.

CONTAINER DISPOSAL

Triple rinse (or equivalent). Then dispose of in a sanitary landfill or by other approved State and local procedures.

GENERAL DIRECTIONS

BIDRIN 8 Insecticide is a water-miscible product which contains 2 pounds of active ingredient per quart. For foliage applications, mix the required amount of BIDRIN in sufficient water to provide uniform coverage. If container is emptied, triple rinse adding rinsate to the spray tank.

USE AND DIRECTIONS FOR APPLICATION ON COTTON

Insect	Dosage Fl. Oz./A	Directions for Application
Aphids, Thrips, Spider Mites, Cotton Fleahopper	For Early Season control 1.6–3.2	Apply to give uniform coverage. Repeat as necessary, but do not apply more than three times per season. Do not apply within 30 days of harvest. Do not graze livestock on treated fields or feed treated gin trash.
Grasshopers, Boll Weevil	4.0	
Aphids, Stinkbugs, Cotton Fleahopper, Spider Mites, Grasshoppers, Boll Weevil, Black Fleahopper, Plant Bugs (Lygus), Salt Marsh Caterpillars, Leaf Perforators	For Mid or Late Season Control 4.0–8.0	

NOTICE OF WARRANTY

SHELL CHEMICAL COMPANY MAKES NO WARRANTY OF MERCHANTABILITY, FITNESS FOR ANY PURPOSE, OR OTHERWISE, EXPRESS OR IMPLIED, concerning this product or its uses which extend beyond the use of the product under normal conditions in accord with the statements made on this label.

BLADEX® 4L

HERBICIDE

Use of BLADEX® 4L in Corn, Cotton and Grain Sorghum

ACTIVE INGREDIENTS	BY WEIGHT
2–[[4–chloro–6–(ethylamino)-s-triazine-2-yl] amino]-2-methylpropionitrile	43%
INERT INGREDIENTS	57%
TOTAL	100%

Contains 4 lbs. BLADEX per Gallon

EPA Reg. No. 201–281

Covered by U.S. Patents Nos. 3,505,325 and 3,620,710.

Keep Out of Reach of Children
WARNING

NOTE CONDITIONS OF SALE AND WARRANTY BEFORE OPENING THE CONTAINER. IF THEY ARE NOT ACCEPTABLE, RETURN UNOPENED PRODUCT.

STATEMENT OF PRACTICAL TREATMENT

IF SWALLOWED, call a physician or poison control center. Drink 1 or 2 glasses of water and induce vomiting by touching back of throat with finger. Do not induce vomiting or give anything by mouth to an unconscious person.

IF IN EYES, flush with plenty of water. If irritation persists call a physician.

IF ON SKIN, wash immediately with plenty of soap and water.

IF INHALED, remove victim to fresh air. If not breathing, give artificial respiration, preferably mouth-to-mouth and get medical attention.

AVISO

PRECAUCION AL USUARIO: Si usted no lee ingles, no use este producto hasta que le etiqueta haya sido explicado ampliamente.

PRECAUTIONARY STATEMENTS

HAZARD TO HUMANS AND DOMESTIC ANIMALS

WARNING

May be fatal if swallowed. Harmful if inhaled or absorbed through the skin. Causes temporary eye injury.

This product may be hazardous to your health. This product has been determined to cause birth defects in laboratory animals. Use of protective clothing and equipment and following the precautions below can reduce risk.

Avoid breathing spray mist. Avoid contact with skin, eyes, or clothing. Do not get in eyes or on clothing. Wear a face shield when mixing and loading. Wash thoroughly with soap and water after handling and before eating or smoking.

Wear long trousers and long-sleeved clothing when applying this product. Wear rubber gloves extending above the wrist, long trousers and long-sleeved clothing when mixing or loading or when adjusting, repairing or cleaning equipment. Remove contaminated clothing and wash before reuse. Heavily contaminated clothing should be taken off immediately and incinerated or burned, if allowed by state and local authorities, or disposed of in a sanitary landfill and not laundered. For 24-hour emergency medical assistance, call: (713) 473-9461.

Do not apply this product in such a manner as to directly, or through drift, expose workers or other persons. The area being treated must be vacated by unprotected persons.

Keep out of reach of domestic animals, particularly cattle. Consumption of this product, spray solutions, or water contaminated with product can result in serious illness or possible death of bovines.

ENVIRONMENTAL HAZARDS

Do not apply directly to water or wetlands.

Do not contaminate water by cleaning of equipment or disposal of wastes.

BLADEX is a pesticide which can move (seep or travel) through soil and can contaminate groundwater which may be used as a drinking water. BLADEX has been found in groundwater as a result of agricultural use. Users are advised not to apply BLADEX where the water table (groundwater) is close to the surface and where the soils are very permeable (i.e. well drained soils such as loamy sands). Your local agricultural agencies can provide further information on the type of soil in your area and the location of groundwater. In case of significant spill, call CHEMTREC (800) 424-9300 or (713) 473-9461.

DIRECTIONS FOR USE

It is a violation of Federal law to use this product in a manner inconsistent with its labeling.

FOR END USE ONLY. DO NOT REPACKAGE OR REFORMULATE WITHOUT MANUFACTURER'S WRITTEN APPROVAL.

STORAGE AND DISPOSAL

STORAGE

Do not contaminate water, food, or feed by storage or disposal. Do not use or store around the home environment. Avoid contact with water. In case of spill or leak, soak up with sand, earth or synthetic absorbent (do not use alkaline abosrbents) and dispose of wastes in compliance with local, State and Health regulations.

PESTICIDE DISPOSAL

Pesticide, spray mixture or rinsate that cannot be used according to label instructions must be disposed of according to applicable Federal, State, or local procedures.

CONTAINER DISPOSAL

Triple rinse (or equivalent). Then offer for recycling or reconditioning, or puncture and dispose of in a sanitary landfill or by incineration or, if allowed by State and local authorities, by burning. If burned, stay out of smoke.

CONDITIONS OF SALE AND WARRANTY

SHELL AND THE SELLER OFFER THIS PRODUCT, AND THE BUYER AND USER ACCEPT THIS PRODUCT, ONLY UNDER THE FOLLOWING AGREED CONDITIONS OF SALE AND WARRANTY: The directions for use of this product are believed to be reliable and should be followed carefully. However, it is impossible to take into account all variables and to eliminate all risks associated with its use. Injury or damage may result because of conditions which are beyond the control of Shell or the seller, including soil texture, organic matter content of soil, weather, presence of other materials, and the manner of use or applicaion. Shell warrants only that this product conforms to the chemical description on the label and is believed to be reasonably fit for the purposes referred to in the Directions for Use when used as directed under normal conditions. SHELL MAKES NO OTHER EXPRESS OR IMPLIED WARRANTY OF FITNESS OR MERCHANTABILITY OR ANY OTHER EXPRESS OR IMPLIED WARRANTY. Any variation or exception from this warranty must be in writing and signed by an authorized Shell representative.

LIMITATION OF LIABILITY

Shell's liability, whether or not based on negligence, breach of expressed or implied warranty, strict liability or any other legal cause, is limited to the cost of this product. In no case shall Shell or the Seller be liable for consequential, incidental, or indirect damages, such as loss of crops, resulting from the use or handling of this product.

GENERAL INFORMATION

BLADEX 4L Herbicide is a selective herbicide for the control of annual grasses and broadleaved weeds in field corn, cotton, and grain sorghum.

Consult your local Agricultural Extension Agent for help in determining soil texture, organic matter content, and the most appropriate herbicide rate for local conditions.

As a preemergence herbicide, BLADEX 4L is active mainly through the roots, and therefore, its effect on weeds is dependent on adequate rainfall or sprinkler irrigation to move the herbicide into the root zone. Moisture should be sufficient to wet the top $1\frac{1}{2}$ to 2 inches of soil or make the soil too wet to cultivate; for most soils $\frac{1}{2}$–$\frac{3}{4}$" of moisture is sufficient. A rotary hoeing or shallow cultivation is recommended if a rainfall or sprinkler irrigation has not occurred within about ten days after application of BLADEX 4L. Under conditions which delay weed germination, such as low temperatures, lack of soil surface moisture, or when germination extends over a long period, the effectiveness of the herbicide may be impaired. Rotary hoeing, a shallow cultivation or a post-emergence herbicide treatment may be of benefit under these circumstances. When applied as a directed post-emergence herbicide on cotton,

BLADEX 4L is also active through foliage as well as through the roots.

OBSERVE ALL CAUTIONS AND LIMITATIONS ON LABELING OF ALL PRODUCTS USED IN MIXTURES.

APPLICATION DIRECTIONS

GENERAL MIXING AND SPRAYING DIRECTIONS

If another pesticide has been used previously, be sure the sprayer tank, boom, and nozzles have been thoroughly flushed with clean water.

Sufficient jet or mechanical agitation should be provided during the filling operation and during application to keep the spray mixture uniformly suspended. Fill the spray tank at least one-half full of clean water and while pump and agitator are running, add the recommended amount of BLADEX 4L to the tank, then fill with water to the desired level. For tank mixes, once the BLADEX 4L is completely suspended in the water, add the recommended amount of other products to the tank while filling with water to the desired level. Add all wettable powders before liquids. For emulsible concentrates (EC), such as Lasso™ or Dual™ , pre-mixing these products in equal amounts of water or pouring the products directly into the water stream as the tank is filling will aid in achieving compatible tank-mixtures. Fluid fertilizers may replace all or part of the water as carrier. Always check the compatibility of BLADEX 4L or BLADEX 4L tank mixes and fluid fertilizer, according to the Jar Test for Campatibility instructions below, before the herbicide and fertilizer are mixed in the spray tank.

For use with a surfactant on vegetable oil, completely suspend BLADEX 4L in water and fill the tank before adding the surfactant on vegetable oil.

Allow the pump to run several minutes before application to ensure proper suspension and mixing of BLADEX 4L or other various mixtures. If BLADEX 4L or tank-mix combinations of BLADEX 4L with atrazine are allowed to stand in a spray tank or nurse tank for an extended period (several hours or overnight), provide good agitation for several minutes before application to ensure proper suspension and mixing. Tank-mix combinations of BLADEX 4L with other products should be continuously agitated and should be mixed and applied as soon as possible.

Apply with flat fan (even flat fan for band application) or flood nozzles in conjunction with 50 mesh nozzle screens and in-line screens. Carefully follow the nozzle manufacturer's recommendation for correct nozzle height, positioning and spacing. Check for uniform distribution of spray droplets. Apply at pressures of 15–35 psi to reduce chemical loss due to fine mist drift.

Apply the spray mixture to the soil surface using at least 15 gals./A. for ground application. Apply by air using a minimum of 4 gals./A. of spray mixture.

Check nozzles, pressure, and output regularly to ensure proper calibration and application.

ALL RATES INDICATED ON THIS LABEL ARE BROADCAST APPLICATION RATES PER ACRE UNLESS OTHERWISE STATED. FOR BAND APPLICATION, USE PROPORTIONATELY LESS.

JAR TEST FOR COMPATIBILITY

A. Add one pint of liquid fertilizer or other carrier to be used to two quart jars. Mark one quart jar "with" and the other jar "without".

B. To the jar marked "with" and ¼ teaspoon of a suitable compatibility agent. Shake gently for 5 to 10 seconds to mix. (¼ teaspoon/pint = 2 pints/100 gallons.)

C. To each jar add the appropriate amount of herbicides corresponding to the herbicide rate and spray volume selected (see following table). If more than one herbicide is used, add them separately; wettable powders first, flowables second, and emulsible concentrates last. Shake gently for 5 to 10 seconds after each addition.

D. Let stand one-half hour and make observations. If any separation, agglomeration, or precipitation has occurred, again shake the jar 10 to 15 seconds.

1. If mixtures appear to be compatible in both jars, the herbicides can probably be used without a compatibility agent.

2. If the mixture in the jar marked "with" is compatible but the one in the jar marked "without" is incompatible, a compatibility agent should be used.

3. If both mixtures are incompatible, the liquid fertilizer and herbicides should not be used in the same spray tank.

Jar Test for BLADEX 4L
Compatibility with Liquid Fertilizer

Gallons Liquid Fertilizer per acre	15	20	25	30	40
Teaspoons of BLADEX 4L* pint of fertilizer	1½	1¼	1	¾	½

* Based on 1 qt. BLADEX 4L (1 pound active ingredients) per acre in fertilizer volumes indicated. Increase volume proportionately to correspond with intended field rate in terms of quarts BLADEX 4L per acre (e.g., for field rate of 4 qts. BLADEX 4L in 20 gallons fertilizer per acre, add 5 teaspoons BLADEX 4L to 1 pint fertilizer for compatibility testing purposes. CALCULATION: At 4 qts BLADEX 4L rate/20 gallons fertilizer, 4 × 1¼ teaspoons = 5 teaspoons BLADEX 4L/pint of fertilizer).

If a test such as outlined indicates that components of a proposed mix are compatible, the applicator still has the responsibility of combining materials in sequence to the spray tank in accordance with directions prescribed on the label of the herbicides or pesticides involved.

Tests have indicated that compatibility agents, noted below by the various tank mix combinations, may give improved compatibility in liquid fertilizers.

Tank Mix Combination Compatibility Agents

BLADEX 4L/Lasso (Liquid Fertilizer Grade)	Probably not needed in 28–0–0, 10–34–0, Compex may help in others.
BLADEX 4L/Sutan + ™6.7E	Probably not needed in 28–0–0. Incompatible in 10–34–0. Unite, Spray-Mate, Kem-Link, may help in others.
BLADEX 4L/Dual 8E.	Probably not needed in 28–0–0. Unite, Spray-Mate, Ivory Liquid may help in others.

APPLICATION IN CENTER PIVOT SPRINKLER IRRIGATION WATER (For use in Midwest and Plains States only).

BLADEX 4L alone or in tank-mix combinations with atrazine or Lasso 4EC, may be applied in center pivot sprinkler irrigation water. Apply any time after planting but prior to crop emergence. Pre-mix the herbicide(s) in water to form a slurry and inject at a carefully calibrated rate into the irrigation water during the entire period of the run. Mix as directed for a spray application. Provide good agitation for the herbicide mixture throughout the application.

Select the correct application rate of BLADEX 4L or the BLADEX tank-mix combinations from the rate tables listed on the label.

Example:

A center pivot sprinkler that is 1320 feet long (¼ mile) will irrigate 125.7 acres. A sandy loam soil containing 2% organic matter will require 2 quarts of BLADEX 4L per acre.

125.7A × 2qts. = 251.4 quarts BLADEX 4L

$\frac{251.4}{4}$ = 62.9 gallons BLADEX 4L

Mixing tank size 600 gallons. Sprinkler takes 20 hours to apply ½ inch of water to the complete circle; inject rate=

$\frac{600}{20}$ = 30 gal./hr/or 0.5 gal./minute

Calibrate injection pump to deliver herbicide mix into the irrigation system at the rate of 30 gallons per hour or 0.5 gallons per minute. Add 62.9 gallons of BLADEX 4L to the tank.

For tank-mix combinations, select the correct dosage rate and follow the same procedures outlined above for each herbicide component.

Apply in ½ to 1 acre inch of water. Use this method of application only with irrigation systems designed for uniform application of water.

ADDITIONAL USE PRECAUTIONS

1. Apply product or tank mixes only through center pivot irrigation systems containing anti-siphon and check valves which will prevent water source contamination and overflow of the slurry tank and containing interlocking controls between metering device and the water pump to ensure simultaneous shut off.
2. Inject the herbicide mix with a positive displacement pump into the main line ahead of a right angle turn to ensure adequate mixing.
3. Application of more than label recommended quantities of irrigation water per acre may result in decreased product performance by removing the chemical from the zone of effectiveness.
4. Do not apply when wind speed favors drift, when system connectors leak, when nozzles do not provide uniform distribution, or when lines containing the product must be drained and dismantled.
5. Greater accuracy in calibration and distribution will be achieved by injecting a large volume or more dilute slurry per hour.
6. Where sprinkler distribution patterns do not overlay sufficiently, unacceptable weed control may result. Where sprinkler distribution patterns overlap excessively, crop injury may result.

CORN

WEEDS CONTROLLED BY BLADEX 4L ALONE AND IN COMBINATION WITH OTHER HERBICIDES ON CORN

Grasses

Annual bluegrass
Annual fescues
Annual (Italian) ryegrass
Barnyardgrass[1]
Bullgrass
Crabgrass
Fall panicum
Giant foxtail
Goosegrass
Green foxtail
Junglerice
Stinkgrass (Indian lovegrass)
Witchgrass
Yellow foxtail

Broadleaves

Annual groundcherry
Annual morningglory
Annual sedge
Black mustard
Buffalobur
Buttercup (annual)
Carpetweed
Cocklebur[2]
Nightshade (annual)
Pigweed[1]
Pineappleweed
Plantain
Poorjoe
Prickly sida (teaweed)
Prostrate knotweed
Prostrate spurge
Common chickweed
Common groundsel
Common mallow
Common purslane
Corn spurry
Curly dock (seedling)
Fiddleneck
Florida pusley (Florida purslane)
Hedge mustard
Jimsonweed[1]
Kochia
Ladysthumb
Lambsquarters
Mayweed
Ragweed
Russian thistle
Shepherdspurse
Smallflower galinsoga
Smartweed (Pennsylvania)
Sunflower[2] (wild, annual, common)
Tarweed cuphea (Gumweed)
Velvetleaf[1]
Wild buckwheat
Wild mustard
Wild radish
Wild turnip

[1] Under conditions such as low temperatures, lack of soil surface moisture or other factors that may cause delay in germination of the seeds, the effectiveness of BLADEX may be impaired against these weeds.

[2] The degree of control will be reduced if soil moisture and temperature conditions cause deep germination of the seed.

PREEMERGENCE-PREPLANT INCORPORATED

Apply BLADEX treatments just before, at or after planting but before crop has emerged. Avoid removal of treated soil from seedrow prior to or during the planting operation.

BLADEX may also be applied early prior to planting or in a split application if pre-season weed control is desired. For split applications, do not exceed the total amount of BLADEX for the soil texture and organic matter shown in Table 1. If BLADEX is applied early, more than 15 days before planting, a split application of BLADEX or some other herbicide treatment may be necessary at or after planting to provide additional length of weed control. For further information see "Early Preplant" recommendations in the Conservation Tillage section of label.

A rotary hoeing is recommended for preemergence applications which do not receive adequate rainfall or sprinkler irrigation to wet the top 1½-2″ of soil within about 10 days after application.

If a BLADEX mixture is to be incorporated, except as noted, single or two pass incorporation is acceptable. Care should be taken to incorporate the BLADEX mixture no deeper than the top two inches of soil.

BLADEX Applied Alone

Use at the proper rate for soil texture and organic matter indicated in Table 1. Any rotational crop may be planted in the fall or following spring following this treatment.

TABLE 1

PREEMERGENCE BROADCAST APPLICATION RATES PER ACRE FOR BLADEX 4L APPLIED ALONE ON CORN

SOIL TEXTURE DESCRIPTION	QUARTS OF BLADEX 4L — PERCENT ORGANIC MATTER IN SOIL*					
	LESS THAN 1%	1%	2%	3%	4%	5% & OVER
Sand, Loamy Sand	DO NOT USE	1.25	1.5	2.25	2.75	3.25
Sandy Loam	1.25	1.75	2.0	2.5	3.0	3.5
Loam, Silt Loam, Silt	1.5	2.0	2.5	3.0	3.5	4.0
Sandy Clay Loam, Clay Loam, Silty Clay Loam	2.0	2.5	3.0	3.5	4.0	4.5
Sandy Clay, Silty Clay, Clay	2.75	3.0	3.5	4.0	4.5	4.75
Peat or Muck	NOT RECOMMENDED					

* For organic matter content between those listed, adjust the rate proportionately.

BLADEX COMBINATIONS

BLADEX plus Atrazine

Use at the proper rate for soil texture and organic matter indicated in Table 2. Table 2 provides rates for generally weedy conditions. The ratio of the amounts of each herbicide may be adjusted as necessary for particular weed conditions as long as the combined rate of the two products does not exceed the combined rate for the soil shown in Table 2. For grassier conditions use a ratio that contains higher levels of BLADEX (3:1). For fields with more broadleaves use a ratio that contains higher levels of atrazine (1:1).

Rotational crops: (1) Plant only corn, peanuts, sorghum, or soybeans the year following the use of this mixture. (2) If soybeans are to be planted, injury may occur due to the carryover of atrazine. (3) If applied after June 10, do not rotate with crops other than corn or sorghum the next year or injury may occur. (4) In the high plains and intermountain areas of the west where rainfall is sparse and erratic or where irrigation is required, use only when corn or sorghum is to be planted the following year, or a crop of corn or sorghum not treated with atrazine is to precede other rotational crops. (5) Small grains may be planted 15 months following treatment. (6) All other crops may be planted 18 months after application.

TABLE 2
PREEMERGENCE BROADCAST APPLICATION RATES PER ACRE FOR TANK-MIX COMBINATIONS OF BLADEX 4L PLUS ATRAZINE 4L ON CORN

SOIL TEXTURE DESCRIPTION	QUARTS OF BLADEX 4L + QUARTS OF ATRAZINE 4L** — PERCENT ORGANIC MATTER IN SOIL* LESS THAN 1%	1%	2%	3%	4%	5% & OVER
Sand, Loamy Sand	DO NOT USE	0.75 + 0.5	1.0 + 0.5	1.5 + 0.5	1.75 + 0.75	2.25 + 1.0
Sandy Loam	0.75 + 0.5	1.0 + 0.5	1.5 + 0.5	1.75 + 0.75	2.25 + 1.0	2.75 + 1.25
Loam, Silt Loam, Silt	1.0 + 0.5	1.5 + 0.5	2.0 + 0.75	2.25 + 1.0	2.75 + 1.25	3.25 + 1.25
Sandy Clay Loam, Clay Loam, Silty Clay Loam	1.5 + 0.5	2.0 + 0.75	2.25 + 1.0	2.75 + 1.25	3.25 + 1.25	3.5 + 1.25
Sandy Clay, Silty Clay, Clay	2.0 + 0.75	2.25 + 1.0	2.75 + 1.25	3.25 + 1.25	3.5 + 1.25	3.75 + 1.5
Peat or Muck	NOT RECOMMENDED					

* For organic matter content between those listed, adjust the rate proportionately.
** If Atrazine 80W is used, multiply rates shown by 1.25 to equal lbs. of Atrazine 80W.

BLADEX plus Lasso 4 EC
Use BLADEX 4L at the proper rate for the soil texture and organic matter shown in Table 3 plus 2 quarts per acre of Lasso. (Use 2.5 quarts Lasso on clay soils containing 5 percent organic matter and over). Any rotational crop may be planted the fall or spring following this treatment.

BLADEX plus Sutan+ 6.7E, or Eradicane™ 6.7E
Use BLADEX 4L at the proper rate for the soil texture and organic matter shown in Table 3 plus 1.8 quarts per acre of Sutan+ or Eradicane for control of many annual grasses and broadleaf weeds. (Use 2.4 quarts of Sutan+ or Eradicane on loam soils containing 5 percent or more organic matter, and clay loams and clays containing 4 percent or more organic matter). Do not use on sands and loamy sands of less than 1 percent organic matter in the light sandy soils of eastern coastal states. Do not use on corn seed stock.

Apply before planting. Incorporate the mixture immediately upon application using power-driven cultivation equipment set for 2-3 inch depth, or tandem disc set to cut to a depth of about 4 inches while operating at 4-6 mph. For thorough mixing, disc in two directions (cross disc), and follow with a harrow, drag, or other leveling device. Prior to the second discing, readjust the disc to prevent cutting deeper than 4 inches. BLADEX 4L may be applied preemergence as an overlay over previous incorporated Sutan+ or Eradicane if desired. Any rotation crop may be planted in the fall or spring following these treatments.

Existing stands of quackgrass, purple and yellow nutsedge must be turned under and thoroughly chopped up prior to chemical treatments.

Additional weeds controlled by Sutan+ or Eradicane combinations:

Grasses
Sandbur
Shattercane (Wild Cane)*
Texas Panicum
Quackgrass (Eradicane only)
Wild Proso Millet*
(Eradicane only)
Perennial Weeds
Yellow Nutsedge (nutgrass)
Purple Nutsedge (nutgrass)

* Suppression only—refer to Sutan + or Eradicane label for appropriate supplement cultural and tillage practices.

For fields with moderate to heavy infestations of these weeds refer to the Sutan+ or Eradicane labels for appropriate higher rates.

BLADEX Plus Dual 8E
Use BLADEX 4L at the proper rate for soil texture and organic matter shown in Table 3. Use Dual as follows:

Soil Texture	Broadcast Rate Per Acre For Dual
Coarse Sand, loamy sand, sandy loam	1.25–1.5 pints
Medium Loam, silt loam, silt	1.5–2.0 pints
Fine Sandy clay loam, silty clay loam, clay loam, sandy clay, silty clay, clay	1.5–2.5 pints

The low end of the rate range should be used for lowest organic matter soils and the rate increased as organic matter increases to a point that soils containing 4 percent organic matter or more require the highest rate shown for that soil texture. Refer to the Dual label for precautions on rotational crops.

Table 3
Preemergence Broadcast Application Rates Per Acre For BLADEX 4L Used In Tank-Mix Combinations With Lasso, Sutan +, Eradicane, Or Dual On Corn

Soil Texture Description	Quarts of BLADEX 4L — Percent Organic Matter in Soil* Less than 1%	1%	2%	3%	4%	5% & Over
Sand, Loamy Sand	0.6**	0.75	1.25	1.5	1.75	2.0
Sandy Loam	0.75	1.25	1.5	1.75	2.0	2.25
Loam, Silt Loam, Silt	1.25	1.5	1.75	2.0	2.25	2.5
Sandy Clay Loam, Clay Loam, Silty Clay Loam	1.5	1.75	2.0	2.25	2.5	2.75
Sandy Clay, Silty Clay, Clay	1.75	2.0	2.25	2.5	2.75	3.0
Peat or Muck	NOT RECOMMENDED					

* For organic matter content between those listed, adjust the rate proportionately.
** Do not use in the Atlantic Coastal Plain.

BLADEX plus Atrazine plus Lasso, Sutan+, Eradicane, or Dual
Use BLADEX 4L plus atrazine at the proper rate for soil texture and organic matter shown in Table 4. Use Lasso, Sutan+, Eradicane, or Dual according to rates shown in "BLADEX Combinations" in this section of the label.

Rotational crops: Refer to Rotational Crops section of "BLADEX plus Atrazine" in this section of the label.

Table 4

Preemergence Broadcast Application Rates Per Acre for BLADEX 4L Plus Atrazine 4L Used in Tank-Mix Combinations With Lasso, Sutan +, Eradicane, Or Dual On Corn

Soil Texture Description	Quarts of BLADEX 4L + Quarts Atrazine 4L*** Percent Organic Matter in Soil* Less than 1%	1%	2%	3%	4%	5% & Over
Sand, Loamy Sand	0.4 + 0.2**	0.5 + 0.25	0.75 + 0.5	1.0 + 0.5	1.25 + 0.5	1.25 + 0.75
Sandy Loam	0.5 + 0.25	0.75 + 0.5	1.0 + 0.5	1.25 + 0.5	1.25 + 0.75	1.5 + 0.75
Loam, Silt Loam, Silt	0.75 + 0.5	1.0 + 0.5	1.25 + 0.5	1.25 + 0.75	1.5 + 0.75	1.75 + 0.75
Sandy Clay Loam, Clay Loam, Silty Clay Loam	1.0 + 0.5	1.25 + 0.5	1.25 + 0.75	1.5 + 0.75	1.75 + 0.75	2.0 + 0.75
Sandy Clay, Silty Clay, Clay	1.25 + 0.5	1.25 + 0.75	1.5 + 0.75	1.75 + 0.75	2.0 + 0.75	2.0 + 1.0
Peat or Muck	NOT RECOMMENDED					

* For organic matter content between those listed, adjust the rate proportionately.
** Do not use in the Atlantic Coast Plain.
*** If Atrazine 80W is used, multiply rates shown by 1.25 to equal lbs. of Atrazine 80W.

Conservation Tillage Weed Control Early Preplant (EPP)
[Prior-to-planting weed control for land going into production of corn.]

Complete any planned early spring tillage prior to application. Apply herbicide treatment before weeds germinate or before weed seedlings are more than 3 inches tall. Tillage after application may reduce the effectiveness of the herbicide treatment.

Apply BLADEX or BLADEX plus atrazine tank mix 15-30 days prior to planting. Use the proper rate for soil texture and organic matter indicated in Table 1 or 2. Where heavy crop residues exist, the rates shown in Tables 1 or 2 should be increased by 25%.

Where broadleaf weeds are present, add 2,4-DLV Ester at 1⅓-2 pts./A (6 lb./gal.) or 2-3 pts./A. (4 lbs./gal.) plus X-77 surfactant at 1 qt./100 gals. of diluted spray, or other suitable surfactant at its recommended rate. If existing weeds exceed 3 inches in height, add 1-2 pts./A. of paraquat to the above mixture. Well established weeds over 6 inches tall may not be well controlled. Use at least 25 gals/A. of spray mixture by ground rig or 5-10 gals./A. for aerial application.

Depending upon weather conditions following the EPP application, a postemergence BLADEX treatment, or some other herbicide treatment may be necessary at or after planting to provide additional length of weed control. If desired, 1½-2 pts./A. of Dual or 2 qts./A. of Lasso may be tank mixed with the BLADEX EPP treatment or applied preemergence at planting.

Rotational Crops: Refer to Rotational Crops section for each treatment in the Preemergence section of the label.

At Planting
BLADEX 4L applied alone or in combination with atrazine and/or Lasso or Dual according to the following directions will kill most existing small weeds and suppress many emerged perennial weeds when corn is planted into no-till stalk ground (corn, sorghum), stubble ground (soybean, small grains), and any minimum-till situation. This treatment then provides residual control of annual weeds as in conventional tillage.

Apply BLADEX 4L alone or with other products according to the directions for those treatments in the Preemergence section of the label. Where heavy crop residues exist, the BLADEX rate shown in Tables 1,2,3 or 4 should be increased by 25%. Add ½–1 pt./A. of 2,4-D Low Volatile 6 lb. Ester (0.75–1.5 pt./A 2,4–D LV 4 lb. Ester). Add the 2,4-D LV to the spray tank last. Use a minimum of 25 gals/A. of carrier. Complete spray coverage of the weeds is essential for best performance. Nitrogen solutions and complete liquid fertilizers are the preferred carriers for this treatment as they help aid in the burndown of existing weeds. Add Ortho X-77™ surfactant at 1-2 qts. per 100 gals. of diluted spray, or other suitable surfactant at its recommended rate. If water is used as a carrier, crop oil concentrate may be used as a surfactant. Apply before weeds exceed 3" in height. For control of existing alfalfa add 1/3–1/2 pt/A. of Banvel™ to the spray mixture. Apply before the alfalfa exceeds 6" in height.

For fields with existing sod grasses such as orchardgrass, bromegrass, rye or timothy, or when very dry conditions exist, or when existing weeds exceed 3" in height add Paraquat CL to the tank-mix. Use 2 pts./A. of Paraquat CL in combination with BLADEX as described above in this section, except the 2,4–DLV may be omitted if desired. Do not apply Paraquat CL in suspension type liquid fertilizer.

COTTON

IDLE SEASON EARLY PREPLANT WEED CONTROL (CALIFORNIA ONLY)

Weeds Controlled

Grasses

Annual Bluegrass	Volunteer Small Grains (suppression)
Annual Ryegrass	Wild Oat*
Barnyardgrass*	Yellow Foxtail
Bristly Foxtail	
Rabbitsfoot Grass	

Broadleaves

Annual Henbit	Miners Lettuce
Black Nightshade	Lambsquarter
Burclover	London Rocket
Cheeseweed*	Pineapple Weed
Chickweed	Prickly Lettuce
Fiddleneck	Shepherdspurse
Groundsel	Sowthistle
Knotweed	Wild Mustard
Marestail	Wild Radish

* Under soil moisture conditions favoring deep germination, these species may not be completely controlled.

BLADEX may be used for burndown of small existing annual weeds and residual control of weeds during the winter and early spring season prior to planting cotton in California only. Complete any planned tillage prior to application. Apply herbicide treatment before weeds germinate or before weed seedlings are more than 3 inches tall. Tillage after application may reduce the effectiveness of the herbicide treatment.

Apply BLADEX at least 30 days prior to planting. Apply the proper rate for the soil texture, organic matter and time interval between application and planting indicated in the following table. Where existing weeds are present, add crop oil concentrate, surfactant, or emulsible vegetable oil at its recommended rate to aid in the burndown of small weeds.

Where existing weeds are greater than 3 inches in height, when very dry conditions exist, or where volunteer grains are a major problem, tank-mix BLADEX with 1-2 pts./A. of paraquat or 2-3 qts./A. of Dinitro 5 Weed Killer. Well established weeds 6-8 inches tall or taller may not be well controlled.

Apply BLADEX in at least 20 gal./A of carrier by ground, 5 gal./A. by air. Add X-77 surfactant at 1-2 qts./100 gals. of diluted spray where paraquat is used. Crop oil concentrate or emulsible vegetable oil are not needed where paraquat or Dinitro Weed Killer are used.

BLADEX can also be tank-mixed with Treflan™ preplant incorporated for fall listed cotton beds instead of surface applied as described above.

Broadcast Application Rates Per Acre for BLADEX 4L
For Idle Season Early Preplant Treatment on Cotton

	DAYS PRIOR TO PLANTING*					
	30 DAYS QTS. OF BLADEX ORGANIC MATTER		60 DAYS QTS. OF BLADEX ORGANIC MATTER		90 DAYS QTS. OF BLADEX ORGANIC MATTER	
SIOL TEXTURE	2% OR LESS	2% OR MORE	2% OR LESS	2% OR MORE	2% OR LESS	2% OR MORE
Sands & Loamy Sand	1.5	2.0	2.5	3.0	3.0	3.5
All other Soils	2.0	2.5	3.0	3.5	3.5	4.0

* For time intervals between those listed, adjust the rate proportionately.

Precautions:

1. Failure to wait the recommended time interval between application and planting may result in crop injury.
2. At least one inch of rainfall or an equivalent irrigation that waters the surface of the soil after application must precede planting.
3. The use of this treatment on calcareous or caliche soil outcroppings may result in crop injury.
4. Do not graze or feed foliage from treated areas to livestock.

PREEMERGENCE USE

Weeds Controlled

Annual morningglory
Cocklebur
Prickly sida (teaweed)
Spurge

BLADEX 4L Herbicide is a selective preemergence herbicide for early season weed control in cotton. Supplemental practices (such as directed postemergence BLADEX 4L application) may be necessary to control late season weeds. BLADEX 4L can be used in a tank-mix combination with Zorial™ 80WP. Apply only in the states of AL, AR, LA and MS.

Carefully match the BLADEX rate with the soil texture. Do not use on fields where the soil texture changes from coarse to fine. Avoid overlapping the spray pattern or overdosing the field with BLADEX 4L. Application rates above those recommended for the soil texture can result in yellowing or stunting of the crop and may result in stand reduction.

While cotton exhibits tolerance to BLADEX 4L, adverse growing conditions such as excessive rains, standing water or cold weather may result in stand reduction.

DO NOT GRAZE OR FEED FOLIAGE FROM TREATED AREAS TO LIVESTOCK.

BLADEX Plus Zorial 80WP

Apply BLADEX 4L plus Zorial 80WP at the proper rate for the soil texture shown in Table 5. The soil must contain at least 1.0 percent organic matter. Seed placement should be $\frac{1}{2}$–$\frac{3}{4}$ inch from the soil surface. Plant only cotton within six months after the last application of Zorial 80WP or injury may occur.

Table 5
Preemergence Broadcast Application Rates Per Acre For BLADEX 4L + Zorial 80WP On Cotton

Soil Texture (Soil Texture must contain at least 1.0% Organic Matter)		BLADEX 4L + Zorial 80WP qts.	lbs.
Coarse Soils	Sandy loam	0.5	0.8
Medium Soils	Silt and silt loam	0.6	1.3
	Loam, clay loam, sandy clay loam and sandy clay	0.9	1.3
Fine Soils	Silty clay loam, silty clay and clay	1.2	1.6

DO NOT USE ON COARSE SOILS (SANDS AND LOAMY SANDS) CONTAINING MORE THAN 70% SAND.

DIRECTED POST EMERGENCE-LAYBY USE

Weeds Controlled

Annual morningglory*	Palmer amaranth
Bristly starbur	Pigweed (redroot and spiny)
Cocklebur	Prickly sida (teaweed)
Crotalaria	Sicklepod
Jimsonweed	Spurge
Lambsquarters	Tropic croton
Nightshade (annual)	Wright groundcherry

* The degree of preemergence control from a layby treatment will be reduced if soil, moisture and temperature conditions cause deep germination of the seed.

BLADEX 4L and tank-mix combinations may be applied directed postemergence/layby to cotton and either preemergence to weeds or postemergence to weeds in all cotton growing states. Apply before weeds are more than 2 inches tall.

Apply the directed postemergence treatment after the cotton has attained the minimum height of 6 inches. For layby treatment, apply BLADEX 4L after the cotton has attained a height of 12 inches or more.

The spray mixture should be directed to the soil around the base of the cotton plants. Care should be taken to prevent the spray from striking the cotton leaves as injury will occur. The use of leaf lifters or shields on application equipment is recommended to help avoid the spray contacting the cotton foliage.

BLADEX may be applied postemergence following a preemergence application of BLADEX. Apply no more than two postemergence and one preemergence application to the same crop in any one year. If BLADEX is not used preemergence, apply no more than three postemergence applications to the same crop in any one year. When applied as a layby treatment before weeds emerge, the effectiveness of BLADEX 4L depends on rainfall or irrigation to move it into the soil. When irrigation water activation is used, every row must be watered and for skip row cotton all treated soil must be irrigated. Any rotational crop may be planted the fall or spring following any of the treatments in this section.

DO NOT GRAZE OR FEED FOLIAGE FROM TREATED AREAS TO LIVESTOCK.

BLADEX Applied Alone

Apply BLADEX 4L directed-postemergence at the rate shown in Table 6. Apply at layby at the rates for the soil texture indicated in Table 7. Add an agricultural surfactant such as X-77 or other surfactant suitable for use on growing cotton at the rate of 2 quarts per 100 gallons of spray mixture (or as directed by the manufacturer).

Table 6
Directed Postemergence Application Rates Per Acre For BLADEX 4L On Cotton

Broadcast	Banded 38" Row	
	12" Band	19" Band
0.6–1.0 qt.	0.2–0.3 qt.	0.3–0.5 qt.

Use the maximum rate when dry or arid conditions exist.

Table 7
Layby Application Rates Per Acre For BLADEX 4L On Cotton

Height of Cotton	Soil Texture		Broadcast Rates
12 inches or more	Coarse	Sandy loam	0.8 qt.
	Medium	Silt and silt loam, loam, clay loam, sandy clay loam and sandy clay	1.2 qt.
	Fine	Silty clay loam silty clay and clay	1.6 qt.

BLADEX Plus MSMA

Apply a tank-mix combination of BLADEX 4L plus MSMA plus surfactant after the cotton is 6" tall but before it reaches the bloom stage. Apply no more than two applications of this mixture before the first bloom stage. Tank-mix BLADEX plus MSMA at the rates indicated in Table 8. Add a surfactant, such as X-77, at the rate of 2 qts./100 gals. of spray mixture (or as directed by the manufacturer).

Table 8
Directed Postemergence Application Rates Per Acre BLADEX 4L + MSMA On Cotton

Broadcast	Banded 38" Row	
	12" Band	19" Band
0.6–1.0 Quart BLADEX 4L	0.2–0.3 qt.	0.3–0.5 qt.
+	+	+
4 pints MSMA (4 lb/gal)	1.3 pints	2 pints
or	or	or
2.4 pints MSMA (6.6 lb/gal)	0.8 pint	1.2 pints

BLADEX plus Dinitro (dinoseb)

Apply a tank-mix combination of BLADEX 4L plus Dinitro 3 plus surfactant after the cotton is 6" tall. Tank-mix BLADEX 4L plus Dinitro 3 at the rates indicated in Table 9. Add a surfactant, such as X-77, at the rate of 2 qts/100 gals. of spray mixture. Do not apply when soil surface is wet or cotton plants are extremely succulent. Do not apply in CA or AZ.

Table 9
Directed Postemergence Application Rates Per Acre BLADEX 4L + Dinitro On Cotton

Broadcast	Banded 38" Row	
	12" Band	19" Band
0.6–1.0 qt. BLADEX 4L	0.2–0.3 qt.	0.3–0.5 qt.
+	+	+
1–2 qts. Dinitro 3	0.3–0.65 qt.	0.5–1.0 qts.

BLADEX Plus MSMA Plus Dinitro

Apply as directed in the "BLADEX Plus MSMA" section of the label. Tank-mix BLADEX plus MSMA as the rates shown in Table 8. Add Dinitro at the rate shown in Table 9. Add a surfactant, such as X-77, at the rate of 2 qts./100 gals. of spray mixture. Do not apply when soil surface is wet or cotton plants are extremely succulent. Do not apply in CA or AZ.

GRAIN SORGHUM (MILO)

Tank-mix combinations of BLADEX 4L plus propachlor (Ramrod™) propazine (Milogard™) may be used for selective preemergence weed control in grain sorghum. Do not use on forage sorghum.

BLADEX 4L in tank-mix combination with propachlor or propazine should be applied only once per crop season. If replanting of grain sorghum is necessary, it may be planted in soil previously treated with these mixtures. Do not make a second application of BLADEX or crop injury may occur. Apply these tank-mix combinations at planting or after planting, but before the crop and weeds have emerged. Heavy rain immediately following application tends to cause excessive concentrations of herbicide in seed furrow, resulting in possible crop injury. Do not apply to furrow-planted sorghum until furrows are leveled (plowed-in). Level deep planter marks or seed furrows before application. Application to sorghum growing under stress caused by minor element deficiency or to sorghum growing on highly calcareous soil may result in crop injury.

BLADEX 4L Plus Propachlor (Ramrod)

WEEDS CONTROLLED

Grasses

Barnyardgrass	Giant foxtail
Crabgrass	Green foxtail
Fall panicum	Yellow foxtail

Broadleaves

Annual morningglory	Pigweed
Carpetweed	Ragweed
Cocklebur*	Smartweed (Pennsylvania)
Common purslane	Velvetleaf*
Lambsquarters	

* Under conditions such as low temperatures, lack of soil surface moisture or other factors that may cause delay in germination of the seeds, the effectiveness of BLADEX may be impaired against these weeds.

Apply BLADEX 4L plus propachlor at the proper rate for soil texture and organic matter shown in Table 10. Only apply this tank-mix to grain sorghum grown in states East of the Rocky Mountains. Any rotational crop may be planted the fall or spring following this treatment.

Table 10
Preemergence Broadcast Application Rates Per Acre For BLADEX 4L + Propachlor (Ramrod) 4L On Grain Sorghum

Soil Texture Description	Percent Organic Matter In Soil* 2% Qts./Acre BLADEX 4L	2% Qts./Acre Propachlor 4L	3% Qts./Acre BLADEX 4L	3% Qts./Acre Propachlor 4L
Sand, Loamy Sand	DO NOT USE			
Sandy Loam	1.0	2.5	1.2	3.0
Loam, Silt Loam, Silt	1.2	3.0	1.4	3.5
Sandy Clay Loam, Clay Loam, Silty Clay Loam	1.4	3.5	1.6	4.0
Sandy Clay, Silty Clay, Clay	1.6	4.0	1.6	4.0
Peat or Muck	NOT RECOMMENDED			

* For organic matter content between those listed, adjust the rate proportionately.

BLADEX Plus Propazine (Milogard)

WEEDS CONTROLLED

Grasses

Crabgrass	Green foxtail
Fall panicum	Yellow foxtail
Giant foxtail	

Broadleaves

Annual morningglory	Pigweed
Carpetweed	Puncturevine
Cocklebur	Ragweed
Common purslane	Smartweed
Lambsquarters	Velvetleaf

Apply BLADEX 4L plus Propazine at the proper rate for soil texture shown in Table 11.

Apply to grain sorghum grown only in the states of Texas, Oklahoma, and Kansas. **Rotational Crops:** In the Texas Gulf Coast and Texas Blacklands, fields treated with this tank-mix may be planted to cotton, soybeans, or corn 12 months after treatment. In Oklahoma and West Texas, fields treated with this tank-mix may be planted to cotton or corn 12 months after application. In Kansas, corn may be planted in rotation 12 months after treatment. Other crops should not be planted for 18 months following treatment.

Table 11
Preemergence Broadcast Application Rates Per Acre For BLADEX 4L + Propazine (Milogard) 4L On Grain Sorghum

Soil Texture	Qts./Acre BLADEX 4L	Qts./Acre Propazine 4L*
Sand, Loamy Sand, Sandy Loam	DO NOT USE	
Loam, silt loam, silt	0.8	0.75
Sandy clay loam, clay loam, silty clay loam	1.0	1.0
Sandy clay, silty clay, clay	1.2	1.25
Peat or Muck	NOT RECOMMENDED	

* If Propazine 80W is used, multiply rate shown by 1.25 to equal pounds of Propazine 80W.

BLADEX®—Trademark of Shell Chemical Co.
Banvel™—Trademark of Velsicol Chemical Co.
Dual™, Milogard™—Trademarks of CIBA-Geigy Corp.
Eradicane™, Sutan ™—Trademarks of Stauffer Chemical Co.
Lasso™, Ramrod™—Trademarks of Monsanto Co.
Ortho X-77™—Trademark of Chevron Chemical Co.
Treflan™—Trademark of Elanco Products Co.
Zorial™—Trademark of Sandoz, Inc.

BLADEX® 80W

HERBICIDE

Use of Bladex 80W in Corn, Cotton, Grain Sorghum, Wheat and Fallow Cropland

ACTIVE INGREDIENTS	BY WEIGHT
2-[[4-chloro-6-(ethylamino)-s triazine-2-yl]amino]-2-methylpropionitrile	80%
INERT INGREDIENTS	20%
TOTAL	100%

EPA Reg No. 201–279
Covered by U.S. Patents Nos. 3,505,325 and 3,620,710.

Keep Out of Reach of Children
WARNING

NOTE CONDITIONS OF SALE AND WARRANTY BEFORE OPENING THE CONTAINER. IF THEY ARE NOT ACCEPTABLE, RETURN UNOPENED PRODUCT.

STATEMENT OF PRACTICAL TREATMENT

IF SWALLOWED, call a physician or poison control center. Drink 1 or 2 glasses of water and induce vomiting by touching back of throat with finger. Do not induce vomiting or give anything by mouth to an unconscious person.

IF IN EYES, flush with plenty of water. If irritation persists call a physician.

IF ON SKIN, wash immediately with plenty of soap and water.

IF INHALED, remove victim to fresh air. If not breathing, give artificial respiration, preferably mouth-to-mouth and get medical attention.

AVISO

PRECAUCION AL USUARIO: Si usted no lee ingles, no use este producto hasta que le etiqueta haya sido explicado ampliamente.

PRECAUTIONARY STATEMENTS

HAZARD TO HUMANS AND DOMESTIC ANIMALS WARNING

WARNING: May be fatal if swallowed. Harmful if inhaled or absorbed through the skin. Causes temporary eye injury.

This product may be hazardous to your health. This product has been determined to cause birth defects in laboratory animals. Use of protective clothing and equipment and following the precautions below can reduce risk.

Avoid breathing spray mist. Avoid contact with skin, eyes, or clothing. Do not get in eyes or on clothing. Wear a face shield when mixing and loading. Wash thoroughly with soap and water after handling and before eating or smoking.

Wear long trousers and long-sleeved clothing when applying this product. Wear rubber gloves extending above the wrist, long trousers and long-sleeved clothing when mixing or loading or when adjusting, repairing or cleaning equipment. Remove contaminated clothing and wash before reuse. Heavily contaminated clothing should be taken off immediately and incinerated or burned, if allowed by the state and local authorities, or disposed of in a sanitary landfill and not laundered.

For 24-hour emergency medical assistance, call (713) 473-9461.

Do not apply this product in such a manner as to directly, or through drift, expose workers or other persons. The area being treated must be vacated by unprotected persons.

Keep out of reach of domestic animals, particularly cattle. Consumption of this product, spray solutions, or water contaminated with product can result in serious illness or possible death of bovines.

ENVIRONMENTAL HAZARDS

Do not apply directly to water or wetlands.
Do not contaminate water by cleaning of equipment or disposal of wastes. In case of significant spill, call CHEM-TREC (800) 424-9300 or (713) 473-9461.

BLADEX is a pesticide which can move (seep or travel) through soil and can contaminate groundwater which may be used as drinking water. BLADEX has been found in groundwater as a result of agricultural use. Users are advised not to apply BLADEX where the water table (groundwater) is close to the surface and where the soils are very permeable (i.e. well drained soils such as loamy sands). Your local agricultural agencies can provide further information on the type of soil in your area and the location of groundwater.

DIRECTIONS FOR USE

It is a violation of Federal law to use this product in a manner inconsistent with its labeling.

FOR END USE ONLY. DO NOT REPACKAGE OR REFORMULATE WITHOUT MANUFACTURER'S WRITTEN APPROVAL.

STORAGE AND DISPOSAL

STORAGE
Do not contaminate water, food, or feed by storage or disposal. Do not use or store around the home environment. Avoid contact with water. In case of spill or leak, soak up with sand, earth, or synthetic absorbent (do not use alkaline absorbents) and dispose of wastes in compliance with local, state and Federal regulations.

PESTICIDE DISPOSAL
Pesticide, spray mixture or rinsate water that cannot be used according to label instructions must be disposed of according to applicable Federal, State, or local procedures.

CONTAINER DISPOSAL
Completely empty bag into application equipment. Empty residue into application equipment. Then dispose of empty bag in a sanitary landfill or by incineration or if allowed by State and local authorities by burning. If burned stay out of smoke.

CONDITIONS OF SALE AND WARRANTY

SHELL AND THE SELLER OFFER THIS PRODUCT, AND THE BUYER AND USER ACCEPT THIS PRODUCT, ONLY UNDER THE FOLLOWING AGREED CONDITIONS OF SALE AND WARRANTY: The directions for use of this product are believed to be reliable and should be followed carefully. However, it is impossible to take into account all variables and to eliminate all risks associated with its use. Injury or damage may result because of conditions which are beyond the control of Shell or the seller, including soil texture, organic matter content of soil, weather, presence of other materials, and the manner of use or application. Shell warrants only that this product conforms to the chemical description on the label and is believed to be reasonably fit for the purposes referred to in the Directions for Use when used as directed under normal conditions. SHELL MAKES NO OTHER EXPRESS OR IMPLIED WARRANTY OF FITNESS OR MERCHANTABILITY OR ANY OTHER EXPRESS OR IMPLIED WARRANTY. Any variation or exception from this warranty must be in writing and signed by an authorized Shell representative.

LIMITATION OF LIABILITY

Shell's liability, whether or not based on negligence, breach of expressed or implied warranty, strict liability or any other legal cause, is limited to the cost of this product. In no case shall Shell or the Seller be liable for consequential, incidental, or indirect damages, such as loss of crops, resulting from the use or handling of this product.

GENERAL INFORMATION

BLADEX 80W Herbicide is a selective herbicide for the control of annual grasses and broadleaved weeds in field corn, cotton, grain sorghum, and fallow cropland.

Consult your local Agricultural Extension Agent for help in determining soil texture, organic matter content, and the most appropriate herbicide rate for local conditions.

Where surfactants or emulsifiable vegetable oils are added to BLADEX 80W for over the top of corn postemergence applications, use on field corn only.

As a preemergence herbicide, BLADEX 80W is active mainly through the roots, and therefore, its effect on weeds is dependent on adequate rainfall or sprinkler irrigation to move the herbicide into the root zone. Moisture should be sufficient to wet the top 1½ to 2 inches of soil or make the soil too wet to cultivate; for most soils ½–¾" of moisture is sufficient. A rotary hoeing or shallow cultivation is recommended if a rainfall or sprinkler irrigation has not occurred within about ten days after application of BLADEX 80W. Under conditions which delay weed germination, such as low temperatures, lack of soil surface moisture, or when germination extends over a long period, the effectiveness of the herbicide may be impaired. Rotary hoeing, a shallow cultivation or a postemergence herbicide treatment may be of benefit under these circumstances. When applied as a postemergence herbicide, BLADEX 80W is also active through foliage as well as through the roots.

OBSERVE ALL CAUTIONS AND LIMITATIONS ON LABELING OF ALL PRODUCTS USED IN MIXTURES.

APPLICATION DIRECTIONS

GENERAL MIXING AND SPRAYING DIRECTIONS

If another pesticide has been used previously, be sure the sprayer tank, boom, and nozzles have been thoroughly flushed with clean water.

Sufficient jet or mechanical agitation should be provided during the filling operation and during application to keep the spray mixture uniformly suspended. Fill the spray tank at least one-half full of clean water and while pump and agitator are running, add the recommended amount of BLADEX 80W to the tank, then fill with water to the desired level. For tank mixes, once the BLADEX 80W is completely suspended in the water, add the recommended amount of other products to the tank while filling with water to the desired level. Add all wettable powders before liquids. For emulsible concentrates (EC), such as Lasso™ or Dual™, pre-mixing these products in equal amounts of water or pouring the products directly into the water stream as the tank is filling will aid in achieving compatible tank-mixtures. Fluid fertilizers may replace all or part of the water as carrier. Always check the compatibility of BLADEX 80W or BLADEX 80W tank mixes and fluid fertilizer, according to the Jar Test for Compatibility instructions below, before the herbicide and fertilizer are mixed in the spray tank. (BLADEX 80W can also be premixed as a slurry before adding to the tank.)

For use with a surfactant or vegetable oil, completely suspend the BLADEX 80W in water and fill the tank before adding the surfactant or vegetable oil.

Allow the pump to run several minutes before application to ensure proper suspension and mixing of BLADEX 80W or other various mixtures. If BLADEX 80W or tank-mix combinations of BLADEX 80W with atrazine are allowed to stand in a spray tank or nurse tank for an extended period (several hours or overnight), provide good agitation for several minutes before application to ensure proper suspension and mixing. Tank-mix combinations of BLADEX 80W with other products should be continuously agitated and should be mixed and applied as soon as possible.

Apply with flat fan (even flat fan for band application) or flood nozzles in conjunction with 50 mesh nozzle screens and in-line screens. Carefully follow the nozzle manufacturer's recommendation for correct nozzle height, positioning and spacing. Check for uniform distribution of spray droplets. Apply at pressures of 15–35 psi to reduce chemical loss due to fine mist drift.

Apply the spray mixture to the soil surface using at least 15 gals./A. for ground application. Apply by air using a minimum of 4 gals./A. of spray mixture.

Check nozzles, pressure, and output regularly to ensure proper calibration and application.

ALL RATES INDICATED ON THIS LABEL ARE BROADCAST APPLICATION RATES PER ACRE UNLESS OTHERWISE STATED. FOR BAND APPLICATION, USE PROPORTIONATELY LESS.

JAR TEST FOR COMPATIBILITY

A. Add one pint of liquid fertilizer or other carrier to be used to two quart jars. Mark one quart jar "with" and the other jar "without."

B. To the jar marked "with" add ¼ teaspoon of a suitable compatibility agent. Shake gently for 5 to 10 seconds to mix. (¼ teaspoon/pint = 2 pints/100 gallons.)

C. To each jar add the appropriate amount of herbicides corresponding to the herbicide rate and spray volume selected (see following table). If more than one herbicide is used, add them separately; wettable powders first, flowables second, and emulsible concentrates last. Shake gently for 5 to 10 seconds after each addition.

D. Let stand one-half hour and make observations. If any separation, agglomeration, or precipitation has occurred, again shake the jar 10 to 15 seconds.

   1. If mixtures appear to be compatible in both jars, the herbicides can probably be used without a compatibility agent.
   2. If the mixture in the jar marked "with" is compatible but the one in the jar marked "without" is incompatible, a compatibility agent should be used.
   3. If both mixtures are incompatible, the liquid fertilizer and herbicide should not be used in the same spray tank.

Jar Test for BLADEX 80W
Compatibility with Liquid Fertilizer

Gallons Liquid Fertilizer per acre	15	20	25	30	40
Teaspoons of BLADEX 80W* per pint of fertilizer	2½	1¾	1½	1¼	1

* Based on 1 pound BLADEX 80W (0.8 pound active ingredients) per acre in fertilizer volumes indicated. Increase volume proportionately to correspond with intended field rate in terms of pounds BLADEX 80W per acre

(e.g., for field rate of 4 pounds BLADEX 80W in 20 gallons fertilizer per acre, add 7 teaspoons BLADEX 80W to 1 pint fertilizer for compatibility testing purposes. CALCULATION: At 4 pounds BLADEX 80W rate/20 gallons fertilizer, 4 × 1¾ teaspoons = 7 teaspoons BLADEX 80W/pint of fertilizer).

If a test such as outlined indicates that components of a proposed mix are compatible, the applicator still has the responsibility of combining materials in sequence to the spray tank in accordance with directions prescribed on the label of the herbicides or pesticides involved.

Tests have indicated that compatibility agents, noted below by the various tank mix combinations, may give improved compatibility in liquid fertilizers.

Tank Mix Combination	Compatibility Agents
BLADEX 80W/Lasso (Liquid Fertilizer Grade)	Probably not needed in 28–0–0, 10–34–0. Unite, Compex, Ortho X-77 may help in others.
BLADEX 80W/ Sutan +™6.7E	Probably not needed in 28–0–0. Unite, Spray-Mate, Kem-Link, Compex may help in others.
BLADEX 80W/Dual 8E	Unite, Spray-Mate, Ivory Liquid may help

CORN

WEEDS CONTROLLED BY BLADEX 80W ALONE AND IN COMBINATION WITH OTHER HERBICIDES ON CORN

Grasses

Annual bluegrass	Goosegrass
Annual fescues	Green foxtail
Annual (Italian) ryegrass	Junglerice
Barnyardgrass[1]	Stinkgrass (Indian lovegrass)
Bullgrass	Witchgrass
Crabgrass	Yellow foxtail
Fall panicum	
Giant foxtail	

Broadleaves

Annual groundcherry	Mayweed
Annual morningglory	Nightshade (annual)
Annual sedge	Pigweed[1]
Black mustard	Pineappleweed
Buffalobur	Plantain
Buttercup (annual)	Poorjoe
Carpetweed	Prickly sida (teaweed)
Cocklebur[2]	Prostrate knotweed
Common chickweed	Prostrate spurge
Common groundsel	Ragweed
Common mallow	Russian thistle
Common purslane	Shepherdspurse
Corn spurry	Smallflower galinsoga
Curly dock (seedling)	Smartweed (Pennsylvania)
Fiddleneck	Sunflower[2] (wild, annual, common)
Florida pusley (Florida purslane)	Tarweed cuphea (Gumweed)
Hedge mustard	Velvetleaf[1]
Jimsonweed[1]	Wild buckwheat
Kochia	Wild mustard
Ladysthumb	Wild radish
Lambsquarters	Wild turnip

[1] Under conditions such as low temperatures, lack of soil surface moisture or other factors that may cause delay in germination of the seeds, the effectiveness of BLADEX may be impaired against these weeds.

[2] The degree of control will be reduced if soil moisture and temperature conditions cause deep germination of the seed.

PREEMERGENCE-PREPLANT INCORPORATED

Apply BLADEX treatments just before, at or after planting but before the crop has emerged. Avoid removal of treated soil from seedrow prior to or during the planting operation.

BLADEX may also be applied early prior to planting or in a split application if pre-season weed control is desired. For split applications, do not exceed the total amount of BLADEX for the soil texture and organic matter shown in Table 1. If BLADEX is applied early, more than 15 days before planting, a split application of BLADEX or some other herbicide treatment may be necessary at or after planting to provide additional length of weed control. For further information see "Early Preplant" recommendation in the Conservation Tillage section of label.

A rotary hoeing is recommended for preemergence applications which do not receive adequate rainfall or sprinkler irrigation to wet the top 1½–2" of soil within about 10 days after application.

If a BLADEX mixture is to be incorporated, except as noted, single or two pass incorporation is acceptable. Care should be taken to incorporate the BLADEX mixture no deeper than the top two inches of soil.

BLADEX Applied Alone

Use at the proper rate for soil texture and organic matter indicated in Table 1. Any rotational crop may be planted in the fall or following spring following this treatment.

Table 1
Preemergence Broadcast Application Rates Per Acre For BLADEX 80W Applied Alone On Corn

Soil Texture Description	Pounds of BLADEX 80W — Percent Organic Matter in Soil*					
	Less than 1%	1%	2%	3%	4%	5% & Over
Sand, Loamy Sand	DO NOT USE	1.5	2.0	2.75	3.5	4.0
Sandy Loam	1.5	2.25	2.5	3.0	3.75	4.5
Loam, Silt Loam, Silt	2.0	2.5	3.0	3.75	4.5	5.0
Sandy Clay Loam, Clay Loam, Silty Clay Loam	2.5	3.0	3.75	4.5	5.0	5.5
Sandy Clay, Silty Clay, Clay	3.5	3.75	4.5	5.0	5.5	6.0
Peat or Muck	NOT RECOMMENDED					

* For organic matter content between those listed, adjust the rate proportionately.

TABLE 2
PREEMERGENCE BROADCAST APPLICATION RATES PER ACRE FOR TANK-MIX COMBINATIONS OF BLADEX 80W PLUS ATRAZINE 80W ON CORN

SOIL TEXTURE DESCRIPTION	POUNDS OF BLADEX 80W + POUNDS OF ATRAZINE 80W** — PERCENT ORGANIC MATTER IN SOIL*					
	LESS THAN 1%	1%	2%	3%	4%	5% & OVER
Sand, Loamy Sand	DO NOT USE	1.0 + 0.5	1.5 + 0.5	1.75 + 0.75	2.0 + 1.0	2.75 + 1.25
Sandy Loam	1.0 + 0.5	1.5 + 0.5	1.75 + 0.75	2.0 + 1.0	2.75 + 1.25	3.5 + 1.5
Loam, Silt Loam, Silt	1.5 + 0.5	1.75 + 0.75	2.5 + 1.0	2.75 + 1.25	3.5 + 1.5	4.0 + 1.5
Sandy Clay Loam, Clay Loam, Silty Clay Loam	1.75 + 0.75	2.5 + 1.0	2.75 + 1.25	3.5 + 1.5	4.0 + 1.5	4.25 + 1.75

SOIL TEXTURE DESCRIPTION	POUNDS OF BLADEX 80W + POUNDS OF ATRAZINE 80W** PERCENT ORGANIC MATTER IN SOIL*					
	LESS THAN 1%	1%	2%	3%	4%	5% & OVER
Sandy Clay, Silty Clay, Clay	2.5 + 1.0	2.75 + 1.25	3.5 + 1.5	4.0 + 1.5	4.25 + 1.75	4.5 + 2.0
Peat or Muck	NOT RECOMMENDED					

* For organic matter content between those listed, adjust the rate proportionately.
** If Atrazine 4L is used, multiply rates shown by 0.8 to equal quarts of Atrazine 4L.

BLADEX COMBINATIONS

BLADEX plus Atrazine

Use at the proper rate for soil texture and organic matter indicated in Table 2. Table 2 provides rates for generally weedy conditions. The ratio of the amounts of each herbicide may be adjusted as necessary for particular weed conditions as long as the combined rate of the two products does not exceed the combined rate for the soil shown in Table 2. For grassier conditions use a ratio that contains higher levels of BLADEX (3:1). For fields with more broadleaves use a ratio that contains higher levels of atrazine (1:1).

Rotational crops: (1) Plant only corn, peanuts, sorghum, or soybeans the year following the use of this mixture. (2) If soybeans are to be planted, injury may occur due to the carryover of atrazine. (3) If applied after June 10, do not rotate with crops other than corn or sorghum the next year or injury may occur. (4) In the high plains and intermountain areas of the west where rainfall is sparse and erratic or where irrigation is required, use only when corn or sorghum is to be planted the following year, or a crop of corn or sorghum not treated with atrazine is to precede other rotational crops. (5) Small grains may be planted 15 months following treatment. (6) All other crops may be planted 18 months after application.

BLADEX plus Lasso 4 EC

Use BLADEX 80W at the proper rate for the soil texture and organic matter shown in Table 3 plus 2 quarts per acre of Lasso. (Use 2.5 quarts Lasso on clay soils containing 5 percent organic matter and over.) Any rotational crop may be planted the fall or spring following this treatment.

BLADEX plus Sutan+ 6.7E, or Eradicane™ 6.7E

Use BLADEX 80W at the proper rate for the soil texture and organic matter shown in Table 3 plus 1.8 quarts per acre of Sutan+ or Eradicane for control of many annual grasses and broadleaf weeds. (Use 2.4 quarts of Sutan+ or Eradicane on loam soils containing 5 percent or more organic matter, and clay loams and clays containing 4 percent or more organic matter.) Do not use on sands and loamy sands of less than 1 percent organic matter in the light sandy soils of eastern coastal states. Do not use on corn seed stock.

Apply before planting. Incorporate the mixture immediately upon application using power-driven cultivation equipment set for 2–3 inch depth, or a tandem disc set to cut to a depth of about 4 inches while operating at 4–6 mph. For thorough mixing, disc in two directions (cross disc), and follow with a harrow, drag, or other leveling device. Prior to the second discing, readjust the disc to prevent cutting deeper than 4 inches. BLADEX 80W may be applied preemergence as an overlay over previous incorporated Sutan+ or Eradicane if desired. Any rotation crop may be planted in the fall or spring following these treatments.

Existing stands of quackgrass, purple and yellow nutsedge must be turned under and thoroughly chopped up prior to chemical treatments.

Additional weeds controlled by Sutan+ or Eradicane combinations:

Grasses
Sandbur
Shattercane (Wild Cane)*
Texas Panicum
Quackgrass (Eradicane only)
Wild Proso Millet* (Eradicane only)

Perennial Weeds
Yellow Nutsedge (nutgrass)
Purple Nutsedge (nutgrass)

* Suppression only—refer to Sutan+ or Eradicane label for appropriate supplemental cultural and tillage practices.

For fields with moderate to heavy infestations of these weeds refer to the Sutan+ or Eradicane labels for appropriate higher rates.

Table 3
Preemergence Broadcast Application Rates Per Acre For BLADEX 80W Used In Tank-Mix Combinations With LASSO, SUTAN+, ERADICANE, Or DUAL On Corn

Soil Texture Description	Pounds of BLADEX 80W Percent Organic Matter in Soil*					
	Less than 1%	1%	2%	3%	4%	5% & Over
Sand, Loamy Sand	0.75**	1.0	1.5	1.75	2.0	2.5
Sandy Loam	1.0	1.5	1.75	2.0	2.5	2.75
Loam, Silt Loam, Silt	1.5	1.75	2.0	2.5	2.75	3.25
Sandy Clay Loam, Clay Loam, Silty Clay Loam	1.75	2.25	2.5	2.75	3.25	3.5
Sandy Clay, Silty Clay, Clay	2.25	2.5	3.0	3.25	3.5	3.75
Peat or Muck	NOT RECOMMENDED					

* For organic matter content between those listed, adjust the rate proportionately.
** Do not use in the Atlantic Coastal Plain.

BLADEX Plus Dual 8E

Use BLADEX 80W at the proper rate for soil texture and organic matter shown in Table 3. Use Dual as follows:

Soil Texture	Broadcast Rate Per Acre
Coarse Sand, loamy sand, sandy loam	1.25–1.5 pints
Medium Loam, silt loam, silt	1.5–2.0 pints
Fine Sandy clay loam, silty clay loam, clay loam, Sandy clay, silty clay, clay	1.5–2.5 pints

The low end of the rate range should be used for lowest organic matter soils and the rate increased as organic matter increases to a point that soils containing 4 percent organic matter or more require the highest rate shown for that soil texture. Refer to the Dual label for precautions on rotational crops.

BLADEX plus Atrazine plus Lasso, Sutan+, Eradicane, or Dual

Use BLADEX 80W plus atrazine at the proper rate for soil texture and organic matter shown in Table 4. Use Lasso, Sutan+, Eradicane, or Dual according to rates shown in "BLADEX Combinations" in this section of the label.

Rotational crops: Refer to Rotational Crops section of "BLADEX plus Atrazine" in this section of the label.

TABLE 4
Preemergence Broadcast Application Rates Per Acre For BLADEX 80W Plus ATRAZINE 80W Used In Tank-Mix Combinations With LASSO, SUTAN+, ERADICANE, Or DUAL On Corn

SOIL TEXTURE DESCRIPTION	POUNDS OF BLADEX 80W + POUNDS OF ATRAZINE 80W*** PERCENT ORGANIC MATTER IN SOIL*					
	LESS THAN 1%	1%	2%	3%	4%	5% & OVER
Sand, Loamy Sand	0.5 + 0.25**	0.75 + 0.5	1.0 + 0.5	1.25 + 0.5	1.25 + 0.75	1.75 + 0.75
Sandy Loam	0.75 + 0.25	1.0 + 0.5	1.25 + 0.5	1.25 + 0.75	1.75 + 0.75	2.0 + 0.75
Loam, Silt Loam, Silt	1.0 + 0.5	1.25 + 0.5	1.5 + 0.5	1.75 + 0.75	2.0 + 0.75	2.25 + 1.0
Sandy Clay Loam, Clay Loam, Silty Clay Loam	1.25 + 0.5	1.5 + 0.75	1.75 + 0.75	2.0 + 0.75	2.25 + 1.0	2.5 + 1.0
Sandy Clay, Silty Clay, Clay	1.5 + 0.75	1.75 + 0.75	2.0 + 1.0	2.25 + 1.0	2.5 + 1.0	2.5 + 1.25
Peat or Muck	NOT RECOMMENDED					

* For organic matter content between those listed, adjust the rate proportionately.
** Do not use in the Atlantic Coast Plain.
*** If Atrazine 4L is used, multiply rates shown by 0.8 to equal quarts of Atrazine 4L.

Conservation Tillage Weed Control Early Preplant (EPP)
[Prior-to-planting weed control for land going into production of corn.]

Complete any planned early spring tillage prior to application. Apply herbicide treatment before weeds germinate or before weed seedlings are more than 3 inches tall. Tillage after application may reduce the effectiveness of the herbicide treatment.

Apply BLADEX or BLADEX plus atrazine tank mix 15-30 days prior to planting. Use the proper rate for soil texture and organic matter indicated in Table 1 or 2. Where heavy crop residues exist, the rates shown in Tables 1 or 2 should be increased by 25 percent.

Where broadleaf weeds are present, add 2,4-D LV Ester at 1⅓–2 pts./A. (6 lb./gal.) or 2–3 pts./A. (4 lbs./gal.) plus X-77 surfactant at 1 qt./100 gals. of diluted spray, or other suitable surfactant at its recommended rate. If existing weeds exceed 3 inches in height, add 1–2 pts./A. of paraquat to the above mixture. Well established weeds over 6 inches tall may not be well controlled. Use at least 25 gals./A. of spray mixture by ground rig or 5–10 gals./A. for aerial application.

Depending upon weather conditions following the EPP application, a postemergence BLADEX treatment, or some other herbicide treatment may be necessary at or after planting to provide additional length of weed control. If desired, 1½–2 pts./A. of Dual or 2 qts./A. of Lasso may be tank-mixed with the BLADEX EPP treatment or applied preemergence at planting.

Rotational Crops: Refer to Rotational Crops section for each treatment in the Preemergence section of the label.

At Planting

BLADEX 80W applied alone or in combination with atrazine and/or Lasso or Dual according to the following directions will kill most existing small weeds and suppress many emerged perennial weeds when corn is planted into no-till stalk ground (corn, sorghum), stubble ground (soybean, small grains), and any minimum-till situation. This treatment then provides residual control of annual weeds as in conventional tillage.

Apply BLADEX 80W alone or with other products according to the directions for those treatments in the Preemergence section of the label. Where heavy crop residues exist, the BLADEX rate shown in Tables 1, 2, or 3 should be increased by 25 percent. Add ½–1 pt./A. of 2, 4-D Low Volatile 6 lb. Ester (0.75–1.5 pt./A. 2,4-D LV 4 lb. Ester). Add the 2,4-D LV to the spray tank last. Use a minimum of 25 gals./A. of carrier. Complete spray coverage of the weeds is essential for best performance. Nitrogen solutions and complete liquid fertilizers are the preferred carriers for this treatment as they aid in the burndown of existing weeds. Add Ortho X-77™ surfactant at 1–2 qts. per 100 gals. of diluted spray, or other suitable surfactant at its recommended rate. If water is used as a carrier, crop oil concentrate may be used as a surfactant. Apply before weeds exceed 3" in height. For control of existing alfalfa add ⅓–½ pt./A. of Banvel™ to the spray mixture. Apply before the alfalfa exceeds 6" in height.

For fields with existing sod grasses such as orchardgrass, bromegrass, rye or timothy, or when very dry conditions exist, or when existing weeds exceed 3 inches in height add paraquat to the tank-mix. Use 1–2 pts./A. of paraquat in combination with BLADEX as described above in this section, except the 2, 4-D LV may be omitted if desired. Do not apply paraquat in suspension type liquid fertilizer.

POSTEMERGENCE

BLADEX Applied Alone

Use BLADEX 80W at the proper rate for the soil texture and organic matter shown in Table 5 or 5A. Use rates shown in Table 5 if BLADEX has not been applied to the soil this season. Use rates shown in Table 5A if BLADEX has been applied to the soil this season. This treatment may be used on peat or muck soils for burndown and suppression of existing weeds but will not provide residual control. Apply from crop emergence through the four-leaf stage of corn growth but before weeds exceed 1½" in height. Do not apply over the top of corn if the fifth leaf is visible. Apply in water only. Do not spray emerged corn plants with BLADEX in a liquid fertilizer carrier.

Under dry, arid conditions of low humidity and the absence of dew formation at night, add a surfactant such as X-77, or an emulsible vegetable (EV) oil suitable for use on growing corn at its recommended rate. Do not use petroleum-based crop oils. Addition of a surfactant or EV oil is not recommended under moist, rainy conditions and when dew forms at night as injury may occur.

Do not apply this treatment under cold, wet, weather conditions or to weather-stressed or storm-damaged corn. Yellowing of the corn may result from this treatment, particularly if cold, adverse growing conditions occur after application. Extended or extreme cold and wet conditions may reduce stands.

Do not use on corn grown for seed. Any rotational crop may be planted the fall or spring following this treatment.

Table 5
Postemergence Broadcast Application Rates Per Acre For BLADEX 80W On Corn

Soil Texture	Pounds of BLADEX 80W Percent Organic Matter in Soil*			
	Less than 1%	1%	2%	Over 2%
Sand, Loamy Sand	DO NOT USE	1.5	2.0	2.5
Sandy Loam	1.5	2.0	2.5	2.5
Loam, Silt Loam, Silt	2.0	2.5	2.5	2.5
All other textures	2.5	2.5	2.5	2.5

* For organic matter content between those listed, adjust the rate proportionately.

Table 5A
Postemergence Broadcast Application Rates Per Acre For BLADEX 80W On Corn BLADEX Used In Prior Application

Soil Texture	Pounds of BLADEX 80W — Percent Organic Matter in Soil* Less than 1%	1%	2%	Over 2%
Sand, Loamy Sand	DO NOT USE	DO NOT USE	1.75	1.75
Sandy Loam	DO NOT USE	DO NOT USE	2.0	2.5
Loam, Silt Loam, Silt	DO NOT USE	1.75	2.5	2.5
All other textures	DO NOT USE	2.25	2.5	2.5

* For organic matter content between those listed, adjust the rate proportionately.

BLADEX COMBINATIONS

BLADEX Plus Atrazine

Apply as directed in "Postemergence—BLADEX Applied Alone" section of this label. Use an amount of BLADEX 80W plus Atrazine 80W equal to the rate shown in Table 5 or 5A for the proper soil texture and organic matter. To determine the amount of BLADEX 80W to use multiply the rate in Table 5 by 0.7. To determine the amount of Atrazine 80W to use, multiply the rate shown in Table 5 by 0.3. (For the 2.5 lb/A. rate shown in Table 5, use 1.75 lbs/A. of BLADEX 80W plus 0.75 lbs/A. of Atrazine 80W.)

Rotational Crops: See Rotational Crops section in the "Preemergence—BLADEX Plus Atrazine" section of this label.

BLADEX Plus Banvel

Apply as directed under "Postemergence—BLADEX Applied Alone" section of this label. Add ½ –⅔ pt/A. of Banvel to the mixture. Do not use with a surfactant or emulsible vegetable oil.

COTTON

IDLE SEASON EARLY PREPLANT WEED CONTROL (CALIFORNIA ONLY)

WEEDS CONTROLLED

Grasses

Annual Bluegrass
Annual Ryegrass
Volunteer Small Grains (suppression)
Barnyardgrass*
Bristly Foxtail
Rabbits Foot Grass
Wild Oat*
Yellow Foxtail

Broadleaves

Annual Henbit
Black Nightshade
Burclover
Cheeseweed*
Chickweed
Fiddleneck
Groundsel
Knotweed
Marestail
Miners Lettuce
Lambsquarter
London Rocket
Pineapple Weed
Prickly Lettuce
Shepherdspurse
Sowthistle
Wild Mustard
Wild Radish

* Under soil moisture conditions favoring deep germination, these species may not be completely controlled.

BLADEX may be used for burndown of small existing annual weeds and residual control of weeds during the winter and early spring season prior to planting cotton in California only. Complete any planned tillage prior to application. Apply herbicide treatment before weeds germinate or before weed seedlings are more than 3 inches tall. Tillage after application may reduce the effectiveness of the herbicide treatment.

Apply BLADEX at least 30 days prior to planting. Apply the proper rate for the soil texture, organic matter and time interval between application and planting indicated in the following table. Where existing weeds are present, add crop oil concentrate, surfactant, or emulsible vegetable oil at its recommended rate to aid in the burndown of small weeds.

Where existing weeds are greater than 3 inches in height, when very dry conditions exist or where volunteer grains are a major problem, tank-mix BLADEX with 1–2 pts./A. of paraquat or 2–3 pts./A. of Dinitro 5 Weed Killer. Well established weeds 6–8 inches tall or taller may not be well controlled.

Apply BLADEX in at least 20 gal./A. of carrier by ground, 5 gal./A. by air. Add X-77 surfactant at 1–2 qts./100 gals. of diluted spray where paraquat is used. Crop oil concentrate or emulsible vegetable oil are not needed where paraquat or Dinitro Weed Killer are used.

BLADEX can also be tank-mixed with Treflan™ preplant incorporated for fall listed cotton beds instead of surface applied as described above.

Broadcast Application Rates Per Acre For BLADEX 80W For Idle Season Early Preplant Treatment On Cotton

	DAYS PRIOR TO PLANTING*					
	30 DAYS POUNDS OF BLADEX 80W ORGANIC MATTER		60 DAYS POUNDS OF BLADEX 80W ORGANIC MATTER		90 DAYS POUNDS OF BLADEX 80W ORGANIC MATTER	
SOIL TEXTURE	2% OR LESS	2% OR MORE	2% OR LESS	2% OR MORE	2% OR LESS	2% OR MORE
Sands & Loamy Sand	2.0	2.5	3.0	3.75	3.75	4.5
All other Soils	2.5	3.0	3.75	4.5	4.5	5.0

*For intervals between those listed, adjust the rate proportionately.

Precautions:

1. Failure to wait the recommended time interval between application and planting may result in crop injury.
2. At least one inch of rainfall or an equivalent irrigation that waters the surface of the soil after application must precede planting.
3. The use of this treatment on calcareous or caliche soil outcroppings may result in crop injury.
4. Do not graze or feed foliage from treated areas to livestock.

PREEMERGENCE USE

Weeds Controlled

Annual morningglory
Cocklebur
Prickly sida (teaweed)
Spurge

BLADEX 80W Herbicide is a selective preemergence herbicide for early season weed control in cotton. Supplemental practices (such as directed postemergence BLADEX 80W application) may be necessary to control late season weeds. BLADEX 80W can be used in a tank-mix combination with Zorial™ 80WP. Apply only in the states of AL, AR, LA, MS.

Carefully match the BLADEX rate with the soil texture. Do not use on fields where the soil texture changes from coarse to fine. Avoid overlapping the spray pattern or overdosing the field with BLADEX 80W. Application rates above those recommended for the soil texture can result in yellowing or stunting of the crop and may result in stand reduction.

While cotton exhibits tolerance to BLADEX 80W, adverse growing conditions such as excessive rains, standing water or cold weather may result in stand reduction.

DO NOT GRAZE OR FEED FOLIAGE FROM TREATED AREAS TO LIVESTOCK.

BLADEX Plus Zorial 80WP

Apply BLADEX 80W plus Zorial 80WP at the proper rate for the soil texture shown in Table 6. The soil must contain at least 1.0 percent organic matter. Seed placement should be ½–¾ inch from the soil surface. Plant only cotton within six months after the last application of Zorial 80WP or injury may occur.

Table 6
Preemergence Broadcast Application Rates Per Acre For BLADEX 80W + Zorial 80WP On Cotton

Soil Texture (Soil Texture must contain at least 1.0% Organic Matter)		BLADEX 80W + Zorial 80WP lbs.	lbs.
Coarse Soils	Sandy loam	.65	.8
Medium Soils	Silt and silt loam	.75	1.3
	Loam, clay loam, sandy clay loam and sandy clay	1.1	1.3
Fine Soils	Silty clay loam, silty clay and clay	1.5	1.6

DO NOT USE ON COARSE SOILS (SANDS AND LOAMY SANDS) CONTAINING MORE THAN 70% SAND.

DIRECTED POSTEMERGENCE—LAYBY USE

Weeds Controlled

Annual morningglory*	Palmer amaranth
Bristly starbur	Pigweed (redroot and spiny)
Cocklebur	Prickly sida (teaweed)
Crotalaria	Sicklepod
Jimsonweed	Spurge
Lambsquarters	Tropic croton
Nightshade (annual)	Wright groundcherry

* The degree of preemergence control from a layby treatment will be reduced if soil moisture and temperature conditions cause deep germination of the seed.

BLADEX 80W and tank-mix combinations may be applied directed postemergence/layby to cotton and either preemergence to weeds or postemergence to weeds in all cotton growing states. Apply before weeds are more than 2 inches tall.

Apply the directed postemergence treatment after the cotton has attained the minimum height of 6 inches. For layby treatment, apply BLADEX 80W after the cotton has attained the height of 12 inches or more.

The spray mixture should be directed to the soil around the base of the cotton plants. Care should be taken to prevent the spray from striking the cotton leaves as injury will occur. The use of leaf lifters or shields on application equipment is recommended to help avoid the spray contacting the cotton foliage.

BLADEX may be applied postemergence following a preemergence application of BLADEX. Apply no more than two postemergence and one preemergence application to the same crop in any one year. If BLADEX is not used preemergence, apply no more than three postemergence applications to the same crop in any one year. (In California apply no more than two postemergence applications.) When applied as a layby treatment before weeds emerge, the effectiveness of BLADEX 80W depends on rainfall or irrigation to move it into the soil. When irrigation water activation is used, every row must be watered and for skip row cotton all treated soil must be irrigated. Any rotational crop may be planted the fall or spring following any of the treatments in this section.

DO NOT GRAZE OR FEED FOLIAGE FROM TREATED AREAS TO LIVESTOCK.

BLADEX Applied Alone

Apply BLADEX 80W directed-postemergence at the rate shown in Table 7. Apply at layby at the rates for the soil texture indicated in Table 8. Add an agricultural surfactant such as X-77 or other surfactant suitable for use on growing cotton at the rate of 2 quarts per 100 gallons of spray mixture (or as directed by the manufacturer).

Table 7
Directed Postemergence Application Rates Per Acre For BLADEX 80W On Cotton

	Banded 38″ Row	
Broadcast	**12″ Band**	**19″ Band**
0.75–1.25 lb.	0.25–0.4 lb.	0.4–0.6 lb.

Use the maximum rate when dry or arid conditions exist.

Table 8
Layby Application Rates Per Acre For BLADEX 80W On Cotton

Height of Cotton	Soil Texture		Broadcast Rates
12 inches or more	Coarse	Sandy loam	1.0 lb.
	Medium	Silt and silt loam, loam, clay loam, sandy clay loam and sandy clay	1.5 lb.
	Fine	Silty clay loam, silty clay and clay	2.0 lb.

BLADEX Plus MSMA

Apply a tank-mix combination of BLADEX 80W plus MSMA plus surfactant after the cotton is 6″ tall but before it reaches the bloom stage. Apply no more than two applications of this mixture before the first bloom stage. Tank-mix BLADEX plus MSMA at the rates indicated in Table 9. Add a surfactant, such as X-77, at the rate of 2 qts./100 gals of spray mixture (or as directed by the manufacturer).

Table 9
Directed Postemergence Application Rates Per Acre BLADEX 80W + MSMA On Cotton

	Banded 38″ Row	
Broadcast	**12″ Band**	**19″ Band**
0.75–1.25 lb. BLADEX 80W	0.25–0.4 lb.	0.4–0.6 lb.
+	+	+
4 pints MSMA (4 lb/gal)	1.3 pints	2 pints
or	or	or
2.4 pints MSMA (6.6 lb/gal.)	0.8 pint	1.2 pints

BLADEX Plus Dinitro (dinoseb)

Apply as a tank-mix combination of BLADEX 80W plus Dinitro 3 plus surfactant after the cotton is 6″ tall. Tank-mix BLADEX 80W plus Dinitro 3 at the rates indicated in Table 10. Add a surfactant, such as X-77, at the rate of 2 qt/100 gals. of spray mixture. Do not apply when soil surface is wet or cotton plants are extremely succulent. Do not apply in California or Arizona.

Table 10
Directed Postemergence Application Rates Per Acre BLADEX 80W + Dinitro On Cotton

	Banded 38″ Row	
Broadcast	**12″ Band**	**19″ Band**
0.75–1.25 lb. BLADEX 80W	0.25–0.4 lb.	0.4–0.6 lb.
+	+	+
1–2 qts. DINITRO 3	0.3–0.65 qt.	0.5–1.0 qts.

BLADEX Plus MSMA Plus Dinitro

Apply as directed in the "BLADEX Plus MSMA" section of the label. Tank-mix BLADEX plus MSMA at the rates shown in Table 9. Add Dinitro at the rate shown in Table 10. Add a surfactant, such as X-77, at the rate of 2 qts./100 gals. of spray mixture. Do not apply when soil surface is wet or cotton plants are extremely succulent. Do not apply in California or Arizona.

GRAIN SORGHUM (MILO)

Tank-mix combinations of BLADEX 80W plus propachlor (Ramrod™) or propazine (Milogard™) may be used for selective preemergence weed control in grain sorghum. Do not use on forage sorghum.

BLADEX 80W in tank-mix combination with propachlor or propazine should be applied only once per crop season. If replanting of grain sorghum is necessary, it may be planted in soil previously treated with these mixtures. Do not make a second application of BLADEX or crop injury may occur. Apply these tank-mix combinations at planting or after planting, but before the crop and weeds have emerged. Heavy rain immediately following application tends to cause excessive concentrations of herbicide in seed furrow, resulting in possible crop injury. Do not apply to furrow-planted sorghum until furrows are leveled (plowed-in). Level deep planter marks or seed furrows before application. Application to sorghum growing under stress caused by minor element deficiency or to sorghum growing on highly calcareous soil may result in crop injury.

BLADEX 80W Plus Propachlor (Ramrod)

WEEDS CONTROLLED

Grasses

Barnyardgrass	Giant foxtail
Crabgrass	Green foxtail
Fall panicum	Yellow foxtail

Broadleaves

Annual morningglory	Pigweed
Carpetweed	Ragweed
Cocklebur*	Smartweed (Pennsylvania)
Common purslane	
Lambsquarters	Velvetleaf*

* Under conditions such as low temperatures, lack of soil surface moisture or other factors that may cause delay in germination of the seeds, the effectiveness of BLADEX may be impaired against these weeds.

Apply BLADEX 80W plus propachlor at the proper rate for soil texture and organic matter shown in Table 11. Only apply this tank-mix to grain sorghum grown in states East of the Rocky Mountains. Any rotational crop may be planted the fall or spring following this treatment.

Table 11
Preemergence Broadcast Application Rates Per Acre For BLADEX 80W + Propachlor (Ramrod) 4L On Grain Sorghum

	Percent Organic Matter in Soil*			
	2%		3%	
Soil Texture Description	**Lbs./Acre BLADEX 80W**	**Qts./Acre Propachlor 4L**	**Lbs./Acre BLADEX 80W**	**Qts./Acre Propachlor 4L**
Sand, Loamy Sand	DO NOT USE			
Sandy Loam	1.2	2.5	1.5	3.0
Loam, Silt Loam, Silt	1.5	3.0	1.7	3.5
Sandy Clay Loam, Clay Loam, Silty Clay Loam	1.7	3.5	2.0	4.0
Sandy Clay, Silty Clay, Clay	2.0	4.0	2.0	4.0
Peat or Muck	NOT RECOMMENDED			

* For organic matter content between those listed, adjust the rate proportionately.

BLADEX Plus Propazine (Milogard)

WEEDS CONTROLLED

Grasses

Crabgrass	Green foxtail
Fall panicum	Yellow foxtail
Giant foxtail	

Broadleaves

Annual morningglory	Pigweed
Carpetweed	Puncturevine
Cocklebur	Ragweed
Common purslane	Smartweed
Lambsquarters	Velvetleaf

Apply BLADEX 80W plus Propazine at the proper rate for soil texture shown in Table 12.

Apply to grain sorghum grown only in the states of Texas, Oklahoma, and Kansas. **Rotational Crops:** In the Texas Gulf Coast and Texas Blacklands, fields treated with this tank-mix may be planted to cotton, soybeans, or corn 12 months after treatment. In Oklahoma and West Texas, fields treated with this tank-mix may be planted to cotton or corn 12 months after application. In Kansas, corn may be planted in rotation 12 months after treatment. Other crops should not be planted for 18 months following treatment.

Table 12
Preemergence Broadcast Application Rates Per Acre For BLADEX 80W + Propazine (Milogard) 4L On Grain Sorghum

Soil Texture	Lbs/Acre BLADEX 80W	Qts/Acre Propazine 4L*
Sand, Loamy Sand, Sandy Loam	DO NOT USE	
Loan, Silt Loam, Silt	1.0	0.75
Sandy Clay Loam, Clay Loam, Silty Clay Loam	1.25	1.0
Sandy Clay, Silty Clay, Clay	1.5	1.25
Peat or Muck	NOT RECOMMENDED	

* If Propazine 80W is used multiply rate shown by 1.25 to equal pounds of Propazine 80W.

WHEAT

Early Preplant—Post-harvest weed control for land going into production of continuous winter wheat.

Weeds controlled: See FALLOW CROPLAND section of this label.

Apply after wheat harvest but 45–60 days prior to planting winter wheat. Use at the proper rate for soil texture and organic matter indicated in Table 13. Add X-77 at 1–2 qts./100 gals. of diluted spray. Use higher rate or surfactant when dry weather conditions prevail. Failure to wait the recommended time interval between application and planting may result in crop injury. Use of this treatment on calcareous or caliche subsoil outcroppings may result in crop injury.

For best results during harvest, use a straw chopper/spreader on the combine that distributes the straw uniformly over the soil surface unless the straw is to be baled and removed prior to treatment. Avoid tillage after application as it may reduce the effectiveness of the herbicide treatment.

Where broadleaf weeds are present at time of application, tank-mix 1⅓–2 pts./A. of 2,4-D low volatile 6 lb./gal. ester (2–3 pts./A. of 2,4-D LV 4 lb./gal). Use the high rate when weeds are over 4" tall or when directed on the 2,4-D label for control of hard-to-kill weed species, such as perennials.

Where grasses and broadleaf weeds are present and exceed 2" in height, tank-mix with 1–2 pts./A. of paraquat. Well established weeds over 6" tall may not be well controlled. Apply at least 25 gals./A. of spray mixture by ground or 5–10 gals./A. by air where paraquat is included in the tank mixture.

No more than two applications of BLADEX may be made prior to planting wheat including applications made under the FALLOW CROPLAND section of this label.

Table 13
Early Preplant Broadcast Application Rates per Acre for BLADEX 80W for Winter Wheat Apply 60 days before planting*

	Pounds of BLADEX 80W	
	Percent Organic Matter in Soil	
Soil Texture Description	**Less than 2%**	**More than 2%**
Sand, Loamy Sand	2.5	3.0
All other textures	3.0	3.5

* For applications made 45 days prior to planting, decrease the BLADEX rate indicated by 0.5 pound/acre.

FALLOW CROPLAND

WEEDS CONTROLLED BY BLADEX 80W ALONE AND IN TANK-MIX COMBINATIONS

Grasses

Annual (Italian) ryegrass	Yellow foxtail
Barnyardgrass*	Indian lovegrass
Crabgrass	Stinkgrass
Cheatgrass	Volunteer wheat
Downy Brome	Witchgrass
Green foxtail	Wild Oat*

Broadleaves

Cocklebur*	Prostrate spurge
Common chickweed	Purslane
Dog fennel	Russian thistle
False flax	Shepherdspurse
Henbit	Smartweed (Pennsylvania)
Horseweed (marestail)	Sunflower* (wild)
Kochia	Purple mustard
Lambsquarters	Tansy mustard
Pennycress	Tumble mustard
Pigweed*	Wild radish
Prickly lettuce	Wild buckwheat*
Prostrate knotweed	

Additional weeds controlled are listed in the Corn section of this label.

* Under soil moisture and temperature conditions favoring deep germination or other factors that may cause delayed germination, these species may not be completely controlled.

BLADEX 80W may be used alone or in tank-mix combination with atrazine for the control of certain annual weeds during a fallow program.

BLADEX 80W or BLADEX 80W plus atrazine should be used in tank-mix combination with Paraquat CL and/or 2,4-D as described below if growing vegetation is present. Should weeds become established before adequate rainfall for herbicide activation occurs, sweep tillage may be employed to destroy them.

APPLICATION DIRECTIONS

Apply fallow cropland herbicide treatments uniformly to the soil surface. Adjust boom height on ground rigs to obtain the correct spray pattern at the top of the stubble rather than the ground. At sprayer speeds over 8 mph and when crop residues are heavy use floodtype nozzles and at least 25 gal./A. of carrier.

BLADEX 80W Applied Alone

Select the appropriate rate of BLADEX 80W from Table 14. Add X-77 surfactant at the rate of 1 qt./100 gals. of diluted spray. Winter wheat may be planted four months or more after treatment. Spring wheat or durum wheat may be planted nine months after treatment. Grain sorghum or field corn may be planted in the spring following a late fall application.

Table 14
Broadcast Application Rates Per Acre For BLADEX 80W On Fallow Cropland*
Lbs. Per Acre

Soil Texture	Post Harvest Treatment for Fall Weed Control	Late Fall Treatment for Spring Weed Control	Spring Treatment for Spring Weed Control
Clay loams, loams, and sandy soils	3.0	4.0	3.0
Clays, loamy clays, silty clays, and sandy clays	3.5	4.0	3.0

* For soils containing more than 2% organic matter, use an additional 0.5 lb/acre of BLADEX 80W.
For soils containing more than 3% organic matter, use an additional 1.0 lb/acre of BLADEX 80W.

BLADEX 80W Plus Atrazine

BLADEX 80W may be used in tank-mix combination with atrazine where a maximum period of weed control is desired in a fallow cropland program. Treatments must be applied before November 15 of the year before planting winter wheat or at least 11 months before planting spring wheat or durum wheat. Select the appropriate rates of BLADEX 80W plus atrazine for a particular location from Table 15. Add X-77 surfactant at 1 qt/100 gals. of diluted spray.

Do not use this treatment on sands or on Rosebud or Canyon series soils, or on calcareous or caliche subsoil outcroppings because of possible atrazine carryover damage to the succeeding crop.

Do not graze or feed foliage from treated areas to livestock within six months after application.

Table 15
Broadcast Application Rates Per Acre For Tank Mix Combinations Of BLADEX 80W Plus Atrazine On Fallow Cropland*

Location/ Treatment		Lbs/Acre BLADEX 80W	+	Qts./Acre Atrazine 4L**
FALL APPLICATION FOR WINTER WHEAT Kansas, Southern Nebraska, Oklahoma, Colorado East Slope between Hwy. I-76 & State Hwy. 96		2.0	+	0.8
FALL APPLICATION FOR WINTER OR SPRING WHEAT Nebraska Panhandle, Wyoming, Utah, Colorado West Slope & East Slope North of Hwy. I-76 & South of Hwy. 96, Idaho, Montana, North Dakota, South Dakota		2.0	+	0.6
SPRING APPLICATION FOR SPRING WHEAT Idaho, Montana, North Dakota, South Dakota, Northern Utah, Northern Wyoming		2.5	+	0.4
FALL APPLICATION FOR WINTER WHEAT Columbia Basin areas of Washington & Oregon	Average Annual Rainfall over 15"	2.5	+	0.4
	10-15"	2.5	+	0.27
	Less than 10"	2.5	+	0.2

* Use an additional 0.5 lb/A. of BLADEX 80W for soils with 2.0 to 3.0 percent organic matter.
Use an additional 1.0 lb/A. of BLADEX 80W for soils containing more than 3.0 percent organic matter.

** If Atrazine 80W is used multiply rate shown by 1.25 to equal pounds of Atrazine 80W.

NOTE: If conditions cause weeds to begin to germinate in the spring or summer following a post-harvest fall treatment of BLADEX or BLADEX plus atrazine, an application of BLADEX 80W may be made for additional weed control in the spring or summer prior to planting winter wheat. Apply as directed in "BLADEX 80W Applied Alone" in this section of the label or in the WHEAT section of this label. Do not make more than two applications of BLADEX 80W prior to planting.

BLADEX Combinations With Paraquat

On fallow crop land having an existing or established weed population, Paraquat CL may be tank mixed at 1–2 pts/A. with either BLADEX 80W or BLADEX 80W/atrazine tank-mix combination as previously described in this section. Apply the recommended rates in at least 25 gallons of spray mixture per acre by ground rig or 5–10 gals./A. for aerial application. Use higher volumes and the high rates of paraquat when weed growth is heavy or when dry weather conditions prevail. Add X-77 surfactant at 1 qt./100 gals. of diluted spray.

BLADEX Combinations With 2,4-D Low Volatile Ester

2,4-D LV Ester may be added to any treatment in this section to help control broadleaf weeds growing at the time of application. Use 1–1/3 to 2 pts/A. of 2,4-D LV 6 lb. Ester (2–3 pts/A. of 2,4-DLV 4 lb. Ester). Use the high rate when weeds are over 4″ tall or when directed on the 2,4-D label for the control of hard-to-kill weed species, such as perennials. When 2,4-D LV is used, it should be added to the spray tank last.

USE OF SUPPLEMENTAL TILLAGE

In fields where established weeds are too large to be effectively controlled with paraquat or 2,4-D, sweep tillage should be employed. Till before applying the herbicide treatment. This type of tillage will preserve a maximum amount of existing stubble on the surface for soil protection. Similar tillage may also be used if weeds become established prior to receiving adequate rainfall for activation of the herbicide treatment. At some point prior to seeding wheat, the herbicide will degrade and no longer be effective. Limited tillage should be employed at this time. This tillage should be kept shallow to preserve as much moisture as possible for the crops.

BLADEX®—Trademark of Shell Chemical Co.
Banvel™ —Trademark of Velsicol Chemical Co.
Dual™, Milogard™ —Trademarks of CIBA-Geigy Corp.
Eradicane™, Sutan+™ —Trademarks of Stauffer Chemical Co.
Lasso™, Ramrod™ —Trademarks of Monsanto Co.
Ortho X-77™ —Trademark of Chevron Chemical Co.
Treflan™— Trademark of Elanco Products Co.
Zorial™ —Trademark of Sandoz, Inc.

BLADEX® 90DF

HERBICIDE

Use of Bladex 90DF in Corn, Cotton, Grain Sorghum, Wheat and Fallow Cropland

BLADEX 90DF is a water dispersible granule.

ACTIVE INGREDIENT	BY WEIGHT
2[[4-chloro-6-(ethylamino)-s-triazin 2-yl]amino]-2-methylpropionitrile	90%
INERT INGREDIENTS	10%
TOTAL	100%

EPA Reg. No. 201-414
Covered by U.S. Patent Nos. 3,505,325 and 3,620,710

NOTE CONDITIONS OF SALE AND WARRANTY BEFORE OPENING THE CONTAINER. IF THEY ARE NOT ACCEPTABLE, RETURN UNOPENED PRODUCT.

Keep Out of Reach of Children
WARNING
AVISO

PRECAUCION AL USUARIO: Si usted no lee ingles, no use este producto hasta que le etiqueta haya sido explicado ampliamente.

STATEMENT OF PRACTICAL TREATMENT

IF SWALLOWED, call a physician or poison control center. Drink 1 or 2 glasses of water and induce vomiting by touching back of throat with finger. Do not induce vomiting or give anything by mouth to an unconscious person.

IF IN EYES, flush with plenty of water. Get medical attention if irritation persists.

IF ON SKIN, wash with plenty of soap and water.

IF INHALED, remove victim to fresh air. If not breathing, give artificial respiration, preferably mouth-to-mouth. Get medical attention.

PRECAUTIONARY STATEMENTS

HAZARD TO HUMANS AND DOMESTIC ANIMALS

WARNING

WARNING: May be fatal is swallowed. Harmful if inhaled or absorbed through the skin. Causes temporary eye injury.

This product may be hazardous to your health. This product has been determined to cause birth defects in laboratory animals. Use of protective clothing and equipment and following the precautions below can reduce risk.

Avoid breathing spray mist. Avoid contact with skin, eyes, or clothing. Do not get in eyes or on clothing. Wear a face shield when mixing and loading. Wash thoroughly with soap and water after handling and before eating or smoking.

Wear long trousers and long-sleeved clothing when applying this product. Wear rubber gloves extending above the wrist, long trousers and long-sleeved clothing when mixing or loading or when adjusting, repairing or cleaning equipment. Remove contaminated clothing and wash before reuse. Heavily contaminated clothing should be taken off immediately and incinerated or burned, if allowed by state and local authorities, or disposed of in a sanitary landfill and not laundered. For 24-hour emergency medical assistance call: (713) 473-9461.

Do not apply this product in such a manner as to directly, or through drift, expose workers or other persons. The area being treated must be vacated by unprotected persons.

Keep out of reach of domestic animals, particularly cattle. Consumption of this product, spray solutions, or water contaminated with product can result in serious illness or possible death of bovines.

ENVIRONMENTAL HAZARDS

Do not apply directly to water or wetlands.

Keep out of lakes, streams or ponds. Do not contaminate water by cleaning equipment or disposal of wastes. BLADEX is a pesticide which can move (seep or travel) through soil and can contaminate groundwater which may be used as drinking water. BLADEX has been found in groundwater as a result of agricultural use. Users are advised not to apply BLADEX where the water table (groundwater) is close to the surface and where the soils are very permeable (i.e., well drained soils such as loamy sands). Your local agricultural agencies can provide further information on the type of soil in your area and the location of groundwater. In case of significant spill, call CHEMTREC (800) 424-9300 or (713) 473-9461.

DIRECTIONS FOR USE

It is a violation of Federal law to use this product in a manner inconsistent with its labeling.

FOR END USE ONLY. DO NOT REPACKAGE OR REFORMULATE WITHOUT MANUFACTURER'S WRITTEN APPROVAL.

STORAGE AND DISPOSAL

STORAGE

Do not contaminate water, food or feed by storage or disposal. Do not use or store around the home environment. Avoid contact with water. In case of spill or leak, avoid breathing dust or vapors, clean up and dispose of wastes in compliance with local, State and Federal regulations.

PESTICIDE DISPOSAL

Pesticide, spray mixture or rinsate that cannot be used according to label instructions must be disposed of according to applicable Federal, State or local procedures.

CONTAINER DISPOSAL

Completely empty bag into application equipment. Then dispose of empty bag in a sanitary landfill or by incineration, or, if allowed by state and local authorities, by burning. If burned, stay out of smoke.

CONDITIONS OF SALE AND WARRANTY

SHELL AND THE SELLER OFFER THIS PRODUCT, AND THE BUYER AND USER ACCEPT THIS PRODUCT, ONLY UNDER THE FOLLOWING AGREED CONDITIONS OF SALE AND WARRANTY: The directions for use of this product are believed to be reliable and should be followed carefully. However, it is impossible to take into account all variables and to eliminate all risks associated with its use. Injury or damage may result because of conditions which are beyond the control of Shell or the seller, including soil texture, organic matter content of soil, weather, presence of other materials, and the manner of use or application. Shell warrants only that this product conforms to the chemical description on the label and is believed to be reasonably fit for the purposes referred to in the Directions for Use when used as directed under normal conditions. SHELL MAKES NO OTHER EXPRESS OR IMPLIED WARRANTY OF FITNESS OR MERCHANTABILITY OR ANY OTHER EXPRESS OR IMPLIED WARRANTY. Any variation or exception from this warranty must be in writing and signed by an authorized Shell representative.

LIMITATION OF LIABILITY

Shell's liability, whether or not based on negligence, breach of expressed or implied warranty, strict liability or any other legal cause, is limited to the cost of this product. In no case shall Shell or the Seller be liable for consequential, incidental, or indirect damages, such as loss of crops, resulting from the use or handling of this product.

GENERAL INFORMATION

BLADEX 90DF Herbicide is a selective herbicide for the control of annual grasses and broadleaved weeds in field corn, cotton, grain sorghum, wheat, and fallow cropland.

Consult your local Agricultural Extension Agent for help in determining soil texture, organic matter content, and the most appropriate herbicide rate for local conditions.

Where surfactants or emulsifiable vegetable oils are added to BLADEX 90DF for over the top of corn postemergence applications, use on field corn only.

As a preemergence herbicide, BLADEX 90DF is active mainly through the roots, and therefore, its effect on weeds is dependent on adequate rainfall or sprinkler irrigation to move the herbicide into the root zone. Moisture should be sufficient to wet the top 1½ to 2 inches of soil or make the soil too wet to cultivate; for most soils ½-¾" of moisture is sufficient. A rotary hoeing or shallow cultivation is recommended if a rainfall or sprinkler irrigation has not occurred within about ten days after application of BLADEX 90DF. Under conditions which delay weed germination, such as low temperatures, lack of soil surface moisture, or when germination extends over a long period, the effectiveness of the herbicide may be impaired. Rotary hoeing, a shallow cultivation or a postemergence herbicide treatment may be of benefit under these circumstances. When applied as a postemergence herbicide, BLADEX 90DF is also active through foliage as well as through the roots.

OBSERVE ALL CAUTIONS AND LIMITATIONS ON LABELING OF ALL PRODUCTS USED IN MIXTURES.

APPLICATION DIRECTIONS

GENERAL MIXING AND SPRAYING INSTRUCTIONS

Use sufficient agitation to ensure that the BLADEX 90DF is completely dispersed (dissolved) and in uniform suspension prior to application or tank mixing with other formulations.

The following general mixing instructions are recommended when using this or any other dry flowable formulation.

I. General

1. Unless otherwise specified, use at least 15 gallons of water per acre for all applications with ground equipment. For aerial application, apply a minimum of four gallons of water per acre.
2. A nitrogen solution or complete liquid fertilizer may replace all or part of the water as a carrier for preemergence or preplant application on corn or sorghum. Do not apply fertilizer mixtures after crop emerges, as injury may occur.
3. Always check the tank mix compatibility (TMC) of this or any other formulation before mixing with liquid fertilizer carriers or other formulations. A simple but generally reliable TMC evaluation procedure has been provided for your use in step II of these mixing instructions.
4. Start with thoroughly clean equipment.
5. Fill tank ¼ **full** with carrier. **Start and maintain consistent agitation through all mixing and spraying procedures.** Make sure that the agitation system is working properly and creates a rippling or rolling action on the liquid surface.
6. Slowly add the dry flowable (DF) to the tank or inductor.
7. Fill tank to 75 percent capacity with carrier. Filling and bypass lines should be kept below liquid surface. Increase tank agitation if necessary to maintain surface action.
8. When desired, appropriate emulsifiable oil, oil concentrate, or other tank mix formulations should be added at this time. Preslurry these added ingredients before addition if the compatibility test shows it to be necessary.
9. Complete filling tank maintaining sufficient agitation at all times to ensure surface action. **This applies to both spray and nurse tanks.**
10. Tank mixtures should always be applied immediately after preparation. If for any reason this is not possible, assure that sufficient agitation has been provided to remix all products and check for complete resuspension prior to application.
11. Empty tank as completely as possible before refilling to prevent buildup of oil or emulsifiable concentrate residues when tank mixing with these formulations. Always maintain agitation to avoid separation.
12. If an oil or emulsifiable concentrate film starts to build up after using these formulations, drain and clean the tank with strong detergent solution or appropriate solvent.
13. It is recommended that the sprayer be thoroughly cleaned by flushing with a detergent solution at the end of each work day when any emulsifiable oil, oil concentrate, or other emulsifiable formulation has been used either alone or in tank mix combinations with other pesticide formulations, even if no obvious problems have been encountered. This precaution will ensure a clean sprayer and continued trouble-free operation.

II. Tank Mix Compatibility Evaluation Procedure

1. Add one pint of carrier liquid to each of two one-quart jars. Mark one quart jar "with" and the other "without."
2. Add ¼ teaspoon of a suitable tank mix compatibility agent (¼ teaspoon/pint = 2 pints/100 gallons of carrier) to the jar marked "with," cap the jar, and shake gently for five to ten seconds to mix.
3. Add the appropriate amount of herbicide to both jars, cap each jar, and shake gently for five to ten seconds to mix. If problems are encountered in mixing wettable powder or dry flowable formulations into a liquid fertilizer, then preslurry these formulations in water prior to their addition to the liquid fertilizer and proceed with the test. The following chart has been provided to assist you in selecting the appropriate BLADEX 90DF use rate for this evaluation. If more than one herbicide is to be used in the tank mixture, each should be added separately as follows: water solubles first, wettable powders or dry flowables second, liquid flowables third, and emulsifiable concentrate or oil formulations last, with each jar capped and gently shaken for five to ten seconds for each addition.

Jar Test For BLADEX 90DF Compatibility

Gallons of Liquid Carrier per Acre	4	10	15	20	25	30	40
Teaspoons of BLADEX 90DF per Pint of Liquid Carrier	6.25	2.5	1.5	1.25	1.0	0.75	0.5

This chart is based on one pound of BLADEX 90DF (0.9 pounds active ingredient) per acre in the indicated carrier volumes. Intended field use rates are achieved by varying the amount of BLADEX

90DF; i.e., for a field use rate of 2.5 pounds of BLADEX 90DF in 15 gallons of carrier per acre, add 3.75 teaspoons of BLADEX to the quart jars containing one pint of carrier. **Calculation:** 2.5 pounds of BLADEX 90DF/15 gallons of carrier per acre = 2.5 X 1.5 teaspoons of BLADEX 90DF per pint of carrier.

4. Let each jar stand one-half hour and make observations. If any separation, agglomeration, or precipitation has occurred, shake the jar again for 10 to 15 seconds, and note whether any of the following occur:
   a. Separated phases do not remix uniformly.
   b. Screen/nozzle plugging lumps do not disperse.
   c. Precipitate does not resuspend readily.
   d. Precipitate sticks tenaciously to the glass.
5. If none of the above problems occur in either jar, then the herbicides can, in most cases, be safely used without a compatibility agent.
6. If problems **a** or **b** occur in the jar marked "without" but do not occur in the jar marked "with," then compatibility agent should be used.
7. If problems **a** or **b** are seen in both jars, then the herbicides and carrier mixture are incompatible and should not be used in the same spray tank. Alternatively, a different tank mix compatibility agent can be evaluated.
8. If problems **c** or **d** occur in the jar marked "without" but do not occur in the jar marked "with," the compatibility agent should be used unless constant, thorough agitation can be maintained and immediate clean-out of spray system is performed.
9. If problems **c** and **d** are seen in the jar marked "with," the user proceeds with mixing and application at his own risk should agitation in the system be insufficient or curtailed.
10. Those mixtures defined as compatible in this test should then be mixed for use as indicated in steps 1-12 of the general mixing instructions listed above.

III. Application Equipment

1. Use application equipment fitted with nozzles that provide accurate and uniform coverage. Be certain that nozzles are uniformly spaced and the same size. Calibrate sprayer before use and recheck frequently during use whenever possible.
2. Use a pump with capacity to:
   a. Maintain 35-40 psi at nozzles.
   b. Provide sufficient agitation in tank to keep mixture in suspension.
   c. Provide a minimum of 20 percent bypass at all times.
3. Use centrifugal pumps which provide sufficient shear action for dispersing and mixing this product. The pump should provide a minimum of 10 gallons/minute/100-gallon tank size circulated through the jets of a correctly-positioned sparger tube.
4. Use screens to protect the pump and to prevent nozzles from clogging. Screens placed on suction side of pump should be 10 to 16 mesh. Do not place a screen in the recirculation line. Use a 40- to 50-mesh screen between the pump and boom and, where required, 50-mesh screens at the nozzles. Check your equipment manufacturer's literature for specific recommendations.

APPLICATION IN CENTER PIVOT SPRINKLER IRRIGATION WATER (FOR USE IN MIDWEST AND PLAINS STATES ONLY)

BLADEX 90DF alone or in tank-mix combinations with atrizine or Lasso™ 4 EC, may be applied in center pivot sprinkler irrigation water. Apply any time after planting but prior to crop emergence. Pre-mix the herbicide(s) in water to form a slurry and inject at a carefully calibrated rate into the irrigation water during the entire period of the run. Mix as directed for a spray application. Provide good agitation for the herbicide mixture throughout the application.

Select the correct application rate of BLADEX 90DF or the BLADEX tank-mix combinations from the rate tables listed on this label.

Example:

A center pivot sprinkler that is 1320 feet long (1/4 mile) will irrigate 125.7 acres. A sandy loam soil containing 2% organic matter will require 2.2 pounds of BLADEX 90DF per acre.

125.7A × 2.2 lbs. = 276.5 lbs. BLADEX 90DF

Mixing tank size 600 gallons. Sprinkler takes 20 hours to apply 1/2 inch of water to the complete circle; inject rate =

$\frac{600}{20}$ = 30 gal./hr. or 0.5 gal./minute

Calibrate injection pump to deliver herbicide mix into the irrigation system at the rate of 30 gallons per hour or 0.5 gallons per minute. Add 276.5 pound of BLADEX 90DF to the tank.

For tank-mix combinations select the correct dosage rate and follow the same procedures outlined above for each herbicide component.

Apply in 1/2 to 1 acre inch of water. Use this method of application only with irrigation systems designed for uniform application of water.

Additional Use Precautions

1. Apply product or tank mixes only through center pivot irrigation systems containing anti-siphon and check valves which will prevent water source contamination and overflow of the slurry tank and containing interlocking controls between metering device and the water pump to ensure simultaneous shutoff.
2. Inject the herbicide mix with a positive displacement pump into the main line ahead of a right angle turn to ensure adequate mixing.
3. Application of more than label recommended quantities of irrigation water per acre may result in decreased product performance by removing the chemical from the zone of effectiveness.
4. Do not apply when wind speed favors drift, when system connectors leak, when nozzles do not provide uniform distribution or when lines containing the product must be drained and dismantled.
5. Greater accuracy in calibration and distribution will be achieved by injecting a large volume or more dilute slurry per hour.
6. Where sprinkler distribution patterns do not overlay sufficiently, unacceptable weed control may result. Where sprinkler distribution patterns overlap excessively, crop injury may result.
7. Make sure the system is designed to shut off the herbicide injection if the traveling sprinkler stalls or stops for any reason, even if the water pump does not shut off.

CORN

WEEDS CONTROLLED BY BLADEX 90DF ALONE AND IN COMBINATION WITH OTHER HERBICIDES ON CORN

Grasses

Annual bluegrass	Giant foxtail
Annual fescues	Goosegrass
Annual (Italian) ryegrass	Green foxtail
Barnyardgrass[1]	Junglerice
Bullgrass	Stinkgrass (Indian lovegrass)
Crabgrass	Witchgrass
Fall panicum	Yellow foxtail

Broadleaves

Annual groundcherry	Mayweed
Annual morningglory	Nightshade (annual)
Annual sedge	Pigweed[1]
Black mustard	Pineappleweed
Buffalobur	Plantain
Buttercup (annual)	Poorjoe
Carpetweed	Prickly sida (teaweed)
Cocklebur[2]	Prostrate knotweed
Common chickweed	Prostrate spurge
Common groundsel	Ragweed
Common mallow	Russian thistle
Common purslane	Shepherdspurse
Corn spurry	Smallflower galinsoga
Curly dock (seedling)	Smartweed (Pennsylvania)

Fiddleneck
Florida pusley
(Florida purslane)
Hedge mustard
Jimsonweed[1]
Kochia
Ladysthumb
Lambsquarters
Sunflower[2]
(wild, annual, common)
Tarweed cuphea (Gumweed)
Velvetleaf[1]
Wild buckwheat
Wild mustard
Wild radish
Wild turnip

[1] Under conditions such as low temperatures, lack of soil surface moisture or other factors that may cause delay in germination of the seeds, the effectiveness of BLADEX may be impaired against these weeds.
[2] The degree of control will be reduced if soil moisture and temperature conditions cause deep germination of the seed.

Table 1
Preemergence Broadcast Application Rates Per Acre For BLADEX 90DF Applied Alone on Corn

Soil Texture Description	Pounds of BLADEX 90DF — Percent Organic Matter in Soil*					
	Less than 1%	1%	2%	3%	4%	5% & Over
Sand, Loamy Sand	Do Not Use	1.3	1.8	2.4	3.1	3.6
Sandy Loam	1.3	2.0	2.2	2.7	3.3	4.0
Loam, Silt Loam, Silt	1.8	2.2	2.7	3.3	4.0	4.4
Sandy Clay Loam, Clay Loam, Silty Clay Loam	2.2	2.7	3.3	4.0	4.4	4.9
Sandy Clay, Silty Clay, Clay	3.1	3.3	4.0	4.4	4.9	5.3
Peat or Muck	NOT RECOMMENDED					

* For organic matter content between those listed, adjust the rate proportionately.

CORN—PREEMERGENCE-PREPLANT INCORPORATED

Apply BLADEX treatments just before, at or after planting but before crop has emerged. Avoid removal of treated soil from seedrow prior to or during the planting operation.

BLADEX may also be applied early prior to planting or in a split application if pre-season weed control is desired. For split applications, do not exceed the total amount of BLADEX for the soil texture and organic matter shown in Table 1. If BLADEX is applied early, more than 15 days before planting, a split application of BLADEX or some other herbicide treatment may be necessary at or after planting to provide additional length of weed control. For further information see "Early Preplant" recommendations in the Conservation Tillage section of label.

A rotary hoeing is recommended for preemergence applications which do not receive adequate rainfall or sprinkler irrigation to wet the top 1½-2" of soil within about 10 days after application.

If a BLADEX mixture is to be incorporated, except as noted, single or two pass incorporation is acceptable. Care should be taken to incorporate the BLADEX mixture no deeper than the top two inches of soil.

BLADEX Applied Alone

Use at the proper rate for soil texture and organic matter indicated in Table 1. Any rotational crop may be planted in the fall or spring following this treatment.

Table 2
Preemergence Broadcast Application Rates Per Acre For Tank-Mix Combinations of Bladex 90DF Plus Atrazine 90DF on Corn

Soil Texture Description	Pounds of BLADEX 90DF + Pounds of Atrazine 90DF** — Percent Organic Matter in Soil*					
	Less than 1%	1%	2%	3%	4%	5% & Over
Sand, Loamy Sand	Do Not Use	0.9+0.4	1.3+0.4	1.6+0.7	1.8+0.9	2.4+1.1
Sandy Loam	0.9+0.4	1.3+0.4	1.6+0.7	1.8+0.9	2.4+1.1	3.1+1.3
Loam, Silt Loam, Silt	1.3+0.4	1.6+0.7	2.2+0.9	2.4+1.1	3.1+1.3	3.6+1.3
Sandy Clay Loam, Clay Loam, Silty Clay Loam	1.6+0.7	2.2+0.9	2.4+1.1	3.1+1.3	3.6+1.3	3.8+1.6
Sandy Clay, Silty Clay, Clay	2.2+0.9	2.4+1.1	3.1+1.3	3.6+1.3	3.8+1.6	4.0+1.8
Peat or Muck	NOT RECOMMENDED					

* For organic matter content between those listed, adjust the rate proportionately.
** If Atrazine 4L is used, multiply rates shown by 0.9 to equal quarts of Atrazine 4L.
If Atrazine 80W is used, multiply rates shown by 1.125 to equal pounds of Atrazine 80W.

BLADEX COMBINATIONS

BLADEX plus Atrazine

Use at the proper rate for soil texture and organic matter indicated in Table 2. Table 2 provides rates for generally weedy conditions. The ratio of the amounts of each herbicide may be adjusted as necessary for particular weed conditions as long as the combined rate of the two products does not exceed the combined rate for the soil shown in Table 2. For grassier conditions use a ratio that contains higher levels of BLADEX (3:1). For fields with more broadleaves use a ratio that contains higher levels of atrazine (1:1).

Rotational Crops: (1) Plant only corn, peanuts, sorghum, or soybeans the year following the use of this mixture. (2) If soybeans are to be planted, injury may occur due to the carryover of atrazine. (3) If applied after June 10, do not rotate with crops other than corn or sorghum the next year or injury may occur. (4) In the high plains and intermountain areas of the West where rainfall is sparse and erratic or where irrigation is required, use only when corn or sorghum is to be planted the following year, or a crop of corn or sorghum not treated with atrazine is to precede other rotational crops. (5) Small grains may be planted 15 months following treatment. (6) All other crops may be planted 18 months after application.

Table 3
Preemergence Broadcast Application Rates Per Acre For Bladex 90DF Used in Tank-Mix Combinations With Lasso, Sutan+, Eradicane, or Dual on Corp

Soil Texture Description	Pounds of Bladex 90DF — Percent Organic Matter in Soil*					
	Less than 1%	1%	2%	3%	4%	5% & Over
Sand, Loamy Sand	0.7**	0.9	1.3	1.6	1.8	2.2
Sandy Loam	0.9	1.3	1.6	1.8	2.2	2.4
Loam, Silt Loam, Silt	1.3	1.6	1.8	2.2	2.4	2.9
Sandy Clay Loam, Clay Loam, Silty Clay Loam	1.6	2.0	2.2	2.4	2.9	3.1
Sandy Clay, Silty Clay, Clay	2.0	2.2	2.7	2.9	3.1	3.3
Peat or Muck	NOT RECOMMENDED					

* For organic matter content between those listed, adjust the rate proportionately.
** Do not use in the Atlantic Coastal Plain.

BLADEX plus Lasso™ 4 EC

Use BLADEX 90DF at the proper rate for the soil texture and organic matter shown in Table 3 plus 2 quarts per acre of Lasso. (Use 2.5 quarts Lasso on clay soils containing 5 percent organic matter and over). Any rotational crop may be planted the fall or spring following this treatment.

BLADEX Plus Sutan™+ 6.7E, or Eradicane™ 6.7E

Use BLADEX 90DF at the proper rate for the soil texture and organic matter shown in Table 3 plus 1.8 quarts per acre of Sutan+ or Eradicane for control of many annual grasses and broadleaf weeds. (Use 2.4 quarts of Sutan+ or Eradicane on loam soils containing 5 percent or more organic matter, and clay loams and clays containing 4 percent or more organic matter). Do not use on sands and loamy sands of less than 1 percent organic matter in the light sandy soils of eastern coastal states. Do not use on corn seed stock.

Apply before planting. Incorporate the mixture immediately upon application using power-driven cultivation equipment set for 2-3 inch depth, or a tandem disc set to cut to a depth of about 4 inches while operating at 4-6 mph. For thorough mixing, disc in two directions (cross disc), and follow with a harrow, drag, or other leveling device. Prior to the second discing, readjust the disc to prevent cutting deeper than 4 inches. BLADEX 90DF may be applied preemergence as an overlay over previous incorporated Sutan+ or Eradicane if desired. Any rotation crop may be planted in the fall or spring following these treatments.

Existing stands of quackgrass, purple and yellow netsedge must be turned under and thoroughly chopped up prior to chemical treatments.

Additional weeds controlled by Sutan+ or Eradicane combinations:

Grasses
Sandbur
Shattercane (Wild Cane)*
Texas Panicum
Quackgrass (Eradicane only)
Wild Proso Millet* (Eradicane only)

Perennial Weeds
Yellow Nutsedge (nutgrass)
Purple Nutsedge (nutgrass)

* Suppression only—refer to Sutan+ or Eradicane label for appropriate supplement cultural and tillage practices.

For fields with moderate to heavy infestations of these weeds refer to the Sutan+ or Eradicane labels for appropriate higher rates.

BLADEX plus Dual™ 8E

Use BLADEX 90DF at the proper rate for soil texture and organic matter shown in Table 3. Use Dual as follows:

Soil Texture	Broadcast Rate Per Acre For Dual
Coarse	
Sand, loamy sand, sandy loam	1.25-1.5 pints
Medium	
Loam, silt loam, silt	1.5-2.0 pints
Fine	
Sandy clay loam, silty clay loam, clay loam, sandy clay, silty clay, clay	1.5-2.5 pints

The low end of the rate range should be used for lowest organic matter soils and the rate increased as organic matter increases to a point that soils containing 4 percent organic matter or more require the highest rate shown for that soil texture. Refer to the Dual label for precautions on rotational crops.

Table 4
Preemergence Broadcast Application Rates Per Acre For Bladex 90DF Plus Atrazine 90DF Used in Tank-Mix Combinations With Lasso, Sutan+, Eradicane, or Dual on Corn

Soil Texture Description	Pounds of BLADEX 90DF + Pounds of Atrazine 90DF***					
	Percent Organic Matter in Soil*					
	Less than 1%	1%	2%	3%	4%	5% & Over
Sand, Loamy Sand	0.5 + 0.2**	0.7 + 0.2	0.9 + 0.4	1.2 + 0.4	1.1 + 0.7	1.6 + 0.6
Sandy Loam	0.7 + 0.2	0.9 + 0.4	1.2 + 0.4	1.1 + 0.7	1.6 + 0.6	1.8 + 0.6
Loam, Silt Loam, Silt	0.9 + 0.4	1.2 + 0.4	1.3 + 0.5	1.6 + 0.6	1.8 + 0.6	2.0 + 0.9
Sandy Clay Loam, Clay Loam, Silty Clay Loam	1.2 + 0.4	1.3 + 0.7	1.6 + 0.6	1.8 + 0.6	2.0 + 0.9	2.2 + 0.9
Sandy Clay, Silty Clay, Clay	1.3 + 0.7	1.6 + 0.6	1.8 + 0.9	2.0 + 0.9	2.2 + 0.9	2.2 + 1.1
Peat or Muck	NOT RECOMMENDED					

* For organic matter content between those listed, adjust the rate proportionately.
** Do not use in the Atlantic Coast Plain.
*** If Atrazine 4L is used, multiply rates shown by 0.9 to equal quarts of Atrazine 4L.

If Atrazine 80W is used, multiply rates shown by 1.125 to equal pounds of Atrazine 80W.

BLADEX plus Atrazine plus Lasso, Sutan+, Eradicane, or Dual

Use BLADEX 90DF plus atrazine at the proper rate for soil texture and organic matter shown in Table 4. Use Lasso, Sutan +, Eradicane, or Dual according to rates shown in "BLADEX Combinations" in this section of the label.

Rotational Crops: Refer to Rotational Crops section of "BLADEX plus Atrazine" in this section of the label.

CONSERVATION TILLAGE WEED CONTROL

Early Preplant (EPP)

[Prior-to-planting weed control for land going into production of corn.]

Complete any planned early spring tillage prior to application. Apply herbicide treatment before weeds germinate or before weed seedlings are more than 3 inches tall. Tillage after application may reduce the effectiveness of the herbicide treatment.

Apply BLADEX or BLADEX plus atrazine tank mix 15-30 days prior to planting. Use the proper rate for soil texture and organic matter indicated in Table 1 or 2. Where heavy crop residues exist, the rates shown in Tables 1 or 2 should be increased by 25%.

Where broadleaf weeds are present, add 2,4-D LV Ester at 1⅓-2 pts./A. (6 lb./gal.) or 2-3 pts./A. (4 lbs./gal.) plus X-77 surfactant at 1 qt./100 gals. of diluted spray, or other suitable surfactant at its recommended rate. If existing weeds exceed 3 inches in height, add 1-2 pts./A. of paraquat to the above mixture. Well established weeds over 6 inches tall may not be well controlled. Use at least 25 gals./A. of spray mixture by ground rig or 5-10 gals./A. for aerial application.

Depending upon weather conditions following the EPP application, a postemergence BLADEX treatment or some other herbicide treatment may be necessary at or after planting to provide additional length of weed control. If desired, 1½-2 pts./A. of Dual or 2 qts./A. of Lasso may be tank-mixed with the BLADEX EPP treatment or applied preemergence at planting.

Rotational Crops: Refer to Rotational Crops section for each treatment in the Preemergence section of the label.

At Planting

BLADEX 90DF applied alone or in combination with atrazine and/or Lasso or Dual according to the following directions will kill most existing small weeds and suppress many emerged perennial weeds when corn is planted into no-till stalk ground (corn, sorghum), stubble ground (soybean, small grains), and any minimum-till situation. This treatment then provides residual control of annual weeds as in conventional tillage.

Apply BLADEX 90DF alone or with other products according to the directions for those treatments in the Preemergence section of the label. Where heavy crop residues exist, the BLADEX rate shown in Tables 1, 2, 3 or 4 should be increased by 25%. Add ½-1 pt./A. of 2,4-D Low Volatile 6 lb. Ester (0.75-1.5 pt./A. 2,4-D LV 4 lb. Ester). Add the 2,4-D LV to the spray tank last. Use a minimum of 25 gals./A. of carrier. Complete spray coverage of the weeds is essential for best performance. Nitrogen solutions and complete liquid fertilizers are the preferred carriers for this treatment as they aid in the burndown of existing weeds. Add Ortho X-77™ surfactant at 1-2 qts. per 100 gals. of diluted spray, or other suitable surfactant at its recommended rate. If water is used as a carrier, crop oil concentrate may be used as a surfactant. Apply before weeds exceed 3" in height. For control of existing alfalfa add ⅓-½ pt./A. of Banvel™ to the spray mixture. Apply before the alfalfa exceeds 6" in height.

For fields with existing sod grasses such as orchardgrass, bromegrass, rye or timothy, or when very dry conditions exist, or when existing weeds exceed 3" in height, add paraquat to the tank-mix.

Use 2 pts./A. of paraquat in combination with BLADEX as described above in this section, except the 2,4-D LV may be omitted if desired. Do not apply paraquat in suspension type liquid fertilizer.

Table 5
Postemergence Broadcast Application Rates
Per Acre For BLADEX 90DF on Corn
No Prior BLADEX Application

Soil Texture	Pounds of BLADEX 90DF — Percent Organic Matter in Soil*: Less than 1%	1%	2%	Over 2%
Sand, Loamy Sand	DO NOT USE	1.3	1.8	2.2
Sandy Loam	1.3	1.8	2.2	2.2
Loam, Silt Loam, Silt	1.8	2.2	2.2	2.2
All other textures	2.2	2.2	2.2	2.2

* For organic matter content between those listed, adjust the rate proportionately.

Table 5A
Postemergence Broadcast Application Rates
Per Acre for Bladex 90DF on Corn
Bladex Used in Prior Application

Soil Texture	Pounds of BLADEX 90DF — Percent Organic Matter in Soil*: Less than 1%	1%	2%	Over 2%
Sandy Loamy Sand	DO NOT USE	DO NOT USE	1.5	1.5
Sandy Loam	DO NOT USE	DO NOT USE	1.75	2.2
Loam, Silt Loam, Silt	DO NOT USE	1.5	2.2	2.2
All other textures	DO NOT USE	2.0	2.2	2.2

* For organic matter content between those listed, adjust the rate proportionately.

POSTEMERGENCE

BLADEX Applied Alone

Use BLADEX 90DF at the proper rate for the soil texture and organic matter shown in Table 5 or 5A. Use rates shown in Table 5 if BLADEX has not been applied to the soil this season. Use rates shown in Table 5A if BLADEX has been applied to the soil this season. This treatment may be used on peat or muck soils for burndown and suppression of existing weeds but will not provide residual control. Apply from crop emergence through the four-leaf stage of corn growth but before weeds exceed about 1½" in height. Do not apply over the top of corn if the fifth leaf is visible. Apply in water only. Do not spray emerged corn plants in a liquid fertilizer carrier.

Under dry, arid conditions of low humidity and the absence of dew formation at night, add a surfactant such as X-77, or an emulsible vegetable (EV) oil suitable for use on growing corn at its recommended rate. Do not use petroleum-based crop oils. Addition of a surfactant or EV oil is not recommended under moist, rainy conditions and when dew forms at night as injury may occur.

DO NOT APPLY THIS TREATMENT UNDER COLD, WET, WEATHER CONDITIONS OR TO WEATHER-STRESSED OR STORM-DAMAGED CORN.

Yellowing of the corn may result from this treatment, particularly if cold, adverse growing conditions occur after application. Extended or extreme cold and wet conditions may reduce stands.

Do not use on corn grown for seed. Any rotational crop may be planted the fall or spring following this treatment.

BLADEX COMBINATIONS

BLADEX plus Atrazine

Apply as directed in "Postemergence-BLADEX Applied Alone" section of this label. Use an amount of BLADEX 90DF plus Atrazine 90DF equal to the rate shown in Table 5 or 5A for the proper soil texture and organic matter. To determine the amount of BLADEX 90DF to use multiply the rate indicated by 0.7. To determine the amount of Atrazine 90DF to use, multiply the rate indicated by 0.3. (For the 2.2 lb/A. rate shown in Table 5, use 1.5 lbs/A. of BLADEX 90DF plus 0.7 lbs/A. of Atrazine 90DF.)

Rotational Crops: See Rotational Crops section in the "Preemergence-BLADEX plus Atrazine" section of this label.

BLADEX plus Banvel™

Apply as directed under "Postemergence—BLADEX Applied Alone" section of this label. Add ½-⅔ pt/A. of Banvel to the mixture. Do not use with a surfactant or emulsible vegetable oil.

COTTON

IDLE SEASON EARLY PREPLANT WEED CONTROL (CALIFORNIA ONLY)

WEEDS CONTROLLED

Grasses

Annual Bluegrass
Annual Ryegrass
Barnyardgrass*
Bristly Foxtail
Rabbitsfoot Grass
Volunteer Small Grains (suppression)
Wild Oat*
Yellow Foxtail

Broadleaves

Annual Henbit
Black Nightshade
Burclover
Cheeseweed*
Fiddleneck
Groundsel
Knotweed
Marestail
Miners Lettuce
Lambsquarter
London Rocket
Pineapple Weed
Prickly Lettuce
Shepherdspurse
Sowthistle
Wild Mustard
Wild Radish

* Under soil moisture conditions favoring deep germination, these species may not be completely controlled.

BLADEX may be used for burndown of small existing annual weeds and residual control of weeds during the winter and early spring season prior to planting cotton in California only. Complete any planned tillage prior to application. Apply herbicide treatment before weeds germinate or before weed seedlings are more than 3 inches tall. Tillage after application may reduce the effectiveness of the herbicide treatment.

Apply BLADEX at least 30 days prior to planting. Apply the proper rate for the soil texture, organic matter and time interval between application and planting indicated in the following table. Where existing weeds are present, add crop oil concentrate, surfactant, or emulsible vegetable oil at its recommended rate to aid in the burndown of small weeds.

Where existing weeds are greater than 3 inches in height, when very dry conditions exist or where volunteer grains are a major problem, tank-mix BLADEX with 1-2 pts./A. of paraquat or 2-3 pts./A. of Dinitro 5 Weed Killer. Well established weeds 6-8 inches tall or taller may not be well controlled.

Apply BLADEX in at least 20 gal./A. of carrier by ground, 5 gal./A. by air. Add X-77 surfactant at 1-2 qts./100 gals. of diluted spray where paraquat is used. Crop oil concentrate or emulsible vegetable oil are not needed where paraquat or Dinitro Weed Killer are used.

BLADEX can also be tank-mixed with Treflan™ preplant incorporated for fall listed cotton beds instead of surface applied as described above.

Precautions: 1). Failure to wait the recommended time interval between application and planting may result in crop injury. 2). At least one inch of rainfall or an equivalent irrigation that waters the surface of the soil after application must precede planting. 3). The use of this treatment on calcereous or caliche soil outcroppings may result in crop injury. 4). Do not graze or feed foliage from treated areas to livestock.

Broadcast Application Rates per Acre for BLADEX 90DF
Idle Season Early Preplant Treatment on Cotton

	DAYS PRIOR TO PLANTING*					
	30 Days		60 Days		90 Days +	
	Pounds of Bladex 90DF		Pounds of BLADEX 90DF		Pounds of BLADEX 90DF	
	Organic Matter		Organic Matter		Organic Matter	
Soil Texture	2% or Less	2% or More	2% or Less	2% or More	2% or Less	2% or More
Sands & Loamy Sands	1.75	2.25	2.75	3.25	3.25	4.0
All other Soils	2.25	2.75	3.25	4.0	4.0	4.5

* For time intervals between those listed, adjust the rate proportionately.

PREEMERGENCE USE

Weeds Controlled

Annual morningglory	Prickly sida (teaweed)
Cocklebur	Spurge

BLADEX 90DF Herbicide is a selective preemergence herbicide for early season weed control in cotton. Supplemental practices (such as directed postemergence BLADEX 90DF application) may be necessary to control late season weeds. BLADEX 90DF can be used in a tank-mix combination with Zorial™ 80WP. Apply only in the states of AL, AR, LA, MS.

Carefully match the BLADEX rate with the soil texture. Do not use on fields where the soil texture changes from coarse to fine. Avoid overlapping the spray pattern or overdosing the field with BLADEX 90DF. Application rates above those recommended for the soil texture can result in yellowing or stunting of the crop and may result in stand reduction.

While cotton exhibits tolerance to BLADEX 90DF, adverse growing conditions such as excessive rains, standing water or cold weather may result in stand reduction.

DO NOT GRAZE OR FEED FOLIAGE FROM TREATED AREAS TO LIVESTOCK.

BLADEX plus Zorial 80WP

Apply BLADEX 90DF plus Zorial 80WP at the proper rate for the soil texture shown in Table 6. The soil must contain at least 1.0 percent organic matter. Seed placement should be $\frac{1}{2}$-$\frac{3}{4}$ inch from the soil surface. Plant only cotton within six months after the last application of Zorial 80WP or injury may occur.

Table 6
Preemergence Broadcast Application Rates Per Acre for BLADEX 90DF + ZORIAL 80WP on Cotton

Soil Texture (Soil Texture must contain at least 1.0% Organic Matter)		BLADEX 90DF + ZORIAL 80WP lbs.	lbs.
Coarse Soils	Sandy loam	0.6	.8
Medium Soils	Silt and silt loam	0.7	1.3
	Loam, clay loam, sandy clay loam and sandy clay	1.0	1.3
Fine Soils	Silty clay loam, silty clay and clay	1.3	1.6

DO NOT USE ON COARSE SOILS (SANDS AND LOAMY SANDS) CONTAINING MORE THAN 70% SAND.

DIRECTED POSTEMERGENCE—LAYBY USE

Weeds Controlled

Annual morningglory*	Palmer amaranth
Bristly starbur	Pigweed (redroot and spiny)
Cocklebur	Prickly sida (teaweed)
Crotalaria	Sicklepod
Jimsonweed	Spurge
Lambsquarters	Tropic croton
Nightshade (annual)	Wright groundcherry

* The degree of preemergence control from a layby treatment will be reduced if soil moisture and temperature conditions cause deep germination of the seed.

BLADEX 90DF and tank-mix combinations may be applied directed postemergence/layby to cotton and either preemergence to weeds or postemergence to weeds in all cotton growing States. Apply before weeds are more than 2 inches tall.

Apply the directed postemergence treatment after the cotton has attained the minimum height of 6 inches. For layby treatment, apply BLADEX 90DF after the cotton has attained the height of 12 inches or more.

The spray mixture should be directed to the soil around the base of the cotton plants. Care should be taken to prevent the spray from striking the cotton leaves as injury will occur. The use of leaf lifters or shields on application equipment is recommended to help avoid the spray contacting the cotton foliage.

BLADEX may be applied postemergence following a preemergence application of BLADEX. Apply no more than two postemergence and one preemergence application to the same crop in any one year. In California, apply no more than two postemergence applications. If BLADEX is not used preemergence, apply no more than three postemergence applications to the same crop in any one year. When applied as a layby treatment before weeds emerge, the effectiveness of BLADEX 90DF depends on rainfall or irrigation to move it into the soil. When irrigation water activation is used, every row must be watered and for skip row cotton all treated soil must be irrigated. Any rotational crop may be planted the fall or spring following any of the treatments in this section.

DO NOT GRAZE OR FEED FOLIAGE FROM TREATED AREAS TO LIVESTOCK.

BLADEX Applied Alone

Apply BLADEX 90DF directed-postemergence at the rate shown in Table 7. Apply at layby at the rates for the soil texture indicated in Table 8. Add an agricultural surfactant such as X-77 or other surfactant suitable for use on growing cotton at the rate of 2 quarts per 100 gallons of spray mixture (or as directed by the manufacturer).

Table 7
Directed Postemergence Application Rates per Acre for BLADEX 90DF on Cotton

	Banded 38″ Row	
Broadcast	12″ Band	19″ Band
0.7-1.1 lb.	0.2-0.4 lb.	0.4-0.5 lb.

Use the maximum rate when dry or arid conditions exist.

Table 8
Layby Application Rates per Acre for BLADEX 90DF on Cotton

Height of Cotton	Soil Texture		Broadcast Rates
12 inches or more	Coarse	Sandy loam	0.9 lb.
	Medium	Silt and silt loam, loam, clay loam, sandy clay loam and sandy clay	1.3 lb.
	Fine	Silty clay loam, silty clay and clay	1.8 lb.

BLADEX plus MSMA

Apply a tank-mix combination of BLADEX 90DF plus MSMA plus surfactant after the cotton is 6″ tall but before it reaches the bloom stage. Apply no more than two applications of this mixture before the first bloom stage. Tank-mix BLADEX plus MSMA at the rates indicated in Table 9. Add a surfactant, such as X-77, at the rate of 2 qts./100 gals of spray mixture (or as directed by the manufacturer).

Table 9
Directed Postemergence
Application Rates per Acre
BLADEX 90DF + MSMA on Cotton

Broadcast	Banded 38" Row 12" Band	Banded 38" Row 19" Band
0.7-1.1 lb. BLADEX 90DF + 4 pints MSMA (4 lb/gal) or 2.4 pints MSMA (6.6 lb/gal)	0.2-0.4 lb. + 1.3 pints or 0.8 pint	0.4-0.5 lb. + 2 pints or 1.2 pints

BLADEX plus Dinitro (dinoseb)

Apply a tank-mix combination of BLADEX 90DF plus Dinitro 3 plus surfactant after the cotton is 6" tall. Tank-mix BLADEX 90DF plus Dinitro 3 at the rates indicated in Table 10. Add a surfactant, such as X-77, at the rate of 2 qt/100 gals. of spray mixture. Do not apply when soil surface is wet or cotton plants are extremely succulent. Do not apply in CA or AZ.

Table 10
Directed Postemergence
Application Rates per Acre
for BLADEX 90DF + Dinitro on Cotton

Broadcast	Banded 38" Row 12" Band	Banded 38" Row 19" Band
0.7-1.1 lb. BLADEX 90DF + 1-2 qts. Dinitro 3	0.2-0.4 lb. + 0.3 -0.65 qt.	0.4-0.5 lb. + 0.5-1.0 qts.

BLADEX Plus MSMA Plus Dinitro (dinoseb)

Apply as directed in the "BLADEX Plus MSMA" section of the label. Tank-mix BLADEX plus MSMA at the rates shown in Table 9. Add Dinitro at the rate shown in Table 10. Add a surfactant, such as X-77, at the rate of 2 qts./700 gals. of spray mixture. Do not apply when soil surface is wet or cotton plants are extremely succulent. Do not apply in CA or AZ.

GRAIN SORGHUM (MILO)

Tank-mix combinations of BLADEX 90DF plus propachlor (Ramrod™) or propazine (Milogard™) may be used for selective preemergence weed control in grain sorghum. Do not use on forage sorghum.

BLADEX 90DF in tank-mix combination with propachlor or propazine should be applied only once per crop season. If replanting of grain sorghum is necessary, it may be planted in soil previously treated with these mixtures. Do not make a second application of BLADEX or crop injury may occur. Apply these tank-mix combinations at planting or after planting, but before the crop and weeds have emerged. Heavy rain immediately following application tends to cause excessive concentrations of herbicide in seed furrow, resulting in possible crop injury. Do not apply to furrow-planted sorghum until furrows are leveled (plowed-in). Level deep planter marks or seed furrows before application. Application to sorghum growing under stress caused by minor element deficiency or to sorghum growing on highly calcareous soil may result in crop injury.

BLADEX 90DF plus Propachlor (Ramrod)

WEEDS CONTROLLED

Grasses

Barnyardgrass
Crabgrass
Fall panicum
Giant foxtail
Green foxtail
Yellow foxtail

Broadleaves

Annual morningglory
Carpetweed
Cocklebur*
Common purslane
Lambsquarters
Pigweed
Ragweed
Smartweed (Pennsylvania)
Velvetleaf*

* Under conditions such as low temperatures, lack of soil surface moisture or other factors that may cause delay in germination of the seeds, the effectiveness of BLADEX may be impaired against these weeds.

Apply BLADEX 90DF plus propachlor at the proper rate for soil texture and organic matter shown in Table 11. Only apply this tank-mix to grain sorghum grown in States East of the Rocky Mountains. Any rotational crop may be planted the fall or spring following this treatment.

Table 11
Preemergence Broadcast Application Rates per Acre for BLADEX 90DF + Propachlor (Ramrod) 4L on Grain Sorghum

Soil Texture Description	Percent Organic Matter in Soil* 2% Lbs./Acre BLADEX 90DF	2% Qts./Acre Propachlor 4L	3% Lbs./Acre BLADEX 90DF	3% Qts./Acre Propachlor 4L
Sand, Loamy Sand	DO NOT USE			
Sandy Loam	1.1	2.5	1.3	3.0
Loam, Silt Loam, Silt	1.3	3.0	1.5	3.5
Sandy Clay Loam, Clay Loam, Silty Clay Loam	1.5	3.5	1.8	4.0
Sandy Clay, Silty Clay, Clay	1.8	4.0	1.8	4.0
Peat or Muck	NOT RECOMMENDED			

* For organic matter content between those listed, adjust the rate proportionately.

BLADEX plus Propazine (Milogard)

WEEDS CONTROLLED

Grasses

Crabgrass
Fall panicum
Giant foxtail
Green foxtail
Yellow foxtail

Broadleaves

Annual morningglory
Carpetweed
Cocklebur
Common purslane
Lambsquarters
Pigweed
Puncturevine
Ragweed
Smartweed
Velvetleaf

Apply BLADEX 90DF plus Propazine at the proper rate for soil texture shown in Table 12.

Apply to grain sorghum grown only in the States of Texas, Oklahoma and Kansas. **Rotational Crops:** in the Texas Gulf Coast and Texas Blacklands, fields treated with this tank-mix may be planted to cotton, soybeans, or corn 12 months after treatment. In Oklahoma and west Texas, fields treated with this tank-mix may be planted to cotton or corn 12 months after application. In Kansas, corn may be planted in rotation 12 months after treatment. Other crops should not be planted for 18 months following treatment.

Table 12
Preemergence Broadcast Application Rates per Acre for BLADEX 90DF + Propazine (Milogard) 4L on Grain Sorghum

Soil Texture	Lbs/Acre BLADEX 90DF	Qts/Acre Propazine 4L*
Sand, Loamy Sand, Sandy Loam	DO NOT USE	
Loam, Silt Loam, Silt	0.9	0.75
Sandy Clay Loam, Clay Loam, Silty Clay Loam	1.1	1.0
Sandy Clay, Silty Clay, Clay	1.3	1.25
Peat or Muck	NOT RECOMMENDED	

* If Propazine 80W is used multiply rate shown by 1.25 to equal pounds of

Propazine 80W.

If Propazine 90DF is used multiply rate shown by 1.125 to equal pounds of Propazine 90DF.

WHEAT

Early Preplant (Post-harvest weed control for land going into production of continuous winter wheat.)

Weeds controlled: See FALLOW CROPLAND section of this label.

Apply after wheat harvest but 45-60 days prior to planting winter wheat. Use at the proper rate for soil texture and organic matter indicated in Table 13. Add X-77 at 1-2 qts./100 gals. of diluted spray. Use higher rate or surfactant when dry weather conditions prevail. Failure to wait the recommended time interval between application and planting may result in crop injury. Use of this treatment on calcareous or caliche subsoil outcroppings may result in crop injury.

For best results during harvest, use a straw chopper/spreader on the combine that distributes the straw uniformly over the soil surface unless the straw is to be baled and removed prior to treatment. Avoid tillage after application as it may reduce the effectiveness of the herbicide treatment.

Where broadleaf weeds are present at time of application, tank-mix 1⅓-2 pts./A. of 2,4-D low volatile 6 lb./gal. ester (2-3 pts./A. of 2,4-D LV 4 lb./gal.). Use the high rate when weeds are over 4″ tall or when directed on the 2,4-D label for control of hard-to-kill weed species, such as perennials.

Where grasses and broadleaf weeds are present and exceed 2″ in height, tank-mix with 1-2 pts./A. of paraquat. Well established weeds over 6″ tall may not be well controlled. Apply at least 25 gals./A. of spray mixture by ground or 5-10 gals./A. by air where paraquat is included in the tank mixture.

No more than two applications of BLADEX may be made prior to planting wheat including applications made under the FALLOW CROPLAND section of this label.

Table 13
Preplant Broadcast Application Rates per Acre for BLADEX 90DF for Winter Wheat 60 days before planting*

Soil Texture Description	Pounds of BLADEX 90DF — Percent Organic Matter in Soil: Less than 2%	More than 2%
Sand, Loamy Sand	2.2	2.7
All other textures	2.7	3.1

* For applications made 45 days prior to planting, decrease the BLADEX rate indicated by 0.4 pound/acre.

FALLOW CROPLAND

WEEDS CONTROLLED BY BLADEX 90DF ALONE AND IN TANK-MIX COMBINATIONS

Grasses

Annual (Italian) ryegrass
Barnyardgrass*
Crabgrass
Cheatgrass
Downy brome
Green foxtail
Yellow foxtail
Indian lovegrass (Stinkgrass)
Volunteer wheat
Witchgrass
Wild oat*

Broadleaves

Cocklebur*
Common chickweed
Dog fennel
False flax
Henbit
Horseweed (marestail)
Kochia
Lambsquarters
Prostrate spurge
Purslane
Russian thistle
Shepherdspurse
Smartweed (Pennsylvania)
Sunflower* (wild)
Purple mustard
Pennycress
Pigweed*
Prickly lettuce
Prostrate knotweed
Tansy mustard
Tumble mustard
Wild radish
Wild buckwheat*

* Under soil moisture and temperature conditions favoring deep germination or other factors that may cause delayed germination, these species may not be completely controlled.

Additional weeds controlled are listed in the CORN section of this label.

BLADEX 90DF may be used alone or in tank-mix combination with atrazine for the control of certain annual weeds during a fallow program.

BLADEX 90DF or BLADEX 90DF plus atrazine should be used in tank-mix combination with Paraquat CL and/or 2,4-D as described below if growing vegetation is present. Should weeds become established before adequate rainfall for herbicide activation occurs, sweep tillage may be employed to destroy them.

Application Directions

Apply fallow cropland herbicide treatments uniformly to the soil surface. Adjust boom height on ground rigs to obtain the correct spray pattern at the top of the stubble rather than the ground. At sprayer speeds over 8 mph and when crop residues are heavy use flood type nozzles and at least 25 gal./A. of carrier.

Table 14
Broadcast Application Rates per Acre for BLADEX 90DF on Fallow Cropland*
Lbs. per Acre

Soil Texture	Post Harvest Treatment for Fall Weed Control	Late Fall Treatment for Spring Weed Control	Spring Treatment for Spring Weed Control
Clay Loam, Loam, and Sandy Soils	2.7	3.6	2.7
Clays, Loamy Clays, Silty Clays, and Sandy Clays	3.1	3.6	2.7

* For soils containing more than 2% organic matter, use an additional 0.4 lb/acre of BLADEX 90DF.

For soils containing more than 3% organic matter, use an additional 0.9 lb/acre of BLADEX 90DF.

BLADEX 90DF Applied Alone

Select the appropriate rate of BLADEX 90DF from Table 14. Add X-77 surfactant at the rate of 1 qt./100 gals. of diluted spray. Winter wheat may be planted four months or more after treatment. Spring wheat or durum wheat may be planted nine months after treatment. Grain sorghum or field corn may be planted in the spring following a late fall application.

Table 15
Broadcast Application Rates per Acre for Tank Mix Combinations of BLADEX 90DF Plus Atrazine on Fallow Cropland*

Location/Treatment	Lbs/Acre BLADEX 90DF	+	Qts/Acre Atrazine 4L**
FALL APPLICATION FOR WINTER WHEAT Kansas, southern Nebraska, Oklahoma, Colorado East Slope between Hwy. I-76 & State Hwy. 96	1.8	+	0.8
FALL APPLICATION FOR WINTER OR SPRING WHEAT Nebraska Panhandle, Wyoming, Utah, Colorado West Slope & East Slope north of Hwy. I-76 & south of Hwy. 96, Idaho, Montana, North Dakota, South Dakota	1.8	+	0.6

Location/Treatment			Lbs/Acre BLADEX 90DF	+	Qts/Acre Atrazine 4L**
SPRING APPLICATION FOR SPRING WHEAT Idaho, Montana, North Dakota, South Dakota, northern Utah, northern Wyoming			2.2	+	0.4
FALL APPLICATION FOR WINTER WHEAT Columbia Basin areas of Washington & Oregon	Average Annual Rainfall	over 15"	2.2	+	0.4
		10-15"	2.2	+	0.27
		Less than 10"	2.2	+	0.2

* Use an additional 0.4 lb./A. of BLADEX 90DF for soils with 2.0 to 3.0 percent organic matter.
Use an additional 0.9 lb./A. of BLADEX 90DF for soils containing more than 3.0 percent organic matter.

** If Atrazine 80W is used multiply rate shown by 1.25 to equal pounds of Atrazine 80W.
If Atrazine 90DF is used, multiply rate shown by 1.1 to equal pounds of Atrazine 90DF.

BLADEX 90DF plus Atrazine

BLADEX 90DF may be used in tank-mix combination with atrazine where a maximum period of weed control is desired in a fallow crop land program. Treatments must be applied before November 15 of the year before planting winter wheat or at least 11 months before planting spring wheat or durum wheat. Select the appropriate rates of BLADEX 90DF plus atrazine for a particular location from Table 15. Add X-77 surfactant at 1 qt/100 gals. of diluted spray.

Do not use this treatment on sands or on Rosebud or Canyon series soils, or on calcareous or caliche subsoil outcroppings because of possible atrazine carryover damage to the succeeding crop.

Do not graze or feed foliage from treated areas to livestock within six months after application.

NOTE: If conditions cause weeds to begin to germinate in the spring or summer following a post-harvest treatment of BLADEX or BLADEX plus atrazine, and atrazine, an application of BLADEX 90DF may be made for additional weed control in the spring or summer prior to planting winter wheat. Apply as directed in "BLADEX 90DF Applied Alone" in this section of the label. Do not make more than two applications of BLADEX 90DF prior to planting.

BLADEX Combinations with Paraquat

On fallow crop land having an existing or established weed population, paraquat may be tank mixed at 1-2 pts/A. with either BLADEX 90DF or BLADEX 90DF/atrazine tank-mix combination as previously described in this section. Apply the recommended rates in at least 25 gallons of spray mixture per acre by ground rig or 5-10 gals./A. for aerial application. Use higher volumes and the high rates of paraquat when weed growth is heavy or when dry weather conditions prevail. Add X-77 surfactant at 1 qt./100 gals. of diluted spray.

BLADEX Combinations with 2,4-D Low Volatile Ester

2,4-D LV Ester may be added to any treatment in this section to help control broadleaf weeds growing at the time of application. Use 1⅓ to 2 pts/A. of 2,4-D LV 6 lb. Ester (2-3 pts/A. of 2,4-D LV 4 lb. Ester). Use the high rate when weeds are over 4" tall or when directed on the 2,4-D label for the control of hard-to-kill weed species, such as perennials. When 2,4-D LV is used, it should be added to the spray tank last.

Use of Supplemental Tillage

In fields where established weeds are too large to be effectively controlled with paraquat or 2,4-D, sweep tillage should be employed. Till before applying the herbicide treatment. This type of tillage will preserve a maximum amount of existing stubble on the surface for soil protection. Similar tillage may also be used if weeds become established prior to receiving adequate rainfall for activation of the herbicide treatment. At some point prior to seeding wheat, the herbicide will degrade and no longer be effective. Limited tillage should be employed at this time. This tillage should be kept shallow to preserve as much moisture as possible for the crops.

BLADEX®—Trademark of Shell Chemical Co.
Banvel™—Trademark of Velsicol Chemical Co.
Dual™, Milogard™—Trademarks of CIBA-Geigy Corp.
Eradicane™, Sutan +™—Trademarks of Stauffer Chemical Co.
Lasso™, Ramrod™—Trademarks of Monsanto Co.
Ortho X-77™—Trademark of Chevron Chemical Co.
Treflan™—Trademark of Elanco Products Co.
Zorial™—Trademark of Sandoz, Inc.

EXTRAZINE® 4L

HERBICIDE

For Field Corn

ACTIVE INGREDIENTS	BY WEIGHT
2-[[4-chloro-6-(ethylamino)-s-triazin-2-yl]amino]-2-methylpropionitrile	28.7%
Atrazine (2-chloro-4-ethylamino-6-isopropylamino-s-triazine)	13.6%
Related compounds	0.7%
INERT INGREDIENTS	57.0%
TOTAL	100.0%

This product contains 4 lbs. active ingredients per gallon.

EPA Reg. No. 201-408

NOTE CONDITIONS OF SALE AND WARRANTY BEFORE OPENING THE CONTAINER. IF THEY ARE NOT ACCEPTABLE, RETURN UNOPENED PRODUCT.

Keep Out of Reach of Children
WARNING
AVISO

PRECAUCION AL USUARIO: Si usted no lee ingles, no use este producto hasta que le etiqueta haya sido explicado ampliamente.

STATEMENT OF PRACTICAL TREATMENT

IF SWALLOWED, drink one or two glasses of water immediately and induce vomiting by sticking finger down back of throat.

IN CASE OF CONTACT, immediately flush eyes or skin with plenty of water. Get medical attention if irritation persists.

PRECAUTIONARY STATEMENTS

HAZARD TO HUMANS AND DOMESTIC ANIMALS

WARNING

May be fatal if swallowed. Harmful if absorbed through the skin. Avoid breathing vapors or spray mist. Avoid contact with eyes, or clothing. Wash thoroughly after handling. Remove and wash contaminated clothing before reuse. For 24-hour emergency medical assistance, call: (713) 473-9461.

ENVIRONMENTAL HAZARDS

Atrazine, an active ingredient of EXTRAZINE 4L, can travel (seep or leach) through soil and can enter groundwater which may be used as drinking water. Atrazine has been found in groundwater. Users are advised not to apply Atrazine to sand and loamy sand soils where the water table (groundwater) is close to the surface and where the soils are very permeable. Your local agriculture agencies can provide further information on the type of soil in your area and the location of groundwater.

Atrazine is toxic to aquatic invertebrates. Do not apply directly to water or wetlands. Do not apply where runoff is likely to occur. Do not apply when weather conditions favor drift from treated areas. Runoff and drift from treated areas may be hazardous to aquatic organisms in neighboring areas. Do not contaminate water by cleaning of equipment or disposal of wastes. In case of significant spill, call CHEMTREC (800) 424-9300 or (713) 473-9461.

DIRECTIONS FOR USE

It is a violation of Federal law to use thir product in a manner inconsistent with its labeling.

Do not apply this product in such a manner as to directly or through drift expose workers or other persons. The area being treated must be vacated by unprotected persons.

Do not enter treated areas without protective clothing until sprays have dried.

FOR END USE ONLY, DO NOT REPACKAGE OR REFORMULATE WITHOUT MANUFACTURER'S WRITTEN APPROVAL.

STORAGE AND DISPOSAL

STORAGE

Do not contaminate water, food, or feed by storage or disposal. Do not use or store around the home environment. Avoid contact with water. In case of spill or leak, soak up with sand, earth or synthetic absorbent (do not use alkaline absorbents) and dispose of wastes in compliance with local, state and Federal regulations.

PESTICIDE DISPOSAL

Wastes resulting from the use of this product may be disposed of on site or at an approved waste disposal facility.

CONTAINER DISPOSAL

Triple rinse (or equivalent). Then offer for recycling or reconditioning, or puncture and dispose of in a sanitary landfill, or incineration, if allowed by state and local authorities, by burning. If burned, stay out of smoke.

WEEDS CONTROLLED BY EXTRAZINE 4L ALONE AND IN COMBINATION WITH OTHER HERBICIDES ON CORN

Grasses

Annual bluegrass	Giant foxtail
Annual fescues	Goosegrass
Annual (Italian) ryegrass	Green foxtail
Barnyardgrass[1]	Junglerice
Bullgrass	Stinkgrass (Indian lovegrass)
Crabgrass	Witchgrass
Fall panicum	Yellow foxtail

Broadleaves

Annual groundcherry	Pigweed[1]
Annual morningglory	Pineappleweed
Annual sedge	Plantain
Black mustard	Poorjoe
Buffalobur	Prickly sida (teaweed)
Buttercup (annual)	Prostrate knotweed
Carpetweed	Prostrate spurge
Cocklebur[2]	Ragweed
Common chickweed	Russian Thistle
Common groundsel	Shepherdspurse
Common mallow	Smallflower galinsoga
Common purslane	Smartweed (Pennsylvania)
Corn spurry	Spinysida
Curly dock (seedling)	Sunflower[2] (wild, annual, common)
Fiddleneck	Tarweed cuphea (Gumweed)
Florida pusley (Florida purslane)	Velvetleaf[1]
Hedge mustard	Wild buckwheat
Jimsonweed[1]	Wild mustard
Kochia	Wild radish
Ladysthumb	Wild turnip
Lambsquarters	
Mayweed	
Nightshade (annual)	

[1] Under conditions such as low temperatures, lack of soil surface, moisture or other factors that may cause delay in germination of the seeds, the effectiveness of EXTRAZINE may be impaired against these weeds.

[2] The degree of control will be reduced if soil moisture and temperature conditions cause deep germination of the seed.

GENERAL INFORMATION

EXTRAZINE 4L Herbicide is a selective preemergence herbicide for the control of annual grasses and broadleaved weeds in field corn.

Consult your local Agricultural Extension Agent for help in determining soil type, organic matter content, and the most appropriate herbicide rate for local conditions.

As a preemergence herbicide, EXTRAZINE 4L is active mainly through the roots, and therefore, its effect on weeds is dependent on adequate rainfall to move the herbicide into the root zone. A rotary hoeing or shallow cultivation is recommended for those applications which are not incorporated at the time of treatment, if a rainfall or sprinkler irrigation has not occurred within about ten days after application of EXTRAZINE 4L. Moisture should be sufficient to wet the top 1½ to 2 inches of soil or make the soil too wet to cultivate for most soils one-half inch of moisture is sufficient. Under conditions which delay weed germination, such as low temperatures, lack of soil surface moisture, or when germination extends over a long period, the effectiveness of the herbicide may be impaired. Rotary hoeing or a shallow cultivation may be of benefit under these circumstances. If the corn is cultivated, tillage should be shallow to minimize herbicide dilution in the soil. Should the corn stand be lost due to adverse weather conditions, the field can be replanted the same season to corn or sorghum.

EXTRAZINE 4L is not effective when used preemergence on peat or muck soils. Do not use EXTRAZINE 4L on sands or loamy sands (soils consisting of more than 70% sand) containing less than 1% organic matter.

Rotational Crops: (1) Plant only corn, sorghum, or soybeans the year following the use of this mixture. (2) If soybeans are to be planted, injury may occur. (3) If applied after June 10, do not rotate with crops other than corn or sorghum the next year or injury may occur. (4) In the high plains and intermountain areas of the West where rainfall is sparse and erratic or where irrigation is required, use only when corn or sorghum is to be planted the following year, or a crop of corn or sorghum not treated with this mixture or atrazine is to precede other rotational crops. (5) Small gains may be planted 15 months following treatment. (6) All other crops may be planted 18 months after application.

APPLICATION DIRECTIONS

General Mixing and Spraying Directions

If another pesticide has been used previously, be sure the sprayer tank, boom, and nozzles have been thoroughly flushed with clean water.

Sufficient jet or mechanical agitation should be provided during the filling operation and during application to keep the spray mixture uniformly suspended. Fill the spray tank at least one-half full of clean water and while pump and agitator are running, add the recommended amount of EXTRAZINE 4L to the tank, then fill with water to the desired level. For tank mixes, once the EXTRAZINE 4L is completely suspended in the water, add the recommended amount of other products to the tank while filling with water to the desired level. Add all wettable powders before liquids. For emulsible concentrates (EC), such as Lasso™ or Dual™, pre-mixing these products in equal amounts of water or pouring the products directly into the water stream as the tank is filling will aid in achieving compatible tank-mixtures. Fluid fertilizers may replace all or part of the water as carrier. always check the compatibility of EXTRAZINE 4L or EXTRAZINE 4L tank mixes and fluid fertilizer, according to the Jar Test for Compatibility instructions below, before the herbicide and fertilizer are mixed in the spray tank.

Allow the pump to run several minutes before application to ensure proper suspension and mixing of EXTRAZINE 4L or other various mixtures. If EXTRAZINE 4L or tank-mix combinations of EXTRAZINE 4L with atrazine are allowed to stand in a spray tank or nurse tank for an extended period (several hours or overnight), provide good agitation for several minutes before application to ensure proper suspension and mixing. Tank-mix combinations of

EXTRAZINE 4L with other products should be continuously agitated and should be mixed and applied as soon as possible.

Apply with flat fan (even flat fan for band application) or flood nozzles in conjunction with 50 mesh nozzle screens and in-line screens. Carefully follow the nozzle manufacturer's recommendation for correct nozzle height, positioning and spacing. Check for uniform distribution of spray droplets. Apply at pressures of 15-35 psi to reduce chemical loss due to fine mist drift.

Apply the spray mixture to the soil surface using at least 15 gals./A for ground application. Apply by air using a minimum of 4 gals./A of spray mixture.

Check nozzles, pressure, and output regularly to ensure proper calibration and application.

ALL RATES INDICATED ON THIS LABEL ARE BROADCAST APPLICATION RATES PER ACRE UNLESS OTHERWISE STATED. FOR BAND APPLICATION, USE PROPORTIONATELY LESS.

Jar Test for Compatibility

A. Add one pint of liquid fertilizer or other carrier to be used to two quart jars. Mark one quart jar "with" and the other jar "without."

B. To the jar marked "with" add 1/4 teaspoon of a suitable compability agent. Shake gently for 5 to 10 seconds to mix. (1/4 teaspoon/pint = 2 pints/100 gallons.)

C. To each jar add the appropriate amount of herbicides corresponding to the herbicide rate and spray volume selected (see following table). If more than one herbicide is used, add them separately; wettable powders first, flowables second, and emulsible concentrates last. Shake gently for 5 to 10 seconds after each addition.

D. Let stand one-half hour and make observations. If any separation, agglomeration, or precipitation has occurred, again shake the jar 10 to 15 seconds.
   1. If mixtures appear to be compatible in both jars, the herbicides can probably be used without a compatibility agent.
   2. If the mixture in the jar marked "with" is compatible but the one in the jar marked "without" is incompatible, a compatibility agent should be used.
   3. If both mixtures are incompatible, the liquid fertilizer and herbicides should not be used in the same spray tank.

Jar Test for EXTRAZINE 4L Compatibility with Liquid Fertilizer

Gallons Liquid Fertilizer per acre	15	20	25	30	40
Teaspoons of EXTRAZINE 4L* per pint of fertilizer	1½	1¼	1	¾	½

* Based on 1 qt. EXTRAZINE 4L (1 pound active ingredients) per acre in fertilizer volumes indicated. Increase volume proportionately to correspond with intended field rate in terms of quarts EXTRAZINE 4L per acre (e.g., for field rate of 4 qts. EXTRAZINE 4L in 20 gallons fertilizer per acre, add 5 teaspoons EXTRAZINE 4L to 1 pint fertilizer for compatibility testing purposes. CALCULATION: At 4 qts. EXTRAZINE 4L rate/20 gallons fertilizer, 4 × 1¼ teaspoons = 5 teaspoons EXTRAZINE 4L/pint of fertilizer).

If a test such as outlined indicates that components of a proposed mix are compatible, the applicator still has the responsibility of combining materials in sequence to the spray tank in accordance with directions prescribed on the label of the herbicides or pesticides involved.

Tests have indicated that compatibility agents, noted below by the various tank mix combinations, may give improved compatibility in liquid fertilizers.

Tank Mix Combination	Compatibility Agents
EXTRAZINE 4L/Lasso (Liquid Fertilizer Grade)	Probably not needed in 28-0-0, 10-34-0. Compex may help in others.
EXTRAZINE 4L/ Sutan+™6.7E	Probably not needed in 28-0-0. Incompatible in 10-34-0. Units, Spray-Mate, Kem-Link, may help in others.
EXTRAZINE 4L/Dual 8E	Probably not needed in 28-0-0. Unite, Spray-Mate, Ivory Liquid may help in others.

PREEMERGENCE-PREPLANT INCORPORATED

Apply EXTRAZINE treatments just before, at or after planting but before crop has emerged. Avoid removal of treated soil from seedrow prior to or during the planting operation.

EXTRAZINE may also be applied early prior to planting or in a split application if pre-season weed control is desired. For split applications, do not exceed the total amount of EXTRAZINE for the soil texture and organic matter shown in Table 1. If EXTRAZINE is applied early, more than 15 days before planting, a split application of EXTRAZINE or some other herbicide treatment may be necessary at or after planting to provide additional length of weed control. For further information, see "Early Preplant" recommendations in the Conservation Tillage section of label.

A rotary hoeing is recommended for preemergence applications which do not receive adequate rainfall or sprinkler irrigation to wet the top 1½-2 inches of soil within about 10 days after application.

If an EXTRAZINE mixture is to be incorporated, except as noted, single or two pass incorporation is acceptable. Care should be taken to incorporate the EXTRAZINE mixture no deeper than the top two inches of soil.

Extrazine Applied Alone

Use at the proper rate for soil texture and organic matter indicated in Table 1.

Table 1
Preemergence Broadcast Application Rates Per Acre For EXTRAZINE 4L Applied Alone on Corn

Soil Texture Description	Quarts of EXTRAZINE 4L — Percent Organic Matter in Soil*					
	Less than 1%	1%	2%	3%	4%	5% & Over
Sand, Loamy Sand	DO NOT USE	1.25	1.5	2.0	2.5	3.25
Sandy Loam	1.25	1.5	2.0	2.5	3.25	4.0
Loam, Silt Loam, Silt	1.5	2.0	2.75	3.25	4.0	4.5
Sandy Clay Loam, Clay Loam, Silty Clay Loam	2.0	2.75	3.25	4.0	4.5	4.75
Sandy Clay, Silty Clay, Clay	2.75	3.25	4.0	4.5	4.75	5.25
Peat or Muck	NOT RECOMMENDED					

* For organic matter content between those listed, adjust the rate proportionately.

EXTRAZINE COMBINATIONS

EXTRAZINE plus Lasso 4EC

Use EXTRAZINE 4L at the proper rate for the soil texture and organic matter shown in Table 2 plus 2 quarts per acre of Lasso. (Use 2.5 quarts Lasso on clay soils containing 5 percent organic matter and over).

EXTRAZINE plus Sutan + 6.7E, or Eradicane™ 6.7E

Use EXTRAZINE 4L at the proper rate for the soil texture and organic matter shown in Table 2 plus 1.8 quarts per acre of Sudan+ or Eradicane for control of many annual grasses and broadleaf weeds. (Use 2.4 quarts of Sutan+ or Eradicane on loam soils containing 5 percent or more organic matter, and clay loams and clays containing 4 percent or more organic matter). Do not use on sands and loamy sands of less than 1 percent organic matter in the light sandy soils of eastern coastal states. Do not use on corn seed stock.

Apply before planting. Incorporate the mixture immediately upon application using power-driven cultivation equipment set for 2-3 inch depth, or a tandem disc set to cut to a depth of about 4 inches while operating at 4-6 mph. For thorough mixing, disc in two directions (cross disc), and follow with a harrow, drag, or other leveling device. Prior to the second discing, readjust the disc to prevent cutting deeper than 4 inches. EXTRAZINE 4L may be applied preemergence as an overlay over previous incorporated Sutan+ or Eradicane if desired.

Existing stands of quackgrass, purple and yellow nutsedge must be turned under and thoroughly chopped up prior to chemical treatments.

Additional weeds controlled by Sutan+ or Eradicane combinations:

Grasses	Sandbur
	Shattercane (Wild Cane)*
	Texas Panicum
	Quackgrass (Eradicane only)
	Wild Proso Millet* (Eradicane only)
Perennial Weeds	Yellow Nutsedge (nutgrass)
	Purple Nutsedge (nutgrass)

* Suppression only—refer to Sutan+ or Eradicane label for appropriate supplement cultural and tillage practices.

For fields with moderate to heavy infestations of these weeds refer to the Sutan+ or Eradicane labels for appropriate higher rates.

EXTRAZINE plus Dual 8E

Use EXTRAZINE 4L at the proper rate for soil texture and organic matter shown in Table 2. Use Dual as follows:

Soil Texture	Broadcast Rate Per Acre
Coarse	
Sand, loamy sand, sandy loam	1.25–1.5 pints
Medium	
Loam, silt loam, silt	1.5–2.0 pints
Fine	
Sandy clay loam, silty clay loam, clay loam, sandy clay, silty clay, clay	1.5–2.5 pints

The low end of the rate range should be used for lowest organic matter soils and the rate increased as organic matter increases to a point that soils containing 4 percent organic matter or more require the highest rate shown for that soil texture. Refer to the Dual label for precautions on rotational crops.

Table 2
Preemergence Broadcast Application Rates Per Acre for EXTRAZINE 4L Used in Tank-Mix Combinations with Lasso, Sutan+, Eradicane, or Dual on Corn

Soil Texture Description	Quarts of EXTRAZINE 4L — Percent Organic Matter in Soil*: Less than 1%	1%	2%	3%	4%	5% & Over
Sand, Loamy Sand	0.6**	0.75	1.25	1.5	1.75	2.0
Sandy Loam	0.75	1.25	1.5	1.75	2.0	2.25
Loam, Silty Loam, Silt	1.25	1.5	1.75	2.0	2.25	2.5
Sandy Clay Loam, Clay Loam, Silty Clay Loam	1.5	1.75	2.0	2.25	2.5	2.75
Sandy Clay, Silty Clay, Clay	1.75	2.0	2.25	2.5	2.75	3.0
Peat or Muck	NOT RECOMMENDED					

* For organic matter content between those listed, adjust the rate proportionately.

** Do not use in the Atlantic Coastal Plain.

CONSERVATION TILLAGE WEED CONTROL

Early Preplant (EPP)

[Prior-to-planting weed control for land going into production of corn.]

Complete any planned early spring tillage prior to application. Apply herbicide treatment before weeds germinate or before weed seedlings are more than 3 inches tall. Tillage after application may reduce the effectiveness of the herbicide treatment.

Apply EXTRAZINE 4L 15–30 days prior to planting. Use the proper rate for soil texture and organic matter indicated in Table 1. Where heavy crop residues exist, the rates shown in Table 1 should be increased by 25 percent.

Where broadleaf weeds are present, add 2, 4-D LV Ester at 1⅓–2 pts./A. (6 lb./gal.) or 2–3 pts./A. (4 lbs./gal.) plus X-77 surfactant at 1 qt./100 gals. of diluted spray, or other suitable surfactant at its recommended rate. If existing weeds exceed 3 inches in height, add 1–2 pts./A. of paraquat to the above mixture. Well established weeds over 6 inches tall may not be well controlled. Use at least 25 gal./A. of spray mixture by ground rig or 5–10 gals./A. for aerial application.

Depending upon weather conditions following the EPP application, a postemergence EXTRAZINE or BLADEX Herbicide treatment, or some other herbicide treatment may be necessary at or after planting to provide additional length of weed control. If desired, 1½–2 pts./A. of Dual or 2 qts./A. of Lasso may be tank mixed with the EXTRAZINE EPP treatment or applied preemergence at planting.

Rotational Crops: Refer to the General Information section of the label.

At Planting

EXTRAZINE 4L applied alone or in combination with Lasso or Dual according to the following directions will kill most existing small weeds and suppress many emerged perennial weeds when corn is planted into no-till stalk ground (corn, sorghum), stubble ground (soybean, small grains), and any minimum-till situation. This treatment then provides residual control of annual weeds as in conventional tillage.

Apply EXTRAZINE 4L alone or with other products according to the directions for those treatments in the Preemergence section of the label. Where heavy crop residues exist, the EXTRAZINE rate shown in Tables 1 or 2 should be increased by 25 percent. Add ½–1 pt./A. of 2,4-D Low Volatile 6 lb. Ester (0.75–1.5 pt./A 2,4-D LV 4 lb. Ester). Add the 2,4-D LV to the spray tank last. Use a minimum of 25 gals./A. of carrier. Complete spray coverage of the weeds is essential for best performance. Nitrogen solutions and complete liquid fertilizers are the preferred carriers for this treatment as they help aid in the burndown of existing weeds. Add Ortho X-77™ surfactant at 1–2 qts. per 100 gals. of diluted spray, or other suitable surfactant at its recommended rate. If water is used as a carrier, crop oil concentrate may be used as a surfactant Apply before weeds exceed 3″ in height. For control of existing alfalfa, add ⅓–½ pt./A. of Banvel™ to the spray mixture. Apply before the alfalfa exceeds 6″ in height.

For fields with existing sod grasses such as orchardgrass, bromegrass, rye or timothy, or when very dry conditions exist, or when existing weeds exceed 3″ in height add Paraquat CL to the tank-mix. Use 2 pts./A. of Paraquat CL in combination with EXTRAZINE as described above in this section, except the 2,4-D LV may be omitted if desired. Do not apply Paraquat CL in suspension type liquid fertilizer.

POSTEMERGENCE

EXTRAZINE Applied Alone

Use EXTRAZINE 4L at the proper rate for the soil texture and organic matter shown in Table 3 or 3A. Use rates shown in Table 3 if EXTRAZINE or BLADEX has not been applied to the soil this season. Use rates shown in Table 3A if EXTRAZINE or BLADEX has been applied to the soil this season. This treatment may be used on peat or muck soils for burndown and suppression of existing weeds but will not provide residual control. Apply from crop emergence through the four-leaf stage of corn growth before weeds exceed about 1½ inches in height. Do not apply over the top of corn if the fifth leaf is visible. Apply in water only. Do not spray emerged corn plants with EXTRAZINE in a liquid fertilizer carrier.

DO NOT APPLY THIS TREATMENT UNDER COLD, WET, WEATHER CONDITIONS OR TO WEATHER STRESSED OR STORM DAMAGED CORN. Yellowing of the corn may result from this treatment, particularly if cold, adverse growing conditions occur after application. Extended or extreme cold and wet conditions may reduce stands. Do not use on corn grown for seed.

Rotational Crops: Refer to the "General Information" section of the label.

Table 3
Postemergence Broadcast Application Rates
Per Acre for EXTRAZINE 4L on Corn
No Prior EXTRAZINE or BLADEX® Application

	Quarts of EXTRAZINE 4L			
	Percent Organic Matter in Soil*			
Soil Texture	**Less than 1%**	**1%**	**2%**	**Over 2%**
Sand, Loamy Sand	DO NOT USE	1.25	1.5	2.0
Sandy Loam	1.25	1.5	2.0	2.0
Loam, Silt Loam, Silt	1.5	2.0	2.0	2.0
All other textures	2.0	2.0	2.0	2.0

* For organic matter content between those listed, adjust the rate proportionately.

Table 3A
Postemergence Broadcast Application Rates
Per Acre For EXTRAZINE 4L on Corn
EXTRAZINE or BLADEX Used in Prior Application

	Quarts of EXTRAZINE 4L			
	Percent Organic Matter in Soil*			
Soil Texture	**Less than 1%**	**1%**	**2%**	**Over 2%**
Sand, Loamy Sand	DO NOT USE	DO NOT USE	1.4	1.4
Sandy Loam	DO NOT USE	DO NOT USE	1.6	2.0
Loam, Silt Loam, Silt	DO NOT USE	1.4	2.0	2.0
All other textures	DO NOT USE	1.8	2.0	2.0

* For organic matter content between those listed, adjust the rate proportionately.

CONDITIONS OF SALE AND WARRANTY

SHELL AND THE SELLER OFFER THIS PRODUCT, AND THE BUYER AND USER ACCEPT THIS PRODUCT, ONLY UNDER THE FOLLOWING AGREED CONDITIONS OF SALE AND WARRANTY. The directions for use of this product are believed to be reliable and should be followed carefully. However, it is impossible to take into account all variables and to eliminate all risks associated with its use. Injury or damage may result because of conditions which are beyond the control of Shell or the seller, including soil texture, organic matter content of soil, weather, presence of other materials, and the manner of use or application.

Shell warrants only that this product conforms to the chemical description on the label and is believed to be reasonably fit for the purposes referred to in the Directions for Use when used as directed under normal conditions. SHELL MAKES NO OTHER EXPRESS OR IMPLIED WARRANTY OF FITNESS OR MERCHANTABILITY OR ANY OTHER EXPRESS OR IMPLIED WARRANTY. Any variation or exception from this warranty must be in writing and signed by an authorized Shell representative.

LIMITATION OF LIABILITY
Shell's liability, whether or not based on negligence, breach of expressed or implied warranty, strict liability or any other legal cause, is limited to the cost of this product. In no case shall Shell or the Seller be liable for consequential, incidental, or indirect damages, such as loss of crops, resulting from the use or handling of this product.

BLADEX®, EXTRAZINE® Trademarks of Shell Chemical Co.

Banvel™ Trademark of Velsicol Chemical Co.
Dual™ Trademark of Ciba-Geigy Corp.
Eradicane™ Trademark of Stauffer Chemical Co.
Lasso™ Trademark of Monsanto Co.
Sutan+™ Trademark of Stauffer Chemical Co.
Ortho X-77™ Trademark of Chevron Chemical Co.

NUDRIN® 1.8
METHOMYL INSECTICIDE SOLUTION

RESTRICTED USE PESTICIDE

For retail sale to and use only by Certified Applicators or persons under their direct supervision and only for those uses covered by the Certified Applicator's certification.

CARBAMATE PESTICIDE, LIQUID, N.O.S. (CONTAINS METHYL ALCOHOL), FLAMMABLE LIQUID (POISON B), UN 2758

ACTIVE INGREDIENT	BY WEIGHT
Methomyl: S-methyl N-[(methyl-carbamoyl)oxy] thioacetimidate	24.1%
INERT INGREDIENTS	75.9%
TOTAL	100.0%

Contains 1.8 lbs. active ingredient per gallon.

EPA Reg. No. 201-347

DANGER–POISON

See Statement of Practical Treatment

Keep Out of Reach of Children
PELIGRO

PRECAUCION AL USUARIO: Si usted no lee ingles, no use este producto hasta que le etiqueta haya sido explicado ampliamente.

STATEMENT OF PRACTICAL TREATMENT
ATROPINE IS AN ANTIDOTE. CALL A PHYSICIAN AT ONCE IN ALL CASES OF SUSPECTED POISONING.

IF SWALLOWED, drink 1 or 2 glasses of water and induce vomiting by touching back of throat with finger. NEVER induce vomiting or give anything by mouth to an unconscious person.

IF INHALED, remove from exposure. Have patient lie down and keep quiet. If patient is not breathing, start artifical respiration, preferably by mouth, immediately. *Get medical attention.*

IN CASE OF CONTACT, wash skin with plenty of soap and water; for eyes, flush with water and *get medical attention;* remove and wash contaminated clothing before re-use.

If warning symptoms appear (see warning symptoms), keep patient prone and quiet. Start artifical respiration immediately if patient is not breathing. Transport the patient immediately to the nearest hospital.

PRECAUTIONARY STATEMENTS

HAZARDS TO HUMANS AND DOMESTIC ANIMALS
DANGER! POISONOUS IF SWALLOWED OR INHALED. FLAMMABLE. CONTAINS METHANOL, MAY BE FATAL OR CAUSE BLINDNESS IF SWALLOWED. POISONOUS IF INHALED. CAUSES EYE DAMAGE. Do not breathe vapor or spray mist. Do not get in eyes, on skin, on clothing.

PRECAUTIONS IN USING
When mixing spray, loading, applying or otherwise handling, wear protective clothing, goggles and a respirator, jointly approved by the Mine Safety and Health Administration and The National Institute for Occupational Safety and Health, suitable for protection against NUDRIN Insecticide vapors or spray mist. Wear clean clothes daily. Wash thoroughly after handling and before eating or smok-

ing. Pilot should not assist in the mixing and loading operation. Do not contaminate or store near feed or foodstuffs. Note Reentry Statements Section.

IMPORTANT: Not for use or storage in or around the home. Keep animals and unprotected persons out of operational areas during treatment and while there is danger of drift.

NOTE TO PHYSICIAN

Warning Symptoms—NUDRIN Insecticide poisoning produces effects associated with anticholinesterase activity which may include weakness, blurred vision, headache, nausea, abdominal cramps, discomfort in the chest, constriction of pupils, sweating, slow pulse, and muscle tremors.

Treatment—Atropine sulfate should be used for treatment. Administer repeated doses, 1.2 to 2.0 mg intravenously every 10 to 30 minutes until full atropinization is achieved. Maintain atropinization until the patient recovers. Artificial respiration or oxygen may be necessary. Allow no further exposure to any cholinesterase inhibitor until recovery is assured. *Do not use morphine or 2-PAM.*

For 24-hour emergency medical assistance, call (713) 473-9461.

ENVIRONMENTAL HAZARDS

This product is toxic to fish, birds and other wildlife. Do not apply directly to water and wetlands. Do not apply where runoff is likely to occur. Do not apply when weather conditions favor drift from areas treated. Do not contaminate water by cleaning of equipment, or disposal of wastes.

This product is toxic to bees and should not be applied when bees are actively visiting the area. Apply late in evening or early morning where honeybees visit fields. In alfalfa frequented by alkali or leafcutting bees, do not apply during bloom or when these bees will be present in fields.

PHYSICAL OR CHEMICAL HAZARDS

FLAMMABLE! Keep away from heat, sparks, and open flame.

DIRECTIONS

It is a violation of Federal law to use this product in a manner inconsistent with its labeling.

FOR END USE ONLY. DO NOT REPACKAGE OR REFORMULATE WITHOUT MANUFACTURER'S WRITTEN APPROVAL.

RE-ENTRY STATEMENTS

Do not apply this product in such a manner as to directly or through drift expose workers or other persons. The area being treated must be vacated by unprotected persons.

Do not enter treated areas without protective clothing until sprays have dried.

Because certain States may require more restrictive reentry intervals for various crops treated with this product, consult your State Department of Agriculture for further information.

Written or oral warnings must be given to workers who are expected to be in a treated area or in an area about to be treated with this product. (Indicate specific oral warnings which inform workers of areas or fields that may not be entered without specific protective clothing, period of time field must be vacated and appropriate actions to take in case of accidental exposure.)

When oral warnings are given, warnings shall be given in a language customarily understood by workers. Oral warnings must be given if there is reason to believe that written warnings cannot be understood by workers.

Written warnings must include the following information: "DANGER! Area treated with NUDRIN® insecticide on (*date of application*). Do not enter without appropriate protective clothing until sprays have dried. (*Insert here actions to take in case of accidental exposure. See* STATEMENT OF PRACTICAL TREATMENT *for details.*)"

STORAGE AND DISPOSAL

Do not contaminate water, food or feed by storage or disposal. Do not reuse empty container. Open dumping is prohibited.

STORAGE

Do not store in or around the home. Store in a secure, dry and temperate area. Store in original container only. Product is flammable. Keep away from heat, sparks or open flame. If spilled avoid exposure. If significant spill call CHEMTREC (800) 424-9300 or (713) 473-9461. Spill can be contained by covering with a sufficient amount of absorbent material, such as clay, diatomaceous earth, sand, sawdust or sweeping compound. Place in drum, label contents and dispose of according to the instructions below in pesticide disposal.

CAUTION: Wear a pesticide respirator and avoid forming dust clouds during clean-up.

PESTICIDE DISPOSAL

Pesticide wastes are acutely hazardous. Improper disposal of excess pesticide, spray mixture, or rinse water is a violation of Federal Law. If the wastes cannot be disposed of according to label instructions, contact your State Pesticide or Environmental Control Agency, or the Hazardous Waste representative at the nearest EPA Regional Office for guidance.

CONTAINER DISPOSAL

Triple rinse (or equivalent). Then offer for recycling, reconditioning, or puncture and dispose of in a sanitary landfill, or, if allowed by State and local authorities, by burning. If burned, stay out of smoke.

GENERAL DIRECTIONS—Shell NUDRIN 1.8 Insecticide Solution is a liquid to be mixed in water for application by ground or air equipment as indicated in the table. Use only in commercial plantings; do not use in home plantings.

Shell NUDRIN Methomyl Insecticide should be used only in accordance with recommendations on the label and booklet, or in separate published Shell recommendations available through local dealers.

SPRAY PREPARATION—Fill spray tank ¼ to ½ full of water. Add the desired amount of Nudrin 1.8 Insecticide to the spray tank and mix thoroughly. If the contents of the product container have been emptied, triple rinse the container and add the rinsate to the spray tank. Fill the tank to the desired level. Agitate until the material is thoroughly mixed; continued agitation is not required.

APPLICATION—Apply at the recommended rates when insects first appear. Unless otherwise noted, continue application at 5- to 7-day intervals or as needed. Use sufficient water to obtain thorough, uniform coverage. For aerial, use a minimum of 5 gals. per acre (gpa) except: 2 gpa for beans (including blackeyed peas and cowpeas), Bermuda grass, carrots, cucumbers, melons (including cantaloupe and watermelon), peanuts, small grains (barley, oats, rye, wheat), sorghum and soybeans; 3 gpa for alfalfa, cabbage, corn, cotton, lettuce, onions, peas, spinach and sugar beets; 10 gpa for grapes, nectarines, and pomegranates; and 15 gpa for citrus.

Apply at low rates on small plants, small insects and light infestations of insects. Use intermediate rates on large insects and heavier infestations. Use 1 to 3 applications of the highest recommended rate for controlling severe infestations. Then, use the lowest rate to maintain control.

NOTICE OF WARRANTY

SHELL CHEMICAL COMPANY MAKES NO WARRANTY OF MERCHANTABILITY, FITNESS FOR ANY PURPOSE, OR OTHERWISE, EXPRESS OR IMPLIED, concerning this product or its uses which extend beyond the use of the product under normal conditions in accord with the statement made in this label.

Sold under U.S. Patent Numbers 3,576,834 and 3,639,633.

CROPS METHOD OF APPLICATION (G MEANS GROUND ONLY)	INSECTS	PINTS PER ACRE	LAST APPLICATION—DAYS TO HARVEST	LAST APPLICATION—DAYS TO LIVESTOCK GRAZING/FEEDING
Alfalfa Do not apply to dormant or semi-dormant alfalfa when min. daily temp. is 50°F or lower	Armyworm†, Loopers, Yellowstripped Armyworm, Western Yellowstriped Armyworm	1–2		7
	Fall Armyworm	2		
	Lygus Bugs, Pea Aphid	2–4		
	Alfalfa Weevil Larvae	4		
	Beet Armyworm, Variegated Cutworm†	1–4		
	Alfalfa Caterpillar	1–⅓–2		
	Aphids, Egyptian Alfalfa Weevil Larvae	2–4		
Anise (Fennel)†	Cabbage Looper	4	7	
	Beet Armyworm	2–4		
Apples†-G (Except varieties Early MacIntosh & Wealthy)	Apple Aphid	½–2 per 100 gals.; up to 400 gpa	8	10
	Rosy Apple Aphid, Tarnished Plant Bug, Tufted Apple Budmoth, Green Fruitworm	1–2 per 100 gals.; up to 400 gpa		
	Codling Moth-10–12 day intervals			
	Leafrollers (Fruittree, Obliquebanded, Redbanded, Variegated), Lesser Appleworm, White Apple Leafhopper, Teniform Leafminer, Cutworm	2 per 100 gals. up to 400 gpa		
Asparagus	Beet Armyworm, Western Yellow Striped Armyworm, Asparagus Beetle, Spotted Asparagus Beetle, White Cutworm†, Redback Cutworm†	2–4	1	
Avocado	Avocado Leafroller, Avocado Looper, Omnivorous Looper	2–4	1	10
Barley†	Armyworms	1–2	7	10
Beans (Includes Blackeyed Peas & Cowpeas)	Fall Armyworm	2	Succulent Beans 1–2 pts.—1 over 2 pts.—3	3 Vines 7 Hay
	European Corn Borer (Initiate when moth flights first appear and continue preventive treatments at 3–4 day intervals)	2–4		
	Leafhopper, Mexican Bean Beetle	1–4		
	Beet Armyworm, Cabbage Looper, Corn Earworm, Lygus Bugs, Loopers† Saltmarsh Caterpillar†, Aphids†, Thrips†, Yellowstriped Armyworm†, Western Yellowstriped Armyworm†	2–4	Dry Beans all rates—25	
	Spotted Cucumber Beetle†	1–2		
Beet (Table)†	Beet Armyworm, Cabbage Looper, Diamondback Moth, Imported Cabbageworm	1–2	7–Roots 30–Tops	
Beet (Tops)	Beet Armyworm, Cabbage Looper, Diamondback Moth, Imported Cabbageworm, Flea Beetles	2–4	10	
Bermuda Grass†	Armyworms	1–2		7—Forage 3—Dehydrated Hay
	Cutworms Heliothis paradoxa	2–4		
Blueberries†	Aphids, Tussock Moth, Weevil, Sharp-Nosed Leafhopper	2	3	
	Cranberry Fruitworm, Cherry Fruitworm	2–4		
	Flea Beetle (larvae), Sawfly (larvae), Blueberry Leafroller	4		
Broccoli	Cabbage Looper, Imported Cabbageworm, Diamondback Moth	1–4*	3	
Brussels Sprouts	Cabbage Looper, Imported Cabbageworm, Diamondback Moth†	2–4*	3	
Cabbage	Cabbage Looper, Imported Cabbageworm, Diamondback Moth	1–4*	1	
	Fall Armyworm†	2*		
Carrots	Beet Armyworm	1⅓–4	1	
	Armyworms†	1–2		
	Aster Leafhopper†	2–4		
Cauliflower	Cabbage Looper, Imported Cabbageworm, Diamondback Moth	1–4*	3	

CROPS METHOD OF APPLICATION (G MEANS GROUND ONLY)	INSECTS	PINTS PER ACRE	LAST APPLICATION—DAYS TO HARVEST	LAST APPLICATION—DAYS TO LIVESTOCK GRAZING/FEEDING
Celery	Beet Armyworm	2–4		
	Loopers	4	7	
	Armyworm†	1–2		
Chicory†	Beet Armyworm, Leafhoppers, Variegated Cutworm	2–4	80	
Chinese Cabbage†	Loopers, Beet Armyworm	2–4	10	
Collards (Fresh Market only—Do not apply when temperature is less than 50°F. Do not apply when crop is less than 10" tall.)	Loopers, Diamondback Moth, Imported Cabbageworm, Beet Armyworm, Flea Beetles	2–4	10	
Corn (Sweet)	Earworm—Whorl as needed	1⅓–2		
	Beet Armyworm†	2		
	Earworm, Fall Armyworm, European Corn Borer, Armyworms†-Ears 1–3 days or as needed Ovicide/Larvicide Earworm, European Corn Borer	1–2	0 Ears	3 Forage
Corn (Field & Popcorn)	Earworm, Armyworms	1–2		
	Fall Armyworm, Beet Armyworm	2	0 Ears	3 Forage
Cotton Use may redden cotton—if excessive, stop or alternate with other insecticides U.S.	Ovicide/Larvicide Bollworm, Tobacco Budworm (Initiate schedule when eggs and significant numbers of newly hatched larvae are present. Continue at 5–7 intervals while eggs are present and larval control adequate. If significant larvae survive, use higher rates below.)	½–1		
	Larvicide Bollworm, Tobacco Budworm, Fall Armyworm, Beet Armyworm, Cotton Leaf Perforator, Lygus Bugs†—Use as occasional spray in regular schedule but not more often than every 10 days	2	15	Do not graze or feed
	Bollworm, Fall Armyworm, Tobacco Budworm, Beet Armyworm, Cotton Leaf Perforator, Lygus Bugs†—After full boll load, use up to 3 sprays at 3–5 day intervals	2–3		
	Cotton Fleahopper†	½–1		
West of Rockies	Beet Armyworm	1⅓–2		
	Cotton Leafperforator	1⅓–3		
Cotton Combination with AZODRIN® 5 WMI.† Note use Restrictions on AZODRIN labeling.	Lygus Bugs, Pink Bollworm, Loopers, Mites, Boll Weevil, Whiteflies	½–3 plus ⅞–1⅝ pt. AZODRIN 5	21	
	Thrips	½–3 plus 3/16–5/16 AZODRIN 5		
Cotton Combination with *Bacillus thuringiensis*†	Ovicide/Larvicide Bollworm, Tobacco Budworm	½–1 plus registered rates of B.T.	15	Do not graze or feed
Cotton Combination with *chlordimeform*†	Ovicide/Larvicide Bollworm, Tobacco Budworm, Lygus Bugs — U.S.	½ plus ⅛ lb. Chlordimeform	21	
	TX	¼–½ plus ⅛ lb. Chlordimeform		
Cotton Combination with EPN† (Do not enter field for 24 hrs. after treatment. Do not hand pick within 3 days of treatment.)	Ovicide/Larvicide Bollworm, Tobacco Budworm, Boll Weevil	½–1 plus 1–2 pts. EPN (4 lb./gal.)	15	

CROPS METHOD OF APPLICATION (G MEANS GROUND ONLY)	INSECTS	PINTS PER ACRE	LAST APPLICATION—DAYS TO HARVEST	LAST APPLICATION—DAYS TO LIVESTOCK GRAZING/FEEDING
Cotton Combination with EPN + methyl parathion† (Do not enter field for 48 hrs. after treatment. Do not hand pick within 7 days of treatment.)	<u>Ovicide/Larvicide</u> Bollworm, Tobacco Budworm, Boll Weevil	½–1 plus ⅔–1⅓ pts. EPN + MP (3 lb. + 3 lb./gal.)	15	
Cotton Combination with methyl parathion Do not enter field for 48 hrs. after treatment. Do not hand pick within 7 days of treatment.	Tobacco Budworm, Bollworm–Budworm control with NUDRIN is obtained through kill of eggs (ovicide) and larvae (larvicide) Bollworm (Ovicide/Larvicide)† (5–7 day intervals)	½–1¼ plus 1–2 pts. methyl parathion (4lb./gal.)	15	Do not graze or feed
Cotton Combination with permethrin.†	<u>Ovicide/Larvicide</u> Bollworm, Tobacco Budworm	½–1 plus 0.1–0.2 lb. permethrin	15	
Cotton	Pink Bollworm <u>Ovicide/Larvicide</u> Bollworm, Tobacco Budworm	½–1 plus 0.05–0.1 Lb ai/ acre PYDRIN		
Combination with PYDRIN® Insecticide	Cotton Leaf Perforator <u>Ovicide/Larvicide</u> Bollworn, Tobacco Budworm	½–1 plus 0.10 Lb ai/ acre PYDRIN		
	Cotton Bollworm, Tobacco Budworm, Lygus Bugs, Cabbage Looper, Beet Armyworm, Boll Weevil <u>Ovicide/Larvicide</u> Bollworm, Tobacco Budworm	½–1 plus 0.10–0.20 Lb ai/ acre PYDRIN	21	
Cucumber	Beet Armyworm, Flea Beetle, Cabbage Looper, Tobacco budworm, Melon Aphid†, Loopers†, Granulate Cutworms†, Yellowstriped Armyworm†, Pickleworm†, Cucumber Beetles†, Melonworm†	2–4	to 2 pts.—1 over 2 pts.—3 3 (CA only)	
	Fall Armyworm†	2		
Dandelions	Beet Armyworm, Cabbage Looper, Diamondback Moth, Imported Cabbageworm, Flea Beetles	2–4	10	
Eggplant†	Green Peach Aphid	1–4	5	
	Tomato Pinworm	2–4		
Garlic	Beet Armyworm	2–4	7	
Grapefruit	Thrips, Spirea Aphid†, Fruittree Leafroller, Citrus Cutworm	2–4	1	10
	Beet Armyworm†	2		
	Whiteflies†, Western Tussock Moth, Orange Tortrix, Omnivorous Leafroller	4		
Grapes East of Rockies G	Grape Berry Moth-Apply pre- & post-bloom & repeat at 10–14 day intervals or as needed Leafhoppers	2–4	1 Fresh & Raisin 14 Wine Grapes	
West of Rockies	Leafhoppers, Grapeleaf Skeletonizer, Saltmarsh Caterpillar			
	Omnivorous Leafroller, Grape Leaffolder, Orange Tortrix, Climbing Cutworm†	4		
Non-bearing Grapevines and Nursery Stock	Thrips, Omnivorous Leafroller, Beet Armyworm	3		
Horseradish†	Aphids, Thrips	2	65	
Kale	Beet Armyworm, Cabbage Looper, Flea Beetles, Diamondback Moth, Imported Cabbageworm	2–4	10	
Lemon	Thrips, Spirea Aphid† Fruittree Leafroller, Citrus Cutworm	2–4	1	10
	Beet Armyworm†	2		
	Whiteflies†, Western Tussock Moth, Orange Tortrix, Omnivorous Leafroller	4		
Lentils†	Western Yellowstriped Armyworm	1–4	25	
Lettuce (Head and Leaf)	Beet Armyworm, Cabbage Looper, Alfalfa Looper, Aphids, Thrips, Corn Earworm	1–4	1–2 pts.—7 over 2 pts.–10	10

CROPS METHOD OF APPLICATION (G MEANS GROUND ONLY)	INSECTS	PINTS PER ACRE	LAST APPLICATION—DAYS TO HARVEST	LAST APPLICATION—DAYS TO LIVESTOCK GRAZING/FEEDING
Melons	Beet Armyworm, Flea Beetle, Cabbage Looper, Tobacco Budworm, Melon Aphid†, Loopers†, Granulate Cutworms†, Yellowstriped Armyworm†, Pickleworm†, Cucumber Beetles†, Melonworm†	2–4	to 2 pts.–1 over 2 pts.–3 3 (CA only)	
	Fall Armyworm†	2		
Mint	Variegated Cutworm, Alfalfa Looper	4	14	
	Flea Beetles†	1–4		
Mustard Greens	Beet Armyworm, Cabbage Looper, Flea Beetles, Diamondback Moth, Imported Cabbageworm	2–4	10	
Nectarines CA & AZ	Thrips–Up to 3 applications	2–4	1	10
Oats†	Armyworms	1–2	7	10
Onions CA & East of Rockies	Beet Armyworm, Thrips, Cutworms– Up to 2 applications maximum	2*	7 Dry 28 Green	
West of Rockies	Beet Armyworm, Thrips, Cutworms– Up to 2 applications maximum	2*	7 Dry & Green	
Oranges	Thrips, Spirea Aphid†, Fruittree Leafroller, Citrus Cutworm	2–4	1	10
	Beet Armyworm†	2		
	Whiteflies†, Western Tussock Moth, Orange Tortrix, Omnivorous Leafroller	4		
Parsley	Beet Armyworm, Cabbage Looper, Diamondback Moth, Imported Cabbageworm, Flea Beetles	2–4	10	
Peaches	Catfacing Plant Bugs†, Stink Bugs†	4–8	4	10
	Oriental Fruitmoth, Twig Borer	4		
	Green Peach Aphid	3–8		
Peanuts	Fall Armyworm	1–2	21	Do not graze or feed
	Corn Earworm, Beet Armyworm, Potato Leafhopper†	1–4		
	Loopers, Green Cloverworm, Velvetbean Caterpillar, Thrips†	2–4		
Pears CT, ME, MA RI, VT, NH NY, NJ	Obliquebanded Leafroller, Green Fruitworm Do not make more than one application per season	2 per 100 gallons spray. 400 gals. of water/acre maximum	7	10
Peas (Succulent)	Pea Aphid, Loopers, Saltmarsh Caterpillar†, Beet Armyworm	2–4	1	5 Forage 14 Hay
	Alfalfa Caterpillar	1–4		
	Armyworm, Green Cloverworm	1–2		
Pecans†	Aphids	2–4	30	10
Peppers	Green Peach Aphid, Beet Armyworm, Cabbage Lopper, Loopers†, Fall Armyworm†	2	3	
	Armyworms†	1–2		
Pomegranates CA	Omnivorous Leafroller	4	14	
Potato	Beet Armyworm, Tuberworm, Cabbage Looper, Ahpids, Loopers†	2–4	6	
	Leafhopper, Variegated Cutworm† Flea Beetle†, Fall Armyworm†	2		
Rye†	Armyworms	1–2	7	10
Sorghum	Corn Earworm (Headworm), Fall Armyworm, Armyworm	1–2	14	14
	Sorghum Webworm	2		
	Sorghum Midge—Apply when 50% bloom then 3–5 days later if needed	1–2		
Soybeans	Green Cloverworm, Beet Armyworm, Velvetbean Caterpillar, Mexican Bean Beetle, Corn Earworm, Saltmarsh Caterpillar†, Fall Armyworm† Thrips†, Bean Leaf Beetle†	1–2	14	less than 2 pts: 3 Forage 7 Hay over 2 pts: 10 forage 12 hay
	Loopers	2–4		
	Velvetbean Caterpillar, Mexican Bean Beetle†, Green Cloverworm†, Corn Earworm; Apply when light to moderate infestations build to economic levels	½–1		

CROPS METHOD OF APPLICATION (G MEANS GROUND ONLY)	INSECTS	PINTS PER ACRE	LAST APPLICATION—DAYS TO HARVEST	TO LIVESTOCK GRAZING/FEEDING
Spinach; Do not apply when min. daily temp. is 32° F or lower. Do not apply to seedlings less than 3 inches in diameter	Loopers, Beet Armyworm	2–4	7	
	Fall Armyworm	2		
Strawberries†	Beet Armyworm, Thrips, Aphids	2–4	3 Fresh	
G	Lygus Bugs, Omnivorous Leaf Tiers	4	10 Processing	
Strawberries (dormant)	Armyworms	3–4	90	
Sugar Beets	Variegated Cutworm†	2		
	Beet Webworm, Flea Beetles, Beet Armyworm, Aphids	1–4	7	30 Tops
Summer Squash	Cabbage Looper, Flea Beetles, Tobacco Budworm, Beet Armyworm, Yellowstriped Armyworm†, Melon Aphid†, Granulate Cutworm†, Melonworm†, Pickleworm†, Loopers†, Cucumber Beetles†	2–4	to 2 pt—1 over 2 pt—3 3 (CA only)	
	Fall Armyworm†	2		
Swiss Chard	Beet Armyworm, Cabbage Looper, Diamondback Moth, Imported Cabbageworm, Flea Beetles	2–4	10	
Tangerines, Tangelo	Thrips, Spirea Aphid†, Fruittree Leafroller, Citrus Cutworm	2–4	1	10
	Beet Armyworm†	2		
	Whiteflies†, Western Tussock Moth, Orange Tortrix, Omnivorous Leafroller	4		
Tobacco (Except Shade)	Flea Beetles, Hornworm	1–2	5 Flue cured	
	Loopers, Aphids, Tobacco Budworm, Fall Armyworm	2	14 Air or fire cured	
Tomato	Fall Armyworm†	2		
	Tomato Fruitworm, Aphids, Cabbage Looper, Beet Armyworm, Pinworm	2–4	1	
	Armyworm†	1–2		
Turf† G	Sod Webworm—After application, sprinkle irrigate for 15 min.	4–8 (1.5–3 fl. oz. per 1000 sq. ft.)		Do not graze or feed
Turnip Greens	Beet Armyworm, Cabbage Looper, Diamondback Moth, Imported Cabbageworm, Flea Beetles	2–4	10	
Watercress† G	Beet Armyworm, Cabbage Looper, Diamondback Moth, Imported Cabbageworm (Use in minimum of 30 gpa in water. Field water should be turned off 24 hours prior to application)	2–4	10	
Wheat†	Armyworms	1–2	7	10
ORNAMENTALS: See also additional Ornamental uses on following pages				
Chrysanthemum G	Cabbage Looper, Corn Earworm, Beet Armyworm, Thrips Omnivorous Leafroller—Weekly	1–2 per 100 gals		
Forest Woodlots, Plantations and Nurseries G	Spruce Budworm (Treat when larvae feeding on buds, retreat after 3–4 days if needed)	1–2		
	Forest Tent Caterpillar†, Western Tent Caterpillar†	2–4		
Gladiolus	Cabbage Looper	1–2 per 100 gals		
Rose	Omnivorous Leafroller	1–2 per 100 gals		

† Claim not registered in California.
* Add wetting agent.

RESTRICTED USE PESTICIDE
For retail sale to and use only by Certified Applicators or persons under their direct supervision and only for those uses covered by the Certified Applicator's Certification.

SUPPLEMENTAL LABELING

NUDRIN 1.8 METHOMYL INSECTICIDE

EPA REG. NO. 201–347

SOLUTION FOR USE ON ORNAMENTALS†

Shell NUDRIN 1.8 Methomyl Insecticide solution is recommended for use on select ornamentals (see list) to control certain insects.

Insects Controlled

Aphids	Flea Beetles
Armyworms	Leafhoppers
Beet Armyworm	Leafrollers
Black Cutworm	Loopers
Corn Earworm	Thrips
Fall Armyworm	Tobacco Budworm

How to Apply: Apply 1 to 2 pts. NUDRIN 1.8 per 100 gals. of water. Spray plants thoroughly (to run-off by ground equipment). Apply when insects first appear and continue at weekly intervals or as needed.

NOTE: Since varieties of ornamentals are numerous, continually changing, and may respond differently to NUDRIN 1.8, test NUD-

RIN 1.8 on a small scale before proceeding to large-scale application.

IMPORTANT

BEFORE USING NUDRIN 1.8, READ AND CAREFULLY OBSERVE THE CAUTIONARY STATEMENTS AND ALL OTHER INFORMATION APPEARING ON THE PRODUCT LABEL.

This bulletin contains new or supplemental instructions for use of this product which may not appear on package label. Follow the instructions carefully.

†Claim not registered in California

ORNAMENTALS FOR WHICH NUDRIN 1.8 IS RECOMMENDED

Scientific Name	Common Name
Aglaonema mandalay 'Aqua'	
Aglaonema 'Silver King'	Silver King Evergreen
Araucaria heterophylla	Norfolk Island Pine
Ardisia crenata	Coral Berry
Areca sp.	Areca Palm
Arecastrum romanzoffianum	Queen Palm
Asparagus densiflorus sprengeri	Asparagus Fern
Bucida buceras	Black Olive
Butia capitata	Jelly Palm
Buxus microphylla	Boxwood
Callistemon viminalis	Weeping Bottlebrush
Carissa grandiflora	Natal Plum
Chamaedorea elegans	Parlor Palm
Chamaedorea elegans 'Bella'	Neanthe Bella Palm
Cocas nucifera	Coconut Palm
Codiaeum variegatum 'Aucubifolium'	Gold Dust Croton
Dieffenbachia amonena	Giant Dumbcane
Dieffenbachia exotica 'Perfection'	Dumbcane
Dieffenbachia maculata 'Perfection'	Perfection Dumbcane
Dieffenbachia maculata	Spotted Dumbcane
Dracaena sanderana	Belgian Evergreen
Epipremnum aureum	Pothos
Epipremnum aureum 'Marble Queen'	Marble Queen Pothos
Eriobotrya japonica	Loquat or Japanese Medlar or Japanese Plum
Euonymus japonica 'Microphylla'	Spindle Tree or Japanese Spindle Tree
Euphorbia pulcherrima	Poinsettia
	'Annette Hegg'
	'Annette Hegg Dark Red'
	'Annette Hegg Diva'
	'Annette Hegg Lady'
	'Annette Hegg Marble'
	'Annette Hegg New Pink'
	'Annette Hegg Super Star'
	'Annette Hegg Supreme'
	'Annette Hegg Top Star'
	'Annette Hegg White'
	'Eckespoint C–1'
	'Eckespoint C–1 White'
Euphorbia pulcherrima	Poinsettia
	'Eckespoint Professor Laurie Pink'
	'Eckespoint Reddy Light'
	'Mikkel Dawn Rochford'
	'Mikkel Fantastic'
	'Mikkel Heritage'
	'Mikkel Imperial'
	'Mikkel Improved Rochford'
	'Mikkel Pink Rochford
	'Mikkel Super Rochford'
	'Mikkel Triumph'
	'Mikkel Vivid Rochford'
	'Mikkel White Rochford'
	'Paul Mikkel'
	'Small Winter Flame'
	'Wonder Star'
Ficus benjamina	Weeping Fig
Ficus robusta	
Ficus triangularis	Triangleleaf Fig
Gardenia jasminoides	Common Gardenia or Cape Jasmine
Hedera helix	English Ivy
Hedera helix	'Needle Point'
Hedera helix	'California'
Hoya carnosa	Wax Plant
Hoya carnosa 'Krinkle Kurl'	'Hindu Rope'
Hoya carnosa 'Tri Color'	Tricolor Wax Plant
Ilex vomitoria	Yaupon or Cassina
Jasminum jasminoides	
Juniperus chinensis 'Bal Harbour'	
Ligustrum japonicum	Wax Leaf Privet or Japanese Privet
Ligustrum lucidium	Glossy Privet or Chinese Privet
Liriope sp	Lily Turf
Magnolia grandiflora	Bull Bay or Southern Magnolia
Maranta leuconeura	Prayer Plant
Maranta leuconeura 'Kerchoviana'	Green Maranta or Rabbit's Foot
Monstera delisciosa	Splitleaf Philodendron
Nephrolepis exaltata 'Bostoniensis'	Boston Swordfern
Nephrolepis exaltata	Florida Ruffles Swordfern
Peperomia obtusifolia	Ovalleaf Peperomia or Baby Rubber Plant
Peperomia obtusifolia 'Variegata'	Variegated Peperomia
Philodendron bipennifolium	Fiddleleaf Philodendron
Philodendron scandens oxycardium	Heart-leaf Philodendron
Phoenix dactylifera	Data Palm
Photinia X 'Frasieri'	
Pinus elliottii	Slash Pine
Pittosporum tobria 'Variegata'	Variegated Japanese Pittosporum
Platycerium bifurcotum 'Netherlands'	Staghorn Fern
Podocarpus macrophyllus	Yew Pine or Southern Yew or Buddhist Pine
Pyracantha angustifolia 'Variegata'	Fire Thorn
Raphiolepis indica	Indian Hawthorn
Raphiolepis umbellata	Yedda Hawthorn
Rhodendron indicum	Southern or Indian Azalea
Rumohra adiantiformis	Leatherleaf Fern
Saintpaulia ionantha	African Violet
Sansevieria trifasciata 'Laurentii'	Variegated Snake Plant
Sansevieria trifasciata	Snake Plant
Schlumbergera bridgesii	Christmas Cactus
Scindapsus pictus	Satin Leaf Pothos
Spathiphyllum cannifolium[1]	Peace Lily

Spathiphyllum clevelandii[1]	
Spathiphyllum X 'Mauna Loa' [1]	Mauna Loa Peace Lily
Syngonium podophyllym	Arrowhead Vine
Syngonium xanthophilum 'Greengold'	
Thuja occidentalis	American Arborvitae or White Cedar
Veitchia merrillii	Christmas Palm
Viburnum odoratissimum	Sweet Viburnum
Viburnum suspensum	

[1] Except seedlings

NUDRIN® 90

METHOMYL INSECTICIDE

Water Soluble Powder

CARBAMATE PESTICIDE, SOLID, N.O.S. POISON B, UN 2757

ACTIVE INGREDIENT	BY WEIGHT
Methomyl: S-methyl N-[(methyl-carbamoyl)oxy] thioacetimidate	90%
INERT INGREDIENTS	10%
TOTAL	100%

EPA Reg. No. 201-324

DANGER–POISON

Keep Out of Reach of Children

PELIGRO

PRECAUCION AL USUARIO: Si usted no lee inglés, no use este producto hasta que le etiqueta haya sido explicada ampliamente.

See Statement of Practical Treatment.

The outer pouch or bag contains NUDRIN 90 in a water soluble bag. Entire inner bag and contents dissolve in water. After opening outer container, immediately dump entire unopened inner bag into spray tank. Do not handle soluble bag or expose it to moisture since this may cause breakage.

STATEMENT OF PRACTICAL TREATMENT

ATROPINE IS AN ANTIDOTE. CALL A PHYSICIAN AT ONCE IN ALL CASES OF SUSPECTED POISONING.

IF SWALLOWED, drink 1 or 2 glasses of water and induce vomiting by touching back of throat with finger. NEVER induce vomiting or give anything by mouth to an unconscious person.

IF INHALED, remove from exposure. Have patient lie down and keep quiet. If patient is not breathing, start artificial respiration immediately. *Get medical attention.*

IN CASE OF CONTACT, wash skin with plenty of soap and water; for eyes, flush with water and *get medical attention;* remove and wash contaminated clothing before re-use.

If warning symptoms appear (see warning symptoms), keep patient prone and quiet. Start artificial respiration immediately if patient is not breathing. Transport the patient immediately to the nearest hospital.

PRECAUTIONARY STATEMENTS

HAZARDS TO HUMANS AND DOMESTIC ANIMALS

DANGER! POISONOUS IF SWALLOWED OR INHALED. Do not breathe dust or spray mist. Do not get in eyes, on skin, on clothing.

PRECAUTIONS IN USING

When mixing spray, loading, applying or otherwise handling, wear protective clothing, goggles and a respirator jointly approved by the Mine Safety and Health Administration and the National Institute for Occupational Safety and Health for protection against NUDRIN Insecticide dust or spray mist. Wear clean clothes daily. Wash thoroughly after handling and before eating or smoking. Pilot should not assist in the mixing and loading operation. Do not contaminate or store near feed or foodstuffs. Note Re-Entry Statements Section.

IMPORTANT: Not for use or storage in or around the home. Keep animals and unprotected persons out of operational areas during treatment and while there is danger of drift.

NOTE TO PHYSICIAN

Warning Symptoms—NUDRIN Insecticide poisoning produces effects associated with anticholinesterase activity which may include weakness, blurred vision, headache, nausea, abdominal cramps, discomfort in the chest, constriction of pupils, sweating, slow pulse, and muscle tremors.

Treatment—Atropine sulfate should be used for treatment. Administer repeated doses, 1.2 to 2.0 mg intravenously every 10 to 30 minutes until full atropinization is achieved. Maintain atropinization until the patient recovers. Artificial respiration or oxygen may be necessary. Allow no further exposure to any cholinesterase inhibitor until recovery is assured. *Do not use morphine or 2-PAM.*

For 24-hour emergency medical assistance, call (713) 473-9461.

ENVIRONMENTAL HAZARDS

This product is toxic to fish, birds and other wildlife. Do not apply directly to water and wetlands. Do not apply where runoff is likely to occur. Do not apply when weather conditions favor drift from areas treated. Do not contaminate water by cleaning of equipment, or disposal of wastes.

This product is toxic to bees and should not be applied when bees are actively visiting the area. Apply late in evening or early morning where honeybees visit fields. In alfalfa frequented by alkali or leaf-cutting bees, do not apply during bloom or when these bees will be present in fields.

DIRECTIONS

It is a violation of Federal law to use this product in a manner inconsistent with its labeling.

FOR END USE ONLY. DO NOT REPACKAGE OR REFORMULATE WITHOUT MANUFACTURER'S WRITTEN APPROVAL.

RE-ENTRY STATEMENTS

Do not apply this product in such a manner as to directly or through drift expose workers or other persons. The area being treated must be vacated by unprotected persons.

Do not enter treated areas without protective clothing until sprays have dried.

Because certain states may require more restrictive reentry intervals for various crops treated with this product, consult your State Department of Agriculture for further information.

Written or oral warnings must be given to workers who are expected to be in a treated area or in an area about to be treated with this product. (*Indicate specific oral warnings which inform workers of areas or fields that may not be entered without specific protective clothing, period of time field must be vacated and appropriate actions to take in case of accidental exposure.*)

When oral warnings are given, warnings shall be given in a language customarily understood by workers. Oral warnings must be given if there is reason to believe that written warnings cannot be understood by workers.

Written warnings must include the following information:

"DANGER! Area treated with NUDRIN® Insecticide on (*date of application*). Do not enter without appropriate protective clothing until sprays have dried. (*Insert here actions to take in case of accidental exposure. See* STATEMENT OF PRACTICAL TREATMENT *for details.*)"

STORAGE AND DISPOSAL

Do not contaminate water, food, or feed by storage or disposal. Do not reuse empty container. Open dumping is prohibited.

STORAGE

Do not store in or around the home. Store in a secure, dry and temperate area. Store in original container only. Handle this envelope carefully to prevent breakage of the inner bag when stored at low temperatures. Allow to warm above 50°F for normal handling. Do not allow inner bag to become exposed to moisture since this may cause breakage. In case of leaking package put leaker inside a heavy plastic bag. To remove spilled powder, cover spill with a sweeping compound or sawdust. Put sweepings in a heavy plastic bag. To dispose of sweepings and leaker put in fiber drum and label contents. Dispose of according to the instructions below in pesticide disposal. In case of a significant spill call CHEMTREC (800) 424-9300 or (713) 473-9461.

CAUTION: Wear a pesticide respirator and avoid forming dust clouds during clean-up.

PESTICIDE DISPOSAL

Pesticide wastes are acutely hazardous. Improper disposal of excess pesticide, spray mixture, or rinse water is a violation of Federal Law. If the wastes cannot be disposed of according to label instructions, contact your State Pesticide or Environmental Control Agency, or the Hazardous Waste representative at the nearest EPA Regional Office for guidance.

CONTAINER DISPOSAL

Completely empty bag into application equipment. Then dispose of empty bag in a sanitary landfill, or, if allowed by State and local authorities, by burning. If burned, stay out of smoke.

GENERAL DIRECTIONS

Shell NUDRIN 90 Methomyl Insecticide is a dry powder packaged in water soluble bags to be dissolved in water for application by ground or air equipment as indicated in the table. Use only in commercial plantings; do not use in home plantings.

Shell NUDRIN 90 Methomyl Insecticide should be used only in accordance with recommendations on the label and booklet, or in separate published Shell recommendations available through local dealers.

SPRAY PREPARATION: Fill spray tank ¼ to ½ full of water. Do not open the NUDRIN 90 water soluble inner bag. Drop this inner bag directly into the spray tank. Then fill the tank to the desired level with water. Mix thoroughly to fully dissolve the insecticide; once dissolved, continued agitation is not required. Use mechanical or hydraulic means; do not use air agitation.

APPLICATION—Apply at the recommended rates when insects first appear. Unless otherwise noted, continue application at 5- to 7-day intervals or as needed. Use sufficient water to obtain thorough, uniform coverage. For aerial, use a minimum of 5 gallons per acre (gpa) except: 2 gpa for beans (including blackeyed peas and cowpeas), Bermuda grass, carrots, cucumbers, melons (including cantaloupe and watermelon), peanuts, small grains (barley, oats, rye, and wheat), sorghum and soybeans; 3 gpa for alfalfa, cabbage, corn, cotton, lettuce, onions, peas, spinach and sugar beets; 10 gpa for grapes, nectarines, pomegranates and 15 gpa for citrus.

Apply at low rates on small plants, small insects and light infestations of insects. Use intermediate rates on large insects and heavier infestations. Use 1 to 3 applications of the highest recommended rate for controlling severe infestations. Then, use the lowest rate to maintain control.

NOTICE OF WARRANTY

SHELL CHEMICAL COMPANY MAKES NO WARRANTY OF MERCHANTABILITY, FITNESS FOR ANY PURPOSE, OR OTHERWISE, EXPRESS OR IMPLIED, concerning this product or its uses which extend beyond the use of the product under normal conditions in accord with the statements made in this label.

Sold under U.S. Patent Numbers 3,576,834 and 3,639,633.

Rate Conversion Tables

Area Treated By One 8 Oz Bag By Application Rate

RATE LB/ACRE	¼	⅓	½	¾	1	Rate Oz/1000 ft^2	.4	.8
ACRES TREATED	2	1.5	1	0.66	0.5	1000 ft^2 treated	20	10

Area Treated By One 2-Pound Bag By Application Rate

RATE LB/ACRE	¼	⅓	½	¾	1	Rate Oz/1000 ft^2	.4	.8
ACRES TREATED	8	6	4	2.64	2	1000 ft^2 treated	80	40

Area Treated By One 10-Pound Bag By Application Rate

RATE LB/ACRE	¼	⅓	½	¾	1	Rate Oz/1000 ft^2	.4	.8
ACRES TREATED	40	30	20	13.2	10	1000 ft^2 treated	400	200

CROP METHOD OF APPLICATION (G MEANS GROUND ONLY)	INSECTS	POUNDS PER ACRE*	LAST APPLICATION—DAYS TO HARVEST	LAST APPLICATION—DAYS TO LIVESTOCK GRAZING/FEEDING
Alfalfa	Fall Armyworm	½		
Do not apply to dormant or semi-dormant alfalfa when min. daily temp. is 50°F or lower	Armyworm†, Loopers, Yellowstriped Armyworm, Western Yellowstriped Armyworm	¼–½		7
	Lygus Bugs, Pea Aphid	½–1		
	Alfalfa Weevil Larvae	1		
	Beet Armyworm, Variegated Cutworm†	¼–1		
	Alfalfa Caterpillar	⅓–½		
	Aphids, Egyptian Alfalfa Weevil Larvae	½–1		
Anise (Fennel)†	Cabbage Looper	1	7	
	Beet Armyworm	½–1		

CROP METHOD OF APPLICATION (G MEANS GROUND ONLY)	INSECTS	POUNDS PER ACRE*	LAST APPLICATION—DAYS TO HARVEST	LAST APPLICATION—DAYS TO LIVESTOCK GRAZING/FEEDING
Apples†-G (Except varieties Early MacIntosh & Wealthy)	Apple Aphid	1/8–1/2 per 100 gals.; up to 400 gpa.	8	10
	Rosy Apple Aphid, Tarnished Plant Bug, Tufted Apple Budmoth, Green Fruitworm	1/4–1/2 per 100 gals.; up to 400 gpa.		
	Codling Moth—10-12 day intervals			
	Leafrollers (Fruittree, Oblique-banded, Redbanded, Variegated), Lesser Appleworm, White Apple Leafhopper, Tentiform Leafminer, Cutworm	1/2 per 100 gals.; up to 400 gpa		
Asparagus	Beet Armyworm, Western Yellow Striped Armyworm, Asparagus Beetle, Spotted Asparagus Beetle, White Cutworm†, Redback Cutworm†	1/2–1	1	
Avocado	Avocado Leafroller, Avocado Looper, Omnivorous Looper	1/2–1	1	10
Barley†	Armyworms	1/4–1/2	7	10
Beans (includes Blackeyed Peas & Cowpeas)	Fall Armyworm	1/2	Succulent Beans 1/4–1/2 lb.—1 over 1/2 lb.—3 Dry Beans all rates—25	3 Vines 7 Hay
	European Corn Borer (Initiate when moth flights first appear and continue preventive treatment at 3–4 day intervals.)	1/2–1		
	Leafhopper, Mexican Bean Beetle	1/4–1		
	Beet Armyworm, Cabbage Looper, Corn Earworm, Lygus Bugs, Loopers†, Saltmarsh Caterpillar†, Aphids†, Thrips†, Yellowstriped Armyworm†, Western Yellowstriped Armyworm†	1/2–1		
	Spotted Cucumber Beetle†	1/4–1/2		
Beet (Table)†	Beet Armyworm, Cabbage Looper, Diamondback Moth, Imported Cabbageworm	1/4–1/2	7—Roots 30—Tops	
Beet (Tops)	Beet Armyworm, Cabbage Looper, Diamondback Moth, Imported Cabbageworm, Flea Beetles	1/2–1	10	
Bermuda Grass†	Armyworms	1/4–1/2		7-Forage 3-Dehydrated Hay
	Cutworms Heliothis paradoxa	1/2–1		
Blueberries†	Aphids, Tussock Moth, Weevil, Sharp-Nosed Leafhopper	1/2	3	
	Cranberry Fruitworm, Cherry Fruitworm	1/2–1		
	Flea Beetle (larvae), Sawfly (larvae), Blueberry Leafroller	1		
Broccoli	Cabbage Looper, Imported Cabbageworm, Diamondback Moth	1/4–1**	3	
Brussels Sprouts	Cabbage Looper, Imported Cabbageworm, Diamondback Moth†	1/2–1**	3	
Cabbage	Cabbage Looper, Imported Cabbageworm, Diamondback Moth	1/4–1**	1	
	Fall Armyworm†	1/2**		
Carrots	Beet Armyworm	1/3–1	1	
	Armyworm†	1/4–1/2		
	Aster Leafhopper†	1/2–1		
Cauliflower	Cabbage Looper, Imported Cabbageworm, Diamondback Moth	1/4–1**	3	
Celery	Beet Armyworm	1/2–1	7	
	Loopers	1		
	Armyworm†	1/4–1/2		
Chicory†	Beet Armyworm, Leafhoppers, Variegated Cutworm	1/2–1	80	
Chinese Cabbage†	Loopers, Beet Armyworm	1/2–1	10	
Collards (Fresh Market only—Do not apply when temperature is less than 50°F. Do not apply when crop is less than 10″ tall.)	Loopers, Diamondback Moth, Imported Cabbageworm, Beet Armyworm, Flea Beetles	1/2–1	10	
Corn (Sweet)	Earworm—Whorl as needed	1/3–1/2	0 Ears	3 Forage
	Beet Armyworm†	1/2		
	Earworm, Fall Armyworm, European Corn Borer, Armyworms†—Ears 1-3 days or as needed **Ovicide/Larvicide** Earworm, European Corn Borer	1/4–1/2		

CROP METHOD OF APPLICATION (G MEANS GROUND ONLY)	INSECTS	POUNDS PER ACRE*	LAST APPLICATION—DAYS TO HARVEST	LAST APPLICATION—DAYS TO LIVESTOCK GRAZING/FEEDING
Corn (Field & Popcorn)	Earworm, Armyworms	1/4–1/2	0 Ears	3 Forage
	Fall Armyworm, Beet Armyworm	1/2		
Cotton Use may redden cotton- if excessive, stop or alternate with other insecticides U.S.	**Ovicide/Larvicide** Bollworm, Tobacco Budworm, (Initiate schedule when eggs and significant numbers of newly hatched larvae are present. Continue at 5–7 day intervals while eggs present and larval control adequate. If significant larvae survive, use higher rates below.)	1/8–1/4		
U.S.	**Larvicide** Bollworm, Tobacco Budworm, Fall Armyworm, Beet Armyworm, Cotton Leaf Perforator, Lygus Bugs†—Use as occasional spray in regular schedule but not more often than every 10 days	1/2	15	Do not graze or feed
U.S.	Bollworm, Fall Armyworm, Tobacco Budworm, Beet Armyworm Cotton Leaf Perforator, Lygus Bugs†—After full boll load, use up to 3 sprays at 3–5 day intervals	1/2–3/4		
U.S.	Cotton Fleahopper†	1/8–1/4		
West of Rockies	Beet Armyworm	1/3–1/2		
West of Rockies	Cotton Leaf Perforator	1/4–3/4		
Cotton Combination with AZODRIN® 5 WMI.† Note use restrictions on AZODRIN labeling.	Lygus Bugs, Pink Bollworm, Loopers, Mites, Boll Weevil, Whiteflies	1/8–3/4 plus 7/8–1 5/8 pt. AZODRIN 5	21	
	Thrips	1/8–3/4 plus 3/16–5/16 AZODRIN 5		
Cotton Combination with *Bacillus thuringiensis*†	**Ovicide/Larvicide** Bollworm, Tobacco Budworm	1/4–1/2 plus registered rates of B.T.	15	Do not graze or feed
Cotton Combination with chlordimeform†	**Ovicide/Larvicide** Bollworm, Tobacco Budworm, Lygus Bugs — U.S.	1/8 plus 1/8 lb. Chlordimeform	21	
	TX	1/16–1/8 plus 1/8 lb. Chlordimeform		
Cotton Combination with EPN† (Do not enter field for 24 hrs. after treatment. Do not hand pick within 3 days of treatment.)	**Ovicide/Larvicide** Bollworm, Tobacco Budworm, Boll Weevil	1/8–1/4 plus 1–2 pts. EPN (4 lb./gal.)	15	
Cotton Combination with EPN + methyl parathion† (Do not enter field for 48 hours after treatment. Do not hand pick within 7 days of treatment.)	**Ovicide/Larvicide** Bollworm, Tobacco Budworm, Boll Weevil	1/8–1/4 plus 2/3–1 1/3 pts. EPN + MP (3 lb. + 3 lb./gal.)	15	
Cotton Combination with methyl parathion (Do not enter field for 48 hours after treatment. Do not hand pick within 7 days of treatment.)	Tobacco Budworm, Bollworm-Budworm control with NUDRIN is obtained through kill of eggs (ovicide) and larvae (larvicide) Bollworm (Ovicide/Larvicide)† (5–7 days intervals)	1/8–1/3 plus 1–2 pts. methyl parathion (4lb./gal.)	15	
Cotton Combination with permethrin.†	**Ovicide/Larvicide** Bollworm, Tobacco Budworm	1/8–1/4 plus 0.1–0.2 lb. permethrin	15	
Cotton Combination with PYDRIN® Insecticide	Pink Bollworm **Ovicide/Larvicide** Bollworm, Tobacco Budworm	1/8–1/4 plus 0.05–0.1 Lb ai/acre PYDRIN	21	

CROP METHOD OF APPLICATION (G MEANS GROUND ONLY)	INSECTS	POUNDS PER ACRE*	LAST APPLICATION—DAYS TO HARVEST	LAST APPLICATION—DAYS TO LIVESTOCK GRAZING/FEEDING
Add adjuvant (X-77, Bio-88, Unite, Triton B-1956 or equivalent) to spray tank, then NUDRIN 90; agitate until thoroughly mixed. Compatibility of equivalent adjuvant should be determined before use.	Cotton Leaf Perforator **Ovicide/Larvicide** Bollworm, Tobacco Budworm	1/8–1/4 plus 0.10 Lb ai/acre PYDRIN		
	Cotton Bollworm, Tobacco Budworm, Lygus Bugs, Cabbage Looper, Beet Armyworm, Boll Weevil **Ovicide/Larvicide** Bollworm, Tobacco Budworm	1/8–1/4 plus 0.10–0.20 Lb ai/acre PYDRIN		
Cucumber	Beet Armyworm, Flea Beetle, Cabbage Looper, Tobacco Budworm, Melon Aphid†, Loopers†, Granulate Cutworms†, Yellowstriped Armyworm†, Pickleworm†, Cucumber Beetles†, Melonworm†	1/2–1	to 1/2 lb.—1 over 1/2 lb.—3 3 (CA only)	
	Fall Armyworm†	1/2		
Dandelions	Beet Armyworm, Cabbage Looper, Diamondback Moth, Imported Cabbageworm, Flea Beetles	1/2–1	10	
Eggplant†	Green Peach Aphid	1/4–1	5	
	Tomato Pinworm	1/2–1		
Garlic	Beet Armyworm	1/2–1	7	
Grapefruit	Thrips, Spirea Aphid†, Fruittree Leafroller, Citrus Cutworm	1/2–1	1	10
	Beet Armyworm†	1/2		
	Whiteflies†, Western Tussock Moth, Orange Tortrix, Omnivorous Leafroller	1		
Grapes East of Rockies G	Grape Berry Moth—Apply pre- & post-bloom & repeat at 10-14 day intervals or as needed intervals or as needed Leafhoppers	1/2–1	1 Fresh & Raisin 14 Wine Grapes	
West of Rockies	Leafhoppers[G], Saltmarsh Caterpillar, Grapeleaf Skeletonizer			
	Omnivorous Leafroller, Grape Leaffolder, Orange Tortrix, Climbing Cutworm†	1		
Grapes—Non-bearing grapevines and nursery stock—West of Rockies	Beet Armyworm, Thrips, Omnivorous Leafroller	3/4		
Horseradish†	Aphids, Thrips	1/2	65	
Kale	Beet Armyworm, Cabbage Looper, Diamondback Moth, Imported Cabbageworm, Flea Beetles	1/2–1	10	
Lemon	Thrips, Spirea Aphid†, Fruittree Leafroller, Citrus Cutworm	1/2–1	1	10
	Beet Armyworm†	1/2		
	Whiteflies†, Western Tussock Moth, Orange Tortrix, Omnivorous Leafroller	1		
Lentils†	Western Yellowstriped Armyworm	1/2–1	25	
Lettuce (Head and Leaf)	Beet Armyworm, Cabbage Looper, Alfalfa Looper, Aphids, Thrips, Corn Earworm	1/4–1	1/4–1/2 lb.—7 over 1/2 lb.—10	10
Melons	Beet Armyworm, Flea Beetle, Cabbage Looper, Tobacco Budworm, Melon Aphid†, Loopers†, Granulate Cutworms†, Yellowstriped Armyworm†, Pickleworm†, Cucumber Beetles†, Melonworm†	1/2–1	to 1/2 lb.—1 over 1/2 lb.—3 3 (CA only)	
	Fall Armyworm†	1/2		
Mint	Variegated Cutworm, Alfalfa Looper	1	14	
	Flea Beetles†	1/4–1		
Mustard Greens	Beet Armyworm, Cabbage Looper, Diamondback Moth, Imported Cabbageworm, Flea Beetles	1/2–1	10	
Nectarines CA & AZ	Thrips—Up to 3 applications	1/2–1	1	10
Oats†	Armyworms	1/4–1/2	7	10
Onions CA & East of Rockies	Beet Armyworm, Thrips, Cutworms—Up to 2 applications maximum	1/2**	7 Dry 28 Green	
West of Rockies	Beet Armyworm, Thrips, Cutworms—Up to 2 applications maximum	1/2**	7 Dry & Green	

CROP METHOD OF APPLICATION (G MEANS GROUND ONLY)	INSECTS	POUNDS PER ACRE*	LAST APPLICATION—DAYS TO HARVEST	LAST APPLICATION—DAYS TO LIVESTOCK GRAZING/FEEDING
Oranges	Thrips, Spirea Aphid†, Fruittree Leafroller, Citrus Cutworm	½–1	1	10
	Beet Armyworm†	½		
	Whiteflies†, Western Tussock Moth, Orange Tortrix, Omnivorous Leafroller,	1		
Parsley	Beet Armyworm, Cabbage Looper, Diamond Moth, Imported Cabbageworm, Flea Beetles	½–1	10	
Peaches	Catfacing Plant Bugs†, Stink Bugs†	1–2	4	10
	Oriental Fruitmoth, Twig Borer	1		
	Green Peach Aphid	¾–2		
Peanuts	Fall Armyworm	¼–½	21	Do not graze or feed
	Corn Earworm, Beet Armyworm Potato Leafhopper†	¼–1		
	Loopers, Green Cloverworm, Velvetbean Caterpillar, Thrips†	½–1		
Pears CT, ME, MA RI, VT, NH NY, NJ	Obliquebanded Leafroller, Green Fruitworm Do not make more than one application per season	½ per 100 gallons spray. 400 gals of water/acre maximum	7	10
Peas (Succulent)	Pea Aphid, Loopers, Saltmarsh Caterpillar†, Beet Armyworm	½–1	1	5-Forage 14 Hay
	Alfalfa Caterpillar	¼–1		
	Armyworm, Green Cloverworm	¼–½		
Pecans†	Aphids	½–1	30	10
Peppers	Green Peach Aphid, Beet Armyworm, Loopers†, Cabbage Looper, Fall Armyworm†	½	3	
	Armyworms†	¼–½		
Pomegranates CA	Omnivorous Leafroller	1	14	
Potato	Beet Armyworm, Tuberworm, Cabbage Looper, Aphids, Loopers†	½–1	6	
	Leafhoppers, Variegated Cutworm†, Flea Beetle†, Fall Armyworm†	½		
Rye†	Armyworms	¼–½	7	10
Sorghum	Corn Earworm (Headworm), Fall Armyworm, Armyworm	¼–½	14	14
	Sorghum Webworm	½		
	Sorghum Midge—Apply when 50% bloom, then 3–5 days later if needed	¼–½		
Soybeans	Green Cloverworm, Beet Armyworm, Velvetbean Caterpillar, Mexican Bean Beetle, Corn Earworm, Saltmarsh Caterpillar†, Fall Armyworm†, Thrips†, Bean Leaf Beetle†	¼–½	14	Less than ½ lb 3 Forage 7 Hay
	Loopers	½–1		over ½ lb 10 forage 12 Hay
	Velvetbean Caterpillar, Mexican Bean Beetle†, Green Cloverworm†, Corn Earworm† Apply when light to moderate infestations build to economic levels	⅛–¼		
Spinach—Do not apply when min. daily temp. is 32°F or lower. Do not apply to seedlings less than 3 inches in diameter.	Loopers, Beet Armyworm	½–1	7	
	Fall Armyworm	½		
Strawberries†	Beet Armyworm, Thrips, Aphids	½–1	3-Fresh 10-Processing	
G	Lygus Bugs, Omnivorous Leaf Tiers	1		
Strawberries (Dormant)	Armyworms	¾–1	90	
Sugar Beets	Variegated Cutworm†	½	7	30 Tops
	Beet Webworm, Flea Beetles, Beet Armyworm, Aphids-	¼–1		

CROP METHOD OF APPLICATION (G MEANS GROUND ONLY)	INSECTS	POUNDS PER ACRE*	LAST APPLICATION—DAYS TO HARVEST	TO LIVESTOCK GRAZING/FEEDING
Summer Squash	Cabbage Looper, Flea Beetles, Tobacco Budworm, Beet Armyworm, Yellowstriped Armyworm†, Granulate Cutworm†, Cucumber Beetles†, Melon Aphid†, Melonworm†, Pickleworm†, Loopers†	½–1	½ lb.—1 over ½ lb.—3 3 (CA only)	
	Fall Armyworm†	½		
Swiss Chard	Beet Armyworm, Cabbage Looper, Diamondback Moth, Imported Cabbageworm, Flea Beetles	½–1	10	
Tangerines, Tangelos	Thrips, Spirea Aphid†, Fruittree Leafroller, Citrus Cutworm	½–1	1	10
	Beet Armyworm†	½		
	Whiteflies†, Western Tussock Moth, Orange Tortrix, Omnivorous Leafroller	1		
Tobacco (except Shade)	Flea Beetle, Hornworm	¼–½	5 Flue cured	
	Loopers, Aphids, Tobacco Budworm, Fall Armyworm	½	14 Air or fire	
Tomato	Fall Armyworm†	½		
	Tomato Fruitworm, Aphids, Cabbage Looper, Beet Armyworm, Pinworm	½–1	1	
	Armyworm†	¼–½		
Turf† G	Sod Webworm—After application, sprinkle irrigate for 15 min.	1–2 (⅖–⅘ ozs. per 100 sq. Ft.)		Do not graze or feed
Turnip Greens	Beet Armyworm, Cabbage Looper, Diamondback Moth, Imported Cabbageworm, Flea Beetles	½–1	10	
Watercress† G	Beet Armyworm, Cabbage Looper, Diamondback Moth, Imported Cabbageworm (Use in minimum of 30 gpa water. Water in field should be turned off 24 hrs. prior to application)	½–1	10	
Wheat†	Armyworms	¼–½	7	10
ORNAMENTALS: See also additional Ornamental uses on following pages.				
Chrysanthemum G	Cabbage Looper, Corn Earworm, Beet Armyworm, Thrips, Omnivorous Leafroller—Weekly	¼–½ per 100 gals.		
Forest Woodlots, Plantations and Nurseries G	Spruce Budworm (treat when larvae feeding on buds, retreat after 3–4 days if needed.)	¼–½		
	Forest Tent Caterpillar†, Western Tent Caterpillar†	½–1		
Gladiolus	Cabbage Looper	¼–½ per 100 gals.		
Rose	Omnivorous Leafroller	¼–½ per 100 gals.		

* Consult the Rate Conversion Tables for acre(s)/bag rates.
Do not open bag; use entire inner package.
** Add wetting agent.
† Claim not registered in California.

SUPPLEMENTAL LABELING

NUDRIN® 90 METHOMYL INSECTICIDE
EPA REG. No. 201-324

WATER SOLUBLE POWDER FOR USE ON ORNAMENTALS†

† Claim not registered in California.

Shell NURDRIN 90 Methomyl Insecticide Water Soluble Powder is reommended for use on select ornamentals (see list) to control certain insects.

Insects Controlled

Aphids	Leafrollers
Beet Armyworm	Loopers
Corn Earworm	Thrips
Fall Armyworm	Tobacco Budworm
Flea Beetles	

How to Apply: Apply ¼ to ½ pts. NUDRIN 90 per 100 gals. of water. Spray plants thoroughly (to run-off by ground equipment).

Apply when insects first appear and continue at weekly intervals or as needed.

Note: Since varieties of ornamentals are numerous, continually changing, and may respond differently to NUDRIN 90, test it on a small scale before proceeding to large-scale application.

IMPORTANT

BEFORE USING NUDRIN 90, READ AND CAREFULLY OBSERVE THE CAUTIONARY STATEMENTS AND ALL OTHER INFORMATION APPEARING ON THE PRODUCT LABEL.

ORNAMENTALS FOR WHICH NUDRIN 90 IS RECOMMENDED

Scientific Name	Common Name
Abies lasiocarpa	Alpine Fir
Asparagus densiflorus 'Sprengeri'	Sprenger or Emerald Fern
Asparagus plumosis	Asparagus Fern or Lace Fern
Araucaria heterophylla	Norfolk Island Pine
Ardisia sp.	Ardisia
Aesculus pavia	Red Buckeye
Arecastrum romanzoffianum	Queen Palm
Bougainvillea glabra	Paper Flower
Bischofia javanica	Bishopwood Tree
Berberis thunbergii cv. Crimson Pygmy	Barberry
Berberis X Media cv. Parkjuweel	Barberry
Buddleja X Pikei cv. Hever	Butterfly Bush
Buxus sempervirens	Boxwood
Calocedrus decurrens	California Incense Cedar
Callicarpa japonica	Beautyberry

Calycanthus floridus	Carolina Allspice or Pineapple Shrub or Strawberry Shrub
Carissa grandiflora	Natal Plum
Carica Papaya	Papaya
Cocos nucifera	Coconut Palm
Codiaeum variegatum 'Apple Leaf'	Apple Leaf Croton
Codiaeum variegatum 'Bravo'	Bravo Croton
Codiaeum variegatum 'Elaine'	Elaine Croton
Codiaeum variegatum	Garden Croton
Chamaecyparis lawsoniana cv. Columnaris Gunica	Lawson Cypress or Port Oxford Cedar
Chamaecyparis lawsoniana cv. Drummondii	Cypress
Chamaecyparis lawsoniana cv. Ellwoodii	Pygmy Cypress
Chamaecyparis lawsoniana cv. Silver Queen	Silver Queen Cypress
Chamaecyparis obtusa cv. Mariesii	Cypress
Chamaecyparis obtusa cv. Pygmaea	Pygmy Cypress
Chamaecyparis pisifera cv. Nana	Cypress
Chamaecyparis pisifera cv. Filifera	Cypress
Chamaecyparis pisifera cv. Plumosa	Cypress
Chamaecyparis pisifera cv. Squarrosa	Cypress
Chamaecyparis pisifera cv. Tsukuma	Cypress
Chlorophytum commosum	Spider Plant or St. Bernards Lily
Chrysanthemum spp.	Chrysanthemums
Cissus rhombifolia	Grape Ivy
Coccolobia uvifera	Sea Grape
Coleus spp.	Flame Nettle, Painted Leaves
Cordyline terminalis 'Baby Doll'	Baby Doll Dracaena
Cotoneaster adpressus	
Cotoneaster atrupurpurea	
Cotoneaster ballatus var. floribundus	
Cotoneaster dammeri cv. Skogsholmen	
Cotoneaster denticulata	
Cotoneaster divaricatus	
Cotoneaster franchetii	
Cotoneaster harrovianus	
Cotoneaster horizontalis cv. Great Dane	
Contoneaster microphyllus var. Repens	
Cotoneaster morrisonensis	
Cotoneaster racemiflorus var. soongoricus	
Cotoneaster salicifolius var. Repens	
Cotoneaster simonsii	
Cotoneaster X watereri cv. Herbstfever	
Cotoneaster zabelii	
Crytomeria japonica	Japanese Cedar
Cytisus purpureus	
Deutzia scabra	
Dieffenbachia amoena	Giant Dumbcane
Dieffenbachia exotica	Exotic Dumbcane
Dracaena fragrans 'Massangeana'	Corn Plant
Epipremnum aureum	Golden Pothos
Epipremnum aureum 'Marble Queen'	Marbel Queen Pothos
Euonymus fortunei var. radicans	
Euonymus japonica cv. Microphylla	
Euphorbia pulcherrima	Poinsettia
Fittonia verschaffeltii	Silver Nerve Plant
Ginkgo biloba cv. Princeton Sentry	Maidenhair Tree
Gladiolus spp.	
Gynura sarmentosa	Purple Passion Velvet Plant
Hoya carnosa	Hindurope or Wax Plant
llex collina	Holly
llex cornuta cv. Bostic	Chinese Holly
llex cornuta cv. Hume	
llex crenata cv. Convexa	Japanese Holly
llex crentata cv. Glory	
llex crenata cv. Helleri	
llex crenata cv. Hetzii	
llex crenata cv. Mariesii	
llex crenata cv. Paludosa	
llex crenata cv. William Jackson	
llex integra	Mochi Tree
llex latifolia	Lusterleaf Holly
llex opaca	American Holly
llex opaca cv. Hedgeholly	
llex pedunculosa	
llexs pernyi	
llex sugerokii	
llex X cv. Autumn Glow	
llex X cv. Foster	
llex X cv. Lydia Morris	
llex X cv. Nellie R. Stevens	
llex X aguiperni cv. Brilliant	
Ixora coccinea 'Superking'	
Juniperus chinensis cv. Hetzii	Juniper
Juniperus chinensis cv. Mas	
Juniperus chinensis cv. Old Gold	
Juniperus chinensis cv. San Jose	
Juniperus chinensis cv. Shimpaki	
Juniperus chinensis var. Sargentii	
Juniperus communis cv. Compressa	
Juniperus communis cv. Effusa	Ground Juniper
Juniperus communis cv. Repanda	
Juniperus conferta	Shore Juniper
Juniperus horizontalis cv. Blue Horizon	Creeping Juniper
Juniperus horizontalis cv. Emerson	Creeping Juniper

Juniperus horizontalis cv. Turquoise Spreader	Creeping Juniper
Juniperus horizontalis cv. Venusta	Creeping Juniper
Juniperus horizontalis cv. Watnong	Creeping Juniper
Juniperus horizontalis cv. Wiltonii	Creeping Juniper
Juniperus scopulorum cv. Blue Heaven	Rocky Mountain Juniper or Colorado Red Cedar
Juniperus squamata cv. Loderi	
Juniperus virginiana cv. Hillii	
Juniperus virginiana cv. Sky Rocket	
Lagerstroemia indica cv. Powhatan	Crape Myrtle
Lagerstroemia indica cv. Catawba	Crape Myrtle
Lagerstroemia X cv. Hardy Red	Crape Myrtle
Lagerstroemia X cv. Hardy Rose	Crape Myrtle
Leucothoe fontanesiana	Dog Hobble or Switch Ivy
Leucothoe fontanesiana cv. Rollisonii	
Ligustrum sinense	Chinese Privet
Magnolia tripetala	Umbrella Tree
Magnolia virginiana	Sweet Bay
Mangifera indica	Mango
Marraya paniculata	Orange Jasmine
Metasequoia glyptostroboides	Dawn Redwood
Monstera deliciosa	Cut Leaf Philodendron
Nephrolepis exaltata 'Florida Ruffles'	Florida Ruffles Swordfern
Osmanthus heterophyllus	Holly Olive or Chinese Holly of False Holly
Osmanthus heterophyllus cv. Myrtifolius	
Osmanthus heterophyllus cv. Rotundifolius	
Peperomia obtusifolia	Baby Rubber Plant or Green Oval Leaf
Peperomia caperata	Emerald ripple Peperomia
Persea americana	Avocado
Philadelphus X cv. Snow White	Mock Orange
Philadelphus X virginalis cv. Enchantment	Mock Orange
Philodendron scandens oxycardium	Heart Leaf Philodendron
Philodendron bipennifolium	Fiddle Leaf Philodendron
Photinia villosa	
Picea abies	Norway Spruce
Picea abies cv. Viminalis	
Picea asperata	
Pieris floribunda	Fetterbush
Pieris japonica cv. Dorthy Wyckoff	Lily of the Valley
Pilea-bronze	
Pinus coulteri	Big Cone Pine
Pinus strobus	White Pine
Plectranthus australis	Swedish Ivy
Podocarpus macrophyllus	Southern Yew or Japanese Yew or Buddhist Pine
Potentilla fruticosa cv. Primrose Beauty	Golden Hardhack
Prunus laurocerasus	Cherry Laurel or English Laurel
Pseudotsuga menzeisii	Douglas Fir
Pyrus calleryana cv. Bradford	Pear
Phoenix dactylifera	Date Palm
Rhododendron arborerum X cv. Pioneer	
Rhododendron beltsville cv. Ping Pong	
Rhododendron catawbiense cv. Powell Glass	Catawba or Purple Laurel
Rhododendron catawbiense X cv. Boule De Neige	
Rhododendron catawbiense X cv. Lee's Dark Purple	
Rhododendron catawbiense X cv. Madame Mason	
Rhododendron catawbiense X cv. Lady Clermont	
Rhododendron caucasicum	
Rhododendron caucasicum X cv. Prince Camilla De Rohan	
Rhododendron cv. P. J. M. Victor	
Rhododendron Den Ouden cv. Van Der Hoop	
Rhododendron dexter X	
Rhododendron dexter A cv. Wheatley	
Rhododendron dexter X/N.Y.-203 Lavender	
Rhododendron gable cv. Caroline	
Rhododendron gable cv. Rose Greenley	
Rhododendron gable X cv. Atroflo	
Rhododendron gable X cv. Caroline Gable	
Rhododendron gable X cv. Louise Gable	
Rhododendron gable X cv. Stewartstonian	
Rhododendron glenn dale cv. Coral Sea	
Rhododendron glenn dale cv. Bishop	
Rhododendron glenn dale X cv. Aphrodite	
Rhododendron glenn dale X cv. Arradia	
Rhododendron glenn dale X cv. Greeting	
Rhododendron glenn dale X cv. Martha Hitchcock	
Rhododendron glenn dale X cv. Treasure	
Rhododendron glenn dale hyb. cv. Sagittaris	
Rhododendron hanceanum var. nanum	
Rhododendron holland X cv. Racil	

Rhododendron kurume X cv. Pink Pearl	
Rhododendron Kurume X cv. Day Break	
Rhododendron indicum X cv. J. T. Lovett	Southern or Indian Azalea
Rhododendron mucronatum X cv. Deseree	
Rhododendron nearing X cv. Windbeam	
Rhododendron noble X cv. The Queen	
Rhododendron obtusum cv. Fedora	
Rhododendron pericot X cv. Sweetheart Supreme	
Rhododendron ponticum X cv. Chionoides	
Rhododendron Robin Hill X cv. Nancy of Robin Hill	
Rhododendron Rufum	
Rhododendron satsuki X cv. Gyoko	
Rhododendron satsuki X cv. Shoshgi-Nohana	
Rhododendron satuski X cv. Tengasa	
Rhododendron slacok X cv. Champagne	
Rhododendron VanVeen X cv. Evening Glow	
Rhododendron Vossberg X cf. Meadowbrook	
Rhododendron Vuykiana X cv. Palestrina	
Rhododendron waterer X cv. Lady Eleanor Cathcart	
Rhododendron waterer X cv. Mrs. Charles S. Sargent	
Rhododendron waterer X cv. Mrs. P. D. Williams	
Rhododendron X cv. Helen Curtis	
Rhododendron X cv. Honey Moon	
Rhododendron X cv. Ice Cube	
Rhododendron X cv. Pink Twins	
Rhododendron X cv. Purple Imp	
Rhododendron X cv. Roslyn	
Rosa spp.	Rose
Saintpaulia ionantha	African Violet
Sansevieria tri-fasciata 'Laurentii'	Variegated Snake Plant
Sansevieria zeylanica	Snake Plant
Sciadopitys verticullata	Umbrella Pine or Japanese Pine
Spiraea thunbergii cv. Compacta	Compact Thunberg Spiraea
Syngonium podophyllum 'Variegatum'	Variegated Nephtuytis
Taxus baccata cv. Repandens	English Yew
Taxus cuspidata cv. Robusta	Japanese Yew
Taxus X media cv., Anthony Wayne	
Taxus X media cv. Grandifolia	
Taxus X media cv. Flushing	
Taxus X media cv. Hatfieldii	
Taxus X media cv. Hicksii	
Thuja occidentalis cv. Fastigata	White Cedar or American Arborvitae
Thuja occidentalis cv. Hetz Midget	
Thuja occidentalis cv. Lutea	
Thuja occidentalis cv. Nigra	
Thuja occidentalis cv. Pendula	
Thuja occidentalis cv. Pygmaea	
Thuja occidentalis cv. Rosenthalii	
Thuja occidentalis cv. Techng	
Thuja occidentalis cv. Umbraculifera	
Thuja occidentalis cv. Woodwardii	
Thuja occidentalis var. Caucasica	
Thuja plicata cv. Zebrina	Giant White Cedar
Tsuga canadensis cv. Rockland	Hemlock or Hemlock Spruce
Veitchia merrilli	Christmas palm
Viburnum cassinoides	Withe Rod or Appalachian Tea or Wild Raisin
Viburnum japonicum	
Zinnia elegans	Zinnia

PHOSDRIN® 4EC

EMULSIBLE CONCENTRATE INSECTICIDE

RESTRICTED USE PESTICIDE

For retail sale to and use only by Certified Applicators or persons under their direct supervision and only for those uses covered by the Certified Applicator's certification.

MEVINPHOS MIXTURE, LIQUID—POISON B, (Flammable Liquid) NA 2783, RQ

Contains 4 lbs. PHOSDRIN Insecticide per gallon.

ACTIVE INGREDIENTS	BY WEIGHT
alpha Isomer of 2-carbomethoxy-1-methylvinyl dimethyl phosphate	28.3%*
Related Compounds	18.8%
Petroleum Hydrocarbons	47.9%
INERT INGREDIENTS	5.0%
TOTAL	100.0%

* Equivalent to 47.1%w PHOSDRIN Insecticide

EPA Reg. No. 201-289 U.S. Patent No. 2,685,552

DANGER–POISON

Keep Out of Reach of Children

STATEMENT OF PRACTICAL TREATMENT

ATROPINE IS AN ANTIDOTE, CALL A PHYSICIAN AT ONCE IN ALL CASES OF SUSPECTED POISONING.

IF SWALLOWED, drink 1 or 2 glasses of water and induce vomiting by touching back of throat with finger. NEVER induce vomiting or give anything by mouth to an unconscious or non-alert person.

IF INHALED, remove from exposure. Have patient lie down and keep quiet. If patient is not breathing, start artifical respiration immediately. Get medical attention.

IN CASE OF CONTACT, wash skin with plenty of soap and water; for eyes, flush with water and get medical attention; remove and wash contaminated clothing before re-use.

IF WARNING SYMPTONS APPEAR: (See WARNING SYMPTOMS Under "NOTE TO PHYSICIAN") Keep patient prone and quiet. Start artificial respiration immediately if patient is not breathing. Transport the patient immediately to the nearest hospital.

See Antidote and Other Precautions

PELIGRO

PRECAUCION AL USUARIO: Si usted no lee ingles, no use este producto hasta que le etiqueta haya sido explicada ampliamente.

PRECAUTIONARY STATEMENTS

HAZARDS TO HUMANS AND DOMESTIC ANIMALS.

DANGER! POISONOUS IF SWALLOWED, INHALED OR ABSORBED THROUGH SKIN. RAPIDLY ABSORBED THROUGH SKIN. REPEATED INHALATION OR SKIN CONTACT MAY, WITHOUT SYMPTOMS, PROGRESSIVELY INCREASE SUSCEPTIBILITY TO PHOSDRIN INSECTICIDE POISONING. DO NOT SWALLOW OR GET IN EYES, ON SKIN, OR ON CLOTHING. DO NOT BREATHE VAPORS. DO NOT CONTAMINATE FOOD OR FEED PRODUCTS. KEEP AWAY FROM HEAT OR OPEN FLAME.

PRECAUTIONS IN USING

Wear clean rubber gloves, clean protective clothing and goggles. Replace gloves frequently and destroy used gloves. Wear a mask or respirator approved for protection against mevinphos (PHOSDRIN) vapors or spray mist. Wear clean clothes daily. Wash thoroughly with soap and water after handling and before eating or smoking. Pilot should not assist in mixing and loading operation. Do not contaminate or store near food or feedstuffs. Note Reentry Statements Section.

IMPORTANT: Not for use or storage in or around the home. Keep animals and unprotected persons out of operational areas and while there is a danger of drift.

In case of spillage, decontaminate areas and equipment by washing with a dilute solution of alkali (less than 5%) and detergent and rinse with water.

NOTE TO PHYSICIAN

WARNING SYMPTOMS: Symptoms incude weakness, headaches, tightness in chest, blurred vision, nonreactive pinpoint pupils, salivation, sweating, nausea, vomiting, diarrhea, and abdominal cramps.

TREATMENT: Atropine is the specific therapeutic antagonist of choice against parasympathetic nervous stimulation. If there are signs of parasympathetic stimulation, atropine sulfate should be injected at 10 minute intervals, in doses of 1 to 2 milligrams, until complete atropinization has occurred.

Pralidoxime chloride (2-PAM chloride) may also be used as an effective antidote in addition to and while maintaining full atropinization. In adults, an initial dose of 1 gram of 2-PAM should be injected, preferably as an infusion in 250 cc of saline over a 15 to 30 minute period. If this is not practical, 2-PAM may be administered slowly by intravenous injection as a 5 percent solution in water over not less than two minutes. After about an hour, a second dose of 1 gram of 2-PAM will be indicated if muscle weakness has not been relieved. For infants and children the dose of 2-PAM is 0.25 grams.

Morphine is an improper treatment.

Clear chest by postural drainage. Oxygen administration may be necessary. Observe patient continuously for 48 hours. Repeated exposure to cholinesterase inhibitors may, without warning, cause prolonged susceptibility to very small doses of any cholinesterase inhibitor. Allow no further exposure until cholinesterase regeneration has been attained as determined by blood test.

For 24-hour emergency medical assistance, call (713) 473-9461.

ENVIRONMENTAL HAZARDS

This product is poisonous to fish and wildlife. Birds feeding on treated areas may be killed. This product is highly toxic to bees exposed to direct treatment or residues on crops. Protective information may be obtained from your Cooperative Agricultural Extension Service. Keep out of lakes, ponds and streams. Do not apply when weather conditions favor drift from areas treated. Do not apply where runoff is likely to occur. Do not contaminate water by cleaning of equipment or disposal of wastes.

In case of a significant spill, call CHEMTREC (800) 424-9300 or (713) 473-9461.

DIRECTIONS FOR USE

It is a violation of Federal law to use this product in a manner inconsistent with its labeling.

FOR END USE ONLY. DO NOT REPACKAGE OR REFORMULATE WITHOUT MANUFACTURER'S WRITTEN APPROVAL.

REENTRY STATEMENTS

Do not apply this product in such a manner as to directly or through drift expose workers or other persons. The area being treated must be vacated by unprotected persons.

Do not enter treated areas without protective clothing until sprays have dried.

Because certain states may require more restrictive reentry intervals for various crops treated with this product, consult your State Department of Agriculture for further information.

Written or oral warnings must be given to workers who are expected to be in a treated area or in an area about to be treated with this product. *(Indicate specific oral warnings which inform workers of areas or fields that may not be entered without specific protective clothing, period of time field must be vacated and appropriate actions to take in case of accidental exposure.)*

When oral warnings are given, warnings shall be given in a language customarily understood by workers. Oral warnings must be given if there is reason to believe that written warnings cannot be understood by workers. Written warnings must include the following information: "DANGER! Area treated with PHOSDRIN® 4EC on (*date of application*). Do not enter without appropriate protective clothing until sprays have dried. (*Insert here actions to take in case of accidental exposure. See* STATEMENT OF PRACTICAL TREATMENT *for details.*)"

STORAGE AND DISPOSAL

Do not contaminate water, food or feed by storage or disposal. Do not reuse empty container. Open dumping is prohibited.

STORAGE

Do not store in or around the home. Store in a secure, dry and temperate area. Store in original container only. Product is flammable. Keep away from heat, sparks or open flame. If spilled avoid exposure. If significant spill call CHEMTREC (800) 424-9300 or (713) 473-9461. Spill can be contained by covering with a sufficient amount of absorbent material, such as clay, diatomaceous earth, sand, sawdust or sweeping compound. Place in drum, label contents and dispose of according to the instructions below in pesticide disposal.

CAUTION

Wear a pesticide respirator and avoid forming dust clouds during clean-up.

PESTICIDE DISPOSAL
Pesticide wastes are acutely hazardous. Improper disposal of excess pesticide, spray mixture, or rinse water is a violation of Federal law. If the wastes cannot be disposed of according to label instructions, contact your State Pesticide or Environmental Control Agency, or the Hazardous Waste representative at the nearest EPA Regional Office for guidance.

CONTAINER DISPOSAL
Triple rinse (or equivalent). Then offer for recycling or reconditioning, or puncture and dispose of in a sanitary landfill, or by other procedures approved by state and local authorities.

USE AND APPLICATION DIRECTIONS

Shell PHOSDRIN® 4EC can be used in conventional hydraulic sprayers, low-volume ground applicators, or airplane sprayers. Use this product only for recommended purposes and at recommended dosages. For best results consult the State Agricultural Extension Service for exact dosage and timing recommendations. Mix with water by vigorous agitation; agitate the mixture while spraying. If allowed to stand, re-agitate before use.

Apply when insects first appear and repeat application as often as necessary to maintain control. Apply in a suitable volume of water to give uniform coverage of foliage. For application by airplane, dilute this product with 3 to 10 gallons of water per acre.

FRUIT AND NUT CROPS: Use recommended dosages in sufficient water to obtain thorough spray coverage, but do not exceed indicated maximum gallonage for that crop. If concentrate ground application is employed, do not apply less than 10 gallons of spray per acre, and adjust dosage of PHOSDRIN Insecticide Emulsible Concentrate to obtain same dosage per acre as recommended for full coverage application.

FIELD AND VEGETABLE CROPS: Use 10 to 125 gallons of spray per acre when using ground application equipment.

Avoid drift of this product to crops for which it is not registered or to crops for which its use may be registered but application would be closer to harvest than recommended.

FRUIT AND NUT CROPS

APPLES, PEACHES, PEARS, PLUMS: For control of aphids and mites use 1/4 to 1/2 pt. per 100 gals. water. Do not exceed 1,200 gallons of finished spray per acre on apples and pears when 1/2 pt. is used. For control of grasshoppers, Lygus bugs and red-banded leafroller use 1/2 to 2/3 pt. per 100 gals. water. Do not exceed 900 gals. of finished spray per acre on apples and pears when 2/3 pt. is used. Do not treat within 1 day of harvest. Do not exceed 750 gals. finished spray per acre on peaches or plums.

CHERRIES (SOUR ONLY): For control of aphids and mites use 1/2 to 2/3 pt. per 100 gals. water. Do not treat within 2 days of harvest. Do not exceed 750 gals. finished spray per acre.

CITRUS (ORANGES, LEMONS, GRAPEFRUIT): For control of aphids use 1 to 2 pt. per acre in 200 or more gals. water. For control of fruit tree leafroller, orange tortrix and omnivorous leafroller use 1 qt. per acre in 500 gals. water. For control of western tussock moth larvae, citrus cutworm, variegated cutworm and pink scavenger caterpillar use 2 qt. per acre in 1,200 gals. water. Allow at least 7 days between applications. Do not treat within 1 day of harvest.

GRAPES: For control of aphids use 1/2 to 1 pt. per acre. Do not treat within 2 days of harvest. For control of leaf folder, leafhoppers, mites, red-banded leafroller and Lygus bugs use 1 to 2 pts per acre. At the 2 pt. dosage do not treat within 5 days of harvest.

RASPBERRIES: For control of aphids use 1/4 to 1/2 pt. per 100 gals. water. For control of mites, leafhoppers, fruit tree leafroller and orange tortrix use 1/2 to 2/3 pt. per 100 gals. water. Apply both dosages in up to 200 gals. water per acre. Do not treat within 3 days of harvest.

STRAWBERRIES: For control of aphids and mites use 1/4 to 1/2 pt. per 100 gals. water. For control of grasshoppers, strawberry leafroller, salt-marsh caterpillar and Lygus bugs use 1/2 to 1 pt. per 100 gals. water. Do not treat within 1 day of harvest. For all hard to kill insects, including aphids and mites, use 1 qt. per 100 gals. water, but do not treat within 2 days of harvest. Do not exceed 100 gals. finished spray per acre.

WALNUTS: For control of aphids use 1/4 to 1/2 pt. per 100 gals. water. For control of mites, omnivorous looper, orange tortrix, fruit tree leafroller, western tussock moth use 1/2 pt. per 100 gals. water. Do not treat within 1 day of harvest. Do not exceed 1,000 gals. finished spray per acre.

FIELD CROPS

ALFALFA, CLOVER: For control of aphids and alfalfa caterpillar use 1/4 to 1/2 pt. per acre. For control of grasshoppers, leafhoppers, cutworms (climbing), mites, alfalfa weevil larvae, Lygus bugs use 1/2 to 1 pt. per acre. Do not treat within 1 day of harvest.

CORN (FIELD, SWEET, AND POPCORN) FOR FORAGE ONLY: For control of aphids use 1/2 to 1 pt. per acre. Do not treat within 1 day of harvest.

MUSTARD GREENS, TURNIP TOPS: For control of aphids use 1/4 to 1/2 pt. per acre. For control of cabbage looper, imported cabbage worm, false chinch bug, dipterous leafminer (adult), grasshoppers, leafhoppers and mites use 1/2 to 1 pt. per acre.

PEA VINES (FOR FORAGE ONLY): For control of aphids use 1/4 to 1/2 pt. per acre. For control of grasshoppers, leafhoppers, mites and cutworms (climbing) use 1/2 to 1 pt. per acre. Do not treat within 1 day of harvest.

SORGHUM (FOR FORAGE AND GRAIN): For control of aphids use 1/4 to 1/2 pt. per acre. For control of corn earworm and webworm use 1/2 to 1 pt. per acre. For control of fall armyworms use 1 pt. per acre. Do not treat within 3 days of harvest.

VEGETABLE CROPS

ARTICHOKES: For control of aphids use 1/4 to 1/2 pt. per acre. For control of plume moth use 1 to 2 pts. per acre. Do not treat within 2 days of harvest.

BEANS: For control of aphids use 1/4 to 1/2 pt. per acre. For control of grasshoppers, leafhoppers, mites and Mexican bean beetle use 1/2 to 1 pt. per acre. Do not treat within 1 day of harvest.

BEETS (INCLUDING TOPS): For control of aphids use 1/4 to 1/2 pt. per acre. For control of cabbage looper, cutworm (climbing), dipterous leafminer (adult), false chinch bug, grasshoppers, imported cabbage worm, leafhopper, mites and salt-marsh caterpillar use 1/2 to 1 pt. per acre. Do not treat within 3 days of harvest.

BROCCOLI, CABBAGE: For control of aphids use 1/4 to 1/2 pt. per acre. For control of cabbage looper, imported cabbage worm, grasshoppers, leafhoppers, salt-marsh caterpillar, mites, cutworms (climbing), dipterous leafminer (adult) and Lygus bugs use 1/2 to 1 pt. per acre. Do not treat within 1 day of harvest. For hard to kill insects, including aphids, use 1 qt. per acre, but do not treat within 3 days of harvest.

BRUSSELS SPROUTS, CAULIFLOWER, COLLARDS, KALE: For control of aphids use 1/4 to 1/2 pt. per acre. For control of cabbage looper, imported cabbage worm, grasshoppers, leafhoppers, salt-marsh caterpillar, mites, cutworms (climbing), dipterous leafminer (adult) and Lygus bugs use 1/2 to 1 qt. per acre. Do not treat within 3 days of harvest. For hard to kill insects, including aphids, use 1 qt. per acre, but do not treat collards and kale within 7 days of harvest.

CARROTS: For control of aphids use 1/4 to 1/2 pt. per acre. For control of leafhoppers, Lygus bugs, mites, cabbage looper, dipterous leafminer (adult), cutworm (climbing) and salt-marsh caterpillar use 1/2 to 1 pt. per acre. Do not treat within 2 days of harvest.

CELERY: For control of aphids use 1/4 to 1/2 pt. per acre. For control of dipterous leafminer (adult), Lygus bug, salt-marsh caterpillar, leafhopper, cabbage looper and mites use 1/2 to 1 pt. per acre. Do not treat within 3 days of harvest. For hard to kill insects, including aphids, use 1 qt. per acre, but do not treat within 5 days of harvest.

CORN (FIELD, SWEET AND POPCORN) FOR GRAIN ONLY: For control of aphids use $\frac{1}{4}$ to 1 pt. per acre. Do not treat within 1 day of harvest.

CUCUMBERS (OUTDOOR): For control of aphids use $\frac{1}{4}$ to $\frac{1}{2}$ pt. per acre. For control of grasshoppers, leafhoppers and mites use $\frac{1}{2}$ to 1 pt. per acre. do not treat within 1 day of harvest.

EGGPLANT, PEPPERS: For control of aphids use $\frac{1}{4}$ to $\frac{1}{2}$ pt. per acre. For control of grasshoppers, leafhoppers and mites use $\frac{1}{2}$ to 1 pt. per acre. Do not treat within 2 days of harvest.

LETTUCE: For control of aphids use $\frac{1}{4}$ to $\frac{1}{2}$ pt. per acre. For control of corn earworm, cutworms (climbing), dipterous leafminer (adult), cabbage looper, imported cabbage worm, grasshoppers, mites, Lygus bugs, salt-marsh caterpillar, false chinch bug and thrips use $\frac{1}{2}$ to 1 pt. per acre. Do not treat within 2 days of harvest. For hard to kill insects, including aphids, use 1 qt. per acre, but do not treat within 4 days of harvest.

MELONS (INCLUDING CANTALOUPES, HONEYDEW MELONS, MUSKMELONS, WATERMELONS): For control of aphids use $\frac{1}{4}$ to $\frac{1}{2}$ pt. per acre. For control of cabbage looper, dipterous leafminer (adult), leafhoppers. Lygus bugs, mites, false chinch bugs, salt-marsh caterpillars, cutworms (climbing) and grasshoppers use $\frac{1}{2}$ to 1 pt. per acre. To control rindworms (cabbage looper, cutworms, salt-marsh caterpillars, tobacco budworms) on watermelons apply $\frac{1}{2}$ pt. per acre. Do not treat cantaloupes, honeydew melons, muskmelons or watermelons within 1 day of harvest.

OKRA: For control of aphids use $\frac{1}{4}$ to $\frac{1}{2}$ pt. per acre. For control of cutworms (climbing), corn earworms, green stink bug, mites and velvet bean caterpillar use $\frac{1}{2}$ to 1 pt. per acre. Do not treat within 1 day of harvest.

ONIONS (INCLUDING GREEN ONIONS): For control of thrips and cutworms (climbing) use $\frac{1}{2}$ to 1 pt. per acre. Do not treat within 1 day of harvest.

PARSLEY: For control of aphids apply 1 to 2 pt. (0.5 lb. to 1.0 lb. a.i./acre) as a foliage application using a minimum of 40 gallons of water per acre by ground application or 10 to 15 gallons of water per acre by air application. For light infestations use lower dosage of 1 pt. (0.5 lb. a.i./acre). Do not treat later than 5 days before harvest. For hard to kill insects and increased populations use 2 pt. (1.0 lb. a.i./acre). Do not treat later than 8 days before harvest for higher dosage. Begin application when insects first appear and repeat as often as necessary to maintain control. Do not treat more than 3 times between harvests.

PEAS (INCLUDING VINES): For control of aphids use $\frac{1}{4}$ to $\frac{1}{2}$ pt. per acre. For control of grasshoppers, leafhoppers, mites and cutworms (climbing) use $\frac{1}{2}$ to 1 pt. per acre. Do not treat within 1 day of harvest.

POTATOES: For control of aphids use $\frac{1}{4}$ to $\frac{1}{2}$ pt. per acre. For control of grasshoppers, leafhoppers and mites use $\frac{1}{2}$ pt. per acre. Do not treat within 1 day of harvest.

SPINACH: For control of aphids use $\frac{1}{4}$ tp $\frac{1}{2}$ pt. per acre. For control of cabbage looper, imported cabbage worm, grasshoppers, leafhoppers, mites, dipterous leafminer (adult), cutworms (climbing), salt-marsh caterpillar and false chinch bug use $\frac{1}{2}$ to 1 pt. per acre. Do not treat within 4 days of harvest. For hard to kill insects, including aphids, use 1 qt. per acre, but do not treat within 4 days of harvest.

SUMMER SQUASH: For control of aphids use $\frac{1}{4}$ to $\frac{1}{2}$ pt. per acre. For control of cabbage looper, dipterous leafminer (adult), leafhoppers. Lygus bugs, mites, false chinch bugs, salt-marsh caterpillar, cutworms (climbing) and grasshoppers use $\frac{1}{2}$ to 1 pt. per acre. Do not treat within 1 day of harvest.

TOMATOES (OUTDOOR): For control of aphids use $\frac{1}{4}$ to $\frac{1}{2}$ pt. per acre. For control of grasshoppers, leafhoppers and mites use $\frac{1}{2}$ to 1 pt. per acre. Do not treat within 1 day of harvest.

TURNIPS: For control of aphids use $\frac{1}{4}$ to $\frac{1}{2}$ pt. per acre. For control of cabbage looper, imported cabbage worm, grasshoppers, leafhoppers, mites, false chinch bug and dipterous leafminer (adult) use $\frac{1}{2}$ to 1 pt. per acre. Do not treat within 3 days of harvest.

GREENHOUSE

Diluted PHOSDRIN 4 EC should be applied in a closed greenhouse (close all doors, windows and ventilators). Lock or barricade all entrances, post warning signs, and take whatever precautions are necessary to prevent unprotected humans and domestic animals from entering the treated area. The operator must wear a full-face mask of a type found adequate for mevinphos (PHOSDRIN Insecticide) protection.

Apply any time the greenhouse ventilators can remain closed for two hours without endangering crops from high temperatures, such as early morning, cloudy days or late afternoons. If application is made during the day, keep the greenhouse closed tightly for at least two hours, then ventilate the enclosure thoroughly for an hour before resuming regular work. If application is made in the late afternoon, the house can be kept closed all night. On the following morning, ventilate for one hour to render the air safe for regular work.

LETTUCE (INCLUDING LEAF, BIBB AND ROMAINE): For control of aphids, corn earworms, cutworms (climbing), dipterous leafminer (adult), cabbage looper, imported cabbage worm, grasshoppers, mites, Lygus bugs, salt-marsh caterpillars, false chinch bugs and thrips use $1\frac{1}{4}$ to $2\frac{1}{3}$ pt. per 50,000 sq. ft. Do not treat within 10 days of harvest.

NOTICE OF WARRANTY

SHELL CHEMICAL COMPANY MAKES NO WARRANTY OF MERCHANTABILITY, FITNESS FOR ANY PURPOSE, OR OTHERWISE, EXPRESS OR IMPLIED concerning this product or its uses which extend beyond the use of the product under normal conditions in accord with the statements made on this label.

PHOSDRIN® 10.3 WS

WATER SOLUBLE INSECTICIDE

RESTRICTED USE PESTICIDE

For retail sale to and use only by Certified Applicators or persons under their direct supervision and only for those uses covered by the Certified Applicator's certification.

MEVINPHOS, POISON B, NA 2783, RQ

ACTIVE INGREDIENT	BY WEIGHT
alpha Isomer of 2-carbomethoxy-1-methylvinyl dimethyl phosphate	60.0%*
Related Compounds	40.0%*
TOTAL	100.0%

Contains 10.3 lbs. Active Ingredients Per Gallon

* Equivalent to 100%w PHOSDRIN Insecticide

EPA Reg. No. 201-291 U.S. Patent No. 2,685,552

DANGER–POISON

Keep Out of Reach of Children

STATEMENT OF PRACTICAL TREATMENT

ATROPINE IS AN ANTIDOTE, CALL A PHYSICIAN AT ONCE IN ALL CASES OF SUSPECTED POISONING.

IF SWALLOWED, drink 1 or 2 glasses of water and induce vomiting by touching back of throat with finger. NEVER induce vomiting or give anything by mouth to an unconscious or non-alert person.

IF INHALED, remove from exposure. Have patient lie down and keep quiet. If patient is not breathing, start artificial respiration immediately. Get medical attention.

IN CASE OF CONTACT, wash skin with plenty of soap and water; for eyes, flush with water and get medical attention; remove and wash contaminated clothing before re-use.

IF WARNING SYMPTOMS APPEAR: (See WARNING SYMPTOMS Under "NOTE TO PHYSICIAN") Keep patient prone and quiet. Start artificial respiration immediately if patient is not breathing. Transport the patient immediately to the nearest hospital.

SEE ANTIDOTE AND OTHER PRECAUTIONS.

PELIGRO

PRECAUCION AL USUARIO: Si usted no lee ingles, no use este producto hasta que le etiqueta haya sido explicado ampliamente.

PRECAUTIONARY STATEMENTS

HAZARDS TO HUMANS AND DOMESTIC ANIMALS

DANGER! POISONOUS IF SWALLOWED, INHALED OR ABSORBED THROUGH SKIN. REPEATED INHALATION OR SKIN CONTACT MAY, WITHOUT SYMPTOMS, PROGRESSIVELY INCREASE SUSCEPTIBILITY TO PHOSDRIN INSECTICIDE POISONING. DO NOT SWALLOW OR GET IN EYES, ON SKIN, OR ON CLOTHING. DO NOT BREATHE VAPORS. DO NOT CONTAMINATE FOOD OR FEED PRODUCTS. KEEP AWAY FROM HEAT OR OPEN FLAME.

PRECAUTIONS IN USING

Wear clean rubber gloves, clean protective clothing and goggles. Replace gloves frequently and destroy used gloves. Wear a mask or respirator approved for protection against mevinphos (PHOSDRIN) vapors or spray mist. Wear clean clothes daily. Wash thoroughly with soap and water after handling and before eating or smoking. Pilot should not assist in mixing and loading operation. Do not contaminate or store near food or feedstuffs. Note reentry statements section. **IMPORTANT:** Not for use or storage in or around the home. Keep animals and unprotected persons out of operational areas during treatment and while there is a danger of drift. Note Reentry Statements Section.

In case of spillage, decontaminate areas and equipment by washing with a dilute solution of alkali (less than 5%) and detergent and rinse with water.

NOTE TO PHYSICIAN

WARNING SYMPTOMS: Symptoms include weakness, headaches, tightness in chest, blurred vision, nonreactive pinpoint pupils, salivation, sweating, nausea, vomiting, diarrhea, and abdominal cramps.

TREATMENT: Atropine is the specific therapeutic antagonist of choice against parasympathetic nervous stimulation. If there are signs of parasympathetic stimulation, atropine sulfate should be injected at 10-minute intervals, in doses of 1 to 2 milligrams, until complete atropinization has occurred.

Pralidoxime chloride (2-PAM chloride) may also be used as an effective antidote in addition to and while maintaining full atropinization. In adults, an initial dose of 1 gram of 2-PAM should be injected, preferably as an infusion in 250 cc of saline over a 15- to 30-minute period. If this is not practical, 2-PAM may be administered slowly by intravenous injection as a 5 percent solution in water over not less than two minutes. After about an hour, a second dose of 1 gram of 2-PAM will be indicated if muscle weakness has not been relieved. For infants and children the dose of 2-PAM is 0.25 grams.

Morphine is an improper treatment.

Clear chest by postural drainage. Oxygen administration may be necessary. Observe patient continuously for 48 hours. Repeated exposure to cholinesterase inhibitors may, without warning, cause prolonged susceptibility to very small doses of any cholinesterase inhibitor. Allow no further exposure until cholinesterase regeneration has been attained as determined by blood test.

For 24-hour emergency medical assistance, call (713) 473-9461.

ENVIRONMENTAL HAZARDS

This product is poisonous to fish and wildlife. Birds feeding on treated areas may be killed. This product is highly toxic to bees exposed to direct treatment or residues on crops. Protective information may be obtained from your Cooperative Agricultural Extension Service. Keep out of lakes, ponds and streams. Do not apply when weather conditions favor drift from areas treated. Do not apply where runoff is likely to occur. Do not contaminate water by cleaning of equipment or disposal of wastes.

In case of a significant spill, call CHEMTREC (800) 424-9300 or (713) 473-9461.

DIRECTIONS FOR USE

It is a violation of Federal law to use this product in a manner inconsistent with its labeling.

RE-ENTRY STATEMENTS

Do not apply this product in such a manner as to directly or through drift expose workers or other persons. The area being treated must be vacated by unprotected persons.

Do not enter treated areas without protective clothing until sprays have dried.

Because certain States may require more restrictive reentry intervals for various crops treated with this product, consult your State Department of Agriculture for further information.

Written or oral warnings must be given to workers who are expected to be in a treated area or in an area about to be treated with this product. *(Indicate specific oral warnings which inform workers of areas or fields that may not be entered without specific protective clothing, period of time field must be vacated and appropriate actions to take in case of accidental exposure.)* When oral warnings are given, warnings shall be given in a language customarily understood by workers. Oral warnings must be given if there is reason to believe that written warnings cannot be understood by workers. Written warnings must include the following information: "DANGER! Area treated with PHOSDRIN® Insecticide on *(date of application)*. Do not enter without appropriate protective clothing until sprays have dried. (*Insert here actions to take in case of accidental exposure. See* STATEMENT OF PRACTICAL TREATMENT *for details.*)"

STORAGE AND DISPOSAL

PROHIBITIONS

Do not contaminate water, food or feed by storage or disposal. Open dumping is prohibited. Do not reuse container.

PESTICIDE DISPOSAL

Pesticide, spray mixture, or rinse water that cannot be used according to label instructions must be disposed of according to Federal or approved State procedures under the hazardous waste provisions (Subtitle C) of the Resource Conservation and Recovery Act.

CONTAINER DISPOSAL

Triple rinse (or equivalent). Dispose of in a sanitary landfill or by other approved State and local procedures.

Read the Directions Carefully and Follow Them at all Times

USE AND APPLICATION DIRECTIONS

Shell PHOSDRIN® 10.3 WS can be used in conventional hydraulic sprayers, low-volume ground applicators, or airplane sprayers. Use this product only for recommended purposes and at recommended dosages. For best results consult the State Agricultural Extension Service for exact dosage and timing recommendations.

Mix with water by vigorous agitation.

Apply when insects first appear and repeat application as often as necessary to maintain control. Apply in a suitable volume of water to give uniform coverage of foliage. For application by airplane, dilute this product with 3 to 10 gallons of water per acre.

FRUIT AND NUT CROPS: Use recommended dosages in sufficient water to obtain thorough spray coverage, but do not exceed indicated maximum gallonage for that crop. If concentrate ground application is employed, do not apply less than 10 gallons of spray per acre, and adjust dosage of PHOSDRIN WS Insecticide to obtain same dosage per acre as recommended for full coverage application.

FIELD AND VEGETABLE CROPS: Use 10 to 125 gallons of spray per acre when using ground application equipment.

Avoid drift of this product to any crops not intended for spraying.

Dosages are expressed in fluid ounces.

FRUIT AND NUT CROPS

APPLES, PEACHES, PEARS, PLUMS: For control of aphids and mites use 1½ to 3 ozs. per 100 gals. water. For control of grasshoppers, Lygus bugs and redbanded leafroller use 3 to 4 ozs. per 100 gals. water. Do not exceed 1,200 gallons of finished spray per acre on apples and pears for the 3 ozs. per 100 gallons of water dosage; do not exceed 900 gallons per acre of finished spray on apples and pears for the 4 ozs. per 100 gallons of water dosage. Do not treat within 1 day of harvest. Do not exceed 750 gals. finished spray per acre on peaches or plums.

CHERRIES (SOUR ONLY): For control of aphids and mites use 3 to 4 ozs. per 100 gals. water. Do not treat within 2 days of harvest. Do not exceed 750 gals. finished spray per acre.

CITRUS (ORANGES, LEMONS, GRAPEFRUIT): For control of aphids use 6 to 12½ ozs. per acre per 200 or more gals. water. For control of fruit tree leafroller, orange tortrix and omnivorous leafroller use 12½ ozs. per acre in 500 gallons water. For control of western tussock moth larvae, citrus cutworm, variegated cutworm and pink scavenger caterpillar use 25 ozs. per acre in 1,200 gals. water. Allow at least 7 days between applications. Do not treat within 1 day of harvest.

GRAPES: For control of aphids use 3 to 6 ozs. per acre. Do not treat within 2 days of harvest. For control of leaf folder, leafhoppers, mites, red-banded leafroller and Lygus bugs use 6 to 12½ ozs. per acre. At 12½ ozs. dosage do not treat within 5 days of harvest.

RASPBERRIES: For control of aphids use 1½ to 3 ozs. per 100 gals water. For control of mites, leafhoppers, fruit tree leafroller and orange tortrix use 3 to 4 ozs. per 100 gals. water. Apply both dosages in up to 200 gals. water per acre. Do not treat within 3 days of harvest.

STRAWBERRIES: For control of aphids and mites use 1½ to 3 ozs. per 100 gals. water. For control of grasshoppers, strawberry leafroller, salt-marsh caterpillar and Lygus bugs use 3 to 6 ozs. per 100 gals. water. Do not treat within 1 day of harvest. In certain areas where these insects, includng aphids and mites, are not adequately controlled at the dosages recommended above, use 12½ ozs. per 100 gals. water, but do not treat within 2 days of harvest. Do not exceed 100 gals. finished spray per acre.

WALNUTS: For control of aphids use 1½ to 3 ozs. per 100 gals. water. For control of mites, omnivorous looper, orange tortrix, fruit tree leafroller and western tussock moth use 3 ozs. per 100 gals. water. Do not treat within 1 day of harvest. Do not exceed 1000 gals. finished sprayer acre.

FIELD CROPS

ALFALFA, CLOVER: For control of aphids and alfalfa caterpillar use 1½ to 3 ozs. per acre. For control of grasshoppers, leafhoppers, cutworms (climbing), mites, alfalfa weevil larvae and Lygus bugs use 3 to 6 ozs. per acre. Do not treat within 1 day of harvest.

CORN (FIELD, SWEET, AND POPCORN) FOR FORAGE ONLY: For control of aphids use 1½ to 6 ozs. per acre. Do not treat within 1 day of harvest.

MUSTARD GREENS, TURNIP TOPS: For control of aphids use 1½ to 3 ozs. per acre. For control of cabbage looper, imported cabbage worm, false chinch bug, dipterous leafminer (adult), grasshopper, leafhoppers, and mites use 3 to 6 ozs. per acre. Do not treat within 3 days of harvest.

PEA VINES (FOR FORAGE ONLY): For control of aphids use 1½ to 3 ozs. per acre. For control of grasshoppers, leafhopper, mites and cutworms (climbing), use 3 to 6 ozs. per acre. Do not treat within 1 day of harvest.

SORGHUM (FOR FORAGE AND GRAIN): For control of aphids use 1½ to 3 ozs. per acre. For control of corn earworm and webworms use 3 to 6 ozs. per acre. For control of fall army worms use 6 ozs. per acre. Do not treat within 3 days of harvest.

VEGETABLE CROPS

ARTICHOKES: For control of aphids use 1½ to 3 ozs. per acre. For control of plume moth use 6 to 12½ ozs. per acre. Do not treat within 2 days of harvest.

BEANS: For control of aphids use 1½ to 3 ozs. per acre. For control of grasshoppers, leafhoppers, mites and Mexican bean beetle use 3 to 6 ozs. per acre. Do not treat within 1 day of harvest.

BEETS (INCLUDING TOPS): For control of aphids use 1½ to 3 ozs. per acre. For control of cabbage looper, cutworm (climbing), dipterous leafminer (adult), false chinch bug, grasshoppers, imported cabbage worm, leafhoppers, mites and salt-marsh caterpillar use 3 to 6 ozs. per acre. Do not treat within 3 days of harvest.

BROCCOLI, CABBAGE: For control of aphids use 1½ to 3 ozs. per acre. For control of cabbage looper, imported cabbage worm, grasshoppers, leafhoppers, salt-marsh caterpillars, mites, cutworms (climbing), dipterous leafminer (adult) and Lygus bugs use 3 to 6 ozs. per acre. Do not treat within 1 day of harvest. In certain areas where these insects, including aphids, are not adequately controlled at the dosages recommended above, use 12½ ozs. per acre, but do not treat within 3 days of harvest.

BRUSSELS SPROUTS, CAULIFLOWER, COLLARDS, KALE: For control of aphids use 1½ to 3 ozs. per acre. For control of cabbage looper, imported cabbage worm, grasshoppers, leafhoppers, salt-marsh caterpillar, mites, cutworms (climbing), dipterous leafminer (adult) and Lygus bugs use 3 to 6 ozs. per acre. Do not treat within 3 days of harvest. In certain areas where these insects, including aphids, are not adequately controlled at the dosages recommended above, use 12½ ozs. per acre, but do not treat collards and kale with 7 days of harvest.

CARROTS: For control of aphids use 1½ to 3 ozs. per acre. For control of leafhoppers, Lygus bugs, mites, dipterous leafminer (adult), cutworms (climbing) and salt-marsh caterpillar use 3 to 6 ozs. per acre. Do not treat within 2 days of harvest.

CELERY: For control of aphids use 1½ to 3 ozs. per acre. For control of dipterous leafminer (adult), Lygus bug, salt-marsh caterpillar, leafhoppers, cabbage loopers and mites use 3 to 6 ozs. per acre. Do not treat within 3 days of harvest. In certain areas where these insects, including aphids, are not adequately controlled at the dosages recommended above, use 12½ ozs. per acre, but do not treat within 5 days of harvest.

CORN (FIELD, SWEET, AND POPCORN) FOR GRAIN ONLY: For control of aphids, use 1½ to 6 ozs. per acre. Do not treat within 1 day of harvest.

CUCUMBERS (OUTDOOR): For control of aphids use 1½ to 3 ozs. per acre. For control of grasshoppers, leafhoppers, and mites use 3 to 6 ozs. per acre. Do not treat within 1 day of harvest.

EGGPLANT, PEPPERS: For control of aphids use 1½ to 3 ozs. per acre. For control of grasshoppers, leafhoppers and mites use 3 to 6 ozs. per acre. Do not treat within 2 days of harvest.

LETTUCE: For control of aphids use 1½ to 3 ozs. per acre. For control of corn earworm, cutworm (climbing), dipterous leafminer (adult), cabbage looper, imported cabbage worm, grasshoppers, mites, Lygus bugs, salt-marsh caterpillar, false chinch bug and thrips use 3 to 6 ozs. per acre. Do not treat within 2 days of harvest. In certain areas where these insects, including aphids, are not adequately controlled at the dosages recommended above, use 12½ ozs. per acre, but do not treat within 4 days of harvest.

MELONS (INCLUDING CANTALOUPES, HONEYDEW MELONS, MUSKMELONS, WATERMELONS): For control of aphids use 1½ to 3 ozs. per acre. For control of cabbage looper, dipterous leafminer (adult), leafhoppers, Lygus bugs, mites, false chinch bugs, salt-marsh caterpillar, cutworms (climbing), and grasshoppers use 3 to 6 ozs. per acre. To control rindworms (cabbage looper, cutworms, saltmarsh caterpillar, tobacco budworms) on watermelons apply 3 ozs. per acre. Do not treat cantaloupes, honey-dew melons, muskmelons or watermelons within 1 day of harvest.

OKRA: For control of aphids use 1½ to 3 ozs. per acre. For cutworms (climbing), corn earworms, green stink bug, mites and velvet bean caterpillar use 3 to 6 ozs. per acre. Do not treat within 1 day of harvest.

ONIONS (INCLUDING GREEN ONIONS): For control of thrips and cutworms (climbing) use 3 to 6 ozs. per acre. Do not treat within 1 day of harvest.

PEAS (INCLUDING VINES): For control of aphids use 1½ to 3 ozs. per acre. For control of grasshoppers, leafhoppers, mites and cutworms (climbing) use 3 to 6 ozs. per acre. Do not treat within 1 day of harvest.

POTATOES: For control of aphids use 1½ to 3 ozs. per acre. For control of grasshoppers, leafhoppers, mites use 3 ozs. per acre. Do not treat within 1 day of harvest.

SPINACH: For control of aphids use 1½ to 3 ozs. per acre. For control of cabbage looper, imported cabbage worm, grasshoppers, leafhoppers, mites, dipterous leafminer (adult), cutworms (climbing), salt-marsh caterpillar and false chinch bug use 3 to 6 ozs. per acre. Do not treat within 4 days of harvest. In certain areas where insects, including aphids, are not adequately controlled at the dosages recommended above, use 12½ ozs. per acre, but do not treat within 4 days of harvest.

SUMMER SQUASH: For control of aphids use 1½ to 3 ozs. per acre. For control of cabbage looper, dipterous leafminer (adult), leafhoppers, Lygus bugs, mites, false chinch bugs, salt-marsh caterpillar, cutworms (climbing) and grasshoppers use 3 to 6 ozs. per acre. Do not treat within 1 day of harvest.

TOMATOES (OUTDOOR): For control of aphids use 1½ to 3 ozs. per acre. For control of grasshoppers, leafhoppers, mites use 3 to 6 ozs. per acre. Do not treat within 1 day of harvest.

TURNIPS: For control of aphids use 1½ to 3 ozs. per acre. For control of cabbage looper, imported cabbage worm, grasshoppers, leafhoppers, mites, false chinch bug and dipterous leafminer (adult) use 3 to 6 ozs. per acre. Do not treat within 3 days of harvest.

GREENHOUSE

Diluted PHOSDRIN 10.3 WS should be applied in a closed greenhouse (close all doors, window, and ventilators). Lock or barricade all entrances, post warning signs, and take whatever precautions are necessary to prevent unprotected humans and domestic animals from entering the treated area. The operator must wear a full-face mask of a type found adequate for mevinphos (PHOSDRIN Insecticide) protection.

Apply any time that the greenhouse ventilators can remain closed for two hours without endangering crops from high temperatures, such as early morning, cloudy days or late afternoons. If application is made during the day, keep the greenhouse closed tightly for at least two hours. Then ventilate the enclosure thoroughly for an hour before resuming regular work. If application is made in the late afternoon, the house can be kept closed all night. On the following morning, ventilate for one hour to render the air safe for regular work.

LETTUCE (INCLUDING LEAF, BIBB AND ROMAINE): For control of aphids, corn earworms, cutworms (climbing), dipterous leafminer (adult) cabbage looper, imported cabbage worm, grasshoppers, mites, Lygus bugs, salt-marsh caterpillars, false chinch bugs, thrips use 6 to 14⅛ ozs. per 50,000 sq. ft. Do not treat within 10 days of harvest.

NOTICE OF WARRANTY

SHELL CHEMICAL COMPANY MAKES NO WARRANTY OF MERCHANTABILITY, FITNESS FOR ANY PURPOSE, OR OTHERWISE, EXPRESS OR IMPLIED concerning this product or its uses which extend beyond the use of the product under normal conditions in accord with the statements made on this label.

PYDRIN®

INSECTICIDE
2.4 EMULSIBLE CONCENTRATE

RESTRICTED USE PESTICIDE

For retail sale to and use only by Certified Applicators or persons under their direct supervision and only for those uses covered by the Certified Applicator's certification.

ACTIVE INGREDIENT	BY WEIGHT
Cyano (3-phenoxyphenyl) methyl-4-chloro-alpha-(1-methylethyl) benzeneacetate	30%
INERT INGREDIENTS	70%
TOTAL	100%

This product contains 2.4 lbs. of PYDRIN per gallon

EPA Reg. No. 201-401

WARNING

Keep Out of Reach of Children

May be irritating to skin, eyes, nose and throat. Causes substantial but temporary eye injury. Harmful if swallowed.

AVISO

PRECAUCION AL USUARIO: Si usted no lee inglés, no use este producto hasta que le etiqueta haya sido explicada ampliamente.

STATEMENT OF PRACTICAL TREATMENT

IF SWALLOWED, do not induce vomiting. Call a physician. Gastric lavage should be supervised by a physician or the professional staff because of the possible pulmonary damage via aspiration of the solvent.

IF ON THE SKIN, wash with plenty of soap and water immediately. After drying apply vitamin E cream or oil. If vitamin E cream or oil is not available, apply vegetable oil over any irritated areas. Get medical attention if irritation persists.

IF IN EYES, flush with plenty of water. Call a physician.

PRECAUTIONARY STATEMENTS

HAZARDS TO HUMANS—WARNING

Causes substantial but temporary eye injury. Wear goggles while handling. Can cause irritation to skin such as burning and tingling. Do not get in eyes, on skin, or on clothing. Wear protective gloves while handling. Harmful if swallowed or absorbed through skin. If material gets on skin, the affected area should be thoroughly washed with plenty of soap and water immediately. After drying area(s) apply vitamin E cream or oil. If vitamin E cream or oil is not available, apply vegetable oil over irritated area(s). Treatment may be performed repeatedly until relief is achieved. Remove and wash clothing before reuse.

For 24-hour emergency medical assistance, call (713) 473-9461.

ENVIRONMENTAL HAZARDS

This pesticide is extremely toxic to fish. Use with care when applying in areas adjacent to any body of water. Do not apply directly to water. Do not apply when weather conditions favor drift from treated areas. Do not contaminate water by cleaning of equipment or disposal of wastes. Apply this product only as specified on this label.

WILDLIFE PROTECTION IS EVERYONE'S RESPONSIBILITY

This product is highly toxic to bees exposed to direct treatment on blooming crops or weeds. Do not apply PYDRIN or allow it to drift to blooming crops or weeds while bees are actively visiting the treatment area. Additional information may be obtained by consulting your Cooperative Extension Service.

DIRECTIONS FOR USE

It is a violation of Federal Law to use this product in a manner inconsistent with the labeling.

FOR END-USE ONLY. DO NOT REPACKAGE OR REFORMULATE WITHOUT MANUFACTURER'S WRITTEN APPROVAL.

RE-ENTRY STATEMENTS

Do not apply this product in such a manner as to directly or through drift expose workers or other persons. The area being treated must be vacated by unprotected persons.

Do not enter treated areas without protective clothing until sprays have dried.

Because certain states may require more restrictive reentry intervals for various crops treated with this product, consult your State Department of Agriculture for further information.

Written or oral warnings must be given to workers who are expected to be in a treated area or in an area about to be treated with this product. *(Indicate specific oral warnings which inform workers of areas or fields that may not be entered without specific protective clothing, period of time field must be vacated and appropriate actions to take in case of accidental exposure).* When oral warnings are given, warnings shall be given in a language customarily understood by workers. Oral warnings must be given if there is reason to believe that written warnings cannot be understood by workers. Written warnings must include the following information: "WARNING! Area treated with PYDRIN Insecticide on *(date of application)*. Do not enter without appropriate protective clothing until sprays have dried. *(Insert here action to take in case of accidental exposure. See front panel STATEMENT OF PRACTICAL TREATMENT for details.)*"

STORAGE AND DISPOSAL

Do not contaminate water, food or feed by storage or disposal.

STORAGE

Store in a secure, dry and temperate area. Keep container closed when not in use. Do not store near food or feed. Do not use or store around the home. Avoid contact with water. In case of spill or leak, soak up with sand, earth or synthetic absorbent (do not use alkaline absorbents) and dispose of wastes in compliance with local, State and Federal regulations.

PESTICIDE DISPOSAL

Pesticide wastes are toxic. Improper disposal of excess pesticide, spray mixture or rinsate is a violation of Federal Law. If these wastes cannot be disposed of by use according to label instructions, contact your State Pesticide or Environment Control Agency or the Hazardous Waste representative at the nearest EPA Regional Office for guidance.

CONTAINER DISPOSAL

Triple rinse (or equivalent). Then offer for recycling or reconditioning if container reuse is permitted. If container reuse is prohibited, puncture metal containers and dispose of in a sanitary landfill, or by other approved State and local procedures. Dispose of plastic containers in a sanitary landfill, or by incineration if allowed by State and local authorities. Dispose of glass containers in a sanitary landfill, or by other approved State and local procedures.

In case of a significant spill, call CHEMTREC (800) 424-9300 or (713) 473-9461.

GENERAL DIRECTIONS

PYDRIN Insecticide 2.4 Emulsible Concentrate contains 2.4 pounds of active ingredient per gallon. For the application given below, mix the required amount of PYDRIN in sufficient diluent to provide uniform coverage (Refer to Use Tables). PYDRIN may be applied by ground or aerial application equipment. Do not enter treated areas before spray is dry. Do not plant root crops other than those in this label within 12 months after last application. Root crops in this label and all other rotation crops may be planted immediately following last application.

NOTICE OF WARRANTY

SHELL CHEMICAL COMPANY MAKES NO WARRANTY OF MERCHANTABILITY, FITNESS FOR ANY PURPOSE, OR OTHERWISE, EXPRESS OR IMPLIED, concerning this product or its uses which extend beyond the use of the product under normal conditions in accord with the statements made on this label.

Sumitomo Chemical Company, Ltd., holds U.S. Patents for the active ingredient. Sales by Shell Chemical Company of products containing the active ingredient are made under agreements with Sumitomo Chemical Company, Ltd.

Federally Registered Uses through June 1, 1985

This booklet contains new and supplemental uses for this product which may not appear in label booklet packed with product and must be in possession of user at time of application.

COTTON

CROP	INSECT	DOSAGE LB AI/ ACRE	DOSAGE FL OZ/ ACRE	NUMBER ACRES TREATED WITH GALLON	FURTHER USE INSTRUCTIONS	DAYS TO HARVEST
Cotton	Pink Bollworm	0.05–0.1	2⅔–5⅓	48–24	May be applied in water or nonvolatile vegetable oils. When using oil, apply a total spray volume of 1 or more quarts/acre. When diluted with water, apply no less than one gallon per acre by air (min. of 3 gal/a in Arizona and 5 gallons per acre in California) or 4 gallons per acre by ground. Repeat as necessary throughout season to maintain control. For the control of Boll Weevil, a 3–5 day interval between sprays may be required. Apply no more than 2 lbs/a to a crop in a growing season. Do not graze livestock on treated fields or feed treated trash. *Low rate to be used only for light infestations. **Aids in the suppression.	21
	Cotton Leaf Perforator	0.10	5⅓	24		
	Cotton Bollworm	0.08*–0.20	4¼–10⅔	30–12		
	Tobacco Budworm, Lygus Bugs, Cabbage Looper, Beet Armyworm, Boll Weevil, Whitefly**, Cutworms	0.1–0.2	5⅓–10⅔	24–12		

COTTON—TANK MIX APPLICATIONS
READ AND FOLLOW ALL PRECAUTIONS AND RESTRICTIONS ON THE PRODUCT LABELING.

TANK MIX COMBINATION	INSECT	DOSAGE	FURTHER USE INSTRUCTIONS	DAYS TO HARVEST
NUDRIN® 1.8 Methomyl Insecticide Solution	Cotton Bollworm and Tobacco Budworm eggs	PYDRIN (from cotton table) with ½ pt/acre NUDRIN 1.8		21
NUDRIN® 90 Methomyl Insecticide Water Soluble Powder (with adjuvant)	Cotton Bollworm and Tobacco Budworm eggs	PYDRIN (from cotton table) with ⅛ lb/acre NUDRIN 90	Add adjuvant (X-77, Bio-88, Unite, Triton B-1956 or equivalent*) to spray tank, then PYDRIN, then NUDRIN 90; agitate until thoroughly mixed.	21
Azodrin® 5 Water Miscible Insecticide	Thrips	PYDRIN (from cotton table) with 3–5 oz/acre AZODRIN 5	Use the higher rates when insect populations are high.	21
	Whiteflies	PYDRIN (from cotton table) with ⅜–1⅝ pt/acre AZODRIN 5		
	Mites	PYDRIN (from cotton table) with ⅞–1⅝ pt/acre AZODRIN 5		
Chlordimeform: Fundal™ 4EC or Galecron™ 4EC	Cotton Bollworm	Pydrin 0.05–0.075 Chlordimeform 0.125		

* Compatibility of alternate should be determined before use.

Fundal™ is a registered trademark of NOR-AM Chemical Co.

Galecron™ is a registered trademark of Ciba-Geigy Corp.

FIELD CROPS

CROP	INSECT	DOSAGE LB AI/ ACRE	DOSAGE FL OZ/ ACRE	NUMBER ACRES TREATED WITH GALLON	FURTHER USE INSTRUCTIONS	DAYS TO HARVEST
Field Corn					Do not exceed 1.0 lb. a.i. per acre per season.	21
	PYDRIN 2.4 EC may be tank-mixed with the following herbicides when insect populations require control concurrent with the need for weed control. Follow all herbicidal label directions regarding proper herbicide usage. Jar-test compatibility should be conducted before field mixtures are made and applications initiated. ATRAZINE 4L, ATRAZINE 80 WP, Banvel™ 4EC, Bicep™ 4.5L, BLADEX® 4L, BLADEX® 80WP, Dual™ 8E, Lasso ™ 4EC, Paraquat CL, Roundup™					
	Cutworm Grashoppers	0.1–0.2	5⅓–10⅔	24–12	Apply as necessary to maintain control.	21
	Western Bean Cutworm	0.05–0.1	2⅔–5⅓	48–24	Apply as necessary to maintain control. Application must be made before larvae enter the ear.	
	Corn Earworm	0.1–0.2	5⅓–10⅔	24–12	First application should be at or before silking. Subsequent applications should be made at 3–5 day intervals until silking is completed. Repeat applications may be applied if economically damaging populations exist.	
	Stalk Borer Flea Beetles	0.1–0.2	5⅓–10⅔	24–12	Application must be made early in migration from grassy areas to corn, before borers enter the plant. More than one application may be necessary.	
	Corn Rootworm (adult control)	0.1–0.2	5⅓–10⅔	24–12	Spray at first sign of silk feeding. Reinfestation may occur. Should reinfestation occur, repeat as necessary to maintain control.	
	Chinch Bugs	0.1–0.2	5⅓–10⅔	24–12	Over the top sprays, as used for control of ear and stalk infesting corn pests, are not adequate for chich bug control. It is very important that the spray be directed at the base of the plant through the use of drop nozzles or some other mechanism. Repeat as necessary to control migrating chinch bugs.	

CROP	INSECT	DOSAGE LB AI/ ACRE	DOSAGE FL OZ/ ACRE	NUMBER ACRES TREATED WITH GALLON	FURTHER USE INSTRUCTIONS	DAYS TO HARVEST
	European Corn Borer	0.15–0.2	8.0–10⅔	16–12	First brood: spray while eggs are in the blackhead stage or before the larvae enter the whorl. Application by ground equipment is suggested. Good coverage of both upper and lower leaf surfaces is essential. This can be accomplished with drop nozzles over the row and on each side of the corn plant. Multiple applications may be required when egg laying is prolonged or where moderate to heavy populations are present. A higher rate is recommended for moderate to heavy populations. Proper coverage by ground equipment usually requires 20–30 gallons of carrier. Once larvae enter the whorl, foliar sprays will not provide adequate control. Second brood: make applications when sufficient egg masses are found. Spray when eggs are in the blackhead stage or starting to hatch. When egg laying is prolonged or a third generation is present, additional sprays may be required. A higher rate is recommended for moderate to heavy populations. Good coverage above, below, and in the ear zone is essential. This usually requires 2–3 gallons of carrier by air. If ground equipment is used, drop nozzles on each side of the plant will provide best coverage.	
	Southwestern Corn Borer	0.1–0.2	5⅓–10⅔	24–12	Apply as necessary to maintain control. Two or more applications may be required. Higher rates (0.15–0.2 lb. a.i.) are recommended for moderate to heavy populations.	
	Armyworm (True Armyworm)	0.1–0.2	5⅓–10⅔	24–12	Apply as necessary to maintain control.	
	Corn Leaf Aphid Oat Bird-Cherry Aphid	0.1–0.2	5⅓–10⅔	24–12	For optimum results, direct the spray at the aphid population so as to achieve maximum coverage of the exposed insects. Aphids not contacted by the spray, such as in whorls and leaf axils, may not be adequately controlled.	21
	Adult Corn Rootworm, European Corn Borer	Pydrin and Parathions from respective corn labels			PYDRIN 2.4 EC Insecticide may be tank-mixed with Parathion (methyl and/or ethyl) and applied on field corn (and seed corn) where supplemental control of Adult Corn Rootworm and/or European Corn Borer is desired in conjunction with insects controlled by PYDRIN insecticide when used alone. For supplemental control of second brood European Corn Borer, make applications when sufficient egg masses are found. Spray when eggs are in the blackhead stage or starting to hatch. When egg laying is prolonged or a third generation is present, additional sprays may be required. A higher rate is recommended for moderate to heavy populations. Good coverage above, below, and in the ear zone is essential. This usually requires 2–3 gallons of carrier by air. If ground equipment is used, drop nozzles on each side of the plant will provide best coverage. For supplemental control of Adult Corn Rootworm, spray at first sign of silk feeding. Reinfestation may occur. Should reinfestation occur, repeat as necessary to maintain control. Refer to the PYDRIN 2.4 EC and Parathion (methyl or ethyl) labels for appropriate rates of the individual products for controlling the respective insects.	
Seed Corn	For specific insect control recommendations refer to Field Corn (above). Follow directions carefully.				Multiple applications and/or shortened intervals between sprays must be used to insure proper insect control. Do not exceed 1.0 lb a.i. per acre per season.	21

CROP	INSECT	DOSAGE LB AI/ ACRE	DOSAGE FL OZ/ ACRE	NUMBER ACRES TREATED WITH GALLON	FURTHER USE INSTRUCTIONS	DAYS TO HARVEST
Pop Corn	For specific insect control recommendations refer to Field Corn (above). Follow directions carefully.				Multiple applications and/or shortened intervals between sprays must be used to insure proper insect control. Do not exceed 2.0 lbs a.i. per acre per season.	1
Peanuts	Corn Earworm, Rednecked Peanut Worm, Potato Leafhopper, Velvetbean Caterpillar	0.05–0.1	2⅔–5⅓	48–24	Repeat as necessary to maintain control. Do not feed or graze livestock on treated vines. Do not exceed 0.6 lb a.i. per acre per season.	21
	Beet Armyworm, Granulate Cutworm, Grasshoppers, Cutworms	0.1–0.2	5⅓–10⅔	24–12		
	Lesser Cornstalk Borer** Fall armyworm***	0.2	10⅔	12	**Aids in the control ***Aids in control. For heavy infestations use of Nudrin Insecticide is recommended. Please refer to the Nudrin label for use instructions.	
Soybean	Mexican Bean Beetle, Green Cloverworm, Velvetbean Caterpillar, Potato Leafhopper, Saltmarsh Caterpillar Woolybear Caterpillar,	0.05–0.1	2⅔–5⅓	48–24	Repeat as necessary to maintain control. Do not feed or graze livestock on treated plants. Do not exceed 0.8 lb a.i. per acre per season.	21
	Cabbage Looper, Corn Earworm, Southern Green Stinkbug, Bean Leaf Bettle, Cutworms, Three-cornered Alfalfa Hopper, Beet Armyworm, Grasshoppers, Japanese Bettle (adult)	0.1–0.2	5⅓–10⅔	24–12	Repeat as necessary to maintain control. Do not feed or graze livestock on treated plants. Do not exceed 0.8 lb a.i. per acre per season. When applying in nonvolatile vegetable oils use a total spray volume of 1 or more quarts.	
Sugarcane	Sugarcane Borer	0.1–0.2	5⅓–10⅔	24–12	Repeat as necessary to maintain control. Do not exceed 0.8 lb. a.i./acre per season.	
Sunflower	Sunflower Beetle	0.05–0.1	2⅔–5⅓	48–24	Repeat as necessary to maintain control. Do not exceed 0.8lb. a.i./acre per season.	28
	Grasshoppers, Cutworms, Heliothis (complex), Sunflower Moth, Sunflower Seed Weevil, Banded Sunflower Moth, Beet Armyworm, Sunflower Maggot, Leafhoppers, Sunflower Stem Weevil	0.1–0.2	5⅓–10⅔	24–12		

BLADEX® is a registered trademark of Shell Chemical Co.

Banvel™ is a registered trademark of Velsicol Chemical Co.

Bicep™ and Dual™ are registered trademarks of Ciba-Geigy Corp.

Lasso™ and Roundup™ are registered trademarks of Monsanto Co.

FRUITS AND NUTS

DILUTE SPRAY: Apply specified dosage *per 100 gallons* of water in a uniform spray applied to the point of drip with conventional ground equipment. Do not exceed maximum number of gallons per acre indicated.

NOTE: In order to apply the correct amount of PYDRIN insecticide to your orchard you must know the number of gallons of water needed to spray one acre of your trees to the point of drip. If you do not already know this gallonage, you should conduct a test to determine it. If you do not know how to conduct such a test with your equipment, you should ask assistance from your equipment dealer or State Extension specialist.

CONCENTRATE SPRAY: Apply specified dosage *per acre* in no less than 30 gals. of water per acre by ground. When applying PYDRIN by air, consult your Cooperative Extension Service for application guidelines.

CROP	INSECT	DOSAGE: CONCENTRATE SPRAY: LB AI/ ACRE	DOSAGE: CONCENTRATE SPRAY: FL OZ/ ACRE	DOSAGE: DILUTE SPRAY: FL OZ/ 100 GALS	NUMBER ACRES TREATED WITH GALLON	FURTHER USE INSTRUCTIONS	DAYS TO HARVEST
Apples	Oblique Banded Leafroller, Codling Month, Variegated Leafroller, White Apple Leafhopper, Tentiform Leaf Miner, San Jose Scale (fruit infestations only), Apple Aphid, Tufted Apple Budmoth, Plum Curculio, Oriental Fruit Moth, Apple Maggot, Red-Banded Leafroller, Lesser Appleworm, Green Fruitworm, Rosy Apple Aphid, Plant Bugs (Tarnished Plant Bug, Stink Bugs), Periodical Cicada	0.1–0.3	5⅓–16	2⅔–8	24–8	Repeat as necessary to maintain control. Do not feed or graze livestock on treated orchard floors. Do not exceed 2.1 lbs. a.i. per acre per season. For dilute spray apply 200–600 gals per acre, but do not exceed 16 fl. oz. of PYDRIN 2.4 EC per acre per treatment **Plant Bug, Rosy Apple Aphid Control—** Time of application is critical in achieving control. Use prebloom and post bloom spray timings recommended by State Extension Services	21
Peach	Peach Twig Borer, Oriental Fruit Month, Peach Tree Borer, Plum Curculio, Lesser Peach Tree Borer, Periodical Cicada Plant Bugs (Tarnished Plant Bug, Stink Bugs)	0.1–0.3	5⅓–16	2⅔–8	24–8	Repeat as necessary to maintain control. Do not exceed 1.5 lbs. a.i. per acre per season with no more than 1.2 lbs. a.i. per acre per season between bloom and harvest. For dilute spray apply 200–400 gals per acre, but do not exceed 16 fl. oz. of PYDRIN 2.4 EC per acre per treatment. Do not graze livestock on treated orchard floors. **Peach Tree Borer, Lesser Peach Tree Borer control—** Apply as directed trunk and scaffold limb spray. Thorough coverage of trunk and scaffold limbs is required. **Peach Twig Borer, Plum Curculio, Oriental Fruit Moth, Periodical Cicada Control—** Apply by ground sprayer to achieve thorough coverage of all aerial portions of the tree. **Plant Bug Control—** Time of application is critical in achieving control. Use prebloom and post bloom spray timings recommended by State Extension Services.	14
Pear (dormant)	Pear Psylla	0.2–0.4	10⅔–21⅓	8–14	12–6	Apply during dormant to prebloom (white bud) stage only. Do not apply more than 0.8 lb a.i. per acre per season. Do not graze orchard floor. For dilute spray apply 150–250 gals per acre, but do not exceed 21⅓ fl. oz. of PYDRIN 2.4 EC per acre per treatment.	
Pear	Pear Psylla, Codling Moth, Green Fruitworm, Plum Curculio, Leafrollers, Pear Slug, Periodical Cicada	0.1–0.3	5⅓–16	2⅔–8	24–8	Repeat as necessary to maintain control. Do not exceed 1.5 lbs. a.i. per acre per season, nor more than 0.9 lb. a.i. per acre between bloom and harvest. Do not feed or graze livestock on treated orchard floors. For dilute spray apply 200-600 gals. per acre, but do not exceed 16 fl. ozs. of Pydrin 2.4 EC per acre per treatment.	28
Filberts	Filbert Worm, Oblique Banded Leafroller	0.2–0.4	10⅔–21⅓	5⅓–10⅔	12–6	Make first application after emergence of filbert worm moths in early summer. Do not apply a second treatment earlier than three weeks after the first. Do not apply more than 0.8 lb. a.i. per acre per season. For dilute spray, apply 200-400 gals. per acre, but do not exceed 21⅓ fl. oz. Pydrin 2.4 EC per acre per treatment. Do not graze livestock in treated orchards.	21

CROP	INSECT	DOSAGE: CONCENTRATE SPRAY: LB AI/ ACRE	DOSAGE: CONCENTRATE SPRAY: FL OZ/ ACRE	DOSAGE: DILUTE SPRAY: FL OZ/ 100 GALS	NUMBER ACRES TREATED WITH GALLON	FURTHER USE INSTRUCTIONS	DAYS TO HARVEST
Pecans	Pecan Weevil, Hickory Shuckworm, Pecan Nut Casebearer, Pecan Aphids, Pecan Spittlebug	0.1–0.3	5⅓–16	2⅔–8	24–8	Repeat as necessary to maintain control. Do not feed or graze livestock on treated orchard floors. Do not exceed 1.2 lbs. a.i. per acre per season. For dilute spray apply 200-600 gals. per acre, but do not exceed 16 fl. oz. of Pydrin 2.4 EC per acre per treatment.	
	Pecan Stem Phylloxera, Pecan Leaf Phylloxera					Correct timing of spray applications is critical in achieving optimum control of leaf and stem phylloxera. Consult local spray recommendations for correct times of application.	
Walnut	Walnut Aphid, Walnut Husk Fly, Codling Moth, Navel Orangeworm	0.2–0.4	10⅔–21⅓	5⅓	12–6	Repeat as necessary to maintain control. Do not exceed 0.8 lb. a.i. per acre per season. For dilute sprays, apply 200-400 gals. per acre. Do not feed or graze livestock on treated crop floors. NOTE: Use of baits in Walnut Husk Fly sprays is recommended where endorsed by local Agricultural Extension Service. CONCENTRATE SPRAY: Apply specified dosage per acre in no less than 30 gals. of water per acre by ground or 5 gals. of water per acre by air.	

VEGETABLE CROPS

CROP	INSECT	DOSAGE: LB AI/ ACRE	DOSAGE: FL OZ/ ACRE	NUMBER ACRES TREATED WITH GALLON	FURTHER USE INSTRUCTIONS	DAYS TO HARVEST
Artichokes	Artichoke Plume Moth	0.1	5⅓	24	Apply no more than three times per crop. Begin application when larvae first appear with subsequent applications at 14 day intervals. Apply in a minimum of 10 gallons per acre by air and 50–200 gallons per acre by ground (use sufficient water to obtain coverage without excessive runoff).	7
Broccoli, Cabbage, Cauliflower	Diamondback Moth	0.05–0.1	2⅔–5⅓	48–24	Repeat as necessary to maintain control. Do not exceed 1.6 lbs. a.i. per acre per season.	3
	Imported Cabbageworm, Cabbage Looper, Beet Armyworm, Alfalfa Looper, Flea Beetle, Grasshopper, Cutworm,	0.1–0.2	5⅓–10⅔	24–12		
Cucumber, Melons (Cantaloupe, Honeydew Melons, Watermelon, Muskmelon), Pumpkin, Squash (Summer, Winter)	Pickleworm, Rindworms, Leafhoppers, Corn Earworm, Cabbage Looper, Squash Bug, Cutworms (seedling spray), Cucumber Beetle (adult control), Grasshoppers, Plant Bugs (Lygus Bugs, Stink Bugs), Squash Vine Borer	0.1–0.2	5⅓–10⅔	24–12	Repeat as necessary to maintain control. Do not exceed 1.0 lb a.i. per acre per season. Do not apply on zucchini squash, as injury may occur.	3
Dry Beans Dry Peas	Mexican Bean Beetle, Saltmarsh Caterpillar	0.05–0.1	2⅔–5⅓	48–24	Repeat as necessary to maintain control. Do not exceed 0.8 lb a.i. per acre per season. Do not feed or graze livestock on treated vines.	21
	Corn Earworm, Potato Leafhopper, Pea Aphid, Grasshoppers, Pea Weevil, Leafhoppers, Cowpea Curculio, Pea Leaf Weevil, Cutworms, Cabbage Looper, Western Bean Cutworm, Velvetbean Caterpillar, Beet Armyworm, Cucumber	0.1–0.2	5⅓–10⅔	24–12		

CROP	INSECT	DOSAGE LB AI/ ACRE	DOSAGE FL OZ/ ACRE	NUMBER ACRES TREATED WITH GALLON	FURTHER USE INSTRUCTIONS	DAYS TO HARVEST
	Beetle, Corn Rootworm (adult control), Painted Lady Butterfly, (larvae), Green Cloverworm					
Eggplant	Colorado Potato Beetle, Flea Beetles, Corn Earworm, Loopers, European Corn Borer	0.1–0.2	5⅓–10⅔	24–12	Begin application when insects or their damage is observed and repeat at 7 to 10 day intervals to achieve control. Do not exceed 1.4 lbs. a.i. per acre per season.	7
Green Peas	Pea Aphid, Green Cloverworm	0.05–0.1	2⅔–5⅓	48–24	Repeat as necessary to maintain control but do not exceed 0.4 lb ai/acre per season. Do not feed treated pea vines to livestock.	3
	Cabbage Looper, Celery Looper, Alfalfa Looper, Cowpea Curculio, Cutworms, Pea Leaf Weevil, Armyworm, Corn Earworm, Imported Cabbageworm, Alfalfa Caterpillar	0.1–0.2	5⅓–10⅔	24–12		
Peppers	European Corn Borer, Fall Armyworm, Beet Armyworm, Flea Beetles, Corn Earworm, Loopers, Colorado Potato Beetle, Pepper Weevil, Cucumber Beetle (adults), Southern Armyworm	0.1–0.2	5⅓–10⅔	24–12	Begin application when insects or their damage is observed and repeat at 7 to 10 day intervals to achieve control. Do not exceed 1.4 lbs. a.i. per acre per season.	7
Potato	Potato Psyllid, Potato Tuberworm	0.05–0.1	2⅔–5⅓	48–24	Repeat as necessary to maintain control. Do not graze livestock on treated vines. Do not exceed 1.4 lbs. a.i. per acre per season.	7
	Fleabeetles, Tarnished Plant Bug, Cabbage Looper, Potato Aphid, Potato Leafhopper, Cucumber Beetle (adult control), Grasshoppers, European Corn Borer, Beet Armyworm, Western Yellow-striped Armyworm, Buckthorn Aphid, Cutworms	0.1–0.2	5⅓–10⅔	24–12		
	Colorado Potato Beetle West of Mississippi River	0.05–0.1	2⅔–5⅓	48–24		
	East of Mississippi River	0.1–0.2	5⅓–10⅔	24–12		
	Long Island, NY	0.2	10⅔	12		
Sweet Corn	Corn Earworm, Southwestern Corn Borer, European Corn Borer, Beet Armyworm, Flea Beetles, Sap Beetles (adults), Banded Cucumber Beetle, Tarnished Plant Bug	0.1–0.2	5⅓–10⅔	24–12	Do not exceed 2.0 lb a.i. per acre per season. Begin applications either before or at time of silking and continue as necessary to maintain control. For protection against corn borers other than at silking time, treat when necessary to maintain control. For additional information consult directions for use under field corn.	1
	Stalk Borer, Cutworms, Corn Rootworms (adult control), Grasshoppers, Chinch Bugs, Armyworm	0.1–0.2	5⅓–10⅔	24–12	For additional information consult directions for use under field corn.	
	Western Bean Cutworm	0.05–0.1	2⅔–5⅓	48–24		
	Corn Leaf Aphid, Oat Bird-Cherry Aphid	0.1–0.2	5⅓–10⅔	24–12	For optimum results, direct the spray at the aphid population so as to achieve maximum coverage of the exposed insects. Aphids not contacted by the spray, such as in whorls and leaf axils, may not be adequately controlled.	

CROP	INSECT	DOSAGE LB AI/ ACRE	DOSAGE FL OZ/ ACRE	NUMBER ACRES TREATED WITH GALLON	FURTHER USE INSTRUCTIONS	DAYS TO HARVEST
Tomato	Tomato Hornworm Tobacco Hornworm	0.05–0.1	2⅔–5⅓	48–24	Repeat as necessary to maintain control. Do not feed or graze livestock on treated vines. Do not exceed 2.0 lbs. a.i. per acre per season. (Note: Some bleaching or spotting may occur on the foliage of young plants. This does not affect yield or fruit quality.)	1
	Tomato Pinworm, Flea Beetle, Tomato Fruitworm, Colorado Potato Beetle, Cutworms, Cabbage Looper, Southern Armyworm, Sugarbeet Armyworm, Western Yellow-Striped Armyworm, Grasshopper, Whitefly	0.1–0.2	5⅓–10⅔	24–12		
	Vegetable Leafminer	0.2	10⅔	12		

SPECIALTY USES

CROP	INSECT	DOSAGE LB AI/ ACRE	DOSAGE FL OZ/ ACRE	NUMBER ACRES TREATED WITH GALLON	FURTHER USE INSTRUCTIONS
Non-Crop Land (excluding public land such as forests, parks, or recreational areas.)	Grasshoppers Saltmarsh Caterpillar	0.05–0.1	2⅔–5⅓	48–24	Spray non-crop land adjacent to tilled areas to control migrating insects (Grasshoppers, Armyworms) which are a threat to crops. Repeat as necessary to maintain control but do not exceed 2 lbs. active ingredient per acre per year. Do not feed treated crop to livestock. Do not spray ditch banks or areas adjacent to water.
	Armyworms, Army Cutworms Chinch Bugs	0.1–0.2	5⅓–10⅔	24–12	

SITE	INSECT	DOSAGE: HIGH VOLUME SPRAYS LB AI/ 100 GALS	DOSAGE: HIGH VOLUME SPRAYS FL OZ/ 100 GALS	DOSAGE: HIGH VOLUME SPRAYS PYDRIN 2.4 EC/ GAL OF SPRAY	DOSAGE: LOW VOLUME SPRAYS (MISTBLOWER) LB AI/A	FURTHER USE DIRECTIONS
Christmas Tree Plantings, Pine Seed Orchards	Nantucket Pine Tip Moth, Spittlebugs, Pineleaf Chermid, Redheaded Pine Sawfly, Balsam Woolly Adelgid, Pine Chafer, Red Pine Sawfly, Pine Conelet Bug, European Pine Sawfly, Cranberry Girdler (adult control), Pine Needle Midge, Balsam Twig Aphid	0.1–0.2	5⅓–10⅔	⅓–⅔ Teaspoon	0.1–0.2	Apply as needed to maintain control. Spray sufficient gallonage to obtain good coverage of entire tree. High volume sprays on Virginia Pine may cause some yellowing of needles, low volume (Mistblower) applications at the rate of 0.1 to 0.2 lb ai/acre are recommended to avoid this problem.
	Pales Weevil, Northern Pine Weevil	0.1–0.2	5⅓–10⅔	⅓–⅔ Teaspoon	Apply as needed to maintain control. Dilute formulation in kerosene, instead of water, and spray tree stumps to control developing larvae. Use sufficient spray volume to thoroughly cover stump.	
	Coneworms, Seedbugs, Seed Chalcid	11 fl oz/100 gals water for high volume sprayers 3.5 pts./100 gals water for low volume sprayers 0.75 lbs ai/ac/application in not less than 10 gals of water for aerial applications			Apply first application within 1 week of female flower closure or peak pollen flight for Webbing Coneworm control. For other Coneworms and Seedbugs, apply first application within 30 days following female flower closure. Repeat applications at intervals of 4 weeks but do not exceed 6.3 lb ai/ac/yr. For Seed Chalcid, apply when all cones are pendant, and repeat at 1-2 week intervals for 2 or more sprays. Apply approximately 5-10 gals of the 11 fl. oz/100 gal dilution per tree with high volume sprayers. With low volume sprayers apply 100 gallons of the 3.5 pts/100 gal dilution per acre. Do not graze or harvest cover crop. Avoid contact with open water. Some foliar chlorosis may occur on Douglas Fir when concentrated (flow volume) sprays are employed. To avoid this problem use high volume spray.	

CROP	INSECT	DOSAGE LB AL/ 100 GALS.	DOSAGE FL OZ/ 100 GALS.	FURTHER USE INSTRUCTIONS
Ornamentals (Outdoor/ Greenhouse)	Whitefly Webworms Cutworms Bagworms Leafrollers Black Vine Weevil Lacebug Oleander Caterpillar Periodical Cicada	0.1–0.2	5⅓–10⅔	Apply as needed to maintain control. Spray sufficient gallonage to obtain good coverage. For best control of leafrollers and webworms, apply sprays before extensive webbing or leafrolling is evident. NOTE: Spotting of leaves may occur on young growth especially when plants are growing rapidly. Caution should be exercised when treating the following: ornamentals, as localized chlorotic leaf spotting or burning has been observed following treatment: Roses, Poinsettias, Chrysanthemums, Hydrangea, Redbud, Maidenhair fern, Crape Myrtle, Chinese Elm, Desert Gum, Scotch Pine, River Willow, Weeping Willow, and Dogwood.

VENDEX® 4L

MITICIDE

Contains 4 lbs. Vendex per gallon.

ACTIVE INGREDIENTS	BY WEIGHT
Hexakis (2-methyl-2-phenylpropyl)-distannoxane	42.0%
INERT INGREDIENTS	58.0%
TOTAL	100%

EPA Reg. No. 201-412 U.S. Patent No. 3,657,451

Keep Out of Reach of Children

DANGER

STATEMENT OF PRACTICAL TREATMENT

In case of contact, immediately flush eyes or skin with plenty of water for at least 15 minutes. For eyes, call a physician.

See additional precautionary statements.

PELIGRO

PRECAUCION AL USUARIO: Si usted no lee ingles, no use este producto hasta que le etiqueta haya sido explicada ampliamente.

PRECAUTIONARY STATEMENTS

HAZARDS TO HUMANS AND DOMESTIC ANIMALS

DANGER! CORROSIVE. THIS PRODUCT MAY CAUSE SKIN IRRITATION, RESPIRATORY IRRITATION, AND EYE DAMAGE. Do not get in eyes, on skin, or on clothing. Wear suitable eye protection, such as goggles or safety glasses with side shields or their equivalent, when pouring product from container. During commercial application or other prolonged exposure, use goggles and a face mask. Wear clean protective clothing that covers the body well.

Do not graze or feed animals on cover crops grown in treated areas. Avoid contamination of food, feedstuffs, and domestic water supplies.

Harmful if swallowed. Avoid contamination of food. Avoid breathing spray mist. Wash thoroughly with soap and water after handling, spraying, and before eating or smoking.

For 24-hour emergency medical assistance, call (713) 473-9461.

NOTE TO PHYSICIAN: Probable mucosal damage may contraindicate the use of gastric lavage. Measures against circulatory shock, respiratory depression, and convulsion may be needed.

ENVIRONMENTAL HAZARDS

This product is toxic to fish. Keep out of lakes, ponds, and streams. Do not contaminate water by cleaning of equipment or disposal of wastes.

DIRECTIONS FOR USE

It is a violation of Federal law to use this product in a manner inconsistent with its labeling.

SHAKE WELL BEFORE USING

FOR END-USE ONLY, DO NOT REPACKAGE OR REFORMULATE WITHOUT MANUFACTURER'S WRITTEN APPROVAL

VENDEX 4L Miticide is recommended for the control of a wide range of plant feeding mites, including strains that are resistant to other miticides. VENDEX 4L may be applied when honeybees are present and when beneficial mites are important in integrated control programs. Shake VENDEX 4L container to mix contents thoroughly before adding to spray tank. VENDEX 4L Miticide can readily be dispersed in water and can be used in conventional dilute sprayers or concentrate sprayers. Agitation is required during mixing and spraying. Thorough and complete coverage of foliage and fruit is necessary for optimum mite control.

In order to apply the correct amount of VENDEX dilute spray to your orchard, you must know the number of gallons of water needed to spray one acre of your trees to the point of drip. If you do not know this gallonage, you should conduct a test to determine it. Equipment dealers may be of some assistance in conducting these tests.

Apply this product only as recommended on this label by ground applicaton.

STORAGE AND DISPOSAL

STORAGE

Store in a secure, dry and temperate area. Keep container closed when not in use. Do not store near food or feed. Do not use or store around the home. Avoid contact with water. In case of spill or leak, soak up with sand, earth or synthetic absorbent and dispose of wastes in compliance with local, State and Federal regulations.

In case of a significant spill, call CHEMTREC: (800) 424-9300 or (713) 473-9461.

PESTICIDE DISPOSAL

Pesticide wastes are acutely hazardous. Improper disposal of excess pesticide, spray mixture or rinsate is a violation of Federal law. If these wastes cannot be disposed of according to label instructions, contact your State Pesticide or Environmental Control Agency or Hazardous Waste Representative at the nearest EPA Regional Office for guidance.

CONTAINER DISPOSAL

Triple rinse (or equivalent) and offer for recycling or reconditioning, or dispose of in a sanitary landfill or by incineration if permitted by State and local authorities.

RE-ENTRY STATEMENTS

Do not apply this product in such a manner as to directly or through drift expose workers or other persons. The area being treated must be vacated by unprotected persons.

Do not enter treated areas without protective clothing until sprays have dried.

Because certain states may require more restrictive reentry intervals for various crops treated with this product, consult your State Department of Agriculture for further information.

Written or oral warnings must be given to workers who are expected to be in a treated area or in an area about to be treated with this product. (Indicate specific oral warnings which inform workers of areas of fields that may not be entered without specific protective clothing, period of time field must be vacated and appropriate actions to take in case of accidental exposure.)

When oral warnings are given, warnings shall be given in a language customarily understood by workers. Oral warnings must be given if there is reason to believe that written warnings cannot be understood by workers.

Written warnings must include the following information: "DANGER! Area treated with VENDEX Miticide on (date of application). **Do not enter without appropriate protective clothing until sprays have dried.** (Insert here actions to take in case of accidental exposure. See front panel **STATEMENT OF PRACTICAL TREATMENT** for details.)"

NOTICE OF WARRANTY

SHELL CHEMICAL COMPANY MAKES NO WARRANTY OF MECHANTABILITY, FITNESS FOR ANY PURPOSE, OR OTHERWISE EXPRESS OR IMPLIED, concerning this product or its use which extends beyond the use of the product under normal conditions in accord with the statements made on this label.

DOSAGE AND CROP RECOMMENDATIONS

CROP	MITES CONTROLLED	FOR CONCENTRATE SPRAYS VENDEX 4L PER ACRE[1]	FOR DILUTE SPRAYS VENDEX 4L PER 100 GAL.	REMARKS
Apples	Apple Rust Carpini Spider European Red McDaniel Spider Two-Spotted Spider	1 to 3 pints	4 to 8 fl. oz. (¼ to ½ pt.)	(1) Do not apply more than 600 gals. dilute spray per acre. (2) Apply when mites first appear. Repeat as necessary to maintain control. (3) Do not apply more than 4 times per season including a pre-bloom spray; apply this product no more than 3 times between petal fall and harvest. (4) Do not apply within 14 days of harvest.
Pears	European Red McDaniel Spider Pear Rust Two-Spotted Spider	1 to 3 pints	4 to 8 fl. oz. (¼ to ½ pt.)	
Citrus[2]	Citrus Red Citrus Rust Texas Citrus Yuma Spider Two-Spotted Spider	2 to 4 pints[3]	4 to 8 fl. oz. (¼ to ½ pt.)	(1) Do not apply more than 800 gals. dilute spray per acre. (2) Apply when mites first appear. Repeat as necessary to maintain control. (3) Do not apply more than 4 sprays in any 12-month period. (4) Do not apply within 7 days of harvest.
Grapes	Pacific Spider European Red Two-Spotted Spider	1 to 2.5 pints	8 to 16 fl. oz. (½ to 1 pt.)	(1) Do not apply more than 250 gals. per acre. (2) Apply when mites first appear. Repeat as necessary to maintain control but not more often than once every 21 days. (3) Do not apply more than 2 times per season. (4) Do not apply within 28 days of harvest.
Peaches Plums Prunes	European Red McDaniel Spider Pacific Spider Two-Spotted Spider Big Beaked Plum Peach Silver Plum Nursery	1 to 2 pints	4 to 8 fl. oz. (¼ to ½ pt.)	(1) Do not apply more than 400 gals. dilute spray per acre. (2) Apply when mites first appear. Repeat as necessary to maintain control. (3) Do not apply more than 2 times per season. (4) Do not apply within 14 days of harvest.
Cherries (including both sweet and sour varieties)	European Red McDaniel Spider Cherry Rust	1.5 to 3 pints	4 to 8 fl. oz. (¼ to ½ pt.)	(1) Do not apply more than 600 gals. dilute spray per acre. (2) Apply when mites first appear. Repeat as necessary to maintain control. (3) Do not apply more than 2 sprays while fruit is on the trees. (4) Do not apply within 14 days of harvest.
Almonds	Pacific Spider European Red Two-Spotted Spider Hickory Spider	1 to 2.5 pints	4 to 8 fl. oz. (¼ to ½ pt.)	(1) Do not apply more than 500 gals. per acre. (2) Apply when mites first appear. Repeat as necessary to maintain control. (3) Do not apply more than 2 times per season. (4) Do not apply within 14 days of harvest.
Pecans	Pecan Hickory			
Walnuts	Two-Spotted Spider European Red Pacific Spider			
Strawberries	Two-Spotted Spider	1.5 to 2 pints	12 to 16 fl. oz. (¾ to 1 pt.)	(1) Apply in 150–200 gal. per acre. (2) Apply when mites first appear. Repeat as necessary to maintain control but not more than 4 times per season. (3) Do not apply within 1 day of harvest. (4) Good coverage is essential to good mite control. Adjust spray volume and nozzle placement to assure good coverage of top and underside of leaves.

DOSAGE AND CROP RECOMMENDATIONS				
CROP	MITES CONTROLLED	FOR CONCENTRATE SPRAYS VENDEX 4L PER ACRE[1]	FOR DILUTE SPRAYS VENDEX 4L PER 100 GAL.	REMARKS
Greenhouse & Outdoor Ornamentals (including nursery stock, flowers, and plants grown for propagation purposes.)	Spruce Spider *Olingonychus* Two-Spotted Spider		8 to 16 fl. oz. (½ to 1 pt.)	(1) Do not add oil to the spray solution. (2) Apply when mites first appear. Repeat as necessary to maintain control, but not more than once every 28 days. (3) Apply to the foliage only of chrysanthemums (pre-bloom) and poinsettias (prebract).

[1] For Concentrate Sprays—Adjust the dosage to apply an amount not exceeding that used in a dilute spray.
[2] Do not use VENDEX 4L on the following citrus: tangerines, tangelos.
[3] Only use dilute spray applications on lemons.

VENDEX® 50WP

WATER SOLUBLE BAG MITICIDE

ACTIVE INGREDIENTS	BY WEIGHT
Hexakis (2-methyl-:2-phenylpropyl)-distannoxane	50%
INERT INGREDIENTS	50%
TOTAL	100%

EPA Reg No. 201–369 U.S. Patent No. 3,657,451

Keep Out of Reach of Children
DANGER

STATEMENT OF PRACTICAL TREATMENT

If on skin or eyes: Immediately flush eyes or skin with plenty of water for at least 15 minutes. For eyes, call a physician. See additional precautionary statements.

PELIGRO

PRECAUCION AL USUARIO: Si usted no lee ingles, no use este producto hasta que le etiqueta haya sido explicado ampliamente.

CAREFULLY OPEN THIS ENVELOPE AND IMMEDIATELY DROP THE INNER BAG INTO THE SPRAY TANK. DO NOT OPEN OR HANDLE THE INNER BAG. MOISTURE WILL CAUSE BREAKAGE.

PRECAUTIONARY STATEMENTS

HAZARDS TO HUMANS AND DOMESTIC ANIMALS

DANGER! CORROSIVE. THIS PRODUCT MAY CAUSE SKIN IRRITATION, RESPIRATORY IRRITATION, AND EYE DAMAGE. Do not get in eyes, on skin, or on clothing. Wear suitable eye protection, such as goggles or safety glasses with side shields or their equivalent, when handling product. During commercial application or other prolonged exposure, wear goggles. Wear clean protective clothing that covers the body well. Avoid handling the inner bag as moisture will cause breakage.

Do not graze or feed animals on cover crops grown in treated areas. Avoid contamination of food, feedstuffs, and domestic water supplies.

Harmful if swallowed. Avoid contamination of food. Avoid breathing spray mist. Wash thoroughly with soap and water after handling, spraying, and before eating or smoking.

For 24 hour emergency medical assistance, call (713) 473-9461.

NOTE TO PHYSICIAN: Probable mucosal damage may contraindicate the use of gastric lavage. Measures against circulatory shock, respiratory depression, and convulsion may be needed.

ENVIRONMENTAL HAZARDS

This product is toxic to fish. Keep out of lakes, ponds, and streams. Do not contaminate water by cleaning of equipment or disposal of wastes.

DIRECTIONS FOR USE

It is a violation of Federal law to use this product in a manner inconsistent with its labeling.

FOR END-USE ONLY, DO NOT REPACKAGE OR REFORMULATE WITHOUT MANUFACTURER'S WRITTEN APPROVAL.

RE-ENTRY STATEMENTS

Do not apply this product in such a manner as to directly or through drift expose workers or other persons. The area being treated must be vacated by unprotected persons.

Do not enter treated areas without protective clothing until sprays have dried.

Because certain states may require more restrictive reentry intervals for various crops treated with this product, consult your State Department of Agriculture for further information.

Written or oral warnings must be given to workers who are expected to be in a treated area or in an area about to be treated with this product. (*Indicate specific oral warnings which inform workers of areas or fields that may not be entered without specific protective clothing, period of time field must be vacated and appropriate actions to take in case of accidental exposure.*)

When oral warnings are given, warnings shall be given in a language customarily understood by workers. Oral warnings must be given if there is reason to believe that written warnings cannot be understood by workers.

Written warnings must include the following information: "DANGER! Area treated with VENDEX *Miticide* **on** (*date of application*) **Do not enter without appropriate protective clothing until sprays have dried.** (*Insert here actions to take in case of accidental exposure. See front panel* **STATEMENT OF PRACTICAL TREATMENT** *for details.*)"

STORAGE AND DISPOSAL

STORAGE

Store in a secure, dry and temperate area. Keep container closed when not in use. Do not store near food or feed. Do not use or store around the home. Avoid contact with water. In case of spill or leak, soak up with sand, earth or synthetic absorbent and dispose of wastes in compliance with local, State and Federal regulations.

In case of a significant spill, call CHEMTREC (800) 424-9300 or (713) 473-9461.

PESTICIDE DISPOSAL

Pesticide wastes are acutely hazardous. Improper disposal of excess pesticide, spray mixture, or rinsate is a violation of Federal law. If these wastes cannot be disposed of according to label instructions, contact your State Pesticide or Environmental Control Agency or Hazardous Waste Representative at the nearest EPA Regional Office for guidance.

CONTAINER DISPOSAL

Completely empty contents of envelope into application equipment. Then dispose of empty envelope in a sanitary landfill, or, if allowed by State and local authorities, by burning. If burned, stay out of smoke.

VENDEX 50WP Miticide is recommended for the control of a wide range of plant feeding mites, including strains that are resistant to other miticides. VENDEX 50WP may be applied when honeybees and beneficial mites are present. VENDEX 50WP Miticide can readily be dispersed in water and can be used in conventional dilute sprayers or concentrate sprayers. Agitation is required during mixing and spraying. Thorough and complete coverage of foliage and fruit is necessary for optimum mite control.

In order to apply the correct amount of VENDEX dilute spray to your orchard, you must know the number of gallons of water needed to spray one acre of your trees to the point of drip. If you do not know this gallonage, you should conduct a test to determine it. Equipment dealers may be of assistance in conducting these tests.

Apply this product only as recommended on this label by ground application.

NOTICE OF WARRANTY

SHELL CHEMICAL COMPANY MAKES NO WARRANTY OF MERCHANTABILITY, FITNESS FOR ANY PURPOSE, OR OTHERWISE EXPRESS OR IMPLIED, concerning this product and its use which extends beyond the use of the product under normal conditions in accord with the statements made on this label.

DOSAGE AND CROP RECOMMENDATIONS[1]

CROP	MITES CONTROLLED	FOR CONCENTRATE SPRAY VENDEX 50[2]		FOR DILUTE SPRAY VENDEX 50		REMARKS
		POUNDS PER ACRE	WATER SOLUBLE BAGS PER ACRE	OZS. PER 100 GALS.	WATER SOLUBLE BAGS PER 400 GALS.	
Apples	Apple Rust Carpini Spider European Red McDaniel Spider Two-Spotted Spider	1 to 3	1 to 3	4 to 8	1 to 2	(1) Do not apply more than 600 gals. dilute spray per acre. (2) Apply when mites appear. Repeat as necessary to maintain control. (3) Do not apply more than 4 times per season including a pre-bloom spray: apply this product no more than 3 times between petal fall and harvest. (4) Do not apply within 14 days of harvest.
Pears	European Red McDaniel Spider Pear Rust Two-Spotted Spider	1 to 3	1 to 3	4 to 8	1 to 2	
Citrus[3]	Citrus Red Citrus Rust Texas Citrus Two-Spotted Spider Yuma Spider	2 to 4[4]	2 to 4[4]	4 to 8	1 to 2	(1) Do not apply more than 800 gals. dilute spray per acre. (2) Apply when mites appear. Repeat as necessary to maintain control. (3) Do not apply more than 4 sprays in any 12-month period. (4) Do not apply within 7 days of harvest.
Grapes	Pacific Spider European Red Two-Spotted Spider	1 to 2.5	1 to 2.5[5]	8 to 16	2 to 4	(1) Do not apply more than 250 gals. per acre. (2) Apply when mites first appear. Repeat as necessary to maintain control but not more often than once every 21 days. (3) Do not apply more than 2 times per season. (4) Do not apply within 28 days of harvest.
Peaches Plums Prunes	European Red McDaniel Spider Pacific Spider Two-Spotted Spider Big-Beaked Plum Peach Silver Plum Nursery	1 to 2	1 to 2	4 to 8	1 to 2	(1) Do not apply more than 400 gals. dilute spray per acre (600 gals. for cherries). (2) Apply when mites first appear. Repeat as necessary to maintain control. (3) Do not apply more than 2 times per season. (4) Do not apply within 14 days of harvest.
Cherries (including both sweet and sour varieties)	European Red McDaniel Spider Cherry Rust	1.5 to 3	1.5[5] to 3	4 to 8	1 to 2	
Strawberries	Two-Spotted Spider	1.5 to 2	1.5[5] to 2	12 to 16	3 to 4	(1) Apply in 150–200 gals. per acre. (2) Apply when mites appear. Repeat as necessary to maintain control but not more than 4 times per season. (3) Do not apply within 1 day of harvest. (4) Good coverage is essential to good mite control. Adjust spray volume and nozzle placement to assure coverage of top and underside of leaves.
Papayas†	Carmine Citrus Red Texas Citrus Broad Red and Black Flat	—	—	4 to 8	1 to 2	(1) Use only as dilute spray. Do not apply more than 400 gals. dilute spray per acre. (2) Apply when mites first appear. Repeat as necessary to maintain control. (3) Do not apply more than once per month to bearing trees. Do not apply more than 9 times per season. (4) Do not apply within 7 days of harvest.

CROP	MITES CONTROLLED	FOR CONCENTRATE SPRAY VENDEX 50[2]: POUNDS PER ACRE	FOR CONCENTRATE SPRAY VENDEX 50[2]: WATER SOLUBLE BAGS PER ACRE	FOR DILUTE SPRAY VENDEX 50: OZS. PER 100 GALS.	FOR DILUTE SPRAY VENDEX 50: WATER SOLUBLE BAGS PER 400 GALS.	REMARKS
Almonds	Pacific Spider European Red Two-Spotted Spider Hickory Spider	1 to 2.5	1 to 2.5[5]	4 to 8	1 to 2	(1) Do not apply more than 500 gals. per acre. (2) Apply when mites first appear. Repeat as necessary to maintain control. (3) Do not apply more than 2 times per season. (4) Do not apply within 14 days of harvest.
Pecans	Pecan Hickory					
Walnuts	Two Spotted Spider European Red Pacific Spider					
Greenhouse & Outdoor Ornamentals (including nursery stock, flowers and plants grown for propagation purposes)	Spruce Spider *Olingonychus* Two-Spotted Spider	—	—	8 to 16	2 to 4	(1) Do not add oil to the spray solution. (2) Apply when first mites appear. Repeat as necessary to maintain control, but not more than once every 28 days. (3) Apply to the foliage only of chrysanthemums (pre-bloom) and poinsettia (prebract).

[1] This product is packaged in WATER SOLUBLE BAGS—drop required number of inner bags—unopened—into the spray tank. Do *not* handle inner bag: moisture will cause breakage.
[2] For Concentrate Sprays—Adjust the dosage to apply an amount not exceeding that used in a dilute spray.
[3] Do not use VENDEX 50WP on the following citrus: Tangerines and tangelos.
[4] Only use dilute spray applications on lemons.
[5] Fractions are used here for the purpose of conversion only. Do *not* use partial bags.
† Not registered for use in California.

Stauffer Chemical Company

Agricultural Chemical Division
Westport, CT 06881
(203) 222-3000

CAPTAN 4 Flowable

WEST OF ROCKIES
AQUEOUS SUSPENSION
AGRICULTURAL FUNGICIDE

ACTIVE INGREDIENTS:	
CAPTAN *N* - [(trichloromethyl) thio] -4-cyclohexene-1,2-dicarboximide	36.4%
Related derivatives	2.1%
INERT INGREDIENTS:	61.5%
TOTAL	100.0%

Contains 4 Lbs. Active Ingredient Per Gallon

EPA Reg. No. 476–1839 A-2R

CAUTION

Keep Out of Reach of Children

HARMFUL IF TAKEN INTERNALLY

Avoid inhaling spray mist
Avoid prolonged or repeated contact with skin
Avoid contamination of feed and food

PRODUCT FREEZING TEMPERATURE 32°F

Satisfactory utility may generally be restored after thawing and mixing a product which has frozen. Repeated freezing-thawing cycles may impair utility.

PRODUCT INFORMATION

CAPTAN 4 FLOWABLE is an aqueous suspension for use in water as a spray for the control of certain fungus diseases of fruit and vegetables.

DIRECTIONS FOR USE

It is a violation of Federal Law to use this product in a manner inconsistent with its labeling.

BEFORE USING MIX CONTENTS THOROUGHLY—STIR OR ROLL!

Pour recommended amount of this material into nearly filled spray tank. Add balance of water. Maintain agitation during filling and spraying operations. Do not allow mixture to stand. Use a screen not finer than 50 mesh in entire system. Do not combine with emulsifiable liquids or wettable powders unless previous experience has proven them to be physically compatible and safe to plants.

USE PRECAUTIONS

This formulation is not suitable for use as a seed protectant.

Observe use limitations given on the label for specific crops.

In order that pesticidal residues on food and forage crops will not exceed tolerances established by the Federal Food and Drug Administration, use only at recommended rates and intervals. Do not apply or allow to drift onto adjoining food, fiber or pasture crops. The grower is responsible for residues on his crops as well as for damage caused by drift from his property to that of others.

Consult State Agricultural Experiment Stations or the State Agricultural Extension Service for additional information, as the timing of applications needed will vary with local conditions.

This product is toxic to fish. Keep out of lakes, streams and ponds.

RECOMMENDATIONS

Unless otherwise stated, dosage rates on this label are given in terms of CAPTAN 4 FLOWABLE per 100 gals. of water.

ALMONDS: Brown rot, scab, shot hole—Apply 1–1½ qts. at popcorn, bloom and petal fall stages.

APPLES (Western Areas) Apple Scab—CAPTAN is most effective when used in a protective schedule where the spray is applied before infection occurs. Primary infection of apple scab may occur at any time during and after the green tip stage if over-wintering scab is mature and there is sufficient moisture to discharge spores. Thorough coverage is necessary for best results, especially in the pre-pink and pink stage spray applications. Danger from primary infection may extend through the time of the first cover spray. In pre-blossom sprays, use 1 qt. per 100 gallons of water in high volume application. In low volume sprays, apply 2½ gals. per acre. In mid-summer, the dosage may be reduced to ½ qt. per 100 gal. in high volume application, or 1¼ gals. per acre in low volume sprays. Bitter rot, black rot, Botryosphaeria, black pox, bull's eye rot: use 1 qt. per 100 gals. Make 1 or 2 applications with late cover sprays and 1 or 2 preharvest applications.

APRICOTS: Brown rot (twig blight), Jacket Rot—Use 1 qt. Apply at bloom, red bud stage and repeat at 75 percent petal fall.

CHERRIES (Western Areas): Brown rot, cherry leaf spot—Use 1 qt. Apply at pre-bloom, bloom and petal fall. For severe infection, use 2 qts. Repeat in cover sprays.

CITRUS (Grapefruit, Lemon, Lime, Orange): Brown Rot—Use 1 to 2 qts. Apply to skirts of trees prior to winter rains. Wet tree thoroughly up to 3 to 4 ft. Repeat application after 10 weeks. CAUTION: Do not feed citrus by-products to dairy animals or animals being finished for slaughter.

GRAPES (Western Areas): Grape Dead Arm Disease in California—Use 2 qts. Apply a minimum of 200 gals. of spray per acre. Start when new shoots are 3–4 inches long and before the first infective rains. Make second spray when shoots are 12–15 inches long, and a third spray before the bunches close. **Black Rot**—Use 1 qt. Make two sprays before bloom and just after bloom. Follow with at least 3 cover sprays before the bunches close.

NECTARINES, PEACHES (Western Areas): Coryneum Blight (peach blight), brown rot, scab—Use 1 qt. Apply at pink bud, full bloom and petal-fall stages. Repeat at 10–14 day intervals after petal-fall. Also, make late applications for Coryneum Blight.

PLUMS, PRUNES (Western Areas): Brown Rot—Use 1 qt. Apply at red bud or popcorn stage. Repeat at bloom and 75% petal-fall stages. Also repeat later in season as conditions warrant.

STRAWBERRIES: BOTRYTIS FRUIT ROT—Use 1–1½ qts. Apply 200 gals. of spray per acre or an equivalent of 2–3 qts. per acre in sufficient water for even coverage. Make first application when new growth starts in the spring before the fruit starts to form. Repeat at weekly intervals. Under severe infection conditions make continuous applications through harvest period, treating immediately after each picking.

ASPARAGUS PREPLANTING ROOT DIP (For Damp-Off Control)—1½ qts. per 100 gal. water (1 tablespoonful per gal.) Dip root sections for 1 minute, drain and plant.

BEANS, PEAS: Damping-off (Beans only)—**Root Rot Fungi**—Apply 5 to 6 qts. per acre as a broadcast treatment, working into top 3 to 4 inches of soil or apply 2½ to 3 qts. per acre in the row at time of planting.

CANTALOUPES, CUCUMBERS, PUMPKINS, SUMMER AND WINTER SQUASH, WATERMELONS: Angular Leaf Spot, Anthracnose, Downy Mildew— Use 1 qt. per 100 gals. water, or apply 2 qts. per acre in sufficient water for good coverage. Apply at first sign of disease and repeat at 5–7 day intervals or as needed to maintain disease control.

CARROTS: Septoria Leaf Spot, Cercospora Blight—Use 1 qt. per 100 gals. water and apply the equivalent of 2 qts. per acre. Apply at first sign of disease and repeat at 6–10 day intervals.

POTATO DISEASES: Early and Late Blight—Use 1 qt. 100 gals. water, or apply 1½–2 qts. per acre in sufficient water for good coverage. Severe Infection Conditions: Use 2 qts. per 100 gals. or apply 2½–3 qts. per acre in sufficient water for good coverage. Apply at first sign of infection and repeat at 5–7 day intervals or as needed to maintain control.

SPINACH (Direct Seeded)—**Seed Furrow Application (For Damping-off Control)**—Use 2½ to 3½ qts. in 30 gal. water per acre. Direct the suspension into the seed furrow over the seed. Close and pack the furrow in the usual manner. May be used throughout the year.

TOMATOES: Anthracnose, Early and Late Blight, Gray Leaf Spot, Septoria Leaf Spot—Use 1 qt. per 100 gals. of water, or apply 2–3 qts. per acre in sufficient water to provide good coverage. Severe Infection Conditions: Use 2 qts. per 100 gals. of water, or apply 3–4 qts. per acre with adequate water. Apply at first sign of infestation and repeat at 5–7 day intervals or as needed to maintain control.

POST-HARVEST DIP: Fruit & Vegetables: CAPTAN can be used as a post-harvest dip or wash on the following crops for prevention of various rots and molds (Rhizopus, Botrytis, Gloeosporium) during storage: Apples, apricots, cantaloupes, cucumbers, grapefruit, lemons, limes, mangoes, nectarines, onions (green and bulb), oranges, peaches, pears, potatoes, tangeloes and tangerines. Dilution Rate: Use 1¼ qts. per 100 gals. water. Apply as a spray or in dip tank. When used as a dip, recharge wash solution periodically as volume in tank is depleted. When tank volume is reduced 25%, bring water back to volume adding ⅜ qt. CAPTAN 4 Flowable for every 25 gals. of water added. At end of every 8-hour shift, empty tank, flush and charge with fresh dilution. Do not allow dip tank solution to stand overnight. Maintain continuous agitation during dipping operations. **CAUTION:** Do not feed mango or citrus by-products to dairy animals or animals being finished for slaughter.

STORAGE AND DISPOSAL

PROHIBITIONS
Do not contaminate water, food or feed by storage or disposal.—Open dumping is prohibited.—Do not reuse empty containers.

PESTICIDE DISPOSAL
Wastes resulting from the use of this product may be disposed of on site or at an approved waste disposal facility.

CONTAINER DISPOSAL
Triple rinse (or equivalent). Then offer for recycling or reconditioning, or puncture and dispose of in a sanitary landfill, or by incineration, or, if allowed by State and local authorities, by burning. If burned, stay out of smoke.

NOTICE—READ CAREFULLY

CONDITIONS OF SALE
Stauffer (and seller) offer(s) this product for sale subject to, and buyer and all users are deemed to have accepted, the following conditions of sale and warranty which may only be varied by written agreement of a duly authorized representative of Stauffer.

WARRANTY LIMITATION
Stauffer warrants that this product conforms to the chemical description on the label and is reasonably fit for the purposes referred to in the directions for use on the label subject to the inherent risks referred to below. Stauffer makes no other express warranties: THERE IS NO IMPLIED WARRANTY OF MERCHANTABILITY and there are no warranties which extend beyond the description on the label hereof.

INHERENT RISKS
The directions for use of this product are believed to be reliable and should be followed carefully. However, it is impossible to eliminate all risks associated with use. Buyer assumes all risks associated with use or application of this product contrary to label instructions or resulting from extraordinary weather conditions.

LIMITATION OF LIABILITY
In no case shall Stauffer be liable for special, indirect, or consequential damages resulting from the use or handling of this product and no claim of any kind shall be greater in amount than the purchase price of the product in respect of which such damages are claimed.

This product is sold only for uses stated on the label. No express or implied license is granted to use or sell this product under any patent in any country except as specified: Country: U.S.A.

This is a specimen label. Stauffer is not responsible for the accuracy of the information contained herein. As labels are subject to revision, always carefully read and follow the label on the product container.

CAPTAN 50-WP

EAST OF ROCKIES FUNGICIDE

For Plant Disease Control

ACTIVE INGREDIENTS:	
CAPTAN: *N*-[(trichloromethyl) thio]-4-cyclohexene-1,2-dicarboximide	47.3%
Related derivatives	2.7%
INERT INGREDIENTS:	50.0%
TOTAL	100.0%

EPA Reg. No. 476-581 A-3RE

CAPTAN 50-WP is a microfine wettable powder for use in water as a spray for the control of certain fungus diseases of fruit, vegetables, and ornamental crops, and as a soil treatment for the control of certain seed rots and damping-off diseases.

CAUTION

Keep Out of Reach of Children

CAUSES SKIN AND EYE IRRITATION.

MAY BE HARMFUL IF SWALLOWED OR INHALED.

Avoid contact with skin, eyes or clothing.
Avoid breathing dust. Use with adequate ventilation.
Wash thoroughly after handling.
Do not store near food or feed.

FIRST AID: In case of eye contact immediately flush eyes with plenty of water for at least 15 minutes. Call a physician. Wash skin with soap and water.

STORE IN COOL, DRY PLACE—PROTECT FROM EXCESSIVE HEAT.

OBSERVE THE FOLLOWING USE PRECAUTIONS

Read all precautions and directions for use before using. Use only for claims listed and only as specified on this label.

In order that pesticide residues on food and forage crops will not exceed federal tolerances, use only at recommended rates and intervals, and do not apply closer to harvest than specified. Do not apply or allow to drift to adjoining food, fiber or pasture crops.

Consult State Agricultural Experiment Stations or State Agricultural Extension Service for additional information, as the timing of applications needed will vary with local conditions.

FISH CAUTIONS: This product is toxic to fish. Do not contaminate any body of water by cleaning of equipment, disposal of wastes or applications where runoff is likely to occur. Do not apply when weather conditions favor drift from areas treated.

COMPATIBILITY AND PLANT SAFETY: CAPTAN 50-WP can be combined safely and effectively at recommended dosage rates with most commonly used fungicides and insecticides, with the exception of oil and strongly alkaline materials. Alkaline materials such as spray lime, lime-sulfur and bordeaux mixture will reduce the fungicidal activity of CAPTAN 50-WP. Do not apply CAPTAN 50-WP in combination with or immediately before or closely following oil sprays. The time factor governing the safe interval between

CAPTAN 50-WP and oil sprays varies due to general climatic conditions, therefore, consult local agricultural spray programs and authorities to determine the proper timing. The use of spreaders which cause excessive wetting is not advised. Combinations with solvent formulations of organic phosphates should not be used. Combinations of CAPTAN 50-WP and sulfur should not be used on crops sensitive to sulfur. Used at high rates or in drenching sprays, CAPTAN 50-WP may cause a necrotic spotting of tender, immature leaves of certain varieties of apples, peaches, plums and cherries. This type of injury is most likely to occur in the early cover sprays during long periods of warm, cloudy, humid weather. To avoid the hazard of leaf spotting under such conditions, use CAPTAN 50-WP and other spray materials at lowest recommended rates and avoid drenching trees.

DIRECTIONS FOR USE

It is a violation of Federal law to use this product in a manner inconsistent with its labeling.

Applications can be made by aircraft or ground power equipment (including concentrate and semi-concentrate equipment). Pour recommended amount of this material into nearly filled spray tank. Add balance of water. Maintain agitation during filling and spraying operations. Do not allow mixture to stand. Do not combine with emulsifiable liquids or wettable powders unless previous experience has proven them to be physically compatible and safe to plants. (Read compatibility and plant safety information given under USE PRECAUTIONS.)

RECOMMENDATIONS

Unless otherwise specified, rates are given in terms of CAPTAN 50-WP per 100 gallons water for fruit and nut crops and ornamentals. For vegetable crops, rates are expressed as CAPTAN 50-WP per acre in sufficient water to provide thorough coverage. Except as specified begin applications before or at first sign of disease and repeat as needed to maintain control but observe use limitations. Unless otherwise specified, application can be made on the day of harvest.

Maximum permissible rate per acre, per application as CAPTAN 50-WP is given in parenthesis, ( ), after crop claim.

FRUIT AND NUT CROPS

IMPORTANT: Read label carefully. Although most of the directions on this label may be followed nationwide, a few are limited to either the eastern or western U.S. Follow those directions for your growing area where specified.

ALMONDS: Brown rot twig and blossom blight, shothole, scab, leaf blight—Use 2 to3 pounds. Apply at popcorn, bloom, and petal fall stages and up to 5 weeks after petal fall. Hulls may be fed to livestock. (32 pounds)

Subsequent applications may be made up to 12 days of harvest in accordance with pest infestations and State Agricultural Extension Service recommendations, provided the hulls are not fed to dairy animals or animals being finished for slaughter. (32 pounds)

APPLES (eastern U.S.): Primary scab, black rot (frogeye), botrytis blossom-end rot—1½ to 2 pounds at 5 to 7 day intervals as needed to maintain cover in prebloom, bloom, petal fall and first cover sprays. Use at 2 pound rate if conditions favor scab or if active scab can be found. Secondary scab, Brooks fruit spot, sooty blotch, fly speck, black rot, black pox, botryosphaeria rot, bitter rot—1 to 2 pounds at 10 to 14 day intervals in second and later cover sprays up to start of harvest. (16 pounds) Powdery mildew—If powdery mildew is a problem add 1⅓ to 3 pounds sulfur or ¼ to ½ pound Karathane to all pre-bloom and bloom sprays and ¼ to ½ pound Karathane to all post bloom sprays until the foliage matures. Use ½ pound Karathane if interval between sprays is longer than 7 days. **Caution**—Do no use CAPTAN in combination with or closely following or in alternation with wettable suflur products on sulfur sensitive varieties of apples such as Red Delicious, Staymen, Baldwin, King, etc., as severe injury and defoliation may occur.

APRICOTS: Brown rot (twig blight), jacket rot—Use 2 pounds. Apply in red bud, bloom, 75% petal fall, and cover stages. (10 pounds)

BLACKBERRY, DEWBERRY, RASPBERRY (eastern U.S.): Anthracnose—Use 4 pounds per acre in sufficient water for thorough coverage in the preblossom (when new canes are 6 to 8 inches long) and first cover (right after petal fall) sprays. **Raspberry**—Botrytis fruit rot—Use 2 pounds per 100 gallons water. Begin application in the first bloom. Make 2 or more repeat applications (as field conditions warrant) at 10 to 15 day intervals. (6 pounds)

CHERRIES (eastern U.S.): Brown rot, leaf spot, Botrytis rot—Use 2 pounds in pre-bloom, bloom, petal fall, shuck, cover and preharvest sprays. Applications at 3 to 4 day intervals may be necessary during bloom to control blossom blight. Repeat applications at 7 to 10 day intervals as needed to maintain cover up to start of harvest. If powdery mildew is a problem add 3 pounds sulfur to the petal fall, shuck and early cover sprays. If sulfur is added, CAPTAN may be reduced to 1 pound in these sprays. Post harvest sprays—Apply 2 pounds immediately after harvest for leaf spot. Repeat application in 10 to 14 days. (20 pounds)

CITRUS (Grapefruit, Lemons, Oranges, Tangelos, Tangerines): Brown rot (Phytophthora)—Use 2 to 4 pounds. Apply to skirts of trees before winter rains. Wet tree thoroughly to height of 3 to 4 feet. Repeat 10 to 12 weeks later (20 pounds). Melanose, scab (oranges and tangelos)—Use 2 pounds plus spray sticker. Apply as necessary from ⅔ petal fall until fruit is ½ inch in diameter. Do not apply after fruit size exceeds ½ inch in diameter. Do not feed raw by-products to livestock. (10 pounds)

GRAPES (eastern U.S.): Black rot, downy mildew—2 pounds per 100 gallons, make first application just before bloom, repeat just after bloom and at 10 to 14 day intervals for 1 to 3 more applications depending upon the susceptibility of the grape variety. **Dead Arm** (current season infection) Use 2 to 4 pounds per 100 gallons dilute spray. Apply 50 gallons per acre when shoots are ½ to 1 inch long. Repeat application when shoots are 4 to 6 inches long. (10 pounds)

MANGOES: Cercospora leaf spot or blotch—Use 2 pounds. Begin at first sign of disease and repeat at weekly intervals, to maintain control. (20 pounds)

PEACHES, NECTARINES: Brown rot, scab—Use 2 pounds in full pink, bloom, petal fall, shuck, cover and preharvest sprays. Applications at 3 to 4 day intervals may be necessary during bloom to control blossom blight. Repeat application at 7 to 14 day intervals as needed to maintain cover. Continue applications through harvest if conditions favor brown rot. If powdery mildew is a problem, add 3 pounds sulfur to the petal fall, shuck and early cover sprays. If sulfur is added, CAPTAN may be reduced to 1 pound in these sprays. Coryneum blight (peach blight, shothole)—Use 2 pounds, apply in pink bud, full bloom, petal fall and cover periods as necessary and as a post harvest spray (but before leaves drop). (10 pounds)

PEARS (eastern U.S.): Scab—Use 2 pounds in preblossom, petal fall and cover sprays. Repeat cover applications at 7 to 14 day intervals under severe scab conditions. Russeting may be reduced on Bosc pears. Do not use on D'Anjou pears. (16 pounds)

PINEAPPLES: Heart rot (Phytophthora parasitica)—Use 10 pounds. Dip slips into solution just prior to planting. Root rot (Phytophthora cinnamonii)—Use 4 pounds. Spray immediately after planting. Repeat at 2 week intervals (20 pounds). **Post Harvest Dip:** Storage and transit rots—Use 10 pounds. Dip or wash harvested fruit.

PLUMS, PRUNES (eastern U.S.): Brown rot—Use 2 pounds in full pink, bloom and petal fall sprays. Repeat applications at 7 to 14 day intervals as needed to maintain cover. Continue applications through harvest if conditions favor brown rot. The addition of a neutral spreader has improved coverage. (10 pounds)

QUINCE: Brown rot, scab—Use 2 pounds. Begin when disease threatens. Repeat at 7 to 10 day intervals. Do not apply within 7 days of harvest. (10 pounds)

STRAWBERRIES: Botrytis (gray mold), leaf spot—Apply 4 to 6 pounds per acre in sufficient water for thorough coverage. Begin applications when new growth starts in the spring and before fruit starts to form. Repeat at weekly intervals. Under conditions favorable to fruit rot continue applications through harvest period treating immediately after each picking. (8 pounds)

VEGETABLE CROPS

ASPARAGUS (PREPLANTING ROOT DIP): Botrytis, root rots, phoma rot, penicillium rot, fusarium—Use 3 pounds per 100 gallons water (4 tablespoonfuls per gallon). Dip root sections for 1 minute, drain and plant.

BEANS, PEAS: Damping off (beans only), fungus root rot—Use 10 to 12 pounds per acre as a broadcast treatment, working into top 3 to 4 inches of soil, or use 5 to 6 pounds per acre in the row at time of planting.

CANTALOUPE, CUCUMBER, PUMPKIN, SUMMER AND WINTER SQUASH, HONEYDEW MELONS, WATERMELON: Angular leaf spot, anthracnose—Use 4 pounds per acre in sufficient water for thorough coverage. Begin applications at first signs of infection and repeat at 5 to 7 day intervals or as needed to maintain control. (7.6 pounds)

CARROTS: Cercospora blights, septoria leaf spot—Use 2 pounds per 100 gallons water and apply the equivalent of 4 pounds CAPTAN 50-WP per acre. Apply at first sign of disease and repeat at 6 to 10 day intervals. (6 pounds)

CELERY: Damping-off, pink rot—Use 2 pounds per 100 gallons water. Apply 450 gallons spray per acre to plant beds when all seeds have germinated. Repeat as necessary. (13 pounds) Late blight (septoria), pink rot (sclerotinia)—Use 2 pounds per 100 gallons water. Apply 200 gallons spray per acre. Spraying should begin in the seedbed and be continued at 7 to 10 day intervals. (4 pounds)

EGGPLANT: Phomoposis blight, fruit rot, anthracnose—Use 2 pounds. Begin when plants emerge in plant bed or when disease first appears in field. Repeat at 3 to 7 day intervals through growing season. **Damping-off**—Use 10 to 12 pounds per acre in sufficient water for thorough coverage. Begin when all seeds have germinated. Repeat at 10 day intervals. (6 pounds as foliar spray; 12 pounds in plant beds before transplanting)

ONIONS (GREEN AND BULB): Downy mildew, purple blotch —Use 2 pounds. Apply at first sign of infection and repeat at 7 to 10 day intervals. (3 pounds)

PEPPERS, PIMENTOS: Anthracnose, cercospora (frogeye leaf spot and stem end rot)—Use 3 pounds. Apply at 3 to 7 day intervals throughout the growing season. (6 pounds) **Damping-off**—Use 2 pounds. Plant bed treatment. Begin when all seeds have germinated. Repeat at 10 day intervals. (12 pounds)

TOMATO: Anthracnose, gray leaf spot, septoria leaf spot—Use 4 to 6 pounds per acre in sufficient water for thorough coverage. Under severe infection conditions use 6 to 8 pounds per acre in sufficient water for thorough coverage. Apply at first signs of infection and repeat at 5 to 7 day intervals or as needed to maintain control. (10 pounds)

SPECIAL USES

PREPLANTING SOIL TREATMENT: For control of damping-off and root rots for the crops listed below, apply spray broadcast to soil surface as a preplanting application and cultivate into upper 3 to 4 inches of soil. Or apply to furrow over the seed when specified. (14,500 linear feet of row equals 1 acre). Close and pack the furrow in usual manner. Apply dosage rate in sufficient water for thorough soil coverage unless otherwise specified.

BEANS, PEAS: Use 10 to 12 pounds broadcast or 5 to 6 pounds in furrow. (12 pounds)

BEETS: Use 2 pounds per 100 gallons water; 15 gallons spray per 1,000 square feet. (13 pounds)

BROCCOLI, BRUSSELS SPROUTS, CABBAGE, CAULIFLOWER, COLLARDS, KALE, KOHLRABI, MUSTARD, RUTABAGAS, TURNIPS: Damping-off only—Use 15 pounds broadcast. (28 pounds)

CANTALOUPES, CUCUMBERS, HONEYDEW MELONS, PUMPKINS, SQUASH (SUMMER AND WINTER), WATERMELONS: Use 2 pounds per 100 gallons water; 600 gallons spray per acre broadcast. (12 pounds)

SPINACH: (Direct seeded) Use 5 to 7 pounds in 30 gallons water in furrow. (7 pounds)

POST HARVEST DIP: For control of various molds and storage rots (Botrytis, Gleosporium, Rhizopus). Use as a post harvest dip or wash on the following fruits and vegetables: **Apples, Apricots, Cantaloupes, Cherries, Cucumbers, Grapefruit, Lemons, Limes, Mangoes, Nectarines, Onions (Green and Bulb), Oranges, Peaches, Pears, Potatoes, Tangelos, Tangarines**—Use 2½ pounds. Apply as a spray or in dip tank. When used as a dip, recharge wash solution periodically when tank volume is reduced by 25%. Bring water back to volume and add 1 pound for each 25 gallons added. At end of every 8 hour shift, empty tank, flush and charge with fresh dilution. Do not allow dip tank solution to stand overnight. Maintain continuous agitation during dipping operation. **Caution:** Do not feed mango or citrus by-products to dairy animals or animals being finished for slaughter.

PACKING BOX TREATMENT: For reduction of storage rots and molds (Rhizopus, Botrytis, Gleosporium) treat packing boxes with a solution of 2 pounds per 100 gallons water. Spray boxes thoroughly or use as a dip. Keep solution agitated continuously. Drain, flush and recharge the dip tank with new solution after each 8 hour shift.

ORNAMENTALS

AZALEAS: Damping-off of cuttings—Use 4 pounds. Dip cuttings before bedding. Petal blight—Use 2 pounds. Apply to soil area around plants and spray flowers just before bloom. Repeat at 7 to 14 day intervals through bloom.

BEGONIAS (TUBEROUS): Damping-off, tuber rot—Use 4 pounds. Dip tubers 30 minutes, drain and plant.

CAMELLIAS: Petal blight—Use 1 pound. Apply to drench soil around plants beginning when flowers start to open. Repeat at 7 to 10 day intervals through bloom.

CARNATIONS: Alternaria leaf spot, rust—Use 2 pounds. Begin at first sign of disease. Repeat at 7 to 10 day intervals. Shorten intervals during frequent rains and heavy dews. Damping-off of cuttings —Use 4 pounds. Dip cuttings before bedding.

CHRYSANTHEMUM: Botrytis flower blight, Septoria leaf spot —Use 2 pounds. Apply at first sign of disease. Repeat at 7 to 10 day intervals. Damping-off of cuttings—Use 4 pounds. Dip cuttings before bedding.

GLADIOLUS (CORMS): Corm rot and decay, damping-off — Use ½ to 1½ pounds per 10 gallons, dip corms 20 to 30 minutes. Drain and plant.

GRASSES (ORNAMENTAL IN NON-PASTURED AREAS ONLY): Leaf spot, damping-off, brown patch, melting out, seedling blights, and brown spot on St. Augustine grass—Use 2 pounds. Apply 10 gallons spray per 1,000 square feet. Begin when growth starts in spring. Repeat at 7 to 14 day intervals throughout season. Do not graze treated areas or feed clippings to livestock.

GRASSES (LAWN SEEDBEDS): Damping-off and other soil borne diseases—Use ¼ to ⅓ pound per 1,000 square feet or 2 pounds in 100 gallons water using 15 gallons spray per 1,000 square feet. Cultivate into upper 3 to 4 inches before planting.

ROSES: Black spot, Botrytis blossom blight—Use 2 pounds. Begin at first growth or first sign of disease. Repeat at 7 to 14 day intervals, and more frequently during frequent rains and heavy dews.

SOIL AND GREENHOUSE BENCH TREATMENT: Preplanting for damping-off and root rot diseases on seedlings or transplants of roses (and other shrubs, trees, flowers). Lawn seedbeds, vegetables (such as beans, celery, crucifers, eggplant, peas, peppers, tomatoes)—Use 2 pounds per 100 gallons water at rate of 15 gallons per 1,000 square feet. Cultivate into upper 3 to 4 inches of soil before planting.

STORAGE AND DISPOSAL

PROHIBITIONS
Do not contaminate water, food, or feed by storage or disposal. Open dumping is prohibited.

PESTICIDE DISPOSALS
Wastes resulting from the use of this product may be disposed of on site or at an approved waste disposal facility.

CONTAINER DISPOSAL
Completely empty bag into application equipment. Then dispose of empty bag in a sanitary landfill or by incineration, or, if allowed by State and local authorities, by burning. If burned, stay out of smoke.

NOTICE —READ CAREFULLY

CONDITIONS OF SALE
Stauffer (and seller) offer(s) this product for sale subject to, and buyer and all users are deemed to have accepted, the following conditions of sale and warranty which may only be varied by written agreement of a duly authorized representative of Stauffer.

WARRANTY LIMITATION
Stauffer warrants that this product conforms to the chemical description on the label and is reasonably fit for the purposes referred to in the directions for use on the label subject to the inherent risks referred to below. Stauffer makes no other express warranties; THERE IS NO IMPLIED WARRANTY OF MERCHANTABILITY and there are no warranties which extend beyond the description on the label hereof.

INHERENT RISKS
The directions for use of this product are believed to be reliable and should be followed carefully. However, it is impossible to eliminate all risks associated with use. Buyer assumes all risks associated with use or application of this product contrary to label instructions or resulting from extraordinary weather conditions.

LIMITATION OF LIABILITY
In no case shall Stauffer be liable for special indirect, or consequential damages resulting from the use or handling of this product and no claim of any kind shall be greater in amount than the purchase price of the product in respect of which such damages are claimed.

This product is sold only for uses stated on the label. No express or implied license is granted to use or sell this product under any patent in any country except as specified: Country: United States of America.

This is a specimen label. Stauffer is not responsible for the accuracy of the information contained herein. As labels are subject to revision, always carefully read and follow the label on the product container.

CAPTAN 50-WP

WEST OF ROCKIES FUNGICIDE

For Plant Disease Control

ACTIVE INGREDIENTS:	
CAPTAN: *N*-[(trichloromethyl) thio]-4-cyclohexene-1,2-dicarboximide	47.3%
Related derivatives	2.7%
INERT INGREDIENTS:	50.0%
TOTAL	100.0%

EPA Reg. No. 476-581 A-3RW

CAPTAN 50-WP is a microfine wettable powder for use in water as a spray for the control of certain fungus diseases of fruit, vegetables, and ornamental crops, and as a soil treatment for the control of certain seed rots and damping-off diseases.

CAUTION

Keep Out of Reach of Children

CAUSES SKIN AND EYE IRRITATION.

MAY BE HARMFUL IF SWALLOWED OR INHALED.

Avoid contact with skin, eyes or clothing.
Avoid breathing dust. Use with adequate ventilation.
Wash thoroughly after handling.
Do not store near food or feed.

FIRST AID: In case of eye contact immediately flush eyes with plenty of water for at least 15 minutes. Call a physician. Wash skin with soap and water.

STORE IN COOL, DRY PLACE—PROTECT FROM EXCESSIVE HEAT.

COMPATIBILITY AND PLANT SAFETY: CAPTAN 50-WP can be combined safely and effectively at recommended dosage rates with most commonly used fungicides and insecticides, with the exception of oil and strongly aklaline materials. Alkaline materials such as spray lime, lime-sulfur and bordeaux mixture will reduce the fungicidal activity of CAPTAN. Do not apply CAPTAN in combination with or immediately before or closely following oil sprays. The time factor governing the safe interval between CAPTAN and oil sprays varies due to general climatic conditions, therefore, consult local agricultural spray programs and authorities to determine the proper timing. The use of spreaders which cause excessive wetting is not advised. Combinations with solvent formulations of organic phosphates should not be used. Combinations of CAPTAN and sulfur should not be used on crops sensitive to sulfur. Used at high rates or in drenching sprays, CAPTAN may cause a necrotic spotting of tender, immature leaves of certain varieties of apples, peaches, plums and cherries. This type of injury is most likely to occur in the early cover sprays during long periods of warm, cloudy, humid weather. To avoid the hazard of leaf spotting under such conditions, use CAPTAN and other spray materials at lowest recommended rates and avoid drenching trees.

OBSERVE THE FOLLOWING USE PRECAUTIONS

Read all precautions and directions for use before using. Use only for claims listed and only as specified on this label.

In order that pesticide residues on food and forage crops will not exceed federal tolerances, use only at recommended rates and intervals, and do not apply closer to harvest than specified. Do not apply or allow to drift to adjoining food, fiber or pasture crops.

Consult State Agricultural Experiment Stations or State Agricultural Extension Service for additional information, as the timing of applications needed will vary with local conditions.

FISH CAUTIONS: This product is toxic to fish. Do not contaminate any body of water by cleaning of equipment, disposal of wastes or application where runoff is likely to occur. Do not apply when weather conditions favor drift from areas treated.

DIRECTIONS FOR USE

It is a violation of Federal law to use this product in a manner inconsistent with its labeling.
Applications can be made by aircraft or ground power equipment (including concentrate and semi-concentrate equipment). Pour recommended amount of this material into nearly filled spray tank. Add balance of water. Maintain agitation during filling and spraying operations. Do not allow mixture to stand. Do not combine with emulsifiable liquids or wettable powders unless previous experience has proven them to be physically compatible and safe to plants. (Read compatibility and plant safety information given under USE PRECAUTIONS).

RECOMMENDATIONS

Unless otherwise specified, rates are given in terms of CAPTAN 50-WP per 100 gallons water for fruit and nut crops and ornamentals. For vegetable crops, rates are expressed as CAPTAN 50-WP per acre in sufficient water to provide thorough coverage. Except as specified begin applications before or at first sign of disease and repeat as needed to maintain control but observe use limitations. Unless otherwise specified, application can be made on the day of harvest.

Maximum permissible rate per acre, per application as CAPTAN 50-WP is given in parenthesis, ( ), after crop claim.

FRUIT AND NUT CROPS

IMPORTANT: Read label carefully. Although most of the directions on this label may be followed nationwide, a few are limited to either the eastern or western U.S. Follow those directions for your growing area where specified.

ALMONDS: Brown rot, twig and blossom blight, shothole, scab, leaf blight—Use 2 to 3 pounds. Apply at popcorn, bloom, and petal fall stages and up to 5 weeks after petal fall. Hulls may be fed to livestock. (32 pounds)

Subsequent applications may be made up to 12 days of harvest in accordance with pest infestations and State Agricultural Extension Service recommendations, provided the hulls are not fed to dairy animals or animals being finished for slaughter. (32 pounds)

APPLES (western U.S.): Apple scab—in preblossom sprays, use 2 lbs. per 100 gallons of water in high volume application. In low volume sprays apply 16 pounds per acre. In mid-summer, the dosage may be reduced to 1 pound per 100 gallons in high volume application or 10 pounds per acre in low volume sprays. Bull's eye rot, Botrytis rot. (Pacific northwest)—2 pounds per 100 gallons spray. Make 1 or 2 applications with late cover sprays and 1 to 2 preharvest applications. (16 pounds)

APRICOTS: Brown rot (twig blight), jacket rot—Use 2 pounds. Apply in red bud, bloom, 75% petal fall, and cover stages. (10 pounds)

BLUEBERRIES (western U.S.): Botrytis gray mold or berry rot, mummy berry—Use 2 pounds in sufficient water for thorough coverage. Begin at mid-bloom, repeat at 7 to 10 day intervals until maturity. (5 pounds)

RASPBERRIES (western U.S.): Anthracnose, Botrytis, spur blight—Apply 2 pounds. Apply at bloom or when new shoots are 8 to 10 inches long. Repeat 2 weeks later, apply fall spray after canes are removed. Fruit rot—Apply 2 pounds. Apply 3 to 5 days before harvest starts. Repeat at mid-harvest and 8 to 10 days later. (6 pounds)

CHERRIES (western U.S.): Brown rot blossom blight, brown rot (fruit), leaf spot— Use 2 pounds in pre-bloom, bloom, petal fall, shuck, cover and preharvest sprays. (20 pounds)

CITRUS (Grapefruit, Lemons, Oranges, Tangelos, Tangerines): Brown rot (Phytophthora)—Use 2 to 4 pounds. Apply to skirts of trees before winter rains. Wet tree thoroughly to height of 3 to 4 feet. Repeat 10 to 12 weeks later (20 pounds). Melanose, scab (oranges and tangelos)—Use 2 pounds plus spray sticker. Apply as necessary from $2/3$ petal fall until fruit is $1/2$ inch in diameter. Do not apply after fruit size exceeds $1/2$ inch in diameter. Do not feed raw by-products to livestock (10 pounds).

CRANBERRIES (western U.S.): Blotch rot (Acanthorynchus vaccinii), Guignardia blight—Apply 6 pounds per acre in 300 gallons of water in early spring. Repeat applications as required. Storage rots, twig blight—Apply 6 pounds per acre in 300 gallons of water during the first week of June for twig blight, and in early and late August for storage rots. (7 pounds)

GRAPES (California): Bunch rot (Botrytis)—Apply 2 pounds per acre. Make 2 applications before bloom and 1 just after bloom. Repeat periodically making 3 cover applications before the bunches close. **Dead Arm** (current season infection)—Apply 4 pounds per 100 gallons spray. Apply 3 to 6 pounds per acre by ground equipment. Apply 4 pounds in 7 to 20 gallons of water per acre by aircraft. Apply first spray when green tissue begins to show but before shoots are 1 inch long. Apply a second spray when shoots are 6 to 8 inches long. (10 pounds)

MANGOES: Cercospora leaf spot or blotch—Use 2 pounds. Begin at first sign of disease and repeat at weekly intervals, to maintain control. (20 pounds)

PEACHES, NECTARINES: Brown rot, scab—Use 2 pounds in full pink, bloom, petal fall, shuck, cover and preharvest sprays. Applications at 3 to 4 day intervals may be necessary during bloom to control blossom blight. Repeat application at 7 to 14 day intervals as needed to maintain cover. Continue applications through harvest if conditions favor brown rot. If powdery mildew is a problem, add 3 pounds sulfur to the petal fall, shuck and early cover sprays. If sulfur is added, CAPTAN may be reduced to 1 pound in these sprays. Coryneum blight (peach blight, shothole)—Use 2 pounds, apply in pink bud, full bloom, petal fall and cover periods as necessary and as a post harvest spray (but before leaves drop). (10 pounds)

PEACHES (NURSERY STOCK) (California): Preventative preplant dip treatment for crown gall. Use 4 pounds CAPTAN 50-WP plus 3.2 pints diluted sodium hypochlorite (5.25% household bleach) per 100 gallons water. Wash nursery trees to remove soil from roots. Cut off all dormant buds and suckers in crown area and prune root system if necessary. Submerge the entire dormant tree for 5 minutes. Recharge dip during operation as necessary to maintain a proper concentration of 200 ppm chlorine. Check the concentration frequently using a chlorine test kit.

PEARS (western U.S.—except Pacific northwest): Pear scab — Use 2 pounds. Apply during early finger stage, petal fall and first cover sprays. Under severe scab infection conditions additional treatments may be necessary. Russeting may be reduced on Bosc pears. Do not use on D'Anjou pears. (16 pounds)

PINEAPPLES: Heart rot (Phytophthora parasitica)—Use 10 pounds. Dip slips into solution just prior to planting. Root rot (Phytophthora cinnamonii)—Use 4 pounds. Spray immediately after planting. Repeat at 2 week intervals (20 pounds). **Post Harvest Dip:** Storage and transit rots—Use 10 pounds. Dip or wash harvested fruit.

PLUMS, PRUNES (western U.S.): Brown rot—use 2 pounds. Apply at green bud, popcorn, bloom and petal fall stages. Repeat in cover sprays as conditions warrant. Prune russet scab (lacy scab) (in California)—Apply 2 pounds per 100 gallons water (8 pounds per acre). Apply at full bloom. (10 pounds)

QUINCE: Brown rot, scab—Use 2 pounds. Begin when disease threatens. Repeat at 7 to 10 day intervals. Do not apply within 7 days of harvest. (10 pounds)

STRAWBERRIES: Botrytis (gray mold), leaf spot—Apply 4 to 6 pounds per acre in sufficient water for thorough coverage. Begin applications when new growth starts in the spring and before fruit starts to form. Repeat at weekly intervals. Under conditions favorable to fruit rot continue applications through harvest period treating immediately after each picking. (8 pounds)

VEGETABLE CROPS

ASPARAGUS (PREPLANTING ROOT DIP): Botrytis, root rots, phoma rot, penicillium rot, fusarium—Use 3 pounds per 100 gallons water (4 tablespoonfuls per gallon). Dip root sections for 1 minute, drain and plant.

BEANS, PEAS: Damping off (beans only), fungus root rot—Use 10 to 12 pounds per acre as a broadcast treatment, working into top 3 to 4 inches of soil, or use 5 to 6 pounds per acre in the row at time of planting.

CANTALOUPE, CUCUMBER, PUMPKIN, SUMMER AND WINTER SQUASH, HONEYDEW MELONS, WATERMELON: Angular leaf spot, anthracnose—Use 4 pounds per acre in sufficient water for thorough coverage. Begin application at first signs of infection and repeat at 5 to 7 day intervals or as needed to maintain control. (7.6 pounds)

CARROTS: Cercospora blights, septoria leaf spot—Use 2 pounds per 100 gallons water and apply the equivalent of 4 pounds CAPTAN 50-WP per acre. Apply at first sign of disease and repeat at 6 to 10 day intervals. (6 pounds)

CELERY: Damping-off, pink rot—Use 2 pounds per 100 gallons water. Apply 450 gallons spray per acre to plant beds when all seeds have germinated. Repeat as necessary. (13 pounds) Late blight (septoria), pink rot (sclerotinia)—Use 2 pounds per 100 gallons water. Apply 200 gallons spray per acre. Spraying should begin in the seedbed and be continued at 7 to 10 day intervals. (4 pounds)

EGGPLANT: Phomoposis blight, fruit rot, anthracnose—Use 2 pounds. Begin when plants emerge in plant bed or when disease first appears in field. Repeat at 3 to 7 day intervals through growing season. Damping-off—Use 10 to 12 pounds per acre in sufficient water for thorough coverage. Begin when all seeds have germinated. Repeat at 10 day intervals. (6 pounds as foliar spray; 12 pounds in plant beds before transplanting)

ONIONS (GREEN): (California)—Downy mildew—Use 3 pounds per acre in 5 to 20 gallons water per acre by air or 20 to 100 gallons water per acre by ground equipment. Apply at first sign of infection and repeat at 5 to 7 day intervals or as needed to maintain disease control. (3 pounds)

ONIONS (GREEN AND BULB): Downy mildew, purple blotch —Use 2 pounds. Apply at first sign of infection and repeat as 7 to 10 day intervals. (3 pounds)

PEPPERS, PIMENTOS: Anthracnose, cercospora (frogeye leaf spot and stem end rot)—Use 3 pounds. Apply at 3 to 7 day intervals throughout the growing season. (6 pounds) Damping off—Use 2 pounds. Plant bed treatment. Begin when all seeds have germinated. Repeat at 10 day intervals. (12 pounds)

TOMATO: Anthracnose, gray leaf spot, septoria leaf spot— Use 4 to 6 pounds per acre in sufficient water for thorough coverage. Under severe infection conditions use 6 to 8 pounds per acre in sufficient water for thorough coverage. Apply at first signs of infection and repeat at 5 to 7 day intervals or as needed to maintain control. (10 pounds)

SPECIAL USES

PREPLANTING SOIL TREATMENT: For control of damping-off and root rots for the crops listed below, apply spray broadcast to soil surface as a preplanting application and cultivate into upper 3 to 4 inches of soil. Or apply to furrow over the seed when specified. (14,500 linear feet of row equals 1 acre.) Close and pack the furrow in usual manner. Apply dosage rate in sufficient water for thorough soil coverage unless otherwise specified.

BEANS, PEAS: Use 10 to 12 pounds broadcast or 5 to 6 pounds in furrow. (12 pounds)

BEETS: Use 2 pounds per 100 gallons water; 15 gallons spray per 1,000 square feet. (13 pounds)

BROCCOLI, BRUSSELS SPROUTS, CABBAGE, CAULIFLOWER, COLLARDS, KALE, KOHLRABI, MUSTARD, RUTABAGAS, TURNIPS: Damping-off only—Use 15 pounds broadcast. (28 pounds)

CANTALOUPES, CUCUMBERS, HONEYDEW MELONS, PUMPKINS, SQUASH (SUMMER AND WINTER), WATERMELONS: Use 2 pounds per 100 gallons water; 600 gallons spray per acre broadcast. (12 pounds)

SPINACH: (Direct seeded) Use 5 to 7 pounds in 30 gallons water in furrow. (7 pounds)

POST HARVEST DIP: For control of various molds and storage rots (Botrytis, Gleosporium, Rhizopus). Use as a post harvest dip or wash on the following fruits and vegetables: **Apples, Apricots, Cantaloupes, Cherries, Cucumbers, Grapefruit, Lemons, Limes, Mangoes, Nectarines, Onions (Green and Bulb), Oranges, Peaches, Pears, Potatoes, Tangelos, Tangerines**—Use 2½ pounds. Apply as a spray or in dip tank. When used as a dip, recharge wash solution periodically when tank volume is reduced by 25%. Bring water back to volume and add 1 pound for each 25 gallons added. At end of every 8 hour shift, empty tank, flush and charge with fresh dilution. Do not allow dip tank solution to stand overnight. Maintain continuous agitation during dipping operation. **Caution:** Do not feed mango or citrus by-products to dairy animals or animals being finished for slaughter.

PACKING BOX TREATMENT: For reduction of storage rots and molds (Rhizopus, Botrytis, Gleosporium) treat packing boxes with a solution of 2 pounds per 100 gallons water. Spray boxes thoroughly or use as a dip. Keep solution agitated continuously. Drain, flush and recharge the dip tank with new solution after each 8 hour shift.

ORNAMENTALS

AZALEAS: Damping-off of cuttings—Use 4 pounds. Dip cuttings before bedding. Petal blight—Use 2 pounds. Apply to soil area around plants and spray flowers just before bloom. Repeat at 7 to 14 day intervals through bloom.

BEGONIAS (TUBEROUS): Damping-off, tuber rot—Use 4 pounds. Dip tubers 30 minutes, drain and plant.

CAMELLIAS: Petal blight—Use 1 pound. Apply to drench soil around plants beginning when flowers start to open. Repeat at 7 to 10 day intervals through bloom.

CARNATIONS: Alternaria leaf spot, rust—Use 2 pounds. Begin at first sign of disease. Repeat at 7 to 10 day intervals. Shorten intervals during frequent rains and heavy dews. Damping-off of cuttings —Use 4 pounds. Dip cuttings before bedding.

CHRYSANTHEMUM: Botrytis flower blight, Septoria leaf spot —Use 2 pounds. Apply at first sign of disease. Repeat at 7 to 10 day intervals. Damping-off of cuttings—Use 4 pounds. Dip cuttings before bedding.

DICHONDRA (CALIFORNIA ONLY): White mold (Sclerotium rolfsii)—Use 2 pounds. Apply 1 gallon spray per 10 square feet. Make 2 to 3 applications at 7 day intervals.

GLADIOLUS (CORMS): Corm rot and decay, damping-off — Use ½ to 1½ pounds per 10 gallons, dip corms 20 to 30 minutes. Drain and plant.

GRASSES (ORNAMENTAL IN NON-PASTURED AREAS ONLY): Leaf spot, damping-off, brown patch, melting out, seedling blights, and brown spot on St. Augustine grass—Use 2 pounds. Apply 10 gallons spray per 1,000 square feet. Begin when growth starts in spring. Repeat at 7 to 14 day intervals throughout season. Do not graze treated areas or feed clippings to livestock.

GRASSES (LAWN SEEDBEDS): Damping-off and other soil borne diseases—Use ¼ to ⅓ pound per 1,000 square feet or 2 pounds in 100 gallons water using 15 gallons spray per 1,000 square feet. Cultivate into upper 3 to 4 inches before planting.

ROSES: Black spot, Botrytis blossom blight—Use 2 pounds. Begin at first growth or first sign of disease. Repeat at 7 to 14 day intervals, and more frequently during frequent rains and heavy dews.

SOIL AND GREENHOUSE BENCH TREATMENT: Preplanting for damping-off and root rot diseases on seedlings or transplants of roses (and other shrubs, trees, flowers). Lawn seedbeds, vegetables (such as beans, celery, crucifers, eggplant, peas, peppers, tomatoes)—Use 2 pounds per 100 gallons water at rate of 15 gallons per 1,000 square feet. Cultivate into upper 3 to 4 inches of soil before planting.

STORAGE AND DISPOSAL

PROHIBITIONS

Do not contaminate water, food, or feed by storage or disposal. Open dumping is prohibited.

PESTICIDE DISPOSAL

Wastes resulting from the use of this product may be disposed of on site or at an approved waste disposal facility.

CONTAINER DISPOSAL

Completely empty bag into application equipment. Then dispose of empty bag in a sanitary landfill or by incineration, or, if allowed by State and local authorities, by burning. If burned, stay out of smoke.

NOTICE —READ CAREFULLY

CONDITIONS OF SALE

Stauffer (and seller) offer(s) this product for sale subject to, and buyer and all users are deemed to have accepted, the following conditions of sale and warranty which may only be varied by written agreement of a duly authorized representative of Stauffer.

WARRANTY LIMITATION

Stauffer warrants that this product conforms to the chemical description on the label and is reasonably fit for the purposes referred to in the directions for use on the label subject to the inherent risks referred to below. Stauffer makes no other express warranties; THERE IS NO IMPLIED WARRANTY OF MERCHANTABILITY and there are no warranties which extend beyond the description on the label hereof.

INHERENT RISKS

The directions for use of this product are believed to be reliable and should be followed carefully. However, it is impossible to eliminate all risks associated with use. Buyer assumes all risks associated with use or application of this product contrary to label instructions or resulting from extraordinary weather conditions.

LIMITATION OF LIABILITY

In no case shall Stauffer be liable for special, indirect, or consequential damages resulting from the use or handling of this product and no claim of any kind shall be greater in amount than the purchase price of the product in respect of which such damages are claimed.

This product is sold only for uses stated on the label. No express or implied license is granted to use or sell this product under any patent in any country except as specified: Country: United States of America

---

This is a specimen label. Stauffer is not responsible for the accuracy of the information contained herein. As labels are subject to revision, always carefully read and follow the label on the product container.

CAPTAN 80-WP

EAST OF ROCKIES FUNGICIDE

For Plant Disease Control

ACTIVE INGREDIENTS:

CAPTAN: *N*-[(trichloromethyl)thio]-4- cyclohexene-1,2-dicarboximide	75.7%
Related derivatives	4.3%
INERT INGREDIENTS:	20.0%
TOTAL	100.0%

EPA Reg. No. 476-1819 A-2RE

CAPTAN 80-WP is a microfine wettable powder for use in water as a spray for the control of certain fungus diseases of fruit, vegetables, and ornamental crops, and as a soil treatment for the control of certain seed rots and damping-off diseases.

CAUTION

Keep Out of Reach of Children

CAUSES SKIN AND EYE IRRITATION.

MAY BE HARMFUL IF SWALLOWED OR INHALED.

Avoid contact with skin, eyes or clothing.
Avoid breathing dust. Use with adequate ventilation.
Wash thoroughly after handling.
Do not store near food or feed.

FIRST AID: *In case of eye contact* immediately flush eyes with plenty of water for at least 15 minutes. Call a physician. Wash skin with soap and water.

STORE IN COOL, DRY PLACE—PROTECT FROM EXCESSIVE HEAT.

OBSERVE THE FOLLOWING USE PRECAUTIONS

Read all precautions and directions for use before using. Use only for claims listed and only as specified on this label.

In order that pesticide residues on food and forage crops will not exceed federal tolerances, use only at recommended rates and intervals, and do not apply closer to harvest than specified. Do not apply or allow to drift to adjoining food, fiber or pasture crops.

Consult State Agricultural Experiment Stations or State Agricultural Extension Service for additional information, as the timing of applications needed will vary with the local conditions.

FISH CAUTIONS: This product is toxic to fish. Do not contaminate any body of water by cleaning of equipment, disposal of wastes or application where runoff is likely to occur. Do not apply when weather conditions favor drift from areas treated.

COMPATIBILITY AND PLANT SAFETY: CAPTAN 80-WP can be combined safely and effectively at recommended dosage rates with most commonly used fungicides and insecticides, with the exception of oil and strongly alkaline materials. Alkaline materials such as spray lime, lime-sulfur and bordeaux mixture will reduce the fungicidal activity of CAPTAN 80-WP. Do not apply CAPTAN 80-WP in combination with or immediately before or closely following oil sprays. The time factor governing the safe interval between CAPTAN 80-WP and oil sprays varies due to general climatic conditions, therefore, consult local agricultural spray programs and authorities to determine the proper timing. The use of spreaders which cause excessive wetting is not advised. Combinations with solvent formulations of organic phosphates should not be used. Combinations of CAPTAN 80-WP and sulfur should not be used on crops sensitive to sulfur. Used at high rates or in drenching sprays, CAPTAN 80-WP may cause a necrotic spotting of tender, immature leaves of certain varieties of apples, peaches, plums and cherries. This type of injury is most likely to occur in the early cover sprays during long periods of warm, cloudy, humid weather. To avoid the hazard of leaf spotting under such conditions, use CAPTAN 80-WP and other spray materials at lowest recommended rates and avoid drenching trees.

DIRECTIONS FOR USE

It is a violation of Federal law to use this product in a manner inconsistent with its labeling.

Applications can be made by aircraft or ground power equipment (including concentrate and semi-concentrate equipment). Pour recommended amount of this material into nearly filled spray tank. Add balance of water. Maintain agitation during filling and spraying operations. Do not allow mixture to stand. Do not combine with emulsifiable liquids or wettable powders unless previous experience has proven them to be physically compatible and safe to plants. (Read compatibility and plant safety information given under USE PRECAUTIONS.)

RECOMMENDATIONS

Unless otherwise specified, rates are given in terms of CAPTAN 80-WP per 100 gallons water for fruit and nut crops and ornamentals. For vegetable crops, rates are expressed as CAPTAN 80-WP per acre in sufficient water to provide thorough coverage. Except as specified begin applications before or at first sign of disease and repeat as needed to maintain control but observe use limitations. Unless otherwise specified, application can be made on the day of harvest.

Maximum permissible rate per acre, per application expressed as CAPTAN 80-WP is given in parentheses ( ), after each crop claim.

FRUIT AND NUT CROPS

IMPORTANT: Read label carefully. Although most of the directions on this label may be followed nationwide, a few are limited to either the eastern or western U.S. Follow those directions for your growing area where specified.

ALMONDS: Brown rot twig and blossom blight, shothole, scab, leaf blight—Use 1¼ to 2 pounds. Apply at popcorn, bloom, and petal fall stages and up to 5 weeks after petal fall. Hulls may be fed to livestock. (20 pounds) Subsequent applications may be made up to 12 days of harvest in accordance with pest infestations and State Agricultural Extension Service recommendations, provided the hulls are not fed to dairy animals or animals being finished for slaughter. (20 pounds)

APPLES (eastern U.S.): Primary scab, black rot (frogeye), Botrytis blossom-end rot—1 to 1¼ pounds at 5 to 7 day intervals as needed to maintain cover in prebloom, bloom, petal fall and first cover sprays. Use at 1¼ pounds rate if conditions favor scab or if active scab can be found. Secondary scab, Brooks fruit spot, sooty blotch, fly speck, black rot, black pox, Botryosphaeria rot, bitter rot—3/4 to 1 1/4 pounds at 10 to 14 day intervals in second and later cover sprays up to start of harvest. (10 pounds) Powdery mildew—If powdery mildew is a problem add 1 1/2 to 3 pounds sulfur or 1/4 to 1/2 pound karathane to all pre-bloom and bloom sprays and 1/4 to 1/2 pound karathane to all post bloom sprays until the foliage matures. Use 1/2 pound karathane if interval between sprays is longer than 7 days. **CAUTION**—Do not use CAPTAN in combination with or closely following or in alternation with wettable sulfur products on sulfur sensitive varieties of apples such as Red Delicious, Staymen, Baldwin, King, etc. as severe injury and defoliation may occur.

APRICOTS: Brown rot (twig blight), jacket rot—1 1/4 pounds. Apply in red bud, bloom, 75% petal fall, and cover stages. (6 1/4 pounds)

BLACKBERRY, DEWBERRY, RASPBERRY (eastern U.S.): Anthracnose—Use 2 1/2 pounds per acre in sufficient water for thorough coverage in the preblossom (when new canes are 6 to 8 inches long) and first cover (right after petal fall) sprays. Raspberry—Botrytis fruit rot—Use 1 1/4 pounds per 100 gallons water. Begin application in the first bloom. Make 2 or more repeat applications (as field conditions warrant) at 10 to 15 day intervals. (3 3/4 pounds)

CHERRIES (eastern U.S.): Brown rot, leaf spot, Botrytis rot — Use 1 1/4 pounds in pre-bloom, bloom, petal fall, shuck, cover and preharvest sprays. Applications at 3 to 4 day intervals may be necessary during bloom to control blossom blight. Repeat applications at 7 to 10 day intervals as needed to maintain cover up to start of harvest. If powdery mildew is a problem add 3 pounds sulfur to the petal fall, shuck and early cover sprays. If sulfur is added, CAPTAN may be reduced to 3/4 pound in these sprays. Post harvest sprays —Use 1 1/4 pounds immediately after harvest for leaf spot. Repeat application in 10 to 14 days. (12 1/2 pounds)

CITRUS (Grapefruit, Lemons, Limes, Oranges, Tangelos, Tangerines): Brown rot (Phytophthora)—Use 1 1/4 to 2 1/2 pounds. Apply to skirts of trees before winter rains. Wet tree thoroughly to height of 3 to 4 feet. Repeat 10 to 12 weeks later (12 1/2 pounds). Melanose, scab (oranges and tangelos)—Use 1 1/4 pounds plus spray sticker. Apply as necessary from 2/3 petal fall until fruit is 1/2 inch in diameter. Do not apply after fruit size exceeds 1/2 inch in diameter. Do not feed raw byproducts to livestock. (6 1/4 pounds)

GRAPES (eastern U.S.): Black rot, downy mildew— 1 1/4 pounds per 100 gallons. Make first application just before bloom repeat just after bloom and at 10 to 14 day intervals for 1 to 3 more applications depending upon the susceptibility of the grape variety. **Dead Arm** (current season infection) use 1 1/4 to 2 1/2 pounds per 100 gallons dilute spray. Apply 50 gallons per acre when shoots are 1/2 to 1 inch long. Repeat application when shoots are 4 to 6 inches long. (6 1/4 pounds)

MANGOES: Cercospora leaf spot or blotch—Use 1 1/4 pounds. Begin at first sign of disease and repeat at regular intervals to maintain control. (12 1/2 pounds)

PEACHES, NECTARINES: Brown rot, scab—Use 1 1/4 pounds in full pink, bloom, petal fall, shuck, cover and preharvest sprays. Applications at 3 to 4 day intervals may be necessary during bloom to control blossom blight. Repeat application at 7 to 14 day intervals as needed to maintain cover. Continue applications through harvest if conditions favor brown rot. If powdery mildew is a problem, add 3 pounds sulfur to the petal fall, shuck and early cover sprays. If sulfur is added, CAPTAN may be reduced to 3/4 pound in these sprays. Coryneum blight (peach blight, shothole)—Use 1 1/4 pounds, apply in pink bud, full bloom, petal fall and cover periods as necessary and as a post harvest spray, but before leaves drop. (6 1/4 pounds)

PEARS (eastern U.S.): Scab, leaf spot—Use 1 1/4 pounds in preblossom, petal fall and cover sprays. Repeat cover applications at 7 to 14 day intervals under severe scab conditions. Russeting may be reduced on Bosc pears. Do not use on D'Anjou pears. (10 pounds)

PINEAPPLES: Heart rot (Phytophthora parasitica)—Use 6 1/4 pounds. Dip slips into solution just prior to planting. Root rot (Phytophthora cinnamonii)—Use 2 1/2 pounds. Spray immediately after planting. Repeat at 2 week intervals (12 1/2 pounds). **Post Harvest Dip:** Storage and transit rots—Use 6 1/4 pounds. Dip or wash harvested fruit.

PLUMS, PRUNES (eastern U.S.): Brown rot—Use 1 1/4 pounds in full pink, bloom and petal fall sprays. Repeat applications at 7 to 14 day intervals as needed to maintain cover. Continue applications through harvest if conditions favor brown rot. The addition of a neutral spreader has improved coverage. (6 1/4 pounds)

QUINCE: Brown rot, scab—Use 1 1/4 pounds. Begin when disease threatens. Repeat at 7 to 10 day intervals. Do not apply within 7 days of harvest. (6 1/4 pounds)

STRAWBERRIES: Botrytis (gray mold), leaf spot—Apply 2 1/2 to 3 3/4 pounds per acre in sufficient water for thorough coverage. Begin applications when new growth starts in the spring and before fruit starts to form. Repeat at weekly intervals. Under conditions favorable to fruit rot continue applications through harvest period treating immediately after each picking. (5 pounds)

VEGETABLE CROPS

ASPARAGUS (Preplanting Root Dip): Botrytis, root rots, phoma rot, penicillium rot, fusarium—Use 1 7/8 pounds per 100 gallons water (2 1/2 tablespoonfuls per gallon). Dip root sections for 1 minute, drain and plant.

BEANS, PEAS: Damping-off (bean only), fungus root rot—Apply 6 1/4 to 7 1/2 pounds per acre as a broadcast treatment, working into top 3 to 4 inches of soil, or apply 3 1/8 to 3 3/4 pounds per acre in the row at time of planting. (7 1/2 pounds)

CANTALOUPE, CUCUMBER, PUMPKIN, SUMMER AND WINTER SQUASH, HONEYDEW MELON, WATERMELON: Angular leaf spot, anthracnose—Use 2 1/2 pounds per acre in sufficient water for thorough coverage. Begin applications at first signs of infection and repeat at 5 to 7 day intervals or as needed to maintain control. (4 3/4 pounds)

CARROTS: Cercospora blights, septoria leaf spot—Use 1 1/4 pounds per 100 gallons water and apply the equivalent of 2 1/2 pounds CAPTAN 80-WP per acre. Apply at first sign of disease and repeat at 6 to 10 day intervals. (3 3/4 pounds)

CELERY: Damping-off, pink rot—Use 1 1/4 pounds per 100 gallons water. Apply 450 gallons spray per acre to plant beds when all seeds have germinated. Repeat as necessary. (8 1/8 pounds) Late blight (Septoria), pink rot (Sclerotinia)—Use 1 1/4 pounds per 100 gallons water. Apply 200 gallons spray per acre. Repeat at 7 to 10 day intervals. (2 1/2 pounds)

EGGPLANT: Phomoposis blight, fruit rot, anthracnose—Use 1 1/4 pounds. Begin when plants emerge in plant bed or when disease first appears in field. Repeat at 3 to 7 day intervals through growing season. Damping-off—Use 6 1/4 to 7 1/2 pounds per acre in sufficient water for thorough coverage. Begin when all seeds have germinated. Repeat at 10 day intervals. (3 3/4 pounds as foliar spray; 7 1/2 pounds in plant beds before transplanting)

ONIONS (Green and Bulb): Downy mildew, purple blotch—Use 1 1/4 pounds. Apply at first sign of infection and repeat at 7 to 10 day intervals. (1 7/8 pounds)

PEPPERS, PIMENTOS: Anthracnose, cercospora (frogeye leaf spot and stem end rot)—Use 1 7/8 pounds. Apply at 3 to 7 day intervals throughout the growing season. (3 3/4 pounds) Damping-off —Use 1 1/4 pounds. Plant bed treatment. Begin when all seeds have germinated. Repeat at 10 day intervals. (7 1/2 pounds)

TOMATO: Anthracnose, gray leaf spot, septoria leaf spot—Use 2 1/2 to 3 3/4 pounds per acre in sufficient water for thorough coverage. Under severe infection conditions use 3 3/4 to 5 pounds per acre in sufficient water for thorough coverage. Apply at first signs of infection and repeat at 5 to 7 day intervals or as needed to maintain control. (6 1/4 pounds)

SPECIAL USES

PREPLANTING SOIL TREATMENT: For control of damping-off and root rots for the crops listed below, apply spray broadcast to soil surface as a preplanting application and cultivate into upper 3 to 4 inches of soil. Or apply to furrow over the seed when specified (14,500 linear feet of row equals 1 acre). Close and pack the furrow in usual manner. Apply dosage rate in sufficient water for thorough soil coverage unless otherwise specified.

BEANS, PEAS: Use $6^1/_4$ to $7^1/_2$ pounds broadcast or $3^1/_8$ to $3^1/_4$ pounds in furrow. ($7^1/_2$ pounds)

BEETS: Use $1^1/_4$ pounds per 100 gallons water: 15 gallons spray per 1,000 square feet. ($8^1/_5$ pounds)

BROCCOLI, BRUSSELS SPROUTS, CABBAGE, CAULIFLOWER, COLLARDS, KALE, KOHLRABI, MUSTARD, RUTABAGAS, TURNIPS: Damping-off only—Use $9^3/_8$ pounds broadcast. ($17^1/_2$ pounds)

CANTALOUPES, CUCUMBERS, HONEYDEW MELONS, PUMPKINS, SQUASH (SUMMER AND WINTER), WATERMELONS: Use $1^1/_4$ pounds per 100 gallons water; 600 gallons spray per acre broadcast. ($7^1/_2$ pounds)

SPINACH: (Direct seeded) Use $3^1/_8$ to $4^3/_8$ pounds in 30 gallons water in furrow. ($4^3/_8$ pounds)

POST HARVEST DIP: For control of various molds and storage rots (Botrytis, Gleosporium, Rhizopus)—Use as a post harvest dip or wash on the following fruits and vegetables: **Apples, Apricots, Cantaloupes, Cherries, Cucumbers, Grapefruit, Lemons, Limes, Mangoes, Nectarines, Onions (Green and Bulb), Oranges, Peaches, Pears, Potatoes, Tangelos, Tangerines**—Use $1^3/_5$ pounds. Apply as a spray or in dip tank. When used as a dip, recharge wash solution periodically when tank volume is reduced by 25%. Bring water back to volume and add $^5/_8$ pound for each 25 gallons added. At end of every 8 hour shift, empty tank, flush and charge with fresh dilution. Do not allow dip tank solution to stand overnight. Maintain continuous agitation during dipping operation. **CAUTION:** Do not feed mango or citrus byproducts to dairy animals or animals being finished for slaughter.

PACKING BOX TREATMENT: For reduction of storage rots and molds (Rhizopus, Botrytis, Gleosporium) treat packing boxes with a solution of $1^1/_4$ pounds per 100 gallons water. Spray boxes thoroughly or use as a dip. Keep solution agitated continuously. Drain, flush and recharge the dip tank with new solution after each 8 hour shift.

ORNAMENTALS

AZALEAS: Damping-off of cuttings—Use $2^1/_2$ pounds. Dip cuttings before bedding. Petal blight—Use $1^1/_4$ pounds. Apply to soil area around plants and spray flowers just before bloom. Repeat at 7 to 14 day intervals through bloom.

BEGONIAS (TUBEROUS): Damping-off, tuber rot—Use $2^1/_2$ pounds. Dip tubers 30 minutes, drain and plant.

CAMELLIAS: Petal blight—Use $1^5/_8$ pounds. Apply to drench soil around plants beginning when flowers start to open. Repeat at 7 to 10 day intervals through bloom.

CARNATIONS: Alternaria leaf spot, rust—Use $1^1/_4$ pounds. Begin at first sign of disease. Repeat at 7 to 10 day intervals. Shorten intervals during frequent rains and heavy dews. Damping-off of cuttings—Use $2^1/_2$ pounds. Dip cuttings before bedding.

GLADIOLUS (CORMS): Corm rot and decay, damping-off—Use $^1/_3$ to 1 pound per 10 gallons, dip corms 20 to 30 minutes. Drain and plant.

GRASSES (ORNAMENTAL IN NON-PASTURED AREAS ONLY): Leaf spot, damping-off, brown patch, melting out, seedling blights, and brown spot on St. Augustine grass—Use $1^1/_4$ pounds. Apply 10 gallons spray per 1,000 square feet. Begin when growth starts in spring. Repeat at 7 to 14 day intervals throughout season. Do not graze treated areas or feed clippings to livestock.

GRASSES (LAWN SEEDBEDS): Damping-off and other soil borne diseases—Use $^1/_5$ to $^1/_4$ pound per 1,000 square feet or $1^1/_4$ pounds in 100 gallons water using 15 gallons spray per 1,000 square feet. Cultivate into upper 3 to 4 inches before planting.

ROSES: Black spot, Botrytis blossom blight—Use $1^1/_4$ pounds. Begin at first growth or first sign of disease. Repeat at 7 to 14 day intervals, and more frequently during frequent rains and heavy dews.

SOIL AND GREENHOUSE BENCH TREATMENT: Preplanting for damping-off and root rot diseases on seedlings or transplants of roses (and other shrubs, trees, flowers). Lawn seedbeds, vegetables (such as beans, celery, crucifers, eggplant, peas, peppers, tomatoes)—Use $1^1/_4$ pounds per 100 gallons water at rate of 15 gallons per 1,000 square feet. Cultivate into upper 3 to 4 inches of soil before planting.

CHRYSANTHEMUM: Botrytis flower blight, Septoria leaf spot—Use $1^1/_4$ pounds. Apply at first sign of disease, repeat at 7 to 10 day intervals. Damping-off of cuttings—Use $2^1/_2$ pounds. Dip cuttings before bedding.

STORAGE AND DISPOSAL

PROHIBITIONS
Do not contaminate water, food, or feed by storage or disposal. Open dumping is prohibited.

PESTICIDE DISPOSAL
Wastes resulting from the use of this product may be disposed of on site or at an approved waste disposal facility.

CONTAINER DISPOSAL
Completely empty bag into application equipment. Then dispose of empty bag in a sanitary landfill or by incineration, or, if allowed by State and local authorities, by burning. If burned, stay out of smoke.

NOTICE—READ CAREFULLY

CONDITIONS OF SALE
Stauffer (and seller) offer(s) this product for sale subject to, and buyer and all users are deemed to have accepted, the following conditions of sale and warranty which may only be varied by written agreement of a duly authorized representative of Stauffer.

WARRANTY LIMITATION
Stauffer warrants that this product conforms to the chemical description on the label and is reasonably fit for the purposes referred to in the directions for use on the label subject to the inherent risks referred to below. Stauffer makes no other express warranties; THERE IS NO IMPLIED WARRANTY OF MERCHANTABILITY and there are no warranties which extend beyond the description on the label hereof.

INHERENT RISKS
The directions for use of this product are believed to be reliable and should be followed carefully. However, it is impossible to eliminate all risks associated with use. Buyer assumes all risks associated with use or application of this product contrary to label instructions or resulting from extraordinary weather conditions.

LIMITATION OF LIABILITY
In no case shall Stauffer be liable for special, indirect, or consequential damages resulting from the use or handling of this product and no claim of any kind shall be greater in amount than the purchase price of the product in respect of which such damages are claimed.

This product is sold only for uses stated on the label. No express or implied license is granted to use or sell this product under any patent in any country except as specified: Country: United States of America

This is a specimen label. Stauffer is not responsible for the accuracy of the information contained herein. As labels are subject to revision, always carefully read and follow the label on the product container.

DEVRINOL® 2-E

SELECTIVE HERBICIDE
EMULSIFIABLE LIQUID

For Use on Tomatoes, Peppers, Strawberries and Tobacco

ACTIVE INGREDIENT:	
N, N-Diethyl-2-(1-napthalenyloxy)-propionamide	21.8%
INERT INGREDIENTS:	78.2%
TOTAL	100.0%

Contains 2 pounds Active Ingredient per Gallon

EPA Reg. No. 476-2150 A6R

Keep Out of Reach of Children
DANGER

Causes eye damage and skin irritation. Corrosive.

Do not get in eyes, on skin, or on clothing.

Wear goggles or face shield and rubber gloves when handling.

Harmful if swallowed or absorbed through the skin.

Avoid contamination of food or feed.

Do not use, store, spill, or pour near heat or open flame.

FIRST AID

CALL A POISON CENTER OR A PHYSICIAN IMMEDIATELY.

If a known exposure occurs or is suspected, immediately start the procedures given below. Simultaneously contact a Poison Center, a physician or the nearest hospital. NOTE: Be sure to advise the person contacted that the product contains a Hydrocarbon Solvent. Describe the type of exposure, the victim's symptoms, and follow the advice given. For additional information, call collect, day or night: (203) 226-6602 Stauffer Chemical Company or (800) 424-9300 Pesticide Team Safety Network.

If swallowed, immediately give several glasses of water but *do not* induce vomiting. If vomiting does occur, give fluids again. Have a physician determine if condition of patient will permit induction of vomiting or evacuation of stomach. Do not give anything by mouth to an unconscious or convulsing person.

In case of contact, immediately flush eyes and skin with large amounts of running water for at least 15 minutes while removing contaminated clothing and shoes. For eyes, hold eyelids apart during the flushing to ensure rinsing of the entire surface of the eye and lids with water. Get medical attention immediately. Wash clothing before reuse.

If inhaled, remove to fresh air. If victim is breathing, oxygen may be given from a demand-type or continuous-flow inhaler, preferably with a physician's advice. If not breathing, clear victim's airway and start mouth-to-mouth artificial respiration. Get medical attention immediately.

PRECAUTIONARY STATEMENTS

HAZARDS TO HUMANS & DOMESTIC ANIMALS

DANGER

Corrosive Causes Eye Damage and Skin Irritation

Do not get in eyes, on skin, or on clothing.

Wear goggles or face shield and rubber gloves when handling.

Harmful if swallowed or absorbed through the skin.

Avoid contamination of food or feed.

Do not use, store, spill, or pour near heat or open flame.

ENVIRONMENTAL HAZARDS

Do not apply directly to water or wetlands.

Do not apply when weather conditions favor drift from areas treated.

Do not contaminate water by cleaning of equipment or disposal of wastes.

Apply this product as specified on this label.

Do not contaminate water used for domestic purposes.

Avoid contamination of Food and Feed.

GENERAL PRECAUTIONS

Read all label directions before using. Use only for recommended purposes at recommended rates. Do not overdose.

Do not contaminate irrigation water or water used for domestic purposes.

Do not store near seeds or fertilizer.

Do not apply more than once per season.

DEVRINOL is recommended for use on mineral soils only. Not recommended for use on soils with over 10% organic matter.

Spraying on ground heavily covered with leaves or trash may adversely affect weed control.

DIRECTIONS FOR USE

It is a violation of Federal law to use this product in a manner inconsistent with its labeling. DEVRINOL 2-E is a selective, surface-applied herbicide for use in strawberries. It is also recommended for weed control in certain vegetables and tobacco, applied and incorporated prior to planting.

DEVRINOL 2-E controls weeds before they emerge. It does not control established weeds. All weed growth and crop stubble in fields intended for vegetables or tobacco must be thoroughly worked into the soil prior to planting.

Spray equipment should be carefully calibrated before use and checked frequently during application to be sure nozzles are free from clogging and are delivering a uniform spray pattern. Apply the recommended rate of DEVRINOL 2-E in 20 to 100 gallons of water per acre.

STRAWBERRIES

WEEDS CONTROLLED

NOTE: DEVRINOL does not control established weeds.

Annual Grasses:	
Annual Bluegrass	*Poa annua*
Barnyardgrass-Watergrass	*Enchinochloa spp.*
Bristly Foxtail	*Setaria verticilla*
Large (Hairy) Crabgrass	*Digitaria sanguinalis*
Ripgut Brome	*Bromus rigidus*
Southwestern Cupgrass	*Eriochloa gracilis*
Wild Barley	*Hordeum leporinum*
Wild Oats	*Avena fatua*
Annual Broadleaf Weeds:	
Chickweed	*Stellaria media*
Common Fiddleneck	*Amsinckia intermedia*
Common Knotweed	*Polygonum aviculare*
Common Purslane	*Portulaca oleracea*
Common Sowthistle	*Sonchus oleraceus*
Filaree	*Erodium spp.*
Groundsel	*Senecio spp.*
Little Mallow*	*Malva parviflora*
Lambsquarters	*Chenopodium album*
Pineappleweed	*Matricaria maricarioides*
Prickley Lettuce	*Lactuca scariola*
Redroot Pigweed	*Amaranthus retroflexus*

* From seed.

RECOMMENDATIONS

NOTE: All rates are given on a broadcast basis. Reduce rates proportionately for band or strip treatment.

STRAWBERRIES (Post Transplant and Established)

For control of all weeds listed, apply 2 gallons DEVRINOL 2-E per broadcast acre to the soil surface. Apply in the fall through early spring, prior to weed emergence. Do not apply DEVRINOL 2-E to frozen ground. Do not apply DEVRINOL from bloom to harvest.

For best results, the treatment must reach the zone of weed seed germination. Treatments made other than in November through February must be irrigated-in within 24 hours (sprinkler or flood) with sufficient water to wet the soil to a depth of at least 2 to 4 inches. Irrigation or shallow incorporation is recommended for treatments made from November through February if no rainfall occurs within one week of application.

VEGETABLES & TOBACCO

WEEDS CONTROLLED

NOTE: DEVRINOL 2-E does not control established weeds.

Annual Grasses:	
Barnyardgrass	*Echinochloa spp.*
Crabgrass	*Digitaria spp.*
Fall Panicum	*Panicum dichotomiflorum*
Goosegrass	*Eleusine indica*
Seedling Johnsongrass	*Sorghum halepense*
Annual Broadleaf Weeds:	
Carpetweed	*Mollugo verticillata*
Common Purslane	*Portulaca oleracea*
Lambsquarters	*Chenopodium album*
Pigweed	*Amaranthus spp.*

DIRECTIONS FOR USE

On flat-planted or bedded crops apply, before planting, to well worked soil that is dry enough to permit thorough incorporation to a depth of 1 to 2 inches. Incorporate the same day as applied, using equipment which will result in uniform incorporation of the herbicide to the desired depth.

After harvest or prior to planting of succeeding crops, a deep moldboard or disc plowing operation must be carried out.

Do not seed to alfalfa, small grains, sorghum, corn, lettuce or sugarbeets for 12 months after DEVRINOL 2-E application.

RECOMMENDATIONS

NOTE: All rates are given on a broadcast basis. Reduce rates proportionately for band or strip treatment.

TOMATOES (Direct Seeded and Transplanted):
Apply $\frac{1}{2}$ to 1 gallon DEVRINOL 2-E per broadcast acre. Use the lower rate on light soil (coarse textured—sandy) and the higher rate on heavy soil (fine textured—clay).

PEPPERS (Direct Seeded and Transplanted):
Apply $\frac{1}{2}$ to 1 gallon DEVRINOL 2-E per broadcast acre. Use the lower rate on light soil (coarse textured—sandy) and the higher rate on heavy soil (fine textured—clay).

TOBACCO (Transplanted):
Before transplanting, apply $\frac{1}{2}$ to 1 gallon DEVRINOL 2-E per broadcast acre. Use the lower rate on light soils (coarse textured—sandy) and the higher rate on heavy soil (fine textured— clay).

DEVRINOL 2-E/TILLAM ® 6-E

TANK MIX COMBINATION

A tank mix combination of DEVRINOL 2-E and TILLAM 6-E can be applied and mixed (incorporated) into the soil before planting for weed control in tomatoes and tobacco.

IMPORTANT:

Observe all cautions and limitations on labeling of all products used in combination.

Make only one application per season.

In each case where a range of rates is given, the lower rates should be used on light (coarse) textured soils and soils low in organic matter. The higher rates should be used on heavy (fine) textured soils.

After harvest or prior to planting of succeeding crops, a deep moldboard or disc plowing operation must be carried out.

Do not seed to alfalfa, small grains, sorghum, corn, lettuce or sugarbeets for 12 months after application of the DEVRINOL 2-E/ TILLAM 6-E tank mix.

WEEDS CONTROLLED

The tank mix is recommended for control of the following weeds:

Annual Grasses:	
Barnyardgrass-watergrass	*Echinochloa spp.*
Bermudagrass	*Cynodon dactylon*
Crabgrass	*Digitaria spp.*
Foxtail-yellow, giant, green	*Setaria spp.*
Goosegrass	*Eleusine indica*
Millet	*Setaria spp.*
Signalgrass	*Brachiaria piligera*
Wild Oats	*Avena fatua*
Annual Broadleaf Weeds:	
Blackeyed Susan, Sweet Suzy	*Rudbeckia serotina*
Common Purslane	*Portilaca oleracea*
Deadnettle, Henbit	*Lamium amplexicaule*
Florida Pusley	*Richardia scabra*
Hairy Nightshade	*Solanum villosum*
Lambsquarters	*Chenopodium album*
Nettleleaf Goosefoot	*Chenopodium murale*
Prostrate Pigweed	*Amaranthus graecizans*
Redroot Pigweed	*Amaranthus retroflexus*
Shepherdspurse	*Capsella bursa-pastoris*
Perennial Weeds:	
Yellow Nutsedge (Yellow Nutgrass)	*Cyperus escultentus*
Purple Nutsedge (Purple Nutgrass)	*Cyperus rotundus*

Existing stands of nutgrass must be turned under and chopped up thoroughly before treatment.

DIRECTIONS FOR USE

The recommended tank mix controls weeds by interfering with normal seed germination and seedling development. It does not control established weeds. All weed growth and crop stubble should be thoroughly worked into the soil before treatment.

TANK MIXING DIRECTIONS

Be sure the sprayer tank, boom and nozzle have been thoroughly flushed with clean water if another pesticide has been used previously.

Sufficient jet or mechanical agitation should be provided during the filling operation and during application. Fill the spray tank at least one-fourth full of clean water and while pump and agitator are running, add the recommended amounts of the herbicides to the tank. Fill the tank with water to the desired level. Do not allow the mixture to stand in the spray tank without agitation.

Allow the pump to run several minutes before application to insure proper suspension and mixing. The tank-mix should not be left in the spray tank for prolonged periods, batches should be mixed and applied the same day.

APPLICATION DIRECTIONS

The soil should be well-worked and dry enough to permit good soil mixing (incorporation). Pour the recommended amounts of DEVRINOL 2-E and TILLAM 6-E into the spray tank during the filling operations. Apply in 25 to 50 gallons of water per acre.

For soil incorporated application, use a properly calibrated, low-pressure, boom-type sprayer having good agitation.

In irrigated areas, do not apply before pre-irrigation. For all band applications, reduce dosage proportionately depending upon row spacing and band to be treated. Do not use on rocky ground because thorough incorporation is not possible.

INCORPORATION DIRECTIONS

The herbicides must be mixed (incorporated) into the soil IMMEDIATELY after application to prevent loss by evaporation. Whenever possible, application and incorporation should be done in the same operation.

INCORPORATION BEFORE PLANTING

Use power-driven cultivation equipment set to cut to a depth two to three inches in coarse soils and three to four inches in heavy soils OR a tandem disc set to cut to a depth of six inches, followed by a spike-tooth harrow or some other leveling device which extends beyond ends of the disc. For thorough mixing, disc in two different directions (cross disc).

INCORPORATION AT PLANTING

Use power-driven cultivation equipment set to cut to a depth of two to three inches OR, on coarse soils only, ground-driven tillers (rolling cultivators, rotary hoe, row wheels, etc.) set to cut to a depth of 1½ inches and operated at 6 to 8 mph. When using ground-driven tillers, precultivate if soil is crusted.

PLANTING

Crops should be planted or seeded immediately after application and incorporation. Do not use a drag behind the planter as it may concentrate the herbicides over the seed row and cause crop injury.

CULTURAL PRACTICES FOLLOWING APPLICATION:

Shallow cultivation may be necessary to control those susceptible weeds that escape control as well as those weeds not susceptible to either herbicide. Do not cultivate deeper than the depth of herbicide incorporation.

When cultivating fields where tank mix has been banded in the row, use shields to prevent the movement of untreated soil into the treated row.

RECOMMENDATIONS

TOMATOES (Direct Seeded (California Only) and Transplant): Apply ½ to 1 gallon DEVRINOL 2-E plus 2⅔ to 4 quarts TILLAM 6-E per broadcast acre. Use the lower rates on light soils (coarse textured—sandy) and the higher rate on heavy soils (fine textured—clay). Do not use hotcaps on transplanted tomatoes.

TOBACCO (Transplanted):

Before transplanting, apply ½ gallon DEVRINOL 2-E plus 2-⅔ quarts TILLAM 6-E per broadcast acre.

STORAGE AND DISPOSAL

PROHIBITIONS

Do not contaminate water, food, or feed by storage or disposal. Open dumping is prohibited. Do not reuse container.

STORAGE

Store in a cool dry place. Do not store below 20°F. Keep container closed when not in use.

PESTICIDE DISPOSAL

Pesticide wastes are acutely hazardous. Improper disposal of excess pesticide, spray mixture or rinsate is a violation of Federal Law. If these wastes cannot be disposed of by use according to label instructions, contact your State Pesticide or Environmental Control Agency, or the Hazardous Waste representative at the nearest EPA Regional Office for guidance.

CONTAINER DISPOSAL

Triple rinse (or equivalent). Then offer for recycling or reconditioning, or puncture and dispose of in a sanitary landfill, or by other procedures approved by State and local authorities.

NOTICE—READ CAREFULLY

CONDITIONS OF SALE

Stauffer (and seller) offer(s) this product for sale subject to, and buyer and all users are deemed to have accepted, the following conditions of sale and warranty which may only be varied by written agreement of a duly authorized representative of Stauffer.

WARRANTY LIMITATION

Stauffer warrants that this product conforms to the chemical description on the label and is reasonably fit for the purposes referred to in the directions for use on the label subject to the inherent risks referred to below. Stauffer makes no other express warranties. THERE IS NO IMPLIED WARRANTY OF MECHANTABILITY and there are no warranties which extend beyond the description on the label hereof.

INHERENT RISKS

The directions for use of this product are believed to be reliable and should be followed carefully. However, it is impossible to eliminate all risks associated with use. Buyer assumes all risks associated with use or application of this product contrary to label instructions or resulting from extraordinary weather conditions.

LIMITATION OF LIABILITY

In no case shall Stauffer be liable for special, indirect or consequential damages resulting from the use or handling of this product and no claim of any kind shall be greater in amount than the purchase price of the product in respect of which such damages are claimed.

This product is sold only for uses stated on its label. No express or implied license is granted to use or sell this product under any patent in any country except as specified: Country: United States of America Patent No.: 3,480,671; 3,718,455

This is a specimen label. Stauffer is not responsible for the accuracy of the information contained herein. As labels are subject to revision, always carefully read and follow the label on the product container.

DEVRINOL® 10-G

SELECTIVE HERBICIDE—GRANULES

For Use on Certain Citrus, Nuts, Pome Fruits, Small Fruits, Stone Fruits, Strawberries and Grapes

ACTIVE INGREDIENT:	
N,N-diethyl-2-(1-naphthalenyloxy)-propionamide	10.0%
INERT INGREDIENTS:	90.0%
TOTAL	100.0%

Contains 10 lbs. of active ingredient per 100 pounds

EPA Reg. No. 476-2218 A-1

CAUTION

Keep Out of Reach of Children

MAY BE HARMFUL IF SWALLOWED

STATEMENT OF PRACTICAL TREATMENT

FIRST AID: If an adverse reaction occurs after exposure, promptly initiate the recommended procedures below. Seek medical attention if further treatment is required.

In case of contact: Flush eyes and skin with plenty of water for several minutes. Get medical attention if eye or skin irritation occurs.

If swallowed: Immediately give several glasses of water and induce vomiting by gagging the victim with a finger or blunt object placed on the back of the victim's tongue. Give fluids until vomitus is clear. If victim is unconscious, do not induce vomiting or give anything by mouth.

If inhaled: Remove to fresh air. Seek medical attention if respiratory irritation occurs or if breathing becomes difficult.

PRECAUTIONARY STATEMENTS

HAZARDS TO HUMANS AND DOMESTIC ANIMALS

CAUTION

May be harmful if swallowed. Avoid contact with eyes, skin and clothing. Avoid inhalation of dust.

ENVIRONMENTAL HAZARDS

Do not apply to any body of water. Do not contaminate water by cleaning of equipment or disposal of wastes. Do not apply when weather conditions favor drift from areas treated. Do not contaminate irrigation water or water used for domestic purposes.

In Case of Emergency—Immediately call (24 hours) (800) 424-9300 Chemtrec, or (203) 226-6602 Stauffer Chemical Company.

GENERAL PRECAUTIONS

Read all label directions before using.

Use only for recommended purposes and at recommended rates. Do not overdose.

DEVRINOL 10-G is recommended for use on mineral soils only. Not recommended for use on soils with over 10% organic matter. Treating ground heavily covered with leaves or trash may adversely affect weed control.
Do not apply to frozen ground.
To avoid injury to crops not specified on this label, do not plant until 12 months after the last DEVRINOL 10-G application. A deep disc plow or moldboard plow operation must be carried out before planting.

GENERAL DIRECTIONS FOR USE

It is a violation of Federal law to use this product in a manner inconsistent with its labeling.

DEVRINOL 10-G is a selective surface-applied herbicide for use in certain types of newly planted or established crops. DEVRINOL 10-G does not control established weeds. All weed growth and crop stubble in the fields must be thoroughly worked into the soil prior to planting. Application equipment should be carefully calibrated before use and checked frequently during application to be sure equipment is free from clogging and is delivering a uniform pattern. Apply the recommended rate of DEVRINOL 10-G by ground or air equipment.

NOTE: All rates are given on a broadcast basis, per acre. Reduce rates proportionately for band or strip treatment.

WEEDS CONTROLLED BY DEVRINOL 10-G

NOTE: DEVRINOL 10-G does not control established weeds.

A surface application of 40 lbs. DEVRINOL 10-G will control the following weeds:

Annual Grasses:	
Barnyardgrass, Junglerice	*Echinochloa* spp.
Wild Barley	*Hordeum Leporinum*
Annual Bluegrass	*Poa Annua*
Bromes	*Bromus*
Cheatgrass	*B. Secalinus*
Downy Brome	*B. Tectorum*
Ripgut Brome	*B. Rigidus*
Soft Chess	*B. Mollis*
Large (Hairy) Crabgrass	*Digitaria Sanguinalis*
Smooth Crabgrass	*Digitaria Ischaemum*
Southwestern Cupgrass	*Eriochloa Gracilis*
Foxtails	*Setaria*
Bristly Foxtail	*S. Verticilla*
Giant Foxtail	*S. Faberi*
Green Foxtail	*S. Viridis*
Yellow Foxtail	*S. Lutescens*
Goosegrass	*Eleusine Indica*
Seedling Johnsongrass	*Sorghum Halepense*
Wild Oats	*Avena Fatua*
Panicums	*Panicum*
Guineagrass	*P. Maximum*
Fall Panicum	*P. Dichotomiflorum*
Texas Panicum	*P. Texanum*
Witchgrass	*P. Capillare*
Italian Ryegrass	*Lolium Multiflorum*
Sandbur	*Cenchrus* spp.
Red Sprangletop	*Leptochloa Filiformis*
Stinkgrass	*Eragrostis Cilianensis*
Annual Broadleaf Weeds:	
Carpetweed	*Mollugo Verticillata*
Common Chickweed	*Stellaria Media*
Coast Fiddleneck	*Amsinckia Intermedia*
Filaree	*Erodium* spp.
Groundsel	*Senecio* spp.
Prostrate Knotweed	*Polygonum Aviculare*
Lambsquarters	*Chenopodium Album*
Prickly Lettuce	*Lactuca Serriola*
Little Mallow*	*Malva Parviflora*
Redroot Pigweed	*Amaranthus Retroflexus*
Pineappleweed	*Matricaria Matricarioides*
Common Purslane	*Portulaca Oleracea*
Horse Purslane	*Trianthema Postullacastrum*
Common Ragweed**	*Ambrosia Artemisiieolia*
Annual Sowthistle	*Sonchus Oleraceus*

* From Seed
** Suppression

NOTE: See Recommendations Section for weeds controlled in Cranberries.

DIRECTIONS FOR USE

For control of all weeds listed, apply 40 pounds of DEVRINOL 10-G per acre to a weed free soil surface. Apply in the fall through early spring prior to weed emergence. For best results, the treatment must reach the zone of weed seed germination. If rainfall does not occur the treatment must be shallowly incorporated or irrigated-in (sprinkler or flood) following application with sufficient water to wet the soil to a depth of 2 to 4 inches. In the Western and Pacific Northwest Regions treatments made in fall through early spring must be incorporated or irrigated-in if no rainfall occurs within two weeks of application. For treatments made other than in fall through early spring in these Regions, and for all treatments applied in the Northern, Southwestern and Southeastern Regions regardless of season, the treatment must be incorporated or irrigated-in within 24 hours if no rainfall occurs.

Avoid contact with fruit or foliage. Do not apply when fruit or nuts are on the ground during the harvest period. Do not graze treated areas.

RECOMMENDATIONS

Application may be made immediately after planting, except where treatment is restricted to established crops only. For these uses, apply to stands that have been established in the field for at least one year.

TREES AND VINES: Apples, Apricots, Cherries, Filberts, Grapefruit, Grapes, Lemons, Nectarines, Oranges, Peaches, Pears, Pecans, Plums, Prunes, Tangerines and Walnuts. Apply 40 pounds DEVRINOL 10-G per acre to a weed free soil surface.

STRAWBERRIES: Apply 40 pounds DEVRINOL 10-G per acre to a weed free soil surface. Delay application until the desired number of daughter plants have become established. Do not apply from bloom through harvest.

BUSH FRUIT: Blueberries and Caneberries (Blackberries, Raspberries, Boysenberries, Loganberries). Apply 40 pounds DEVRINOL 10-G per acre to a weed free soil surface.

CRANBERRIES: CAUTION: Do not apply in the springtime to bogs that do not have sprinkler systems for frost control.

Massachusetts (Established Bogs Only): For control of nutsedge (Cyperus Dentatus) and Rice Cutgrass (Leersia Oryzoides), apply 60-90 pounds DEVRINOL 10-G per acre to a weed free soil surface before spring growth begins or in the fall after harvest.

New Jersey (Established Bogs Only): For control of nutsedge (Cyperus Dentatus), Rice Cutgrass (Leersia Oryzoides) and Purplestem Beggarticks (Bidens Cannata), apply 60-90 pounds DEVRINOL 10-G per acre to a weed free soil surface before spring growth begins or in the fall after harvest.

Oregon and Washington (Established Bogs Only): For control of Clover (Trifolium Fimbriatum), Aster (Aster Douglasii), Silverweed Cinquefoil* (Silverleaf) (Potentilla Anserina), Red Sorrel (Sheep Sorrel) (Rumex Acetosella) and Birdsfoot Trefoil (Lotus Corniculatus):

* Suppression only

In peat beds: Apply 150 pounds DEVRINOL 10-G per acre to a weed free soil surface before spring growth begins in Mid-February or before the end of March. Do not apply after buds start to swell.

In sand beds: Apply 40-90 pounds DEVRINOL 10-G per acre to a weed free soil surface before spring growth begins in Mid-February or before the end of March. Use the lower rate on one-year old beds only. Do not apply after buds start to swell. For further information, consult your Local Ag Extension Service.

Wisconsin: For control of Purplestem Beggarticks (Bidens Connata) and Rice Cutgrass (Leersia Oryzoides) in:

New Plantings: Apply 30 pounds DEVRINOL 10-G per acre to a weed free soil surface in the spring after setting vines.

Established Plantings: Apply 60 pounds DEVRINOL 10-G per acre to a weed free soil surface before spring growth begins or in the fall after harvest. Do not apply when bogs are under winter flood.

MINT: Established mint (Pacific Northwest only). Apply 40 pounds DEVRINOL 10-G per acre to a weed free soil surface.

STORAGE AND DISPOSAL

PROHIBITIONS
Do not contaminate water, food or feed by storage or disposal.

STORAGE
Keep container closed when not in use. Store in a cool dry place. Do not store near seeds, fertilizer or foodstuffs.

PESTICIDE DISPOSAL
Pesticide or rinsate that cannot be used according to label instructions must be disposed of according to applicable Federal, state or local procedures.

CONTAINER DISPOSAL
Completely empty bag by shaking and tapping sides and bottom to loosen clinging particles. Empty residue into application equipment. Then dispose of bags in a sanitary landfill or by incineration if allowed by state and local authorities.

NOTICE—READ CAREFULLY

CONDITIONS OF SALE
Stauffer (and seller) offer(s) this product for sale subject to and buyer and all users are deemed to have accepted, the following conditions of sale and warranty which may only be varied by written agreement of a duly authorized representative of Stauffer.

WARRANTY LIMITATIONS
Stauffer warrants that this product conforms to the chemical description on the label and is reasonably fit for the purposes referred to in the directions for use on the label subject to the inherent risks referred to below. Stauffer makes no other express warranties; THERE IS NO IMPLIED WARRANTY OF MERCHANTABILITY and there are no warranties which extend beyond the descripton on the label thereof.

INHERENT RISKS
The directions for use of this product are believed to be reliable and should be followed carefully. However, it is impossible to eliminate all risks associated with use. Buyer assumes all risks associated with use or application of this product contrary to label instructions or resulting from extraordinary weather conditions.

LIMITATION OF LIABILITY
In no case shall Stauffer be liable for special, indirect or consequential damages resulting from the use or handling of this product and no claim of any kind shall be greater in amount than the purchase price of the product in respect of which such damages are claimed.

This product is sold only for uses stated on its label. No express or implied license is granted to use or sell this product under any patent in any country except as specified: Country: United States Patent No.: 3,480,671; 3,718,455.

This is a specimen label. Stauffer is not responsible for the accuracy of the information contained herein. As labels are subject to revision, always carefully read and follow the label on the product container.

DEVRINOL® 50-WP

SELECTIVE HERBICIDE WETTABLE POWDER

For use on certain Citrus, Nuts, Pome fruits, Small fruits, Stone fruits, Vegetables and Tobacco

Complete Directions for Use, Use Precautions and Recommendations

ACTIVE INGREDIENT:
N,N-Diethyl-2-(1-naphthalenyloxy)-
propionamide 50%
INERT INGREDIENTS 50%
TOTAL 100%

Contains 50 lbs. of active ingredient per 100 lbs.

EPA Reg. No. 476-2108 A-13R

CAUTION

Keep Out of Reach of Children

STATEMENT OF PRACTICAL TREATMENT

FIRST AID
Immediately start the procedures given below. If further treatment is required, contact a Poison Center, a physician, or the nearest hospital, or call collect, day or night: (203) 226–6602 Stauffer Chemical Company or (800) 424-9300 Pesticide Team Safety Network.

If swallowed: Immediately give several glasses of water and induce vomiting by gagging the victim with a finger placed on the back of the victim's tongue. Give fluids until vomitus is clear. If victim is unconscious or convulsing, do not induce vomiting or give anything by mouth.

For skin contact: Flush all affected areas with plenty of water for several minutes. For eye contact, hold eyelids apart and flush with plenty of water for 15 minutes. Remove and clean contaminated clothing and shoes before reuse. Seek medical attention if skin or eye irritation occurs.

If inhaled: Remove to fresh air. Seek medical attention if respiratory irritation occurs or if breathing becomes difficult.

PRECAUTIONARY STATEMENTS

HAZARDS TO HUMANS AND DOMESTIC ANIMALS

CAUTION

MAY BE HARMFUL IF SWALLOWED
Do not breathe spray mist. Avoid contact with skin, eyes and clothing. Wash thoroughly with soap and water after use.

ENVIRONMENTAL HAZARDS
Do not apply directly to water or wetlands.
Do not contaminate water by cleaning of equipment or disposal of wastes. Apply this product as specified on this label.
Do not contaminate water used for domestic purposes.
Avoid Contamination of Food and Feed.

GENERAL PRECAUTIONS

Read all instructions before using.
Use only for recommended purposes and at recommended rates. Do not overdose.
Except where specifically noted, DEVRINOL 50-WP is not recommended for use on soil containing more than 10% organic matter.
Spraying on ground heavily covered with leaves or trash may adversely affect weed control.
Do not apply to frozen ground.
To minimize injury to succeeding crops not listed on this label, a deep disc plow or moldboard plow operation (minimum 10 inches) must be carried out prior to planting. To avoid injury to crops not

specified on the label, do not plant until 12 months after the last application of DEVRINOL 50-WP.

When applied at rates no greater than 4 lbs. DEVRINOL 50-WP per acre, rotational crops which are not registered on this label may be planted no sooner than 12 months after application when harvest and use of mature crop(s) is only for food or feed purposes. Crops are not to be grazed or harvested for food or feed purposes earlier than maturity.

GENERAL DIRECTIONS FOR USE

It is a violation of Federal law to use this product in a manner inconsistent with its labeling.

DEVRINOL 50-WP is a selective surface-applied or incorporated herbicide for use in certain types of newly planted or established crops. DEVRINOL 50-WP does not control established weeds. All weed growth and crop stubble in the fields must be thoroughly worked into the soil prior to planting. Spray equipment should be carefully calibrated before use and checked frequently during application to be sure nozzles are operating properly and are delivering a uniform spray pattern. Apply the recommended rate of DEVRINOL 50-WP in 20 to 100 gallons of water per acre in a low pressure boom-type sprayer.

Be sure the spray tank, boom and nozzles have been thoroughly flushed with clean water before use. Sufficient jet or mechanical agitation should be provided during the filling operation and during application.

When mixing DEVRINOL 50-WP alone, or in combination with other recommended wettable powders, make a slurry and add to a nearly filled spray tank with adequate agitation. This will enhance initial mixing. Recommended contact herbicides and spreaders should always be added to a nearly filled spray tank. Never add them to the slurry.

Whenever a rate range is recommended for DEVRINOL 50-WP use the lower rate on light soil (coarse textured-sandy) and the higher rate on heavy soil (fine textured-clay).

NOTE: All rates are given on a broadcast basis, per acre. Reduce rates proportionately for band or strip treatment.

WEEDS CONTROLLED BY 8 LBS. DEVRINOL 50-WP PER ACRE

SURFACE APPLICATION

NOTE: DEVRINOL 50-WP does not control established weeds.

Annual Grasses	
Barnyardgrass, Junglerice	*Echinochloa* spp.
Barley	*Hordeum jubatum*
Wild Barley	*Hordeum leporinum*
Annual Bluegrass	*Poa annua*
Bromes	*Bromus*
Cheatgrass	*B. secalinus*
Downy Brome	*B. tectorum*
Ripgut Brome	*B. rigidus*
Soft Chess	*B. mollis*
Canarygrass	*Phalaris canariensis*
Large (Hairy) Crabgrass	*Digitaria sanguinalis*
Smooth Crabgrass	*Digitaria ischaemum*
Southwestern Cupgrass	*Eriochloa gracilis*
Foxtails	*Setaria*
Bristly Foxtail	*S. verticilla*
Giant Foxtail	*S. faberi*
Green Foxtail	*S. viridis*
Yellow Foxtail	*S. lutescens*
Goosegrass	*Eleusine indica*
Seedling Johnsongrass	*Sorghum halepense*
Wild Oats	*Avena fatua*
Panicums	*Panicum*
Guineagrass	*P. Paximum*
Fall Panicum	*P. dichotomiflorum*
Texas Panicum	*P. texanum*
Witchgrass	*P. capillare*
Italian Ryegrass	*Lolium multiflorum*
Sandbur	*Cenchrus* spp.
Mexican Sprangletop	*Leptochloa uninervia*
Red Sprangletop	*Leptochloa filiformis*
Stinkgrass	*Eragrostis cilianesis*
Annual Broadleaf Weeds	
Carpetweed	*Mullugo verticillata*
Common Chickweed	*Stellaria media*
Coast Fiddleneck	*Amsinckia intermedia*
Cudweed, Purple	*Gnaphalium purpureum*
Redstem Filaree	*Erodium fiarcutorium*
Common Groundsel	*Senecio vulgaris*
Prostrate Knotweed	*Polygonum aviculare*
Lambsquarters	*Chenopodium album*
Prickly Lettuce	*Lactuca serriola*
Little Mallow*	*Malva parviflora*
Redroot Pigweed	*Amaranthus retroflexus*
Pineappleweed	*Matricaria matricarioides*
Common Purslane	*Portulaca oleracea*
Horse Purslane	*Trianthema portullacastrum*
Common Ragweed**	*Ambrosia artemisiifolia*
Annual Sowthistle	*Sonchus oleraceus*

* From seed
** Suppression

REGIONAL USE MAP

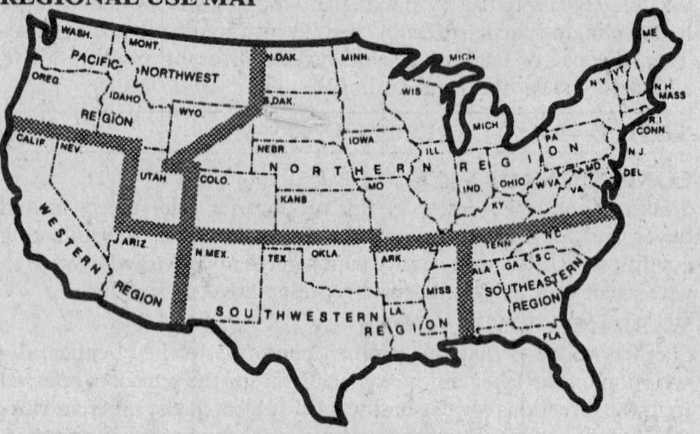

DIRECTIONS FOR USE

NOTE: All rates are given on a broadcast basis. Reduce rates proportionately for band or strip treatment.

For control of all weeds listed, apply 8 pounds DEVRINOL 50-WP per broadcast acre to the soil surface. Apply in the fall through early spring prior to weed emergence. Do not apply DEVRINOL 50-WP to frozen ground. For best results, the treatment must reach the zone of weed seed germination. If rainfall does not occur the treatment must be shallowly incorporated or irrigated in following application (sprinkler or flood) with sufficient water to wet the soil to a depth of 2 to 4 inches.

ALLOWABLE TIME FOR TREATMENT TO REMAIN ON SOIL SURFACE (By Region):

WEST: Three weeks (November through February) or 24 hours (remainder of year).

PACIFIC NORTHWEST: Three weeks (Fall through Early Spring) or 24 hours (remainder of year).

OTHER REGIONS: 24 hours.

Use as a directed spray and avoid contact with fruit or foliage. Do not apply when fruit or nuts are on the ground during the harvest period. Do not graze treated areas.

FRUIT AND NUT CROPS

SURFACE APPLICATION RECOMMENDATIONS

DEVRINOL 50-WP is recommended for use on the following crops. Refer to regional map under "DIRECTIONS FOR USE".

ALMONDS: (All Regions)
Rate: 8 pounds DEVRINOL 50-WP per broadcast acre. In the Western Region an additional 8 pounds may be applied during the growing season. Do not apply within 35 days of harvest.
Use Recommendations: Apply to a weed free soil surface. May be applied to newly planted and established crop.

APPLES: (All Regions)
Rate: 8 pounds DEVRINOL 50-WP per broadcast acre. In the Western Region an additional 8 pounds may be applied during the growing season. Do not apply within 35 days of harvest.
Use Recommendations: Apply to a weed free soil surface. May be applied to newly planted and established crop.

APRICOTS: (All Regions)
Rate: 8 pounds DEVRINOL 50-WP per broadcast acre. In the Western Region an additional 8 pounds may be applied during the growing season. Do not apply within 35 days of harvest.
Use Recommendations: Apply to a weed free soil surface. May be applied to newly planted and established crop.

AVOCADOS: (Western Region Only)
Rate: 8 pounds DEVRINOL 50-WP per broadcast acre. An additional 8 pounds may be applied during the growing season. Do not apply within 35 days of harvest.
Use Recommendations: Apply to a weed free soil surface. May be applied to established crop only.

BLUEBERRIES: (All Regions)
Rate: 8 pounds DEVRINOL 50-WP per broadcast acre.
Use Recommendations: Apply to a weed free soil surface. May be applied to newly planted and established crop.

CANEBERRIES (blackberries, raspberries, boysenberries and loganberries): (All Regions)
Rate: 8 pounds DEVRINOL 50-WP per broadcast acre.
Use Recommendations: Apply to a weed free soil surface. May be applied to newly planted and established crop.

CHERRIES: (All Regions)
Rate: 8 pounds DEVRINOL 50-WP per broadcast acre. In the Western Region an additional 8 pounds may be applied during the growing season. Do not apply within 35 days of harvest.
Use Recommendations: Apply to a weed free soil surface. May be applied to newly planted and established crop.

FIGS: (All Regions)
Rate: 8 pounds DEVRINOL 50-WP per broadcast acre. In the Western Region an additional 8 pounds may be applied during the growing season. Do not apply within 35 days of harvest.
Use Recommendations: Apply to a weed free soil surface. May be applied to newly planted and established crop.

FILBERTS: (All Regions)
Rate: 8 pounds DEVRINOL 50-WP per broadcast acre. In the Western Region an additional 8 pounds may be applied during the growing season. Do not apply within 35 days of harvest.
Use Recommendations: Apply to a weed free soil surface. May be applied to newly planted and established crop.

GRAPEFRUIT: (All Regions)
Rate: 8 pounds DEVRINOL 50-WP per broadcast acre. In the Western Region an additional 8 pounds may be applied during the growing season. Do not apply within 35 days of harvest.
Use Recommendations: Apply to a weed free soil surface. May be applied to newly planted and established crop.

GRAPES: (All Regions)
Rate: 8 pounds DEVRINOL 50-WP per broadcast acre. In the Western Region an additional 8 pounds may be applied during the growing season. Do not apply within 35 days of harvest.
Use Recommendations: Apply to a weed free soil surface. May be applied to a newly planted and established crop.

KIWIFRUIT: (Western Region Only)
Rate: 8 pounds DEVRINOL 50-WP per broadcast acre. An additional 8 pounds may be applied during the growing season. Do not apply within 35 days of harvest.
Use Recommendations: Apply to a weed free soil surface. May be applied to newly planted and established crop.

LEMONS: (All Regions)
Rate: 8 pounds DEVRINOL 50-WP per broadcast acre. In the Western Region an additional 8 pounds may be applied during the growing season. Do not apply within 35 days of harvest.
Use Recommendations: Apply to a weed free soil surface. May be applied to newly planted and established crop.

NECTARINES: (All Regions)
Rate: 8 pounds DEVRINOL 50-WP per broadcast acre. In the Western Region an additional 8 pounds may be applied during the growing season. Do not apply within 35 days of harvest.
Use Recommendations: Apply to a weed free soil surface. May be applied to newly planted and established crop.

OLIVES: (Western Region Only)
Rate: 8 pounds DEVRINOL 50-WP per broadcast acre. An additional 8 pounds may be applied during the growing season. Do not apply within 35 days of harvest.
Use Recommendations: Apply to a weed free soil surface. May be applied to newly planted and established crop.

ORANGES: (All Regions)
Rate: 8 pounds DEVRINOL 50-WP per broadcast acre. In the Western Region an additional 8 pounds may be applied during the growing season. Do not apply within 35 days of harvest.
Use Recommendations: Apply to a weed free soil surface. May be applied to newly planted and established crop.

PEACHES: (All Regions)
Rate: 8 pounds DEVRINOL 50-WP per broadcast acre. In the Western Region an additional 8 pounds may be applied during the growing seaseon. Do not apply within 35 days of harvest.
Use Recommendations: Apply to a weed free soil surface. May be applied to newly planted and established crop.

PEARS: (All Regions)
Rate: 8 pounds DEVRINOL 50-WP per broadcast acre. In the Western Region an additional 8 pounds may be applied during the growing season. Do not apply within 35 days of harvest.
Use Recommendations: Apply to a weed free soil surface. May be applied to newly planted and established crop.

PECANS: (All Regions)
Rate: 8 pounds DEVRINOL 50-WP per broadcast acre. In the Western Region an additional 8 pounds may be applied during the growing season. Do not apply within 35 days of harvest.
Use Recommendations: Apply to a weed free soil surface. May be applied to a newly planted and established crop.

PERSIMMONS: (Western Region Only)
Rate: 8 pounds DEVRINOL 50-WP per broadcast acre. An additional 8 pounds may be applied during growing season. Do not apply within 35 days of harvest.
Use Recommendations: Apply to a weed free soil surface. May be applied to established crop only.

PISTACHIOS: (All Regions)
Rate: 8 pounds DEVRINOL 50-WP per broadcast acre.
Use Recommendations: Apply to a weed free soil surface. May be applied to newly planted and established crop.

PLUMS: (All Regions)
Rate: 8 pounds DEVRINOL 50-WP per broadcast acre. In the Western Region an additional 8 pounds may be applied during the growing season. Do not apply within 35 days of harvest.
Use Recommendations: Apply to a weed free soil surface. May be applied to newly planted and established crop.

PRUNES: (All Regions)
Rate: 8 pounds DEVRINOL 50-WP per broadcast acre. In the Western Region an additional 8 pounds may be applied during the growing season. Do not apply within 35 days of harvest.
Use Recommendations: Apply to a weed free soil surface. May be applied to newly planted and established crop.

RHUBARB: (Pacific Northwest Only)
Rate: 8 pounds DEVRINOL 50-WP per broadcast acre.
Use Recommendations: Apply to a weed free soil surface in Winter when rhubarb is dormant.

STRAWBERRIES: (All Regions)
Rate: 8 pounds DEVRINOL 50-WP per broadcast acre.
Use Recommendations: Apply to a weed free soil surface. May be applied to newly transplanted and established crop. Delay application until desired number of daughter plants have become established. Do not apply from bloom through harvest.

TANGERINES: (All Regions)
Rate: 8 pounds DEVRINOL 50-WP per broadcast acre. In the Western Region an additional 8 pounds may be applied during the growing season. Do not apply within 35 days of harvest.
Use Recommendations: Apply to a weed free soil surface. May be applied to newly planted and established crop.

WALNUTS: All Regions)
Rate: 8 pounds DEVRINOL 50-WP per broadcast acre. In the Western Region an additional 8 pounds may be applied during the growing season. Do not apply within 35 days of harvest.
Use Recommendations: Apply to a weed free soil surface. May be applied to newly planted and established crop.

OTHER CROPS SURFACE APPLICATION

RECOMMENDATIONS

ARTICHOKES: (Western Region Only)
Rate: 8 pounds DEVRINOL 50-WP per broadcast acre.
Use Recommendations: Apply to a weed free soil surface. May be applied to established crop only.

ASPARAGUS: (All Regions)
Rate: 8 pounds DEVRINOL 50-WP per broadcast acre.
Use Recommendations: Apply to a weed free soil surface. May be applied to established crop only.

MINT: (Pacific Northwest Only)
Rate: 8 pounds DEVRINOL 50-WP per broadcast acre.
Use Recommendations: Apply to a weed free soil surface. May be applied to newly planted and established crop.

Where recommended for use on established crops, apply DEVRINOL 50-WP only to stands that have been established for at least one growing season. Where recommended for use on newly planted crops, application may be made immediately after planting young trees, vines, and bushes.

DEVRINOL 50-WP may be applied at the recommended rates through irrigation water for pre-emergence control of listed weeds in grapes and tree crops (Western Region Only). Do not allow treatment to contact foliage or crop. Remove existing weed growth before application. Apply at the start of irrigation set and with sufficient water to wet the soil to a depth of 2–4 inches. Avoid runoff. These irrigation systems should be equipped with automatic shutoff devices and valves to prevent back flow into the water source.

WEEDS CONTROLLED BY 2 to 4 LBS. DEVRINOL 50-WP PREPLANT INCORPORATED APPLICATION

NOTE: DEVRINOL 50-WP does not control established weeds.

Annual Grasses	
Barnyardgrass, Junglerice	*Echinochloa* spp.
Crabgrass	*Digitaria* spp.
Foxtails	*Setaria*
Giant Foxtail	*S. faberi*
Green Foxtail	*S. viridis*
Yellow Foxtail	*S. lutescens*
Goosegrass	*Eleusine indica*
Seedling Johnsongrass	*Sorghum halepense*
Panicums	*Panicum*
Fall Panicum	*P. dichotomiflorum*
Witchgrass	*P. capillare*
Stinkgrass	*Eragrostis cilianensis*
Annual Broadleaf Weeds	
Carpetweed	*Molugo verticillata*
Purple Cudweed	*Gnaphalium purpureum*
Galinsoga*	*Galinsoga* spp.
Lambsquarters	*Chenopodium album*
Redroot Pigweed	*Amaranthus retroflexus*
Smooth Pigweed	*Amaranthus hybridus*
Common Purslane	*Portulaca oleracea*
Florida Pusley	*Richardia scabra*
Common Ragweed**	*Ambrosia artemisiifolia*

* Controlled at 4 lb. DEVRINOL 50-WP.
** Suppression

DIRECTIONS FOR USE OF DEVRINOL 50-WP PRE-PLANT INCORPORATED

On flat-planted or bedded crops apply, before planting, to well-worked soil that is dry enough to permit thorough incorporation to a depth of 1 to 2 inches. Incorporate the same day as applied, using equipment which will result in uniform incorporation of the herbicide to the desired depth.
After harvest or prior to planting of succeeding crops, a deep moldboard or disc plowing operation must be carried out.

When applied at rates no greater than 4 lbs. DEVRINOL 50-WP per acre, rotational crops which are not registered on this label may be planted no sooner than 12 months after application when harvest and use of the mature crop(s) only is for food or feed purposes. Crops are not to be grazed or harvested for food or feed purposes earlier than maturity.

MELONS, TOBACCO AND VEGETABLES PRE-PLANT INCORPORATED APPLICATION

RECOMMENDATIONS

NOTE: All rates are given on a broadcast basis. Reduce rates proportionately for band or strip treatment.

COLE CROPS—Broccoli, Brussels Sprouts, Cauliflower: (California: To be used in Monterey, San Benito, San Mateo, Santa Cruz Counties Only)
Rate: 1-2 pounds DEVRINOL 50-WP per broadcast acre.
Use Recommendations: For suppression of Annual Bluegrass, Barnyardgrass, Common Groundsel, Common Purslane and Redroot Pigweed. Apply to a weed free soil surface. May be applied to direct seeded and transplanted crop, pre-plant incorporated or as a surface spray after planting. Shallow incorporate to a depth of 1-2 inches or sprinkler irrigate within 24 hours using sufficient water to wet the soil to a depth of 2-4 inches. Use the lower rate on light soil (coarse textured-sandy) and the higher rate on heavy soil (fine textured-clay).

Cole Crops—Broccoli, Brussels Sprouts, Cabbage, Cauliflower: (Northern Region Only)
Rate: 2 pounds DEVRINOL 50-WP per broadcast acre.
Use Recommendations: Apply to a weed free surface. May be applied to direct seeded and transplanted crop, pre-plant incorporated or as a surface spray after planting. Shallow incorporate no deeper than seeding depth or sprinkler irrigate within 24 hours using sufficient water to wet the soil to a depth of 2-4 inches.

Cole Crops—Broccoli, Brussels Sprouts, Cabbage, Cauliflower: (Pacific Northwest Only)
Rate: 2 pounds DEVRINOL 50-WP per broadcast acre.
Use Recommendations: Apply to a weed free surface. May be applied to direct seeded and transplanted crop, pre-plant incorporated or as a surface spray after planting. Shallow incorporate no deeper than seeding depth or sprinkler irrigate within 24 hours using sufficient water to wet the soil to a depth of 2-4 inches. Use the lower rate on light soil (coarse textured-sandy) and the higher rate on heavy soil (fine textured-clay).

Cole Crops—Broccoli, Brussels Sprouts, Cabbage, Cauliflower: (Southeast Region Only)
Rate: 4 pounds DEVRINOL 50-WP per broadcast acre.
Use Recommendations: Apply to a weed free surface. May be applied to direct seeded (except in Florida) and transplanted crop, as a surface spray after planting. Sprinkler irrigate within 24 hours using sufficient water to wet the soil to a depth of 2-4 inches.

Cole Crops—Broccoli, Brussels Sprouts, Cabbage, Cauliflower: (Southwest Region)
Rate: 2-4 pounds DEVRINOL 50-WP per broadcast acre.

Use Recommendations: Apply to a weed free surface. May be applied to direct seeded and transplanted crop, pre-plant incorporated or as a surface spray after planting. Incorporate the surface spray to a shallow depth of one inch, avoiding the seed placement when mixing the soil. Use the lower rate on light soil (coarse textured-sandy) and the higher rate on heavy soil (fine textured-clay).

EGGPLANTS: (All Regions)

Rate: Before transplanting, apply 2-4 pounds DEVRINOL 50-WP per broadcast acre.

Use Recommendations: Apply to a weed free soil surface. May be applied to transplanted crop only. Use the lower rate on light soil (coarse textured-sandy) and the higher rate on heavy soil (fine textured-clay).

MELONS (except cantaloupes): (Texas, South of Highway I-10 Only)

Rate: 2-4 pounds DEVRINOL 50-WP per broadcast acre.

Use Recommendations: Apply to a weed free soil surface. May be applied to direct seeded crop as a preemergence surface treatment only. Incorporate the surface spray to a shallow depth of one inch, avoiding the seed placement when mixing the soil. Use the lower rate on light soil (coarse textured-sandy) and the higher rate on heavy soil (fine textured-clay).

PEPPERS: (All Regions)

Rate: 2-4 pounds DEVRINOL 50-WP per broadcast acre.

Use Recommendations: Apply to a weed free soil surface. May be applied to direct seeded and transplanted crop. Use the lower rate on light soil (coarse textured-sandy) and the higher rate on heavy soil (fine textured-clay).

TOMATOES: (All Regions)

Rate: 2-4 pounds DEVRINOL 50-WP per broadcast acre.

Use Recommendations: Apply to a weed free soil surface. May be applied to direct seeded and transplanted crop. Use the lower rate on light soil (coarse textured-sandy) and the higher rate on heavy soil (fine textured-clay).

TOBACCO: (All Regions)

Rate: Before transplanting, apply 2-4 pounds DEVRINOL 50-WP per broadcast acre.

Use Recommendations: Apply to a weed free soil surface. Use the lower rate on light soil (coarse textured-sandy) and the higher rate on heavy soil (fine textured-clay). In Southeast only: Apply up to 3 weeks before transplanting. No waiting period is necessary.

POSTTRANSPLANT SURFACE AND LAYBY SURFACE APPLICATIONS FOR USE IN TRANSPLANTED TOBACCO (Kentucky, Maryland, Virginia and Southeastern Region Only)

WEEDS CONTROLLED

See list of weeds controlled under DEVRINOL 50-WP preplant incorporated application.

DIRECTIONS FOR USE.

At Transplanting: Apply the recommended amount of DEVRINOL 50-WP to the soil surface immediately after transplanting. May be applied over the top of transplants.

At Layby: Apply the recommended amount of DEVRINOL 50-WP in a band immediately after the layby cultivation. Apply as a directed spray to the row middles.

Use 20 to 50 gallons of spray per acre. For best results the treatment must reach the zone of weed seed germination. If rainfall does not occur within 5 days after application, the treatment must be shallowly incorporated or irrigated-in with sufficient overhead irrigation to wet the soil to a depth of 2 to 4 inches.

RECOMMENDATIONS

Range: Apply 2 to 4 pounds DEVRINOL 50-WP per broadcast acre.

Use Recommendations: Apply to a weed free soil surface. Reduce rates proportionately for band application. Treatments may follow a preplant incorporated application of DEVRINOL 50-WP or DEVRINOL 50-WP/TILLAM® 6-E. Do not apply more than 4 pounds DEVRINOL 50-WP in any one season.

Rotational Crop Restrictions

When DEVRINOL 50-WP is used on tobacco posttransplant or layby at 2 to 4 pounds, surface applied, small grains may be seeded in rotation in the fall to prevent soil erosion. These grains may be stunted but not otherwise adversely affected.

No tolerance is established for residues of DEVRINOL 50-WP in small grains. Rotational cover crops must be destroyed by plowing under or otherwise prevented from reaching food or feed production.

If rotating out of tobacco the following spring, to avoid injury to crops not specified on this label do not plant until 12 months after the last DEVRINOL 50-WP application.

When applied at rates no greater than 4 lbs. DEVRINOL 50-WP per acre, rotational crops which are not registered on this label may be planted no sooner than 12 months after application when harvest and use of the mature crop(s) only is for food or feed purposes. Crops are not to be grazed or harvested for food or feed purposes earlier than maturity.

TANK MIXTURES FRUIT AND NUT CROPS

IMPORTANT

Observe all cautions and limitations on labeling of all products used in combination.

Make only one application per season.

Do not contaminate water by cleaning of equipment or disposal of wastes. Apply these combinations only as specified on this label.

For a tank mix application, apply the recommended dosage (see directions for use) in 20 to 100 gallons of water per acre.

In any tank mix including PARAQUAT CL, add an approved nonionic spreader to the tank water, such as ORTHO*X-77 spreader, at the rate of 8 oz. per 100 gallons.

* ORTHO is a registered trademark of Chevron Chemical Company.

DEVRINOL 50-WP/PARAQUAT CL COMBINATION

DEVRINOL 50-WP will not control emerged weeds. PARAQUAT CL may be used to control existing weeds when applied prior to, or in combination with DEVRINOL 50-WP.

Refer to the section on DEVRINOL 50-WP surface application for a list of weeds controlled.

DIRECTIONS FOR USE

Follow use directions for DEVRINOL 50-WP surface application. Read the PARAQUAT CL label for timing with respect to stage of weed growth and for application methods recommended to minimize crop injury.

For control of all weeds listed, apply 8 pounds DEVRINOL 50-WP and 1 to 2 quarts PARAQUAT CL per broadcast acre.

RECOMMENDATIONS

Consult the PARAQUAT CL label for recommended uses in orchards and vineyards. Then read the crop recommendations for DEVRINOL 50-WP surface application to insure that the combination may be used for a particular crop in your state. Recommended *only* for the uses specified on the PARAQUAT CL and DEVRINOL 50-WP labels. Do not apply this combination to caneberries.

DEVRINOL 50-WP/DINOSEB COMBINATION (California Only)

For use on the same crops listed under DEVRINOL 50-WP/PARAQUAT CL. Apply 8 pounds DEVRINOL 50-WP and the recommended amount of the DINOSEB product chosen. Follow use directions for DEVRINOL 50-WP surface application.

NOTE: DINOSEB products vary considerably in recommendations, read the manufacturer's label carefully to determine the following:

- Crops on which the product can be used.
- Stage of weed growth for timing of application.
- Requirements for adding petroleum oil or spray adjuvants.
- Specific application methods needed to minimize crop injury.

Follow all restrictions, cautions and limitations on the DINOSEB label.

DEVRINOL 50-WP/GOAL* 1.6-E COMBINATION (California Only)

DEVRINOL 50-WP may be used in combination with GOAL 1.6-E for control of additional weeds not controlled by DEVRINOL 50-WP alone.

Additional weeds controlled preemergence by DEVRINOL 50-WP/GOAL 1.6-E tank mix, applied preemergence to a relatively weed-free established berm or soil surface, at recommended rates:

Burclover	*Medicago hipide*
Filaree, (Redstem,	*Erodium cicutarium*
Broadleaf)	*Erodium botrys*
Henbit	*Lamium amplexicaule*
London rocket	*Sisymbrium irio*
Redmaids	*Calandrinia caulescens*
Shepherdspurse	*Capsella bursa-pastoris*

* GOAL is a registered trademark of Rohm and Haas Company

DIRECTIONS FOR USE—DORMANT APPLICATION (California Only)

For control of all weeds listed, apply 8 pounds of DEVRINOL 50-WP and 5.0 quarts GOAL 1.6-E (2.0 lb active) per broadcast acre. Spray should be directed to the soil and the base of dormant trees or vines. Use of a low pressure sprayer equipped with a break-away boom and flat fan or off-center (OC) nozzle is recommended. For best results, the treatment must reach the zone of weed germination. Irrigation sufficient to wet the soil to a minimum depth of 2 to 4 inches is recommended for treatments made from November through February 15 if no rainfall occurs within three weeks of application.

NOTE: Applications made in October must be irrigated-in (sprinkler or flood) the same day as applied with sufficient water to wet the soil to a minimum depth of 2 to 4 inches.

Cultural practices which result in redistribution or disturbance of the soil surface after treatment will decrease the herbicidal effectiveness of GOAL 1.6-E. For best results from the DEVRINOL 50-WP/GOAL 1.6-E tank mix combination, application should be made to established berms or soil surfaces which are left undisturbed during the time period for which weed control is desired.

Read the GOAL 1.6-E label for specific use directions and precautions.

RECOMMENDATIONS

For use on almond, apricot, cherry, fig, grape, nectarine, peach, pear, pistachio, plum, prune and walnut in California only.

NOTE: To avoid possible crop injury follow these precautions:

- Apply only to dormant fruit or nut trees or grape vines. Do not apply when foliage, buds or fruit/nuts are present.
- Grapes: Do not apply to grapes established less than 3 years. Do not apply to grapes which are not staked or trellised.

DEVRINOL 50-WP/GOAL 1.6-E/PARAQUAT CL COMBINATION (California Only)

DEVRINOL 50-WP/GOAL 1.6-E/PARAQUAT CL is recommended for residual preemergence control of susceptible broadleaf and grassy weeds. PARAQUAT CL is included in the tank mix to control existing emerged weeds.

For weeds controlled refer to the weed lists for DEVRINOL 50-WP and DEVRINOL 50-WP/GOAL 1.6-E tank mix surface applied.

DIRECTIONS FOR USE

For control of all weeds listed, apply 8 pounds DEVRINOL 50-WP, 5 quarts GOAL 1.6-E (2.0 lb active) and 1 quart PARAQUAT CL per broadcast acre.

Follow the use directions and recommendations for the DEVRINOL 50-WP/GOAL 1.6-E combination. Consult the PARAQUAT CL label for timing with respect to stage of weed growth and for application methods recommended to minimize crop injury.

RECOMMENDATIONS

For use on almond, apricot, cherry, fig, grape, nectarine, peach, pear, plum, prune and walnut in California only.

DEVRINOL 50-WP/KARMEX* 80-WP COMBINATION

DEVRINOL 50-WP may be used in combination with KARMEX 80-WP for improved control of annual broadleaf weeds. This tank mix does not control emerged weeds.

* KARMEX is a registered trademark of E.I. DuPont de Nemours & Company

DIRECTIONS FOR USE

NOTE: All rates are given on a broadcast basis. Reduce rates proportionately for band or strip treatment.

For control of all weeds listed, apply 8 pounds DEVRINOL 50-WP and 1.25 pounds KARMEX 80-WP per broadcast acre to the soil surface. Apply in the fall through early spring prior to weed emergence. Do not apply to frozen ground. For best results, the treatment must reach the zone of weed seed germination. If rainfall does not occur the treatment must be shallowly incorporated or irrigated-in following application (sprinkler or flood) with sufficient water to wet the soil to a depth of 2 to 4 inches. In the Western and Pacific Northwest Regions treatment made in November through February must be incorporated-in or irrigated-in if no rainfall occurs within three weeks of application.

For treatments made other than in November through February in these regions, and for all treatments applied in the Northern, Southwestern, and Southeastern Regions regardless of season, the treatment must be incorporated-in or irrigated-in within 24 hours if no rainfall occurs.

Use as a directed spray and avoid contact with fruit or foliage. Do not apply when fruit is on the ground during the harvest period. Do not graze treated areas.

RECOMMENDATIONS

For use on the following crops: Apples and peaches (All Regions). Citrus (California and Arizona only). Grapes, pears and walnuts (California only).

NOTE: To avoid possible crop injury follow these precautions:

- **Apples and Pears:** Use only under trees established in the orchard for at least 1 year; do not treat varieties grafted on full-dwarf root stocks.
- **Citrus:** Use only under trees established in the grove for at least 1 year.
- **Grapes:** Apply only to established vineyards (at least 3 years old) as a band treatment to grape rows. Do not apply to vines with trunks less than 1½" in diameter as injury may result. On soils low in clay or organic matter (1 to 2%), severe plant injury may result if heavy rainfall or more than one inch of irrigation occurs soon after treatment. This risk must be assumed by the user.
- **Peaches:** Use only under trees established in the orchard for at least 2 years.
- **Walnuts:** Use only under trees established in the orchard for at least 1 year.

DEVRINOL 50-WP/ROUNDUP* COMBINATION

ROUNDUP may be used to control existing weeds when applied prior to or in combination with DEVRINOL 50-WP.

*ROUNDUP is a registered trademark on Monsanto Company.

DIRECTIONS FOR USE

Follow use directions for DEVRINOL 50-WP surface application.

Read the ROUNDUP label for timing with respect to stage of weed growth and for application methods. For control of all weeds listed, apply 8 pounds of DEVRINOL 50-WP in 20 to 100 gallons of water per broadcast acre. For application rates of ROUNDUP, refer to the ROUNDUP label.

RECOMMENDATIONS

Consult the ROUNDUP label for recommended uses in orchards and vineyards. Then read the crop recommendations for DEVRI-

NOL 50-WP surface application to insure that the combination may be used for a particular crop in your state. Recommended only for the uses specified on the ROUNDUP and DEVRINOL 50-WP labels.

DEVRINOL 50-WP/SIMAZINE 80-WP COMBINATION (California and Pacific Northwest Only)

DEVRINOL 50-WP may be used in combination with SIMAZINE 80-WP for control of additional weeds not controlled by DEVRINOL 50-WP alone. This tank mix does not control emerged weeds.

Additional weeds controlled by DEVRINOL 50-WP/SIMAZINE 80-WP tank mix, surface applied, at recommended rates:

California burclover	*Medicago polymorpha*
Henbit	*Lamium amplexicaule*
London rocket	*Sisymbrium irio*
Marestail (horsewood)	*Conyza canadensis*
Miners lettuce	*Montia perfoliata*
Panicle willoweed	*Epilobium paniculatum*
Redmaids (rockpurslane)	*Calandrinia caulescens*

DIRECTIONS FOR USE

For control of all weeds listed, apply 8 pounds DEVRINOL 50-WP and 0.6 to 1.25 pounds SIMAZINE 80-WP per broadcast acre. Limit treatment of almonds, grapes and peaches to a 4 to 8 foot band centered on the tree row, reducing the dosage proportionately. Apply in the fall through early spring, prior to weed emergence. For best results, the treatment must reach the zone of weed seed germination. Treatments made other than in November through February must be shallowly incorporated or irrigated-in the same day as applied (sprinkler or flood) with sufficient water to wet the soil to a depth of 2 to 4 inches. Irrigation or shallow incorporation is recommended for treatments made from November through February if no rainfall occurs within three weeks of application.

Read directions on the SIMAZINE 80-WP label concerning nozzle and screen sizes.

Use as a directed spray and avoid contact with fruit or foliage. Do not apply when fruit or nuts are on the ground during the harvest period. Do not graze treated areas.

RECOMMENDATIONS

For use on almonds, apples, avocados, grapefruit, grapes, lemons, olives, oranges, peaches, pears, and walnuts in California and apples, filberts, grapes and walnuts in the Pacific Northwest.

NOTE: To avoid possible crop injury follow these precautions.

- **Apples and Pears.** Make application only to orchards where trees have been established one year or more.
- **Almonds and Peaches** (California Only). Do not treat trees established in the orchard less than 3 years. Do not treat the Mission variety of almonds. Do not apply to almond trees propagated on plum rootstocks. Do not replant almonds or peaches in treated soil. Do not apply on gravelly, sandy, or loamy sand soil, or on soil with less than 1% organic matter. Do not treat areas where water will accumulate (flood or furrow irrigation water should not be applied to any treated areas or allow to stand on treated areas).
- **Filberts** (Oregon and Washington Only). Do not apply on gravelly, sandy or loamy sand soil.
- **Grapes.** Do not use in vineyards established less than three years.
- **Grapefruit, Lemons, and Oranges** (California Only). Do not use in the Imperial, Coachella, or Palo Verde Valleys. Do not use in nurseries. Do not apply to bedded grapefruit, lemons, or oranges. Do not apply to trees under stress from freeze damage for one year after the freeze.
- **Walnuts.** Do not apply to sandy soil. Leveling and furrowing operations after application will lessen effectiveness of weed control.

DEVRINOL 50-WP/SIMAZINE 80-WP/PARAQUAT CL COMBINATION (California and Pacific Northwest Only)

DEVRINOL 50-WP/SIMAZINE 80-WP will not control emerged weeds. PARAQUAT CL may be used to control existing weeds when applied prior to, or in combination with DEVRINOL 50-WP/SIMAZINE 80-WP.

For weeds controlled refer to the weed lists for DEVRINOL 50-WP and DEVRINOL 50-WP/SIMAZINE 80-WP tank mix surface applied.

For control of all weeds listed, apply 8 pounds DEVRINOL 50-WP, 0.6 to 1.25 pounds SIMAZINE 80-WP, and 1 to 2 quarts PARAQUAT CL per broadcast acre.

Follow the use directions and recommendations for the DEVRINOL 50-WP/SIMAZINE 80-WP combination. Consult the PARAQUAT CL label for timing with respect to stage of weed growth and for application methods recommended to minimize crop injury.

RECOMMENDATIONS

For crop recommendations, read the section on DEVRINOL 50-WP/SIMAZINE 80-WP.

DEVRINOL 50-WP/SINBAR* 80-WP COMBINATION

DEVRINOL 50-WP may be used in combination with SINBAR 80-WP for improved control of annual broadleaf weeds. This tank mix does not control emerged weeds.

* SINBAR is a registered trademark of E.I. DuPont de Nemours & Company

DIRECTIONS FOR USE

NOTE: All rates are given on a broadcast basis. Reduce rates proportionately for band or strip treatment.

For control of all weeds listed, apply 8 pounds DEVRINOL 50-WP and 1.25 pounds SINBAR 80-WP per broadcast acre to the soil surface. Apply in the fall through early spring prior to weed emergence. Do not apply to frozen ground. For best results, the treatment must reach the zone of weed seed germination. If rainfall does not occur the treatment must be shallowly incorporated or irrigated-in following application (sprinkler or flood) with sufficient water to wet the soil to a depth of 2 to 4 inches. In the Western and Pacific Northwest Regions treatment made in November through February must be incorporated-in or irrigated-in if no rainfall occurs within three weeks of application. For treatments made other than in November through February in these regions, and for all treatments applied in the Northern, Southwestern, and Southeastern Regions regardless of season, the treatment must be incorporated-in or irrigated-in within 24 hours if no rainfall occurs.

Use as a directed spray and avoid contact with fruit or foliage. Do not apply when fruit is on the ground during the harvest period. Do not graze treated areas.

RECOMMENDATIONS

For use on apples, peaches and citrus (restrict use on citrus to California only) and mint (restrict use on mint to Pacific Northwest only).

NOTE: To avoid possible crop injury follow these precautions:

- **Apples and Peaches:** Where crop is grown under furrow irrigation or under raised-berm flood irrigation (trees 4" to 6" above waterline), apply only as a band treatment. Do not treat trees planted in the bottom of irrigation furrows or trees grown under flat flood or basin irrigation, and do not use on eroded areas where subsoil or tree roots are exposed, as injury to trees may result. Use only on trees established in the orchard for at least two years.
- **Citrus:** Temporary yellowing of citrus leaves may occur following treatment. Treatment may be used on coarse soils but is not to be used on soils low in organic matter (less than 1%), on poorly drained soils, nor on trees planted in irrigation furrows, as injury to trees may result. Treated areas may be planted to citrus trees one year after last application. Use only on trees established in the grove for at least one year.
- **Mint:** Apply in the Fall through early Spring as post-emergence harvest treatment after last cultivation.

TANK MIXTURES
VEGETABLES AND TOBACCO

IMPORTANT

Observe all cautions and limitations on labeling of all products used in combination.

Make only one application per season.

Do not contaminate water by cleaning of equipment or disposal of wastes.
In each case where a range of rates is given, the lower rates should be used on light (coarse) textured soils and soils low in organic matter. The higher rates should be used on heavy (fine) textured soils.

After harvest or prior to planting of succeeding crops, a deep moldboard or disc plowing operation must be carried out. To avoid injury to crops not specified on the label, do not plant until 12 months after the last DEVRINOL 50-WP application.

When applied at rates no greater than 4 lbs. DEVRINOL 50-WP per acre, rotational crops which are not registered on this label may be planted no sooner than 12 months after application when harvest and use of the mature crop(s) only is for food or feed purposes. Crops are not to be grazed or harvested for food or feed purposes earlier than maturity.

DIRECTIONS FOR USE
The recommended tank mix controls weeds by interfering with normal seed germination and seedling development. It does not control established weeds. All weed growth and crop stubble should be thoroughly worked into the soil before treatment.

TANK MIX DIRECTIONS
Follow directions given in the sections DEVRINOL 50-WP GENERAL DIRECTIONS FOR USE AND TANK MIXTURES — IMPORTANT.

At labeled spray volumes of 20–50 gallons per acre on transplanted tobacco, DEVRINOL 50-WP is physically compatible with Dyfonate®, Dasanit*, Diazinon**, Nemacur*, Ridomil** and Tillam®.

* Dasanit and Nemacur are registered trademarks of Mobay Chemical Corporation.

** Diazinon and Ridomil are registered trademarks of Ciba-Geigy Corporation.

DEVRINOL 50-WP/TILLAM 6-E TANK MIX COMBINATION APPLIED PRE-PLANT INCORPORATED

A tank mix combination of DEVRINOL 50-WP and TILLAM 6-E can be applied and mixed (incorporated) into the soil before planting for weed control in tomatoes and tobacco.

Additional weeds controlled by DEVRINOL 50-WP/TILLAM 6-E tank mix at recommended rates.

Annual Grasses	
Bermudagrass	*Cynodon dactylon*
Narrowleaf Signalgrass	*Brachiaria piligera*
Wild Oats	*Avena fatua*
Millet, foxtail	*Setaria italica*
Annual Broadleaf Weeds	
Henbit (Deadnettle)	*Lamium amplexicaule*
Nettleleaf goosefoot	*Chenopodium murale*
Hairy nightshade*	*Solanum villosum*
Blackeyedsusan	*Rudbeckia serotina*
Shepherdspurse	*Capsella bursa-pastoris*

* Suppression (partial control)

Perennial Weeds	
Purple nutsedge (purple nutgrass)	*Cyperus rotundus*
Yellow nutsedge (yellow nutgrass)	*Cyperus esculentus*

Existing stands of nutgrass must be turned under and chopped up thoroughly before treatment.

APPLICATION DIRECTIONS
The soil should be well-worked and dry enough to permit good soil mixing (incorporation). Pour the recommended amounts of DEVRINOL 50-WP and TILLAM 6-E into spray tank during the filling operations. Apply in 25 to 50 gallons of water per acre.

For soil incorporated applications, use a properly calibrated, low-pressure, boom-type sprayer having good agitation.

In irrigated areas, do not apply before pre-irrigation. For all band applications, reduce dosage proportionately depending upon row spacing and band to be treated. Do not band application on rocky ground because thorough incorporation is not possible.

INCORPORATION DIRECTIONS
The herbicides must be mixed (incorporated) into the soil IMMEDIATELY after application to prevent loss by evaporation. Whenever possible, application and incorporation should be done in the same operation.

INCORPORATION BEFORE PLANTING
Use power-driven cultivation equipment set to cut to a depth of two to three inches in coarse soils and three to four inches in heavy soils OR a tandem disc set to cut to a depth of six inches, followed by a spike-tooth harrow or some other leveling device which extends beyond ends of the disc. For thorough mixing, disc in two different directions (cross disc).

INCORPORATION AT PLANTING
Use power-driven cultivation equipment set to cut to a depth of two to three inches.

PLANTING
Crops should be planted or seeded immediately after application and incorporation. Do not use a drag behind the planter as it may concentrate the herbicides over the seed row and cause crop injury.

CULTURAL PRACTICES FOLLOWING APPLICATION
Shallow cultivation may be necessary to control those susceptible weeds that escape control as well as those weeds not susceptible to either herbicide. Do not cultivate deeper than the depth of herbicide incorporation.

When cultivating fields where the tank mix has been banded in the row, use shields to prevent the movement of untreated soil into the treated row.

RECOMMENDATIONS
TOMATOES (Direct Seeded [California Only] and Transplanted): Apply 2 to 4 pounds DEVRINOL 50-WP plus 2⅔ to 4 quarts TILLAM 6-E per broadcast acre. Use the lower rates on light soils (coarse textured–sandy) and the higher rate on heavy soils (fine textured-clay). Do not use hotcaps on transplanted tomatoes.

TOBACCO (Transplanted): Before transplanting, apply 2 pounds DEVRINOL 50-WP plus 2⅔ quarts TILLAM 6-E. This combination may be tank mixed with Furadan* 4-F or Mocap** 6-E at recommended rates. Read the Furadan 4-F and Mocap 6-E labels for recommended uses, directions for use, precautions and limitations. When filling the spray tank follow these orders of mixing: DEVRINOL 50-WP slurry, Furadan 4-F, TILLAM 6-E, OR DEVRINOL 50-WP slurry, TILLAM 6-E, Mocap 6-E.

* Furadan is a registered trademark of FMC Corporation

** Mocap is a registered trademark of Rhone-Poulenc Chemical Company

DEVRINOL 50-WP/TILLAM 6-E/DYFONATE [4-E, 4-EC or 4-ED] TANK MIX COMBINATION

A tank mix combination of DEVRINOL 50-WP, DYFONATE 4-E, 4-EC or 4-ED and TILLAM 6-E can be applied and mixed into the soil before transplanting tomatoes and tobacco, for control of weeds and soil insects. Refer to DEVRINOL 50-WP/TILLAM 6-E Tank Mix Combination for list of weeds controlled.

NOTE: Dyfonate 4-E, 4-EC and 4-ED are Restricted Use Pesticides. Observe all cautions and limitations on labeling of all products used in combinations. Make only one application per season. Do not contaminate water by cleaning equipment or disposal of wastes.

DIRECTIONS FOR USE
The recommended tank mix controls weeds by interfering with normal seed germination and seedling development. It does not control established weeds. All weed growth and crop stubble should be thoroughly worked into the soil before treatment.

For GENERAL USE PRECAUTIONS and DIRECTIONS FOR USE REFER TO DEVRINOL 50-WP/TILLAM 6-E Tank Mix Recommendations for methods of tank mixing, application and incorporation. When using DEVRINOL 50-WP, make a slurry and add to a nearly filled spray tank with adequate agitation.

RECOMMENDATIONS
TOMATOES (Direct Seeded [California Only] and Transplanted): Apply 2 to 4 pounds DEVRINOL 50-WP plus 2⅔ to 4 qts. TILLAM 6-E plus 2 qts. DYFONATE 4-E, 4-EC or 4-ED for

control of symphylans or 4 qts. for control of symphylans and wireworms in 25 to 50 gallons of water per broadcast acre. Use the lower DEVRINOL 50-WP and TILLAM 6-E rates on light soils (coarse texture-sandy) and the higher rates on heavy soils (fine textured-clay). Do not use hotcaps on transplanted tomatoes.

TOBACCO (Transplanted): Before transplanting apply 2 pounds DEVRINOL 50-WP plus 2⅔ qts. TILLAM 6-E plus 2 qts. DYFONATE 4-E, 4-EC or 4-ED for control of wireworms in 25 to 50 gallons of water per broadcast acre.

STORAGE AND DISPOSAL

Do not contaminate water, food or feed by storage or disposal.

STORAGE
Keep container closed when not in use. Store in a cool dry place. Do not store near seeds, fertilizer or food stuffs.

PESTICIDE DISPOSAL
Wastes resulting from the use of this product may be disposed of on site or at an approved waste disposal facility.

CONTAINER DISPOSAL
Do not reuse. Completely empty bag into application equipment. Then dispose of empty bag in a sanitary landfill or by incineration, or, if allowed by State and local authorities, by burning. If burned, stay out of smoke.

NOTICE—READ CAREFULLY

CONDITIONS OF SALE
Stauffer (and seller) offer(s) this product for sale subject to, and buyer and all users are deemed to have accepted, the following conditions of sale and warranty which may only be varied by written agreement of a duly authorized representative of Stauffer.

WARRANTY LIMITATION
Stauffer warrants that this product conforms to the chemical description on the label and is reasonably fit for the purposes referred to in the directions for use on the label subject to the inherent risks referred to below. Stauffer makes no other express warranties; THERE IS NO IMPLIED WARRANTY OF MERCHANTABILITY and there are no warranties which extend beyond the description on the label hereof.

INHERENT RISKS
The directions for use of this product are believed to be reliable and should be followed carefully. However, it is impossible to eliminate all risks associated with use. Buyer assumes all risks associated with use or application of this product contrary to label instructions or resulting from extraordinary weather conditions.

LIMITATION OF LIABILITY
In no case shall Stauffer be liable for special, indirect or consequential damages resulting from the use or handling of this product and no claim of any kind shall be greater in amount than the purchase price of the product in respect of which such damages are claimed.

This product is sold only for uses stated on its label. No express or implied license is granted to use or sell this product under any patent in any country except as specified: Country: United States of America Patent No.: 3,480,671; 3,718,455

This is a specimen label. Stauffer is not responsible for the accuracy of the information contained herein. As labels are subject to revision, always carefully read and follow the label on the product container.

DYFONATE® 4-EC

EMULSIFIABLE LIQUID INSECTICIDE

Complete Directions For Use, Use Precautions and Recommendations

RESTRICTED USE PESTICIDE

Due to ACUTE DERMAL TOXICITY

For retail sale to and use only by Certified Applicators or personnel under their direct supervision.

ACTIVE INGREDIENTS:	
O-Ethyl *S*-phenyl ethylphosphonodithioate*	47.3%
Petroleum hydrocarbon solvent	39.1%
INERT INGREDIENTS:	13.6%
TOTAL	100.0%

*DYFONATE
Contains 4 pounds DYFONATE per gallon

EPA Reg. No. 476-2134 A-11R

Keep Out of Reach of Children
CAUTION

DANGER–POISON

PELIGRO

PRECAUCION AL USUARIO: SI USTED NO LEE INGLES, NO USE ESTE PRODUCTO HASTA QUE LA ETIQUETA LE HAYA SIDO EXPLICADA AMPLIAMENTE.

READ ALL LABEL DIRECTIONS BEFORE USING

STATEMENT OF PRACTICAL TREATMENT

FOR PESTICIDE EMERGENCY

CALL A PHYSICIAN IMMEDIATELY: If a known exposure occurs, or if poisoning is suspected, do not wait for symptoms to develop. Immediately initiate the recommended procedures below. Simultaneously contact a physician, or the nearest hospital, or the nearest Poison Control Center. Inform the person contacted of the type and extent of exposure, describe the victim's symptoms, and follow the advice given. For additional information, call collect, day or night: Stauffer Chemical Company (203) 226-6602 or Chemtrec (800) 424-9300. Note: Be sure to advise the physician that the compound is a cholinesterase inhibitor, and follow the physician's advice.

POISON SYMPTOMS

DYFONATE 4-EC is a highly toxic cholinesterase inhibitor. Symptoms include headache, blurred vision, weakness, nausea, discomfort in the chest, abdominal cramps, vomiting, diarrhea, salivation, sweating and pinpoint pupils.

ALWAYS CALL A PHYSICIAN IN CASE OF SUSPECTED POISONING. ATROPINE BY INJECTION IS ANTIDOTAL.

FIRST AID
If patient has stopped breathing **immediately** remove from contaminated area, then start artificial respiration, preferably by mechanical means.

If swallowed, immediately dilute the swallowed material by giving large quanities of water and induce vomiting by gagging the victim with the finger or blunt object placed on the back of the victim's tongue. Give fluids until vomitus is clear. Never give anything by mouth to an unconscious person. Call a physician or the nearest Poison Control Center.

If inhaled, remove to fresh air. Get medical attention if respiratory irritation occurs.

For eye contact, flush with large amounts of running water for at least 15 minutes. Hold eye-lids apart to insure rinsing of the entire surface of the eyes. Get medical attention.

For skin contact, remove contaminated clothing and immediately wash skin thoroughly for 15 minutes with soap and water. Avoid abrading the skin.

INFORMATION FOR PHYSICIAN

Initial measures include removal of secretions, maintenance of a patent airway and, if necessary, artificial respiration. *When cyanosis is relieved,* atropine may be administered in large therapeutic doses, repeated as necessary to the point of tolerance. If symptoms warrant further treatment, protopam chloride (pralidoxime chloride, 2-PAM chloride) has shown utility as *adjunctive* therapy. Never use morphine. Continued absorption of the poison may occur, resulting in a fatal relapse after initial improvement; **very close supervision of the patient is indicated for at least 48 to 72 hours.**

IN CASE OF SPILL

ISOLATE the spill.

Hold this package, other cargo and vehicles involved. IMMEDIATELY TELEPHONE (24-hour service) for emergency assistance, 800-424-9300.

PRECAUTIONARY STATEMENTS

HAZARDS TO HUMANS AND DOMESTIC ANIMALS
POISON

DANGER–POISON

DANGER: POISONOUS IF SWALLOWED, INHALED OR ABSORBED THROUGH THE SKIN. RAPIDLY ABSORBED THROUGH THE SKIN. DO NOT GET IN THE EYES, ON SKIN, OR ON CLOTHING. DO NOT BREATHE SPRAY MIST.

Use a respirator approved by the National Institute for Occupational Safety and Health (30 CFR Part 11—Subpart M) for protection against pesticide inhalation.

Wear rubber gloves, protective clothing, and goggles when handling or applying.

Wash hands, arms, and face thoroughly with soap and water after handling and before eating or smoking. Routinely shower or bathe after work and wash all clothing with soap and hot water before reusing. In case of contact, immediately remove contaminated clothing and wash thoroughly with soap and water.

Do not transport or store DYFONATE 4-EC with any food or feed intended for human or animal consumption.

Not for use or storage in or around the home.

Do not use, pour, spill or store near heat or open flame.

Do not apply or allow to drift to areas occupied by unprotected humans or beneficial animals or onto adjoining food, fiber or pasture crops. Do not contaminate food or feed.

REENTRY STATEMENT

- Because certain states may require more restrictive reentry intervals for various crops treated with this product, consult your State Department of Agriculture for further information.
- Do not apply this product in such a manner as to directly or through drift expose workers or other persons. The area being treated must be vacated by unprotected persons.
- Do not enter treated areas without protective clothing until sprays have dried.
- Do not apply or allow to drift to areas occupied by beneficial animals or onto adjoining food, fiber or pasture crops. Do not contaminate food or feed.
- Written or oral warnings must be given to workers who are expected to be in a treated area or in an area about to be treated with this product. Oral warnings must be given if there is reason to believe that written warnings cannot be understood by workers.

When oral warnings are given, warnings shall be given in a language customarily understood by workers and must indicate the specific period of time that treated areas may not be entered without protective clothing. Oral warnings must also include appropriate first aid instructions in case of accidental exposure. These instructions are given in the STATEMENT OF PRACTICAL TREATMENT at the beginning of this label.

Written warnings must include the following information, "WARNING. Area treated with DYFONATE® Agricultural Insecticide on (date of application). Do not enter treated areas without appropriate protective clothing until sprays have dried (or other reentry interval if your state has a more restrictive interval for this product). If accidental exposure occurs, follow the instructions below". (Written warnings must include the STATEMENT OF PRACTICAL TREATMENT given at the beginning of this label.)

ENVIRONMENTAL HAZARDS

This pesticide is toxic to fish and wildlife. Birds feeding in treated areas may be killed. Remove or cover if spills occur. Runoff from treated areas may be hazardous to fish in neighboring areas. Do not apply directly to water or wetlands. Do not contaminate water by cleaning of equipment or disposal of waste. For guidance contact your regional office of the EPA. Apply this product only as specified on the label.

PHYSICAL AND CHEMCIAL HAZARDS

Do not use or store near heat or open flame.

GENERAL PRECAUTIONS

Read all directions before using.

Do not apply to any crop, or in any manner not specified on this label.

Do not rotate with carrots.

INSECTS CONTROLLED

Black Cutworm	*Agrotis ipsilon**
Cabbage Maggot	*Hylema brassicae*
Corn Leaf Aphid	*Rhopalosiphum maidis*
Garden Symphylan	*Scutigerella immaculata*
Onion Maggot	*Hylema antiqua*
Greenbug	*Schizaphis graminum*
Seed Corn Maggot	*Hylema platura*

* (Suppression only)

FLEA BEETLES:	
Elongate Flea Beetle	*Systena elongata*
Palestriped Flea Beetle	*Systena blanda*
Threespotted Flea Beetle	*Disonycha triangularis*
ROOTWORMS:	
Northern Corn Rootworm	*Diabrotica longicornis*
Southern Corn Rootworm	*Diabrotica undecimpunctata howardi*
Western Corn Rootworm	*Diabrotica virgifera*
WIREWORMS:	
Corn Wireworm	*Melanotus communis*
Pacific Coast Wireworm	*Limonius canus*
Southern Potato Wireworm	*Conoderus falli*
Sugar Beet Wireworm	*Limonius californicus*
Tobacco Wireworm	*Conoderus vespertinus*
Western Field Wireworm	*Limonius infuscatus*
Wheat Wireworm	*Agriotes mancus*
Ctenicera lobata	
Dalopius pallidus	

DIRECTIONS FOR USE

It is a violation of federal law to use this product in a manner inconsistent with its labeling.

Application should be made only with tractor mounted spray equipment or larger, custom applicator type vehicles. Application should be made by trained personnel only, using approved protective equipment. Pour the recommended amount of DYFONATE 4-EC into nearly filled spray tank. Add balance of water to fill tank. Keep agitator running during filling and spraying operations. Apply in 20 to 50 gallons of water per broadcast acre using a properly calibrated, low-pressure sprayer having good agitation.

DYFONATE 4-EC may be combined with solution, slurry, and suspension fertilizers. In addition, certain herbicides may be combined with DYFONATE 4-EC, both with and without fertilizer.

Read the recommendations for specific uses. Before combining in the spray tank, physical compatibility with these fluid fertilizers must be determined. See Appendix 1 for special directions regarding these combinations. Even though found to be compatible, agitation must be maintained during filling and application operations to insure uniform distribution of material.

Preplant broadcast applications should be mixed (incorporated) into the soil to a depth of 2 to 3 inches.

RECOMMENDATIONS

BEANS (GREEN AND DRY): For control of garden symphylans and wireworms, apply 2 quarts (for symphylans) or 1 gallon (for symphylans and wireworms) DYFONATE 4-EC per acre, broadcast prior to planting and incorporate into the soil by discing.

For control of seed corn maggot: **Northwestern U.S.:** Apply 2 quarts to 1 gallon DYFONATE 4-EC per acre, broadcast on soil surface, prior to planting and incorporate to a depth of 2 to 4 inches using suitable tillage equipment. Use higher rate where population pressure is severe. **Northeastern U.S.:** Apply 1 gallon DYFONATE 4-EC per acre, broadcast on soil surface, prior to planting and incorporate to a depth of 2 to 4 inches using suitable tillage equipment.

All treatments may be applied as a tank mix with EPTAM® 7-E Selective Herbicide in fluid fertilizer and/or water. Follow the appropriate regional recommendations in the crop section of the EPTAM 7-E label for broadcast, preplant incorporated applications. Use only fluid fertilizers compatible with EPTAM 7-E and DYFONATE 4-EC (see Appendix 1). Observe all use precautions and limitations on the EPTAM 7-E label.

BEANS, (LIMA): For control of garden symphylans and wireworms, apply 2 quarts (for symphylans) or 1 gallon (for symphylans and wireworms) DYFONATE 4-EC per acre, broadcast, prior to planting and incorporate into the soil by discing. The higher rate is recommended on lima beans grown in California only.

COLE CROPS (BROCCOLI, BRUSSELS SPROUTS, CABBAGE, CAULIFLOWER):

Garden Symphylans—Prior to or at time of seeding or transplanting, apply 2 quarts DYFONATE 4-EC per acre, broadcast, and incorporate into the soil by discing.

Cabbage Maggot—(Northeastern U.S. only)—For transplant cole crops: use 1 to 2 quarts DYFONATE 4-EC in 200 to 400 gallons of water per acre. Apply with a sprayer equipped with drop nozzles to direct a drenching spray to the base of the plants. Apply immediately after transplanting.

CORN (FIELD, SWEET, AND POPCORN):

NOTE: May be used in corn grown for grain or silage.

A. Band application for control of northern, southern and western corn rootworms—

Apply, at planting, 3/4 to 1 quart DYFONATE 4-EC per acre in a 6 to 8 inch band over corn rows 38 to 40 inches apart.

Incorporate into the top 1/2 to 1 inch of soil by making application ahead of the press wheels or by use of special covering devices or by dragging a short length of chain behind the press wheels.

Do not place DYFONATE 4-EC in direct contact with seed.

B. Broadcast application for control of symphylans, northern, southern and western corn rootworms, and wireworms, and for suppression of black cutworm: Apply 2 quarts (for symphylans), 3 quarts (for symphylans and corn rootworms), or 4 quarts (for symphylans, corn rootworms, wireworms and black cutworm) DYFONATE 4-EC per acre prior to planting and incorporate into the soil by discing. This treatment may be applied as a tank mix with ERADICANE® 6.7-E or SUTAN® + 6.7-E, selective herbicides or their recommended tank mixes with ATRAZINE 80W or 4L or BLADEX* 80W or 4L. A compatible fluid fertilizer (see Appendix 1) or water may be used as carrier.

*BLADEX is a registered trademark of Shell Chemical Company.

Follow all directions, limitations, and use precautions on the ERADICANE 6.7-E, SUTAN + 6.7-E, ATRAZINE, and BLADEX labels. Do not tank mix DYFONATE with ERADICANE 6.7-E on sweet corn or popcorn.

C. Split boot application for control of northern and western corn rootworms (recommended in the Northcentral and Midwestern corn-growing areas only)—

Apply, at planting, 1 quart DYFONATE 4-EC in 5 to 20 gallons of compatible fluid fertilizer (see Appendix 1) per acre (40 inch row spacing) in subsurface split bands on each side of the corn row, 1 1/4 to 1 1/2 inches to each side of the seed and 1 to 2 inches deep. **Do not place DYFONATE 4-EC in direct contact with the seed.**

MINT (PEPPERMINT, SPEARMINT): For control of garden symphylans apply 2 quarts DYFONATE 4-EC per acre, broadcast, and incorporate into the soil by discing prior to planting or on established plantings prior to Spring growth. (Do not apply terbacil for at least 3 weeks before or after the DYFONATE 4-EC application.)

BULB ONIONS: For control of onion maggot, apply as a furrow treatment at time of planting. Based on a 20 inch row spacing, apply at a rate of 1 quart DYFONATE 4-EC per acre (1 quart per 26,135 feet of row; 3 1/4 fluid ounces per 2,613 feet of row; 1 1/4 fluid ounces per 1,000 feet of row). Use on organic soils only (soils containing 10% or more organic matter).

PEPPERS: For control of garden symphylans and wireworms, apply 2 quarts (for symphylans) or 4 quarts (for symphylans and wireworms) DYFONATE 4-EC per acre, broadcast, prior to planting and incorporate into the soil by discing.

POTATOES (IRISH):

A. Broadcast preplant application for control of garden symphylans and wireworms:

Apply 2 quarts (for symphylans) or 4 quarts (for symphylans and wireworms) DYFONATE 4-EC per acre prior to planting and incorporate into the soil by discing. This treatment may also be applied in certain areas as a tank mix with EPTAM 7-E Selective Herbicide in a compatible fluid fertilizer and/or water (see Appendix 1). Follow recommendations in the crop section of the EPTAM 7-E label for broadcast, preplant incorporated applications. For use in the Northern, Western, and Pacific Northwest Regions only. Observe all use precautions and limitations on the EPTAM 7-E label.

B. Banded preplant soil injection with fluid fertilizer for control of wireworms (Northwestern U.S. only):

Apply through fluid fertilizer shank immediately ahead of the planter. Set the shank to run 8 to 10 inches deep. Apply 2 quarts DYFONATE 4-EC per crop acre, based on 36" row spacing, in the recommended quantity of compatible liquid fertilizer (see Appendix 1).

SORGHUM (MILO): For control of corn leaf aphid and greenbug, apply 3/4 to 1 quart DYFONATE 4-EC in 5 to 30 gallons of water per acre over the top using aerial or ground application equipment. Do not apply within 14 days of harvest nor feed or graze to livestock within 14 days of application. Limit use to 2 applications per crop.

STRAWBERRIES: For control of symphylans apply 2 quarts DYFONATE 4-EC per acre, broadcast, and incorporate into the soil by discing prior to planting.

SUGAR BEETS: For control of garden symphylans and wireworms, apply 2 quarts (for symphylans) or 1 gallon (for symphylans and wireworms) DYFONATE 4-EC per acre, broadcast, prior to planting and incorporate into the soil by discing.

SUGARCANE: At Planting—For wireworm control, apply 4 quarts DYFONATE 4-EC per acre in a 14 inch band on cane rows 60 inches apart. This is equivalent to 11 to 14 1/2 fluid ounces of DYFONATE 4-EC per 1,000 linear feet of row. Apply over the top of sugarcane seed pieces immediately ahead of covering discs.

TABLE BEETS: For control of cabbage maggot and garden symphylans, apply 2 quarts DYFONATE 4-EC per acre, broadcast, prior to planting. Incorporate into the soil by discing.

TOBACCO: For control of wireworms, apply 1 to 2 quarts DYFONATE 4-EC per acre, broadcast, prior to bedding and transplanting and incorporate into the soil by discing.

FLUE CURED TOBACCOS ONLY: DYFONATE 4-EC in combination with TILLAM® 6-E controls wireworms and certain weeds (see TILLAM 6-E label) in tobacco. Add to the spray tank and apply 1 quart DYFONATE 4-EC and 2⅔ quarts TILLAM 6-E in 20 to 50 gallons of water per acre (see Appendix 1). Use a properly calibrated, low-pressure, boom-type sprayer having good agitation. IMPORTANT: Refer to both DYFONATE 4-EC and TILLAM 6-E labels for pertinent directions for use, and use precautions and restrictions. The soil should be well worked and dry enough to permit good soil incorporation. The tank mix must be incorporated (mixed) into the soil immediately after application to prevent loss of the herbicide by evaporation. Whenever possible, application and incorporation should be done in the same operation. Use power-driven cultivation equipment set to cut to a depth of 3" in light soil and 3" to 4" in heavy soil, or a tandem disc set to cut to a depth of 4" to 6". Cross disc at right angles for thorough mixing. The bedding operation should follow the incorporation as soon as practical. Planting may be done immediately after bedding or at a later date; however, to insure long term weed control, the crop should be planted as soon as possible after application and bedding.

TOMATOES: For control of garden symphylans, apply 2 quarts of DYFONATE 4-EC per acre, broadcast, prior to planting and incorporate into the soil by discing.

APPENDIX I

PROCEDURE FOR TESTING THE TANK-MIX COMPATIBILITY OF DYFONATE 4-EC WITH FLUID FERTILIZERS AND DYFONATE 4-EC WITH HERBICIDES AND WATER OR FLUID FERTILIZERS.

This procedure is suggested for determining whether DYFONATE 4-EC and a specific fluid fertilizer and/or the specifically approved herbicide(s) may be combined for spray-tank application.

Materials Required:

1. a. DYFONATE 4-EC.
   b. DYFONATE 4-EC herbicide combinations (EPTAM 7-E, ERADICANE 6.7-E, SUTAN + 6.7-E, TILLAM 6-E, ATRAZINE 80W or 4L, or BLADEX 80W or 4L).
2. Water or fluid fertilizer to be used.
3. Adjuvant for tank mix—Compex*, Sponto* 168-D, Unite* or equivalent. The adjuvant which performs best depends on the specific fertilizer being considered.

* Compex is a trademark of Kalo Laboratories, Inc., Sponto 168-D is a trademark of Witco Chemical Corporation; Unite is a trademark of Hopkins Agricultural Chemical Company.

4. Two one-quart, wide-mouth glass jars with tops.
5. Measuring spoons (a graduated cylinder or pipette will provide more accurate measurement).
6. Measuring cup, 8 oz. (237 mL).

Procedure:

1. Add one pint (2 cups or approximately 473 mL) of the test water or fluid fertilizer to each of the quart jars.
2. Add adjuvant to one of the jars and mix (see rate tables).
3. Carefully add the pesticides to each jar (see rate tables) as follows:
   a. If only DYFONATE 4-EC is to be used, add it directly to each jar.
   b. If a tank mix with a herbicide emulsifiable concentrate is to be used, add the herbicide first and mix, and then add the DYFONATE 4-EC.
   c. If a tank mix with a herbicide emulsifiable concentrate *and* a wettable powder or flowable formulation is to be used, add and mix the pesticides in the following order:
      1. Add and mix the wettable powder or flowable formulation first (premixing these in water prior to addition to fluid fertilizer often provides a more compatible mix with fluid fertilizer),
      2. Add and mix the herbicide emulsifiable concentrate second,
      3. And finally, add the DYFONATE 4-EC.
4. Securely close each jar using the top and mix the contents by inverting the jar ten times.
5. Inspect the mixtures immediately and after standing for thirty minutes without mixing.
6. Dispose of jar mixes according to disposal statements contained within this label.

A mixture may be incompatible if either of the following conditions is observed:

a. an oil layer or large oil globules are observed, or
b. sludge or clumps are present.

If either mixture remains uniform for thirty minutes, the combination may be used. Should either mixture separate after thirty minutes, but remix readily with ten jar inversions, the combination can be used if good agitation is maintained in the spray tank. If nondispersible oil, sludge or clumps persist, the combination must not be used.

If the mixture with the adjuvant stays mixed, or remixes readily to form a uniform dispersion, but the one without adjuvant does not, be sure to use the adjuvant in the spray tank. Add the adjuvant first to the water or fluid fertilizer at the recommended rate. Minimize foaming by using moderate agitation.

RATE TABLE FOR DYFONATE® 4-EC/EPTAM® 7-E AND ADJUVANT WITH THE FLUID FERTILIZER**

Gallons of Fluid Fertilizer To Be Applied Per Acre	Milliliters or Teaspoons of DYFONATE 4-EC* to Be added to 1 Pint of Fertilizer		Milliliters or Teaspoons of EPTAM 7-E* to Be Added to 1 Pint of Fertilizer	
	mL	Tsp.	mL	Tsp.
10	12	2½	6⅔	1⅓
15	8	1⅔	5	1
20	6	1¼	3⅓	⅔
25	5	1	2½	½
40	3	⅔	1⅔	⅓

* Based on a field rate of one pound active ingredient per acre in the fertilizer volumes indicated. Adjust volumes proportionately to correspond with intended field rates in terms of pounds active ingredients per acre (e.g. for a field rate of 3 pounds actual EPTAM 7-E in 20 gallons of fertilizer per acre, add 10 mL or 2 tsp. EPTAM 7-E to each jar for compatibility testing purposes).

** A typical adjuvant rate of use in 3 pints per 100 gallons of fluid fertilizer. Two (2) mL or one-half (½) tsp. of adjuvant should be added to 1 pint of fertilizer in order to equal a rate of 3 pints of adjuvant per 100 gallons of fluid fertilizer.

RATE TABLE FOR DYFONATE 4-EC AND ADJUVANT WITH THE FLUID FERTILIZER**

Gallons of Fluid Fertilizer To Be Applied Per Acre	Milliliters or Teaspoons of DYFONATE 4-EC* To Be Added To 1 Pint of Fertilizer	
	mL	Tsp.
10	12	2½
15	8	1⅔
20	6	1¼
25	5	1
40	3	⅔

* Based on a fluid rate of one pound active ingredient per acre in the fertilizer volumes indicated. Adjust volume proportionately to correspond with intended field rate in terms of pounds active ingredient per acre (e.g. for a field rate of *2* pounds active ingredient in *10* gallons of fertilizer per acre, add 24 mL or 5 tsp. DYFONATE 4-EC to each jar for compatibility testing purposes.

** A typical adjuvant rate of use is 3 pints per 100 gallons of fluid fertilizer. Two (2) milliliters or one-half (½) teaspoon of adjuvant should be added to 1 pint of fertilizer in order to equal a rate of 3 pints of adjuvant per 100 gallons of fluid fertilizer.

RATE TABLE FOR DYFONATE 4-EC/ERADICANE 6.7-E AND DYFONATE 4-EC/SUTAN + 6.7-E AND ADJUVANT** WITH THE FLUID FERTILIZER

Gallons of Fluid Fertilizer To Be Applied Per Acre	Milliliters or Teaspoons of DYFONATE 4-EC* to Be added to 1 Pint of Fertilizer		Milliliters or Teaspoons of ERADICANE 6.7-E* or SUTAN + 6.7-E To Be Added to 1 Pint of Fertilizer	
	mL	Tsp.	mL	Tsp.
10	12	2½	7	1½
15	8	1⅔	5	1
20	6	1¼	3½	¾
25	5	1	2½	½
40	3	⅔	1¾	⅓

* Based on a field rate of one pound active ingredient per acre in the fertilizer volumes indicated. Adjust volumes proportionately to correspond with intended field rates in terms of pounds active ingredients per acre (e.g. for a field rate of 4 pounds actual SUTAN + 6.7-E in 20 gallons of fertilizer per acre, add 14 mL or 3 tsp. SUTAN + 6.7-E to each jar for compatibility testing purposes).

** A typical adjuvant rate of use is 3 pints per 100 gallons of fluid fertilizer. Two (2) mL or one-half (½) tsp. of adjuvant should be added to 1 pint of fertilizer in order to equal a rate of 3 pints of adjuvant per 100 gallons of fluid fertilizer.

TANK MIX DIRECTIONS

1. Fill the spray tank at least ⅔ full of the appropriate carrier fluid. Be sure tank is clean and not contaminated with 2,4-D type herbicides, oil residue or other materials.
2. Start agitation and continue moderate agitation throughout mixing and application procedures.
3. Add the proper amount of adjuvant and mix (if adjuvant is to be used).
4. Next add the ATRAZINE 80W or 4L or BLADEX 80W or 4L (if it is to be combined in the tank) and mix to disperse it thoroughly.
5. Then add the herbicide emulsifiable concentrate (EPTAM 7-E, ERADICANE 6.7-E, SUTAN + 6.7-E or TILLAM 6-E) and mix thoroughly.
6. Lastly, add the DYFONATE 4-EC.
7. Complete filling the remaining portion of the tank with carrier fluid; mix thoroughly.
8. The tank-mix combination should not be left in the spray tank for prolonged periods, batches should be mixed and applied the same day.

NOTE: Constant moderate agitation is required for successful tank-mix preparation and application. Avoid excessive agitation which can cause tank-mix incompatibility. All return lines to the tank must discharge below liquid level.

RATE TABLE FOR DYFONATE 4-EC ERADICANE 6.7-E/ATRAZINE OR BLADEX AND DYFONATE 4-EC SUTAN + 6.7-E/ATRAZINE OR BLADEX AND ADJUVANT** WITH WATER OR FLUID FERTILIZER

Gallons of Water or Fluid Fertilizer To Be Applied/Acre	10		15		20		25		40	
Measure	mL	Tsp.	mL	Tsp.	mL	Tsp.	mL	Tsp.	mL	Tsp.
Milliliters or Teaspoons of DYFONATE 4-EC* To Be Added to 1 Pint of Water or Fertilizer	12	2½	8	1⅔	6	1¼	5	1	3	⅔
Milliliters or Teaspoons of ERADICANE 6.7-E* or SUTAN + 6.7-E* To Be Added to 1 Pint of Water or Fertilizer	7	1½	5	1	3½	¾	2½	½	1¾	⅓
Milliliters or Teaspoons of ATRAZINE 80W* or BLADEX 80W* To Be Added to 1 Pint of Water or Fertilizer	22	4½	15	3	11	2¼	8⅔	1¾	5	1
Milliliters or Teaspoons of ATRAZINE 4L* or BLADEX 4L* To Be Added to 1 Pint of Water or Fertilizer	12	2½	8	1⅔	6	1¼	5	1	3	⅔

* Based on a field rate of one pound active ingredient per acre in the water or fertilizer volumes indicated. Adjust volumes proportionately to correspond with intended field rates in terms of pounds active ingredients per acre (e.g. for a field rate of 4 pounds of actual SUTAN + 6.7-E in 40 gallons of fertilizer per acre, add 7 mL or 1½ tsp. SUTAN + 6.7-E to each jar for compatibility testing purposes, i.e., 4 times the 10 gallon rate for 1 pound equals 28 mL or 6 tsp.; then dividing by the dilution factor four [40÷10=4] gives 7 mL or 1½ tsp. per pint, which is equivalent to 4 pounds actual SUTAN + 6.7-E/40 gallons).

** A typical adjuvant rate of use is 3 pints per 100 gallons of fluid fertilizer. Two (2) mL or one-half (½) tsp. of adjuvant should be added to 1 pint of fertilizer in order to equal a rate of 3 pints of adjuvant per 100 gallons of fluid fertilizer.

RATE TABLE FOR DYFONATE 4-EC/TILLAM 6-E AND ADJUVANT**

Gallons of Fluid Fertilizer To Be Applied Per Acre	Milliliter or Teaspoons of DYFONATE 4-EC* to Be Added to 1 Pint of Fertilizer		Milliliters or Teaspoons of TILLAM 6-E* to Be Added to 1 Pint of Fertilizer	
	mL	Tsp.	mL	Tsp
10	12	2½	8	1⅔
15	8	1⅔	5¼	1
20	6	1¼	4	¾
25	5	1	3¼	⅔
40	3	⅔	2	⅓

* Based on a field rate of one pound active ingredient per acre in the fertilizer volumes indicated. Adjust volumes proportionately to correspond with intended field rates in terms of pounds active ingredients per acre (e.g. for a field rate of 4 pounds actual TILLAM 6-E in 20 gallons of fertilizer per acre, add 10 mL or 2 tsp. TILLAM 6-E to each jar for compatibility testing purposes).

** A typical adjuvant rate of use in 3 pints per 100 gallons of fluid fertilizer. Two (2) mL or one-half (½) tsp. of adjuvant should be added to 1 pint of fertilizer in order to equal a rate of 3 pints of adjuvant per 100 gallons of fluid fertilizer.

STORAGE AND DISPOSAL

STORAGE

Minimum storage temperature minus 40°F. Any steel container may be corroded by its contents. Dents or damage may promote corrosion. Use damaged and older containers FIRST. Inspect rusted cans for leaks.

PROHIBITIONS

Do not contaminate water, food, or feed by storage or disposal. Open dumping is prohibited.

PESTICIDE DISPOSAL
Pesticide wastes are acutely hazardous. Improper disposal of excess pesticide, spray mixture, or rinsate is a violation of Federal law. If these wastes cannot be disposed of by use according to label instructions, contact your State Pesticide or Environmental Control Agency, or the Hazardous Waste representative at the nearest EPA Regional Office for guidance.

CONTAINER DISPOSAL
Triple rinse (or equivalent). Then offer for recycling or reconditioning, or puncture and dispose of in a sanitary landfill, or by other procedures approved by State and local authorities.

NOTICE—READ CAREFULLY

CONDITIONS OF SALE
Stauffer (and seller) offer(s) this product for sale subject to, and buyer and all users are deemed to have accepted, the following conditions of sale and warranty which may only be varied by written agreement of a duly authorized representative of Stauffer.

WARRANTY LIMITATION
Stauffer warrants that this product conforms to the chemical description on the label and is reasonably fit for the purposes referred to in the directions for use on the label subject to the inherent risks referred to below. Stauffer makes no other express warranties. THERE IS NO IMPLIED WARRANTY OF MERCHANTABILITY and there are no warranties which extend beyond the description of the label hereof.

INHERENT RISKS
The directions for use of this product are believed to be reliable and should be followed carefully. However, it is impossible to eliminate all risks associated with use. Buyer assumes all risks associated with use or application of this product contrary to label instructions or resulting from extraordinary weather conditions.

LIMITATION OF LIABILITY
In no case shall Stauffer be liable for special, indirect or consequential damages resulting from the use or handling of this product and no claim of any kind shall be greater in amount than the purchase price of the product in respect of which such damages are claimed.

This product is sold only for uses stated on its label. No express or implied license is granted to use or sell this product under any patent in any country except as specified: Country: United States of America.

This is a specimen label. Stauffer is not responsible for the accuracy of the information contained herein. As labels are subject to revision, always carefully read and follow the label on the product container.

DYFONATE® 10-G

GRANULAR INSECTICIDE

ACTIVE INGREDIENT:	
O-Ethyl *S*-phenyl ethylphosphonodithioate*	10.0%
INERT INGREDIENTS	90.0%
TOTAL	100.0%

*DYFONATE
Contains 10 Lbs. Active Ingredient Per 100 Pounds
EPA Reg. No. 476-1995 A-7R

Keep Out of Reach of Children
WARNING

Read all directions before using.

Do not transport or store DYFONATE 10-G with food or feed intended for human or animal consumption.

STATEMENT OF PRACTICAL TREATMENT FOR PESTICIDE EMERGENCY
CALL A PHYSICIAN IMMEDIATELY. If a known exposure occurs, or if poisoning is suspected, do not wait for symptoms to develop. Immediately initiate the recommended procedures below. Simultaneously contact a physician, or the nearest hospital, or the nearest Poison Control Center. Inform the person contacted of the type and extent of exposure, describe the victim's symptoms, and follow the advice given. For additional information, call collect, day or night, Stauffer Chemical Company (203) 226-6602 or Chemtrec (800) 424-9300. Note: Be sure to advise the physician that the compound is a cholinesterase inhibitor, and follow the physician's advice.

FIRST AID
If patient has stopped breathing, *immediately* remove from contaminated area, then start artificial respiration, preferably by mechanical means.

If swallowed—immediately dilute the swallowed material by rapidly giving large quantities of water and induce vomiting by gagging the victim with a finger or blunt object placed on the back of the victim's tongue. Give fluids until vomitus is clear. Never give anything by mouth to an unconscious person. Call a physician or the nearest Poison Control Center.

If inhaled remove to fresh air. Get medical attention if respiratory irritation occurs.

For eye contact, flush eyes with large amounts of running water for at least 15 minutes. Hold eyelids apart to insure rinsing of the entire surface of the eyes. Get medical attention.

For skin contact, remove contaminated clothing and immediately wash skin thoroughly for 15 minutes with soap and water. Avoid abrading the skin.

NOTE TO PHYSICIAN: Exposure may cause cholinesterase inhibition. Atropine by injection is antidotal. 2-PAM also is antidotal when administered early, and in conjunction with atropine.

IN CASE OF SPILL, ISOLATE THE SPILL.
HOLD this package, other cargo and vehicles involved. IMMEDIATELY TELEPHONE (24-hour service) for emergency assistance: (800) 424-9300.

PRECAUTIONARY STATEMENTS

HAZARDS TO HUMANS AND DOMESTIC ANIMALS
WARNING
Harmful or fatal if swallowed. Skin contact or inhalation of dusts may be harmful. Do not get on skin, in eyes or on clothing. Avoid breathing dust. Wear gloves when handling. In case of contact, wash immediately with soap and water. Always wash thoroughly after handling.

- Do not transport or store DYFONATE 10-G with any food or feed intended for human or animal consumption.
- Not for use or storage in or around the home.
- Do not apply or allow to drift to areas occupied by unprotected humans or beneficial animals or onto adjoining food, fiber or pasture crops. Do not contaminate food or feed.

ENVIRONMENTAL PRECAUTIONS
- This product will kill fish and wildlife. Keep out of lakes, ponds and streams. Birds feeding on treated areas may be killed.
- Do not contaminate water by cleaning of equipment, or disposal of wastes. Apply this product only as specified on this label.

GENERAL PRECAUTIONS
- Read all directions before using.
- Do not apply on any crop, or in any manner not specified on this label.
- Do not rotate with carrots.

INSECTS CONTROLLED

Black cutworm	*Agrotis ipsilon**
Burrowing bug	*Pangaeus bilineatus*
Cabbage maggot	*Hylemya brassicae*
Corn leaf aphid	*Rhopalosiphum maidis*
European corn borer	*Ostrinia nubilalis*
Garden symphylan	*Scutigerella immaculata*
Greenbug	*Schizaphis graminum*
Lesser cornstalk borer	*Elasmopalpus lignosellus*
Maize billbug	*Spenophorus maidis*
Onion maggot	*Hylemya antiqua*
Seed corn beetle	*Agonodersus lecontei*

Seed corn maggot	*Hylemya cilicrura*
Southwestern Corn Borer	*Diatraea grandiosella*
Sugarbeet root maggot	*Tetanops myopaeformis*
*(Suppression only)	
Flea Beetles:	
Elongate flea beetle	*Systenta elongata*
Palestriped flea beetle	*Systena blanda*
Threespotted flea beetle	*Disonycha triangularis*
Rootworms:	
Northern corn rootworm*	*Diabrotica longicornis*
Southern corn rootworm	*Diabrotica undecimpunctata howardi*
Western corn rootworm	*Diabrotica virgifera*
Wireworms:	
Corn wireworm	*Melanotus communis*
Pacific coast wireworm	*Limonius canus*
Southern potato wireworm	*Conoderus falli*
Sugar beet wireworm	*Limonius californicus*
Tobacco wireworm	*Conoderus vespertinus*
Western field wireworm	*Limonius infuscatus*
Wheat wireworm	*Agriotes mancus*
Ctenicera lobata	
Dalopius pallidus	

DIRECTIONS FOR USE AND RECOMMENDATIONS

It is a violation of Federal law to use this product in a manner inconsistent with its labeling.

NOTE: Preplant broadcast applications should be mixed (incorporated) into the soil to a depth of 2 to 3 inches.

ASPARAGUS: (California only) For control of garden symphylans, apply 30 to 100 pounds per acre, broadcast. Incorporate into the soil with standard cultivation equipment. Do not apply within 30 days of harvest.

BEANS, GREEN & DRY: For control of garden symphylans and wireworms, apply 20 pounds (for symphylans) or 40 pounds (for symphylans and wireworms) DYFONATE 10-G per acre, broadcast, prior to planting and incorporate into the soil by discing. For control of seed corn maggot. (Northwestern U.S. only) apply 20 to 40 pounds DYFONATE 10-G per acre, broadcast on soil surface, prior to planting, and incorporate to a depth of 2 to 4 inches using suitable tillage equipment. Use the higher rate where population pressure is severe.

BEANS, LIMA: For control of garden symphylans, apply 20 pounds DYFONATE 10-G per acre, broadcast, prior to planting and incorporate into the soil by discing.

COLE CROPS (broccoli, Brussels sprouts, cabbage, cauliflower): For control of garden symphylans prior to and at time of transplanting, apply 20 pounds per acre, broadcast, and incorporate into the soil by discing.

CORN (field, sweet and popcorn):

NOTE: May be used in corn grown for grain or silage.

A. Band application for control of northern, southern and western corn rootworms, lesser cornstalk borer, seed corn beetle and seed corn maggot, and for suppression of wireworms and black cutworm (recommended for seed corn beetle in the Central and Midwestern corn growing areas only).

Apply the rates recommended in the following table in a 6 to 8 inch band over corn rows 38 to 40 inches apart.

Insect	Rate Per Acre	Rate Per 1000 Linear Feet
Corn rootworms	7½ to 10 pounds	9 to 12 ounces
Seed corn beetle		
Seed corn maggot		
Wireworms	10 pounds	12 ounces
Black cutworms		
Lesser cornstalk borer	20 pounds	22 ounces

1. APPLICATION AT PLANTING: Incorporate into the top ½ to 1 inch of soil by making application ahead of the press wheels, or by dragging a short length of chain behind the press wheels. For John Deere 7000 series planters, place DYFONATE 10-G behind firming wheel. Do not place DYFONATE 10-G in direct contact with the seed.

OR

2. APPLICATION AT CULTIVATION: Cover the treated band with 2 to 3 inches of soil by making application immediately ahead of disc hillers or cultivation equipment. Make application in May or June when activity of rootworm larvae is first observed. Not recommended for control of lesser cornstalk borer or for suppression of wireworms or black cutworm. Do not apply within 45 days of harvest nor feed or graze to livestock within 30 days of treatment.

B. Broadcast application for control of symphylans, northern, southern, and western corn rootworms, wireworms and maize billbugs:

Apply 20 pounds (for symphylans) or 40 pounds (for symphylans, wireworms and maize billbugs) DYFONATE 10-G per acre prior to planting and incorporate into the soil by discing. Recommended for billbugs in the Southeastern U.S. only.

C. Band or broadcast application for control of European corn borer and southwestern corn borer: Apply 10 pounds DYFONATE 10-G per acre over the corn, so granules fall into whorls. May be applied broadcast using air or ground equipment, or in 6 to 8 inch bands over corn rows 30 inches or more apart.

Do not apply within 30 days of harvest nor feed or graze to livestock within 30 days of treatment.

1. European corn borer. Make application when 50% of the plants show first generation borer feeding. To control second generation corn borer, apply when field counts show 100 egg masses per 100 plants.
2. Southwestern corn borer. Apply when newly hatched larvae first appear. Repeat application as necessary to control later generations.

D. Band application for control of wireworms in sweet corn grown on muck soils (Florida only).

Based on 36 inch row spacing, apply 20 pounds DYFONATE 10-G per acre in a 2 inch band over the furrow at planting. This is equivalent to 20 pounds DYFONATE 10-G per 14,520 linear feet of row (22 ounces per 1,000 linear feet of row).

MINT (PEPPERMINT, SPEARMINT): For control of garden symphylans apply 20 pounds per acre, broadcast, and incorporate into the soil by discing prior to planting or on established plantings prior to spring growth. Do not apply terbacil for at least 3 weeks before or after the DYFONATE 10-G application.

BULB ONIONS: For control of onion maggot, apply as a furrow treatment at time of planting. Based on a 20 inch row spacing apply at a rate of 10 pounds per acre (10 lbs. per 26,135 ft. of row; 1 lb. per 2,613 ft. of row; ⅜ lb. per 1,000 ft. of row). Use on organic soils only (soils containing 10% or more organic matter).

PEANUTS: Southeastern U.S.—For control of lesser cornstalk borer, use 20 lbs. DYFONATE 10-G per acre, in an 18 inch band on 36 inch rows.

OR

40 lbs. DYFONATE 10-G per acre, broadcast, by air or ground equipment. Apply at first signs of borer infestation after pegging begins. May be used up to 30 days after pegging begins. For control of southern corn rootworm, use 20 pounds DYFONATE 10-G per acre in 18 inch bands on 36 inch rows —apply at pegging.

Southwestern U.S.—For control of lesser cornstalk borer and southern corn rootworm, apply 20 pounds DYFONATE 10-G per acre in 18 inch bands on 36 inch rows. For lesser cornstalk borer only, 10 to 14 pounds DYFONATE 10-G per acre are recommended, applied in 14 inch bands on rows planted 32 or more inches apart. All rates may be applied up to 30 days after pegging begins. For burrowing bug control, apply 10 to 20 pounds DYFONATE 10-G per acre in 18 inch bands on 36 inch rows.

OR

Broadcast 20 pounds DYFONATE 10-G per acre. Lightly incorporate into the soil with 1 to 2 inches of water. Recommended for control of burrowing bug in Texas only.

POTATOES (IRISH): For control of garden symphylans and wireworms, apply 20 pounds (for symphylans) or 40 pounds (for symphylans and wireworms) DYFONATE 10-G per acre, broadcast, prior to planting and incorporate into the soil by discing.

OR

For wireworm control in the Pacific Northwest only, apply at planting 20 pounds DYFONATE 10-G per acre (36 inch row spacing) in subsurface bands 3 to 4 inches on both sides of and 2 inches below the seed pieces.

RADISH: For control of cabbage maggot and garden symphylans prior to and at time of seeding, apply 20 pounds per acre, broadcast, and incorporate into the soil by discing.

SORGHUM (MILO): (Texas only)—For control of corn leaf aphid and greenbug, apply 7.5 pounds DYFONATE 10-G per acre over the top using aerial application equipment. Use only on irrigated sorghum. Do not apply within 14 days of harvest nor feed or graze to livestock within 14 days of application. Limit post-emergence use to 2 applications per crop.

STRAWBERRIES: For control of symphylans apply 20 pounds per acre, broadcast, and incorporate into the soil by discing prior to planting or transplanting.

SUGAR BEETS: BAND APPLICATION FOR CONTROL OF SUGAR BEET ROOT MAGGOTS: Apply 10 to 15 pounds DYFONATE 10-G per acre in a 7 inch band over sugar beet rows 22 or more inches apart. Apply at planting and lightly incorporate into the soil. Do not place DYFONATE 10-G in direct contact with the seed. Recommended in Colorado and the Northwest sugar beet growing areas only.

SUBSURFACE APPLICATION FOR CONTROL OF SUGAR BEET ROOT MAGGOTS: (Recommended in the Big Horn Basin of Wyoming only)—apply 15 pounds DYFONATE 10-G per acre 2 inches from the seed furrow and 2 inches below the soil surface on the irrigated side of sugar beet rows spaced 22 or more inches apart (equivalent to 10½ ounces DYFONATE 10-G per 1,000 linear feet of row). Do not place DYFONATE 10-G in direct contact with the seed.

BROADCAST APPLICATION FOR CONTROL OF SYMPHYLANS AND WIREWORMS: Apply 20 pounds (for symphylans) or 40 pounds (for symphylans and wireworms) DYFONATE 10-G per acre, broadcast, prior to planting and incorporate into the soil by discing.

SUGARCANE: For wireworm control apply 30 to 40 pounds (40 pounds outside of Florida) DYFONATE 10-G per acre in a 14 inch band on cane rows 60 inches apart. This is equivalent to 3.4–4.6 pounds of DYFONATE 10-G per 1,000 linear feet of row.

Apply over the tops of sugar cane seed pieces immediately ahead of covering discs.

TABLE BEETS: For control of cabbage maggot and garden symphylans, apply 20 pounds per acre, broadcast, prior to planting. Incorporate into the soil by discing.

TOBACCO: For control of wireworms listed apply 10 to 20 pounds per acre, broadcast, prior to bedding and transplanting and incorporate into the soil by discing.

TOMATOES: For control of garden symphylans, apply 20 pounds of DYFONATE 10-G per acre, broadcast, prior to planting and incorporate into the soil by discing.

GAUGE SETTINGS AND CALIBRATION INSTRUCTIONS

	7.50 Pounds Per Acre					10 Pounds Per Acre				
speed in mph	**3**	**4**	**5**	**6**	**7**	**3**	**4**	**5**	**6**	**7**
AC			7					9		
Gandy	20	23	25	28	30	23	26	29	32	34
IH	1/8.5	2/2.0	2/6.0	3/0.0	3/5.0	2/1.5	2/7.0	3/3.5	3/9.0	4/5.5
*John Deere (old)	1/29	2/5	2/11	2/17	2/24	2/5	2/13	2/21	2/31	3/4
*John Deere (new)	15	20	23	26	30	20	24	28	32	34
*Noble (old)	14	19	23	27	30	19	24	29	32	35
*Noble (new)	11	13	16	18	20	13	16	19	22	25

* John Deere and Noble have introduced new metering units into their late model machinery. Be sure to use the appropriate chart.

These are suggested beginning settings for application with one hopper-opening per row. Be sure to check your actual application rate, under your operating conditions. Fill hoppers level full and plant a known acreage such as 0.8 acre—one round with four-row planter on 80-rod (¼-mile) rows spaced 40 inches apart. Weigh a full bag, refill hoppers, weigh again, subtract to see how much was applied. Divide amount applied by acreage covered to get actual rate. EXAMPLE: If desired rate is 10#/A, and you apply 8 pounds on 0.8 A, your actual rate is 10#/A because 8 divided by 0.8 = 10. If rate is not quite as desired, adjust up or down and try again. OR Fasten a container (a bag or calibration tube) over the opening of your applicator, then plant a known area. Divide the weight of the granules (in pounds) by the area covered (in acres) to get the actual rate per acre. EXAMPLE: If you recover 8 pounds of granules in the container after covering 0.8 acres, divide 8 pounds by 0.8, which equals 10#/A.

STORAGE AND DISPOSAL

PROHIBITIONS

Do not contaminate water, food or feed by storage or disposal. Open dumping is prohibited.

STORAGE

Keep container closed when not in use. Do not store near seeds, fertilizers, or foodstuffs.

PESTICIDE DISPOSAL

Wastes resulting from the use of this product may be disposed of on site or at an approved waste disposal facility.

CONTAINER DISPOSAL

Completely empty bag into application equipment. Then dispose of empty bag in a sanitary landfill or by incineration, or, if allowed by State and local authorities, by burning. If burned, stay out of smoke.

NOTICE—READ CAREFULLY

CONDITIONS OF SALE

Stauffer (and seller) offer(s) this product for sale subject to, and buyer and all users are deemed to have accepted, the following conditions of sale and warranty which may only be varied by written agreement of a duly authorized representative of Stauffer.

WARRANTY LIMITATION

Stauffer warrants that this product conforms to the chemical description on the label and is reasonably fit for the purposes referred to in the directions for use on the label subject to the inherent risks referred to below. Stauffer makes no other express warranties. THERE IS NO IMPLIED WARRANTY OF MERCHANTIBILITY and there are no warranties which extend beyond the description on the label hereof.

INHERENT RISKS

The directions for use of this product are believed to be reliable and should be followed carefully. However, it is impossible to eliminate all risks associated with use. Buyer assumes all risks associated with the use or application of this product contrary to label instructions or resulting from extraordinary weather conditions.

LIMITATION OF LIABILITY

In no case shall Stauffer be liable for special, indirect or consequential damages resulting from the use or handling of this product and no claim of any kind shall be greater in amount than the purchase price of the product in respect of which such damages are claimed.

This product is sold only for uses stated on its label. No express or implied license is granted to use or sell this product under any patent in any country except as specified: Country: United States of America.

This is a specimen label. Stauffer is not responsible for the accuracy of the the information contained herein. As labels are subject to revision, always carefully read and follow the label on the product container.

DYFONATE® 20-G

GRANULAR INSECTICIDE

Soil Insecticide for Protection Against Corn Rootworm and Other Major Pests

ACTIVE INGREDIENT:
O-Ethyl *S*-phenylethylphosphonodithioate* 20.0%
INERT INGREDIENTS: 80.0%
TOTAL 100.0%

* DYFONATE
Contains 20 Pounds Active Ingredient per 100 Pounds
EPA Reg. No. 476-2028 A-9R

Keep Out of Reach of Children

DANGER–POISON

PELIGRO

PRECAUCION AL USUARIO: Si usted no lee ingles, no use este producto hasta que la etiqueta le haya sido explicado ampliamente.

HAZARDOUS TO LIVESTOCK

STATEMENT OF PRACTICAL TREATMENT FOR PESTICIDE EMERGENCY

CALL A PHYSICIAN IMMEDIATELY: If a known exposure occurs, or if poisoning is suspected, *do not wait for symptoms to develop.* Immediately start the procedures given below and simultaneously contact a Poison Control Center, a physician, or the nearest hospital, or call collect, day or night, Stauffer Chemical Company (203) 226-6602 or Chemtrec (800) 424-9300.

Note: Be sure to advise the person contacted that the compound is a cholinesterase inhibitor. Describe the type of exposure, the victim's symptoms, and follow the advice given.

POISON SYMPTOMS: DYFONATE is a highly toxic cholinesterase inhibitor. Symptoms of poisoning include headache, blurred vision, weakness, nausea, discomfort in the chest, abdominal cramps, vomiting, diarrhea, salivation, sweating and pinpoint pupils.

ALWAYS CALL A PHYSICIAN IN CASE OF SUSPECTED POISONING.
ATROPINE BY INJECTION IS ANTIDOTAL.

If swallowed, immediately give several glasses of water and induce vomiting by placing a finger on the back of the victim's tongue. Give fluids until vomitus is clear. Do not induce vomiting or give anything by mouth to an unconscious or convulsing person.

For skin contact, flush with large amounts of running water for 15 minutes. Remove contaminated clothing and shoes. Get medical attention. Wash clothing before reuse.

For eye contact, immediately flush eyes with large amounts of running water for at least 15 minutes. Hold eyelids apart to insure rinsing of the entire surface of the eye and lids with water. Get medical attention immediately!

If inhaled, remove to fresh air. If not breathing, clear victim's airway and start mouth-to-mouth artificial respiration.

If breathing is difficult, give oxygen, preferably with a physician's advice. Get medical attention immediately.

INFORMATION FOR MEDICAL PERSONNEL: Initial measures include removal of secretions, maintenance of patent airway, and if necessary, artificial respiration. *When cyanosis is relieved,* atropine by injection is antidotal and may be administered in large therapeutic doses, repeated as necessary to the point of tolerance. If symptoms warrant further treatment, 2-PAM (Protopam Chloride) has shown utility as *adjunctive* therapy. Never use morphine. Continued absorption of the poison may occur, resulting in fatal relapse after initial improvement; *very close supervision of the patient is indicated for at least 48 to 72 hours.*

Do not transport or store DYFONATE 20-G with food or feed intended for human or animal consumption.
See back panel for additional precautionary statements

PRECAUTIONARY STATEMENTS

HAZARDS TO HUMANS AND DOMESTIC ANIMALS
DANGER—POISON

Hazardous to livestock—do not contaminate feed used for livestock.
Poisonous if swallowed, inhaled or absorbed through the skin. Rapidly absorbed through the skin. Do not get in the eyes, on skin, or on clothing. Avoid breathing of dust.
Wear a respirator approved by the National Institute for Occupational Safety and Health, (30 CFR Part 11—Subpart M) for protection against pesticide inhalation.

Wear rubber gloves, protective clothing and goggles.

Wash hands, arms, and face thoroughly with soap and water after handling and before eating or smoking. Routinely shower or bathe after work and wash all clothing with soap and hot water before reusing. In case of contact, immediately remove contaminated clothing and wash thoroughly with soap and water.

- Do not transport or store DYFONATE 20-G with any food or feed intended for human or animal consumption.
- Not for use or storage in or around the home.

ENVIRONMENTAL HAZARDS
This product is toxic to fish, birds and other wildlife. Keep out of lakes, streams or ponds. Do not contaminate water by cleaning of equipment or disposal of wastes. Apply this product only as specified on this label.

IN CASE OF SPILL, ISOLATE THE SPILL.
HOLD this package, other cargo and vehicles involved. IMMEDIATELY TELEPHONE (24-hour service) for emergency assistance (800) 424-9300.

GENERAL PRECAUTIONS
Read all directions before using.
Do not apply on any crop, or in any manner not specified on this label.
Do not rotate with carrots.

INSECTS CONTROLLED

Black Cutworm	*Agrotis ipsilon*
Cabbage Maggot	*Hylemy abrassicae*
European Corn Borer	*Ostrinia nubilalis*
Garden Symphylan	*Scutigerella immaculata*
Lesser Cornstalk Borer	*Elasmopalpus lignosellus*
Maize Billbug	*Sphenophorus maidis*
Seed Corn Beetle	*Agonoderus lecontei*
Seed Corn Maggot	*Hylemya cilicrura*
Southwestern Corn Borer	*Diatraea grandiosella*
Sugarbeet Root Maggot	*Tetanops myopaeformis*
White Grub	*Phyllophaga crinata*

* (Control of light and moderate infestation)

Flea Beetles	
Elongate Flea Beetle	*Systena elongata*
Palestriped Flea Beetle	*Systena blanda*
Sweetpotato Flea Beetle	*Chaetocnema confinis*
Threespotted Flea Beetle	*Disonycha triangularis*
Rootworms	
Northern Corn Rootworm	*Diabrotica longicornis*
Western Corn Rootworm	*Diabrotica virgifera*
Southern Corn Rootworm	*Diabrotica undecimpunctata howardi*
Wireworms	
Corn Wireworm	*Melanotus communis*
Pacific Coast Wireworm	*Limonius canus*
Southern Potato Wireworm	*Conoderus falli*
Sugarbeet Wireworm	*Limonius californicus*
Tobacco Wireworm	*Conoderus vespertinus*
Western Field Wireworm	*Limonius infuscatus*

Wheat Wireworm	*Agriotes mancus*
Ctenicera lobata	
Dalopius pallidus	

REENTRY STATEMENT

Do not apply this product in such a manner as to directly or through drift expose workers or other persons. The area being treated must be vacated by unprotected persons.

- Do not enter treated areas without protective clothing until dusts have settled.
- Do not apply or allow to drift to areas occupied by beneficial animals or onto adjoining food, fiber or pasture crops. Do not contaminate food or feed.
- Because certain states may require more restrictive reentry intervals for various crops treated with this product, consult your State Department of Agriculture for further information.
- Written or oral warnings must be given to workers who are expected to be in a treated area or in an area about to be treated with this product. Oral warnings must be given if there is reason to believe that written warnings cannot be understood by workers.

When oral warnings are given, warnings shall be given in a language customarily understood by workers and must indicate the specific period of time that treated areas may not be entered without protective clothing. Oral warnings must also include appropriate first aid instructions in case of accidental exposure. These instructions are given in the STATEMENT OF PRACTICAL TREATMENT at the beginning of this label.

Written warnings must include the following information, "WARNING. Area treated with DYFONATE® Agricultural Insecticide on (date of application). Do not enter treated areas without appropriate protective clothing until dusts have settled (or other reentry interval if your state has a more restrictive interval for this product). If accidental exposure occurs, follow the instructions below." (Written warnings must include the STATEMENT OF PRACTICAL TREATMENT given at the beginning of this label).

DIRECTIONS FOR USE AND RECOMMENDATIONS

It is a violation of Federal law to use this product in a manner inconsistent with its labeling.

NOTE: Preplant broadcast applications should be mixed (incorporated) into the soil to a depth of 2 to 3 inches.

COLE CROPS (Broccoli, Brussels Sprouts, Cabbage and Cauliflower):

For control of cabbage maggot prior to and at time of seeding or transplanting, apply 20 pounds of DYFONATE 20-G per acre, broadcast and incorporate into the soil by discing. Not recommended in the Northwestern U.S.

CORN (Field, Sweet and Popcorn):

NOTE: May be used in corn grown for grain or silage.

A. Band application for control of northern, southern and western corn rootworms, black cutworm, lesser cornstalk borer, seed corn beetle and seed corn maggot, and for suppression of wireworms (recommended for seed corn beetle in the central and midwestern corn growing areas only): Apply the rates recommended in the following table in a 6 to 8 inch band over corn rows 38 to 40 inches apart.

Insect	Rate per acre	Rate per 1000 linear feet
Corn rootworms	3¾ to 5 pounds	4½ to 6 ounces
Seed corn beetle		
Seed corn maggot		
Wireworms	5 pounds	6 ounces
Black cutworm		
Lesser cornstalk borer	10 pounds	12 ounces

* With heavy infestations, a sufficient number of black cutworm larvae may survive the DYFONATE treatment to cause damage to the corn. If this occurs, an insecticide rescue treatment with another insecticide registered for this use can be made to control these surviving larvae.)

1. **Application at planting:** Incorporate into the top ½ to 1 inch of soil by making application ahead of the press wheels, or by dragging a short length of chain behind the press wheels. For John Deere 7000 series planters place DYFONATE 20-G behind the firming wheels. *Do not place DYFONATE 20-G in direct contact with the seed.* **. . . or**
2. **Application at cultivation:** Cover the treated band with 2 to 3 inches of soil by making application immediately ahead of disc hillers or cultivation equipment. Make application in May or June when activity of rootworm larvae is first observed. NOT recommended for control of lesser cornstalk borer or black cutworm or for suppression of wireworms. Do not apply within 45 days of harvest nor feed or graze livestock within 30 days of treatment.

B. Broadcast application for control of symphylans, northern, southern and western corn rootworms, wireworms and maize billbugs.

Apply 10 pounds (for symphylans) or 20 pounds (for symphylans, rootworms, wireworms and maize billbugs) DYFONATE 20-G per acre, prior to planting and incorporate into the soil by discing. Recommended for billbugs in the southeastern U.S. only.

C. Band or broadcast application for control of European corn borer and southwestern corn borer.

Apply 5 pounds DYFONATE 20-G per acre over the corn, so granules fall into whorls. May be applied broadcast, using air or ground equipment or in 6 to 8 inch bands over corn rows 30 inches or more apart.

Do not apply within 30 days of harvest nor feed or graze livestock within 30 days of treatment.

1. **European corn borer:** Make application when 50% of the plants show first generation borer feeding. To control second generation corn borer, apply when field counts show 100 egg masses per 100 plants.
2. **Southwestern corn borer:** Apply when newly hatched larvae first appear. Repeat application as necessary to control later generations.

D. Band application for control of wireworms in sweet corn grown on muck soils (Florida only):

Based on 36 inch row spacing, apply 10 pounds DYFONATE 20-G per acre in a 2 inch band over the furrows at planting. This is equivalent to 10 pounds DYFONATE 20-G per 14,520 linear feet of row.

PEANUTS:

Southeastern U.S.: For control of southern corn rootworm, use 10 pounds DYFONATE 20-G per acre in an 18 inch band on 36 inch rows—apply at pegging. For control of lesser cornstalk borer, use 10 pounds DYFONATE 20-G per acre in an 18 inch band on 36 inch rows—apply at first signs of borer infestation after pegging begins. May be used up to 30 days after pegging begins.

SUGAR BEETS:

Band application for control of sugar beet root maggots: Apply 5 to 7½ pounds DYFONATE 20-G per acre in a 7 inch band over sugar beet rows 22 or more inches apart. Apply at planting and lightly incorporate into the soil. Do not place DYFONATE 20-G in direct contact with the seed. Recommended in Colorado and the northwest sugar beet growing areas only.

SUGARCANE:

For control of the white grub, *Phyllophaga crinata,* apply 20 to 30 pounds DYFONATE 20-G per acre in a 20 inch band on cane rows 60 inches apart. This is equivalent to 2.3–3.4 pounds of DYFONATE 20-G per 1,000 linear feet of row. Make application over the top of the cane row and lightly incorporate as close to the plants as possible. For best results, irrigate after application. Application should be timed according to the flight of the adult beetles. Apply at any time from the peak of the adult flight up until two weeks after flight peak. Recommended in the southwestern U.S. only.

SWEET POTATOES:

Southeastern U.S.: For control of wireworms and flea beetle larvae, apply 20 pounds DYFONATE 20-G per acre, broadcast, at planting and incorporate into the soil by discing, or, apply 15 pounds DYFONATE 20-G over the top at root swelling.

Other Growing Regions: For control of wireworms and flea beetle larvae, apply 20 pounds DYFONATE 20-G per acre, broadcast, at planting and incorporate into the soil by discing, or, over the top at root swelling.

GAUGE SETTINGS AND CALIBRATION INSTRUCTIONS

	3.75 Pounds Per Acre					5 Pounds Per Acre				
	Speed in mph									
Equipment	**3**	**4**	**5**	**6**	**7**	**3**	**4**	**5**	**6**	**7**
AC			3					5		
Gandy	13	15	17	19	20	15	17	20	22	23
IH	1/3.0	1/4.0	1/5.3	1/6.5	1/8.0	1/4.2	1/5.5	1/7.2	1/9.0	2/1.0
*John Deere (old)	1/21	1/23	1/25	1/27	1/29	1/23	1/26	1/28	1/33	2/3
*John Deere (new)	8	10	12	14	16	9	12	15	18	20
*Noble (old)	6	9	11	12	14	9	11	14	17	19
*Noble (new)	5	6	7	9	10	6	8	9	11	12

* John Deere and Noble have introduced new metering units into their late model machinery. Be sure to use the appropriate chart.

These are suggested beginning settings for application with one hopper-opening per row. Be sure to check your actual application rate, under your operation conditions. Fill hoppers (level full and plant a known acreage such as 0.8 acre—one round with four—row planter on 80-rod (1/4 mile) rows spaced 40 inches apart. Weigh a full bag, refill hoppers, weigh again, subtract to see how much was applied. Divide amount applied by acreage covered to get actual rate. EXAMPLE: If desired rate is 5 lbs./A, and you apply 4 pounds on 0.8 A, your actual rate is 5 lbs./A because 4 divided by 0.8 = 5. If rate is not quite as desired, adjust up or down and try again. Or fasten a container (a bag or calibration tube) over the opening of your applicator, then plant a known area. Divide the weight of the granules (in pounds) by the area covered (in acres) to get the actual rate per acre. EXAMPLE: If you recover 4 pounds of granules in the container after covering 0.8 acre, divide 4 pounds by 0.8, which equals 5 lbs./A.

STORAGE AND DISPOSAL

PROHIBITIONS

Do not contaminate water, food, or feed by storage or disposal. Open dumping is prohibited.

STORAGE

Do not transport or store with any food or feed intended for human or animal consumption. Not for use or storage in or around the home.

PESTICIDE DISPOSAL

Pesticide wastes are acutely hazardous. Improper disposal of excess pesticide, or rinsate is a violation of Federal law. If these wastes cannot be disposed of by use according to label instructions, contact your State Pesticide or Environmental Control Agency, or the Hazardous Waste representative at the nearest EPA Regional office for guidance.

CONTAINER DISPOSAL

Completely empty bag into application equipment. Then dispose of empty bag in a sanitary landfill or by incineration, or, if allowed by State and local authorities, by burning. If burned, stay out of smoke.

NOTICE—READ CAREFULLY

CONDITIONS OF SALE

Stauffer (and seller) offer(s) this product for sale subject to, and buyer and all users are deemed to have accepted, the following conditions of sale and warranty which may only be varied by written agreement of a duly authorized representative of Stauffer.

WARRANTY LIMITATION

Stauffer warrants that this product conforms to the chemical description on the label and is reasonably fit for the purposes referred to in the directions for use on the label subject to the inherent risks referred to below. Stauffer makes no other express warranties. THERE IS NO IMPLIED WARRANTY OF MERCHANTABILITY and there are no warranties which extend beyond the description on the label hereof.

INHERENT RISKS

The directions for use of this product are believed to be reliable and should be followed carefully. However, it is impossible to eliminate all risks associated with use. Buyer assumes all risks associated with use or application of this product contrary to label instructions or resulting from extraordinary weather conditions.

LIMITATION OF LIABILITY

In no case shall Stauffer be liable for special, indirect or consequential damages resulting from the use or handling of this product and no claim of any kind shall be greater in amount than the purchase price of the product in respect of which such damages are claimed.

This product is sold only for uses stated on its label. No express or implied license is granted to use or sell this product under any patent in any country except as specified. Country: United States of America

This is a specimen label. Stauffer is not responsible for the accuracy of the information contained herein. As labels are subject to revision, always carefully read and follow the label on the product container.

EPTAM® 7-E

SELECTIVE HERBICIDE EMULSIFIABLE LIQUID

Complete Directions For Use, Use Precautions Regional Crop Recommendations, National Ornamental Recommendations

ACTIVE INGREDIENT:

S-ethyl dipropylthiocarbamate 87.8%

INERT INGREDIENTS: 12.2%

TOTAL 100.0%

Contains 7 pounds of active ingredient per gallon

EPA Reg. No. 476-2154 A-8R

CAUTION

Keep Out of Reach of Children

HARMFUL IF SWALLOWED

STATEMENT OF PRACTICAL TREATMENT

FIRST AID

Immediately start the procedures given below. If further treatment is required, contact a Poison Center, a physician, or the nearest hospital.

If swallowed:

Immediately give several glasses of water but *do not* induce vomiting. If vomiting does occur, give fluids again. Have a physician determine if condition of patient will permit induction of vomiting or evacuation of stomach. Do not give anything by mouth to an unconscious or convulsing person.

For eye contact:

Immediately flush eyes with large amounts of running water for at least 15 minutes. Hold eyelids apart to ensure rinsing of the entire surface of the eye and lids with water. Get medical attention immediately.

For skin contact:

Flush all affected areas with plenty of water for several minutes. Remove and clean contaminated clothing and shoes. Seek medical attention if irritation occurs.

If inhaled:

Remove to fresh air. Seek medical attention if respiratory irritation occurs or if breathing becomes difficult.

PRECAUTIONARY STATEMENTS

HAZARDS TO HUMANS AND DOMESTIC ANIMALS

CAUTION

HARMFUL IF SWALLOWED

Avoid contact with skin, eyes and clothing—Avoid inhalation of spray mist—Wear rubber gloves and fresh clothing—Wear goggles to avoid contact with eyes—Wash with soap and water immediately after use or contact—Do not contaminate food or feed.

Do not apply this product in such a manner as to directly or through drift expose workers or other persons. When applied through sprinkler irrigation method of application, the area being treated must be vacated by unprotected persons.

ENVIRONMENTAL HAZARDS

Do not contaminate water to be used on susceptible crops and ornamentals, or to be used for domestic purposes. This product is toxic to shrimp. Keep out of tidal marshes and estuaries. Do not apply where runoff is likely to occur. Do not apply when weather conditions favor drift from areas treated. Do not contaminate water by cleaning of equipment or disposal of wastes.

IN CASE OF EMERGENCY: Immediately call (24 hours) (800) 424-9300 Pesticide Team Safety Network, or (203) 226-6602 Stauffer Chemical Company.

GENERAL USE PRECAUTIONS

Read all label directions before using.

EPTAM 7-E should be used only for recommended purposes and recommended rates. DO NOT OVERDOSE.

EPTAM 7-E is recommended for use on mineral soils only (soils containing less than 10% organic matter).

Keep container tightly closed when not in use. Do not store near seeds or fertilizers. Store out of reach of children, pets and domestic animals.

Rinse spray equipment and empty container.

Apply this product only as specified on this label.

SPECIAL PRECAUTIONS FOR CROP USES

For incorporated applications, use equipment which has been proven to incorporate thoroughly to the recommended depth.

In irrigated areas, do not apply EPTAM 7-E prior to pre-irrigation.

Tank-mix this product with fungicides, insecticides or herbicides only as recommended.

When properly applied and weather conditions exist for normal plant growth through the season, EPTAM 7-E will not harm the treated crop nor should harmful soil residues remain beyond harvest. However, during germination and early growth, extended periods of unusually cold and wet or hot and dry weather, insect, nematode, or plant disease attack, carry-over soil residues of certain persistent herbicides, the use of certain soil applied systemic insecticides, highly saline or alkaline soil conditions, improperly placed fertilizers or soil insecticides may create abnormal conditions that weaken crop seedlings. Also some of these abnormal conditions may weaken established crops: alfalfa, almonds, etc. EPTAM 7-E used under these abnormal conditions could result in crop injury.

SPECIAL PRECAUTIONS FOR ORNAMENTAL USES

EPTAM 7-E must be thoroughly mixed into the soil for all ornamental uses.

EPTAM 7-E may cause injury to ornamentals under certain soil and climatic conditions or if directions are not followed.

WEEDS CONTROLLED

EPTAM 7-E will not control established weeds.

Annual Grasses:

Annual Bluegrass	*Poa annua*
Annual Ryegrass (Italian Ryegrass)	*Lolium multiflorum*
Barnyardgrass (Watergrass Junglerice)	*Echinochloa* spp.
Bermudagrass Seedlings	*Cynodon dactylon*
Crabgrass	*Digitaria* spp.
Giant Foxtail	*Setaria faberii*
Goosegrass	*Eleusine indica*
Green Foxtail	*Setaria viridis*
Johnsongrass Seedlings	*Sorghum halepense*
Lovegrass (Stinkgrass)	*Eragrostis cilianensis*
Panicum, Fall	*Panicum dichotomiflorum*
Panicum*, Texas	*Panicum texanum*
Rescuegrass	*Bromus willdenowii*
Sandbur, Field	*Chenchrus pauciflorus*
Wild Cane (Shattercane)	*Sorghum bicolor*
Signalgrass	*Brachiaria* spp.
Volunteer grains (Barley, Oats, Wheat)	
Wild Oats	*Avena fatua*
Witchgrass*	*Panicum capillare*
Yellow Foxtail	*Setaria glauca*

* May not be controlled at less than 3½ pints of EPTAM 7-E.

Annual Broadleaf Weeds:

Annual Morningglory	*Ipomoea purpurea*
Black Nightshade	*Solanum nigrum*
Carpetweed	*Mollugo verticillata*
Chickweed, Common	*Stellaria media*
Corn Spurry	*Spergula arvensis*
Deadnettle (Henbit)	*Lamium amplexicaule*
Fiddleneck	*Amsinckia* spp.
Florida Purslane	*Richardia scabra*
Hairy Nightshade	*Solanum villosum*
Lambsquarters*, Common	*Chenopodium album*
Nettleleaf, Goosefoot	*Chenopodium murale*
Purslane, Common	*Portulaca oleracea*
Prostrate Pigweed	*Amaranthus Graecizans*
Prickly Sida*	*Sida spinosa*
Redroot Pigweed* (Common Pigweed)	*Amaranthus retroflexus*
Shepherdspurse	*Capsella bursa-pastoris*
Sicklepod*	*Cassia obtusifolia*
Tumble Pigweed	*Amaranthus albus*

The annual broadleaf weeds listed above will be controlled only if treatment is made when conditions are favorable for weed germination and growth. Broadleaf weeds may only be suppressed at less than 3½ pints EPTAM 7-E per acre in heavier soils or under very cold soil conditions.

* May not be controlled at less than 3½ pints of EPTAM 7-E.

Perennial Weeds:

Bermudagrass	*Cynodon dactylon*
Purple Nutsedge* (Nutgrass)	*Cyperus rotundus*
Quackgrass (Couchgrass, Quitchgrass)	*Agropyron repens*
Yellow Nutsedge* (Nutgrass)	*Cyperus esculentus*
Mugwort** (Chrysanthemumweed)	*Artemisia vulgaris**

* May not be controlled at less than 3½ pints of EPTAM 7-E per acre.
** Controlled by high EPTAM 7-E rates recommended for use on certain ornamentals only. See ornamental recommendations for specific uses.

Perennial weeds must be turned under and chopped up thoroughly prior to treatment. The underground rhizomes of quackgrass and the rhizomes and stolons of bermudagrass must be cut up thoroughly so that four or less nodes remain on a strand. For the suppression or control of quackgrass and bermudagrass the disc must be set to cut 6 inches deep. Use 4½ to 7 pints EPTAM 7-E for quackgrass and 3½ to 7 pints for bermudagrass. The EPTAM 7-E should be incorporated by discing or applied in the irrigation water after the rhizomes and stolons have been cut up. **Consult recommendations for crops on which these higher rates may be used.** Nutsedge may not be controlled by water-run applications in heavier soils.

TANK MIX COMBINATIONS
For broader spectrum weed control and increased contol of certain broadleaf weeds EPTAM 7-E may be tank mixed with the following herbicides. On dry beans, EPTAM 7-E may be tank mixed with Treflan* or with a combination of Treflan* and Amiben**. On Irish potatoes, EPTAM 7-E may be tank mixed with metribuzin (Lexone† or Sencor‡). Consult supplemental literature for exact rates and application directions.

* Treflan—Registered Trademark of Elanco Products Incorporated.

** Amiben chloramben herbicide—Registered Trademark of Union Carbide Agricultural Products Co., Inc.

† Lexone—Registered Trademark of DuPont Company.

‡ Sencor-Registered Trademark of Mobay Chemical Corp.

DIRECTIONS FOR USE
CROP SECTION

It is a violation of Federal law to use this product in a manner inconsistent with its labeling.

EPTAM 7-E is a selective herbicide which for most uses must be mixed (incorporated), or injected subsurface into the soil, or applied in the irrigation water for control of weeds listed on this label. EPTAM 7-E controls weeds by interfering with normal germination and seedling development. It does not control established weeds.

APPLICATION DIRECTIONS
Pour the recommended amount of EPTAM 7-E into the spray tank during the filling operations. Apply in 10 to 50 gallons of water per acre using a properly calibrated, low-pressure sprayer having good agitation. The soil should be well worked and dry enough to permit good soil mixing (incorporation).

EPTAM 7-E may be combined with solution, slurry or suspension fertilizers. However, physical compatibility with these fluid fertilizers must be determined before combining in the spray tank. See Appendix 1 for special direction regarding these combinations. Even though found to be compatible, constant agitation is necessary to keep the EPTAM 7-E uniformly mixed with the fluid fertilizer.

For all band applications, reduce dosage proportionately depending upon row spacing and band width to be treated.

IMPREGNATION ON DRY FERTILIZER
EPTAM 7-E may be impregnated on dry fertilizer for use on registered crops. However, uniform distribution of the EPTAM 7-E on fertilizer particles and uniform application are necessary to assure good results. See special instructions elsewhere on this label giving directions for impregnation and use.

INCORPORATION DIRECTIONS
EPTAM 7-E must be incorporated into the soil immediately to prevent loss of the herbicide. Whenever possible, application and incorporation should be done in the same operation.

SOIL MIXING (INCORPORATION) BEFORE PLANTING: The following equipment commonly is used for soil mixing (incorporation) before planting:

Power Driven Cultivation Equipment (recommended on all soil types) set to cut to a depth of 2 to 3 inches.

Tandem Discs (recommended on all soil types) set to cut to a depth of 4 to 6 inches, operated at 4 to 6 mph followed by a spiked-tooth harrow or some other leveling device which extends beyond the ends of the discs. For more thorough mixing (for perennial grasses and in heavier soils) disc in two different directions (cross disc).

Field Cultivators (recommended for spring application on coarse textured soils, and for fall application on all soils. Use only on soils in good tilth). Use 3 to 4 rows of sweeps, spaced at 7 inch or less intervals and staggered so that no soil is left unturned, followed by a spiked-tooth harrow pulled behind the cultivator. Do not use chisel plows to incorporate. Set the cultivator to cut 4 inches deep, operated at 5 mph or more. Run the equipment over the field twice, the second run at an angle to the first.

Rotary Ground Driven or Spring-Tooth Cultivators (recommended on coarse textured soils in good tilth only). Set to penetrate to a depth of 4 to 6 inches and operated at 5 to 8 mph in two different directions.

INCORPORATION AT OR AFTER PLANTING: Use power driven cultivation equipment set to cut to a depth of 2 to 3 inches OR on coarse textured soils only, ground driven tillers (rolling cultivators, rotary hoe, row wheels, etc.) set to cut to a depth of $1\frac{1}{2}$ inches and operated at 6 to 8 mph. When incorporating after planting, care must be taken not to disturb the crop seed or seedling.

SUBSURFACE APPLICATION—
AT PLANTING OR POST EMERGENCE
Apply EPTAM 7-E in 10 or more gallons of water per acre.

Special equipment designed for subsurface application MUST be used. Injector and sweep units must be rigidly mounted on the planter or cultivation unit. When using sweeps at planting they must be mounted ahead of the planters.

SOIL INJECTION—Injector shanks must be spaced $2\frac{1}{2}$ to 3 inches apart and mounted in staggered positions to avoid trash buildup. Set shanks to inject EPTAM 7-E 2 to 3 inches below the soil surface. The width of the band in which weed control is desired will determine the number and spacing of injector shanks required per row. (Example: Four injector shanks spaced 3 inches apart give a 12-inch band). A broadcast application can be made by increasing the number of shanks. The two shanks adjacent to the drill row must be $1\frac{1}{4}$ to $1\frac{1}{2}$ inches on either side of it, EXCEPT IN COTTON WHERE THE DISTANCE MUST BE 4 INCHES ON EITHER SIDE OF THE DRILL ROW, AND SUGAR BEETS WHERE THE DISTANCE MUST BE $2\frac{3}{4}$ INCHES ON EITHER SIDE OF THE DRILL ROW.

COVERED SWEEPS—Set the sweeps to run below the soil suface deep enough to cover the EPTAM 7-E with 2 to 3 inches of soil. Calibrate by measuring the spray band width at the back of the sweep, not the sweep width. For broadcast applications stagger sweeps on double tool bar so they overlap sufficiently to allow spray bands to meet.

NOTE: When applying with either injectors or sweeps, EPTAM 7-E must be applied deep enough to allow 2 to 3 inches of soil to remain over the treatment after the planting operations.

PLANTING DIRECTIONS
For pre-plant applications, seeding should be done as soon as possible after treatment to obtain a maximum period of weed control.

IRRIGATION APPLICATION—
POST-PLANTING AND ESTABLISHED CROPS
Meter EPTAM 7-E into the irrigation water by using a metering device that will introduce a constant flow into the water. For flood, furrow, or sprinkler irrigation meter the EPTAM 7-E into the water during the entire period OR, for sprinkler irrigation, the EPTAM 7-E may be metered into sufficient water to penetrate to a depth of 3 to 4 inches. Time this EPTAM 7-E application to insure that proper penetration of the herbicide corresponds with the end of the irrigation period. Flush the lines and then turn the water off promptly. Consult "RECOMMENDATIONS" on this label for proper timing of application for each crop for which irrigation application is recommended. A flow rate chart for water run applications will be found in Appendix 3 of this booklet.

USE PRECAUTIONS FOR SPRINKLER IRRIGATION SYSTEMS
A. Apply the product only through irrigation systems containing anti-siphon and check valves which will prevent water source contamination and overflow of the slurry tank and containing interlocking controls between the metering device and the water pump to insure simultaneous shut-off.
B. Inject the product with a positive displacement pump into the main line ahead of a right angle turn to insure adequate mixing.
C. Application of more than label recommended quantities of irrigation water per acre may result in decreased product performance by removing the chemical from the zone of effectiveness.
D. Do not apply when wind speed favors drift, when system connections or fittings leak, when nozzles do not provide uniform distribution or when lines containing the product must be dismantled and drained.

USE PRECAUTIONS FOR FLOOD OR FURROW IRRIGATION
Tailwater (runoff water) from flood or furrow irrigation should be recirculated or used only on other crops which are registered for this type of application.

CULTURAL PRACTICES FOLLOWING APPLICATION
Should weeds develop, a shallow cultivation or rotary hoeing will generally result in better weed control. When cultivating for any reason, it should be shallow, i.e., no more than ½ the depth the herbicide was incorporated or injected. Pre-emergence or post-emergence herbicides may be necessary to control weeds resistant to EPTAM 7-E.

REGIONAL USE MAP

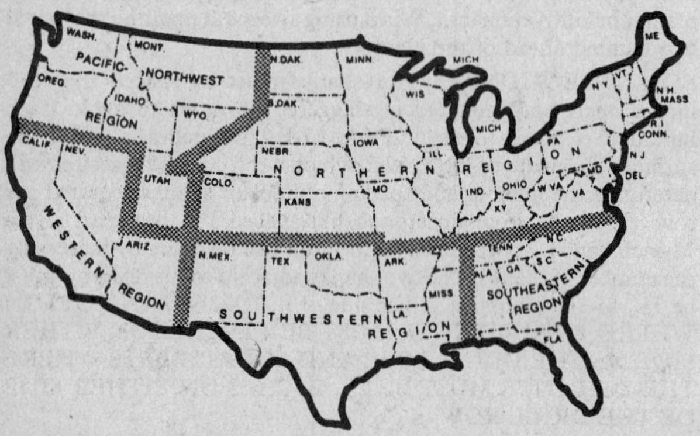

CROP RECOMMENDATIONS
All crop recommendations are given on a regional basis. There are five regions, as delineated on the U.S. map printed above. USE THE RECOMMENDATIONS IN YOUR REGION ONLY.

RATE CONVERSION TABLE
Dosage rates in this booklet are expressed as pints EPTAM 7-E per acre. The following table shows pints EPTAM 7-E per acre in the left column and the equivalent amount of active ingredient per acre in the center column.

Pints EPTAM 7-E Per Acre	Lb. Active Ingredient Per Acre	Acres Treated By One Gallon EPTAM 7-E
1¼	1	7
1¾	1½	4⅔
2¼	2	3½
3½	3	2⅓
4½	4	1¾
5¼	4½	1½
5¾	5	1⅖
7	6	1⅙
8½	7½	1
17	15	½

RECOMMENDATIONS

NORTHERN REGION

These recommendations are given as the broadcast (overall) rates of EPTAM 7-E per acre. For band treatment, use proportionately less material per acre depending upon the width of the band to be treated and the crop row spacing. Do not use band application on rocky ground because thorough incorporation is not possible.

ALFALFA*, BIRDSFOOT TREFOIL, CLOVERS, LESPEDEZA: Do not use EPTAM 7-E if a grass or grain nurse crop is to be planted with the legume. Do not use on white dutch clover. Apply and incorporate 3½ to 4½ pints EPTAM 7-E per acre just before planting. Temporary crop stunting and sealing of the first leaves will occur if conditions for germination and growth are not optimum (e.g. lack of moisture), and will be relieved by irrigation or adequate rainfall.

OR ALFALFA* (FOR CONTROL OF ANNUAL GRASSES GROWING FROM SEED ONLY): Apply and incorporate 2¼ pints EPTAM 7-E per acre just before planting. Temporary crop stunting and sealing of the first leaves will occur if conditions for germination and growth are not optimum.

* Alfalfa is sensitive to soil residues of Atrazine. Do not use EPTAM 7-E on alfalfa if Atrazine was applied within the previous 12 months.

AND/OR ALFALFA (ESTABLISHED STANDS): Meter 2¼ to 3½ pints EPTAM 7-E per acre into the irrigation water applied to established stands prior to weed emergence. Use the lower rate on very coarse textured soils. Do not apply within 14 days of harvesting or grazing alfalfa.

LADINO CLOVER (ESTABLISHED STANDS): Meter 2¼ to 3½ pints EPTAM 7-E per acre into the irrigation water applied to established stands prior to weed emergence. Use the lower rate on very coarse textured soils. Do not apply within 45 days of harvesting or grazing.

BEANS, GREEN OR DRY: Do not use EPTAM 7-E on Adzuki beans, cowpeas (blackeye peas, blackeye beans), soybeans, lima beans, Mung beans, Garbanzo beans or other flat-podded beans except Romano. Under abnormal weather conditions, stunting may occur on Gratiot, Michilite, Sanilac, Seafarer, and Seaway varieties. Do not exceed 3½ pints EPTAM 7-E per acre on small white beans or green beans grown on coarse textured soils.

FALL APPLICATION: (dry beans Minnesota, North Dakota only): Apply and incorporate in the late fall before the ground freezes. Use 4½ pints EPTAM 7-E per acre on coarse textured soils and 5¼ pints EPTAM 7-E per acre on medium and fine textured soils.

APPLICATION AT PLANTING: Apply and incorporate just before planting or meter into the irrigation water before or immediately after planting, 3½ to 4½ pints EPTAM 7-E per acre. Rotary hoe lightly during or shortly after emergence of the beans to break any crust which occurs.

OR

LAY-BY APPLICATION: At time of last cultivation for the season apply and incorporate 3½ to 4½ pints EPTAM 7-E per acre. Apply as a directed spray to the soil at the base of the plants before bean pods start to form. Do not feed or pasture vines to livestock within 45 days after application.

For control of smartweed and ragweed and for increased control of labeled broadleaf weeds, EPTAM 7-E may be followed by a pre-emergence application of Amiben* 2E, 2-EC. Apply according to label directions and observe all use precautions. Slight phytotoxicity may be noted following sequential application.

* Chloramben herbicide—Registered trademark of Union Carbide Agricultural Products Company, Inc.

CASTOR BEANS: Apply and incorporate 2¼ pints EPTAM 7-E per acre immediately after planting. Use a rotary hoe for incorporation. Early cultivation after EPTAM 7-E application enhances weed control.

FLAX:

FALL APPLICATION (Minnesota, North Dakota): Apply and incorporate in the late fall before the ground freezes. Use 4½ pints EPTAM 7-E per acre on coarse textured soils and 5¼ pints EPTAM 7-E per acre on medium and fine textured soils.

POTATOES, IRISH: Do not exceed 7 pints EPTAM 7-E per acre per crop. The Superior variety potato is sensitive to EPTAM 7-E and under stress conditions, early season stunting may occur.

FALL APPLICATION (Minnesota, North Dakota). Apply and incorporate in the late fall before the ground freezes. Use 5¼ pints EPTAM 7-E per acre on coarse textured soils and 7 pints EPTAM 7-E per acre on medium and fine textured soils.

PRE-PLANT

INCORPORATION: Apply and incorporate 3½ to 7 pints EPTAM 7-E per acre. For quackgrass and nutgrass control use the higher rate.

OR

DRAG-OFF (COME UP, WEEDING TIME)

INCORPORATION: Apply and incorporate 3½ to 7 pints EPTAM 7-E per acre. For nutgrass control use the higher rate. The field first must be "dragged-off", followed by EPTAM 7-E application and incorporation. Use spiked-tooth harrows or cultivation equipment for incorporation.

AND/OR LAY-BY

INCORPORATION: Apply and incorporate 3½ to 4½ pints EPTAM 7-E per acre to clean cultivated soil after potato plants have emerged from the soil. Apply as directed spray to the soil. Do not apply within 45 days of harvest.

OR

IRRIGATION: Meter up to 3½ pints EPTAM 7-E per acre into the irrigation water after clean cultivation. Do not apply within 45 days of harvest.

SAFFLOWER: Apply and incorporate 3½ pints EPTAM 7-E per acre just before planting.

SUGAR BEETS:

FALL APPLICATION (Minnesota, North Dakota): Apply and incorporate in the late fall before the ground freezes. Use 4½ pints EPTAM 7-E per acre on coarse textured soils and 5¼ pints EPTAM 7-E per acre on medium and fine textured soils.

PRE-PLANT (Iowa, Eastern Nebraska, North Dakota, South Dakota, Minnesota, Michigan): Apply and incorporate 2¼ pints EPTAM 7-E per acre on coarse textured soils, or 3½ pints per acre on medium and fine textured soils just before planting. Injury will occur if conditions for germination and growth are not optimum.

OR

POST EMERGENCE

(After the first true leaves have formed)

IRRIGATION WATER: Meter 2¼ to 3½ pints EPTAM 7-E per acre into the first irrigation applied after the last cultivation for the season.

INCORPORATION: Apply 3½ pints EPTAM 7-E per acre after thinning and clean cultivation and incorporate to a depth of 2 to 3 inches. Treatment may be used following a fall application of EPTAM 7-E at recommended rates.

OR

SUBSURFACE INJECTION: Apply 3½ pints EPTAM 7-E per broadcast acre, or in band treatment (using 2 shanks per row 5½ inches apart centered on the drill row with rows 22 inches apart) use 1¾ pints EPTAM 7-E per acre. Prior to application, a clean cultivation must be made for all existing weed growth to be destroyed.

SUNFLOWER (Minnesota, North Dakota, South Dakota): Apply and incorporate 2½ to 3½ pints EPTAM 7-E per acre just before planting. Use lower rates on lighter soil.

FALL APPLICATION (Minnesota, North Dakota): Apply and incorporate in the late fall before the ground freezes. Use 4½ pints EPTAM 7-E per acre on coarse textured soils and 5¼ pints EPTAM 7-E per acre on medium and fine textured soils.

RECOMMENDATIONS SOUTHEASTERN REGION

These recommendations are given as the broadcast (overall) rate of EPTAM 7-E per acre. For band treatment, use proportionately less material per acre depending on the width of band to be treated and the crop row spacing. Do not use band application on rocky ground because thorough incorporation is not possible.

ALFALFA*, BIRDSFOOT TREFOIL, CLOVERS, LESPEDEZA: Do not use EPTAM 7-E if a grass or grain nurse crop is to be planted with the legume. Do not use on white dutch clover. Apply and incorporate 3½ pints EPTAM 7-E per acre just before planting. (For fall seeded alfalfa in South Carolina only, apply and incorporate 1¾ pints EPTAM 7-E per acre just before planting.) Temporary crop stunting and sealing of the first leaves will occur if conditions for germination and growth are not optimum (e.g. lack of moisture), and will be relieved by irrigation or adequate rainfall.

* Alfalfa is sensitive to soil residues of Atrazine. Do not use EPTAM 7-E on alfalfa if Atrazine was applied within the previous 12 months.

BEANS, GREEN OR DRY: Do not use EPTAM 7-E on Adzuki beans, cowpeas (blackeye peas, blackeye beans), soybeans, lima beans, Mung beans, Garbanzo beans or other flat-podded beans except Romano. Under abnormal weather conditions stunting may occur on Gratiot, Michilite, Sanilac, Seafarer and Seaway varieties.

AT PLANTING

PRE-PLANT (FLAT-PLANTED): Use 3½ pints EPTAM 7-E per acre incorporated just before planting on dry, snap and pole beans. Rotary hoe lightly during or shortly after emergence of the beans to break any crust which occurs.

OR

SUBSURFACE APPLICATION: Apply 2¼ pints EPTAM 7-E per acre pre-plant or at planting. See DIRECTIONS FOR USE.

OR

BED TREATMENTS

METHOD A—Apply 3½ pints EPTAM 7-E per acre broadcast and disc in 6 inches deep prior to forming beds and planting.

METHOD B—Apply 1¾ pints EPTAM 7-E per acre broadcast (do not disc in) immediately ahead of bedding discs. Plant 7 days after treatment.

METHOD C—Apply as a band treatment (do not disc in) immediately ahead of bedding discs, or as a band treatment to partially formed beds or bed tops immediately in front of the rebedding operation. Use a band rate equivalent to 2¼ pints per acre broadcast. Care should be taken not to fold in treatment.

EXAMPLE: To apply EPTAM 7-E as an 18 inch band on 36-inch rows, use 1¼ pints per crop acre. Plant 7 days after application.

NOTE: With Methods B and C, if bed shapers (levelers) are used, the bedding up and shaping should be done so that 3 to 4 inches of soil remain over the EPTAM 7-E.

OR

LAY-BY

At the time of last cultivation apply and incorporate 3½ pints EPTAM 7-E per acre. Apply as a directed spray to the soil at the base of the plants before bean pods start to form. Do not feed or pasture vines to livestock until 45 days after application.

CITRUS NURSERY STOCK AND YOUNG FIELD PLANTINGS (Non-bearing Orange and Grapefruit Groves): After lining out, apply 3½ to 7 pints EPTAM 7-E per acre as a directed spray to the soil. Incorporate with cultivation equipment, i.e., tree hoes and rotary hoes.

CITRUS (Oranges, Tangerines, Grapefruit): AFTER CLEAN CULTIVATION OR PRIOR TO WEED EMERGENCE in bearing citrus, apply 3½ pints EPTAM 7-E per acre by flood or furrow irrigation. Meter EPTAM 7-E into the water during the entire irrigation period. Do not apply within 15 days of harvest.

COTTON: NON-IRRIGATED AREAS ONLY—

APPLICATION AFTER STAND IS ESTABLISHED: Apply 2¼ pints EPTAM 7-E per broadcast acre. Use specially designed injector units or sweeps for application. If incorporated application is to be made, use power driven rotary tillers set to a depth of 2 to 3 inches. Apply after cotton has 2 to 4 leaves. Do not apply after first bolls open. DO NOT APPLY CLOSER THAN 4 INCHES EITHER SIDE OF THE COTTON DRILL.

NOTE: Tandem discs may be used for incorporation in the skips of skip row cotton.

Cotton is susceptible to injury from EPTAM 7-E. Follow directions for use carefully to avoid crop injury.

PINE SEEDLING NURSERIES (LOBLOLLY, SLASH, LONGLEAF, SHORTLEAF): Apply and incorporate 7 pints EPTAM 7-E per acre 14 days prior to seeding.

POTATOES, IRISH: Do not exceed 3½ pints EPTAM 7-E per acre per crop.

CAUTION: In Florida on winter and early spring potatoes apply only after potatoes have emerged and true leaves have formed.

BEFORE OR AT PLANTING

PRE-PLANT: Apply and incorporate 3½ pints EPTAM 7-E per broadcast acre just before planting. For incorporated applications to beds, apply as a band application and incorporate with ground or power driven tillers.

EXAMPLE: In 18-inch bands on 36-inch rows, use 1¾ pints per crop acre. See DIRECTIONS FOR USE: Incorporation.

OR

BEFORE PLANTING AND BEFORE BED FORMATION: Band application: Apply as a band, equivalent to 3½ pints per acre broadcast basis. Cover with 3 to 4 inches of soil with bedding discs, middle busters or other suitable bed making equipment. Care should be taken not to fold in the band treatment.

OR

AFTER PLANTING BUT BEFORE BED FORMATION: Apply 1¾ pints EPTAM 7-E per broadcast acre over planted crop and bed up immediately with bedding discs set to cover 3 to 4 inches of soil.

OR

AFTER PLANTING AND AFTER BED FORMATION: Apply EPTAM 7-E as a band at a rate equivalent to 3½ pints per acre, broadcast basis. Re-bed immediately after application with bedding discs set to cover with 3 to 4 inches of soil. Care should be taken not to fold in the band treatment.

OR

AFTER PLANTING AND AFTER BED FORMATION: Apply 1¾ pints EPTAM 7-E per broadcast acre. Re-bed immediately after application with bedding discs set to cover with 3 to 4 inches of soil.

OR

DRAG-OFF (COME UP, WEEDING TIME)

Apply EPTAM 7-E as a band treatment after drag-off, at a rate equivalent to 3½ pints per acre (broadcast basis) and cover with bedding discs set to cover with 3 to 4 inches of soil. Care should be taken not to fold in the band treatment.

OR

POST-EMERGENCE—LAY-BY

INCORPORATION: Apply and incorporate EPTAM 7-E at a rate equivalent to 3½ pints per acre (broadcast basis) after potato plants have emerged from the soil. Apply as a directed spray to the soil in bands on both sides of the row.

Immediately cover the EPTAM 7-E with 3 to 4 inches of soil by re-bedding with bedding discs. Care should be taken not to fold in the band treatment. *Example:* Apply 2¼ pints EPTAM 7-E per crop acre as a directed spray to the soil in 12-inch bands on both sides of 36-inch rows. Do not apply within 45 days of harvest.

OR

IRRIGATION: Meter 3½ pints EPTAM 7-E per acre into the irrigation water after clean cultivation. Do not apply within 45 days of harvest.

RECOMMENDATIONS SOUTHWESTERN REGION

These recommendations are given as the broadcast (overall) rate of EPTAM 7-E per acre. For band treatment, use proportionately less material per acre depending on the width of band to be treated and the crop row spacing. Do not use band application on rocky ground because thorough incorporation is not possible.

ALFALFA*, BIRDSFOOT TREFOIL, CLOVERS, LESPEDEZA: Do not use EPTAM 7-E if a grass or grain nurse crop is to be planted with the legume. Do not use on white dutch clover. Apply and incorporate 3½ pints EPTAM 7-E per acre just before planting. Temporary crop stunting and sealing of the first leaves will occur if conditions for germination and growth are not optimum (e.g. lack of moisture), and will be relieved by irrigation or adequate rainfall.

* Alfalfa is sensitive to soil residues of Atrazine. Do not use EPTAM 7-E on alfalfa if Atrazine was applied within the previous 12 months.

AND/OR ALFALFA (ESTABLISHED STANDS): Meter 2¼ to 3½ pints EPTAM 7-E per acre into the irrigation water applied to established stands prior to weed emergence. Use the lower rate on very coarse textured soils. Do not apply within 14 days of harvesting or grazing alfalfa.

LADINO CLOVER (ESTABLISHED STANDS): Meter 2¼ to 3½ pints EPTAM 7-E per acre into the irrigation water applied to established stands prior to weed emergence. Use the lower rate or very coarse textured soils. Do not apply within 45 days of harvesting or grazing.

BEANS, GREEN OR DRY: Do not use EPTAM 7-E on Adzuki beans, cowpeas (blackeye peas, blackeye beans), soybeans, lima beans, Mung beans, Garbanzo beans or other flat-podded beans except Romano. Under abnormal weather conditions, stunting may occur on Gratiot, Michilite, Sanilac, Seafarer and Seaway varieties.

BEFORE OR AT PLANTING

PRE-PLANT (FLAT-PLANTED): Apply and incorporate 3½ pints EPTAM 7-E per acre just before planting. Rotary hoe lightly during or shortly after emergence of the beans to break any crust which occurs.

OR

SUBSURFACE APPLICATION: Apply 3½ pints EPTAM 7-E per acre pre-plant, or at planting. See DIRECTIONS FOR USE.

OR LAY-BY

At the time of the last cultivation, apply and incorporate 3½ pints EPTAM 7-E per acre, apply as a directed spray to the soil at the base of the plants before bean pods start to form. Do not feed or pasture vines to livestock until 45 days after application.

CITRUS NURSERY STOCK AND YOUNG FIELD PLANTINGS (Non-bearing Orange and Grapefruit Groves): After lining out, apply 3½ to 7 pints EPTAM 7-E per acre as a directed spray to the soil. Incorporate with cultivation equipment, i.e., tree hoes and rotary hoes.

CITRUS (Oranges, Tangerines, Grapefruit): AFTER CLEAN CULTIVATION OR PRIOR TO WEED EMERGENCE in bearing citrus, apply 3½ pints EPTAM 7-E per acre by flood or furrow irrigation. Meter EPTAM 7-E into the water during the entire irrigation period. Do not apply within 15 days of harvest.

COTTON: NON-IRRIGATED AREAS ONLY—

APPLICATION AFTER STAND IS ESTABLISHED: Apply 2¼ pints EPTAM 7-E per broadcast acre. Use specially designed injector units or sweeps for application. If incorporated application is to be made, use power driven rotary tillers set to a depth of 2 to 3 inches. Apply after cotton has 2 to 4 leaves. Do not apply after first bolls open. DO NOT APPLY CLOSER THAN 4 INCHES EITHER SIDE OF THE COTTON DRILL.

NOTE: Tandem discs may be used for incorporation in the skips of skip row cotton.

Cotton is susceptible to injury from EPTAM 7-E. Follow directions for use carefully to avoid crop injury.

PINE SEEDLING NURSERIES (LOBLOLLY, SLASH, LONGLEAF, SHORTLEAF): Apply and incorporate 7 pints EPTAM 7-E per acre 14 days prior to seeding.

POTATOES, IRISH: Do not exceed 7 pints EPTAM 7-E per acre per crop.

PRE-PLANT: Apply and incorporate 3½ to 7 pints EPTAM 7-E per acre just before planting.

OR

DRAG-OFF (COME UP, WEEDING TIME)

INCORPORATION: Apply and incorporate 3½ to 7 pints EPTAM 7-E per acre. For nutgrass control, use the higher rate. The field first must be "dragged-off", followed by EPTAM 7-E application and incorporation. Use spiked-tooth harrows or cultivation equipment for incorporation.

AND/OR LAY-BY

INCORPORATION: Apply and incorporate 3½ to 7 pints EPTAM 7-E per acre after potato plants have emerged from the soil. Apply as a directed spray to the soil. Do not apply within 45 days before harvest.

OR

IRRIGATION: Meter up to 3½ pints EPTAM 7-E per acre into the irrigation water after clean cultivation. Do not apply within 45 days of harvest.

SUGAR BEETS—POST THINNING:

IRRIGATION WATER: Meter $2^1/_4$ to $3^1/_2$ pints EPTAM 7-E per acre into the first irrigation applied after the last cultivation for the season.

OR

INCORPORATION: Apply and incorporate $2^1/_4$ pints EPTAM 7-E per acre after thinning and clean cultivation and incorporate to a depth of 2 to 3 inches.

SWEET POTATOES:

PRE-PLANT: Apply and incorporate $2^1/_4$ pints EPTAM 7-E per acre on coarse textured soils or $3^1/_2$ pints per acre on medium and fine textured soils just before planting. Incorporate to maximum depth of 3 inches.

OR

PRE-PLANT—BED-OVER: Apply $1^3/_4$ pints EPTAM 7-E per acre on coarse textured soils or $2^1/_4$ pints per acre on medium and fine textured soils just before planting. Treat a band width equal to $^1/_3$ of the total distance between rows.

Soil from areas adjacent to the band that is not treated is then placed on top of the treated band with bed shaping equipment forming the bed. Band depth in finished and planted bed should be 2 to 4 inches below the bed surface. Bed over immediately after application.

OR

PRE-PLANT—BED-UP: Apply $1^3/_4$ pints EPTAM 7-E per acre on coarse textured soils or $2^1/_4$ pints per acre on medium and fine textured soils just before planting. After preshaped beds have been dragged down, EPTAM 7-E is applied broadcast. Soil is then shaped into beds with bed shaping equipment so that the undisturbed EPTAM 7-E layer in the finished bed is 2 to 4 inches below the bed surface. Bed-up immediately after application.

OR

POST-PLANT. Apply $8^1/_2$ pints EPTAM 7-E per acre immediately after planting or within 2 days after planting slips or vine cuttings. Apply to a dry soil surface and do not mix into the soil. If sweet potatoes are irrigated, EPTAM 7-E should be applied prior to irrigation. Apply as a solid overall spray.

RECOMMENDATIONS PACIFIC NORTHWEST REGION

These recommendations are given as the broadcast (overall) rate of EPTAM 7-E per acre. For band treatment, use proportionately less material per acre depending on the width of band to be treated and the crop row spacing. Do not use band application on rocky ground because thorough incorporation is not possible.

ALFALFA*, BIRDSFOOT TREFOIL, CLOVERS, LESPEDEZA: Do not use EPTAM 7-E if grass or grain nurse crop is to be planted with the legume. Do not use on white dutch clover. Apply and incorporate $2^1/_4$ to $4^1/_2$ pints EPTAM 7-E per acre just before planting. (Use the lower rate on very coarse textured soils.) Temporary crop stunting and sealing of the first leaves will occur if conditions for germination and growth are not optimum (e.g. lack of moisture), and will be relieved by irrigation or adequate rainfall.

OR ALFALFA*: Meter $2^1/_4$ to $3^1/_2$ pints EPTAM 7-E per acre into the irrigation water that is applied immediately after planting. (Use the lower rate on very coarse textured soils.)

AND/OR ALFALFA (ESTABLISHED STANDS): Meter $2^1/_4$ to $3^1/_2$ pints EPTAM 7-E per acre into the irrigation water applied to established stands prior to weed emergence. Use the lower rate on very coarse textured soils. Do not apply within 14 days of harvesting or grazing alfalfa.

* Alfalfa is sensitive to soil residues of Atrazine. Do not use EPTAM 7-E on alfalfa if Atrazine was applied within the previous 12 months.

LADINO CLOVER (ESTABLISHED STANDS): Meter $2^1/_4$ to $3^1/_2$ pints EPTAM 7-E per acre into the irrigation water applied to established stands prior to weed emergence. Use the lower rate on very coarse textured soils. Do not apply within 45 days of harvesting or grazing.

BEANS, GREEN OR DRY: Do not use EPTAM 7-E on Adzuki beans, cowpeas (blackeye peas, blackeye beans), soybeans, lima beans, Mung beans, Garbanzo beans or other flat-podded beans except Romano. Under abnormal weather conditions, stunting may occur on Gratiot, Michilite, Sanilac, Seafarer and Seaway varieties.

Apply and incorporate $3^1/_2$ to $4^1/_2$ pints EPTAM 7-E per acre just before planting. Rotary hoe lightly during or shortly after emergence of the beans to break any crust which occurs.

OR

SUBSURFACE APPLICATION: Apply $3^1/_2$ pints EPTAM 7-E per acre pre-plant, just before planting or at planting. See DIRECTIONS FOR USE.

OR

LAY-BY INCORPORATION: At time of last cultivation for the season, apply and incorporate $3^1/_2$ to $4^1/_2$ pints EPTAM 7-E per acre for grass and broadleaf control. Apply as a directed spray to the soil at the base of the plants before bean pods start to form. Do not feed or pasture vines to livestock until 45 days after application.

OR

LAY-BY SUBSURFACE: Prior to application, a clean cultivation must be made for all existing weed growth to be destroyed. Apply $3^1/_2$ pints EPTAM 7-E per broadcast acre or in a band treatment (using 2 shanks per row $5^1/_2$ inches apart, centered on the drill row with rows 38 inches apart) use $1^3/_4$ pints per acre. See DIRECTIONS FOR USE.

PEAS, GREEN PROCESSING (WESTERN WASHINGTON ONLY): Apply and incorporate $2^1/_4$ pints EPTAM 7-E per acre just before planting. Early stunting of crop may occur.

POTATOES, IRISH: Do not exceed 14 pints EPTAM 7-E per acre per crop.

PRE-PLANT: Apply and incorporate just before planting $3^1/_2$ pints EPTAM 7-E per acre; $4^1/_2$ pints per acre for quackgrass control.

LAY-BY: Apply and incorporate $3^1/_2$ pints EPTAM 7-E per acre after potato plants have emerged from the soil. Apply as a directed spray to the soil. Do not apply within 45 days of harvest.

IRRIGATION: Meter $3^1/_2$ pints EPTAM 7-E per acre into the irrigation water after clean cultivation. Do not apply within 45 days of harvest.

SAFFLOWER: Apply and incorporate $3^1/_2$ pints EPTAM 7-E per acre per crop.

SUGAR BEETS: Post Emergence (after the first true leaves have formed). Do not exceed $3^1/_2$ pints EPTAM 7-E per acre per crop except for irrigation applications where 2 applications of $3^1/_2$ pints may be made.

INCORPORATION: Apply $3^1/_2$ pints EPTAM 7-E per acre after thinning and clean clutivation and incorporate to a depth of 2 to 3 inches.

OR

IRRIGATION WATER: Meter $2^1/_4$ to $3^1/_2$ pints EPTAM 7-E per acre into the irrigation water after clean cultivation. Do not exceed 7 total pints EPTAM 7-E per acre per crop. Do not apply within 49 days of harvest.

OR

SUBSURFACE INJECTION: Apply $3^1/_2$ pints EPTAM 7-E per broadcast acre or in band treatment (using 2 shanks per row $5^1/_2$ inches apart, centered on the drill row with rows 22 inches apart) use $1^3/_4$ pints EPTAM 7-E per acre. Prior to application, a clean cultivation must be made for all existing weed growth to be destroyed.

TABLE BEETS: Apply and incorporate $2^1/_4$ pints EPTAM 7-E per acre just before planting. (Under normal use table beets are susceptible to injury from EPTAM 7-E and when EPTAM 7-E is used the seeding rate should be increased 10 percent.)

WALNUTS: After clean cultivation or prior to weed emergence on well established trees, meter $3^1/_2$ pints EPTAM 7-E per acre into the irrigation water during the entire irrigation period.

RECOMMENDATIONS WESTERN REGION

These recommendations are given as the broadcast (overall) rate of EPTAM 7-E per acre. For band treatment, use proportionately less material per acre depending on the width of band to be treated and the crop row spacing. Do not use band application on rocky ground because thorough incorporation is not possible.

ALFALFA*, BIRDSFOOT TREFOIL, CLOVERS, LESPEDEZA: Do not use EPTAM 7-E if a grass or grain nurse crop is to be planted with the legume. Do not use on white dutch clover. Apply and incorporate 2¼ to 4½ pints EPTAM 7-E per acre just before planting. (Use the lower rate on very coarse textured soils.) Temporary crop stunting and sealing of the first leaves will occur if conditions for germination and growth are not optimum (e.g. lack of moisture), and will be relieved by irrigation or adequate rainfall.

OR ALFALFA*: Meter 2¼ to 3½ pints EPTAM 7-E per acre into the irrigation water that is applied immediately after planting. (Use the lower rate on very coarse textured soils.)

Do not use EPTAM 7-E pre-emergence on rill irrigated (corrugated) alfalfa.

AND/OR ALFALFA (ESTABLISHED STANDS): Meter 2¼ to 3½ pints EPTAM 7-E per acre into the irrigation water applied to established stands prior to weed emergence. Use the lower rate on very coarse textured soils. Limit use to one application per cutting. Up to 14 pints EPTAM 7-E per acre per year may be used if applied into the irrigation water. Do not apply within 14 days of harvesting or grazing alfalfa.

* Alfalfa is sensitive to soil residues of Atrazine. Do not use EPTAM 7-E on alfalfa if Atrazine was applied within the previous 12 months.

LADINO CLOVER (ESTABLISHED STANDS): Meter 2¼ to 3½ pints EPTAM 7-E per acre into the irrigation water applied to established stands prior to weed emergence. Use the lower rate on very coarse textured soils. Do not apply within 45 days of harvesting or grazing.

ALMONDS: After making the last cultivation for the season, meter 2½ to 3½ pints EPTAM 7-E per acre into the irrigation water. Do not make more than 2 applications per season or exceed 7 pints per acre. Do not apply within 16 days of harvest.

BEANS, GREEN OR DRY: Do not use EPTAM 7-E on Adzuki beans, cowpeas (blackeye peas, blackeye beans), soybeans, lima beans, Mung beans, Garbanzo beans or other flat-podded beans except Romano. Under abnormal weather conditions stunting may occur on Gratiot, Michilite, Sanilac, Seafarer and Seaway varieties.

PRE-PLANT

Apply and incorporate 3½ pints EPTAM 7-E per acre just before planting. Rotary hoe lightly during or shortly after emergence of the beans to break any crust which occurs.

OR

SUBSURFACE APPLICATION: Apply 3½ pints EPTAM 7-E per acre per plant, just before planting or at planting. See DIRECTIONS FOR USE.

OR

LAY-BY INCORPORATION: At time of last cultivation for the season, apply and incorporate 3½ to 4½ pints EPTAM 7-E per acre for grass and broadleaf control. Apply as a directed spray to the soil at the base of the plants before bean pods start to form. Do not feed or pasture vines to livestock until 45 days after application.

OR

LAY-BY SUBSURFACE: Prior to application, a clean cultivation must be made for all existing weed growth to be destroyed. Apply 3½ pints EPTAM 7-E per broadcast acre or in a band treatment (using 2 shanks per row 5½ inches apart, centered on the drill row with rows 38 inches apart) use 1¾ pints per acre. See DIRECTIONS FOR USE.

CITRUS NURSERY STOCK AND YOUNG FIELD PLANTINGS (Non-bearing Orange, Grapefruit and Lemon Groves): After lining out, apply 3½ to 7 pints EPTAM 7-E per acre as a directed spray to the soil. Incorporate with cultivation equipment, i.e., tree hoes and rotary hoes.

CITRUS (Oranges, Tangerines, Grapefruit, Lemons): After clean cultivation or prior to weed emergence in bearing citrus, apply 3½ pints EPTAM 7-E per acre by flood or furrow irrigation. Do not exceed 10½ pints total EPTAM 7-E per acre per year when multiple applications are made. Do not apply within 14 days of harvest.

POTATO, IRISH: Do not exceed 14 pints EPTAM 7-E per acre per crop.

PRE-PLANT: Apply and incorporate just before planting 3½ pints EPTAM 7-E per acre.

DRAG-OFF (COME UP, WEEDING TIME)

Incorporation: Apply and incorporate 3½ pints EPTAM 7-E per acre. The field first must be "dragged off," followed by EPTAM 7-E application and incorporation. Use spiked-tooth harrows or cultivation equipment for incorporation.

AND/OR LAY-BY

Incorporation: Apply and incorporate 3½ to 4½ pints EPTAM 7-E per acre after potato plants have emerged from the soil. (Use lower rate on coarse textured soils). Apply as a directed spray to the soil. Do not apply within 30 days of harvest.

OR

Irrigation: Meter 3½ pints EPTAM 7-E per acre into the irrigation water after clean cultivation. Do not apply within 30 days of harvest.

SAFFLOWER: Apply and incorporate 3½ pints EPTAM 7-E per acre just before planting.

SUGAR BEETS: Post Emergence (after first true leaves have formed).

INCORPORATION: Apply 3½ pints EPTAM 7-E per acre after thinning and clean cultivation and incorporate to a depth of 2 to 3 inches.

OR

IRRIGATION WATER: Meter 2¼ to 3½ pints EPTAM 7-E per acre into the first irrigation applied after the last cultivation for the season. Two applications of 2¼ pints each should be made when beets are to be carried in the ground longer than the normal growing season.

OR

SUBSURFACE INJECTION: Apply 3½ pints EPTAM 7-E per broadcast acre, or in band treatment (using 2 shanks per row 5½ inches apart, centered on the drill row) use 1¾ pints EPTAM 7-E per acre. Prior to application, a clean cultivation must be made for all existing weed growth to be destroyed. See DIRECTIONS FOR USE.

TOMATOES: LAY-BY APPLICATION

(Northern California Counties only, i.e., Butte, Colusa, Contra Costa, Glenn, Merced (North of Highway 152), Sacramento, San Joaquin, Solano, Stanislaus, Sutter, Yolo and Yuba.)

For use on tomatoes at least 3–4 inches tall; on clay and clay loam soils only. DO NOT USE ON SANDY SOILS.

Apply EPTAM 7-E as a spray to the soil surface at a rate of 3½ pints per treated acre. Incorporate immediately. For band applications reduce rates proportionately. DO NOT APPLY WITHIN 2 INCHES OF THE CROP ROW. Do not use where grain will be planted within 90 days. Do not irrigate for at least 5 days after application. Do not apply within 21 days of harvest.

WALNUTS: After clean cultivation or prior to weed emergence on well established trees, meter 3½ pints EPTAM 7-E per acre into the irrigation water during the entire irrigation period.

DIRECTIONS FOR USE ORNAMENTAL SECTION

SOIL PREPARATION: The soil to be treated should be loose and free of clods. All weed growth should be removed or thoroughly worked into the soil before application.

APPLICATION: The recommended rate of EPTAM 7-E should be applied as uniformly as possible. Apply to well-worked soil that is dry enough to permit thorough mixing with incorporation equipment. When treating around established plants, direct spray to soil surface for maximum coverage, use one of the following appropriate means of application:

LOW PRESSURE HERBICIDE SPRAYER: For broadcast application use 10 to 50 gallons of water per acre. For band application (in front of power tiller), use less water depending upon row spacing and width of band desired. Check pressure and nozzles frequently to assure uniform application.

HOSE PROPORTIONER: Make sure proportioner is working properly. A more uniform application can be made by applying half the required amount of EPTAM 7-E over the area to be treated, then applying the remainder at right angles or crosswise.

KNAPSACK SPRAYER: Apply as suggested for the hose proportioner.

SOIL (MIXING) INCORPORATION: Immediately after application, thoroughly mix EPTAM 7-E into the soil to a depth of 2 to 3 inches. Mix to a depth of 6 inches for nutgrass, quackgrass, bermudagrass and chrysanthemumweed (mugwort) control. Thorough soil mixing is necessary for good weed control.

Use the following equipment or other equipment which has proven satisfactory under local conditions.

COMMERCIAL NURSERY:

Use nursery cultivator or rototillers for pre-plant broadcast (overall) applications, pre-plant band applications and post-plant applications.

HOME GARDEN:

PRE-PLANT APPLICATION—Rototiller.

POST-PLANT APPLICATION—For annual weeds use hand rake or hoe or water in immediately after application to a depth of 2 to 3 inches. Where incorporation by hand raking is done, light watering after raking is recommended. For perennial weeds, incorporate to a depth of 6 inches with a rototiller.

EPTAM 7-E CAN BE USED ON THESE ORNAMENTALS

HERBACEOUS PLANTS

Ageratum	Daylilies
Alyssum	Dianthus
Amaranthus	Marigold
Asters	Nasturtium
Balsam	Pansy
Begonia	Petunia
Chrysanthemum	Zinnia
Dahlia	

GROUND COVERS

Ajuga	Pachysandra
Gazania	Periwinkle *(Vinca Minor)*
Hypericum	Sedum
Ice Plant	Strawberry *(ornamental)*
Ivy	

EVERGREEN AND DECIDUOUS TREES AND SHRUBS

Azalea	Leucothoe
Berberis	Lilac
Boxwood	Linden
Camellia	Magnolia
Chamaecyparis	Maple
Citrus (Non-bearing)	Oak
Dogwood	Pieris
Euonymus	Pine
Fir	Podocarpus
Hemlock	Rhododendron
Holly (American and Japanese)	Spruce
	Viburnum
Juniper	Yew *(Texas)*

NOTE: All flowering bulbs, salvia, phlox, snapdragon and ornamental pepper are susceptible to injury from an application of EPTAM 7-E.

RECOMMENDATIONS

FOR ANNUAL WEED CONTROL—Use EPTAM 7-E at the rate of $5\frac{3}{4}$ pints in 10 to 50 gallons of water per acre (2 fl. oz. per 1,000 square feet).

FOR QUACKGRASS, NUTGRASS AND BERMUDAGRASS CONTROL IN TREES AND SHRUBS ONLY—existing stands of these perennial grasses must be turned under and chopped up thoroughly before treatment. Use EPTAM 7-E at the rate of 7 pints in 10 to 50 gallons of water per acre. (2.5 fl. oz. per 1,000 square feet).

FOR MUGWORT (CHRYSANTHEMUMWEED) CONTROL IN THE FOLLOWING PLANTS: JUNIPER, JAPANESE HOLLY, IVY, PACHYSANDRA, PETUNIAS—use 17 pints of EPTAM 7-E in 10 to 50 gallons of water per acre (6 fl. oz. per 1,000 square feet), mix thoroughly into the top 6 inches of soil. Apply 4 weeks before desired planting date.

WHEN TO USE EPTAM 7-E

HERBACEOUS PLANTS AND GROUND COVERS: Apply 2 weeks after transplanting or after growth starts in the spring.

TREES AND SHRUBS: Apply 2 weeks before transplanting balled and canned stock (only) and anytime after transplanting. Around established plants apply after growth starts in the spring.

APPENDIX I

EPTAM 7-E WITH FLUID FERTILIZERS

The following procedure is suggested for determining whether EPTAM-7E may be combined with a specific fluid fertilizer for spray tank application.

MATERIAL REQUIRED:

1. EPTAM 7-E.
2. Fluid fertilizer to be used.
3. Adjuvant for fertilizer tank mix: Compex*, Sponto 168-D*, Unite*, or equivalent. The adjuvant which provides the best emulsification depends on the specific fertilizer under consideration.

* Compex, Colloidal Products Incorp., Petaluma, Calif. Sponto 168-D, Retzloff Chemical Company, Houston, Texas. Unite, Hopkins Agricultural Chemesol Co., Madison, Wisconsin.

4. Two one-quart, wide-mouth glass jars with lid or stopper.
5. Measuring spoons (a 25 mL pipette or graduated cylinder provides more accurate measurement).
6. Measuring cup, 8 oz. (238 mL).

PROCEDURE:

1. Pour a pint (about 473 mL) of the fluid fertilizer into each of the quart jars.
2. Add adjuvant to one of the jars and mix (see next rate table).
3. Add the EPTAM 7-E to both jars (see next rate table).
4. Close both jars with lid or stopper and mix the contents by turning the jars upside down ten times.
5. Inspect the surface and body of the mixtures—
   (A) Immediately after completing the jar inversions.
   (B) After allowing the jars to stand quietly for 30 minutes.
   (C) And then again after turning the jars upside down ten times.

If a uniform mix cannot be made, the mixture should not be used. If either mixture remains uniform for 30 minutes, the combination may be used. Should either mixture separate after 30 minutes, but readily remix uniformly with ten jar inversions, the mixture can be used if adequate agitation is maintained in the tank. If the mixture with adjuvant is satisfactory, but the one without adjuvant is not, be sure to use the adjuvant in the spray tank. Add the adjuvant first at a rate of 3 pints per 100 gallons of fluid fertilizer; foaming can be minimized by using moderate agitation.

If nondispersible oil, sludge or clumps of solids form in the mixtures, the combination should not be used.

NOTE 1: For some combinations, premixing wettable powders in a little water in a pail before adding them to the spray tank will improve the compatibility of the final mixtures with EPTAM 7-E. This technique can be tested in the small-scale jar test by premixing the wettable powder in one-eighth (1/8) cup of water prior to addition to the pint of fluid fertilizer.

RATE TABLE FOR EPTAM 7-E AND ADJUVANT WITH THE FLUID FERTILIZER**

Gallons of fluid fertilizer to be applied per acre	mL or Tsp. of EPTAM 7-E to be added to 1 pint of fertilizer	
	7-E	
	mL	Tsp.
10	7	1 1/3
15	4	3/4
20	3	2/3
25	3	2/3
30	2	1/2
40	2	1/2

* Based on field rate of 1 pound active ingredient per acre in the fertilizer volumes indicated. Increase volume proportionately to correspond with intended field rate in terms of pounds active ingredient per acre (e.g., for field rate of 4 pounds actual EPTAM 7-E in 40 gallons fertilizer per acre, add 8 mL or 2 tsp. EPTAM 7-E to each jar for compatibility testing purposes).

** Two (2) milliliters or one-half (1/2) teaspoon of adjuvant to be added to 1 pint of fluid fertilizer in order to equal the rate of 3 pints of adjuvant per 100 gallons of fluid fertilizer.

APPENDIX II

EPTAM 7-E IMPREGNATION ON DRY BULK FERTILIZERS

CAUTION: EPTAM 7-E alone or in combination with other herbicides must not be impregnated on ammonium nitrate, sodium nitrate or potassium nitrate. Such mixtures may cause explosion and fire.

All individual state regulations relating to bulk dry fertilizer blending, registration, labeling and application are the responsibility of the individual and/or company selling the fertilizer and EPTAM 7-E mixture.

EPTAM 7-E may be impregnated on many dry bulk fertilizers and applied and incorporated in the soil before planting for the control of grass and broadleaf weeds.

All EPTAM 7-E supplementary literature instructions and label recommendations regarding rates per acre, soil incorporation, application, cautions, general use precautions and other directions must be followed.

Test results have shown that EPTAM 7-E on bulk dry fertilizers gives weed control equal to EPTAM 7-E applied as a spray in water or liquid fertilizer. However, uniform impregnation of EPTAM 7-E on dry fertilizer particles and uniform application in the field are necessary to assure good results.

A minimum of 200 pounds and a maximum of 700 pounds of approved ingredients impregnated with EPTAM 7-E at the recommended rate must be applied per acre.

For impregnating EPTAM 7-E on dry fertilizers, use a closed rotary-drum mixer or similar type of closed blender equipped with suitable spray equipment. The spray nozzle (or nozzles) should be positioned inside of the mixer to provide *uniform* spray coverage of the tumbling fertilizer.

The EPTAM 7-E should be sprayed uniformly onto the fertilizer using a fine spray pattern.

The physical properties of fertilizers vary, particularly in liquid absorptive capacity. When absorptivity is sufficient, simple spray impregnation of the fertilizer with EPTAM 7-E provides a satisfactory, dry mixture.

If the absorptive capacity is inadequate, use of a highly absorptive powder is required to provide a dry, free-flowing mixture.

Microcel E (Johns-Manville Products Corp.) is the recommended absorbent powder. It should be added separately and uniformly to the prepared EPTAM 7-E fertilizer mixture, in a quantity that is sufficient to provide a suitably free-flowing mixture. Generally less than 2% by weight of Microcel E is required.

The amount of EPTAM 7-E actually required in the manufacture of individual fertilizer mixtures should be determined carefully for each production operation. This is necessary to ensure that the amount of EPTAM 7-E actually contained in the mixture applied to the soil represents the correct rate of use.

Bulk fertilizer impregnated with EPTAM 7-E *should be applied immediately,* NOT STORED. All bulk containers must be tightly covered while the product is being transported and applied to reduce chances of EPTAM 7-E loss via volatilization.

EPTAM 7-E Physical Data

Specific Gravity (20/20°C): 0.954 (typical)
Pounds/Gallon (20°C): 7.94 (typical)
Flashpoint: 190°F (Tag. Closed Cup)
Viscosity: Sprayable down to minus 20°F.

Approved Dry Fertilizer Ingredients

	N	P	K
Ammonium sulfate	21	0	0
Diammonium phosphate	18	46	0
Potassium chloride	0	0	60
Potassium sulfate	0	0	52
Super-phosphate (single)	0	20	0
Triple super-phosphate	0	46	0
Urea	45	0	0
Ammonium phosphate-sulfate	16	20	0
11-48-0	11	48	0

NOTE: K-Mag has been shown to be compatible with EPTAM 7-E and is approved for use.

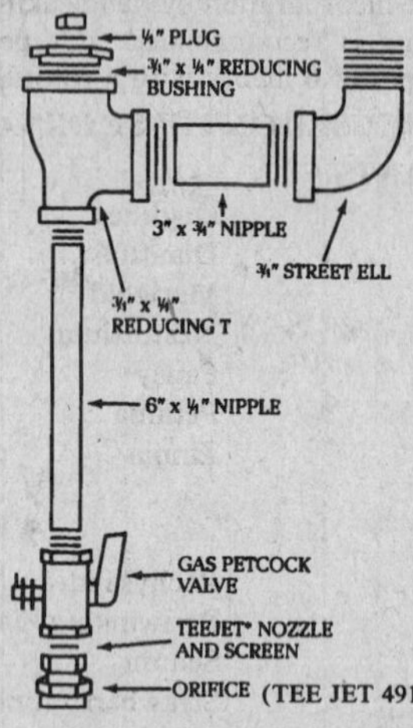

RATE CHART FOR THE IMPREGNATION OF DRY BULK FERTILIZERS WITH EPTAM 7-E

Fertilizer Rate Per Acre	EPTAM 7-E Rate per acre		
	3 1/2 pts. per acre	4 1/2 pts. per acre	7 pts. per acre
200 lbs.	17 1/2 qts./ton	22 1/4 qts./ton	35 qts./ton
250 lbs.	14 qts./ton	18 qts./ton	28 qts./ton
300 lbs.	11 2/3 qts./ton	15 qts./ton	23 1/3 qts./ton
350 lbs.	10 qts./ton	12 7/8 qts./ton	20 qts./ton
400 lbs.	8 3/4 qts./ton	11 1/4 qts./ton	17 1/5 qts./ton
450 lbs.	7 3/4 qts./ton	10 qts./ton	15 1/5 qts./ton
500 lbs.	7 qts./ton	9 qts./ton	14 qts./ton
550 lbs.	6 1/3 qts./ton	8 1/5 qts./ton	12 2/3 qts./ton
600 lbs.	5 7/8 qts./ton	7 1/2 qts./ton	11 3/4 qts./ton
650 lbs.	5 2/5 qts./ton	7 qts./ton	10 4/5 qts./ton
700 lbs.	5 qts./ton	6 2/5 qts./ton	10 qts./ton

APPENDIX III

FLOW RATES FOR EPTAM 7-E USING VARIOUS TEE JET* ORIFICES (4916)**

Tee Jet Orifice	Ounces Per Minute	cc Per Minute	Gallons Per Hour	Pounds Per Hour
.012	0.215	6.37	0.101	0.707
.014	0.286	8.45	0.134	0.938
.015	0.324	9.59	0.152	1.064
.016	0.375	11.10	0.176	1.232
.018	0.523	15.46	0.245	1.715
.020	0.610	18.04	0.286	2.002
.022	0.796	23.53	0.373	2.611
.024	0.896	26.50	0.420	2.940
.025	0.996	29.46	0.467	3.269
.026	1.111	32.87	0.521	3.647
.027	1.269	37.54	0.595	4.165
.029	1.284	37.98	0.602	4.214
.030	1.502	44.42	0.704	4.928
.032	1.641	48.52	0.769	5.383
.034	1.871	55.33	0.877	6.139
.035	2.091	61.83	0.980	6.860
.037	2.223	65.74	1.042	7.294
.039	2.539	75.08	1.190	8.330
.040	2.603	76.97	1.220	8.540
.041	2.807	83.03	1.316	9.212
.043	2.882	85.24	1.351	9.457
.045	3.334	98.61	1.563	10.941
.046	3.441	101.77	1.613	11.291
.047	3.678	108.77	1.724	12.068
.048	3.951	116.84	1.852	12.965
.051	4.102	121.32	1.923	13.461
.052	4.437	131.42	2.083	14.581
.054	4.849	143.41	2.273	15.911
.055	5.079	150.22	2.381	16.667
.057	5.333	157.73	2.500	17.500
.059	5.926	175.27	2.788	19.446
.063	6.272	185.49	2.940	20.580
.067	7.110	210.28	3.333	23.331
.070	8.205	242.65	3.846	26.922

* Registered trademark of Spraying Systems Co.

** Figures were taken at 70°F. and are approximate. Be sure occasionally to measure flow in the field to make certain you have the correct orifice and because rates vary with temperature. (Flow on an .037 orifice increases from 2.2 ounces at 70°F. to 2.4 ounces at 92°F.). Use a 300 mesh screen on orifice sizes below .014 and a 200 mesh screen on all others.

STORAGE AND DISPOSAL

PROHIBITIONS

Do not contaminate water, food or feed by storage or disposal. Open dumping is prohibited.

STORAGE

Keep container tightly closed when not in use. Do not store near seeds, fertilizers or foodstuffs. Can be stored at temperatures as low as minus 50°F.

PESTICIDE DISPOSAL

Wastes resulting from the use of this product may be disposed of on site or at an approved waste disposal facility.

CONTAINER DISPOSAL

Triple rinse (or equivalent). Then offer for recycling or reconditioning, or puncture and dispose of in a sanitary landfill, or by other procedures approved by State and local authorities.

NOTICE—READ CAREFULLY

CONDITIONS OF SALE

Stauffer (and seller) offer(s) this product for sale subject to, and buyer and all users are deemed to have accepted, the following conditions of sale and warranty which may only be varied by written agreement of a duly authorized representative of Stauffer.

WARRANTY LIMITATION

Stauffer warrants that this product conforms to the chemical description on the label and is reasonably fit for the purposes referred to in the directions for use on the label subject to the inherent risks referred to below. Stauffer makes no other express warranties; THERE IS NO IMPLIED WARRANTY OF MERCHANTABILITY and there are no warranties which extend beyond the description on the label hereof.

INHERENT RISKS

The directions for use of this product are believed to be reliable and should be followed carefully. However, it is impossible to eliminate all risks associated with use. Buyer assumes all risks associated with use or application of this product contrary to label instructions or resulting from extraordinary weather conditions.

LIMITATION OF LIABILITY

In no case shall Stauffer be liable for special, indirect, or consequential damages resulting from the use or handling of this product and no claim of any kind shall be greater in amount than the purchase price of the product in respect of which such damages are claimed.

This product is sold only for uses stated on its label. No express or implied license is granted to use or sell this product under any patent in any country except as specified: Country: United States of America.

This is a specimen label. Stauffer is not responsible for the accuracy of the information contained herein. As labels are subject to revision, always carefully read and follow the label on the product container.

SUPPLEMENTAL LABELING

EPTAM® 7-E Selective Herbicide

EPA Reg. No. 476–2154

EPTAM 7-E/TREFLAN E.C.* TANK MIX

For Weed Control in Beans (Green and Dry), Sugar Beets and Sunflowers

A tank mix combination of EPTAM 7-E plus TREFLAN E.C. will give a broader spectrum of weed control than either product used separately.

CAUTION: Do not graze or feed forage from treated fields to livestock. Read both the EPTAM 7-E and TREFLAN E.C. labels carefully before using. Observe all cautions and limitations on labeling of both products.

Do not contaminate water by cleaning of equipment or disposal of wastes.

* TREFLAN is the registered trademark of Elanco Products Incorporated

WEEDS CONTROLLED BY THE COMBINATION

Annual Grasses

Crabgrass	Brachiaria
Barnyardgrass (Watergrass)	Sandbur
Foxtails (including giant foxtail)	Junglerice
Johnsongrass (from seed)	Annual Bluegrass
Goosegrass	Sprangletop
Wild Cane (Shattercane)	Cheat
Stinkgrass	Wild Oat
Bromegrass	Texas Panicum**

Annual Broadleaves

Annual Morningglory	Knotweed
Pigweeds (Spiny, Redroot)	Stinging Nettle
Lambsquarters	Puncturevine
Carpetweed	Nettleleaf Goosefoot
Russian Thistle	Chickweed
Kochia	Corn Spurry
Purslane	Henbit
Florida Purslane	Nightshade

Perennial Weed

Nutsedge** (Nutgrass)

** May not be controlled at less than $3\frac{1}{2}$ pints per acre of EPTAM 7-E.

DIRECTIONS FOR USE
It is a violation of Federal law to use this product in a manner inconsistent with its labeling.

MIXING: Add the recommended rates of both EPTAM 7-E and TREFLAN E.C. to the spray tank during filling and mix thoroughly. Apply in 10 to 40 gallons of water per acre.

SPRAY EQUIPMENT: Use any properly calibrated low pressure boom type herbicide sprayer which will apply the material uniformly. Check calibration frequently during application and observe the nozzles to be sure they are delivering a uniform spray pattern.

SOIL INCORPORATION: The EPTAM 7-E and TREFLAN E.C. combination must be incorporated (mixed) thoroughly into the top 2 to 3 inches of soil immediately after spraying. Spraying and incorporation should be accomplished in the same operation if possible. This can be done by mounting the tank and boom right on the incorporation rig.

Thorough incorporation can be achieved with any of the following equipment.

PREPLANT: Power-driven rotary cultivators—set to cut 2 to 3 inches deep. Double disc (or double disc with spike-tooth harrow in tandem)—set to cut 3 to 6 inches deep and operate in two directions (cross disc) at 4 to 6 mph.

IN ESTABLISHED ROW CROPS: Power-driven rotary cultivators—set to cut 2 to 3 inches deep. Rolling cultivators—set to cut 2 to 4 inches and operated 2 times at 6 to 8 mph. Use on coarse and medium textured soils only.

Adjust equipment to throw treated soil toward the crop. Avoid mechanically damaging the crop.

Shallow incorporation with implements set to cut less than 2 inches may result in erratic weed control.

RECOMMENDATIONS

BEANS, GREEN AND DRY: The combination of EPTAM 7-E and TREFLAN E.C. should not be used on Adzuki beans, cowpeas (black-eyed peas, black-eyed beans) soybeans, lima beans, Mung beans, Garbanzo beans or other flat-podded beans, except Romano.

Make application before planting, using the rates listed in the following table.

APPLICATION RATES PER BROADCAST ACRE

EPTAM 7-E	TREFLAN E.C. Soil Type	Organic Matter Content	Rate
	Coarse (sand)	0–2%	1 Pint
	Coarse (sand)	2–5%	1–1½ Pints
	Medium (loam)	0–5%	1½ Pints
2½ to 3½ pints†	Fine (clay)	0–5%	1½ Pints
	All soil types	5.1–10%	1½ Pints

† Use the higher rate for nutsedge control.

PLANTING INSTRUCTIONS

DRY BEANS—Plant within 48 hours after incorporation. In the lighter soils under sprinkler irrigation, when it is necessary to irrigate beans after planting and before emergence, sufficient water should be applied to wet the soil well below the depth of planted seed.

GREEN BEANS—Plant soon after incorporation to provide the maximum period of weed control.

SUNFLOWERS (Minnesota, North Dakota, South Dakota)—Apply and incorporate just before planting, using the rates listed below.

APPLICATION RATES PER BROADCAST ACRE

EPTAM 7-E	Soil Type	Organic Matter Content	TREFLAN E.C. Rate: Minnesota Eastern Dakotas	Western Dakotas
2¼ pints	Coarse (sand)	0–2%	1 Pint	1 Pint
	Coarse (sand)	2–5%	1½–2 Pints	1½–2 Pints
	Medium (loam)	0–5%	1½ Pints	1¼–1½
	Fine (clay)	0–5%	2 Pints	1½ Pints
	All soil types	5.1–10%	2 Pints	2 Pints

SUGAR BEETS (California)—Apply as a broadcast spray over the top when plants are 2 to 6 inches tall, using the rates listed below.

APPLICATION RATES PER BROADCAST ACRE

EPTAM 7-E	TREFLAN E.C. Soil Type	Rate
3½ Pints	Coarse (sand)	1 Pint
	Medium (loam)	1¼–1½ Pints
	Fine (clay)	1½ Pints

Exposed beet roots should be covered with soil before application to reduce possibility of girdling. Set incorporation machinery to throw treated soil toward the plants in the row. Care should be taken that incorporation machinery does not damage the sugar beet taproot.

GENERAL USE PRECAUTIONS
Before using EPTAM 7-E plus TREFLAN E.C. as a tank mix, read and observe the cautionary statements, and all other information appearing in both product labels.

SUPPLEMENTAL LABELING

EPTAM 7-E Selective Herbicide
EPA Reg. No. 476-2154

EPTAM 7-E/AMIBEN* TANK MIX

FOR WEED CONTROL IN SUNFLOWERS in the States of Minnesota, North Dakota and South Dakota.

A tank mix combination of EPTAM 7-E and AMIBEN will give improved control of many of the broadleaf weeds listed on the EPTAM 7-E label. The combination will also control wild mustard and kochia.

CAUTION: Do not graze treated fields of sunflowers or use these plants as livestock feeds.

Do not contaminate water by cleaning of equipment or disposal of wastes.

* AMIBEN is the registered trademark of Union Carbide Agricultural Products Company, Inc.

DIRECTIONS FOR USE
It is a violation of Federal law to use this product in a manner inconsistent with its labeling.

MIXING: Pour the recommended amount of EPTAM 7-E into the spray tank during the filling operation. Then add the recommeded amount of AMIBEN. Apply in 10 to 50 gallons of water per acre.

SPRAY EQUIPMENT: Use any properly calibrated low pressure boom type herbicide sprayer which will apply the material uniformly. Check calibration frequently during application and observe the nozzles to be sure they are delivering a uniform spray pattern.

SOIL INCORPORATION: The EPTAM 7-E and AMIBEN combination must be incorporated (mixed) thoroughly into the top 2 to 3 inches of soil immediately after spraying. Spraying and incorporation should be accomplished in the same operation if possible. This can be done by mounting the tank and boom right on the incorporation rig.

Thorough incorporation can be achieved with any of the following equipment.

- Power Driven Cultivation Equipment (recommended on all soil types) set to cut to a depth of 2 to 3 inches.
- Tandem Discs (recommended on all soil types) set to cut to a depth of 4 to 6 inches, operated at 4 to 6 mph followed by a spike-tooth harrow or some other leveling device which extends beyond the ends of the discs. For more thorough mixing in heavier soils, disc in two different directions (cross disc).
- Field Cultivators (recommended on coarse textured soils. Use only on soils in good tilth). Use 3 to 4 rows of sweeps, spaced at 7 inch or less intervals and staggered so that no soil is left unturned, followed by a spike-tooth harrow pulled behind the

cultivator. Do not use chisels or points. Set the cultivator to cut 4 inches deep, operated at 5 mph or more. Run the equipment over the field twice, the second run at an angle to the first.

- Rotary Ground Driven or Spring-Tooth Cultivators (recommended on coarse textured soils in good tilth only). Set to penetrate to a depth of 4 to 6 inches and operated at 5 to 8 mph in two different directions.

RECOMMENDATIONS

Apply and incorporate just before planting, using the rates given below.

BROADCAST RATES PER ACRE

Soil Texture	EPTAM 7-E	AMIBEN
Coarse (sandy)	2¼ Pints	4 Pints
Medium (loams)*	2¼–3½ Pints	4–8 Pints
Fine (clays)	3½ Pints	4 Pints
All soil types with organic matter 3%	3⅓ Pints	8 Pints
Peat or Muck	NOT RECOMMENDED	

* Use the higher rates with greater amounts of clay and organic matter

GENERAL USE PRECAUTIONS

Before using EPTAM 7-E plus AMIBEN as a tank mix, read and observe the cautionary statements, and all other information appearing in both product labels.

SUPPLEMENTAL LABELING

EPTAM 7-E Selective Herbicide

EPA Reg. No. 476-2154

EPTAM 7-E AND METRIBUZIN TANK MIX FOR CONTROL OF WEEDS IN IRISH POTATOES

A tank mix combination of EPTAM 7-E and Metribuzin (Sencor **4, Sencor 50W, Sencor Sprayule 75%, Lexone ‡4, Lexone 50W or Lexone Dry Flowable 75%) can be applied to Irish potatoes in the Pacific Northwestern and Northern regions of the U.S. to give a broader spectrum of weed control than either product used separately.

** SENCOR is the TM of Metribuzin, manufactured by Mobay Chemical Corp.

‡ LEXONE is the TM of Metribuzin, manufactured by DuPont Company.

CAUTION: It is a violation of Federal law to use this product in a manner inconsistent with its labeling.

Before using EPTAM 7-E and Metribuzin (Sencor, Lexone) as a tank mix, read both the EPTAM 7-E and Metribuzin (Sencor, Lexone) labels carefully. Observe all cautions and limitations on labeling of both products.

Do not contaminate water by cleaning of equipment or disposal of wastes.

WEEDS CONTROLLED

Annual Grasses

Annual Bluegrass
Annual Ryegrass (Italian Ryegrass)
Barnyardgrass (Watergrass, Junglerice)
Bermudagrass seedlings
Crabgrass
Foxtail (Giant, Green, Yellow)
Goosegrass
Johnsongrass seedlings
Lovegrass (Stinkgrass)
Panicum, Fall
Panicum, Texas
Rescuegrass
Sandbur, Field
Signalgrass
Volunteer Grains (Barley, Oat, Wheat)
Wild Cane
Wild Oats
Witchgrass

Annual Broadleaf Weeds

Annual Morningglory
Carpetweed
Chickweed, Common
Cocklebur
Corn Spurry
Deadnettle (Henbit)
Nightshade, Hairy
Pennsylvania Smartweed
Prickly Sida
Prostrate Pigweed
Purslane, Common
Ragweed, Common
Redroot Pigweed
Fiddleneck
Florida Pusley
Jimsonweed
Lambsquarters, Common
Nettleleaf Goosefoot
Shepherdspurse
Sicklepod
Tumble Pigweed
Wild Mustard

Perennial Grasses

Bermudagrass
Mugwort (Chrysanthemum weed)

DIRECTION FOR USE

MIXING: Add the recommended rates of EPTAM 7-E and Metribuzin (Sencor, Lexone) to the spray tank during the filling operation. Add EPTAM 7-E last. Mix thoroughly. Apply in 20 to 30 gallons of water per acre.

SPRAY EQUIPMENT: Use any properly calibrated, low pressure, boom type herbicide sprayer which will apply the material uniformly. Check calibration frequently during application and observe to be sure that the nozzles are delivering a uniform spray pattern.

SOIL INCORPORATION: The EPTAM 7-E and Metribuzin (Sencor, Lexone) combination must be incorporated (mixed) thoroughly into the top 2 to 3 inches of soil immediately after spraying. Spraying and incorporation should be accomplished in the same operation if possible. See the EPTAM 7-E label for complete incorporation directions.

SPRINKLER APPLICATION (Pacific Northwest only): Apply through irrigation sprinkler system after planting as a pre-emergence application and/or as early post emergence application until potatoes are 4 to 6 inches tall. Regulate post emergence dosage rates according to soil type on rate chart.

Pre-mix Metribuzin (Sencor, Lexone) first with water in the holding tank at desired rate in 4 to 5 parts water per 1 part chemical. Add EPTAM 7-E last. Meter chemical—water mixture into sprinkling system at rate proportionate to acreage covered. For center pivot systems, apply ½ to ¾ inch of water per acre. For solid set, wheel lines or hand lines, moisten soil surface lightly first, then apply chemical during first hour of irrigation set.

All irrigation systems for this purpose must have check valves and antisyphon device to prevent back-flow into water source or supply tank and must be equipped with a power interlock to insure simultaneous shut-off of water pumps and injector.

RECOMMENDATIONS

Broadcast the combination of EPTAM 7-E and Metribuzin (Sencor, Lexone) according to the following rates:

	Pints EPTAM 7-E	Pints Sencor 4 to Lexone 4 or Pounds Sencor 50 or Lexone 50	Pounds Sencor Sprayule 75% or Lexone Dry Flowable 75%
Coarse Soils			
Sand			
Sandy loam	3½	½	⅓
Loamy sand			
Fine Soils			
Loam			
Silt loam	4½	½–1	⅓–⅔
Sandy clay			
Clay loam			

GENERAL USE PRECAUTIONS

Before using Eptam 7-E plus Metribuzin (Sencor, Lexone) as a tank mix, read and observe the cautionary statements, and all other information appearing in both product labels.

SUPPLEMENTAL LABELING

EPTAM® 7-E Selective Herbicide

EPA Reg. No. 476-2154

EPTAM 7-E/DUAL 8E* TANK MIX FOR WEED CONTROL IN DRY BEANS IN THE NORTHERN AND PACIFIC NORTHWEST REGIONS

A tank mix combination of EPTAM 7-E plus DUAL 8E will give better weed control than either product used separately.

CAUTION: Read both the EPTAM 7-E and DUAL 8E labels carefully before using this tank mix. Observe all cautions and limitations on each label.

* DUAL 8E is the registered trademark of Ciba-Geigy Corp.

DIRECTIONS FOR USE

It is a violation of Federal law to use these products in a manner inconsistent with their labeling.

Apply the EPTAM 7-E/DUAL 8E tank mix as a pre-plant soil incorporated treatment.

MIXING: Add the recommended rates of both EPTAM 7-E and DUAL 8E to the spray tank during filling and mix thoroughly. Apply in 10 to 40 gallons of water per acre.

SPRAY EQUIPMENT: Use any properly calibrated low pressure boom type herbicide sprayer which will apply the material uniformly. Check calibration frequently during application and observe the nozzles to be sure they are delivering a uniform spray pattern.

SOIL INCORPORATION: Immediately after spraying, the EPTAM 7-E and DUAL 8E combination must be incorporated thoroughly into the top 2 to 3 inches of soil. Spraying and incorporation should be accomplished in the same operation if possible by mounting the tank and boom directly on the incorporation equipment. Incorporate the herbicide into the soil using any of the following equipment:

- **Power Driven Cultivation Equipment** (recommended on all soil types) set to cut to a depth of 2 to 3 inches.
- **Tandem Discs** (recommended on all soil types) set to cut to a depth of 4 to 6 inches, operated at 4 to 6 mph followed by a spike-tooth harrow or some other leveling device which extends beyond the ends of the discs. For more thorough mixing (for perennial grasses and in heavier soils) disc in two different directions (cross disc).
- **Field Cultivators** (recommended for spring application on coarse textured soils, and for fall application on all soils. Use only on soils in good tilth). Use 3 to 4 rows of sweeps, spaced at 7 inch or less intervals and staggered so that no soil is left unturned, followed by a spike-tooth harrow pulled behind the cultivator. Do not use chisel plows to incorporate. Set the cultivator to cut 4 inches deep, operated at 5 mph or more. Run the equipment over the field twice, the second run at an angle to the first.

APPLICATION RATES: Use the appropriate rates of EPTAM 7-E and DUAL 8E as indicated in the following table:

APPLICATION RATES PER BROADCAST ACRE

		DUAL 8E	
Soil Type	**EPTAM 7–E**	**Less than 3% Organic Matter**	**3% or Greater Organic Matter**
Coarse (sand)	3½-4½ pints	1¼ pints	1½ pints
Medium (loam)	3½–4½ pints	1½ pints	2 pints
Fine (clay)	3½–4½ pints	2 pints	2–2½ pints

PLANTING: Seeding should be done as soon as possible after treatment to obtain a maximum period of weed control.

GENERAL USE PRECAUTIONS

Before using EPTAM 7-E plus DUAL 8E as a tank mix, read and observe the cautionary statements, and all other information appearing in both product labels.

SUPPLEMENTAL LABELING

EPTAM® 7-E Selective Herbicide

EPA Reg. No. 476-2154

EPTAM 7-E/SONALAN E.C.* TANK MIX FOR WEED CONTROL IN DRY BEANS IN THE PACIFIC NORTHWEST REGION

A tank mix combination of EPTAM 7-E plus SONALAN E.C. will give a broader spectrum of weed control than either product used separately.

CAUTION: Read both the EPTAM 7-E and SONALAN E.C. labels carefully before using this tank mix. Observe all cautions and limitations on each label.

In addition to the weeds listed on the label for EPTAM 7-E alone, the following annual broadleaf weeds can be controlled with an EPTAM 7-E/SONALAN E.C. tank mix:
Kochia (*Kochia scoparia*)
Russian Thistle (*Salsoa kali*)
Wild Buckwheat (*Polygonum convoluvlus*)

* SONALAN is the registered trademark of Elanco Products Inc.

DIRECTIONS FOR USE

It is a violation of Federal law to use these products in a manner inconsistent with their labeling.

Apply the EPTAM 7-E/SONALAN E.C. tank mix as a pre-plant soil incorporated treatment.

MIXING: Add the recommended rates of both EPTAM 7-E and SONALAN E.C. to the spray tank during filling and mix thoroughly. Apply in 10 to 40 gallons of water per acre.

SPRAY EQUIPMENT: Use any properly calibrated low pressure boom type herbicide sprayer which will apply the material uniformly. Check calibration frequently during application and observe the nozzles to be sure they are delivering a uniform spray pattern.

SOIL INCORPORATION: Immediately after spraying, the EPTAM 7-E and SONALAN E.C. combination must be incorporated thoroughly in the top 2 to 3 inches of soil. Spraying and incorporation should be accomplished in the same operation if possible by mounting the tank and boom directly on the incorporation equipment. Incorporate the herbicides into the soil using any of the following equipment:

- **Power Driven Cultivation Equipment** (recommended on all soil types) set to cut to a depth of 2 to 3 inches.
- **Tandem Discs** (recommended on all soil types) set to cut to a depth of 4 to 6 inches, operated at 4 to 6 mph followed by a spike-tooth harrow or some other leveling device which extends beyond the ends of the discs. For more thorough mixing (for perennial grasses and in heavier soils) disc in two different directions (cross disc).
- **Field Cultivators** (recommended for spring application on coarse textured soils, and for fall application on all soils. Use only on soils in good tilth). Use 3 to 4 rows of sweeps, spaced at 7 inch or less intervals and staggered so that no soil is left unturned, followed by a spike-tooth harrow pulled behind the cultivator. Do not use chisel plows to incorporate. Set the cultivator to cut 4 inches deep, operated at 5 mph or more. Run the equipment over the field twice, the second run at an angle to the first.

APPLICATION RATES: Use the appropriate rates of EPTAM 7-E and SONALAN E.C. as indicated in the following table:

APPLICATION RATES PER BROADCAST ACRE

Soil Type	EPTAM 7–E	SONALAN E.C.
Coarse (sand)	3½–4½ pints	1½–2 pints
Medium (loam)	3½–4½ pints	2–2½ pints
Fine (clay)	3½–4½ pints	2½–3 pints

PLANTING: Seeding should be done as soon as possible after treatment to obtain a maximum period of weed control.

GENERAL USE PRECAUTIONS

Before using EPTAM 7-E plus SONALAN E.C. as a tank mix, read and observe the cautionary statements, and all other information appearing in both product labels.

SUPPLEMENTAL LABELING

EPTAM® 7-E Selective Herbicide

EPA Reg. No. 476-2154

EPTAM 7-E/LASSO 4-E* TANK MIX FOR WEED CONTROL IN DRY BEANS IN THE NORTHERN REGION

A tank mix combination of EPTAM 7-E plus LASSO 4-E will give a broader spectrum of weed control than either product used separately.

CAUTION: Read both the EPTAM 7-E and LASSO 4-E labels carefully before using this mix. Observe all cautions and limitations on each label.

In addition to the weeds listed on the label for EPTAM 7-E alone, the following annual broadleaf weeds can be controlled with an EPTAM 7-E/LASSO 4-E tank mix.

Common Ragweed *(Ambrosia artemisiifolia)*

Pennsylvania Smartweed (*Polygonum pensylvanicum*)

* LASSO is the registered trademark of Monsanto Agricultural Products Co.

DIRECTIONS FOR USE

It is a violation of Federal law to use these products in a manner inconsistent with their labeling.

Apply the EPTAM 7-E/LASSO 4-E tank mix as a pre-plant soil incorporated treatment.

MIXING: Add the recommended rates of both EPTAM 7-E and LASSO 4-E to the spray tank during filling and mix thoroughly. Apply in 10 to 40 gallons of water per acre.

SPRAY EQUIPMENT: Use any properly calibrated low pressure boom type sprayer which will apply the material uniformly. Check calibration frequently during operation and observe nozzles to be sure they are delivering a uniform spray pattern.

SOIL INCORPORATION: Immediately after spraying, the EPTAM 7-E and LASSO 4-E combination must be incorporated thoroughly into the top 2 to 3 inches of soil. Spraying and incorporation should be accomplished in the same operation if possible by mounting the tank and boom directly on the incorporation equipment. Incorporate the herbicides into the soil using any of the following equipment:

- **Power Driven Cultivation Equipment** (recommended on all soil types) set to cut to a depth of 2 to 3 inches.
- **Tandem Discs** (recommended on all soil types) set to cut to a depth of 4 to 6 inches, operated at 4 to 6 mph followed by a spike-tooth harrow or some other leveling device which extends beyond the ends of the discs. For more thorough mixing (for perennial grasses and in heavier soils) disc in two different directions (cross disc).
- **Field Cultivators** (recommended for spring application on coarse textured soils, and for fall application on all soils. Use only on soils in good tilth). Use 3 to 4 rows of sweeps, spaced at 7 inch or less intervals and staggered so that no soil is left unturned, followed by a spike-tooth harrow pulled behind the cultivator. Do not use chisel plows to incorporate. Set the cultivator to cut 4 inches deep, operated at 5 mph or more. Run the equipment over the field twice, the second run at an angle to the first.

APPLICATION RATES: Use 2 to 3 pints of EPTAM 7-E plus 4 to 6 pints of LASSO 4-E. Use only the 4 pint rate of LASSO 4-E in Michigan. Use the higher rates of herbicides for heavy weed infestations and hard-to-control weeds.

PLANTING: Seeding should be done as soon as possible after treatment to obtain a maximum period of weed control.

GENERAL USE PRECAUTIONS

Before using EPTAM 7-E plus LASSO 4-E as a tank mix, read and observe the cautionary statements, and all other information appearing in both product labels.

SUPPLEMENTAL LABELING

EPTAM® 7-E Selective Herbicide

EPA Reg. No 476-2154

EPTAM 7-E/TREFLAN*/AMIBEN** ON DRY FIELD BEANS

EPTAM herbicide may be applied as a preplant soil incorporated tank mixture with Amiben and Treflan herbicides, or a tank mix of EPTAM and Treflan may be applied preplant soil incorporated and followed by a pre-emergence application of Amiben for weed control in dry beans in the Northern region of the U.S. When Amiben is applied pre-emergence, effectiveness depends on adequate rainfall after treatment. When applied according to directions, this three-way combination will give increased control of broadleaf weeds listed on the EPTAM label as well as control of additional broadleaf weeds and labeled grasses.

WEEDS CONTROLLED

Annual Grasses:

Annual bluegrass	Johnsongrass seedlings
Barnyardgrass (Watergrass, Junglerice)	Lovegrass (Stinkgrass)
	Sandbur, Field
Bromegrass	Signalgrass, Brachiaria
Crabgrass	Sprangletop
Cheat	Texas Panicum†
Foxtails (Giant, Green, Yellow)	Wild Cane
Goosegrass	Wild Oats

Broadleaf weeds:

Annual Morningglory	Nettleleaf Goosefoot
Bindweed, Field	Nightshade (Black, Hairy)
Carpetweed	Pigweed
Chickweed	Puncturevine
Corn Spurry	Purslane
Florida Pusley	Ragweed
Henbit	Russian Thistle
Knotweed	Shepherdspurse
Kochia	Smartweed
Lambsquarters	Stinging Nettle

Perennial Weeds:

Nutsedge† (Purple, Yellow)

† Suppression (partial control)

DIRECTIONS FOR USE

It is a violation of Federal law to use this product in a manner inconsistent with its labeling.

DRY FIELD BEANS

Make treatments according to the following table:

Chemical	Timing	Amounts of Formulation Per Acre
EPTAM 7-E	Preplant Soil Incorporated	2.5 pints
+		
Treflan* 4–E	Preplant Soil Incorporated	1 pint
+		
Amiben**	Preplant Soil Incorporated or Pre-emergence (surface)	6–8 pints

EPTAM/Amiben**/Treflan* or the EPTAM/Treflan tank mixes should be incorporated immediately after application with a field cultivator, tandem disc, mulch treaders, rolling cultivator or a power driven tiller. Thoroughly mix into the top 2–3 inches by incorporating twice in different directions. Refer to complete product labels for further information on incorporation procedures.

Apply in ten or more gallons of water per acre. Prepare the spray mixture by first filling the spray tank 1/2 to 2/3 full of water. With agitator running slowly add Amiben, followed by EPTAM and Treflan. Do not mix the concentrate before adding to water. Fill tank with remaining required water.

If sufficient rain does not occur within five (5) days after pre-emergence application of Amiben, a light cultivation or incorporation with a rotary hoe or similar tool is recommended.

** Amiben is the registered trademark of Union Carbide Agricultural Products Co.

* Treflan is the registered trademark of Elanco Products Co.

GENERAL USE PRECAUTIONS

Before using the tank mix of EPTAM/Amiben/Treflan, read and observe the cautionary statements and all other information on the EPA registered EPTAM 7-E, Treflan 4-E and Amiben labels. This labeling must be in the possession of the user at time of application.

This is a specimen label. Stauffer is not responsible for the accuracy of the information contained herein. As labels are subject to revision, always carefully read and follow the label on the product container.

EPTAM® 10-G

GRANULES—SELECTIVE HERBICIDE

ACTIVE INGREDIENT:

S-ethyl dipropylthiocarbamate	10.0%
INERT INGREDIENT:	90.0%
TOTAL	100.0%

Contains 10 Pounds Active Ingredient Per 100 Pounds

EPA Reg. No. 476-1620 A-3MR

IMPORTANT
RECALIBRATE YOUR EQUIPMENT FOR THIS PRODUCT BEFORE USING

CAUTION

Keep Out of Reach of Children

HARMFUL IF SWALLOWED

PRECAUTIONARY STATEMENTS

HAZARDS TO HUMANS AND DOMESTIC ANIMALS

Harmful If Swallowed

Avoid contact with skin and clothing. Avoid breathing dust. Wash thoroughly after use. Avoid contamination of feed and food.

ENVIRONMENTAL HAZARDS

Do not apply to water or wetlands. Do not contaminate water, food or feed by storage or disposal.

USE PRECAUTIONS

READ ALL LABEL DIRECTIONS BEFORE USING.

EPTAM 10-G should be used only for recommended purposes and at recommended rates (DO NOT OVERDOSE). Recalibrate your application equipment for this product before using.

Applied according to directions and under normal growing conditions, EPTAM 10-G will not harm the treated crop. However, during germination and early growth, extended periods of unusually cold and wet or hot and dry weather, insect, nematode, or plant disease attack, carryover pesticide residues, the use of certain soil applied systemic insecticides, highly saline or alkaline soil conditions, improperly placed fertilizers or soil insecticides may create abnormal conditions that weaken crop seedlings. Used under these abnormal conditions, EPTAM 10-G could result in crop injury.

Use equipment which has proven to incorporate thoroughly to the recommended depth.

Do not contaminate water to be used on susceptible crops or for domestic purposes.

EPTAM 10-G is recommended for use on mineral soils only.

When applied according to directions and under conditions for normal crop growth, no harmful residues should remain beyond harvest.

Do not apply prior to pre-irrigation.

Keep container closed when not in use. Do not store near seeds or fertilizers.

DIRECTIONS FOR USE

It is a violation of Federal law to use this product in a manner inconsistent with its labeling. EPTAM 10-G is a selective herbicide which is mixed (incorporated) into the soil for control of weeds listed on this label. EPTAM 10-G controls weeds as their seeds germinate by interfering with normal germination and seedling development. It does not control established weeds.

APPLICATION DIRECTIONS

Apply only to well-worked soil that is dry enough to permit thorough mixing with incorporation equipment. Uniformly apply the recommended rate using equipment designed for application of granular herbicides. Equipment should be carefully calibrated before use and checked frequently during application to be sure it is working properly and delivering a uniform distribution pattern. Avoid overlaps that will increase EPTAM 10-G dosage above recommended rates as plant injury may occur.

INCORPORATION DIRECTIONS

EPTAM 10-G must be incorporated into the soil immediately to prevent loss of the herbicide. Whenever possible, application and incorporation should be done in the same operation.

SOIL MIXING (INCORPORATION) BEFORE PLANTING: Use power-driven cultivation equipment set to cut to a depth of 2 to 3 inches *or* tandem disc set to cut to a depth of 4 to 6 inches, operated at 4 to 6 mph, followed by a spike-tooth harrow or some other leveling device which extends beyond the ends of the disc. For more thorough mixing (for perennial grasses and in heavier soils) disc in two different directions (cross disc).

INCORPORATION AT OR AFTER PLANTING: Use power-driven cultivation equipment set to cut to a depth of 2 to 3 inches or, on light soils only, ground-driven tillers (rolling cultivators, rotary hoe, row wheels, etc.) set to cut to a depth of 1½ inches and operated at 6 to 8 mph. When using ground-driven tillers, precultivate if soil is crusted. When incorporating after planting, care must be taken not to disturb the crop seed or seeding.

PLANTING DIRECTIONS

For pre-plant applications, seeding should be done as soon as possible after treatment to obtain a maximum period of weed control.

CULTURAL PRACTICES FOLLOWING APPLICATION

EPTAM 10-G is not persistent in the soil and susceptible weeds germinating later during the growing season may not be controlled. Shallow cultivation may be necessary to control these susceptible weeds that escape control by EPTAM 10-G as well as those weeds not susceptible to EPTAM 10-G. Do not cultivate deeper than the depth of the herbicide incorporation.

WEEDS CONTROLLED

EPTAM 10-G will not control established weeds.

Annual Grasses:

Annual Bluegrass	*Poa annua*
Annual Ryegrass (Italian ryegrass)	*Lolium multiflorum*
Barnyardgrass (Watergrass, Junglerice)	*Echinochloa spp.*
Crabgrass	*Digitaria spp.*
Field Sandbur	*Cenchrus pauciflorus*
Giant Foxtail	*Setaria faberii*
Goosegrass	*Eleusine indica*
Green Foxtail	*Setaria viridis*
Johnsongrass Seedlings	*Sorghum halepense*
Wildcane (Shattercane, Black ambercane, Chicken-corn, Scattercane)	*Sorghum bicolor*
Volunteer Grains (Barley, Oats, Wheat)	
Wild oats	*Avena fatua*
Yellow Foxtail	*Setaria glauca*

Annual Broadleaf Weeds:

Black Nightshade	*Solanum nigrum*
Common Chickweed	*Stellaria media*
Common Lambsquarters	*Chenopodium album*
Corn Spurry	*Spergula arvensis*
Deadnettle (Henbit)	*Lamium amplexicaule*
Florida Purslane	*Richardia scabra*
Hairy Nightshade	*Solanum villosum*
Nettleleaf Goosefoot	*Chenopodium murale*
Common Purslane	*Portulaca oleracea*
Prostrate Pigweed	*Amaranthus graecizans*
Redroot Pigweed (Common pigweed)	*Amaranthus retroflexus*
Tumble Pigweed	*Amaranthus albus*

The annual broadleaf weeds listed above will be controlled only if treatment is made when conditions are favorable for weed germination and growth.

Perennial Weeds:

Bermudagrass	*Cynodon dactylon*
Purple Nutsedge (Nutgrass)	*Cyperus rotundus*
Quackgrass (Couchgrass, Witchgrass)	*Agropyron repens*
Yellow Nutsedge (Nutgrass)	*Cyperus esculentus*

Perennial weeds must be turned under and chopped up thoroughly prior to treatment. The underground rhizomes of quackgrass and the rhizomes and stolons of bermudagrass must be cut up thoroughly so that four or less nodes remain on a strand. For the control of quackgrass and bermudagrass, the disc must be set to cut 6 inches deep and the higher rates of EPTAM 10-G must be used. Consult recommendations for crops on which the higher rates may be used.

RECOMMENDATIONS

These recommendations are given as broadcast (overall) rates of EPTAM 10-G per acre. For band treatment, use proportionately less material per acre depending on the width of band to be treated and the crop row spacing. Do not apply EPTAM 10-G on rocky ground because thorough incorporation is not possible.

IMPORTANT: The following recommendations are national or regional. National recommendations are for all regions. Regional recommendations are limited to the regions specified. Refer to the regional map below to determine your EPTAM 10-G region.

REGIONAL USE MAP

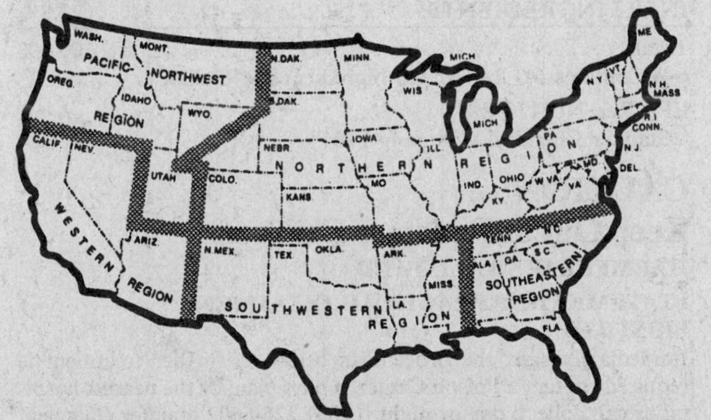

ALFALFA, BIRDSFOOT TREFOIL, CLOVER, LESPEDEZA

Temporary crop stunting and sealing of first leaves will occur if conditions for germination and crop growth are not good. Do not use EPTAM 10-G if a grass or grain nurse crop is to be planted with the legume.

NATIONAL: For control of annual grasses and broadleaf weeds and nutsedge apply and incorporate 30 pounds of EPTAM 10-G per acre just before planting.

NORTHERN, PACIFIC NORTHWEST AND WESTERN REGIONS ONLY: Apply 40 pounds EPTAM 10-G per acre for quackgrass suppression, and for control of nutsedge and annual broadleaf weeds and annual grasses in medium and fine textured soils.

Do not exceed 30 pounds EPTAM 10-G per acre in coarse textured soils.

BEANS, GREEN OR DRY

Do not use EPTAM 10-G on cowpeas (blackeye peas, blackeye beans), soybeans, Adzuki beans, lima beans, or other flat-podded beans except Romano. Under abnormal weather conditions, stunting may occur on gratiot, michilite, sanilac and seaway varieties. Do not exceed the 30 pound per acre rate on small white beans, or on green beans grown on coarse textured soils.

Apply and incorporate EPTAM 10-G just before planting, immediately after planting or at the time of last cultivation. If applied before planting, rotary hoe during or shortly after emergence of the beans to break any crust which occurs. If applied at the time of last cultivation apply to the soil before bean pods start to form, and do not feed or pasture vines to livestock within 45 days after application.

NATIONAL: Use 30 pounds EPTAM 10-G per acre for nutsedge, annual grass and broadleaf weed control.

NORTHERN AND PACIFIC NORTHWEST REGIONS ONLY: Use 40 pounds of EPTAM 10-G per acre for quackgrass suppression, and for control of seedling johnsongrass, nutsedge, annual broadleaf weeds and annual grasses in medium and fine textured soils.

FALL APPLICATION (DRY BEANS: MINNESOTA, NORTH DAKOTA ONLY): Apply and incorporate in the late Fall before the ground freezes. Use 40 pounds EPTAM 10-G per acre on coarse textured soils and 45 pounds per acre on medium and fine textured soils.

CITRUS, NURSERY STOCK AND YOUNG FIELD PLANTINGS (NON-BEARING ORANGE AND GRAPEFRUIT GROVES)

NATIONAL: After lining out, apply 30 to 60 pounds of EPTAM 10-G per acre to the soil and incorporate with cultivation equipment such as a tree hoe or rotary hoe.

CONIFER NURSERY STOCK (TWO-YEAR OLD LINING OUT STOCK OF AUSTRIAN PINE, EASTERN HEMLOCK, NORWAY PINE, RED PINE, SCOTCH PINE, WHITE PINE)

NATIONAL: Apply and incorporate 30 pounds of EPTAM 10-G per acre just before planting or over well established plants. For control of quackgrass, heavy infestations of nutsedge, and annual broadleaf weeds apply and incorporate 60 pounds of EPTAM 10-G per acre.

CORN (FIELD, SILAGE AND SWEET)

Corn is susceptible to injury from EPTAM 10-G. Do not use on milo or sorghum. Do not use on corn seed stock. Plant corn to a maximum depth of 2 inches.

PACIFIC NORTHWEST REGION ONLY: Pre-Plant Incorporation: Apply and incorporate 20 pounds of EPTAM 10-G per acre just before planting.

NORTHERN AND SOUTHEASTERN REGIONS ONLY: Pre-Emergence Incorporation: (For use only in areas where seedling johnsongrass and wildcane are problems). Plant corn. Apply 20 to 30 pounds EPTAM 10-G per acre over the planted crop. Use the higher rate if broadleaf weeds are a problem. Use rotary hoe type equipment to incorporate EPTAM 10-G into the soil to a maximum depth of 1½ inches.

FLAX

FALL APPLICATION (MINNESOTA, NORTH DAKOTA): Apply and incorporate in the late Fall before the ground freezes. Use 40 pounds EPTAM 10-G per acre on coarse textured soils and 45 pounds per acre on medium and fine textured soils.

POTATOES, IRISH

PRE-PLANT

SOUTHEAST, SOUTHWEST AND WESTERN REGIONS ONLY: Apply and incorporate 30 pounds EPTAM 10-G per acre.

NORTHERN AND PACIFIC NORTHWEST REGIONS ONLY: For annual grasses and broadleaf weeds use 30 to 60 pounds EPTAM 10-G per acre. For quackgrass suppression (or control) and control of nutsedge, annual broadleaf weeds and annual grasses in fine textured soils, use 40 to 60 pounds per acre.

OR

DRAG-OFF (COME UP, WEEDING TIME)

The field must first be "dragged-off" followed by EPTAM 10-G application and incorporation. Use spike-tooth harrows or cultivation equipment for incorporation.

NATIONAL: For annual grasses and broadleaf weeds use 30 pounds EPTAM 10-G per acre.

NORTHERN AND PACIFIC NORTHWEST REGIONS ONLY: For control of nutsedge, annual grasses and broadleaf weeds in medium and fine textured soils use 40 to 60 pounds EPTAM 10-G per acre.

LAYBY

Apply after a clean cultivation. Apply when vines are dry and use cloth drag to knock granules off plant. Do not apply within 45 days of harvest.

SOUTHEAST, SOUTHWEST AND WESTERN REGIONS ONLY: Apply and incorporate 30 pounds EPTAM 10-G per acre.

NORTHERN AND PACIFIC NORTHWEST REGIONS ONLY: Apply to the soil and incorporate 30 to 40 pounds EPTAM 10-G per acre after the potato plants have emerged.

FALL APPLICATION (MINNESOTA, NORTH DAKOTA): Apply and incorporate in the late Fall before the ground freezes. Use 40 pounds EPTAM 10-G per acre on coarse textured soils and 45 pounds per acre on medium and fine textured soils.

SAFFLOWER

NATIONAL: Apply and incorporate 30 pounds EPTAM 10-G per acre just before planting.

SUGAR BEETS

NATIONAL (POST-THINNING): Apply and incorporate 30 pounds of EPTAM 10-G per acre after thinning and clean cultivation and incorporate to a depth of 2 to 3 inches.

FALL APPLICATION (MINNESOTA, NORTH DAKOTA): Apply and incorporate in the late Fall before the ground freezes. Use 40 pounds EPTAM 10-G per acre on coarse textured soils and 45 pounds per acre on medium and fine textured soils.

TABLE BEETS

PACIFIC NORTHWEST REGION ONLY: Apply and incorporate 20 pounds EPTAM 10-G per acre just before planting (under normal use table beets are susceptible to injury and when EPTAM 10-G is used the seeding rate should be increased 10 percent).

SUNFLOWERS

FALL APPLICATION (MINNESOTA, NORTH DAKOTA): Apply and incorporate in the late Fall before the ground freezes. Use 40 pounds EPTAM 10-G per acre on coarse textured soils and 45 pounds per acre on medium and fine textured soils.

STORAGE AND DISPOSAL

PROHIBITIONS

Do not contaminate water, food, or feed by storage or disposal. Open dumping is prohibited.

STORAGE

Keep container closed when not in use. Store in a cool, dry place. Do not store near seeds, fertilizers or foodstuffs.

PESTICIDE DISPOSAL

Wastes resulting from the use of this product may be disposed of on site or at an approved waste disposal facility.

CONTAINER DISPOSAL

Completely empty bag into application equipment. Then dispose of empty bag in a sanitary landfill or by incineration, or, if allowed by State and local authorities, by burning. If burned, stay out of smoke.

NOTICE—READ CAREFULLY

CONDITIONS OF SALE

Stauffer (and seller) offer(s) this product for sale subject to, and buyer and all users are deemed to have accepted, the following conditions of sale and warranty which may only be varied by written agreement of a duly authorized representative of Stauffer.

WARRANTY LIMITATION

Stauffer warrants that this product conforms to the chemical description on the label and is reasonably fit for the purposes referred to in the directions for use on the label subject to the inherent risks referred to below. Stauffer makes no other express warranties; THERE IS NO IMPLIED WARRANTY OF MERCHANTABILITY and there are no warranties which extend beyond the description on the label hereof.

INHERENT RISKS

The directions for use of this product are believed to be reliable and should be followed carefully. However, it is impossible to eliminate all risks associated with use. Buyer assumes all risks associated with use or application of this product contrary to label instructions or resulting from extraordinary weather conditions.

LIMITATION OF LIABILITY

In no case shall Stauffer be liable for special, indirect, or consequential damages resulting from the use or handling of this product and no claim of any kind shall be greater in amount than the purchase price of the product in respect of which such damages are claimed.

This product is sold only for uses stated on its label. No express or implied license is granted to use or sell this product under any patent in any country except as specified: Country: United States of America.

This is a specimen label. Stauffer is not responsible for the accuracy of the information contained herein. As labels are subject to revision, always carefully read and follow the label on the product container.

ERADICANE® EXTRA

SELECTIVE HERBICIDE

Pre-plant Incorporated Liquid Herbicide for Tough Weed Control in Corn

ACTIVE INGREDIENT:	
S-ethyl dipropylthiocarbamate	72.2%
INERT INGREDIENTS:	27.8%
Total	100.0%

6-E Contains 6.0 lbs. Active Ingredient Per Gallon

EPA Reg. No. 476–2207 A-6R

CAUTION

Keep Out of Reach of Children

HARMFUL IF SWALLOWED

STATEMENT OF PRACTICAL TREATMENT

FIRST AID

Immediately start the procedures below. If further treatment is required contact a Poison Center, a physician, or the nearest hospital or call collect, day or night, (203) 226-6602 Stauffer Chemical Company, or (800) 424-9300 Pesticide Team Safety Network.

If swallowed: Immediately give several glasses of water but *do not* induce vomiting. If vomiting does occur, give fluids again. Have a physician determine if condition of patient will permit induction of vomiting or evacuation of stomach. Do not give anything by mouth to an unconscious or convulsing person.

For eye contact: Immediately flush eyes with large amounts of running water for at least 15 minutes. Hold eyelids apart to ensure rinsing of the entire surface of the eye and lids with water. Get medical attention immediately.

For skin contact: Flush all affected areas with plenty of water for several minutes. Remove and clean contaminated clothing and shoes. Seek medical attention if irritation occurs.

If inhaled: Remove to fresh air. Seek medical attention if respiratory irritation occurs or if breathing becomes difficult.

PRECAUTIONARY STATEMENTS

HAZARDS TO HUMANS AND DOMESTIC ANIMALS

CAUTION

HARMFUL IF SWALLOWED—Avoid contact with skin, eyes and clothing—Avoid inhalation of spray mist—Wear rubber gloves and fresh clothing—Wear goggles to avoid contact with eyes—Wash with soap and water immediately after use or contact—Do not contaminate food or feed.

Do not apply this product in such a manner as to directly or through drift, expose workers or other persons. When applied through the sprinkler irrigation method of application, the area being treated must be vacated by unprotected persons.

ENVIRONMENTAL HAZARDS

Keep out of any body of water—Do not contaminate water by cleaning of equipment or disposal of wastes.

PHYSICAL AND CHEMICAL HAZARDS

Keep away from heat and flame.

Can be stored at temperatures as low as minus 50°F.

GENERAL USE PRECAUTIONS

READ ALL LABEL DIRECTIONS BEFORE USING.

ERADICANE EXTRA should be used for recommended purposes and at recommended rates. (DO NOT OVERDOSE.)

ERADICANE EXTRA is recommended for use on mineral soils only (soils containing less than 10% organic matter).

Do not use on milo or sorghum. Do not use on corn seed stock (Breeders, Foundation or Increase).

Do not contaminate water to be used on susceptible crops and ornamentals, or for domestic purposes. Do not contaminate feed or food. Keep container tightly closed when not in use. Do not store near seeds or fertilizers.

When properly applied and weather conditions exist for normal plant growth through the season, ERADICANE EXTRA will not harm the treated crop nor should harmful soil residues remain beyond harvest. However, during germination and early growth, extended periods of unusually cold and wet or hot and dry weather, insect, nematode, or plant disease attack, carry-over soil residues of certain persistent herbicides, the use of certain soil applied systemic insecticides may create abnormal conditions that weaken crop seedlings. ERADICANE EXTRA used under these abnormal conditions could result in crop injury.

WEEDS CONTROLLED BY ERADICANE EXTRA

ERADICANE EXTRA will not control established weeds.

Annual Grasses:

Annual Bluegrass	*Poa annua*
Annual Ryegrass (Italian Ryegrass)	*Lolium multiflorum*
Barnyardgrass (Watergrass, Junglerice)	*Echinochloa* spp.
Bermudagrass Seedlings	*Cynodon dactylon*
Crabgrass	*Digitaria* spp.
Giant Foxtail	*Setaria faberi*
Goosegrass	*Eleusine indica*
Green Foxtail	*Setaria viridis*
Johnsongrass Seedlings	*Sorghum halepense*
Lovegrass (Stinkgrass)	*Eragrostis cilianensis*
Millet, Wild Proso*‡	*Panicum miliaceum*
Panicum, Fall	*Panicum dichotomiflorum*
Red Rice	*Oryza sativa*
Rescuegrass	*Bromus willdenowii*
Sandbur, Field*	*Cenchrus pauciflorus*
Signalgrass	*Brachiaria* spp.
Texas Panicum (Buffalograss, Coloradograss)*†	*Panicum texanum*
Volunteer Grains (Barley, Oats, Wheat)	
Wild Cane (Shattercane)*¶	*Sorghum bicolor*
Wild Oats	*Avena fatua*
Woolly Cupgrass	*Eriochloa villosa*
Yellow Foxtail	*Setaria glauca*

¶ Partial Control (suppression) See Wild Cane Section

* Cultivation is suggested in addition to ERADICANE EXTRA treatment in fields with moderate to heavy infestations of wild cane, sandbur, Texas panicum, or wild proso millet.

† Southeastern U.S. Only.

‡ Central Region Only—Partial Control (Suppression)

Annual Broadleaf Weeds:

Annual Morningglory	*Ipomoea purpurea*
Black Nightshade	*Solanum nigrum*
Carpetweed	*Mollugo verticillata*
Chickweed, Common	*Stellaria media*
Corn Spurry	*Spergula arvensis*
Deadnettle (Henbit)	*Lamium amplexicaule*
Fiddleneck	*Amsinckia* spp.
Florida Purslane	*Richardia scabra*
Hairy Nightshade	*Solanum sarachoides*
Lambsquarters, Common	*Chenopodium album*
Nettleleaf Goosefoot	*Chenopodium murale*
Puncturevine**	*Tribulus terrestris*
Purslane, Common	*Portulaca oleracea*
Prostrate Pigweed	*Amaranthus graecizans*
Redroot Pigweed (Common Pigweed)	*Amaranthus retroflexus*
Shepherdspurse	*Capsella bursa-pastoris*
Sickepod	*Cassia obtusifolia*
Teaweed (Spiny Sida)	*Sida spinosa*
Tumble Pigweed	*Amaranthus albus*
Velvetleaf	*Abutilon theophrasti*

** Western Region only.

The annual broadleaf weeds listed above will be controlled only if treatment is made when conditions are favorable for weed seed germination and growth. Under prolonged conditions of very cold soil, control of the above broadleaf weeds may not be adequate.

Perennial Grasses:

Bermudagrass	*Cynodon dactylon*
Purple Nutsedge (Nutgrass)	*Cyperus rotundus*
Quackgrass	*Agropyron repens*
Rhizome Johnsongrass	*Sorghum halepense*
Yellow Nutsedge (Nutgrass)	*Cyperus esculentus*

Perennial weeds must be turned under and chopped up thoroughly prior to treatment. The underground rhizomes of quackgrass and johnsongrass and the rhizomes and stolons of bermudagrass must be cut up thoroughly into small pieces so that 4 or less nodes remain on a strand. Consult RECOMMENDATIONS for ERADICANE EXTRA rates necessary to control the above listed weeds.

DIRECTIONS FOR USE

It is a violation of Federal law to use this product in a manner inconsistent with its labeling.

ERADICANE EXTRA is a selective herbicide which must be mixed (incorporated) into the soil for control of weeds listed on this label. ERADICANE EXTRA controls weeds by interfering with normal germination and seedling development. ERADICANE EXTRA does not control established weeds.

APPLICATION DIRECTIONS

Pour the recommended amount of ERADICANE EXTRA into the spray tank during the filling operation. Apply in 10 to 50 gallons of water per acre using a properly calibrated, low-pressure sprayer having good agitation. The soil should be well-worked and dry enough to permit good soil mixing (incorporation). See Appendix I for soil mixing directions.

For band applications, calculate the amount to be applied per acre as follows:

$$\frac{\text{Band Width in inches}}{\text{Row Width in Inches}} \times \text{Rate/A for Broadcast Treatment} = \text{Amount Needed for Band Treatment}$$

COMBINATION WITH FLUID FERTILIZERS

ERADICANE EXTRA may be combined with fluid (solution, slurry or suspension) fertilizers. However, physical compatibility with these fertilizers must be verified before combining in the spray tank. See Appendix II for special directions regarding these combinations. Even though found to be compatible, constant agitation is necessary to keep the ERADICANE EXTRA uniformly mixed with fluid fertilizers during application.

IMPREGNATION ON DRY BULK FERTILIZERS

ERADICANE EXTRA may be impregnated on dry fertilizer and applied as the fertilizer is spread. See Appendix III for information and directions regarding impregnation and use of these combinations.

CENTER PIVOT SPRINKLER APPLICATION

Apply and soil incorporate ERADICANE EXTRA, ERADICANE EXTRA/ATRAZINE, or ERADICANE EXTRA/ BLADEX* (see TANK MIX RECOMMENDATIONS) with irrigation water before planting (see PLANTING), or immediately after planting before either corn or weeds emerge. It is important that application and incorporation be done immediately after the last tillage operation since poor results will occur if the weeds have germinated before application of the herbicide. Meter the ERADICANE EXTRA or tank mix combinations with ATRAZINE or BLADEX into the irrigation water during the entire period. Apply in ½ to ¾ inches of water.

* Registered trademark of Shell Chemical Company

USE PRECAUTIONS

A. Apply the product only through central pivot irrigation systems containing anti-siphon and check valves which will prevent water source contamination and overflow of the slurry tank and containing interlocking controls between the metering device and the water pump to insure simultaneous shut-off.

B. Inject the product with a positive displacement pump into the main line ahead of a right angle turn to insure adequate mixing.

C. Application of more than recommended quantities of irrigation water per acre may result in decreased product performance by removing the chemical from the zone of effectiveness.

D. Do not apply when wind speed favors drift, when system connections or fittings leak, when nozzles do not provide uniform distribution or when lines containing the product must be dismantled and drained.

E. Greater accuracy in calibration and distribution will be achieved by injecting a larger volume of a more dilute slurry per hour.

EXAMPLE:

110 Acres to be covered by 5⅓ pints (4 lbs. active) ERADICANE EXTRA

110 A. × 4 lbs. = 440 lbs. active ERADICANE EXTRA

$$\frac{440}{6.0} = 73.3 \text{ gals. ERADICANE EXTRA}$$

Mixing tank size 750 gallons. Sprinkler takes 30 hours to apply ½ inch of water

$$\text{Injection rate} = \frac{750}{30} = 25 \text{ gal./hr. or } 0.4 \text{ gal./min.}$$

Calibrate injection pump to deliver the herbicide mix into the irrigation system at the rate of 25 gal. per hr. or 0.4 gal. per minute.

PLANTING

Plant seed to a maximum depth of 2 inches within 2 weeks after treatment. To avoid removing ERADICANE EXTRA from the row (with loss of weed control in the row), do not move or shape soil after treatment such as by planting in deep furrows.

CULTURAL PRACTICES FOLLOWING APPLICATION

Should weeds develop, a shallow cultivation or rotary hoeing will generally result in better weed control. When cultivating for any reason, it should be shallow, i.e. no more than ½ the depth the herbicide was incorporated. A sequential application of 2,4-D, BANVEL* 4E or a pre-emergence or post-emergence herbicides may be necessary to control weeds resistant to ERADICANE EXTRA.

* BANVEL is a registered trademark of Velsicol Chemical Corporation.

RECOMMENDATIONS

Consult the U.S. regional map below and use the recommendations in your region only.

IMPORTANT: Limit use on sweet corn to 5⅓ pints ERADICANE EXTRA per acre.

EASTERN U.S.

(Note: In the Southeastern U.S., when ERADICANE EXTRA is used for weed control in silage corn, do not seed small seeded grains after corn harvest until September.)

CENTRAL U.S.

(Note: Western slope of Colorado: Where furrow irrigation will be used when corn is flat planted, do not form furrows until corn is in the three-leaf stage.)

WESTERN U.S.

(Note: ERADICANE EXTRA is not recommended on corn grown in Arizona or in the 10 Southeastern-most California counties. In California, do not use ERADICANE EXTRA on corn planted after May 1.)

NOTE: Soils lacking sufficient moisture for seed germination must be preirrigated prior to ERADICANE EXTRA application.

NOTE: Do not use on "Golden Jubilee" sweet corn west of the Cascade Mountains in Oregon or Washington.

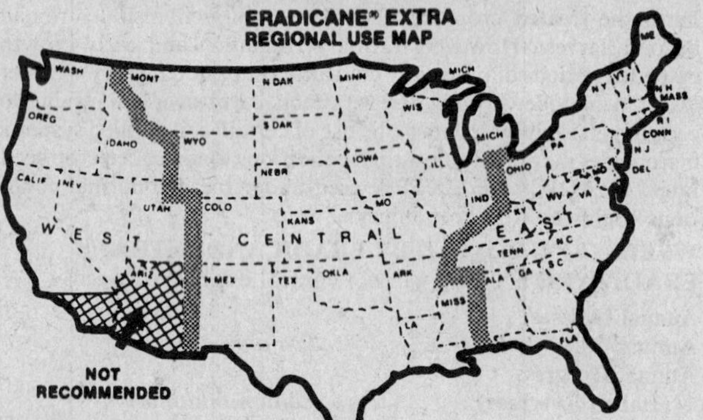

PREPLANT INCORPORATION

Weed	ERADICANE EXTRA Broadcast Rate Per Acre In Pints — Eastern U.S.	Central U.S.	Western U.S.	Restrictions
Annual Grasses & Annual Broadleaf Weeds	4 to 5⅓	4 to 8	4 to 5⅓	Use the lower rate for light infestations and in sand and light (coarse textured) soils.
Bermudagrass	5⅓ to 8	5⅓ to 8	4 to 5⅓	Use the higher rate for heavy infestations.
Nutsedge	5⅓ to 8	5⅓ to 8	4 to 8	
Quackgrass	8	5⅓ to 8	5⅓	For control of moderate infestations and suppression of heavy infestations.

Rhizome Johnsongrass	8	8	8	Application should follow a program of plowing and discing for good control. See special instructions
Sandbur, Field	4	5⅓ to 8	5⅓ to 8	
Wild Cane	5⅓ to 8	5⅓ to 8	—	Partial control (suppression) only in central U.S. See special instructions.
Wild Oat	4	4	4 to 5⅓	
Wild Proso Millet (suppression)	—	5⅓ to 8	—	Control may be improved with a combination treatment of ERADICANE EXTRA/ ATRAZINE or ERADICANE EXTRA/BLADEX
Velvetleaf	8	8	5⅓ to 8	See tank mix section.

Use the lower rates on sand and light (coarse textured) soils. Use the higher rates on heavy (fine textured) soils.

Rhizome Johnsongrass—For control, apply and incorporate 8 pints ERADICANE EXTRA per acre. Using a program of plowing and discing, followed by ERADICANE EXTRA will give good control of seedlings and rhizomes.

In some areas discing 2 to 3 times spaced at 4 to 5 week intervals during fall and winter can be done. If this cannot be done, start with fall plowing followed by several discings in the spring before planting. The fall plowing and discing cuts rhizomes into small pieces and brings them to the surface, where both weather and ERADICANE EXTRA will kill or weaken them.

Wild Cane—For partial control (suppression)—Apply 5⅓ to 8 pints ERADICANE EXTRA per acre. Use the higher rate on heavy infestations or on soils with higher organic content. For enhanced control of wild cane, especially with heavy infestations, ERADICANE EXTRA may be used in combination with ATRAZINE, BLADEX or PRINCEP* (see TANK MIX COMBINATIONS).

*PRINCEP is a registered trademark of Ciba Geigy Corporation.

Suppression of wild cane populations may be improved if all the following practices are followed:

1. Fall or spring moldboard plow.
2. Before application, the soil should be well worked with a disc and dry enough to permit good soil mixing of the herbicide (incorporation). All trash should be worked into the soil.
3. Delay application until soil warms to a constant 55°F or greater temperature. Plant immediately.
4. Incorporate thoroughly. Use recommended incorporation equipment and procedures found on this label.
5. Cultivate immediately if wild cane emerges. Cultivation should be shallow, no more than ½ the depth of incorporation of the herbicide.

CENTER PIVOT SPRINKLER APPLICATION

Apply 4 to 8 pints of ERADICANE EXTRA per acre for control of annual grasses and annual broadleaf weeds. ATRAZINE or BLADEX may be tank mixed with ERADICANE EXTRA for sprinkler application (see TANK MIX RECOMMENDATIONS).

SUBSURFACE INJECTION

(Southeastern U.S. Only)

For control of light infestations of annual grasses (except wild cane) annual broadleaf weeds and nutsedge, apply 4 pints of ERADICANE EXTRA per broadcast acre, adjusting the dosage proportionately depending on the row spacing and width to be treated. Recommended on lighter (coarse textured) soils only.

TANK MIX COMBINATIONS

For control of additional weeds and for increased control of labelled broadleaf weeds, ATRAZINE, BLADEX, PRINCEP, or a combination of ATRAZINE and BLADEX may be tank mixed with the recommended rates of ERADICANE EXTRA and applied preplant incorporated. ERADICANE EXTRA/ATRAZINE or ERADICANE EXTRA/BLADEX may be applied through a center pivot irrigation system.

As an alternative, ERADICANE EXTRA can be applied preplant incorporated at recommended rates followed by ATRAZINE applied pre-emergence to soil surface. Consult the pre-emergence application directions on the ATRAZINE label. A rotary hoeing or shallow cultivation is recommended if rainfall or sprinkler irrigation has not occurred within 10 days of the surface application. For enhanced or additional broadleaf weed control, an overlay treatment of 2,4-D, or BANVEL 4E may be made.

With the combination of ERADICANE EXTRA/BLADEX/ ATRAZINE, the potential hazard of ATRAZINE residues affecting certain crops the following year is reduced because a lower rate of ATRAZINE can be used.

WEEDS CONTROLLED BY ERADICANE EXTRA/ ATRAZINE TANK MIX

Annual Grasses:

Annual Bluegrass	*Poa annua*
Annual Ryegrass (Italian Ryegrass)	*Lolium multiflorum*
Barnyardgrass (Watergrass, Junglerice)	*Echinochloa* spp.
Bermudagrass Seedlings	*Cynodon dactylon*
Crabgrass	*Digitaria* spp.
Giant Foxtail	*Setaria faberi*
Goosegrass	*Eleusine indica*
Green Foxtail	*Setaria viridis*
Johnsongrass Seedlings	*Sorghum halepense*
Lovegrass (Stinkgrass)	*Eragrostis cilianensis*
Millet, Wild Proso*‡	*Panicum miliaceum*
Panicum, Fall	*Panicum dichotomiflorum*
Rescuegrass	*Bromus willdenowii*
Sandbur, Field*	*Cenchrus pauciflorus*
Signalgrass	*Brachiaria* spp.
Texas Panicum (Buffalograss, Coloradograss)*†	*Panicum texanum*
Volunteer Grains (Barley, Oats, Wheat)	
Wild Cane (Shattercane)*¶	*Sorghum bicolor*
Wild Oats	*Avena fatua*
Witchgrass	*Panicum capillare*
Woolly Cupgrass	*Eriochloa villosa*
Yellow Foxtail	*Setaria glauca*

¶ See Wild Cane Section.
* Cultivation is suggested in addition to ERADICANE EXTRA/ATRAZINE treatment in fields with moderate to heavy infestation of wild cane, sandbur, seedling johnsongrass, Texas panicum, or wild proso millet.
† Southeastern U.S. Only.
‡ Central Region Only—Partial Control (Suppression)

Annual Broadleaf Weeds:

Annual Morningglory	*Impomoea purpurea*
Black Nightshade	*Solanum nigrum*
Carpetweed	*Mollugo verticillata*
Chickweed, Common	*Stellaria media*
Cocklebur, Common	*Xanthium pensylvanicum*
Corn Spurry	*Spergula arvensis*
Deadnettle (Henbit)	*Lamium amplexicaule*
Fiddleneck	*Amsinckia* spp.
Florida Purslane	*Richardia scabra*
Hairy Nightshade	*Solanum sarachoides*

Jimsonweed	*Datura stramonium*
Kochia	*Kochia scoparia*
Lambsquarters, Common	*Chenopodium album*
Mustard	*Brassica* spp.
Nettleleaf Goosefoot	*Chenopodium murale*
Puncturevine	*Tribulus terrestris*
Purslane, Common	*Portulaca oleracea*
Prostrate Pigweed	*Amaranthus graecizans*
Ragweed	*Ambrosia* spp.
Redroot Pigweed (Common Pigweed)	*Amaranthus retroflexus*
Shepherdspurse	*Capsella bursa-pastoris*
Sicklepod	*Cassia obtusifolia*
Smartweed	*Polygonum* spp.
Teaweed (Spiny Sida)	*Sida spinosa*
Thistle, Russian	*Salsola kali*
Tumble Pigweed	*Amaranthus albus*
Velvetleaf	*Abutilon theophrasti*
Venice Mallow	*Hibiscus trionum*
Perennial Grasses:	
Bermudagrass	*Cynodon dactylon*
Purple Nutsedge (Nutgrass)	*Cyperus rotundus*
Quackgrass	*Agropyron repens*
Rhizome Johnsongrass	*Sorghum halepense*
Yellow Nutsedge (Nutgrass)	*Cyperus esculentus*

Perennial weeds must be turned under and chopped up thoroughly prior to treatment. The underground rhizomes of quackgrass and johnsongrass and the rhizomes and stolons of bermudagrass must be cut up thoroughly into small pieces so that 4 or less nodes remain on a strand.

WEEDS CONTROLLED BY ERADICANE EXTRA/BLADEX TANK MIX

Annual Grasses:	
Annual Bluegrass	*Poa annua*
Annual Ryegrass (Italian Ryegrass)	*Lolium multiflorum*
Barnyardgrass (Watergrass, Junglerice)	*Echinochloa* spp.
Bermudagrass Seedlings	*Cynodon dactylon*
Crabgrass	*Digitaria* spp.
Giant Foxtail	*Setaria faberi*
Goosegrass	*Eleusine indica*
Green Foxtail	*Setaria viridis*
Johnsongrass Seedlings	*Sorghum halepense*
Lovegrass (Stinkgrass)	*Eragrostis cilianensis*
Millet, Wild Proso*‡	*Panicum miliaceum*
Panicum, Fall	*Panicum dichotomiflorum*
Rescuegrass	*Bromus willdenowii*
Sandbur, Field*	*Cenchrus pauciflorus*
Signalgrass	*Brachiaria* spp.
Texas Panicum (Buffalograss, Coloradograss)*†	*Panicum texanum*
Volunteer Grains (Barley, Oats, Wheat)	
Wild Cane (Shattercane)*¶	*Sorghum bicolor*
Wild Oats	*Avena fatua*
Witchgrass	*Panicum capillare*
Woolly Cupgrass	*Eriochloa villosa*
Yellow Foxtail	*Setaria glauca*

¶ See Wild Cane Section
* Cultivation is suggested in addition to the ERADICANE EXTRA/BLADEX treatment in fields with moderate to heavy infestations of wild cane, sandbur, seedling johnsongrass, Texas panicum, or wild proso millet.
† Southeastern U.S. Only.
‡ Central region Only—Partial Control (Suppression)

Annual Broadleaf Weeds:	
Annual Morningglory	*Ipomoea purpurea*
Black Nightshade	*Solanum nigrum*
Carpetweed	*Mollugo verticillata*
Chickweed, Common	*Stellaria media*
Cocklebur, Common	*Xanthium pensylvanicum*
Corn Spurry	*Spergula arvensis*
Deadnettle (Henbit)	*Lamium amplexicaule*
Fiddleneck	*Amsinckia* spp.
Florida Purslane	*Richardia scabra*
Hairy Nightshade	*Solanum sarachoides*
Jimsonweed	*Datura stramonium*
Kochia	*Kochia scoparia*
Lambsquarters, Common	*Chenopodium album*
Mustard	*Brassica* spp.
Nettleleaf Goosefoot	*Chenopodium murale*
Puncturevine	*Tribulus terrestris*
Purslane, Common	*Portulaca oleracea*
Prostrate Pigweed	*Amaranthus graecizans*
Ragweed	*Ambrosia* spp.
Redroot Pigweed (Common Pigweed)	*Amaranthus retroflexus*
Shepherdspurse	*Capsella bursa-pastoris*
Sicklepod	*Cassia obtusifolia*
Smartweed	*Polygonum* spp.
Teaweed (Spiny Sida)	*Sida spinosa*
Tumble Pigweed	*Amaranthus albus*
Venice Mallow	*Hibiscus trionum*
Velvetleaf	*Abutilon theophrasti*
Perennial Grasses:	
Bermudagrass	*Cynodon dactylon*
Purple Nutsedge (Nutgrass)	*Cyperus rotundus*
Quackgrass	*Agropyron repens*
Rhizome Johnsongrass	*Sorghum halepense*
Yellow Nutsedge (Nutgrass)	*Cyperus esculentus*

Perennial weeds must be turned and chopped up thoroughly prior to treatment. The underground rhizomes of quackgrass and johnsongrass and the rhizomes and stolons of bermudagrass must be cut up thoroughly into small pieces so that 4 or less nodes remain on a strand.

WEEDS CONTROLLED BY ERADICANE EXTRA/PRINCEP TANK MIX

Annual Grasses:	
Annual Bluegrass	*Poa annua*
Annual Ryegrass (Italian Ryegrass)	*Lolium multiflorum*
Barnyardgrass (Watergrass, Junglerice)	*Echinochloa* spp.
Bermudagrass Seedlings	*Cynodon dactylon*
Crabgrass	*Digitaria* spp.
Giant Foxtail	*Setaria faberi*
Goosegrass	*Eleusine indica*
Green Foxtail	*Setaria viridis*
Johnsongrass Seedlings	*Sorghum halepense*
Lovegrass (Stinkgrass)	*Eragrostis cilianensis*
Millet, Wild Proso*‡	*Panicum miliaceum*
Panicum, Fall	*Panicum dichotomiflorum*
Rescuegrass	*Bromus willdenowii*
Sandbur, Field*	*Cenchrus pauciflorus*
Signalgrass	*Brachiaria* spp.
Texas Panicum (Buffalograss, Coloradograss)*†	*Panicum texanum*
Volunteer Grains (Barley, Oats, Wheat)	

Wild Cane (Shattercane)*¶	*Sorghum bicolor*
Wild Oats	*Avena fatua*
Woolly Cupgrass	*Ericochloa villosa*
Yellow Foxtail	*Setaria glauca*

¶ See Wild Cane Section.
* Cultivation is suggested in addition to ERADICANE EXTRA/PRINCEP treatment in fields with moderate to heavy infestation of wild cane, sandbur, seedling johnsongrass, Texas panicum, or wild proso millet.
† Southeastern U.S. Only.
‡ Central Region Only—Partial Control (Suppression)

Annual Broadleaf Weeds:

Annual Morningglory	*Ipomoea purpurea*
Black Nightshade	*Solanum nigrum*
Carpetweed	*Mollugo verticillata*
Chickweed, Common	*Stellaria media*
Cocklebur, Common	*Xanthium pensylvanicum*
Corn Spurry	*Spergula arvensis*
Deadnettle (Henbit)	*Lamium amplexicaule*
Fiddleneck	*Amsinckia* spp.
Florida Purslane	*Richardia scabra*
Hairy Nightshade	*Solanum sarachoides*
Jimsonweed	*Datura stramonium*
Lambsquarters, Common	*Chenopodium album*
Nettleleaf Goosefoot	*Chenopodium murale*
Puncturevine**	*Tribulus terrestris*
Purslane, Common	*Portulaca oleracea*
Prostrate Pigweed	*Amaranthus graecizans*
Ragweed	*Ambrosia* spp.
Redroot Pigweed (Common Pigweed)	*Amaranthus retroflexus*
Shepherdspurse	*Capsella bursa-pastoris*
Sicklepod	*Cassia obtusifolia*
Smartweed	*Polygonum* spp.
Teaweed (Spiny Sida)	*Sida spinosa*
Tumble Pigweed	*Amaranthus albus*
Velvetleaf	*Abutilon theophrasti*

** Western Region only.

Perennial Grasses:

Bermudagrass	*Cynodon dactylon*
Purple Nutsedge (Nutgrass)	*Cyperus rotundus*
Rhizome Johnsongrass*	*Sorghum halepense*
Yellow Nutsedge (Nutgrass)	*Cyperus esculentus*

* Partial control (suppression) only.

Perennial weeds must be turned under and chopped up thoroughly prior to treatment. The underground rhizomes of quackgrass and johnsongrass and the rhizomes and stolons of bermudagrass must be cut up thoroughly into small pieces so that 4 or less nodes remain on a strand.

APPLICATION DIRECTIONS

Apply ERADICANE EXTRA/ATRAZINE, ERADICANE EXTRA/BLADEX, ERADICANE EXTRA/PRINCEP or ERADICANE EXTRA/BLADEX/ATRAZINE in 20-50 gallons of water per acre. It is recommended that the compatibility of tank mix combinations be tested on a small scale, for example, a jar test before actual tank mixing. See Appendix II which describes the procedure for such tests.

RECOMMENDATIONS

BROADCAST RATES PER ACRE

Atrazine: Use the recommended rates of ERADICANE EXTRA plus 1¼ to 2 pounds of ATRAZINE 80W or 2 to 3 pints of ATRAZINE 4L per acre for both tank mix (preplant incorporation or center pivot sprinkler application) and sequential application. Use the higher rate of ATRAZINE in soils with over 5% organic matter.

Bladex: Use the recommended rates of ERADICANE EXTRA plus 2 to 2½ pounds of BLADEX 80W or 3 to 4 pints BLADEX 4L per acre for preplant incorporation of application through center pivot sprinkler system. Use the lower rate of BLADEX on sand or loam soils. Use the higher rate of BLADEX on fine textured (clay) soils or soils with 3% organic matter and above.

Princep: Use 1¼ to 3¾ pounds of PRINCEP 80W, 1⅒ to 3⅓ pounds of PRINCEP CALIBER®* 90 or 2 to 6 pints of PRINCEP 4L per acre with recommended rates of ERADICANE EXTRA preplant incorporated. Use the higher rate of PRINCEP on fine-textured soils and where heavy broadleaf weed infestations are expected.

* CALIBER is a registered trademark of Ciba Geigy Corporation.

Bladex/Atrazine: Use the recommended rate of ERADICANE EXTRA plus 1¼ to 2½ pounds of BLADEX 80W or 2 to 4 pints of BLADEX 4L and ¾ to 1¼ pounds of ATRAZINE 80W or 1 to 2 pints of ATRAZINE 4L per acre. BLADEX and ATRAZINE should be used in a two to one proportion.

2,4-D or Banvel 4E: Use the amount of 2,4-D or BANVEL 4E recommended on the manufacturer's label.

NOTE: BLADEX is not recommended for use on sandy soils consisting of more than 70% sand and containing less than 1% organic matter. Consult the BLADEX label for more specific recommended rates on sandy soils.

Follow all the use precautions and warnings that appear on the ATRAZINE, BLADEX, PRINCEP, BANVEL and 2,4-D labels and supplemental literature.

Make only one application per crop. After a treatment including ATRAZINE, do not plant to any crop except corn until the following year or injury may occur.

Do not plant to alfalfa where more than 1.2 pounds actual ATRAZINE was used the previous year.

PRECAUTIONS TO FOLLOW WHEN USING ERADICANE EXTRA/PRINCEP

Do not rotate to any crop except corn until the following year or injury may occur.

After harvest, plow and thoroughly till the soil in fall or spring to minimize possible injury to spring-seeded rotational crops, regardless of rate used.

Do not apply ERADICANE EXTRA/PRINCEP 80W, ERADICANE EXTRA/PRINCEP CALIBER 90 or ERADICANE EXTRA/PRINCEP 4L in the High Plains and Intermountain areas of the WEST (including central and western KS, western NE, western OK, and the Panhandle of TX) where rainfall is sparse and erratic or where irrigation is required.

Injury may occur to soybeans planted in north-central and northwest IA, south-central and southwest MN, northeast NE, southeast SD, and other areas the year following application on soils having a calcareous surface layer.

Do not plant sugar beets, tobacco, vegetables (including dry beans), spring-seeded small grains, or small-seeded legumes and grasses the year after ERADICANE EXTRA/PRINCEP 80W, ERADICANE EXTRA/PRINCEP CALIBER 90 or ERADICANE EXTRA/PRINCEP 4L application or injury may occur.

Quackgrass: For improved quackgrass control ERADICANE EXTRA can be preceded by a preplow application of ATRAZINE. Broadcast 2½ lbs. ATRAZINE 80W or 2 quarts ATRAZINE 4L per acre in the fall or spring and plow 1-3 weeks later. In the spring, follow with the recommended rate of ERADICANE EXTRA.

Velvetleaf: For control of velvetleaf use 5⅓ to 8 pints of ERADICANE EXTRA plus 2 pounds of ATRAZINE 80W or 3 pints of ATRAZINE 4L per broadcast acre.

TANK MIXING DIRECTIONS

1. Fill the spray tank at least ⅔ full of clean water. Be sure tank is clean and not contaminated.
2. Start agitation and continue moderate agitation throughout mixing.

3. Mix and disperse wettable powders first, flowable products next and ERADICANE EXTRA last.
4. Premix the BLADEX, ATRAZINE or PRINCEP products individually in proper order (step 3 above) making sure the products are thoroughly wetted and dispersed in the tank prior to adding ERADICANE EXTRA.
5. Add the ERADICANE EXTRA to the tank ⅔ full of water already mixed with BLADEX and ATRAZINE or PRINCEP.
6. Complete filling the remaining portion of the tank with water; mix thoroughly. Maintain moderate agitation to provide a uniform mix during application.
7. The tank-mix combination should not be left in the spray tank for prolonged periods; batches should be mixed and applied the same day.

Do not use excessive agitation as it may cause the BLADEX, ATRAZINE or PRINCEP to settle out and set up in the spray tank. All return lines to the tank must discharge below liquid level.

APPENDIX I

SOIL MIXING (INCORPORATION) DIRECTIONS

ERADICANE EXTRA must be incorporated into the soil to prevent loss of the herbicide. Whenever possible, application and incorporation should be done in the same operation. If there must be a delay (less than four hours) in incorporation of the herbicide when application and incorporation are done in separate operations, ERADICANE EXTRA must be applied to a dry soil surface (at least ½ inch deep) and free from dew and incidental moisture.

Pacific Northwest Region (Semi-arid areas of Eastern Washington, Eastern Oregon and Idaho only): When application and incorporation are done in separate operations. ERADICANE EXTRA must be incorporated on the same day as applied. Application must be made on a dry soil surface (at least ½ inch deep) free from dew and incidental moisture.

DELAYED INCORPORATION OF DRY BULK FERTILIZER: The application and incorporation of ERADICANE EXTRA must be carried out on the same day and delayed only if soil moisture conditions permit. See Appendix III for directions on the delay in soil incorporation.

IMPORTANT: Thorough mixing of ERADICANE EXTRA in the soil is necessary. The soil must be in good tilth and cross discing is recommended. Improper discing or poor tilth (large clods) may result in unsatisfactory control.

INCORPORATION DIRECTIONS:

The following equipment commonly is used for soil mixing (incorporation) before planting.

Power-Driven Cultivation Equipment: (Recommended on all soil types previously primary tilled): *Ground speed should be adjusted with all PTO driven cultivation tools to insure thorough soil mixing.*

a. Horizontal Axis of Rotation: Set to cut 3 to 4 inches deep.
b. Vertical Axis of Rotation: Set to cut 4 to 6 inches deep.

Tandem Discs (Recommended on all soil types): Set to cut to a depth of 4 to 6 inches, operated at 4 to 6 mph followed by a spike-tooth harrow or some other leveling device which extends beyond the ends of the discs. For more thorough mixing (for control or suppression of deep-germinating grasses—wild cane, seedling johnsongrass, quackgrass and nutsedge), disc in two different directions. When discing the second time, the disc should be reset to cut no deeper than 4 to 6 inches.

Field Cultivators (Recommended on lighter soils in good tilth): Use 3 to 4 rows of sweeps spaced at 7 inch or less intervals and staggered so that no soil is left unturned, followed by a spike-tooth harrow pulled behind the cultivator. *Do not use chisel plows.* Set the cultivator to cut 4 to 6 inches deep, operated at 5 mph or more. The field cultivator should not be used alone as an incorporation tool where control or suppression of deep-germinating weeds such as nutgrass, quackgrass, johnsongrass seedlings or wild cane is desired. For control of these weeds, use a disc for at least one of the passes over the field.

Sprinkler Incorporation Pacific Northwest Region (Semi-arid areas of Eastern Washington, Eastern Oregon and Idaho only): Surface apply ERADICANE EXTRA immediately after planting. The soil surface should be dry (at least ½ inch deep) and free from dew and incidental moisture. Incorporate using ½ to ¾ inch of water within 36 hours following application. The application and incorporation must be done within 5 days after the last tillage operation since poor results will occur if weeds have germinated.

APPENDIX II

PROCEDURE FOR TESTING THE COMPATABILITY OF ERADICANE EXTRA WITH FLUID FERTILIZERS

The following procedure is suggested for determining whether ERADICANE EXTRA may be combined with a specific fluid fertilizer for spray tank application.

MATERIALS REQUIRED

1. ERADICANE EXTRA
2. Fluid fertilizer to be used.
3. Adjuvant for fertilizer tank mix: Compex*, Sponto 168-D*, Unite*, or equivalent. The adjuvant which provides the best emulsification depends on the specific fertilizer under consideration.
4. Two one-quart, wide-mouth glass jars with lid or stopper.
5. Measuring spoons (a 25 mL pipette or graduated cylinder provides more accurate measurement).
6. Measuring cup, 8 oz. (237 mL).

* Compex, Kalo Laboratories Inc., Kansas City, Missouri; Sponto 168-D, Witco Chemical Company, Houston, Texas; Unite, Hopkins Agricultural Chemical Co., Madison, Wisconsin.

PROCEDURE

1. Pour a pint (about 473 mL) of the fluid fertilizer into each of the quart jars.
2. Add adjuvant to one of the jars and mix (see rate table on reverse side).
3. (Add the ERADICANE EXTRA to both jars (see rate table on reverse side).
4. Close both jars with lid or stopper and mix the contents by turning the jars upside down ten times.
5. Inspect the surface and body of the mixtures—
   (A) Immediately after completing the jar inversions.
   (B) After allowing the jars to stand quietly for 30 minutes.
   (C) And then again after turning the jars upside down 10 times.

If a uniform mix cannot be made, the mixture should not be used. If either mixture remains uniform for 30 minutes, the combination may be used. Should either mixture separate after 30 minutes, but readily remix uniformly with ten jar inversions, the mixture can be used if adequate agitation is maintained in the tank. If the mixture with adjuvant is satisfactory, but the one without adjuvant is not, be sure to use the adjuvant in the spray tank. Add the adjuvant first at a rate of 3 pints per 100 gallons of fluid fertilizer; foaming can be minimized by using moderate agitation.

If nondispersible oil, sludge or clumps of solids form in the mixtures, the combination should not be used.

NOTE 1: For some combinations, premixing wettable powders in a little water in a pail before adding them to the spray tank will improve the compatibility of the final mixtures with ERADICANE EXTRA. This technique can be tested in the small-scale jar test by premixing the wettable powder in one-eighth (⅛) cup of water prior to addition to the pint of fluid fertilizer.

Rate Table For ERADICANE EXTRA and Adjuvant With The Fluid Fertilizer**

Gallons of fluid fertilizer to be applied per acre	mL or tsp. of ERADICANE EXTRA* to be added to 1 pint of fertilizer	
	mL	tsp.
10	8	1⅗
15	5	1
20	4	⅘

25	3	2/3
30	2.5	1/2
40	2	2/5

* Based on field rate of 1 pound active ingredient per acre in the fertilizer volumes indicated. Increase volume proportionately to correspond with intended field rate in terms of pounds active ingredient per acre (e.g. for field rate of 4 pounds actual active ingredient in 40 gallons fertilizer per acre, add 8 mL, or 1 3/5 tsp. ERADICANE EXTRA to each jar for compatibility testing purposes).

** Two (2) milliliters or two-fifths (2/5) teaspoon of adjuvant to be added to 1 pint of fluid fertilizer in order to equal the rate of 3 pints of adjuvant per 100 gallons of fluid fertilizer.

APPENDIX III

IMPREGNATION ON DRY BULK FERTILIZERS

ERADICANE EXTRA, ERADICANE EXTRA/ATRAZINE (4L OR 80W) OR ERADICANE EXTRA/BLADEX (4L OR 80W)

CAUTION: ERADICANE EXTRA alone or in combination with other herbicides must not be impregnated on ammonium nitrate, potassium nitrate or sodium nitrate fertilizers. Such mixtures may cause explosion and fire.

All individual state regulations relating to bulk dry fertilizer blending, registration, labeling and application are the responsibility of the individual and/or company selling the ERADICANE EXTRA or ERADICANE EXTRA/ATRAZINE (4L or 80W) or ERADICANE EXTRA/BLADEX (4L or 80W) and fertilizer mixtures.

ERADICANE EXTRA alone and in combination with ATRAZINE or BLADEX may be impregnated on many dry bulk fertilizers (1) (2) and applied and incorporated in the soil before planting for the control of grass and broadleaf weeds in corn.

Test results have shown that ERADICANE EXTRA, ERADICANE EXTRA/ATRAZINE (4L or 80W) and ERADICANE EXTRA/BLADEX (4L or 80W) on many dry bulk fertilizers give weed control equal to ERADICANE EXTRA or a tank mix of ERADICANE EXTRA/ATRAZINE (4L or 80W) or ERADICANE EXTRA/BLADEX (4L or 80W) applied as a spray in water or liquid fertilizer. However, uniform impregnation of the pesticides on the dry fertilizer particles and uniform application in the field are necessary to assure good results.

A minimum of 200 pounds and a maximum of 700 pounds of approved ingredients (1) (2) impregnated with the appropriate amount of ERADICANE EXTRA alone or in combination with ATRAZINE (4L or 80W) or BLADEX (4L or 80W) must be applied per acre.

For impregnating the pesticides on dry fertilizers, use a closed rotary-drum mixer or a similar type of closed blender equipped with suitable spray equipment. The spray nozzle should be positioned inside of the mixer to provide uniform spray coverage of the tumbling fertilizer. The ERADICANE EXTRA should be sprayed uniformly onto the fertilizer using a fine spray pattern. For the ERADICANE EXTRA/ATRAZINE or ERADICANE EXTRA/ BLADEX combination product, the ATRAZINE (4L or 80W) or BLADEX (4L or 80W) can be added and mixed uniformly with the fertilizer prior to the ERADICANE EXTRA addition, or alternately, the ERADICANE EXTRA/ATRAZINE (4L or 80W) or ERADICANE EXTRA/BLADEX (4L or 80W) can be mixed in the proposed use ratio to prepare a uniform slurry for joint spray impregnation.

The physical properties of fertilizers vary, particularly in liquid absorptive capacity. When absorptivity is sufficient, simple spray impregnation of the fertilizer with ERADICANE EXTRA, ERADICANE EXTRA/ATRAZINE (4L or 80W), or ERADICANE EXTRA/BLADEX (4L or 80W) provides a satisfactory, dry mixture.

If the absorptive capacity is inadequate, use of a highly absorptive powder is required to provide a dry, flowable mixture.

Microcel E (Johns-Manville Products Corp.) is the recommended absorbent powder. It should be added separately and uniformly to the prepared ERADICANE EXTRA, ERADICANE EXTRA/ATRAZINE (4L or 80W), or ERADICANE EXTRA/BLADEX (4L or 80W) fertilizer mixture, in a quantity that is sufficient to provide a suitably flowable mixture. Generally less than 2% by weight of Microcel E is required.

The amount of ERADICANE EXTRA, ERADICANE EXTRA/ATRAZINE (4L or 80W) or ERADICANE EXTRA/ BLADEX (4L or 80W) actually required in the manufacture of individual fertilizer mixtures should be determined carefully for each production operation. This is necessary to ensure that the amount of pesticides actually contained in the mixture applied to the soil represents the correct rate of use.

Bulk fertilizer impregnated with ERADICANE EXTRA, ERADICANE EXTRA/ATRAZINE (4L or 80W) or ERADICANE EXTRA/BLADEX (4L or 80W) should be *applied immediately,* not stored. It is recommended that all bulk containers be tightly covered while the product is being transported and applied to reduce chances of ERADICANE EXTRA loss via volatilization.

When application and incorporation are done in separate operations. ERADICANE EXTRA impregnated on the dry fertilizer must be incorporated on the same day as applied. Application must be made on a dry soil surface (at least 1/2 inch deep) free from dew and incidental moisture.

ERADICANE EXTRA PHYSICAL DATA

Specific Gravity 68/68°F: 0.997 (typical)
Pounds/Gallon 68°F: 8.30 (typical)
Flashpoint: 194°F (Setaflash closed cup)
Viscosity: Sprayable down to minus 20°F

(1) APPROVED DRY FERTILIZER INGREDIENTS FOR USE WITH ERADICANE EXTRA

	N	P	K
Ammonium Sulfate	21	0	0
Diammonium Phosphate	18	46	0
Potassium Chloride	0	0	60
Potassium Sulfate	0	0	52
Single Super-phosphate	0	20	0
Treble Super-phosphate	0	46	0
Urea*	45	0	0
Ammonium Phosphate-Sulfate	16	20	0

* Some ureas may be phytotoxic when high rates are applied to corn. Use only urea rates known to be safe for corn application.

NOTE: K-Mag and 11–48–0 are also approved for use.

(2) APPROVED DRY FERTILIZER INGREDIENTS FOR USE WITH ERADICANE EXTRA/ATRAZINE (4L or 80W) OR ERADICANE EXTRA/BLADEX (4L OR 80W)

	N	P	K
Ammonium Sulfate	21	0	0
Diammonium Phosphate	18	46	0
Potassium Chloride	0	0	60
Potassium Sulfate	0	0	52
Urea*	45	0	0
Ammonium Phosphate-Sulfate	16	20	0

* Some ureas may be phytotoxic when high rates are applied to corn. Use only urea rates known to be safe for corn application.

NOTE: K-Mag and 11–48–0 are also approved for use.

NOTE: Do not combine ERADICANE EXTRA plus ATRAZINE (4L or 80W) or BLADEX (4L or 80W) with single superphosphate (0–20–0) or treble super-phosphate (0–46–0).

(3) RATE CHART FOR IMPREGNATION OF DRY BULK FERTILIZERS (1) (2) WITH ERADICANE EXTRA OR ERADICANE EXTRA/ATRAZINE 4L or 80W

FERTILIZER RATE PER ACRE	ERADICANE EXTRA RATE PER ACRE			ATRAZINE 4L		ATRAZINE 80W	
	4pts.	5 1/3 pts.	8pts.	2 pts.	3pts.	1 1/4 lbs.	2lbs.
200 lbs.	20 qts/ton	26 2/3 qts/ton	40 qts/ton	10 qts/ton	15 qts/ton	12 1/2 lbs/ton	20 lbs/ton
250 lbs.	16 qts/ton	21 1/3 qts/ton	32 qts/ton	8 qts/ton	12 qts/ton	10 lbs/ton	16 lbs/ton
300 lbs.	13 1/3 qts/ton	17 3/4 qts/ton	26 2/3 qts/ton	6 2/3 qts/ton	10 qts/ton	8 1/3 lbs/ton	13 1/3 lbs/ton
350 lbs.	11 1/2 qts/ton	15 1/4 qts/ton	23 qts/ton	5 3/4 qts/ton	8 3/5 qts/ton	7 1/5 lbs/ton	11 2/5 lbs/ton
400 lbs.	10 qts/ton	13 1/3 qts/ton	20 qts/ton	5 qts/ton	7 1/2 qts/ton	6 1/4 lbs/ton	10 lbs/ton
450 lbs.	9 qts/ton	11 4/5 qts/ton	17 4/5 qts/ton	4 1/2 qts/ton	6 2/3 qts/ton	5 2/3 lbs/ton	8 7/8 lbs/ton
500 lbs.	8 qts/ton	10 2/3 qts/ton	16 qts/ton	4 qts/ton	6 qts/ton	5 lbs/ton	8 lbs/ton
550 lbs.	7 1/4 qts/ton	9 3/4 qts/ton	14 1/2 qts/ton	3 2/3 qts/ton	5 1/2 qts/ton	4 2/3 lbs/ton	7 1/4 lbs/ton
600 lbs.	6 2/3 qts/ton	9 qts/ton	13 1/3 qts/ton	3 1/3 qts/ton	5 qts/ton	4 1/5 lbs/ton	6 2/3 lbs/ton
650 lbs.	6 1/4 qts/ton	8 1/4 qts/ton	12 1/3 qts/ton	3 1/5 qts/ton	4 3/5 qts/ton	3 7/8 lbs/ton	6 1/5 lbs/ton
700 lbs.	5 3/4 qts/ton	7 2/3 qts/ton	11 1/2 qts/ton	2 7/8 qts/ton	4 1/4 qts/ton	3 2/3 lbs/ton	5 3/4 lbs/ton

(4) RATE CHART FOR IMPREGNATION OF DRY BULK FERTILIZERS (1) (2) WITH ERADICANE EXTRA OR ERADICANE EXTRA/BLADEX 4L or 80W

FERTILIZER RATE PER ACRE	ERADICANE EXTRA RATE PER ACRE			BLADEX 4L RATE PER ACRE		BLADEX 80W RATE PER ACRE	
	4pts.	5 1/3 pts.	8pts.	3 pts.	4pts.	2 lbs.	2 1/2lbs.
200 lbs.	20 qts/ton	26 2/3 qts/ton	40 qts/ton	15 qts/ton	20 qts/ton	20 lbs/ton	25 lbs/ton
250 lbs.	16 qts/ton	21 1/3 qts/ton	32 qts/ton	12 qts/ton	16 qts/ton	16 lbs/ton	20 lbs/ton
300 lbs.	13 1/3 qts/ton	17 3/4 qts/ton	26 2/3 qts/ton	10 qts/ton	13 1/3 qts/ton	13 1/3 lbs/ton	16 2/3 lbs/ton
350 lbs.	11 1/2 qts/ton	15 1/4 qts/ton	23 qts/ton	8 3/5 qts/ton	11 1/2 qts/ton	11 2/5 lbs/ton	14 1/3 lbs/ton
400 lbs.	10 qts/ton	13 1/3 qts/ton	20 qts/ton	7 1/2 qts/ton	10 qts/ton	10 lbs/ton	12 1/2 lbs/ton
450 lbs.	9 qts/ton	11 4/5 qts/ton	17 4/5 qts/ton	6 2/3 qts/ton	9 qts/ton	8 7/8 lbs/ton	11 1/5 lbs/ton
500 lbs.	8 qts/ton	10 2/3 qts/ton	16 qts/ton	6 qts/ton	8 qts/ton	8 lbs/ton	10 lbs/ton
550 lbs.	7 1/4 qts/ton	9 3/4 qts/ton	14 1/2 qts/ton	5 1/2 qts/ton	7 1/4 qts/ton	7 1/4 lbs/ton	9 1/5 lbs/ton
600 lbs.	6 2/3 qts/ton	9 qts/ton	13 1/3 qts/ton	5 qts/ton	6 2/3 qts/ton	6 2/3 lbs/ton	8 1/3 lbs/ton
650 lbs.	6 1/4 qts/ton	8 1/4 qts/ton	12 1/3 qts/ton	4 3/5 qts/ton	6 1/4 qts/ton	6 1/5 lbs/ton	7 2/3 lbs/ton
700 lbs.	5 3/4 qts/ton	7 2/3 qts/ton	11 1/2 qts/ton	4 1/4 qts/ton	5 3/4 qts/ton	5 3/4 lbs/ton	7 1/5 lbs/ton

STORAGE AND DISPOSAL

PROHIBITIONS
Do not contaminate water, food, or feed by storage or disposal. Open dumping is prohibited. Keep away from heat and flame.

STORAGE
Keep container tightly closed when not in use. Do not store near seeds, fertilizers, or foodstuffs. Can be stored at temperatures as low as minus 50°F.

PESTICIDE DISPOSAL
Waste resulting from the use of this product may be disposed of on site or at an approved waste disposal facility.

CONTAINER DISPOSAL
METAL—Triple rinse (or equivalent), then offer for recycling or reconditioning, or puncture and dispose of in a sanitary landfill, or by other procedures approved by State or local authorities.
PLASTIC—Triple rinse (or equivalent), then offer for recycling or reconditioning, or puncture and dispose of in a sanitary landfill, or by incineration, or, if allowed by State and local authorities, by burning. If burned, stay out of smoke.

NOTICE—READ CAREFULLY

CONDITIONS OF SALE
Stauffer (and seller) offer(s) this product for sale subject to, and buyer and all users are deemed to have accepted, the following conditions of sale and warranty which may only be varied by written agreement of a duly authorized representative of Stauffer.

WARRANTY LIMITATION
Stauffer warrants that this product conforms to the chemical description on the label and is reasonably fit for the purposes referred to in the directions for use on the label subject to the inherent risks referred to below. Stauffer makes no other express warranties; THERE IS NO IMPLIED WARRANTY OF MERCHANTABILITY and there are no warranties which extend beyond the description on the label hereof.

INHERENT RISKS
The directions for use of this product are believed to be reliable and should be followed carefully. However, it is impossible to eliminate all risks associated with use. Buyer assumes all risks associated with use or application of this product contrary to label instructions or resulting from extraordinary weather conditions.

LIMITATION OF LIABILITY
In no case shall Stauffer be liable for special, indirect, or consequential damages resulting from the use or handling of this product and no claim of any kind shall be greater in amount than the purchase price of the product in respect of which such damages are claimed.

This is a specimen label. Stauffer is not responsible for the accuracy of the information contained herein. As labels are subject to revision, always carefully read and follow the label on the product container.

This product is sold only for uses stated on its label. No express or implied license is granted to use or sell this product under any patent in any country except as specified: Country: United States of America. Patent Nos. 4,021,224; 4,137,070 and other patents pending.

SUPPLEMENTAL LABELING

ERADICANE® EXTRA SELECTIVE HERBICIDE
EPA Reg. No. 476–2207

DIRECTIONS FOR INJECTION WITH ANHYDROUS AMMONIA

INJECTION DIRECTIONS FOR ERADICANE EXTRA WITH ANHYDROUS AMMONIA
It is a violation of Federal law to use this product in a manner inconsistent with its labeling.

For pre-plant soil injection applications of anhydrous ammonia, on conservation or conventional tilled acreage, ERADICANE EXTRA can be introduced into the anhydrous line using a metering pump which offers a constant rate of ERADICANE EXTRA output per acre independent of the anhydrous ammonia output rate. The ERADICANE EXTRA discharge hose should tee into the ammonia discharge line between the meter and the manifold. For best results, ERADICANE EXTRA must be used at rates of 2 to 4 quarts per acre, injection shanks must be spaced at 8 to 10 inch intervals, and injection depth must be 4 to 5 inches. Spacings greater than 10 inches up to 15 inches will require a cross discing to achieve acceptable weed control. Do not inject at spacings greater than 15 inches as weed control may be unacceptable. A minimum rate of 125 pounds per acre of anhydrous ammonia is recommended.

Weed control will be inadequate if the ERADICANE EXTRA is injected more than 4 to 5 inches below the soil surface. Soil injections at less than a 4 inch depth may result in some loss of the anhydrous ammonia through the soil surface. Injection should only be performed in soils dry enough to permit good soil mixing.

NOTE: Atrazine and Bladex* have little lateral movement in the soil and injections with anhydrous ammonia will give only a narrow band of broadleaf weed control. These products may be surface applied and incorporated at the same time the ERADICANE EXTRA and anhydrous ammonia are injected or post-emergent applications of recommended broadleaf herbicides may be used.

*BLADEX is a registered trademark of Shell Chemical Company

All applicable recommendations, directions, restrictions, and precautions on the EPA-registered label are to be followed.

This labeling must be in the possession of the user at the time of pesticide application.

This is a specimen label. Stauffer is not responsible for the accuracy of the information contained herein. As labels are subject to revision, always carefully read and follow the label on the product container.

ERADICANE® 6.7–E

SELECTIVE HERBICIDE

For Control of Weeds in Corn

(Field, Sweet, Silage and Popcorn)

Including Wild Cane, Seedling Johnsongrass, Quackgrass, Wild Proso Millet, Nutsedge, Bermudagrass, and Suppression of Rhizome Johnsongrass

ACTIVE INGREDIENT:	
S-ethyl dipropylthiocarbamate	82.6%
INERT INGREDIENTS	17.4%
TOTAL	100.0%

Contains 6.7 lbs. Active Ingredient Per Gallon

EPA Reg. No. 476-2157 A-10R

CAUTION

Keep Out of Reach of Children

HARMFUL IF SWALLOWED

STATEMENT OF PRACTICAL TREATMENT

FIRST AID

Immediately start the procedures below. If further treatment is required contact a Poison Center, a physician, or the nearest hospital or call collect, day or night, (203) 226–6602 Stauffer Chemical Company, or (800) 424–9300 Pesticide Team Safety Network.

If swallowed: Immediately give several glasses of water but *do not* induce vomiting. If vomiting does occur, give fluids again. Have a physician determine if condition of patient will permit induction of vomiting or evacuation of stomach. Do not give anything by mouth to an unconscious or convulsing person.

For eye contact: Immediately flush eyes with large amounts of running water for at least 15 minutes. Hold eyelids apart to ensure rinsing of the entire surface of the eye and lids with water. Get medical attention immediately.

For skin contact: Flush all affected areas with plenty of water for several minutes. Remove and clean contaminated clothing and shoes. Seek medical attention if irritation occurs.

If inhaled: Remove to fresh air. Seek medical attention if respiratory irritation occurs or if breathing becomes difficult.

PRECAUTIONARY STATEMENTS

HAZARDS TO HUMANS AND DOMESTIC ANIMALS
CAUTION

HARMFUL IF SWALLOWED—Avoid contact with skin, eyes and clothing—Avoid inhalation of spray mist—Wear rubber gloves and fresh clothing—Wear goggles to avoid contact with eyes—Wash with soap and water immediately after use or contact—Do not contaminate food or feed.

Do not apply this product in such a manner as to directly or through drift, expose workers or other persons. When applied through the sprinkler irrigation method of application, the area being treated must be vacated by unprotected persons.

ENVIRONMENTAL HAZARDS

Keep out of any body of water—Do not contaminate water by cleaning of equipment or disposal of wastes.

PHYSICAL AND CHEMICAL HAZARDS

Keep away from heat and flame.

Can be stored at temperatures as low as minus 50°F.

GENERAL USE PRECAUTIONS

READ ALL LABEL DIRECTIONS BEFORE USING.

ERADICANE 6.7-E should be used for recommended purposes and at recommended rates. (DO NOT OVERDOSE.)

ERADICANE 6.7-E is recommended for use on mineral soils only (soils containing less than 10% organic matter).

Do not use on milo or sorghum. Do not use on corn seed stock. (Breeders, Foundation or Increase).

Do not contaminate water to be used on susceptible crops and ornamentals, or for domestic purposes. Do not contaminate feed or food. Keep container tightly closed when not in use. Do not store near seeds or fertilizers.

When properly applied and weather conditions exist for normal plant growth through the season, ERADICANE 6.7-E will not harm the treated crop nor should harmful soil residues remain beyond harvest. However, during germination and early growth, extended periods of unusually cold and wet or hot and dry weather, insect, nematode, or plant disease attack, carry-over soil residues of certain persistent herbicides, the use of certain soil applied systemic insecticides may create abnormal conditions that weaken crop seedlings. ERADICANE 6.7-E used under these abnormal conditions could result in crop injury.

WEEDS CONTROLLED BY ERADICANE 6.7-E

ERADICANE 6.7-E will not control established weeds.

Annual Grasses:	
Annual Bluegrass	*Poa annua*
Annual Ryegrass (Italian Ryegrass)	*Lolium multiflorum*
Barnyardgrass (Watergrass, Junglerice)	*Echinochloa* spp.
Bermudagrass Seedlings	*Cynodon dactylon*
Crabgrass	*Digitaria* spp.
Giant Foxtail	*Setaria faberi*
Goosegrass	*Eleusine indica*
Green Foxtail	*Setaria viridis*
Johnsongrass Seedlings	*Sorghum halepense*
Lovegrass (Stinkgrass)	*Eragrostis cilianensis*
Millet, Wild Proso*‡	*Panicum miliaceum*
Panicum, Fall	*Panicum dichotomiflorum*
Red Rice	*Oryza sativa*
Rescuegrass	*Bromus willdenowii*
Sandbur, Field*	*Cenchrus pauciflorus*
Signalgrass	*Brachiaria* spp.
Texas Panicum (Buffalograss, Coloradograss)*†	*Panicum texanum*
Volunteer Grains (Barley, Oats, Wheat)	

Wild Cane (Shattercane)*¶	*Sorghum bicolor*
Wild Oats	*Avena fatua*
Woolly Cupgrass	*Eriochloa villosa*
Yellow Foxtail	*Setaria glauca*

¶ Partial Control (Suppression) See Wild Cane Section
* Cultivation is suggested in addition to ERADICANE 6.7–E treatment in fields with moderate to heavy infestations of wild cane, sandbur, Texas panicum, or wild proso millet.
† Southeastern U.S Only.
‡ Central Region Only—Partial Control (Suppression)

Annual Broadleaf Weeds:

Annual Morningglory	*Ipomoea purpurea*
Black Nightshade	*Solanum nigrum*
Carpetweed	*Mollugo verticillata*
Chickweed, Common	*Stellaria media*
Corn Spurry	*Spergula arvensis*
Deadnettle (Henbit)	*Lamium amplexicaule*
Fiddleneck	*Amsinckia* spp.
Florida Purslane	*Richardia scabra*
Hairy Nightshade	*Solanum sarachoides*
Lambsquarters, Common	*Chenopodium album*
Nettleleaf Goosefoot	*Chenopodium murale*
Puncturevine**	*Tribulus terrestris*
Purslane, Common	*Portulaca oleracea*
Prostrate Pigweed	*Amaranthus graecizans*
Redroot Pigweed (Common Pigweed)	*Amaranthus retroflexus*
Shepherdspurse	*Capsella bursa-pastoris*
Sicklepod	*Cassia obtusifolia*
Teaweed (Spiny Sida)	*Sida spinosa*
Tumble Pigweed	*Amaranthus albus*
Velvetleaf	*Abutilon theophrasti*

** Western Region Only

The annual broadleaf weeds listed above will be controlled only if treatment is made when conditions are favorable for weed seed germination and growth. Under prolonged conditions of very cold soil, control of the above broadleaf weeds may not be adequate.

Perennial Grasses:

Bermudagrass	*Cynodon dactylon*
Purple Nutsedge (Nutgrass)	*Cyperus rotundus*
Quackgrass	*Agropyron repens*
Rhizome Johnsongrass*	*Sorghum halepense*
Yellow Nutsedge (Nutgrass)	*Cyperus esculentus*

* Partial Control (Suppression) Only.

Perennial weeds must be turned under and chopped up thoroughly prior to treatment. The underground rhizomes of quackgrass and johnsongrass and the rhizomes and stolons of bermudagrass must be cup up thoroughly into small pieces so that 4 or less nodes remain on a strand. Consult RECOMMENDATIONS for ERADICANE 6.7-E rates necessary to control the above listed weeds.

DIRECTIONS FOR USE

It is a violation of Federal law to use this product in a manner inconsistent with its labeling.

ERADICANE 6.7-E is a selective herbicide which must be mixed (incorporated) into the soil for control of weeds listed on this label. ERADICANE 6.7-E controls weeds by interfering with normal germination and seedling development. ERADICANE 6.7-E does not control established weeds .

APPLICATION DIRECTIONS

Pour the recommended amount of ERADICANE 6.7-E into the spray tank during the filling operation. Apply in 10 to 50 gallons of water per acre using a properly calibrated, low-pressure sprayer having good agitation. The soil should be well-worked and dry enough to permit good soil mixing (incorporation). See Appendix 1 for soil mixing directions.

For band application, calculate the amount to be applied per acre as follows:

$$\frac{\text{Band Width in Inches}}{\text{Row Width in Inches}} \times \text{Rate/A for Broadcast Treatment} = \text{Amount Needed for Band Treatment}$$

COMBINATION WITH FLUID FERTILIZERS

ERADICANE 6.7-E may be combined with fluid (solution, slurry or suspension) fertilizers. However, physical compatibility with these fertilizers must be verified before combining in the spray tank. See Appendix II for special directions regarding these combinations. Even though found to be compatible, constant agitation is necessary to keep the ERADICANE 6.7-E uniformly mixed with fluid fertilizers during application.

IMPREGNATION ON DRY BULK FERTILIZERS

ERADICANE 6.7-E may be impregnated on dry fertilizer and applied as the fertilizer is spread. See Appendix III for information and directions regarding impregnation and use of these combinations.

CENTER PIVOT SPRINKLER APPLICATION

Apply and soil incorporate ERADICANE 6.7-E, ERADICANE 6.7-E/ATRAZINE, or ERADICANE 6.7-E/BLADEX* (see TANK MIX RECOMMENDATIONS) with irrigation water before planting (see PLANTING), or immediately after planting before either corn or weeds emerge. It is important that application and incorporation be done immediately after the last tillage operation since poor results will occur if the weeds have germinated before application of the herbicide. Meter the ERADICANE 6.7-E or tank mix combinations with ATRAZINE or BLADEX into the irrigation water during the entire period. Apply in ½ to ¾ inches of water.

* BLADEX is a registered trademark of Shell Chemical Company.

USE PRECAUTIONS

A. Apply the product only through central pivot irrigation systems containing anti-siphon and check valves which will prevent water source contamination and overflow of the slurry tank and containing interlocking controls between the metering device and the water pump to insure simultaneous shut-off.
B. Inject the product with a positive displacement pump into the main line ahead of a right angle turn to insure adequate mixing.
C. Application of more than recommended quantities of irrigation water per acre may result in decreased product performance by removing the chemical from the zone of effectiveness.
D. Do not apply when wind speed favors drift, when system connections or fittings leak, when nozzles do not provide uniform distribution or when lines containing the product must be dismantled and drained.
E. Greater accuracy in calibration and distribution will be achieved by injecting a larger volume of a more dilute slurry per hour.

EXAMPLE

110 Acres to be covered by 4¾ pints (4 lbs. active) ERADICANE 6.7-E per acre.

110 A. × 4 lbs. = *440* lbs. active ERADICANE 6.7-E

$$\frac{440}{6.7} = 65.6 \text{ gals. ERADICANE 6.7–E}$$

Mixing tank size 750 gallons. Sprinkler takes 30 hours to apply ½ inch of water.

$$\text{Injection rate} = \frac{750}{30} = 25 \text{ gal./hr. or } 0.4 \text{ gal./min.}$$

Calibrate injection pump to deliver the herbicide mix into the irrigation system at the rate of *25* gal. per hr. or *0.4* gal. per minute.

PLANTING

Plant seed to a maximum depth of 2 inches within 2 weeks after treatment. To avoid removing ERADICANE 6.7-E from the row (with loss of weed control in the row), do not move or shape soil after treatment such as by planting in deep furrows.

CULTURAL PRACTICES FOLLOWING APPLICATION

Should weeds develop, a shallow cultivation or rotary hoeing will generally result in better weed control. When cultivating for any reason, it should be shallow, i.e., no more than ½ the depth the herbicide was incorporated. A sequential application of 2,4-D, BANVEL* 4E or a pre-emergence or post-emergence herbicides may be necessary to control weeds resistant to ERADICANE 6.7-E.

* BANVEL is a registered trademark of Velsicol Chemical Corporation.

RECOMMENDATIONS

Consult the U.S. regional map below and use the recommendations in your region only.

IMPORTANT: Limit use on sweet corn to 4¾ pints ERADICANE 6.7-E per acre.

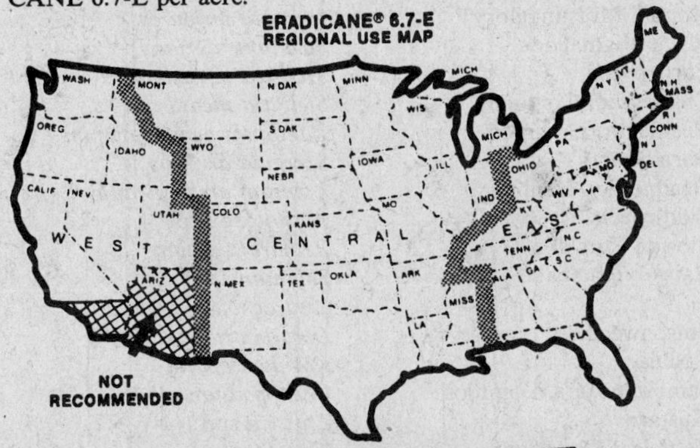

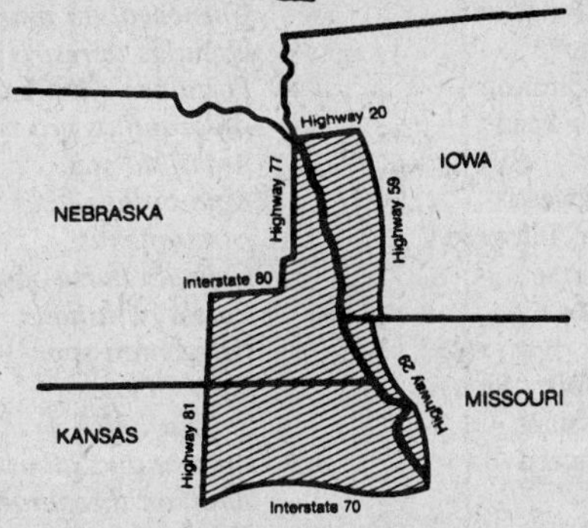

EASTERN U.S.

(Note: In the Southeastern U.S., when ERADICANE 6.7-E is used for weed control in silage corn, do not seed small seeded grains after corn harvest until September.)

CENTRAL U.S.

(Note: Western slope of Colorado: Where furrow irrigation will be used when corn is flat planted, do not form furrows until corn is in the three-leaf stage.)

ERADICANE 6.7-E is not recommended for wild cane control or suppression in the following areas of Nebraska, Iowa, Missouri, and Kansas.

NEBRASKA—east of Highway 81 and south of Interstate 80.

NEBRASKA—north of Interstate 80 and east of Highway 77.

IOWA—south of Highway 20 and west of Highway 59.

MISSOURI—west of Interstate 29 and north of Interstate 70.

KANSAS—north of Interstate 70 and east of Highway 81.

WESTERN U.S.

(Note: ERADICANE 6.7-E is not recommended on corn grown in Arizona or in the 10 Southernmost California counties. In California, do not use ERADICANE 6.7-E on corn planted after May 1.)

NOTE: Soils lacking sufficient moisture for seed germination must be preirrigated prior to ERADICANE 6.7-E application.

NOTE: Do not use on "Golden Jubilee" sweet corn west of the Cascade Mountains in Oregon or Washington.

PREPLANT INCORPORATION:

ERADICANE 6.7–E BROADCAST RATE PER ACRE IN PINTS

Weed	Eastern U.S.	Central U.S.	Western U.S.	Restrictions
Annual Grasses & Annual Broadleaf Weeds	3¾ to 4¾	4¾ to 7⅓	3¾ to 4¾	Use the lower rate for light infestations and in sand and light (coarse textured) soils.
Bermudagrass	4¾ to 7⅓	4¾ to 7⅓	3¾ to 4¾	Use the higher rate for heavy infestations and in heavier soils.
Nutsedge	4¾ to 7⅓	4¾ to 7⅓	3¾ to 7⅓	
Quackgrass	7⅓	4¾ to 7⅓	4¾	For control of moderate infestations and suppression of heavy infestations. See tank mix section.
Rhizome Johnsongrass (suppression)	7⅓	7⅓	7⅓	Application should follow a program of plowing and discing for good control. See special instructions.
Sandbur, Field	3¾	4¾ to 7⅓	4¾ to 7⅓	
Wild Cane	4¾ to 7⅓	7⅓	—	Partial control (suppression) only in central U.S. See special instructions.
Wild Oat	3¾	3¾	3¾ to 4¾	
Wild Proso Millet (suppression)	—	7⅓	—	Control may be improved with a combination treatment of ERADICANE 6.7–E/ATRAZINE or ERADICANE 6.7–E/BLADEX.
Velvetleaf	7⅓	7⅓	4¾ to 7⅓	See tank mix section.

Use the lower rates on sand and light (coarse textured) soils. Use the higher rates on heavy (fine textured) soils.

Rhizome Johnsongrass—For partial control, apply and incorporate 7⅓ pints ERADICANE 6.7-E per acre. Using a program of plowing and discing, followed by ERADICANE 6.7-E will give good control of seedlings and greatly reduce competition from plants coming from rhizomes.

In some areas discing 2 to 3 times spaced at 4 to 5 week intervals during fall and winter can be done. If this cannot be done, start with fall plowing followed by several discings in the spring before planting. The fall plowing and discing cuts rhizomes into small pieces and brings them to the surface, where both weather and ERADICANE 6.7-E will kill or weaken them.

Wild Cane—For partial control (suppression)—When applied and incorporated *immediately* before planting. ERADICANE 6.7-E at the rate of 7⅓ pints per acre on a broadcast basis will reduce competition from wild cane. On continuous corn acreage with high cane seed population, rotation to another crop is recommended. For enhanced control of wild cane, especially with heavy infestations. ERADICANE 6.7-E may be used in combination with ATRAZINE BLADEX or PRINCEP* (see TANK MIX COMBINATIONS).

* PRINCEP is a registered trademark of Ciba Geigy Corporation.

Suppression of wild cane populations may be improved if all the following practices are followed:

1. Fall or spring moldboard plow.
2. Before application, the soil should be well worked with a disc and

dry enough to permit good soil mixing of the herbicide (incorporation). All trash should be worked into the soil.
3. Delay application until soil warms to a constant 55°F or greater temperature. Plant immediately.
4. Incorporate thoroughly. Use recommended incorporation equipment and procedures found on this label.
5. Cultivate immediately if wild cane emerges. Cultivation should be shallow, no more than 1/2 the depth of incorporation of the herbicide.

CENTER PIVOT SPRINKLER APPLICATION

Apply 4 3/4 to 7 1/3 pints of ERADICANE 6.7-E per acre for control of annual grasses and annual broadleaf weeds. ATRAZINE or BLADEX may be tank mixed with ERADICANE 6.7-E for sprinkler application (see TANK MIX RECOMMENDATIONS).

SUBSURFACE INJECTION

(Southeastern U.S. Only)

For control of light infestations of annual grasses (except wild cane) annual broadleaf weeds and nutsedge, apply 3 3/4 pints of ERADICANE 6.7-E per broadcast acre, adjusting the dosage proportionately depending on the row spacing and width to be treated. Recommended on lighter (coarse textured) soils only.

TANK MIX COMBINATIONS

For control of additional weeds and for increased control of labelled broadleaf weeds, ATRAZINE, BLADEX, PRINCEP, or a combination of ATRAZINE and BLADEX may be tank mixed with the recommended rates of ERADICANE 6.7-E and applied preplant incorporated. ERADICANE 6.7-E/ATRAZINE or ERADICANE 6.7-E/BLADEX may be applied through a center pivot irrigation system.

As an alternative, ERADICANE 6.7-E can be applied preplant incorporated at recommended rates followed by ATRAZINE applied pre-emergence to soil surface. Consult the pre-emergence application directions on the ATRAZINE label. A rotary hoeing or shallow cultivation is recommended if rainfall or sprinkler irrigation has not occurred within 10 days of the surface application. For enhanced or additional broadleaf weed control, an overlay treatment of 2,4-D may be made.

With the combination of ERADICANE 6.7-E/BLADEX/ATRAZINE, the potential hazard of ATRAZINE residues affecting certain crops the following year is reduced because a lower rate of ATRAZINE can be used.

WEEDS CONTROLLED BY ERADICANE 6.7–E/ATRAZINE TANK MIX

Annual Grasses:	
Annual Bluegrass	*Poa annua*
Annual Ryegrass (Italian Ryegrass)	*Lolium multiflorum*
Barnyardgrass (Watergrass, Junglerice)	*Echinochloa* spp.
Bermudagrass Seedlings	*Cynodon dactylon*
Crabgrass	*Digitaria* spp.
Giant Foxtail	*Setaria faberi*
Goosegrass	*Eleusine indica*
Green Foxtail	*Setaria viridis*
Johnsongrass Seedlings	*Sorghum halepense*
Lovegrass (Stinkgrass)	*Eragrostis cilianensis*
Millet, Wild Proso*‡	*Panicum miliaceum*
Panicum, Fall	*Panicum dichotomiflorum*
Rescuegrass	*Bromus willdenowii*
Sandbur, Field*	*Cenchrus pauciflorus*
Signalgrass	*Brachiaria* spp.
Texas Panicum (Buffalograss, Coloradograss)*†	*Panicum texanum*
Volunteer Grains (Barley, Oats, Wheat)	
Wild Cane (Shattercane)*¶	*Sorghum bicolor*
Wild Oats	*Avena fatua*
Witchgrass	*Panicum capillare*
Woolly Cupgrass	*Eriochloa villosa*
Yellow Foxtail	*Setaria glauca*

¶ See Wild Cane Section.
* Cultivation is suggested in addition to ERADICANE 6.7–E/ATRAZINE treatment in fields with moderate to heavy infestation of wild cane, sandbur, seedling johnsongrass, Texas panicum or wild proso millet.
† Southeastern U.S. only.
‡ Central Region Only—Partial Control (Suppression).

Annual Broadleaf Weeds:	
Annual Morningglory	*Ipomoea purpurea*
Black Nightshade	*Solanum nigrum*
Carpetweed	*Mollugo verticillata*
Chickweed, Common	*Stellaria media*
Cocklebur, Common	*Xanthium pensylvanicum*
Corn Spurry	*Spergula arvensis*
Deadnettle (Henbit)	*Lamium amplexicaule*
Fiddleneck	*Amsinckia* spp.
Florida Purslane	*Richardia scabra*
Hairy Nightshade	*Solanum sarachoides*
Jimsonweed	*Datura stramonium*
Kochia	*Kochia scoparia*
Lambsquarters, Common	*Chenopodium album*
Mustard	*Brassica* spp.
Nettleleaf Goosefoot	*Chenopodium murale*
Puncturevine**	*Tribulus terrestris*
Purslane, Common	*Portulaca oleracea*
Prostrate Pigweed	*Amaranthus graecizans*
Ragweed	*Ambrosia* spp.
Redroot Pigweed (Common Pigweed)	*Amaranthus retroflexus*
Shepherdspurse	*Capsella bursa-pastoris*
Sicklepod	*Cassia obtusifolia*
Smartweed	*Polygonum* spp.
Teaweed (Spiny Sida)	*Sida spinosa*
Thistle, Russian	*Salsola kali*
Tumble Pigweed	*Amaranthus albus*
Velvetleaf	*Abutilon theophrasti*
Venice Mallow	*Hibiscus trionum*

** Western Region only.

Perennial Grasses:	
Bermudagrass	*Cynodon dactylon*
Purple Nutsedge (Nutgrass)	*Cyperus rotundus*
Quackgrass	*Agropyron repens*
Rhizome Johnsongrass*	*Sorghum halepense*
Yellow Nutsedge (Nutgrass)	*Cyperus esculentus*

Perennial weeds must be turned under and chopped up thoroughly prior to treatment. The underground rhizomes of quackgrass and johnsongrass and the rhizomes and stolons of bermudagrass must be cut up thoroughly into small pieces so that 4 or less nodes remain on a strand.

* Partial Control (Suppression) Only.

WEEDS CONTROLLED BY ERADICANE 6.7–E/BLADEX TANK MIX

Annual Grasses:	
Annual Bluegrass	*Poa annua*
Annual Ryegrass (Italian Ryegrass)	*Lolium multiflorum*
Barnyardgrass (Watergrass, Junglerice)	*Echinochloa* spp.
Bermudagrass Seedlings	*Cynodon dactylon*
Crabgrass	*Digitaria* spp.
Giant Foxtail	*Setaria faberi*

Goosegrass	*Eleusine indica*
Green Foxtail	*Setaria viridis*
Johnsongrass Seedlings	*Sorghum halepense*
Lovegrass (Stinkgrass)	*Eragrostis cilianensis*
Millet, Wild Proso*‡	*Panicum miliaceum*
Panicum, Fall	*Panicum dichotomiflorum*
Rescuegrass	*Bromus willdenowii*
Sandbur, Field*	*Cenchrus pauciflorus*
Signalgrass	*Brachiaria* spp.
Texas Panicum (Buffalograss, Coloradograss)*†	*Panicum texanum*
Volunteer Grains (Barley, Oats, Wheat)	
Wild Cane (Shattercane)*¶	*Sorghum bicolor*
Wild Oats	*Avena fatua*
Witchgrass	*Panicum capillare*
Woolly Cupgrass	*Eriochloa villosa*
Yellow Foxtail	*Setaria glauca*

¶ See Wild Cane Section
* Cultivation is suggested in addition to the ERADICANE 6.7–E/BLADEX treatment in fields with moderate to heavy infestations of wild cane, sandbur, seedling johnsongrass, Texas panicum or wild proso millet.
Southeastern U.S. Only.
Central Region Only—Partial Control (Suppression).

Annual Broadleaf Weeds:

Annual Morningglory	*Ipomoea purpurea*
Black Nightshade	*Solanum nigrum*
Carpetweed	*Mollugo verticillata*
Chickweed, Common	*Stellaria media*
Cocklebur, Common	*Xanthium pensylvanicum*
Corn Spurry	*Spergula arvensis*
Deadnettle (Henbit)	*Lamium amplexicaule*
Fiddleneck	*Amsinckia* spp.
Florida Purslane	*Richardia scabra*
Hairy Nightshade	*Solanum sarachoides*
Jimsonweed	*Datura stramonium*
Kochia	*Kochia scoparia*
Lambsquarters, Common	*Chenopodium album*
Mustard	*Brassica* spp.
Nettleleaf Goosefoot	*Chenopodium murale*
Puncturevine**	*Tribulus terrestris*
Purslane, Common	*Portulaca oleracea*
Prostrate Pigweed	*Amaranthus graecizans*
Ragweed	*Ambrosia* spp.
Redroot Pigweed (Common Pigweed)	*Amaranthus retroflexus*
Shepherdspurse	*Capsella bursa-pastoris*
Sicklepod	*Cassia obtusifolia*
Smartweed	*Polygonum* spp.
Teaweed (Spiny Sida)	*Sida spinosa*
Tumble Pigweed	*Amaranthus albus*
Velvetleaf	*Abutilon theophrasti*
Venice Mallow	*Hibiscus trionum*

** Western Region Only.

Perennial Grasses:

Bermudagrass	*Cynodon dactylon*
Purple Nutsedge (Nutgrass)	*Cyperus rotundus*
Quackgrass	*Agropyron repens*
Rhizome Johnsongrass*	*Sorghum halepense*
Yellow Nutsedge (Nutgrass)	*Cyperus esculentus*

Perennial weeds must be turned and chopped up thoroughly prior to treatment. The underground rhizomes of quackgrass and johnsongrass and the rhizomes and stolons of bermudagrass must be cut up thoroughly into small pieces so that 4 or less nodes remain on a strand.

* Partial Control (Suppression) Only.

WEEDS CONTROLLED BY ERADICANE 6.7–E/PRINCEP TANK MIX

Annual Grasses:

Annual Bluegrass	*Poa annua*
Annual Ryegrass (Italian Ryegrass)	*Lolium multiflorum*
Barnyardgrass (Watergrass, Junglerice)	*Echinochloa* spp.
Bermudagrass Seedlings	*Cynodon dactylon*
Crabgrass	*Digitaria* spp.
Giant Foxtail	*Setaria faberi*
Goosegrass	*Eleusine indica*
Green Foxtail	*Setaria viridis*
Johnsongrass Seedlings	*Sorghum halepense*
Lovegrass (Stinkgrass)	*Eragrostis cilianensis*
Millet, Wild Proso*‡	*Panicum miliaceum*
Panicum, Fall	*Panicum dichotomiflorum*
Rescuegrass	*Bromus willdenowii*
Sandbur, Field*	*Cenchrus pauciflorus*
Signalgrass	*Brachiaria* spp.
Texas Panicum (Buffalograss, Coloradograss)*†	*Panicum texanum*
Volunteer Grains (Barley, Oats, Wheat)	
Wild Cane (Shattercane)*¶	*Sorghum bicolor*
Wild Oats	*Avena fatua*
Woolly Cupgrass	*Ericochloa villosa*
Yellow Foxtail	*Setaria glauca*

¶ See Wild Cane Section.
* Cultivation is suggested in addition to ERADICANE 6.7–E/PRINCEP treatment in fields with moderate to heavy infestation of wild cane, sandbur, seedling johnsongrass, Texas panicum or wild proso millet.
† Southeastern U.S. Only.
‡ Central Region Only—Partial Control (Suppression).

Annual Broadleaf Weeds:

Annual Morningglory	*Ipomoea purpurea*
Black Nightshade	*Solanum nigrum*
Carpetweed	*Mollugo verticillata*
Chickweed, Common	*Stellaria media*
Cocklebur, Common	*Xanthium pensylvanicum*
Corn Spurry	*Spergula arvensis*
Deadnettle (Henbit)	*Lamium amplexicaule*
Fiddleneck	*Amsinckia* spp.
Florida Purslane	*Richardia scabra*
Hairy Nightshade	*Solanum sarachoides*
Jimsonweed	*Datura stramonium*
Lambsquarters, Common	*Chenopodium album*
Nettleleaf Goosefoot	*Chenopodium murale*
Puncturevine**	*Tribulus terrestris*
Purslane, Common	*Portulaca oleracea*
Prostrate Pigweed	*Amaranthus graecizans*
Ragweed	*Ambrosia* spp.
Redroot Pigweed (Common Pigweed)	*Amaranthus retroflexus*
Shepherdspurse	*Capsella bursa-pastoris*
Sicklepod	*Cassia obtusifolia*
Smartweed	*Polygonum* spp.
Teaweed (Spiny Sida)	*Sida spinosa*
Tumble Pigweed	*Amaranthus albus*
Velvetleaf	*Abutilon theophrasti*

** Western Region Only.

Perennial Grasses:

Bermudagrass	*Cynodon dactylon*
Purple Nutsedge (Nutgrass)	*Cyperus rotundus*
Rhizome Johnsongrass*	*Sorghum halepense*

Yellow Nutsedge (Nutgrass)	*Cyperus esculentus*

* Partial Control (Suppression) Only.

Perennial weeds must be turned under and chopped up thoroughly prior to treatment. The underground rhizomes of quackgrass and johnsongrass and the rhizomes and stolons of bermudagrass must be cut up thoroughly into small pieces so that 4 or less nodes remain on a strand.

APPLICATION DIRECTIONS

Apply ERADICANE 6.7-E/ATRAZINE, ERADICANE 6.7-E/BLADEX, ERADICANE 6.7-E/PRINCEP or ERADICANE 6.7-E/BLADEX/ATRAZINE in 20–50 gallons of water per acre. It is recommended that the compatibility of tank mix combinations be tested on a small scale, for example, a jar test before actual tank mixing. See Appendix II which describes the procedure for such tests.

RECOMMENDATIONS

BROADCAST RATES PER ACRE

Atrazine: Use the recommended rates of ERADICANE 6.7-E (in the Central U.S. use 3¾ to 7⅓ pints) plus 1¼ to 2 pounds of ATRAZINE 80W or 2 to 3 pints of ATRAZINE 4L per acre for both tank mix (preplant incorporation or center pivot sprinkler application) and sequential application. Use the higher rate of ATRAZINE in soils with over 5% organic matter.

Bladex: Use the recommended rates of ERADICANE 6.7-E (in the Central U.S. use 3¾ to 7⅓ pints) plus 2 to 2½ pounds of BLADEX 80W or 3 to 4 pints BLADEX 4L per acre for preplant incorporation or application through center pivot sprinkler system. Use the lower rate of BLADEX on sand or loam soils. Use the higher rate of BLADEX on fine textured (clay) soils or soils with 3% organic matter and above.

Princep: Use 1¼ to 3¾ pounds of PRINCEP 80W, 1–1/10 to 3⅓ pounds of PRINCEP CALIBER* 90 or 2 to 6 pints of PRINCEP 4L per acre with recommended rates of ERADICANE 6.7-E preplant incorporated. Use the higher rate of PRINCEP on fine-textured soils and where heavy broadleaf weed infestations are expected.

* CALIBER is a registered trademark of Ciba Geigy Corporation.

Bladex/Atrazine: Use the recommended rate of ERADICANE 6.7-E (in the Central U.S. use 3¾ to 7⅓ pints) plus 1¼ to 2½ pounds of BLADEX 80W or 2 to 4 pints of BLADEX 4L and ¾ to 1¼ pounds of ATRAZINE 80W or 1 to 2 pints of ATRAZINE 4L per acre. BLADEX and ATRAZINE should be used in a two to one proportion.

2,4-D or Banvel 4E: Use the amount of 2,4-D or BANVEL 4E recommended on the manufacturer's label.

NOTE: BLADEX is not recommended for use on sandy soils consisting of more than 70% sand and containing less than 1% organic matter. Consult the BLADEX label for more specific recommended rates on sandy soils.

Follow all the use precautions and warnings that appear on the ATRAZINE, BLADEX, PRINCEP, BANVEL and 2,4-D labels and supplemental literature.

Make only one application per crop. After a treatment including ATRAZINE, do not plant to any crop except corn until the following year or injury may occur.

Do not plant to alfalfa where more than 1.2 pounds actual ATRAZINE was used the previous year.

PRECAUTIONS TO FOLLOW WHEN USING ERADICANE 6.7-E/PRINCEP:

Do not rotate to any crop except corn until the following year or injury may occur.

After harvest, plow and thoroughly till the soil in fall or spring to minimize possible injury to spring-seeded rotational crops, regardless of rate used.

Do not apply ERADICANE 6.7-E/PRINCEP 80W, ERADICANE 6.7-E/PRINCEP CALIBER 90 or ERADICANE 6.7-E/PRINCEP 4L in the High Plains and Intermountain areas of the WEST (including central and western KS, western NE, western OK, and the Panhandle of TX) where rainfall is sparse and erratic or where irrigation is required.

Injury may occur to soybeans planted in north-central and northwest IA, south-central and southwest MN, northeast NE, southeast SD, and other areas the year following application on soils having a calcareous surface layer.

Do not plant sugar beets, tobacco, vegetables (including dry beans), spring-seeded small grains, or small-seeded legumes and grasses the year after ERADICANE 6.7-E/PRINCEP 80W, ERADICANE 6.7-E/PRINCEP CALIBER 90 or ERADICANE 6.7-E/PRINCEP 4L application or injury may occur.

Quackgrass: For improved quackgrass control ERADICANE 6.7-E can be preceded by a preplow application of ATRAZINE. Broadcast 2½ lbs. ATRAZINE 80W or 2 quarts ATRAZINE 4L per acre in the fall or spring and plow 1-3 weeks later. In the spring, follow with the recommended rate of ERADICANE 6.7-E.

Velvetleaf: For control of velvetleaf use 4¾ to 7⅓ pints of ERADICANE 6.7-E plus 2 pounds of ATRAZINE 80W or 3 pints of ATRAZINE 4L per broadcast acre.

TANK MIXING DIRECTIONS

1. Fill the spray tank at least ⅔ full of clean water. Be sure tank is clean and not contaminated.
2. Start agitation and continue moderate agitation throughout mixing.
3. Mix and disperse wettable powders first, flowable products next and ERADICANE 6.7-E last.
4. Premix the BLADEX, ATRAZINE or PRINCEP products individually in proper order (step 3 above) making sure the products are thoroughly wetted and dispersed in the tank prior to adding ERADICANE 6.7-E.
5. Add the ERADICANE 6.7-E to the tank ⅔ full of water already mixed with BLADEX and ATRAZINE or PRINCEP.
6. Complete filling the remaining portion of the tank with water; mix thoroughly. Maintain moderate agitation to provide a uniform mix during application.
7. The tank-mix combination should not be left in the spray tank for prolonged periods; batches should be mixed and applied the same day.

Do not use excessive agitation as it may cause the BLADEX, ATRAZINE or PRINCEP to settle out and set up in the spray tank. All return lines to the tank must discharge below liquid level.

APPENDIX I

SOIL MIXING (INCORPORATION) DIRECTIONS

ERADICANE 6.7-E must be incorporated into the soil to prevent loss of the herbicide. Whenever possible, application and incorporation should be done in the same operation. If there must be a delay (less than four hours) in incorporation of the herbicide when application and incorporation are done in separate operations, ERADICANE 6.7-E must be applied to a dry soil surface (at least ½ inch deep) and free from dew and incidental moisture.

Pacific Northwest Region (Semi-arid areas of Eastern Washington, Eastern Oregon and Idaho only): When application and incorporation are done in separate operations, ERADICANE 6.7-E must be incorporated on the same day as applied. Application must be made on a dry soil surface (at least ½ inch deep) free from dew and incidental moisture.

DELAYED INCORPORATION OF DRY BULK FERTILIZERS: The application and incorporation of ERADICANE 6.7-E must be carried out on the same day and delayed only if soil moisture conditions permit. See Appendix III for directions on the delay in soil incorporation.

IMPORTANT: Thorough mixing of ERADICANE 6.7-E in the soil is necessary. The soil must be in good tilth and cross discing is recommended. Improper discing or poor tilth (large clods) may result in unsatisfactory control.

INCORPORATION DIRECTIONS

The following equipment commonly is used for soil mixing (incorporation) before planting.

Power-Driven Cultivation Equipment: (Recommended on all soil types previously primary tilled): *Ground speed should be adjusted with all PTO driven cultivation tools to insure thorough soil mixing.*

a. Horizontal axis of rotation: Set to cut 3 to 4 inches deep.
b. Vertical axis of rotation: Set to cut 4 to 6 inches deep.

Tandem Discs (Recommended on all soil types): Set to cut to a depth of 4 to 6 inches, operated at 4 to 6 mph followed by a spike-tooth harrow or some other leveling device which extends beyond the ends of the discs. For more thorough mixing (for control or suppression of deep-germinating grasses—wild cane, seedling johnsongrass, quackgrass and nutsedge), disc in two different directions. When discing the second time, the disc should be reset to cut no deeper than 4 to 6 inches.

Field Cultivators (Recommended on lighter soils in good tilth): Use 3 to 4 rows of sweeps spaced at 7 inch or less intervals and staggered so that no soil is left unturned, followed by a spike-tooth harrow pulled behind the cultivator. *Do not use chisel plows.* Set the cultivator to cut 4 to 6 inches deep, operated at 5 mph or more. The field cultivator should not be used alone as an incorporation tool where control or suppression of deep-germinating weeds such as nutgrass, quackgrass, johnsongrass seedlings or wild cane is desired. For control of these weeds, use a disc for at least one of the passes over the field.

Sprinkler Incorporation Pacific Northwest Region (Semi-arid areas of Eastern Washington, Eastern Oregon and Idaho only): Surface apply ERADICANE 6.7-E immediately after planting. The soil surface should be dry (at least ½ inch deep) and free from dew and incidental moisture. Incorporate using ½ to ¾ inch of water within 36 hours following application. The application and incorporation must be done within 5 days after the last tillage operation since poor results will occur if weeds have germinated.

APPENDIX II

PROCEDURE FOR TESTING THE COMPATIBILITY OF ERADICANE 6.7-E WITH FLUID FERTILIZERS

The following procedure is suggested for determining whether ERADICANE 6.7-E may be combined with a specific fluid fertilizer for spray tank application.

MATERIALS REQUIRED

1. ERADICANE 6.7-E.
2. Fluid fertilizer to be used.
3. Adjuvant for fertilizer tank mix: Compex*, Sponto 168-D*, Unite*, or equivalent. The adjuvant which provides the best emulsification depends on the specific fertilizer under consideration.
4. Two one-quart, wide-mouth glass jars with lid or stopper.
5. Measuring spoons (a 25 mL pipette or graduated cylinder provides more accurate measurement).
6. Measuring cup, 8 oz. (237 mL).

* Compex, Kalo Laboratories Inc., Kansas City, Missouri; Sponto 168-D, Witco Chemical Company, Houston, Texas; Unite, Hopkins Agricultural Chemical Co., Madison, Wisconsin.

PROCEDURE

1. Pour a pint (about 473 mL) of the fluid fertilizer into each of the quart jars.
2. Add adjuvant to one of the jars and mix (see rate table on reverse side).
3. Add the ERADICANE 6.7-E to both jars (see rate table on reverse side).
4. Close both jars with lid or stopper and mix the contents by turning the jars upside down ten times.
5. Inspect the surface and body of the mixtures—
   (A) Immediately after completing the jar inversions.
   (B) After allowing the jars to stand quietly for 30 minutes.
   (C) And then again after turning the jars upside down 10 times.

If a uniform mix cannot be made, the mixture should not be used. If either mixture remains uniform for 30 minutes, the combination may be used. Should either mixture separate after 30 minutes, but readily remix uniformly with ten jar inversions, the mixture can be used if adequate agitation is maintained in the tank. If the mixture with adjuvant is satisfactory, but the one without adjuvant is not, be sure to use the adjuvant in the spray tank. Add the adjuvant first at a rate of 3 pints per 100 gallons of fluid fertilizer; foaming can be minimized by using moderate agitation.

If nondispersible oil, sludge or clumps of solids form in the mixtures, the combination should not be used.

NOTE 1: For some combinations, premixing wettable powders in a little water in a pail before adding them to the spray tank will improve the compatibility of the final mixtures with ERADICANE 6.7-E. This technique can be tested in the small-scale jar test by premixing the wettable powder in one-eighth (⅛) cup of water prior to addition to the pint of fluid fertilizer.

Rate Table for ERADICANE 6.7–E and Adjuvant With The Fluid Fertilizer**

Gallons of fluid fertilizer to be applied per acre	mL or tsp. of ERADICANE 6.7–E* to be added to 1 pint of fertilizer	
	mL	tsp.
10	7	1⅓
15	5	1
20	4	¾
25	3	⅔
30	2	½
40	2	½

* Based on field rate of 1 pound active ingredient per acre in the fertilizer volumes indicated. Increase volume proportionately to correspond with intended field rate in terms of pounds active ingredient per acre (e.g. for field rate of 4 pounds actual active ingredient in 40 gallons fertilizer per acre, add 8 mL or 1⅗ tsp. ERADICANE 6.7–E to each jar for compatibility testing purposes).

** Two (2) milliliters or two-fifths (⅖) teaspoon of adjuvant to be added to 1 pint of fluid fertilizer in order to equal the rate of 3 pints of adjuvant per 100 gallons of fluid fertilizer.

APPENDIX III

IMPREGNATION ON DRY BULK FERTILIZERS

ERADICANE 6.7-E, ERADICANE 6.7-E/ATRAZINE (4L OR 80W) OR ERADICANE 6.7-E/BLADEX (4L OR 80W)

CAUTION: ERADICANE 6.7-E alone or in combination with other herbicides must not be impregnated on ammonium nitrate, potassium nitrate or sodium nitrate fertilizers. Such mixtures may cause explosion and fire.

All individual state regulations relating to bulk dry fertilizer blending, registration, labeling and application are the responsibility of the individual and/or company selling the ERADICANE 6.7-E or ERADICANE 6.7-E/ATRAZINE (4L or 80W) or ERADICANE 6.7-E/BLADEX (4L or 80W) and fertilizer mixtures.

ERADICANE 6.7-E alone and in combination with ATRAZINE or BLADEX may be impregnated on many dry bulk fertilizers (1) (2) and applied and incorporated in the soil before planting for the control of grass and broadleaf weeds in corn.

Test results have shown that ERADICANE 6.7-E, ERADICANE 6.7-E/ATRAZINE (4L or 80W) and ERADICANE 6.7-E/ BLADEX (4L or 80W) on many dry bulk fertilizers give weed control equal to ERADICANE 6.7-E or a tank mix of ERADICANE 6.7-E/ATRAZINE (4L or 80W) or ERADICANE 6.7-E/ BLADEX (4L or 80W) applied as a spray in water or liquid fertilizer. However, uniform impregnation of the pesticides on the dry fertilizer particles and uniform application in the field are necessary to assure good results.

A minimum of 200 pounds and a maximum of 700 pounds of approved ingredients (1) (2) impregnated with the appropriate amount of ERADICANE 6.7-E alone or in combination with

ATRAZINE (4L or 80W) or BLADEX (4L or 80W) must be applied per acre.

For impregnating the pesticides on dry fertilizers, use a closed rotary-drum mixer or a similar type of closed blender equipped with suitable spray equipment. The spray nozzle should be positioned inside of the mixer to provide uniform spray coverage of the tumbling fertilizer. The ERADICANE 6.7-E should be sprayed uniformly onto the fertilizer using a fine spray pattern. For the ERADICANE 6.7-E/ATRAZINE or ERADICANE 6.7-E/BLADEX combination product, the ATRAZINE (4L or 80W) or BLADEX (4L or 80W) can be added and mixed uniformly with the fertilizer prior to the ERADICANE 6.7-E addition, or alternately, the ERADICANE 6.7-E/ATRAZINE (4L or 80W) or ERADICANE 6.7-E/BLADEX (4L or 80W) can be mixed in the proposed use ratio to prepare a uniform slurry for joint spray impregnation.

The physical properties of fertilizers vary, particularly in liquid absorptive capacity. When absorptivity is sufficient, simple spray impregnation of the fertilizer with ERADICANE 6.7-E, ERADICANE 6.7-E/ATRAZINE (4L or 80W), or ERADICANE 6.7-E/BLADEX (4L or 80W) provides a satisfactory, dry mixture.

If the absorptive capacity is inadequate, use of a highly absorptive powder is required to provide a dry, flowable mixture.

Microcel E (Johns-Manville Products Corp.) is the recommended absorbent powder. It should be added separately and uniformly to the prepared ERADICANE 6.7-E, ERADICANE 6.7-E/ATRAZINE (4L or 80W), or ERADICANE 6.7-E/BLADEX (4L or 80W) fertilizer mixture, in a quantity that is sufficient to provide a suitably flowable mixture. Generally less than 2% by weight of Microcel E is required.

The amount of ERADICANE 6.7-E, ERADICANE 6.7-E/ATRAZINE (4L or 80W) or ERADICANE 6.7-E/BLADEX (4L or 80W) actually required in the manufacture of individual fertilizer mixtures should be determined carefully for each production operation. This is necessary to ensure that the amount of pesticides actually contained in the mixture applied to the soil represents the correct rate of use.

Bulk fertilizer impregnated with ERADICANE 6.7-E, ERADICANE 6.7-E/ATRAZINE (4L or 80W) or ERADICANE 6.7-E/BLADEX (4L or 80W) should be *applied immediately,* not stored. It is recommended that all bulk containers be tightly covered while the product is being transported and applied to reduce chances of ERADICANE 6.7-E loss via volatilization.

When application and incorporation are done in separate operations, ERADICANE 6.7-E impregnated on the dry fertilizer should be incorporated on the same day as applied. Application must be made on a dry soil surface (at least ½ inch deep) free from dew and incidental moisture.

ERADICANE 6.7-E PHYSICAL DATA

Specific Gravity 68/68°F: 0.972 (typical)
Pounds/Gallon 68°F: 8.09 (typical)
Flashpoint: 186°F (Tagliabue closed cup)
Viscosity: Sprayable down to minus 20°F

(1) APPROVED DRY FERTILIZER INGREDIENTS FOR USE WITH ERADICANE 6.7–E

	N	P	K
Ammonium Sulfate	21	0	0
Diammonium Phosphate	18	46	0
Potassium Chloride	0	0	60
Potassium Sulfate	0	0	52
Single Super-phosphate	0	20	0
Treble Super-phosphate	0	46	0
Urea*	45	0	0
Ammonium Phosphate-Sulfate	16	20	0

* Some ureas may be phytotoxic when high rates are applied to corn. Use only urea rates known to be safe for corn application.

NOTE: K-Mag and 11-48-0 are also approved for use.

(2) APPROVED DRY FERTILIZER INGREDIENTS FOR USE WITH ERADICANE 6.7–E/ATRAZINE (4L or 80W) OR ERADICANE 6.7–E/BLADEX (4L or 80W)

	N	P	K
Ammonium Sulfate	21	0	0
Diammonium Phosphate	18	46	0
Potassium Chloride	0	0	60
Potassium Sulfate	0	0	52
Urea*	45	0	0
Ammonium Phosphate-Sulfate	16	20	0

* Some ureas may be phytotoxic when high rates are applied to corn. Use only urea rates known to be safe for corn application.

NOTE: K-Mag and 11-48-0 are also approved for use.

NOTE: Do not combine ERADICANE 6.7-E plus ATRAZINE (4L or 80W) or BLADEX (4L or 80W) with single super-phosphate (0-20-0) or treble super-phosphate (0-46-0).

(3) RATE CHART FOR IMPREGNATION OF DRY BULK FERTILIZERS (1) (2) WITH ERADICANE 6.7–E OR ERADICANE 6.7–E PLUS ATRAZINE 4L OR 80W

FERTILIZER RATE/A	ERADICANE 6.7–E RATE/A			ATRAZINE 4L		ATRAZINE 80W	
	3¾ pts.	4¾ pts.	7⅓ pts.	2 pts.	3 pts.	1¼ lbs.	2 lbs.
200 lbs.	18¾ qts/ton	23¾ qts/ton	36⅔ qts/ton	10 qts/ton	15 qts/ton	12½ lbs/ton	20 lbs/ton
250 lbs.	15 qts/ton	19 qts/ton	29⅓ qts/ton	8 qts/ton	12 qts/ton	10 lbs/ton	16 lbs/ton
300 lbs.	12½ qts/ton	15⅘ qts/ton	24⅖ qts/ton	6⅔ qts/ton	10 qts/ton	8⅓ lbs/ton	13⅓ lbs/ton
350 lbs.	10¾ qts/ton	13⅗ qts/ton	21 qts/ton	5¾ qts/ton	8⅗ qts/ton	7⅕ lbs/ton	11⅖ lbs/ton
400 lbs.	9⅖ qts/ton	11⅞ qts/ton	18⅓ qts/ton	5 qts/ton	7½ qts/ton	6¼ lbs/ton	10 lbs/ton
450 lbs.	8⅓ qts/ton	10½ qts/ton	16⅓ qts/ton	4½ qts/ton	6⅔ qts/ton	5⅔ lbs/ton	8⅞ lbs/ton
500 lbs.	7½ qts/ton	9½ qts/ton	14⅔ qts/ton	4 qts/ton	6 qts/ton	5 lbs/ton	8 lbs/ton
550 lbs.	6⅘ qts/ton	8⅔ qts/ton	13⅓ qts/ton	3⅔ qts/ton	5½ qts/ton	4⅔ lbs/ton	7⅓ lbs/ton
600 lbs.	6¼ qts/ton	8 qts/ton	12⅕ qts/ton	3⅓ qts/ton	5 qts/ton	4⅕ lbs/ton	6⅔ lbs/ton
650 lbs.	5¾ qts/ton	7⅓ qts/ton	11¼ qts/ton	3⅕ qts/ton	4⅗ qts/ton	3⅞ lbs/ton	6⅕ lbs/ton
700 lbs.	5⅓ qts/ton	6⅘ qts/ton	10½ qts/ton	2⅞ qts/ton	4¼ qts/ton	3⅔ lbs/ton	5¾ lbs/ton

(4) RATE CHART FOR IMPREGNATION OF DRY BULK FERTILIZERS (1) (2) WITH ERADICANE 6.7–E OR ERADICANE 6.7–E PLUS BLADEX 4L OR 80W

FERTILIZER RATE/A	ERADICANE 6.7–E RATE/A			BLADEX 4L RATE/A		BLADEX 80W RATE/A	
	3¾ pts.	4¾ pts.	7⅓ pts.	3 pts.	4 pts.	2 lbs.	2½ lbs.
200 lbs.	18¾ qts/ton	23¾ qts/ton	36⅔ qts/ton	15 qts/ton	20 qts/ton	20 lbs/ton	25 lbs/ton
250 lbs.	15 qts/ton	19 qts/ton	29⅓ qts/ton	12 qts/ton	16 qts/ton	16 lbs/ton	20 lbs/ton
300 lbs.	12½ qts/ton	15⅘ qts/ton	24⅖ qts/ton	10 qts/ton	13⅓ qts/ton	13⅓ lbs/ton	16⅔ lbs/ton
350 lbs.	10¾ qts/ton	13⅗ qts/ton	21 qts/ton	8⅗ qts/ton	11⅖ qts/ton	11⅖ lbs/ton	14⅓ lbs/ton
400 lbs.	9⅖ qts/ton	11⅞ qts/ton	18⅓ qts/ton	7½ qts/ton	10 qts/ton	10 lbs/ton	12½ lbs/ton
450 lbs.	8⅓ qts/ton	10½ qts/ton	16⅓ qts/ton	6⅔ qts/ton	8⅞ qts/ton	8⅞ lbs/ton	11⅕ lbs/ton
500 lbs.	7½ qts/ton	9½ qts/ton	14⅔ qts/ton	6 qts/ton	8 qts/ton	8 lbs/ton	10 lbs/ton
550 lbs.	6⅘ qts/ton	8⅔ qts/ton	13⅓ qts/ton	5½ qts/ton	7⅓ qts/ton	7⅓ lbs/ton	9⅕ lbs/ton
600 lbs.	6¼ qts/ton	8 qts/ton	12⅕ qts/ton	5 qts/ton	6⅔ qts/ton	6⅔ lbs/ton	8⅓ lbs/ton
650 lbs.	5¾ qts/ton	7⅓ qts/ton	11¼ qts/ton	4⅗ qts/ton	6⅕ qts/ton	6⅕ lbs/ton	7¾ lbs/ton
700 lbs.	5⅓ qts/ton	6⅘ qts/ton	10½ qts/ton	4¼ qts/ton	5¾ qts/ton	5¾ lbs/ton	7⅕ lbs/ton

STORAGE AND DISPOSAL

PROHIBITIONS
Do not contaminate water, food or feed by storage or disposal. Open dumping is prohibited. Keep away from heat and flame.

STORAGE
Keep container tightly closed when not in use. Do not store near seeds, fertilizers or foodstuffs. Can be stored at temperatures as low as minus 50°F.

PESTICIDE DISPOSAL
Waste resulting from the use of this product may be disposed of on site or at an approved waste disposal facility.

CONTAINER DISPOSAL
METAL—Triple rinse (or equivalent), then offer for recycling or reconditioning, or puncture and dispose of in a sanitary landfill, or by other procedures approved by State and local authorities.
PLASTIC—Triple rinse (or equivalent), then offer for recycling or reconditioning, or puncture and dispose of in a sanitary landfill, or by incineration, or, if allowed by State and local authorities, by burning. If burned stay out of smoke.

NOTICE—READ CAREFULLY

CONDITIONS OF SALE
Stauffer (and seller) offer(s) this product for sale subject to, and buyer and all users are deemed to have accepted, the following conditions of sale and warranty which may only be varied by written agreement of a duly authorized representative of Stauffer.

WARRANTY LIMITATION
Stauffer warrants that this product conforms to the chemical description on the label and is reasonably fit for the purposes referred to in the directions for use on the label subject to the inherent risks referred to below. Stauffer makes no other express warranties; THERE IS NO IMPLIED WARRANTY OF MERCHANTABILITY and there are no warranties which extend beyond the description on the label hereof.

INHERENT RISKS
The directions for use of this product are believed to be reliable and should be followed carefully. However, it is impossible to eliminate all risks associated with use. Buyer assumes all risks associated with use or application of this product contrary to label instructions or resulting from extraordinary weather conditions.

LIMITATION OF LIABILITY
In no case shall Stauffer be liable for special, indirect, or consequential damages resulting from the use or handling of this product and no claim of any kind shall be greater in amount than the purchase price of the product in respect of which such damages are claimed.

This product is sold only for uses stated on its label. No express or implied license is granted to use or sell this product under any patent in any country except as specified: Country: United States of America, Patent No. 4,021,224; 4,137,070 and other patents pending.

This is a specimen label. Stauffer is not responsible for the accuracy of the information contained herein. As labels are subject to revision, always carefully read and follow the label on the product container.

SUPPLEMENTAL LABELING

ERADICANE® 6.7-E Selective Herbicide
EPA Reg. No. 476–2157

DIRECTIONS FOR INJECTION WITH ANHYDROUS AMMONIA

INJECTION DIRECTIONS FOR ERADICANE 6.7-E WITH ANHYDROUS AMMONIA

It is a violation of Federal law to use this product in a manner inconsistent with its labeling.

For pre-plant soil injection applications of anhydrous ammonia, on conservation or conventional tilled acreage, ERADICANE 6.7-E can be introduced into the anhydrous line using a metering pump which offers a constant rate of ERADICANE 6.7-E output per acre independent of the anhydrous ammonia output rate. The ERADICANE 6.7-E discharge hose should tee into the ammonia discharge line between the meter and the manifold. For best results, ERADICANE 6.7-E must be used at rates of $4\frac{3}{4}$ to $7\frac{1}{3}$ pints per acre, injection shanks must be spaced at 8 to 10 inch intervals, and injection depth must be 4 to 5 inches. Spacings greater than 10 inches up to 15 inches will require a cross discing to achieve acceptable weed control. Do not inject at spacings greater than 15 inches as weed control may be unacceptable. A minimum rate of 125 pounds per acre of anhydrous ammonia is recommended.

Weed control will be inadequate if the ERADICANE 6.7-E is injected more than 4 to 5 inches below the soil surface. Soil injections at less than a 4 inch depth may result in some loss of the anhydrous ammonia through the soil surface. Injection should only be performed in soils dry enough to permit good soil mixing.

NOTE: Atrazine and BLADEX* have little lateral movement in the soil and injections with anhydrous ammonia will give ony a narrow band of broadleaf weed control. These products may be surface applied and incorporated at the same time the ERADICANE 6.7-E and anhydrous ammonia are injected or post-emergent applications of recommended broadleaf herbicides may be used.

* BLADEX is a registered trademark of Shell Chemical Company.

All applicable recommendations, diections, restrictions, and precautions on the EPA-registered label are to be followed.

This labeling must be in the possession of the user at the time of pesticide application.

This is a specimen label. Stauffer is not responsible for the accuracy of the information contained herein. As labels are subject to revision, always carefully read and follow the label on the product container.

FOLPET 50-WP (Phaltan*)

AGRICULTURAL FUNGICIDE

For Plant Disease Control
Wettable Powder For Use in Water

ACTIVE INGREDIENT:

FOLPET: *N*-(Trichloromethylthio) phthalimide	50%
INERT INGREDIENTS	50%
TOTAL	100%

* PHALTAN is a registered trademark for phthalimide fungicide.

EPA Reg. No. 476-1609 A-1R

CAUTION

Keep Out of Reach of Children

STATEMENT OF PRACTICAL TREATMENT

FIRST AID

Immediately start the procedures given below and contact a Poison Center, a physician or the nearest hospital for further assistance or call Collect, Day or Night, (203) 226-6602 Stauffer Chemical Company or (800) 424-9300 Pesticide Team Safety Network.

If swallowed: Immediately give several glasses of water and induce vomiting by gagging the victim with a finger placed at the back of the victim's tongue. Give fluids until vomitus is clear. If victim is unconscious or convulsing, do not induce vomiting or give anything by mouth.

For skin contact: Flush all affected areas with plenty of water for several minutes. Remove and clean contaminated clothing and shoes. Seek medical attention if irritation occurs.

For eye contact: Immediately flush eyes with large amounts of running water for at least 15 minutes. Hold eyelids apart during the flushing to ensure rinsing entire surface of the eye and lids with water. Get medical attention immediately.

If inhaled: Remove to fresh air. Seek medical attention if respiratory irritation occurs or if breathing becomes difficult.

PRECAUTIONARY STATEMENTS

HAZARDS TO HUMANS AND DOMESTIC ANIMALS

CAUTION

CAUSES EYE IRRITATION,
MAY BE HARMFUL IF SWALLOWED OR INHALED.

Avoid contact with eyes. Avoid Breathing Dust. Use with adequate ventilation. Wear respirator and goggles. Wash thoroughly after handling. Do not use or store near food or feed.

ENVIRONMENTAL HAZARDS

This product is toxic to fish. Do not apply directly to water. Do not contaminate any body of water by cleaning of equipment, disposal of wastes or application where runoff is likely to occur.

OBSERVE THE FOLLOWING USE PRECAUTIONS

Use only for claims listed and only as specified on this label. FOLPET (Phaltan) 50-WP may be combined with most commonly used insecticides, adjuvants and fungicides. Fungicidal activity is reduced if combined with strongly alkaline materials such as hydrated lime. This material should not be used in combination with or closely following oil sprays. Do not use with TEPP when a concentrate mixture is being applied.

In order that pesticidal residues on food and forage crops will not exceed tolerances established by the Environmental Protection Agency, use only at recommended rates and intervals, and do not apply closer to harvest than specified. Do not apply or allow to drift to areas occupied by unprotected humans or beneficial animals or onto adjoining food, fiber or pasture crops. **THE GROWER IS RESPONSIBLE** for residues on his crop as well as for damages caused by drift from his property to that of others. Consult State Agricultural Experiment Stations or State Agricultural Extension Service for additional information, as the timing, number and rate of applications needed will vary with the local conditions.

DIRECTIONS FOR USE

It is a violation of Federal law to use this product in a manner inconsistent with its labeling. Pour recommended amount of this material on surface of water in nearly filled spray tank or pre-mix in a bucket half filled with water and pour mixture through screen into nearly filled spray tank. Add balance of water to fill tank. Keep agitator running during filling and spraying operations. Do not allow mixture to stand. Do not use in low volume gear-type spray equipment. Do not combine wettable powders with emulsifiable liquids in the same spray tank unless previous use of the materials being combined has proven them to be physically compatible.

RECOMMENDATIONS
FRUITS AND VEGETABLES

Unless otherwise specified dosage rates are in terms of pounds of FOLPET (Phaltan) 50-WP per 100 gallons of water.

Maximum permissible rate per acre, expressed as FOLPET (Phaltan) 50-WP is given in parenthesis, ( ), after each crop claim.

APPLES, CRABAPPLES: Alternaria Leaf Spot, Bitter Rot, Black Rot, Botryosphaeria Rot, Brooks Spot, Fly Speck, Scab, Sooty Blotch. Use 1½ to 2 lbs. Apply up to 800 gals. per acre. Make first application at delayed-dormant and repeat at 7 to 14 day intervals through cover sprays (16 lbs.)

Note to User:

EASTERN STATES: May cause russeting on sensitive varieties such as Golden Delicious, Red Delicious and Stayman Winesap when used from just before bloom to 30 days after petal fall.

WESTERN STATES: Do not apply between bloom and July 1 to avoid injury.

AVOCADOS: Scab. Use 3 lbs. Apply up to 1000 gals. per acre. Apply at 2 to 3 week intervals as needed. (36 lbs.)

BLACKBERRIES, BOYSENBERRIES, DEWBERRIES, LOGANBERRIES, RASPBERRIES: Fruit Rot, Gray Mold, Spur Blight. Use 2 lbs. Apply 250 gals. per acre. Make first application when berries have set and repeat at 7 day intervals until harvest. (5 lbs.)

BLUEBERRIES, HUCKLEBERRIES: Gleosporium Leaf Spot, Gray Mold, Powdery Mildew. Use 2 lbs. Apply 250 gals. per acre. Make first application before first infection and repeat at 7 day intervals until harvest. (5 lbs.)

CANTALOUPE, HONEYDEW MELONS, MUSKMELONS, PUMPKINS, SUMMER SQUASH, WATERMELONS, WINTER SQUASH: Anthracnose, Downy and Powdery Mildews. Use 2 to 4 lbs. Apply up to 200 gals. per acre depending on size of vines. Make first application when first true leaves appear and repeat at weekly intervals until harvest. (8 lbs.)

CELERY: Cercospora Leaf Spot, Early and Late Blights. Use 2 to 4 lbs. per acre in sufficient water for thorough coverage. Make first application after plants are set in field and established. Repeat at 7 day intervals. Do not apply within 7 days of harvest. (4 lbs.)

CHERRIES (RED TART): Brown Rot. Use 1 to 2 lbs. Apply in delayed dormant, popcorn, and bloom stages and at 7 to 14 day intervals; thereafter, as necessary. **Leaf Spot.** Use 1 to 2 lbs. Apply in cover period at 7 to 14 day intervals as necessary and repeat once or twice as post-harvest foliage applications. Do not use on Sweet Cherries as foliage injury may occur. (12 lbs.)

CITRUS (GRAPEFRUIT, KUMQUATS, LEMONS, LIMES, ORANGES, TANGELOES, TANGERINES): Greasy Spot. Use 2 lbs. plus sticker. Apply in August and September, 3 to 10 weeks after first growth. **Melanose.** Use 2 to 4 lbs. Make first application 1 week after bloom and complete as soon as possible. **Scab.** Use 2 lbs. Apply in dormant and ⅔ petal fall stages, 2 weeks later, and on fall flush in August and September. (40 lbs.)

CRANBERRIES: Storage Rots. Use 9 lbs. per acre, in sufficient water for thorough coverage. Apply during bloom and repeat at 10 to 14 day intervals. Do not apply within 30 days of harvest. (18 lbs.)

CUCUMBERS: Anthracnose, Downy Mildew, Fruit Rots (Pythium and Rhizoctonia), Powdery Mildew. Use 2 to 4 lbs. Apply 200 gals. per acre. Make first application when first true leaves appear. Repeat at weekly intervals. (8 lbs.)

CURRANTS, GOOSEBERRIES: Leaf Spot. Use 4 lbs. Apply 150 gals. per acre. Make first application before infection or before bloom. Repeat at 7 to 10 day intervals in the post-bloom period. (6 lbs.)

GRAPES: Black Rot, Downy Mildew. Use 2 lbs. Apply 250 to 300 gals. per acre. Make first application just before bloom. Repeat just after bloom and at 7 to 10 day intervals for 1 to 3 more applications. **Dead-Arm.** Use 2 lbs. In **Eastern States,** begin when new growth is 1 to 2 inches long. Repeat when 8 to 10 inches long. In **Western States,** begin at bud break. Repeat 2 weeks later or when new growth is 4 to 8 inches long. **Powdery Mildew.** Use 2 lbs. Begin 2 to 3 weeks after bloom. Repeat at 2 week intervals. **NOTE: During periods of extended hot, dry weather foliage injury may occur.** (6 lbs.)

LETTUCE: Downy Mildew. Use 2 lbs. Apply up to 200 gals. per acre. Make first application when plants are established in field. Repeat at 7 day intervals until harvest. (4 lbs.)

ONIONS, GARLIC, LEEKS, SHALLOTS: Downy Mildew, Purple Blotch. Use 4 lbs. Apply 200 gals. per acre. Make first application before infection. Repeat at 7 day intervals until tops begin to dry before harvest. (8 lbs.)

STRAWBERRIES: Fruit Rot, Leaf Spot. Use 2 lbs. Apply up to 200 gals. per acre. Make first application before bloom and repeat at 7 day intervals until after harvest. (4 lbs.)

TOMATOES: Anthracnose. Use 4 lbs. Apply up to 200 gals. per acre. Make first application during bloom and repeat at 7 to 10 day intervals through harvest. (8 lbs.)

ORNAMENTALS

Unless otherwise specified dosage rates are in terms of FOLPET (Phaltan) 50-WP per 100 gallons of water. Thorough coverage is required.

ASTERS, PHLOX: Powdery Mildew. Use 2 lbs. (2 tablespoonfuls per 1 gal. water). Make first application before disease appears and repeat at 7 to 10 day intervals. Increase frequency during periods of rapid new growth and severe infection caused by rains or high humidity.

AZALEAS (Cuttings): Damping-Off. Use 1½ to 2 lbs. Soak cuttings for 15 to 30 minutes before placing in rooting media.

CARNATIONS: Alternaria Leaf Spot, Rust. Use 1 to 2 lbs. (1 to 2 tablespoonfuls per 1 gal. water). Make first application 2 weeks after planting and repeat at 7 to 10 day intervals as necessary.

CHRYSANTHEMUMS: Powdery Mildew, Septoria Leaf Spot. Use 2 lbs. (2 tablespoonfuls per 1 gal. water). Make first application immediately after planting and repeat at 7 to 10 day intervals. Increase frequency during periods of rapid new growth and severe infection caused by rains or high humidity.

GLADIOLUS (CORMS): Corm Rot, Damping Off. Use 15 teaspoonfuls per 1 gal. water. Apply to Corms as a pre-planting soak for 20 to 30 minutes. Keep mixture stirred. Drain and plant immediately.

GRASS (BLUE GRASS IN NON-PASTURED SITES): Melting Out (Helminthosporium). Use 6 ozs. in 5 gal. water per 1000 sq. ft. Make first application before disease is expected to appear and repeat at 7 day intervals as necessary. Do not graze treated areas. Do not feed clippings to livestock.

IRIS: Didymellina Leaf Spot. Use 2 lbs. (2 tablespoonfuls per 1 gal. water). Make first application when plants emerge and repeat at 7 to 10 day intervals.

MARIGOLDS, ZINNIAS: Alternaria Leaf Spot. Use 2 lbs. (2 tablespoonfuls per 1 gal. water). Make first application when plants emerge and repeat at 3 to 7 day intervals. Use higher frequency during periods of rapid new growth and severe infection caused by rains or high humidity.

POINSETTIAS (GREENHOUSE): Pythium, Root Rot. Use 2 lbs. plus a spreader and use 1 pint of suspension per 8 inch pan. Make application at time of panning and repeat 10 days later.

ROSES: Black Spot, Powdery Mildew. Use 1 to 2 lbs. (1 to 2 tablespoonfuls per 1 gal. water). Make first application when first leaves unfold and repeat at 7 to 10 day intervals throughout the season. Do not use wetting agents on roses as injury may result.

SNAPDRAGONS: Anthracnose, Powdery Mildew, Rust. Use 2 lbs. (2 tablespoonfuls per 1 gal. water). Make first application when plants emerge and repeat at 3 to 7 day intervals. Use higher frequency during periods of rapid new growth and severe infection caused by rains or high humidity.

STORAGE AND DISPOSAL

PROHIBITIONS

Do not contaminate water, food, or feed by storage or disposal. Open dumping is prohibited.

STORAGE

Store in cool, dry place. Protect from excessive heat.

PESTICIDE DISPOSAL

Wastes resulting from the use of this product may be disposed of on site or at an approved waste disposal facility.

CONTAINER DISPOSAL

Completely empty bag into application equipment. Then dispose of empty bag in a sanitary landfill or by incineration, or, if allowed by State and local authorities, by burning. If burned, stay out of smoke.

NOTICE—READ CAREFULLY

CONDITIONS OF SALE

Stauffer (and seller) offer(s) this product for sale subject to, and buyer and all users are deemed to have accepted, the following conditions of sale and warranty which may only be varied by written agreement of a duly authorized representative of Stauffer.

WARRANTY LIMITATION

Stauffer warrants that this product conforms to the chemical description on the label and is reasonably fit for the purposes referred to in the directions for use on the label subject to the inherent risks referred to below. Stauffer makes no other express warranties; THERE IS NO IMPLIED WARRANTY OF MERCHANTABILITY and there are no warranties which extend beyond the description on the label hereof.

INHERENT RISKS

The directions for use of this product are believed to be reliable and should be followed carefully. However, it is impossible to eliminate all risks associated with use. Buyer assumes all risks associated with use or application of this product contrary to label instructions or resulting from extraordinary weather conditions.

LIMITATION OF LIABILITY

In no case shall Stauffer be liable for special, indirect, or consequential damages resulting from the use or handling of this product and no claim of any kind shall be greater in amount than the purchase price of the product in respect of which such damages are claimed.

This product is sold only for uses stated on its label. No express or implied license is granted to use or sell this product under any patent in any country except as specified: Country: United States of America.

This is a specimen label. Stauffer is not responsible for the accuracy of the information contained herein. As labels are subject to revision, always carefully read and follow the label on the product container.

IMIDAN® 50 WP

AGRICULTURAL INSECTICIDE WETTABLE POWDER

FOR AGRICULTURAL USE ONLY

Complete Directions For Use, Use Precautions and Recommendations

ACTIVE INGREDIENT:
N-(Mercaptomethyl) phthalimide,
S-(*O,O*-dimethyl phosphorodithioate) 50%
INERT INGREDIENTS: 50%
TOTAL 100%

EPA Reg. No. 476-1917 A-10R

Keep Out of Reach of Children

WARNING

AVISO

PRECAUCION AL USUARIO: Si usted no lee ingles, no use este producto hasta que la etiqueta le haya sido explicada ampliamente.

STATEMENT OF PRACTICAL TREATMENT

Call a Poison Center or a Physician Immediately! If a known exposure occurs or is suspected, immediately start the recommended procedures below. Simultaneously contact a Poison Center, a physician, or the nearest hospital, or call collect, day or night:
STAUFFER CHEMICAL COMPANY (203) 226-6602
OR
PESTICIDE TEAM SAFETY NETWORK (800) 424-9300

Inform the person contacted of the type and extent of exposure, describe the victim's symptoms and follow the advice given.

NOTE: Be sure to advise the person contacted that the compound is a cholinesterase inhibitor.

NOTE TO MEDICAL PERSONNEL: Exposure may cause cholinesterase inhibition. Atropine by injection is antidotal. 2-PAM (Protopam Chloride) is also antidotal when administered early and in conjunction with atropine.

If Swallowed: Immediately give several glasses of water and induce vomiting by gagging the victim with a finger placed on the back of the victim's tongue. Give fluids until vomitus is clear. If victim is unconscious or convulsing, do not induce vomiting or give anything by mouth.

If Inhaled: Remove to fresh air. If not breathing, clear victim's airway and start mouth-to-mouth artificial respiration. If breathing is difficult, give oxygen, preferably with a physician's advice.

For Skin Contact: Flush all affected areas with plenty of water for several minutes. Remove and clean contaminated clothing and shoes. Seek medical attention if skin irritation occurs.

For Eye Contact: Hold eyelids apart and flush eyes with large amounts of running water for at least 15 minutes. Seek medical attention if eye irritation occurs.

PRECAUTIONARY STATEMENTS

HAZARDS TO HUMANS AND DOMESTIC ANIMALS WARNING

MAY BE HARMFUL OR FATAL IF SWALLOWED, INHALED OR ABSORBED THROUGH THE SKIN.

Do not get in eyes or on skin, or on clothing.

Do not breathe dust or spray mist.

Wash hands thoroughly after handling and before eating or smoking.

Wash all contaminated clothing with soap and hot water before reuse.

ENVIRONMENTAL HAZARDS

This product is toxic to fish, birds, and other wildlife. Birds and other wildlife feeding on treated areas may be killed.

Keep out of lakes, streams, and ponds. Do not contaminate water by cleaning of equipment or disposal of wastes.

Apply this product only as specified on this label.

This pesticide is highly toxic to bees exposed to direct treatment or to residues remaining on the treated area. Do not apply when bees are actively visiting the crop, cover crop, or weeds blooming in the treatment area. Applications should be timed to provide the maximum possible interval between treatment and the next period of bee activity.

USE PRECAUTIONS

Read all precautions and directions before using. Apply this product only as specified on this label.

IMIDAN 50-WP is compatible with most commonly used insecticides and fungicides, but is incompatible with alkaline materials such as spray lime, lime sulfur, and bordeaux mixtures. These materials will reduce the insecticidal activity of IMIDAN 50-WP.

Insecticidal activity may also be reduced when the water in the spray solution has a pH of 7 or higher. The pH of the spray solution may be corrected by the addition of a suitable buffering or acidifying agent for optimum insecticidal activity.

DIRECTIONS FOR USE

It is a violation of Federal law to use this product in a manner inconsistent with its labeling.

Pour recommended amount of this material on surface of water in nearly filled spray tank. Add balance of water to fill tank. Keep agitator running during filling and spraying operations. Do not allow mixture to stand. Do not use in gear-type spray equipment.

Applications may be made by aircraft or ground power equipment including concentrate and semiconcentrate equipment. The recommended dosages, except as specified for alfalfa, corn, cotton, peas, and pine seedlings are calculated for high volume conventional hydraulic sprayers.. When concentrate and semiconcentrate sprays are used, apply the recommended rate of IMIDAN 50-WP per acre in proportionately lower volumes of water according to equipment manufacturer's directions and State spray schedules. Do not exceed the maximum rate of IMIDAN 50-WP per acre or the time limitations specified for the individual crops. [In California, use a minimum of 4 pounds IMIDAN 50-WP per acre on mature Pome and Stone Fruit Trees (3½ pounds on cherries)].

REENTRY STATEMENT

Do not apply this product in such a manner as to directly or through drift expose workers or other persons. The area being treated must be vacated by unprotected persons.

Do not enter treated areas without protective clothing until sprays have dried.

Because certain states may require more restrictive reentry intervals for various crops treated with this product, consult your State Department of Agriculture for further information.

Written or oral warnings must be given to workers who are expected to be in a treated area or in an area about to be treated with this product. Oral warnings must be given if there is reason to believe that written warnings cannot be understood by workers.

When oral warnings are given, warnings shall be given in a language customarily understood by workers and must indicate the specific period of time that treated areas may not be entered without protective clothing. Oral warnings must also include appropriate first aid instructions in case of accidental exposure. These instructions are given in the STATEMENT OF PRACTICAL TREATMENT at the beginning of this label.

Written warnings must include the following information, "WARNING. Area treated with IMIDAN® Agricultural Insecticide on (date of application). Do not enter treated areas without appropriate protective clothing until sprays have dried (or other reentry interval if your State has a more restrictive interval for this product). If accidental exposure occurs, follow the instructions below." (Written warnings must include the STATEMENT OF PRACTICAL TREATMENT given at the beginning of this label.)

RECOMMENDATIONS: FRUIT AND NUT CROPS

NOTE: Directions for applications to **dormant** fruit and nut crops are located at the end of this section.

ALMONDS: For control of Peach Twig Borer, apply 1 pound IMIDAN 50-WP per 100 gallons of water (up to a maximum of 6 pounds IMIDAN 50-WP per acre). Apply as a full cover spray when infestation starts. Limit use on bearing almonds to one foliar application per season. Do not apply within 30 days of harvest.

APPLES: For control of Apple Maggot, Codling Moth, Elm Spanworm, Fruit Tree Leafroller, Green Apple Aphid, Green Fruitworm, Gypsy Moth, Japanese Beetle, Obliquebanded Leafroller, Orange Tortix*, Plum Curculio, Red Banded Leafroller, Redhumped Caterpillar, Rosy Apple Aphid and Tarnished Plant Bug, apply 1 to 1½ pounds per 100 gallons water (or a maximum of 8 pounds per acre per application) as a full cover spray. For heavy insect infestations use the higher dosage rate. Repeat as necessary in accordance with insect infestations and local or State spray programs. Usually suppresses European Red and Two Spotted Mites when used in a seasonal program. Do not apply within 7 days of harvest. Recommended for control of Elm Spanworm, Green Fruitworm, Gypsy Moth, Fruit Tree Leafroller, and Obliquebanded Leafroller in the Northeast only.

* California only. Consult your local County Agent or Extension Service Representative regarding proper timing of application.

Tank Mix (Northeast Only)

For control of the insects listed above for APPLES plus Tentiform Leafminer, Tufted Apple Budmoth and White Apple Leafhopper, a tank mix of IMIDAN 50-WP and methomyl (LANNATE* or NUDRIN**) may be used as follows: Mix ½ to 1 pound of IMI-

DAN 50-WP plus $1/2$ to 1 pint of methomyl 1.8-L or 2 to 4 oz. of methomyl 90%-SP per 100 gallons of water. Apply as a full cover spray (350-400 gallons per acre). Repeat as necessary in accordance with insect infestation and local or State spray programs. Do not use on Early MacIntosh or Wealthy varieties. Do not apply within 8 days of harvest. Do not graze/feed livestock under treated trees for 10 days after application.

* LANNATE is a registered trademark of E.I. duPont de Nemours and Company, Inc.

** NUDRIN is a registered trademark of Shell Oil Company.

APRICOTS: For control of Apple Maggot, Oriental Fruit Moth, Peach Twig Borer, Plum Curculio, Red Banded Leafroller, apply 1 to $1\frac{1}{2}$ pounds per 100 gallons water (or maximum of 6 pounds per acre per application) as a full cover spray. For heavy insect infestations use the higher dosage rate. Repeat as necessary in accordance with insect infestations and local or State spray programs. Usually suppresses European Red and Two Spotted Mites when used in a seasonal program. Do not apply within 14 days of harvest.

BLUEBERRIES (Northeastern U.S. only): For control of Blueberry Maggot, Cranberry Fruitworm, and Plum Curculio, apply 2 pounds IMIDAN 50-WP in 150 to 200 gallons of water per acre as a foliar spray. A second application may be made when indicated by insect infestations and local or State spray programs. May be applied up to 3 days of harvest.

CHERRIES, SOUR (TART): For control of Cherry Fruit Fly, Fruit Tree Leafroller, Japanese Beetle, Peach Twig Borer, and Plum Curculio, apply 1 pound per 100 gallons water (or maximum of $3\frac{1}{2}$ pounds per acre per application) as a full cover spray. Repeat as necessary in accordance with insect infestations and local or State spray programs. Usually suppresses European Red and Two Spotted Mites when used in a seasonal program. Do not apply within 7 days of harvest. Recommended for Cherry Fruit Fly, Fruit Tree Leafroller and Plum Curculio in the Northeast only.

CITRUS (Lemons, Oranges—California, Arizona, and Texas only). For control of California Red Scale and Brown Soft Scale, apply 1 pound IMIDAN 50-WP per 100 gallons of water plus a minimum of one quart (per 100 gallons) of a suitable spray oil according to oil manufacturer's specifications. Apply as a full cover spray using ground application equipment (up to a maximum of 30 pounds IMIDAN 50-WP per acre). Repeat as necessary in accordance with scale infestations and local or State spray programs, up to three applications per season. Do not reapply within 30 days of the previous application. Do not apply within 7 days of harvest.

GRAPES (West): For control of Grape Leaffolder and Omnivorous Leafroller, apply 2 pounds IMIDAN 50-WP in 200 gallons of water per acre. Spray both sides of each row, also tops of vines to assure adequate coverage of fruit and foliage. Applications may be made at any time between egg hatch and pupation for Leaffolder and Leafroller. Do not apply within 7 days of harvest.

GRAPES (East): For control of Grape Berrymoth and Grape Leafhopper and Japanese Beetle, apply 2 to 3 pounds IMIDAN 50-WP in 200 gallons of water per acre. For Grape Berrymoth, apply prebloom, postbloom, first and late cover sprays as needed. For Grape Leafhopper apply when most nymphs hatch (generally coincides with Berrymoth). Use the higher rate of control for Japanese Beetle. Spray both sides of each row, also tops of vines to assure adequate coverage of fruit and foliage. Make applications as necessary to maintain control in accordance with insect infestations and local or state spray programs. Do not apply within 14 days of harvest.

NECTARINES: For control of Apple Maggot, Oriental Fruit Moth, Peach Twig Borer, Plum Curculio, and Red Banded Leafroller, apply 1 to $1\frac{1}{2}$ pounds per 100 gallons water (or maximum of 6 pounds per acre per application) as a full cover spray. For heavy insect infestations use the higher dosage rate. Repeat as necessary in accordance with insect infestations and local or State spray programs. Usually suppresses European Red and Two Spotted Mites when used in a seasonal program. Do not apply within 14 days of harvest. Do not use OMITE* in combination with IMIDAN 50-WP on late maturing nectarine varieties as fruit injury may result.

* OMITE is a registered trademark of Uniroyal Chemical.

PEACHES: For control of Japanese Beetle, Oriental Fruit Moth, Peach Twig Borer, Plum Curculio and Tarnished Plant Bug, apply 1 to $1\frac{1}{2}$ pounds per 100 gallons water (or a maximum of 6 pounds per acre per application) as a full cover spray. For heavy insect infestations use the higher dosage rate. Repeat as necessary in accordance with insect infestations and local or State spray programs. Usually suppresses European Red and Two Spotted Mites when used in a seasonal program. Do not apply within 14 days of harvest.

PEARS: For control of Codling Moth, Elm Spanworm, Fruit Tree Leafroller, Green Apple Aphid, Gypsy Moth, Pear Psylla, Plum Curculio, Red Banded Leafroller, Rosy Apple Aphid and Tarnished Plant Bug, apply 1 to $1\frac{1}{2}$ pounds per 100 gallons of water (or maximum of 10 pounds per acre per application) as a full cover spray. For heavy insect infestations use the higher dosage rate. Repeat as necessary in accordance with insect infestations and local or State programs. Usually suppresses European Red and Two Spotted Mites when used in a seasonal program. Do not apply within 7 days of harvest. Recommended for control of Elm Spanworm, Gypsy Moth and Fruit Tree Leafroller in Northeast only.

PECANS: For control of Pecan Nut Casebearer*, Pecan Weevil and Hickory Shuckworm, apply $1\frac{1}{2}$ pounds IMIDAN 50-WP per 100 gallons of water as a full cover spray when infestation starts. Repeat as necessary in accordance with insect infestations and local or State programs. For Pecan Weevil, repeat applications should be made at 7 day intervals. Do not apply within 14 days of harvest. Do not graze livestock on cover crops grown in treated pecan groves.

* Recommended for Pecan Nut Casebearer in Texas only.

PLUMS, PRUNES: For control of Apple Maggot, Codling Moth*, Oriental Fruit Moth, Peach Twig Borer, Plum Curculio, Red Banded Leafroller and Redhumped Caterpillar, apply 1 to $1\frac{1}{2}$ pounds per 100 gallons water (or maximum of 6 pounds per acre per application) as a full cover spray. For heavy insect infestations use the higher dosage rate. Repeat as necessary in accordance with insect infestations and local or State spray programs. Usually suppresses Eropean Red and Two Spotted Mites when used in a seasonal program. Do not apply within 7 days of harvest.

*For control of Codling Moth in California only, apply 1 pound per 100 gallons water (or maximum of 6 pounds per acre per application) by ground equipment.

FRUIT AND NUT CROPS—DORMANT APPLICATIONS (Pacific Coast States Only)

Almonds, Apples, Apricots, Cherries, Nectarines, Peaches, Pears, Plums, Prunes .... San Jose Scale

Almonds, Apricots, Cherries, Nectarines, Peaches, Plums, Prunes .... Peach Twig Borer

Use 1 pound IMIDAN 50-WP per 100 gallons of water plus suitable spray oil according to oil manufacturer's specifications. Make thorough coverage using conventional ground sprayers.

FIELD, FORAGE, AND VEGETABLE CROPS

ALFALFA (Arizona, California, Nevada Only): For control of Common Alfalfa Weevil Larvae and Egyptian Alfalfa Weevil Larvae, apply $1\frac{1}{2}$ pounds IMIDAN 50-WP in 10 to 50 gallons of water per acre (20 to 50 gallons of water for dense stands) by ground equipment, or in 5 to 10 gallons of water per acre by aircraft. Consult your local Farm Advisor regarding the proper timing of application. Larvae should be sprayed when they are actively feeding. Do not apply more than once per cutting. Do not graze or cut for hay within 14 days of application. Do not use a spreader, sticker, or surfactant with IMIDAN 50-WP on alfalfa in Arizona, California, or Nevada. Do not apply to alfalfa in the bloom period.

FOR PEA APHIDS, apply the recommended rate of IMIDAN 50-WP in combination with $1/3$ to $2/3$ pint DIMETHOATE 2.67-E or $1/4$ to $1/2$ pint DIMETHOATE 4-E. Do not apply more than once per cutting. Do not graze or cut for hay within 14 days of application. Do not apply to alfalfa in the bloom period. Wear the protective clothing specified on the DIMETHOATE label when preparing and using this combination.

ALFALFA (Except Arizona, California, Nevada): For control of Common and Egyptian Alfalfa Weevil Larvae and Adults, Potato Leafhopper*, Meadow Spittlebug and Alfalfa Blotch Leafminer, apply 2 pounds IMIDAN 50-WP in 10 to 50 gallons of water per acre (20 to 50 gallons of water for dense stands) by ground equipment or 2 pounds IMIDAN 50-WP in 2 to 10 gallons of water per acre by aircraft. Consult your Local County Agent or Extension Service Representative regarding the proper timing of application. Larvae should be sprayed when they are actively feeding. Application for adult weevils should be timed to periods of visible activity. For Alfalfa Blotch Leafminer and Meadow Spittlebug, application should be made when first signs of infestation are visible. Do not apply more than once per cutting. Do not graze or cut for hay within 7 days of application.

* Recommended for control of Potato Leafhopper in Northeast and North Central States Only.

FOR PEA APHIDS, apply the recommended rate of IMIDAN 50-WP in combination with 1/3 to 2/3 pint DIMETHOATE 2.67-E or 1/4 to 1/2 pint DIMETHOATE 4-E. Do not apply more than once per cutting. Do not graze or cut for hay within 10 days of application. Do not apply to alfalfa in the bloom period. Wear the protective clothing specified on the DIMETHOATE label when preparing and using this combination.

CORN (Corn Belt Only): For control of Corn Rootworm adult populations apply 1/2 to 1 pound IMIDAN 50-WP in 2 to 5 gallons of water per acre by aircraft, or in 20 to 50 gallons of water per acre by ground equipment. Apply when beetles are present in sufficient numbers to warrant treatment for protection of the silk from adult feeding. Consult your Local County Agent or Extension Service Representative regarding proper timing of application. Do not apply within 14 days of harvest.

COTTON (Except San Joaquin Valley, CA) For control of the overwintering generation of boll weevils, apply 1/2 to 1 pound of IMIDAN 5-WP per acre in 5 to 20 gallons of water by ground equipment or 1 pound IMIDAN 50-WP per acre in 3 to 5 gallons of water by aircraft. Make two applications. The first application should be made at the 1/3 grown square stage and the second application 5 to 7 days later. Use the higher rate under heavy infestations. For control of first, second, or third generation Boll Weevils, apply 1 1/2 to 2 pounds IMIDAN 50-WP in 3 to 5 gallons of water per acre by aircraft or in 5 to 20 gallons of water per acre by ground equipment. Make applications at intervals of 3 to 7 days depending upon weevil population and weevil migration in fields. Use the higher rate for heavy infestations. Check infestations regularly. Do not exceed 20 pounds IMIDAN 50-WP per acre per season. Do not apply within 21 days of harvest. Do not graze or feed forage to livestock. Use on cotton prohibited in Copiah and Claiborne Counties, MS; Lauderdale and Madison Counties, AL; and Lawrence County, TN. Do not apply within one mile of any coastal or estuarine waters. Do not apply within 100 feet of aquatic habitants.

PEAS, FRESH AND DRY (Pacific Northwest Only): For control of Pea Weevil and Pea Leaf Weevil, apply 1 1/2 to 2 pounds IMIDAN 50-WP in 5 to 10 gallons of water per acre by aircraft, or in 20 to 50 gallons of water per acre by ground equipment. Apply between emergence and early pod formation when adult populations are present but before eggs are laid. Consult your Local County Agent or Extension Service Representative regarding proper timing of application. Do not apply within 7 days of harvest. Do not graze or feed forage to livestock within 7 days of application. Do not cut treated fresh pea forage for hay within 10 days of application.

POTATOES: For control of Colorado Potato Beetle, Potato Flea Beetle, Potato Leafhopper, apply 2 pounds per acre (1 1/2 to 2 lb. per acre for Colorado Potato Beetle and Potato Flea Beetle in the Northeast) in sufficient water to provide good coverage. Repeat applications as necessary throughout the growing season. Do not apply within 7 days of harvest. Not recommended for use on potatoes in California.

DECIDUOUS SHADE AND ORNAMENTAL TREES AND WOODY EVERGREENS

IMIDAN 50-WP is recommended for use by commercial applicators on Deciduous Shade and Ornamental Trees and Woody Evergreens in parks, residental and recreational areas, along thoroughfares and other localized areas where infestations of Gypsy Moth, Spring Cankerworm, Elm Spanworm, Birch Leaf Miner, Eastern Tent Caterpillar, Elm Leaf Beetle Larvae, Japanese Beetle, or Red-humped Caterpillar occur. When such insects or their damage occur, apply at a rate of 1 to 1 1/2 pounds IMIDAN 50-WP per 100 gallons of water, thoroughly wetting all parts of the affected plants to the point of runoff. For heavily insect infestations use the higher dosage rate. Repeat application as necessary to maintain insect control.

- Recommended on **Deciduous Shade and Ornamental Trees** (such as Ash, Beech, Oak, Dogwood, Willow, Hickory, Locust, Liquidambar, Hawthorne, Birch, Elm, Maple) to control: Gypsy Moth, Spring Cankerworm, Elm Spanworm, Eastern Tent Caterpillar, Elm Leaf Beetle Larvae, Japanese Beetle, and Red-humped Caterpillar.
- Recommended on **Woody Evergreens** (such as Arborvitae, Cedar, Fir, Hemlock, Juniper, Pine, Spruce, Yew) to control: Gypsy Moth.
- Recommended on **Birch Trees** to control: Birch Leaf Miner.

Choose a cool, calm period, preferably in early morning or evening. Do not apply if rain is expected or before leaf surfaces are dry.

PINE TREES (Pacific Northwest Only): For control of European, Pine Shoot Moth (*Rhyacionia buoliana*), apply 2 pounds of IMIDAN 50-WP per acre or for individual trees at a rate of 2 pounds per 100 gallons of water. Spray trees for thorough coverage of foliage. Repeat at 2 week intervals throughout flight period of the moths.

PINE SEEDLINGS

White, Slash and Loblolly: For control of Pales Weevil *(Hylobius pales)* and Pitch-eating Weevil *(Pachylobius picivorus),* use IMIDAN 50-WP as a 4% top dip, dipping down to and including the root collar only. Dip in bundles loose enough to allow solution to penetrate the bundles. Avoid coverage of roots. Swish tops in solution for 10-15 seconds to assure adequate coverage of all top growth. Drain and allow seedlings to dry before planting. The addition of an extender such as Nu-Film* 17 or Plantgard** may help to retain the IMIDAN 50-WP on the foliage under high rainfall conditions.

* Nu-Film is a registered trademark of Miller Chemical & Fertilizer Corp.

** Plantgard is a registered trademark of Polymetrics International Inc.

Some slight needle burn and first year growth reduction may occur following treatment of loblolly pine.

Wear rubber gloves during treating and planting.

DILUTION TABLE FOR A 4% TOP DIP

TO MAKE	5 gallons	30 gallons	50 gallons	100 gallons
Use this amount of IMIDAN 50-WP	3 1/2 lbs.	20 lbs.	33 1/2 lbs.	67 lbs.

Agitate frequently to keep IMIDAN 50-WP in suspension. Five gallons of solution should be enough to treat 10,000 seedlings. Do not keep IMIDAN solution overnight. Make up a fresh batch each day that seedlings will be dipped.

STORAGE AND DISPOSAL

Do not contaminate water, food or feed by storage or disposal.

STORAGE
Store in a cool dry place.

PESTICIDE DISPOSAL
Wastes resulting from the use of this product may be disposed of on site or at an approved waste disposal facility.

CONTAINER DISPOSAL

Completely empty bag into application equipment. Then dispose of empty bag in a sanitary landfill or by incineration, or, if allowed by State and local authorities, by burning. If burned, stay out of smoke.

SPILL OR LEAK

A small spill can be handled routinely. Use adequate ventilation and wear an air-supplied respirator to prevent inhalation. Wear suitable protective clothing and eye protection to prevent skin and eye contact. Use the following procedures:

1. Sweep up spilled material being careful not to create dust.
2. Place sweepings in an open drum.
3. Generously cover the contaminated areas with a common household detergent. Using a stiff brush and small amounts of water, work the detergent into the spill material forming a slurry. Do not splatter on one's self or bystanders, and completely avoid skin or eye contact with this material. Brush the slurry into cracks and crevices and allow to stand for 2 to 3 minutes.
4. Spread a suitable absorbent such as clay, sawdust or kitty litter on the slurried liquid. Shovel absorbed material into an open drum.
5. Repeat if necessary.
6. Flush the area with water while observing proper environmental considerations.
7. Seal drum and dispose of contaminated material in an approved pesticide landfill.

Large spills must be handled according to a predetermined plan. For assistance in developing a plan, contact Stauffer's Agricultural Chemical Division, Westport, CT 06881-0850.

NOTICE—READ CAREFULLY

CONDITIONS OF SALE

Stauffer (and seller) offer(s) this product for sale subject to, and buyer and all users are deemed to have accepted, the following conditions of sale and warranty which may only be varied by written agreement of a duly authorized representative of Stauffer.

WARRANTY LIMITATION

Stauffer warrants that this product conforms to the chemical description on the label and is reasonably fit for the purposes referred to in the directions for use on the label subject to the inherent risks referred to below. Stauffer makes no other express warranties: THERE IS NO IMPLIED WARRANTY OF MERCHANTABILITY and there are no warranties which extend beyond the description on the label hereof.

INHERENT RISKS

The directions for use of this product are believed to be reliable and should be followed carefully. However, it is impossible to eliminate all risks associated with use. Buyer assumes all risks associated with use or application of this product contrary to label instructions or resulting from extraordinary weather conditions.

LIMITATION OF LIABILITY

In no case shall Stauffer be liable for special, indirect or consequential damages resulting from the use or handling of this product and no claim of any kind shall be greater in amount than the purchase price of the product in respect of which such damages are claimed.

This product is sold only for uses stated on its label. No express or implied license is granted to use or sell this product under any patent in any country except as specified: Country: United States of America.

This is a specimen label. Stauffer is not responsible for the accuracy of the information contained herein. As labels are subject to revision, always carefully read and follow the label on the product container.

ORDRAM® 8-E

SOUTHWEST
EMULSIFIABLE LIQUID HERBICIDE

For Use on Rice

This Label Contains Directions For Use on Rice Grown in the Southwest Only

ACTIVE INGREDIENT:

S-ethyl hexahydro-1*H*-azepine-1-carbothioate	90.9%
INERT INGREDIENTS	9.1%
TOTAL	100.0%

Contains 8 Pounds Active Ingredient Per Gallon.

EPA Reg. No. 476-2107 A-2R

WARNING
Keep Out of Reach of Children

MAY BE FATAL IF SWALLOWED.

Avoid contact with skin, eyes and clothing. Wash with soap and water after use. Avoid breathing spray mist. Avoid contamination of feed and food. Repeated skin contact may cause cumulative effects.

GENERAL USE PRECAUTIONS

Read all label instructions before using. Do not overdose.

ORDRAM should be used only for recommended purposes and at recommended rates. ORDRAM may cause crop injury under extreme soil or climatic conditions or if directions are not followed.

Except as specified on this label, do not apply other herbicides in conjunction with ORDRAM as injury may occur.

ORDRAM may be combined with fluid (solution, slurry, or suspension) fertilizers. However, physical compatibility with these fluid fertilizers must be determined before combining in the spray tank. See special literature giving directions for these combinations. Even though found to be compatible, constant agitation is necessary to keep the ORDRAM uniformly mixed with the fluid fertilizer.

DO NOT STORE NEAR SEEDS OR FERTILIZER.

KEEP CONTAINER CLOSED WHEN NOT IN USE.

DO NOT CONTAMINATE WATER USED FOR DOMESTIC PURPOSES.

GENERAL APPLICATION DIRECTIONS

All equipment used in the application of ORDRAM should be carefully calibrated before use and checked frequently during application to be sure the application is uniform. Avoid overlaps that will increase ORDRAM dosages above recommended limits because crop injury may occur. Apply ORDRAM 8-E in 10 to 50 gallons of water by ground equipment and 5 to 10 gallons of water by airplane except as specified below.

To avoid uneven distribution of ORDRAM in flooded fields the water flow through the field should be at a minimum at the time of application and for at least two hours after application.

Do not apply more than 6 pounds actual ORDRAM (6 pints ORDRAM 8-E) per acre per season where more than one method of application is used. Allow 30 days between applications.

DIRECTIONS FOR USE AND RECOMMENDATIONS

It is a violation of Federal law to use this product in a manner inconsistent with its labeling.

1. PRE-PLANT, PRE-FLOOD: For use on water-seeded rice ONLY.

For control of Watergrass (Barnyardgrass) *Echinochloa* spp. ONLY.

SOIL PREPARATION: Prepare the soil for seeding according to good agricultural practices. All weed growth must be thoroughly worked into the soil before treatment.

FOR GROUND APPLICATION ONLY: Use standard low-pressure (20–50 psi) boom-spray equipment. DO NOT USE SINGLE NOZZLE BOOM-JET TYPE SPRAYERS.

APPLICATION AND INCORPORATION: Apply to dry soil in good tilth. ORDRAM must be incorporated immediately (within minutes) to prevent loss of the herbicide. Whenever possible, application and incorporation should be done in the same operation. Incorporation should be done by at least one discing or spike-tooth harrowing. Flood the field for seeding and hold the flood for 4 to 6 days following seeding.

RECOMMENDED RATE: Apply 3 pints ORDRAM 8-E per acre.

2. POST-EMERGENCE, PRE-FLOOD: For use on dry-seeded and water-seeded rice.

For control of Dayflower *(Commelina communis)* in the two leaf stage or less and small Sprangletop *(Leptochloa* spp.) emerged less than one inch, apply ORDRAM 8-E as a tank mix combination with a registered propanil brand in 10–15 gallons of water per acre by airplane or in 20–50 gallons of water per acre by ground equipment. Flood the rice within 5 days after treatment. Follow all directions and use precautions on the propanil label. Avoid drift or accidental application to all crops, shrubs or ornamentals not in the rice field being treated.

RECOMMENDED RATE: 2 to 3 pints ORDRAM 8-E (3 pints for Sprangletop) plus 3 pounds actual propanil per acre.

3. POST-EMERGENCE AT FLOODING: For use on dry-seeded rice ONLY.

For control of Watergrass (Barnyardgrass) and Dayflower under 5 inches tall.

When establishing the permanent flood, meter the recommended rate of ORDRAM 8-E per acre into the irrigation water at the header ditch or canal where water enters the treated field. Meter the ORDRAM during the entire irrigation period using a constant head metering device.

RECOMMENDED RATE: 3 pints ORDRAM 8-E per acre.

The rice must be established at the time of application. Dayflower must be completely submerged and Barnyardgrass ⅔ submerged by the flooding operation.

Flood should be maintained at the prescribed depth until Barnyardgrass and/or Dayflower is controlled.

4. POST-EMERGENCE, POST-FLOOD: For use on dry-seeded and water seeded rice.

For control of Watergrass (Barnyardgrass) ONLY.

APPLICATION: Apply the recomended rate directly to the flooded field. Apply by airplane ONLY. Apply only when wind velocity is low, 0 to 5 mph.

RECOMMENDED RATE: 3 pints ORDRAM 8-E per acre. The Watergrass must be a minimum of 2 inches high but not higher than 5 inches, and it must be at least ⅔ submerged at time of application. This water level (not less than 2 inches) must be maintained until the Watergrass dies. Application should be made before rice is 2 inches above surface of water.

STORAGE AND DISPOSAL

PROHIBITIONS

Do not contaminate water, food or feed by storage or disposal. Open dumping is prohibited.

STORAGE

Keep container tightly closed when not in use. Do not store near seeds, fertilizers or foodstuffs. Can be stored at temperatures as low as minus 50 F.

PESTICIDE DISPOSAL

Wastes resulting from the use of this product may be disposed of on site or at an approved waste disposal facility.

CONTAINER DISPOSAL

Triple rinse (or equivalent). Then offer for recycling or reconditioning, or puncture and dispose of in a sanitary landfill, or by other procedures approved by State and local authorities.

NOTICE—READ CAREFULLY

CONDITIONS OF SALE

Stauffer (and seller) offer(s) this product for sale subject to, and buyer and all users are deemed to have accepted, the following conditions of sale and warranty which may only be varied by written agreement of a duly authorized representative of Stauffer.

WARRANTY LIMITATION

Stauffer warrants that this product conforms to the chemical description on the label and is reasonably fit for the purposes referred to in the directions for use on the label subject to the inherent risks referred to below. Stauffer makes no other express warranties; THERE IS NO IMPLIED WARRANTY OF MERCHANTABILITY and there are no warranties which extend beyond the description on the label hereof.

INHERENT RISKS

The directions for use of this product are believed to be reliable and should be followed carefully. However, it is impossible to eliminate all risks associated with use. Buyer assumes all risks associated with use or application of this product contrary to label instructions or resulting from extraordinary weather conditions.

LIMITATION OF LIABILITY

In no case shall Stauffer be liable for special, indirect or consequential damages resulting from the use or handling of this product and no claim of any kind shall be greater in amount than the purchase price of the product in respect of which such damages are claimed.

This product is sold only for uses stated on its label. No express or implied license is granted to use or sell this product under any patent in any country except as specified: Country: United States of America. Patent No.: 3,198,786; 3,573,031.

This is a specimen label. Stauffer is not responsible for the accuracy of the information contained herein. As labels are subject to revision, always carefully read and follow the label on the product container.

SUPPLEMENTAL LABELING

ORDRAM® 8–E

EPA Reg. No. 476-2107

FOR DISTRIBUTION AND USE IN THE SOUTHWEST ONLY

DIRECTIONS FOR USE

It is a violation of federal law to use this product in a manner inconsistent with its labeling.

PRE-PLANT, INCORPORATED GROUND APPLICATION

FOR USE ON DRY-SEEDED RICE ONLY

For partial control of early season grasses (watergrass [barnyardgrass], sprangletop, broadleaf signalgrass and crabgrass) and nutsedge (*Cyperus* spp.)

On a thoroughly clean and prepared seed bed, apply 3–4 pints of ORDRAM 8-E. Application must be made to a dry soil surface. Use 3 pints on light soil, 3–4 pints on medium soil, and 4 pints on heavy soil. Use ground application equipment only. Incorporate immediately to a depth of 1–2″. Whenever possible, application and incorporation should be done in the same operation. When applied in fluid fertilizer, incorporation can be delayed up to 2 hours. Early flushing or flooding may improve the control received by this treatment method.

CAUTION: Other control methods may be needed for escaped weeds or for late season infestations.

All applicable directions, restrictions, and precautions on the EPA-registered label are to be followed.

This labeling must be in the posession of the user at the time of pesticide application.

NOTICE—READ CAREFULLY

CONDITIONS OF SALE

Stauffer (and seller) offer(s) this product for sale subject to, and buyer and all users are deemed to have accepted, the following conditions of sale and warranty which may only be varied by written agreement of a duly authorized representative of Stauffer.

WARRANTY LIMITATION

Stauffer warrants that this product conforms to the chemical description on the label and is reasonably fit for the purposes referred to in the directions for use on the label subject to the inherent risks referred to below. Stauffer makes no other express warranties; THERE IS NO IMPLIED WARRANTY OF MERCHANTABILITY and there are no warranties which extend beyond the description on the label hereof.

INHERENT RISKS

The directions for use of this product are believed to be reliable and should be followed carefully. However, it is impossible to eliminate all risks associated with use. Buyer assumes all risks associated with use or application of this product contrary to label instructions or resulting from extraordinary weather conditions.

LIMITATION OF LIABILITY

In no case shall Stauffer be liable for special, indirect, or consequential damages resulting from the use or handling of this product and no claim of any kind shall be greater in amount than the purchase price of the product in respect of which such damages are claimed.

This is a specimen label. Stauffer is not responsible for the accuracy of the information contained herein. As labels are subject to revision, always carefully read and follow the label on the product container.

ORDRAM® 8-E

CALIFORNIA EMULSIFIABLE LIQUID HERBICIDE

For Use on Rice Grown in California Only

ACTIVE INGREDIENT:

S-ethyl hexahydro-1*H*-azepine-1-carbothioate	90.9%
INERT INGREDIENTS	9.1%
TOTAL	100.0%

Contains 8 Pounds ACTIVE INGREDIENT per Gallon

EPA Reg. No. 476-2107 AA A-1R

WARNING

Keep Out of Reach of Children

MAY BE FATAL IF SWALLOWED.

Avoid contact with skin, eyes and clothing. Wash with soap and water after use. Avoid breathing spray mist. Avoid contamination of feed and food. Repeated skin contact may cause cumulative effects.

GENERAL USE PRECAUTIONS

READ ALL LABEL INSTRUCTIONS BEFORE USING. ORDRAM 8-E should be used only for recommended purposes and at recommended rates. ORDRAM 8-E may cause crop injury under extreme soil or climatic conditions or if directions are not followed.

Do not apply any other herbicide in conjunction with ORDRAM 8-E as injury may occur.

ORDRAM 8-E may be combined with fluid (solution, slurry, or suspension) fertilizers. However, physical compatibility with these fluid fertilizers must be determined before combining in the spray tank. See special literature giving directions for these combinations. Even though found to be compatible, constant agitation is necessary to keep the ORDRAM 8-E uniformly mixed with the fluid fertilizer.

DO NOT STORE NEAR SEEDS OR FERTILIZER.

DO NOT CONTAMINATE WATER USED FOR DOMESTIC PURPOSES.

Keep container closed when not in use.

DIRECTIONS FOR USE AND RECOMMENDATIONS

It is a violation of Federal law to use this product in a manner inconsistent with its labeling.

For use on rice to control WATERGRASS (Barnyardgrass) *Echinochloa* spp.

PRE-PLANT, PRE-FLOOD: (Water-seeded rice only)

SOIL PREPARATION: Prepare the soil for seeding according to good agricultural practices. All weed growth must be thoroughly worked into the soil before treatment.

FOR GROUND APPLICATION ONLY: Use standard low-pressure (20–50 psi) boom-spray equipment. DO NOT USE SINGLE-NOZZLE BOOM-JET TYPE SPRAYERS. Spray equipment should be carefully calibrated before use and checked frequently during application to be sure nozzles are free from clogging and are delivering a uniform spray pattern. Avoid overlaps that will increase ORDRAM 8-E dosage above recommended limits because crop injury may occur. Apply the recommended rate of ORDRAM 8-E in 20 to 50 gallons of water per acre.

RATE: Apply 3 pints per acre.

SOIL INCORPORATION: Apply to dry soil in good tilth. ORDRAM 8-E must be incorporated immediately (within minutes) to prevent loss of the herbicide. Whenever possible, application and incorporation should be done in the same operation. Incorporation should be done by at least one discing or spike-tooth harrowing. Flood the field for seeding and hold the flood for 4 to 6 days following seeding.

OR

POST-FLOOD, POST-EMERGENCE: (Water-seeded or drilled rice)

Rice should be in the seedling stage at the time of application.

FOR AIR APPLICATION ONLY: Watergrass must be a minimum of 2″ high but not higher than 5″, and it must be at least ⅔ submerged at time of application. This water level (not less than 2″) must be maintained until the watergrass dies.

Equipment should be carefully calibrated before use and checked frequently during application to be sure application is uniform. Avoid overlaps that will increase ORDRAM 8-E dosage above recommended limits because crop injury may occur. Apply the recommended rate of ORDRAM 8-E in 5 to 10 gallons of water per acre. Aerial application should be made only when the wind velocity is low, 0–5 mph.

To avoid uneven distribution of ORDRAM 8-E, the water flow through the field should be at a minimum at the time of application and for at least 2 hours after application.

Hold flood water a minimum of 4 days after application.

RATE: Apply 3 pints per acre.

STORAGE AND DISPOSAL

PROHIBITIONS

Do not contaminate water, food or feed by storage or disposal. Open dumping is prohibited.

STORAGE

Keep container tightly closed when not in use. Do not store near seeds, fertilizers or foodstuffs. Can be stored at temperatures as low as minus 50 F.

PESTICIDE DISPOSAL

Wastes resulting from the use of this product may be disposed of on site or at an approved waste disposal facility.

CONTAINER DISPOSAL

Triple rinse (or equivalent). Then offer for recycling or reconditioning, or puncture and dispose of in a sanitary landfill, or by other procedures approved by State and local authorities.

NOTICE—READ CAREFULLY

CONDITIONS OF SALE
Stauffer (and seller) offer(s) this product for sale subject to, and buyer and all users are deemed to have accepted, the following conditions of sale and warranty which may only be varied by written agreement of a duly authorized representative of Stauffer.

WARRANTY LIMITATION
Stauffer warrants that this product conforms to the chemical description on the label and is reasonably fit for the purposes referred to in the directions for use on the label subject to the inherent risks referred to below. Stauffer makes no other express warranties; THERE IS NO IMPLIED WARRANTY OF MERCHANTABILITY and there are no warranties which extend beyond the description on the label hereof.

INHERENT RISKS
The directions for use of this product are believed to be reliable and should be followed carefully. However, it is impossible to eliminate all risks associated with use. Buyer assumes all risks associated with use or application of this product contrary to label instructions or resulting from extraordinary weather conditions.

LIMITATION OF LIABILITY
In no case shall Stauffer be liable for special, indirect, or consequential damages resulting from the use or handling of this product and no claim of any kind shall be greater in amount than the purchase price of the product in respect of which such damages are claimed.

This is a specimen label. Stauffer is not responsible for the accuracy of the information contained herein. As labels are subject to revision, always carefully read and follow the label on the product container.

ORDRAM® 10-G

SOUTHWEST
GRANULES—SELECTIVE HERBICIDE

For Use on Rice Only

ACTIVE INGREDIENT:

S-ethyl hexahydro-1*H*-azepine-1-carbothioate	10%
INERT INGREDIENTS	90%
TOTAL	100%

Contains 10 Pounds of Active Ingredient Per 100 Pounds

EPA Reg. No. 476–1932 A-2R

CAUTION

Keep Out of Reach of Children

HARMFUL IF SWALLOWED.

STATEMENT OF PRACTICAL TREATMENT
If a known exposure occurs or is suspected, **immediately** start the procedures given below. If further treatment is required, contact a Poison Center, a physician, or the nearest hospital, or call collect, day or night: (203) 226-6602 Stauffer Chemical Company, or (800) 424-9300 Pesticide Team Safety Network.

If Swallowed: Immediately give several glasses of water and induce vomiting by gagging the victim with a finger placed on the back of the victim's tongue. Give fluids until vomitus is clear. Do not induce vomiting or give anything by mouth to an unconscious or convulsing person. Call a physician immediately.

For Eye Contact: Hold eyelids apart and flush eyes with large amounts of running water for at least 15 minutes. Get medical attention.

For Skin Contact: Wash all affected areas with plenty of soap and water. Remove and clean contaminated clothing and shoes before reuse. Seek medical attention if irritation occurs.

If Inhaled: Remove to fresh air. Seek medical attention if respiratory irritation occurs or if breathing becomes difficult.

PRECAUTIONARY STATEMENTS

HAZARDS TO HUMANS AND DOMESTIC ANIMALS

CAUTION

HARMFUL IF SWALLOWED: Avoid contact with skin, eyes and clothing. Avoid inhalation of dust. Wear rubber gloves and fresh clothing. Wear goggles to avoid contact with eyes. Wash with soap and water immediately after use or contact. Do not contaminate food and feed.

ENVIRONMENTAL HAZARDS
Do not contaminate water used for domestic purposes. Do not contaminate water by cleaning of equipment or disposal of wastes. Do not apply directly to water except as stated on label.

USE PRECAUTIONS

READ ALL LABEL INSTRUCTIONS BEFORE USING. ORDRAM 10-G should be used only for recommended purposes and at recommended rates. ORDRAM 10-G may cause crop injury under extreme soil or climatic conditions or if directions are not followed. Do not apply any other herbicide in conjunction with ORDRAM 10-G as injury may occur.

DIRECTIONS FOR USE

It is a violation of Federal law to use this product in a manner inconsistent with its labeling.

DIRECTIONS AND RECOMMENDATIONS FOR PRE-PLANT, PRE-FLOOD USE ON RICE

(Water Seeded Rice only)
For use on RICE to control **Barnyardgrass** (watergrass) (*Echinochloa* spp.)

SOIL PREPARATION: Use only on soils in good tilth. Prepare the soil for seeding according to good agricultural practices. All weed growth must be thoroughly worked into the soil before treatment.

APPLICATION: When ORDRAM 10-G is applied before flooding and planting, application may be made by either air or ground equipment. Application equipment should be carefully calibrated before use and checked frequently during use to be sure application is uniform. Avoid overlaps that will increase ORDRAM dosage above recommended limits because crop injury may occur. Aerial application should be made only when the wind velocity is low, 0–10 mph.

SOIL INCORPORATION: All applications of ORDRAM 10-G must be incorporated into the soil within 6 hours by mechanical incorporation. Incorporation should be done by at least one discing and spike-tooth harrowing. Do not set disc to cut deeper than 4 inches.

RATE: Apply 30 pounds of ORDRAM 10-G per acre.

FLOOD THE FIELD FOR SEEDING AND HOLD THE WATER FOUR TO SIX DAYS FOLLOWING SEEDING.

DIRECTIONS AND RECOMMENDATIONS FOR POST-FLOOD, POST-EMERGENCE USE ON RICE

(Water Seeded or Drilled Rice):
For use on RICE to control **Barneyardgrass** (watergrass) (*Echinochloa* spp.), **Dayflower** (*Commelina* spp.), and for suppression of **Annual Sedges** (*Cyperus* spp.) and **Spikerush** (*Eleocharis* spp.)

STANDARD TREATMENT: Rice must be established at the time of application. Barnyardgrass (watergrass), Dayflower, Annual Sedges and Spikerush must not be more than 5 inches high. Barnyardgrass, Annual Sedges and Spikerush must be at least ⅔ submerged. Dayflower must be completely submerged. Floods should be maintained at these depths until the Dayflower or Watergrass dies or until the Annual Sedges and Spikerush are suppressed.

RATE: Apply 20–30 pounds of ORDRAM 10-G per acre depending on size and degree of infestation of Watergrass. Use 25–30 pounds per acre on Annual Sedges, Spikerush and heavy infestations of Dayflower.

IN EMERGENCY SITUATIONS: ORDRAM may be used for suppression (stopping normal development) of **Barnyardgrass** over 5 inches tall and under 12 inches tall. Flood water must be maintained at a 4 to 6 inch depth.

RATE: Apply at 30 pounds per acre.

STORAGE AND DISPOSAL

STORAGE
Do not contaminate water, food, feed, fertilizer or seeds by storage or disposal. Open dumping is prohibited. Keep container closed when not in use.

PESTICIDE DISPOSAL
Wastes resulting from the use of this product may be disposed of on site or at an approved waste disposal facility.

CONTAINER DISPOSAL
Completely empty bag into application equipment. Then dispose of empty bag in a sanitary landfill or by incineration, or, if allowed by State and local authorities, by burning. If burned, stay out of smoke.

NOTICE—READ CAREFULLY

CONDITIONS OF SALE
Stauffer (and seller) offer(s) this product for sale subject to, and buyer and all users are deemed to have accepted, the following conditions of sale and warranty which may only be varied by written agreement of a duly authorized representative of Stauffer.

WARRANTY LIMITATION
Stauffer warrants that this product conforms to the chemical description on the label and is reasonably fit for the purposes referred to in the directions for use on the label subject to the inherent risks referred to below. Stauffer makes no other express warranties; THERE IS NO IMPLIED WARRANTY OF MERCHANTABILITY and there are no warranties which extend beyond the description on the label hereof.

INHERENT RISKS
The directions for use of this product are believed to be reliable and should be followed carefully. However, it is impossible to eliminate all risks associated with use. Buyer assumes all risks associated with use or application of this product contrary to label instructions or resulting from extraordinary weather conditions.

LIMITATION OF LIABILITY
In no case shall Stauffer be liable for special, indirect, or consequential damages resulting from the use or handling of this product and no claim of any kind shall be greater in amount than the purchase price of the product in respect of which such damages are claimed.

This product is sold only for uses stated on its label. No express or implied license is granted to use or sell this product under any patent in any country except as specified: Country: United States of America Patent No: 3,573,031.

This is a specimen label. Stauffer is not responsible for the accuracy of the information contained herein. As labels are subject to revision, always carefully read and follow the label on the product container.

ORDRAM® 10-G

CALIFORNIA GRANULES SELECTIVE HERBICIDE

For Use on Rice Only

ACTIVE INGREDIENT:	
S-ethyl hexahydro-1*H*-azepine-1-carbothioate	10.0%
INERT INGREDIENTS:	90.0%
TOTAL	100.0%

Contains 10 Pounds of Active Ingredient Per 100 Pounds.

EPA Reg. No. 476-1932 A-3R

CAUTION

Keep Out of Reach of Children

HARMFUL IF SWALLOWED

STATEMENT OF PRACTICAL TREATMENT
If a known exposure occurs or is suspected, **immediately** start the procedures given below. If further treatment is required, contact a Poison Center, a physician, or the nearest hospital, or call collect, day or night: (203) 226-6602 Stauffer Chemical Company, or (800) 424-9300 Pesticide Team Safety Network.

If Swallowed: Immediately give several glasses of water and induce vomiting by gagging the victim with a finger placed on the back of the victim's tongue. Give fluids until vomitus is clear. Do not induce vomiting or give anything by mouth to an unconscious or convulsing person. Call a physician immediately.

For Eye Contact: Hold eyelids apart and flush eyes with large amounts of running water for at least 15 minutes. Get medical attention.

For Skin Contact: Wash all affected areas with plenty of soap and water. Remove and clean contaminated clothing and shoes before reuse. Seek medical attention if irritation occurs.

If Inhaled: Remove to fresh air. Seek medical attention if respiratory irritation occurs or if breathing becomes difficult.

PRECAUTIONARY STATEMENTS

HAZARDS TO HUMANS AND DOMESTIC ANIMALS

CAUTION

Harmful If Swallowed. Avoid contact with skin, eyes and clothing. Avoid inhalation of dust. Wear rubber gloves and fresh clothing. Wear goggles to avoid contact with eyes. Wash with soap and water immediately after use or contact. Do not contaminate food and feed.

ENVIRONMENTAL HAZARDS
Do not contaminate water used for domestic purposes. Do not contaminate water by cleaning of equipment or disposal of wastes. Do not apply directly to water except as stated on label.

USE PRECAUTIONS

READ ALL LABEL INSTRUCTIONS BEFORE USING. ORDRAM 10-G should be used only for recommended purposes and at recommended rates. ORDRAM 10-G may cause crop injury under extreme soil or climatic conditions or if directions are not followed. Do not apply any other herbicide in conjunction with ORDRAM 10-G as injury may occur.
Do not mix with fertilizer.

DIRECTIONS FOR USE

It is a violation of Federal law to use this product in a manner inconsistent with its labeling.

DIRECTIONS AND RECOMMENDATIONS FOR PRE-PLANT, PRE-FLOOD USE ON RICE (Water Seeded Rice only):
For use on Rice to control Baryardgrass (watergrass) (*Echinochloa* spp.)

SOIL PREPARATION: Use only on soils in good tilth. Prepare the soil for seeding according to good agricultural practices. All weed growth must be thoroughly worked into the soil before treatment.

APPLICATION: When ORDRAM 10-G is applied before flooding and planting, application may be made by either air or ground equipment. Application equipment should be carefully calibrated before use and checked frequently during use to be sure application is uniform. Avoid overlaps that will increase ORDRAM 10-G dosage above recommended limits because crop injury may occur. Aerial application should be made only when the wind velocity is low, 0–10 mph.

SOIL INCORPORATION: All applications of ORDRAM 10-G must be incorporated into the soil to a depth of 2 to 3 inches within 6 hours by mechanical incorporation. Incorporation should be done by at least two discings or heavy spike-tooth harrowings. Do not set disc to cut deeper than 4 inches.

RATE: Apply 30 pounds of ORDRAM 10-G per acre.

FLOOD THE FIELD FOR SEEDING AND HOLD THE WATER FOUR TO SIX DAYS FOLLOWING SEEDING.

DIRECTIONS AND RECOMMENDATIONS FOR POST-FLOOD, POST-EMERGENCE USE ON RICE (Water Seeded or Drilled Rice):

For use on Rice to control Barnyardgrass (watergrass) (*Echinochloa* spp.). Rice should be in the seedling stage at the time of application.

FOR AIR APPLICATION ONLY: Barnyardgrass must be a minimum of 2″ high but not higher than 5″, and it must be at least 2/3 submerged at time of application. This water level (not less than 2″) must be maintained until the Barnyardgrass dies.

Equipment should be carefully calibrated before use and checked frequently during use to be sure application is uniform. Avoid overlaps that will increase ORDRAM 10-G dosage above recommended limits because crop injury may occur. Aerial application should be made only when the wind velocity is low, 0–10 mph. To avoid uneven distribution of ORDRAM 10-G, the water flow through the field should be at a minimum at the time of application and for at least 2 hours after application.

HOLD FLOOD WATER A MINIMUM OF 4 DAYS AFTER APPLICATION.

RATE: Apply 30 pounds of ORDRAM 10-G per acre.

STORAGE AND DISPOSAL

STORAGE

Do not contaminate water, food, feed, fertilizer or seeds by storage or disposal. Open dumping is prohibited. Keep container closed when not in use.

PESTICIDE DISPOSAL

Wastes resulting from the use of this product may be disposed of on site or at an approved waste disposal facility.

CONTAINER DISPOSAL

Completely empty bag into application equipment. Then dispose of empty bag in a sanitary landfill or by incineration, or, if allowed by State and local authorities, by burning. If burned, stay out of smoke.

NOTICE—READ CAREFULLY

CONDITIONS OF SALE

Stauffer (and seller) offer(s) this product for sale subject to, and buyer and all users are deemed to have accepted, the following conditions of sale and warranty which may only be varied by written agreement of a duly authorized representative of Stauffer.

WARRANTY LIMITATION

Stauffer warrants that this produce conforms to the chemical description on the label and is reasonably fit for the purposes referred to in the directions for use on the label subject to the inherent risks referred to below. Stauffer makes no other express warranties; THERE IS NO IMPLIED WARRANTY OF MERCHANTABILITY and there are no warranties which extend beyond the description on the label hereof.

INHERENT RISKS

The directions for use of this product are believed to be reliable and should be followed carefully. However, it is impossible to eliminate all risks associated with use. Buyer assumes all risks associated with use or application of this product contrary to label instructions or resulting from extraordinary weather conditions.

LIMITATION OF LIABILITY

In no case shall Stauffer be liable for special, indirect, or consequential damages resulting from the use or handling of this product and no claim of any kind shall be greater in amount than the purchase price of the product in respect of which such damages are claimed.

This product is sold only for uses stated on its label. No express or implied license is granted to use or sell this product under any patent in any country except as specified: Country: United States of America Patent No. 3,573,031.

This is a specimen label. Stauffer is not responsible for the accuracy of the information contained herein. As labels are subject to revision, always carefully read and follow the label on the product container.

PREFAR® 4-E

EMULSIFIABLE LIQUID SELECTIVE HERBICIDE

For Control of Annual Grasses and Broadleaf Weeds

ACTIVE INGREDIENT:	
Bensulide *S*-(*O,O*-diisopropyl phosphorodithioate) ester of *N*-(2-mercaptoethyl)benzenesulfonamide	46.0%
INERT INGREDIENTS:	54.0%
TOTAL	100.0%

Contains 4 Pounds Active Ingredient Per Gallon.

EPA Reg. No. 476-2004 A-2

CAUTION

Keep Out of Reach of Children

STATEMENT OF PRACTICAL TREATMENT

If a known exposure occurs or is suspected, immediately start the procedures given below and simultaneously contact a Poison Center, a physician, or the nearest hospital, or call collect, day or night, (203) 226-6602 Stauffer Chemical Company or (800) 424-9300 Pesticide Team Safety Network.

NOTE: Be sure to advise the person contacted that the active ingredient in this compound is a cholinesterase inhibitor. Describe the type of exposure, the victim's symptoms, and follow the advice given.

Note to Medical Personnel: Should cholinesterase inhibition occur atropine by injection is antidotal. 2-PAM (Protopam Chloride) is also antidotal when administered early and in conjunction with atropine.

If swallowed, immediately give several glasses of water but *do not* induce vomiting. If vomiting does occur, give fluids again. Have a physician determine if condition of patient will permit induction of vomiting or evacuation of stomach. Do not give anything by mouth to an unconscious or convulsing person.

For skin contact, flush all affected areas with plenty of water for several minutes. Remove and clean contaminated clothing and shoes. Seek medical attention if irritation occurs.

For eye contact, flush eyes with large amounts of running water for at least 15 minutes. Hold eyelids apart to ensure rinsing of the entire surface of the eye and lids with water. Get medical attention immediately.

If inhaled, remove to fresh air. Seek medical attention if respiratory irritation occurs or if breathing becomes difficult.

READ ALL LABEL DIRECTIONS BEFORE USING

PRECAUTIONARY STATEMENTS

HAZARDS TO HUMANS AND DOMESTIC ANIMALS

CAUTION

Harmful if swallowed or absorbed through the skin. Avoid breathing spray mist. Causes moderate eye irritation. Do not get in eyes, on skin, or on clothing. Wash thoroughly after handling and before eating or smoking. Wash all contaminated clothing with soap and hot water before re-use.

ENVIRONMENTAL HAZARDS

This product is toxic to fish. Do not apply directly to any body of water. Do not contaminate water by cleaning of equipment or disposal of wastes. Do not apply when weather conditions favor drift from area treated.

PHYSICAL HAZARDS

Combustible. Do not use, store, pour, or spill near heat or open flame. Keep container closed when not in use.

GENERAL USE PRECAUTIONS

READ ALL DIRECTIONS BEFORE USING. MIX WELL BEFORE DILUTING WITH WATER. PREFAR 4-E should be used only for recommended purposes and at recommended rates. (DO NOT OVERDOSE.) Do not apply in combination with fluid fertilizers. Use on mineral soils only. Do not contaminate irrigation water.

All crops specified on the label may be planted in rotation but do not apply PREFAR 4-E more often than once every 12 months. To avoid injury to crops not specified on the label do not plant until 18 months after the last PREFAR 4-E application. Soybeans may be planted 12 months after PREFAR 4-E application.

WEEDS CONTROLLED

Annual Grasses

Crabgrass	*Digitaria* spp.
Foxtail	*Setaria* spp.
Jungle Rice	*Echinochloa colonum*
Watergrass	*Echinochloa crusgali*
Goosegrass	*Eleusine Indica*
Fall Panicum	*Panicum dichotomiflorum*
Sprangletop	*Leptochloa* spp.

Annual Broadleaf Weeds (Arizona, New Mexico, Texas and ten southernmost California counties only)

Lambsquarters	*Chenopodium album*
Redroot Pigweed	*Amaranthus retroflexus*
Purslane	*Portulaca oleracea*

DIRECTIONS FOR USE

It is a violation of Federal law to use this product in a manner inconsistent with its labeling.

PREFAR 4-E is a selective herbicide which is applied pre-plant and mixed (incorporated) into the soil. On certain crops which are irrigated up, PREFAR 4-E may be applied to the soil surface after planting. PREFAR 4-E controls weeds as their seeds germinate by interfering with normal germination and seedling development.

SOIL PREPARATION: Prepare the soil according to good agricultural practice. PREFAR 4-E does not control emerged weeds and all weed growth and crop stubble must be thoroughly worked into the soil prior to application.

APPLICATION AND SOIL INCORPORATION: Spray equipment should be carefully calibrated before use and checked frequently during application to be sure nozzles are free from clogging and are delivering a uniform spray pattern. Apply uniformly the recommended rate of PREFAR 4-E in 10 to 50 gallons of water per acre.

Pre-plant Application: Apply to well-worked soil that is dry enough to permit thorough incorporation. Incorporate to a depth of 1 to 2 inches before planting. Use power driven rotary cultivators or other equipment which will result in satisfactory incorporation.

Pre-emergence Application: (For use only on crops that will be irrigated up). After planting apply PREFAR 4-E as a surface application. Follow immediately with irrigation (sprinkler or furrow). For sprinkler irrigation apply 1 inch of water, wetting the soil at least 2 to 4 inches deep. For furrow irrigation thoroughly saturate the entire bed top (completely black out bed top).

RECOMMENDATIONS

All rates are given on a broadcast basis. Reduce rate proportionately for band treatment.

COTTON

Pre-plant: Apply 5 to 6 quarts PREFAR 4-E per broadcast acre. Apply to the flat soil surface before furrowing (listing) for the pre-planting irrigation and then continue with normal practices.

OR

Apply after furrowing (listing) but before the pre-planting irrigation. Incorporate with a sectional, rolling cultivator.

OR

Apply after the pre-planting irrigation but before the final seed bed preparation.

Pre-plant: (Arizona, New Mexico & Texas only) For control of watergrass, jungle rice, foxtail and crabgrass only, apply 2 quarts PREFAR 4-E per broadcast acre. Use only on light, sandy soils.

Apply to the flat soil surface and thoroughly disc the PREFAR 4-E into the soil before furrowing (listing) for the pre-planting irrigation and then continue with normal practices.

OR

Apply after furrowing (listing). Incorporate with a sectioned, rolling cultivator. Follow with pre-plant irrigation.

Pre-emergence: (Use only on cotton that will be irrigated up). Apply 5 to 6 quarts PREFAR 4-E per broadcast acre as a surface application.

CUCURBITS (cucumbers, summer squash, winter squash, cantaloupes, crenshaw melons, muskmelons, persian melons, pumpkins, watermelons).

Apply, pre-plant, 5 to 6 quarts PREFAR 4-E per broadcast acre to flat-planted or bedded crops. Follow directions for APPLICATION AND SOIL INCORPORATION given on this label.

COLE CROPS (broccoli, brussels sprouts, cabbage, cauliflower).

Apply, pre-plant or pre-emergence, 5 to 6 quarts PREFAR 4-E per broadcast acre. Follow directions for APPLICATION AND SOIL INCORPORATION given on this label. Use pre-emergence only on crops to be irrigated up.

(NOTE: Limit use on cabbage to Southern and California/Arizona growing regions.)

FRUITING VEGETABLES (tomatoes [direct seeded and transplants], seeded bell peppers, chili peppers).

Apply, pre-plant or pre-emergence, 5 to 6 quarts (4 to 5 quarts for tomatoes) PREFAR 4-E per broadcast acre. Follow directions for APPLICATIONS AND SOIL INCORPORATION given on this label. Use pre-emergence only on crops to be irrigated up.

NOTE: Limit use on bell peppers to the Southern and California/Arizona growing regions; on chili peppers to the Southwestern U.S.)

LETTUCE (head or leaf)

Apply, pre-plant or pre-emergence, 5 to 6 quarts PREFAR 4-E per broadcast acre. Follow directions for APPLICATION AND SOIL INCORPORATION given on this label. Use pre-emergence only on lettuce to be irrigated up.

ROOT CROP VEGETABLES

NOTE: Limit use on carrots to Texas; on bulb onions to Idaho, New Mexico, Oregon, and Texas. Use on onions in Idaho and Oregon is restricted to fall application followed by spring planting).

CARROTS (Texas Only): Apply pre-plant or pre-emergence, 5 to 6 quarts PREFAR 4-E per broadcast acre. Follow directions for APPLICATION AND SOIL INCORPORATION given on this label. Use pre-emergence only on carrots to be irrigated up. Do not feed treated carrots to livestock.

BULB ONIONS (New Mexico and Texas): Apply pre-plant or pre-emergence, 5 to 6 quarts PREFAR 4-E per broadcast acre. Follow directions for APPLICATION AND SOIL INCORPORATION given on this label. Use pre-emergence only on onions to be irrigated up.

BULB ONIONS (Idaho and Oregon): For control of watergrass, redroot pigweed and common lambsquarters, prepare a good, flat seed bed by standard cultural practices. Apply in the fall 3 quarts of PREFAR 4-E per acre in 10 to 50 gallons of water in a 10 to 12 inch band and bed-up, throwing soil from the furrows over the sprayed band of PREFAR 4-E. Allow to remain undistributed until spring. Prior to planting in spring, drag-off the bed tops with a spike-tooth harrow being careful not to drag off soil below the level of PREFAR 4-E application. Plant in the center of the beds.

GRASS SEED CROPS (bentgrass, bluegrass)

For *Poa annua* control in bentgrass and bluegrass seed fields: Apply 2 to 2½ gallons PREFAR 4-E per acre to the soil surface after straw removal but before weed seed germination. If field has been burned, ashes should be completely dispersed before application of PREFAR 4-E.

Do not feed straw to livestock. Do not graze livestock in treated fields.

(NOTE: Limit use to Oregon, Washington and Idaho.)

PREFAR 4-E/ALANAP* TANK MIX DIRECTIONS

* ALANAP is a registered trademark of Uniroyal Chemical

A tank mix combination of PREFAR 4-E and ALANAP can be mixed (incorporated) into the soil before planting for weed control in cantaloupes, cucumbers, muskmelons and watermelons. This combination will control a broader spectrum of weeds than will either material used alone.

IMPORTANT—Follow all general directions for use and use precautions that appear on the ALANAP label and elsewhere on this label.

Weeds controlled by PREFAR 4-E/ALANAP combinations:

Annual Grasses

Watergrass (Barnyardgrass)	Foxtails
Crabgrass	Goosegrass
Fall Panicum	Sprangletop

Annual Broadleaf Weeds

Bindweed	Lambsquarters
Carpetweed	Mustards
Chickweed	Pigweed
Cocklebur	Purslane
Galinsoga	Ragweed
Groundcherry	Shepherdspurse
	Velvetleaf

DIRECTIONS FOR USE

Soil Preparation: Prepare the soil according to good agricultural practice. All weed growth and crop stubble must be thoroughly worked into the soil.

Tank Mixing Directions: Combine PREFAR 4-E and ALANAP in the spray tank in the recommended dosage proportions using 10 to 40 gallons of water per acre of area to be treated.

Spray equipment should be carefully calibrated before use and checked frequently during application to be sure nozzles are free from clogging and are delivering a uniform spray pattern.

Pre-plant Application and Soil Incorporation: Apply 4 to 6 quarts per acre of PREFAR 4-E plus 4 to 8 quarts of ALANAP in sufficient water (10 to 40 gallons) to provide thorough coverage. Incorporate lightly (½ to 1 inch) into the soil using suitable equipment just prior to planting. Use lower rates on light (sandy) soils and higher rates on heavier (clay) soils. Reduce the rate proportionately for band applications. Incorporation is not necessary if the application is followed by immediate irrigation.

STORAGE AND DISPOSAL

Do not contaminate water, food, or feed by storage or disposal.

STORAGE

Protect from temperatures below 42°F. At temperatures below 42 °F, product may crystalize. Dissolve crystals with quiet storage at warm temperatures (75 °F). Redissolution rate: 40 °-48 °F (40-60 days) or 60 °F (1-3 weeks); 70 °-80 °F (1-3 days). Roll container and agitate contents to provide a uniform product prior to use.

PESTICIDE DISPOSAL

Wastes resulting from the use of this product may be disposed of on site or at an approved waste disposal facility.

CONTAINER DISPOSAL

Triple rinse (or equivalent). Then offer for recycling or reconditioning, or puncture and dispose of in a sanitary landfill, or by other approved State and local procedures.

SPILL OR LEAK: Small spills can be handled routinely. Use adequate ventilation and wear an air-supplied respirator to prevent inhalation. Wear suitable protective clothing to prevent skin and eye contact. Use the following procedures:

1. Spread a suitable absorbent such as clay, sawdust or kitty litter on the spilled material.
2. Place sweepings in an appropriate chemical waste container.
3. Generously cover the contaminated areas with a common household detergent. Using a stiff brush and small amounts of water, work the detergent into the spill material forming a slurry. Do not splatter on one's self or bystanders and completely avoid skin or eye contact with this material. Brush the slurry into cracks and crevices and allow to stand for 2-3 minutes.
4. Spread absorbents on the slurried liquid and shovel the absorbed material into the container.
5. Repeat if necessary.
6. Flush the area with water while observing proper environmental considerations.
7. Seal container and dispose of in approved pesticide landfill.

Large spills should be handled according to a predetermined plan. For assistance in developing a plan, contact the Agricultural Chemical Division, Stauffer Chemical Company, Westport, CT 06881-0850.

NOTICE—READ CAREFULLY

CONDITIONS OF SALE

Stauffer (and seller) offer(s) this product for sale subject to, and buyer and all users are deemed to have accepted, the following conditions of sale and warranty which may only be varied by written agreement of a duly authorized repesentative of Stauffer.

WARRANTY LIMITATIONS

Stauffer warrants that this product conforms to the chemical description on the label and is reasonably fit for the purposes referred to in the directions for use on the label subject to the inherent risks referred to below. Stauffer makes no other express warranties: THERE IS NO IMPLIED WARRANTY OF MERCHANTABILITY and there are no warranties which extend beyond the description on the label hereof.

INHERENT RISKS

The directions for use of this product are believed to be reliable and should be followed carefully. However, it is impossible to eliminate all risks associated with use. Buyer assumes all risks associated with use or application of this product contrary to label instructions or resulting from extraordinary weather conditions.

LIMITATION OF LIABILITY

In no case shall Stauffer be liable for special, indirect or consequential damages resulting from the use or handling of this product and no claim of any kind shall be greater in amount than the purchase price of the product in respect of which such damages are claimed.

This product is sold only for uses stated on its label. No express or implied license is granted to use or sell this product under any patent in any country except as specified. Country: United States of America.

PRO-DRONE®

INSECT GROWTH REGULATOR

Insect Growth Regulator for Control of the Red Imported Fire Ant

ACTIVE INGREDIENT:
1-(8-methoxy-4,8-dimethylnonyl)-4-(1-methylethyl) benzene 1.2%
INERT INGREDIENTS: 98.8%

TOTAL 100.0%
Contains 1.2 lbs. active ingredient per 100 pounds

EPA Reg. No. 476-2211 A-3MR

STATEMENT OF PRACTICAL TREATMENT

If a known exposure occurs or is suspected, immediately initiate the recommended procedures below. If further treatment is required, contact a Poison Center, a physician, or the nearest hospital.

If swallowed: Immediately give large quantities of water and induce vomiting by gagging the victim with a finger or blunt object placed on the back of the victim's tongue. Give fluids until vomitus is clear. If victim is unconscious, do not induce vomiting or give anything by mouth.

For skin contact: Flush all affected areas with plenty of water for 15 minutes. Remove contaminated clothing and shoes. Wash clothing before reuse. Seek medical attention if skin irritation occurs.

For eye contact: Flush eyes with plenty of water for several minutes. Seek medical attention if eye irritation occurs.

If inhaled: Remove to fresh air. Seek medical attention if respiratory irritation occurs or if breathing becomes difficult.

PRECAUTIONARY STATEMENTS

HAZARDS TO HUMANS AND DOMESTIC ANIMALS

CAUTION

Harmful if swallowed or absorbed through the skin. Avoid contact with skin, eyes or clothing. Wash with soap and water after use.

For further information, call collect, day or night: Stauffer Chemical Company (203) 226-6602 or Pesticide Safety Team Network, (800) 424-9300.

PRODUCT INFORMATION

PRO-DRONE® is an insect growth regulator which acts slowly to reduce fire ant colonies. The bait is passed throughout the ant colony and causes a disappearance of worker brood within 12-16 weeks. This disappearance of workers leads to a decline in food gathering, brood tending, and eventual death of the colonies.

DIRECTIONS FOR USE

It is a violation of Federal law to use this product in a manner inconsistent with its labeling.

Apply broadcast to pasture and range grass, lawns, turf, and non-agricultural land for control of the red imported fire ant, *Solenopsis invicta*. Make two applications per year: (1) spring and fall, or (2) fall and spring. Application should be made when the soil temperature is above 60 °F and when ants are actively foraging. Do not apply until dew has evaporated. Do not apply if heavy rainfall is expected within 4-6 hours.

Large Area Application

Use at 0.88 lbs (400 grams) per acre per application. Apply by air when wind is not more than 12 mph. Apply by ground equipment when wind is not more than 15 mph. Apply with conventional agricultural aircraft properly calibrated for small quantities of granular pesticides or with conventional ground granular applicator properly calibrated for small quantities.

Small Area Application

Apply at the rate of 3 teaspoons per 250 square feet. Do not disturb the fire ant mounds. The contents of this container (6 ozs net) is sufficient to treat $^2/_5$ acre (18,500 square feet). Note: Do not contaminate kitchen utensils by storage or use.

STORAGE AND DISPOSAL

Do not contaminate water, food or feed by storage or disposal.

CONTAINER FOR LARGE AREA APPLICATION

Storage: Store in a cool, dry place and keep container tightly closed.

Pesticide Disposal: Wastes resulting from the use of this product may be disposed of on site or at an approved waste disposal facility.

Container Disposal: Completely empty bag into application equipment. Then dispose of empty bag in a sanitary landfill or by incineration, or if allowed by State and local authorities, by burning. If burned, stay out of smoke.

CONTAINER FOR SMALL AREA APPLICATION

Storage: Store in a cool, dry place and keep container tightly closed.

Disposal: Do not reuse container. Securely wrap original container in several layers of newspaper and discard in trash.

NOTICE—READ CAREFULLY

CONDITIONS OF SALE

Stauffer (and seller) offer(s) this product for sale subject to, and buyer and all users are deemed to have accepted, the following conditions of sale and warranty which may only be varied by written agreement of a duly authorized representative of Stauffer.

WARRANTY LIMITATION

Stauffer warrants that this product conforms to the chemical description on the label and is reasonably fit for the purposes referred to in the directions for use on the label subject to the inherent risks referred to below. Stauffer makes no other express warranties: THERE IS NO IMPLIED WARRANTY OF MERCHANTABILITY and there are no warranties which extend beyond the description on the label hereof.

INHERENT RISKS

The directions for use of this product are believed to be reliable and should be followed carefully. However, it is impossible to eliminate all risks associated with use. Buyer assumes all risks associated with use or application of this product contrary to label instructions or resulting from extraordinary weather conditions.

LIMITATION OF LIABILITY

In no case shall Stauffer be liable for special, indirect or consequential damages resulting from the use or handling of this product and no claim of any kind shall be greater in amount than the purchase price of the product in respect of which such damages are claimed.

This product is sold only for uses stated on its label. No express or implied license is granted to use or sell this product under any patent in any country except as specified.
Country: United States of America.

This is a specimen label. Stauffer is not responsible for the accuracy of the information contained herein. As labels are subject to revision, always carefully read and follow the label on the product container.

REWARD™

SELECTIVE HERBICIDE

Liquid Herbicide for Weed Control in Soybeans

ACTIVE INGREDIENT:

S-Propyl dipropylthiocarbamate	74.1%
INERT INGREDIENTS:	25.9%
TOTAL	100.0%

6-E Contains 6.0 pounds of active ingredient per gallon

EPA Reg. No. 476-2219

 A-1R

CAUTION

Keep Out of Reach of Children

HARMFUL IF SWALLOWED

STATEMENT OF PRACTICAL TREATMENT

If a known exposure occurs or is suspected, immediately initiate the recommended procedures below. If further treatment is required, contact a poison center, a physician, or the nearest hospital.

If swallowed: Immediately give several glasses of water but *do not* induce vomiting. If vomiting does occur, give fluids again. Have a physician determine if condition of patient will permit evacuation of stomach. Do not give anything by mouth to an unconscious person.

For eye contact: Hold eyelids apart and flush for at least 15 minutes. Get medical attention immediately.

For skin contact: Immediately flush all affected areas with plenty of water for at least 15 minutes. Remove contaminated clothing and shoes. Seek medical attention if skin irritation occurs. Wash clothing before reuse.

If inhaled: Remove to fresh air. Seek medical attention if respiratory irritation occurs or if breathing becomes difficult.

PRECAUTIONARY STATEMENTS

HAZARDS TO HUMANS AND DOMESTIC ANIMALS

CAUTION

Harmful if swallowed. Avoid contact with skin, eyes and clothing. Avoid inhalation of spray mist. Wear rubber gloves, goggles and protective clothing. Wash with soap and water immediately after use or contact. Do not contaminate food and feed. Keep away from heat and flame.

CHEMICAL AND PHYSICAL HAZARDS

Keep away from heat and flame.

ENVIRONMENTAL HAZARDS

Do not apply directly to water. Do not contaminate water by cleaning of equipment or disposal of wastes.

In case of emergency call (day or night) Chemtrec 800–424-9300 or Stauffer Chemical Company 203–226-6602.

WEEDS CONTROLLED BY REWARD

REWARD will not control established weeds:

Annual Grasses:

Barnyardgrass, Watergrass	*Echinochloa crus-galli*
Crabgrass	*Digitaria* spp.
Fall Panicum	*Panicum dichotomiflorum*
Foxtail millet	*Setaria italica*
Giant foxtail	*Setaria faberi*
Goosegrass	*Eleusine indica*
Green foxtail	*Setaria viridis*
Johnsongrass seedling	*Sorghum halepense*
Wild cane*	*Sorghum bicolor*
Yellow foxtail	*Setaria glauca*

* Cultivation is suggested in addition to the REWARD treatment in fields with moderate or heavy infestations of wild cane.

Annual Broadleaf Weeds:

Annual morningglory	*Ipomoea purpurea*
Balloonvine*	*Cardiospermun halicacabum*
Carpetweed	*Mollugo verticillata*
Common lambsquarters	*Chenopodium album*
Florida purslane—Florida pusley	*Richardia scabra*
Purslane	*Portulaca oleracea*
Redroot pigweed	*Amaranthus retroflexus*
Sicklepod—coffeeweed	*Cassia obtusifolia*
Smooth pigweed	*Amaranthus hybridus*
Velvetleaf—buttonweed	*Abutilon theophrasti*

* In Alabama only

The above listed broadleaf weeds will be controlled only if treatment is made when conditions are favorable for weed seed germination and growth.

Perennial Grasses:

Purple nutsedge—nutgrass	*Cyperus rotundus*
Yellow nutsedge—nutgrass	*Cyperus esculentus*

Existing stands of nutgrass must be turned under and chopped up thoroughly before treatment. For nutgrass control REWARD must be thoroughly mixed into the soil prior to planting or applied subsurface to the soil with injectors or sweeps. Cultivation is suggested in addition to REWARD treatment in fields with moderate to heavy nutgrass infestations.

GENERAL USE PRECAUTIONS

Read entire label before using.

REWARD should be used only for recommended purposes and at recommended rates (do not overdose).

Failure to follow label instructions, or use on crops not specifically listed on the label may result in crop injury or destruction.

Applied according to directions and under normal growing conditions, REWARD will not harm the treated crop. Malformation of the primary leaves is not unusual. This condition is only temporary and will not affect yields. During germination and early stages of growth, extended periods of unusually cold and wet or hot and dry weather, insect or plant disease attack, the use of certain soil-applied systemic insecticides, improperly placed fertilizers or soil insecticides may create abnormal conditions that weaken crop seedlings. REWARD used under these abnormal conditions could result in crop injury.

REWARD is recommended for use on mineral soils only.

Do not tank mix REWARD with insecticides or fungicides.

Do not contaminate irrigation water or water used for domestic purposes. In irrigated areas, do not apply REWARD before pre-irrigation.

DIRECTIONS FOR USE

IT IS A VIOLATION OF FEDERAL LAW TO USE THIS PRODUCT IN A MANNER INCONSISTENT WITH ITS LABELING.

REWARD is a selective herbicide which is mixed (incorporated) into the soil or applied below the surface of the soil for control of weeds listed on this label. REWARD controls annual weeds as their seeds germinate be interfering with normal germination and seedling development. It does not control established weeds. All weed growth and crop stubble should be thoroughly worked into the soil before treatment.

APPLICATION DIRECTIONS

Pour the recommended amount of REWARD into the spray tank during the filling operation. Apply in 10 to 50 gallons of water per acre using a properly calibrated, low-pressure (20 to 40 PSI) sprayer having good agitation. The soil should be well-worked and dry enough to permit good soil mixing (incorporation) or subsurface application. Soil should be as free as possible of trash for subsurface application with sweeps or injectors. Avoid overlaps that will increase dosage above recommended limits.

FLUID FERTILIZER STATEMENT

REWARD may be combined with fluid (solution, slurry, or suspension) fertilizers. However, physical compatibility with these fluid fertilizers must be determined before combining in the spray tank. See Appendix I for directions regarding these combinations.

DRY FERTILIZER STATEMENT

REWARD may be impregnated on many dry bulk fertilizers for use on soybeans. However, uniform distribution of REWARD on fertilizer particles and uniform application are necessary to assure good results. See Appendix II for information and directions for impregnation and use.

INCORPORATION DIRECTIONS

REWARD must be incorporated into the soil immediately (within minutes) to prevent loss of the herbicide. Whenever possible, application and incorporation should be done in the same operation.

SOIL MIXING (INCORPORATION) BEFORE PLANTING: Use power-driven cultivation equipment set to cut to a depth of 2 to 3 inches or tandem discs set to cut to a depth of 4 to 6 inches, operated at 4 to 6 mph, followed by a spike-tooth harrow or some other leveling device which extends beyond ends of the disc. For thorough mixing, disc in two different directions (cross disc). Prior to the second discing, raise the disc to prevent cutting deeper than 4 to 6 inches.

NOTE: In the southern lower Delta of the Mississippi Valley, disc incorporation is not recommended on the heavy (gumbo) clay soils.

SOIL MIXING (INCORPORATION) AFTER PLANTING: Use power-driven cultivation equipment set to cut to a depth of 2 inches, or on coarse textured (light) soils only, ground-driven tillers (rolling cultivators, rotary hoe, row wheels, etc.) set to cut to a depth of 1$^1/_2$ inches and operated at 6 to 8 mph. When incorporating after planting, care must be taken not to disturb the crop seed or seedlings by adjusting tines or blades of incorporation equipment so they are not directly over the drill row.

SUBSURFACE APPLICATION BEFORE OR AT PLANTING: Special equipment designed for subsurface application MUST be used. Injector units must be rigidly mounted on or IMMEDIATELY ahead or IMMEDIATELY behind the planter unit. Sweeps must be mounted IMMEDIATELY ahead of the planter unit.

(1) **Soil Injection**—Injector shanks must be spaced 2$^1/_2$ to 3 inches apart and mounted in staggered positions to avoid trash buildup. Set shanks to inject REWARD 1$^1/_2$ to 3 inches below the soil surface. The width of the band in which weed control is desired will determine the number and spacing of injector shanks per row. (Example: Four injector shanks spaced 3 inches apart gives a 12-inch band). A broadcast application can be made by increasing the number of shanks. The two shanks adjacent to the drill row must be 1$^1/_4$ to 1$^1/_2$ inches on either side of it.

(2) **Covered Sweeps**—Set the sweeps to run 3 to 4 inches below the soil surface. Calibrate by measuring the spray band width at the back edge of the sweep, not the sweep width. For BROADCAST APPLICATIONS, stagger sweeps on double tool bar so that they overlap sufficiently to allow spray bands to meet.

NOTE: When applying with sweeps, REWARD must be applied deep enough to allow 3 to 4 inches of soil to remain over the treatment after the planting operation.

PLANTING

Do not plant seed deeper than 2 inches. For REWARD application made before planting, seeding can be done immediately, or it may be delayed up to 10 days. If it is necessary to work the soil following application and incorporation, but before planting do not work the soil deeper than the depth the herbicide was incorporated. To avoid removing REWARD from the row (with loss of weed control in the row) do not move or shape soil after treatment, such as by planting in deep furrows.

SOYBEAN RECOMMENDATIONS

These recommendations are given in terms of broadcast (overall) rates of REWARD per acre. For band or row application, reduce broadcast rates proportionately, depending upon band width to be treated and row spacing.

The following recommendations are for soybeans grown in all areas except in the Southeastern States:

REWARD may be applied before planting, at planting, or immediately after planting. When applying at or after planting, care must be taken not to disturb the crop seed or seedlings. Use 2$^2/_3$ pints on coarse textured (sandy-light) soils, 3$^1/_3$ pints on medium to fine textured (sandy-loam to clay loam-medium to heavy) soils, 4 pints on very fine textured (clay-very heavy) soils.

For soybeans grown in the Southeastern States including North Carolina, South Carolina, Georgia, Alabama, Florida and Eastern Tennessee (Tenn. River East):

REWARD may be applied before planting, at planting, or immediately after planting. For incorporated applications made before planting, do not plant in deep furrows, or use shanks or other device ahead of planter which bring up untreated soil to the surface. For furrow planted soybeans, apply REWARD at or immediately after planting in bands over the row and incorporate with row wheels only. When applying at or after planting, care must be taken not to disturb the crop seed or seedlings. Use 2$^2/_3$ pints on coarse textured (sandy-light) soils; 3$^1/_3$ pints on medium to fine textured (sandy loam to clay loam—medium to heavy) soils.

NOTE: In areas where beds are made prior to planting soybeans, REWARD may be applied in a band immediately ahead of the bedding discs, set to cover the REWARD with 3 to 4 inches of soil. If bed levelers are used before planting soybeans, make the beds higher to compensate for leveling, so 2 to 3 inches of soil remain over the treatment after the planting operation is done.

SEQUENTIAL APPLICATIONS

IMPORTANT: Always consult exact label of each herbicide used for specific recommendations and limitations.

For control of weeds not listed on the REWARD label, and for increased control of labeled broadleaf weeds, the following herbicides are suggested as overlay treatments after preplant applications of REWARD:

AMIBEN* 2 lbs./gal. liquid

Broadcast rates per acre:

Coarse textured (light) soils—1 Gallon
Medium textured (medium) soils—1$^1/_8$ Gallons
Fine textured (heavy) soils—1$^1/_2$ Gallons

BASAGRAN* 4 lbs./gal. liquid

Broadcast rates per acre: Apply $^3/_4$ to 1 quart per acre. Rates depend upon weeds to be controlled. Consult and the follow recommended rates on the BASAGRAN label.

DYANAP* 3 lbs./gal. liquid

Broadcast rates per acre: Use 6 quarts as a broadcast spray just after the soybeans are planted but before the soybeans plants emerge from the soil.

LOROX* 50W

Broadcast rates per acre: Use 1 to 2$^1/_2$ lbs. LOROX 50W per acre depending on the soil type. Use on coarse to medium textured soils. Apply after planting before soybeans emerge. Plant soybeans at least 1$^3/_4$" deep on flat seed bed only to avoid possible crop injury. Do not use on sand or loamy sand or soil with less than 1% organic matter since crop injury may result.

*AMIBEN—Reg. TM Union Carbide Agricultural Products Co., Inc.

*BASAGRAN—Trademark BASF Aktiegesellschaft

*DYANAP—Reg. TM Uniroyal Chemical, Inc.

*LORAX—Reg. TM E.I. Du Pont de Nemours & Co., Inc.

TANK MIX DIRECTIONS FOR USE ON SOYBEANS

REWARD may be combined with the compounds listed in this section for tank mix applications on soybeans. Recommended rates for each tank mix are listed in the sections below. The mixing, spray equipment, soil incorporation, and planting directions for the REWARD tank mixes on soybeans are listed at the end of the soybean tank mix section.

REWARD/TREFLAN* 4-EC TANK MIX DIRECTIONS

*TREFLAN® Registered trademark of Elanco Products Company.

How to use REWARD and TREFLAN 4-EC as a spray tank mix for weed control on soybeans.

A tank mix combination of REWARD (6 pounds active ingredient per gallon) and TREFLAN E.C. (4 pounds of active ingredient per gallon) can be used prior to planting for weed control in soybeans.

The combination will control a broader spectrum of weeds than either material used alone. This includes tough weeds such as nutgrass, velvetleaf, foxtails, johnsongrass, redroot pigweed and wild cane.

Dependable performance—because the herbicides are mixed into the soil, rainfall is not needed to move them into the weed germinating zone. They are already there to kill grass and broadleaf weeds when they sprout.

WEEDS CONTROLLED

Annual Grasses:

Annual bluegrass	Goosegrass
Barnyardgrass, Watergrass	Junglerice
Brachiaria	Johnsongrass, seedling
Bromegrass	Sprangletop
Cheat	Stinkgrass
Crabgrass	Texas panicum—(buffalograss) (coloradograss)
Fall Panicum	
Field sandbur	
Foxtails	Wild cane

Annual Broadleaf Weeds:

Balloonvine**	Palmer armaranth—(carelessweed)
Carpetweed	
Chickweed	Pigweed
Common lambsquarters	Purslane
	Russian thistle
Florida pusley	Sicklepod—(coffeeweed)
Goosefoot	
Knotweed	Stinging nettle
Kochia	Velvetleaf—(buttonweed)

Perennial Grasses:

Nutsedge—(nutgrass)
Rhizome johnsongrass*

* Suppression
** In Alabama Only

APPLICATION DIRECTIONS

Read and follow all directions, use limitations, precautions and warnings that apper on the REWARD and TREFLAN E.C. labels. Do not contaminate water by cleaning of equipment or disposal of wastes.

Application Rates

Soil Texture	Broadcast Rates/Acre REWARD*	TREFLAN 4-EC*
Coarse (light-sandy soils)	2⅔ pints	1 pint
Medium (silt loam & sandy clay soils)	2⅔ to 3⅓ pints	1 to 1½ pints
Fine (heavy—clay soils)	3⅓ to 4 pints	1 to 2 pints

* For rhizome johnsongrass suppression use the high rate of TREFLAN 4-EC and REWARD.

** For nutgrass, wild cane and velvetleaf control, use the high rate of 3⅓ pints REWARD on medium textured soils.

For Rhizome Johnsongrass—Use mold board plow, chisel plow or similar implement to bring rhizomes to the top of the soil. Then follow with a disc set to cut 4–6 inches deep and operate in two different directions at 4–6 mph before application to cut the rhizomes into small (2–3 inch) pieces. This should also destroy any emerged johnsongrass.

Cultivation is suggested in addition to REWARD TREFLAN 4-EC treatment in fields with moderate to heavy infestations of nutgrass, wild cane, velvetleaf, and rhizome johnsongrass.

REWARD/AMIBEN TANK MIX DIRECTIONS

How to use REWARD and AMIBEN as a spray tank mix for weed control on soybeans.

A tank mix combination of REWARD (6 pounds active ingredient per gallon) and AMIBEN (2 pounds active ingredient per gallon) can be used prior to planting, for weed control in soybeans.

The combination will control a broader spectrum of weeds than either material used alone.

WEEDS CONTROLLED

Annual Grasses:

Barnyardgrass, Watergrass	German millet
Crabgrass	Goosegrass
Foxtail (giant, green, yellow)	Johnsongrass, seedling
	Stinkgrass

Perennial Grasses:

Purple nutsedge—nutgrass
Yellow nutsedge—nutgrass

Annual Broadleaves:

Carpetweed	Nightshade, black
Chickweed	Pigweed (redroot, smooth)
Coffeeweed	Prickly sida—teaweed
Florida pusley	Purslane
Kochia	Ragweed, common
Lambsquarters, common	Sicklepod
Morningglory, annual	Smartweed, Pennsylvania
Mustard, wild	Spurge, annual
	Velvetleaf—buttonweed

APPLICATION DIRECTIONS

Read and follow all directions, use limitations, precautions and warnings that appear on the REWARD and AMIBEN labels. Do not contaminate water by cleaning of equipment or disposal of wastes.

Application Rates

Soil Texture	Broadcast Rates/Acre REWARD	AMIBEN
Medium (silt loam & sandy clay soils)	2⅔ pints	3 quarts

For mixing, spray equipment, soil incorporation, and planting directions, refer to the directions at the end of the REWARD tank mix on soybeans section of this label.

REWARD/BASALIN* 4-E TANK MIX DIRECTIONS

* BASALIN Registered trademark of BASF Wyandotte Corp.

How to use REWARD and BASALIN 4-E as a spray tank mix on soybeans.

A tank mix combination of REWARD (6 pounds active ingredient per gallon) and BASALIN (4 pounds active ingredient per gallon) can be used prior to planting for weed control in soybeans.

The combination will provide better weed control than either herbicide used alone.

WEEDS CONTROLLED

Annual Grasses:

Barnyardgrass, Watergrass	German millet
Crabgrass	Goosegrass
Foxtail (giant, green yellow)	Johnsongrass, seedling

Annual Broadleaves:

Carpetweed	Pigweed; redroot, smooth
Florida pusley	Purslane
Lambsquarters, common	Sicklepod
Morningglory, annual	Velvetleaf—buttonweed

Perennial Grasses

Purple nutsedge—nutgrass
Yellow nutsedge—nutgrass

APPLICATION DIRECTIONS

Read and follow all directions, use limitations, precautions and warnings that appear on the REWARD and BASALIN 4-E labels. Do not graze treated plant parts to livestock. Do not contaminate water by cleaning of equipment or disposal of wastes.

Application Rates

Soil Texture	Broadcast Rates/Acre REWARD	BASALIN 4-E
Coarse (light-sandy soils)	2 pints	3/4 pint
Medium (silt loam and sandy clay soils)	2⅔ pints	1 pint

For mixing, spray equipment, soil incorporation, and planting directions, refer to the directions at the end of the REWARD tank mix on soybean section of this label.

REWARD/LASSO* 4-EC TANK MIX DIRECTIONS

* LASSO Registered trademark of Monsanto Company

How to use REWARD and LASSO 4-EC as a spray tank mix for weed control on soybeans.

A tank mix combination of REWARD (6 pounds active ingredient per gallon) and LASSO 4-EC (4 pounds active ingredient per gallon) can be used prior to planting for weed control in soybeans grown in North Central States of Michigan, Indiana, and Illinois. Do not use on sand or loamy sandy soils as injury may occur.

The combination will provide better weed control than either herbicide used alone.

WEEDS CONTROLLED

Annual Grasses:

Barnyardgrass, Watergrass
Crabgrass
Fall panicum
Foxtail (giant, green yellow)
Goosegrass
Johnsongrass, seedling

Annual Broadleaves:

Carpetweed
Florida pusley
Lambsquarters, common
Morningglory, annual
Pennsylvania smartweed
Pigweed; redroot, smooth
Purslane
Sicklepod
Velvetleaf—buttonweed

Perennial Grasses:

Purple nutsedge—nutgrass
Yellow nutsedge—nutgrass

APPLICATION DIRECTIONS

Read and follow all directions, use limitations, precautions and warnings that appear on the REWARD and LASSO 4-EC labels. Do not contaminate water by cleaning of equipment or disposal of wastes.

Application Rates

Soil Texture	Broadcast Rates/Acre REWARD	LASSO 4-EC
Coarse (light-sandy soils)	2 pints	2 pints
Medium (silt loam and sandy clay soils)	2⅔ pints**	3 pints
Fine (heavy—clay soils)	3⅓ pnts	4 pints

** For nutgrass and velvetleaf control, use the rate of 3⅓ pints of REWARD on medium and fine textured (medium and heavy—clay) soils.

REWARD/PROWL* 4-E TANK MIX DIRECTIONS

* PROWL Registered trademark of American Cyanamid Company.

How to use REWARD and PROWL 4-E as a spray tank mix for weed control on soybeans.

A tank mix combination of REWARD (6 pounds active ingredient per gallon) and PROWL 4-E (4 pounds active ingredient per gallon) can be used prior to planting for weed control in soybeans.

The combination will provide better weed control than either herbicide used alone.

WEEDS CONTROLLED

Annual Grasses:

Barnyardgrass, Watergrass
Crabgrass
Foxtail millet
Foxtail (giant, green, yellow)
Goosegrass
Johnsongrass, seedling

Annual Broadleaves:

Carpetweed
Florida pusley
Lambsquarters, common
Morningglory, annual
Pigweed; redroot, smooth
Purslane
Sicklepod
Velvetleaf—buttonweed

Perennial Grasses:

Purple nutsedge—nutgrass
Yellow nutsedge—nutgrass

APPLICATION DIRECTIONS

Read and follow all directions, use limitations, precautions and warnings that appear on the REWARD and PROWL 4-E labels. Do not contaminate water by cleaning of equipment or disposal of wastes.

Application Rates

Soil Texture	Broadcast Rates/Acre: REWARD	PROWL 4-E
Coarse (light-sandy soils)	2 pints	3/4 pint
Medium (silt loam and sandy clay soils)	2⅔ pints**	1 pint
Fine (heavy—clay soils)	3⅓ pints	1½ pints

** For nutgrass and velvetleaf control, use the rate of 3⅓ pints of REWARD on medium textured (medium and heavy—clay) soils.

For mixing, spray equipment, soil incorporation, and planting directions, refer to the directions at the end of the REWARD tank mix on soybean section of this label.

USE THE FOLLOWING DIRECTIONS FOR THE REWARD TANK MIXES ON SOYBEANS.

Mixing:

Add the recommended rates of both REWARD and TREFLAN 4-EC to the spray tank during filling and mix thoroughly. Begin adding the herbicides when spray tank is half full. Apply in 10-50 gallons of water per acre.

Spray Equipment:

Use a properly calibrated, low pressure, boom-type sprayer having good agitation. Check frequently during application to be sure nozzles are free from clogging and delivering a uniform spray pattern.

DRY FERTILIZER STATEMENT

REWARD/TREFLAN 4-EC tank mix may be impregnated on many dry bulk fertilizers for use on soybeans. However, uniform distribution of REWARD/TREFLAN 4-EC on fertilizer particles and uniform application are necessary to assure good results. See Appendix II giving directions for impregnation and use.

Soil Incorporation:

The tank mix combinations with REWARD must be mixed (incorporated) into the soil immediately after application. Whenever possible, application and incorporation should be done in the same operation. This can be done by mounting the tanks and boom right on the incorporation rig.

For thorough incorporation, use the following equipment:

Tandem Disc—Set to cut to a depth of 4 inches, followed by a spike-tooth harrow or some other leveling device which extends beyond the ends of the disc. For thorough mixing, disc in two different directions (cross disc).

OR: Power-driven cultivation equipment set to cut to a depth of 2–3 inches.

Planting:
Do not plant seed deeper than 2 inches. Seeding can be done immediately, or it may be delayed up to 10 days. If it is necessary to work the soil following application and incorporation, but before planting, do not work the soil deeper than the depth the herbicide was incorporated i.e., no more than ½ the depth the herbicides were incorporated. To avoid removing the herbicide from the row (with loss of weed control in the row) do not move or shape soil after treatment, such as by planting in deep furrows.

APPENDIX I

PROCEDURE FOR TESTING THE COMPATIBILITY OF REWARD AND COMBINATIONS WITH FLUID FERTILIZERS

The following procedure is suggested for determining whether REWARD may be combined with a specific fluid fertilizer for spray tank application.

MATERIALS REQUIRED

1. REWARD and/or combination.
2. Fluid fertilizer to be used.
3. Adjuvant for fertilizer tank mix: Compex*, Sponto 168-D*, Unite* or equivalent. The adjuvant which provides the best compatibility depends on the specific fertilizer under consideration.

* Compex+up,TM , Kalo Laboratories, Inc., Kansas City, Missouri; Sponto 168-D+up,TM , Witco Chemical Company, Houston, Texas; Unite, Hopkins Agricultural Chemical Company, Madison, Wisconsin.

4. Two one-quart, wide mouth glass jars with lid.
5. Measuring spoons; a 25 ml pipette or graduated cylinder provides more accurate measurement; 1 tablespoon equals 3 teaspooons, 1 teaspoon equals approximately 5 milliliters.
6. Measuring cup, 8 oz. (237 ml).

PROCEDURE

1. Pour a pint (about 473 ml) of the fluid fertilizer into each of the quart jars.
2. Add adjuvant to one of the jars and mix (see rate table).
3. Add the REWARD to the desired combinations of herbicide to both jars (see rate table for REWARD).
4. Close both jars with lid and mix the contents by turning the jars upside down ten times.
5. Inspect the surface and body of the mixtures—
   (A) Immediately after completing the jar inversions.
   (B) After allowing the jars to stand quietly for 30 minutes.
   (C) And then again after turning the jars upside down 10 times.

If a uniform mix cannot be made, the mixture should not be used. If either mixture remains uniform for 30 minutes, the combination may be used. Should either mixture separate after 30 minutes, but readily remix uniformly with the jar inversions, the mixture can be used if adequate agitation is maintained in the tank. If the mixture with adjuvant is satisfactory, but the one without adjuvant is not, be sure to use the adjuvant in the spray tank. Add the adjuvant first at a rate of 3 pints per 100 gallons of fluid fertilizer, foaming can be minimized by using moderate agitation.

If nondispersible oil, sludge or clumps of solids form in the mixtures, the combinations should not be used.

Rate Table for REWARD and Adjuvant**
With the Fluid Fertilizer

Gallons of Fluid Fertilizer To Be Applied Per Acre	mL or Tsp. of REWARD* To Be Added to 1 Pint of Fertilizer	
	REWARD	
	mL.	Tsp.
10	8	1–3/5
15	5	1
20	4	4/5
25	3	2/3
30	2.5	1/2
40	2	2/5

* Based on field rate of 1 pound active ingredient per acre in the fertilizer volumes indicated. Increase volume proportionately to correspond with intended field rate in terms of pounds active ingredient per acre (e.g. for field rate of 4 pounds actual active ingredient in 40 gallons fertilizer per acre, add 8 ml. of 2 tsp. REWARD to each jar for compatibility testing purposes).

** Two (2) milliliters or one-half (½) teaspoon of adjuvant should be added to 1 pint of fluid fertilizer in order to equal the rate of 3 pints of adjuvant per 100 gallons of fluid fertilizer.

APPENDIX II

REWARD and REWARD/TREFLAN* 4-EC IMPREGNATION ON DRY BULK FERTILIZERS

CAUTION: REWARD alone or in combination with other herbicides must not be impregnated on ammonium nitrate, sodium nitrate or potassium nitrate. Such mixtures may cause explosion and fire.

REWARD alone may be impregnated on many* dry bulk fertilizers for use *on soybeans.* REWARD in combination with TREFLAN may be impregnated on many* dry bulk fertilizers and applied and incorporated in the soil before planting for the control of labeled grasses and broadleaf weeds.

Field results have shown that REWARD and REWARD/TREFLAN 4-EC on bulk dry fertilizers give weed control equal to REWARD or REWARD/TREFLAN 4-EC applied as a spray in water or liquid fertilizer. However, uniform impregnation of herbicide solution on the dry fertilizer particles and uniform application in the field are necessary to assure good results.

For impregnating pesticides on dry fertilizers, use a closed rotary-drum mixer or a similar type of closed blender equipped with suitable spray equipment. The spray nozzle (or nozzles) should be positioned inside of the mixer to provide *uniform* spray coverage of the tumbling fertilizer.

For the REWARD/TREFLAN 4-EC combination, premix in the proper use ratio to prepare a uniform solution for joint spray impregnation. The physical properties of fertilizers vary, particularly in liquid absorptive capacity. When absorptivity is sufficient, simple spray impregnation of the fertilizer with REWARD or REWARD/TREFLAN 4-EC provides a satisfactory, dry mixture. If the absorptive capacity is inadequate, use of a highly absorptive powder is required to provide a dry, flowable mixture. Microcel E (Johns-Manville Products Corp.) is the recommended absorbent powder. It should be added separately and uniformly to the prepared REWARD or REWARD/TREFLAN 4-EC fertilizer mixture, in a quantity that is sufficient to provide a suitably flowable mixture. Generally less than 2% by weight of Microcel E is required. The amount of product actually required in the manufacture of individual fertilizer mixture should be determined carefully for each production operation. This is necessary to ensure that the amount of REWARD/TREFLAN 4-EC actually contained in the mixture and applied to the soil represents the correct rate of use.

All REWARD and TREFLAN 4-EC label recommendations regarding rates per acre, soil incorporation, application, cautions, general use precautions and other directions must be followed.

All individual state regulations relating to bulk dry fertilizer blending, registration, labeling and application are the responsibility of the individual and/or company selling the fertilizer and REWARD and/or REWARD/TREFLAN 4-EC mixture.

A minimum of 200 pounds and a maximum of 450 pounds of approved ingredients* impregnated with REWARD/TREFLAN 4-EC at the recommended rate must be applied per acre.

Bulk fertilizer impregnated with REWARD should be applied immediately, *not stored.* It is recommended that all bulk containers be tightly covered while the product is being transported and applied to reduce chances of REWARD loss via volatilization.

***Approved Dry Fertilizer Ingredients**

	N	P	K
Ammonium Sulfate	21	0	0
Diammonium Phosphate	18	46	0
Potassium Chloride	0	0	60
(Single) Superphosphate	0	20	0

Triple Superphosphate	0	46	0
Urea	45	0	0
Ammonium Phosphate-Sulfate	16	20	0
Potassium Sulfate	0	0	52

NOTE: Sul-PO-MAG, K-MAG, and 11-48-0 have been shown to be compatible with REWARD plus TREFLAN 4-EC and are approved for use.

RATE CHART FOR THE IMPREGNATION OF DRY BULK FERTILIZERS WITH REWARD AND REWARD/TREFLAN 4-EC ADDED TO A TON OF FERTILIZER*

Fertilizer Rate Per Acre	REWARD Rate/Acre				TREFLAN 4-EC Rate/Acre	
	2 pts.	2⅔ pts.	3⅓ pts.	4 pts.	1 pt.	2 pts.
200 lbs.	10 qts./ton	13⅓ qts./ton	16⅔ qts./ton	20 qts./ton	5 qts./ton	10 qts./ton
250 lbs.	8 qts./ton	10⅔ qts./ton	13⅓ qts./ton	16 qts./ton	4 qts./ton	8 qts./ton
300 lbs.	6⅔ qts./ton	9 qts./ton	11 qts./ton	13⅓ qts./ton	3⅓ qts./ton	7 qts./ton
350 lbs.	5¾ qts./ton	7⅗ qts./ton	9½ qts./ton	11½ qts./ton	2⅘ qts./ton	6 qts./ton
400 lbs.	5 qts./ton	6⅔ qts./ton	8⅓ qts./ton	10 qts./ton	2½ qts./ton	5 qts./ton
450 lbs.	4½ qts./ton	6 qts./ton	7⅖ qts./ton	9 qts./ton	2¼ qts./ton	4½ qts./ton

* Refer to application rates under "APPLICATION DIRECTIONS"

REWARD Typical Physical Data

Specific Gravity, 20/20 °C	0.972 (typical)
Pounds/Gallon, 20 °C	8.091 (typical)
Flashpoint:	187 °F (Setaflash closed cup)
Viscosity:	Sprayable down to minus 20 °F

STORAGE AND DISPOSAL

PROHIBITIONS

Do not contaminate water, food, feed, fertilizer or seeds by storage or disposal.

STORAGE

Keep container tightly closed when not in use. Can be stored at temperatures as low as minus 50°F.

PESTICIDE DISPOSAL

Rinse spray equipment. Wastes resulting from the use of this product may be disposed of on site or at an approved waste disposal facility.

CONTAINER DISPOSAL

METAL: Triple rinse (or equivalent), then offer for recycling or reconditioning, or puncture and dispose of in a sanitary landfill, or by other approved State and local procedures.

PLASTIC: Triple rinse (or equivalent), then offer for recycling or reconditioning or puncture and dispose of in a sanitary landfill, or by incineration, or, if allowed by State and local authorities, by burning. If burned, stay out of smoke.

NOTICE—READ CAREFULLY

CONDITIONS OF SALE

Stauffer (and seller) offer(s) this product for sale subject to, and buyer and all users are deemed to have accepted, the following conditions of sale and warranty which may only be varied by written agreement of a duly authorized representative of Stauffer.

WARRANTY LIMITATION

Stauffer warrants that this product conforms to the chemical description on the label and is reasonably fit for the purposes referred to in the directions for use on the label subject to the inherent risks referred to below. Stauffer makes no other express warranties: THERE IS NO IMPLIED WARRANTY OF MERCHANTABILITY and there are no warranties which extend beyond the description on the label hereof.

INHERENT RISKS

The directions for use of this product are believed to be reliable and should be followed carefully. However, it is impossible to eliminate all risks associated with use. Buyer assumes all risks associated with use or application of this product contrary to label instructions or resulting from extraordinary weather conditions.

LIMITATION OF LIABILITY

In no case shall Stauffer be liable for special, indirect or consequential damages resulting from the use or handling of this product and no claim of any kind shall be greater in amount than the purchase price of the product in respect of which such damages are claimed.

This product is sold only for uses stated on its label. No express or implied license is granted to use or sell this product under any patent in any Country except as specified: Country: United States of America. Patent No. 4,004,001; 4,035,490. Other Patents Pending.

SUPPLEMENTAL LABELING

REWARD,™ Selective Herbicide

EPA Reg. No. 476-2219

FOR USE ON SOYBEANS

REWARD/TREFLAN* 4E/METRIBUZIN TANK MIX DIRECTIONS

* TREFLAN is the registered trademark of Elanco Products Company.

This label must be in the possession of the user at time of pesticide application.

It is a violation of Federal law to use this product in a manner inconsistent with its labeling.

How to use REWARD and TREFLAN 4-EC and METRIBUZIN (4-L or 50-WP) as a spray tank mix.

A tank mix of REWARD (6 pounds active ingredient per gallon) and TREFLAN 4-EC (4 pounds active ingredient per gallon) and METRIBUZIN 4-L (4 pounds active ingredient per gallon) or METRIBUZIN 50-WP (50% active ingredient by weight) can be used prior to planting for weed control in soybeans.

The combination will control a broader spectrum of weeds than the REWARD/TREFLAN 4-EC tank mix.

WEEDS CONTROLLED

Annual Grasses:

Annual bluegrass	Foxtails
Barnyardgrass—watergrass	Goosegrass
Brachiaria	Junglerice
Bromegrass	Johnsongrass, seedling
Cheat	Sprangletop
Crabgrass	Stinkgrass
Fall panicum	Texas panicum—
Field sandbur	(buffalograss)
	(coloradograss)
	Wild Cane

Annual Broadleaf Weeds:

Balloonvine**	Palmer amaranth—
Carpetweed	(carelessweed)
Chickweed	Pennsylvania smartweed
Common lambsquarters	Pigweed
Common ragweed	Prickly sida
Florida pusley	Purslane
Goosefoot	Russian thistle
Knotweed	Sicklepod-coffeeweed
Kochia	Stinging nettle
	Velvetleaf—buttonweed
	Venice mallow

** In Alabama only.

Perennial Grasses:

Purple nutsedge—nutgrass

Yellow nutsedge—nutgrass

APPLICATION DIRECTIONS

Read and follow all directions, use limitations, precautions and warnings that appear on the REWARD, TREFLAN 4-C and appropriate METRIBUZIN (4-L or 50-WP) labels. Do not contaminate water by cleaning of equipment or disposal of wastes.

APPLICATION RATES

		Broadcast Rates/Acre	
Soil Texture	REWARD	TREFLAN 4-EC	METRIBUZIN 4L or 50-WP
Medium (silt loam & sandy clay soils)	2⅔ to 3⅓ pints	1 pint	½ pint or ½ pound
Fine (Heavy-clay soils)	3⅓ pints	1 pint	½ pint or ½ pound

For nutsedge, wild cane, and velvetleaf control, use the high rate of 3⅓ pints of REWARD on medium-textured soils.

Do not use the REWARD/TREFLAN/METRIBUZIN tank mix on coarse (light sandy) soils.

Cultivation is suggested in addition to REWARD/TREFLAN 4-EC/METRIBUZIN (4-L or 50-WP) treatment in fields with moderate to heavy infestations of nutgrass, wild cane, and velvetleaf.

This is a specimen label. Stauffer is not responsible for the accuracy of the information contained herein. As labels are subject to revision, always carfully read and follow the label on the product container.

RO-NEET® 6-E

EMULSIFIABLE LIQUID SELECTIVE HERBICIDE

For Sugar Beets, Table Beets & Spinach

ACTIVE INGREDIENT:

S-Ethyl cyclohexylethylthiocarbamate 73.9%

INERT INGREDIENTS: 26.1%

TOTAL 100.0%

Contains six pounds active ingredient per gallon.

EPA Reg. No. 476-1979 A-4R

CAUTION

Keep Out of Reach of Children

HARMFUL IF SWALLOWED

READ ENTIRE LABEL BEFORE USING

Protect from temperatures below 20°F.
Product crystallizes at lower temperatures.
Warm or store at higher temperatures, and mix to redissolve crystals and assure uniformity before use.

STATEMENT OF PRACTICAL TREATMENT

FIRST AID: If a known exposure occurs or is suspected, immediately start the recommended procedures below. Simultaneously contact a Poison Control Center, a physician or the nearest hospital. Describe the situation and follow the advice given.

If swallowed: Immediately give large quantities of water but DO NOT induce vomiting. *This product contains hydrocarbon solvent.* If vomiting does occur, give fluids again. Have a physician determine if condition of patient will permit evacuation of stomach. DO NOT give anything by mouth to an unconscious person. Call a physician immediately.

In case of contact: Hold eyelids apart and flush eyes and skin with plenty of water for at least 15 minutes while removing contaminated clothing and shoes. Get medical attention. Wash clothing before re-use.

If inhaled: Remove to fresh air. Seek medical attention if respiratory irritation occurs.

PRECAUTIONARY STATEMENTS

HAZARDS TO HUMANS AND DOMESTIC ANIMALS

CAUTION

HARMFUL IF SWALLOWED

Avoid contact with skin, eyes and clothing.
Wear rubber gloves, and protective clothing.
Wash hands, arms and face thoroughly with soap and water after handling and before eating or smoking.
Routinely shower or bathe after work and wash all clothing with soap and hot water before re-using.
In case of contact, immediately remove contaminated clothing.
Avoid contamination of feed or food.

ENVIRONMENTAL HAZARDS

Do not apply directly to water. Do not contaminate water by cleaning of equipment or disposal of wastes.

In case of emergency call (day or night) Pesticide Team Safety Network 800-424-9300 or Stauffer Chemical Company 203-226-6602.

WEEDS CONTROLLED

Annual Grasses:	
Annual Bluegrass	*Poa annua*
Annual Ryegrass	*Lolium multiflorum*
Barley, Volunteer	*Hordeum* spp.
Barnyardgrass (Watergrass)	*Echinochloa* spp.
Crabgrass	*Digitaria* spp.
Foxtails	*Setaria* spp.
Wild Oats	*Avena fatua*
Annual Broadleaf Weeds:	
Black Nightshade	*Solanum nigrum*
Hairy Nightshade	*Solanum villosum*
Henbit, Deadnettle	*Lamium* spp.
Lambsquarters	*Chenopodium album*
Nettleleaf Goosefoot*	*Chenopodium murale*
Common Purslane	*Portulaca oleracea*
Redroot Pigweed	*Amaranthus retroflexus*
Shepherdspurse	*Capsella bursa-pastoris*
Small Stinging Nettle, Burning Nettle	*Urtica urens*

* Arizona and California excluded

The above broadleaf weeds will be controlled only if the application is made when conditions are favorable for germination. RO-NEET 6-E WILL NOT CONTROL ESTABLISHED WEEDS.

Perennial Weeds:	
Yellow Nutgrass (Nutsedge)	*Cyperus esculentus*
Purple Nutgrass (Nutsedge)	*Cyperus rotundus*

Existing stands of nutgrass must be turned under and chopped thoroughly before treatment. Prior to planting, RO-NEET 6-E must be thoroughly mixed into the soil, or applied sub-surface to the soil with injectors for nutgrass control.

GENERAL USE PRECAUTIONS

Not for use or storage in or around the home.

RO-NEET 6-E should be used only for recommended purposes and at recommended rates (DO NOT OVERDOSE). Applied according to directions and under normal growing conditions RO-NEET 6-E will not harm the treated crop. During germination and early growth, extended-periods of unusually cold and wet or hot and dry weather, insect, nematode, or plant disease attack, carryover pesticide residues, the use of certain soil applied systemic insecticides, highly saline or alkaline soil conditions, improperly placed fertilizers or soil insecticides may create abnormal conditions that weaken crop seedlings. RO-NEET 6-E used under these abnormal conditions could result in crop injury.

RO-NEET 6-E may cause crop injury on very light sandy soil.

RO-NEET 6-E may be combined with fluid (solution, slurry, or suspension) fertilizers. However, physical compatibility with these fluid fertilizers must be determined before combining in the spray tank. See special literature giving directions for these combinations. Even though found to be compatible, constant agitation is necessary to keep the RO-NEET 6-E evenly mixed with the fluid fertilizer.

Tank mix this product with fungicides, insecticides or herbicides only as recommended.

Do not contaminate irrigation water or water used for domestic purposes.

DIRECTIONS FOR USE

It is a violation of Federal law to use this product in a manner inconsistent with its labeling.

RO-NEET 6-E is a selective herbicide which is mixed (incorporated) into the soil or applied below the surface of the soil for control of weeds listed on this label. RO-NEET 6-E controls weeds by interfering with normal seed germination and seedling development. It does not control established weeds. All weed growth and crop stubble should be thoroughly worked into the soil before treatment.

SOIL PREPARATION: Prepare the soil for seeding according to good agricultural practice. All weed growth should be thoroughly worked into the soil before treatment. DO NOT APPLY BEFORE PRE-IRRIGATION.

APPLICATION: Apply RO-NEET 6-E to well-worked soil which is dry enough to permit thorough mixing with incorporation equipment, or proper injection. Use standard low-pressure (20–50 psi) boom spray equipment. DO NOT USE SINGLE NOZZLE BOOM-JET TYPE SPRAYERS. Application equipment should be carefully calibrated before use and checked frequently during application to be sure nozzles are free from clogging and delivering a uniform pattern. Apply UNIFORMLY the recommended rate of RO-NEET 6-E in 20 to 50 gallons of water per acre. Avoid overlaps that will increase RO-NEET 6-E dosage above recommended limits because plant injury will occur.

RO-NEET 6-E WITH FLUID FERTILIZER: RO-NEET 6-E may be combined with fluid (solution, slurry, or suspension) fertilizers. However, physical compatibility with these fluid fertilizers must be determined before combining in the spray tank. See Appendix I giving directions for these combinations. Even though found to be compatible, constant agitation is necessary to keep the RO-NEET 6-E evenly mixed with the fluid fertilizer.

IMPREGNATION ON DRY BULK FERTILIZERS: RO-NEET 6-E may be impregnated on many dry bulk fertilizers for use on sugar beets. RO-NEET 6-E impregnated on dry bulk fertilizers may be applied and incorporated into the soil either in the fall (in states where fall application for RO-NEET 6-E is recommended) before the ground freezes or before planting. However, uniform distribution of RO-NEET 6-E on fertilizer particles and uniform application are necessary to assure good results. See Appendix II for special instructions regarding directions for impregnation and use.

INCORPORATION DIRECTIONS

Immediately (within minutes) after application thoroughly mix RO-NEET 6-E into the soil to a depth of 2 to 3 inches. (California only: Incorporate to a depth of 3 inches.) Thorough soil mixing is necessary for good weed control.

RO-NEET 6-E must be incorporated into the soil immediately to prevent loss of the herbicide. Whenever possible, application and incorporation should be done in the same operation. Use equipment which has been proven to incorporate thoroughly to the recommended depth.

SOIL MIXING (INCORPORATION) BEFORE PLANTING: The following equipment commonly is used for soil mixing (incorporation) before planting:

For Broadcast (Overall) Application:

Power Driven Cultivation Equipment (recommended on all soil types) set to cut to a depth of 2 to 3 inches.

Tandem Discs (recommended on all soil types) set to cut to a depth of 4 to 6 inches, operated at 4 to 6 mph followed by a spike-tooth harrow or some other leveling device which extends beyond the ends of the discs. For more thorough mixing (for nutgrass, heavier soils) disc in two different directions (cross disc).

Field Cultivators (recommended on lighter soils in good tilth only). Not recommended in California or Arizona. Use 3 to 4 rows of sweeps, spaced at 7 inch or less intervals and staggered so that no soil is left unturned, followed by a spike-tooth harrow pulled behind the cultivator. Do not use chisels or points. Set the cultivator to cut 4 inches deep, operated at 5 mph or more. Run the equipment over the field twice, the second run at an angle to the first.

Rotary Ground Driven or Spring-Tooth Cultivators (recommended on lighter soils in good tilth only). Not recommended in California or Arizona. Set to penetrate to a depth of 4 to 6 inches and operated at 5 to 8 mph in two different directions.

For Band (Row) Application:

Uniformly mix to a depth of 2 to 3 inches.

Hooded power-driven rotary tillers.

Hooded ground-driven rotary tillers.

Rolling:

Under most conditions, the treated area should be firmly rolled between the incorporation unit and the planters.

PLANTING

For maximum weed control benefits, crops should be planted or seeded immediately after application. Do not use a drag behind the planter as it may concentrate RO-NEET 6-E over the seed row and cause crop injury.

DOSAGE RECOMMENDATIONS

Solid Broadcast Treatments

Crop	When to Apply	Gallons RO-NEET 6-E Per Acre	Remarks
Sugar Beets*	Pre-plant (soil incorporation) or at planting (soil injection)	½ To ⅔	Use on mineral soils ONLY. Use lower rate on sandy soils. Use higher dosage rate on heavier soil. Injury may result in highly saline or alkaline soils.
	Fall application	⅔	Recommended only in the states of Idaho, Minnesota, Montana, North Dakota, Oregon, Wyoming and Washington. Apply and incorporate in late fall before the ground freezes.
Table Beets	Pre-plant (soil incorporation)	½ To ⅔	Use on mineral soils ONLY. Use lower rate on sandy soils. Use higher dosage on heavier dosage on heavier soils. When applying RO-NEET 6-E in combination with fluid fertilizer, do not apply over 150 lbs. actual nitrogen per acre.
Spinach**	Pre-plant (soil incorporation)	½	Use on sandy mineral soils only.

* Make only one (1) application per growing season. If RO-NEET 6-E is applied in the fall, do not reapply RO-NEET 6-E the following spring.

** Recommended on spinach in Arkansas, Connecticut, Delaware, Maine, Maryland, Massachusetts, Mississippi, New Hampshire, New Jersey, New York, Ohio, Oklahoma, Pennsylvania, Texas, Vermont, Virginia and Western Tennessee only.

CULTURAL PRACTICES FOLLOWING APPLICATION

RO-NEET 6-E is not persistent in the soil and susceptible weeds germinating later during the growing season may not be controlled. Shallow cultivation or approved or post emergence herbicides may be necessary to control these susceptible weeds that escape control by RO-NEET 6-E as well as those weeds not susceptible to RO-NEET 6-E. Do not cultivate deeper than the depth of the herbicide incorporation.

When cultivating fields where RO-NEET 6-E has been banded in the row, use shields to prevent the movement of untreated soil into the treated row.

BAND TREATMENT: The rates given above are for solid coverage basis. For band treatment, reduce the amount of RO-NEET 6-E proportionately depending on the row spacing and band width to be treated. DO NOT USE BAND APPLICATION ON ROCKY GROUND BECAUSE THOROUGH INCORPORATION IS NOT POSSIBLE.

Soil injection application on sugar beets—Special equipment designed for soil injection must be used. The injector units must be rigidly mounted immediately ahead or immediately behind the planter unit.

Soil Injection—Injector shanks must be spaced 2½ to 3 inches apart and mounted in staggered positions to avoid trash building up on them. Injector shanks must be set to inject RO-NEET 6-E 1½ to 2 inches below the soil surface. The width of the band in which weed control is desired will determine the number of injector shanks required per row (that is, four injector shanks for a 10 to 12-inch band, six injector shanks for a 15 to 18-inch band). The two injector shanks adjacent to the drill row must be 1¼ to 1½ inches on either side of it.

APPENDIX I

RO-NEET 6-E WITH FLUID FERTILIZERS

The following procedure is suggested for determining whether RO-NEET 6-E may be combined with a specific fluid fertilizer for spray tank application.

MATERIAL REQUIRED

1. RO-NEET 6-E.
2. Fluid fertilizer to be used.
3. Adjuvant for fertilizer tank mix: Compex*, Sponto 168-D*, Unite*, or equivalent. The adjuvant which provides the best emulsification depends on the specific fertilizer under consideration.
4. Two one-quart, wide-mouth glass jars with lid or stopper.
5. Measuring spoons (a 25 mL pipette or graduated cylinder provides more accurate measurement).
6. Measuring cup, 8 oz. (237 mL).

* Compex, Kalo Laboratories Inc., Kansas City, Missouri, Sponto 168-D, Witco Chemical Company, Houston, Texas, Unite, Hopkins Agricultural Chemical Company, Madison, Wisconsin.

PROCEDURE

1. Pour a pint (about 473 mL) of the fluid fertilizer into each of the quart jars.
2. Add adjuvant to one of the jars and mix (see rate table).
3. Add the RO-NEET 6-E to both jars (see rate table).
4. Close both jars with lid or stopper and mix the contents by turning the jars upside down ten times.
5. Inspect the surface and body of the mixtures—
   (A) Immediately after completing the jar inversions.
   (B) After allowing the jars to stand quietly for 30 minutes.
   (C) And then again after turning the jars upside down 10 times.

If the uniform mix cannot be made, the mixture should not be used. If either mixture remains uniform for 30 minutes, the combination may be used. Should either mixture separate after 30 minutes, but readily remix uniformly with the jar inversions, the mixture can be used if adequate agitation is maintained in the tank. If the mixture with adjuvant is satisfactory, but the one without adjuvant is not, be sure to use the adjuvant in the spray tank. Add the adjuvant first at a rate of 3 pints per 100 gallons of fluid fertilizer; foaming can be minimized by using moderate agitation.

If nondispersible oil, sludge or clumps of solids form in the mixtures, the combinations should not be used.

RATE TABLE FOR RO-NEET 6-E AND ADJUVANT WITH THE FLUID FERTILIZER**

Gallons of fluid fertilizer to be applied per acre	mL or Tsp. of RO-NEET 6-E* to be added to 1 pint of fertilizer	
	6-E	
	mL	Tsp.
10	8	1⅔
15	5	1
20	4	¾
25	3	⅔
30	2½	½
40	2	⅖

* Based on field rate of 1 pound active ingredient per acre in the fertilizer volumes indicated. Increase volume proportionately to correspond with intended field rate in terms of pounds active ingredient per acre (e.g., for field rate of 4 pounds actual RO-NEET 6-E in 40 gallons fertilizer per acre, add 8 mL or 2 tsp. RO-NEET 6-E to each jar for compatibility testing purposes).

** Two (2) milliliters or one-half (½) teaspoon of adjuvant to be added to 1 pint of fluid fertilizer in order to equal the rate of 3 pints of adjuvant per 100 gallons of fluid fertilizer.

APPENDIX II

RO-NEET 6-E IMPREGNATION ON DRY BULK FERTILIZERS

CAUTION: Nitrate fertilizers represent a potential explosive hazard, particularly in contact with organic substances such as RO-NEET 6-E. Do not impregnate RO-NEET 6-E on ammonium nitrate. Do not use fertilizers containing ammonium, potassium or sodium nitrate. Such mixtures may cause explosion.

All individual state regulations, relating to dry bulk fertilizer blending, registration, labeling and application are the responsibility of the individual and/or company selling the RO-NEET fertilizer mixtures.

RO-NEET 6-E may be impregnated on many dry bulk fertilizers (1) and applied and incorporated into the soil before planting for the control of labeled grasses and broadleaf weeds in sugar beets.

All RO-NEET 6-E label and supplementary literature instructions and precautions regarding rates per acre, soil type and soil incorporation, application and other directions must be followed.

Test results have shown RO-NEET 6-E on many dry bulk fertilizers gives weed control equal to RO-NEET 6-E applied as a spray in water or liquid fertilizer. However, uniform impregnation of the pesticide on the dry fertilizer particles and uniform application in the field are necessary to assure good results.

A minimum of 200 pounds and a maximum of 700 pounds of approved fertilizer ingredients (1) impregnated with the appropriate amount of RO-NEET 6-E must be applied per acre.

For impregnating the pesticide on dry fertilizers, use a closed rotary-drum type mixer equipped with suitable spraying equipment. The spray nozzles should be positioned inside of the mixer to provide uniform spray coverage of the tumbling fertilizer. The RO-NEET 6-E should be sprayed uniformly onto the fertilizer using a fine spray pattern.

The physical properties of fertilizers vary, particularly in liquid absorptive capacity. When absorptivity is sufficient, simple spray impregnation of the fertilizer with RO-NEET 6-E provides a satisfactory dry mixture.

If the absorptivity is not adequate, use of a highly absorptive powder is required to provide a dry, free-flowing mixture. Microcel E (Johns-Manville Products Corporation) is the recommended absorbent powder. It should be added separately and uniformly to the prepared pesticide-fertilizer mixture, in a quantity that is sufficient to provide a suitably free-flowing mixture. Generally, less than 2% by weight of Microcel E is required.

The amount of RO-NEET 6-E actually required in the manufacture of individual fertilizer mixtures should be determined carefully for each production operation. This is necessary to ensure that the amounts of RO-NEET actually contained in the mixture applied to the soil represent the correct rates of use.

Bulk fertilizers impregnated with RO-NEET 6-E should be **applied immediately**, NOT STORED.

All bulk containers should be tightly covered while the products are being transported and applied to reduce chances of RO-NEET loss via volatilization.

(1) APPROVED DRY FERTILIZER INGREDIENTS FOR USE WITH RO-NEET 6-E

	N	P	K
Ammonium Sulfate	21	0	0
Ammonium Phosphate-sulfate	16	20	0
Diammonium Phosphate	18	46	0
Monoammonium Phosphate	11	56	0
Potassium Chloride	0	0	60
Potassium Sulfate	0	0	52
Single Superphosphate	0	20	0
Treble Superphosphate	0	46	0
Urea	45	0	0

NOTE: SUL-PO-MAG, K-MAG and 11-48-0 have been shown to be compatible with RO-NEET 6-E and are approved for use.

RO-NEET 6-E Physical Data

Specific Gravity 20/20°C:	0.976
Pounds/gallon, 20/20°C:	8.12
Flashpoint:	168°F (Tagliabue Closed Cup)
Viscosity:	Sprayable down to 25°F
Freeze point:	20°F (seeded)

(2) RATE CHART FOR THE IMPREGNATION OF DRY BULK FERTILIZERS WITH RO-NEET 6-E:

Fertilizer Rate Per Acre	RO-NEET 6-E Rate Per Acre: 2 qts. per acre	RO-NEET 6-E Rate Per Acre: 2⅔ qts. per acre
200 lbs.	20 quarts/ton	26⅔ quarts/ton
250 lbs.	16 quarts/ton	21⅓ quarts/ton
300 lbs.	13⅓ quarts/ton	17¾ quarts/ton
350 lbs.	11⅖ quarts/ton	15¼ quarts/ton
400 lbs.	10 quarts/ton	13⅓ quarts/ton
450 lbs.	8⅞ quarts/ton	11⅞ quarts/ton
500 lbs.	8 quarts/ton	10⅔ quarts/ton
550 lbs.	7⅓ quarts/ton	9¾ quarts/ton
600 lbs.	6⅔ quarts/ton	8⅞ quarts/ton
650 lbs.	6⅕ quarts/ton	8¼ quarts/ton
700 lbs.	5¾ quarts/ton	7⅝ quarts/ton

STORAGE AND DISPOSAL

PROHIBITIONS
Do not contaminate water, food, or feed by storage or disposal. Open dumping is prohibited.

STORAGE
Keep containers closed when not in use. Do not store near seeds, fertilizers or foodstuffs. Protect from temperatures below 20°F, but do not store near sources of heat or open flame.

PESTICIDE DISPOSAL
Waste resulting from the use of this product may be disposed of on site or at an approved waste disposal facility.

CONTAINER DISPOSAL
Triple rinse (or equivalent), then offer for recycling or reconditioning, or puncture and dispose of in a sanitary landfill, or by other procedures approved by State and local authorities.

NOTICE READ CAREFULLY

CONDITIONS OF SALE
Stauffer (and seller) offer(s) this product for sale subject to, and buyer and all users are deemed to have accepted, the following conditions of sale and warranty which may only be varied by written agreement of a duly authorized representative of Stauffer.

WARRANTY LIMITATION
Stauffer warrants that this product conforms to the chemical description on the label and is reasonably fit for the purposes referred to below. Stauffer makes no other express warranties; THERE IS NO IMPLIED WARRANTY OF MERCHANTABILITY and there are no warranties which extend beyond the description on the label hereof.

INHERENT RISKS
The directions for use of this product are believed to be reliable and should be followed carefully. However, it is impossible to eliminate all risks associated with use. Buyer assumes all risks associated with use or application of this product contrary to label instructions or resulting from extraordinary weather conditions.

LIMITATION OF LIABILITY
In no case shall Stauffer be liable for special, indirect or consequential damages resulting from the use or handling of this product and no claim of any kind shall be greater in amount than the purchase price of the product in respect of which such damages are claimed.

This is a specimen label. Stauffer is not responsible for the accuracy of the information contained herein. As labels are subject to revision, always carefully read and follow the label on the product container.

RO-NEET® 10-G

GRANULES
SELECTIVE HERBICIDE

For Sugar Beets

ACTIVE INGREDIENT:
S-ethyl cyclohexylethylthiocarbamate 10%
INERT INGREDIENTS: 90%
TOTAL 100%
Contains 10 lbs. active ingredient per 100 lbs.

EPA Reg. No. 476-1978 A-1R

IMPORTANT

RECALIBRATE YOUR EQUIPMENT FOR THIS PRODUCT BEFORE USING

CAUTION
Keep Out of Reach of Children

STATEMENT OF PRACTICAL TREATMENT
If a known exposure occurs or is suspected, immediately initiate the recommended procedures below. If further treatment is required, contact a Poison Center, a physician, or the nearest hospital.

If swallowed: Immediately give several glasses of water and induce vomiting by gagging the victim with a finger placed on the back of the victim's tongue. Give fluids until vomitus is clear. If victim is unconscious, do not induce vomiting or give anything by mouth.

For eye contact: Hold eyelids apart and flush eyes with large amounts of running water for 15 minutes. Seek medical attention if eye irritation occurs.

For skin contact: Flush affected areas with plenty of water for several minutes. Seek medical attention if skin irritation occurs.

If inhaled: Remove to fresh air. Seek medical attention if respiratory irritation occurs or if breathing becomes difficult.

PRECAUTIONARY STATEMENTS

HAZARDS TO HUMANS AND DOMESTIC ANIMALS

HARMFUL IF SWALLOWED
Avoid contact with skin, eyes and clothing. Avoid breathing dust. Wear goggles, rubber gloves and protective clothing. Wash thoroughly after handling.

ENVIRONMENTAL HAZARDS
Keep out of lakes, ponds and streams. Do not contaminate water by cleaning of equipment or disposal of wastes. Do not apply directly to water or wetlands.

In case of emergency, call collect, day or night, Pesticide Team Safety Network (800)424-9300 or Stauffer Chemical Company (203)226-6602.

GENERAL USE PRECAUTIONS

READ ALL LABEL DIRECTIONS BEFORE USING.

RO-NEET 10-G should be used only for recommended purposes and at recommended rates (DO NOT OVERDOSE). Applied according to directions and under normal growing conditions RO-NEET 10-G will not harm the treated crop. During germination and early stages of growth, unusually cold and wet or hot and dry weather, insect or plant disease attack, the use of certain soil applied systemic insecticides, improperly placed fertilizers or soil insecticides may create abnormal conditions that will weaken crop seedlings. RO-NEET 10-G used under these abnormal conditions could result in crop injury. RO-NEET 10-G may cause crop injury on very light sandy soil.

Do not apply in combination with fertilizer, insecticide or fungicide.

Use equipment which has been proven to incorporate thoroughly to the recommended depth.

WEEDS CONTROLLED

Annual Grasses

Annual Bluegrass	*Poa annua*
Annual Ryegrass	*Lolium multiflorum*
Barley, volunteer	*Hordeum* spp.
Barnyardgrass (Watergrass)	*Echinochloa* spp.
Crabgrass	*Digitaria* spp.
Foxtails (Yellow, Green, Giant)	*Setaria* spp.
Wild Oats	*Avena fatua*

Annual Broadleaf Weeds

Black Nightshade	*Solanum nigrum*
Hairy Nightshade (Ground Cherry)	*Solanum villosum*
Henbit (Deadnettle)	*Lamium* spp.
Lambsquarter	*Chenopodium album*
Nettleleaf Goosefoot	*Chenopodium murale*
Purslane	*Portulaca oleracea*
Redroot Pigweed	*Amaranthus retroflexus*
Shepherdspurse	*Capsella bursa-pastoris*
Small Stinging Nettle (Burning Nettle)	*Urtica urens*

The above listed annual grasses and broadleaf weeds will be controlled only if the application is made when conditions are favorable for germination.

RO-NEET 10-G will not control established weeds.

Perennial Weeds

Yellow Nutgrass (Nutsedge)	*Cyperus esculentus*
Purple Nutgrass (Nutsedge)	*Cyperus rotundus*

Existing stands of nutgrass must be turned under and chopped thoroughly before treatment. Prior to planting, RO-NEET 10-G must be thoroughly mixed into the soil.

DIRECTIONS FOR USE

It is a violation of Federal law to use this product in a manner inconsistent with its labeling.

SOIL PREPARATION: Prepare the soil for seeding according to good agricultural practice. All weed growth should be thoroughly worked into the soil before treatment. DO NOT APPLY BEFORE PRE-IRRIGATION.

APPLICATION: Apply RO-NEET 10-G to well-worked soil which is dry enough to permit thorough incorporation. Apply uniformly the recommended rate of RO-NEET 10 Granular using equipment designed for application of granular herbicides. Equipment should be carefully calibrated before use and checked frequently during application to be sure equipment is working properly and delivering a uniform distribution pattern. Avoid overlaps that will increase RO-NEET 10-G dosage above recommended limits because plant injury may occur.

SOIL (MIXING) INCORPORATION APPLICATION: Immediately (within minutes) after application, thoroughly mix RO-NEET 10-G into the soil to a depth of two to three (2-3) inches. Thorough soil mixing is necessary for good weed control.

INCORPORATION EQUIPMENT

For Broadcast (Over-All) Application

A. HOODED POWER-DRIVEN ROTARY TILLER—Set teeth close enough together so all soil is picked up. Uniformly mix to a depth of two to three inches.

B. DISC—Mix to a depth of two to three inches by setting tandem or offset discs at four to six-inch depth. On heavy and medium soils, cross disc (disc twice at right angles). On light soils, disc once followed by harrowing for additional mixing and to level the seed bed.

For Band (Row) Application: Uniformly mix to a depth of two to three inches.

A. HOODED POWER-DRIVEN ROTARY TILLERS.

B. HOODED GROUND-DRIVEN ROTARY TILLERS.

ROLLING—Under most conditions, the treated area should be firmly rolled between the incorporation unit and the planters.

PLANTING

For maximum weed control benefits, crops should be planted or seeded immediately after application. Do not use a drag behind the planter as it may concentrate RO-NEET 10-G over the seed row and cause crop injury.

CULTURAL PRACTICES FOLLOWING APPLICATION

RO-NEET 10-G is not a persistent chemical in the soil and susceptible weeds germinating late in the season may not be controlled. Cultivate to control those susceptible weeds that escape control by treatment as well as those weeds not susceptible to RO-NEET 10-G.

Should cultivation be desirable at any time following application, maximum weed control will be obtained by cultivating shallow, i.e., no deeper than depth of treatment. For banded treatments, use shields to prevent the movement of untreated soil into the treated row.

RECOMMENDATIONS

FOR CONTROL OF WEEDS LISTED IN CROP SHOWN BELOW

SOLID BROADCAST TREATMENT

Crop	When to Apply	Dosage lbs./Acre	Remarks
SUGAR BEETS	Pre-Plant (Soil Incorporation)	30 to 40	Use on mineral soils ONLY. Use lower rate on sandy soils. Use higher dosage rate in heavier soils.

BAND APPLICATION: The rates given above are for solid coverage basis. For band application, reduce the amount of RO-NEET 10-G proportionately dependent upon the row spacing and band width to be treated. Do not use band application on rocky ground because thorough incorporation is not possible.

STORAGE AND DISPOSAL

PROHIBITIONS

Do not contaminate water, food or feed by storage or disposal. Open dumping is prohibited.

PESTICIDE DISPOSAL

Waste resulting from the use of this product may be disposed of on site or at an approved waste disposal facility.

CONTAINER DISPOSAL

Completely empty bag into application equipment. Then dispose of empty bag(s) in a sanitary landfill or by incineration, or, if allowed by State and local authorities, by burning. If burned, stay out of smoke.

NOTICE—READ CAREFULLY

CONDITIONS OF SALE

Stauffer (and seller) offer(s) this product for sale subject to, and buyer and all users are deemed to have accepted, the following conditions of sale and warranty which may only be varied by written agreement of a duly authorized representative of Stauffer.

WARRANTY LIMITATION

Stauffer warrants that this product conforms to the chemical description on the label and is reasonably fit for the purposes referred to in the directions for use on the label subject to the inherent risks referred to below. Stauffer makes no other express warranties: THERE IS NO IMPLIED WARRANTY OF MERCHANTABILITY and there are no warranties which extend beyond the description on the label hereof.

INHERENT RISKS

The directions for use of this product are believed to be reliable and should be followed carefully. However, it is impossible to eliminate all risks associated with use. Buyer assumes all risks associated with use or application of this product contrary to label instructions or resulting from extraordinary weather conditions.

LIMITATION OF LIABILITY

In no case shall Stauffer be liable for special, indirect, or consequential damages resulting from the use or handling of this product and no claim of any kind shall be greater in amount than the purchase price of the product in respect of which such damages are claimed.

This is a specimen label. Stauffer is not responsible for the accuracy of the information contained herein. As labels are subject to revision, always carefully read and follow the label on the product container.

SURPASS® 6.7-E

SELECTIVE HERBICIDE EMULSIFIABLE LIQUID

For Corn (Field, Sweet and Silage)

Complete Directions for Use, Use Precautions and Recommendations

ACTIVE INGREDIENT:	
S-propyl dipropylthiocarbamate	83.4%
INERT INGREDIENTS:	16.6%
TOTAL	100.0%

Contains 6.7 Pounds of Active Ingredient Per Gallon

EPA Reg. No. 476-2179 A-7R

CAUTION

Keep Out of Reach of Children

HARMFUL IF SWALLOWED

STATEMENT OF PRACTICAL TREATMENT

If a known exposure occurs or is suspected, please start the procedures given below. If further treatment is required, contact a Poison Center, a physician, or the nearest hospital.

If swallowed: Immediately give several glasses of water but *do not* induce vomiting. If vomiting does occur, give fluids again. Have a physician determine if condition of patient will permit induction of vomiting or evacuation of stomach. Do not give anything by mouth to an unconscious or convulsing person.

For eye contact: Immediately flush eyes with large amounts of running water for at least 15 minutes. Hold eyelids apart to ensure rinsing of the entire surface of the eye and lids with water. Get medical attention immediately.

For skin contact: Flush all affected areas with plenty of water for several minutes. Remove and clean contaminated clothing and shoes. Seek medical attention if irritation occurs.

If inhaled: Remove to fresh air. Seek medical attention if respiratory irritation occurs or if breathing becomes difficult.

CAN BE STORED AT TEMPERATURES AS LOW AS MINUS 50° F.

PRECAUTIONARY STATEMENTS

HAZARDS TO HUMANS AND DOMESTIC ANIMALS

CAUTION

HARMFUL IF SWALLOWED. Avoid contact with skin, eyes and clothing. Avoid inhalation of spray mist. Wear rubber gloves, goggles and protective clothing. Wash with soap and water immediately after use or contact. Do not contaminate food and feed. Keep away from heat or flame.

Do not apply this product in such a manner as to directly or through drift, expose workers or other persons. When using the sprinkler method of application, the area being treated must be vacated by unprotected persons.

ENVIRONMENTAL HAZARDS

Do not apply directly to water. Do not contaminate water by cleaning of equipment or disposal of wastes.

WEEDS CONTROLLED BY SURPASS 6.7-E IN CORN

SURPASS 6.7-E will not control established weeds.

Annual Grasses:

Barnyardgrass; Watergrass	*Echinochloa crus-galli*
Crabgrass	*Digitaria* spp.
Fall Panicum*	*Panicum dichotomiflorum*
Field Sandbur*	*Cenchrus pauciflorus*
German Millet	*Setaria italica*
Giant Foxtail	*Setaria faberi*
Goosegrass	*Eleusine indica*
Green Foxtail	*Setari viridis*
Seedling Johnsongrass*	*Sorghum halepense*
Wild Cane*† Shattercane; Black Amber Cane; Chicken Corn	*Sorghum bicolor*
Wild Oats	*Avena fatua*
Wild Proso Millet*	*Panicum miliaceum*
Witchgrass	*Panicum capillare*
Yellow Foxtail	*Setaria glauca*

* Cultivation is suggested in addition to the SURPASS 6.7-E treatment in fields with moderate to heavy infestations of wild cane, sandbur, johnsongrass seedlings, fall panicum or wild proso millet.

† Partial control (suppression) only.

Annual Broadleaf Weeds:

The broadleaf weeds listed below will be controlled only if treatment is made when conditions are favorable for weed seed germination and growth.

Annual Morningglory	*Ipomoea purpurea*
Carpetweed	*Mollugo verticillata*
Common Purslane	*Portulaca oleracea*
Florida Pusley	*Richardia scabra*
Lambsquarters	*Chenopodium album*
Redroot Pigweed	*Amaranthus retroflexus*
Sicklepod	*Cassia obtusifolia*
Smooth Pigweed	*Amaranthus hybridus*
Velvetleaf (Buttonweed)	*Abutilon theophrasti*
Perennial Weeds:	
Purple Nutsedge, Nutgrass	*Cyperus rotundus*
Yellow Nutsedge, Nutgrass	*Cyperus esculentus*
Rhizome Johnsongrass	*Sorghum halepense*
Quackgrass*	*Agropyron repens*

* Partial control (suppression) only.

Existing stands of perennial weeds must be turned under and chopped up thoroughly before treatment. Cultivation is suggested in addition to SURPASS 6.7-E treatment in fields with moderate to heavy infestations.

GENERAL PRECAUTIONS

Read all label directions before using.

Applied according to directions and under normal growing conditions, SURPASS 6.7-E will not harm the treated crop. During the germination and early stages of growth, extended periods of unusually cold and wet or hot and dry weather, insect or plant disease attack, carry-over pesticide residues, the use of certain soil-applied systemic insecticides, improperly placed fertilizers or soil insecticides may create abnormal conditions that weaken crop seedlings. SURPASS 6.7-E used under these abnormal conditions could result in crop injury.

When applied according to directions and when conditions exist for normal plant growth through the season, no harmful residues of SURPASS 6.7-E should remain beyond harvest. In the Southeastern U.S., when SURPASS 6.7-E is used for weed control in silage corn, do not seed small seeded grains after corn harvest until September.

Do not use on milo or sorghum. Do not use on corn seed stock (Breeders, Foundation, or Increase.)

SURPASS 6.7-E is recommended for use on mineral soils only (soils containing less than 10% organic matter).

Do not contaminate irrigation water (used for crops other than corn) or water used for domestic purposes.

DIRECTIONS FOR USE

It is a violation of federal law to use this product in a manner inconsistent with its label.

SURPASS 6.7-E is a selective herbicide which is mixed (incorporated) or injected into the soil for control of weeds listed on this label. SURPASS 6.7-E controls annual weeds as their seeds germinate by interfering with normal germination and seedling development. It does not control established weeds. All weed growth and crop stubble should be thoroughly worked into the soil before treatment.

IMPREGNATION OF DRY BULK FERTILIZERS

Dry bulk fertilizer may be impregnated with SURPASS 6.7-E (except nitrate fertilizers) and applied as the fertilizer is spread.

See Appendix III for information and directions regarding impregnation and use of these combinations.

COMBINATION WITH FLUID FERTILIZERS

SURPASS 6.7-E may be combined with fluid (solution, slurry or suspension) fertilizers. *However, physical compatibility with these fluid fertilizers must be determined before combining in the spray tank.* See Appendix IV for directions regarding these combinations. Even though found to be compatible, constant agitation is necessary to keep the SURPASS 6.7-E evenly mixed with the fluid fertilizer during application.

APPLICATION AND SOIL INCORPORATION DIRECTIONS

APPLICATION DIRECTIONS

Pour the recommended amount of SURPASS 6.7-E into the spray tank during the filling operations. Apply in 10 to 50 gallons of water per acre using a properly calibrated, low-pressure (20–40 psi) boom type sprayer having good agitation (use the higher gallonage in arid areas). The soil should be well-worked and dry enough to permit good soil mixing (incorporation).

INCORPORATION DIRECTIONS

SURPASS 6.7-E must be incorporated into the soil immediately to prevent loss of the herbicide. Whenever possible, application and incorporation should be done in the same operation.

See Appendix I for soil mixing (incorporation) before planting directions.

IMPORTANT: Thorough mixing for SURPASS 6.7-E in the soil is necessary. The soil must be in good tilth. Improper discing or poor tilth (large clods) may result in unsatisfactory control.

INCORPORATION DIRECTIONS (PACIFIC NORTHWEST ONLY)

Mechanical Incorporation: SURPASS 6.7-E must be incorporated into the soil within 24 hours after application. The application should be made just prior to planting. The soil surface should be dry (at least ½ inch deep) and free from dew and incidental moisture.

IMPORTANT: Thorough mixing of SURPASS 6.7-E in the soil is necessary. The soil must be in good tilth. Improper discing or poor tilth (large clods) may result in unsatisfactory control.

Sprinkler Incorporation: Surface apply SURPASS 6.7-E immmediately after planting. The soil surface should be dry (at least ½ inch deep) and free from dew and incidental moisture. Incorporate using ½ to ¾ inch of water within 36 hours following application. The application and incorporation must be done within 5 days after the last tillage operation since poor results will occur if weeds have germinated.

SOIL INJECTION

Coulters and injector shanks must be spaced 3½ inches apart. The two shanks adjacent to the drill row must be 1¾ to 2 inches of either side of it. Mount injectors in staggered positions to avoid trash buildup. Set injectors to inject SURPASS 6.7-E 3 inches below the soil surface. The width of the band in which weed control is desired will determine the number of injectors required per row. (Example: Four injections spaced 3½ inches apart give a 14 inch band). A broadcast application can be made by increasing the number of shanks. Seal injector openings by using a roller or chain drag behind the injectors. Apply the recommended rate of SURPASS 6.7-E in 20 to 50 gallons of water per acre.

For band applications, calculate the amount to be applied per acre as follows:

$$\frac{\text{Band width in inches}}{\text{distance between rows in inches}} \times \text{Rate/A for broadcast treatment} = \text{Amount needed for band treatment}$$

CENTER PIVOT SPRINKLER APPLICATION

Apply SURPASS 6.7-E with irrigation water before planting (see planting), or immediately after planting before corn or weeds emerge. It is important that application and incorporation be done immediately after the last tillage operation since poor results will occur if the weeds have germinated before application of the herbicide. Meter the SURPASS 6.7-E into the irrigation water during the entire period.

Apply in ½ to 1 inch of water or sufficient water to penetrate the soil to a depth of 5 to 7 inches. If diluted, SURPASS 6.7-E should be mixed at a volume ratio of no more than 1 part SURPASS 6.7-E to 4 parts of water. Good agitation should be maintained during the entire application period.

USE PRECAUTIONS

A. Apply the product only through center pivot irrigation systems containing anti-siphon and check valves which will prevent water source contamination and overflow of the slurry tank and containing interlocking controls between the metering device and the water pump to insure simultaneous shut-off.

B. Inject the product with a positive displacement pump into the main line ahead of a right angle turn to insure adequate mixing.

C. Application of more than label recommended quantities of irrigation water per acre may result in decreased product performance by removing the chemical from the zone of effectiveness.

D. Do not apply when wind speed favors drift, when system connections or fittings leak, when nozzles do not provide uniform distribution or when lines containing the product must be dismantled and drained.

PLANTING

Plant corn seed to a maximum depth of 2 inches. Ideally, planting should be done within 2 weeks after preplant application. If planting must be delayed, do not delay more than 4 weeks after preplant application. To avoid removing SURPASS 6.7-E from the row (with loss of weed control in the row), do not move or shape soil after treatment, such as by planting in deep furrows. If it is necessary to work the soil following application and incorporation, but before planting, do not work the soil deeper than the depth SURPASS 6.7-E was incorporated.

CULTURAL PRACTICES FOLLOWING APPLICATION

Should weeds develop, a shallow cultivation or rotary hoeing will generally result in better weed control. When cultivating for any reason, it should be shallow, i.e., no more than ½ the depth the herbicide was incorporated.

Under circumstances when a poor stand of corn is obtained after SURPASS 6.7-E treatment (caused by flooding, disease, etc.) corn can be planted again without working the field. If it is too late to plant corn again, soybeans or peanuts may be planted providing no other herbicide was used with the SURPASS 6.7-E. An interval of 4 weeks should elapse between when SURPASS 6.7-E was applied and the soybeans or peanuts are planted. Disc cultivate the soil thoroughly prior to planting the soybeans.

DOSAGE RECOMMENDATIONS

All rates are expressed as pints of formulation per broadcast acre. For all uses cultivation is recommended in addition to the SURPASS 6.7-E treatment in fields with moderate to heavy infestations of wild cane, nutgrass, quackgrass, sandbur, johnsongrass seedlings, fall panicum or wild proso millet. NOTE: SURPASS 6.7-E must be applied prior to May 1st on corn grown in Arizona or in the ten southern California counties.

For control of most annual grasses and broadleaf weeds: Use 3¾ to 4¾ pints of SURPASS 6.7-E. Use the higher rate on medium (silt loam, clay loam, etc.) to fine (clay) soils and for control of sandbur, johnsongrass seedlings, fall panicum and witchgrass.

For partial control (suppression) of wild cane and for control of nutsedge (nutgrass) and wild proso millet: Use 4¾ to 7⅓ pints of SURPASS 6.7-E, use the higher rates on medium (silt loam, clay loam, etc.) soils, fine (clay) soils and for heavy infestations.

For partial control (suppression) of quackgrass: Use 7⅓ pints of SURPASS 6.7-E.

Johnsongrass: Using a program of plowing and discing followed by SURPASS 6.7-E will give good control of seedlings and greatly reduce competition from plants coming from rhizomes.

In some areas discing 2 to 3 times spaced at 4 to 6 week intervals during fall and winter can be done. If this cannot be done, start with fall plowing followed by several discings in the spring before planting. The discing cuts rhizomes into small pieces and brings them to the surface, where the weather will kill or weaken them.

Next, apply and disc-incorporate SURPASS 6.7-E just before planting at the rate of 7⅓ pints per acre. Disc a second time at right angles to the first, cultivate at least once, preferably twice.

SURPASS 6.7 PLUS ATRAZINE (80W, 4L or AATREX* NINE-0) TANK MIX COMBINATION

* AATREX NINE-0 is a registered trademark of Ciba-Geigy Corporation

A tank mix combination of SURPASS 6.7-E and ATRAZINE can be applied and mixed into the soil before planting for weed control in corn. This combination will control a broader spectrum of weeds than will either material used alone. In addition, potential hazard of ATRAZINE residues affecting certain succeeding crops is reduced because a lower rate of ATRAZINE can be used when it is combined with SURPASS 6.7-E than when it is used alone.

Follow all use precautions and warnings that appear on the ATRAZINE label, supplementary ATRAZINE literature, and elsewhere on this label.

Make only one application per crop. After treatment with the SURPASS 6.7-E/ATRAZINE combination, do not plant any crop except corn until the following year or injury may occur.

When applied as directed, the SURPASS 6.7-E/ATRAZINE tank mixture will control these weeds:

Annual Grasses:	
Barnyardgrass; Watergrass	*Echinochloa crus-galli*
Crabgrass	*Digitaria* spp.
Fall Panicum*	*Panicum dichotomiflorum*
Field Sandbur*	*Cenchrus pauciflorus*
German Millet	*Setaria italica*
Giant Foxtail	*Setaria faberi*
Goosegrass	*Eleusine indica*
Green Foxtail	*Setaria viridis*
Seedling Johnsongrass*	*Sorghum halepense*
Wild Cane*†	*Sorghum bicolor*
Wild Oats	*Avena fatua*
Wild Proso Millet*	*Panicum miliaceum*
Witchgrass	*Panicum capillare*
Yellow Foxtail	*Setaria glauca*

* Cultivation is suggested in addition to the SURPASS 6.7-E/ATRAZINE treatment in fields with moderate to heavy infestations of wild cane, sandbur, johnsongrass seedlings, fall panicum or wild proso millet.

† Partial control (suppression) only.

Annual Broadleaf Weeds:	
Annual Morningglory	*Ipomoea purpurea*
Buffalobur	*Solanum rostratum*
Common Cocklebur	*Xanthium pensylvanicum*
Common Lambsquarters	*Chenopodium album*
Common Purslane	*Portulaca oleracea*
Florida Pusley	*Richardia scabra*
Horsenettle	*Solanum carolinense*
Jimsonweed	*Datura stramonium*
Ladysthumb	*Polygonum persicaria*
Pennsylvania Smartweed	*Polygonum pensylvanicum*
Prostrate Pigweed	*Amaranthus graecizans*
Ragweed	*Ambrosia* spp.
Redroot Pigweed	*Amaranthus retroflexus*
Sicklepod	*Cassia obtusifolia*
Smooth Pigweed	*Amaranthus hybridus*
Velvetleaf	*Abutilon theophrasti*
Wild Mustard	*Brassica kaber*
Perennial Grasses:	
Purple Nutsedge, Nutgrass	*Cyperus rotundus*
Yellow Nutsedge, Nutgrass	*Cyperus esculentus*
Rhizome Johnsongrass	*Sorghum halepense*
Quackgrass**	*Agropyron repens*

** For partial control (suppression, of quackgrass: use 7⅓ pints of SURPASS 6.7-E and 1¼ to 2* pounds ATRAZINE 80W, 1 to 1½ quarts ATRAZINE 4L, or 1 to 1¾* pounds AATREX NINE-O per acre.

Existing stands of perennial weeds must be turned under and chopped up thoroughly before treatment. Cultivation is suggested in addition to SURPASS 6.7-E/ATRAZINE treatment in fields with moderate or heavy infestations.

* Use the higher rate of ATRAZINE on soils with over 5% organic matter.

DIRECTIONS FOR USE

The SURPASS 6.7-E/ATRAZINE (80W or 4L) combination can be mixed with water or combined with fluid (solution, slurry, or suspension) fertilizers and applied as a spray, or be impregnated on dry bulk fertilizers. The SURPASS 6.7-E/AATREX NINE-O combination can be mixed with water and applied as a spray.

Dry bulk fertilizer may be impregnated with SURPASS 6.7-E/ATRAZINE (80W or 4L) (except nitrate fertilizers) and applied as the fertilizer is spread. See Appendix III for information and directions regarding impregnation and use of these combinations.

SURPASS 6.7-E/ATRAZINE (80W or 4L) combination can be combined with fluid fertilizers for applications made before planting. However, the physical compatibility with these fluid fertilizers must be determined before combining in the spray tank, and even though found to be compatible, constant agitation is necessary to keep the SURPASS 6.7-E/ATRAZINE evenly mixed with the fluid fertilizer during application. Refer to Appendix II regarding the procedure for testing the compatiblity of SURPASS 6.7-E/ATRAZINE combination with fluid fertilizers.

See Appendix II for Tank Mix Directions.

APPLICATION AND SOIL INCORPORATION DIRECTIONS

SPRAY APPLICATION AND MECHANICAL INCORPORATION

Apply the recommended dosage in 10 to 50 gallons of water per acre (use the higher gallonage in arid areas), using a properly calibrated, low-pressure (20-40 psi) boom-type sprayer having good, but not excessive agitation. Check frequently during application to be sure nozzles are free from clogging and are delivering a uniform spray pattern. Read directions on the ATRAZINE label concerning nozzle and screen sizes.

The herbicides must be incorporated into the soil immediately after application to prevent loss of the SURPASS. Whenever possible, application and incorporation should be done in the same operation. Apply the herbicides to well-worked soil, dry enough to permit good soil mixing (incorporation).

See Appendix I for soil mixing (incorporation) before planting directions.

IMPORTANT: Thorough mixing of SURPASS 6.7-E in the soil is necessary. The soil must be in good tilth. Improper discing or poor tilth (large clods) may result in unsatisfactory control.

INCORPORATION DIRECTIONS (PACIFIC NORTHWEST ONLY)

Mechanical Incorporation: The tank mix must be incorporated into the soil within 24 hours after application. The application should be made just prior to planting. The soil surface should be dry (at least ½ inch deep) and free from dew and incidental moisture.

IMPORTANT: Thorough mixing of SURPASS 6.7-E in the soil is necessary. The soil must be in good tilth. Improper discing or poor tilth (large clods) may result in unsatisfactory control.

Sprinkler Incorporation: Surface apply SURPASS 6.7-E/ATRAZINE immediately after planting. The soil surface should be dry (at least ½ inch deep) and free from dew and incidental moisture. Incorporate using ½ to ¾ inch of water within 36 hours following application. The application and incorporation must be done within 5 days after the last tillage operation since poor results will occur if weeds have germinated.

PLANTING

Do not plant seed deeper than 2 inches. For SURPASS 6.7-E/ATRAZINE application made before planting, seeding can be done immediately, or it may be delayed up to 4 weeks. To avoid removing SURPASS 6.7-E/ATRAZINE from the row (with loss of weed control in the row), do not move or shape soil after treatment, such as by planting in deep furrows. If it is necessary to work the soil following application and incorporation, but before planting, do not work the soil deeper than the depth the hebicides were incorporated. Do not use on milo or sorghum.

Do not use on corn seed stock (Breeders, Foundation, or Increase).

DOSAGE RECOMMENDATIONS

All rates are expressed as pints of SURPASS 3.7-E combined with quarts of ATRAZINE 4L or pounds of ATRAZINE 80W or AATREX NINE-O per broadcast acre.

NOTE: The SURPASS 6.7-E/ATRAZINE tank mix combination is not recommended on corn grown in Arizona or in the ten southern California counties.

For all uses cultivation is recommended in addition to the SURPASS 6.7-E/ATRAZINE treatment in fields with moderate to heavy infestations of wild cane, nutgrass, quackgrsss, sandbur, johnsongrass seedlings, fall panicum or wild proso millet.

For control of annual grasses and broadleaf weeds: Use 3¾ to 4¾ pints SURPASS 6.7-E and 1 to 2* pounds of ATRAZINE 80W, ¾ to 1½* quarts ATRAZINE 4L, or 8/10 to 1¾ pounds AATREX NINE-O per acre. Use the higher rate of SURPASS 6.7-E on medium (silt loam, clay loam, etc.) to fine (clay) soils and for control of sandbur, johnsongrass seedlings, fall panicum and witchgrass.

For partial control (suppression) of wild cane and for control of nutsedge (nutgrass) and wild proso millet (in addition to the broadleaf weeds and annual grasses listed): Use 4¾ to 7⅓ pints of SURPASS 6.7-E and 1 to 2* pounds ATRAZINE 80W or ¾ to 1½* quarts ATRAZINE 4L or 8/10 to 1¾* pounds AATREX NINE-O per acre. Use the higher rates of SURPASS 6.7-E on medium (silt loam, clay loam, etc.) soils, fine (clay) soils, and for heavy infestations.

For partial control (suppression) of quackgrass: Use 7⅓ pints of SURPASS 6.7-E and 1¼ to 2* pounds ATRAZINE 80W or 1 to 1½ quarts ATRAZINE 4L, or 1 to 1¾* pounds AATREX NINE-O per acre.

Johnsongrass: Using a program of plowing and discing followed by SURPASS 6.7-E and ATRAZINE will give good control of seedlings and greatly reduce competition from plants coming from rhizomes.

In some areas discing 2 to 3 times spaced at 4 to 6 week intervals during fall and winter is recommended. If this cannot be done, start with fall plowing followed by several discings in the spring before planting. The discing cuts rhizomes into small pieces and brings them to the surface, where both the weather and SURPASS 6.7-E will kill or weaken them.

NEXT, apply and disc-incorporate SURPASS 6.7-E just before planting at the rate of 7⅓ pints plus 1 to 2* pounds of ATRAZINE 80W or ¾ to 1½* quarts of ATRAZINE 4L, or 8/10 to 1¾* pounds AATREX NINE-O per acre. Disc a second time at right angles to the first. Cultivate at least once, preferably twice.

* Use the higher rate of ATRAZINE on soils with over 5% organic matter.

CULTURAL PRACTICES FOLLOWING APPLICATION

Should weeds develop, a shallow cultivation or rotary hoeing will generally result in better weed control. When cultivating for any reason, it should be shallow, i.e., no more than ½ the depth the herbicides were incorporated.

For all uses cultivation is recommended in addition to the SURPASS 6.7-E/ATRAZINE treatment in fields with moderate to heavy infestations of wild cane, nutgrass, quackgrass, sandbur, johnsongrass seedings, fall panicum or wild proso millet.

SURPASS 6.7-E/BLADEX* (80W or 4L) COMBINATION

* BLADEX is a registered trademark of Shell Chemical Company

SURPASS 6.7- E may be used in combination with BLADEX (80W or 4L) for weed control in corn. These combinations will control a broader spectrum of weeds than will either of the products used alone.

These combinations may be applied as a tank mix, preplant and soil-incorporated; or the BLADEX(80W or 4L) may be applied pre-emergence following a preplant incorporated application of SURPASS 6.7-E.

IMPORTANT: Observe all cautions and imitations on the BLADEX (80W or 4L) and SURPASS 6.7-E labels.

Make only one application per crop.

Do not use on milo or sorghum. Do not use on corn seed stock (Breeders, Foundation, or Increase).

Do not contaminate water by cleaning of equipment or disposal of wastes.

WEEDS CONTROLLED BY SURPASS 6.7-E/BLADEX

Annual Grasses:

Annual Bluegrass	*Poa annua*
Annual Fescues	*Festuca* spp.
Annual (Italian) Ryegrass	*Lolium multiflorum*
Barnyardgrass, Watergrass	*Echinochloa crus-galli*
Bullgrass	*Paspalum boscianum*
Crabgrass	*Digitaria* spp.
Fall Panicum*	*Panicum dichotomiflorum*

Field Sandbur*	*Cenchrus pauciflorus*
German Millet	*Setaria italica*
Giant Foxtail	*Setaria faberi*
Goosegrass	*Eleusine indica*
Green Foxtail	*Setaria viridis*
Indian Lovegrass	*Eragrostis pilosa*
Junglerice	*Echinochloa colonum*
Seedling Johnsongrass*	*Sorghum halepense*
Stinkgrass	*Eragrostis cilianesis*
Wild Cane*†	
Shattercane,	
Black Amber Cane	
Chicken Corn	
Scatter Cane	*Sorghum bicolor*
Wild Oats	*Avena fatua*
Wild Proso Millet*	*Panicum miliaceum*
Witchgrass	*Panicum capillare*
Yellow Foxtail	*Setaria glauca*

* Cultivation is suggested in addition to the SURPASS 6.7-E/BLADEX treatment in fields with moderate to heavy infestations of fall panicum, sandbur, johnsongrass seedlings, wild cane or wild proso millet.

† Partial control (suppression) only.

Annual Broadleaf Weeds:	
Annual Buttercup	*Ranunculus* spp.
Annual Groundcherry	*Physalis lanceifolia*
Annual Morningglory	*Ipomoea purpurea*
Annual Sedge	*Carex* spp.
Black Mustard	*Brassica nigra*
Buffalobur	*Solanum rostratum*
Carpetweed	*Mollugo verticillata*
Common Chickweed	*Stellaria media*
Common Cocklebur	*Xanthium pensylvanicum*
Common Groundsel	*Senecio vulgaris*
Common Lambsquarters	*Chenopodium album*
Common Mallow	*Malva neglecta*
Common Purslane	*Portulaca oleracea*
Corn Spurry	*Spergula arvensis*
Floras Paintbrush	*Emilia sagittata*
Florida Pusley	*Richardia scabra*
Hedge Mustard	*Sisymbrium officinale*
Jimsonweed	*Datura stramonium*
Kochia	*Kochia scoparia*
Ladysthumb	*Polygonum persicaria*
Mayweed	*Anthemis cotula*
Pennsylvania Smartweed	*Polygonum pensylvanicum*
Pigweed	*Amaranthus* spp.
Pineappleweed	*Matricaria matricarioides*
Plantain	*Plantago* spp.
Poorjoe	*Diodia teres*
Prostrate Knotweed	*Polygonum aviculare*
Prostrate Spurge	*Euphorbia supina*
Ragweed	*Ambrosia* spp.
Seedling Curly Duck	*Rumex crispus*
Shepherdspurse	*Capsella bursa-pastoris*
Sicklepod	*Cassia obtusifolia*
Smallflower Galinsoga	*Galinsoga parviflora*
Prickly sida	*Sida spinosa*
Sunflower	*Helianthus annuus*
Tarweed	*Hemizonia congesta*
Velvetleaf	*Abutilon theophrasti*
Wild Buckwheat	*Polygonum convolvulus*
Wild Mustard	*Brassica kaber*
Wild Radish	*Raphanus raphanistrum*
Wild Turnip	*Brassica campestris*
Perennial Weeds:	
Rhizome Johnsongrass	*Sorghum halepense*
Purple Nutgrass, Nutsedge	*Cyperus rotundus*
Yellow Nutgrass, Nutsedge	*Cyperus esculentus*
Quackgrass*	*Agropyron repens*

* Partial control (suppression) only

Existing stands of perennial weeds must be turned under and chopped up thoroughly before treatment. Cultivation is suggested in addition to SURPASS 6.7-E/BLADEX treatment in fields with moderate to heavy infestations.

DIRECTIONS FOR USE

The SURPASS 6.7-E/BLADEX (80W or 4L) combination can be mixed with water or combined with fluid (solution, slurry, or suspension) fertilizers and applied as a spray.

TANK MIX DIRECTION: See Appendix II

APPLICATION AND INCORPORATION DIRECTIONS:

For a tank mix application, apply the recommended dosage (see RECOMMENDATIONS) in 10 to 50 gallons of water per acre, (use the higher gallonage in arid areas) using a properly calibrated, low-pressure type nozzle (flood, fan or swirl chamber) on boom-type sprayer having good agitation. Check frequently during application to be sure the nozzles are free from clogging and are delivering a uniform spray pattern. Read directions on the BLADEX (80W or 4L) label concerning nozzle and screen sizes.

The tank mix must be incorporated into the soil immediately after application to prevent loss of the SURPASS 6.7-E. Whenever possible, application and incorporation should be done in the same operation. Apply the herbicides to worked soil dry enough to permit good mixing.

See Appendix I for soil mixing incorporation before planting directions.

IMPORTANT: Thorough mixing of SURPASS 6.7-E in the soil is necessary. The soil must be in good tilth. Improper discing or poor tilth (large clods) may result in unsatisfactory control.

INCORPORATION DIRECTIONS (PACIFIC NORTHWEST ONLY)

Mechanical Incorporation: The tank mix must be incorporated into the soil within 24 hours after application. The application should be made just prior to planting. The soil surface should be dry (at least ½ inch deep) and free from dew and incidental moisture.

IMPORTANT: Thorough mixing of SURPASS 6.7-E in the soil is necessary. The soil must be in good tilth. Improper discing or poor tilth (large clods) may result in unsatisfactory control.

Sprinkler Incorporation: Surface apply SURPASS 6.7-E/ BLADEX immediately after planting. The soil surface should be dry (at least ½ inch deep) and free from dew and incidental moisture. Incorporate using ½ to ¾ inch of water within 36 hours following application. The application and incorporation must be done within 5 days after the last tillage operation since poor results will occur if weeds have germinated.

SURPASS 6.7-E/BLADEX (80W or 4L)
FLUID FERTILIZER TANK MIX

The SURPASS 6.7-E/BLADEX (80W or 4L) combination can be combined with fluid fertilizers for applications made before planting. However, the physical compatiblity with these fluid fertilizers must be determined before combining in the spray tank, and even though found to be compatible, constant agitation is necessary to keep SURPASS 6.7-E/BLADEX evenly mixed with the fluid fertilizer during application.

Refer to Appendix IV regarding the procedure for testing the compatibility of SURPASS 6.7-E/BLADEX combination with fluid fertilizers.

For sequential application of SURPASS 6.7-E preplant incorporated followed by BLADEX (80W or 4L) pre-emergence, consult the appropriate application directions on the respective product labels. Follow the dosage recommendations given below.

DOSAGE RECOMMENDATIONS

For all uses, cultivation is recommended in addition to the SURPASS 6.7-E/BLADEX treatment in fields with moderate to heavy infestations of wild cane, nutgrass, quackgrass, sandbur, johnsongrass seedlings, fall panicum or wild proso millet.

Use the lower rate of BLADEX on coarse to medium textured soils. Use the higher rate of BLADEX on fine (fine texture-clay) soils.

Consult the BLADEX label for more specific recommended rates on various soil types.

SURPASS 6.7-E/BLADEX can be applied either as a tank mix combination and incorporated into the soil or apply SURPASS 6.7-E preplant incorporated followed by BLADEX pre-emergence. If BLADEX is surface-applied on top of a preplant application of SURPASS 6.7-E, a rotary hoeing or shallow cultivation is recommended if rainfall or sprinkler irrigation has not occurred within about 10 days after the application of BLADEX.

NOTE: SURPASS 6.7-E must be applied prior to May 1st on corn grown in Arizona or the ten southern California counties.

Broadcast Rate Per Acre (Conventional Sprays)

For control of annual grasses and broadleaf weeds: Use 3¾ to 4¾ pints of SURPASS 6.7-E and 2 to 2½ pounds of BLADEX 80W or 3 to 4 pints BLADEX 4L per acre. Use the higher rates of SURPASS 6.7-E on medium (medium fine texture-silt loam, clay loam, etc.) to fine (fine textured-clay) soils and for control of sandbur, johnsongrass seedlings, fall panicum and witchgrass.

For partial control (suppression) of wild cane and for control of nutsedge (nutgrass) and wild proso millet (in addition to the broadleaf weeds and annual grasses listed): Use 4¾ to 7⅓ pints of SURPASS 6.7-E and 2 to 2½ pounds BLADEX 80W or 3 to 4 pints BLADEX 4L per acre. Use the higher rates of SURPASS 6.7-E on medium (silt loam, clay loam, etc.) soils, fine (clay) soils, and for heavy infestations.

For partial control (suppression) of quackgrass: Use 7⅓ pints SURPASS 6.7-E and 2 to 2½ pounds of BLADEX 80W or 3 to 4 pints BLADEX 4L per acre.

Johnsongrass: Using a program of plowing and discing followed by SURPASS 6.7-E and BLADEX will give good control of seedlings and greatly reduce competition from plants coming from rhizomes.

In some areas discing 2 to 3 times spaced at 4 to 6 week intervals during fall and winter is recommended. If this cannot be done, start with fall plowing followed by several discings in the spring before planting. The discing cuts rhizomes into small pieces and brings them to the surface, where both the weather and SURPASS 6.7-E will kill or weaken them.

NEXT, apply and disc-incorporate SURPASS 6.7-E just before planting at the rate of 7⅓ pints plus 2 to 2½ pounds of BLADEX 80W or 3 to 4 pints of BLADEX 4L per acre. Disc a second time at right angles to the first. Cultivate at least once, preferably twice.

PLANTING

Do not plant seed deeper than 2 inches. For a SURPASS 6.7-E/BLADEX application made before planting, seeding can be done immediately or it may be delayed up to 4 weeks. To avoid removing SURPASS 6.7-E/BLADEX from the row (with loss of weed control in the row), do not move or shape soil after treatment, such as by planting in deep furrows. If it is necessary to work the soil following application and incorporation, but before planting, do not work the soil deeper than the depth the herbicides were incorporated.

CULTURAL PRACTICES FOLLOWING APPLICATION

Should weeds develop, a shallow cultivation or rotary hoeing will generally result in better weed control. When cultivating for any reason, it should be shallow, i.e., no more than ½ the depth the herbicide was incorporated.

APPENDIX I

SOIL MIXING (incorporation) BEFORE PLANTING DIRECTIONS

The following equipment commonly is used for soil mixing (incorporation) before planting.

Power-Driven Cultivation Equipment (Recommended on all soil types previously primary tilled): Ground speed should be adjusted with all PTO driven cultivation tools to insure thorough soils mixing.

a. Horizontal action: set to cut 4 to 6 inches deep.
b. Vertical action: set to cut 3 to 4 inches deep.

Tandem Discs (Recommended on all soil types): Set to cut to a depth of 4 to 6 inches, operated at 4 to 6 mph followed by a spike-tooth harrow or some other leveling device which extends beyond the ends of the discs. For more thorough mixing (for control or suppression of deep-germinating grasses—wild cane, wild oats, seedling johnsongrass, nutsedge and wild proso millet), disc in two different directions. When discing the second time, the disc should be reset to cut no deeper than 4 to 6 inches.

Field Cultivator (Recommended on lighter soils in good tilth): Use 3 to 4 rows of sweeps spaced at 7 inch or less intervals and staggered so that no soil is left unturned, followed by a spike-tooth harrow pulled behind the cultivator. Do not use chisel plows for incorporation. Set the cultivator to cut 4 to 6 inches deep, operated at 5 mph or more. The field cultivator should not be used alone as an incorporation tool where control or suppression of deep-germinating weeds such as nutgrass, wild proso millet, johnsongrass seedlings, wild cane or wild oats is desired. For control of these weeds, use a disc alone for at least one of the passes over the field.

APPENDIX II

TANK MIX DIRECTIONS

1) Fill the spray tank at least ⅔ full of appropriate carrier. Be sure tank is clean and not contaminated with 2,4-D, oil residues or other materials.
2) Start agitation and continue moderate agitation throughout mixing.
3) Premix the BLADEX or ATRAZINE making sure it is thoroughly wetted and dispersed in the tank prior to adding SURPASS 6.7-E.
4) Add the SURPASS 6.7-E to the tank ⅔ full of carrier already mixed with BLADEX or ATRAZINE.
5) Complete filling the remaining portion of the tank with carrier; mix thoroughly, maintain moderate agitation to provide a uniform mix during application.
6) The tank mix combination should not be left in the spray tank for prolonged periods; batches should be mixed and applied the same day.

Do not use excessive agitation as it may cause the BLADEX or ATRAZINE to settle out and set up in the spray tank. All return lines to the tank must discharge below liquid level.

APPENDIX III

SURPASS 6.7-E or SURPASS 6.7-E/ATRAZINE (80W or 4L) IMPREGNATION ON DRY BULK FERTILIZERS

CAUTION: SURPASS 6.7-E alone or in combination with other herbicides must not be impregnated on ammonium nitrate, potassium nitrate or sodium nitrate fertilizers. Such mixtures may cause explosion and fire.

All individual state regulations relating to dry bulk fertilizer blending, registration, labeling and application are the responsibility of the individual and/or company selling the SURPASS 6.7-E or SURPASS 6.7-E/ATRAZINE fertilizer mixtures.

SURPASS 6.7-E alone and in combination with ATRAZINE may be impregnated on many dry bulk fertilizers (1) (2) and applied and incorporated into the soil either in the Fall before the ground freezes or before planting for the control of certain grassy weeds by SURPASS 6.7-E and many grassy and broadleaf weeds by the SURPASS 6.7-E/ATRAZINE combination in corn.

All SURPASS 6.7-E and ATRAZINE (80W or 4L) label and supplementary literature instructions and precautions regarding rates per acre, soil type and soil incorporation, application and other directions must be followed.

Test results have shown that SURPASS 6.7-E and SURPASS 6.7-E/ATRAZINE (80W or 4L) on many dry bulk fertilizers give weed control equal to SURPASS 6.7-E or tank mixtures of SURPASS 6.7-E/ATRAZINE (80W or 4L) applied as a spray in water or liquid fertilizer. However, uniform impregnation of the pesticides on the dry fertilizer particles and uniform application in the field are necessary to assure good results.

A minimum of 200 pounds and a maximum of 700 pounds of approved fertilizer ingredients (1) (2) impregnated with the appropriate amount of SURPASS 6.7-E alone or in combination with ATRAZINE (80W or 4L) must be applied per acre.

For impregnating the pesticides on dry fertilizers, use a closed rotary-drum type mixer equipped with suitable spraying equipment. The spray nozzles should be positioned inside of the mixer to provide uniform spray coverage of the tumbling fertilizer. The SURPASS 6.7-E should be sprayed uniformly onto the fertilizer using a fine spray pattern. For the SURPASS 6.7-E/ATRAZINE combination product, the ATRAZINE (80W or 4L) can be added and mixed uniformly with the fertilizer prior to the SURPASS 6.7-E addition, or alternatively, the SURPASS 6.7-E/ATRAZINE (80W or 4L) can be mixed in the proper use ratio to prepare a uniform slurry for joint spray impregnation.

The physical properties of fertilizers vary, particularly in liquid absorptive capacity. When absorptivity is sufficient, simple spray impregnation of the fertilizer with SURPASS 6.7-E provides a satisfactory dry mixture.

If the absorptive capacity is not adequate, use of a drying agent is required to provide a dry, free-flowing mixture.

For mixtures to be used in spinning-disc applicators, Microcel E calcium silicate powder (Manville, Filtration & Minerals) is recommended for use as a drying agent.

For mixtures to be used in pneumatic applicators, Microcel E or Agsorb 16/30 RVM-MS granular-clay (Oil-Dri Corporation) is recommended for use as the drying agent.

The drying agent should be added separately and uniformly to the prepared pesticide-fertilizer mixture, in a quantity that is sufficient to provide a suitably free-flowing mixture. Generally, less than 2% Microcel E or 5% Agsorb 16/30 RVM-MS by weight is required.

The amount of SURPASS 6.7-E or SURPASS 6.7-E/ATRAZINE (80W or 4L) actually required in the manufacture of individual fertilizer mixtures should be determined carefully for each production operation (3). This is necessary to ensure that the amounts of SURPASS 6.7-E and ATRAZINE actually contained in the mixture applied to the soil represent the correct rates of use.

Bulk fertilizers impregnated with SURPASS 6.7-E or SURPASS 6.7-E/ATRAZINE (80W or 4L) should be *applied immediately,* NOT STORED. All bulk containers should be tightly covered while the products are being transported and applied to reduce chances of SURPASS 6.7-E loss via volatilization.

Do not combine SURPASS 6.7-E plus ATRAZINE (80W or 4L) with single superphosphate (0–20–0) or treble superphosphate (0–46–0).

SURPASS 6.7-E Physical Data

Specific Gravity 20/20°C:	0.966 (typical)
Pounds/Gallon, 20/20°C:	8.04 (typical)
Flashpoint:	>212°F (Setaflash closed cup)
Viscosity:	Sprayable down to minus 20°F.

(1) APPROVED DRY FERTILIZER INGREDIENTS FOR USE WITH SURPASS 6.7-E

	N	P	K
Ammonium Sulfate	21	0	0
Ammonium Phosphate-Sulfate	16	20	0
Diammonium Phosphate	18	46	0
Monoammonium Phosphate	11	56	0
Potassium Chloride	0	0	60
Potassium Sulfate	0	0	52
Single Superphosphate	0	20	0
Treble Superphosphate	0	46	0
Urea*	45	0	0

* Some ureas may be phytotoxic when high rates are applied to corn. Use only urea rates known to be safe for corn application.

NOTE: Sul-Po-Mag, K-Mag and 11-48-0 have been shown to be compatible with SURPASS 6.7-E and are approved for use.

(2) APPROVED DRY FERTILIZER INGREDIENTS FOR USE WITH SURPASS 6.7-E/ATRAZINE (80W or 4L)

	N	P	K
Ammonium Sulfate	21	0	0
Ammonium Phosphate-Sulfate	16	20	0
Diammonium Phosphate	18	46	0
Monoammonium Phosphate	11	56	0
Potassium Chloride	0	0	60
Potassium Sulfate	0	0	52
Urea*	45	0	0

* Some ureas may be phytotoxic when high rates are applied to corn. Use only urea rates known to be safe for corn application.

NOTE: Sul-Po-Mag, K-Mag and 11-48-0 have been shown to be compatible with SURPASS 6.7-E/ATRAZINE 80W and are approved for use.

(3) RATE CHART FOR THE IMPREGNATION OF DRY BULK FERTILIZERS WITH SURPASS 6.7-E OR SURPASS 6.7-E PLUS ATRAZINE (80W OR 4L)

FERTILIZER RATE PER ACRE	SURPASS 6.7-E RATE/ACRE			ATRAZINE 80WP RATE/ACRE		ATRAZINE 4L RATE/ACRE	
	3¾ PTS. PER ACRE	4¾ PTS. PER ACRE	7⅓ PTS. PER ACRE	1¼ LBS. PER ACRE	2 LBS. PER ACRE	1 QT. PER ACRE	1½ QTS. PER ACRE
200 lbs.	18¾ qts./ton	23¾ qts./ton	36⅔ qts./ton	12½ lbs./ton	20 lbs./ton	10 qts./ton	15 qts./ton
250 lbs.	15 qts./ton	19 qts./ton	29⅓ qts./ton	10 lbs./ton	16 lbs./ton	8 qts./ton	12 qts./ton
300 lbs.	12½ qts./ton	15⅘ qts./ton	24⅖ qts./ton	8⅓ lbs./ton	13⅓ lbs./ton	6⅗ qts./ton	10 qts./ton
350 lbs.	10¾ qts./ton	13⅗ qts./ton	21 qts./ton	7⅛ lbs./ton	11⅖ lbs./ton	5⅘ qts./ton	8⅗ qts./ton
400 lbs.	9⅖ qts./ton	11⅞ qts./ton	18⅓ qts./ton	6¼ lbs./ton	10 lbs./ton	5 qts./ton	7½ qts./ton
450 lbs.	8⅓ qts./ton	10½ qts./ton	16⅓ qts./ton	5½ lbs./ton	8⅞ lbs./ton	4⅖ qts./ton	6⅗ qts./ton
500 lbs.	7½ qts./ton	9½ qts./ton	14⅔ qts./ton	5 lbs./ton	8 lbs./ton	4 qts./ton	6 qts./ton
550 lbs.	6⅘ qts./ton	8⅔ qts./ton	13⅓ qts./ton	4⅔ lbs./ton	7⅓ lbs./ton	3⅔ qts./ton	5½ qts./ton
600 lbs.	6¼ qts./ton	8 qts./ton	12⅕ qts./ton	4⅕ lbs./ton	6⅔ lbs./ton	3⅓ qts./ton	5 qts./ton
650 lbs.	5¾ qts./ton	7⅓ qts./ton	11¼ qts./ton	3⅞ lbs./ton	6⅕ lbs./ton	3⅛ qts./ton	4⅗ qts./ton
700 lbs.	5⅓ qts./ton	6⅘ qts./ton	10½ qts./ton	3⅔ lbs./ton	5¾ lbs./ton	2⅞ qts./ton	4¼ qts./ton

APPENDIX IV

PROCEDURE FOR TESTING THE COMPATIBILITY OF SURPASS 6.7-E or SURPASS 6.7-E/ATRAZINE (80W or 4L) or SURPASS 6.7-E/BLADEX (80W or 4L) WITH FLUID FERTILIZERS.

The following procedure is suggested for determining whether the above herbicide combinations may be combined with a specific fluid fertilizer for spray tank application.

MATERIALS REQUIRED

1. A. SURPASS 6.7-E

   B. or SURPASS 6.7-E/ATRAZINE (80W or 4L) combination

   C. or SURPASS 6.7-E/BLADEX (80W or 4L) combination.

2. Fluid fertilizer to be used.

3. Tank mix adjuvants for fluid fertilizer combinations: Compex*, Sponto 168-D*, Uni-Mix*, Unite* or equivalent. The adjuvant which provides the best compatibility depends on the specific fertilizer involved.

* Compex, Kalo Laboratories Inc., Kansas City, Missouri; Sponto 168-D, Witco Chemical Company, Houston, Texas; Uni-Mix, Thompson-Hayward Chemical Co., Kansas City, Kansas; Unite, Hopkins Agricultural Chemical Company, Madison, Wisconsin.

4. Two one-quart, wide-mouth glass jars with lids.
5. Measuring spoons; a graduated pipette or cylinder will provide more accurate measurement; 1 tablespoon equals 3 teaspoons, 1 teaspoon equals approximately 5 milliliters.
6. Measuring cup, 8 oz. (237 mL).

PROCEDURE (See Note 1):

1. Pour a pint (473 mL) of the fluid fertilizer into each of the quart jars.
2. Add adjuvant to one of the jars and mix (see next rate table).
3. Add the SURPASS 6.7-E or the desired combination of herbicides to both jars (see next four rate tables). The order of addition should be: wettable powders first with mixing, flowables next with mixing, and the SURPASS 6.7-E last.
4. Close both jars with lids and mix the contents by turning the jars upside down ten times.
5. Inspect the surface and body of the mixtures—
   (A) Immediately after completing the jar inversions.
   (B) After allowing the jars to stand quietly for 30 minutes.
   (C) And then again after turning the jars upside down 10 times.

If a uniform mix cannot be made, the mixture should not be used. If either mixture remains uniform for 30 minutes, the combination may be used. Should either mixture separate after 30 minutes, but readily remix uniformly with ten jar inversions, the mixture can be used if adequate agitation is maintained in the tank. If the mixture with adjuvant is satisfactory, but the one without adjuvant is not, be sure to use the adjuvant in the spray tank. Add the adjuvant first at a rate of 3 pints per 100 gallons of fluid fertilizer; foaming can be minimized by using moderate agitation.

If nondispersible oil, sludge or clumps of solids form in the mixtures, the combination should not be used.

NOTE 1: For some combinations, premixing wettable powders in a little water in a pail before adding them to the spray tank will improve the compatibility of the final mixtures with SURPASS 6.7-E. This technique can be tested in the small-scale jar test by premixing the wettable powder in one-eighth (1/8) cup of water prior to addition to the pint of fluid fertilizer.

Rate Table for SURPASS 6.7-E and Adjuvant with the Fluid Fertilizer**

Gallons of Fluid Fertilizer to be applied per acre	mL. or tsp. of SURPASS 6.7-E* to be added to 1 pint of Fertilizer	
	mL.	tsp.
10	7	1 1/3
15	5	1
20	4	3/4
25	3	2/3
30	2	1/2
40	2	1/2

* Based on field rate of 1 pound active ingredient per acre in the fertilizer volumes indicated. Increase volume proportionately to correspond with intended field rate in terms of pounds active ingredient per acre (e.g. for field rate of 4 pounds actual SURPASS 6.7-E in 40 gallons fertilizer per acre, add 8 mL. or 2 tsp. SURPASS 6.7-E to each jar for compatibility testing purposes).

** Two (2) milliliters or one-half (1/2) teaspoon of adjuvant should be added to 1 pint of fluid fertilizer in order to equal the rate of 3 pints of adjuvant per 100 gallons of fluid fertilizers.

Rate Table* for SURPASS 6.7-E/ATRAZINE (80W or 4L) and Adjuvant with Fluid Fertilizer**

Gallons of Fluid Fertilizer to be applied per acre	10	15	20	25	30	40
tsp. of SURPASS 6.7-E	5 3/4	3 3/4	2 3/4	2 1/3	2	1 1/2
tsp. of ATRAZINE 80W	7	4 2/3	3 1/2	2 3/4	2 1/3	1 3/4
tsp. of ATRAZINE 4L	3 2/3	2 1/2	1 3/4	1 1/2	1 1/4	1

* Based on field rates of 4 pounds actual SURPASS and 1.6 pounds actual ATRAZINE (80W) or 1.5 pounds actual ATRAZINE (4L) per acre in the fertilizer volumes indicated.

** Two (2) milliliters or one-half (1/2) teaspoon of adjuvant should be added to 1 pint of fluid fertilizer in order to equal the rate of 3 pints of adjuvant per 100 gallons of fluid fertilizer.

Rate Table* for SURPASS 6.7-E/BLADEX (80W or 4L) and Adjuvant with Fluid Fertilizer**

Gallons of Fluid Fertilizer to be applied per acre	10	15	20	25	30	40
tsp. of SURPASS 6.7-E	5 3/4	3 3/4	2 3/4	2 1/3	2	1 1/2
tsp. of BLADEX 80W	7 3/4	5 1/4	4	3	2 2/3	2
tsp. of BLADEX 4L	4 3/4	3 1/4	2 1/2	2	1 2/3	1 1/4

* Based on field rates of 4 pounds actual SURPASS and 2 pounds actual BLADEX (80W or 4L) per acre in the fertilizer volumes indicated.

** Two (2) milliliters or one-half (1/2) teaspoon of adjuvant should be added to 1 pint of fluid fertilizer in order to equal the rate of 3 pints of adjuvant per 100 gallons of fluid fertilizer.

STORAGE AND DISPOSAL

PROHIBITIONS

Do not contaminate water, food or feed by storage and disposal.

PESTICIDE DISPOSAL

Rinse spray equipment. Wastes resulting from the use of this product may be disposed of on site or at an approved waste disposal facility.

CONTAINER DISPOSAL

Triple rinse (or equivalent), then offer for recycling, reconditioning, or puncture and dispose of in a sanitary landfill, or by other approved State or local procedures.

NOTICE—READ CAREFULLY

CONDITIONS OF SALE

Stauffer (and seller) offer(s) this product for sale subject to, and buyer and all users are deemed to have accepted, the following conditions of sale and warranty which may only be varied by written agreement of a duly authorized representative of Stauffer.

WARRANTY LIMITATION

Stauffer warrants that this product conforms to the chemical description on the label and is reasonably fit for the purposes referred to in the directions for use on the label subject to the inherent risk referred to below. Stauffer makes no other express warranties; THERE IS NO IMPLIED WARRANTY OF MERCHANTABILITY and there are no warranties which extend beyond the description on the label hereof.

INHERENT RISKS

The directions for use of this product are believed to be reliable and should be followed carefully. However, it is impossible to eliminate all risks associated with use. Buyer assumes all risks associated with use or application of this product contrary to label instructions or resulting from extraordinary weather conditions.

LIMITATION OF LIABILITY

In no case shall Stauffer be liable for special, indirect or consequential damages resulting from the use or handling of this product and no claim of any kind shall be greater in amount than the purchase price of the product in respect of which such damages are claimed.

This product is sold only for uses stated on the label. No express or implied license is granted to use or sell this product under any patent in any country except as specified: Country: United States of America.
Patent No. 4,021,224; 4,137,070.
SURPASS is a registered trademark of Stauffer Chemical Company.

This is a specimen label. Stauffer is not responsible for the accuracy of the information contained herein. As labels are subject to revision, always carefully read and follow the label on the product container.

SUTAN® 6.7-E

SELECTIVE HERBICIDE EMULSIFIABLE LIQUID

Corn (Field, Sweet, Silage and Popcorn) Herbicide for Control of Annual Grasses

ACTIVE INGREDIENT:

S-Ethyl diisobutylthiocarbamate	85.1%
INERT INGREDIENTS:	14.9%
TOTAL	100.0%

Contains 6.7 Pounds of Active Ingredient Per Gallon

EPA Reg. No. 476-2156

 A-12R

CAUTION

Keep Out of Reach of Children

HARMFUL IF SWALLOWED

STATEMENT OF PRACTICAL TREATMENT

If a known exposure occurs or is suspected, immediately initiate the recommended procedures below. If further treatment is required, contact a Poison Center, a physician, or the nearest hospital, or call collect, day or night: (203) 226-6602 Stauffer Chemical Company or (800) 424-9300 Chemtrec Pesticide Team Safety Network.

If swallowed: Immediately give several glasses of water but *do not* induce vomiting. If vomiting does occur, give fluids again. Have a physician determine if condition of patient will permit evacuation of stomach. Do not give anything by mouth to an unconscious or convulsing person.

For eye contact: Immediately flush eyes with large amounts of running water for at least 15 minutes. Hold eyelids apart during flushing to ensure rinsing of the entire surface of the eye and lids with water. Get medical attention immediately.

For skin contact: Immediately flush all affected areas with plenty of water for at least 15 minutes. Remove contaminated clothing and shoes. Seek medical attention if skin irritation occurs. Wash clothing before reuse.

If inhaled: Remove to fresh air. Seek medical attention if respiratory irritation occurs or if breathing becomes difficult.

PRECAUTIONARY STATEMENTS

HAZARDS TO HUMANS AND DOMESTIC ANIMALS
CAUTION
HARMFUL IF SWALLOWED

Avoid contact with skin, eyes and clothing. Avoid breathing spray mist. Wear goggles, rubber gloves and protective clothing. Wash skin with soap and water immediately after contact. Flush eyes with water.

Do not apply this product in such a manner as to directly or through drift expose workers or other persons. When applied through sprinkler irrigation method of application, the area being treated must be vacated by unprotected persons.

ENVIRONMENTAL HAZARDS

Do not apply directly to water or wetlands. Do not contaminate water by cleaning of equipment or disposal of wastes. Cover or incorporate spills.

CAN BE STORED AT TEMPERATURES AS LOW AS MINUS 50°F

READ ALL LABEL DIRECTIONS BEFORE USING

WEEDS CONTROLLED BY SUTAN + 6.7-E

SUTAN + 6.7-E will not control established weeds.

Annual Grasses:

Barnyardgrass, Watergrass	*Echinochloa crus-gaili*
Crabgrass	*Digitaria* spp.
Fall Panicum	*Panicum dichotomiflorum*
Field Sandbur*	*Cenchrus pauciflorus*
Giant Foxtail	*Setaria faberi*
Green Foxtail	*Setaria viridis*
Goosegrass	*Eleusine indica*
Seedling Bermudagrass	*Cynodon dactylon*
Seedling Johnsongrass*	*Sorghum halepense*
Texas Panicum*†, Buffalograss, Coloradograss	*Panicum texanum*
Volunteer Sorghum*	*Sorghum* spp.
Wild Cane*‡, Shattercane, Black Amber Cane, Chicken Corn, Scatter Cane	*Sorghum bicolor*
See Wild Cane Rate and Cultural Recommendations	
Yellow Foxtail	*Setaria glauca*

‡ Partial control (suppression) only.
* Cultivation is suggested in addition to the SUTAN + 6.7-E treatment in fields with moderate to heavy infestations of wild cane, sandbur, johnsongrass seedlings, Texas panicum, or volunteer sorghum.
† Southeast Region only.

Perennial Weeds:

Purple Nutgrass, Nutsedge	Cyperus rotundus
Rhizome Bermudagrass†	*Cynodon dactylon*
Rhizome Johnsongrass‡	*Sorghum halepense*
Yellow Nutgrass, Nutsedge	Cyperus esculentus

‡ Partial control (suppression) only.

† Southeast Region only.

Existing stands of nutgrass must be turned under and chopped up thoroughly before treatment. Cultivation is suggested in addition to SUTAN + 6.7-E treatment in fields with moderate to heavy nutgrass infestations.

NOTE: See special instructions under "RECOMMENDATIONS" for Johnsongrass and Rhizome Bermudagrass.

GENERAL USE PRECAUTIONS

Read all label directions before using.

Applied according to directions and under normal growing conditions, SUTAN + 6.7-E will not harm the treated crop. During germination and early stages of growth, extended periods of unusually cold and wet or hot and dry weather, insect or plant disease attack, carryover pesticide residues, the use of certain soil-applied systemic insecticides, improperly placed fertilizers or soil insecticides may create abnormal conditions that weaken crop seedlings. SUTAN + 6.7-E used under these abnormal conditions could result in crop injury.

When applied according to directions and when conditions exist for normal plant growth through the season, no harmful residues of SUTAN + 6.7-E should remain beyond harvest. In the Southeastern U.S., when SUTAN + 6.7-E is used for weed control in silage corn, do not seed small seeded grains after corn harvest until September.

Do not use on milo or sorghum. Do not use on corn seed stock (Breeders, Foundation, or Increase). SUTAN + 6.7-E may be used on Production seed corn.

SUTAN + 6.7-E is recommended for use on mineral soils only (soils containing less than 10% organic matter).

Do not contaminate irrigation water (used for crops other than corn) or water used for domestic purposes.

Do not contaminate food or feed.

Keep container closed when not in use. Do not store near seeds, fertilizers, or foodstuffs.

Do not use in Arizona (except at elevations of 2500 feet or higher, where field and silage corn may be treated).

Do not use in the ten Southernmost California counties (except Kern County where field and silage corn may be treated).

In irrigated areas, do not apply SUTAN + 6.7-E before pre-irrigation.

DIRECTIONS FOR USE ON CORN
(Field, Sweet, Silage and Popcorn, including Production Seed Corn)

It is a violation of Federal law to use this product in a manner inconsistent with its labeling.

SUTAN + 6.7-E is a selective herbicide which is mixed (incorporated) or injected into the soil for control of weeds listed on this label. SUTAN + 6.7-E controls annual grasses as their seeds germinate by interfering with normal germination and seedling development. It does not control established weeds.

COMBINATION WITH FLUID FERTILIZERS
SUTAN + 6.7-E may be combined with fluid (solution, slurry, or suspension) fertilizers. *However, physical compatibility with these fluid fertilizers must be determined before combining in the spray tank.* See Appendix II for directions regarding these combinations. Even though found to be compatible, constant agitation is necessary to keep the SUTAN + 6.7-E evenly mixed with the fluid fertilizer during application.

IMPREGNATION ON DRY BULK FERTILIZERS
Dry bulk fertilizers (except nitrate fertilizers) may be impregnated or coated with SUTAN + 6.7-E. However, uniform distribution of SUTAN + 6.7-E on fertilizer particles and uniform application are necessary to assure good results. See Appendix IV for information and directions regarding impregnation and use for these combinations.

APPLICATION AND SOIL INCORPORATION DIRECTIONS

SPRAY APPLICATION AND MECHANICAL INCORPORATION
Pour the recommended amount of SUTAN + 6.7-E into the spray tank during the filling operations. Apply in 10 to 50 gallons of water per acre using a properly calibrated, low-pressure type nozzle (flood, fan or swirl chamber) on boom type sprayer having good agitation (use the higher gallonage in arid areas). The soil should be dry enough to permit good soil mixing (incorporation).

INCORPORATION DIRECTIONS
See Appendix I for soil mixing (incorporation) before planting directions.

SUTAN + 6.7-E must be incorporated into the soil within 4 hours (within 1 hour in Western Region) to prevent loss of the herbicide. Whenever possible, application and incorporation should be done in the same operation.

If there must be a delay (less than 4 hours, except less than 1 hour in Western Region) in incorporation of the herbicide when application and incorporation are done in separate operations, SUTAN + 6.7-E must be applied to a dry soil surface (at least ½ inch deep), free from dew and incidental moisture.

Loss of weed control will result from any delay in incorporation if SUTAN + 6.7-E is applied to a moist soil surface.

IMPORTANT: Thorough mixing of SUTAN + 6.7-E in the soil is necessary. The soil must be in good tilth. Improper discing or poor tilth (large clods) may result in unsatisfactory control.

INCORPORATION OF IMPREGNATED DRY BULK FERTILIZER
The application and incorporation of SUTAN + 6.7-E must be carried out on the same day and delayed only if soil moisture conditions permit. See Appendix IV for directions on the delay in soil incorporation.

CENTER PIVOT SPRINKLER APPLICATION
Apply SUTAN + 6.7-E with irrigation water before planting (see PLANTING), or immediately after planting before either corn or weeds emerge. It is important that application be done immediately after the last tillage operation since poor results will occur if the weeds have germinated before application of the herbicide. Meter the SUTAN + 6.7-E into the irrigation water during the entire period.

Apply in ½ to 1 inch of water or sufficient water to penetrate the soil to a depth of 5 to 7 inches. If diluted, SUTAN + 6.7-E should be mixed at a volume ratio of no more than 1 part SUTAN + 6.7-E to 4 parts of water or fluid fertilizer. Good agitation should be maintained during the entire application period.

USE PRECAUTIONS
A. Apply the product only through center pivot irrigation systems containing anti-siphon and check valves which will prevent water source contamination and overflow of the slurry tank and containing interlocking controls between the metering device and the water pump to insure simultaneous shut-off.

B. Inject the product with a positive displacement pump into the main line ahead of a right angle turn to insure adequate mixing.

C. Application of more than recommended quantities of irrigation water may result in decreased product performance by removing the chemical from the zone of effectiveness.

D. Do not apply when wind speed favors drift, when system connections or fittings leak, when nozzles do not provide uniform distribution or when lines containing the product must be dismantled and drained.

Subsurface Application Before or At Planting (Limit use to the Southeastern U.S. only).

Special equipment designed for subsurface application *must* be used. Injector units must be rigidly mounted on, *immediately* ahead of, or *immediately* behind the planter unit.

Soil Injection: Coulters and injector shanks must be spaced 3½ inches apart. The two shanks adjacent to the drill row must be 1¾ to 2 inches on either side of it. Mount injectors in staggered positions to avoid trash buildup. Set injectors to inject SUTAN + 6.7-E 3 inches below the soil surface. The width of the band in which weed control is desired will determine the number of injectors required per row. (Example: Four injectors spaced 3½ inches apart give a 14 inch band). A broadcast application can be made by increasing the number of shanks. Seal injector openings by using a roller or chain drag behind the injectors. Apply the recommended rate of SUTAN + 6.7-E in 20 to 50 gallons of water per acre.

For band applications, calculate the amount to be applied per acre as follows:

$$\frac{\text{Band width In Inches}}{\text{Distance Between Rows In Inches}} \times \text{Rate/A for Broadcast Treatment} = \text{Amount Needed For Band Treatment}$$

PLANTING

Plant corn seed to a maximum depth of 2 inches. Seeding can be done immediately or it may be delayed up to 2 weeks in the Southeast, Southwest, and Western regions or up to 4 weeks in the Northeast, North Central, Midwest, and Pacific Northwest regions.* To avoid removing SUTAN + 6.7-E from the row (with loss of weed control in the row), do not move or shape soil after treatment, such as by planting in deep furrows. If it is necessary to work the soil following application and incorporation, but before planting, do not work the soil deeper than the depth SUTAN + 6.7-E was incorporated.

* For Wild Cane—See Special Recommendations.

CULTURAL PRACTICES FOLLOWING APPLICATION

Should weeds develop, a shallow cultivation or rotary hoeing will generally result in better weed control. When cultivating for any reason, it should be shallow, i.e., no more than ½ the depth the herbicide was incorporated or injected. A sequential application of 2,4-D or BANVEL* 4E may be necessary to control weeds resistant to SUTAN + 6.7-E.

* BANVEL is a registered trademark of Velsicol Chemical Corporation

Under circumstances when a poor stand of corn is obtained after SUTAN + 6.7-E treatment (caused by flooding, disease, etc.) corn can be planted again without working the field.

RECOMMENDATIONS

For all uses, cultivation is recommended in addition to SUTAN + 6.7-E treatment in fields with moderate to heavy infestations of wild cane, nutgrass, sandbur, johnsongrass seedlings, rhizome bermudagrass, Texas panicum or volunteer sorghum.

Broadcast Rates Per Acre (Conventional Sprays and Center Pivot Sprinklers).

For control of most annual grasses: Use 4¾ pints of SUTAN + 6.7-E per acre.

For control of Nutsedge and Texas Panicum (in addition to the other annual grasses listed): Use 4¾ to 7⅓ pints of SUTAN + 6.7-E per acre. Use the higher rates on medium (medium fine textured-silt loam, clay loam, etc.) soils, fine (fine textured-clay) soils and for heavy infestations.

For partial control (suppression) of Wild Cane (and for control of the annual grasses listed): When applied and incorporated *immediately* before planting, SUTAN + 6.7-E at the rate of 7⅓ pints per acre on a broadcast basis will reduce competition from wild cane. On continuous corn acreage with high cane seed population, rotation to another crop is recommended.

Suppression of wild cane may be improved if all the following practices are followed:

1. Fall or Spring Moldboard Plow.
2. Before application, the soil should be well worked with a disc and dry enough to permit good soil mixing of the herbicide (incorporation). All trash should be worked into the soil.
3. Delay application until soil warms to a constant 55 degrees or greater temperature. Plant immediately.
4. Incorporate thoroughly. Use recommended incorporation equipment and procedures found on this label.
5. Cultivate immediately if wild cane emerges. Cultivation should be shallower than the depth used when incorporating the herbicide.

For control of Johnsongrass and Bermudagrass (in addition to the annual grasses listed): Apply and disc-incorporate SUTAN + 6.7-E just before planting at the rate of 4¾ to 7⅓ pints per acre. Use the higher rates of SUTAN + 6.7-E on medium (medium fine textured-silt loam, clay loam, etc.) soils, fine (fine textured-clay) soils, for rhizome johnsongrass suppression, rhizome bermudagrass, and for heavy infestations. Disc a second time at right angles to the first. Cultivate at least once, preferably twice.

Johnsongrass: Using a program of plowing and discing followed by SUTAN + 6.7-E will give good control of seedlings and greatly reduce competition from plants coming from rhizomes.

In some areas discing 2 to 3 times spaced at 4 to 6 week intervals during Fall and Winter can be done. If this cannot be done, start with Fall plowing followed by several discings in the Spring before planting. The discing cuts rhizomes into small pieces and brings them to the surface, where both the weather and SUTAN + 6.7-E will kill or weaken them.

Rhizome Bermudagrass: The rhizomes and stolons of bermudagrass must be cut up thoroughly so that four or less nodes remain on a strand. For control of bermudagrass, the disc must be set to cut 6 inches deep.

Subsurface Injection Application: (Limit use to the Southeastern U.S. only) Use 3¾ pints SUTAN + 6.7-E per broadcast acre, reducing the dosage proportionately depending upon the row spacing and band width to be treated.

Fall Application: For use only in Minnesota, Northern Iowa (North of Interstate 80), Northeast Nebraska (North of Interstate 80 and East of State Highway 14), South Dakota (East of Missouri River), Wisconsin and Northern Illinois (North of Interstate 80). Apply and incorporate 7⅓ pints SUTAN + 6.7-E per acre in the late Fall before the ground freezes.

See Appendix IV for directions on use of dry bulk fertilizer applied in the Fall with SUTAN + 6.7-E.

NOTE: Do not use Fall application for suppression of wild cane.

Arizona: For control of most annual grasses in field and silage corn grown at elevations of 2500 feet or higher, use 3¾ to 4¾ pints of SUTAN + 6.7-E per acre. Use the higher rate on medium (medium fine textured silt loam, clay loam, etc.) to heavy (fine textured clay) soils. For nutsedge control, use 4¾ pints of SUTAN + 6.7-E per acre.

SUTAN + 6.7-E/ATRAZINE (80W, 4L or AATREX* NINE-0*)

TANK MIX DIRECTIONS

* AATREX trademark of Ciba-Geigy Corporation for atrazine

* NINE-0 trademark of Ciba-Geigy Corporation

A tank mix combination of SUTAN + 6.7-E and ATRAZINE can be applied and mixed into the soil before planting for weed control in corn. This combination will control a broader spectrum of weeds than will either material used alone. In addition, potential hazard of ATRAZINE residues affecting certain succeeding crops is reduced because a lower rate of ATRAZINE can be used when it is combined with SUTAN + 6.7-E than when it is used alone.

- FOLLOW ALL USE PRECAUTIONS AND WARNINGS THAT APPEAR ON THE ATRAZINE LABEL, SUPPLEMENTARY ATRAZINE LITERATURE AND ELSEWHERE ON THIS LABEL.
- MAKE ONLY ONE APPLICATION PER CROP. AFTER TREATMENT WITH SUTAN + 6.7-E/ATRAZINE COMBINATION, DO NOT PLANT ANY CROP EXCEPT CORN UNTIL THE FOLLOWING YEAR OR INJURY MAY OCCUR.
- DO NOT USE ON MILO OR SORGHUM.
- DO NOT USE ON CORN SEED STOCK (BREEDERS, FOUNDATION, OR INCREASE). SUTAN + 6.7-E/ATRAZINE MAY BE USED ON PRODUCTION SEED CORN.

WEEDS CONTROLLED BY SUTAN + 6.7-E/ATRAZINE

Annual Grasses:

Broadleaf Signalgrass‡*	*Brachiaria platyphylla*
Crabgrass	*Digitaria* spp.
Fall Panicum	*Panicum dichotomiflorum*
Field Sandbur*	*Cenchrus pauciflorus*
Giant Foxtail	*Setaria faberi*
Green Foxtail	*Setaria viridis*
Goosegrass	*Eleusine indica*
Seedling Bermudagrass	*Cynodon dactylon*
Seedling Johnsongrass*	*Sorghum halepense*
Texas Panicum*†, Buffalograss, Coloradograss	*Panicum texanum*
Volunteer Sorghum*	*Sorghum* spp.
Wild Cane*†, Shattercane, Black Amber Cane, Chicken Corn, Scatter Cane	*Sorghum bicolor*
See Wild Cane Rate and Cultural Recommendations	
Wild Oats	*Avena fatua*
Witchgrass	*Panicum capillare*
Yellow Foxtail	*Setaria glauca*

‡ Partial control (suppression) only.

* Cultivation is suggested in addition to the SUTAN + 6.7-E/ATRAZINE treatment in fields with moderate to heavy infestations of wild cane, sandbur, johnsongrass, Texas panicum, volunteer sorghum, or broadleaf signalgrass.

† Southeast Region only.

Annual Broadleaf Weeds:

Annual Morningglory	*Ipomoea purpurea*
Carpetweed	*Mollugo verticillata*
Common Cocklebur	*Xanthium pensylvanicum*
Common Lambsquarters	*Chenopodium album*
Common Purslane	*Portulaca oleracea*

Florida Purslane	*Richardia scabra*
Jimsonweed	*Datura stramonium*
Mustards	*Brassica* spp.
Nightshade	*Solanum* spp.
Prostrate Pigweed	*Amaranthus graecizans*
Ragweed	*Ambrosia* spp.
Redroot Pigweed	*Amaranthus retroflexus*
Sicklepod†, Coffeeweed	*Cassia obtusifolia*
Smartweed	*Polygonum* spp.
Smooth Pigweed	*Amaranthus hybridus*
Velvetleaf	*Abutilon theophrasti*

† Southeast Region only.

Perennial Weeds:	
Hemp Dogbane‡	*Apocynum cannabinum*
Purple Nutgrass, Nutsedge	*Cyperus rotundus*
Rhizome Bermudagrass†	*Cynodon dactylon*
Rhizome Johnsongrass‡	*Sorghum halepense*
Yellow Nutgrass, Nutsedge	*Cyperus esculentus*

‡ Partial control (suppression) only.
† Southeast Region only.

Existing stands of nutgrass must be turned under and chopped up thoroughly before treatment. Cultivation is suggested in addition to SUTAN + 6.7-E/ATRAZINE treatment in fields with moderate to heavy nutgrass infestations.

DIRECTIONS FOR USE

The SUTAN + 6.7-E/ATRAZINE (80W or 4L) combination can be mixed with water or combined with fluid (solution, slurry, or suspension) fertilizers and applied as a spray, or can be impregnated on dry bulk fertilizers. The SUTAN + 6.7-E/AATREX NINE-O combination can be mixed with water and applied as a spray, or can be impregnated on dry bulk fertilizers.

SUTAN + 6.7-E/ATRAZINE (80W, 4L or AATREX NINE-O) TANK MIX.

See Appendix III For Tank Mix Directions.

SUTAN + 6.7-E/ATRAZINE (80W or 4L) FLUID FERTILIZER TANK MIX.

The SUTAN + 6.7-E/ATRAZINE (80W or 4L) combination can be combined with fluid fertilizers for applications made before planting. However, *the physical compatibility with these fluid fertilizers must be determined before combining in the spray tank, and even though found to be compatible, constant agitation is necessary to keep the SUTAN + 6.7-E/ATRAZINE evenly mixed with the fluid fertilizer during application.*

Refer to Appendix II regarding the procedure for testing the compatibility of SUTAN + 6.7-E/ATRAZINE combination with fluid fertilizers.

IMPREGNATION ON DRY BULK FERTILIZERS

Dry bulk fertilizers (except single superphosphate, treble superphosphate and nitrate fertilizers) may be impregnated or coated with SUTAN + 6.7-E/ATRAZINE (80W, 4L or AATREX NINE-O). However, uniform distribution of SUTAN + 6.7-E/ATRAZINE (80W, 4L or AATREX NINE-O) on fertilizer particles and uniform application are necessary to assure good results. See Appendix IV for information and directions for impregnation and use for these combinations.

APPLICATION AND SOIL INCORPORATION DIRECTIONS

SPRAY APPLICATION AND MECHANICAL INCORPORATION

Apply the recommended dosage in 10 to 50 gallons of water per acre (use the higher gallonage in arid areas) using a properly calibrated, low-pressure type nozzle (flood, fan or swirl chamber) on boom-type sprayer having good, but not excessive agitation. Check frequently during application to be sure nozzles are free from clogging and are delivering a uniform spray pattern. *Read directions on ATRAZINE label concerning nozzle and screen sizes.*

See Appendix I for Soil Mixing (incorporation) before Planting Directions.

The herbicides must be incorporated into the soil within 4 hours (within 1 hour in Western Region) after application to prevent loss of the SUTAN + 6.7-E. Whenever possible, application and incorporation should be done in the same operation. Apply the herbicides to soil dry enough to permit good soil mixing.

If there must be a delay (less than 4 hours, except less than 1 hour in Western Region) in incorporation of the herbicides when application and incorporation are done in separate operations, SUTAN + 6.7-E must be applied to a dry soil surface (at least ½ inch deep), free from dew and incidental moisture.

Loss of weed control will result from any delay in incorporation if SUTAN + 6.7-E is applied to a moist soil surface.

IMPORTANT: Thorough mixing of SUTAN + 6.7-E in the soil is necessary. The soil must be in good tilth. Improper discing or poor tilth (large clods) may result in unsatisfactory control.

INCORPORATION OF IMPREGNATED DRY BULK FERTILIZER

The application and incorporation of SUTAN + 6.7-E must be carried out on the same day and delayed only if soil moisture conditions permit. See Appendix IV for directions on the delay in soil incorporation.

CENTER PIVOT SPRINKLER APPLICATION

Apply SUTAN + 6.7-E/ATRAZINE (80W or 4L) with irrigation water before planting (see PLANTING), or immediately after planting before either corn or weeds emerge. It is important that application be done immediately after the last tillage operation since poor results will occur if the weeds have germinated before application of the herbicide. Meter the pre-mixed SUTAN + 6.7-E/ATRAZINE into the irrigation water during the entire period. Apply in ½ to 1 inch of water or sufficient water to penetrate the soil to a depth of 5 to 7 inches. Good agitation of the pre-mixed SUTAN + 6.7-E/ ATRAZINE should be maintained during the entire application period. See Appendix III for tank mixing directions for SUTAN + 6.7-E with ATRAZINE.

USE PRECAUTIONS

A. Apply the product only through center pivot irrigation systems containing anti-siphon and check valves which will prevent water source contamination and overflow of the slurry tank and containing interlocking controls between the metering device and the water pump to insure simultaneous shut-off.
B. Inject the product with a positive displacement pump into the main line ahead of a right angle turn to insure adequate mixing.
C. Application of more than recommended quantities of irrigation water may result in decreased product performance by removing the chemical from the zone of effectiveness.
D. Do not apply when wind speed favors drift, when system connections or fittings leak, when nozzles do not provide uniform distributions or when lines containing the product must be dismantled and drained.

PLANTING

Do not plant seed deeper than 2 inches. For SUTAN + 6.7-E/ATRAZINE application made before planting, seeding can be done immediately, or it may be delayed up to 2 weeks in the Southeast, Southwest, and Western regions or up to 4 weeks in the Northeast, North Central, Midwest, and Pacific Northwest regions. To avoid removing SUTAN + 6.7-E/ATRAZINE from the row (with loss of weed control in the row), do not move or shape soil after treatment, such as by planting in deep furrows. If it is necessary to work the soil following application and incorporation, but before planting, do not work the soil deeper than the depth the herbicides were incorporated.

RECOMMENDATIONS

NOTE: The SUTAN + 6.7-E/ATRAZINE tank mix combination is not recommended on corn grown in Arizona or in the ten Southernmost California counties.

For all uses, cultivation is recommended in addition to the SUTAN + 6.7-E/ATRAZINE treatment in fields with moderate to heavy infestations of wild cane, nutgrass, sandbur, johnsongrass seedlings, rhizome bermudagrass, Texas panicum or volunteer sorghum.

Broadcast Rates Per Acre [Conventional sprays—SUTAN + 6.7-E/ATRAZINE (80W, 4L, AATREX NINE-O); center pivot sprinklers—SUTAN + 6.7-E/ATRAZINE (80W, 4L)].

For control of annual grasses and broadleaf weeds: Use 3¾ to 4¾ pints SUTAN + 6.7-E and 1 to 2* pounds of ATRAZINE 80W, ¾ to 1½* quarts ATRAZINE 4L, or ⅘ to 1¾* pounds of AATREX NINE-O per acre. Use the higher rate of SUTAN + 6.7-E on medium (medium fine textured-silt loam, clay loam, etc.) to fine (fine textured-clay) soils and for sandbur and johnsongrass seedling control.

* Use the higher rate of ATRAZINE on soils with over 5% organic matter.

For control of Nutsedge and Texas Panicum (in addition to the broadleaf weeds and other annual grasses listed): Use 4¾ to 7⅓ pints of SUTAN + 6.7-E and 1 to 2* pounds ATRAZINE 80W, ¾ to 1½* quarts ATRAZINE 4L, or ⅘ to 1¾* pounds AATREX NINE-O per acre. Use the higher rates of SUTAN + 6.7-E on medium (medium fine textured-silt loam, clay loam, etc.) soils, fine (fine textured-clay) soils and for heavy infestations.

For partial control (suppression) of Wild Cane (and for control of the broadleaf weeds and annual grasses listed): When applied and incorporated *immediately* before planting, SUTAN + 6.7-E at the rate of 7⅓ pints and 1 to 2* pounds ATRAZINE 80W, ¾ to 1½* quarts ATRAZINE 4L, or ⅘ to 1¾* pounds AATREX NINE-O per acre will reduce competition from wild cane. On continuous corn acreage with high cane seed population, rotation to another crop is recommended.

NOTE: See preceding SUTAN + 6.7-E section for wild cane supplemental cultural and tillage practices.

For control of Johnsongrass and Bermudagrass (in addition to the broadleaf weeds and annual grasses listed): Apply and disc-incorporate SUTAN + 6.7-E just before planting at the rate of 4¾ to 7⅓ pints plus 1 to 2* pounds of ATRAZINE 80W, or ¾ to 1½* quarts of ATRAZINE 4L, or ⅘ to 1¾* pounds AATREX NINE-O per acre. Use the higher rate of SUTAN + 6.7-E on medium (medium fine textured-silt loam, clay loam, etc.) soils, fine (fine textured-clay) soils, for rhizome johnsongrass suppression and for heavy infestations. Disc a second time at right angles to the first. Cultivate at least once, preferably twice.

* Use the higher rate of ATRAZINE on soils with over 5% organic matter.

Johnsongrass: Using a program of plowing and discing followed by SUTAN + 6.7-E and ATRAZINE will give good control of seedlings and greatly reduce competition from plants coming from rhizomes.

In some areas discing 2 to 3 times spaced at 4 to 6 week intervals during Fall and Winter can be done. If this cannot be done, start with Fall plowing followed by several discings in the Spring before planting. The discing cuts rhizomes into small pieces and brings them to the surface, where both the weather and SUTAN + 6.7-E will kill or weaken them.

Rhizome Bermudagrass: The rhizomes and stolons of bermudagrass must be cut up thoroughly so that four or less nodes remain on a strand. For control of bermudagrass, the disc must be set to cut 6 inches deep.

For suppression of Broadleaf Signalgrass: For early and mid-season suppression, use 7⅓ pints of SUTAN + 6.7-E and 1¼ to 2* pounds ATRAZINE 80W, 1 to 1½* quarts ATRAZINE 4L, or 1 to 1¾* pounds AATREX NINE-O per acre. Cultivation or application of a post-emergence herbicide recommended for broadleaf signalgrass will aid in extending control to late season.

For suppression of Hemp Dogbane: For early and mid-season suppression, use 7⅓ pints of SUTAN + 6.7-E and 1¼ to 2* pounds ATRAZINE 80W, 1 to 1½* quarts ATRAZINE 4L, or 1 to 1¾* pounds AATREX NINE-O per acre. Cultivation or application of a post-emergence herbicide recommended for hemp dogbane may aid in extending control beyond mid-season.

* Use the higher rate of ATRAZINE on soils with over 5% organic matter.

Fall Application: For use only in Minnesota, Northern Iowa (North of Interstate 80), Northeast Nebraska (North of Interstate 80 and East of State Highway 14), South Dakota (East of Missouri River), Wisconsin and Northern Illinois (North of Interstate 80). Apply and incorporate 7⅓ pints SUTAN + 6.7-E and 2 pounds ATRAZINE 80W or 1½ quarts ATRAZINE 4L per acre in the late fall before the ground freezes.

See Appendix IV for directions on use of dry bulk fertilizer applied in the Fall with this combination.

NOTE: Do not use Fall application for suppression of wild cane.

CULTURAL PRACTICES FOLLOWING APPLICATION

Should weeds develop, a shallow cultivation or rotary hoeing will generally result in better weed control. When cultivating for any reason, it should be shallow; i.e., no more than ½ the depth the herbicide was incorporated.

SUTAN + 6.7-E/BLADEX* (80W or 4L) COMBINATION

* BLADEX is the registered trademark of Shell Chemical Company

GENERAL INFORMATION

SUTAN + 6.7-E may be used in combination with BLADEX (80W or 4L) for weed control in corn. These combinations will control a broader spectrum of weeds than will either of the products used alone.

These combinations may be applied as a tank mix, preplant and soil-incorporated; or the BLADEX (80W or 4L) may be applied pre-emergence following a pre-plant incorporated application of SUTAN + 6.7-E.

IMPORTANT: Observe all cautions and limitations on the BLADEX (80W or 4L) and SUTAN + 6.7-E labels.

Make only one application per crop.

Do not use on milo or sorghum. Do not use on corn seed stock (Breeders, Foundation, or Increase). SUTAN + /BLADEX may be used on Production seed corn.

WEEDS CONTROLLED BY SUTAN + 6.7-E/BLADEX

Annual Grasses:

Annual Bluegrass	*Roa annua*
Annual Fescues	*Festuca* spp.
Annual (Italian) Ryegrass	*Lolium multiflorum*
Barnyardgrass, Watergrass	*Echinochloa crus-galli*
Bullgrass	*Paspalum boscianum*
Crabgrass	*Digitaria* spp.
Fall Panicum	*Panicum dichotomiflorum*
Field Sandbur*	*Cenchrus pauciflorus*
Giant Foxtail	*Setaria faberi*
Goosegrass	*Eleusine indica*
Green Foxtail	*Setaria viridis*
Indian Lovegrass	*Eragrostis pilosa*
Junglerice	*Echinochloa colonum*
Seedling Bermudagrass	*Cynodon dactylon*
Seedling Johnsongrass*	*Sorghum halepense*
Stinkgrass	*Eragrostis cilianensis*
Texas Panicum*†, Buffalograss, Coloradograss	*Panicum texanum*
Volunteer Sorghum*	*Sorghum* spp.
Wild Cane*‡, Shattercane, Black Amber Cane, Chicken Corn, Scatter Cane	*Sorghum bicolor*
See Wild Cane Rate and Cultural Recommendations.	
Wild Oats	*Avena fatua*
Witchgrass	*Panicum capillare*
Yellow Foxtail	*Setaria glauca*

‡ Partial control (suppression) only.
* Cultivation is suggested in addition to the SUTAN + 6.7-E/ BLADEX treatment in fields with moderate to heavy infestations of wild cane, sandbur, Texas panicum, johnsongrass seedlings, or volunteer sorghum.
† Southeast Region only.

Annual Broadleaf Weeds:	
Annual Buttercup	*Ranunculus* spp.
Annual Groundcherry	*Physalis lanceifolia*
Annual Morningglory	*Ipomoea purpurea*
Annual Sedge	*Carex* spp.
Black Mustard	*Brassica nigra*
Buffalobur	*Solanum rostratum*
Carpetweed	*Mollugo verticillata*
Common Chickweed	*Stellaria media*
Common Cocklebur	*Xanthium pensylvanicum*
Common Groundsel	*Senecio vulgaris*
Common Lambsquarters	*Chenopodium album*
Common Mallow	*Malva neglecta*
Common Purslane	*Portulaca oleracea*
Corn Spurry	*Spergula arvensis*
Floras Paintbrush	*Emilia sagittata*
Florida Pusley	*Richardia scabra*
Hedge Mustard	*Sisymbrium officinale*
Jimsonweed	*Datura stramonium*
Kochia	*Kochia scoparia*
Mayweed	*Anthemis cotula*
Pennsylvania Smartweed	*Polygonum pensylvanicum*
Pigweed	*Amaranthus* spp.
Pineappleweed	*Matricaria matricarioides*
Plantain	*Plantago* spp.
Poorjoe	*Diodia teres*
Prostrate Knotweed	*Polygonum aviculare*
Prostrate Spurge	*Euphorbia supina*
Ragweed	*Ambrosia* spp.
Seedling Curly Dock	*Rumex crispus*
Shepherdspurse	*Capsella bursa-pastoris*
Sicklepod†, Coffeeweed	*Cassia obtusifolia*
Smallflower Galinsoga	*Galinsoga parviflora*
Prickly Sida	*Sida spinosa*
Tarweed	*Hemizonia congesta*
Velvetleaf	*Abutilon theophrasti*
Wild Buckwheat	*Polygonum convolvulus*
Wild Mustard	*Brassica kaber*
Wild Radish	*Raphanus raphanistrum*
Wild Turnip	*Brassica campestris*

† Southeast Region only.

Perennial Weeds:	
Hemp Dogbane‡	*Apocynum cannabinum*
Rhizome Bermudagrass†	*Cynodon dactylon*
Rhizome Johnsongrass‡	*Sorghum halepense*
Purple Nutgrass, Nutsedge	*Cyperus rotundus*
Yellow Nutgrass, Nutsedge	*Cyperus esculentus*

‡ Partial Control (suppression only)
† Southeast Region only

Existing strands of nutgrass must be turned under and chopped up thoroughly before treatment. Cultivation is suggested in addition to SUTAN + 6.7-E/BLADEX treatment in fields with moderate to heavy nutgrass infestations.

DIRECTIONS FOR USE

The SUTAN + 6.7-E/BLADEX (80W or 4L) combination can be mixed with water or combined with fluid (solution, slurry, or suspension) fertilizers and applied as a spray, or it can be impregnated on dry bulk fertilizers.

TANK MIX DIRECTIONS: See Appendix III

APPLICATION AND INCORPORATION DIRECTIONS

For a tank mix application, apply the recommended dosage (see RECOMMENDATIONS) in 10 to 50 gallons of water per acre (use the higher gallonage in arid areas), using a properly calibrated, low-pressure type nozzle (flood, fan or swirl chamber) or boom-type sprayer having good agitation. Check frequently during application to be sure the nozzles are free from clogging and are delivering a uniform spray pattern. Read directions on the BLADEX (80W or 4L) label concerning nozzle and screen sizes.

See Appendix 1 for Soil Mixing (incorporation) before Planting Directions.

The tank mix must be incorporated into the soil within 4 hours (within 1 hour in Western Region) after application to prevent loss of the SUTAN + 6.7-E. Whenever possible, application and incorporation should be done in the same operation. Apply the herbicides to soil dry enough to permit good soil mixing.

If there must be a delay (less than 4 hours, except less than 1 hour in Western Region) in incorporation of the herbicides when application and incorporation are done in separate operations, SUTAN + 6.7-E must be applied to a dry soil surface (at least ½ inch deep), free from dew and incidental moisture.

Loss of weed control will result from any delay in incorporation if SUTAN + 6.7-E is applied to a moist soil surface.

IMPORTANT: Thorough mixing of SUTAN + 6.7-E in the soil is necessary. The soil must be in good tilth. Improper discing or poor tilth (large clods) may result in unsatisfactory control.

INCORPORATION OF IMPREGNATED DRY BULK FERTILIZER

The application and incorporation of SUTAN + 6.7-E must be carried out on the same day and delayed only if soil moisture conditions permit. See Appendix IV for directions on the delay in soil incorporation.

CENTER PIVOT SPRINKLER APPLICATION

Apply SUTAN + 6.7-E/BLADEX (80W or 4L) with irrigation water before planting (see PLANTING), or immediately after planting before either corn or weeds emerge. It is important that application be done immediately after the last tillage operation since poor results will occur if the weeds have germinated before application of the herbicide. Meter the pre-mixed SUTAN + 6.7-E/BLADEX into the irrigation water during the entire period. Apply in ½ to 1 inch of water or sufficient water to penetrate the soil to a depth of 5 to 7 inches. Good agitation of the pre-mixed SUTAN + 6.7-E/BLADEX should be maintained during the entire application period. See Appendix III for tank mixing directions for SUTAN + 6.7-E with BLADEX.

USE PRECAUTIONS

A. Apply the product only through center pivot irrigation systems containing anti-siphon and check valves which will prevent water source contamination and overflow of the slurry tank and containing interlocking controls between the metering device and the water pump to insure simultaneous shut-off.
B. Inject the product with a positive displacement pump into the main line ahead of a right angle turn to insure adequate mixing.
C. Application of more than recommended quantities of irrigation water may result in decreased product performance by removing the chemical from the zone of effectiveness.
D. Do not apply when wind speed favors drift, when system connections or fittings leak, when nozzles do not provide uniform distribution or when lines containing the product must be dismantled and drained.

SUTAN + 6.7-E/BLADEX (80W or 4L) FLUID FERTILIZER TANK MIX

The SUTAN + 6.7-E/BLADEX (80W or 4L) combination can be combined with fluid fertilizers for applications made before planting. However, the physical compatibility with these fluid fertlizers must be determined before combining in the spray tank, and even though found to be compatible, constant agitation is necessary to keep the SUTAN + 6.7-E/BLADEX evenly mixed with the fluid fertilizer during application.

Refer to Appendix II regarding the procedure for testing the compatibility of SUTAN + 6.7-E/BLADEX combination with fluid fertilizers.

IMPREGNATION ON DRY BULK FERTILIZERS

Dry bulk fertilizers (except single superphosphate, treble superphosphate and nitrate fertilizers) may be impregnated or coated with SUTAN + 6.7-E/BLADEX (80W or 4L). However, uniform distribution of SUTAN + 6.7-E/BLADEX (80W or 4L) on fertilizers particles and uniform application are necessary to assure good results. See Appendix IV for information and directions for impregnation and use for these combinations.

SUTAN + 6.7-E/BLADEX (80W or 4L) Sequential Application
For sequential application of SUTAN + 6.7-E preplant incorporated followed by BLADEX (80W or 4L) pre-emergence, consult the appropriate application directions on the respective product labels. Follow the dosage recommendations given below.

RECOMMENDATIONS

NOTE: The SUTAN + 6.7-E/BLADEX tank mix combination is not recommended on corn grown in Arizona or in the ten Southernmost California counties.

For all uses, cultivation is recommended in addition to the SUTAN + 6.7-E/BLADEX treatment in fields with moderate to heavy infestations of wild cane, nutgrass, sandbur, johnsongrass seedlings, rhizome bermudagrass, Texas panicum, or volunteer sorghum.

Broadcast Rates Per Acre (Conventional Sprays and Center Pivot Sprinklers)

For control of annual grasses and broadleaf weeds: Use 3¾ to 4¾ pints of SUTAN + 6.7-E and 2 to 2½ pounds BLADEX 80W or 3 to 4 pints BLADEX 4L per acre. Use the higher rates of SUTAN + 6.7-E on medium (medium fine textured-silt loam, clay loam, etc.) to fine (fine textured-clay) soils and for sandbur and johnsongrass seedling control.

For control of Nutsedge and Texas Panicum (in addition to the broadleaf weeds and annual grasses listed): Use 4¾ to 7⅓ pints of SUTAN + 6.7-E and 2 to 2½ pounds BLADEX 80W or 3 to 4 pints BLADEX 4L per acre.

For partial control (suppression) or Wild Cane (and for control of the broadleaf weeds and annual grasses listed): When applied and incorporated immediately before planting, SUTAN + 6.7-E at the rate of 7⅓ pints and 2 to 2½ pounds BLADEX 80W or 3 to 4 pints BLADEX 4L per acre will reduce competition from wild cane. On continous corn acreage with high cane seed population, rotation to another crop is recommended.

NOTE: See preceding SUTAN + 6.7-E section for wild cane supplemental cultural and tillage practices.

For control of Johnsongrass and Bermudagrass (in addition to the broadleaf weeds and annual grasses listed): Apply and disc-incorporate SUTAN + 6.7-E just before planting at the rate of 4¾ to 7⅓ pints plus 2 to 2½ pounds of BLADEX 80W or 3 to 4 pints BLADEX 4L per acre. Use the higher rate of SUTAN + 6.7-E on medium (medium fine textured-silt loam, clay loam, etc.) soils, fine (fine textured-clay) soils, for rhizome johnsongrass suppression and for heavy infestations. Disc a second time at right angles to the first. Cultivate at least once, preferably twice.

Johnsongrass: Using a program of plowing and discing followed by SUTAN + 6.7-E will give good control of seedlings and greatly reduce competition from plants coming from rhizomes.

In some areas discing 2 to 3 times spaced at 4 to 6 week intervals during Fall and Winter can be done. If this cannot be done, start with Fall plowing followed by several discings in the Spring before planting. The discing cuts rhizomes into small pieces and brings them to the surface, where both the weather and SUTAN + 6.7-E will kill or weaken them.

Rhizome Bermudagrass: The rhizomes and stolons of bermudagrass must be cut up thoroughly so that four or less nodes remain on a strand. For control of bermudagrass, the disc must be set to cut 6 inches deep.

For suppression of Hemp Dogbane: For early and mid-season suppression use 7⅓ pints of SUTAN + 6.7-E and 2 to 2½ pounds BLADEX 80W or 3 to 4 pints BLADEX 4L per acre. Cultivation or application of a post-emergence herbicide recommended for hemp dogbane may aid in extending control beyond mid-season.

NOTE: BLADEX in combination with SUTAN + 6.7-E is not recommended for use on coarse or medium textured soils containing less than 1% organic matter. Use the lower rate of BLADEX on coarse to medium textured soils containing 1% organic matter and above. Use the higher rate of BLADEX on fine (fine textured-clay) soils with 3% organic matter and above.

SUTAN + 6.7-E/BLADEX can be applied either as a tank mix combination and incorporated into the soil or apply SUTAN + 6.7-E preplant incorporated followed by BLADEX pre-emergence. If BLADEX is surface-applied on top of a preplant application of SUTAN + 6.7-E, a rotary hoeing or shallow cultivation is recommended if rainfall or sprinkler irrigation had not occurred within about 10 days after the application of BLADEX.

Fall Application: For use only in Minnesota, Northern Iowa (North of Interstate 80), Northeast Nebraska (North of Interstate 80 and East of State Highway 14), South Dakota (East of Missouri River), Wisconsin, and Northern Illinois (North of Interstate 80). Apply an incorporate 7⅓ pints SUTAN + 6.7-E and 2½ pounds BLADEX 80W or 2 quarts BLADEX 4L per acre in the late Fall before the ground freezes.

See Appendix IV for directions on use of dry bulk fertilizer applied in the Fall with this combination.

NOTE: Do not use Fall application for suppression of wild cane.

PLANTING
Do not plant seed deeper than 2 inches. For SUTAN + 6.7-E/BLADEX application made before planting, seeding can be done immediately, or it may be delayed up to 2 weeks in the Southeast, Southwest, and Western regions or up to 4 weeks in the Northeast, North Central, Midwest, and Pacific Northwest regions. To avoid removing SUTAN + 6.7-E/BLADEX from the row (with loss of weed control in the row), do not move or shape soil after treatment, such as by planting in deep furrows. If it is necessary to work the soil following application and incorporation, but before planting, do not work the soil deeper than the depth the herbicides were incorporated.

CULTURAL PRACTICES FOLLOWING APPLICATION
Should weeds develop, a shallow cultivation or rotary hoeing will generally result in better weed control. When cultivating for any reason, it should be shallow, i.e., no more than ½ the depth the herbicide was incorporated.

SUTAN + 6.7-E/ATRAZINE (80W or 4L)/BLADEX (80W or 4L)

3-WAY TANK MIX DIRECTIONS
A tank mix combination of SUTAN + 6.7-E, ATRAZINE (80W or 4L) and BLADEX (80W or 4L) can be applied and mixed into the soil before planting for weed control in corn. This combination will control a broader spectrum of weeds than will any of these materials used alone. In addition, potential hazards of ATRAZINE residues affecting certain succeeding crops is reduced because a lower rate of ATRAZINE can be use when it is combined with SUTAN + 6.7-E/BLADEX than when it is used alone.

IMPORTANT: Observe all cautions and limitations on the ATRAZINE (80W or 4L), BLADEX (80W or 4L), and SUTAN + 6.7-E labels.

Make only one application per crop. After treatment with SUTAN + 6.7-E/ATRAZINE/BLADEX combination, do not plant any crop except corn until the following year or injury may occur.

Do not use on milo or sorghum. Do not use on corn seed stock (Breeders, Foundation, or Increase). SUTAN + 6.7-E/ ATRAZINE/BLADEX may be used on Production seed corn.

WEEDS CONTROLLED BY SUTAN + 6.7-E/ATRAZINE/BLADEX TANK MIX

Annual Grasses:

Annual Bluegrass	*Poa annua*
Annual Fescues	*Festuca* spp.
Annual (Italian) Ryegrass	*Lolium multiflorum*
Barnyardgrass, Watergrass	*Echinochloa crus-galli*
Bullgrass	*Paspalum boscianum*
Crabgrass	*Digitaria* spp.
Fall Panicum	*Panicum dichotomiflorum*
Field Sandbur*	*Cenchrus pauciflorus*

Giant Foxtail	*Setaria faberi*
Goosegrass	*Eleusine indica*
Green Foxtail	*Setaria viridis*
Indian Lovegrass	*Eragrostis pilosa*
Junglerice	*Echinochloa colonum*
Seedling Bermudagrass	*Cynodon dactylon*
Seedling Johnsongrass*	*Sorghum halepense*
Stinkgrass	*Eragrostic cilianensis*
Texas Panicum*† Buffalograss, Coloradograss	*Panicum texanum*
Volunteer Sorghum*	*Sorghum* spp.
Wild Cane*‡, Shattercane, Black Amber Cane, Chicken Corn, Scatter Cane	*Sorghum bicolor*
See Wild Cane Rate and Cultural Recommendations.	
Wild Oats	*Avena fatua*
Witchgrass	*Panicum capillare*
Yellow Foxtail	*Setaria glauca*

‡ Partial control (suppression) only.
* Cultivation is suggested in addition to the SUTAN + 6.7-E/ATRAZINE/BLADEX treatment in fields with moderate to heavy infestations of wild cane, sandbur, Texas panicum, johnsongrass seedlings or volunteer sorghum.
† Southeast Region Only.

Annual Broadleaf Weeds:

Annual Buttercup	*Ranunculus* spp.
Annual Groundcherry	*Physalis lanceifolia*
Annual Morningglory	*Ipomoea purpurea*
Annual Sedge	*Carex* spp.
Buffalobur	*Solanum rostratum*
Carpetweed	*Mollugo verticillata*
Common Chickweed	*Stellaria media*
Common Cocklebur	*Xanthium pensylvanicum*
Common Groundsel	*Senecio vulgaris*
Common Lambsquarters	*Chenopodium album*
Common Mallow	*Malva neglecta*
Common Purslane	*Portulaca oleracea*
Corn Spurry	*Spergula arvensis*
Floras Paintbrush	*Emilia sagittata*
Florida Pusley	*Richardia scabra*
Hedge Mustard	*Sisymbrium officinale*
Jimsonweed	*Datura stramonium*
Kochia	*Kochia scorparia*
Mayweed	*Anthemis cotula*
Mustards	*Brassica* spp.
Nightshade	*Solanum* spp.
Pigweed	*Amaranthus* spp.
Pineappleweed	*Matricaria matricarioides*
Plantain	*Plantago* spp.
Poorjoe	*Diodia teres*
Prostrate Knotweed	*Polygonum aviculare*
Prostrate Spurge	*Euphorbia supina*
Ragweed	*Ambrosia* spp.
Seedling Curly Dock	*Rumex crispus*
Shepherdspurse	*Capsella burse-pastoris*
Smallflower Galinsoga	*Galinsoga parviflora*
Smartweed	*Polygonum* spp.
Pickly Sida	*Sida spinosa*
Tarweed	*Hemizonia congesta*
Velvetleaf	*Abutilon theophrasti*
Wild Buckwheat	*Polygonum convolvulus*
Wild Radish	*Raphanus raphanistrum*
Wild turnip	*Brassica campestris*

Perennial Weeds:

Rhizome Bermudagrass†	*Cynodon dactylon*
Purple Nutgrass, Nutsedge	*Cyperus rotundus*
Yellow Nutgrass, Nutsedge	*Cyperus esculentus*
Rhizome Johnsongrass‡	*Sorghum halespense*

Existing stands of nutgrass must be turned under and chopped up thoroughly before treatment. Cultivation is suggested in addition to SUTAN + 6.7-E/AZTRAZINE/BLADEX treatment in fields with moderate to heavy nutgrass infestations.

‡ Partial Control (Suppression) Only.
† Southeast Region Only.

DIRECTIONS FOR USE

The SUTAN + 6.7-E/ATRAZINE (80W or 4L)/BLADEX (80W or 4L) combination can be mixed with water or combined with fluid (solution, slurry or suspension) fertilizers and applied as a spray, or can be impregnated on dry bulk fertilizers.

TANK MIX DIRECTIONS: See Appendix III.

APPLICATION AND INCORPORATION DIRECTIONS:

For a tank mix application, apply the recommended dosage (see RECOMMENDATIONS) in 10 to 50 gallons of water per acre (use the higher gallonage in arid areas) using a properly calibrated, low-pressure type nozzle (flood, fan or swirl chamber), on boom-type sprayer having good, but not excessive agitation. Check frequently during application to be sure the nozzles are free from clogging and are delivering a uniform pattern. Read directions on the ATRAZINE (80W or 4L) and BLADEX (80W or 4L) labels concerning nozzle and screen sizes.

See Appendix 1 for Soil Mixing (incorporation) before Planting Directions.

The tank mix must be incorporated into the soil within 4 hours (within 1 hour in Western Region) after application to prevent loss of the SUTAN + 6.7-E. Whenever possible, application and incorporation should be done in the same operation. Apply the herbicides to soil dry enough to permit good soil mixing.

Loss of weed control will result from any delay in incorporation if SUTAN + 6.7-E is applied to a moist soil surface.

If there must be a delay (less than 4 hours, except less than 1 hour in Western Region) in incorporation of the herbicides when application and incorporation are done in separate operations, SUTAN + 6.7-E must be applied to a dry soil surface (at least ½ inch deep), and free from dew and incidental moisture.

IMPORTANT: Thorough mixing of SUTAN + 6.7-E in the soil is necessary. The soil must be in good tilth. Improper discing or poor tilth (large clods) may result in unsatisfactory control.

INCORPORATION OF IMPREGNATED DRY BULK FERTILIZER

The application and incorporation of SUTAN + 6.7-E must be carried out on the same day and delayed only if soil moisture conditions permit. See Appendix IV for directions on the delay in soil incorporation.

SUTAN + 6.7-E/ATRAZINE (80W or 4L)/BLADEX (80W or 4L) FLUID FERTILIZER TANK MIX

The SUTAN + 6.7-E/ATRAZINE (80W or 4L)/BLADEX (80W or 4L) 3-way combination can be combined with fluid fertilizers for applications made before planting. However, the physical compatibility with these fertilizers must be determined before combining in the spray tank, and even though found to be compatible, constant agitation is necessary to keep the SUTAN + 6.7-E/ATRAZINE/ BLADEX evenly mixed with the fluid fertilizer during application.

Refer to Appendix II regarding the procedure for testing the compatibility of SUTAN + 6.7-E/ATRZAINE/BLADEX combination with fluid fertilizers.

IMPREGNATION ON DRY BULK FERTILIZERS

Dry bulk fertilizers (except single superphosphate, treble superphosphate and nitrate fertilizers) may be impregnated or coated with SUTAN + 6.7-E/ATRZAINE (80W or 4L)/BLADEX (80W or 4L). However, uniform distribution of SUTAN + 6.7-E/ATRAZINE (80W or 4L)/BLADEX (80W or 4L) on fertilizer particles and uniform application are necessary to assure good results. See Appendix IV for information and directions for impregnation and use of these combinations.

PLANTING

Do not plant seed deeper than 2 inches. For SUTAN + 6.7-E/ATRAZINE/BLADEX application made before planting, seeding can be done immediately, or it may be delayed up to 2 weeks in the Southeast, Southwest, and Western regions or up to 4 weeks in the Northeast, North Central, Midwest, and Pacific Northwest regions. To avoid removing SUTAN + 6.7-E/ATRAZINE/BLADEX from the row (with loss of weed control in the row), do not move or shape soil after treatment, such as by planting in deep furrows. If it is necessary to work the soil following application and incorporation, but before planting, do not work the soil deeper than the depth the herbicides were incorporated.

RECOMMENDATIONS

NOTE: The SUTAN + 6.7-E/ATRAZINE/BLADEX tank mix combination is not recommended on corn grown in Arizona or in the ten Southernmost California counties.

For all uses, cultivation is recommended in addition to the SUTAN + 6.7-E/ATRAZINE/BLADEX treatment in fields with moderate to heavy infestations of wild cane, nutgrass, sandbur, johnsongrass seedlings, rhizome bermudagrass, Texas panicum, or volunteer sorghum.

Broadcast Rates Per Acre (Conventional sprays)

For control of annual grasses and broadleaf weeds: Use 3¾ to 4¾ pints of SUTAN + 6.7-E plus ¾ to 1¼ pounds of ATRAZINE 80W or 1 to 2 pints ATRAZINE 4L plus 1¼ to 2½ pounds BLADEX 80W or 1 to 2 quarts of BLADEX 4L per acre. ATRAZINE and BLADEX should be used in a one to two proportion. Use the higher rate of SUTAN + 6.7-E on medium (medium fine texture-silt loam, clay loam, etc.) to fine (fine textured clay) soils and for sandbur and johnsongrass seedling control.

For control of Nutgrass and Texas Panicum (In addition to the broadleaf weeds and annual grasses listed): Use 4¾ to 7⅓ pints of SUTAN + 6.7-E plus ¾ to 1¼ pounds ATRAZINE 80W or 1 to 2 pints ATRAZINE 4L plus 1¼ to 2½ pounds BLADEX 80W or 1 to 2 quarts of BLADEX 4L per acre. ATRAZINE and BLADEX should be used in a one to two proportion. Use the higher rates of SUTAN + 6.7-E on medium (medium fine textured-silt loam, clay loam, etc.) soils, fine (fine textured-clay) soils and for heavy infestations.

For partial control (suppression) of Wild Cane (and for control of the broadleaf weeds and annual grasses listed): When applied and incorporated immediately before planting, SUTAN + 6.7-E at the rate of 7⅓ pints plus ¾ to 1¼ pounds ATRAZINE 80W or 1 to 2 pints ATRAZINE 4-L plus 1¼ to 2½ pounds BLADEX 80W or 1 to 2 quarts BLADEX 4-L per acre will reduce competition from wild cane. ATRAZINE and BLADEX should be used in a one to two proportion. On continuous corn acreage with high cane seed population, rotation to another crop is recommended.

NOTE: See preceding SUTAN + 6.7-E section for wild cane supplemental cultural and tillage practices.

For control of Johnsongrass and Bermudagrass (in addition to the broadleaf weeds and annual grasses listed): Apply and disc-incorporate SUTAN + 6.7-E just before planting at the rate of 4¾ to 7⅓ pints plus ¾ to 1¼ pounds of ATRAZINE 80W or 1 to 2 pints ATRAZINE 4L plus 1¼ to 2½ pounds of BLADEX 80W or 1 to 2 quarts BLADEX 4L per acre. ATRAZINE and BLADEX should be used in a one to two proportion. Use the higher rates of SUTAN + 6.7-E on medium (medium fine textured-silt loam, clay loam, etc.) soils, fine (fine textured-clay) soils, for rhizome johnsongrass suppression, and for heavy infestations. Disc a second time at right angles to the first. Cultivate at least once, preferably twice.

Rhizome Bermudagrass: The rhizomes and stolons of bermudagrass must be cut up thoroughly so that four or less nodes remain on a strand. For control of bermudagrass, the disc must be set to cut 6 inches deep.

Johnsongrass: Using a program of plowing and discing followed by SUTAN + 6.7-E/ATRAZINE/BLADEX will give good control of seedlings and greatly reduce competition from plants coming from rhizomes.

In some areas discing 2 to 3 times spaced at 4 to 6 week intervals during Fall and Winter can be done. If this cannot be done, start with Fall plowing followed by several discings in the Spring before planting. The discing cuts rhizomes into small pieces and brings them to the surface, where both the weather and SUTAN + 6.7-E will kill or weaken them.

NOTE: BLADEX in combination with SUTAN + 6.7-E and ATRAZINE is not recommended for use on coarse or medium textured soils containing less than 1% organic matter.

Fall Application: For use only in Minnesota, Northern Iowa (North of Interstate 80), Northeast Nebraska (North of Interstate 80 and East of State Highway 14), South Dakota (East of Missouri River), Wisconsin, and Northern Illinois (North of Interstate 80). Apply and incorporate 7⅓ pints SUTAN + 6.7-E plus 1¼ pounds ATRAZINE 80W or 2 pints ATRAZINE 4L plus 2½ pounds BLADEX 80W or 2 quarts BLADEX 4L per acre in the late Fall before the ground freezes.

See Appendix IV for directions on use of dry bulk fertilizer applied in the Fall wth this combination.

NOTE: Do not use Fall application for suppression of wild cane.

CULTURAL PRACTICES FOLLOWING APPLICATION

Should weeds develop, a shallow cultivation or rotary hoeing will generally result in better weed control. When cultivating for any reason, it should be shallow; i.e., no more than ½ the depth the herbicide was incorporated.

SUTAN + 6.7-E/PRINCEP* (4L, 80W or CALIBER*90) COMBINATION

* PRINCEP is the registered trademark of Ciba-Geigy Corporation for simazine.

* CALIBER is the registered trademark of Ciba-Geigy Corporation.

GENERAL INFORMATION

SUTAN + 6.7-E may be used in a preplant incorporated tank mixture with PRINCEP (4L, 80W or CALIBER 90). These combinations will control a broader spectrum of weeds than will either of the products used alone.

IMPORTANT: Observe all cautions and limitations on the SUTAN + 6.7-E and PRINCEP (4L, 80W or CALIBER 90) labels.

Make only one application per crop.

Do not use on milo or sorghum.

Do not use on corn seed stock (Breeders, Foundations, or Increase). SUTAN + 6.7-E/PRINCEP may be used on Production seed corn.

WEEDS CONTROLLED BY SUTAN + 6.7-E/PRINCEP

Annual Grasses:

Barnyardgrass, Watergrass	*Echinochloa crus-galli*
Crabgrass	*Digitaria* spp.
Fall Panicum	*Panicum dichotomiflorum*
Field Sandbur*	*Cenchrus pauciflorus*
Giant Foxtail	*Setaria faberi*
Green Foxtail	*Setaria viridis*
Goosegrass	*Eleusine indica*
Seedling Bermudagrass	*Cynodon dactylon*
Seedling Johnsongrass*	*Sorghum halepense*
Texas Panicum*†, Buffalograss, Coloradograss	*Panicum texanum*
Volunteer Sorghum*	*Sorghum* spp.
Wild Cane*‡, Shattercane, Black Amber Cane, Chicken Corn, Scatter Cane	*Sorghum bicolor*

See Wild Cane Rate and Cultural Recommendations

Witchgrass	*Panicum capillare*
Yellow Foxtail	*Setaria glauca*

† Southeast Region only.
‡ Partial control (suppression) only.
* Cultivation is suggested in addition to the SUTAN + 6.7-E/PRINCEP treatment in fields with moderate to heavy infestations of johnsongrass, sandbur, Texas panicum, volunteer sorghum and wild cane.

Annual Broadleaf Weeds:	
Common Lambsquarters	*Chenopodium album*
Common Ragweed	*Ambrosia artemisiifolia*
Florida Pusley	*Richardia scabra*
Redroot Pigweed	*Amaranthus retroflexus*
Velvetleaf	*Abutilon theophrasti*
Perennial Weeds:	
Purple Nutgrass, Nutsedge	*Cyperus rotundus*
Rhizome Bermudagrass†	*Cynodon dactylon*
Rhizome Johnsongrass‡	*Sorghum halepense*
Yellow Nutgrass, Nutsedge	*Cyperus esculentus*

† Southeast Region only.
‡ Partial control (suppression) only.
Existing stands of nutgrass must be turned under and chopped up thoroughly before treatment. Cultivation is suggested in addition to SUTAN + 6.7-E/PRINCEP treatment in fields with moderate to heavy nutgrass infestations.

DIRECTIONS FOR USE

TANK MIX DIRECTIONS: See Appendix III

APPLICATION AND INCORPORATION DIRECTIONS

For a tank mix application, apply the recommended dosage (see RECOMMENDATIONS) in 20 to 50 gallons of water per acre (use the higher gallonage in arid areas), using a properly calibrated, low-pressure type nozzle (flood, fan or swirl chamber) on boom-type sprayer having good agitation. Check frequently during application to be sure the nozzles are free from clogging and are delivering a uniform spray pattern. Read directions on the PRINCEP (4L, 80W or CALIBER 90) label concerning nozzle and screen sizes.

See Appendix I for Soil Mixing (incorporation) before Planting Directions.

The tank mix must be incorporated into the soil within 4 hours (within 1 hour in Western Region) after application to prevent loss of the SUTAN + 6.7-E. Whenever possible, application and incorporation should be done in the same operation. Apply the herbicides to soil dry enough to permit good soil mixing.

If there must be a delay (less than 4 hours, except less than 1 hour in Western Region) in incorporation of the herbicides when application and incorporation are done in separate operations, SUTAN + 6.7-E must be applied to a soil surface that is dry (at least $1/2$ inch deep), free from dew and incidental moisture.

Loss of weed control will result from any delay in incorporation if SUTAN + 6.7-E is applied to a moist soil surface.

IMPORTANT: Thorough mixing of SUTAN + 6.7-E in the soil is necessary. The soil must be in good tilth. Improper discing or poor tilth (large clods) may result in unsatisfactory control.

RECOMMENDATIONS

NOTE: The SUTAN + 6.7-E/PRINCEP tank mix combination is not recommended on corn grown in Arizona or in the ten Southernmost California counties.

For all uses, cultivation is recommended in addition to the SUTAN + 6.7-E/PRINCEP treatment in fields with moderate to heavy infestations of johnsongrass seedlings, nutgrass, rhizome bermudagrass, sandbur, Texas panicum, volunteer sorghum, or wild cane.

Broadcast Rates Per Acre (Conventional sprays)

* Use the higher rate of PRINCEP on fine-textured soils and where heavy broadleaf weed infestations are expected.

For control of annual grasses and broadleaf weeds: Use $3^3/_4$ to $4^3/_4$ pints of SUTAN + 6.7-E and 2 to 6* pints of PRINCEP 4L, $1^1/_4$ to $3^3/_4$* pounds PRINCEP 80W, or $1^1/_{10}$ to $3^1/_3$* pounds of PRINCEP CALIBER 90 per acre. Use the higher rates of SUTAN + 6.7-E on medium (medium fine textured silt loam, clay loam, etc.) to fine (fine textured-clay) soils and for sandbur and johnsongrass seedling control.

For control of Nutsedge, Texas Panicum and Velvetleaf (in addition to the broadleaf weeds and annual grasses listed): Use $4^3/_4$ to $7^1/_3$ pints of SUTAN + 6.7-E and 2 to 6* pints of PRINCEP 4L, $1^1/_4$ to $3^3/_4$* pounds PRINCEP 80W, or $1^1/_{10}$ to $3^1/_3$* pounds of PRINCEP CALIBER 90 per acre.

For partial control (suppression) of Wild Cane (and for control of the broadleaf weeds and the annual grasses listed): When applied and incorporated immediately before planting SUTAN + 6.7-E at the rate of $7^1/_3$ pints and 2 to 6* pints PRINCEP 4L, $1^1/_4$ to $3^3/_4$* pounds PRINCEP 80W, or $1^1/_{10}$ to $3^1/_3$* pounds of PRINCEP CALIBER 90 per acre will reduce competition from wild cane.

On continuous corn acreage with high cane seed population, rotation to another crop is recommended.

NOTE: See preceding SUTAN + 6.7-E section for wild cane supplemental cultural and tillage practices.

For control of Johnsongrass and Bermudagrass (in addition to the broadleaf weeds and annual grasses listed): Apply and disc-incorporate SUTAN + 6.7-E just before planting at the rate of $4^3/_4$ to $7^1/_3$ pints plus 2 to 6* pints PRINCEP 4L, $1^1/_4$ to $3^3/_4$* pounds PRINCEP 80W, or $1^1/_{10}$ to $3^1/_3$* pounds of PRINCEP CALIBER 90 per acre. Use the higher rate of SUTAN + 6.7-E on medium (medium fine textured-silt loam, clay loam, etc.) soils, fine (fine textured-clay) soils, for rhizome johnsongrass suppression and for heavy infestations. Disc a second time at right angles to the first. Cultivate at least once, preferable twice.

Johnsongrass: Using a program of plowing and discing followed by SUTAN + 6.7-E will give good control of seedlings and greatly reduce competition from plants coming from rhizomes.
In some areas discing 2 to 3 times spaced at 4 to 6 week intervals during fall and winter can be done. If this cannot be done, start with fall plowing followed by several discings in the spring before planting. The discing cuts rhizomes into small pieces and brings them to the surface, where both the weather and SUTAN + 6.7-E will kill or weaken them.

Rhizome Bermudagrass: The rhizomes and stolons of bermudagrass must be cut up thoroughly so that four or less nodes remain on a strand. For control of bermudagrass, the disc must be set to cut 6 inches deep.

PLANTING

Do not plant seed deeper than 2 inches. For a SUTAN + 6.7-E/PRINCEP application made before planting, seeding can be done immediately or it may be delayed up to 2 weeks in the Southeast, Southwest, and Western regions or up to 4 weeks in the Northeast, North Central, Midwest and Pacific Northwest regions. To avoid removing SUTAN + 6.7-E/PRINCEP from the row (with loss of weed control in the row), do not remove or shape soil after treatment, such as by planting in deep furrows. If it is necessary to work the soil following application and incorporation, but before planting, do not work the soil deeper than the depth the herbicides were incorporated.

CULTURAL PRACTICES FOLLOWING APPLICATION

Should weeds develop, a shallow cultivation or rotary hoeing will generally result in better weed control. When cultivating for any reason, it should be shallow, i.e., no more than $1/2$ the depth the herbicide was incorporated.

APPENDIX I

SOIL MIXING (INCORPORATION) BEFORE PLANTING DIRECTIONS

The following equipment commonly is used for soil mixing (incorporation) before planting.

Power-Driven Cultivation Equipment (Recommended on all soil types previously primary tilled): Ground speed should be adjusted with all PTO driven cultivation tools to insure thorough soil mixing.

a. Horizontal action: set to cut 4 to 6 inches deep.

b. Vertical action: set to cut 3 to 4 inches deep.

Tandem Discs (Recommended on all soil types): Set to cut to a depth of 4 to 6 inches, operated at 4 to 6 mph followed by a spike-tooth harrow or some other leveling device which extends beyond the ends of the discs. For more thorough mixing (or control of deep-germinating grasses—wild cane, seedling johnsongrass and nutsedge), disc in two different directions. When discing the second time, the disc should be reset to cut no deeper than 4 to 6 inches.

Field Cultivator (Recommended on lighter soils in good tilth): Use 3 to 4 rows of sweeps spaced at 7 inch or less intervals and staggered so that no soil is left unturned, followed by a spike-tooth harrow pulled behind the cultivator. Do not use chisel plows to incorporate. Set the cultivator to cut 4 to 6 inches deep, operated at 5 mph or more. The field cultivator should not be used alone as an incorporation tool where control of deep-germinating weeds such as nutgrass, johnsongrass seedlings or wild cane is desired. For control of these weeds, use a disc alone for at least one of the passes over the field.

Till 'N Hill Bed Conditioner (Recommended for bedded fields of all soil types in good tilth): Set tillers (choppers) to a depth of 4 to 6 inches. Operate at a speed of 5 to 7 mph. To insure moisture conservation and proper mixing, set the disc hippers to run only in the upper loose soil of the bed. Do not use in fields with excessive trash and/or moisture. For johnsongrass, wild cane and nutgrass control: In winter or prior to the SUTAN+ application, run a row disc to chop up rhizomes and to destroy vegetating nutgrass tubers and early germinating seeds. Till 'N Hill equipment is most effective on well-peaked beds. On low beds, two passes are necessary for effective incorporation.

APPENDIX II

PROCEDURE FOR TESTING THE COMPATIBILITY OF SUTAN + 6.7-E or SUTAN + 6.7-E/ATRAZINE (80W or 4L) or SUTAN + 6.7-E/BLADEX (80W or 4L) or SUTAN + 6.7-E/ATRAZINE (80W or 4L)/BLADEX (80W or 4L) WITH FLUID FERTILIZERS.

The following procedure is suggested for determining whether the above herbicide combinations may be combined with a specific fluid fertilizer for spray tank application.

MATERIALS REQUIRED:

1. A. SUTAN + 6.7-E
   B. or SUTAN + 6.7-E/ATRAZINE (80W or 4L) combination
   C. or SUTAN + 6.7-E/BLADEX (80W or 4L) combination.
   D. or SUTAN + 6.7-E/ATRAZINE (80W or 4L) BLADEX (80W or 4L) combination.
2. Fluid fertilizer to be used.
3. Tank mix adjuvants for fluid fertilizer combinations: Compex*, Sponto 168-D*, Unite* or equivalent. The adjuvant which provides the best compatibility depends on the specific fertilizer involved.

* Compex, Kalo Laboratories Inc., Kansas City, Missouri; Sponto 168-D, Witco Chemical Company, Houston, Texas; Unite, Hopkins Agricultural Chemical Company, Madison, Wisconsin.

4. Two one-quart, wide-mouth glass jars with lids.
5. Measuring spoons; a graduate pipette or cylinder will provide more accurate measurement; 1 tablespoon equals 3 teaspoons, 1 teaspoon equals approximately 5 milliliters.
6. Measuring cup, 8 oz. (237 mL).

PROCEDURE: (See Note 1):

1. Pour a pint (473 mL) of the fluid fertilizer into each of the quart jars.
2. Add adjuvant to one of the jars and mix (see next rate table).
3. Add the SUTAN + 6.7-E or the desired combination of herbicides to both jars (see next four rate tables). The order of addition should be: wettable powders first with mixing, flowables next with mixing, and the SUTAN + 6.7-E last.
4. Close both jars with lids and mix the contents by turning the jars upside down ten times.
5. Inspect the surface and body of the mixtures—
   (A) Immediately after completing the jar inversions.
   (B) After allowing the jars to stand quietly for 30 minutes.
   (C) And then again after turning the jars upside down 10 times.

If a uniform mix cannot be made, the mixture should not be used. If either mixture remains uniform for 30 minutes, the combination may be used. Should either mixture separate after 30 minutes, but readily remix uniformly with ten jar inversions, the mixture can be used if adequate agitation is maintained in the tank. If the mixture with adjuvant is satisfactory, but the one without adjuvant is not, be sure to use the adjuvant in the spray tank. Add the adjuvant first at a rate of 3 pints per 100 gallons of fluid fertilizer; foaming can be minimized by using moderate agitation.

If nondispersible oil, sludge or clumps of solids form in the mixtures, the combination should not be used.

NOTE 1: For some combinations, premixing wettable powders in a little water in a pail before adding them to the spray tank will improve the compatibility of the final mixtures with SUTAN + 6.7-E. This technique can be tested in the small-scale jar test by premixing the wettable powder in one-eighth (⅛) cup of water prior to addition to the pint of fluid fertilizer.

Rate Table For SUTAN + 6.7-E and Adjuvant with the Fluid Fertilizer**

Gallons of Fluid Fertilizer to be applied per acre	mL or tsp. of SUTAN + 6.7-E* to be added to 1 pint of Fertilizer	
	mL	tsp.
10	7	1⅓
15	5	1
20	4	¾
25	3	⅔
30	2	½
40	2	½

* Based on field rate of 1 pound active ingredient per acre in the fertilizer volumes indicated. Increase volume proportionately to correspond with intended field rate in terms of pounds active ingredient per acre (e.g. for field rate of 4 pounds actual SUTAN in 40 gallons fertilizer per acre, add 8 mL or 2 tsp. SUTAN + 6.7-E to each jar for compatibility testing purposes).

** Two (2) milliliters or one-half (½) teaspoon of adjuvant should be added to 1 pint of fluid fertilizer in order to equal the rate of 3 pints of adjuvant per 100 gallons of fluid fertilizer.

Rate Table* For SUTAN + 6.7-E/ATRAZINE (80W or 4L) and Adjuvant with Fluid Fertilizer**

Gallons of Fluid Fertilizer to be applied per acre	10	15	20	25	30	40
tsp. of SUTAN + 6.7-E	5¾	3¾	2¾	2⅓	2	1½
tsp. of ATRAZINE 80W	7	4⅔	3½	2¾	2⅓	1¾
tsp. of ATRAZINE 4L	3⅔	2½	1¾	1½	1¼	1

* Based on field rates of 4 pounds actual SUTAN and 1.6 pounds actual ATRAZINE (80W) or 1.5 pounds actual ATRAZINE (4L) per acre in the fertilizer volumes indicated.

** Two (2) milliliters or one-half (½) teaspoon of adjuvant should be added to 1 pint of fluid fertilizer in order to equal the rate of 3 pints of adjuvant per 100 gallons of fluid fertilizer.

Rate Table* For SUTAN + 6.7-E/BLADEX (80W or 4L) and Adjuvant with Fluid Fertilizer**

Gallons of Fluid Fertilizer to be applied per acre	10	15	20	25	30	40
tsp. of SUTAN + 6.7-E	$5\frac{3}{4}$	$3\frac{3}{4}$	$2\frac{3}{4}$	$2\frac{1}{3}$	2	$1\frac{1}{2}$
tsp. of BLADEX 80W	$7\frac{3}{4}$	$5\frac{1}{4}$	4	3	$2\frac{1}{3}$	2
tsp. of BLADEX 4L	$4\frac{3}{4}$	$3\frac{1}{4}$	$2\frac{1}{2}$	2	$1\frac{2}{3}$	$1\frac{1}{4}$

* Based on field rates of 4 pounds actual SUTAN and 2 pounds actual BLADEX (80W or 4L) per acre in the fertilizer volumes indicated.

** Two (2) milliliters or one-half ($\frac{1}{2}$) teaspoon of adjuvant should be added to 1 pint of fluid fertilizer in order to equal the rate of 3 pints of adjuvant per 100 gallons of fluid fertilizer.

Rate Table* for SUTAN + 6.7-E/ATRAZINE (80W or 4L)/ BLADEX (80W or 4L) and Adjuvant with Fluid Fertilizer**

Gallons of Fluid Fertilizer to be applied per acre	10	15	20	25	30	40
tsp. of SUTAN + 6.7-E	$5\frac{3}{4}$	$3\frac{3}{4}$	$2\frac{3}{4}$	$2\frac{1}{3}$	2	$1\frac{1}{2}$
tsp. of ATRAZINE 80W	$2\frac{1}{4}$	$1\frac{1}{2}$	1	$\frac{3}{4}$	$\frac{3}{4}$	$\frac{1}{2}$
tsp. of ATRAZINE 4L	$1\frac{1}{4}$	$\frac{3}{4}$	$\frac{2}{3}$	$\frac{1}{2}$	$\frac{1}{3}$	$\frac{1}{3}$
tsp. of BLADEX 80W	4	$2\frac{2}{3}$	2	$1\frac{1}{2}$	$1\frac{1}{4}$	1
tsp. of BLADEX 4L	$2\frac{1}{2}$	$1\frac{2}{3}$	$1\frac{1}{4}$	1	$\frac{3}{4}$	$\frac{2}{3}$

* Based on field rates of 4 pounds actual SUTAN, 0.5 pounds actual ATRAZINE and 1 pound actual BLADEX per acre in the fertilizer volumes indicated.

** Two (2) milliliters or one-half ($\frac{1}{2}$) teaspoon of adjuvant should be added to 1 pint of fluid fertilizer in order to equal the rate of 3 pints of adjuvant per 100 gallons of fluid fertilizer.

APPENDIX III

TANK MIX DIRECTIONS

Proper care must be used in preparing combination tank mixtures to avoid incompatibility problems. Small-scale tests should be made to establish that the mixtures are compatible (refer to Appendix II for general test procedures).

1) Fill the spray tank at least $\frac{2}{3}$ full of the appropriate carrier, except when dry flowable products are used. These products should be thoroughly wet out, mixed and predispersed first in $\frac{1}{4}$ to $\frac{1}{2}$ of a tank of carrier liquid. Then the tank should be filled $\frac{2}{3}$ full with additional carrier. Be sure the tank is clean and not contaminated with 2,4-D, oil residue or other materials.
2) Continue moderate agitation throughout mixing and application.
3) Add, mix and disperse dry flowables first, wettable powders second, liquid flowables third and SUTAN + 6.7-E last. For some combinations, premixing dry flowables and wettable powders in a little water in a pail before adding them to the spray tank will improve the compatibility of the final mixtures with SUTAN + 6.7-E.
4) Add the ATRAZINE, BLADEX and PRINCEP products individually to the spray tank in proper order (steps 1 and 3). Make sure each product is thoroughly wetted and dispersed prior to addition of the next product. In particular, dry flowables should be thoroughly predispersed prior to addition of other pesticides.
5) Add the SUTAN + 6.7-E to the tank at least $\frac{2}{3}$ full of carrier and mix; if ATRAZINE, BLADEX or PRINCEP are being used in the tank mix, they should be thoroughly mixed in the carrier *prior* to adding SUTAN + 6.7-E.
6) Complete filling the tank with carrier; mix thoroughly, and maintain moderate agitation to provide a uniform mix during application.
7) The tank-mix combination should not be left in the spray tank for prolonged periods; batches should be mixed and applied the same day.

Do not use excessive agitation because it may cause the ATRAZINE, BLADEX or PRINCEP to settle out and set up in the spray tank. All return lines to the tank must discharge below liquid level.

APPENDIX IV

SUTAN + 6.7-E, SUTAN + 6.7-E/ATRAZINE (80W, 4L or AATREX NINE-0), SUTAN + 6.7-E/BLADEX (80W or 4L) OR SUTAN + 6.7-E/ATRAZINE (80W or 4L)/BLADEX (80W or 4L) IMPREGNATION ON DRY BULK FERTILIZERS

CAUTION: SUTAN + 6.7-E alone or in combination with other herbicides must not be impregnated on ammonium nitrate, potassium nitrate or sodium nitrate fertilizers. Such mixtures may cause explosion and fire.

All individual state regulations relating to dry bulk fertilizer blending, registration, labeling and application are the responsibility of the individual and/or company selling the SUTAN + 6.7-E, SUTAN + 6.7-E/ATRAZINE, SUTAN + 6.7-E/BLADEX or SUTAN + 6.7-E/ATRAZINE/BLADEX fertilizer mixtures.

SUTAN + 6.7-E alone and in combination with ATRAZINE and/ or BLADEX may be impregnated on many dry bulk fertilizers (1)(2) and applied and incorporated into the soil either in the Fall before the ground freezes or before planting for the control of certain grassy weeds by SUTAN + 6.7-E and many grassy and broadleaf weeds by the SUTAN + 6.7-E/ATRAZINE, SUTAN + 6.7-E/ BLADEX or SUTAN + 6.7-E/ATRAZINE/BLADEX combinations in corn.

All SUTAN + 6.7-E, ATRAZINE (80W, 4L or AATREX NINE-O) and BLADEX (80W or 4L) label and supplemental literature instructions and precautions regarding rates per acre, soil type and soil incorporation, application and other directions must be followed.

Test results have shown that SUTAN + 6.7-E, SUTAN + 6.7-E/ ATRAZINE (80W, 4L or AATREX NINE-O), SUTAN + 6.7-E/BLADEX (80W or 4L) and SUTAN + 6.7-E/ATRAZINE (80W or 4L)/BLADEX (80W or 4L) on many dry bulk fertilizers give weed control equal to SUTAN + 6.7-E or tank mixtures of SUTAN + 6.7-E/ATRAZINE (80W, 4L or AATREX NINE-O), SUTAN + 6.7-E/BLADEX (80W or 4L) and SUTAN + 6.7-E/ ATRAZINE (80W or 4L)/BLADEX (80W or 4L) applied as a spray in water or liquid fertilizer. However, uniform impregnation of the pesticides on the dry fertilizer particles and uniform application in the field are necessary to assure good results.

A minimum of 200 pounds and a maximum of 700 pounds of approved fertilizer ingredients (1) (2) impregnated with the appropriate amount of SUTAN + 6.7-E alone or in combination with ATRAZINE (80W, 4L or AATREX NINE-O) and/or BLADEX (80W or 4L) must be applied per acre.

For impregnating the pesticides on dry fertilizers, use a closed rotary-drum type mixer equipped with suitable spraying equipment. The spray nozzles should be positioned inside of the mixer to provide uniform spray coverage of the tumbling fertilizer. The SUTAN + 6.7-E should be sprayed uniformly onto the fertilizer using a fine spray pattern. For the SUTAN + 6.7-E/ATRAZINE (80W or 4L), SUTAN + 6.7-E/BLADEX or SUTAN + 6.7-E/ ATRAZINE (80W or 4L)/BLADEX combination product, the ATRAZINE (80W or 4L) and/or BLADEX (80W or 4L) can be added and mixed uniformly with the fertilizer prior to the SUTAN + 6.7-E addition, or alternatively, the SUTAN + 6.7-E/ATRAZINE (80W or 4L), SUTAN + 6.7-E/BLADEX (80W or 4L) or SUTAN + 6.7-E/ ATRAZINE (80W or 4L)/BLADEX (80W or 4L) can be mixed in the proper use ratio to prepare a uniform slurry for joint spray impregnation. For the SUTAN + 6.7-E/AATREX NINE-O combination, the AATREX NINE-O should be mixed and thoroughly dispersed in the smallest amount of water feasible, and then uniformly sprayed onto the dry fertilizer. Following this, the SUTAN 6.7-E can be sprayed onto the dry fertilizer.

The physical properties of fertilizers vary, particularly in liquid absorptive capacity. When absorptivity is sufficient, simple spray impregnation of the fertilizer with SUTAN + 6.7-E provides a satisfactory dry mixture.

If the absorptivity is not adequate, use of a highly absorptive powder is required to provide a dry, free-flowing mixture.

Microcel E (Johns-Manville Products Corporation) is the recommended absorbent powder. It should be added separately and uniformly to the prepared pesticide-fertilizer mixture, in a quantity that is sufficient to provide a suitably free-flowing mixture. Generally, less than 2% by weight of Microcel E is required.

This amount of SUTAN + 6.7-E, SUTAN + 6.7-E/ATRAZINE (80W, 4L or AATREX NINE-O), SUTAN + 6.7-E/BLADEX (80W or 4L) or SUTAN + 6.7-E/ATRAZINE (80W or 4L)/BLADEX (80W or 4L) actually required in the manufacture of individual fertilizer mixtures should be determined carefully for each production operation (3) (4) (5). This is necessary to ensure that the amounts of SUTAN + 6.7-E, ATRAZINE and BLADEX actually contained in the mixture applied to the soil represent the correct rates of use.

Bulk fertilizers impregnated with SUTAN + 6.7-E, SUTAN + 6.7-E/ATRAZINE (80W, 4L or AATREX NINE-O), SUTAN + 6.7-E/BLADEX (80W or 4L) or SUTAN + 6.7-E/ATRAZINE (80W or 4L)/BLADEX (80W or 4L) should be **applied immediately, NOT STORED.** All bulk containers should be tightly covered while the products are being transported and applied to reduce chances of SUTAN + 6.7-E loss via volatilization.

INCORPORATION OF IMPREGNATED FERTILIZERS

When application and incorporation are done in separate operations, SUTAN + 6.7-E impregnated on dry fertilizer must be incorporated on the same day as applied. Application must be made on a dry soil surface (at least 1/2 inch deep) free from dew and incidental moisture.

Do not combine SUTAN + 6.7-E plus ATRAZINE (80W, 4L or AATREX NINE-O) and/or BLADEX (80W or 4L) with single superphosphate (0-20-0) or treble superphosphate (0-46-0).

SUTAN + 6.7-E Physical Data

Specific Gravity 20/20°C: 0.946 (typical)
Pounds/Gallon 20/20°C: 7.87 (typical)
Flashpoint: 197°F (Tagliabue closed cup)
Viscosity: Sprayable down to minus 20°F.

(1) Approved Dry Fertilizer Ingredients For Use With SUTAN + 6.7-E

	N	P	K
Ammonium Sulfate	21	0	0
Ammonium Phosphate-Sulfate	16	20	0
Diammonium Phosphate	18	46	0
Monoammonium Phosphate	11	56	0
Potassium Chloride	0	0	60
Potassium Sulfate	0	0	52
Single Superphosphate	0	20	0
Treble Superphosphate	0	46	0
Urea*	45	0	0

* Some ureas may be phytotoxic when high rates are applied to corn. Use only urea rates known to be safe for corn application.

NOTE: Sul-po-mag, K-mag and 11-48-0 have been shown to be compatible with SUTAN + 6.7-E and are approved for use.

(2) Approved Dry Fertilizer Ingredients For Use with SUTAN + 6.7-E/ATRAZINE (80W, 4L or AATREX NINE-O), SUTAN + 6.7-E/BLADEX (80W or 4L) or SUTAN + 6.7-E/ATRAZINE (80W or 4L)/BLADEX (80W or 4L)

	N	P	K
Ammonium Sulfate	21	0	0
Ammonium Phosphate-Sulfate	16	20	0
Diammonium Phosphate	18	46	0
Monoammonium Phosphate	11	56	0
Potassium Chloride	0	0	60
Potassium Sulfate	0	0	52
Urea*	45	0	0

* Some ureas may be phytotoxic when high rates are applied to corn. Use only urea rates known to be safe for corn application.

(3) RATE CHART FOR THE IMPREGNATION OF DRY BULK FERTILIZERS WITH SUTAN + 6.7-E or SUTAN + 6.7-E PLUS ATRAZINE (80W, 4L or AATREX NINE-O)

FERTILIZER RATE PER ACRE	SUTAN + 6.7-E RATE/ACRE			ATRAZINE 80W RATE/ACRE			ATRAZINE 4L RATE/ACRE			AATREX NINE-O RATE/ACRE		
	3 3/4 pts. per acre	4 3/4 pts. per acre	7 1/3 pts. per acre	1 lb. per acre	1 1/4 lbs. per acre	2 lbs. per acre	3/4 qt. per acre	1 qt. per acre	1 1/2 qts. per acre	4/5 lb. per acre	1 lb. per acre	1 3/4 lbs. per acre
200 lbs.	18 3/4 qts./ton	23 3/4 qts./ton	36 2/3 qts./ton	10 lbs./ton	12 1/2 lbs./ton	20 lbs./ton	7 1/2 qts./ton	10 qts./ton	15 qts./ton	8 lbs./ton	10 lbs./ton	17 1/2 lbs./ton
250 lbs.	15 qts./ton	19 qts./ton	29 1/3 qts./ton	8 lbs./ton	10 lbs./ton	16 lbs./ton	6 qts./ton	8 qts./ton	12 qts./ton	6 2/5 lbs./ton	8 lbs./ton	14 lbs./ton
300 lbs.	12 1/2 qts./ton	15 4/5 qts./ton	24 2/5 qts./ton	6 2/3 lbs./ton	8 1/3 lbs./ton	13 1/3 lbs./ton	5 qts./ton	6 3/5 qts./ton	10 qts./ton	5 1/3 lbs./ton	6 2/3 lbs./ton	11 2/3 lbs./ton
350 lbs.	10 3/4 qts./ton	13 3/5 qts./ton	21 qts./ton	5 3/4 lbs./ton	7 1/8 lbs./ton	11 2/5 lbs./ton	4 1/4 qts./ton	5 4/5 qts./ton	8 3/5 qts./ton	4 2/5 lbs./ton	5 3/4 lbs./ton	10 lbs./ton
350 lbs.	10 3/4 qts./ton	13 3/5 qts./ton	21 qts./ton	5 3/4 lbs./ton	7 1/8 lbs./ton	11 2/5 lbs./ton	4 1/4 qts./ton	5 4/5 qts./ton	8 3/5 qts./ton	4 2/5 lbs./ton	5 3/4 lbs./ton	10 lbs./ton
400 lbs.	9 2/5 qts./ton	11 7/8 qts./ton	18 1/3 qts./ton	5 lbs./ton	6 1/4 lbs./ton	10 lbs./ton	3 3/4 qts./ton	5 qts./ton	7 1/2 qts./ton	4 lbs./ton	5 lbs./ton	8 3/4 lbs./ton
450 lbs.	8 1/3 qts./ton	10 1/2 qts./ton	16 1/3 qts./ton	4 2/5 qts./ton	5 1/2 lbs./ton	8 7/8 lbs./ton	3 1/3 qts./ton	4 2/5 qts./ton	6 3/5 qts./ton	3 3/5 lbs./ton	4 2/5 lbs./ton	7 3/4 lbs./ton
500 lbs.	7 1/2 qts./ton	9 1/2 qts./ton	14 2/3 qts./ton	4 lbs./ton	5 lbs./ton	8 lbs./ton	3 qts./ton	4 qts./ton	6 qts./ton	3 1/5 lbs./ton	4 lbs./ton	7 lbs./ton
550 lbs.	6 4/5 qts./ton	8 2/3 qts./ton	13 1/3 qts./ton	3 2/3 lbs./ton	4 2/3 lbs./ton	7 1/3 lbs./ton	2 3/4 qts./ton	3 2/3 qts./ton	5 1/2 qts./ton	2 7/8 lbs./ton	3 2/3 lbs./ton	6 3/8 lbs./ton
600 lbs.	6 1/4 qts./ton	8 qts./ton	12 1/5 qts./ton	3 1/3 lbs./ton	4 1/5 lbs./ton	6 2/3 lbs./ton	2 1/2 qts./ton	3 1/3 qts./ton	5 qts./ton	2 2/3 lbs./ton	3 1/3 lbs./ton	5 4/5 lbs./ton
650 lbs.	5 3/4 qts./ton	7 1/3 qts./ton	11 1/4 qts./ton	3 1/8 lbs./ton	3 7/8 lbs./ton	6 1/5 lbs./ton	2 1/3 qts./ton	3 1/8 qts./ton	4 3/5 qts./ton	2 1/2 lbs./ton	3 1/8 lbs./ton	5 3/8 lbs./ton
700 lbs.	5 1/3 qts./ton	6 4/5 qts./ton	10 1/2 qts./ton	2 7/8 lbs./ton	3 2/3 lbs./ton	5 3/4 lbs./ton	2 1/8 qts./ton	2 7/8 qts./ton	4 1/4 qts./ton	2 1/4 lbs./ton	2 7/8 lbs./ton	5 lbs./ton

STORAGE AND DISPOSAL

PROHIBITIONS

Do not contaminate water, food, or feed by storage of disposal. Open dumping is prohibited.

STORAGE

Keep container tightly closed when not in use. Do not store near seeds, fertilizers or foodstuffs. Store out of reach of children, pets and domestic animals. Can be stored at temperatures as low as minus 50°F.

PESTICIDE DISPOSAL

Rinse spray equipment. Waste resulting from the use of this product may be disposed of on site or at an approved waste disposal facility.

CONTAINER DISPOSAL
METAL—Triple rinse (or equivalent). Then offer for recycling or reconditioning, or puncture and dispose of in a sanitary landfill, or by other procedures approved by State and local authorities.
PLASTIC—Triple rinse (or equivalent). Then offer for recycling or reconditioning, or puncture and dispose of in a sanitary landfill, or by incineration, or, if allowed by State and local authorities, by burning. If burned, stay out of smoke.

NOTICE—READ CAREFULLY

CONDITIONS OF SALE
Stauffer (and seller) offer(s) this product for sale subject to, and buyer and all users are deemed to have accepted, the following conditions of sale and warranty which may only be varied by written agreement of a duly authorized representative of Stauffer.

WARRANTY LIMITATION
Stauffer warrants that this product conforms to the chemical description on the label and is reasonably fit for the purposes referred to in the directions for use on the label subject to the inherent risks referred to below. Stauffer makes no other express warranties; THERE IS NO IMPLIED WARRANTY OF MERCHANTABILITY and there are no warranties which extend beyond the description on the label hereof.

INHERENT RISKS
The directions for use of this product are believed to be reliable and should be followed carefully. However, it is impossible to eliminate all risks associated with use. Buyer assumes all risks associated with use or application of this product contrary to label instructions or resulting from extraordinary weather conditions.

LIMITATION OF LIABILITY
In no case shall Stauffer be liable for special, indirect, or consequential damages resulting from the use or handling of this product and no claim of any kind shall be greater in amount than the purchase price of the product in respect of which such damages are claimed.

This product is sold only for uses stated on its label. No express or implied license is granted to use or sell this product under any patent in any country except as specified: Country: United States of America. Patent No.: 4,021,224; 4,137,070.

This is a specimen label. Stauffer is not responsible for the accuracy of the information contained herein. As labels are subject to revision, always carefully read and follow the label on the product container.

SUPPLEMENTAL LABELING

SUTAN® + 6.7-E Selective Herbicide
EPA Reg. No. 476–2156

DIRECTIONS FOR INJECTION WITH ANHYDROUS AMMONIA

INJECTION DIRECTIONS FOR SUTAN + 6.7-E WITH ANHYDROUS AMMONIA
It is a violation of Federal law to use this product in a manner inconsistent with its labeling.

For pre-plant soil injection applications of anhydrous ammonia, on conservation or conventional tilled acreage, SUTAN + 6.7-E can be introduced into the anhydrous line using a metering pump which offers a constant rate of SUTAN + 6.7-E output per acre independent of the anhydrous ammonia output rate. The SUTAN + 6.7-E discharge hose should tee into the ammonia discharge line between the meter and the manifold. For best results, SUTAN + 6.7-E must be used at rates of $4^3/_4$ to $7^1/_3$ pints per acre, injection shanks must be spaced at 8 to 10 inch intervals, and injection depth must be 4 to 5 inches. Spacings greater than 10 inches up to 15 inches will require a cross discing to achieve acceptable weed control. Do not inject at spacings greater than 15 inches as weed control may be unacceptable. A minimum rate of 125 pounds per acre of anhydrous ammonia is recommended.

Weed control will be inadequate if the SUTAN + 6.7-E is injected more than 4 to 5 inches below the soil surface. Soil injections at less than a 4 inch depth may result in some loss of the anhydrous ammonia through the soil surface. Injection should only be performed in soils dry enough to permit good soil mixing.

NOTE: Atrazine, BLADEX* and PRINCEP* have little lateral movement in the soil and injections with anhydrous ammonia will give only a narrow band of broadleaf weed control. These products may be surface applied and incorporated at the same time the SUTAN + 6.7-E and anhydrous ammonia are injected or post-emergent applications of recommended broadleaf herbicides may be used.

All applicable recommendations, directions, restrictions, and precautions on the EPA-registered label are to be followed.

This labeling must be in the possession of the user at the time of pesticide application.

* BLADEX is a Registered Trademark of Shell Chemical Company.

* PRINCEP is a Registered Trademark of Ciba-Geigy Corporation.

This is a specimen label. Stauffer is not responsible for the accuracy of the information contained herein. As labels are subject to revision, always carefully read and follow the label on the product container.

SUTAN®+ 10-G

SELECTIVE HERBICIDE GRANULES

For Corn (Field, Sweet and Silage)

ACTIVE INGREDIENT:	
S-ethyl diisobutylthiocarbamate	10.0%
INERT INGREDIENTS:	90.0%
TOTAL	100.0%

Contains 10 pounds of active ingredient per 100 pounds.
EPA Reg. No. 476–2001 A-2R

CAUTION

Keep Out of Reach of Children

STATEMENT OF PRACTICAL TREATMENT
If a known exposure occurs or is suspected, immediately start the procedures given below. If further treatment is required, contact a Poison Center, a physician, or the nearest hospital, or call Collect day or night: (203) 226-6602 Stauffer Chemical Company, or (800) 424-9300 Pesticide Team Safety Network.

If swallowed: Immediately give several glasses of water and induce vomiting by gagging the victim with a finger placed on the back of the victim's tongue. Give fluids until vomitus is clear. If victim is unconscious or convulsing, do not induce vomiting or give anything by mouth. Call a physician or nearest Poison Center immediately.

For skin contact: Flush all affected areas with plenty of water for several minutes. Remove and clean contaminated clothing and shoes before reuse.

For eye contact: Hold eyelids apart and flush with plenty of water for 15 minutes. Seek medical attention if irritation occurs.

If inhaled: Remove to fresh air. Seek medical attention if respiratory irritation occurs or if breathing becomes difficult.

PRECAUTIONARY STATEMENTS

HAZARDS TO HUMANS AND DOMESTIC ANIMALS

CAUTION

Harmful if Swallowed
Avoid contact with skin, eyes and clothing. Avoid inhalation of dust. Wash thoroughly with soap and water after use.

ENVIRONMENTAL HAZARDS
Do not apply directly to water or wetlands. Do not contaminate water by cleaning of equipment or disposal of wastes. Cover or incorporate spills.

READ ALL LABEL DIRECTIONS BEFORE USING

GENERAL USE PRECAUTIONS

Read all label directions before using. Apply this product as specified on this label.

Applied according to directions and under normal growing conditions, SUTAN+ 10-G will not harm the treated crop. During germination and early stages of growth, extended periods of unusually cold and wet or hot and dry weather, insect or plant disease attack, the use of certain soil-applied systemic insecticides, improperly placed fertilizers or soil insecticides may create abnormal conditions that weaken crop seedlings. SUTAN+ 10-G used under these abnormal conditions could result in crop injury.

When applied according to directions and when conditions exist for normal plant growth through the season, no harmful residues of SUTAN+ 10-G should remain beyond harvest. In the **Southeastern U.S.**, when SUTAN+ 10-G is used for weed control in silage corn, do not seed small seeded grains after corn harvest until September.

- Do not use on milo or sorghum. Do not use on corn seed stock. (Breeders, Foundation, or Increase). SUTAN+ 10-G may be used on Production seed corn.
- This product is recommended for use on mineral soils only.
- Do not apply in combination with fertilizers, insecticides or fungicides.
- Do not contaminate food or feed.
- Do not store near seeds, fertilizers, or foodstuffs.
- Do not contaminate irrigation water or water used for domestic purposes.

WEEDS CONTROLLED

SUTAN+ 10-G will not control established weeds.

Annual Grasses:	
Barnyardgrass; (Watergrass)	*Echinochloa crusgalli*
Crabgrass	*Digitaria* spp.
Giant Foxtail	*Setaria faberi*
Green Foxtail	*Setaria viridis*
Yellow Foxtail	*Setaria glauca*
Goosegrass	*Eleusine indica*
Seedling Johnsongrass	*Sorghum halepense*
Perennial Weeds:	
Purple Nutgrass (Nutsedge)	*Cyperus rotundus*
Yellow Nutgrass (Nutsedge)	*Cyperus esculentus*

Existing stands of nutgrass must be turned under and chopped up thoroughly before treatment. Cultivation is suggested in addition to SUTAN+ 10-G treatment in fields with moderate or heavy nutgrass infestations.

DIRECTIONS FOR USE

It is a violation of Federal law to use this product in a manner inconsistent with its labeling.

SUTAN+ 10-G is a selective herbicide which is mixed (incorporated) into the soil for control of weeds listed on this label. SUTAN+ 10-G controls annual grasses as their seeds germinate by interfering with normal germination and seedling development. It does not control established weeds. All weed growth and crop stubble should be thoroughly worked into the soil before treatment.

APPLICATION DIRECTIONS

Apply only to well-worked soil that is dry enough to permit thorough mixing with incorporation equipment. Uniformly apply the recommended rate of SUTAN+ 10-G using equipment designed for application of granular herbicides. Equipment should be carefully calibrated before use and checked frequently during application to be sure it is working properly and delivering a uniform distribution pattern. Avoid overlaps that will increase SUTAN+ 10-G dosage above recommended rates as plant injury may occur.

In irrigated areas, do not apply SUTAN+ 10-G before pre-irrigation.

INCORPORATION DIRECTIONS

SUTAN+ 10-G must be incorporated into the soil immediately after application to prevent loss of the herbicide. Whenever possible, application and incorporation should be done in the same operation.

SOIL MIXING (INCORPORATION) BEFORE PLANTING: Use power-driven cultivation equipment set to cut to a depth of 2 to 3 inches; OR tandem discs set to cut to a depth of 4 to 6 inches operated at 4 to 6 mph, followed by a spike-tooth harrow or some other leveling device which extends beyond the ends of the disc. For thorough mixing, disc in two different directions (cross disc).

PLANTING

Plant seed to a maximum depth of 2 inches within 2 weeks after treatment. To avoid removing SUTAN+ 10-G from the row (with loss of weed control in the row) do not move or shape soil after treatment, such as by planting in deep furrows.

CULTURAL PRACTICES FOLLOWING APPLICATION

SUTAN+ 10-G is not persistent in the soil and susceptible weeds germinating late in the season may not be controlled. Shallow cultivation may be necessary to control these late-germinating weeds. Do not cultivate deeper than the depth of herbicide incorporation. Pre-emergence or post-emergence herbicides may be necessary to control weeds resistant to SUTAN+ 10-G.

RECOMMENDATIONS

Do not use on any type corn in Arizona or in the ten southern-most California counties (except in Kern County where field and silage corn may be treated).

BROADCAST RATES PER ACRE: Apply and incorporate 30 lbs. SUTAN+ 10-G per acre on light soils (coarse textured-sandy); apply 40 lbs. SUTAN+ 10-G on medium (medium textured-sandy loam) to heavy (fine textured-clay) soils. In the north central and midwest corn growing regions, 40 lbs. of SUTAN+ 10-G per acre also is recommended on all soils with moderate to heavy infestations of nutgrass. Use 30 lbs. SUTAN+ 10-G per acre on blow sands in these regions. Cultivation is suggested in addition to the SUTAN+ 10-G treatment in fields with moderate or heavy infestations of nutgrass.

STORAGE AND DISPOSAL

PROHIBITIONS: Do not contaminate water, food, or feed by storage or disposal. Open dumping is prohibited.

STORAGE: Keep container closed when not in use. Do not store near seeds, fertilizers, or foodstuffs.

PESTICIDE DISPOSAL: Wastes resulting from the use of this product may be disposed of on site or at an approved waste disposal facility.

CONTAINER DISPOSAL: Completely empty bag into application equipment. Then dispose of empty bag in a sanitary landfill or by incineration, or, if allowed by State and local authorities, by burning. If burned, stay out of smoke.

NOTICE—READ CAREFULLY

CONDITIONS OF SALE

Stauffer (and seller) offer(s) this product for sale subject to, and buyer and all users are deemed to have accepted, the following conditions of sale and warranty which may only be varied by written agreement of a duly authorized representative of Stauffer.

WARRANTY LIMITATION

Stauffer warrants that this product conforms to the chemical description on the label and is reasonably fit for the purposes referred to in the directions for use on the label subject to the inherent risks referred to below. Stauffer makes no other express warranties; THERE IS NO IMPLIED WARRANTY OF MERCHANTABILITY and there are no warranties which extend beyond the description on the label hereof.

INHERENT RISKS
The directions for use of this product are believed to be reliable and should be followed carefully. However, it is impossible to eliminate all risks associated with use. Buyer assumes all risks associated with use or application of this product contrary to label instructions or resulting from extraordinary weather conditions.

LIMITATION OF LIABILITY
In no case shall Stauffer be liable for special, indirect, or consequential damages resulting from the use or handling of this product and no claim of any kind shall be greater in amount than the purchase price of the product in respect of which such damages are claimed.

This product is sold only for uses stated on its label. No express or implied license is granted to use or sell this product under any patent in any country except as specified. Country: United States of America. Patent No. 4,021,224; 4,137,070.

This is a specimen label. Stauffer is not responsible for the accuracy of the information contained herein. As labels are subject to revision, always carefully read and follow the label on the product container.

SUTAZINE™+ 18-6.G

SELECTIVE HERBICIDE GRANULES

For Selective Weed Control in Field, Sweet, and Silage Corn

ACTIVE INGREDIENTS:	
S-ethyl diisobutylthiocarbamate	18.0%
Atrazine: 2-chloro-4-(ethylamino)-6-(isopropylamino)-*s*-triazine	5.7%
Related triazines	0.3%
INERT INGREDIENTS:	76.0%
TOTAL	100.0%

Contains 24 pounds of active ingredients per 100 pounds.

EPA Reg. No. 476-2103 A-4R

CAUTION

Keep Out of Reach of Children

HARMFUL IF SWALLOWED

STATEMENT OF PRACTICAL TREATMENT
If a known exposure occurs or is suspected, immediately start the procedures given below. If further treatment is required contact a Poison Center, a physician, or the nearest hospital, or call collect day or night: (203) 226-6602 Stauffer Chemical Company, or (800) 424-9300 Pesticide Team Safety Network.

If swallowed: Immediately give several glasses of water and induce vomiting by gagging the victim with a finger placed on the back of the victim's tongue. Give fluids until vomitus is clear. If victim is unconscious or convulsing, do not induce vomiting or give anything by mouth.

For skin contact: Flush all affected areas with plenty of water for several minutes.

For eye contact: Hold eyelids apart and flush with plenty of water for 15 minutes. Remove and clean contaminated clothing and shoes before reuse. Seek medical attention if irritation occurs.

If inhaled: Remove to fresh air. Seek medical attention if respiratory irritation occurs or if breathing becomes difficult.

READ ALL LABEL DIRECTIONS BEFORE USING

PRECAUTIONARY STATEMENTS

HAZARDS TO HUMANS AND DOMESTIC ANIMALS
CAUTION
Harmful If Swallowed

Avoid contact with skin, eyes and clothing. Avoid inhalation of dust. Wash thoroughly with soap and water after use.

ENVIRONMENTAL HAZARDS
Do not apply directly to water or wetlands. Do not contaminate water by cleaning of equipment or disposal of wastes. Cover or incorporate spills.

GENERAL USE PRECAUTIONS

Read all label directions before using. Apply this product as specified on this label. Make only one application per crop.

After treatment with SUTAZINE+18-6·G, do not plant any crop except corn the following year or injury may occur.

Do not use on milo or sorghum. Do not use on corn seed stock (Breeders, Foundation, or Increase). SUTAZINE+18-6·G may be used on Production seed corn.

This product is recommended for use on mineral soils only.

Applied according to directions and under normal growing conditions, SUTAZINE+18-6·G will not harm the treated crop. During germination and early stages of growth, extended periods of unusually cold and wet or hot and dry weather, insect or plant disease attack, the use of certain soil-applied systemic insecticides, improperly placed fertilizers or soil insecticides, may create abnormal conditions that weaken crop seedlings. SUTAZINE+18-6·G used under these abnormal conditions could result in crop injury.

Care should be taken to avoid use where adjacent desirable trees, shrubs, or plants may be injured as a result of the product leaching through the soil.

Do not contaminate food or feed.

Do not store near seeds, fertilizers, or foodstuffs.

Do not contaminate irrigation water or water used for domestic purposes.

WEEDS CONTROLLED
This product will not control established weeds.

Annual Grasses	
Barnyardgrass	*Echinochloa crus-galli*
Crabgrass	*Digitaria* spp.
Fall Panicum	*Panicum dichotomiflorum*
Giant Foxtail	*Setaria faberi*
Goosegrass	*Eleusine indica*
Green Foxtail	*Setaria viridis*
Seedling Bermudagrass	*Cynodon dactylon*
Seedling Johnsongrass	*Sorghum halepense*
*Wild Cane	*Sorghum bicolor*
Wild Oats	*Avena fatua*
Witchgrass	*Panicum capillare*
Yellow Foxtail	*Setaria glauca*
*Texas Panicum; Buffalo-grass—Coloradograss	*Panicum Texanum*

* Cultivation is suggested in addition to the SUTAZINE + 18-6·G treatment in fields with heavy infestation of wild cane or buffalograss (Texas panicum).

Annual Broadleaf Weeds	
Annual Morningglory	*Ipomoea purpurea*
Carpetweed	*Mollugo verticillata*
Cocklebur Common	*Xanthium pensylvanicum*
Jimsonweed	*Datura stramonium*
Lambsquarters	*Chenopodium album*
Mustards	*Brassica* spp.
Nightshade	*Solanum* spp.
Common Purslane	*Portulaca oleracea*
Ragweed	*Ambrosia* spp.
Redroot Pigweed	*Amaranthus retroflexus*
Prostrate Pigweed	*Amaranthus graecizans*
Smooth Pigweed	*Amaranthus hybridus*
Smartweed	*Polygonum* spp.
Velvetleaf	*Abutilon theophrasti*
Florida Purslane	*Richardia scabra*

Perennial Grasses

Purple Nutgrass (Nutsedge)	*Cyperus rotundus*
Yellow Nutgrass (Nutsedge)	*Cyperus esculentus*

Existing stands of nutgrass must be turned under and chopped up thoroughly before treatment Cultivation is suggested in addition to the SUTAZINE+18-6·G treatment in fields with moderate to heavy nutgrass infestations.

DIRECTIONS FOR USE

It is a violation of Federal law to use this product in a manner inconsistent with its labeling.

SUTAZINE+18-6·G is a selective herbicide which is mixed (incorporated) into the soil for control of weeds listed on this label by interfering with normal germination and seedling development. It does not control established weeds. All weed growth and crop stubble should be thoroughly worked into the soil before treatment.

APPLICATION DIRECTIONS

Apply only to well-worked soil that is dry enough to permit thorough mixing with incorporation equipment. Uniformly apply the recommended rate of SUTAZINE+18-6·G using equipment designed for application of granular herbicides. Equipment should be carefully calibrated before use and checked frequently during application to be sure it is working properly and delivering a uniform distribution pattern. Avoid overlaps that will increase dosage above recommended rates as plant injury may occur.

In irrigated areas, do not apply before pre-irrigation.

INCORPORATION DIRECTIONS

SUTAZINE+18-6·G must be incorporated into the soil after application to prevent loss of the Butylate. Whenever possible, application and incorporation should be done in the same operation. When application and incorporation are done in separate operations, SUTAZINE+18-6·G must be incorporated on the same day as applied. Application must be made on a dry soil surface (at least ½ inch deep) free from dew and incidental moisture.

SOIL MIXING (INCORPORATION) BEFORE PLANTING

Incorporate using power-driven cultivation equipment set to cut to a depth of 2 to 3 inches, or a tandem disc set to cut to a depth of 4 to 6 inches, operated at 4 to 6 mph, followed by a spike-tooth harrow or some other leveling device which extends beyond the ends of the disc. For thorough mixing, disc in two different directions (cross disc). Prior to the second discing adjust the disc to prevent cutting deeper than 4 to 6 inches.

PLANTING

Do not plant seed deeper than 2 inches. For SUTAZINE+18-6·G application made before planting, seeding can be done immediately, or it may be delayed up to 2 weeks. To avoid removing SUTAZINE+18-6·G from the row (with loss of weed control in the row), do not move or shape soil after treatment, such as by planting in deep furrows. If it is necessary to work the soil following application and incorporation, but before planting, do not work the soil deeper than the depth the herbicides were incorporated.

RECOMMENDATIONS

Do not use on any type corn in Arizona or in the ten southernmost California counties (except Kern County where field and silage corn may be treated).

NOTE: All doses expressed as broadcast rates per acre. For band applications, reduce dosage proportionately depending upon the row spacing and band width to be treated.

BEFORE PLANTING (PREPLANT) APPLICATION

In The North Central, Northeastern And Midwestern Corn Growing Areas

Apply and incorporate 16⅔ to 22⅔ pounds of SUTAZINE+18-6·G per broadcast acre. Use the higher rate on extremely heavy soils (fine textured-clay) with over 5% organic matter. On fields with moderate to heavy infestations of wild cane or nutgrass use the higher rate regardless of soil texture, except on blow sands and the light soils (coarse textured-sandy) of the Eastern Coastal States. Cultivation is suggested in addition to the SUTAZINE+18-6·G treatment in fields with moderate to heavy infestations of wild cane or nutgrass.

In All Other Corn Growing Areas

Apply and incorporate 16⅔ pounds of SUTAZINE+18-6·G per broadcast acre. Cultivation is suggested in addition to the SUTAZINE+18-6·G treatment in fields with moderate to heavy infestations of wild cane, nutgrass or buffalograss (Texas panicum).

AFTER PLANTING APPLICATION NOTE: Recommended in the following states *only*—North and South Dakota, Nebraska, Kansas, Minnesota and Iowa. If *local* conditions in the above-named states make preplant SUTAZINE+18-6·G application unfeasible, the product may be applied and mixed into the soil at any time after planting before emergence of corn. Apply and incorporate 16⅔ pounds SUTAZINE+18-6·G per broadcast acre. For band applications, use standard planter attached equipment designed for band application of granular herbicides. Normally, application and incorporation are done in the same operation using equipment that attaches to the back of the applicator, i.e., row wheels, vibro fillers and similar equipment. Incorporate to a depth of 1 to 1½ inches. A rotary hoe can also be used set to cut to a depth of 1½ inches operated at 6 to 8 mph.

NOTE: After-planting application is not recommended for control of nutgrass, Texas panicum, or wild cane. See "BEFORE PLANTING (PREPLANT) APPLICATION" for control of these weeds.

CULTURAL PRACTICES FOLLOWING APPLICATION

Should weeds develop, a shallow cultivation or rotary hoeing will generally result in better weed control. When cultivating for any reason, it should be shallow, i.e., no more than ½ the depth the herbicide was incorporated.

STORAGE AND DISPOSAL

PROHIBITIONS

Do not contaminate water, food, or feed by storage or disposal. Open dumping is prohibited.

STORAGE

Keep container closed when not in use. Do not store near seeds, fertilizers, or foodstuffs.

PESTICIDE DISPOSAL

Wastes resulting from the use of this product may be disposed of on site or at an approved waste disposal facility.

CONTAINER DISPOSAL

Completely empty bag into application equipment. Then dispose of empty bag in a sanitary landfill or by incineration, or, if allowed by State and local authorities, by burning. If burned, stay out of smoke.

NOTICE—READ CAREFULLY

CONDITIONS OF SALE

Stauffer (and seller) offer(s) this product for sale subject to, and buyer and all users are deemed to have accepted, the following conditions of sale and warranty which may only be varied by written agreement of a duly authorized representative of Stauffer.

WARRANTY LIMITATION

Stauffer warrants that this product conforms to the chemical description on the label and is reasonably fit for the purposes referred to in the directions for use on the label subject to the inherent risks referred to below. Stauffer makes no other express warranties. THERE IS NO IMPLIED WARRANTY OF MERCHANTABILITY and there are no warranties which extend beyond the description on the label hereof.

INHERENT RISKS

The directions for use of this product are believed to be reliable and should be followed carefully. However, it is impossible to eliminate all risks associated with use. Buyer assumes all risks associated with use or application of this product contrary to label instructions or resulting from extraordinary weather conditions.

LIMITATION OF LIABILITY

In no case shall Stauffer be liable for special, indirect, or consequential damages resulting from the use or handling of this product and no claim of any kind shall be greater in amount than the purchase price of the product in respect of which such damages are claimed.

This product is sold only for uses stated on its label. No express or implied license is granted to use or sell this product under any patent in any country except as specified. Country: United States of America.
Patent No. 4,021,224; 4,137,070.

This is a specimen label. Stauffer is not responsible for the accuracy of the information contained herein. As labels are subject to revision, always carefully read and follow the label on the product container.

SUTAZINE™+

SELECTIVE HERBICIDE

Preplant Incorporated Herbicide for Annual Grass and Broadleaf Weed Control in Corn (Field, Sweet, Silage and Popcorn)

ACTIVE INGREDIENTS:	
S-ethyl diisobutylthiocarbamate	56.80%
Atrazine: 2-chloro-4-(ethylamino)-6-(isopropylamino)-s-triazine	13.90%
Related Triazines	0.30%
INERT INGREDIENTS:	29.00%
TOTAL	100.00%

6-ME Contains 6.0 pounds active ingredients per gallon.

EPA Reg. No. 476-2213 A-3R

Keep Out of Reach of Children
DANGER

CORROSIVE, CAUSES IRREVERSIBLE EYE DAMAGE HARMFUL IF SWALLOWED

STATEMENT OF PRACTICAL TREATMENT

First Aid: If a known exposure occurs or is suspected, immediately start the procedures given below. If further treatment is required, CONTACT A POISON CENTER, A PHYSICIAN, OR THE NEAREST HOSPITAL, or call collect, day or night: (203) 226-6602 Stauffer Chemical Company, or (800) 424-9300 Pesticide Team Safety Network. Describe the type and extent of exposure, the victim's symptoms, and follow the advice given.

If swallowed: Immediately give several glasses of water and induce vomiting by gagging the victim with a finger placed on the back of the victim's tongue. Give fluids until vomitus is clear. If victim is unconscious or convulsing, do not induce vomiting or give anything by mouth.

Eye contact: Immediately flush the eyes with plenty of running water for at least 15 minutes. Hold the eyelids apart during the flushing to ensure rinsing of the entire surface of the eye and lids with water. Call a physician.

Skin contact: Flush affected areas with plenty of water for several minutes. Remove and clean any contaminated clothing and shoes. Seek medical attention if skin irritation occurs.

Inhalation: If inhaled, remove to fresh air. Seek medical attention if respiratory irritation occurs or breathing becomes difficult.

PRECAUTIONARY STATEMENTS

HAZARDS TO HUMANS AND DOMESTIC ANIMALS DANGER

CORROSIVE, CAUSES IRREVERSIBLE EYE DAMAGE. HARMFUL IF SWALLOWED. Avoid contact with skin, eyes and clothing. Avoid breathing spray mist. Wear goggles, rubber gloves and protective clothing when handling.

Wash skin with soap and water immediately after contact. Flush eyes with water. Do not apply this product in such a manner as to directly or through drift expose workers or other persons. When applied through sprinkler irrigation method of application, the area being treated must be vacated by unprotected persons.

ENVIRONMENTAL HAZARDS

Do not apply directly to water or wetlands. Do not contaminate water by cleaning of equipment or disposal of wastes.

WEEDS CONTROLLED BY SUTAZINE+

Barnyardgrass, Watergrass	*Echinochloa crus-galli*
Broadleaf Signalgrass†*‡	*Brachiaria platyphylla*
Crabgrass	*Digitaria* spp.
Fall Panicum	*Panicum dichotomiflorum*
Field Sandbur*	*Cenchrus pauciflorus*
Giant Foxtail	*Setaria faberi*
Green Foxtail	*Setaria viridis*
Goosegrass	*Eleusine indica*
Seedling Bermudagrass	*Cynodon dactylon*
Seedling Johnsongrass*	*Sorghum halepense*
Texas Panicum*‡, Buffalograss, Coloradograss	*Panicum texanum*
Volunteer Sorghum*	*Sorghum* spp.
Wild Cane*†, Shattercane, Black Amber Cane, Chicken Corn, Scatter Cane	*Sorghum bicolor*
See Wild Cane Rate and Cultural Recommendations	
Wild Oats	*Avena fatua*
Witchgrass	*Panicum capillare*
Yellow Foxtail	*Setaria glauca*

† Partial control (suppression) only.
* Cultivation is suggested in addition to the SUTAZINE+ treatment in fields with moderate to heavy infestations of wild cane, sandbur, johnsongrass, Texas panicum, volunteer sorghum, or broadleaf signalgrass.
‡ Southeast Region only.

Annual Broadleaf Weeds:

Annual Morningglory	*Ipomoea purpurea*
Carpetweed	*Mollugo verticillata*
Common Cocklebur	*Xanthium pensylvanicum*
Common Lambsquarters	*Chenopodium album*
Common Purslane	*Portulaca oleracea*
Florida purslane	*Richardia scabra*
Jimsonweed	*Datura stramonium*
Mustards	*Brassica* spp.
Nightshade	*Solanum* spp.
Prostrate Pigweed	*Amaranthus graecizans*
Ragweed	*Ambrosia* spp.
Redroot Pigweed	*Amaranthus retroflexus*
Sicklepod‡, Coffeeweed	*Cassia obtusifolia*
Smartweed	*Polygonum* spp.
Smooth Pigweed	*Amaranthus hybridus*
Velvetleaf	*Abutilon theophrasti*

‡ Southeast Region Only

Perennial Weeds:

Hemp Dogbane†	*Apocynum cannabinum*
Purple Nutgrass, Nutsedge	*Cyperus rotundus*
Rhizome Bermudagrass‡	*Cynodon dactylon*
Rhizome Johnsongrass†	*Sorghum halepense*
Yellow Nutgrass, Nutsedge	*Cyperus esculentus*

† Partial control (suppression) only.
‡ Southeast Region only.

Existing stands of nutgrass must be turned under and chopped up thoroughly before treatment. Cultivation is suggested in addition to SUTAZINE+ treatment in fields with moderate to heavy nutsedge infestations.

GENERAL USE PRECAUTIONS

Read all label directions before using.

Applied according to directions and under normal growing conditions, SUTAZINE+ will not harm the treated crop. During germination and early stages of growth, extended periods of unusually cold and wet or hot and dry weather, insect or plant disease attack, carryover pesticide residues, the use of certain soil-applied systemic insecticides, improperly placed fertilizers or soil insecticides may create abnormal conditions that weaken crop seedlings. SUTAZINE+ used under these abnormal conditions could result in crop injury.

Make only one application per crop. After treatment with SUTAZINE+, do not plant any crop except corn until the following year or injury may occur. Do not apply after June 10.

Following corn harvest, thoroughly invert soil (deep disk or plow) in the fall or early spring to minimize injury potential to rotational spring-seeded crops. Corn or soybeans may be planted the spring season following treatment. Injury may occur to soybeans planted in North Central and Northwest Iowa, South Central and Southwest Minnesota, Northern Nebraska and Southeast South Dakota on soils having a calcareous surface layer. Small grains may be planted 15 months following treatment. All other atrazine-sensitive crops may be planted 18 months after application.

Do not use on milo or sorghum. Do not use on corn seed stock (Breeders, Foundation, or Increase). SUTAZINE+ may be used on Production seed corn.

SUTAZINE+ is recommended for use on mineral soils only (soils containing less than 10% organic matter).

Do not contaminate irrigation water (used for crops other than corn) or water used for domestic purposes.

Do not contaminate food or feed.

Keep container closed when not in use. Do not store near seeds, fertilizers, or foodstuffs.

Do not use in Arizona or in the ten Southernmost California counties.

In irrigated areas, do not apply SUTAZINE+ before pre-irrigation.

DIRECTIONS FOR USE

It is a violation of Federal law to use this product in a manner inconsistent with its labeling.

SUTAZINE+ is a selective herbicide which is mixed (incorporated) into the soil for control of weeds listed on this label. SUTAZINE+ controls annual grasses as their seeds germinate by interfering with normal germination and seedling development. It does not control established weeds.

SUTAZINE + can be mixed with water or combined with fluid (solution, slurry, or suspension) fertilizers and applied as a spray, or can be impregnated on dry bulk fertilizers.

SUTAZINE+ FLUID FERTILIZER TANK MIX

SUTAZINE+ can be combined with fluid fertilizers for applications made before planting. However, *the physical compatibility with these fluid fertilizers must be determined before combining in the spray tank, and even though found to be compatible, constant agitation is necessary to keep the SUTAZINE + evenly mixed with the fluid fertilizer during application.*

Refer to Appendix II regarding the procedure for testing the compatibility of SUTAZINE+ with fluid fertilizers.

IMPREGNATION ON DRY BULK FERTILIZERS

Dry bulk fertilizers (except single superphosphate, treble superphosphate and nitrate fertilizers) may be impregnated or coated with SUTAZINE+. However, uniform distribution of SUTAZINE+ on fertilizer particles and uniform application are necessary to assure good results. See Appendix III for information and directions for impregnation and use for these combinations.

APPLICATION AND SOIL INCORPORATION DIRECTIONS

SPRAY APPLICATION AND MECHANICAL INCORPORATION

Apply the recommended dosage in 10 to 50 gallons of water per acre (use the higher volume in arid areas) using a properly calibrated, low-pressure type nozzle (flood, fan or swirl chamber) on boom-type sprayer having good, but not excessive agitation. Check frequently during application to be sure nozzles are free from clogging and are delivering a uniform spray pattern.

The herbicides must be incorporated into the soil within 4 hours (within 1 hour in Western Region) to prevent loss of the SUTAZINE+. Whenever possible, application and incorporation should be done in the same operation. Apply the herbicides to soil dry enough to permit good soil mixing.

See Appendix 1 for Soil Mixing (incorporation) before Planting Directions.

Loss of weed control will result from any delay in incorporation if SUTAZINE+ is applied to a moist soil surface.

If there must be a delay (less than four hours, except less than one hour in Western Region) in incorporation of the herbicides when application and incorporation are done in separate operations, SUTAZINE+ must be applied to a dry soil surface (at least ½ inch deep) and free from dew and incidental moisture.

IMPORTANT: Thorough mixing of SUTAZINE+ in the soil is necessary. The soil must be in good tilth. Improper discing or poor tilth (large clods) may result in unsatisfactory control.

INCORPORATION OF IMPREGNATED DRY BULK FERTILIZER

The application and incorporation of SUTAZINE+ must be carried out on the same day and delayed only if soil moisture conditions permit. See Appendix III for directions on the delay in soil incorporation.

CENTER PIVOT SPRINKLER APPLICATION

Apply SUTAZINE+ with irrigation water before planting (see PLANTING), or immediately after planting before either corn or weeds emerge. It is important that application be done immediately after the last tillage operation since poor results will occur if the weeds have germinated before application of the herbicide. Meter the pre-mixed SUTAZINE+ into the irrigation water during the entire period. Apply in ½ to 1 inch of water or sufficient water to penetrate the soil to a depth of 5 to 7 inches. Good agitation should be maintained during the entire application period.

USE PRECAUTIONS

A. Apply the product only through center pivot irrigation systems containing anti-siphon and check valves which will prevent water source contamination and overflow of the slurry tank and containing interlocking controls between the metering device and the water pump to insure simultaneous shut-off.

B. Inject the product with a positive displacement pump into the main line ahead of a right angle turn to insure adequate mixing.

C. Application of more than recommended quantities of irrigation water may result in decreased product performance by removing the chemical from the zone of effectiveness.

D. Do not apply when wind speed favors drift, when system connections or fittings leak, when nozzles do not provide uniform distribution or when lines containing the product must be dismantled and drained.

PLANTING

Do not plant seed deeper than 2 inches. For a SUTAZINE+ application made before planting, seeding can be done immediately, or it may be delayed up to 2 weeks*. To avoid removing SUTAZINE+ from the row (with loss of weed control in the row), do not move or shape soil after treatment, such as by planting in deep furrows. If it is necessary to work the soil following application and incorporation but before planting, do not work the soil deeper than the depth the herbicides were incorporated.

* For Wild Cane—See Special Recommendations.

RECOMMENDATIONS FOR USE ON CORN
(Field, Sweet, Silage and Popcorn)

NOTE: SUTAZINE+ is not recommended on corn grown in Arizona or in the ten Southernmost California counties.

For all uses, cultivation is recommended in addition to the SUTAZINE+ treatment in fields with moderate to heavy infestations of wild cane, nutgrass, sandbur, johnsongrass seedlings, rhizome bermudagrass, Texas panicum or volunteer sorghum.

Broadcast Rates Per Acre Conventional sprays and center pivot sprinklers.

For control of annual grasses and broadleaf weeds: Use 5.25 to 7 pints of SUTAZINE+ per acre. Use the higher rate of SUTAZINE+ on medium (medium fine textured-silt loam, clay loam, etc.) to fine (fine textured-clay) soils and for sandbur and johnsongrass seedling control.

For control of nutsedge and Texas panicum (in addition to the broadleaf weeds and other annual grasses listed): Use 7 to 10.5 pints of SUTAZINE+ per acre. Use the higher rates of SUTAZINE+ on medium (medium fine textured-silt loam, clay loam, etc.) soils, fine (fine textured-clay) soils and for heavy infestations.

For partial control (suppression) of Wild Cane (and for control of the broadleaf weeds and annual grasses listed): When applied and incorporated *immediately* before planting, SUTAZINE+ at the rate of 10.5 pints per acre on a broadcast basis will reduce competition from wild cane. On continuous corn acreage with high cane seed population, rotation to another crop is recommended.

Suppression of wild cane may be improved if all the following practices are followed:
1. Fall or Spring Moldboard Plow.
2. Before application, the soil should be well worked with a disc and dry enough to permit good soil mixing of the herbicide (incorporation). All trash should be worked into the soil.
3. Delay application until soil warms to a constant 55 degrees or greater temperature. Plant immediately.
4. Incorporate thoroughly. Use recommended incorporation equipment and procedures found on this label.
5. Cultivate immediately if wild cane emerges. Cultivation should be shallower than the depth used when incorporating the herbicide.

For control of Johnsongrass and Bermudagrass (in addition to the broadleaf weeds and annual grasses listed): Apply and disc-incorporate SUTAZINE+ just before planting at the rate of 7 to 10.5 pints per acre. Use the higher rate of SUTAZINE+ on medium (medium fine textured silt loam, clay loam, etc.) soils, fine (fine textured-clay) soils, for rhizome johnsongrass suppression and for heavy infestations. Disc a second time at right angles to the first. Cultivate at least once, preferably twice.

Johnsongrass: Using a program of plowing and discing followed by SUTAZINE+ will give good control of seedlings and greatly reduce competition from plants coming from rhizomes.

In some areas discing 2 to 3 times spaced at 4 to 6 week intervals during fall and winter can be done. If this cannot be done, start with fall plowing followed by several discings in the spring before planting. The discing cuts rhizomes into small pieces and brings them to the surface, where both the weather and SUTAZINE+ will kill or weaken them.

Rhizome Bermudagrass: The rhizomes and stolons of bermudagrass must be cut up thoroughly so that four or less nodes remain on a strand. For control of bermudagrass, the disc must be set to cut 6 inches deep.

For Suppression of Broadleaf Signalgrass: For early and mid-season suppression, use 10.5 pints of SUTAZINE+ per acre. Cultivation or application of a post-emergence herbicide recommended for broadleaf signalgrass will aid in extending control to late season.

For Suppression of Hemp Dogbane: For early and mid-season suppression, use 10.5 pints of SUTAZINE+ per acre. Cultivation or application of a post-emergence herbicide recommended for hemp dogbane may aid in extending control beyond mid-season.

Fall Application: For use only in Minnesota, Northern Iowa (North of Interstate 80), Northeast Nebraska (North of Interstate 80 and East of State Highway 14), South Dakota (East of Missouri River), Wisconsin and Northern Illinois (North of Interstate 80). Apply and incorporate 10.5 pints SUTAZINE+ per acre in the late fall before the ground freezes.

NOTE: Do not use Fall application for suppression of wild cane.

CULTURAL PRACTICES FOLLOWING APPLICATION

Should weeds develop, a shallow cultivation or rotary hoeing will generally result in better weed control. When cultivating for any reason, it should be shallow; i.e., no more than ½ the depth the herbicide was incorporated.

APPENDIX I

SOIL MIXING (INCORPORATION) BEFORE PLANTING DIRECTIONS

The following equipment commonly is used for soil mixing (incorporation) before planting.

Power-Driven Cultivation Equipment (Recommended on all soil types previously primary tilled): All PTO driven cultivation tools must be adjusted so as to have a tillage action speed (originating from internal gears and/or tractor gear) synchronized to the tractor's forward speed so that thorough soil mixing is obtained.

a. Horizontal action: set to cut 4 to 6 inches deep.

b. Vertical action: set to cut 3 to 4 inches deep.

Tandem discs (Recommended on all soil types): Set to cut to a depth of 4 to 6 inches, operated at 4 to 6 mph followed by a spike-tooth harrow or some other leveling device which extends beyond the ends of the discs. For more thorough mixing (for control of deep-germinating grasses—wild cane, seedling johnsongrass and nutsedge), disc in two different directions. When discing the second time, the disc should be reset to cut no deeper than 4 to 6 inches.

Field cultivator (Recommended on lighter soils in good tilth): Use 3 to 4 rows of sweeps spaced at 7 inch or less intervals and staggered so that no soil is left unturned, followed by a spike-tooth harrow pulled behind the cultivator. Do not use chisel plows to incorporate. Set the cultivator to cut 4 to 6 inches deep, operated at 5 mph or more. The field cultivator should not be used alone as an incorporation tool where control of deep-germinating weeds such as nutgrass, johnsongrass seedlings or wild cane is desired. For control of these weeds, use a disc alone for at least one of the passes over the field.

APPENDIX II

PROCEDURE FOR TESTING THE COMPATIBILITY OF SUTAZINE+ WITH FLUID FERTILIZERS.

The following procedure is suggested for determining whether SUTAZINE+ may be combined with a specific fluid fertilizer for spray tank application.

MATERIALS REQUIRED

1. SUTAZINE+
2. Fluid fertilizer to be used.
3. Tank mix adjuvants for fluid fertilizer combinations: Compex*, Sponto 168-D*, Unite* or equivalent. The adjuvant which provides the best compatibility depends on the specific fertilizer involved.
4. Two one-quart, wide-mouth glass jars with lids.
5. Measuring spoons; a graduated pipette or cylinder will provide more accurate measurement; 1 tablespoon equals 3 teaspoons, 1 teaspoon equals approximately 5 milliliters.
6. Measuring cup, 8 oz. (237 mL).

* Compex, Kalo Laboratories Inc., Kansas City, Missouri; Sponto 168-D, Witco Chemical Company, Houston, Texas; Unite, Hopkins Agricultural Chemical Company, Madison, Wisconsin.

PROCEDURE

1. Pour a pint (473 mL) of the fluid fertilizer into each of the quart jars.
2. Add adjuvant to one of the jars and mix (see next rate table).
3. Add the SUTAZINE+ to both jars.
4. Close both jars with lids and mix the contents by turning the jars upside down ten times.
5. Inspect the surface and body of the mixtures:
   (A) Immediately after completing the jar inversions.
   (B) After allowing the jars to stand quietly for 30 minutes.
   (C) And then again after turning the jars upside down 10 times.

If a uniform mix cannot be made, the mixture should not be used. If either mixture remains uniform for 30 minutes, the combination may be used. Should either mixture separate after 30 minutes, but readily remix uniformly with ten jar inversions, the mixture can be used if adequate agitation is maintained in the tank. If the mixture with adjuvant is satisfactory, but the one without adjuvant is not, be sure to use the adjuvant in the spray tank. Add the adjuvant at a rate of 3 pints per 100 gallons of fluid fertilizer; foaming can be minimized by using moderate agitation.

If nondispersible oil, sludge or clumps of solids form in the mixtures, the combination should not be used.

Rate Table* for SUTAZINE + and Adjuvant with the Fluid Fertilizer**

Gallons of Fluid Fertilizer to be applied per acre	mL of SUTAZINE+ to be added to 1 pint of Fertilizer	
	mL	tsp.
10	10½	2
15	7	1⅓
20	5¼	1
25	4¼	¾
30	3½	⅔
40	2½	½

* Based on field rate of 1 pound SUTAN® and ¼ pound atrazine per acre in the fertilizer volumes indicated. Increase volume proportionately to correspond with intended field rate in terms of pounds active ingredients per acre (e.g. for field rate of 4 pounds actual SUTAN® and 1 pound atrazine in 40 gallons fertilizer per acre, add 10 mL or 2 tsp. SUTAZINE+ to each jar for compatibility testing purposes).

** Two (2) milliliters or one-half (½) teaspoon of adjuvant should be added to 1 pint of fluid fertilizer in order to equal the rate of 3 pints of adjuvant per 100 gallons of fluid fertilizer.

APPENDIX III

SUTAZINE+ IMPREGNATION ON DRY BULK FERTILIZERS

CAUTION: SUTAZINE+ alone or in combination with other herbicides must not be impregnated on ammonium nitrate, potassium nitrate or sodium nitrate fertilizers. Such mixtures may cause explosion and fire.

All individual state regulations relating to dry bulk fertilizer blending, registration, labeling and application are the responsibility of the individual and/or company selling the SUTAZINE+ /fertilizer mixtures.

SUTAZINE+ may be impregnated on many dry bulk fertilizers (1) and applied and incorporated into the soil either in the Fall before the ground freezes or before planting for the control of many grassy and broadleaf weeds.

All SUTAZINE+ label and supplementary literature instructions and precautions regarding rates per acre, soil type and soil incorporation, application and other directions must be followed.

Test results have shown that SUTAZINE+ on many dry bulk fertilizers gives weed control equal to SUTAN + 6.7-E or tank mixtures of SUTAN + 6.7-E/ATRAZINE (80W or 4L) applied as a spray in water or liquid fertilizer, or impregnated on dry bulk fertilizer. However, uniform impregnation of the SUTAZINE+ on the dry fertilizer particles and uniform application in the field are necessary to assure good results.

A minimum of 200 pounds and a maximum of 700 pounds of approved fertilizer ingredients (1) impregnated with the appropriate amount of SUTAZINE+ must be applied per acre.

For impregnating the SUTAZINE+ on dry fertilizers, use a closed rotary-drum type mixer equipped with suitable spraying equipment. The spray nozzles should be positioned inside of the mixer to provide uniform spray coverage of the tumbling fertilizer. The SUTAZINE+ should be sprayed uniformly onto the fertilizer using a fine spray pattern. SUTAZINE+ is a relatively viscous material, particularly at low temperatures (20°F). At warmer temperatures (45–80°F) the viscosity is significantly lower (see viscosity data later in this appendix).

The physical properties of fertilizers vary, particularly in liquid absorptive capacity. When absorptivity is sufficient, simple spray impregnation of the fertilizer with SUTAZINE+ provides a satisfactory dry mixture.

If the absorptivity is not adequate, use of a highly absorptive powder is required to provide a dry, free-flowing mixture.

For mixtures to be used in spinning-disc applicators, Microcel E calcium silicate powder (Manville, Filtration & Minerals) is recommended for use as a drying agent.

For mixtures to be used in pneumatic applicators, Microcel E or Agsorb 16/30 RVM-MS granular-clay (Oil-Dri Corporation) is recommended for use as the drying agent.

The drying agent should be added separately and uniformly to the prepared pesticide-fertilizer mixture, in a quantity that is sufficient to provide a suitably free-flowing mixture. Generally, less than 2% Microcel E or 5% Agsorb 16/30 RVM-MS by weight is required.

The amount of SUTAZINE+ actually required in the manufacture of individual fertilizer mixtures should be determined carefully for each production operation (2). This is necessary to ensure that the amount of SUTAZINE+ actually contained in the mixture applied to the soil represents the correct rates of use.

Bulk fertilizers impregnated with SUTAZINE+ should be *applied immediately,* NOT STORED. All bulk containers should be tightly covered while the products are being transported and applied to reduce chances of SUTAN loss via volatilization.

Do not combine SUTAZINE+ with single superphosphate (0–20–0) or treble superphosphate (0–46–0).

INCORPORATION OF IMPREGNATED FERTILIZERS

When application and incorporation are done in separate operations, SUTAZINE+ impregnated on dry fertilizer must be incorporated on the same day as applied. Application must be made on a dry soil surface (at least ½ inch deep) free from dew and incidental moisture.

SUTAZINE+ Physical Data

Specific Gravity 20/20°C:	1.019 (typical)
Pounds/Gallon (20°C):	8.480 (typical)
Flashpoint:	>216°F (Setaflash closed cup)
Viscosity: (typical)	1120 centipoise at 20°F 660 centipoise at 32°F 430 centipoise at 45°F 275 centipoise at 60°F 175 centipoise at 80°F 145 centipoise at 95°F

(1) APPROVED DRY FERTILIZER INGREDIENTS FOR USE WITH SUTAZINE+

	N	P	K
Ammonium Sulfate	21	0	0
Ammonium Phosphate-Sulfate	16	20	0
Diammonium Phosphate	18	46	0
Monoammonium Phosphate	11	56	0
Potassium Chloride	0	0	60
Potassium Sulfate	0	0	52
Urea*	45	0	0

* Some ureas may be phytotoxic when high rates are applied to corn. Use only urea rates known to be safe for corn application.

NOTE: Sul-Po-Mag, K-Mag and 11-48-0 have been shown to be compatible with SUTAZINE+ and are approved for use.

(2) RATE CHART FOR THE IMPREGNATION OF DRY BULK FERTILIZERS WITH SUTAZINE+

FERTILIZER Rate/Acre	SUTAZINE+ Rate/Acre		
	5¼ pts. per acre	7 pts. per acre	10½ pts. per acre
200 lbs.	26¼ qts./ton	35 qts./ton	52½ qts./ton
250 lbs.	21 qts./ton	28 qts./ton	42 qts./ton
300 lbs.	17½ qts./ton	23⅓ qts./ton	35 qts./ton
350 lbs.	14½ qts./ton	20 qts./ton	30 qts./ton
400 lbs.	13¼ qts./ton	17½ qts./ton	26¼ qts./ton
450 lbs.	11⅔ qts./ton	15½ qts./ton	23⅓ qts./ton
500 lbs.	10½ qts./ton	14 qts./ton	21 qts./ton
550 lbs.	9½ qts./ton	12¾ qts./ton	19 qts./ton
600 lbs.	8¾ qts./ton	11⅔ qts./ton	17½ qts./ton
650 lbs.	8 qts./ton	10¾ qts./ton	16¼ qts./ton
700 lbs.	7½ qts./ton	10 qts./ton	15 qts./ton

STORAGE AND DISPOSAL

PROHIBITIONS
Do not contaminate water, food or feed by storage or disposal. Open dumping is prohibited. Do not reuse empty container.

STORAGE
Keep container tightly closed when not in use. Do not store near seeds, fertilizers or foodstuffs. Can be stored at temperatures as low as minus 30°F.

PESTICIDE DISPOSAL
Rinse spray equipment. Pesticide wastes are acutely hazardous. Improper disposal of excess pesticide, spray mixture, or rinsate is a violation of Federal Law. If these wastes cannot be disposed of by use according to label instructions, contact your State Pesticide or Environmental Control Agency, or the Hazardous Waste representative at the nearest EPA Regional Office for guidance.

CONTAINER DISPOSAL
METAL—Triple rinse (or equivalent) then offer for recycling or reconditioning, or puncture and dispose of in a sanitary landfill or by other procedures approved State or local procedures.
PLASTIC—Triple rinse (or equivalent). Then offer for recycling or reconditioning, or puncture and dispose of in a sanitary landfill, or by incineration, or, if allowed by State and local authorities, by burning. If burned, stay out of smoke.

NOTICE—READ CAREFULLY

CONDITIONS OF SALE
Stauffer (and seller) offer(s) this product for sale subject to, and buyer and all users are deemed to have accepted, the following conditions of sale and warranty which may only be varied by written agreement of a duly authorized representative of Stauffer.

WARRANTY LIMITATION
Stauffer warrants that this product conforms to the chemical description on the label and is reasonably fit for the purposes referred to in the directions for use on the label, subject to the inherent risks referred to below. Stauffer makes no other express warranties; THERE IS NO IMPLIED WARRANTY OF MERCHANTABILITY and there are no warranties which extend beyond the description on the label hereof.

INHERENT RISKS
The directions for use of this product are believed to be reliable and should be followed carefully. However, it is impossile to eliminate all risks associated with use. Buyer assumes all risks associated with use or application of this product contrary to label instructions or resulting from extraordinary weather conditions.

LIMITATION OF LIABILITY
In no case shall Stauffer be liable for special, indirect, or consequential damages resulting from the use or handling of this product and no claim of any kind shall be greater in amount than the purchase price of the product in respect of which such damages are claimed.

This product is sold only for uses stated on its label. No express or implied license is granted to use or sell this product under any patent in any country except as specified: Country: United States of America. Patent No. 4,021,224; 4,137,070.

This is a specimen label. Stauffer is not responsible for the accuracy of the information contained herein. As labels are subject to revision, always carefully read and follow the label on the product container.

TILLAM® 6-E

SELECTIVE HERBICIDE

Liquid Herbicide for Weed Control in Tobacco, Tomatoes and Sugar Beets

ACTIVE INGREDIENT:

S-propyl butylethylthiocarbamate	77.32%
INERT INGREDIENTS:	22.68%
TOTAL	100.00%

EPA Reg. No. 476-1615 A-7MR
Contains 6 Lbs. Active Ingredient Per Gallon

CAUTION

Keep Out of Reach of Children

HARMFUL IF SWALLOWED

STATEMENT OF PRACTICAL TREATMENT

If swallowed: **DO NOT** induce vomiting. Immediately give large quantities of water and get medical attention. If vomiting occurs, give fluids again. Never give anything by mouth to an unconscious person.

If inhaled: Remove from contaminated atmosphere. Get medical attention if respiratory irritation occurs.

For eye contact: Hold eyelids apart and flush eyes with large amounts of running water for at least 15 minutes. Call a physician.

For skin contact: Immediately flush with large amounts of running water for 15 minutes. Remove any contaminated clothing and shoes. Call a physician. Wash clothing before re-use.

PRECAUTIONARY STATEMENTS

HAZARDS TO HUMANS AND DOMESTIC ANIMALS
CAUTION
HARMFUL IF SWALLOWED
Avoid contact with skin, eyes and clothing. Avoid inhalation of spray mist. Wear rubber gloves and fresh clothing. Wear goggles to avoid contact with eyes. Wash with soap and water immediately after use or contact. Keep away from heat and flame. Do not contaminate food and feed.

Do not apply this product in such a manner as to directly or through drift, expose workers or other persons. When using the solid set sprinkler method of application, the area being treated must be vacated by unprotected persons.

ENVIRONMENTAL HAZARDS
Keep out of lakes, ponds or streams. Do not contaminate water by cleaning of equipment or disposal of wastes.

WEEDS CONTROLLED

Annual Grasses:

Barnyardgrass (Watergrass)	*Echinochloa* Spp.
Bermudagrass, seedling	*Cynodon dactylon*
Crabgrass	*Digitaria* Spp.
Foxtails—Yellow, Giant, Green	*Setaria* Spp.
Goosegrass	*Eleusine indica*
Millet	*Setaria* Spp.
Narrow Leaf Signalgrass	*Brachiaria piligera*
Wild Oats	*Avena fatua*

Annual Broadleaf Weeds:

Black-eyed-susan; (Sweet Suzy)	*Rudbeckia serotina*
Deadnettle (Henbit)	*Lamium amplexicaule*
Florida pusley	*Richardia scabra*
Hairy nightshade*	*Solanum villosum*
Lambsquarters	*Chenopodium album*
Nettleleaf goosefoot	*Chenopodium murale*
Prostrate pigweed	*Amaranthus graecizans*
Purslane; Common	*Portulaca oleracea*
Redroot pigweed	*Amaranthus retroflexus*
Shepherdspurse	*Capsella bursa-pastoris*

The above broadleaf weeds will be controlled only if TILLAM 6-E is applied when conditions are favorable for germination.

* Suppression (partial control)

Perennial Weeds:

Yellow nutsedge (Yellow nutgrass) *Cyperus esculentus*
Purple nutsedge (Purple nutgrass) *Cyperus rotundus*

Existing stands of Nutgrass must be turned under and chopped up thoroughly before treatment.

GENERAL USE PRECAUTIONS

Read all label directions before using.
TILLAM 6-E should be used only for recommended purposes and at recommended rates (do not overdose).

Applied according to directions and under normal growing conditions, TILLAM 6-E will not harm the treated crop. During germination and early growth, extended periods of unusually cold and wet or hot and dry weather, insect, nematode, or plant disease attack, the use of certain soil applied systemic insecticides, highly saline or alkaline soil conditions, improperly placed fertilizers or soil insecticides may create abnormal conditions that weaken crop seedlings. TILLAM 6-E used under these abnormal conditions could result in crop injury.

TILLAM 6-E WITH FLUID FERTILIZER

TILLAM 6-E may be combined with fluid (solution, slurry or suspension) fertilizers. However, physical compatibility with these fertilizers must be determined before combining in the spray tank. See Appendix I giving directions for these combinations. Even though found to be compatible constant agitation is necessary to keep the TILLAM 6-E evenly mixed with the liquid fertilizer.

IMPREGNATION ON DRY FERTILIZER

TILLAM 6-E may be impregnated on many dry bulk fertilizers for incorporation prior to transplanting of tobacco to control weeds listed on this label. To insure optimum results, TILLAM 6-E must be uniformly distributed on the fertilizer particles. See Appendix II for special instructions regarding directions for impregnation and use.

DIRECTIONS FOR USE

It is a violation of Federal law to use this product in a manner inconsistent with its labeling.

TILLAM 6-E is a selective herbicide which is mixed (incorporated) into the soil or applied below the surface of the soil, or applied through solid set sprinklers for control of weeds listed on this label. TILLAM 6-E controls weeds by interfering with normal seed germination and seedling development. It does not control established weeds. All weed growth and crop stubble should be thoroughly worked into the soil before treatment.

APPLICATION DIRECTIONS

The soil should be well-worked and dry enough to permit good soil mixing (incorporation) or sub-surface application. Pour the recommended amount of TILLAM 6-E into the spray tank during the filling operations. Apply in 10 to 50 gallons of water per acre.

For soil incorporated applications, use a properly calibrated, low-pressure, boom-type sprayer having good agitation. For sub-surface applications, soil should be free of trash. Special equipment designed for sub-surface application must be used. Sweeps must be mounted immediately ahead of the planter unit.

In irrigated areas, do not apply TILLAM 6-E before pre-irrigation.

For all band applications, use proportionately less material per acre depending upon row spacing and band width to be treated. Do not use band application on rocky ground because thorough incorporation is not possible.

INCORPORATION DIRECTIONS

TILLAM 6-E must be mixed (incorporated) into the soil IMMEDIATELY after application to prevent loss by evaporation of the herbicide. Whenever possible, application and incorporation should be done in the same operation.

INCORPORATION BEFORE PLANTING: Use power-driven cultivation equipment set to cut to a depth of two to three inches in light soils and three to four inches in heavy soils OR a tandem disc set to cut to a depth of six inches, followed by a spike-tooth harrow or some other leveling device which extends beyond ends of the disc. For thorough mixing, disc in two different directions (cross disc).

INCORPORATION AT OR AFTER PLANTING: Use power-driven cultivation equipment set to cut to a depth of two to three inches, OR on light soils only, ground-driven tillers (rolling cultivators, rotary hoe, row wheels, et cetera) set to cut to a depth of 1½ inches and operated at 6 to 8 mph. When using ground-driven tillers pre-cultivate if soil is crusted. When incorporating after planting, care must be taken not to disturb the crop seedling.

SUB–SURFACE DIRECTIONS

(Tobacco only); Special equipment designed for sub-surface sweep application MUST be used. Sweeps must be mounted immediately ahead of the planter unit.

Covered Sweeps: Set the sweeps to run below the soil surface deep enough to cover the TILLAM 6-E with 5 to 6 inches of soil. Calibrate by measuring the spray band width at the back of the sweep, not the sweep width. NOTE: When applying with sweeps, TILLAM 6-E must be applied deep enough to allow 3–4 inches of soil to remain over the treatment after the transplanting operation.

BED–OVER METHOD OF APPLICATION

Apply TILLAM 6-E in a band immediately in front of bedding discs. Set discs so that TILLAM 6-E is covered with 6 inches of soil. Set bedding discs so that treated soil is not folded in. If bed levelers are used at transplanting, beds should be made higher to compensate for leveling, i.e., so that 3 to 4 inches of soil are left above the TILLAM 6-E treatment.

SOLID SET SPRINKLER APPLICATION (Tomatoes only, Western Region): Meter the TILLAM 6-E into the irrigation water during the entire period. Apply in ¼ to ½ acre inch of water or sufficient water to penetrate the soil to a depth of 2 to 4 inches.

Time the TILLAM application to insure that proper penetration of the herbicide corresponds with the end of the irrigation period. Flush the lines then turn the water off promptly. Tailwater (run-off water) from irrigation should be recirculated.

Application of more than recommended quantities of irrigation water may result in decreased product performance by removing the chemical from the zone of effectiveness.

PLANTING

For maximum weed control benefits, crops should be planted or seeded immediately after application. Do not use a drag behind the planter as it may concentrate TILLAM 6-E over the seed row and cause crop injury.

CULTURAL PRACTICES FOLLOWING APPLICATION

TILLAM 6-E is not persistent in the soil and susceptible weeds germinating later in the season may not be controlled. Shallow cultivation or approved post-emergence herbicides may be necessary to control susceptible weeds that escape control by TILLAM 6-E as well as those weeds not susceptible to TILLAM 6-E. Do not cultivate deeper than the depth of herbicide incorporation.

When cultivating fields where TILLAM 6-E has been banded in the row, use shields to prevent the movement of untreated soil into the treated row.

RECOMMENDATIONS

SUGAR BEETS, PRE-PLANT: Apply just prior to planting; see DIRECTIONS FOR USE for methods of application and incorporation. Recommended dosages are as follows:

For light textured soils and soils with less than 10% organic matter, apply 2⅔ qts. TILLAM 6-E/acre.

For heavy textured soils and soils with over 10% organic matter, apply 4 qts. TILLAM 6-E/acre.

TOBACCO, FLUE-CURED

Before bedding and transplanting: Apply TILLAM 6-E broadcast at 2⅔ qts. per acre. Incorporate immediately with a tandem disc and bed up immediately prior to transplanting.

OR

After bedding but before transplanting: Apply TILLAM 6-E at a broadcast rate of 2⅔ qts. per acre prior to transplanting, reducing amount per acre proportionately depending upon band width and row spacing. Incorporate immediately using a power-driven rotary hoe with variable length blades (i.e., one with short blades on the inside and longer blades on the outside). Bed levelers, if used should be kept shallow as excessive removal of treated soil will destroy weed control.

OR

Sub-surface sweeps after bedding but before transplanting: Run at a depth so that when bed is in final shape, 3 to 4 inches of soil will cover the TILLAM 6-E. Apply TILLAM 6-E at 2 qts. per acre, just prior to transplanting. For band treatment, reduce amount per acre proportionately depending upon band width and row spacing.

TOBACCO, LIGHT AIR CURED

Includes Type 31 (Burley) & Type 32 Tobaccos, Type 41 Cigar Filler. Types 54 and 55 Cigar Binder:

Before transplanting: Apply TILLAM 6-E broadcast at 2⅔ qts. per acre. Incorporate immediately, only with a tandem disc on flat ground followed by transplanting. For band treatment reduce amount per acre proportionately depending upon band width and row spacing. Incorporate immediately, using a power-driven rotary hoe.

TOMATOES

SEE DIRECTIONS FOR USE FOR METHODS OF APPLICATION AND INCORPORATION.

DIRECT SEEDED: (Western Region Only)

Apply TILLAM 6-E just prior to planting.

For light textured soils and soils with less than 10% organic matter, apply 2⅔ qts. TILLAM 6-E/acre.

For heavy textured soils and soils with more than 10% organic matter, apply 4 qts. TILLAM 6-E/acre.

OR

DIRECT SEEDED, LAYBY: (Not pre-plant)

Apply TILLAM 6-E as a directed spray to clean cultivated soil. Do not apply within eight (8) days of harvest.

For light textured soils and soils with less than 10% organic matter, apply 2⅔ qts. TILLAM 6-E/acre.

For heavy textured soils and soils with more than 10% organic matter, apply 4 qts. TILLAM 6-E/acre.

OR

DIRECT SEEDED, SPRINKLER APPLICATION (Western Region only): Apply TILLAM 6-E through solid set sprinklers prior to or immediately after planting, before weeds emerge.

For light textured soils and soils with less than 10% organic matter, apply 2⅔ qts. TILLAM 6-E/acre.

For heavy textured soils and soils with more than 10% organic matter, apply 4 qts. TILLAM 6-E/acre.

OR

PLUG PLANTED, PRE-PLANT (Western Region only): Prior to plug planting tomatoes, apply and soil incorporate up to 6⅔ quarts of TILLAM 6-E per acre. Use higher rate for hairy nightshade.

TOMATO TRANSPLANTS (DO NOT USE HOTCAPS)

Before transplanting—Apply TILLAM 6-E prior to transplanting.

OR

After transplanting—Apply TILLAM 6-E as a directed spray to clean, cultivated soil. Do not apply within eight (8) days of harvest.

For light textured soils and soils with less than 10% organic matter, apply 2⅔ qts. TILLAM 6-E/acre.

For heavy textured soils and soils with more than 10% organic matter, apply 4 qts. TILLAM 6-E/acre.

TOMATO TRANSPLANTS (South Carolina only)

Before bedding and transplanting: Apply TILLAM 6-E at a broadcast rate of 2 qts. per acre just prior to bedding and transplanting, reducing the amount per acre proportionately depending upon band width and row spacing. Use bed-over method of application.

OR

After bedding but before transplanting: Apply TILLAM 6-E at a broadcast rate of 2 qts. per acre just prior to transplanting, reducing amount per acre proportionately depending upon band width and row spacing. Apply to the top of the bed and incorporate four to six inches into the bed top.

OR

Post-emergence, or after transplanting: Apply TILLAM 6-E at a broadcast rate of 2 qts. per acre reducing amount per acre proportionately depending upon band width and row spacing. Apply in a directed spray to soil around base of tomato plants using bands on either side of the row. Immediately after treatment, sweep out row centers to effectively cover TILLAM 6-E treatment with a layer of soil two inches deep. Do not apply within eight (8) days of harvest.

TILLAM 6-E/DASANIT* SPRAY CONCENTRATE TANK MIX COMBINATION

* DASANIT is a Registered Trademark of the Parent Company of Farbenfabriken Bayer, GMBH, Leverkusen.

TILLAM 6-E IN COMBINATION WITH DASANIT SPRAY CONCENTRATE controls nematodes, wireworms and weeds controlled by TILLAM 6-E in tobacco (Flue Cured and Light Air Cured).

NOTE: DASANIT is a Restricted Use Pesticide. Refer to the label for DASANIT SPRAY CONCENTRATE for pertinent directions and precautions for use, and restrictions.

APPLICATION DIRECTIONS AND RECOMMENDATIONS

The soil should be well worked and dry enough to permit good soil incorporation (mixing into the soil). Apply 2⅔ qts. TILLAM 6-E and 4 qts. DASANIT SPRAY CONCENTRATE in 20 to 50 gallons water per acre. Use a properly calibrated low-pressure boom-type sprayer having good agitation.

SOIL INCORPORATION

The tank mix must be incorporated (mixed) into the soil immediately after application to prevent loss of the herbicide by evaporation. Whenever possible, application and incorporation should be done in the same operation.

Use power-driven cultivation equipment set to cut to a depth of 3″ in light soil and 3 to 4″ in heavy soil or a tandem disc set to cut to a depth of 4 to 6″.

Cross disc at right angles for thorough mixing. The bedding operation should follow the incorporation as soon as is practical. Planting may be done immediately after bedding or at a later date; however, to insure long term weed control, the crop should be planted as soon as possible after application and bedding.

TILLAM 6-E/MOCAP* EC TANK MIX COMBINATIONS

* MOCAP is a Registered Trademark of Mobil Chemical Company.

A tank mix of TILLAM 6-E and MOCAP EC can be applied and mixed into the soil before transplanting for control of weeds, wireworms and nematodes in tobacco.

NOTE: MOCAP is a Restricted Use Pesticide. Refer to the label of MOCAP EC for pertinent directions and precautions for use, and restrictions.

TANK MIXING DIRECTIONS

The TILLAM 6-E/MOCAP EC tank mix combination can be mixed with water and applied as a spray. Pour the recommended amount of TILLAM 6-E directly into the nearly filled spray tank while agitating. Next add the recommended amount of MOCAP EC. Fill the remaining portion of the tank with water; mix thoroughly. Maintain a moderate agitation to provide a uniform mix during application.

APPLICATION DIRECTIONS AND RECOMMENDATIONS

The soil should be well-worked and dry enough to permit good soil mixing (incorporation). Apply 2⅔ qts. TILLAM 6-E plus 1 gallon MOCAP EC in 10 to 50 gallons water per acre. use a properly calibrated low-pressure, boom-type sprayer having good agitation.

INCORPORATION DIRECTIONS

The TILLAM 6-E/MOCAP EC tank mix combination must be mixed (incorporated) into the soil immediately after application to prevent loss by evaporation. Whenever possible, application and incorporation should be done in the same operation. Use a tandem disc set to cut to a depth of six inches, followed by a spike-tooth harrow or some other leveling device which extends beyond the ends of the disc. For thorough mixing, disc in two different directions (cross disc).

TILLAM 6-E/PAARLAN* EC TANK MIX COMBINATIONS

* PAARLAN is a Registered Trademark of Elanco Products Co., Div. of Eli Lilly and Co.

A tank mix of TILLAM 6-E and PAARLAN EC can be applied and mixed into the soil before transplanting for additional weed control and longer season grass control in tobacco.

NOTE: Refer to the label of PAARLAN EC for pertinent directions and precautions for use, and restrictions.

TANK MIXING DIRECTIONS

The TILLAM 6-E/PAARLAN EC tank mix combination can be mixed with water and applied as a spray. Pour the recommended amount of TILLAM 6-E directly into the nearly filled spray tank while agitating. Next add the recommended amount of PAARLAN EC. Fill the remaining portion of the tank with water; mix thoroughly. Maintain a moderate agitation to provide a uniform mix during application.

APPLICATION DIRECTIONS AND RECOMMENDATIONS

The soil should be well-worked and dry enough to permit good soil mixing (incorporation). Apply 2⅔ qts. TILLAM 6-E plus 2 pints PAARLAN EC in 10 to 50 gallons of water per acre. Use a properly calibrated low-pressure, boom-type sprayer having good agitation.

INCORPORATION DIRECTIONS

The TILLAM 6-E/PAARLAN EC tank mix combination must be mixed (incorporated) into the soil immediately after application to prevent loss by evaporation. Whenever possible, application and incorporation should be done in the same operation. Use a tandem disc set to cut to a depth of six inches, followed by a spike-tooth harrow or some other leveling device which extends beyond the ends of the disc. For thorough mixing, disc in two different directions (cross disc).

TILLAM 6-E/DEVRINOL ® 50-WP TANK MIX COMBINATION APPLIED PRE-PLANT INCORPORATED

A tank mix combination of TILLAM 6-E and Devrinol 50-WP can be applied and mixed (incorporated) into the soil before planting for weed control in tomatoes and tobacco.

WEEDS CONTROLLED

Annual Grasses:	
Barnyardgrass (Watergrass)	*Echinochloa crus-galli*
Bermudagrass, seedling	*Cynodon dactylon*
Crabgrass	*Digitaria* Spp.
Foxtails (Giant, Green, Yellow)	*Setaria* Spp.
Goosegrass	*Eleusine indica*
Millet	*Setaria* Spp.
Narrow Leaf Signalgrass	*Brachiaria piligera*
Wild Oats	*Avena fatua*
Annual Broadleaf Weeds:	
Black-eyed-Susan (Sweet Suzy)	*Rudbeckia serotina*
Deadnettle (Henbit)	*Lamium amplexicaule*
Florida pusley	*Richardia scabra*
Hairy Nightshade*	*Solanum villosum*
Lambsquarters	*Chenopodium album*
Nettleleaf Goosefoot	*Chenopodium murale*
Prostrate pigweed	*Amaranthus graecizans*
Purslane; Common	*Portulaca oleracea*
Redroot Pigweed	*Amaranthus retroflexus*
Shepherdspurse	*Capsella bursa-pastoris*

* Suppression (partial control)

The above broadleaf weeds will be the controlled only when the TILLAM/DEVINROL tank mix is applied when conditions are favorable for germination.

Perennial Weeds:	
Purple Nutsedge (Purple Nutgrass)	*Cyperus rotundus*
Yellow Nutsedge (Yellow Nutgrass)	*Cyperus esculentus*

Existing stand of nutgrass must be turned under and chopped up thoroughly before treatment.

Note: Observe all cautions and limitations on labeling of all products used in combination.

Make only one application per season.

In each case where a range of rates is given, the lower rates should be used on light (coarse) textured soils and soils low in organic matter. The higher rates should be used on heavy (fine) textured soils.

After harvest or prior to planting of succeeding crops, a deep moldboard or disc plowing operation must be carried out.

TANK MIXING DIRECTIONS

The TILLAM 6-E/Devrinol 50-WP tank mix combination can be mixed with water and applied as a spray. Pour the recommended amounts of TILLAM 6-E and Devrinol 50-WP into a spray tank during the filling operations. When using Devrinol 50-WP, add first to the spray tank, then add the Tillam 6-E. Maintain adequate agitation of the tank mix. Apply in 25 to 50 gallons of water per acre.

APPLICATION DIRECTIONS AND RECOMMENDATIONS

The soil should be well-worked and dry enough to permit good soil mixing (incorporation).

For soil incorporated applications, use a properly calibrated, low-pressure, boom-type sprayer having good agitation.

In irrigated areas do not apply before pre-irrigation. For all band applications, reduce dosage proportionately depending upon row spacing and band to be treated. Do not use band application on rocky ground because thorough incorporation is not possible.

INCORPORATION DIRECTIONS

The herbicides must be mixed (incorporated) into the soil immediately after application to prevent loss by evaporation. Wherever possible, application and incorporation should be done in the same operation.

INCORPORATION BEFORE PLANTING

Use power-driven cultivation equipment set to cut to a depth of two to three inches in coarse soils and three to four inches, in heavy soils OR a tandem disc set to cut to a depth of six inches, followed by a spike-tooth harrow or some other leveling device which extends beyond ends of the disc. For thorough mixing, disc in two different directions (cross disc).

INCORPORATION AT PLANTING

Use power-driven cultivation equipment set to cut to a depth of two to three inches OR, on coarse soils only, ground-driven tillers (rolling cultivators, rotary hoe, row wheels, et cetera) set to cut to a depth of 1½ inches and operated at 6 to 8 mph. When using ground-driven tillers, precultivate if soil is crusted.

PLANTING

Crops should be planted or seeded immediately after application and incorporation. Do not use a drag behind the planter as it may concentrate the herbicides over the seed row and cause crop injury.

CULTURAL PRACTICES FOLLOWING APPLICATION

Shallow cultivation may be necessary to control those susceptible weeds that escape control as well as those weeds not susceptible to either herbicide. Do not cultivate deeper than the depth of herbicide incorporation.

When cultivating fields where the tank mix has been banded in the row, use shields to prevent the movement of untreated soil into the treated row.

RECOMMENDATIONS

TOMATOES (Direct Seeded [California Only] and Transplanted): Apply 2 to 4 pounds DEVRINOL 50-WP plus 2⅔ to 4 qts. TILLAM 6-E per broadcast acre. Use the lower rates on light soils (coarse textured-sandy) and the higher rate on heavy soils (fine textured-clay). Do not use hotcaps on transplanted tomatoes.

TOBACCO (Transplanted): Before transplanting, apply 2 pounds DEVRINOL 50-WP plus 2⅔ qts. TILLAM 6-E per broadcast acre.

To avoid injury to crops not specified on the label, do not plant until 12 months after the last DEVRINOL application.

TILLAM 6-E/DYFONATE ® (4-E, 4-EC OR 4-ED)/ DEVRINOL 50-WP TANK MIX COMBINATION

A tank mix combination of TILLAM 6-E, DYFONATE 4-E, 4-EC or 4-ED and DEVRINOL 50-WP can be applied and mixed into the soil before transplanting tomatoes and tobacco, for control of weeds and soil insects. Refer to Tillam/Devrinol Tank Mix Combination for list of weeds controlled.

NOTE: Dyfonate 4-E, 4-EC and 4-ED are Restricted Use Pesticides. Observe all cautions and limitations on labeling of all products used in combinations. Make only one application per season. Do not contaminate water by cleaning of equipment or disposal of wastes.

DIRECTIONS FOR USE

The recommended tank mix controls weeds by interfering with normal seed germination and seedling development. It does not control established weeds. All weed growth and crop stubble should be thoroughly worked into the soil before treatment.

For GENERAL USE PRECAUTIONS and DIRECTIONS FOR USE REFER TO TILLAM/DEVRINOL Tank Mix Recommendations for methods of tank mixing, application and incorporation.

RECOMMENDATIONS

TOMATOES (Direct Seeded [California Only] and Transplanted): Apply 2⅔ to 4 qts. TILLAM 6-E plus 2 to 4 pounds DEVRINOL 50-WP plus 2 qts. DYFONATE 4-E, 4-EC, or 4-ED for control of symphylans or 4 qts. for control of symphylans and wireworms in 25 to 50 gallons of water per broadcast acre. Use the lower DEVRINOL and TILLAM rates on light soils (coarse textured-sandy) and the higher rates on heavy soils (fine textured-clay). Do not use hotcaps on transplanted tomatoes.

TOBACCO (Transplanted): Before transplanting, apply 2⅔ qts. TILLAM 6-E plus 2 pounds DEVRINOL 50-WP plus 2 qts. DYFONATE 4E, 4-EC or 4-ED for control of wireworms in 25 to 50 gallons of water per broadcast acre.

To avoid injury to crops not specified on the label, do not plant until 12 months after the last DEVRINOL application.

APPENDIX I

TILLAM 6-E WITH FLUID FERTILIZERS

The following procedure is suggested for determining whether TILLAM 6-E may be combined with a specific fluid fertilizer for spray tank application.

MATERIALS REQUIRED

1. TILLAM 6-E.
2. Fluid fertilizer to be used.
3. Adjuvant for fertilizer tank mix: Compex*, Sponto 168-D*, Uni-Mix*, or equivalent. The adjuvant which provides the best emulsification depends on the specific fertilizer under consideration.
4. Two one-quart, wide-mouth glass jars with lid or stopper.
5. Measuring spoons (a 25 mL pipette or graduated cylinder provides more accurate measurement).
6. Measuring cup, 8 oz. (237 mL).

* Compex, Colloidal Products Incorp., Petaluma, Calif. Sponto 168-D, Retzloff Chemical Company, Houston, Texas, Uni-Mix, Thompson-Hayward Chemical Co., Kansas City, Kansas.

PROCEDURE

1. Pour a pint (about 473 mL) of the fluid fertilizer into each of the quart jars.
2. Add adjuvant to one of the jars and mix (see rate table).
3. Add the TILLAM 6-E to both jars (see rate table).
4. Close both jars with lid or stopper and mix the contents by turning the jars upside down ten times.
5. Inspect the surface and body of the mixtures—
   (A) Immediately after completing the jar inversions.
   (B) After allowing the jars to stand quietly for 30 minutes.
   (C) And then again after turning the jars upside down 10 times.

If a uniform mix cannot be made, the mixture should not be used. If either mixture remains uniform for 30 minutes, the combination may be used. Should either mixture separate after 30 minutes, but readily remix uniformly with the jar inversions, the mixture can be used if adequate agitation is maintained in the tank. If the mixture with adjuvant is satisfactory, but the one without adjuvant is not, be sure to use the adjuvant in the spray tank. Add the adjuvant first at a rate of 3 pints per 100 gallons of fluid fertilizer; foaming can be minimized by using moderate agitation.

If nondispersible oil, sludge or clumps of solids form in the mixtures, the combinations should not be used.

RATE TABLE FOR TILLAM 6-E AND ADJUVANT WITH THE FLUID FERTILIZER**

Gallons of fluid fertilizer to be applied per acre	mL or Tsp. of TILLAM 6-E* to be added to 1 pint of fertilizer 6-E	
	mL	Tsp.
10	8	1⅔
15	5	1
20	4	¾
25	3	⅔
30	2½	½
40	2	⅖

* Based on field rate of 1 pound active ingredient per acre in the fertilizer volumes indicated. Increase volume proportionately to correspond with intended field rate in terms of pounds active ingredient per acre (e.g., for field rate of 4 pounds actual TILLAM 6-E in 40 gallons fertilizer per acre, add 8 mL or 2 tsp. TILLAM 6-E to each jar for compatability testing purposes).

** Two (2) milliliters or one-half (½) teaspoon of adjuvant to be added to 1 pint of fluid fertilizer in order to equal the rate of 3 pints of adjuvant per 100 gallons for fluid fertilizer.

APPENDIX II

TILLAM® 6-E IMPREGNATION ON DRY BULK FERTILIZERS FOR USE ON TOBACCO

CAUTION: TILLAM 6-E must not be impregnated on ammonium nitrate, sodium nitrate or potassium nitrate. Such mixtures may cause explosion and fire.

TILLAM 6-E may be impregnated on many dry bulk fertilizers* and applied and incorporated in the soil before planting for the control of grasses and broadleaf weeds in tobacco.

Field results have shown that TILLAM 6-E on bulk dry fertilizers gives weed control equal to TILLAM 6-E applied as a spray in water. However, uniform impregnation of TILLAM 6-E on the dry fertilizer particles and uniform application in the field are necessary to assure good results.

For impregnating TILLAM 6-E on dry fertilizer, use a closed rotary-drum mixer or a similar type of closed blender equipped with suitable spray equipment. The spray nozzle (or nozzles) should be positioned inside of the mixer to provide *uniform* spray coverage of the tumbling fertilizer. The physical properties of fertilizers vary, particularly in liquid absorptive capacity. When absorptivity is sufficient, simple spray impregnation of the fertilizer with TILLAM 6-E provides a satisfactory, dry mixture. If the absorptive capacity is inadequate, use of a highly absorptive powder is required to provide a dry, flowable mixture. Microcel E (Johns-Manville Products Corp.) is the recommended absorbent powder. It should be added separately and uniformly to the prepared TILLAM 6-E fertilizer mixture, in a quantity that is sufficient to provide a suitably flowable mixture. Generally less than 2% by weight of Microcel E is required. The amount of TILLAM 6-E actually required in the manufacture of individual fertilizer mixtures should be determined carefully for each production operation. This is necessary to ensure that the amount of TILLAM 6-E actually contained in the mixture applied to the soil represents the correct rate of use.

All TILLAM 6-E supplementary literature instructions and label recommendations regarding rates per acre, soil incorporation, application, cautions, general use precautions and other directions must be followed.

All individual state regulations relating to bulk dry fertilizer blending, registration, labeling and application are the responsibility of the individual and/or company selling the fertilizer and TILLAM 6-E mixture.

A minimum of 200 pounds and a maximum of 2,000 pounds of approved ingredients* impregnated with TILLAM 6-E at the recommended rate must be applied per acre.

Bulk fertilizer impregnated with TILLAM 6-E should be applied immediately, not stored. It is recommended that all bulk containers be tightly covered while the product is being transported and applied to reduce chances of TILLAM 6-E loss via volatilization.

***APPROVED DRY FERTILIZER INGREDIENTS**

	N	P	K
Ammonium sulfate	21	0	0
Diammonium phosphate	18	46	0
Potassium chloride	0	0	60
Potassium sulfate	0	0	52
Super-phosphate (single)	0	20	0
Super-phosphate (triple)	0	46	0
Urea	45	0	0
Ammonium phosphate sulfate	16	20	0
11-48-0	11	48	0

NOTE: K-Mag and two MAP fertilizers (12.5-50-0, 11-54-0) have been shown to be compatible with TILLAM 6-E and are approved for use.

RATE CHART FOR IMPREGNATION OF DRY BULK FERTILIZERS WITH TILLAM 6-E

Fertilizer Rate Per Acre	Tillam 6-E Rate Per Acre (2⅔ qts./acre)
200	26⅔ qts./ton
300	17⅘ qts./ton
400	13⅓ qts./ton
500	10⅔ qts./ton
600	8⅞ qts./ton
800	6⅔ qts./ton
1000	5⅓ qts./ton
1500	3⅗ qts./ton
2000	2⅔ qts./ton

STORAGE AND DISPOSAL

PROHIBITIONS
Do not contaminate water, food, feed or fertilizer by storage or disposal. Open dumping is prohibited.

PESTICIDE DISPOSAL
Rinse spray equipment. Wastes resulting from the use of this product may be disposed of on site or at an approved waste disposal facility.

CONTAINER DISPOSAL
METAL—Triple rinse (or equivalent), then, offer for recycling, reconditioning, or puncture and dispose of in a sanitary landfill, or by other approved State and local procedures.

PLASTIC—Triple rinse (or equivalent), then offer for recycling, or reconditioning, or puncture and dispose of in a sanitary landfill or incineration if allowed by State and local authorities, by burning. If burned, stay out of smoke.

NOTICE—READ CAREFULLY

CONDITIONS OF SALE
Stauffer (and seller) offer(s) this product for sale subject to, and buyer and all users are deemed to have accepted, the following conditions of sale and warranty which may only be varied by written agreement of a duly authorized representative of Stauffer.

WARRANTY LIMITATION
Stauffer warrants that this product conforms to the chemical description on the label and is reasonably fit for the purposes referred to in the directions for use on the label, subject to the inherent risks referred to below. Stauffer makes no other express warranties; THERE IS NO IMPLIED WARRANTY OF MERCHANTABILITY and there are no warranties which extend beyond the description on the label hereof.

INHERENT RISKS
The directions for use of this product are believed to be reliable and should be followed carefully. However, it is impossible to eliminate all risks associated with use. Buyer assumes all risks associated with use or application of this product contrary to label instructions or resulting from extraordinary weather conditions.

LIMITATION OF LIABILITY
In no case shall Stauffer be liable for special, indirect or consequential damages resulting from the use or handling of this product and no claim of any kind shall be greater in amount than the purchase price of the product in respect of which such damages are claimed.

This product is sold only for uses stated on the label. No express or implied license is granted to use or sell this product under any patent in any country except as specified: Country: United States of America.

This is a specimen label. Stauffer is not responsible for the accuracy of the information contained herein. As labels are subject to revision, always carefully read and follow the label on the product container.

TILLAM® 10-G

SELECTIVE HERBICIDE—GRANULES

ACTIVE INGREDIENT:

S-propyl butylethylthiocarbamate	10%
INERT INGREDIENTS:	90%
TOTAL	100%

Contains 10 Pounds Active Ingredient per 100 lbs.

EPA Reg. No. 476-1616 A-3R

CAUTION

Keep Out of Reach of Children

HARMFUL IF SWALLOWED

STATEMENT OF PRACTICAL TREATMENT
If a known exposure occurs or is suspected, immediately start the procedures given below. If further treatment is required, contact a Poison Control Center, a physician, or the nearest hospital, or call

collect day or night: (203) 226-6602 Stauffer Chemical Company, or (800) 424-9300 Pesticide Team Safety Network.

If ingested: Immediately give large amounts of water and induce vomiting by gagging the victim with a finger placed on the back of the victim's tongue. Give fluids until vomitus is clear. *Do not* induce vomiting or give anything by mouth to an unconscious or convulsing person. Call a physician immediately.

For eye contact: Immediately flush eyes with large amounts of running water for at least 15 minutes. Hold eyelids apart during flushing to ensure rinsing of the entire surface of the eye and lids with water. Call a physician.

For skin contact: Wash all affected areas with plenty of soap and water. Seek medical attention if irritation occurs.

If inhaled: Remove to fresh air. If not breathing, give artificial respiration, preferably mouth-to-mouth. If breathing is difficult, give oxygen, preferably with a physician's advice.

PRECAUTIONARY STATEMENTS

HAZARDS TO HUMANS AND DOMESTIC ANIMALS

CAUTION

Harmful if swallowed.

Avoid contact with skin, eyes and clothing. Avoid inhalation of dust. Wear rubber gloves and fresh clothing. Wear goggles to avoid contact with eyes. Wash with soap and water immediately after use or contact. Do not contaminate food and feed.

ENVIRONMENTAL HAZARDS

Do not apply directly to water. Do not contaminate water by cleaning of equipment or disposal of wastes.

WEEDS CONTROLLED

Annual Grasses:

Barnyard grass (Watergrass)	*Echinochloa* spp.
Crabgrass	*Digitaria* spp.
Foxtails—yellow, green, giant	*Setaria* spp.
Millet	*Setaria* spp.
Wild Oats	*Avena fatua*

Annual Broadleaf Weeds:

Black-Eyed Susan; Sweet Suzy	*Rubeckia serotina*
Common Purslane	*Portulaca oleracea*
Deadnettle, Henbit	*Lamium amplexicaule*
Florida Purslane	*Richardia scabra*
Hairy Nightshade	*Solanum villosum*
Lambsquarters	*Chenopodium album*
Nettleleaf Goosefoot	*Chenopodium murale*
Prostrate Pigweed	*A. graecizans*
Redroot Pigweed	*Amaranthus retroflexus*

The above broadleaf weeds will be controlled only when TILLAM 10-G is applied when conditions are favorable for germination.

Perennial Weeds:

Yellow Nutsedge (Yellow Nutgrass)	*Cyperus esculentus*
Purple Nutsedge (Purple Nutgrass)	*Cyperus rotundus*

Existing stands of nutgrass must be turned under and chopped up thoroughly before treatment.

GENERAL USE PRECAUTIONS

READ ALL LABEL DIRECTIONS BEFORE USING

TILLAM 10-G should be used only for recommended purposes and at recommended rates (DO NOT OVERDOSE): Applied according to directions and under normal growing conditions, TILLAM 10-G will not harm the treated crop. During germination and early growth, extended periods of unusually cold and wet or hot and dry weather, insect, nematode, or plant disease attack, the use of certain soil applied systemic insecticides, highly saline or alkaline soil conditions, improperly placed fertilizers or soil insecticides may create abnormal conditions that weaken crop seedlings. TILLAM 10-G used under these abnormal conditions could result in crop injury.

Do not apply in combination with fertilizer, insecticide or fungicide.

DIRECTIONS FOR USE

It is a violation of Federal law to use this product in a manner inconsistent with its labeling.

TILLAM 10-G is a selective herbicide which is mixed (incorporated) into the soil for control of weeds listed on this label. TILLAM 10-G controls weeds by interfering with normal seed germination and seedling development. It does not control established weeds. All weed growth and crop stubble should be thoroughly worked into the soil before treatment.

APPLICATION DIRECTIONS: Apply uniformly the recommended rate of TILLAM 10-G using equipment designed for application of granular herbicides. Equipment should be carefully calibrated before use and checked frequently during application to be sure equipment is working properly and delivering a uniform distribution pattern. Avoid overlaps that will increase TILLAM 10-G dosage above recommended rates as plant injury may occur. Apply only to well-worked soil that is dry enough to permit good soil mixing (incorporation). In irrigated areas, do not apply TILLAM 10-G before pre-irrigation.

For all band applications, reduce dosage proportionately depending upon row spacing and band width. Do not use band application on rocky ground because thorough incorporation is not possible.

INCORPORATION DIRECTIONS: TILLAM 10-G must be mixed (incorporated) into the soil IMMEDIATELY after application to prevent loss by evaporation of the herbicide. Whenever possible, application and incorporation should be done in the same operation.

INCORPORATION BEFORE PLANTING: Use power-driven cultivation equipment set to cut to a depth of 2 to 3 inches in light soils and 3 to 4 inches in heavy soils, OR tandem discs set to cut to a depth of 6 inches, followed by a spike-tooth harrow or some other leveling device which extends beyond ends of the disc. For thorough mixing, disc in two different directions (cross disc).

INCORPORATION AT OR AFTER PLANTING: Use power-driven cultivation equipment set to cut to a depth of 2–3 inches OR, on light soils only, ground-driven tillers (rolling cultivators, rotary hoe, row wheels, etcetera) set to cut to a depth of 1½ inches and operated at 6 to 8 mph. When using ground driven tillers, pre-cultivate if soil is crusted. When incorporating after planting, care must be taken not to disturb the crop seedling.

PLANTING

For maximum weed control benefits, crops should be planted or seeded immediately after application. DO NOT use a drag behind the planter as it may concentrate TILLAM 10-G over the seed row and cause crop injury.

CULTURAL PRACTICES FOLLOWING APPLICATION

TILLAM 10-G is not persistent in the soil and susceptible weeds germinating late in the season may not be controlled. Shallow cultivation or approved post emergence herbicides may be necessary to control these susceptible weeds that escape control by TILLAM 10-G as well as those weeds not susceptible to TILLAM 10-G. Do not cultivate deeper than the depth of herbicide incorporation.

When cultivating fields where TILLAM 10-G has been banded in the row, use shields to prevent the movement of untreated soil into the treated row.

RECOMMENDATIONS

SUGAR BEETS, PRE-PLANT—Apply just prior to planting. See DIRECTIONS FOR USE for methods of application and incorporation. Recommended dosages are as follows: For light textured soils and soils with less than 10% organic matter, apply 40 lbs. TILLAM 10-G per acre. For heavy textured soils and soils with over 10% organic matter, apply 60 lbs. TILLAM 10-G per acre.

TOBACCO, FLUE-CURED: BEFORE BEDDING AND TRANSPLANTING: Apply TILLAM 10-G broadcast of 40 lbs. per acre. Incorporate immediately with a tandem disc and bed up immediately prior to transplanting, OR

AFTER BEDDING BUT BEFORE TRANSPLANTING: Apply TILLAM 10-G at a broadcast rate of 40 lbs. per acre prior to transplanting, reducing rates proportionately depending upon band width and row spacing. Incorporate immediately, using a power-driven rotary hoe with variable length blades (i.e., one with short blades on the inside and longer blades on the outside). Bed levelers, if used, should be kept shallow, as excessive removal of treated soil will destroy weed control.

TOBACCO, LIGHT AIR CURED: INCLUDES TYPE 31 (Burley) AND TYPE 32 TOBACCOS

BEFORE TRANSPLANTING: Apply TILLAM 10-G broadcast at 40 lbs. per acre. Incorporate immediately, only with a tandem disc on flat ground followed by transplanting. For band treatment, reduce rates proportionately depending upon band width and row spacing. Incorporate immediately, using a power-driven rotary hoe.

TOMATOES: See DIRECTIONS FOR USE for methods of application and incorporation.

DIRECT SEEDED, LAYBY: Apply TILLAM 10-G to clean cultivated soils. Do not apply within eight (8) days of harvest. For light textured soils and soils with less than 10% organic matter, apply 40 lbs. TILLAM 10-G per acre. For heavy textured soils and soils with more than 10% organic matter, apply 60 lbs. TILLAM 10-G per acre.

DIRECT SEEDED, PRE-PLANT (CALIFORNIA only): Apply TILLAM 10-G prior to planting. For light textured soils and soils with less than 10% organic matter, apply 40 lbs. TILLAM 10-G per acre. For heavy textured soils and soils with more than 10% organic matter, apply 60 lbs. TILLAM 10-G per acre.

TOMATO TRANSPLANTS (Do not use hotcaps): BEFORE TRANSPLANTS—Apply TILLAM 10-G prior to transplanting. OR, AFTER TRANSPLANTING—Apply TILLAM 10-G to clean cultivated soil. Do not apply within eight (8) days of harvest. For light textured soils and soils with less than 10% organic matter, apply 40 lbs. TILLAM 10-G per acre. For heavy texured soils and soils with more than 10% organic matter, apply 60 lbs. TILLAM 10-G per acre.

STORAGE AND DISPOSAL

STORAGE
Do not contaminate water, food, or feed by storage or disposal. Open dumping is prohibited.

PESTICIDE DISPOSAL
Wastes resulting from the use of this product may be disposed of on site or at an approved waste disposal facility.

CONTAINER DISPOSAL
Completely empty bag into application equipment. Then dispose of empty bags in a sanitary landfill or by incineration, or, if allowed by State or local authorities, by burning. If burned, stay out of smoke.

NOTICE—READ CAREFULLY

CONDITIONS OF SALE
Stauffer (and seller) offer(s) this product for sale subject to, and buyer and all users are deemed to have accepted, the following conditions of sale and warranty which may only be varied by written agreement of a duly authorized representative of Stauffer.

WARRANTY LIMITATION
Stauffer warrants that this product conforms to the chemical description in the directions for use on the label subject to the inherent risks referred to below. Stauffer makes no other express warranties; THERE IS NO IMPLIED WARRANTY OF MERCHANTABILITY and there are no warranties which extend beyond the description of the label hereof.

INHERENT RISK
The directions for use of this product are believed to be reliable and should be followed carefully. However, it is impossible to eliminate all risks associated with use. Buyer assumes all risks associated with use or application of this product contrary to label instructions or resulting from extraordinary weather conditions.

LIMITATION OF LIABILITY
In no case shall Stauffer be liable for special, indirect or consequential damages resulting from the use or handling of this product and no claim of any kind shall be greater in amount than the purchase price of the product in respect of which such damages are claimed.

This product is sold only for uses stated on its label. No express or implied license is granted to use or sell this product under any patent in any country except as specified: Country: United States of America. Patent Nos. 3,175,897; 3,185,720.

This is a specimen label. Stauffer is not responsible for the accuracy of the information contained herein. As labels are subject to revision, always carefully read and follow the label on the product container.

VAPAM®

SOIL FUMIGANT SOLUTION

For All Crops

ACTIVE INGREDIENTS:

Sodium methyl dithiocarbamate (anhydrous)	32.7%
INERT INGREDIENTS:	67.3%
TOTAL	100.0%

Contains 3.18 lbs. active ingredient per gallon

EPA Reg. No. 476-859 A-1R

2 ARC 720 428-0

CAUTION

Keep Out of Reach of Children

STATEMENT OF PRACTICAL TREATMENT

FIRST AID
Immediately start the procedures given below and contact a Poison Center, a physician, or the nearest hospital. Report the type and extent of exposure, describe the victim's symptoms, and follow the advice given.

If on skin: Immediately flush skin with large amounts of running water for at least 15 minutes while removing contaminated clothing and shoes. Get medical attention immediately.

If in eyes: Immediately flush eyes with large amounts of running water for at least 15 minutes. Hold eyelids apart to ensure rinsing of the entire surface of the eye and lids with water. Get medical attention immediately.

If inhaled: Remove to fresh air. If not breathing, clear the victim's airway and start mouth-to-mouth artificial respiration. If breathing is difficult, give oxygen, preferably with a physician's advice. Get medical attention immediately.

If swallowed: Immediately give several glasses of water but *do not* induce vomiting. If vomiting does occur, give fluids again. Have a physician determine if condition of patient will permit induction of vomiting or evacuation of stomach. Do not give anything by mouth to an unconscious or convulsing person.

See Side Panel for Additional Precautionary Statements.

PRECAUTIONARY STATEMENTS
HAZARDS TO HUMANS AND DOMESTIC ANIMALS

ENVIRONMENTAL HAZARDS
This product is toxic to fish. Do not apply directly to water or wetlands. Do not apply where runoff is likely to occur. Do not contaminate water by cleaning of equipment of disposal of waste. Apply this product only as specified on the label.

In case of Emergency—Immediately call (24 hours) (800) 424-9300 Chemtrec, or (203) 226-6602 Stauffer Chemical Company.

HARMFUL IF INHALED OR SWALLOWED
IRRITATING TO EYES, NOSE, THROAT AND SKIN

Avoid breathing vapor or spray mist.

Do not get in eyes, on skin, on clothing.

In case of contact, immediately remove contaminated clothing or shoes and flush skin with plenty of water, and apply soothing lotion; for eyes, flush with water for at least 15 minutes and get medical attention.

Wash and dry clothing and shoes before reuse.

When applying in enclosed areas wear a mask or respirator of a type approved by the U. S. Bureau of Mines for VAPAM protection. Keep children and pets out of treated areas.

PRODUCT INFORMATION

VAPAM is a water soluble liquid. When applied to properly prepared soil, the liquid is converted into a gaseous fumigant. After sufficient interval of time, the gas dissipates leaving the soil ready for planting. VAPAM is recommended for the control of the following soil-borne pests that attack ornamental, food and fibre crops: Weeds and germinating weed seeds [annual bluegrass, Bermudagrass, chickweed, dandelion, ragweed, henbit, lambsquarters, Amaranthus sp. (pigweed, careless weed) watergrass, Johnsongrass, nutgrass, wild morningglory and purslane] nematodes, symphylids (garden centipede); and soil-borne diseases (Rhizoctonia, Pythium, Phytophora, Verticillium, Sclerotinia, Oak root fungus and club root of crucifers).

USE PRECAUTIONS

Keep children and pets out of treated area. Keep off desirable lawns and plants. Do not apply within 3 feet of the drop line of desirable plants, shrubs or trees. Do not use in confined areas without adequate ventilation OR where fumes may enter nearby houses containing growing plants. Do not use in greenhouses where desirable plants are present. Keep container tightly closed when not in use. Do not store near feed or food.

DIRECTIONS FOR USE

It is a violation of Federal Law to use this product in a manner inconsistent with its labeling.

Before applying VAPAM always cultivate thoroughly area to be treated, breaking up clods and loosening soil deeply and thoroughly. A week before treatment, moisten soil after cultivation to the desired depth; sprinkle or flood irrigate. **This step is essential for all methods of use.** Immediately before application, cultivate lightly if the soil has crusted.

To prevent loss from evaporation, use only at times when air temperature is moderate and there is little wind movement. Soil temperature must be in the range of 60°–90°F. at a 3 inch depth. For other conditions, see section, "Cultivation & Planting After Application". Do not apply to the soil surface, as in the sprinkler method, when air temperature is over 90°F. or when low humidity or high winds would cause loss of VAPAM before it can be drenched into the soil with additional water. If fumes become unpleasant during treatment, apply more water to seal the fumes into the soil where they should be confined to achieve maximum fumigant benefit.

The activity of VAPAM is increased by the use of a tarp (plastic, paper or fabric) spread loosely over the treated area and secured to prevent removal by wind. Keep covered for a minimum period of 48 hours. Seven days after treatment cultivate area to depth of 2 inches to aerate the soil. Do not seed earlier than 21 days after application when tarping method is used.

USE PROMPTLY AFTER MIXING WITH WATER. DO NOT ALLOW SOLUTION TO STAND. Flush all equipment with water after each day's use. Disassemble valves and clean carefully.

CULTIVATION AND PLANTING AFTER APPLICATION

On well drained soils of light to medium texture which are not excessively wet or cold following application, planting may take place 14 to 21 days after treatment. If soils are heavy or especially high in organic matter or remain wet and/or cold (below 60°F.) following application of VAPAM, a minimum interval of 30 days should be observed. Where dosages are greater than 100 gal. per acre, wait at least 60 days. On heavy, wet soils, light surface cultivation to break up crusting and promote drying of the soil should be done 5 to 7 days after application. This cultivation may be repeated as necessary. To avoid reinfesting treated soils, cultural practices should be such that untreated soils are not mixed with treated soils.

SPECIAL INSTRUCTIONS: When treating potting soil, or heavier field soils, including soils high in clay or organic matter, it is **especially important** that the soils be allowed to aerate and dry thoroughly after using VAPAM. During cold and/or wet weather, frequent shallow cultivations may aid the escape of VAPAM from the soil. If in doubt, transplant a seedling plant and examine for injury before planting crop.

USES, APPLICATION METHODS & RATES

FOR SHALLOW PESTS IN SEED BEDS, PLANT BEDS, LAWNS AND OTHER LIMITED AREAS

SPRINKLING CAN METHOD: Place 1 pt. VAPAM (1½ pt. on very heavy soils or for deep-rooted weeds) in a sprinkling can, fill with water, and sprinkle uniformly over 50 sq. ft. of well-prepared soil. Sprinkle immediately with water until soil is sealed, or tarp for 48 hours.

HOSE PROPORTIONER METHOD: Add 1 qt. VAPAM to 3 qt. water in a bucket or other container and apply, using a hose proportioner to an area of 100 sq. ft. Sprinkle with water until soil is sealed, or tarp for 48 hours.

SOIL INJECTION: Space injection shanks 5 in. apart and inject VAPAM 4 in. into well prepared soil. Follow immediately with a roller to smooth and compact the soil surface. Light watering or a tarp after rolling helps prevent gas escape. For seedbeds a dosage of 75 to 100 gal. per acre (1½ pt. to 2 pt. per 100 sq. ft.) is recommended.

ROTARY TILLER: Spray or sprinkle diluted VAPAM immediately in front of tiller. Use 1 qt. VAPAM in 2½ gal. water per each 100 sq. ft. Follow immediately with a roller to smooth and compact the soil surface. Light watering or a tarp after rolling will help prevent gas escape.

FIELD APPLICATION—WHERE ENTIRE AREA IS BEING TREATED

SOIL INJECTION: Space thin injection shanks 5 in. apart and inject VAPAM 4 in. deep into well prepared soil. Follow immediately with a roller to smooth and compact surface. Light watering or a tarp after rolling helps prevent gas escape. For field use, 40 to 100 gal. VAPAM per acre is recommended.

SPRINKLER SYSTEM: Use only those sprinkler systems which give large water droplets to prevent excessive loss. Use 75 to 100 gal. VAPAM per acre. For control of shallow pests (top 1 ft. or less of soil), run sprinklers 5 to 10 min. In next 10 to 20 min. inject into the sprinkler system all VAPAM needed for the area covered. On very light soils keep surface moist by sprinkling for 2 or 3 days. For control of pests deeper than 18 in. in the soil, divide VAPAM into 3 or more equal parts and apply at intervals during the sprinkling period.

CHECK OR FLOOD IRRIGATION: Meter VAPAM at a steady rate into water during irrigation. Use 50 to 100 gal. VAPAM per acre, depending upon the kind of pest and depth desired, in 3 to 18 inches of water per acre.

FIELD APPLICATION TO BEDS OR ROWS

SOIL INJECTION: VAPAM may be injected into pre-formed plant beds following the directions given above under soil injection. If a wider treated band is desired, space 2 or more shanks at intervals of 5 in. to cover the desired treating width. Roll immediately.

SOIL COVERING METHODS: (Bed-over methods). VAPAM may be sprayed or dripped onto the soil immediately ahead of bed-shaping equipment. Cover the VAPAM with soil to a depth of 3 to 6 inches. The soil should be rolled and compacted immediately. The recommended rate of VAPAM is 50 to 100 gal. per acre, approximately equivalent to 1 to 2 pints per 100 ft. of 12-inch row.

TREATMENT OF TREE REPLANT SITES: After removing dead or diseased tree and as much of the root system as possible, make a shallow basin over the planting site. Add VAPAM to the stream of water while filling the basin. Use 1 qt. VAPAM per 100 sq. ft. in sufficient water (depending on soil type) to penetrate at least 6 ft. For control of oak root fungus, use a basin at least 20 x 20 ft. square. Increase dosage to 2 qt. VAPAM per 100 sq. ft. in sufficient water to penetrate to the depth of root system. If water is tanked to the planting site, add VAPAM to the water and mix before filling basin.

TREATMENT OF POTTING SOIL

A. SPRINKLE METHOD:

1. Spread soil in a smooth layer 4 inches high on concrete or on pre-treated soil.
2. Sprinkle VAPAM at rate of 1 pint in 5 gallons of water per 100 sq. ft. of surface area.
3. Layers can be treated one on top of another.
4. Sprinkle top layer with sufficient additional water to seal the surface, or cover with tarp (plastic, kraft paper, etc.).

B. CEMENT MIXER:

1. Add VAPAM to soil mix at rate of 1 fl. oz. VAPAM per 2 cu. ft. of soil, in cement or similar mixer. Mix thoroughly.
2. After soil is treated and piled, sprinkle water over entire surface to seal in gas. Or, cover with tarp (plastic, kraft paper, etc.).

C. SHREDDER:

1. Dilute VAPAM in sufficient water to obtain even distribution. As soil is ejected from shredder, spray uniformly on soil stream at rate of 1 fl. oz. VAPAM per 2 cu. ft. of soil.
2. After all soil is treated and piled, apply light water seal to entire surface or cover with tarp.

TOBACCO PLANT BEDS: Fall applications are recommended wherever possible. Read and follow DIRECTIONS FOR USE carefully. Treatment in the South should generally be made before November 30.

A. TARP METHOD: Prepare the bed 5 to 7 days before application to insure best conditions for weed seed germination and fumigant action of VAPAM. The bed should be free of clods, level and in good tilth. Apply 1 to 1½ gal. of VAPAM in a minimum of 40 gal. of water per 100 sq. yds. Apply uniformly over the entire bed. Cover the bed immediately with a plastic cover. Keep covered no less than one day, but no more than two days. The cover need not be tented, but should be secured to prevent wind from uncovering the treated area. Seven days after date of VAPAM application, loosen the treated soil to a depth of 2 in. Do not seed tobacco earlier than 21 days after VAPAM application.

B. DRENCH METHOD: Apply 2½ gal. VAPAM in 150 to 200 gal. of water per 100 sq. yd. Application may be made with sprinklers, sprayers with nozzles or any suitable equipment. Follow directions given above for seed bed treatment.

SYMPHYLID CONTROL: Soil should be in good seed bed condition to a depth of 8 to 10 inches. Maintain adequate moisture during spring season. Treat during July-August when symphylids are in the upper soil surface. Apply 20 gal. VAPAM per acre using blade or chisel injector. Inject below level of symphylid concentration, usually 6 to 8 inches. Pack soil immediately after application.

EARLY MATURITY DISEASES OF POTATOES IN OREGON. Apply 40 gals. VAPAM per acre using thin shank injector rig.

PEPPERMINT—Verticilium wilt control. When infestation is limited to small spots in a field, spread can be reduced by treating the soil with 100 gals. VAPAM per acre using injector blade or thin shank injector rig.

PREVENTION OF ROOT GRAFT TRANSMISSION OF DUTCH ELM AND OAK WILT DISEASE

Immediately after a tree is diagnosed as having Dutch Elm or Oak Wilt disease, isolate the diseased tree from healthy trees with the VAPAM treatment. If a diseased tree is less than 20 feet from a healthy tree or has advanced wilt symptoms, it may be necessary to treat at two sites—one between the diseased and the first healthy-appearing tree and one between the first and the second healthy-appearing trees. This measure is advisable because the causal fungus may have already passed from the diseased to the first healthy-appearing tree before VAPAM was applied.

Use VAPAM diluted one part to three parts water for Dutch Elm disease and diluted one part to ten parts of water for Oak Wilt disease. Drill holes approximately ¾ to 1 inch in diameter, 15 inches deep to 6 to 9 inches apart. Fill each hole with diluted VAPAM to within 2 inches of the soil surface.

Make the line of treatment sufficiently long to kill all roots of the two adjacent trees that are likely to be root-grafted. Apply the chemical slowly and carefully to avoid overflowing the drilled holes, this will reduce grass kill. Tamp each hole closed with the heel. Allow at least two weeks after treatment before removing the diseased tree.

STORAGE AND DISPOSAL

PROHIBITION

Do not contaminate water, food or feed by storage or disposal.

STORAGE

Do Not Store Below 0°F. Product crystallizes at lower temperatures. Warm or store at higher temperatures and mix to redissolve crystals and assure uniformity before use.

PESTICIDE DISPOSAL

Wastes resulting from the use of this product may be disposed of on site or at an approved waste disposal facility.

CONTAINER DISPOSAL:

METAL—Triple rinse (or equivalent). Then offer for recycling or reconditioning, or puncture and dispose of in a sanitary landfill or by other procedures approved by State and local authorities.

PLASTIC—Triple rinse (or equivalent). Puncture and dispose of in a sanitary landfill, or by incineration, or, if allowed by State and local authorities, by burning. If burned, stay out of smoke.

NOTICE—READ CAREFULLY

CONDITIONS OF SALE

Stauffer (and seller) offer(s) this product for sale subject to, and buyer and all users are deemed to have accepted, the following conditions of sale and warranty which may only be varied by written agreement of a duly authorized representative of Stauffer.

WARRANTY LIMITATION

Stauffer warrants that this product conforms to the chemical description on the label and is reasonably fit for the purposes referred to in the directions for use on the label subject to the inherent risks referred to below. Stauffer makes no other express warranties. THERE IS NO IMPLIED WARRANTY OF MERCHANTABILITY and there are no warranties which extend beyond the description on the label hereof.

INHERENT RISKS

The directions for use of this product are believed to be reliable and should be followed carefully. However, it is impossible to eliminate all risks associated with use. Buyer assumes all risks associated with use or application of this product contrary to label instructions or resulting from extraordinary weather conditions.

LIMITATION OF LIABILITY

In no case shall Stauffer be liable for special, indirect, or consequential damages resulting from the use or handling of this product and no claim of any kind shall be greater in amount than the purchase price of the product in respect of which such damages are claimed.

This product is sold only for uses stated on its label. No express or implied license is granted to use or sell this product under any patent in any country except as specified: Country: United States of America.

This is a specimen label. Stauffer is not responsible for the accuracy of the information contained herein. As labels are subject to revision, always carefully read and follow the label on the product container.

VERNAM® 7-E

SELECTIVE HERBICIDE

Liquid Herbicide For Weed Control In Soybeans And Peanuts

Complete Directions For Use, Use Precautions and Recommendations

ACTIVE INGREDIENT:

S-propyl dipropylthiocarbamate	88.5%
INERT INGREDIENTS	11.5%
TOTAL	100.0%

Contains Seven Pounds of Active Ingredient Per Gallon

EPA Reg. No. 476-2155 A-12MR

CAUTION

Keep Out of Reach of Children

HARMFUL IF SWALLOWED

STATEMENT OF PRACTICAL TREATMENT

If swallowed: **DO NOT** induce vomiting. Immediately give large quantities of water. If vomiting does occur, give fluids again. Never give anything by mouth to an unconscious person. Call a physician or the nearest Poison Control Center immediately.

For eye contact: Hold eyelids apart and flush with large amounts of running water for at least 15 minutes. Call a physician.

For skin contact: Flush with large amounts of running water for 15 minutes. Remove contaminated clothing and shoes. Get medical attention if skin irritation occurs. Wash clothing before reuse.

If inhaled: Remove to fresh air. Get medical attention if respiratory irritation occurs.

PRECAUTIONARY STATEMENTS

HAZARDS TO HUMANS AND DOMESTIC ANIMALS

CAUTION

Harmful if swallowed. Avoid contact with skin, eyes and clothing. Avoid inhalation of spray mist. Wear rubber gloves, goggles and protective clothing. Wash with soap and water immediately after use or contact. Do not contaminate food and feed. Keep away from heat and flame.

ENVIRONMENTAL HAZARDS

Do not apply directly to water. Do not contaminate water by cleaning of equipment or disposal of wastes.

Do not apply this product in such a manner as to directly or through drift, expose workers or other persons. When applied through the sprinkler irrigation method of application, the area being treated must be vacated by unprotected persons.

In case of emergency call (day or night) Chemtrec (800) 424-9300 or Stauffer Chemical Company (203) 226-6602.

WEEDS CONTROLLED BY VERNAM 7-E

VERNAM 7-E will not control established weeds.

Annual Grasses:

Barnyardgrass (watergrass)	*Echinochloa crusgalli*
Bermudagrass seedlings**	*Cynodon dactylon*
Crabgrass	*Digitaria* spp.
Fall Panicum	*Panicum dichotomiflorum*
Foxtail millet	*Setaria italica*
Giant foxtail	*Setaria faberi*
Goosegrass	*Eleusine indica*
Green foxtail	*Setaria viridis*
Johnsongrass seedling	*Sorghum halepense*
Wild cane*	*Sorghum bicolor*
Yellow foxtail	*Setaria glauca*

* Cultivation is suggested in addition to the VERNAM 7-E treatment in fields with moderate or heavy infestations of wild cane.

** Southeastern U.S. Only.

Annual Broadleaf Weeds:

Annual morningglory	*Ipomoea purpurea*
Balloonvine*	*Cardiospermun halicacabum*
Carpetweed	*Mollugo verticillata*
Common lambsquarters	*Chenopodium album*
Florida purslane Florida pusley	*Richardia scabra*
Purslane	*Portulaca oleracea*
Redroot pigweed	*Amaranthus retroflexus*
Sicklepod (coffeeweed)	*Cassia obtusifolia*
Smooth pigweed	*Amaranthus hybridus*
Velvetleaf (buttonweed)	*Abutilon theophrasti*

* In Alabama Only.

The above listed broadleaf weeds will be controlled only if treatment is made when conditions are favorable for weed seed germination and growth.

Perennial Grasses:

Bermudagrass*	*Cynodon dactylon*
Purple nutsedge (nutgrass)	*Cyperus rotundus*
Yellow nutsedge (nutgrass)	*Cyperus esculentus*

* Southeastern U.S. Only.

Existing stands of nutgrass must be turned under and chopped up thoroughly before treatment. For nutgrass control VERNAM 7-E must be thoroughly mixed into the soil prior to planting or applied subsurface to the soil with injectors or sweeps. Cultivation is suggested in addition to VERNAM 7-E treatment in fields with moderate to heavy nutgrass infestations.

GENERAL USE PRECAUTIONS

Read entire label before using.

VERNAM 7-E should be used only for recommended purposes and at recommended rates (do not overdose).

Failure to follow label instructions, or use on crops not specifically listed on the label may result in crop injury or destruction.

Applied according to directions and under normal growing conditions, VERNAM 7-E will not harm the treated crop. Malformation of the primary leaves is not unusual. This condition is only temporary and will not affect yields. During germination and early stages of growth, extended periods of unusually cold and wet or hot and dry weather, insect or plant disease attack, the use of certain soil-applied systemic insecticides, improperly placed fertilizers or soil insecticides may create abnormal conditions that weaken crop seedlings. VERNAM 7-E used under these abnormal conditions could result in crop injury.

VERNAM 7-E is recommended for use on mineral soils only.

Do not tank mix VERNAM 7-E with insecticides or fungicides.

Do not contaminate irrigation water or water used for domestic purposes.

In irrigated areas, do not apply VERNAM 7-E before pre-irrigation.

DIRECTIONS FOR USE

It is a violation of federal law to use this product in a manner inconsistent with its labeling.

VERNAM 7-E is a selective herbicide which is mixed (incorporated) into the soil or applied below the surface of the soil for control of weeds listed on this label. VERNAM 7-E controls annual weeds as their seeds germinate by interfering with normal germination and seedling development. It does not control established weeds. All weed growth and crop stubble should be thoroughly worked into the soil before treatment.

APPLICATION DIRECTIONS

Pour the recommended amount of VERNAM 7-E into the spray tank during the filling operation. Apply in 10 to 50 gallons of water per acre using a properly calibrated, low-pressure (20 to 40 PSI) sprayer having good agitation. The soil should be well-worked and dry enough to permit good soil mixing (incorporation) or subsurface application. Soil should be as free as possible of trash for subsurface application with sweeps or injectors. Avoid overlaps that will increase dosage above recommended limits.

FLUID FERTILIZER STATEMENT

VERNAM 7-E may be combined with fluid (solution, slurry, or suspension) fertilizers. However, physical compatibility with these fluid fertilizers must be determined before combining in the spray tank. See Appendix I for directions regarding these combinations.

DRY FERTILIZER STATEMENT

VERNAM 7-E may be impregnated on many dry bulk fertilizers for use on *registered crops.* However, uniform distribution of VERNAM 7-E on fertilizer particles and uniform application are necessary to assure good results. See Appendix II for information and directions for impregnation and use.

INCORPORATION DIRECTIONS

VERNAM 7-E must be incorporated into the soil immediately (within minutes) to prevent loss of the herbicide. Whenever possible, application and incorporation should be done in the same operation.

SOIL MIXING (INCORPORATION) BEFORE PLANTING: Use power-driven cultivation equipment set to cut to a depth of 2 to 3 inches or tandem discs set to cut to a depth of 4 to 6 inches, operated at 4 to 6 mph, followed by a spike-tooth harrow or some other leveling device which extends beyond ends of the disc. For thorough mixing, disc in two different directions (cross disc). Prior to the second discing, raise the disc to prevent cutting deeper than 4 to 6 inches.

NOTE: In the southern lower Delta of the Mississippi Valley, disc incorporation is not recommended on the heavy (gumbo) clay soils.

SOIL MIXING (INCORPORATION) AFTER PLANTING: Use power-driven cultivation equipment set to cut to a depth of 2 inches, or on coarse textured (light) soils only, ground-driven tillers (rolling cultivators, rotary hoe, row wheels, etc.) set to cut to a depth of 1½ inches and operated at 6 to 8 mph. When incorporating after planting, care must be taken not to disturb the crop seed or seedlings by adjusting tines or blades of incorporation equipment so they are not directly over the drill row.

SUBSURFACE APPLICATION BEFORE OR AT PLANTING (for peanuts and soybeans only): Special equipment designed for subsurface application MUST be used. Injector units must be rigidly mounted on or IMMEDIATELY ahead or IMMEDIATELY behind the planter unit. Sweeps must be mounted IMMEDIATELY ahead of the planter unit.

(1) Soil Injection—Injector shanks must be spaced 2½ to 3 inches apart and mounted in staggered positions to avoid trash buildup. Set shanks to inject VERNAM 7-E 1½ to 3 inches below the soil surface. The width of the band in which weed control is desired will determine the number and spacing of injector shanks required per row. (Example: Four injector shanks spaced 3 inches apart gives a 12-inch band.) A broadcast application can be made by increasing the number of shanks. The two shanks adjacent to the drill row must be 1¼ to 1½ inches on either side of it.

(2) Covered Sweeps—Set the sweeps to run 3 to 4 inches below the soil surface. Calibrate by measuring the spray band width at the back edge of the sweep, not the sweep width. For BROADCAST APPLICATIONS, stagger sweeps on double tool bar so that they overlap sufficiently to allow spray bands to meet.

NOTE: When applying with sweeps, VERNAM 7-E must be applied deep enough to allow 3 to 4 inches of soil to remain over the treatment after the planting operation.

PLANTING—SOYBEANS, PEANUTS

Do not plant seed deeper than 2 inches. For VERNAM 7-E application made before planting, seeding can be done immediately, or it may be delayed up to 10 days. If it is necessary to work the soil following application and incorporation, but before planting, do not work the soil deeper than the depth the herbicide was incorporated. To avoid removing VERNAM 7-E from the row (with loss of weed control in the row) do not move or shape soil after treatment, such as by planting in deep furrows.

RECOMMENDATIONS

These recommendations are given in terms of broadcast (overall) rates of VERNAM 7-E per acre. For band or row application, reduce broadcast rates proportionately, depending upon band width to be treated and row spacing.

PEANUTS

VERNAM 7-E may be applied before planting, at planting or immediately after planting. When applying after planting, care must be taken not to disturb the seed. Use 2⅓ pints on coarse textured (sandy—light) soils; 3 pints on medium to fine textured (sandy loam to clay loam—medium to heavy) soils.

SOYBEANS

The following recommendations are for soybeans grown in all areas except in the Southeastern States:

VERNAM 7-E may be applied before planting, at planting, or immediately after planting. When applying at or after planting, care must be taken not to disturb the crop seed or seedlings. Use 2⅓ pints on coarse textured (sandy—light) soils, 3 pints on medium to fine textured (sandy loam to clay loam—medium to heavy) soils, 3½ pints on very fine textured (clay—very heavy) soils.

For soybeans grown in the Southeastern States including North Carolina, South Carolina, Georgia, Alabama, Florida and Eastern Tennessee (Tenn. River East):

VERNAM 7-E may be applied before planting, at planting, or immediately after planting. For incorporated applications made before planting, do not plant in deep furrows, or use shanks or other device ahead of planter which bring up untreated soil to the surface. For furrow planted soybeans, apply VERNAM 7-E at or immediately after planting in bands over the row and incorporate with row wheels only. When applying at or after planting, care must be taken not to disturb the crop seed or seedlings. Use 2⅓ pints on coarse textured (sandy—light) soils; 3 pints on medium to fine textured (sandy loam to clay loam—medium to heavy) soils.

NOTE: In areas where beds are made prior to planting soybeans, VERNAM 7-E may be applied in a band immediately ahead of the bedding discs, set to cover the VERNAM 7-E with 3 to 4 inches of soil. If bed levelers are used before planting soybeans, make the beds higher to compensate for leveling, so 2 to 3 inches of soil remain over the treatment after the planting operation is done.

SEQUENTIAL APPLICATIONS

IMPORTANT: Always consult exact label of each herbicide used for specific recommendations and limitations.

For control of weeds not listed on the VERNAM 7-E label, and for increased control of labeled broadleaf weeds, the following herbicides are suggested as overlay treatments after preplant applications of VERNAM 7-E.

PEANUTS

DYANAP* 3 lbs./gal. liquid

Broadcast rates per acre:

Apply at cracking at the rate of 6 quarts per acre.

ENIDE* 50W + DINITRO (PREMERGE 3 lbs./gal.)

Broadcast rates per acre:

Apply at cracking or up to 7 days after at the rate of 4 lbs. ENIDE 50W and 2 quarts PREMERGE per acre.

LASSO* 4-EC (4 lbs. a.i./gal.) + DINITRO (PREMERGE 3 lbs./gal.)

Broadcast rates per acre:

Apply at cracking at the rate of 2 to 3 quarts of LASSO 4-EC and 2 quarts of PREMERGE per acre.

SOYBEANS

AMIBEN* 2 lbs./gal. liquid

Broadcast rates per acre:

Coarse textured (light) soils—1 Gallon
Medium textured (medium) soils—1⅛ Gallons
Fine textured (heavy) soils—1½ Gallons

BASAGRAN* 4 lbs./gal. liquid

Broadcast rates per acre: Apply 3/4 to 1 quart per acre. Rates depend upon weeds to be controlled. Consult and follow the recommended rates on the BASAGRAN label.

DYANAP* 3 lbs./gal. liquid

Broadcast rates per acre: Use 6 quarts as a broadcast spray just after the soybeans are planted but before the soybean plants emerge from the soil.

LOROX* 50W

Broadcast rates per acre: Use 1 to 2 1/2 lbs. LOROX 50W per acre depending on soil type. Use on coarse to medium textured soils. Apply after planting before soybeans emerge. Plant soybeans at least 1 3/4" deep on flat seed bed only to avoid possible crop injury. Do not use on sand or loamy sand or soil with less than 1% organic matter since crop injury may result.

*AMIBEN chloramben herbicide is a registered trademark of Union Carbide Agricultural Products Co., Inc.

*BASAGRAN is a registered trademark of BASF Aktiegesellschaft

*DYANAP is a registered trademark of Uniroyal Chemical, Inc.

*ENIDE is a registered trademark of Tuco, Division of the Upjohn Company

*LASSO is a registered trademark of Monsanto Company

*LOROX is a registered trademark of E.I. Du Pont de Nemours & Co., Inc.

*PREMERGE is a registered trademark of Dow Chemical Company

TANK MIX DIRECTIONS FOR USE ON PEANUTS

VERNAM 7-E may be combined with the compounds listed in this section for tank mix application on peanuts. Recommended rates for each tank mix are listed below. Mixing, spray equipment, soil incorporation, and planting directions for VERNAM 7-E tank mixes on peanuts are listed at the end of the peanut tank mix section.

VERNAM 7-E/BALAN* 1.5 EC FOR PEANUTS

How to use VERNAM 7-E and BALAN 1.5 EC for weed control on peanuts.

A tank mix combination of VERNAM 7-E and BALAN E.C. can be used prior to planting for weed control in peanuts.

NOTE: Follow all use precautions and warnings that appear on the BALAN 1.5 EC label and on the VERNAM 7-E label. Do not contaminate water by cleaning of equipment or disposal of wastes.

WEEDS CONTROLLED

Annual Grasses:

Annual bluegrass	Goosegrass
Barnyardgrass (watergrass)	Junglerice
Crabgrass	Ryegrass
Crowfootgrass	Johnsongrass, seedling
Fall panicum	Texas panicum (buffalograss) (coloradograss)
Field sandbur	
Foxtails	

Annual Broadleaves:

Balloonvine*	Knotweed
Carpetweed	Palmer amaranth (carelessweed)
Chickweed	Pigweed
Lambsquarters, common	Purslane
Florida purslane (pusley)	Redmaids

* In Alabama Only

Perennial Grasses:

Nutsedge (nutgrass)

APPLICATION DIRECTIONS

The following directions give broadcast rates of VERNAM 7-E and BALAN 1.5 EC for application prior to planting. Do not apply this combination after planting. BALAN 1.5 EC is not recommended for use on muck or other soil high in organic matter content.

*BALAN is a registered trademark of Elanco Products Company

Broadcast Rates Per Acre (Conventional spray and center pivot sprinklers):

Apply 2 1/3 pints VERNAM 7-E plus 3 quarts BALAN 1.5-EC on coarse and medium textured (sandy and loam—light and medium) soils.

Apply 3 pints VERNAM 7-E plus 4 quarts BALAN 1.5 EC on fine textured (silt and clay—heavy) soils.

CENTER PIVOT SPRINKLER APPLICATION

(Limit use to the Southeastern U.S. only.)

Apply VERNAM 7-E/BALAN 1.5–EC with irrigation water before planting or immediately after planting before either peanuts or weeds emerge. It is important that application be done immediately after the last tillage operation since poor results will occur if the weeds have germinated before application of the herbicides. Meter the premixed VERNAM 7-E/BALAN 1.5-EC into the irrigation water during the entire period. Apply in 1/2 inch of water.

USE PRECAUTIONS

A. Apply the product only through center pivot irrigation systems containing anti-siphon and check valves which will prevent water source contamination and overflow of the slurry tank and containing interlocking controls between the metering device and the water pump to insure simultaneous shut-off.

B. Inject the product with a positive displacement pump into the main line ahead of a right angle turn to insure adequate mixing.

C. Application of more than recommended quantities of irrigation water may result in decreased product performance by removing the chemical from the zone of effectiveness.

D. Do not apply when wind speed favors drift, when system connections or fittings leak, when nozzles do not provide uniform distribution or when lines containing the product must be dismantled and drained.

VERNAM 7-E/TREFLAN* 4-EC FOR PEANUTS

Selective Herbicide—Tank Mix Directions

For TEXAS and OKLAHOMA ONLY

How to use VERNAM 7-E and TREFLAN 4-EC as a spray tank mix for weed control on peanuts.

A tank mix combination of VERNAM 7-E and TREFLAN 4-EC can be used prior to planting for weed control in peanuts. The combination will control a broader spectrum of weeds than either material used alone. It is especially useful where nutgrass and Texas panicum are a problem.

* TREFLAN is a registered trademark of Elanco Products Company.

WEEDS CONTROLLED

Annual Grasses:

Annual bluegrass	Goosegrass
Barnyardgrass (watergrass)	Junglerice
Brachiaria	Johnsongrass, seedling
Bromegrass	Sandbur
Cheat	Sprangletop
Crabgrass	Stinkgrass
Fall panicum	Texas panicum (coloradograss)
Foxtails (including giant foxtail)	

Annual Broadleaves:

Balloonvine*	Kochia
Carpetweed	Lambsquarters
Carelessweed	Pigweeds (spiny, redroot)
Chickweed	Puncturevine
Florida purslane (pusley)	Purslane
Goosefoot	Russian thistle
Knotweed	Stinging nettle

* In Alabama Only

Perennial Grasses:

Nutsedge (Nutgrass)

APPLICATION DIRECTIONS

The following directions give broadcast rates for VERNAM 7-E and TREFLAN 4-EC as a spray tank mix for application prior to planting. Read and follow all directions, use limitations, precautions and warnings that appear on the TREFLAN 4-EC and VERNAM 7-E labels.

Broadcast rates per Acre:

Apply and incorporate before planting 2⅓ pints of VERNAM 7-E and 1 pint TREFLAN 4-EC per acre. For band applications, reduce dosage proportionately, depending on row spacing and band width to be treated.

Use the following directions for the VERNAM 7-E tank mixes on peanuts.

MIXING

Add the recommended rates of the herbicides to the spray tank during filling and mix thoroughly. Begin adding the herbicides when spray tank is half full. Apply in 10 to 50 gallons of water per acre.

SPRAY EQUIPMENT

Use a properly calibrated, low pressure, boom-type sprayer having good agitation. Check frequently during application to be sure nozzles are free from clogging and are delivering a uniform spray pattern.

SOIL INCORPORATION

The tank mix combinations must be mixed (incorporated) into the soil immediately after application. Whenever possible, application and incorporation should be done in the same operation. This can be done by mounting the tanks and boom right on the incorporation rig.

For thorough incorporation, use the following equipment: TANDEM DISC: Set to cut 3 to 6 inches operated at 4-6 mph, followed by a spike-toothed harrow or some other leveling device, which extends beyond the ends of the disc. For thorough mixing, disc in two different directions (cross-disc).

OR: POWER DRIVEN cultivation equipment (tillers, cultivators, hoes) set to cut 2 to 3 inches deep.

Shallow incorporation of less than 2 inches may result in erratic weed control.

PLANTING

Peanuts should be planted within 48 hours after the combination has been applied to obtain the maximum period of weed control during the growing season.

TANK MIX DIRECTIONS FOR USE ON SOYBEANS

VERNAM 7-E may be combined with the compounds listed in this section for tank mix applications on soybeans. Recommended rates for each tank mix are listed in the sections below. The mixing, spray equipment, soil incorporation, and planting directions for the VERNAM 7-E tank mixes on soybeans are listed at the end of the soybean tank mix section.

VERNAM 7-E/TREFLAN 4-EC TANK MIX DIRECTIONS

How to use VERNAM* 7-E and TREFLAN* 4-EC as a spray tank mix for weed control on soybeans.

A tank mix combination of VERNAM 7-E (7 pounds active ingredient per gallon) and TREFLAN 4-E.C. (4 pounds of active ingredient per gallon) can be used prior to planting for weed control in soybeans.

The combination will control a broader spectrum of weeds than either material used alone. This includes tough weeds such as nutgrass, velvetleaf, foxtails, johnsongrass, redroot pigweed and wild cane.

Dependable performance—because the herbicides are mixed into the soil, rainfall is not needed to move them into the weed germinating zone. They are already there to kill grass and broadleaf weeds when they sprout.

WEEDS CONTROLLED:

Annual Grasses:

Annual bluegrass
Barnyardgrass
(watergrass)
Goosegrass
Junglerice
Johnsongrass, seedling
Brachiaria
Bromegrass
Cheat
Crabgrass
Fall panicum
Field sandbur
Foxtails
Sprangletop
Stinkgrass
Texas panicum
(buffalograss)
(coloradograss)
Wild cane

Annual Broadleaves:

Annual morningglory
Balloonvine**
Carpetweed
Chickweed
Common
lambsquarters
Florida pusley
Goosefoot
Knotweed
Kochia
Palmer amaranth
(carelessweed)
Pigweed
Purslane
Russian thistle
Sicklepod
(coffeeweed)
Stinging nettle
Velvetleaf
(buttonweed)

Perennial Grasses:

Purple nutsedge (nutgrass)
Yellow nutsedge (nutgrass)
Rhizome johnsongrass*

* Suppression
** In Alabama Only

APPLICATION DIRECTIONS

Read and follow all directions, use limitations, precautions and warnings that appear on the VERNAM 7-E and TREFLAN EC labels. Do not contaminate water by cleaning of equipment or disposal of wastes.

Application Rates

	Broadcast Rates/Acre	
Soil Texture	VERNAM VERNAM 7-E*	TREFLAN TREFLAN 4-EC*
Coarse (light—sandy soils)	2⅓ pints	1 pint
Medium (silt loam & sandy clay soils)	2⅓ to 3** pints	1 to 1½ pints
Fine (heavy—clay soils)	3 to 3½ pints	1 to 2 pints

* For rhizome johnsongrass suppression use the high rates of TREFLAN 4-EC and VERNAM 7-E.
** For nutgrass, wild cane and velvetleaf control, use the high rate of 3 pints VERNAM 7-E on medium textured soils.

For Rhizome Johnsongrass—Use mold board plow, chisel plow or similar implement to bring rhizomes to the top of the soil. Then follow with a disc set to cut 4–6 inches deep and operate in two different directions at 4–6 mph before application to cut the rhizomes into small (2–3 inch) pieces. This should also destroy any emerged johnsongrass.

Cultivation is suggested in addition to VERNAM 7-E/TREFLAN 4-EC treatment in fields with moderate to heavy infestations of nutgrass, wild cane, velvetleaf, and johnsongrass rhizomes.

VERNAM 7-E/TREFLAN 4-E/METRIBUZIN TANK MIX DIRECTIONS

How to use VERNAM 7-E and TREFLAN 4-EC and METRIBUZIN (4-L or 50-WP) as a spray tank mix.

A tank mix of VERNAM 7-E (7 pounds active ingredient per gallon) and TREFLAN 4-EC (4 pounds active ingredient per gallon) and METRIBUZIN 4-L (4 pounds active ingredient per gallon) or METRIBUZIN 50-WP (50% active ingredient by weight) can be used prior to planting for weed control in soybeans.

The combination will control a broader spectrum of weeds than the VERNAM 7-E/TREFLAN 4-EC tank mix.

WEEDS CONTROLLED

Annual Grasses:

Annual bluegrass
Barnyardgrass
(watergrass)
Goosegrass
Junglerice
Johnsongrass, seedling

Brachiaria	Sprangletop
Bromegrass	Stinkgrass
Cheat	Texas panicum (buffalograss) (coloradograss)
Crabgrass	Wildcane
Fall panicum	
Field sandbur	
Foxtails	

Annual Broadleaves:

Balloonvine**	Pennsylvania smartweed
Carpetweed	Pigweed
Chickweed	Prickly sida
Cocklebur	Purslane
Common lambsquarters	Russian thistle
Common ragweed	Sicklepod (coffeeweed)
Florida pusley	Stinging nettle
Goosefoot	Velvetleaf (buttonweed)
Knotweed	Venice mallow
Kochia	
Palmer amaranth (carelessweed)	

** In Alabama Only

Perennial Grasses:

Purple nutsedge (nutgrass)	Rhizome johnsongrass*
Yellow nutsedge (nutgrass)	

* Suppression

APPLICATION DIRECTIONS

Read and follow all directions, use limitations, precautions and warnings that appear on the VERNAM 7-E, TREFLAN 4-EC and appropriate METRIBUZIN (4-L or 50-WP) labels. Do not contaminate water by cleaning of equipment or disposal of wastes.

Application Rates

	Broadcast Rates/Acre		
Soil Texture	VERNAM 7-E*	TREFLAN 4-EC	METRIBUZIN 4L or 50WP
Medium (silt loam & sandy clay soils)	2⅓ to 3** pints	1 pint	½ pint or ½ pound
Fine (heavy-clay soils)	3 to 3½ pints	1 pint	½–¼ pint or ½–¾ pound

* For Rhizome johnsongrass suppression use the high rate of VERNAM 7-E.

** For nutsedge, wild cane, and velvetleaf control, use the high rate of 3 pints of VERNAM on medium-textured soils.

Do not use the VERNAM/TREFLAN/METRIBUZIN tank mix on coarse (light sandy) soils.

For Rhizome johnsongrass—use mold board plow, chisel plow or similar implement to bring rhizomes to the top of the soil. Then follow with a disc set to cut 4–6 inches deep and operate to two different directions at 4–6 mph before application to cut the rhizomes into small (2–3 inch) pieces. This should also destroy any emerged johnsongrass.

Cultivation is suggested in addition to VERNAM 7-E/TREFLAN 4-EC/METRIBUZIN (4-L or 50-WP) treatment in fields with moderate to heavy infestations of nutgrass, wild cane, velvetleaf, and johnsongrass rhizomes.

VERNAM 7-E/AMIBEN* TANK MIX DIRECTIONS

* AMIBEN chloramben herbicide is a registered trademark of Union Carbide Agricultural Products Co., Inc.

How to use VERNAM 7-E and AMIBEN as a spray tank mix for weed control on soybeans.

A tank mix combination of VERNAM 7-E (7 pounds active ingredient per gallon) and AMIBEN (2 pounds active ingredient per gallon) can be used prior to planting, for weed control in soybeans.

The combination will control a broader spectrum of weeds than either material used alone.

WEEDS CONTROLLED

Annual Grasses:

Barnyardgrass (watergrass)	German millet
Crabgrass	Goosegrass
Foxtail (giant, green, yellow)	Johnsongrass, seedling
	Stinkgrass

Annual Broadleaves:

Carpetweed	Nightshade, black
Chickweed	Pigweed (redroot, smooth)
Coffeeweed	Prickly sida (teaweed)
Florida pusley	Purslane
Kochia	Ragweed, common
Lambsquarters, common	Sicklepod
Morningglory, annual	Smartweed, Pennsylvania
Mustard, wild	Spurge, annual
	Velvetleaf (buttonweed)

Perennial Grasses:

Purple nutsedge (nutgrass)	Yellow nutsedge (nutgrass)

APPLICATION DIRECTIONS

Read and follow all directions, use limitations, precautions and warnings that appear on the VERNAM 7-E and AMIBEN labels. Do not contaminate water by cleaning of equipment or disposal of wastes.

Application Rates

	Broadcast Rates/Acre	
Soil Texture	VERNAM 7-E	AMIBEN
Medium (silt loam & sandy clay soils)	2⅓ pints	3 quarts

For mixing, spray equipment, soil incorporation, and planting directions, refer to the directions at the end of the VERNAM 7-E tank mix on soybeans section of this label.

VERNAM 7-E/BASALIN* 4-E TANK MIX DIRECTIONS

*BASALIN is a registered trademark of BASF Wyandotte Corp.

How to use VERNAM 7-E and BASALIN 4-E as a spray tank mix on soybeans.

A tank mix combination of VERNAM 7-E (7 pounds active ingredient per gallon) and BASALIN (4 pounds active ingredient per gallon) can be used prior to planting for weed control in soybeans.

The combination will provide better weed control than either herbicide used alone.

WEEDS CONTROLLED

Annual Grasses:

Barnyardgrass (watergrass)	German millet
Crabgrass	Goosegrass
Foxtail (giant, green, yellow)	Johnsongrass, seedling

Annual Broadleaves:

Carpetweed	Pigweed (redroot, smooth)
Florida pusley	Purslane
Lambsquarters, common	Sicklepod
Morningglory, annual	Velvetleaf (buttonweed)

Perennial Grasses:

Purple nutsedge (nutgrass)

Yellow nutsedge (nutgrass)

APPLICATION DIRECTIONS

Read and follow all directions, use limitations, precautions and warnings that appear on the VERNAM 7-E and BASALIN 4-E labels. Do not graze treated plant parts to livestock. Do not contaminate water by cleaning of equipment or disposal of wastes.

Application Rates

Soil Texture	Broadcast Rates/Acre VERNAM 7-E	BASALIN 4-E
Coarse (light-sandy soils)	$1\frac{3}{4}$ pints	$\frac{3}{4}$ pint
Medium (silt loam and sandy clay soils)	$2\frac{1}{3}$ pints	1 pint

For mixing, spray equipment, soil incorporation, and planting directions, refer to the directions at the end of the VERNAM 7-E tank mix on soybeans section of this label.

VERNAM 7-E/LASSO* 4-EC TANK MIX DIRECTIONS

*LASSO is a registered trademark of Monsanto Company

How to use VERNAM 7-E and LASSO 4-EC as a spray tank mix for weed control on soybeans.

A tank mix combination of VERNAM 7-E (7 pounds active ingredient per gallon) and LASSO 4-EC (4 pounds active ingredient per gallon) can be used prior to planting for weed control in soybeans grown in North Central States of Michigan, Indiana, and Illinois. Do not use on sand or loamy sandy soils as injury may occur.

The combination will provide better weed control than either herbicide used alone.

WEEDS CONTROLLED

Annual Grasses:

- Barnyardgrass (watergrass)
- Crabgrass
- Fall panicum
- Foxtail (giant, green, yellow)
- Goosegrass
- Johnsongrass, seedling

Annual Broadleaves:

- Carpetwood
- Florida pusley
- Lambsquarters, common
- Morningglory, annual
- Pennsylvania smartweed
- Pigweed (redroot, smooth)
- Purslane
- Sicklepod
- Velvetleaf (buttonweed)

Perennial Grasses:

- Purple nutsedge (nutgrass)
- Yellow nutsedge (nutgrass)

APPLICATION DIRECTIONS

Read and follow all directions, use limitations, precautions and warnings that appear on the VERNAM 7-E and LASSO 4-EC labels. Do not contaminate water by cleaning of equipment or disposal of wastes.

Applications Rates

Soil Texture	Broadcast Rates/Acre VERNAM 7-E	LASSO 4-EC
Coarse (light-sandy soils)	$1\frac{3}{4}$ pints	2 pints
Medium (slit loan and sandy clay soils)	$2\frac{1}{3}$ pints**	3 pints
Fine (heavy-clay soils)	3 pints	4 pints

** For nutgrass and velvetleaf control, use the rate of 3 pints of VERNAM 7-E on medium and fine textured (medium and heavy-clay) soils.

For mixing, spray equipment, soil incorporation, and planting directions, refer to the directions at the end of the VERNAM 7-E tank mix on soybean section of this label.

VERNAM 7-E/PROWL* 4-E TANK MIX DIRECTIONS

*PROWL is a registered trademark of American Cyanamid Company.

How to use VERNAM 7-E and PROWL 4-E as a spray tank mix for weed control on soybeans.

A tank mix combination of VERNAM 7-E (7 pounds active ingredient per gallon) and PROWL 4-E (4 pounds active ingredient per gallon) can be used prior to planting for weed control in soybeans.

The combination will provide better weed control than either herbicide used alone.

WEEDS CONTROLLED

Annual Grasses:

- Barnyardgrass (watergrass)
- Crabgrass
- Foxtail millet
- Foxtail (giant, green, yellow)
- Goosegrass
- Johnsongrass, seedling

Annual Broadleaves:

- Carpetweed
- Florida pusley
- Lambsquarters, common
- Morningglory, annual
- Pigweed (redroot, smooth)
- Purslane
- Sicklepod
- Velvetleaf (buttonweed)

Perennial Grasses:

- Purple nutsedge (nutgrass)
- Yellow nutsedge (nutgrass)

APPLICATION DIRECTIONS

Read and follow all directions, use limitations, precautions and warnings that appear on the VERNAM 7-E and PROWL 4-E labels. Do not contaminate water by cleaning of equipment or disposal of wastes.

Application Rates

Soil Texture	Broadcast Rates/Acre VERNAM 7-E	PROWL 4-E
Coarse (light-sandy soils)	$1\frac{3}{4}$ pint	$\frac{3}{4}$ pint
Medium (silt loam and sandy clay soils)	$2\frac{1}{3}$ pints**	1 pint
Fine (heavy—clay soils)	3 pints	$1\frac{1}{2}$ pints

** For nutgrass and velvetleaf control, use the rate of 3 pints of VERNAM 7-E on medium textured (medium and heavy—clay) soils.

For mixing, spray equipment, soil incorporation, and planting directions, refer to the directions at the end of the VERNAM 7-E tank mix on soybean section of this label.

VERNAM 7-E/TOLBAN* 4-E TANK MIX DIRECTIONS

*TOLBAN is a registered trademark of Ciba-Geigy Corp.

How to use VERNAM 7-E and TOLBAN 4-E as a spray tank mix for weed control on soybeans.

A tank mix combination of VERNAM 7-E (7 pounds active ingredient per gallon) and TOLBAN 4-E (4 pounds active ingredient per gallon) can be used prior to plantng for weed control in soybeans.

The combination will provide better weed control than either herbicide used alone.

WEEDS CONTROLLED

Annual Grasses:

- Barnyardgrass (watergrass)
- Crabgrass
- Foxtail (giant, green, yellow)
- German millet
- Goosegrass
- Johnsongrass, seedling

Annual Broadleaves:

- Carpetweed
- Florida pusley
- Pigweed (redroot, smooth)

Lambsquarters, common
Morningglory, annual
Purslane
Sicklepod
Velvetleaf (buttonweed)

Perennial Grasses:
Purple nutsedge (nutgrass)
Yellow nutsedge (nutgrass)

APPLICATION DIRECTIONS

Read and follow all directions, use limitations, precautions and warnings that appear on the VERNAM 7-E and TOLBAN 4-E labels. Do not contaminate water by cleaning of equipment or disposal of wastes.

Application Rates

Soil Texture	Broadcast Rates/Acre VERNAM 7-E	TOLBAN 4-E
Coarse (light-sandy soils)	1¾ pints	¾ pints
Medium (silt loam and sandy clay soils)	2⅓ pints**	1 pint
Fine (heavy—clay soils)	3 pints	1½ pints

** For nutgrass and velvetleaf control, use the rate of 3 pints of VERNAM 7-E on medium textured soils.

For mixing, spray equipment, soil incorporation, and planting directions, refer to the directions at the end of the VERNAM 7-E tank mix on soybean section of this label.

USE THE FOLLOWING DIRECTIONS FOR THE VERNAM 7-E TANK MIXES ON SOYBEANS

Mixing:
Add the recommended rates of both VERNAM 7-E and TREFLAN 4-EC to the spray tank during filling and mix thoroughly. Begin adding the herbicides when spray tank is half full. Apply in 10–50 gallons of water per acre.

Spray Equipment:
Use a properly calibrated, low pressure, boom-type sprayer having good agitation. Check frequently during application to be sure nozzles are free from clogging and delivering a uniform spray pattern.

DRY FERTILIZER STATEMENT:
VERNAM 7-E/TREFLAN 4-EC tank mix may be impregnated on many dry bulk fertilizers for use on soybeans. However, uniform distribution of VERNAM 7-E/TREFLAN 4-EC on fertilizer particles and uniform application are necessary to assure good results. See Appendix II giving directions for impregnation and use.

Soil Incorporation:
The tank mix combinations with VERNAM 7-E must be mixed (incorporated) into the soil immediately after application. Whenever possible, application and incorporation should be done in the same operation. This can be done by mounting the tanks and boom right on the incorporation rig.

For thorough incorporation, use the following equipment:

Tandem Disc—Set to cut to a depth of 4 inches, followed by a spike-tooth harrow or some other leveling device which extends beyond the ends of the disc. For thorough mixing, disc in two different directions (cross disc).
OR:
Power-driven cultivation equipment set to cut to a depth of 2–3 inches.

Planting:
Do not plant seed deeper than 2 inches. Seeding can be done immediately, or it may be delayed up to 10 days. If it is necessary to work the soil following application and incorporation, but before planting, do not work the soil deeper than the depth the herbicide was incorporated i.e., no more than ½ the depth the herbicides were incorporated. To avoid removing the herbicide from the row (with loss of weed control in the row) do not move or shape soil after treatment, such as by planting in deep furrows.

APPENDIX I

PROCEDURE FOR TESTING THE COMPATIBILITY OF VERNAM 7-E AND COMBINATIONS WITH FLUID FERTILIZERS

The following procedure is suggested for determining whether VERNAM 7-E may be combined with a specific fluid fertilizer for spray tank application.

MATERIALS REQUIRED

1. VERNAM 7-E and/or combination.
2. Fluid fertilizer to be used.
3. Adjuvant for fertilizer tank mix: Compex*, Sponto 168-D*, Unite* or equivalent. The adjuvant which provides the best compatibility depends on the specific fertilizer under consideraton.
4. Two one-quart, wide-mouth glass jars with lid.
5. Measuring spoons; a 25 mL pipette or graduated cylinder provides more accurate measurement; 1 tablespoon equals 3 teaspoons, 1 teaspoon equals approximately 5 milliliters.
6. Measuring cup, 8 oz. (237 mL).

* Compex is a trademark of Kalo Laboratories, Inc., Kansas City, Missouri; Sponto 168-D is a trademark of Witco Chemical Company, Houston, Texas; Unite is a trademark of Hopkins Agricultural Chemical Company, Madison, Wisconsin.

PROCEDURE

1. Pour a pint (about 473 mL) of the fluid fertilizer into each of the quart jars.
2. Add adjuvant to one of the jars and mix (see rate table).
3. Add the VERNAM 7-E to the desired combinations of herbicide to both jars (see rate table for VERNAM 7-E).
4. Close both jars with lid and mix the contents by turning the jars upside down ten times.
5. Inspect the surface and body of the mixtures—
(A) Immediately after completing the jar inversions.
(B) After allowing the jars to stand quietly for 30 minutes.
(C) And then again after turning the jars upside down 10 times.

If a uniform mix cannot be made, the mixture should not be used. If either mixture remains uniform for 30 minutes, the combination may be used. Should either mixture separate after 30 minutes, but readily remix uniformly with the jar inversions, the mixture can be used if adequate agitation is maintained in the tank. If the mixture with adjuvant is satisfactory, but the one without adjuvant is not, be sure to use the adjuvant in the spray tank. Add the adjuvant first at a rate of 3 pints per 100 gallons of fluid fertilizer; foaming can be minimized by using moderate agitation.

If nondispersible oil, sludge or clumps of solids form in the mixtures, the combinations should not be used.

RATE TABLE FOR VERNAM 7-E AND ADJUVANT** WITH THE FLUID FERTILIZER

Gallons of Fluid Fertilizer To Be Applied Per Acre	mL or Tsp. of VERNAM 7-E* To Be Added To 1 Pint of Fertilizer VERNAM 7-E mL	Tsp.
10	7	1⅓
15	5	1
20	4	¾
25	3	⅔
30	2	½
40	2	½

* Based on field rate of 1 pound active ingredient per acre in the fertilizer volumes indicated. Increase volume proportionately to correspond with intended field rate in terms of pounds active ingredient per acre (e.g. for field rate of 4 pounds actual VERNAM 7-E in 40 gallons fertilizer per acre, add 8 ml. or 2 tsp. VERNAM 7-E to each jar for compatibility testing purposes).

** Two (2) milliliters or one-half (½) teaspoon of adjuvant should be added to 1 pint of fluid fertilizer in order to equal the rate of 3 pints of adjuvant per 100 gallons of fluid fertilizer.

APPENDIX II

VERNAM 7-E and VERNAM/TREFLAN 4-EC IMPREGNATION ON DRY BULK FERTILIZERS

CAUTION: VERNAM 7-E alone or in combination with other herbicides must not be impregnated on ammonium nitrate, sodium nitrate or potassium nitrate. Such mixtures may cause explosion and fire.

VERNAM 7-E alone may be impregnated on many* dry bulk fertilizers for use *on registered crops.* VERNAM 7-E in combination with TREFLAN may be impregnated on many* dry bulk fertilizers and applied and incorporated in the soil before planting for the control of *labeled grasses* and broadleaf weeds *in soybeans.*

Field results have shown that VERNAM 7-E and VERNAM/TREFLAN 4-EC on bulk dry fertilizers give weed control equal to VERNAM 7-E or VERNAM 7-E/TREFLAN 4-EC applied as a spray in water or liquid fertilizer. However, uniform impregnation of herbicide solution on the dry fertilizer particles and uniform application in the field are necessary to assure good results.

For impregnating pesticides on dry fertilizers, use a closed rotary-drum mixer or a similar type of closed blender equipped with suitable spray equipment. The spray nozzle (or nozzles) should be positioned inside of the mixer to provide *uniform* spray coverage of the tumbling fertilizer. For the VERNAM 7-E/TREFLAN 4-EC combination, premix in the proper use ratio to prepare a uniform solution for joint spray impregnation. The physical properties of fertilizers vary, particularly in liquid absorptive capacity. When absorptivity is sufficient, simple spray impregnation of the fertilizer with VERNAM 7-E or VERNAM 7-E/TREFLAN 4-EC provides a satisfactory, dry mixture. If the absorptive capacity is inadequate, use of a highly absorptive powder is required to provide a dry, flowable mixture. Microcel E (Johns-Manville Products Corp.) is the recommended absorbent powder. It should be added separately and uniformly to the prepared VERNAM 7-E or VERNAM 7-E/TREFLAN 4-EC fertilizer mixture, in a quantity that is sufficient to provide a suitably flowable mixture. Generally less than 2% by weight of Microcel E is required. The amount of product actually required in the manufacture of individual fertilizer mixture should be determined carefully for each production operation. This is necessary to ensure that the amount of VERNAM 7-E/TREFLAN 4-EC actually contained in the mixture applied to the soil represents the correct rate of use.

All VERNAM 7-E and TREFLAN 4-EC label recommendations regarding rates per acre, soil incorporation, application, cautions, general use precautions and other directions must be followed.

All individual state regulations relating to bulk dry fertilizer blending, registration, labeling and application are the responsibility of the individual and/or company selling the fertilizer and VERNAM 7-E and/or VERNAM/TREFLAN 4-EC mixture.

A minimum of 200 pounds and a maximum of 450 pounds of approved ingredients* impregnated with VERNAM 7-E/TREFLAN 4-EC at the recommended rate must be applied per acre.

Bulk fertilizer impregnated with VERNAM 7-E should be applied immediately, *not stored.* It is recommended that all bulk containers be tightly covered while the product is being transported and applied to reduce chances of VERNAM 7-E loss via volatilization.

***APPROVED DRY FERTILIZER INGREDIENTS**

	n	P	K
Ammonium Sulfate	21	0	0
Diammonium Phosphate	18	46	0
Potassium Chloride	0	0	60
(Single) Superphosphate	0	20	0
Triple Superphosphate	0	46	0
Urea	45	0	0
Ammonium Phosphate-Sulfate	16	20	0
Potassium Sulfate	0	0	52

NOTE: Sul-PO-MAG, K-MAG, and 11-48-0 have been shown to be compatible with VERNAM 7-E plus TREFLAN 4-EC and are approved for use.

RATE CHART FOR THE IMPREGNATION OF DRY BULK FERTILIZERS WITH VERNAM 7-E and VERNAM/TREFLAN 4-EC ADDED TO A TON OF FERTILIZER*

Fertilizer Rate Per Acre	VERNAM 7-E Rate/Acre $1\frac{3}{4}$ pts.	$2\frac{1}{3}$ pts.	3 pts.	$3\frac{1}{2}$ pts.	TREFLAN 4-EC Rate/Acre 1 pt.	2 pts.
200 lbs.	$8\frac{3}{4}$ qts./ton	$11\frac{2}{3}$ qts./ton	15 qts./ton	$17\frac{1}{2}$ qts./ton	5 qts./ton	10 qts./ton
250 lbs.	7 qts./ton	$9\frac{1}{3}$ qts./ton	12 qts./ton	14 qts./ton	4 qts./ton	8 qts./ton
300 lbs.	$5\frac{4}{5}$ qts./ton	$7\frac{3}{4}$ qts./ton	10 qts./ton	$11\frac{2}{3}$ qts./ton	$3\frac{1}{3}$ qts./ton	7 qts./ton
350 lbs.	5 qts./ton	$6\frac{2}{3}$ qts./ton	$8\frac{3}{5}$ qts./ton	10 qts./ton	$2\frac{4}{5}$ qts./ton	6 qts./ton
400 lbs.	$4\frac{1}{3}$ qts./ton	$5\frac{4}{5}$ qts./ton	$7\frac{1}{2}$ qts./ton	$8\frac{3}{4}$ qts./ton	$2\frac{1}{2}$ qts./ton	5 qts./ton
450 lbs.	$3\frac{4}{5}$ qts./ton	$5\frac{1}{5}$ qts./ton	$6\frac{2}{3}$ qts./ton	$7\frac{4}{5}$ qts./ton	$2\frac{1}{4}$ qts./ton	$4\frac{1}{2}$ qts./ton

* Refer to application rates under "APPLICATION DIRECTIONS".

VERNAM 7-E Typical Physical Data

Pounds/Gallon, 20/20°C	7.88 Approximate (varies with temperature
Flashpoint:	186°F (tagliabue closed cup)
Viscosity:	Sprayable down to minus 20°F

STORAGE AND DISPOSAL

PROHIBITIONS
Do not contaminate water, food, feed or fertilizer by storage or disposal.

STORAGE
Keep container tightly closed when not in use. Can be stored at temperatures as low as minus 50° F.

PESTICIDE DISPOSAL
Rinse spray equipment. Wastes resulting from the use of this product may be disposed of on site or at an approved waste disposal facility.

CONTAINER DISPOSAL
METAL—Triple rinse (or equivalent), then offer for recycling or reconditioning, or dispose of in a sanitary landfill, or by other approved State and local procedures.
PLASTIC—Triple rinse (or equivalent), then offer for recycling or reconditioning or puncture and dispose of in a sanitary landfill, or by incineration, or, if allowed by State and local authorities, by burning. If burned, stay out of smoke.

NOTICE—READ CAREFULLY

CONDITIONS OF SALE
Stauffer (and seller) offer(s) this product for sale subject to, and buyer and all users are deemed to have accepted, the following conditions of sale and warranty which may only be varied by written agreement of a duly authorized representative of Stauffer.

WARRANTY LIMITATION
Stauffer warrants that this product conforms to the chemical description on the label and is reasonably fit for the purposes referred to in the directions for use on the label subject to the inherent risks referred to below. Stauffer makes no other express warranties; THERE IS NO IMPLIED WARRANTY OF MERCHANTABILITY and there are no warranties which extend beyond the description on the label hereof.

INHERENT RISKS
The directions for use of this product are believed to be reliable and should be followed carefully. However, it is impossible to eliminate all risks associated with use. Buyer assumes all risks associated with use or application of this product contrary to label instructions or resulting from extraordinary weather conditions.

LIMITATION OF LIABILITY

In no case shall Stauffer be liable for special, indirect or consequential damages resulting from the use or handling of this product and no claim of any kind shall be greater in amount than the purchase price of the product in respect of which such damages are claimed. This product is sold only for uses stated on its label. No express or implied license is granted to use or sell this product under any patent in any Country except as specified: Country: United States of America.

This is a specimen label. Stauffer is not responsible for the accuracy of the information contained herein. As labels are subject to revision, always carefully read and follow the label on the product container.

VERNAM® 10-G

SELECTIVE HERBICIDE GRANULES

IMPORTANT

RECALIBRATE YOUR EQUIPMENT FOR THIS PRODUCT BEFORE USING

ACTIVE INGREDIENT:

S-propyl dipropylthiocarbamate	10.0%
INERT INGREDIENTS:	90.0%
TOTAL	100.0%

Contains 10 Pounds Active Ingredient Per 100 Pounds

EPA Reg. No. 476-1872 A-2R

Keep Out of Reach of Children

CAUTION

HARMFUL IF SWALLOWED

STATEMENT OF PRACTICAL TREATMENT

If a known exposure occurs or is suspected, immediately start the procedures given below. If further treatment is required, contact a Poison Center, a physician, or the nearest hospital, or call collect day or night: (203) 226-6602 Stauffer Chemical Company, or (800) 424-9300 Pesticide Team Safety Network.

If ingested: Immediately give large amounts of water and induce vomiting by gagging the victim with a finger placed on the back of the victim's tongue. Give fluids until vomitus is clear. Do not induce vomiting or give anything by mouth to an unconscious or convulsing person. Call a physician immediately.

For eye contact: Immediately flush eyes with large amounts of running water for at least 15 minutes. Hold eyelids apart during flushing to ensure rinsing of the entire surface of the eye and lids with water. Call a physician.

For skin contact: Wash all affected areas with plenty of soap and water. Seek medical attention if irritation occurs.

If inhaled: Remove to fresh air. If not breathing, give artificial respiration, preferably mouth-to-mouth. If breathing is difficult, give oxygen, preferably with a physician's advice.

PRECAUTIONARY STATEMENTS

HAZARDOUS TO HUMANS AND DOMESTIC ANIMALS

CAUTION

HARMFUL IF SWALLOWED. Avoid contact with skin, eyes and clothing. Avoid breathing dust. Wash with soap and water immediately after use or contact. Do not contaminate food and feed.

ENVIRONMENTAL HAZARDS

Keep out of lakes, ponds or streams. Do not contaminate water by cleaning of equipment or disposal of wastes.

READ ENTIRE LABEL BEFORE USING.

WEEDS CONTROLLED

VERNAM 10-G will not control established weeds.

Annual Grasses:

Barnyardgrass (Watergrass)	*Echinochloa crus-galli*
Crabgrass	*Digitaria* spp.
Giant Foxtail	*Setaria faberi*
Green Foxtail	*Setaria viridis*
Yellow Foxtail	*Setaria glauca*
Foxtail Millet	*Setaria italica*
Goosegrass	*Eleusine indica*
Wild Cane*	*Sorghum bicolor*
Johnsongrass Seedlings	*Sorghum halepense*

* Cultivation is suggested in addition to the VERNAM 10-G treatment in fields with moderate or heavy infestations of wild cane.

Annual Broadleaf Weeds:

Carpetweed	*Mollugo verticillata*
Sicklepod (Coffeeweed)	*Cassia tora*
Florida Pusley	*Richardia scabra*
Common Lambsquarters	*Chenopodium album*
Annual Morningglory	*Ipomoea purpurea*
Smooth Pigweed	*Amaranthus hybridus*
Redroot Pigweed	*Amaranthus retroflexus*
Purslane	*Portulaca oleracea*
Velvetleaf (Buttonweed)	*Abutilon theophrasti*

The above listed broadleaf weeds will be controlled only if treatment is made when conditions are favorable for weed seed germination and growth.

Perennial Weeds:

Yellow Nutgrass (Nutsedge)	*Cyperus esculentus*
Purple Nutgrass (Nutsedge)	*Cyperus rotundus*

Existing stands of nutgrass must be turned under and chopped up thoroughly before treatment. For nutgrass control VERNAM 10-G must be thoroughly mixed into the soil prior to planting. Cultivation is suggested in addition to VERNAM 10-G treatment in fields with moderate or heavy nutgrass infestations.

GENERAL USE PRECAUTIONS

VERNAM 10-G should be used only for recommended purposes and at recommended rates (do not overdose).

Applied according to directions and under normal growing conditions, VERNAM 10-G will not harm the treated crop. Malformation of the primary leaves is not unusual. This condition is only temporary and will not affect yields. During germination and early stages of growth, extended periods of unusually cold and wet or hot and dry weather, insect or plant disease attack, the use of certain soil-applied systemic insecticides, improperly placed fertilizers or soil insecticides may create abnormal conditions that weaken crop seedlings. VERNAM 10-G used under these abnormal conditions could result in crop injury.

VERNAM 10-G is recommended for use on mineral soils only.

Do not mix VERNAM 10-G with insecticides or fungicides.

Do not contaminate irrigation water or water used for domestic purposes.

DIRECTIONS FOR USE

It is a violation of Federal law to use this product in a manner inconsistent with its labeling.

VERNAM 10-G is a selective herbicide which is mixed (incorporated) into the soil, except where noted under **RECOMMENDATIONS,** for control of weeds listed on this label. VERNAM 10-G controls annual weeds as their seeds germinate by interfering with normal germination and seedling development. It does not control established weeds. All weed growth and crop stubble should be thoroughly worked into the soil before treatment.

APPLICATION DIRECTIONS

The soil should be well worked and dry enough to permit good soil mixing (incorporation). Use granular applicators designed to handle herbicide granules. Callibrate the applicator according to the manufacturer's directions prior to use and check frequently during application to be sure equipment is working properly and distributing the granules uniformly. Avoid overlaps that will increase dosage above recommended limits. When using certain broadcast applicators, a splash pan is needed to spread the granules and avoid concentration of the material in narrow bands. Broadcast applications before planting may be made by airplane.

INCORPORATION DIRECTIONS

With soil incorporation applications, VERNAM 10-G must be mixed (incorporated) into the soil immediately (within minutes) after application to prevent loss of herbicide. Whenever possible, application and incorporation should be done in the same operation.

SOIL MIXING (INCORPORATION) BEFORE PLANTING: Use power-driven cultivation equipment set to cut to a depth of 2 to 3 inches OR a tandem disc set to cut to a depth of 4 to 6 inches operated at 4 to 6 mph, followed by a spike-tooth harrow or some other leveling device which extends beyond the ends of the disc. For thorough mixing, disc in two different directions (cross disc). Prior to the second discing, raise the disc to prevent cutting deeper than 4 to 6 inches.

SOIL MIXING (INCORPORATION) AFTER PLANTING: Use power-driven cultivation equipment set to cut to a depth of 2 inches OR on light soils only, ground-driven tillers (rolling cultivators, rotary hoe, row wheels, etc.) set to a depth of $1\frac{1}{2}$ inches and operated at 6 to 8 mph. When incorporating after planting, care must be taken not to disturb the crop seed or seedlings. Adjust tines or blades of incorporation equipment so they are not directly over the drill row.

PLANTING—SOYBEANS, PEANUTS

Do not plant seed deeper than 2 inches. For VERNAM 10-G applications made before planting, seeding can be done immediately or it may be delayed up to 10 days. If it is necessary to work the soil following application and incorporation, but before planting, do not work the soil deeper than the depth the herbicide was incorporated. To avoid removing VERNAM 10-G from the row (with loss of weed control in the row) do not move or shape soil after treatment, such as planting in deep furrows.

RECOMMENDATIONS

These recommendations are given in terms of broadcast (over-all) rates of VERNAM 10-G per acre. For band or row applications, reduce broadcast rates proportionately depending on band width to be treated and row spacing.

PEANUTS

VERNAM 10-G may be applied before planting, or immediately after planting. When incorporating after planting, care must be taken not to disturb the seed. **Broadcast Rates Per Acre**—20 pounds on sandy (light) soils; 25 pounds on sandy loam to clay loam (medium to heavy) soils.

SOYBEANS

For Soybeans Grown in Midwestern, North Central and Northeastern States: VERNAM 10-G may be applied before planting by broadcasting 20 pounds on sandy (light) soils, or 25 pounds on sandy loam to clay loam (medium to heavy) soils.

VERNAM 10-G may also be applied at planting or immediately after planting by broadcasting 20 pounds on sandy (light) soils, 25 pounds on sandy loam to clay loam (medium to heavy) soils, and 30 pounds on clay (heavy) soils. When incorporating after planting, care must be taken not to disturb the crop seed or seedlings.

For Soybeans Grown in the Southeastern States Including North Carolina, South Carolina, Georgia, Alabama, Florida and Eastern Tennessee (Tenn. River East): VERNAM 10-G may be applied before planting or immediately after planting. For applications made before planting, do not plant in deep furrows or use shanks or other devices ahead of planter which bring up untreated soil to the surface. For furrow planted soybeans, apply VERNAM 10-G immediately after planting in bands over the row and incorporate with row wheels only. When incorporating after planting, care must be taken not to disturb the crop seed or seedlings.

Broadcast Rates per Acre—20 pounds on sandy (light) soils and 25 pounds on sandy loam to clay loam (medium to heavy) soils.

In the Midwest and North Central States Only, VERNAM 10-G may also be applied to the soil surface immediately after planting without mixing it into the soil by broadcasting 30 pounds on sandy (light) to clay (heavy) soils. This application method will not control nutgrass, Johnsongrass from seed, wild cane or velvetleaf. Also, rainfall of at least one-half inch in one rain is necessary to carry the chemical down to the germinating weed seeds. One or more light rains are not sufficient. Poor weed control will result if sufficient rain does not occur within 7 days after application. If sufficient rainfall does not occur, OR if weeds have emerged, rotary hoeing or standard cultivation should be done before weeds become well established.

For Soybeans Grown in the Southwestern States Including Arkansas, Louisiana, New Mexico, Texas, Oklahoma, Mississippi and Western Tennessee (Tenn. River West): VERNAM 10-G may be applied before planting or immediately after planting. When incorporating at or after planting, care must be taken not to disturb the crop seed or seedlings. **Broadcast Rates Per Acre**—20 pounds on sandy (light) soils, 25 pounds on sandy loam to clay loam (medium to heavy) soils and 30 pounds on clay (very heavy) soils.

NOTE: If beds are made prior to planting soybeans, VERNAM 10-G may be applied in a band immediately ahead of the bedding discs set to cover the VERNAM 10-G with 3 to 4 inches of soil. If bed levelers are used before planting soybeans, make the beds higher to compensate for leveling, so 2 to 3 inches of soil remains over the treatment after the planting operation is done.

CULTURAL PRACTICES FOLLOWING APPLICATION

VERNAM 10-G is not persistent in the soil, and susceptible weeds germinating later during the growing season may not be controlled. Shallow cultivation may be necessary to control these late-germinating weeds. Do not cultivate deeper than the depth of the herbicide incorporation. Recommended pre-emergence or post-emergence herbicides may be applied over the top of preplant incorporated applications of VERNAM 10-G when necessary to control weeds resistant to VERNAM 10-G.

STORAGE AND DISPOSAL

PROHIBITIONS

Do not contaminate water, food, or feed by storage or disposal. Open dumping is prohibited.

PESTICIDE DISPOSAL

Waste resulting from the use of this product may be disposed of on site or at an approved waste disposal facility.

CONTAINER DISPOSAL

Completely empty bag into application equipment, then dispose of empty bags in a sanitary landfill or by incineration, or, if allowed by State and local authorities, by burning. If burned, stay out of smoke.

NOTICE—READ CAREFULLY

CONDITIONS OF SALE

Stauffer (and seller) offer(s) this product for sale subject to, and buyer and all users are deemed to have accepted, the following conditions of sale and warranty which may only be varied by written agreement of a duly authorized representative of Stauffer.

WARRANTY LIMITATION

Stauffer warrants that this product conforms to the chemical description on the label and is reasonably fit for the purposes referred to in the directions for use on the label subject to the inherent risks referred to below. Stauffer makes no other express warranties; THERE IS NO IMPLIED WARRANTY OF MERCHANTABILITY and there are no warranties which extend beyond the description on the label hereof.

INHERENT RISKS

The directions for use of this product are believed to be reliable and should be followed carefully. However, it is impossible to eliminate all risks associated with use. Buyer assumes all risks associated with use or application of this product contrary to label instructions or resulting from extraordinary weather conditions.

LIMITATION OF LIABILITY

In no case shall Stauffer be liable for special, indirect, or consequential damages resulting from the use or handling of this product and no claim of any kind shall be greater in amount than the purchase price of the product in respect of which such damages are claimed.

This product is sold only for uses stated on its label. No express or implied license is granted to use or sell this product under any patent in any country except as specified: Country: United States of America.

This is a specimen label. Stauffer is not responsible for the accuracy of the information contained herein. As labels are subject to revision, always carefully read and follow the label on the product container.

Union Carbide Agricultural Products Company, Inc.

P.O. Box 12014
T.W. Alexander Drive
Research Triangle Park, NC 27709
(919) 549-2000

AMIBEN® GRANULAR

CHLORAMBEN HERBICIDE

Controls Annual Broadleaf Weeds and Annual Grasses in Soybeans and Other Crops

ACTIVE INGREDIENTS:

Ammonium Salt of Chloramben (3-amino,-2,5-dichlorobenzoic acid)*	9.7%
Ammonium salts or related aminodichlorobenzoic acids	1.1%
INERT INGREDIENTS:	89.2%
Total	100.0%

*Total aminodichlorobenzoic acids 10.0% by weight. Chloramben equivalent by weight. Related aminodichlorobenzoic acid equivalents 1.0% by weight 9.0%

E.P.A. Reg. No. 264-167AA E.P.A. Est. No. 264-IA-01

Keep Out of Reach of Children

CAUTION

IN CASE OF EMERGENCY TELEPHONE (24 HOURS A DAY) IN U.S.A. 1-800-UCC-HELP.

FOR GENERAL PRODUCT INFORMATION CALL 1-800-334-9745.

GENERAL INFORMATION

AMIBEN® granular chloramben herbicide has many desirable features: Outstanding control of annual broadleaf weeds and annual grasses. Excellent crop tolerance.

No harmful residue in soil—no change in normal crop rotations. One of the most consistent preemergence herbicides available. Tested and recommended by state colleges and extension agents. Proven effective on millions of acres of soybeans.

HOW AMIBEN® granular WORKS

AMIBEN® granular chloramben herbicide is most effective applied at planting time, when spreading and planting are done in the same operation. Subsequent moisture moves the chemical into the soil where germinating weed seeds take it up and are controlled.

AMIBEN® granular remains active and effective for several weeks ... long enough generally for crops to fill middles and shade out late season weed growth.

AMIBEN® granular controls the following important Annual Grass and Broadleaf Weeds:

barnyardgrass (water grass)	*Echinochloa crusgalli*
carpetweed	*Mollugo verticillata*
chickweed, annual	*Stellaria media*
coffeeweed (sesbania)	*Sesbania exaltata*
crabgrass, hairy (large)	*Digitaria sanguinalis*
crabgrass, smooth	*Digitaria ischaemum*
fall panicgrass	*Panicum dichotomiflorum*
Florida beggarweed	*Desmodium tortuosum*
foxtail, giant	*Setaria faberii*
foxtail, green	*Setaria viridis*
foxtail, yellow	*Setaria glauca*
goosegrass (silver crabgrass)	*Eleusine indica*
johnsongrass-seedling	*Sorghum halepense*
kochia	*Kochia scoparia*
lambsquarter, common	*Chenopodium album*
mustard, wild	*Brassica kaber*
nightshade, black	*Solanum nigrum*
pigweeds	*Amaranthus sp.*
prickly sida (teaweed)	*Sida spinosa*
purslane (pusley)	*Portulaca oleracea*
ragweed, common	*Ambrosia artemisiifolia*
Russian thistle	*Salsola kali*
smartweed, Pennsylvania	*Polygonum pennsylvanicum*
spurge, annual	*Euphorbia maculata*
stinkgrass	*Eragrostis cilianesis*
velvetleaf (buttonweed)	*Abutilon theophrasti*
Venice mallow	*Hibiscus trionum*

Velvetleaf, jimsonweed, buffalobur, wild oats, and cocklebur often germinate in soil at depths below the chemical penetration. Even though populations of these weeds are reduced under normal rainfall conditions, weed seeds which germinate outside the chemical zone are not affected.

For control of cocklebur and suppression of annual morningglory, follow AMIBEN® granular herbicide treatment with an early, directed, postemergence application of BUTYRAC 200 broadleaf herbicide on soybeans.

Sprouts coming from roots of established perennial weeds such as horsenettle, hedge birdweed, nutsedge (nutgrass), milkweed, dogbane, quackgrass and johnsongrass are not controlled by any surface preemergence chemical treatment, including AMIBEN® granular.

Band or broadcast applications of AMIBEN® granular may be made for broadleaf weed control over the top of preplant soil-incorporated grass herbicides.

DIRECTIONS

SOYBEANS, DRY BEANS (including white, navy, kidney, pinto, lima), PEANUTS (Oklahoma, North Texas), SUNFLOWERS

Plant and treat the same operation. Apply at the rate of 20 to 30 pounds AMIBEN® granular chloramben herbicide per broadcast acre. The higher rate should be used on clay loams, clay, or high organic (3 to 8%) mineral soils.

When Band Applying: Mount the spreader behind the last packer or row wheel so that planting and spreading are done in the same operation. Proper calibration and maintenance of desired spreader height above the ground are very important. Weeds growing in the middles between the treated rows should be cultivated as soon as necessary. Use care while cultivating to avoid throwing fresh soil on treated bands.

If sufficient rain does not fall within 7 days after spreading AMIBEN® granular but there is enough soil moisture to germinate weeds and grasses, a light cultivation with a rotary hoe or similar tool will uproot these small broadleaf weeds and grasses. The shallow mixing of AMIBEN® granular in the surface soil will not interfere with the action of AMIBEN® granular when rains come.

If heavy rain packs the soil surface, rotary hoe the field to aid bean emergence. Do not disrupt the chemical band with deep cultivation.

For dry beans grown in low rainfall areas, incorporate AMIBEN® granular into the top 1 to 2 inches of soil with a disk or harrow. Plant beans immediately after incorporation.

AMIBEN® granular split application to soybeans: Apply 15 pounds of AMIBEN® granular per acre as a broadcast treatment just prior to planting. Incorporate AMIBEN® **granular by shallow cultivation. The same day, if possible, plant beans and apply AMIBEN® granular in a band over the row at a rate equivalent to 10 to 15 pounds AMIBEN® granular per broadcast acre.**

PROGRAMMED WEED CONTROL FOR SOYBEANS AND SUNFLOWERS

AMIBEN® granular applied over preplant "TREFLAN"

AMIBEN® granular chloramben herbicide may be applied in a band over the soybean row at planting time in fields where Treflan has been applied as a preplant soil incorporated herbicide. Apply AMIBEN® granular on a band at a rate equivalent to 20 to 30 pounds per broadcast acre. Use the higher rate of AMIBEN® granular herbicide where heavy populations of smartweed, velvetleaf, ragweed, wild mustard or black nightshade are expected. Refer to Treflan label for exact directions, for rate of application, method of incorporation and timing for your area. Observe all directions and cautions on the AMIBEN® granular label.

CORN

Use only on heavy soils in Missouri, Minnesota, Iowa, Ohio, Illinois, Kansas, Nebraska and Indiana. **If rains and extended cold weather occur after planting, corn injury may result.** Plant corn at least 2 to 2½ inches deep. Do not use AMIBEN® granular on corn grown in light textured sandy soils. Apply 20 pounds of AMIBEN® granular chloramben herbicide per acre. For best results, plant and treat at the same time.

SWEET POTATOES

Apply AMIBEN® granular chloramben herbicide at the rate of 40 pounds per broadcast acre at planting time. For the first few cultivations after treatment, set sweeps so that soil in treated bands is not disturbed. For the final cultivation, set sweeps closer to pull soil as high as possible over the sweet potato row.

ESTABLISHED TOMATOES AND PEPPERS (From Transplants or Seed)

Use 30 to 40 pounds AMIBEN® granular chloramben herbicide per acre as an overall treatment. Apply AMIBEN® granular once during the season, using the timing recommended below for your soil type and cultural practice. **Apply AMIBEN® granular herbicide only when tomato and pepper foliage is dry.**

TIMES TO APPLY: (Use Only One.)

A. Transplants on Medium to Heavy Soils.

On medium to heavy soils apply AMIBEN® granular after plants are established (recovered from wilting that follows transplanting). Do not use AMIBEN® granular at this time for tomatoes or peppers growing on light sandy soil as injury may result.

B. Transplants on Light Sandy Soils.

On light textured sandy soils cultivate or hoe to control weeds after transplanting. Apply AMIBEN® granular immediately after the last cultivation and before weeds emerge. Weeds in the crop row should be removed prior to application.

C. Direct Seeded Tomatoes or Peppers.

On direct seeded tomatoes or peppers, do not apply until plants have reached the 5 to 6 leaf stage. Cultivate before application of AMIBEN® granular herbicide to remove existing weeds. Weeds in the crop row should be removed prior to application of AMIBEN® granular. **Apply AMIBEN® granular herbicide only when tomato or pepper foliage is dry.**

SEEDED SQUASH AND PUMPKIN

Treat at planting time or as soon after planting as possible, before weeds and crop emerge. Mount the spreader on the planter so that spreading and seeding are carried out in the same operation. Apply at a rate of 30 pounds AMIBEN® granular chloramben herbicide per broadcast acre on light soils and 40 pounds per broadcast acre on heavy or high organic soils. If band applications are used, use the table below to determine the rate of AMIBEN® granular to use per banded acre on your soil type. If band applications are used, use normal cultivations after squash or pumpkin emerge to control weeds growing in the untreated strips between the rows.

POUNDS AMIBEN® GRANULAR PER BANDED ACRE

Row Width	Band Width	Light Soils	High Organic Soils
72 inch	14 width	5.9 lbs	7.8 lbs
76 inch	14 width	5.6 lbs	7.4 lbs
80 inch	14 width	5.3 lbs	7.0 lbs

After squash and pumpkin emerge, use normal cultivations to control weeds growing in the untreated strips between the rows.

LIMA BEANS—Overall Treatment

Apply 20 to 40 pounds of AMIBEN® granular per broadcast acre. Treat at planting time or as soon after planting as possible, before weeds and lima beans emerge.

SEEDLING ASPARAGUS—Overall Treatment

Apply 30 pounds of AMIBEN® granular per acre as soon as possible after asparagus is planted and before seeds emerge. Irrigation with ½ inch of water following application is suggested to produce maximum weed control.

BAND TREATMENTS

POUNDS AMIBEN® GRANULAR CHLORAMBEN HERBICIDE PER BANDED ACRE

ROW WIDTH	BAND WIDTH	@20 LB/A*	@25 LB/A*	@30 LB/A*	@40 LB/A*
21 in	10 in	9 lbs 8 oz	11 lbs 14 oz	14 lbs 5 oz	19 lbs
28 in	12 in	8 lbs 10 oz	10 lbs 11 oz	12 lbs 14 oz	17 lbs 2 oz
	14 in	10 lbs	12 lbs 8 oz	15 lbs	20 lbs
30 in	12 in	8 lbs	10 lbs	12 lbs	16 lbs
	14 in	9 lbs 5 oz	11 lbs 11 oz	14 lbs	18 lbs 11 oz
36 in	12 in	6 lbs 11 oz	8 lbs 5 oz	10 lbs	13 lbs 5 oz
	14 in	7 lbs 13 oz	9 lbs 11 oz	11 lbs 11 oz	15 lbs 10 oz
38 in	12 in	6 lbs 5 oz	7 lbs 14 oz	9 lbs 8 oz	12 lbs 10 oz
	14 in	7 lbs 6 oz	9 lbs 3 oz	11 lbs	14 lbs 11 oz
40 in	12 in	6 lbs	7 lbs 8 oz	9 lbs	12 lbs
	14 in	7 lbs	8 lbs 13 oz	10 lbs 8 oz	14 lbs

* Pounds AMIBEN® granular per **broadcast** acre.

To determine amount of AMIBEN granular to purchase follow these 4 steps and use the above table.

1. Select row width (Example 28 inch)
2. Select band width (Example 24 inch)
3. Select broadcast rate * that fits your weed problems (soybean example—use 30 lb/A where velvetleaf and black nightshade are problems).
4. Formula

15 lbs. per acre X number of acres equals Amount to Purchase.

How to use the granular equipment charts: Step (1). Identify your granular equipment. Note differences between "old" and "new" equipment. (2). Decide ground speed within a range of 2 to 10 mph. (3) Decide broadcast rate within a range of 20 to 30 lbs. per broadcast acre. Note that each rate range is in a column "below" the miles per hour. (Example): if you have the new Gandy, and if you decide 5 mph and a broadcast rate of 30 lbs. AMIBEN granular per acre your gauge setting is 20.

OLD INTERNATIONAL HARVESTER

(prior to 1981)
MILES PER HOUR

GA	2	3	4	5	6	7	8	9	10
1/4	17.1								
1/5	20.7								
1/6	24.2								
1/7	27.8	18.5							
1/8	31.4	20.9							
1/9		23.3							
2/0		25.7	19.3						
2/1		27.7	20.8						
2/2		29.7	22.2						
2/3		31.7	23.7	19.0					
2/4			25.2	20.2					
2/5			26.7	21.4					
2/6			28.2	22.6					
2/7			29.7	23.8	19.8				
2/8			31.2	24.6	20.8				
2/9				26.1	21.8				
3/0				27.3	22.8	19.5			
3/1				28.5	23.8	20.4			
3/2				29.7	24.8	21.2			
3/3				30.9	25.8	22.1	19.3		
3/4					26.8	22.9	20.1		
3/5					27.8	23.8	20.8		
3/6					28.7	24.6	21.6		
3/7					29.7	25.5	22.3	19.8	
3/8					30.7	26.3	23.1	20.5	
3/9						27.2	23.8	21.2	
4/0						28.0	24.5	21.8	19.6
4/1						29.0	25.4	22.6	20.3
4/2						30.0	26.3	23.3	21.0
4/3							27.1	24.1	21.7
4/4							28.0	24.9	22.4
4/5							28.8	25.6	23.1
4/6							29.7	26.4	23.7
4/7							30.5	27.1	24.4
4/8								27.9	25.1
4/9								28.7	25.8
5/0								29.4	26.5
5/1								30.2	27.2
5/2									27.9
5/3									28.5
5/4									29.2
5/5									29.9
5/6									30.6

NEW INTERNATIONAL HARVESTER

(1981 and after)
MILES PER HOUR

GA	2	3	4	5	6	7	8	9	10
1/5	16.4								
1/6	20.5								
1/7	24.5								
1/8	28.9	19.3							
1/9	32.6	21.8							
2/0		24.5							
2/1		26.5	19.9						
2/2		28.5	21.4						
2/3		30.5	22.9						
2/4			24.4	19.5					
2/5			25.9	20.7					
2/6			27.4	21.9					
2/7			28.9	23.1	19.3				
2/8			30.4	24.3	20.3				
2/9				25.5	21.3				
3/0				26.7	22.3				
3/1				27.9	23.3	19.9			
3/2				29.1	24.3	20.8			
3/3				30.3	25.3	21.7			
3/4					26.3	22.5	19.7		
3/5					27.3	23.4	20.5		
3/6					28.3	24.2	21.2		
3/7					29.3	25.1	21.9	19.5	
3/8					30.3	26.0	22.7	20.2	
3/9						26.8	23.5	20.9	
4/0						27.7	24.2	21.5	
4/1						28.5	25.0	22.2	20.0
4/2						29.4	25.7	22.9	20.6
4/3						30.3	26.5	23.5	21.2
4/4							27.2	24.2	21.8
4/5							28.0	24.9	22.4
4/6							28.7	25.5	23.0
4/7							29.5	26.2	23.6
4/8							30.2	26.9	24.2
4/9								27.5	24.8
5/0								28.2	25.4
5/1								28.9	26.0
5/2								29.5	26.6
5/3								30.2	27.2
5/4									27.8
5/5									28.4
5/6									29.0
5/7									29.6
5/8									30.2

ALLIS CHALMERS

GA	RATE IS UNIFORM FOR ALL SPEEDS	#/Br. A
3		18.1
4		22.2
5		29.0
6		34.7

OLD JOHN DEERE

(metal hoppers 3 range gauge)
MILES PER HOUR

GAUGE SETTINGS

GA	2	3	4	5	6	7	8	9	10
1/22	18.7								
1/23	21.5								
1/24	24.3								
1/25	27.0								
1/26	29.8	19.9							
1/27	32.5	21.7							
1/28		23.5							
1/29		25.4	19.0						
1/30		27.2	20.4						
1/31		27.2	20.4						
1/32		27.2	20.4						
1/33		27.2	20.4						
2/1		28.6	21.4						
2/2		29.8	22.3						
2/3		31.0	23.2						
2/4			24.1						
2/5			25.0	20.0					
2/6			25.9	20.7					
2/7			26.8	21.5					
2/8			27.7	22.2					
2/9			28.6	22.9					
2/10			29.5	23.6	19.7				
2/11			30.4	24.3	20.3				
2/12				25.0	20.9				
2/13				25.7	21.5				
2/14				26.5	22.1				
2/15				27.2	22.6				
2/16				27.6	23.0				
2/17				28.0	23.4	20.0			
2/18				28.5	23.7	20.3			
2/19				28.9	24.1	20.6			
2/20				29.3	24.4	20.9			
2/21				29.7	24.8	21.2			
2/22				30.2	25.1	21.5			
2/23					25.5	21.8			
2/24					25.8	22.1			
2/25					26.2	22.5			
2/26					26.5	22.8	19.9		
2/27					26.9	23.1	20.2		
2/28					27.3	23.4	20.4		
2/29					27.6	23.4	20.7		
2/30					28.0	24.0	21.0		
2/31					28.8	24.7	21.6		
2/32					29.6	25.4	22.2	19.7	
2/33					30.4	26.1	22.8	20.3	
3/1						27.9	24.4	21.7	19.5
3/2						28.9	25.3	22.5	20.2
3/3						29.9	26.2	23.3	21.0
3/4						31.0	27.1	24.1	21.7
3/5							28.1	24.9	22.4
3/6							28.9	25.7	23.1
3/7							29.8	26.5	23.9
3/8							30.8	27.3	24.6
3/9								28.1	25.3
3/10								29.0	26.1
3/11								29.8	26.8
3/12								30.6	27.5
3/13									28.2
3/14									29.0
3/15									29.7
3/16									30.4

NEW JOHN DEERE

(plastic hoppers 10-20-30 gauge)
MILES PER HOUR

GAUGE SETTINGS

GA	2	3	4	5	6	7	8	9	10
8	19.3								
9	22.4								
10	25.5								
11	28.0	18.7							
12	30.5	20.4							
13		22.0							
14		23.7							
15		25.4	19.1						
16		27.1	20.3						
17		28.8	21.6						
18		30.5	22.9						
19			24.1	19.3					
20			25.4	20.3					
21			26.9	21.6					
22			28.5	22.8	19.0				
23			30.3	24.2	20.2				
24				25.6	21.3				
25				27.0	22.5	19.3			
26				28.4	23.7	20.3			
27				30.1	25.1	21.5			
28					26.5	22.7	19.9		
29					27.9	23.9	21.0		
30					29.4	25.2	22.0	19.6	
31					31.3	26.8	23.4	20.8	18.8
32						28.8	25.2	22.4	20.2
33						31.1	27.2	24.2	21.8
34							29.4	26.1	23.5
35							31.6	28.1	25.3
36								30.0	27.0
37									29.3
38									31.8

NEWEST JOHN DEERE

(5-15-25 plastic hopper)
MILES PER HOUR

GAUGE SETTINGS

GA	2	3	4	5	6	7	8	9	10
7	16.0								
8	20.3								
9	24.6								
10	28.9	19.3							
11	33.2	22.1							
12		25.0	18.8						
13		28.3	21.2						
14		31.5	23.6	18.9					
15			26.1	20.9					
16			28.6	22.9	19.1				
17			31.2	25.0	20.8				
18				27.0	22.5	19.3			
19				29.5	24.6	21.1			
20				32.0	26.7	22.9	20.0		
21					28.8	24.7	21.6	19.2	
22					30.9	26.5	23.2	20.6	18.6
23						28.7	25.1	22.3	20.1
24						30.9	27.0	24.0	21.6
25							29.2	25.9	23.4
26							31.4	27.9	25.1
27								29.8	26.9
28								31.8	28.6
29									30.7

OLD NOBLE

(round opening in bottom of plastic hopper)
MILES PER HOUR

GAUGE SETTINGS

GA	2	3	4	5	6	7	8	9	10
6	17.4								
7	20.3								
8	23.3								
9	26.4								
10	29.6	19.8							
11	32.4	21.6	16.2						
12		23.4	17.6						
13		25.3	19.0						
14		27.2	20.4						
15		29.1	21.8						
16		31.0	23.3						
17			24.8	19.8					
18			26.3	21.0					
19			27.6	22.1					
20			29.5	23.6	19.6				
21			31.2	24.9	20.8				
22				26.3	21.9				
23				27.7	23.1	19.8			
24				29.1	24.3	20.8			
25				30.7	25.6	21.9	19.2		
26					26.9	23.1	20.2		
27					24.2	21.2	18.9		
28					29.6	25.4	22.2	19.8	
29					31.0	26.6	23.2	20.7	
30						27.7	24.3	21.6	19.4
31						29.4	25.7	22.9	20.6
32						31.1	27.2	24.2	21.8
33							28.7	25.5	23.0
34							30.2	26.8	24.2
35								28.4	25.6
36								30.1	27.1
37									28.9
38									30.6

OLD GANDY

(bolt on bottom)
MILES PER HOUR

GAUGE SETTINGS

GA	2	3	4	5	6	7	8	9	10
10	20.0								
11	25.0								
12	30.0	20.0							
13		23.5	17.6						
14		27.5	20.6						
15		31.5	23.6	18.9					
16			26.6	21.3					
17			29.9	23.9	19.9				
18			33.1	26.5	22.1	18.9			
19				29.2	24.3	20.9	18.3		
20				32.1	26.7	22.9	20.1		
21					29.4	25.2	22.1	19.6	
22					32.5	27.9	24.4	21.7	19.5
23						30.5	26.7	23.7	21.4
24							29.3	26.1	23.5
25							32.0	28.4	25.6
26								31.0	27.9
27									30.4

NEW NOBLE

(snap on bottom)
MILES PER HOUR

GAUGE SETTINGS

GA	2	3	4	5	6	7	8	9	10
4	18.3								
5	21.5								
6	24.8								
7	28.0	18.7							
8	31.3	20.9							
9		23.0							
10		25.2	18.9						
11		28.7	21.5						
12		32.2	24.2	19.3					
13			26.8	21.5					
14			29.5	23.6	19.6				
15			32.1	25.7	21.4				
16				27.8	23.2	19.9			
17				29.9	24.9	21.4	18.7		
18				32.0	26.7	22.9	20.0		
19					28.5	24.4	21.3	19.0	
20					30.2	25.9	22.7	20.1	
21						27.4	24.0	21.3	19.2
22						28.9	25.3	22.5	20.2
23						30.4	26.6	23.7	21.3
24							27.9	24.8	22.4
25							29.3	26.0	23.4
26							30.6	27.2	24.5
27								28.4	25.5
28								29.5	26.6
29								30.7	27.6
30									28.7
31									29.7
32									30.8

NEW GANDY

(snap on bottom)
MILES PER HOUR

GAUGE SETTINGS

GA	2	3	4	5	6	7	8	9	10
10	17.0								
11	20.8								
12	24.8								
13	29.3	19.5							
14	34.8	23.2	17.4						
15		27.5	20.6						
16		31.8	23.9	19.1					
17			27.4	21.9	18.3				
18			31.1	24.9	20.8				
19				28.0	23.3	20.0			
20				30.9	25.7	22.1	19.3		
21					28.3	24.3	21.3	18.9	
22					31.3	26.9	23.5	20.9	18.8
23						29.8	26.1	23.2	20.9
24						32.6	28.6	25.4	22.9
25							31.2	27.7	25.0
26								30.5	27.5
27									30.0

PRECAUTIONARY STATEMENTS

CAUTION

HAZARD TO HUMANS AND DOMESTIC ANIMALS.

Harmful if swallowed. Avoid contact with skin, eyes, clothing.

STATEMENT OF PRACTICAL TREATMENT

In case of eye contact, rinse immediately with water and get medical attention.

In case of skin contact, wash immediately with soap and water and remove contaminated clothing.

ENVIRONMENTAL HAZARDS
Do not contaminate water used for irrigation or domestic purposes by cleaning of equipment or disposal of wastes.

DIRECTIONS FOR USE
IT IS A VIOLATION OF FEDERAL LAW TO USE THIS PRODUCT IN A MANNER INCONSISTENT WITH ITS LABELING.

Thoroughly clean hoppers and hoses of granular applicators before using to apply AMIBEN® Granular Chloramben Herbicide.

Do not graze treated field of sunflowers or use these plants as livestock feed.

STORAGE AND DISPOSAL

Do not store near fertilizers, seed, insecticides, or fungicides. Do not contaminate water, food, or feed by storage or disposal. Open dumping is prohibited.

PESTICIDE DISPOSAL
Wastes resulting from the use of this product may be disposed of on site or at an approved waste disposal facility.

CONTAINER DISPOSAL
Completely empty bag into application equipment. Then dispose of empty bag in a sanitary landfill or by incineration, or if allowed by State and local authorities, by burning. If burned, stay out of smoke.

LIMITED WARRANTY AND DISCLAIMER

The manufacturer warrants (a) that this product conforms to the chemical description on the label; (b) that this product is reasonably fit for the purposes set forth in the directions for use when it is used in accordance with such directions; and (c) that the directions, warnings, and other statements on this label are based upon responsible experts' evaluation of reasonable tests of effectiveness, of toxicity to laboratory animals and to plants, and of residues on food crops, and upon reports of field experience. Tests have not been made on all varieties or in all states or under all conditions. THE MANUFACTURER NEITHER MAKES NOR INTENDS, NOR DOES IT AUTHORIZE ANY AGENT OR REPRESENTATIVE TO MAKE ANY OTHER WARRANTIES, EXPRESS OR IMPLIED, AND IT EXPRESSLY EXCLUDES AND DISCLAIMS ALL IMPLIED WARRANTIES OF MERCHANTABILITY OR FITNESS FOR A PARTICULAR PURPOSE.

THIS WARRANTY DOES NOT EXTEND TO, AND THE BUYER SHALL BE SOLELY RESPONSIBLE FOR, ANY AND ALL LOSS OR DAMAGE WHICH RESULTS FROM THE USE OF THIS PRODUCT IN ANY MANNER WHICH IS INCONSISTENT WITH THE LABEL DIRECTIONS, WARNINGS OR CAUTIONS.

BUYER'S EXCLUSIVE REMEDY AND MANUFACTURER'S OR SELLER'S EXCLUSIVE LIABILITY FOR ANY AND ALL CLAIMS, LOSSES, DAMAGES, OR INJURIES RESULTING FROM THE USE OR HANDLING OF THIS PRODUCT, WHETHER OR NOT BASED ON CONTRACT, NEGLIGENCE, STRICT LIABILITY IN TORT OR OTHERWISE, SHALL BE LIMITED, AT THE MANUFACTURER'S OPTION, TO REPLACEMENT OF, OR THE REPAYMENT OF THE PURCHASE PRICE FOR, THE QUANTITY OF PRODUCT WITH RESPECT TO WHICH DAMAGES ARE CLAIMED. IN NO EVENT SHALL MANUFACTURER OR SELLER BE LIABLE FOR SPECIAL, INDIRECT OR CONSEQUENTIAL DAMAGES RESULTING FROM THE USE OR HANDLING OF THIS PRODUCT.

THIS SPECIMEN LABEL IS INTENDED FOR USE ONLY AS A GUIDE IN PROVIDING GENERAL INFORMATION REGARDING THE DIRECTIONS, WARNINGS AND CAUTIONS ASSOCIATED WITH THE USE OF THIS PRODUCT. AS WITH ANY AGRICULTURAL CHEMICAL, ALWAYS FOLLOW THE LABEL INSTRUCTIONS ON THE PACKAGE BEFORE USING.

AMIBEN®

CHLORAMBEN HERBICIDE

Controls Annual Broadleaf Weeds and Annual Grasses in Soybeans and Other Crops

ACTIVE INGREDIENTS:
Ammonium salt of chloramben
(3-amino-2,5-dichlorobenzoic acid)* 21.1%
Ammonium salts of related aminodichlorobenzoic acids* 2.3%
INERT INGREDIENTS: 76.6%
TOTAL 100.0%
*Total aminodichlorobenzoic acids 21.6% by weight or 2 pounds per gallon.
Chloramben equivalent 19.5% by weight or 1.8 pounds per gallon.
Related aminodichlorobenzoic acid equivalent 2.1% by weight or 0.2 pounds per gallon.

E.P.A. Reg. No. 264-138AA E.P.A. Est. No. 264-IA-01

Keep Out of Reach of Children
CAUTION

IN CASE OF EMERGENCY—TELEPHONE DAY OR NIGHT (UCC HELP) 800-822-4357

FOR GENERAL PRODUCT INFORMATION CALL 1-800-334-9745

- Outstanding control of annual broadleaf weeds and grasses.
- Excellent crop tolerance.
- No harmful residue in soil—no changes in normal crop rotations. Completely water soluble . . . easy to mix and handle.
- One of the most consistent preemergence herbicides available.
- Tested and recommended by state colleges and extension agents.
- Proven effective on millions of acres of soybeans.

GENERAL INFORMATION

AMIBEN® Chloramben Herbicide is effective applied preplant incorporated, preemergence and early postemergence. When used preemergence it is most effective applied at planting time, when spraying and planting are done in the same operation. Rain, irrigation, water, or mechanical incorporation moves AMIBEN® into the soil where weed seeds sprout.

Uniform spray application over the soil surface is important. About 10–20 gallons of water or sprayable fluid fertilizer for ground and about 3–5 gallons for air application is enough spray solution for good coverage. See specific sections of this label for any special spray volume requirements.

AMIBEN® Herbicide remains active and effective for several weeks . . . long enough generally for crops to fill middles and shade out the late season weed growth. Performance is improved when moisture is received within 5 days after application and prior to weed emergence. High intensity or excessive rainfall or excessive irrigation after application may reduce control.

In coarse textured soils with less than 1% organic matter heavy rains may move AMIBEN® Herbicide below the weed seed germination zone and produce erratic weed control.

AMIBEN® controls the following important Annual Grass and Broadleaf Weeds:

barnyardgrass (water grass)	*Echinochloa crusgalli*
carpetweed	*Mullugo verticillata*
chickweed, annual	*Stellaria media*
coffeeweed (sesbania)	*Sesbania exaltata*
crabgrass, hairy (large)	*Digitaria sanguinalis*
crabgrass, smooth	*Digitaria ischaemum*
fall panicgrass	*Panicum dichotomiflorum*
Florida beggarweed	*Desmodium tortuosum*
foxtail, giant	*Setaria faberii*
foxtail, green	*Setaria viridis*
foxtail, yellow	*Setaria glauca*
goosegrass (silver crabgrass)	*Eleusine indica*

johnsongrass-seedling	*Sorghum halepense*
kochia	*Kochia scoparia*
lambsquarters, common	*Chenopodium album*
mustard, wild	*Brassica kaber*
nightshade, black	*Solanum nigrum*
pigweeds	*Amaranthus sp.*
prickly sida (teaweed)	*Sida spinosa*
purlane (pusley)	*Portulaca oleracea*
ragweed, common	*Ambrosia artemisiifolia*
Russian thistle	*Salsola kali*
smartweed, Pennsylvania	*Polygonum pensylvanicum*
spurge, annual	*Euphorbia maculata*
stinkgrass	*Eragrostis cilianensis*
velvetleaf (buttonweed)	*Abutilon theophrasti*
Venice mallow	*Hibiscus trionum*
wild proso millet	*Panicum miliaceum*

Band application of AMIBEN® may be made for broadleaf weed control over the top of preplant soil incorporated grass killers.

Deep germinating annual weeds and sprouts from established perennial weeds are not controlled. Sprouts coming from roots of established perennial weeds such as horsenettle, hedge bindweed, nutsedge (nutgrass), milkweed, dogbane, quackgrass, and johnsongrass are not controlled by any selective, surface preemergence chemical treatment including AMIBEN®.

Only the following recommendations referenced in this label are registered for use in California.

CROPS

Asparagus, seedling
Drybeans (White, Navy, Kidney, Pinto)
Lima beans
Pumpkin
Squash
Sunflowers
Sweet Potatoes
Soybeans
- AMIBEN® with fertilizer
- AMIBEN® preemergence on preplant incorporated
- AMIBEN® delayed preemergence
- AMIBEN® plus Lasso
- AMIBEN® plus Treflan
- AMIBEN® plus Lorox
- AMIBEN® plus Prowl
- AMIBEN® plus Preemerge 3

AMIBEN® is not registered for control of Venice mallow (*Hibiscus trionum*) in California.

DIRECTIONS FOR INCORPORATION OF AMIBEN®

AMIBEN® Herbicide applied alone, and several AMIBEN® tank mixtures as specified in the crops section of this label, may be incorporated into the top 1 to 3 inches of the final seedbed before planting. When preplant incorporation is recommended in the crops section of this label the following equipment and operating procedures should be used. Be sure to use machinery that mixes AMIBEN® thoroughly with the soil. Incorporation with implements set to cut less than 2 inches deep may result in erratic weed control.

Recommended Equipment Includes:

1. Two passes with a disc set to cut 4 inches deep at 4 to 6 mph with the second pass running at an angle to the first. A tandem or double-disc operated one time does not provide adequate incorporation.
2. One pass with a field cultivator set to cut 3 to 4 inches deep and operated at 6 mph or more. The field cultivator should have 3 to 4 rows of sweeps, spaced at intervals of 7 inches or less and staggered so that no soil is left unturned. Chisel points should not be used. Do not set cultivator deeper than 4 inches. The field cultivator used alone or in combination with the double-disc will provide effective incorporation. If used in combination with a double-disc the spray boom should be mounted behind the disc blades but ahead of the cultivator sweeps.
3. One pass with a rolling cultivator set to cut 2 to 3 inches deep and operated at 6 to 8 mph. Rolling cultivators are adequate for use on coarse and medium textured soils only.
4. One pass with a mulch treader or other similar disc type implement set to cut 2 to 3 inches deep and operated at 5 to 8 mph.
5. PTO-driven equipment (tillers, cultivators, hoes) set to cut 2 to 3 inches deep with rotors spaced to provide a clean sweep of the soil operated one time. PTO-driven equipment should not be operated at a speed greater than 4 mph.

MIXING INSTRUCTIONS

AMIBEN® in water or fluid fertilizer: first fill the spray tank 1/4 to 1/2 full with clean water. Start agitation and add AMIBEN® while adding the remaining required water or fluid fertilizer.

AMIBEN® Plus Emulsifiable Concentrates in Water: first fill tank 1/4 to 1/2 full with clean water, start agitation and add AMIBEN®. Mix thoroughly and add water until tank is 3/4 full, continue agitation and add the emulsifiable concentrate material. Fill tank with remaining required water and continue agitation during spraying until tank is empty. Agitation should sweep bottom contents of tank up into the main body of the spray mixture for thorough mixing. If a compatibility agent is required it should be thoroughly mixed with the water before adding the emulsifiable concentrate.

To prevent foaming, avoid stirring or splashing air into the mixture by placing the end of the fill pipe below the surface of the water in the spray tank.

AMIBEN® Plus Wettable Powders or Flowables: Fill tank 1/2 full with clean water, start sprayer agitation and add the recommended amount of the flowable or wettable powder. Mix thoroughly and add AMIBEN® while filling tank with remaining required water. Continue agitation during entire spraying operation until tank has been emptied. Agitation should sweep bottom contents of the tank up into the main body of the spray mixture for thorough mixing and uniform suspension.

DIRECTIONS FOR USING AMIBEN® THROUGH OVERHEAD IRRIGATION SYSTEMS

AMIBEN® may be applied as a preemergence treatment in irrigation water alone, or in tank mixtures with any of the other herbicides on this label which are registered for center pivot or overhead irrigation application, after planting but before weeds or crop emerge. Use the broadcast rates per acre for each product as specified elsewhere on this label for each of the tank mixtures, and follow all use precautions below and on respective labels for any other products applied in this manner. Use only through automated (center pivot or continuous lateral move) equipment which is properly maintained and operated to apply water uniformly.

DIRECTIONS FOR USE

Remove scale, pesticide residues and other foreign matter from the chemical tank and the entire injector system. Flush with clean water. In the chemical tank prepare a mixture with a minimum of 1 part of water to 1 part of herbicide(s). Maintain continuous agitation in the chemical tank while injecting into the irrigation system with a positive displacement pump. Injecting a larger volume of a more dilute mixture will usually provide more accurate calibration of metering equipment.

Set sprinkler to deliver 0.5 inch or less of water per acre. Application of more than 0.5 inch of water per acre may reduce weed control by moving the herbicide below the weed germination zone.

USER PRECAUTIONS

Apply AMIBEN® only through systems containing antisiphon and check valves and having interlocking controls between the metering device and the water pump to insure their simultaneous shutoff to prevent water source contamination and overflow of the mixing tank.

To insure adequate mixing inject the chemical into the main line at least 8 feet ahead of a right angle turn.

Do not apply when conditions favor drift, such as from windy conditions, or when system joints and connections are leaking, or when nozzles do not provide uniform distribution, as crop injury may result.

Where sprinkler distribution patterns do not overlap sufficiently, unacceptable weed control may result. Where sprinkler distribution patterns overlap excessively, crop injury may result.

DIRECTIONS FOR USING AMIBEN® IN CONSERVATION OR REDUCED TILLAGE SYSTEMS

When applied as directed below, AMIBEN® tank mixtures will control many emerged annual weeds, suppress many emerged perennial weeds and will give preemergence control of many annual grasses and broadleaf weeds in conservation tillage or reduced tillage systems.

When used in tillage systems with practices which eliminate all existing vegetation before planting, AMIBEN® may be used alone or in tank mixture with any of the other products on this label which are labeled for use in reduced tillage operations. AMIBEN® and AMIBEN® combinations with other preemergence or preplant incorporated herbicides will *not* control emerged weeds. The presence of previous crop residue (trash) which is uniformly distributed on the soil surface will not affect the performance of AMIBEN®, providing sufficient rainfall (0.375 to 0.5 inch) is received before weeds begin to germinate. IF WEEDS HAVE BEGUN TO EMERGE AT TIME OF APPLICATION FOLLOW THE DIRECTIONS BELOW:

When used in tillage systems where emerged weeds are present at planting, either Roundup, or Paraquat plus Ortho X-77 spreader should be added to the AMIBEN® tank mixture. Refer to the respective Roundup, Paraquat and Ortho X77 labels for crops labeled, weeds controlled, use rates, and specific precautions to be taken. Use a minimum of 15 to 40 gallons of water per acre and apply the tank mixture during or after planting, but BEFORE CROP *EMERGENCE.* As density of stubble, crop residue or weeds increase, spray gallonage and rate should be increased, within recommended ranges, to assure good coverage.

MIXING INSTRUCTIONS FOR USE OF AMIBEN® WITH SPRAYABLE FLUID FERTILIZERS

In most cases, AMIBEN® Chloramben Herbicide can be mixed and applied with liquid nitrogen, blended clear liquids, or suspension type fertilizers.

Be sure the spray tank is clean and uncontaminated. First fill the spray tank 1/4 to 1/2 full with clean water. Start agitation and add AMIBEN® Chloramben Herbicide. Add required fluid fertilizer and water to fill tank. After the tank is full, no further agitation is necessary.

The compatibility of tank mixtures of AMIBEN® with other herbicides in combination with liquid fertilizers should be evaluated on a small scale before mixing spray tank quantities. Do not use fertilizer mixes with combinations of AMIBEN® and Premerge 3.

DIRECTIONS FOR USE ON SOYBEANS

AMIBEN® PREEMERGENCE OR PREPLANT INCORPORATED

AMIBEN® Chloramben Herbicide may be used as a preemergence or preplant incorporated herbicide. AMIBEN® may be applied and incorporated at planting or several days prior to planting. Uneven spray application or improper soil incorporation of AMIBEN® can produce erratic weed control. Incorporate chemical in the top 0 to 2 inch level as most weeds sprout in this zone. Refer to the "Directions for Incorporation" section of this label for approved incorporation equipment and operating procedures.

In coarse textured soils heavy rains on incorporated AMIBEN® Herbicide may move it below the weed seed germination zone and produce erratic weed control.

Make all broadcast treatments according to the following table:

Soil Texture	AMIBEN® Qts/A
Loamy sand, sandy loam	4 to 5
Loam, silt loam, silt	5 to 6
Silty clay loam, silty clay, sandy clay, clay loam, clay	6

Use the higher rate of AMIBEN® when heavy populations of black nightshade, velvetleaf, wild mustard, common ragweed or Pennsylvania smartweed are expected. To insure good spray coverage use 10 to 20 gallons of water per acre with ground equipment or a minimum of 5 gallons of water per acre by aircraft.

When band spraying, mount the spray nozzle on the packer or row wheel so that planting and spraying are done in the same operation. Proper calibration and maintenance of desired nozzle height above the ground are very important. Weeds growing in the middles between the treated rows should be cultivated as soon as necessary.

Use care while cultivating to avoid throwing fresh soil on treated bands.

If sufficient rain does not fall within 5 days after spraying AMIBEN® herbicide, but there is enough soil moisture to germinate weeds and grasses, a light cultivation with a rotary hoe or similar tool will uproot these small broadleaf weeds and grasses. The shallow mixing of AMIBEN® in the surface soil will not interfere with the action of AMIBEN® when rains come. If heavy rain packs the soil surface, rotary hoe the field to aid bean emergence. Do not disrupt the chemical band with deep cultivation.

"AMIBEN®" AND "LASSO" FOR SOYBEANS

A tank mixture of AMIBEN® Chloramben Herbicide and Lasso may be applied as a preplant soil incorporated herbicide treatment. AMIBEN® Herbicide may be applied as a preemergence broadcast tank mix treatment with Lasso during planting or after planting before weeds and soybeans emerge. Following a preplant soil incorporated application of Lasso, AMIBEN® may be applied preemergence during planting or after planting before weeds and soybeans emerge.

A tank mixture or the sequential treatment of AMIBEN® Herbicide and Lasso will control the following annual weeds: barnyardgrass, crabgrass, foxtail (green, yellow, giant), fall panicum, lambsquarters, pigweed, smartweed, ragweed, black nightshade, velvetleaf and wild mustard, as well as all other weeds listed on the AMIBEN® and Lasso labels.

For uniform application of the tank mixture of AMIBEN® Herbicide and Lasso use 10 to 20 gallons of water per acre for ground application. To prepare the tank mixture refer to the "Mixing Instructions" portion of this label. DO NOT mix AMIBEN® Herbicide and Lasso concentrates before the addition of water as an emulsion will not be formed.

Apply AMIBEN® and Lasso preplant incorporated or preemergence using the appropriate rates from the table below. For preplant incorporated treatments apply the tank mixture to the soil and incorporate into the top two inches of soil as directed in the "Directions for Incorporation" portion of this label. For preemergence treatments, apply the tank mixture during planting (behind the planter) or after planting before weeds or soybeans emerge. For sequential applications apply Lasso alone preplant incorporated as specified on the Lasso label. Follow with a preemergence application of AMIBEN® herbicide during planting (behind the planter) or after planting before weeds or soybeans emerge.

With preemergence treatments rain is necessary to move the chemicals into the soil and make them available to control germinating weed seeds. If no rainfall occurs within 5 days following application, a shallow incorporation no deeper than 2 inches is recommended. High intensity or excessive rainfall after application may reduce control.

Make all broadcast treatments according to the following table:

Soil Texture	AMIBEN® Qts/A	Lasso Qts/A
Loamy sand, sandy loam	4	1.5 to 2.0
Loam, silt loam, silt	4 to 5	2.0 to 2.5
Silty clay loam, silty clay, sandy clay, clay loam, clay	5 to 6	2.5 to 3.0

Use the higher rate of AMIBEN® when heavy populations of black nightshade, velvetleaf, mustard or smartweed are expected. Use the higher rate of Lasso when treatment is incorporated or when high populations of black nightshade, witchgrass or foxtail are expected.

For band treatments divide the band width by the row width and multiply by the appropriate rates indicated above to determine the amount of chemical to use per acre.

Read thoroughly the cautionary statements and all other information appearing on labels for both Lasso and AMIBEN®.

AMIBEN® CHLORAMBEN HERBICIDE TANK MIXTURE WITH "LASSO" AND "SENCOR" OR "LEXONE" FOR SOYBEANS

GENERAL INFORMATION

A tank of mixture of AMIBEN® Chloramben Herbicide plus Lasso plus Sencor or Lexone when applied in accordance with the following directions, either preplant soil incorporated or preemergence, effectively controls most weeds listed on the current AMIBEN®, Lasso, and Lexone WP/4L/DF or Sencor WP/4/DF labels.

DIRECTIONS FOR USE

Except as directed in the "Conservation or Reduced Tillage" portion of this label, the seedbed should be free of existing weeds before application. Surface trash (plant residue from previous crop) should be uniformly distributed, and the seedbed should be free of clods. The tank mixture should be applied in from 10 to 20 gallons of water per acre using any properly calibrated low pressure sprayer equipped with herbicide tips and in-line strained screens no finer than 50 mesh. To prepare the tank mixture refer to the "Mixing Instructions" portion of this label. Apply the entire contents of the spray tank within 4 hours of initial mixing.

PREPLANT SOIL INCORPORATED USE

Incorporation should place the AMIBEN® Herbicide plus Lasso plus Sencor or Lexone into the top 1 to 3 inches of the final seedbed. Refer to the "Directions for Incorporation" portion of this label for recommended incorporation equipment and operating procedures.

Apply the tank mixture using the table below to determine the proper rate of each chemical for your soil type.

PREEMERGENCE USE

The tank mixture of AMIBEN® plus Lasso plus Lexone or Sencor may be applied as a preemergence broadcast treatment at time of planting. Use the table below to determine the proper rate of each chemical for your soil type.

Make all broadcast treatments according to the following table:

Soil Texture	AMIBEN® Qts/A	Lasso 4E Qts/A	Sencor 50WP or Lexone WP Lbs/A	Sencor 4 or Lexone 4L Pts/A	Sencor DF or Lexone DF Lbs/A
Coarse (loamy sand, sandy loam)	4	1.5 to 2.0	0.5	0.5	0.33
Medium (loam, silt loam, silt)	4 to 5	2.0 to 2.5	0.5 to 0.75*	0.5 to 0.75*	0.33 to 0.5*
Fine (clay, clay loam, silty clay, sandy clay, sandy clay loam, silty clay loam)	5 to 6	2.5 to 3.0	0.75 to 1.0*	0.75 to 1.0*	0.5 to 0.67*

* On Clarion Webster soils in Minnesota and Iowa or on similar alkaline (calcareous) soils a pH of 7.5 or above, apply Sencor WP of Lexone WP at 0.5 pound/A, Sencor 4 or Lexone 4L at 0.5 pint/A, or Lexone DF or Sencor DF at 0.33 pound/A on medium soils. On fine soils use 0.5 to 0.75 pound/A Sencor WP or Lexone WP, 0.5 to 0.75 pint Sencor 4 or Lexone 4L or 0.33 to 0.5 pound/A Lexone DF or Sencor DF. Use the higher rate only where soil pH is less than 7.5 and where pressure is heavy.

Use the higher rate of AMIBEN® where heavy stands of Pennsylvania smartweed, velvetleaf, ragweed, wild mustard, or black nightshade are expected. Use the higher rate of Lasso when treatments are incorporated or where heavy stands of black nightshade, foxtail, witchgrass, panicum, or yellow nutsedge are expected. Use the high rate of Sencor or Lexone where heavy stands of Florida pusley, sesbania, velvetleaf or Jimsonweed are expected except when soil pH is over 7.5.

CULTIVATION AFTER PLANTING

Soil treated with AMIBEN® Chloramben Herbicide plus Lasso plus Sencor or Lexone may be shallow-cultivated, rotary-hoed, or hand-hoed without reducing the weed control activity of the tank mixture. With preplant soil incorporated treatments, do not cultivate deeper than the AMIBEN® plus Lasso plus Sencor or Lexone treated layer of soil since this may bring untreated soil to the surface and poor weed control may result. If sufficient rain does not fall within 5 days after preemergence applications of the tank mixture, but there is enough soil moisture to germinate weeds and grasses, a light cultivation with a rotary hoe or similar tool will uproot these small broadleaf weeds and grasses.

SPECIAL PRECAUTIONS

Applied according to directions and under normal growing conditions, the tank mix of AMIBEN® plus Lasso plus Sencor or Lexone will not harm the treated crop. Over-application may result in crop injury. Uneven application or improper soil incorporation of AMIBEN® plus Lasso plus Sencor or Lexone can result in erratic weed control or crop injury. Seedling disease, cold weather, deep planting, excessive moisture, soil pH over 7.5, high salt concentration, or drought may weaken crop seedlings and increase the possibility of damage from the tank mixtures. Under these conditions, delayed crop development or reduced yields may result.

Do not use on sand or on medium to fine soils with less than 1% organic matter. This mixture applied preplant incorporated may cause slight soybean injury early, but the soybeans should recover and the subsequent growth or yields should not be affected.

Do not use the foliage from soybeans treated with the tank mix of AMIBEN® and Lasso and Sencor or AMIBEN® and Lasso and Lexone for feed or forage.

Caution: Observe all cautions and limitations on labeling of all products used in mixtures.

"AMIBEN®" AND "LASSO" AND "LOROX" FOR SOYBEANS

A tank mixture of AMIBEN® Chloramben Herbicide and Lasso and Lorox L or Lorox 50W may be applied as a preemergence broadcast tank mix treatment during planting or after planting before weeds and soybeans emerge.

A tank mixture of AMIBEN® Herbicide and Lasso and Lorox will control the following annual weeds: barnyardgrass, carpetweed, crabgrass, foxtail (green, yellow, giant), fall panicum, lambsquarters, pigweed, smartweed, ragweed, black nightshade, velvetleaf and wild mustard, as well as all other weeds listed on the AMIBEN®, Lasso and Lorox labels. Velvetleaf and cocklebur often germinate in soil at depths below the chemical penetration. Even though populations of these weeds are reduced under normal rainfall conditions, weed seeds which germinate outside the chemical zone are not affected.

Except as directed in the "Conservation or Reduced Tillage" portion of this label, the seedbed should be free of existing weeds before application. Surface trash (plant residue from previous crop) should be uniformly distributed, and the seedbed should be free of clods. For uniform application of the tank mixture of AMIBEN® Herbicide and Lasso and Lorox use 10 to 20 gallons of water per acre for ground application. To prepare the tank mixture refer to the "Mixing Instructions" portion of this label. DO NOT mix AMIBEN® Herbicide and Lasso and Lorox concentrates before the addition of water as an emulsion will not be formed.

Apply AMIBEN® and Lasso and Lorox as a preemergence treatment using the appropriate rates from the table below. Apply the tank mixture during planting (behind the planter) or after planting before weeds or soybeans emerge.

Rain is necessary to move the chemicals into the soil and make them available to control germinating weed seeds. If no rainfall occurs within 5 days following application, a shallow incorporation no deeper than 2 inches is recommended. High intensity or excessive rainfall after application may reduce control.

Make all broadcast treatments according to the following table:

3% or Less Organic Matter

Soil Texture	AMIBEN® Qts/A	Lasso Qts/A	Lorox 50W or Lorox L Lbs or pts/A
Loamy sand, sandy loam	4	1.5 to 2.0	0.67 to 1.5
Loam, silt loam, silt	4 to 5	2.0 to 2.5	1.0 to 1.67
Silty clay loam, silty clay, sandy clay, clay loam, clay	5 to 6	2.5 to 3.0	1.25 to 2.0

Greater than 3% Organic Matter

Soil Texture	AMIBEN® Qts/A	Lasso Qts/A	Lorox 50W or Lorox L Lbs or pts/A
Loamy sand, sandy loam	4	1.5 to 2.0	1.5 to 1.0
Loam, silt loam, silt	4 to 5	2.0 to 2.5	1.67 to 2.5
Silty clay loam, silty clay, sandy clay, clay loam, clay	5 to 6	2.5 to 3.0	2.0 to 3.0

Use the higher rate of AMIBEN® when heavy populations of black nightshade, velvetshade, mustard or smartweed are expected. Use the higher rate of Lasso when high populations of black nightshade, witchgrass or foxtail are expected. Use the higher rate of Lorox on soils with organic matter contents which are toward the high end of the ranges given in the table.

For band treatments divide the band width by the row width and multiply by the appropriate rates indicated above to determine the amount of chemical to use per acre.

Plant seeds at least 1.75 inches deep and do not treat if beans have begun to emerge.

Read thoroughly the cautionary statements and all other information appearing on labels for Lasso, Lorox L or Lorox 50W and AMIBEN®.

"AMIBEN®" AND "DUAL" FOR SOYBEANS

A tank mixture of AMIBEN® Chloramben Herbicide and Dual 8E may be applied as a preplant soil incorporated herbicide treatment. AMIBEN® Herbicide may be applied as a preemergence broadcast tank mix treatment with Dual 8E during planting or after planting before weeds and soybeans emerge. Following a preplant soil incorporated application of Dual 8E, AMIBEN® Herbicide may be applied preemergence during planting or after planting before weeds and soybeans emerge.

When applied according to label directions, tank mixtures of AMIBEN® and Dual will control all weeds controlled by AMIBEN® alone or by Dual alone including fall panicum, foxtail millet, red rice, signalgrass, witchgrass, yellow nutsedge and Florida pusley. Refer to the general information section of this label and the Dual label for additional weeds controlled by AMIBEN® and Dual 8E.

For uniform application of the tank mixture of AMIBEN® and Dual, use 10 to 20 gallons of water with ground equipment. To prepare the tank mixture refer to the "Mixing Instructions" portion of this label.

Apply AMIBEN® and Dual preplant incorporated or preemergence using the appropriate rates from the table below. For preplant incorporated treatments apply the tank mixture to the soil and incorporate into the top two inches of soil following the procedures in the "Directions for Incorporation" portion of this label. Use a preplant incorporate application if furrow irrigation is used or when a period of dry weather is expected. If soybeans are planted on beds apply and incorporate the tank mixture after bed formation. For preemergence treatments, apply the tank mixture during planting (behind the planter) or after planting before weeds or soybeans emerge. For sequential applications apply Dual 8E alone preplant incorporated as specified on the Dual label. Follow with a preemergence application of AMIBEN® herbicide during planting (behind the planter) or after planting before weeds or soybeans emerge.

Make all broadcast treatments according to the following table:

Soil Texture	AMIBEN® Qts/A	Dual 8E Pts/A
Loamy sand, sandy loam	4	1.5 to 2.0
Loam, silt loam, silt	4 to 5	2.0 to 2.5
Silty clay loam, silty clay, sandy clay, clay loam, clay	5 to 6	2.5 to 3.0

Do not use on muck or peat soils.

Use the higher rate of AMIBEN® when heavy populations of black nightshade, velvetleaf, mustard or smartweed are expected. Use the higher rate of Dual when treatment is incorporated, when organic matter is over 3%, or when high populations of black nightshade, witchgrass, panicum, foxtail, or yellow nutsedge are expected.

Before applying this combination read thoroughly all cautionary statements and all other information appearing on or attached to the containers of AMIBEN® Herbicide and Dual 8E.

"AMIBEN®" TANK MIXTURE WITH "DUAL" AND "SENCOR" OR "LEXONE" FOR SOYBEANS

GENERAL INFORMATION

A tank mixture of AMIBEN® Chloramben Herbicide plus Dual 8E plus Sencor or Lexone when applied in accordance with the following directions, either preplant soil incorporated or preemergence, effectively controls most weeds listed on the current AMIBEN®, Dual 8E, and Lexone WP/4L/DF or Sencor WP/4/DF labels.

DIRECTIONS FOR USE

Except as directed in the "Conservation or Reduced Tillage" portion of this label, the seedbed should be free of existing weeds before application. Surface trash (plant residue from previous crop) should be uniformly distributed, and the seedbed should be free of clods.

The mixture should be applied in from 10 to 20 gallons of water per acre using any properly calibrated low pressure sprayer equipped with herbicide tips and in-line strained screens no finer than 50 mesh. To prepare the tank mixture refer to the "Mixing Instructions" portion of this label. Apply the entire contents of the spray tank within 4 hours of initial mixing.

PREPLANT SOIL INCORPORATED USE

Incorporation should place the AMIBEN® plus Dual 8E Sencor or Lexone into the top 1 to 3 inches of the final seedbed as directed in the "Directions for Incorporation" portion of this label. Apply the tank mixture using the table below to determine the proper rate of each chemical for your soil type.

PREEMERGENCE USE

The tank mixtures of AMIBEN® plus Dual 8E plus Lexone or Sencor may be applied as a preemergence broadcast treatment at time of planting. Use the table below to determine the proper rate of each chemical for your soil type.

Make all broadcast treatments according to the following table:

Soil Texture	AMIBEN® Qts	Dual 8E Pts/A	Sencor 50WP or Lexone WP Lbs/A	Sencor 4 or Lexone 4L Pts/A	Sencor DF or Lexone DF Pts/A
Coarse (loamy sand, sandy loam)	4	1.5 to 2.0	0.5	0.5	0.33
Medium (loam, silt loam, silt)	4 to 5	2.0 to 2.5	0.5 to 0.75*	0.5 to 0.75*	0.33 to 0.5*
Fine (clay, clay loam, silty clay, sandy clay, sandy clay loam, silty clay loam)	5 to 6	2.5 to 3.0	0.75 to 1.0*	0.75 to 1.0*	0.5 to 0.67*

* On Clarion Webster soils in Minnesota and Iowa or on similar alkaline (calcareous) soils a pH of 7.5 or above, apply Sencor WP or Lexone WP at 0.5 pound/A, Sencor 4 or Lexone 4L at 0.5 pint/A, or Lexone DF or Sencor DF at 0.33 pound/A on medium soils. On fine soils use 0.5 to 0.75 pound/A Sencor WP or Lexone WP, 0.5 to 0.75 pint Sencor 4 or Lexone 4L or 0.33 to 0.5 pound/A Lexone DF or Sencor DF. Use the higher rate only where soil pH is less than 7.5 and where weed pressure is heavy.

Use the higher rate of AMIBEN® where heavy stands of Pennsylvania smartweed, velvetleaf, ragweed, wild mustard, or black nightshade are expected. Use the higher rate of Dual 8E when treatments are incorporated, when organic matter is greater than 3%, or where heavy stands of black nightshade, foxtail, witchgrass, panicum, or yellow nutsedge are expected.

Use the high rate of Sencor or Lexone where heavy stands of Florida pusley, sesbania, velvetleaf or Jimsonweed are expected except when soil pH is over 7.5.

CULTIVATION AFTER PLANTING

Soil treated with AMIBEN® plus Dual plus Sencor or Lexone may be shallow-cultivated, rotary-hoed, or hand-hoed without reducing the weed control activity of the tank mixture. With preplant soil incorporated treatments, do not cultivate deeper than the AMIBEN® plus Dual plus Sencor or Lexone treated layer of soil since this may bring untreated soil to the surface and poor weed control may result. If sufficient rain does not fall within 5 days after preemergence applications of the tank mixture, but there is enough soil moisture to germinate weeds and grasses, a light cultivation with a rotary hoe or similar tool will uproot these small broadleaf weeds and grasses.

SPECIAL PRECAUTIONS

Applied according to directions and under normal growing conditions, the tank mix of AMIBEN® plus Dual plus Sencor or Lexone will not harm the treated crop. Over-application may result in crop injury. Uneven application or improper soil incorporation of AMIBEN® plus Dual plus Sencor or Lexone can result in erratic weed control or crop injury. Seedling disease, cold weather, deep planting, excessive moisture, soil pH over 7.5, high salt concentration, or drought may weaken crop seedlings and increase the possibility of damage from the tank mixture. Under these conditions, delayed crop development or reduced yields may result.

Do not use on sand or on medium to fine soils with less than 1% organic matter. This mixture applied preplant incorporated may cause slight soybean injury early, but the soybeans should recover and the subsequent growth or yields should not be affected.

Do not use the foliage from soybeans treated with the tank mix of AMIBEN® and Dual and Sencor or AMIBEN® and Dual and Lexone for feed for forage.

Caution: Observe all cautions and limitations on labeling of all products used in mixtures.

"AMIBEN®" AND "DUAL" AND "LOROX" FOR SOYBEANS

A tank mixture of AMIBEN® Chloramben Herbicide and Dual and Lorox L or Lorox 50W may be applied as a preemergence broadcast tank mix treatment during planting or after planting before weeds and soybeans emerge.

A tank mixture of AMIBEN® Herbicide and Dual and Lorox will control the following annual weeds: barnyardgrass, carpetweed, crabgrass, foxtail (green, yellow, giant), fall panicum, lambsquarters, pigweed, smartweed, ragweed, black nightshade, velvetleaf and wild mustard, as well as all other weeds listed on the AMIBEN®, Dual and Lorox labels. Velvetleaf and cocklebur often germinate in soil at depths below the chemical penetration. Even though populations of these weeds are reduced under normal rainfall conditions, weed seeds which germinate outside the chemical zone are not affected.

Except as directed in the "Conservation or Reduced Tillage" portion of this label, the seedbed should be free of existing weeds before application. Surface trash (plant residue from previous crop) should be uniformly distributed, and the seedbed should be free of clods. For uniform application of the tank mixture of AMIBEN® Herbicide and Dual and Lorox use 10 to 20 gallons of water per acre for ground application. To prepare the tank mixture refer to the "Mixing Instructions" portion of this label.

Apply AMIBEN® and Dual and Lorox as a preemergence treatment using the appropriate rates from the table below. Apply the tank mixture during planting (behind the planter) or after planting before weeds or soybeans emerge.

Rain is necessary to move the chemicals into the soil and make them available to control germinating weed seeds. If no rainfall occurs within 5 days following application, a shallow incorporation no deeper than 2 inches is recommended. High intensity or excessive rainfall after application may reduce control.

Make all broadcast treatments according to the following table:

3% or Less Organic Matter

Soil Texture	AMIBEN® Qts/A	Dual Pts/A	Lorox 50W or Lorox L Lbs or pts/A
Loamy sand, sandy loam	4	1.5 to 2.0	0.67 to 1.5
Loam, silt loam, silt	4 to 5	2.0 to 2.5	1.0 to 1.67
Silty clay loam, silty clay, sandy clay, clay loam, clay	5 to 6	2.5 to 3.0	1.25 to 2.0

Greater than 3% Organic Matter

Soil Texture	AMIBEN® Qts/A	Dual Pts/A	Lorox 50W or Lorox L Lbs or pts/A
Loamy sand, sandy loam	4	1.5 to 2.0	1.5 to 2.0
Loam, silt loam, silt	4 to 5	2.0 to 2.5	1.67 to 2.5
Silty clay loam, silty clay, sandy clay, clay loam, clay	5 to 6	2.5 to 3.0	2.0 to 3.0

Use the higher rate of AMIBEN® when heavy populations of black nightshade, velvetleaf, mustard or smartweed are expected. Use the higher rate of Dual when high populations of black nightshade, witchgrass or foxtail are expected. Use the higher rate of Lorox on soils with organic matter contents which are toward the high end of the ranges given in the table.

For band treatments divide the band width by the row width and multiply by the appropriate rates indicated above to determine the amount of chemical to use per acre.

Plant seeds at least 1.75 inches deep and do not treat if beans have begun to emerge.

Read thoroughly the cautionary statements and all other information appearing on labels for Dual, Lorox L or Lorox 50W and AMIBEN®.

"AMIBEN®" AND "TREFLAN" FOR SOYBEANS

AMIBEN® Chloramben Herbicide may be applied in a band over the soybean row at planting time in fields where Treflan has been applied as a preplant soil incorporated herbicide. Or AMIBEN® Herbicide may be applied several days prior to planting as a broadcast tank mix with Treflan. The tank mixture should be used as a spring preplant soil incorporated treatment. The combination of AMIBEN® Herbicide with Treflan improves broadleaf weed control of species such as Pennsylvania smartweed, velvetleaf and ragweed. For broadcast treatments, incorporate chemicals immediately and thoroughly to an approximate depth of 2 inches with a disc, field cultivator or similar tool set to cut a depth of 4 to 6 inches. Refer to the Treflan label for additional incorporation information.

Apply AMIBEN® at a rate of 4 to 6 quarts per broadcast acre. Use the higher rate where heavy populations of Pennsylvania smartweed, velvetleaf, ragweed, wild mustard or black nightshade are expected.

Apply Treflan at a rate of 1½ pints for medium soils and 2 pints for fine soils. Do not use on muck or charcoal soils. Read and observe all directions and cautions on labels for AMIBEN® and Treflan.

"AMIBEN®" TANK MIXTURES WITH "TREFLAN" AND "SENCOR" OR "LEXONE" FOR SOYBEANS

GENERAL INFORMATION

A tank mix of AMIBEN® Chloramben Herbicide plus Treflan plus Sencor or Lexone when applied preplant soil incorporated as recommended herein for use on soybeans, effectively controls the weeds listed on the current AMIBEN®, Treflan and Sencor WP/4/DF or Lexone WP/4L/DF labels.

DIRECTIONS FOR USE

Except as directed in the "Conservation or Reduced Tillage" portion of this label, the seedbed should be free of existing weeds before application. Surface trash (plant residue from previous crop) should be uniformly distributed, and the seedbed should be free of clods.

To prepare the tank mixture refer to the "Mixing Instructions" portion of this label. Apply the entire contents of the spray tank within 4 hours of initial mixing. Apply the tank mixture in from 10 to 20 gallons of water per acre using any properly calibrated low pressure sprayer equipped with herbicide tips and in-line strained screens no finer than 50 mesh.

For best results AMIBEN® plus Treflan plus Sencor or Lexone should be incorporated as soon as possible after application. The three way tank mixture must be incorporated within 24 hours after application. If AMIBEN® plus Treflan plus Sencor or Lexone is applied to a wet, warm soil surface or if the wind velocity is 10 mph or higher, variable weed control may result from delaying incorporation beyond 24 hours.

Incorporation should place the AMIBEN® Herbicide plus Treflan plus Sencor or Lexone into the top 2 to 3 inches of the final seedbed. Refer to the incorporation portion of the Treflan label for approved incorporation equipment and operating procedures.

CULTIVATION AFTER PLANTING

Soil treated with AMIBEN® plus Treflan plus Sencor or Lexone may be shallow-cultivated, rotary-hoed or hand-hoed without reducing the weed control activity of the tank mixture. Do not cultivate deeper than the AMIBEN® plus Treflan plus Sencor or Lexone treated layer or soil since this may bring untreated soil to the surface and poor weed control may result.

Make all broadcast treatments according to the following table:

Soil Texture	AMIBEN® Qts/A	Treflan 4EC Pts/A	Sencor 50WP or Lexone WP Lbs/A	Sencor 4 or Lexone 4L Pts/A	Sencor DF or Lexone DF Lbs/A
Coarse (sand, loamy sand, sandy loam) [a]	4	1.0	0.5	0.5	0.33
Medium (loam, silt loam, silt)	4 to 5*	1.5	0.5 to 0.75**	0.5 to 0.75**	0.33 to 0.5**
Fine (clay, clay loam, silty clay, sandy clay, sandy clay loam, silty clay loam)	5 to 6*	2.0	0.75 to 1.0**	0.75 to 1.0**	0.5 to 0.67**

[a] Do not use Sencor or Lexone on coarse soils with less than 1% organic matter.

* Use the higher rate of AMIBEN® chloramben herbicide when velvetleaf or black nightshade is a problem.

** On Clarion Webster soils in Minnesota and Iowa or on similar alkaline (calcareous) soils with a pH of 7.5 or above apply Sencor WP or Lexone WP at 0.5 pound/A, Sencor 4 or Lexone 4L at 0.5 pint/A, or Lexone DF or Sencor DF at 0.33 pound/A on medium soils and at 0.5 to 0.75 pound/A Sencor WP or Lexone WP, 0.5 to 0.75 pint/A Sencor 4 or Lexone 4L or 0.33 to 0.5 pound/A Lexone DF or Sencor DF on fine soils. Use the higher rate only where soil pH is less than 7.5 and where weed pressure is heavy.

SPECIAL PRECAUTIONS

Applied according to directions and under normal growing conditions, the AMIBEN® and Treflan and Sencor or AMIBEN® and Treflan and Lexone tank mix will not harm the treated crop. Over-application may result in crop injury. Uneven application or improper soil incorporation of the AMIBEN® and Treflan and Sencor or AMIBEN® and Treflan and Lexone tank mix can result in erratic weed control or crop injury. Seedling disease, cold weather, deep planting, excessive moisture, soil pH over 7.5, high salt concentration, or drought may weaken crop seedlings and increase possibility of damage from the tank mixtures. Under these conditions, delayed crop development or reduced yields may result.

Caution: Observe all cautions and limitations on labeling of all products used in mixtures.

Do not use the foliage from soybeans treated with AMIBEN® and Treflan and Sencor or AMIBEN® and Treflan and Lexone tank mix for feed or forage.

"AMIBEN®" AND "SONALAN" FOR SOYBEANS

AMIBEN® Chloramben Herbicide may be applied as a preemergence band or broadcast treatment in soybeans at planting time in fields where Sonalan has been applied as a preplant soil incorporated herbicide. Or AMIBEN® Herbicide may be applied prior to planting as a broadcast tank mix with Sonalan. The tank mixture should be used as a spring preplant soil incorporated treatment. The combination of AMIBEN® Herbicide with Sonalan improves broadleaf weed control of species such as Pennsylvania smartweed, velvetleaf, common ragweed, black nightshade and wild mustard. For broadcast treatments, incorporate chemicals immediately and thoroughly to an approximate depth of 2 inches as described on the Sonalan label.

Make all broadcast treatments according to the following table:

Soil Texture	AMIBEN® Qts/A	Sonalan Pts/A
Loamy sand, sandy loam	4	1.25 to 2.0
Loam, silt loam, silt	4 to 5	1.75 to 2.5
Silty clay loam, silty clay, sandy clay, clay loam, clay	5 to 6	2.25 to 3.0

Use the higher rate of AMIBEN® where a heavy population of smartweed, velvetleaf, ragweed, wild mustard or black nightshade are expected. Use the higher rate of Sonalan on finer textured soils and on soils with organic matter content over 3%.

"AMIBEN®" AND "PROWL" FOR SOYBEANS

A tank mixture of AMIBEN® Chloramben Herbicide and Prowl may be applied as a preplant soil incorporated herbicide treatment or AMIBEN® Herbicide may be applied as a preemergence broadcast treatment at planting time in soybean fields where Prowl has been applied as a preplant soil incorporated herbicide treatment. On soybeans grown south of Interstate 80 the tank mixture of AMIBEN® plus Prowl may also be applied as a preemergence broadcast treatment. The combination may be used only in Illinois, Indiana, Iowa, Kansas, Kentucky, Michigan, Minnesota, Nebraska, New York, North Dakota, Ohio, Pennsylvania, South Dakota, Wisconsin and Missouri except in the "Bootheel Region."

When applied according to directions, tank mixtures of AMIBEN® Herbicide and Prowl herbicide either preplant incorporated or preemergence, or the sequential treatment of AMIBEN® preemergence over Prowl preplant incorporated will control fall panicum, signalgrass, Venice mallow and velvetleaf in addition to all other weeds listed on the AMIBEN® and Prowl labels.

For uniform application of the tank mixture of AMIBEN® and Prowl, use 10 to 20 gallons of water per acre with ground equipment or 3 to 5 or more gallons of water per acre by aircraft. To prepare the tank mixture refer to the "Mixing Instructions" portion of this label.

Except as directed in the "Conservation or Reduced Tillage" portion of this label, the seedbed should be free of existing weeds before application. Surface trash (plant residue from previous crop) should be uniformly distributed, and the seedbed should be free of clods. When using the tank mixture preplant soil incorporated or when using Prowl preplant incorporated followed by a preemergence application of AMIBEN® the herbicides should be applied within 7 days prior to planting. When applied as a preplant soil incorporated treatment the tank mixture of AMIBEN® plus Prowl should be thoroughly incorporated into the top 0 to 2 inches of soil using equipment and techniques described elsewhere on this label and on the Prowl label. When using the tank mixture preemergence, treatments should be made at planting or within 2 days after planting before weeds and soybeans emerge.

Make all broadcast treatments according to the following table:

Soil Texture	AMIBEN® Qts/A	Prowl Pts/A
Loamy sand, sandy loam	4	1.0 to 2.0
Loam, silt loam, silt	4 to 5	1.5 to 2.5
Silty clay loam, silty clay, sandy clay, clay loam, clay	5 to 6	2.0 to 3.0

Use the higher rate of AMIBEN® when heavy populations of black nightshade, velvetleaf, wild mustard, common ragweed or Pennsylvania smartweed are expected. Use the higher rate of Prowl when treatments are incorporated on soils with more than 3% organic matter, or where heavy stands of foxtail, witchgrass or panicum are expected. Do not use on peat or muck soils.

If sufficient rain does not fall within 5 days after preemergence treatment, but there is enough soil moisture to germinate weeds and soybeans a light cultivation with a rotary hoe or similar tool will uproot these small weeds. The shallow mixing of the herbicides into the soil surface will not interfere with their action when rains come. If soybean crop loss occurs due to weather conditions, soybeans may be replanted. If replanting is necessary, DO NOT rework the soil.

Before applying this combination, read thoroughly the cautionary statements and all other information appearing on or attached to the containers of both the Prowl and AMIBEN® herbicides.

"AMIBEN®" TANK MIXTURE WITH "PROWL" AND "SENCOR" OR "LEXONE" FOR SOYBEANS

GENERAL INFORMATION

A tank mixture of AMIBEN® Chloramben Herbicide plus Prowl plus Sencor or Lexone may be applied either preplant soil incorporated or preemergence in soybeans grown in the northeastern and north central states (Illinois, Indiana, Iowa, Kansas, Kentucky, Michigan, Minnesota, Nebraska, New York, North Dakota, Ohio, Pennsylvania, South Dakota, Wisconsin, Missouri (except the "Bootheel Region"), all areas of Delaware, Maryland, New Jersey, and Virginia (except the Coastal Plains of these states). Do not make preemergence applications of Prowl north of Interstate 80. When applied according to the following directions the tank mixture of AMIBEN® plus Prowl plus Sencor or Lexone effectively controls the weeds listed on the current AMIBEN®, Prowl and Sencor WP/4/DF or Lexone WP/4L/DF labels.

DIRECTIONS FOR USE

Except as directed in the "Conversion or Reduced Tillage" portion of this label, the seedbed should be free of existing weeds before application. Surface trash (plant residue from previous crop) should be uniformly distributed, and the seedbed should be free of clods. Uniformly apply the recommended AMIBEN® plus Prowl plus Sencor or Lexone tank mixture in 10 to 20 gallons of water per acre with ground equipment or in 3 to 5 or more gallons of water per acre by aircraft using herbicide nozzle tips and in line strained screens no finer than 50 mesh. To prepare the tank mixture refer to the "Mixing Instructions" portion of this label. Apply the entire contents of the spray tank within 4 hours of initial mixing.

PREPLANT INCORPORATED USE

The tank mixture of AMIBEN® plus Prowl plus Sencor or Lexone must be incorporated into the soil and soybeans must be planted within 7 days after application. Uneven spray application or improper soil incorporation can produce erratic weed control. The tank mixture should be thoroughly incorporated into the top 1 to 2 inches of soil using equipment and techniques described elsewhere on this label.

PREEMERGENCE USE

Apply the tank mixture of AMIBEN® plus Prowl plus Sencor or Lexone as a broadcast treatment at time of planting or within 2 days after planting before weeds and soybeans emerge. This treatment is most effective when adequate rainfall or overhead irrigation is received within 5 days after application. If sufficient rain does not fall within 5 days after spraying the tank mixture of AMIBEN® plus Prowl plus Sencor or Lexone, but there is enough soil moisture to germinate weeds and grasses a light cultivation with a rotary hoe or similar tool will up-root these small broadleaf weeds and grasses. The shallow mixing of the tank mixture in the surface soil will improve consistency of performance and reduce dependency on rainfall to activate the herbicides, but will not interfere with their action when rains come.

APPLICATION RATES

Make all broadcast treatments according to the following table:

SOIL TEXTURE	AMIBEN® QTS/A	PROWL PTS/A	SENCOR 50WP or LEXONE WP LBS./A	SENCOR 4 or LEXONE 4L PTS./A	SENCOR DF or LEXONE DF LBS./A
Coarse (loamy sand, sandy loam)	4	1.5 to 2.0	0.5	0.5	0.33
Medium (loam, silt loam, silt)	4 to 5	1.5 to 2.5	0.5 to 0.75*	0.5 to 0.75*	0.33 to 0.5*
Fine (clay, clay loam, silty clay, sandy clay, sandy clay loam, silty clay loam)	5 to 6	2.0 to 3.0	0.75 to 1.0*	0.75 to 1.0*	0.5 to 0.67*

* On Clarion Webster soils in Minnesota and Iowa or on similar alkaline (calcareous) soils with a pH of 7.5 or above, apply Sencor WP or Lexone WP at 0.5 pound/A, Sencor 4 or Lexone 4L at 0.5 pint/A, or Lexone DF or Sencor DF at 0.33 pound/A on medium soils. On fine soils use 0.5 to 0.75 pound/A Sencor WP or Lexone WP, 0.5 to 0.75 pint Sencor 4 or Lexone 4L or 0.33 to 0.5 pound/A Lexone DF or Sencor DF. Use the higher rate only where soil pH is less than 7.5 and where pressure is heavy.

Use the higher rate of AMIBEN® where heavy stands of Pennsylvania smartweed, velvetleaf, ragweed, wild mustard, or black nightshade are expected. Use the higher rate of Prowl when treatments are incorporated on soils with more than 3% organic matter, or where heavy stands of foxtail, witchgrass or panicum are expected. Use the high rate of Sencor or Lexone where heavy stands of Florida pusley, sesbania, velvetleaf or Jimsonweed are expected except when soil pH is over 7.5.

SPECIAL PRECAUTIONS

Soybean seed should be planted at least 1½ inches below soil surface. Soil incorporation deeper than 2 inches will reduce weed control and can result in crop injury. Seedling disease, cold weather, deep planting, excessive soil moisture, soil pH over 7.5, high salt concentration, or drought may weaken crop seedlings and increase possibility of damage from the tank mixtures. Under these conditions delayed crop development or reduced yields may result..

Do not use on sensitive soybean varieties such as Tracy, Semmes, Altona, Vansoy or Coker 102.

Do not use the foliage from soybeans treated with the AMIBEN® plus Prowl plus Sencor or Lexone tank mix for feed or forage.

Caution: Observe all cautions and limitations on labeling of all products used in mixtures.

"AMIBEN®" AND METRIBUZIN ("SENCOR" OR "LEXONE") TANK MIXTURE FOR SOYBEANS

AMIBEN® Chloramben Herbicide and metribuzin may be applied in a tank mixture combination as a preemergence band or broadcast application. Use standard spray equipment, and 10 to 20 gallons of spray solution per broadcast acre. To prepare the tank mixture refer to the "Mixing Instructions" portion of this label.

This tank mixture controls Florida pusley, sesbania (coffeebean or coffeeweed), shepherdspurse and sicklepod in addition to those weeds controlled by AMIBEN® Chloramben Herbicide. Performance on cocklebur, fall panicum and black nightshade may be erratic depending on soil temperature, depth of weed seeds in soil and amount of moisture. Control may be improved with timely cultivation.

Make all broadcast treatments according to the following table:

	3% or Less Organic Matter			
Soil Texture	AMIBEN® Qts/A	Sencor 50W or Lexone WP Lbs/A	Sencor 4 or Lexone 4L Pts/A	Sencor DF or Lexone DF Lbs/A
Loamy sand to sandy loam (greater than 2% organic matter only)	4	0.5	0.5	0.33
Loam to sandy clay loam	4 to 5	0.5	0.75	0.5
Silty clay loam to clay	5 to 6	1.0	1.0	0.67

	Greater than 3% Organic Matter			
Soil Texture	AMIBEN® Qts/A	Sencor 50W or Lexone WP Lbs/A	Sencor 4 or Lexone 4L Pts/A	Sencor DF or Lexone DF Lbs/A
Loamy sand to sandy loam (greater than 2% organic matter only)	4	0.75	0.75	0.5
Loam to sandy clay loam	4 to 5	1.0	1.0	0.67
Silty clay loam to clay	5 to 6	1.0	1.0	0.67

In the Mississippi Delta only, on silty clay to heavy clay textured soil use 6 quarts of AMIBEN® plus Lexone 50 WP—1.5 lb., Lexone DF—1 lb. or 2 lbs. of Sencor 50 WP, 2 pts. of Sencor 4, or 1 lb. of Sencor DF. Do not use on muck soils. Use the high rates when heavy weed stands are expected.

For band application rates divide Band width by the Row spacing and multiply by rate/acre selected from the chart above.

SPECIAL PRECAUTIONS

Injury may occur to soybeans following application of metribuzin on soils having a calcareous surface area or a pH of 7.5 or higher. Sensitive soybean varieties such as Tracy, Semmes, Altona, Vansoy or Coker 102, should not be treated with metribuzin or injury may result.

If initial planting does not produce a satisfactory stand, fields treated with AMIBEN® Herbicide and metribuzin tank mixtures may be replanted to soybeans. If replanting is necessary, do not rework soil. Do not apply a second treatment over the first treatment as injury to soybeans may occur. Read Sencor or Lexone labels and literature carefully to understand conditions to insure crop safety. For additional precautions and information concerning spray equipment, mixing directions, and restrictions refer to the appropriate sections of the AMIBEN® Herbicide and the Lexone/Sencor labels. Do not use vines treated with the AMIBEN®, metribuzin tank mix for feed or forage.

"AMIBEN®" AND "LOROX" TANK MIXTURE FOR SOYBEANS

A tank mixture of AMIBEN® Herbicide and Lorox 50W or Lorox L will control annual weeds such as barnyardgrass, carpetweed, crabgrass, fall panicum, foxtail, lambsquarters, pigweed, ragweed and Pennsylvania smartweed. Velvetleaf and cocklebur often germinate in soil at depths below the chemical penetration. Even though populations of these weeds are reduced under normal rainfall conditions, weed seeds which germinate outside the chemical zone are not affected.

Treatments should be made at planting time as a simple preemergence application using 10 to 20 gallons of water per acre. For air application use 3 to 5 gallons per acre. To prepare the tank mixture refer to the "Mixing Instructions" portion of this label.

Plant seeds at least 1.75 inches deep and do not treat if beans have begun to emerge.

Make all broadcast treatments according to the following table:

	3% or Less Organic Matter		Greater than 3% Organic Matter	
Soil Texture	AMIBEN® Qts/A	Lorox 50W or Lorox L Lbs or pts/A	AMIBEN® Qts/A	Lorox 50W or Lorox L Lbs or pts/A
Sandy loam	4	0.67 to 1.5	4	1.5 to 2.0
Silt loam	5	1.0 to 1.67	5	1.67 to 2.5
Clay loam	6	1.25 to 2.0	6	2.0 to 3.0

For band treatments, use proportionally less AMIBEN® and Lorox. Before applying this tank mixture as a broadcast or band treatment, read thoroughly the cautionary statements and all other information appearing on both product labels.

"AMIBEN®" AND "BASALIN" FOR SOYBEANS

AMIBEN® Chloramben Herbicide may be applied over soybeans at planting time in fields where Basalin herbicide has been applied as a preplant soil incorporated herbicide. Or AMIBEN® Herbicide may be applied as a preplant incorporated tank mix with Basalin.

When applied according to direction, tank mixtures of AMIBEN® Herbicide and Basalin or the sequential treatment will control fall panicum, signalgrass, and common purslane in addition to those weeds controlled with AMIBEN® Herbicide alone.

For uniform application of the tank mixture of AMIBEN® and Basalin use 10 to 20 gallons of water per acre with ground equipment. To prepare the tank mixture refer to the "Mixing Instructions" portion of this label. The tank mixture may be applied and incorporated at planting or several days prior to planting.

When using Basalin herbicide followed by preemergence application of AMIBEN®, Basalin may be applied as many as 6 to 8 weeks before planting or up to planting. Except as directed in the "Conservation or Reduced Tillage" portion of this label, the seedbed should be free of existing weeds before application. Surface trash (plant residue from previous crop) should be uniformly distributed, and the seedbed should be free of clods. Basalin is to be incorporated into the soil within eight hours of application prior to planting. Follow the incorporation instructions on the

Basalin label. Read all instructions and cautions on both the AMIBEN® and Basalin labels.

Apply AMIBEN® herbicide at a rate of 4 to 6 quarts per broadcast acre. Apply Basalin herbicide at 1.5 to 3 pints per broadcast acre. Use the higher rate of the range on heavier, finer textured soils or on soils with over 4% organic mattter. Do not use on peat or muck soils.

Before applying this combination read thoroughly the cautionary statements and all other information appearing on or attached to the containers of both Basalin and AMIBEN® herbicides.

"AMIBEN®" TANK MIXTURES WITH "BASALIN" AND "SENCOR" OR "LEXONE" FOR SOYBEANS

GENERAL INFORMATION

A tank mix of AMIBEN® Chloramben Herbicide plus Basalin plus Sencor or Lexone when applied preplant soil incorporated as recommended herein for use on soybeans, effectively controls the weeds listed on the current AMIBEN® Basalin and Sencor WP/4/DF or Lexone WP/4L/DF labels.

DIRECTIONS FOR USE

Except as directed in the "Conservation or Reduced Tillage" portion of this label, the seedbed should be free of existing weeds before application. Surface trash (plant residue from previous crop) should be uniformly distributed, and the seedbed should be free of clods.

To prepare the tank mixture refer to the "Mixing Instructions" portion of this label. Apply the entire contents of the spray tank within 4 hours of initial mixing. Apply the tank mixture in from 10 to 20 gallons of water per acre using any properly calibrated low pressure sprayer equipped with herbicide tips and in-line strained screens no finer than 50 mesh.

For best results AMIBEN® plus Basalin plus Sencor or Lexone should be incorporated as soon as possible after application. The three way tank mixture must be incorporated within 24 hours after application. If AMIBEN® plus Basalin plus Sencor or Lexone is applied to a wet, warm soil surface or if the wind velocity is 10 mph or higher, variable weed control may result from delaying incorporation beyond 24 hours.

Incorporation should place the AMIBEN® Herbicide plus Basalin plus Sencor or Lexone into the top 2 to 3 inches of final seedbed. Refer to the incorporation portion of the Basalin label for approved incorporation equipment and operating procedures.

CULTIVATION AFTER PLANTING

Soil treated with AMIBEN® plus Basalin plus Sencor or Lexone may be shallow-cultivated, rotary-hoed or hand-hoed without reducing the weed control activity of the tank mixture. Do not cultivate deeper than the AMIBEN® plus Basalin plus Sencor or Lexone treated layer of soil since this may bring untreated soil to the surface and poor weed control may result.

Make all broadcast treatments according to the following table:

Soil Texture	AMIBEN® Qts/A	Basalin Pts/A	Sencor 50WP or Lexone WP Lbs/A	Sencor 4 or Lexone 4L Pts/A	Sencor DF or Lexone DF Lbs/A
Coarse (sand, loamy sand, sandy loam)[a]	4	1.0	0.5	0.5	0.33
Medium (loam, silt loam, silt)	4 to 5*	0.5 to 0.75**	0.5 to 0.75**		0.33 to 0.5**
Fine (clay, clay loam, silty clay, sandy clay, sandy clay loam, silty clay loam)	5 to 6*	2.0	0.75**	0.75**	0.5**

[a] Do not use Sencor or Lexone on coarse soils with less than 1% organic matter.

* Use the higher rate of AMIBEN® Chloramben Herbicide when velvetleaf or black nightshade is a problem.

** On Clarion Webster soils in Minnesota and Iowa or on similar alkaline (calcareous) soils with a pH of 7.5 of above apply Sencor WP or Lexone WP at 0.5 pound/A, Sencor 4 or Lexone 4L at 0.5 pint/A or Lexone DF or Sencor DF at 0.33 pound/A on medium soils and at 0.5 to 0.75 pound/A Sencor WP or Lexone WP, 0.5 to 0.75 pint/A Sencor 4 or Lexone 4L or 0.33 to 0.5 pound/A Lexone DF or Sencor DF on fine soils. Use the higher rate only where soil pH is less than 7.5 and where weed pressure is heavy.

SPECIAL PRECAUTIONS

Applied according to directions and under normal growing conditions, the AMIBEN® and Basalin and Sencor or AMIBEN® and Basalin and Lexone tank mix will not harm the treated crop. Over-application may result in crop injury. Uneven application or improper soil incorporation of the AMIBEN® and Basalin and Sencor or AMIBEN® and Basalin and Lexone tank mix can result in erratic weed control or crop injury. Seedling disease, cold weather, deep planting, excessive moisture, soil pH over 7.5, high salt concentration, or drought may weaken crop seedlings and increase possibility of damage from the tank mixtures. Under these conditions, delayed crop development or reduced yields may result.

Caution: Observe all cautions and limitations on labeling of all products used in mixtures.

Do not use the foliage from soybeans treated with the AMIBEN® and Basalin and Sencor or AMIBEN® and Basalin and Lexone tank mix for feed for forage.

"AMIBEN®" AND "REWARD" FOR SOYBEANS

GENERAL INFORMATION

A tank mixture of AMIBEN® Chloramben Herbicide and Reward can be used prior to planting for weed control in soybeans or AMIBEN® may be applied over soybeans at planting time when Reward has been applied as a preplant soil incorporated herbicide.

The combination will control a broader spectrum of weeds than either material alone.

WEEDS CONTROLLED:

Nutsedge (yellow & purple)	Velvetleaf-buttonweed
Barnyardgrass (watergrass)	Stinkgrass
Crabgrass (hairy & smooth)	Carpetweed
Foxtail (giant, green, yellow)	Chickweed
German millet	Coffeeweed
Goosegrass	Florida pusley
Johnsongrass (seedling)	Kochia
Mustard	Lambsquarters (common)
Nightshade	Morningglory (annual)
Prickly sida (teaweed)	Pigweed (redroot & smooth)
Ragweed (common)	Purslane
Smartweed (Pennsylvania)	Sicklepod
	Spurge (annual)

DIRECTIONS FOR USE

Reward must be incorporated into the soil immediately (within minutes) to prevent loss of the herbicide. Whenever possible application and incorporation of Reward or a tank mixture of AMIBEN® and Reward should be done in the same operation. Refer to the Reward label for equipment and operating instructions to best incorporate the tank mixture.

Sequential Applications:

Before planting, apply Reward at the rates listed in the table below for your soil type. Incorporate into the soil as described above and on the Reward label. At or after planting apply AMIBEN® Herbicide at the rates listed in the table below as a broadcast treatment.

Tank Mix Application:

On medium textured soils a tank mixture of AMIBEN® and Reward may be applied and incorporated prior to planting. Use the rates of AMIBEN® and Reward listed in the table below. Follow the incorporation directions described elsewhere on this label. To prepare the tank mixture refer to the "Mixing Instructions" portion of this label. Apply in 10 to 20 gallons of water per acre.

Make all broadcast treatments according to the following table:

Soil Type	AMIBEN® Qts/Acre	Reward Pts/Acre
Coarse (sandy-light)	4	2.5
Medium to Fine (sandy loam to clay loam)	5	3.0
Fine to Very Fine (clay)	6	3.5

Read and follow all directions, use limitations, precautions and warnings that appear on the AMIBEN® and Reward labels.

"AMIBEN®" TANK MIX WITH "SURFLAN" FOR WEED CONTROL IN SOYBEANS

GENERAL INFORMATION

The AMIBEN® Chloramben Herbicide and Surfla tank mixture effectively controls all annual grass and broadleaf weeds listed on the Surflan and AMIBEN® Herbicide labels including chickweed, coffeeweed, kochia, wild mustard, black nightshade, prickly sida (teaweed), common ragweed, Russian thistle, Pennsylvania smartweed, annual spurge, stinkgrass and velvetleaf (buttonweed).

AMIBEN® and Surflan tank mixtures may be applied within two days after planting, but before crop emergence using a properly calibrated, low-pressure herbicide sprayer that will provide uniform soil coverage. Agitate thoroughly before and during application. Follow normal AMIBEN® and Surflan procedures for soil preparation. Do not spray over top of emerged soybeans.

DIRECTIONS FOR USE

Make all broadcast application according to the following table:

Soil Texture	AMIBEN® Qts/A	Surflan 4AS Qts/A	Surflan 75W Lbs/A
Coarse Soils	4 to 6	0.5 to 0.75	0.67 to 1.0
Medium Soils	4 to 6	0.75 to 1.0	1 to 1.33
Fine Soils	4 to 6	1.0 to 1.25	1.33 to 1.67

The higher rate of AMIBEN® Herbicide should be used on clay loams, clay or high organic (3 to 5%) mineral soils. Do not use the AMIBEN® and Surflan tank mix on soils containing more than 5% organic mattter.

SPECIAL PRECAUTION

Poor weed control and/or crop injury may result if directions are not followed. Over-application may result in crop injury and/or soil residue of Surflan. Do not use treated vines for feed or forage. Observe all precautions on both the AMIBEN® and Surflan labels.

"AMIBEN®" AND "PREMERGE 3" TANK MIXTURE FOR SOYBEANS

Preemergence—Use a tank mix combination of AMIBEN®

Chloramben Herbicide and Premerge 3 consisting of 4 to 6 quarts of AMIBEN® plus 4 to 6 quarts of Premerge 3. Apply at time of planting or before crop emergence. Use the higher rate of each product on clay and clay loam soils.

Early postemergence (when soybeans have emerged and are still in the cotyledon stage before first leaves open to expose the terminal bud). Combine 4 to 6 quarts of AMIBEN® with 1 to 2 quarts of Premerge 3 to control emerged weeds including pigweed, cocklebur, annual morningglory, jimsonweed, common ragweed, velvetleaf, annual Pennsylvania smartweed, small seedling grasses and many others that emerge before or with the crop. AMIBEN® Herbicide provides residual control of annual weed seedlings that may germinate after spraying. Use 7 gallons of water when applied by aircraft or about 30 gallons of water applied by ground equipment. The amount of Premerge 3 to be used per acre depends on the maximum air temperature expected within 2 hours; use 2 quarts below 75 degrees F, 1 quart from 75 to 95 degrees F and do not apply above 95 degrees F.

Some injury to soybean plants may occur but yields usually are not reduced. Note: Do not spray if soil surface is wet or if first leaves have opened. Do not use this tank mix if AMIBEN® has been applied preemergence.

AMIBEN® PROGRAMMED CONTROL OF WILD PROSO MILLET IN SOYBEANS

When used as directed below, a weed control program of sequential AMIBEN® Chloramben Herbicide applications will control wild proso millet in addition to all other weed species listed on the AMIBEN® label.

DIRECTIONS FOR USE

1 to 4 days before planting apply a preplant incorporated tank mixture of 4 to 6 quarts AMIBEN® plus the full labeled rate of another grass control material such as Treflan, Basalin, Prowl, Lasso or Dual as directed elsewhere on this label. After soybean emergence, but before emergence of wild proso millet seedlings make a second broadcast application of 4 to 6 quarts AMIBEN® per acre. This application should be made when soybeans are in the cotyledon to 2nd trifoliate growth stage, but before emergence of wild proso millet seedlings. Under normal conditions this will be 10 to 21 days after planting. If wild proso millet or other weeds have begun to emerge use a rotary hoe to remove emerged seedling weeds.

Do not feed straw or green fodder from soybeans treated with AMIBEN® postemergence. Observe caution statements on this and all other herbicide labels.

AMIBEN® FOR POSTEMERGENCE WEED CONTROL IN SOYBEANS

"AMIBEN®" AND "CROP OIL" FOR POSTEMERGENCE WEED CONTROL IN SOYBEANS

For programmed weed control apply a preemergence or preplant incorporated grass herbicide at planting followed by a later application of AMIBEN® plus crop oil concentrate when soybeans are in the cracking to 4th trifoliate stage of growth. Apply a tank mixture of 6 quarts of AMIBEN® plus 1 quart per acre of crop oil concentrate to control or suppress 1 to 8 inch velvetleaf and 1 to 3 inch Pennsylvania smartweed, common ragweed and pigweed causing weeds to either die or cease growth and become noncompetitive with the crop. Application at this timing provides both postemergence activity on the above species and extended preemergence activity on those late germinating annual grass and broadleaf weeds listed in the general information section of the AMIBEN® label which have not yet germinated at the time of treatment. Make applications in 10 to 20 gallons water per acre by ground or 3 to 5 gallons water per acre by air. Good coverage is essential for maximum weed control. Wide spaced, low pressure, flood type nozzles do not give satisfactory coverage. Nozzle pressure should be at least 30 pounds per square inch. Width of swath must not exceed length of boom. Under good soil moisture conditions this treatment will control small grassy weeds (1 to 2 leaf) if rainfall occurs within 48 hours of application.

Caution: Do not use soybean straw or green fodder which has been treated postemergence with AMIBEN® for livestock feed. Do not apply later than 33 days after planting.

Do not apply more than 6 quarts per acre of AMIBEN® alone or in combination as a postemergence treatment. If a preemergence or preplant soil incorporated application of AMIBEN® is made, do not apply more than a total of 12 quarts of AMIBEN® per acre per growing season (6 quarts preemergence plus 6 quarts postemergence).

AMIBEN® CULTIVATION FOR SOYBEANS

If heavy rainfall has prevented application of preplant incorporated or preemergence herbicides AMIBEN® Chloramben Herbicide may be applied when soybeans are in the cracking to second trifoliate stage of growth at the rate of 5 to 6 quarts of AMIBEN® Herbicide per acre. Grass and broadleaf weeds which have emerged must be removed. To remove these germinated weeds, use shallow cultivation with a tool such as a rotary hoe before or after AMIBEN® Herbicide has been applied. Avoid ridging the soil if a cultivator is used after AMIBEN® has been applied. Use the higher rate on fine textured soils or when heavy weed populations are expected.

Caution: Do not use soybean straw or green fodder which has been treated postemergence with AMIBEN® for livestock feed. Do not apply later than 33 days after planting.

Do not apply more than 6 quarts per acre of AMIBEN® alone or in combination as a preplant soil incorporated or preemergence treatment. If a postemergence application of AMIBEN® is made, do not apply more than a total of 12 quarts of AMIBEN® per acre per growing season (6 quarts preemergence plus 6 quarts postemergence).

AMIBEN® AND BLAZER FOR POSTEMERGENCE WEED CONTROL IN SOYBEANS

GENERAL INFORMATION

When applied to young actively growing broadleaf weed seedlings the combination of AMIBEN® Chloramben Herbicide plus Blazer will control all weeds listed on the Blazer label including cocklebur, jimsonweed, morningglory, Pennsylvania smartweed, pigweed, Eastern black nightshade, and common ragweed and will also provide preemergence control of those grass and broadleaf weeds listed in the general information section of the AMIBEN® label which have not yet germinated at time of treatment. While the combination of AMIBEN® plus Blazer will provide suppression of small (2 to 3 leaf) foxtail and fall panicum it must not be used as a basic grass control program and should only be applied to fields which have received a preemergence or preplant incorporated grass herbicide treatment as directed on the AMIBEN® label. In addition to providing preemergence control of late germinating weeds the combination of AMIBEN® plus Blazer will provide control of velvetleaf and increased control of Pennsylvania smartweed, and pigweed which are approaching the maximim weed size listed on the Blazer label.

USE DIRECTIONS

Apply a tank mixture of 5 to 6 quarts AMIBEN® plus 1.5 to 2 pints Blazer when broadleaf weeds have 2 to 4 leaves as directed on the Blazer label (usually 14 to 21 days after planting). Use the higher rate on larger weeds. Use 10 to 20 gallons of water per acre with ground equipment or 5 to 10 gallons of water per acre by air to insure good coverage. Wide spaced, low pressure, flood type nozzles do not give satisfactory coverage. Width of swath must not exceed length of boom. Do not make applications later than 33 days after planting. Refer to the Blazer label for specific directions regarding the addition of a surfactant or crop oil. Follow these same directions when using the tank mixture of AMIBEN® plus Blazer to determine when to add surfactant or crop oil except that 1 quart per acre of crop oil concentrate must be added to provide acceptable control of velvetleaf.

Caution: Do not use soybean straw or green fodder which has been treated postemergence with AMIBEN® plus Blazer for livestock feed. Read and observe all directions and cautions on labels for Blazer and AMIBEN®. Do not apply later than 33 days after planting.

"AMIBEN®" PLUS BUTYRAC® 200 FOR POSTEMERGENCE WEED CONTROL IN SOYBEANS

GENERAL INFORMATION

When applied as directed below a tank mixture of AMIBEN® Chloramben Herbicide plus BUTYRAC® 200 will control or suppress pigweed, velvetleaf, cocklebur and Pennsylvanis smartweed causing them to either die or become noncompetitive with the crop. In addition, this tank mixture will provide preemergence control of those grass and broadleaf weeds listed in the general information section of the AMIBEN® label which have not yet germinated at time of treatment. The tank mixture of AMIBEN® plus BUTYRAC® 200 will not control emerged grassy weeds and should only be used on fields which have received a preemergence or preplant incorporated grass control herbicide.

This combination may cause temporary twisting, curling or malformation of soybean leaves and stems. Soybeans will usually outgrow these effects with no reduction in yield. Do not apply to fields which are under stress from moisture or disease. Good coverage is essential for maximum weed control. Wide spaced, low pressure, flood type nozzles do not give satisfactory coverage. Nozzle pressure should be at least 30 pounds per square inch. Width of swath must not exceed length of boom.

USE DIRECTIONS

Apply a tank mixture of 5 to 6 quarts AMIBEN® plus 2 fluid ounces Butyrac® 200 in 10 to 20 gallons of water per acre by ground equipment when soybeans have 2 to 4 trifoliate leaves but before weeds are 4 inches tall. Use the higher rate when weeds are over 2 inches tall. Do not make application later than 33 days after planting. When weed pressure is heavy addition of 1 pint per acre of crop oil concentrate has improved weed control, but may cause more soybean malformation as described above.

Caution: Do not use soybean straw or green fodder which has been treated postemergence with AMIBEN® plus BUTYRAC® 200 for livestock feed. Read and observe directions and cautions on labels for BUTYRAC® 200 and AMIBEN®. Do not make applications later than 33 days after planting.

"AMIBEN®" PLUS "ALANAP" FOR POSTEMERGENCE WEED CONTROL IN SOYBEANS

GENERAL INFORMATION

When applied according to the directions below a tank mixture of AMIBEN® Chloramben Herbicide plus Alanap will control or suppress velvetleaf, pigweed, cocklebur, and Pennsylvania smartweed causing them to either die or become noncompetitive with the crop. While the combination of AMIBEN® plus Alanap often provides control or suppression of small (2 to 3 leaf) annual grasses it should only be applied to fields which have received a preemergence or preplant incorporated application of a grass control herbicide. The tank mixture of AMIBEN® plus Alanap may cause temporary stunting of soybeans. Under normal conditions the soybeans will recover and yields will not be affected.

USE DIRECTIONS

Apply a tank mixture of 5 to 6 quarts AMIBEN® plus 2 to 3 quarts Alanap in 10 to 20 gallons per acre of water by ground equipment or in 5 or more gallons per acre of water by aerial equipment. To assure good coverage use a minimum of 30 pounds per square inch of nozzle pressure. Apply when soybeans have 3 to 6 trifoliate leaves, but no later than 33 days after planting. Use the higher rates when weeds are larger (over 4 inches). Addition of 1 quart per acre of a surfactant or crop oil concentrate has shown increased control of these large weeds.

Caution: Do not use soybean straw or green fodder which has been treated postemergence with AMIBEN® plus Alanap for livestock feed. Read and observe all directions and cautions on labels for Alanap and AMIBEN®. Do not make applications later than 33 days after planting.

DIRECTIONS FOR USE ON DRYBEANS (INCLUDING WHITE, NAVY, KIDNEY, PINTO, LIMA AND ADZUKI BEANS).

AMIBEN® PREEMERGENCE OR PREPLANT INCORPORATED

AMIBEN® Chloramben Herbicide may be used on drybeans as a preemergence or preplant incorporated herbicide. AMIBEN® may be applied and incorporated at planting or several days prior to planting. Uneven spray application or improper soil incorporation of AMIBEN® can produce erratic weed control. Refer to the "Directions for Incorporation" portion of this label for approved incorporation equipment and recommended operating procedures.

Make all broadcast treatments according to the following table:

Soil Texture	AMIBEN® Qts/A
Loamy sand, sandy loam	4 to 5
Loam, silt loam, silt	5 to 6
Silty clay loam, silty clay, sandy clay, clay loam, clay	6

Use the higher rate of AMIBEN® when heavy populations of black nightshade, velvetleaf, wild mustard, common ragweed or Pennsylvania smartweed are expected. To insure good spray coverage use 10 to 20 gallons of water per acre with ground equipment or a minimum of 5 gallons of water per acre by aircraft.

When band spraying, mount the spray nozzle on the packer or row wheel so that planting and spraying are done in the same operation. Proper calibration and maintenance of desired nozzle height above the ground are very important. Weeds growing in the middles between the treated rows should be cultivated as soon as necessary.

Use care while cultivating to avoid throwing fresh soil on treated bands.

If sufficient rain does not fall within 5 days after spraying AMIBEN® herbicide as a preemergence treatment, but there is enough soil moisture to germinate weeds and grasses, a light cultivation with a rotary hoe or similar tool will up-root these small broadleaf weeds and grasses. The shallow mixing of AMIBEN® in the surface soil will not interfere with the action of AMIBEN® when rains come.

If heavy rain packs the soil surface, rotary hoe in the field to aid bean emergence. Do not disrupt the chemical band with deep cultivation. In coarse textured soils heavy rains on incorporated AMIBEN® herbicide may move it below the weed seed germination zone and produce erratic weed control.

DO NOT apply more than 6 quarts per acre of AMIBEN® herbicide to the same field of drybeans in the same season.

"AMIBEN®" AND "TREFLAN" FOR DRYBEANS

AMIBEN® Chloramben Herbicide may be applied as a preemergence band or broadcast treatment in drybeans at planting time in fields where Treflan has been applied as a preplant soil incorporated herbicide. Or AMIBEN® Herbicide may be applied several days prior to planting as a broadcast tank mix with Treflan. The tank mixture should be used as a spring preplant soil incorporated treatment. The combination of AMIBEN® Herbicide with Treflan improves broadleaf weed control of species such as Pennsylvania smartweed, velvetleaf and ragweed. For broadcast treatments, incorporate chemicals immediately and thoroughly to an approximate depth of 2 inches as described on the Treflan label.

Apply AMIBEN® Herbicide at a rate of 4 to 6 quarts per broadcast acre. Use the higher rate where heavy populations of Pennsylvania smartweed, velvetleaf, ragweed, wild mustard or black nightshade are expected.

Apply Treflan at a rate of $1\frac{1}{2}$ pints for medium soils and 2 pints for fine soils. Do not use on muck or charcoal soils. Read and observe all directions and cautions on labels for AMIBEN® and Treflan.

"AMIBEN®" AND "EPTAM" FOR DRYBEANS

AMIBEN® Chloramben Herbicide may be applied as a preplant soil incorporated tank mixture with "Eptam" herbicide. Or AMIBEN® may be applied preemergence over drybeans at planting time where Eptam has been applied as preplant soil incorporated treatment. When applied preemergence, effectiveness depends on adequate rainfall after treatment. When applied according to directions, this combination will control annual broadleaves including black nightshade, ragweed, Pennsylvania smartweed, wild mustard and annual grasses.

DIRECTIONS FOR USE

Apply AMIBEN® Herbicide either preplant soil incorporated or preemergence at a rate of 4 to 6 quarts per acre. Apply Eptam 7E preplant soil incorporated at a rate of 2.5 to 3.5 pints per acre. Use the higher rates on fine textured soils or where weed pressure is heavy.

The AMIBEN® plus Eptam tank mix should be incorporated immediately after application with a field cultivator, tandem disc, mulch treader, rolling cultivator or power driven tiller as described on the Eptam label. Thoroughly mix in the top 2-3" of the soil surface by incorporating twice in different directions. Refer to the Eptam label for further information on incorporation procedures.

Rainfall is not as critical for activation of AMIBEN® Herbicide when it is applied preplant incorporated as when surface applied.

Apply in 10 to 20 gallons of water per acre. Prepare the spray mixture refer to the "Mixing Instructions" portion of this label. Do not mix the concentrates before adding to water.

If sufficient rain does not occur within 5 days after spraying AMIBEN® preemergence, a light cultivation or incorporation with a rotary hoe or similar tool is recommended.

Do not use Eptam or an AMIBEN® plus Eptam tank mixture on adzuki beans, cowpeas (blackeye peas, blackeye beans), soybeans, lima beans or other flat podded beans except Romano. Under abnormal weather conditions stunting may occur on Gratiot, Michilite, Sanilac, Seafarer, and Seaway varieties. Do not use on muck or charcoal soils. Read and observe all directions and cautions on labels for AMIBEN® and Eptam.

"AMIBEN®" PLUS "EPTAM" AND "TREFLAN" FOR DRYBEANS

AMIBEN® Chloramben Herbicide may be applied as a preplant soil incorporated tank mixture with "Eptam" and "Treflan" herbicides. Or AMIBEN® may be applied preemergence over drybeans at planting time where Eptam has been applied as preplant soil incorporated treatment. When applied preemergence, effectiveness depends on adequate rainfall after treatment. When applied according to directions, this combination will control annual broadleaves including black nightshade, ragweed, Pennsylvania smartweed, wild mustard annual grasses.

DIRECTIONS FOR USE

Apply AMIBEN® Herbicide either preplant soil incorporated or preemergence at a rate of 4 to 6 quarts per acre. Apply Eptam preplant soil incorporated at a rate of 2.5 pints per acre and Treflan preplant soil incorporated at a rate of 1 pint per acre. Use the higher rate of AMIBEN® on fine textured soils or where weed pressure is heavy.

The AMIBEN® plus Eptam plus Treflan or the Eptam plus Treflan tank mixes should be incorporated immediately after application following the equipment and operating procedures described on the Eptam and Treflan labels.

Rainfall is not as critical for activation of AMIBEN® Herbicide when it is applied preplant incorporated as when surface applied.

Apply in 10 to 20 gallons of water per acre. To prepare the spray mixture refer to the "Mixing Instructions" portion of this label. Do not mix the concentrates before adding to water.

If sufficient rain does not occur within 5 days after spraying AMIBEN® preemergence, light cultivation or incorporation with a rotary hoe or similar tool is recommended.

Do not use Eptam or an AMIBEN® plus Eptam plus Treflan tank mixture on adzuki beans, cowpeas (blackeye peas, blackeye beans), soybeans, lima beans or other flat podded beans except Romano. Under abnormal weather conditions stunting may occur on Gratiot, Michilite, Sanilac, Seafarer, and Seaway varieties. Do not use on muck or charcoal soils. Read and observe all directions and cautions on labels for AMIBEN®, Eptam and Treflan.

"AMIBEN®" AND "SONALAN" FOR DRYBEANS

AMIBEN® Chloramben Herbicide may be applied as a preemergence band or broadcast treatment in drybeans at planting time in fields where Sonalan has been applied as a preplant soil incorporated herbicide. Or AMIBEN® Herbicide may be applied prior to planting as a broadcast tank mix with Sonalan. The tank mixture should be used as spring preplant soil incorporated treatment. The combination of AMIBEN® Herbicide with Sonalan improves broadleaf weed control of species such as Pennsylvania smartweed, velvetleaf, common ragweed, black nightshade and wild mustard. For broadcast treatments, incorporate chemicals immediately and thoroughly to an approximate depth of 2 inches as described on the Sonalan label.

Make all broadcast treatments according to the following table:

Soil Texture	AMIBEN® Qts/A	Sonalan Pts/A
Loamy sand, sandy loam	4	1.25 to 2.0
Loam, silt loam, silt	4 to 5	1.75 to 2.5
Silty clay loam, silty clay, sandy clay, clay loam, clay	5 to 6	2.25 to 3.0

Use the higher rate of AMIBEN® where a heavy population of smartweed, velvetleaf, ragweed, wild mustard or black nightshade are expected. Use the higher rate of Sonalan on finer textured soils and on soils with organic matter content over 3%.

"AMIBEN®" AND "PROWL" FOR DRYBEANS (except Adzuki beans)

A tank mixture of AMIBEN® Chloramben Herbicide and Prowl may be applied as a preplant soil incorporated herbicide treatment or AMIBEN® Herbicide may be applied as a preemergence broadcast treatment at planting time in drybean fields where Prowl has been applied as a preplant soil incorporated herbicide treatment.

When applied according to directions, tank mixtures of AMIBEN® Herbicide and Prowl herbicide either preplant incorporated or the sequential treatment of AMIBEN® preemergence over Prowl preplant incorporated will control fall panicum, signalgrass, Venice mallow and velvetleaf in addition to all other weeds listed on the AMIBEN® and Prowl labels.

For uniform application of the tank mixture of AMIBEN® and Prowl, use 10 to 20 gallons of water per acre with ground equipment or 3 to 5 or more gallons of water per acre by aircraft. To prepare the tank mixture refer to the "Mixing Instructions" portion of this label.

Except as directed in the "Conservation or Reduced Tillage" portion of this label, the seedbed should be free of existing weeds before application. Surface trash (plant residue from previous crop) should be uniformly distributed, and the seedbed should be free of clods. When using the tank mixture preplant soil incorporated or when using Prowl preplant incorporated followed by a preemergence application of AMIBEN® the herbicides should be applied within 7 days prior to planting. When applied as a preplant soil incorporated treatment the tank mixture of AMIBEN® plus Prowl should be thoroughly incorporated into the top 0 to 2 inches of soil using equipment and techniques described elsewhere on this label and on the Prowl label. When using the sequential treatment AMIBEN® should be applied at planting or within 2 days after planting before weeds and drybeans emerge.

Make all broadcast treatments according to the following table:

Soil Texture	AMIBEN® Qts/A	Prowl Pts/A
Loamy sand, sandy loam	4	1.0 to 2.0
Loam, silt loam, silt	4 to 5	1.5 to 2.5
Silty clay loam, silty clay, sandy clay, clay loam, clay	5 to 6	2.0 to 3.0

Use the higher rate of AMIBEN® when heavy populations of black nightshade, velvetleaf, wild mustard, common ragweed or Pennsylvania smartweed are expected. Use the higher rate of Prowl when treatments are incorporated on soils with more than 3% organic matter, or where heavy stands of foxtail, witchgrass or panicum are expected. Do not use on peat or muck soils.

If sufficient rain does not fall within 5 days after preemergence AMIBEN® treatment, but there is enough soil moisture to germinate weeds and soybeans a light cultivation with a rotary hoe or similar tool will uproot these small weeds. The shallow mixing of the herbicides into the soil surface will not interfere with their action when rains come. If drybean crop loss occurs due to weather conditions, drybeans may be replanted. If replanting is necessary, DO NOT rework the soil.

Before applying this combination, read thoroughly the cautionary statements and all other information appearing on or attached to the containers of both the Prowl and AMIBEN® herbicides.

"AMIBEN®" AND "PROWL" AND "EPTAM 7E" FOR DRYBEANS (except Aduzuki beans)

A tank mixture of AMIBEN® Chloramben Herbicide and Prowl and Eptam 7E may be applied as a preplant soil incorporated herbicide treatment or AMIBEN® Herbicide may be applied as a preemergence broadcast treatment at planting time in drybean fields where Prowl and Eptam 7E have been applied as a preplant soil incorporated herbicide treatment.

When applied according to directions, tank mixtures of AMIBEN® Herbicide and Prowl herbicide and Eptam 7E either preplant incorporated or the sequential treatment of AMIBEN® preemergence over Prowl and Eptam 7E preplant incorporated will control fall panicum, signalgrass, Venice mallow, nightshade and velvetleaf in addition to all other weeds listed on the AMIBEN®, Prowl and Eptam 7E labels.

For uniform application of the tank mixture of AMIBEN®, Prowl and Eptam 7E, use 10 to 20 gallons of water per acre with ground equipment or 3 to 5 or more gallons of water per acre by aircraft. To prepare the tank mixture refer to the "Mixing Instructions" portion of this label.

Except as directed in the "Conservation or Reduced Tillage" portion of this label, the seedbed should be free of existing weeds before application. Surface trash (plant residue from previous crop) should be uniformly distributed, and the seedbed should be free of clods. When using the tank mixture preplant soil incorporated or when using Prowl and Eptam preplant incorporated followed by a preemergence application of AMIBEN® the herbicides should be applied within 7 days prior to planting. When applied as a preplant soil incorporated treatment the tank mixture of AMIBEN® plus Prowl and Eptam 7E should be thoroughly incorporated into the top 0 to 2 inches of soil using equipment and techniques described elsewhere on this label and on the Prowl and Eptam 7E label. When using the sequential treatment AMIBEN® should be applied at planting or within 2 days after planting before weeds and drybeans emerge.

Make all broadcast treatments according to the following table:

Soil Texture	AMIBEN® Qts/A	Prowl Pts/A	Eptam 7E Pts/A
Loamy sand, sandy loam	4	1.0 to 2.0	2.5
Loam, silt loam, silt	4 to 5	1.5 to 2.5	2.5 to 3.0
Silty clay loam, silty clay, sandy clay, clay loam, clay	5 to 6	2.0 to 3.0	3.0 to 3.5

Use the higher rate of AMIBEN® when heavy populations of black nightshade, velvetleaf, wild mustard, common ragweed or Pennsylvania smartweed are expected. Use the higher rate of Prowl when treatments are incorporated on soils with more than 3% organic matter, or where heavy stands of foxtail, witchgrass or panicum are expected. Use the higher rate of Eptam 7E on finer textured soils or where weed pressure is heavy. Do not use on peat or muck soils.

If sufficient rain does not fall within 5 days after preemergence AMIBEN® treatment, but there is enough soil moisture to germinate weeds and soybeans a light cultivation with a rotary hoe or similar tool will uproot these small weeds. The shallow mixing of the herbicides into the soil surface will not interfere with their action when rains come. If drybean crop loss occurs due to weather conditions, drybeans may be replanted. If replanting is necessary, DO NOT rework the soil.

Before applying this combination, read thoroughly the cautionary statements and all other information appearing on or attached to the containers of Prowl, Eptam 7E and AMIBEN® herbicides.

"AMIBEN®" AND "DUAL" FOR DRYBEANS

A tank mixture of AMIBEN® Chloramben Herbicide and Dual 8E may be applied as a preplant soil incorporated herbicide treatment. AMIBEN® Herbicide may be applied as a preemergence broadcast tank mix treatment with Dual 8E during planting or after planting before weeds and drybeans emerge. Following a preplant soil incorporated application of Dual 8E, AMIBEN® Herbicide may be applied preemergence during planting or after planting before weeds and drybeans emerge.

When applied according to label directions, tank mixtures of AMIBEN® Herbicide and Dual will control all weeds controlled by AMIBEN® alone or by Dual alone including fall panicum, foxtail millet, red rice, signalgrass, witchgrass, yellow nutsedge and black nightshade. Refer to the general information section of this label and the Dual label for additional weeds controlled by AMIBEN® and Dual 8E.

For uniform application of the tank mixture of AMIBEN® and Dual, use 10 to 20 gallons of water with ground equipment. To prepare the tank mixture, refer to the "Mixing Instructions" portion of this label.

Apply AMIBEN® and Dual preplant incorporated or preemergence using the appropriate rates from the table below. For preplant incorporated treatments apply the tank mixture to the soil and incorporate into the top two inches of soil just before planting following the procedures in the "Directions for Incorporation" portion of this label. Use a preplant incorporated application if furrow irrigation is used or when a period of dry weather is expected. If drybeans are planted on beds apply and incorporate the tank mixture after bed formation. For preemergence treatments, apply the tank mixture during planting (behind the planter) or after planting before weeds or drybeans emerge. For sequential applications apply Dual 8E alone preplant incorporated and follow with a preemergence application of AMIBEN® Herbicide during planting (behind the planter) or after planting before weeds or drybeans emerge.

Make all broadcast treatments according to the following table:

Soil Texture	AMIBEN® Qts/A	Dual 8E Pts/A
Loamy sand, sandy loam	4	1.5
Loam, silt loam, silt	4 to 5	1.5 to 2.0
Silty clay loam, silty clay, sandy clay, clay loam, clay	5 to 6	2.0 to 2.5

Do not use on muck or peat soils.

Use the higher rate of AMIBEN® when heavy populations of black nightshade, velvetleaf, mustard or smartweed are expected. Use the higher rate of Dual when treatment is incorporated, when organic matter is over 3% or when high populations of black nightshade, witchgrass, panicum, foxtail, or yellow nutsedge are expected.

Before applying this combination read thoroughly all cautionary statements and all other information appearing on or attached to the containers of AMIBEN® Herbicide and Dual 8E.

"AMIBEN®" AND "LASSO" FOR DRYBEANS

AMIBEN® Chloramben Herbicide may be applied as a preemergence broadcast treatment at planting or after planting before weeds and drybeans emerge in fields where Lasso has been applied as a preplant soil incorporated treatment or a tank mixture of AMIBEN® and Lasso may be applied as a preplant soil incorporated treatment.

When applied according to label directions the tank mixture or the sequential treatment of AMIBEN® plus Lasso will control all weeds listed on both the AMIBEN® and Lasso labels including nightshade, ragweed, Pennsylvania smartweed, lambsquarters, foxtails and barnyardgrass when applied within 7 days prior to planting.

To prepare the tank mixture refer to the "Mixing Instructions" portion of this label. DO NOT mix AMIBEN® Herbicide and Lasso concentrates before the addition of water as an emulsion will not be formed.

When used as a tank mixture preplant soil incorporated the AMIBEN® Herbicide and Lasso must be uniformly incorporated within 7 days prior to planting, no deeper than the top 2 inches of soil. Refer to the incorporation directions given elsewhere on this label and the Lasso label. Use the rates indicated in the table below.

When AMIBEN® is applied following a preplant soil incorporated application of Lasso follow the incorporation directions on the Lasso label. Apply as a broadcast treatment 10 to 20 gallons of water per acre at the rates indicated in the table below.

Make all broadcast treatments according to the following table:

Soil Texture	AMIBEN® Qts/A	Lasso Qts/A
Loamy sand, sandy loam	DO NOT USE	DO NOT USE
Loam, silt loam, silt	4 to 5	1.5 to 2.0
Silty clay loam, silty clay, sandy clay, clay loam, clay	5 to 6	2.0 to 2.5

Do not use on muck or peat soils.

Before applying the tank mixture of AMIBEN® plus Lasso or AMIBEN® preemergence following Lasso preplant incorporated, read all directions and cautions on both the AMIBEN® and Lasso labels.

DIRECTIONS FOR USE ON SUNFLOWERS

AMIBEN® CHLORAMBEN HERBICIDE PREEMERGENCE OR PREPLANT INCORPORATED

AMIBEN® Chloramben Herbicide may be used on sunflowers as a preemergence or preplant incorporated herbicide. AMIBEN® may be applied and incorporated at planting or several days prior to planting. Uneven spray application or improper soil incorporation of AMIBEN® can produce erratic weed control. Incorporate the chemical in the top 0 to 2 inch level as most weeds sprout in this zone. Refer to the "Directions for Incorporation" portion of this label for equipment and operating procedures to be used for proper incorporation.

Make all broadcast treatments according to the following table:

Soil Texture	AMIBEN® Qts/A
Loamy sand, sandy loam	4 to 5
Loam, silt loam, silt	5 to 6
Silty clay loam, silty clay, sandy clay, clay loam, clay	6

Use the higher rate of AMIBEN® DS when heavy populations of black nightshade, velvetleaf, wild mustard, common ragweed or Pennsylvania smartweed are expected. To insure good spray coverage use 10 to 20 gallons of water per acre with ground equipment or a minimum of 5 gallons of water per acre by aircraft.

When band spraying, mount the spray nozzle on the packer or row wheel so that planting and spraying are done in the same operation. Proper calibration and maintenance of desired nozzle height above the ground are very important. Weeds growing in the middles between the treated rows should be cultivated as soon as necessary.

Use care while cultivating to avoid throwing fresh soil on treated bands.

If sufficient rain does not fall within 5 days after spraying AMIBEN® herbicide as a preemergence treatment, but there is enough soil moisture to germinate weeds and grasses, a light cultivation with a rotary hoe or similar tool will up-root these small broadleaf weeds and grasses. The shallow mixing of AMIBEN® in the surface soil will not interfere with the action of AMIBEN® when rains come.

In coarse textured soils heavy rains on incorporated AMIBEN® Herbicide may move it below the weed seed germination zone and produce erratic weed control.

If heavy rain packs the soil surface, rotary hoe in the field to aid sunflower emergence. Do not disrupt the chemical band with deep cultivation.

DO NOT apply more than 6 quarts per acre of AMIBEN® Herbicide to the same field of sunflowers in the same season.

"AMIBEN®" AND "PROWL" FOR SUNFLOWERS

A tank mixture of AMIBEN® Chloramben Herbicide and Prowl may be applied as a preplant soil incorporated herbicide treatment or AMIBEN® Herbicide may be applied as a preemergence broadcast treatment at planting time in sunflower fields where Prowls had been applied as a preplant soil incorporated herbicide treatment. The tank mixture of AMIBEN® plus Prowl may also be applied as a preemergence broadcast treatment.

When applied according to directions, tank mixtures of AMIBEN® Herbicide and Prowl herbicide either preplant incorporated or preemergence, or the sequential treatment of AMIBEN® preemergence over Prowl preplant incorporated will control fall panicum, signalgrass, Venice mallow and velvetleaf in addition to all other weeds listed on the AMIBEN® and Prowl labels.

For uniform application of the tank mixture of AMIBEN® and Prowl, use 10 to 20 gallons of water per acre with ground equipment or 3 to 5 or more gallons of water per acre by aircraft. To prepare the tank mixture refer to the "Mixing Instructions" portion of this label.

Except as directed in the "Conservation or Reduced Tillage" portion of this label, the seedbed should be free of existing weeds before application. Surface trash (plant residue from previous crop) should be uniformly distributed, and the seedbed should be free of clods. When using the tank mixture preplant soil incorporated or when using Prowl preplant incorporated followed by a preemergence application of AMIBEN® the herbicides should be applied within 7 days prior to planting. When applied as a preplant soil incorporated treatment the tank mixture of AMIBEN® plus Prowl should be thoroughly incorporated into the top 0 to 2 inches of soil using equipment and techniques described elsewhere on this label and on the Prowl label. When using the tank mixture preemergence, treatments should be made at planting or within 2 days after planting before weeds and sunflowers emerge.

Make all broadcast treatments according to the following table:

Soil Texture	AMIBEN® Qts/A	Prowl Pts/A
Loamy sand, sandy loam	4	1.0 to 2.0
Loam, silt loam, silt	4 to 5	1.5 to 2.5
Silty clay loam, silty clay, sandy clay, clay loam, clay	5 to 6	2.0 to 3.0

Use the higher rate of AMIBEN® when heavy populations of black nightshade, velvetleaf, wild mustard, common ragweed or Pennsylvania smartweed are expected. Use the higher rate of Prowl when treatments are incorporated on soils with more than 3% organic matter, or where heavy stands of foxtail, witchgrass or panicum are expected. Do not use on peat or muck soils.

If sufficient rain does not fall within 5 days after preemergence treatment, but there is enough soil moisture to germinate weeds and sunflowers a light cultivation with a rotary hoe or similar tool will uproot these small weeds. The shallow mixing of the herbicides into the soil surface will not interfere with their action when rains come. If sunflower crop loss occurs due to weather conditions, sunflowers or soybeans may be replanted. If replanting is necessary, DO NOT rework the soil.

Before applying this combination, read thoroughly the cautionary statements and all other information appearing on or attached to the containers of both the Prowl and AMIBEN® herbicides.

"AMIBEN®" AND "TREFLAN" FOR WEED CONTROL IN SUNFLOWERS

GENERAL INFORMATION

AMIBEN® Chloramben Herbicide may be applied in a band or broadcast over sunflowers at planting time in fields where Treflan has been applied as a preplant soil incorporated herbicide, or AMIBEN® may be applied several days prior to planting as a broadcast tank mix with Treflan. The tank mixture should be used as a spring preplant soil incorporated treatment.

The combination of AMIBEN® Herbicide with Treflan improves broadleaf weed control of species such as mustard, Pennsylvania smartweed, velvetleaf and ragweed.

For broadcast treatments, incorporate chemicals immediately and thoroughly in the top 2 to 3 inches as most weeds sprout in this zone. Refer to the Treflan label for equipment and operating directions for incorporation of the tank mixture.

In coarse textured soils heavy rains on incorporated AMIBEN® may move it below the weed seed germination zone and cause erratic weed control.

If sufficient rain does not fall within 5 days after a preemergence application of AMIBEN®, but there is enough soil moisture to germinate weeds and grasses, a light cultivation with a rotary hoe or similar tool will uproot these small broadleaf weeds and grasses. The shallow mixing of AMIBEN® Herbicide in the surface soil will not interfere with the action of AMIBEN® when rains come.

DIRECTIONS FOR USE

Apply AMIBEN® Herbicide and Treflan at the rates indicated in the table below. If mustard, common ragweed or black nightshade are present, use higher rates of AMIBEN® for best control. Do not use on muck or charcoal soils. Read and observe all directions and precautions on the AMIBEN® chloramben and Treflan labels.

Make all broadcast treatments according to the following table:

Soil Texture	AMIBEN® Qts/A	TREFLAN 4EC Pts/A
Coarse	4	1.0
Medium	4 to 6*	1.5
Fine	4 to 6*	2.0

* Use the 6 quart rate if mustard, common ragweed or black nightshade is present for best control.

"AMIBEN®" AND "EPTAM" FOR SUNFLOWERS

For use only in the states of Minnesota, North Dakota and South Dakota

AMIBEN® Chloramben Herbicide may be applied as a preplant soil incorporated tank mixture with "Eptam" herbicide. Or AMIBEN® may be applied preemergence over sunflowers at planting time where Eptam has been applied as preplant soil incorporated treatment. When applied preemergence, effectiveness depends on adequate rainfall after treatment. When applied according to directions, this combination will control annual broadleaves including black nightshade, ragweed, Pennsylvania smartweed, wild mustard and annual grasses.

DIRECTIONS FOR USE

Apply AMIBEN® Herbicide either preplant soil incorporated or preemergence at a rate of 4 to 6 quarts per acre. Apply Eptam 7E preplant soil incorporated at a rate of 2.5 to 3.5 pints per acre. Use the higher rates on fine textured or high organic matter soils, or where weed pressure is heavy.

The AMIBEN® plus Eptam tank mix should be incorporated immediately after application with a field cultivator, tandem disc, mulch treader, rolling cultivator or power driven tiller as described on the Eptam label. Thoroughly mix in the top 2-3" of the soil surface by incorporating twice in different directions. Refer to the Eptam label for further information on incorporation procedures.

Rainfall is not as critical for activation of AMIBEN® Herbicide when it is applied preplant incorporated as when surface applied.

Apply in 10 to 20 gallons of water per acre. To prepare the spray mixture refer to the "Mixing Instructions" portion of this label. Do not mix the concentrates before adding to water.

If sufficient rain does not occur within 5 days after spraying AMIBEN® preemergence, a light cultivation or incorporation with a rotary hoe or similar tool is recommended.

"AMIBEN®" AND "LASSO" FOR SUNFLOWERS

A tank mixture of AMIBEN® Chloramben Herbicide and Lasso may be applied as a preplant soil incorporated herbicide treatment. AMIBEN® Herbicide may be applied as a preemergence broadcast tank mix treatment with Lasso during planting or after planting before weeds and sunflowers emerge. Following a preplant soil incorporated application of Lasso, AMIBEN® may be applied preemergence during planting or after planting before weeds and sunflowers emerge.

For uniform application of the tank mixture of AMIBEN® Herbicide and Lasso use 10 to 20 gallons of water per acre for ground application. To prepare the tank mixture refer to the "Mixing Instructions" portion of this label. DO NOT mix AMIBEN® Herbicide and Lasso concentrates before the addition of water as an emulsion will not be formed.

Apply AMIBEN® and Lasso preplant incorporated or preemergence using the appropriate rates from the table below. For preplant incorporated treatments apply the tank mixture to the soil and incorporate into the top two inches of soil as directed in the "Directions for Incorporation" portion of this label. For preemergence treatments, apply the tank mixture during planting (behind the planter) or after planting before weeds or sunflowers emerge. For sequential applications apply Lasso alone preplant incorporated as specified on the Lasso label. Follow with a preemergence application of AMIBEN® Herbicide during planting (behind the planter) or after planting before weeds or sunflowers emerge.

With preemergence treatments rain is necessary to move the chemicals into the soil and make them available to control germinating weed seeds. If no rainfall occurs within 5 days following application, a shallow incorporation no deeper than 2 inches is recommended. High intensity or excessive rainfall after application may reduce control.

Make all broadcast treatments according to the following table:

Soil Texture	AMIBEN® Qts/A	Lasso Qts/A
Loamy sand, sandy loam	4	1.5 to 2.0
Loam, silt loam, silt	4 to 5	2.0 to 2.5
Silty clay loam, silty clay, sandy clay, clay loam, clay	5 to 6	2.5 to 3.0

Use the higher rate of AMIBEN® when heavy populations of black nightshade, velvetleaf, mustard or smartweed are expected. Use the higher rate of Lasso when treatment is incorporated or when high populations of Black nightshade, witchgrass or foxtail are expected.

Read thoroughly the cautionary statements and all other information appearing on labels for both Lasso and AMIBEN®.

DIRECTIONS FOR USE ON PEANUTS

"AMIBEN®" AND "PREMERGE" 3 ON PEANUTS

Tank mixtures of AMIBEN® Chloramben Herbicide and Premerge 3 control weeds in two ways. AMIBEN® controls germinating annual grasses and broadleaf weeds while Premerge 3 is a contact herbicide and quickly burns weeds that have germinated before treatment. This tank mixture is especially effective in controlling Florida beggarweed (*Desmodium tortuosum*).

DIRECTIONS FOR USE

Apply a tank mixture of AMIBEN® and Premerge 3 at a rate of 4 to 6 quarts of AMIBEN® and 4 pints of Premerge 3 per broadcast acre. When the air temperature during the day of application is expected to exceed 90 ° F, reduce the Premerge 3 rate to 3 pints. Apply as a ground cracking treatment when the oldest peanut plants reach the two leaf stage or when the first weeds become visible. Use enough water for thorough coverage, generally 15 or more gallons per acre.

Caution: Read thoroughly the cautionary statements and all other information appearing on labels for both AMIBEN® and Premerge 3.

DO NOT apply more than 6 quarts per acre of AMIBEN® Herbicide to the same field of peanuts in the same season.

"AMIBEN®" AND "DYANAP" ON PEANUTS

Tank mixtures of AMIBEN® Chloramben Herbicide and Dyanap control weeds in two ways. AMIBEN® plus Dyanap controls many germinating and emerged seedling annual grasses and broadleaf weeds. This tank mixture is especially effective in controlling Florida beggarweed (*Desmodium tortuosum*) and smallflower morningglory (*Jacquemontia tamnifolia*).

DIRECTIONS FOR USE:

Apply a tank mixture of AMIBEN® and Dyanap at the rate of 4 to 6 quarts of AMIBEN® and 6 quarts of Dyanap per broadcast acre. Apply as a ground cracking treatment when the first weeds become visible and up to the time most of the peanuts are in the "Cracking stage." Use enough water for thorough coverage, generally 10 to 20 gallons per acre.

Caution: Read thoroughly the cautionary statements and all other information appearing on labels for both AMIBEN® and Dyanap.

DO NOT apply more than 6 quarts per acre of AMIBEN® Herbicide to the same field of peanuts in the same season.

FOR PEANUTS IN OKLAHOMA AND NORTH TEXAS

Plant and treat in the same operation. Apply at the rate of 4 to 6 quarts AMIBEN® Chloramben Herbicide per broadcast acre. The higher rate should be used on clay loams, clay or high organic (3 to 8 percent) mineral soils. Use enough water for good spray coverage.

When band spraying, mount the spray nozzle on the packer or row wheel so that planting and spraying are done in the same operation. Proper calibration and maintenance of desired nozzle height above the ground are very important. Weeds growing in the middles between the treated rows should be cultivated as soon as necessary. Use care while cultivating to avoid throwing fresh soil on treated bands.

If sufficient rain does not fall within 5 days after spraying AMIBEN® Herbicide, but there is enough soil moisture to germinate weeds and grasses, a light cultivation with a rotary hoe or similar tool will up-root these small broadleaf weeds and grasses. The shallow mixing of AMIBEN® in the surface soil will not interfere with the action of AMIBEN® when rains come.

DO NOT apply more than 6 quarts per acre of AMIBEN® Herbicide to the same field of peanuts in the same season.

DIRECTIONS FOR USE ON SWEET POTATOES

Apply AMIBEN® Chloramben Herbicide at the rate of 8 quarts per broadcast acre at planting time. For the first few cultivations after treatment, set sweeps so that soil in treated bands is not disturbed. For the final cultivation, set sweeps closer to pull soil as high as possible over the sweet potato row and treated band. For best results, plant and treat at the same time.

DIRECTIONS FOR USE ON SQUASH AND PUMPKIN

Treat preplant incorporated or preemergence at planting time or as soon after planting as possible, before weeds and crops emerge. Follow incorporation instructions under "Directions for use for Soybeans."

In coarse textured soils, heavy rains on incorporated AMIBEN® Chloramben Herbicide may move it below the weed seed germination zone and produce erratic weed control.

For preemergence application, the sprayer may be mounted on the planter so that spraying and seeding are carried out in the same operation. Apply 6 quarts of AMIBEN® on lighter soils and 8 quarts of AMIBEN® on heavier or high organic soils per broadcast acre in at least 10 to 20 gallons of water. If band applications are used, calculate the rate by dividing the band width by the row width and multiplying by 6 or 8 quarts of AMIBEN® per broadcast acre whichever is appropriate for your soil type. After squash or pumpkin emerge, use normal cultivations to control weeds that grow in strips unprotected by any chemical.

BAND TREATMENTS

Row Width	Band Width	Qts AMIBEN® Per Band Acre: Lighter Soils	Qts AMIBEN® Per Band Acre: Heavier or High Organic Soils
72 inch	14 inch	1.2	1.6
	20 inch	1.7	2.2
76 inch	14 inch	1.1	1.5
	20 inch	1.6	2.1
80 inch	14 inch	1.1	1.4
	20 inch	1.5	2.0

DIRECTIONS FOR USE ON SEEDLING ASPARAGUS

Apply 6 quarts of AMIBEN® Chloramben Herbicide per broadcast acre in about 10 to 20 gallons of water as soon as possible after asparagus seed is planted and before weeds emerge. Irrigation with about ½ inch of water after applying AMIBEN® is suggested to produce maximum weed control.

DIRECTIONS FOR USE ON LIMA BEANS

Apply 4 to 8 quarts of AMIBEN® Chloramben Herbicide per broadcast acre in about 10 to 20 gallons of water.

Treat preplant incorporated or preemergence at planting time or as soon after planting as possible, before weeds and lima beans emerge. Follow incorporation instructions under the appropriate section of this label.

In coarse textured soils, heavy rains on incorporated AMIBEN® Herbicide may move it below the weed seed germination zone and produce erratic weed control.

If band applications are used, calculate the rate by dividing the band width by the row width and multiplying by 4, 5, 6, or 8 quarts of AMIBEN® per broadcast acre whichever is appropriate for your weed problem and soil type. Mount band sprayer on planter so that spraying and seeding are carried out in the same operation. Make final adjustment on proper nozzle height when actually spraying in the field.

DIRECTIONS FOR USE ON SNAPBEANS (GREEN AND YELLOW)

Apply AMIBEN® at the rate of 4 to 5 quarts per broadcast acre as a preemergence treatment in sufficient water for good spray coverage (10 to 20 gallons of water per acre). Use the higher rate of AMIBEN® on heavy or high organic mineral soils or where black nightshade is a problem. Use the lower rate on loam soils. Do not use on sandy soils low in clay content or with less than 1% organic matter. Attach the applicator to the planter so that planting and spraying are done in the same operation. Rain or sprinkler irrigation carries AMIBEN® into the soil where weed seeds sprout. If weeds begin to emerge after AMIBEN® application, but prior to rainfall or irrigation, a light cultivation is recommended to up-root the small broadleaf weeds and grasses. The shallow mixing of AMIBEN® in the soil surface will not interfere with its weed killing action when rain comes.

WARNING: Stunting of beans has been observed when rains wash the chemical into contact with the germinating seed or the root system of young plants. Under good growing conditions plants normally recover and yields are generally not affected.

Heavy or prolonged rainfall may move AMIBEN® below the weed seed germination zone and produce erractic weed control.

"AMIBEN®" AND "DUAL" FOR SNAPBEANS (GREEN AND YELLOW)

AMIBEN® Herbicide may be applied as a preemergence broadcast tank mix treatment with Dual 8E during planting or after planting before weeds and snapbeans emerge. Following a preplant soil incorporated application of Dual 8E, AMIBEN® Herbicide may be applied preemergence during planting or after planting before weeds and snapbeans emerge.

When applied according to label directions, tank mixtures of AMIBEN® Herbicide and Dual will control all weeds controlled by AMIBEN® alone or by Dual alone including fall panicum, foxtail millet, red rice, signalgrass, witchgrass, yellow nutsedge and black nightshade. Refer to the general informaion section of this label and the Dual label for additional weeds controlled by AMIBEN® and Dual 8E.

For uniform application of the tank mixture of AMIBEN® and Dual, use 10 to 20 gallons of water with ground equipment. To prepare the tank mixture, refer to the "Mixing Instructions" portion of this label.

Apply AMIBEN® and Dual using the appropriate rates from the table below. For preemergence treatments, apply the tank mixture during planting (behind the planter) or after planting before weeds or snapbeans emerge. For sequential applications apply Dual 8E alone preplant incorporated and follow with a preemergence application of AMIBEN® Herbicide during planting (behind the planter) or after planting before weeds or snapbeans emerge.

Make all broadcast treatments according to the following table:

Soil Texture	AMIBEN® Qts/A	Dual 8E Pts/A
Loamy sand, sandy loam	DO NOT USE	1.5 to 2.0
Loam, silt loam, silt	4 to 5	2.0 to 2.5
Silty clay loam, silty clay, sandy clay, clay loam, clay	4 to 5	2.0 to 3.0

Do not use on muck or peat soils.

Use the higher rate of AMIBEN® when heavy populations of black nightshade, velvetleaf, mustard or smartweed are expected. Use the higher rate of Dual when treatment is incorporated, when organic matter is over 3%, or when high populations of black nightshade, witchgrass, panicum, foxtail, or yellow nutsedge are expected.

Before applying this combination read thoroughly all cautionary statements and all other information appearing on or attached to the containers of AMIBEN® Herbicide and Dual 8E.

DIRECTIONS FOR USE ON FIELD SEEDED TOMATOES PROTECTED WITH ACTIVATED CARBON

AMIBEN® Chloramben Herbicide may be applied as a broadcast preemergence herbicide treatment on field seeded tomatoes which have been protected by the use of an activated carbon plus vermiculite anticrustant system as described below. Do not use AMIBEN® on field seeded tomatoes without the use of the activated carbon plus vermiculite protection system as crop injury will result.

Thoroughly mix 1 pound of activated carbon with each cubic foot of number 2 or number 3 horticultural grade vermiculite. Apply 1 cubic foot of this mixture per 600 feet of seeded row (15 cubic feet per acre) by filling the seed furrow with the mixture. Use a positive feed applicator to deliver the activated carbon plus vermiculite anticrustant mixture to the seed furrow directly ahead of the planter press wheel.

Immediately after seeding and before weeds or tomatoes emerge apply 4 to 6 quarts per acre of AMIBEN® in 10 to 40 gallons of water per acre as a broadcast treatment. Use only on medium or heavy soils. Use the higher rate of AMIBEN® on heavier soils or where high populations of black nightshade, velvetleaf, mustard or smartweed are expected.

The use of the activated carbon plus vermiculite anticrustant system in the seed furrow will protect the tomato seedling from the phytotoxic effects of AMIBEN® When used in this manner AMIBEN® *will not control* weed seeds germinating within the vermiculite treated row.

DIRECTIONS FOR USE ON FIELD SEEDED CUCUMBERS PROTECTED WITH ACTIVATED CARBON

AMIBEN® Chloramben Herbicide may be applied as a broadcast preemergence herbicide treatment on field seeded cucumbers which have been protected by the use of an activated carbon plus vermiculite anticrustant system as described below. Do not use AMIBEN® on field seeded cucumbers without the use of the activated carbon plus vermiculite protection system as crop injury will result.

Thoroughly mix 1 pound of activated carbon with each cubic foot of number 2 or number 3 horticultural grade vermiculite. Apply 1 cubic foot of this mixture per 600 feet of seeded row (15 cubic feet per acre) by filling the seed furrow with the mixture. Use a positive feed applicator to deliver the activated carbon plus vermiculite anticrustant mixture to the seed furrow directly ahead of the planter press wheel.

Immediately after seeding and before weeds or cucumbers emerge apply 4 to 6 quarts per acre of AMIBEN® in 10 to 40 gallons of water per acre as a broadcast treatment. Use only on medium or heavy soils. Use the higher rate of AMIBEN® on heavier soils or where high populations of black nightshade, velvetleaf, mustard or smartweed are expected.

The use of the activated carbon plus vermiculite anticrustant system in the seed furrow will protect the cucumber seedling from the phytotoxic effects of AMIBEN®. When used in this manner AMIBEN® **will not control** weed seeds germinating within the vermiculite treated row.

DIRECTIONS FOR USE ON FIELD SEEDED CANTALOUPE PROTECTED WITH ACTIVATED CARBON

AMIBEN® Chloramben Herbicide may be applied as a broadcast preemergence herbicide treatment on field seeded cantaloupe which have been protected by the use of an activated carbon plus vermiculite anticrustant system as described below. Do not use AMIBEN® on field seeded cantaloupe without the use of the activated carbon plus vermiculite protection system as crop injury will result.

Thoroughly mix 1 pound of activated carbon with each cubic foot of number 2 or number 3 horticultural grade vermiculite. Apply 1 cubic foot of this mixture per 600 feet of seeded row (15 cubic feet per acre) by filling the seed furrow with the mixture. Use a positive feed applicator to deliver the activated carbon plus vermiculite anticrustant mixture to the seed furrow directly ahead of the planter press wheel.

Immediately after seeding and before weeds or cantaloupe emerge apply 4 to 6 quarts per acre of AMIBEN® in 10 to 40 gallons of water per acre as a broadcast treatment. Use only on medium or heavy soils. Use the higher rate of AMIBEN® on heavier soils or where high populations of black nightshade, velvetleaf, mustard or smartweed are expected.

The use of the activated carbon plus vermiculite anticrustant system in the seed furrow will protect the cantaloupe seedling from the phytotoxic effects of AMIBEN®. When used in this manner AMIBEN® **will not control** weed seeds germinating within the vermiculite treated row.

DIRECTIONS FOR USE ON CORN

Use only on heavy soils in Missouri, Minnesota, Iowa, Ohio, Illinois, Kansas, Nebraska and Indiana. If rains and extended cold weather occur after planting, corn injury may result.

Apply 4 quarts of AMIBEN® Chloramben Herbicide per acre in at least 10 to 20 gallons of water. Plant and treat at the same time or as soon after planting as possible before weeds and crop emerge. Plant corn at least 2 to $2\frac{1}{2}$ inches deep. Do not use AMIBEN® on corn grown in light textured sandy soils.

AMIBEN® AND "ATRAZINE" 80W TANK MIXTURE

Application: Apply a tank mixture of AMIBEN® Chloramben Herbicide and Atrazine 80W during or immediately after planting. A light incorporation with a rotary hoe or similar equipment may be desirable, especially if there is no rainfall soon after application. Do not use AMIBEN® and Atrazine for weed control in crops other than corn. Observe all directions on the Atrazine label, especially those relating to crop rotation.

Use only in the states of Missouri, Minnesota, Iowa, Ohio, Illinois, Kansas, Nebraska and Indiana.

Amounts of use: For silty loam soils apply 2.0 quarts of AMIBEN® and 1.25 pounds of Atrazine. For heavier soils, apply 2 quarts of AMIBEN® and 2.5 pounds of Atrazine or 3 quarts of AMIBEN® and 1.87 pounds of Atrazine. This mixture is not recommended for use on very light soils such as sands, or sandy loams.

PRECAUTIONARY STATEMENTS

CAUTION

HAZARDS TO HUMANS AND DOMESTIC ANIMALS

Harmful if swallowed. Avoid contact with skin, eyes or clothing.

STATEMENT OF PRACTICAL TREATMENT

If swallowed, call a physician or Poison Control Center. Drink 1 to 2 glasses of water and induce vomiting by touching back of throat with finger. Do not induce vomiting or give anything by mouth to an unconscious person.

In case of skin contact, wash immediately with soap and water and remove contaminated clothing.

In case of eye contact flush with plenty of water. Get medical attention.

If inhaled, remove victim to fresh air.

ENVIRONMENTAL HAZARDS

Do not apply directly to water. Do not contaminate water used for irrigation or domestic purposes by cleaning of equipment, or disposal of waste.

DIRECTIONS FOR USE

IT IS A VIOLATION OF FEDERAL LAW TO USE THIS PRODUCT IN A MANNER INCONSISTENT WITH ITS LABELING.

Clean sprayers thoroughly before using to apply AMIBEN®. To clean sprayer after using AMIBEN®, rinse all parts of sprayer with water several times.

Avoid spray drift to adjacent crops as injury may result. Coarse sprays are less likely to drift.

Do not graze treated fields of sunflowers or use these plants as livestock feed.

STORAGE AND DISPOSAL

Do not store near fertilizers, seed, insecticides, or fungicides. Open dumping is prohibited.

PESTICIDE DISPOSAL

Do not contaminate water, food, or feed by storage or disposal. Wastes resulting from the use of this product may be disposed of on site or at an approved waste disposal facility.

CONTAINER DISPOSAL

Do not reuse empty container. Triple rinse (or equivalent). Then offer for recycling or reconditioning, or puncture and dispose of in a sanitary landfill or by other procedures approved by state and local authorities.

Consult Federal, state, or local disposal authorities for approved alternative procedures such as limited open burning.

LIMITED WARRANTY AND DISCLAIMER

The manufacturer warrants (a) that this product conforms to the chemical description on the label; (b) that this product is reasonably fit for the purposes set forth in the directions for use when it is used in accordance with such directions; and (c) that the directions, warning and other statements on this label are based upon responsible experts' evaluation of reasonable tests of effectiveness, of toxicity to laboratory animals and to plants, and of residues on food crops and upon reports of field experience. Tests have not been made on all varieties or in all states or under all conditions. THE MANUFACTURER NEITHER MAKES NOR INTENDS, NOR DOES IT AUTHORIZE ANY AGENT OR REPRESENTATIVE TO MAKE, ANY OTHER WARRANTIES, EXPRESS OR IMPLIED, AND IT EXPRESSLY EXCLUDES AND DISCLAIMS ALL IMPLIED WARRANTIES OR MERCHANTABILITY OR FITNESS FOR A PARTICULAR PURPOSE.

THIS WARRANTY DOES NOT EXTEND TO, AND THE BUYER SHALL BE SOLELY RESPONSIBLE FOR, ANY AND ALL LOSS OR DAMAGE WHICH RESULTS FROM THE USE OF THIS PRODUCT IN ANY MANNER WHICH IS INCONSISTENT WITH THE LABEL DIRECTION, WARNINGS OR CAUTIONS.

BUYER'S EXCLUSIVE REMEDY AND MANUFACTURER'S OR SELLER'S EXCLUSIVE LIABILITY FOR ANY AND ALL CLAIMS, LOSSES, DAMAGES, OR INJURIES RESULTING FROM THE USE OR HANDLING OF THIS PRODUCT, WHETHER OR NOT BASED ON CONTRACT, NEGLIGENCE, STRICT LIABILITY IN TORT OR OTHERWISE SHALL BE LIMITED, AT THE MANUFACTURER'S OPTION, TO REPLACEMENT OF, OR THE REPAYMENT OF THE PURCHASE PRICE FOR, THE QUANTITY OF PRODUCT WITH RESPECT TO WHICH DAMAGES ARE CLAIMED. IN NO EVENT SHALL MANUFACTURER OR SELLER BE LIABLE FOR SPECIAL, INDIRECT OR CONSEQUENTIAL DAMAGES RESULTING FROM THE USE OR HANDLING OF THIS PRODUCT.

AMIBEN® and BUTYRAC® 200 are registered trademarks of Union Carbide Agricultural Products, Company, Inc.

Blazer is a registered trademark of Rohm and Haas Company.

Treflan and Surflan are registered trademarks of Elanco Products Company.

Lasso and Roundup are registered trademarks of Monsanto Company.

Dual is a registered trademark of Ciba-Geigy Corporation.

Sencor is a registered trademark of Farbenfabrik Bayer GmbH, Leverkusen.

Lorox and Lexone are registered trademarks of E.I. du Pont de Nemours and Company.

Prowl is a registered trademark of American Cyanamid Company.

Alanap and Dyanap are registered trademarks of Uniroyal, Inc.

Premerge is a registered trademark of Dow Chemical Company.

Reward and Eptam are registered trademarks of Stauffer Chemical Company.

Basalin is a registered trademark of BASF Wyandotte Corporation.

THIS SPECIMEN LABEL IS INTENDED FOR USE ONLY AS A GUIDE IN PROVIDING GENERAL INFORMATION REGARDING THE DIRECTIONS, WARNINGS AND CAUTIONS ASSOCIATED WITH THE USE OF THIS PRODUCT. AS WITH ANY AGRICULTURAL CHEMICAL, ALWAYS FOLLOW THE LABEL INSTRUCTIONS ON THE PACKAGE BEFORE USING.

AMIBEN® DS

CHLORAMBEN HERBICIDE

Controls Annual Broadleaf Weeds and Annual Grasses in Soybeans and Other Crops

ACTIVE INGREDIENTS:
Sodium salt of chloramben
(3-amino-2,5-dichlorobenzoic acid)* 83%
INERT INGREDIENTS: 17%

TOTAL 100%

*Chloramben acid equivalent: 75% by weight

E.P.A. Reg. No. 264-305 E.P.A. Est. No. 10352-GA-01

Keep Out of Reach of Children

CAUTION

IN CASE OF EMERGENCY—TELEPHONE DAY OR NIGHT (UCC HELP) 800-822-4357

FOR GENERAL PRODUCT INFORMATION CALL 1-800-334-9745

GENERAL INFORMATION

AMIBEN® DS Chloramben Herbicide is effective applied preplant incorporated, preemergence and early postemergence. When used preemergence it is most effective applied at planting time, when spraying and planting are done in the same operation. Rain, irrigation, water, or mechanical incorporation moves AMIBEN® DS into the soil where weed seeds sprout.

Uniform spray application over the soil surface is important. About 10-20 gallons of water or sprayable fluid fertilizer for ground and about 3-5 gallons for air application is enough spray solution for good coverage. See specific sections of this label for any special spray volume requirements.

AMIBEN® DS Herbicide remains active and effective for several weeks ... long enough generally for crops to fill middles and shade out the late season weed growth. Performance is improved when moisture is received within 5 days after application and prior to weed emergence. High intensity or excessive rainfall or excessive irrigation after application may reduce control.

In coarse textured soils with less than 1% organic matter heavy rains may move AMIBEN® DS Herbicide below the weed seed germination zone and produce erratic weed control.

AMIBEN® DS controls the following important Annual Grass and Broadleaf Weeds:

barnyardgrass (water grass)	*Echinochloa crusgalli*
carpetweed	*Mullugo verticillata*
chickweed, annual	*Stellaria media*
coffeeweed (sesbania)	*Sesbania exaltata*
crabgrass, hairy (large)	*Digitaria sanguinalis*
crabgrass, smooth	*Digitaria ischaemum*
fall panicgrass	*Panicum dichotomiflorum*
Florida beggarweed	*Desmodium tortuosum*
foxtail, giant	*Setaria faberii*
foxtail, green	*Setaria viridis*
foxtail, yellow	*Setaria glauca*
goosegrass (silver crabgrass)	*Eleusine indica*
johnsongrass-seedling	*Sorghum halepense*
kochia	*Kochia scoparia*
lambsquarters, common	*Chenopodium album*
mustard, wild	*Brassica kaber*
nightshade, black	*Solanum nigrum*
pigweeds	*Amaranthus sp.*
prickly sida (teaweed)	*Sida spinosa*
purslane (pusley)	*Portulaca oleracea*
ragweed, common	*Ambrosia artemisiifolia*
Russian thistle	*Salsola kali*
smartweed, Pennsylvania	*Polygonum pensylvanicum*

spurge, annual	*Euphorbia maculata*
stinkgrass	*Eragrostis cilianensis*
velvetleaf (buttonweed)	*Abutilon theophrasti*
Venice mallow	*Hibiscus trionum*
wild proso millet	*Panicum miliaceum*

Band application of AMIBEN® DS may be made for broadleaf weed control over the top of preplant soil incorporated grass killers.

Deep germinating annual weeds and sprouts from established perennial weeds are not controlled. Sprouts coming from roots of established perennial weeds such as horsenettle, hedge bindweed, nutsedge (nutgrass), milkweed, dogbane, quackgrass, and johnsongrass are not controlled by any selective, surface preemergence chemical treatment including AMIBEN® DS.

Only the following recommendations referenced in this label are registered for use in California.

CROPS

Asparagus, seedling
Drybeans (White, Navy, Kidney, Pinto)
Lima beans
Pumpkin
Squash
Sunflowers
Sweet Potatoes
Soybeans
- AMIBEN® DS with fertilizer
- AMIBEN® DS preemergence on preplant incorporated
- AMIBEN® DS delayed preemergence
- AMIBEN® DS plus Lasso
- AMIBEN® DS plus Treflan
- AMIBEN® DS plus Lorox
- AMIBEN® DS plus Prowl
- AMIBEN® DS plus Preemerge 3

AMIBEN® DS is not registered for control of Venice mallow *(Hibiscus trionum)* in California.

DIRECTIONS FOR INCORPORATION OF AMIBEN® DS

AMIBEN® DS Herbicide applied alone, and several AMIBEN® DS tank mixtures as specified in the crops section of this label, may be incorporated into the top 1 to 3 inches of the final seedbed before planting. When preplant incorporation is recommended in the crops section of this label the following equipment and operating procedures should be used. Be sure to use machinery that mixes AMIBEN® DS thoroughly with the soil. Incorporation with implements set to cut less than 2 inches deep may result in erratic weed control.

Recommended Equipment Includes:

1. Two passes with a disc set to cut 4 inches deep at 4 to 6 mph with the second pass running at an angle to the first. A tandem or double-disc operated one time does not provide adequate incorporation.
2. One pass with a field cultivator set to cut 3 to 4 inches deep and operated at 6 mph or more. The field cultivator should have 3 to 4 rows of sweeps, spaced at intervals of 7 inches or less and staggered so that no soil is left unturned. Chisel points should not be used. Do not set cultivator deeper than 4 inches. The field cultivator used alone or in combination with the double-disc will provide effective incorporation. If used in combination with a double-disc the spray boom should be mounted behind the disc blades but ahead of the cultivator sweeps.
3. One pass with a rolling cultivator set to cut 2 to 3 inches deep and operated at 6 to 8 mph. Rolling cultivators are adequate for use on coarse and medium textured soils only.
4. One pass with a mulch treader or other similar disc type implement set to cut 2 to 3 inches deep and operated at 5 to 8 MPH.
5. PTO-driven equipment (tillers, cultivators, hoes) set to cut 2 to 3 inches deep with rotors spaced to provide a clean sweep of the soil operated one time. PTO-driven equipment should not be operated at a speed greater than 4 mph.

MIXING INSTRUCTIONS

AMIBEN® DS in water or fluid fertilizer: first fill the spray tank ¼ to ½ full with clean water. Start agitation and add AMIBEN® DS while adding the remaining required water or fluid fertilizer.

AMIBEN® DS Plus Emulsifiable Concentrates in Water: first fill tank ¼ to ½ full with clean water, start agitation and add AMIBEN® DS. Mix thoroughly and add water until tank is ¾ full, continue agitation and add the emulsifiable concentrate material. Fill tank with remaining required water and continue agitation during spraying until tank is empty. Agitation should sweep bottom contents of the tank up into the main body of the spray mixture for thorough mixing. If a compatibility agent is required it should be thoroughly mixed with the water before adding the emulsifiable concentrate. To prevent foaming, avoid stirring or splashing air into the mixture by placing the end of the fill pipe below the surface of the water in the spray tank.

AMIBEN® DS Plus Wettable Powders or Flowables: Fill tank ½ full with clean water, start sprayer agitation and add the recommended amount of the flowable or wettable powder. Mix thoroughly and add AMIBEN® DS while filling tank with remaining required water. Continue agitation during entire spraying operation until tank has been emptied. Agitation should sweep bottom contents of the tank up into the main body of the spray mixture for thorough mixing and uniform suspension.

DIRECTIONS FOR USING AMIBEN® DS THROUGH OVERHEAD IRRIGATION SYSTEMS

AMIBEN® DS may be applied as a premergence treatment in irrigation water alone, or in tank mixtures with any of the other herbicides on this label which are registered for center pivot or overhead irrigation application, after planting but before weeds or crop emerge. Use the broadcast rates per acre for each product as specified elsewhere on this label for each of the tank mixtures, and follow all use precautions below and on respective labels for any other products applied in this manner. Use only through automated (center pivot or continuous lateral move) equipment which is properly maintained and operated to apply water uniformly.

DIRECTIONS FOR USE

Remove scale, pesticide residues and other foreign matter from the chemical tank and the entire injector system. Flush with clean water. In the chemical tank prepare a mixture with a minimum of 1 part of water to 1 part of herbicide(s). Maintain continuous agitation in the chemical tank while injecting into the irrigation system with a positive displacement pump. Injecting a larger volume of a more dilute mixture will usually provide more accurate calibration of metering equipment.

Set sprinkler to deliver 0.5 inch or less of water per acre. Application of more than 0.5 inch of water per acre may reduce weed control by moving the herbicide below the weed germination zone.

USER PRECAUTIONS

Apply AMIBEN® DS only through systems containing antisiphon and check valves and having interlocking controls between the metering device and the water pump to insure their simultaneous shutoff to prevent water source contamination and overflow of the mixing tank.

To insure adequate mixing inject the chemical into the main line at least 8 feet ahead of a right angle turn.

Do not apply when conditions favor drift, such as from windy conditions, or when system joints and connections are leaking, or when nozzles do not provide uniform distribution, as crop injury may result.

Where sprinkler distribution patterns do not overlap sufficiently, unacceptable weed control may result. Where sprinkler distribution patterns overlap excessively, crop injury may result.

DIRECTIONS FOR USING AMIBEN® DS IN CONSERVATION OR REDUCED TILLAGE SYSTEMS

When applied as directed below, AMIBEN® DS tank mixtures will control many emerged annual weeds, suppress many emerged perennial weeds and will give preemergence control of many annual grasses and broadleaf weeds in conservation tillage or reduced tillage systems.

When used in tillage systems with practices which eliminate all existing vegetation before planting, AMIBEN® DS may be used alone or in tank mixture with any of the other products on this label which are labeled for use in reduced tillage operations. AMIBEN® DS and AMIBEN® DS combinations with other preemergence or preplant incorporated herbicides will not control emerged weeds. The presence of previous crop residue (trash) which is uniformly distributed on the soil surface will not affect the performance of AMIBEN® DS, providing sufficient rainfall (0.375 to 0.5 inch) is received before weeds begin to germinate. IF WEEDS HAVE BEGUN TO EMERGE AT TIME OF APPLICATION FOLLOW THE DIRECTIONS BELOW.

When used in tillage systems where emerged weeds are present at planting, either Roundup, or Paraquat plus Ortho X-77 spreader should be added to the AMIBEN® DS tank mixture. Refer to the respective Roundup, Paraquat and Ortho X77 labels for crops labeled, weeds controlled, use rates, and specific precautions to be taken. Use a minimum of 15 to 40 gallons of water per acre and apply the tank mixture during or after planting, but BEFORE CROP EMERGENCE. As density of stubble, crop residue or weeds increase, spray gallonage and rate should be increased, within recommended ranges, to assure good coverage.

MIXING INSTRUCTIONS FOR USE OF AMIBEN® DS WITH SPRAYABLE FLUID FERTILIZERS

In most cases, AMIBEN® DS Chloramben Herbicide can be mixed and applied with liquid nitrogen, blended clear liquids, or suspension type fertilizers.

Be sure the spray tank is clean and uncontaminated. First fill the spray tank ¼ to ½ full with clean water. Start agitation and add AMIBEN® DS Chloramben Herbicide. Add required fluid fertilizer and water to fill tank. After the tank is full, no further agitation is necessary.

The compatibility of tank mixtures of AMIBEN® DS with other herbicides in combination with liquid fertilizers should be evaluated on a small scale before mixing spray tank quantities. Do not use fertilizer mixes with combinations of AMIBEN® DS and Premerge 3.

DIRECTIONS FOR AMIBEN® DS IMPREGNATION ON DRY FERTILIZER

Determine rate of fertilizer and AMIBEN® DS needed per acre. Higher rates of AMIBEN® DS should be used when heavy infestations of velvetleaf, black nightshade, mustard or Pennsylvania smartweed are expected.

Determine number of gallons of AMIBEN® DS Solution needed per ton of fertilizer from Table 1 using the appropriate rates of fertilizer and AMIBEN® DS. Example: At a fertilizer rate of 350 lbs/A and an AMIBEN® DS rate per acre of 3.0 lbs it would take 3.5 gallons of AMIBEN® DS solution per ton of fertilizer.

To prepare the AMIBEN® DS Solution mix two 12 pound bags of AMIBEN® DS in 5 gallons of water. It is necessary to maintain good agitation to keep AMIBEN® DS in suspension.

Apply up to 5 gallons of the AMIBEN® DS Solution to each ton of fertilizer. Provide sufficient mixing during impregnation to assure even distribution of all fertilizer particles.

TABLE 1

GALLONS OF AMIBEN® DS SOLUTION[1] PER TON OF FERTILIZER

Fertilizer LBS/A	AMIBEN DS Rates Per Acre 2.4 lbs	3.0 lbs	3.6 lbs
150	*	*	*
200	5.0 Gal.	*	*
250	4.0 Gal.	5.0 Gal.	*
300	3.3 Gal.	4.2 Gal.	5.0 Gal.
350	3.0 Gal.	3.5 Gal.	4.3 Gal.
400	2.5 Gal.	3.1 Gal.	3.8 Gal.
450	2.3 Gal.	2.8 Gal.	3.3 Gal.
500	2.0 Gal.	2.5 Gal.	3.0 Gal.

[1] Solution = 24 lbs AMIBEN® DS Per 5 Gal. Water.

* DO NOT USE = Fertilizer will become too wet resulting in poor handling and spreading.

Note: For fertilizer rates not in Table 1 apply proportional amounts of the standard AMIBEN® DS solution per ton of fertilizer from the table above.

Example: For 375 lb fertilizer per acre and 2.4 lb AMIBEN® DS per acre apply 2.75 gallons of AMIBEN® DS solution per ton of fertilizer.

DIRECTIONS FOR USE ON SOYBEANS

AMIBEN® DS PREEMERGENCE OR PREPLANT INCORPORATED

AMIBEN® DS Chloramben Herbicide may be used as a preemergence or preplant incorporated herbicide. AMIBEN® DS may be applied and incorporated at planting or several days prior to planting. Uneven spray application or improper soil incorporation of AMIBEN® DS can produce erratic weed control. Incorporate chemical in the top 0 to 2 inch level as most weeds sprout in this zone. Refer to the "Directions for Incorporation" section of this label for approved incorporation equipment and operating procedures.

In coarse textured soils heavy rains on incorporated AMIBEN® DS Herbicide may move it below the weed seed germination zone and produce erratic weed control.

Make all broadcast treatments according to the following table:

Soil Texture	AMIBEN® DS Lbs/A
Loamy sand, sandy loam	2.4 to 3.0
Loam, silt loam, silt	3.0 to 3.6
Silty clay loam, silty clay, sandy clay, clay loam, clay	3.6

Use the higher rate of AMIBEN® DS when heavy populations of black nightshade, velvetleaf, wild mustard, common ragweed or Pennsylvania smartweed are expected. To insure good spray coverage use 10 to 20 gallons of water per acre with ground equipment or a minimum of 5 gallons of water per acre by aircraft.

When band spraying, mount the spray nozzle on the packer or row wheel so that planting and spraying are done in the same operation. Proper calibration and maintenance of desired nozzle height above the ground are very important. Weeds growing in the middles between the treated rows should be cultivated as soon as necessary.

Use care while cultivating to avoid throwing fresh soil on treated bands.

If sufficient rain does not fall within 5 days after spraying AMIBEN® DS Herbicide, but there is enough soil moisture to germinate weeds and grasses, a light cultivation with a rotary hoe or similar tool will up-root these small broadleaf weeds and grasses. The shallow mixing of AMIBEN® DS in the surface soil will not interfere with the action of AMIBEN® DS when rains come.

If heavy rain packs the soil surface, rotary hoe the field to aid bean emergence. Do not disrupt the chemical band with deep cultivation.

AMIBEN® DS AND "LASSO" FOR SOYBEANS

A tank mixture of AMIBEN® DS Chloramben Herbicide and Lasso may be applied as a preplant soil incorporated herbicide treatment. AMIBEN® DS Herbicide may be applied as a preemergence broadcast tank mix treatment with Lasso during planting or after planting before weeds and soybeans emerge. Following a preplant soil incorporated application of Lasso, AMIBEN® DS may be applied preemergence during planting or after planting before weeds and soybeans emerge.

A tank mixture or the sequential treatment of AMIBEN® DS Herbicide and Lasso will control the following annual weeds: barnyardgrass, crabgrass, foxtail (green, yellow, giant), fall panicum, lambsquarters, pigweed, smartweed, ragweed, black nightshade, velvetleaf and wild mustard, as well as all other weeds listed on the AMIBEN® DS and Lasso labels.

For uniform application of the tank mixture of AMIBEN® DS Herbicide and Lasso use 10 to 20 gallons of water per acre for ground application. To prepare the tank mixture refer to the "Mixing Instructions" portion of this label. DO NOT mix AMIBEN® DS Herbicide and Lasso concentrates before the addition of water as an emulsion will not be formed.

Apply AMIBEN® DS and Lasso preplant incorporated or preemergence using the appropriate rates from the table below. For preplant incorporated treatments apply the tank mixture to the soil and incorporate into the top two inches of soil as directed in the "Directions for Incorporation" portion of this label. For preemergence treatments, apply the tank mixture during planting (behind the planter) or after planting before weeds or soybeans emerge. For sequential applications apply Lasso alone preplant incorporated as specified on the Lasso label. Follow with a preemergence application of AMIBEN® DS Herbicide during planting (behind the planter) or after planting before weeds or soybeans emerge.

With preemergence treatments rain is necessary to move the chemicals into the soil and make them available to control germinating weed seeds. If no rainfall occurs within 5 days following application, a shallow incorporation no deeper than 2 inches is recommended. High intensity or excessive rainfall after application may reduce control.

Make all broadcast treatments according to the following table:

Soil Texture	AMIBEN® DS Lbs/A	Lasso Qts/A
Loamy sand, sandy loam	2.4	1.5 to 2.0
Loam, silt loam, silt	2.4 to 3.0	2.0 to 2.5
Silty clay loam, silty clay, sandy clay, clay loam, clay	3.0 to 3.6	2.5 to 3.0

Use the higher rate of AMIBEN® DS when heavy populations of black nightshade, velvetleaf, mustard or smartweed are expected. Use the higher rate of Lasso when treatment is incorporated or when high populations of black nightshade, witchgrass or foxtail are expected.

For band treatments divide the band width by the row width and multiply by the appropriate rates indicated above to determine the amount of chemical to use per acre.

Read thoroughly the cautionary statements and all other information appearing on labels for both Lasso and AMIBEN® DS.

"AMIBEN®" DS TANK MIXTURE WITH "LASSO" AND "SENCOR" OR "LEXONE" FOR SOYBEANS

GENERAL INFORMATION

A tank mixture of AMIBEN® DS Chloramben Herbicide plus Lasso plus Sencor or Lexone when applied in accordance with the following directions, either preplant soil incorporated or preemergence, effectively controls most weeds listed on the current AMIBEN® DS, Lasso, and Lexone WP/4L/DF or Sencor WP/4/DF labels.

DIRECTIONS FOR USE

Except as directed in the "Conservation or Reduced Tillage" portion of this label, the seedbed should be free of existing weeds before application. Surface trash (plant residue from previous crop) should be uniformly distributed, and the seedbed should be free of clods. The tank mixture should be applied in from 10 to 20 gallons of water per acre using any properly calibrated low pressure sprayer equipped with herbicide tips and in-line strained screens no finer than 50 mesh. To prepare the tank mixture refer to the "Mixing Instructions" portion of this label. Apply the entire contents of the spray tank within 4 hours of initial mixing.

PREPLANT SOIL INCORPORATED USE

Incorporation should place the AMIBEN® DS Herbicide plus Lasso plus Sencor or Lexone into the top 1 to 3 inches of the final seedbed. Refer to the "Directions for Incorporation" portion of this label for recommended incorporation equipment and operating procedures.

Apply the tank mixture using the table below to determine the proper rate of each chemical for your soil type.

PREEMERGENCE USE

The tank mixture of AMIBEN® DS plus Lasso plus Lexone or Sencor may be applied as a preemergence broadcast treatment at time of planting. Use the table below to determine the proper rate of each chemical for your soil type.

Make all broadcast treatments according to the following table:

Soil Texture	AMIBEN® DS Lbs/A	Lasso 4E Qts/A	Sencor 50WP or Lexone WP Lbs/A	Sencor 4 or Lexone 4L Pts/A	Sencor DF or Lexone DF Lbs/A
Coarse (loamy sand, sandy loam)	2.4	1.5 to 2.0	0.5	0.5	0.33
Medium (loam, silt loam, silt)	2.4 to 3.0	2.0 to 2.5	0.5 to 0.75*	0.5 to 0.75*	0.33 to 0.5*
Fine (clay, clay loam, silty clay, sandy clay, sandy clay loam, silty clay loam)	3.0 to 3.6	2.5 to 3.0	0.75 to 1.0*	0.75 to 1.0*	0.5 to 0.67*

* On Clarion Webster soils in Minnesota and Iowa or on similar alkaline (calcareous) soils a pH of 7.5 or above, apply Sencor WP or Lexone WP at 0.5 pound/A, Sencor 4 or Lexone 4L at 0.5 pint/A, or Lexone DF or Sencor DF at 0.33 pound/A on medium soils. On fine soils use 0.5 to 0.75 pound/A Sencor WP or Lexone WP, 0.5 to 0.75 pint Sencor 4 or Lexone 4L or 0.33 to 0.5 pound/A Lexone DF or Sencor DF. Use the higher rate only where soil pH is less than 7.5 and where pressure is heavy.

Use the higher rate of AMIBEN® DS where heavy stands of Pennsylvania smartweed, velvetleaf, ragweed, wild mustard, or black nightshade are expected. Use the higher rate of Lasso when treatments are incorporated or where heavy stands of black nightshade, foxtail, witchgrass, panicum, or yellow nutsedge are expected. Use the high rate of Sencor or Lexone where heavy stands of Florida pusley, sesbania, velvetleaf or Jimsonweed are expected except when soil pH is over 7.5.

CULTIVATION AFTER PLANTING

Soil treated with AMIBEN® DS Chloramben Herbicide plus Lasso plus Sencor or Lexone may be shallow-cultivated, rotary-hoed, or hand-hoed without reducing the weed control activity of the tank mixture. With preplant soil incorporated treatments, do not cultivate deeper than the AMIBEN® DS plus Lasso plus Sencor or Lexone treated layer of soil since this may bring untreated soil to the surface and poor weed control may result. If sufficient rain does not fall within 5 days after preemergence applications of the tank mixture, but there is enough soil moisture to germinate weeds and grasses, a light cultivation with a rotary hoe or similar tool will uproot these small broadleaf weeds and grasses.

SPECIAL PRECAUTIONS

Applied according to directions and under normal growing conditions, the tank mix of AMIBEN® DS plus Lasso plus Sencor or Lexone will not harm the treated crop. Over-application may result in crop injury. Uneven application or improper soil incorporation of AMIBEN® DS plus Lasso plus Sencor or Lexone can result in erratic weed control or crop injury. Seedling disease, cold weather, deep planting, excessive moisture, soil pH over 7.5, high salt concentration, or drought may weaken crop seedlings and increase the possibility of damage from the tank mixtures. Under these conditions, delayed crop development or reduced yields may result.

Do not use on sand or on medium to fine soils with less than 1% organic matter. This mixture applied preplant incorporated may cause slight soybean injury early, but the soybeans should recover and the subsequent growth or yields should not be affected.

Do not use the foliage from soybeans treated with the tank mix of AMIBEN® DS and Lasso and Sencor or AMIBEN® DS and Lasso and Lexone for feed or forage.

Caution: Observe all cautions and limitations on labeling of all products used in mixtures.

"AMIBEN®" DS AND "LASSO" AND "LOROX" FOR SOYBEANS

A tank mixture of AMIBEN® DS Chloramben Herbicide and Lasso and Lorox L or Lorox 50W may be applied as a preemergence broadcast tank mix treatment during planting or after planting before weeds and soybeans emerge.

A tank mixture of AMIBEN® DS Herbicide and Lasso and Lorox will control the following annual weeds: barnyardgrass, carpetweed, crabgrass, foxtail (green, yellow, giant), fall panicum, lambsquarters, pigweed, smartweed, ragweed, black nightshade, velvetleaf and wild mustard, as well as all other weeds listed on the AMIBEN® DS, Lasso and Lorox labels. Velvetleaf and cocklebur often germinate in soil at depths below the chemical penetration. Even though populations of these weeds are reduced under normal rainfall conditions, weed seeds which germinate outside the chemical zone are not affected.

Except as directed in the "Conservation or Reduced Tillage" portion of this label, the seedbed should be free of existing weeds before application. Surface trash (plant residue from previous crop) should be uniformly distributed, and the seedbed should be free of clods. For uniform application of the tank mixture of AMIBEN® DS Herbicide and Lasso and Lorox use 10 to 20 gallons of water per acre for ground application. To prepare the tank mixture refer to the "Mixing Instructions" portion of this label. DO NOT mix AMIBEN® DS Herbicide and Lasso and Lorox concentrates before the addition of water as an emulsion will not be formed.

Apply AMIBEN® DS and Lasso and Lorox as a preemergence treatment using the appropriate rates from the table below. Apply the tank mixture during planting (behind the planter) or after planting before weeds or soybeans emerge.

Rain is necessary to move the chemicals into the soil and make them available to control germinating weed seeds. If no rainfall occurs within 5 days following application, a shallow incorporation not deeper than 2 inches is recommended. High intensity or excessive rainfall after application may reduce control.

Make all broadcast treatments according to the following table:

3% or Less Organic Matter

Soil Texture	AMIBEN® DS Lbs/A	Lasso Qts/A	Lorox 50W or Lorox L Lbs or pts/A
Loamy sand, sandy loam	2.4	1.5 to 2.0	0.67 to 1.5
Loam, silt loam, silt	2.4 to 3.0	2.0 to 2.5	1.0 to 1.67
Silty clay loam, silty clay, sandy clay, clay loam, clay	3.0 to 3.6	2.5 to 3.0	1.25 to 2.0

Greater than 3% Organic Matter

Soil Texture	AMIBEN® DS Lbs/A	Lasso Qts/A	Lorox 50W or Lorox L Lbs or pts/A
Loamy sand, sandy loam	2.4	1.5 to 2.0	1.5 to 2.0
Loam, silt loam, silt	2.4 to 3.0	2.0 to 2.5	1.67 to 2.5
Silty clay loam, silty clay, sandy clay, clay loam, clay	3.0 to 3.6	2.5 to 3.0	2.0 to 3.0

Use the higher rate of AMIBEN® DS when heavy population of black nightshade, velvetleaf, mustard or smartweed are expected. Use the higher rate of Lasso when high populations of black nightshade, witchgrass or foxtail are expected. Use the higher rate of Lorox on soils with organic matter contents which are toward the high end of the ranges given in the table.

For band treatments divide the band width by the row width and multiply by the appropriate rates indicated above to determine the amount of chemical to use per acre.

Plant seeds at least 1.75 inches deep and do not treat if beans have begun to emerge.

Read thoroughly the cautionary statements and all other information appearing on labels for Lasso, Lorox L or Lorox 50W and AMIBEN® DS.

"AMIBEN®" DS and "DUAL" FOR SOYBEANS

A tank mixture of AMIBEN® DS Chloramben Herbicide and Dual 8E may be applied as a preplant soil incorporated herbicide treatment. AMIBEN® DS Herbicide may be applied as a preemergence broadcast tank mix treatment with Dual 8E during planting or after planting before weeds and soybeans emerge. Following a preplant soil incorporated application of Dual 8E, AMIBEN® DS Herbicide may be applied preemergence during planting or after planting before weeds and soybeans emerge.

When applied according to label directions, tank mixtures of AMIBEN® DS and Dual will control all weeds controlled by AMIBEM® DS alone or by Dual alone including fall panicum, foxtail millet, red rice, signalgrass, witchgrass, yellow nutsedge and Florida pusley. Refer to the general information section of this label and the Dual label for additional weeds controlled by AMIBEN® DS and Dual 8E.

For uniform application of the tank mixture of AMIBEN® DS and Dual, use 10 to 20 gallons of water with ground equipment. To prepare the tank mixture, refer to the "Mixing Instructions" portion of this label.

Apply AMIBEN® DS and Dual preplant incorporated or preemergence using the appropriate rates from the table below. For preplant incorporated treatments apply the tank mixture to the soil and incorporate into the top two inches of soil following the procedures in the "Directions for Incorporation" portion of this label. Use a preplant incorporate application if furrow irrigation is used or when a period of dry weather is expected. If soybeans are planted on beds apply and incorporate the tank mixture after bed formation. For preemergence treatments, apply the tank mixture during planting (behind the planter) or after planting before weeds or soybeans emerge. For sequential applications apply Dual 8E alone preplant incorporated as specified on the Dual label. Follow with a preemergence application of AMIBEN® DS Herbicide during planting (behind the planter) or after planting before weeds or soybeans emerge.

Make all broadcast treatments according to the following table:

Soil Texture	AMIBEN® DS Lbs/A	Dual 8E Pts/A
Loamy sand, sandy loam	2.4	1.5 to 2.0
Loam, silt loam, silt	2.4 to 3.0	2.0 to 2.5
Silty clay loam, silty clay, sandy clay, clay loam, clay	3.0 to 3.6	2.5 to 3.0

Do not use on muck or peat soils.

Use the higher rate of AMIBEN® DS when heavy populations of black nightshade, velvetleaf, mustard or smartweed are expected. Use the higher rate of Dual when treatment is incorporated, when organic matter is over 3%, or when high populations of black nightshade, witchgrass, panicum, foxtail, or yellow nutsedge are expected.

Before applying this combination read thoroughly all cautionary statements and all other information appearing on or attached to the containers of AMIBEN® DS Herbicide and Dual 8E.

"AMIBEN®" DS TANK MIXTURE WITH "DUAL" AND "SENCOR" OR "LEXONE" FOR SOYBEANS

GENERAL INFORMATION

A tank mixture of AMIBEN® DS Chloramben Herbicide plus Dual 8E plus Sencor or Lexone when applied in accordance with the following directions, either preplant soil incorporated or preemergence, effectively controls most weeds listed on the current AMIBEN® DS, Dual 8E, and Lexone WP/4L/DF or Sencor WP/4/DF labels.

DIRECTIONS FOR USE

Except as directed in the "Conservation or Reduced Tillage" portion of this label, the seedbed should be free of existing weeds before application. Surface trash (plant residue from previous crop) should be uniformly distributed, and the seedbed should be free of clods. The mixture should be applied in from 10 to 20 gallons of water per acre using any properly calibrated low pressure sprayer equipped with herbicide tips and in-line strained screens no finer than 50 mesh. To prepare the tank mixture refer to the "Mixing Instructions" portion of this label. Apply the entire contents of the spray tank within 4 hours of initial mixing.

PREPLANT SOIL INCORPORATED USE

Incorporation should place the AMIBEN® DS plus Dual 8E plus Sencor or Lexone into the top 1 to 3 inches of the final seedbed as directed in the "Directions for Incorporation" portion of this label. Apply the tank mixture using the table below to determine the proper rate of each chemical for your soil type.

PREEMERGENCE USE

The tank mixtures of AMIBEN® DS plus Dual 8E plus Lexone or Sencor may be applied as a preemergence broadcast treatment at time of planting. Use the table below to determine the proper rate of each chemical for your soil type.

Make all broadcast treatments according to the following table:

Soil Texture	AMIBEN® DS Lbs/A	Dual 8E Pts/A	Sencor 50WP or Lexone WP Lbs/A	Sencor 4 or Lexone 4L Pts/A	Sencor DF or Lexone DF Lbs/A
Coarse (loamy sand, sandy loam)	2.4	1.5 to 2.0	0.5	0.5	0.33
Medium (loam, silt loam, silt)	2.4 to 3.0	2.0 to 2.5	0.5 to 0.75*	0.5 to 0.75*	0.33 to 0.5*
Fine (clay, clay loam, silty clay, sandy clay, sandy clay loam, silty clay loam)	3.0 to 3.6	2.5 to 3.0	0.75 to 1.0*	0.75 to 1.0*	0.5 to 0.67*

* On Clarion Webster soils in Minnesota and Iowa or on similar alkaline (calcareous) soils a pH of 7.5 or above, apply Sencor WP or Lexone WP at 0.5 pound/A, Sencor 4 or Lexone 4L at 0.5 pint/A, or Lexone DF or Sencor DF at 0.33 pound/A on medium soils. On fine soils use 0.5 to 0.75 pound A/Sencor WP or Lexone WP, 0.5 to 0.75 pint Sencor 4 or Lexone 4L or 0.33 to 0.5 pound/A Lexone DF or Sencor DF. Use the higher rate only where soil pH is less than 7.5 and where weed pressure is heavy.

Use the higher rate of AMIBEN® DS where heavy stands of Pennsylvania smartweed, velvetleaf, ragweed, wild mustard, or black nightshade are expected. Use the higher rate of Dual 8E when treatments are incorporated, when organic matter is greater than 3%, or where heavy stands of black nightshade, foxtail, witchgrass, panicum, or yellow nutsedge are expected. Use the high rate of Sencor or Lexone where heavy stands of Florida pusley, sesbania, velvetleaf of Jimsonweed are expected except when soil pH is over 7.5.

CULTIVATION AFTER PLANTING

Soil treated with AMIBEN® DS plus Dual plus Sencor or Lexone may be shallow-cultivated, rotary-hoed, or hand-hoed without reducing the weed control activity of the tank mixture. With preplant soil incorporated treatments, do not cultivate deeper than the AMIBEN® DS plus Dual plus Sencor or Lexone treated layer of soil since this may bring untreated soil to the surface and poor weed control may result. If sufficient rain does not fall within 5 days after preemergence applications of the tank mixture, but there is enough soil moisture to germinate weeds and grasses, a light cultivation with a rotary hoe or similar tool will uproot these small broadleaf weeds and grasses.

SPECIAL PRECAUTIONS

Applied according to directions and under normal growing conditions, the tank mix of AMIBEN® DS plus Dual plus Sencor or Lexone will not harm the treated crop. Over-application may result in crop injury. Uneven application or improper soil incorporation of AMIBEN® DS plus Dual plus Sencor or Lexone can result in erratic weed control or crop injury. Seedling disease, cold weather, deep planting, excessive moisture, soil pH over 7.5, high salt concentration, or drought may weaken crop seedlings and increase the possibility of damage from the tank mixture. Under these conditions, delayed crop development or reduced yields may result.

Do not use on sand or on medium to fine soils with less than 1% organic matter. This mixture applied preplant incorporated may cause slight soybean injury early, but the soybeans should recover and the subsequent growth or yields should not be affected.

Do not use the foliage from soybeans treated with the tank mix of AMIBEN® DS and Dual and Sencor or AMIBEN® DS and Dual and Lexone for feed or forage.

Caution: Observe all cautions and limitations on labeling of all products used in mixtures.

"AMIBEN®" DS AND "DUAL" AND "LOROX" FOR SOYBEANS

A tank mixture of AMIBEN® DS Chloramben Herbicide and Dual and Lorox L or Lorox 50W may be applied as a preemergence broadcast tank mix treatment during planting or after planting before weeds and soybeans emerge.

A tank mixture of AMIBEN® DS Herbicide and Dual and Lorox will control the following annual weeds: barnyardgrass, carpetweed, crabgrass, foxtail (green, yellow, giant), fall panicum, lambsquarters, pigweed, smartweed, ragweed, black nightshade, velvetleaf and wild mustard, as well as all other weeds listed in the AMIBEN® DS, Dual and Lorox labels. Velvetleaf and cocklebur often germinate in soil at depths below the chemical penetration. Even though populations of these weeds are reduced under normal rainfall conditions, weed seeds which germinate outside the chemical zone are not affected.

Except as directed in the "Conservation or Reduced Tillage" portion of this label, the seedbed should be free of existing weeds before application. Surface trash (plant residue from previous crop) should be uniformly distributed, and the seedbed should be free of clods. For uniform application of the tank mixture of AMIBEN® DS Herbicide and Dual and Lorox use 10 to 20 gallons of water per acre for ground application. To prepare the tank mixture refer to the "Mixing Instructions" portion of this label.

Apply AMIBEN® DS and Dual and Lorox as a preemergence treatment using the appropriate rates from the table below. Apply the tank mixture during planting (behind the planter) or after planting before weeds or soybeans emerge.

Rain is necessary to move the chemicals into the soil and make them available to control germinating weed seeds. If no rainfall occurs within 5 days following application, a shallow incorporation no deeper than 2 inches is recommended. High intensity or excessive rainfall after application may reduce control.

Make all broadcast treatments according to the following table:

3% or Less Organic Matter

Soil Texture	AMIBEN® DS Lbs/A	Dual Pts/A	Lorox 50W or Lorox L Lbs or pts/A
Loamy sand, sandy loam	2.4	1.5 to 2.0	0.67 to 1.5
Loam, silt loam, silt	2.4 to 3.0	2.0 to 2.5	1.0 to 1.67
Silty clay loam, silty clay, sandy clay, clay loam, clay	3.0 to 3.6	2.5 to 3.0	1.25 to 2.0

Greater than 3% Organic Matter

Soil Texture	AMIBEN® DS Lbs/A	Dual Pts/A	Lorox 50W or Lorox L Lbs or pts/A
Loamy sand, sandy loam	2.4	1.5 to 2.0	1.5 to 2.0
Loam, silt loam, silt	2.4 to 3.0	2.0 to 2.5	1.67 to 2.5
Silty clay loam, silty clay, sandy clay, clay loam, clay	3.0 to 3.6	2.5 to 3.0	2.0 to 3.0

Use the higher rate of AMIBEN® DS when heavy populations of black nightshade, velvetleaf, mustard or smartweed are expected. Use the higher rate of Dual when high populations of black nightshade, witchgrass or foxtail are expected. Use the higher rate of Lorox on soils with organic matter contents which are toward the high end of the ranges given in the table.

For band treatments divide the band width by the row width and multiply by the appropriate rates indicated above to determine the amount of chemical to use per acre.

Plant seeds at least 1.75 inches deep and do not treat if beans have begun to emerge.

Read thoroughly the cautionary statements and all other information appearing on labels for Dual, Lorox L or Lorox 50W and AMIBEN® DS.

"AMIBEN®" DS AND "TREFLAN" FOR SOYBEANS

AMIBEN® DS Chloramben Herbicide may be applied in a band over the soybean row at planting time in fields where Treflan has been applied as a preplant soil incorporated herbicide. Or AMIBEN® DS Herbicide may be applied several days prior to planting as a broadcast tank mix with Treflan. The tank mixture should be used as a spring preplant soil incorporated treatment. The combination of AMIBEN® DS Herbicide with Treflan improves broadleaf weed control of species such as Pennsylvania smartweed, velvetleaf and ragweed. For broadcast treatments, incorporate chemicals immediately and thoroughly to an approximate depth of 2 inches with a disc, field cultivator or similar tool set to cut a depth of 4 to 6 inches. Refer to the Treflan label for additional incorporation information.

Apply AMIBEN® DS at a rate of 2.4 to 3.6 pounds per broadcast acre. Use the higher rate where heavy populations of Pennsylvania smartweed, velvetleaf, ragweed, wild mustard or black nightshade are expected.

Apply Treflan at a rate of 1½ pints for medium soils and 2 pints for fine soils. Do not use on muck or charcoal soils. Read and observe all directions and cautions on labels for AMIBEN® DS and Treflan.

"AMIBEN®" DS TANK MIXTURES WITH "TREFLAN" AND "SENCOR" OR "LEXONE" FOR SOYBEANS

GENERAL INFORMATION

A tank mix of AMIBEN® DS Chloramben Herbicide plus Treflan plus Sencor or Lexone when applied preplant soil incorporated as recommended herein for use on soybeans, effectively controls the weeds listed on the current AMIBEN® DS, Treflan and Sencor WP/4/DF or Lexone WP/4L/DF labels.

DIRECTIONS FOR USE

Except as directed in the "Conservation or Reduced Tillage" portion of this label, the seedbed should be free of existing weeds before application. Surface trash (plant residue from previous crop) should be uniformly distributed, and the seedbed should be free of clods.

To prepare the tank mixture refer to the "Mixing Instructions" portion of this label. Apply the entire contents of the spray tank within 4 hours of initial mixing. Apply the tank mixture in from 10 to 20 gallons of water per acre using any properly calibrated low pressure sprayer equipped with herbicide tips and in-line strained screens no finer than 50 mesh.

For best results AMIBEN® DS plus Treflan plus Sencor or Lexone should be incorporated as soon as possible after application. The three way tank mixture must be incorporated within 24 hours after application. If AMIBEN® DS plus Treflan plus Sencor or Lexone is applied to wet, warm soil surface or if the wind velocity is 10 mph or higher, variable weed control may result from delaying incorporation beyond 24 hours.

Incorporation should place the AMIBEN® DS Herbicide plus Treflan plus Sencor or Lexone into the top 2 to 3 inches of the final seedbed. Refer to the incorporation portion of the Treflan label for approved incorporation equipment and operating procedures.

CULTIVATION AFTER PLANTING

Soil treated with AMIBEN® DS plus Treflan plus Sencor or Lexone may be shallow-cultivated, rotary-hoed or hand-hoed without reducing the weed control activity of the tank mixture. Do not cultivate deeper than the AMIBEN® DS plus Treflan plus Sencor or Lexone treated layer of soil since this may bring untreated soil to the surface and poor weed control may result.

Make all broadcast treatments according to the following table:

Soil Texture	AMIBEN® DS Lbs/A	Treflan 4EC Pts/A	Sencor 50WP or Lexone WP Lbs/A	Sencor 4 or Lexone 4L Pts/A	Sencor DF or Lexone DF Lbs/A
Course (sand, loamy sand, sandy loam) a/	2.4	1.0	0.5	0.5	0.33
Medium (loam, silt loam, silt)	2.4 to 3.0*	1.5	0.5 to 0.75**	0.5 to 0.75**	0.33 to 0.5**
Fine (clay, clay loam, silty clay, sandy clay, sandy clay loam, silty clay loam)	3.0 to 3.6*	2.0	0.75 to 1.0**	0.75 to 1.0**	0.5 to 0.67**

a/ Do not use Sencor or Lexone on coarse soils with less than 1% organic matter.

* Use the higher rate of AMIBEN® DS Chloramben Herbicide when velvetleaf or black nightshade is a problem.

** On Clarion Webster soils in Minnesota and Iowa or on similar alkaline (calcareous) soils with a pH of 7.5 or above apply Sencor WP or Lexone WP at 0.5 pound/A, Sencor 4 or Lexone 4L at 0.5 pint/A or Lexone DF or Sencor DF at 0.33 pound/A on medium soils and at 0.5 to 0.75 pound/A Sencor WP or Lexone WP, 0.5 to 0.75 pint/A Sencor 4 or Lexone 4L or 0.33 to 0.5 pound/A Lexone DF or Sencor DF on fine soils. Use the higher rate only where soil pH is less than 7.5 and where weed pressure is heavy.

SPECIAL PRECAUTIONS
Applied according to directions and under normal growing conditions, the AMIBEN® DS and Treflan and Sencor or AMIBEN® DS and Treflan and Lexone tank mix will not harm the treated crop. Over-application may result in crop injury. Uneven application or improper soil incorporation of the AMIBEN® DS and Treflan and Sencor or AMIBEN® DS and Treflan and Lexone tank mix can result in erratic weed control or crop injury. Seedling disease, cold weather, deep planting, excessive moisture, soil pH over 7.5, high salt concentration, or drought may weaken crop seedlings and increase possibility of damage from the tank mixtures. Under these conditions, delayed crop development or reduced yields may result.

Caution: Observe all cautions and limitations on labeling of all products used in mixtures.

Do not use the foliage from soybeans treated with the AMIBEN® DS and Treflan and Sencor or AMIBEN® DS and Treflan and Lexone tank mix for feed or forage.

"AMIBEN®" DS AND "SONALAN" FOR SOYBEANS

AMIBEN® DS Chloramben Herbicide may be applied as a preemergence band or braodcast treatment in soybeans at planting time in fields where Sonalan has been applied as a preplant soil incorporated herbicide. Or AMIBEN® DS Herbicide may be applied prior to planting as a broadcast tank mix with Sonalon. The tank mixture should be used as a spring preplant soil incorporated treatment. The combination of AMIBEN® DS Herbicide with Sonalan improves broadleaf weed control of species such as Pennsylvania smartweed, velvetleaf, common ragweed, black nightshade and wild mustard. For broadcast treatments, incorporate chemicals immediately and thoroughly to an appropriate depth of 2 inches as described on the Sonalan label.

Make all broadcast treatments according to the following table:

Soil Texture	AMIBEN® DS Lbs/A	Sonalan Pts/A
Loamy sand, sandy loam	2.4	1.25 to 2.0
Loam, silt loam, silt	2.4 to 3.0	1.75 to 2.5
Silty clay loam, silty clay, sandy clay, clay loam, clay	3.0 to 3.6	2.25 to 3.0

Use the higher rate of AMIBEN® DS where heavy populations of smartweed, velvetleaf, ragweed, wild mustard or black nightshade are expected. Use the higher rate of Sonalan on finer textured soils and on soils with organic matter content over 3%.

"AMIBEN®" DS AND "PROWL" FOR SOYBEANS

A tank mixture of AMIBEN® DS Chloramben Herbicide and Prowl may be applied as a preplant soil incorporated herbicide treatment or AMIBEN® DS Herbicide may be applied as a preemergence broadcast treatment at planting time in soybean fields where Prowl has been applied as a preplant soil incorporated herbicide treatment. On soybeans grown south of Interstate 80 the tank mixture of AMIBEN® DS plus Prowl may also be applied as a preemergence broadcast treatment. The combination may be used only in Illinois, Indiana, Iowa, Kansas, Kentucky, Michigan, Minnesota, Nebraska, New York, North Dakota, Ohio, Pennsylvania, South Dakota, Wisconsin and Missouri except in the "Bootheel Region".

When applied according to directions, tank mixtures of AMIBEN® DS Herbicide and Prowl herbicide either preplant incorporated or preemergence, or the sequential treatment of AMIBEN® DS preemergence over Prowl preplant incorporated will control fall panicum, signalgrass, Venice mallow and velvetleaf in addition to all other weeds listed on the AMIBEN® DS and Prowl labels.

For uniform application of the tank mixture of AMIBEN® DS and Prowl, use 10 to 20 gallons of water per acre with ground equipment or 3 to 5 or more gallons of water per acre by aircraft. To prepare the tank mixture refer to the "Mixing Instructions" portion of this label.

Except as directed in the "Conservation or Reduced Tillage" portion of this label, the seedbed should be free of existing weeds before application. Surface trash (plant residue from previous crop) should be uniformly distributed, and the seedbed should be free of clods. When using the tank mixture preplant soil incorporated or when using Prowl preplant incorporated followed by a preemergence application of AMIBEN® DS the herbicides should be applied within 7 days prior to planting. When applied as a preplant soil incorporated treatment the tank mixture of AMIBEN® DS plus Prowl should be thoroughly incorporated into the top 0 to 2 inches of soil using equipment and techniques described elsewhere on this label and on the Prowl label. When using the tank mixture preemergence, treatments should be made at planting or within 2 days after planting before weeds and soybeans emerge.

Make all broadcast treatments according to the following table:

Soil Texture	AMIBEN® DS Lbs/A	Prowl Pts/A
Loamy sand, sandy loam	2.4	1.0 to 2.0
Loam, silt loam, silt	2.4 to 3.0	1.5 to 2.5
Silty clay loam, silty clay, sandy clay, clay loam, clay	3.0 to 3.6	2.0 to 3.0

Use the higher rate of AMIBEN® DS when heavy populations of black nightshade, velvetleaf, wild mustard, common ragweed or Pennsylvania smartweed are expected. Use the higher rate of Prowl when treatments are incorporated on soils with more than 3% organic matter, or where heavy stands of foxtail, witchgrass or panicum are expected. Do not use on peat or muck soils.

If sufficient rain does not fall within 5 days after preemergence treatment, but there is enough soil moisture to germinate weeds and soybeans a light cultivation with a rotary hoe or similar tool will uproot these small weeds. The shallow mixing of the herbicides into the soil surface will not interfere with their action when rains come. If soybean crop loss occurs due to weather conditions, soybeans may be replanted. If replanting is necessary, DO NOT rework the soil.

Before applying this combination, read thoroughly the cautionary statements and all other information appearing on or attached to the containers of both the Prowl and AMIBEN® DS herbicides.

"AMIBEN®" DS TANK MIXTURE WITH "PROWL" AND "SENCOR" OR "LEXONE" FOR SOYBEANS

GENERAL INFORMATION
A tank mixture of AMIBEN® DS Chloramben Herbicide plus Prowl plus Sencor or Lexone may be applied either preplant soil incorporated or preemergence in soybeans grown in the northeastern and north central states (Illinois, Indiana, Iowa, Kansas, Kentucky, Michigan, Minnesota, Nebraska, New York, North Dakota, Ohio, Pennsylvania, South Dakota, Wisconsin, Missouri (except the "Bootheel Region"), all areas of Delaware, Maryland, New Jersey, and Virginia (except the Coastal Plains of these states). Do not make preemergence applications of Prowl north of Interstate 80. When applied according to the following directions the tank mixture of AMIBEN® DS plus Prowl plus Sencor or Lexone effectively controls the weeds listed on the current AMIBEN®, Prowl and Sencor WP/4/DF or Lexone WP/4L/DF labels.

DIRECTIONS FOR USE
Except as directed in the "Conservation or Reduced Tillage" portion of this label, the seedbed should be free of existing weeds before application. Surface trash (plant residue from previous crop) should be uniformly distributed, and the seedbed should be free of clods. Uniformly apply the recommended AMIBEN® DS plus Prowl plus Sencor or Lexone tank mixture in 10 to 20 gallons of water per acre with ground equipment or in 3 to 5 or more gallons of water per acre by aircraft using herbicide nozzle tips and in line strained screens no finer than 50 mesh. To prepare the tank mixture refer to the "Mixing Instructions" portion of this label. Apply the entire contents of the spray tank within 4 hours of initial mixing.

PREPLANT INCORPORATED USE

The tank mixture of AMIBEN® DS plus Prowl plus Sencor or Lexone must be incorporated into the soil and soybeans must be planted within 7 days after application. Uneven spray application or improper soil incorporation can produce erratic weed control. The tank mixture should be thoroughly incorporated into the top 1 to 2 inches of soil using equipment and techniques described elsewhere on this label.

PREEMERGENCE USE

Apply the tank mixture of AMIBEN® DS plus Prowl plus Sencor or Lexone as a broadcast treatment at time of planting or within 2 days after planting before weeds and soybeans emerge. This treatment is most effective when adequate rainfall or overhead irrigation is received within 5 days after application. If sufficient rain does not fall within 5 days after spraying the tank mixture of AMIBEN® DS plus Prowl plus Sencor or Lexone, but there is enough soil moisture to germinate weeds and grasses a light cultivation with a rotary hoe or similar tool will up-root these small broadleaf weeds and grasses. The shallow mixing of the tank mixture in the surface soil will improve consistency of performance and reduce dependency on rainfall to activate the herbicides, but will not interfere with their action when rains come.

APPLICATION RATES

Make all broadcast treatments according to the following table:

Soil Texture	AMIBEN® DS Lbs/A	Prowl Pts/A	Sencor 50WP or Lexone WP Lbs/A	Sencor 4 or Lexone 4L Pts/A	Sencor DF or Lexone DF Lbs/A
Coarse (loamy sand, sandy loam)	2.4	1.5 to 2.0	0.5	0.5	0.33
Medium (loam, silt loam, silt)	2.4 to 3.0	1.5 to 2.5	0.5 to 0.75*	0.5 to 0.75*	0.33 to 0.5*
Fine (clay, clay loam, silty clay, sandy clay, sandy clay loam, silty clay loam)	3.0 to 3.6	2.0 to 3.0	0.75 to 1.0*	0.75 to 1.0*	0.5 to 0.67*

* On Clarion Webster soils in Minnesota and Iowa or on similar alkaline (calcareous) soils with a pH of 7.5 or above, apply Sencor WP or Lexone WP at 0.5 pound/A, Sencor 4 or Lexone 4L at 0.5 pint/A, or Lexone DF or Sencor DF at 0.33 pound/A on medium soils. On fine soils use 0.5 to 0.75 pound/A Sencor WP or Lexone WP, 0.5 to 0.75 pint Sencor 4 or Lexone 4L or 0.33 to 0.5 pound/A Lexone DF or Sencor DF. Use the higher rate only where soil pH is less than 7.5 and where pressure is heavy.

Use the higher rate of AMIBEN® DS where heavy stands of Pennsylvania smartweed, velvetleaf, ragweed, wild mustard, or black nightshade are expected. Use the higher rate of Prowl when treatments are incorporated on soils with more than 3% organic matter, or where heavy stands of foxtail, witchgrass or panicum are expected. Use the high rate of Sencor or Lexone where heavy stands of Florida pusley, sesbania, velvetleaf or Jimsonweed are expected except when soil pH is over 7.5.

SPECIAL PRECAUTIONS

Soybean seed should be planted at least 1½ inches below soil surface. Soil incorporation deeper than 2 inches will reduce weed control and can result in crop injury. Seedling disease, cold weather, deep planting, excessive soil moisture, soil pH over 7.5, high salt concentration, or drought may weaken crop seedlings and increase possibility of damage from the tank mixtures. Under these conditions delayed crop development or reduced yields may result.

Do not use on sensitive soybean varieties such as Tracy, Semmes, Altona, Vansoy or Coker 102.

Do not use the foliage from soybeans treated with the AMIBEN® DS plus Prowl plus Sencor or Lexone tank mix for feed or forage.

Caution: Observe all cautions and limitations on labeling of all products used in mixtures.

"AMIBEN®" DS AND METRIBUZIN ("SENCOR" OR "LEXONE") TANK MIXTURE FOR SOYBEANS

AMIBEN® DS Chloramben Herbicide and metribuzin may be applied in a tank mixture combination as a preemergence band or broadcast application. Use standard spray equipment, and 10 to 20 gallons of spray solution per broadcast acre. To prepare the tank mixture refer to the "Mixing Instructions" portion of this label.

This tank mixture controls Florida pusley, sesbania (coffeebean or coffeeweed), shepherdspurse and sicklepod in addition to those weeds controlled by AMIBEN® DS Herbicide. Performance on cocklebur, fall panicum and black nightshade may be erratic depending on soil temperature, depth of weed seeds in soil and amount of moisture. Control may be improved with timely cultivation.

Make all broadcast treatments according to the following table:

3% or Less Organic Matter

Soil Texture	AMIBEN® DS Lbs/A	Sencor 50W or Lexone WP Lbs/A	Sencor 4 or Lexone 4L Pts/A	Sencor DF or Lexone DF Lbs/A
Loamy sand to sandy loam (greater than 2% organic matter only)	2.4	0.5	0.5	0.33
Loam to sandy clay loam	2.4 to 3.0	0.5	0.75	0.5
Silty clay loam to clay	3.0 to 3.6	1.0	1.0	0.67

Greater than 3% Organic Matter

Soil Texture	AMIBEN® DS Lbs/A	Sencor 50W or Lexone WP Lbs/A	Sencor 4 or Lexone 4L Pts/A	Sencor DF or Lexone DF Lbs/A
Loamy sand to sandy loam (greater than 2% organic matter only)	2.4	0.75	0.75	0.5
Loam to sandy clay loam	2.4 to 3.0	1.0	1.0	0.67
Silty clay loam to clay	3.0 to 3.6	1.0	1.0	0.67

In the Mississippi Delta only, on silty clay to heavy clay textured soil use 3.6 pounds of AMIBEN® DS plus Lexone 50 WP—1.5 lb., Lexone DF—1 lb. or 2 lbs. of Sencor 50 WP, 2 pts. of Sencor 4, or 1 lb. of Sencor DF. Do not use on muck soils. Use the high rates when heavy weed stands are expected.

For band application rates divide Band width by the Row spacing and multiply by rate/acre selected from the chart above.

SPECIAL PRECAUTIONS

Injury may occur to soybeans following application of metribuzin on soils having a calcareous surface area or a pH of 7.5 or higher.

Sensitive soybean varieties such as Tracy, Semmes, Altona, Vansoy or Coker 102, should not be treated with metribuzin or injury may result.

If initial planting does not produce a satisfactory stand, fields treated with AMIBEN® DS Herbicide and metribuzin tank mixtures may be replanted to soybeans. If replanting is necessary, do not rework soil. Do not apply a second treatment over the first treatment as injury to soybeans may occur. Read Sencor or Lexone labels and literature carefully to understand conditions to insure crop safety. For additional precautions and information concerning spray equipment, mixing directions, and restrictions refer to the appropriate sections of the AMIBEN® DS Herbicide and the Lexone/Sencor labels. Do not use vines treated with the AMIBEN® DS, metribuzin tank mix for feed or forage.

"AMIBEN®" DS AND "LOROX" TANK MIXTURE FOR SOYBEANS

A tank mixture of AMIBEN® DS Herbicide and Lorox 50W or Lorox L will control annual weeds such as barnyardgrass, carpetweed, crabgrass, fall panicum, foxtail, lambsquarters, pigweed, ragweed and Pennsylvania smartweed. Velvetleaf and cocklebur often germinate in soil at depths below the chemical penetration. Even though populations of these weeds are reduced under normal rainfall conditions, weed seeds which germinate outside the chemical zone are not affected.

Treatments should be made at planting time as a simple preemergence application using 10 to 20 gallons of water per acre. For air application use 3 to 5 gallons per acre. To prepare the tank mixture refer to the "Mixing Instructions" portion of this label.

Plant seeds at least 1.75 inches deep and do not treat if beans have begun to emerge.

Make all broadcast treatments according to the following table.

	3% or Less Organic Matter		Greater than 3% Organic Matter	
Soil Texture	AMIBEN® DS Lbs/A	Lorox 50W or Lorox L Lbs or pts/A	AMIBEN® DS Lbs/A	Lorox 50W or Lorox L Lbs or pts/A
Sandy Loam	2.4	0.67 to 1.5	2.4	1.5 to 2.0
Silt Loam	3.0	1.0 to 1.67	3.0	1.67 to 2.5
Clay Loam	3.6	1.25 to 2.0	3.6	2.0 to 3.0

For band treatments, use proportionally less AMIBEN® DS and Lorox. Before applying this tank mixture as a broadcast or band treatment, read thoroughly the cautionary statements and all other information appearing on both product labels.

"AMIBEN®" DS AND "BASALIN" FOR SOYBEANS

AMIBEN® DS Chloramben Herbicide may be applied over soybeans at planting time in fields where Basalin herbicide has been applied as a preplant soil incorporated herbicide. Or AMIBEN® DS Herbicide may be applied as a preplant incorporated tank mix with Basalin.

When applied according to direction, tank mixtures of AMIBEN® DS Herbicide and Basalin or the sequential treatment will control fall panicum, signalgrass, and common purslane in addition to those weeds controlled with AMIBEN® DS Herbicide alone.

For uniform application of the tank mixture of AMIBEN® DS and Basalin use 10 to 20 gallons of water per acre with ground equipment. To prepare the tank mixture refer to the "Mixing Instructions" portion of this label. The tank mixture may be applied and incorporated at planting or several days prior to planting.

When using Basalin herbicide followed by preemergence application of AMIBEN® DS, Basalin may be applied as many as 6 to 8 weeks before planting or up to planting. Except as directed in the "Conservation or Reduced Tillage" portion of this label, the seedbed should be free of existing weeds before application. Surface trash (plant residue from previous crop) should be uniformly distributed, and the seedbed should be free of clods. Basalin is to be incorporated into the soil within eight hours of application prior to planting. Follow the incorporation instructions on the Basalin label. Read all instructions and cautions on both the AMIBEN® DS and Basalin labels.

Apply AMIBEN® DS Herbicide at a rate of 2.4 to 3.6 pounds per broadcast acre. Apply Basalin herbicide at 1.5 to 3 pints per broadcast acre. Use the higher rate of the rate range on heavier, finer textured soils or on soils with over 4% organic matter. Do not use on peat or muck soils.

Before applying this combination read thoroughly the cautionary statements and all other information appearing on or attached to the containers of both Basalin and AMIBEN® DS herbicides.

"AMIBEN®" DS TANK MIXTURES WITH "BASALIN" AND "SENCOR" OR "LEXONE" FOR SOYBEANS

GENERAL INFORMATION

A tank mix of AMIBEN® DS Chloramben Herbicide plus Basalin plus Sencor or Lexone when applied preplant soil incorporated as recommended herein for use on soybeans, effectively controls the weeds listed on the current AMIBEN® DS, Basalin and Sencor WP/4/DF or Lexone WP/4L/DF labels.

DIRECTIONS FOR USE

Except as directed in the "Conservation or Reduced Tillage" portion of this label, the seedbed should be free of existing weeds before application. Surface trash (plant residue from previous crop) should be uniformly distributed, and the seedbed should be free of clods.

To prepare the tank mixture refer to the "Mixing Instructions" portion of this label. Apply the entire contents of the spray tank within 4 hours of initial mixing. Apply the tank mixture in from 10 to 20 gallons of water per acre using any properly calibrated low pressure sprayer equipped with herbicide tips and in-line strained screens no finer than 50 mesh.

For best results AMIBEN® DS plus Basalin plus Sencor or Lexone should be incorporated as soon as possible after application. The three way tank mixture must be incorporated within 24 hours after application. If AMIBEN® DS plus Basalin plus Sencor or Lexone is applied to a wet, warm soil surface or if the wind velocity is 10 mph or higher, variable weed control may result from delaying incorporation beyond 24 hours.

Incorporation should place the AMIBEN® DS Herbicide plus Basalin plus Sencor or Lexone into the top 2 to 3 inches of the final seedbed. Refer to the incorporation portion of the Basalin label for approved incorporation equipment and operating procedures.

CULTIVATION AFTER PLANTING

Soil treated with AMIBEN® DS plus Basalin plus Sencor or Lexone may be shallow-cultivated, rotary-hoed or hand-hoed without reducing the weed control activity of the tank mixture. Do not cultivate deeper than the AMIBEN® DS plus Basalin plus Sencor or Lexone treated layer of soil since this may bring untreated soil to the surface and poor weed control may result.

Make all broadcast treatments according to the following table:

Soil Texture	AMIBEN® DS Lbs/A	Basalin Pts/A	Sencor 50WP or Lexone WP Lbs/A	Sencor 4 or Lexone 4L Pts/A	Sencor DF or Lexone DF Lbs/A
Coarse (sand, loamy sand, sandy loam) a/	2.4	1.0	0.5	0.5	0.33
Medium (loam, silt loam, silt)	2.4 to 3.0*	1.5	0.5 to 0.75**	0.5 to 0.75**	0.33 to 0.5**
Fine (clay, clay loam, silty clay, sandy clay, sandy clay loam, silty clay loam)	3.0 to 3.6*	2.0	0.75**	0.75**	0.5**

a/ Do not use Sencor or Lexone on coarse soils with less than 1% organic matter.

* Use the higher rate of AMIBEN® DS Chloramben Herbicide when velvetleaf or black nightshade is a problem.

** On Clarion Webster soils in Minnesota and Iowa or on similar alkaline (calcareous) soils with a pH of 7.5 or above apply Sencor WP or Lexone WP at 0.5 pound/A, Sencor 4 or Lexone 4L at 0.5 pint/A or Lexone DF or Sencor DF at 0.33 pound/A on medium soils and at 0.5 to 0.75 pound/A Sencor WP or Lexone WP, 0.5 to 0.75 pint/A Sencor 4 or Lexone 4L or 0.33 to 0.5 pound/A Lexone DF or Sencor DF on fine soils. Use the higher rate only where soil pH is less than 7.5 and where weed pressure is heavy.

SPECIAL PRECAUTIONS

Applied according to directions and under normal growing conditions, the AMIBEN® DS and Basalin and Sencor or AMIBEN® DS and Basalin and Lexone tank mix will not harm the treated crop. Over-application may result in crop injury. Uneven application or improper soil incorporation of the AMIBEN® DS and Basalin and Sencor or AMIBEN® DS and Basalin and Lexone tank mix can result in erratic weed control or crop injury. Seedling disease, cold weather, deep planting, excessive moisture, soil pH over 7.5, high salt concentration, or drought may weaken crop seedlings and increase possibility of damage from the tank mixtures. Under these conditions, delayed crop development or reduced yields may result.

Caution: Observe all cautions and limitations on labeling of all products used in mixtures.

Do not use the foliage from soybeans treated with the AMIBEN® DS and Basalin and Sencor or AMIBEN® DS and Basalin and Lexone tank mix for feed or forage.

"AMIBEN®" DS AND "REWARD" FOR SOYBEANS

GENERAL INFORMATION

A tank mixture of AMIBEN® DS Chloramben Herbicide and Reward can be used prior to planting for weed control in soybeans or AMIBEN® DS may be applied over soybeans at planting time when Reward has been applied as a preplant soil incorporated herbicide.

The combination will control a broader spectrum of weeds than either material alone.

WEEDS CONTROLLED:

Nutsedge (yellow & purple)	Velvetleaf—buttonweed
Barnyardgrass (watergrass)	Stinkgrass
Crabgrass (hairy & smooth)	Carpetweed
Foxtail (giant, green, yellow)	Chickweed
German millet	Coffeeweed
Goosegrass	Florida pusley
Johnsongrass (seedling)	Kochia
Mustard	Lambsquarters (common)
Nightshade	Morningglory (annual)
Prickly sida (teaweed)	Pigweed (redroot & smooth)
Ragweed (common)	Purslane
Smartweed (Pennsylvania)	Sicklepod
	Spurge (annual)

DIRECTIONS FOR USE

Reward must be incorporated into the soil immediately (within minutes) to prevent loss of the herbicide. Whenever possible application and incorporation of Reward or a tank mixture of AMIBEN® DS and Reward should be done in the same operation. Refer to the Reward label for equipment and operating instructions to best incorporate the tank mixture.

Sequential Applications:
Before planting, apply Reward at the rates listed in the table below for your soil type. Incorporate into the soil as described above and on the Reward label. At or after planting apply AMIBEN® DS Herbicide at the rates listed in the table below as a broadcast treatment.

Tank Mix Application:
On medium textured soils a tank mixture of AMIBEN® DS and Reward may be applied and incorporated prior to planting. Use the rates of AMIBEN® DS and Reward listed in the table below. Follow the incorporation directions described elsewhere on this label. To prepare the tank mixture refer to the "Mixing Instructions" portion of this label. Apply in 10 to 20 gallons of water per acre.

Make all broadcast treatments according to the following table:

Soil Type	AMIBEN® DS Lbs/Acre	Reward Pts/Acre
Coarse (sandy-light)	2.4	2.5
Medium to Fine (sandy loam to clay loam)	3.0	3.0
Fine to Very Fine (clay)	3.6	3.5

Read and follow all directions, use limitations, precautions and warnings that appear on the AMIBEN® DS and Reward labels.

"AMIBEN®" DS TANK MIX WITH "SURFLAN" FOR WEED CONTROL IN SOYBEANS

GENERAL INFORMATION

The AMIBEN® DS Chloramben Herbicide and Surflan tank mixture effectively controls all annual grass and broadleaf weeds listed on the Surflan and AMIBEN® DS Herbicide labels including chickweed, coffeeweed, kochia, wild mustard, black nightshade, prickly sida (teaweed), common ragweed, Russian thistle, Pennsylvania smartweed, annual spurge, stinkgrass and velvetleaf (buttonweed).

AMIBEN® DS and Surflan tank mixtures may be applied within two days after planting, but before crop emergence using a properly calibrated, low-pressure herbicide sprayer that will provide uniform soil coverage. Agitate thoroughly before and during application. Follow normal AMIBEN® DS and Surflan procedures for soil preparation. Do not spray over top of emerged soybeans.

DIRECTIONS FOR USE

Make all broadcast application according to the following table:

Soil Texture	AMIBEN® DS Lbs/A	Surflan 4AS Qts/A	Surflan 75W Lbs/A
Coarse Soils	2.4 to 3.6	0.5 to 0.75	0.67 to 1.0
Medium Soils	2.4 to 3.6	0.75 to 1.0	1 to 1.33
Fine Soils	2.4 to 3.6	1.0 to 1.25	1.33 to 1.67

The higher rate of AMIBEN® DS Herbicide should be used on clay loams, clay or high organic (3 to 5%) mineral soils. Do not use the AMIBEN® DS and Surflan tank mix on soils containing more than 5% organic matter.

SPECIAL PRECAUTION

Poor weed control and/or crop injury may result if directions are not followed. Over-application may result in crop injury and/or soil residue of Surflan. Do not use treated vines for feed or forage. Observe all precautions on both the AMIBEN® DS and Surflan labels.

"AMIBEN®" DS AND "PREMERGE 3" TANK MIXTURE FOR SOYBEANS

Preemergence—Use a tank mix combination of AMIBEN® DS Chloramben Herbicide and Premerge 3 consisting of 2.4 to 3.6 pounds of AMIBEN® DS plus 4 to 6 quarts of Premerge 3. Apply at time of planting or before crop emergence. Use the higher rate of each product on clay and clay loam soils.

Early postemergence (when soybeans have emerged and are still in the cotyledon stage before first leaves open to expose the terminal bud). Combine 2.4 to 3.6 pounds of AMIBEN® DS with 1 to 2 quarts of Premerge 3 to control emerged weeds including pigweed, cocklebur, annual morningglory, jimsonweed, common ragweed, velvetleaf, annual Pennsylvania smartweed, small seedling grasses and many others that emerge before or with the crop. AMIBEN® DS Herbicide provides residual control of annual weed seedlings that may germinate after spraying. Use 7 gallons of water when applied by aircraft or about 30 gallons of water applied by ground equipment. The amount of Premerge 3 to be used per acre depends on the maximum air temperature expected within 2 hours; use 2 quarts below 75 degrees F, 1 quart from 75 to 95 degrees F and do not apply above 95 degrees F.

Some injury to soybean plants may occur but yields usually are not reduced. Note: Do not spray if soil surface is wet or if first leaves have opened. Do not use this tank mix if AMIBEN® DS has been applied preemergence.

AMIBEN® DS PROGRAMMED CONTROL OF WILD PROSO MILLET IN SOYBEANS

When used as directed below, a weed control program of sequential AMIBEN® DS Chloramben Herbicide applications will control wild proso millet in addition to all other weed species listed on the AMIBEN® DS label.

DIRECTIONS FOR USE

1 to 4 days before planting apply a preplant incorporated tank mixture of 2.4 to 3.6 pounds AMIBEN® DS plus the full labeled rate of another grass control material such as Treflan, Basalin, Prowl, Lasso or Dual as directed elsewhere on this label. After soybean emergence, but before emergence of wild proso millet seedlings make a second broadcast application of 2.4 to 3.6 pounds AMIBEN® DS per acre. This application should be made when soybeans are in the cotyledon to 2nd trifoliate growth stage, but before emergence of wild proso millet seedlings. Under normal conditions this will be 10 to 21 days after planting. If wild proso millet or other weeds have begun to emerge use a rotary hoe to remove emerged seedling weeds.

Do not feed straw or green fodder from soybeans treated with AMIBEN® DS postemergence. Observe caution statements on this and all other herbicide labels.

AMIBEN® DS FOR POSTEMERGENCE WEED CONTROL IN SOYBEANS

"AMIBEN®" DS AND "CROP OIL" FOR POSTEMERGENCE WEED CONTROL IN SOYBEANS

For programmed weed control apply a preemergence or preplant incorporated grass herbicide at planting followed by a later application of AMIBEN® DS plus crop oil concentrate when soybeans are in the cracking to 4th trifoliate stage of growth. Apply a tank mixture of 3.6 pounds of AMIBEN® DS plus 1 quart per acre of crop oil concentrate to control or suppress 1 to 8 inch velvetleaf and 1 to 3 inch Pennsylvania smartweed, common ragweed and pigweed causing weeds to either die or cease growth and become noncompetitive with the crop. Application at this timing provides both postemergence activity on the above species and extended preemergence activity on those late germinating annual grass and broadleaf weeds listed in the general information section of the AMIBEN® DS label which have not yet germinated at the time of treatment. Do not apply later than 33 days after planting. Make applications in 10 to 20 gallons water per acre by ground or 3 to 5 gallons water per acre by air. Good coverage is essential for maximum weed control. Wide spaced, low pressure, flood type nozzles do not give satisfactory coverage. Nozzle pressure should be at least 30 pounds per square inch. Width of swatch must not exceed length of boom. Under good soil moisture conditions this treatment will control small grassy weeds (1 to 2 leaf) if rainfall occurs within 48 hours of application.

Caution: Do not use soybean straw or green fodder which has been teated postemergence with AMIBEN® DS for livestock feed. Do not apply later than 33 days after planting.

Do not apply more than 3.6 pounds per acre of AMIBEN® DS alone or in combination as a postemergence treatment. If a preemergence or preplant soil incorporated application of AMIBEN® DS is made, do not apply more than a total of 7.2 pounds of AMIBEN® DS per acre per growing season (3.6 pounds preemergence plus 3.6 pounds postemergence).

AMIBEN® DS CULTIVATION FOR SOYBEANS

If heavy rainfall has prevented application of preplant incorporated or preemergence herbicides AMIBEN® DS Herbicide may be applied when soybeans are in the cracking to second trifoliate stage of growth at the rate of 3.0 to 3.6 pounds AMIBEN® DS Herbicide per acre. Grass and broadleaf weeds which have emerged **must** be removed. To remove these germinated weeds, use shallow cultivation with a tool such as a rotary hoe before or after AMIBEN® DS Herbicide has been applied. Avoid ridging the soil if a cultivator is used after AMIBEN® DS has been applied. Use the higher rate on fine textured soils or when heavy weed populations are expected.

Caution: Do not use soybean straw or green fodder which has been treated postemergence with AMIBEN® DS for livestock feed. Do not apply later than 33 days after planting.

Do not apply more than 3.6 pounds per acre of AMIBEN® DS alone or in combination as a preplant soil incorporated or preemergence treatment. If a postemergence application of AMIBEN® DS is made, do not apply more than a total of 7.2 pounds of AMIBEN® DS per acre per growing season (3.6 pounds preemergence plus 3.6 pounds postemergence).

"AMIBEN®" DS AND "BLAZER" FOR POSTEMERGENCE WEED CONTROL IN SOYBEANS

GENERAL INFORMATION

When applied to young actively growing broadleaf weed seedlings the combination of AMIBEN® DS Chloramben Herbicide plus Blazer will control all weeds listed on the Blazer label including cocklebur, jimsonweed, morninglory, Pennsylvania smartweed, pigweed, Eastern black nightshade, and common ragweed and will also provide preemergence control of those grass and broadleaf weeds listed in the general information section of the AMIBEN® DS label which have not yet germinated at time of treatment. While the combination of AMIBEN® DS plus Blazer will provide suppression of small (2 to 3 leaf) foxtail and fall panicum it must not be used as a basic grass control program and should only be applied to fields which have received a preemergence or preplant incorporated grass herbicide treatment as directed on the AMIBEN® DS label. In addition to providing preemergence control of late germinating weeds the combination of AMIBEN® DS plus Blazer will provide control of velvetleaf and increased control of Pennsylvania smartweed, and pigweed which are approaching the maximum weed size listed on the Blazer label.

USE DIRECTIONS

Apply a tank mixture of 3.0 to 3.6 pounds AMIBEN® DS plus 1.5 to 2 pints Blazer when broadleaf weeds have 2 to 4 leaves as directed on the Blazer label (usually 14 to 21 days after planting). Use the higher rate on larger weeds. Use 10 to 20 gallons of water per acre with ground equipment or 5 to 10 gallons of water per acre by air to insure good coverage. Wide spaced, low pressure, flood type nozzles do not give satisfactory coverage. Nozzle pressure should be at least 30 pounds per square inch. Width of swath must not exceed length of boom. Do not make applications later than 33 days after planting. Refer to the Blazer label for specific directions regarding the addition of a surfactant or crop oil. Follow these same directions when using the tank mixture of AMIBEN® plus Blazer to determine when to add surfactant or crop oil except that 1 quart per acre of crop oil concentrate must be added to provide acceptable control of velvetleaf.

Caution: Do not use soybean straw or green fodder which has been treated postemergence with AMIBEN® DS plus Blazer for livestock feed. Read and observe all directions and cautions on labels for Blazer and AMIBEN®.

"AMIBEN®" DS AND "BUTYRAC®" 200 FOR POSTEMERGENCE WEED CONTROL IN SOYBEANS

GENERAL INFORMATION

When applied as directed below a tank mixture of AMIBEN® DS Chloramben Herbicide plus Butyrac® 200 will control or suppress pigweed, velvetleaf, cocklebur and Pennsylvania smartweed causing them to either die or become noncompetitive with the crop. In addition, this tank mixture will provide preemergence control of those grass and broadleaf weeds listed in the general information section of the AMIBEN® DS label which have not yet germinated at time of treatment. The tank mixture of AMIBEN® DS plus Butyrac® 200 will not control emerged grassy weeds and should only be used on fields which have received a preemergence or preplant incorporated grass control herbicide.

This combination may cause temporary twisting, curling or malformation of soybean leaves and stems. Soybeans will usually outgrow these effects with no reduction in yield. Do not apply to fields which are under stress from moisture or disease. Good coverage is essential for maximum weed control. Wide spaced, low pressure, flood type nozzles do not give satisfactory coverage. Nozzle pressure should be at least 30 pounds per square inch. Width of swath must not exceed length of boom.

USE DIRECTIONS
Apply a tank mixture of 3.0 to 3.6 pounds AMIBEN® DS plus 2 fluid ounces Butyrac® 200 in 10 to 20 gallons of water per acre by ground equipment when soybeans have 2 to 4 trifoliate leaves but before weeds are 4 inches tall. Use the higher rate when weeds are over 2 inches tall. Do not make application later than 33 days after planting. When weed pressure is heavy addition of 1 pint per acre of crop oil concentrate has improved weed control, but may cause more soybean malformation as described above.

Caution: Do not use soybean straw or green fodder which has been treated postemergence with AMIBEN® DS plus Butyrac® 200 for livestock feed. Read and observe directions and cautions on labels for Butyrac® 200 and AMIBEN®. Do not make applications later than 33 days after planting.

"AMIBEN®" DS PLUS "ALANAP" FOR POSTEMERGENCE WEED CONTROL IN SOYBEANS

GENERAL INFORMATION
When applied according to the directions below a tank mixture of AMIBEN® DS Chloramben Herbicide plus Alanap will control or suppress velvetleaf, pigweed, cocklebur, and Pennsylvania smartweed causing them to either die or become noncompetitive with the crop. While the combination of AMIBEN® DS plus Alanap often provides control or suppression of small (2 to 3 leaf) annual grasses it should only be applied to fields which have received a preemergence or preplant incorporated application of a grass control herbicide. The tank mixture of AMIBEN® DS plus Alanap may cause temporary stunting of soybeans. Under normal conditions the soybeans will recover and yields will not be affected.

USE DIRECTIONS
Apply a tank mixture of 3.0 to 3.6 pounds AMIBEN® DS plus 2 to 3 quarts Alanap in 10 to 20 gallons per acre of water by ground equipment or in 5 or more gallons per acre of water by aerial equipment. To assure good coverage use a minimum of 30 pounds per square inch of nozzle pressure. Apply when soybeans have 3 to 6 trifoliate leaves, but no later than 33 days after planting. Use the higher rates when weeds are larger (over 4 inches). Addition of 1 quart per acre of a surfactant or crop oil concentrate has shown increased control of these large weeds.

Caution: Do not use soybean straw or green fodder which has been treated postemergence with AMIBEN® DS plus Alanap for livestock feed. Read and observe all directions and cautions on labels for Alanap and AMIBEN® DS. Do not make applications later than 33 days after planting.

DIRECTIONS FOR USE ON DRY BEANS (INCLUDING WHITE, NAVY, KIDNEY, PINTO, LIMA AND ADZUKI BEANS)

AMIBEN® DS PREEMERGENCE OR PREPLANT INCORPORATED

AMIBEN® DS Chloramben Herbicide may be used on drybeans as a preemergence or preplant incorporated herbicide. AMIBEN® DS may be applied and incorporated at planting or several days prior to planting. Uneven spray application or improper soil incorporation of AMIBEN® DS can produce erratic weed control. Refer to the "Directions for Incorporation" portion of this label for approved incorporation equipment and recommended operating procedures.

Make all broadcast treatments according to the following table:

Soil Texture	AMIBEN® DS Lbs/A
Loamy sand, sandy loam	2.4 to 3.0
Loam, silt loam, silt	3.0 to 3.6
Silty clay loam, silty clay, sandy clay, clay loam, clay	3.6

Use the higher rate of AMIBEN® DS when heavy populations of black nightshade, velvetleaf, wild mustard, common ragweed or Pennsylvania smartweed are expected. To insure good spray coverage use 10 to 20 gallons of water per acre with ground equipment or a minimum of 5 gallons of water per acre by aircraft.

When band spraying, mount the spray nozzle on the packer or row wheel so that planting and spraying are done in the same operation. Proper calibration and maintenance of desired nozzle height above the ground are very important. Weeds growing in the middles between the treated rows should be cultivated as soon as necessary.

Use care while cultivating to avoid throwing fresh soil on treated bands.

If sufficient rain does not fall within 5 days after spraying AMIBEN® DS Herbicide as a preemergence treatment, but there is enough soil moisture to germinate weeds and grasses, a light cultivation with a rotary hoe or similar tool will up-root these small broadleaf weeds and grasses. The shallow mixing of AMIBEN® DS in the surface soil will not interfere with the action of AMIBEN® DS when rains come.

If heavy rain packs the soil surface, rotary hoe in the field to aid bean emergence. Do not disrupt the chemical band with deep cultivation. In coarse textured soils heavy rains on incorporated AMIBEN® DS Herbicide may move it below the weed seed germination zone and produce erratic weed control.

DO NOT apply more than 3.6 pounds per acre of AMIBEN® DS Herbicide to the same field of drybeans in the same season.

"AMIBEN®" DS AND "TREFLAN" FOR DRYBEANS

AMIBEN® DS Chloramben Herbicide may be applied as a preemergence band or broadcast treatment in drybeans at planting time in fields where Treflan has been applied as a preplant soil incorporated herbicide. Or AMIBEN® DS Herbicide may be applied several days prior to planting as a broadcast tank mix with Treflan. The tank mixture should be used as a spring preplant soil incorporated treatment. The combination of AMIBEN® DS Herbicide with Treflan improves broadleaf weed control of species such as Pennsylvania smartweed, velvetleaf and ragweed. For broadcast treatments, incorporate chemicals immediately and thoroughly to an approximate depth of 2 inches as described on the Treflan label.

Apply AMIBEN® DS Herbicide at a rate of 2.4 to 3.6 pounds per broadcast acre. Use the higher rate where heavy populations of Pennsylvania smartweed, velvetleaf, ragweed, wild mustard or black nightshade are expected.

Apply Treflan at a rate of 1½ pints for medium soils and 2 pints for fine soils. Do not use on muck or charcoal soils. Read and observe all directions and cautions on labels for AMIBEN® DS and Treflan.

"AMIBEN®" DS AND "EPTAM" FOR DRYBEANS

AMIBEN® DS Chloramben Herbicide may be applied as a preplant soil incorporated tank mixture with "Eptam" herbicide. Or AMIBEN® DS may be applied preemergence over drybeans at planting time where Eptam has been applied as preplant soil incorporated treatment. When applied preemergence, effectiveness depends on adequate rainfall after treatment. When applied according to directions, this combination will control annual broadleaves including black nightshade, ragweed, Pennsylvania smartweed, wild mustard and annual grasses.

DIRECTIONS FOR USE
Apply AMIBEN® DS Herbicide either preplant soil incorporated or preemergence at a rate of 2.4 to 3.6 pounds per acre. Apply Eptam 7E preplant soil incorporated at a rate of 2.5 to 3.5 pints

per acre. Use the higher rates on fine textured soils or where weed pressure is heavy.

The AMIBEN® DS plus Eptam tank mix should be incorporated immediately after application with a field cultivator, tandem disc, mulch treader, rolling cultivator or power driven tiller as described on the Eptam label. Thoroughly mix in the top 2-3" of the soil surface by incorporating twice in different directions. Refer to the Eptam label for further information on incorporation procedures.

Rainfall is not as critical for activation of AMIBEN® DS Herbicide when it is applied preplant incorporated as when surface applied.

Apply in 10 to 20 gallons of water per acre. To prepare the spray mixture refer to the "Mixing Instructions" portion of this label. Do not mix the concentrates before adding to water.

If sufficient rain does not occur within 5 days after spraying AMIBEN® DS preemergence, a light cultivation or incorporation with a rotary hoe or similar tool is recommended.

Do not use Eptam or an AMIBEN® DS plus Eptam tank mixture on adzuki beans, cowpeas (blackeye peas, blackeye beans), soybeans, lima beans or other flat podded beans except Romano. Under abnormal weather conditions stunting may occur on Gratiot, Michilite, Sanilac, Seafarer, and Seaway varieties. Do not use on muck or charcoal soils. Read and observe all directions and cautions on labels for both AMIBEN® DS and Eptam.

"AMIBEN®" DS PLUS "EPTAM" AND "TREFLAN" FOR DRYBEANS

AMIBEN® DS Chloramben Herbicide may be applied as a preplant soil incorporated tank mixture with "Eptam" and "Treflan" herbicides. Or AMIBEN® DS may be applied preemergence over drybeans at planting time where Eptam has been applied as preplant soil incorporated treatment. When applied preemergence, effectiveness depends on adequate rainfall after treatment. When applied according to directions, this combination will control annual broadleaves including black nightshade, ragweed, Pennsylvania smartweed, wild mustard and annual grasses.

DIRECTIONS FOR USE

Apply AMIBEN® DS Herbicide either preplant soil incorporated or preemergence at a rate of 2.4 to 3.6 pounds per acre. Apply Eptam preplant soil incorporated at a rate of 2.5 pints per acre and Treflan preplant soil incorporated at a rate of 1 pint per acre. Use the higher rate of AMIBEN® DS on fine textured soils or where weed pressure is heavy.

The AMIBEN® DS plus Eptam plus Treflan or the Eptam plus Treflan tank mixes should be incorporated immediately after application following the equipment and operating procedures described on the Eptam and Treflan labels.

Rainfall is not as critical for activation of AMIBEN® DS Herbicide when it is applied preplant incorporated as when surface applied.

Apply the tank mixture in 10 to 20 gallons of water per acre. To prepare the spray mixture refer to the "Mixing Instructions" portion of this label. Do not mix the concentrates before adding to water.

If sufficient rain does not occur within 5 days after spraying AMIBEN® DS preemergence, light cultivation or incorporation with a rotary hoe or similar tool is recommended.

Do not use Eptam or an AMIBEN® DS plus Eptam plus Treflan tank mixture on adzuki beans, cowpeas (blackeye peas, blackeye beans), soybeans, lima beans or other flat podded beans except Romano. Under abnormal weather conditions stunting may occur on Gratiot, Michilite, Sanilac, Seafarer, and Seaway varieties. Do not use on muck or charcoal soils. Read and observe all directions and cautions on labels for AMIBEN® DS, Eptam and Treflan.

"AMIBEN®" DS AND "SONALAN" FOR DRYBEANS

AMIBEN® DS Chloramben Herbicide may be applied as a preemergence band or broadcast treatment in drybeans at planting time in fields where Sonalan has been applied as a preplant soil incorporated herbicide. Or AMIBEN® DS Herbicide may be applied prior to planting as a broadcast tank mix with Sonalan. The tank mixture should be used as a spring preplant soil incorporated treatment. The combination of AMIBEN® DS Herbicide with Sonalan improves broadleaf weed control of species such as Pennsylvania smartweed, velvetleaf, common ragweed, black nightshade and wild mustard. For broadcast treatments, incorporate chemicals immediately and thoroughly to an approximate depth of 2 inches as described on the Sonalan label.

Make all broadcast treatments according to the following table:

Soil Texture	AMIBEN® DS Lbs/A	Sonalan Pts/A
Loamy sand, sandy loam	2.4	1.25 to 2.0
Loam, silt loam, silt	2.4 to 3.0	1.75 to 2.5
Silty clay loam, silty clay, sandy clay, clay loam, clay	3.0 to 3.6	2.25 to 3.0

Use the higher rate of AMIBEN® DS where a heavy population of smartweed, velvetleaf, ragweed, wild mustard or black nightshade are expected. Use the higher rate of Sonalan on finer textured soils and on soils with organic matter content over 3%.

"AMIBEN®" DS AND "PROWL" FOR DRYBEANS (except Adzuki beans)

A tank mixture of AMIBEN® DS Chloramben Herbicide and Prowl may be applied as a preplant soil incorporated herbicide treatment or AMIBEN® DS Herbicide may be applied as a preemergence broadcast treatment at planting time in drybean fields where Prowl has been applied as a preplant soil incorporated herbicide treatment.

When applied according to directions, tank mixtures of AMIBEN® DS Herbicide and Prowl herbicide either preplant incorporated or the sequential treatment of AMIBEN® DS preemergence over Prowl preplant incorporated will control fall panicum, signalgrass, Venice mallow and velvetleaf in addition to all other weeds listed on the AMIBEN® DS and Prowl labels. For uniform application of the tank mixture of AMIBEN® DS and Prowl, use 10 to 20 gallons of water per acre with ground equipment or 3 to 5 or more gallons of water per acre by aircraft. To prepare the tank mixture refer to the "Mixing Instructions" portion of this label.

Except as directed in the "Conservation or Reduced Tillage" portion of this label, the seedbed should be free of existing weeds before application. Surface trash (plant residue from previous crop) should be uniformly distributed, and the seedbed should be free of clods. When using the tank mixture preplant soil incorporated or when using Prowl preplant incorporated followed by a preemergence application of AMIBEN® DS the herbicides should be applied with 7 days prior to planting. When applied as a preplant soil incorporated treatment the tank mixture of AMIBEN® DS plus Prowl should be thoroughly incorporated into the top 0 to 2 inches of soil using equipment and techniques described elsewhere on this label and on the Prowl label. When using the sequential treatment AMIBEN® DS should be applied at planting or within 2 days after planting before weeds and drybeans emerge.

Make all broadcast treatments according to the following table:

Soil Texture	AMIBEN® DS Lbs/A	Prowl Pts/A
Loamy sand, sandy loam	2.4	1.0 to 2.0
Loam, silt loam, silt	2.4 to 3.0	1.5 to 2.5
Silty clay loam, silty clay, sandy clay, clay loam, clay	3.0 to 3.6	2.0 to 3.0

Use the higher rate of AMIBEN® DS when heavy populations of black nightshade, velvetleaf, wild mustard, common ragweed or Pennsylvania smartweed are expected. Use the higher rate of Prowl when treatments are incorporated on soils with more than 3% organic matter, or where heavy stands of foxtail, witchgrass or panicum are expected. Do not use on peat or muck soils.

If sufficient rain does not fall within 5 days after preemergence AMIBEN® DS treatment, but there is enough soil moisture to germinate weeds and soybeans a light cultivation with a rotary hoe or similar tool will uproot these small weeds. The shallow mixing of the herbicides into the soil surface will not interfere with their action when rains come. If drybean crop loss occurs due to weather conditions, drybeans may be replanted. If replanting is necessary, DO NOT rework the soil.

Before applying this combination, read thoroughly the cautionary statements and all other information appearing on or attached to the containers of both the Prowl and AMIBEN® DS herbicides.

"AMIBEN®" DS AND "PROWL" AND "EPTAM 7E" FOR DRYBEANS (except Adzuki beans)

A tank mixture of AMIBEN® DS Chloramben Herbicide and Prowl and Eptam 7E may be applied as a preplant soil incorporated herbicide treatment or AMIBEN® DS Herbicide may be applied as a preemergence broadcast treatment at planting time in drybean fields where Prowl and Eptam 7E have been applied as a preplant soil incorporated herbicide treatment.

When applied according to directions, tank mixtures of AMIBEN® DS Herbicide and Prowl herbicide and Eptam 7E either preplant incorporated or the sequential treatment of AMIBEN® DS preemergence over Prowl and Eptam 7E preplant incorporated will control fall panicum, signalgrass, Venice mallow nightshade and velvetleaf in addition to all other weeds listed on the AMIBEN® DS, Prowl and Eptam 7E labels.

For uniform application of the tank mixture of AMIBEN® DS, Prowl and Eptam 7E, use 10 to 20 gallons of water per acre with ground equipment or 3 to 5 or more gallons of water per acre by aircraft. To prepare the tank mixture refer to the "Mixing Instructions" portion of this label.

Except as directed in the "Conservation or Reduced Tillage" portion of this label, the seedbed should be free of existing weeds before application. Surface trash (plant residue from previous crop) should be uniformly distributed, and the seedbed should be free of clods. When using the tank mixture preplant soil incorporated or when using Prowl and Eptam preplant incorporated followed by a preemergence application of AMIBEN® DS the herbicides should be applied within 7 days prior to planting. When applied as a preplant soil incorporated treatment the tank mixture of AMIBEN® DS plus Prowl and Eptam 7E should be thoroughly incorporated into the top 0 to 2 inches of soil using equipment and techniques described elsewhere on this label and on the Prowl and Eptam 7E label.

When using the sequential treatment AMIBEN® DS should be applied at planting or within 2 days after planting before weeds and drybeans emerge.

Make all broadcast treatments according to the following table:

Soil Texture	AMIBEN® DS Lbs/A	Prowl Pts/A	Eptam 7E Pts/A
Loamy sand, sandy loam	2.4	1.0 to 2.0	2.5
Loam, silt loam, silt	2.4 to 3.0	1.5 to 2.5	2.5 to 3.0
Silty clay loam, silty clay, sandy clay, clay loam, clay	3.0 to 3.6	2.0 to 3.0	3.0 to 3.5

Use the higher rate of AMIBEN® DS when heavy populations of black nightshade, velvetleaf, wild mustard, common ragweed or Pennsylvania smartweed are expected. Use the higher rate of Prowl when treatments are incorporated on soils with more than 3% organic matter, or where heavy stands of foxtail, witchgrass or panicum are expected. Use the higher rate of Eptam 7E on finer textured soils or where weed pressure is heavy. Do not use on peat or muck soils.

If sufficient rain does not fall within 5 days after preemergence AMIBEN® DS treatment, but there is enough soil moisture to germinate weeds and soybeans a light cultivation with a rotary hoe or similar tool will uproot these small weeds. The shallow mixing of the herbicides into the soil surface will not interfere with their action when rains come. If drybean crop loss occurs due to weather conditions, drybeans may be replanted. If replanting is necessary, DO NOT rework the soil.

Before applying this combination, read thoroughly the cautionary statements and all other information appearing on or attached to the containers of Prowl, Eptam 7E and AMIBEN® DS herbicides.

"AMIBEN®" DS AND "DUAL" FOR DRYBEANS

A tank mixture of AMIBEN® DS Chloramben Herbicide and Dual 8E may be applied as a preplant soil incorporated herbicide treatment. AMIBEN® DS Herbicide may be applied as a preemergence broadcast tank mix treatment with Dual 8E during planting or after planting before weeds and drybeans emerge. Following a preplant soil incorporated application of Dual 8E, AMIBEN® DS Herbicide may be applied preemergence during planting or after planting before weeds and drybeans emerge.

When applied according to label directions, tank mixtures of AMIBEN® DS Herbicide and Dual will control all weeds controlled by AMIBEN® DS alone or by Dual alone including fall panicum, foxtail millet, red rice, signalgrass, witchgrass, yellow nutsedge and black nightshade. Refer to the general information section of this label and the Dual label for additional weeds controlled by AMIBEN® DS and Dual 8E.

For uniform application of the tank mixture of AMIBEN® DS and Dual, use 10 to 20 gallons of water with ground equipment. To prepare the tank mixture, refer to the "Mixing Instructions" portion of this label.

Apply AMIBEN® DS and Dual preplant incorporated or preemergence using the appropriate rates from the table below. For preplant incorporated treatments apply the tank mixture to the soil and incorporate into the top two inches of soil just before planting following the procedures in the "Directions for Incorporation" portion of this label. Use a preplant incorporated application if furrow irrigation is used or when a period of dry weather is expected. If drybeans are planted on beds apply and incorporate the tank mixture after bed formation. For preemergence treatments, apply the tank mixture during planting (behind the planter) or after planting before weeds or drybeans emerge. For sequential applications apply Dual 8E alone preplant incorporated and follow with a preemergence application of AMIBEN® DS Herbicide during planting (behind the planter) or after planting before weeds or drybeans emerge.

Make all broadcast treatments according to the following table:

Soil Texture	AMIBEN® DS Lbs/A	Dual 8E Pts/A
Loamy sand, sandy loam	2.4	1.5
Loam, silt loam, silt	2.4 to 3.0	1.5 to 2.0
Silty clay loam, silty clay, sandy clay, clay loam, clay	3.0 to 3.6	2.0 to 2.5

Do not use on muck or peat soils.

Use the higher rate of AMIBEN® DS when heavy populations of black nightshade, velvetleaf, mustard or smartweed are expected. Use the higher rate of Dual when treatment is incorporated, when organic matter is over 3%, or when high populations of black nightshade, witchgrass, panicum, foxtail, or yellow nutsedge are expected.

Before applying this combination read thoroughly all cautionary statements and all other information appearing on or attached to the containers of AMIBEN® DS Herbicide and Dual 8E.

"AMIBEN®" DS AND "LASSO" FOR DRYBEANS

AMIBEN® DS Chloramben Herbicide may be applied as a preemergence broadcast treatment at planting or after planting before weeds and drybeans emerge in fields where Lasso has been applied as a preplant soil incorporated treatment or a tank mixture of AMIBEN® DS and Lasso may be applied as a preplant soil incorporated treatment.

When applied according to label directions the tank mixture or the sequential treatment of AMIBEN® DS plus Lasso will control all weeds listed on both the AMIBEN® DS and Lasso labels including nightshade, ragweed, Pennsylvania smartweed, lambsquarters, foxtails and barnyardgrass when applied within 7 days prior to planting.

To prepare the tank mixture refer to the "Mixing Instructions" portion of this label. DO NOT mix AMIBEN® DS Herbicide and Lasso concentrates before the addition of water as an emulsion will not be formed.

When used as a tank mixture preplant soil incorporated the AMIBEN® DS Herbicide and Lasso must be uniformly incorporated within 7 days prior to planting, no deeper than the top 2 inches of soil. Refer to the incorporation directions given elsewhere on this label and the Lasso label. Use the rates indicated in the table below.

When AMIBEN® DS is applied following a preplant soil incorporated application of Lasso follow the incorporation directions on the Lasso label. Apply as a broadcast treatment in 10 to 20 gallons of water per acre at the rates indicated in the table below.

Make all broadcast treatments according to the following table:

Soil Texture	AMIBEN® DS Lbs/A	Lasso Qts/A
Loamy sand, sandy loam	DO NOT USE	DO NOT USE
Loam, silt loam, silt	2.4 to 3.0	1.5 to 2.0
Silty clay loam, silty clay sandy clay, clay loam, clay	3.0 to 3.6	2.0 to 2.5

Do not use on muck or peat soils

Before applying the tank mixture of AMIBEN® DS plus Lasso or AMIBEN® DS preemergence following Lasso preplant incorporated, read all directions and cautions on both the AMIBEN® DS and Lasso labels.

DIRECTIONS FOR USE ON SUNFLOWERS

AMIBEN® DS PREEMERGENCE OR PREPLANT INCORPORATED

AMIBEN® DS Chloramben Herbicide may be used on sunflowers as a preemergence or preplant incorporated herbicide. AMIBEN® DS may be applied and incorporated at planting or several days prior to planting.Uneven spray application or improper soil incorporation of AMIBEN® DS can produce erratic weed control. Incorporate the chemical in the top 0 to 2 inch level as most weeds sprout in this zone. Refer to the "Directions for Incorporation" portion of this label for equipment and operating procedures to be used for proper incorporation.

Make all broadcast treatments according to the following table:

Soil Texture	AMIBEN® DS Lbs/A
Loamy sand, sandy loam	2.4 to 3.0
Loam, silt loam, silt	3.0 to 3.6
Silty clay loam, silty clay, sandy clay, clay loam, clay	3.6

Use the higher rate of AMIBEN® DS when heavy populations of black nightshade, velvetleaf, wild mustard, common ragweed or Pennsylvania smartweed are expected. To insure good spray coverage use 10 to 20 gallons of water per acre with ground equipment or a minimum of 5 gallons of water per acre by aircraft.

When band spraying, mount the spray nozzle on the packer or row wheel so that planting and spraying are done in the same operation. Proper calibration and maintenance of desired nozzle height above the ground are very important. Weeds growing in the middles between the treated rows should be cultivated as soon as necessary. Use care while cultivating to avoid throwing fresh soil on treated bands.

If sufficient rain does not fall within 5 days after spraying AMIBEN® DS Herbicide as a preemergence treatment, but there is enough soil moisture to germinate weeds and grasses, a light cultivation with a rotary hoe or similar tool will up-root these small broadleaf weeds and grasses. The shallow mixing of AMIBEN® DS in the surface soil will not interfere with the action of AMIBEN® DS when rains come.

In coarse textured soils heavy rains on incorporated AMIBEN® DS Herbicide may move it below the weed seed germination zone and produce erratic weed control.

If heavy rain packs the soil surface, rotary hoe in the field to aid sunflower emergence. Do not disrupt the chemical band with deep cultivation.

DO NOT apply more than 3.6 pounds per acre of AMIBEN® DS Herbicide to the same field of sunflowers in the same season.

"AMIBEN®" DS AND "PROWL" FOR SUNFLOWERS

A tank mixture of AMIBEN® DS Chloramben Herbicide and Prowl may be applied as a preplant soil incorporated herbicide treatment or AMIBEN® DS Herbicide may be applied as a preemergence broadcast treatment at planting time in sunflowers fields where Prowl has been applied as a preplant soil incorporated herbicide treatment. The tank mixture of AMIBEN® DS plus Prowl may also be applied as a preemergence broadcast treatment.

When applied according to directions, tank mixtures of AMIBEN® DS Herbicide and Prowl herbicide either preplant incorporated or preemergence, or the sequential treatment of AMIBEN® DS preemergence over Prowl preplant incorporated will control fall panicum, signalgrass, Venice mallow and velvetleaf in addition to all other weeds listed on the AMIBEN® DS and Prowl labels.

For uniform application of the tank mixture of AMIBEN® DS and Prowl, use 10 to 20 gallons of water per acre with ground equipment or 3 to 5 or more gallons of water per acre by aircraft. To prepare the tank mixture refer to the "Mixing Instructions" portion of this label.

Except as directed in the "Conservation or Reduced Tillage" portion of this label, the seedbed should be free of existing weeds before application. Surface trash (plant residue from previous crop) should be uniformly distributed, and the seedbed should be free of clods. When using the tank mixture preplant soil incorporated or when using Prowl preplant incorporated followed by a preemergence application of AMIBEN® DS the herbicides should be applied within 7 days prior to planting. When applied as a preplant soil incorporated treatment the tank mixture of AMIBEN® DS plus Prowl should be thoroughly incorporated into the top 0 to 2 inches of soil using equipment and techniques described elsewhere on this label and on the Prowl label. When using the tank mixture preemergence, treatments should be made at planting or within 2 days after planting before weeds and sunflowers emerge.

Make broadcast treatments according to the following table:

Soil Texture	AMIBEN® DS Lbs/A	Prowl Pts/A
Loamy sand, sandy loam	2.4	1.0 to 2.0
Loam, silt loam, silt	2.4 to 3.0	1.5 to 2.5
Silty clay loam, silty clay, sandy clay, clay loam, clay	3.0 to 3.6	2.0 to 3.0

Use the higher rate of AMIBEN® DS when heavy populations of black nightshade, velvetleaf, wild mustard, common ragweed or Pennsylvania smartweed are expected. Use the higher rate of Prowl when treatments are incorporated on soils with more than 3% organic matter, or where heavy stands of foxtail, witchgrass or panicum are expected. Do not use on peat or muck soils.

If sufficient rain does not fall within 5 days after preemergence treatment, but there is enough soil moisture to germinate weeds and sunflowers a light cultivation with a rotary hoe or similar tool will uproot these small weeds. The shallow mixing of the herbicides into the soil surface will not interfere with their action when rains come. If sunflower crop loss occurs due to weather conditions, sunflowers or soybeans may be replanted. If replanting is necessary, DO NOT rework the soil.

Before applying this combination, read thoroghly the cautionary statements and all other information appearing on or attached to the containers of both the Prowl and AMIBEN® DS herbicides.

"AMIBEN®" DS AND "TREFLAN" FOR WEED CONTROL IN SUNFLOWERS

GENERAL INFORMATION

AMIBEN® DS Chloramben Herbicide may be applied in a band or broadcast over sunflowers at planting time in fields where Treflan has been applied as a preplant soil incorporated herbicide, or AMIBEN® DS may be applied several days prior to planting as a broadcast tank mix with Treflan. The tank mixture should be used as a spring preplant soil incorporated treatment.

The combination of AMIBEN® DS Herbicide with Treflan improves broadleaf weed control of species such as mustard, Pennsylvania smartweed, velvetleaf and ragweed.

For broadcast treatments, incorporate chemicals immediately and thoroughly in the top 2 to 3 inches as most weeds sprout in this zone. Refer to the Treflan label for equipment and operating directions for incorporation of the tank mixture.

In coarse textured soils heavy rains on incorporated AMIBEN® DS may move it below the weed seed germination zone and cause erratic weed control.

If sufficient rain does not fall within 5 days after a preemergence application of AMIBEN® DS, but there is enough soil moisture to germinate weeds and grasses, a light cultivation with a rotary hoe or similar tool will uproot these small broadleaf weeds and grasses. The shallow mixing of AMIBEN® DS Herbicide in the surface soil will not interfere with the action of AMIBEN® DS when rains come.

DIRECTIONS FOR USE

Apply AMIBEN® DS Herbicide and Treflan at the rates indicated in the table below. If mustard, common ragweed or black nightshade are present, use higher rates of AMIBEN® DS for best control. Do not use on muck or charcoal soils. Read and observe all directions and precautions on the AMIBEN® DS chloramben and Treflan labels.

Make all broadcast treatments according to the following table.

Soil Texture	AMIBEN® DS Lbs/A	TREFLAN 4EC Pts/A
Coarse	2.4	1.0
Medium	2.4 to 3.6*	1.5
Fine	2.4 to 3.6*	2.0

* Use 3.6 pound rate if mustard, common ragweed or black nightshade is present for best control.

"AMIBEN®" DS AND "EPTAM" FOR SUNFLOWERS

For use only in the states of Minnesota, North Dakota and South Dakota

AMIBEN® DS Chloramben Herbicide may be applied as a preplant soil incorporated tank mixture with "Eptam" herbicide. Or AMIBEN® DS may be applied preemergence over sunflowers at planting time where EPTAM has been applied as preplant soil incorporated treatment. When applied preemergence, effectiveness depends on adequate rainfall after treatment. When applied according to directions, this combination will control annual broadleaves including black nightshade, ragweed, Pennsylvania smartweed, wild mustard and annual grasses.

DIRECTIONS FOR USE

Apply AMIBEN® DS Herbicide either preplant soil incorporated or preemergence at a rate of 2.4 to 3.6 pounds per acre. Apply Eptam 7E preplant soil incorporated at a rate of 2.5 to 3.5 pints per acre. Use the higher rates on fine textured or high organic matter soils, or where weed pressure is heavy.

The AMIBEN® DS plus Eptam tank mix should be incorporated immediately apply application with a field cultivator, tandem disc, mulch treader, rolling cultivator or power driven tiller as described on the Eptam label. Thoroughly mix in the top 2-3" of the soil surface by incorporating twice in different directions. Refer to the Eptam label for further information on incorporation procedures.

Rainfall is not as critical for activation of AMIBEN® DS Herbicide when it is applied preplant incorporated as when surface applied.

Apply in 10 to 20 gallons of water per acre. To prepare the spray mixture refer to the "Mixing Instructions" portion of this label Do not mix the concentrates before adding to water.

If sufficient rain does not occur within 5 days after spraying AMIBEN® DS preemergence, a light cultivation or incorporation with a rotary hoe or similar tool is recommended

"AMIBEN®" DS AND "LASSO" FOR SUNFLOWERS

A tank mixture of AMIBEN® DS Chloramben Herbicide and Lasso may be applied as a preplant soil incorporated herbicide treatment. AMIBEN® DS Herbicide may be applied as a preemergence broadcast tank mix treatment with Lasso during planting or after planting before weeds and sunflowers emerge. Following a preplant soil incorporated application of Lasso, AMIBEN® DS may be applied preemergence during planting or after planting before weeds and sunflowers emerge.

For uniform application of the tank mixture of AMIBEN® DS Herbicide and Lasso use 10 to 20 gallons of water per acre for ground application. To prepare the tank mixture refer to the "Mixing Instructions" portion of this label. DO NOT mix AMIBEN® DS Herbicide and Lasso concentrates before the addition of water as an emulsion will not be formed.

Apply AMIBEN® DS and Lasso preplant incorporated or preemergence using the appropriate rates from the table below. For preplant incorporated treatments apply the tank mixture to the soil and incorporate into the top two inches of soil as directed in the "Direction for Incorporation" portion of this label. For preemergence treatments, apply the tank mixture during planting (behind the planter) or after planting before weeds or sunflowers emerge. For sequential applications apply Lasso alone preplant incorporated as specified on the Lasso label. Follow with a preemergence application of AMIBEN® DS Herbicide during planting (behind the planter) or after planting before weeds or sunflowers emerge.

With preemergence treatments rain is necessary to move the chemicals into the soil and make them available to control germinating weed seeds. If no rainfall occurs within 5 days following application, a shallow incorporation no deeper than 2 inches is recommended. High intensity or excessive rainfall after applicatin may reduce control.

Make all broadcast treatments according to the following table:

Soil Texture	AMIBEN® DS Lbs/A	Lasso QtsA
Loamy sand, sandy loam	2.4	1.5 to 2.0
Loam, silt loam, silt	2.4 to 3.0	2.0 to 2.5
Silty clay loam, silty clay, sandy clay, clay loam, clay	3.0 to 3.6	2.5 to 3.0

Use the higher rate of AMIBEN® DS when heavy populations of black nightshade, velvetleaf, mustard or smartweed are expected. Use the higher rate of Lasso when treatment is incorporated or when high populations of Black nightshade, witchgrass or foxtail are expected.

Read thoroughly the cautionary statements and all other information appearing on labels for both Lasso and AMIBEN® DS.

DIRECTIONS FOR USE ON PEANUTS

"AMIBEN®" DS AND "PREMERGE" 3 on PEANUTS

Tank mixtures of AMIBEN® DS Chloramben Herbicide and Premerge 3 control weeds in two ways. AMIBEN® DS controls germinating annual grasses and broadleaf weeds while Premerge 3 is a contact herbicide and quickly burns weeds that have germinated before treatment. This tank mixture is especially effective in controlling Florida beggarweed (*Desmodium tortuosum).*

DIRECTIONS FOR USE

Apply a tank mixture of AMIBEN® DS and Premerge 3 at a rate of 2.4 to 3.6 pounds of AMIBEN® DS and 4 pints of Premerge 3 per broadcast acre. When the air temperature during the day of application is expected to exceed 90°F, reduce the Premerge 3 rate to 3 pints. Apply as a ground cracking treatment when the oldest peanut plants reach the two leaf stage or when the first weeds become visible. Use enough water for thorough coverage, generally 15 or more gallons per acre.

Caution: Read thoroughly the cautionary statements and all other information appearing on labels for both AMIBEN® DS and Premerge 3.

DO NOT apply more than 3.6 pounds per acre of AMIBEN® DS Herbicide to the same field of peanuts in the same season.

"AMIBEN®" DS AND "DYANAP" ON PEANUTS

Tank mixtures of AMIBEN® DS Chloramben Herbicide and Dyanap control weeds in two ways. Amiben plus Dyanap controls many germinating and emerged seeding annual grasses and broadleaf weeds. This tank mixture is especially effective in controlling Florida beggarweed (*Desmodiou tortuosum*) and smallflower morningglory (*Jacquemontia tamnifolia*).

DIRECTIONS FOR USE

Apply a tank mixture of AMIBEN® DS and Dyanap at the rate of 2.4 to 3.6 pounds of AMIBEN® DS and 6 quarts of Dyanap per broadcast acre. Apply as a ground cracking treatment when the first weeds become visible and up to the time most of the peanuts are in the "Cracking stage." Use enough water for thorough coverage, generally 10 to 20 gallons per acre.

Caution: Read thoroughly the cautionary statements and all other information appearing on labels for both AMIBEN® DS and Dyanap.

DO NOT apply more than 3.6 pounds per acre of AMIBEN® DS Herbicide to the same field of peanuts in the same season.

FOR PEANUTS IN OKLAHOMA AND NORTH TEXAS

Plant and treat in the same operation. Apply at the rate of 2.4 to 3.6 pounds AMIBEN® DS Chloramben Herbicide per broadcast acre. The higher rate should be used on clay loams, clay or high organic (3 to 8 percent) mineral soils. Use enough water for good spray coverage.

When band spraying, mount the spray nozzle on the packer or row wheel so that planting and spraying are done in the same operation. Proper calibration and maintenance of desired nozzle height above the ground are very important. Weeds growing in the middles between the treated rows should be cultivated as soon as necessary.

Use care while cultivating to above throwing fresh soil on treated bands.

If sufficient rain does not fall within 5 days after spraying AMIBEN® DS Herbicide, but there is enough soil moisture to germinate weeds and grasses, a light cultivation with a rotary hoe or similar tool will up-root these small broadleaf weeds and greases. The shallow mixing of AMIBEN® DS in the surface soil will not interfere with the action of AMIBEN® DS when rains come.

DO NOT apply more than 3.6 pounds per acre of AMIBEN® DS Herbicide to the same field of peanuts in the same season.

DIRECTIONS FOR USE ON SWEET POTATOES

Apply AMIBEN® DS Chloramben Herbicide at the rate of 4.8 pounds per broadcast acre at planting time. For the first few cultivations after treatment, set sweeps so that soil in treated bands is not disturbed. For the final cultivation, set sweeps closer or pull soil as high as possible over the sweet potato row and treated band. For best results, plant and treat at the same time.

DIRECTIONS FOR USE ON SQUASH AND PUMPKIN

Treat preplant incorporated or preemergence at planting time or as soon after planting as possible, before weeds and crops emerge. Follow incorporation instructions the appropriate section of this label.

In coarse textured soil, heavy rains on incorporated AMIBEN® DS Chloramben Herbicide may move it below the weed seed germination zone and produce erratic weed control.

For preemergence application, the sprayer may be mounted on the planter so that spraying and seeding are carried out in the same operation. Apply 3.6 pounds of AMIBEN® DS on ligher soils and 4.8 pounds of AMIBEN® DS on heavier or high organic soils per broadcast acre in at least 10 to 20 gallons of water. If band applications are used, calculate the rate by dividing the band width by the row width and multiplying by 3.6 or 4.8 pounds of AMIBEN® DS per broadcast acre whichever is appropriate for your soil type. After squash or pumpkin emerge, use normal cultivations to control weeds that grow in strips unprotected by any chemical.

BAND TREATMENTS

Row Width	Band Width	Lbs AMIBEN® DS Per Band Acre: Lighter Soils	Lbs AMIBEN® DS Per Band Acre: Heavier or High Organic Soils
72 inch	14 inch	0.72	0.96
	20 inch	1.02	1.32
76 inch	14 inch	0.66	0.90
	20 inch	0.96	1.26
80 inch	14 inch	0.66	0.84
	20 inch	0.90	1.20

DIRECTIONS FOR USE ON SEEDLING ASPARAGUS

Apply 3.6 pounds of AMIBEN® DS Chloramben Herbicide per broadcast acre in about 10 to 20 gallons of water as soon as possible after asparagus seed is planted and before weeds emerge. Irrigation with about ½ inch of water after applying AMIBEN® DS is suggested to produce maximum weed control.

DIRECTIONS FOR USE ON LIMA BEANS

Apply 2.4 to 4.8 pounds of AMIBEN® DS Chloramben Herbicide per broadcast acre in about 10 to 20 gallons of water.

Treat preplant incorporated or preemergence at planting time or as soon after planting as possible, before weeds and lima beans emerge. Follow incorporation instructions under the appropriate section of this label.

In coarse textured soils, heavy rains on incorporated AMIBEN® DS Herbicide may move it below the weed germination zone and produce erratic weed control.

If band applications are used, calculate the rate by dividing the band width by the row width and multiplying by 2.4, 3.0, 3.6, or 4.8 pounds of AMIBEN® DS per broadcast acre whichever is appropriate for your weed problem and soil type. Mount band sprayer on planter so that spraying and seeding are carried out in the same operation. Make final adjustment on proper nozzle height when actually spraying in the field.

DIRECTIONS FOR USE ON SNAPBEANS (GREEN AND YELLOW)

Apply AMIBEN® DS at the rate of 2.4 to 3.0 pounds per broadcast acre as a preemergence treatment in sufficient water for good spray coverage (10 to 20 gallons of water per acre). Use the higher rate of AMIBEN® DS on heavy or high organic mineral soils or where Black Nightshade is a problem. Use the lower rate on loam soils. Do not use on sandy soils low in clay content or with less than 1% organic matter. Attach the applicator to the planter so that planting and spraying are done in the same operation. Rain or sprinkler irrigation carries AMIBEN® DS into the soil where weed seeds sprout. If weeds begin to emerge after AMIBEN® DS application, but prior to rainfall or irrigation, a light cultivation is recommended to up-root the small broadleaf weeds and grasses. The shallow mixing of AMIBEN® DS in the soil surface will not interfere with its weed killing action when rain comes.

WARNING: Stunting of beans has been observed when rains wash the chemical into contact with the germinating seed or the root system of young plants. Under good growing conditions plants normally recover and yields are generally not affected.

Heavy or prolonged rainfall may move AMIBEN® DS below the weed seed germination zone and produce erractic weed control.

"AMIBEN®" DS AND "DUAL" FOR SNAPBEANS (GREEN AND YELLOW)

AMIBEN® DS Herbicide may be applied as a preemergence broadcast tank mix treatment with Dual 8E during planting or after planting before weeds and snapbeans emerge. Following a preplant soil incorporated application of Dual 8E, AMIBEN® DS Herbicide may be applied preemergence during planting or after planting before weeds and snapbeans emerge.

When applied according to label directions, tank mixtures of AMIBEN® DS Herbicide and Dual will control all weeds controlled by AMIBEN® DS alone or by Dual alone including fall panicum, foxtail millet, red rice, signalgrass, witchgrass, yellow nutsedge and black nightshade. Refer to the general information section of this label and the Dual label for additional weeds controlled by AMIBEN® DS and Dual 8E.

For uniform application of the tank mixture of AMIBEN® DS and Dual, use 10 to 20 gallons of water with ground equipment. To prepare the tank mixture, refer to the "Mixing Instructions" portion of this label.

Apply AMIBEN® DS and Dual using the appropriate rates from the table below. For preemergence treatments, apply the tank mixture during planting (behind the planter) or after planting before weeds or snapbeans emerge. For sequential applications apply Dual 8E alone preplant incorporated and follow with a preemergence application of AMIBEN® DS Herbicide during planting (behind the planter) or after planting before weeds or snapbeans emerge.

Make all broadcast treatments according to the following table:

Soil Texture	AMIBEN® DS Lbs/A	Dual 8E Pts/A
Loamy sand, sandy loam	DO NOT USE	1.5 to 2.0
Loam, silt loam, silt	2.4 to 3.0	2.0 to 2.5
Silty clay loam, silty clay, sandy clay, clay loam, clay	2.4 to 3.0	2.0 to 3.0

Do not use on muck or peat soils.

Use the higher rate of AMIBEN® DS when heavy populations of black nightshade, velvetleaf, mustard or smartweed are expected. Use the higher rate of Dual when treatment is incorporated, when organic matter is over 3% or when high populations of black nightshade, witchgrass, panicum, foxtail, or yellow nutsedge are expected.

Before applying this combination read thoroughly all cautionary statements and all other information appearing on or attached to the containers of AMIBEN® DS Herbicide and Dual 8E.

DIRECTIONS FOR USE ON FIELD SEEDED TOMATOES PROTECTED WITH ACTIVATED CARBON

AMIBEN® DS Chloramben Herbicide may be applied as a broadcast preemergence herbicide treatment on field seeded tomatoes which have been protected by the use of an activated carbon plus vermiculite anticrustant system as described below. Do not use AMIBEN® DS on field seeded tomatoes without the use of the activated carbon plus vermiculite protection system as crop injury will result.

Thoroughly mix 1 pound of activated carbon with each cubic foot of number 2 or number 3 horticultural grade vermiculite. Apply 1 cubic foot of this mixture per 600 feet of seeded row (15 cubic feet per acre) by filling the seed furrow with the mixture. Use a positive feed applicator to deliver the activated carbon plus vermiculite anticrustant mixture to the seed furrow directly ahead of the planter press wheel.

Immediately after seeding and before weeds or tomatoes emerge apply 2.4 to 3.6 pounds per acre of AMIBEN® DS in 10 to 40 gallons of water per acre as a broadcast treatment. Use only on medium or heavy soils. Use the higher rate of AMIBEN® DS on heavier soils or where high populations of black nightshade, velvetleaf, mustard or smartweed are expected.

The use of the activated carbon plus vermiculite anticrustant system in the seed furrow will protect the tomato seedling from the phytotoxic effects of AMIBEN® DS. When used in this manner AMIBEN® DS **will not control** weed seeds germinating within the vermiculite treated row.

DIRECTIONS FOR USE ON FIELD SEEDED CUCUMBERS PROTECTED WITH ACTIVATED CARBON

AMIBEN® DS Chloramben Herbicide may be applied as a broadcast preemergence herbicide treatment on field seeded cucumbers which have been protected by the use of an activated carbon plus vermiculite anticrustant system as described below. Do not use AMIBEN® DS on field seeded cucumbers without the use of the activated carbon plus vermiculite protection system as crop injury will result.

Thoroughly mix 1 pound of activated carbon with each cubic foot of number 2 or number 3 horticultural grade vermiculite. Apply 1 cubic foot of this mixture per 600 feet of seeded row (15 cubic feet per acre) by filling the seed furrow with the mixture. Use a positive feed applicator to deliver the activated carbon plus vermiculite anticrustant mixture to the seed furrow directly ahead of the planter press wheel.

Immediately after seeding and before weeds or cucumbers emerge apply 2.4 to 3.6 pounds per acre of AMIBEN® DS in 10 to 40 gallons of water per acre as a broadcast treatment. Use only on medium or heavy soils. Use the higher rate of AMIBEN® DS on heavier soils or where high populations of black nightshade, velvetleaf, mustard or smartweed are expected.

The use of the activated carbon plus vermiculite anticrustant system in the seed furrow will protect the cucumber seedling from the phytotoxic effects of AMIBEN® DS. When used in this manner AMIBEN® DS **will not control** weed seeds germinating within the vermiculite treated row.

DIRECTIONS FOR USE ON FIELD SEEDED CANTALOUPE PROTECTED WITH ACTIVATED CARBON

AMIBEN® DS Chloramben Herbicide may be applied as a broadcast preemergence herbicide treatment on field seeded cantaloupe which have been protected by the use of an activated carbon plus vermiculite anticrustant system as described below. Do not use AMIBEN® DS on field seeded cantaloupe without the use of the activated carbon plus vermiculite protection system as crop injury will result.

Thoroughly mix 1 pound of activated carbon with each cubic foot of number 2 or number 3 horticultural grade vermiculite. Apply 1 cubic foot of this mixture per 600 feet of seeded row (15 cubic feet per acre) by filling the seed furrow with the mixture. Use a positive feed applicator to deliver the activated carbon plus vermiculite anticrustant mixture to the seed furrow directly ahead of the planter press wheel.

Immediately after seeding and before weeds or cantaloupe emerge apply 2.4 to 3.6 pounds per acre of AMIBEN® DS in 10 to 40 gallons of water per acre as a broadcast treatment. Use only on medium or heavy soils. Use the higher rate of AMIBEN® DS on heavier soils or where high populations of black nightshade, velvetleaf, mustard or smartweed are expected.

The use of the activated carbon plus vermiculite anticrustant system in the seed furrow will protect the cantaloupe seedling from the phytotoxic effects of AMIBEN® DS. When used in this manner AMIBEN® DS **will not control** weed seeds germinating within the vermiculite treated row.

DIRECTIONS FOR USE ON CORN

Use only on heavy soils in Missouri, Minnesota, Iowa, Ohio, Illinois, Kansas, Nebraska and Indiana. If rains and extended cold weather occur after planting, corn injury may result.

Apply 2.4 pounds of AMIBEN® DS Chloramben Herbicide per acre in at least 10 to 20 gallons of water. Plant and treat at the same time or as soon after planting as possible before weeds and crop emerge.

Plant corn at least 2 to 2½ inches deep. Do not use AMIBEN® DS on corn grown in light textured sandy soils.

AMIBEN® DS AND "ATRAZINE" 80W TANK MIXTURE

Application: Apply a tank mixture of AMIBEN® DS Chloramben Herbicide and Atrazine 80W during or immediately after planting. A light incorporation with a rotary hoe or similar equipment may be desirable, especially if there is no rainfall soon after application. Do not use AMIBEN® DS and Atrazine for weed control in crops other than corn. Observe all directions on the Atrazine label, especially those relating to crop rotation.

Use only in the states of Missouri, Minnesota, Iowa, Ohio, Illinois, Kansas, Nebraska and Indiana.

Amounts of use: For silty loam soils apply 1.2 pounds of AMIBEN® DS and 1.25 pounds of Atrazine. For heavier soils, apply 1.2 pounds of AMIBEN® DS and 2.5 pounds of Atrazine or 1.8 pounds of AMIBEN® DS and 1.87 pounds of Atrazine. This mixture is not recommended for use on very light soils such as sands, or sandy loams.

PRECAUTIONARY STATEMENTS

CAUTION

HAZARDS TO HUMANS AND DOMESTIC ANIMALS

Harmful if swallowed. Avoid contact with skin, eyes or clothing.

STATEMENT OF PRACTICAL TREATMENT

If swallowed, call a physician or Poison Control Center. Drink 1 to 2 glasses of water and induce vomiting by touching back of throat with finger. Do not induce vomiting or give anything by mouth to an unconscious person.

In case of skin contact, wash immediately with soap and water and remove contaminated clothing.

In case of eye contact flush with plenty of water. Get medical attention.

If inhaled, remove victim to fresh air.

ENVIRONMENTAL HAZARDS

Do not apply directly to water. Do not contaminate water used for irrigation or domestic purposes by cleaning of equipment or disposal of waste.

DIRECTIONS FOR USE

IT IS A VIOLATION OF FEDERAL LAW TO USE THIS PRODUCT IN A MANNER INCONSISTENT WITH ITS LABELING.

Clean sprayers thoroughly before using to apply AMIBEN® DS. To clean sprayer after using AMIBEN® DS, rinse all parts of sprayer with water several times.

Avoid spray drift to adjacent crops as injury may result. Coarse sprays are likely to drift.

Do not graze treated field of sunflowers or use these plants as livestock feed.

STORAGE AND DISPOSAL

Do not store near fertilizers, seed, insecticides, or fungicides. Open dumping is prohibited.

PESTICIDE DISPOSAL

Do not contaminate water, food, or feed by storage or disposal. Wastes resulting from the use of this product may be disposed of on site or at an approved waste disposal facility.

CONTAINER DISPOSAL

Do not reuse empty container. Completely empty bag into application equipment. Then dispose of empty bag in a sanitary landfill or by incineration, or if allowed by state and local authorities, by burning. If burned, stay out of smoke.

Consult Federal, state, or local disposal authorities for approved alternative procedures such as limited open burning.

LIMITED WARRANTY AND DISCLAIMER

The manufacturer warrants (a) that this product conforms to the chemical description on the label; (b) that this product is reasonably fit for the purposes set forth in the directions for use when it is used in accordance with such directions; and (c) that the directions, warnings and other statements on this label are based upon responsible experts' evaluations of reasonable tests of effectiveness, of toxicity to laboratory animals and to plants, and of residues on food crops, and upon reports of field experience. Tests have not been made on all varieties or in all states or under all conditions.

THE MANUFACTURER NEITHER MAKES NOT INTENDS, NOR DOES IT AUTHORIZE ANY AGENT OR REPRESENTATIVE TO MAKE, ANY OTHER WARRANTIES, EXPRESS OR IMPLIED, AND IT EXPRESSLY EXCLUDES AND DISCLAIMS ALL IMPLIED WARRANTIES OR MERCHANTABILITY OR FITNESS FOR A PARTICULAR PURPOSE.

THIS WARRANTY DOES NOT EXTEND TO, AND THE BUYER SHALL NOT BE SOLELY RESPONSIBLE FOR, ANY AND ALL LOSS OR DAMAGE WHICH RESULTS FROM THE USE OF THIS PRODUCT IN A MANNER WHICH IS INCONSISTENT WITH THE LABEL DIRECTIONS, WARNINGS OR CAUTIONS.

BUYER'S EXCLUSIVE REMEDY AND MANUFACTURER'S OR SELLER'S EXCLUSIVE LIABILITY FOR ANY AND ALL CLAIMS, LOSSES, DAMAGES, OR INJURIES RESULTING FROM THE USE OR HANDLING OF THIS PRODUCT, WHETHER OR NOT BASED ON CONTRACT, NEGLIGENCE, STRICT LIABILITY IN TORT OR OTHERWISE SHALL BE LIMITED, AT THE MANUFACTURER'S OPTION, TO REPLACEMENT OF, OR THE REPAYMENT OF THE PURCHASE PRICE FOR, THE QUANTITY OF PRODUCT WITH RESPECT TO WHICH DAMAGES ARE CLAIMED. IN NO EVENT SHALL MANUFACTURER OR SELLER BE LIABLE FOR SPECIAL, INDIRECT OR CONSEQUENTIAL DAMAGES RESULTING FROM THE USE OR HANDLING OF THIS PRODUCT.

AMIBEN® DS and BUTYRAC® 200 are registered trademarks of Union Carbide Agricultural Products, Company, Inc.

Blazer is a registered trademark of Rohm and Haas Company.

Treflan and Surflan are registered trademarks of Elanco Products Company.

Lasso and Roundup are registered trademarks of Monsanto Company.

Dual is a registered trademark of Ciba-Geigy Corporation.

Sencor is a registered trademark of Farbenfabrik Bayer GmbH, Leverkusen.

Lorox and Lexone are registered trademarks of E.I. duPont de Nemours and Company.

Prowl is a registered trademark of American Cyanamid Company.

Alanap and Dyanap are registered trademarks of Uniroyal, Inc.

Premerge is a registered trademark of Dow Chemical Company.

Vernam and Eptam are registered trademarks of Stauffer Chemical Company.

Basalin is a registered trademark of BASF Wyandotte Corporation.

THIS SPECIMEN LABEL IS INTENDED FOR USE ONLY AS A GUIDE IN PROVIDING GENERAL INFORMATION REGARDING THE DIRECTIONS, WARNINGS AND CAUTIONS ASSOCIATED WITH THE USE OF THIS PRODUCT. AS WITH ANY AGRICULTURAL CHEMICAL, ALWAYS FOLLOW THE LABEL INSTRUCTIONS ON THE PACKAGE BEFORE USING.

AMID-THIN® W

For thinning apples and pears

ACTIVE INGREDIENT:

1-Naphthaleneacetamide	8.4%
INERT INGREDIENTS:	91.6%
TOTAL	100.0%

E.P.A. Reg. No. 264-137 AA E.P.A. Est. No. 477-MD-01

Keep Out of Reach of Children

CAUTION

IN CASE OF EMERGENCY TELEPHONE (24 HOURS A DAY) IN U.S.A. 1-800-UCC-HELP.

FOR GENERAL PRODUCT INFORMATION CALL 1-800-334-9745.

PRECAUTIONARY STATEMENTS

CAUTION

HAZARDS TO HUMANS AND DOMESTIC ANIMALS

Harmful if swallowed. Avoid contact with skin, eyes, and clothing. Avoid breathing spray mist.

STATEMENT OF PRACTICAL TREATMENT

In case of eye contact flush with water for at least 15 minutes. In case of skin contact wash with soap and water. Wash contaminated clothing before reuse.

ENVIRONMENTAL HAZARDS

Do not apply directly to water. Do not contaminate water by cleaning of equipment or disposal of waste. Do not contaminate irrigation ditches or water used for irrigation or domestic purposes. Do not apply when weather conditions favor drift from treated areas or use in a greenhouse.

REENTRY STATEMENT

Do not apply this product in such a manner as to directly or through drift expose workers or other persons. The area treated must be vacated by unprotected persons.

Do not enter treated areas without protective clothing until sprays have dried.

Because certain states may require more restrictive reentry intervals for various crops treated with this product, consult your State Department of Agriculture for further information.

Written or oral warnings must be given to workers who are expected to be in a treated area or in an area about to be treated with this product. Advise workers to stay out of fields during application and until sprays have dried. Regular long-sleeved work clothing should be worn when working in treated fields. See PRECAUTIONARY STATEMENTS; STATEMENT OF PRACTICAL TREATMENT for information on accidental exposures. When oral warnings are given, warnings shall be given a language customarily understood by workers. Oral warnings must be given if there is reason to believe that written warnings cannot be understood by workers. Written warnings must include the following information: appropriate signal word (CAUTION), area treated with AMID-THIN W, date of application, appropriate protective clothing, and reentry interval (i.e., until sprays have dried).

STORAGE AND DISPOSAL

STORAGE

Do not store near fertilizers, seeds, insecticides, or fungicides.

PESTICIDE DISPOSAL

Do not contaminate water, food, or feed by storage or disposal. Open dumping is prohibited. Wastes resulting from the use of this product may be disposed of on site or at an approved waste disposal facility.

CONTAINER DISPOSAL

Do not reuse empty container. Completely empty container into application equipment. Then dispose of container in a sanitary landfill or by incineration, or, if allowed by State and local authorities, by burning. If container is burned, stay out of smoke.

DIRECTIONS FOR USE

IT IS A VIOLATION OF FEDERAL LAW TO USE THIS PRODUCT IN A MANNER INCONSISTENT WITH ITS LABELING.

AMID-THIN W contains naphthaleneacetamide and is used to thin many varieties of apples as a petal fall spray.

DIRECTIONS FOR MIXING

When using AMID-THIN W fill the tank with water and then add AMID-THIN W with agitator running and keep under constant agitation during spray operations. To insure proper mixing of AMID-THIN W, allow a few minutes of agitation after material has been added to water before beginning spray operation. One-quarter pound AMID-THIN W in 100 gallons water will give 25 ppm of naphthaleneacetamide. Use up to 400 gallons of spray solution per acre (1 lb. per acre AMID-THIN W).

FACTORS AFFECTING CONCENTRATIONS OF AMID-THIN W

The proper concentration for a given variety depends on many factors: vigor of tree, provisions for cross-pollination, weather conditions favorable to bee flight and fruit setting, healthy foliage free from insect and disease injury and the past history of the trees in relation to food reserves based on the previous crop and conditions of foliage the preceding year. Higher concentrations given in table should be used on vigorous trees having ample provisions for pollinations for good set. Lower concentrations for weaker, less vigorous trees with less food reserves. One of the outstanding advantages of AMID-THIN W from the standpoint of overthinning is its wide range of safety at the concentrations suggested.

It is important that early summer varieties are sprayed no later than petal fall since applications two to three weeks later often give insufficient thinning and premature fruit.

APPLE THINNING TABLE—PPM TO USE

(1/4 lb. package in 100 gallons water gives 25 ppm)

(2 1/2 lb. package in 1000 gallons water gives 25 ppm)

25 to 40 ppm at petal fall in normal years* Williams Early Red, Yellow Transparent, Oldenburg (Dutchess), Red Astrachan

50 ppm at 2-2 1/2 weeks after full bloom Early McIntosh, Wealthy, Golden Delicious, Baldwin, Rome Beauty, Jonathan, Stayman

40 to 50 ppm at 2-2 1/2 weeks after full bloom McIntosh, Grimes Golden, Gravenstein, York

40 ppm at 2-2 1/2 weeks after full bloom Cortland, Macoun, Northern Spy

25 to 40 ppm at 2-2 1/2 weeks after full bloom Rhode Island Greening

50 ppm at petal fall Red Delicious

40 to 50 ppm at petal fall Yellow Newton

*Spraying could be delayed to 3 weeks for the 2-2 1/2 weeks after full bloom varieties if unusually cool weather has slowed growth.

Petal fall—approximately 90%-95% of petals off.

Full bloom—90% of the blossoms open with petals beginning to fall from king flowers.

This table is offered only as a guide.

Weather conditions, amount of bloom and other factors that make for good set will vary in different localities. Contact your local extension pomologists as to time of application and rates to be used for your specific thinning problems.

Do not apply to red delicious apples grown in the Pacific Northwest as "pigmy" fruit may develop.

DIRECTIONS FOR THINNING BARTLETT AND BOSC PEARS

Mix ¼ pound of AMID-THIN W in 50 gallons of water. This makes a spray concentration of 50 ppm of naphthaleneacetamide.

Spray at petal fall or within 5 to 7 days after petal fall. Use up to 400 gallons of spray solution per acre (2 lbs. per acre AMID-THIN W).

Applying AMID-THIN W in this manner reduces the time and hand labor required for supplemental thinning of pears to the desired level.

LIMITED WARRANTY AND DISCLAIMER

The manufacturer warrants that this product conforms to the chemical description on the label; that this product is reasonably fit for the purposes set forth in the directions for use when it is used in accordance with such directions; and that the directions, warnings and other statements on this label are based upon responsible experts' evaluation of reasonable tests of effectiveness, of toxicity to laboratory animals and to plants, and of residues on food crops and upon reports of field experience. Test have not been made on all varieties or in all states or under all conditions. THE MANUFACTURER NEITHER MAKES NOR INTENDS, NOR DOES IT AUTHORIZE ANY AGENT OR REPRESENTATIVE TO MAKE, ANY OTHER WARRANTIES, EXPRESS OR IMPLIED, AND IT EXPRESSLY EXCLUDES AND DISCLAIMS ALL IMPLIED WARRANTIES OF MERCHANTABILITY OR FITNESS FOR A PARTICULAR PURPOSE.

THIS WARRANTY DOES NOT EXTEND TO, AND THE BUYER SHALL BE SOLELY RESPONSIBLE FOR, ANY AND ALL LOSS OR DAMAGE WHICH RESULTS FROM THE USE OF THIS PRODUCT IN ANY MANNER WHICH IS INCONSISTENT WITH THE LABEL DIRECTIONS, WARNINGS, OR CAUTIONS.

BUYER'S EXCLUSIVE REMEDY AND MANUFACTURER'S OR SELLER'S EXCLUSIVE LIABILITY FOR ANY AND ALL CLAIMS, LOSSES, DAMAGES OR INJURIES RESULTING FROM THE USE OR HANDLING OF THIS PRODUCT, WHETHER OR NOT BASED IN CONTRACT, NEGLIGENCE, STRICT LIABILITY IN TORT OR OTHERWISE SHALL BE LIMITED, AT THE MANUFACTURER'S OPTION, TO REPLACEMENT OF, OR THE REPAYMENT OF THE PURCHASE PRICE FOR, THE QUANTITY OF PRODUCT WITH RESPECT TO WHICH DAMAGES ARE CLAIMED. IN NO EVENT SHALL MANUFACTURER OR SELLER BE LIABLE FOR SPECIAL, INDIRECT OR CONSEQUENTIAL DAMAGES RESULTING FROM THE USE OR HANDLING OF THIS PRODUCT.

THIS SPECIMEN LABEL IS INTENDED FOR USE ONLY AS A GUIDE IN PROVIDING GENERAL INFORMATION REGARDING THE DIRECTIONS, WARNINGS AND CAUTIONS ASSOCIATED WITH THE USE OF THIS PRODUCT. AS WITH ANY AGRICULTURAL CHEMICAL, ALWAYS FOLLOW THE LABEL INSTRUCTIONS ON THE PACKAGE BEFORE USING.

AMITROL T™

LIQUID HERBICIDE

RESTRICTED USE PESTICIDE

For retail sale to and application only by certified applicators or personnel under their direct supervision

Controls quackgrass, Canada thistle and other tough annual and perennial weeds.

ACTIVE INGREDIENT:
Amitrole (3-amino-1,2,4-triazole)* 21.6%
INERT INGREDIENTS: 78.4%
TOTAL 100.0%
*Equivalent to 2 lbs. per gal.

EPA Reg. No. 264-135 EPA Est. No. 477-MD-01

Keep Out of Reach of Children

CAUTION

IN CASE OF EMERGENCY PHONE COLLECT (24 HOURS A DAY) IN U.S.A. 1-800-UCC-HELP.

FOR GENERAL PRODUCT INFORMATION CALL 1-800-334-9745.

PRECAUTIONARY STATEMENTS

CAUTION

HAZARDS TO HUMANS AND DOMESTIC ANIMALS

Harmful if swallowed. Avoid contact with skin, eyes, or clothing.

The use of this product may be hazardous to your health. This product contains amitrole which has been determined to cause cancer in laboratory animals. Wear a respirator during mixing and loading of this product.

Wear lightweight waterproof clothing (jumpsuit or coverall, boots or shoes, gloves, and a wide-brimmed plastic hardhat) when applying all products to control dense, tall vegetation. Workers applying this product in all other situations must wear lightweight waterproof gloves and boots (or shoes).

STATEMENT OF PRACTICAL TREATMENT

IF SWALLOWED: Induce vomiting immediately. Get medical attention.

IF ON SKIN: Wash thoroughly with soap and water. Remove contaminated clothing and wash before reuse.

IF IN EYES: Flush eyes with plenty of water.

NOTE TO PHYSICIAN

This product may affect thyroid function. Overexposure can be determined by monitoring thyroid function. If worker is under treatment for hypo- or hyperthyroidism, he should be removed from spray crew until the euthyroid state is achieved.

ENVIRONMENTAL HAZARDS

Do not allow spray or spray drift to contaminate edible crops or water which will be used for irrigation, drinking or domestic purposes as no chemical residue in crops is permitted. Spray drift from this product may injure cotton, tomatoes, lawns, ornamentals and other desirable vegetation. Coarse sprays are less likely to drift.

Do not use spray equipment for other purposes.

Do not use in greenhouse.

Do not allow livestock to graze or feed in treated non-crop areas.

Do not apply directly to water and wetlands. Do not contaminate water by cleaning of equipment or disposal of wastes.

Do not allow spray solution to remain in a sprayer any longer than necessary. Rinse all spray equipment with clean water immediately after each use to prevent corrosion of metal parts.

STORAGE AND DISPOSAL

STORAGE

Do not store near fertilizers, seeds, insecticides, or fungicides.

PESTICIDE DISPOSAL

Do not contaminate water, food, or feed by storage or disposal. Open dumping is prohibited.

Pesticide wastes are toxic. Improper disposal of excess pesticide, spray mixture, or rinsate is a violation of Federal law. If these wastes cannot be disposed of by use according to label instructions, contact your State Pesticide or Environmental Control Agency, or the Hazardous Waste Representative at the nearest EPA Regional Office for guidance.

CONTAINER DISPOSAL

Do not reuse empty container. Triple rinse or equivalent. Then offer for recycling or reconditioning or puncture and dispose of in a sanitary landfill or by incineration, or if allowed by state and local authorities, by burning. If burned, stay out of smoke.

LIMITED WARRANTY AND DISCLAIMER

The manufacturer warrants (a) that this product conforms to the chemical description on the label; (b) that this product is reasonably fit for the purposes set forth in the directions for use when it is used in accordance with such directions; and (c) that the directions, warnings and other statements on this label are based upon responsible experts' evaluation of reasonable tests of effectiveness, of toxicity to laboratory animals and to plants, and upon reports of field experience. Tests have not been made on all varieties or in all states or under all conditions. THE MANUFACTURER NEITHER MAKES NOR INTENDS, NOR DOES IT AUTHORIZE ANY AGENT OR REPRESENTATIVE TO MAKE ANY OTHER WARRANTIES, EXPRESS OR IMPLIED, AND IT EXPRESSLY EXCLUDES AND DISCLAIMS ALL IMPLIED WARRANTIES OF MERCHANTABILITY OR FITNESS FOR A PARTICULAR PURPOSE. THIS WARRANTY DOES NOT EXTEND TO, AND THE BUYER SHALL BE SOLELY RESPONSIBLE FOR, ANY AND ALL LOSS OR DAMAGE WHICH RESULTS FROM THE USE OF THIS PRODUCT IN ANY MANNER WHICH IS INCONSISTENT WITH THE LABEL DIRECTIONS, WARNINGS OR CAUTIONS.

BUYER'S EXCLUSIVE REMEDY AND MANUFACTURER'S OR SELLER'S EXCLUSIVE LIABILITY FOR ANY AND ALL CLAIMS, LOSSES, DAMAGES, OR INJURIES RESULTING FROM THE USE OR HANDLING OF THIS PRODUCT, WHETHER OR NOT SUCH LIABILITY IS BASED ON CONTRACT, NEGLIGENCE, STRICT LIABILITY IN TORT OR OTHERWISE, SHALL BE LIMITED, AT THE MANUFACTURER'S OPTION, TO REPLACEMENT OF, OR THE REPAYMENT OF THE PURCHASE PRICE FOR, THE QUANTITY OF PRODUCT WITH RESPECT TO WHICH DAMAGES ARE CLAIMED. IN NO EVENT SHALL MANUFACTURER OR SELLER BE LIABLE FOR SPECIAL, INDIRECT OR CONSEQUENTIAL DAMAGES RESULTING FROM THE USE OR HANDLING OF THIS PRODUCT.

DIRECTIONS FOR USE

IT IS A VIOLATION OF FEDERAL LAW TO USE THIS PRODUCT IN A MANNER INCONSISTENT WITH ITS LABELING.

WOODY PLANTS AND VINES

	GALS. AMITROL T	GALS. WATER	WHEN TO APPLY
Poison ivy — poison oak —	1	100	Treat from the time foliage is fully developed until plants begin to go dormant. For effective control, all leaves, stems and suckers must be thoroughly wet to groundline. *If regrowth of honeysuckle appears retreat in August or the following spring.
honeysuckle* — kudzu —	2	100	
salmonberry — western dewberry (blackberry) — big leaf maple	2	100	
Wild cherry — ash — locust —sumac	1½	100	

WEEDS IN NON-CROP AREAS
SUCH AS RAILROADS, ROADSIDES, FENCE ROWS AND HARDWOOD NURSERIES

WEEDS	SPOT TREATMENTS [1] SPRAY MIXTURE		ACRE TREATMENTS [2] AMOUNTS PER ACRE		WHEN TO APPLY
	GALLONS AMITROL T	GALLONS WATER	GALLONS AMITROL T	MIN. GALS. WATER	
CONTROL annual grasses and broadleaf weeds (foxtail, ryegrass, cheatgrass, barnyardgrass, annual bluegrass, wild hemp or marijuana, pigweed, kochia, sunflower, wild barley)	½	50	1	100	Spray when most weeds are 3-4 inches tall.
CONTROL annual grasses, annual and perennial broadleaf weeds (whitetop, Canada thistle, volunteer alfalfa, milkweed). SUPPRESS perennial grasses (ripgut grass, reed canarygrass, quackgrass).	1	50	2	100	Spray when most weeds are 6-10 inches tall.
Canada thistle sowthistle	1	50	2	40	Spray when most thistle are up and in the bud to bloom stage. Do not mow treated plants. In the far West, treat during the bloom stage.
quackgrass (couch, witch)	1	50	2	40	Spray when vigorous young growth is 6-8 inches tall.
whitetop (hoary cress)	1	50	2	40	Spray in spring when weeds are in the bud to bloom stage. Do not mow treated plants.
leafy spurge (Western states)	1	50	2	40	
milkweed	1	50	1	50	

WEEDS	SPOT TREATMENTS [1] SPRAY MIXTURE		ACRE TREATMENTS [2] AMOUNTS PER ACRE		WHEN TO APPLY
	GALLONS AMITROL T	GALLONS WATER	GALLONS AMITROL T	MIN. GALS. WATER	
Bermudagrass	2	50	4	50	Treat from mid-summer to early fall when shoots are growing vigorously. Spot treat any green regrowth with a solution of 1 gal. AMITROL T per 50 gals. of water.
nutsedge (nut grass)	1	50	2	50	Spray when shoots are 4-5 inches tall and growing actively.
horsenettle	1	50	2	100	In East and Midwest, spray when horsenettle is in the bud to bloom stage. In the far West, treat from mid-summer to early frost.
horsetail rush	1½	50	3	100	Spray when 6-8 inches tall and growing actively.
dock	½	50	1	50	Apply anytime when dock is actively growing, preferably before development of bloomstalk.
wild chrysanthemum in hardwood nurseries.	2	50	4	100	Apply as a directed spray on young weed growth 4-6 inches tall. Keep spray off foliage of hardwood nursery stock.

[1] A second treatment may be necessary to control weeds not thoroughly sprayed or weeds coming from dormant seeds and nuts or old roots. Apply when regrowth is young and growing actively.

[2] INTERMOUNTAIN STATES: Canada thistle—use 4 gals. AMITROL T Liquid Herbicide in 100 gals. water per acre. All other weeds—use the rates of AMITROL T Liquid Herbicide per acre as listed in the preceding chart, applied in a **minimum of 100 gallons of water per acre.**

KNAPSACK SPRAYERS

Spray weeds thoroughly, wetting all leaves and stems to ground line. Mix ½ cup (4 fluid ounces) of AMITROL T per gallon of water. One gallon of spray solution is usually enough to cover 1 square rod (272 square feet) infested with weeds. In intermountain states or other dryland areas, mix ¼ cup (2 fluid ounces) AMITROL T per gallon of water and apply 2 gallons of spray solution per square rod.

POWER SPRAYERS

Apply enough of the spray mixture to wet thoroughly all leaves and stems to ground line.

THIS SPECIMEN LABEL IS INTENDED FOR USE ONLY AS A GUIDE IN PROVIDING GENERAL INFORMATION REGARDING THE DIRECTIONS, WARNINGS AND CAUTIONS ASSOCIATED WITH THE USE OF THIS PRODUCT. AS WITH ANY AGRICULTURAL CHEMICAL, ALWAYS FOLLOW THE LABEL INSTRUCTIONS ON THE PACKAGE BEFORE USING.

AQUA-KLEEN®

GRANULAR AQUATIC HERBICIDE

For Controlling Water Weeds.

ACTIVE INGREDIENT:
2,4-Dichlorophenoxyacetic acid,
butoxyethyl ester* 27.6%
INERT INGREDIENTS: 72.4%
TOTAL 100.0%

*2,4-Dichlorophenoxyacetic acid equivalent 19% by weight. Isomer specific by AOAC method No. 6.D01-5

EPA Reg. No. 264-109

Keep Out of Reach of Children
CAUTION

IN CASE OF EMERGENCY PHONE (24 HOURS A DAY) IN U.S.A. 1-800-UCC-HELP

FOR GENERAL PRODUCT INFORMATION CALL 1-800-334-9745.

GENERAL INFORMATION

AQUA-KLEEN Granular Aquatic Herbicide contains 2,4-D formulated as the low-volatile butoxyethyl ester. This has been formulated on special heat treated attaclay granules that resist rapid decomposition in water. AQUA-KLEEN sinks quickly to lake or pond bottoms and releases the weed killing chemical in the critical root zone area.

OXYGEN RATIO

Fish breathe oxygen in the water and a water-oxygen ratio must be maintained. Decaying weeds use up oxygen, but during the period when AQUA-KLEEN Granular Aquatic Herbicide should be used the weed mass is fairly sparse and the weed decomposition rate is slow enough so that the water-oxygen ratio is not disturbed by treating the entire area at one time.

If treatments must be applied later in the season when the weed mass is dense and repeat treatments are needed, spread granules in lanes, leaving buffer strips which can then be treated when vegetation in treated lanes has disintegrated. During the growing season, weeds decompose in a 2 to 3 week period following treatment. Buffer lanes should be 50 to 100 feet wide. Treated lanes should be as wide as the buffer strips.

PERMIT TO USE CHEMICALS IN WATER

In many states, permits are required to control weeds by chemical means in public water. If permits are required, they may be obtained from the Chief, Fish Division, State Department of Conservation, or the State Department of Public Health.

DIRECTIONS FOR USE

IT IS A VIOLATION OF FEDERAL LAW TO USE THIS PRODUCT IN A MANNER INCONSISTENT WITH ITS LABELING.

When to Apply

For best results, spread AQUA-KLEEN Granular Aquatic Herbicide in the spring and early summer, during the time weeds start to grow. If desired, this timing can be checked by sampling the lake bottom in areas heavily infested with weeds the year before.

If treatments are delayed until weeds form a dense mat or reach the surface, two treatments may be necessary. Make the second treatment when weeds show signs of recovery, but no later than mid-August.

Occasionally, a second application will be necessary if heavy regrowth occurs or weeds move in from untreated areas.

How to Apply

For large areas, use a fertilizer spreader or mechanical seeder such as the Gerber or Gandy. Before spreading any chemical, calibrate your method of application to be sure of spreading the proper amount. When using boats and power equipment, you must determine the proper combination of (1) boat speed (2) rate of delivery from the spreader, and (3) width of swath covered by the granules.

For small areas-around docks or isolated patches of weeds-use a portable spreader such as the Cyclone seeder. Estimate or measure out the area you want to treat. Weigh out the amount of material needed and spread this uniformly over the area. More uniform coverage is obtained by dividing the required amount in two and covering the area twice, applying the second half at right angles to the first.

When spreading AQUA-KLEEN from a boat, we suggest setting out guide posts at frequent intervals marking out the area. This helps to steer a straight course.

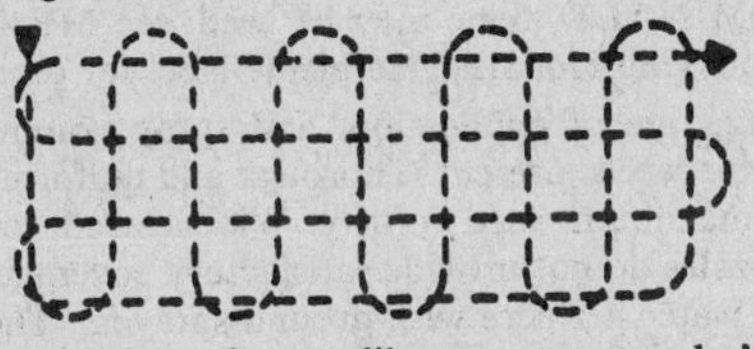

Use the following formula to calibrate your spreader's delivery in pounds of AQUA-KLEEN Granular Aquatic Herbicide per minute.

$$\frac{\textbf{miles per hour} \times \textbf{spreader width} \times \textbf{pounds per acre}}{\textbf{495}} = \textbf{pounds per minute}$$

Example: to apply 100 pounds of AQUA-KLEEN per acre using a spreader that covers a 20-foot wide swath from a boat traveling at 4 miles per hour, set the spreader to deliver 16 pounds of AQUA-KLEEN granules per minute.

$$\frac{\textbf{4 mph} \times \textbf{20 ft.} \times \textbf{100 lbs./A}}{\textbf{495}} = \textbf{16 lbs./min.}$$

Amounts to use

Rates of application vary with the resistance of weed species to the chemical, density of weed mass at time of treatment, water depth, and rate of water flow through the treated area. Use the higher rate for dense weeds, when water is more than 8 feet deep and where there is a large volume turnover.

	AQUA-KLEEN	
	lbs./A	**lbs./2000 sq. ft.**
Susceptible weeds	**100**	**5**
water milfoil (*Myriophyllum* spp.)		
water stargrass *(Heteranthera dubia)*		
Slightly to moderately resistant weeds	150-200	7½-10
bladderwort (*Utricularia* spp.)		
white water lily (*Nymphaea* spp.)		
yellow water lily or spatterdock* *(Nuphar spp.)*		
water shield (*Brasenia* spp.)		
water chestnut *(Trapa natans)*		
coontail* *(Ceratophyllum demersum)*		

*Repeat treatments may be needed.

This product is designed to control the weeds listed on the label. Control of other weeds may not be satisfactory.

Generally, weeds are difficult to control in lakes where water replacement comes from bottom springs.

Water pH

Lake water that is extremely acid or alkaline may influence the effectiveness of AQUA-KLEEN Granular Aquatic Herbicide. A pH on the acid side (pH 6.0 or below) generally favors the action of the chemical while a pH on the alkaline side (pH 8.0 or above) may reduce the action. If regrowth occurs within a period of 6 to 8 weeks, a second application may be needed.

CAUTION

Avoid contact with skin, eyes, or clothing.

Do not use in or near a greenhouse. Clean spreader equipment thoroughly before using it for any other purpose. Vapors from this product may injure susceptible plants in the immediate vicinity.

Do not apply to waters used for irrigation, agricultural sprays, watering dairy animals, or domestic water supplies.

Avoid drift of dust to susceptible plants.

Do not use this product for purposes other than those recommended on the label.

STORAGE AND DISPOSAL

STORAGE

Do not store near fertilizers, seeds, insecticides, or fungicides.

PESTICIDE DISPOSAL

Do not contaminate water, food, or feed by storage or disposal. Pesticide wastes are toxic. Improper disposal of excess pesticide, spray mixture, or rinsate is a violation of Federal Law. If these wastes cannot be disposed of by use according to label instruction, contact your State Pesticide or Environmental Control Agency, of the Hazardous Waste representative at the nearest EPA Regional Office for guidance.

CONTAINER DISPOSAL

Do not reuse empty bag. Completely bag into application equipment. Then dispose of empty bag in a sanitary landfill or by incineration, or, if allowed by State and local authorities, by burning. If bag is burned, stay out of smoke.

LIMITED WARRANTY AND DISCLAIMER

The manufacturer warrants (a) that this product conforms to the chemical description on the label; (b) that this product is reasonably fit for the purposes set forth in the directions for use when it is used in accordance with such direction; and (c) that the directions, warnings and other statements on this label are based upon responsible experts' evaluation of reasonable tests of effectiveness, of toxicity to laboratory animals and to plants, and of residues, and upon reports of field experience. Tests have not been made on all varieties or in all states or under all conditions. THE MANUFACTURER NEITHER MAKES NOR INTENDS, NOR DOES IT AUTHORIZE ANY AGENT OR REPRESENTATIVE TO MAKE, ANY OTHER WARRANTIES, EXPRESS OR IMPLIED, AND IT EXPRESSLY EXCLUDES AND DISCLAIMS ALL IMPLIED WARRANTIES OF MERCHANTABILITY OR FITNESS FOR A PARTICULAR PURPOSE.

THIS WARRANTY DOES NOT EXTEND TO, AND THE BUYER SHALL BE SOLELY RESPONSIBLE FOR, ANY AND ALL LOSS OR DAMAGE WHICH RESULTS FROM THE USE OF THIS PRODUCT IN ANY MANNER WHICH IS INCONSISTENT WITH THE LABEL DIRECTIONS, WARNINGS OR CAUTIONS.

BUYER'S EXCLUSIVE REMEDY AND MANUFACTURER'S OR SELLER'S EXCLUSIVE LIABILITY FOR ANY AND ALL CLAIMS, LOSSES, DAMAGES, OR INJURIES RESULTING FROM THE USE OR HANDLING OF THIS PRODUCT, WHETHER OR NOT SUCH LIABILITY IS BASED IN CONTRACT, NEGLIGENCE, STRICT LIABILITY IN TORT OR OTHERWISE, SHALL BE LIMITED, AT THE MANUFACTURER'S OPTION, TO REPLACEMENT OF, OR THE REPAYMENT OF THE PURCHASE PRICE FOR, THE QUANTITY OF PRODUCT WITH RESPECT TO WHICH DAMAGES ARE CLAIMED. IN NO EVENT SHALL MANUFACTURER OR SELLER BE LIABLE FOR SPECIAL, INDIRECT OR CONSEQUENTIAL DAMAGES RESULTING FROM THE USE OR HANDLING OF THIS PRODUCT.

THIS SPECIMEN LABEL IS INTENDED FOR USE ONLY AS A GUIDE IN PROVIDING GENERAL INFORMATION RE-

GARDING THE DIRECTIONS, WARNINGS AND CAUTIONS ASSOCIATED WITH THE USE OF THIS PRODUCT. AS WITH ANY AGRICULTURAL CHEMICAL, ALWAYS FOLLOW THE LABEL INSTRUCTIONS ON THE PACKAGE BEFORE USING.

3 + 3® BROMINAL®

BROMOXYNIL MCPA BROADLEAF HERBICIDE

Controls annual broadleaf weeds in wheat, barley, oats, rye, and annual canarygrass.

ACTIVE INGREDIENTS:

Octanoic acid, 3,5-dibromo-4-hydroxybenzonitrile ester*	20.7%
Butyric acid, 3,5-dibromo-4-hydroxybenzonitrile ester*	17.8%
2-Methyl-4-chlorophenoxyacetic acid, butoxyethyl ester**	42.7%
INERT INGREDIENTS:	18.8%
TOTAL	100.0%

* Total Bromoxynil (3,5-Dibromo-4-hydroxybenzonitrile) equivalent 28.5% by weight or 3 pounds per gallon.

** MCPA (2-Methyl-4-chlorophenoxyacetic acid) equivalent 28.5% by weight or 3 pounds per gallon.

E.P.A. Reg. No. 264-339 E.P.A. Est. No. 264-MO-01

Keep Out of Reach of Children
WARNING

IN CASE OF EMERGENCY, TELEPHONE (24 HOURS A DAY) IN U.S.A. 1-800-UCC-HELP.

FOR GENERAL PRODUCT INFORMATION CALL 1-800-334-9745

GENERAL INFORMATION—WHEAT, BARLEY, OATS, RYE, AND ANNUAL CANARYGRASS

3 + 3® BROMINAL® Bromoxynil MCPA Broadleaf Herbicide applied as directed, controls young broadleaf weeds in small grain cereals, thus reducing competition with crops during their early growth.

3 + 3® BROMINAL® Bromoxynil MCPA Broadleaf Herbicide can be applied in the winter or early spring depending on the area. 3 + 3® BROMINAL® is not registered for applications to small grains undersown with legumes.

The most important factors for effective weed control are:

1. Treat while weeds are small.
2. Use the recommended rate.
3. Thorough and uniform spray coverage.

3 + 3® BROMINAL® should be applied in late winter, or early spring before broadleaf weeds have more than four true leaves. Treat after small grain has three to four leaves or are tillering. Earlier applications may affect crop vigor during poor growing conditions. If severe winter damage has occurred, apply only after surviving plants have become re-established with new roots and leaves.

Most weeds should be treated before the four leaf stage. Weeds forming rosettes such as mustards, pennycress, tarweed, gromwell, and fiddleneck should be treated when they are less than 2 inches across.

Where wild buckwheat and/or cow cockle are the principal weeds ME4® BROMINAL® alone may be used; see ME4® BROMINAL® Bromoxynil Broadleaf Herbicide label for directions.

Proper timing, rate of application, and spray volume will vary according to the weeds present. Thorough and uniform coverage is essential for maximum weed control. Wide-spaced, low pressure, flood type nozzles do not provide satisfactory coverage. Use 10 or more gallons water per acre with ground sprayers. The minimum aerial spray volumes in the direction for use tables may not provide sufficient spray coverage when weeds are dense or crop growth is vigorous. Nozzle pressure should be at least 30 psi. Swath width must not exceed the length of the boom for ground application.

Short-straw varieties, such as Yoccora Rojo wheat, may not develop the canopy fast enough to shade subsequent weed flushes. Initial treatment of 3 + 3® BROMINAL® Bromoxynil MCPA Broadleaf Herbicide will not control weeds from subsequent flushes.

Do not apply to small grains during or after the boot stage.

Do not apply when the crop is under moisture stress, or when crop forms a canopy over weeds.

Do not graze treated fields for 30 days after application.

Vapors from this product may injure susceptible plants in the immediate vicinity.

DO NOT APPLY TO ONIONS, GARLIC, OR FLAX.

Do not graze livestock in treated annual canarygrass fields.

Do not tank mix 3 + 3® BROMINAL® with Hoelon because the MCPA in this formulation will interfere with the action of Hoelon.

If tank mixtures other than those listed on this label are chosen, check the compatibility by mixing small quantities of the components in the same proportions and dilutions planned for the application. Check product labels to see if tank mixture is prohibited or produces undesirable response.

DIRECTIONS FOR CROP USE—WHEAT, BARLEY, OATS, RYE, ANNUAL CANARYGRASS

IT IS A VIOLATION OF FEDERAL LAW TO USE THIS PRODUCT IN A MANNER INCONSISTENT WITH ITS LABELING.

AREA & CROP GROWTH STAGE	PRODUCT RATE/A	WEEDS & WEED GROWTH STAGE	SPRAY VOLUME/ACRE
Washington, Oregon, Idaho, Western Montana Winter or early spring application; grains 3–4 leaf stage or tillering.	3 + 3 BROMINAL 1–1.33 (1–1⅓) pint	Gromwell, tarweed, fiddleneck, mustards, & other broadleaf weeds; apply before 4-leaf stage or rosettes are less than 2 inches across.	Use 10 to 20 gallons for ground and 5 to 10 gallons for aerial application. Use higher volumes for larger weeds or when weed growth is slow due to dry or cold weather.
Montana, North & South Dakota, Minnesota, Colorado Spring application; grains 3-4 leaf stage or tillering	3 + 3 BROMINAL 0.67 (⅔) pint	Russian thistle & pennycress mixed with wild buckwheat and/or cow cockle, apply before 4-leaf stage. Where wild buckwheat and/or cow cockle are the principal weeds. ME 4® BROMINAL® alone may be used; see ME 4® BROMINAL® broadleaf herbicide label for directions.	Use 10 to 20 gallons for ground, and 2 to 5 gallons for aerial application.
	3 + 3 BROMINAL 1 pint	Weeds listed above applied after 4-leaf stage but before 6 inches high.	Use the higher spray volumes indicated above on larger weeds.

AREA & CROP GROWTH STAGE	PRODUCT RATE/A	WEEDS & WEED GROWTH STAGE	SPRAY VOLUME/ACRE
All Other Areas Fall, Winter, or Spring application; grains 3–4 leaf or after tillering.	3 + 3 BROMINAL 0.67–1.33 (⅔–1⅓) pint	Annual broadleaf weeds as listed. Fall and Winter applications are more effective on winter annuals such as henbit in winter wheat.	Use 10 to 20 gallons for ground and 5 to 10 gallons for aerial application.
Throughout the United States Annual Canarygrass	3 + 3 BROMINAL 0.67–1.33 (⅔–1⅓) pint	Weeds 3–4 leaf up to but before 6 inches high.	Use 10 to 20 gallons for ground and 5 to 10 gallons for aerial application.

3 + 3® BROMINAL® AND MCPA* TANK MIXTURE FOR WHEAT, BARLEY, RYE, AND OATS

AREA & CROP GROWTH STAGE	PRODUCT RATE/A	WEEDS & WEED GROWTH STAGE	SPRAY VOLUME/ACRE
Throughout the United States Wheat, barley, rye, and oats. Treat at 3 to 4 leaf stage or tillering but before boot stage.	3 + 3 BROMINAL 0.67–1.33 (⅔–1⅓) pint + MCPA* 0.25 (¼) lb/Active	Broadleaf Weeds—dense and susceptible to MCPA, 3–4 leaf stage and not larger than 6 inches.	Use 10 to 20 gallons for ground and 5 to 10 gallons for aerial application.

* Weedone® MCPA Ester or Weedar® Concentrate or equivalent.

3 + 3® BROMINAL® AND BANVEL TANK MIXTURE FOR WHEAT

AREA & CROP GROWTH STAGE	PRODUCT RATE/A	WEEDS & WEED GROWTH STAGE	SPRAY VOLUME/ACRE
Throughout the United States Winter wheat, Apply fall through early spring when wheat is from 3 leaf up to joint stage.	3 + 3 BROMINAL 0.67–1.33 (⅔–1⅓) pint + Banvel (dicamba) 0.125 (⅛) pint	Treat when weeds are 3–4 leaf stage of growth.	Use 10 or more gallons water for ground and 5 to 10 gallons for aerial application.

3 + 3® BROMINAL® and Avenge Tank Mixture for Wheat and Barley

When using a tank mix of 3 + 3® BROMINAL® Bromoxynil MCPA Broadleaf Herbicide and Avenge, add about ⅔ of the required water to the tank and start agitation. Add 3 + 3® BROMINAL® and mix thoroughly, then add Avenge and the remainder of the required water. Agitate and recycle for at least five minutes before spraying, and continue agitation until spraying is completed. Do not apply Avenge tank-mix under stress conditions or if the plants are wet with rain or dew or if rain is predicted within 6 hours.

Use 2.5 (2½) pints of Avenge for 1–10 wild oat plants per sq. ft. or 3 pints for 11–25 and 4 pints for more than 25.

Do not use this tank mixture on Mexican durum wheat varieties Jori, Mexicali, Crane or Modoc; spring type Borah wheat, spring-seeded durum wheat varieties Lakota and Wascana. Injury can occur to some wheat varieties including Bonanza, Bounty, Lark, Waldron, and WS 1809. Do not graze Avenge-treated fields or cut treated forage for silage, although grain and straw from treated fields can be fed to livestock.

AREA & CROP GROWTH STAGE	PRODUCT RATE/A	WEEDS & WEED GROWTH STAGE	SPRAY VOLUME/ACRE
Throughout the United States winter wheat 4-leaf to tillering stage. Spring wheat ID, OR, WA, and Era variety in MN and ND. Durum in MN, MT, ND & SD. Wheat 5–6 leaf stage.	3 + 3 BROMINAL 0.67–1.33 (⅔–1⅓) pint + Avenge (difenzoquat) 2.5 (2½) 3 or 4 pints	Wild oats, fiddleneck, wild buckwheat and other broadleaf weeds, apply to wild oats in the 3–5 leaf stage and to emerged susceptible broadleaf weeds in the 2–4 leaf stage or when rosettes are less than 1½ inches across.	Use 5 to 20 gal. for ground and 3 to 10 gal. for aerial application. If spray vol. exceeds 10 gal., add 0.6 fluid ounces of surfactant such as Tergitol NPX for each gal. in excess.

3 + 3® BROMINAL® AND KARMEX TANK MIXTURE FOR WHEAT AND BARLEY

AREA & CROP GROWTH STAGE	PRODUCT RATE/A	WEEDS & WEED GROWTH STAGE	SPRAY VOLUME/ACRE
Washington, Oregon and *Idaho* where annual rainfall exceeds 16 inches on winter wheat and winter barley. Apply a single application in the spring after wheat has initiated spring growth and has 3 to 4 leaves.	3 + 3 BROMINAL 0.67 (⅔) pint + Karmex (diuron) 0.5 (½) lb/A tank mixture	Henbit, gromwell, fiddleneck, other weeds. Apply before weeds are 2 inches tall or across.	Use a minimum of 25 gal. for ground and 5 to 10 gal. for aerial application. Read and observe all applicable precautions and restrictions on the 3 + 3 BROMINAL and Karmex labels before applying this tank mixture

3 + 3® BROMINAL® and Glean Tank Mixture for Use in Winter Wheat

The tank mixture of 3 + 3® BROMINAL® Bromoxynil MCPA Broadleaf Herbicide and Glean provides winter wheat producers with an effective contact and soil active herbicide combination. This tank mixture improves control of henbit, tansy mustard, gromwell, Chinese lettuce, bedstraw, lambsquarters, chickweed, wild buckwheat, and miners lettuce.

Apply 3 + 3® BROMINAL® Broadleaf Herbicide and Glean tank mixture to actively growing weeds since weeds hardened off by low fertility, drought stress, or cold weather may not be controlled.

Thorough and uniform spray coverage is essential for maximum weed control. Use 10 or more gallons water per acre for ground application or 4 to 10 gallons water per acre for aerial application. Wide spaced, low pressure flood type nozzles do not provide satisfactory coverage. Nozzle pressure should be at least 30 psi.

When preparing 3 + 3® BROMINAL® bromoxynil herbicide and Glean tank mix, add ⅔ the required amount of water to the spray tank and mix in Glean with agitation. With Glean in suspension, and the agitator running, add 3 + 3® BROMINAL® and remainder of the required water. Agitation is required for uniform mixing and application.

Carefully follow all precautions on the Glean label, particularly those regarding crop rotation, recropping interval guidelines and spray cleanup procedures.

Do not apply this tank mixture when conditions favor spray drift.

AREA & CROP GROWTH STAGE	PRODUCT RATE/A	WEEDS & WEED GROWTH STAGE	SPRAY VOLUME/ ACRE
Throughout the United States Winter Wheat 3–4 leaf or tillering but before boot stage.	3 + 3 BROMINAL 0.5–1 pint + Glean (75% active ingredient) 0.13 to 0.33 ounce	Apply to the most susceptible seedling weeds in the 1–4 leaf stage but before they are 6" tall. On less susceptible weeds apply before rosettes are 1.5" in diameter.	Use 10 to 20 gal. for ground or 4 to 10 gal. for aerial application.

3 + 3® BROMINAL® and Liquid Fertilizer Tank Mixture

3 + 3® BROMINAL® Broadleaf Herbicide may be applied with liquid fertilizer. Maintain agitation until all the mixture is applied. Do not allow mixture to stand overnight. NOTE: Herbicides can increase contact burn of fertilizers on plant foliage. A reduction in the fertilizer rate and concentration may reduce the hazard of leaf burn.

Different fertilizer sources and ratios may influence the stability or compatibility of these tank mixtures. If in doubt, test a small sample in the dilution ratio planned for application.

APPLICATION OF 3 + 3® BROMINAL® THROUGH IRRIGATION SYSTEMS—BROMIGATION® POST-EMERGENCE WEED CONTROL SERVICE

General Information: 3 + 3® BROMINAL® Bromoxynil MCPA Broadleaf Herbicide may be applied through automated (center point or continuous lateral move) or wheel line, hand line, and solid set overhead sprinkler irrigation systems for broadleaf weed control. 3 + 3® BROMINAL® controls fiddleneck, wild buckwheat, sunflower, fanweed, kochia, Russian thistle and other susceptible broadleaf weeds as listed on the 3 + 3® BROMINAL® label. For best results, apply 3 + 3® BROMINAL® when weeds are small and before they have exerted yield reducing competition on the crop.

Directions for use: Remove scale, pesticide residues and other foreign matter from the chemical tank and the entire injector system. Flush with clean water. Add 3 + 3® BROMINAL® Broadleaf Herbicide or 3 + 3® BROMINAL® and water dilution to a mix tank. For greater accuracy in calibration and distribution, inject a larger quantity of a more dilute solution.

For best performance, set the sprinklers to deliver approximately 0.5 inch or less of water per acre. Start sprinklers and then inject 3 + 3® BROMINAL® into the irrigation line. 3 + 3® BROMINAL® Broadleaf Herbicide should be injected with a positive displacement pump into the main line at least 8 feet ahead of a right angle turn to insure adequate mixing.

For cereals, use the 1.33 pint rate on weeds with 4 leaves. Follow 3 + 3® BROMINAL® label for detailed information on application timing.

User precautions: Application of more than 0.5 inch/acre of irrigation water may result in decreased product performance on certain soils.

Apply 3 + 3® BROMINAL® only through systems containing antisiphon, check valves and interlocking controls between the metering device and the water pump to insure simultaneous shutoff to prevent water source contamination and overflow of the mixing tank.

Do not apply when conditions favor drift, since this formulation contains MCPA which can produce damage symptoms on sensitive adjacent crops. Do not apply when system connections or fittings leak, or when nozzles do not provide uniform distribution.

Allow sufficient time for pesticide to be flushed through all the lines and nozzles before turning off irrigation water.

WEEDS CONTROLLED

Most Susceptible

Use 0.67 pint 3 + 3® BROMINAL® Broadleaf Herbicide per Acre (0.25 lb. bromoxynil and 0.25 lb. MCPA active):

black mustard	*(Brassica nigra)*
cocklebur	*(Xanthium pensylvanicum)*
cow cockle	*(Saponaria vaccaria)*
Eastern black nightshade	*(Solanum ptycanthum)*
field pennycress, fanweed	*(Thalaspi arvense)*
green smartweed	*(Polygonum scabrum)*
Hairy nightshade	*(Solanum sarachoides)*
kochia	*(Kochia scoparia)*
lambsquarters	*(Chenopodium album)*
London rocket	*(Sisymbrium irio)*
morningglory	*(Ipomoea hederacea)*
nettleleaf goosefoot	*(Chenopodium murale)*
nightshade (annual)	*(Solanum nigrum, villosum)*
parthenium ragweed	*(Parthenium hysterophorus)*
pigweed	*(Amaranthus retroflexus)*
Russian thistle	*(Salsola kali)*
shepherdspurse	*(Capsella bursa-pastoris)*
sow thistle (annual)	*(Sonchus oleraceus)*
sunflower	*(Helianthus annuus)*
treacle mustard	*(Erysimum repandum)*
tumble (Jim Hill) mustard	*(Sisymbrium altissimum)*
velvetleaf (buttonweed)	*(Abutilon theophrasti)*
wild buckwheat	*(Polygonum convolvulus)*
wild radish	*(Raphanus raphanistrum)*

Susceptible

Use 1.0 pint 3 + 3® BROMINAL® per acre (0.375 lb bromoxynil and 0.375 lb MCPA active):

blue (purple) mustard	*(Chorispora tenella)*
bullthistle	*(Cirsium vulgare)*
corn chamomile	*(Anthemis arvensis)*
corn gromwell	*(Lithospermum arvense)*
dogfennel (Mayweed)	*(Anthemis coluta)*
dyers woad	*(Isatis tineforia)*
fiddleneck	*(Amsinchia intermedia)*
fivehooked bassia	*(Bassia hyssopifolia)*
jimsonweed	*(Datura stramonium)*
knawel (German moss)	*(Scleranthas annus)*
Ladysthumb	*(Polygonum persicaria)*
mare's-tail	*(Conyza canadensis)*
minerslettuce	*(Montia perfoliata)*
nodding thistle	*(Carduus nutans)*
pepperweed	*(Lepidium campestre)*
popcorn flower	*(Allocarya figurata)*
prickly (Chinese) lettuce	*(Lactuca scariola)*
puncture vine	*(Tribulus terrestris)*
ragweed	*(Ambrosia artemisiifolia)*
redmaids rockpurslane	*(Colandrinia caulescens)*
silver leaf nightshade	*(Solanum elaeagnifolium)*
small seeded false flax	*(Carmelina microcarpa)*
smartweed (annual)	*(Polygonum scabrum, pensylvanicum)*
southern giant curl mustard	*(Brassica japonica)*
tansymustard	*(Descurainia pinnata)*
tartary buckwheat	*(Fagopyrum tataricum)*
tarweed	*(Hemizonia spp.)*
wild mustard	*(Brassica kaber)*
winter vetch	*(Vicia sativa)*
yellow rocket	*(Barbarea vulgaris)*
yellow star thistle	*(Centaurea solstitialis)*

Less Susceptible

Use 1.33 pint 3 + 3® BROMINAL® per acre (0.5 lb. bromoxynil and 0.5 MCPA active):

bur buttercup	*(Ranunculus testiculatus)*
fumitory	*(Fumaria officnalis)*
Giant ragweed	*(Ambrosia trifida)*
groundsel	*(Senecio vulgaris)*

henbit	*(Lamium amplexicaule)*
Prostrate knotweed	*(Polygonum aviculare)*

This treatment will not control established perennial broadleaf weeds such as Canada thistle, bindweed, or horsetail rush.

PRECAUTIONARY STATEMENTS

WARNING

HAZARDS TO HUMANS AND DOMESTIC ANIMALS

Harmful if swallowed. Avoid contact with skin, eyes, or clothing. Avoid breathing spray mist.

STATEMENT OF PRACTICAL TREATMENT

In case of eye contact, rinse thoroughly with water and get medical attention. In case of skin contact, wash immediately with soap and water and remove contaminated clothing.

If swallowed, do not induce vomiting. Drink promptly large quantities of milk, egg white, gelatin solution or if these are not available, drink large quantities of water. Avoid alcohol. Get medical attention.

NOTE TO PHYSICIAN

There is no specific antidote. This product contains petroleum distillates. Gastric lavage may be contraindicated because of possibility of aspiration. Treatment of overexposure should be directed at the control of the symptoms and clinical condition.

ENVIRONMENTAL HAZARDS

This product is toxic to fish and wildlife. Do not apply directly to water. Do not contaminate water use for irrigation or domestic purposes by cleaning of equipment or disposal of waste. Do not apply when weather conditions favor drift from areas treated.

Clean sprayers thoroughly before using to apply 3 + 3® BROMINAL® Broadleaf Herbicide. To clean sprayer after using 3 + 3® BROMINAL® Broadleaf Herbicide, rinse all parts of sprayer with water several times.

Avoid spray drift to adjacent crops as injury may result. Coarse sprays are less likely to drift.

STORAGE AND DISPOSAL

Freezing does not affect the quality of 3 + 3® BROMINAL®.

Do not store near fertilizers, seeds, insecticides, or fungicides.

PESTICIDE DISPOSAL

Do not contaminate water, food, or feed by storage or disposal. Open dumping is prohibited.

Wastes resulting from the use of this product may be disposed of on site or at an approved waste disposal facility.

CONTAINER DISPOSAL

Do not reuse empty container. Triple rinse (or equivalent). Then offer for recycling or reconditioning, or puncture and dispose of in a sanitary landfill, or by incineration, or if allowed by state and local authorities, by burning. If burned, stay out of smoke.

LIMITED WARRANTY AND DISCLAIMER

The manufacturer warrants that (a) this product conforms to the chemical description on the label; (b) that this product is reasonably fit for the purposes set forth in the directions for use when it is used in accordance with such directions; (c) and that the directions, warnings and other statements on this label are based upon responsible experts' evaluation of reasonable tests of effectiveness, of toxicity to laboratory animals and to plants, and of residues on food crops and upon reports of field experience. Tests have not been made on all varieties or in all states or under all conditions. THE MANUFACTURER NEITHER MAKES NOR INTENDS, NOR DOES IT AUTHORIZE ANY AGENT OR REPRESENTATIVE TO MAKE, ANY OTHER WARRANTIES, EXPRESS OR IMPLIED, AND IT EXPRESSLY EXCLUDES AND DISCLAIMS ALL IMPLIED WARRANTIES OF MERCHANTABILITY OR FITNESS FOR A PARTICULAR PURPOSE.

THIS WARRANTY DOES NOT EXTEND TO, AND THE BUYER SHALL BE SOLELY RESPONSIBLE FOR, ANY AND ALL LOSS OR DAMAGE WHICH RESULTS FROM THE USE OF THIS PRODUCT IN ANY MANNER WHICH IS INCONSISTENT WITH THE LABEL DIRECTIONS, WARNINGS OR CAUTIONS.

BUYER'S EXCLUSIVE REMEDY AND MANUFACTURER'S OR SELLER'S EXCLUSIVE LIABILITY FOR ANY AND ALL CLAIMS, LOSSES, DAMAGES, OR INJURIES RESULTING FROM THE USE OR HANDLING OF THIS PRODUCT, WHETHER OR NOT BASED ON CONTRACT, NEGLIGENCE, STRICT LIABILITY IN TORT OR OTHERWISE, SHALL BE LIMITED, AT THE MANUFACTURER'S OPTION, TO REPLACEMENT OF, OR THE REPAYMENT OF THE PURCHASE PRICE FOR, THE QUANTITY OF PRODUCT WITH RESPECT TO WHICH DAMAGES ARE CLAIMED. IN NO EVENT SHALL MANUFACTURER OR SELLER BE LIABLE FOR SPECIAL, INDIRECT OR CONSEQUENTIAL DAMAGES RESULTING FROM THE USE OR HANDLING OF THIS PRODUCT.

3 + 3® BROMINAL®, WEEDONE® MCPA Ester and WEEDAR® MCPA Concentrate are trademarks of Union Carbide

Karmex® is a registered trademark of E.I. DuPont de Nemours & Company, Inc.

Avenge® is a registered trademark of American Cyanamid Company.

Banvel® is a registered trademark of Velsicol Chemical Corporation.

THIS SPECIMEN LABEL IS INTENDED FOR USE ONLY AS A GUIDE IN PROVIDING GENERAL INFORMATION REGARDING THE DIRECTIONS, WARNINGS AND CAUTIONS ASSOCIATED WITH THE USE OF THIS PRODUCT. AS WITH ANY AGRICULTURAL CHEMICAL, ALWAYS FOLLOW THE LABEL INSTRUCTIONS ON THE PACKAGE BEFORE USING.

ME4® BROMINAL®

BROMOXYNIL BROADLEAF HERBICIDE

CROP USE

Controls annual broadleaf weeds in wheat, barley, oat, rye, field corn, popcorn, grain sorghum, flax, garlic, onions, and newly seeded and established grasses.

INDUSTRIAL USE

Controls Russian thistle and other annual broadleaf weeds on roadsides, vacant lots, dry drainage beds, railroads and industrial sites.

For Commercial or Agricultural Use Only

ACTIVE INGREDIENTS:	
Octanoic acid, 3,5-dibromo-4-hydroxybenzonitrile ester*	29.0%
Butyric acid, 3,5-dibromo-4-hydroxybenzonitrile ester*	24.9%
INERT INGREDIENTS:	46.1%
TOTAL	100.0%

* Total Bromoxynil (3,5-Dibromo-4-hydroxybenzonitrile) equivalent 39.4% by weight or 4 pounds per gallon.

E.P.A. Reg. No. 264-340 E.P.A. Est. No. 264-MO-01

Keep Out of Reach of Children

WARNING AVISO

PRECAUCION AL USARIO: Si usted no lee ingles no use este producto hasta que la informacion en la etiqueta le haya sido explicada ampliamente.

IN CASE OF EMERGENCY—TELEPHONE DAY OR NIGHT (UCC HELP) 800-822-4357

FOR GENERAL PRODUCT INFORMATION CALL 1-800-334-9745.

GENERAL INFORMATION CROP USE

Postemergence application of ME4® BROMINAL® Bromoxynil Broadleaf Herbicide effectively controls seedling broadleaf weeds.

ME4® BROMINAL® Bromoxynil Broadleaf Herbicide applied early controls seedling broadleaf weeds that emerge in small grains, corn, sorghum, onions, garlic and flax, thus reducing weed competition with these crops during their early growth.

ME4® BROMINAL® Broadleaf Herbicide may be applied in the fall, winter or spring of the year, as long as the weeds are in the seedling stage. Application of ME4® BROMINAL® is not registered for use on small grains undersown with legumes.

Apply 0.5 to 1 pint ME4® BROMINAL® Broadleaf Herbicide per acre after crops emerge, but before weeds are past the 3 to 4 leaf stage. Weeds that form rosettes should be treated before they reach $1\frac{1}{2}$ inches across.

Thorough and uniform spray coverage is essential for maximum weed control. Use 10 or more gallons of water per acre with ground sprayers with at least 30 psi nozzle pressure. Wide-spaced low pressure flood type nozzles do not give satisfactory coverage. Swath width must not exceed the length of spray boom for ground application. Use adequate spray volume for aerial application according to the label directions. The minimum spray volume may not provide sufficient coverage when weeds are dense or crop growth is vigorous.

The most important factors for effective weed control are: 1. spraying weeds in the seedling stage; 2. Thorough and uniform spray coverage.

Do not apply when crop canopy covers the weeds. Do not apply when crops are under moisture stress. Do not apply to wheat, barley, or rye that will be exposed to cold, desiccating winds in the absence of snow cover. If high crop mortality from winds has occurred, apply ME4® BROMINAL® Broadleaf Herbicide only after plants have developed new roots and begin new top growth. Do not apply to small grains during or after the boot stage. Do not graze treated fields for 30 days after treatment.

Occasional leaf burn may occur on cereals. Moderate temperature conditions, dry foliage, and at least 10 gallons of spray solution per acre without soluble fertilizer reduce the incidence of cereal leaf burn. When cereal foliage is sensitive, weeds are also easier to control. Since ME4® BROMINAL® Bromoxynil is not systemic, recovery of the crop is rapid with no lasting effect.

Short-straw varieties, for example Yaccora Rojo wheat, may not develop the crop canopy fast enough to shade the subsequent flushes of weeds. Since ME4® BROMINAL® has little residual activity, subsequent flushes of weeds will not be controlled by the initial treatment.

A wide range of tank mixes are recommended for ME4® BROMINAL® to handle specific weed problems. Use ME4® BROMINAL® Bromoxynil Broadleaf Herbicide alone or in combination with liquid fertilizers or other herbicides according to the label directions. Follow all cautions and limitations on the labels of the products you select to mix with ME4® BROMINAL® Bromoxynil.

If tank mixtures other than those listed on this label are chosen, check compatibility by mixing small quantities of the components in the same proportions and dilutions planned for the application. Do not combine ME4® BROMINAL®/2,4-D or MCPA with Hoelon as the 2,4-D or MCPA will interfere with the control of grass by Hoelon.

ME4® BROMINAL® contains 4 lb active bromoxynil per gallon. For easy reference, 0.5 pint contains 0.25 lb active ingredient; 0.75 pint contains 0.375 lb active ingredient; 1.0 pint contains 0.5 lb active ingredient.

Only the following recommendations referenced in this label are registered in California:

- ME4® BROMINAL® Wheat, Barley, Oats & Rye
- ME4® BROMINAL® MCPA, 2,4-D Tank Mixtures
- ME4® BROMINAL® Avenge Tank Mixture
- ME4® BROMINAL® Hoelon Tank Mixture
- ME4® BROMINAL® Flax
- ME4® BROMINAL® Field Corn
- ME4® BROMINAL® Grain Sorghum
- Garlic
- Onions
- ME4® BROMINAL®—BROMIGATION®
- Newly Planted Grasses
- Industrial Use
- Weeds Susceptibility

DIRECTIONS FOR USE

DO NOT USE THIS PRODUCT FOR PURPOSES OTHER THAN THOSE RECOMMENDED ON THIS LABEL. IT IS A VIOLATION OF FEDERAL LAW TO USE THIS PRODUCT IN A MANNER INCONSISTENT WITH ITS LABELING.

ME4® BROMINAL® on Wheat, Barley, Oats and Rye

AREA/PRODUCT	RATE/ACRE	APPLICATION TIMING CROP	APPLICATION TIMING WEEDS	SPRAY VOLUME/ACRE
Cereal Areas except WA, OR, ID ME4® BROMINAL®	0.5-1 pint	*Winter cereals:* Fall through early spring applications before crop has reached the boot stage (not underseeded with legumes). *Spring cereals:* One leaf to boot stage.	Weeds in the 2-4 leaf stage or when rosettes are less than 1.5 inches across.	Use 10 to 20 gal. for ground or 5 to 10 gal. for aerial application. Use higher volumes for larger weeds or under poor growing conditions. Use 2 or more gal. with midwest air application.
Washington, Oregon, Idaho ME4® BROMINAL®	0.75-1 pint		Gromwell, tarweed, fiddleneck, mustards, and other broadleaf weeds apply before the 4 leaf stage, or before rosettes are 1.5 inches across.	Use 10 to 20 gal. for ground or 5 to 10 gal. for aerial application. Use higher rates and volumes for larger weeds or where weed growth is slow due to dry or cold weather.

ME4® BROMINAL® and MCPA-2,4-D Tank Mixtures on Wheat, Barley, Oats and Rye

When using ME4® BROMINAL® Bromoxynil Broadleaf Herbicide in combination with 2,4-D or MCPA, first add 2,4-D or MCPA to the water, then add ME4® BROMINAL®. Use the higher rates and volumes for larger weeds and where weed growth is slow due to dry or cold weather. Do not include Hoelon in this tank mixture, since the effectiveness of Hoelon will be reduced.

AREA/PRODUCT	RATE/ACRE	APPLICATION TIMING CROP	APPLICATION TIMING WEEDS	SPRAY VOLUME/ACRE
Cereal Areas except Washington, Oregon, Idaho ME4® BROMINAL® + 2,4-D* or MCPA*	0.5-0.75 pint + 0.25-0.5 lb. active ingredient	Apply after tillering but before jointing stage.	All weeds including wild buckwheat, cow cockle, fiddleneck, tarweed; apply at 1-4 leaf stage but before 6 inches high.	Use 10 to 20 gal. for ground or 4 to 10 gal. water for aerial application. Use higher volume on larger weeds.
Washington, Oregon, Idaho ME4® BROMINAL® + 2.4-D* or MCPA*	0.75-1 pint + 0.25-0.5 lb. active ingredient		Gromwell, tarweed, fiddleneck, mustard, other broadleaf weeds; apply before weeds are 6 inches high.	Use 10 to 20 gal. for ground or 5 to 10 gal. water for aerial application. Use higher volume on larger weeds.

* 2,4-D amine - WEEDAR 64® Herbicide, *2,4-D esters - WEEDONE® LV 4 Herbicide, *MCPA amine - WEEDAR® MCPA Concentrate, *MCPA Ester - WEEDONE® MCPA Ester Broadleaf Herbicide.

ME4® BROMINAL® and Banvel Tank Mixture on Wheat
When Banvel is tank mixed with ME4® BROMINAL® Bromoxynil Broadleaf Herbicide, water should be added to the spray tank first, followed by ME4® BROMINAL® and then Banvel.

AREA/PRODUCT	RATE/ACRE	APPLICATION TIMING CROP	APPLICATION TIMING WEEDS	SPRAY VOLUME/ACRE
Throughout the United States ME4® BROMINAL® + Banvel (dicamba)	0.5-0.75 pint + 0.0625-0.125 pint	*Spring Wheat:* Apply when spring wheat is from 3-leaf up to joint stage.	Bachelor's button, knotweed, mustards, nightshade, Russian thistle, wild buckwheat, lambsquarters, fanweed, dogfennel, cow cockle; apply when weeds are small (early seedling).	Use 10 or more gal. of water per acre for ground application.
Throughout the United States ME4® BROMINAL® + Banvel (dicamba)	0.5-0.75 pint + 0.125 pint	*Winter Wheat:* Fall through early spring application when winter wheat is from 3-leaf up to joint stage.	Kochia, fiddleneck, minerslettuce, knotweed, wild buckwheat, prickley lettuce, jacobsladder, alkanet, burchevil, corn cockle, corn gromwell, cow cockle, dogfennel, fanweed, falseflax, fixweed, henbit, lambsquarters, catchweed; apply when weeds are small (early seedling).	

Do not graze Banvel-treated areas or harvest for dairy feed prior to crop maturity.

Banvel is a registered trademark of Velsicol Chemical Corporation.

ME4® BROMINAL® and Karmex Tank Mixture on Wheat and Barley
When Karmex is tank mixed with ME4® BROMINAL® Bromoxynil Broadleaf Herbicide for henbit, gromwell and other weed control, observe all cautions on the Karmex label and especially those regarding winter stress, varieties, soil texture, organic matter, planting depth, plant vigor and replanting. Use where average annual rainfall exceeds 16 inches on winter wheat and winter barley.

AREA/PRODUCT	RATE/ACRE	APPLICATION TIMING CROP	APPLICATION TIMING WEEDS	SPRAY VOLUME/ACRE
Washington, Oregon, Idaho ME4® BROMINAL® + Karmex (diuron)	0.5 pint + 0.5 lb.	*Winter Wheat and Barley:* Make a single application in the fall after wheat has emerged, but before soil freezes or in the spring as soon as soil thaws.	Henbit, gromwell, fiddleneck, other weeds. Apply before weeds are 2 inches tall or across.	Use a minimum of 25 gal. for ground or 5 to 10 gal. for aerial application.

Karmex is a registered trademark of E.I. DuPont de Nemours and Company.

ME4® BROMINAL® and Avenge Tank Mixture for Broadleaf and Wild Oat Control in Wheat and Barley

When using a tank mix of ME® BROMINAL® Bromoxynil Broadleaf Herbicide and Avenge, fill the tank with about ⅔ of the required water and start agitation. Add ME4® BROMINAL® and mix thoroughly and with the agitation still running add the Avenge and the remainder of the required water. Agitate and recycle for at least five minutes before spraying and continue agitation until spraying is completed. Do not apply Avenge tank-mix under stress conditions or if the plants are wet with rain or dew or if rain is predicted within 6 hours.

AREA/PRODUCT	RATE/ACRE	APPLICATION TIMING CROP	APPLICATION TIMING WEEDS	SPRAY VOLUME/ACRE
Throughout the United States ME4® BROMINAL® + Avenge (difenzoquat)	0.75-1 pint + 2.5 pints (1-10 oats per sq. ft.) 3 pints (11-25 oats per sq. ft.) 4 pints (more than 25 oats per sq. ft.)	*Winter Wheat:* 4-leaf to tillering stage. *Spring Wheat:* Idaho, Oregon, Washington and Era variety in Minnesota and North Dakota, 5-6 leaf stage. *Spring Durum Wheat:* 5-6 leaf stage. *Barley:* 2-7 leaf stage.	Wild oats, fiddleneck, wild buckwheat and other broadleaf weeds; apply to wild oat in the 3-5 leaf stage, to emerged susceptible broadleaf weeds in the 2-4 leaf stage or when rosettes are less than 1.5 inches across.	Use 5 to 20 gal. with ground or 3 to 10 gal. for aerial application. If spray vol. exceeds 10 gallons, add 0.6 fluid ounce of a surfactant such as Tergitol NPX for each gallon in excess of 10 gal.

Do not use this tank mixture on Mexican durum wheat varieties Jori, Mexicali, Crane or Modoc; spring type Borah wheat; spring-seeded durum wheat varieties Lokota and Wascana. Injury can occur to some wheat varieties including Bonanza, Bounty, Lark, Waldron, and WS1809. Do not graze Avenge-treated fields or cut treated forage for silage, although grain and straw from treated fields can be fed to livestock. Avenge is a registered trademark of American Cyanamid Company.

ME4® BROMINAL® and Carbyne (barban) Tank Mixture for Broadleaf and Wild Oat Control in Wheat and Barley

When applying a tank mix of ME4® BROMINAL® Bromoxynil Broadleaf Herbicide and Carbyne follow the directions for spray nozzle arrangements on the Carbyne label. Maintain vigorous agitation in the spray tank until spraying is completed. Sequential treatments have been most effective.

Do not spray this tank mixture when plants are wet with heavy dew or rain.

Do not let spray drift into adjacent fields, particularly fields of oats, rye or buckwheat.

Do not allow livestock to graze on fields sprayed with the Carbyne 2EC tank mixed with ME4® BROMINAL® until after harvest of the crop.

AREA/PRODUCT	RATE/ACRE	APPLICATION TIMING CROP	APPLICATION TIMING WEEDS	SPRAY VOLUME/ACRE
Throughout the United States Sequential Carbyne treatment ME4® BROMINAL® + Carbyne 2EC 1st application + Carbyne 2EC 2nd application	0.5-0.75 pint + 1-1.5 pints + 1-1.5 pints	*Winter and Spring Wheat and Spring Barley:* For the sequential treatment, ME4® BROMINAL® can be combined with 1st or 2nd Carbyne application.	Broadleaf weeds 2-4 leaves. Wild oats, canary grass in 2 leaf stage. Apply the ME4® BROMINAL® tank mix only once, if the split Carbyne application is used. Use the high rate of Carbyne when the grassy weed population is over 50 plants per sq. ft.	Use 5 to 10 gal. of water and sufficient pressure (minimum 45 psi) to break spray into small droplets with ground equipment, and 3 to 5 gal. for aerial application.
ME4® BROMINAL® + Carbyne 2EC	0.5-0.75 pint + 1-1.5 pints	*Winter and Spring Wheat and Spring Barley.*	Broadleaf weeds 2-4 leaves. Wild oats, canary grass in 2 leaf stage. Use the high rate of Carbyne when the grassy weed population is over 50 plants per sq. ft.	
ME4® BROMINAL® + Carbyne 2EC	½-¾ (0.5-0.75) pint + 2 pints	*Winter and Spring Wheat and Spring Barley except Durum Wheat.*	Broadleaf weeds 2-4 leaves. Wild oats, canary grass in 2.5-3.5 leaf stage, do not apply if wild oats or canary grass have 4 or more leaves.	

Carbyne is a registered trademark of Velsicol Chemical Corporation

ME4® BROMINAL® and Hoelon Tank Mixture For Broadleaf, Wild Oat, and Foxtail (pigeon grass) Control in Wheat and Barley

When using the ME4® BROMINAL® Bromoxynil Broadleaf Herbicide and Hoelon tank mixture, cold (lower than 40 ° F) and/or prolonged wet weather conditions increase the sensitivity of barley to Hoelon, and Hoelon should not be used under these adverse weather conditions. Use the low rate of Hoelon when weather conditions are cool and moist. Do not use 3+3® BROMINAL® (Bromoxynil plus MCPA) Broadleaf Herbicide with Hoelon as it will reduce the effectiveness of Hoelon.

AREA/PRODUCT	RATE/ACRE	APPLICATION TIMING CROP	APPLICATION TIMING WEEDS	SPRAY VOLUME/ACRE
Throughout the United States ME4® BROMINAL® + Hoelon	0.75-1 pint + 2.66 pints	*Winter and Spring Wheat and Spring Barley:* Before or after tillering.	Grasses, wild oats, foxtail, (pigeon grass), annual ryegrass, barnyardgrass, 1-3 leaves. Fiddleneck and other broadleaf weeds 2-4 leaves. Use 1 pint ME4® BROMINAL® for 3-4 leaf broadleaf weeds.	Use at least 10 gal. with minimum 30 psi pressure for ground, or at least 5 gal. for aerial application.
ME4® BROMINAL® + Hoelon	0.75-1 pint + 2.66-3.33 pints	*Winter and Spring Wheat:* Before or after tillering.	Grasses, 3-4 leaves; Broadleaves, 3-4 leaves.	

Hoelon is a registered trademark of American Hoechst Corporation. Refer to the Hoelon label for list of states where applicable.

ME4® BROMINAL® and Glean on Barley and Wheat

The tank mixture of ME4® BROMINAL® Bromoxynil Broadleaf Herbicide and Glean provides an effective combination of contact and soil active herbicides. This tank mixture improves control of henbit, tansy mustard, gromwell and miners lettuce.

Apply this tank mixture to actively growing weeds, since weeds hardened off by low fertility, drought stress or cold weather may not be controlled.

Thorough and uniform spray coverage is essential for maximum weed control. Use 10 or more gallons of water per acre with ground sprayers or 4 to 10 gallons of water per acre for aerial application. Wide spaced, low pressure flood-type nozzles do not provide satisfactory coverage. Nozzle pressure should be at least 30 psi.

When preparing this tank mix, fill the spray tank with ⅔ of the required water and add proper amount of Glean to water with agitation. With Glean in suspension and the agitator running, and ME4® BROMINAL® Bromoxynil Broadleaf Herbicide and the remainder of required water. Agitation is required for uniform mixing and application.

Refer to Glean label for information on crop rotation and other guidelines.

AREA/PRODUCT	RATE/ACRE	APPLICATION TIMING CROP	 WEEDS	SPRAY VOLUME/ACRE
Throughout the United States ME4® BROMINAL® + Glean (75% active ingredient)	 0.38-1 pint 0.13-0.33 ounce	*Winter and Spring Wheat and Barley:* 2-3 leaf but before boot stage.	Apply to the most susceptible seedling weeds in the 1-4 leaf growth stage but before they are 6" tall. On less susceptible weeds apply before rosettes are 1.5" in diameter.	Use 10 to 20 gal. for ground or 4 to 10 gal. water for aerial application.

Glean is a registered trademark of E.I. DuPont de Nemours & Company.

ME4® BROMINAL® and Metribuzin (Lexone or Sencor) on Winter Wheat

A tank mixture of ME4® BROMINAL® Bromoxynil Broadleaf Herbicide and Metribuzin will control chickweed, filaree, henbit and dogfennel in addition to all the weeds listed on the ME4® BROMINAL® Bromoxynil label.

Crop injury can occur from this combination if used in wheat that is too small and/or stressed from winter injury. The higher rates of ME4® BROMINAL® in combination with higher rates of Metribuzin should be used only in an emergency weed situation and if some minor crop injury is acceptable. When using the higher rates the wheat should be fully tillered and have a well-established secondary root system.

To prepare the spray mixture, add ⅔ to ¾ the required amount of water to the tank. Start agitation and add the proper amount of Metribuzin as determined from the table below, followed by ME4® BROMINAL® Bromoxynil and finally the remaining water.

A recognized authority in your area should be consulted concerning proper rates and timings for your conditions.

Do not use Metribuzin on irrigated wheat as crop injury may result. Do not exceed the rates on this label. Avoid applications when crop is under stress from winter kill, frost damage, disease or drought.

AREA/PRODUCT	RATE/ACRE	APPLICATION TIMING CROP	 WEEDS	SPRAY VOLUME/ACRE
States of Idaho, Montana, Washington, Oregon ME4® BROMINAL® + Metribuzin (active ingredient)	 0.5-0.625 pint 0.125-0.250 lb. active ingredient	*Winter Wheat:* Spring after wheat has started to grow and has established secondary roots with a minimum of 3 to 4 tillers but before boot stage.	Apply before weeds are 2 inches tall or less than 2 inches across.	Use 10 to 40 gallons water for ground or 5 to 10 gallons for aerial application.

DETERMINE RATE OF FORMULATED PRODUCT PER ACRE

METRIBUZIN ACTIVE INGREDIENT	SENCOR 4 OR LEXONE 4L	SENCOR DF OR LEXONE DF
0.125 lb ai/A	0.25 pt/A	0.167 lb/A
0.188 lb ai/A	0.38 pt/A	0.250 lb/A
0.250 lb ai/A	0.50 pt/A	0.333 lb/A

ME4® BROMINAL® on Flax

Application of ME4® BROMINAL® Bromoxynil Broadleaf Herbicide under high humidity conditions can injure Flax. Temperatures at the time of treatment, or for the three days after treatment, must be below 85 ° F to avoid serious Flax injury.

AREA/PRODUCT	RATE/ACRE	APPLICATION TIMING CROP	 WEEDS	SPRAY VOLUME/ACRE
Throughout the United States ME4® BROMINAL®	 0.5-1 pint	 Flax 2-4 inches tall	 Weeds 2-4 leaf stage.	 Use 10 to 20 gal for ground or 5 to 10 gal. for aerial application.

ME4® BROMINAL® on Corn (Field and Pop)

ME4® BROMINAL® Bromoxynil Broadleaf Herbicide controls seedling broadleaf weeds such as velvetleaf (buttonweed), cocklebur, smartweed, sunflower, jinsonweed, wild cucumber, wild buckwheat, kochia, and other weeds, some of which are resistant to 2,4-D.

ME4® BROMINAL® Broadleaf Herbicide requires good spray coverage to provide effective weed control. Coarse sprays such as those produced by flood jet nozzles may not provide adequate coverage and should be avoided.

Warm temperatures increase the speed of action of ME4® BROMINAL® Bromoxynil and increases the chance of some corn foliage burn on the uppermost leaves. New growth will be normal and no adverse effect has been observed on yields. Spray when corn foliage is dry since rain, irrigation moisture or dew increases the chance of leaf burn.

Pigweed is somewhat resistant to ME4® BROMINAL® and must be no larger than the 4 leaf stage of growth for effective control. If pigweed is the primary weed, use the high rate of ME4® BROMINAL® Bromoxynil or tank mix with 0.25 lb, 2,4-D/A. Observe all precautions necessary with 2,4-D.

Tank mixtures with 1 lb. active atrazine/A have given broad spectrum postemergence broadleaf weed control with soil activity. This tank mixture has been effective in controlling atrazine-resistant lambsquarters, wild turnip, groundsel and common ragweed. The option of reducing the atrazine rate to 1 lb. per active ingredient (1.25 lbs. of 80 W) reduces the risk of carryover. As ME4® BROMINAL® Bromoxynil enhances the postemergence action of atrazines the addition of oil is not necessary in this tank mixture.

Do not cut for feed or graze within 30 days of treatment. Do not plant a rotational crop until the following use season.

		APPLICATION TIMING		
AREA/PRODUCT	RATE/ACRE	CROP	WEEDS	SPRAY VOLUME/ACRE
Throughout the United States ME4® BROMINAL®	0.5-1 pint	Corn (Field, Pop) 2 leaf to 14" height	Velvetleaf (buttonweed), Cocklebur, Smartweed, Sunflower, Jimsonweed, Wild cucumber, Wild buckwheat, and Kochia; 2-4 leaf stage.	Use 10 or more gal. water for ground or 5 or more gal. water for aerial application.
ME4® BROMINAL® + Atrazine	0.5-1 pint + 1.0 lb (active ingredient)			

ME4® BROMINAL® on Grain Sorghum

ME4® BROMINAL® Bromoxynil Broadleaf Herbicide controls seedling broadleaf weeds such as cocklebur, velvetleaf (buttonweed), sunflower, morningglory, kochia, smartweed, burcucumber, some of which are resistant to 2,4-D.

ME4® BROMINAL® Bromoxynil Broadleaf Herbicide requires thorough and uniform spray coverage to provide effective weed control. Coarse sprays such as those produced by flood jet nozzles may not provide adequate coverage and should be avoided. Although ME4® BROMINAL® Bromoxynil spray drift does not harm plants like 2,4-D, high concentrations may desiccate adjacent crop foliage.

Warm temperatures increase the speed of ME4® BROMINAL® Broadleaf Herbicide action on weeds and may also increase the chance of some grain sorghum foliage burn on the uppermost leaves. New growth will be normal and yields will not be reduced. Do not spray when grain sorghum foliage is wet from dew, rain or irrigation.

Pigweed is somewhat resistant to ME4® BROMINAL® and should be no larger than the 4 leaf stage of growth for effective control. Use 1 pt. ME4® BROMINAL® Bromoxynil per acre if pigweed is the primary weed, or use a 2,4-D (0.25 lb/A) tank mix. Observe all precautions necessary with 2,4-D.

Do not cut for feed or graze within 30 days of treatment. Do not plant a rotational crop until the following use season.

		APPLICATION TIMING		
AREA/PRODUCT	RATE/ACRE	CROP	WEEDS	SPRAY VOLUME/ACRE
Throughout the United States				
ME4® BROMINAL®	0.5-1 pint	Grain Sorghum 2 leaf to 14" height	Cocklebur, Velvetleaf (buttonweed), Sunflower, Morningglory, Pigweed, Kochia, Smartweed and Burcucumber; 2-4 leaf stage.	Use 10 or more gal. water for ground, or 5 or more gal. for aerial applications.
ME4® BROMINAL® + Atrazine	0.5-1 pint + 1.0 lb (active ingredient)			

DIRECTIONS FOR USE

GARLIC

Apply 1 to 2 pints (0.5 to 1.0 lb active ingredient) ME4® BROMINAL® Bromoxynil Broadleaf Herbicide per broadcast acre when the weeds are 1 to 4 inches tall. Use the lower rate on small weeds and the higher rate on 4 inch weeds. Treat when garlic has 1 to 2 true leaves and is up to 12 inches tall.

For ground application use 40 to 100 gallons spray solution per broadcast acre. Thorough and uniform coverage of weeds is essential for good control. Higher spray volumes are preferred. For aerial application, use 5 gallons of spray solution per acre.

Do not apply within 112 days of anticipated harvest.

ONIONS

Apply 0.5 to 0.75 pint (0.25 to 0.38 lb active ingredient) ME4® BROMINAL® Bromoxynil Broadleaf Herbicide per acre on a broadcast basis. Water volume is *important*—use 50-70 gallons water/acre and apply when onions have 2-5 true leaves. *Concentrated sprays kill onions.* Thorough and uniform coverage is necessary for good weed control. Do not treat thrips or sand damaged crop. *Do not add surfactants.*

In onion-producing areas, certain environmental conditions reduce development of waxy coating on the onion leaves, thus increasing the possibility of injury. Dry soil, dry onion foliage, high light intensity, low humidity, and high temperatures tend to increase the waxy coating on onion leaves and thus reducing chances for injury.

It is essential that the soil and onion foilage be dry at the time of application. Humidity should be low and dew should be off the plants.

Applications should not be made 24 hours before or after cultivation, irrigation, or rain; longer than 24 hours if possible. It is best to spray on a bright, sunny day after dew has completely dried from onion foliage and before it has formed.

Do not use ME4® BROMINAL® Bromoxynil Broadleaf Herbicide on onions grown under low light intensity, in areas such as Oregon, west of the Cascades.

The sensitivity of onions varies with the variety and environmental conditions. Therefore, even if all the label directions are followed,

ME4® BROMINAL® application may still kill onions under certain circumstances.

APPLICATION OF ME4® BROMINAL® THROUGH IRRIGATION SYSTEMS—BROMIGATION® POSTEMERGENCE WEED CONTROL SERVICE

GENERAL INFORMATION

ME4® BROMINAL® Bromoxynil Broadleaf Herbicide may be applied through automated (center point or continuous lateral move) or wheel lines, hand lines or solid set overhead sprinkler irrigation systems for broadleaf weed control. Properly applied ME4® BROMINAL® Broadleaf Herbicide will control fiddleneck, wild buckwheat, sunflower, fanweed, kochia, Russian thistle and other susceptible broadleaf weeds listed on the ME4® BROMINAL® Bromoxynil label. For best results, apply when weeds are small (3-4 leaf stage), and before they have exerted yield reducing competition on the crop.

DIRECTIONS FOR USE

Remove scale, pesticide residues and other foreign matter from the chemical tank and entire injector system. Flush with clean water. Add ME4® BROMINAL® Broadleaf Herbicide or a water dilution of it to a mix tank. Application accuracy is increased with increased dilution.

For best performance, set sprinkler to deliver approximately 0.5 inch or less of water per acre. Start the sprinklers and then inject ME4® BROMINAL® Broadleaf Herbicide into the irrigation line. ME4® BROMINAL® Broadleaf Herbicide should be injected with a positive displacement pump into the main line at least 8 feet ahead of a right angle turn to insure adequate mixing. Use 1 pint rate if pigweed or lambsquarters are part of the weed complex. Refer to the ME4® BROMINAL® Bromoxynil Broadleaf Herbicide label for detailed information on application rates and timings.

ME4® BROMINAL® can be applied through overhead sprinkler irrigation systems to small grain cereals, field corn, flax, garlic, onions and sorghum.

USER PRECAUTIONS

Application of more than 0.5 inch/acre of irrigation water may result in decreased product performance on certain soils.

Apply ME4® BROMINAL® Bromoxynil Broadleaf Herbicide only through systems containing antisiphon, check valves and interlocking controls between a metering device and the water pump to insure simultaneous shutoff to prevent water source contamination and overflow of the mixing tank.

Do not apply when conditions favor drift, when system connections or fittings leak, or when nozzles do not provide uniform distribution.

Allow sufficient time for pesticide to be flushed through all the lines and nozzles before turning off irrigation water.

ME4® BROMINAL® on Established Lawn, Ornamental Turf, Sod and Grass Seed Production Areas

ME4® BROMINAL® Bromoxynil Broadleaf Herbicide may be used on established Bermudagrass, St. Augustinegrass, centipedegrass, Kentucky bluegrass, fescue, orchardgrass, bentgrass, and ryegrass. For weeds controlled see "Weed Susceptibility" list in addition to the weeds listed below. Do not use on bentgrass putting greens or tees. Seedling weeds must be actively growing. Good coverage is essential for effective control of weeds.

Temporary turfgrass discoloration may occur particularly if excessive overlapping of spray solution occurs, or if ME4® BROMINAL® Bromoxynil Broadleaf Herbicide is applied at temperatures of 85 °F or above. The grass will recover after two mowings leaving no permanent damage.

AREA/PRODUCT	RATE/ACRE	APPLICATION TIMING CROP	APPLICATION TIMING WEEDS	SPRAY VOLUME/ACRE
Throughout the United States				
ME4® BROMINAL®	0.75-1 pint Use high rate on weeds larger than 3-4 leaf stage but less than 6 inches tall or wide.	Established turf lawn, sod and grass seed production areas.	Weeds 3-4 leaf stage to 6 inches tall or wide.	20 or more gallons water per acre or 2 to 4 gallons per 1000 sq. ft.
ME4® BROMINAL®	2 pints		Seedling, burweed, yellow woodsorrel, black medic, bur clover, spiny fruited crowfoot, hop clover.	
ME4® BROMINAL®	4 pints		Seedling prostrate spurge	

ME4® BROMINAL® Tank Mixtures for Broadleaf Weed Control in Established Lawns and Turf Areas

For weeds controlled see "Weed Susceptibility" list in addition to the weeds listed below.

AREA/PRODUCT	RATE/ACRE	APPLICATION TIMING CROP	APPLICATION TIMING WEEDS	SPRAY VOLUME/ACRE
ME4® BROMINAL® + WEEDONE® DPC (2,4-D + 2,4-DP)	1-2 pints + 3-4 pints Use low rates on annual weeds and high rates on perennial and biennial broadleaf weeds	Established turf except bent, St. Augustine, centipede and carpetgrasses.	Actively growing dandelion, plantain, ground ivy, red clover, white clover, hop clover, bur clover, prostrate spurge, knotweed, and oxalis.	25 or more gallons water per acre or 2 to 4 gallons water per 1000 sq. ft.

AREA/PRODUCT	RATE/ACRE	APPLICATION TIMING CROP	WEEDS	SPRAY VOLUME/ACRE
ME4® BROMINAL® + WEEDONE® DPC (2,4-D + 2,4-DP)	1 pint + 1.5-2 pints	Established bent, St. Augustine, and centipedegrasses. Retreat in 2 to 3 weeks. These grasses may be slightly injured. Do not use on bentgrass putting greens or tees.		
ME4® BROMINAL® + WEEDAR 64® (2,4-D amine) or WEEDONE® LV4 (2,4-D ester) Use ester for best control of wild garlic and onion.	1 pint + 2-4 pints (0.95-1.9 lb active ingredient) or 2-4 pints (0.95-1.9 lb active ingredient) Use the 2 pt. rate on annual broadleaf weeds and 3-4 pts on perennial and biennial broadleaf weeds.	Established turf except bent, St. Augustine, and centipedegrasses.	Seedling and established weeds such as ragweed, dandelion, plantain, wild garlic or onion. Treat wild garlic or onion for three successive seasons (example: spring, fall, spring).	20 or more gallons water per acre or 2 to 4 gallons of water per 1000 sq. ft.
ME4® BROMINAL® + dicamba (Banvel)	1-2 pints + 0.2-0.5 pint (0.1-0.25 lb active ingredient) Use high rates for control of yellow woodsorrel and prostrate spurge. Repeat treatments may be made. Do not exceed 1 lb ai Banvel per acre per growing season.	Established turf except Bentgrass. Banvel in excess of 0.5 lb ai per acre may cause noticeable stunting or discoloration of sensitive grass species such as carpetgrass, buffalograss, and St. Augustine grass.	Seedling weeds such as knotweed, chickweed, red clover, and white clover.	20 or more gallons water per acre or 2 to 4 gallons of water per 1000 sq. ft.
ME4® BROMINAL® + MCPP	1-2 pints + 1 lb active ingredient Use high rate on established weeds.	Established turf except bentgrass.	Seedling and established weeds such as plantain, dandelion, red clover, ground ivy, prostrate spurge, white clover, chickweed, knotweed, and starwort.	20 or more gallons water per acre or 2 to 4 gallons water per 1000 sq. ft.
ME4® BROMINAL® + MCPP + dicamba (Banvel)	1-2 pints + 0.5-1 lb active ingredient + 0.25-0.5 pts (0.125-0.25 lb active ingredient) Use high rate of ME4® BROMINAL® Bromoxynil when spurweed and/or prostrate spurge is present.	Established turf except bentgrass.	Seedling and established weeds such as white clover, dandelion, and plantain.	20 or more gallons water per acre or 2 to 4 gallons water per 1000 sq. ft.

AREA/PRODUCT	RATE/ACRE	CROP	APPLICATION TIMING WEEDS	SPRAY VOLUME
ME4® BROMINAL® + MCPP + 2.4-D	1-2 pints + 0.5-1 lb active ingredient) + 0.5-1 lb (active ingredient) Use lower rate on St. Augustine and Carpetgrasses.	Established turf except bentgrass.	Seedling and established weeds such as chickweed, knotweed, red sorrel, dandelion and plantain.	20 or more gallons water per acre or 2 to 4 gallons per 1000 sq. ft.

ME4® BROMINAL® Tank Mixtures for Broadleaf and Grass Weed Control in Established Lawns and Turf Areas
For broadleaf weeds controlled see "Weed Susceptibility" list.

AREA/PRODUCT	RATE/ACRE	CROP	APPLICATION TIMING WEEDS	SPRAY VOLUME
ME4® BROMINAL® + MSMA or DSMA	1 pint + 2-3 lb active ingredient Refer to MSMA or DSMA label directions for proper rates.	Established bluegrass, bermudagrass, tall fescue and zoysia. Do not apply to bentgrass, carpetgrass, centipedegrass, or St. Augustine grass. Grasses may show temporary discoloration.	Seedling broadleaf weeds, crabgrass, goosegrass, nutsedge and sandbur. Apply during warm weather 80-90 degrees F. Begin treatment when grassy weeds have 3-4 leaves. More mature plants require additional applications 7-10 days apart. Three to four applications of MSMA or DSMA may be necessary for nutsedge and sandbur control. Spray under adequate soil moisture.	20 or more gallons water per acre or 2 to 4 gallons water per 1000 sq. ft.

CONVERSION CHART

RATE (PINTS/ACRE)	RATE (FLUID OUNCES/10,000 SQ. FT.)
0.125	0.46
0.25	0.92
0.50	1.84
0.75	2.75
1.0	3.67
2.0	7.35
3.0	11.02
4.0	14.70

ME4® BROMINAL® FOR BROADLEAF WEED CONTROL IN NEWLY SEEDED, NEWLY SPRIGGED, ESTABLISHED LAWN AND ORNAMENTAL TURF AREAS.

Newly Seeded and Sprigged Grasses

ME4® BROMINAL® Bromoxynil Broadleaf Herbicide may be used on seedling grasses such as Merion, Park, Delta or common Kentucky bluegrass, Pennlawn, Chewing's, Illahee or Alta fescues, orchardgrass, Penncross, Highland, Seaside or Astoria, bentgrass, Excalibre and perennial ryegrasses. ME4® BROMINAL® Broadleaf Herbicide may be used on newly sprigged Bermudagrass, centipedegrass or St. Augustinegrass. For weeds controlled see "Weed Susceptibility" list.

AREA/PRODUCT	RATE/ACRE	CROP	APPLICATION TIMING WEEDS	SPRAY VOLUME
Throughout the United States				
ME4® BROMINAL®	0.75 pint	Seedling grasses or newly sprigged Bermuda, centipede, or St. Augustine	Seedling broadleaf weeds 3–4 leaf stage.	20 or more gallons water per acre or 2 to 4 gallons per 1000 sq. ft.
ME4® BROMINAL®	1.0 pint		Seedling broadleaf weeds larger than 3–4 leaf stage but less than 6 inches tall.	

Read carefully the labels of any product used in a tank mixture. Note cautions and limitations on this and products used in tank mixtures to assure maximum weed control and turf and applicator safety.

DO NOT FEED TURF GRASSES TREATED WITH ME4® BROMINAL® BROMOXYNIL BROADLEAF HERBICIDE TO LIVESTOCK.

WEED SUSCEPTIBILITY IN SEEDLING STAGE

Most Susceptible Weeds: Use 0.5 point ME4® BROMINAL® Bromoxynil Broadleaf Herbicide (0.25 lb active ingredient) per acre:

Black mustard	*(Brassica nigra)*
Bristly starbur	*(Acanthospermum hispidum)*
Cocklebur	*(Xanthium pensylvanicum)*
Eastern black nightshade	*(Solanum ptycanthum)*
Field pennycress, fanweed	*(Thlaspi arvense)*
Green smartweed	*(Polygonum scabrum)*
Hairy nightshade	*(Solanum sarachoides)*
Kochia	*(Kochia scoparia)*
Lambsquarters	*(Chenopodium album)*
London rocket	*(Sisymbrium irio)*
Morningglory (annual)	*(Ipomoea hederacea)*
Nettleleaf goosefoot	*(Chenopodium murale)*
Nightshade	*(Solanum nigrum, villosum)*
Parthenium ragweed	*(Parthenium hysterophorus)*
Ragweed	*(Ambrosia artemisiifolia)*

Shepherdspurse	*(Capsella bursa-pastoris)*
Silver leaf nightshade	*(Solanum elaeagnifolium)*
Sowthistle (annual)	*(Sonchus oleraceus)*
Sunflower	*(Helianthus annuus)*
Treacle mustard	*(Erysimum repandum)*
Tumble (Jim Hill) mustard	*(Sisymbrium altissimum)*
Velvetleaf (or Buttonweed)	*(Abutilon theophrasti)*
Wild buckwheat	*(Polygonum convolvulus)*
Wild radish	*(Raphanus raphanistrum)*

Susceptible Weeds: Use 0.75 pint ME4® BROMINAL® Bromoxynil Broadleaf Herbicide (0.375 lb active ingredient) per acre:

Blue (purple) mustard	*(Chorispora tenella)*
Bullthistle	*(Cirsium vulgare)*
Corn chamomile	*(Anthemis arvensis)*
Corn gromwell	*(Lithospermum arvense)*
Cow cockle	*(Saponaria vaccaria)*
Dyers woad	*(Isatis tinctoria)*
Fiddleneck	*(Amsinckia intermedia)*
Fivehooked bassia	*(Bassia hyssopifolia)*
Jimsonweed	*(Datura stramonium)*
Knawel (German moss)	*(Scleranthus annuus)*
Laydysthumb	*(Polygonum persicaria)*
Mare's-tail	*(Conyza canadensis)*
Nodding thistle	*(Carduus nutans)*
Pennsylvania smartweed	*(Polygonum pensylvanicum)*
Popcorn flower	*(Allocarya figurata)*
Prickly (Chinese) lettuce	*(Lactuca scariola)*
Puncture vine	*(Tribulus terrestris)*
Redmaids rockpurslane	*(Calandrinia caulescens)*
Russian thistle	*(Salsola kali)*
Smallseeded falseflax	*(Camelina microcarpa)*
Southern giant curl mustard	*(Brassica japonica)*
Tartary buckwheat	*(Fagopyrum tataricum)*
Tarweed	*(Hemizonia spp.)*
Wild mustard	*(Brassica kaber)*
Winter vetch	*(Vicia sativa)*
Yellow rocket	*(Barbarea vulgaris)*
Yellow star thistle	*(Centaurea solstitialis)*

Less Susceptible Weeds: Use 1.0 pint ME4® BROMINAL® Bromoxynil Broadleaf Herbicide (0.5 lb active ingredient) per acre:

Bur buttercup	*(Ranunculus testiculatus)*
Dogfennel (Mayweed)	*(Anthemis cotula)*
Fumitory	*(Fumaria officinalis)*
Giant ragweed	*(Ambrosia trifida)*
Groundsel	*(Senecia vulgaris)*
Henbit	*(Lamium amplexicaule)*
Minerslettuce	*(Montia perfoliata)*
Pepperweed	*(Lepidium campestre)*
Pigweed	*(Amaranthus retroflexus)*
Prickly (Chinese) lettuce	*(Lactuca scariola)*
Prostrate knotweed	*(Polygonum aviculare)*
Spiny pigweed*	*(Amaranthus spinosus)*
Tall waterhemp*	*(Amaranthus tuberculatus)*
Tansymustard	*(Descurainia pinnata)*

* Treat before 4 leaf stage or less than 4 inches tall.

INDUSTRIAL USE

For controlling Russian Thistle and Annual Broadleaf weeds on roadsides, vacant lots, dry drainage beds, railroads and industrial sites.

Directions for Industrial Use

ME4® BROMINAL® Broadleaf Herbicide effectively controls Russian thistle and certain other summer annual broadleaf weeds such as saltbrush *(Atriplex)*, mare's-tail, puncture vine, prostrate spurge, and Bassia.

ME4® BROMINAL® Broadleaf Herbicide should be applied in early summer when Russian thistle is young and growing vigorously.

For broadcast treatment of small weeds: Apply 1 quart ME4® BROMINAL® Broadleaf Herbicide with 2 quarts SURFEL Spray Adjuvant or equivalent in 150 gallons water per acre. Thorough and uniform coverage is important. Make sure spray boom is high enough to contact tops of all plants. Use adequate spray pressure to contact all leaf surfaces.

For spot treatments of scattered infestations or large weeds along fencerows or buildings: Mix 0.5 quart ME4® BROMINAL® Bromoxynil Broadleaf Herbicide with 1 quart of surfactant such as SURFEL per 100 gallons spray solution. Cover all the weed foliage thoroughly. Use at least 200 gallons of spray solution for each acre sprayed. Best industrial results with the ME4® BROMINAL® Broadleaf Herbicide and surfactant mixture are obtained where temperatures are 85 ° F or higher.

PRECAUTIONARY STATEMENTS

WARNING

HAZARD TO HUMANS AND DOMESTIC ANIMALS

Harmful if swallowed. Avoid contact with skin, eyes, clothing. Avoid breathing spray mist.

STATEMENT OF PRACTICAL TREATMENT

In case of eye contact, rinse thoroughly with water and get medical attention. In case of skin contact, wash immediately with soap and water and remove contaminated clothing.

If swallowed, do not induce vomiting. Drink promptly large quantities of milk, egg white, gelatin solution or if these are not available, drink large quantities of water. Avoid alcohol. Get medical attention.

NOTE TO PHYSICIAN

There is no specific antidote. This product contains petroleum distillates. Gastric lavage may be contraindicated because of possibility of aspiration. Treatment of overexposure should be directed at the control of the symptoms and clinical condition.

GENERAL WORKER PROTECTION STATEMENTS

Do not apply this product in such a manner as to directly or through drift expose workers or other persons. The area being treated must be vacated by unprotected persons.

Do not enter treated areas without protective clothing until sprays have dried. Because certain states may require more restrictive reentry intervals for various crops treated with this product, consult your State Department of Agriculture for further information. Written or oral warnings must be given to workers who are expected to be in a treated area or in an area about to be treated with this product. Advise workers to stay out of fields during application and until sprays have dried. Regular long-sleeved work clothing should be worn when working in treated fields. See Precautionary Statement, Statement of Practical Treatment and Note to Physician for information on accidental exposures. When oral warnings are given warnings shall be given in a language customarily understood by workers. Oral warnings must be given if there is reason to believe that written warnings cannot be understood by workers. Written warnings must include the following information: appropriate signal word (WARNING), area treated with ME4® BROMINAL® Bromoxynil Broadleaf Herbicide, date of application, appropriate protective clothing, and reentry interval if applicable.

ENVIRONMENTAL HAZARDS

This product is toxic to fish. Do not apply directly to water. Do not apply when weather conditions favor drift from treated areas.

Do not contaminate water used for irrigation or domestic purposes by cleaning of equipment or disposal of waste.

Avoid spray drift into neighboring crops or pastures. Clean sprayer thoroughly before using for other purposes.

STORAGE AND DISPOSAL

Freezing does not affect the quality of this product. Do not store near fertilizers, seed, insecticides or fungicides. Do not contaminate water, food, or feed by storage or disposal. Open dumping is prohibited.

PESTICIDE DISPOSAL

Wastes resulting from the use of this product may be disposed of on site or at an approved waste disposal facility.

CONTAINER DISPOSAL

Do not reuse empty container. Triple rinse (or equivalent). Then offer for recycling or reconditioning, or puncture and dispose of in a sanitary landfill, or by incineration or if allowed by state and local authorities, by burning. If burned, stay out of smoke.

LIMITED WARRANTY AND DISCLAIMER

The manufacturer warrants (a) that this product conforms to the chemical description on the label; (b) that this product is reasonably fit for the purposes set forth in the directions for use when it is used in accordance with such directions; and (c) that the directions, warnings and other statements on this label are based upon responsible experts' evaluation of reasonable tests of effectiveness of toxicity to laboratory animals and to plants, and of residues on food crops, and upon reports of field experience. Tests have not been made on all varieties or in all states or under all conditions. THE MANUFACTURER NEITHER MAKES NOR INTENDS, NOR DOES IT AUTHORIZE ANY AGENT OR REPRESENTATIVE TO MAKE ANY OTHER WARRANTIES, EXPRESS OR IMPLIED, AND IT EXPRESSLY EXCLUDES AND DISCLAIMS ALL IMPLIED WARRANTIES OF MERCHANTABILITY OR FITNESS FOR A PARTICULAR PURPOSE.

THIS WARRANTY DOES NOT EXTEND TO, AND THE BUYER SHALL BE SOLELY RESPONSIBLE FOR, ANY AND ALL LOSS OR DAMAGE WHICH RESULTS FROM THE USE OF THIS PRODUCT IN ANY MANNER WHICH IS INCONSISTENT WITH THE LABEL DIRECTIONS, WARNINGS OR CAUTIONS.

BUYER'S EXCLUSIVE REMEDY AND MANUFACTURER'S OR SELLER'S EXCLUSIVE LIABILITY FOR ANY AND ALL CLAIMS, LOSSES, DAMAGES, OR INJURIES RESULTING FROM THE USE OR HANDLING OF THIS PRODUCT, WHETHER OR NOT BASED IN CONTRACT, NEGLIGENCE, STRICT LIABILITY IN TORT, OR OTHERWISE, SHALL BE LIMITED AT THE MANUFACTURER'S OPTION, TO REPLACEMENT OF, OR THE REPAYMENT OF THE PURCHASE PRICE FOR, THE QUANTITY OR PRODUCT WITH RESPECT TO WHICH DAMAGES ARE CLAIMED. IN NO EVENT SHALL MANUFACTURER OR SELLER BE LIABLE FOR SPECIAL, INDIRECT OR CONSEQUENTIAL DAMAGES RESULTING FROM THE USE OR HANDLING OF THIS PRODUCT.

WEEDONE® LV4, WEEDONE® MCPA Ester, WEEDAR 64®, WEEDAR® MCPA Concentrate, ME4® BROMINAL® and SURFEL® Spray Adjuvant are registered trade marks and BROMIGATION® is a service mark of Union Carbide Agricultural Products Company, Inc.

THIS SPECIMEN LABEL IS INTENDED FOR USE ONLY AS A GUIDE IN PROVIDING GENERAL INFORMATION REGARDING THE DIRECTIONS, WARNINGS AND CAUTIONS ASSOCIATED WITH THE USE OF THIS PRODUCT. AS WITH ANY AGRICULTURAL CHEMICAL, ALWAYS FOLLOW THE LABEL INSTRUCTIONS ON THE PACKAGE BEFORE USING.

BROMINAL®

BROADLEAF HERBICIDE

CROP USE

Controls annual broadleaf weeds in wheat, barley, oats, rye, flax, garlic, onions, established turf areas and newly seeded or sprigged grasses.

INDUSTRIAL USE

Controls Russian thistle and other annual broadleaf weeds on roadsides, vacant lots, railroads, and industrial sites.

For Commercial or Agricultural Use Only.

ACTIVE INGREDIENT:	% BY WEIGHT
Octanoic acid ester of bromoxynil* (3,5-dibromo-4-hydroxybenzonitrile)	31.7%
INERT INGREDIENTS:	68.3%
TOTAL	100.0%

*Bromoxynil equivalent 21.8% by weight or 2 pounds per gallon.

EPA Reg. No. 264-204-AA EPA Est. No. 264-MO-01

Keep Out of Reach of Children

WARNING—AVISO

PRECAUCION AL USARIO: Si usted no lee ingles, no use este producto hasta que la etiqueta le haya sido explicada ampliamente.

IN CASE OF EMERGENCY, TELEPHONE (24 HOURS A DAY) IN U.S.A. 1-800-UCC-HELP.

FOR GENERAL PRODUCT INFORMATION CALL 1-800-334-9745.

DIRECTIONS FOR USE

IT IS A VIOLATION OF FEDERAL LAW TO USE THIS PRODUCT IN A MANNER INCONSISTENT WITH THE LABELING.

WHEAT, BARLEY, OATS AND RYE

Apply 1 to 2 pints of BROMINAL broadleaf herbicide per acre after small grains emerge, but before weeds are past the 3 to 4 leaf stage. Weeds that form rosettes should be treated when these weeds are less than $1\frac{1}{2}$ inches across.

Proper timing, rate of application, and spray volume will vary according to the weed species present. See the table below.

Good coverage is essential for maximum weed control. Use 10 or more gallons of water per acre with ground sprayers. Nozzle pressure should be at least 30 psi. Use 2 or more gallons of water per acre with aerial sprayers according to the table below. Width of swath must not exceed length of spray boom.

Do not apply to small grains during or after the boot stage. Do not apply when crop is under stress for lack of moisture or when crop forms a canopy over weeds.

BROMINAL broadleaf herbicide may be applied in liquid fertilizer or combinations of liquid fertilizer and water. Use constant agitation until the spray mixture has been completely applied.

Tank mixes permit choosing a herbicide combination most effective for the weed problems present on your farm. Use BROMINAL broadleaf herbicide alone or in a combination according to the table below. If the amine of 2,4-D or MCPA is added, mix these into the water first, then add BROMINAL. If Banvel is tank mixed with BROMINAL, water should be added to the spray tank first, followed by BROMINAL and Banvel. If Karmex is tank mixed with BROMINAL for henbit, gromwell, and other weed control, observe all cautions on the Karmex label and especially those regarding winter stress, varieties, soil texture, organic matter, planting depth, plant vigor, and replanting. Follow all cautions and limitations on the labels of the products you select to mix with BROMINAL.

When using a tank mix of BROMINAL broadleaf herbicide and Avenge fill the tank with about $\frac{2}{3}$'s of the required water and start agitation. Add BROMINAL and mix thoroughly and with the agitation still running add the Avenge and the remainder of the required water. Agitate and recycle for at least five minutes before spraying and continue agitation until spraying is completed. Do not apply Avenge tank-mix under stress conditions or if the plants are wet with rain or dew or if rain is predicted within 6 hours.

When using the BROMINAL and Hoelon tank mixture, cold (lower than 40°F) and/or prolonged wet weather conditions increase the sensitivity of barley to Hoelon and Hoelon should not be used under these adverse weather conditions. Use the low rate of Hoelon when weather conditions are cool and moist.

When applying a tank mix of BROMINAL and Carbyne follow the directions for spray nozzle arrangements on the Carbyne label.

Do not graze treated fields for 30 days after application. When treating with Banvel do not graze treated areas or harvest for dairy feed prior to crop maturity. When treating with Avenge do not graze treated fields or cut treated forage for silage although grain and straw from treated fields can be fed to livestock.

The tank mixture of BROMINAL® bromoxynil broadleaf herbicide and Glean provides winter wheat producers with an effective contact and soil active herbicide combination. This tank mixture improves control of henbit, tansy mustard, gromwell, Chinese lettuce, bedstraw, lambsquarters, chickweed, wild buckwheat and miners lettuce.

Apply the tank mixture BROMINAL® and Glean to actively growing weeds as weeds hardened off by low fertility, drought stress or cold weather may not be controlled.

Good spray coverage is essential for maximum weed control. Use 10 or more gallons of water per acre with ground sprayers or 4-10 gallons of water per acre for air application. Wide spaced low pressure flood type nozzles do not give satisfactory coverage. Nozzle pressure should be at least 30 psi.

When preparing this tank mix of BROMINAL® bromoxynil and Glean fill the spray tank with ⅔'s of the required water and mix the proper amount of Glean in the water with the agitator running. With the Glean in suspension and the agitator running add the BROMINAL® and the remainder of the required water. Agitation is required for uniform mixing and application.

Carefully follow all precautions on the label for Glean. Be particularly observant of crop rotation, recropping interval guidelines and spray cleanup procedures. Minimum recropping interval guidelines are described in part as follows on the Glean label: "Crop rotation plans are determined by soil pH, recropping interval, and for crops other than wheat or barley, results of a field bioassay. Minimum recropping interval is from time of last application to anticipated date of planting next crop." Read the complete label for Glean weed killer dry flowable and "Glean-Field Bioassay" for information to provide maximum flexibility for crop rotation.

Do not apply this tank mixture when conditions favor spray drift.

A tank mixture of BROMINAL® and metribuzin will control chickweed, filaree, henbit and dogfennel in addition to all weeds listed on the BROMINAL® label in non-irrigated winter wheat.

Crop injury can occur from this combination if used in wheat that is too small and/or stressed from winter injury. The higher rates of BROMINAL® in combination with the higher rates of Metribuzin should be used only in an emergency weed situation and if some minor crop injury is acceptable. When using the higher rates the wheat should be fully tillered and have a well established secondary root system.

CONVERSION TABLE FOR DETERMINING RATE OF PRODUCT PER ACRE TO USE

Metribuzin	Sencor or Lexone 50 WP	Sencor 4 or Lexone 4L	Sencor 75 Spray or Lexone 75 DP
⅛ lb ai/A	¼ lb/A	¼ pt/A	⅙ lb/A
3/16 lb ai/A	⅜ lb/A	⅜ pt/A	⅕ lb/A
¼ lb ai/A	½ lb/A	½ pt/A	⅓ lb/A

To prepare the spray mixture fill the tank ⅔ to ¾ full of water. With agitation running, add the proper amount of Metribuzin as determined from the table above, add the required amount of BROMINAL® and finally add the remaining water.

A recognized authority in your area should be consulted concerning proper rates and timings for your conditions.

Do not use Metribuzin on irrigated wheat or crop injury may result. Do not exceed the rates on this label. Avoid applications when crop is under stress from winter kill, frost damage, disease or drought.

WHEAT—BARLEY—OATS—RYE When to Spray, Amounts to Use

AREA & CROP STAGE	CHEMICAL RATE/ACRE	SIZE OF WEEDS	AMOUNT OF WATER NEEDED PER ACRE
CEREAL AREAS Fall through early spring applications before or after tillering (not underseeded) with legumes.	1-2 pints BROMINAL	Apply to emerged susceptible weeds in the 2-4 leaf stage or when rosettes are less than 1.5 inches across.	Use 10 gal. to 20 gal. with ground or 5-10 gal. air application. Use higher volumes for larger weeds or when growing conditions are poor. Use 2 or more gal. with midwest air application.
Application after tillering and before the forming of joints in the stem.	1-1.5 pints BROMINAL + ¼ lb/A of 2,4-D or MCPA*	All weeds including wild buckwheat, cow cockle, fiddleneck, tarweed; apply with 4-leaf stage but before 6 inches high.	Use 10-20 gal. with ground or 4-10 gal. with air application. Use higher volume on larger weeds.
Apply when spring wheat is from 3-leaf up to joint stage.	1-1.5 pints BROMINAL + 1/16-⅛ pints of Banvel	Bachelor's button, knotweed, mustard, nightshade, Russian thistle, wild buckwheat, lambsquarters, fanweed, dogfennel, cow cockle; apply when weeds are small (early seeding).	Use 10 or more gal of water per acre for ground application.
Fall through early spring application when wheat is from 3-leaf up to joint stage.	1-1.5 pints BROMINAL + ⅛ pint of Banvel	Kochia, fiddleneck, minerslettuce, knotweed, wild buckwheat, prickly lettuce, jacobsladder, alkanet, burchevil, corn cockle, corn gromwell, cow cockle, dogfennel, fanweed, falseflax, flixweed, henbit, lambsquarters, catchweed; apply when weeds are small (early seeding).	Use 10 or more gal. of water per acre for ground application.
Winter type wheat 4-leaf to tillered stage and barley 2 to 7 leaf stage throughout the United States. Spring type wheat Idaho, Oregon, Washington and Era variety in Minnesota and North Dakota 5 to 6 leaf stage.	1.5-2 pints BROMINAL + Avenge at 2.5 pints for 1-10 oats per sq. ft. 3 pints for 11-25 per sq. ft. and 4 pints for more than 25 per sq. ft.	Wild oats, fiddleneck, wild buckwheat and other broadleaf weeds; apply to wild oats in the 3 to 5 leaf stage to emerge susceptible broadleaf weeds in the 2-4 leaf stage or when rosettes are less than 1.5 inches across.	Use 5 to 20 gal. with ground or 3 to 10 gal. with air application. If spray vol. exceeds 10 gal. adopt 0.6 fl. oz. of a surfactant such as Tergitol NPX for each gal. in excess.

AREA & CROP STAGE	CHEMICAL RATE/ACRE	SIZE OF WEEDS	AMOUNT OF WATER NEEDED PER ACRE
Winter type wheat and springtype wheat and barley	1-1.5 pints BROMINAL + 2 pints of Carbyne applied at one or two times when wild oats are in the 2 leaf stage or 4 pints at the 2.5 to 3.5 leaf stage.	Broadleaf weeds 2-4 leaves. Wild oats canarygrass in 2 to 3.5 leaf stage. Apply the BROMINAL tank mix only once if the split Carbyne application is used.	Use 5-10 gal. of water and sufficient pressure (minimum 45 lb) to break spray into small droplets with ground equipment and 3-5 gallon for air application.
Winter type wheat and spring type wheat and barley before tillering or after tillering	1.5-2 pints BROMINAL + Use 2 pints for 3-4 leaf broadleaf weeds + 2.66 pints Hoelon	Grasses, wild oats, foxtails, annual ryegrass, barnyardgrass, 1-3 leaves. Fiddleneck and other broadleaf weeds 2-4 leaves.	Use at least 10 gallons with minimum 30 psi for ground or at least 5 gal. with air application.
Winter and spring type wheat before or after tillering	1.5 pints BROMINAL + 2.66-3.33 pints Hoelon	Grasses 3-4 leaves Broadleaves, 3-4 leaves.	Use at least 10 gal. with minimum 30 psi for ground or at least 5 gal. with air application.
Washington, Oregon, Idaho. Fall through early spring application before or after tillering.	1.5-2 pints BROMINAL	Gromwell, tarweed, fiddleneck, mustard and other broadleaf weeds; apply before the 4 leaf stage, before rosettes are 1.5 inches across.	Use 10 to 20 gal. with ground or 5-10 gal. with application. Use higher rates and volume for larger weeds or where weed growth is slow during dry or cold weather.
Application after tillering and before the forming of joints in the stem.	1.5-2 pints BROMINAL + ¼ lb/A of 2,4-D or MCPA*	Gromwell, tarweed, fiddleneck, mustard and other broadleaf weeds; apply before weeds are 6 inches high	
Use where average annual rainfall exceeds 16 inches winter barley. Apply a single application in the fall after wheat has emerged but before soil freezes or in the spring as soon as soil thaws.	1 pint BROMINAL + ½ lb/A Karmex	Henbit, gromwell, fiddleneck, other weeds. Apply before weeds are 2 inches tall or across.	Use a minimum of 25 gal. with ground or 5-10 gal. with air application for thorough coverage of weed foliage.
Washington, Oregon, Idaho, Montana. Non irrigated winter wheat. Apply only in the spring after wheat has started to grow and has a well established secondary root system with a minimum of 3 to 4 tillers but before the boot stage.	1-1.25 pts. BROMINAL + 0.125-0.25 lbs. ai/A metribuzin. See label of selected metribuzin product for amount of formulated product.	Chickweed, filaree, henbit, dogfennel. Apply before weeds are 2 inches tall or have 2 inch rosettes.	Use 10-40 gal. with ground or 5-10 gal. of water with air application.
FLAX spring application	1-2 pints BROMINAL	Weeds in 2-4 leaf stage.	Use 10-20 gal/A with ground or 5-10 gal/acre with air application.
Washington, Oregon, Idaho Montana Winter wheat 2-3 leaf but before boot stage	1-2 pints BROMINAL + 0.17 + 0.33 ounce Glean (75% active ingredient)	Apply to the most susceptible seedling weeds in the 1-4 leaf growth stage but before they are 6" tall. On less susceptible weeds apply before rosettes are 1.5" in diameter.	Use 10-20 gal/A with ground or 4-10 gal. water with air application.

* 2,4-D amine—WEEDAR 64 herbicide, *2,4-D esters—WEEDONE LV4 herbicide, MCPA amine—WEEDAR MCPA broadleaf herbicide concentrate, MCPA Ester—WEEDONE* MCPA Ester broadleaf herbicide.

GENERAL INFORMATION CROP USE

BROMINAL broadleaf herbicide effectively controls broadleaf weeds germinating from seed, applied with these weeds are in the seedling stage. BROMINAL applied early as directed kills seedling broadleaf weeds that emerge in small grains and flax, thus reducing competition with these crops during their early growth.

BROMINAL broadleaf herbicide may be applied in the fall, winter, or spring of the year depending on the area. Spraying must be done while weeds are in the seedling stage. Application of BROMINAL is not registered for use on small grains undersown with legumes.

The most important factors for effective weed control are: 1. spraying weeds in the seedling stage; 2. Uniform spray coverage.

Do not apply when crops form a canopy over weeds. Do not apply when crops are under moisture stress. Do not apply to wheat, barley, or rye that will be exposed to cold desiccating winds in the absence of snow cover. If high crop mortality from winds has occurred, apply BROMINAL only after plants have sent out new roots and begin new top growth. Do not apply to FLAX when the humidity is high or the temperature at the time of treatment or for the next three days after treatment is or is expected to exceed 85°F.

APPLICATION OF BROMINAL THROUGH IRRIGATION SYSTEMS

GENERAL INFORMATION

BROMINAL bromoxynil broadleaf herbicide may be applied through automated (center point or continuous lateral move) or hand move and wheel line irrigation equipment for broadleaf weed controlling wheat and barley. Properly applied BROMINAL will control fiddleneck, wild buckwheat, sunflower, fanweed, kochia, Russian thistle, and other susceptible broadleaf weeds as listed on the BROMINAL label. For most efficient use of BROMINAL apply when weeds are small, before they have exerted yield reducing competition on the small grain.

DIRECTION FOR USE

Remove scale, pesticide residues, and other foreign matter from the chemical tank and entire injector system. Flush with clean water. Add BROMINAL broadleaf herbicide or BROMINAL and water dilution to mix tank. For greater accuracy in calibration and distribution, inject a larger quantity of a more dilute solution per unit time.

For best performance set sprinkler to deliver approximately 0.5 inch of water per acre. Start sprinkler and then uniformly inject BROMINAL into the irrigation line so as to deliver a rate of 1.5 to 2 pints BROMINAL broadleaf herbicide (0.25 to 0.5 lb bromoxynil) per acre. Weeds should not be larger than 3-4 leaf growth stage. Use the 2 pint rate on weeds with 4 leaves. Use the 2 pint rate if pigweed or lambsquarters are part of the weed complex. Refer to BROMINAL label for detailed information on application timing. BROMINAL broadleaf herbicide should be injected with a positive displacement pump into the main line at least 8 feet ahead of a right angle turn to insure adequate mixing.

USER PRECAUTIONS

Application of more than 0.5 inch/acre of irrigation water may result in decreased product performance on certain soils.

Apply BROMINAL only through systems containing anti-siphon, check valves, and interlocking controls between the metering device and the water pump to insure simultaneous shutoff to prevent water source contamination and overflow of the mixing tank.

Do not apply when conditions favor drift, when system connections or fittings leak, or when nozzles do not provide uniform distribution.

Allow sufficient time for pesticides to be flushed through all lines and all nozzles before turning off irrigation water.

GARLIC

For broadleaf weed control in garlic apply 2 to 4 pints of BROMINAL per broadcast acre when the weeds are 1 to 4 inches tall. Use the lower rate on small weeds and the higher rate on 4 inch weeds. The garlic will generally have 1 to 2 true leaves and treatment should be completed before the garlic is 12 inches tall.

For ground application use 40 to 100 gallons of spray solution per broadcast acre. Good coverage of weeds is essential for good control and the higher volumes are preferred for complete coverage. For aerial application use 5 gallons of spray solution per acre.

Do not apply within 112 days of anticipated harvest.

ONIONS

Apply 1 to 1.5 pints (0.25 to 0.38 lbs. ai.) of BROMINAL per acre on a broadcast basis. Water volume is important—use 50-70 gallons of water/acre and apply when onions have 2-5 true leaves. Some crop damage may occur even under ideal conditions. Concentrate spray kills onions. Good coverage is necessary for good weed control. Do not treat if thrips or sand damage have occurred to the onion foliage. BROMINAL can injure onions. Do not add surfactants.

In onion-producing areas, it is essential for greatest safety that the soil and onion foliage be dry. Humidity should be low and dew should be off the plants. The suggested temperature for spraying is 80°F. Applications should not be made 24 hours before or after cultivation, irrigation, or rain; longer than 24 hours if possible. It is best to spray between 11:00 and 4:30 p.m. on a bright sunny day after dew has completely dried from onion foliage and before it has formed.

BROMINAL—DIRECTIONS FOR BROADLEAF WEED CONTROL IN NEWLY SEEDED, NEWLY SPRIGGED, ESTABLISHED LAWN AND ORNAMENTAL TURF AREAS.

Newly Seeded and Sprigged Grasses

BROMINAL may be used on seedling grasses such as Merion, Park, Delta or common Kentucky bluegrass, Pennlawn, Chewing's, Illahee or Alta fescues, orchardgrass, Penncross, Highland, Seaside or Astoria, bentgrass, Excalibre *(Paspalum vagintum)* and perennial ryegrasses. BROMINAL may be used on newly sprigged Bermudagrass, centipedegrass or St. Augustinegrass. For weeds controlled see "Susceptible Weeds" list.

PRODUCT/AREA	RATE/ACRE	CROP	APPLICATION TIMING WEED	SPRAY VOLUME
BROMINAL Throughout the United States	1.5 pt. (0.375 lb ai*)	Seedling grasses or newly sprigged Bermuda, centipede or St. Augustine grasses.	Seedling broadleaf weeds 3-4 leaf stage.	20 or more gallons of water per acre for large areas. 2-4 gallons per 1000 sq. ft. for small areas.
BROMINAL	2.0 pt. (0.5 lb ai)		Seedling broadleaf weeds larger than 3-4 leaf stage but less than 6 inches.	

* ai = active ingredient

Established Lawn, Ornamental Turf Areas, Sod and Grass Seed Production Areas

BROMINAL may be used on established Bermudagrass, St. Augustinegrass, centipedegrass, Kentucky bluegrass, fescue, orchardgrass, bentgrass, and ryegrass. For weeds controlled see "Susceptible Weeds" list and additional weeds listed below in direction tables. Do not use on bentgrass putting greens or tees. Seedling weeds must be actively growing. Good coverage is essential for effective control of weeds.

Temporary turfgrass discoloration may occur particularly if excessive overlapping of spray solution occurs, or if BROMINAL is applied at temperatures of 85 degrees F or above. This will usually disappear following two mowings leaving no permanent damage.

PRODUCT/AREA	RATE/ACRE	CROP	APPLICATION TIMING WEED	SPRAY VOLUME
BROMINAL Throughout the United States	1.5-2pt/A (0.375-0.5 lb ai*/A) Use higher rate on weeds larger than 3-4 stage but less than 6 inches tall or wide	Established turf lawn, sod and grass seed production.	Seedling 3-4 leaf to 6 inches tall or wide.	20 or more gallons of water per acre for large area 2-4 gallons per 1000 sq. ft. for small areas.
BROMINAL	4 pts/A (1 lb. ai/A)		Seedling spurweed (*Soliva* sp.) yellow woodsprrel (*Oxalis* sp.) black medic, bur clover, spiny fruited crowfoot, hop clover.	
BROMINAL	8 pts (2 lb ai)		Seedling prostrate spurge (*Euphorbia supina*)	

* ai = active ingredient

Tank Mixture for Broadleaf Weed Control in Established Lawns and Turf Areas
For weeds controlled see "Susceptible Weeds" list and weeds as listed below with tank mixtures.

PRODUCT/AREA	RATE/ACRE	APPLICATION TIMING CROP	APPLICATION TIMING WEED	SPRAY VOLUME
BROMINAL + WEEDONE DPC (2,4-D + 2,4-DP)	2-4 pts/A (0.5-1 lb ai) + 3-4 pts. Use low rates on annual weeds and high rates on perennial and biennial broadleaf weeds.	Established turf except bent, St. Augustine, centipede and carpetgrasses.	Actively growing dandelion, plantain, ground ivy, red clover, bur clover chickweed, prostrate spurge, knotweed and oxalis.	25 or more gallons of water per acre for large area; 2-4 gallons of water per 1000 sq. ft. for small area.
BROMINAL + WEEDONE DPC (2,4-D + 2,4-DP)	2 pt (0.5 lb ai) + 1.5-2 pts.	Established bent, St. Augustine and centipedegrasses. Retreat in 2 to 3 weeks. These grasses may be slightly injured. Do not use on bentgrass putting greens or tees.		
BROMINAL + WEEDAR 64 (2,4-D amine) or WEEDONE LV4 (2,4-D ester) User ester for best control of wild garlic and onion.	2 pt (0.5 lb ai) + 2-4 pts. (0.95-1.9 lb ai) or 2-4 pts. (0.95-1.9 lb ai) Use the 2 pt. rate on annual broadleaf weeds and 3-4 pts. on perennial and biennial broadleaf weeds.	Established turf except bent, St. Augustine, and centipedegrasses	Seedling and established weeds such as ragweed, dandelion, plantain, wild garlic or onions. Treat wild garlic or onion for three successive seasons (example: spring, fall, spring)	20 or more gallons of water per acre for large area 2-4 gallons of water per 1000 sq. ft. for small area.
BROMINAL + dicamba (Banvel)	2-4 pts (0.5-1 lb ai) + 0.2-0.5 pts. (0.1-0.25 lb ai) Use high rates for control of yellow woodsorrel and prostrate spurge. Repeat treatment may be made. Do not exceed 1 lb ai Banvel acre per growing season.	Established turf except bentgrass. Rates of Banvel in excess of 0.5 lb ai per treated area may cause noticeable stunting or discoloration of sensitive grass species such as carpetgrass, buffalograss, and St. Augustinegrass.	Seedling actively growing weeds such as chickweed, red clover, and white clover.	20 or more gallons of water per acre for large areas. 2-4 gallons of water per 1000 sq. ft. for small areas.
BROMINAL MCPP	2-4 pts (0.5-1 lb ai) + 1 lb ai Use high rate on established weeds.	Established turf except bentgrass.	Seedling and established weeds such as plantain. dandelion, red clover, ground ivy, prostrate spurge, white clover, chickweed, knotweed and stitchwort.	20 or more gallons of water per acre for larger areas. 2-4 gallons of water per 1000 sq. ft. for small areas.
BROMINAL + MCPP + dicamba (Banvel)	2-4 pts (0.5-1 lb ai) + 0.5-1 lb ai + 0.25-0.5 pts (0.125-0.25 lb ai) Use higher rate of BROMINAL when spurweed and/or prostrate spurge is present.	Established turf except bentgrass	Seedling and established weeds such as white clover, dandelion, and plantain.	20 or more gallons of water per acre for larger areas. 2-4 gallons of water per 1000 sq. ft. for small areas.
BROMINAL + MCPP + 2,4-D	2-4 pts (0.5-1 lb ai) + 0.5-1 lb ai + 0.5-1 lb ai use lower rates on St. Augustine and carpetgrasses.	Established turf except bentgrass	Seedling and established broadleaf weeds such as chickweed, knotweed, red sorrel, dandelion and plantain.	20 or more gallons of water per acre for larger areas. 2-4 gallons of water per 1000 sq. ft. for small areas.

* ai = active ingredient

Tank Mixture for Broadleaf and Grass Weed Control in Established Lawns and Turf Areas

For broadleaf weeds controlled see "Susceptible Weeds" list.

PRODUCT/AREA	RATE/ACRE	APPLICATION TIMING CROP	APPLICATION TIMING WEED	SPRAY VOLUME
BROMINAL + MSMA or DSMA	2 pts (0.5 lb ai*) + 2-3.0 lb ai Refer to manufacturer's label for proper rate per acre of MSMA or DSMA.	Established bluegrass, bermudagrass, tall fescue and zoysia. Do not apply to bentgrass, carpetgrass, centipedegrass or St. Augustinegrass. Grasses may show temporary discoloration.	Seedling broadleaf weeds, crabgrass, goosegrass, nutsedge and sandbur. Apply during warm weather 80-90 degrees F. Begin treatment when weedy grasses have 3-4 leaves. More mature plants require additional applications 7-10 days apart. Three to four applications of MSMA or DSMA may be necessary for nutsedge and sandbur control. Spray when soil moisture is good.	20 or more gallons of water per acre for large areas. 2-4 gallons of water per 1000 sq. ft. for small areas.

* ai = active ingredient

Read carefully the labels of any product used as a tank mixture. Note cautions and limitations on this and products used in tank mixtures to assure maximum weed control and turf and applicator safety.

DO NOT FEED TURF GRASSES TREATED WITH BROMINAL TO LIVESTOCK.

Conversion Chart	Rate Pints Per Acre	Rate Fluid Ounces per 10,000 sq. ft.
	0.125	0.46
	0.25	0.92
	0.50	1.84
	0.75	2.75
	1.0	3.67
	2.0	7.35
	3.0	11.02
	4.0	14.70
	8.0	29.40

Susceptible Weeds When Treated at the Proper Stage:

Annual sowthistle	*(Sonchus oleraceus)*
Black mustard	*(Brassica nigra)*
Blue (purple) mustard	*(Chorispora tenella)*
Bristly starbur	*(Acanthospermum hispidum)*
Bullthistle	*(Cirsium vulgare)*
Bur buttercup	*(Ranunculus testiculatus)*
Cocklebur	*(Xanthium pensylvanicum)*
Corn chamomile	*(Anthemis arvensis)*
Corn gromwell	*(Lithospermum arvense)*
Cow cockle	*(Saponaria vaccaria)*
Dogfennel (Mayweed)	*(Anthemis cotula)*
Dyers woad	*(Isatis tinctoria)*
Fiddleneck	*(Amsinckia intermedia)*
Field pennycress, fanweed	*(Thlaspi arvense)*
Fivehooked bassia	*(Bassia hyssopifolia)*
Fumitory	*(Fumaria officinalis)*
Green smartweed	*(Polygonum scabrum)*
Groundsel	*(Senecio vulgaris)*
Henbit	*(Lamium amplexicaula)*
Jimsonweed	*(Datura stramonium)*
Knawel (German moss)	*(Scleranthus annuus)*
Kochia	*(Kochia scoparia)*
Lambsquarters	*(Chenopodium album)*
London rocket	*(Sisymbrium irio)*
Mare's-tail	*(Conyza canadensis)*
Minerslettuce	*(Montia perfoliata)*
Morningglory (annual)	*(Ipomoea hederacea)*
Nettleleaf goosefoot	*(Chenopodium murale)*
Nightshade	*(Solanum nigrum, villosum)*
Nodding thistle	*(Carduus nutans)*
Parthenium ragweed	*(Parthenium hysterophorus)*
Pennsylvania smartweed	*(Polygonum pensylvanicum)*
Pepperweed	*(Lepidium campestre)*
Pigweed	*(Amaranthus retroflexus)*
Popcorn flower	*(Allocarya figurata)*
Prickly (Chinese) lettuce	*(Lactuca scariola)*
Puncture vine	*(Tribulus terrestris)*
Ragweed	*(Ambrosia artemisiifolia)*
Redmaids rockpurslane	*(Calandrinia caulescens)*
Russian thistle	*(Salsola kali)*
Shepherdspurse	*(Capsella bursa-pastoris)*
Silver leaf nightshade	*(Solanum elaeagnifolium)*
Smallseeded falseflax	*(Camelina microcarpa)*
Southern giant curl mustard	*(Brassica japonica)*
Sunflower	*(Helianthus annuus)*
Tansymustard	*(Descurainia pinnata)*
Tartary buckwheat	*(Fagopyrum tataricum)*
Tarweed	*(Hemizonia spp.)*
Treacle mustard	*(Erysimum repandum)*
Tumble (Jim Hill) mustard	*(Sisymbrium altissimum)*
Velvetleaf	*(Abutilon theophrasti)*
Wild buckwheat	*(Polygonum convolvulus)*
Wild mustard	*(Brassica kaber)*
Wild radish	*(Raphanus raphanistrum)*
Winter vetch	*(Vicia sativa)*
Yellow rocket	*(Barbarea vulgaris)*
Yellow starthistle	*(Centaurea solstitialis)*

This treatment will not control established perennial broadleaf weeds such as Canada thistle, bindweed, or horsetail rush.

DIRECTIONS FOR INDUSTRIAL USE

Russian Thistle and Annual Broadleaf weeds on roadsides, vacant lots, railroads, and industrial sites.

For broadleaf treatment of small weeds.

Apply 2 quarts of BROMINAL broadleaf herbicide with either 2 quarts of surfactant (such as X-77) or 2 gallons of diesel oil in 150 gallons of water for each acre sprayed. Thorough coverage is important. Make sure spray boom is high enough to contact tops of all plants. Use adequate spray pressure to contact all leaf surfaces.

For spot treatments of scattered infestations or large weeds along fencerows or buildings:

Mix 1 quart BROMINAL with either 1 quart surfactant (such as X-77 or 1 gallon diesel oil per 100 gallons of spray solution. Cover all the weed foliage thoroughly. Use at least 200 gallons of spray solution for each acre sprayed.

When using oil, add chemical to oil then add this mixture to water agitating vigorously.

GENERAL INFORMATION INDUSTRIAL USE
BROMINAL broadleaf herbicide effectively controls Russian thistle and certain other summer annual broadleaf weeds such as saltbru*** *(Atriplex)*, mare's tail, puncture vine, prostrate spurge, and Bassia.

PRECAUTIONARY STATEMENTS

WARNING

HAZARDS TO HUMANS AND DOMESTIC ANIMALS
Harmful if swallowed. Avoid contact with skin, eyes, clothing. Avoid breathing spray mist.

STATEMENT OF PRACTICAL TREATMENT
In case of eye contact, rinse thoroughly with water and get medical attention. In case of skin contact, wash immediately with soap and water and remove contaminated clothing. If swallowed, drink promptly large quantities of milk, egg white, gelatin solution or if these are not available, drink large quantities of water. Avoid alcohol. Get medical attention.

GENERAL WORKER PROTECTION STATEMENT
Do not apply this product in such a manner as to directly or through drift expose workers or other persons. The area being treated must be vacated by unprotected persons.

Do not enter treated areas without protective clothing until sprays have dried. Because certain states may require more restrictive reentry intervals for various crops treated with this product, consult your State Department of Agriculture for further information. Written or oral warnings must be given to workers who are expected to be in a treated area or in an area about to be treated with this product. Advise workers to stay out of fields during application and until sprays have dried. Regular long-sleeved work clothing should be worn when working in treated fields. See Precautionary Statement and Statement of Practical Treatment for information on accidental exposures. When oral warnings are given, warnings shall be given in a language customarily understood by workers. Oral warnings must be given if there is reason to believe that written warnings cannot be understood by workers. Written warnings must include the following information: appropriate signal word (WARNING), area treated with BROMINAL Broadleaf Herbicide, date of application, appropriate protective clothing, and reentry interval if applicable.

ENVIRONMENTAL HAZARDS
This product is toxic to fish. Do not apply directly to water. Do apply when weather conditions favor drift from treated areas.

Do not contaminate water used for irrigation or domestic purposes by cleaning of equipment or disposal of waste.

Do not use this product for purposes other than those recommended on this label. Open dumping is prohibited.

Clean sprayers thoroughly before using to apply BROMINAL broadleaf herbicide. To clean sprayers after using BROMINAL broadleaf herbicide, rinse all parts of sprayer with water several times.

Avoid spray drift to adjacent crops as injury may result. Coarse sprays are less likely to drift.

STORAGE AND DISPOSAL

Do not store near fertilizers, seed, insecticide or fungicides. Do not contaminate water, food, or feed by storage or disposal.

PESTICIDE DISPOSAL
Wastes resulting from the use of this product may be disposed of on site or at an approved waste disposal facility.

CONTAINER DISPOSAL
Triple rinse (or equivalent). Then offer for recycling or reconditioning or puncture and dispose of in a sanitary landfill, or by incineration, if allowed by state and local authorities, by burning. If burned, stay out of smoke.

LIMITED WARRANTY AND DISCLAIMER

The manufacturer warrants (a) that this product conforms to the chemical description on the label; (b) that this product is reasonably fit for the purposes set forth in the directions for use when it is used in accordance with such directions; and (c) that the directions, warnings, and other statements on this label are based upon responsible experts' evaluation of reasonable tests of effectiveness, of toxicity to laboratory animals and to plants, and of residues on food crops, and upon reports of field experience. Tests have not been made on all varieties or in all states or under all conditions. THE MANUFACTURER NEITHER MAKES NOR INTENDS, NOR DOES IT AUTHORIZE ANY AGENT OR REPRESENTATIVE TO MAKE, ANY OTHER WARRANTIES, EXPRESS OR IMPLIED, AND IT EXPRESSLY EXCLUDES AND DISCLAIMS ALL IMPLIED WARRANTIES OF MERCHANTABILITY OR FITNESS FOR A PARTICULAR PURPOSE.

THIS WARRANTY DOES NOT EXTEND TO, AND THE BUYER SHALL BE SOLELY RESPONSIBLE FOR, ANY AND ALL LOSS OR DAMAGE WHICH RESULTS FROM THE USE OF THIS PRODUCT IN ANY MANNER WHICH IS INCONSISTENT WITH THE LABEL DIRECTIONS, WARNINGS OR CAUTIONS.

BUYER'S EXCLUSIVE REMEDY AND MANUFACTURER'S OR SELLER'S EXCLUSIVE LIABILITY FOR ANY AND ALL CLAIMS, LOSSES, DAMAGES, OR INJURIES RESULTING FROM THE USE OR HANDLING OF THIS PRODUCT, WHETHER OR NOT BASED IN CONTRACT, NEGLIGENCE, STRICT LIABILITY IN TORT OR OTHERWISE, SHALL BE LIMITED, AT THE MANUFACTURER'S OPTION, TO REPLACEMENT OF, OR THE REPAYMENT OF THE PURPOSE PRICE FOR, THE QUANTITY OF PRODUCT WITH RESPECT TO WHICH DAMAGES ARE CLAIMED. IN NO EVENT SHALL MANUFACTURER OR SELLER BE LIABLE FOR SPECIAL, INDIRECT OR CONSEQUENTIAL DAMAGES RESULTING FROM THE USE OR HANDLING OF THIS PRODUCT.

THIS SPECIMEN LABEL IS INTENDED FOR USE ONLY AS A GUIDE IN PROVIDING GENERAL INFORMATION REGARDING THE DIRECTIONS, WARNINGS AND CAUTIONS ASSOCIATED WITH THE USE OF THIS PRODUCT. AS WITH ANY AGRICULTURAL CHEMICAL, ALWAYS FOLLOW THE LABEL INSTRUCTIONS ON THE PACKAGE BEFORE USING.

BROOT® 15GX

GRANULAR TRIMETHACARB INSECTICIDE

ACTIVE INGREDIENTS:	% BY WEIGHT
Trimethacarb [3,4,5-trimethylphenyl methylcarbamate, mixture with 2,3,5-trimethylphenyl methylcarbamate (approximately 4:1)]	15%
INERT INGREDIENTS:	85%
TOTAL	100%

E.P.A. Reg. No. 264-381 E.P.A. Est. No. 10352-GA-01

Keep Out of Reach of Children

CAUTION

IN CASE OF EMERGENCY, TELEPHONE (24 HOURS A DAY) IN U.S.A. 1-800-UCC-HELP.

FOR GENERAL PRODUCT INFORMATION CALL 1-800-334-9745.

DIRECTIONS FOR USE

IT IS A VIOLATION OF FEDERAL LAW TO USE THIS PRODUCT IN A MANNER INCONSISTENT WITH ITS LABELING.

Do not use BROOT® 15GX except as directed on this label or on other Union Carbide Agricultural Products Company supplemental labeling for this product.

GENERAL WORKER PROTECTION STATEMENTS

Do not apply this product in such a manner as to directly or through drift expose workers or other persons. The area being treated must be vacated by unprotected persons.

Do not enter treated areas without protective clothing until dusts have settled. Because certain states may require more restrictive reentry intervals for various crops treated with this product, consult your State Department of Agriculture for further information. Written or oral warnings must be given to workers who are expected to be in a treated area or in an area about to be treated with this product. Advise workers to stay out of fields during application and until dusts have settled. Regular long-sleeved work clothing should be worn when working in treated fields. See Precautionary Statement, Statement of Practical Treatment and Note to Physician for information on accidental exposures. When oral warnings are given, warnings shall be given in a language customarily understood by workers. Oral warnings must be given if there is reason to believe that written warnings cannot be understood by workers. Written warnings must include the following information: appropriate signal word (CAUTION), area treated with BROOT 15GX, date of application, appropriate protective clothing, and reentry interval if applicable.

STORAGE AND DISPOSAL

STORAGE

Do not store in or around the home. Store unused product in a cool, dry, locked area. Do not allow prolonged storage in areas where temperatures frequently exceed 115°F (46°C). Do not contaminate food, water or feed by storage or disposal. NEVER TRANSFER THIS PRODUCT TO ANOTHER CONTAINER FOR STORAGE.

PESTICIDE DISPOSAL

Wastes resulting from the use of this product may be disposed of on site or at an approved waste disposal facility.

CONTAINER DISPOSAL

NEVER REUSE EMPTY CONTAINERS. Completely empty bag into application equipment. Then dispose of bag in a sanitary landfill or by incineration, or if allowed by state and local authorities, by open burning. When burning bag, stay out of smoke.

RECOMMENDED RATES AND APPLICATION DIRECTIONS

CROP	PEST	DOSAGE BROOT®15GX	SPECIFIC DIRECTIONS
Corn (Field and Pop)	Corn rootworm	8 oz. per 1000 feet of row	APPLICATION AT PLANTING: Apply the granules in a 6-8 inch band over the row behind the planter shoe and in front of the press wheel and incorporate into the top 1 inch of soil using suitable equipment or devices. Conservation Tillage Systems: Under NO-TILL conditions where heavy crop residues occur, apply the granules in a 6-8 inch band over the row behind the planter shoe and ahead of the press wheel. DO NOT USE A SEED FURROW TREATMENT. APPLICATION AT CULTIVATION: Apply the granules to the base of plants on both sides of the row or over the top of plants just ahead of cultivation shovels and cover with soil. Do not apply at cultivation if BROOT® 15GX was used at planting. DO NOT HARVEST WITHIN 90 DAYS OF APPLICATION.

Row Spacing Chart

Recommended Rates of Broot® 15GX According to Row Spacing

Row Spacing (inches)	Amount of BROOT® 15GX Per Acre Row Treatment at Planting Time
40	6.7 lbs.
38	7.0 lbs.
36	7.3 lbs.
34	7.7 lbs.
32	8.2 lbs.
30	8.7 lbs.

Following is a table of suggested initial calibration settings for BROOT® 15GX trimethacarb granular insecticide applied at 8 ounces per 1000 row feet. Confirm accuracy of gauge settings under your operating conditions by calibrating each unit before application.

Applicator Type	SPEED (MPH) 4	5	6	7	8
	Gauge Setting[1]				
John Deere Max Emerge Even Numbered Housing[2]	12	15	17	20	22
John Deere Max Emerge Odd Numbered Housing[3]	10	12	14	15	17
John Deere 71[4]	1/293	2/2	2/5	2/8	2/11
Gandy[5]	20	22	24	26	28
International Harvester[6]					
400 and 500 Series Planters Plate Type and Cyclo Air	1/9.0	2/2.5	2/4.5	2/8.5	3/0
800 Series Planters Early Riser and Cyclo Air	1/9.0	2/1.0	2/4.0	2/7.5	3/0.5
Noble	10	11	13	14	16

1 Based upon one hopper outlet per row.
2 Models prior to 1979
3 Models after 1979
4 Gauge Setting—range/notch
5 Rubber Rotor
6 Refers to gate/dial settings

PRECAUTIONARY STATEMENTS

CAUTION

HAZARDS TO HUMANS AND DOMESTIC ANIMALS

Harmful if swallowed. Avoid contact with skin and eyes. Avoid breathing dust.

Wear long-sleeved clothing while using. Remove contaminated clothing daily; wash before reuse. Wash exposed skin thoroughly with soap and water after use and before eating or smoking.

SIGNS AND SYMPTOMS OF OVEREXPOSURE

Salivation, Watery eyes, Pinpoint eye pupils, Blurred vision, Muscle tremors, Difficult breathing, Excessive sweating, Abdominal cramps, Nausea, Vomiting, Diarrhea, Weakness, Headache. In severe cases, convulsions, unconsciousness and respiratory failure may occur.

ANTIDOTE STATEMENT

ATROPINE SULFATE IS HIGHLY EFFECTIVE AS AN ANTIDOTE. See NOTE TO PHYSICIAN below.

STATEMENT OF PRACTICAL TREATMENT

GENERAL

Contact a physician immediately in all cases of suspected poisoning. If breathing stops, start artificial respiration, establish an airway and provide oxygen. Make certain to remove all sources of continuing contamination. Remove clothing and wash skin and hair immediately with large amounts of water. Transport the patient to a physician or hospital immediately and SHOW A COPY OF THIS LABEL TO THE PHYSICIAN.

IF SWALLOWED: Drink 1 or 2 glasses of water and induce vomiting by touching the back of the throat with finger. Never give anything by mouth to an unconscious or convulsing person. Call a physician and follow general advice listed above.

IF IN EYES: Flush eyes with plenty of water. Get medical attention if irritation persists.

IF ON SKIN: Wash thoroughly with soap and water.

IF INHALED: Call a physician and follow General advice listed above.

NOTE TO PHYSICIAN

BROOT® trimethacarb is a methyl carbamate insecticide, which is a cholinesterase inhibitor. Overexposure to this substance may cause toxic signs and symptoms due to stimulation of the cholinergic nervous system. These effects of overexposure are spontaneously and rapidly reversible.

Specific treatment consists of parenteral atropine sulfate. Caution should be maintained to prevent overatropinization. Mild cases may be given 1 to 2 mg intramuscularly every 10 minutes until full atropinization has been achieved and repeated thereafter whenever symptoms reappear. Severe cases should be given 2 to 4 mg intravenously every 10 minutes until fully atropinized, then intramuscularly every 30 to 60 minutes as needed to maintain the effect for at least 12 hours. Dosages for children should be appropriately reduced. Complete recovery from overexposure is to be expected within 24 hours.

Narcotics and other sedatives should not be used. Further, drugs like 2-PAM (pyridine-2-aldoxime methiodide) are NOT recommended unless organophosphate intoxication is also suggested.

To aid in the confirmation of a diagnosis, urine samples should be obtained within 24 hours of exposure and immediately frozen. Analyses will be arranged by Union Carbide Agricultural Products Company, Inc.

Consultation on therapy can be obtained at all hours by calling the Union Carbide emergency number: (304) 744-3487.

ENVIRONMENTAL HAZARDS

This product is toxic to fish, birds and other wildlife. Treated granules exposed on soil surface may be hazardous to birds and other wildlife. Do not apply directly to water. Do not apply where runoff is likely to occur. Do not contaminate water by cleaning of equipment or disposal of wastes. Apply this product only as specified on this label.

LIMITED WARRANTY AND DISCLAIMER

The manufacturer warrants (a) that this product conforms to the chemical description on the label; (b) that this product is reasonably fit for the purposes set forth in the directions for use when it is used in accordance with such directions; and (c) that the directions, warnings and other statements on this label are based upon responsible experts evaluation of reasonable tests of effectiveness, of toxicity to laboratory animals and to plants, and of residues on food crops, and upon reports of field experiences. Tests have not been made on all varieties or in all states or under all conditions. THE MANUFACTURER NEITHER MAKES NOR INTENDS, NOR DOES IT AUTHORIZE ANY AGENT OR REPRESENTATIVE TO MAKE, ANY OTHER WARRANTIES, EXPRESS OR IMPLIED, AND IT EXPRESSLY EXCLUDED AND DISCLAIMS ALL IMPLIED WARRANTIES OR MERCHANTABILITY OR FITNESS FOR A PARTICULAR PURPOSE.

THIS WARRANTY DOES NOT EXTEND TO, AND THE BUYER SHALL BE SOLELY RESPONSIBLE FOR, ANY AND ALL LOSS OR DAMAGE WHICH RESULTS FROM THE USE OF THIS PRODUCT IN ANY MANNER WHICH IS INCONSISTENT WITH THE LABEL DIRECTIONS, WARNINGS OR CAUTIONS.

BUYER'S EXCLUSIVE REMEDY AND MANUFACTURER'S OR SELLER'S EXCLUSIVE LIABILITY FOR ANY AND ALL CLAIMS, LOSSES, DAMAGES, OR INJURIES RESULTING FROM THE USE OR HANDLING OF THIS PRODUCT, WHETHER OR NOT BASED IN CONTRACT, NEGLIGENCE, STRICT LIABILITY IN TORT OR OTHERWISE, SHALL BE LIMITED, AT THE MANUFACTURER'S OPTION, TO REPLACEMENT OF, OR THE REPAYMENT OF THE PURCHASE PRICE FOR, THE QUANTITY OF PRODUCT WITH RESPECT TO WHICH DAMAGES ARE CLAIMED. IN NO EVENT SHALL MANUFACTURER OR SELLER BE LIABLE FOR SPECIAL, INDIRECT OR CONSEQUENTIAL DAMAGES RESULTING FROM THE USE OR HANDLING OF THIS PRODUCT.

THIS SPECIMEN LABEL IS INTENDED FOR USE ONLY AS A GUIDE IN PROVIDING GENERAL INFORMATION REGARDING THE DIRECTIONS, WARNINGS AND CAUTIONS ASSOCIATED WITH THE USE OF THIS PRODUCT. AS WITH ANY AGRICULTURAL CHEMICAL, ALWAYS FOLLOW THE LABEL INSTRUCTIONS ON THE PACKAGE BEFORE USING.

BUTYRAC®175

Controls Cocklebur and Other Broadleaf Weeds in Soybeans and Peanuts

ACTIVE INGREDIENT:

*4-(2,4-Dichlorophenoxy) Butyric acid, dimethylamine salt	23.0%
INERT INGREDIENTS	77.0%
TOTAL	100.0%

*4-(2,4-Dichlorophenoxy) Butyric acid equivalent 19.5% by weight.

EPA Reg. No. 264-164 EPA Est. No. 264-MO-01

Keep Out of Reach of Children

CAUTION

IN CASE OF EMERGENCY TELEPHONE (24 HOURS A DAY) IN U.S.A. 1-800-UCC-HELP.

FOR GENERAL PRODUCT INFORMATION CALL 1-800-334-9745.

DIRECTIONS

IT IS A VIOLATION OF FEDERAL LAW TO USE THIS PRODUCT IN A MANNER INCONSISTENT WITH ITS LABELING.

SOYBEANS

OVER-ALL APPLICATIONS: For control of cocklebur and morningglory. Use 0.8 to 1.0 pint of BUTYRAC 175 per acre (equivalent to 0.175 to 0.22 pounds active 2,4-DB acid per acre). It is very important to calibrate sprayers before application to be sure you are applying the right amount.

Apply BUTYRAC 175 from 7 to 10 days before bloom up to mid-bloom, when soybeans are about knee-high and growing actively. Soybean foliage should be dark green indicating nodulation and nitrification are already underway. Do not apply over-all sprays earlier than this stage of growth.

MIDWEST IMPORTANT: Cocklebur must form a protective canopy over the bean plants before spraying or injury to the soybeans may result. Application after mid-bloom may affect flowering and reduce soybean yields.

Apply as a broadcast spray. Use a boom sprayer with flat fan-type nozzles. Space nozzles so that one nozzle is over each row and over each middle. Adjust the height of the boom above the crop to give complete coverage of all weeds. Use low pressure (20-40 psi). Do not use boom jet nozzles.

For a time, soybeans will show some effects from the chemical in the form of temporary twisting or curling of the stems and leaves. However, these effects disappear in about 3 to 5 days. Cocklebur will die out in from 3 days to 3 weeks.

DIRECTED BAND APPLICATIONS

For control of cocklebur and morningglory, apply the equivalent of 0.8 to 1.0 pint of BUTYRAC 175 per broadcast acre (equivalent to 0.175 to 0.22 pounds active 2,4-DB acid per acre). Use sufficient water for good coverage; at least 10 gallons per acre.

Treat when beans are at least 8 inches tall. Adjust nozzle height to allow spray to contact no more than the bottom one-third of soybean plants. Do not allow spray or spray drift to contact growing terminals of beans, as excessive crop injury will result.

For best results treat when the cocklebur and morningglory are no more than 3 inches tall. All leaves and growing terminals of weeds should be contacted by the spray.

For control of weeds one inch or less in height and/or suppression of weeds greater than about one inch in height such as velvetleaf, jimsonweed, common ragweed and lambsquarters, apply 1.6 to 1.8 pints of BUTYRAC 175 per acre (equivalent to 0.35 to 0.4 lb. active 2,4-DB acid per acre). Do not use more than two applications per season. Stunting of the soybeans may be observed when two applications of the higher rates are used for weed control.

To insure proper placement of spray on lower one-third of soybean plants, apply with sprayer nozzles mounted on skid shoes, oiling shoes, or cultivators with gauge wheels. Do not mount on booms with drop nozzles, or on cultivators without gauge wheels. Use fan type nozzle tips #8001 or larger, or their equivalent. Nozzle pressure should be less than 40 psi to reduce spray drift.

DIRECTED BAND APPLICATIONS OF MIXTURES OF BUTYRAC 175 AND LOROX 50W*

For control of cocklebur, morningglory, sesbania, velvetleaf, lambsquarters, prickly sida (teaweed), jimsonweed, ragweed and crabgrass (2 leaf), apply 1.0 pint of BUTYRAC 175 plus 1 lb. Lorox per broadcast acre (equivalent to 0.22 pounds active 2,4-DB and ½ lb. linuron active per acre). Use a proportional amount of the two chemicals on the band (for example, on a 12 inch band in 36 inch rows use ⅓ pt. of BUTYRAC 175 and ⅓ lb. of Lorox per acre). Use an 8002 T-Jet (or equivalent) or larger with a minimum of 25 gals. per broadcast acre and spray pressure of 20 to 25 psi, to avoid spray drift. The addition of 1 pint of surfactant such as DuPont WK per 2.5 gallons of spray mix. Continuous agitation in the spray tank is required to keep the material in suspension. Agitate by mechanical or hydraulic means; if by-pass or return line is used, it should terminate at the bottom of the tank to minimize foaming. Openings in nozzle screens should be equal to or larger than 50 mesh.

Apply directed spray when beans are at least 8 inches tall and when weeds do not exceed 2 inches in height. Adjust nozzle height to spray the lower one-third of the soybean plant. Apply with sprayer nozzles mounted on skid shoes, oiling shoes, or on cultivators with gauge wheels. Do not mount on booms with drop nozzles or on cultivators without gauge wheels. If a new flush of weeds occurs after initial treatment, make a second application but do not apply within 60 days of harvest or make more than 2 applications per growing season. Do not allow spray or spray drift to contact growing terminals of beans, as excessive crop injury will result. Do not feed soybean forage or hay to livestock from fields treated with this mixture. Harvested soybeans may be used for feed or oil purposes.

NOTE: Do not use on sands, loamy sands or gravelly soils or on exposed subsoils. Do not use on soils containing less than ½% organic matter as crop injury may result.

* Lorox 50W—Registered trademark of E.I. DuPont de Nemours & Co (Inc.)

IMPORTANT

Drought stressed beans should not be sprayed.

Do not harvest beans within 60 days after application.

Use either the over-all or the directed application. Do not use both applications on the same field in the same season.

Do not use this product on soybeans that show symptoms of Phytophthora root rot disease.

Do not add any wetting agents or detergents to the spray solution. Be sure sprayer is clean before using BUTYRAC 175 in it.

PEANUTS

BUTRYAC 175 has given control or suppression of the following weeds depending on the size of the weed.

Common Name	Botanical Name	Suppression only if larger than:
Pigweed or carelessweed	*Amaranthus sp.*	3 inches
morningglory	*Ipomea sp.*	3 foot
cocklebur	*Xanthium sp.*	3 foot
velvetleaf	*Abutilon theophrasti*	1 inch
Virginia copperleaf	*Acalypha virginica*	1 inch
teaweed	*Sida spinosa*	1 inch
common ragweed	*Ambrosia artemisiifolia*	1 inch
jimsonweed	*Datura stramonium*	1½ inch
devils claw	*Proboscidea louisianica*	2 inches
goatweed	*Croton lindheimeri*	1 inch
lambsquarters	*Chenopodium album*	1 inch
silverleaf nightshade	*Sodanum elaeagnifolium*	4 inches

For best results apply BUTYRAC 175 when the weeds are one to two inches in size. For ground application use 10 to 30 gallons of spray solution per acre. For air application use 5 gallons of spray solution per acre.

Georgia, Virginia, North Carolina, Alabama, Florida, Arkansas
Apply 0.9 pints to 1.1 pints (equivalent to 0.2 to 0.25 pounds active 2,4-DB acid) of BUTYRAC 175 per acre on peanuts 2 to 12 weeks after planting. For late germinating cocklebur and morningglory, apply a second application about 3 weeks after the first application.

In the Virginia-Carolina area, do not apply later than 45 days before harvest.

Oklahoma, Texas, New Mexico
Apply 0.9 to 1.8 pints (0.2 to 0.4 pounds active 2,4-DB acid) of BUTYRAC 175 per acre. Use the low rate on morningglory and cocklebur up to 12 inches in size. Use the higher rate on other weeds. BUTYRAC 175 may be applied twice during the season. For best suppression of Sida, space the treatments 14 days apart. The second application should not be made later than 100 days after planting or 30 days before harvest.

AMOUNT OF BUTYRAC 175 TO USE

Acreage to Spray	Rate Desired @0.2lb/A Active	@0.4lb/A Active
1	0.9 pt. or 14.6 oz	1.8 pt. or 29.0 oz.
8.75	1 gal.	2 gal.

Do not feed treated peanut vines and peanut hay to livestock. Do not apply BUTYRAC 175 if peanuts are under stress from drought as injury to peanuts may occur.

CAUTION
Harmful if swallowed. Avoid contact with skin, eyes or clothing.

Do not contaminate water used for irrigation or domestic purposes.

Avoid spray drift to susceptible plants such as cotton, beans, tomatoes and ornamentals as this product may cause injury. Coarse sprays are less likely to drift. Do not use same spray equipment for other purposes.

Under very high temperature, vapors from this product may injure susceptible plants in the immediate vicinity.

Do not use in a greenhouse.

Do not use this product for purposes other than those recommended on this label.

A small amount of 2,4-D, MCPA, silvex or 2,4,5-T mixed with BUTYRAC 175 can cause serious injury to soybeans. A sprayer previously used to apply these chemicals must be thoroughly cleaned before applying BUTYRAC 175.

This material is not recommended as a preemergence treatment. Do not use on sweet clover.

STORAGE AND DISPOSAL

Do not store near fertilizers, seeds, insecticides, or fungicides. Keep from freezing. If allowed to freeze, thaw and mix well before using.

PESTICIDE DISPOSAL
Do not contaminate water, food, or feed by storage or disposal. Wastes resulting from the use of this product may be disposed of on site or at an approved waste disposal facility.

CONTAINER DISPOSAL
Triple rinse or equivalent. Then offer for recycling, or reconditioning or puncture and dispose of in a sanitary landfill or incineration or if allowed by state and local authorities, by burning. If burned, stay out of smoke.

LIMITED WARRANTY AND DISCLAIMER

The manufacturer warrants that this product conforms to the chemical description on the label; that this product is reasonably fit for the purposes set forth in the directions for use when it is used in accordance with such directions; and that the directions, warnings and other statements on this label are based upon responsible experts' evaluation of reasonable tests of effectiveness, of toxicity to laboratory animals and to plants, and of residues on food crops, and upon reports of field experience. Tests have not been made on all varieties or in all states or under all conditions. THE MANUFACTURER NEITHER MAKES NOR INTENDS, NOR DOES IT AUTHORIZE ANY AGENT OR REPRESENTATIVE TO MAKE, ANY OTHER WARRANTIES, EXPRESS OR IMPLIED, AND IT EXPRESSLY EXCLUDES AND DISCLAIMS ALL IMPLIED WARRANTIES OF MERCHANTABILITY OR FITNESS FOR A PARTICULAR PURPOSE.

THIS WARRANTY DOES NOT EXTEND TO, AND THE BUYER SHALL BE SOLELY RESPONSIBLE FOR, ANY AND ALL LOSS OR DAMAGE WHICH RESULTS FROM THE USE OF THIS PRODUCT IN ANY MANNER WHICH IS INCONSISTENT WITH THE LABEL DIRECTIONS, WARNINGS OR CAUTIONS.

BUYER'S EXCLUSIVE REMEDY AND MANUFACTURER'S OR SELLER'S EXCLUSIVE LIABILITY FOR ANY AND ALL CLAIMS, LOSSES, DAMAGES, OR INJURIES RESULTING FROM THE USE OR HANDLING OF THIS PRODUCT, WHETHER OR NOT BASED IN CONTRACT, NEGLIGENCE, STRICT LIABILITY IN TORT OR OTHERWISE, SHALL BE LIMITED, AT THE MANUFACTURER'S OPTION TO REPLACEMENT OF, OR THE REPAYMENT OF THE PURCHASE PRICE FOR, THE QUANTITY OF PRODUCT WITH RESPECT TO WHICH DAMAGES ARE CLAIMED. IN NO EVENT SHALL MANUFACTURER OR SELLER BE LIABLE FOR SPECIAL, INDIRECT OR CONSEQUENTIAL DAMAGES RESULTING FROM THE USE OR HANDLING OF THIS PRODUCT.

THIS SPECIMEN LABEL IS INTENDED FOR USE ONLY AS A GUIDE IN PROVIDING GENERAL INFORMATION REGARDING THE DIRECTIONS, WARNINGS AND CAUTIONS ASSOCIATED WITH THE USE OF THIS PRODUCT. AS WITH ANY AGRICULTURAL CHEMICAL, ALWAYS FOLLOW THE LABEL INSTRUCTIONS ON THE PACKAGE BEFORE USING.

BUTYRAC®200

BROADLEAF HERBICIDE

Controls Broadleaf Weeds in Soybeans, Peanuts, Alfalfa, Trefoil and Clovers.

ACTIVE INGREDIENT:
4-(2,4-Dichlorophenoxy) butyric acid,
dimethylamine salt* 25.9%
INERT INGREDIENTS: 74.1%

TOTAL 100.0%
*4-(2,4-Dichlorophenoxy) butyric acid equivalent 22.0% by weight or 2 per gallon.

E.P.A. Reg. No. 264-105ZA E.P.A. Est. No. 264-MO-01

Keep Out of Reach of Children
DANGER

IN CASE OF EMERGENCY, TELEPHONE (24 HOURS A DAY) IN U.S.A. 1-800-UCC-HELP.

FOR GENERAL PRODUCT INFORMATION CALL 1-800-334-9745.

DIRECTIONS FOR USE

IT IS A VIOLATION OF FEDERAL LAW TO USE THIS PRODUCT IN A MANNER INCONSISTENT WITH ITS LABELING.

WEED CONTROL IN SOYBEANS

OVER-THE-TOP APPLICATION: For control of cockleburs and annual morningglories in Southern states.

MIDWEST IMPORTANT: Over-the-top application in the Midwest can reduce yields. When BUTYRAC 200 is applied alone at the 0.7 to 0.9 pint rates, use it only as a directed band application in IA, IL, IN, KS, KY (except the KY Purchase area), MO (except the MO Bootheel), MI, MN, NE, ND, OH, SD and WI. See directions for band application.

Apply 0.7 to 0.9 pint (11.2 to 14.4 fluid ounces) of BUTYRAC® 200 broadleaf herbicide per acre as a broadcast spray. It is very important to calibrate sprayers before application to be sure the right amount is applied.

Apply BUTYRAC 200 broadleaf herbicide from 7 to 10 days before bloom up to mid-bloom, when soybeans are about knee-high and growing actively. Soybean foliage should be dark green indicating that nodulation and nitrification are already underway. Do not apply over-the-top sprays at these rates earlier than this stage of growth. Application after mid-bloom may affect flowering and reduce soybean yields.

Use a boom sprayer with hollow cone type nozzles. Adjust the height of the boom so nozzles are above the crop to give complete coverage of all weeds. Use low pressure (20-40 psi). Do not use flood type nozzles operating with extremely low pressures as coverage will not be adequate.

For a time, soybeans will show some temporary effects from the chemical in the form of twisted or curled stems and leaves. However, these effects disappear in about 3 to 5 days and yield may be reduced but is not usually affected. Cocklebur will die in 3 days to 3 weeks.

DIRECTED BAND APPLICATION

To control cockleburs and annual morningglories when soybeans are 8 or more inches tall apply the equivalent of 0.7 to 0.9 pint (11.2 to 14.4 fluid ounces) of BUTYRAC 200 broadleaf herbicide per broadcast acre as a directed band treatment. For best results, spray when the cockleburs and morningglories are no more than 3 inches tall.

To control other broadleaf weeds one inch or less in height and/or suppress weeds including velvetleaf, jimsonweed, common ragweed and lambsquarters taller than about one inch, apply as a directed treatment of 1.4 to 1.6 pints (22.4 to 25.6 fluid ounces) of BUTYRAC 200 broadleaf herbicide per broadcast acre. Use enough water for good coverage, at least 10 gallons per acre. Apply no more than two treatments per season. Soybean plants may be stunted when two applications of the higher rate are used for weed control.

Apply with sprayer nozzles mounted on skid shoes, oiling shoes or cultivators with gauge wheels to insure proper placement of spray on only the lower one-third of the soybean plants. *Do not mount nozzles on booms with drop pipes or on cultivators without gauge wheels.* Use flat fan type nozzles, 8001 or larger or the equivalent. *Do not allow spray to contact growing terminals of bean plants,* as excessive crop injury will result. Nozzle pressure should be less than 40 psi to control spray drift.

IMPORTANT: *Beans stressed by drought or other influences should not be sprayed. Do not use this product on soybeans that show symptoms of disease such as Phytophthora root rot.*

Be sure sprayer is clean before applying BUTYRAC 200 broadleaf herbicide.

Do not harvest beans within 60 days after spray application.

TANK MIXTURES OF BUTYRAC® 200 BROADLEAF HERBICIDE AND OTHER HERBICIDES

Applying tank mixtures of low rates of BUTYRAC 200 broadleaf herbicide with other soybean herbicides as directed below can reduce competition from early weed populations and can improve weed control or control mid- to late-season weed populations, thus minimizing the likelihood of yield reduction in soybean fields with heavy broadleaf weed infestations. However, treating soybeans under stress (as from drought or disease such as Phytophthora root rot) or in any other manner not directed on this label can cause crop injury and yield reduction. The advantage of treating weeds with an appropriate tank mixture should be balanced against any potential for loss. Base your decision on an evaluation of each factor. Follow the local recommendations of your state, Cooperative Extension Service, or other agricultural weed control authority. The only BUTYRAC 200 tank mix registered for use in California is with Lorox L.

BROADCAST OVERALL TREATMENT

BUTYRAC® 200 BROADLEAF HERBICIDE PLUS ALANAP®-L FOR SOYBEANS

For postemergence control of mid-season cockleburs, giant ragweed, wild mustard, volunteer sunflower.

Ground Application

Spray 3 to 4 ounces of BUTYRAC 200 broadleaf herbicide with 2 to 3 quarts of Alanap-L in 10 to 20 gallons of water per acre (broadcast rate) on soybeans that are about 18 inches tall (7 to 10 days before bloom through mid-bloom).

The spray boom should be positioned 18-24" above the tops of the soybean plants or weeds and be equipped with hollow cone nozzles. Maintain a high spray pressure (40-50 psi) during application. This insures a fine spray mist for better overall coverage.

Aerial Application

Apply 3 to 4 ounces of BUTYRAC 200 broadleaf herbicide with 2 to 3 quarts of Alanap-L in a minimum of 5 gallons of water per acre to soybeans that are about 18 inches tall (7 to 10 days before bloom through mid-bloom). Nozzle size and orientation should provide maximum droplet breakup without causing excessive drift. Do not apply by air near susceptible crops such as cotton, tomatoes and tobacco.

For best results use 3 ounces of BUTYRAC 200 broadleaf herbicide with 2 quarts of Alanap-L on weeds that are 10-12 inches tall such as common cocklebur, giant ragweed, wild mustard or sunflower. However, if these weeds are taller than 12 inches use 4 ounces of BUTYRAC 200 with 3 quarts of Alanap-L. If applied according to directions BUTYRAC 200 herbicide with Alanap-L will control or suppress growth of above weeds, reducing their competition with the soybean crop.

NOTE: Observe all precautions and limitations on the labels for BUTYRAC 200 and Alanap-L.

BUTYRAC® 200 BROADLEAF HERBICIDE PLUS BASAGRAN® HERBICIDE FOR SOYBEANS

For improved postemergence control of annual morningglories and cocklebur.

To the label rate of Basagran herbicide selected according to the other weed species present, add 2 fluid ounces per acre of BUTYRAC 200 broadleaf herbicide for annual morningglories and cockleburs.

Broadcast the tank mix of BUTYRAC 200 broadleaf herbicide and Basagran herbicide when weeds are actively growing and before annual morningglory vines are a maximum of 6 inches long in the South (AL, AR, FL, GA, LA, MS, NC, OK, SC, TN, TX, VA) or a maximum of 10 inches long in all other states. This is normally up to the time when soybeans have two expanded trifoliate leaves. If application is delayed until weeds are larger than maximum size stated on the Basagran herbicide label, control will be inadequate.

Use a minimum of 20 gallons of total spray mixture per acre with ground equipment and 40 psi pressure with flat fan or hollow cone nozzles spaced 20 inches apart.

Fill the spray tank half way with water, add the recommended amount of BUTYRAC 200 and Basagran herbicides while the agitator is running, then finish filling.

The tank mix is effective partly through contact action, so weeds must be thoroughly covered with spray. Canopies of large crop and weed leaves shelter smaller weeds and prevent adequate spray coverage. Soybeans tolerate the tank mix of BUTYRAC 200 broadleaf herbicide and Basagran; however, under certain conditions soybean leaves may burn, crinkle and bronze.

NOTE: Observe all precautions and limitations on the labels for BUTYRAC 200 and Basagran.

BUTYRAC® 200 BROADLEAF HERBICIDE PLUS BLAZER® 2S HERBICIDE FOR SOYBEANS

For controlling morningglories, cockleburs, common ragweed, jimsonweed and pigweed when weed size exceeds that specified on the Blazer 2S label.

Mix 2 fluid ounces of BUTYRAC® 200 with 1.5 to 2 pints of Blazer 2S for each acre being treated.

For optimum control, apply the tank mixture of BUTYRAC 200 herbicide and Blazer 2S to weeds growing actively, before they are 12 inches tall or long. Applications at later stages will result in partial control or suppression. The use of this tank mix will cause soybean foliage injury and may reduce yields. Control with this mixture may decrease with increasing weed size, weed density or heavy soybean canopy, due to poor spray coverage. Treatment at late stages will only partially control or merely suppress large weeds.

For best results, the tank mixture of BUTYRAC 200 broadleaf herbicide and Blazer 2S herbicide should be applied with ground equipment. Always use at least 20 gallons of water per acre to assure maximum coverage of weeds partially covered by soybean foliage. In extremely weedy fields, increasing the spray volume above 20 gallons of water will usually assist in activity. When using this tank mixture, nozzle pressures must not exceed 40 psi to avoid excessive spray drift. *The addition of surfactant is not recommended since hormonal crop response will be intensified.*

NOTE: Observe all precautions and limitations on the labels for BUTYRAC 200 and Blazer 2S herbicides.

BUTYRAC® 200 BROADLEAF HERBICIDE PLUS DYANAP FOR SOYBEANS

The addition of $1^3/_4$ fluid ounces of BUTYRAC 200 broadleaf herbicide to the label rate of Dyanap selected improves the control of larger and hard-to-kill weeds.

Follow all directions and observe all cautions and limitations on the labels for using BUTYRAC 200 and Dyanap herbicides.

DIRECTED TREATMENT

BUTYRAC® 200 BROADLEAF HERBICIDE PLUS LOROX® L ON SOYBEANS

For directed postemergence control of morningglory, sicklepod, teaweed, cocklebur, velvetleaf, sesbania and other problem weeds.

Apply BUTYRAC 200 broadleaf herbicide plus Lorox L as a directed spray to cover weed foliage with minimum contact of the soybean plant. Do not spray higher than 3" on the soybean stem or crop injury may result. *Do not spray over top of soybean plants.*

For soybeans at least 8" high and when weeds do not exceed 4" in height, apply 13 fluid ounces of BUTYRAC 200 broadleaf herbicide with 1 pint of Lorox L per acre. A second application may be made if needed, but do not make more than two applications per season.

For directed broadcast application, use a single flood-type spray nozzle (K series or equivalent) per middle mounted on an oiling shoe or gauge wheel. *For directed band treatment,* use two nozzles per row mounted on oiling shoes or gauge wheels, one on each side of row. To avoid spray drift, which may cause crop injury, do not exceed nozzle pressure of 25 psi or use nozzle tips smaller than 8002 TeeJet (or equivalent) and do not spray when it is windy. For best results, use a preemergence treatment or cultivation to control early weed growth and to increase the differential between height of soybeans and weeds.

NOTE: Do not use on sands, loamy sands or gravelly soils, on exposed subsoils, or on soils with less than 1/2% organic matter. Do not apply within 60 days of harvest. Do not feed soybean forage or hay to livestock from fields treated postemergence. Harvested soybeans may be used for food, feed, or oil purposes.

Observe all precautions and limitations on labels for BUTYRAC 200 and Lorox L.

BUTYRAC® 200 BROADLEAF HERBICIDE PLUS SENCOR® FOR SOYBEANS

For controlling weeds that escape preplant or preemergence herbicides or for new flushes of weeds germinating after soybeans have emerged.

Apply as a *directed spray* in a 6- to 8-inch band on each side of the row after soybeans are 8 inches tall and before broadleaf weeds are 3 inches tall and grasses and common ragweed are 1 inch tall.

Add 0.9 pint of BUTYRAC 200 broadleaf herbicide to $1/3$-$2/3$ pound Sencor DF per broadcast acre, depending on the weed species. Use the lowest rate for Florida beggarweed, redroot pigweed (carelessweed), carpetweed, common cocklebur, dayflower, Mexicanweed, common purslane, sicklepod (coffeeweed), velvetleaf, and large and smooth crabgrass. Use $1/2$ to $2/3$ pound Sencor DF per broadcast acre for hemp sesbania, prickly sida (spiny sida or teaweed), and $2/3$ pound per acre for burcucumber, common ragweed, morningglory species, wild gourd, and wild poinsettia. At the rate of $2/3$ pounds Sencor DF per broadcast acre horsenettle, Florida purslane (Florida pusley) and spotted spurge are suppressed when Sencor is applied before these weeds are 3 inches tall. This rate will suppress broadleaf signalgrass up to 1 inch tall. Equivalent rates of Sencor 4 or Sencor 50% wettable powder may be used.

To prepare the mixture, fill the sprayer tank half full of water and add the required amount of Sencor. Agitate thoroughly, then add the required amount of BUTYRAC 200 broadleaf herbicide and add more water to fill sprayer tank to desired level.

For maximum contact activity plus residual weed control, a directed broadcast application between the rows is recommended. For best results the spray must cover weed foliage with minimum or no contact of soybean foliage. Add Ortho® X-77 non-ionic spreader or similar surfactant/oil combination to the spray mixture to wet the weed leaf surfaces better.

Do not apply Sencor directly to soybeans or serious crop injury will occur. Soybean leaves contacted by the spray will be killed. Limit spray to only the lower $1/4$ to $1/3$ of soybean plants. Soybean leaves contacted by the spray will be killed.

Do not apply Sencor postemergence to Tracy, Semmes, Altona, Vansoy, or Coker 102 soybeans.

Observe all precautions and limitations on the labels for BUTYRAC 200 and Sencor.

PEANUTS

BUTYRAC 200 broadleaf herbicide will control or suppress the following weeds depending on weed size:

Common name	Botanical name	Maximum size controlled*
pigweed or carelessweed	*Amaranthus* spp.	3 inches
morningglory	*Ipomoea* spp.	3 feet
cocklebur	*Xanthium* spp.	3 feet
velvetleaf	*Abutilon theophrasti*	1 inch
Virginia copperleaf	*Acalypha virginica*	1 inch
teaweed	*Sida spinosa*	1 inch
common ragweed	*Ambrosia artemisiifolia*	1 inch
jimsonweed	*Datura stramonium*	$1^1/_2$ inches
devils claw	*Proboscidea louisianica*	2 inches
goatweed	*Croton lindheimeri*	1 inch
lambsquarters	*Chenopodium album*	1 inch
silverleaf nightshade	*Solanum elaeagnifolium*	4 inches

* Growth of larger weeds will be only suppressed.

For best results apply BUTYRAC 200 broadleaf herbicide to peanuts when the weeds are 1" to 2" in size. For ground application, use 10 to 30 gallons of spray solution per acre. For air application use a minimum of 5 gallons of spray solution per acre.

Alabama, Arkansas, Florida, Louisiana, North Carolina, South Carolina, Tennessee

Apply 0.8 pint to 1.0 pint (12.8 to 16 fluid ounces) of BUTYRAC 200 broadleaf herbicide per acre on peanuts 2 to 12 weeks after planting. For late germinating cocklebures and morningglories, make a second application about 3 weeks after the first one.

In the Virginia-Carolina area, do not apply later than 45 days before harvest.

Oklahoma, Texas, New Mexico

Apply 0.8 to 1.6 pint (12.8 to 25.6 fluid ounces) of BUTYRAC 200 broadleaf herbicide per acre. Use the low rate on morningglories and cockleburs up to 12" in size. Use the higher rate on other weeds as listed above. BUTYRAC 200 broadleaf herbicide may be applied twice during the season. For best suppression of *Sida* make applications 14 days apart. The second application should be made not more than 100 days after planting or 30 days before harvest.

Do not feed treated peanut vines or peanut hay to livestock. Do not apply BUTYRAC 200 broadleaf herbicide if peanut plants are under stress from drought as injury may occur.

WEED CONTROL IN SEEDLING ALFALFA, BIRDSFOOT TREFOIL, ALSIKE, LADINO OR RED CLOVER

Best results will be obtained by applying BUTYRAC 200 broadleaf herbicide to spring or fall-seeded legumes after their emergence *when seedling weeds are small,* no more than 2-3 inches high or rosettes less than 2 inches across and not bolting. Later application to more mature weeds or to new growth from old established weeds will not be satisfactory.

Weed and crop seedlings should be growing strongly for greatest selectivity. Control is difficult under cold or droughty conditions. Rainfall within 7-10 days after treatment can cause injury. For irrigated crops, apply BUTYRAC 200 broadleaf herbicide as soon as possible after irrigation. Delay the next irrigation for at least 7 to 10 days after spraying.

Amounts of BUTYRAC® 200 broadleaf herbicide to use on seedling alfalfa, birdsfoot trefoil, alsike, Ladino or red clover. Use the higher rate in dry, low humidity areas.

Seedling Broadleaf Weeds:

Lambsquarters, pigweed, pennycress, common mustard, ragweed, cocklebur, yellow rocket, kochia, nightshade, Russian thistle, annual morningglory:

Less than 1 inch high .... 1 to 2 quarts per acre
1 to 3 inches high .... 2 to 3 quarts per acre

Smartweed, Curled Dock:

Up to 3 inches high .... 3 quarts per acre

This material does not adequately control wild radish, corn chamomile, wild buckwheat, winter vetch, chickweeds or overwintered pennycress and mustards.

Apply BUTYRAC 200 broadleaf herbicide in 10 or more gallons of water per acre when using ground equipment and in at least 5 gallons per acre with aerial equipment. The volume of spray applied should provide good coverage. SEE ADDITIONAL USE PRECAUTIONS BELOW.

WEED CONTROL IN ESTABLISHED ALFALFA

Certain weeds will emerge in the fall and over-winter in the rosette stage. Best control of these weeds will result from postemergence treatment in late fall or early winter rather than in spring. Follow state recommendations for timing and rates to be used under local conditions.

Overhead sprinkler irrigation should be delayed as long as possible (10 days or more) following application of BUTYRAC 200 broadleaf herbicide to avoid washing the chemical into the root zone. In established alfalfa there may be some twisting of stems and malformation of leaves. This condition is usually outgrown. SEE ADDITIONAL USE PRECAUTIONS BELOW.

USE PRECAUTIONS

1. Do not graze established alfalfa or feed straw or hay from treated crops to livestock within 30 days after application. Do not graze or feed seedling alfalfa, clover or birdsfoot trefoil within 60 days after application.
2. Do not use on sweet clover, peas or other legumes not mentioned, nor on established clovers grown for seed.
3. Do not apply when crop is stressed from lack of moisture.
4. Do not spray when daytime temperatures are expected to exceed 90° F within the next two or three days or are likely to fall below 40° during or shortly after treatment.

PRECAUTIONARY STATEMENTS

DANGER

HAZARD TO HUMANS AND DOMESTIC ANIMALS

Corrosive. Causes eye damage. Harmful if swallowed. May be fatal if absorbed through the skin. Do not get in eyes, on skin or on clothing. Avoid breathing spray mist. Wear goggles or face shield when handling concentrate. In case of contact, immediately flush eyes or skin with plenty of water. Get medical attention if irritation persists. Remove and wash contaminated clothing before reuse.

STATEMENT OF PRACTICAL TREATMENT

If swallowed, call a physician or poison control center. Drink 1 to 2 glasses of water and induce vomiting by touching back of throat with finger. Do not induce vomiting or give anything by mouth to an unconscious person. In case of skin contact, wash skin with plenty of soap and water. In case of eye contact, flush eyes with large amounts of water. Get medical attention.

ENVIRONMENTAL HAZARDS

This product is toxic to fish. Do not apply directly to water. Do not contaminate irrigation ditches or water used for irrigation or domestic purposes. Do not use in or near a greenhouse.

PROTECTION STATEMENTS

Do not apply BUTYRAC 200 broadleaf herbicide directly to or otherwise permit it to come in contact with cotton, okra, grapes, tomatoes, fruit trees, vegetables, flowers or other desirable crop or ornamental plants. Do not permit spray mist to drift onto susceptible plants since very small quantities of the BUTYRAC 200 broadleaf herbicide can cause severe injury during the growing or dormant periods.

Use coarse sprays to minimize drift. Do not apply with nozzles that produce fine spray droplets. *Drift from ground application* may be reduced by: (1) keeping the spray boom as near to the crop as possible in order to obtain complete coverage; (2) by applying 10 gallons or more of spray per acre; (3) by using no more than 20 pounds of pressure at the nozzle tips; and (4) by not spraying when wind exceeds 5 miles per hour.

Drift from aerial application may be reduced by: (1) applying as near to the target as possible to obtain adequate coverage; (2) by applying 5 or more gallons of spray per acre; (3) by using 20 pounds pressure or less at the nozzle tips; (4) by using nozzles which produce a coarse spray pattern; (5) by spraying when there is no possibility for a temperature inversion at time of spraying.

Applications by aircraft, ground rig and hand sprayers should be carried out only where there is no hazard from spray drift.

Spray tank residues of 2,4-D or MCPA mixed with BUTYRAC 200 broadleaf herbicide can cause serious crop or ornamental plant injury. A sprayer previously used to apply these chemicals must be thoroughly cleaned before applying BUTYRAC 200.

Do not treat soybeans with a tank mixture of BUTYRAC 200 broadleaf herbicide and SEVIN® carbaryl insecticide as severe injury may result.

STORAGE AND DISPOSAL

Do not store near food, feedstuffs, fertilizers, seeds, insecticides or fungicides.

PESTICIDE DISPOSAL

Do not contaminate water, food or feed by storage or disposal. Open dumping is prohibited. Pesticide wastes are acutely hazardous. Improper disposal of excess pesticide, spray mixture, or rinsate is a violation of Federal law. If these wastes cannot be disposed of by use according to label instructions, contact your State Hazardous Waste representative at the nearest EPA Regional Office for guidance.

CONTAINER DISPOSAL

Triple rinse (or equivalent). Then offer for recycling or reconditioning, or puncture and dispose of in a sanitary landfill, or by incineration or if allowed by state and local authorities by burning. If burned, stay out of smoke.

LIMITED WARRANTY AND DISCLAIMER

The manufacturer warrants that this product conforms to the chemical description on the label; that this product is reasonably fit for the purposes set forth in the directions for use when it is used in accordance with such direction; and that the directions, warnings and other statements on this label are based upon responsible experts' evaluation of reasonable tests of effectiveness, of toxicity to laboratory animals and to plants, and of residues on food crops, and upon reports of field experience. Tests have not been made on all varieties or in all states or under all conditions. THE MANUFACTURER NEITHER MAKES NOR INTENDS NOR DOES IT AUTHORIZE ANY AGENT OR REPRESENTATIVE TO MAKE ANY OTHER WARRANTIES, EXPRESS OR IMPLIED, AND IT EXPRESSLY EXCLUDES AND DISCLAIMS

ALL IMPLIED WARRANTIES OF MERCHANTABILITY OR FITNESS FOR A PARTICULAR PURPOSE.

THIS WARRANTY DOES NOT EXTEND TO, AND THE BUYER SHALL BE SOLELY RESPONSIBLE FOR, ANY AND ALL LOSS OR DAMAGE WHICH RESULTS FROM THE USE OF THIS PRODUCT IN ANY MANNER WHICH IS INCONSISTENT WITH THE LABEL DIRECTIONS, WARNINGS OR CAUTIONS.

BUYER'S EXCLUSIVE REMEDY AND MANUFACTURER'S OR SELLER'S EXCLUSIVE LIABILITY FOR ANY AND ALL CLAIMS, LOSSES, DAMAGES, OR INJURIES RESULTING FROM THE USE OR HANDLING OF THIS PRODUCT, WHETHER OR NOT SUCH LIABILITY IS BASED IN CONTRACT, NEGLIGENCE, STRICT LIABILITY IN TORT OR OTHERWISE, SHALL BE LIMITED, AT THE MANUFACTURER'S OPTION, TO REPLACEMENT OF, OR THE REPAYMENT OF THE PURCHASE PRICE FOR THE QUANTITY OF PRODUCT WITH RESPECT TO WHICH DAMAGES ARE CLAIMED. IN NO EVENT SHALL MANUFACTURER OR SELLER BE LIABLE FOR SPECIAL, INDIRECT OR CONSEQUENTIAL DAMAGES RESULTING FROM THE USE OR HANDLING OF THIS PRODUCT.

BUTYRAC is a registered trademark of Union Carbide Agricultural Products Company, Inc.

SEVIN is a registered trademark of Union Carbide Corporation

Alanap is a registered trademark of Uniroyal Chemical Division of Uniroyal, Inc.

Basagran is a registered trademark of BASF Wyandotte Corporation

Blazer is a trademark of Rohm and Haas Company

Lorox is a registered trademark of E. I. du Pont de Nemours & Company, Inc.

Sencor is a registered trademark of Farbenfabrik Bayer GmbH, Leverkusen

THIS SPECIMEN LABEL IS INTENDED FOR USE ONLY AS A GUIDE IN PROVIDING GENERAL INFORMATION REGARDING THE DIRECTIONS, WARNINGS AND CAUTIONS ASSOCIATED WITH THE USE OF THIS PRODUCT. AS WITH ANY AGRICULTURAL CHEMICAL, ALWAYS FOLLOW THE LABEL INSTRUCTIONS ON THE PACKAGE BEFORE USING.

BUTYRAC®ESTER

BROADLEAF HERBICIDE

Controls broadleaved weeds in alfalfa and birdsfoot trefoil.

ACTIVE INGREDIENT:

4-(2,4-Dichlorophenoxy) butyric acid, butoxyethanol ester*	34.4%
INERT INGREDIENTS:	65.6%
TOTAL	100.0%

*4-(2,4-Dichlorophenoxy) butyric acid equivalent 24.6% by weight or 2 pounds per gallon.

EPA Reg. No. 264-145 EPA Est. No. 264-MO-01

Keep Out of Reach of Children

WARNING

IN CASE OF EMERGENCY PHONE (24 HOURS A DAY) IN U.S.A. 1-800-UCC-HELP.

FOR GENERAL PRODUCT INFORMATION CALL 1-800-334-9745.

DIRECTIONS FOR USE

IT IS A VIOLATION OF FEDERAL LAW TO USE THIS PRODUCT IN A MANNER INCONSISTENT WITH ITS LABELING.

Apply BUTYRAC ESTER Broadleaf Herbicide in spring or fall, when seedling legumes have 2 to 4 trifoliate leaves and when weeds are less than 3 inches high. Applications made later to more mature weeds will give less satisfactory control. Weeds such as dock and fanweed that form rosettes should be treated when these rosettes are less than 3 inches across.

Apply BUTYRAC ESTER Broadleaf Herbicide in 5 or more gallons of water per acre using air or ground equipment. The total volume of spray applied should be enough for good coverage, but not so much that spray will run off weed leaves.

For irrigated crops, apply BUTYRAC ESTER Broadleaf Herbicide as soon as possible after irrigation. Delay the next irrigation for at least 7 to 10 days after spraying to avoid washing the chemical into the root zone.

Do not use BUTYRAC ESTER Broadleaf Herbicide as a preemergence spray. Do not use on a sweet clover or other legumes not mentioned.

WEED CONTROL IN ALFALFA AND BIRDSFOOT TREFOIL

Weeds Controlled:

lambsquarters	groundsel	bull thistle
pigweed	stinkweed	filaree
kochia	fanweed	shepherdspurse
nightshade	smartweed	dock
Russian thistle	annual ragweed	fiddleneck
prickly (Chinese) lettuce		

Fiddleneck with more than 2-3 leaves will only be suppressed.

Amount BUTYRAC ESTER Broadleaf Herbicide per acre: 2 to 4 pints. Use this higher rate only if weeds have grown to a height of 4 to 6 inches before spraying. Poorer weed control and crop stunting can be expected when using this higher rate on older weeds.

NOTE: Do not graze treated crops nor feed hay from treated crops to livestock within 60 days after application.

BAND TREATMENTS

Use BUTYRAC ESTER Broadleaf Herbicide in proportion to the area actually sprayed. For example, if a 12-inch band is sprayed on 36-inch rows, the amount used per banded acre will be ⅓ of the amount used per broadcast acre. Calibrate sprayers before making band applications. Keep boom at the same height while calibrating and spraying.

MIXING DIRECTIONS

Add one-half the required amount of water to the spray tank, then add this product with agitation, and finally the balance of the water with continued agitation. The material forms an emulsion in water—not a solution. This tends to separate on standing. Provide agitation to prevent such separation.

Do not add any wetting agents or detergents to the spray solution.

NOTE: A sprayer or other container previously used to apply 2,4-D and MCPA must be thoroughly cleaned with alkali and water before using for this product. Even a small amount of one of these chemicals mixed with this product can cause serious injury to legumes. Contact your supplier for special instructions on cleaning out spray equipment.

PRECAUTIONARY STATEMENTS

WARNING

HAZARDS TO HUMANS AND DOMESTIC ANIMALS

Do not get in eyes, on skin or on clothing. May be fatal if absorbed through skin. Harmful if swallowed. Avoid breathing vapors or spray mist. Remove and wash contaminated clothing before reuse. Wash thoroughly before eating or smoking.

STATEMENT OF PRACTICAL TREATMENT

In case of contact wash skin with soap and water; for eyes flush with water for at least 15 minutes and get medical attention. If swallowed, do not induce vomiting. Never give anything by mouth to an unconscious person. Get medical attention.

ENVIRONMENTAL HAZARDS
Do not apply directly to any body of water. Do not contaminate water by cleaning of equipment or disposal of wastes. Do not contaminate irrigation ditches or water used for irrigation or domestic purposes. This product is toxic to fish. Do not use in a greenhouse.

PROTECTION STATEMENTS
Do not apply BUTYRAC ESTER Broadleaf Herbicide directly to, or otherwise permit it to come in contact with cotton, okra, grapes, tomatoes, fruit trees, vegetables, flowers or other desirable crop or ornamental plants which are susceptible to BUTYRAC ESTER Broadleaf Herbicide. Do not permit spray mist to drift onto susceptible plants since very small quantities of the BUTYRAC ESTER Broadleaf Herbicide can cause severe injury during the growing or dormant periods.

Use coarse sprays to minimize drift. Do not apply with nozzles that produce fine spray droplets. Drift from **ground application** may be reduced by: (1) keeping the spray boom as near to the crop as possible in order to obtain complete coverage; (2) by applying 10 gallons or more of spray per acre; (3) by using no more than 20 pounds of pressure at the nozzle tips; and (4) by not spraying when the wind exceeds 5 miles per hour.

Drift from *aerial application* may be reduced by: (1) applying as near to the crop as possible in order to obtain coverage; (2) by applying 5 or more gallons of spray per acre; (3) by using 20 pounds pressure or less at the nozzle tips; (4) by using nozzles which produce a coarse spray pattern; (5) by spraying when there is no possibility for a temperature inversion at time of spraying.

Applications by aircraft, ground rig and hand sprayers should be carried out only when there is no hazard from spray drift. Do not apply near cotton, grapes, tomatoes or near desirable BUTYRAC ESTER Broadleaf Herbicide susceptible crop or ornamental vegetation.

Spray tank residues of 2,4-D and MCPA mixed with BUTYRAC ESTER Broadleaf Herbicide can cause serious crop or ornamental plant injury. A sprayer previously used to apply these chemicals must be thoroughly cleaned with alkali and water before applying BUTYRAC ESTER Broadleaf Herbicide. Do not treat with a tank mixture of BUTYRAC ESTER Broadleaf Herbicide and SEVIN® carbaryl insecticide as severe injury may result.

STORAGE AND DISPOSAL

STORAGE
Do not store near food, feedstuffs, fertilizers, seeds, insecticides, or fungicides.

PESTICIDE DISPOSAL
Do not contaminate water, food, or feed by storage or disposal. Open dumping is prohibited. Wastes resulting from the use of this product may be disposed of on site or at an approved waste disposal facility.

CONTAINER DISPOSAL
Triple rinse (or equivalent). Then offer for recycling, or reconditioning, or puncture and dispose of in a sanitary landfill, or incineration, or, if allowed by state and local authorities, by burning. If container is burned, stay out of smoke.

LIMITED WARRANTY AND DISCLAIMER

The manufacturer warrants (a) that this product conforms to the chemical description on the label; (b) that this product is reasonably fit for the purposes set forth in the directions for use when it is used in accordance with such direction; and (c) that the directions, warnings and other statements on this label are based upon responsible experts' evaluation of reasonable tests of effectiveness, of toxicity to laboratory animals and to plants, and of residues, and upon reports of field experience. Tests have not been made on all varieties or in all states or under all conditions. THE MANUFACTURER NEITHER MAKES NOR INTENDS, NOR DOES IT AUTHORIZE ANY AGENT OR REPRESENTATIVE TO MAKE, ANY OTHER WARRANTIES, EXPRESS OR IMPLIED, AND IT EXPRESSLY EXCLUDES AND DISCLAIMS ALL IMPLIED WARRANTIES OF MERCHANTABILITY OR FITNESS FOR A PARTICULAR PURPOSE. THIS WARRANTY DOES NOT EXTEND TO, AND THE BUYER SHALL BE SOLELY RESPONSIBLE FOR, ANY AND ALL LOSS OR DAMAGE WHICH RESULTS FROM THE USE OF THIS PRODUCT IN ANY MANNER WHICH IS INCONSISTENT WITH THE LABEL DIRECTIONS, WARNINGS OR CAUTIONS.

BUYER'S EXCLUSIVE REMEDY AND MANUFACTURER'S OR SELLER'S EXCLUSIVE LIABILITY FOR ANY AND ALL CLAIMS, LOSSES, DAMAGES, OR INJURIES RESULTING FROM THE USE OR HANDLING OF THIS PRODUCT, WHETHER OR NOT SUCH LIABILITY IS BASED IN CONTRACT, NEGLIGENCE, STRICT LIABILITY IN TORT OR OTHERWISE, SHALL BE LIMITED, AT THE MANUFACTURER'S OPTION, TO REPLACEMENT OF, OR THE REPAYMENT OF THE PURCHASE PRICE FOR, THE QUANTITY OF PRODUCT WITH RESPECT TO WHICH DAMAGES ARE CLAIMED. IN NO EVENT SHALL MANUFACTURER OR SELLER BE LIABLE FOR SPECIAL, INDIRECT OR CONSEQUENTIAL DAMAGES RESULTING FROM THE USE OR HANDLING OF THIS PRODUCT.

THIS SPECIMEN LABEL IS INTENDED FOR USE ONLY AS A GUIDE IN PROVIDING GENERAL INFORMATION REGARDING THE DIRECTIONS, WARNINGS AND CAUTIONS ASSOCIATED WITH THE USE OF THIS PRODUCT. AS WITH ANY AGRICULTURAL CHEMICAL, ALWAYS FOLLOW THE LABEL INSTRUCTIONS ON THE PACKAGE BEFORE USING.

CERONE®

PLANT REGULATOR

For Reducing Lodging in Barley and Wheat

For Agricultural or Commercial Use Only. Not For Residential Use

ACTIVE INGREDIENT:
Ethephon
(2-chloroethyl) phosphonic acid* 39.9%
INERT INGREDIENT: 60.1%
TOTAL 100.0%

*Contains four pounds ethephon per gallon

EPA Reg. No. 264-377 EPA EST NO. 264-PA-01

Keep Out of Reach of Children

DANGER

IN CASE OF EMERGENCY—TELEPHONE DAY OR NIGHT (UCC HELP) 800-822-4357

FOR GENERAL PRODUCT INFORMATION CALL 1-800-334-9745.

GENERAL INFORMATION

CERONE® Plant Regulator is intended to be used only as an antilodging agent on intensively managed barley and wheat crops that are likely to lodge.

Vigorously growing small grain cereal crops in fertile soils are likely to lodge. Lodging often decreases recoverable grain yields, and increases harvesting costs. Lodging also contributes to uneven maturity, higher moisture content, loss of grain quality due to sprouting and possible molding.

CERONE® Plant Regulator reduces the plant height and increases the straw strength, thus providing greater resistance to lodging. A standing crop would allow for more efficient harvesting and improved grain recovery. CERONE® Plant Regulator also reduces "necking" in barley and the resultant yield decrease due to ear loss prior to harvest.

IMPORTANT NOTES

CERONE® Plant Regulator should only be used on vigorously growing grain crops that are likely to lodge, and which have maximum yield potential.

Since CERONE® Plant Regulator treatment provides greater resistance to lodging, higher than normal rates of supplemental nitrogen and other inputs for intensive management can be used. Consult your local extension service for fertilizer recommendations.

Management practices aimed at maximizing grain yields can produce conditions conducive to disease development. If rust, septoria, or mildew disease infestations are present, CERONE® treatment may enhance the severity of the disease. Growers planning to use CERONE® Plant Regulator should plant disease-resistant varieties or monitor their crop for disease and use effective fungicides.

Do not apply CERONE® Plant Regulator to crops grown under low fertility conditions, or seeded later than local recommendation.

Do not apply CERONE® Plant Regulator to crops already lodged, severely infected with rust or other diseases, or damaged by insects.

Dwarf wheat varieties with known good standability should not be treated with CERONE® Plant Regulator.

CERONE® Plant Regulator treatment may produce undesirable secondary tillers in some spring barley varieties especially under moisture stress. Treatment is not recommended for Azure barley.

Do not apply CERONE® Plant Regulator to crops under severe moisture stress.

Irrigation should be continued following CERONE® Plant Regulator treatment until grain head filling is completed.

Do not add surfactants or wetting agents to CERONE® Plant Regulator other than those recommended on this label.

Do not apply CERONE® Plant Regulator if rain is expected within 5 hours.

For CERONE® Plant Regulator to be effective, the average daytime temperature following application should reach at least 60°F. Antilodging protection begins 7 days or more after treatment.

DO NOT GRAZE OR FORAGE TREATED CROP OR CUT FOR HAY OR SILAGE.

SPRAY PREPARATION

Add ½ to ¾ of the required amount of water to the spray tank and begin agitation. Add the required amount of CERONE® Plant Regulator, followed by the remaining amount of water. Prepare only as much solution as can be used on the day of mixing. Do not allow spray solution to stand overnight.

Do not spill the concentrated product on spray equipment, or any airplane parts. ANY SPILLS SHOULD BE RINSED IMMEDIATELY WITH PLENTY OF WATER AS CERONE® PLANT REGULATOR IS CORROSIVE. Use of a nurse tank is highly recommended for avoiding possible spills of concentrated formulation on spray equipment or airplane parts.

EQUIPMENT CLEANING

Because of the acidic nature of this product, prolonged exposure to spray deposit will damage acrylic plastics, certain paints, and metals.

Rinse thoroughly with a detergent and water all exposed acrylic plastic-type materials (e.g., aircraft windshields), and painted surfaces **within an hour** after exposure to spray deposits.

At the end of each, rinse thoroughly with a detergent and water the spray system and all the metal parts of the aircraft exposed to the spray deposits.

COMPATIBILITY

CERONE® Plant Regulator can be tank-mixed with most currently registered wheat and barley fungicides. The addition of Triton CS-7 to Mancozeb wettable powder fungicide can be used in MN, ND and SD. Consult your local Union Carbide field representative for specific recommendation.

DIRECTIONS FOR USE

IT IS A VIOLATION OF FEDERAL LAW TO USE THIS PRODUCT IN A MANNER INCONSISTENT WITH ITS LABELING

BARLEY AND WHEAT

Apply 0.5 to 1.0 pint (0.25 to 0.5 lb. active ingredient) CERONE® Plant Regulator in 3 to 25 gallons water per acre by air or ground equipment. With low volume controlled droplet ground equipment, use a lower rate of CERONE® Plant Regulator. Use the higher rate (1.0 pint) for extremely vigorous crops. Apply between early boot to late boot stages, but before the awns or spikes have begun to emerge. The spike or head should never be exposed to spray solution as damage and reduced yield will result from the treatment applied too late.

DO NOT APPLY MORE THAN A TOTAL OF 1.0 PINT (0.5 LB ACTIVE) OF CERONE® PLANT REGULATOR PER ACRE PER YEAR.

See the illustration below for growth stages and proper time for CERONE® Plant Regulator application.

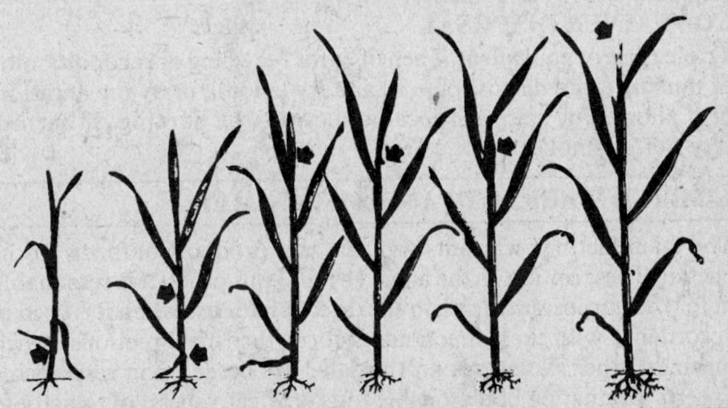

Growth Description	1st node detectable	2nd node detectable	Flag leaf just visible	Flag leaf ligule visible	Boot swollen	First spikelet visible
Feekes—Large Scale	6	7	8	9	10	10.1
			Recommended Cerone Treatment			

PRECAUTIONARY STATEMENTS

DANGER

HAZARDS TO HUMANS AND DOMESTIC ANIMALS

Corrosive. Causes irreversible eye damage. Causes skin irritation. Do not get in eyes, on skin or on clothing. Wear protective clothing and goggles or face shield when handling.

Harmful or fatal if swallowed. Avoid contamination of food. Avoid inhalation of spray mist. Wash thoroughly with soap and water after handling and before eating or smoking. Remove and wash contaminated clothing before reuse.

STATEMENT OF PRACTICAL TREATMENT

IF ON SKIN: Flush immediately with plenty of water for at least 15 minutes.

IF INHALED: Stop spraying and move to uncontaminated area.

IF IN EYES: Flush immediately with plenty of water for at least 15 minutes and call a physician.

IF SWALLOWED: Promptly drink a large quantity of milk, egg whites, gelatin solution or if these are not available, drink large quantities of water. Avoid alcohol. Call a physician immediately.

NOTE TO PHYSICIAN

This material possesses the characteristics of a strong acid, and may cause mucosal damage if swallowed. Do not induce vomiting. Gastric lavage should be undertaken with care if necessary. Appropriate convention treatment for circulatory shock, respiratoty depression and convulsions may be needed.

ENVIRONMENTAL HAZARDS

Do not contaminate water used for irrigation or domestic purposes.

Do not apply directly to water.

Avoid spray drift to nearby crops as this product will cause modifications in plant growth. Plant injury or reduced yields will result. Do not plant another crop within 30 days after treatment.

IMPORTANT

Do not use this product for purposes other than those listed on the label.

Mix only the amount of spray you expect to use each day. Do not allow mixed solution to stand overnight.

STORAGE AND DISPOSAL

Do not contaminate water, food, or feed by storage or disposal.

PESTICIDE DISPOSAL

Pesticide wastes are acutely hazardous. Improper disposal of excess pesticide, spray mixture, or rinsate is a violation of Federal law. If these wastes cannot be disposed of by use according to label instructions, contact your State Pesticide or Environmental Control Agency or the Hazardous Waste representative at the nearest EPA Regional Office for guidance.

CONTAINER DISPOSAL

Triple rinse or equivalent. Then offer for recycling or reconditioning or puncture and dispose of in a sanitary landfill, or by incineration, or if allowed by state and local authorities, by burning. If burned, stay out of smoke.

LIMITED WARRANTY AND DISCLAIMER

The manufacturer warrants (a) that this product conforms to the chemical description on the label; (b) that this product is reasonably fit for the purposes set forth in the direction for use when it is used in accordance with such directions; and (c) that the directions, warnings and other statements on this label are based upon responsible experts' evaluation of reasonable tests of effectiveness, of toxicity to laboratory animals and to plants, and of residues on food crops and upon reports of field experience. Tests have not been made on all varieties or in all states or under all conditions. THE MANUFACTURER NEITHER MAKES NOR INTENDS, NOT DOES IT AUTHORIZE ANY AGENT OR REPRESENTATIVE TO MAKE, ANY OTHER WARRANTIES, EXPRESS OR IMPLIED, AND IT EXPRESSLY EXCLUDES AND DISCLAIMS ALL IMPLIED WARRANTIES OF MERCHANTABILITY OR FITNESS FOR A PARTICULAR PURPOSE.

THIS WARRANTY DOES NOT EXTEND TO, AND THE BUYER SHALL BE SOLELY RESPONSIBLE FOR, ANY AND ALL LOSS OR DAMAGE WHICH RESULTS FROM THE USE OF THIS PRODUCT IN ANY MANNER WHICH IS INCONSISTENT WITH THE LABEL DIRECTIONS, WARNINGS OR CAUTIONS.

BUYER'S EXCLUSIVE REMEDY AND MANUFACTURER'S OR SELLER'S EXCLUSIVE LIABILITY FOR ANY AND ALL CLAIMS, LOSSES, DAMAGES OR INJURIES RESULTING FROM THE USE OR HANDLING OF THIS PRODUCT, WHETHER OR NOT BASED IN CONTRACT, NEGLIGENCE, STRICT LIABILITY IN TORT OR OTHERWISE SHALL BE LIMITED, AT THE MANUFACTURER'S OPTION, TO REPLACEMENT OF, OR THE REPAYMENT OF THE PURCHASE PRICE FOR, THE QUANTITY OF PRODUCT WITH RESPECT TO WHICH DAMAGES ARE CLAIMED. IN NO EVENT SHALL MANUFACTURER OR SELLER BE LIABLE FOR SPECIAL, INDIRECT OR CONSEQUENTIAL DAMAGES RESULTING FROM THE USE OR HANDLING OF THIS PRODUCT.

THIS SPECIMEN LABEL IS INTENDED FOR USE ONLY AS A GUIDE IN PROVIDING GENERAL INFORMATION REGARDING THE DIRECTIONS, WARNINGS AND CAUTIONS ASSOCIATED WITH THE USE OF THIS PRODUCT. AS WITH ANY AGRICULTURAL CHEMICAL, ALWAYS FOLLOW THE LABEL INSTRUCTIONS ON THE PACKAGE BEFORE USING.

ETHREL®

PLANT REGULATOR

For Flue Cured Tobacco

Aids Harvesting and Curing Efficiency

ACTIVE INGREDIENT:
Ethephon [(2-chloroethyl) phosphonic acid]* 21.7%
INERT INGREDIENTS: 78.3%
TOTAL 100.0%

* Contains 2 pounds ethephon per gallon.

EPA Reg. No. 264-292 EPA Est. No. 264-PA-01

Keep Out of Reach of Children

WARNING—AVISO

PRECAUCION AL USARIO: Si usted no lee ingles, no use este producto hasta que la etiqueta le haya sido explicada ampliamente.

IN CASE OF EMERGENCY TELEPHONE (24 HOURS A DAY) IN U.S.A. 1-800-UCC-HELP.

FOR GENERAL PRODUCT INFORMATION CALL 1-800-334-9745.

SUPPLEMENTAL INFORMATION

NUMBER OF PLANTS PER ACRE AT DIFFERENT ROW WIDTHS AND DRILL SPACINGS

Row Width	Spacing in Drill						
	16 In.	18 In.	20 In.	22 In.	24 In.	26 In.	28 In.
36 in	10,890	9,680	8,712	7,920	7,260	6,701	6,222
38 in	10,317	9,170	8,255	7,503	6,878	6,349	5,896
40 in	9,801	8,712	7,841	7,128	6,535	6,032	5,601
42 in	9,334	8,297	7,467	6,788	6,222	5,744	5,333
44 in	8,912	7,920	7,128	6,480	5,940	5,483	5,091
46 in	8,523	7,576	6,818	6,198	5,682	5,245	4,870
48 in	8,167	7,260	6,534	5,940	5,445	5,026	4,667

NUMBER OF LEAVES PER ACRE AT DIFFERENT PLANT POPULATIONS

Plants per Acre	Height of Topping (Number of Leaves)					
	16	18	20	22	24	26
	Leaves per Acre (To nearest 100)					
9800	156,800	176,400	196,000	215,600	235,200	254,800
9300	148,800	167,400	186,000	204,600	223,200	241,800
8900	142,400	160,200	178,000	195,800	213,600	231,400
8700	139,200	156,600	174,000	191,400	208,800	226,200
8500	136,000	153,000	170,000	187,000	204,000	221,000
8300	132,800	149,400	166,000	182,600	199,200	215,800
8200	131,200	147,600	164,000	180,400	196,800	213,200
7900	126,400	142,200	158,000	173,800	189,600	205,400
7800	124,800	140,400	156,000	171,600	187,200	202,800
7600	121,600	136,800	152,000	167,200	182,400	197,600
7500	120,000	135,000	150,000	165,000	180,000	195,000
7300	116,800	131,400	146,000	160,600	175,200	189,800
7100	113,600	127,800	142,000	156,200	170,400	184,600
6800	108,800	122,400	136,000	149,600	163,200	176,800
6500	104,000	117,000	130,000	143,000	156,000	169,000
6200	99,200	111,600	124,000	136,400	148,800	161,200
6000	96,000	108,000	120,000	132,000	144,000	156,000
5900	94,400	106,200	118,000	129,800	141,600	153,400
5700	91,200	102,600	114,000	125,400	136,800	148,200
5500	88,000	99,000	110,000	121,000	132,000	143,000
5400	86,400	97,200	103,000	118,800	129,600	140,400
5200	83,200	93,600	104,000	114,400	124,800	135,200
5000	80,000	90,000	100,000	110,000	120,000	130,000

For further help in determining the time of application, changes in curing procedure, and/or the required number of treated acres needed to fill one barn, contact your County Extension Agent, Area Tobacco Specialist, or your Union Carbide Representative.

See supplemental information for guidance on the amount of tobacco needed to fill a barn.

USE PRECAUTIONS

Treat only the acreage of tobacco which can be harvested and barned in one day. Application made to immature leaves may result in no or slight partial yellowing. Do not allow crop to over ripen in the field after using ETHREL, since this may cause some reduction in yield and quality.

Do not plant another crop within 30 days after treatment.

Do not treat before anticipated major storm which could prevent harvest and resultant crop loss.

GENERAL INFORMATION

A foliar spray of ETHREL plant regulator hastens "yellowing" of mature tobacco and reduces curing time. ETHREL can reduce the number of primings required to harvest a crop of flue cured tobacco. ETHREL increases the capacity of the curing barn by shortening the curing time and allows adjustments in harvest schedules. ETHREL can be used as a directed spray to the bottom or middle portion of the tobacco plant or as an over-the-top spray.

PROPER TIME FOR TREATMENT

Successful results with ETHREL calls for treatment when leaves are mature . . . not overly rank green when sprayed. To easily determine the proper time for treatment and to determine the number of leaves per stalk that are ready for harvest, test spray several plants in more than one location in each field and observe their response. Mature leaves will begin yellowing in 24 to 72 hours. If the test plants fail to yellow the number of leaves desired after 72 hours wait a few days to permit further natural maturing, then make another test spray or "maturity" check.

When you have determined the number of leaves per plant that will color, you can determine the number of acres to treat in order to fill a barn. (See Supplemental Information).

Prepare your test spray by mixing four teaspoons of ETHREL in one quart of water. Spray each test plant with about one ounce of this mixture, covering all leaves with a fine mist. ETHREL will not color immature sprayed leaves.

DIRECTIONS FOR USE

DIRECTED SPRAY APPLICATION

For a directed spray to the bottom or middle leaves of a tobacco plant, apply 2 quarts of ETHREL in 50 to 60 gallons of water per acre. Drop nozzles similar to those used in corn can be used for directed application of ETHREL to tobacco. Nozzles mounted on drop pipes plus slow tractor speeds will keep the boom from bouncing and insure proper placement of the directed spray. Use TG or OC spray tips designed to apply 50-60 gpa at 35-40 psi and at tractor speed of 2-3 mph. Direct ETHREL in a fine spray to the mid or lower leaves to be treated using 2 nozzles per row; one on each side of the row dropped low enough to direct the spray to the leaves to be ripened and harvested. Thorough spray coverage is essential.

OVER-THE-TOP APPLICATION

Mix 2 to 4 quarts of ETHREL with 40 to 60 gallons of water for each acre of tobacco to be treated.

Treat only when leaves remaining on the stalk are mature, after the second or third priming. To ensure remaining leaves are mature, test spray several tobacco plants as described under the section on Proper Time of Treatment. Use the lower rate in a normally matured crop when experience indicates that minimum ripening inducement is required. Use the higher rate when the crop is heavy and has a tendency to be more rank or when temperatures are lower than normal. Test spraying, however, should still indicate whether the tobacco will respond to treatment with ETHREL.

Apply the over-the-top ETHREL spray as a fine mist using three nozzles, (one nozzle tip over the center of the plant, and one on each side) so all leaves are covered thoroughly . . . similar to the application pattern of systemic sucker control agents. Use a spray pressure of 40 to 60 psi.

HARVESTING AND CURING

All mature, sprayed leaves will color 48 to 72 hours after ETHREL application. The yellowing process is weather dependent . . . cool days will delay, while hot sunny weather can speed up the process. Harvest treated tobacco when leaves have reached the desired color intensity.

With a directed spray be sure to harvest all leaves with 20% or more yellowing. When using the over-the-top application harvesting can commence 24 hours after the ETHREL application depending on weather conditions or to avoid overripe tobacco. Tobacco harvested 24 hours after treatment will require more time in the coloring phase of the curing process.

Do not allow tobacco to become overripe before starting harvest. Leaf drop can occur when tobacco is overripe and left on the stalk. If a number of the bottom leaves have yellowed at the time of treatment, they should be harvested prior to spraying with ETHREL.

Curing procedures are as much an art as a science and each cure must be judged on the basis of tobacco condition, interval between treatment and harvest, weather and type of curing facility before prescription temperature and ventilation schedules can be established. To obtain maximum quality, care must be taken to observe and control the curing process closely, especially during the late "coloring" and early "drying" stages of the leaf.

ETHREL treated tobacco will have started the coloring process when harvested, reducing the time required in the coloring phase of curing. Treated tobacco should be dried faster. If tobacco leaves are green or contain some green when harvested, it may be necessary to color them for a few hours. If the leaves are completely yellow, the temperature and ventilation must be adjusted in a manner to dry the tobacco as fast as possible without scalding. Once the leaf is dried ($^3/_4$ dry), the grower should follow normal procedures for curing. Since ETHREL treated leaves cure faster, treated and untreated leaves should not be cured together in the same barn.

PRECAUTIONARY STATEMENTS

WARNING

HAZARDS TO HUMANS AND DOMESTIC ANIMALS

HARMFUL IF SWALLOWED. May cause eye damage. Wear safety goggles when handling. Avoid contact with skin or clothing and do not inhale vapors as this product will irritate skin and mucous membranes.

STATEMENT OF PRACTICAL TREATMENT

In case of splashing into eyes, flood eyes immediately with plenty of water for at least 15 minutes and get medical attention.

In case of skin contact, wash immediately with plenty of soap and water and remove contaminated clothing. Avoid contamination of food. If taken internally call a physician and promptly drink large quantities of milk, egg white, gelatin solution or if these are not available, large quantities of water.

NOTE TO PHYSICIAN

Probable mucosal damage may contraindicate the use of gastric lavage.

ENVIRONMENTAL HAZARDS

Do not contaminate water used for irrigation or domestic purposes.

GENERAL WORKER PROTECTION STATEMENTS

Do not apply this product in such a manner as to directly or through drift expose workers or other persons. The area being treated must be vacated by unprotected persons.

Do not enter treated areas without protective clothing until sprays have dried. Because certain states may require more restrictive reentry intervals for various crops treated with this product, consult your State Department of Agriculture for further information. Written or oral warnings must be given to workers who are expected to be in a treated area or in an area about to be treated with this product. When oral warnings are given, warnings shall be given in a language customarily understood by workers. Oral warnings must be given if there is reason to believe that written warnings cannot be understood by workers. Written warnings must include the following information: appropriate signal word (WARNING), area treated, date of application, appropriate protective clothing, and reentry interval if applicable.

DIRECTIONS FOR USE

IT IS A VIOLATION OF FEDERAL LAW TO USE THIS PRODUCT IN A MANNER INCONSISTENT WITH ITS LABELING.

Avoid spray drift to nearby crops as this product will cause modifications in plant growth. Plant injury or reduced yields will result.

IMPORTANT:
Do not use ETHREL with any additives other than those specifically recommended on this label.

Mix only the amount of spray you expect to use each day. Do not allow mixed solution to stand overnight.

STORAGE AND DISPOSAL

STORAGE
Do not store near fertilizers, seeds, insecticides or fungicides.

PESTICIDE DISPOSAL
Do not contaminate water, food, or feed by storage or disposal. Wastes resulting from the use of this product may be disposed of on site or at an approved waste disposal facility.

CONTAINER DISPOSAL
Do not reuse container. Triple rinse (or equivalent). Then offer for recycling or reconditioning, or puncture or dispose of in a sanitary landfill or incineration, or if allowed by state and local authorities, by burning. If burned, stay out of smoke.

LIMITED WARRANTY AND DISCLAIMER

The manufacturer warrants (a) that this product conforms to the chemical description on the label; (b) that this product is reasonably fit for the purposes set forth in the directions for use when it is used in accordance with such directions; and (c) that the directions, warnings and other statements on this label are based upon responsible experts' evaluation of reasonable tests of effectiveness, of toxicity to laboratory animals and to plants, and of residues on food crops and upon reports of field experience. Tests have not been made on all varieties or in all states or under all conditions. THE MANUFACTURER NEITHER MAKES NOR INTENDS, NOR DOES IT AUTHORIZE ANY AGENT OR REPRESENTATIVE TO MAKE, ANY OTHER WARRANTIES, EXPRESS OR IMPLIED, AND IT EXPRESSLY EXCLUDES AND DISCLAIMS ALL IMPLIED WARRANTIES OF MERCHANTABILITY OR FITNESS FOR A PARTICULAR PURPOSE.

THIS WARRANTY DOES NOT EXTEND TO, AND THE BUYER SHALL BE SOLELY RESPONSIBLE FOR, ANY AND ALL LOSS OR DAMAGE WHICH RESULTS FROM THE USE OF THIS PRODUCT IN ANY MANNER WHICH IS INCONSISTENT WITH THE LABEL DIRECTIONS, WARNINGS OR CAUTIONS.

BUYER'S EXCLUSIVE REMEDY AND MANUFACTURER'S OR SELLER'S EXCLUSIVE LIABILITY FOR ANY AND ALL CLAIMS, LOSSES, DAMAGES, OR INJURIES RESULTING FROM THE USE OR HANDLING OF THIS PRODUCT, WHETHER OR NOT BASED IN CONTRACT, NEGLIGENCE, STRICT LIABILITY IN TORT OR OTHERWISE SHALL BE LIMITED, AT THE MANUFACTURER'S OPTION, TO REPLACEMENT OF, OR THE REPAYMENT OF THE PURCHASE PRICE FOR, THE QUANTITY OF PRODUCT WITH RESPECT TO WHICH DAMAGES ARE CLAIMED. IN NO EVENT SHALL MANUFACTURER OR SELLER BE LIABLE FOR SPECIAL, INDIRECT OR CONSEQUENTIAL DAMAGES RESULTING FROM THE USE OR HANDLING OF THIS PRODUCT.

THIS SPECIMEN LABEL IS INTENDED FOR USE ONLY AS A GUIDE IN PROVIDING GENERAL INFORMATION REGARDING THE DIRECTIONS, WARNINGS AND CAUTIONS ASSOCIATED WITH THE USE OF THIS PRODUCT. AS WITH ANY AGRICULTURAL CHEMICAL, ALWAYS FOLLOW THE LABEL INSTRUCTIONS ON THE PACKAGE BEFORE USING.

ETHREL®
PLANT REGULATOR

For Commercial Or Agricultural Use Only. Not For Residential Use. For use on apples, blackberries, blueberries, cantaloupes, cherries, cranberries, figs, filberts, grapes, lemons, peppers, tangerine, tangerine hybrids, tomato (fresh and processed), walnuts

ACTIVE INGREDIENT:	
Ethephon (2-chloroethyl) phosphonic acid*	21.7%
INERT INGREDIENTS:	78.3%
TOTAL	100.0%

*Contains 2 pounds ethephon per gallon.

E.P.A. Reg. No. 264-267AA E.P.A. Est. No. 264-PA-01

Keep Out of Reach of Children
WARNING AVISO

PRECAUCION AL USARIO: Si usted no lee ingles, no use este producto hasta que la etiqueta le haya sido explicada ampliamente.

IN CASE OF EMERGENCY TELEPHONE COLLECT (24 HOURS A DAY) IN U.S.A. (304) 744-3487

FOR GENERAL PRODUCT INFORMATION CALL 1-800-334-9745.

GENERAL INFORMATION

DIRECTIONS FOR USE
For local recommendations on rates, spray volumes, and spray equipment under varying temperature and rainfall conditions consult your Extension Pomologist, Horticultural Specialist, Union Carbide Representative or Farm Advisor for his experience with this product in your area. Do not exceed the rate of ETHREL Plant Regulator per acre recommended on this label.

APPLES—A foliar spray of ETHREL Plant Regulator promotes fruit maturity and loosens apples, making harvest by hand or machine easier and more efficient. ETHREL applied in combination with a chemical to control pre-harvest drop stimulates early development of red color and ripening without loosening fruit. Applied to young trees, ETHREL will suppress vegetative growth and promote flower bud formation.

BLACKBERRIES IN OREGON AND WASHINGTON—A foliar spray of ETHREL Plant Regulator will concentrate maturity and loosen fruit improving the harvesting efficiency while reducing cane injury from mechanical harvest.

BLUEBERRIES—A foliar spray of ETHREL Plant Regulator concentrates maturity of blueberries for easier and more efficient harvest. A foliar spray of ETHREL will abort black barrenberry *(Aronia melanocarpa)* flowers and/or fruit growing in Maine lowbush blueberry fields reducing the number of undesirable barrenberry fruit harvested with blueberries.

CANTALOUPES IN CALIFORNIA, ARIZONA AND TEXAS—A foliar spray of ETHREL Plant Regulator promotes abscission (slipping) of fruit allowing more efficient and economical harvesting.

CHERRIES—A foliar spray of ETHREL Plant Regulator loosens fruit and promotes early uniform ripening. These effects are important for increasing efficiency, recoverable yields, and for maintaining fruit quality from a once-over mechanical harvest. DO NOT USE ON CHERRIES IN CALIFORNIA. A fall application of ETHREL to sweet cherries in the Pacific Northwest area increases dormant fruit and hardness and delays bloom the following spring.

CRANBERRIES IN MASSACHUSETTS, NEW JERSEY AND WISCONSIN—A foliar spray of ETHREL Plant Regulator can hasten maturity and intensify cranberry color development, allowing earlier harvest of more uniform high quality fruit.

FIGS IN CALIFORNIA—A foliar spray of ETHREL Plant Regulator accelerates early uniform fig ripening, allowing earlier harvest of high quality fruit and improving the efficiency of scheduled harvest.

FILBERTS IN OREGON—A foliar spray of ETHREL Plant Regulator will hasten bush maturity in an increased recoverable yield from an earlier once-over harvest.

GRAPES IN CALIFORNIA—An application of ETHREL Plant Regulator promotes early uniform color development on table grape varieties Cardinal, Emperor, Flame Seedless, Red Malaga, Queen, and Tokay allowing more efficient harvests of quality fruit.

A foliar spray of ETHREL Plant Regulator will hasten maturity of Thompson Seedless grapes resulting in reduced acids, increased sugars and increased raisin quality.

LEMONS POSTHARVEST DEGREENING, FLORIDA—A postharvest spray of ETHREL Plant Regulator will shorten the degreening time for mature lemons.

PEPPERS—A foliar spray of ETHREL Plant Regulator promotes early, uniform ripening and coloring for more efficient packing and handling of fruit.

TANGERINES AND TANGERINE HYBRIDS, FLORIDA—A foliar spray of ETHREL Plant Regulator will loosen and degreen tangerines and tangerine hybrids for reduced fruit plugging and improved color, while reducing or stimulating ethylene chamber degreening.

TOMATOES—FRESH MARKET IN CALIFORNIA—A foliar spray of ETHREL Plant Regulator accelerates and concentrates tomato ripening increasing recoverable fruit yield from a once-over harvest.

TOMATOES—PROCESSED—A foliar spray of ETHREL Plant Regulator accelerates tomato ripening increasing the early yields of marketable ripe fruit. ETHREL application starts the normal ripening process earlier and increases the rate of ripening of mature green fruit concentrating maturity. Concentrated fruit maturity enables high recoverable yields of ripe tomatoes in once-over field harvest. Early concentrated maturity extends the normal harvest season and will help growers to schedule harvesting sequence and handling of fruit more efficiently.

WALNUTS IN CALIFORNIA—A foliar spray of ETHREL Plant Regulator will loosen walnuts for increased efficiency of mechanical harvest, promote earlier harvest, improve hull removal, and increase recoverable yields from a once-over harvest.

PRECAUTIONARY STATEMENTS

HAZARDS TO HUMANS AND DOMESTIC ANIMALS

WARNING: May cause eye damage. Do not get in eyes. Wear goggles or face shield when handling in case of eye contact, immediately flush eyes with plenty of warm water for at least 15 minutes. Call a physician.

HARMFUL IF SWALLOWED: Avoid contamination of food. If taken internally call a physician and promptly drink large quantities of milk, egg white, gelatin solution or if these are not available, large quantities of water.

Avoid contact with skin or clothing and do not inhale vapors as this product may irritate the skin or mucous membranes. In case of skin contact, wash immediately with plenty of soap and water. Remove contaminated clothing.

NOTE TO PHYSICIAN

Probable mucosal damage may contraindicate the use of gastric lavage.

GENERAL WORKER PROTECTION STATEMENTS

Do not apply this product in such a manner as to directly or through drift expose workers or other persons. The area being treated must be vacated by unprotected persons.

Do not enter treated areas without protective clothing until sprays have dried. Because certain states may require more restrictive reentry intervals for various crops treated with this product, consult your State Department of Agriculture for further information. Written or oral warnings must be given to workers who are expected to be in a treated area or in an area about to be treated with this product. When oral warnings are given, warnings shall be given in a language customarily understood by workers. Oral warnings must be given if there is reason to believe that written warnings cannot be understood by workers. Written warnings must include the following information: appropriate signal word (WARNING), area treated, date of application, appropriate protective clothing, and reentry interval if applicable.

ENVIRONMENTAL HAZARDS

Keep out of lakes, ponds or streams. Do not contaminate water used for irrigation or domestic purpose.

DIRECTIONS FOR USE

Avoid spray drift to nearby crops as this product will cause modification in plant growth. Plant injury or reduced yields will result.

Do not plant another crop within 30 days after treatment.

IMPORTANT:

Do not use ETHREL Plant Regulator with any additives other than those specifically recommended on this label.

Do not use this product for purposes other than those listed on this label.

Mix only the amount of spray you expect to use each day. Do not allow mixed solution to stand overnight.

IT IS A VIOLATION OF FEDERAL LAW TO USE THIS PRODUCT IN A MANNER INCONSISTENT WITH ITS LABELING.

STORAGE AND DISPOSAL

Do not contaminate water, food, or feed by storage and disposal.

PESTICIDE DISPOSAL

Wastes resulting from the use of this product may be disposed of on site or at an approved waste disposal facility.

CONTAINER DISPOSAL

Triple rinse (or equivalent). Then offer for recycling or reconditioning or puncture and dispose of in a sanitary landfill or by incineration, or if allowed by state and local authorities, by burning. If burned, stay out of smoke.

LIMITED WARRANTY AND DISCLAIMER

The manufacturer warrants (a) that this product conforms to the chemical description on the label; (b) that this product is reasonably fit for the purposes set forth in the directions for use when it is used in accordance with such directions; and (c) that the directions, warnings and other statements on this label are based upon responsible experts' evaluation of reasonable tests of effectiveness, of toxicity to laboratory animals and to plants and of residues and upon reports of field experience. Tests have not been made on all varieties or in all states or under all conditions. THE MANUFACTURER NEITHER MAKES NOR INTENDS, NOR DOES IT AUTHORIZE ANY AGENT OR REPRESENTATIVE TO MAKE, ANY OTHER WARRANTIES, EXPRESS OR IMPLIED, AND IT EXPRESSLY EXCLUDES AND DISCLAIMS ALL IMPLIED WARRANTIES OF MERCHANTABILITY OR FITNESS FOR A PARTICULAR PURPOSE.

THIS WARRANTY DOES NOT EXTEND TO, AND THE BUYER SHALL BE SOLELY RESPONSIBLE FOR, ANY AND ALL LOSS OR DAMAGE WHICH RESULTS FROM THE USE OF THIS PRODUCT IN ANY MANNER WHICH IS INCONSISTENT WITH THE LABEL DIRECTIONS, WARNINGS OR CAUTIONS.

BUYER'S EXCLUSIVE REMEDY AND MANUFACTURER'S OR SELLER'S EXCLUSIVE LIABILITY FOR ANY AND ALL CLAIMS, LOSSES, DAMAGES, OR INJURIES RESULTING FROM THE USE OR HANDLING OF THIS PRODUCT, WHETHER OR NOT BASED IN CONTRACT, NEGLIGENCE, STRICT LIABILITY IN TORT OR OTHERWISE SHALL BE LIMITED AT THE MANUFACTURER'S OPTION, TO REPLACEMENT OF, OR THE REPAYMENT OF THE PURCHASE PRICE FOR, THE QUANTITY OF PRODUCT WITH RESPECT TO WHICH DAMAGES ARE CLAIMED. IN NO EVENT SHALL MANUFACTURER OR SELLER BE LIABLE FOR SPECIAL, INDIRECT OR CONSEQUENTIAL DAMAGES RESULTING FROM THE USE OR HANDLING OF THIS PRODUCT.

ETHREL® Helps Nature Help You

ETHREL® Plant Regulator

- Promotes earlier coloration and maturity of APPLES, CRANBERRIES, GRAPES, TOMATO, and PEPPER
- Loosens CHERRIES, WALNUTS, and APPLES for earlier, more efficient harvest
- Promotes fruit abscission (slipping) in CANTALOUPES
- Increases dormant fruit bud hardiness and delayed spring bloom of SWEET CHERRIES in the Pacific Northwest
- Reduces degreening time for LEMONS. Loosens and degreens TANGERINE and TANGERINE HYBRIDS
- Accelerates BLACKBERRY ripening and loosening
- Concentrates BLUEBERRY maturity and aids in barrenberry control
- Accelerates early uniform FIG ripening
- Hastens FILBERT husk maturity for earlier more efficient harvest
- Promotes early uniform color development of EMPEROR and TOKAY GRAPES. Hastens maturity of THOMPSON SEEDLESS GRAPES

DIRECTIONS

APPLES

For best results, a thorough uniform spray coverage of leaves and fruit is needed. A wetting agent may improve spray coverage. Fruit size reduction may occur when used to obtain early maturity, especially if fruit is small at the time of treatment. Use no more than one application per season on bearing trees.

For Fruit Loosening

Apply a foliar spray to apple trees 7 to 14 days prior to normal anticipated harvest. Pick apples when they reach the degree of loosening desired for mechanical or hand harvesting. Observe treated fruit closely for abscission response since temperatures above 75°F accelerate fruit loosening and promote preharvest drop.

I. Early and Midseason Maturing Varieties *(such as Tydeman, Milton, Wealthy and McIntosh).*

Apply 2½ pints of ETHREL Plant Regulator per acre. Use sufficient water for thorough uniform spray coverage, generally 50 to 500 gallons per acre depending on the type and delivery of sprayer used and size of trees.

II. Late Maturing Varieties *(such as Cortland, Red Delicious, Rome, Idared and Jonathan).*

Apply 5 pints of ETHREL Plant Regulator per acre. Use sufficient water for thorough uniform spray coverage, generally 50 to 500 gallons per acre depending on the type and delivery of sprayer used and size of trees.

For Promoting Uniform Ripening and Coloring of Red Varieties Without Loosening Fruit

Apply a foliar spray of ETHREL Plant Regulator and a preharvest drop control chemical such as FRUITONE®-N. Begin spraying 2 to 3 weeks before normal harvest period and about 1 to 2 weeks before desired harvest date.

I. Early or Midseason Maturing Varieties *(varieties maturing with McIntosh or earlier).*

Mix ½ to 1 pint of ETHREL Plant Regulator in 100 gallons of water. Use the lower rate on apples intended for storage. Apply as a normal dilute spray using sufficient water for thorough uniform coverage, generally 200 to 400 gallons per acre depending on tree size.

II. Late Maturing Varieties *(varieties maturing later than McIntosh).*

Mix 1 pint of ETHREL Plant Regulator in 100 gallons of water. Apply as a normal dilute spray using sufficient water for thorough uniform coverage, generally 200 to 400 gallons per acre depending on size of trees. Since recommendations for chemical control of preharvest drop vary according to location and variety, follow directions on the FRUITONE®-N label or consult your local extension Pomologist or Horticulturist for his recommendations.

Treat when air temperatures are between 60°F and 90°F. However, application may be made at 50°F under rising temperature conditions. If daytime temperatures are high and nighttime temperatures are warm, color response from ETHREL treatment will be reduced, but ripening and loosening effects will be accelerated. Cool weather may extend the interval between treatment and harvest.

Important: Harvest fruit at proper maturity and put into cold storage immediately. Observe fruit daily as the proper picking period is shorter with ETHREL treated fruit than with untreated fruit. Do not allow fruit to become overripe on trees. Fruit intended for fresh market must be checked for quality and maturity, not just color. Color alone is not an adequate guide for fruit maturity. Use a pressure gauge or other suitable methods for determining internal maturity. If the fruit is over-mature when harvested and stored, the fruit may soften sooner than untreated fruit. ETHREL applied earlier than 3 weeks before normal anticipated harvest may result in reduced fruit quality and size. Do not apply ETHREL to more acreage than can be harvested in 1 to 2 days. Treated fruit can be satisfactorily stored in CA or regular cold air storage provided fruit is in good condition when put into storage.

For Increased Flower Bud Development on Young Trees

To increase flower bud development on young non-bearing trees apply a foliar spray of ETHREL Plant Regulator 3 to 4 weeks after bloom. On young trees just beginning to bear fruit, apply ETHREL 4 to 6 weeks after full bloom (immediately after June drop period) to minimize overthinning. Reduced vegetative growth and increased bud development during the season of application should increase flowering the following spring. Trees should be large enough to support a crop of apples before they are treated to initiate flower buds. Treat only vigorous trees since excessive growth reduction will result on low weak trees. Treat when air temperatures are between 60°F and 90°F.

I. Spur Type Trees: Mix 1⅔ pints of ETHREL Plant Regulator in 100 gallons of water (5 pints in 300 gallons). Spray trees thoroughly and uniformly to the point of runoff.

II. Non-spur Type Trees: Mix 3⅓ pints of ETHREL Plant Regulator in 100 gallons of water (10 pints in 300 gallons). Spray trees thoroughly and uniformly to the point of runoff. This rate may completely defruit trees.

BLACKBERRIES (Cultivars Chehalem, Thornless Evergreen and Boysenberry) in Oregon and Washington

To loosen fruit and concentrate maturity, apply a foliar spray of ETHREL Plant Regulator at least 3 days prior to the expected harvest when the air temperature is between 60° and 90°F and rain is not predicted within 24 hours. Treat early in the harvest season to reduce the number of pickings OR treat later in the harvest season for a once-over final picking.

Mix 5 pints of ETHREL per 100 gallons water and apply 100-175 gallons of spray mixture per acre.

Do not spray more than can be harvested on the third day following treatment as considerable fruit may drop. Do not apply more than once per season.

Use only at recommended times and rates on healthy, vigorous plants, as a reduction in berry size may occur. Do not use on damaged or diseased plants.

BLUEBERRIES

For concentration of maturity and earlier fruit coloring on cultivars Bluecrop, Weymouth, Jersey, Rancocas, Rubel, Bluetta, Erliblue, Wolcott, Croatan, Murphy, Angela, Morrow, Garden Blue, Trifblue and NC 901. Application to other cultivars may cause excessive steminess and/or premature crops.

Apply a foliar spray of ETHREL Plant Regulator at a rate of 4 to 8 pints in 150 to 200 gallons per acre. Use the higher spray volume and rate on large dense foliage bushes or when temperatures are cool. Treat when air temperatures are 60° to 90°F. Thorough uniform spray coverage of foliage and fruit is essential for maximum fruit maturation and coloration. To aid uniform wetting of plants, use a wetting agent such as Triton B-1956 at 0.5 pints per 100 gallons or X-77 at 2.0 pints per 100 gallons.

To concentrate first harvest, apply ETHREL when 15 to 20% of the berries are blue. To concentrate final harvest, treat after the first or second picking. Harvest blueberries at proper maturity when colored, generally one to two weeks after treatment. Observe fruit daily as the proper picking period for optimum internal and external quality is shorter with ETHREL treated fruit than with untreated fruit. High temperatures (90°F) accelerate fruit ripening.

ETHREL treatments may increase fruit with stems, slightly decrease fruit size, soluble solids and acidity. ETHREL may not induce internal ripening as rapidly as fruit color. Make no more than one application per season. If used for barrenberry control in Maine, do not make a second application for blueberry concentration of maturity or coloring.

Black Barrenberry Control in Low Bush Blueberries in Maine

Apply 4 to 8 pints per acre of ETHREL Plant Regulator in approximately 100 to 200 gallons per acre for ground sprayers and approximately 10 gallons per acre for aerial application. Thorough uniform spray coverage is essential. To aid in uniform wetting of plants, use a wetting agent such as X-77 at 0.1% of the spray volume. Use the lower rates of ETHREL when black barrenberries are between 90 to 100% petal fall and use the higher rates of ETHREL when black barrenberry fruit is 1/8 to 3/16 inch in diameter. This is generally 7 to 10 days after blueberries are in the same stages of development. Harvest blueberry fruit when ripe, generally 6 to 8 weeks after application.

Do not treat blueberry plants under drought stress or excessively high temperatures as defoliation and reduced yield may result.

CANTALOUPES in California, Arizona and Texas

ETHREL Plant Regulator promotes abscission (slipping) so fruit separates more easily from vines. This allows more efficient and economical harvesting. Abscission generally will occur two to five days after ETHREL treatment. The effect will be faster at higher temperatures. Proper timing of ETHREL treatment is critical, but may vary from season to season. Fruit quality, in terms of soluble solids and flesh color, does not improve following treatment. Therefore, ETHREL should not be used until after fruit has developed marketable levels of soluble solids and flesh color. ETHREL promotes abscission of immature fruit as well as marketable fruit. Thus, using ETHREL too soon will result in poorly colored fruit with substandard soluble solids.

Observe treated fields frequently and pick fruit when market quality is reached. Fruit allowed to remain in the field too long will lose quality. Some yellowing or rapid aging of vines will be seen following treatment. Plants with low vigor will not respond properly. Treat each field only once with ETHREL. Do not plant another crop within 30 days after treatment. For more specific directions under varying temperature and moisture conditions, consult your Extension Horticulturist or Farm Advisor for his experience with ETHREL in your area.

CALIFORNIA AND ARIZONA

Treat only those fruit that have a fairly uniform fruit set, have vines in good condition, and have fruit with marketable soluble solids and internal flesh color. Do not treat fields where soluble solids are running less than 10%.

Apply 3 pints of ETHREL Plant Regulator per acre in 40 to 80 gallons of water. Good spray coverage is important for producing uniform abscission. Be prepared to harvest abscised fruit two to five days after treatment. The exact pre-harvest interval will vary with temperature. Be sure to coordinate your harvesting schedule with your packer/shipper.

TEXAS

Treat only those fields that have a fairly uniform fruit set, have vines in good condition, and have fruit with marketable soluble solids and internal flesh color. Apply 3 pints of ETHREL Plant Regulator per acre in 40 to 80 gallons of water with ground sprayers, or 10 to 12 gallons of water with aerial applications. ETHREL should not be applied when night temperatures are 60°F or below.

CHERRIES

For Fruit Loosening and Uniform Ripening

To loosen fruit, apply a foliage spray to cherry trees 7 to 14 days before anticipated harvest date. Treat when air temperatures are between 60°F and 85°F. To promote earlier fruit ripening, apply 7 to 14 days prior to normal harvest. Observe trees closely to determined application date.

Fruit clusters on shaded inside branches should be of marketable size and have some color development or signs of ripening prior to treatment. Applying earlier than two weeks prior to the normal peak harvest period may loosen immature or poorly colored fruit, may prevent fruit from reaching normal size, and may inhibit accumulation of sugars.

Inspect treated fruit to evaluate crop conditions. Harvest loosened fruit at proper maturity, generally 7 to 14 days after treatment. For programming harvest, spray a different orchard block each day and harvest blocks daily in the same sequence.

Be certain your spray equipment provides good penetration and uniform coverage in top and middle (interior) portions of trees. Concentrate applications will promote adequate loosening, but may cause adverse effect on trees such as gummosis and tip dieback.

ETHREL Plant Regulator works most effectively when applied to healthy trees. Some gummosis of cherry trees is associated with treatment. Do not apply to trees that are of low vigor or have experienced severe stress (such as winter injury, drought, disease), since gummosis will be excessive, particularly when high temperatures and/or drought follow treatment. If possible, irrigate orchards to prevent drought stress resulting in gummosis. Do not treat trees that had severe gummosis the previous year as tree damage will result. Use no more than one application per season.

NOT FOR USE ON CHERRIES IN CALIFORNIA

FOR USE ON TART CHERRIES—Varieties such as Montmorency: Mix 2/3 pints of ETHREL Plant Regulator in 100 gallons of water. Apply as a dilute spray with sufficient water for thorough, uniform coverage, generally 300 gallons per acre depending on the type and delivery of sprayer used and size of trees.

For concentrate sprayers, apply 2 1/2 pints of ETHREL per acre in 50 to 100 gallons of water. Uniform coverage is essential for optimum response.

FOR USE ON SWEET CHERRIES—Varieties such as Windsor, Napoleon (Royal Anne), Emperor Francis: Mix 5/6 to 1 2/3 pints of ETHREL Plant Regulator in 100 gallons of water. Apply as a dilute spray with sufficient water for thorough uniform coverage, generally 300 gallons per acre depending on the type and delivery of sprayer used and size of trees. Use the lower rates on the light or yellow colored varieties. A longer treatment to harvest interval will occur with the lower rate.

For concentrate sprayers, apply 5 pints of ETHREL per acre in 50 to 100 gallons of water per acre. Uniform coverage is essential for optimum response.

INCREASED DORMANT FRUIT BUD HARDINESS AND DELAYED SPRING BLOOM OF SWEET CHERRIES IN THE PACIFIC NORTHWEST

A fall application of ETHREL Plant Regulator will increase fruit bud hardiness by decreasing the chance for winter injury and delaying bloom by 3 to 5 days which may help avoid frost injury. Treatment of early flowering varieties may delay bloom to better coincide with pollenization from other varieties. Apply 3 pints of ETHREL per acre in a dilute spray (300 to 400 gallons per acre). Treat during the first two weeks of September. Some early leaf yellowing and drop may be noted following application. Slight gummosis and individual fruit size reduction may be noted.

CRANBERRIES—Grown in Massachusetts, New Jersey and Wisconsin

In Wisconsin apply ETHREL Plant Regulator during the first two weeks in September. In Massachusetts and New Jersey apply ETHREL after 1 to 5% of the total fruit is pink and red. To determine percent colored fruit, take a scoop and harvest fruit from several plants. Separate and count pink and red versus white and green fruit. Do not look at top fruit only. Use only one ETHREL treatment per season. Harvest treated berries when the desired intensity of maturity and coloration is reached and before bogs freeze over. An increase in maturity and color may occur in 10 to 14 days.

For best results it is essential to thoroughly cover all leaves and fruit. Earlier coloration can be produced on berries that have been directly contacted by ETHREL spray.

Apply when daytime temperature above 65°F is expected for several days after treatment. To aid in uniform wetting of plants add a surfactant such as X-77 at 2 pints per 100 gallons or Triton B-1956 at 0.5 pints per 100 gallons to the spray mixture.

Apply 4 to 6 pints of ETHREL per acre in 200 to 300 gallons of water for ground sprayers and 5 to 30 gallons of water for aerial sprayers. Use the higher rates and volumes when foliage is dense and temperatures are cool.

FIGS in California

A foliar spray of ETHREL Plant Regulator accelerates early uniform fig ripening, allowing earlier harvest of high quality fruit and involving the efficiency of scheduled harvest.

ETHREL should be used only on Calimyrna and second crop Black Mission figs. Do not use on other varieties as crop loss may result. Apply 4 pints of ETHREL per acre using 250 gallons of water. For best results, a thorough uniform spray coverage of leaves and fruit is required. Treat when 70 to 90% of the basal figs (closest to branch fork) are yellow.

Do not treat earlier as reduced fruit quality, increased incidence of "puff" balls, or premature fruit drop may result.

Figs generally start to drop 14 to 21 days following treatment. Pick up fruit promptly to avoid loss of quality due to insect damage or adverse weather. Dry and process figs as usual.

Treat trees only once per season. For more information regarding local experience with this product on figs, consult your Farm Advisor or Extension Horticulturist.

FILBERTS in Oregon

Apply a foliar spray of ETHREL Plant Regulator to trees when the first nuts have loosened, will spin within their husk and begin to fall naturally. At this time some husks have turned brown at the tips but many green husks are still present. Mix 3½ pints of ETHREL per 100 gallons of water and apply approximately 75 to 100 gallons of spray mixture per acre for average size trees. To assure uniform coverage of large trees, apply 150 gallons of spray mixture per acre. Use only one ETHREL spray per season.

For best results, a thorough, uniform spray coverage of leaves and husks is needed. Some premature leaf drop will be associated with treatment. Do not apply to diseased or low vigor trees as leaf drop may be excessive. For maximum benefit, harvest should begin when 90% or more of the filberts have dropped.

TABLE GRAPES in California

An application of ETHREL Plant Regulator promotes early uniform color development of colored table grape varieties including Cardinal, Flame Seedless, Red Malaga, Queen and Tokay allowing more efficient harvest of quality fruit.

Apply 1 to 2 pints of ETHREL per acre in sufficient water to wet vines and fruit clusters uniformly, generally 50 to 300 gallons using conventional ground sprayers. Rates higher than 1 pint/A may increase incidence of cracked fruit and should only be used in areas or during weather conditions where grapes have been especially hard to color in past seasons. Except Tokay treat all varieties when 5 to 30% of the berries are showing color. Treat Tokays when 5 to 15% of the berries are showing color. Aerial application of 1 to 2 pints of ETHREL in 20 gallons of spray volume per acre of concentrate ground applications of 1 to 2 pints of ETHREL in 25 to 50 gallons per acre may be used to treat Tokays.

Harvest fruit at desired maturity and quality, as indicated by sugar content, acidity and color. This is generally two weeks or more after treatment. Watch treated areas closely and harvest before berries become too dark.

Some berry softening is associated with ETHREL treatment of some varieties which may limit or influence storage considerations. Do not store Tokays.

Treat each vine only once per season. For more information regarding local experiences with this product on grapes consult your Farm Advisor or Extension Viticulturist.

THOMPSON SEEDLESS GRAPES FOR RAISIN PRODUCTION

A foliar spray of ETHREL Plant Regulator will hasten maturity of Thompson Seedless grapes resulting in reduced acids, increased sugars and increased raisin quality.

Apply a foliar spray of 1 to 2 pints (0.25-0.5 lb.) per acre ETHREL Plant Regulator in 30 to 300 gpa of water at 5% berry softening. Do not apply to grapes which are under stress from insect damage or which are under moisture stress.

Observe treated vineyards closely and harvest at proper maturity as determined by sugar and acid levels.

Treat each vine only once per season. For more information regarding local experience with this product on grapes consult your Farm Advisor or Extension Viticulturist.

LEMONS—Postharvest Degreening (Florida)

Thoroughly mix 3⅓ pints of ETHREL Plant Regulator per 100 gallons of water and apply a postharvest spray mist to lemons prior to placement in the curing room. One hundred gallons of spray mix will treat approximately 100 to 130 tons of fruit. Complete, uniform coverage of fruit is essential for desired response. If ETHREL is to be used in conjunction with a fungicide such as TBZ, follow all directions and cautions on the fungicide label.

An ETHREL treatment will shorten the degreening time for lemons by up to ⅓ over untreated fruit. Cure treated lemons following normal practices, but check frequently to determine when desirable color is obtained. If ETHREL is used for postharvest degreening do not place fruit in ethylene chambers.

Do not make preharvest applications to lemons or treat more than once after harvest. Do not pick lemons until the minimum juice requirement (30% v/v) has been obtained.

PEPPERS

A foliar spray of ETHREL Plant Regulator promotes early, uniform ripening and coloring for more efficient picking and handling of fruit. Apply ETHREL to bell peppers when 10% of the fruit are red and chocolate and to pepper varieties, chili and pimento when 10 to 30% of the fruit are red and chocolate and there is sufficient mature green fruit to produce the desired tonnage per acre. Check several field locations to determine the crop stage and degree of maturity. ETHREL treatments may reduce total yields if applied too early or there is a lack of uniform, mature green fruit resulting from split fruit set or variable soil cultural practices. Do not apply when prolonged average high temperatures of 95°F after treatment are expected. Do not treat when temperatures exceed 100°F. Applications made under high temperatures will accentuate fruit ripening, yellowing of foliage, defoliation and immature fruit abscission while average low temperatures, 60°F after treatment, have reduced or negated the effects of ETHREL.

Apply 3 to 4 pints of ETHREL in 40 to 100 gallons of water per acre with ground spray equipment. Use the higher rate when cool temperatures are anticipated (less than 65°F), when plants are growing vigorously or when foliage is dense to assure thorough wetting of all leaves and fruit. Some yellowing and general aging of leaves will be noted after treatment. Do not make more than one application of ETHREL per season. Harvest fruit when the desired color and maturity have developed, generally 14 or more days after treatment.

ETHREL will not ripen immature green fruit. Do not plant any crop until 30 days after ETHREL application.

TANGERINES and TANGERINE HYBRIDS in Florida

ETHREL promotes loosening of tangerines for easier harvest and degreening for improved marketability. Apply a preharvest spray of ETHREL Plant Regulator to leaves and fruit when 10 to 20% of the mature fruit are showing color break. Observe treated fruit to determine optimum harvest response which generally occurs 5 to 10 days after treatment depending on percent color and maturity at time of spraying and temperature after application.

For treatment of tangerine and tangerine hybrids such as Dancy, Robinson, Nova, Murcott and Lee, mix and apply 3⅓ to 5 pints of ETHREL in 400 to 600 gallons of water per acre, respectively. For Orlando, apply ⅔ pints of ETHREL in 100 gallons and apply 400 to 600 gallons of spray mixture per acre.

Use a higher rate and spray volume on large dense vigorous trees. Uniform spray coverage is essential for maximum response.

Some defoliation or yellowing of older leaves may occur from ETHREL treatment. Excessive defoliation can result from treatment of trees with low vigor or treating trees exhibiting freeze injury, drought or other stress symptoms. Do not use a surfactant or wetting agent since this promotes defoliation. Do not make more than one application of ETHREL per harvest season. Do not treat trees with new vegetative growth or during flowering.

TOMATOES

Some yellowing and general aging of foliage may be observed following treatment. Do not treat plants with poor root systems or growing under stress due to poor soil conditions, drought, disease or insect damage. Treatment of weak plants will result in rapid loss of foliage cover increasing sunburn and sun scald potential, especially under high temperature conditions. Do not treat when sustained temperatures above 105°F are anticipated. Foliar damage is more severe on certain varieties such as VF 10, VF 315, VF 145, 21-4, and 13L. Do not treat sensitive varieties during period when temperatures exceed 100°F.

ETHREL Plant Regulator will not ripen immature green fruit. Check fields closely for stage and degree of uniform maturity prior to application. Sample several plants throughout the field to determine proper spray date. Sort, weigh and calculate the percent fruit which is red and pink, including breakers. Fruit size alone is not an adequate indicator.

Timely harvest is essential for maximum return of high quality fruit. Consult processor representatives concerning delivery schedules and quotas prior to application of ETHREL.

Apply recommended amount of ETHREL in 20 to 100 gallons of water per acre for ground boom sprayers to 10 to 12 gallons of water per acre for aerial application. Uniform ripening requires uniform spray coverage on plants. Do not tank mix with sun protection whiteners. Use only one application per season. Do not apply ETHREL to more acreage than can be harvested in 2 to 3 days. Do not apply ETHREL to varieties known to soften rapidly or shatter when ripe. Rates listed below are for broadcast coverage of the total area.

If bands are sprayed over the row reduce the amount of ETHREL used in proportion to the area actually treated. Overdosing from overlapping swaths or spray nozzle patterns can cause severe foliage injury.

FOR USE IN CALIFORNIA, TEXAS AND OTHER WESTERN AREAS HAVING SIMILAR VARIETIES AND CULTURES

I. Early and Midseason Tomato Crops

Apply 3¼ pints of ETHREL Plant Regulator per acre (0.8 lb. ethephon) when 5% to 15% of the fruit in the field are red and pink (including breakers) and there is sufficient mature green fruit to produce the desired tonnage. During the midseason, particularly under high temperatures, rates as low as 1¼ pints ETHREL per acre have been effective for increasing recoverable yields and reducing sorting times with less vine damage. There will be little change in the harvest date, compared to untreated tomatoes. Walk treated fields frequently to evaluate crop condition. Maintain normal cultural practice between treatment and harvest. Harvest fruit at proper maturity generally 14 to 21 days after treatment.

II. Late Season or Coastal Tomato Crops

Apply 3¼ to 6½ pints ETHREL Plant Regulator per acre (0.8 to 1.6 lb. ethephon) when 5% to 30% of fruit in the field are red and pink (including breakers) and there is sufficient mature green fruit to produce the desired tonnage. Maximum response will be when treated fruit is 5 to 15% pink and red. Use the higher rate of ETHREL when nighttime temperatures are cool (below 65°F) or vegetative growth is dense.

Walk treated fields frequently to evaluate condition of crop. Maintain normal cultural practice between treatment and harvest. Harvest maturity of late season cooler coastal crops can be expected 18 to 24 days after treatment. However, nighttime temperatures below 65°F retard natural color development and may extend the interval between treatment and harvest.

FOR USE IN TOMATO PRODUCING AREAS OF THE MIDWEST AND EAST

Apply 3¼ pints of ETHREL Plant Regulator per acre (0.8 lb. ethephon) when 5% to 30% of the fruit in the field are red and pink (including breakers) and there is sufficient mature green fruit to produce the desired tonnage. When average temperatures are cool (below 65°F) and vegetative growth is dense, apply 6½ pints of ETHREL per acre (1.6 lbs. ethephon) when 5% to 30% of the fruit in the field are red and pink (including breakers).

Walk treated fields frequently to evaluate condition of crop. Maintain normal cultural practice between treatment and harvest. Harvest fruit at proper maturity, generally 14 to 21 days after treatment. Temperatures above 90°F may cause fruit to ripen sooner, whereas temperatures below 65°F retard natural color development and may extend the interval between treatment and harvest.

TOMATOES—Fresh Market—in California

A foliar spray of ETHREL Plant Regulator accelerates and concentrates tomato ripening, increasing recoverable fruit yield from a once-over harvest. Apply ETHREL when desired tonnage of fruit has reached marketable size and maturity, generally 3 to 6 days before desired harvest date. Expect only one harvest from treated areas. ETHREL will not ripen immature green fruit and accelerates aging of tomato plants. Check fields closely for stage and degree of uniform maturity prior to application. Sample several locations throughout the field to determine proper spray date.

Do not treat when temperatures above 100°F are anticipated. Do not use on greenhouse tomatoes. Do not tank mix with sun protection whiteners. Use only one application per season.

Consult your local Union Carbide Representative, Extension Horticultural Specialist or Farm Advisor for suggestions on the proper use of ETHREL when fields have variable plant vigor due to differences in soil conditions or cultural practices and for rates of fruit ripening as affected by temperature, within the rate and timing limitations shown on the label. Do not plant to any other crop until 30 days after ETHREL treatment.

Apply a foliar spray of 3¼ to 5 pints of ETHREL per acre in 40 or more gallons of water for ground sprayers or 10 to 12 gallons of water for aerial sprayers. Use the higher rate and longer preharvest interval on late season crops, when temperatures are cool (generally below 85°F) and foliage is dense. Uniform ripening requires uniform spray coverage of plants. Rates listed are for broadcast coverage of the total area. If bands are sprayed over the row reduce the amount of ETHREL used in proportion to the area actually treated. Overdosing from overlapping swaths or spray nozzle patterns can cause severe foliage injury. Walk treated fields frequently to evaluate conditions of crop. Maintain normal cultural practice between treatment and harvest.

Timely harvest is essential for maximum return of high quality fruit.

WALNUTS in California

ETHREL Plant Regulator application promotes uniform hull split providing greater potential for increased recoverable yields of optimum quality walnuts with early harvest. Maximum benefit is obtained with application immediately after walnut packing tissue has turned brown.

Apply 3 to 5 pints of ETHREL per acre in ground spray volumes of 100 to 500 gallons per acre depending on spray equipment and size of trees. Treat when air temperatures are between 60°F and 90°F and after the packing tissue of all nuts inspected are completely brown. Use 5 pints per acre when minimum time interval between treatment and harvest is desired and with cool temperature harvest conditions.

Uniform spray coverage of walnut hulls is essential for maximum hull split and loosening response. Be sure equipment provides good penetration of spray to all parts of the tree. Check nozzle systems frequently to assure uniform distribution of spray droplets. Application equipment should not be operated in excess of 3 mph and directional spray volute equipment should be utilized when treating tall dense foliage trees.

Inspect treated orchards frequently and harvest as soon as adequate hull split occurs, generally 5 to 10 days after treatment. Cool temperatures increase the time requirement for optimum harvest response. As with early harvest, longer drying time may be required.

Treatment prior to brown coloration of packing tissue may reduce nut quality. Some leaf drop is associated with treatment. Do not treat diseased, low vigor or moisture stressed trees as excessive leaf drop may occur. Use only one spray application per season.

Do not use rates above that recommended since tree injury (such as excessive defoliation, reduced catkin formation, twig dieback) may result.

For a programmed harvest, spray a different orchard block each day and harvest blocks daily in the same sequence starting 5 to 10 days after treatment.

THIS SPECIMEN LABEL IS INTENDED FOR USE ONLY AS A GUIDE IN PROVIDING GENERAL INFORMATION REGARDING THE DIRECTIONS, WARNINGS AND CAUTIONS ASSOCIATED WITH THE USE OF THIS PRODUCT. AS WITH ANY AGRICULTURAL CHEMICAL, ALWAYS FOLLOW THE LABEL INSTRUCTIONS ON THE PACKAGE BEFORE USING.

FLOREL™

PLANT GROWTH REGULATOR

For use on ornamentals, for production of cucumber, pumpkin and squash hybrid seed, for increasing pickling cucumber fruit set and yields, for inducing flowering of ornamental bromelaids and for earlier ripening of home garden tomatoes, reducing plant height of potted daffodils and stem topple of potted hyacinths.

ACTIVE INGREDIENT:
Ethephon (2-chloroethyl)
phosphonic acid 3.9%
INERT INGREDIENTS: 96.1%
TOTAL 100.0%
Contains 0.33 lbs. ethephon per gallon.

E.P.A. Reg. No. 264-263-AA E.P.A. Est. No. 264-PA-01

Keep Out of Reach of Children
CAUTION

IN CASE OF EMERGENCY TELEPHONE (24 HOURS A DAY) IN U.S.A. 1-800-UCC-HELP.

FOR GENERAL PRODUCT INFORMATION CALL 1-800-334-9745.

PRECAUTIONARY STATEMENTS

HAZARDS TO HUMANS AND DOMESTIC ANIMALS
CAUTION
Harmful if swallowed. Avoid contact with skin, eyes or clothing and do not inhale vapors.

STATEMENT OF PRACTICAL TREATMENT

If swallowed, call a physician and promptly drink large quantities of milk, egg white, gelation solution or if these are not available, large quantities of water.

In case of eye contact, flood eyes immediately with plenty of water for at least 15 minutes and get medical attention.

In case of skin contact, wash with plenty of soap and water. Remove and wash contaminated clothing before reuse.

NOTE TO PHYSICIAN
Florel possesses characteristics of a strong acid. Probable mucosal damage may contraindicate the use of gastric lavage. Measures against circulatory shock, respiratory depression and convulsion may be needed.

GENERAL WORKER PROTECTION STATEMENTS
Do not apply this product in such a manner as to directly or through drift expose workers or other persons. The area being treated must be vacated by unprotected persons.

Do not enter treated areas without protective clothing until sprays have dried. Because certain states may require more restrictive reentry intervals for various crops treated with this product, consult your State Department of Agriculture for further information. Written or oral warnings must be given to workers who are expected to be in a treated area or in an area about to be treated with this product. When oral warnings are given, warnings shall be given in a language customarily understood by workers. Oral warnings must be given if there is reason to believe that written warnings cannot be understood by workers. Written warnings must include the following information: appropriate signal word (CAUTION), area treated, date of application, appropriate protective clothing, and reentry interval if applicable.

ENVIRONMENTAL HAZARDS
Do not apply directly to lakes, streams or ponds. Do not contaminate water by cleaning of equipment or disposal of wastes. Avoid spray drift to nearby crops as this product may cause temporary modifications in plant growth. Do not contaminate water used for irrigation or domestic purposes.

STORAGE AND DISPOSAL

Do not contaminate water, food, or feed by storage or disposal.

PESTICIDE DISPOSAL

Wastes resulting from the use of this product may be disposed of on site or at an approved waste disposal facility.

CONTAINER DISPOSAL

Triple rinse (or equivalent). The offer for recycling or reconditioning, or puncture and dispose of in a sanitary landfill, or incineration or if allowed by state and local authorities, by open burning. If container is burned, stay out of smoke.

LIMITED WARRANTY AND DISCLAIMER

The manufacturer warrants (a) that this product conforms to the chemical description on the label; (b) that this product is reasonably fit for the purposes set forth in the direction for use when it is used in accordance with such directions; and (c) that the directions, warnings and other statements on this label are based upon responsible experts' evaluation of reasonable test of effectiveness, of toxicity to laboratory animals and to plants, and of residues on food crops, and upon reports of field experience. Tests have not been made on all varieties or in all states or under all conditions. THE MANUFACTURER NEITHER MAKES NOR INTENDS, NOR DOES IT AUTHORIZE ANY AGENT OR REPRESENTATIVE TO MAKE, ANY OTHER WARRANTIES, EXPRESS OR IMPLIED, AND IT EXPRESSLY EXCLUDES AND DISCLAIMS ALL IMPLIED WARRANTIES OF MERCHANTABILITY OR FITNESS FOR A PARTICULAR PURPOSE.

THIS WARRANTY DOES NOT EXTEND TO, AND THE BUYER SHALL BE SOLELY RESPONSIBLE FOR, ANY AND ALL LOSS OR DAMAGE WHICH RESULTS FROM THE USE OF THIS PRODUCT IN ANY MANNER WHICH IS INCONSISTENT WITH THE LABEL DIRECTIONS, WARNINGS, OR CAUTIONS.

BUYER'S EXCLUSIVE REMEDY AND MANUFACTURER'S OR SELLER'S EXCLUSIVE LIABILITY FOR ANY AND ALL CLAIMS, LOSSES, DAMAGES OR INJURIES RESULTING FROM THE USE OR HANDLING OF THIS PRODUCT, WHETHER OR NOT BASED IN CONTRACT, NEGLIGENCE, STRICT LIABILITY IN TORT OR OTHERWISE, SHALL BE LIMITED, AT THE MANUFACTURER'S OPTION, TO REPLACEMENT OF, OR THE REPAYMENT OF THE PURCHASE PRICE FOR, THE QUANTITY OF PRODUCT WITH RESPECT TO WHICH DAMAGES ARE CLAIMED. IN NO EVENT SHALL MANUFACTURER OR SELLER BE LIABLE FOR SPECIAL, INDIRECT OR CONSEQUENTIAL DAMAGES RESULTING FROM THE USE OR HANDLING OF THIS PRODUCT.

DIRECTIONS FOR USE

IT IS A VIOLATION OF FEDERAL LAW TO USE THIS PRODUCT IN A MANNER INCONSISTENT WITH ITS LABELING.

HYBRID SEED PRODUCTION

FLOREL™ Plant Growth Regulator modifies sex expression and flowering pattern of cucumbers, squash and pumpkins to facilitate hybrid seed production. Application of FLOREL™ increases the number of pistillate (female) and decreases the number of staminate (male) flowers. FLOREL™ brings about earlier formation of female flowers at lower nodes where normally only male flowers are formed in standard (monoecious) cucumber, squash and pumpkins varieties.

Sexual modification towards femaleness in treated plants is temporary (5 to 15 nodes). Variety, location, climate and cultural practices may influence the best rate for treatment. New breeding lines will require trial applications before extensive treatments are made.

HYBRID SEED PRODUCTION IN CUCUMBERS AND SQUASH

Amounts To Use: Apply one quart of FLOREL™ Plant Growth Regulator in 40 to 100 gallons of water per acre. Spray plants thoroughly when they have developed two true leaves.

When germination is variable under field conditions, repeat application of one quart of FLOREL™ plant growth regulator per acre when the first true leaf is the size of a quarter and the third true leaf is expanding.

HYBRID SEED PRODUCTION IN PUMPKINS (ILLINOIS ONLY)

Amounts To Use: Apply 3 quarts of FLOREL™ Plant Growth Regulator in 40 to 100 gallons of water per acre. Apply no more than 6 spray applications per year at 7-10 day intervals beginning at the 2-4 leaf stage.

Do not harvest within 42 days of last application.

GENERAL COMMENTS:

Apply FLOREL™ Plant Growth regulator spray solution within 4 hours after mixing. Mix only the amount of spray you plan to use immediately. DO NOT save unused diluted spray solutions as they will not be as effective as fresh solutions.

Plants sprayed with FLOREL™ often flower 7 to 10 days earlier than untreated. Therefore pollen source plants should be seeded prior to seed parent to insure adequate pollen availability for hybrid seed production.

Temporary growth inhibition resulting from shortened internodes may be observed.

NOTE: For home garden use follow spray guide at end of pamphlet.

PICKLING CUCUMBERS YIELD INCREASE

FLOREL™ Plant Growth Regulator modifies sex expression and flowering pattern of pickling-type cucumbers and induces femaleness and the potential for increased yield.

Amounts to Use: Mix 1.25 to 2.5 pints of FLOREL Plant Growth Regulator in 50 gallons of water. Use the lower rate for gynoecious varieties to maintain femaleness under adverse environmental conditions, and the higher rate for monoecious varieties to induce femaleness.

Make two applications, the first at the 2 to 4 leaf stage and a second 5 to 7 days later. Apply to point of run-off with low pressure ground sprayer at 25 to 50 gallons per acre.

Leave one untreated row in every 10 rows to ensure sufficient male flowers for pollination.

For use on pickling-type cucumbers only. Do not apply when fruit is present.

CAUSES BROMELIADS TO FLOWER

FLOREL™ Plant Growth Regulator initiates flowering of ornamental bromeliads such as ANANAS, AECHMEA, NEOREGELIA, VRIESIA and BILLBERGIA.

AMOUNTS TO USE:

For most bromeliad varieties, mix 8 fluid ounces of FLOREL™ Plant Growth Regulator per gallon of water. This prepares a spray concentration of approximately 2,500 ppm of ethephon.

Spray all surfaces of the plant to "wet". Avoid overspraying to "runoff" which may cause damage to leaves or the growing points. For treating groups of plants, use approximately ½ pint of spray solution per 10 square feet of greenhouse bench or outdoor bed area.

Apply FLOREL™ Plant Growth Regulator spray solutions within 4 hours after mixing. Mix up only the amount of spray you plan to use immediately. DO NOT save unused diluted spray solutions as they will not be as effective as fresh solutions.

The degree of flower induction with a given rate of FLOREL is influenced by the plant age, variety, growth rate, climate and cultural conditions. Lower rates may effectively force flowering or produce desirable foliage coloring on certain varieties. Trial applications at lower rates are suggested before making extensive treatments.

GUIDELINES FOR CONSISTENT FLOWER FORCING

1. Grow plants on Photoperiods regulated to maintain plants vega-

tatively active prior to treatment; long days for ANANAS, BILLBERGIA, NEOREGELIA, and short days for AECHMEA, and VRIESIA.

2. Treat mature plants that have well established root systems. Treatments too early in the development of the plant will cause erratic flower initiation and the few flowers formed will be small.
3. Remove water at base of leaves. Allow foliage to dry prior to treatment. Water may be replaced 24 hours after treatment.
4. A minimum night temperature of 65-70°F or higher should be maintained throughout the forcing period.
5. Do not apply fertilizer two weeks prior or for two weeks after treatment.

NOTE: Inconsistent results may be obtained if the leaf surface is covered with algae.

HOME GARDEN TOMATO PRODUCTION USE

FLOREL™ Plant Growth Regulator can be used as an overall, foliar spray to concentrate and accelerate tomato ripening. Apply FLOREL™ at the end of the growing season to hasten ripening of mature green tomatoes and allow harvesting of fruit before frost. FLOREL™ spray will be most effective on healthy plants. Harvest fruit at proper maturity, generally 14 to 21 days after treatment. Some defoliation or yellowing of foliage following treatment will be observed.

AMOUNTS TO USE:
WHEN TREATING SEVERAL PLANTS MIX 1 TEASPOON OF FLOREL™ in 3 ounces (6 tablespoons) of water. When treating a larger number of plants (10 to 15) mix 1 tablespoon of FLOREL™ in 9 ounces (18 tablespoons) of water. Apply FLOREL™ with a clean hand sprayer. Thoroughly wet all leaves and fruit.

DEFOLIATION:
A foliar spray of FLOREL™ Plant Growth Regulator will cause earlier leaf drop of roses, Tallhedge and apple nursery stock. Apply only after the buds are mature or some injury may result.

FOR ROSES: Mix 1 quart FLOREL™ in 10 gallons of water and apply to thoroughly wet foliage. Amount of spray used will depend on the size of the rose bush. Amount of defoliation obtained will depend on the variety and temperature. The addition of 1 pint of nonionic surfactant such as Tween 20 or X-77 per 100 gallons of spray solution will improve defoliation. Do not treat sensitive varieties such as Red American Beauty as bud injury may result.

FOR TALLHEDGE BUCKTHORN: Mix 5 to 10 quarts FLOREL™ in 10 gallons water and apply to thoroughly wet foliage. Amount of spray used will depend on size of tallhedge. Use the higher rate when temperatures are cool or earlier defoliation is desired.

FOR APPLE NURSERYS STOCK IN WASHINGTON: Mix 1-2 pints FLOREL™ plus 3 quarts Dupont Surfactant WK in 50 gallons water and apply no more solution than is necessary to moisten foliage without runoff. A second treatment 3-7 days later using the above rates may be applied. Apply no more than 4 pints FLOREL™ per season. Do not use on Rome apples as defoliation will not be satisfactory. Amount of defoliation obtained will depend on variety and temperature.

FRUIT ELIMINATION:
A foliar spray of FLOREL™ Plant Growth Regulator will reduce or eliminate undesirable fruit from apple trees, crabapple trees, carob trees, and olive trees. Applications should be made at the flower bud to full bloom stage, prior to fruit set.

FOR APPLES AND CRABAPPLES: Mix ¾ to 1 quart FLOREL™ in 10 gallons of water and apply to thoroughly wet foliage. Amount of spray used will depend on tree size. Use the higher rate when temperatures are cool. Do not use on small, red fruited varieties of crabapples as fruit elimination will not be satisfactory.

FOR CAROB (Ceratonia siliqua): Mix ½ quart FLOREL™ in 10 gallons of water and apply to thoroughly wet foliage. Amount of spray used will depend on tree size. Some leaf drop of older leaves may occur after treatment.

FOR OLIVE (Olea europaea): Mix 1 quart FLOREL™ in 10 gallons water and apply to thoroughly wet foliage. Amount of spray used will depend on tree size. Some temporary leaf yellowing may occur. Applications must be made before fruit set for best results.

NOTE: This recommendation is not intended to be used in commercial fruit production as a thinning agent.

INCREASED LATERAL BRANCHING:
A foliar spray of FLOREL™ Plant Growth Regulator will increase lateral branching in Azaleas and Geraniums.

FOR AZALEAS: Mix 2½ to 5 quarts of FLOREL™ in 10 gallons of spray solution and apply to thoroughly wet foliage. Amount of spray used will depend upon plant size. Apply at normal pinching time in conjunction with either hand or chemical pinching agents. Use the higher rate on vigorous tolerant varieties as determined by experience. Do not treat sensitive varieties such as Sweetheart as growth the flower inhibition can occur.

FOR GERANIUMS: Mix 1 quart of FLOREL™ in 10 gallons of water and spray to thoroughly wet foliage. Apply two weeks prior to normal pinching. Use only on stock plants as delay in flowering can occur.

REDUCTION OF HYACINTH STEM TOPPLE AND DAFFODIL PLANT HEIGHT

To reduce potted HYACINTH stem topple at time of full flower, apply a foliar spray of FLOREL before florets show any color. Most cultivars will respond to 1000 ppm (½ pint in 2½ gallons of water) of FLOREL. Bismarck, Jan Bes, Blue Giant, Delft Blue, and Madame Kruger may benefit from a second spray 2 days after the first.

To reduce total plant height of potted daffodils, apply a foliar spray of FLOREL when the shoots are 4 to 5 inches tall. Most cultivars will respond to 2000 ppm (1 pint in 2½ gallons) sprays of FLOREL. For earlier forcings, Dutch Marter, Joseph MacLeod, Flower Record, and Barrett Browning will benefit from a second spray 2 or 3 days after the first. Bridal Crown and Geranium require only 1000 ppm (½ pint in 2½ gallons of water). Gold Medal, Van Sien, February Gold, and Tete-a-tete do not require FLOREL treatment.

FLOREL HOME SPRAY GUIDE

Read Detailed Instructions in This Pamphlet First.

APPLICATION	Amounts to Use: FLOREL	WATER
Flower Induction		
Bromeliads	1 Oz.	1 Pt.
Tomato Ripening		
Fall Season	1 Tsp.	3 Oz.
Fruit Elimination		
Apples/Crabapples	3 Oz.	1 Gal.
Carob	1½ Oz.	1 Gal.
Olive	3 Oz.	1 Gal.
Lateral Branching		
Azaleas	2-4 Oz.	1 Qt.
Geraniums	3 Oz.	1 Gal.

FLOREL™ PLANT GROWTH REGULATOR

- Modifies sex expression and flowering pattern of CUCUMBERS and SQUASH to facilitate hybrid seed production
- Increases PICKLING CUCUMBER fruit set and yield
- Causes BROMELIADS to flower
- Concentrates and accelerates TOMATO ripening for home garden tomato production
- Promotes earlier defoliation of ROSES, TALLHEDGE BUCKTHORN, APPLE NURSERY STOCK
- Eliminates undesirable fruit on APPLE and CRABAPPLES, CAROB and OLIVE
- Stimulates lateral branching of AZALEAS and GERANIUMS
- Reduces plant height of potted DAFFODILS and stem topple of potted HYACINTHS

GENERAL INFORMATION (See attached Pamphlet for Detailed Instructions)

Hybrid Seed Production

A foliar spray of FLOREL™ Plant Growth Regulator will modify sex expression and flowering pattern of cucumbers and squash to facilitate hybrid seed production. Do not use on cucumbers or squash to be harvested for fresh or processed markets.

INCREASED YIELDS OF PICKLING CUCUMBERS

A foliar spray of FLOREL™ Plant Growth Regulator will modify sex expression and flowering patterns of pickling-type cucumbers, increasing fruit set and yields.

INHIBITS GROWTH AND PREMATURE FLOWERING OF SEEDBED TOBACCO

A foliar spray of FLOREL™ Plant Growth Regulator will temporarily inhibit growth and premature flowering of seedbed Tobacco for easier transplanting.

Flower Induction of Bromeliads

A foliar spray of FLOREL™ Plant Growth Regulator initiates flowering of ornamental bromeliads such as ANANAS, AECHMEA, NEOREGELIA, VRIESIA, and BILLBERGIA.

Home Garden Tomato Ripening

FLOREL™ Plant Growth Regulator can be used as an overall, foliar spray to concentrate and accelerate tomato ripening. Applied at the end of the growing season, FLOREL™ will hasten ripening of mature green tomatoes and allow harvesting of fruit before frost.

DeFoliation

A foliar spray of FLOREL™ Plant Growth Regulator applied after buds are matured, will initiate earlier leaf drop of roses, tall-hedge, and apple nursery stock (in Washington only). This will allow digging of stock plants prior to the onset of unfavorable weather.

Undesirable Fruit Elimination

A foliar spray of FLOREL™ will reduce or eliminate undesirable fruit development on Apple, Crabapple, Carob, and Olive trees.

Increasing Lateral Branching

A foliar spray of FLOREL™ Plant Growth Regulator in conjunction with chemical or mechanical pinching, will increase the number of lateral branches on Azaleas. When used on Geranium stock plants, prior to normal pinching time, an increase in the number of lateral branches to be used as cutting material can be expected.

Reduction of Plant Height and Stem Topple

A foliar spray of FLOREL Plant Growth Regulator will aid in reducing total plant height of potted daffodils and stem topple of potted hyacinths at time of full flower.

THIS SPECIMEN LABEL IS INTENDED FOR USE ONLY AS A GUIDE IN PROVIDING GENERAL INFORMATION REGARDING THE DIRECTIONS, WARNINGS AND CAUTIONS ASSOCIATED WITH THE USE OF THIS PRODUCT. AS WITH ANY AGRICULTURAL CHEMICAL, ALWAYS FOLLOW THE LABEL DIRECTIONS ON THE PACKAGE BEFORE USING.

FRUITONE®N

To control pre-harvest drop of apples and pears.

Also useful in thinning apples.

ACTIVE INGREDIENTS:	
1-Naphthaleneacetic Acid, Sodium Salt*	3.5%
INERT INGREDIENTS:	96.5%
TOTAL	100.0%

*1-Naphthaleneacetic acid equivalent 3.1%.

EPA Reg. No. 264-141AA EPA Est. No. 477-MD-01

Keep Out of Reach of Children
CAUTION

IN CASE OF EMERGENCY TELEPHONE (24 HOURS A DAY) IN U.S.A. 1-800-UCC-HELP.

FOR GENERAL PRODUCT INFORMATION CALL 1-800-334-9745.

DIRECTIONS FOR USE

IT IS A VIOLATION OF FEDERAL LAW TO USE THIS PRODUCT IN A MANNER INCONSISTENT WITH ITS LABELING.

PRE-HARVEST DROP:

Spraying just before harvest time (before pre-harvest drop commences) permits a longer harvest period, reduces wind losses, decreases the need for spot picking. Be sure applications are completed in time to become effective before actual pre-harvest drop begins. With summer and early fall varieties during periods of warm weather, FRUITONE-N usually becomes effective in one to two days. For later varieties during cool weather, a slightly longer period is necessary. Healthy foliage, freedom from mites and frost damage is necessary for the most effective action from this material. If possible, sprays should be applied when temperature is 70°F or higher.

TIMING

For summer varieties such as Yellow Transparent and Wealthy, sprays should be applied when the first sound, uninjured fruit begin to drop. The effectiveness of this material is usually much longer with these varieties in hot weather. For summer varieties one half the concentration (5 ppm) may be effective, especially in hot weather.

For late fall varieties such as McIntosh, Baldwin and Red Delicious, application should be made as the first sound, uninjured fruit begin to drop. With a variety such as McIntosh, apply at 10 ppm as the first sound, uninjured fruit begin to drop. If the area treated is not to be picked within 7 to 10 days after first treatment, we suggest a second application be made to the remaining unpicked fruit 5 to 6 days after the first application. With later varieties such as Baldwin and Red Delicious, one application of a 10 ppm spray at the time the first sound, uninjured fruit begin to drop is usually sufficient.

For Bartlett pears, apply a spray of one half concentration (5 ppm) as the first sound, uninjured fruit begin to drop.

IMPORTANT:

FRUITONE-N inhibits fruit drop but does not affect normal ripening. The fruit must be picked at the proper stage of maturity even though the hormone-like material is still effectively delaying excessive drop. Do not make more than two applications, and do not apply closer than 5 days before harvest.

CHEMICAL THINNING

FRUITONE-N should not be used on early summer varieties such as Yellow Transparent, Williams Early Red and Oldenburg, since severe injury or overthinning may occur. For later fall varieties such as Red Delicious, Golden Delicious and Romes, application should be made 14 to 18 days after full bloom. Because of wide variations in varieties, climatic factors, tree vigor and fruit set, a range of rates of 5 to 20 ppm can be used. For rates to be used for your specific variety and location, contact your local extension pomologist.

One and one-quarter pounds of FRUITONE-N give 10 ppm of naphthaleneacetic acid in 500 gallons of water at 1X concentration.

PRECAUTIONARY STATEMENTS

CAUTION

HAZARD TO HUMANS AND DOMESTIC ANIMALS

Harmful if swallowed. Avoid contact with skin, eyes, and clothing. Avoid breathing spray mist.

STATEMENT OF PRACTICAL TREATMENT

In case of eye contact flush with water for at least 15 minutes. In case of skin contact wash with soap and plenty of water. Remove and wash contaminated clothing before reuse.

ENVIRONMENTAL HAZARDS

Do not apply directly to water. Do not contaminate water by cleaning of equipment or disposal of wastes. Do not contaminate irrigation ditches or water used for irrigation or domestic purposes. Do not apply when weather conditions favor drift from treated areas or use in a greenhouse.

STORAGE AND DISPOSAL

STORAGE

Do not store near fertilizers, seeds, insecticides, or fungicides.

PESTICIDE DISPOSAL

Do not contaminate water, food, or feed by storage or disposal. Open dumping is prohibited. Wastes resulting from the use of this product may be disposed of on site or at an approved waste disposal facility.

CONTAINER DISPOSAL

Do not reuse empty container. Triple rinse (or equivalent). Then offer for recycling, or reconditioning, or puncture and dispose of in a sanitary landfill, or by other procedures approved by state and local authorities.

LIMITED WARRANTY AND DISCLAIMER

The manufacturer warrants (a) that this product conforms to the chemical description of the label; (b) that this product is reasonably fit for the purposes set forth in the directions for use when it is used in accordance with such direction; and (c) that the directions, warnings and other statements on this label are based upon responsible experts' evaluation of reasonable tests of effectiveness, of toxicity to laboratory animals and to plants, and of residues, and upon reports of field experience. Tests have not been made on all varieties or in all states or under all conditions. THE MANUFACTURER NEITHER MAKES NOR INTENDS, NOR DOES IT AUTHORIZE ANY AGENT OR REPRESENTATIVE TO MAKE, ANY OTHER WARRANTIES, EXPRESS OR IMPLIED, AND IT EXPRESSLY EXCLUDES AND DISCLAIMS ALL IMPLIED WARRANTIES OF MERCHANTABILITY OR FITNESS FOR A PARTICULAR PURPOSE, THIS WARRANTY DOES NOT EXTEND TO, AND THE BUYER SHALL BE SOLELY RESPONSIBLE FOR, ANY AND ALL LOSS OR DAMAGE WHICH RESULTS FROM THE USE OF THIS PRODUCT IN ANY MANNER WHICH IS INCONSISTENT WITH THE LABEL DIRECTIONS, WARNINGS OR CAUTIONS.

BUYER'S EXCLUSIVE REMEDY AND MANUFACTURER'S OR SELLER'S EXCLUSIVE LIABILITY FOR ANY AND ALL CLAIMS, LOSSES, DAMAGES, OR INJURIES RESULTING FROM THE USE OR HANDLING OF THIS PRODUCT, WHETHER OR NOT SUCH LIABILITY IS BASED IN CONTRACT, NEGLIGENCE, STRICT LIABILITY IN TORT OR OTHERWISE, SHALL BE LIMITED, AT THE MANUFACTURER'S OPTION, TO REPLACEMENT OF, OR THE REPAYMENT OF THE PURCHASE PRICE FOR, THE QUANTITY OF PRODUCT WITH RESPECT TO WHICH DAMAGES ARE CLAIMED. IN NO EVENT SHALL MANUFACTURER OR SELLER BE LIABLE FOR SPECIAL, INDIRECT OR CONSEQUENTIAL DAMAGES RESULTING FROM THE USE OR HANDLING OF THIS PRODUCT.

THIS SPECIMEN LABEL IS INTENDED FOR USE ONLY AS A GUIDE IN PROVIDING GENERAL INFORMATION REGARDING THE DIRECTIONS, WARNINGS AND CAUTIONS ASSOCIATED WITH THE USE OF THIS PRODUCT. AS WITH ANY AGRICULTURAL CHEMICAL, ALWAYS FOLLOW THE LABEL INSTRUCTIONS ON THE PACKAGE BEFORE USING.

LARVIN® 3.2

THIODICARB INSECTICIDE

Aqueous Flowable

ACTIVE INGREDIENT:
THIODICARB: Dimethyl N, N' [thiobis [(methylimino) carbonyloxy]] bis [ethanimidothioate] 34%
INERT INGREDIENTS: 66%

TOTAL 100%

Contains 3.2 pounds active thiodicarb per U.S. gallon

EPA Registration No. 264-379 EPA Est. No. 264-MO-02

Keep Out of Reach of Children

WARNING
AVISO

PRECAUCION AL USARIO: Si usted no lee ingles, no use este producto hasta que la etiqueta le haya sido explicada ampliamente. (TO THE USER: If you cannot read English do not use this product until the information on the label has been fully explained to you.)

IN CASE OF EMERGENCY—TELEPHONE DAY OR NIGHT (UCC HELP) 1-800-822-4357

FOR GENERAL PRODUCT INFORMATION CALL 1-800-334-9745.

GENERAL INFORMATION

LARVIN® 3.2 Thiodicarb Insecticide is an aqueous flowable formulation that readily disperses in water for spraying by ground or air equipment. Directions on this label are based on tests and field experience relating to effectiveness, impact on environment and residues remaining in food and feed. READ THE ENTIRE LABEL AND OBSERVE ALL LABEL DIRECTIONS AND PRECAUTIONS BEFORE USE.

DIRECTIONS FOR USE

NOTE: IT IS A VIOLATION OF FEDERAL LAW TO USE THIS PRODUCT IN A MANNER INCONSISTENT WITH ITS LABELING.

Do not use this product except as directed on this label or on other Union Carbide Agricultural Products Company, Inc. supplemental labeling for this product.

GENERAL WORKER PROTECTION STATEMENTS

Workers may reenter treated field once spray deposits are dry on foliage.

Do not apply this product in such a manner as to directly or through drift expose workers or other persons. The area being treated must be vacated by unprotected persons.

Do not enter treated areas without protective clothing until sprays have dried.

Because certain states may require more restrictive reentry intervals for various crops treated with this product, consult your State Department of Agriculture for further information.

Written or oral warnings must be given to workers who are expected to be in a treated area or in an area about to be treated with this product. Advise workers to stay out of fields during application and until sprays have dried. Regular long-sleeved work clothing should be worn when working in treated fields. See Precautionary Statement, Statement of Practical Treatment and Note to Physician for information on accidental exposures. When oral warnings are given, warnings shall be given in a language customarily understood by workers. Oral warnings must be given if there is reason to believe that written warnings cannot be understood by workers. Written warnings must include the following information: appropriate signal word (WARNING), area treated with LARVIN 3.2 Thiodicarb Insecticide, date of application, appropriate protective clothing, appropriate reentry interval, if applicable.

STORAGE AND DISPOSAL

STORAGE
Do not store in or around the home. Store unused product in a cool, ventilated, dry, locked area. Do not allow prolonged storage in areas where temperatures frequently exceed 115°F (46°C). Do not contaminate food, water or feed by storage or disposal. NEVER TRANSFER THIS PRODUCT TO ANOTHER CONTAINER FOR STORAGE.

PESTICIDE DISPOSAL
Wastes resulting from the use of this product may be disposed of on site or at an approved waste disposal facility.

CONTAINER DISPOSAL
NEVER REUSE EMPTY CONTAINERS. Triple rinse or equivalent. Then offer for recycling or reconditioning or puncture and dispose of in a sanitary landfill or by incineration, or if allowed by state or local authorities, by burning. If container is burned, stay out of smoke.

ACCIDENTS
In case of a major spill of LARVIN® 3.2 Insecticide:

TELEPHONE COLLECT (24 HOURS A DAY) IN THE U.S.A. 1-800 (UCC-HELP)822-4357.

In case of minor spills or leaks follow all precautions indicated above and cleanup immediately. Soak up with sand, earth or other suitable material and dispose of wastes and broken or empty containers in a landfill approved for pesticide use.

GENERAL DIRECTIONS

SPRAY PREPARATION: TO ASSURE A UNIFORM PRODUCT, AGITATE OR SHAKE ALL CONTAINERS OF THIS PRODUCT PRIOR TO USE. Remove oil, rust, scale, pesticide residues and other foreign matter from sprayer and strainer. Flush with clean water. Use 50 mesh screens or slotted strainers in spray system. To prepare for spraying, fill tank approximately one-half full of water. Add LARVIN 3.2 insecticide and mix thoroughly by mechanical or hydraulic agitation. Finish filling tank with water to desired volume and thoroughly mix. Do not store spray mixture for prolonged periods. If tank-mixes are to be used, LARVIN 3.2 insecticide must be fully dispersed in water first, followed by the addition of the intended tank-mix material.

COMPATIBILITY: LARVIN 3.2 thiodicarb insecticide, when diluted with an equal volume of water, is physically compatible with a wide range of commonly used spray products, but the full range of compatibilities under local conditions is not known. Therefore, it is essential that before using any tank mixture of LARVIN 3.2 insecticide, the compatibility of the mixture be established. Add a small amount of LARVIN 3.2 insecticide to an equal volume of water in a small container and then add the other pesticide or spray product and mix thoroughly. DO NOT USE MIXTURES THAT CURDLE, PRECIPITATE, OR GREASE. THIS PRODUCT IS NOT MISCIBLE WITH VEGETABLE OIL DILUENTS. TANK MIXTURES SHOULD BE USED WITHIN 6 HOURS AFTER MIXING.

NOTE: LARVIN Thiodicarb Insecticide is hydrolytically sensitive to degradation of active ingredient by strong acids, strong bases, and certain heavy metal oxides and salts of certain fungicides such as Maneb, Mancozeb, Copper Count N-8, and Bordeaux mixture. DO NOT USE LARVIN 3.2 INSECTICIDE IN SPRAY MIXTURES WITH THE ABOVE METAL PRODUCTS. DO NOT ADD LARVIN 3.2 INSECTICIDE TO WATER WITH pH VALUES BELOW 3.0 OR ABOVE 8.5. If necessary, water should be buffered within this range before adding LARVIN thiodicarb insecticide.

APPLICATION: Begin application when insect populations reach recognized economic threshold levels. Consult the Cooperative Extension Service, Professional Consultants or other qualified authorities to determine appropriate threshold levels for treatment in your area. Apply a minimum finished spray volume of 3 gallons per acre by air or 15 gallons per acre by ground for sweet corn. Repeat application as needed to maintain control. Use higher dosage rates for heavier infestations or larger larvae. Use lower rates for light to moderate infestations.

To clean the sprayer after use, drain and flush with water.

RECOMMENDED USES

COTTON

INTEGRATED PEST MANAGEMENT (IPM) USE OF THIS PRODUCT: LARVIN® Thiodicarb Insecticide has both contact and residual ovicidal activity which can be effectively utilized in IPM programs (See ovicide section). LARVIN® Thiodicarb Insecticide has been shown to leave populations of beneficial insects and spiders after use. The lower rates allow for maximum beneficial survival and faster rebound of beneficial populations. Larvicidal and ovicidal control coupled with retention of beneficial insects and spiders can offer significant benefits to those producers utilizing integrated pest control programs.

SPRAY VOLUME FOR COTTON: LARVIN® brand 3.2 Thiodicarb Insecticide should be applied in a minimum finished spray volume of 2 gallons per acre by aircraft and 5 gallons per acre by ground equipment (minimum of 3 gal/A in Arizona and 5 gal/A in California with aerial application).

SITE	PEST	DOSAGE PER ACRE POUNDS ACTIVE	FLUID OUNCES	SPECIFIC DIRECTIONS	USE RESTRICTIONS
COTTON	Ovicidal/Larvicidal Cotton bollworm Tobacco budworm	0.6	24.0	LARVIN® Thiodicarb Insecticide is a direct contact and residual contact ovicide. Thorough spray coverage is necessary as eggs must be contacted by LARVIN© Thiodicarb Insecticide in order for activity to occur. Eggs laid on untreated foliage or not contacted by LARVIN® Thiodicarb Insecticide will not be controlled. Initiate sprays when eggs and sufficient newly hatched larvae are present.	Do not allow livestock to graze treated fields. Do not apply less than 28 days before harvest.
	Cotton Bollworm Tobacco budworm Armyworms (such as: Beet, Fall, etc.) Cabbage looper Soybean looper Cotton leaf-perforator Cotton leafworm Pink bollworm	0.6 to 0.9	24.0 to 36.0	As pest pressure increases, reduce intervals between applications and/or increase dosage rate as indicated in specific directions below. **A rate of 24 fluid ounces (0.60 lbs. active) should be used when:** Maximum survival of beneficial insects and spiders is desired. **Rates of 30 to 36 fluid ounces (0.75 to 0.90 lbs. active) should be used when:**	

SITE	PEST	DOSAGE PER ACRE POUNDS ACTIVE	FLUID OUNCES	SPECIFIC DIRECTIONS	USE RESTRICTIONS
	Fleahopper (Suppression) Plantbugs (Suppression)			Insect populations are heavy or infestation is predominately large larvae. Larvae present in bolls will not be controlled. Applications for control of Pink Bollworm should be timed for the presence of adult moths.	
	Boll weevil (Suppression)	0.6 to 0.9	24.0 to 36.0	**LIGHT BOLL WEEVIL POPULATIONS:** When spraying LARVIN® Thiodicarb Insecticide on a 5 to 7 day worm control schedule, control of boll weevil will usually be adequate to main-tain boll weevil populations below established economical thresholds. **MODERATE TO HEAVY BOLL WEEVIL POPULATIONS:** When spraying LARVIN® Thiodicarb Insecticide on a worm control schedule, intersprays of a registered boll weevil control product and/or tank mix of LARVIN plus a registered boll weevil product should be used. A 3-5 day interval between sprays may be required to maintain boll weevil control under heavy infestations.	
	Cutworms	See below.		See "SPECIAL DIRECTIONS ON CUTWORMS."	Do not allow livestock to graze treated fields. Do not apply less than 28 days before harvest.
SOYBEANS	Podworms (Bollworm, Corn earworm, Tobacco budworm) Armyworms (Such as: Beet, Fall, Southern, Yellow-striped, etc.) Green cloverworm Velvetbean caterpillar Wooly Bear	0.25 to 0.4	10.0 to 16.0	Use the lower rates for low to moderate populations and maximum protection of beneficials. Apply in a minimum finished spray volume of 2 gallons per acre by air or 5 gallons per acre by ground.	Do not feed forage, hay, or straw to livestock. Do not apply less than 28 days before harvest.
	Soybean looper Cabbage looper Mexican bean beetle Bean leaf beetle Stinkbugs (Suppression) Three-Cornered alfalfa hopper (Suppression)	0.45 to 0.75	18.0 to 30.0		
	Cutworms	See below.		See "SPECIAL DIRECTIONS ON CUTWORMS."	
SWEET CORN (Fresh Market Only)	Corn earworm European corn borer Armyworms Western bean cutworm	0.5 to 0.75	20 to 30 (Florida only)	**Insects attacking silks and ears:** Apply at 1-7 day intervals starting when silks first appear and continuing until silks begin to dry or infestation potential ceases. Follow local recommendations for proper timing. **Whorl or foliage feeders:** For best results use sprays of 15 gallons or more per acre applied by ground equipment using solid or hollow cone nozzles directed into the plant whorl. Also refer to general directions for use.	After silk initiation, do not exceed 7.5 pounds of active ingredient (300 fluid ounces of LARVIN 3.2) per acre per season. Do not allow livestock to graze treated field. Do not feed treated corn silage (green plant) or fodder to livestock. Ears may be harvested on day of last application.

SPECIAL DIRECTIONS ON CUTWORMS

SITE	PEST	DOSAGE PER ACRE POUNDS ACTIVE	FLUID OUNCES	SPECIFIC DIRECTIONS	USE RESTRICTIONS
CROPS INDICATED ABOVE	Cutworms	0.5 to 0.75	20.0 to 30.0	**Postemergence rescue treatments:** For best results, apply as a broadcast spray when cutworms are actively feeding. If cutworms are feeding below the ground, control will be more difficult. Light cultivation with a rotary hoe before spraying may enhance control. If banding, use a minimum width of 10 inches over the row. When banding, determine the amount of chemical to use per acre, by dividing the bank width by the row width and multiplying by the appropriate broadcast rate. Apply in a minimum finished spray volume of 3 gallons per acre by air or 15 gallons per acre by ground.	Refer to each specific crop for use restrictions.

FIELD USE CONVERSION TABLE

DOSAGE PER ACRE Pounds Active	Fluid Ounces	Acres treated per gallon LARVIN® 3.2
0.25	10.0	12.8
0.33	13.2	9.7
0.40	16.0	8.0
0.45	18.0	7.1
0.50	20.0	6.4
0.55	22.0	5.9
0.60	24.0	5.3
0.65	26.0	4.9
0.70	28.0	4.6
0.75	30.0	4.2
0.90	36.0	3.6

1U.S. Gallon Equals 128 Fluid Ounces

PRECAUTIONARY STATEMENTS

WARNING

HAZARDS TO HUMANS AND DOMESTIC ANIMALS

May be harmful or fatal if swallowed. May be harmful if inhaled or absorbed through the skin. May be irritating to eyes and may cause mild skin sensitization. Do not take internally. Avoid contact with skin, eyes or clothing. Avoid breathing spray mist.

Wear long-sleeved clothing while using. Remove contaminated clothing daily; wash before reuse. Wash exposed skin thoroughly with soap and water after use and before eating or smoking. Discontinue use if allergic reaction occurs.

SIGNS AND SYMPTONS OF OVEREXPOSURE:

Salivation, Watery eyes, Pinpoint eye pupils, Blurred vision, Muscle tremors, Difficult breathing, Excessive sweating, Abdominal cramps, Nausea, Vomiting, Diarrhea, Weakness, Headache. In severe cases, convulsions, unconsciousness and respiratory failure may occur.

ANTIDOTE STATEMENT

ATROPINE SULFATE HIGHLY EFFECTIVE AS AN ANTIDOTE. See NOTE TO PHYSICIAN below.

STATEMENT OF PRACTICAL TREATMENT

GENERAL

Contact a physician immediately in all cases of suspected poisoning. If breathing stops, establish an airway, start artificial respiration and provide oxygen. Make certain to remove all sources of continuing contamination. Remove clothing and wash skin and hair immediately with large amounts of water. Transport the patient to a physician or hospital immediately and SHOW A COPY OF THIS LABEL TO THE PHYSICIAN.

IF SWALLOWED: Drink 1 or 2 glasses of water and induce vomiting by touching back of throat with finger. Never give anything by mouth to an unconscious or convulsing person. Call a physician and follow General advice listed above.

IF IN EYES: Flush eyes with plenty of water. Get medical attention if irritation persists.

IF ON SKIN: Wash thoroughly with soap and water.

IF INHALED: Call a physician and follow advice listed above.

NOTE TO PHYSICIAN

LARVIN® thiodicarb is an oxime carbamate insecticide, which is a cholinesterase inhibitor. Overexposure to this substance may cause toxic signs and symptoms due to stimulation of the cholinergic nervous system. These effects of overexposure are spontaneously and rapidly reversible.

Specific treatment consists of parenteral atropine sulfate. Caution should be maintained to prevent overatropinization. Mild cases may be given 1 to 2 mg intramuscularly every 10 minutes until full atropinization has been achieved and repeated thereafter whenever symptoms reappear. Severe cases should be given 2 to 4 mg intravenously every 10 minutes until fully atropinized, then intramuscularly every 30 to 60 minutes as needed to maintain the effect for at least 12 hours. Dosages for children should be appropriately reduced. Complete recovery from overexposure is to be expected within 24 hours.

Narcotics and other sedatives should not be used. Further, drugs like 2-PAM (pyridine-2-aldoxime methiodide) are NOT recommended unless organophosphate intoxication is also suggested.

To aid in confirmation of a diagnosis, urine samples should be obtained within 24 hours of exposure and immediately frozen. Analyses will be arranged by Union Carbide Agricultural Products Company, Inc.

Consultation on therapy can be obtained at all hours by calling the Union Carbide emergency number 1-800-UCC-HELP.

ENVIRONMENTAL HAZARDS

Avoid direct applications to lakes, streams and ponds. Do not apply when weather conditions favor drift from area treated. This product is toxic to bees exposed to direct treatment. Do not apply when bees are actively visiting the treatment area.

LIMITED WARRANTY AND DISCLAIMER

The manufacturer warrants (a) that this product conforms to the chemical description on the label; (b) that this product is reasonably fit for the purposes set forth in the directions for use when it is used in accordance with such directions; and (c) that the directions, warnings and other statements on this label are based upon responsible experts' evaluation of reasonable tests of effectiveness, of toxicity to laboratory animals and to plants, and of residues on food crops, and upon reports of field experiences. Tests have not been made on all varieties or in all states or under all conditions. THE MANUFACTURER NEITHER MAKES NOR INTENDS, NOR DOES IT AUTHORIZE ANY AGENT OR REPRESENTATIVE TO MAKE, ANY OTHER WARRANTIES, EXPRESS OR IMPLIED, AND IT EXPRESSLY EXCLUDES AND DISCLAIMS

ALL IMPLIED WARRANTIES OF MERCHANTABILITY OR FITNESS FOR A PARTICULAR PURPOSE.

THIS WARRANTY DOES NOT EXTEND TO, AND THE BUYER SHALL BE SOLELY RESPONSIBLE FOR, ANY AND ALL LOSS OR DAMAGE WHICH RESULTS FROM THE USE OF THIS PRODUCT IN ANY MANNER WHICH IS INCONSISTENT WITH THE LABEL DIRECTIONS, WARNINGS OR CAUTIONS.

BUYER'S EXCLUSIVE REMEDY AND MANUFACTURER'S OR SELLER'S EXCLUSIVE LIABILITY FOR ANY AND ALL CLAIMS, LOSSES, DAMAGES, OR INJURIES RESULTING FROM THE USE OF HANDLING OF THIS PRODUCT, WHETHER OR NOT BASED IN CONTRACT, NEGLIGENCE, STRICT LIABILITY IN TORT OR OTHERWISE, SHALL BE LIMITED, AT THE MANUFACTURER'S OPTION, TO REPLACEMENT OF, OR THE REPAYMENT OF THE PURCHASE PRICE FOR, THE QUANTITY OF PRODUCT WITH RESPECT TO WHICH DAMAGES ARE CLAIMED. IN NO EVENT SHALL MANUFACTURER OR SELLER BE LIABLE FOR SPECIAL, INDIRECT OR CONSEQUENTIAL DAMAGES RESULTING FROM THE USE OR HANDLING OF THIS PRODUCT.

NOTICE TO THE BUYER

Purchase of this material does not confer any rights under patents governing this product or the use thereof in countries outside of the United States.

THIS SPECIMEN LABEL IS INTENDED FOR USE ONLY AS A GUIDE IN PROVIDING GENERAL INFORMATION REGARDING THE DIRECTIONS, WARNINGS AND CAUTIONS ASSOCIATED WITH THE USE OF THIS PRODUCT. AS WITH ANY AGRICULTURAL CHEMICAL, ALWAYS FOLLOW THE LABEL INSTRUCTIONS ON THE PACKAGE BEFORE USING.

NAA-800

For Control of Pre-harvest Drop of Apples and Pears

ACTIVE INGREDIENT:
*Potassium salt of
1-Naphthaleneacetic acid 24.2%
INERT INGREDIENTS: 75.8%
TOTAL 100.0%

*1-Naphthaleneacetic acid equivalent 20.1% by weight or 800 grams per gallon.

E.P.A. Reg. No. 264-142AA E.P.A. Est. No. 264-MO-01

Keep Out of Reach of Children
CAUTION

IN CASE OF EMERGENCY TELEPHONE (24 HOURS A DAY) IN U.S.A. 1-800-UCC-HELP.

FOR GENERAL PRODUCT INFORMATION CALL 1-800-334-9745.

DIRECTIONS FOR USE

IT IS A VIOLATION OF FEDERAL LAW TO USE THIS PRODUCT IN A MANNER INCONSISTENT WITH ITS LABELING.

Do not use same spray equipment for other purpose unless thoroughly cleaned.

FOR CONTROL OF PRE-HARVEST DROP OF APPLES & PEARS

DIRECTIONS—PRE-HARVEST DROP

NAA-800 contains 800 grams of naphthaleneacetic acid equivalent per gallon and has been formulated for aerial and ground applications to control pre-harvest drop of apples and pears.

It is important that aerial operators apply NAA-800 under conditions where drift can be minimized so that material is equally distributed over the area being sprayed.

Applications should be completed in time to become effective before actual pre-harvest drop begins. NAA-800 becomes effective in one or two days when applied to summer and early fall maturing varieties. For later varieties ripening during cooler weather, a slightly longer period is necessary. Healthy foliage, freedom from mites and frost damage is necessary for the most effective action from this material. For maximum effectiveness, sprays should be applied when the temperature is 70°F or higher.

For aerial application use at least 5 gallons of water per acre and for ground applications use sufficient amounts of water to insure good coverage.

AMOUNTS PER ACRE

APPLES

Proper timing of sprays to reduce pre-harvest drop is extremely important. Make applications as the first few sound, uninjured fruit begin to drop. With summer varieties, 1/4 pint NAA-800 per acre should be used, especially in hot weather.

For a variety such as McIntosh, 1/3 pint NAA-800 per acre is suggested. A second application of the same amount can be applied 5 to 6 days after the first application to an area that will not be picked within 7 to 10 days of first application.

With later varieties, such as Delicious, 1/3 pint NAA-800 for one acre of apples is suggested.

PEARS

Make applications as the first few sound uninjured fruit begin to drop. Use 1/8 to 1/4 pint of NAA-800 per acre.

IMPORTANT: NAA-800 inhibits fruit drop, but does not affect normal ripening. The fruit must be picked at the proper stage of maturity even though NAA-800 is still effectively delaying excessive drop. Do not apply within 2 days of harvest. Do not make more than 2 applications per growing season.

PRECAUTIONARY STATEMENTS

CAUTION

HAZARD TO HUMANS AND DOMESTIC ANIMALS

Harmful if swallowed. Avoid contact with skin, eyes and clothing. Avoid breathing spray mist.

STATEMENT OF PRACTICAL TREATMENT

IN CASE OF EYE CONTACT: Flush with water for at least 15 minutes.

IN CASE OF SKIN CONTACT: Wash with soap and plenty of water. Remove and wash contaminated clothing before reuse.

ENVIRONMENTAL HAZARDS

Do not apply directly to water. Do not contaminate water by cleaning of equipment or disposal of wastes. Do not contaminate irrigation ditches or water used for irrigation or domestic purposes. Do not apply when weather conditions favor drift from treated areas. Do not use in a greenhouse.

REENTRY STATEMENT

Do not apply this product in such a manner as to directly or through drift expose workers or other persons. The area treated must be vacated by unprotected persons.

Do not enter treated areas without protective clothing until sprays have dried.

Because certain states may require more restrictive reentry intervals for various crops treated with this product, consult your State Department of Agriculture for further information.

Written or oral warnings must be given to workers who are expected to be in a treated area or in an area about to be treated with this product. Advise workers to stay out of fields during application and until sprays have dried. Regular long-sleeved work clothing should be worn when working in treated fields. See PRECAUTIONARY STATEMENTS; STATEMENT OF PRACTICAL TREATMENT for information on accidental exposures. When oral warnings are given, warnings shall be given in a language customarily understood by workers. Oral warnings must be given if there is reason to believe that written warnings cannot be understood by workers. Written warnings must include the following information: appropriate signal word (CAUTION), area treated with NAA-800, date of application, appropriate protective clothing, and reentry interval (i.e., until sprays have dried).

STORAGE AND DISPOSAL

STORAGE

Do not store near fertilizers, seeds, insecticides or fungicides.

PESTICIDE DISPOSAL

Do not contaminate water, food or feed by storage or disposal. Wastes resulting from the use of this product may be disposed of on site or at an approved waste disposal facility.

CONTAINER DISPOSAL

Triple rinse (or equivalent). Then offer for recycling or reconditioning, or puncture and dispose of in a sanitary landfill, or incineration, or if allowed by state and local authorities, by burning. If container is burned, stay out of smoke.

LIMITED WARRANTY AND DISCLAIMER

The manufacturer warrants (a) that this product conforms to the chemical description on the label; (b) that this product is reasonably fit for the purposes set forth in the directions for use when it is used in accordance with such directions; and (c) that the directions, warnings and other statements on this label are based upon responsible experts' evaluation of reasonable tests of effectiveness, of toxicity to laboratory animals and to plants, and of residues on food crops and upon reports of field experience. Tests have not been made on all varieties or in all states or under all conditions. THE MANUFACTURER NEITHER MAKES NOR INTENDS, NOR DOES IT AUTHORIZE ANY AGENT OR REPRESENTATIVE TO MAKE, ANY OTHER WARRANTIES, EXPRESS OR IMPLIED, AND IT EXPRESSLY EXCLUDES AND DISCLAIMS ALL IMPLIED WARRANTIES OF MERCHANTABILITY OR FITNESS FOR A PARTICULAR PURPOSE.

THIS WARRANTY DOES NOT EXTEND TO, AND THE BUYER SHALL BE SOLELY RESPONSIBLE FOR, ANY AND ALL LOSS OR DAMAGE WHICH RESULTS FROM THE USE OF THIS PRODUCT IN ANY MANNER WHICH IS INCONSISTENT WITH THE LABEL DIRECTIONS, WARNINGS, OR CAUTIONS.

BUYER'S EXCLUSIVE REMEDY AND MANUFACTURER'S OR SELLER'S EXCLUSIVE LIABILITY FOR ANY AND ALL CLAIMS, LOSSES, DAMAGES, OR INJURIES RESULTING FROM THE USE OR HANDLING OF THIS PRODUCT, WHETHER OR NOT BASED IN CONTRACT, NEGLIGENCE, STRICT LIABILITY IN TORT OR OTHERWISE SHALL BE LIMITED, AT THE MANUFACTURER'S OPTION, TO REPLACEMENT OF, OR THE REPAYMENT OF THE PURCHASE PRICE FOR, THE QUANTITY OF PRODUCT WITH RESPECT TO WHICH DAMAGES ARE CLAIMED, IN NO EVENT SHALL MANUFACTURER OR SELLER BE LIABLE FOR SPECIAL, INDIRECT OR CONSEQUENTIAL DAMAGES RESULTING FROM THE USE OR HANDLING OF THIS PRODUCT.

THIS SPECIMEN LABEL IS INTENDED FOR USE ONLY AS A GUIDE IN PROVIDING GENERAL INFORMATION REGARDING THE DIRECTIONS, WARNINGS AND CAUTIONS ASSOCIATED WITH THE USE OF THIS PRODUCT. AS WITH ANY AGRICULTURAL CHEMICAL, ALWAYS FOLLOW THE LABEL INSTRUCTIONS ON THE PACKAGE BEFORE USING.

PREP™

PLANT REGULATOR FOR COTTON

For Commercial Use or Agricultural Use Only.
Not for Residential Use.

ACTIVE INGREDIENT:	
Ethephon (2-chloroethyl) phosphonic acid*	39.9%
INERT INGREDIENTS:	60.1%
TOTAL	100.0%

This product contains 4 pounds ethephon per gallon.

E.P.A. Reg. No. 264-380 E.P.A. Est. No. 264-PA-01

Keep Out of Reach of Children

DANGER

IN CASE OF EMERGENCY TELEPHONE COLLECT (24 HOURS A DAY) IN U.S.A. (UCC-HELP) 1-800-822-4357.

FOR GENERAL PRODUCT INFORMATION CALL 1-800-334-9745.

GENERAL INFORMATION

A foliar spray of PREP™ Plant Regulator will accelerate opening of mature unopened cotton bolls and enhance defoliation which can result in earlier harvest with an increased recoverable yield. PREP™ Plant Regulator treatment allows increased efficiency from a once-over harvest.

Some premature drop of small, immature bolls may be associated with treatment. Do not apply before sufficient, mature, unopened bolls have developed to produce the desired yield of cotton.

A boll is mature when it is too hard to be dented when squeezed between thumb and fingers, too hard to be sliced with a sharp knife, and when the seed coat becomes light brown in color.

For best results, thorough and uniform spray coverage is required. It is essential that the bolls are contacted by the spray to obtain optimum boll opening. Observe treated fields closely. Harvest at optimum boll opening and defoliation, generally 14 to 21 days after treatment. Sustained cold weather (65°F or lower) after treatment with PREP™ Plant Regulator will delay the boll opening response.

DO NOT APPLY PREP™ PLANT REGULATOR IF RAIN IS EXPECTED WITHIN 6 TO 8 HOURS. Rainfall within 6 to 8 hours of application may reduce product performance.

Certified cotton seed growers should observe boll and seed maturity before treatment with PREP™ Plant Regulator. Bolls not completely mature might open with treatment of PREP™ Plant Regulator giving lower quality seed.

Spray deposits of PREP™ Plant Regulator, when left on aircraft acrylic plastic type windshields for longer than 1 hour, may react with this material, forming an opaque film or spots on the surface which cannot be removed by washing with water or any other cleaner. Rinse windshields with water immediately after each tankload of PREP™ Plant Regulator has been applied. Thoroughly rinse the aircraft with clean water after each day PREP™ Plant Regulator is applied.

PREP™ Plant Regulator may also affect certain paints, and if accidental deposits on painted objects such as cars, trucks, agricultural equipment, etc. occurs, these objects should be rinsed with water as soon as possible (within 1 hour) to prevent permanent staining of the painted surface.

SPRAY PREPARATION

Add $\frac{1}{2}$ to $\frac{3}{4}$ of the required amount of water to the spray tank. Start agitation. Add the required amount of PREP™ Plant Regulator, and the remaining amount of water. Prepare only as much spray solution as can be used on the day of mixing. Do not allow spray solution to stand overnight.

Do not spill the concentrated product on spray equipment, or any airplane parts. ANY SPILLS SHOULD BE RINSED IMMEDIATELY WITH PLENTY OF WATER.

Use of a nurse tank is highly recommended for avoiding possible spills of concentrated formulation on spray equipment or any airplane parts.

EQUIPMENT CLEANING
Because of the acidic nature of this product, prolonged exposure to spray deposit will damage acrylic plastics, certain paints, and metals.

Rinse thoroughly with a detergent and water all exposed acrylic plastic-type materials (e.g., aircraft windshields), and painted surfaces **within an hour** after exposure to spray deposits.

At the end of each day, rinse thoroughly with a detergent and water all the metal parts of the aircraft and the associated spray equipment exposed to the spray deposits.

DIRECTIONS FOR USE

IT IS A VIOLATION OF FEDERAL LAW TO USE THIS PRODUCT IN A MANNER INCONSISTENT WITH ITS LABELING.

BOLL OPENING

Rates For Boll Opening

Conditions Expected 5-7 Days After Treatment	PREP™ Quarts/Acre	One Gallon Of PREP™ Treats	Spray Volume Gal./Acre Ground	Spray Volume Gal./Acre Air
Hot and dry 80°F or higher	1.0 (1.0 lb a.i.)	4 Acres	15 to 50	2 to 5*
Dry with temperature 75°F to 80°F	1.5 (1.5 lb a.i.)	2.5 Acres	15 to 50	2 to 5*
Cool but higher than 65°F	2.0 (2.0 lb a.i.)	2 Acres	15 to 50	2 to 5*
Rank cotton	2.0 (2.0 lb a.i.)	2 Acres	25 to 50	5

* NOTE: For California and Arizona use a volume of no less than 5 gallons per acre for aerial applications

Time of Application (Boll Opening)
Apply when 30% to 60% of the bolls are open and/or there are sufficient mature, unopened bolls to produce the desired yield (see General Information section). IN CALIFORNIA AND ARIZONA, DO NOT TANK MIX PREP™ PLANT REGULATOR WITH A DEFOLIANT OR DESICCANT.

PRECONDITIONING FOR DEFOLIATION

Rates For Preconditioning For Defoliation

Conditions Expected 5-7 Days After Treatment	PREP™ Quarts/Acre	One Gallon Of PREP™ Treats	Spray Volume Gal./Acre Ground	Spray Volume Gal./Acre Air
Hot and dry 80°F or higher	0.5 (0.5 lb a.i.)	8 Acres	15 to 50	2 to 5*
Cool but higher than 65° and also for rank cotton	1.0 (1.0 lb a.i.)	4 Acres	15 to 50	2 to 5*

* NOTE: For California and Arizona use a volume of no less than 5 G.P.A. for area applications.

Time of Application (Preconditioning)
Apply PREP™ 4 to 7 days prior to application of defoliant or tank mixed with the defoliant (see section on compatibility). IN CALIFORNIA AND ARIZONA, DO NOT TANK MIX PREP™ PLANT REGULATOR WITH A DEFOLIANT OR DESICCANT.

Pre-treatment With Defoliants Prior to PREP™ Plant Regulator Treatment (Not in California and Arizona).
If the cotton is overly rank or laying down in the middles and good spray coverage of the bolls with PREP™ Plant Regulator is difficult, a pre-treatment with defoliants will be useful to improve boll coverage with PREP™ Plant Regulator. Use dosage rates of PREP™ Plant Regulator recommended for boll opening. Read and observe all appropriate label use directions and precautions for the defoliant used.

COMPATIBILITY
PREP™ Plant Regulator is compatible with DEF, FOLEX, DROPP, HARVADE, METHYL PARATHION, GUTHION AND MALATHION, and may be applied in sequence or as a tank mixture (DO NOT TANK MIX WITH DEFOLIANTS OR DESICCANTS IN CALIFORNIA AND ARIZONA). In some cases slight reduction in boll opening response has been observed when tank mixes with defoliants were used.

Good agitation in the spray tank is essential and a tank mixture should not be allowed to stand without agitation for more than 5 to 10 minutes. Read and observe all appropriate label use directions and precautions for the defoliants and insecticides used.

PRECAUTION: DO NOT TANK MIX PREP™ PLANT REGULATOR WITH DEFOLIANTS CONTAINING SODIUM CHLORATE BECAUSE THIS RESULTS IN THE FORMATION OF HYPOCHLOROUS ACIDS WHICH UPON HEATING EMIT TOXIC CHLORINE FUMES.

WHEN TO HARVEST: Do not harvest cotton sooner than 7 days after a treatment with PREP™ Plant Regulator. Observe the treated crop and harvest when optimum boll opening has been reached. Too early harvest might reduce the full advantage of the treatment and too late a harvest may result in reduced quality and loss of lint which will drop from the plant.

USE PRECAUTIONS
Avoid spray drift to nearby crops as this product will cause modifications in plant growth. Plant injury or reduced yields will result.

Do not plant another crop within 30 days after treatment.

Mix only the amount of spray you expect to use each day. Do not allow mixed solution to stand overnight.

PRECAUTIONARY STATEMENTS

DANGER

HAZARDS TO HUMANS AND DOMESTIC ANIMALS

CORROSIVE. Causes eye and skin damage. May be fatal if absorbed through skin. Harmful if swallowed or inhaled. Avoid breathing spray mist. Do not get in eyes, on skin or on clothing. Wear protective clothing and goggles or face shield when handling. Remove and wash contaminated clothing before reuse.

STATEMENT OF PRACTICAL TREATMENT

If in eyes, flush with water for at least 15 minutes and call a physician.

If on skin or clothing, remove clothing and wash skin thoroughly with soap and water, get medical attention.

If swallowed, promptly drink a large quantity of milk, egg whites, gelatin solution or if these are not available, drink large quantities of water. Do not induce vomiting. Avoid alcohol. Call a physician immediately.

If inhaled, remove victim to fresh air. If not breathing, give artificial respiration, preferably mouth to mouth.

NOTE TO PHYSICIAN
PREP™ Plant Regulator possesses characteristics of a strong acid. Probable mucosal damage may contraindicate the use of gastric lavage or induction of vomiting. Measures against circulatory shock, respiratory depression and convulsion may be needed.

ENVIRONMENTAL HAZARDS
Do not apply directly to water. Do not contaminate water used for irrigation or domestic purposes.

STORAGE AND DISPOSAL

STORAGE

Do not contaminate water, food, or feed by storage or disposal.

PESTICIDE DISPOSAL

Pesticide wastes are acutely hazardous. Improper disposal of excess pesticide, spray mixture, or rinsate is a violation of Federal law. If these wastes cannot be disposed of by use according to label instructions, contact your State Pesticide or Environmental Control Agency or the Hazardous Waste representative at the nearest EPA Regional Office for guidance.

CONTAINER DISPOSAL

Triple rinse (or equivalent) the empty containers. Then offer for recycling or reconditioning, or puncture and dispose of in a sanitary landfill, or by incineration or if allowed by state and local authorities, by burning. If container is burned, stay out of smoke.

LIMITED WARRANTY AND DISCLAIMER

The manufacturer warrants (a) that this product conforms to the chemical description on the label; (b) that this product is reasonably fit for the purposes set forth in the directions for use when it is used in accordance with such directions; and (c) that the directions, warning and other statements on this label are based upon responsible experts' evaluation of reasonable tests of effectiveness, of toxicity to laboratory animals and to plants, and of residues on food crops and upon reports of field experience. Tests have not been made on all varieties or in all states or under all conditions. THE MANUFACTURER NEITHER MAKES NOR INTENDS, NOR DOES IT AUTHORIZE ANY AGENT OR REPRESENTATIVE TO MAKE, ANY OTHER WARRANTIES, EXPRESS OR IMPLIED, AND IT EXPRESSLY EXCLUDES AND DISCLAIMS ALL IMPLIED WARRANTIES OF MERCHANTABILITY OR FITNESS FOR A PARTICULAR PURPOSE.

THIS WARRANTY DOES NOT EXTEND TO, AND THE BUYER SHALL BE SOLELY RESPONSIBLE FOR, ANY AND ALL LOSS OR DAMAGE WHICH RESULTS FROM THE USE OF THIS PRODUCT IN ANY MANNER WHICH IS INCONSISTENT WITH THE LABEL DIRECTIONS, WARNINGS OR CAUTIONS.

BUYER'S EXCLUSIVE REMEDY AND MANUFACTURER'S OR SELLER'S EXCLUSIVE LIABILITY FOR ANY AND ALL CLAIMS, LOSSES, DAMAGES, OR INJURIES RESULTING FROM THE USE OR HANDLING OF THIS PRODUCT, WHETHER OR NOT BASED IN CONTRACT, NEGLIGENCE, STRICT LIABILITY IN TORT OR OTHERWISE SHALL BE LIMITED, AT THE MANUFACTURER'S OPTION, TO REPLACEMENT OF, OR THE REPAYMENT OF THE PURCHASE PRICE FOR, THE QUANTITY OF PRODUCT WITH RESPECT TO WHICH DAMAGES ARE CLAIMED. IN NO EVENT SHALL MANUFACTURER OR SELLER BE LIABLE FOR SPECIAL, INDIRECT OR CONSEQUENTIAL DAMAGES RESULTING FROM THE USE OR HANDLING OF THIS PRODUCT.

THIS SPECIMEN LABEL IS INTENDED FOR USE ONLY AS A GUIDE IN PROVIDING GENERAL INFORMATION REGARDING THE DIRECTIONS, WARNINGS AND CAUTIONS ASSOCIATED WITH THE USE OF THIS PRODUCT. AS WITH ANY AGRICULTURAL CHEMICAL, ALWAYS FOLLOW THE LABEL INSTRUCTIONS ON THE PACKAGE BEFORE USING.

SEVIN® BRAND XLR PLUS

CARBARYL INSECTICIDE

For Agricultural or Commercial Use Only

ACTIVE INGREDIENT:	% BY WEIGHT
Carbaryl (1-naphthyl N-methylcarbamate)	43.4%
INERT INGREDIENTS:	56.6%
TOTAL	100.0%

Contains 4 pounds carbaryl per gallon.

E.P.A. Reg. No. 264-333 E.P.A. Est. No. 264-MO-02

Keep Out of Reach of Children

CAUTION

IN CASE OF EMERGENCY TELEPHONE (24 HOURS A DAY) IN U.S.A. 1-800-UCC-HELP.

FOR GENERAL PRODUCT INFORMATION CALL 1-800-334-9745.

PRECAUTIONARY STATEMENTS

CAUTION

HAZARDS TO HUMANS AND DOMESTIC ANIMALS

MAY BE HARMFUL IF SWALLOWED. Avoid breathing spray mist. Do not take internally. Avoid contact with eyes, skin or clothing. Wear regular long-sleeved work clothing. Change to clean clothing daily. Wash hands and face before eating. Wash thoroughly after handling.

ANTIDOTE STATEMENT

ATROPINE SULFATE IS HIGHLY EFFECTIVE AS AN ANTIDOTE. See NOTE TO PHYSICIAN below.

STATEMENT OF PRACTICAL TREATMENT

GENERAL

Contact a physician immediately in all cases of suspected poisoning. Transport to a physician or hospital immediately and SHOW A COPY OF THIS LABEL TO THE PHYSICIAN.

IF SWALLOWED: Never give anything by mouth to an unconscious or convulsing person. If conscious and not convulsing, drink 1 to 2 glasses of water and induce vomiting by touching the back of the throat with finger.

IF IN EYES: Flush eyes with plenty of water. Get medical attention if irritation persists.

IF ON SKIN: Wash thoroughly with soap and water.

IF INHALED: Move from contaminated atmosphere and call a physician.

NOTE TO PHYSICIAN

Carbaryl is a carbamate insecticide, which is a chloinesterase inhibitor. Overexposure to this substance may cause toxic signs and symptoms due to stimulation of the cholinergic nervous system. These effects of overexposure are spontaneously and rapidly reversible.

Specific treatment consists of parenteral atropine sulfate. Caution should be maintained to prevent overatropinization. Mild cases may be given 1 to 2 mg intramuscularly every 10 minutes until full atropinization has been achieved and repeated thereafter whenever symptoms reappear. Severe cases should be given 2 to 4 mg intravenously every 10 minutes until fully atropinized, then intramuscularly every 30 to 60 minutes to maintain the effect for at least 12 hours. Dosages for children should be appropriately reduced. Complete recovery from overexposure is to be expected within 24 hours.

Narcotics and other sedatives should not be used. Further, drugs like 2-PAM (pyridine-2-aldoxime methiodide) are NOT recommended unless organophosphate intoxication is also suggested.

To aid in confirmation of a diagnosis, urine samples should be obtained within 24 hours of exposure and immediately frozen. Analyses will be arranged by Union Carbide Agricultural Products Company, Inc.

Consultation on therapy can be obtained at all hours by calling the Union Carbide emergency number 1-800-UCC-HELP.

ENVIRONMENTAL HAZARDS

This product is extremely toxic to aquatic and estuarine invertebrates. Do not apply directly to water and wetlands, except under the forest canopy and use on rice. Discharge from rice fields may kill aquatic and estaurine invertebrates. Do not apply when weather conditions favor drift from area treated. Do not contaminate water by cleaning equipment or disposal of wastes.

BEE CAUTION

This product is highly toxic to bees exposed to direct treatment on blooming crops or weeds. However, field studies have shown that SEVIN® brand XLR PLUS Carbaryl Insecticide is less hazardous to honey bees than other carbaryl products when direct application to bees is avoided and the spray residues have dried. For maximum honey bee hazard reduction, apply SEVIN® brand XLR PLUS Carbaryl Insecticide from late evening to early morning or when bees are not foraging. Do not apply this product or allow it to drift to blooming crops or weeds if bees are foraging in the treatment area. However, applications may be made during foraging periods if the beekeeper takes one of the following precautionary measures prior to bee flight activity on the day of treatment: (1) Confine the honey bees to the hive by covering the colony or screening the entrance or; (2) locate hives beyond bee flight range from the treated area. Precautionary measures may be discontinued after spray residues have dried. Contact your cooperative Agricultural Extension Service or your local Union Carbide Agricultural Products Company representative for further information.

DIRECTIONS FOR USE

IT IS A VIOLATION OF FEDERAL LAW TO USE THIS PRODUCT IN A MANNER INCONSISTENT WITH ITS LABELING.

GENERAL INFORMATION

SEVIN® brand XLR PLUS Carbaryl Insecticide is a suspension of microfine SEVIN carbaryl insecticide in an aqueous medium. It readily disperses in water to form a spray which may be applied by air or ground. READ THIS LABEL BEFORE USE. STRICTLY OBSERVE LABEL DIRECTIONS AND CAUTIONS, AND APPLICABLE FEDERAL AND STATE REGULATIONS.

GENERAL WORKER PROTECTION STATEMENT

Do not apply this product in such a manner as to directly or through drift expose workers or other persons. The area being treated must be vacated by unprotected persons.

Do not enter treated areas without protective clothing until sprays have dried.

Because certain states may require more restrictive reentry intervals for various crops treated with this product, consult your State Department of Agriculture for further information.

Written or oral warnings must be given to workers who are expected to be in a treated area or in an area about to be treated with this product. Advise workers to stay out of fields during application and until sprays have dried. Regular long-sleeved work clothing should be worn when working in treated fields. See PRECAUTIONARY STATEMENTS, STATEMENT OF PRACTICAL TREATMENT and NOTE TO PHYSICIAN for information on accidental exposures. When oral warnings are given, warnings shall be given in a language customarily understood by workers. Oral warnings must be given if there is reason to believe that written warnings cannot be understood by workers. Written warnings must include the following information: appropriate signal word (CAUTION), area treated with SEVIN® brand XLR PLUS Carbaryl Insecticide, date of application, appropriate clothing, and re-entry interval (i.e., until sprays have dried).

PREHARVEST AND GRAZING USE INFORMATION AND LIMITATIONS

Tolerances established under the Federal Food, Drug and Cosmetic Act permit the sale of crops bearing probably carbaryl residues when this product is used in accordance with the label directions. If used as directed, treated forage may be grazed or used as feed for dairy and meat animals without causing illegal residues in meat or milk. This product may be applied up to and including the day of harvest or grazing of forage crops. Application may be made without removing livestock from area being treated.

Do not plant rotational food and feed crops not listed on this or other carbaryl labels in carbaryl treated soil.

Do not use reclaimed irrigation water from crops treated with carbaryl on upland crops for which carbaryl tolerances are not established.

PLANT RESPONSE PRECAUTIONS

To avoid possible injury to tender foliage, do not apply to wet foliage or when rain or high humidity is expected during the next two days.

Do not use on Boston Ivy, Virginia creeper and maindenhair fern as injury will result. During early season, it may also injure Virginia and sand pines.

Observe label instructions on apple thinning and on combinations with certain herbicides on rice and soybeans.

SPRAY PREPARATION

TO ASSURE A UNIFORM PRODUCT, AGITATE, STIR OR RECIRCULATE ALL CONTAINERS OF SEVIN® BRAND XLR PLUS CARBARYL INSECTICIDE PRIOR TO USE. Remove oil, rust, scale, pesticide residues and other foreign matter from mix tanks and entire spray system. Flush with clean water. Fill spray or mix tank with $\frac{1}{2}$ to $\frac{3}{4}$ the desired amount of water. Start mechanical or hydraulic agitation. Slowly add the required amount of SEVIN® brand XLR PLUS Carbaryl Insecticide, and then the remaining volume of water. Include rinse water from container. Prepare only as much spray mixture as can be applied on the day of mixing. MAINTAIN CONTINUOUS AGITATION DURING MIXING AND APPLICATION TO ASSURE A UNIFORM SUSPENSION. DO NOT STORE SPRAY MIXTURE FOR PROLONGED PERIODS.

WASHOFF RESISTANCE AND COVERAGE

Dilution of 1 volume of SEVIN® brand XLR PLUS Carbaryl Insecticide with 1 volume of water will allow maximum resistance to washoff by rainfall or overhead irrigation. Dilutions higher than 1:39 are not recommended when washoff resistance is desired.

SEVIN® brand XLR PLUS Carbaryl Insecticide must be diluted as stated above, and droplets must dry on the foliage before they become washoff resistant. When atmospheric humidity is low, a drying time of at least two hours is generally adequate. Under high humidity a longer drying time may be required. Washoff resistance cannot be expected if SEVIN® brand XLR PLUS Carbaryl Insecticide is applied to wet foliage which does not dry adequately prior to rainfall or overhead irrigation.

On all crops use sufficient spray volume to obtain thorough coverage. Optimum pest control under certain crop, pest or climatic conditions may require spray gallonages higher than the 1:39 dilution. For example in hot, dry weather, or when humidity is low, higher spray gallonage per area may be required to minimize loss from evaporation and insure thorough coverage. The total spray volume required for effective pest control can best be determined by previous experience, pest and crop conditions and local recommendations.

COMPATABILITY

SEVIN® brand XLR PLUS Carbaryl Insecticide, when diluted with an equal volume of water, is compatible with a wide range of pesticides. It is not compatible with diesel fuel, kerosene, fuel oil or aromatic solvents. If compatibility of SEVIN® brand XLR PLUS Carbaryl Insecticide with another product and the resulting crop response is unknown, it should be tested on a small scale. Curdling, precipitation, greasing, layer formation or increased viscosity are symptoms of incompatibility. Incompatibility will reduce insect control and may cause mechanical difficulties or plant injury. Observe all cautions and limitations on labeling of all products used in mixtures. WHEN PREPARING COMBINATION SPRAYS, FIRST ADD "SEVIN® BRAND XLR PLUS CARBARYL INSECTICIDE" TO AT LEAST AN EQUAL VOLUME OF WATER, MIX THOROUGHLY, AND THEN ADD COMBINATION PRODUCTS TO THE MIXTURE. DO NOT APPLY TANK MIX COMBINATIONS UNLESS YOUR PREVIOUS EXPERIENCE INDICATES THE MIXTURE IS EFFECTIVE AND WILL NOT RESULT IN APPLICATION PROBLEMS, EXCESSIVE RESIDUES OR PLANT INJURY.

Unstable under highly alkaline conditions. Not effective if used with alkaline materials such as Bordeaux, lime-sulfur and casein-lime spreaders.

APPLICATION

On all crops use sufficient gallonage to obtain thorough and uniform coverage.

Calibrate spray equipment to deliver the required volume. The flow rate of SEVIN® brand XLR PLUS Carbaryl Insecticide diluted 1:1 with water is similar to water. Use 50 mesh slotted strainers in spray system and 25 mesh slotted strainers behind nozzles.

To clean spray system after use, drain and flush with a water and detergent mixture. Rinse thoroughly with clean water. Refer to the Storage and Disposal directions for proper disposal of this water and detergent mixture and rinse water.

INSECT CONTROL

Apply when insects or their damage appear. To maintain control repeat at 7 to 14 day intervals or as necessary unless a shorter interval is specified below. Where a dosage range is indicated, use lower rate on young plants and early instars and higher rate on mature plants, advanced instars and adults. Thorough and uniform spray coverage is essential for effective control.

SEVIN® brand XLR PLUS Carbaryl Insecticide does not control spider mites. If spider mites are a problem, use a registered miticide. Regional differences have been noted in the susceptibility of certain strains of fall armyworm to SEVIN® brand XLR PLUS Carbaryl Insecticide. If local experience indicates inadequate control, use an alternative pesticide.

FORAGE, FIELD AND VEGETABLE CROPS

Apply in sufficient spray volume to obtain thorough coverage. Where maximum washoff resistance is needed, use up to a 1:1 (vol/vol) dilution with water. Do not dilute greater than 1:39 where washoff resistance is desired. To prepare small volumes of spray, use 1 tablespoonful (½ fluid ounce) of SEVIN® brand XLR PLUS Carbaryl Insecticide per gallon of water where rates of 1 quart per acre or 1 quart per 100 gallons are indicated in the tables below.

CROP	INSECT		QUARTS OF "SEVIN®" BRAND XLR PLUS/ACRE	PRE-HARVEST INTERVAL (DAYS)	SPECIFIC DIRECTIONS
All Forage, Field and Vegetable Crops in This Section	Grasshoppers		½ to 1½	See specific Forage, Field or Vegetable Crop	Use ½ to 1 quart for nymphs on small plants or sparse vegetation in wasteland, rangeland, ditch banks and borders. Use 1 to 1½ quarts for adult grasshoppers or applications to dense vegetation.
Alfalfa Clovers	Blister beetles Mexican bean beetle		½ to 1	0 (clovers) 3 (alfalfa)	Observe plant response precautions.
	Alfalfa caterpillar Bean leaf beetle Cucumber beetles Green cloverworm Japanese beetle	Leafhoppers Three cornered alfalfa hopper Thrips Velvetbean caterpillar	1		
	Alfalfa-weevil larvae Armyworm Cloverhead weevil Corn earworm Cutworms Egyptian alfalfa weevil larvae Essex skipper	European alfalfa beetle Fall armyworm Lygus bugs Stink bugs Webworms Yellowstriped armyworm	1 to 1½		For alfalfa weevil larvae, if pretreatment damage is extensive, cut alfalfa and treat the stubble. Use higher rate in areas east of the Rocky Mountains. On dense growth use 25 to 40 gallons per acre with ground equipment to ensure adequate coverage.
Asparagus	Asparagus beetle		1 to 2	1	Treat ferns or brush growth. Do not treat more than once every 3 days.
	Apache cicada Asparagus beetle		2 to 4	Post harvest application only.	
Beans (including blackeyed peas, cowpeas, crowder or southern peas, dry beans, green beans, lima beans, navy beans and snap beans)	Blister beetles Mexican bean beetle		½ to 1	0 (except cowpeas) 3 (cowpeas)	Observe plant response precautions.
	Alfalfa caterpillar Bean leaf beetle Cucumber beetles Flea beetles Green cloverworm Japanese beetle	Leafhoppers Three cornered alfalfa hopper Thrips Velvetbean caterpillar Western bean cutworm	1		
	Armyworm Corn earworm Cutworms European corn borer	Fall armyworm Stink bugs Tarnished plant bug Webworms	1 to 1½		
	Cowpea curculio		2		
	Corn earworm Limabean pod borer	Lygus bugs Stink bugs	2		CALIFORNIA ONLY

CROP	INSECT		QUARTS OF "SEVIN®" BRAND XLR PLUS/ACRE	PRE-HARVEST INTERVAL (DAYS)	SPECIFIC DIRECTIONS
Cabbage Broccoli Brussels Sprouts Cauliflower Kohlrabi	Flea beetles	Harlequin bug	½ to 1	3	
	Armyworm Corn earworm Fall armyworm	Imported cabbage-worm	1 to 2		
Chinese cabbage Collards Hanover salad Horseradish Kale Mustard greens Radishes Rutabagas Turnips	Flea beetles Harlequin bug	Leafhoppers	½ to 1	3 (horseradish, radishes, rutabagas and turnip roots)	
	Aster leafhopper		1 to 1½	14 (Chinese cabbage, collards, Hanover salad, kale mustard greens, and turnip tops)	
	Armyworm Corn earworm Fall armyworm	Imported cabbage-worm Stink bugs Tarnished plant bug	1 to 2		
Carrots Celery Parsnips Parsley	Flea beetles	Leafhoppers	½ to 1	0 (carrots)	NOT FOR USE ON CELERY IN CALIFORNIA Treat on a 5 to 7 day schedule.
	Aster leafhopper Lygus bugs	Spittlebugs	1 to 1½	3 (parsnips)	
	Armyworm Corn earworm Fall armyworm	Stink bugs Tarnished plant bug	1 to 2	14 (celery and parsley)	
Corn (field, sweet, pop)	Armyworm Chinch bugs Corn earworm Corn rootworm adults European corn borer Fall armyworm	Flea beetles Japanese beetle Sap beetles Southwestern corn borer Leafhopper	1 to 2	0	OBSERVE BEE CAUTION. For insects attacking silks and ears apply at 1 to 6 day intervals starting when first silks appear and continuing until silks begin to dry. For larvae in whorl and foliage feeders, apply as necessary. Optimum timing and good coverage are essential for effective control. For optimum chinch bug control, apply 20 gallons of water per acre by ground and direct spray toward stalk to provide thorough coverage.
	Western bean cutworm		3		Treat when infestation averages 15% and at 90% to 100% tassel emergence. Treatment after 100% silk emergence will reduce effectiveness.
	Cutworms		2 to 6½		Apply in a 12 inch band, using 6 fluid ounces per 1000 linear feet of row, in at least 15 gallons of water per acre. For broadcast application use 6.5 quarts in at least 20 gallons (ground) or 5 gallons (air) of water per acre.
Cotton	Cotton fleahopper Cotton leafworm Flea beetles	Striped blister beetle Thrips	½ to 1	7	Early season insect control.
	Boll weevil Bollworms Cotton leafperforator Fall armyworm Leafrollers Leafhoppers	Tarnished plant bug Yellowstriped armyworm (cotton cutworm)	1 to 2		Treat on a 5 to 7 day schedule for as long as control is necessary. Mid and late season insect control. May be applied after bolls open.
	Lygus bugs		1 to 2		For light to moderate populations in Western irrigated cotton.
	Pink bollworm		1½ to 2½		Aphid populations will be suppressed by repeated applications of this insecticide.
	Stink bugs	Saltmarsh caterpillar	2		
Cucumber Melons Pumpkin Squash	Pickleworm	Melonworm	½ to 1	0	Observe plant response precautions.
	Cucumber beetles Flea beetles	Leafhoppers Squash bugs	1		Avoid excessive applications.

CROP	INSECT		QUARTS OF "SEVIN®" BRAND XLR PLUS/ACRE	PRE-HARVEST INTERVAL (DAYS)	SPECIFIC DIRECTIONS
Dandelion Endive (Escarole) Lettuce Salsify	Flea beetles Harlequin bug	Leafhoppers	½ to 1	3 (head lettuce & salsify roots) 14 (dandelion, endive (escarole), leaf lettuce & salsify tops)	Observe plant response precautions. Treat on a 5 to 7 day schedule after heads begin to form.
	Aster leafhopper Lygus bugs	Spittlebugs	1 to 1½		
	Armyworm Corn earworm Fall armyworm	Imported cabbage worm Stink bugs Tarnished plant bug	1 to 2		
Forage Grasses Pasture Rangeland	Armyworm Black grass bugs Chinch bugs Essex skipper Fall armyworm	Range caterpillars Range crane fly Striped grass looper Thrips	1 to 1½	0	To control thrips in grasses grown for seed use high spray pressure to improve penetration into boot.
	White grubs (green June beetle)		1½ to 2		
Garden beet Spinach Swiss chard	Flea beetles Harlequin bug	Leafhoppers	½ to 1	3 (garden beet roots) 14 (garden beet tops, spinach, Swiss chard)	Treat on a 5 to 7 day schedule.
	Aster leafhopper		1 to 1½		
	Armyworm Corn earworm Fall armyworm	Stink bugs Tarnished plant bug	1 to 2		
Okra	Corn earworm	Stink bugs	1 to 2	0	Treat on a 5 to 7 day schedule.
Peanuts	Blister beetles	Mexican bean beetle	½ to 1	0	Observe plant response precautions.
	Alfalfa caterpillar Bean leaf beetle Cucumber beetles Green cloverworm Japanese beetle Leafhoppers	Rednecked peanutworm Three cornered alfalfa hopper Thrips Velvetbean caterpillar	1		
	Armyworm Corn earworm Fall armyworm	Stink bugs Webworms	1 to 1½		
	Whitefringed beetle adults Cutworms		2		
Peas	Colorado potato beetle Leafhoppers		1	3	
	Armyworms		1 to 1½		
	Alfalfa caterpillar Cutworms Pea leaf weevil	Pea weevil Yellowstriped armyworm	1½		
	Alfalfa looper		2½		WASHINGTON STATE ONLY
Potato Tomato Eggplant Pepper	Colorado potato beetle Flea beetles Leafhoppers		½ to 1	0	
	European corn borer Fall armyworm Lace bugs Stink bugs	Tarnished plant bug Tomato fruitworm Tomato hornworm Tomato pinworm	1 to 2		
	Cutworms		2		
Rice	Armyworm Chinch bugs	Fall armyworm Stink bugs	1 to 1½	14	CAUTION: May kill shrimp and crabs. Do not use in areas where these are important resources. Do not use on rice fields in which crayfish and catfish farming are included in the cultural practice. DO NOT APPLY PROPANIL HERBICIDES WITHIN 15 DAYS BEFORE OR AFTER APPLICATION OF THIS PRODUCT OR PLANT INJURY WILL RESULT. MISSISSIPPI DELTA & TEXAS.

CROP	INSECT	QUARTS OF "SEVIN®" BRAND XLR PLUS/ACRE	PRE-HARVEST INTERVAL (DAYS)	SPECIFIC DIRECTIONS
	Armyworm, Leafhoppers, Tadpole shrimp	2		CALIFORNIA ONLY. For optimum tadpole shrimp control apply to water when pest first appears.
Sorghums (milo, grain sorghum sweet sorghum and hybrids)	Armyworm, Chinch bugs, Corn earworm, Fall armyworm, Stink bugs, Webworms	1 to 2	21 (grain) 0 (forage)	Direct spray into forming heads for optimum insect control. Treat for sorghum midge when 25 to 30 percent of heads have emerged from boot and are in bloom. Repeat application 3 to 5 days later if adults are still active. A third application may be necessary in late planted sorghum or if midge are abundant. For chinch bugs use high gallonage ground application directed at the base of plants.
	Sorghum midge, Southwestern corn borer	1½		
	Cutworms	2		
Soybeans	Bean leaf beetle, Cucumber beetles, Green cloverworm, Mexican bean beetle, Velvetbean caterpillar	½	0	DO NOT APPLY A COMBINATION OF THIS PRODUCT AND 2,4 DB HERBICIDES TO SOYBEANS AS CROP INJURY MAY RESULT. Use lower rates for light to moderate populations and smaller instars. Use the higher rates for heavy populations and larger instars.
	Corn earworm	½ to ¾ 1 to 1½		
	Blister beetles, Grape colaspis, Mexican bean beetle	½ to 1		
	Alfalfa caterpillar, Bean leaf beetle, Cucumber beetles, Green cloverworm, Japanese beetle, Leafhoppers, Three cornered alfalfa hopper, Thrips, Velvetbean caterpillar	1		
	Armyworm, Cutworms, Fall armyworm, Stink bugs, Webworms	1 to 1½		
	Painted lady (Thistle caterpillar), Saltmarsh caterpillar, Yellowstriped armyworm	2 to 2½		
Sugar beets	Armyworm, Beet leaf beetle, Fall armyworm, Flea beetles, Leafhoppers, Webworms	1 to 1½	14	
	Cutworms	1½		
Sunflower	Cutworms	1½	60	DO NOT USE IN CALIFORNIA.
	Armyworm, Fall armyworm	1½ to 2		
	Stem Weevil, Sunflower Beetle	1 to 2		
Sweet Potato	Corn earworm, Cucumber beetles, Flea beetles, Sweet potato hornworm, Tortoise beetles	1 to 2	0	DO NOT USE IN CALIFORNIA. Apply as a foliar spray as needed.
	Yellowstriped armyworm	2		
	Sweet potato weevil	1 to 2		FOLIAR APPLICATION Full coverage of plants is essential. Use lower rate on young plants and higher rate on mature plants.
Tobacco	Tobacco flea beetle	2 Tbs./gallon or 1 qt./50 gal. and apply 6 gal./100 sq. yards	0	IN PLANT BEDS To prevent plant injury, avoid excessive applications.
	Green June beetle grubs	½ qt./100 gallons		Apply only to areas that larvae have uprooted by sprinkling as a drench with 50 to 100 gallons per 100 square yards.
	Budworms, Fall armyworm, Flea beetles, Hornworms, Japanese beetle, June beetle, Suckfly	1 to 2		IN FIELDS Use lower rate on young plants (up to knee height). Use at least 10 gallons of prepared spray per acre. Begin treatments when worms are small. If late season applications are necessary, allow three days before priming or cutting.

CROP	INSECT	QUARTS OF "SEVIN®" BRAND XLR PLUS/ACRE	PRE-HARVEST INTERVAL (DAYS)	SPECIFIC DIRECTIONS
Wheat	Cereal leaf beetle	1	0 (forage)	DO NOT USE IN CALIFORNIA. Application is effective against eggs, larvae and adults.
	Armyworm Fall armyworm	1 to 1½	21 (grain)	

TREE FRUIT AND NUT CROPS

For dilute sprays apply the specified dosage per 100 gallons of water. For concentrate and aerial sprays increase the concentration of SEVIN® brand XLR PLUS Carbaryl Insecticide in the spray mixture to apply an amount per acre equivalent to that in a dilute spray. The optimum spray gallonage will depend on tree size, density and stage of growth. Typical spray gallonages per acre range from 200-300 gallons for dilute sprays, 30-100 gallons for concentrate sprays and 10-25 gallons for aerial sprays. Do not exceed maximum label rate per acre per application.

CROP	QUARTS OF "SEVIN®" BRAND XLR PLUS/ 100 GAL.	PRE-HARVEST INTERVAL (DAYS)	SPECIFIC DIRECTIONS
APPLE THINNING Apples only	¼ to ½	1	Apply 1 full coverage dilute spray between 10 and 25 days after full bloom. If factors such as tree age, variety, nutrition, previous crop, pruning, bloom and degree of set favor excessive fruit thinning with this product, exercise caution to avoid possible yield reduction. Consult with your County Extension Service or other experts for advice on the proper use of this product. For easily thinned varieties including Cortland, Grimes, Jonathan, McIntosh, Orleans, Rome Beauty, Puritan, Red Delicious, Winesap, Yellow Newton.
	½ to 1	1	For difficult to thin varieties including Baldwin, Ben Davis, Duchess, Early McIntosh, Golden Delicious, Lady Apple, Northern Spy, Rhode Island Greening, Steele Red, Turley, Wealthy, Yellow Transparent and York Imperial.

CROP	INSECT	QUARTS OF "SEVIN®" BRAND XLR PLUS/ACRE	PRE-HARVEST INTERVAL (DAYS)	SPECIFIC DIRECTIONS
Almond	Peach twig borer San Jose scale Fruittree leafroller	1	28	Apply in "popcorn" or petal fall stages and again when the May brood of the peach twig borer begins to hatch or thereafter as needed.
	Navel orangeworm			Time early and mid season applications to correspond to moth flight peaks. Make a late season application at initiation of hull split or up to 10% hull split. Do not apply more than 5 quarts per acre.
Apples Pears	Apple aphid Apple rust mite Apple sucker Bagworms California peaslug (pear sawfly) Codling moth Eyespotted bud moth Green fruitworm Lecanium scales Lesser appleworm Lygus bugs Orange tortrix Oystershell scale Pear leaf blister mite Pear psylla Pear rust mite San Jose scale Tentiform leafminers Woolly apple aphid	¾ to 1	1	WEST OF THE ROCKY MOUNTAINS To avoid undesired apple thinning, delay use until at least 30 days after full bloom. For psylla control apply when eggs hatch or young nymphs are present.
	Apple mealybug Apple aphid Codling moth White apple leafhopper	½	1	EAST OF THE ROCKY MOUNTAINS To avoid undesired apply thinning, delay use until at least 30 days after full bloom. For optimum scale control apply when crawlers are present.
	Apple maggot Apple rust mite Bagworms Eastern tent caterpillar European apple sawfly Eyespotted bud moth Forbes scale Fruittree leafroller Green fruitworm Japanese beetle Lecanium scales Lesser appleworm Oystershell scale Pear psylla Pear leaf blister mite Pear rust mite Periodical cicada Plum curculio Redbanded leafroller Rosy apple aphid San Jose scale Tarnished plant bug Tentiform leafminers Woolly apple aphid Yellowheaded fireworm	1		

CROP	INSECT	QUARTS OF "SEVIN®" BRAND XLR PLUS/ACRE	PRE-HARVEST INTERVAL (DAYS)	SPECIFIC DIRECTIONS
Chestnuts	Chestnut weevil	2 to 3	0	DO NOT USE IN CALIFORNIA. Make 4 applications at weekly intervals beginning in late July for adult chestnut weevil control. Last application should be made prior to shuck split.
Citrus Fruits (such as grapefruit, lemons, limes, oranges, tangelos tangerines, citrus citron, kumquats and hybrids)	Avocado leafroller California orangedog Citrus cutworm Citrus root weevil Fruittree leafroller Orange tortrix Western tussock moth West Indian sugarcane borer (adults)	1	5	Do not apply more than 20 quarts of this product per acre per application. Do not apply less than 10 gallons of dilute spray mixture per mature tree. May be mixed with petroleum oils commonly used on citrus.
	Black scale Brown soft scale California red scale Citricola scale Citrus snow scale Yellow scale	¾ to 1		
Filbert	Filbert aphid Filbert leafroller Filbertworm	1	0	Apply when leafroller eggs are hatching. Repeat on first appearance of adult fibert moths and again 3 to 4 weeks later.
Olives	Olive scale	¾ to 1	0	For optimum scale control add 1½ gallons of summer oil and apply mixture when crawlers are present. Do not exceed 2 applications per year. Do not apply more than 15 quarts of this product per acre per application.
Peaches Apricots Nectarines	Apple pandemis Codling moth Cucumber beetles European earwig Fruittree leafroller Japanese beetle June beetles Lecanium scales Lesser peachtree borer Olive scale Orange tortrix Oriental fruit moth Peach twig borer Periodical cicada Plum curculio Redbanded leafroller San Jose scale Tarnished plant bug Tussock moths Variegated leafroller	1	1 (peaches) 3 (apricots & nectarines)	Do not apply more than 6 quarts of this product per acre per application to apricots. For optimum scale control apply when crawlers are present. Spray limbs and trunk thoroughly, weekly during moth flight.
Pecans	Black margined aphid Fall webworm Hickory shuckworm Lesser webworm Pecan leaf phylloxera Pecan nut casebearer Pecan spittlebug Pecan stem phylloxera Pecan weevil Twig girdler Walnut caterpillar	1 to 2½	0	Do not apply more than 7.2 quarts of this product per acre per application.
Pistachios	Navel orangeworm	½ to 2	14	Apply dilute volumes of 150 to 300 gallons of mixed spray per acre for full coverage at onset of hull split.
Plums Prunes Cherries	Black cherry aphid Brown soft scale Cherry fruitworm Cherry maggot European earwig Eyespotted bud moth Forbes scale Fruittree leafroller Green fruitworm Japanese beetle Lecanium scales Lesser peachtree borer Mealy plum aphid Oystershell scale Peach twig borer Plum curculio Prune leafhopper Redbanded leafroller Rose chafer San Jose scale Variegated leafroller	1	1	Do not apply more than 6 quarts of this product per acre per application. For optimum scale control apply when crawlers are present. For lesser peachtree borer control, spray limbs and tree trunks thoroughly, weekly during moth flight.
	Codling moth Eastern tent caterpillar Orange tortrix Tussock moth	¾		
Walnut	Calico scale European fruit lecanium Filbertworm Fruittree leafroller Frosted scale	½	0	Apply 1000 gallons of dilute spray per acre for mature trees.
	Codling moth	½		For codling moth apply first spray when average cross-sectional diameters of developing nuts are ½ to ¾ inch. Repeat during middle or late June as needed.
	European earwig	2		Spray tree trunks to point of run-off.

SMALL FRUIT CROP

Recommended dosages refer to quarts of SEVIN® brand XLR PLUS Carbaryl Insecticide per acre. The optimum spray gallonage will depend on plant size, density and stage of growth. Typical spray gallonages per acre range from 100-300 gallons for dilute sprays, 30-100 gallons for concentrate sprays and 10-25 gallons for aerial sprays. Do not exceed maximum label rate per acre per application.

CROP	INSECT		QUARTS OF "SEVIN®" BRAND XLR PLUS/ACRE	PRE-HARVEST INTERVAL (DAYS)	SPECIFIC DIRECTIONS
Blackberries Raspberries Dewberries (including boysenberries and loganberries)	European raspberry aphid Japanese beetle	Leafhoppers Leafrollers Rose chafer Snowy tree cricket	1 to 2	7	
	Omnivorous leafroller Raspberry sawfly		2		CALIFORNIA ONLY.
Blueberries	Blueberry maggot Cherry fruitworm Cranberry fruitworm	European fruit lecanium Japanese beetle	1½ to 2	0	Apply 3 weeks before harvest and repeat as necessary.
Cranberries	Cutworms Cranberry fireworms Cranberry fruitworms	Japanese beetle Leafhoppers Rose chafer	1½ to 3	1	CAUTION! May kill shrimp and crabs. Do not use in areas where these are important resources. Apply in late bloom and as needed at 7 to 10 day intervals.
Grapes	European fruit lecanium Grape leaffolder Grape leafhopper	Western grapeleaf skeletonizer Western yellowstriped armyworm	1 to 2	0	Apply before first brood leaffolder larvae emerge from rolls.
	Cutworms Eight-spotted forester Grape berry moth Japanese beetle June beetles	Orange tortrix Omnivorous leafroller Redbanded leafroller Saltmarsh caterpillar	2		
Strawberries	Flea beetles Meadow spittlebug Omnivorous leaftier (strawberry fruitworm)	Strawberry leafroller Strawberry weevil	1 to 2	1	Carbaryl may injure Early Dawn and Sunrise varieties on the Delmarva Peninsula.

TREES AND ORNAMENTALS

For dilute-spray ground applications to trees (including shade trees, shelter belts, forests, plantations, parks and recreational areas), ornamentals, woody plants and shrubs, apply the specified dosage per 100 gallons of water. For concentrate-spray ground applications, apply the specified dosage per acre in sufficient spray volume to provide thorough coverage. To prepare small volumes of spray, use 1 tablespoon (½ fluid ounce) of SEVIN® brand XLR PLUS Carbaryl Insecticide per gallon of water, where rates of 1 quart are indicated. For aerial applications to forest trees (including shade trees, shelter belts, plantations, parks and recreational areas) and commercially grown ornamentals, woody plants and shrubs, apply the specified dosage per acre in sufficient spray volume to provide thorough coverage. Avoid direct application to lakes, streams and ponds.

CROP	INSECT			QUARTS OF "SEVIN®" BRAND XLR PRE-HARVEST PLUS/ACRE or 100 GAL.	SPECIFIC DIRECTIONS
GROUND APPLICATION: Trees (including shade trees, shelter belts, forests, plantations, parks and recreational areas). Ornamentals (including roses, flowers and other herbaceous plants), Woody Plants and Shrubs.	Ants Apple aphid Armyworm Azalea leafminer Bagworms Birch leafminer Blister beetle Boxelder bug Boxwood leafminer Brown tail moth Cankerworms Catalpa sphinx Chiggers Cooley spruce gall aphid Cutworms	Grasshoppers Gypsy moth Hackberry nipplegall maker Holly bud moth Holly leafminer Jackpine budworm Japanese beetle Jeffrey pine needleminer June beetles Lace bugs Leafhoppers Leafrollers Locust borer Maple leafcutter	Plant bugs Poinsettia hornworm Psyllids Puss caterpillar Redhumped oakworm Rose aphid Rose chafer Roseslug Saddled prominent Sawflies (exposed) Scale insects Sowbugs Spiny elm caterpillar Springtails Spruce budworm	1	Observe plant response precautions. Apply dilute sprays to obtain thorough coverage of upper and lower leaf surfaces. To control scale insects, treat trunks, stems and twigs in addition to plant foliage. For optimum worm control, treat when in early instars. Addition of a sticker may improve residual control.

CROP	INSECT			QUARTS OF "SEVIN®" BRAND XLR PRE-HARVEST PLUS/ACRE or 100 GAL.	SPECIFIC DIRECTIONS
AERIAL APPLICATION: Forest Trees (including shade trees, shelter belts, forests, plantations, parks and recreational areas) and Commercially Grown Ornamentals (including roses, flowers and other herbaceous plants), Woody Plants and Shrubs.	Cypress tip moth Douglas-fir tussock moth Eastern spruce gall aphid Elm leaf aphid Elm leaf beetle Elm spanworm Eriophyid mites European pine shoot moth Fall armyworm Flea beetles Fuller rose beetle Gall midges Gall wasps Green striped mapleworm	Mealybugs Mimosa webworm Nantucket pine tip moth Oak leafminers Oak leaf skeletonizer Oakworm complex Oleander caterpillar Olive ash borer Orange-striped oakworm Orange tortrix Periodical cicada Pine sawfly Pine spittlebug Pitch pine tip moth	Spruce needleminer Subtropical pine tip moth Tent caterpillars Thorn bug Thrips (exposed) Ticks Walnut caterpillar Webworms Western hemlock looper Western spruce budworm Willow leaf beetles Yellow poplar weevil		
	Ips engraver beetles Mountain pine beetle Roundheaded pine beetle Western pine beetle			20	Effective as a preventive treatment only. Repeat annually as required to prevent beetle attacks. Apply 1 gallon of spray per 50 square feet of bark in late May to early July, or prior to beetle attack. Treat tree trunks from ground level up, until trunk diameter is less than 5 inches.
	Elm bark beetle				Apply approximately 20-30 gallons of spray mixture for each 50 feet of elm tree for thorough coverage of all bark surfaces on trunks, limbs and twigs.

LAWNS AND RECREATIONAL AREAS

CROP	INSECT		SPECIFIC DIRECTIONS
Lawns and Recreational Areas (including turf, golf courses and parks)	Ants Armyworm Bluegrass billbug Centipedes Chiggers Cutworms Earwigs Essex skipper European chafer Fall armyworm Fleas Grasshoppers	June beetles Leafhoppers Lucerne moth Millipedes Mosquitoes Sod webworm (lawn moths) Sowbugs Springtails Ticks Yellowstriped armyworm	Use 1 quart of this product in 150 to 200 gallons of spray mixture per 5000 square feet of established law, recreational area, roadway or path border. For best results, mow lawn and apply after watering or rain. Insure good penetration of turf. Garden hose or pressure applications may be used. Repeat 2 to 3 weeks later if necessary.
	Chinch bugs White grubs		Use 1½ quarts of this product in 150 to 200 gallons of spray per 5000 square feet.

IMPORTED FIRE ANT CONTROL

To prepare small amounts, use 1 tablespoon (½ fluid ounce) of SEVIN® brand XLR PLUS Carbaryl Insecticide per gallon of water.

SITE	QUARTS OF "SEVIN®" BRAND XLR PLUS/100 GALLONS OF WATER	SPECIFIC DIRECTIONS
Lawns, cemeteries and Recreational Areas (including turf, golf courses, and parks) Pastures Rangeland Forested Lands, and Wastelands	1½	Apply a total of 2 gallons of the diluted solution over the surface of each mound or at least 1 quart per 6 inches of mound diameter using a bucket, can or other appropriate equipment. Thoroughly wet mound and surrounding area to a 4 ft. diameter (12 sq. ft.). Do not disturb mounds prior to treatment. Pour solution from a height of about three feet to give sufficient force to break mound apex and flow into ant tunnels. For best results apply in cool weather. 65-80°F, or in early morning or late evening hours. Repeat application if mound activity resumes after 10 days. Treat new mounds as they appear. Pressurized sprays may disturb the ants and cause migration, reducing product effectiveness.

MOSQUITO CONTROL

For dilute-spray ground applications to trees (including shade trees, shelter belts, forests, plantation, parks and recreational areas), ornamentals, woody plants and shrubs, apply the specified dosage per 100 gallons of water. For concentrate-spray ground applications, apply the specified dosage per acre in sufficient spray volume to provide thorough coverage. To prepare small volumes of spray, use 1 tablespoon (½ fluid ounce) of SEVIN® brand XLR PLUS Carbaryl Insecticide per gallon of water, where rates of 1 quart are indicated.

For aerial applications to forest trees (including shade trees, shelter belts, plantations, parks and recreational areas) and commercially grown ornamentals, woody plants and shrubs, apply the specified dosage per acre in sufficient spray volume to provide thorough coverage. Avoid direct application to lakes, streams, and points.

CROP	QUARTS OF "SEVIN®" BRAND XLR PLUS/100 GALLONS OF WATER	SPECIFIC DIRECTIONS
Pastures, Rangelands Parks, Recreational Areas, Logging Camps, Military Posts and Adjacent Forested lands or Wastelands.	¼ to 1 (½ to 1 tablespoonful per gallon)	ADULT MOSQUITOES CAUTION! May kill shrimp and crabs. Do not use in areas where these are important resources. OBSERVE BEE CAUTION. Treat shrubbery and areas where adult mosquitoes congregate. Treat when adult mosquitoes are active in early mornings or late evenings. Repeat at 7 to 10 days intervals. Use ¼ to ½ quart per 100 gallons in mistblowers, ½ to 1 quart per acre in aerial sprays and 1 quart per acre in low pressure ground sprayers.
	25	For residual control in subtropical regions apply 4 gallons prepared spray per 2000 square feet of surface area. Repeat in 3 to 6 months or when necessary.

PEST CONTROL AROUND BUILDINGS LIMITED TO PEST CONTROL OPERATORS ONLY

INSECT	QUARTS OF "SEVIN®" BRAND XLR PLUS/100 GALLONS OF WATER	SPECIFIC DIRECTIONS
Ants Cockroaches	20 (20 tablespoonsful/ gallon)	For external use around buildings only. Do not treat animals. Apply as a course wet spray or with a paint brush to outdoor sleeping quarters of pets outside perimeter of dwellings and other areas where insects tend to congregate.
Brown dog tick Earwigs Fleas Millipedes Silverfish Spiders	10 (10 tablespoonsful/ gallon)	Do not use this product in commercial food areas of food handling establishments, restaurants or other places where food is prepared or processed. Do not use in serving areas while food is exposed. Avoid application to surfaces where visible spray residues are objectionable.

POULTRY ROOST AND BUILDINGS ONLY

POULTRY QUARTERS	QUARTS OF "SEVIN®" BRAND XLR PLUS/100 GALLONS OF WATER	SPECIFIC DIRECTIONS
Bed bugs Chicken mite Fleas Lice Northern fowl mite	4	DO NOT TREAT POULTRY OR GAME BIRDS. Apply 1 to 2 gallons of spray mixture per 1000 square feet of wall, litter or roost surface. Force spray into cracks. Repeat as needed.
Fowl tick	16	Avoid spraying nests, eggs and feeding and watering troughs. Ventilate while spraying. Do not treat premises within 7 days of slaughter.
Lesser mealworms	50	Apply 2 gallons of spray mixture per 1000 square feet of floor space or litter surface. Repeat as needed. Ventilate while spraying. Do not treat premises within 7 days of slaughter.

STORAGE AND DISPOSAL

STORAGE

Store unused SEVIN® brand XLR PLUS Carbaryl Insecticide in original container only, in cool, dry area out of reach of children and animals. Do not store in areas where temperatures frequently exceed 100°F.

PESTICIDE DISPOSAL

Do not contaminate water, food or feed by storage or disposal. Open dumping is prohibited. Wastes resulting from the use of this product may be disposed of on site or at an approved waste disposal facility.

CONTAINER DISPOSAL

Triple rinse (or equivalent). Then offer for recycling or reconditioning, or dispose of in a sanitary landfill, or incineration, or if allowed by state and local authorities, by burning. If container is burned, stay out of smoke.

LIMITED WARRANTY AND DISCLAIMER

The manufacturer warrants (a) that this product conforms to the chemical description on the label; (b) that this product is reasonably fit for the purposes set forth in the directions for use when it is used in accordance with such directions; and (c) that the directions, warnings and other statements on this label are based upon responsible experts' evaluation of reasonable tests of effectiveness, of toxicity to laboratory animals and to plants, and of residues on food crops, and upon reports of field experience. Tests have not been made on all varieties or in all states or under all conditions. THE MANUFACTURER NEITHER MAKES NOR INTENDS, NOR DOES IT AUTHORIZE ANY AGENT OR REPRESENTATIVE TO MAKE, ANY OTHER WARRANTIES, EXPRESS OR IMPLIED, AND IT EXPRESSLY EXCLUDES AND DISCLAIMS ALL IMPLIED WARRANTIES OF MERCHANTABILITY OR FITNESS FOR A PARTICULAR PURPOSE.

THIS WARRANTY DOES NOT EXTEND TO, AND THE BUYER SHALL BE SOLELY RESPONSIBLE FOR, ANY AND ALL LOSS OR DAMAGE WHICH RESULTS FROM THE USE OF THIS PRODUCT IN ANY MANNER WHICH IS INCONSISTENT WITH THE LABEL DIRECTIONS, WARNINGS OR CAUTIONS.

BUYER'S EXCLUSIVE REMEDY AND MANUFACTURER'S OR SELLER'S EXCLUSIVE LIABILITY FOR ANY AND ALL CLAIMS, LOSSES, DAMAGES, OR INJURIES RESULTING FROM THE USE OR HANDLING OF THIS PRODUCT, WHETHER OR NOT BASED IN CONTRACT, NEGLIGENCE, STRICT LIABILITY IN TORT OR OTHERWISE, SHALL BE LIMITED, AT THE MANUFACTURER'S OPTION, TO REPLACEMENT OF, OR THE REPAYMENT OF THE PURCHASE PRICE FOR, THE QUANTITY OF PRODUCT WITH RESPECT TO WHICH DAMAGES ARE CLAIMED. IN NO EVENT SHALL MANUFACTURER OR SELLER BE LIABLE FOR SPECIAL, INDIRECT OR CONSEQUENTIAL DAMAGES RESULTING FROM THE USE OR HANDLING OF THIS PRODUCT.

THIS SPECIMEN LABEL IS INTENDED FOR USE ONLY AS A GUIDE IN PROVIDING GENERAL INFORMATION REGARDING THE DIRECTIONS, WARNINGS AND CAUTIONS ASSOCIATED WITH THE USE OF THIS PRODUCT. AS WITH ANY AGRICULTURAL CHEMICAL, ALWAYS FOLLOW THE LABEL INSTRUCTIONS ON THE PACKAGE BEFORE USING.

SEVIN® BRAND 4-OIL

CARBARYL INSECTICIDE

For Agricultural or Commercial Use Only.

ACTIVE INGREDIENTS:

Carbaryl (1-naphthyl N-methylcarbamate)	49.0%
Petroleum oils	46.3%
INERT INGREDIENTS:	4.7%
TOTAL	100.0%

Contains 4 pounds carbaryl per gallon.

E.P.A. Reg. No. 264-323 E.P.A. Est. No. 264-MO-02

Keep Out of Reach of Children

CAUTION

IN CASE OF EMERGENCY TELEPHONE (24 HOURS A DAY) IN U.S.A. 1-800-UCC-HELP.

FOR GENERAL PRODUCT INFORMATION CALL 1-800-334-9745.

PRECAUTIONARY STATEMENTS

CAUTION

HAZARDS TO HUMANS AND DOMESTIC ANIMALS

MAY BE HARMFUL IF SWALLOWED. Avoid breathing of spray mist. Do not take internally. Avoid contact with eyes, skin or clothing. Wear regular long-sleeved work clothing. Change to clean clothing daily. Wash hands and face before eating. Wash thoroughly after handling.

ANTIDOTE STATEMENT

ATROPINE SULFATE IS HIGHLY EFFECTIVE AS AN ANTIDOTE. See NOTE TO PHYSICIAN below.

STATEMENT OF PRACTICAL TREATMENT

GENERAL

Contact a physician immediately in all cases of suspected poisoning. Transport to a physician or hospital immediately and SHOW A COPY OF THIS LABEL TO THE PHYSICIAN.

IF SWALLOWED: Never give anything by mouth to an unconscious or convulsing person. Do not induce vomiting.

IF IN EYES: Flush eyes with plenty of water. Get medical attention if irritation persists.

IF ON SKIN: Wash thoroughly with soap and water.

IF INHALED: Move from contaminated atmosphere and call a physician.

NOTE TO PHYSICIAN

Carbaryl is a carbamate insecticide, which is a chloinesterase inhibitor. Overexposure to this substance may cause toxic signs and symptoms due to stimulation of the cholinergic nervous system. These effects of overexposure are spontaneously and rapidly reversible.

Specific treatment consists of parenteral atropine sulfate. Caution should be maintained to prevent overatropinization. Mild cases may be given 1 to 2 mg intramuscularly every 10 minutes until full atropinization has been achieved and repeated thereafter whenever symptoms reappear. Severe cases should be given 2 to 4 mg intravenously every 10 minutes until fully atropinized, then intramuscularly every 30 to 60 minutes to maintain the effect for at least 12 hours. Dosages for children should be appropriately reduced. Complete recovery from overexposure is to be expected within 24 hours.

Narcotics and other sedatives should not be used. Further, drugs like 2-PAM (pyridine-2-aldoxime methiodide) are NOT recommended.

To aid in confirmation of a diagnosis, urine samples should be obtained within 24 hours of exposure and immediately frozen. Analyses will be arranged by Union Carbide Agricultural Products Company, Inc.

Consultation on therapy can be obtained at all hours by calling the Union Carbide emergency number 1-800-UCC-HELP.

ENVIRONMENTAL HAZARDS

This product is extremely toxic to aquatic and estuarine invertebrates. Do not apply directly to water and wetlands, except under the forest canopy. Do not contaminate water by cleaning equipment or disposal of wastes.

BEE CAUTION

This product is highly toxic to bees exposed to direct treatment or residues on blooming crops and weeds. Do not apply this product or allow it to drift to blooming crops or weeds if bees are visiting the treatment area.

DIRECTIONS FOR USE

IT IS A VIOLATION OF FEDERAL LAW TO USE THIS PRODUCT IN A MANNER INCONSISTENT WITH ITS LABELING.

GENERAL INFORMATION

SEVIN® brand 4-OIL Carbaryl Insecticide is a dispersion of finely ground technical carbaryl in a non-aromatic, low volatile oil. It is designed for air application as a low volume or ultra low volume spray. It may also be applied by ground mist blower or cold fogging, or diluted with diesel fuel, kerosene or #2 fuel oil. READ THIS LABEL BEFORE USE. USE IN STRICT ACCORDANCE WITH DIRECTIONS AND CAUTIONS.

GENERAL WORKER PROTECTION STATEMENT

Do not apply this product in such a manner as to directly or through drift expose workers or other persons. The area treated must be vacated by unprotected persons.

Do not enter treated areas without protective clothing until sprays have dried.

Because certain states may require more restrictive reentry intervals for various crops treated with this product, consult your State Department of Agriculture for further information.

Written or oral warnings must be given to workers who are expected to be in treated area or in an area about to be treated with this product. Advise workers to stay out of fields during application and until sprays have dried. Regular long-sleeved work clothing should be worn when working in treated fields. See PRECAUTIONARY STATEMENTS, STATEMENT OF PRACTICAL TREATMENT AND NOTE TO PHYSICIAN for information on accidental exposures. When oral warnings are given, warnings shall be given in a language customarily understood by workers. Oral warnings must be given if there is reason to believe that written warnings cannot be understood by workers. Written warnings must include the following information: Appropriate signal work (CAUTION), area treated with SEVIN® brand 4-OIL Carbaryl Insecticide, date of application, appropriate clothing, and re-entry interval (i.e., until sprays have dried).

PRE-HARVEST AND GRAZING USE INFORMATION AND LIMITATIONS

Tolerances established under the Federal Food, Drug and Cosmetic Act permit the sale of crops bearing probable carbaryl residues when this product is used in accordance with label directions. If used as directed, treated forage may be grazed or used as feed for dairy and meat animals without causing illegal residues in meat or milk. This product may be applied up to and including the day of harvest or grazing of forage crops.

Do not plant rotational food and feed crops not listed on this or other carbaryl labels in carbaryl treated soil.

Do not use reclaimed irrigation water from crops treated with carbaryl on upland crops for which carbaryl tolerances are not established.

PLANT RESPONSE PRECAUTION

To avoid possible injury to tender foliage, do not apply to wet foliage or when rain or high humidity is expected during the next two days.

SPRAY PREPARATION

SEVIN® brand 4-OIL Carbaryl Insecticide thickens during storage. Roll or agitate drums or recirculate product in bulk tanks before use to reduce viscosity and restore homogeneity. SEVIN® brand 4-OIL Carbaryl Insecticide may be held prior to use and transferred in any materials of construction suitable for use with fuel oil, kerosene or diesel oil. Protect from water. Before and after use, flush nurse tanks, pumps, lines, hoses and entire spray systems with diesel fuel, kerosene or #2 fuel oil until clean. Water, scale, rust and other residue must be removed from pumps, mixing and spray systems before use.

COMPATIBILITY

"SEVIN®" BRAND 4-OIL CARBARYL INSECTICIDE IS NOT COMPATIBLE WITH AND SHOULD NOT BE MIXED WITH WATER, ALCOHOL OR AROMATIC SOLVENTS, but can be diluted with aliphatic oils (diesel fuel, kerosene or #2 fuel oil). Compatibility with other pesticides has not been thoroughly tested. DO NOT USE IN TANK MIX COMBINATIONS UNLESS YOUR EXPERIENCE INDICATES THAT THE MIXTURE IS EFFECTIVE AND WILL NOT RESULT IN PLANT INJURY OR MECHANICAL DIFFICULTY. For best mixing when preparing diluted SEVIN® brand 4-OIL Carbaryl Insecticide, add diluent oil to mix tank before adding proper volume of SEVIN® brand 4-OIL Carbaryl Insecticide. Provide complete recirculation of diluted suspensions once every 24 hours to maintain uniformity.

APPLICATION

On all crops use sufficient gallonage to obtain thorough and uniform coverage. Calibrate equipment to deliver the desired spray volume. Flat fan nozzles may be used to apply both undiluted and diluted SEVIN® brand 4-OIL Carbaryl Insecticide. Rotary atomizing and hollow cone nozzles are not recommended for applying undiluted SEVIN® brand 4-OIL Carbaryl Insecticide but can be used if 1 to 1 dilution is made. Use of a high-volume 50-mesh in-line strainer is suggested. Use of screens behind nozzles is not recommended. Apply by air undiluted, or diluted with kerosene, diesel fuel or #2 fuel oil. At temperatures below 50°F (10°C) dilution of 4 volumes of SEVIN® brand 4-OIL Carbaryl Insecticide with 1 volume of diluent oil is recommended to assure uniform flow and spray distribution. Dilutions greater than 1 to 1 by volume are not recommended; higher dilutions will reduce residual properties. Dilute 1 to 1 by volume when applying with ground equipment. Apply only when weather conditions are favorable. Wind and rising air currents may cause undesirable spray drift and reduce insect control.

INSECT CONTROL

Apply when insects or their damage appear. To maintain control repeat at 7 to 14 day intervals or as necessary unless a shorter interval is specified below. Where a dosage range is indicated, use lower rate on young plants and early instars and higher rate on mature plants, advanced instars and adults. Thorough and uniform spray coverage is essential for effective control.

SEVIN® brand 4-Oil Carbaryl Insecticide does not control spider mites. If spider mites are a problem, use a registered miticide.

Regional differences have been noted in the susceptibility of certain strains of fall armyworm to SEVIN® brand 4-Oil Carbaryl Insecticide. If local experience indicates inadequate control, use an alternative pesticide.

Follow directions for most effective use. DOSAGES LISTED REFER TO QUARTS OF "SEVIN®" BRAND 4-OIL CARBARYL INSECTICIDE PER ACRE, NOT TO VOLUME OF FINISHED SPRAY PER ACRE. Refer to Steps 4 and 6 under DIRECTIONS FOR USE for instructions on dilution.

ALFALFA AND CLOVERS INSECT CONTROL

For grasshoppers: Apply ½ to 1½ quarts of this product per acre. To avoid possible injury to tender alfalfa foliage, do not apply to wet foliage or when rain or high humidity is expected during the next two days.

Do not apply to alfalfa within 3 days of harvest.

CORN (field, pop) INSECT CONTROL

For corn rootworm beetles: Apply 1 quart of this product per acre as needed at first silking if populations are at economic levels and/or apply later when needed to control adult population buildup to reduce larval damage to next year's crop. Do not make more than two applications.

For western bean cutworm: Apply a single application of 2 quarts of this product plus 2 quarts diesel fuel, kerosene or #2 fuel oil per acre when an average of 1 in 7 plants have western bean cutworm egg masses or newly hatched larvae and 90 to 100% of the tassels have emerged. Application after 100% of the silks have emerged will reduce effectiveness.

For Japanese beetle: Apply 1 quart of this product per acre when adult beetles are present.

For grasshoppers: Apply ½ to 1½ quarts of this product per acre. Lower rate is suggested for nymphs on small plants. Use 1½ quarts when grasshoppers of foliage are mature and greater coverage is required.

FOREST INSECT CONTROL

For control of gypsy moth, fall and spring cankerworms, saddled prominent, forest tent caterpillar, elm spanworm, oak leafroller complex and Japanese beetle: Apply 1 quart of this product per acre when larvae are in early instars and leaves are at least ⅓ grown, or when adult Japanese beetles are present.

For spruce budworm and western spruce budworm: East of the Rocky Mountains apply ½ to 1 quart of this product per acre when a majority of the larvae are in the third through sixth instars. Lower rate is suggested for heavy spruce budworm infestations where 2 applications are necessary for adequate foliage protection. In the Rocky Mountains and west of the Rocky Mountains apply ½ to 1 quart of this product per acre when a majority of the larvae are in the third through fifth instars.

For Douglas-fir tussock moth: For maximum foliage protection apply 2 quarts of this product per acre when larvae are in first and second instars. Application to later instars will provide control but with reduced foliage protection.

For population control and foliage protection of high value trees using ground mist blower application: Apply 0.8 pint of spray mixture containing equal volumes of this product and diesel fuel, kerosene or #2 fuel oil per 20 to 30 foot tree (0.2 lb. ai/tree) when larvae are in early instars. Cover foliage thoroughly, but avoid runoff.

PASTURE, RANGELAND AND WASTELAND INSECT CONTROL

For grasshoppers: Apply ½ to 1½ quarts of SEVIN® brand 4-OIL Carbaryl Insecticide per acre. Lower rate is suggested for nymphs on small plants or sparse vegetation. Use higher rate when grasshoppers are mature or vegetation is thick.

For range caterpillar: Apply ½ to 1 quart of this product per acre. Use the lower rate when larvae are in early instars and the higher rate for later instars.

For Japanese beetle: Apply 1 quart of this product per acre when adult beetles are present.

For black grass bugs: Apply ½ to 1½ quarts of this product per acre. Use lower rate for control of immature bugs in sparse vegetation. Use higher rate for mature bugs or when vegetation is thick.

WHEAT INSECT CONTROL—(DO NOT USE IN CALIFORNIA.)

Do not apply within 21 days of harvest of grain. No time limitation on green wheat used as pasture or forage.

For grasshoppers: Apply ½ to 1½ quarts of this product per acre. Lower rate is suggested for nymphs on small plants. Use higher rate when grasshoppers are mature or vegetation is thick.

For cereal leaf beetle: Apply 1 quart of this product per acre. Application is effective against eggs, larvae and adults.

For armyworm and fall armyworm: Apply 1 to 1½ quarts of this product per acre.

STORAGE AND DISPOSAL

STORAGE

Store unused SEVIN® brand 4-OIL Carbaryl Insecticide in original container only, in cool, dry area out of reach of children and animals. Do not store in areas where temperatures frequently exceed 100°F.

PESTICIDE DISPOSAL

Do not contaminate water, food or feed by storage or disposal. Wastes resulting from the use of this product may be disposed of on site or at an approved waste disposal facility.

CONTAINER DISPOSAL

Decontaminate empty bulk tanks. For drums: Triple rinse (or equivalent). Then offer for recycling or reconditioning, or puncture and dispose of in a sanitary landfill, or by other procedures approved by state and local authorities.

Consult Federal, State or local disposal authorities for approved alternative procedures.

LIMITED WARRANTY AND DISCLAIMER

The manufacturer warrants (a) that this product conforms to the chemical description on the label; (b) that this product is reasonably fit for the purposes set forth in the directions for use when it is used in accordance with such directions; and (c) that the directions, warnings and other statements on this label are based upon responsible experts' evaluation of reasonable tests of effectiveness, of toxicity to laboratory animals and to plants, and of residues on food crops, and upon reports of field experience. Tests have not been made on all varieties or in all states or under all conditions. THE MANUFACTURER NEITHER MAKES, NOR INTENDS, NOR DOES IT AUTHORIZE ANY AGENT OR REPRESENTATIVE TO MAKE, ANY OTHER WARRANTIES, EXPRESS OR IMPLIED, AND IT EXPRESSLY EXCLUDES AND DISCLAIMS ALL IMPLIED WARRANTIES OF MERCHANTABILITY OR FITNESS FOR A PARTICULAR PURPOSE.

THIS WARRANTY DOES NOT EXTEND TO, AND THE BUYER SHALL BE SOLELY RESPONSIBLE FOR, ANY AND ALL LOSS OR DAMAGE WHICH RESULTS FROM THE USE OF THIS PRODUCT IN ANY MANNER WHICH IS INCONSISTENT WITH THE LABEL DIRECTIONS, WARNINGS OR CAUTIONS.

BUYER'S EXCLUSIVE REMEDY AND MANUFACTURER'S OR SELLER'S EXCLUSIVE LIABILITY FOR ANY AND ALL CLAIMS, LOSSES, DAMAGES, OR INJURIES RESULTING FROM THE USE OR HANDLING OF THIS PRODUCT, WHETHER OR NOT BASED IN CONTRACT, NEGLIGENCE, STRICT LIABILITY IN TORT OR OTHERWISE, SHALL BE LIMITED, AT THE MANUFACTURER'S OPTION, TO REPLACEMENT OF, OR THE REPAYMENT OF THE PURCHASE PRICE FOR, THE QUANTITY OF PRODUCT WITH RESPECT TO WHICH DAMAGES ARE CLAIMED. IN NO EVENT SHALL MANUFACTURER OR SELLER BE LIABLE FOR SPECIAL, INDIRECT OR CONSEQUENTIAL DAMAGES RESULTING FROM THE USE OR HANDLING OF THIS PRODUCT.

THIS SPECIMEN LABEL IS INTENDED FOR USE ONLY AS A GUIDE IN PROVIDING GENERAL INFORMATION REGARDING THE DIRECTIONS, WARNINGS AND CAUTIONS ASSOCIATED WITH THE USE OF THIS PRODUCT. AS WITH ANY AGRICULTURAL CHEMICAL, ALWAYS FOLLOW THE LABEL INSTRUCTIONS ON THE PACKAGE BEFORE USING.

SEVIN® BRAND 80S

CARBARYL INSECTICIDE

For Agricultural or Commercial Use Only

ACTIVE INGREDIENT:	% BY WEIGHT
Carbaryl (1-naphthyl N-methylcarbamate)	80%
INERT INGREDIENTS:	20%
TOTAL	100%

E.P.A. Reg. No. 264-316 E.P.A. Est. No. 10352-GA-01

Keep Out of Reach of Children

WARNING—AVISO

PRECAUCION AL USARIO: Si usted no lee ingles, no use este producto hasta que la etiqueta le haya sido explicada ampliamente.

IN CASE OF EMERGENCY TELEPHONE (24 HOURS A DAY) IN U.S.A. 1-800-UCC-HELP

FOR GENERAL PRODUCT INFORMATION CALL 1-800-334-9745.

PRECAUTIONARY STATEMENTS

WARNING

HAZARDS TO HUMANS AND DOMESTIC ANIMALS

MAY BE FATAL IF SWALLOWED. Avoid breathing of dust or spray mist. Do not take internally. Avoid contact with eyes, skin or clothing. Wear regular long-sleeved work clothing. Change to clean clothing daily. Wash hands and face before eating. Wash thoroughly after handling.

ANTIDOTE STATEMENT

ATROPINE SULFATE IS HIGHLY EFFECTIVE AS AN ANTIDOTE. See NOTE TO PHYSICIAN below.

STATEMENT OF PRACTICAL TREATMENT

GENERAL

Contact a physician immediately in all cases of suspected poisoning. Transport to a physician or hospital immediately and SHOW A COPY OF THIS LABEL TO THE PHYSICIAN.

IF SWALLOWED: Never give anything by mouth to an unconscious or convulsing person. If conscious and not convulsing, drink 1 to 2 glasses of water and induce vomiting by touching the back of the throat with finger.

IF IN EYES: Flush eyes with plenty of water. Get medical attention if irritation persists.

IF ON SKIN: Wash thoroughly with soap and water.

IF INHALED: Move from contaminated atmosphere and call a physician.

NOTE TO PHYSICIAN

Carbaryl is a carbamate insecticide, which is a chloinesterase inhibitor. Overexposure to this substance may cause toxic signs and symptoms due to stimulation of the cholinergic nervous system. These effects of overexposure are spontaneously and rapidly reversible.

Specific treatment consists of parenteral atropine sulfate. Caution should be maintained to prevent overatropinization. Mild cases may be given 1 to 2 mg intramuscularly every 10 minutes until full atropinization has been achieved and repeated thereafter whenever symptoms reappear. Severe cases should be given 2 to 4 mg intravenously every 10 minutes until fully atropinized, then intramuscularly every 30 to 60 minutes to maintain the effect for at least 12 hours. Dosages for children should be appropriately reduced. Complete recovery from overexposure is to be expected within 24 hours.

Narcotics and other sedatives should not be used. Further, drugs like 2-PAM (pyridine-2-aldoxime methiodide) are NOT recommended.

To aid in confirmation of a diagnosis, urine samples should be obtained within 24 hours of exposure and immediately frozen. Analyses will be arranged by Union Carbide Agricultural Products Company, Inc.

Consultation on therapy can be obtained at all hours by calling the Union Carbide emergency number 1-800-UCC-HELP.

ENVIRONMENTAL HAZARDS

This product is extremely toxic to aquatic and estuarine invertebrates. Do not apply directly to water and wetlands, except under the forest canopy and use on rice. Discharge from rice fields may kill aquatic and estuarine invertebrates. Do not contaminate water by cleaning equipment or disposal of wastes.

BEE CAUTION

This product is highly toxic to bees exposed to direct treatment or residues on blooming crops and weeds. Do not apply this product or allow it to drift to blooming crops or weeds if bees are visiting the treatment area.

DIRECTIONS FOR USE

IT IS A VIOLATION OF FEDERAL LAW TO USE THIS PRODUCT IN A MANNER INCONSISTENT WITH ITS LABELING.

GENERAL INFORMATION

SEVIN® brand 80S Carbaryl Insecticide is a dry powder for dispersion in water and application as an insecticidal spray in hydraulic sprayers, mist blowers, low gallonage ground equipment and aircraft. READ THIS LABEL BEFORE USE. STRICTLY OBSERVE LABEL DIRECTIONS AND CAUTIONS, AND APPLICABLE FEDERAL AND STATE REGULATIONS.

GENERAL WORKER PROTECTION STATEMENTS

Do not apply this product in such a manner as to directly or through drift expose workers or other persons. The area treated must be vacated by unprotected persons.

Do not enter treated areas without protective clothing until sprays have dried.

Because certain states may require more restrictive reentry intervals for various crops treated with this product, consult your State Department of Agriculture for further information.

Written or oral warnings must be given to workers who are expected to be in treated area or in an area about to be treated with this product. Advise workers to stay out of fields during application and until sprays have dried. Regular long-sleeved work clothing should be worn when working in treated fields. See PRECAUTIONARY STATEMENTS, STATEMENT OF PRACTICAL TREATMENT and NOTE TO PHYSICIAN for information on accidental exposures. When oral warnings are given, warnings shall be given in a language customarily understood by workers. Oral warnings must be given if there is reason to believe that written warnings cannot be understood by workers. Written warnings must include the following information: appropriate signal word (WARNING), area treated with SEVIN® brand 80S Carbaryl Insecticide, date of application, appropriate clothing, and re-entry interval (i.e., until sprays have dried).

PREHARVEST AND GRAZING USE INFORMATION AND LIMITATIONS

Tolerances established under the Federal Food, Drug and Cosmetic Act permit the sale of crops bearing probable carbaryl residues when this product is used in accordance with label directions. If used as directed, treated forage may be grazed or used as feed for dairy and meat animals without causing illegal residues in meat or milk. This product may be applied up to and including the day of harvest or grazing of forage crops. Application may be made without removing livestock from area being treated.

Do not plant rotational food and feed crops not listed on this or other carbaryl labels in carbaryl treated soil.

Do not use reclaimed irrigation water from crops treated with carbaryl on upland crops for which carbaryl tolerances are not established.

PLANT RESPONSE PRECAUTIONS

To avoid possible injury to tender foliage, do not apply to wet foliage or when rain or high humidity is expected during the next two days.

Do not use on Boston ivy, Virginia creeper and maidenhair fern as injury will result. During early season, it may also injure Virginia and sand pines.

Observe label instructions on apply thinning and on combinations with certain herbicides on rice and soybeans.

SPRAY PREPARATION

TO ASSURE A UNIFORM SUSPENSION, AGITATE, STIR OR RECIRCULATE ALL "SEVIN®" BRAND 80S CARBARYL INSECTICIDE SPRAY MIXTURES PRIOR TO AND DURING USE. Remove oil, rust, scale, pesticide residues and other foreign matter from mix tanks and entire spray system. Flush with clean water. Fill spray or mix tank with ½ to ¾ the desired amount of water. Start mechanical or hydraulic agitation. Slowly add the required amount of SEVIN® brand 80S Carbaryl Insecticide, and then the remaining volume of water. Prepare only as much spray mixture as can be applied on the day of mixing. MAINTAIN CONTINUOUS AGITATION DURING MIXING AND APPLICATION TO ASSURE A UNIFORM SUSPENSION. DO NOT STORE SPRAY MIXTURES FOR PROLONGED PERIODS.

COMPATIBILITY

SEVIN® brand 80S Carbaryl Insecticide has been used without plant injury with most common insecticides, miticides, fungicides,nutrients, adjuvants and petroleum oil as used on citrus. If compatibility of SEVIN® brand 80S Carbaryl Insecticide with another product and the resulting crop response is unknown, it should be tested on a small scale. Curdling, precipitation, greasing, layer formation or increased viscosity are symptoms of incompatibility. Incompatibility will reduce insect control and may cause mechanical difficulties or plant injury. Observe all cautions and limitations on labeling of all products used in mixtures. WHEN PREPARING COMBINATION SPRAYS, FIRST ADD "SEVIN®" BRAND 80S CARBARYL INSECTICIDE TO AT LEAST ½ THE DESIRED AMOUNT OF WATER, MIX THOROUGHLY, AND THEN ADD COMBINATION PRODUCTS TO THE MIXTURE, AND THEN THE REMAINING VOLUME OF WATER. DO NOT APPLY TANK MIX COMBINATIONS UNLESS YOUR PREVIOUS EXPERIENCE INDICATES THE MIXTURE IS EFFECTIVE AND WILL NOT RESULT IN APPLICATION PROBLEMS, EXCESSIVE RESIDUES OR PLANT INJURY.

Unstable under highly alkaline conditions. Not effective if used with alkaline materials such as Bordeaux, lime-sulfur and casein-lime spreaders.

APPLICATION

On all crops, use sufficient gallonage to obtain thorough and uniform coverage. Calibrate spray equipment to deliver the required volume. Use 50 mesh slotted strainers in spray system and 25 mesh slotted strainers behind nozzles. Avoid applications just before rainfall as poor insect control may result. To clean spray system after use, drain and flush with water. Refer to the Storage and Disposal directions for proper disposal of rinse water.

INSECT CONTROL

Apply when insects or their damage appear. To maintain control repeat at 7 to 14 day intervals or as necessary unless a shorter interval is specified below. Where a dosage range is indicated, use lower rate on young plants and early instars and higher rate on mature plants, advanced instars and adults. Thorough and uniform spray coverage is essential for effective control.

SEVIN® brand 80S Carbaryl Insecticide does not control spider mites. If spider mites are a problem, use a registered miticide.

Regional differences have been noted in the susceptibility of certain strains of fall armyworm to SEVIN® brand 80S Carbaryl Insecticide. If your experience indicates inadequate control, use an alternative pesticide.

FORAGE, FIELD AND VEGETABLE CROPS

Use at least 1 gallon of finished spray per acre for aerial application and at least 3 gallons of finished spray per acre for concentrate ground application. To prepare small volumes of spray, use 1¼ tablespoonfuls of SEVIN® brand 80S Carbaryl Insecticide per gallon of water where rates of 1¼ pounds per acre or 1½ pounds per 100 gallons are indicated in the tables below.

CROP	INSECT		POUNDS OF "SEVIN®" BRAND 80S/ACRE	PRE-HARVEST INTERVAL (DAYS)	SPECIFIC DIRECTIONS
All Forage, Field and Vegetable Crops	Grasshoppers		⅔ to 1⅞	See specific Forage, Field or Vegetable Crop	Use ⅔ to 1¼ pounds for nymphs on small plants or sparse vegetation in wasteland, rangeland, ditch banks and borders. Use 1¼ to 1⅞ pounds for adult grasshoppers or applications to dense vegetation.
Alfalfa* Clovers	Blister Beetles Mexican bean beetle		⅔ to 1¼	0 (clovers) 3 (alfalfa)	Observe plant response precautions
	Alfalfa caterpillar Bean leaf beetle Cucumber beetles Green cloverworm Japanese beetle	Leafhoppers Three cornered alfalfa hopper Thrips Velvetbean caterpillar	1¼		
	Alfalfa weevil larvae Armyworm Cloverhead weevil Corn earworm Cutworms Egyptian alfalfa weevil larvae Essex skipper	European alfalfa beetle Fall armyworm Lygus bugs Webworms Yellowstriped armyworm	1¼ to 1⅞		For alfalfa weevil larvae, if pretreatment damage is extensive, cut alfalfa and treat the stubble. Use higher rate in areas east of the Rocky Mountains. On dense growth use 25 to 40 gallons per acre with ground equipment to ensure adequate coverage.
* . . . For application to Alfalfa via Center Pivot Irrigation Systems, refer to "Directions for Use Through Center Pivot Irrigation Systems."					
Asparagus	Asparagus beetle		1¼ to 2½	1	
	Apache cicada Asparagus beetle		2½ to 5	Post harvest application only	Treat ferns or brush growth. Do not treat more than once every 3 days.
Beans (including blackeyed peas cowpeas, crowder or southern peas, dry beans, green beans, lima beans, navy beans and snap beans)	Blister beetles Mexican bean beetle		⅔ to 1¼	0 (except cowpeas) 3 (cowpeas)	Observe plant response precautions
	Alfalfa caterpillar Bean leaf beetle Cucumber beetles Flea beetles Green cloverworm Japanese beetle	Leafhoppers Three cornered alfalfa hopper Thrips Velvetbean caterpillar Western bean cutworm	1¼		

CROP	INSECT	POUNDS OF "SEVIN®" BRAND 80S/ACRE	PRE-HARVEST INTERVAL (DAYS)	SPECIFIC DIRECTIONS
	Armyworm Corn earworm Cutworms European corn borer Fall armyworm Stink bugs Tranished plant bug Webworms	1¼ to 1⅞		
	Cowpea curculia	2½		
	Corn earworm Limabean pod borer Lygus bugs Stink bugs	2½		CALIFORNIA ONLY
Cabbage Broccoli Brussels Sprouts Cauliflower Kohlrabi	Flea beetles Harlequin bug	⅔ to 1¼	3	
	Armyworm Corn earworm Fall armyworm Imported cabbageworm	1¼ to 2½		
Chinese cabbage Collards Hanover salad Horseradish Kale Mustard greens Radishes Rutabagas Turnips	Flea beetles Harlequin bug Leafhoppers	⅔ to 1¼	3 (Horseradish, radishes, rutabagas and turnip roots) 14 (Chinese cabbage, collards, Hanover salad, kale, mustard greens, and turnip tops)	
	Aster leafhopper	1¼ to 1⅞		
	Armyworm Corn earworm Fall armyworm Imported cabbageworm Stink bugs Tarnished plant bug	1¼ to 2½		
Carrots Celery Parsnips Parsley	Flea beetles Leafhoppers	⅔ to 1¼	0 (carrots) 3 (parsnips) 14 (celery and parsley)	DO NOT USE ON CELERY IN CALIFORNIA
	Aster leafhopper Lygus bugs Spittlebugs	1¼ to 1⅞		Treat on a 5 to 7 day schedule.
	Armyworm Corn earworm Fall armyworm Stink bugs Tarnished plant bugs	1¼ to 2½		
Corn* (field, sweet, pop)	Armyworm Chinch bugs Corn earworm Corn rootworm adults European corn borer Fall armyworm Flea beetles Japanese beetle Sap beetles Southwestern corn borer Leafhoppers	1¼ to 2½	0	OBSERVE BEE CAUTION: For insects attacking silks and ears apply at 1 to 6 day intervals starting when first silks appear and continuing until silks begin to dry. For larvae in whorl and foliage feeders, apply as necessary. For chinch bugs use high gallonage ground application directed at the base of plants. Optimum timing and good coverage are essential for effective control.
	Western bean cutworm	2½		Treat when infestation averages 15% and at 90% to 100% tassel emergence. Treatment after 100% silk emergence will reduce effectiveness.
	Cutworms	2½ to 8		Apply in a 12 inch band, using ¼ pound (4 ounces) per 1000 linear feet of row, in at least 15 gallons of water per acre. For broadcast application use up to 8 pounds in at least 20 gallons (ground) or 5 gallons (air) of water per acre.

* . . .For application to Corn via Center Pivot Irrigation Systems, refer to "Directions for Use Through Center Pivot Irrigation Systems."

CROP	INSECT	POUNDS OF "SEVIN®" BRAND 80S/ACRE	PRE-HARVEST INTERVAL (DAYS)	SPECIFIC DIRECTIONS
Cotton	Cotton fleahopper Cotton leafworm Flea beetles Striped blister beetle Thrips	⅔ to 1¼	7	Early season insect control.
	Boll weevil Bollworms Cotton leafperforator Fall armyworm Leafrollers Leafhoppers Tarnished plant bug Yellowstriped armyworm (cotton cutworm)	1¼ to 2½		Treat on a 5 to 7 day schedule for as long as control is necessary. Mid and late season insect control. May be applied after bolls open.

CROP	INSECT		POUNDS OF "SEVIN®" BRAND 80S/ACRE	PRE-HARVEST INTERVAL (DAYS)	SPECIFIC DIRECTIONS
	Lygus bugs		1¼ to 2½		For light to moderate populations in Western irrigated cotton.
	Pink bollworm		1⅞ to 3		Aphid populations will be suppressed by repeated applications of this insecticide.
	Stink bugs	Saltmarsh caterpillar	2½		
Cucumber Melons Pumpkins Squash	Pickleworm	Melonworm	⅔ to 1¼	0	Observe plant response precautions.
	Cucumber beetles Flea beetles	Leafhoppers Squash bugs	1¼		Avoid excessive applications.
Dandelion Endive (Escarole) Lettuce Salsify	Flea beetles Harlequin bug	Leafhoppers	⅔ to 1¼	3 (head lettuce & salsify roots) 14 (dandelion, endive (escarole) leaf lettuce & salsify tops)	Observe plant response precautions.
	Aster leafhopper Lygus bugs	Spittlebugs	1¼ to 1⅞		Treat on a 5 to 7 day schedule after heads begin to form.
	Armyworm Corn earworm Fall armyworm	Imported cabbageworm Stink bugs Tarnished plant bug	1¼ to 2½		
Forage grasses Pasture	Armyworm Black grass bugs Chinch bugs Essex skipper Fall armyworm	Range caterpillars Range crane fly Striped grass looper Thrips	1¼ to 1⅞	0	To control thrips in grasses grown for seed use high spray pressure to improve penetration into boot.
	White grubs (green June beetle)		1⅞ to 2½		
Garden beet Spinach Swiss chard	Flea beetles Harlequin bug	Leafhoppers	⅔ to 1¼	3 (garden beet roots) 14 (garden beet tops, spinach, Swiss chard)	Treat on a 5 to 7 day schedule.
	Aster leafhopper		1¼ to 1⅞		
	Armyworm Corn earworm Fall armyworm	Stink bugs Tarnished plant bug	1¼ to 2½		
Okra	Corn earworm	Stink bugs	1¼ to 2½	0	Treat on a 5 to 7 day schedule.
Peanuts	Blister beetles	Mexican bean beetle	⅔ to 1¼	0	Observe plant response precautions.
	Alfalfa caterpillar Bean leaf beetle Cucumber beetles Green cloverworm Japanese beetle Leafhoppers	Red-necked peanut worm Three cornered alfalfa hopper Thrips Velvetbean caterpillar	1¼		
	Armyworm Corn earworm Fall armyworm	Stink bugs Webworms	1¼ to 1⅞		
	Whitefringed beetle adults Cutworms		2½		
Peas	Colorado potato beetle Leafhoppers		1¼	3	
	Armyworm		1¼ to 1⅞		
	Alfalfa caterpillar Cutworms Pea leaf weevil	Pea weevil Yellowstriped armyworm	1⅞		
Potato* Tomato Eggplant Pepper	Colorado potato beetle Flea beetles Leafhoppers		⅔ to 1¼	0	
	European corn borer Fall armyworm Lace bugs Stink bugs	Tarnished plant bug Tomato fruit worm Tomato hornworm Tomato pinworm	1½ to 2½		
	Cutworms		2½		

* . . . For application to Potatoes via Center Pivot Irrigation Systems, refer to "Directions for Use Through Center Pivot Irrigation Systems."

CROP	INSECT	INSECT	POUNDS OF "SEVIN®" BRAND 80S/ACRE	PRE-HARVEST INTERVAL (DAYS)
Rice	Armyworm Chinch bugs	Fall armyworm Stink bugs	1¼ to 1⅞	14
	Armyworm Leafhoppers	Tadpole shrimp	2½	
Sorghums (milo, grain sorghum, sweet sorghum and hybrids)	Armyworm Chinch bugs Corn earworm	Fall armyworm Stink bugs Webworms	1¼ to 2½	21 (grain) 0 (forage)
	Sorghum midge, Southwestern corn borer		1⅞	
	Cutworms		2½	
Soybeans	Bean leaf beetle Cucumber beetles Green cloverworm	Mexican bean beetle Velvetbean caterpillar	⅔	0
	Corn earworm		⅔ to 1 1¼ to 1⅞	
	Blister beetles Grape colaspis Mexican bean beetle		⅔ to 1¼	
	Alfalfa caterpillar Bean leaf beetle Cucumber beetles Green cloverworm Japanese beetle	Leafhoppers Three cornered alfalfa hopper Thrips Velvetbean caterpillar	1¼	
	Armyworm Cutworms	Fall armyworm Stink bugs Webworms	1¼ to 1⅞	
	Painted lady (Thistle caterpillar)	Saltmarsh caterpillar Yellowstriped armyworm	2½ to 3	
Sugar beets	Armyworm Beet leaf beetle Fall armyworm	Flea beetles Leafhoppers Webworms	1¼ to 1⅞	14
	Cutworms		1⅞	
Sunflower	Cutworms		1⅞	60
	Armyworm Fall armyworm		1⅞ to 2½	
	Stem weevil Sunflower beetle		1¼ to 2½	

CROP	SPECIFIC DIRECTIONS
Rice	CAUTION! May kill shrimp and crabs. Do not use in areas where these are important resources. Do not use on rice fields in which crayfish and/or catfish farming are included in the cultural practice. DO NOT APPLY PROPANIL HERBICIDES WITHIN 15 DAYS BEFORE OR AFTER APPLICATION OF THIS PRODUCT OR PLANT INJURY WILL RESULT. MISSISSIPPI DELTA & TEXAS. CALIFORNIA ONLY. For optimum tadpole shrimp control apply to water when pest first appears.
Sorghums (milo, grain sorghum, sweet sorghum and hybrids)	Direct spray into forming heads for optimum insect control. Treat for sorghum midge when 25 to 30 percent of heads have emerged from boot and are in bloom. Repeat application 3 to 5 days later if adults are still active. A third application may be necessary in late planted sorghum or if midge are abundant. For chinch bugs use high gallonage ground application directed at base of plants.
Soybeans	DO NOT APPLY A COMBINATION OF THIS PRODUCT AND 2,4-DB HERBICIDES TO SOYBEANS AS CROP INJURY MAY RESULT. Use lower rates for light to moderate populations and smaller instars. Use the higher rate for heavy infestations and larger instars.
Sugar beets	
Sunflower	DO NOT USE IN CALIFORNIA. Do not allow animals to graze on treated crops.

CROP	INSECT	POUNDS OF "SEVIN®" BRAND 80S/ACRE	PRE-HARVEST INTERVAL (DAYS)	SPECIFIC DIRECTIONS
Sweet potato	Sweet potato weevil	9 Tbsps./gal.	0	DO NOT USE IN CALIFORNIA PREPLANT DIP Just prior to planting, dip sweet potato cuttings in a suspension of this product (9 Tbsps./gal. of water).
	Corn earworm Cucumber beetles Flea beetles Sweet potato hornworm Tortoise beetles	1¼ to 2½		Apply as a foliar spray as needed.
	Sweet potato weevil	1¼ to 2½		FOLIAR APPLICATION Full coverage of plants is essential. Use lower rate on young plants and higher rate on mature plants
Tobacco	Tobacco flea beetle	2½ Tbs./gal. or 1¼ lbs./50 gal. and apply 6 gal./100 sq. yards	0	IN PLANT BEDS. To prevent plant injury, avoid excessive applications.
	Green June beetle grubs	⅔ lb./100 gallons		Apply only to areas that larvae have uprooted by sprinkling as a drench with 50 to 100 gallons per 100 square yards.
	Budworms Fall armyworm Flea beetles Hornworms Japanese beetle June beetles Suckfly	1¼ to 2½		IN FIELDS Use lower rate on young plants (up to knee height). Use at least 10 gallons of prepared spray per acre. Begin treatments when worms are small. If late season applications are necessary, allow three days before priming or cutting
Wheat	Cereal leaf beetle	1¼	0 (Forage) 21 (grain)	DO NOT USE IN CALIFORNIA Do not make more than two applications after grain heads emerge from boot. Application is effective against eggs larvae and adults.
	Armyworm Fall armyworm	1¼ to 1⅞		

TREE FRUIT AND NUT CROPS

For dilute sprays apply the specified dosage per 100 gallons of water. For concentrate and aerial sprays increase the concentration of SEVIN brand 80S Carbaryl Insecticide in the spray mixture to apply an amount per acre equivalent to that in a dilute spray. The optimum spray gallonage will depend on tree size, density and stage of growth. Typical spray gallonages per acre range from 200-300 gallons for dilute sprays, 30-100 gallons for concentrate sprays and 10-25 gallons for aerial sprays. Do not exceed maximum label rate per acre per application.

CROP	POUNDS "SEVIN®" BRAND 80S/100 GAL	PRE-HARVEST INTERVAL (DAYS)	SPECIFIC DIRECTIONS
APPLE THINNING Apples only	⅓ to ⅔		Apply 1 full coverage dilute spray between 10 and 25 days after full bloom. If factors such as tree age, variety, nutrition, previous crop, pruning, bloom and degree of set favor excessive fruit thinning with this product, exercise caution to avoid possible yield reduction. Consult with your County Extension Service or other experts for advice on the proper use of this product. For easily thinned varieties including Cortland, Grimes, Jonathan, McIntosh, Orleans, Rome Beauty, Puritan, Red Delicious, Winesap, and Yellow Newton.

CROP	POUNDS "SEVIN®" BRAND 80S/100 GAL	PRE-HARVEST INTERVAL (DAYS)	SPECIFIC DIRECTIONS
	⅔ to 1¼	1	For difficult to thin varieties including Baldwin, Ben Davis, Duchess, Early McIntosh, Golden Delicious, Lady Apple, Northern Spy, Rhode Island Greening, Steele Red, Turley, Wealthy, Yellow Transparent and York Imperial.

CROP	INSECT		POUNDS OF "SEVIN®" BRAND 80S/100 GAL	PRE-HARVEST INTERVAL (DAYS)	SPECIFIC DIRECTIONS
Almond	Peach twig borer San Jose scale	Fruittree leafroller	1¼	28	Apply in "popcorn" or petal fall stages and again when the May brood of the peach twig borer begins to hatch or thereafter as needed.
	Naval orangeworm				Time early and mid season applications to correspond to moth flight peaks. Make a late season application at initiation of hull split or up to 10% hull split. Do not apply more than 6¼ pounds per acre.
Apples Pears	Apple aphid Apple rust mite Apple sucker Bagworms California pearslug (pear sawfly) Codling moth Eyespotted bud moth Green fruitworm Lecanium scales Lesser appleworm	Lygus bugs Orange tortrix Oystershell scale Pear leaf blister mite Pear psylla Pear rust mite San Jose scale Tentiform leafminers Woolly apple aphid	1 to 1¼	1	WEST OF THE ROCKY MOUNTAINS To avoid undesired apple thinning, delay use until at least 30 days after full bloom. For psylla control apply when eggs hatch or young nymphs are present.
	Apple mealybug Apple aphid Codling moth	White apple leafhopper	⅔		EAST OF THE ROCKY MOUNTAINS To avoid undesired apple thinning, delay use until at least 30 days after full bloom. For optimum scale control apply when crawlers are present.
	Apple maggot Apple rust mite Bagworms Eastern tent caterpillar European apple sawfly Eyespotted bud moth Forbes scale Fruittree leafroller Green fruitworm Japanese beetle Lecanium scales Lesser appleworm Oystershell scale	Pear psylla Pear leaf blister mite Pear rust mite Periodical cicada Plum curculio Redbanded leafroller Rosy apple aphid San Jose scale Tarnished plant bug Tentiform leafminers Woolly apple aphid Yellowheaded fireworm	1¼		
Chestnuts	Chestnut weevil		2½ to 3¾	0	DO NOT USE IN CALIFORNIA. Make 4 applications at weekly intervals beginning in late July for adult chestnut weevil control. Last application should be made prior to shuck split.
Citrus Fruits (such as grapefruit, lemons, lime, oranges tangelos, tangerines, citrus citron, kumquats and hybrids)	Avocado leafroller California orangedog Citrus cutworm Citrus root weevil Fruittree leafroller	Orange tortrix Western tussock moth West Indian sugarcane borer (adults)	1¼	5	Do not apply more than 25 pounds of this product per acre per application. Do not apply less than 10 gallons of dilute spray mixture per mature tree. May be mixed with petroleum oils commonly used on citrus.
	Black scale Brown soft scale California red scale	Citricola scale Citrus snow scale Yellow scale	1 to 1¼		
Filbert	Filbert aphid Filbert leafroller Filbertworm		1¼	0	Apply when leafroller eggs are hatching. Repeat on first appearance of adult filbert moths and again 3 to 4 weeks later.
Olives	Olive scale		1 to 1¼	0	For optimum scale control add 1½ gallons of summer oil and apply mixture when crawlers are present. Do not exceed 2 applications per year. Do not apply more than 18¾ pounds of this product per acre per application.

CROP	INSECT		POUNDS OF "SEVIN®" BRAND 80S/100 GAL	PRE-HARVEST INTERVAL (DAYS)	SPECIFIC DIRECTIONS
Peaches Apricots Nectarines	Apple pandemis Codling moth Cucumber beetles European earwig Fruittree leafroller Japanese beetle June beetles Lecanium scales Lesser peachtree borer Olive scale	Orange tortrix Oriental fruit moth Peach twig borer Periodical cicada Plum curculio Redbanded leafroller San Jose scale Tarnished plant bug Tussock moths Variegated leafroller	1¼	1 (peaches) 3 (apricots & nectarines)	Do not apply more than 7½ pounds of this product per acre per application to apricots. For optimum scale control apply when crawlers are present. Spray limbs and trunk thoroughly, weekly during moth flight.
Pecans	Black margined aphid Fall webworm Hickory shuckworm Lesser webworm Pecan leaf phylloxera	Pecan nut casebearer Pecan spittlebug Pecan stem phylloxera Pecan weevil Twig girdler Walnut caterpillar	1¼ to 3	0	Do not apply more than 9 pounds of this product per acre per application.
Pistachios	Navel orangeworm		⅔ to 2½	14	Apply dilute volumes of 150 to 300 gallons of mixed spray per acre for full coverage at onset of hull split.
Plums Prunes Cherries	Black cherry aphid Brown soft scale Cherry fruitworm Cherry maggot European earwig Eyespotted bud moth Forbes scale Fruittree leafroller Green fruitworm Japanese beetle Lecanium scales Lesser peach tree borer	Mealy plum aphid Oystershell scale Peach twig borer Plum curculio Prune leafhopper Redbanded leafroller Rose chafer San Jose scale Variegated leafroller	1¼	1	Do not apply more than 7½ pounds of this product per acre per application. For optimum scale control apply when crawlers are present. For lesser peachtree borer control, spray limbs and tree trunks thoroughly, weekly during moth flight.
	Codling moth Eastern tent caterpillar	Orange tortrix Tussock moth	1		
Walnut	Calico scale European fruit lecanium	Filbertworm Fruittree leafroller Frosted scale	⅔	0	Apply 1000 gallons of dilute spray per acre for mature trees.
	Codling moth				For codling moth apply first spray when average cross sectional diameters of developing nuts are ½ to ¾ inch. Repeat during middle or late June as needed.
	European earwig		2½		Spray tree trunks to point of run-off

SMALL FRUIT CROPS

Recommended dosages refer to pounds of SEVIN® brand 80S Carbaryl Insecticide per acre. The optimum spray gallonage will depend on plant size, density and stage of growth. Typical spray gallonages per acre range from 100-300 gallons for dilute sprays, 30-100 gallons for concentrate sprays and 10-25 gallons for aerial sprays. Do not exceed maximum label rate per acre per application.

CROP	INSECT		POUNDS OF "SEVIN®" BRAND 80S/ACRE	PRE-HARVEST INTERVAL (DAYS)	SPECIFIC DIRECTIONS
Blackberries Raspberries Dewberries (including boysenberries and loganberries)	European raspberry aphid Japanese beetle	Leafhoppers Leafrollers Rose chafer Snowy tree cricket	1¼ to 2½	7	
	Omnivorous leafroller Raspberry sawfly		2½		CALIFORNIA ONLY.
Blueberries	Blueberry maggot Cherry fruitworm Cranberry fruitworm	European fruit lecanium Japanese beetle	1⅞	0	Apply 3 weeks before harvest and repeat as necessary.
Cranberries	Cutworms Cranberry fireworms Cranberry fruitworms	Japanese beetle Leafhoppers Rose chafer	1⅞ to 3⅗	1	CAUTION! May kill shrimp and crabs. Do not use in areas where these are important resources. Apply in late bloom and as needed at 7 to 10 day intervals.
Grapes	European fruit lecanium Grape leaffolder Grape leafhopper	Western grapeleaf skeletonizer Western yellow-striped armyworm	1¼ to 2½	0	Apply before first brood leaffolder larvae emerge from rolls.

CROP	INSECT		POUNDS OF "SEVIN®" BRAND 80S/100 GAL	PRE-HARVEST INTERVAL (DAYS)	SPECIFIC DIRECTIONS
	Cutworms Eight spotted forester Grape berry moth Japanese beetle June beetles	Orange tortrix Omnivorous leafroller Redbanded leafroller Saltmarsh caterpillar	2½		
Strawberries	Flea beetles Meadow spittlebug Omnivorous leaftier (strawberry fruitworm)	Strawberry leafroller Strawberry weevil	1½ to 2½	1	Carbaryl may injure Early Dawn and Sunrise varieties on the Delmarva Peninsula.

TREES AND ORNAMENTALS

For dilute-spray ground applications to trees (including shade trees, shelter belts, forests, plantations, parks and recreational areas), ornamentals, woody plants and shrubs, apply the specified dosage per 100 gallons of water. For concentrate-spray ground applications, apply the specified dosage per acre in sufficient spray volume to provide thorough coverage. To prepare small volumes of spray, use 1¼ tablespoonsfuls of SEVIN® brand 80S Carbaryl Insecticide per gallon of water where rates of 1¼ pounds are indicated.

For aerial applications to forest trees (including shade trees, shelter belts, plantations, parks, and recreational areas) and commercially grown ornamentals, woody plants and shrubs, apply the specified dosage per acre in sufficient spray volume to provide thorough coverage. Avoid direct application to lakes, streams and ponds.

CROP	INSECT			POUNDS OF "SEVIN®" BRAND 80S	SPECIFIC DIRECTIONS
GROUND APPLICATION Trees (including shade trees, shelter belts, forests, plantations, parks and recreational areas). Ornamentals (including roses, flowers and other herbaceous plants), Woody plants and Shrubs. AERIAL APPLICATION Forest Trees (including shade trees, shelter belts, forests, plantations, parks and recreational areas) and Commercially Grown Ornamentals (including roses, flowers and other herbaceous plants), Woody Plants and Shrubs.	Ants Apple aphid Armyworm Azalea leafminer Bagworms Birch leafminer Blister beetle Boxelder bug Boxwood leafminer Brown tail moth Cankerworms Catalpa sphinx Chiggers Cooley spruce gall aphid Cutworms Cypress tip moth Douglas fir tussock moth Eastern spruce gall aphid Elm leaf aphid Elm leaf beetle Elm spanworm Eriophyid mites European pine shoot moth Fall armyworm Flea beetles Fuller rose beetle Gall midges Gall wasps	Green striped mapleworm Grasshoppers Gypsy moth Hackberry nipplegall maker Holly bud moth Holly leafminer Jackpine budworm Japanese beetle Jeffrey pine needleminer June beetles Lace bugs Leafhoppers Leafrollers Locust borer Maple leafcutter Mealybugs Mimosa webworm Nantucket pine tip moth Oak leafminers Oak leaf skeletonizer Oakworm complex Oleander caterpillar Olive ash borer Orange-striped oakworm Orange tortrix Periodical cicada Pine sawfly Pine spittlebug	Pitch pine tip moth Plant bugs Poinsettia hornworm Psyllids Puss caterpillar Redhumped oakworm Rose aphid Rose chafer Roseslug Saddled prominent Sawflies (exposed) Scale insects Sowbugs Spiny elm caterpillar Springtails Spruce budworm Spruce needleminer Subtropical pine tip moth Tent caterpillars Thorn bug Thrips (exposed) Ticks Walnut caterpillar Webworms Western hemlock looper Western spruce budworm Willow leaf beetles Yellow poplar weevil	1¼	Observe plant response precautions. Apply dilute sprays to obtain thorough coverage of upper and lower leaf surfaces. To control scale insects, treat trunks, stems and twigs in addition to plant foliage. For optimum worm control, treat when in early instars. Addition of a sticker may improve residual control. Applications for control of Maple leafcutter on sugar maple should be made when larvae are in 2nd instar, after mining, and as cases are being formed.
	Ips engraver beetles Mountain pine beetle Roundheaded pine beetle Western pine beetle			25	Effective as a preventive treatment only. Repeat annually as required to prevent beetle attacks. Apply 1 gallon of spray per 50 square feet of bark in late May to early July, or prior to beetle attack. Treat tree trunks from ground level up, until trunk diameter is less than 5 inches.
	Elm bark beetle				Apply approximately 20 to 30 gallons of spray mixture for each 50 feet of elm tree for thorough coverage of all bark surfaces on trunk, limbs and twigs.

LAWNS AND RECREATIONAL AREAS

CROP	INSECT		SPECIFIC DIRECTIONS
Turf grasses	Ants Armyworm Bluegrass billbug Centipedes Chiggers Chinch bugs Cutworms Earwigs Essex skipper European chafer Fall armyworm Fiery skipper Fleas Grasshoppers	June beetles Leafhoppers Lucerne moth Millipedes Mosquitoes Sod webworm (lawn moths) Sowbugs Springtails Ticks White grubs Yellowstriped armyworm	Use 4 ounces of SEVIN brand Carbaryl Insecticide per 1000 square feet ($10\frac{3}{4}$ pounds per acre) of turf grass. Make applications in sufficient spray volume for thorough coverage and turf thatch penetration. Repeat treatment as necessary. For Armyworm, Cutworms, Fall Armyworm and Sod Webworm Control: Do not irrigate treated areas following insecticide application. For Chinch Bug Control: Treat entire turf grass area rather than just damaged areas. Irrigation of turf grass area before insecticide application will aid in penetration into turf grass. For White Grub Control: Applications should be made when grubs are feeding near the soil surface, usually during late March through May, or July to early September, or as recommended by local Agricultural Extension Service agents. Water or irrigate turf grasses thoroughly soon after treatment.

IMPORTED FIRE ANT CONTROL

To prepare small amounts, use $\frac{1}{3}$ ounce of SEVIN® brand 80S Carbaryl Insecticide per gallon of water.

SITE	QUARTS OF "SEVIN®" BRAND 80S/ 100 GALLONS OF WATER	SPECIFIC INSTRUCTIONS
Lawns, cemeteries and Recreational Areas (Including turf, golf courses, and parks) Pastures Rangeland Forested Lands and Wastelands	2	Apply a total of 2 gallons of the diluted solution over the surface of each mound or at least 1 quart per 6 inches of mound diameter using a bucket, can or other appropriate equipment. Thoroughly wet mound and surrounding area to a 4 ft. diameter (12 sq. ft.). Do not disturb mounds prior to treatment. Pour solution from a height of about three feet to give sufficient force to break mound apex and flow into ant tunnels. For best results apply in cool weather, 65-80°F, or in early morning or late evening hours. Repeat application if mound activity resumes after 10 days. Treat new mounds as they appear. Pressurized sprays may disturb the ants and cause migration, reducing product effectiveness.

MOSQUITO CONTROL

For dilute-spray ground applications to trees (including shade trees, shelter belts, forests, plantations, parks and recreational areas), ornamentals, woody plants and shrubs, apply the specified dosage per 100 gallons of water. For concentrate-spray ground applications, apply the specified dosage per acre in sufficient spray volume to provide thorough coverage. To prepare small volumes of spray, use $1\frac{1}{4}$ tablespoonsful of SEVIN® brand 80S Carbaryl Insecticide per gallon of water, where rates of $1\frac{1}{4}$ pounds are indicated. For aerial applications to forest trees (including shade trees, shelter belts, plantations, parks and recreational areas) and commercially grown ornamentals, woody plants and shrubs, apply the specified dosage per acre in sufficient spray volume to provide thorough coverage. Avoid direct application to lakes, streams and ponds.

CROP	POUNDS OF "SEVIN®" BRAND 80S/ 100 GAL.	SPECIFIC DIRECTIONS
Pastures Rangelands, Parks, Recreational areas Logging Camps, Military Posts and Adjacent Forested lands or Wastelands.	$\frac{1}{3}$ to $1\frac{1}{4}$ ($\frac{1}{3}$ to $1\frac{1}{4}$ tablespoonful per gallon)	ADULT MOSQUITOES CAUTION! May kill shrimp and crabs. Do not use in areas where these are important resources. OBSERVE BEE CAUTION. Treat shrubbery and areas where adult mosquitoes congregate. Treat when adult mosquitoes are active in early morning or late evening. Repeat at 7 to 10 day intervals. Use $\frac{1}{3}$ to $\frac{2}{3}$ pound per 100 gallons in mistblowers, $\frac{2}{3}$ to $1\frac{1}{4}$ pounds per acre in aerial sprays and $1\frac{1}{4}$ pounds per acred in low pressure ground sprayers.
	$31\frac{1}{4}$	For residual control in subtropical regions apply 4 gallons of prepared spray per 2000 square feet of surface area. Repeat in 3 to 6 months or when necessary.

PEST CONTROL IN AND AROUND BUILDINGS LIMITED TO PEST CONTROL OPERATORS ONLY

INSECT	POUNDS OF "SEVIN®" BRAND 80S/100 GAL.	SPECIFIC DIRECTIONS
Ants Cockroaches	25 (25 tablespoonfuls/ gallon)	For use in and around buildings such as homes, apartments, warehouses, barns, and municipal and recreational areas.

INSECT	POUNDS OF "SEVIN®" BRAND 80S/100 GAL.	SPECIFIC DIRECTIONS
Brown dog tick Earwigs Fleas Millipedes Silverfish Spiders	12½ (12½ tablespoonfuls gallon)	Apply as a coarse wet spray or with a paint brush, and thoroughly treat cracks and crevices, baseboards, corners, walls, door sills, and window frames, quarters of household pets, ant trails, outside perimeter of dwellings and other areas where insects tend to congregate. Do not use as a space spray. Repeat when necssary but not more often than twice per week. Do not use this product in commercial food areas of food handling establishments, restaurants, or other places where food is prepared or processed. Do not use in serving areas while food is exposed. Avoid application to surfaces where visible spray residues are objectionable.

POULTRY INSECT CONTROL

POULTRY	INSECT	SPECIFIC DIRECTIONS
Chickens, Ducks, Geese Gamebirds, Pigeons, and Turkeys	Chicken mite Fleas Lice Northern Fowl mite	For use as a direct spray on birds by: 1. Misting with Electric Fog Machine: Mix ⅜ pounds (6 ounces) of this product in 1 gallon of spray. Use 1½ gallons per 1000 hens in cages, on litter or on slatted floors. Repeat in 4 weeks if necessary. or 2. Spraying with knapsack or cylinder type compressed air sprayers: Mix ¼ pound (4 ounces) of this product in 5 gallons of spray. Use 1 gallon per 100 hens in cages, on litter or on slatted floors. Repeat in 4 weeks if necessary. Direct mist spraying for chicken mite and fleas is a supplement to spraying roosts and buildings for control of these pests. Do not apply to poultry and gamebirds within 7 days of slaughter.

POULTRY QUARTERS	INSECT	SPECIFIC DIRECTIONS
Roosts and Buildings	Bedbugs Chicken mite Fleas Fowl tick Lesser mealworms Lice Northern fowl mite	Spray roosts and buildings with conventional power spray or knapsack equipment. For chicken mite, fleas and bedbugs, use 5 pounds of this product per 100 gallons of water. For fowl ticks, use 20 pounds of this product per 100 gallons of water. For lesser mealworms, use 62½ pounds per 100 gallons. Spray 1 to 2 gallons per 1000 square feet of wall, bedding, litter or roost surface. Force spraying into cracks. Repeat as needed. Avoid spraying of nests, eggs and feeding and watering troughs. Ventilate while spraying. Do not treat premises within 7 days of slaughter.

CENTER PIVOT IRRIGATION APPLICATIONS FOR ALFALFA, CORN AND POTATO

General Precautions for Applications Through Center Pivot Irrigation Systems

MAINTAIN CONTINUOUS AGITATION IN MIX TANK DURING MIXING AND APPLICATION TO ASSURE A UNIFORM SUSPENSION.

GREATER ACCURACY IN CALIBRATION AND DISTRIBUTION WILL BE ACHIEVED BY INJECTING A LARGER VOLUME OF A MORE DILUTE SOLUTION PER UNIT TIME.

APPLICATION OF MORE THAN LABEL RECOMMENDED QUANTITIES OF IRRIGATION WATER PER ACRE MAY RESULT IN DECREASED PRODUCT PERFORMANCE.

APPLY "SEVIN®" BRAND 80S CARBARYL INSECTICIDE ONLY THROUGH SYSTEMS CONTAINING ANTI-SIPHON AND CHECK VALVES WHICH WILL PREVENT WATER SOURCE CONTAMINATION AND OVERFLOW OF THE MIX TANK, AND CONTAINING INTERLOCKING CONTROLS BETWEEN THE METERING DEVICE AND THE WATER PUMP TO INSURE SIMULTANEOUS SHUT-OFF.

DO NOT APPLY WHEN WIND SPEED FAVORS DRIFT, WHEN SYSTEM CONNECTIONS OR FITTINGS LEAK, WHEN NOZZLES DO NOT PROVIDE UNIFORM DISTRIBUTION OR WHEN LINES CONTAINING THE PRODUCT MUST BE DISMANTLED AND DRAINED.

WHERE SPRINKLER DISTRIBUTION PATTERNS DO NOT OVERLAP SUFFICIENTLY, UNACCEPTABLE INSECT CONTROL MAY RESULT. WHERE SPRINKLER DISTRIBUTION PATTERNS OVERLAP EXCESSIVELY, PLANT INJURY MAY RESULT ON SENSITIVE CROPS.

ALLOW SUFFICIENT TIME FOR PESTICIDE TO BE FLUSHED THROUGH ALL LINES AND ALL NOZZLES BEFORE TURNING OFF IRRIGATION WATER.

DIRECTIONS FOR USE THROUGH CENTER PIVOT IRRIGATION SYSTEMS

Spray Preparation: Remove scale, pesticide residues, and other foreign matter from the chemical tank and entire injector system. Flush with clean water.

APPLICATION INSTRUCTIONS:

Prepare a suspension of SEVIN® brand 80S Carbaryl Insecticide in a mix tank. Fill tank with ½ to ¾ the desired amounts of water. Start mechanical or hydraulic agitation. Slowly add the required amount of this product, and then the remaining volume of water. Suspension concentrations of 1¼ to 2½ pounds of this product per 1 to 4 gallons of water are recommended.

Set sprinkler to deliver 0.1 to 0.3 inches of water per acre. Start sprinkler and then uniformly inject the suspension of this product into the irrigation water line so as to deliver the desired rate per acre. The suspension of this product should be injected with a positive displacement pump into the main line ahead of a right angle turn to insure adequate mixing.

Note: When treatment with this product has been completed, further field irrigation over the treated area should be avoided for 24 to 48 hours to prevent washing the chemical off the crop.

FOR CONTROL THROUGH CENTER PIVOT IRRIGATION SYSTEMS

Recommended dosages refer to pounds of SEVIN® brand 80S Carbaryl Insecticide per acre. Consult the Directions for Use and General Precautions Through Center Pivot Irrigation Systems before application.

CROP	INSECT	POUNDS OF "SEVIN®" BRAND 80S/ACRE	PRE-HARVEST INTERVAL (DAYS)	SPECIFIC DIRECTIONS
Alfalfa	Alfalfa caterpillar	1¼ to 2½	0	Observe plant response precautions.
	Grasshoppers	2½		
	Leafhoppers	1¼ to 2½		
Corn (field sweet, pop)	Corn earworm	2½	0	OBSERVE BEE CAUTION. Apply when first silks appear. If local populations warrant, a second application should be made 5 to 7 days after the first. Optimum timing is essential for effective control.
	European corn borer Southwestern corn borer			Apply as necessary. Optimum timing is essential for effective control.
	Western bean cutworm			Make 2 applications. First applications should be made at 60% egg hatch. Second application should be made at 98% egg hatch. Optimum timing is essential for effective control.
	Corn rootworm beetles Grasshoppers			Treat as necessary.
Potato	Colorado potato beetles European corn borer Flea beetle	2½	0	Treat as necessary.

STORAGE AND DISPOSAL

STORAGE

Store unused SEVIN® brand 80S Carbaryl Insecticide in original container only, in cool, dry area out of reach of children and animals. Do not store in areas where temperatures frequently exceed 100°F.

PESTICIDE DISPOSAL

Do not contaminate water, food or feed by storage or disposal. Open dumping is prohibited. Wastes resulting from the use of this product may be disposed of on site or at an approved waste disposal facility.

CONTAINER DISPOSAL

Completely empty bag into application equipment. Then dispose of empty bag in a sanitary landfill or by incineration, or, if allowed by State and local authorities, by burning. If bag is burned, stay out of smoke.

LIMITED WARRANTY AND DISCLAIMER

The manufacturer warrants (a) that this product conforms to the chemical description on the label; (b) that this product is reasonably fit for the purposes set forth in the directions for use when it is used in accordance with such directions; and (c) that the directions, warnings and other statements on this label are based upon responsible experts' evaluation of reasonable tests of effectiveness, of toxicity to laboratory animals and to plants, and of residues on food crops, and upon reports of field experience. Tests have not been made on all varieties or in all states or under all conditions. THE MANUFACTURER NEITHER MAKES NOR INTENDS, NOR DOES IT AUTHORIZE ANY AGENT OR REPRESENTATIVE TO MAKE, ANY OTHER WARRANTIES, EXPRESS OR IMPLIED, AND IT EXPRESSLY EXCLUDES AND DISCLAIMS ALL IMPLIED WARRANTIES OF MERCHANTABILITY OR FITNESS FOR A PARTICULAR PURPOSE.

THIS WARRANTY DOES NOT EXTEND TO, AND THE BUYER SHALL BE SOLELY RESPONSIBLE FOR, ANY AND ALL LOSS OR DAMAGE WHICH RESULTS FROM THE USE OF THIS PRODUCT IN ANY MANNER WHICH IS INCONSISTENT WITH THE LABEL DIRECTIONS, WARNINGS OR CAUTIONS.

BUYER'S EXCLUSIVE REMEDY AND MANUFACTURER'S OR SELLER'S EXCLUSIVE LIABILITY FOR ANY AND ALL CLAIMS, LOSSES, DAMAGES, OR INJURIES RESULTING FROM THE USE OR HANDLING OF THIS PRODUCT, WHETHER OR NOT BASED IN CONTRACT, NEGLIGENCE, STRICT LIABILITY IN TORT OR OTHERWISE, SHALL BE LIMITED, AT THE MANUFACTURER'S OPTION, TO REPALCEMENT OF, OR THE REPLACEMENT OF THE PURCHASE PRICE FOR, THE QUANTITY OF PRODUCT WITH RESPECT TO WHICH DAMAGES ARE CLAIMED. IN NO EVENT SHALL MANUFACTURER OR SELLER BE LIABLE FOR SPECIAL, INDIRECT OR CONSEQUENTIAL DAMAGES RESULTING FROM THE USE OR HANDLING OF THIS PRODUCT.

THIS SPECIMEN LABEL IS INTENDED FOR USE ONLY AS A GUIDE IN PROVIDING GENERAL INFORMATION REGARDING THE DIRECTIONS, WARNINGS AND CAUTIONS ASSOCIATED WITH THE USE OF THIS PRODUCT. AS WITH ANY AGRICULTURAL CHEMICAL, ALWAYS FOLLOW THE LABEL INSTRUCTIONS ON THE PACKAGE BEFORE USING.

SEVIN® BRAND 50-W

CARBARYL INSECTICIDE

For Agricultural Or Commercial Use Only

ACTIVE INGREDIENT:	% BY WEIGHT
Carbaryl (1-naphthyl N-methylcarbamate)	50%
INERT INGREDIENTS:	50%
TOTAL:	100%

EPA Reg. No. 264-314 E.P.A. Est. No. 477-MD-01

Keep Out of Reach of Children

CAUTION

IN CASE OF EMERGENCY TELEPHONE (24 HOURS A DAY) IN U.S.A. 1-800-UCC-HELP.

FOR GENERAL PRODUCT INFORMATION CALL 1-800-334-9745.

PRECAUTIONARY STATEMENTS

CAUTION

HAZARDS TO HUMANS AND DOMESTIC ANIMALS

MAY BE HARMFUL IF SWALLOWED. Avoid breathing of dust and spray mist. Do not take internally. Avoid contact with eyes, skin or clothing. Wear regular long-sleeved work clothing. Change to clean clothing daily. Wash hands and face before eating. Wash thoroughly after handling.

ANTIDOTE STATEMENT

ATROPINE SULFATE IS HIGHLY EFFECTIVE AS AN ANTIDOTE.

See NOTE TO PHYSICIAN below.

STATEMENT OF PRACTICAL TREATMENT

GENERAL

Contact a physician immediately in all cases of suspected poisoning. Transport to a physician or hospital immediately and SHOW A COPY OF THIS LABEL TO THE PHYSICIAN.

IF SWALLOWED: Never give anything by mouth to an unconscious or convulsing person. If conscious and not convulsing, drink 1 to 2 glasses of water and induce vomiting by touching the back of the throat with finger.

IF IN EYES: Flush eyes with plenty of water. Get medical attention if irritation persists.

IF ON SKIN: Wash thoroughly with soap and water.

IF INHALED: Move from contaminated atmosphere and call a physician.

NOTE TO PHYSICIAN

Carbaryl is a carbamate insecticide, which is a chloinesterase inhibitor. Overexposure to this substance may cause toxic signs and symptoms due to stimulation of the cholinergic nervous system. These effects of overexposure are spontaneously and rapidly reversible.

Specific treatment consists of parenteral atropine sulfate. Caution should be maintained to prevent overatropinization. Mild cases may be given 1 to 2 mg intramuscularly every 10 minutes until full atropinization has been achieved and repeated thereafter whenever symptoms reappear. Severe cases should be given 2 to 4 mg intravenously every 10 minutes until fully atropinized, then intramuscularly every 30 to 60 minutes to maintain the effect for at least 12 hours. Dosages for children should be appropriately reduced. Complete recovery from overexposure is to be expected within 24 hours.

Narcotics and other sedatives should not be used. Further, drugs like 2-PAM (pyridine-2-aldoxime methiodide) are NOT recommended.

To aid in confirmation of a diagnosis, urine samples should be obtained within 24 hours of exposure and immediately frozen. Analyses will be arranged by Union Carbide Agricultural Products Company, Inc.

Consultation on therapy can be obtained at all hours by calling the Union Carbide emergency phone number 1-800-UCC-HELP.

ENVIRONMENTAL HAZARDS

This product is extremely toxic to aquatic and estuarine invertebrates. Do not apply directly to water and wetlands, except under the forest canopy and use on rice. Discharge from rice fields may kill aquatic and estaurine invertebrates. Do not contaminate water by cleaning equipment or disposal of wastes.

BEE CAUTION

This product is highly toxic to bees exposed to direct treatment or residues on blooming crops and weeds. Do not apply this product or allow it to drift to blooming crops or weeds if bees are visiting the treatment area.

DIRECTIONS FOR USE

IT IS A VIOLATION OF FEDERAL LAW TO USE THIS PRODUCT IN A MANNER INCONSISTENT WITH ITS LABELING.

GENERAL INFORMATION

SEVIN® brand 50-W Carbaryl Insecticide is a dry powder for dispersion in water and application as an insecticide spray. READ THIS LABEL BEFORE USE. STRICTLY OBSERVE LABEL DIRECTIONS AND CAUTIONS, AND APPLICABLE FEDERAL AND STATE REGULATIONS.

GENERAL WORKER PROTECTION STATEMENTS

Do not apply this product in such a manner as to directly or through drift expose workers or other persons. The area treated must be vacated by unprotected persons.

Do not enter treated areas without protective clothing until sprays have dried.

Because certain states may require more restrictive reentry intervals for various crops treated with this product, consult your State Department of Agriculture for further information.

Written or oral warnings must be given to workers who are expected to be in treated area or in an area about to be treated with this product. Advise workers to stay out of fields during application and until sprays have dried. Regular long-sleeved work clothing should be worn when working in treated fields. See PRECAUTIONARY STATEMENTS, STATEMENT OF PRACTICAL TREATMENT and NOTE TO PHYSICIAN for information on accidental exposures. When oral warnings are given, warnings shall be given in a language customarily understood by workers. Oral warnings must be given if there is reason to believe that written warnings cannot be understood by workers. Written warnings must include the following information: Appropriate signal word (CAUTION), area treated with SEVIN® brand 50-W Carbaryl Insecticide, date of application, appropriate clothing, and reentry interval (i.e., until sprays have dried).

PRE-HARVEST AND GRAZING USE INFORMATION AND LIMITATIONS

Tolerances established under the Federal Food, Drug and Cosmetic Act permit the sale of crops bearing probable carbaryl residues when carbaryl is used in accordance with label directions. Pre-harvest and grazing use limitations are listed at the end of the directions for each crop use. Where no pre-harvest or grazing time limitations are stated, application can be made to the day of harvest and crops can be grazed without time delay if directions for use on the crop are followed. If SEVIN® brand 50-W Carbaryl Insecticide is used in accordance with label directions, forage and feed crops, including almond hulls, bean vines, carrot tops, citrus pulp, cowpea hay, peanut hay, pea vines, rice straw, soybean hay, and sugar beet tops, may be grazed or harvested for use as feed for dairy and meat animals without resulting in residues in milk or meat.

Do not plant rotational food and feed crops not listed on this or other carbaryl labels in carbaryl treated soil.

Do not use reclaimed irrigation water from crops treated with carbaryl on upland crops for which carbaryl tolerances are not established.

PLANT RESPONSE PRECAUTIONS

To avoid possible injury on tender foliage, do not apply to wet foliage or when rain or high humidity is expected during the next two days.

Do not use on Boston Ivy, Virginia creeper and maidenhair fern as injury will result. During early season, it may also injure Virginia and sand pines.

Observe label instructions on apple thinning and on combinations with certain herbicides on rice and soybeans.

SPRAY PREPARATION

TO ASSURE A UNIFORM SUSPENSION, AGITATE, STIR OR RECIRCULATE ALL "SEVIN®" BRAND 50-W CARBARYL INSECTICIDE SPRAY MIXTURES PRIOR TO AND DURING USE. Remove oil, rust, scale, pesticide residues and water. Fill spray or mix tank with ½ to ¾ the desired amount of water. Start mechanical or hydraulic agitation. Slowly add the required amount of SEVIN® brand 50-W Carbaryl Insecticide and then the remaining volume of water. Prepare only as much spray mixture as can be applied on the day of mixing. MAINTAIN CONTINUOUS AGITATION DURING MIXING AND APPLICATION TO ASSURE A UNIFORM SUSPENSION. DO NOT STORE SPRAY MIXTURE FOR PROLONGED PERIODS.

COMPATIBILITY

SEVIN brand 50-W Carbaryl Insecticide has been effectively used without plant injury with most common insecticides, miticides, fungicides, nutrients, adjuvants, and with petroleum oil as used on citrus. If compatibility with another product and the resulting crop response is unknown, it should be tested on a small scale. Curdling, precipitation, greasing, layer formation or increased viscosity are symptoms of incompatibility. Incompatibility will reduce insect control and may cause mechanical difficulties or plant injury. Observe all cautions and limitations on labeling of all products used in mixtures. WHEN PREPARING COMBINATION SPRAYS, FIRST ADD "SEVIN®" BRAND 50-W CARBARYL INSECTICIDE TO ½ THE DESIRED AMOUNT OF WATER, MIX THOROUGHLY, AND THEN ADD COMBINATION PRODUCTS TO THE MIXTURE. DO NOT APPLY TANK MIX COMBINATIONS UNLESS YOUR PREVIOUS EXPERIENCE INDICATES THE MIXTURE IS EFFECTIVE AND WILL NOT RESULT IN APPLICATION PROBLEMS, EXCESSIVE RESIDUES OR PLANT INJURY.

Unstable under highly alkaline conditions. Not effective if used with alkaline materials such as Bordeaux, lime-sulfur and casein-lime spreaders.

Do not use with summer oils on deciduous fruits.

APPLICATION

On all crops use sufficient gallonage to obtain thorough and uniform coverage. Calibrate spray equipment to deliver the required volume. Use 50 mesh slotted strainers in spray system and 25 mesh slotted strainers behind nozzles. Avoid applications just before rainfall as poor insect control may result. To clean spray system after use, drain and flush with water. Refer to the Storage and Disposal directions for proper disposal of rinse water.

INSECT CONTROL

Apply when insects or their damage appear. To maintain control repeat at 7 to 14 day intervals or as necessary unless a shorter interval is specified below. Where a dosage range is indicated, use lower rate on young plants and early instars and higher rate on mature plants, advanced instars and adults. Thorough and uniform spray coverage is essential for effective control.

SEVIN® brand 50-W Carbaryl Insecticide does not control spider mites. If spider mites are a problem, use a registered miticide.

Regional differences have been noted in the susceptibility of certain strains of fall armyworm to SEVIN® brand 50-W Carbaryl Insecticide. If local experience indicates inadequate control, use an alternative pesticide.

FORAGE, FIELD AND VEGETABLE CROP INSECT CONTROL

Recommended dosages refer to pounds of SEVIN® brand 50-W Carbaryl Insecticide per acre. Where a dosage range is indicated, use lower rate on young plants, higher rate on mature plants. Use sufficient spray gallonage to obtain full coverage. This will usually vary from 3 to 40 gallons per acre depending on equipment, degree of insect infestation, and stage of crop growth. Keep spray mixture well agitated. Apply when insects or their damage appear and repeat at 7 to 10 day intervals or as necessary unless shorter interval is specified below. See specific directions below for grasshopper control.

ALFALFA AND CLOVERS

Use 1 to 2 pounds for blister beetles and Mexican bean beetle.

Use 2 pounds for alfalfa caterpillar, bean leaf beetle, cucumber beetles, green cloverworm, Japanese beetle, leafhoppers, three-cornered alfalfa hopper, thrips and velvetbean caterpillar.

Use 2 to 3 pounds for alfalfa weevil larvae, armyworm, cloverhead weevil, corn earworm, cutworms, Egyptian alfalfa weevil larvae, essex skipper, European alfalfa beetle, fall armyworm, lygus bugs, stink bugs, webworms, and yellowstriped armyworm. Use higher rate in areas east of Rocky Mountains. On dense growth, use 25 to 40 gallons per acre with ground equipment to ensure adequate coverage. For alfalfa weevil larvae, if pretreatment damage is extensive, cut alfalfa and treat the stubble.

To avoid possible injury on tender foliage, do not apply when foliage is wet or when rain or excessive humidity is expected during the next 2 days.

Do not use on alfalfa within 3 days of harvest.

ASPARAGUS

Use 2 to 4 pounds for asparagus beetle on seedlings or spears.

Use 4 to 8 pounds for asparagus beetle and Apache cicada on ferns or brush growth in the post-harvest period.

Do not apply more often than once every 3 days.

Do not apply within 1 day of harvest.

BEANS (including blackeyed peas, cowpeas, crowder or southern peas, dry beans, green beans, lima beans, navy beans and snap beans)

Use 1 to 2 pounds for blister beetles and Mexican bean beetle.

Use 2 pounds for alfalfa caterpillar, bean leaf beetle, cucumber beetles, flea beetles, green cloverworm, Japanese beetle, leafhoppers, three-cornered alfalfa hopper, thrips, velvetbean caterpillar, and western bean cutworm.

Use 2 to 3 pounds for armyworm, corn earworm, cutworms, European corn borer, fall armyworm, stink bugs, tarnished plant bug, and webworms.

Use 4 pounds for cowpea curculio.

In California, use 4 pounds for corn earworm, limabean pod borer, lygus bugs, and stink bugs.

Do not apply to cowpeas within 3 days of harvest.

CABBAGE, BROCCOLI, BRUSSELS SPROUTS, CAULIFLOWER AND KOHLRABI

Use 1 to 2 pounds for flea beetles and harlequin bug.

Use 2 to 4 pounds for armyworm, fall armyworm, imported cabbageworm and corn earworm.

Do not apply within 3 days of harvest.

CHINESE CABBAGE, COLLARDS, HANOVER SALAD, HORSERADISH, KALE, MUSTARD GREENS, RADISHES, RUTABAGAS AND TURNIPS.

Use 1 to 2 pounds for flea beetles, harlequin bug and leafhoppers.

Use 2 to 3 pounds in a 5 to 7 day schedule for aster leafhopper.

Use 2 to 4 pounds for armyworm, fall armyworm, imported cabbageworm, corn earworm, tarnished plant bug and stink bugs.

Do not apply within 3 days of harvest of horseradish, radishes, rutabagas and turnips (roots).

Do not apply within 14 days of harvest of Chinese cabbage, collards, hanover salad, kale, mustard greens and turnips (tops).

CARROTS, CELERY, PARSNIPS AND PARSLEY (DO NOT USE ON CELERY IN CALIFORNIA)

Use 1 to 2 pounds for flea beetles and leafhoppers.

Use 2 to 3 pounds in a 5 to 7 day schedule for aster leafhopper, lygus bugs, and spittlebugs.

Use 2 to 4 pounds for armyworm, fall armyworm, corn earworm, tarnished plant bug and stink bugs.

No time limitation on carrots.

Do not apply within 3 days of harvest of parsnips.

Do not apply within 14 days of harvest of parsley and celery.

CORN (FIELD, SWEET AND POP)
Use 2 to 4 pounds for armyworm, chinch bugs, corn earworm, corn rootworm adults, European corn borer, fall armyworm, flea beetles, Japanese beetle, sap beetles, Southwestern corn borer, and leafhoppers. For larvae in whorl and foliage feeders, apply to entire plant. Repeat as necessary. For insects attacking silks and ears, apply at 1 to 6 day intervals starting when first silks appear and continuing until silks begin to dry. For chinch bugs use high gallonage ground application directed at the base of plants. Three or more applications may be required depending on severity of infestation. Timing and good coverage are absolutely essential for effective control. OBSERVE BEE CAUTION.

Use 4 pounds for Western bean cutworm. Treat when infestation averages 15% and at 90% to 100% tassel emergence. Treatment after 100% silk emergence will reduce effectiveness.

Use 4 to 13 pounds in at least 15 gallons of water per acre for cutworm control. Spray in 12-inch band using $^3/_8$ pound (6 ounces) per 1000 linear feet of row. For broadcast ground application, use up to 13 pounds in at least 20 gallons of water per acre.

COTTON
Use 1 to 2 pounds for early season control of cotton fleahopper, cotton leafworm, flea beetles, striped blister beetle, and thrips.

Use 2 to 4 pounds for mid and late season control of boll weevil, bollworms, cotton leafperforator, fall armyworm, leafrollers, leafhoppers, tarnished plant bug, and yellowstriped armyworm (cotton cutworm). Treat on a 5 to 7 day schedule for as long as necessary. May be applied after bolls open.

Use 2 to 4 pounds for light to moderate populations of lygus bugs in Western irrigated cotton.

Use 3 to 5 pounds for pink bollworm.

Use 4 pounds for stink bugs and saltmarsh caterpillar.

Aphid populations will be suppressed by repeated applications of this insecticide.

Do not use SEVIN® brand 50-W Carbaryl Insecticide within 7 days of harvest.

CUCUMBER, MELONS, PUMPKIN AND SQUASH
Use 1 to 2 pounds for pickleworm and melonworm; 2 pounds for cucumber beetles, flea beetles, leafhoppers and squash bug. Avoid excessive applications.

To avoid possible injury on tender foliage, do not apply when foliage is wet or when rain or excessive humidity is expected during the next 2 days.

DANDELION, ENDIVE (ESCAROLE), LETTUCE AND SALSIFY
Use 1 to 2 pounds for flea beetles, harlequin bug and leafhoppers.

Use 2 to 3 pounds in a 5 to 7 day schedule for aster leafhopper, lygus bugs, and spittlebugs.

Use 2 to 4 pounds for armyworm, fall armyworm, imported cabbageworm, corn earworm, tarnished plant bug and stink bugs. For optimum control of corn earworm after lettuce heads begin to form, use 4 pounds in a 5 to 7 day schedule.

To avoid possible injury on lettuce, do not apply when foliage is wet or when rain or excessive humidity is expected during the next 2 days.

Do not apply within 3 days of harvest of head lettuce and salsify (roots).

Do not apply within 14 days of harvest of dandelion, endive (escarole), leaf lettuce and salsify (tops).

FORAGE GRASSES AND PASTURE
Use 2 to 3 pounds for armyworm, blackgrass bugs, chinch bugs, essex skipper, fall armyworm, range caterpillar, range cranefly, striped grass looper, and thrips. For thrips control in grasses grown for seed, use high spray pressure to improve penetration into boot.

Use 3 to 4 pounds for white grubs (green June beetle).

GARDEN BEET, SPINACH AND SWISS CHARD
Use 1 to 2 pounds for the flea beetles, harlequin bug and leafhoppers.

Use 2 to 3 pounds in a 5 to 7 day schedule for aster leafhopper.

Use 2 to 4 pounds for armyworm, corn earworm, fall armyworm, tarnished plant bug and stink bugs.

Do not apply within 3 days of harvest of garden beets (roots).

Do not apply within 14 days of harvest of garden beets (tops), spinach and Swiss chard.

OKRA
Use 2 to 4 pounds for corn earworm and stink bugs. Repeat at 5 to 7 day intervals or as necessary.

PEANUTS
Use 1 to 2 pounds for blister beetles and Mexican bean beetle.

Use 2 pounds for alfalfa caterpillar, bean leaf beetle, cucumber beetles, green cloverworm, Japanese beetle, leafhoppers, red-necked peanutworm, three-cornered alfalfa hopper, thrips and velvetbean caterpillar.

Use 2 to 3 pounds for armyworm, fall armyworm, corn earworm, stink bugs and webworms.

Use 4 pounds for white fringed beetle adults and cutworms.

To avoid possible injury on tender foliage, do not apply when foliage is wet or when rain or excessive humidity is expected during the next 2 days.

PEAS
Use 2 pounds for leafhoppers and Colorado potato beetle.

Use 2 to 3 pounds for armyworm.

Use 3 pounds for alfalfa caterpillar, cutworms, pea leaf weevil, pea weevil, and yellow-striped armyworm.

Use 5 pounds for alfalfa looper in Washington State only. Repeat as necessary.

Do not apply within 3 days of harvest.

POTATO, TOMATO, EGGPLANT AND PEPPER
Use 1 to 2 pounds for Colorado potato beetle, flea beetles and leafhoppers.

Use 2 to 4 pounds for European corn borer, fall armyworm, lace bugs, tomato fruitworm, tomato hornworm, tomato pinworm, tarnished plant bug and stink bugs.

Use 4 pounds for cutworm control.

RICE
CAUTION: May kill shrimp and crabs. Do not use in areas where these are important resources. Do not use on rice fields in which crayfish and/or catfish farming are included in the cultural practice.

In the Mississippi Delta and Texas use 2 to 3 pounds for armyworm, fall armyworm, chinch bugs, and stink bugs.

DO NOT APPLY PROPANIL HERBICIDES WITHIN 15 DAYS BEFORE OR AFTER APPLICATION OF THIS PRODUCT OR PLANT INJURY WILL RESULT.

In California use 4 pounds for armyworm, leafhoppers and tadpole shrimp. For optimum tadpole shrimp control apply to the water when the pest first appears.

Do not apply within 14 days of harvest.

SORGHUMS (milo, grain sorghum, sweet sorghum, and hybrids)
Use 2 to 4 pounds for armyworm, chinch bugs, corn earworm, fall armyworm, stink bugs, and webworms. For chinch bugs use high gallonage ground application directed to base of plants.

Use 3 pounds for sorghum midge and Southwestern corn borer. Treat for sorghum midge when 25 to 30 percent of heads have emerged from boot and are in bloom. Repeat 3 to 5 days later if adults are still active. A third application may be necessary.

For optimum insect control, direct spray into forming heads.

Do not apply within 21 days of harvest of grain. No time limitations on forage.

Use 4 pounds for cutworm control.

SOYBEANS
DO NOT APPLY A COMBINATION OF THIS PRODUCT AND 2,4-DB HERBICIDES TO SOYBEANS AS CROP INJURY MAY RESULT.

Use 1 pound for light to moderate populations or 2 pounds for heavy infestations of bean leaf beetle, cucumber beetle, green cloverworm, and velvetbean caterpillar.

Use 1 to $1\frac{1}{2}$ pounds for light to moderate populations or 2 to 3 pounds for heavy infestations of corn earworm.

Use 1 to 2 pounds for blister beetles, grape colaspis, and Mexican bean beetle.

Use 2 pounds for alfalfa caterpillar, Japanese beetle, leafhoppers, three cornered alfalfa hopper, and thrips.

Use 2 to 3 pounds for armyworm, cutworm, fall armyworms, stink bugs, and webworms.

Use 4 to 5 pounds for painted lady (thistle caterpillar), saltmarsh caterpillar, and yellowstriped armyworm.

Use higher rates for heavy infestations.

SUGAR BEETS

Use 2 to 3 pounds for armyworms, beet leaf beetle, fall armyworm, flea beetles, leafhoppers and webworms.

Use 3 pounds for cutworms.

Do not apply within 14 days of harvest.

SWEET POTATOES—(DO NOT USE IN CALIFORNIA.)

Use 13 Tbsps. per gallon of water as a preplant dip for sweet potato weevil. Just prior to planting, dip sweet potato cuttings in a suspension of SEVIN® brand 50-W Carbaryl Insecticide.

Use 2 to 4 pounds for corn earworm, cucumber beetles, flea beetles, sweet potato hornworm, sweet potato weevil, and tortoise beetles.

TOBACCO

IN PLANT BEDS

For tobacco flea beetle, use 4 level tablespoonfuls per gallon or 2 pounds per 50 gallons, and apply 6 gallons per 100 square yards.

For green June beetle grubs, use 1 pound per 100 gallons of water when insects or their damage appear. Apply only to areas that larvae have uprooted by sprinkling as a drench with 50 to 100 gallons per 100 square yards. May be applied before or after seeding. Avoid excessive application as plant injury may result.

IN FIELDS:

For budworms, fall armyworm, flea beetles, Japanese beetle, June beetles, hornworms and suckfly, use 2 to 4 pounds in 50 to 100 gallons of water per acre. Full coverage of plants is essential. Use lower rate on young plants up to knee height, higher rate on mature plants. Begin treatments when worms are small. Avoid excessive application in bud of plant as injury may result.

Treated fields may be entered immediately after foliage has dried. If late applications are necessary, allow 3 days before priming or cutting.

GRASSHOPPERS

Apply 1 to 3 pounds as often as necessary. TO CONTROL GRASSHOPPERS ON THE ABOVE FORAGE FIELD AND VEGETABLE CROPS, FOLLOW PREHARVEST LIMITATIONS GIVEN AFTER EACH CROP. A 1 to 2 pounds rate is suggested for nymphs on small plants or sparse vegetation in wasteland, rangeland, ditchbanks and borders. A 2 to 3 pounds dosage should be used when grasshoppers are mature or material is applied to crops requiring greater coverage.

TREE FRUIT AND NUT INSECT CONTROL

Recommended dosages refer to pounds of SEVIN® brand 50-W Carbaryl Insecticide per 100 gallons of dilute spray in conventional hydraulic-type sprayers. Apply dilute sprays for full coverage. Spray gallonage will vary according to tree size, density, planting distance and stage of growth. With concentrate spray, where lower volumes of spray are applied per acre, the concentration of SEVIN® brand 50-W Carbaryl Insecticide should be increased in the spray mixture so that the amount of SEVIN® brand 50-W Carbaryl Insecticide applied per acre is equivalent to a full-coverage dilute spray. Keep spray mixtures well agitated. Additional instructions appear with the crop categories below.

ALMOND

Use 2 pounds for peach twig borer, San Jose scale and fruittree leafroller. Apply in "popcorn" or petal fall stages and again when the May brood of the peach twig borer begins to hatch or thereafter as needed.

Use 2 pounds for navel orangeworm. Time early and mid season applications to correspond to moth flight peaks. Make late season application at initiation of or up to 10% hull split.

Do not apply more than 10 pounds per acre.

Do not apply within 28 days of harvest.

APPLES AND PEARS

Application of SEVIN® brand 50-W Carbaryl Insecticide within 30 days after full bloom may also provide apple thinning; to avoid this, delay use until at least 30 days after bloom. Foliar injury may result from combination with summer oils west of the Rocky Mountains, or when used before second cover on York and McIntosh apples east of the Rocky Mountains.

West of the Rocky Mountains: Use $1\frac{1}{2}$ to 2 pounds for apple sucker, apple aphid, woolly apple aphid, bagworms, California pearslug (pear sawfly), codling moth, apple rust mite, pearleaf blister mite, pear rust mite, eyespotted bud moth, green fruitworm, lesser appleworm, lygus bugs, orange tortrix, pear psylla, tentiform leafminers, lecanium scales, oystershell scale, and San Jose scale. For optimum scale control apply when crawlers are present. For psylla control apply when eggs hatch or when young nymphs are present. When heavy aphid infestations are present, repeated applications may be necessary for optimum control.

East of the Rocky Mountains: Use 1 pound for apple mealybug, apple aphid, codling moth and white apple leafhopper. When heavy aphid infestations are present, repeated applications may be necessary for optimum control.

Use 2 pounds for apple maggot, bagworms, eastern tent caterpillar, European apple sawfly, eyespotted bud moth, fruittree leafroller, green fruitworm, Japanese beetle, lesser appleworm, pear psylla, periodical cicada, plum curculio, redbanded leafroller, rosy apple aphid, woolly apple aphid, apple rust mite, pearleaf blister mite, pear rust mite, tarnished plant bug, tentiform leafminers, yellow headed fireworm, forbes scale, lecanium scales, oystershell scale, and San Jose scale. For optimum scale control, apply when crawlers are present. To control rosy apple aphid, apply before leaves are curled.

Do not apply within 1 day of harvest of apples and pears.

APPLE THINNING

Many factors influence the degree of apple thinning obtained with SEVIN brand 50-W Carbaryl Insecticide. These include tree age, variety, nutrition, previous crop, pruning, degree of set, bloom, weather and use of other chemical thinners. When any of these favor reduced fruit set, caution should be observed in applying SEVIN® brand 50-W Carbaryl Insecticide so that excessive thinning and possible yield reduction will not occur. Recommended dosages refer to pounds of SEVIN® brand 50-W Carbaryl Insecticide per 100 gallons dilute spray.

Consult with your County Extension Service or other experts for advice on the proper use of this product.

Concentration: Use $\frac{1}{2}$ to 1 pound on easily thinned varieties. Apply for full coverage. On hard-to-thin varieties, use 1 to 2 pounds.

Timing: Apply in one spray timed between 10 and 25 days after full bloom.

Varietal Response: Easily-thinned varieties include Cortland, Grimes, Jonathan, McIntosh, Orleans, Rome Beauty, Puritan, Red Delicious, Winesap, and Yellow Newton. Difficult-to-thin varieties include Baldwin, Ben Davis, Dutchess, Early McIntosh, Golden Delicious, Lady Apple, Northern Spy, Rhode Island Greening, Steele Red, Turley, Wealthy, Yellow Transparent, and York Imperial.

CITRUS FRUITS (such as grapefruit, lemons, limes, oranges, tangelos, tangerines, citrus citron, kumquats and hybrids).
Use 2 pounds for avocado leafroller, California orangedog, citrus cutworm, citrus root weevil, fruittree leafroller, orange tortrix, Western tussock moth and West Indian sugarcane borer (adults).

Use $1\frac{1}{2}$ to 2 pounds for black scale, brown soft scale, California red scale, citricola scale, citrus scale, and yellow scale.

Apply lower dosage early in the season or against light infestation and the higher dosage late in the season or against heavy infestation and situations involving infested fruit. May be used with petroleum oil as used in common practice on citrus.

Do not apply more than 40 pounds of SEVIN® brand 50-W Carbaryl Insecticide per acre per application.

Do not apply less than 10 gallons of dilute spray mixture per mature tree.

Do not apply within 5 days of harvest of these crops.

CHESTNUTS—DO NOT USE IN CALIFORNIA.
Use 6 pounds for chestnut weevil.

Make 4 applications at weekly intervals beginning in late July for adult chestnut weevil control. Last application should be made prior to shuck split.

FILBERT
Use 2 pounds for filbert aphid, filbert leafroller and fibertworm. Apply when leafroller eggs are hatching. Repeat on first appearance of adult filbert moths and again 3 to 4 weeks later. With dilute sprays, 400 gallons per acre is suggested.

OLIVES
Use $1\frac{1}{2}$ to 2 pounds with $1\frac{1}{2}$ gallons of summer oil, in sprays for olive scale control. For optimum scale control, apply when crawlers are present. A maximum of 2 applications is permitted.

Do not apply more than 30 pounds of this product per acre per application.

PEACHES APRICOTS AND NECTARINES
Use 2 pounds for codling moth, cucumber beetles, European earwig, Japanese beetle, apple pandemis, June beetles, lesser peachtree borer, orange tortrix, oriental fruit moth, peach twig borer, periodical cicada, plum curculio, fruittree leafroller, redbanded leafroller, variegated leafroller, tarnished plant bug, tussock moths, lecanium scale, San Jose scale and olive scale. For optimum scale control, apply when crawlers are present. For lesser peachtree borer control, spray limbs and trunk thoroughly, weekly during moth flight.

Do not apply more than 12 pounds of this product per acre per application on apricots.

Do not apply within 1 day of harvest of peaches.

Do not apply within 3 days harvest of apricots and nectarines.

PECANS
Use 2 to 5 pounds for black margined aphid, fall webworm, hickory shuckworm, lesser webworm, pecan leaf phylloxera, pecan nut casebearer, pecan spittlebug, pecan weevil, twig girdler, walnut caterpillar.

Do not apply more than $14\frac{2}{5}$ pounds of SEVIN® brand 50-W Carbaryl Insecticide per acre per application.

PLUMS, PRUNES, AND CHERRIES
Use 2 pounds for black cherry aphid, mealy plum aphid, cherry maggots, cherry fruitworm, eyespotted bud moth, fruittree leafroller, redbanded leafroller, Japanese beetle, lesser peachtree borer, peach twig borer, plum curculio, prune leafhopper, European earwig, green fruitworm, rose chafer, variegated leafroller, brown soft scale, Forbes scale, lecanium scales, oystershell scale, and San Jose scale. For optimum scale control, apply when crawlers are present. For lesser peachtree borer control spray limbs and trunk thoroughly, weekly during moth flight.

Use $1\frac{1}{2}$ pounds for eastern tent caterpillar, codling moth, orange tortrix, and tussock moths.

Do not apply more than 12 pounds of this product per acre per application.

Do not apply within 1 day of harvest of these crops.

PISTACHIOS
Use 1 to 4 pounds per 100 gallons of water for navel orangeworm control. Apply 150 to 300 gallons of mixed spray per acre for full coverage at onset of hull split. Do not use within 14 days of harvest.

WALNUT
Use 1 pound for calico scale, codling moth, European fruit lecanium, filbertworm, fruittree leafroller, and frosted scale. Apply 1000 gallons of diluted spray per acre for mature trees. For codling moth, apply first spray when average cross-sectional diameters of developing nuts are $\frac{1}{2}$ to $\frac{3}{4}$ inch. Repeat during middle or late June as needed.

Use 4 pounds for European earwig. Spray tree trunks to point of runoff.

SMALL FRUIT INSECT CONTROL

Recommended dosage refers to pounds of SEVIN® brand 50-W Carbaryl Insecticide per acre. Keep spray mixtures well agitated. Apply when insects or the damage appear. Repeat at 7 to 10 day intervals or as necessary.

Typical spray gallonage per acre ranges from 100-300 gallons for dilute sprays, and 30-100 gallons for concentrate sprays. Do not exceed maximum label rate per acre per application.

BLACKBERRIES, RASPBERRIES AND DEWBERRIES (including Boysenberries and Loganberries)
Use 2 to 4 pounds for Japanese beetle, leafrollers, leafhoppers, European raspberry aphid, rose chafer, snowy tree cricket.

In California, use 4 pounds for omnivorous leafroller and raspberry sawfly.

Do not apply within 7 days of harvest of these crops.

BLUEBERRIES
Use 3 pounds for blueberry maggot, cherry and cranberry fruitworms, European fruit lecanium and Japanese beetle. Apply 3 weeks before harvest and repeat as necessary.

CRANBERRIES
Use 3 to 6 pounds for cutworms, cranberry fireworms, cranberry fruitworm, Japanese beetle, leafhoppers, and rose chafer. Apply in late bloom and as needed at 7 to 10 day intervals.

Do not apply within 1 day of harvest.

CAUTION: May kill shrimp and crabs. Do not use in areas where these are important resources.

GRAPES
Use 2 to 4 pounds for European fruit lecanium, grape leaffolder, grape leafhopper, western grapeleaf skeletonizer, and Western yellowstriped armyworm.

Apply just before first brood leaffolder larvae emerge from rolls and as needed for leafhoppers.

Use 4 pounds for grape berry moth, Japanese beetle, June beetles, orange tortrix, omnivorous leafroller, redbanded leafroller, cutworms, eight spotted forester, and saltmarsh caterpillar.

STRAWBERRIES
Use 2 to 4 pounds for flea beetles, meadow spittlebug, omnivorous leaftier (strawberry fruitworm), strawberry leafroller and strawberry weevil. Carbaryl may injure Early Dawn and Sunrise varieties on Delmarva Peninsula.

Do not apply within 1 day of harvest.

TREES AND ORNAMENTALS INSECT CONTROL

For dilute spray ground applications to trees (including shade trees, sugar maple, shelter belts, forests, plantations, parks, and recreational areas), ornamentals (including roses, flowers and other herbaceous plants), woody plants and shrubs, apply the specified dosage per 100 gallons of water. For concentrate-spray ground applications, apply the specified dosage per acre in sufficient spray volume to provide thorough coverage. To prepare small volumes of spray, use 2 tablespoonsful of SEVIN® brand 50-W Carbaryl Insecticide per gallon of water where rates of 2 pounds are indicated.

Use 2 pounds for control of:

Ants	Green striped mapleworm	Pine spittlebug
Apple aphid	Grasshoppers	Pitch pine tip moth
Armyworm	Gypsy moth	Plant bugs
Azalea leafminer	Hackberry nipplegall maker	Poinsettia hornworm
Bagworms	Holly bud moth	Psyllids
Birch leafminer	Holly leafminer	Puss caterpillar
Blister beetle	Jackpine budworm	Redhumped oakworm
Boxelder bug	Japanese beetle	Rose aphid
Boxwood leafminer	Jeffrey pine needleminer	Rose chafer
Brown tail moth	June beetles	Roseslug
Cankerworms	Lace bugs	Saddled prominent
Catalpa sphinx	Leafhoppers	Sawflies (exposed)
Chiggers	Leafrollers	Scale insects
Cooley spruce gall aphid	Locust borer	Sowbugs
Cutworms	Maple leafcutter	Spiny elm caterpillar
Cypress tip moth	Mealybugs	Springtails
Douglas-fir tussock moth	Mimosa webworm	Spruce budworm
Eastern spruce gall aphid	Nantucket pine tip moth	Spruce needleminer
Elm leaf aphid	Oak leafminer	Subtropical pine tip moth
Elm leaf beetle	Oak leaf skeletonizer	Tent caterpillars
Elm spanworm	Oakworm complex	Thorn bug
Eriophyid mites	Oleander caterpillar	Thrips (exposed)
European pine shoot moth	Olive ash borer	Ticks
Fall armyworm	Orange-striped oakworm	Walnut caterpillar
Flea beetles	Orange tortrix	Webworms
Fuller rose beetle	Periodical cicada	Western hemlock looper
Gall midges	Pine sawfly	Western spruce budworm
Gall wasps		Willow leaf beetles
		Yellow poplar weevil

Apply dilute sprays to obtain thorough coverage of upper and lower leaf surfaces. To control scale insects, treat trunks, stems, and twigs in addition to plant foliage. For optimum worm control, treat when in early instars. Addition of a sticker may improve residual control.

Applications for control of Maple leafcutter on sugar maple should be made when larvae are in 2nd instar, after mining, and as cases are being formed.

Use 40 pounds for control of Ips engraver beetles, mountain pine beetle, roundheaded pine beetle, and western pine beetle. Effective as a preventive treatment only. Repeat annually as required to prevent beetle attacks. Apply 1 gallon of spray per 50 square feet of bark in late May to early July, or prior to beetle attack. Treat tree trunks from ground level up, until trunk diameter is less than 5 inches.

LAWN AND RECREATIONAL AREA INSECT CONTROL

Use 6 ounces of SEVIN® brand 50-W Carbaryl Insecticide per 1000 square feet (17 pounds per acre) of turf grass for control of:

Ants	Cutworms	Fleas	Mosquitoes
Armyworm	Earwigs	Grasshoppers	Sod webworms (lawn moths)
Bluegrass billbug	Essex skipper	June beetles	Sowbugs
Centipedes	European chafer	Leafhoppers	Springtails
Chiggers	Fall armyworm	Lucerne moth	Ticks
Chinch bugs	Fiery skipper	Millipedes	White grubs
			Yellowstriped armyworm

Make applications using hose-end sprayers or pressure type equipment in sufficient spray volume for thorough coverage and turf thatch penetration. Repeat treatment as necessary.

For armyworm, cutworm, fall armyworm and sod webworm control, do not irrigate treated area following insecticide application.

For chinch bug control, treat entire turf grass area rather than just damaged areas. Irrigation of turf grass area before insecticide application will aid in penetration into turf grass.

For white grub control, treatments should be applied when grubs are feeding near the soil surface, usually during late March through May, or July to early September, or as recommended by local Agricultural Extension Service agents. Water or irrigate turf grasses thoroughly soon after treatment.

IMPORTED FIRE ANT CONTROL

To prepare small volumes of spray, use ½ ounce of SEVIN® brand 50-W Carbaryl Insecticide per gallon water.

Site	Pounds of "SEVIN® Brand 50-W	Specific Instructions
Lawns, cemeteries and Recreational Areas (including turf, golf courses, and parks) Pastures Rangeland Forested Lands and Wastelands	3	Apply a total of 2 gallons of the diluted solution over the surface of each mound or at least 1 quart per 6 inches of mound diameter using a bucket, can or other appropriate equipment. Thoroughly wet mound and surrounding area to a 4 ft. diameter (12 sq. ft.). Do not disturb mounds prior to treatment. Pour solution from a height of about 3 ft. to give sufficient force to break mound apex and flow into ant tunnels. For best results apply in cool weather, 65-80°F, or in early morning or late evening hours. Repeat application if mound activity resumes after 10 days. Treat new mounds as they appear. Pressurized spray may disturb the ants and causes migration, reducing product effectiveness.

ADULT MOSQUITOES

On pasture, rangelands and non-agricultural lands such as yards, parks, recreation areas, logging camps, military posts and adjacent forested lands or wastelands, use ½ to 1 pound per acre in mist blowers, 2 pounds per acre in suitable low pressure ground equipment. Apply in sufficient spray gallonage to effectively cover areas to be treated. For optimum control, apply in early morning or evening when air is calm and adult mosquitoes are active.

Use 2 pounds per 100 gallons in hydraulic sprayers or 2 tablespoonfuls per gallon in hand sprayers for treating backyards or other limited areas. Thoroughly cover lower shade tree foliage, shrubbery, flower beds, grass and shaded areas around buildings where adult mosquitoes congregate. Repeat at 7 to 10 day intervals or as necessary. Do not use on Boston ivy, Virginia creeper and maidenhair fern as injury will result.

OBSERVE BEE CAUTION.

CAUTION: May kill shrimp and crabs. Do not use in areas where these are important resources.

Use 50 pounds of SEVIN® brand 50-W Carbaryl Insecticide in 100 gallons of water for residual control in subtropical regions. Apply 4 gallons of prepared spray per 2000 square feet of surface area. Repeat in 3 to 6 months or when necessary.

PEST CONTROL IN AND AROUND BUILDINGS

LIMITED TO PEST CONTROL OPERATORS ONLY

For use in and around buildings such as homes, apartments, warehouses, barns, and municipal and recreation areas.

Use 40 pounds per 100 gallons of water (40 tablespoons per gallon of water) for ants and cockroaches.

Use 20 pounds per 100 gallons of water (20 tablespoons per gallon of water) for brown dog tick, earwigs, fleas, millipeds, and silverfish.

Apply as a coarse wet spray or with a paint brush, and thoroughly treat cracks and crevices, base boards, corners, walls, door sills and window frames, quarters of household pets, ant trails, outside perimeter of dwellings and other areas where insects tend to congregate.

Do not use as a space spray. Repeat when necessary, but not more than twice per week.

Do not use this product in commercial food areas of food handling establishments, restaurants, or other places where food is prepared or processed. Do not use in serving areas while food is exposed.

Avoid application to surfaces where visible spray residues are objectionable.

POULTRY INSECT CONTROL
CHICKENS, DUCKS, GAME BIRDS, GEESE, PIGEONS AND TURKEYS

Direct Mist Spray on Birds: Control northern fowl mite, chicken mite, lice and fleas by:

1. Misting with Electric Fog Machine: Mix $^5/_8$ pound (10 ounces) of SEVIN® brand 50-W Carbaryl Insecticide in 1 gallon of spray. Use $1^1/_2$ gallons per 1000 hens in cages, on litter or slatted floor. Repeat in 4 weeks if necessary, or
2. Spraying with knapsack or cylinder type compressed air sprayers. Mix $^3/_8$ pound (6 ounces) of this product in 5 gallons of spray. Use 1 gallon per 100 hens in cages on litter or on slatted floor. Repeat in 4 weeks if necessary.

Direct mist spraying for chicken mite and fleas, as a supplement to spraying roosts and buildings for control of these pests.

For bedbugs, chicken mite, fleas, foul tick, lesser mealworms, lice and northern fowl mite; spray roosts and buildings with conventional power spray or knapsack equipment. For chicken mite, fleas and bedbugs, use 2 pounds per 25 gallons of water. For fowl ticks, use 8 pounds per 25 gallons of water. Spray 1 to 2 gallons per 1000 square feet of wall, bedding, litter or roost surfaces. Force spray into cracks and crevices. Repeat as needed. Ventilate while spraying. Avoid spraying of nests, eggs and feeding and watering troughs.

Do not apply to poultry and game birds or to their premises within 7 days of slaughter.

STORAGE AND DISPOSAL

STORAGE
Store unused SEVIN® brand 50-W Carbaryl Insecticide in original container only, in cool, dry areas out of reach of children and animals. Do not store in areas where temperatures frequently exceed 100°F.

PESTICIDE DISPOSAL
Do not contaminate water, food or feed by storage or disposal. Open dumping is prohibited. Wastes resulting from the use of this product may be disposed of on site or at an approved waste disposal facility.

CONTAINER DISPOSAL
Completely empty bag into application equipment. Then dispose of bag in a sanitary landfill or by incineration, or, if allowed by State and local authorities, by burning. If bag is burned, stay out of smoke.

LIMITED WARRANTY AND DISCLAIMER

The manufacturer warrants (a) that this product conforms to the chemical description on the label; (b) that this product is reasonably fit for the purposes set forth in the directions for use when it is used in accordance with such directions; and (c) that the directions, warnings and other statements on this label are based upon responsible experts' evaluation of reasonable tests of effectiveness, of toxicity to laboratory animals and to plants, and of residues on food crops, and upon reports of field experience. Tests have not been made on all varieties or in all states or under all conditions. THE MANUFACTURER NEITHER MAKES NOR INTENDS, NOR DOES IT AUTHORIZE ANY AGENT OR REPRESENTATIVE TO MAKE, ANY OTHER WARRANTIES, EXPRESS OR IMPLIED, AND IT EXPRESSLY EXCLUDES AND DISCLAIMS ALL IMPLIED WARRANTIES OF MERCHANTABILITY OR FITNESS FOR A PARTICULAR PURPOSE.

THIS WARRANTY DOES NOT EXTEND TO, AND THE BUYER SHALL BE SOLELY RESPONSIBLE FOR, ANY AND ALL LOSS OR DAMAGE WHICH RESULTS FROM THE USE OF THIS PRODUCT IN ANY MANNER WHICH IS INCONSISTENT WITH THE LABEL DIRECTIONS, WARNINGS OR CAUTIONS.

BUYER'S EXCLUSIVE REMEDY AND MANUFACTURER'S OR SELLER'S EXCLUSIVE LIABILITY FOR ANY AND ALL CLAIMS, LOSSES, DAMAGES, OR INJURIES RESULTING FROM THE USE OR HANDLING OF THIS PRODUCT, WHETHER OR NOT BASED IN CONTACT, NEGLIGENCE, STRICT LIABILITY IN TORT OR OTHERWISE, SHALL BE LIMITED, AT THE MANUFACTURER'S OPTION, TO REPLACEMENT OF, OR THE REPAYMENT OF THE PURCHASE PRICE FOR, THE QUANTITY OF PRODUCT WITH RESPECT TO WHICH DAMAGES ARE CLAIMED. IN NO EVENT SHALL MANUFACTURER OR SELLER BE LIABLE FOR SPECIAL, INDIRECT OR CONSEQUENTIAL DAMAGES RESULTING FROM THE USE OR HANDLING OF THIS PRODUCT.

THIS SPECIMEN LABEL IS INTENDED FOR USE ONLY AS A GUIDE IN PROVIDING GENERAL INFORMATION REGARDING THE DIRECTIONS, WARNINGS AND CAUTIONS ASSOCIATED WITH THE USE OF THIS PRODUCT. AS WITH ANY AGRICULTURAL CHEMICAL, ALWAYS FOLLOW THE LABEL INSTRUCTIONS ON THE PACKAGE BEFORE USING.

SEVIN®BRAND 20% BAIT

CARBARYL INSECTICIDE

For Agricultural or Commercial Use Only

For Control of Insect Pests

ACTIVE INGREDIENT:	% BY WEIGHT
Carbaryl (1-naphthyl N-methyl-carbamate)	20%
INERT INGREDIENT:	80%
TOTAL	100%

EPA Reg. No. 264-311 E.P.A. Est. No. 10352-GA-01

Keep Out of Reach of Children
CAUTION

IN CASE OF EMERGENCY TELEPHONE (24 HOURS A DAY) IN U.S.A. 1-800-UCC-HELP.

FOR GENERAL PRODUCT INFORMATION CALL 1-800-334-9745.

PRECAUTIONARY STATEMENTS

CAUTION

HAZARDS TO HUMANS AND DOMESTIC ANIMALS
MAY BE HARMFUL IF SWALLOWED. Avoid breathing of dust. Do not take internally. Avoid contact with eyes, skin, or clothing. Wear regular long-sleeved work clothing. Change to clean clothing daily. Wash hands and face before eating. Wash thoroughly after handling.

ANTIDOTE STATEMENT
ATROPINE SULFATE IS HIGHLY EFFECTIVE AS AN ANTIDOTE. See NOTE TO PHYSICIAN below.

STATEMENT OF PRACTICAL TREATMENT

GENERAL

Contact a physician immediately in all cases of suspected poisoning. Transport to a physician or hospital immediately and SHOW A COPY OF THIS LABEL TO THE PHYSICIAN.

IF SWALLOWED: Never give anything by mouth to an unconscious or convulsing person. If conscious and not convulsing, drink 1 to 2 glasses of water and induce vomiting by touching the back of the throat with finger.

IF IN EYES: Flush eyes with plenty of water. Get medical attention if irritation persists.

IF ON SKIN: Wash thoroughly with soap and water.

IF INHALED: Move from contaminated atmosphere and call a physician.

NOTE TO PHYSICIAN

Carbaryl is a carbamate insecticide, which is a chloinesterase inhibitor. Overexposure to this substance may cause toxic signs and symptoms due to stimulation of the cholinergic nervous system. These effects of overexposure are spontaneously and rapidly reversible.

Specific treatment consists of parenteral atropine sulfate. Caution should be maintained to prevent overatropinization. Mild cases may be given 1 to 2 mg intramuscularly every 10 minutes until full atropinization has been achieved and repeated thereafter whenever symptoms reappear. Severe cases should be given 2 to 4 mg intravenously every 10 minutes until fully atropinized, then intramuscularly every 30 to 60 minutes to maintain the effect for at least 12 hours. Dosages for children should be appropriately reduced. Complete recovery from overexposure is to be expected within 24 hours.

Narcotics and other sedatives should not be used. Further, drugs like 2-PAM (pyridine-2-aldoxime methiodide) are NOT recommended.

To aid in confirmation of a diagnosis, urine samples should be obtained within 24 hours of exposure and immediately frozen. Analyses will be arranged by Union Carbide Agricultural Products Company, Inc.

Consultation on therapy can be obtained at all hours by calling the Union Carbide emergency number 1-800-UCC-HELP.

ENVIRONMENTAL HAZARDS

This product is extremely toxic to aquatic and estuarine invertebrates. Do not apply directly to water or wetlands. Do not contaminate water by cleaning equipment or disposal of wastes.

DIRECTIONS FOR USE

IT IS A VIOLATION OF FEDERAL LAW TO USE THIS PRODUCT IN A MANNER INCONSISTENT WITH ITS LABELING.

IMPORTANT PRE-HARVEST AND GRAZING USE INFORMATION

Tolerances established under the Federal Food, Drug and Cosmetic Act permit the sale of crops bearing probable carbaryl residues when carbaryl is used in accordance with label directions. If SEVIN carbaryl insecticide is used in accordance with label directions, these crops, including bean vines, carrot tops, pea vines, sugar beet tops, corn forage and cotton forage, may be grazed or harvested for use as feed for dairy and meat animals without resulting in illegal residues in milk or meat.

Do not plant rotational food and feed crops not listed on this or carbaryl labels in carbaryl treated soil.

Do not use reclaimed irrigation water on upland crops for which carbaryl tolerances are not established.

FIELD AND VEGETABLE CROPS, LAWN, TURF, AND ORNAMENTAL PLANTS

Broadcast applications may be made with either ground equipment or aircraft. Treatment may be repeated as necessary.

CROP	INSECT	POUNDS OF "SEVIN" BRAND 20% BAIT/ACRE	PRE-HARVEST INTERVAL (DAYS)	SPECIFIC DIRECTIONS
Cucumbers Squash Melons Peanuts	Cutworms Armyworms Crickets Darkling ground beetle Grasshoppers Sowbugs	5	0	
Alfalfa Peas Cotton	Cutworms Armyworms Crickets Darkling ground beetle Grasshoppers Sowbugs	7.5	3 (peas, alfalfa) 7 (cotton)	
Beans Carrots Corn Eggplants Okra Peppers Potatoes Tomatoes Sweet corn	Cutworms Armyworms Crickets Darkling ground beetles Grasshoppers Sowbugs	5-10	0	
Cowpeas	Cutworms Armyworms Crickets Darkling ground beetles Grasshoppers Sowbugs	5-10	3	
Asparagus Strawberries	Cutworms Armyworms Crickets Darkling ground beetle Grasshoppers Sowbugs	5-10	1	

CROP	INSECT	POUNDS OF "SEVIN" BRAND 20% BAIT/ACRE	PRE-HARVEST INTERVAL (DAYS)	SPECIFIC DIRECTIONS
Broccoli Brussels sprouts Cabbage Cauliflower Head lettuce Garden beets (roots) Horseradish Parsnips Radishes Rutabagas Turnips (roots)	Cutworms Armyworms Crickets Darkling ground beetles Grasshoppers Sowbugs	5-10	3	
Sugarbeets Collards Endive (escarole) Garden beets (tops) Kale Leaf lettuce Parsley Spinach Swiss chard Turnips (tops)	Armyworms Crickets Darkling ground beetles Grasshoppers Sowbugs	5-10	14	
Sunflowers	Cutworms	5-10	60	Do not allow animals to graze on treated crops.
Wheat	Grasshoppers	2½ to 7½	0 (Forage) 21 (Grain)	Do not make more than 2 applications after grain heads emerge from boot. Do not apply within 21 days of harvest of grain. No time limitation on green wheat used as pasture or forage.
	Armyworm Fall Armyworm	5 to 7½		
Pastures Rangeland Wasteland	Grasshoppers Crickets Armyworm Fall Armyworm	5-10	0	Treatment may be repeated as necessary. Lower rate is suggested for early instar grasshoppers, cricket and armyworms on small plants or sparse vegetation. Use higher rate when grasshoppers and crickets are mature and armyworms are in late instars or vegetation is dense.
Lawn and Turf	Cutworms Armyworms Crickets Darkling ground beetle Grasshoppers Mole crickets Sowbugs	5-10 (2-4 oz. per 1000 sq. ft.)		Treat entire area evenly. Apply with a rotary broadcast spreader to obtain uniform coverage, giving particular attention to edges of sidewalks and along porch and patio foundations. For best results, cut lawn, remove clippings and water thoroughly before treatment.
Ornamental plants	Cutworms Armyworms Crickets Darkling ground beetle Mole crickets Sowbugs	5-10 (2-4 oz. per 1000 sq. ft.)		Apply evenly to soil around plants, shrubs and trees. For best results, area to be treated should be damp.

SITE	INSECT	OUNCES OF "SEVIN®" BRAND 20% BAIT PER 1000 SQUARE FEET	SPECIFIC DIRECTIONS
Poultry Houses (chickens, turkeys, ducks, geese, gamebirds, and pigeons)	Darkling Beetle	8-12	Apply bait uniformly to floor or litter area using appropriate equipment such as chest-carried fertilizer/seed spreader. It is not necessary to remove birds during application, but uniform coverage may be difficult to achieve with birds present. Repeat application after 5 weeks if necessary.

STORAGE AND DISPOSAL

STORAGE

Store unused SEVIN® brand 20% Bait Carbaryl Insecticide in original container only, in cool, dry area out of reach of children and animals. Do not store in areas where temperatures frequently exceed 100°F.

PESTICIDE DISPOSAL

Do not contaminate water, food or feed by storage or disposal. Open dumping is prohibited. Wastes resulting from the use of this product may be disposed of on site or at an approved waste facility.

CONTAINER DISPOSAL

Completely empty bag into application equipment. Then dispose of empty bag in a sanitary landfill or by incineration, or, if allowed by State and local authorities, by burning. If bag is burned, stay out of smoke.

LIMITED WARRANTY AND DISCLAIMER

The manufacturer warrants (a) that this product conforms to the chemical description on the label; (b) that this product is reasonably fit for the purposes set forth in the directions for use when it is used in accordance with such directions, and (c) that the directions, warnings and other statements on this level are based upon responsible experts' evaluation of reasonable tests of effectiveness, of toxicity to laboratory animals and to plants, and of residues on food crops, and upon reports of field experience. Tests have not been made on all varieties or in all states or under all conditions. THE MANUFACTURER NEITHER MAKES NOR INTENDS, NOR DOES IT AUTHORIZE ANY AGENT OR REPRESENTATIVE TO MAKE, ANY OTHER WARRANTIES, EXPRESS OR IM-

PLIED WARRANTIES OF MERCHANTABILITY OR FITNESS FOR A PARTICULAR PURPOSE.

THIS WARRANTY DOES NOT EXTEND TO, AND THE BUYER SHALL BE SOLELY RESPONSIBLE FOR, ANY AND ALL LOSS OR DAMAGE WHICH RESULTS FROM THE USE OF THIS PRODUCT IN ANY MANNER WHICH IS INCONSISTENT WITH THE LABEL DIRECTIONS, WARNINGS OR CAUTIONS.

BUYER'S EXCLUSIVE REMEDY AND MANUFACTURER'S OR SELLER'S EXCLUSIVE LIABILITY FOR ANY AND ALL CLAIMS, LOSSES, DAMAGES, OR INJURIES RESULTING FROM THE USE OR HANDLING OF THIS PRODUCT, WHETHER OR NOT BASED IN CONTRACT, NEGLIGENCE, STRICT LIABILITY IN TORT OR OTHERWISE, SHALL BE LIMITED, AT THE MANUFACTURER'S OPTION, TO REPLACEMENT OF, OR THE REPAYMENT OF THE PURCHASE PRICE FOR, THE QUANTITY OF PRODUCT WITH RESPECT TO WHICH DAMAGES ARE CLAIMED. IN NO EVENT SHALL MANUFACTURER OR SELLER BE LIABLE FOR SPECIAL, INDIRECT OR CONSEQUENTIAL DAMAGES RESULTING FROM THE USE OR HANDLING OF THIS PRODUCT.

THIS SPECIMEN LABEL IS INTENDED FOR USE ONLY AS A GUIDE IN PROVIDING GENERAL INFORMATION REGARDING THE DIRECTIONS, WARNINGS AND CAUTIONS ASSOCIATED WITH THE USE OF THIS PRODUCT. AS WITH ANY AGRICULTURAL CHEMICAL, ALWAYS FOLLOW THE LABEL INSTRUCTIONS ON THE PACKAGE BEFORE USING.

SEVIN® BRAND 4F

CARBARYL INSECTICIDE

For Agricultural Or Commercial Use Only

ACTIVE INGREDIENTS:	% BY WEIGHT
Carbaryl (1-naphthyl N-methylcarbamate)	43%
INERT INGREDIENTS:	57%
TOTAL	100%

(Contains 4 pounds Carbaryl per Gallon)

EPA Reg. No. 264-349 EPA Est. No. 264-MO-01

Keep Out of Reach of Children

CAUTION

IN CASE OF EMERGENCY TELEPHONE (24 HOURS A DAY) IN U.S.A. 1-800-UCC-HELP.

FOR GENERAL PRODUCT INFORMATION CALL 1-800-334-9745.

PRECAUTIONARY STATEMENTS

CAUTION

HAZARDS TO HUMANS AND DOMESTIC ANIMALS

MAY BE HARMFUL IF SWALLOWED. Avoid breathing spray mist. Do not take internally. Avoid contact with eyes, skin or clothing. Wear regular long-sleeved work clothing. Change to clean clothing daily. Wash hands and face before eating. Wash thoroughly after handling.

ANTIDOTE STATEMENT

ATROPINE SULFATE IS HIGHLY EFFECTIVE AS AN ANTIDOTE. See NOTE TO PHYSICIAN below.

STATEMENT OF PRACTICAL TREATMENT

GENERAL

Contact a physician immediately in all cases of suspected poisoning. Transport to a physician or hospital immediately and SHOW A COPY OF THIS LABEL TO THE PHYSICIAN.

IF SWALLOWED: Never give anything by mouth to an unconscious or convulsing person. If conscious and not convulsing, drink 1 to 2 glasses of water and induce vomiting by touching the back of the throat with finger.

IF IN EYES: Flush eyes with plenty of water. Get medical attention if irritation persists.

IF ON SKIN: Wash thoroughly with soap and water.

IF INHALED: Move from contaminated atmosphere and call a physician.

NOTE TO PHYSICIAN

Carbaryl is a carbamate insecticide, which is a chloinesterase inhibitor. Overexposure to this substance may cause toxic signs and symptoms due to stimulation of the cholinergic nervous system. These effects of overexposure are spontaneously and rapidly reversible.

Specific treatment consists of parenteral atropine sulfate. Caution should be maintained to prevent overatropinization. Mild cases may be given 1 to 2 mg intramuscularly every 10 minutes until full atropinization has been achieved and repeated thereafter whenever symptoms reappear. Severe cases should be given 2 to 4 mg intravenously every 10 minutes until fully atropinized, then intramuscularly every 30 to 60 minutes to maintain the effect for at least 12 hours. Dosages for children should be appropriately reduced. Complete recovery from overexposure is to be expected within 24 hours.

Narcotics and other sedatives should not be used. Further, drugs like 2-PAM (pyridine-2-aldoxime methiodide) are NOT recommended.

To aid in confirmation of a diagnosis, urine samples should be obtained within 24 hours of exposure and immediately frozen. Analyses will be arranged by Union Carbide Agricultural Products Company, Inc.

Consultation of therapy can be obtained at all hours by calling the Union Carbide emergency number 1-800-UCC-HELP.

ENVIRONMENTAL HAZARDS

This product is extremely toxic to aquatic and estuarine invertebrates. Do not apply directly to water and wetlands, except for rice use. Discharge from rice fields may kill aquatic and estaurine invertebrates. Do not contaminate water by cleaning equipment or disposal of wastes.

BEE CAUTION

This product is highly toxic to bees exposed to direct treatment or residues on blooming crops and weeds. Do not apply this product or allow it to drift to blooming crops or weeds if bees are visiting the treatment area.

DIRECTIONS FOR USE

IT IS A VIOLATION OF FEDERAL LAW TO USE THIS PRODUCT IN A MANNER INCONSISTENT WITH ITS LABELING.

GENERAL INFORMATION

SEVIN® brand 4F Carbaryl Insecticide is a suspension of microfine SEVIN carbaryl insecticide in an aqueous medium. It readily disperses in water to form a spray which may be applied by air or ground. READ THIS LABEL BEFORE USE. STRICTLY OBSERVE LABEL DIRECTIONS AND CAUTIONS, AND APPLICABLE FEDERAL AND STATE REGULATIONS.

GENERAL WORKER PROTECTION STATEMENTS

Do not apply this product in such a manner as to directly or through drift expose workers or other persons. The area treated must be vacated by unprotected persons.

Do not enter treated areas without protective clothing until sprays have dried.

Because certain states may require more restrictive reentry intervals for various crops treated with this product, consult your State Department of Agriculture for further information.

Written or oral warnings must be given to workers who are expected to be in treated area or in an area about to be treated with this product. Advise workers to stay out of fields during application and until sprays have dried. Regular long-sleeved work clothing should be worn when working in treated fields. See PRECAUTIONARY STATEMENTS, STATEMENT OF PRACTICAL TREATMENT AND NOTE TO PHYSICIAN for information on accidental exposures. When oral warnings are given, warnings shall be given in a language customarily understood by workers. Oral warnings must be given if there is reason to believe that written warnings cannot be understood by workers. Written warnings must include the following information: appropriate signal word (CAUTION), area treated with SEVIN® **brand 4F Carbaryl Insecticide, date of application, appropriate clothing, and re-entry interval (i.e., until sprays have dried).**

PREHARVEST AND GRAZING USE INFORMATION AND LIMITATIONS

Tolerances established under the Federal Food, Drug and Cosmetic Act permit the sale of crops bearing probable carbaryl residues when this product is used in accordance with label directions. If used as directed, treated forage may be grazed or used as feed for dairy and meat animals without causing illegal residues in meat or milk. This product may be applied up to and including the day of harvest or grazing of forage crops. Application may be made without removing livestock from area being treated.

Do not plant rotational food and feed crops not listed on this or other carbaryl labels in carbaryl treated soil.

Do not use reclaimed irrigation water from crops treated with carbaryl on upland crops for which carbaryl tolerances are not established.

PLANT RESPONSE PRECAUTIONS

To avoid possible injury to tender foliage, do not apply to wet foliage or when rain or high humidity is expected during the next two days.

Do not use on Boston ivy, Virginia creeper and maidenhair fern as injury will result. During early season, it may also injure Virginia and sand pines.

Observe label instructions on apple thinning and on combinations with certain herbicides on rice and soybeans.

SPRAY PREPARATION

TO ASSURE A UNIFORM PRODUCT, AGITATE, STIR OR RECIRCULATE ALL CONTAINERS OF "SEVIN®" BRAND 4F CARBARYL INSECTICIDE CONTAINERS PRIOR TO USE. Remove oil, rust, scale, pesticide residues or other foreign matter from mix tanks and entire spray system. Flush with clean water. Fill spray or mix tank with ½ to ¾ the desired amount of water. Start mechanical or hydraulic agitation. Slowly add the required amount of SEVIN® brand 4F Carbaryl Insecticide, and then the remaining volume of water. Include rinse water from container. Prepare only as much spray mixture as can be applied on the day of mixing. MAINTAIN CONTINUOUS AGITATION DURING MIXING AND APPLICATION TO ASSURE A UNIFORM SUSPENSION. DO NOT STORE SPRAY MIXTURES FOR PROLONGED PERIODS.

COMPATIBILITY

SEVIN® brand 4F Carbaryl Insecticide, when diluted with an equal volume of water, is compatible with a wide range of pesticides. It is not compatible with diesel fuel, kerosene, fuel oil or aromatic solvents. If compatibility of SEVIN® brand 4F Carbaryl Insecticide with another product and the resulting crop response is unknown, it should be tested on a small scale. Curdling, precipitation, greasing, layer formation or increased viscosity are symptoms of incompatibility.

Incompatibility will reduce insect control and may cause mechanical difficulties or plant injury. Observe all cautions and limitations on labeling of all products used in mixtures. WHEN PREPARING COMBINATIONS SPRAYS, FIRST ADD "SEVIN®" BRAND 4F CARBARYL INSECTICIDE TO AT LEAST AN EQUAL VOLUME OF WATER, MIX THOROUGHLY, AND THEN ADD COMBINATION PRODUCTS TO THE MIXTURE. DO NOT APPLY TANK MIX COMBINATIONS UNLESS YOUR PREVIOUS EXPERIENCE INDICATES THE MIXTURE IS EFFECTIVE AND WILL NOT RESULT IN APPLICATION PROBLEMS, EXCESSIVE RESIDUES OR PLANT INJURY.

Unstable under highly alkaline conditions. Not effective if used with alkaline materials such as Bordeaux, lime-sulfur and casein-lime spreaders.

APPLICATION

On all crops, use sufficient gallonage to obtain thorough and uniform coverage. Calibrate spray equipment to deliver the required volume. The flow rate of SEVIN® brand 4F Carbaryl Insecticide diluted 1:1 with water is similar to water. Use 50 mesh strainers in spray system and 25 mesh slotted strainers behind nozzles. Avoid applications just before rainfall as poor insect control may result. To clean spray system after use, drain and flush with water. Refer to the Storage and Disposal directions for proper disposal of this rinse water.

INSECT CONTROL

Apply when insects or their damage appear. To maintain control repeat at 7 to 14 day intervals or as necessary unless a shorter interval is specified below. Where a dosage range is indicated, use lower rate on young plants and early instars and higher rate on mature plants, advanced instars and adults. Thorough and uniform spray coverage is essential for effective control.

SEVIN® brand 4F Carbaryl Insecticide does not control spider mites. If spider mites are a problem, use a registered miticide.

Regional differences have been noted in the susceptibility of certain strains of fall armyworm of SEVIN® brand 4F Carbaryl Insecticide. If local experience indicates inadequate control, use an alternative pesticide.

FORAGE, FIELD AND VEGETABLE CROPS

Use at least 1 gallon of finished spray per acre for aerial application and at least 3 gallons of finished spray per acre for concentrate ground application. To prepare small volumes of spray, use 1 tablespoonful (½ fluid ounce) of SEVIN® brand 4F Carbaryl Insecticide per gallon of water where rates of 1 quart per acre or 1 quart per 100 gallons are indicated in the tables below.

CROP	INSECT	QUARTS OF "SEVIN" BRAND 4F/ACRE	PRE-HARVEST INTERVAL (DAYS)	SPECIFIC DIRECTIONS
All Forage, Field and Vegetable Crops in this Section	Grasshoppers	½ to 1½	See specific Field or Vegetable Crop	Use ½ to 1 quart for nymphs on small plants or sparse vegetation in wasteland, rangeland, ditch banks and borders. Use 1 to 1½ quarts for adult grasshoppers or applications to dense vegetation.
Alfalfa, Clovers	Blister beetles Mexican bean beetle	½ to 1	0 (clovers) 3 (alfalfa)	Observe plant response precautions.
	Alfalfa caterpillar Bean leaf beetle Cucumber beetles Green cloverworm Japanese beetle Leafhoppers Three cornered alfalfa hopper Thrips Velvetbean caterpillar	1		

CROP	INSECT		QUARTS OF "SEVIN" BRAND 4F/ACRE	PRE-HARVEST INTERVAL (DAYS)	SPECIFIC DIRECTIONS
	Alfalfa weevil larvae Armyworm Cloverhead weevil Corn earworm Cutworms Egyptian alfalfa weevil larvae Essex skipper	European alfalfa beetle Fall armyworm Lygus bugs Stink bugs Webworms Yellowstriped armyworm	1 to 1½		For alfalfa weevil larvae, if pretreatment damage is extensive, cut alfalfa and treat the stubble. Use higher rate in areas east of the Rocky Mountains. On dense growth use 25 to 40 gallons per acre with ground equipment to ensure adequate coverage.
Asparagus	Asparagus beetle		1 to 2	1	
	Apache cicada Asparagus beetle		2 to 4	Post harvest application only	Treat ferns or brush growth. Do not treat more than once every 3 days.
Beans (including blackeyed peas, cowpeas, crowder or southern peas, dry beans, green beans, lima beans, navy beans and snap beans)	Blister beetles Mexican bean beetle		½ to 1	0 (except cowpeas) 3 (cowpeas)	Observe plant response precautions.
	Alfalfa caterpillar Bean leaf beetle Cucumber beetles Flea beetles Green cloverworm Japanese beetle	Leafhoppers Three cornered alfalfa hopper Thrips Velvetbean caterpillar	1		
	Western bean cutworm		1 to 2		
	Armyworm Corn earworm Cutworms European corn borer	Fall armyworm Stink bugs Tarnished plant bug Webworms	1 to 1½		
	Cowpea curculio		2		
	Corn earworm Limabean pod borer	Lygus bugs Stink bugs	2		CALIFORNIA ONLY
Cabbage Broccoli Brussels Sprouts Cauliflower Kohlrabi	Flea beetles	Harlequin bug	½ to 1	3	
	Armyworm Corn earworm Fall armyworm	Imported cabbageworm	1 to 2		
Chinese cabbage Collards Hanover salad Horseradish Kale Mustard greens Radishes Rutabagas Turnips	Flea beetles Harlequin bug	Leafhoppers	½ to 1	3 (Horseradish, radishes, rutabagas and turnip roots) 14 (Chinese cabbage, collards, Hanover salad, kale, mustard greens, and turnip tops)	
	Aster leafhopper		1 to 1½		
	Armyworm Corn earworm Fall armyworm	Imported cabbageworm Stink bugs Tarnished plant bug	1 to 2		
Carrots Parsnips Parsley Celery	Flea beetles	Leafhoppers	½ to 1	0 (carrots) 3 (parsnips) 14 (celery and parsley)	DO NOT USE ON CELERY IN CALIFORNIA. Treat on a 5 to 7 day schedule.
	Aster leafhopper Lygus bugs	Spittlebugs	1 to 1½		
	Armyworm Corn earworm Fall armyworm	Stink bugs Tarnished plant bug	1 to 2		
Corn (field, sweet, pop)	Armyworm Chinch bugs Corn earworm Corn rootworm adults European corn borer Fall armyworm	Flea beetles Japanese beetle Sap beetles Southwestern corn borer Leafhoppers	1 to 2	0	OBSERVE BEE CAUTION. For insects attacking silks and ears apply at 1 to 6 day intervals starting when first silks appear and continuing until silks begin to dry. For larvae in whorl and foliage feeders, apply as necessary. For Chinch bugs use high gallonage ground application directed at the base of plants. Optimum timing and good coverage are essential for effective control.
	Western bean cutworm		2		Treat when infestation averages 15% and at 90 to 100% tassel emergence. Treatment after 100% silk emergence will reduce effectiveness.

CROP	INSECT		QUARTS OF "SEVIN" BRAND 4F/ACRE	PRE-HARVEST INTERVAL (DAYS)	SPECIFIC DIRECTIONS
	Cutworms		2 to 6½		Apply in a 12 inch band, using 6 fluid ounces per 1000 linear feet of row, in at least 15 gallons of water per acre. For broadcast application use 6.5 quarts in at least 20 gallons (ground) or 5 gallons (air) of water per acre.
Cotton	Cotton fleahopper Cotton leafworm Flea beetles	Striped blister beetle Thrips	½ to 1	7	Early season insect control.
	Boll weevil Bollworms Cotton leafperforator Fall armyworm Leafrollers	Leafhoppers Tarnished plant bug Yellowstriped armyworm (cotton cutworm)	1 to 2		Treat on a 5 to 7 day schedule for as long as control is necessary. Mid and late season insect control. May be applied after bolls open.
	Lygus bugs		1 to 2		For light to moderate populations in Western irrigated cotton.
	Pink bollworm		1½ to 2½		Aphid populations will be suppressed by repeated applications of this insecticide.
	Stink bugs	Saltmarsh caterpillar	2		
Cucumber Melons Pumpkins Squash	Pickleworm	Melonworm	½ to 1	0	Observe plant response precautions. Avoid excessive applications.
	Cucumber beetles Flea beetles	Leafhoppers Squash bugs	1		
Dandelion Endive (Escarole) Lettuce Salsify	Flea beetles Harlequin bug	Leafhoppers	½ to 1	3 (head lettuce & salsify roots) 14 (dandelion, endive (escarole) leaf lettuce & salsify tops)	Observe plant response precautions. Treat on a 5 to 7 day schedule after heads begin to form.
	Aster leafhopper Lygus bugs	Spittlebugs	1 to 1½		
	Armyworm Corn earworm Fall armyworm	Imported cabbageworm Stink bugs Tarnished plant bug	1 to 2		
Forage Grasses Pasture	Armyworm Black grass bugs Chinch bugs Essex skipper Fall armyworm	Range caterpillars Range crane fly Striped grass looper Thrips	1 to 1½	0	To control thrips in grasses grown for seed use high spray pressure to improve penetration into boot.
	White grubs (green June beetle)		1½ to 2		
Garden beet Spinach Swiss chard	Flea beetles Harlequin bug	Leafhoppers	½ to 1	3 (garden beet roots) 14 (garden beet tops, spinach, Swiss chard)	Treat on a 5 to 7 day schedule
	Aster leafhopper		1 to 1½		
	Armyworm Corn earworm Fall armyworm	Stink bugs Tarnished plant bug	1 to 2		
Okra	Corn earworm	Stink bugs	1 to 2	0	Treat on a 5 to 7 day schedule.
Peanuts	Blister beetles	Mexican bean beetle	½ to 1	0	Observe plant response precautions.
	Alfalfa caterpillar Bean leaf beetle Cucumber beetles Green cloverworm Japanese beetle Leafhoppers	Rednecked peanut worm Three cornered alfalfa hopper Velvetbean caterpillar	1		
	Armyworm Corn earworm Fall armyworm	Stink bugs Webworms	1 to 1½		
	Whitefringed beetle adults	Cutworms	2		
Peas	Colorado potato beetle Leafhoppers		1	3	
	Armyworm		1 to 1½		
	Alfalfa caterpillar Cutworms Pea leaf weevil	Pea weevil Yellowstriped armyworm	1½		

CROP	INSECT		QUARTS OF "SEVIN" BRAND 4F/ACRE	PRE-HARVEST INTERVAL (DAYS)	SPECIFIC DIRECTIONS
Potato Tomato Eggplant Pepper	Colorado potato beetle Flea beetles Leafhoppers		½ to 1		
	European corn borer Fall armyworm Lace bugs Stink bugs	Tarnished plant bug Tomato fruit worm Tomato hornworm Tomato pinworm	1 to 2	0	
	Cutworms		2		
Rice	Armyworm Chinch bugs	Fall armyworm Stink bugs	1 to 1½		CAUTION: May kill shrimp and crabs. Do not use in areas where these are important resources. Do not use on rice fields in which crayfish and/or catfish farming are included in the cultural practice. DO NOT APPLY PROPANIL HERBICIDES WITHIN 15 DAYS BEFORE OR AFTER APPLICATION OF THIS PRODUCT OR PLANT INJURY WILL RESULT. MISSISSIPPI DELTA & TEXAS.
	Armyworm Leafhoppers	Tadpole shrimp	2	14	CALIFORNIA ONLY. For optimum tadpole shrimp control apply to water when pest first appears.
Sorghums (milo, grain sorghum, sweet sorghum and hybrids)	Armyworm Chinch bugs Corn earworm	Fall armyworm Stink bugs Webworms	1 to 2	21 (grain) 0 (forage)	Direct spray into forming heads for optimum insect control.
	Sorghum midge Southwestern corn borer		1½		Treat for sorghum midge when 25 to 30 percent of heads have emerged from boot and are in bloom. Repeat application 3 to 5 days later if adults are still active. A third application may be necessary in late planted sorghum or if midge are abundant. For chinch bugs use high gallonage ground application directed at the base of plants
	Cutworms		2		
Soybeans	Bean leaf beetle Cucumber beetles Green cloverworm	Mexican bean beetle Velvetbean caterpillar	½	0	DO NOT APPLY A COMBINATION OF THIS PRODUCT AND 2,4-DB HERBICIDES TO SOYBEANS AS CROP INJURY MAY RESULT. For light to moderate populations and smaller instars. Use the higher rates for heavy populations and larger instars.
	Corn earworm		½ to ¾ 1-1½		
	Blister beetles Grape colaspis Mexican bean beetle		½ to 1		
	Alfalfa caterpillar Bean leaf beetle Cucumber beetles Green cloverworm Japanese beetle	Leafhoppers Three cornered alfalfa hopper Thrips Velvetbean caterpillar	1		
	Armyworm Cutworms Fall armyworm	Stink bugs Webworms	1 to 1½		
	Painted lady (Thistle caterpillar)	Saltmarsh caterpillar Yellowstriped Armyworm	2 to 2½		
Sugar beets	Armyworm Beet leaf beetle Fall armyworm	Flea beetles Leafhoppers Webworms	1 to 1½	14	
	Cutworms		1½		
Sunflower	Cutworms		1½		DO NOT USE IN CALIFORNIA.
	Armyworm Fall armyworm		1½ to 2	60	
	Stem Weevil Sunflower beetle		1 to 2		

CROP	INSECT	QUARTS OF "SEVIN" BRAND 4F/ACRE	PRE-HARVEST INTERVAL (DAYS)	SPECIFIC DIRECTIONS
Sweet potato	Corn earworm Cucumber beetles Flea beetles Sweet potato hornworm Tortoise beetles	1 to 2	0	Apply as a foliar spray as needed. DO NOT USE IN CALIFORNIA.
	Yellowstriped armyworm	2		
	Sweet potato weevil	1 to 2		FOLIAR APPLICATION Full coverage of plants is essential. Use lower rate on young plants and higher rate on mature plants.
Tobacco	Tobacco flea beetle	2 Tbs./gal or 1 qt/50 gal. and apply 6 gal/100 sq. yards	0	IN PLANT BEDS. To prevent plant injury, avoid excessive applications.
	Green june beetle grubs	½ qt/100 gallons		Apply only to areas that larvae have uprooted by sprinkling as a drench with 50 to 100 gallons per 100 square yards.
	Budworms Fall armyworm Flea beetles Hornworms Japanese beetle June beetle Suckfly	1 to 2		IN FIELDS Use lower rate on young plants (up to knee height). Use at least 10 gallons of prepared spray per acre. Begin treatments when worms are small. If late season applications are necessary, allow three days before priming or cutting.
Wheat	Cereal leaf beetle	1	0 (forage) 21 (grain)	DO NOT USE IN CALIFORNIA. Do not make more than two applications after grain head emerge from boot. Application is effective against eggs, larvae and adults.
	Armyworm Fall armyworm	1 to 1½		

TREE FRUIT AND NUT CROPS

For dilute sprays apply the specified dosage per 100 gallons of water. For concentrate and aerial sprays increase the concentration of SEVIN brand 4F Carbaryl Insecticide in the spray mixture to apply an amount per acre equivalent to that in a dilute spray. The optimum spray gallonage will depend on tree size, density and stage of growth. Typical spray gallonages per acre range from 200-300 gallons for dilute sprays, 30-100 gallons for concentrate sprays and 10-25 gallons for aerial sprays. Do not exceed maximum label rate per acre per application.

CROP	QUARTS OF "SEVIN" BRAND 4F/100 GAL.	PRE-HARVEST INTERVAL (DAYS)	SPECIFIC DIRECTIONS
APPLE THINNING			Apply 1 full coverage dilute spray between 10 and 25 days after full bloom. If factors such as tree age, variety, nutrition, previous crop, pruning, bloom and degree of set favor excessive fruit thinning with this product, exercise caution to avoid possible yield reduction. Consult with your County Extension Service or other experts for advise on the proper use of this product.
Apples only	¼ to ½	1	For easily thinned varieties including Cortland, Grimes, Jonathan, Early McIntosh, Orleans, Rome Beauty, Puritan, Red Delicious, Winesap, Yellow Newton.
	½ to 1	1	For difficult to thin varieties including Baldwin, Ben Davis, Duchess, Early McIntosh, Golden Delicious, Lady Apple, Northern Spy, Rhode Island Greening, Steele Red, Turley, Wealthy, Yellow Transparent and York Imperial

CROP	INSECT	QUARTS OF "SEVIN" BRAND 4F/100 GAL.	PRE-HARVEST INTERVAL (DAYS)	SPECIFIC DIRECTIONS
Almond	Peach twig borer San Jose Scale Fruittree leafroller	1	28	Apply in "popcorn" or petal fall stages and again when the May brood of the peach twig borer begins to hatch or thereafter as needed.
	Naval orangeworm			Time early and mid season applications to correspond to moth fight peaks. Make a late season application at initiation of hull split or up to 10% hull split. Do not apply more than 5 quarts per acre.
Apples Pears	Apple aphid Apple rust mite Apple sucker Bagworms California pearslug (pear sawfly) Codling moth Eyespotted bud moth Green fruitworm Lecanium scales Lesser appleworm Lygus bugs Orange tortrix Oystershell scale Pear leaf blister mite Pear psylia Pear rust mite San Jose scale Tentiform leafminers Woolly apple aphid	¾ to 1	1	WEST OF ROCKY MOUNTAINS To avoid undesired apple thinning, delay use until at least 30 days after full bloom. For psylla control apply when eggs hatch or young nymphs are present.
	Apple mealybug Apple aphid Codling moth White apple leafhopper	½		EAST OF ROCKY MOUNTAINS
	Apple maggot Apple rust mite Bagworms Eastern tent caterpillar European apple sawfly Eyespotted bud moth Forbes scale Fruittree leafroller Green fruitworm Japanese beetle Tarnished plant bug Tentiform leafminers Lecanium scales Lesser appleworm Oystershell scale Pear psylla Pear leaf blister mite Pear rust mite Periodical cicada Plum curculio Redbanded leafroller Rosy apple aphid San Jose scale Woolly apple aphid Yellowheaded fireworm	1	1	To avoid undesired apple thinning, delay use until at least 30 days after full bloom. For optimum scale control apply when crawlers are present.
Chestnuts	Chestnut weevil	2 to 3	0	DO NOT USE IN CALIFORNIA. Make 4 applications at weekly intervals beginning in late July for adult chestnut weevil control. Last application should be made prior to shuck split.
Pistachios	Navel orangeworm	½ to 2	14	Apply dilute volumes of 150 to 300 gallons of mixed spray per acre for full coverage at onset of hull split.
Citrus Fruits (such as grapefruit, lemons, limes, oranges, tangelos, tangerines, citrus citron, kumquats and hybrids)	Avocado leafroller California orangedog Citrus cutworm Citrus root weevil Fruittree leafroller Orange tortrix Western tussock moth West Indian sugarcane borer (adults)	1	5	Do not apply more than 20 quarts of this product per acre per application. Do not apply less than 10 gallons of dilute spray mixture per mature tree. May be mixed with petroleum oils commonly used on citrus.
	Black scale Brown soft scale California red scale Citricola scale Citrus snow scale Yellow scale	¾ to 1		
Filbert	Filbert aphid Filbert leafroller Filbertworm	1	0	Apply when leafroller eggs are hatching. Repeat on first appearance of adult filbert moths and again 3 to 4 weeks later.
Olives	Olive scale	¾ to 1	0	For optimum scale control add 1½ gallons of summer oil and apply mixture when crawlers are present. Do not exceed 2 applications per year. Do not apply more than 15 quarts of this product per acre per application.

CROP	INSECT		QUARTS OF "SEVIN" BRAND 4F/100 GAL.	PRE-HARVEST INTERVAL (DAYS)	SPECIFIC DIRECTIONS
Peaches Apricots Nectarines	Apple pandemis Codling moth Cucumber beetles European earwig Fruittree leafroller Japanese beetle June beetles Lecanium scales Lesser peachtree borer Olive scale Orange tortrix	Oriental fruit moth Peach twig borer Periodical cicada Plum curculio Redbanded leaf-roller San Jose scale Tarnished plant bug Tussock moths Variegated leafroller	1	1 (peaches) 3 (apricots & nectarines)	Do not apply more than 6 quarts of this product per acre per application to apricots. For optimum scale control apply when crawlers are present. Spray limbs and trunk thoroughly, weekly during moth flight.
Pecans	Balck margined aphid Fall webworm Hickory shuck-worm Lesser webworm Pecan leaf phylloxera	Pecan nut case-bearer Pecan spittlebug Pecan weevil Twig girdler Walnut caterpillar	1 to 2½	0	Do not apply more than 7.2 quarts of this products per acre per application.
Plums Prunes Cherries	Black cherry aphid Brown soft scale Cherry fruitworm Cherry maggot European earwig Eyespotted bud moth Forbes scale Fruittree leafroller Green fruitworm Japanese beetle Lecanium scales Lesser peachtree borer	Mealy plum aphid Oystershell scale Peach twig borer Plum curcuilio Prune leafhopper Redbanded leaf-roller Rose chafer San Jose scale Variegated leaf-roller	1	1	Do not apply more than 6 quarts of this product per acre per application. For optimum scale control apply when crawlers are present. For lesser peachtree borer control, spray limbs and tree trunks thoroughly, weekly during moth flight.
	Codling moth Eastern tent caterpillar	Orange tortrix Tussock moth	¾		
Walnut	Calico scale European fruit lecanium	Filbertworm Fruittree leafroller Frosted scale	½	0	Apply 1000 gallons of dilute spray per acre for mature trees.
	Codling moth				For codling moth apply when average cross sectional diameters of developing nuts are ½ to ¾ inch. Repeat during middle or late June as needed.
	European earwig		2		Spray tree trunks to point of run-off.

SMALL FRUIT CROPS

Recommended dosages refer to quarts of SEVIN brand 4F Carbaryl Insecticide per acre. The optimum spray gallonage will depend on plant size, density and stage of growth. Typical spray gallonages per acre range from 100-300 gallons for dilute sprays, 30-100 gallons for concentrate sprays and 10-25 gallons for aerial sprays. Do not exceed maximum label rate per acre per application.

CROP	INSECT		QUARTS OF "SEVIN" BRAND 4F/ACRE	PRE-HARVEST INTERVAL (DAYS)	SPECIFIC DIRECTIONS
Blackberries Raspberries Dewberries (including boysen-berries and loganberries)	European rasp-berry aphid Japanese beetle	Leafhoppers Leafrollers Rose chafer Snowy tree cricket	1 to 2	7	
	Omnivorous leafroller Raspberry sawfly		2		CALIFORNIA ONLY
Blueberries	Blueberry maggot Cherry fruitworm Cranberry fruit-worm	European fruit lecanium Japanese beetle	1½	0	Apply 3 weeks before harvest and repeat as necessary.
Cranberries	Cutworms Cranberry fire-worms Cranberry fruit-worm	Japanese beetle Leafhoppers Rose chafer	1½ to 3	1	Apply in late bloom and as needed at 7 to 10 day intervals. **Caution:** May kill shrimp and crabs. Do not use in areas where these are important resources.

CROP	INSECT		QUARTS OF "SEVIN" BRAND 4F/ACRE	PRE-HARVEST INTERVAL (DAYS)	SPECIFIC DIRECTIONS
Grapes	European fruit lecanium Grape leaffolder Grape leafhopper	Western grapeleaf skeletonizer Western yellow-striped armyworm	1 to 2	0	Apply before first brood leaffolder larvae emerge from rolls.
	Cutworms Eight spotted forester Grape berry moth Japanese beetle June beetles	Orange tortrix Omnivorous leaf-roller Redbanded leaf-roller Saltmarsh cater-pillar	2		
Strawberries	Flea beetles Meadow spittlebug Omnivorous leaftier (strawberry fruitworm)	Strawberry leaf-roller Strawberry weevil	1 to 2	1	Carbaryl may injure Early Dawn and Sunrise varieties on the Delmarva Peninsula.

POULTRY INSECT CONTROL

POULTRY	INSECT	SPECIFIC DIRECTIONS
Chickens, Ducks, Geese, Gamebirds, Pigeons, and Turkeys	Chicken mite Fleas Lice Northern Fowl mite	For use as a direct mist spray on birds by: 1. Misting with Electric Fog Machine: Mix 10 ounces of this product in 1 gallon of spray. Use $1\frac{1}{2}$ gallons per 1000 hens in cages, on litter or on slatted floor. Repeat in 4 weeks if necessary. or 2. Spraying with knapsack or cylinder type compressed air sprayers: Mix 6 ounces of this product in 5 gallons of spray. Use 1 gallon per 100 hens in cages, on litter or on slatted floor. Repeat in 4 weeks if necessary. Direct mist spraying for chicken mite and fleas is a supplement to spraying roosts and buildings for control of these pests. Do not apply to poultry and game birds within 7 days of slaughter.

POULTRY QUARTERS	INSECT	SPECIFIC DIRECTIONS
Roosts and Buildings	Bedbugs Chicken mite Fleas Fowl tick Lesser mealworms Lice Northern fowl mite	Spray roosts and buildings with conventional power spray or knapsack equipment. For chicken mite, fleas and bedbugs, use 4 quarts of this product per 100 gallons of water. For lesser mealworms, use 50 quarts per 100 gallons. Spray 1 to 2 gallons per 1000 square feet of wall, bedding, litter or roost surface. Force spray into cracks. Repeat as needed. Avoid spraying nests, eggs and feeding and watering troughs. Ventilate while spraying. Do not treat premises within 7 days of slaughter.

STORAGE AND DISPOSAL

STORAGE

Store unused SEVIN® brand 4F Carbaryl Insecticide in original container only, in cool, dry area out of reach of children and animals. Do not store in areas where temperatures frequently exceed 100°F.

PESTICIDE DISPOSAL

Do not contaminate water, food or feed by storage or disposal. Open dumping is prohibited. Wastes resulting from the use of this product may be disposed of on site or at an approved waste disposal facility.

CONTAINER DISPOSAL

Triple rinse (or equivalent). Then offer for recycling or reconditioning, or dispose of in a sanitary landfill, or by incineration, or, if allowed by State and local authorities, by burning. If container is burned, stay out of smoke.

LIMITED WARRANTY AND DISCLAIMER

The manufacturer warrants (a) that this product conforms to the chemical description on the label; (b) that this product is reasonably fit for the purposes set forth in the directions for use when it is used in accordance with such directions; and (c) that the directions, warnings and other statements on this label are based upon responsible experts' evaluation of reasonable tests of effectiveness, of toxicity to laboratory animals and to plants, and of residues on food crops, and upon reports of field experience. Tests have not been made on all varieties or in all states or under all conditions. THE MANUFACTURER NEITHER MAKES NOR INTENDS, NOR DOES IT AUTHORIZE ANY AGENT OR REPRESENTATIVE TO MAKE, ANY OTHER WARRANTIES, EXPRESS OR IMPLIED, AND IT EXPRESSLY EXCLUDES AND DISCLAIMS ALL IMPLIED WARRANTIES OF MERCHANTABILITY OR FITNESS FOR A PARTICULAR PURPOSE.

THIS WARRANTY DOES NOT EXTEND TO, AND THE BUYER SHALL BE SOLELY RESPONSIBLE FOR, ANY AND ALL LOSS OR DAMAGE WHICH RESULTS FROM THE USE OF THIS PRODUCT IN ANY MANNER WHICH IS INCONSISTENT WITH THE LABEL DIRECTIONS, WARNINGS OR CAUTIONS.

BUYER'S EXCLUSIVE REMEDY AND MANUFACTURER'S OR SELLER'S EXCLUSIVE LIABILITY FOR ANY AND ALL CLAIMS, LOSSES, DAMAGES, OR INJURIES RESULTING FROM THE USE OR HANDLING OF THIS PRODUCT, WHETHER OR NOT BASED IN CONTRACT, NEGLIGENCE, STRICT LIABILITY IN TORT OR OTHERWISE, SHALL BE LIMITED, AT THE MANUFACTURER'S OPTION, TO REPLACEMENT OF, OR THE REPAYMENT OF THE PURCHASE PRICE FOR, THE QUANTITY OF PRODUCT WITH RESPECT TO WHICH DAMAGES ARE CLAIMED. IN NO EVENT SHALL MANUFACTURER OR SELLER BE LIABLE FOR SPECIAL, INDIRECT OR CONSEQUENTIAL DAMAGES RESULTING FROM THE USE OR HANDLING OF THIS PRODUCT.

THIS SPECIMEN LABEL IS INTENDED FOR USE ONLY AS A GUIDE IN PROVIDING GENERAL INFORMATION REGARDING THE DIRECTIONS, WARNINGS AND CAUTIONS ASSOCIATED WITH THE USE OF THIS PRODUCT. AS WITH ANY AGRICULTURAL CHEMICAL, ALWAYS FOLLOW THE LABEL INSTRUCTIONS ON THE PACKAGE BEFORE USING.

TEMIK® BRAND 15G

ALDICARB PESTICIDE

RESTRICTED USE PESTICIDE

For Retail Sale To And Use Only By Certified Applicators Or Persons Under Their Direct Supervision And Only For Those Uses Covered By The Certified Applicators Certification

For Control of Certain Insects, Mites and Nematodes

ACTIVE INGREDIENT: Aldicarb [2-methyl-2-(methylthio) propionaldehyde 0-(methylcarbamoyl) oxime]	15%
INERT INGREDIENTS	85%
TOTAL	100%

EPA Reg. No. 264-330 EPA Est. No. 10352-GA-01

IN CASE OF EMERGENCY TELEPHONE COLLECT (24 HOURS A DAY) IN U.S.A. (304) 744-3487

FOR GENERAL PRODUCT INFORMATION CALL 1-800-334-9745.

Keep Out of Reach of Children

DANGER–POISON

PELIGRO

PRECAUCION AL USUARIO: Si usted no lee ingles, use este producto hasta que la informacion en la etiqueta le haya sido explicada ampliamente.

PRECAUTIONARY STATEMENTS

DANGER

HAZARDS TO HUMANS AND DOMESTIC ANIMALS

POISONOUS IF SWALLOWED. CAUSES CHOLINESTERASE INHIBITION. MAY BE FATAL OR HARMFUL BY SKIN OR EYE CONTACT OR BY BREATHING DUST. RAPIDLY ABSORBED THROUGH SKIN OR EYES. DO NOT GET ON SKIN OR EYES. DO NOT BREATHE THE DUST. KEEP AWAY FROM DOMESTIC ANIMALS.

WEAR LONG-SLEEVED CLOTHING, FULL-LENGTH TROUSERS, HEAD COVERINGS, AND PROTECTIVE GLOVES WHEN HANDLING. ALWAYS STAND UP-WIND FROM HOPPER WHEN LOADING. WASH HANDS AND FACE BEFORE EATING OR USING TOBACCO. BATHE AT THE END OF WORK DAY, WASHING ENTIRE BODY AND HAIR WITH SOAP AND WATER. CHANGE CLOTHING DAILY. WASH CONTAMINATED CLOTHING IN STRONG WASHING SODA SOLUTION AND RINSE THOROUGHLY BEFORE REUSING.

SIGNS AND SYMPTOMS OF OVEREXPOSURE:

Salivation	Muscle tremor	Nausea
Watery eyes	Difficult breathing	Vomiting
Pinpoint eye pupils	Excessive sweating	Diarrhea
Blurred vision	Abdominal cramps	Weakness
		Headache

In severe cases, convulsions, unconsciousness and respiratory failure may occur.

ANTIDOTE STATEMENT

ATROPINE SULFATE IS HIGHLY EFFECTIVE AS AN ANTIDOTE.

See NOTE TO PHYSICIAN below.

STATEMENT OF PRACTICAL TREATMENT

GENERAL

Contact a physician immediately in all cases of suspected poisoning. Illness may be produced rapidly following overexposure to TEMIK-Aldicarb. If breathing stops, establish an airway, start artificial respiration, and provide oxygen. Make certain to remove all sources of continuing contamination. Remove clothing and wash skin and hair immediately with large amounts of water. Transport the patient to a physician or hospital immediately and SHOW A COPY OF THIS LABEL TO THE PHYSICIAN. If poisoning is suspected in animals, contact a veterinarian.

IF SWALLOWED: Drink 1 to 2 glasses of water and induce vomiting by touching back of throat with finger. Do not induce vomiting or give anything orally to an unconscious or convulsing person. Call a physician and follow General advice listed above.

IF IN EYES: Flush eyes with plenty of water and get medical attention.

IF ON SKIN: Wash thoroughly with soap and water. Remove contaminated clothing and wash before reuse.

IF INHALED: Call a physician and follow General advice listed above.

NOTE TO PHYSICIAN

TEMIK® brand aldicarb is a methyl carbamate insecticide, which is a cholinesterase inhibitor. Overexposure to this substance may cause toxic signs and symptoms due to stimulation of the parasympathetic nervous system. These effects of overexposure are spontaneously and rapidly reversible.

Gastric lavage may be used if this product has been swallowed. TEMIK brand aldicarb poisoning may occur rapidly after ingestion and prompt removal of stomach contents is indicated.

Specific treatment consists of parenteral atropine sulfate. Caution should be maintained to prevent overatropinization. Mild cases may be given 1 to 2 mg intramuscularly every 10 minutes until full atropinization has been achieved and repeated thereafter whenever symptoms reappear. Severe cases should be given 2 to 4 mg intravenously every 10 minutes until fully atropinized, then intramuscularly every 30 to 60 minutes as needed to maintain the effect for at least 12 hours. Dosages for children should be appropriately reduced. Complete recovery from overexposure is to be expected within 24 hours.

Narcotics and other sedatives should not be used. Further, drugs like 2-PAM (pyridine-2-aldoxime methiodide) are NOT recommended unless organophosphate intoxication is also suggested.

To aid in confirmation of a diagnosis, urine samples should be obtained within 24 hours of exposure and immediately frozen. Analyses will be arranged by Union Carbide Agricultural Products Company, Inc.

Consultation on therapy can be obtained at all hours by calling the Union Carbide emergency number: (304) 744-3487.

ENVIRONMENTAL HAZARDS

DECOMPOSITION AND MOVEMENT IN SOIL

This product is readily decomposed into harmless residues under most use conditions. However, a combination of permeable and acidic soil conditions, moderate to heavy irrigation and/or rainfall, use of 20 or more pounds per acre, and soil temperature below 50°F at application time, tend to reduce degradation and promote movement of residues to groundwater. If this describes your local use conditions and groundwater in your area is used for drinking, do not use this product without first contacting Union Carbide at 1-800-334-8577.

TOXIC TO FISH, BIRDS, AND WILDLIFE

This product is toxic to fish, birds and other wildlife. Birds feeding on treated areas may be killed. Cover or incorporate granules in spill areas. Run-off from treated areas may be hazardous to fish in neighboring areas. Do not apply directly to water or wetlands. Keep out of lakes, streams and ponds. Do not contaminate water by cleaning of equipment or disposal of wastes. Apply this product only as specified on this label.

NOTICE: Under the Endangered Species Act, it is a Federal Offense to use any pesticide in a manner that results in the death of a member of an endangered species.

This Act protects Attwater's Greater Prairie Chicken in the Texas counties of Aransas, Austin, Brazoria, Colorado, Galveston, Goliad, Harris, Refugio, and Victoria.

Prior to making application in these counties, the user must determine that this species is not located in or immediately adjacent to the area to be treated. If the user is in doubt whether or not the above named endangered species may be affected, he should contact either the regional U.S. Fish & Wildlife Service office (Endangered Species Specialist) or personnel of the State Fish and Game Office.

GENERAL DIRECTIONS FOR USE

GENERAL INFORMATION

READ ENTIRE LABEL BEFORE USING THIS PRODUCT.

TEMIK® 15G Aldicarb Pesticide controls certain insects, mites and nematodes. When applied into moist soil at-planting and/or postemergence the active ingredient is rapidly absorbed by roots and translocated throughout all parts of the plant. Control often lasts more than six weeks varying with growing conditions, rate of use and pests. Crop yields are usually increased with treatments of TEMIK® BRAND 15G Aldicarb Pesticide.

GENERAL CAUTIONS

Use TEMIK® BRAND 15G Aldicarb Pesticide only in accordance with label directions, warnings and cautions. Do not use on any crop not listed on this label or supplemental labeling as any residues remaining may be illegal or harmful. DO NOT STORE OR USE IN OR AROUND THE HOME OR HOME GARDEN. Use high rate on heavy organic or clay soils. Do not exceed maximum label rate. Applications at higher rates or at more frequent intervals than stated on the label may result in illegal residues in crop, meat and milk. Treatments in excess of 7 pounds per acre made directly in the seed furrow of cotton, sugar beets, soybeans or drybeans may delay emergence and reduce plant stand.

Make side-dress applications close enough to plants to allow good uptake by the roots without injury to the plants from root pruning. In irrigated areas, follow application with irrigation within one week. If alternate furrows are irrigated after side-dress application, the TEMIK® and water must be on the same side of the plant row.

Calibrate and adjust application equipment to insure proper rate and accurate placement. Clean application equipment thoroughly after use. Bury any excess material in soil (see instructions for STORAGE AND DISPOSAL in this booklet).

Deep disc any spills at row ends immediately to prevent birds from feeding on exposed granules.

DO NOT APPLY THIS PRODUCT IN SUFFOLK COUNTY, LONG ISLAND, NEW YORK OR DEL NORTE COUNTY, CALIFORNIA.

COMPATIBILITY

Pesticidal activity of TEMIK® 15G Aldicarb Pesticide is not affected by normal applications of fertilizers or other pesticides. Its effectiveness may be reduced or lost if applied with alkaline materials such as lime. To minimize potential exposure hazards, do not mix TEMIK® Aldicarb Pesticide with other materials before application.

GENERAL WORKER PROTECTION STATEMENTS

Do not apply this product in such a manner as to directly expose workers or other persons. The area being treated must be vacated by unprotected persons.

Do not enter treated area without protective footgear. Persons coming in direct contact with wet treated soil after the initial irrigation or rainfall following treatment shall wear coveralls and rubber or neoprene boots and gloves. Because certain states may require more restrictive re-entry intervals for various crops treated with this product, consult your State Department of Agriculture for further information. Written or oral warnings must be given to workers who are expected to be in a treated area or in an area about to be treated with this product. Advise workers to stay out of fields during application. Regular long-sleeved work clothing should be worn when working in treated fields. See Precautionary Statement, Statement of Practical Treatment and Note To Physician for information on accidental exposures. When oral warnings are given, warnings shall be given in a language customarily understood by workers. Oral warnings must be given if there is reason to believe that written warnings cannot be understood by workers. Written warnings must include the following information: appropriate signal word (DANGER-POISON), area treated with TEMIK® 15G Aldicarb Pesticide, date of application, appropriate protective clothing, and re-entry requirements.

STORAGE AND DISPOSAL

STORAGE

STORE UNUSED TEMIK® BRAND ALDICARB PESTICIDE IN ORIGINAL CONTAINER ONLY, IN WELL VENTILATED CLEAN DRY AREA OUT OF REACH OF CHILDREN AND ANIMALS. DO NOT STORE IN AREAS WHERE TEMPERATURE AVERAGES 115°F OR GREATER. DO NOT STORE IN OR AROUND THE HOME OR HOME GARDEN.

IF CONTAINER IS BROKEN, BEFORE CLEANING UP PUT ON LONG-SLEEVED SHIRT, FULL LENGTH TROUSERS, HEAD COVERING AND RUBBER GLOVES. DO NOT GET DUST OR GRANULES ON SKIN OR EYES. DO NOT BREATHE THE DUST. SWEEP UP AND BURY ANY SMALL SPILLS OR EXCESS TEMIK® AT LEAST 18 INCHES DEEP IN SOIL ISOLATED FROM WATER SUPPLIES AND FOOD CROPS.

PESTICIDE DISPOSAL

DO NOT CONTAMINATE WATER, FOOD, OR FEED BY STORAGE OR DISPOSAL.

OPEN DUMPING IS PROHIBITED.

PESTICIDE WASTES ARE ACUTELY HAZARDOUS. IMPROPER DISPOSAL OF EXCESS PESTICIDE IS A VIOLATION OF FEDERAL LAW. IF THESE WASTES CANNOT BE DISPOSED OF BY USE ACCORDING TO LABEL INSTRUCTIONS, CONTACT YOUR STATE PESTICIDE OR ENVIRONMENTAL CONTROL AGENCY OR THE HAZARDOUS WASTE REPRESENTATIVE AT THE NEAREST EPA REGIONAL OFFICE FOR GUIDANCE.

CONTAINER DISPOSAL

COMPLETELY EMPTY CONTAINER INTO APPLICATION EQUIPMENT. THEN DISPOSE OF EMPTY CONTAINER IN A SANITARY LANDFILL OR BY INCINERATION, OR IF ALLOWED BY STATE AND LOCAL AUTHORITIES, BY OPEN BURNING. IF CONTAINER IS BURNED, STAY AWAY FROM AND DO NOT BREATHE OR CONTACT SMOKE.

IN CASE OF TRANSPORTATION OR WAREHOUSE EMERGENCY INVOLVING A SPILL, FIRE, OR EXPOSURE, CALL COLLECT (304) 744-3487 TWENTY-FOUR HOURS DAY IN THE USA.

PRE-HARVEST, GRAZING, AND ENVIRONMENTAL LIMITATIONS

To avoid excessive residues in or on:

COTTONSEED

- Do not make more than one at-planting application and one postemergence application per crop.
- Do not apply within 90 days of harvest.
- Do not feed cotton forage to livestock or allow livestock to graze in treated area.

POTATOES
- Do not make more than one application per crop.
- Do not harvest within 90 days of an at-planting application or within 50 days of an 'after-planting' application.

PEANUTS
- Do not make more than one application per crop.
- Do not harvest within 90 days of application.
- Do not hog off treated fields.
- Do not feed hay or vines to livestock.

SUGAR BEETS
- Do not make more than one at-planting application and two postemergence applications per crop.
- Do not exceed a total of 40 pounds per acre.
- Do not apply within 90 days of harvest.
- If tops are to be fed to livestock, do not apply within 120 days of harvest.
- Do not use tops as food for humans.

SOYBEANS
- Do not make more than one application per crop.
- Do not harvest within 90 days of application.
- Do not feed green forage, hay or straw to livestock.

DRYBEANS
- Do not make more than one application per crop.
- Do not harvest within 90 days of application.
- Do not feed green forage, hay or straw to livestock.
- Do not use green pods as food for humans.

SUGARCANE AND SWEET POTATOES
- Do not make more than one application per crop.
- Do not harvest within 120 days of application.
- Do not feed sweet potato vines to livestock.

PECANS
- Do not make more than one application per year.
- Do not allow livestock to graze in treated areas.
- Do not harvest foliage or hay from treated areas.

SORGHUM
- Do not make more than one application per year.
- Do not harvest within 90 days of application.
- Do not feed green forage to livestock.

OTHER CROPS
- Do not plant any crop other than those with registered aldicarb uses in aldicarb treated soil.
- Do not apply this product in Suffolk County, Long Island, New York or Del Norte County, California.

MEAT AND MILK
- Do not allow livestock to graze in treated areas before harvest.

GROUNDWATER
- Observe Environmental Hazard precautions regarding Decomposition and Movement in Soil, and carefully follow Directions For Use.
- Do Not apply in Suffolk County, Long Island, New York or Del Norte County, California.
- Do not apply more than 33 lbs/A in the state of Florida.

DIRECTIONS FOR USE AND RECOMMENDED APPLICATIONS

IT IS A VIOLATION OF FEDERAL LAW TO USE THIS PRODUCT IN A MANNER INCONSISTENT WITH ITS LABELING. Do not apply in Del Norte County, California. To provide maximum performance and to minimize hazard to birds, work granules into or cover with soil to a depth of 2 inches or more. Deep disc any spills at row ends immediately to prevent birds from feeding on granules.

CROP & TIME OF APPLICATION	PESTS CONTROLLED	POUNDS/ACRE	OUNCES/1000 FEET OF ROW	RECOMMENDED APPLICATION
COTTON		40-inch row spacing		
At-planting	Aphids, Thrips	2 to 3.5	2.5 to 4.5	Drill granules just below seed line or place in seed furrow and cover with soil. If seeds and TEMIK are hill-dropped, TEMIK rates may be reduced by one-half.
	Fleahoppers, Leafminers, Mites, Overwintering boll weevil (adults feeding on foliage), Plant bugs, including Lygus	4 to 7	5 to 8.5	
	Nematodes	3.5 to 10 (Except Far West)	4.5 to 12	Drill granules just below seed line or place in seed furrow and cover with soil. If rate exceeds 7 pounds per acre apply granules in a 4 to 6-inch band and work into the soil or cover with soil; plant seed in or above treated zone.
	Nematodes	14 to 27 (Far West only)	17 to 33	Apply granules in a 4 to 6-inch band 3 to 6 inches below seed line or cover with 3 to 6 inches of soil. Plant seed above treated zone.
At first squaring	Leafhoppers Fleahoppers	7 to 14	8.5 to 17	
	Mites, Boll weevil, Plant bugs, including Lygus	14 to 20	17 to 24.5	Side-dress granules 8 to 16 inches to one side of the plant row and 2 to 6 inches deep (usually at or below bottom of water furrow).
From squaring through early bloom	Cotton leaf perforator	14	17	
	Whiteflies	14 to 20	17 to 24.5	
GRAIN SORGHUM	Nematodes	36-inch row spacing		Apply granules in seed furrow and cover with soil.
		7	7.5	
		3.5 to 7 (Texas and Oklahoma only)	4 to 7.5	
At Planting	Greenbug And Chinch Bug (early season control)	7	7.5	

CROP & TIME OF APPLICATION	PESTS CONTROLLED	POUNDS/ACRE	OUNCES/1000 FEET OF ROW	RECOMMENDED APPLICATION
POTATOES		34-inch row spacing		
DO NOT APPLY IN SUFFOLK COUNTY, LONG ISLAND, NEW YORK In the following areas apply at emergence only: **STATES OF:** MAINE and WISCONSIN **COUNTIES OF:** Hartford, CT Kent and New Castle, DE Franklin and Hampshire, MA Worchester, MD Atlantic, Burlington, Cumberland, Monmouth, and Salem, NJ Newport and Washington, RI Accomack and Northampton, VA	Aphids	7	7 (Maine Only)	Apply granules in 4- to 6-inch band over row on top of hill. Cover granules with soil by cultivators or hilling discs. DO NOT APPLY THIS PRODUCT TO ANY FIELD MORE THAN ONCE EVERY TWO YEARS.
	Colorado potato beetle, Leafhoppers Aphids (except in Maine)	14	14.5	
IN ALL OTHER AREAS: **At planting**	Flea beetles	20	21	Apply granules with seed pieces in planting furrow and cover with soil.
	Aphid, Colorado potato beetle, Leafhoppers	14 to 20	14.5 to 21	Apply as above, OR drill granules 2 to 4 inches on both sides of seed row and 3 to 8 inches deep (usually 1 to 2 inches below seed pieces).
	Psyllids	14	14.5 (Arizona only)	Apply as above
	Nematodes (Except Golden Nematode. Root-knot nematodes are suppressed. Verticillium wilt has been suppressed where nematode populations have been reduced)	20	21	Apply granules with seed pieces in planting furrow and cover with soil, OR apply granules in an 8-inch band and work into the soil or cover with soil to a depth of 4 inches. Plant seed pieces in treated zone.
After Planting At 75% emergence or before plants are 6 inches high.	Aphids, Colorado potato beetle	14 to 20	14.5 to 21	Side-dress granules 4 to 5 inches on both sides of plant row and 4 inches deep where pest populations are expected 2-3 weeks after plant emergence. Do not apply if at planting treatment was made.
PEANUTS	Thrips	36-inch row spacing		Apply granules in seed furrow and cover with soil. In Southwest use high rate only.
		7 to 14	7.5 to 15	
At-planting	Nematodes	14 to 20	15 to 22	Apply granules in a 6 to 12-inch band and work into the soil or cover with soil to a depth of 2 to 4 inches. Plant seed in treated zone.
SUGAR BEETS		22-inch row spacing		
At-planting or within one week before planting	Nematodes	27 to 33	18 to 22	Apply granules in a 4 to 6-inch band and immediately work into the soil or cover with soil to a depth of 2 to 4 inches. Plant seed in or above treated zone, OR drill granules 1.5 to 3 inches to one side of seed row and 2 to 4 inches deep.
At-planting	Aphids	7 to 14	4.5 to 9.5	Drill granules 1 to 3 inches below seed line. Granules can be placed in seed furrow if rate does not exceed 7 pounds per acre.
	Leafminers Leafhoppers	14 to 20	9.5 to 13.5	
	Sugar beet root maggot	7 to 14	4.5 to 9.5	Apply granules in a 2 to 4-inch band over seed row and immediately work into the soil or cover with soil, OR where furrow irrigation is employed for seed germination, drill granules 2 inches deep and 2 inches from seed row on water furrow side. For the 7 lb. rate, apply granules in a 1 to 2-inch band in front of the press wheel as the furrow is closing.

CROP & TIME OF APPLICATION	PESTS CONTROLLED	POUNDS/ACRE	OUNCES/1000 FEET OF ROW	RECOMMENDED APPLICATION
At-planting plus post-emergence (split application)	Nematodes	14 at-planting and 14 postemergence	9.5 at-planting and 9.5 post-emergence	AT PLANTING: Apply granules in a 4- to 6-inch band and immediately work into the soil or cover with soil to a depth of 2 to 4 inches. Plant seed in or above treated zone, OR drill granules 1.5 to 3 inches to one side of seed row and 2 to 4 inches deep. POST-EMERGENCE: Side-dress granules 4 to 8 inches to one side of plant row and 3 to 6 inches deep (furrow depth). Apply 40 to 60 days after planting.
Postemergence	Sugar beet root maggot	7 to 14	4.5 to 9.5	Apply granules to both sides of plant row and immediately work into the soil or cover with soil.
(If at planting plus postemergence applications are made, do not exceed a total of 40 lbs/acre.	Aphids	7 to 14	4.5 to 9.5	Side-dress granules 4 to 8 inches to one side of plant row and 2 to 6 inches deep. A repeat application may be required for continued protection against virus vectors (aphids, leafhoppers).
	Leafminers, Leafhoppers	14 to 20	9.5 to 13.5	
	Nematodes	27	18	Side-dress granules 4 to 8 inches to one side of plant row and 3 to 6 inches deep (furrow depth). Apply 40 to 60 days after planting.
SOYBEANS At Planting		30 to 48 inch row spacing		
	Mexican bean beetle Thrips Suppression of three-cornered alfalfa hopper	5 to 10 (Based on 36 inch row spacing)	5.5 to 11	Drill granules 2 to 3 inches below seed line, OR 2 to 3 inches to the side of the seed row and 2 to 3 inches deep. Granules can be placed in seed furrows if rate does not exceed 5 pounds per acre.
	Nematodes	10 to 20 (Based on 36 inch row spacing	11 to 22	Apply granules in a 6 to 8-inch band and immediately work into soil or cover with soil. Plant seed in treated zone.
BEANS (Dried only) At-planting		22 to 48 inch row spacing		
	Seedcorn maggot	3.5 to 5 (Based on 36 inch row spacing) (Michigan and New York Only)	4.0 to 5.5	Apply granules in seed furrow and cover with soil.
	Aphids	5 to 7 (Based on 36-inch row spacing)	5.5 to 7.5	Drill granules 2 to 3 inches below seed line, OR 2 to 3 inches to side of seed row and 2 to 3 inches deep. OR Granules can be placed in seed furrow if rate does not exceed 5 pounds per acre.
	Leafhoppers, Mexican bean beetle, Mites	7 to 14 (Based on 36 inch row spacing)	7.5 to 15	
	Nematodes	7 to 14 (Based on 36 inch row spacing)	7.5 to 15	Apply granules in a 6 to 8-inch band and immediately work into the soil or cover with soil. Plant seed in treated zone.
SWEET POTATOES At-planting	Nematodes	10 to 20 (Based on 48 inch row spacing)	15 to 30	Apply granules in a 12-inch band in opened row. Cover immediately with soil by hilling up 8 to 10 inches. Plant in center of treated zone.
SUGAR CANE At-planting		60-inch row spacing		
	Nematodes	14 to 20 (Louisiana Only)	26 to 37	Apply granules in open row on top of newly planted cane and cover immediately with at least 6 inches of soil.
PECANS (Southeast Only) Producing trees During period from bud break to nut set, or about April 15 to May 15	Aphids, Mites	33-67 (except Florida)	Not Applicable	Apply as a 4 to 6-foot band along dripline on both sides of tree row by spreading granules uniformly and immediately working into soil, OR by shanking 2-3 inches into soil on 12 inch centers.
One to two weeks before bud break.	Suppression of pecan leaf phylloxera	33 (Florida only)		In Florida, apply between January 1 and April 30 only.

CROP & TIME OF APPLICATION	PESTS CONTROLLED	POUNDS/ACRE	OUNCES/1000 FEET OF ROW	RECOMMENDED APPLICATION
Newly Transplanted Trees (1-5 years old) During period of bud break, or about April 15 to May 15 (except Florida)	Aphids, Mites	Dosage Ounces TEMIK 15G/ Tree	Not Applicable	Apply as a side-dress to individual trees by spreading the granules uniformly around the tree and immediately working into the soil to a depth of 2-3 inches.
One to two weeks before bud bread	Budmoth, Suppression of pecan leaf phylloxera	5-20		In Florida, do not exceed 33 lbs/A and apply between January 1 and April 30 only.

SEE PREHARVEST, GRAZING AND ENVIRONMENTAL LIMITATIONS

PLANT GROWTH STIMULATION: In the absence of recognizable target pests and under certain growing conditions, stimulation of plant growth by TEMIK aldicarb on certain crops has been demonstrated under laboratory, greenhouse, and field conditions. Plants may be visibly taller, greener, and denser. Faster grow-off and an increase in fruiting rate and size often are results of the growth enhancement by TEMIK brand aldicarb. However, such effects do not occur with all crops or under all conditions. Therefore, this product should not be used solely as a plant growth regulator, and its effect as one, should be considered a side benefit which may or may not occur as a consequence of its use.

LIMITED WARRANTY AND DISCLAIMER

The manufacturer warrants (a) that this product conforms to the chemical description on the label; (b) that this product is reasonably fit for the purposes set forth in the directions for use when it is used in accordance with such directions; and (c) that the directions, warnings and other statements on this label are based upon responsible experts' evaluation of reasonable tests of effectiveness, or toxicity to laboratory animals and to plants, and of residues on food crops, and upon reports of field experience. Tests have not been made on all varieties or in all states or under all conditions. THE MANUFACTURER NEITHER MAKES NOR INTENDS, NOR DOES IT AUTHORIZE ANY AGENT OR REPRESENTATIVE TO MAKE, ANY OTHER WARRANTIES, EXPRESS OR IMPLIED, AND IT EXPRESSLY EXCLUDES AND DISCLAIMS ALL IMPLIED WARRANTIES OF MERCHANTABILITY OR FITNESS FOR A PARTICULAR PURPOSE.

THIS WARRANTY DOES NOT EXTEND TO, AND THE BUYER SHALL BE SOLELY RESPONSIBLE FOR, ANY AND ALL LOSS OR DAMAGE WHICH RESULTS FROM THE USE OF THIS PRODUCT IN ANY MANNER WHICH IS INCONSISTENT WITH THE LABEL DIRECTIONS, WARNINGS OR CAUTIONS.

BUYER'S EXCLUSIVE REMEDY AND MANUFACTURER'S OR SELLER'S EXCLUSIVE LIABILITY FOR ANY CLAIMS, LOSSES, DAMAGES, OR INJURIES RESULTING FROM THE USE OR HANDLING OF THIS PRODUCT WHETHER OR NOT SUCH LIABILITY IS BASED IN CONTRACT, NEGLIGENCE, STRICT LIABILITY IN TORT OR OTHERWISE SHALL BE LIMITED, AT THE MANUFACTURER'S OPTION, TO REPLACEMENT OF, OR THE REPAYMENT OF THE PURCHASE PRICE FOR, THE QUANTITY OF PRODUCT WITH RESPECT TO WHICH DAMAGES ARE CLAIMED. IN NO EVENT SHALL MANUFACTURER OR SELLER BE LIABLE FOR SPECIAL, INDIRECT OR CONSEQUENTIAL DAMAGES RESULTING FROM THE USE OR HANDLING OF THIS PRODUCT.

THIS SPECIMEN LABEL IS INTENDED FOR USE ONLY AS A GUIDE IN PROVIDING GENERAL INFORMATION REGARDING THE DIRECTIONS, WARNINGS AND CAUTIONS ASSOCIATED WITH THE USE OF THIS PRODUCT. AS WITH ANY AGRICULTURAL CHEMICAL, ALWAYS FOLLOW THE LABEL INSTRUCTIONS ON THE PACKAGE BEFORE USING.

TEMIK® 15G

ALDICARB PESTICIDE

RESTRICTED USE PESTICIDE

For Retail Sale To And Use Only By Certified Applicators Or Persons Under Their Direct Supervision And Only For Those Uses Covered By The Certified Applicators Certification.

For use on CITRUS ONLY

For Control of Certain Insects, Mites and Nematodes on Oranges, Grapefruit, Lemons and Limes Only

ACTIVE INGREDIENT:
Aldicarb [2-methyl-2-(methylthio) propionaldehyde O-(methylcarbamoyl)oxime] 15%
INERT INGREDIENTS 85%
TOTAL 100%

EPA Reg. No. 264-417 EPA Est. No. 10352-GA-01

DANGER–POISON

Keep Out of Reach of Children

IN CASE OF EMERGENCY TELEPHONE COLLECT (24 HOURS A DAY) IN U.S.A. (304) 744-3487

FOR GENERAL PRODUCT INFORMATION CALL 1-800-334-9745.

PRECAUTIONARY STATEMENTS

DANGER

HAZARDS TO HUMANS AND DOMESTIC ANIMALS

POISONOUS IF SWALLOWED. MAY BE FATAL OR HARMFUL BY SKIN OR EYE CONTACT OR BY BREATHING DUST. RAPIDLY ABSORBED THROUGH SKIN OR EYES. DO NOT GET ON SKIN OR EYES. DO NOT BREATHE DUST.

WEAR LONG-SLEEVED CLOTHING AND PROTECTIVE GLOVES WHEN HANDLING. WASH HANDS AND FACE BEFORE EATING OR SMOKING. BATHE AT THE END OF WORK DAY, WASHING ENTIRE BODY AND HAIR WITH SOAP AND WATER. CHANGE CLOTHING DAILY. WASH CONTAMINATED CLOTHING IN STRONG WASHING SODA SOLUTION AND RINSE THOROUGHLY BEFORE REUSING.

SIGNS AND SYMPTOMS OF OVEREXPOSURE:

Salivation	Muscle tremor	Nausea
Watery eyes	Difficult breathing	Vomiting

Pinpoint eye pupils	Excessive sweating	Diarrhea
Blurred vision	Abdominal cramps	Weakness
	Headache	

IN SEVERE CASES, CONVULSIONS, UNCONSCIOUSNESS AND RESPIRATORY FAILURE MAY OCCUR.

ANTIDOTE STATEMENT

ATROPINE SULFATE IS HIGHLY EFFECTIVE AS AN ANTIDOTE. SEE NOTE TO PHYSICIAN BELOW.

STATEMENT OF PRACTICAL TREATMENT

GENERAL

Contact a physician immediately in all cases of suspected poisoning. If breathing stops, establish an airway, start artificial respiration and provide oxygen. Make certain to remove all sources of continuing contamination. Remove clothing and wash skin and hair immediately with large amounts of water. Transport the patient to a physician or hospital immediately and SHOW A COPY OF THIS LABEL TO THE PHYSICIAN.

IF SWALLOWED: Drink 1 to 2 glasses of water and induce vomiting by touching back of throat with finger. Do not induce vomiting or give anything by mouth to an unconscious or convulsing person. Call a physician and follow advice listed above.

IF IN EYES: Flush eyes with plenty of water and get medical attention.

IF ON SKIN: Wash thoroughly with soap and water. Remove contaminated clothing and wash before reuse.

IF INHALED: Call physician and follow advice listed above.

NOTE TO PHYSICIAN

TEMIK® aldicarb is a methyl carbamate insecticide, which is a cholinesterase inhibitor. Overexposure to this substance may cause toxic signs and symptoms due to stimulation of the cholinergic nervous system. These effects of overexposure are spontaneously and rapidly reversible.

Specific treatment consists of parenteral atropine sulfate. Caution should be maintained to prevent overatropinization. Mild cases may be given 1 to 2 mg intramuscularly every 10 minutes until full atropinization has been achieved and repeated thereafter whenever symptoms reappear. Severe cases should be given 2 to 4 mg intravenously every 10 minutes until fully atropinized, then intramuscularly every 30 to 60 minutes as needed to maintain the effect for at least 12 hours. Dosages for children should be appropriately reduced. Complete recovery from overexposure is to be expected within 24 hours.

Narcotics and other sedatives should not be used. Further, drugs like 2-PAM (pyridine-2-aldoxime methiodide) are NOT recommended unless organophosphate intoxication is also suggested.

To aid in confirmation of a diagnosis, urine samples should be obtained within 24 hours of exposure and immediately frozen. Analyses will be arranged by Union Carbide Agricultural Products Company, Inc.

Consultation on therapy can be obtained at all hours by calling the Union Carbide emergency number 1-800-UCC-HELP.

ENVIRONMENTAL HAZARDS

DECOMPOSITION AND MOVEMENT IN SOIL: This product is readily decomposed into harmless residues under most use conditions. However, a combination of sandy and acidic soil conditions, moderate to heavy irrigation and/or rainfall, use of 20 or more pounds per acre, and soil temperature below 50°F at application time, tend to reduce degradation and promote movement of residues to groundwater. If this describes your local use conditions and groundwater in your area is used for drinking, do not use this product without first contacting Union Carbide at 1 (800) 334-8577.

TOXIC TO FISH, BIRDS, AND WILDLIFE

This product is toxic to fish, birds and other wildlife. Birds feeding on treated areas may be killed. Cover or incorporate granules in spill areas. Runoff from treated areas may be hazardous to fish in neighboring areas. Do not apply directly to water or wetlands. Do not contaminate water by cleaning or equipment or disposal of wastes. Apply this product only as specified on this label.

GENERAL DIRECTIONS FOR USE

READ ENTIRE LABEL BEFORE USING THIS PRODUCT.

GENERAL INFORMATION

This product controls certain insects, mites and nematodes. When applied into moist soil the active ingredient is rapidly absorbed by roots and translocated throughout all parts of the plant. Rainfall or irrigation soon after application will insure prompt uptake or aldicarb. Control often lasts more than six weeks varying with growing conditions, rate of use and pests. Crop yields are usually increased with treatments of TEMIK®.

GENERAL CAUTIONS

Use this product only in accordance with label directions, warnings and cautions. Do not use on any crop not listed on this label or supplemental labeling as any residues remaining may be illegal or harmful. DO NOT STORE OR USE IN OR AROUND THE HOME OR HOME GARDEN. Use high rate on heavy organic or clay soils. Do not over-apply.

Calibrate and adjust application equipment to insure proper rate and accurate placement. Clean application equipment thoroughly after use. For any left over material see instructions for STORAGE AND DISPOSAL in this booklet.

Deep disc and spills at row ends immediately to prevent birds from feeding on exposed granules.

COMPATIBILITY

Pesticidal activity of this product is not affected by normal applications of fertilizers or other pesticides. It may be reduced or lost if applied with alkaline materials such as lime. To minimize potential exposure hazards, do not mix this product with other materials before application.

GENERAL WORKER PROTECTION STATEMENTS

Do not apply this product in such a manner as to directly expose workers or other persons. The area being treated must be vacated by unprotected persons.

Do not enter treated areas without protective footgear. Persons coming in direct contact with wet treated soil after the initial irrigation or rainfall following treatment shall wear coveralls and rubber or neoprene boots and gloves. Because certain states may require more restrictive reentry intervals for various crops treated with this product, consult your State Department of Agriculture for further information. Written or oral warnings must be given to workers who are expected to be in a treated area or in an area about to be treated with this product. Advise workers to stay out of fields during application. Regular long-sleeved work clothing should be worn when working in treated fields. See Precautionary Statements, Statement of Practical Treatment and Note to Physician for information on accidental exposures. When oral warnings are given, warnings shall be given in a language customarily understood by workers. Oral warnings must be given if there is reason to believe that written warnings cannot be understood by workers. Written warnings must include the following information: Appropriate signal word (DANGER-POISON), area treated with aldicarb, date of application, appropriate protective clothing, and reentry requirements.

STORAGE AND DISPOSAL

STORAGE

Store unused TEMIK® Aldicarb Pesticide in original container only, in well ventilated clean dry area out of reach of children and animals. Do not store in areas where temperature averages 115°F or greater. Do not store in or around the home or home garden.

If container is broken, handle with rubber gloves. Do not get dust or granules on skin or eyes. Do not breathe dust. Sweep up and bury any small spills or excess TEMIK® Aldicarb at least 18 inches deep in soil isolated from water supplies and food crops.

PESTICIDE DISPOSAL

Do not contaminate water, food, or feed by storage or disposal. Open dumping is prohibited. Pesticide wastes are acutely hazardous. Improper disposal of excess pesticide is a violation of Federal law. If these wastes cannot be disposed of by use according to label instructions, contact your State Pesticide or Environmental Control Agency, or the Hazardous Waste representative at the nearest EPA Regional Office for guidance.

CONTAINER DISPOSAL

Completely empty container into the application equipment. Then dispose of container in a sanitary landfill or by incineration, or if alllowed by State and local authorities, open burning. When burning the container, stay away from and do not breathe or contact smoke.

IN CASE OF TRANSPORTATION OR WAREHOUSE EMERGENCY INVOLVING A SPILL, FIRE, OR EXPOSURE, CALL COLLECT 1-800-UCC-HELP TWENTY-FOUR HOURS A DAY IN THE USA.

LIMITED WARRANTY AND DISCLAIMER

The manufacturer warrants (a) that this product conforms to the chemical description on the label; (b) that this product is reasonably fit for the purposes set forth in the directions for use when it is used in accordance with such directions; and (c) that the directions, warning and other statements on this label are based upon responsible experts' evaluation of reasonable tests of effectiveness, of toxicity to laboratory animals and to plants, and of residues on food crops and upon reports of field experience. Tests have not been made on all varieties or in all states or under all conditions. THE MANUFACTURER NEITHER MAKES NOR INTENDS, NOR DOES IT AUTHORIZE ANY AGENT OR REPRESENTATIVE TO MAKE, ANY OTHER WARANTIES, EXPRESS OR IMPLIED, AND IT EXPRESSLY EXCLUDES AND DISCLAIMS ALL IMPLIED WARRANTIES OF MERCHANTABILITY OR FITNESS FOR A PARTICULAR PURPOSE.

THIS WARRANTY DOES NOT EXTEND TO, AND THE BUYER SHALL BE SOLELY RESPONSIBLE FOR, ANY AND ALL LOSS OR DAMAGE WHICH RESULTS FROM THE USE OF THIS PRODUCT IN ANY MANNER WHICH IS INCONSISTENT WITH THE LABEL DIRECTIONS, WARNINGS, OR CAUTIONS.

BUYER'S EXCLUSIVE REMEDY AND MANUFACTURER'S OR SELLER'S EXCLUSIVE LIABILITY FOR ANY AND ALL CLAIMS, LOSSES, DAMAGES, OR INJURIES RESULTING FROM THE USE OR HANDLING OF THIS PRODUCT, WHETHER OR NOT BASED IN CONTRACT, NEGLIGENCE, STRICT LIABILITY IN TORT OR OTHERWISE SHALL BE LIMITED, AT THE MANUFACTURER'S OPTION, TO REPLACEMENT OF, OR THE REPAYMENT OF THE PURCHASE PRICE FOR, THE QUANTITY OF PRODUCT WITH RESPECT TO WHICH DAMAGES ARE CLAIMED. IN NO EVENT SHALL MANUFACTURER OR SELLER BE LIABLE FOR SPECIAL, INDIRECT OR CONSEQUENTIAL DAMAGES RESULTING FROM THE USE OR HANDLING OF THIS PRODUCT.

DIRECTIONS FOR USE ON CITRUS (Grapefruit, Oranges, Lemons, and Limes only)

IT IS A VIOLATION OF FEDERAL LAW TO USE THIS PRODUCT IN A MANNER INCONSISTENT WITH ITS LABELING. To provide maximum performance and to minimize hazard to birds, work granules into or cover with soil to a depth of 2 inches or more. Deep disc any spills at row ends immediately to prevent birds from feeding on granules.

PESTS CONTROLLED	POUNDS/ACRE (18 to 25 FT ROW SPACING)	RECOMMENDED APPLICATION
Citrus Nematode (Except Florida)	67 33-67 (Texas only)	Apply just prior to or during the spring flush of foliage growth in a band along the dripline on both sides of tree row. Spread granules uniformly and immedately work into the soil (preferred method); OR, shank granules 2 to 3 inches deep using 4 to 6 shanks on 12 inch centers. The width of each band should equal one fourth the tree row spacing.
Aphids Mites: Citrus Red Citrus Rust Texas Citrus	33-67 (Except Florida)	Apply as above, or apply in irrigation furrows using two shanks per furrow.
Aphids Mites: Citrus Rust Texas Citrus	33 (Florida only)	Apply as above between January 1 and April 30 only.

PRE-HARVEST AND GRAZING USE INFORMATION AND OTHER LIMITATIONS

To avoid illegal residues:

- Do not make more than one application per year.
- Do not allow livestock to graze in treated areas.
- Do not harvest lemons within 30 days of application.

THIS SPECIMEN LABEL IS INTENDED FOR USE ONLY AS A GUIDE IN PROVIDING GENERAL INFORMATION REGARDING THE DIRECTIONS, WARNINGS AND CAUTIONS ASSOCIATED WITH THE USE OF THIS PRODUCT. AS WITH ANY AGRICULTURAL CHEMICAL, ALWAYS FOLLOW THE LABEL INSTRUCTIONS ON THE PACKAGE BEFORE USING.

TEMIK® TSX

GRANULAR PESTICIDE

RESTRICTED USE PESTICIDE

For retail sale to and use only by certified applicators or persons under their direct supervision and only for those uses covered by the certified applicator's certification.

For control of certain insects, mites, nematodes, and soil-borne seedling diseases of cotton.

ACTIVE INGREDIENTS:

Aldicarb [2-methyl-2-(methylthio) propionalde hyde-O-(methylcarbamoyl)oxime]	5.0%
Pentachloronitrobenzene	10.0%
5-Ethoxy-3-(trichloromethyl) 1,2,4-thiadiazole	2.5%
INERT INGREDIENTS:	82.5%
TOTAL	100.0%

EPA Reg. No. 264-319 EPA Est. No. 10352-GA-01

Keep Out of Reach of Children

DANGER–POISON

IN CASE OF EMERGENCY TELEPHONE COLLECT (24 HOURS A DAY) IN U.S.A. 1-800-UCC-HELP.

FOR GENERAL PRODUCT INFORMATION CALL 1-800-334-9745.

PRECAUTIONARY STATEMENTS

HAZARDS TO HUMANS AND DOMESTIC ANIMALS

POISONOUS IF SWALLOWED. MAY BE FATAL OR HARMFUL BY SKIN OR EYE CONTACT OR BY BREATHING DUST. RAPIDLY ABSORBED THROUGH SKIN OR EYES. DO NOT GET ON SKIN OR EYES. DO NOT BREATHE DUST. ANTIDOTE IS ATROPINE SULFATE. SEE STATEMENT OF PRACTICAL TREATMENT AND OTHER DETAILED WARNINGS ON THIS PACKAGE. READ ENTIRE LABEL BEFORE USING THIS PESTICIDE.

WEAR LONG-SLEEVED CLOTHING AND PROTECTIVE GLOVES WHEN HANDLING. WASH HANDS AND FACE BEFORE EATING OR SMOKING. BATHE AT THE END OF WORK DAY, WASHING ENTIRE BODY AND HAIR WITH SOAP AND WATER. CHANGE CLOTHING DAILY. WASH CONTAMINATED CLOTHING IN STRONG WASHING SODA SOLUTION AND RINSE THOROUGHLY BEFORE REUSING.

SYMPTOMS OF POISONING

May be one or more of the following:

Salivation	Difficult Breathing	Vomiting
Watery Eyes	Excessive Sweating	Diarrhea
Pinpoint Eye Pupils	Abdominal Cramps	Weakness
Blurred Vision	Headache	
Muscle Tremor	Nausea	

CONTACT A PHYSICIAN IMMEDIATELY IN ALL CASES OF SUSPECTED POISONING.

STATEMENT OF PRACTICAL TREATMENT

ANTIDOTE

ANTIDOTE IS ATROPINE SULFATE.

FIRST AID TREATMENT

If symptoms are apparent, give atropine sulfate tablets as directed by the physician. Do not give antidote unless symptoms of poisoning have occurred. Transport patient to physician or hospital whichever is faster. Keep patient warm and quiet. If breathing is difficult give oxygen. Start artificial respiration immediately if breathing stops. SHOW A COPY OF THIS LABEL TO THE PHYSICIAN.

IN CASE OF SWALLOWING: Drink 1 or 2 glasses of water and induce vomiting by touching back of throat with finger. Do not induce vomiting or give anything by mouth to an unconscious person. Get medical attention.

IN CASE OF SKIN OR EYE CONTACT: Wash skin immediately and thoroughly with soap and water and rinse well. Flush eyes with plenty of water for 15 minutes. Remove contaminated clothing and wash before reuse.

INFORMATION FOR PHYSICIAN

This product is a carbamate pesticide containing 2-methyl-2-(methylthio)propionaldehyde O-(methylcarbamoyl) oxime. It is a spontaneously reversible cholinesterase inhibitor causing parasympathetic nerve stimulation. Preferred treatment of poisoning in adults is atropine sulfate given intravenously. As much as 2 to 4 mg may be needed every 10 to 12 minutes until patient is fully atropinized. Dosage for children is appropriately reduced. Atropinization should be maintained for 12 hours by intramuscular administration of atropine in lower doses given at appropriate time intervals. Do not administer opiates or cholinesterase inhibiting drugs. Artificial respiration or oxygen may be necessary. Observe patient continuously for at least 24 hours. Allow no further exposure to any cholinesterase inhibitors until cholinesterase level is normal by blood test.

ENVIRONMENTAL HAZARDS

TOXIC TO FISH, BIRDS, AND WILDLIFE

This product is toxic to fish, birds, and other wildlife. Birds feeding on treated areas may be killed. Keep out of lakes, streams and ponds. Do not contaminate water by cleaning of equipment or disposal of wastes. Apply this product only as specified on this label.

STORAGE AND DISPOSAL

STORAGE

Store unused TEMIK® TSX Granular Pesticide in original container only in well ventilated clean dry area out of reach of children and animals. Do not store in areas where temperature averages 115°F or greater. Do not store in or around the home or home garden.

If container is broken, handle with rubber gloves. Do not get dust or granules on skin or eyes. Do not breathe dust. Sweep up and bury any small spills or excess TEMIK® TSX at least 18 inches deep in soil isolated from water supplies and food crops.

PESTICIDE DISPOSAL

Do not contaminate water, food, or feed by storage or disposal. Open dumping is prohibited. Pesticide wastes are acutely hazardous. Improper disposal of excess pesticide is a violation of Federal law. If these wastes cannot be disposed of by use according to label instructions, contact your State Pesticide or Environmental Control Agency, or the Hazardous Waste representative at the nearest EPA Regional Office for guidance.

CONTAINER DISPOSAL

Completely empty container into application equipment. Then dispose of empty container in a sanitary landfill, or by incineration, or if allowed by state and local authorities, by open burning. If container is burned, stay away from and do not breathe or contact smoke.

IN CASE OF A TRANSPORTATION OR WAREHOUSE EMERGENCY INVOLVING A SPILL, FIRE, OR EXPOSURE CALL COLLECT (304) 744-3487 TWENTY-FOUR HOURS A DAY IN THE USA.

GENERAL DIRECTIONS FOR USE

TEMIK® TSX Granular Pesticide controls certain nematodes, insects, mites, and soil-borne seedling diseases of cotton. When applied into moist soil at planting the active fungicide ingredients (Terraclor and Terrazole fungicides) protect the germinating seed and seedling from Rhizoctonia, Pythium, and Fusarium diseases. At the same time, the active nematacide ingredient (aldicarb) protects the developing root system from nematodes which damage the seedling directly and promote seedling disease. When the germinating plant emerges from the soil, the active insecticide ingredient (aldicarb) has moved upward from the roots to protect the developing stem and leaves from attack by early season mites and insects. Control of insects and mites often lasts more than six weeks, depending upon growing conditions, rate of use and pests. Cotton yields are usually increased with treatments of TEMIK® TSX Granular Pesticide.

GENERAL CAUTIONS

Use TEMIK® TSX Granular Pesticide only in accordance with label directions, warnings and cautions. Do not use on any crop other than cotton, and any residues remaining may be illegal or harmful. DO NOT STORE OR USE IN OR AROUND THE HOME OR HOME GARDEN.

Use high rate on heavy organic or clay soils. Do not overapply. Treatments in excess of 15 pounds per acre may delay emergence and reduce plant stand.

In irrigated areas, follow application with irrigation within one week.

Do not drill TEMIK® TSX Granular Pesticide if seed is hill-dropped.

Calibrate and adjust application equipment to insure proper rate and accurate placement. Clean application equipment thoroughly after use. Bury any excess material in soil (see STORAGE AND DISPOSAL instructions on FRONT PANEL of the bag).

Disc any spills at row ends immediately to prevent birds from feeding on exposed granules.

COMPATIBILITY

Pesticide activity of TEMIK® TSX Granular Pesticide is not affected by normal applications of fertilizers or other pesticides. It may be reduced or lost if applied with alkaline materials such as lime or fertilizers containing lime.

PRE-HARVEST AND GRAZING USE INFORMATION AND LIMITATIONS

Do not apply within 90 days of harvest.

Do not plant any crop other than cotton in soil treated with TEMIK® TSX Granular Pesticide within 100 days of application.

Do not allow livestock to graze in treated areas before harvest.

FOR CONTROL OF CERTAIN INSECTS, MITES, NEMATODES, AND SOIL-BORNE SEEDLING DISEASE OF COTTON.

DIRECTIONS FOR USE AND RECOMMENDED APPLICATIONS

IT IS A VIOLATION OF FEDERAL LAW TO USE THIS PRODUCT IN A MANNER INCONSISTENT WITH ITS LABELING. To provide maximum performance and to minimize hazard to birds, work granules into or cover with soil to a depth of 2 inches or more. Deep disc any spills at row ends immediately to prevent birds from feeding on granules.

CROP AND TIME OF APPLICATION	PESTS CONTROLLED	POUNDS/ACRE	OUNCES/1000 FEET OF ROW	RECOMMENDED APPLICATION
COTTON At-planting	Aphids, mites, thrips, boll-weevil (adults feeding on foliage); Nematodes (Except Far West); Seedling Disease complex (Rhizoctonia, Pythium, Fusarium)	For example: 40-inch row spacing 10-15	12.3-18.5	Place granules in seed furrow and cover with soil. If seeds and TEMIK® TSX are hill-dropped, reduce rates of application of TEMIK® TSX are indicated.

Hill Drop Rates

Hill Spacing	Pounds/ Acre	Ounces/1000 feet of row
8-inch	10	12.5
12-inch	8.5	10.5
16-inch	7.5	9
20-inch	6	7

LIMITED WARRANTY AND DISCLAIMER

The manufacturer warrants (a) that this product conforms to the chemical description on the label; (b) that this product is reasonably fit for the purposes set forth in the directions for use when it is used in accordance with such directions and; (c) that the directions, warnings, and other statements on this label are based upon responsible experts' evaluation of reasonable tests of effectiveness, of toxicity to laboratory animals and to plants, and of residues on food crops, and upon reports of field experience. Tests have not been made on all varieties or in all states or under all conditions. THE MANUFACTURER NEITHER MAKES NOR INTENDS, NOR DOES IT AUTHORIZE ANY AGENT OR REPRESENTATIVE TO MAKE, ANY OTHER WARRANTIES, EXPRESS OR IMPLIED, AND IT EXPRESSLY EXCLUDES AND DISCLAIMS ALL IMPLIED WARRANTIES OF MERCHANTABILITY OR FITNESS FOR A PARTICULAR PURPOSE.

THIS WARRANTY DOES NOT EXTEND TO, AND THE BUYER SHALL BE SOLELY RESPONSIBLE FOR ANY AND ALL LOSS OR DAMAGE WHICH RESULTS FROM THE USE OF THIS PRODUCT IN ANY MANNER WHICH IS INCONSISTENT WITH THE LABEL DIRECTIONS, WARNINGS OR CAUTIONS.

BUYER'S EXCLUSIVE REMEDY AND MANUFACTURER'S OR SELLER'S EXCLUSIVE LIABILITY FOR ANY AND ALL CLAIMS LOSSES, DAMAGES, OR INJURIES RESULTING FROM THE USE OR HANDLING OF THIS PRODUCT WHETHER OR NOT BASED IN CONTRACT, NEGLIGENCE, STRICT LIABILITY IN TORT OR OTHERWISE SHALL BE LIMITED, AT THE MANUFACTURER'S OPTION, TO REPLACEMENT OF, OR THE REPAYMENT OF THE PURCHASE PRICE FOR, THE QUANTITY OF PRODUCT WITH RESPECT TO WHICH DAMAGES ARE CLAIMED. IN NO EVENT SHALL MANUFACTURER OR SELLER BE LIABLE FOR SPECIAL, INDIRECT OR CONSEQUENTIAL DAMAGES RESULTING FROM THE USE OR HANDLING OF THIS PRODUCT.

THIS SPECIMEN LABEL IS INTENDED FOR USE ONLY AS A GUIDE IN PROVIDING GENERAL INFORMATION REGARDING THE DIRECTIONS, WARNINGS AND CAUTIONS ASSOCIATED WITH THE USE OF THIS PRODUCT.

AS WITH ANY AGRICULTURAL CHEMICAL, ALWAYS FOLLOW THE LABEL INSTRUCTIONS ON THE PACKAGE BEFORE USING.

THISTROL®

THISTLE HERBICIDE

Contains 4-(MCPB). Controls Canada thistle, other broadleaf weeds in peas.

ACTIVE INGREDIENT:
Sodium salt of 4-(2-methyl-4-chlorophenoxy)butyric acid* 23.5%
INERT INGREDIENTS: 76.5%

TOTAL 100.0%

* 4-(2-Methyl-4-chlorophenoxy) butyric acid equivalent 21.4% by weight or 2 pounds per gallon.

EPA Reg. No. 264-181

Keep Out of Reach of Children

CAUTION

IN CASE OF EMERGENCY PHONE COLLECT (24 HOURS A DAY) IN U.S.A. 1-800-UCC-HELP.

FOR GENERAL PRODUCT INFORMATION CALL 1-800-334-9745.

DIRECTIONS FOR USE

NOTE: IT IS A VIOLATION OF FEDERAL LAW TO USE THIS PRODUCT IN A MANNER INCONSISTENT WITH ITS LABELING.

Canada thistle growing in pea fields is a serious problem. It can reduce pea yields and the Canada thistle flower buds can contaminate the harvested peas, reducing market value and cash return to the grower. At rates used to control Canada thistle, THISTROL Thistle Herbicide will control or suppress other broadleaf weeds such as lambsquarters, pigweed, smartweed, sow thistle, and annual morningglory.

WHEN TO APPLY

Spray THISTROL Thistle Herbicide on thistles as soon as good shoot emergence has developed and when annual broadleaf weeds are less than 3 inches high. During this period, peas will have developed 6 to 12 nodes depending on variety. Do not apply THISTROL later than 3 nodes before pea flowering, as pea yields may be affected adversely. Do not spray peas under stress from lack of moisture, or when temperatures are over 90°F. Do not apply after pea flower buds appear.

AMOUNTS OF THISTROL TO USE

Weeds and stage of growth	Thistle Herbicide per acre
Young susceptible weeds less than 3 inches high (2 to 3 leaves)	2-3 pints
Canada thistle 4 to 10 inches high, or before thistle buds appear. Other susceptible weeds over 3 inches high.	3-4 pints
Susceptible weeds requiring higher rates due to local conditions	up to 6 pints

For ground equipment apply in 20 to 40 gallons of water per acre as a uniform, overall spray, using 40 psi maximum pressure. For aerial application use 5 to 10 gallons of spray solution per acre. Do not exceed the recommended dosage. Most pea varieties may be safely sprayed with THISTROL Thistle Herbicide; however, temporary twisting of some pea varieties may follow application. Check with state weed specialists or with local canning company representative.

Do not feed treated peas or vines to livestock.

GENERAL WORKER PROTECTION STATEMENTS

Do not apply this product in such a manner as to directly or through drift expose workers or other persons. The area treated must be vacated by unprotected persons.

Do not enter treated areas without protective clothing until sprays have dried.

Because certain states may require more restrictive reentry intervals for various crops treated with this product, consult your State Department of Agriculture for further information.

Written or oral warnings must be given to workers who have expected to be in treated area or in an area about to be treated with this product. Advise workers to stay out of fields during application and until sprays have dried. Regular long-sleeved work clothing should be worn when working in treated fields. See CAUTION Statements for information on accidental exposures. When oral warnings are given, warnings shall be given in a language customarily understood by workers. Oral warnings must be given if there is reason to believe that written warnings cannot be understood by workers. Written warnings must include the following information: Appropriate signal word (CAUTION), area treated with Thistrol® Thistle Herbicide, date of application, appropriate protective clothing, and reentry interval (i.e., until sprays have dried).

CAUTION

Harmful if swallowed. Avoid contact with skin, eyes or clothing. In case of contact, wash with plenty of water.

Avoid spray drift as this product may injure susceptible crops and plants such as cotton, beans, grape, tomatoes and ornamentals. (Coarse sprays are less likely to drift.) Do not use same spray equipment for other purposes unless thoroughly cleaned prior to use.

STORAGE AND DISPOSAL

STORAGE

Do not store near fertilizers, seeds, insecticides, or fungicides. Protect from freezing. If allowed to freeze, hold in warm temperature for several days and remix before using.

PESTICIDE DISPOSAL

Do not contaminate water, food, or feed by storage or disposal. Wastes resulting from the use of this product may be disposed of on site or at an approved waste disposal facility.

CONTAINER DISPOSAL

Triple rinse (or equivalent). Then offer for recycling, or reconditioning, or puncture and dispose of in a sanitary landfill, or, incineration, or if allowed by state and local authorities, by burning. If burned, stay out of smoke.

Do not use this product for purposes other than those recommended on the label.

LIMITED WARRANTY AND DISCLAIMER

The manufacturer warrants (a) that this product conforms to the chemical description on the label; (b) that this product is reasonably fit for the purposes set forth in the directions for use when it is used in accordance with such directions; and (c) that the directions, warnings and other statements on this label are based upon responsible experts' evaluation of reasonable tests of effectiveness, or toxicity to laboratory animals and to plants, and of residues on food crops, and upon reports of field experience. Tests have not been made on all varieties or in all states or under all conditions. THE MANUFACTURER NEITHER MAKES NOR INTENDS, NOR DOES IT AUTHORIZE ANY AGENT OR REPRESENTATIVE TO MAKE, ANY OTHER WARRANTIES, EXPRESS OR IMPLIED, AND IT EXPRESSLY EXCLUDES AND DISCLAIMS ALL IMPLIED WARRANTIES OF MERCHANTABILITY OR FITNESS FOR A PARTICULAR PURPOSE.

THIS WARRANTY DOES NOT EXTEND TO, AND THE BUYER SHALL BE SOLELY RESPONSIBLE FOR, ANY AND ALL LOSS OR DAMAGE WHICH RESULTS FROM THE USE OF THIS PRODUCT IN ANY MANNER WHICH IS INCONSISTENT WITH THE LABEL DIRECTIONS, WARNINGS OR CAUTIONS.

BUYER'S EXCLUSIVE REMEDY AND MANUFACTURER'S OR SELLER'S EXCLUSIVE LIABILITY FOR ANY AND ALL CLAIMS, LOSSES, DAMAGES, OR INJURIES RESULTING FROM THE USE OR HANDLING OF THIS PRODUCT, WHETHER OR NOT BASED IN CONTRACT, NEGLIGENCE, STRICT LIABILITY IN TORT OR OTHERWISE, SHALL BE LIMITED, AT THE MANUFACTURER'S OPTION, TO REPLACEMENT OF, OR THE REPAYMENT OF THE PURCHASE PRICE FOR, THE QUANTITY OF PRODUCT WITH RESPECT TO WHICH DAMAGES OF THE PURCHASE PRICE FOR, THE QUANTITY OF PRODUCT WITH RESPECT TO WHICH DAMAGES ARE CLAIMED. IN NO EVENT SHALL MANUFACTURER OR SELLER BE LIABLE FOR SPECIAL, INDIRECT OR CONSEQUENTIAL DAMAGES RESULTING FROM THE USE OR HANDLING OF THIS PRODUCT.

THIS SPECIMEN LABEL IS INTENDED FOR USE ONLY AS A GUIDE IN PROVIDING GENERAL INFORMATION REGARDING THE DIRECTIONS, WARNINGS AND CAUTIONS ASSOCIATED WITH THE USE OF THIS PRODUCT. AS WITH ANY AGRICULTURAL CHEMICAL, ALWAYS FOLLOW THE LABEL INSTRUCTIONS ON THE PACKAGE BEFORE USING.

TRE-HOLD®

TREE WOUND DRESSING

Controls regrowth of tree sprouts following proper pruning.

ACTIVE INGREDIENT:

1-Naphthaleneacetic acid, ethyl ester	1%
INERT INGREDIENTS:	99%
TOTAL	100%

E.P.A. Reg. No. 264-248AA E.P.A. Est. No. 34631-PA-01

Keep Out of Reach of Children

CAUTION

IN CASE OF EMERGENCY TELEPHONE COLLECT (24 HOURS A DAY) IN U.S.A. 1-800-UCC-HELP.

FOR GENERAL PRODUCT INFORMATION CALL 1-800-334-9745.

GENERAL INFORMATION

Many trees sprout prolifically after they have been pruned. Periodic pruning to maintain line clearance is one of the major expenses of many public utilities. TRE-HOLD® Tree Wound Dressing, in combination with proper pruning techniques, reduces the number of new spouts on many tree species. This provides a longer period before they must be trimmed again and reduces the number of sprouts for later trimming.

DIRECTIONS

Spray TRE-HOLD® Tree Wound Dressing on the pruning cut when the cuts are made.

SHAKE CAN WELL. Hold the can upright 10" to 20" from the pruned surface and depress button. Thoroughly cover the pruned surface and in the bark around the cut.

For most effective control of resprouts, prune to a lateral branch (drop-crotch pruning). Do not leave stubs and avoid severe trimming. Treatment of all cuts is important.

TRE-HOLD® Tree Wound Dressing is effective on the following:

silver maple	American elm
sugar maple	Chinese elm
red maple	water oak
sycamore	

PRECAUTIONARY STATEMENTS

CAUTION

HAZARD TO HUMANS AND DOMESTIC ANIMALS

Harmful if swallowed. Avoid contact with skin, eyes, or clothing. Avoid breathing spray mist.

STATEMENT OF PRACTICAL TREATMENT

In case of eye contact flush with water for at least 15 minutes.

In case of skin contact wash with soap and water.

DIRECTIONS FOR USE

It is a violation of Federal law to use this product in a manner inconsistent with its labeling.

STORAGE AND DISPOSAL

STORAGE

Extremely flammable. Contents under pressure. Keep away from fire, sparks, or heated surfaces. Do not puncture or incinerate container. Exposed to temperatures above 130°F may cause bursting.

PESTICIDE DISPOSAL

Do not contaminate water, food or feed by storage and disposal. Wastes resulting from the use of this product may be disposed of on site or at an approved waste disposal facility.

CONTAINER DISPOSAL

Replace cap and discard containers in trash. Do not incinerate or puncture.

LIMITED WARRANTY AND DISCLAIMER

The manufacturer warrants (a) that this product conforms to the chemical description on the label; (b) that this product is reasonably fit for the purposes set forth in the directions for use when it is used in accordance with such direction; and (c) that the directions, warnings and other statements on this label are based upon responsible experts' evaluation of reasonable tests of effectiveness, of toxicity to laboratory animals and to plants, and of residues, and upon reports of field experience. Tests have not been made on all varieties or in all states or under all conditions. THE MANUFACTURER NEITHER MAKES NOR INTENDS, NOR DOES IT AUTHORIZE ANY AGENT OR REPRESENTATIVE TO MAKE, ANY OTHER WARRANTIES, EXPRESS OR IMPLIED, AND IT EXPRESSLY EXCLUDES AND DISCLAIMS ALL IMPLIED WARRANTIES OF MERCHANTABILITY OR FITNESS FOR A PARTICULAR PURPOSE.

THIS WARRANTY DOES NOT EXTEND TO, AND THE BUYER SHALL BE SOLELY RESPONSIBLE FOR, ANY AND ALL LOSS OR DAMAGE WHICH RESULTS FROM THE USE OF THIS PRODUCT IN ANY MANNER WHICH IS INCONSISTENT WITH THE LABEL DIRECTIONS, WARNINGS OR CAUTIONS.

BUYER'S EXCLUSIVE REMEDY AND MANUFACTURER'S OR SELLER'S EXCLUSIVE LIABILITY FOR ANY AND ALL CLAIMS, LOSSES, DAMAGES, OR INJURIES RESULTING FROM THE USE OR HANDLING OF THIS PRODUCT, WHETHER OR NOT SUCH LIABILITY IS BASED IN CONTRACT, NEGLIGENCE, STRICT LIABILITY IN TORT OR OTHERWISE, SHALL BE LIMITED, AT THE MANUFACTURER'S OPTION, TO REPLACEMENT OF, OR THE REPAYMENT OF THE PURCHASE PRICE FOR, THE QUANTITY OF PRODUCT WITH RESPECT TO WHICH DAMAGES ARE CLAIMED IN NO EVENT SHALL MANUFACTURER OR SELLER BE LIABLE FOR SPECIAL INDIRECT OF CONSEQUENTIAL DAMAGES RESULTING FROM THE USE OR HANDLING OF THIS PRODUCT.

TRE-HOLD®

SPROUT INHIBITOR A112

Controls sprout or sucker growth on apple, pear and olive trees.

For commercial or agricultural use. Not for residential use.

ACTIVE INGREDIENT:

1-Naphthaleneacetic acid, ethyl ester*	15.1%
INERT INGREDIENTS:	84.9%
TOTAL	100.0%

* 1-Naphthaleneacetic acid equivalent
1 lb./gal. or 13.2% by weight

EPA Reg. No. 264-336 EPA Est. No. 264-MO-01

Keep Out of Reach of Children

DANGER—PELIGRO

PRECAUCIÓN AL USARIO: Si usted no lee ingles, no use este producto hasta que la etiqueta le haya sido explicada ampliamente.

IN CASE OF EMERGENCY TELEPHONE COLLECT (24 HOURS A DAY) IN U.S.A. 1-800-UCC-HELP.

FOR GENERAL PRODUCT INFORMATION CALL 1-800-334-9745.

PRECAUTIONARY STATEMENTS

DANGER

HAZARD TO HUMANS AND DOMESTIC ANIMALS

Corrosive. Causes eye and skin irritation. Do not get in eyes, on skin or on clothing. Wear goggles or face shield and rubber gloves when handling. Harmful or fatal if swallowed. Avoid contamination of food. Flammable, keep away from heat and open flame.

STATEMENT OF PRACTICAL TREATMENT

In case of contact, immediately flush eyes or skin with plenty of water for at least 15 minutes. For eyes, call a physician. Remove and wash contaminated clothing before reuse. "NOTE TO PHYSICIAN" If ingested: Probable mucosal damage may contraindicate the use of gastric lavage. Measures against circulatory shock, respiratory depression and convulsion may be needed.

ENVIRONMENTAL HAZARDS

Do not apply directly to water. Do not contaminate water by cleaning of equipment or disposal of waste. Do not contaminate irrigation ditches or water used for irrigation or domestic purposes. Do not apply when weather conditions favor drift from treated areas. Do not use in greenhouse.

DIRECTIONS FOR USE

IT IS A VIOLATION OF FEDERAL LAW TO USE THIS PRODUCT IN A MANNER INCONSISTENT WITH ITS LABELING. Do not use the same spray equipment for other purpose unless thoroughly cleaned.

GENERAL WORKER PROTECTION STATEMENTS

Do not apply this product in such a manner as to directly or through drift expose workers or other persons. The area being treated must be vacated by unprotected persons.

Do not enter treated areas without protective clothing until sprays have dried. Because certain states may require more restrictive reentry intervals for various crops treated with this product, consult your State Department of Agriculture for further information. Written or oral warnings must be given to workers who are expected to be in a treated area or in an area about to be treated with this product. Advise workers to stay out of fields during application and until sprays have dried. Regular long-sleeved work clothing should be worn when working in treated fields. See PRECAUTIONARY STATEMENTS and STATEMENT OF PRACTICAL TREATMENT for information on accidental exposures. When oral warnings are given, warnings shall be given in a language customarily understood by workers. Oral warnings must be given if there is reason to believe that written warnings cannot be understood by workers. Written warnings must include the following information: appropriate signal word (DANGER), area treated, date of application, appropriate protective clothing, and reentry interval if applicable.

STORAGE AND DISPOSAL

STORAGE

Do not use or store near heat or open flame. Do not store near fertilizers, seeds, insecticides or fungicides.

PESTICIDE DISPOSAL

Do not contaminate water, food, or feed by storage or disposal. Pesticide wastes are acutely hazardous. Improper disposal of excess pesticide, spray mixture, or rinsate is a violation of Federal law. If these wastes cannot be disposed of by use according to label instructions, contact your State Pesticide or Environmental Control Agency or the Hazardous Waste representative at the nearest EPA Regional Office for guidance.

CONTAINER DISPOSAL

Triple rinse (or equivalent). Then offer for recycling or reconditioning, or puncture and dispose of in a sanitary landfill, or by other procedures approved by state and local authorities.

GENERAL INFORMATION

Many trees sprout vigorously from the rootstock, base of the trunks, scaffold limbs, and around pruning cuts on limbs. The sprouts must be removed for best tree form and to allow light penetration for good fruit color.

TRE-HOLD Sprout Inhibitor A112 induces bud dormancy and helps prevent resprouting on bearing and non-bearing apple, olive and pear trees and on ornamental olives, pears and crabapples.

TRE-HOLD Sprout Inhibitor A112 is formulated for dilution in water and interior latex paint. After mixing, the solution should be applied to pruning cuts on the scaffold limbs, base of trunks and on rootstock suckers to control resprouting. Thorough coverage of treated area is essential for most effective sprout control.

MIXING DIRECTIONS

TRE-HOLD Sprout Inhibitor A112 MUST BE DILUTED BEFORE USE.

To make one gallon of dilute spray sufficient to treat 50-100 trees, add 10 fluid ounces of TRE-HOLD Sprout Inhibitor A112 to sufficient water to make one gallon of spray mixture. One to four pints of white interior latex paint may be substituted for an equal volume of water for marking purposes. Be sure the paint used is not harmful to your tree variety. The number of trees treated will vary with tree size and extent of tree surface treated.

To make 100 gallons of spray mixture, add 8 gallons of TRE-HOLD Sprout Inhibitor A112 to 92 gallons of water. Five to 20 gallons of white interior latex paint may be substituted for an equal volume of water for marking and shading purposes.

APPLICATION INSTRUCTIONS—APPLES AND PEARS

Scaffold Limbs

Prune existing sprouts and treat during the dormant season. Thoroughly cover area where existing sprouts were removed but do not treat any more than upper ⅓ to ½ of the circumference of the scaffold limb. Apply TRE-HOLD Sprout Inhibitor A112 with a brush or small hand held pump up type sprayer. Use low pressure (not to exceed 25 psi) to avoid fine spray particles from drifting or splashing on surrounding fruiting wood or buds. The use with interior white latex paint is required to help reflect sun light and cut down on high bark temperature.

Do not treat scaffold limbs after bud activity starts in the Spring. There has been some fruit size reduction on Golden Delicious and Anjou pears from excessive treatments to scaffold limbs on older, weaker trees. To minimize the chance of injury cut down on the area treated, and Do Not treat weak trees. Do Not repeat applications on the same tree for at least one year.

Trunk Basal Sprouts and Root Suckers

Prune existing sprouts and apply Tre-Hold Sprout Inhibitor A112 during the dormant season prior to resprouting "or" applications can be made when the new shoots are 6 to 12 inches in height. The later applications have been the most effective. On bearing trees do not treat suckers during bud swell, bloom or fruit set. This period is from the start of growth to 4 weeks after petal fall. Application during this time can result in excessive thinning.

Apply TRE-HOLD Sprout Inhibitor A112 with a brush, small hand held sprayer or back pack sprayer. For large projects a power operated herbicide type spray equipment with attached hand gun or trunk directed nozzle can be used. Maintain low spray pressure (not to exceed 25 psi) to avoid spray drift of fine spray particles.

Pruning Cuts on Large Lateral or Upright Limbs.

To control sprouts in the vicinity of pruning cuts, paint the 1% dilute mixture of TRE-HOLD Sprout Inhibitor A112 on the bark in the immediate vicinity of the cut. Treat several inches of bark around the cut. DO NOT apply with a sprayer since the chance of hitting near-by fruit buds is too great. DO NOT allow the mixture to splash or drip on to other parts of the tree. Treat only in the dormant season prior to bud activity.

Olives—Trunk Basal Sprouts and Root Suckers.

(Do not treat scaffold branches or pruning cuts on olives.) Prune off existing sprouts or suckers and allow new sprouts to regrow. Treat with a dilute spray of TRE-HOLD Sprout Inhibitor A112 when sprouts are no more than 10" tall. TRE-HOLD Sprout Inhibitor A112 will control new sprout growth of olives, but will not control old woody sprouts. Do not treat during olive bloom or fruit set.

LIMITED WARRANTY AND DISCLAIMER

The manufacturer warrants that this product conforms to the chemical description on the label; that this product is reasonably fit for the purposes set forth in the directions for use when it is used in accordance with such directions; and that the directions, warnings and other statements on this label are based upon responsible experts' evaluation of reasonable tests of effectiveness, of toxicity to laboratory animals and to plants, and of residues on food crops and upon reports of field experience. Tests have not been made on all varieties or in all states or under all conditions. THE MANUFACTURER NEITHER MAKES NOR INTENDS, NOR DOES IT AUTHORIZE ANY AGENT OR REPRESENTATIVE TO MAKE, ANY OTHER WARRANTIES, EXPRESS OR IMPLIED, AND IT EXPRESSLY EXCLUDES AND DISCLAIMS ALL IMPLIED WARRANTIES OF MERCHANTABILITY OR FITNESS FOR A PARTICULAR PURPOSE.

THIS WARRANTY DOES NOT EXTEND TO, AND THE BUYER SHALL BE SOLELY RESPONSIBLE FOR, ANY AND ALL LOSS OR DAMAGE WHICH RESULTS FROM THE USE OF THIS PRODUCT IN ANY MANNER WHICH IS INCONSISTENT WITH THE LABEL DIRECTIONS, WARNINGS OR CAUTIONS.

BUYER'S EXCLUSIVE REMEDY AND MANUFACTURER'S OR SELLER'S EXCLUSIVE LIABILITY FOR ANY AND ALL CLAIMS, LOSSES, DAMAGES OR INJURIES RESULTING FROM THE USE OR HANDLING OF THIS PRODUCT, WHETHER OR NOT BASED IN CONTRACT, NEGLIGENCE, STRICT LIABILITY IN TORT OR OTHERWISE SHALL BE LIMITED, AT THE MANUFACTURER'S OPTION, TO REPLACEMENT OF, OR THE REPAYMENT OF THE PURCHASE PRICE FOR, THE QUANTITY OF PRODUCT WITH RESPECT TO WHICH DAMAGES ARE CLAIMED. IN NO EVENT SHALL MANUFACTURER OR SELLER BE LIABLE FOR SPECIAL, INDIRECT OR CONSEQUENTIAL DAMAGES RESULTING FROM THE USE OR HANDLING OF THIS PRODUCT.

THIS SPECIMEN LABEL IS INTENDED FOR USE ONLY AS A GUIDE IN PROVIDING GENERAL INFORMATION REGARDING THE DIRECTIONS, WARNINGS AND CAUTIONS ASSOCIATED WITH THE USE OF THIS PRODUCT. AS WITH ANY AGRICULTURAL CHEMICAL, ALWAYS FOLLOW THE LABEL INSTRUCTIONS ON THE PACKAGE BEFORE USING.

WEEDAR 64®

BROADLEAF HERBICIDE

The 2,4-D Amine Weedkiller

ACTIVE INGREDIENT:
Dimethylamine salt of
2,4-Dichlorophenoxyacetic acid* 46.8%
INERT INGREDIENTS: 53.2%

TOTAL 100.0%

*2,4-Dichlorophenoxyacetic acid equivalent 38.9% by weight or 3.8 pounds per gallon.
*Isomer specific by AOAC method No. 6.D01-5

E.P.A. Reg. No. 264-2AA E.P.A. Est. No. 264-MO-01

Keep Out of Reach of Children

DANGER—PELIGRO

PRECAUCION AL USARIO: Si usted no lee ingles, no use este producto hasta que la etiqueta le haya sido explicada ampliamente.

IN CASE OF EMERGENCY TELEPHONE COLLECT (24 HOURS A DAY) IN U.S.A. 1-800-UCC-HELP.

FOR GENERAL PRODUCT INFORMATION CALL 1-800-334-9745.

DIRECTIONS FOR USE

IT IS A VIOLATION OF FEDERAL LAW TO USE THIS PRODUCT IN A MANNER INCONSISTENT WITH ITS LABELING.

WEEDS IN CROP	PINTS WEEDAR 64 PER ACRE	GALLONS WATER PER ACRE	DIRECTIONS
Wheat, barley, rye, oats,			
Winter grain			
Annual and biennial weeds	½-2*	8 or more	Apply after grain is fully tillered (about 4″-8″ high) but not forming joints in the stem.
Perennial broadleaf weeds	1-2*	8 or more	Apply when weeds are near bud stage. Do not spray grain in boot to dough stage.
Spring grain			
Annual broadleaf weeds	½-2*	8 or more	Apply after grain is fully tillered (about 4″-8″ high) but not forming joints in the stem.
Perennial broadleaf weeds	1-2*	8 or more	Apply when weeds are near bud stage. Do not spray grain in boot to dough stage.
Grain underseeded with legumes	¼-½*	5-8	Apply after grains are about 8″ tall. **Do not spray grain in boot to dough stage.** Do not spray alfalfa or sweet clover unless the infestation is severe and injury to these legumes can be tolerated.

*Use the lower rate if small annual and biennial weeds are the major problem. Use the higher rate if perennial weeds or annual and biennial weeds are present which are in the hard-to-kill categories as determined by local experience. The higher rates increase the risk of grain injury and should be used only where the weed control problem justifies the grain damage risk. For aerial application on grain, apply WEEDAR 64 Broadleaf Herbicide in 3-10 gallons of water per acre.

Do not forage or graze treated grain fields within 2 weeks after treatment with 2,4-D.

Do not feed treated straw to livestock.

WEEDS IN CROP	PINTS WEEDAR 64 PER ACRE	GALLONS WATER PER ACRE	DIRECTIONS
Corn (field and sweet)			
Preemergence (for annual grasses and broadleaf weeds)	2-3	15-30	Apply just before corn emerges. Do not use on light sandy soil, or where soil moisture is low.
Postemergence annual broadleaf weeds	½-1	8-15	Apply when most weeds have germinated. Spray after corn emerges and until it is 5″-8″ tall.
Perennial broadleaf weeds	1-1½	8-15	Apply when weeds are in bud to bloom stage. Use drop pipes after corn is 10″ tall. Do not spray corn in tassel stage. 2,4-D may cause brittleness to corn. Winds or cultivations may cause stalk breakage while brittle. Certain single cross corn hybrids may be more susceptible to 2,4-D injury than others.
Sorghum—postemergence	1	6-10	Apply when sorghum is 4″-12″ tall. Use drop pipes to keep spray off sorghum plants.
Rice	1½-2½	5-10	To control curly indigo and other broadleaf weeds, apply 7-10 weeks after planting when rice is fully tillered. **Do not spray rice in boot stage.**

WEEDS IN CROP	PINTS WEEDAR 64 PER ACRE	GALLONS WATER PER ACRE	DIRECTIONS
Sugarcane			
Preemergence	4	15-20	
Postemergence	1½-2	10-30	Apply when cane is 1-2 feet tall.
Non-crop land (Fencerows, roadsides, drainage ditchbanks)			
Annual broadleaf weeds	2-3	15-30	Apply when weeds are young and growing vigorously.
Perennial broadleaf weeds	4-8	15-30	Apply when weeds are growing rapidly—generally near the bud stage. Repeated applications may be necessary.
Pastures & other turf areas (golf courses, parks, cemeteries)	2-4*	15-30	Do not apply to newly seeded turf until grass has been cut several times. Where bentgrass predominates, make 2 applications of 1 pint per acre at 3-week intervals. Do not graze dairy animals for 7 days after treatment. Do not use on susceptible Southern grasses such as St. Augustine. Bentgrass, dichondra and clover may be injured by this treatment.

* Use the lower rate if annual and biennial weeds are the major problem, the higher rate for perennial weeds.

For Emergency Weed Control in Wheat

Perennial broadleaf weeds: Apply 3 pints per acre when weeds are approaching bud stage, but do not spray grain in the boot to dough stage.

The 3 pint per acre application can produce injury to wheat. Balance the severity of your weed problem against the possibility of crop damage. Where perennial weeds are scattered spot treatment is suggested to minimize the extent of crop injury.

WEEDS AND BRUSH ON IRRIGATION CANAL DITCHBANKS—SEVENTEEN WESTERN STATES

Arizona, California, Colorado, Idaho, Kansas, Montana, Nebraska, New Mexico, Nevada, North Dakota, Oklahoma, Oregon, South Dakota, Texas, Utah, Washington and Wyoming.

For control of annual and perennial broadleaf weeds, apply 1 to 2 quarts of WEEDAR 64 Broadleaf Herbicide per acre in approximately 20 to 100 gallons per acre. Treat when weeds are young and actively growing before the bud or early bloom stage. For harder-to-control weeds a repeat spray after 3 to 4 weeks using the same rates may be needed for maximum results. Apply no more than two treatments per season.

For woody brush and patches of perennial broadleaf weeds, mix one gallon of WEEDAR 64 Broadleaf Herbicide in 150 gallons of water. Wet foliage thoroughly using about one gallon of solution per square yd.

Spraying Instructions

Apply with low pressure (10 to 40 psi) power spray equipment mounted on a truck, tractor or boat. Apply while traveling upstream to avoid accidental concentration of chemical into water. Spray when the air is fairly calm; 5 mph or less. Do not use on small canals (less than 10 cfs) where water will be used for drinking purposes.

Boom spraying onto water surface must be held to a minimum and no cross-stream spraying to opposite banks should be permitted. When spraying shoreline weeds, allow no more than two-foot overspray onto water with an average of less than one-foot overspray to prevent introduction of greater than negligible amounts of chemical into the water.

Do not allow dairy animals to graze on treated areas for at least 7 days after spraying. Water within treated banks should not be fished.

For spot spraying general weeds in non-cropped areas such as fencerows, roadsides and ditchbanks, mix 2 to 3 fluid ounces of WEEDAR 64 Broadleaf Herbicide in 3 gallons of water. Wet all weeds and stems thoroughly. For best results, treat when weeds are growing actively.

FORESTRY—TREE INJECTION

Make injections as near the root collar as possible, using one injection per inch of trunk dbh (4½ feet). For resistant species such as hickory, injections should overlap. For best results, injections should be made during the growing season, May 15—October 15.

For Dilute Injection

Mix 1 gallon of WEEDAR 64 Broadleaf Herbicide in 19 gallons of water for dilute injections.

For Concentrate Injections

Use 1 to 2 ml of concentrate WEEDAR 64 Broadleaf Herbicide per injection. The injector bit must penetrate the inner bark.

CONTROL THESE WEEDS

ANNUAL AND BIENNIAL WEEDS

Beggarticks, bull thistle, burdock, cockle, cocklebur, coffeeweed, fleabane (daisy), frenchweed, galinsoga, goatsbeard, jimsonweed, kochia, knotweed, lambsquarters, lettuce (wild), mallow, marshelder, morningglory, mustard, parsnip, peppergrass, pigweed, prickly lettuce, primrose, radish (wild), ragweed (common), Russian thistle, smartweed, sowthistle (common), sunflower, tumbleweed, vervains, vetch, wild carrot.

PERENNIAL WEEDS

Artichoke, aster, Austrian fieldcress, bindweed, blue lettuce, Canada thistle, catnip, chicory, dandelion, docks, dogbane, goldenrod, ground ivy, healall, hoary cress, horsetail, ironweed, loco weed, nettles, orange hawkweed, plantains, poverty weed, ragweed, sowthistle, stinging nettles, strawberry (wild), tanweed, toadflax, vervains, wild garlic, wild onion, wild sweet potato.

PRECAUTIONARY STATEMENTS

DANGER

HAZARDS TO HUMANS AND DOMESTIC ANIMALS

Harmful if swallowed. Do not get in eyes, on skin, or on clothing. Causes eye damage. Avoid breathing vapors or spray mist. Remove and wash contaminated clothing before reuse. Wash thoroughly before eating or smoking.

STATEMENT OF PRACTICAL TREATMENT

In case of contact wash skin with soap and water; for eyes flush with water for at least 15 minutes and get medical attention. If swallowed induce vomiting by touching the back of the throat or give strong salt water to drink. Repeat until vomit is clear. Never give anything by mouth to an unconscious person. Get medical attention.

ENVIRONMENTAL HAZARDS

Do not contaminate water by cleaning equipment or disposal of wastes.

Do not apply when weather conditions favor drift from treated areas.

Do not use the same spray equipment for other purposes unless thoroughly cleaned.

Do not use in or near a greenhouse.

Do not contaminate water used for irrigation or domestic purposes (except as specifically recommended on this label) especially in areas where grapes, cotton, tomatoes or other susceptible plants are grown.

Do not treat irrigation ditches in areas where water will be used to overhead (sprinkler) irrigate susceptible crops especially grapes, tomatoes, tobacco and cotton.

Do not apply WEEDAR 64 Broadleaf Herbicide directly to, or permit mist to drift onto cotton, okra, grapes, tomatoes, fruit trees, vegetables, flowers or other desirable crop or ornamental plants which are susceptible to 2,4-D herbicide. Do not apply near susceptible plants since very small quantities of the 2,4-D will cause severe injury during the growing or dormant periods. Crops contacted by WEEDAR 64 Broadleaf Herbicide sprays or spray drift may be killed or suffer significant stand loss with extensive quality and yield reduction.

Do not apply when a temperature air inversion exists. Such a condition is characterized by little or no air movement and an increase in air temperature with an increase in height. In humid regions a fog or mist may form. An inversion may be detected by producing a smoke column and checking for a layering effect. If questions exist pertaining to the existence of an inversion, consult with local weather services before making an application.

Use coarse sprays to minimize drift. Do not apply with hollow cone type insecticide or other nozzles that produce fine spray droplets.

Drift from aerial or ground application may be reduced by: (1) applying as near the target as possible in order to obtain coverage; (2) by increasing the volume of spray mix per acre; (3) by decreasing the pounds of pressure at the nozzle tips; (4) by using nozzles which produce a coarse spray pattern; (5) by not applying when wind is blowing toward susceptible crops or valuable plants.

GENERAL WORKER PROTECTION STATEMENTS

Do not apply this product in such a manner as to directly or through drift expose workers or other persons. The area treated must be vacated by unprotected persons.

Do not enter treated areas without protective clothing until sprays have dried.

Because certain states may require more restrictive reentry intervals for various crops treated with this product, consult your State Department of Agriculture for further information.

Written or oral warnings must be given to workers who are expected to be in a treated area or in an area about to be treated with this product. Advise workers to stay out of fields during application and until sprays have dried. Regular long-sleeved work clothing should be worn when working in treated fields. See Precautionary Statements, Statement of Practical Treatment and Note to Physician for information on accidental exposures. When oral warnings are given, warnings shall be given in a language customarily understood by workers. Oral warnings must be given if there is reason to believe that written warnings cannot be understood by workers. Written warnings must include the following information: appropriate signal word (DANGER—PELIGRO), area treated with WEEDAR 64® Broadleaf Herbicide, date of application, appropriate protective clothing, and reentry interval (i.e., until sprays have dried).

STORAGE AND DISPOSAL

STORAGE

Do not store near food, feedstuff, fertilizers, seeds, insecticides, fungicides or other pesticides.

PESTICIDE DISPOSAL

Do not contaminate water, food or feed by storage or disposal. Open dumping is prohibited. Pesticide wastes are acutely hazardous. Improper disposal of excess pesticide, spray mixture, or rinsate is a violation of Federal law. If these wastes cannot be disposed of by use according to label instructions, contact your State Pesticide or Environmental Control Agency, or the Hazardous Waste representative at the nearest EPA Regional Office for guidance.

CONTAINER DISPOSAL

Triple rinse or (equivalent). Then offer for recycling or reconditioning, or puncture and dispose of in a sanitary landfill, or by other procedures approved by state and local authorities.

Local conditions may affect the use of herbicides. Consult your State Agriculture Experiment Station, Farm Advisors, or Extension Weed Specialists for advice in selecting treatment from this label to best fit local conditions.

Be sure that use of this product conforms to all applicable laws, rules and regulations. Certain states have restrictions pertaining to application distances from susceptible crops. The applicator should become familiar with these laws, rules or regulations and follow them exactly. Apply this product only as specified on this label.

LIMITED WARRANTY AND DISCLAIMER

The manufacturer warrants (a) that this product conforms to the chemical description on the label; (b) that this product is reasonably fit for the purposes set forth in the directions for use when it is used in accordance with such directions; (c) and that the directions, warnings and other statements on this label are based upon responsible experts' evaluation of reasonable tests of effectiveness, of toxicity to laboratory animals and to plants, and of residues on food crops, and upon reports of field experience. Tests have not been made on all varieties or in all states or under all conditions. THE MANUFACTURER NEITHER MAKES NOR INTENDS, NOR DOES IT AUTHORIZE ANY AGENT OR REPRESENTATIVE TO MAKE, ANY OTHER WARRANTIES, EXPRESS OR IMPLIED, AND IT EXPRESSLY EXCLUDES AND DISCLAIMS ALL IMPLIED WARRANTIES OF MERCHANTABILITY OR FITNESS FOR A PARTICULAR PURPOSE.

THIS WARRANTY DOES NOT EXTEND TO, AND THE BUYER SHALL BE SOLELY RESPONSIBLE FOR, ANY AND ALL LOSS OR DAMAGE WHICH RESULTS FROM THE USE OF THIS PRODUCT IN ANY MANNER WHICH IS INCONSISTENT WITH THE LABEL DIRECTIONS, WARNINGS OR CAUTIONS.

BUYER'S EXCLUSIVE REMEDY AND MANUFACTURER'S OR SELLER'S EXCLUSIVE LIABILITY FOR ANY AND ALL CLAIMS, LOSSES, DAMAGES, OR INJURIES RESULTING FROM THE USE OR HANDLING OF THIS PRODUCT, WHETHER OR NOT BASED IN CONTRACT, NEGLIGENCE, STRICT LIABILITY IN TORT OR OTHERWISE SHALL BE LIMITED, AT THE MANUFACTURER'S OPTION, TO REPLACEMENT OF, OR THE REPAYMENT OF THE PURCHASE PRICE FOR, THE QUANTITY OF PRODUCT WITH RESPECT TO WHICH DAMAGES ARE CLAIMED. IN NO EVENT SHALL MANUFACTURER OR SELLER BE LIABLE FOR SPECIAL, INDIRECT OR CONSEQUENTIAL DAMAGES RESULTING FROM THE USE OR HANDLING OF THIS PRODUCT.

THIS SPECIMEN LABEL IS INTENDED FOR USE ONLY AS A GUIDE IN PROVIDING GENERAL INFORMATION REGARDING THE DIRECTIONS, WARNINGS AND CAUTIONS ASSOCIATED WITH THE USE OF THIS PRODUCT. AS WITH ANY AGRICULTURAL CHEMICAL, ALWAYS FOLLOW THE LABEL INSTRUCTIONS ON THE PACKAGE BEFORE USING.

WEEDAR 64®A

HERBICIDE

2,4-D amine for weed control.

ACTIVE INGREDIENT:
2,4-Dichlorophenoxyacetic acid,
diethanolamine salt* 54.6%
INERT INGREDIENTS: 45.4%

TOTAL 100.0%

*Isomer specific by AOAC method No. 6.D01-5.

*2,4-Dichlorophenoxyacetic acid equivalent 37.0% by weight or 3.8 pounds per gallon.

E.P.A. Reg. No. 264-143AA   E.P.A. Est. No. 264-MO-01

Keep Out of Reach of Children
DANGER—PELIGRO

PRECAUCION AL USARIO: Si usted no lee ingles, no use este producto hasta que la etiqueta le haya sido explicada ampliamente.

IN CASE OF EMERGENCY TELEPHONE COLLECT (24 HOURS A DAY) IN U.S.A. 1-800-UCC-HELP.

FOR GENERAL PRODUCT INFORMATION CALL 800-334-9745.

PRECAUTIONARY STATEMENTS

DANGER

HAZARD TO HUMANS AND DOMESTIC ANIMALS.

Harmful if swallowed. Avoid inhalation of spraymist and contact with skin, eyes, or clothing. Causes eye damage.

STATEMENT OF PRACTICAL TREATMENT

In case of eye contact, flush with water for at least 15 minutes. In case of skin contact wash with soap and plenty of water.

ENVIRONMENTAL HAZARDS

Do not apply where runoff is likely to occur. Do not contaminate water by cleaning of equipment or disposal of wastes. Do not contaminate irrigation ditches of water used for irrigation or domestic purposes. Do not use in or near a greenhouse.

PROTECTION STATEMENTS

Do not apply WEEDAR 64A Herbicide directly to, or otherwise permit it to come in contact with cotton, okra, grapes, tomatoes, fruit trees, vegetables, flowers or other desirable crop or ornamental plants which are susceptible to 2,4-D herbicide. Do not permit spray mist to drift onto susceptible plants since very small quantities of the 2,4-D can cause severe injury during the growing or dormant periods. Use coarse sprays to minimize drift. Do not apply with hollow cone-type insecticide or other nozzles that produce fine spray droplets. Drift from ground application may be reduced by: (1) keeping the spray boom as near to the crop as possible in order to obtain complete coverage; (2) by applying 5 gallons or more of spray per acre; (3) by using no more than 20 pounds of pressure at the nozzle tips; and (4) by not spraying when the wind exceeds 5 miles per hour.

Drift from aerial application may be reduced by: (1) applying as near to the crop as possible in order to obtain coverage; (2) by applying 3 or more gallons of spray per acre; (3) by using 20 pounds pressure or less at the nozzle tips; (4) by using nozzles which produce a coarse spray pattern; (5) by spraying only when the wind velocity is less than 5 miles per hour; and (6) by spraying when there is no possibility for a temperature inversion at time of spraying.

Applications by aircraft, ground rig and hand sprayers should be carried out only when there is no hazard from spray drift. Do not apply near cotton, grapes, tomatoes or near desirable 2,4-D susceptible crop or ornamental vegetation.

At air temperatures above 95°F do not apply WEEDAR 64A as vapors from this product may injure susceptible plants. Do not use the same spray equipment for other purposes.

GENERAL WORKER PROTECTION STATEMENTS

Do not apply this product in such a manner as to directly or through drift expose workers or other persons. The area being treated must be vacated by unprotected persons.

Do not enter treated areas without protective clothing until sprays have dried.

Because certain states may require more restrictive reentry intervals for various crops treated with this product, consult your State Department of Agriculture for further information.

Written or oral warnings must be given to workers who are expected to be in a treated area or in an area about to be treated with this product. Advise workers to stay out of fields during application and until sprays have dried. Regular long-sleeved work clothing should be worn when working in treated fields. See PRECAUTIONARY STATEMENTS and STATEMENT OF PRACTICAL TREATMENT for information on accidental exposures. When oral warnings are given, warnings shall be given in a language customarily understood by workers. Oral warnings must be given if there is reason to believe that written warnings cannot be understood by workers. Written warnings must include the following information: appropriate signal word (DANGER), area treated with WEEDAR 64® A Herbicide, date of application, appropriate protective clothing, and reentry interval (i.e. until sprays have dried).

STORAGE AND DISPOSAL

STORAGE

Do not store near fertilizers, seeds, insecticides or fungicides.

PESTICIDE DISPOSAL

Do not contaminate water, food, or feed by storage or disposal.

Pesticide wastes are acutely hazardous. Improper disposal of excess pesticide, spray mixture, or rinsate is a violation of Federal law. If these wastes cannot be disposed of by use according to label instructions, contact your State Pesticide or Environmental Control Agency or the Hazardous Waste representative at the nearest EPA Regional Office for guidance.

CONTAINER DISPOSAL

Triple rinse (or equivalent). Then offer for recycling or reconditioning, or puncture and dispose of in a sanitary landfill, or by incineration, or if allowed by state and local authorities, by burning. If burned stay out of smoke.

Local conditions may affect the use of herbicides. Consult your State Agriculture Experiment Station, Farm Advisors, or Extension Weed Specialists for advice in selecting treatment from this label to best fit local conditions.

Be sure that use of this product conforms to all applicable laws, rules and regulations. Certain states have restrictions pertaining to application distances from susceptible crops. The applicator should become familiar with these laws, rules or regulations and follow them exactly. Apply this product only as specified on this label.

DIRECTIONS

IT IS A VIOLATION OF FEDERAL LAW TO USE THIS PRODUCT IN A MANNER INCONSISTENT WITH ITS LABELING.

For ground application apply WEEDAR 64A Herbicide in 5 to 40 gallons of water per acre with the higher volume used to provide good coverage when weeds are dense.

For aerial application on grain apply WEEDAR 64A in 3 to 10 gallons of water per acre.

FOR EMERGENCY WEED CONTROL IN WHEAT

Perennial broadleaf weeds:

Apply 3 pints per acre when weeds are approaching bud stage, but do not spray grain in the seedling or boot to dough stage.

The 3 pint per acre application can produce injury to wheat. Balance the severity of your weed problem against the possibility of crop damage. Where perennial weeds are scattered, spot treatment is suggested to minimize the extent of crop injury.

WHEAT, BARLEY, RYE, OATS	PINTS WEEDAR 64® A PER ACRE	GALLONS WATER PER ACRE	DIRECTIONS
Winter grain			
Annual and biennial weeds	½-2*	8 or more	Apply after grain is fully tillered (about 4"-8" high) but not forming joints in the stem.
Perennial broadleaf weeds	1-2*	8 or more	Apply when weeds are near bud stage. Do not spray grain in boot to dough stage.
Spring grain			
Annual broadleaf weeds	½-2"	8 or more	Apply after grain is fully tillered (about 4"-8" high) but not forming joints in the stem.
Perennial broadleaf weeds	1-2"	8 or more	Apply when weeds are near bud stage. Do not spray grain in boot to dough stage.
Grain Underseeded with Legumes	¼-½	5-8	Apply after grains are about 8" tall. **Do not spray grain in boot to dough stage.** Do not spray alfalfa or sweet clover unless the infestation is severe and injury to these legumes can be tolerated.

* Use the lower rate if small annual and biennial weeds are the major problem. Use the higher rate if perennial weeds or annual and biennial weeds are present which are in the hard to kill categories as determined by local experience. The higher rates increase the risk of grain injury and should be used only where the weed control problem justifies the grain damage risk. Do not apply WEEDAR 64® A to grain in the seedling stage.

Do not forage or graze treated grain fields within two weeks after treatment with this product. Do not feed treated straw to livestock.

CORN (Field and sweet)

Preemergence: To control annual grasses and broadleaf weeds, apply 2 to 3 pints per acre just before corn emerges. Do not use on light sandy soil or when soil moisture is low.

Postemergence: For annual broadleaf weeds, apply ½ to 1 pint per acre as soon as most weeds have germinated. Spray after corn emerges and until corn is 5" to 8" tall. For perennial broadleaf weeds, use 1 to 1½ pints per acre when weeds are in the bud to bloom stage. Use drop pipes after corn is 12" tall. Do not spray corn in tassel.

2,4-D may cause brittleness to corn. Winds or cultivations may cause stalk breakage while brittle. Certain single cross corn hybrids may be more susceptible to 2,4-D injury than others.

SORGHUM—Apply 1 pint per acre when sorghum is 4" to 12" tall, according to state recommendations. Use drop pipes to keep off sorghum plants.

RICE—To control curly indigo and other broadleaf weeds in rice, use 1½ to 2½ pints per acre 7 to 10 weeks after planting when rice is fully tillered. Do not spray rice in the boot stage.

CONTROLLING WEEDS ON NON-CROPLAND (drainage ditchbanks, highway rights of way)

Annual broadleaf weeds: Apply 2 to 3 pints per acre when weeds are young and growing vigorously. Perennial broadleaf weeds: Apply 4 to 8 pints per acre when weeds are growing rapidly—generally near the bud stage. Repeated applications are often necessary for complete control.

For spot spraying general weeds in non-crop areas such as highway rights of way and drainage ditchbanks, mix 2 to 3 fluid ounces of WEEDAR 64® A in 3 gallons of water. Wet all weeds and stems thoroughly. For best results, treat when weeds are growing actively.

CONTROLLING WEEDS IN PASTURES, OTHER TURF AREAS—(golf courses, parks, cemeteries)

Apply 2 to 4 pints per acre in spring or fall when weeds are growing actively. Use higher rate where perennial broadleaf weeds are the major problem. Do not apply to newly seeded areas until grass has been cut several times. Where bentgrass predominates, make 2 applications of 1 pint per acre at 3-week intervals.

Do not use for susceptible Southern grasses such as St. Augustine. Bentgrasses and clover may be injured by this treatment.

Do not graze dairy animals on treated areas within 7 days after treatment.

WEEDS CONTROLLED

Annual and Biennial Weeds

beggarticks
bull thistle
burdock
cockle
cocklebur
coffeeweed
fleabane (daisy)
frenchweed
galinsoga
goatsbeard
jimsonweed
kochia
knotweed
lambsquarters
lettuce (wild)
mallow
marshelder
morningglory
mustard
parsnip
peppergrass
pigweed
prickly lettuce
primrose
radish (wild)
ragweed (common)
Russian thistle
smartweed
sowthistle (common)
sunflower
tumbleweed
vervains
vetch
wild carrot

Perennial Weeds

artichoke
aster
Austrian fieldcress
bindweed
blue lettuce
Canada thistle
catnip
chicory
dandelion
docks
dogbane
goldenrod
ground ivy
healall
hoary cress
horsetail
ironweed
locoweed
nettles
orange hawkweed
plantains
povertyweed
ragweed
sowthistle
stinging neetles
strawberry (wild)
tanweed
toadflax
vervains
wild garlic
wild onion
wild sweet potato

LIMITED WARRANTY AND DISCLAIMER

The manufacturer warrants (a) that this product conforms to the chemical description on the label; (b) that this product is reasonably fit for the purposes set forth in the directions for use when it is used in accordance with such directions; and (c) that the directions, warnings, and other statements on this label are based upon responsible experts evaluation of reasonable tests of effectiveness, of toxicity to laboratory animals and to plants, and of residues on food crops and upon reports of field experience. Tests have not been made on all varieties or in all states or under all conditions. THE MANUFACTURER NEITHER MAKES NOR INTENDS, NOR DOES IT AUTHORIZE ANY AGENT OR REPRESENTATIVE TO MAKE, ANY OTHER WARRANTIES, EXPRESS OR IMPLIED, AND IT EXPRESSLY EXCLUDES AND DISCLAIMS ALL IMPLIED WARRANTIES OF MERCHANTABILITY OR FITNESS FOR A PARTICULAR PURPOSE.

THIS WARRANTY DOES NOT EXTEND TO, AND THE BUYER SHALL BE SOLELY RESPONSIBLE FOR, ANY AND ALL LOSS OR DAMAGE WHICH RESULTS FROM THE USE

OF THIS PRODUCT IN ANY MANNER WHICH IS INCONSISTENT WITH THE LABEL DIRECTIONS, WARNINGS OR CAUTIONS.

BUYER'S EXCLUSIVE REMEDY AND MANUFACTURER'S OR SELLER'S EXCLUSIVE LIABILITY FOR ANY AND ALL CLAIMS, LOSSES, DAMAGES, OR INJURIES RESULTING FROM THE USE OR HANDLING OF THIS PRODUCT, WHETHER OR NOT BASED IN CONTRACT, NEGLIGENCE, STRICT LIABILITY IN TORT OR OTHERWISE SHALL BE LIMITED, AT THE MANUFACTURER'S OPTION, TO REPLACEMENT OF, OR THE REPAYMENT OF THE PURCHASE PRICE FOR, THE QUANTITY OF PRODUCT WITH RESPECT TO WHICH DAMAGES ARE CLAIMED. IN NO EVENT SHALL MANUFACTURER OR SELLER BE LIABLE FOR SPECIAL, INDIRECT OR CONSEQUENTIAL DAMAGES RESULTING FROM THE USE OR HANDLING OF THIS PRODUCT.

THIS SPECIMEN LABEL IS INTENDED FOR USE ONLY AS A GUIDE IN PROVIDING GENERAL INFORMATION REGARDING THE DIRECTIONS, WARNINGS AND CAUTIONS ASSOCIATED WITH THE USE OF THIS PRODUCT. AS WITH ANY AGRICULTURAL CHEMICAL, ALWAYS FOLLOW THE LABEL INSTRUCTIONS ON THE PACKAGE BEFORE USING.

WEEDAR® MCPA CONCENTRATE

BROADLEAF POSTEMERGENCE HERBICIDE

Controls weeds in flax, rice, peas, and small grains

ACTIVE INGREDIENT:
Dimethylamine salt of
2-methyl-4-chlorophenoxyacetic acid* 52.1%
INERT INGREDIENTS: 47.9%

TOTAL 100.0%

*2-Methyl-4-chlorophenoxyacetic acid equivalent 42.6% by weight or four pounds per gallon.

E.P.A. Reg. No. 264-47AA E.P.A. Est. No. 264-MO-01

Keep Out of Reach of Children
DANGER—PELIGRO

PRECAUCION AL USARIO: Si usted no lee ingles, no use este producto hasta que la etiqueta le haya sido explicada ampliamente.

IN CASE OF EMERGENCY, TELEPHONE COLLECT (24 HOURS A DAY) IN U.S.A. 1-800-UCC-HELP.

FOR GENERAL PRODUCT INFORMATION CALL 1-800-334-9745.

GENERAL INFORMATION

Tests in this country and in Europe show that several crops such as flax, rice and small grains underseeded with legumes are more tolerant of MCPA than they are of 2,4-D.

WEEDAR® MCPA Concentrate Broadleaf Postemergence Herbicide is effective on a large number of broadleaf weeds (mustards, thistles, pigweed, ragweed, burdock, plantain, dandelion, cocklebur, shepherdspurse, stinkweed, purslane and many others) and is useful for controlling these weeds in certain crops. Use MCPA on oats because oats are more tolerant of MCPA than they are of 2,4-D.

WEEDAR® MCPA Concentrate Broadleaf Postemergence Herbicide is the amine form of MCPA that is easily dissolved in water, even in the hard waters found in some parts of this country. Mix thoroughly while diluting this material and mix again if allowed to stand long before spraying.

To convert local recommendations into terms of WEEDAR® MCPA Concentrate, use the following table

MCPA acid	1¼ lb.	1 lb.	¾ lb.	½ lb.	⅜ lb.	¼ lb.
WEEDAR® MCPA Concentrate	2½ pt.	2 pt.	1½ pt.	1 pt.	¾ pt.	½ pt.

DIRECTIONS FOR USE

IT IS A VIOLATION OF FEDERAL LAW TO USE THIS PRODUCT IN A MANNER INCONSISTENT WITH ITS LABELING.

GENERAL WORKER PROTECTION STATEMENTS

Do not apply this product in such a manner as to directly or through drift expose workers or other persons. The area treated must be vacated by unprotected persons.

Do not enter treated areas without protective clothing until sprays have dried.

Because certain states may require more restrictive reentry intervals for various crops treated with this product, consult your State Department of Agriculture for further information.

Written or oral warnings must be given to workers who are expected to be in treated area or in an area about to be treated with this product. Advise workers to stay out of fields during application and until sprays have dried. Regular long-sleeved work clothing should be worn when working in treated fields. See Precautionary Statements, Statement of Practical Treatment and Note to Physician for information on accidental exposures. When oral warnings are given, warnings shall be given in a language customarily understood by workers. Oral warnings must be given if there is reason to believe that written warnings cannot be understood by workers. Written warnings must include the following information: Appropriate signal word (DANGER), area treated with WEEDAR® MCPA Concentrate Broadleaf Postemergence Herbicide, date of application, appropriate protective clothing, and reentry interval (ie. until sprays have dried).

CROP	AMOUNT WEEDAR® MCPA CONCENTRATE PER ACRE	DIRECTIONS
SMALL GRAINS (wheat, barley, rye not seeded to legumes)		
Spring planted	½ pint	Treat in spring when grain is fully tillered and 8 to 10 inches tall. Do not spray grain in the boot to dough stage. Use about 8 gallons of water per acre.
Fall planted	½ pint	Apply in early spring when annual broadleaf weeds are small and grain is fully tillered, but before grain is in the jointed stage. Use about 8 gallons of water per acre.

CROP	AMOUNT WEEDAR® MCPA CONCENTRATE PER ACRE	DIRECTIONS
Emergency treatment for perennial broadleaf weeds in wheat (all producing areas)	3 pints	Treat in spring after grain is tillered and at least 8 to 10 inches tall. Apply when weeds are approaching bud stage, but do not spray grain in the boot to dough stage. The 3 pint (1.5 pound acid equivalent) per acre application of any MCPA product can produce injury to wheat. Balance the severity of your weed problem against the possibility of crop damage. Where perennial weeds are scattered, spot treatment is suggested to minimize the effect of crop injury. Use at least 8 to 10 gallons of water per acre for ground application and at least 3 to 5 gallons of water per acre for aerial application.
SMALL GRAINS		
Underseeded with mixed legumes (alfalfa, alsike, red and ladino clovers, trefoil) (all producing areas)	½ pint	Treat in early spring when grain is in the tiller to boot stage and 8 to 10 inches tall, and when legume seedlings are 2 to 3 inches tall. Do not spray when grain is in the boot to dough stage. Use 5 to 6 gallons of water per acre since higher volumes may result in injury to legumes.
Do not forage or graze meat animals on treated areas within seven days of slaughter.		
OATS	½ to 1 pint	Spray after oats are fully tillered (4 to 6 inches tall) but before they are in the boot to dough stage. Apply the lower rate where annual weeds are the major problem, the higher rate for perennial weeds. Use at least 8 to 10 gallons of water per acre for ground application and at least 3 to 5 gallons of water per acre for aerial application.
FLAX	⅓-⅜ pint	Where susceptible weeds such as mustard are the main problem. Where weeds such as pigweed, lambsquarters, stinkweed, ragweed, cocklebur are the problem.
	½ pint	Apply when flax is 2 to 8 inches tall but before bud stage. Do not spray after flax is in the bud stage. Use at least 8 to 10 gallons of water per acre for ground application and at least 3 to 5 gallons of water per acre for aerial application.
RICE	1½ to 2½ pints	Treat 7 to 10 weeks after planting when rice is fully tillered and 6 to 8 inches above water. Do not spray after rice is in the boot stage. Use the higher rate where curly indigo, sedge or bulrush are the major problems. Use lower rate for arrowhead, water plantain and redstem. Do not spray when temperatures are over 90°F. Use at least 3 to 5 gallons of water per acre for aerial application.
Do not forage or graze livestock animals on treated areas within seven days of slaughter. Do not grow crayfish or catfish in treated rice fields.		
PEAS	½ to ¾ pint	Treat when peas are 4 to 6 inches tall using 15 to 30 gallons of water per acre for ground application and 10 gallons of water per acre for aerial application. Do not spray while peas are in blossom. Peas may show slight injury, but they usually recover a few days after treatment. Use the higher rate before buds form on Canada thistle to prevent from going to seed. Do not spray peas that are stressed from lack of moisture or when temperatures are over 90°F.
Do not forage or graze treated pea fields. Do not feed treated vines to livestock.		

PRECAUTIONARY STATEMENTS

DANGER

HAZARDS TO HUMANS AND DOMESTIC ANIMALS

Corrosive. Causes irreversible eye damage. Harmful if swallowed. May be fatal if absorbed through skin. Avoid contact with eyes, skin or clothing. Avoid breathing vapors or spray mist. Remove and wash contaminated clothing before reuse. Wash thoroughly before eating or smoking.

STATEMENT OF PRACTICAL TREATMENT

If swallowed, drink large quantities of water. Do not induce vomiting. Get medical attention. In case of skin contact, wash skin with plenty of soap and water. In case of eye contact, flush eyes with large amounts of water for 15 minutes. Get medical attention. Never give anything by mouth to an unconscious person.

Note to physician: aspiration hazard may contraindicate the use of gastric lavage.

ENVIRONMENTAL HAZARDS

Do not apply directly to water except as specified on this label. Do not contaminate water by cleaning of equipment or disposal of wastes.

Avoid spray drift to susceptible plants such as cotton, beans, tomatoes, and ornamentals (coarse sprays are less likely to drift).

Do not apply when weather conditions favor drift from treated areas. Avoid use of small diameter spray nozzles. At high air or ground surface temperatures, vapors from this product may injure susceptible plants.

Do not contaminate water used for irrigation or domestic purposes.

Do not use the same spray equipment for other purposes unless thoroughly cleaned.

Do not use in or near a greenhouse.

STORAGE AND DISPOSAL

Do not store near food, feedstuffs, fertilizers, seeds, insecticides or fungicides.

PESTICIDE DISPOSAL

Do not contaminate water, food or feed by storage or disposal. Open dumping is prohibited.

Pesticide wastes are acutely hazardous. Improper disposal of excess pesticide, spray mixture or rinsate is a violation of Federal law. If these wastes cannot be disposed of by use according to label instructions, contact your State Pesticide or Environmental Control Agency, or the Hazardous Waste representative at the nearest EPA Regional Office for guidance.

CONTAINER DISPOSAL

Triple rinse (or equivalent). Then offer for recycling or reconditioning, or puncture and dispose of in a sanitary landfill, or by incineration, or if allowed by state and local authorities, by burning. If burned, stay out of smoke.

LIMITED WARRANTY AND DISCLAIMER

The manufacturer warrants that this product conforms to the chemical description on the label, that this product is reasonably fit for the purposes set forth in the directions for use when it is used in accordance with such directions, and that the directions, warnings and other statements on this label are based upon responsible experts' evaluation of reasonable tests of effectiveness, of toxicity to laboratory animals and to plants, and of residues on food crops, and upon reports of field experience. Tests have not been made on all varieties or in all states or under all conditions. THE MANUFACTURER NEITHER MAKES NOR INTENDS, NOR DOES IT AUTHORIZE ANY AGENT OR REPRESENTATIVE TO MAKE, ANY OTHER WARRANTIES, EXPRESS OR IMPLIED, AND IT EXPRESSLY EXCLUDES AND DISCLAIMS ALL IMPLIED WARRANTIES OF MERCHANTABILITY OR FITNESS FOR A PARTICULAR PURPOSE.

THIS WARRANTY DOES NOT EXTEND TO, AND THE BUYER SHALL BE SOLELY RESPONSIBLE FOR, ANY AND ALL LOSS OR DAMAGE WHICH RESULTS FROM THE USE OF THIS PRODUCT IN ANY MANNER WHICH IS INCONSISTENT WITH THE LABEL DIRECTIONS, WARNINGS OR CAUTIONS.

BUYER'S EXCLUSIVE REMEDY AND MANUFACTURER'S OR SELLER'S EXCLUSIVE LIABILITY FOR ANY AND ALL CLAIMS, LOSSES, DAMAGES, OR INJURIES RESULTING FROM THE USE OR HANDLING OF THIS PRODUCT, WHETHER OR NOT BASED IN CONTRACT, NEGLIGENCE, STRICT LIABILITY IN TORT OR OTHERWISE, SHALL BE LIMITED, AT THE MANUFACTURER'S OPTION, TO REPLACEMENT OF, OR THE REPAYMENT OF THE PURCHASE PRICE FOR, THE QUANTITY OF PRODUCT WITH RESPECT TO WHICH DAMAGES ARE CLAIMED. IN NO EVENT SHALL MANUFACTURER OR SELLER BE LIABLE FOR SPECIAL, INDIRECT OR CONSEQUENTIAL DAMAGES RESULTING FROM THE USE OR HANDLING OF THIS PRODUCT.

THIS SPECIMEN LABEL IS INTENDED FOR USE ONLY AS A GUIDE IN PROVIDING GENERAL INFORMATION REGARDING THE DIRECTIONS, WARNINGS AND CAUTIONS ASSOCIATED WITH THE USE OF THIS PRODUCT.

AS WITH ANY AGRICULTURAL CHEMICAL, ALWAYS FOLLOW THE LABEL INSTRUCTIONS ON THE PACKAGE BEFORE USING.

WEEDAR® SODIUM MCPA

Controls weeds in rice, small grains, and peas

ACTIVE INGREDIENT:
*Sodium salt of 2-methyl-4-chlorophenoxyacetic acid 24.0%
INERT INGREDIENTS: 76.0%
TOTAL 100.0%

* 2-Methyl-4-chlorophenoxyacetic acid equivalent 21.7% or 2 pounds per gallon.

E.P.A. Reg. No. 264-55AA E.P.A. Est. No. 264-MO-01

Keep Out of Reach of Children

DANGER—PELIGRO

PRECAUCION AL USARIO: Si usted no lee ingles, no use este producto hasta que la etiqueta le haya sido explicada ampliamente.

IN CASE OF EMERGENCY TELEPHONE (24 HOURS A DAY) IN U.S.A. 1-800-UCC-HELP.

FOR GENERAL PRODUCT INFORMATION CALL 1-800-334-9745.

DIRECTIONS

IT IS A VIOLATION OF FEDERAL LAW TO USE THIS PRODUCT IN A MANNER INCONSISTENT WITH ITS LABELING.

CROP	AMOUNT PER ACRE	REMARKS
SMALL GRAINS (wheat, barley, rye, oats) Not underseeded with mixed legumes Broadleaf weeds	1 to 3 pints	Treat in the spring when grain is fully tillered and 8 to 10 inches tall. Do not spray when grain is in the boot to dough stage.
Emergency treatment of perennial broadleaf weeds in wheat	6 pints	Apply 6 pints per acre when weeds are approaching bud stage, but do not spray grain in the boot to dough stage. The 6 pint (1.5 pounds acid equivalent) per acre application of any MCPA product can produce injury to wheat. Balance the severity of your weed problem against the possibility of crop damage. Where perennial weeds are scattered, spot treatment is suggested to minimize the extent of crop injury. Treat in the spring when grain is fully tillered and 8 to 10 inches tall.
Underseeded with mixed legumes (alfalfa, alsike, red and ladino clovers, trefoil)	½ to 1 pint	Treat in early spring when legumes are 2 to 3 inches tall, when grain is fully tillered and 8 to 10 inches tall. Do not spray when grain is in the boot to dough stage. Use 5 to 6 gallons of water per acre since higher volumes may cause injury to legumes.
RICE	3 to 5 pints	Controls arrowhead, waterplantain and red stem. Controls sedge and bulrush. Treat 55 to 65 days after planting when rice is fully tillered and 6-8 inches above water. Water should be at least 2-3 inches deep. Do not spray after rice is in the boot stage, or when temperatures are over 90°F.
For airplane applications use 10 gallons of water per acre. For ground application use a conventional low pressure, low volume sprayer that delivers 10 to 40 gallons of water per acre. When diluting, the material must be mixed thoroughly before using. If the material is allowed to stand for any length of time after dilution, it should be stirred again before using.		
PEAS	½ to 1 ½ pint	Treat when peas are 4-6 inches tall using 15 to 30 gallons of water per acre. Do not spray while peas are in blossom. Peas may show slight injury but they usually recover a few days after treatment. Use the higher rate before buds form on Canada thistle to prevent from going to seed.
Do not feed treated vines to livestock. Do not forage or graze treated pea fields. Do not spray peas that are stressed from lack of moisture or when temperatures are over 90°F.		

GENERAL INFORMATION

Among the weeds controlled are:

Susceptible

lambsquarters	*wild radish*	*purslane*
mustard (annual)	*yellow rocket*	*peppergrass*
ragweed	*wintercress*	*cocklebur*
stinkweed		

Moderately Resistant

buttercup	*hempnettle*	*kochia*
Canada thistle	*stinging nettle*	*shepherdspurse*
dandelion	*sowthistle*	*whitetop (hoary cress)*
plantain	*vetch*	

PRECAUTIONARY STATEMENTS

DANGER

HAZARDS TO HUMANS AND DOMESTIC ANIMALS

Corrosive. Causes eye damage. Harmful if swallowed. May be fatal if absorbed through the skin. Do not get in eyes, on skin or on clothing. Avoid breathing spray mist. Wear goggles or face shield when handling concentrate. In case of contact immediately flush eyes or skin with plenty of water. Get medical attention if irritation persists. Remove and wash contaminated clothing before reuse.

STATEMENT OF PRACTICAL TREATMENT

If swallowed, call a physician or poison control center. Drink 1 or 2 glasses of water and induce vomiting by touching back of throat with finger. Do not induce vomiting or give anything by mouth to an unconscious person. In case of eye contact, rinse thoroughly with water. Get medical attention. In case of skin contact, wash with plenty of soap and water.

ENVIRONMENTAL HAZARDS

Do not apply directly to water except as specified on this label. Do not contaminate water used for irrigation or domestic purposes by cleaning equipment or disposal of wastes. Do not use in a greenhouse. Avoid spray drift to susceptible plants such as cotton, beans, peas, and ornamentals as injury may result. Coarse sprays are less likely to drift. Do not use same spray equipment for other purposes.

STORAGE AND DISPOSAL

STORAGE

Do not store near fertilizers, seeds, insecticides, or fungicides.

PESTICIDE DISPOSAL

Do not contaminate water, food, or feed by storage or disposal. Pesticide wastes are acutely hazardous. Improper disposal of excess pesticide, spray mixture, or rinsate is a violation of Federal Law. If these wastes cannot be disposed of by use according to label instructions, contact your State Pesticide or Environmental Control Agency, or the Hazardous Waste representative at the nearest EPA Regional Office for guidance.

CONTAINER DISPOSAL

Triple rinse (or equivalent). Then offer for recycling or reconditioning, or puncture and dispose of in a sanitary landfill, or by other procedures approved by state and local authorities.

LIMITED WARRANTY AND DISCLAIMER

The manufacturer warrants that this product conforms to the chemical description on the label; that this product is reasonably fit for the purposes set forth in the directions for use when it is used in accordance with such directions; and that the directions, warning and other statements on this label are based upon responsible experts' evaluation of reasonable tests of effectiveness, of toxicity to laboratory animals and to plants, and of residues on food crops and upon reports of field experience. Tests have not been made on all varieties or in all states or under all conditions. THE MANUFACTURER NEITHER MAKES NOR INTENDS, NOR DOES IT AUTHORIZE ANY AGENT OR REPRESENTATIVE TO MAKE, ANY OTHER WARRANTIES, EXPRESS OR IMPLIED, AND IT EXPRESSLY EXCLUDES AND DISCLAIMS ALL IMPLIED WARRANTIES OF MERCHANTIBILITY OR FITNESS FOR A PARTICULAR PURPOSE.

THIS WARRANTY DOES NOT EXTEND TO, AND THE BUYER SHALL BE SOLELY RESPONSIBLE FOR, ANY AND ALL LOSS OR DAMAGE WHICH RESULTS FROM THE USE OF THIS PRODUCT IN ANY MANNER WHICH IS INCONSISTENT WITH THE LABEL DIRECTIONS, WARNINGS OR CAUTIONS.

BUYER'S EXCLUSIVE REMEDY AND MANUFACTURER'S OR SELLER'S EXCLUSIVE LIABILITY FOR ANY AND ALL CLAIMS, LOSSES, DAMAGES OR INJURIES RESULTING FROM THE USE OR HANDLING OF THIS PRODUCT, WHETHER OR NOT BASED IN CONTRACT, NEGLIGENCE, STRICT LIABILITY IN TORT OR OTHERWISE SHALL BE LIMITED, AT THE MANUFACTURER'S OPTION, TO REPLACEMENT OF, OR THE REPAYMENT OF THE PURCHASE PRICE FOR, THE QUANTITY OF PRODUCT WITH RESPECT TO WHICH DAMAGES ARE CLAIMED. IN NO EVENT SHALL MANUFACTURER OR SELLER BE LIABLE FOR SPECIAL, INDIRECT OR CONSEQUENTIAL DAMAGES RESULTING FROM THE USE OR HANDLING OF THIS PRODUCT.

THIS SPECIMEN LABEL IS INTENDED FOR USE ONLY AS A GUIDE IN PROVIDING GENERAL INFORMATION REGARDING THE DIRECTIONS, WARNINGS AND CAUTIONS ASSOCIATED WITH THE USE OF THIS PRODUCT. AS WITH ANY AGRICULTURAL CHEMICAL, ALWAYS FOLLOW THE LABEL INSTRUCTIONS ON THE PACKAGE BEFORE USING.

WEEDONE LV4®

BROADLEAF POSTEMERGENCE HERBICIDE

The 2,4-D low volatile ester for agricultural weed control, conifer release and site preparation

ACTIVE INGREDIENT:	
2,4-Dichlorophenoxyacetic acid, butoxyethyl ester*	60.8%
INERT INGREDIENTS:	39.2%
TOTAL	100.0%

*2,4-Dichlorophenoxyacetic acid equivalent 41.9% by weight or 3.8 pounds per gallon.

*Isomer specific by AOAC method No. 6.D01-5

E.P.A. Reg. No. 264-20ZA E.P.A. Est. No. 264-MO-01

Keep Out of Reach of Children

CAUTION

IN CASE OF EMERGENCY TELEPHONE (24 HOURS A DAY) IN U.S.A. 1-800-UCC-HELP

FOR GENERAL PRODUCT INFORMATION CALL 1-800-334-9745.

DIRECTIONS FOR USE

IT IS A VIOLATION OF FEDERAL LAW TO USE THIS PRODUCT IN A MANNER INCONSISTENT WITH ITS LABELING.

GENERAL INFORMATION

WEEDONE LV4 Broadleaf Postemergence Herbicide contains the butoxyethyl ester of 2,4-D, the original and one of the best low volatile esters. In cropland, WEEDONE LV4 is more effective than amines for controlling hard-to-kill weeds such as bindweed, thistle, smartweed, wild garlic, curled dock, musk thistle, tansy ragwort, and wild onions.

TO PREPARE A SPRAY: Add one-half the required amount of water to the spray tank, then add the WEEDONE LV4 with agitation, and finally, the balance of the water with continued agitation. This material forms an emulsion in water—not a solution. This tends to separate on standing. Provide agitation to prevent such separation and insure uniform spray mixtures.

NOTICE: Use WEEDONE LV6 Emulsifiable Broadleaf Herbicide to make oil solutions for airplane spraying.

RECOMMENDATIONS FOR APPLYING WEEDONE LV4 BROADLEAF POSTEMERGENCE HERBICIDE

WEEDS IN CROPS

	AMOUNT OF WEEDONE LV4 PER ACRE	DIRECTIONS
WHEAT, BARLEY, RYE (not underseeded with legumes) Annual and biennial weeds Perennial broadleaf weeds	 ½-2 pints* 1-2 pints*	Apply after grain is fully tillered (usually about 4-8″ high) but not forming joints in the stem. Apply when weeds are near bud stage. **Do not spray grain in boot to dough stage.**
WEEDONE LV4 and ME4® BROMINAL® tank mixture on wheat, barley, oats and rye Cereal areas *except* Washington, Oregon, Idaho Washington, Oregon, Idaho	 ½-1 pint plus ½-¾ pint ME4 BROMINAL ½-1 pint plus ¾-1 pint ME4 BROMINAL	ME4® BROMINAL® Broadleaf Herbicide will control some annual weeds that are resistant to WEEDONE LV4 and may be tank mixed with WEEDONE LV4 for broader spectrum weed control on small grains. To prepare the spray, first mix WEEDONE LV4 in water then add the ME4 BROMINAL. Use the higher rates for larger weeds or where weed growth is slow due to dry or cold weather. Apply before weeds are 6 inches high. Use 10-20 gallons total spray volume per acre with ground equipment or use 5-10 gallons total spray volume with air application. Use higher volume on larger weeds. Observe all directions, precautions, and restrictions on both products' labeling.
Wild onion or garlic	1½-2 pints	Apply 1½ pints when grain is fully tillered and wild garlic or onion plants are small. Apply 2 pints after harvest in the crop stubble.
Emergency weed control in wheat Perennial broadleaf weeds	 3 pints	Apply when weeds are approaching bud stage, after the dough stage, but **do not spray during the boot to dough stage.** The 3 pint per acre application can produce injury to wheat. Balance the severity of your weed problem against the possibility of crop damage. Where perennial weeds are scattered, spot treatment is suggested to minimize the extent of crop injury.
Preharvest	1-2 pints*	Apply when grains are in the hard dough stage to suppress large weeds that may interfere with harvest. Best results will be obtained when soil moisture is sufficient to induce succulent weed growth.

* Use the lower rate if small annual and biennial weeds are the major problem. Use the higher rate if perennial weeds or annual and biennial weeds are present which are in the hard-to-kill categories as determined by local experience. The higher rates increase the risk of grain injury and should be used only where the weed control problem justifies the risk of grain damage.

- For aerial application on grain, apply WEEDONE LV4 in 2 to 5 gallons of water per acre. For ground application, use a minimum of 10 gallons of water per acre.
- Do not mix with oil for crop uses.
- Do not forage or graze treated fields for 2 weeks after application.
- Do not feed treated straw to livestock.

FIELD CORN		
Preemergence	2-3 pints	To control broadleaf weeds and suppress annual grasses, apply 3 to 5 days after planting, but before corn emerges. Use high rate on soil high in organic matter. Do not use on light, sandy soil, or where soil moisture is inadequate for normal weed growth.
Postemergence Annual broadleaf weeds	 ½ pint	Avoid spraying just after corn leaves unfold, as injury may occur. Apply when weeds are small and corn is less than 10″ tall. If corn more than 10″ tall must be sprayed, use drop pipes.
Perennial broadleaf weeds	¾ pint	Spray when weeds are in the bud to bloom stage. If corn is more than 10″ tall, use drop pipes to keep spray off corn leaves. **Do not spray in tassel to dough stage.** 2,4-D may make corn brittle. Winds or cultivation may cause stalk breakage while brittle. Certain single cross hybrids may be more susceptible to 2,4-D injury than other varieties.
Preharvest (brown silk)	1-2 pints	After the hard dough (denting) stage, to suppress weeds that interfere with harvest such as bindweed, cocklebur, dogbane, jimsonweed, ragweed, sunflower, and velvetleaf, and to decrease production of weed seeds, spray with air or ground equipment. The high rate will be needed for tough weeds under stress.
GRAIN SORGHUM		
Annual broadleaf weeds Perennial broadleaf weeds	¾ pint 1¼ pints	Treat when sorghum is 6 to 10″ tall.
GRASSES FOR SEED PRODUCTION	1-1½ pints	Apply in established stands in spring from tiller to early boot stage. **Do not spray in boot stage.** New spring seedlings may be treated with the lower rate after grass seedlings have at least 5 leaves. Perennial weed regrowth may be treated in the fall.
WEEDS AND BRUSH IN PASTURES, RANGELAND, TURF AREAS, AND FALLOW LAND		
RANGELAND, PASTURES, GOLF COURSES, CEMETERIES, PARKS, AND OTHER TURF AREAS		
Annual broadleaf weeds	2 pints	Apply in spring or fall when annual weeds such as knotweed or common chickweed are up but still small and in or near the seedling stage, and growing vigorously. Also controls cocklebur, galinsoga, lambsquarters, mustards, pigweed, ragweed, sunflower.
Perennial and biennial broadleaf weeds	3-4 pints	To control dandelion, plantain, and most other common broadleaf weeds in turf, apply in spring or fall when these weeds are growing actively. Treat wild garlic or onion in three successive seasons. (Example: spring, fall, spring.) Do not apply to newly seeded areas until sod is formed and grass has been cut a few times. Do not use WEEDONE LV4 for susceptible Southern grasses such as St. Augustine. Bentgrass and clover may be injured by this treatment. Also controls bindweed, thistles, (Canada, musk, bull, and Scotch), smartweed, tansy ragwort, wild onion, wild garlic.

	AMOUNT OF WEEDONE LV4 PER ACRE	DIRECTIONS
Sand sage, sagebrush, shinnery oak, and other 2,4-D susceptible woody species	1-4 quarts	Apply aerially to brush in 2-10 gallons water per acre. One gallon fuel oil may be included in the mixture. Consult state or local brush control specialists for most effective rate, volume, and timing of spray applications.
FALLOW LAND		
Annual broadleaf weeds	1-2 pints	Use the lower rate when weeds are small (2-3" tall) and growing actively. Use the higher rate on older and drought-stressed plants.
Biennial broadleaf weeds	2-4 pints	Spray while musk thistles or other biennial species are in the seedling to rosette stage, and before flower stalks are initiated. The lower rate can be used in spring during rosette stage. In fall or after flower stalks have developed, use the highest rate.
Perennial broadleaf weeds	2-4 pints	Spray weeds in bud to bloom stage, or in good vegetative growth. Do not disturb treated area for at least 2 weeks after treatment, or until weed tops are dead.

- For ground spraying on non-crop areas, pastures, rangeland, and turf areas, use enough water for good wetting (minimum 10 gallons per acre).
- Do not graze dairy animals on treated areas within 7 days after treatment.
- Do not plant treated fallow land until three months after treatment or until chemical has disappeared from soil.

COMBINATION WITH LIQUID NITROGEN FERTILIZER: Use $\frac{1}{2}$ - $\frac{3}{4}$ pint WEEDONE LV4 per acre for weeding and feeding corn, small grains, grasses for seed production, or grass pastures as directed on this label. Use the fertilizer at rates recommended by supplier or Extension Service Specialist. Fill the spray tank about half full with the liquid fertilizer, then add the WEEDONE LV4 with vigorous agitation, and complete filling the tank with fertilizer. Apply immediately and continue agitation in the spray tank during application. Application during very cold weather (near freezing) is not advisable. Do not allow mixture to stand overnight. Incompatibility may be encountered with some fertilizer brands or under some environmental conditions. If in doubt, test a small sample in the dilution ratio planned for application. NOTE: Herbicides can increase foliage contact burn of fertilizers. Reducing the fertilizer rate and concentration will reduce the hazard of leaf burn.

WEEDS AND BRUSH IN NON-CROP AREAS		
FENCEROWS, ROADSIDES, DRAINAGE DITCHBANKS, VACANT LOTS, AIRFIELDS, RAILROAD RIGHTS-OF-WAY, AND OTHER NON-CROP AREAS.		
Annual broadleaf weeds	2-4 pints	Apply when most weeds are still young and growing vigorously.
Perennial and biennial broadleaf weeds	3-6 pints	Apply when weeds are actively growing and near the bud stage, but before flowering. For best results on tansy ragwort and musk thistle, treat in rosette stage, before bolting. A second application is usually needed for best results on thistle, nettle, and bindweed. Treat wild onion or garlic in early spring and in fall when they are young and growing actively.
Cattails and tules	$1\frac{1}{2}$-$2\frac{1}{2}$ gallons	Mix 2 quarts WEEDONE LV4 in 2 quarts kerosene or diesel oil, then add this mixture to 100 gallons of water. Apply 300 to 500 gallons of spray per acre, depending on the stand. The addition of a wetting agent such as SURFEL® Spray Adjuvant is suggested.
Woody plants	1-2 gallons	Apply in 30-100 gallons of spray per acre, preferably through low-volume equipment such as DIRECTA-SPRA®, WOBBLER®, MINI-WOBBLER®, or SPIROMETER®. Treat when 2,4-D-susceptible species are in full leaf and growing actively. Repeated applications, oil or wetting agent, and/or high volume spraying (300 to 500 gallons of spray per acre) may be needed for acceptable control of some species or if brush growth is dense.
WEEDONE LV4 and AMDON® K Herbicide tank mixtures for non-crop areas		To prepare the spray, add the total required amount of water to the spray tank. When using NALCO-TROL® drift control additive or LO-DRIFT® Spray Additive, add at rates specified on its container label. Next add the required amount of WEEDONE LV4 while agitating the mixture in the spray tank then add the required amount of AMDON K while continuing agitation.
Perennial broadleaf weeds and susceptible woody species	$\frac{1}{2}$-$1\frac{1}{2}$ gallons plus 1-3 quarts AMDON K	Apply mixture of WEEDONE LV4 and AMDON K in enough spray volume to obtain adequate coverage. Usually 15 to 25 gallons per acre.
Difficult to control woody species such as balsam fir, black spruce, bracken fern, gums, hickory, maple, oaks, salmonberry and sourwood	$1\frac{1}{2}$-2 gallons plus 3-4 quarts AMDON K	Observe all directions, precautions, and restrictions on both products labeling.

- For aerial application to solid stands of susceptible brush, use 2-4 quarts in 3-12 gallons volume per acre. Two to 4 quarts of fuel oil may be included in this mixture.
- Applications will be less effective when conditions such as deficient soil moisture reduce brush growth.

FOREST MANAGEMENT

CONIFER RELEASE

To control alder, apply $1\frac{1}{2}$ -2 quarts WEEDONE LV4 Broadleaf Postemergence Herbicide in $9\frac{1}{2}$ -15 gallons water per acre as a foliage spray. Treat when $\frac{3}{4}$ of the brush foliage has attained full size leaves and before new conifer growth reaches 2" in length. This is usually between early May and mid-June. Adjust treatment date depending on stage of growth of conifers and brush species. This may cause leader deformation on exposed firs, but they should overcome this during the second year after spraying.

To control tanoak, madrone, ceanothus, canyon live oak, and manzanita, and to release Douglas fir, hemlock, Sitka spruce or grand fir, apply 3 quarts WEEDONE LV4 plus 3 quarts fuel oil in $8\frac{1}{2}$ -15 gallons water per acre before new growth on Douglas fir is 2" long. To control manzanita and ceanothus in ponderosa pine, apply 3 quarts WEEDONE LV4 before pine growth begins in spring.

To control hazel brush in the Lake states, apply 2 quarts WEEDONE LV4 in 1½ -4½ gallons water per acre when new shoot growth of hazel is complete (usually mid-July).

In northern areas, if possible conifer injury can be tolerated, 1½ -3 quarts WEEDONE LV4 applied aerially in 8-25 gallons of water per acre after conifers such as jack pine, red pine, black spruce, and white spruce have hardened off (usually mid-July) will provide control of competing hardwoods including alder, aspen, birch, hazel, and willow.

SITE PREPARATION

Budbreak spray: To control alder before planting forest seedlings, apply 2-4 quarts WEEDONE LV4 in 9-15 gallons fuel oil per acre after alder buds break, but before foliage is ¼ full size.

Foliage spray: To control alder before planting forest tree seedlings, apply 2 quarts WEEDONE LV4 plus 2 quarts fuel oil in 9-15 gallons water per acre after most alder leaves are full size. One-tenth gallon TRANS-VERT® Herbicide Additive may be added for additional alder control.

Consult your regional or extension forester or state herbicide specialist for recommendation to fit local conditions.

WEEDONE, ME4 BROMINAL, SURFEL, DIRECTA-SPRA, WOBBLER, MINI WOBBLER, SPIROMETER, AMDON, LO-DRIFT and TRANSVERT are registered trademarks of Union Carbide Agricultural Products Company, Inc.

NALCO-TROL is a trademark of NALCO Chemical Company.

PRECAUTIONARY STATEMENTS

CAUTION
HAZARD TO HUMANS AND DOMESTIC ANIMALS

Harmful if swallowed or absorbed through skin. Do not get in eyes on skin or on clothing. Avoid breathing vapors or spray mist. Remove and wash contaminated clothing before reuse. Wash thoroughly before eating or smoking.

STATEMENT OF PRACTICAL TREATMENT

In case of contact wash skin with soap and water; for eyes flush with water for at least 15 minutes and get medical attention. If swallowed, do not induce vomiting. Get medical attention.

Note to Physician: Aspiration hazards may contraindicate the use of gastric lavage.

Never induce vomiting or give anything by mouth to an unconscious person. Get medical attention immediately.

ENVIRONMENTAL HAZARDS

This product is toxic to fish. Do not apply directly to water. Do not contaminate water by cleaning of equipment or disposal of wastes.

USE PRECAUTIONS

Do not apply when weather conditions favor drift from treated areas.

Do not use the same spray equipment for other purposes unless thoroughly cleaned.

Do not use in or near a greenhouse.

Do not contaminate water used for irrigation or domestic purposes.

Do not apply WEEDONE LV4 Broadleaf Postemergence Herbicide directly to, or permit spray mist to drift onto cotton, okra, grapes, tomatoes, fruit trees, vegetables, flowers or other desirable crop or ornamental plants which are susceptible to 2,4-D herbicide. Do not apply near susceptible plants since very small quantities of the 2,4-D will cause severe injury during the growing or dormant periods. Crops contacted by WEEDONE LV4 Broadleaf Postemergence Herbicide sprays or spray drift may be killed or suffer significant stand loss with extensive quality and yield reduction.

Do not apply when a temperature air inversion exists. Such a condition is characterized by little or no air movement and an increase in air temperature with an increase in height. In humid regions a fog or mist may form. An inversion may be detected by producing a smoke column and checking for a layering effect. If questions exist pertaining to the existence of an inversion, consult with local weather services before making an application.

Use coarse sprays to minimize drift. Do not apply with hollow cone type insecticide or other nozzles that produce fine spray droplets.

Drift from **aerial or ground application** may be reduced by: (1) applying as near to the target as possible in order to obtain coverage; (2) by increasing the volume of spray mix per acre; (3) by decreasing the pounds of pressure at the nozzle tips; (4) by using nozzles which produce a coarse spray pattern; (5) by not applying when wind is blowing toward susceptible valuable plants.

At high air or ground surface temperatures, vapors from this product may injure susceptible plants.

STORAGE AND DISPOSAL

STORAGE

Do not store near food, feedstuff, fertilizers, seeds, insecticides, fungicides or other pesticides.

PESTICIDE DISPOSAL

Do not contaminate water, food or feed by storage or disposal. Pesticide wastes are toxic. Improper disposal of excess pesticide, spray mixture, or rinsate is a violation of Federal law. If these wastes cannot be disposed of by use according to label instructions, contact your State Pesticide or Environmental Control Agency or the Hazardous Waste representative at the nearest EPA Regional Office for guidance.

CONTAINER DISPOSAL

Triple rinse (or equivalent). Then offer for recycling or reconditioning, or puncture and dispose of in a sanitary landfill or by incineration, or if allowed by state and local authorities, by burning. If burned, stay out of smoke.

LIMITED WARRANTY AND DISCLAIMER

The manufacturer warrants that this product conforms to the chemical description on the label; that this product is reasonably fit for the purposes set forth in the directions for use when it is used in accordance with such directions; and that the directions, warnings and other statements on this label are based upon responsible experts' evaluation of reasonable tests of effectiveness, of toxicity to laboratory animals and to plants, and of residues on food crops and upon reports of field experience. Tests have not been made on all varieties or in all states or under all conditions. THE MANUFACTURER NEITHER MAKES NOR INTENDS, NOR DOES IT AUTHORIZE ANY AGENT OR REPRESENTATIVE TO MAKE, ANY OTHER WARRANTIES, EXPRESS OR IMPLIED, AND IT EXPRESSLY EXCLUDES AND DISCLAIMS ALL IMPLIED WARRANTIES OF MERCHANTABILITY OR FITNESS FOR A PARTICULAR PURPOSE.

THIS WARRANTY DOES NOT EXTEND TO, AND THE BUYER SHALL BE SOLELY RESPONSIBLE FOR, ANY AND ALL LOSS OR DAMAGE WHICH RESULTS FROM THE USE OF THIS PRODUCT IN ANY MANNER WHICH IS INCONSISTENT WITH THE LABEL DIRECTIONS, WARNINGS OR CAUTIONS.

BUYER'S EXCLUSIVE REMEDY AND MANUFACTURER'S OR SELLER'S EXCLUSIVE LIABILITY FOR ANY AND ALL CLAIMS, LOSSES, DAMAGES, OR INJURIES RESULTING FROM THE USE OR HANDLING OF THIS PRODUCT, WHETHER OR NOT BASED IN CONTRACT, NEGLIGENCE, STRICT LIABILITY IN TORT OR OTHERWISE SHALL BE LIMITED, AT THE MANUFACTURER'S OPTION, TO REPLACEMENT OF, OR THE REPAYMENT OF THE PURCHASE PRICE FOR, THE QUANTITY OF PRODUCT WITH RESPECT TO WHICH DAMAGES ARE CLAIMED. IN NO EVENT SHALL MANUFACTURER OR SELLER BE LIABLE FOR SPECIAL, INDIRECT OR CONSEQUENTIAL DAMAGES RESULTING FROM THE USE OR HANDLING OF THIS PRODUCT.

THIS SPECIMEN LABEL IS INTENDED FOR USE ONLY AS A GUIDE IN PROVIDING GENERAL INFORMATION REGARDING THE DIRECTIONS, WARNINGS AND CAUTIONS ASSOCIATED WITH THE USE OF THIS PRODUCT. AS WITH ANY AGRICULTURAL CHEMICAL, ALWAYS FOLLOW THE LABEL INSTRUCTIONS ON THE PACKAGE BEFORE USING.

WEEDONE®LV6

EMULSIFIABLE BROADLEAF HERBICIDE

Weed and brush control in small grains, field corn, rangeland, pastures, roadsides and fencerows.

ACTIVE INGREDIENT:
2,4-Dichlorophenoxyacetic acid,
butoxyethyl ester* 83.5%
INERT INGREDIENTS: 16.5%
TOTAL 100.0%

*2,4-Dichlorophenoxyacetic acid equivalent 5.7 lb./gal. or 57.5%/wt.
Isomer specific by AOAC method No. 6.D01-5
EPA Reg. No. 264-271 EPA Est. No. 264-MO-01

Keep Out of Reach of Children
CAUTION

IN CASE OF EMERGENCY PHONE (24 HOURS A DAY) IN U.S.A. 1-800-UCC-HELP.

FOR GENERAL PRODUCT INFORMATION CALL 1-800-334-9745.

DIRECTIONS FOR USE

IT IS A VIOLATION OF FEDERAL LAW TO USE THIS PRODUCT IN A MANNER INCONSISTENT WITH ITS LABELING.

WEEDS IN CROPS	AMOUNT PER ACRE	DIRECTIONS
Field corn		
Preemergence	1⅓-2 pints	To control broadleaf weeds and suppress annual grasses, apply 3 to 5 days after planting but before corn emerges. Use high rate on soil high in organic matter or clay. Do not use on light sandy soil or when soil moisture is inadequate for normal weed growth.
At emergence (spike)	⅔ pint	Treat just as plants are breaking ground.
Postemergence		
Annual broadleaf weeds	⅓ pint	Apply when weeds are small and corn is less than 10″ tall. Avoid spraying just after corn leaves unfold as injury may occur. If corn more than 10″ tall must be sprayed, use drop pipes.
Perennial broadleaf weeds	1 pint	Spray when weeds are in the bud to bloom stage. If corn is more than 10″ tall use drop pipes to keep spray off corn leaves. Do not spray corn from tassel to dough stage. 2,4-D may cause brittleness to corn. Winds or cultivation may cause stalk breakage while brittle. Certain single cross hybrids may be more susceptible to 2,4-D injury than other varieties.
Preharvest (brown silk)	⅔-1⅓ pint	Apply with air or ground equipment after the hard dough (denting) stage to suppress weeds that interfere with harvest such as bindweed, cocklebur, dogbane, jimsonweed, ragweed, sunflower and vevetleaf, and to decrease production of weed seeds. The high rate will be needed for tough weeds under stress.
Spring seeded wheat, barley, rye (not underseeded with legumes)		
Annual broadleaf weeds	⅓ pint	Spray after grain is fully tillered (about 4 to 6 inches tall) but before it is in the boot stage.
Perennial broadleaf weeds	⅓-1⅓ pint*	Spray after grain is fully tillered (about 4 to 6 inches tall) and when weeds are nearing the bud stage. Do not spray grain in the boot to dough stage.
Fall seeded wheat, rye (not underseeded with legumes)	⅓-1⅓ pint*	Spray in the spring before grain is in the boot to dough stage.

* Use the lower rate if small annual and biennial weeds are the major problem. Use the higher rate if perennial weeds or annual and biennial weeds are present which are in the hard to kill categories as determined by local experience. The higher rates increase the risk of injury and should be used only where the weed control problem justifies the grain damage risk. Do not forage or graze treated grain fields within two weeks after treatment with this product.

For aerial application to small grains, preharvest corn, pastures or rangeland, use the recommended amount of this product in 2 to 10 gallons of water per acre.

WEEDS AND BRUSH IN FALLOW LAND, RANGELAND, PASTURES, ROADSIDES, FENCEROWS

WEEDS IN CROPS	AMOUNT PER ACRE	DIRECTIONS
Annual broadleaf weeds	1⅓-2⅔ pints	Spray when weeds are young and growing vigorously. Controls cocklebur, galinsoga, lambsquarters, mustards, pigweed, ragweed, sunflower.
Biennial and perennial broadleaf weeds	2⅔-4 pints	Spray when weeds are actively growing and near the bud stage. Repeat applications may be needed for complete control. Controls bindweed, Canada thistle, chicory, dandelion, dock, musk thistle, plantain, smartweed, tansy mustard, wild garlic, wild onion.
Brush	2⅔-4 pints per 100 gal water	Spray to thoroughly wet plants when they are in full leaf and growing actively. Where practical cut tall woody plants and spray sucker growth when 2 to 4 feet tall. Retreatment may be needed for some species. Controls Cherokee rose, Japanese honeysuckle, Virginia creeper, wild grape, willow.
Aerial brush spraying	1⅓-2⅔ pints in 1 gallon diesel oil 2 to 4 gallons water	Aerial spraying is a specialized job. Secure qualified technical guidance and employ a competent reliable applicator. Become familiar with state laws governing the use of herbicides. Treat in spring when brush is fully leaved and growing actively. Controls big sagebrush, buckbrush, rabbitbrush, sandsage, shinnery oak.

Do not graze dairy animals on treated areas within 7 days after treatment. Do not plant treated fallow land until 3 months after treatment, or until chemical has disappeared from soil. Do not apply when grass is in the early boot to milk stage where grass seed production is desired. Do not apply to newly seeded areas or on alfalfa, clover, bent, or susceptible grass pastures as injury may result.

TO PREPARE A SPRAY: Add one-half the required amount of water to the spray tank, then add this product with agitation and finally the balance of water with continued agitation. This product forms an emulsion in water, not a solution.

Local conditions may affect the use of herbicides. Consult your State Agricultural Experiment Station, Farm Advisor, or Extension Weed Specialist for advice in selecting treatment from this label to best fit local conditions. Be sure that use of this product conforms to all applicable laws, rules and regulations. Certain states have restrictions pertaining to application distances from susceptible crops. The applicator should become familiar with these laws, rules or regulations and follow them exactly.

PRECAUTIONARY STATEMENTS

CAUTION

HAZARD TO HUMANS AND DOMESTIC ANIMALS

Harmful if swallowed or absorbed through skin. Do not get in eyes, on skin or on clothing. Avoid breathing vapors or spray mist. Wash thoroughly before eating or smoking. Remove and wash contaminated clothing before reuse.

STATEMENT OF PRACTICAL TREATMENT

In case of contact wash skin with soap and water; for eyes flush with water for at least 15 minutes and get medical attention. If swallowed, drink 1 or 2 glasses of water and induce vomiting by finger in back of throat. Never induce vomiting or give anything by mouth to an unconscious person. Get medical attention immediately.

ENVIRONMENTAL HAZARDS

This product is toxic to fish. Do not apply directly to water. Do not contaminate water by cleaning of equipment or disposal of wastes.

USE PRECAUTIONS

Do not apply when weather conditions favor drift from treated areas.

Do not use the same spray equipment for other purposes unless thoroughly cleaned.

Do not use in or near a greenhouse.

Do not contaminate water used for irrigation or domestic purposes.

Do not apply WEEDONE LV6 Emulsifiable Broadleaf Herbicide directly to, or permit spray mist to drift onto cotton, okra, grapes, tomatoes, fruit trees, vegetables, flowers or other desirable crop or ornamental plants which are susceptible to 2,4-D herbicide. Do not apply near susceptible plants since very small quantities of the 2,4-D will cause severe injury during the growing or dormant periods. Crops contacted by WEEDONE LV6 Emulsifiable Broadleaf Herbicide sprays or spray drift may be killed or suffer significant stand loss with extensive quality and yield reduction.

Do not apply when a temperature air inversion exists. Such a condition is characterized by little or no air movement and an increase in air temperature with an increase in height. In humid regions a fog or mist may form. An inversion may be detected by producing a smoke column and checking for a layering effect. If questions exist pertaining to the existence of an inversion, consult with local weather services before making an application.

Use coarse sprays to minimize drift. Do not apply with hollow cone type insecticide or other nozzles that produce fine spray droplets. Drift from **serial** or **ground application** may be reduced by: (1) applying as near to the target as possible in order to obtain coverage; (2) by increasing the volume of spray mix per acre; (3) by decreasing the pounds of pressure at the nozzle tips; (4) by using nozzles which produce a coarse spray pattern; and (5) by not applying when wind is blowing toward susceptible valuable plants.

At high air or ground surface temperatures, vapors from this product may injure susceptible plants.

STORAGE AND DISPOSAL

STORAGE

Do not store near food, feedstuff, fertilizers, seeds, insecticides, fungicides or other pesticides.

PESTICIDE DISPOSAL

Do not contaminate water, food, or feed by storage or disposal.

Pesticide wastes are toxic. Improper disposal of excess pesticide, spray mixture, or rinsate is a violation of Federal Law. If these wastes cannot be disposed of by use according to label instructions, contact your State Pesticide or Environmental Control Agency, or the Hazardous Waste representative at the nearest EPA Regional Office for guidance.

CONTAINER DISPOSAL

Triple rinse (or equivalent). Then offer for recycling, or reconditioning, or puncture and dispose of in a sanitary landfill, or incineration, or, if allowed by state and local authorities, by burning. If burned, stay out of smoke.

LIMITED WARRANTY AND DISCLAIMER

The manufacturer warrants that this product conforms to the chemical description on the label; that this product is reasonably fit for the purposes set forth in the directions for use when it is used in accordance with such directions; and that the directions, warnings and other statements on this label are based upon responsible experts' evaluation of reasonable tests of effectiveness, of toxicity to laboratory animals and to plants, and of residues on food crops, and upon reports of field experience. Tests have not been made on all varieties or in all states or under all conditions. THE MANUFACTURER NEITHER MAKES NOR INTENDS TO MAKE, ANY OTHER WARRANTIES, EXPRESS OR IMPLIED, AND IT EXPRESSLY EXCLUDES AND DISCLAIMS ALL IMPLIED WARRANTIES OF MERCHANTABILITY OR FITNESS FOR A PARTICULAR PURPOSE.

THIS WARRANTY DOES NOT EXTEND TO, AND THE BUYER SHALL BE SOLELY RESPONSIBLE FOR, ANY AND ALL LOSS OR DAMAGE WHICH RESULTS FROM THE USE OF THIS PRODUCT IN ANY MANNER WHICH IS INCONSISTENT WITH THE LABEL DIRECTIONS, WARNINGS OR CAUTIONS.

BUYER'S EXCLUSIVE REMEDY AND MANUFACTURER'S OR SELLER'S EXCLUSIVE LIABILITY FOR ANY AND ALL CLAIMS, LOSSES, DAMAGES, OR INJURIES RESULTING FROM THE USE OR HANDLING OF THIS PRODUCT, WHETHER OR NOT BASED IN CONTRACT, NEGLIGENCE, STRICT LIABILITY IN TORT OR OTHERWISE, SHALL BE LIMITED, AT THE MANUFACTURER'S OPTION, TO REPLACEMENT OF, OR THE REPAYMENT OF THE PURCHASE PRICE FOR, THE QUANTITY OF PRODUCT WITH RESPECT TO WHICH DAMAGES ARE CLAIMED. IN NO EVENT SHALL MANUFACTURER OR SELLER BE LIABLE FOR SPECIAL, INDIRECT OR CONSEQUENTIAL DAMAGES RESULTING FROM THE USE OR HANDLING OF THIS PRODUCT.

THIS SPECIMEN LABEL IS INTENDED FOR USE ONLY AS A GUIDE IN PROVIDING GENERAL INFORMATION REGARDING THE DIRECTIONS, WARNINGS AND CAUTIONS ASSOCIATED WITH THE USE OF THIS PRODUCT. AS WITH ANY AGRICULTURAL CHEMICAL, ALWAYS FOLLOW THE LABEL INSTRUCTIONS ON THE PACKAGE BEFORE USING.

WEEDONE® MCPA ESTER

BROADLEAF HERBICIDE

Controls Broadleaf weeds in small grains.

ACTIVE INGREDIENT:
2-Butoxyethyl ester of
2-methyl-4-chlorophenoxyacetic acid* 66.7%
INERT INGREDIENTS: 33.3%

TOTAL 100.0%

This product contains Petroleum Distillate.
*2-Methyl-4-chlorophenoxyacetic acid equivalent 44.5% by weight or 4.0 pounds per gallon.

E.P.A. Reg. No. 264-240 E.P.A. Est. No. 264-MO-01

Keep Out of Reach of Children

WARNING

IN CASE OF EMERGENCY, TELEPHONE (24 HOURS A DAY) IN U.S.A. 1-800-UCC-HELP.

FOR GENERAL PRODUCT INFORMATION CALL 1-800-334-9745.

DIRECTIONS FOR USE

IT IS A VIOLATION OF FEDERAL LAW TO USE THIS PRODUCT IN A MANNER INCONSISTENT WITH ITS LABELING.

Use the lower rate where susceptible weeds such as lambsquarters, mustards, ragweed, stinkweed, wild radish, yellow rocket, purslane, peppergrass, shepherdspurse, hempnettle and annual sowthistle are a problem. Use the higher rate where less susceptible weeds such as dandelion, plantain, pigweed, pennycress, kochia and perennial sowthistle are a problem.

CROP	AMOUNT WEEDONE® MCPA Ester Per Acre	DIRECTIONS
SMALL GRAINS (Wheat, Barley, Rye) Spring Planted (not seeded to legumes)	½ to 1½ pints	Treat when grain is fully tillered and 8 to 10 inches tall. Do not spray grain in the boot to dough stage. Use at least 8 to 10 gallons of water per acre for ground application and at least 3 to 5 gallons of water per acre for aerial application.
Fall Planted (not seeded to legumes) Fall application (From 3 to 4 leaf stage to the boot stage)	½ to 1 pint	The safest time to treat fall planted small grains is when they are fully tillered and 8 to 10 inches tall. Infestations of broadleaf weeds germinating in fall and winter may, however, justify treatment at an earlier stage. Fall planted wheat, barley, or rye can be sprayed with ½ to 1 pint of WEEDONE® MCPA Ester Broadleaf Herbicide per acre from the 3 to 4 leaf stage to the boot stage. Use at least 8 to 10 gallons of water per acre for ground application and at least 3 to 5 gallons of water per acre for aerial application.
Spring application (After tillering but prior to jointed stage)	½ to 1½ pint	Apply in early spring when annual broadleaf weeds are small and grain is fully tillered but before grain is in the jointed stage. Use at least 8 to 10 gallons of water per acre for ground applications and at least 3 to 5 gallons of water per acre for aerial applications.
DO NOT FORAGE OR GRAZE MEAT ANIMALS ON TREATED AREAS WITHIN SEVEN DAYS OF SLAUGHTER.		
FLAX	¼ to ⅜ pint	Where susceptible weeds such as mustard are the main problem.
	½ pint	Where weeds such as pigweed, lambsquarters, stinkweed, ragweed, cocklebur are the problem. Apply when flax is 2 to 8 inches tall but before bud stage. Do not spray after flax is in the bud stage or severe damage may result. Use at least 8 to 10 gallons of water per acre for ground applications and at least 3 to 5 gallons of water per acre for aerial application.

CROP	AMOUNT WEEDONE® MCPA Ester Per Acre	DIRECTIONS
OATS	½ to 1 pint	Spray after oats are fully tillered (4 to 6 inches tall) but before they are in the boot to dough stage. Apply the lower rate where annual weeds are the problem the higher rate for perennial weeds. Use at least 8 to 10 gallons of water per acre for ground application and at least 3 to 5 gallons of water per acre for aerial application.
RICE	1½ to 2½ pints	Treat 7 to 10 weeks after planting when rice is fully tillered and 6 to 8 inches above water. Do not spray after rice is in the boot stage. Use the higher rate where curly indigo, sedge, or bulrush are the major problem. Use lower rate for arrowhead, water plantain and redstem. Use at least 8 to 10 gallons of water per acre for ground application and at least 3 to 5 gallons of water per acre for aerial applications. Do not spray when temperatures are over 90° F.

DO NOT GROW CRAYFISH OR CATFISH IN TREATED RICE FIELDS.

DO NOT FORAGE OR GRAZE MEAT ANIMALS ON TREATED AREAS WITHIN SEVEN DAYS OF SLAUGHTER.

TANK MIXTURES OF WEEDONE® MCPA Ester Broadleaf Herbicide and BROMINAL® Bromoxynil Broadleaf Herbicide or ME4® BROMINAL® Bromoxynil Broadleaf Herbicide

When using WEEDONE® MCPA Ester Broadleaf Herbicide in combination with BROMINAL® or ME4® BROMINAL® Bromoxynil Broadleaf Herbicides, use the higher rates and volumes for larger weeds or where weed growth is slow due to dry or cold weather.

Read and follow all restrictions, precautions and use limitations on this label and the labels for BROMINAL® and ME4® BROMINAL®.

CROP	AMOUNT OF HERBICIDE PER ACRE	DIRECTIONS
SMALL GRAINS (Wheat, Barley, Oats and Rye) Washington, Oregon and Idaho	½ to 1 pint WEEDONE MCPA Ester + .375 to .5 lbs active* bromoxynil	Apply after grain is fully tillered and before the forming of joints in the stems. Use 10 to 20 gallons of water per acre for ground application and 4 to 10 gallons of water per acre for aerial application. Use higher volume on larger weeds. (Follow timing and use directions above.)
Cereal Areas except Washington, Oregon & Idaho	½ to 1 pint WEEDONE MCPA Ester + .25 to .375 lbs active* bromoxynil	

DO NOT FORAGE OR GRAZE MEAT ANIMALS ON TREATED AREAS WITHIN SEVEN DAYS OF SLAUGHTER.

* BROMINAL® contains 2 lbs active bromoxynil per gallon.

ME4® BROMINAL® contains 4 lbs active bromoxynil per gallon.

To Convert local recommendation into terms of WEEDONE® MCPA Ester Broadleaf Herbicide use the following table:

MCPA acid	1¼ lb.	1 lb.	¾ lb.	½ lb.	⅜ lb.	¼ lb.
WEEDONE® MCPA ESTER	2½ pt.	2 pt.	1½ pt.	1 pt.	¾ pt.	½ pt.

IMPORTANT

CLOVERS AND OTHER LEGUMES MAY BE INJURED OR KILLED BY THESE RATES.

PRECAUTIONARY STATEMENTS

WARNING

HAZARDS TO HUMANS AND DOMESTIC ANIMALS

Do not get in eyes, on skin or on clothing. May be fatal if absorbed through skin. Harmful if swallowed. Avoid breathing vapors or spray mist. Remove and wash contaminated clothing before reuse. Wash thoroughly before eating or smoking.

STATEMENT OF PRACTICAL TREATMENT

In case of contact with skin wash with soap and water. Flush eyes with water for at least 15 minutes. Get medical attention.

If swallowed, drink large quantities of water. Do not induce vomiting. Never give anything by mouth to an unconscious person. Get medical attention.

ENVIRONMENTAL HAZARDS

Do not apply directly to water except as specified on this label. Do not contaminate water by cleaning of equipment or disposal of wastes. This product is toxic to fish.

Avoid spray drift to susceptible plants such as cotton, beans, tomatoes, and ornamentals. (Coarse sprays are less likely to drift). Do not apply when weather conditions favor drift from treated areas. Avoid use of small diameter spray nozzles. At high air or ground surface temperatures, vapors from this product may injure susceptible plants.

Do not contaminate water used for irrigation or domestic purposes.

Do not use the same spray equipment for other purposes unless thoroughly cleaned.

Do not use in or near a greenhouse.

STORAGE AND DISPOSAL

Do not store near food, feedstuffs, fertilizers, seeds, insecticides or fungicides.

PESTICIDE DISPOSAL

Do not contaminate water, food or feed by storage or disposal. Open dumping is prohibited.

Wastes resulting from the use of this product may be disposed of on site or at an approved waste disposal facility.

CONTAINER DISPOSAL

Triple rinse (or equivalent). Then offer for recycling or reconditioning, or puncture and dispose of in a sanitary landfill, or by incineration or if allowed by state and local authorities by burning. If burned, stay out of smoke.

LIMITED WARRANTY AND DISCLAIMER

The manufacturer warrants that this product conforms to the chemical description on the label, that this product is reasonably fit for the purposes set forth in the directions for use when it is used in accordance with such directions, and that the directions, warnings and other statements on this label are based upon responsible experts' evaluation of reasonable tests of effectiveness of toxicity to laboratory animals and to plants, and of residues on food crops, and upon reports of field experience. Tests have not been made on all varieties or in all states or under all conditions. THE MANUFACTURER NEITHER MAKES NOR INTENDS, NOR DOES IT AUTHORIZE ANY AGENT OR REPRESENTATIVE TO MAKE ANY OTHER WARRANTIES, EXPRESS OR IMPLIED, AND IT EXPRESSLY EXCLUDES AND DISCLAIMS ALL IMPLIED WARRANTIES OF MERCHANTABILITY OR FITNESS FOR A PARTICULAR PURPOSE.

THIS WARRANTY DOES NOT EXTEND TO, AND THE BUYER SHALL BE SOLELY RESPONSIBLE FOR, ANY AND ALL LOSS OR DAMAGE WHICH RESULTS FROM THE USE OF THIS PRODUCT IN ANY MANNER WHICH IS INCONSISTENT WITH THE LABEL DIRECTIONS, WARNINGS OR CAUTIONS.

BUYER'S EXCLUSIVE REMEDY AND MANUFACTURER'S OR SELLER'S EXCLUSIVE LIABILITY FOR ANY AND ALL CLAIMS, LOSSES, DAMAGES, OR INJURIES RESULTING FROM THE USE OR HANDLING OF THIS PRODUCT, WHETHER OR NOT BASED IN CONTRACT, NEGLIGENCE, STRICT LIABILITY IN TORT OR OTHERWISE, SHALL BE LIMITED, AT THE MANUFACTURER'S OPTION, TO THE REPLACEMENT OF OR THE REPAYMENT OF THE PURCHASE PRICE FOR, THE QUANTITY OF PRODUCT WITH RESPECT TO WHICH DAMAGES ARE CLAIMED. IN NO EVENT SHALL MANUFACTURER OR SELLER BE LIABLE FOR SPECIAL, INDIRECT OR CONSEQUENTIAL DAMAGES RESULTING FROM THE USE OR HANDLING OF THIS PRODUCT.

THIS SPECIMEN LABEL IS INTENDED FOR USE ONLY AS A GUIDE IN PROVIDING GENERAL INFORMATION REGARDING THE DIRECTIONS, WARNINGS AND CAUTIONS ASSOCIATED WITH THE USE OF THIS PRODUCT. AS WITH ANY AGRICULTURAL CHEMICAL, ALWAYS FOLLOW THE LABEL INSTRUCTIONS ON THE PACKAGE BEFORE USING.

WEEDONE® 170

WOODY PLANT HERBICIDE

Contains the butoxyethanol ester of 2,4-DP and 2,4-D. For controlling mixed brush on utility rights-of-way, along highways and solid stands of oak or elm.

ACTIVE INGREDIENTS:

*2,4-Dichlorophenoxypropionic acid,
butoxyethanol ester 29.3%

**2,4-Dichlorophenoxyacetic acid,
butoxyethanol ester*** 29.9%

INERT INGREDIENTS: 40.8%

TOTAL 100.0%

*2,4-Dichlorophenoxypropionic acid equivalent 20.6% by weight or 1.85 pounds per gallon.

**2,4-Dichlorophenoxyacetic acid equivalent 20.6% by weight or 1.85 pounds per gallon.

Isomer specific by AOAC method No. 6.D01-5

EPA Reg. No. 264-222 ZB EPA Est. No. 264-MO-01

Keep Out of Reach of Children

CAUTION

IN CASE OF EMERGENCY PHONE (24 HOURS A DAY) IN U.S.A. 1-800-UCC-HELP

FOR GENERAL PRODUCT INFORMATION CALL 1-800-334-9745.

GENERAL INFORMATION

WEEDONE 170 Woody Plant Herbicide is a concentrated 50-50 combination of 2,4-D and 2,4-DP as the low-volatile butoxyethanol ester.

WEEDONE 170 Woody Plant Herbicide is specifically designed for utilities and other industrial users to control woody plants.

WEEDONE 170 Woody Plant Herbicide mixes with oil or kerosene and emulsifies in water.

WEEDONE 170 Woody Plant Herbicide controls many noxious perennial weeds on uncropped land, such as along highways and drainage ditchbanks and fence rows.

WEEDONE 170 Woody Plant Herbicide is effective for controlling:

alder	currant	red elm
aspen	fir	red maple
bigleaf maple	gooseberry	salmonberry
birch	hemlock	sand shinnery oak
blackberry	honeysuckle	serviceberry
black cherry	manzanita	snowberry
blackjack oak	multiflora rose	spruce
buckbrush	oaks	sycamore
black locust	pine	tulip poplar
ceanothus	poison ivy	willow
chamise	poison oak	winged elm
coffeeberry	poplar	yerba santa
		red elder

(and many other species)

WEEDONE 170 Woody Plant Herbicide in combination with TRANSVERT is not registered for use in California.

TO PREPARE A SPRAY: Add one-half the required amount of oil or water to the spray tank, then add this product with agitation and finally the balance of the water or oil with continued agitation.

If this material is to be used in straight oil mixtures, do not let water get into it or the finished mixture.

This material forms an emulsion in water—not a solution. This tends to separate on standing. Provide agitation to prevent such separation and ensure a uniform mixture.

DIRECTIONS FOR USE

IT IS A VIOLATION OF FEDERAL LAW TO USE THIS PRODUCT IN A MANNER INCONSISTENT WITH ITS LABELING.

FOLIAGE STEM METHOD

This is the standard method for high volume sprays along fencerows, highways and rights-of-way. Use it as a "first spray" on thick brush composed of mixed species.

Apply spray to both foliage and stems of all plants from the time foliage is fully developed until plants begin to go dormant. *For effective control, all leaves, stems and suckers must be thoroughly wet to ground line.* Some regrowth may be expected on resistant species, such as ash, maple, oak and persimmon.

Add 1 to 1½ gallons of WEEDONE® 170 Woody Plant Herbicide to 100 gallons of water. Mix thoroughly before spraying. Use 200 to 600 gallons of spray mixture per acre depending on height and density of brush.

SPOT TREATMENT

For spot spraying with backpack sprayers mix 1 cup (8 oz.) WEEDONE® 170 Woody Plant Herbicide with 5 gallons of water. Wet brush stems and foliage thoroughly.

BASAL BARK METHOD

Thoroughly wet the base and root collar of all stems until the spray collects around the root collar at ground line. This spray may be applied in any season. Apply WEEDONE 170 Woody Plant Herbicide with the basal bark method on scattered brush or as a "second spray" on species resistant to a first foliage spray.

Mix 3 to 4 gallons of WEEDONE 170 Woody Plant Herbicide in 100 gallons of oil. Apply with low volume knapsack sprayer or power equipment. Concentration will depend on species present, volume of spray used, and season applied. Use a coarse spray to avoid drift.

MODIFIED BASAL METHOD

Drench the base of plants, then wet the lower ⅘ of remaining stems and leaves to run-off. Treat when brush is in full foliage. This method may be used when susceptible species have been controlled by prior sprays and resistant species such as maple and oak remain. *Drenching the base of the plant and wetting all stems to run-off is essential for complete control.*

Early season spraying: Add 1 to 1½ gallons of WEEDONE 170 Woody Plant Herbicide to 10 gallons of diesel oil and mix thoroughly. Add this mixture to 89 gallons of water.

During dry weather or the latter part of spray season: Add 1½ gallons of WEEDONE 170 Woody Plant Herbicide to 15 gallons of diesel oil and mix thoroughly. Add this mixture to 83 ½ gallons of water. Agitate thoroughly before spraying, particularly if material is allowed to stand for more than one hour after mixing.

CUT SURFACE

Stump: This method can be used at any time of year, but it is more effective when applied as soon as possible after trees are cut. Spray the entire stump, particularly bark and exposed roots. *A thorough drenching is essential for complete control.*

Use this method after original or capital clearing. It is the first step toward a chemical brush control program on newly cleared highways and rights-of-way. Spray is most effective and economical on stumps 3 to 4 inches and larger.

Mix 3 to 4 gallons of WEEDONE 170 Woody Plant Herbicide in 100 gallons of oil. Apply with low volume knapsack sprayer using solid cone nozzle of medium orifice.

CONCENTRATE STUMP TREATMENT: For small (up to 3-inch diameter) stems, cut them as close to the ground as possible and apply undiluted WEEDONE® 170 Woody Plant Herbicide directly from the can to the surface of the freshly cut stump.

Frill: Make a frill by using an axe to cut overlapping notches in a continuous ring around the trunk near its base. Cut through the bark, but do not remove chips. This method is recommended for cull trees 5 to 6 inches in diameter and larger. Treat freshly cut frills at any time of the year. Mix 3 to 4 gallons of WEEDONE 170 Woody Plant Herbicide in 100 gallons of oil. Pour in as much of this mixture as the frill will hold without wasting chemical.

POWERED KNAPSACK BASAL SPRAY

Mix 8 to 14 gallons of WEEDONE 170 with fuel oil or kerosene to make 100 gallons of spray solution. Apply with a portable knapsack mistblower to lower brush stems. Apply spray to all sides of stems; good root collar coverage is essential. Run knapsack mistblower at ¼ to ⅓ throttle for best spray delivery and coverage. For maximum drift control use a basal nozzle attachment. Do not raise spray nozzle above horizontal position.

FENCE ROW APPLICATION

To control mixed brush, perennial and annual broadleaf weeds use one of the application methods such as the foliage stem method described on this label. Some regrowth may be expected on resistant species such as ash, maple, oak and persimmon.

RAILROAD ON-TRACK APPLICATION

Weedone 170 Woody Plant Herbicide in combination with TRANSVERT is not registered for use in California.

Use 1 gallon of WEEDONE 170 Woody Plant Herbicide and 1 quart of TRANSVERT Herbicide Additive in each 100 gallons of water and apply 100 to 300 gallons per acre to control mixed brush growing along railroad rights-of-way and under railroad communication lines. Apply from spray trains or Hi-Rail equipment only. Avoid contact with skin, eyes, or clothing.

LOW VOLUME STEM FOLIAGE SPRAY—AERIAL OR GROUND

Weedone® 170 Woody Plant Herbicide in combination with TRANSVERT® is not registered for use in California.

Mixed Brush—Utility Rights-of-Way: Use 3 gallons of WEEDONE® 170 and ½ gallon TRANSVERT® mixed in water to make 12 to 15 gallons total volume per acre to control mixed brush growing on utility rights-of-way. Where red maple, conifers, ash, hickory or oak are the major problems, use 3 gallons of WEEDONE 170 and ⅔ gallon TRANSVERT in 15 gallons total volume per acre.

Apply the spray only through equipment designed to provide effective drift control. A helicopter mounted MICROFOIL® applicator or other equipment that provides equivalent drift control is recommended for aerial sprays. A DIRECTA-SPRA® or mini-wobbler is recommended for ground sprays.

Solid Stands of Oak or Elm: Apply with fixed-wing or helicopter aircraft. Apply in the spring after hardwoods have just developed full sized leaves. The spray season normally runs from early May to mid-June in Texas and California, and from early May to early July in Oklahoma and northward. Spray when wind velocities are less than 5 miles per hour.

Brush	WEEDONE 170	Oil	Add water to Make Total
post, blackjack oak and winged elm	⅔ gallon	1 gallon	5 gallon
sand shinnery oak	½-1 quart	1 gallon	3 gallon

If necessary, repeat spray the second year with 1 quart WEEDONE 170 Woody Plant Herbicide per acre. For maximum control, use higher rate and repeat spray the second year using 1 quart WEEDONE 170 Woody Plant Herbicide per acre.

For aerial application to solid stands of brush in California use 2 to 4 quarts in 3 to 12 gallons volume per acre. Two to 4 quarts of fuel oil may be included in this mixture.

FORESTRY SITE PREPARATION—To reduce competition from mixed trees and brush before planting forest trees. Apply one gallon WEEDONE® 170 for easy to control species such as aspen, hazel, alder, sassafrass, sumac, black locust willow, and other similar species. Apply 2 gallons WEEDONE® 170 to control more difficult species such as oaks, sweet gum, black gum, hickory, and tulip-poplar.

To control annual grasses and provide additional foliage brownout for prescribed foliage burning, add one quart of TRANSVERT® per acre to the above spray mixtures.

Mix with water to make 10 gallons total solution when applying as an aerial spray or 20 gallons total solution when applying as a ground spray. Do not apply to established plantations as this spray mix will injure planted conifers.

PINE RELEASE—To control hardwood brush and release Northern conifers such as red pine, Jack pine, white pine and white spruce.

Apply 2-4 qts. WEEDONE® 170 as a broadcast spray in 10 gallons of water per acre. Make applications in midsummer after pine height growth is complete and the conifer buds are set. WEEDONE® 170 will control aspen, birch, cherry, alder, hazel, oaks and similar species. WEEDONE® 170 will not, at these rates, provide satisfactory control of red maple, sugar maple or ash.

AERIAL OR GROUND APPLICATION TO CONTROL MIXED BRUSH INCLUDING CONIFERS AND ROOT SUCKERING SPECIES ON UTILITY, RAILROAD AND HIGHWAY RIGHTS-OF-WAY, FENCEROWS AND OTHER NON CROP AREAS.

WEEDONE® 170 AND TORDON* MIXTURE

Where red maple, conifers and root-suckering species such as sassafras, sumac, black locust or persimmon are the major problem, use 1 to 2½ gallons of WEEDONE 170 Woody Plant Herbicide in combination with 1 to 2½ gallons of Tordon* 101 herbicide mixture or in combination with 3 to 6 pints of Tordon* K herbicide. Dilute with water and apply in a total spray volume of 15 to 50 gallons per acre. Use lower rates to control brush on sandy soils or coarse soils with susceptible species such as seedling aspen, cherry, willow and sumac. Use higher rates to control brush on clay soils or fine soils or rocky terrain with more resistant species such as maple, oak, pine and red cedar. When making the spray mixture, add WEEDONE® 170 to the required amount of water while agitating the mixture in the spray tank. Then add the TORDON* 101 or TORDON* K while continuing agitation. (Do not mix concentrates of WEEDONE® 170 with concentrates of TORDON* 101 or TORDON* K .) Aerial applications of the tank mixture should be made only with a helicopter mounted MICROFOIL® applicator or an equipment system providing equivalent drift control. Do not use these tank mixtures on drainage ditchbanks or for site preparation in forest plantations. Observe all restrictions, precautions and limitations on the labeling of each product used in tank mixtures.

* Trademark of the Dow Chemical Company.

WEEDON® 170 and Banvel Herbicide* Mixtures

HIGH VOLUME STEM FOLIAGE SPRAY

Mix 1 to 1½ gallons WEEDON® 170 with 1 to 2 qts. Banvel 4WS per 100 gallons total spray mix. Use the low rate for easy to kill brush such as alder, aspen, cherry, sycamore, tulip-poplar, and willows. Use the high rate to control species such as oak, ash, elm, pines, spruce, and fir.

LOW VOLUME STEM FOLIAGE SPRAY—Aerial or Ground

Mix 2-3 gallons of WEEDONE® 170 with 1-1½ gallons of Banvel herbicide in 15-50 gallons of water per acre to control mixed brush including conifers and ash. Observe all restrictions, precautions and limitations on the labeling of each product used in tank mixtures.

* Trademark of the Velsicol Chemical Co.

PRECAUTIONARY STATEMENTS

CAUTION

HAZARD TO HUMANS AND DOMESTIC ANIMALS

Harmful if swallowed or absorbed through skin. Do not get in eyes, on skin, or clothing. Avoid breathing vapors or spray mist. Wash thoroughly before eating or smoking. Remove and wash contaminated clothing before reuse.

STATEMENT OF PRACTICAL TREATMENT

In case of contact wash skin with soap and water; for eyes flush with water for at least 15 minutes and get medical attention. Never give anything by mouth to an unconscious person. Get medical attention.

ENVIRONMENTAL HAZARDS

This product is toxic to fish. Do not apply directly to water. Do not contaminate water by cleaning of equipment or disposal of wastes.

Do not apply when weather conditions favor drift from treated areas.

Do not graze dairy animals on treated areas within 6 weeks after application.

Do not graze meat animals on treated areas within 2 weeks of slaughter.

Do not use the same spray equipment for other purposes unless thoroughly cleaned. Do not contaminate water used for domestic or irrigation purposes. Do not use in or near a greenhouse.

Do not apply WEEDONE® 170 Woody Plant Herbicide directly to, or permit mist to drift onto cotton, okra, grapes, tomatoes, fruit trees, vegetables, flowers or other desirable crop or ornamental plants which are susceptible to 2,4-D or 2,4-DP herbicides. Do not apply near susceptible plants since very small quantities of the 2,4-D or 2,4-DP will cause severe injury during the growing or dormant periods. Crops contacted by WEEDONE® 170 Woody Plant Herbicide sprays or spray drift may be killed or suffer significant stand loss with extensive quality and yield reduction.

Do not apply when a temperature air inversion exists. Such a condition is characterized by little or no air movement and an increase in air temperature with an increase in height. In humid regions a fog or mist may form. An inversion may be detected by producing a smoke column and checking for a layering effect. If questions exist pertaining to the existence of an inversion consult with local weather services before making an application.

Use coarse sprays to minimize drift. Do not apply with hollow cone type insecticide or other nozzles that produce fine spray droplets.

Drift from **aerial or ground application** may be reduced by: (1) applying as near to the target as possible in order to obtain coverage; (2) by increasing the volume of spray mix per acre; (3) by decreasing the pounds of pressure at the nozzle tips; (4) by using nozzles which produce a coarse spray pattern; (5) by not applying when wind is blowing toward susceptible crops or valuable plants.

At high air or ground surface temperatures, vapors from this product may injure susceptible plants.

STORAGE AND DISPOSAL

STORAGE

Do not contaminate or store near food, feedstuff, fertilizers, seeds, insecticides, fungicides or other pesticides.

PESTICIDE DISPOSAL

Do not contaminate water, food, or feed by storage or disposal. Open dumping is prohibited. Pesticide wastes are toxic. Improper disposal of excess pesticide, spray mixture, or rinsate is a violation of Federal Law. If these wastes cannot be disposed of by use according to label instructions, contact your State Pesticide or Environmental Control Agency, of the Hazardous Waste representative at the nearest EPA Regional Office for guidance.

CONTAINER DISPOSAL

Triple rinse or (equivalent). Then offer for recycling, or reconditioning, or puncture and dispose of in a sanitary landfill, or by other procedures approved by state and local authorities.

Local conditions may affect the use of herbicides. Consult your State Agricultural Experiment Station, Farm Advisors, or Extension Weed Specialists for advice in selecting treatment from this label to best fit local conditions.

Be sure that use of this product conforms to all applicable laws, rules and regulations. Certain states have restrictions pertaining to application distances from susceptible crops. The applicator should become familiar with these laws, rules or regulations and follow them exactly. Apply this product only as specified on this label.

LIMITED WARRANTY AND DISCLAIMER

The manufacturer warrants (a) that this product conforms to the chemical description on the label; (b) that this product is reasonably fit for the purposes set forth in the directions for use when it is used in accordance with such direction; and (c) that the directions, warnings and other statements on this label are based upon responsible experts' evaluation of reasonable tests of effectiveness, of toxicity to laboratory animals and to plants, and of residues, and upon reports of field experience. Tests have not been made on all varieties or in all states or under all conditions. THE MANUFACTURER NEITHER MAKES NOR INTENDS, NOR DOES IT AUTHORIZE ANY AGENT OR REPRESENTATIVE TO MAKE, ANY OTHER WARRANTIES, EXPRESS OR IMPLIED, AND IT EXPRESSLY EXCLUDES AND DISCLAIMS ALL IMPLIED WARRANTIES OF MERCHANTABILITY OR FITNESS FOR A PARTICULAR PURPOSE. THIS WARRANTY DOES NOT EXTEND TO, AND THE BUYER SHALL BE SOLELY RESPONSIBLE FOR, ANY AND ALL LOSS OR DAMAGE WHICH RESULTS FROM THE USE OF THIS PRODUCT IN ANY MANNER WHICH IS INCONSISTENT WITH THE LABEL DIRECTIONS, WARNINGS OR CAUTIONS.

BUYER'S EXCLUSIVE REMEDY AND MANUFACTURER'S OR SELLER'S EXCLUSIVE LIABILITY FOR ANY AND ALL CLAIMS, LOSSES, DAMAGES, OR INJURIES RESULTING FROM THE USE OR HANDLING OF THIS PRODUCT, WHETHER OR NOT SUCH LIABILITY IS BASED IN CONTRACT, NEGLIGENCE, STRICT LIABILITY IN TORT OR OTHERWISE, SHALL BE LIMITED, AT THE MANUFACTURER'S OPTION, TO REPLACEMENT OF, OR THE REPAYMENT OF THE PURCHASE PRICE FOR, THE QUANTITY OF PRODUCT WITH RESPECT TO WHICH DAMAGES ARE CLAIMED. IN NO EVENT SHALL MANUFACTURER OR SELLER BE LIABLE FOR SPECIAL, INDIRECT OR CONSEQUENTIAL DAMAGES RESULTING FROM THE USE OR HANDLING OF THIS PRODUCT.

WEEDONE, TRANSVERT, DIRECTA-SPRA, MICROFOIL are registered trademarks of Union Carbide Agricultural Products Company, Inc.

THIS SPECIMEN LABEL IS INTENDED FOR USE ONLY AS A GUIDE IN PROVIDING GENERAL INFORMATION REGARDING THE DIRECTIONS, WARNINGS AND CAUTIONS ASSOCIATED WITH THE USE OF THIS PRODUCT. AS WITH ANY AGRICULTURAL CHEMICAL, ALWAYS FOLLOW THE LABEL INSTRUCTIONS ON THE PACKAGE BEFORE USING.

WEEDONE®638

BROADLEAF HERBICIDE

A special formulation containing 2,4-D acid.

ACTIVE INGREDIENTS:

*2,4-dichlorophenoxyacetic acid 13.8%

*2,4-dichlorophenoxyacetic acid, butoxyethyl ester 24.5%

INERT INGREDIENTS: 61.7%

TOTAL 100.0%

*Isomer specific by AOAC method No. 6.D01-5 Total 2,4-dichlorophenoxyacetic acid equivalent 30.8% by weight or 2.8 pounds per gallon.

E.P.A. Reg. No. 264-37AA E.P.A. Est. No. 264-MO-01

Keep Out of Reach of Children

DANGER—PELIGRO

PRECAUCION AL USARIO: Si usted no lee ingles, no use este producto hasta que la etiqueta le haya sido explicada ampliamente.

IN CASE OF EMERGENCY TELEPHONE (24 HOURS A DAY) IN U.S.A. 1-800-UCC-HELP.

FOR GENERAL PRODUCT INFORMATION CALL 1-800-334-9745.

DIRECTIONS

WEEDONE 638 Broadleaf Herbicide is a special formulation containing 2,4-D acid. WEEDONE 638 is more effective than 2,4-D amines for controlling hard-to-kill weeds such as field bindweed (morningglory), Russian knapweed, Canada thistle, leafy spurge, cattails, tules and nutgrass.

ROADSIDES, FENCEROWS, DRAINAGE DITCH BANKS

Spray when daytime temperatures are lower than 80 to 85 degrees. Spraying when temperatures are higher may cause rapid foliage burn and decrease the effectiveness of WEEDONE 638 Broadleaf Herbicide. Treat at the state of growth recommended below. Wet weeds thoroughly, applying about 150 to 300 gallons of spray mixture per acre, depending on the stand of weeds and spray equipment used. Where necessary, repeat applications should be made on active regrowth when it develops.

WEEDS	QUARTS WEEDONE 638 PER ACRE	WHEN TO SPRAY
Field bindweed (morningglory), hoary cress	2	Bud to full bloom
Leafy spurge, Russian knapweed	2	Full bloom
Canada thistle, camelthorn, lotus	2	Bud to early bloom
Cattails, tules	3	Late spring to early summer when 4 to 6 feet tall; repeat in fall and following spring.
Tansy ragwort	3	Rosette stage
Nutgrass	3	6 to 8 inches high, before bloom; repeat when regrowth is 6 to 8 inches high.

PASTURES, FALLOW AREAS, TURF

WEEDS	AMOUNTS PER ACRE QUARTS WEEDONE 638	GALLONS WATER	REMARKS
Field bindweed (morningglory)	1½	50	For boom spraying*
	1½	10	For aerial spraying*
			*Treat during early to full bloom.

Fallow land treated with WEEDONE 638 Broadleaf Herbicide in the spring or summer can be returned to crops in the fall or following spring. Do not graze livestock for seven days after spraying pastures with this product. Do not use on newly seeded turf until grass has been cut several times. WEEDONE 638 Broadleaf Herbicide may injure bentgrass, St. Augustinegrass, dichondra, carpetgrass, and clover lawns.

CROPPED AREAS

CROP	AMOUNT WEEDONE 638 PER ACRE	WHEN TO SPRAY
Wheat*, Barley		
Annual broadleaf weeds	1 pint	Apply after grains are fully tillered (4 to 6 inches tall) but before they are in the boot stage.
Perennial broadleaf weeds	2 pints	Apply when weeds are in the bud stage if possible, but do not spray after grains are in the boot to dough stage.
* Treat fall planted wheat in the spring after grains are fully tillered but before they are in the boot stage.		
Grain Sorghum		
Annual broadleaf weeds Perennial broadleaf weeds	1 pint $1\frac{1}{2}$ pints	Treat when sorghum is 6 to 10 inches tall.

Do not forage or graze treated grain fields within two weeks after treatment with this product. Do not feed treated straw to livestock.

Among the weeds controlled are:

Annuals: lambsquarters, pigweed, mustard, kochia and Russian thistle.

Perennials: bindweeds (morningglory), Canada thistle, blueweed, bur ragweed, alkali mallow.

Use low pressure sprayers (25 to 40 psi) and apply in a minimum of 5 to 10 gallons of water per acre.

GENERAL INFORMATION

WEED CONTROL IN CORN

One thorough shallow cultivation is recommended when corn is 10-12 inches high.

Apply spray when corn is 18 to 24 inches high.

Use drop pipes to direct spray away from corn leaves and onto weeds at the base of the corn.

Use 1 pint WEEDONE 638 Broadleaf Herbicide in 30 gallons of water for each acre sprayed.

2,4-D may cause brittleness to corn. Winds or cultivations may cause stalk breakage while brittle.

Certain single cross corn hybrids may be more susceptible to 2,4-D injury than others.

Combination with nitrogen solutions:

Use 14 fluid ounces ($\frac{7}{8}$ pint) WEEDONE 638 Broadleaf Herbicide per acre, mixed with the amount of nitrogen solution required to apply the pounds of nitrogen you want per acre.

Pour liquid nitrogen into the spray tank, then add WEEDONE 638 Broadleaf Herbicide. Mix thoroughly before spraying and maintain good bypass agitation while spraying.

MIXING SPRAYS

Add the required amount of water to the spray tank. Start agitation. While agitating, slowly add WEEDONE 638 Broadleaf Herbicide. Mix thoroughly. Whenever practical, maintain continuous agitation and use spray material as soon as possible after mixing.

PRECAUTIONARY STATEMENTS

DANGER

HAZARD TO HUMANS AND DOMESTIC ANIMALS

Harmful if swallowed. Avoid contact with skin, eyes or clothing. Causes eye damage.

STATEMENT OF PRACTICAL TREATMENT

In case of contact wash skin with soap and water; for eyes flush with water for at least 15 minutes and get medical attention.

GENERAL WORKER PROTECTION STATEMENTS

Do not apply this product in such a manner as to directly or through drift expose workers or other persons. The area being treated must be vacated by unprotected persons.

Do not enter treated areas without protective clothing until sprays have dried. Because certain states may require more restrictive reentry intervals for various crops treated with this product, consult your State Department of Agriculture for further information. Written or oral warnings must be given to workers who are expected to be in a treated area or in an area about to be treated with this product. Advise workers to stay out of fields during application and until sprays have dried. Regular long-sleeved work clothing should be worn when working in treated fields. See PRECAUTIONARY STATEMENTS and STATEMENT OF PRACTICAL TREATMENT for information on accidental exposures. When oral warnings are given, warnings shall be given in a language customarily understood by workers. Oral warnings must be given if there is reason to believe that written warnings cannot be understood by workers. Written warnings must include the following information: appropriate signal word (DANGER), area treated with WEEDONE® 638 Broadleaf Herbicide, date of application, appropriate protective clothing, and reentry interval (i.e. until sprays have dried).

ENVIRONMENTAL HAZARDS

This product is toxic to fish. Do not apply directly to water. Do not contaminate water by cleaning of equipment or disposal of wastes.

DIRECTIONS FOR USE

IT IS A VIOLATION OF FEDERAL LAW TO USE THIS PRODUCT IN A MANNER INCONSISTENT WITH ITS LABELING.

Do not contaminate water used for irrigation or domestic purposes. Do not apply where runoff is likely to occur. Do not apply when weather conditions favor drift from treated areas. Avoid spray drift to susceptible plants such as cotton, beans, peas, ornamentals, and most vegetables, as injury may occur. Coarse sprays are less likely to drift.

Under very high temperature, vapors from this product may injure susceptible plants in the immediate vicinity. Do not use in or near a greenhouse.

Do not use this product for purposes other than those recommended on this label.

STORAGE AND DISPOSAL

STORAGE

Do not store near fertilizers, seeds, insecticides, or fungicides.

PESTICIDE DISPOSAL

Do not contaminate water, food, or feed by storage or disposal. Open dumping is prohibited. Pesticide wastes are acutely hazardous. Improper disposal of excess pesticide, spray mixture or rinsate is a violation of Federal law. If these wastes cannot be disposed of by use according to label instructions, contact your State Pesticide or Environmental Control Agency or the Hazardous Waste representative at the nearest EPA Regional office for guidance.

CONTAINER DISPOSAL

Triple rinse (or equivalent). Then offer for recycling or reconditioning, or puncture and dispose of in a sanitary landfill, or incineration, or, if allowed by state and local authorities, by burning. If container is burned, stay out of smoke.

LIMITED WARRANTY AND DISCLAIMER

The manufacturer warrants (a) that this product conforms to the chemical description on the label; (b) that this product is reasonably fit for the purposes set forth in the directions for use when it is used in accordance with such directions; and (c) that the directions, warnings and other statements on this label are based upon responsible experts' evaluation of reasonable tests of effectiveness, of toxicity to laboratory animals and to plants, and of residues on food crops and upon reports of field experience. Tests have not been made on all varieties or in all states or under all conditions. THE MANUFACTURER NEITHER MAKES NOR INTENDS, NOR DOES IT AUTHORIZE ANY AGENT OR REPRESENTATIVE TO MAKE, ANY OTHER WARRANTIES, EXPRESS OR IMPLIED, AND IT EXPRESSLY EXCLUDES AND DISCLAIMS ALL IMPLIED WARRANTIES OF MERCHANTABILITY OR FITNESS FOR A PARTICULAR PURPOSE.

THIS WARRANTY DOES NOT EXTEND TO, AND THE BUYER SHALL BE SOLELY RESPONSIBLE FOR, ANY AND ALL LOSS OR DAMAGE WHICH RESULTS FROM THE USE OF THIS PRODUCT IN ANY MANNER WHICH IS INCONSISTENT WITH THE LABEL DIRECTIONS, WARNINGS OR CAUTIONS.

BUYER'S EXCLUSIVE REMEDY AND MANUFACTURER'S OR SELLER'S EXCLUSIVE LIABILITY FOR ANY AND ALL CLAIMS, LOSSES, DAMAGES, OR INJURIES RESULTING FROM THE USE OR HANDLING OF THIS PRODUCT, WHETHER OR NOT BASED IN CONTRACT, NEGLIGENCE, STRICT LIABILITY IN TORT OR OTHERWISE SHALL BE LIMITED, AT THE MANUFACTURER'S OPTION, TO REPLACEMENT OF, OR THE REPAYMENT OF THE PURCHASE PRICE FOR, THE QUANTITY OF PRODUCT WITH RESPECT TO WHICH DAMAGES ARE CLAIMED. IN NO EVENT SHALL MANUFACTURER OR SELLER BE LIABLE FOR SPECIAL, INDIRECT OR CONSEQUENTIAL DAMAGES RESULTING FROM THE USE OR HANDLING OF THIS PRODUCT.

THIS SPECIMEN LABEL IS INTENDED FOR USE ONLY AS A GUIDE IN PROVIDING GENERAL INFORMATION REGARDING THE DIRECTIONS, WARNINGS AND CAUTIONS ASSOCIATED WITH THE USE OF THIS PRODUCT. AS WITH ANY AGRICULTURAL CHEMICAL, ALWAYS FOLLOW THE LABEL INSTRUCTIONS ON THE PACKAGE BEFORE USING.

WEEDONE® CB

WOODY PLANT HERBICIDE

For controlling woody plants in forest management programs and on fencerows, roadsides, drainage ditchbanks, railroads, utility right-of-ways and other non-crop areas.

ACTIVE INGREDIENTS:

*2-(2,4-Dichlorophenoxy) propionic acid, butoxyethanol ester	11.7%
2,4-Dichlorophenoxyacetic acid, butoxyethanol ester*	11.9%
INERT INGREDIENTS:	76.4%
TOTAL	100.0%

*2-(2,4-Dichlorophenoxy) propionic acid equivalent 8.2% by weight or 0.66 pounds per gallon.***

2,4-Dichlorophenoxyacetic acid equivalent 8.2% by weight or 0.66 pounds per gallon.*

***Isomer specific by AOAC method No. 6.D01-5.

E.P.A. Reg. No. 264-393 E.P.A. Est. No. 264-MO-01

Keep Out of Reach of Children

WARNING

IN CASE OF EMERGENCY TELEPHONE (24 HOURS A DAY) IN U.S.A. 1-800-UCC-HELP.

FOR GENERAL PRODUCT INFORMATION CALL 1-800-334-9745.

GENERAL INFORMATION

WEEDONE® CB Woody Plant Herbicide is a ready-to-apply formulation for controlling unwanted woody plants by basal, stump or stem application. WEEDONE® CB Woody Plant Herbicide is applied directly to the woody plant stem; it is not necessary to cut or frill the bark. WEEDONE® CB Woody Plant Herbicide will not freeze under normal storage or application conditions, so it can be applied during the growth period or in the dormant season.

For most efficient treatment with WEEDONE® CB Woody Plant Herbicide use the nozzles and spray equipment suggested in the Directions For Use.

During the growing season, vegetation near treated stems may be discolored and even killed. WEEDONE® CB Woody Plant Herbicide normally allows ground cover plants to become reestablished in the treated area.

Species controlled by WEEDONE® CB Woody Plant Herbicide include:

OAK SPECIES
- Bear (scrub) oak
- Black oak
- Bur oak
- Chestnut oak
- Pin oak
- Red oak
- Water oak
- White oak

CONIFER SPECIES
- Balsam fir
- Red cedar
- Red pine
- Southern pine*
- Virginia pine*
- White pine
- Spruce

ELM SPECIES
- American elm
- Red elm
- Winged elm

CHERRY SPECIES
- Black Cherry*
- Choke cherry
- Mazzard cherry
- Pin cherry

OTHER SPECIES
- Ailanthus
- Ash*
- Aspen
- Birch
- Blackgum
- Black locust
- Boxelder
- Camphorwood
- Hackberry
- Hickory
- Ironwood (Ostrya)
- Ironwood (Carpinus) blue beech
- Multiflora rose
- Osageorange
- Persimmon
- Poison ivy

Pricklypear
Red Maple
Sassafras
Sour orange
Sumac
Sweetgum
Sycamore
Tallowtree
Titi (swamp cyrilla)
Tulip Poplar
Wax myrtle

* Treat at least lower 36 inches of stem as specified in Directions For Use.

DIRECTIONS FOR USE

IT IS A VIOLATION OF FEDERAL LAW TO USE THIS PRODUCT IN A MANNER INCONSISTENT WITH ITS LABELING.

Be sure that use of this product conforms to all applicable laws, rules and regulations. Certain states have restrictions pertaining to application distances from susceptible crops. The applicator should become familiar with these laws, rules or regulations and follow them exactly. Apply this product only as specified on this label.

Do not dilute WEEDONE® CB Woody Plant Herbicide with water. Do not apply WEEDONE® CB Woody Plant Herbicide when stems are wet. Rainfall which follows application of WEEDONE® CB to dry stems does not normally reduce brush control.

Do not graze meat or dairy animals on treated area.

FOR CONTROLLING WOODY PLANTS IN FOREST MANAGEMENT PROGRAMS AND ON FENCEROWS, ROADSIDES, DRAINAGE DITCHBANKS, RAILROADS, UTILITY RIGHT-OF-WAYS AND OTHER NON-CROP AREAS.

Average application volume of WEEDONE® CB Woody Plant Herbicide is 3 gallons per acre, although actual amount used will vary from 1-5 gallons depending on the amount and size of brush in the treated area. Total spray volume should be approximately 10% of that required for a conventional basal treatment. Do not exceed a total application volume of 20 gallons WEEDONE® CB Woody Plant Herbicide per acre.

BASAL BARK TREATMENT

To control general mixed brush up to 4 inches DBH (trunk diameter at breast height) including oak, hickory, sweetgum, blackgum, and many others, apply WEEDONE® CB Woody Plant Herbicide undiluted to the lower 18-24 inches of the plant stem. Treat all around the stem including the root collar area and any exposed roots. It is not necessary for WEEDONE® CB Woody Plant Herbicide to run down or puddle in the root collar area.

To control pines larger than 2 inches DBH, or cherry, ash and maple, increase treatment height to cover the lower 36 inches of stem bark.

WEEDONE® CB Woody Plant Herbicide basal bark treatment will control top growth of root-suckering species such as aspen, sumac, Sassafras, persimmon and black locust, but regrowth from lateral roots may occur the following growing season. Fall and winter treated stems may briefly leaf out in the following spring before final kill.

STUMP TREATMENT

Control stems larger than 4 inches DBH, by cutting the tree and thoroughly treating the bark and root collar area of the remaining stump. The stump can be cut at any height—low or high as long as the remaining wood, bark and root collar area is thoroughly treated. Stumps do not require immediate treatment permitting separate cut and spray operations for greater efficiency.

If other oil-miscible woody plant herbicides are mixed with WEEDONE CB Woody Plant Herbicide to increase the spectrum of species controlled, do not exceed the ratio of 1 quart other herbicide per 3 quarts WEEDONE CB Woody Plant Herbicide. Mix other herbicide with WEEDONE CB thoroughly before application. When mixing WEEDONE CB with other herbicides, follow all directions, precautions and limitations on both product labels.

FENCEROW TREATMENT

To control brush in fencerows apply WEEDONE CB to brush or stumps as directed above. Do not allow WEEDONE CB spray or spray drift to contact crops, forage plants or ornamentals.

APPLICATION EQUIPMENT

WEEDONE® CB Woody Plant Herbicide can be applied from a hand pump back pack sprayer or a hand-carried pump-up pressurized sprayer such as Chapin Model 152, Sears 1443, or Birchmeier Iris 10.

For most efficient use of the spray mixture, apply with a gun such as the Spraying Systems Model 23L-7676 equipped with a Spraying System 5500 Y series adjustable ConeJet nozzle. Equip all nozzles with a 200 mesh screen.

An alternate system which will reduce but not eliminate dripping between trigger releases is a 200 mesh TeeJet strainer and check valve number 4193A-SS having a spring opening that operates at 10 psi with the Spraying Systems 5500 Y series adjustable ConeJet nozzle mounted on a conventional basal spray wand.

Since WEEDONE® CB Woody Plant Herbicide is applied as a concentrate, the most efficient nozzle size will vary with the diameter of the brush stems to be treated:

—For average brush of 1-3 inches DBH, or for heavy stands of brush with most stems less than 1 inch DBH use the Spraying Systems 5500 adjustable ConJet nozzle with a Y2 orifice.

—For scattered stands or large trees or brush or to treat large stumps, use the Spraying Systems 5500 adjustable ConeJet nozzle with a Y3 orifice.

—For small stems less than 1 inch diameter or to limit amount of WEEDONE® CB applied, use the Spraying Systems 5500 adjustable ConeJet nozzle with a Y1 orifice.

EQUIPMENT MAINTENANCE

WEEDONE CB Woody Plant Herbicide may affect 0 rings and other seals in certain sprayer units. The most effective method of minimizing this problem is to drain the sprayer completely every evening. WEEDONE CB is a concentrate and can be returned to the original container.

To prevent time loss in the field, have available replacement parts whenever you start out to spray.

Sprayers can be cleaned of WEEDONE CB at the end of the season, or prior to shifting to other herbicides, by first rinsing with kerosene, then following with a triple rinse of hot water and detergent. At the end of the spray season replace all 0 rings and/or seals showing signs of deterioration. Such a maintenance program will prolong the life of your equipment.

PRECAUTIONARY STATEMENTS

WARNING

HAZARD TO HUMANS AND DOMESTIC ANIMALS

Harmful if swallowed, inhaled or absorbed through skin. Causes skin irritation. Causes eye irritation. Do not get on skin or clothing. Avoid breathing spray mist. Remove and wash contaminated clothing before reuse. Wash thoroughly before eating or smoking.

STATEMENT OF PRACTICAL TREATMENT

In case of skin contact, wash with plenty of soap and water. Get medical attention if irritation develops.

In case of eye contact, flush with plenty of water.

If swallowed, drink promptly large quantities of milk, egg whites, gelatin solution or if these are not available, drink large quantities of water.

Avoid alcohol. Get medical attention.

Do not induce vomiting.

Do not give anything by mouth to an unconscious person.

ENVIRONMENTAL HAZARDS

This product is toxic to fish. Do not apply directly to water.

Do not contaminate water by cleaning of equipment or disposal of wastes.

Do not graze meat or dairy animals on treated area.

Do not apply when weather conditions favor drift. Do not use the same spray equipment for other purposes unless thoroughly cleaned. Do not contaminate water used for domestic or irrigation purposes. Do not use in or near a greenhouse. Do not apply WEEDONE® CB Woody Plant Herbicide directly to or permit spray mist to drift onto cotton, okra, grapes, tomatoes, fruit trees, vegetables, flowers, or other desirable crop or ornamental plants which are susceptible to 2,4-D or 2,4-DP herbicides. Do not allow spray to contact susceptible plants since very small quantities of the 2,4-D or 2,4-DP will cause severe injury during the growing or dormant periods. Crops contacted by WEEDONE® CB Woody Plant Herbicide sprays or spray drift may be killed or suffer significant stand loss with extensive quality and yield reduction.

Drift may be reduced by: (1) applying as near to the target as possible; (2) by decreasing the pounds of pressure at the nozzle tips; (3) by using nozzles which produce a coarse spray pattern; (4) by not applying when wind is blowing toward susceptible crops or valuable plants.

STORAGE AND DISPOSAL

STORAGE

Do not store near fertilizers, seeds, insecticides, or fungicides.

PESTICIDE DISPOSAL

Do not contaminate water, food, or feed by storage or disposal. Pesticide wastes are toxic. Improper disposal of excess pesticide, spray mixture, or rinsate is a violation of Federal law. If these wastes cannot be disposed of by use according to label instructions, contact your State Pesticide or Environmental Control Agency or the Hazardous Wastes representative at the nearest EPA Regional Office for guidance.

CONTAINER DISPOSAL

Triple rinse (or equivalent). Then offer for recycling or reconditioning, or puncture and dispose of in a sanitary landfill or by other procedures approved by state and local procedures.

LIMITED WARRANTY AND DISCLAIMER

The manufacturer warrants (a) that this product conforms to the chemical description on the label; (b) that this product is reasonably fit for the purposes set forth in the directions for use when it is used in accordance with such directions; and (c) that the directions, warning and other statements on this label are based upon responsible experts' evaluation of reasonable tests of effectiveness, of toxicity to laboratory animals and to plants, and of residues on food crops and upon reports of field experience.

Tests have not been made on all varieties or in all states or under all conditions. THE MANUFACTURER NEITHER MAKES NOR INTENDS, NOR DOES IT AUTHORIZE ANY AGENT OR REPRESENTATIVE TO MAKE, ANY OTHER WARRANTIES, EXPRESS OR IMPLIED, AND IT EXPRESSLY EXCLUDES AND DISCLAIMS ALL IMPLIED WARRANTIES OF MERCHANTABILITY OR FITNESS FOR A PARTICULAR PURPOSE.

THIS WARRANTY DOES NOT EXTEND TO, AND THE BUYER SHALL BE SOLELY RESPONSIBLE FOR, ANY AND ALL LOSS OR DAMAGE WHICH RESULTS FROM THE USE OF THIS PRODUCT IN ANY MANNER WHICH IS INCONSISTENT WITH THE LABEL DIRECTIONS, WARNINGS OR CAUTIONS.

BUYER'S EXCLUSIVE REMEDY AND MANUFACTURER'S OR SELLER'S EXCLUSIVE LIABILITY FOR ANY AND ALL CLAIMS, LOSSES, DAMAGES, OR INJURIES RESULTING FROM THE USE OR HANDLING OF THIS PRODUCT, WHETHER OR NOT BASED IN CONTRACT, NEGLIGENCE, STRICT LIABILITY IN TORT OR OTHERWISE SHALL BE LIMITED, AT THE MANUFACTURER'S OPTION, TO REPLACEMENT OF, OR THE REPAYMENT OF THE PURCHASE PRICE FOR, THE QUANTITY OF PRODUCT WITH RESPECT TO WHICH DAMAGES ARE CLAIMED. IN NO EVENT SHALL MANUFACTURER OR SELLER BE LIABLE FOR SPECIAL, INDIRECT OR CONSEQUENTIAL DAMAGES RESULTING FROM THE USE OR HANDLING OF THIS PRODUCT.

WEEDONE® is a registered trademark of Union Carbide Agricultural Products Company, Inc.

ConeJet and TeeJet are registered trademarks of Spraying Systems Company.

WEEDONE® AERO CONCENTRATE E

BROADLEAF HERBICIDE

2,4-D Butyl ester for weed control in small grains.

ACTIVE INGREDIENT:
2,4-Dichlorophenoxyacetic acid,
butyl ester* 73.9%
INERT INGREDIENTS 26.1%
TOTAL 100.0%

*Total 2,4-dichlorophenoxyacetic acid equivalent 59.0% by weight or 5.7 pounds per gallon.

Isomer specific by AOAC method No. 6.D01-5

EPA Reg. No. 264-52 EPA Est. No. 264-MO-01

Keep Out of Reach of Children

CAUTION

IN CASE OF EMERGENCY PHONE (24 HOURS A DAY) IN U.S.A. 1-800-UCC-HELP.

FOR GENERAL PRODUCT INFORMATION CALL 1-800-334-9745.

DIRECTIONS FOR USE

IT IS A VIOLATION OF FEDERAL LAW TO USE THIS PRODUCT IN A MANNER INCONSISTENT WITH ITS LABELING.

	PINTS PER ACRE	GALLONS WATER PER ACRE	
WHEAT, BARLEY, RYE			
Winter grain			
Annual and biennial weeds	½-2*	8 or more	Apply after grain is fully tillered (about 4-8" high) but not forming joints in the stem.
Perennial broadleaf weeds	1-2*	8 or more	Apply when weeds are near bud stage. Do not spray grain in boot to dough stage.
Spring grain			
Annual broadleaf weeds	½-2*	8 or more	Apply after grain is fully tillered (about 4-8" high) but not forming joints in the stem.
Perennial broadleaf weeds	1-2*	8 or more	Apply when weeds are near bud stage. Do not spray grain in boot to dough stage.

* Use the lower rate if small annual and biennial weeds are the major problem. Use the higher rate if perennial weeds or annual and biennial weeds are present which are in the hard-to-kill categories as determined by local experience. The higher rates increase the risk of grain injury and should be used only where the weed control problem justifies the grain damage risk.

Do not forage or graze treated grain fields within 2 weeks after treatment with 2,4-D. Do not feed treated straw to livestock.

WEEDS IN FALLOW LAND, GRASS PASTURES		
Annual broadleaf weeds	$1\frac{1}{3}$ to $2\frac{2}{3}$ pints	Spray when these weeds are young and growing vigorously.
Perennial broadleaf weeds.	$2\frac{2}{3}$ to 4 pints	Spray when these weeds are actively growing and near the bud stage. Repeat applications may be needed for complete control. Do not graze dairy animals for seven days after spraying pastures with this product.

Do not plant any crop in fallow land until 3 months after treatment or until the chemical has disappeared from the soil.

Aerial spraying is a specialized job. Secure qualified technical guidance and employ a competent, reliable applicator. Become familiar with state laws governing the use of herbicides.

TO PREPARE A SPRAY

Add one-half of the required amount of water to the spray tank, then add this product with agitation and finally the balance of the water with continued agitation. This product forms an emulsion in water—not a solution. This tends to separate on standing. Provide agitation to prevent such separation and insure a uniform spray mixture. Use enough water per acre to provide adequate coverage.

GENERAL DIRECTIONS

WEEDONE AERO-CONCENTRATE E can be used in either airplane or ground equipment. For best results, uniform coverage is important.

CONTROL THESE WEEDS:

Annual and biennial weeds

beggarticks	knotweed	Russian thistle
bullthistle	lambsquarters	smartweed
burdock	mallow	sowthistle
cockle	marshelder	sunflower
cocklebur	morningglory	tansy ragwort
coffeeweed	mustard	tumbleweed
daisy fleabane	parsnip	vervain
frenchweed	peppergrass	vetch
galinsoga	pigweed	wild carrot
goatsbeard	prickly lettuce	wild lettuce
jimsonweed	primrose	wild radish
kochia	ragweed	

Perennial weeds

artichoke	goldenrod	ragweed
aster	ground ivy	sowthistle
Austrian fieldcress	healall	stinging nettles
	hoary cress	tarweed
bindweed	horsetail	toadflax
blue lettuce	ironweed	vervain
Canada thistle	loco weed	wild garlic
catnip	nettles	wild onion
chicory	orange hawkweed	wild strawberry
dandelion		wild sweet potato
dock	plantains	
dogbane	povertyweed	

PRECAUTIONARY STATEMENTS

CAUTION

HAZARD TO HUMANS AND DOMESTIC ANIMALS.

Harmful if swallowed. Avoid contact with skin, eyes or clothing.

STATEMENT OF PRACTICAL TREATMENT

In case of contact wash skin with soap and water; for eyes flush with water for at least 15 minutes and get medical attention.

ENVIRONMENTAL HAZARDS

This product is toxic to fish. Keep out of lakes, streams or ponds. Do not contaminate water by cleaning of equipment or disposal of wastes.

Do not contaminate water used for irrigation or domestic purposes. Do not apply when weather conditions favor drifts of spray from areas treated. Spray drift may injure cotton, beans, grapes, peas, tomatoes, ornamentals and many other desirable plants. Coarse sprays are less likely to drift. Do not use same spray equipment for other purposes.

Vapors from this product may injure highly sensitive plants already growing in the immediate vicinity or planted there two to three weeks after spraying. Do not use in or near a greenhouse.

Do not use this product for purposes other than those recommended on this label.

STORAGE AND DISPOSAL

STORAGE

Do not store near fertilizers, seeds, insecticides, or fungicides.

PESTICIDE DISPOSAL

Do not contaminate water, food, or feed by storage or disposal. Open dumping is prohibited. Pesticide wastes are toxic. Improper disposal of excess pesticide, spray mixture, or rinsate is a violation of Federal Law. If these wastes cannot be disposed of by use according to label instructions, contact your State Pesticide or Environmental Control Agency, of the Hazardous Waste representative at the nearest EPA Regional Office for guidance.

CONTAINER DISPOSAL

Triple rinse (or equivalent). Then offer for recycling, or reconditioning, or puncture and dispose of in a sanitary landfill, or by other procedures approved by state and local authorities.

LIMITED WARRANTY AND DISCLAIMER

The manufacturer warrants (a) that this product conforms to the chemical description on the label; (b) that this product is reasonably fit for the purposes set forth in the directions for use when it is used in accordance with such direction; and (c) that the directions, warnings and other statements on this label are based upon responsible experts' evaluation of reasonable tests of effectiveness, of toxicity to laboratory animals and to plants, and of residues, and upon reports of field experience. Tests have not been made on all varieties or in all states or under all conditions. THE MANUFACTURER NEITHER MAKES NOR INTENDS, NOR DOES IT AUTHORIZE ANY AGENT OR REPRESENTATIVE TO MAKE, ANY OTHER WARRANTIES, EXPRESS OR IMPLIED, AND IT EXPRESSLY EXCLUDES AND DISCLAIMS ALL IMPLIED WARRANTIES OF MERCHANTABILITY OR FITNESS FOR A PARTICULAR PURPOSE.

THIS WARRANTY DOES NOT EXTEND TO, AND THE BUYER SHALL BE SOLELY RESPONSIBLE FOR, ANY AND ALL LOSS OR DAMAGE WHICH RESULTS FROM THE USE OF THIS PRODUCT IN ANY MANNER WHICH IS INCONSISTENT WITH THE LABEL DIRECTIONS, WARNINGS OR CAUTIONS.

BUYER'S EXCLUSIVE REMEDY AND MANUFACTURER'S OR SELLER'S EXCLUSIVE LIABILITY FOR ANY AND ALL CLAIMS, LOSSES, DAMAGES, OR INJURIES RESULTING FROM THE USE OR HANDLING OF THIS PRODUCT, WHETHER OR NOT SUCH LIABILITY IS BASED IN CONTRACT, NEGLIGENCE, STRICT LIABILITY IN TORT OR OTHERWISE, SHALL BE LIMITED, AT THE MANUFACTURER'S OPTION, TO REPLACEMENT OF, OR THE REPAYMENT OF THE PURCHASE PRICE FOR, THE QUANTITY OF PRODUCT WITH RESPECT TO WHICH DAMAGES ARE CLAIMED. IN NO EVENT SHALL MANUFACTURER OR SELLER BE LIABLE FOR SPECIAL, INDIRECT OR CONSEQUENTIAL DAMAGES RESULTING FROM THE USE OR HANDLING OF THIS PRODUCT.

THIS SPECIMEN LABEL IS INTENDED FOR USE ONLY AS A GUIDE IN PROVIDING GENERAL INFORMATION REGARDING THE DIRECTIONS, WARNINGS AND CAUTIONS ASSOCIATED WITH THE USE OF THIS PRODUCT. AS WITH ANY AGRICULTURAL CHEMICAL, ALWAYS FOLLOW THE LABEL INSTRUCTIONS ON THE PACKAGE BEFORE USING.

Uniroyal Chemical

Crop Protection
World Headquarters
Middlebury, CT 06749
(203) 573-2000

ALANAP®-L

KILLS BROADLEAF WEEDS

COMPOSITION

ACTIVE INGREDIENTS:	% BY WEIGHT
Sodium salt of naptalam (N-1-Naphthylphthalamic acid)*	23.7%
INERT INGREDIENTS:	76.3%
TOTAL:	100.0%

* Each gallon contains two pounds of active ingredient.

EPA Reg. No. 400-49-AA EPA Est. No.

WARNING

Keep Out of Reach of Children

STATEMENT OF PRACTICAL TREATMENT

IF SWALLOWED: Drink promptly a large quantity of milk, egg white or gelatin solution; if not available drink large amounts of water. Avoid alcohol. Call a physician immediately.

IF INHALED: If abnormal symptoms develop, see a physician.

IF IN EYES: Immediately wash with plenty of water for at least 15 minutes. Call a physician. Remove and wash contaminated clothing before use.

See left side panel for additional precautionary statement.

PRECAUTIONARY STATEMENTS

HAZARDS TO HUMANS

WARNING

Causes eye irritation. Do not get in eyes, on skin or on clothing. Harmful if swallowed.

ENVIRONMENTAL HAZARDS

Do not apply directly to water. Do not contaminate water by cleaning of equipment or disposal of wastes.

DIRECTIONS FOR USE

It is a violation of Federal law to use this product in a manner inconsistent with its labeling.

STORAGE AND DISPOSAL

Do not contaminate water, food or feed by storage or disposal.

STORAGE

Store in a dry location. ALANAP-L is normally a solution but it may crystallize in the cold (32°F.). Gentle warming (60°F.) will redissolve any crystals. Mild agitation, such as rolling, shaking, or stirring before use is recommended.

PESTICIDE DISPOSAL

Wastes resulting from the use of this product may be disposed of on site or at an approved disposal facility.

CONTAINER DISPOSAL

Triple rinse (or equivalent). Then offer for recycling or reconditioning, or puncture and dispose of in a sanitary landfill, or by other procedures approved by state and local authorities.

EQUIPMENT CLEANING

Thoroughly rinse spray tank, hose and nozzles with water after application. Use ALANAP-L in metal or fiberglass tanks. ALANAP-L can be removed from metal or fiberglass tanks by washing thoroughly with ammonia water and rinsing several times with water containing a good detergent.

GENERAL INSTRUCTIONS

1. PREPARATION OF SOIL

Area to be treated must be free of all weeds at time of application. ALANAP-L usually will not control established weeds. To insure maximum weed control, it is important to have a relatively smooth soil surface. If the area to be treated has been plowed, the soil clods should be broken up. If weather factors make replanting necessary after ALANAP-L has been applied, thoroughly disk the area prior to replanting.

2. METHOD OF APPLICATION

PREEMERGENCE

a. Uniform coverage of soil surface is essential.
b. Apply ALANAP-L within 48 hours of planting.
c. Rotary hoe may be used to destroy small weeds without impairing herbicide activity if dry weather follows ALANAP-L treatment.
d. If irrigation is used best results are obtained when water is applied after spraying ALANAP-L.
e. Avoid spray drift on areas planted to such sensitive crops as beets, tomatoes and spinach.

POSTEMERGENCE

a. Uniform coverage of foliage is essential.
b. Do not apply if rain is expected within 6 hours of application.

NOTES

1. **If heavy rains immediately follow ALANAP-L application, weed control normally will not be impaired. However, some stunting of growth or even some stand reduction may be noted, especially in the lighter soils temperature may accentuate this condition. With the return of good growing conditions the stunted plants will usually recover and maintain a normal growth pattern.**
2. **Do not use ALANAP-L on silt loam soils of extremely fine texture.**
3. **Do not mix ALANAP-L with liquid fertilizer solutions since solid precipitates may form.**

3. WEEDS CONTROLLED: (Preemergence)

BROADLEAF WEEDS: *Common chickweed, common lambsquarters, common ragweed, common purslane, velvetleaf, common cocklebur, field and hedge bindweed, cutleaf groundcherry, white mustard, shepherdspurse, redroot pigweed, hairy galinsoga and carpetweed.*

4. DIRECTIONS FOR USE: Put one-half the recommended amount of water in the spray tank; add the required amount of ALANAP-L then add the remainder of water to the tank while mixing. Use a bypass agitation until completely mixed. A line strainer should be placed between the spray tank and the pump. Each nozzle should be equipped with a 50 mesh screen. Calibration of herbicide sprayers is essential, since a successful chemical weed control program requires that known and correct amounts of herbicides be applied.

CUCURBITS

Cantaloupe, Cucumber, Muskmelon and Watermelon.

Apply ALANAP-L as a preemergence spray immediately after seeding. A second application postemergence to the crop may be made about a month after planting just before the crop is starting to vine, but before weeds have emerged. ALANAP-L can also be used as a postemergence spray immediately after transplanting crops. DO NOT USE POSTEMERGENCE IN WESTERN IRRIGATION AREAS. ALANAP-L may be used as a tank mix with bensulfide [PREFAR (T)]. See specific directions.

1. PREEMERGENCE

ALANAP-L: Use 6–8 quarts of ALANAP-L in 20–40 gallons of water per acre (broadcast). On light sandy soils reduce the rate to

4–6 quarts of ALANAP-L. On silt or clay soils the rate should be increased to 12 quarts.

2. IRRIGATED MELONS (Preplant or Preemergence)
Prepare and shape seed beds. Band treat on the bed with the recommended rate (see above use directions for ALANAP-L alone or use directions for Prefar plus ALANAP-L) incorporate to 1–2½ inches. Plant, then irrigate as needed. On sandy loam or lighter soils ALANAP-L should not be used with sprinkler irrigation.

3. POSTEMERGENCE
ALANAP-L may be used postemergence in areas outside the western irrigated areas. Reduce dosage in silt or clay soils to 8 qts.

NOTES

1. **Do not use ALANAP-L on vine crops of any kind when growing conditions are adverse namely in early spring when weather is cold and wet. Under such conditions ALANAP-L may cause stunting.**
2. **Do not apply postemergence to vine crops when temperature is 100°F. or above.**
3. **Do not use ALANAP-L postemergence to cucumbers, cantaloupe, muskmelon, and watermelon until plants are ready to vine.**

ALANAP PREFAR (T)

DIRECTIONS FOR USE PRE-PLANT OR PREEMERGENCE:
Cantaloupe, cucumber, muskmelon and watermelon—Apply 1–1½ gallons per acre of PREFAR 4E (4 to 6 lbs. a.i.) plus 1 to 2 gallons per acre of ALANAP-L (2 to 4 lbs. a.i.) in sufficient water (10 to 40 gal./acre) to provide uniform distribution. Incorporate lightly (½ to 1 inch) into the soil using suitable equipment just prior to planting. Use lower rates on light (sandy) soils and higher rates on heavier (clay) soils. Reduce the rate proportionately for band applications. **For irrigated melons see treating sequence given above.**

BROADLEAF WEEDS CONTROLLED

PREFAR plus ALANAP-L in tank mix combination will control: *common chickweed, common lambsquarters, common ragweed, common purslane, velvetleaf, common cocklebur, field and hedge bindweed, cutleaf groundcherry, white mustard, shepherdspurse, redroot pigweed, hairy galinsoga, carpetweed.*

ANNUAL GRASSES CONTROLLED

Smooth and large crabgrass, foxtails, barnyard grass, goosegrass, fall panicum and *red sprangletop.*

IMPORTANT—Follow all general directions for use and precautions listed above on the ALANAP-L label, as well as those listed on the PREFAR label.

NURSERY STOCK

A. For established Woody Plants (e.g. elm, arborvitae, taxus, pine, spruce, maple, cedar, cottonwood, plum, thuja, picea, lilac, privet, philadelphus, azalea, holly, rhododendron, juniper.)

1. Apply ALANAP-L on a broadcast basis or as a band spray centering on the row. Additional applications may be made as required.
2. Direct spray so as to strike the nursery stock about 2–3 inches or less above the ground.

B. For small Woody Transplants (2 methods suggested).

1. Apply ALANAP-L prior to transplanting. Care should be taken not to disturb much of the treated area.
2. Apply ALANAP-L as directed spray after transplanting.

Use 12 quarts of ALANAP-L in 20–40 gallons of water per acre (broadcast). On light sandy soils the rate should be reduced to 8 quarts. If applying on silt or clay soils increase the rate to 16 quarts.

NOTE: Avoid application to foliage or nursery plantings. Do not use on herbaceous (non-woody) plants.

PEANUTS

Use ALANAP-L as preemergence spray immediately after seeding. Do not apply ALANAP-L to soil area where crop seeds are exposed or shallow planted. The usual depth of planting to insure a stand is sufficient providing that the crop seeds are well covered with soil.

Use 8 quarts of ALANAP-L in 20–40 gallons of water per acre (broadcast). On light sandy soils the rate should be reduced to 4 quarts. When applying on silt or clay soils increase the rate to 12 quarts.

ALANAP DINOSEB: For use at "Cracking Time". Use 1½ gallons of ALANAP-L and ½ gallon (1.5 lbs. active) Dinoseb (2-sec-butyl-4,6-dinitrophenol) (alkanolamine salts) in 20–30 gallons of water per acre on broadcast basis. Spraying a 12 inch band on a 36 inch row requires only ⅓ of the above recommended dosage.

SOYBEANS

Use ALANAP-L as a preemergence spray immediately after seeding, or as an overtop application when soybeans are about 18" tall and are growing actively.

1. PREEMERGENCE THROUGH CRACKING
Use 8 quarts of ALANAP-L in 20–40 gallons of water per acre (broadcast). On light sandy soils the rate should be reduced to 4 quarts. When applying on silt or clay soils increase the rate to 12 quarts. A band treatment may also be used.

ALANAP DINOSEB: For use at preemergence through cracking. Use 1½ gallons of ALANAP-L and ⅔ gallons (2 lbs. active) Dinoseb (2-sec-butyl-4,6-dinitrophenol) (alkanolamine salts) in 30–40 gallons of water per acre on broadcast basis.

NOTES:

1. **Do not apply ALANAP-L to soil area where crop seeds are exposed or shallow planted. The usual depth of planting to insure a stand is sufficient providing that the crop seeds are well covered with soil.**

2. POSTEMERGENCE
ALANAP-L Butyrac 200 or Butoxone

Ground Application:
Apply 2–3 quarts of ALANAP-L plus 3–4 ounces of Butyrac* 200 or Butoxone** in 10–20 gallons of water per acre (broadcast rate) on soybeans that are about 18 inches tall (7–10 days before bloom through mid bloom). The spray boom should be positioned 18"–24" above the tops of the soybean plants or weeds and be equipped with hollow cone nozzles. Maintain a high spray pressure (40–50 psi) during application. This insures a fine spray mist for better overall coverage.

Aerial Application:
Apply 2–3 quarts of ALANAP-L plus 3–4 ounces of Butyrac 200 or Butoxone in a minimum of 5 gallons of water per acre on soybeans that are about 18 inches tall (7 to 10 days before bloom through mid bloom). Nozzle size and orientation should provide maximum droplet breakup without causing excessive drift. Do not apply by air within 50 feet of susceptible crops such as cotton, tomatoes and tobacco.

For best results, use 2 quarts of ALANAP-L plus 3 ounces of Butyrac 200 or Butoxone on weeds that are 10–12 inches tall such as *common cocklebur, giant ragweed, wild mustard or sunflower.* However, if these weeds are taller than 12 inches, use 3 quarts of ALANAP-L plus 4 ounces of Butyrac 200 or Butoxone. ALANAP-L plus Butyrac 200 or Butoxone if applied according to directions will control or suppress growth of above weeds reducing their competition to the soybean crop.

NOTES

1. **Use surfactant (nonionic 80–20 etc.) at ½% by volume with this tank mix.**
2. **Avoid application of this tank mix to drought stressed soybeans.**
3. **Do not apply the tank mix if rain is expected within 6 hours after application.**
4. **Applications of this tank mix should not be made when weather conditions favor drift.**
5. **Observe all cautions and limitations on the Butyrac 200 or Butoxone labels.**

* Butyrac 200 contains 2 lbs/ gallon of 4-(2,4-dicholorophenoxy) butyric acid and is a registered trademark of Union Carbide.

** Butoxone contains 1.75 lbs/gallon of 4-(2,4-dicholorophenoxy) butyric acid and is a registered trademark of Rhone-Poulenc, Inc.

DO NOT GRAZE OR FEED FORAGE FROM TREATED SOYBEANS TO LIVESTOCK.

OBSERVE ALL CAUTIONS AND LIMITATIONS ON LABELING OF ALL PRODUCTS USED IN MIXTURES.

IMPORTANT NOTICE

Seller warrants that this product conforms to its chemical description and is reasonably fit for the purposes stated on the label when used in accordance with the directions and instructions specified on the label under normal conditions of use, **but neither this warranty nor any other warranty of merchantability or fitness for a particular purpose, express or implied, extends to the use of this product, contrary to label instructions. or under abnormal conditions, or under conditions not reasonably foreseeable to seller, and buyer assumes the risk of any such use.**

(T) PREFAR is a registered trademark of Stauffer Chemical.

ALAR®-85

A PLANT GROWTH REGULANT

COMPOSITION

ACTIVE INGREDIENT	% BY WEIGHT
Daminozide[butanedioic acid mono (2,2-dimethylhydrazide)]	85.0%
INERT INGREDIENTS	15.0%
TOTAL:	100.0%

EPA Reg. No. 400-79 EPA Est. No.

ALAR is a Registered Trademark of UNIROYAL CHEMICAL COMPANY, INC.

CAUTION

Keep Out of Reach of Children

STATEMENT OF PRACTICAL TREATMENT

IF SWALLOWED: Drink promptly a large quantity of milk, egg white or gelatin solution; if not available drink large amounts of water. Avoid alcohol. Call a physician immediately.

IF ON EYES OR SKIN: Wash with plenty of water for at least 15 minutes. For eyes, call a physician. Remove and wash contaminated clothing before use.

IF INHALED: and abnormal symptoms develop see a physician.

See Left Side Panel For Additional Precautionary Statement.

PRECAUTIONARY STATEMENTS

HAZARDS TO HUMANS

CAUTION

Harmful if absorbed through skin. Avoid contact with skin, eyes or clothing. Do not allow spray solution to touch skin. Wear protective clothing to protect hands, arms and legs, both when diluting and applying this product. Avoid breathing dusts. Wear a protective mask during mixing/loading operations. Wash hands and face thoroughly with soap and water after use or before eating or smoking.

ENVIRONMENTAL HAZARDS

Do not apply directly to water or wetlands. Do not discharge into lakes, streams, ponds, or public waters unless in accordance with NPDES permit. For guidance, contact your Regional Office of the EPA. Do not contaminate water by cleaning of equipment or disposal of wastes.

PHYSICAL OR CHEMICAL HAZARDS

Do not use or store near heat or open flame.

DIRECTIONS FOR USE

Each formulator is responsible for obtaining EPA registrations for its formulated product(s).

It is a violation of Federal Law to use this product in a manner inconsistent with its labeling.

REENTRY STATEMENT

Do not reenter treated areas until 24 hours after application, unless wearing protective clothing.

STORAGE AND DISPOSAL

Do not contaminate water, food or feed by storage or disposal.

STORAGE

Store in a dry location.

Use spray and stock solutions within 24 hours. Do not store solutions. Immediate use is appropriate, if another component is added to spray mixture.

PESTICIDE DISPOSAL

Wastes resulting from the use of this product may be disposed of on site or at an approved waste disposal facility.

CONTAINER DISPOSAL

Completely empty bag into application equipment. Then dispose of empty bag in a sanitary landfill or by incineration, or if allowed by State and local authorities, by burning. If burned, stay out of smoke.

NOT FOR REFORMULATION

GENERAL INSTRUCTIONS AND PRECAUTIONS

ALAR-85 is a water soluble plant growth regulant. It is effective in reducing vegetative growth, promoting flower bud initiation, increasing fruit set, controlling time of fruit maturity, preventing preharvest fruit drop, and general **improving market quality** of fruit **at harvest** and/or **after storage.** See specific crop instructions.

1. ALAR-85 should be applied as a foliar spray in a manner that insures uniform and complete coverage.
2. Since ALAR-85 must be absorbed by the plant to be effective, best performance can be expected with good absorption conditions (a healthy well watered plant and slow drying conditions) and a rain free period of about 6 to 12 hours following application. Concentrate sprays (50 to 100 gals./A) of ALAR-85 may be used.
3. Do not spray ALAR-85 on any low vigor trees or trees under stress as from drought, low fertility, winter injury etc. Such treatment may accentuate reduction in vegetative growth and fruit size. ALAR-85 is most effective when used on healthy vigorously growing plants.
4. On fruit trees the earlier ALAR-85 is applied and the higher the rate used the greater the reduction of both vegetative growth and fruit size. Less reductions will occur on high vigor trees.
5. Spray ALAR-85 only once per year on fruit trees. Split applications are not recommended.
6. ALAR-85 is compatible with most commonly used spray materials. East of the Rockies it is generally recommended as a separate spray. Consult your advisor for local recommendations. Do not tank mix with highly alkaline materials (such as lime) or copper sprays or apply within 28 days (before or after) a copper spray application. Observe all precautions and limitations on labels of all products used. When ALAR-85 is mixed with water for spraying, use the mixture on the same day.
7. Do not graze animals on cover crops grown among sprayed trees.
8. Do not handle treated plants until sprays have dried.

FOLLOWING SEASON EFFECTS

ALAR-85 may promote an increase in fruit bud development and set the year following treatment. To maximize this advantage and assure optimum fruit size and quality, improve tree vigor through fertilization, pruning and uniform spray application. If excessive fruit set occurs, follow your normal thinning program. If excessive fruit set occurs accompanied by excessive size reduction, short thickened stems, shallow stem cavities and misshapen fruit, skip a year of ALAR-85 treatment and bring the trees to proper vigor through fertilization, pruning, thinning, etc. Always leave some untreated trees in each orchard for comparison.

APPLES

The effect of ALAR-85 will vary depending on time and rate of application. The higher the rate the greater the magnitude of effect during the year of treatment and the higher the possibility for following season effects (See Following Season Effects). Only one ALAR-85 spray may be used on apple trees during a growing season. From the following table select Spray Schedule Timing A "Spring", Timing B "Mid-Season" or Timing C "Summer" and follow down the column for Time and Rate to Apply and Expected Desirable Effects.

APPLE

SPRAY SCHEDULE	TIMING A "SPRING"		TIMING B "MID SEASON"		TIMING C "SUMMER"	
Time to Apply						
General U.S.	10 to 21 days after petal fall when new vegetative growth is about 4-6 inches long		21 days after petal fall to 70 days before harvest		60 to 70 days before normal harvest or approximately half way between full bloom and maturity (see footnote 1)	
Northwest U.S.	10 to 14 days after full bloom		14 to 70 days after full bloom		70 to 80 days after full bloom	
Rate to Apply relative to Tree Vigor[2]	lbs/100 gal.[3]	lbs/A	lbs/100 gal.[3]	lbs/A	lbs/100 gal.[3]	lbs/A
Low Vigor	Do not use		Do not use		Do not use	
Moderate Vigor	1–1½	4–6	¾	3	½–¾	2–3
Good to High Vigor[4]	1–1½	4–6	1	4	1	4
Very High Vigor	2	8	1	4	1	4
Young Non Bearing Trees	(See Special Instructions *1)					
Heavily Pruned Trees	(See Special Instructions *2)					
Expected Desirable Effects						
1. Maturity delayed[5]	Yes		Yes		Yes	
2. Red Color increased	Yes		Yes		Yes	
3. Preharvest fruit drop prevented	Yes		Yes		Yes	
4. Fruit firmness maintained	Yes		Yes		Yes	
5. Handling bruises reduced	Yes		Yes		Yes	
6. Scald reduced	Yes		Yes		Yes	
7. Water core delayed	Yes		Yes		Yes	
8. Fruit cracking reduced	Yes		Yes		Yes	
9. Weather check reduced	Yes		Yes		Yes	
10. Bitterpit & cork spot reduced	Yes		Yes		Yes	
11. Fruit size controlled	Greatest		Moderate		Least	
12. Vegetative growth reduced	Greatest		Some to None		None	
13. Flower bud initiation promoted, increasing return bloom, set and yield	Greatest		Some to None		Some to None	
ALAR-85 maintains or improves quality while preventing drop.	Use this timing where vegetative growth control and increased return bloom are primary objectives.		Effects 11, 12, and 13 will be pronounced with an early mid season ALAR-85 spray and reduced with later applications.		Use this timing where fruit quality and drop control only are primary objectives.	

FOOTNOTES

1. Example. If Red Delicious are 150 days between bloom and maturity the ALAR-85 should be applied about 75 days after full bloom. (75 days before maturity).
2. Maintenance of vigorous trees is essential for best results. Annual pruning resulting in removal of weak wood, proper fertilization, and other cultural practices which contribute to high tree vigor should be used. Use leaf analysis services where available to assist in determining degree of tree vigor.
3. This dilution rate is based on applications of 400 gallons per acre (to drip point) on mature bearing trees. Most field experience has been with dilute sprays but concentrate sprays can be used if care is taken to get uniform and complete coverage. Do not exceed the per acre recommended rate.
4. In areas with short growing seasons treat good to very high vigor trees as moderate.
5. The higher rates achieve the longest delays in maturity.

APPLES—*SPECIAL INSTRUCTIONS

1. To Bring Young Trees Into Bearing and Control Vegetative Growth:

Rate: 1½ to 2 pounds ALAR-85 per 100 gal. water.

Time: 10 to 21 days after full bloom when new growth is about 4 to 6 inches long. Spray to drip.

2. Vegetative Growth Control on Heavily Pruned Trees:

Rate: 1 to 2 pounds ALAR-85 per 100 gal. water (Do not exceed 8 lbs. per acre).

Time: When regrowth (terminal or sucker) is about 4 to 6 inches long.

Use sufficient spray solution to insure uniform and complete coverage. Use this spray where excessive vigor and associated pruning practices make it desirable to control regrowth throughout the entire tree.

On Heavily Topped Trees—Use 2 to 4 pounds ALAR-85 per 100 gal. water (Do not exceed 8 pounds per acre) and direct the spray to the heavily pruned area in the tops of trees with lighter application on the remainder.

PEARS
(WASHINGTON and OREGON Only)

To control premature ripening and aid in preventing preharvest fruit drop:

Rate: 4 to 5 pounds ALAR-85 per acre on mature bearing trees.

Time: 18 to 24 days before normal harvest.

Apply in sufficient water to give uniform and complete coverage.

To promote flower bud initiation and control vegetative growth:

Rate: 4 to 5 pounds per acre on mature trees.

Time: 7 to 30 days after full bloom.

Apply in sufficient water to give uniform and complete coverage. Use the higher rate on large and high vigor trees.

To bring trees into bearing:

Rate: 1½ to 2 pounds ALAR-85 per 100 gallons water and spray to drip point.

Time: 7 to 30 days after full bloom.

Apply only one ALAR-85 spray per growing season. NOTE—ALAR-85 will control preharvest fruit drop through 7 to 10 day normal harvest period or about 28 to 35 days after spray application. If it is desired to leave the fruit on the tree for a longer period of time, an additional approved "Stop-drop" material should be applied at the recommended dosage and timing.

PEACHES AND NECTARINES

To improve fruit color, concentrate maturity and hasten ripening:
Rate: 4 pounds ALAR-85 per acre on mature bearing trees. Apply in a minimum of 400 gals. per acre.
Time: From early to late pit hardening stage.
Use only once per growing season as a separate dilute spray in a thoroughly cleaned sprayer. Injury may occur as "shotholing" and possibly leaf drop especially on early varieties. Do not spray low vigor trees as fruit size reduction may occur. Do not use ALAR-85 on "Cherokee" or "Arm King" nectarines as it may cause excessive injury on these varieties.

SOUR CHERRIES

To improve color, hasten ripening, concentrate maturity, decrease fruit removal force thus reducing shaker damage, prevent late maturity fruit swell and produce a uniform firm cherry for improved pitting and grade. To reduce vegetative growth, lessening wind whip damage and promoting fruit buds and set for the following season.
Rate: 4 to 8 pounds ALAR-85 per acre on mature bearing trees.
Time: Approximately two weeks after full bloom.
Apply in sufficient water to give uniform and complete coverage. Use the higher rate for earliest maturity and on first time sprayed high vigor trees. Use only once per growing season.

SWEET CHERRIES

To hasten ripening, concentrate maturity and improve fruit color:
Rate: 4 to 8 pounds ALAR-85 per acre on mature bearing trees.
Time: Two weeks after full bloom.
Apply in sufficient water to give uniform and complete coverage. Use the higher rate for earliest maturity and on very high vigor trees. Use only once per growing season.

WARNING: Premature "green" drop has been claimed to be sporadically associated with the use of ALAR-85 on sweet cherry trees under apparently unusual and undetermined physiological and/or physical stress conditions.

GRAPES

To increase yield of varieties benefitting from increased berry set (such as Concord)—To reduce rate of vine growth:
Rate: See Following Table

Vine Growth	Annual Pruning Weight per Vine (8 × 9 feet spacing)	Pounds ALAR-85 per acre
Low Vigor	Less than 2 pounds	Do not use
Medium Vigor	2–3 pounds	1–2
High Vigor	3 pounds or more	2–2½

Time: First bloom to full bloom.
Apply in sufficient water to give uniform and complete coverage.

TOMATO TRANSPLANTS
(Do not use in California)

FOR USE BEFORE TRANSPLANTING
To retard stem elongation, delay early fruiting and aid in concentrating maturity:
For Field Grown Transplants: Apply once or twice as growth demands.
Rate: 5 pounds ALAR-85 per 100 gallons water.
Time: First spray—when plants are in the 1 to 4 true leaf stage.
Second spray—2 weeks after first.
Apply to runoff using 25 to 50 gallons per acre.

For Greenhouse Grown Transplants: Apply once or twice as growth demands.
Rate: 2½ pounds per 100 gallons water.
Time: First spray—when plants are in the 1 to 4 true leaf stage.
Second spray—2 weeks after first.
Apply to runoff using 25 to 50 gallons per acre.

CAUTION: Do not tank mix ALAR-85 with copper sprays, or apply ALAR-85 within 7 days before or after a copper spray application.

IMPORTANT NOTICE

Seller warrants that this product conforms to its chemical description and is reasonably fit for the purposes stated on the label when used in accordance with the directions and instructions specified on the label under normal conditions of use, **but neither this warranty nor any other warranty of merchantability or fitness for a particular purpose, express or implied, extends to the use of this product, contrary to label instructions, or under abnormal conditions, or under conditions not reasonably foreseeable to seller, and buyer assumes the risk of any such use.**

CASORON® 4G

DICHLOBENIL WEED AND GRASS KILLER

COMPOSITION

ACTIVE INGREDIENT:	% BY WEIGHT
Dichlobenil*	4.0%
INERT INGREDIENTS:	96.0%
TOTAL:	100.0%

* 2,6-dichlorobenzonitrile, U.S. Patent 3,027,248

EPA Reg. No. 400-168 EPA Est. No.
CASORON is a Registered Trademark of UNIROYAL CHEMICAL COMPANY, INC.

CAUTION

Keep Out of Reach of Children

STATEMENT OF PRACTICAL TREATMENT
IF SWALLOWED: Call a physician or poison control center immediately.
IF INHALED: Remove to fresh air; if abnormal symptoms develop, contact a physician.
IF ON SKIN: Wash affected area with plenty of soap and water.
IF IN EYES: Flush eyes with plenty of water for at least 15 minutes. Contact a physician.
See left side panel for additional precautionary statement.

PRECAUTIONARY STATEMENTS

HAZARDS TO HUMANS

CAUTION
Harmful if swallowed. Avoid breathing of dust. Avoid contact with eyes or skin.

ENVIRONMENTAL HAZARDS
Do not apply directly to water. Do not contaminate water by cleaning of equipment or disposal of wastes.

DIRECTIONS FOR USE

It is a violation of Federal Law to use this product in a manner inconsistent with its labeling.

STORAGE AND DISPOSAL

Do not contaminate water, food or feed by storage or disposal.

STORAGE
Store in a dry location. Do not store with propagative structures such as seed, bulbs, tubers, nursery stock, etc., or with food or feed products.

PESTICIDE DISPOSAL
Wastes resulting from the use of this product may be disposed of on site or at an approved wastes disposal facility.

CONTAINER DISPOSAL

Completely empty bag into application equipment. Then dispose of empty bag in a sanitary landfill or by incineration, or if allowed by State and local authorites, by burning. If burned, stay out of smoke.

GENERAL DIRECTIONS

PERENNIAL WEED CONTROL

CASORON 4G is recommended for control of certain perennial weeds such as *Artemisia, Canada Thistle, Curly Dock, Falsedandelion (Catsear), Fescue, Leafy Spurge, Orchardgrass, Quackgrass, Russian Knapweed, Smooth Brome, Timothy, Wild Artichoke, Wild Aster, Wild Carrot, and Yellow Rocket* in established woody ornamentals, nuts and fruit and for control in non-crop areas such as roadways, around buildings and fences, railroad and utility rights-of-way and installations.

SOIL SURFACE TREATMENT: Apply CASORON 4G as a soil surface treatment in late fall from November 15 to February 15. Old weed growth should not be removed before application. Do not make surface application to areas which have been cultivated during the fall or summer prior to application.

INCORPORATED TREATMENT: Apply CASORON 4G and incorporate immediately after application in late fall or very early spring before May 1.

ANNUAL WEED CONTROL

CASORON 4G is recommended for control of annual grassy and broadleaved weeds and certain perennial weeds such as *Annual Bluegrass (Poa annua), Bluegrass, Bull Thistle, Camphorweed, Carpetweed, Chickweed, Citron Melon, Coffeeweed, Crabgrass, Cudweed, Dandelion, Dogfennel, Evening Primrose, Fiddleneck, Florida Purslane (Pusley), Foxtail, Gisekia, Groundsel, Henbit, Horsetail, Jerusalemoak, Goosefoot, Knotweed, Lambsquarter, Maypops, Milkweed Vine (Morrenia odorata), Miners Lettuce, Natalgrass, Old Witchgrass, Peppergrass, Pineapple Weed, Plantain, Purslane, Ragweed, Red Deadnettle, Redroot Pigweed, Rosarypea, Russian Thistle, Shepherdspurse, Smartweed, Spanishneedles, Spurge, Teaweed, Texas Panicum (Hurrah grass), Wild Barley, Wild Mustard, Wild Radish, and Yellow Woodsorrel.*

Apply CASORON 4G either in early spring before seeds of annual weeds germinate or after cultivation has removed the growing weeds. Distribute granules uniformly over the soil surface. If treatment is followed by application of $1/2$ to 1 inch of water by overhead irrigation, the lower rates are recommended. Shallow incorporation immediately after application is recommended in southern areas. In other areas, shallow incorporation or sprinkler irrigation is suggested when applications are made during periods of high temperature.

NOTE: Use of CASORON 4G on plants not listed on the label may result in damage.

FRUIT AND NUT ORCHARDS AND NURSERIES

CASORON 4G may be used on the following crops.

A. **Almond, Apple, Blueberry, Cherry, Grape, Mango, Nectarine, Peach, Pear, Prune, Plum, Filbert, Pecan** (do not use on pecan seedings within 6 months of planting), **English Walnut:** May be used on bearing, non-bearing and nursery stock. Do not apply until 4 weeks after transplanting. Do not make application within one month of harvest of nuts or stone fruits. Apply 100 to 150 pounds per acre for annual weed control based on area actually treated. For control of perennial weeds apply 150 pounds per acre. Follow general directions above.

B. **Blackberry and Raspberry:** For control of annual weeds follow general directions above except use only on established plantings and do not apply more than 100 pounds per acre. Do not apply during new shoot emergence.

C. **Citrus** (bearing and non-bearing oranges, grapefruit, tangerines, lemons and limes) in groves: Apply 75 to 150 pounds per acre and incorporate immediately to a depth of 1 to 2 inches by use of power driven rotary hoe. Do not apply less than one year after transplanting; do not apply before water ring is removed.

D. **Citrus Nurseries.** CASORON 4G may be used for control of annual weeds in the following citrus root stocks: Cleopatra mandarin, rough lemon, sour orange, and sweet orange. Apply CASORON 4G to weed-free soil at 100 to 150 pounds per acre. Make first application in early spring after removal of banks, and second in early summer, not less than 90 days after the first. Apply broadcast, and unless soil is wet irrigate lightly with overhead system ($1/4$ to $1/2$ inch).

Do not apply CASORON 4G during 6 months following possible freeze damage on any citrus. Do not allow livestock to graze treated orchards.

WOODY ORNAMENTALS, SHELTERBELTS AND FOREST PLANTINGS (ESTABLISHED AND IN NURSERIES)

For control of annual weeds CASORON 4G may be used on the following plants. Apply 100 to 150 pounds of CASORON 4G per acre, based on area actually treated. Do not apply until 4 weeks after transplanting. Follow general directions above.

For control of perennial weeds listed, apply 150 to 200 pounds per acre. Do not apply more than 150 pounds per acre on Azalea, Boxwood, Euonymus, Forsythia, Heather, Holly, Ivy, Lilac, Leucothoe. Do not apply more than 150 pounds per acre on plantings less than 1 year old. Follow general directions listed above.

Amur Corktree	Ivy (Hedera helix)
Arbor Vitae	Juniper
Ash	Kinnikinnick
Azalea[1]	Laurel
Barberry	Leucothoe
Beauty Bush	Lilac
Birch	Linden
Boxelder	Locust
Boxwood	Magnolia
Camellia	Maple
Caragana	Mock Orange
Cleyera japonica	Mountain Ash
Cotoneaster	Nandina
Cottonwood	Oak
Crabapple	Osmanthus (Holly-olive)
Deutzia	Pachistima
Dogwood	Pecan
Eastern Red Cedar	Photinia
Elm	Pittosporum
Elwood Cypress	Poplar
English Walnut	Privet
Euonymus	Pyracantha
Filbert	Rhododendron
Flowering Almond	Rockrose
Flowering Quince	Rose
Forsythia	Russian Olive
Gardenia	Spirea
Goldenraintree	Squawcarpet
Hackberry	Weigela
Heather	Willow
Holly (Illex spp)[2]	Yew (Taxus)
Honeysuckle	

[1] Kurume, Mollis, Hardy hybrid types and hardy native Azalea species only.

[2] Except I. crenata, I. rotunda, I. vomitoria.

USE PRECAUTIONS

Do not use in seed beds, transplant or cutting beds or greenhouses. Do not apply until 6 months after rooting of cuttings in the field, or after planting rooted cuttings. Do not apply CASORON 4G to fruit stock within 3 months prior to or following grafting or budding of root stocks or planting of new grafts. Do not apply more than the maximum recommended amount of CASORON 4G per growing season. Do not use on "St. Lucie fine sand," "Arzell fine sand" or other light sand soil. Do not graze livestock in treated areas.

WEED CONTROL IN ESTABLISHED ALFALFA SEED AND LADINO SEED PRODUCING FIELDS

(For Seed Alfalfa in Northwest Area, Seed Ladino Clover in Pacific Coast Area)

For control of all species of dodder and certain grassy broadleaved weeds such as *lambsquarters, pigweed, green foxtail, buckhorn plantain, prickly oxtongue thistle, annual ryegrass, chickweed, shepherdspurse, knotweed, prickly lettuce, sow thistle,* and germinating seeds of these weeds, apply CASORON 4G at 35 to 50 pounds per acre. Use the lower rate on light soils and the higher rate on heavier soils. Distribute uniformly by use of either ground or aerial granule distribution equipment. Do not skip or overlap.

ALFALFA: Apply CASORON 4G in spring as a surface application just prior to dodder and weed seed germination. To insure continued control of dodder, make a second application four to six weeks after the first, using 50 pounds per acre in the standing alfalfa without incorporation, just before irrigating. CASORON 4G can be used for dodder control under either sprinkler or furrow irrigation.

LADINO CLOVER: Make application after harvest when adequate rainfall or supplemental irrigation is assured, while crop is semidormant to dormant.

USE PRECAUTIONS

Use only in fields devoted exclusively to seed production. Do not apply CASORON 4G if any part of the crop or subsequent cuttings are used for food or feed purposes. Do not graze or use forage from treated areas. Do not use seed for food or feed purposes. Apply only to established stands. Do not apply more than 50 pounds of CASORON 4G at any one time or more than 100 pounds per season.

WEED CONTROL IN CRANBERRIES

Control of perennial and annual, broadleaved and grassy weeds (including rushes and sedges) in cranberries can be obtained by use of CASORON 4G. The following weeds are controlled: *Sensitive Fern, Bracken Fern, Royal Fern, Hair Cap Moss, Common Horsetail, Water Horsetail (pipes), Rush (Juncus spp.), Dodder, Bluejoint grass, Rattlesnake grass (Manna grass), Summer grass, Velvetgrass, Bent Grasses, Little Hairgrass, Crabgrass, Rice Cutgrass, Bunch grass, Muskrat grass, Nutsedge (Nutgrass), Short Wiregrass, Wideleaf grass, Stargrass, Woolgrass, Cotton grass, Needle grass, Oniongrass, Arrowleaved Tear Thumb, Beggarsticks, Knotweed, Loosestrife, Marsh St. Johnswort, Tideland clover, Ragweed, Sorrell, Wild Strawberry, Aster, Buckbean, Hawkweed, Western Lilaeopsis, Marsh Pea, Plantain, Smartweed (Marsh pepper Pennsylvania, Spotted, Swamp and Water).*

Application should be made in early spring while perennial weeds are still dormant and annual weeds have not started to germinate or in late fall after crop has been harvested. Apply pre-bloom or post-harvest.

Spring Treatment

AREAS OTHER THAN PACIFIC NORTHWEST: Apply 100 pounds CASORON 4G per acre. Do not make spring application if treatment was made the previous fall. See specific directions for spring treatment in the Pacific Northwest.

PACIFIC NORTHWEST: Two equal applications may be made in early spring with an interval of 3 to 6 weeks between treatments. Total application should be 100 pounds per acre.

Fall Treatment

In all growing areas apply 100 to 150 pounds of CASORON 4G per acre in the fall after crop has been harvested. Do not make fall application if treatment was made the previous spring. Use high rate only on very weedy bogs, or on bogs abandoned because of weeds.

USE PRECAUTIONS

Distribute CASORON 4G uniformly. Avoid over-application which may result from over-lapping during treatment period. Temporary reddening of cranberry plants may occur especially with late spring application or when applied on sandy bogs. Do not apply after cranberry plants start to bloom as yields may be reduced. Do not use on young beds or newly sanded beds or on bogs prior to or immediately after mowing for vines.

OTHER USES

NUTSEDGE CONTROL: To control nutsedge (nutgrass) on mineral soils in areas on which containers of nursery stock are to be grown or held—and also for control in non-crop areas—apply 250 to 500 pounds of CASORON 4G per acre and incorporate immediately to a depth of 4 to 6 inches. Seasonal control of annual weeds will also be obtained. Containers of plants listed above may be placed on treated soil immediately. Do not transplant into treated soil.

UNDER ASPHALT: For general weed control under asphalt (in areas such as roadways, parking lots, recreational areas), after final grade is achieved, apply 250 to 300 lbs. per acre, ($5^3/_4$ to 7 lbs. per 1,000 sq. ft.). Treated area should be covered with asphalt as soon as possible. For re-surfacing work, existing weeds should be sprayed with a contact herbicide prior to CASORON 4G application.

IMPORTANT NOTICE

Seller warrants that this product conforms to its chemical description and is reasonably fit for the purpose stated on the label when used in accordance with the directions and instructions specified on the label under normal conditions of use. **but neither this warranty nor any other warranty of merchantability or fitness for a particular purpose, express or implied, extends to the use of this product, contrary to label instructions, or under abnormal conditions, or under conditions not reasonably foreseeable to seller, and buyer assumes the risk of any such use.**

CASORON® 10G

DICHLOBENIL WEED AND GRASS KILLER AND AQUATIC HERBICIDE

COMPOSITION

ACTIVE INGREDIENT:	% BY WEIGHT
Dichlobenil*	10.0%
INERT INGREDIENTS:	90.0%
TOTAL:	100.0%

* 2,6-dichlorobenzonitrile

EPA Reg. No. 400-178 EPA Est. No.

CAUTION

Keep Out of Reach of Children

STATEMENT OF PRACTICAL TREATMENT

IF SWALLOWED: Call a physician or poison control center immediately.

IF INHALED: Remove to fresh air; if abnormal symptoms develop, contact a physician.

IF ON SKIN: Wash affected area with plenty of soap and water.

IF IN EYES: Flush eyes with plenty of water for at least 15 minutes. Contact a physician.

See left side panel for additional precautionary statement.

PRECAUTIONARY STATEMENTS

HAZARDS TO HUMANS

CAUTION

Harmful if swallowed. Avoid breathing of dust. Avoid contact with eyes or skin.

ENVIRONMENTAL HAZARDS

Do not apply directly to any body of water, except as specified on this label. Do not apply to water which will be used for crop irrigation, for livestock water or for human consumption. Do not use fish from treated water for food or feed for a period of 90 days after application. Do not use in commercial fish or shellfish water. Do not contaminate water by cleaning of equipment or disposal of wastes.

DIRECTIONS FOR USE

It is a violation of Federal Law to use this product in a manner inconsistent with its labeling.

STORAGE AND DISPOSAL

Do not contaminate water, food or feed by storage or disposal.

STORAGE

Store in a dry location. Do not store with propagative structures such as seed, bulbs, tubers, nursery stock, etc., or with food or feed products.

PESTICIDE DISPOSAL

Wastes resulting from the use of this product may be disposed of on site or at an approved waste disposal facility.

CONTAINER DISPOSAL

Completely empty bag into application equipment. Then dispose of empty bag in a sanitary landfill or by incineration, or if allowed by State and local authorities, by burning. If burned, stay out of smoke.

USE DIRECTIONS
AQUATIC WEED CONTROL

CASORON 10G is recommended for control of submerged aquatic weeds in non-flowing water, such as ponds, reservoirs and lakes. CASORON 10G applied before weed growth begins will give season long control of several species of weeds including the following: *Elodea, Naiads, Chara, Northern watermilfoil, Coontail, Pondweeds (Potamogeton spp.)*

Apply CASORON 10G during early spring either to exposed pond bottoms or through the water. CASORON 10G is active through the soil rather than through the water, therefore uniform distribution of the granules over the entire area to be treated is essential. It is particularly suited for use as a partial pond treatment around boat docks and swimming areas.

Treatment of Exposed Bottoms and Shorelines: Apply CASORON 10G at a rate of 70 to 100 pounds per acre. Complete and uniform coverage of the area, whether it be applied to the soil or into the water is necessary for control. Use the lower rate if the soil is wet at the time of application, or if the water is less than 1 foot deep. In the case of treatment after drawdown, the pond should be refilled promptly.

Treatment Through Water: Distribute CASORON 10G uniformly over the surface of the water using a rate of 100 to 150 pounds per acre of surface area. The lower rate should be used in shallow water less than 2 feet deep, the higher rate in deep water over 5 to 6 feet. Application should be made during periods of calm when wave action is at a minimum. Use of a non-motorized boat is recommended to avoid excessive water turbulence.

CASORON 10G should be spread using a hand cyclone type spreader. If areas to be treated are small, satisfactory application can be made by hand, throwing granules with a small scoop. CASORON 10G is designed to sink rapidly in water.

NOTE: Consult your State Fish Game Department before applying CASORON 10G to public waters.

NON-SELECTIVE WEED CONTROL

CASORON 10G is recommended for industrial weed control. Annual and perennial broadleaf and grass weeds listed below are controlled in a wide range of non-cultivated situations. When applied as directed this product wll give non-selective weed control for an extended period of time. Length of residual action is dependent upon rain-fall, soil type and season of application.

CASORON 10G is a free-flowing granule, non-flammable and easy to use. Best results are obtained by distributing uniformly over the soil surface by hand operated or tractor mounted granule applicator.

CASORON 10G is recommended for control of certain perennial weeds such as: *Artemisia, Bindweed, Canada Thistle, Curly Dock, Falsedandelion (Catsear), Fescue, Leafy Spurge, Orchardgrass, Quackgrass, Russian Knapweed, Sheep Sorrel, Timothy, Wild Artichoke, Wild Aster, Wild Carrot and Yellow Rocket.*

CASORON 10G is recommended for control of annual grassy and broadleaved weeds and certain perennial weeds such as: *Annual Bluegrass, Bluegrass, Bull Thistle, Camphorweed, Carpetweed, Chickweed, Citron Melon, Coffeeweed, Crabgrass, Cudweed, Dandelion, Dogfennel, Evening-primrose, Fiddleneck, Florida Purslane (Pusley), Foxtail, Gisekia, Groundsel, Henbit, Horsetail, Jerusalem-oak Goosefoot, Knotweed, Lambsquarters, Maypops, Milkweed Vine (Morrenia odorata), Minerslettuce, Natalgrass, Old Witchgrass, Peppergrass, Pineappleweed, Plantain, Purslane, Ragweed, Red Deadnettle, Redroot Pigweed, Rosarypea, Russian Thistle, Shepherdspurse, Smartweed, Spanishneedles, Spurge, Teaweed, Texas Panicum (Hurrahgrass), Wild Barley, Wild Mustard, Wild Radish and Yellow Woodsorrel.*

For annual weed control CASORON 10G should be applied prior to germination or when new plants are very small. For perennial weeds, make application during coldest weather when plants are fully dormant, or in the winter rosette stage.

The weeds listed above can be controlled in industrial and non-crop areas such as:

Electric substations—Equipment sales and storage areas.
Lumber yards—Around above ground pipes and tanks.
Non-surfaced roadways—Railroad and highway right-of-way.
Petroleum installations—Around buildings.
Along fence rows—Farm fuel storage tanks and out buildings.

For non-selective weed control CASORON 10G should be applied at the rate of 120 to 200 pounds per acre ($2^3/_4$ to $4^1/_2$ pounds per thousand square feet, $4^1/_2$ to $7^1/_2$ ounces per 100 square feet or $^3/_4$ to $1^1/_4$ pounds per square rod). The lower rates may be used if the weed infestation is primarily annuals. For perennial weeds and in dry soil areas the higher rates should be used. Dead weeds from previous seasons growth do not need to be removed and soil should not be cultivated prior to application.

CASORON 10G should be applied during the season of lowest temperatures. Apply November 1 to March 1 in most areas or until April 1 in the northern tier of states. Do not apply if air temperature is expected to rise above 70°F within the following week. In areas of limited rainfall apply just prior to or during rainy season. For prompt and effective results rainfall or other precipitation should occur within a few days of application. Do not seed or transplant into treated area for 24 months.

USE PRECAUTIONS

This product is not selective in action and may destroy all types of vegetation. The treated area may be totally or partially nonproductive for one or more years. Apply only to areas where complete plant control is desired. Do not apply to areas on slope above lawns or susceptible crops as run-off water from treated area could cause injury. Do not graze livestock in treated areas.

OTHER USES

UNDER ASPHALT: For general weed control under asphalt (in areas such as roadways, parking lots, recreational areas), after final grade is achieved, apply 100 to 120 lbs. per acre. (2.3 to 2.8 lbs. per 1,000 sq. ft.). Treated area should be covered with asphalt as soon as possible. For re-surfacing work, existing weeds should be sprayed with a contact herbicide prior to CASORON 10G application.

IMPORTANT NOTICE

Seller warrants that this product conforms to its chemical description and is reasonably fit for the purposes stated on the label when used in accordance with the directions and instructions specified on the label under normal conditions of use, **but neither this warranty nor any other warranty of merchantability or fitness for a particular purpose, express or implied, extends to the use of this product, contrary to label instructions, or under abnormal conditions, or under conditions not reasonably foreseeable to seller, and buyer assumes the risk of any such use.**

CASORON® 50W

DICHLOBENIL WEED AND GRASS KILLER

COMPOSITION

ACTIVE INGREDIENT:	% BY WEIGHT
Dichlobenil*	50.0%
INERT INGREDIENTS:	50.0%
TOTAL	100.0%

* 2,6-dichlorobenzonitrile, U.S. Patent 3,027,248

EPA Reg. No. 400-169 EPA Est. No.

CAUTION

Keep Out of Reach of Children

STATEMENT OF PRACTICAL TREATMENT

IF SWALLOWED: Call a physician or poison control center immediately.

IF INHALED: Remove to fresh air; if abnormal symptoms develop, contact a physician.

IF ON SKIN: Wash affected area with plenty of soap and water.

IF IN EYES: Flush eyes with plenty of water for at least 15 minutes. Contact a physician.

See left side panel for additional precautionary statement.

PRECAUTIONARY STATEMENTS

HAZARDS TO HUMANS

CAUTION

Harmful if swallowed. Avoid breathing of dust. Avoid contact with eyes or skin.

ENVIRONMENTAL HAZARDS

Do not apply directly to water. Do not contaminate water by cleaning of equipment or disposal of wastes.

DIRECTIONS FOR USE

It is a violation of Federal Law to use this product in a manner inconsistent with is labeling.

STORAGE AND DISPOSAL

Do not contaminate water, food or feed by storage or disposal.

STORAGE

Store in a dry location. Do not store with propagative structures such as seed, bulbs, tubers, nursery stock, etc., or with food or feed products.

PESTICIDE DISPOSAL

Wastes resulting from the use of this product may be disposed of on site or at an approved wastes disposal facility.

CONTAINER DISPOSAL

Completely empty bag into application equipment. Then dispose of empty bag in a sanitary landfill or by incineration, or if allowed by State and local authorities, by burning. If burned, stay out of smoke.

GENERAL DIRECTIONS

ANNUAL WEED CONTROL

CASORON 50W is recommended for control of annual grassy and broadleaved weeds and certain perennial weeds such as *Annual Bluegrass (Poa annua), Bluegrass, Bull Thistle, Camphorweed, Carpetweed, Chickweed, Citron Melon, Coffeeweed, Crabgrass, Cudweed, Dandelion, Dogfennel, Evening Primrose, Fiddleneck, Florida Purslane (Pusley), Foxtail, Gisekia, Groundsel, Henbit, Horsetail, Jerusalemoak, Goosefoot, Knotweed, Lambsquarters, Maypops, Milkweed Vine (Morrenia odorata), Minerslettuce, Natalgrass, Old Witchgrass, Peppergrass, Pineapple Weed, Plantain, Purslane, Ragweed, Red Deadnettle, Redroot Pigweed, Rosarypea, Russian Thistle, Sheperdspurse, Smartweed, Spanishneedles, Spurge, Teaweed, Texas Panicum (Hurrahgrass), Wild Barley, Wild Mustard, Wild Radish, and Yellow Woodsorrel.* Apply CASORON 50W over-all or as a directed spray either in early spring before seeds of annual weeds germinate or after cultivation has removed the growing weeds. Apply in sufficient water for uniform coverage (at least 50 gallons per acre). If treatment is followed by application of ½ to 1 inch of water by overhead irrigation, the lower rates are recommended. Shallow incorporation immediately after application is recommended in southern areas. In other areas, shallow incorporation or sprinkler irrigation is suggested when applications are made during periods of high temperature. CASORON 50W may be used for weed control on the plants listed below.

USE PRECAUTIONS

Do not use in seed beds, transplant or cutting beds or greenhouses. Do not apply until 6 months after rooting of cuttings in the field, or after planting rooted cuttings. Do not apply on fruit stock within 3 months prior to or following grafting or budding of root stocks or planting of new grafts. Do not apply more than the maximum recommended amount of CASORON 50W per growing season. Do not use on "St. Lucie fine sand," "Arzell fine sand" or other light sand soil. Do not graze livestock in treated areas.

FRUIT AND NUT ORCHARDS AND NURSERIES

CASORON 50W may be used on the following crops.

A. **Apple, Almond, Blueberry, Cherry, Grape, Mango, Nectarine, Peach, Pear, Prune, Plum, Filbert, Pecan** (do not use on pecan seedlings within 6 months of planting), **English Walnut:** May be used on bearing, non-bearing and nursery stock. Do not apply until 4 weeks after transplanting. Do not make application within one month of harvest of nuts or stone fruits. Apply 8 to 12 pounds per acre based on area actually treated. Follow general directions above.

B. **Blackberry and Raspberry.** Follow application directions above except use only on established planting and do not apply more than 8 pounds per acre. Do not apply during new shoot emergence.

C. **Citrus** (bearing and non-bearing oranges, grapefruit, tangerines, lemons and limes) in groves: Apply 6 to 12 pounds per acre and incorporate immediately to a depth of 1 to 2 inches by use of power driven rotary hoe. Do not apply less than one year after transplanting; do not apply before water ring is removed.

D. **Citrus Nurseries.** CASORON 50W may be used for control of annual weeds in the following citrus root stocks: Cleopatra, mandarin, rough lemon, sour orange, and sweet orange. Apply CASORON 50W to weed-free soil at 8 to 12 pounds per acre. Make first application in early spring after removal of banks, and second in early summer, not less than 90 days after the first. Apply broadcast, and unless soil is wet irrigate lightly with overhead system (¼ to ½ inch).

Do not apply CASORON 50W during 6 months following possible freeze damage on any citrus.

WOODY ORNAMENTALS, SHELTERBELTS AND FOREST PLANTINGS (ESTABLISHED AND IN NURSERIES)

CASORON 50W may be used for weed control on the plants listed below. Apply 8 to 12 pounds per acre, based on area actually treated. Do not apply until 4 weeks after transplanting. Following general directions above.

Amur Corktree	Ivy (Hedera helix)
Arbor Vitae	Juniper
Ash	Kinnikinnick
Azalea[1]	Laurel
Barberry	Leucothoe

Beauty Bush	Lilac
Birch	Linden
Boxelder	Locust
Boxwood	Magnolia
Camellia	Maple
Caragana	Mock Orange
Cleyera japonica	Mountain Ash
Cotoneaster	Nandina
Cottonwood	Oak
Crabapple	Osmanthus (Holly-olive)
Deutzia	Pachistima
Dogwood	Pecan
Eastern Red Cedar	Photinia
Elm	Pittosporum
Elwood Cypress	Poplar
English Walnut	Privet
Euonymus	Pyracantha
Filbert	Rhododendron
Flowering Almond	Rockrose
Flowering Quince	Rose
Forsythia	Russian Olive
Gardenia	Spirea
Goldenraintree	Squawcarpet
Hackberry	Weigela
Heather	Willow
Holly (Illex spp.)[2]	Yew (Taxus)
Honeysuckle	

[1] Kurume, Mollis, Hardy hybrid types and hardy native Azalea species only.
[2] Except I. crenata, I. rotunda, I. vomitoria.

OTHER USES

UNDER ASPHALT: For general weed control under asphalt (in areas such as roadways, parking lots, recreational areas); after final grade is achieved, apply 20–24 lbs. in 75 to 100 gals. of water per acre. (7½ to 9 oz. per 1,000 sq. ft. in 1½ to 2½ gals. water). Treated area should be covered with asphalt as soon as possible. For re-surfacing work, existing weeds should be sprayed with a contact herbicide to CASORON 50W application.

IMPORTANT NOTICE

Seller warrants that this product conforms to its chemical description and is reasonably fit for the purposes stated on the label when used in accordance with the directions and instructions specified on the label under normal conditions of use, **but neither this warranty nor any other warranty of merchantability or fitness for a particular purpose, express or implied, extends to the use of this product, contrary to label instructions, or under abnormal conditions, or under conditions not reasonably foreseeable to seller, and buyer assumes the risk of any such use.**

COMITE®

AGRICULTURAL MITICIDE

COMPOSITION

ACTIVE INGREDIENTS:	% BY WEIGHT
Propargite [2-(p-tert-butylphenoxy) cyclohexyl 2-propynyl sulfite]*	73.0%
INERT INGREDIENTS:	27.0%
TOTAL:	100.0%

* U.S. Patent No. 3,463,859

** Contains 6.55 pounds OMITE per gallon

EPA Reg. No. 400–104 EPA Est. No.

DANGER
Keep Out of Reach of Children
PELIGRO

PRECAUCION AL USUARIO: Si usted no lee ingles, no use este producto hasta que le etiqueta haya sido explicado ampliamente.

STATEMENT OF PRACTICAL TREATMENT

IF SWALLOWED: Drink promptly a large quantity of milk, egg whites or gelatin solution. If these are not available, drink large quantities of water. Avoid alcohol. Call a physician or Poison Control Center immediately.

IF INHALED: See a physician if an abnormal reaction occurs.

IF ON SKIN: Wash thoroughly with soap and water.

IF IN EYES: Immediately flush eyes with plenty of water. See a physician.

See left side panel for additional precautionary statement.

PRECAUTIONARY STATEMENTS

HAZARDS TO HUMANS

DANGER

Corrosive, causes eye damage. May be fatal if inhaled. Harmful if swallowed or absorbed through skin. Do not breathe vapors or spray mist. Do not get in eyes, on skin or on clothing. Wear goggles or face shield when handling. Wash hands and face thoroughly with soap and water after use and before eating or smoking.

ENVIRONMENTAL HAZARDS

This product is toxic to fish and shrimp. Keep out of lakes, streams or ponds. Do not apply where runoff is likely to occur. Do not apply when weather conditions favor drift from areas treated. Do not contaminate water by cleaning of equipment or disposal of wastes.

PHYSICAL OR CHEMICAL HAZARDS

Flammable, keep away from heat and open flame.

DIRECTIONS FOR USE

It is a violation of Federal law to use this product in a manner inconsistent with its labeling.

STORAGE AND DISPOSAL

Do not contaminate water, food or feed by storage or disposal

PESTICIDE DISPOSAL

Pesticide wastes are acutely hazardous. Improper disposal of excess pesticide, spray mixture or rinsate is a violation of Federal Law. If these wastes cannot be disposed of by use according to label instructions, contact your State Pesticide or Environmental Control Agency, or the Hazardous Waste representative at the nearest EPA Regional Office for guidance.

CONTAINER DISPOSAL

Triple rinse (or equivalent). Then offer for recycling or reconditioning, or puncture and dispose of in a sanitary landfill, or by other procedures approved by state and local authorities.

REENTRY STATEMENT

Do not apply this product in such a manner as to directly or through drift expose workers or other persons. The area being treated must be vacated by unprotected persons.

Do not enter treated areas without protective clothing until sprays have dried.

Because certain states may require more restrictive reentry intervals for various crops treated with this product, consult your State Department of Agriculture for further information.

Written or oral warnings must be given to the workers who are expected to be in a treated area or in an area about to be treated with this product. When oral warnings are given, warnings shall be given in a language customarily understood by workers. Oral warnings must be given if there is reason to believe that written warnings cannot be understood by workers. Written or oral warnings must include the following information:

"DANGER. Area to be treated with Propargite on (date of application). Do not enter without appropriate protective clothing until sprays have dried. If accidentally exposed, consult the Statement of Practical Treatment portion of the pesticide label for first aid recommendations."

GENERAL INSTRUCTIONS

COMITE is a liquid emulsifiable concentrate for preparing sprays to control the banks grass, citrus red, citrus rust, false spider, Pacific spider, six-spotted, strawberry spider, Texas citrus, and two-spotted spider mites.

COMITE is not systemic in action: therefore complete coverage of both upper and lower leaf surfaces and of fruit is necessary for effective control.

After three-quarters full of water, add recommended dosage (from table) of COMITE to the spray tank. Fill tank, agitate and spray thoroughly to cover foliage and fruit for best results.

DOSAGE INSTRUCTIONS

CROP	MITES CONTROLLED	TIMING OF APPLICATION	AMOUNT OF COMITE PER ACRE	GALLONS OF SPRAY SOLUTIONS PER ACRE		TOTAL NUMBER OF SPRAYS PER YEAR	DAYS BEFORE HARVEST (LAST SPRAY)
				GROUND APPLICATION	AERIAL APPLICATION		
Cotton	two-spotted spider Pacific spider strawberry spider	Early[1]	1 pint	15[2] to 30 gallons	Not recommended	3	Before bolls open
		Mid Season to Layby	1½ to 2 pints	25[2] to 40 gallons	—		
			2 pints	—	5 to 15 gallons		
		Layby	2 pints	25[2] to 50 gallons	5 to 15 gallons		
		Layby to Boll Opening	2 pints		5 to 15 gallons		
Cotton (East of the Rocky Mts.)	two-spotted spider Pacific spider Strawberry spider	Early[1]	1 pint	15[2] to 30 gallons	Not recommended	3	Before bolls open
		Mid Season to Layby	1 to 2 pints	25[2] to 40 gallons	5 to 15 gallons		
		Layby to Boll Opening	2 pints	25[2] to 50 gallons	5 to 15 gallons		
Field Corn (grain & silage)	two-spotted spider Banks grass		2–3 pints	minimum of 20 gallons	minimum of 2 gallons[4]	1	30
Sweet Corn[3] (California only)	two-spotted spider Banks grass		2–3 pints	minimum of 20 gallons	minimum of 2 gallons	1	30
Grain Sorghum[5] East of Rocky Mountains)	two-spotted spider Banks grass		1½ to 2 pints	minimum of 20 gallons	minimum of 5 gallons	1	30[6]
Peanuts	two-spotted spider		2 pints	minimum of 20 gallons	minimum of 5 gallons	1	14
Potatoes (Pacific North-west Only)	two-spotted spider		1½ to 2½ pints[7]	20 to 50 gallons	minimum of 10 gallons	2	14
Beans[8] (green, lima & dry) (West of Rocky Mts.)	two-spotted spider		2–3 pints	minimum of 20 gallons	minimum of 5 gallons	2	succulent-7 dry-28
Oranges & Grapefruit[9] (Florida and Texas only)	citrus red Texas citrus six-spotted citrus rust false spider		2–3 pints	minimum of 25 gallons	minimum of 10 gallons	2	7

APPLICATION NOTES

COTTON

1. COMITE may be phytotoxic to young cotton under certain conditions. Age and climatic conditions result in a physiological hardening when plants are generally between 6 and 12 inches tall. Applications can begin after plants are hardened.
2. COMITE may be used in concentrate sprayers designed for concentrate row crop applications. Maintain COMITE dosage rate.

CORN

3. Treat early. Best results are achieved when mite populations are

low, before the mites have damaged the plants. Treatment is recommended when mite populations are just beginning to build and the corn is still small enough (usually 3 to 4 feet) to allow good coverage. In California, ensure applications are made only when corn leaves are dry. This is especially true for white corn varieties which are the most susceptible to plant injury.

4. In Kansas, Colorado and New Mexico use a minimum of 2 gallons spray solution per acre. In California, best results are obtained from applying 10-20 gallons of spray solution per acre. In Texas and New Mexico and other states use a minimum of 5 gallons spray solution per acre.

GRAIN SORGHUM

5. Due to the large number of varieties of sorghum, it is recommended that a small test plot be treated to check phytotoxicity before using COMITE on a large area.
6. SILAGE STAGE: Grain should not be harvested within 60 days of spraying COMITE.

POTATOES

7. The $2\frac{1}{2}$ pints per acre rate should be used when the crop is under extreme mite pressure.

BEANS

8. Do not use on varieties grown for fresh market or on varieties whose value is linked to pod appearance as pods may be spotted.

ORANGES AND GRAPEFRUIT

9. Apply in sufficient water for thorough coverage. On citrus, leaf distortion and/or fruit spotting may occur when used in the spring or it may occur if COMITE is tank mixed with oil or applied within two weeks prior to or following an oil application.

Do not use in spray solution above pH 10.

USE RESTRICTIONS

Cotton—Apply only before bolls open. Do not feed treated foliage or cotton trash to livestock.

Beans—Do not feed or forage treated bean vines or bean trash after harvesting.

Peanuts—Do not graze of feed livestock on treated areas or cut treated forage for hay.

Do not graze or feed livestock on cover crops grown among the trees listed on this label.

General—Do not mix COMITE with alkaline materials (such as lime, Bordeaux mixture or lime sulfur) or with materials containing a large amount of petroleum type solvents.

IMPORTANT NOTICE

Seller warrants that this product conforms to its chemical description and is reasonably fit for the purposes stated on the label when used in accordance with the directions and instructions specified on the label under normal conditions of use, **but neither this warranty nor any other warranty of merchantability or fitness for a particular purpose, express or implied, extends to the use of this product, contrary to label instructions, or under abnormal conditions, or under conditions not reasonably foreseeable to seller, and buyer assumes the risk of any such use.**

DIMILIN® 25W

COTTON AND SOYBEANS

RESTRICTED USE PESTICIDE

For retail sale to and use only by Certified Applicators, or persons under their direct supervision, and only for those uses covered by the Certified applicator's certification.

COMPOSITION

ACTIVE INGREDIENT:	% BY WEIGHT
N-[[(4-Chlorophenyl)amino]carbonyl]-2,6-difluorobenzamide	25.0%
INERT INGREDIENTS:	75.0%
TOTAL:	100.0%

U.S. Patent Numbers 3,748,356; 3,933,908; 3,989,842; 4,013,717; 4,110,469; 4,399,152 and other patents pending.

EPA Reg. No. 37100-8-400 EPA Est. No.

DIMILIN® is a Registered Trademark of Duphar B.V. Amsterdam, Netherlands

CAUTION

Keep Out of Reach of Children

See left side panel for additional precautionary statement.

PRECAUTIONARY STATEMENTS

HAZARDS TO HUMANS

CAUTION

Harmful if swallowed. Avoid breathing of dust. Avoid contact with eyes or skin. Wear one piece overalls which have long sleeves and long pants constructed of finely-woven fabric as specified in the USDA/EPA Guide for Commercial Applicators during mixing, loading or application of DIMILIN. Remove contaminated clothing and wash before reuse. Wear a pesticide respirator jointly approved by the Mining Enforcement and Safety Administration (formerly the U.S. Bureau of Mines) and by the National Institute for Occupational Safety and Health while mixing, loading, or making ground application of DIMILIN.

Avoid contamination of feed and foodstuffs.

ENVIRONMENTAL HAZARDS

Do not apply to lakes, streams, ponds or other bodies of water. Do not apply when weather conditions favor significant drift from target area to lakes, streams, ponds or other bodies of water. Do not apply DIMILIN near any person or dwelling. It is recommended that the product be applied at least 300 feet from dwellings when they are downwind. Flagmen may not be used for marking the spray swath or pattern, unless protected in an enclosed air conditioned vehicle. Do not enter the treated field until the day following application or the spray deposit has dried. Do not contaminate water by cleaning of equipment or disposal of wastes. Do not graze or feed treated foliage.

DIRECTIONS FOR USE

It is a violation of Federal Law to use this product in a manner inconsistent with its labeling.

STORAGE AND DISPOSAL

Do not contaminate water, food or feed by storage or disposal.

STORAGE

Store in a dry location.

PESTICIDE DISPOSAL

Wastes resulting from the use of this product may be disposed of on site or at an approved waste disposal facility.

CONTAINER DISPOSAL

Completely empty bag into application equipment. Then dispose of empty bag in a sanitary landfill or by incineration, or if allowed by State and local authorities, by burning. If burned, stay out of smoke.

DIRECTIONS FOR USE

COTTON

To control *boll weevil* through reproduction suppression, apply $^1/_4$ to $^1/_2$ pound of DIMILIN 25W per acre per treatment in combination with 2 to 4 quarts of emulsifiable paraffinic crop oil, such as DI-MOIL®. Consult your supplier or Uniroyal representative for crop oil specifications.

MIXING:

Paraffinic oil use:

1. Fill tank with $^1/_2$ of required water.
2. Begin agitation and add required amount of DIMILIN.
3. Continue agitation while adding remainder of water.
4. Add proper quantity of oil slowly.

To avoid formation of an invert emulsion use at least 2 parts water for each part oil.

For **aerial application** apply in 2 to 3 gallons total volume per acre. For **ground application** apply in 5 to 30 gallons of water per acre to give uniform coverage.

For maximum control of weevil reproduction, make first application at pinhead square stage of cotton growth when overwintering weevils are entering the fields. Repeat treatments should allow at least 5 to 7 day intervals between applications. Do not make more than six applications and do not apply more than $1^1/_2$ pounds of DIMILIN 25W per acre during one growing season. Do not apply after bolls begin to open. Do not rotate crops other than cotton or soybeans until six (6) months following last application of DIMILIN.

DIMILIN does not kill the adult weevils, however, eggs deposited by treated female weevils will not hatch, thus limiting reproduction. The control of egg hatch and larval development within the square prevents shedding and will allow normal boll development. After initial contact, 7–10 days are required before infertile eggs are laid.

In use for boll weevil control, DIMILIN allows normal build-up of beneficial insects which may aid in control of bollworm and budworm. Emulsifiable concentrate insecticide formulations used in tank mixes, in the presence of oil, may result in phytotoxicity. Care should be taken where such mixtures are used.

SOYBEANS

To control *velvetbean caterpillar. Mexican bean beetle larvae* and *green cloverworm*, apply DIMILIN 25W at the rate of 2–4 ounces ($^1/_8$–$^1/_4$ pound) per acre. Make application when larvae are small (less than $^1/_2$ inch) to give greater control and minimum insect damage to leaves. Repeat application if damaging numbers reappear. DIMILIN 25W may be applied at the lower rate (2 ounces) to prevent a velvetbean caterpillar build-up when the vegetative growth of soybeans is completed and as pod formation begins. Consult local Extension Service regarding infestation levels requiring treatment.

Aerial Application

Apply recommended amount of DIMILIN 25W in sufficient water (1–3 gallons per acre) to achieve uniform coverage of foliage.

Ground Application

Apply recommended amount of DIMILIN 25W in 9–35 gallons of water per acre to give uniform coverage.

Do not make more than two applications per season. Do not apply within 21 days of harvest. Do not cut for hay nor allow milk or meat animals to graze treated fields.

DIMILIN inhibits the moulting process of larvae, therefore it does not provide immediate kill. From 3 to 5 days may be required before populations are reduced.

Do not rotate crops other than soybeans or cotton until six (6) months following last application.

IMPORTANT NOTICE

Seller warrants that this product conforms to its chemical description and is reasonably fit for the purposes stated on the label when used in accordance with the directions and instructions specified on the label under normal conditions of use, **but neither this warranty nor any other warranty of merchantability or fitness for a particular purpose, express or implied, extends to the use of this product, contrary to label instructions, or under abnormal conditions, or under conditions not reasonably foreseeable to seller, and buyer assumes the risk of any such use.**

DIMILIN® 25W

MUSHROOMS AND CHRYSANTHEMUMS

RESTRICTED USE PESTICIDE

For retail sale to and use only by Certified Applicators, or persons under their direct supervision, and only for those uses covered by the Certified applicator's certification.

COMPOSITION

ACTIVE INGREDIENT:	% BY WEIGHT
N-[[(4-Chlorophenyl)amino]carbonyl] -2.6-difluorobenzamide*	25.0%
INERT INGREDIENTS:	75.0%
TOTAL:	100.0%

* Diflubenzuron; U.S. Patent Numbers 3,748,356; 3,933,908; 3,989,842; 4,013,717; 4,110,469; 4,399,152

EPA Reg. No. 37100-16-400 EPA Est. No.

DIMILIN® is a Registered Trademark of Duphar B.V. Amsterdam, Netherlands

CAUTION

Keep Out of Reach of Children

See left side panel for additional precautionary statement.

PRECAUTIONARY STATEMENTS

HAZARDS TO HUMANS

CAUTION

Harmful if swallowed. Avoid breathing of dust. Avoid contact with eyes or skin. Avoid contamination of feed and foodstuffs.

ENVIRONMENTAL HAZARDS

Do not contaminate water by cleaning of equipment or disposal of wastes.

DIRECTIONS FOR USE

It is a violation of Federal Law to use this product in a manner inconsistent with its labeling.

STORAGE AND DISPOSAL

Do not contaminate water, food or feed by storage or disposal.

STORAGE

Store in a dry location.

PESTICIDE DISPOSAL

Wastes resulting from the use of this product may be disposed of on site or at an approved waste disposal facility.

CONTAINER DISPOSAL

Completely empty bag into application equipment. Then dispose of empty bag in a sanitary landfill or by incineration, or if allowed by State and local authorities, by burning. If burned, stay out of smoke.

DIRECTIONS FOR USE

MUSHROOMS

To control sciarid fly larvae in mushroom growing facilities. The Insect Growth Regulator DIMILIN in the mushroom growing media will prevent the development of the larval stages of the sciarid fly. This effectively stops reproduction in the growing medium and prevents damage to the mushrooms.

Because of its unique type of activity, do not expect immediate reductions in adult fly populations. DIMILIN does not directly affect adults, but kills the larva in the growing medium.

COMPOST TREATMENT: Apply 2.4 to 4 lbs. DIMILIN 25W per 1000 sq. feet to the compost at spawning time by thorough incorporation such as with a spawning machine. This is equivalent to 30–50 ppm active assuming a compost wet weight of 40 lbs. per cu. ft.

CASING TREATMENT: Apply 13.5 oz of DIMILIN 25W per 1000 sq. feet at the time of casing by thorough incorporation into the casing or in sufficient water to obtain a thorough drench. This is equivalent to a rate of 30 ppm active assuming a casing weight of 6700 pounds per 1000 sq. ft.

CHRYSANTHEMUMS

For control of beet armyworm, apply ½ to 1 lb. of DIMILIN 25W per acre in a dilute spray not to exceed 200 gallons of water per acre. Begin applications when larvae appear and repeat at weekly intervals as required. DIMILIN interferes with the molting process of insects and is, therefore, slow in its action. The insect dies during molting following contact, therefore, full effect will not be seen for 3 to 5 days following application.

IMPORTANT NOTICE

Seller warrants that this product conforms to its chemical description and is reasonably fit for the purposes stated on the label when used in accordance with the directions and instructions specified on the label under normal conditions of use, **but neither this warranty nor any other warranty of merchantability or fitness for a particular purpose, express or implied, extends to the use of this product, contrary to label instructions, or under abnormal conditions, or under conditions not reasonably foreseeable to seller, and buyer assumes the risk of any such use.**

DYANAP®

Kills Annual Broadleaf Weeds in Peanuts and Soybeans

ACTIVE INGREDIENTS:	% BY WEIGHT
Sodium salt of naptalam (N-1-Naphthylphthalamic acid)*	22.3%
Sodium Salt of dinoseb (2-sec-butyl-4,6-dinitrophenol)**	11.5%
INERT INGREDIENTS:	66.2%
TOTAL:	100%

* Equivalent to 2 lbs. of the sodium salt of naptalam per gallon.

** Equivalent to 1 lb. of the sodium salt of dinoseb per gallon.

EPA Reg. No. 400-75 EPA Est. No.

DYANAP is a Registered Trademark of UNIROYAL CHEMICAL COMPANY, INC.

DANGER
Keep Out of Reach of Children
PELIGRO

PRECAUCION AL USUARIO: Si usted no lee inglés, no use este producto hasta que le etiqueta haya sido explicado ampliamente.

STATEMENT OF PRACTICAL TREATMENT

SYMPTOMS OF POISONING: Excessive Fatigue, Sweating, Thirst and Fever. If symptoms of poisoning develop from any type of exposure to DYANAP **SEND FOR A PHYSICIAN.**

FIRST AID: Have patient lie quiet in coolest spot available. If feverish, cool with cold compresses or by immersion in cool water.

IF SWALLOWED: Call a physician or Poison Control Center. Drink 1 or 2 glasses of water and induce vomiting by touching back of throat with finger or blunt object. Do not induce vomiting or give anything by mouth to an unconscious person.

IF SPLASHED IN EYES: Immediately flush eyes with plenty of water for at least 15 minutes and get medical attention.

IF SPILLED ON SKIN: Immediately remove contaminated clothing, INCLUDING SHOES, and wash skin with soap and plenty of water. If symptoms of poisoning develop, send for a physician and treat as in FIRST AID, above. Discard contaminated clothing and shoes, or clean them thoroughly before reuse. Staining may result if contaminated articles are allowed to contact carpets, upholstery, etc.

NOTE TO ATTENDING PHYSICIAN: Active ingredient is a metabolic stimulant. Treat symptomatically. Do not administer atropine, aspirin, or other antipyretics to control fever.

SEE LEFT SIDE PANEL FOR ADDITIONAL PRECAUTIONARY STATEMENT.

PRECAUTIONARY STATEMENTS

HAZARDS TO HUMANS

DANGER

Corrosive, causes eye damage. May be fatal if swallowed or absorbed through skin. Harmful if inhaled. Do not breathe vapors or spray mist. Do not get in eyes, on skin or on clothing. Wear goggles or face shield and impermeable gloves and apron while pouring and transferring the concentrated product and at other times when contact with concentrate or spray is likely. Wear a long-sleeved shirt, long-legged pants or coveralls, and shoes and socks, while mixing concentrate and while spraying. Do not wear contaminated clothing or shoes. Remove contaminated clothing immediately and wash thoroughly before reuse. Wash hands and face thoroughly with soap and water after use and before eating or smoking.

ENVIRONMENTAL HAZARDS

This product is toxic to fish and wildlife. Birds and other wildlife in treated areas may be killed. Do not apply directly to water. Do not apply where runoff is likely to occur. Do not apply when weather conditions favor drift from areas treated. Do not contaminate water by cleaning of equipment, or disposal of wastes. Apply this product only as specified on this label.

PHYSICAL OR CHEMICAL HAZARDS

Do not use, pour, spill or store near heat or open flame.

DIRECTIONS FOR USE

It is a violation of Federal law to use this product in a manner inconsistent with its labeling.

REENTRY STATEMENT

Do not apply this product in such a manner as to directly or through drift expose workers or other persons. The area being treated must be vacated by unprotected persons.

Do not enter treated areas without protective clothing until sprays have dried.

STORAGE AND DISPOSAL

Do not contaminate water, food or feed by storage or disposal.

STORAGE

DYANAP is normally a solution but it may crystallize in the cold (32°F.). Gentle warming (60°F.) will redissolve any crystals. Mild agitation, such as rolling, shaking, or stirring before use is recommended. Avoid contamination of foodstuffs, feeds, fertilizers, seeds, plants and pesticides. Store in a dry location.

PESTICIDE DISPOSAL

Pesticide wastes are acutely hazardous. Improper disposal of excess pesticide spray mixture, or rinsate is a violation of Federal Law. If these wastes cannot be disposed of by use according to label instructions, contact your State Pesticide or Environmental Control Agency, or the Hazardous Waste representative at the nearest EPA Regional Office for guidance.

CONTAINER DISPOSAL
Triple rinse (or equivalent). Then offer for recycling or reconditioning, or puncture and dispose of in a sanitary landfill, or by other procedures approved by state and local authorities.

EQUIPMENT CLEANING
Thoroughly rinse spray tank, hose and nozzles with water after application. Use DYANAP in metal, plastic or fiberglass tanks. DYANAP can be removed from tanks by washing thoroughly with ammonia water and rinsing several times with water containing a good detergent. Denatured alcohol is an effective solvent for removing caked residues of DYANAP. Acetone and lacquer thinner are equally effective but must be used with caution. These solvents may affect certain nonmetal tanks and pump parts.

DIRECTIONS FOR USE

GENERAL INSTRUCTIONS
DYANAP® is a herbicide for controlling certain broadleaf weeds in soybeans and peanuts. DYANAP may be applied preemergence or postemergence to either crop. For broader spectrum weed control DYANAP may be tank mixed with certain grass and broadleaf herbicides. DYANAP can also be applied after using a preplant incorporated herbicide.

MIXING INSTRUCTIONS
Put one half the recommended amount of water in the spray tank, add the required amount of DYANAP and/or other herbicides, then add the remainder of the water to the tank while mixing. Stir or use bypass agitation until completely mixed. Calibrate sprayer accurately prior to use on crop.
Before tank mixing DYANAP with other recommmended herbicides, read and carefully observe the precautionary statements and restrictions on each label. Add powdered formulations to spray mix first.

BE SURE SPRAY TANK HAS BEEN CLEANED OF ALL PREVIOUSLY USED PESTICIDES BEFORE FILLING WITH DYANAP SOLUTION.

DO NOT MIX DYANAP WITH LIQUID FERTILIZER SOLUTIONS SINCE SOLID PRECIPITATES MAY FORM.
DO NOT ADD SURFACTANTS OR ADJUVANTS TO DYANAP SPRAY SOLUTIONS.

GRAZING RESTRICTION
DO NOT GRAZE OR FEED TREATED HAY OR FORAGE FROM ANY TREATED CROP.

SOYBEANS
DIRECTIONS FOR USE

WEEDS CONTROLLED
Consult the chart for the weeds you need to control, then go to the DYANAP RATE & TIMING TABLE to find out how to use DYANAP to control them. Read the PREEMERGENCE and POSTEMERGENCE USE notes for additional information.

WEED CONTROL EXPECTED

Weed	Preemergence	Postemergence 1-3 inches	Postemergence 3-6 inches
annual morningglories	partial	yes[2]	yes[2,5]
black nightshade	partial[1]	yes	partial
bristly starbur	yes	yes	yes
Canada thistle	no	partial[4]	partial[4]
carpetweed	yes[1]	no	no
chickweed	yes	no	no
cocklebur	partial	yes	yes[2]
common ragweed	yes[1]	partial	no
Florida beggarweed	yes	no	no
Florida pusley	yes[1]	no	no
galinsoga	yes	no	no
giant ragweed (horseweed)	partial	yes[2]	partial[2]
jimsonweed	yes	yes	yes
lambsquarters	yes[1]	no	no
mustards	yes	yes	partial
pigweed	yes[1]	partial	no
purslane	yes[1]	no	no
shepherdspurse	yes	no	no
smartweed (seedling)	partial[1]	partial[2]	no
velvetleaf	partial	no	no
wild buckwheat	partial	yes[2]	yes[2]
wild sunflower	partial	yes[2]	partial[2,3]

1. LASSO® or DUAL® tank mix improves control.
2. Control is improved with tank mixes of BUTYRAC® 200 or BUTOXONE®.
3. Apply before 3-4 leaves fully open.
4. Burnback.
5. Before vining begins.

Partial control refers to reducing competition from the weed so that the soybean may favorably compete with and outgrow the weed.

Note that good control of heavy infestations of large-seeded broadleaf weeds including cocklebur, giant ragweed and morningglory may require more management than a single preemergence to cracking stage treatment. Cultivation and or post emergence treatments may be required.

RATE & TIMING TABLE

Stage of Soybean Weed Height	Preemergence	Postemergence 1-3 inches	Postemergence 3-6 inches
Rate/acre broadcast	3-6 quarts[1]	1.5-3 quarts[2,3]	3-4 quarts[2,3]
GPA[4]	20-40	8-10	8-10
PSI at nozzle	20-30	40-60	40-60
SUGGESTED NOZZLES	flat fan, flood and whirl-chamber	hollow cone (TX) or disc and core	hollow cone (TX) or disc and core

1. A 3 quart rate of DYANAP may be used only on light sandy soil or when a supplemental postemergence DYANAP application is part of the weed control program. Soil temperatures of 60°F enhance activity.
2. Use the lower rate when temperatures are greater than 85°F and the higher rate when temperatures are 75°-85°F. If temperatures are below 75°F or above 95°F, delay application until appropriate temperatures are reached.
3. When low rates are used, weeds larger than 3 inches may not be killed and new weeds may emerge. A repeat application with 2 quarts DYANAP per acre may be made after 10-14 days. Follow temperature guidelines in note 2.
4. DYANAP can be applied aerially. Use 5 gallons of water per acre and fly at 10-15 feet above tops of soybeans. Application of fine spray particles is essential. Do not apply when conditions favor drift from target area.

PREEMERGENCE THROUGH CROOK TREATMENTS
DYANAP Preemergence Treatments may be made at Planting through crook stage.

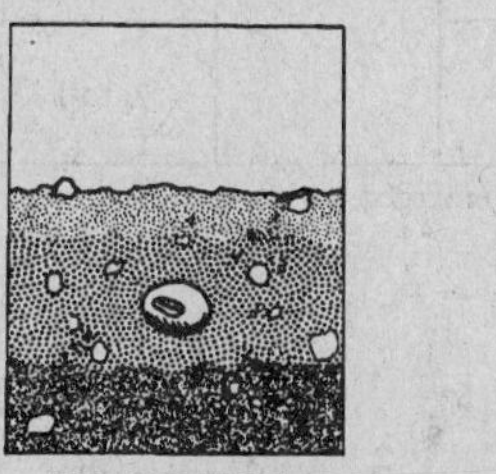

PLANTING
(immediately after seed is planted.)

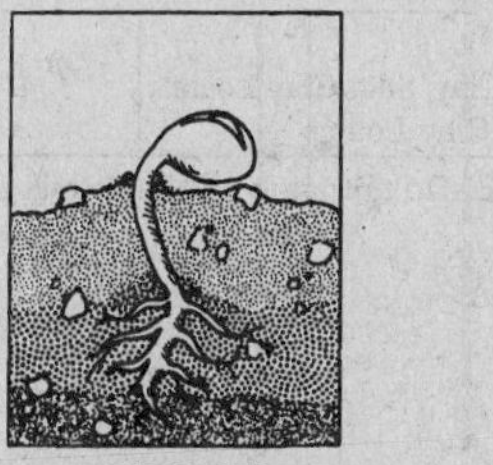

CROOK
(from cracking to before the soybean cotyledons open.)

Preemergence Tank Mix Treatments

DYANAP plus LASSO Use Directions:
In addition to broadleaf weeds that DYANAP® will control, DYANAP plus LASSO® will control the following annual grasses and broadleaf weeds:

Crabgrass, barnyardgrass, goosegrass, Johnsongrass (seedling), fall panicum, witchgrass, broadleaf signalgrass, giant foxtail, green foxtail, yellow foxtail, black nightshade, hairy nightshade, red rice, and yellow nutsedge.

Use 6 quarts of DYANAP and 2 quarts of LASSO as tank mixture in 20 to 40 gallons of water per acre. On light sandy soils reduce rates by 25%. Reduce rates proportionally for band treatments.

DYANAP plus SURFLAN Use Directions:
In addition to broadleaf weeds that DYANAP® will control, DYANAP plus SURFLAN 75W or SURFLAN A.S. will control the following annual grasses:

Crabgrass, barnyardgrass, goosegrass, Johnsongrass (seedling), fall panicum, witchgrass, broadleaf signalgrass, giant foxtail, green foxtail, sandbur, cupgrass, sprangletop, stinkgrass, and windmillgrass. Apply in 20–40 gallons of spray mixture per acre as a band or broadcast application. Tank mixture rates are given in the following table. Reduce rates proportionally for band treatments. Apply before crop emerges.

DYANAP/SURFLAN
Broadcast Rates Per Acre

Soil Texture	DYANAP [Qts.]		SURFLAN 75W [Lbs.]		SURFLAN A.S. [Qts.]
Light Sandy Soils	$4\frac{1}{2}$	PLUS	$\frac{2}{3}$–1	OR	$\frac{1}{2}$–$\frac{3}{4}$
Medium Soils	6	PLUS	1–$1\frac{1}{3}$	OR	$\frac{3}{4}$–1
Clay Loam Soils	6	PLUS	$1\frac{1}{3}$–$1\frac{2}{3}$	OR	1–$1\frac{1}{4}$

NOTE: Do not apply this tank mix to emerged soybeans. Do not apply to any soils with over 5% organic matter. Do not use on silt loam soils of extremely fine textures.

DYANAP plus DUAL Use Directions:
In addition to broadleaf weeds that DYANAP will control, DYANAP plus DUAL® 8E will control the following grasses and broadleaf weeds:

Crabgrass, barnyardgrass, goosegrass, fall panicum, giant foxtail, green foxtail, Johnsongrass (seedling), sandbur,* yellow foxtail, yellow nutsedge and witchgrass and black nightshade*.*

* DYANAP/DUAL will provide partial control.

Apply Dual 8E + Dyanap in 20–40 gals. of spray mixture per acre during the interval preemergence during planting (behind the planter) through crookstage (from soil cracking to before soybeans true leaves open) according to the rates specified in the following table. Do not apply this tank mixture in liquid fertilizer.

	DUAL/DYANAP Broadcast Rates Per Acre			
	Less Than 3% Organic Matter		3% Organic Matter or Greater	
Soil Texture	DUAL 8E	DYANAP	DUAL 8E	DYANAP
Coarse Sand, Loamy Sand, Sandy Loam	$1\frac{1}{4}$ pts.	4–5 qts.	$1\frac{1}{2}$ pts.	4–5 qts.
Medium Loam, Silt Loam, Silt	$1\frac{1}{2}$ pts.	6 qts.	2 pts.	6 qts.
Fine Silty Clay Loam, Sandy Clay Loam, Silty Clay, Sandy Clay, Clay Loam, Clay	2 pts.	6 qts.	2–$2\frac{1}{2}$ pts.	6 qts.
Muck or Peat Soils	DO NOT USE			

If $\frac{1}{2}$–1 inch of rainfall does not occur within 7 days after application, shallow incorporation ($\frac{1}{2}$–1 inch) is recommended. Cultivate if weeds develop.

The following tank mix may be applied UP TO THE CRACKING STAGE of soybeans.

DYANAP plus SENCOR (after a preplant incorporated herbicide) OR **DYANAP plus SENCOR plus LASSO.**

This tank mix may be applied UP TO THE SOIL CRACKING STAGE of soybeans.

Use Directions:
In addition to the broadleaf weeds and grasses that DYANAP/LASSO will control a three way tank mix will provide added control of:

Crowfootgrass, spurred anoda, buffalobur, Venice mallow. Sesbania (coffeebean or coffeeweed), sicklepod (cassia), prickly sida (teaweed), and spotted spurge.

Apply in 20–40 gallons of spray mixture per acre as a band or broadcast application. Tank mixture rates are given in the following table. Reduce rates proportionally for band treatments.

DYANAP/SENCOR/LASSO
Broadcast Rates Per Acre

SOIL TEXTURE	DYANAP QUARTS		SENCOR 50W POUNDS		SENCOR 4F PINTS		SENCOR DF POUNDS		LASSO 4EC QUARTS
Coarse[a] Sandy Loam, Loamy Sand	2 to 4		$\frac{1}{2}$		$\frac{1}{2}$		$\frac{1}{3}$		$1\frac{1}{2}$
Medium Loam, Silt Loam, Silt, Sandy Clay, Sandy Clay Loam	3 to 4	PLUS	$\frac{3}{4}$ to 1	OR	$\frac{3}{4}$ to 1	OR	$\frac{1}{2}$ to $\frac{2}{3}$	PLUS	$1\frac{1}{2}$ to 2
Fine Silty Clay, Silty Clay Loam[b], Clay, Clay Loam	4		$\frac{3}{4}$ to 1		$\frac{3}{4}$ to 1		$\frac{1}{2}$ to $\frac{2}{3}$		2

NOTE: Do not apply this tank mix to emerged soybeans.

DYANAP/LEXONE/LASSO Use Directions:
May also be used as a preemergence treatment to control weeds in soybeans.

DYANAP/SENCOR/LASSO
Broadcast Rates Per Acre

	DYANAP		SENCOR 50W		SENCOR 4L		SENCOR DF		LASSO 4EC
SOIL TEXTURE	QUARTS		POUNDS		PINTS		POUNDS		QUARTS
Coarse Sandy Loam, Loamy Sand	2	PLUS	¼ to ½	OR	¼ to ½	OR	⅛ to ⅓	PLUS	1½ to 2
Medium Loam, Silt Loam, Silt, Sandy Clay, Sandy Clay Loam	3	PLUS	½ to ¾*	OR	½ to ¾*	OR	⅓ to ½	PLUS	2 to 2½
Fine Silty Clay, Silty Clay Loam[b], Clay, Clay Loam	4	PLUS	¾ to 1*	OR	¾ to 1*	OR	½ to ⅔*	PLUS	2-3

* On Clarion Webster soils in Minnesota and Iowa or on similar alkaline (calcareous) soils with a pH of 7.5 or above, apply LEXONE® at ½ pound/acre on medium soils. On fine soils, use ½ to ¾ pound/acre LEXONE. Use the higher rate only where soil is less than 7.5 and where weed pressure is heavy.

[a] Not recommended on sandy soil with less than 1% organic matter.
[b] Silty clay loam soils are transitional soils and may be classified as medium textured soils in some regions of the U.S.

APPLICATION NOTES

NOTE 1. Soybean seed should be well covered with soil. Do not use on shallow planted or exposed seed.

NOTE 2. Accurate sprayer calibration and thorough, uniform coverage of the soil is essential for good weed control.

NOTE 3. Do not apply where water is standing in the field.

NOTE 4. If heavy rains immediately follow a preemergence application of DYANAP, weed control normally will not be impaired. However, some stunting of growth or even stand reduction may be noted, especially in lighter soils. Cool soil temperatures may accentuate this condition. With the return of good growing conditions, the stunted plants will usually recover and maintain a normal growth pattern.

NOTE 5. If replanting is necessary after DYANAP has been applied, thoroughly disc area prior to replanting.

NOTE 6. Some rainfall or irrigation is needed to get DYANAP in contact with the germinating weed seed. Rainfall or sprinkler irrigation amounting to ½-¾ inches within 5 days after application is ideal. If neither has occurred within 7 days, shallow incorporation such as with a rotary hoe (½-1 inch) is recommended.

POSTEMERGENCE TREATMENTS

When weeds have emerged and soybean true leaves (unifoliate) have opened, DYANAP should be applied with high pressure hollow cone type nozzles to obtain small spray droplets. Small droplets provide good coverage on weeds and normally are held off the soybean leaf surface by leaf hairs. Movement through soybean leaf hairs could result in increased leaf injury. Leaf injury is minimized by following these precautions:

1. Use high pressure (40-60 psi).
2. Use low volume (8-10 gpa) apply by ground or 5 gpa by air.
3. Do not apply to wet soybean foliage. Wait until dew has evaporated.
4. Do not spray if rain is expected within 6 hours.
5. Do not use adjuvants or surfactants in the spray solution. Be sure all previous spray chemicals and spray adjuvants have been cleaned from sprayer.
6. Apply to actively growing soybeans.
7. Be sure spray boom is positioned above foliage to give complete coverage. This is usually 18-35″ above the soybean tops.

POSTEMERGENCE TANK MIX: The addition of 1-2 fl. oz. of BUTYRAC® 200 or BUTOXONE® to the labeled rate of DYANAP may be used on soybeans after the first trifoliate stage is reached. This tank mix will improve the control of larger and hard to kill weeds. Refer to the weed chart. Use lower rates on smaller weeds and soybeans. Use no more than 2 oz. BUTYRAC 200 or BUTOXONE during growing season.

The tank mix is very effective for the control of cocklebur from the time it emerges until it is 6 inches tall. DYANAP will suppress the growth of cocklebur and morningglory taller than 6 inches reducing their competition to the soybean crop. For best results, spray when weeds are small and succulent.

POSTEMERGENCE TREATMENT APPLICATION NOTES:

NOTE 1. DYANAP may also be used as a directed spray. Adjust rates appropriately. Caution—soybean leaves contacted by the direct spray may be burned.

NOTE 2. Do not apply when wind conditions favor drift to non-target areas.

NOTE 3. Do not apply to soybeans that have been stressed by herbicide injury, draught, excessive water, diseases or nematodes. If applied under these conditions, soybean tolerance to DYANAP will be low and weed control will be impaired.

NOTE 4. Do not use BUTYRAC 200 or BUTOXONE on stressed soybeans.

PEANUTS
DIRECTIONS FOR USE

Preemergence Treatments

Read and follow all General Instructions

DYANAP will control the following broadleaf weeds:

Florida pusley, chickweed, lambsquarters, ragweed, purslane, velvetleaf, cocklebur, bindweed, groundcherry, mustards, shepherdspurse, pigweed, galinsoga, carpetweed, morningglory, beggarweed (beggar lice), teaweed (prickly sida), jimsonweed, smartweed, wild sunflower and bristly starbur.

Ground Application:

Use 6 quarts of DYANAP in 20 to 40 gallons of water per acre broadcast. DYANAP may be also applied as a band treatment.

Preemergence treatments may be made at planting through soil cracking.

Preemergence Tank Mix Treatments

DYANAP® plus LASSO® Use Directions:

In addition to broadleaf weeds that DYANAP will control, DYANAP® plus LASSO® will control the following annual grasses:

Crabgrass, barnyardgrass, goosegrass, Johnsongrass (seedling), Fall panicum, witchgrass, broadleaf signalgrass, giant foxtail, green foxtail and yellow foxtail.

Use 6 quarts of DYANAP plus 2 quarts of LASSO as a tank mixture in 20 to 40 gallons of water per acre broadcast. Application may be made immediately after planting through cracking. Can be used as a sequential application after Vernam® or Balan®.

DYANAP® plus DUAL® Use Directions:

In addition to broadleaf weeds that DYANAP will control, DYANAP plus DUAL will control:

Crabgrass, barnyardgrass, goosegrass, fall panicum, witchgrass, broadleaf signalgrass, giant foxtail, green foxtail, yellow foxtail, foxtail millet, southwestern cupgrass and yellow nutsedge.

This tank mix will partially control:

Black nightshade, coffeeweed, sandbur, johnsongrass (seedling), sicklepod, volunteer sorghum.

Use 6 quarts of DYANAP plus 1½–3 pints of DUAL®8E as a tank mixture in 15–40 gallons of water per acre broadcast. Application may be made immediately after planting through cracking. Can be used as a sequential application after Vernam®, Balan®, or Treflan®.

Postemergence Treatments

An application of DYANAP by ground or air made over the top when the peanuts are 2 to 3 inches across the crown will control the following broadleaf weeds:

Morningglory, giant ragweed, pigweed, bristly starbur, prickly sida, common cocklebur, jimsonweed.

For best results, spray when the weeds are small and succulent.

Ground Application:

Apply 2 to 6 quarts of DYANAP in 8 to 10 gallons of water per acre (broadcast rate). The spray boom should be positioned 18″–36″ above the tops of the peanuts and be equipped with hollow cone nozzles. Maintain a high spray pressure (i.e. 40 to 50 PSI) during application. This insures a fine spray mist for better overall coverage. Use 2 to 4 quarts of DYANAP on broadleaf weeds up to 3 inches tall, such as jimsonweed, morningglory, and pigweed or use 6 quarts if the weed pressure is heavy. DYANAP is very effective for the control of cocklebur from the time it emerges until it is 6 inches tall. DYANAP will suppress the growth of cocklebur and morningglory taller than 6 inches, reducing their competition to the peanut crop. For best results in Texas and Oklahoma, use 4–6 quarts per acre.

Aerial Application:

Apply 2 to 6 quarts of DYANAP in a minimum amount of 5 gallons of water per acre and follow instructions given for ground application above.

AERIAL APPLICATION NOTES

NOTE 1. Fly plane 10 to 15 feet above the tops of the peanuts.

NOTE 2. Do not apply when weather conditions favor drift from area treated.

NOTE 3. DO NOT fly closer than 50 feet to any cotton crop.

Postemergence Tank Mix Treatment

A tank mix of DYANAP plus Butyrac* 200 or Butoxone** will improve the control of larger and hard-to-kill weeds.

For peanuts grown in the **Southeast** add 1¾ to 3½ fl. oz. of Butyrac 200 or Butoxone and DYANAP.

* Butyrac 200 contains 2 lbs/gallon of 4-(2,4-dichlorophenoxy) butyric acid and is a registered trademark of Union Carbide.

** Butoxone contains 1.75 lbs/gallon of 4-(2,4-dichlorophenoxy) butyric acid and is registered trademark of Rhone-Poulenc, Inc.

POSTEMERGENCE TREATMENT APPLICATION NOTES

NOTE 1. Applications should not be made to wet peanut foliage.

NOTE 2. Do not use surfactants with DYANAP in overtop applications.

NOTE 3. Delay spray application if rain is expected within 6 hours of spraying.

NOTE 4. Under extreme drought conditions and low humidity weeds are harder to control. DYANAP effect on weeds sprayed under these conditions is lessened and weed control will be impaired.

NOTE 5. Do not mix DYANAP with liquid fertilizer solutions.

IMPORTANT NOTICE

Seller warrants that this product conforms to its chemical description and is reasonably fit for the purposes stated on the label when used in accordance with the directions and instructions specified on the label under normal conditions of use, **but neither this warranty nor any other warranty of merchantability or fitness for a particular purpose, express or implied, extends to the use of this product, contrary to label instructions, or under abnormal conditions, or under conditions not reasonably forseeable to seller, and buyer assumes the risk of any such use.**

BUTYRAC 200 contains 2 lbs/gallons of 4-[2-4-dichlorophenoxy] butyrac acid and is a registered trademark of Union Carbide.

BUTOXONE contains 1.75 lbs/gallon of 4-[2,4-dichlorophenoxy] butyrac acid and is a registered trademark of Rhone-Poulenc, Inc.

DUAL is a registered trademark of Ciba Geigy Corp.

LASSO is a registered trademark of Monsanto Agricultural Products Co.

LEXONE is a registered trademark of E.I. duPont de Nemours & Co., Inc.

SENCOR is a registered trademark of Mobay Chemical Corp.

TREFLAN, SURFLAN & BALAN are registered trademarks of Elanco Products Co., Div. of Eli Lilly Co.

VERNAM is a registered trademark of Stauffer Chemical Co.

HARVADE®-5F

HARVEST GROWTH REGULANT

For Cotton

COMPOSITION

ACTIVE INGREDIENT:	% BY WEIGHT
dimethipin (2,3-Dihydro-5,6-dimethyl-1, 4-dithiin 1,1,4,4-tetraoxide*)	48.0%**
INERT INGREDIENTS:	52.0%
TOTAL:	100.0%

* U.S. Patent No. 3,920,438

** Contains 4.9 pounds of active ingredient per gallon.

EPA Reg. No. 400-155 EPA Est. No.

HARVADE is a Registered Trademark of UNIROYAL CHEMICAL COMPANY, INC.

CAUTION

Keep Out of Reach of Children

SHAKE WELL BEFORE USING

STATEMENT OF PRACTICAL TREATMENT

IF SWALLOWED: Drink two large glasses of water and induce vomiting. Contact a physician.

IF ON SKIN: Wash affected area with plenty of soap and water.

IF IN EYES: Flush eyes with plenty of water for at least 15 minutes. Contact a physician.

IF INHALED: Remove to fresh air; if abnormal symptoms develop, contact a physician.

See left panel for additional precautionary statements.

PRECAUTIONARY STATEMENTS

HAZARDS TO HUMANS

CAUTION

Harmful if swallowed or inhaled. Do not breathe vapors or spray mist. Avoid contact with skin, eyes or clothing. Wear a long sleeve shirt and impermeable gloves during mixing, loading, flagging, or application of HARVADE-5F. In case of contact, immediately flush eyes or skin with plenty of water. Get medical attention if irritation persists. Keep away from food or food products.

Wash hands and face thoroughly with soap and water after use and before eating or smoking.

ENVIRONMENTAL HAZARDS

Do not apply directly to water. Do not contaminate water by cleaning of equipment or disposal of wastes. Do not apply when weather conditions favor drift from treated area.

DIRECTIONS FOR USE

It is a violation of Federal law to use this product in a manner inconsistent with its labeling.

STORAGE AND DISPOSAL

Do not contaminate water, food or feed by storage or disposal.

PESTICIDE DISPOSAL
Wastes resulting from the use of this product may be disposed of on site or at an approved waste disposal facility.

CONTAINER DISPOSAL
Triple rinse (or equivalent). Then offer for recycling or reconditioning, or puncture and dispose of in a sanitary landfill, or incineration, or if allowed by State and local authorities, by burning. If burned, stay out of smoke.

MIXING INSTRUCTIONS

Invert container several times before using to insure that the contents are uniformly mixed. Fill the spray tank with 2/3 of the required amount of water. Add the desired amount of HARVADE-5F, the selected spray adjuvant and remaining required amount of water. The HARVADE-5F and spray adjuvant will suspend easily with minimum agitation.

DIRECTIONS FOR USE

General Information
HARVADE-5F is a plant growth regulant that effects certain plant processes which lead to defoliation in cotton. Since HARVADE-5F must enter the plant through the leaves for optimum activity, it must be applied in a manner that will allow maximum penetration through the leaf surface. Complete coverage of the entire plant is essential for optimum effectiveness. **HARVADE-5F MUST BE USED WITH ONE OF THE DESIGNATED SPRAY ADJUVANTS.**

Application Equipment
Nearly all spray application equipment is suitable for applying HARVADE-5F if fine spray droplets can be deposited. If aerial equipment is used, it should be fitted with a fixed boom set at 45–90° relative to the trailing edge of wing and flown 10–12 ft. above the crop canopy.

Compatibility
HARVADE-5F is compatible with most commonly used cotton desiccants, defoliants, and insecticides. Consult your supplier or Uniroyal for details.

COTTON—FOR DEFOLIATION

HARVADE-5F should be applied to mature cotton plants when 70% or more of the bolls are open and 7 to 10 days prior to anticipated harvest.

Air Application:
Mix 6½-8 fl. oz.[1] of HARVADE-5F plus one of the spray adjuvants designated below in a minimum of 5 gallons of spray solution per acre. If cotton is rank or conditions favor regrowth, the higher rate is desirable. A second application at rates of 6 fluid oz/A of HARVADE-5F may be necessary 5–7 days later. See application note[2].

Ground Application:
Mix 6½-8 fl. oz.[1] of HARVADE-5F plus one of the spray adjuvants designated below in 10–20 gallons of spray solution per acre. If cotton is rank or conditions favor regrowth, the higher rate is desirable. A second application at rates of 6 fluid oz/A of HARVADE-5F may be necessary 5–7 days later. See application note[2].

Application Notes:
[1]One gallon of HARVADE-5F will treat 20 acres of cotton, when applied at the 6½ fl. oz. rate.
[2]Partial defoliation from the first application will most likely result from one or more of the following conditions, requiring a second application:
- Rank cotton foliage is dense enough to inhibit spray droplets from reaching bottom leaves.
- Bolls are not 70% open
- Night temperatures are below 55°F for 3–4 nights after application.
- Rainfall occurs within 6 hours after application.

TANK MIX COMBINATIONS

HARVADE-5F plus PREP®:*
HARVADE-5F may be applied with 1–2 quarts of PREP® to enhance boll opening and control regrowth. Apply this tank mixture when the bolls are 60–70% open.

HARVADE-5F plus PARAQUAT:
On rain-belt cotton if a high percentage of cotton leaves are still actively growing, then 2–4 fl. oz./A of PARAQUAT should be added to the labeled rate of HARVADE-5F for more complete defoliation.

* PREP® is the registered trademark of Union Carbide Corporation

SPRAY ADJUVANTS REQUIRED WITH HARVADE-5F

Any oil based surfactant similar to those listed below should provide equal results when applied at a minimum rate of 1 pint/A: Miller-Aide, MorAct, Western Farm Defoliant Spreader Sticker, Agridex, Prime Oil, Super Oil, Surf Oil.

USE PRECAUTIONS

1. Do not plant rotational crops within 6 months after application.
2. Spray drift.
   a. Do not apply HARVADE-5F directly to, or permit mist to drift onto citrus, broccoli, almonds, grapes, peppers, tomatoes, lettuce, alfalfa, sugarbeets or winter wheat. These crops may sustain significant injury if contacted with HARVADE-5F.
   b. Do not apply HARVADE-5F to, or permit mist to drift onto soil that will subsequently be seeded to peppers, tomatoes, lettuce, broccoli, alfalfa or sugarbeets. Significant crop injury may occur from such an application.
   c. Do not apply HARVADE-5F to, or permit mist to drift onto irrigation water that will subsequently be applied to peppers, tomatoes or lettuce.

IMPORTANT NOTICE

Seller warrants that this product conforms to its chemical description and is reasonably fit for the purposes stated on the label when used in accordance with the directions and instructions specified on the label under normal conditions of use, **but neither this warranty nor any other warranty of merchantability or fitness for a particular purpose, express or implied, extends to the use of this product, contrary to label instructions, or under abnormal conditions, or under conditions not reasonably foreseeable to seller, and buyer assumes the risk of any such use.**

KYLAR®-85

A PLANT GROWTH REGULANT

COMPOSITION

ACTIVE INGREDIENT:	% BY WEIGHT
Daminozide[butanedioic acid mono (2,2-dimethylhydrazide)]	85.0%
INERT INGREDIENTS:	15.0%
TOTAL:	100.0%

EPA Reg. No. 400-103 EPA Est. No.
KYLAR is a Registered Trademark of UNIROYAL CHEMICAL COMPANY, INC.

CAUTION

Keep Out of Reach of Children

STATEMENT OF PRACTICAL TREATMENT

IF SWALLOWED: Drink promptly a large quantity of milk, egg white or gelatin solution; if not available drink large amounts of water. Avoid alcohol. Call a physician immediately.

IF ON EYES OR SKIN: Wash with plenty of water for at least 15 minutes. For eyes, call a physician. Remove and wash contaminated clothing before use.

IF INHALED: and abnormal symptoms develop see a physician.

See Left Side Panel For Additional Precautionary Statement.

PRECAUTIONARY STATEMENTS

HAZARDS TO HUMANS

CAUTION

Harmful if absorbed through skin. Avoid contact with skin, eyes or clothing. Do not allow spray solution to touch skin. Wear protective clothing to protect hands, arms and legs, both when diluting and applying this product. Avoid breathing dusts. Wear a protective mask during mixing/loading operations. Wash hands and face thoroughly with soap and water after use or before eating or smoking.

ENVIRONMENTAL HAZARDS

Do not apply directly to water or wetlands. Do not discharge into lakes, streams, ponds, or public waters unless in accordance with NPDES permit. For guidance, contact your Regional Office of the EPA. Do not contaminate water by cleaning of equipment or disposal of wastes.

PHYSICAL OR CHEMICAL HAZARDS

Do not use or store near heat or open flame.

DIRECTIONS FOR USE

Each formulator is responsible for obtaining EPA registrations for its formulated product(s).

It is a violation of Federal Law to use this product in a manner inconsistent with its labeling.

REENTRY STATEMENT

Do not reenter treated areas until 24 hours after application, unless wearing protective clothing.

STORAGE AND DISPOSAL

Do not contaminate water, food or feed by storage or disposal.

STORAGE

Store in a dry location. Use spray and stock solutions within 24 hours. Do not store solutions. Immediate use is appropriate, if another component is added to spray mixture.

PESTICIDE DISPOSAL

Wastes resulting from the use of this product may be disposed of on site or at an approved waste disposal facility.

CONTAINER DISPOSAL

Completely empty bag into application equipment. Then dispose of empty bag in a sanitary landfill or by incineration, or if allowed by State and local authorities, by burning. If burned, stay out of smoke.

GENERAL

KYLAR-85 is a water soluble growth regulant for peanuts. When the solution is sprayed on plants, the chemical moves into the leaves than moves freely inside the plant. Since six hours may be required for the chemical to get inside the plant, application should be delayed if rain is expected within six hours. Peanut vines are shorter, more erect, greener in color and yields are usually increased as a result of KYLAR-85 treatment.

DOSAGE AND TIMING

KYLAR-85 can be applied from the time peanut plants are at least 12 inches across until 30 days before harvest.

A. An application of **1 lb. per acre** in a minimum of **10 gallons** of water by ground sprayer, **or 5 gallons** of water by aerial sprayer when peanut plants are **12–24 inches** across encourages fruiting closer to the tap root. This timing or later application just before "lapping" also suppresses vine growth, improves yields, and makes harvesting easier. These application times usually correspond to the **second, third or fourth** fungicide application. In years favorable for rapid vine growth, an additional **½ lb. per acre** may be needed about six weeks after the first to continue suppression of vine growth. This additional application should be made at least 30 days before harvest.

or

B. KYLAR-85 can also be applied in **three split applications of ½ lb. per acre** each. This gives results as discussed above. The first **½ lb. per acre** should be applied when peanuts are 12–24 inches across to lapping and the additional two **½ lb. per acre** application made at approximately 14 day intervals. The last application should be made at least 30 days before harvest.

or

C. KYLAR-85 can also be applied in each of **four to six split applications of ¼ lb. per acre** with Fungicide applications. Six split applications of **¼ lb. per acre** is recommended when conditions are favorable for rapid vine growth. The first **¼ lb. per acre** should be applied when peanuts are 12–24 inches across and additional **¼ lb. per acre** applications made at approximately 14 day intervals. The last application should be made at least 30 days before harvest.

For best results on Southeastern peanuts use options A or B.

For best results on Southwestern peanuts use options B or C.

NOTES

1. KYLAR-85 works best when plants are actively growing. Applications should be not be made when plants are wilted from drought stress.
2. KYLAR-85 may be applied any time of the day, but best results can be expected from morning applications. Leaves should be thoroughly covered with spray solution.
3. Application should be delayed if rain is expected within six hours.
4. KYLAR-85 may be applied in a tank mixture with one of the following commercial fungicides: Benlate, Bravo, Dithane M-45, Du-Ter or Polyram. Always observe all cautions and limitations on labels of all products used in mixtures. When KYLAR-85 is mixed with water for spraying, use the mixture on the same day.
5. Water soluble KYLAR-85 can be easily rinsed from equipment with water.
6. Do not feed livestock on peanut hay prior to crop maturity.

IMPORTANT NOTICE

Seller warrants that this product conforms to its chemical description and is reasonably fit for the purposes stated on the label when used in accordance with the directions and instructions specified on the label under normal conditions of use, **but neither this warranty nor any other warranty of merchantability or fitness for a particular purpose, express or implied, extends to the use of this product, contrary to label instructions, or under abnormal conditions, or under conditions not reasonably foreseeable to seller, and buyer assumes the risk of any such use.**

OMITE® 30W

AGRICULTURAL MITICIDE
EAST OF THE ROCKY MOUNTAINS

COMPOSITION

ACTIVE INGREDIENTS:	% BY WEIGHT
Propargite [2-(p-tert-butylphenoxy) cyclohexyl 2-propynyl sulfite]*	30.0%
INERT INGREDIENTS:	70.0%
TOTAL:	100.0%

* U.S. Patent No. 3,463,859

EPA Reg. No. 400-82-AA EPA Est. No.

OMITE is a Registered Trademark of UNIROYAL CHEMICAL COMPANY, INC.

DANGER
Keep Out of Reach of Children
PELIGRO

PRECAUCION AL USUARIO: Si usted no lee ingles, no use este producto hasta que le etiqueta haya sido explicado ampliamente.

STATEMENT OF PRACTICAL TREATMENT

IF SWALLOWED: Call a physician or Poison Control Center immediately. Drink promptly a large quantity of milk, egg whites or gelatin solution. If these are not available, drink large quantities of water. Avoid alcohol.

IF INHALED: See a physician if an abnormal reaction occurs.

IF ON SKIN: Wash thoroughly with soap and water.

IF IN EYES: Immediately flush eyes with plenty of water. See a physician.

See left side panel for additional precautionary statement.

PRECAUTIONARY STATEMENTS

HAZARDS TO HUMANS

DANGER

Corrosive, causes eye damage. Harmful if swallowed, inhaled or absorbed through skin. Do not breathe vapors or spray mist. Do not get in eyes, on skin or on clothing. Wear goggles or face shield when handling. Wash hands and face thoroughly with soap and water after use and before eating or smoking.

ENVIRONMENTAL HAZARDS

This product is toxic to fish and shrimp. Keep out of lakes, streams or ponds. Do not apply where runoff is likely to occur. Do not apply when weather conditions favor drift from areas treated. Do not contaminate water by cleaning of equipment or disposal of wastes.

PHYSICAL OR CHEMICAL HAZARDS

Flammable, keep away from heat and open flame.

DIRECTIONS FOR USE

It is a violation of Federal law to use this product in a manner inconsistent with its labeling.

STORAGE AND DISPOSAL

Do not contaminate water, food or feed by storage or disposal.

STORAGE

Store in a dry location.

PESTICIDE DISPOSAL

Pesticide wastes are acutely hazardous. Improper disposal of excess pesticide, spray mixture or rinsate is a violation of Federal Law. If these wastes cannot be disposed of by use according to label instructions, contact your State Pesticide or Environmental Control Agency, or the Hazardous Waste representative at the nearest EPA Regional Office for guidance.

CONTAINER DISPOSAL

Completely empty bag into application equipment. Then dispose of empty bag in a sanitary landfill or by incineration, or if allowed by State and local authorities, by burning. If burned, stay out of smoke.

REENTRY STATEMENT

Do not apply this product in such a manner as to directly or through drift expose workers or other persons. The area being treated must be vacated by unprotected persons.

Do not enter treated areas without protective clothing until sprays have dried.

Because certain states may require more restrictive reentry intervals for various crops treated with this product, consult your State Department of Agriculture for further information.

Written or oral warnings must be given to the workers who are expected to be in a treated area or in an area to be treated with this product. When oral warnings are given, warnings shall be given in a language customarily understood by workers. Oral warnings must be given if there is reason to believe that written warnings cannot be understood by workers. Written or oral warnings must include the following information:

"**DANGER,** Area to be treated with Propargite on (date of application). Do not enter without appropriate protective clothing until sprays have dried." Or in the case of avocados or grape vineyards, "Do not enter treated areas to perform activities that involve substantial or prolonged contact with treated foliage for 7 days. If accidentally exposed, consult the Statement of Practical Treatment portion of this label for first aid recommendations."

GENERAL INSTRUCTIONS

A dry wettable powder for preparing sprays to control brown almond, citrus red, clover, European red, McDaniel, Pacific spider, peach silver, strawberry spider, two-spotted spider and Willamette mites.

OMITE-30W is not systemic in action; therefore complete coverage of both upper and lower leaf surfaces and of fruit is necessary for effective control. Always use at least the minimum recommended amount of OMITE-30W per acre. OMITE-30W can be used in concentrate sprays.

AL USUARIO: Si usted no puede leer o entender ingles no use este producto hasta que la etiqueta le haya sido explicada ampliamente.

TO THE USER: If you cannot read or understand English, do not use this product until the label has been fully explained to you.

DOSAGE INSTRUCTIONS

CROP	MITES CONTROLLED	AMOUNT OMITE-30W PER ACRE PER TREATMENT	GALLONS SPRAY SOLUTION PER ACRE	TOTAL NUMBER OF SPRAYS PER YEAR	LAST SPRAY DAYS BEFORE HARVEST
APPLES	European red two-spotted spider Pacific Spider McDaniel	5–12 lbs.	400–800	3	7
CARNATIONS CHRYSANTHEMUMS ROSES	two-spotted spider	1–1½ lbs. per 100 gallons of water	Spray to wet	—	—

DOSAGE INSTRUCTIONS					
CROP	MITES CONTROLLED	AMOUNT OMITE-30W PER ACRE PER TREATMENT	GALLONS SPRAY SOLUTION PER ACRE	TOTAL NUMBER OF SPRAYS PER YEAR	LAST SPRAY DAYS BEFORE HARVEST
ORNAMENTALS* (field and nursery)	two-spotted spider	1 lb. per 100 gallons of water	Spray to wet	—	—
PEACHES	European red Pacific spider two-spotted spider strawberry spider McDaniel peach silver	5–15 lbs.	400–1000	2	14
PEANUTS	two-spotted spider	3–5 lbs.	See application note 7	2	14
PLUMS PRUNES	European red two-spotted spider Pacific spider	5–15 lbs.	400–1000	2	14
STRAWBERRIES (Florida and Louisiana only)	two-spotted spider	2½ lbs.	100	3** (per crop season)	3

* At end of booklet, see list of ornamentals which may show signs of phytotoxicity.
** Do not apply more than 3 applications of Omite in any 21 day period during the harvesting season.

FOR POST HARVEST AND NON-BEARING TREES ONLY

Non-bearing trees are those that will not bear fruit within one year of Omite Application.

DOSAGE INSTRUCTIONS (EAST OF THE ROCKY MOUNTAINS)

DOSAGE INSTRUCTIONS					
CROP	MITES CONTROLLED	AMOUNT OMITE-30W PER ACRE PER TREATMENT	GALLONS SPRAY SOLUTION PER ACRE	TOTAL NUMBER OF SPRAYS PER YEAR	LAST SPRAY DAYS BEFORE HARVEST
PEARS	European red two-spotted spider	5–7½ lbs.	400–500	—	—

APPLICATION NOTES

1. Complete and uniform coverage is necessary for optimum results.
2. Do not reduce the recommended rate per acre when applying OMITE-30W in concentrate sprays.
3. Do not apply OMITE-30W under the following conditions as leaf injury and fruit spotting may occur:
   —when air temperatures are expected to exceed 90°F.
   —in combination with Guthion emulsifiable concentrate.
   —in combination with foliar spray oils.
4. Do not use additional spray adjuvants.
5. OMITE-30W is not compatible with alkaline materials such as lime, lime sulfur or Bordeaux. The effectiveness of OMITE-30W will be reduced.
6. Ornamentals—1 lb. of OMITE-30W per 100 gallons of water equals 1 ounce (28.4 grams) per 6.25 gallons. The user should initially treat only a small portion of his crop to confirm plant safety under his conditions.
7. Peanuts—Premix OMITE-30W with a small amount of water to form a slurry before adding to spray tank. Apply in sufficient water to provide uniform and complete coverage of foliage when using low gallonage equipment. A minimum of 20 gallons of water per acre is recommended for best results.
8. Roses—Use is limited to greenhouse, slat and saran grown. Because cultivar reaction and cultural practices differ widely, the user should initially treat only a small portion of his crop to confirm plant safety under his conditions. Repeat applications of OMITE-30W are normally necessary for extended control.

USE RESTRICTIONS

Do not graze or feed livestock on cover crops grown among the trees listed on this label.

COMPATIBILITIES

Except as noted above OMITE-30W is compatible with most pesticides. (See application notes 3 & 4). Check with the OMITE-30W use booklet or your supplier for details.

The following ornamentals may show some signs of phytotoxicity.

PLANTS SHOWING SLIGHT NECROSIS (leaf burn)

Abelia grandiflora
Acer palmatum
Alectryon subcinereus
Aralia seiboldi
Aucuba japonica picturata
Aucuba japonica variegata
Beaumontia grandiflora
Berberis thunbergi atropurpurea
Berberis thunbergi atropurpurea nana "Crimson Pygmy"
Bougainvillea rhodamine
Buxus microphylla "Richardi"
Cissus mandianum
Cocculus laurifolius
Cotoneaster horizontalis
Cotoneaster salicifolia repandens
Cotoneaster "Skogsholmen"
Cupania anacardioides
Eucalyptus cinerea
Eucalyptus globulus compacta
Eucalyptus niphophila
Eucalyptus rostrata
Eucalyptus sideroxylon rosea
Eucalyptus torquata
Eucalyptus viminalis
Hedera canariensis
Hedera "Ghost Tree"
Hydrangea macrophylla "Hamburg"
Hydrangea macrophylla hortensis compacta
Hydrangea macrophylla hortensis "Excelsior"
Hydrangea macrophylla "Merrit's Beauty"
Hydrangea macrophylla "Revelation"
Hydrangea macrophylla "Trophee"
Hydrangea macrophylla "White"
Ilex "Nellie Stevens"
Koelreuteria paniculata
Lagerstroemia "Petite Orchid"
Lagerstroemia "Petite Pinkie"
Mahonia repens
Nyssa sylvatica
Pachysandra terminalis
Philadelphus "Glacier"
Podocarpus gracilior
Prunus integrifolia
Rosa Banksiae alba

Gardenia jasminoides "Mystery"	Schinus terebinthifolius
Gardenia jasminoides "Veitchi"	Tecoma stans
	Weigela "Bristol Snowflake"

PLANTS SHOWING MODERATE TO SEVERE NECROSIS (leaf burn)

Acacia podalyriaefolia	Hedera helix "Goldheart"
Alnus rhombifolia	Leptospermum scoparium "Keatley"
Aucucba japonica "Dwarf Female"	Liquidambar orientalis
Beloperone tomentosa	Magnolia grandiflora "Majestic Beauty"
Berberis "Mentorensis"	Philadelphus "Minnesota Snowflake"
Ceanothus "Ray Hartman"	Philadelphus virginalis
Cissus capensis	Potentilla fruticose "Moon light"
Cissus hypoglauca	Potentilla grandiflora
Cotoneaster bullata floribunda	Quercus agrifolia
Dendromecon harfordi rhamnoides	Ternstroemia japonica
Eucalyptus ficifolia	Tristania conferta
Gelsemium sempervirens	
Grevillea "Noelli"	

IMPORTANT NOTICE

Seller warrants that this product conforms to its chemical description and is reasonably fit for the purposes stated on the label when used in accordance with the directions and instructions specified on the label under normal conditions of use, **but neither this warranty nor any other warranty of merchantability or fitness for a particular purpose, express or implied, extends to the use of this product, contrary to label instructions, or under abnormal conditions, or under conditions not reasonably foreseeable to seller, and buyer assumes the risk of any such use.**

OMITE®-30W

AGRICULTURAL MITICIDE
WEST OF THE ROCKY MOUNTAINS

COMPOSITION

ACTIVE INGREDIENTS:	% BY WEIGHT
Propargite 2-(p-tert-butylphenoxy) cyclohexyl 2-propynyl sulfite*	30.0%
INERT INGREDIENTS:	70.0%
TOTAL:	100.0%

* U.S. Patent No. 3,463,859

EPA Reg. No. 400-82-AA EPA Est. No.

OMITE is a Registered Trademark of UNIROYAL CHEMICAL COMPANY, INC.

DANGER
Keep Out of Reach of Children
PELIGRO

PRECAUCION AL USUARIO: Si usted no lee ingles, no use este producto hasta que le etiqueta haya sido explicado ampliamente.

STATEMENT OF PRACTICAL TREATMENT

IF SWALLOWED: Call a physician or Poison Control Center immediately. Drink promptly a large quantity of milk, egg whites or gelatin solution. If these are not available, drink large quantities of water. Avoid alcohol.

IF INHALED: See a physician if an abnormal reaction occurs.

IF ON SKIN: Wash thoroughly with soap and water.

IF IN EYES: Immediately flush eyes with plenty of water. See a physician.

See left side panel for additional precautionary statement.

PRECAUTIONARY STATEMENTS

HAZARDS TO HUMANS

DANGER

Corrosive, causes eye damage. Harmful if swallowed, inhaled or absorbed through skin. Do not breathe vapors or spray mist. Do not get in eyes, on skin or on clothing. Wear goggles or face shield when handling. Wash hands and face thoroughly with soap and water after use and before eating or smoking.

ENVIRONMENTAL HAZARDS

This product is toxic to fish and shrimp. Keep out of lakes, streams or ponds. Do not apply where runoff is likely to occur. Do not apply when weather conditions favor drift from areas treated. Do not contaminate water by cleaning of equipment or disposal of wastes.

PHYSICAL OR CHEMICAL HAZARDS

Flammable, keep away from heat and open flame.

DIRECTIONS FOR USE

It is a violation of Federal law to use this product in a manner inconsistent with its labeling.

STORAGE AND DISPOSAL

Do not contaminate water, food or feed by storage or disposal.

STORAGE

Store in a dry location.

PESTICIDE DISPOSAL

Pesticide wastes are acutely hazardous. Improper disposal of excess pesticide, spray mixture or rinsate is a violation of Federal Law. If these wastes cannot be disposed of by use according to label instructions, contact your State Pesticide or Environmental Control Agency, or the Hazardous Waste representative at the nearest EPA Regional Office for guidance.

CONTAINER DISPOSAL

Completely empty bag into application equipment. Then dispose of empty bag in a sanitary landfill or by incineration, or if allowed by State and local authorities, by burning. If burned, stay out of smoke.

REENTRY STATEMENT

Do not apply this product in such a manner as to directly or through drift expose workers or other persons. The area being treated must be vacated by unprotected persons.

Do not enter treated areas without protective clothing until sprays have dried.

Because certain states may require more restrictive reentry intervals for various crops treated with this product, consult your State Department of Agriculture for further information.

Written or oral warnings must be given to the workers who are expected to be in a treated area or in an area to be treated with this product. When oral warnings are given, warnings shall be given in a language customarily understood by workers. Oral warnings must be given if there is reason to believe that written warnings cannot be understood by workers. Written or oral warnings must include the following information:

"DANGER, Area to be treated with Propargite on (date of application). Do not enter without appropriate protective clothing until sprays have dried." Or in the case of avocados or grape vineyards, "Do not enter treated areas to perform activities that involve substantial or prolonged contact with treated foliage for 7 days. If accidentally exposed, consult the Statement of Practical Treatment portion of this label for first aid recommendations."

GENERAL INSTRUCTIONS

A dry wettable powder for preparing sprays to control brown almond, citrus red, clover, European red, McDaniel, Pacific spider, peach silver, strawberry spider, two-spotted spider and Willamette mites.

OMITE-30W is not systemic in action; therefore complete coverage of both upper and lower leaf surfaces and of fruit is necessary for

effective control. Always use at least the minimum recommended amount of OMITE-30W per acre. OMITE-30W can be used in concentrate sprays.

OMITE-30W performance is best when day temperatures average above 70°F.

PRECAUCION AL USUARIO: Si usted no puede lee ingles no use este producto hasta que la etiqueta le haya sido explicada ampliamente.

DOSAGE INSTRUCTIONS

CROP	MITES CONTROLLED	AMOUNT OMITE-30W PER ACRE	GALLONS SPRAY SOLUTION PER ACRE		TOTAL NUMBER OF SPRAYS PER YEAR	LAST SPRAY DAYS BEFORE HARVEST
			GROUND	AERIAL		
Almonds (CA & AZ)	clover European red two-spotted spider strawberry spider Pacific spider	5–22 lbs.	500–1500	—	2	28
Apples	European red two-spotted spider Pacific spider McDaniel	5–12 lbs.	500–800	—	3	7
Apricots	Pacific spider two-spotted spider	5–9 lbs.	300–600	—	2	14
Figs	two-spotted spider Pacific spider strawberry spider	3–6 lbs.	200–400	—	2	14
Grapes 7-Day reentry Interval (see note 10)	Pacific spider strawberry spider Willamette two-spotted spider	5–9 lbs.	350–600	—	2	21
Hops	two-spotted spider	5 lbs.	200	—	2	14
Lemons (CA & AZ)	citrus red	10–15 lbs.	600–1500	—	2	7
Nectarines	Pacific spider brown almond	5–9 lbs.*	300–600	Minimum 20 gal.	2	14
Ornamentals (field and nursery)†	two-spotted spider	1 lb. per 100 gallons of water	Spray to wet	—	—	—
Peaches	European red Pacific spider two-spotted spider strawberry spider McDaniel peach silver	5–15 lbs.**	300–1000	Minimum 20 gal.	2	14
Plums Prunes	European red two-spotted spider Pacific spider	5–15 lbs.**	300–1000	Minimum 20 gal.	2	14
Roses (greenhouse grown)	two-spotted spider Pacific spider strawberry spider	1 lb. per 100 gallons of water	—	—	—	—
Roses (field grown)	two-spotted spider Pacific spider	2–5 lbs.	50–300	—	—	—
Strawberries (See application notes below)	Pacific spider strawberry spider two-spotted spider	5–10 lbs.	100–200	—	3†† (per crop season)	3
Walnuts (CA only)	European red two-spotted spider	7–22 lbs.	500–1500	—	2	14

† At end of booklet, see list of ornamentals which may show signs of phytotoxicity.
†† Do not apply more than 3 applications of Omite in any 40 day period during the harvesting season.
* Use $7\frac{1}{2}$-9 pounds of OMITE-30W per acre when applied by air.
** Use $7\frac{1}{2}$-10 pounds of OMITE-30W per acre when applied by air.

APPLICATION NOTES:

1. Complete and uniform coverage is necessary for optimum results.

2. Do not reduce the recommended rate per acre when applying OMITE-30W in concentrate sprays.

3. Do not combine with other chemicals or spray adjuvants, when using on strawberries, as leaf and fruit injury may occur.

4. California: Do not apply at day temperatures above 80°F as fruit burn may occur. (Strawberries only).

5. OMITE-30W may cause some leaf burn on new growth (feather growth) of lemons.

6. Omite performance is best when daily temperatures average above 70°F.

7. Leaf injury and/or fruit spotting may occur, if the following conditions are not observed:

—Do not apply OMITE-30W with DIAZINON-WP if air temperatures are expected to exceed 95°F the day of spraying apples or hops.

—OMITE-30W is not compatible with foliar spray oils. (Further information from supplier or Uniroyal).

8. OMITE-30W is not compatible with alkaline materials such as lime, lime sulfur or Bordeaux. The effectiveness of OMITE-30W will be reduced.

9. Ornamentals - 1 lb. of OMITE-30W per 100 gallons of water equals 1 ounce (28.4 grams) per 6.25 gallons. The user should initially treat only a small portion of his crop to confirm plant safety under his conditions.

10. Keep workers out for seven (7) days after application. There is a 7-day safety reentry interval for grapes. Workers shall not enter treated areas to perform activities that involve substantial or prolonged contact with treated foliage until the 7-day safety reentry has expired.

POST HARVEST ONLY

CROP	MITES CONTROLLED	AMOUNT OMITE-30W PER ACRE PER TREATMENT	GALLONS SPRAY SOLUTION PER ACRE	TOTAL NUMBER OF SPRAYS PER YEAR	LAST SPRAY DAYS BEFORE HARVEST
Navel Oranges	citrus red	10–15 lbs.	600–1500	1	SEE CAUTION BELOW
Grapefruit	citrus red	10–15 lbs.	600–1500	1	SEE CAUTION BELOW

CAUTION: DO NOT USE ON NAVEL ORANGES OR GRAPEFRUIT AFTER NEW FRUIT IS EXPOSED IN THE SPRING AS FRUIT SPOTTING MAY OCCUR.

FOR POST HARVEST AND NON-BEARING TREES ONLY

Non-bearing trees are those that will not bear fruit within one year of Omite Application.

DOSAGE INSTRUCTIONS (WEST OF THE ROCKY MOUNTAINS)

CROP	MITES CONTROLLED	AMOUNT OMITE-30W PER ACRE TREATMENT	GALLONS SPRAY SOLUTION PER ACRE	TOTAL NUMBER OF SPRAYS PER YEAR	LAST SPRAY DAYS BEFORE HARVEST
Cherries	European red two-spotted spider Pacific spider McDaniel	6 lbs.	400	—	—
Pears	Eurpoean red two-spotted spider	5–7½ lbs.	FOR POST HARVEST TREES ONLY 300–500	—	—

USE RESTRICTIONS—Do not feed livestock on cover crops grown among the trees listed on this label.

COMPATIBILITIES—Except as noted above OMITE-30W is compatible with many pesticides. (See application note 7). Check with your supplier or Uniroyal for details. Observe all cautions and limitations of labeling of all products used in mixtures.

The following ornamentals may show some signs of phytotoxicity.

PLANTS SHOWING SLIGHT NECROSIS (leaf burn)

Abelia grandiflora
Acer palmatum
Alectryon subcinereus
Aralia seiboldi
Aucuba japonica picturata
Aucuba japonica variegata
Beaumontia grandiflora
Berberis thunbergi atropurpurea
Berberis thunbergi atropurpurea nana "Crimson Pygmy"
Bougainvillea rhodamine
Buxus microphylla "Richardi"
Cissus mandianum
Cocculus laurifolius
Cotoneaster horizontalis
Cotoneaster salicifolia repandens
Cotoneaster "Skogsholmen"
Cupania anacardioides
Eucalyptus cinerea
Eucalyptus globulus compacta
Gardenia jasminoides "Veitchi"
Hedera canariensis
Hedera "Ghost Tree"
Hydrangea macrohphylla "Hamburg"
Hydrangea macrophylla hortensis compacta
Hydrangea macrophylla hortensis "Excelsior"
Hydrangea macrophylla "Merritt's Beauty"
Hydrangea macrophylla "Revelation"
Hydrangea macrophylla "Trophee"
Hydrangea macrophylla "White"
Ilex "Nellie Stevens"
Koelreuteria paniculata
Lagerstroemia "Petite Orchid"
Lagerstroemia "Petite Pinkie"
Mahonia repens
Nyssa sylvatica
Pachysandra terminalis
Philadelphus "Glacier"
Podocarpus gracilior
Eucalyptus niphophila
Eucalyptus rostrata
Eucalyptus sideroxylon rosea
Eucalyptus torquata
Eucalyptus viminalis
Gardenia jasminoides "Mystery"
Prunus integrifolia
Rosa Banksiae alba
Schinus terebinthifolius
Tecoma stans
Weigela "Bristol Snowflake"

PLANTS SHOWING MODERATE TO SEVERE NECROSIS (leaf burn)

Acacia podalyriaefolia
Alnus rhombifolia
Aucuba japonica "Dwarf Female"
Beloperone tomentosa
Berberis "Mentorensis"
Ceanothus "Ray Hartman"
Ciccus capensis
Cissus hypoglauca
Cotoneaster bullata floribunda
Dendromecon harfordi rhamnoides
Eucalyptus ficifolia
Gelsemium sempervirens
Grevillea "Noelli"
Hedera helix "Goldheart"
Leptospermum scoparium "Keatley"
Liquidambar orientalis
Magnolia grandiflora "Majestic Beauty"
Philadelphus "Minnesota Snowflake"
Philadelphus virginalis
Potentilla fruitcosa "Moonlight"
Potentilla grandiflora
Quercus agrifolia
Ternstroemia japonica
Tristania conferta

IMPORTANT NOTICE

Seller warrants that this product conforms to its chemical description and is reasonably fit for the purposes stated on the label when used in accordance with the directions and instructions specified on the label under normal conditions of use, **but neither this warranty nor any other warranty of merchantability or fitness for a particular purpose, express or implied, extends to the use of this product, contrary to label instructions, or under abnormal conditions, or under conditions not reasonably foreseeable to seller, and buyer assumes the risk of any such use.**

OMITE®-CR

AGRICULTURAL MITICIDE

COMPOSITION

ACTIVE INGREDIENTS:	% BY WEIGHT
Propargite [2-(p-tert-butylphenoxy) cyclohexyl 2-propynyl sulfite]*	30.0%
INERT INGREDIENTS:	70.0%
TOTAL:	100.0%

* U.S. Patent No. 3,463,859

EPA Reg. No. 400-82-AA EPA Est. No.

OMITE is a Registered Trademark of UNIROYAL CHEMICAL COMPANY, INC.

DANGER

Keep Out of Reach of Children

PELIGRO

PRECAUCION AL USUARIO: Si usted no lee ingles, no use este producto hasta que le etiqueta haya sido explicado ampliamente.

STATEMENT OF PRACTICAL TREATMENT

IF SWALLOWED: Drink promptly a large quantity of milk, egg whites or gelatin solution. If these are not available, drink large quantities of water. Avoid alcohol. Call a physician or Poison Control Center immediately.

IF INHALED: See a physician if an abnormal reaction occurs.

IF ON SKIN: Wash thoroughly with soap and water.

IF IN EYES: Immediately flush eyes with plenty of water. See a physician.

AVOID EXPOSURE TO DUST

See left side panel for additional precautionary statement.

PRECAUTIONARY STATEMENTS

HAZARDS TO HUMANS

DANGER

Corrosive, causes eye damage. May be fatal if inhaled. Harmful if swallowed or absorbed through skin. Do not breathe vapors or spray mist. Do not get in eyes, on skin or on clothing. Wear goggles or face shield when handling. Wash hands and face thoroughly with soap and water after use and before eating or smoking.

ENVIRONMENTAL HAZARDS

This product is toxic to fish and shrimp. Keep out of lakes, streams or ponds. Do not apply where runoff is likely to occur. Do not apply when weather conditions favor drift from areas treated. Do not contaminate water by cleaning of equipment or disposal of wastes.

PHYSICAL OR CHEMICAL HAZARDS

Flammable, keep away from heat and open flame.

DIRECTIONS FOR USE

It is a violation of Federal law to use this product in a manner inconsistent with its labeling.

STORAGE AND DISPOSAL

Do not contaminate water, food or feed by storage or disposal.

STORAGE

Store in a dry location.

PESTICIDE DISPOSAL

Pesticide wastes are acutely hazardous. Improper disposal of excess pesticide, spray mixture or rinsate is a violation of Federal Law. If these wastes cannot be disposed of by use according to label instructions, contact your State Pesticide or Environmental Control Agency, or the Hazardous Waste representative at the nearest EPA Regional Office for guidance.

CONTAINER DISPOSAL

Completely empty bag into application equipment. Then dispose of empty bag in a sanitary landfill or by incineration, or if allowed by State and local authorities, by burning. If burned, stay out of smoke.

REENTRY STATEMENT

Do not apply this product in such a manner as to directly or through drift expose workers or other persons. The area being treated must be vacated by unprotected persons.

Do not enter treated areas without protective clothing until sprays have dried.

Because certain states may require more restrictive reentry intervals for various crops treated with this product, consult your State Department of Agriculture for further information.

Written or oral warnings must be given to the workers who are expected to be in a treated area or in an area to be treated with this product. When oral warnings are given, warnings shall be given in a language customarily understood by workers. Oral warnings must be given if there is reason to believe that written warnings cannot be understood by workers. Written or oral warnings must include the following information:

"DANGER. Area to be treated with Propargite on (date of application). Do not enter without appropriate protective clothing until sprays have dried." Or in the case of avocados or grape vineyards, "Do not enter treated areas to perform activities that involve substantial or prolonged contact with treated foliage for 7 days. If accidentally exposed, consult the Statement of Practical Treatment portion of this label for first aid recommendations."

GENERAL INSTRUCTIONS

OMITE-CR is a dry wettable powder for preparing sprays to control brown almond, citrus red, clover, European red, McDaniel, Pacific spider, peach silver, strawberry spider, two-spotted spider and Willamette mites.

OMITE-CR is not systemic in action; therefore complete coverage of both upper and lower leaf surfaces and of fruit is necessary for effective control. Always use at least the minimum recommended amount of OMITE-CR per acre. OMITE-CR can be used in concentrate sprays.

AL USUARIO: Si usted no puede leer o entender ingles no use este producto hasta que la etiqueta le haya sido explicada ampliamente.

TO THE USER: If you cannot read or understand English, do not use this product until the label has been fully explained to you.

MIXING INSTRUCTIONS

This package contains $2 \times 1\frac{1}{2}$ pound water soluble bags of OMITE-CR. After opening package:

1. Do not open or perforate the inner water soluble bags of OMITE-CR. The water soluble bags and contents will dissolve in water.
2. Immediately dump entire contents into spray tank.
3. Do not touch water soluble bags as handling may cause breakage.
4. If only one water soluble bag is used, store the remaining bag in the original package in a dry location away from heat or cold.

DOSAGE INSTRUCTIONS

CROP	MITES CONTROLLED	NUMBER OF 1½ POUND WATER SOLUBLE BAGS PER ACRE	GALLONS SPRAY SOLUTION PER ACRE		TOTAL NUMBER OF SPRAYS PER YEAR	LAST SPRAY DAYS BEFORE HARVEST
			GROUND	AERIAL		
Almonds (CA only)	clover European red two-spotted spider strawberry spider Pacific spider	3–15	500–1500	—	2	28
Apples	European red two-spotted spider Pacific spider McDaniel	3–8	500–800	—	3	7
Apricots	Pacific spider two-spotted spider	3–6	300–600	—	2	14
Figs	two-spotted spider Pacific spider strawberry spider	2–3	200–400	—	2	14
Grapes 7-Day reentry Interval (see note 6)	Pacific spider strawberry spider Willamette two-spotted spider	3–6	350–600	—	2	21
Hops	two-spotted spider	3	200	—	2	14
Lemons (AZ only)	citrus red	7–10	600–1500	—	2	7
Nectarines	Pacific spider brown almond	3–6*	300–600	Minimum 20 gal.	2	14
Ornamentals (field and nursery)†	two-spotted spider	1 See footnote 5	Spray to wet	—	—	—
Oranges Grapefruit Lemons (CA only)	citrus red two-spotted spider	7–10	1000	—	2	7
Peaches	European red Pacific spider two-spotted spider strawberry spider McDaniel peach silver	3–10**	300–1000	Minimum 20 gal.	2	14
Plums Prunes	European red two-spotted spider Pacific spider	3–10**	300–1000	Minimum 20 gal.	2	14
Roses (greenhouse grown)	two-spotted spider Pacific spider strawberry spider	1 See footnote 5	150	—	—	—
Roses (field grown)	two-spotted spider Pacific spider	1–3 See footnote 5	50–300	—	—	—
Strawberries (See application notes below)	Pacific spider strawberry spider two-spotted spider	3–7	100–200	—	3†	3
Walnuts (CA only)	European red two-spotted spider	5–15	500–1500	—	2	14

† Do not apply more than 3 applications of OMITE-CR in any 40 day period during the harvesting season.
* Use 5–6 water soluble bags per acre when applied by air.
** Use 5–7 water soluble bags per acre when applied by air.

APPLICATION NOTES

1. Do not combine with other chemicals or spray adjuvants, when using on strawberries, as leaf and fruit injury may occur.

2. California: Do not apply at day temperatures above 80°F as fruit burn may occur. (Strawberries only).

3. OMITE-CR may cause some leaf burn on new growth (feather growth) of lemons.

4. OMITE performance is best when daily temperatures average above 70°F.

5. Ornamentals—1½ lb. of OMITE-CR per 150 gallons of water equals 1 ounce (28.4 grams) per 6.25 gallons. The user should initially treat only a small portion of his crop to confirm plant safety under his conditions.

6. Keep workers out for seven (7) days after application. There is a 7-day safety reentry interval for grapes. Workers shall not enter treated areas to perform activities that involve substantial or prolonged contact with treated foliage until the 7-day safety reentry has expired.

7. OMITE-CR when used alone may follow foliar oils at an interval of greater than 40 days after an oil application.

8. Foliar oil applicatons must not follow OMITE-CR treatments within 30 days.

FOR POST HARVEST AND NON-BEARING TREES ONLY

Non-bearing trees are those that will not bear fruit within one year of OMITE Application.

DOSAGE INSTRUCTIONS

CROP	MITES CONTROLLED	NUMBER OF 1½ POUND WATER SOLUBLE BAGS PER ACRE	GALLONS SPRAY SOLUTION PER ACRE	TOTAL NUMBER OF SPRAYS PER YEAR	LAST SPRAY DAYS BEFORE HARVEST
Cherries	European red two-spotted spider Pacific spider McDaniel	4	400	—	—

USE RESTRICTIONS—Do not feed livestock on cover crops grown among the trees listed on this label.

CAUTION:

A. Leaf injury and/or fruit spotting may occur if the following conditions are not observed.

1. Do not apply OMITE-CR with Diazinon-WP if air temperatures are expected to exceed 95°F the day of spraying apples and hops.
2. OMITE-CR is not compatible with foliar spray oils. (Further information from supplier or Uniroyal).

B. OMITE-CR is not compatible with alkaline materials such as lime, lime sulfur or Bordeaux. The effectiveness of OMITE-CR will be reduced.

COMPATIBILITIES—Except as noted above OMITE-CR is compatible with many pesticides. (See caution statements). Check with your supplier or Uniroyal for details. Observe all cautions and limitations of labeling of all products used in mixtures.

IMPORTANT NOTICE

Seller warrants that this product conforms to its chemical description and is reasonably fit for the purposes stated on the label when used in accordance with the directions and instructions specified on the label under normal conditions of use, **but neither this warranty nor any other warranty of merchantability or fitness for a particular purpose, express or implied, extends to the use of this product, contrary to label instructions, or under abnormal conditions, or under conditions not reasonably foreseeable to seller, and buyer assumes the risk of any such use.**

OMITE®-6E

AGRICULTURAL MITICIDE

COMPOSITION

ACTIVE INGREDIENTS:	% BY WEIGHT
Propargite 2-(p-tert-butylphenoxy) cyclohexyl 2-propynyl sulfite*	68.1%**
INERT INGREDIENTS:	31.9%
TOTAL:	100.0%

* U.S. Patent No. 3,463,859

** 6 lbs. Active per Gallon

EPA Reg. No. 400-89-AA EPA Est. No.

OMITE® is a Registered Trademark of UNIROYAL CHEMICAL COMPANY, INC.

WARNING
Keep Out of Reach of Children
AVISO

PRECAUCION AL USUARIO: Si usted no lee ingles, no use este producto hasta que le etiqueta haya sido explicado ampliamente.

STATEMENT OF PRACTICAL TREATMENT

IF SWALLOWED: Call a physician or poison control center immediately. Drink a large quantity of milk, egg whites or gelatin solution. If these are not available, drink large quantities of water. Avoid alcohol.

IF INHALED: See a physician if an abnormal reaction occurs.

IF ON SKIN: Wash thoroughly with soap and water.

IF IN EYES: Immediately flush eyes with plenty of water. See a physician.

See left side panel for additional precautionary statement.

PRECAUTIONARY STATEMENTS

HAZARDS TO HUMANS AND DOMESTIC ANIMALS

WARNING

Causes eye irritation. Harmful if swallowed or absorbed through skin. Do not breathe vapors or spray mist. Do not get in eyes, on skin or on clothing. Wear goggles or face shield and rubber gloves when handling. Wash hands and face thoroughly with soap and water after use and before eating or smoking. Remove contaminated clothing and wash before reuse.

ENVIRONMENTAL HAZARDS

This product is toxic to fish, shrimp and oysters. Keep out of lakes, streams or ponds. Do not apply where runoff is likely to occur. Do not apply when weather conditions favor drift from areas treated. Do not contaminate water by cleaning of equipment or disposal of wastes.

PHYSICAL OR CHEMICAL HAZARDS

Do not use or store near heat or open flame.

DIRECTIONS FOR USE

It is a violation of Federal law to use this product in a manner inconsistent with its labeling.

STORAGE AND DISPOSAL

Do not contaminate water, food or feed by storage or disposal.

PESTICIDE DISPOSAL

Wastes resulting from the use of this product may be disposed of on site or at an approved waste disposal facility.

CONTAINER DISPOSAL

Triple rinse (or equivalent). Then offer for recycling or reconditioning, or puncture and dispose of in a sanitary landfill, or by other procedures approved by state and local authorities.

REENTRY STATEMENT

Do not apply this product in such a manner as to directly or through drift expose workers or other persons. The area being treated must be vacated by unprotected persons.

Do not enter treated areas without protective clothing until sprays have dried.

Because certain states may require more restrictive reentry intervals for various crops treated with this product, consult your State Department of Agriculture for further information.

Written or oral warnings must be given to workers who are expected to be in a treated area or in an area about to be treated with this product. When oral warnings are given, warnings shall be given in a language customarily understood by workers. Oral warnings must be given if there is reason to believe that written warnings cannot be understood by workers. Written or oral warnings must include the following information:

"WARNING" Area to be treated with propargite on (date of application). Do not enter without appropriate protective clothing until sprays have dried. If accidentally exposed, consult the Statement of Practical Treatment portion of the pesticide label for first aid recommendations.

DIRECTIONS FOR USE

OMITE®-6E is a liquid emulsifiable concentrate for preparing sprays to control European red, Pacific spider, southern red, strawberry spider, and two-spotted spider mites.

OMITE®-6E is not systemic in action; therefore complete coverage of both upper and lower leaf surfaces and of fruit is necessary for effective control.

Add recommended dosage (from table) of OMITE®-6E to the spray tank, agitate and spray thoroughly to cover foliage and fruit for best results.

DOSAGE INSTRUCTIONS

See dosage instructions chart and application notes.

AL USUARIO: Si usted no puede leer o entender ingles no use este producto hasta que la etiqueta le haya sido explicada ampliamente.

TO THE USER: If you cannot read or understand English, do not use this product until the label has been fully explained to you.

USE RESTRICTIONS: Do not graze or feed livestock on cover crops grown among the trees and vines listed on this label. Do not feed hop refuse or cones to livestock. However, spent hops from the extract operation may be fed. Do not feed mint spent hay to livestock.

COMPATIBILITIES: OMITE®-6E is compatible with most pesticides (see caution statements). Check with your supplier for details. Observe all cautions and limitations on labeling of all products used in mixtures.

CAUTION:

A. Leaf injury may occur if the following conditions are not observed.
   1. Do not apply OMITE®-6E with Diazinon WP if air temperatures are expected to exceed 95°F. the day of spraying hops.
   2. OMITE®-6E is not compatible with foliar spray oils. (Further information available from supplier).
   3. Do not apply OMITE®-6E to cranberries if air temperatures are above 85°F.

B. OMITE®-6E is not compatible with alkaline materials such as lime, lime sulfur or Bordeaux. The effectiveness of OMITE will be reduced.

DOSAGE INSTRUCTIONS

CROP	MITES CONTROLLED	AMOUNT PER ACRE	GALLONS OF SPRAY SOLUTION PER ACRE		NUMBER OF SPRAYS PER YEAR	DAYS BEFORE HARVEST (LAST SPRAY)
			GROUND APPLICATION	AERIAL APPLICATION		
Almonds (California & Arizona only)	Clover European red strawberry spider two-spotted spider pacific spider	3–4 pints	minimum of 50 gallons	15[1]	2	28
Apricots (California only)	two-spotted spider Pacific spider	2–3 pints		minimum of 10 gallons	2	post-harvest only
Cranberries[2] (Massachusetts only)	southern red mite	30–60[3,4] fluid ounces	300		2	14
Field Corn (California only)	two-spotted spider	2 pints	20–50	minimum of 10 gallons	1	56
Hops	two-spotted spider	2 pints	200		2	14
Mint	two-spotted spider	2–3 pints	20–50	minimum of 10 gallons	2	14
Potatoes (Pacific Northwest only)	two-spotted spider	2–3 pints	20–50	minimum of 10 gallons	2	14
Walnuts	two-spotted spider European red	2–9 pints	500–1500	minimum of 20 gallons[5]	2	14

APPLICATION NOTES—ALMONDS

1. One application of OMITE-6E may be applied by air to almonds over one application of OMITE-30W.

APPLICATION NOTES—CRANBERRIES

2. OMITE-6E may be applied through a sprinkler irrigation system to cranberries.
3. 30–60 ounces OMITE-6E may be applied to cranberries once prior to bloom followed by a second application of 30 ounces anytime prior to 14 days harvest.
4. Regardless of the size of the irrigation system, if OMITE-6E is used at the recommended rate per acre, the ratio of diluent to product in the slurry tank should be approximately 4:1.

APPLICATION NOTES—WALNUTS

5. Use 3 to 4 pints of OMITE-6E in a minimum of 20 gallons of water per acre when applied by air.

IRRIGATION USE PRECAUTIONS

A. Apply the product only through systems containing anti-siphon and check valves which will prevent water source contamination and overflow of the slurry tank, and containing interlocking controls between the metering device and the water pump to insure simultaneous shut-off.

B. Constant agitation must be maintained in the chemical supply tank during the entire period of insecticide application.

C. Inject the product with a positive displacement pump into the main line ahead of a right angle turn, to insure adequate mixing.

D. Application of more than label recommended quantities of irrigation water per acre may result in decreased product performance by removing the chemical from the zone of effectiveness.

E. Do not apply when wind speed favors drift, when system connections or fittings leak, when nozzles do not provide uniform distribution or when lines containing the product must be dismantled and drained.

F. Greater accuracy in calibration and distribution will be achieved by injecting a larger volume of more dilute slurry per hour.

G. Where sprinkler distribution patterns do not overlap sufficiently unacceptable mite control may result. Where sprinkler distribution patterns overlap excessively crop injury may result.

IMPORTANT NOTICE

Seller warrants that this product conforms to its chemical description and is reasonably fit for the purposes stated on the label when used in accordance with the directions and instructions specified on the label under normal conditions of use, **but neither this warranty nor any other warranty of merchantability or fitness for a particular purpose, express or implied, extends to the use of this product, contrary to label instructions, or under abnormal conditions, or under conditions not reasonably foreseeable to seller, and buyer assumes the risk of any such use.**

RESCUE

POSTEMERGENT HERBICIDE

For Soybeans

COMPOSITION

ACTIVE INGREDIENT	% BY WEIGHT
Sodium salt of naptalam (N-1-Naphthylphthalamic acid)*	23.7%
Sodium salt of 4-(2,4-dichlorophenoxy) butyric acid**	.6%
INERT INGREDIENTS	75.7%
TOTAL	100.0%

* Equivalent to 2 lbs. of the sodium salt of naptalam per gallon.

** Equivalent to 0.06 lbs 4-(2,4-DB) acid per gallon

EPA Reg. No. 400-166 EPA Est. No.

RESCUE is a Trademark of UNIROYAL CHEMICAL COMPANY, INC.

WARNING

Keep Out of Reach of Children

STATEMENT OF PRACTICAL TREATMENT

IF SWALLOWED: Drink two large glasses of water and induce vomiting. Contact a physician.

IF ON SKIN: Wash affected area with plenty of soap and water.

IF IN EYES: Flush eyes with plenty of water for at least 15 minutes. Contact a physician.

IF INHALED: Remove to fresh air; if abnormal symptoms develop, contact a physician.

See left side panel for additional precautionary statements.

PRECAUTIONARY STATEMENTS

HAZARDS TO HUMANS

WARNING

Causes eye irritation. Do not get in eyes. Harmful if inhaled or absorbed through the skin. Avoid breathing spray mist. Avoid contact with skin or clothing.

ENVIRONMENTAL HAZARDS

Do not apply directly to water. Do not contaminate water by cleaning of equipment or disposal of wastes. Do not apply when weather favors drift from treated area.

DIRECTIONS FOR USE

It is a violation of Federal Law to use this product in a manner inconsistent with its labeling.

STORAGE AND DISPOSAL

Do not contaminate water, food or feed by storage and disposal.

STORAGE

RESCUE is normally a solution but it may crystallize in the cold (32° F.). Gentle warming (60° F.) will redissolve crystals. Mild agitation, such as rolling, shaking, or stirring before use is recommended.

PESTICIDE DISPOSAL

Wastes resulting from the use of this product may be disposed of on site or at an approved waste disposal facility.

CONTAINER DISPOSAL

Triple rinse (or equivalent). Then offer for recycling or reconditioning, or puncture and dispose of in a sanitary landfill, or by other procedures approved by state and local authorities.

EQUIPMENT CLEANING

Thoroughly rinse spray tank, hose and nozzles with water after application. Use RESCUE in metal or fiberglass tanks. RESCUE can be removed from metal or fiberglass tanks by washing thoroughly with ammonia water and rinsing several times with water containing a good detergent.

USE DIRECTIONS

GENERAL

RESCUE is a selective herbicide recommended for mid-to-late season postemergence weed control in soybeans.

RESCUE is a systemic herbicide and therefore requires 20-30 days to obtain maximum affect. The reduction of weed competition to the soybeans begins immediately after RESCUE is applied even though the weeds are not killed immediately. DO NOT APPLY RESCUE TO STRESSED SOYBEAN PLANTS.

WEEDS CONTROLLED: When applied according to label instructions, RESCUE will control or suppress the following broadleaf weeds, reducing their competition to the soybean crop:

Weed	Growth Stage For Optimum Results	Expected Results
Cocklebur	8 inches thru 24 inches but before flowering begins	—Stem twisting —Growth termination —Reduction of viable seed —Eventual desiccation will occur if soybeans canopy over weeds —Cocklebur from 24-36" tall will be suppressed and normally will not set seed or interfere with harvest operations
Giant Ragweed	10 inches thru 36 inches, but before flowering begins	—Stem twisting —Eventual desiccation will occur if soybeans canopy over weeds Growth termination
Sunflower	12 inches thru early flower bud	—Stem twisting —Terminal growth inhibition —Eventual desiccation will occur if soybeans canopy over weeds NOTE: Maximum control will be obtained from applications made prior to 24" height.
Marshelder	12 inches thru early flower bud	—Stem twisting —Eventual desiccation will occur if soybeans canopy over weeds

APPLICATION TIMING

Determinate Soybeans—Apply when soybeans are at least 14 inches or taller.

Indeterminate Soybeans—Apply when soybeans are approximately 14 inches tall or after first bloom.

GROUND APPLICATION

Apply 3 quarts of RESCUE in 10–25 gallons of water per acre (broadcast rate). A NON-IONIC SPRAY ADJUVANT OR CROP OIL CONCENTRATE SHOULD BE USED AT THE MANUFACTURERS RECOMMENDED RATE. The spray boom should be positioned 18–24" above the tops of the soybean plants or weeds and be equipped with standard high pressure hollow cone or flat fan nozzles that provide thorough, uniform coverage of the weed foliage. Maintain a high pressure (40–50 p.s.i.) during application to insure better overall coverage. When making band applications, reduce rates and water volume according to width of band row spacing.

SPOT SPRAYERS RESCUE may be applied through spot spraying equipment to control or suppress the above labeled weeds. The use of a spray solution containing 7.5% RESCUE (1½ gal. RESCUE in 20 gallons of water plus surfactant/adjuvant) has obtained the best results. Applications should be made when the weeds are actively growing with additional recommendations stated in the broadcast section. A flat fan or hollow cone nozzle tip should be used with the handgun sprayer to ensure complete coverage of the target weed.

AERIAL APPLICATION

Apply 3 quarts of RESCUE in a minimum of 5 gallons of water per acre. A NON-IONIC SPRAY ADJUVANT OR CROP OIL CONCENTRATE SHOULD BE USED AT THE MANUFACTURERS RECOMMENDED RATE. Nozzle size and orientation should provide maximum droplet breakup without causing excessive drift.

ATTENTION: APPLICATION NOTES

1. Avoid applications to drought, heat or disease stressed soybeans.
2. Less activity should be expected if applied on stressed weeds.
3. If rain occurs within 6 hours after application effectiveness may be reduced.
4. Drift-Applications should not be made when weather conditions favor drift. Do not apply by air within 500 feet of susceptible crops such as cotton, tomatoes or tobacco.
5. Soybean tolerance— Most soybeans are tolerant to RESCUE when used according to label instructions. Some leaf wrinkling or epinasty may occur under certain conditions, however, soybeans will usually outgrow this condition and continue to develop normally. Mitchell and Forrest varieties are more sensitive to RESCUE than other varieties.
6. DO NOT GRAZE OR FEED FORAGE FROM TREATED SOYBEANS TO LIVESTOCK.

IMPORTANT NOTICE

Seller warrants that this product conforms to its chemical description and is reasonably fit for the purposes stated on the label when used in accordance with the directions and instructions specified on the label under normal conditions of use, **but neither this warranty nor any other warranty of merchantability or fitness for a particular purpose, express or implied, extends to the use of this product, contrary to label instructions, or under abnormal conditions, or under conditions not reasonably foreseeable to seller, and buyer assumes the risk of any such use.**

ROYAL MH-30®

GROWTH RETARDANT WITH SORBATRAN*

Controls Tobacco Sucker Growth

COMPOSITION

ACTIVE INGREDIENTS:	% BY WEIGHT
Potassium salt of 1,2-dihydro-3, 6-pyridazinedione	21.7%
INERT INGREDIENTS:	78.3%
TOTAL:	100.0%

(One gallon contains 1.5 pounds as maleic hydrazide)

U.S. Patent No. 3,503,729

EPA Reg. No. 400-84 EPA Est. No.

* Trademark of UNIROYAL, Inc.

MH-30 is a Registered Trademark of UNIROYAL CHEMICAL COMPANY, INC.

CAUTION

Keep Out of Reach of Children

For precautionary statements see left panel.

PRECAUTIONARY STATEMENTS

HAZARDS TO HUMANS

CAUTION

Avoid contact with skin, eyes or clothing. In case of contact immediately flush eyes or skin with plenty of water. Get medical attention if irritation persists.

ENVIRONMENTAL HAZARDS

Do not apply directly to water. Do not contaminate water by cleaning of equipment or disposal of wastes.

DIRECTIONS FOR USE

It is a violation of Federal law to use this product in a manner inconsistent with its labeling.

STORAGE AND DISPOSAL

Do not contaminate water, food or feed by storage and disposal.

STORAGE

The active ingredient in ROYAL MH-30 will separate from the surfactant at temperatures above 134°F. If containers are exposed to excessive heating, the contents can be remixed by lowering the temperature of the entire formulation below 134°F. and applying agitation. Always shake or stir well before use.

PESTICIDE DISPOSAL

Pesticide wastes are toxic. Improper disposal of excess pesticide, spray mixture or rinsate is a violation of Federal Law. If these wastes cannot be disposed of by use according to label instructions, contact your State Pesticide or Environmental Control Agency, or the Hazardous Waste representative at the nearest EPA Regional Office for guidance.

CONTAINER DISPOSAL

Triple rinse (or equivalent). Then offer for recycling or reconditioning, or puncture and dispose of in a sanitary landfill, or incineration, or if allowed by State and local authorities, by burning. If burned stay out of smoke.

EQUIPMENT CLEANING

ROYAL MH-30 is not corrosive to spray equipment. Following applications, all spray equipment should be thoroughly rinsed with water.

REENTRY STATEMENT

Do not apply this product in such a manner as to directly or through drift expose workers or other persons. The area being treated must be vacated by unprotected persons.

Do not enter treated areas without protective clothing until sprays have dried.

Because certain states may require more restrictive reentry intervals for various crops treated with this product, consult your State Department of Agriculture for further information.

Written or oral warnings must be given to workers who are expected to be in a treated area or in an area about to be treated with this product. When oral warnings are given, warnings shall be given in a language customarily understood by workers. Oral warnings must be given if there is reason to believe that written warnings cannot be understood by workers. Written or oral warnings must include the following information:

"CAUTION" "Area to be treated with ROYAL MH-30 on (date of application). Do not enter without appropriate protective clothing until sprays have dried. If accidentally exposed, consult the Statement of Practical Treatment portion of the pesticide label for first aid recommendations."

NOTES

1. To control tobacco suckers apply ROYAL MH-30 at the proper time. Too early spraying may prevent normal development of top leaves. Too late spraying will allow some sucker growth. FULL FLOWER is the stage when 90% of your plants have begun to shed their first flowers.
2. Rains occurring within 6 hours after spraying may reduce effectiveness. Should wash off occur within 6 hours, a single repeat application may be made.
3. Follow dosage recommendations carefully and spray field uniformly.
4. Compatibility studies with ROYAL MH-30 and tobacco insecticides have not been completed. Separate application is recommended.
5. Do not use on tobacco plants grown for seed production.
6. Do not spray tobacco in the middle of hot, dry days when leaves are severely wilted. For best results spray during the early morning, all day on cloudy days, avoiding the late afternoon or night applications. (Use a nozzle delivering a fine mist spray.)
7. Closely hand-suckered tobacco or tobacco treated with ROYAL MH-30 will not ripen as rapidly as poorly suckered tobacco. Also, under some conditions ROYAL MH-30 treated tips have a tendency to turn yellow prematurely. This is not harmful and does not mean that your tobacco is ripe. Therefore, do not harvest tobacco treated with ROYAL MH-30 by using time and color of the leaves as your only guide to ripeness. Use other characteristics indicative of ripeness such as:
   a. Fading of color from midrib.
   b. Change in apparent leaf texture.
8. If tobacco plants have been grown under drought conditions, absorption of ROYAL MH-30 will be slow and results may not be as good as plants treated under normal conditions.
9. Too little ROYAL MH-30 will result in less than the desired control, too much reduces quality of tips.
10. Use only one application of ROYAL MH-30 per season.
11. Allow at least 7 days after application before harvest.
12. Even though ROYAL MH-30 is non-volatile no drift should be permitted onto nearby crops.

TOBACCO—DIRECTIONS FOR USE

To prevent sucker development on flue cured, burley, dark fire, Maryland and cigar (Pennsylvania, and Wisconsin) tobacco.

ROYAL MH-30 is a potassium salt formulation of maleic hydrazide used to prevent sucker development in tobacco. Since ROYAL MH-30 is absorbed and moves inside the tobacco plant, spraying the whole plant is not necessary. Best results are obtained by spraying the upper ⅓ to ½ of the plant. It is very important to apply equal amounts of ROYAL MH-30 to each plant at the correct time to obtain satisfactory sucker control.

Under favorable conditions, ROYAL MH-30 kills tobacco suckers in a manner similar to contact agents through systemic action.

TIMING

For flue cured, burley, dark fire and Maryland tobacco

1. If all plants flower at about the same time, proceed as follows: As soon as the plants are in FULL FLOWER, top, pull all suckers from early maturing plants and spray ROYAL MH-30 uniformly onto the entire field within 24 hours after topping. (See notes 1–12).
2. If there is a wide range in the time of flowering, the following procedure should be used:
   a. Go through the field and top the few early flowering plants.
   b. Wait until the remaining plants are in FULL FLOWER, top them, remove all suckers from the entire field, and apply ROYAL MH-30 within 24 hours after topping. (See notes 1–12). Tip leaves to be harvested should be at least 6 to 8 inches long at time of spraying.

For Pennsylvania cigar tobacco

1. If your field flowers uniformly:
   a. Top when about ¼ to ½ of your plants show one or more open flowers on each head.
   b. Wait three days after topping.
   c. Spray ROYAL MH-30. For best results ROYAL MH-30 should be sprayed on dry or nearly dry leaves. Avoid spraying on leaves covered with dew.
2. If your field doesn't flower uniformly:
   a. When flowering in your field is particularly uneven you should top when about 50 to 60 percent of the plants show one or more open flowers per head.
   b. Wait three days after topping and remove by hand any suckers more than two inches long.
   c. Spray ROYAL MH-30 over your entire field.

For Wisconsin cigar tobacco

1. Top the field when most of the plants in your field are in full flower.
2. Remove, by hand, any suckers larger than one or two inches.
3. Spray within 24 hours after topping and removal of suckers.

HOW TO SPRAY

ROYAL MH-30 may be applied with either a conventional sprayer, or with a compressed air shoulder sprayer.

1. ROYAL MH-30 can be applied by spraying one or more rows at a time, wetting upper ½ to ⅓ of plant.
2. Compressed air shoulder sprayer with nozzle spraying a wide cone. Spray one side of each tobacco row while walking at a constant rate. Use up and down motion of spray nozzle to moisten upper 6 to 8 leaves of each plant. Apply equal amounts of spray to each plant.

RATE OF APPLICATION

Apply the indicated amounts of ROYAL MH-30 in 40 to 50 gallons of water per acre; Apply at 40 to 50 psi.

FOR flue cured, burley, cigar (Pennsylvania, Wisconsin) and Maryland Tobacco

Apply 1.5 to 2 pints of ROYAL MH-30 per 1000 plants. Example: use 1.5 to 2 gallons of ROYAL MH-30 for 8000 plants per acre.

For dark tobacco

Apply 3.0 pints of ROYAL MH-30 per 1000 plants. Example: use 3.0 gallons ROYAL MH-30 for 8000 plants per acre.

NOTE: ALWAYS SHAKE OR STIR WELL BEFORE USE.

IMPORTANT NOTICE

Seller warrants that this product conforms to its chemical description and is reasonably fit for the purposes stated on the label when used in accordance with the directions and instructions specified on the label under normal conditions of use, **but neither this warranty nor any other warranty of merchantability or fitness for a particular purpose, express or implied, extends to the use of this product, contrary to label instructions, or under abnormal conditions, or under conditions not reasonably foreseeable to seller, and buyer assumes the risk of any such use.**

ROYAL MH-30®SG

WATER SOLUBLE
GRANULE GROWTH RETARDANT

For Potatoes and Onions

COMPOSITION

ACTIVE INGREDIENTS:	% BY WEIGHT
Potassium salt of 1,2-dihydro-3, 6-pyridazinedione	80.0%*
INERT INGREDIENTS:	20.0%
TOTAL:	100.0%

*Contains 60% by weight of maleic hydrazide

U.S. Patent No. 3,503,729

EPA Reg. No. 400-165 EPA Est. No.

MH-30 is a Registered Trademark of UNIROYAL CHEMICAL COMPANY, INC.

WARNING
Keep Out of Reach of Children
AVISO

PRECAUCION AL USUARIO: Si usted no lee ingles, no use este producto hasta que le etiqueta haya sido explicado ampliamente.

STATEMENT OF PRACTICAL TREATMENT

IF INHALED: Remove victim to fresh air. If not breathing, give artificial respiration, preferably mouth-to-mouth. Get medical attention.

IF SWALLOWED: Drink large quantities of water and induce vomiting by placing finger in back of throat.

IF IN EYES: Flush with plenty of water and call a physician.

IF ON SKIN: Wash thoroughly with soap and water.

See Left Panel For Additional Precautionary Statement

PRECAUTIONARY STATEMENTS

HAZARDS TO HUMANS

WARNING

Causes eye irritation. Do not get in eyes, on skin, or on clothing. Harmful if swallowed.

ENVIRONMENTAL HAZARDS

Do not apply directly to water. Do not contaminate water by cleaning of equipment or disposal of wastes.

DIRECTIONS FOR USE

It is a violation of Federal law to use this product in a manner inconsistent with its labeling.

STORAGE AND DISPOSAL

Do not contaminate water, food or feed by storage and disposal.

STORAGE

Store in a dry location.

PESTICIDE DISPOSAL

Pesticide wastes are toxic. Improper disposal of excess pesticide, spray mixture or rinsate is a violation of Federal Law. If these wastes cannot be disposed of by use according to label instructions, contact your State Pesticide or Environmental Control Agency, or the Hazardous Waste representative at the nearest EPA Regional Office for guidance.

EQUIPMENT CLEANING

ROYAL MH-30SG is not corrosive to spray equipment. Following applications, all spray equipment should be thoroughly rinsed with water.

CONTAINER DISPOSAL

Fiber carton with liners or bags: Completely empty liner or bag by shaking and tapping sides and bottom to loosen clinging particles. Empty residue into application equipment. Then dispose of liner or bag in a sanitary landfill, or by incineration if permitted by State and local authorities. Dispose of carton in the same manner.

REENTRY STATEMENT

Do not apply this product in such a manner as to directly or through drift expose workers or other persons. The area being treated must be vacated by unprotected persons.

Do not enter treated areas without protective clothing until sprays have dried.

Because certain states may require more restrictive reentry intervals for various crops treated with this product, consult your State Department of Agriculture for further information.

Written or oral warnings must be given to workers who are expected to be in a treated area or in an area about to be treated with this product. When oral warnings are given, warnings shall be given in a language customarily understood by workers. Oral warnings must be given if there is reason to believe that written warnings cannot be understood by workers. Written or oral warnings must include the following information:

"WARNING" "Area to be treated with ROYAL MH-30SG on (date of application). Do not enter without appropriate protective clothing until sprays have dried. If accidentally exposed, consult the Statement of Practical Treatment portion of the pesticide label for first aid recommendations."

NOTES

1. Follow dosage recommendations carefully and spray field uniformly.
2. Do not use on potato plants grown for seed production.
3. Even though ROYAL MH-30SG is non-volatile no drift should be permitted onto nearby crops. Particular care should be taken to prevent drift when aerial applications are being made.
4. Spray potatoes or onions at least 24 hours before rain or applying overhead irrigation. Rain or overhead irrigation occurring within 24 hours after application will reduce effectiveness.
5. At least a 2-week interval should be allowed between a ROYAL MH-30SG application to potatoes and the use of a vine killer.
6. Occasional injury to the terminal growth of potato foliage has been observed following the application of ROYAL MG-30SG.
7. To obtain desired results plants should not be stressed from either too little or excessive water before or after application.

MIXING INSTRUCTIONS

NOTE: ROYAL MH-30SG must be dissolved first if it is to be tank mixed with an insecticide or fungicide.

1. Be sure sprayer is clean and not contaminated with any material.
2. Fill tank $^3/_4$ full with clean water, and start agitation.
3. Pour ROYAL MH-30SG slowly from bag into tank. Avoid dumping all contents at once.
4. Allow 2–3 minutes agitation for dissolving.
5. Add compatible tank mix insecticide or fungicide if needed.
6. Complete filling tank.

DIRECTIONS FOR USE

ONIONS

TO CONTROL SPROUTING OF ONIONS IN STORAGE

Timing

ROYAL MH-30SG should be applied when the bulbs are fully mature, have at least 5–8 green leaves, and necks soft enough for the tops to fall if they have not already done so. The spray is often applied when 50 percent of the tops have fallen but all tops are still green. This condition usually occurs about two weeks to 10 days before harvest time. Green tops are essential for the absorption of ROYAL MG-30SG.

RATE OF APPLICATION

NOTE: Uniform and complete coverage is essential for good sprout control.

Ground Application

Use $3^1/_3$ pounds of ROYAL MH-30SG in a minimum of 30 gallons of water per acre.

Aerial Application
Use 3⅓ pounds of ROYAL MH-30SG in a minimum of 10 gallons of water per acre. Do not fly at heights of less than 8–12 feet. Maintaining this height will give more uniform coverage and aid in preventing an excessive build-up of chemical on the left side of the treated swath. These heights should be reduced as cross-wind velocities exceed 2–4 MPH.

Spray Adjuvants
West of Rocky Mountains: Use Regulaid, X-77 or R11 at the rate of 1 quart per 100 gallons of spray solution in arid sections of the U.S. These products will help provide better coverage and aid in translocation of the ROYAL MH-30SG.
East of Rocky Mountains: use of surfactants is normally not recommended.

Temperature
ROYAL MH-30SG should be applied at temperatures below 85°F or at temperatures below 80°F if the temperature is expected to exceed 85°F. ROYAL MH-30SG is water soluble and under high temperature and low humidity conditions will crystalize on the leaf surface and not be absorbed into the plant.

CAUTION
Avoid spraying edible onions too early. Spongy bulbs may result if spraying is done earlier than two weeks before maturity.

Varieties
Use of ROYAL MH-30SG should be combined with selection of a good-keeping variety. Do not apply to onion varieties with poor-keeping quality. ROYAL MH-30SG will not make poor-keeping varieties equal to good-keeping varieties.

POTATOES—BENEFITS

Sprout Control
ROYAL MH-30SG will control sprouting of potatoes in storage. However, ROYAL MH-30SG will not make a poor-keeping potato such as Norgold or Norchip into a good-keeping potato.

Reduction of Volunteer Potatoes
ROYAL MH-30SG will reduce the number of volunteer potatoes the following spring because of its ability to control sprouting. If applied at the proper time it will also reduce the number of undersize potatoes left in the field, which also reduces the potential for volunteers.

May Improve Grade
ROYAL MH-30SG may improve the grade of potatoes, especially Russet Burbank variety, if the application is made prior to any secondary tuber formation.

Undersize Potatoes
ROYAL MH-30SG will help reduce any additional "sets" of potatoes, thus decreasing the number of undersize potatoes in the field at harvest.

Reduces Shrinkage
ROYAL MH-30SG applications can reduce shrinkage in storage up to 15 percent.

DIRECTIONS FOR USE

Timing
Timing is important. To allow translocation of chemical down to the tubers ROYAL MH-30SG should be applied after tubers are formed and while vines are still green and are free from insects, diseases or other damage. Time of application may occur as early as 2–3 weeks past full bloom when the primary and secondary inflorescences have lost most of their flowers and a few third order inflorescences and buds are present. This occurs when a few lower leaves show a yellow color. Application prior to this time may result in reduced size of potatoes and phytotoxicity. The following tuber sizes may also be used as a guide, provided the above conditions have been met.

Russet Burbank
The smallest tubers that are required to reach a minimum size of 4–6 oz. must be 1½–2" in diameter (and elongating) at time of spraying.

Norgold or Norchip
The smallest tubers that are required to reach maximum size must be 2" or more in diameter at time of spraying.

Other Varieties
The smallest tubers that are required to reach maximum size, should be 1½" in diameter at time of spraying.

Rate of Application
NOTE: Uniform and complete coverage is essential for satisfactory results.

Ground
Use 5 pounds of ROYAL MH-30SG in a minimum of 30 gallons of water per acre.

Aerial
Use 5 pounds of ROYAL MH-30SG per acre.
West of the Rocky Mountains: Use in a minimum of 10 gallons of water per acre.
East of the Rocky Mountains: Use in a sufficient quantity of water to gain complete uniform coverage.
Do not fly at heights of less than 8–12 feet. Maintaining this height will give more uniform coverage and aid in preventing an excessive build up of chemical on the left side of the treated swath. These heights should be reduced as cross-wind velocities exceed 2–4 MPH.

Spray Adjuvants
West of Rocky Mountains: Use Regulaid, X-77 or R11 at the rate of 1 quart per 100 gallons of spray solution in arid sections of the U.S. These products will help provide better coverage and aid in translocation of the ROYAL MH-30SG.
East of the Rocky Mountains: Use of surfactants is normally not recommended.

Temperature
ROYAL MH-30SG should be applied at temperatures below 85°F or at temperatures below 80°F if the temperature is expected to exceed 85°F. ROYAL MH-30SG is water soluble and under high temperature and low humidity conditions will crystalize on the leaf surface and not be absorbed into the plant.

IMPORTANT NOTICE

Seller warrants that this product conforms to its chemical description and is reasonably fit for the purposes stated on the label when used in accordance with the directions and instructions specified on the label under normal conditions of use, **but neither this warranty nor any other warranty of merchantability or fitness for a particular purpose, express or implied, extends to the use of this product, contrary to label instructions, or under abnormal conditions, or under conditions not reasonably foreseeable to seller, and buyer assumes the risk of any such use.**

TERRACLOR®

2 LB. EMULSIFIABLE SOIL FUNGICIDE

COMPOSITION

ACTIVE INGREDIENT:	% BY WEIGHT
Pentachloronitrobenzene	23.8%
INERT INGREDIENTS:	76.2%
TOTAL	100.0%

Contains 2 lbs. of Pentachloronitrobenzene per U.S. Gallon at 68°F.
EPA Reg. No. 400-400 EPA Est. No.
TERRACLOR is a Registered Trademark of UNIROYAL CHEMICAL COMPANY, INC.

CAUTION

Keep Out of Reach of Children

STATEMENT OF PRACTICAL TREATMENT
IF IN EYES: Flush with plenty of water. Get medical attention.
IF ON SKIN: Wash with plenty of soap and water. Get medical attention.
IF SWALLOWED: Drink promptly a large quantity of milk, egg whites, gelatin solution or, if these are not available, drink large quantities of water. Avoid alcohol. Get medical attention.

See left panel for additional precautionary statement.

PRECAUTIONARY STATEMENTS

HAZARDS TO HUMANS AND DOMESTIC ANIMALS
CAUTION

This material may cause skin irritation. Avoid contact with eyes or clothing. Wash thoroughly with soap and water after handling. Use with adequate ventilation. Do not contaminate feed or foodstuffs. The following protective clothing are required during mixing/loading operations: respirator, gloves, long-sleeve shirt, long pants, socks and shoes.

ENVIRONMENTAL HAZARDS

This product is toxic to fish. Do not apply directly to water or wetlands. Do not apply where runoff is likely to occur. Do not contaminate water by cleaning of equipment or disposal of wastes.

PHYSICAL OR CHEMICAL HAZARDS

Do not use, pour, spill or store near heat or open flame. Protect from temperatures below 0°F.

DIRECTIONS FOR USE

It is a violation of Federal law to use this product in a manner inconsistent with its labeling.

STORAGE AND DISPOSAL

Do not contaminate water, food or feed by storage or disposal.

PESTICIDE DISPOSAL

Pesticide wastes are toxic. Improper disposal of excess pesticide, spray mixture, or rinsate is a violation of Federal Law. If these wastes cannot be disposed of by use according to label instructions, contact your State Pesticide or Environmental Control Agency, or the Hazardous Waste representative at the nearest EPA Regional Office for guidance.

CONTAINER DISPOSAL

Triple rinse (or equivalent). Then offer for recycling or reconditioning, or puncture and dispose of in a sanitary landfill, or by other procedures approved by State and local authorities.

DIRECTIONS FOR USE

BEANS (Root and Stem Rot—Rhizoctonia solani): FURROW—Application rates are 14,500 linear feet of row for bush type beans and to 8,430 linear feet of row for pole beans. Apply 2–3 qts. in 8 to 10 gallons of water. Spray planting furrow and covering soil at planting time. Avoid applying directly to seed; otherwise delayed emergence may occur. **In California and Arizona** apply 3–4 pints in 8 to 10 gallons of water. Apply in-furrow and incorporate with covering soil at planting. Use lower rate for lighter soils. Avoid application to bare seed.

SNAP AND DRY BEANS (White Mold—Sclerotinia sclerotiorum): Use 1 gallon mixed in 8 to 10 gals. of water and apply to 14,500 linear feet of row for bush type beans or to 8,430 linear feet of row for pole beans. Spray an 8″ wide band, centered on the row, immediately after or at time of seeding. Avoid applying directly to seed; otherwise delayed emergence may occur. (Where dinitro type pre-emergence herbicides are used, add 1 gallon to the final herbicide spray mix and apply simultaneously). If disease is severe, application may be repeated at 2 to 3 week intervals, using 1 or 2 nozzles per row and directing spray at base of plants. Do not apply after first bloom. Soil should remain undisturbed after treatment. Do not feed treated vines to livestock.

COTTON (Damping Off—Rhizoctonia solani): Use ½ to 1 gallon per acre in at least 15 gallons of water (per 13,000 linear feet of row based on 40″ spacings). Spray on seed and surrounding soil at planting time. Do not allow the feeding or grazing of cotton foliage by livestock.

GARLIC (White Rot—Sclerotium cepivorum): Use 10 gals. mixed in 100 gals. of water and apply to 21,800 linear feet of row at planting time. Attach a suitable rig to the machine planter, using 2 nozzles per planting furrow. Direct the front nozzle to spray the bottom of the open furrow and the "seed" as it is dropped. Direct the rear nozzle to spray the covering soil.

POTATOES (Rhizoctonia solani): BROADCAST APPLICATION—Apply 9 to 12½ gals. to one acre mixed with sufficient water to total 25 gallons total volume. Apply to the soil surface prior to planting and thoroughly mix (disc and cross disc) to a depth of 4 inches. A drag harrow is usually attached behind the discs on the last discing. It is very important to thoroughly mix the TERRACLOR into the tuber-forming zone of the soil. It is desirable to make treatment when the soil is slightly moist.

POTATOES (Rhizoctonia solani): ROW (in furrow) APPLICATION—Use 5 gals. per 12,400 linear feet of row mixed in sufficient water to total 20 gallons of finished spray. Direct the spray solution to several points in the row to insure optimum treatment of the stem, stolen and tuber-forming zones, using three nozzles per row. Place one fan nozzle ahead of the furrow opening discs and two cone spray nozzles above and slightly ahead of the closing discs. Select nozzles that will deliver approximately ⅓ of the recommended spray volume per acre through each nozzle at the desired planting speed and pressure.

NOTE: Do not plant any root crop not registered for PCNB in rotation on PCNB treated soil.

IMPORTANT NOTICE

Seller warrants that this product conforms to its chemical description and is reasonably fit for the purposes stated on the label when used in accordance with the directions and instructions specified on the label under normal conditions of use, **but neither this warranty nor any other warranty of merchantability or fitness for a particular purpose, express or implied, extends to the use of this product, contrary to label instructions, or under abnormal conditions, or under conditions not reasonably foreseeable to seller, and buyer assumes the risk of any such use.**

TERRACLOR®

10% GRANULAR
SOIL FUNGICIDE

Controls Certain Soil-Borne Diseases of Cotton, Beans, Crucifers Peanuts and Potatoes

COMPOSITION

ACTIVE INGREDIENT:	% BY WEIGHT
Pentachloronitrobenzene	10%
INERT INGREDIENTS:	90%
TOTAL:	100%

EPA Reg. No. 400-402 EPA Est. No.

Terraclor is a Registered Trademark of UNIROYAL CHEMICAL COMPANY, INC.

CAUTION
Keep Out of Reach of Children

See left panel for additional precautionary statement.

PRECAUTIONARY STATEMENTS

HAZARDS TO HUMANS
CAUTION

Harmful if swallowed. Avoid inhaling dust. This material may cause skin irritation. Wash hands and face thoroughly after using. The following protective clothing are required during mixing/loading operations: gloves, a respirator, long-sleeved shirt, long pants, socks and shoes.

ENVIRONMENTAL HAZARDS

This product is toxic to fish. Do not apply directly to water or wetlands. Do not contaminate water by cleaning of equipment or disposal of wastes. Do not plant any root crop not registered for PCNB in rotation on PCNB treated soil.

DIRECTIONS FOR USE

It is a violation of Federal law to use this product in a manner inconsistent with its labeling.

STORAGE AND DISPOSAL

Do not contaminate water, food or feed by storage or disposal.

STORAGE

Store in a dry location.

PESTICIDE DISPOSAL

Pesticide wastes are toxic. Improper disposal of excess pesticide, spray mixture, or rinsate is a violation of Federal Law. If these wastes cannot be disposed of by use according to label instructions, contact your State Pesticide or Environmental Control Agency, or the Hazardous Waste representative at the nearest EPA Regional Office for guidance.

CONTAINER DISPOSAL

Completely empty bag into application equipment. Then dispose of empty bag in a sanitary landfill or by incineration, or if allowed by State and local authorities, by burning. If burned, stay out of smoke.

DIRECTIONS FOR USE

NOTE: **Do not plant any root crop not registered for PCNB in rotation on PCNB treated soil.**

COTTON **(Damping-off, Seed and Seedling Rots—Rhizoctonia solani):** For in-furrow application apply 10 to 20 lbs per acre to the planting furrow. (1 to 2 lbs. per 1300 linear feet of row, based on 40 in. rows). Apply over seed and covering soil at planting time, using suitable application equipment. Do not allow the feeding or grazing of cotton foliage by livestock.

BEANS **(Root and Stem Rot—Rhizoctonia solani) FURROW TREATMENT:** Apply 10 to 15 lbs. per 14,500 linear feet of row for bush beans, or per 8,430 linear feet of row for pole beans. Direct the material into the planting furrow and covering soil at planting time. NOTE: Do not feed treated vines to livestock.

CABBAGE, CAULIFLOWER, BRUSSELS SPROUTS, AND BROCCOLI **(Clubroot—Plasmodiophora brassicae):** Row Application: Apply 200 lbs. per 13,100 feet of row for cabbage, broccoli and brussels sprouts or to 10,900 linear feet of row for cauliflower, in a 12 to 15 inch band, by means of a suitable applicator, and rototil to a depth of 4 to 6 inches, immediately prior to transplanting.

BROADCAST: Apply 300 lbs. per acre to the soil surface prior to transplanting. Disc and cross disc to a depth of 4 to 6 inches.

NOTE: For clubroot of cabbage, cauliflower, broccoli and brussels sprouts, it is very important to thoroughly mix it with the soil.

CAULIFLOWER **(Wire Stem or Black root—Corticium solani):** Apply 110 to 150 lbs. per 10,900 feet in an 8-inch band, centered on the row, immediately prior to seeding. For smaller areas apply 12 lbs. per 1000 sq. ft. Rototil to a depth of 2 inches.

PEANUTS **(Southern blight—Sclerotium rolfsii):** Surface Band Treatment: Apply 100 lbs. per 12,400 linear feet of row, in an 8 to 12 inch band, centered on the row, at pegging time.

PRE-PLANT SOIL MIX: Apply 100 lbs. per 12,400 linear feet of row, in an 8 to 12 inch band, centered on the row, and mix to a depth of 2 to 4 inches. Application may be made simultaneously with planting, using suitable equipment.

SOIL MIX DURING CULTIVATION: Apply in split applications during three cultivations at the rate of 33 lbs. per 12,400 linear feet of row per application. Attach two delivery tubes per row ahead of inside cultivator sweeps (one on each side) and adjust so as to treat a total band of 12 inches wide, centered on the row. Do not pile untreated soil on the crown of plant.

SPLIT PEGGING APPLICATIONS: Apply TERRACLOR 10% Granular at the rate of 50 lbs. of granules per 14,000 linear feet of row (based on a 36 inch row spacing) as a 12 to 18 inch band centered over the row followed by shallow incorporation if possible. The initial application should be timed to coincide with early pegging. A second application at the same rate may be made if disease symptoms reappear. Depending on weather, this usually occurs 4 to 5 weeks after the first application.

NOTE: Do not exceed 10 lbs. actual PCNB per acre in any one season, regardless of the application methods used. Do not feed treated vines to livestock. Do not allow hogging down of peanuts.

POTATOES **(Rhizoctonia solani): BROADCAST TREATMENT:** Use 180 to 250 lbs. per acre. Spread evenly over the soil surface prior to planting. Disc and cross disc to a depth of 4 to 6 inches.

ROW (in-furrow) APPLICATION: Apply 100 lbs. per 12,400 linear feet of row (36 inch spacings) and work the fungicide thoroughly into the top 4 to 6 inches of the soil (it is essential that the fungicide is worked into the tuber forming zone of the soil). It is desirable to make treatment when the soil is slightly moist.

IMPORTANT NOTICE

Seller warrants that this product conforms to its chemical description and is reasonably fit for the purposes stated on the label when used in accordance with the directions and instructions specified on the label under normal conditions of use, **but neither this warranty nor any other warranty of merchantability or fitness for a particular purpose, express or implied, extends to the use of this product, contrary to label instructions, or under abnormal conditions, or under conditions not reasonably foreseeable to seller, and buyer assumes the risk of any such use.**

TERRACLOR®

75% WETTABLE POWDER SOIL FUNGICIDE

Controls Certain Soil-Borne Diseases of Cotton, Crucifers, Peanuts, Peppers, Beans, Potatoes, Garlic, Ornamentals and Turf

COMPOSITION

ACTIVE INGREDIENT	% BY WEIGHT
Pentachloronitrobenzene	75%
INERT INGREDIENTS:	25%
TOTAL:	100%

EPA Reg. No. 400-399 EPA Est. No.

Terraclor is a Registered Trademark of UNIROYAL CHEMICAL COMPANY, INC.

CAUTION

Keep Out of Reach of Children

STATEMENT OF PRACTICAL TREATMENT

IF SWALLOWED: Call a physician or Poison Control Center. Drink 1 or 2 glasses of water and induce vomiting by touching back of throat with finger, or if available by administering Syrup of Ipecac. Do not induce vomiting or give anything by mouth to an unconscious person.

IF IN EYES: Flush with plenty of water. Get medical attention if irritation persists.

IF ON SKIN: Wash with plenty of soap and water. Get medical attention if irritation persists.

IF INHALED: Remove to fresh air.

See Left Panel For Additional Precautionary Statement

PRECAUTIONARY STATEMENTS

HAZARDS TO HUMANS AND DOMESTIC ANIMALS

CAUTION

Harmful if swallowed, Avoid inhaling dust. May cause skin irritation. Wash hands and face thoroughly after using. The following protective clothing are required during mixing/loading operations: gloves, long-sleeved shirt, long pants, socks and shoes. Wash thoroughly all contaminated clothing before reuse.

ENVIRONMENTAL HAZARDS

This product is toxic to fish. Do not apply directly to water or wetlands. Do not contaminate water by cleaning of equipment or disposal of wastes. Do not apply directly adjacent to potable water supplies.

DIRECTIONS FOR USE

It is a violation of Federal law to use this product in a manner inconsistent with its labeling.

STORAGE AND DISPOSAL

Do not contaminate water, food or feed by storage or disposal.

STORAGE

Store in a dry location.

PESTICIDE DISPOSAL

Pesticide wastes are toxic. Improper disposal of excess pesticide, spray mixture, or rinsate is a violation of Federal Law. If these wastes cannot be disposed of by use according to label instructions, contact your State Pesticide or Environmental Control Agency, or the Hazardous Waste representative at the nearest EPA Regional Office for guidance.

CONTAINER DISPOSAL

Paper Bags: Completely empty bag into application equipment. Then dispose of bag in a sanitary landfill or by incineration, of if allowed by State and local authorities, by burning. If burned, stay out of smoke.

Cardboard Canisters: **(Commercial)** Completely empty canisters into application equipment. Then dispose of empty container in a sanitary landfill or by incineration, or if allowed by State and local authorities, by burning. If burned, stay out of smoke.

Household: Do not reuse empty container. Securely wrap original container in several layers of newspaper and discard in trash.

RECOMMENDED APPLICATIONS

CROP	DISEASE	POUNDS TERRACLOR 75% W.P.	REMARKS
FIELD CROPS COTTON	Damping off (Rhizoctonia solani)	1⅓ to 2¾	Apply specified dosage per acre in at least 15 gallons of water (per 13,100 linear feet of row based on 40 inch row spacing). Apply as a directed in-furrow spray to the seed and surrounding soil at planting time. Do not allow the feeding or grazing of cotton foliage by livestock.
PEANUTS	Southern stem and root rot (Sclerotium rolfsii)	13	**Pre-Plant Soil Mix:** Apply specified dosage per 12,400 linear feet of row as a spray in 8 to 10 gallons of water in an 8 to 12 inch band centered on the row and incorporate to a depth of 2 to 4 inches. Application may be made simultaneously with planting using suitable application equipment.
			At-Pegging Band: Apply specified dosage per 12,400 linear feet of row as a spray in 8 to 10 gallons of water in a 12 inch band centered on the row so as to cover the crown and pegging zones of the plant. If the row middles are cultivated at the same time, attach two spray nozzles (one to each side)—ahead of inside cultivator sweeps. Do not move any untreated soil into the area (if necessary remove inside sweep).
		4	**At Cultivation:** Apply specified dosage per 12,400 linear feet of row at cultivation. Do not apply more than 3 times per crop season. Attach two spray nozzles per row of inside cultivator sweeps (one on each side) and adjust so as to treat a total band of 12 inches centered on the row. Do not pile untreated soil on crown of plant.
	NOTE: Do not exceed 10 lbs. actual PCNB per acre in any one season, regardless of application methods used. Do not apply within 45 days of harvest. Do not feed treated vines to livestock. Do not allow hogging down of peanuts.		
VEGETABLES BEANS	Root and Stem rot (Rhizoctonia solani)	1⅓ to 2	Apply specified dosage in 8 to 10 gallons of water (per 14,500 linear feet of row for bush beans or 8,430 feet of row for pole beans). Apply as a directed spray in the seed furrow and to the covering soil at planting time.
BEANS (Snap and Dry)	White Mold (Sclerotinia sclerotiorum)	2¾	Apply specified dosage in 8 to 10 gallons of water (per 14,500 linear feet of row for bush beans, or 8,430 linear feet of row for pole beans). Apply as an 8 inch band centered on the row at or immediately after planting. If the disease problem is severe, applications may be repeated at 2 to 3 week intervals using directed spray application with one or two nozzles per row aimed at the soil at the base of the plants. Soil should remain undisturbed after application. Do not apply after first bloom. Do not exceed 10 lbs. actual PCNB per acre in any one season, regardless of method of application. Do not feed treated vines to livestock.
BROCCOLI BRUSSELS SPROUTS CABBAGE CAULIFLOWER	Clubroot (Plasmodiophora brassicae)	2 to 6	**Transplant Solution:** Suspend specified dosage in 100 gallons of water and apply ½ to ¾ pint of this solution per plant. The applied dosage of Terraclor should be based on the severity of the disease infection. Maintain agitation to hold the powder in suspension.
		30	**Band Application:** Apply specified dosage per acre in at least 25 gallons of water (per 13,100 linear feet of row for broccoli, brussels sprouts and cabbage, and 10,900 linear feet of row for cauliflower). Spray as a 12 to 15 inch band centered on the row and incorporate to a depth of 4 to 6 inches immediately prior to transplanting.
		40	**Broadcast Application:** Apply specified dosage per acre in at least 30 gallons of water as a pre-plant broadcast spray prior to transplanting. Thoroughly incorporate to a depth of 4 to 6 inches using a disc or other suitable equipment.
	NOTE: For effective control of clubroot in broccoli, brussels sprouts, cabbage and cauliflower, it is extremely important to thoroughly mix Terraclor with the soil.		
	Wire Stem or Black Root (Corticium solani)	15 to 20	**Broadcast Drench Application:** Apply specified dosage in not less than 50 gallons of water per acre as a soil drench at the time of or immediately after seeding. For smaller areas use 1 level tablespoon per gallon of water as a soil drench per 50 sq. ft. of seed bed. Apply with water can or similar equipment.
		10 to 15	**Row Drench Application:** Apply specified dosage per acre in at least 35 gallons of water (per 13,100 linear feet of row for broccoli, brussels sprouts and cabbage and 10,900 linear feet of row for cauliflower). Spray as an 8 inch band centered on the row at the time of or immediately after seeding.

CROP	DISEASE	POUNDS TERRACLOR 75% W.P.	REMARKS
GARLIC	White rot (Sclerotium capivorum)	27½	**In-Furrow Application:** Mix specified dosage in 100 gallons of water and apply as an in-furrow spray at planting time. Apply at the rate of 1 gallon of solution per 218 linear feet of row based on 24 inch row spacing. Attach a suitable spray rig to the machine planter using 2 nozzles per furrow. Direct front nozzle to spray the "seed" as it is dropped; direct the rear nozzle to spray the covering soil. Use good mechanical agitation in the spray tank to maintain a uniform suspension.
		13½	**Clove Treatment:** Use specified dosage per 1000 lbs. of "seed" cloves. Mix in sufficient water to apply dosage as a uniform mist-spray for complete coverage of treated cloves. For optimum results add a commercial sticker to the spray solution.
PEPPERS TOMATOES	Southern blight (Sclerotium rolfsii)	3 to 5	**Transplant Solution** (Field Use Only): Mix specified dosage in 100 gallons of water and apply ½ pint of solution per plant at the time of transplanting. The solution should be agitated often to maintain a uniform suspension and assure proper suspension and assure proper dosage.
		10	**In-Furrow Application:** Apply specified dosage per acre in 100 gallons of water (per 14,500 linear feet for peppers, 7300 linear feet for flat tomatoes, or 10,900 linear feet for staked tomatoes). Apply as an in-furrow spray to the open "V" trench, just prior to setting transplants. Set plants in bottom of trench, then press walls of trench against stems of the young transplants. When cultivating, set plows as flat as possible to avoid getting untreated soil against stems of plants.
POTATOES	Rhizoctonia solani	24 to 33	**Broadcast Application:** Apply specified dosage per acre in approximately 25 gallons of water as a pre-plant broadcast spray prior to planting. Thoroughly incorporate to a depth of 4 to 6 inches using a disc or other suitable equipment.
		13⅓	**In-Furrow Application:** Apply specified dosage per acre in approximately 20 gallons of water (per 12,400 linear feet of row). Spray several points in the planting furrow so as to evenly treat all the soil that surrounds and covers the seed piece and potato forming zone. For best results, attach 3 nozzles to the planter; one fan nozzle ahead of the opening discs (approximately 5 gallons of spray per acre); one cone nozzle placed above and slightly ahead of the closing discs (approximately 10 gallons of spray per acre); one fan nozzle to rear of closing discs (approximately 5 gallons of spray per acre). Spray pattern should treat a 12 to 14 inch band centered on the row.
ORNAMENTALS AFRICAN VIOLETS BEGONIAS CARNATIONS CHRYSANTHE-MUMS POINSETTIAS SNAPDRAGONS and similar BEDDING PLANTS	Stem rot (Rhizoctonia)	2½	Apply specified dosage as a soil drench in 300 gallons of water per 1000 sq. ft. of bed or bench area. For best results, apply to a well prepared seed bed.
AZALEAS	Azalea petal blight (Ovulinia azalea)	1	Apply specified dosage per 150 sq. ft. in sufficient water to assure uniform coverage. Apply uniformly to the ground beneath the bushes and surrounding area. Begin application prior to opening of buds and 3 to 4 weeks during the bloom period.
IRIS HYASCINTH NARCISSUS TULIPS	Crown rot Black rot	3¼ to 6½	**Broadcast Application:** Apply specified dosage per 1000 sq. ft. to the soil surface in sufficient water to insure uniform coverage. Mix thoroughly in upper 6 to 7 inches by double discing or other suitable means. See note below.
		4½ to 6	**In-Furrow Application:** Apply specified dosage per 1000 linear feet of row in 5 to 10 gallons of water. Direct spray into open furrow so as to thoroughly cover the entire furrow and the bulbs. Close furrow in normal manner. **Note:** Use the higher rate for heavier rate for heavier soils and for Black rot control.
		See Remarks	**Bulb Dip:** Dip bulbs for 5 minutes in a 7.5% concentration which is obtained by adding 3 lbs. Terraclor 75% W.P. to 4.85 gallons of water. Add 1% sticker. Maintain good agitation in the dip tank.
CALENDUAL LARKSPUR SNAPDRAGON SWEET PEA	Stem rot (Rhizoctonia)	⅓	Apply specified dosage per 100 sq. ft. in sufficient water to insure complete coverage. Applications should be made to the soil surface prior to planting and thoroughly incorporate to a depth of 4 inches.
CAMELLIA	Camellia flower blight	1	Apply specified dosage per 150 sq. ft. in sufficient water to insure complete coverage to the surface of the soil and leaf litter beneath the bushes. Begin applications prior to opening of buds and repeat every 3 to 4 weeks during the bloom season.

CROP	DISEASE	POUNDS TERRACLOR 75% W.P.	REMARKS
DUTCH IRIS	Bulb rot Stem rot (Sclerotium rolfsii)	3¼ to 4½	**Broadcast Application:** Apply specified dosage per 1000 sq. ft. in sufficient water for complete coverage. Application should be made to the soil surface prior to planting and thoroughly incorporated to a depth of 2 to 3 inches.
		½	**Band Application:** Apply specified dosage per 1000 linear feet of row as a 12 inch band spray centered on the row at planting time. Follow with 3 additional applications at the same rate at 15 days intervals.
EASTER LILLIES	Rhizoctonia root rot	3¼ to 6½	**Broadcast Application:** Apply specified dosage per 1000 sq. ft. in sufficient water to insure complete coverage. Apply to soil surface prior to planting and thoroughly incorporate to a depth of 6 to 7 inches. See note below.
		4½ to 6	**In-Furrow Application:** Apply specified dosage per 1000 linear feet of row in 5 to 10 gallons of water. Direct spray into open furrow so as to thoroughly cover the entire furrow and the bulbs. Close furrow in the normal manner. **Note:** Use the lower rates for light soils and the higher for heavy soils.
		4 to 6	**Bulb Dip:** Mix specified dosage in 100 gallons of water and dip bulbs for 5 to 15 minutes. A sticking agent may be added. Maintain good agitation in dip tanks.
GLADIOLUS	Neck or Bulb Dry rot (Stomatinia or Sclerotina gladioli)	3⅔	**Broadcast Application:** Apply specified dosage per 1000 sq. ft. in sufficient water to insure complete coverage. Apply to soil surface prior to planting and thoroughly incorporate to a depth of 4 to 6 inches. Bulbs may be planted immediately after incorporation.
		5½	**Band Application:** Apply specified dosage per 1000 linear feet of row as a 12 inch band spray centered on the row at planting time. Incorporate thoroughly to a depth of 4 to 6 inches. Bulb may be planted immediately after incorporation.
MAGNOLIA GRANDIFLORA	Magnolia Leaf Spot (Phyllosticta cookeri)	2	Apply specified dosage in 100 gallons of water as a broadcast foliar spray. Add a spreader-sticker at the rate of 1 pint per 100 gallons. Begin application approximately 1 week before disease normally appears. Apply at least 4 sprays at 2 week intervals. **Note:** Do not use on Magnoli fuscata as injury may result.
ROSES	Botrytis Storage Rot	1 to 2	**Dip Application:** Mix specified dosage in 100 gallons of water and dip dormant roses prior to storage. Maintain good agitation in dip tank.
		2	**Spray Application:** Mix specified dosage in 100 gallons of water. Thoroughly spray dormant roses prior to storage.
SOUTHERN PINE SEEDLINGS	Rhizoctonia Needle Blight	50	Apply specified dosage per acre as a broadcast spray in 100 gallons of water. An additional ½ inch of water per acre should be applied by irrigation. Seed may be planted immediately after irrigation or within one week of this date.
LAWNS and TURF	Brown Patch (Rhizoctonia solani)	See Remarks	**For Southern or warm season lawns such as St. Augustine, Bermuda and similar grasses:** Make one application in the fall or spring at first indication of infection. Mix 1 lb. in sufficient water (10–15 gals. to cover 1000 sq. ft.). Apply by means of a sprinkler can, sprayer or any other convenient method of obtaining even coverage. Following application, the treated area should be lightly watered to insure movement through the grass to soil level. If the treated area is subject to unusually heavy rainfall, or flooded, or if the disease is severe or reappears, the area should be retreated 3–4 weeks later. **For Dichondra:** Treat as above, using 1 lb. mixed in 40 gallons of water per 1000 sq. ft. **For Northern or cool season lawns such as Rye Grass, Fescue, Bluegrass, Bent and similar grasses:** Mix 3–4 ozs. in 3 to 6 gallons of water to cover 1000 sq. ft. Apply at first sign of disease. Repeat every 7–10 days during warm, damp weather. Water treated area lightly to wash chemical down to soil. **Note:** Under certain growing conditions, a temporary discoloration of the grass occasionally may occur. This causes no harm and will disappear in a short time. Do graze treated areas or feed clippings to livestock.
	Snow Mold (Typhula incarnata, Fusarium nivale)	½	Apply specified dosage per 1000 sq. ft. of treated area in sufficient water (5–10 gallons) to insure complete coverage. Application should be made prior to first snowfall. Following application, water treated area lightly to insure movement through turf to soil level. **Note:** Do not graze treated areas or feed clippings to livestock.

USE PRECAUTIONS

1. Do not plant any root crop not registered for PCNB in rotation on PCNB treated soil.

IMPORTANT NOTICE

Seller warrants that this product conforms to its chemical description and is reasonably fit for the purposes stated on the label when used in accordance with the directions and instructions specified on the label under normal conditions of use, **but neither this warranty nor any other warranty of merchantability or fitness for a particular purpose, express or implied, extends to the use of this product, contrary to label instructions, or under abnormal conditions, or under conditions not reasonably foreseeable to seller, and buyer assumes the risk of any such use.**

TERRACLOR SUPER X®
EMULSIFIABLE SOIL FUNGICIDE

Controls Rhizoctonia, Pythium, Thielaviopsis and Fusarium Soil-Borne Seedling Diseases of Cotton

COMPOSITION

ACTIVE INGREDIENT:	% BY WEIGHT
Pentachloronitrobenzene	23.2%
5-Ethoxy-3-(trichloromethyl)-1,2,4-thiadiazole	5.8%
INERT INGREDIENTS	71.0%
TOTAL:	100.0%

Contains 2 lbs. of Pentachloronitrobenzene and 0.5 lb. of 5-Ethoxy-3-trichloromethyl-1,2,4-thiadiazole per U.S. gallon at 68° F.

EPA Reg. No. 400-405 EPA Est. No.

TERRACLOR SUPER X is a Registered Trademark of UNIROYAL CHEMICAL COMPANY, INC.

WARNING
Keep Out of Reach of Children

See left panel for additional precautionary statement

PRECAUTIONARY STATEMENTS

HAZARDS TO HUMANS

WARNING

Harmful if swallowed. Do not inhale spray mist. Do not get on skin or in eyes. May cause skin irritation. Skin contact may be harmful. The following protective clothing are required during mixing/loading operations: respirator, gloves, long-sleeved shirt, long pants, socks and shoes. Wash thoroughly all exposed skin. In case of eye contact, flush with plenty of water. Wash thoroughly all contaminated cothing before reuse.

ENVIRONMENTAL HAZARDS

This product is toxic to fish. Do not apply directly to water or wetlands. Do not apply where runoff is likely to occur. Do not apply when weather conditions favor drift of spray from the area being treated. Do not contaminate water by cleaning of equipment or disposal of wastes. Apply this product only as specified on this label.

PHYSICAL OR CHEMICAL HAZARDOUS

COMBUSTIBLE LIQUID. Do not use, pour, spill or store near heat or open flame.

DIRECTIONS FOR USE

It is a violation of Federal law to use this product in a manner inconsistent with its labeling.

STORAGE AND DISPOSAL

Do not contaminate water, food or feed by storage or disposal.

STORAGE

Keep container closed. Protect from temperatures below 20° F.

PESTICIDE DISPOSAL

Pesticide wastes are toxic. Improper disposal of excess pesticide, spray mixture, or rinsate is a violation of Federal Law. If these wastes cannot be disposed of by use according to label instructions, contact your State Pesticide or Environmental Control Agency, or the Hazardous Waste representative at the nearest EPA Regional Office for guidance.

CONTAINER DISPOSAL

Triple rinse (or equivalent). Then offer for recycling or reconditioning, or puncture and dispose of in a sanitary landfill, or by other procedures approved by State and local authorities.

DIRECTIONS FOR USE

COTTON

(Seedling Disease Complex—Rhizoctonia, Pythium, Thielaviopsis and Fusarium)

SOIL IN-FURROW APPLICATION: At planting time, use 2 to 3 quarts per acre (13,000 linear ft. based on 40 in. rows) mixed with 5 to 15 gallons of water. Spray all the soil that surrounds and covers the seed. Two nozzles work best; center the front nozzle over the open furrow to spray soil around the seed and the rear nozzle to spray the covering soil as it fills the furrow. Do not allow the feeding or grazing of cotton foliage by livestock. Do not feed gin waste to livestock.

USE PRECAUTIONS

Do not plant treated fields to any crop other than cotton.

IMPORTANT NOTICE

Seller warrants that this product conforms to its chemical description and is reasonably fit for the purposes stated on the label when used in accordance with the directions and instructions specified on the label under normal conditions of use, **but neither this warranty nor any other warranty of merchantability or fitness for a particular purpose, express or implied, extends to the use of this product, contrary to label instructions, or under abnormal conditions, or under conditions not reasonably foreseeable to seller, and buyer assumes the risk of any such use.**

TERRACLOR SUPER X®
GRANULAR SOIL FUNGICIDE

For Cotton

COMPOSITION

ACTIVE INGREDIENT:	% BY WEIGHT
Pentachloronitrobenzene	10.0%
5-Ethoxy-3-(trichloromethyl)-1,2,4-thiadiazole	2.5%
INERT INGREDIENTS:	87.5%
TOTAL:	100.0%

EPA Reg. No. 400-406 EPA Est. No.

Terraclor Super X is a Registered Trademark of UNIROYAL CHEMICAL COMPANY, INC.

WARNING
Keep Out of Reach of Children

See left panel for additional precautionary statement.

PRECAUTIONARY STATEMENTS

HAZARDS TO HUMANS

WARNING

Harmful if swallowed. Do not inhale dust. Do not get on skin or in eyes. May cause skin irritation. Contact may be harmful. The following protective clothing are required during mixing/loading operations: respirator, gloves, long-sleeved shirt, long pants, socks and shoes. Wash thoroughly all exposed skin. In case of eye contact, flush with plenty of water. Wash thoroughly all contaminated clothing before reuse.

ENVIRONMENTAL HAZARDS

This product is toxic to fish. Do not apply directly to water or wetlands. Do not contaminate water by cleaning of equipment or disposal of wastes. Do not apply product in areas which would permit runoff to contaminate water supplies.

DIRECTIONS FOR USE

It is a violation of Federal law to use this product in a manner inconsistent with its labeling.

STORAGE AND DISPOSAL

Do not contaminate water, food or feed by storage or disposal.

STORAGE

Store in a dry location.

PESTICIDE DISPOSAL

Pesticide wastes are toxic. Improper disposal of excess pesticide, spray mixture, or rinsate is a violation of Federal Law. If these wastes cannot be disposed of by use according to label instructions, contact your State Pesticide or Environmental Control Agency, or the Hazardous Waste representative at the nearest EPA Regional Office for guidance.

CONTAINER DISPOSAL

Completely empty bag into application equipment. Then dispose of empty bag in a sanitary landfill or by incineration, or if allowed by State and local authorities, by burning. If burned, stay out of smoke.

DIRECTIONS FOR USE

COTTON

(Seedling Disease Complex—Rhizoctonia, Pythium and Fusarium)

SOIL IN-FURROW APPLICATION: At planting time, use 10 to 15 lbs. per acre (per 13,000 linear feet based on 40 in. rows). Apply over the seed and to the covering soil, using suitable equipment. Do not allow the feeding or grazing of cotton foliage by livestock.

USE CAUTIONS

Do not plant treated fields to any crop other than cotton.

IMPORTANT NOTICE

Seller warrants that this product conforms to its chemical description and is reasonably fit for the purposes stated on the label when used in accordance with the directions and instructions specified on the label under normal conditions of use, **but neither this warranty nor any other warranty of merchantability or fitness for a particular purpose, express or implied, extends to the use of this product, contrary to label instructions, or under abnormal conditions, or under conditions not reasonably foreseeable to seller, and buyer assumes the risk of any such use.**

TERRACLOR SUPER X®

GRANULAR SOIL FUNGICIDE

With Di-Syston®

SYSTEMIC INSECTICIDE

RESTRICTED USE PESTICIDE

For retail sale to and application only by certified applicators or personnel under their direct supervision.

COMPOSITION

ACTIVE INGREDIENT:	% BY WEIGHT
Pentachloronitrobenzene	6.50%
5-Ethoxy-3-(trichloromethyl)-1,2,4-thiadiazole	1.63%
0,0-Diethyl S-[2-(ethylthio) ethyl] phosphorodithioate*	6.50%
INERT INGREDIENTS:	85.37%
TOTAL:	100.00%

*U.S. Patent No. 2,759,010.

EPA Reg. No. 400-408 EPA Est. No.

Keep Out of Reach of Children

DANGER

POISONOUS IF SWALLOWED, INHALED OR ABSORBED THROUGH THE SKIN. RAPIDLY ABSORBED THROUGH SKIN.

ATROPINE IS AN ANTIDOTE FOR DI-SYSTON.

CALL A PHYSICIAN IMMEDIATELY IN ALL CASES OF SUSPECTED POISONING.

DANGER–POISON

STATEMENT OF PRACTICAL TREATMENT

IF SWALLOWED: Call a physician or Poison Control Center immediately. Drink one or two glasses of water and induce vomiting by touching back of throat with finger. Repeat until vomit fluid is clear. Do not induce vomiting or give anything by mouth to an unconscious or convulsing person.

IF INHALED: Remove victim to fresh air. Apply artificial respiration if indicated.

IF ON SKIN: Remove contaminated clothing and wash affected areas with soap and water.

IF IN EYES: Flush with plenty of water. Call a physician immediately.

NOTE TO PHYSICIAN: ANTIDOTE: Administer atropine sulfate in large therapeutic doses. Repeat as necessary to the point of tolerance. 2-PAM is also antidotal and may be administered in conjunction with atropine.

Compound inhibits cholinesterase resulting in stimulation of the central nervous system, the parasympathetic nervous system, and the somatic motor nerves. Do not give morphine. Watch for pulmonary edema, which may develop in serious cases of poisoning even after 12 hours. At first sign of pulmonary edema, the patient should be placed in an oxygen tent and treated symptomatically.

See left side panel for additional precautionary statement.

PRECAUTIONARY STATEMENTS

HAZARDS TO HUMANS

DANGER

Poisonous if swallowed, inhaled, or absorbed through the skin. Rapidly absorbed through the skin. Do Not Breathe Dust. Do Not Get In Eyes, On Skin, Or Clothing. In case of contact, immediately remove contaminated clothing and wash skin thoroughly with soap and water. Wear freshly laundered long-sleeve work clothes daily. Wash thoroughly with soap and water before eating or smoking. Bathe at the end of the work day and change outer clothing.

While transferring material from package to equipment, wear a clean cap, goggles, gloves (cotton or rubber) and a pesticide respirator jointly approved by the Mining Enforcement and Safety Administration (formerly the U.S. Bureau of Mines) and by the National Institute for Occupational Safety and Health under the provisions of 30 CFR Part 11. If cotton gloves are used, they must be laundered or discarded after each day■s use. Rubber gloves should be washed with soap and water after each use. Do not wear the same gloves for other work. Destroy and replace gloves frequently. Keep all unprotected persons out of the operating area. The following protective clothing are required during mixing/loading operations: gloves, long sleeved shirt, long pants, socks and shoes.

ENVIRONMENTAL HAZARDS

This pesticide is toxic to fish, birds and other wildlife. Do not apply directly to water or wetlands. Drift and run-off from treated areas may be hazardous to fish in neighboring areas. Do not contaminate water by cleaning of equipment or disposal of wastes. Cover or incorporate spills. This pesticide is toxic to bees exposed to direct application. Applications should be timed to coincide with periods of minimum bee activity, usually between late evening and early morning.

REENTRY STATEMENT

Do not allow reentry into treated fields within 24 hours of application unless appropriate protective clothing is worn.

DIRECTIONS FOR USE

It is a violation of Federal law to use this product in a manner inconsistent with its labeling.

STORAGE AND DISPOSAL

Do not contaminate water, food or feed by storage or disposal.

STORAGE

Store in a dry location.

PESTICIDE DISPOSAL

Pesticide wastes are acutely hazardous. Improper disposal of excess pesticide, spray mixture, or rinsate is a violation of Federal Law. If these wastes cannot be disposed of by use according to label instructions, contact your State Pesticide or Environmental Control Agency, or the Hazardous Waste representative at the nearest EPA Regional Office for guidance.

CONTAINER DISPOSAL

Completely empty bag into application equipment. Then dispose of empty bag in a sanitary landfill or by incineration, or if allowed by State and local authorities, by burning. If burned, stay out of smoke.

DIRECTIONS FOR USE ON COTTON

For control of Seedling Disease Complex - Rhizoctonia, Pythium and Fusarium and for early season control of thrips, aphids and mites.

Apply as a soil in-furrow application at planting time with a granular pesticide applicator properly calibrated to assure accurate placement and proper dosage.

Application Method	Lbs./drill row per acre	Lbs./1300 linear feet of row based on 40 inch spacing
DRILL PLANTING	15	1.5
HILL-DROP PLANTING	5	0.5

Soil temperature, moisture and tilth should be favorable for good germination and emergence. Systemic insecticides can possibly damage seed germination when placed on or very near the seed, however, the addition of a fungicide such as Terraclor Super X, considerably lessens this hazard. Do not allow the feeding or grazing of cotton foliage by livestock. Do not plant treated fields to any crop other than cotton.

IMPORTANT NOTICE

Seller warrants that this product conforms to its chemical description and is reasonably fit for the purposes stated on the label when used in accordance with the directions and instructions specified on the label under normal conditions of use, **but neither this warranty nor any other warranty of merchantability or fitness for a particular purpose, express or implied. extends to the use of this product, contrary to label instructions, or under abnormal conditions, or under conditions not reasonably foreseeable to seller, and buyer assumes the risk of any such use.**

TERRACLOR SUPER X®

GRANULAR
SOIL FUNGICIDE

With Thimet®

SYSTEMIC INSECTICIDE

RESTRICTED USE PESTICIDE

Due to acute oral and dermal toxicity; and bird toxicity.

For retail sale to and application only by certified applicators or personnel under their direct supervision.

NOT FOR USE OR STORAGE IN OR AROUND THE HOME.

COMPOSITION

ACTIVE INGREDIENT:	% BY WEIGHT
Pentachloronitrobenzene	6.50%
5-Ethoxy-3-(trichloromethyl)-1,2,4-thiadiazole	1.63%
Phorate [O,O-Diethyl S-(ethylthio) methyl phosphorodithioate]	6.50%
INERT INGREDIENTS:	85.37%
TOTAL:	100.0%

EPA Reg. No. 400-409 EPA Est. No.

TERRACLOR SUPER X is a Registered Trademark of UNIROYAL CHEMICAL COMPANY, INC.

®Thimet is a Trademark of American Cyanamid Company

DANGER

Keep Out of Reach of Children

POISONOUS BY SKIN CONTACT, INHALATION OR SWALLOWING ABSORBED THROUGH SKIN. REPEATED INHALATION OR SKIN CONTACT MAY, WITHOUT SYMPTOMS PROGRESSIVELY INCREASE SUSCEPTIBILITY TO THIMET POISONING.

ATROPINE IS AN ANTIDOTE FOR THIMET.

CALL A PHYSICIAN IMMEDIATELY IN ALL CASES OF SUSPECTED POISONING.

DANGER–POISON

STATEMENT OF PRACTICAL TREATMENT

IF SWALLOWED: Call a physician or Poison Control Center. Drink 1 or 2 glasses of water and induce vomiting by touching back of throat with finger or blunt object. Do not induce vomiting or give anything by mouth to an unconscious person.

IF SPLASHED IN EYES: Immediately flush eyes with plenty of water for at least 15 minutes and get medical attention.

IF SPILLED ON SKIN: Immediately remove contaminated clothing. Wash with soap and plenty of water. If symptoms of poisoning develop, send for a physician.

IF INHALED: Remove to fresh air; if abnormal symptoms develop, contact a physician.

PHYSICIAN'S NOTE: Warning symptoms include weakness, headache, tightness in chest, blurred vision, non-reactive pinpoint pupils, salivation, sweating, nausea, vomiting, diarrhea and adbominal cramps.

TREATMENT: Give atropine intramuscularly or intravenously, 2 to 4 milligrams (3 to 6 tablets. 1/100 grain each) at once and every hour as required until pupils dilate. Twenty to 30 milligrams, or more, may be required during the first 24 hours. Never give morphine. Clear chest by postural drainage. Artificial respiration or oxygen administration may be necessary. Observe patient continuously for 48 hours. Repeated exposure to cholinesterase inhibitors may, without warning, cause increasing susceptibility to very small doses of any cholineserase inhibitor. Allow no further exposure until cholinesterase regeneration has taken place as determined by blood tests.

See left side panel for additional precautionary statement.

PRECAUTIONARY STATEMENTS

HAZARDS TO HUMANS

DANGER

Do Not Breathe Dust. Do Not Get In Eyes, On Skin, Or Clothing. In case of contact, immediately remove contaminated clothing and wash skin thoroughly with soap and water. Wash thoroughly with soap and water before eating or smoking. Bathe at the end of the work day and change outer clothing.

Wear freshly laundered, long-sleeved work clothing daily. While transferring from package to equipment, wear a clean cap and gloves (rubber or cotton) and a mask or respirator of a type to give adequate protection against this material. If cotton gloves are used, they must be laundered or discarded after each day's use. Rubber gloves should be washed with soap and water after each use. Do not wear the same gloves for other work. Destroy and replace gloves frequently. The following protective clothing are required during mixing/loading operations: gloves, long-sleeved shirt, long pants, socks and shoes. While emptying bags into equipment, pour downwind and allow as little free fall as possible. Do not pour at face level and do not allow dust to reach the breathing zone. Remove unprotected persons and domestic animals from operating and adjacent areas where this product may drift.

When applying product, wear protective clothing, gloves and goggles.

ENVIRONMENTAL HAZARDS

This pesticide is toxic to fish and wildlife. Birds and other wildlife feeding in treated areas may be killed. Do not apply directly to water or wetlands. Runoff and drift from treated areas may be hazardous to aquatic organisms in adjacent sites. Do not contaminate water by cleaning of equipment or disposal of wastes. Cover or incorporate spills.

DIRECTIONS FOR USE

It is a violation of Federal law to use this product in a manner inconsistent with its labeling.

STORAGE AND DISPOSAL

Do not contaminate water, food or feed by storage or disposal.

STORAGE

Store in a dry location.

Once a bag has been opened, use it completely.

PESTICIDE DISPOSAL

Pesticide wastes are acutely hazardous. Improper disposal of excess pesticide, spray mixture, or rinsate is a violation of Federal Law. If these wastes cannot be disposed of by use according to label instructions, contact your State Pesticide or Environmental Control Agency or the Hazardous Waste representative at the nearest EPA Regional Office for guidance.

CONTAINER DISPOSAL

Completely empty bag into application equipment. Then dispose of empty bag in a sanitary landfill or by incineration, or if allowed by State and local authorities, by burning. If burned, stay out of smoke.

REENTRY STATEMENT

Do not reenter treated fields before 24 hours after application. If reentry prior to 24 hours is required, take the same precautions as an applicator of the product including use of protective clothing, gloves and goggles.

DIRECTIONS FOR USE ON COTTON

For control of Seedling Disease Complex-Rhizoctonia, Pythium and Fusarium and for early season control of thrips, aphids and mites.

Apply as a soil in-furrow application at planting time with a granular pesticide applicator properly calibrated to assure accurate placement and proper dosage.

Application Method	Lbs./drill row per acre	Lbs./1300 linear feet of row based on 40 inch spacing
Drill Planting	15	1.5
Hill-Drop Planting	5	0.5

Soil temperature, moisture and tilth should be favorable for good germination and emergence. Systemic insecticides can possibly damage seed germination when placed on or very near the seed, however, the addition of a fungicide such as Terraclor Super-X, considerably lessens this hazard. Do not allow the feeding or grazing of cotton foliage by livestock. Do not plant treated fields to any crop other than cotton.

IMPORTANT NOTICE

Seller warrants that this product conforms to its chemical description and is reasonably fit for the purposes stated on the label when used in accordance with the directions and instructions specified on the label under normal conditions of use, **but neither this warranty nor any other warranty of merchantability or fitness for a particular purpose, express or implied, extends to the use of this product, contrary to label instructions, or under abnormal conditions, or under conditions not reasonably foreseeable to seller, and buyer assumes the risk of any such use.**

TERRAZOLE®

5% GRANULAR FUNGICIDE

COMPOSITION

ACTIVE INGREDIENT:	% BY WEIGHT
5-Ethoxy-3-(trichloromethyl)-1,2,4-thiadiazole	5%
INERT INGREDIENTS:	95%
TOTAL:	100.0%

EPA Reg. No. 400-419 EPA Est. No.

CAUTION

Keep Out of Reach of Children

STATEMENT OF PRACTICAL TREATMENT

IF SWALLOWED: Induce vomiting.

IF ON SKIN: Remove contaminated clothing and wash skin with soap and warm water.

IF INHALED: Remove to fresh air.

IF IN EYES: Flush with water for 15 minutes.

For Further Treatment, Get Medical Treatment

See Left Panel For Additional Precautionary Statement

PRECAUTIONARY STATEMENTS

HAZARDS TO HUMANS AND DOMESTIC ANIMALS

CAUTION

Harmful if swallowed, inhaled or absorbed through skin. Avoid breathing dust. Avoid contact with eyes, skin or clothing. Wash thoroughly with soap and water after handling. Wash contaminated clothing before reuse. Do not handle or apply product with bare hands.

ENVIRONMENTAL HAZARDS

Do not contaminate water by cleaning of equipment or disposal of wastes. Do not apply directly to water or wetlands.

DIRECTIONS FOR USE

It is a violation of Federal law to use this product in a manner inconsistent with its labeling.

STORAGE AND DISPOSAL

Do not contaminate water, food or feed by storage or disposal.

STORAGE

Store in a dry location.

PESTICIDE DISPOSAL

Wastes resulting from the use of this product may be disposed of on site or at an approved wastes disposal facility.

CONTAINER DISPOSAL

Completely empty bag into application equipment. Then dispose of empty bag in a sanitary landfill, or by incineration, or if allowed by State and local authorities, by burning. If burned, stay out of smoke.

HOUSEHOLD

Do not reuse empty container. Securely wrap original container in several layers of newspaper and discard in trash.

DIRECTIONS FOR USE

ORNAMENTALS—For control of damping-off and root rot diseases caused by Pythium and Phytophthora at the time for seeding or transplanting.

Plant Type		Recommendations
Foliage Plants		**Dry Soil Mix**
Aglaonema Anthurium Brassaia Caladium Chinese Evergreen Dieffenbachia Dracaena	Echeveria Epipremnum Nephthytis Peperomia Pilea Philodendron Zygocacus	To mix with potting, greenhouse or other transplanting soils, add TERRAZOLE 5G at the rate of 10 ozs. per cubic yard of soil and mix thoroughly. Soil to be mixed should be free of large lumps and dry enough to prevent agglomeration during mixing. Where **Rhizoctonia** control is desired, add recommended quantities of Terraclor. Retreat as needed for continued control with soil drench applications of Terrazole 35% Wettable Powder or Terrazole 25% Emulsifiable.
Container & Bedgrown		
Herbaceous	**Woody**	
African Violet Begonia Calceolaria Chrysanthemum Easter Lily Geranium Gerbera Gloxinia Gypsophilia Orchids Sinningia	Arctostaphylos Azalea Camellia Ceanothus Ficus Fir Hemlock Ilex Juniper Leucothoe Magnolia Poinsettia Rhamnus Rhododendron Southern Pine Western Pine	NOTE: Overdosage may result in phytotoxicity. If plants are suspected of being sensitive, treat only a few plants and observe for 4 to 5 days prior to treating the entire planting.

AVOCADOS (California Only): For control of Avocado root rot (Phytophthora cinnamomi). Apply TERRAZOLE 5% Granular evenly to the soil surface surrounding the Avocado tree, in a circular area extending out to the drip line. The amount of material to be applied to each tree will be governed by the area to be covered, as per chart below:

Diameter of Circular Surface	Approximate Square Feet Covered	Amount of TERRAZOLE 5% Granular per tree
20′ diameter	314	86 ounces
15′ diameter	176	48 ounces
10′ diameter	79	22 ounces
5′ diameter	20	6 ounces

Following application apply approx. one gallon of water per square foot of area treated. Repeat application once a month for 9 to 10 months during the growing season.

Do not treat within 30 days of harvest. During application, care should be excerised so that the treatment does not come in contact with the fruit.

IMPORTANT NOTICE

Seller warrants that this product conforms to its chemical description and is reasonably fit for the purposes stated on the label when used in accordance with the directions and instructions specified on the label under normal conditions of use, **but neither this warranty nor any other warranty of merchantability or fitness for a particular purpose, express or implied, extends to the use of this product, contrary to label instructions, or under abnormal conditions, or under conditions not reasonably foreseeable to seller, and buyer assumes the risk of any such use.**

TERRAZOLE®

4 LB. EMULSIFIABLE FUNGICIDE

Controls Certain Soil-Borne Diseases

COMPOSITION

ACTIVE INGREDIENT:	% BY WEIGHT
5-Ethoyx-3-(trichloromethyl)-1,2,4-thiadiazole	45.8%
INERT INGREDIENTS:	54.2%
TOTAL:	100.0%

Contains 4 lbs. of 5-Ethoxy-3-(trichloromethyl)-1,2,4-thiadiazole per U.S. Gallon at 68°F (20°C).

EPA Reg. No. 400–422 EPA Est. No.

Terrazole is a Registered Trademark of UNIROYAL CHEMICAL COMPANY, INC.

WARNING

Keep Out of Reach of Children

STATEMENT OF PRACTICAL TREATMENT

IF SWALLOWED: Drink promptly a large quantity of water. Do not induce vomiting. Contains petroleum distilates. Call a physician.

IF ON SKIN: Remove contaminated clothing and wash skin with soap and warm water.

IF IN EYES: Flush with plenty of water for at least 15 minutes.

IF INHALED: Remove to fresh air.

If patient is unconscious, maintain breathing and heartbeat [CPR: cardiopulmonary resuscitation, if necessary]. Contact your local Poison Control Center, hospital or physician immediately.

NOTE TO PHYSICIAN: Probable mucosal damage may contraindicate the use of gastric lavage.

See Left Panel For Additional Precautionary Statement

PRECAUTIONARY STATEMENTS

HAZARDS TO HUMANS AND DOMESTIC ANIMALS

WARNING

May be fatal if absorbed through skin. Causes skin irritation. Harmful if swallowed or inhaled. Do not breathe vapors. Do not get on skin or on clothing. Wash contaminated clothing prior to reuse. When handling or using the concentrate, wear natural rubber gloves. Wash thoroughly with soap and water after handling and before eating or smoking.

ENVIRONMENTAL HAZARDS

Do not apply directly to water or wetlands.

Do not contaminate water by cleaning of equipment or disposal of wastes.

PHYSICAL OR CHEMICAL HAZARDS

Do not use, pour, spill or store near heat or open flame.

DIRECTIONS FOR USE

It is a violation of Federal law to use this product in a manner inconsistent with its labeling.

STORAGE AND DISPOSAL

Do not contaminate water, food or feed by storage or disposal.

PESTICIDE DISPOSAL

Wastes resulting from the use of this product may be disposed of on site or at an approved waste disposal facility.

CONTAINER DISPOSAL

Triple rinse (or equivalent). Then offer for recycling or reconditioning, or puncture and dispose of in a sanitary landfill, or by other procedures approved by State and local authorities.

DIRECTIONS FOR USE

Terrazole is useful as a soil fungicide for control of damping-off, root and stem diseases caused by Phythium and Phytophthora. Use only at recommended rates as overdosage may be harmful to sensitive plants. Do not mix with chemicals other than Terraclor®, unless prior use has proven compatibility.

AVOCADO (California Only): For control of avocado root rot (Phytophthora cinnamomi). Using drip emitter or low-profile star sprinkler irrigation systems, equipped with automatic shutoff devices and anti-siphoning devices to prevent backflow into the water source, meter Terrazole-4E into irrigation water lines at the rate of 2½ fl. oz./100 gallons of water. Meter this solution evenly to the soil surface out to the drip line at the rate of 1 gallon of solution per square foot of treated area.

For treatment of individual trees, add Terrazole 4E at the rate indicated in the chart below to sufficient water (3 to 5 gallons) and apply evenly to the soil surface under the tree extending out to the drip line. Water the treatment into soil immediately following application using approximately 1 gallon of water per square foot of treated area.

Diameter of Circular Surface	Approximate Square Feet Covered	Amount of Terrazole 4E per tree
20′ diameter	314	8½ fluid ounces
15′ diameter	176	5 fluid ounces
10′ diameter	79	2 fluid ounces
5′ diameter	20	½ fluid ounce

Repeat applications monthly during the entire 9 to 10 month active growing season. Do not apply within 30 days of harvest. During application, exercise care to prevent treatment from coming in contact with the fruit. If the irrigation system has PVC or other plastic fittings, flush with clean water after application.

STRAWBERRIES (California Only): For prevention of fungi causing root rot (Phytophthora cactorum, Ceratobasidium spp. Pythium ultimum). Using in-bed drip irrigation or furrow irrigation systems, meter Terrazole 4E into the irrigation water at the rate of 1 pint per acre beginning with the first irrigation. Repeat as needed through the growing and harvest seasons for a maximum of 15 applications. Allow at least 10 days between applications. Applications may be made on the day of harvest. Do not apply through overhead irrigation systems where treated water could come into direct contact with foliage, flowers or fruit.

FESCUE & RYEGRASS SEED: To inhibit the development of soil-borne Phythium fungi. Apply Terrazole 4E at the rate of 3 fl. oz./100 lbs. of seed using suitable seed treating equipment. Terrazole 4E may be diluted with sufficient water to insure optimum coverage of seed. Seed should be sound, well cured and free from trash prior to treatment. Do not use treated seed for feed purposes.

ORNAMENTAL PLANTS

For bedding bench plantings use with water, as a soil drench at time of seeding and transplanting. The drench solution should be continually agitated to assure uniform application of the fungicidal material. To improve penetration of the fungicide, the drenching should be followed with an equal quantity of water applied either by sprinkling or by irrigation.

Plants		Directions
BEDDING PLANTS		
Alyssum	Pansies	**BEDDING PLANTS:** (Soil 2" to 3" deep)—Add $\frac{3}{8}$ to $\frac{1}{2}$ fl. oz. to 25 gallons of water and apply to 100 sq. ft. of bed area. Repeat application may be made at one of two month intervals, if necessary.
Calendula	Petunia	
Celosia	Phlox	
Chrysanthemum	Salvia	
Coleus	Snapdragon	
Dahlia	Verbena	
Dianthus	Vinca	
Larkspur	Zinnia	
Marigold		
FOLIAGE PLANTS		
Aglaonema	Hedera	**TROPICAL FOLIAGE HERBACEOUS AND WOODY PLANTS** (Peat and other high organic media 6" deep).
Anthurium	Hoya	
Catus	Maranta	
Caladium	Nephrotepis	
Calathea	Nephthytis	**Bed and Bench Plantings:** Add 0.5 to 1 fl. oz. in 25 gallons of water and apply to 100 sq. ft. of bed or bench area. Repeat application at two to three month intervals to maintain disease control.
Chamadorea	Peperomia	
Chinese Evergreen	Pilea	
Cassula	Pothos	
Dieffenbachi	Stromanthe	
Gynura	Zygocatus	
HERBACEOUS AND WOODY		
African Violet	Gerbera	**Pot Plantings:** Use $\frac{1}{2}$ pint solution per 6" pot or in sufficient amount to saturate the soil.
Azalea	Juniper	
Boxwood	Kalanchoe	
Chrysanthemum	Pieris	
Easter Lily	Poinsettia	
Euonymous	Red Pine	
Ficus	Rhododendron	
Geranium	Sinningia	

IMPORTANT NOTICE

Seller warrants that this product conforms to its chemical description and is reasonably fit for the purposes stated on the label when used in accordance with the directions and instructions specified on the label under normal conditions of use, **but neither this warranty nor any other warranty of merchantability or fitness for a particular purpose, express or implied, extends to the use of this product, contrary to label instructions, or under abnormal conditions, or under conditions not reasonably foreseeable to seller, and buyer assumes the risk of any such use.**

Velsicol Chemical Corporation

341 East Ohio Street
Chicago, IL 60611
(312) 670-4500

BANVEL®

HERBICIDE

ACTIVE INGREDIENTS:

Dimethylamine Salt of dicamba (3,6-dichloro-o-anisic acid)*	48.2%
Dimethylamine Salts of related acids	12.0%
INERT INGREDIENTS	39.8%
TOTAL	100.0%

* Contains 40.0% 3,6-dichloro-o-anisic acid (dicamba) or 4 pounds per gallon

EPA Reg.No. 876-25-AA

WARNING

Keep Out of Reach of Children

This labeling must be in the possession of the user at the time of pesticide application.

DIRECTIONS FOR USE

It is a violation of Federal Law to use this product in a manner inconsistent with its labeling.

BEFORE USING BANVEL HERBICIDE READ AND FOLLOW THE PRECAUTIONS APPEARING ON THE CONTAINER.

LIMITED WARRANTY AND LIABILITY

NOTICE: Read this Limited Warranty and Liability before buying or using this product. If the terms are not acceptable, return it at once unopened.

It is critical that this product be used and mixed only as specified on the label. The laws of a State may make some or all of this paragraph inapplicable or may give you rights in addition to your rights hereunder. Except to the extent prohibited by applicable law, the exclusive remedy of the user or buyer and the limit of liability of this Company or any other Seller for any and all losses, personal injuries or damages resulting from the use of this product, shall be the purchase price paid by the user or buyer for the quantity of product involved. Except to the extent prohibited by State law, there is no warranty, and this Company and other Sellers disclaim all liability for losses, personal injury or damages: (i) arising from any use of this product in a manner or for a purpose not recommended in its label directions, or from mixing this product before use with any substance except as recommended by the product's label, (ii) arising from handling or storage in violation of label instructions, (iii) for all indirect special or consequential damages, (iv) when not reported to this company within one year of discovery, and (v) arising from product not used within the label-designated shelf life of four years from date of purchase, whichever first occurs. THERE ARE NO IMPLIED WARRANTIES AND NO WARRANTIES OF MERCHANTABILITY OR FITNESS.

PRECAUTIONARY STATEMENTS

IMPORTANT

The following directions apply to all uses of BANVEL herbicide. Additional precautions and restrictions will be found in each specific use section.

Do not contaminate irrigation ditches or water used for domestic purposes.

SENSITIVE CROP PRECAUTIONS: BANVEL herbicide may cause injury to desirable trees and plants, particularly beans, cotton, flowers, fruit trees, grapes, ornamentals, peas, potatoes, soybeans, sunflowers, tobacco, tomatoes and other broadleaf plants when contacting their roots, stems or foliage. These plants are most sensitive to BANVEL herbicide during their development or growing stage. FOLLOW THE PRECAUTIONS LISTED BELOW WHEN USING BANVEL HERBICIDE.

- Do not treat areas where either possible downward movement into the soil or surface washing may cause contact of BANVEL herbicide with the roots of desirable plants such as trees and shrubs.
- Avoid making applications when spray particles may be carried by air currents to areas where sensitive crops and plants are growing. Do not spray near sensitive plants if wind is gusty or in excess of 5 mph and moving in the direction of nearby sensitive crops. However, always make applications when there is some movement to determine the direction and distance of possible spray drift. Leave an adequate buffer zone between area to be treated and sensitive plants. Coarse sprays are less likely to drift out of the target area than fine sprays. Agriculturally approved drift-reducing additives may be used.
- Do not apply BANVEL herbicide in the vicinity of sensitive crops when the temperature on the day of application is expected to exceed 85°F as drift is more likely to occur.
- Do not use aerial equipment to apply BANVEL herbicide when sensitive crops and plants are growing in the vicinity of area to be treated.
- To avoid injury to desirable plants, equipment used to apply BANVEL herbicide should be thoroughly cleaned (see PROCEDURE FOR CLEANING SPRAY EQUIPMENT) before reusing to apply any other chemicals.

All crop uses of BANVEL herbicide are intended for a normal growing interval between planting and harvest. If this interval is shortened, such as in cover crops that will be plowed under, do not follow up with the planting of a sensitive crop.

Crops growing under stress conditions such as drought, poor fertility, or foliar damage due to hail, wind or insects, can exhibit various injury symptoms that may be more pronounced if herbicides are applied.

Consult your local or state authorities for possible application restrictions and advice concerning these and other special local use situations.

PROCEDURE FOR CLEANING SPRAY EQUIPMENT

The steps listed below are suggested for thorough cleaning of spray equipment following applications of BANVEL herbicide or tank mixes of BANVEL herbicide plus 2,4-D amine.

1) Hose down thoroughly the inside as well as outside surfaces of equipment while filling the spray tank half full of water. Flush by operating sprayer until the system is purged of the rinse water.
2) Fill tank with water while adding 1 quart of household ammonia for every 25 gallons of water. Operate the pump to circulate the ammonia solution through the sprayer system for 15 to 20 minutes and discharge a small amount of the ammonia solution through the boom and nozzles. Let the solution stand for several hours, preferably overnight.
3) Flush the solution out of the spray tank through the boom.
4) Remove the nozzles and screens and flush the system with two full tanks of water.

The steps listed below are suggested for thorough cleaning of spray equipment used to apply BANVEL herbicide as a tank mix with wettable powders (WP), emulsifiable concentrates (EC), or other types of water-dispersible formulations. BANVEL herbicide tank mixes with water-dispersible formulations require the use of a water/detergent rinse.

5) Complete step 1.

6) Fill tank with water while adding 2 lbs. of detergent for every 40 gallons of water. Operate the pump to circulate the detergent solution through the sprayer system for 5 to 10 minutes and discharge a small amount of the solution through the boom and nozzles. Let the solution stand for several hours, preferably overnight.

7) Flush the detergent solution out of the spray tank through the boom.

8) Repeat step 1, and follow with steps 2, 3, and 4.

REFER TO THE CONTAINER LABEL FOR INSTRUCTIONS CONCERNING DISPOSAL OF WASTE AND CLEANING RINSES.

MIXING AND APPLICATION

UNLESS OTHERWISE SPECIFIED UNDER THE INDIVIDUAL USE HEADINGS OF THIS BOOKLET, THE FOLLOWING DIRECTIONS APPLY TO ALL CROP AND NON-CROP USES OF BANVEL HERBICIDE. REFER TO INDIVIDUAL USE SECTIONS FOR ADDITIONAL PRECAUTIONS, RESTRICTIONS, APPLICATION RATES AND TIMINGS.

BANVEL herbicide is a water-soluble formulation that can be applied using water or sprayable fluid fertilizer as the carrier. If a fluid fertilizer is to be used, a compatibility test (see COMPATIBILITY TEST) should be made prior to tank mixing.

Ground or aerial application equipment which will give good spray coverage of weed foliage should be used. HOWEVER, DO NOT USE AERIAL APPLICATION EQUIPMENT IF SENSITIVE CROPS ARE GROWING IN THE VICINITY OF THE AREA TO BE TREATED.

Apply 5 to 50 gallons of diluted spray per treated acre when using ground application equipment, or 3 to 10 gallons of diluted spray per treated acre when using aerial application equipment. Use the higher level of the listed spray volumes when treating dense or tall vegetation. Use coarse sprays.

Select nozzles designed to produce minimal amounts of fine spray particles. Spray with nozzles as close to the weeds as is practical for good weed coverage.

BANVEL herbicide should not be applied during periods of gusty wind or when wind is in excess of 15 mph as uneven spray coverage may occur.

Avoid disturbing (e.g. cultivating or mowing) treated areas for at least 7 days following application.

BAND TREATMENTS

BANVEL herbicide may be applied as a band treatment. Use the formulas below to determine the appropriate rate and volume per treated acre.

$$\frac{\text{Band width in inches}}{\text{Row width in inches}} \times \text{Broadcast RATE per treated acre} = \text{Band RATE per treated acre}$$

$$\frac{\text{Band width in inches}}{\text{Row width in inches}} \times \text{Broadcast VOLUME per treated acre} = \text{Band VOLUME per treated acre}$$

COMPATIBILITY TEST

Before mixing in the spray tank, it is advisable to test compatibility by mixing all components in a small container in proportionate quantities (see following table).

Amount of Herbicide to Add to One Pint of Spray Carrier (Assuming Volume is 25 Gallons per Acre)

Herbicide Formulations	Rate Per Acre	Level Teaspoons
Dry	1 lb.	$1\frac{1}{2}$
Liquid	1 pt.	$\frac{1}{2}$

If herbicide(s) do not ball-up or form flakes, sludge, gels, oily films or layers, or other precipitates, then the tested spray mix is compatible. Usually incompatibility in any of the above described forms will occur within 5 minutes after mixing.
If components are incompatible, the use of a compatibility agent is recommended. Rerun the above COMPATIBILITY TEST with a suitable compatibility agent ($\frac{1}{4}$ teaspoon is equivalent to 2 pints per 100 gallons of fluid fertilizer).

GENERAL WEED LIST

This is a general list of weeds which may be treated with BANVEL herbicide in accordance with this label as recommended under the rates and timing sections of the individual use headings. Proper usage of this product will give control or growth suppression of many ANNUAL, BIENNIAL, and PERENNIAL broadleaf weeds, and many WOODY brush and vine species including:

ANNUALS

amaranth, spiny
(spiny pigweed)
aster, slender
beggarweed, Florida
broomweed, common
buckwheat, tartary
buckwheat, wild
buffalobur
burclover, California
burcucumber
buttercup, roughseed
carpetweed
catchfly, nightflowering
chamomile, corn
chickweed, common
clovers *(annual)*
cockle, corn
cockle, cow
cocklebur, common
croton, tropic
croton, woolly
cucumber, wild
daisy, English
eveningprimrose, cutleaf
fleabane, annual
henbit
jimsonweed
knawel *(German moss)*
knotweed
kochia
ladysthumb
lambsquarters, common
lambsquarters *(triazine resistant)*
mallow, common
mallow, Venice
mayweed
morningglory, ivyleaf
morningglory, tall
mustard, tansy
mustard, wild
mustards *(yellowtops)*
nightshade, black
pennycress, field
(fanweed, frenchweed, stinkweed)
pepperweed, Virginia
(peppergrass)
pigweed, prostrate
pigweed, redroot
(carelessweed)
pigweed, rough
pigweed, smooth
pigweed
(triazine resistant)
pigweed, tumble
poorjoe
puncturevine
purslane, common
pusley, Florida
ragweed, common
ragweed, giant
(buffaloweed)
rubberweed, bitter
(bitterweed)
sesbania, hemp
shepherdspurse
sicklepod
sida, prickly
(teaweed)
smartweed, green
smartweed, Pennsylvania
sneezeweed, bitter
sowthistle, annual
sowthistle, spiny
spanishneedles
spikeweed, common
spurge, prostrate
spurry, corn
starbur, bristly
sumpweed, rough
sunflower, common *(wild)*
sunflower, volunteer
thistle, Russian
velvetleaf
waterhemp
waterprimrose, winged
wormwood, annual

BIENNIALS

burdock, common
carrot, wild
(Queen Anne's lace)
cockle, white
eveningprimrose, common
geranium, Carolina
knapweed, diffuse
knapweed, spotted
mallow, drawf
plantain, bracted
ragwort, tansy
starthistle, yellow
sweetclover
teasel
thistle, bull
thistle, musk
thistle, plumeless

PERENNIALS

* alfalfa
artichoke, Jerusalem
aster, spiny
aster, whiteheath
bedstraw, smooth
bindweed, field
bindweed, hedge
blueweed, Texas
* bursage
(bur ragweed, lakeweed, povertyweed)
bursage, woollyleaf
(lakeweed)
* buttercup, tall
campion, bladder
chickweed, field
chickweed, mouseear
chicory
* clover, hop
cress, hoary
(whitetop)
* dandelion, common
* dock, broadleaf
(bitter dock)
dock, curly
dogbane, hemp
* dogfennel
(cypressweed)
eupatorium, late
(thoroughwort)
fern, bracken
garlic, wild
goldenrod, Canada
goldenrod, Missouri
goldenweed, common
hawkweed
horsenettle, Carolina
ironweed
knapweed, black
knapweed, Russian
lespedeza
lupine, silvery
mare's tail
(horseweed)
milkweed, climbing
milkweed, common
milkweed, honeyvine
milkweed, western whorled
nettle, stinging
nightshade, silverleaf
(white horsenettle)
onion, wild
* plantain, broadleaf
* plantain, buckhorn
pokeweed
ragweed, western
redvine
smartweed, swamp
snakeweed, broom
* sorrel, red
(sheep sorrel)
sowthistle
sowthistle, perennial
spurge, leafy
sundrop, halfshrub
(eveningprimrose)
thistle, Canada
toadflax, Dalmatian
trumpetcreeper
(buckvine)
vetch
waterhemlock
waterprimrose, creeping
* woodsorrel, common yellow
wormwood, common
wormwood, Louisiana
* yankeeweed
yarrow, common

* Noted perennials may be controlled using BANVEL herbicide at rates lower than those recommended for other listed perennial weeds. (See application rates and timings under headings: PASTURE, RANGELAND & NON-CROPLAND, TURF AND LAWNS, and BETWEEN CROPPING APPLICATIONS.)

WOODY

alder
ash
aspen
basswood
beech
birch
blackberry
blackgum
cedar
cherry
chinquapin
condalia, lotebrush *(lote)*
cottonwood
mesquite
oak
oak, poison
olive, autumn
olive, Russian
peppertree, Brazil
(Florida holly, Schinus)
persimmon, eastern
pine
plum, sand *(wild plum)*
poplar
rabbitbrush
redcedar, eastern
creosotebush
cucumbertree
dewberry
dogwood
elm
grape
guava
hawthorn *(thornapple)*
hemlock
hickory
honeysuckle
hornbeam
huckleberry
huisache
ivy, poison
kudzu
locust, black
maple
melaleuca
rose, McCartney
rose, multiflora
sagebrush, fringed
sassafras
serviceberry
snowberry, western
(buckbrush)
spicebush
spruce
sumac
sweetgum
sycamore
tarbush
waxmyrtle
willow
witchhazel
yaupon
yucca

FIELD AND SILAGE CORN
(Conventional, Minimum, and No Tillage)

IMPORTANT
OBSERVE ALL PRECAUTIONS
UNDER PRECAUTIONARY STATEMENTS HEADING
READ AND FOLLOW MIXING AND APPLICATION INSTRUCTIONS

Direct chemical contact with corn seed must be avoided. Corn seed should be placed 1½ inches or more below the soil surface if applications are to be made prior to corn emergence. If the corn seeds are less than 1½ inches below the surface, delay application until the corn has spiked.

BANVEL herbicide preemergence treatments do not require mechanical incorporation to become active. To move this herbicide into the weed germination zone, a shallow mechanical incorporation is recommended for applications which are not followed by adequate rainfall or sprinkler irrigation. Cultivations made due to soil crusting should be shallow.

A PREPLANT OR A PREEMERGENCE TO EARLY POSTEMERGENCE application may be followed by one POSTEMERGENCE application of BANVEL herbicide during a growing season.

Applications of BANVEL herbicide to corn during periods of rapid growth may result in temporary leaning. Corn will usually become erect within 3 to 7 days. Cultivation should be delayed until after corn is growing normally to avoid breakage.

Prior to the ensilage (milk) stage of the crop, do not harvest or graze corn for dairy or beef feed.

AVOID SPRAY DRIFT: Observe the following precautions in addition to those appearing under the Precautionary Statements heading when applying POSTEMERGENCE sprays of BANVEL herbicide to corn if sensitive crops are growing nearby:

- Use coarse sprays. Select nozzles which are designed to produce minimal amounts of fine spray particles. Keep the spray pressure below 20 psi and the spray volume at or above 20 gpa unless otherwise required by the manufacturer of drift-reducing nozzles.
- Use drop pipes (drop nozzles) when corn is 8 inches or more in height to direct the spray beneath the lower leaves of the corn and onto the weeds and soil.
- Do not apply BANVEL herbicide sprays when soybeans are growing nearby if:

1) corn is more than 24 inches tall,
2) soybeans are more than 10 inches tall,
3) soybeans have begun to bloom, whichever comes first.

WEEDS CONTROLLED

BANVEL herbicide, when applied at recommended rates, will control many ANNUAL broadleaf weeds or give growth suppression of many PERENNIAL broadleaf weeds commonly found in corn. (Refer to GENERAL WEED LIST.)

Control of cocklebur, velvetleaf and jimsonweed resulting from a pre-emergence application may be reduced if conditions such as low temperature or lack of soil moisture cause delayed or deep germination of weeds.

If wild mustards, common sunflower, or velvetleaf are over 5 inches tall or are growing under drought stress at time of application, adding an agriculturally approved surfactant to BANVEL herbicide or tank mixing BANVEL herbicide plus 2,4-D herbicide will improve control. Refer to tank mix treatments in this section.

PREPLANT USES

For reduced tillage situations, application of BANVEL herbicide may be made before planting to emerged and actively growing broadleaf weeds. Application rate depends on soil texture and organic matter content. Refer to table below. Mechanical incorporation is not required for activation. Avoid disturbing (e.g. cultivating or mowing) treated areas for 7 days.

PREEMERGENCE TO EARLY POSTEMERGENCE USES

Application of BANVEL herbicide may be made immediately after planting until corn is no more than 5 inches tall, with the application rate depending on soil texture and organic matter content. For best performance, make application when ANNUAL broadleaf weeds are emerging from the soil. Refer to table below.

Soil Texture	Broadcast Rate Per Treated Acre: 2½% or less organic matter	more than 2½% organic matter
COARSE SOILS		
sand, sandy loam and loamy sand	*½ pt. (¼ lb. a.i.)	*¾ pt. (⅜ lb. a.i.)
MEDIUM SOILS		
loam, silt loam, silt, sandy clay and sandy clay loam . .	*¾ pt. (⅜ lb. a.i.)	1 pt. (½ lb. a.i.)
FINE SOILS		
silty clay, silty clay loam, clay loam and clay	*1 pt. (½ lb. a.i.)	1 pt. (½ lb. a.i.)
ALL SOILS CONTAINING GREATER THAN 8% ORGANIC MATTER—USE 1 PINT (½ lb. a.i.) PER TREATED ACRE		

* Make application after corn and weeds have begun to emerge.

POSTEMERGENCE USES

Application of BANVEL herbicide may be made any time after weeds have emerged and are actively growing, but before corn is 36 inches tall or 15 days before tassel emergence, whichever comes first. For best performance, make application when weeds are small, less than 3 inches tall. Drop nozzles should be used to direct spray beneath the corn canopy if weeds are covered by the corn leaves. Poor control of some weed species may result if weeds are greater than 12 inches tall at time of application.

BROADCAST RATE PER TREATED ACRE:
½ pint (¼ lb. a.i.)
The addition of an agriculturally approved surfactant will improve wetting and coverage of weed foliage, and improve control of drought stressed weeds. Use drop pipes (drop nozzles) if the crop is taller than 8 inches. Keeping the spray off the corn leaves and out of the whorl will reduce the likelihood of crop injury and improve spray coverage of weed foliage.

OVERLAY (SEQUENTIAL) TREATMENTS

BANVEL herbicide may be applied to ground previously treated with one or more of the following herbicides:

Herbicide	Maximum rate per treated acre (lbs. a.i.)
alachlor (Lasso®)	4
atrazine	4
Bicep® (atrazine + metolachlor premix)	4.5
Bronco® (alachlor + glyphosate premix)	5
butylate (Sutan+®)	6
cyanazine (Bladex®)	4
EPTC (Eradicane®)	6
glyphosate (Roundup®)	5
metolachlor (Dual®)	3
paraquat	1
pendimethalin (Prowl®)	2
propachlor (Bexton®, Ramrod®)	6
simazine (Princep®)	4

Read and follow the label of each of the above products for precautionary statements, directions for use and other restrictions.

TANK MIX TREATMENTS

BANVEL herbicide may be tank mixed with one or more of the following herbicides for control of grasses or additional broadleaf weeds. Read and follow the label of each tank mix product used for precautionary statements, directions for use, rates and timings, and other restrictions.

BANVEL PLUS ATRAZINE: Application may be made before grasses are 1½ inches tall. Use 1.25–4.0 lbs. a.i. atrazine per treated acre.

BANVEL PLUS BLADEX: Application may be made before grasses are 1½ inches tall and the corn is not beyond the four true leaf stage. Use 1.25–4.0 lbs. a.i. Bladex per treated acre for preemergence treatments or 1.25–2.0 lbs. a.i. Bladex per treated acre for postemergence treatments. AFTER CORN EMERGENCE, USE ONLY THE BLADEX 80W FORMULATION.

BANVEL PLUS DUAL: Application may be made immediately after planting, but before weeds and corn emerge. Make applications only on medium or fine textured soils containing greater than 2–½% organic matter. Use 1.5–3.0 lbs. a.i. Dual per treated acre.

BANVEL PLUS LASSO: Application may be made until grasses reach the two-leaf stage and before corn is greater than 3 inches tall. Applications prior to crop emergence should only be made to fine textured soils containing 3% or more organic matter. Use 1.5–4.0 lbs. a.i. Lasso per treated acre.

BANVEL PLUS PARAQUAT: Application may be made to emerged weeds, but before corn emerges. Use 0.25–1.0 lb. a.i. paraquat per treated acre.

BANVEL PLUS PRINCEP: Application may be made before weeds and corn emerges. Use 2.0-3.0 lbs. a.i. Princep per treated acre.

BANVEL PLUS PROWL: Application may be made immediately after planting, but before grasses and corn emerge. Make applications only on medium or fine textured soils containing greater than 2–½% organic matter. Use 1.0–2.0 lbs. a.i. Prowl per treated acre.

BANVEL PLUS ROUNDUP: Application may be made to emerged weeds, but before corn emerges. Use 1.0–3.0 lbs. a.i. Roundup per treated acre.

BANVEL PLUS 2,4-D: Tank mix ⅛–¼ lb. 2,4-D acid equivalent per treated acre. Use drop pipes (drop nozzles) if the crop is taller than 8 inches. Keeping the spray off the corn leaves and out of the whorl will reduce likelihood of crop injury and improve spray coverage of weed foliage.

GRAIN SORGHUM (Milo)

IMPORTANT

OBSERVE ALL PRECAUTIONS

UNDER PRECAUTIONARY STATEMENTS HEADING

(Including the Reference to Crops Growing Under Stress)

READ AND FOLLOW MIXING AND APPLICATION INSTRUCTIONS

Applications of BANVEL herbicide to sorghum during periods of rapid growth may result in temporary leaning of plants or rolling of leaves. These effects are usually outgrown within 10 to 14 days.

Do not graze or feed treated sorghum forage or silage prior to mature grain stage.

Do not apply BANVEL herbicide to sorghum grown for seed production.

Make no more than one application per growing season.

Delay harvest until 30 days after treatment.

WEEDS CONTROLLED

BANVEL herbicide, when applied at the recommended rate for grain sorghum, will control many actively growing ANNUAL broadleaf weeds and will reduce competition from established PERENNIAL broadleaf weeds as well as control their seedlings. (Refer to GENERAL WEED LIST.)

RATES AND TIMINGS

BANVEL herbicide may be applied to emerged and actively growing weeds up to 15 days prior to planting. Postemergence application of BANVEL herbicide must be made after sorghum is in the 3 leaf stage but before sorghum is 15 inches tall. For best performance, make applications when sorghum is in the 3–5 leaf stage and weeds are small (less than 3 inches tall). Use drop pipes (drop nozzles) if crop is taller than 8 inches. Keeping the spray off the sorghum leaves and out of the whorl will reduce the likelihood of crop injury and improve spray coverage of weed foliage.

BROADCAST RATE PER TREATED ACRE: ½ pint (¼ lb. a.i.)

OVERLAY (SEQUENTIAL) TREATMENTS

BANVEL herbicide may be applied to ground previously treated with one or more of the following herbicides:

Herbicide	Maximum Rate per Treated Acre (lbs. a.i.)
alachlor (Lasso®)	4
(Screen®-treated seed)	
atrazine	3
metolachlor (Dual®)	2.5
(Concep®-treated seed)	
propachlor (Ramrod®)	5
propazine (Milogard®)	3.2

PREHARVEST USES

For Use Only in the States of Texas and Oklahoma

Application of BANVEL herbicide may be made any time after the sorghum has reached the soft dough stage of development for suppression of weeds. An agriculturally approved surfactant may be used to improve performance.

BROADCAST RATE PER TREATED ACRE: ½ pint (¼ lb. a.i.)

SMALL GRAINS
(Not Underseeded to Legumes)

IMPORTANT

OBSERVE ALL PRECAUTIONS

UNDER PRECAUTIONARY STATEMENTS HEADING

READ AND FOLLOW MIXING AND APPLICATION INSTRUCTIONS

If small grains are grown for pasture only, refer to the PASTURE, RANGELAND and NON-CROPLAND section.

Do not graze or harvest for livestock feed prior to crop maturity.

WEEDS CONTROLLED

BANVEL herbicide, when applied at recommended rates, will control ANNUAL broadleaf weeds commonly found in small grains, such as:

buckwheat, tartary	lambsquarters, common	ragweed, giant *(buffaloweed)*
buckwheat, wild	mallow, common	smartweed, green
chamomile, corn	mustard, tansy	smartweed, Pennsylvania
cockle, corn	nightshade, black	
cockle, cow		
cocklebur, common	pigweed, redroot *(carelessweed)*	sowthistle, annual
knawel *(German moss)*	pigweed, rough	sunflower, common *(wild)*
knotweed	pigweed, tumble	sunflower, volunteer
kochia	purslane, common	thistle, Russian
ladysthumb	ragweed, common	velvetleaf

BANVEL herbicide and BANVEL herbicide tank mixes will reduce competition from established PERENNIAL broadleaf weeds and control their seedlings. (Refer to GENERAL WEED LIST.)

THE SPECIAL USE TANK MIX FOR FALL SEEDED WHEAT ONLY allows a higher rate of 2,4-D to be used in combination with BANVEL herbicide. This tank mix treatment may be used for improved performance of difficult-to-control weeds including:

fiddleneck *(tarweed)*	henbit
*garlic, wild	*onion, wild
gromwell	

* Spring applications may not control weeds that develop in the fall. For fall applications, refer to the BETWEEN CROPPING APPLICATIONS section.

RATES AND TIMINGS

Application of BANVEL herbicide may be made before, during or after planting to emerged and actively growing weeds. See specific crop for timing restrictions. For best performance, make application when weeds are in the 2-3 leaf stage and rosettes are less than 2 inches across. Use the higher level of listed rate ranges when treating more mature weeds or dense vegetative growth.

FALL SEEDED WHEAT

BANVEL HERBICIDE MUST BE APPLIED TO FALL SEEDED WHEAT PRIOR TO THE JOINTING STAGE.

BROADCAST RATE PER TREATED ACRE: ¼ pint (⅛ lb. a.i.)

TANK MIX TREATMENTS

For control of grasses or additional broadleaf weeds, BANVEL herbicide may be tank mixed with the following herbicides. Read and follow the label of each tank mix product used for precautionary statements, directions for use, weeds controlled and geographic and other restrictions.

BROADCAST RATE PER TREATED ACRE:

Apply ⅛–¼ pint (1/16–⅛ lb. a.i.) BANVEL herbicide with:

Herbicide	amount product*	lb. a.i.
2,4-D	½–¾ pint	¼–⅜
MCPA	½–¾ pint	¼–⅜
bromoxynil	1–1½ pint	¼–⅜
bromoxynil + MCPA	1 + ½ pint	¼ + ¼
metribuzin		
(Sencor® 75 DF, Lexone® DF™)	⅓ lb.	¼ lb.
(Sencor® 4, Lexone® 4L)	½ pint	¼
(Lexone® 50WP)	½ lb.	¼
chlorsulfuron		
(Glean® 75 DF)	⅙–½ wt. oz.	0.008–0.024
diuron (Karmex® 80WP)	½–2 lbs.	⅖–1⅗
terbutryn (Igran® 80W)	1½–2¾ lbs.	1⅕–2⅕

* based on 4 pounds per gallon formulations of MCPA and 2,4-D and 2 pounds per gallon formulation bromoxynil.

SPECIAL USE TANK MIX FOR FALL SEEDED WHEAT ONLY

BANVEL	¼ pint	⅛ lb. a.i.
plus	plus	plus
2,4-D amine	1 to 2 pints	½ to 1 lb. a.i.
or	or	or
2,4-D ester	1 to 1½ pints	½ to ¾ lb. a.i.

Note: Do not use unless possible crop injury will be acceptable.

SPRING SEEDED WHEAT

BANVEL HERBICIDE MUST BE APPLIED BEFORE SPRING SEEDED WHEAT EXCEEDS THE 5 LEAF STAGE.

BROADCAST RATE PER TREATED ACRE: 1/4 pint (1/8 lb. a.i.)

TANK MIX TREATMENTS

For control of additional broadleaf weeds, BANVEL herbicide may be tank mixed with the following herbicides. Read and follow the label of each tank mix product used for precautionary statements, directions for use, weeds controlled and geographic and other restrictions.

BROADCAST RATE PER TREATED ACRE:

Apply 1/8–1/4 pint (1/16–1/8 lb. a.i.) BANVEL herbicide with:

Herbicide	amount product*	lb. a.i.
2,4-D	1/2–3/4 pint	1/4–3/8
MCPA	1/2–3/4 pint	1/4–3/8
bromoxynil	1–1 1/2 pint	1/4–3/8
chlorsulfuron (Glean® 75 DF)	1/6–1/2 wt. oz.	0.008–0.024

* based on 4 pounds per gallon formulations of MCPA, 2,4-D and 2 pounds per gallon formulation of bromoxynil.

FALL SEEDED BARLEY

BANVEL HERBICIDE MUST BE APPLIED TO FALL SEEDED BARLEY PRIOR TO THE JOINTING STAGE. Note: For fall barley varieties that are seeded during the winter months or later, follow the rates and timings given for Spring Seeded Barley.

BROADCAST RATE PER TREATED ACRE: 1/4 pint (1/8 lb. a.i.)

TANK MIX TREATMENTS

For control of additional broadleaf weeds, BANVEL herbicide may be tank mixed with the following herbicides. Read and follow the label of each tank mix product used for precautionary statements, directions for use, weeds controlled and geographic and other restrictions.

BROADCAST RATE PER TREATED ACRE:

Apply 1/8–1/4 pint (1/16–1/8 lb. a.i.) BANVEL herbicide with:

Herbicide	amount product*	lb. a.i.
2,4-D	1/2 pint	1/4
MCPA	1/2–3/4 pint	1/4–3/8
chlorsulfuron (Glean® 75 DF)	1/6–1/2 wt. oz.	0.008–0.024

* based on 4 pounds per gallon formulations of MCPA and 2,4-D.

SPRING SEEDED BARLEY

BANVEL HERBICIDE MUST BE APPLIED BEFORE SPRING SEEDED BARLEY EXCEEDS THE 3 LEAF STAGE.

BROADCAST RATE PER TREATED ACRE: 3/16 pint (3/32 lb. a.i.)

TANK MIX TREATMENTS

For control of additional broadleaf weeds, BANVEL herbicide may be tank mixed with the following herbicides. Read and follow the label of each tank mix product used for precautionary statements, directions for use, weeds controlled and geographic and other restrictions.

BROADCAST RATE PER TREATED ACRE:

Apply 1/8–3/16 pint (1/16–3/32 lb. a.i.) BANVEL herbicide with:

Herbicide	amount product*	lb. a.i.
MCPA	1/2 pint	1/4
chlorsulfuron (Glean® 75 DF)	1/6–1/2 wt. oz.	0.008–0.024

* based on 4 pounds per gallon formulations of MCPA.

FALL AND SPRING SEEDED OATS

BANVEL HERBICIDE MUST BE APPLIED BEFORE SPRING SEEDED OATS EXCEED THE 5 LEAF STAGE. APPLICATIONS TO FALL SEEDED OATS MUST BE MADE PRIOR TO THE JOINTING STAGE.

BROADCAST RATE PER TREATED ACRE: 1/4 pint (1/8 lb. a.i.)

TANK MIX TREATMENTS

For control of additional broadleaf weeds, BANVEL herbicide may be tank mixed with the following herbicides. Read and follow the label of each tank mix product used for precautionary statements, directions for use, weeds controlled and geographic and other restrictions.

BROADCAST RATE PER TREATED ACRE:

Apply 1/8–1/4 pint (1/16–1/8 lb. a.i.) BANVEL herbicide with:

Herbicide	amount product*	lb. a.i.
MCPA	1/2–3/4 pint	1/4–3/8

* based on 4 pounds per gallon formulations of MCPA.

SUGARCANE

IMPORTANT
OBSERVE ALL PRECAUTIONS
READ AND FOLLOW MIXING AND APPLICATION INSTRUCTIONS

Consult your local or state authorities for possible application restrictions, especially concerning aerial applications and advice concerning special local use situations.

WEEDS CONTROLLED

BANVEL herbicide, when applied at recommended rates, will control many ANNUAL, BIENNIAL and PERENNIAL broadleaf weeds commonly found in sugarcane. (Refer to GENERAL WEED LIST).

RATES AND TIMINGS

Application of BANVEL herbicide may be made any time after weeds have emerged and are actively growing but before the close-in stage of sugarcane. Application rates and timings of BANVEL herbicide are given below. Use the higher level of listed rate ranges when treating dense vegetative growth.

Weed Stage & Type	Broadcast rate per treated acre	
	amount	lbs. a.i.
Annual		
Small, actively growing	1/2–1 pt.	1/4–1/2
Established weed growth	1–1 1/2 pts.	1/2–3/4
Biennial	1–2 pts.	1/2–1
Perennial		
Noted (*) Perennials	2–4 pts.	1–2
Other Perennials	4–6 pts.	2–3†

† Application made over the top of actively growing sugarcane may result in crop injury.

When possible, direct the spray beneath the sugarcane canopy in order to minimize the likelihood of crop injury. The use of directed sprays will also aid in maximizing spray coverage of weed foliage.

Retreatments may be made as needed, however, do not exceed a total of 6 pints (3 lbs. a.i.) of BANVEL herbicide per treated acre during a growing season.

TANK MIX TREATMENTS

BANVEL herbicide may be tank mixed with one or more of the following herbicides for control of grasses or additional broadleaf weeds. Read and follow the label of each tank mix product used for precautionary statements, directions for use, rates and timings, weeds controlled, geographic and other restrictions.

Herbicide	Rates per treated acre (lbs. a.i.)
ametryn (Evik®)	2/5 to 8
asulam (Asulox®)	2 to 3 1/3
atrazine	2/5 to 4
dalapon (Dalapon®)	3 1/2 to 8 1/2
2,4-D	1/2 to 3*

* Application of BANVEL herbicide plus 2,4-D tank mix at the higher listed rate ranges may result in crop injury.

PASTURE, RANGELAND AND NON-CROPLAND AREAS

BANVEL herbicide is recommended for use for pasture, rangeland, general farmstead weed and brush control and for use on non-cropland areas such as fence rows, roadways, rights-of-way (utility, railroad, highway, pipeline), nonselective forest brush control (including site preparation), drainage ditch banks, wasteland and other non-cropland areas.

IMPORTANT

OBSERVE ALL PRECAUTIONS

UNDER PRECAUTIONARY STATEMENTS HEADING

READ AND FOLLOW MIXING AND APPLICATION INSTRUCTIONS

BANVEL herbicide uses described in this section also pertain to small grains such as barley, oats, rye or wheat grown for pasture use only.

NEWLY SEEDED AREAS, including small grains such as barley, oats, rye or wheat grown for pasture, may be severely injured if rates of BANVEL herbicide are applied in excess of those listed for control of ANNUAL weeds.

ESTABLISHED GRASS CROPS growing under stress can exhibit various injury symptoms that may be more pronounced if herbicides are applied. Furthermore, rates of BANVEL herbicide in excess of 2 quarts (2 lbs. a.i.) per treated acre may cause temporary injury to many grass species.

Bentgrass, carpetgrass, buffalograss and St. Augustine grass may be injured at rates exceeding 1 pint BANVEL herbicide (½ lb. a.i.) per treated acre. Usually colonial bentgrasses are more tolerant than creeping types. Velvetgrasses are most easily injured. Treatments will kill or injure alfalfa, clovers, lespedeza, wild winter peas, vetch and other legumes.

REMOVE MEAT ANIMALS FROM TREATED AREAS 30 DAYS PRIOR TO SLAUGHTER.

THERE IS NO WAITING PERIOD BETWEEN TREATMENT AND GRAZING FOR NON-LACTATING ANIMALS.

Timing Restrictions For Lactating Dairy Animals Following Treatment

BANVEL herbicide Rate Per Treated Acre	Days Before Grazing	Days Before Hay Harvest
Up to 1 pint (½ lb. a.i.)	7 days	37 days
Up to 1 quart (1 lb. a.i.)	21 days	51 days
Up to 2 quarts (2 lbs. a.i.)	40 days	70 days
Up to 8 quarts (8 lbs. a.i.)	60 days	90 days

NOTE: Observe all precautions and restrictions on labels of products used in tank mixtures.

MIXING AND APPLICATION

BANVEL herbicide can be applied using water, oil-water emulsions (including invert systems), or sprayable fluid fertilizer as the carrier.

BANVEL herbicide may be applied broadcast using either ground or aerial application equipment. When using ground equipment, apply 5 to 600 gallons of diluted spray per treated acre. Volume of spray applied will depend on the height, density, and type of weeds or brush being treated and on the type of equipment being used. When using aerial equipment, apply 3 to 40 gallons of diluted spray per treated acre.

BANVEL herbicide may be applied to individual clumps or small areas (SPOT TREATMENT) of undesirable vegetation using handgun or similar types of application equipment. Apply diluted sprays to allow complete wetting (up to runoff) of foliage and stems.

ACCUTROL® herbicide adjuvant or other spray additives (emulsifiers, surfactants, wetting agents, drift control agents, or penetrants) may be used for wetting, penetration, or drift control. Spray additives must be agriculturally approved when used in pasture applications. If spray additives are used, read and follow all use recommendations and precautions on product label.

WEEDS CONTROLLED

BANVEL herbicide, when applied at recommended rates, will give control of many ANNUAL, BIENNIAL, and PERENNIAL broadleaf weeds, and many WOODY brush and vine species commonly found in Pasture, Rangeland and Non-Cropland areas. (Refer to GENERAL WEED LIST.) Noted (*) PERENNIAL weeds may be controlled with lower rates of either BANVEL herbicide or BANVEL herbicide plus 2,4-D. See RATES AND TIMINGS.

RATES AND TIMINGS

Application rates and timing of BANVEL herbicide are given below. Use the higher level of listed rate ranges when treating dense or tall vegetative growth.

Weed Stage & Type	Broadcast Rate per Treated Acre	
	amount	lbs. a.i.
Annual		
Small, actively growing	½–1 pt.	¼–½
Established weed growth	1–1½ pts.	½–¾
†Biennial		
Rosette diameter		
Less than 3 inches	½–1 pt.	¼–½
3 inches or more	1–2 pts.	½–1
Bolting	2–3 pts.	1–1½
Perennial		
Suppression or top growth control	½–1 qt.	½–1
Noted (*) Perennials	1–2 qts.	1–2
Other perennials	2–4 qts.	2–4
Dense stands	4–6 qts.	4–6
Woody Brush & Vines		
Foliage Suppression	½–1 qt.	½–1
Stems	1–2 qts.	1–2
Stems and Stem Sprouts	½–1 gal.	2–4
Stems and Root Sprouts	1–2 gals.	4–8

† For best performance, make application when BIENNIAL WEEDS are in the rosette stage.

Retreatments may be made as needed; however, do not exceed a total of 2 gallons (8 lbs. a.i.) of BANVEL herbicide per treated acre during a growing season.

TANK MIX TREATMENTS

Read and follow the label of each tank mix product used for precautionary statements, directions for use, application rates and other restrictions. BANVEL herbicide may be tank mixed with one or more of the following herbicides for control of grasses, additional broadleaf weeds, and woody brush and vines.

Herbicide	Rate per Treated Acre (lbs. a.i.)
Pasture, rangeland, and non-cropland use:	
diuron (Karmex®)	4 to 48
glyphosate (Roundup®)	¾ to 3¾
simazine (Princep®)	5 to 40
paraquat	½ to 1
2,4-D	¼ to 6
Rangeland and non-cropland use only:	
picloram (Tordon®)	¼ to 3
2,4,5-T	¼ to 6
Non-cropland use only:	
amitrole	2 to 8
atratol (Atratol®)	4⅘ to 40
bromacil (Hyvar®)	1½ to 24
dalapon (Dalapon®)	4¼ to 12¾
diquat	½
fosamine ammonium (Krenite®)	6 to 12
hexazinone (Velpar®)	2 to 12
MSMA	2
prometon (Pramitol®)	10 to 60
sulfometuron methyl (Oust™)	0.14 to 0.56

tebuthiuron (Spike®)	1 to 16
triclopyr (Garlon®)	3/4 to 9
2,4-DP (Weedone®)	1/2 to 11

Due to the variations that may occur in formulated products and specific use ingredients (e.g. water supplies), a COMPATIBILITY TEST as described earlier in this label is recommended prior to actual tank mixing.

CUT SURFACE TREE TREATMENTS

BANVEL herbicide may be applied as a cut surface treatment for control of unwanted trees and prevention of sprouts of cut trees. A mix of 1 part BANVEL herbicide with 1 to 3 parts water should be used in application. Use the lower dilution when treating difficult-to-control species.

- TREE INJECTIONS*: Injector cuts must penetrate the bark and the sapwood, and should be made completely around tree trunk at intervals of 1 to 2 inches. Apply 1 milliliter of BANVEL herbicide/water mix to each cut.
- FRILL OR GIRDLE TREATMENTS*: Make a continuous cut or a series of overlapping cuts using an axe to girdle tree trunk. Spray or paint cut surface with the BANVEL herbicide/water mix.
- STUMP TREATMENTS: Spray or paint freshly cut surface with the BANVEL herbicide/water mix. The area adjacent to the bark should be thoroughly wet.

* Note: For more rapid foliar effects, 2,4-D may be added to the BANVEL herbicide/water mix.

MULTIFLORA ROSE CONTROL
Dormant Applications

BANVEL herbicide can be applied when plants are dormant as an undiluted SPOT-CONCENTRATE directly to the soil or as a LO-OIL BASAL BARK treatment using an oil-water emulsion solution.

MIXING AND APPLICATION

SPOT CONCENTRATE applications of BANVEL herbicide should be applied directly to the soil as close as possible to the root crown but within 6–8 inches of the crown. On sloping terrain, application should be made to the up-hill side of the crown. Do not make application when snow or water prevents applying BANVEL herbicide directly to the soil.

LO-OIL BASAL BARK applications of BANVEL herbicide should be applied to the basal stem region from the ground line up to a height of 12–18 inches. Spray until runoff, with special emphasis on covering the root crown. For best results, make application when plants are dormant. Do not make application after bud break or when plants are showing signs of active growth. Do not make application when snow or water prevents applying BANVEL herbicide to the ground line.

NOTE: To prepare oil in water emulsions, half fill spray tank with water plus appropriate amount of herbicide. With continuous agitation, slowly add a premix of oil plus a suitable emulsifier to the spray tank. Complete filling of spray tank with water. Maintain vigorous agitation during spray operation to prevent oil and water from forming separate layers.

RATES AND TIMINGS

Application rates of BANVEL herbicide are given below:

SPOT CONCENTRATE TREATMENT

Canopy diameter (feet)	BANVEL herbicide (ounces)
5	1/4
10	1
15	2 1/4

Do not exceed a total of 2 gallons (8 lbs. a.i.) of BANVEL herbicide per acre per year.

LO-OIL BASAL BARK TREATMENT

Mix the appropriate amount of BANVEL herbicide with the appropriate amount of water, emulsifier and No. 2 diesel fuel to obtain the volume of spray desired. See table below. Refer to MIXING and APPLICATION section before mixing. Do not exceed 30 gallons of spray solution per acre per year.

Volume of spray solution desired (gal.)	Ounces			
	Water	Emulsifier	BANVEL herbicide	No. 2 Diesel Fuel
1	100* (3 qts.)	1/2	8	20
2	200 (6 qts.)	1	16	40
5	500 (3.75 gals.)	2 1/2	40	100
10	1000 (7.5 gals.)	5	80	200

* Conversion: 100 ounces = 3 quarts

ASPARAGUS
For Use Only in the States of California, Oregon and Washington

IMPORTANT

OBSERVE ALL PRECAUTIONS UNDER PRECAUTIONARY STATEMENTS HEADING

READ AND FOLLOW MIXING AND APPLICATION INSTRUCTIONS

If spray contacts emerged spears, crooking (twisting) of some spears may result. If such crooking occurs, discard affected spears.

Do not harvest prior to 24 hours after treatment.

Make only one application per season.

RATES AND TIMINGS

Apply BANVEL herbicide to emerged and actively growing weeds in 40 to 60 gallons of diluted spray per treated acre immediately after cutting the field, but at least 24 hours before the next cutting.

Weeds	Rate Per Treated Acre
mustard, black pigweed, redroot (*carelessweed*) sowthistle, annual *thistle, Canada thistle, Russian	1/2 to 1 pt (1/4–1/2 lb. a.i.)
*bindweed, field chickweed, common goosefoot, nettleleaf radish, wild thistle, milk	1 pt. (1/2 lb. a.i.)

BANVEL herbicide plus 2,4-D herbicide tank mixture may be used for improved control of noted (*) weeds. Read and follow 2,4-D product labeling for precautionary statements, directions for use, application rates and timings, and other restrictions.

TURF AND LAWNS
Including Golf Course Fairways, Aprons, Tees and Rough.

IMPORTANT

OBSERVE ALL PRECAUTIONS

UNDER PRECAUTIONARY STATEMENTS HEADING

READ AND FOLLOW MIXING AND APPLICATION INSTRUCTIONS

To avoid injury to newly seeded grasses, application of BANVEL herbicide should be delayed until after the second mowing. Furthermore, application rates in excess of 1 pint (1/2 lb. a.i.) per treated acre may cause noticeable stunting or discoloration of sensitive grass species such as bentgrass, carpetgrass, buffalograss, and St. Augustine grass.

In areas where roots of sensitive plants extend, do not apply in excess of 1/4 pint (1/8 lb. a.i.) of BANVEL herbicide per treated acre on coarse textured (sandy-type) soils, or in excess of 1/2 pint (1/4 lb. a.i.) per treated acre on fine textured (clayey-type) soils. Do not make repeat applications in these areas for 30 days and until previous applications of BANVEL herbicide have been activated in the soil by rain or irrigation.

WEEDS CONTROLLED

BANVEL herbicide, when applied at recommended rates, will give control of many ANNUAL, BIENNIAL, and noted (*) PERENNIAL broadleaf weeds commonly found in turf. BANVEL herbicide will also give growth suppression of many other listed PERENNIAL broadleaf weeds and WOODY brush and vine species. (Refer to GENERAL WEED LIST.)

MIXING AND APPLICATION

Apply 30 to 200 gallons of diluted spray per treated acre (3 qts. to 4¼ gals. per 1,000 sq. ft.), depending on density or height of weeds treated and on the type of equipment used.

RATES AND TIMINGS

Use the higher level of listed rate ranges when treating dense vegetative growth.

Weed Stage & Type	BANVEL herbicide pints per treated acre	lbs. a.i. per treated acre	teaspoons per 1000 sq. ft.
Annual			
Small, actively growing	½–1	¼–½	1–2¼
Established weed growth	1–1½	½–¾	2¼–3¼
Biennial			
Rosette diameter			
Less than 3 inches	½–1	¼–½	1–2¼
3 inches or more	1–2	½–1	2¼–4½
Perennials and Woody Brush and Vines	1–2	½–1	2¼–4½

For best performance, apply when weeds are emerged and actively growing.

Retreatments may be made as needed; however, do not exceed a total of 2 pints (1 lb. a.i.) BANVEL herbicide per treated acre during a growing season.

TANK MIX TREATMENTS

READ AND FOLLOW THE LABEL OF EACH TANK MIX PRODUCT USED FOR PRECAUTIONARY STATEMENTS, DIRECTIONS FOR USE, APPLICATION RATES AND TIMINGS AND OTHER RESTRICTIONS.

Tank mix treatments of BANVEL herbicide may be made with 2,4-D, MCPA, MCPP, or bromoxynil for control of additional weeds listed on the tank mix product label.

Apply ⅕ to ½ pint (¹⁄₁₀–¼ lb. a.i.) of BANVEL herbicide per treated acre with ½ to 1½ lb. acid equivalent of 2,4-D, MCPA, or MCPP, or with ⅜ to ½ lb. a.i. of bromoxynil. Use the higher level of the listed rate ranges when treating established weeds. Repeat treatments may be made as needed; however, do not exceed 2 pints (1 lb. a.i.) of BANVEL herbicide per treated acre during the growing season.

GRASS SEED CROPS

Perennial Grasses such as Bluegrass, Lawntype Fescue and Ryegrass

IMPORTANT

OBSERVE ALL PRECAUTIONS

UNDER PRECAUTIONARY STATEMENTS HEADING

READ AND FOLLOW MIXING AND APPLICATION INSTRUCTIONS

Refer to the PASTURE, RANGELAND, AND NONCROPLAND IMPORTANT section for possible grazing and feeding restrictions.

Do not use on bentgrass unless possible crop injury can be tolerated.

RATES AND TIMINGS

Apply ½ to 2 pints (¼–1 lb. a.i.) of BANVEL herbicide in 5 to 40 gallons of diluted spray per treated acre after weeds have emerged and are actively growing for control of broadleaf weeds such as:

alfalfa	cockle, white
* bindweed, field	dock, curly
catchfly, nightflowering	* knapweed, Russian
chamomile, corn	knotweed
chickweed, common	sorrel, red *(sheep sorrel)*
chickweed, mouseear	starwort, little
clover	* thistle, Canada

* Top growth control only

Use ½ to 1 pint (¼–½ lb. a.i.) of BANVEL herbicide per treated acre on SEEDLING GRASS after the crop reaches the 3–5 leaf stage. Up to 2 pints (1 lb. a.i.) of BANVEL herbicide per treated acre may be used on well-established perennial grass. DO NOT APPLY AFTER THE GRASS SEED CROP BEGINS TO JOINT.

For control of ANNUAL GRASS WEEDS such as:

brome, downy *(cheatgrass)*	hairgrass
brome, ripgut	oats, wild
fescue, rattail	windgrass

Apply 2 to 4 quarts (2–4 lbs. a.i.) of BANVEL herbicide per treated acre in the fall or late summer after harvest and burning of established grass seed crops. Applications should be made within 3–14 days following first irrigation and before weeds have more than 2 leaves.

BETWEEN CROPPING APPLICATIONS (BCA) FOR BROADLEAF WEED CONTROL

IMPORTANT

OBSERVE ALL PRECAUTIONS

UNDER PRECAUTIONARY STATEMENTS HEADING

READ AND FOLLOW MIXING AND APPLICATION INSTRUCTIONS

WEEDS CONTROLLED

BANVEL herbicide, when applied at the recommended rates, will control many ANNUAL and BIENNIAL broadleaf weeds. (Refer to GENERAL WEED LIST.) In addition, BANVEL herbicide will control the PERENNIAL broadleaf weeds listed below.

* alfalfa	dogbane, hemp
artichoke, Jerusalem	garlic, wild
bindweed, field	horsenettle, Carolina
bindweed, hedge	nightshade, silverleaf
blueweed, Texas	redvine
* bursage *(bur ragweed, povertyweed, lakeweed)*	smartweed, swamp
	* sowthistle, perennial
* dandelion, common	thistle, Canada
* dock, curly	trumpetcreeper *(buckvine)*

* Note: perennials may be controlled using BANVEL herbicide at rates lower than those recommended for other listed perennial weeds. (See RATES AND TIMINGS.)

** See SPECIAL TANK MIX TREATMENTS, for specific control program.

RATES AND TIMINGS

Apply BANVEL herbicide as a broadcast or spot treatment to emerged and actively growing weeds after crop harvest and before a killing frost. Agriculturally approved spray additives, such as surfactants or oils, may be used to enhance spray coverage and the herbicide's penetration of weed foliage. See ROTATIONAL CROPS for recommended interval between application and planting to prevent crop injury.

For best performance, make application when ANNUAL weeds are less than 6 inches tall, when BIENNIAL weeds are in the rosette stage, and to PERENNIAL weed regrowth in late summer or fall following a mowing or tillage treatment. Most effective control of upright perennial broadleaf weeds such as Canada thistle and Jerusalem artichoke occurs if application is made when the majority of weeds is 8 inches or taller. Viney perennial broadleaf weeds, such as field bindweed and hedge bindweed, are best controlled when weeds are in or beyond the full bloom stage.

Avoid disturbing treated areas for at least 7 days following application. Treatments may not kill weeds which develop from seed or underground plant parts, such as rhizomes or bulbets, after the effective period for BANVEL herbicide. For seedling control, a follow-up program or other cultural practices could be instituted (refer to corn, sorghum and wheat in-crop uses of BANVEL herbicide).

Weed Stage & Type	BANVEL Herbicide per Treated Acre	
	amount	lbs. a.i.
Annual	½-1 pt.	¼-½
Biennial	1-2 pts.	½-1
Perennial		
Noted (*) Perennials	2-4 pts.	1-2
Other Perennials	4 pts.	2

Retreatments may be made as needed; however, do not exceed a total of 4 pints (2 lbs. a.i.) of BANVEL herbicide per treated acre during any given fallow period.

TANK MIX TREATMENTS

BANVEL herbicide may be tank mixed with one or more of the following herbicides for control of grasses or additional broadleaf weeds. Read and follow the label of each tank mix product used for precautionary statements, directions for use, rates and timings, weeds controlled and geographic or other restrictions.

Herbicide	Rate per Treated Acre (lbs. a.i.)
ANNUAL WEED CONTROL	
atrazine	½ to 3
chlorsulfuron (Glean®)	0.016 to 0.024 (⅓-½ wt. oz. product)
cyanazine (Bladex®)	1⅗ to 3⅕
glyphosate (Roundup®)	¼ to ½
metribuzin (Sencor®, Lexone®)	⅓ to ¾
paraquat	½ to 1
2,4-D	¼ to ½
PERENNIAL WEED CONTROL	
glyphosate	1 to 2
2,4-D	1 to 2

SPECIAL TANK MIX TREATMENTS

For wild garlic control, apply 1 pint (½ lb. a.i.) BANVEL herbicide with 1½ lbs. acid equivalent 2,4-D low volatile ester per treated acre. Apply when wild garlic is 4 to 8 inches tall.

For Canada thistle control, use BANVEL herbicide or BANVEL plus Roundup herbicide tank mix treatments.

ROTATIONAL CROPS

The following recommendations are based on BANVEL herbicide use rates up to 4 pints (2 lbs. a.i.) per treated acre.

CORN and SORGHUM may be planted in the spring following applications made during the previous year.

SOYBEANS may be planted in the spring following applications made during the previous year. If less than 1 inch of rainfall occurs between application and first killing frost, treated areas should be cultivated to allow herbicide to come in contact with moist soil. Cultivation may take place before or immediately after ground thaw.

WHEAT may be planted in the fall or spring following applications. Also, spot applications may be made anytime prior to crop emergence if crop injury can be tolerated in treated areas. Wheat injury may occur if the interval between application and planting is less than specified.

East of the Mississippi River, the interval is 20 days per pint of BANVEL herbicide per treated acre. Exclude days when ground is frozen.

West of the Mississippi River, the interval is 45 days per pint of BANVEL herbicide per treated acre. Exclude days when ground is frozen.

Following a normal harvest of corn, sorghum, soybeans or wheat, any rotational crop may be planted. If the interval before harvest is shortened, such as when cover crops will be plowed under, do not follow up with the planting of a sensitive crop.

CONTROL OF PERENNIAL BROADLEAF WEEDS IN CROPLAND (SPOT APPLICATION ONLY)

For Use Only in the States of Idaho, Montana, Nevada, Oregon, Utah, and Washington.

IMPORTANT

OBSERVE ALL PRECAUTIONS

UNDER PRECAUTIONARY STATEMENTS HEADING

READ AND FOLLOW MIXING AND APPLICATION INSTRUCTIONS

Do not treat subirrigated cropland or areas where the soil remains saturated with water throughout the year.

Make only one application of BANVEL herbicide per year.

WEEDS CONTROLLED

BANVEL herbicide, when applied at recommended rates, will control many broadleaf weeds including:

bindweed, field	knapweed, Russian
dock, broadleaf *(bitterdock)*	ragwort, tansy
dock, curly	spurge, leafy
knapweed, black	thistle, Canada

RATES AND TIMINGS

BANVEL herbicide may be applied at any time following a crop harvest to stubble fallow or other cropland. Application should be made when weeds are actively growing and prior to a killing frost.

Apply 4 to 6 quarts (4–6 lbs. a.i.) of BANVEL herbicide per treated acre. Application may be be made up to one month prior to the planting of wheat. NOTE: Do not use unless injury to wheat or rotated barley will be acceptable.

Barley, oats, corn, sorghum (milo), annual or perennial grass crops may be planted into treated areas one year after application. Crops grown for seed (other than perennial grass seed) should not be planted into treated areas until three years after application. Do not plant broadleaf crops such as alfalfa, beans, peas, potatoes, or sugarbeets into treated areas until two years after application.

In most cases, treatments will not kill perennial weed seedlings which germinate from seed one or two years after treatment. Once the effect of the chemical has been lost, a follow-up program for seedling control or other cultural practices should be instituted.

WIPER APPLICATION USES

IMPORTANT

OBSERVE ALL PRECAUTIONS

UNDER PRECAUTIONARY STATEMENTS HEADING

BANVEL herbicide may be applied through wiper application equipment to control or suppress actively growing broadleaf weeds, brush and vines. Use a solution containing 1 part BANVEL herbicide to 1 part water. Do not contact desirable vegetation with herbicide solution. Wiper application should only be made to crops (including pastures) and non-cropland areas described in this label with the exception of Grain Sorghum (Milo).

REGISTERED TRADEMARKS

PROWL® is a registered trademark of American Cyanamid Company.

ATRATOL®, BICEP®, CONCEP®, DUAL®, EVIK®, IGRAN®, MILOGARD®, PRAMITOL®, and PRINCEP® are registered trademarks of Ciba-Geigy Corporation, Agricultural Division.

BEXTON®, DOWPON®, GARLON®, and TORDON® are registered trademarks of Dow Chemical Company.

DALAPON® is a registered trademark of Diamond Shamrock Corporation.

GLEAN®, HYVAR®, KARMEX®, KRENITE®, LEXONE®, and VELPAR® are registered trademarks of E.I. duPont de Nemours & Co., Inc.

OUST™ is a trademark of E.I. duPont de Nemours & Co., Inc.

SPIKE® is a registered trademark of Elanco Products Company.

SENCOR® is a registered trademark of Mobay Chemical Corporation.

BRONCO®, LASSO®, RAMROD®, and ROUNDUP® and SCREEN® are registered trademarks of Monsanto Agricultural Products Company.

ASULOX® is a registered trademark of Rhone-Poulenc Chemical Company.

BLADEX® is a registered trademark of Shell Chemical Company.

ERADICANE® and SUTAN+® are registered trademarks of Stauffer Chemical Company.

BROMOXYNIL is trademarked as BROMINAL® by Union Carbide Corporation, and as BUCTRIL® by Rhone-Poulenc Chemical Company.

BANVEL® herbicide and ACCUTROL® herbicide adjuvant are registered trademarks of Velsicol Chemical Corporation.

24(c) SPECIAL LOCAL NEED LABELING

BANVEL® HERBICIDE

PREHARVEST USES FOR CONTROL OF HEMP DOGBANE IN FIELD CORN AND SORGHUM

For Use Only in the States of Illinois, Iowa, Kansas, Michigan, Minnesota, Missouri, Nebraska, Ohio and South Dakota
EPA SLN Nos. IL-840005, IA-760001, KS-790015, MI-830020, MN-830002, MO-810020, NE-790017, OH-810003, SD-790013.

IMPORTANT
OBSERVE ALL PRECAUTIONS

To prevent possible injury to soybeans growing in the vicinity where applications to corn or sorghum will be made, soybeans must be fully podded and leaves should be turning yellow. Soybeans must be past the bloom stage.

Do not graze or feed fodder from treated areas until 35 days after treatment.

APPLICATION DIRECTIONS

CORN: Apply ¼ pound per acre of BANVEL (equivalent to ½ pint BANVEL herbicide) as a tank mixture with 1 pound per acre 2,4-D acid equivalent basis, as the amine or LV ester formulation after corn is in the brown silk stage. Do not harvest treated corn for at least 7 days after treatment.

SORGHUM: Apply ¼ pound per acre of BANVEL (equivalent to ½ pint BANVEL HERBICIDE) as a tank mixture with 1 pound per acre 2,4-D acid equivalent basis, as the amine or LV ester formulation after sorghum is in soft dough stage. Do not harvest treated sorghum for at least 30 days after treatment.

For best results make application when hemp dogbane is green and underground rhizomes have swollen pink buds. A non-ionic surfactant may be added to spray mixture. Observe all applicable directions, restrictions, and precautionary statements on the 2,4-D product label.

For Use Only in the State of Indiana
EPA SLN No. IN-800012

Refer to the precautionary statements on previous page and note application directions above. In Indiana, make application from September 15 until the first killing frost. Do not apply near homesteads or residential areas.

24(c) SPECIAL LOCAL NEED LABELING

BANVEL® HERBICIDE

PREHARVEST APPLICATION IN FALL SEEDED WHEAT

For Distribution and Use Only Within the States of Kansas and Oklahoma
EPA SLN Nos. KS-840003, OK-840010

IMPORTANT
OBSERVE ALL APPLICATION DIRECTIONS, RESTRICTIONS AND PRECAUTIONS

Also follow all applicable directions, restrictions, and precautions on the 2,4-D product label.

SENSITIVE CROP PRECAUTIONS: BANVEL herbicide may cause injury to desirable trees and plants, particularly beans, cotton, flowers, fruit trees, grapes, or ornamentals, peas, potatoes, soybeans, sunflowers, tobacco, tomatoes and other broadleaf plants when contacting their roots, stems or foliage. These plants are most sensitive to BANVEL herbicide during their development or growing stage.

WEED LIST: BANVEL herbicide, when applied at recommended rates, will control or suppress the following annual weeds that may interfere with harvest of fall seeded wheat:

buckwheat, wild	mustard spp.
kochia	pigweed spp.
lambsquarters, common	sunflower
	thistle, Russian

RATES AND TIMINGS

Apply BANVEL herbicide as a broadcast or spot treatment to annual broadleaf weeds when wheat is in the hard dough stage and green color is gone from the nodes (joints) of the stem. Best results will be obtained if application can be made when weeds are actively growing but before weeds canopy over the wheat. A WAITING INTERVAL OF 10-14 DAYS IS REQUIRED BEFORE HARVEST.

For control or suppression of listed broadleaf weeds, apply ½ pint BANVEL herbicide (¼ lb. a.i.) with ½-1 lb. a.i. 2,4-D amine or ester per treated acre. Check with local agricultural extension agent or 2,4-D label for specific use rates.

For aerial applications, apply a minimum of 3 gallons of diluted spray per treated acre. Agriculturally approved spray additives, such as surfactants, or oils, may be used to enhance spray coverage and the herbicide's penetration of weed foliage. Refer to pages 1-2 for additional directions for proper application.

NOTE: Do not graze or feed fodder from treated area. Do not use treated wheat for seed.

24(c) SPECIAL LOCAL NEED LABELING

BANVEL® HERBICIDE

SMALL GRAINS
AERIAL APPLICATION

For Use Only in the States of Colorado, Montana, Wyoming, Minnesota, North Dakota and South Dakota
EPA SLN Nos. CO-780032, MN-800008, MT-790004, ND-800002, WY-780011, SD-800001

Refer to the SMALL GRAINS section of this label for DIRECTIONS FOR USE information.

BANVEL herbicide may be applied using 1-5 gallons of diluted spray per treated acre when using aerial application equipment in the above listed states only.

Product continued on following page

24(c) SPECIAL LOCAL NEED LABELING

BANVEL® HERBICIDE

WILD BUCKWHEAT CONTROL IN FALL SEEDED WHEAT

For Use Only Within the State of Oklahoma

EPA SLN No. OK-810004

IMPORTANT

OBSERVE ALL RESTRICTIONS, PRECAUTIONS AND APPLICATION DIRECTIONS

TIMING OF APPLICATION: Application of BANVEL herbicide tank mixtures given below can be made on wheat up to the development of a second node. Apply this mixture after the wild buckwheat has emerged and is actively growing. For good control apply before wild buckwheat plants are larger than two true leaves. Spray before a crop canopy is formed which would protect the weeds from the spray.

TANK MIX TREATMENTS: Tank mix $1/16$ pound active ($1/8$ pint) of BANVEL herbicide with $1/8$ pound active Bromoxynil plus $1/8$ pound active MCPA or with $3/16$ pound active Bromoxynil per treated acre. Bromoxynil plus MCPA is sold as Bronate® (trademark of Rhone-Poulenc Inc.) or Brominal® Plus (trademark of Union Carbide Corp.). Bromoxynil is available as Brominal® (trademark of Union Carbide Corp.) or Buctril® (trademark of Rhone-Poulenc Inc.). Read and follow all applicable directions, restrictions, and precautions on the Bromoxynil and MCPA labels.

Dilute with water a minimum of 3 gallons total spray per acre for aerial application and a minimum of 10 gallons total spray for ground application.

24(c) SPECIAL LOCAL NEED LABELING

BANVEL® HERBICIDE

ANNUAL WEED CONTROL IN MILLET

For Use Only within the State of Nebraska

EPA SLN No. NE-830009

IMPORTANT—OBSERVE ALL PRECAUTIONS

WEEDS CONTROLLED

ANNUALS

buckwheat, tartary	pigweed, redroot *(carelessweed)*
buckwheat, wild	pigweed, rough
carpetweed	pigweed, tumble
chamomile, corn	pineappleweed
chickweed, common	puncturevine
cockle, corn	purslane, common
cockle, cow	radish, wild
cocklebur, common	ragweed, common
dragonhead, American	ragweed, giant *(buffaloweed)*
hempnettle	salsify *(goatsbeard)*
knawel *(German Moss)*	shepherdspurse
knotweed	smartweed, green
kochia	smartweed, Pennsylvania
ladysthumb	sowthistle, annual
lambsquarters, common	starthistle, yellow
lettuce, prickly	sunflower, common *(wild)*
mallow, common	sunflower, volunteer
mayweed	thistle, Russian
mustard, annuals	velvetleaf
mustard, tansy	vetch
nightshade, black	
pennycress, field *(fanweed, frenchweed, stinkweed)*	

APPLICATION DIRECTIONS

Apply BANVEL herbicide as a broadcast or spot treatment to emerged and actively growing weeds and when millet is in the 2 to 5 leaf stage.

- For control of listed Annual broadleaf weeds, apply $1/4$ pint BANVEL herbicide ($1/8$ lb. a.i.) with $3/8$ lb. a.i. 2,4-D.

5 to 50 gallons of diluted spray per treated acre may be applied when using ground application equipment, or 3 to 10 gallons of diluted spray per treated acre when using aerial application equipment.

24(c) SPECIAL LOCAL NEED LABELING

BANVEL® HERBICIDE

BETWEEN CROPPING SYSTEM USES FOR BROADLEAF WEED CONTROL

For Use Only Within the State of South Dakota

EPA SLN No. SD-810024

IMPORTANT

OBSERVE ALL PRECAUTIONS

WEEDS CONTROLLED

ANNUALS

buckwheat	morningglory, tall
buffalobur	mustards, yellow tops
catchfly, nightflowering	pennycress, field
chickweed, common	pigweed, redroot
cocklebur, common	pigweed, tumble
cockle, corn	ragweed, common
cockle, cow	ragweed, giant *(buffaloweed)*
henbit	smartweed, Pennsylvania
knotweed	sowthistle, annual
kochia	sowthistle, spiny
ladysthumb	sunflower
lambsquarters, common	thistle, Russian *(tumbleweed)*
mayweed	
morningglory ivyleaf	

BIENNIALS

starthistle, yellow	thistle, bull
teasel	thistle, musk

PERENNIALS

bindweed, field	nightshade, silverleaf *(whiteweed)*
blueweed, Texas	redvine
bursage, skeletonleaf *(bur ragweed, povertyweed)*	sowthistle, perennial
bursage, woollyleaf *(lakeweed)*	thistle, Canada
dock, curly	trumpetcreeper *(buckvine)*
garlic, wild	

APPLICATION DIRECTIONS

Apply BANVEL herbicide as a broadcast or spot treatment to emerged and actively growing weeds after crop harvest and before a killing frost. Ground or aerial application equipment which will give good spray coverage of weed foliage may be used. Apply 5 to 40 gallons of diluted spray per treated acre when using ground application equipment, or 3 to 10 gallons of diluted spray per treated acre when using aerial application equipment. Spray additives, such as surfactants or oils, may be used to enhance spray coverage and herbicide's penetration of weed foliage.

- For control of ANNUAL broadleaf weeds, apply $1/2$ to 1 pint per acre of BANVEL herbicide. For best results, make application when weeds are less than six inches tall.
- For control of BIENNIAL broadleaf weeds, apply 1 to 2 pints

per acre of BANVEL herbicide. For best results, make application when weeds are in the rosette stage.

- For control of PERENNIAL broadleaf weeds, apply 2 to 4 pints per acre of BANVEL herbicide. (See ROTATIONAL CROPS for appropriate interval between application and planting to prevent crop injury.)

Use the higher level of the listed rate ranges when treating dense weed growth. Avoid disturbing treated areas for 7 days following application. These treatments may not kill weeds which develop from seed or underground plant parts, such as rhizomes or bulblets, after the effective period for BANVEL herbicide. A follow-up program for seedling control or other cultural practices should be instituted. Repeat treatments with BANVEL herbicide may be made during the season to control regrowth, however, do not exceed 4 pints per acre during any given fallow period.

TANK MIX TREATMENTS

BANVEL herbicide may be applied as a tank mix treatment with 2,4-D amine or LV ester formulations for control of additional broadleaf weed species. Tank mix treatments with Roundup® Herbicide (registered trademark of Monsanto Agricultural Products Co.) may be made for control of grasses and additional broadleaf weeds. Read and follow all applicable directions, restrictions, and precautionary statements on the 2,4-D or Roundup Herbicide product labels.

ANNUAL GRASS AND ANNUAL BROADLEAF WEED CONTROL

For **ANNUAL** grass and **ANNUAL** broadleaf weed control during fallow periods, tank mix 1/4 to 1 pint per acre of BANVEL herbicide with 1/4 to 1 pint per acre of Roundup Herbicide. Add 0.5% by volume of agriculturally approved NONIONIC surfactant, such as X-77® (registered trademark of Kalo Laboratories), to the spray solution. Use 1/4 pint per acre of BANVEL plus 1/4 pint per acre of Roundup for control of foxtails, kochia, redroot pigweed, Russian thistle, lambsquarter, wild and tame sunflower when these weeds are actively growing and less thab 4 inches tall. Use higher rates for larger weeds, weeds under drought stress or other weed species. For ground and aerial applications, use 3 to 10 gallons of water per acre.

PERENNIAL BROADLEAF WEED CONTROL

For control of PERENNIAL broadleaf weeds such as field bindweed and Canada thistle, tank mix 1 to 4 pints per acre of BANVEL herbicide with 2 to 4 pints per acre of Roundup Herbicide. In situations where a short waiting interval requires the low rates of BANVEL herbicide (See ROTATIONAL CROPS), tank mix with the higher listed rates of Roundup Herbicide.

ROTATIONAL CROPS

- CORN may be planted in areas treated with up to 1 pint per acre of BANVEL anytime after spray application. CORN may be planted in areas treated with higher rates (up to 2 quarts per acre) of BANVEL if spray applications were made during the previous year.
- SORGHUM may be planted in areas treated with up to 1/2 pint per acre of BANVEL anytime after spray application. SORGHUM may be planted in areas treated with higher rates (up to 2 quarts per acre) of BANVEL if spray applications were made during the previous year.
- WINTER WHEAT, SPRING WHEAT, BARLEY and OATS may be planted in areas treated with up to 1/4 pint of BANVEL anytime prior to planting. If BANVEL rates exceed 1/4 pint per acre, delay planting for 20 days for each 1/2 pint per acre of BANVEL used to avoid crop injury. Exclude days when ground is frozen.

NOTE: SPOT APPLICATIONS MAY BE MADE ANYTIME PRIOR TO CROP EMERGENCE IF CROP INJURY CAN BE TOLERATED IN TREATED AREAS.

CALIFORNIA: Special Directions for Use

BANVEL herbicide has been approved by the U.S. Environmental Protection Agency for the uses, crops and sites listed in this label. Approval of the items listed below is pending under the State of California registration requirements.

BANVEL herbicide may not be used for control of the following weeds in California until approval is received:

buckwheat, tartary	melaleuca
burdock, common	oats, wild
condalia, lotebush	olive, autumn
cress, hoary	olive, Russian
dock, broadleaf *(bitterdock)*	peppertree, Brazil
	rubberweed, bitter
eupatorium, late *(thoroughwort)*	sagebrush, fringed
	snowberry, Western
guava	spanishneedles
hairgrass	waxmyrtle
lespedeza	windgrass
lupine, silvery	

MIXING DIRECTIONS

For Oil in Water Emulsions

BANVEL herbicide can be applied using water, oil in water emulsions, or sprayable fluid fertilizer as a carrier. A COMPATIBILITY TEST should be made prior to tank mixing.

To prepare oil in water emulsions, half-fill spray tank with water plus appropriate amount of herbicide. With continuous agitation, slowly add a premix of oil (such as diesel oil or fuel oil) plus a suitable emulsifier to spray tank. Complete filling of spray tank with water. Maintain vigorous agitation during spray operation to prevent oil and water from forming separate layers.

SUPPLEMENTAL LABELING

BANVEL® HERBICIDE

EPA Reg. No. 876-25

TANK MIX WITH ATRAZINE FOR USE ON GRAIN SORGHUM (Milo)

DIRECTIONS FOR USE

It is a violation of Federal law to use this product in a manner inconsistent with its labeling.

IMPORTANT

Before using this product, read and carefully observe all applicable directions, restrictions, and precautionary statements on the EPA-registered container label and the Banvel Herbicide Directions For Use booklet attached to the container. ALSO, READ AND FOLLOW THE LABEL OF EACH TANK MIXED PRODUCT USED FOR PRECAUTIONARY STATEMENTS DIRECTIONS FOR USE, WEEDS CONTROLLED, GEOGRAPHIC AND OTHER RESTRICTIONS.

APPLICATION DIRECTIONS

BANVEL PLUS ATRAZINE: For improved control of emerged, actively growing broadleaf weeds including triazine resistant species and added suppression of perennial broadleaf weeds, tank mix 1/2 pint Banvel® Herbicide with 0.5 to 1.25 lbs. a.i. atrazine per treated acre. For control of grasses (less than 1.5 inches tall), tankmix 1/2 pint Banvel® Herbicide with 2-3 lbs. a.i. atrazine per treated acre. For best performance and minimal crop injury, make application when sorghum is 3-8″ tall and when broadleaf weeds are small (less than 6 inches tall).

The atrazine rate will depend upon soil texture and length of residual weed control desired.

THIS LABELING MUST BE IN THE POSSESSION OF THE USER AT THE TIME OF PESTICIDE APPLICATION.

SUPPLEMENTAL LABELING

BANVEL® HERBICIDE

EPA Reg. No. 876-25

BETWEEN CROPPING APPLICATIONS (BCA) BANVEL® PLUS KERB® TANK MIX FOR USE IN FALLOW ROTATED TO WHEAT

DIRECTIONS FOR USE

It is a violation of Federal law to use this product in a manner inconsistent with its labeling.

IMPORTANT

Before using this product, read and carefully observe all applicable directions, restrictions, and precautionary statements on the EPA-registered container label and the Banvel Herbicide Directions For Use booklet attached to the container. ALSO, READ AND FOLLOW THE LABEL OF EACH TANK MIXED PRODUCT USED FOR PRECAUTIONARY STATEMENTS, DIRECTIONS FOR USE, WEEDS CONTROLLED, GEOGRAPHIC AND OTHER RESTRICTIONS

APPLICATION DIRECTIONS

A tankmix of Banvel Herbicide plus Kerb 50 W Herbicide will control volunteer barley, bolbous bluegrass, downy brome, jointed goatgrass, common rye and volunteer wheat when they are actively growing.

During fallow periods, tankmix 1 pint of Banvel Herbicide ($\frac{1}{2}$ lb.a.i.) with $\frac{1}{2}$-$\frac{3}{4}$ lb. of Kerb 50 W Herbicide ($\frac{1}{4}$-$\frac{3}{8}$ lb.a.i. pronamide) per treated acre. Do not exceed 4 pints of Banvel Herbicide (2 lbs. a.i.) per treated acre. Fall seeded wheat may be planted into fallow land treated with kerb 50 W Herbicide 9 months (270 days or more after application.

For best performance, make application between mid-October and mid-December, prior to soil freeze up.

THIS LABELING MUST BE IN THE POSSESSION OF THE USER AT THE TIME OF PESTICIDE APPLICATION.

Kerb® is a registered trademark of Rohm and Hass Company.

SUPPLEMENTAL LABELING

BANVEL® HERBICIDE

EPA Reg. No. 876-25

BANVEL® PLUS SENCOR® TANK MIX ON FALL SEEDED WHEAT

DIRECTIONS FOR USE

It is a violation of Federal law to use this product in a manner inconsistent with its labeling.

IMPORTANT

Before using this product, read and carefully observe all applicable directions, restrictions, and precautionary statements on the EPA-registered container label and the Banvel Herbicide Directions For Use booklet attached to the container. Do not graze or harvest for livestock feed prior to crop maturity. ALSO, READ AND FOLLOW THE LABEL OF EACH TANK MIXED PRODUCT USED FOR PRECAUTIONARY STATEMENTS, DIRECTIONS FOR USE, WEEDS CONTROLLED, GEOGRAPHIC AND OTHER RESTRICTIONS.

APPLICATION DIRECTIONS

A single postemergence application of Banvel Herbicide plus Sencor Herbicide at the recommended rates will only control the listed broadleaf weeds. Apply only after fall seeded wheat has started to grow and has a well established secondary root system or is beginning to tiller but prior to the jointing stage. Best performance is achieved when weeds are small (less than 2 inches in diameter or height).

WEEDS CONTROLLED

chickweed, common	knotweed, prostrate
cockle, corn	mayweed (dog fennel)
fiddleneck	mustard, blue
henbit	mustard, Jim Hill
lambsquarters	mustard, tansy
lettuce, miners	speedwell, ivyleaf
lettuce, prickly	

Broadcast Rate Per Treated Acre: Apply $\frac{1}{8}$-$\frac{1}{4}$ pint ($\frac{1}{16}$-$\frac{1}{8}$ lb. a.i.) Banvel Herbicide with:

	Amount of product	lb. a.i.
Sencor® 75 DF	$\frac{1}{6}$ lb.	0.125
Sencor® 4	$\frac{1}{4}$ pint	0.125

When using Sencor Herbicide, refer to product label registered for this use for additional directions, rates, and weed species controlled. Observe all precautions and limitations on labeling of all products used in the mixture.

THIS LABELING MUST BE IN THE POSSESSION OF THE USER AT THE TIME OF PESTICIDE APPLICATION.

Sencor® is a registered trademark of Mobay Chemical Corporation.

SUPPLEMENTAL LABELING

BANVEL® HERBICIDE

EPA Reg. No. 876-25

BETWEEN CROPPING APPLICATIONS (BCA) BANVEL® PLUS LANDMASTER® TANK MIX FOR USE IN FALLOW ROTATED TO WHEAT

For Use in Colorado, Idaho, Kansas, Montana, North Dakota, Nebraska, New Mexico, Oklahoma, Oregon, South Dakota, Texas, Utah, Washington and Wyoming.

DIRECTIONS FOR USE

It is a violation of Federal law to use this product in a manner inconsistent with its labeling.

IMPORTANT: Before using this product, read and carefully observe all applicable directions, restrictions, and precautionary statements on the EPA registered container label and the BANVEL Herbicide Directions For Use booklet attached to the container. ALSO, READ AND FOLLOW THE LABEL OF EACH TANK MIXED PRODUCT USED FOR PRECAUTIONARY STATEMENTS, DIRECTIONS FOR USE, WEEDS CONTROLLED, GEOGRAPHIC AND OTHER RESTRICTIONS.

APPLICATION DIRECTIONS

A tank mix of BANVEL Herbicide and LANDMASTER will improve control of annual broadleaf weeds such as kochia, wild buckwheat, prickly lettuce and labeled perennial broadleaf weeds such as field bindweed and Canada thistle.

During fallow periods, tank mix 4-8 fluid ounces of BANVEL Herbicide (1/8-1/4 lb. a.i.) with 40-54 fluid ounces of LANDMASTER (0.38 lb. a.i. glyphosate + 0.5 lb. a.i. 2,4-D to 0.5 lb. a.i. glyphosate + 0.675 lb. a.i. 2,4-D) per treated acre for above listed annual weed control. For perennial weed suppression, tank mix 8-16 fluid ounces of BANVEL Herbicide (1.4-1.2 lb. a.i.) with 40-54 fluid ounces of LANDMASTER (0.38 lb. a.i. glyphosate + 0.5 lb. a.i. 2,4-D to 0.5 lb. a.i. glyphosate + 0.675 lb. a.i. 2,4-D).

Do not exceed 4 pints of BANVEL Herbicide (2 lbs. a.i.) per treated acre during any given fallow period.

Wheat may be planted into fallow land treated with BANVEL Herbicide, however, injury may occur if interval between application and planting is less than specified. West of the Mississippi River the interval is 45 days per pint of BANVEL Herbicide per treated acre. Exclude all days when ground is frozen.

THIS LABELING MUST BE IN THE POSSESSION OF THE USER AT THE TIME OF PESTICIDE APPLICATION

LANDMASTER® is a registered trademark of Monsanto.d

BANVEL® 520

HERBICIDE

OIL SOLUBLE INDUSTRIAL HERBICIDE

For Brush and Broadleaf Weed Control

ACTIVE INGREDIENTS:

Dicamba (3,6-dichloro-o-anisic acid)	11.9%
Related acids	2.1%
Isooctyl ester of 2,4-dichlorophenoxy-acetic acid*	34.0%
INERT INGREDIENTS:	52.0%
TOTAL	100.0%

* Equivalent to 22.6%, 2,4-dichlorophenoxyacetic acid. Isomer specific by AOAC method 6.275-6.279, 13th edition.

Contains 1 Pound Dicamba and 1.9* Pounds 2,4-D Per Gallon

E.P.A. Reg. No. 876-168-AA E.P.A. Est. No. 876-TX-2

ROLL OR SHAKE DRUM BEFORE USING

Keep Out of Reach of Children
DANGER

SEE LEFT PANEL FOR ADDITIONAL PRECAUTIONARY STATEMENTS

BEFORE USING BANVEL-520 HERBICIDE READ AND FOLLOW THE PRECAUTIONARY STATEMENTS APPEARING ON THE CONTAINER

BANVEL-520 Oil Soluble Industrial Herbicide is recommended for use for rights-of-way (utility, railroad, highway, pipeline), nonselective forest brush control, fence rows, drainage ditch banks, wasteland, and similar non-cropland areas.

LIMITED WARRANTY AND LIABILITY

NOTICE: Read this Limited Warranty and Liability before buying or using this product. If the terms are not acceptable, return it at once unopened.

It is critical that this product be used and mixed only as specified on the label. The laws of a State may make some or all of this paragraph inapplicable or may give you rights in addition to your rights hereunder. Except to the extent prohibited by applicable law, the exclusive remedy of the user or buyer and the limit of liability of this Company or any other Seller for any and all losses, personal injuries or damages resulting from the use of this product, shall be the purchase price paid by the user or buyer for the quantity of product involved. Except to the extent prohibited by State law, there is no warranty, and this Company and other Sellers disclaim all liability for losses, personal injury or damages: (i) arising from any use of this product in a manner or for a purpose not recommended in its label directions, or from mixing this product before use with any substance except as recommended by the product's label, (ii) arising from handling or storage in violation of label instructions, (iii) for all indirect special or consequential damages, (iv) when not reported to this company within one year of discovery, and (v) arising from product not used within the label-designated shelf life or four years from date of purchase, whichever first occurs. THERE ARE NO IMPLIED WARRANTIES AND NO WARRANTIES OF MERCHANTABILITY OR FITNESS.

IMPORTANT

Do not contaminate irrigation ditches or water used for domestic purposes.

BANVEL-520 herbicide may cause injury to desirable trees or plants, particularly beans, cotton, flowers, fruit trees, grapes, ornamentals, peas, potatoes, soybeans, sunflowers, tobacco, tomatoes, and other broadleaf plants when contacting their roots, stems, or foliage. Plants are most sensitive to BANVEL-520 herbicide during their development or growing stage. FOLLOW THE PRECAUTIONS LISTED ON PAGES 2-3 WHEN USING BANVEL-520 HERBICIDE.

- Do not apply to areas where possible downward soil movement or surface washing may cause contact of BANVEL-520 herbicide with the roots of desirable plants, such as trees and shrubs.
- Avoid making application when spray particles may be carried by air currents to areas where sensitive crops and other desirable plants are growing. Always make application when there is some air movement in order to determine the direction and distance of possible spray drift. Leave an adequate buffer zone between the area to be treated and desirable sensitive plants. Coarse sprays are less likely to drift out of target area.
- To avoid injury to desirable plants, equipment used for BANVEL-520 herbicide should be thoroughly cleaned (see PROCEDURE FOR CLEANING SPRAY EQUIPMENT) before handling or applying any other chemical.
- Aerial application should be made with extreme care to prevent off-target drift. An adequate buffer zone must be maintained between the area to be treated and desirable sensitive plants. Observe all drift precautionary statements described in this label.
- DO NOT APPLY IN AREAS WHICH MAY BE GRAZED BY LIVESTOCK OR CUT FOR HAY WITHIN 1 YEAR OF APPLICATION.

Application of BANVEL-520 herbicide may injure newly seeded grasses. Furthermore, rates in excess of 1½ pints per treated acre may cause stunting of sensitive established grass species such as bentgrass and St. Augustine grass. Treatments will kill or injure alfalfa, clovers, lespedeza, wild winter peas, vetch and other legumes.

Consult your local or state authorities for possible application restrictions and advice concerning special local use situations.

PROCEDURE FOR CLEANING SPRAY EQUIPMENT

The steps listed below are suggested for thorough cleaning of spray equipment following applications of BANVEL-520 herbicide or tank mixtures with BANVEL-520 herbicide and other products.

1. Rinse with a small quantity of kerosene or diesel oil. Circulate the liquid throughout the sprayer system, pumping it through the hose and nozzles for at least two minutes. Flush by operating sprayer until the system is purged of liquid.
2. Fill tank with water while adding 2 lbs. of detergent for every 40 gallons of water. Operate the pump to circulate the detergent solution through the sprayer system for 5 to 10 minutes and discharge a small amount of the solution through the boom and nozzles. Let the solution stand for several hours, preferably overnight.
3. Flush the detergent solution out of the spray tank through the boom.
4. Hose down thoroughly the inside as well as outside surfaces of equipment while filling the spray tank half full of water. Flush by operating sprayer until the system is purged of the rinse water.
5. Fill tank with water while adding 1 quart of household ammonia for every 25 gallons of water. Operate the pump to circulate the ammonia solution through the sprayer system for 15 to 20 minutes and discharge a small amount of the ammonia solution through the boom and nozzles. Let the solution stand for several hours, preferably overnight.
6. Flush the solution out of the spray tank through the boom.
7. Remove the nozzles and screens and flush the system with two full tanks of water.

REFER TO THE CONTAINER LABEL FOR INSTRUCTIONS CONCERNING DISPOSAL OF WASTE AND CLEANING RINSES.

BRUSH AND VINES CONTROLLED

BANVEL-520 herbicide, when applied in accordance with this label, will give control or growth suppression of many woody brush and vine species including:

ailanthus	fir	poplar
alder	grape	*rabbitbrush

apple	guava	*redcedar, eastern
ash	hawthorn (thornapple)	rose, McCartney
aspen	hemlock	rose, multiflora
basswood	hickory	sagebrush, fringed
beech	honeysuckle	sassafras
*birch	hornbeam	serviceberry
*blackberry	*huckleberry	snowberry, western (buckbrush)
blackgum	huisache	spicebush
cascara	ivy, poison	spruce
cedar	kudzu	sweetgum
*cherry	locust, black	sumac
chinquapin	*maple	sycamore
*condalia, lotebush (lote)	*melaleuca (cajeput tree)	tarbush
cottonwood	*mesquite	trumpetcreeper (buckvine)
creosotebush	oak, poison	waxmyrtle
cucumbertree	oak	willow
*dewberry	*peppertree, Brazil (Florida holly, Schinus)	witchhazel
dogwood	persimmon, eastern	*yaupon
*elderberry	pine	yucca
elm	plum, sand (wild plum)	

* When treating noted species, use the more concentrated of the spray mixtures recommended.

MIXING

BANVEL-520 herbicide is an oil soluble formulation which can be applied using oil or oil-water emulsions as the carrier. To prepare oil in water emulsions, follow the W A C O Mixing Sequence described under LOW OIL SPRAY TREATMENTS on page 8. When tank mixing with other herbicides, read and follow the instructions on the label of all products used concerning mixing and application. A COMPATIBILITY TEST should be made prior to any tank mixing.

MAINTAIN AGITATION DURING SPRAY OPERATION TO PREVENT OIL AND WATER FROM FORMING SEPARATE LAYERS.

Spray additives such as ACCUTROL® spray adjuvant (Trademark of Velsicol Chemical Corporation), may be used for improved wetting and penetration. Read and follow all use recommendations and precautions on each product label.

APPLICATION

BANVEL-520 herbicide may be applied as a stump or basal treatment or as a dormant stem treatment using high volume ground, low volume ground, or aerial application equipment in an oil or low oil spray. Volume of spray applied will depend on the height, density, and type of weeds or brush being treated and the type of equipment being used. Apply 10 to 40 gallons of diluted spray per treated acre when using aerial application equipment. Refer to the IMPORTANT section for drift precautionary statements.

- STUMP or BASAL treatments may be made by applying BANVEL-520 herbicide at any time during the year, including winter (except when snow or water prevents spraying to the ground line or when extreme cold causes the spray mixture to freeze). A drench treatment may be made by spraying the entire stump until runoff or, if making a basal application, by spraying the basal parts of the brush and tree trunk (particularly bark and exposed roots) from the ground line up to a height of two feet. The soil surface around the stump should also be sprayed. Application may be made either as a hydraulic high volume or as a manual or powered low volume backpack treatment. See pages 7-8 for more information.
- DORMANT STEM treatments may be made by applying BANVEL-520 herbicide when most of the foliage has dropped off or at any time of the year when foliage does not inhibit spray coverage of the stem. Spray the collar at ground line and each stem until runoff, with special emphasis on covering the root crown. Hydraulic high volume, manual or powered low volume backpack or aerial treatments may be made. See pages 7-8 for more information.

OIL SPRAY TREATMENTS

Hydraulic High Volume: Mix 3 to 6 gallons of BANVEL-520 herbicide in sufficient oil or in an oil-water emulsion (see LOW OIL SPRAY TREATMENTS) to make 100 gallons of spray mixture.

Manual Backpack Low Volume: Mix 4 to 6 gallons of BANVEl-520 herbicide in sufficient oil to make 100 gallons of spray mixture. To mix in an oil in water emulsion, see LOW OIL SPRAY TREATMENTS.

Powered Backpack Low Volume: Mix 9 to 18 gallons of BANVEL-520 herbicide in sufficient oil to make 100 gallons of spray mixture. To mix in an oil in water emulsion, see LOW OIL SPRAY TREATMENTS.

LOW OIL SPRAY TREATMENTS

A low oil treatment may be applied at any time of the year for brush control (except when snow or water prevents spraying to the ground line or when extreme cold causes freezing of the spray mixture) as a hydraulic high volume or an air or ground low volume spray treatment. When using low oil treatments, follow the steps of the **W A C O Mixing Sequence listed below:**

1. W – With continuous agitation, add half the desired amount of **water** to the spray tank.
2. A – Add the amount of **ACCUTROL** spray adjuvant required for the spray volume being mixed.
3. C – Add the amount(s) of **chemical(s)** (BANVEL-520 herbicide or BANVEL-520 herbicide and a tank mix product recommended by this label) necessary for the total volume of spray being mixed.
4. O – Add the amount of **oil** required. Finally, add the remaining amount of water needed to bring the spray mixture to the desired volume, maintaining constant agitation in the spray tank throughout mixing. (Note: A premix of BANVEL-520 herbicide and oil may, with constant agitation, be added to the water-Accutrol mix in place of the last two steps of this sequence.)

Low Oil Mixing Table

Means of Application	Gallons of Mix*				
	Water	ACCUTROL®	Chemical	Oil	Total
Hydraulic High Volume	89.4 to 78.5	0.5 0.5	3 to 6	10 to 15	— 100
Air or Ground Low Volume	7.5 28.5	0.5 0.5	1 6	1 5	10 40

* refer to page 9.
* Adjust volumes of spray components proportionately when mixing volumes other than those listed on page 8. Do not apply more than 6 gallons BANVEL-520 herbicide per treated acre per growing season.

TANK MIX TREATMENTS

Read the label of each tank mix product and follow the more restrictive of the precautionary statements, limitations, and directions for use. No label dosage rate may be exceeded. This product must not be mixed with any product whose label prohibits such mixing. Water soluble products may be applied as a low oil spray treatment only. BANVEL-520 herbicide may be tank mixed with the following herbicides:

Compound	Rate lbs. a.i./treated acre
picloram (TORDON®)	2 to 3
triclopyr (GARLON®)	1 to 8
*2, 4-D ester	2 to 6
2, 4-DP ester	2 to 4

* Do not exceed 12 lbs. total 2,4-D acid equivalent per treated acre.

(BANVEL-520 herbicide contains the equivalent of 1.9 lbs./gal. 2,4-D acid.)

Due to the variations that may occur in formulated products and specific use ingredients (e.g. water supplies), a COMPATIBILITY TEST as described on page 10 is recommended prior to actual tank mixing.

TORDON® and GARLON® are registered trademarks of Dow Chemical Co.

COMPATIBILITY TEST

Before mixing in the spray tank, it is advisable to test compatibility by mixing all components in a small container in proportionate quantities.

First, determine the actual application method (low oil or oil treatment). Then, in the following table, find the appropriate vertical column and add the necessary spray components in the order listed. Shake the mixture after each component is added.

Compatibility Mixing Chart for BANVEL-520 Herbicide (Assuming component volume is per 100 gallons spray mix)

	Low Oil Treatment		Oil Treatment	
Spray Component	Gallons	Level Teaspoons	Gallons	Level Teaspoons
Carrier	one quart water		one quart oil	
Herbicide(s)	3 6 18	3 6 18	3 6 18	3 6 18
Accutrol*	½	1	—	—
Oil	10 15	10 15	— —	— —

* or other suitable emulsifier.

The complete test mixture should not separate into layers for at least 5 minutes. Should separation occur earlier, repeat the test by decreasing the amount of oil, by increasing the amount of Accutrol spray adjuvant or by adding a compatibility agent such as Surfel® spray adjuvant (trademark of Union Carbide Agricultural Products Co., Inc.). ¼ teaspoon of compatibility agent is equivalent to 2 pints per 100 gallons of spray solution.

THIS LABELING MUST BE IN THE POSSESSION OF THE USER AT THE TIME OF THE PESTICIDE APPLICATION.

NOTICE: READ "LIMITED WARRANTY AND LIABILITY" ON THE CONTAINER BEFORE BUYING OR USING. IF TERMS ARE NOT ACCEPTABLE, RETURN AT ONCE UNOPENED.

BANVEL®-720

WATER SOLUBLE INDUSTRIAL HERBICIDE

For Brush and Broadleaf Weed Control

ACTIVE INGREDIENTS:

Dimethylamine salt of dicamba (3,6-dichloro-o-anisic acid)* 12.82%

Dimethylamine salts of related acids 3.19%

Dimethylamine salt of 2,4 dichlorophenoxyacetic acid** 24.58%

INERT INGREDIENTS: 59.41%

TOTAL 100.00%

*Equivalent to 10.6% 3,6-dichloro-o-anisic acid

**Equivalent to 20.4% 2,4-dichlorophenoxyacetic acid

**Isomer specific by AOAC method 6 DO1-5

Contains 1 Pound Dicamba and 1.9** Pounds 2,4-D Per Gallon

E.P.A. Reg. No. 876-177-AA E.P.A. Est. No. 876-TX-2

CAUTION

Keep Out of Reach of Children

See Below for Additional Precautionary Statements

BEFORE USING BANVEL-720 WATER SOLUBLE INDUSTRIAL HERBICIDE READ AND FOLLOW THE PRECAUTIONS APPEARING ON THE CONTAINER.

LIMITED WARRANTY AND LIABILITY

NOTICE: Read this Limited Warranty and Liability before buying or using this product. If the terms are not acceptable, return it at once unopened.

It is critical that this product be used and mixed only as specified on the label. The laws of a State may make some or all of this paragraph inapplicable or may give you rights in addition to your rights hereunder. Except to the extent prohibited by applicable law, the exclusive remedy of the user or buyer and the limit of liability of this Company or any other Seller for any and all losses, personal injuries or damages resulting from the use of this product, shall be the purchase price paid by the user or buyer for the quantity of product involved. Except to the extent prohibited by State law, there is no warranty, and this Company and other Sellers disclaim all liability for losses, personal injury or damages: (i) arising from any use of this product in a manner or for a purpose not recommended in its label directions, or from mixing this product before use with any substance except as recommended by the product's label, (ii) arising from handling or storage in violation of label instructions, (ii) for all indirect special or consequential damages, (iv) when not reported to this company within one year of discovery, and (v) arising from product not used within the label-designated shelf life or four years from date of purchase, whichever first occurs. THERE ARE NO IMPLIED WARRANTIES AND NO WARRANTIES OF MERCHANTABILITY OF FITNESS.

PRECAUTIONARY STATEMENTS

HAZARDS TO HUMANS AND TO DOMESTIC ANIMALS

CAUTION

Harmful if swallowed. Avoid contact with skin, eyes, and clothing. In case of contact, wash exposed skin with plenty of soap and water; flush eyes with plenty of water for at least 15 minutes and get medical attention. Keep out of reach of children.

If swallowed, drink 1 or 2 glasses of water and induce vomiting by sticking finger down back of throat. Never induce vomiting or give anything by mouth to an unconscious person.

ENVIRONMENTAL HAZARDS

Do not apply directly to water. Do not contaminate water by cleaning of equipment or disposal of wastes.

STORAGE AND DISPOSAL

Do not contaminate water, food or feed by storage or disposal.

PESTICIDE DISPOSAL

Pesticide, spray mixture or rinsate that cannot be used according to label instructions must be disposed of according to Federal, State or local procedures under Subtitle C of the Resource Conservation and Recovery Act.

CONTAINER DISPOSAL

METAL Triple rinse (or equivalent). Then offer for recycling or reconditioning, or puncture and dispose of in a sanitary landfill, or by other approved state and local procedures.

PLASTIC Triple rinse (or equivalent). Then offer for recycling or reconditioning, or dispose of in a sanitary landfill, or by incineration, if allowed by state and local authorities.

IMPORTANT

The following directions apply to all uses of BANVEL-720 herbicide.

Do not contaminate irrigation ditches or water used for domestic purposes.

SENSITIVE CROP PRECAUTIONS: BANVEL-720 herbicide may cause injury to desirable trees and plants, particularly beans, cotton, flowers, fruit trees, grapes, ornamentals, peas, potatoes, soybeans, sunflowers, tobacco, tomatoes and other broadleaf plants when contacting their roots, stems or foliage. These plants are most sensitive to BANVEL-720 herbicide during their development or growing stage. FOLLOW THE PRECAUTIONS LISTED BELOW WHEN USING BANVEL-720 HERBICIDE.

Do not treat areas where either possible downward movement into the soil or surface washing may cause contact of BANVEL-720 herbicide with the roots of desirable plants such as trees and shrubs.

Avoid making application when spray particles may be carried by air currents to areas where sensitive crops and other desirable plants are growing. Always make application when there is some air movement in order to determine the direction and distance of possible spray drift. Leave an adequate buffer zone between the area to be treated and desirable sensitive plants. Coarse sprays are less likely to drift out of target area.

Do not use aerial equipment to apply BANVEL-720 herbicide when sensitive crops and plants are growing in the vicinity of area to be treated. Broadcast applications of BANVEL-720 herbicide made to conifer seedlings being grown as a crop or to established conifer plantations may result in injury.

Select nozzles designed to produce minimal amounts of fine spray particles. Drift control agents and equipment designed to minimize fine spray particles may be used.

Do not graze dairy animals on treated areas within 7 days after application.

To avoid injury to desirable plants, equipment used to apply BANVEL-720 herbicide should be thoroughly cleaned (see PROCEDURE FOR CLEANING SPRAY EQUIPMENT) before reusing to apply any other chemicals.

Consult your local or state authorities for possible application restrictions and advice concerning these and other special local use situations.

BANVEL-720 Water Soluble Industrial Herbicide For Brush and Broadleaf Weed Control is recommended for use on rights-of-way (utility, railroad, highway, pipeline), nonselective forest brush control (including site preparation), fence rows, drainage ditch banks, wasteland and similar non-cropland areas.

MIXING AND APPLICATION

BANVEL-720 herbicide is a water soluble formulation which can be applied using water or oil/water emulsions as the carrier. Herbicide adjuvants or other spray additives (emulsifiers, surfactants, wetting agents, drift control agents, or penetrants) may be used for wetting, penetration, or drift control. Read and follow all use recommendations and precautions on the label(s) of product(s) used. A COMPATIBILITY TEST should be made prior to tank mixing.

BANVEL-720 herbicide may be applied with ground or aerial equipment for broadcast treatments, or with handgun, backpack, or similar types of equipment for treatment of individual clumps or small areas (spot treatments).

NOTE: To prepare oil/water emulsions, half fill spray tank with water. Add a suitable emulsifier. Then, with constant agitation, slowly add the appropriate amount of herbicide followed by the necessary oil (or slowly add a premix of emulsifier and oil). Finally, fill the rest of the tank with water. Maintain vigorous agitation during spray operation to prevent oil and water from forming separate layers.

SPECIES CONTROLLED

When used as directed, BANVEL-720 herbicide will control or suppress many herbacous broadleaf weeds (annual, biennial, and perennial) as well as many unwanted woody plant and vine species. Species controlled include:

ANNUALS

Buckwheat, wild	Knawel
Carpetweed	Lambsquarters
Chickweed	Morningglory
Clover	Mustard
Cocklebur	Pigweed
Daisy, English	Purslane
Henbit	Ragweed
	Smartweed
	Velvetleaf

BIENNIALS

Ragwort, tansy	Thistle, musk

PERENNIALS

Bindweed, field	Ragweed, perennial
Carrot, wild (Queen Anne's lace)	Sorrel, sheep
Dock, curly	Spurge, leafy
Dogfennel	Thistle, Canada
Knapweed, Russian	Toadflax, Dalmatian
Milkweed	

WOODY BRUSH AND VINES

Alder	Oak
Ash	Olive, Russian
Aspen	Persimmon
Basswood	Pine
Beech	Plum, wild
Birch	Poplar
Blackberry	Puncturevine
Cherry	Raspberry
Creeper, Virginia	Redcedar, Eastern
*Creosotebush	Redvine
Cucumbertree	Rose, multiflora
Dogwood	Sagebrush
Elderberry	Sassafras
Elm	Schinus (Florida holly, Brazil peppertree, Christmas-berry)
Gum	Serviceberry
Hawthorn	Snowberry
Hemlock	Spruce
Honeysuckle	Sumac
Ivy, poison	Sycamore
Kudzu	Trumpetcreeper
Locust	Waxmyrtle
Maple	Willow
	Witchhazel
	Yaupon

* Suppression

DIRECTIONS FOR USE

Regardless of the species to be controlled, spray volumes should be high enough to allow for good spray coverage. Make applications when weeds and brush are actively growing. The addition of surfactants can increase control. Biennials are best controlled when treated in the rosette stage. Regrowth may occur on resistant species. Retreatments may be made as needed; however, do not exceed a total of 6 gallons (6 lbs. a.i. dicamba + 12 lbs. a.i. 2,4-D) of BANVEL-720 herbicide per treated acre during a growing season. To control additional weed species, BANVEL-720 herbicide may be tankmixed with any of the products below.

HERBACEOUS BROADLEAF WEED CONTROL: Apply 1 to 4 quarts BANVEL-720 herbicide in 20 to 100 gallons of water per treated acre. One to two quarts of BANVEL-720 herbicide is recommended for annuals, two to four quarts for biennials and easy-to-kill perennials, and four quarts for established perennials. Do not apply more than 3 gallons of product per treated acre.

BRUSH AND VINE CONTROL—High Volume Foliar Spot Applications:

Mix 1 to 3 gallons of BANVEL-720 herbicide in enough water to make 100 gallons of spray mix. Spray volume applied will depend on the size and density of the brush to be treated, but do not apply more than 3 gallons of product per treated acre. Direct the spray to treat all foliage, stems, and root collars to wet.

BRUSH AND VINE CONTROL—Broadcast Applications with Ground Equipment:
Apply 1 to 3 gallons BANVEL-720 herbicide in 20 to 100 gallons of water per treated acre. Spray volume applied will depend on the size and density of the brush to be treated, but do not apply more than 3 gallons of product per treated acre. Spray all foliage, stems, and root collars to wet.

AERIAL APPLICATIONS: Aerial applications may be made to control either herbaceous or woody plants. Apply 1 to 4 quarts BANVEL-720 herbicide (for herbaceous weeds) or 1 to 3 gallons BANVEL-720 herbicide (for woody brush and vines) in 5 to 40 gallons of water per treated acre. Coverage is important, so increase spray volume when treating dense stands of brush or weeds. Do not apply more than 3 gallons of product per treated acre.

TANK MIX TREATMENTS
READ AND FOLLOW THE LABEL OF EACH TANK MIX PRODUCT USED FOR PRECAUTIONARY STATEMENTS, DIRECTIONS FOR USE, AND OTHER RESTRICTIONS.
For broader spectrum control, BANVEL-720 herbicide may be tank mixed with one or more of the following herbicides for non-cropland uses (e.g. railroad, highway, pipeline, etc.) including forest management applications, if permitted by product labeling (e.g. 2,4-D). Add water to the spray tank prior to the addition of the tank mix products. Do not premix concentrates.

Herbicide	Rate (lbs. a.i./treated acre)
Amitrol*	1.8 to 8.0
Asulam (Asulox)	2.92 to 6.68
Atratol	4.77 to 39.96
Bromacil (Hyvar)	1.6 to 24.0
Chlorflurecol (Maintain)	2.0 to 3.0
Chlorsulfuron (Telar)	0.1875 to 2.25 oz***
Dalapon	4.25 to 12.75
Dicamba (Banvel)	0.25 to 8.0****
Diquat	0.5 to 1.0
Diuron (Karmex)	4.0 to 48.0
Fenac (Fenatrol)	4.5 to 18.0
Fosamine ammonium (Krenite)	6.0 to 12.0
Glyphosate (Roundup)	0.75 to 3.75
Hexazinone (Velpar)	0.675 to 10.8
Imazapyr** (Arsenal)	0.5 to 1.5
Limit**	0.625
Maleic hydrazide (Royal Slo-Gro)	2.25 to 4.5
Mefluidide (Embark)	0.25 to 1.0
MSMA	1.0 to 2.475
Paraquat*	0.5
Picloram* (Tordon)	2.0 to 3.0
Simazine* (Princep)	4.8 to 40.0
Sulfometuron methyl (Oust)	0.75 to 9.0 oz***
Tebuthiuron (Spike)	1.0 to 16.0
Triclopyr (Garlon)	0.75 to 12.0
2,4-D	0.475 to 16.92*****
2,4-DP	0.5 to 11.1

Due to the variations that may occur in formulated products and specific use ingredients (e.g. water supplies), a COMPATIBILITY TEST is recommended prior to actual tank mixing.

* Restricted use pesticides limited to certified applicators.
** Imazapyr is the proposed common name for Arsenal. Limit does not have a common name.
*** Telar = 0.1875 to 2.25 oz. ai/A (0.25 to 3.0 oz product/A) Oust = 0.75 to 9.0 oz. ai/A (1.0 to 12.0 oz. product/A)
**** Do not exceed a total of 8 lbs. a.i. dicamba per treated acre per growing season. (BANVEL-720 herbicide contains 1 lb. a.i. dicamba per gallon.)
***** Use of extremely hard water (500 ppm) may cause this product to form a precipitate when used in a tankmix with BANVEL-720 herbicide. Do not exceed 16 lbs. total 2,4-D acid equivalent per treated acre per growing season. (BANVEL-720 herbicide contains the equivalent of 1.9 lbs./gal. 2,4-D acid.)

COMPATIBILITY TEST
Before mixing in the spray tank, it is advisable to test compatibility by mixing all components in a small container in proportionate quantities (see following table).
Amount of Herbicide to Add to One Pint of Spray Carrier
(Assuming Volume is 25 Gallons per Acre)

HERBICIDE FORMULATIONS	RATE PER ACRE	LEVEL TEASPOONS
Dry	1 lb.	1½
Liquid	1 pt.	½

If herbicide(s) do not ball-up or form flakes, sludge, gels, oily films or layers, or other precipitates, then the tested components are compatible. Usually incompatibility in any of the above described forms will occur within 5 minutes after mixing.
If components are incompatible, the use of a compatibility agent is recommended. Rerun the above COMPATIBILITY TEST with a suitable compatibility agent (¼ teaspoon is equivalent to 2 pints per 100 gallons of spray solution).

PROCEDURE FOR CLEANING SPRAY EQUIPMENT

The steps listed below are suggested for thorough cleaning of spray equipment following applications of BANVEL-720 herbicide or tank mixes of BANVEL-720 herbicide plus 2,4-D amine.

1) Hose down thoroughly the inside as well as outside surfaces of equipment while filling the spray tank half full of water. Flush by operating sprayer until the system is purged of the rinse water.
2) Fill tank with water while adding 1 quart of household ammonia for every 25 gallons of water. Operate the pump to circulate the ammonia solution through the sprayer system for 15 to 20 minutes and discharge a small amount of the ammonia solution through the boom and nozzles. Let the solution stand for several hours, preferably overnight.
3) Flush the solution out of the spray tank through the boom.
4) Remove the nozzles and screens and flush the system with two full tanks of water.

The steps listed below are suggested for thorough cleaning of spray equipment used to apply BANVEL-720 herbicide as a tank mix with wettable powders (WP), emulsifiable concentrates (EC), or other types of water-dispersible formulations. BANVEL-720 herbicide tank mixes with water-dispersible formulations require the use of a water/detergent rinse.

5) Complete step 1.
6) Fill tank with water while adding 2 lbs. of detergent for every 40 gallons of water. Operate the pump to circulate the detergent solution through the boom and nozzles. Let the solution stand for several hours, preferably overnight.
7) Flush the detergent solution out of the spray tank through the boom.
8) Repeat step 1, and follow with steps 2, 3, and 4.

REFER TO THE CONTAINER LABEL FOR INSTRUCTIONS CONCERNING DISPOSAL OF WASTE AND CLEANING RINSES.

REGISTERED TRADEMARKS
Arsenal is a registered trademark of American Cyanamid Company.
Princep is a registered trademark of Ciba-Geigy Corporation.
Garlon and Tordon are registered trademarks of Dow Chemical Company.
Hyvar, Karmex, Krenite, Oust, Telar, and Velpar are registered trademarks of E.I. duPont de Nemours & Co., Inc.
Spike is a registered trademark of Elanco Products Company.
Embark is a registered trademark of 3M.
Limit and Roundup are registered trademarks of Monsanto Agricultural Products Company.
Asulox is a registered trademark of Rhone Poulenc Chemical Company.
Fenatrol is a registered trademark of Union Carbide Agricultural Products Co., Inc.
Royal Slo-Gro is a registered trademark of Uniroyal, Inc.
Maintain is a registered trademark of U.S. Borax & Chemical Corporation.
Banvel is a registered trademark of Velsicol Chemical Corporation.

MARKSMAN®

HERBICIDE

ACTIVE INGREDIENTS:

Potassium salt of dicamba*	13.42%
Potassium salt of related acid	3.35%
Atrazine**	22.23%
INERT INGREDIENTS	61.00%
TOTAL	100.00%

* Equivalent to 11.45% 3,6-dichloro-o-anisic acid (dicamba) or 1.1 pounds/gallon.

** Equivalent to 22.23% 2-chloro-4-ethylamino-6-isopropyl/amino-s-triazine (atrazine) or 2.1 pounds/gallon.

E.P.A. Reg. No. 876-451

CAUTION

Keep Out of Reach of Children

DIRECTIONS FOR USE

IT IS A VIOLATION OF FEDERAL LAW TO USE THIS PRODUCT IN A MANNER INCONSISTENT WITH ITS LABELING.

BEFORE USING MARKSMAN HERBICIDE READ AND FOLLOW THE PRECAUTIONS APPEARING ON THE CONTAINER.

LIMITED WARRANTY AND LIABILITY

NOTICE: Read this Limited Warranty and Liability before buying or using this product. If the terms are not acceptable, return it at once unopened.

It is critical that this product be used and mixed only as specified on the label. The laws of a State may make some or all of this paragraph inapplicable or may give you rights in addition to your rights hereunder. Except to the extent prohibited by applicable law, the exclusive remedy of the user or buyer and the limit of liability of this Company or any other Seller for any and all losses, personal injuries or damages resulting from the use of this product, shall be the purchase price paid by the user or buyer for the quantity of product involved. Except to the extent prohibited by State law, there is no warranty, and this Company and other Sellers disclaim all liability for losses, personal injury or damages: (i) arising from any use of this product in a manner or for a purpose not recommended in its label directions, or from mixing this product before use with any substance except as recommended by the product's label, (ii) arising from handling or storage in violation of label instructions, (iii) for all indirect special or consequential damages, (iv) when not reported to this company within one year of discovery, and (v) arising from product not used within the label-designated shelf life or four years from date of purchase, whichever first occurs. THERE ARE NO IMPLIED WARRANTIES AND NO WARRANTIES OF MERCHANTABILITY OR FITNESS.

PRECAUTIONARY STATEMENTS

IMPORTANT

Do not contaminate irrigation ditches or water used for domestic purposes.

Atrazine leaches readily and accepted label rates have been found to result in contamination of water supplies by way of ground-water. Therefore users are advised to avoid use of atrazine in well drained soils, particularly areas having high ground-water tables.

Corn growing under stress conditions such as drought, poor fertility, or foliar damage due to hail, wind or insects, can exhibit various injury symptoms that may be more pronounced if herbicides are applied.

Broadleaf crops are sensitive to Marksman herbicide and contact may result in crop injury. Do not apply when weather conditions favor drift from target areas.

Prior to the ensilage (milk) stage of the crop, do not harvest or graze corn for dairy or beef feed.

Consult your local or state authorities for possible application restrictions and advice concerning these and other special local use situations.

REFER TO THE CONTAINER LABEL FOR INSTRUCTIONS CONCERNING DISPOSAL OF WASTE AND CLEANING RINSES.

MIXING AND APPLICATION

Marksman herbicide is a water-dispersible formulation that can be applied using water or sprayable fluid fertilizer as the carrier. Use of a sprayable fluid fertilizer carrier after the corn emerges may result in crop injury. If a fluid fertilizer is to be used, a compatibility test (see COMPATIBILITY TEST) should be made prior to tank mixing.

Ground or aerial application equipment which will give good spray coverage of weed foliage should be used. However, do not use aerial application equipment if broadleaf crops are growing in the vicinity of the area to be treated.

Before making application, be sure all equipment is clean and not contaminated with any other materials or crop injury or sprayer clogging may result.

Apply 10 to 50 gallons of diluted spray per treated acre when using ground application equipment, or 3 to 10 gallons of diluted spray per treated acre when using aerial application equipment. Use the higher level of the listed spray volumes when treating dense or tall vegetation. Use coarse sprays.

Thoroughly clean equipment following applications of Marksman herbicide. Any tank mix with a water-dispersible formulation will require the use of a water/detergent rinse.

Avoid cultivating treated areas for at least seven days following postemergence application.

BAND TREATMENTS

Marksman herbicide may be applied as a band treatment. Use the formulas below to determine the appropriate rate and volume per treated acre.

$$\frac{\text{Band width in inches}}{\text{Row width in inches}} \times \text{Broadcast RATE per treated acre} = \text{Band RATE per treated acre}$$

$$\frac{\text{Band width in inches}}{\text{Row width in inches}} \times \text{Broadcast VOLUME per treated acre} = \text{Band VOLUME per treated acre}$$

COMPATIBILITY TEST

Before mixing in the spray tank, it is advisable to test compatibility by mixing all components in a small container in proportionate quantities (see following table).

Amount of Herbicide to Add to One Pint of Spray Carrier (Assuming Volume is 25 Gallons per Acre)

Herbicide Formulations	Rate Per Acre	Level Teaspoons
Dry	1 lb.	1½
Liquid	1 pt.	½

If herbicide(s) do not ball-up or form flakes, sludge, gels, oily films or layers, or other precipitates, then the tested spray-mix is compatible. Usually incompatibility in any of the above described forms will occur within 5 minutes after mixing.

If components are incompatible, the use of a compatibility agent is recommended. Rerun the above COMPATIBILITY TEST with a suitable compatibility agent (¼ teaspoon is equivalent to 2 pints per 100 gallons of fluid fertilizer).

WEEDS CONTROLLED

Marksman herbicide, when applied at recommended rates, will control many ANNUAL broadleaf weeds and give growth suppression of many PERENNIAL broadleaf weeds commonly found in corn.

ANNUALS

buckwheat, wild
burcucumber
chickweed, common
clovers *(annual)*
cocklebur, common
cucumber, wild
jimsonweed
knotweed
kochia
ladysthumb
lambsquarters, common
lambsquarters (triazine resistant)
mallow, common
mallow, Venice
morningglory, ivyleaf
morningglory, tall
mustard, wild
mustards, *(yellowtops)*
nightshade, black
pigweed, prostrate
pigweed, redroot *(carelessweed)*
pigweed, rough
pigweed, smooth
pigweed (triazine resistant)
pigweed, tumble
puncturevine
purslane, common
ragweed, common
ragweed, giant *(buffaloweed)*
sicklepod
sida, prickly *(teaweed)*
smartweed, green
smartweed, Pennsylvania
Spanishneedles
spurge, prostrate
sunflower, common (wild)
sunflower, volunteer
tansymustard
thistle, Russian
velvetleaf

PERENNIALS

alfalfa
artichoke, Jerusalem
bindweed, field
bindweed, hedge
clovers *(perennials)*
dandelion, common
dock, broadleaf *(bitterdock)*
dock, curly
dogbane, hemp
horsenettle, Carolina
lespedeza
milkweed, common
smartweed, swamp
thistle, Canada
vetch

RATES AND TIMINGS

Application of Marksman herbicide may be made prior to, during, or after planting, but before the corn exceeds the 5 leaf stage. For best performance, make application when broadleaf weeds are emerged and actively growing. Refer to the table below for specific application rates, based on soil texture and organic matter content.

SOIL TEXTURE	Broadcast Rate Per Treated Acre: 2% or less organic matter	Broadcast Rate Per Treated Acre: more than 2% organic matter
COARSE COILS		
sand, sandy loam and loamy sand	2 pints	3 pints
MEDIUM SOILS		
loam, silt loam, silt, sandy clay and sandy clay loam	3 pints	$3\frac{1}{2}$ pints
FINE SOILS		
silty clay, silty clay loam, clay loam and clay	$3\frac{1}{2}$ pints	$3\frac{1}{2}$ pints

ALL SOILS CONTAINING GREATER THAN 8% ORGANIC MATTER—USE $3\frac{1}{2}$ PINTS PER TREATED ACRE.

The addition of an agriculturally approved surfactant, crop oil, or petroleum oil to the spray mixture may improve weed control. A surfactant or oil may be used with all preplant and preemergence applications of Marksman herbicide. However, do not use oil when making post emergence applications.

OVERLAY (SEQUENTIAL TREATMENTS)

Marksman herbicide may be applied to ground previously treated with one or more of the following herbicides:

Herbicide	Maximum rate per treated acre (lbs. a.i.)
alachlor (Lasso®)	4
Bronco® (alachlor + glyphosate premix)	5
butylate (Sutan +®)	6
cyanazine (Bladex®)	4
EPTC (Eradicane®)	6
glyphosate (Roundup®)	5
metolachlor (Dual®)	3
paraquat	1
pendimethalin (Prowl®)	2
propachlor (Bexton,® Ramrod®)	6

Read and follow the label of each of the above products for precautionary statements, directions for use and other restrictions.

TANK MIX TREATMENTS

Marksman herbicide will control emerged and actively growing broadleaf weeds prior to, during, or after planting but before the corn exceeds the 5 leaf stage. Marksman herbicide may be tank mixed with one or more of the following herbicides for control of grasses or additional broadleaf weeds. Read and follow the label of each tank mix product used for precautionary statements, directions for use, rates and timings, and other restrictions.

Marksman PLUS BLADEX: Application may be made before grasses are $1\frac{1}{2}$ inches tall and the corn is not beyond the four true leaf stage. Use 1.25–4.0 lbs. a.i. Bladex per treated acre for preemergence treatments or 1.25–2.0 lbs. a.i. Bladex per treated acre for postemergence treatments. AFTER CORN EMERGENCE, USE ONLY THE BLADEX 80W FORMULATION.

Marksman PLUS DUAL: Application may be made until grasses reach the two-leaf stage and before corn is greater than 3 inches tall. Applications prior to crop emergence should only be made to fine textured soils containing 2.5% or more organic matter. Use 1.5–2.5 lbs. a.i. Dual per treated acre.

Marksman PLUS LASSO: Application may be made until grasses reach the two-leaf stage and before corn is greater than 3 inches tall. Applications prior to crop emergence should only be made to fine textured soils containing 3% or more organic matter. Use a 1.5–4.0 lbs. a.i. Lasso per treated acre.

Marksman PLUS PARAQUAT: Application may be made to emerged weeds, but before corn emerges. Use 0.25–1.0 lb. a.i. paraquat per treated acre.

Marksman PLUS PROWL: Application may be made immediately after planting, but before grasses and corn emerge. Make applications only on medium or fine textured soils containing greater than $2\frac{1}{2}$% organic matter. Use the 1.0–2.0 lbs. a.i. Prowl per treated acre.

Marksman PLUS ROUNDUP: Applications may be made to emerged weeds, but before corn emerges. Use 1.0–3.0 lbs. a.i. Roundup per treated acre.

Marksman PLUS 2,4–D: Application may be made to emerged and actively growing weeds prior to, during, and after planting but before corn exceeds the 5 leaf stage. Use $\frac{1}{4}$–$\frac{1}{2}$ lb. 2,4–D acid equivalent per treated acre before corn emerges and $\frac{1}{8}$–$\frac{1}{4}$ lb. 2,4–D acid equivalent per treated acre after corn emerges.

REGISTERED TRADEMARKS

Bexton® is a registered trademark of Dow Chemical Company.
Bladex® is a registered trademark of Shell Chemical Company.
Bronco®, Lasso®, Ramrod®, and Roundup® are registered trademarks of Monsanto Agricultural Products Company.
Dual® is a registered trademark of Ciba-Geigy Corporation, Agricultural Division.
Eradicane® and Sutant +® are registered trademarks of Stauffer Chemical Company.
Prowl® is a registered trademark of American Cyanamid Company.

THIS LABELING MUST BE IN THE POSSESSION OF THE USER AT THE TIME OF THE PESTICIDE APPLICATION.

USE PESTICIDES PROPERLY.

PROSPER WITH PESTICIDES BY USING THEM PROPERLY! READ AND FOLLOW LABEL DIRECTIONS.

NOTICE: READ "LIMITED WARRANTY AND LIABILITY" ON THE CONTAINER BEFORE BUYING OR USING. IF TERMS ARE NOT ACCEPTABLE, RETURN AT ONCE UNOPENED.

WEEDMASTER®

HERBICIDE

For Use on Pasture and Rangeland Grasses and Noncropland Areas such as Fencerows, Roadways and Around Farm Buildings

ACTIVE INGREDIENTS:

Dimethylamine salt of dicamba (3,6-dichloro-o-anisic acid)*	12.4%
Dimethylamine salts of related acids	3.1%
Dimethylamine salt of 2,4-dichlorophenoxyacetic acid**	35.7%
INERT INGREDIENTS	48.8%
TOTAL	100.0%

Contains 1 lb. Dicamba and 2.87** lbs. 2,4-D per gallon

* Equivalent to 10.03% 3,6-dichloro-o-anisic acid.

** Equivalent to 29.6% 2,4-dichlorophenoxyacetic acid.

** Isomer specific by AOAC method 6.DO1-5.

E.P.A. Reg. No. 876-203AA E.P.A. Est. No. 876-TX-2

CAUTION

Keep Out of Reach of Children

See below for additional cautions

DO NOT USE IN ANY MANNER NOT SPECIFIED ON THE LABEL.

CONTAINER DISPOSAL

Do not reuse empty container. Wash thoroughly with water and detergent. Discard in a safe place. Do not burn.

CAUTION

Caution: Harmful if swallowed. Avoid breathing spray mist. Avoid contact with skin, eyes or clothing. Wash thoroughly after handling. In case of contact, wash skin with soap and water; for eyes, flush with water for 15 minutes and get medical attention.

LIMITED WARRANTY AND LIABILITY

NOTICE: Read this Limited Warranty and Liability before buying or using this product. If the terms are not acceptable, return it at once unopened.

It is critical that this product be used and mixed only as specified on the label. The laws of a State may make some or all of this paragraph inapplicable or may give you rights in addition to your rights hereunder. Except to the extent prohibited by applicable law, the exclusive remedy of the user or buyer and the limit of liability of this Company or any other Seller for any and all losses, personal injuries or damages resulting from the use of this product, shall be the purchase price paid by the user or buyer for the quantity of product involved. Except to the extent prohibited by State law, there is no warranty, and this Company and other Sellers disclaim all liability for losses, personal injury or damages: (i) arising from any use of this product in a manner or for a purpose not recommended in its label directions, or from mixing this product before use with any substance except as recommended by the product's label, (ii) arising from handling or storage in violation of label instructions, (ii) for all indirect special or consequential damages, (iv) when not reported to this company within one year of discovery, and (v) arising from product not used within the label-designated shelf life or four years from date of purchase, whichever first occurs. THERE ARE NO IMPLIED WARRANTIES AND NO WARRANTIES OF MERCHANTABILITY OR FITNESS.

BEFORE USING WEEDMASTER HERBICIDE READ AND FOLLOW THE PRECAUTIONS APPEARING ON THE CONTAINER

IMPORTANT

The following directions apply to all uses of WEEDMASTER herbicide.

Do not contaminate irrigation ditches or water used for domestic purposes.

SENSITIVE CROP PRECAUTIONS: WEEDMASTER herbicide may cause injury to desirable trees and plants, particularly beans, cotton, flowers, fruit trees, grapes, ornamentals, peas, potatoes, soybeans, sunflowers, tobacco, tomatoes and other broadleaf plants when contacting their roots, stems or foliage. These plants are most sensitive to WEEDMASTER herbicide during their development or growing stage. FOLLOW THE PRECAUTIONS LISTED BELOW WHEN USING WEEDMASTER HERBICIDE.

- Do not treat areas where either possible downward movement into the soil or surface washing may cause contact of WEEDMASTER herbicide with the roots of desirable plants such as trees and shrubs.
- Avoid making applications when spray particles may be carried by air currents to areas where sensitive crops and plants are growing. Do not spray near sensitive plants if wind is gusty or in excess of 5 mph and moving in the direction of nearby sensitive crops. However, always make applications when there is some air movement to determine the direction and distance of possible spray drift. Leave an adequte buffer zone between area to be treated and sensitive plants. Coarse sprays are less likely to drift out of the target area than fine sprays. Agriculturally approved drift-reducing additives may be used.
- Do not use aerial equipment to apply WEEDMASTER herbicide when sensitive crops and plants are growing in the vicinity of area to be treated.
- To avoid injury to desirable plants, equipment used to apply WEEDMASTER herbicide should be thoroughly cleaned (see PROCEDURE FOR CLEANING SPRAY EQUIPMENT) before reusing to apply any other chemicals.

Crops growing under stress conditions such as drought, poor fertility, or foliar damage due to hail, wind or insects, can exhibit various injury symptoms that may be more pronounced if herbicides are applied.

Consult your local or state authorities for possible application restrictions and advice concerning these and other special local use situations.

MIXING AND APPLICATION

WEEDMASTER herbicide is a water-soluble formulation that may be applied using water.

WEEDMASTER herbicide may be applied broadcast using either ground or aerial equipment. When using ground equipment, apply 5 to 40 gallons of diluted spray per treated acre. Volume of spray applied will depend on the height, density, and type of weeds or brush being treated and on the equipment being used. When using aerial equipment, apply 3 to 10 gallons of diluted spray per treated acre.

WEEDMASTER herbicide may be applied to individual clumps or small areas (SPOT TREATMENT) of undesirable vegetation using handgun or similar types of application equipment. Apply diluted sprays to allow complete wetting (up to runoff) of foliage and stems.

Select nozzles designed to produce minimal amounts of fine spray particles. Spray with nozzles as close to the weeds as is practical for good weed coverage.

WEEDMASTER herbicide should not be applied during periods of gusty wind or when wind is in excess of 15 mph as uneven spray coverage may occur.

Avoid disturbing (e.g. cultivating or mowing) treated areas for at least 7 days following application.

PROCEDURE FOR CLEANING SPRAY EQUIPMENT

The steps listed below are suggested for thorough cleaning of spray equipment following applications of WEEDMASTER herbicide.

1) Hose down thoroughly the inside as well as outside surfaces of equipment while filling the spray tank half full of water. Flush by operating sprayer until the system is purged of the rinse water.
2) Fill tank with water while adding 1 quart of household ammonia for every 25 gallons of water. Operate the pump to circulate the ammonia solution through the sprayer system for 15 to 20 minutes and discharge a small amount of the ammonia solution through the boom and nozzles. Let the solution stand for several hours, preferably overnight.
3) Flush the solution out of the spray tank through the boom.
4) Remove the nozzles and screens and flush the system with two full tanks of water.

REFER TO THE CONTAINER LABEL FOR INSTRUCTIONS CONCERNING DISPOSAL OF WASTE AND CLEANING RINSES.

PASTURE RANGELAND AND NONCROPLAND

IMPORTANT

Do not use on bent, susceptible grass pastures, lespedeza, wild winter peas, vetch, clover and alfalfa pastures as injury will occur. For pasture renovation, wait 3 weeks per quart per acre of WEEDMASTER herbicide used before interseeding or injury may occur.

If grasses are grown for seed or for seed-down purposes, do not apply when grass is in the boot to milk stage. Do not apply to newly seeded areas or after heading begins.

Do not graze meat animals in treated areas within 30 days of slaughter.

Do not graze lactating dairy animals within 7 days of treatment.

Treated-grasses may be harvested for dry hay, however, do not harvest within 37 days of treatment.

Do not use seed from treated grasses for feed or food purposes.

Adjuvants or other spray additives (emulsifiers, surfactants, wetting agents, drift control agents, or penetrants) may be used for wetting, penetration, or drift control. Spray additives must be agriculturally approved when used in pasture applications. If spray additives are used, read and follow all use recommendations and precautions on product label.

WEEDS CONTROLLED, RATES AND TIMINGS

Application rates and timings of WEEDMASTER herbicide are given below. When perennial weeds are reaching maturity, mowing and allowing some regrowth will enhance control.

	WEED STAGE/APPLICATION STAGE					
WEED	WEEDMASTER RATE (PINTS/ACRE)					
	0.5	1.0	1.5	2.0	3.0	4.0
beebalm, spotted				pre-bloom	post-bloom	
bindweed, field						actively growing
broomweed	1-3"	3"-branching		branching		after branching
buckeye spp.*					full leaf	
buffalobur				first flower	late flower	mature
burdock spp.		pre-flower				
buttercup spp.		pre-flower		post- flower		
chickweed, common		seedling	1-3"	over 3"		
chicory				rosette	early bolting	
clover, bur			prior to flowering	mature		
cocklebur, common		1-6"	6-12"	12-30"		
coreopsis, plains		1-6"		prior to flowering		
croton, woolly (doveweed, goatweed)	1-4"	4-12"	12-30"			
dandelion, common		rosette		bolting		
dewberry, southern*						spring or fall
dock, curly			prior to bolting		after bolting	
dogfennel (cypressweed, rosinweed)				10-15"		
elderberry**						actively growing
eupatorium, late (thoroughwort)				10-15"		
fleabane, annual		1-4"	4-8"	>8"		
goldenrod, Missouri				3" to flower		
goldenweed, common						actively growing
groundsel, Texas (yellowtop)		rosette	post bolting			
henbit			pre-flower		flower	
honeysuckle, hairy					spring or fall	
horsenettle, Carolina*						full flower
ivy, poison				spring or fall		
knapweed, black**						actively growing
knapweed, Russian**						actively growing
knapweed, spotted						actively growing
knotweed spp.		<3" runners		>3" runners		
kochia		1-6"	6-10"	10-20"		
lambsquarters, common		1-6"	6-10"	10-20"		
morningglory, tall		pre-flower		post-flower		
mustards, annual			rosette	early bolt		

WEED STAGE/APPLICATION STAGE

WEED	WEEDMASTER RATE (PINTS/ACRE) 0.5	1.0	1.5	2.0	3.0	4.0
nightshade, silverleaf*						full flower early berry
pennycress, field				rosette		
pepperweed, Virginia		1-3"	3-6"	after branching		full leaf
persimmon, Eastern*						actively growing
pigweed, redroot (carelessweed)		< 3"	3-10"	> 10"		
poorjoe		prior to flowering		mature		
privet**						actively growing
rabbitbrush**						actively growing
ragweed, common	1-3"	3-6"	6-10"	> 10"		
ragwort, tansy				rosette		
redvine**						actively growing
sage, fringed**						actively growing
sneezeweed, bitter (bitterweed)		1-4"	prior to flowering	flower		
smartweed, Pennsylvania		< 4"				
smartweed, perennial						actively growing
sorrel, red (sour dock, sheep sorrel)		rosette	bolting	flower		
sowthistle, perennial**						actively growing
spurge, leafy**						actively growing
sunflower		1-3"	3-6"	6-24"		
thistle, bull			rosette	bolting		
thistle, Canada**						actively growing
thistle, musk			rosette	bolting		
thistle, plumeless			rosette	bolting		
velvetleaf		< 6"	6-20"	> 20"		
vetch, hairy		1-4"	4-8"	8" full flower		
yankeeweed (rosinweed)				10-18"		

* may require repeat applications.
** recommended rate will provide Top Growth Suppression only.

BETWEEN CROPPING APPLICATION (BCA) FOR BROADLEAF WEED CONTROL IN CROPLAND ROTATED TO WHEAT

IMPORTANT

Before using this product, read and carefully observe all applicable directions, restrictions, and precautionary statements on the EPA-registered container label and the Weedmaster Herbicide Directions For Use booklet.

WEEDS CONTROLLED

Weedmaster herbicide, when applied at the recommended rates, will control the ANNUAL weeds and suppress the BIENNIAL and PERENNIAL weeds listed below.

ANNUALS
buckwheat, wild
cockle, cow
cocklebur, common
knotweed
kochia
lambsquarters, common
mallow, common
mustards
nightshade, black
pigweed, redroot (careless-weed)
pigweed, rough
purslane, common
ragweed, common
smartweed, Pennsylvania
sowthistle, annual
sunflower
tansymustard
thistle, Russian
velvetleaf

BIENNIALS
carrot, wild
ragwort, tansy
thistle, bull
thistle, musk
thistle, plumeless
star thistle, yellow

PERENNIALS
bindweed, field
dock, curly
thistle, Canada

RATES AND TIMINGS

Application may be made to fallow land, wheat stubble or land to be rotated to wheat. Application should be made to emerged and actively growing weeds. Use higher rate when treating dense vegetative growth. Avoid disturbing treated areas for seven days following application.

Wheat injury may occur if the interval between application and planting is less than 10 days for each pint per acre of Weedmaster herbicide used. Exclude days when ground is frozen.

Weed Type & Stage	Broadcast Rate per Treated Acre Amount
Annual	
small, actively growing (less than 4 inches)	1–1½ pints
established weed growth (greater than 4 inches)	1½–3 pints
Biennial	
rosette diameter	
3 inches or less	1½–2 pints
3 inches or more	2–4 pints
greater than 4 inches, tillering, bolted or flowering	4 pints
Perennial	
suppression or top growth control	2–4 pints
seasonal control	4–8 pints

Add 0.5% v/v of an agriculturally approved surfactant to Weedmaster herbicide when used alone or in a tank mix. The addition of surfactant will enhance spray coverage and the herbicide's penetration and weed foliage.

Retreatments may be made as needed; however, do not exceed a total of 8 pints of WEEDMASTER herbicide per treated acre.

TANK MIX TREATMENTS

Weedmaster herbicide may be tank mixed with one or more of the following herbicides for control of grasses or additional broadleaf weeds. Read and follow the label of each tank mix product used for precautionary statements, directions for use, rates and timings, weeds controlled, geographic or other restrictions. Add 0.5% v/v of an agriculturally approved surfactant to all tank mixes.

Herbicide	Rate per Treated Acre (lbs. a.i.)
atrazine	½ to 3
chlorsulfuron (Glean®)	0.016 to 0.024
cyanazine	1⅗ to 3⅕
glyphosate (Roundup®)	¼ to 2
metribuzin	⅓ to ¾
paraquat	½ to 1

SMALL GRAINS (NOT UNDERSEEDED TO LEGUMES)

IMPORTANT

If small grains are grown for pasture only, refer to the PASTURE, RANGELAND and NONÇROPLAND section.

Do not graze or harvest for livestock feed prior to crop maturity.

WEEDS CONTROLLED

Weedmaster herbicide, when applied at recommended rates, will control ANNUAL broadleaf weeds commonly found in small grains, such as:

buckwheat, wild	pigweed, rough
*cockle, cow	*purslane, common
*cocklebur, common	ragweed, common
*knotweed	*smartweed, Pennsylvania
kochia	*sowthistle, annual
lambsquarters, common	sunflower
*mallow, common	tansymustard
mustards	thistle, Russian
*nightshade, black	*velvetleaf
pigweed, redroot (careless-weed)	

* Use higher rate for control of noted (*) annuals.

RATES AND TIMINGS

Application of BANVEL herbicide may be made before, during or after planting to emerged and actively growing weeds. See specific crop for timing restrictions. For best performance, make application when weeds are in the 2-3 leaf stage and rosettes are less than 2 inches across. Use the higher level of listed rate ranges when treating more mature weeds or dense vegetative growth.

FALL SEEDED WHEAT

WEEDMASTER HERBICIDE MUST BE APPLIED TO FALL SEEDED WHEAT PRIOR TO THE JOINTING STAGE.

BROADCAST RATE PER TREATED ACRE:

¾–1 pint

TANK MIX TREATMENTS

For control of grasses or additional broadleaf weeds, Weedmaster herbicide may be tank mixed with the following herbicides. Read and follow the label of each tank mix product used for precautionary statements, directions for use, weeds controlled, geographic and other restrictions.

BROADCAST RATE PER TREATED ACRE:

Apply ½–1 pint WEEDMASTER herbicide with:

Herbicide	Amount Product*	lb. a.i.
MCPA	½–¾ pint	¼–⅜
bromoxynil	1–1½ pint	¼–⅜
bromoxynil + MCPA	1 + ½ pint	¼ + ¼
metribuzin		
(Sencor® 75DF, Lexone® DF™)	⅓ lb.	¼
(Sencor® 4, Lexone® 4L)	½ pint	¼
(Lexone® 50WP)	½ lb.	¼
chlorsulfuron		
(Glean® 75 DF)	⅙–½ wt. oz.	0.008–0.024
diuron (Karmex® 80WP)	½–2 lbs.	⅖–1⅗
terbutryn (Igran® 80W)	1½–2¾ lbs.	1⅕–2⅕

* based on 4 pounds per gallon formulations of MCPA and 2 pounds per gallon formulation bromoxynil.

SPRING SEEDED WHEAT

WEEDMASTER HERBICIDE MUST BE APPLIED BEFORE SPRING SEEDED WHEAT EXCEEDS THE 5 LEAF STAGE.

BROADCAST RATE PER TREATED ACRE:

¾–1 pint

TANK MIX TREATMENTS

For control of additional broadleaf weeds, Weedmaster herbicide may be tankmixed with the following herbicides. Read and follow the label of each tank mix product used for precautionary statements, directions for use, weeds controlled, geographic and other restrictions.

BROADCAST RATE PER TREATED ACRE:

Apply ½–1 pint WEEDMASTER herbicide with:

Herbicide	Amount Product*	lb. a.i.
MCPA	½–¾ pint	¼–⅜
bromoxynil	1–1½ pint	¼–⅜
chlorsulfuron (Glean® 75DF)	⅙–½ wt. oz.	0.008–0.024

* based on 4 pounds per gallon formulations of MCPA and 2 pounds per gallon formulation of bromoxynil.

SUGARCANE

IMPORTANT

Consult your local or state authorities for possible application restrictions, especially concerning aerial applications and advice concerning special local use situations.

WEEDS CONTROLLED

Weedmaster herbicide, when applied at recommended rates, will control the ANNUAL and PERENNIAL broadleaf weeds commonly found in sugarcane, such as:

ANNUALS

burclover, California	pigweed, smooth
buttercup (Ranunculus spp.)	purslane, common
chickweed, common	ragweed, common
clovers, annual	shepherdspurse
cocklebur, common	sida, prickly (teaweed)
eveningprimrose, cutleaf	sowthistle, annual
fleabane, annual	spurge, prostrate
geranium, Carolina	
henbit	
lettuce, prickly	
pepperweed, Virginia	
pigweed, prostrate	
pigweed, redroot (careless-weed)	

PERENNIALS

aster spp.	garlic, wild
bindweed, field	goldenrod spp.
chickweed, mouseear	groundsel, cressleaf
dandelion, common	onion, wild
dock, curly	sowthistle, perennial
dogfennel	

APPLICATION DIRECTIONS

DO NOT USE AERIAL APPLICATION EQUIPMENT IF SENSITIVE CROPS ARE GROWING IN THE VICINITY OF THE

AREA TO BE TREATED. When band treatments are desired apply an equivalent rate and volume per acre by using the formula below:

$$\frac{\text{Band width in inches}}{\text{Row width in inches}} \times \text{Broadcast RATE per treated acre} = \text{Band RATE per treated acre}$$

$$\frac{\text{Band width in inches}}{\text{Row width in inches}} \times \text{Broadcast VOLUME per treated acre} = \text{Band VOLUME per treated acre}$$

Application of Weedmaster herbicide can be made any time after the weeds have emerged and are actively growing but prior to the close-in stage of sugarcane. When possible, direct the spray beneath the sugarcane canopy in order to minimize the likelihood of crop injury. The use of directed sprays will also aid in maximizing spray coverage of weed foliage. Application rates and timings are given below. Use the higher level of listed rate ranges when treating dense vegetative growth.

- For control of listed ANNUAL broadleaf weeds, apply 1 quart of Weedmaster herbicide per treated acre.
- For suppression of listed PERENNIALS, apply 1-2 quarts of Weedmaster herbicide per treated acre.
- For control of listed PERENNIAL broadleaf weeds, apply 2-3 quarts of Weedmaster herbicide per treated acre.

Retreatments may be made as needed, however, do not exceed 8 quarts of Weedmaster herbicide per treated acre during a growing season.

TANK MIX TREATMENTS

Weedmaster herbicide may be tank mixed with one or more of the following herbicides for control of grasses or additional broadleaf weeds. Read and follow the label of each tank mix product used for precautionary statements, directions for use, rates and timings, weeds controlled, geographic and other restrictions.

Herbicide	Rates Per Treated Acre (lbs. a.i.)
ametryn (Evik®)	⅖ to 8
asulam (Asulox®)	2 to 3⅓
atrazine	⅖ to 4
dalapon (Dalapon®)	3½ to 8½
fenac (Fenatrol® Sugarcane)	2.7 to 6.0
metribuzin (Sencor®, Lexone®)	0.75 to 3.0
terbacil (Sinbar®)	1.6 to 3.2

REGISTERED TRADEMARKS

Glean®, Lexon®, Karmex®, and Sinbar® are registered trademarks of E.I. duPont de Nemours & Co., Inc.

Roundup® is a registered trademark of Monsanto Agricultural Products Company.

Sencor® is a registered trademark of Mobay Chemical Corporation.

Evik® and Igran® are registered trademarks of Ciba Geigy Corporation, Agricultural Division.

Asulox® is a registered trademark of Rhone-Poulenc Inc.

Dalapan® is a registered trademark of SDS Biotech Corporation.

Fenatrol® Sugarcane is a registered trademark of Union Carbide Agricultural Products Co., Inc.

Vertac Chemical Corporation

Suite 2414
5100 Poplar Avenue
Memphis, TN 38137
(901) 767-6851

BOLLS-EYE®

SODIUM CACODYLATE AND CACODYLIC ACID LIQUID PLUS SURFACTANT

For Use As A Defoliant For Irrigated and Dryland Cotton

ACTIVE INGREDIENTS:	
Sodium Cacodylate	27.38%
Dimethylarsinic Acid (Cacodylic Acid)	4.67%
INERT INGREDIENTS:	67.95%
TOTAL	100.00%

Total Arsenic (as elemental) all in water soluble form 15.41%
Product contains 3.10 lbs. cacodylic acid equivalent per gallon.

EPA Reg. No. 39511-111 EPA Est. No. 39511-MS-1[A], 39511-AR-2[B]

Superscript used corresponds to letter in lot number.

CAUTION

Keep Out of the Reach of Children

Harmful if swallowed. Avoid contact with skin. Avoid breathing spray mist. Wash hands after using. Avoid storage near feed or food products. Keep children and domestic animals off treated areas until this material has been washed into the soil.

STATEMENT OF PRACTICAL TREATMENT
If taken internally, induce vomiting and call physician at once.

ENVIRONMENTAL HAZARDS
Do not contaminate waters used for domestic consumption or by animals, wildlife, and aquatic life or for irrigation purposes. Do not feed treated foliage to livestock or graze treated areas.

For chemical spill, leak, fire, or exposure call toll free 1-800-424-9300

AGRICULTURAL CHEMICAL
Do Not Ship or Store with Food, Feeds, Drugs or Clothing

READ ENTIRE LABEL BEFORE USING THIS PRODUCT.

GENERAL INFORMATION: BOLLS-EYE Cotton Defoliant is used as a harvest aid for the defoliation of cotton. Its phytotoxic properties are quickly inactivated on contact with the soil. It is a combination of a defoliating agent and surfactant. It is desirable to add additional surfactant when using ground application equipment (see directions). None need be added for aerial application.

MIXING INSTRUCTIONS: BOLLS-EYE Cotton Defoliant is completely water soluble. Fill the spray equipment reservoir about half with water and add the required amount of defoliant with agitation. Finish filling the reservoir with water and apply. After use clean equipment thoroughly by flushing with water. Do not store spray solution in tank for a prolonged period.

DIRECTIONS FOR USE

BOLLS-EYE Cotton Defoliant is useful as a harvest aid for the defoliation of irrigated and dryland cotton with aerial or ground application equipment. For the most effective defoliation, good coverage of all leaves is essential.

Nozzles must be arranged to provide thorough coverage of the foliage. If applied by airplane use 5 to 10 gallons of water per acre. This range will generally provide adequate coverage. If applied by ground equipment, a spray volume of 15 to 25 gallons of water per acre is preferred. With ground applicator equipment and at spray volumes of 15 to 25 gallons per acre, add $\frac{1}{3}$ to $\frac{2}{3}$ pints of MULTI-FILM X-77 or similar approved nonionic surfactant.

CONTAINER DISPOSAL: Do not reuse empty container. Wash thoroughly with water and detergent. Crush if possible and discard in a safe place.

CARE OF EQUIPMENT: Although BOLLS-EYE Cotton Defoliant is only moderately corrosive do not apply with any applicator that is lined with zinc, tin, or aluminum.

DEFOLIATION OF DRYLAND COTTON: Apply 2 to 2.5 pints of BOLLS-EYE Cotton Defoliant per acre when 50% or more of the bolls are open (7 to 10 days prior to anticipated picking).

DEFOLIATION OF IRRIGATED COTTON: Apply 2.5 to 3 pints of BOLLS-EYE Cotton Defoliant per acre when 50% or more of the bolls are open (7 to 10 days prior to anticipated picking).

TANK-MIX RECOMMENDATIONS: BOLLS-EYE Cotton Defoliant may be tank mixed with DEF Cotton Defoliant for more consistent defoliation over a wide range of environmental conditions. The combination should be applied when 50% or more of the bolls are open (7 to 10 days prior to anticipated picking). Apply at the rate of $1\frac{1}{2}$–2 pints/A of BOLLS-EYE plus 1–$1\frac{1}{2}$ pints/A of DEF.

WARRANTY —CONDITION OF SALE

DIRECTIONS FOR USE of this product are based on field use and tests believed reliable and should be followed carefully. It is, however, impossible to eliminate all risks associated with use of this product. Because such factors as weather conditions, foreign material and manner of use for application are all beyond the control of Vertac Chemical Corporation or the Seller of this product, such things as crop injury, ineffectiveness or other unintended consequences may result. ALL SUCH RISKS ARE ASSUMED BY THE BUYER.

Vertac Chemical Corporation warrants that this product conforms to the chemical description on the label and is reasonably fit for the purposes referred to in the directions for use as modified by the above. Vertac Chemical Corporation makes no other warranties express or implied including FITNESS or MERCHANTABILITY. In no case shall Vertac Chemical Corporation or the Seller be liable for consequential, special or indirect damages resulting from the use or handling of this product. The foregoing is a condition of sale by Vertac Chemical Corporation and is accepted as such by the Buyer.

PATENT NOTICE: The price of this product includes a royalty for a license to use the contents thereof under the claims of U.S. Patent No. 3,378,364. A license under the said patent is available to anyone who wishes to practice the patented method with materials obtained from other sources.

BROADSIDE®

MSMA, SODIUM CACODYLATE AND CACODYLIC ACID LIQUID PLUS SURFACTANT FOR GENERAL POST EMERGENT WEED CONTROL

ACTIVE INGREDIENTS:	
Monosodium Acid Methanearsonate	26.02%
Sodium Cacodylate	10.47%
Dimethylarsinic Acid (Cacodylic Acid)	1.80%
INERT INGREDIENTS:	61.71%
TOTAL	100.00%

Total Arsenic (as elemental) all in water soluble form 17.92%
Product contains 3.00 lbs. MSMA and 1.25 lbs. cacodylic acid equivalent per gallon.

EPA Reg. No. 39511-109 EPA Est. No. 39511-MS-1[A], 39511-AR-2[B]

Superscript used corresponds to letter in lot number.

CAUTION

Keep Out of the Reach of Children

Harmful if swallowed. Avoid contact with skin. Avoid breathing spray mist. Wash hands after using. Avoid storage near feed or food products. Keep children and domestic animals off treated areas until this material has been washed into the soil.

STATEMENT OF PRACTICAL TREATMENT

If taken internally, induce vomiting and call physician at once.

ENVIRONMENTAL HAZARDS

Do not contaminate waters used for domestic consumption, or by animals, wildlife and aquatic life, or for irrigation purposes. Avoid spray contact with foliage on fruit or food crops and ornamentals. Do not pasture livestock on treated areas.

For chemical spill, leak, fire, or exposure call toll free 1-800-424-9300

AGRICULTURAL CHEMICAL

Do Not Ship or Store with Food, Feeds, Drugs or Clothing

Read entire label before using this product.

GENERAL INFORMATION

BROADSIDE Herbicide is a special formulation of herbicides in combination with a surface-active (wetting) agent, for general post-emergent weed control. It is unnecessary to add any other surfactant to the spray solution. This product is effective on both broadleaf weeds and grasses. The phytotoxic properties of these materials are quickly inactivated on contact with the soil. Best results are obtained on young, actively growing weeds at air temperatures above 70°F.

MIXING INSTRUCTIONS

BROADSIDE Herbicide is completely water soluble. Fill the spray equipment reservoir about half full with water and add the required amount of herbicide with agitation. Finish filling the tank with water and apply. After use, clean equipment thoroughly by flushing with water. Do not store spray solution in tank for a prolonged period. Although BROADSIDE Herbicide is only moderately corrosive do not use in galvanized or aluminum equipment.

DIRECTIONS FOR USE

BROADSIDE Herbicide is a post-emergent useful for controlling broadleaf weeds and grasses, such as puncturevine, wild mustard, wild oats, chickweed, sandbur, common ragweed, pigweed, crabgrass, lambsquarters, common plantain, prostrate spurge, giant foxtail and yellow foxtail, and to top-kill certain perennials such as johnsongrass, dallisgrass and nutsedge.

CONTAINER DISPOSAL

Do not reuse empty container. Perforate, crush and bury in a safe place. Drums may be returned to a drum reconditioner.

BROADSIDE Herbicide is useful for general weed control on drainage ditchbanks, rights-of-way, fence rows, and storage yards, along highways, utility lines and pipe lines, around power plants and buildings and on similar non crop areas.

Use BROADSIDE Herbicide at the rate of 1 to 2 gallons per acre in sufficient water to get full coverage. Ground application normally requires 40 to 100 gallons of spray solution per acre. Any spray equipment that gives good coverage may be used. If regrowth occurs, reapply as required.

BINDWEED (MORNINGGLORY) CONTROL

Use 1½ to 2 gallons BROADSIDE Herbicide in 50 to 100 gallons of water per acre. Make first application at first bloom, wetting foliage just short of run-off. Repeat application as needed for top-kill, but limit to three applications per year.

WARRANTY —CONDITION OF SALE

DIRECTIONS FOR USE of this product are based on field use and tests believed reliable and should be followed carefully. It is, however, impossible to eliminate all risks associated with use of this product. Because such factors as weather conditions, foreign material and manner of use for application are all beyond the control of Vertac Chemical Corporation or the Seller of this product such things as crop injury, ineffectiveness or other unintended consequences may result. ALL SUCH RISKS ARE ASSUMED BY THE BUYER.

Vertac Chemical Corporation warrants that this product conforms to the chemical description on the label and is reasonably fit for the purposes referred to in the directions for use as modified by the above. Vertac Chemical Corporation makes no other warranties, express or implied, including FITNESS OR MERCHANTABILITY. In no case shall Vertac Chemical Corporation or the Seller be liable for consequential, special or indirect damages resulting from the use or handling of this product. The foregoing is a condition of sale by Vertac Chemical Corporation and is accepted as such by the Buyer.

BUTOXONE

HERBICIDE

For the Control of Certain Seedling Broadleaf Weeds in Peanuts, Soybeans and seedling stands of Alfalfa, Birdsfoot Trefoil, Red Clover, Alsike, White and Ladino Clovers, and for use on Established Stands of Alfalfa.

ACTIVE INGREDIENT:	
Dimethylamine salt of 4-(2,4-dichlorophenoxy) butyric acid*	23.0%
INERT INGREDIENTS:	77.0%
TOTAL	100.0%

* Equivalent to 19.5% by weight of 4-(2,4-dichlorophenoxy) butyric acid, or not less than 1.75 lbs. 4-(2,4-DB) acid per gal. at 68°F.

E.P.A. Reg. No. 39511-122 E.P.A. Est. No. 359-MO-1

CAUTION

Keep Out of Reach of Children

STATEMENT OF PRACTICAL TREATMENT

IF SWALLOWED: Drink 1 or 2 glasses of water and induce vomiting by touching back of throat with finger. Do not give anything by mouth to an unconscious person. Get medical attention.

IF ON SKIN: Wash with plenty of soap and water. Get medical attention if irritation persists.

IF IN EYES: Flush with water for 15 minutes. Get medical attention.

See Side Panel for Additional Precautionary Statements

PRECAUTIONARY STATEMENTS

HAZARDS TO HUMANS AND DOMESTIC ANIMALS

CAUTION

Harmful if swallowed. Wash thoroughly after handling. Avoid contact with skin, eyes, or clothing. Avoid breathing spray mist.

ENVIRONMENTAL HAZARDS

Keep out of lakes, streams, or ponds. Do not apply when weather conditions favor drift from target area.

DIRECTIONS FOR USE

It is a violation of Federal Law to use this product in a manner inconsistent with its labeling.

STORAGE AND DISPOSAL

STORAGE

Do not contaminate water, food or feed by storage or disposal, or cleaning of equipment. Open dumping is prohibited. Do not reuse empty container. Do not store near fertilizers, other pesticides, or seeds. Store at temperatures above 32°F. If allowed to freeze remix before using.

PESTICIDE DISPOSAL

Pesticide, spray mixture, or rinsate that cannot be used or chemically reprocessed should be disposed of in a landfill approved for pesticides or buried in a safe place away from water supplies.

CONTAINER DISPOSAL

Triple rinse (or equivalent) or dispose in approved landfill or bury in a safe place. Do not burn.

Consult Federal, State or Local Disposal Authorities for approved alternative procedures.

Apply this product only as specified on this label.

NOTE: Avoid spray drift to susceptible plants and crops; cotton, tobacco, tomatoes, ornamentals, etc. Coarse sprays are less likely to drift.

Spray equipment previously used to apply another phenoxy must be thoroughly cleaned with alkali and water prior to use with BUTOXONE. Similarly after using BUTOXONE wash sprayer thoroughly before using again to spray susceptible crops. Do not use near greenhouses.

Observe all caution and limitations on labeling of all products used in mixtures.

Use agitation to keep emulsion well mixed, especially if emulsion has been allowed to stand.

Calibrate sprayer accurately.

DO NOT APPLY WHEN CROP IS STRESSED FROM LACK OF MOISTURE.

ALFALFA, BIRDSFOOT TREFOIL AND CLOVERS

APPLICATION

For use in seedling alfalfa, birdsfoot trefoil, and clovers spray when the crop has reached the 1 to 2 trifoliate leaf stage and growing conditions are good.

In established alfalfa certain weeds will emerge in the fall and over winter in a rosette stage. Best control of these weeds will result from application in late fall or early winter rather than in the spring. Do not apply after flowering.

WHEN TO APPLY

Weeds must be in the young seedling stage and actively growing to achieve satisfactory results. For best results spray weeds in the 2 to 5 leaf stage of growth.

AMOUNT TO USE

Apply BUTOXONE as an overall spray by ground sprayer or airplane. Apply at rates listed below according to weed problems. Use the higher rate if weeds are past the seedling stage. Aerial applications should be made in 10 gallons of water per acre, while ground sprayer applications should be in 20 to 30 gallons of water per acre. Spray gallonage should give adequate coverage of the weeds without run-off. The use of an approved non-ionic surfactant for agricultural uses (such as X-77®, Surfactant WK, etc.) at their label rates will usually result in better weed control, especially if weeds are beyond the seedling stage. Do not use with surfactant in the desert areas of Arizona and California.

X-77® Registered trademark of Kalo Laboratories Inc.

Broadleaf Weeds	Rate of BUTOXONE per Acre
Annual Morningglory (*Ipomoea* spp.) Cocklebur (*Xanthium* spp.) Common Lambsquarters (*Chenopodium album*) Jimsonweed (*Datura stramonium*) Kochia or Mexican Fireweed (*Kochia scoparia*) Pigweed (*Amaranthus* spp.) Velvetleaf (*Abutilon theophrasti*) Wild Turnip (*Brassica campestris*)	4⅓ pints
Black Mustard (*Brassica nigra*) Buckhorn Plantain (*Plantago lanceolata*) Common Ragweed (*Ambrosia artemisiifolia*)	6½ pints
Curly Dock (*Rumex cripus*) Field Pennycress (Fanweed or Stinkweed) (*Thiaspi arvense*) Hedge Smartweed (*polygonum scandens*) Ladysthumb (*Polygonum persicaria*) Prickly Lettuce (*Lactuca serriola*) Shepherdspurse (*Capsella bursa-pastoris*) Sweet Clover (Volunteer plants) (*Melitotus* spp.) Wild Beet (*Beta maritina*) Wild Mustard (*Brassica Kaber*)	6½ pints

Do not graze established alfalfa, nor feed straw or hay from established alfalfa to livestock within 30 days after application. Do not graze or feed SEEDLING Alfalfa, Clover, and Birdsfoot Trefoil within 60 days after application. Do not use on Sweet Clover and other legumes not mentioned, nor on established clovers grown for seed. Irrigation, in particular, overhead sprinkler irrigation, should be delayed as long as possible (10 days or more) following application of BUTOXONE, to avoid washing the chemical into the root zone. Do not apply when crop is stressed from lack of moisture. Do NOT use on peas.

Apply as a postemergence spray. When properly timed there is little or no effect on the crop. In established Alfalfa there may be some twisting of stems and malformation of leaves. This condition is usually outgrown.

Do not spray when daytime temperatures are expected to exceed 90°F. within the next 2 or 3 days. Do not apply if temperatures are likely to fall below 40°F. during or shortly after treatment.

PEANUTS

APPLICATION

Apply BUTOXONE postemergence as an overall spray by ground sprayer or airplane. With ground applications use a boom sprayer with flat fan-type nozzles. Adjust the height of the boom above the crop so the growing terminals of all weeds are sprayed. Use low pressure (about 20 to 40 lbs. per square inch). Do not use boom jet nozzles. With airplane applications BUTOXONE has been successfully applied using 5 gallons of total solution per acre. The height of the application should insure **thorough coverage** of the tops of the weeds. Best control has resulted from airplane application with coarse spray when cocklebur is at least crop high and actively growing. Other weed species should be in the seedling stage and actively growing. Airplane application is not recommended in the vicinity of susceptible crops.

Do not apply if peanuts are suffering from lack of moisture.

AMOUNT TO USE

Apply the recommended rates as indicated in the table below.

Broadleaf Weeds	Rate of BUTOXONE per Acre
Cocklebur (*Xanthium* spp.)	1 pint
Annual Morningglory (*Ipomoea spp.*) Common Lambsquarters (*Chenopodlum album*) Common Ragweed (*Ambrosia artemisiifolia*) Jimsonweed (*Datura stramonium*) Sicklepod (Coffeeweed) (*Cassia obtusifolia*) Velvetleaf (*Abutilon theophrasti*)	1$^3/_4$ pints

Do not apply more than two applications per season; the second application should be made no later than the late bloom stage of peanuts (about 90 to 100 days after planting). Do not feed treated peanut vines and peanut hay to livestock. Do not apply to peanuts within 30 days of harvest.

SOYBEANS

APPLICATION

Apply BUTOXONE postemergence as an over-all spray by ground sprayer or airplane. With ground applications use a boom sprayer with flat fan-type nozzles. Adjust the height of the boom above the crop so the growing terminals of all weeds are sprayed. Use low pressure (about 20 to 40 lbs. per square inch). Do not use boom jet nozzles. With airplane applications BUTOXONE has been successfully applied using 5 gallons of total spray solution per acre. The height of the application should insure **thorough coverage** of the tops of the weeds. Best control has resulted when cocklebur is at least crop high and actively growing. Airplane application is not recommended in the vicinity of susceptible crops. Do not apply if soybeans are suffering from lack of moisture.

TOPICAL (OVERHEAD) APPLICATIONS

Apply from 7 to 10 days before soybeans bloom through mid-bloom. A good indication that the beans are about to bloom is when the soybean plants have turned a dark green color. If application is made 7 to 10 days before bloom, apply 1 gallon of BUTOXONE per 10 acres; if application is made from early-bloom through mid-bloom, apply 1 gallon of BUTOXONE per 8 acres. Use sufficient water to obtain adequate coverage. **The rates recommended are for the control of cocklebur.** These rates will also stunt or partially control certain other weeds including annual morningglory, velvetleaf, and jimsonweed. In the Midwest use only as a directed application since topical application may reduce soybean yields.

DIRECTED APPLICATIONS

For best results apply when weeds do not exceed 3 inches in height. Good coverage of the growing terminals of weeds is essential for effective control. Do not allow spray pattern to contact more than the lower one-third of the soybean plants.

Do not apply more than two applications per season. **Precise application is essential to prevent damage to the crop.** In order to maintain the correct spraying height, nozzles must be mounted on oiling shoes, skid shoes, or on cultivators with gauge wheels. Do not use booms with drop nozzles.

First Application: Apply as a directed spray when soybeans are 8 to 12 inches tall and weeds have emerged.

Second Application: Apply as a directed spray no later than mid-bloom stage.

AMOUNT TO USE

Apply the recommended rates as indicated in the table below.

Broadleaf Weeds	Rate of BUTOXONE per Acre
Cocklebur (*Xanthium* spp.)	1 pint
Annual Morningglory (*Ipomoea* spp.) Common Lambsquarters (*Chenopodium album*) Common Ragweed (*Ambrosia artemisiifolia*) Jimsonweed (*Datura stramonium*) Sicklepod (Coffeeweed) (*Cassia obtusifolia*) Velvetleaf (*Abutilon theophrasti*)	1$^3/_4$ pints

While some temporary twisting of soybean plants may follow an application, this will have little or no permanent effect on the crop. Applications made under adverse growing conditions, such as periods of drought, will result in injury to the beans; under such conditions the crop should not be sprayed.

Do not spray BUTOXONE on soybeans showing a general infestation of **Phytophthora.**

Do not apply more than the recommended amount.

Do not harvest soybeans within 60 days after application.

Do not use on soybeans grown West of the Rocky Mountains.

DIRECTED BAND APPLICATIONS OF MIXTURES OF BUTOXONE AND LOROX 50W*

APPLICATION

Apply directed spray when beans are at least 8 inches tall and when weeds do not exceed 2 inches in height. Direct spray to cover weed foliage with minimum contact of the soybean plant. Do not spray higher than 3 inches on the soybean stem or crop injury may result.

Use an 8002 T-Jet (or equivalent) or larger with a minimum of 25 gals. per broadcast acre and spray pressure of 20 to 25 psi, to avoid spray drift. For each 25 gals. of spray mixture a pint of surfactant such as Surfactant WK may be added. Continuous agitation in the spray tank is required to keep the material in suspension. Agitate by mechanical or hydraulic means; if by-pass or return line is used, it should terminate at the bottom of the tank to minimize foaming. Openings in nozzle screens should be equal to or larger than 50 mesh. Apply with sprayer nozzles mounted on skid shoes, oiling shoes, or on cultivators with gauge wheels. Do not mount on booms with drop nozzles or on cultivators without gauge wheels.

AMOUNT TO USE:

Apply the recommended rates as indicated in the table below.

Broadleaf Weeds	Rates of BUTOXONE and LOROX per Broadcast Acre
Annual Morningglory (*Ipomoea* spp.) Cocklebur (*Xanthium* spp.) Common Lambsquarters (*Chenopodium album*) Common Ragweed (*Ambrosia artemisiifolia*) Crabgrass (2 leaf) (*Digitaria* spp.) Jimsonweed (*Datura stramonium*) Prickly Sida (Teaweed) (*Sida spinosa*) Sesbania (*Sesbania* spp.) Velvetleaf (*Abutilon theophrasti*)	1 pint + 1 pound

Use a proportional amount of the two chemicals on the band (for example, on a 12 inch band in 36 inch rows use ⅓ pint of BUTOXONE and ⅓ lb. of Lorox per acre).

If a new flush of weeds occurs after initial treatment, make a second application but do not apply within 60 days of harvest or make more than 2 applications per growing season. **Do not allow spray or spray drift to contact growing terminals of beans, as excessive crop injury will result.** Do not feed soybean forage or hay to livestock from fields treated with this mixture. Harvested soybeans may be used for feed or oil purposes.

NOTE: Do not use on sands, loamy sands, gravelly soils, or on exposed sub-soils. Do not use on soils containing less than ½% organic matter as crop injury may result. Observe all cautions and limitations on labeling of all products used in mixtures.

*Lorox 50W—registered trademark of E.I. duPont de Nemours & Co. (Inc.)

CONDITION OF SALE, WARRANTY, LIMITATION OF LIABILITY

This product conforms to the chemical description on the label thereof and is reasonably fit for the purpose stated on such label only when used in accordance with directions under normal use conditions. Follow directions carefully. Timing and method of application, weather and crop conditions, mixture with other chemicals not specifically recommended and other influencing factors in the use of this product are beyond the control of Vertac or the seller. Buyer assumes all risks of use, storage, or handling of this material, not in strict accordance with directions given herewith. In no case shall Vertac or the seller be liable for consequential, special, or indirect damages such as loss of profits or values resulting from the use or handling of this product.

BUTOXONE® Ester

SELECTIVE WEED KILLER

For the control of certain seedling broadleaf weeds in seedling stands of Alfalfa, Birdsfoot Trefoil, Red Clover, Alsike, White and Ladino Clovers.

ACTIVE INGREDIENTS:

Isooctyl ester of 4-(2,4-dichlorophenoxy) butyric acid*	38.6%
INERT INGREDIENTS:	61.4%
TOTAL	100.0%

* Equivalent to 26.64% by weight of 4-(2,4-dichlorophenoxy) butyric acid or 2 pounds of 4-(2,4-dichlorophenoxy) butyric acid per gallon at 68° F.

E.P.A. Reg. No. 39511-118 E.P.A. Est. No. 359-MO-1
U.S. Patent No. 2,863,754

CAUTION

Keep Out of Reach of Children

IF SWALLOWED: Call a physician or Poison Control Center. Drink 1 or 2 glasses of water and induce vomiting by touching back of throat with finger. Do not induce vomiting or give anything to an unconscious person.

IF ON SKIN: Wash with plenty of soap and water. Get medical attention if irritation persists.

IF IN EYES: Flush with plenty of water. Get medical attention if irritation persists.

PRECAUTIONARY STATEMENTS

CAUTION

HAZARDS TO HUMANS AND DOMESTIC ANIMALS

Harmful if swallowed or absorbed through skin. Avoid contact with skin, eyes, or clothing. Wash thoroughly with soap and water after handling.

ENVIRONMENTAL HAZARDS

Do not apply directly to lakes, streams or ponds. Do not apply when weather conditions favor drift from treated area. Do not contaminate water by cleaning of equipment or disposal of wastes.

PHYSICAL AND CHEMICAL HAZARDS

Do not use or store near heat or open flame.

DIRECTIONS FOR USE

It is a violation of Federal Law to use this product in a manner inconsistent with its labeling.

STORAGE AND DISPOSAL

Do not contaminate water, food, or feed by storage or disposal.

PESTICIDE DISPOSAL

Pesticide, spray mixture, or rinsate that cannot be used according to label instructions must be disposed of according to Federal, State or local procedures under the Resource Conservation and Recovery Act.

CONTAINER DISPOSAL

Triple rinse (or equivalent), puncture and dispose of in a sanitary landfill or by other approved State and local procedures.

Do not apply when crop is stressed from lack of moisture.

Do not graze or feed SEEDLING Alfalfa, Clover, and Birdsfoot Trefoil within 60 days after application. Do not use on Sweet Clover and other legumes not mentioned, nor on established clovers grown for seed. Irrigation, in particular, overhead sprinkler irrigation, should be delayed as long as possible (10 days or more) following application of BUTOXONE ESTER, to avoid washing the chemical into the root zone. Avoid spray drift to susceptible plants and crops as it may injure cotton, tomatoes and ornamentals (coarse sprays are less likely to drift). Vapors from this product may injure susceptible plants in the vicinity. Do not use near greenhouses. Do not use same spray equipment for other purposes without first thoroughly cleaning with alkali and water. Small amount of 2,4-D or MCPA may damage legumes; sprayers previously used to apply these chemicals must be thoroughly cleaned with alkali and water before being used for applying BUTOXONE ESTER. Contact your supplier for special instructions on cleaning spray equipment.

GENERAL INFORMATION

Apply as a postemergence spray (read restrictions under "when to apply"). When properly timed there is little or no effect on the crop.

CONTROLS THESE WEEDS

At the recommended rates BUTOXONE ESTER will control many seedling broadleaf weeds including those in the following list.

Lambsquarters	*Chenopodium album*
Jimsonweed	*Datura* spp.
Ragweed	*Ambrosia artemisiifolia*
Groundsel	*Senecio vulgaris*
Sweet Clover	*Volunteer plants*
Russian Thistle	*Salsola kali*
Annual Morningglory	*Ipomoea* spp.
Sugar Beet, Wild Beet	*Volunteer plants*
Velvetleaf	*Abutilon theophrasti*
Wild Turnip	*Brassica campestris*
Stinkweed (Fanweed or Penny cress)	*Thlaspi arvense*
Curly Dock	*Rumex crispus*
Bullthistle	*Cirsium lanceolatum*
London Rocket	*Sisymbrium irio*
Hairy Cats Ear	*Hypochoeris radicata*
Narrowleaf Plantain	*Plantago lanceolatum*
Pigweed	*Amaranthus* spp.
Kochia or Mexican Fireweed	*Kochia scoparia*
Shepherdspurse	*Capsella bursa-pastoris*
Wild Mustards	*Brassica kaber, B. nigra*
Prickly Lettuce	*Lactuca scariola*
Tansy Mustard	*Descurainia pinnata*
Cocklebur	*Xanthium* spp.
Smartweed	*Polygonum persicaria*

DIRECTION FOR USE (First Read Cautions)

WHEN TO APPLY

Weeds must be in the young seedling stage and actively growing to achieve satisfactory results. The older the weeds become the more difficult they are to control. For best results spray weeds in the 2 to 5 leaf stage of growth. Do not spray when daytime temperatures are expected to exceed 90°F. within the next 2 or 3 days. Do not apply when crop is stressed for lack of moisture. Do not apply if temperatures are likely to fall below 40°F. during or shortly after treatment.

Seedling Legumes—Spray when legume crop has reached the 1 to 2 trifoliate leaf stage and growing conditions are good. DO NOT use on established alfalfa. DO NOT use on peas.

HOW TO APPLY

Read caution regarding spray drift. Apply as an overall spray by ground sprayer or airplane. Use agitation to keep emulsion well mixed, especially if emulsion has been allowed to stand. Calibrate sprayer accurately.

AMOUNT TO USE

Mix in water and apply at the rates shown in the table. Use the higher rate if weeds have passed the seedling stage.

Broadleaf Weeds	Amt. Butoxone Ester Per Acre	4-(2,4-DB) ACID EQUIX Per Acre	Suggested Dilution Per Acre
For Lambsquarters, Pigweed, Kochia, Russian Thistle, Prickly Lettuce, Groundsel, Wild Mustard, Hairy Cats Ear, Cocklebur, Velvetleaf, Wild Beet, Annual Morningglory, Wild Turnip	3 pts.	3/4 lbs.	Airplane Applications: 10 gals. of water or Ground Sprayer Applications: 20 to 30 gals. of water
For Stickweed, Smartweed, Ragweed, Bullthistle, Shepherdspurse, Curly Dock, Tansy Mustard, London Rocket, Narrowleaf Plantain, Volunteer Sweet Clover, Jimsonweed	4 pts.*	1 lb.	

* On Red Clover, do not use more than 3 pints per acre.

CONDITION OF SALE, WARRANTY, LIMITATION OF LIABILITY

This product conforms to the chemical description on the label thereof and is reasonably fit for the purpose stated on such label only when used in accordance with directions under normal use conditions. Follow directions carefully. Timing and method of application, weather and crop conditions, mixture with other chemicals not specifically recommended and other influencing factors in the use of this product are beyond the control of Vertac or the seller. Buyer assumes all risks of use, storage, or handling of this material, not in strict accordance with directions given herewith. In no case shall Vertac or the seller be liable for consequential, special, or indirect damages such as loss of profits or values resulting from the use or handling of this product.

A method for weed control in crop areas is covered by U.S. Patent 2,863,754, under which this Company is authorized to extend licenses to others at a royalty rate equal to Ten Cents per pound of active ingredient used. This royalty is included in the price of the product sold under this label. A license to practice the invention with material from any source at the same rate of royalty may be acquired upon written request to the Company.

BUTOXONE® 200

HERBICIDE

For the Control of Certain Seedling Broadleaf Weeds in Peanuts, Soybeans and seedling stands of Alfalfa, Birdsfoot Trefoil, Red Clover, Alsike, White and Ladino Clovers, and for use on Established Stands of Alfalfa.

ACTIVE INGREDIENT:
Dimethylamine salt of 4-(2,4-dichlorophenoxy) butyric acid* 26.2%
INERT INGREDIENTS: 73.8%
TOTAL 100.0%

* Equivalent to 22.19% by weight of 4-(2,4-dichlorophenoxy) butyric acid, or 2 pounds of 4-(2,4-dichlorophenoxy) butyric acid per gallon at 68° F.

E.P.A. Reg. No. 39511-117 E.P.A. Est. No. 359-MO-1

CAUTION

Keep Out of Reach of Children

STATEMENT OF PRACTICAL TREATMENT

IF SWALLOWED: Drink 1 or 2 glasses of water and induce vomiting by touching back of throat with finger. Do not give anything by mouth to an unconscious person. Get medical attention.

IF ON SKIN: Wash with plenty of soap and water. Get medical attention if irritation persists.

IF IN EYES: Flush with water for 15 minutes. Get medical attention.

PEANUTS

APPLICATION: Apply BUTOXONE 200 postemergence as an overall spray by ground sprayer or airplane. With ground applications use a boom sprayer with flat fan-type nozzles. Adjust the height of the boom above the crop so the growing terminals of all weeds are sprayed. Use low pressure (about 20 to 40 lbs. per square inch). Do not use boom jet nozzles. With airplane applications BUTOXONE 200 has been succesfully applied using 5 gallons of total solution per acre. The height of the application should insure **thorough coverage** of the tops of the weeds. Best control has resulted from airplane applications with coarse spray when cocklebur is at least crop high and actively growing. Other weed species should be in the seedling stage and actively growing. Airplane application is not recommended in the vicinity of susceptible crops.

Do not apply if peanuts are suffering from lack of moisture.

AMOUNT TO USE: Apply the recommended rates as indicated in the table below.

Broadleaf Weeds	Rate of BUTOXONE 200 per Acre
Cocklebur (*Xanthium* spp.)	0.9 pint (1 gal/9 acres)
Annual Morningglory (*Ipomoea* spp.) Common Lambsquarters (*Chenopodium album*) Common Ragweed (*Ambrosia artemisiifolia*) Jimsonweed (*Datura stramonium*) Sicklepod (Coffeeweed) (*Cassia obtusifolia*) Velvetleaf (*Abutilon theophrasti*)	1.6 pints (1 gal/5 acres)

Do not apply more than two applications per season; the second application should be made no later than the late bloom stage of peanuts (about 90 to 100 days after planting). Do not feed treated peanut vines and peanut hay to livestock. Do not apply to peanuts within 30 days of harvest.

ALFALFA, BIRDSFOOT TREFOIL AND CLOVERS

APPLICATION: For use in seedling alfalfa, birdsfoot trefoil and clovers spray when the crop has reached the 1 to 2 trifoliate leaf stage and growing conditions are good.

In established alfalfa certain weeds will emerge in the fall and over winter in a rosette stage. Best control of these weeds will result from application in late fall or early winter rather than in the spring. Do not apply after flowering.

WHEN TO APPLY: Weeds must be in the young seedling stage and actively growing to achieve satisfactory results. For best results spray weeds in the 2 to 5 leaf stage of growth.

AMOUNT TO USE: Apply BUTOXONE 200 as an overall spray by ground sprayer or airplane. Apply at rates listed below according to weed problems. Use the higher rate if weeds are past the seedling stage. Aerial applications should be made in 10 gallons of water per acre, while ground sprayer applications should be in 20 to 30 gallons of water per acre. Spray gallonage should give adequate coverage of the weeds without run-off. The use of an approved non-ionic surfactant for agricultural uses (such as X-77®, Surfactant WK, etc.) at their label rates will usually result in better weed control, especially if weeds are beyond the seedling stage. Do not use with surfactant in the desert areas of Arizona and California.

X-77® Registered trademark of Kalo Laboratories Inc.

Broadleaf Weeds	Rate of BUTOXONE 200 per Acre
Annual Morningglory (*Ipomoea* spp.) Cocklebur (*Xanthium* spp.) Common Lambsquarters (*Chenopodium album*) Jimsonweed (*Datura stramonium*) Kochia or Mexican Fireweed (*Kochia scoparia*) Pigweed (*Amaranthus* spp.) Velvetleaf (*Abutilon theophrasti*) Wild Turnip (*Brassica campesris*)	4 pints
Black Mustard (*Brassica nigra*) Buckhorn Plaintain (*Plantago lanceolata*) Common Ragweed (*Ambrosia artemisiifolia*) Curly Dock (*Rumex crispus*) Field Pennycress (Fanweed or Stinkweed) (*Thlaspi arvense*) Hedge Smartweed (*Polygonum scandens*) Ladysthumb (*Polygonum persicaria*) Prickly Lettuce (*Lactuca serriola*) Shepherdspurse (*Capsella buras-pastoris*) Sweet Clover (Volunteer plants) (*Melilotus* spp.) Wild Beet (*Beta maritina*) Wild Mustard (*Brassica kaber*)	6 pints

Do not graze established alfalfa, nor feed straw or hay from established alfalfa to livestock within 30 days after application. Do not graze or feed SEEDLING Alfalfa, Clover, and Birdsfoot Trefoil within 60 days after application. Do not use Sweet Clover and other legumes not mentioned, nor on established clovers grown for seed. Irrigation, in particular, overhead sprinkler irrigation, should be delayed as long as possible (10 days or more) following application of BUTOXONE 200, to avoid washing the chemical into the root zone. Do not apply when crop is stressed from lack of moisture. Do NOT use on peas.

Apply as a postemergence spray. When properly timed there is little or no effect on the crop. In established Alfalfa there may be some twisting of stems and malformation of leaves. This condition is usually outgrown.

Do not spray when daytime temperatures are expected to exceed 90°F, within the next 2 or 3 days. Do not apply if temperatures are likely to fall below 40°F, during or shortly after treatment.

SOYBEANS

APPLICATION: Apply BUTOXONE 200 postemergence as an over all spray by ground sprayer or airplane. With ground applications use a boom sprayer with flat fan-type nozzles. Adjust the height of the boom above the crop so the growing terminals of all weeds are sprayed. Use low pressure (about 20 to 40 lbs. per square inch). Do not use boom jet nozzles. With airplane applications BUTOXONE 200 has been successfully applied using 5 gallons of total spray solution per acre. The height of the application should insure **thorough coverage** of the tops of the weeds. Best control has resulted when cocklebur is at least crop high and actively growing. Airplane application is not recommended in the vicinity of susceptible crops. Do not apply if soybeans are suffering from lack of moisture.

TOPICAL (OVERHEAD) APPLICATIONS

Apply from 7 to 10 days before soybeans bloom through mid-bloom. A good indication that the beans are about to bloom is when the soybean plants have turned a dark green color. If application is made 7 to 10 days before bloom, apply 1 gallon of BUTOXONE 200 per 11 acres; if application is made from early-bloom through mid-bloom, apply 1 gallon BUTOXONE 200 per 9 acres. Use sufficient water to obtain adequate coverage. **The rates recommended are for the control of cocklebur.** These rates will also stunt or partially control certin other weeds including annual morningglory, velvetleaf, and jimsonweed. In the Midwest use only as a directed application since topical application may reduce soybean yields.

DIRECTED APPLICATIONS

For best results apply when weeds do not exceed 3 inches in height. Good coverage of the growing terminals of weeds is essential for effective control. Do not allow spray pattern to contact more than the lower one-third of the soybean plants.

Do not apply more than two applications per season. **Precise application is essential to prevent damage to the crop.** In order to maintain the correct spraying height, nozzles must be mounted on oiling shoes, skid shoes, or on cultivators with gauge wheels. Do not use booms with drop nozzles.

First Application: Apply as a directed spray when soybeans are 8 to 12 inches tall and weeds have emerged.

Second Application: Apply as a directed spray no later than mid-bloom stage.

AMOUNT TO USE: Apply the recommended rates as indicated in the table below.

Broadleaf Weeds	Rate of BUTOXONE 200 per Acre
Cocklebur (*Xanthium* spp.)	.9 pint
Annual Morningglory (*Ipomoea* spp.) Common Lambsquarters (*Chenopodium album*) Common Ragweed (*Ambrosia artemisiifolia*) Jimsonweed (*Datura stramonium*) Sicklepod (Coffeeweed) (*Cassia obtusifolia*) Velvetleaf (*Abutilon theophrasti*)	1.5 pint

While some temporary twisting of soybean plants may follow an application, this will have little or no permanent effect on the crop.

Applications made under adverse growing conditions, such as periods of drought, will result in injury to the beans; under such conditions the crop should not be sprayed.

Do not spray BUTOXONE 200 on soybeans showing a general infestation of **Phytophthora.**

Do not apply more than the recommended amount.

Do not harvest soybeans within 60 days after application.

Do not use soybeans grown West of the Rocky Mountains.

DIRECTED BAND APPLICATIONS OF MIXTURES OF BUTOXONE 200 AND LOROX 50W*

APPLICATION: Apply directed spray when beans are at least 8 inches tall and when weeds do not exceed 2 inches in height. Direct spray to cover weed foliage with minimum contact of the soybean plant do not spray higher than 3 inches on the soybean stem or crop injury may result.

Use an 8002 T-Jet (or equivalent) or larger with a minimum of 25 gals. per broadcast acre and spray pressure of 20 to 25 psi, to avoid spray drift. For each 25 gals. of spray mixture a pint of surfactant such as Surfactant WK may be added. Continuous agitation in the spray tank is required to keep the material in suspension. Agitate by mechanical or hydraulic means; if by-pass or return line is used, it should terminate at the bottom of the tank to minimize foaming. Openings in nozzle screens should be equal to or larger than 50 mesh. Apply with sprayer nozzles mounted on skid shoes, oiling shoes, or on cultivators with gauge wheels. Do not mount on booms with drop nozzles or on cultivators without gauge wheels.

AMOUNT TO USE: Apply the recommended rates as indicated in the table below.

Broadleaf Weeds	Rates BUTOXONE 200 and LOROX per broadcast Acre
Annual Morningglory (*Ipomoea* spp.) Cocklebur (*Xanthium* spp.) Common Lambsquarters (*Chenopodium album*) Common Ragweed (*Ambrosia artemisiifolia*) Crabgrass (*Digitaria* spp.) (2 leaf) Jimsonweed (*Datura stramonium*) Pricky Sida (Teaweed) (*Sida spinosa*) Sesbania (*Sesbania* spp.) Velvetleaf (*Abutilon theophrasti*)	0.9 pint + 1 pound

Use a proportional amount of the two chemicals on the band (for example, on a 12 inch band in 36-inch rows use ⅓ pint of BUTOXONE 200 and ⅓ lb. of Lorox per acre).

If a new flush of weeds occur after initial treatment, make a second application but do not apply within 60 days of harvest or make more than 2 applications per growing season. **Do not allow spray or spray drift to contact growing terminals of beans, as excessive crop injury will result.** Do not feed soybean forage or hay to livestock from fields treated with this mixture. Harvested soybeans may be used for feed or oil purposes.

NOTE: Do not use on sands, loamy sands, gravelly soils, or on exposed sub-soils. Do not use on soils containing less than ½% organic matter as crop injury may result. Observe all cautions and limitations on labeling of all products used in mixtures.

Lorox 50W-registered trademark of E.I. duPont de Nemours & Co. (Inc.)

CONDITION OF SALE, WARRANTY, LIMITATION OF LIABILITY

This product conforms to the chemical description on the label thereof and is reasonably fit for the purpose stated on such label only when used in accordance with directions under normal use conditions. Follow directions carefully. Timing and method of application, weather and crop conditions, mixture with other chemicals not specifically recommended and other influencing factors in the use of this product are beyond the control of Vertac or the seller. Buyer assumes all risks of use, storage, or handling of this material, not in strict accordance with directions given herewith. In no case shall Vertac or the seller be liable for consequential, special, or indirect damages such as loss of profits or values resulting from the use or handling of this product.

DMA* 4 HERBICIDE

CONTAINS DIMETHYLAMINE SALT† of 2,4-D

For Selective Control of Many Broadleaf Weeds in Non-Crop Areas, Grass Pastures, Rangelands, and in Certain Crops

Also for Control of Trees by Injection

ACTIVE INGREDIENT:

Dimethylamine Salt of 2,4-Dichlorophenoxyacetic Acid‡	46.3%
INERT INGREDIENTS:	53.7%
TOTAL	100.0%

2,4-Dichlorophenoxyacetic Acid‡

Equivalent: 38.4% –3.8 lb/gal

‡Isomer Specific by AOAC Method No. 6.275–6.279 (13th Ed.)

EPA Reg. No. 464-196-39511 EPA Est. 464-MI-1

† Salts are the least volatile forms of 2,4-D and do not release enough vapors from treated areas to reduce yield of adjacent susceptible crops.

WARNING

Keep Out of Reach of Children

PRECAUCION AL USUARIO: Si usted no lee inglés, no use este producto hasta que la etiqueta le haya sido explicada ampliamente.

TRANSLATION: **(TO THE USER:** If you cannot read English, do not use this product until the label has been fully explained to you.)

PRECAUTIONARY STATEMENTS

HAZARDS TO HUMANS AND DOMESTIC ANIMALS

INJURIOUS TO EYES • HARMFUL IF SWALLOWED
MAY CAUSE SKIN IRRITATION

Do Not Get In Eyes Or On Clothing • Wear Goggles Or Face Shield When Handling • Avoid Contact With Skin • Wash Thoroughly After Handling

STATEMENTS OF PRACTICAL TREATMENT

In case of eye contact, immediately flush eyes with plenty of water for 15 minutes. Call a physician.

In case of skin contact, wash with soap and plenty of water. Remove and wash contaminated clothing before reuse. Discard contaminated shoes.

If swallowed, induce vomiting immediately by giving two glasses of water and sticking finger down throat. Call a physician. Do not induce vomiting or give anything by mouth to an unconscious person.

PHYSICAL OR CHEMICAL HAZARDS

DO NOT CUT OR WELD CONTAINER

ENVIRONMENTAL HAZARDS

Do not contaminate water by cleaning of equipment or disposal of wastes.
Do not contaminate water used for irrigation or domestic purposes.

For chemical spill, leak, fire or exposure call toll free 1-800-424-9300

AGRICULTURAL CHEMICAL

Do Not Ship or Store with Foods, Feeds, Drugs or Clothing

WEED LIST

bitterweed	poorjoe
broomweed	Florida pusley
burdock	wild radish
carpetweed	common ragweed
wild carrot	wild rape
chicory	yellow rocket
cocklebur	
coffeeweed	

croton
dandelion
dock
flixweed
galinsoga
wild hemp
jewelweed
jimsonweed
kochia
lambsquarters
bigbend loco
lupines
Venice mallow
marshelder
annual morningglory
mustards
pennycress
pennwort
pepperweed
pigweed
plantains
shepherdspurse
sicklepod
smartweed
bitter sneezeweed
annual sowthistle
spanishneedles
sunflower
sweetclover
tansymustard
bull thistle
musk thistle
Russian thistle
tumbleweed
velvetleaf
vetch
waterplantain
witchweed
wormwood

DIRECTIONS FOR USE

It is a violation of Federal law to use this product in manner inconsistent with its labeling.

Generally, the lower dosages given will be satisfactory for young, succulent growth of sensitive weed species. For less sensitive species and under conditions where control is more difficult, the higher dosages will be needed. Apply DMA 4 during warm weather when weeds are young and growing actively. Use enough spray volume for uniform coverage by ground or air application. If only bands or rows are treated, leaving middles unsprayed, the dosage per crop acre is reduced proportionately. **Do not apply where drift may be a problem due to proximity of susceptible crops or other desirable plants. Read and follow all Use Precautions given on this label.**

To Prepare the Spray, mix DMA 4 only with water, unless otherwise directed on this label. Add about half the water to the mixing tank, then add the DMA 4 Herbicide with agitation, and finally the rest of the water with continuing agitation. **NOTE:** Adding oil, wetting agent, or other surfactant to the spray may increase effectiveness on weeds, but also may reduce selectivity to crops resulting in crop damage.

WEED CONTROL IN SMALL GRAINS NOT UNDERSEEDED WITH A LEGUME (Barley, Oats, Rye, Wheat): See Table for recommended use rates. Spray after grain begins tillering and before the boot stage (usually 4 to 8 inches tall) and weeds are small. **Do not apply before the tiller stage nor from early boot through the milk stage.** To control large weeds that will interfere with harvest or to suppress perennial weeds, preharvest treatment can be applied when the grain is in the dough stage. Best results will be obtained when soil moisture is adequate for plant growth and weeds are growing well. **Note:** Do not permit dairy animals or meat animals being finished for slaughter to forage or graze treated grain fields within 2 weeks after treatment. Do not feed treated straw to livestock.

WEED CONTROL IN CORN: See Table for recommended use rates. **Preemergence**—Apply to soil anytime after planting but before corn emerges. Do not use on very light, sandy soil. **Emergence**—Apply just as corn plants are breaking ground. **Post-emergence**—Apply to emerged corn. When corn is over 8 inches tall, use drop nozzles to keep spray off corn foliage. Do not apply from tasseling to dough stage. Injury to corn is most likely to occur if DMA 4 is applied when corn is growing rapidly under high temperatures and high soil moisture conditions. In such situations, use the low rate of ½ pint per acre. After application, delay cultivation for 8 to 10 days to allow the corn to overcome any temporary brittleness. **NOTE:** Hybrids vary in tolerance to 2,4-D. Some are easily injured. Spray only varieties known to be tolerant to 2,4-D. Consult the seed company or your Agricultural Experiment Station or Extension Service Weed Specialist for this information.

WEED CONTROL IN SORGHUM (MILO): See Table for recommended use rates. Treat only after the sorghum is 6 inches high and preferably before it is 15 inches high. Do not treat during the boot, tasseling, or early dough stages. Reduce spray drift by keeping the boom and spray nozzles as low as possible. If crop is taller than 8 inches, use drop nozzles to keep the spray off the leaves. Temporary crop injury can be expected under conditions of high soil moisture and high air temperatures. If it is necessary to apply DMA 4 under these conditions, use no more than ⅔ pint per acre.

NOTE: Hybrids vary in tolerance to 2,4-D. Some are easily injured. Spray only varieties known to be tolerant to 2,4-D. Consult the seed company or your Agricultural Experiment Station or Extension Service Weed Specialist for this information.

WEED CONTROL IN RICE: See Table for recommended use rates. Apply in the late tillering stage of rice development, at the time of first joint development (first to second green ring), usually 6 to 9 weeks after emergence. Do not apply after panicle initiation, after rice internodes exceed ½ inch, at early seedling, early panicle, boot, flowering, or early heading growth stages. **NOTE:** Some rice varieties under certain conditions can be injured by 2,4-D. Therefore, before spraying consult local Extension Service or University specialists for appropriate rates and timing of 2,4-D sprays.

WEED CONTROL IN SUGARCANE: See table for recommended use rates. Apply as a preemergence or post-emergence spray in accordance with State recommendations. For grass control, use DOWPON† M grass herbicide in addition to DMA 4. Always read the label directions and precautions for the use of these products before using them with DMA 4.

AMOUNT OF DMA 4 HERBICIDE TO USE IN CROPS

By Air or Ground Application

NOTE: Do not apply when weather conditions favor drift from treated areas. Read complete directions and precautions before using.

Crop	Dosage Per Acre: Normal rates (usually safe to crops)	Higher rates for special situations[2] (more likely to injure crop)
Small Grains		
Spring postemergence wheat, barley, rye oats	⅔ to 1⅓ pints	2 to 3 pints
Preharvest (dough stage)	½ to 1 pint	½ to 2 pints
wheat, barley, oats	1 to 2 pints	2 to 3 pints
Corn[1]		
Preemergence	2 to 4 pints	
Emergence[1]	1 pint	1½ pints
Postemergence[1]		
up to 8 inches	½ to 1 pint	
tall 8 inches to tasseling (use only directed spray)	1 pint	1½ to 2½ pints
Sorghum (Milo)[1]		
Postemergence		
6 to 8 inches tall	⅔ to 1 pint	
8 to 15 inches tall (use only directed spray)	1 pint	1½ to 2 pints
Rice	1 to 2½ pints	2 to 3 pints
Sugarcane	2 to 4 pints	

[1] Corn and sorghum varieties vary in tolerance to 2,4-D; some are easily injured. Before spraying, get information on 2,4-D tolerance of specific varieties and spray only those known to be resistant to 2,4-D injury. If plants are more than 8 inches tall, use directed spray and keep spray off corn and sorghum foliage.

[2] These higher rates may be needed to handle difficult weed problems in certain areas such as under dry conditions especially in western areas. However, do not use unless possible crop injury will be acceptable. Consult State Agricultural Experiment Station or Extension Service weed specialists for recommendations or suggestions to fit local conditions.

WEED CONTROL ON FALLOW LAND: Use 1 to 2 quarts of DMA 4 per acre on annual broadleaf weeds and up to 3 quarts per acre on established perennial species, such as Canada thistle and field bindweed. Apply to weeds actively growing. Do not plant any crop for 3 months after treatment or until chemical has disappeared from soil.

WEED CONTROL IN ESTABLISHED GRASS PASTURES AND RANGELANDS: Use at 2 to 4 pints per acre. Apply preferably when weeds are small and growing actively before the bud stage. Do not use on bentgrass, alfalfa, clover, or other legumes. Do not use on newly seeded areas until grass is well established. Do not use from early boot to milk stage where grass seed production is desired. Do not graze dairy animals on treated areas within 7 days after application.

CONTROL OF SOUTHERN WILD ROSE: On rangelands, roadsides, and fencerows, use 1 gallon of DMA 4 plus 4 to 8 fluid ounces of an agricultural surfactant per 100 gallons of water and spray thoroughly as soon as foliage is well developed. Two or more treatments may be required. On rangeland, apply a maximum of 6 quarts of DMA 4 per acre per application. Do not graze dairy animals on treated areas within 7 days after application.

GRASS SEED CROPS: Use 1 to 4 pints per acre in spring or fall to control broadleaf weeds in grass being grown for seed. **Do not apply from early boot to the milk stage.** Spray seedling grass only after the five-leaf stage, using $^3/_4$ to 1 pint per acre to control small seedling weeds. After the grass is well established, higher rates of up to 4 pints can be used to control hard to kill annual or perennial weeds. For best results, apply when soil moisture is adequate for good growth.

NOTE: Do not use on bentgrass unless grass injury can be tolerated. Do not graze dairy animals nor cut forage for hay within 7 days after application.

BROADLEAF WEED CONTROL IN NON-CROPLAND GRASS AREAS SUCH AS LAWNS, GOLF COURSES, CEMETERIES AND PARKS, AIRFIELDS, ROADSIDES, VACANT LOTS, DRAINAGE DITCH BANKS: Use 1 to 3 quarts of DMA 4 per acre in the amount of water needed for uniform application. Treat when weeds are young and growing well. Usually 2 quarts per acre will provide adequate weed control. Do not use on dichondra or other herbaceous ground covers. Do not use on creeping grasses such as bent except for spot treating nor on freshly seeded turf until grass is well established. Reseeding of lawns should be delayed following treatment. With spring application, reseed in the fall; with fall application, reseed in the spring. Legumes are usually damaged or killed. Deeprooted perennial weeds such as bindweed and Canada thistle may require repeated applications.

SPOT TREATMENT IN NON-CROP AREAS: To control broadleaf weeds in small areas with a hand sprayer, use $^1/_4$ pint of DMA 4 in 3 gallons of water and spray to thoroughly wet all foliage.

TREE INJECTION TREATMENT: To control unwanted hardwood trees such as elm, hickory, oak, and sweetgum in forest and other non-crop areas, apply DMA 4 by injecting 1 ml of the undiluted product through the bark around the trunk at intervals of 1 to 3 inches between edges of the injector wounds. For harder to control species such as ash, maple, and dogwood use 2 ml of undiluted DMA 4 per injection site. Continuous cuts around the trunk often provide improved control. Also, cuts near the ground level may be more effective than at higher levels. Treatments can be made at any season; however, effectiveness may be less during winter months. Maples should not be treated during the spring sap flow.

FOR RIGHTS-OF-WAY: For perennial broadleaf weeds and susceptible woody species, use up to 2 gal DMA 4 per acre. For difficult to control perennial broadleaf weeds and woody species, use up to 2 gal DMA 4 and 1 to 4 qts. GARLON* 3A herbicide per acre. **For Ground application:** (High volume) apply a total of 100 to 400 gal per acre; (low volume) apply a total of 20 to 100 gal per acre. **For helicopter:** Apply a total of 10 to 30 gal per acre spray volume.

STORAGE AND DISPOSAL

Do not contaminate water, food or feed by storage or disposal.

STORAGE

Keep container tightly closed when not in use. If exposed to subfreezing temperatures, the product should be warmed to at least 40F and mixed thoroughly before using.

PESTICIDE DISPOSAL

Pesticide wastes are acutely hazardous. Improper disposal of excess pesticide, spray mixture, or rinsate is a violation of Federal Law. If these wastes cannot be disposed of by use according to label instructions, contact your State Pesticide or Evironmental Control Agency, or the Hazardous Waste representative at the nearest EPA Regional Office for guidance.

CONTAINER DISPOSAL

Triple rinse (or equivalent). Then offer for recycling or reconditioning, or puncture and dispose of in a sanitary landfill, or by other procedures approved by state and local authorities.

USE PRECAUTIONS

Do not apply DMA 4 directly to, or otherwise permit it to come into contact with cotton, flowers, fruit trees, grapes, ornamentals, vegetables, or other desirable plants which are sensitive to 2,4-D herbicides and do not use in a greenhouse. Do not permit spray mist containing it to drift onto them, since even very small quantities of the spray, which may not be visible, can cause severe injury during both growing and dormant periods. Use coarse sprays to minimize drift. With ground equipment, spray drift can also be minimized by keeping the spray boom as low as possible, by applying 20 gallons or more of spray per acre, by using no more than 20 pounds per square inch spraying pressure, by using flat fan or flood fan nozzle tips, and by stopping all spraying when wind velocity exceeds 8 miles per hour. Do not apply using cone-type insecticide or other nozzles that produce a fine-droplet spray. With aircraft application, drift can be lessened by using no more than 20 pounds spray pressure at the nozzles; by using nozzles which produce a coarse spray pattern; and by spraying only when the wind velocity is less than 5 miles per hour. **Applications by airplane, ground rigs, and hand dispensers should be carried out only when there is no hazard from drift. Do not apply in the vicinity of cotton, grapes, tomatoes, or other desirable vegetation susceptible to 2,4-D. Do not spray when the wind is blowing across the area to be sprayed towards susceptible crops or ornamental plants.** Violent windstorms may move soil particles. If 2,4-D is on these particles and they are blown onto susceptible plants, visible symptoms may appear. Serious injury is unlikely. The hazard of movement of 2,4-D on dust during violent windstorms is reduced if treated fields are irrigated or if rain occurs shortly after application. Do not contaminate irrigation ditches or water used for irrigation or domestic purposes.

To avoid injury to desirable plants, do not store, handle, or apply other agricultural chemicals with the same containers or equipment used for DMA 4 except as specified on this label. Excessive amounts of 2,4-D in the soil may temporarily inhibit seed germination or plant growth.

Consult your State Agricultural Experiment Stations or Extension Service weed specialists in many states for recommendations from this label that best fit local conditions. Be sure that use of this product conforms to all applicable regulations. **Apply this product only as specified on this label.**

CONDITIONS OF SALE AND WARRANTY

VERTAC AND SELLER OFFER THIS PRODUCT AND THE BUYER AND USER ACCEPTS THIS PRODUCT ONLY UNDER THE FOLLOWING AGREED CONDITIONS OF SALE AND WARRANTY.

The directions for use of this product are believed to be reliable and should be followed carefully. However, it is impossible to take into account all variables and to eliminate all risks associated with its use. Injury or damage may result because of conditions which are beyond the control of Vertac or the Seller. Vertac warrants only that this product conforms to the chemical description on the label and is believed to be reasonably fit for the purposes referred to in the Directions for Use when used as directed under normal conditions.

VERTAC MAKES NO OTHER EXPRESS OR IMPLIED WARRANTY OF FITNESS OR MERCHANTABILITY OR ANY OTHER EXPRESS OR IMPLIED WARRANTY. In no case shall Vertac or the Seller be liable for consequential, special or indirect damages resulting from the use or handling of this product. Any variation or exception from this warranty must be in writing and signed by an authorized Vertac representative.

*Trademark of VERTAC CHEMICAL CORPORATION

† Trademark of THE DOW CHEMICAL COMPANY

DOWPON† M

GRASS KILLER

FREE FLOWING FORMULATION

For Controlling Annual Grasses Perennial Grasses and Cattails

ACTIVE INGREDIENTS:	
Sodium salt of dalapon‡	72.5%
Magnesium salt of dalapon‡	12.0%
INERT INGREDIENTS:	15.5%
TOTAL	100.0%

‡ Total active ingredients equivalent to 74.0% dalapon (2,2-dichloropropionic acid).

EPA Reg. No. 464-402-39511 EPA Est. 464-MI-1

U.S. Patent No. 3,671,652

† Trademark of The Dow Chemical Company

CAUTION

Keep Out of Reach of Children

PRECAUCION AL USUARIO: Si usted no lee inglés, no use este producto hasta que la etiqueta le haya sido explicada ampliamente.

TRANSLATION: (TO THE USER: If you cannot read English, do not use this product until the label has been fully explained to you.)

PRECAUTIONARY STATEMENTS

HAZARDS TO HUMANS

CAUSES EYE IRRITATION
MAY BE HARMFUL IF SWALLOWED
MAY CAUSE SKIN IRRITATION
AVOID CONTACT WITH EYES, SKIN OR CLOTHING

STATEMENT OF PRACTICAL TREATMENT

In case of contact, flush eyes with plenty of water for a least 15 minutes and get medical attention. For skin, remove contaminated clothing and wash skin with plenty of water. Wash clothing before re-use.

AGRICULTURAL CHEMICAL

Do Not Ship or Store with Food, Feeds, Drugs or Clothing

For chemical spill, leak, fire or exposure call toll free 1-800-424-9300

USE PRECAUTIONS

Apply this product only as specified on this label or on supplementary registered labeling. Be sure that user of this product conforms to all applicable regulations.

To avoid injury to desirable plants, do not allow spray or spray drift containing DOWPON M to contact foliage or roots of any such plants, including lawns, and other desirable turf. Flush spraying equipment with plenty of water immediately after use to avoid contamination of later sprays and possible corrosion.

Local conditions affect the use of herbicides. State Agricultural Experiment Station and Extension Service weed specialists in many states issue recommendations to fit local conditions.

READ ENTIRE LABEL BEFORE USING DOWPON M

DIRECTIONS FOR USE

It is a violation of Federal law to use this product in a manner inconsistent with its labeling.

STORAGE AND DISPOSAL

Do not reuse container. Destroy by burning or burying in non-cropland away from water supplies or follow official local container disposal procedures.

Use DOWPON M only as directed on this label. Before using this product, read GENERAL INFORMATION, USE PRECAUTIONS, TREATMENT DETAILS AND FOOTNOTES for each intended use.

GENERAL INFORMATION

DOWPON M grass killer controls cattails; annual grasses such as foxtails, barnyardgrass, crabgrass, panicums, wild oats; and perennial grasses such as bermudagrass, bluegrass, guineagrass, johnsongrass, maidencane, paragrass, phragmites, quackgrass, raoulgrass, torpedograss, vaseygrass.

MODE OF ACTON: DOWPON M is absorbed through foliage and roots and translocates through the plant system. Grass growth is rapidly inhibited but kill is often delayed. If sprayed grass produces new growth, another treatment is suggested.

SOIL PERSISTENCE: DOWPON M is readily decomposed in soil by micro-organisms. In warm, moist soils breakdown normally occurs in 3 to 5 weeks. In cold or dry soil, and soils with little or no organic matter, slower degradation can be expected.

WHEN TO APPLY: Apply when grasses are growing well and before heading. Adequate soil moisture favors good growth and effective herbicidal action. Application during drought or cold weather or when grass is maturing will be less effective.

HOW TO APPLY: Apply by aircraft or ground equipment as a water spray to foliage which should be wet uniformly. Use enough water for adequate wetting of foliage, usually 5 to 10 gallons per acre by aircraft and 10 to 100 gallons by ground equipment. Dense grass stands may require up to 500 gallons for uniform coverage. For spot treatment, use 10 pounds of DOWPON M per 100 gallons of water. Per acre rates are for overall broadcast treatment. With band, row, or spot treatment the rates per crop acre should be reduced in proportion to the areas actually sprayed. Note surfactant directions.

SURFACTANT: Use any approved agricultural surfactant in all applications of DOWPON M to improve wetting of foliage and grass control. For annual grasses, add ½ to 2 pints per 100 gallons of spray. For perennial grasses, add 4 pints per 100 gallons of spray. **NOTE: For some crops, note that less surfactant is recommended. Adding more surfactant to sprays applied to crop foliage may increase the risk of crop injury.** For non-crop uses, control of perennial grasses may be improved by using more than 4 pints of surfactant per 100 gallons of spray.

COMBINATIONS: DOWPON M can be used in combination sprays with phenoxy or other herbicides as inidicated on this label. Always note label directions and precautions for other products before using with DOWPON M. Do not use combination sprays containing 2,4,5-T on ditch banks, around the home or recreation areas.

MIXING INSTRUCTIONS

DOWPON M grass killer is soluble in water. Add it slowly to the water while stirring or agitating until completely dissolved. Then add desired amount of surfactant. In combination sprays with other herbicides, dissolve the DOWPON M first and use no more than 8 pounds in 10 gallons of water.

NON-CROPLAND USES

Railroad ballast and berm areas, drainage ditches, fence rows, industrial sites, and other non-cropland areas including forest planting sites. Apply by air or ground equipment. Tank mix combinations with FORMULA 40* or ESTERON* herbicides may be used for controlling both broadleaf weeds and grasses.

OTHER USES NOT ON THIS LABEL

Consult VERTAC CHEMICAL CORPORATION for registered use directions for DOWPON M in the following crops: asparagus, bananas, birdsfoot trefoil, coffee, cranberries, macadamia nuts and peas.

CONDITIONS OF SALE AND WARRANTY

VERTAC AND SELLER OFFER THIS PRODUCT AND THE BUYER AND USER ACCEPTS THIS PRODUCT ONLY UNDER THE FOLLOWING AGREED CONDITIONS OF SALE AND WARRANTY.

The directions for use of this product are believed to be reliable and should be followed carefully. However, it is impossible to take into account all variables and to eliminate all risks associated with its use. Injury or damage may result because of conditions which are beyond the control of Vertac or the Seller. Vertac warrants only that this product conforms to the chemical description on the label and is believed to be reasonably fit for the purposes referred to in the Directions for Use when used as directed under normal conditions. VERTAC MAKES NO OTHER EXPRESS OR IMPLIED WARRANTY OF FITNESS OR MERCHANTABILITY OR ANY OTHER EXPRESS OR IMPLIED WARRANTY. In no case shall Vertac or the Seller be liable for consequential, special or indirect damages resulting from the use or handling of this product. Any variation or exception from this warranty must be in writing and signed by an authorized Vertac representative.

Use Directions On Non-Cropland

Problem	DOWPON M lb per acre	Treatment Details (note surfactant section above)
Perennial grasses and cattails	13.5-30	Good spray coverage of healthy foliage essential and retreatment may be necessary. With spray boom—use at least 100 gallons spray per acre. With hand gun—use up to 300 gallons spray per acre. Spray cattails when 2 to 3 feet tall or during catkin development.
Annual grasses	4-10	Treat when growing well and prior to heading. Low rates will retard growth of most grasses and can kill certain species such as foxtails.
Control of grasses on forest planting sites	4-15	Use 4 to 5 pounds per acre on seedling grasses. Use 10 to 15 pounds on perennial grasses. Wait at least 2 weeks after application before planting forest seedlings.
Spot treatment		Use 13.5 pounds of DOWPON M in 100 gallons of water and spray to wet grass without runoff. Repeat as needed.

Use Directions On Irrigation Ditchbanks—Western United States

Problem	DOWPON M lb/A	Treatment Details (note surfactant section above)
Control of grasses such as reed canary, bermuda, Johnson, phragmites, cattails, and tule on banks of irrigation systems	13.5	Good spray coverage of healthy foliage is essential (a,b,c,d). If browning and dieback does not occur within three to six weeks, retreatment may be necessary. With spray boom use at least 100 gallons spray per acre. No waiting period is necessary before this water can be used for irrigation purposes.
	spot treatment	Use 13.5 lb in 100 gallons of water and apply by hand spray gun to control intermittent infestations (not to exceed fifty percent of the total area). Spray to wet all foliage thoroughly but without runoff. This may go up to 300 gallons of spray per treated acre. (a,b,c,d). Spray to wet all foliage thoroughly but without runoff. This may go up to 300 gallons of spray per treated acre. (a,b,c,d).

a. Always begin spraying at downstream end of the section to be treated and work upstream. Do not spray across stream, minimize contact of spray on water surface, and do not spray both banks simultaneously. Do not spray more than 2 miles of the same ditchbank during any 24-hour period for ditches carrying less than 50 cubic feet of water per second.

b. Cattails and phragmites (giant plume grass) are controlled best when new growth is 2 to 3 feet tall and plants are growing well. Fall spraying of cattails after catkin development is effective.

c. Do not permit fishing in treated waterways within the calendar year of treatment.

d. For grazing restrictions see footnotes 20 and 21.

USE DIRECTIONS ON CROPLAND
(BEFORE MAKING ANY APPLICATION, READ APPROPRIATE FOOTNOTES)

CROP	APPLICATION	DOWPON M LB PER ACRE	TREATMENT DETAILS	FOOTNOTES
APPLES, PEARS	directed under trees	5–10	Use low rate for trees less than 4 years old. Apply as spot, band or broadcast application.	1, 2
APRICOTS	directed under trees	6–7	**Western areas**—use spot, band or broadcast application.	1, 2
BEANS (Field, kidney, lima, snap)	Spring preplant	8	Quackgrass—apply in spring on 4 to 6 inch grass. Wait 4 days. Plow or deep disc. Wait 4 to 5 weeks after plowing or deep discing before planting.	2, 3, 4
	Fall preplant	15	Quackgrass—apply in fall (usually September or October) on growing grass. Plow or deep disc 10 days after application or in the spring before planting.	2, 3
CITRUS groves	directed under trees at least 4 years old	2–4	Use 1 pound in 20 gallons water and spray to wet without runoff. **Florida and Texas**—apply broadcast at 2 to 10 day intervals. **California and Arizona**—spot spray grass at 4 to 8 week intervals.	2, 6
ditches, borders	spot or broadcast	5–10	Use 5 pounds per 100 gallons water and spray to wet grass without runoff. If needed, make a second application in 1 to 2 weeks.	2, 3, 7

CROP	APPLICATION	DOWPON M LB PER ACRE	TREATMENT DETAILS	FOOTNOTES
CORN	spring preplant	8	Quackgrass—same as for BEANS	2, 3, 4
	spring preplant	5–7	Johnsongrass—spray when 8 to 12 inches tall. Wait 3 days. Plow or deep disc. Wait at least 5 days after plowing or deep discing before planting.	2,3 4,5
	fall preplant	15	Quackgrass—same as for BEANS	2, 3
COTTON	spring preplant	5–7	Johnsongrass—spray when 8 to 12 inches tall. Wait 3 days. Plow and plant.	2, 3, 5
	preplant (repeat treatments)	5–7	Bermudagrass and johnsongrass on stubble or fallow land.	2, 3 5, 12
	spot spray		Perennial grasses—use 1 pound in 10 gallons water. Direct spray to grass and keep off cotton plants. Spray up to 3 times per season.	
	lay-by	5	Use directed spray under the cotton after last tillage in western areas.	
FLAX	broadcast (for foxtail control)	1	Apply in 5 to 10 gallons water per acre on 1 to 6 inch flax and 1 to 2 inch foxtails. Use with MCP Amine Weed Killer to add broadleaf weed control.	3, 8
GRAPES	directed under vines	10	Use in bearing vineyards in Arizona and California. Apply on 30 inch band centered on the vine row. Keep spray out of irrigation furrows.	1, 2
PASTURE (established perennial grasses such as Pangola-, Bermuda-, Bahia-, Dallis-, and Fescue grasses)	when grasses are growing well	4–6	Smutgrass control in Gulf Coast States including Georgia. Apply broadcast at 4–6 lb per acre.	3, 4, 10, 19, 21, 22, 23, 25
			Spot treat with 6 lb/100 gal water to thoroughly wet smutgrass foliage.	4, 18, 19, 20, 21, 23, 25
PEACHES	directed under trees	6–7	**Western areas**—use spot, band or broadcast application.	1, 2
		2–3½	**Other areas**—use 1 pound in 20 gallons water. Apply up to 70 gallons spray per acre.	1, 2
PLUMS (PRUNES)	directed under trees	6–7	**Western areas**—use spot, band or broadcast application.	1, 2
POTATOES	spring preplant	8	Quackgrass—same for BEANS except potatoes can be planted right after plowing.	2, 3, 11
	fall preplant	15	Quackgrass—same as for BEANS.	2, 3, 11
	pre-emergence	3	Grass seedlings—spray grass before emergence of potatoes. Can be used in combination spray with PREMERGE* 3 Dinitro Amine Herbicide.	3, 11
		10	Quackgrass—spray to wet grass if it comes up after planting and **before** emergence of potatoes.	3, 11
SORGHUM	spring preplant	5–7	Johnsongrass—spray when 8 to 12 inches tall. Wait 3 days. Plow. Wait at least 2½ weeks before planting.	2, 3, 4, 5
SOYBEANS	spring preplant	5–7	Johnsongrass—spray when 8 to 12 inches tall. Wait 3 days. Plow or deep disc. Wait at least 5 days after plowing or deep discing before planting.	2, 3 4, 5
SUGAR BEET	spring preplant	8	Quackgrass—same as for BEANS except beets can be planted right after plowing.	2, 3, 24
	fall preplant	15	Quackgrass—same as for BEANS.	2, 3, 12, 24
	broadcast (post-emergence)	2–4	Seedling grasses—apply overall from emergence to 6-leaf stage of sugarbeets.	3, 9, 10, 24
	directed	3–5	Apply under beet foliage using leaf lifters from 6-leaf stage of sugarbeets until they are 14 inches tall.	9, 10, 24

CROP	APPLICATION	DOWPON M LB PER ACRE	TREATMENT DETAILS	FOOTNOTES
SUGARCANE Florida, Puerto Rico	directed	4	Spray grass in spring before close-in using 20 to 40 gallons water per acre. Repeat 2 or 3 times as needed. Do not spray cane foliage.	13
	spot spray		Use 10 pounds DOWNPON M in 100 gallons water and spray grass infested areas before close-in. Repeat as needed. Do not spray cane foliage.	13
SUGARCANE Hawaii	directed	3–4	Spray grass after cane planting or ratooning. Repeat 2 or 3 times as needed prior to close-in. Do not spray cane foliage.	14
	spot spray		Same as for Florida cane.	14
SUGARCANE Louisiana	ground	5½	**Stubble cane**—Apply 2¼ pounds on 30 inch bands on actively growing johnsongrass or bermudagrass in the spring. Repeat in 2 to 3 weeks if needed.	15, 16, 17
	aerial	2¼	**Stubble cane**—Apply on actively growing johnsongrass or bermudagrass in the spring using 3 to 5 gallons of water per acre. Repeat in 2 to 3 weeks if needed.	16, 17
	ground	2⅖	**Plant cane**—Apply on actively growing seedling grasses such as johnsongrass and raoulgrass. Repeat in 2 to 3 weeks if needed.	15, 16, 17
	aerial	1	**Plant cane**—Apply in spring in 3 to 5 gallons of water per acre on actively growing grass seedlings. Repeat in 2 to 3 weeks if needed.	16, 17
FALLOW OR STUBBLE LAND	preplant or repeat treatments	5–7	Use in grain stubble or on fallow land to control quackgrass, bermudagrass and johnsongrass during the growing season prior to planting beans, corn, cotton, potatoes, sorghum, soybeans and sugarbeets.	2, 3, 5, 11, 12

THESE FOOTNOTES ARE AN IMPORTANT PART OF THE USE DIRECTIONS

1. Apply when grasses have begun growing well in spring, but before heading. Spray to wet the grass without runoff. Apply when soil moisture is high such as after a rain or irrigation. Use one or two applications per season. For band or spot treatment, the amount of spray, and DOWPON M, should be reduced in proportion to the area actually treated. Do not spray bare soil under fruit trees or vines. Keep spray off crop foliage and fruit. Do not apply within 30 days of harvest.
2. Do not graze livestock on treated areas during application season.
3. Apply by air or ground equipment.
4. If soil remains unusually dry after preplant treatment, a longer waiting period may be necessary to avoid residue injury to the crop. A rain of half an inch or more will help to move the herbicide into the soil so that microbial degradation (breakdown) will take place. Wait 5 days after the rain before planting susceptible crops.
5. This treatment will not control unemerged seedling grasses.
6. Keep spray off crop foliage and fruit and do not spray bare soil under the trees. Apply after rain or irrigation so that adequate soil moisture is present. Use up to 3 applications per season in orange, grapefruit, tangerine and lime groves. Use low rate on very light soil. Do not use in groves under drought stress, diseased, in decline, or suffering from cold damage. In Arizona and California, spray only up to half the area over the rootzone of the trees.
7. DOWPON M may be used in combination sprays with ESTERON* products or FORMULA 40 to control weeds and brush in addition to grass.
8. Use no more than ½ pint of surfactant per 100 gallons of spray. Do not expect rapid grass kill. CAUTION: Poor grass control and injury to flax may occur if applied during periods of drought. Do not apply under drought conditions unless crop injury can be tolerated. Flax varieties vary in tolerance to DOWPON M and MCPA. Consult local Extension Service or University Specialists for further information.
9. Do not apply when temperature is above 85°F. Using more than 1 pint of surfactant per 100 gallons of spray may improve grass control but also may increase the risk of crop injury. As much as 8 pounds of DOWPON M per acre may be used for better control of wild oats and quackgrass but also may injure the beets. The 5 to 8 lb. rate should be applied only from ground equipment and as a spray directed away from the sugar beet plants. For aerial application, use at least 5 gallons of spray per acre. DOWPON M at 2 to 3 pounds per acre may be used in tank mix combination with Pyramin W herbicide at its recommended rates and times to control broadleaf weeds as well as grasses.
10. Use no more than 8 pounds of DOWPON M per acre postemergence per season. Beets and tops may be fed to livestock.
11. Do not use on land planted to or to be planted to red-skinned varieties or White Rose. Do not spray potato plants.
12. To control established bermudagrass or johnsongrass on grain stubble or fallow land, use 5 to 7 pounds per acre in 2 or 3 applications 2 to 3 weeks apart, or as needed. Use tillage prior to planting, the following spring on such crops as beans, corn, cotton, potatoes, sorghum, soybeans and sugarbeets.
13. Do not apply more than 12 pounds per acre per season. Do not apply within 5 months of harvest. Do not spray cane foliage.
14. Do not apply more than 20 pounds per acre during 24–36 month crop cycle. Do not spray cane foliage.
15. One pound per acre on 30 inch bands is equivalent to 2.4 pounds per acre on broadcast treatment basis.
16. Apply before May 20 to avoid injury to cane. DOWPON M can be used with KURON or 2,4-D in accordance with label and local recommendations.
17. Varieties C.P. 48-103 and C.P. 47-193 are less tolerant to DOWPON M than other varieties. Consult Extension Specialist concerning response of new varieties.
18. Best results obtained if mowed 2–4 weeks after treatment and fertilized to hasten pasture recovery.
19. Pastures recover most rapidly if mowed, adequately fertilized, ample moisture is made available and other good management practices are followed. Desirable perennial grasses and forbs (such as white clover) may be temporarily reduced in stand or vigor but spring treated pastures usually recover within 2 to 4 months under favorable conditions. Fall treated pastures usually show excellent recovery the following spring under good management and favorable conditions. Any annual grasses present may be seriously injured or killed.
20. When less than 10% of the pasture area is treated, grazing need not be discontinued.
21. Grazing Restrictions

Pastures unmowed—do not graze meat animals on treated pastures until 8 weeks after treatment and withhold from such grazing for 1 week before slaughter. Fourteen weeks after treatment no withholding period is necessary. Do not graze lactating dairy animals on treated pastures until 14 weeks after treatment.

Pastures mowed—mow 2 to 4 weeks after treatment if desired. Do not graze lactating or meat animals on treated pastures until 4 weeks after mowing. Do not harvest hay from treated pastures during the season of application.

22. Use only the lower rate of DOWPON M on bahiagrass pastures since it is more sensitive than other desirable pasture grasses and may suffer considerable reduction under unfavorable conditions.
23. Do not sow grass seed until 2 to 4 weeks after treatment. Species such as annual ryegrass can be fall seeded to provide winter pasture. Defer grazing for 10 weeks or longer after treatment allowing adequate time for sufficient growth to sustain grazing.
24. CAUTION: Use of thiocarbamates such as RO-NEET, TILLAM, EPTAM and PRE-BETA herbicides may increase risk of injury to sugar beets from the postemergence application of DOWPON M. Also, environmental conditions which cause poor growth or reduced sugar of sugar beets such as unusually cold or wet weather, drought stress, insect or disease attack, poor nutrition and high concentrations of salts in the soil increase risk of injury. Therefore, DO NOT APPLY DOWPON M postemergence to sugar beets if thiocarbamate herbicide was applied preplanting or if crop is growing under adverse environmental conditions, unless crop injury can be tolerated.
25. Apply only once per year.

MISCELLANEOUS USES

ASPARAGUS: TO control quackgrass, apply 10 pounds of DOWPON M per acre before cutting and 3 to 4 weeks later as needed. With a single treatment, use 20 pounds of DOWPON M either before or after the cutting season. For bermudagrass control, apply 20 pounds of DOWPON M per acre either during or after the cutting season, and a second spray of 10 pounds as needed. **NOTE:** During the cutting season, spray immediately after cutting and use no more than 1 pint of surfactant per 100 gallons of spray. Postharvest sprays must be directed under the fern growth to avoid crop injury. Use enough water to provide uniform wetting of the grass. Spot spray patches of grass with 1 pound of DOWPON M in 10 gallons of water.

BANANA PLANTINGS: To control perennial grasses such as guinea (**Panicum maximum**), amargo (**Axonopus** sp.), and gamalote (**Paspalum** sp.), use 10 to 15 pounds of DOWPON M per acre in enough water to obtain good coverage. Often rates of 6 to 8 pounds per acre applied in the same manner give very satisfactory results. A first treatment should be made when grass is growing well, but before heading. Repeat treatments, usually spot applications, should be made where needed in about a month to six weeks and later on as required. Use no more than a total of 30 pounds of DOWPON M per acre per year. Keep spray off banana fruits, foliage or young plants. Use sufficient spray to wet grass foliage thoroughly, but avoid run-off.

Birdsfoot Trefoil, for Seed Production Only: For use in Vermont and New York states only. To control or suppress quackgrass, and other perennial grasses, annual grasses and clovers in birdsfoot trefoil seed fields apply 3 to 5 pounds DOWPON M in 20 to 30 gallons of water per acre in fall or early spring when clovers and grasses and 4 to 6 inches tall. Some temporary crop injury may occur but seed yield and quality are usually increased. **NOTE: Do not graze treated fields to livestock or feed forage, hay, or chaff.**

COFFEE: To control crabgrass, bermudagrass, kikuyugrass, para (California) grass and other grasses in established coffee plantations in Hawaii, dissolve one pound DOWPON M in 10 gallons water. Apply as a directed spray to just wet the grass without run-off but not to exceed 100 gallons per acre. **Keep the spray off the coffee foliage.** Make initial application to grasses which are growing well and prior to seed head formation. Respray 4 to 6 weeks later or when regrowth warrants, but do not apply more than two sprays in any one crop year and do not apply within 8 months of harvest.

CRANBERRIES: Ditch and shore treatment: Use 20 pounds of DOWPON M in 200 to 300 gallons of water per acre to control cattails, cutgrass, povertygrass, bur-reed and common rushes and sedges in irrigation and drainage ditches and along shore areas of cranberry bags. Make the first application usually in early June, when weeds are growing well but before seed heads form. Make second application as needed but not later than July 31. Apply the spray to thoroughly wet the weedy vegetation. **Keep spray off cranberry foliage.**

Vine Treatment: Use 10 pounds of DOWPON M in about 300 gallons of water per acre on cranberry bags to control summergrass, cutgrass, povertygrass, wamsy, warty panicgrass, bunchgrass, wide leaf grasses, round rush and sedge weeds.

Use no more than ½ pint of surfactant per 100 gallons of spray. Treatment should be made only after harvest in October and early November. **CAUTION:** This treatment may reduce the next season's crop. For example, on the Howes variety the crop may be reduced only slightly whereas on Early Blacks it may be reduced up to 75% and on certain varieties as much as 90%. Follow recommendations of agricultural experiment station or extension weed specialists.

Swab Treatment: Apply DOWPON M (to the grass and not to the cranberry foliage) at the rate of 2 pounds in 5 gallons of water (up to 5 gallons mixture per acre) as a swab treatment as soon as the grassy weeds appear above the vines. Two applications at two-week intervals are suggested. Swab applications will suppress growth of grassy weeds, and may result in crop reduction in the year following application if the concentrated solution is applied carelessly or drips onto the cranberry vines. **Do not apply within 60 days of harvest. Do not us more than two applications per season.** Follow recommendations of agricultural experiment station or extension weed specialists.

MACADAMIA NUTS: To control crabgrass, bermudagrass, kikuyagrass, para (California) grass, dallisgrass in macadamia orchards in Hawaii, use DOWPON M in concentrations of ½ to 1 pound per 10 gallons of water. Spray to just wet the grass without run-off, but not to exceed 100 gallons per acre. Make first application before harvest and respray at intervals as regrowth warrants. Do not exceed 24 pounds of DOWPON M per acre per year. **CAUTION:** During harvest periods, spray only immediately after gleaning nuts from the ground. **Do not spray directly on fallen nuts.**

PEAS: To control seedling grasses, giant, green and yellow foxtail (**Setaria**) use one pound of DOWPON M per acre in enough water for good spray coverage. This treatment may not give satisfactory control of species other than these sensitive foxtails. Use no more than ½ pint surfactant per 100 gallons of spray. Apply at least 25 days prior to harvest when the grass seedlings are young and the peas are 2 to 6 inches tall. Fresh pea vines or pea silage may be fed to livestock if desired, but do not feed pea vine hay. DOWPON M can also be used in tank-mix combination sprays with PREMERGE* 3 Dinitro Amine Herbicide. (Read PREMERGE 3 label for directions and precautions).

* Esteron, Formula 40, Premerge 3 are trademarks of VERTAC Chemical Corporation.

ESTERON* 6E

HERBICIDE
CONCENTRATED—EFFECTIVE
LOW VOLATILE

Contains Isooctyl Esters of 2,4-D

For the Control of Many Broadleaf Weeds, Herbaceous Perennials and Woody Plants Susceptible to 2,4-D in Grass Pastures, Certain Crops and Non-Crop Areas.

ACTIVE INGREDIENT:

2,4-Dichlorophenoxyacetic Acid, Isooctyl Esters‡ 89.4%

INERT INGREDIENTS: 10.6%

TOTAL 100.0%

2,4-Dichlorophenoxyacetic Acid Equivalent 59.4%—5.6 lb/gal

‡ Isomer Specific by AOAC Method No. 6.275–6.279 (13th Ed.)

EPA Reg. No. 464-347-39511 EPA Est. 464-MI-1

*Trademark of VERTAC Chemical Corporation

CAUTION

Keep Out of Reach of Children

PRECAUCION AL USUARIO: Si usted no lee ingles, no use este products hasta que la etiqueta le haya sido explicada ampliamente.

TRANSLATION: (TO THE USER: If you cannot read English, do not use this product until the label has been fully explained to you.)

PRECAUTIONARY STATEMENTS

HAZARDS TO HUMANS AND DOMESTIC ANIMALS

HARMFUL IF SWALLOWED. MAY CAUSE IRRITATION

Avoid Contact with Skin, Eyes, or Clothing

Wash Thoroughly After Handling

STATEMENTS OF PRACTICAL TREATMENT

In case of contact, immediately flush eyes or skin with plenty of water. Get medical attention if irritation persists.

If swallowed, induce vomiting immediately by giving two glasses of water and sticking finger down throat. Call a physician. Do not induce vomiting or give anything by mouth to an unconscious person.

ENVIRONMENTAL HAZARDS

Do not apply directly to water. Do not apply where runoff is likely to occur. Do not contaminate water by cleaning of equipment or disposal of wastes. Do not contaminate irrigation ditches or water used for irrigation or domestic purposes.

For chemical spill, leak, fire or exposure call toll free 1-800-424-9300

AGRICULTURAL CHEMICAL

Do Not Ship or Store with Food, Feeds, Drugs or clothing

WEED LIST

ESTERON 6E herbicide is recommended for control of numerous broadleaf weeds and certain 2,4-D susceptible woody plants without injury to most established grasses. Species controlled include the following, plus many others:

beggarticks
bitterweed
blueweed, Texas
broomweed
buckbrush
buckwheat, wild
burdock
burhead
carpetweed
catnip
chamise
chicory
cocklebur
onion, wild
pennycress
pepperweed, field
pigweed
plantains
poorjoe
rabbitbrush
radish, wild
ragweed
rape, wild
redstem
sage, coastal
sagebrush, big
coffeeweed
cornflower
coyotebrush
croton
dandelion
docks
dogfennel
elderberry
fanweed
galinsoga
garlic, wild
goatsbeard
halogeton
hemp, wild
jewelweed
jimsonweed
kochia
lambsquarters
loco, bigbend
mallow, Venice
manzanita
marshelder
milkvetch
morningglory, annual
mustards
nettles
sagebrush, sand
salsify
sand shinnery oak
shepherdspurse
sicklepod
smartweed
sneezeweed, bitter
sowthistle, annual
spanishneedles
sumac
sunflower
sweetclover
tansymustard
tansyragwort
thistle, bull
thistle, musk
thistle, Russian
tumbleweed
velvetleaf
vervains
vetch
water plantain
willow
witchweed
wormwood
yellow rocket
yellow starthistle

DIRECTIONS FOR USE

It is a violation of Federal law to use this product in a manner inconsistent with its labeling.

Apply ESTERON 6E as water or oil spray during warm weather when weeds or brush are actively growing. Application under drought conditions often will give poor results. Use low spray pressure to minimize spray drift. On cropland and along roadsides, do not exceed 20 psi pressure. Apply enough spray volume to provide uniform coverage of weeds and brush, usually 5 to 20 gallons per acre by ground equipment and 1 to 5 gallons by aircraft. Higher gallonage may be used if desired to improve spray coverage. Generally, the lower dosages recommended on this label will be satisfactory for young, succulent growth of sensitive weed species. For less sensitive species and under conditions where control is more difficult, the higher dosages will be needed. For crop uses, do not mix with oil or other adjuvants unless specifically recommended on this label. Deep-rooted perennial weeds such as Canada thistle and field bindweed and many woody plants usually require repeated applications for maximum control. **Do not apply ESTERON 6E where spray drift may contact nearby susceptible crops or other desirable plants or may contaminate water for irrigation or domestic use. Read and follow all Use Precautions given on this label.**

NOTE: If there are uncertainties concerning special local use situations or specific crop variety tolerances to 2,4-D, consult your agricultural Experiment Station or Extension Service weed specialists for advice.

TO PREPARE THE SPRAY: (1) Fill the spray tank about half full with water, then add the required amount of ESTERON 6E, with agitation, and finally the rest of the water. NOTE: ESTERON 6E in water forms an emulsion which tends to separate unless the mixture is kept agitated. (2) If oil is added, first mix the ESTERON 6E and the oil and then add this mixture to the water. However, with adequate agitation, the oil can be added after the ESTERON 6E is mixed in the water. (3) If straight oil is used, a solution is formed and separation does not occur. Do not allow any water to get into the oil-herbicide mixture to avoid formation of an invert emulsion.

WEED CONTROL IN SMALL GRAINS NOT UNDERSEEDED WITH A LEGUME

NOTE: Do not permit dairy animals or meat animals being finished for slaughter to forage or graze treated grain fields within 2 weeks after treatment.

Spring and winter wheat, barley and rye: Apply $1/3$ to $2/3$ pint per acre. Spray when grain is in full tiller stage (usually 4 to 8 inches tall) but before the boot stage and when weeds are small. Do not apply before the tiller stage nor from early boot to the dough stage. Higher rates, up to $1\frac{1}{3}$ pints per acre, may be needed to handle difficult weed problems in certain areas such as under dry conditions especially in western areas. **However, do not use unless possible crop injury will be acceptable.**

Spring Seeded Oats: Apply $1/3$ pint per acre at the full tiller stage but before the early boot stage. Oats are less tolerant to 2,4-D than wheat or barley and are more likely to suffer some injury.

Fall Seeded Oats (Southern) Grown for Grain: Apply $1/2$ to 1 pint per acre after full tillering but before the early boot stage. Some difficult weeds may require higher rates for maximum control but crop injury may result. Do not spray during or immediately following cold weather.

Preharvest Treatment: Apply $2/3$ to $1\frac{1}{3}$ pints per acre when grains are in the hard dough stage to control large weeds that may interfere with harvest. Best results will be obtained when soil moisture is sufficient to cause succulent weed growth.

NOTE: Do not feed treated straw to livestock.

WEED CONTROL IN CORN: Use one of the following three programs: **Preemergence:** Apply 1 to $2\frac{2}{3}$ pints per acre to soil anytime after planting but before corn emerges. Do not use on light, sandy soil. **Emergence:** Apply $2/3$ pint per acre just as corn plants are breaking ground. **Postemergence:** After emergence of corn, use $1/3$ pint per acre. Application of $1/2$ to $2/3$ pint per acre may be needed for maximum control of some weeds but such rates are more likely to injure the corn. If corn is over 8 inches tall, use drop nozzles to keep the spray off the corn foliage as much as possible. Do not apply from the tasseling to dough stage. Do not use with oil, atrazine or other adjuvants. Crop injury is more likely to occur if corn is growing rapidly under high temperature and high soil moisture conditions. To reduce breakage of stalks from temporary brittleness caused by 2,4-D, delay cultivation for 8 to 10 days after treatment. **NOTE:** Hybrids vary in response to 2,4-D and some are easily injured. Spray only varieties known to be tolerant to 2,4-D. Contact seed company, Agricultural Experiment Station or Extension Service weed specialists for this information.

Preharvest Corn Treatment: After the hard dough or denting stage, apply $2/3$ to $1\frac{1}{3}$ pints per acre by air or ground equipment to suppress perennial weeds, decrease weed seed production, and control tall weeds such as bindweed, cocklebur, dogbane, jimsonweed, ragweed, sunflower, velvetleaf and vines that interfere with harvesting. Do not forage or feed corn fodder for 7 days following application.

WEED CONTROL IN SORGHUM (MILO): Apply $1/3$ pint per acre when sorghum is 5 to 15 inches tall. A higher rate of $1/2$ to $2/3$ pint per acre may be needed to control some weeds but the chance for crop injury is likewise increased. Do not use with oil. Do not treat before the sorghum is 5 inches tall nor during the boot, flowering or early dough stages. If sorghum is taller than 8 inches, use drop nozzles to keep the spray off the foliage as much as possible. Temporary crop injury may occur under conditions of high soil moisture and high air temperatures. Varieties vary in tolerance to 2,4-D and some hybrids are quite sensitive. Spray only varieties known to be tolerant to 2,4-D. Contact seed company and Extension Service authorities for this information.

WEED CONTROL IN GRASS SEED CROPS: Use $2/3$ to 1 pint per acre in the amount of water required for uniform application by air or ground equipment. Apply to established stands in spring from the tiller to early boot stage. Do not spray in boot stage. New spring seedings may be treated with the lower rate after the grasses have at least five leaves. Perennial weed regrowth may be treated in the fall.

WEED AND BRUSH CONTROL IN RANGELAND AND GRASS PASTURES

NOTE: Do not graze dairy animals on treated areas within 7 days after application. Do not use on bentgrasses, alfalfa, clover or other legumes or on newly seeded pastures. Do not apply after heading begins or when grass is in the boot to milk stage where grass seed production is desired.

Bitterweed, Broomweed, Croton, Docks, Kochia, Marshelder, Muskthistle and Other Broadleaf Weeds: Use $2\frac{2}{3}$ pints of ESTERON 6E per acre in the amount of water needed for uniform application. If the weeds are young and growing actively, $1\frac{1}{3}$ pints per acre will provide control of some species. Deep-rooted perennial weeds may require repeated treatments in the same year or in subsequent years.

Wild Garlic and Wild Onion: Apply $2\frac{2}{3}$ to 4 pints per acre, making three applications (fall-spring-fall or spring-fall-spring) starting in late fall or early spring.

Weed Control in Newly Sprigged Coastal Bermudagrass: Apply $1\frac{1}{3}$ to $2\frac{2}{3}$ pints per acre preemergence and/or postemergence.

Sand Shinnery Oak and Sand Sagebrush: On the oak, use $1\frac{1}{3}$ pints in 5 gallons of oil or in 4 gallons of water plus 1 gallon of oil per acre. Apply by aircraft between May 15 and June 15. On the sagebrush, use $1\frac{1}{3}$ pints in 3 gallons of oil per acre and apply by aircraft when foliage is fully expanded and the brush is actively growing.

Big Sagebrush and Rabbitbrush: Use $2\frac{2}{3}$ to 4 pints per acre in 2 to 3 gallons of oil or in 3 to 5 gallons of oil-water emulsion spray. For rabbitbrush, the 4 pint rate is usually required. Brush should be leafed out and growing actively when treated. Retreatment may be needed.

Chamise, Manzanita, Buckbrush, Coastal Sage, Coyotebrush and Certain Other Chaparral Species: Use $2\frac{2}{3}$ to 4 pints per acre in 5 to 10 gallons of water. One gallon of fuel oil may be included in the spray mixture for added effectiveness. Make applications by aircraft or ground equipment to obtain uniform spray coverage. For effective control, the brush **must** be fully leafed out and growing actively when sprayed. Retreatment may be needed.

WOODY PLANT CONTROL IN NON-CROP AREAS: To control species susceptible to 2,4-D in right-of-ways, fencerows, roadsides, and along drainage ditchbanks, spray brush up to 5 to 8 feet tall after spring foliage is well developed, using 4 to 5 pints of ESTERON 6E in 100 gallons of water and wetting all parts of the brush including foliage, stems and bark. This may require up to 400 gallons of spray per acre for adequate coverage of solid stands of brush. Make application in such a way as to prevent drift of the spray off the area being treated. Spraying can be effective at any time up to 3 weeks before frost as long as soil moisture is sufficient for active growth of the brush. Control will be less effective in midsummer during hot dry weather when soil moisture is deficient and plants are not actively growing. Oil or wetting agent may be added to the spray, if needed for increased effectiveness.

WEED CONTROL IN NON-CROP AREAS SUCH AS LAWNS, GOLF COURSES, CEMETERIES, PARKS, AIRFIELDS, ROADSIDES, VACANT LOTS, DRAINAGE DITCH BANKS: Apply $1\frac{1}{3}$ to 4 pints of ESTERON 6E per acre in the amount of water needed for uniform application. Usually $2\frac{2}{3}$ pints per acre provides good weed control under average conditions. Treat when weeds are young and growing well. Do not use on creeping grasses such as bent and St. Augustine except for spot treating, nor on newly seeded turf until grass is well established. Reseeding of treated areas should be delayed following treatment. With spring application, reseed in the fall; with fall application, reseed in the spring. Legumes are usually damaged or killed so do not treat areas where the legumes are desired. Deep-rooted perennial weeds may require repeated treatments in the same season or in subsequent years.

TOLE (BULRUSH) AND OTHER RUSHES: Mix $2\frac{2}{3}$ pints of ESTERON 6E and 1 gallon of diesel oil or kerosene, then add this mixture to 100 gallons of water. Spray to wet all foliage (400–800 gallons per acre). Addition of a wetting agent may be advisable. Apply in the spring during flower head emergence. Respray if needed when regrowth is 3 to 5 feet tall.

SPOT TREATMENT: To control broadleaf weeds in small noncropland areas with a hand sprayer, use $1/6$ pint of ESTERON 6E in

3 gallons of water and spray to thoroughly wet all weed foliage. Keep spray mixture agitated to prevent separation.

STORAGE AND DISPOSAL

Do not contaminate water, food, or feed by storage or disposal.

STORAGE

Keep container tightly closed when not in use. This product can be stored in an unheated building.

PESTICIDE DISPOSAL

Pesticide wastes are toxic. Improper disposal of excess pesticide, spray mixture, or rinsate is a violation of Federal Law. If these wastes cannot be disposed of by use according to label instructions, contact your State Pesticide or Environment Control Agency, or the Hazardous Waste representative at the nearest EPA Regional Office for guidance.

CONTAINER DISPOSAL

Triple rinse (or equivalent). Then offer for recycling or reconditioning, or puncture and dispose of in a sanitary landfill, or by other procedures approved by state and local authorities.

USE PRECAUTIONS

Do not APPLY ESTRON 6E herbicide directly to, or otherwise permit it to come into contact with cotton, grapes, fruit trees, vegetables, flowers, or other desirable crops or ornamental plants which are sensitive to 2,4-D herbicide. Do not permit spray mist containing it to drift onto them, since even small quantities of the spray, which may not be visible, can cause severe injury during both growing and dormant periods. Use coarse sprays to minimize drift. With ground equipment, spray drift can be lessened by keeping the spray boom as low as possible; by applying 20 gallons or more of spray per acre; by using no more than 20 pounds spraying pressure with flat fan or flooding fan nozzle tips; by spraying when wind velocity is low; and by stopping all spraying when wind exceeds 6 to 7 miles per hour. Do not apply with hollow cone-type insecticide or other nozzles that produce a fine-droplet spray.

With aircraft application, drift can be lessened by applying not less than 5 gallons of spray per acre; by using no more than 20 pounds spray pressure at the nozzles; by using nozzles which produce a coarse spray pattern; and by spraying only when the wind velocity is less than 5 miles per hour.

Applications by aircraft, ground rig and hand dispenser should be carried out only when there is no hazard from spray drift. Do not apply in the vicinity of cotton, grapes, tomatoes, or other desirable 2,4-D susceptible crop or ornamental vegetation. Do not spray when the wind is blowing towards susceptible crops or ornamental plants.

At high temperatures vapors from this product may injure susceptible plants growing nearby. Do not use in a greenhouse. Excessive amounts of this herbicide in the soil may temporarily inhibit seed germination or plant growth.

To avoid injury to desirable plants, do not handle or apply other agricultural chemicals with the same equipment used for ESTERON 6E except as specified on this label.

Local conditions may affect the use of herbicides. Consult your State Agricultural Experiment Station or Extension Service weed specialists for advice in selecting treatments from this label to best fit local conditions. **Be sure that use of this product conforms to all applicable regulations. Apply this product only as specified on the label.**

CONDITIONS OF SALE AND WARRANTY

VERTAC AND SELLER OFFER THIS PRODUCT AND THE BUYER AND USER ACCEPTS THIS PRODUCT ONLY UNDER THE FOLLOWING AGREED CONDITIONS OF SALE AND WARRANTY.

The directions for use of this product are believed to be reliable and should be followed carefully. However, it is impossible to take into account all variables and to eliminate all risks associated with its use. Injury or damage may result because of conditions which are beyond the control of Vertac or the Seller. Vertac warrants only that this product conforms to the chemical description on the label and is believed to be reasonably fit for the purposes referred to in the Directions for Use when used as directed under normal conditions. VERTAC MAKES NO OTHER EXPRESS OR IMPLIED WARRANTY OF FITNESS OR MERCHANTABILITY OR ANY OTHER EXPRESS OR IMPLIED WARRANTY. In no case shall Vertac or the Seller be liable for consequential, special or indirect damages resulting from the use or handling of this product. Any variation or exception from this warranty must be in writing and signed by an authorized Vertac representative.

ESTERON* 76BE

HERBICIDE

For the Control of Many Broadleaf Weeds, Herbaceous Perennials and Woody Plants Susceptible to 2,4-D in Grass Pastures, Certain Crops and Non-Crop Areas.

Contains Butyl Esters of 2,4-D

ACTIVE INGREDIENT:

2,4-Dichlorophenoxyacetic Acid Butyl Esters††	75.2%
INERT INGREDIENTS	24.8%
TOTAL	100.0%

†† Isomer Specific by AOAC Method No. 6.275–6.279 (13 Ed.)

EPA Reg. No. 464-279-39511 EPA Est. 464-MI-1

* Trademark of VERTAC Chemical Corporation

CAUTION

Keep Out of Reach of Children

PRECAUCION AL USUARIO: Si usted no lee inglés, no use este producto hasta que la etiqueta le haya sido explicada ampliamente.

TRANSLATION: (**TO THE USER:** If you cannot read English, do not use this product until the label has been fully explained to you.)

PRECAUTIONARY STATEMENTS

HAZARDS TO HUMANS AND DOMESTIC ANIMALS

HARMFUL IF SWALLOWED

MAY CAUSE IRRITATION

Avoid Contact with Skin, Eyes, or Clothing
Wash Thoroughly After Handling

STATEMENTS OF PRACTICAL TREATMENT

In case of contact, immediately flush eyes or skin with plenty of water. Get medical attention if irritation persists.

If swallowed, induce vomiting immediately by giving two glasses of water and sticking finger down throat. Call a physician. Do not induce vomiting or give anything by mouth to an unconscious person.

PHYSICAL OR CHEMICAL HAZARDS

Do not Cut or Weld Container

ENVIRONMENTAL HAZARDS

This product is toxic to fish. Do not apply directly to water. Do not apply where runoff is likely to occur. Do not contaminate water by cleaning of equipment or disposal of wastes. Do not contaminate irrigation ditches or water used for irrigation or domestic purposes.

For chemical spill, leak, fire or exposure call toll free 1-800-424-9200

AGRICULTURAL CHEMICAL

Do Not Ship or Store with Foods, Feeds, Drugs or Clothing

DIRECTIONS FOR USE

It is a violation of Federal law to use this product in a manner inconsistent with its labeling

Use ESTERON 76 BE herbicide to control bitterweed, croton, dandelion, docks, galinsoga, henbit, kochia, lambsquarters, marshelder, mustards, peppergrass, pigweed, plantains, ragweed, shepherdspurse, thistles, vetch, wild radish, and many other broadleaf weeds, without injury to most established grasses; also for control of

2,4-D susceptible woody plants such as coastal sage, sandsage, elderberry, hazel, locust, poison oak, sumac, and willow. Use in small grains, corn, sorghum, grass seed crops, pastures, rangeland, and in non-crop areas.

Apply ESTERON 76 BE as water or oil spray during warm weather when weeds or brush are actively growing. Application under drought conditions often will give poor results. Use low spray pressure to minimize spray drift. On cropland and along roadsides, do not exceed 20 psi pressure. Apply enough spray volume to provide uniform coverage of weeds and brush, usually 5 to 20 gallons per acre by ground equipment and 3 to 5 gallons by aircraft. Higher gallonage may be used if desired to improve spray coverage and to reduce the hazard from spray drift.

Generally, the lower dosages recommended on this label will be satisfactory for young, succulent growth of sensitive weed species. For less sensitive species and under conditions where control is more difficult, the higher dosages will be needed. For crop uses, do not mix with oil or other adjuvants unless specifically recommended on this label. Deep-rooted perennial weeds such as Canada thistle and field bindweed and many woody plants usually require repeated applications for maximum control. **Do not apply ESTERON 76 BE where spray drift may contact nearby susceptible crops or other desirable plants or may contaminate water for irrigation or domestic use. Do not apply in the vicinity of 2,4-D sensitive crops or ornamental plants since vapors from this product may cause injury to such crops or plants. Read and follow all Use Precautions given on this label.**

NOTE: If there are uncertainties concerning special local use situations or specific crop variety tolerances to 2,4-D, consult local Extension Service or University Specialists for advice.

TO PREPARE THE SPRAY

(1) Fill the spray tank about half full with water, then add the required amount of ESTERON 76 BE, with agitation, and finally the rest of the water. NOTE: ESTERON 76 BE in water forms an emulsion which tends to separate unless the mixture is kept agitated. (2) If oil is added, first mix the ESTERON 76 BE and the oil and then add this mixture to the water. However, with adequate agitation, the oil can be added **after** the ESTERON 76 BE is mixed in the water if **strong agitation** is provided. (3) If straight oil is used, a solution is formed and separation does not occur. Do not allow any water to get into the oil-herbicide mixture to avoid formation of an invert emulsion.

WEED CONTROL IN SMALL GRAINS NOT UNDERSEEDED WITH A LEGUME

NOTE: Water is recommended to make up the spray. If oil is used, there is a greater risk of crop injury and of spray drift. Do not permit dairy animals or meat animals being finished for slaughter to forage or graze treated grain fields within 2 weeks after treatments.

Spring Wheat and Barley: Apply $\frac{1}{3}$ to $\frac{2}{3}$ pint per acre by air or ground equipment. A $\frac{1}{2}$ pint per acre rate of ESTERON 76 BE is an average dosage, effective on many weeds. Spray when grain is in full tiller stage (usually 4 to 8 inches tall) but before the boot stage and when weeds are small. Do not apply before the tiller stage nor from early boot to the dough stage. Higher rates (up to $1\frac{1}{3}$ pints per acre) may be required to control certain weeds but crop injury may result.

Winter Wheat and Rye: Apply $\frac{1}{3}$ to $\frac{1}{2}$ pint per acre in the spring at the full tiller stage but before the early boot stage. See more complete use directions under **Spring Wheat and Barley.**

Spring Seeded Oats: Apply $\frac{1}{3}$ pint per acre at the full tiller stage but before the early boot stage. Oats are less tolerant to 2,4-D than wheat or barley and are more likely to suffer some injury, especially if higher rate ($\frac{1}{2}$ to $\frac{2}{3}$ pint) are used to control difficult weeds.

Preharvest Treatment: Apply $\frac{2}{3}$ to $1\frac{1}{3}$ pints per acre when grains are in the hard dough stage to control large weeds that may interfere with harvest. Best results will be obtained when soil moisture is sufficient to cause succulent weed growth. NOTE: Do not feed treated straw to livestock.

WEED CONTROL IN CORN

Use one of the following three programs:

Preemergence: Apply $1\frac{1}{3}$ to $2\frac{2}{3}$ pints per acre to soil any time after planting but before corn emerges. Do not use on light, sandy soil.

Emergence: Apply $\frac{2}{3}$ pint per acre just as corn plants are breaking ground.

Postemergence: After emergence of corn, use $\frac{1}{3}$ pint per acre. Application of $\frac{1}{2}$ to $\frac{2}{3}$ pint per acre may be needed for maximum control of some weeds but such rates are more likely to injure the corn. If corn is over 8 inches tall, use drop nozzles to keep the spray off the corn foliage as much as possible. Do not apply from the tasseling to dough stage. Do not use with oil, atrazine, or adjuvants. Crop injury is more likely to occur if corn is growing rapidly under high temperature and high soil moisture conditions. To reduce breakage of stalks from temporary brittleness caused by 2,4-D, delay cultivation for 8 to 10 days after treatment. NOTE: Hybrids vary in response to 2,4-D and some are easily injured. Spray only varieties known to be tolerant to 2,4-D. Contact seed company and Extension Service Weed Specialists for this information.

WEED CONTROL IN SORGHUM (MILO)

Apply $\frac{1}{3}$ pint per acre when sorghum is 5 to 15 inches tall. A higher rate of $\frac{1}{2}$ to $\frac{2}{3}$ pint per acre may be needed to control some weeds but the chance for crop injury is likewise increased. Do not use with oil or other adjuvants. Do not treat before the sorghum is 5 inches tall nor during the boot, flowering, or early dough stages. If sorghum is taller than 8 inches, use drop nozzles to keep the spray off the foliage as much as possible. Temporary crop injury may occur under conditions of high soil moisture and high air temperatures. Varieties vary in tolerance to 2,4-D and some hybrids are quite sensitive. Spray only varieties known to be tolerant to 2,4-D. Contact seed company and Extenstion Service weed specialists for this information. Do not apply in the vicinity of cotton, soybeans, and other 2,4-D susceptible plants.

WEED CONTROL IN GRASS SEED CROPS

Use $\frac{2}{3}$ to 1 pint per acre in the amount of water required for uniform application by air or ground equipment. Apply to established stands in spring from the tiller to early boot stage. Do not spray in boot stage. New spring seedings may be treated with the lower rate after the grasses have at least five leaves. Perennial weed regrowth may be treated in the fall.

WEED AND BRUSH CONTROL IN RANGELAND AND GRASS PASTURES BY AIR OR GROUND EQUIPMENT

NOTE: Do not graze dairy animals on treated areas within 7 days after application. The following treatments will injure or kill legumes so use only where loss of legumes can be tolerated. Do not apply on newly seeded areas until grass is well established. Do not apply from early boot to milk stage when grass seed production is desired.

Broadleaf Weeds: To control bitterweed, broomweed, croton, docks, kochia, marshelder, muskthistle, and others, use $2\frac{1}{2}$ to 3 pints of ESTERON 76 BE per acre in the amount of water needed for uniform application. If the weeds are young and growing actively, $1\frac{1}{2}$ to 2 pints per acre will provide control of many species. Deep-rooted perennial weeds may require repeated treatments in the same year or in subsequent years.

Chaparral Brush Species: To control chamise, manzanita, buckbrush, coastal sage, and certain other chaparral species, use $1\frac{1}{2}$ to 2 quarts per acre in 8 to 10 gallons of water. A gallon of oil per acre may be included in the spray mixture for added effectiveness. For effective control, the brush must be fully leafed out and growing actively when sprayed. Retreatment may be needed.

Big Sagebrush: Use $\frac{1}{2}$ to 2 quarts per acre in 2 to 3 gallons of oil or in 3 to 5 gallons of oil-water emulsion spray. For effective control the sagebrush should be in full foliage and growing actively when sprayed.

WEED CONTROL IN NON-CROP AREAS SUCH AS LAWNS, AIRFIELDS, ROADSIDES, VACANT LOTS, DRAINAGE DITCH BANKS

Apply 1 to 2 quarts of ESTERON 76 BE per acre in the amount of water needed for uniform application. Usually $1\frac{1}{2}$ quarts per acre provides good weed control under average conditions. Treat when weeds are young and growing well. Do not use on golf greens nor on dichondra or other broadleaf herbaceous ground covers. Do not use on creeping grasses such as bent and St. Augustine except for spot treating, nor on newly seeded turf until grass is well established. Reseeding of treated areas should be delayed following treatment. With spring application, reseed in the fall; with fall application, reseed in the spring. Legumes are usually damaged or killed so do not treat areas where the legumes are desired. Deep-rooted perennial weeds may require repeated treatments in the same season or in subsequent years.

WOODY PLANT CONTROL IN NON-CROP AREAS

To control species susceptible to 2,4-D in right-of-ways, fencerows, roadsides, and along drainage ditchbanks, spray brush up to 5 to 8 feet tall after spring foliage is well developed, using 2 to 3 quarts of ESTERON 76 BE in 100 gallons of water and wetting all parts of the brush including foliage, stems, and bark. This may require up to 400 gallons of spray for adequate coverage of solid stands of brush. Make application in such a way as to prevent drift of the spray off the area being treated. Spraying can be effective at any time up to 3 weeks before frost as long as soil moisture is sufficient for active growth of the brush. Less effective control may be obtained during hot dry weather when soil moisture is deficient and plants are not actively growing. Oil or wetting agent may be added to the spray, if needed for increased effectiveness.

SPOT TREATMENT

To control broadleaf weeds in small non-cropland areas with a hand sprayer, use 1/6 pint of ESTERON 76 BE in 3 gallons of water and spray to thoroughly wet all weed foliage. Keep spray mixture agitated to prevent separation.

STORAGE AND DISPOSAL

Do not contaminate water, food or feed by storage or disposal.

STORAGE

Keep container tightly closed when not in use. This product can be stored in an unheated building. If exposed to subfreezing temperatures, it should be warmed to at least 40°F and mixed thoroughly before using.

PESTICIDE DISPOSAL

Pesticide wastes are toxic. Improper disosal of excess pesticide, spray mixture, or rinsate is a violation of Federal Law. If these wastes cannot be disposed of by use according to label instructions, contact you State Pesticide or Environmental Control Agency, or the Hazardous Waste representative at the nearest EPA Regional Office for guidance.

CONTAINER DISPOSAL

Triple rinse (or equivalent). Then offer for recycling or reconditioning, or puncture and dispose of in a sanitary landfill, or incineration, or, if allowed by state and local authorities, by burning. If burned, stay out of smoke.

USE PRECAUTIONS

Do not apply ESTERON 76 BE herbicide directly to, or otherwise permit it to come into contact with cotton, grapes, fruit trees, vegatables, flowers, or desirable crop or ornamental plants which are sensitive to 2,4-D herbicide. Do not permit spray mist containing it to drift onto them, since even very small quantities of the spray, which may not be visible, can cause severe injury during both growing and dormant periods. Use coarse sprays to minimize drift. With ground equipment, spray drift can be lessened by keeping the spray boom as low as possible; by applying 20 gallons or more of spray per acre; by using no more than 20 pounds spraying pressure with flat fan or flooding flat fan nozzle tips; by spraying when wind velocity is low; and by stopping all spraying when wind exceeds 6 to 7 miles per hour. Do not apply with hollow cone-type insecticide or other nozzles that produce a fine-droplet spray. With aircraft application, drift can be reduced by applying a minimum of 5 gallons of spray per acre; by using no more than 20 pounds spray pressure at the nozzles; by using nozzles which produce a coarse spray pattern; and by spraying only when the wind velocity is less than 5 miles per hour.

Applications by aircraft, ground rig and hand dispenser should be carried out only when there is no hazard from spray drift. Do not apply in the vicinity of cotton, grapes, tomatoes, or other desirable 2,4-D susceptible crop or ornamental vegetation. Do not spray when the wind is blowing towards suscptible crops or ornamental plants.

This is a high-volatile 2,4-D ester formulation. Vapors from this product may injure susceptible plants growing nearby. Do not use in or near a greenhouse. Excessive amounts of this herbicide in the soil may temporarily inhibit seed germination or plant growth.

To avoid inury to desirable plants, do not handle or apply other agricultural chemicals with the same equipment used for ESTERON 76 BE except as specified on this label.

Local conditions may affect the use of herbicides. Consult your State Agricultural Experiment Station or Extension Service weed specialists for advice in selecting treatments from this label to best fit local conditions. **Be sure that use of this product conforms to all applicable regulations. Apply this product only as specified on this label.**

CONDITIONS OF SALE AND WARRANTY

VERTAC AND SELLER OFFER THIS PRODUCT AND THE BUYER AND USER ACCEPTS THIS PRODUCT ONLY UNDER THE FOLLOWING AGREED CONDITIONS OF SALE AND WARRANTY.

The directions for use of this product are believed to be reliable and should be followed carefully. However, it is impossible to take into account all variables and to eliminate all risks associated with its use. Injury or damage may result because of conditions which are beyond the control of Vertac or the Seller. Vertac warrants only that this product conforms to the chemical description on the label and is believed to be reasonably fit for the purposes referred to in the Directions for Use when used as directed under normal conditions. VERTAC MAKES NO OTHER EXPRESS OR IMPLIED WARRANTY OF FITNESS OR MERCHANTABLILITY OR ANY OTHER EXPRESS OR IMPLIED WARRANTY. In no case shall Vertac or the Seller be liable for consequential, special or indirect damages resulting from the use or handling of this product. Any variation or exception from this warranty must be in writing and signed by an authorized Vertac representative.

ESTERON* 99* Concentrate

2,4-D LOW VOLATILE HERBICIDE

ACTIVE INGREDIENT:

Isooctyl Ester of 2,4-Dichlorophenoxyacetic Acid	65.4%
INERT INGREDIENTS	34.6%
TOTAL	100%

Equivalent to 43.4% of 2,4-Dichlorophenoxyacetic Acid or 3.8 lbs./gal.

Isomer specific by AOAC Method 6.275, 13th Ed., 1980.
E.P.A. Registration No. 39511-75
E.P.A. Est. 464-MI-1[MM], 39511-AR-1[AR]
Superscript used corresponds to letters in Lot Number.

CAUTION

Keep Out of Reach of Children

CAUCION: PRECAUCION AL USUARIO

Si usted no lee ingles, no use este producto hasta que la etiqueta le haya sido explicada ampliamente.

PRECAUTIONARY STATEMENTS

HAZARDS TO HUMANS AND DOMESTIC ANIMALS

HARMFUL IF SWALLOWED

Avoid Breathing Spray Mist • Avoid Contact with Skin, Eyes, and Clothing • In Case of Contact Immediately Flush Eyes or Skin with Plenty of Water • Get Medical Attention If Irritation Persists • Do Not Apply this Product in Such a Manner as to Directly or Through Drift Expose Workers or Other Persons • The Area Being Treated Must be Vacated by Unprotected Persons

STATEMENT OF PRACTICAL TREATMENT

If Swallowed: DO NOT induce vomiting. Get medical attention. Contains petroleum distillate. **If On Skin:** Wash contaminated areas thoroughly with soap and water. Do not reuse contaminated clothing until washed. Get medical attention if irritation persists. **If In Eyes:** Flush with copious amounts of clean water for 10-15 minutes. Get medical attention.

NOTE TO ATTENDING PHYSICIAN: May cause chemical pneumenitis if aspirated. If lavage is performed, suggest endotracheal and or esophagoscopic control.

Physical or Chemical Hazards

Do Not Use or Store Near Heat or Open Flame

ENVIRONMENTAL HAZARDS

Do not apply directly to water except as specified on this label for aquatic weed control. Do not contaminate water by cleaning of equipment or disposal of wastes. Do not contaminate water intended for irrigation or domestic purposes. Do not apply when weather conditions favor drift from target area.

For chemical spill, leak, fire or exposure call toll free 1-800-424-9300

AGRICULTURAL CHEMICAL

Do Not Ship or Store with Foods, Feeds, Drugs or Clothing

DIRECTIONS FOR USE

It is a violation of Federal law to use this product in a manner inconsistent with its labeling.

REENTRY STATEMENT

Do not enter treated areas without protective clothing until sprays have dried. Protective clothing should include: hat or other suitable head covering, long sleeved shirt and long legged trousers, or a coverall type garment, shoes, and socks.

Because certain states may require more restrictive reentry intervals for various crops treated with this product, consult your State Department of Agriculture for further information.

Written or oral warnings must be given to workers who are expected to be in a treated area or in an area about to be treated with this product. The front panel PRECAUTIONARY STATEMENTS should be read to workers as well as the instruction not to enter until sprays have dried. When oral warnings are given, warnings shall be given in a language customarily understood by workers. Oral warnings must be given if there is reason to believe that written warnings cannot be understood by workers. Written warnings must include the following information.

"CAUTION: Area treated with Esteron 99 Concentrate on (date of application). Do not enter without appropriate protective clothing until sprays have dried. (Insert here Statements of Practical Treatment as on front panel.)"

STORAGE AND DISPOSAL

STORAGE

Do not contaminate water, food or feed by storage or disposal. Open dumping is prohibited. Do not store this product near fertilizers, seeds, insecticides, or fungicides. Do not store near heat or open flame. Containers should not be stacked more than six (6) high. Reclose all partially used containers by thoroughly tightening screw cap. Damaged or leaking containers which cannot be used immediately should be transferred to suitable sound containers and properly marked. Absorb any spills with a suitable clay absorbant and dispose of as indicated under "Pesticide Disposal."

For safety and prevention of unauthorized use, all pesticides should be stored in locked facilities.

To prevent accidental misuse, different pesticides should be stored in separate areas with enough distance between to provide clear identification.

Opened, partially used pesticides should be stored in original labeled containers when possible. When transfer to another container is necessary because of leakage or damage, carefully mark and identify contents of the new container.

PESTICIDE DISPOSAL

Pesticide wastes are toxic. Improper disposal of excess pesticide, spray mixture, or rinsate is a violation of Federal Law. If these wastes cannot be disposed of by use according to label instructions, contact your State Pesticide or Environmental Control Agency, or the Hazardous Waste representative at the nearest EPA Regional Office for guidance.

CONTAINER DISPOSAL

Triple rinse (or equivalent), adding rinsate to spray tank. Then offer for recycling or reconditioning, or puncture and dispose of in a sanitary landfill, or by other procedures approved by state and local authorities.

GENERAL INFORMATION

Performance of this product may be affected by local conditions, crop varieties, and application method. User should consult local extension service, agricultural experiment, or university weed specialists, and state regulatory agencies for recommendations in your area.

Best results are obtained when product is applied to young succulent weeds that are actively growing. Application rates lower than recommended will be satisfactory on susceptible annual weeds. For perennial weeds and conditions such as the very dry areas of the western states, where control is difficult, the higher recommended rates should be used.

When product is used for weed control in crops, the growth stage of the crop must be considered.

Some plants and weeds, especially woody varieties, are difficult to control and may require repeat applications. Application rates should be 1 to 5 gallons of total spray by air or 5 to 25 gallons by ground equipment unless otherwise directed. In either case, use the same amount of 2,4-D recommended per acre. For crop uses, do not mix with oil, surfactants, or other adjuvants unless specifically recommended on label. To do so may reduce herbicides selectivity and could result in crop damage.

Aerial application should be used only when there is no danger of drift to susceptible crops. Many states have regulations concerning aerial application of 2,4-D formulations. Consult local regulatory authorities before making applications. Although Esteron 99 Concentrate is a low volatile formulation, at temperatures above 95° F vapors may damage susceptible plants nearby.

Because coarse sprays are less likely to drift than fine, do not use equipment (such as hollow cone small orifice nozzles) or conditions (such as high pressure) that produce such sprays.

Product should not be allowed to come into contact with desirable susceptible plants such as beans, cotton, fruit trees, grapes, legumes, ornamentals, peas, tomatoes, and other vegetables. Product should not be used in greenhouses. Excessive amounts of this product in the soil may temporarily inhibit seed germination and all plant growth.

Users should note that herbicide treatment of public water requires a permit from appropriate state agencies in most states. Your state Conservation Department, or Game and Fish Commission will aid you in securing a permit in your state.

If stored below freezing, it may be necessary to warm product to 40° F and agitate before using. This does not affect the efficiency of the product.

Spray equipment used to apply 2,4-D should not be used for any other purpose until thoroughly cleaned with a suitable chemical cleaner.

Spray Preparation: Add the recommended amount of product to approximately one-half the volume of water to be used for spraying.

Agitate well, then add the remainder of the water. Continue agitation during application until spray tank is empty.

Use in Liquid Nitrogen Fertilizer: Product may be combined with liquid nitrogen fertilizer suitable for foliar application on corn, grass, pastures, or small grains in one operation. Use Esteron 99 Concentrate according to directions on this label for those crops. Use liquid fertilizer at rates recommended by supplier or extension service specialist. Mix the Esteron 99 Concentrate and fertilizer according to the following instructions:

Fill the spray tank approximately ½ full with the liquid fertilizer. Add the Esteron 99 Concentrate while agitating the tank. Add the remainder of the liquid fertilizer while continuing to agitate. Apply immediately maintaining agitation during application until tank is empty. DO NOT APPLY DURING COLD (NEAR FREEZING) WEATHER. Spray mixture must be used immediately and may not be stored.

WHERE TO USE

Esteron 99 Concentrate is used to control broad-leaved weeds in cereal crops, corn, sorghum, weeds and brush in rangeland, pastures, rights-of-way, and similar noncrop uses; and for aquatic weed control.

PLANTS CONTROLLED

Esteron 99 Concentrate will kill or control the following in addition to many other noxious plants susceptible to 2,4-D:

alligatorweed
arrowhead
artichoke
beggarticks
bindweed, hedge
bindweed, field
bindweed, European
bitterweed
bitter wintercress
blueweed, Texas
boxelder
broomweed
buckbrush
buckhorn
buckwheat, wild
bull thistle
bulrush
burdock
burhead
bur ragweed
buttercup
carpetweed
catnip
chamise
chickweed
chickory
cocklebur
coffeebean
coffeeweed
cornflower
coyotebrush
creeping jenny
croton
curly indigo
dandelion
docks
dogfennel
duckweed
elderberry
galinsoga
goatsbeard
goldenrod
ground ivy
halogeton
hemp
hoary cress
honeysuckle
milkvetch
morningglory
mustard
nettles
parrotfeather
pennycress (fanweed)
pennywort
pepperweed
pigweed
plantains
poison ivy
pokeweed
poorjo
povertyweed
puncturevine
purslane
rabbitbrush
ragweed
rape, wild
redstem sage
rush
Russian thistle
sagebrush, big coast
sagebrush, sand
salsify
sand shinnery oak
shepherdspurse
sicklepod
smartweed, annual
sneezeweed, bitter
sowthistle
spanishneedles, annual
stinkweed
sumac
sunflower
sweet clover
tansyragwort
thistle, musk
tumbleweed
velvetweed
vervains
vetch
Virginia creeper
waterhyacinth
waterlily
water plantain
waterprimrose
indigo
ironweed
jewelweed
jimsonweed
ladysthumb
lambsquarters
loco weed
loco, big bend
mallow, venice
manzanita
marshelder
Mexican weed
wild garlic
wild lettuce
wild mustard
wild onion
wild radish
willow
witchweed
wormwood
yellow rocket
yellow starthistle

NOTICE

The control of "hybrid" pigweeds appears to be less satisfactory from 2,4-D products than formerly experienced on "non-hybrid" varieties. Since 2,4-D herbicides are not as effective on the "hybrid" pigweeds, it is necessary to apply higher recommended rates of 2,4-D for control, especially later in the growing season. Higher rates injure some crops, so less than satisfactory pigweed control may be experienced by the highest tolerated crop dosages.

Therefore, the Vertac Chemical Corporation does not include pigweed among the species covered by the performance guarantee statements on the labels for the product. At this time, this disclaimer applies only to the High Plains of Texas and western Oklahoma, including the Panhandles. All other guarantees on this product label are unchanged by this disclaimer.

CROPS

SMALL GRAINS NOT UNDERSEEDED WITH A LEGUME (barley, oats, wheat, rye): See table for recommended use rates.

Spray when weeds are small after grain begins tillering but before boot stage (usually 4 to 8 inches tall). Do not apply before the tiller stage nor from early boot through milk stage. To control large weeds that will interfere with harvest or to suppress perennial weeds, preharvest treatment can be applied when the grain is in the dough stage. Best results will be obtained when soil moisture is adequate for plant growth and weeds are growing well.

Spring Planted Oats: Use ½ pint per acre in sufficient water to give good coverage. Apply after the fully tillered stage, except during the boot to dough stage.

Fall Planted Oats: Apply ¼ to 1¼ pints per acre after full tillering but before early boot stage. Some difficult weeds may require higher rates of ¾ to 1¼ pints per acre for maximum control, but crop injury may result. Do not spray during or immediately following cold weather.

Note: Oats are less tolerant to 2,4-D than wheat or barley and more likely to be injured. Do not forage or graze treated grain fields within 2 weeks after treatment with 2,4-D. Do not feed treated straw to livestock.

CORN: See table for recommended use rates.

Preemergence: Apply Esteron 99 Concentrate from 3 to 5 days after planting but before corn emerges. Do not use on very light, sandy soils. Use the higher rates on heavy soils. Plant corn as deep as practical.

Post Emergence: Best results are usually obtained when weeds are small and corn is 4 to 18 inches tall. When corn is over 8 inches tall, use drop nozzles. Do not apply from tasseling to dough stage. If corn is growing rapidly and temperature and soil moisture content is high, use ½ pint per acre rate to reduce possibility of crop damage. Delay cultivation for 8 to 10 days to prevent stalk breakage due to temporary brittleness caused by 2,4-D. Application rates of up to 1 pint per acre may be used to control some hard-to-control weeds. However, the possibility of injury to the corn is increased.

If corn is over 8 inches tall, use drop nozzles to keep spray off corn foliage as much as possible. Do not use with oil, atrazine, or other adjuvants. Since the tolerance to 2,4-D of individual hybrids varies, consult your local Extension Service, Agricultural Experiment Station, or University Weed Specialist for information.

Pre-Harvest: After the hard dough or denting stage, apply 1 to 2 pints of Esteron 99 Concentrate per acre by air or ground equipment to suppress perennial weeds, decrease weed seed production, and control tall weeds such as Bindweed, Cocklebur, Dogbane, Jimson-weed, Ragweed, Sunflower, Velvetleaf, and vines that interfere with harvesting. Do not forage or feed corn fodder to livestock for 7 days following application.

SORGHUM (Milo): See table for recommended rate.

Apply to sorghum when crop is 4 to 12 inches high with secondary roots well established. Use drop nozzles when crop is over 10 inches high. Do not apply from flowering to dough stage. Rates of up to 1 pint per acre may be used to control some hard-to-control weeds. However, the chance of crop injury is increased with the higher rates. Do not use with oil. Use lower rate if conditions of high temperature and high soil moisture exist.

Recommended Rates of Esteron 99 Concentrate Per Acre[2]

Crop (See Detailed Instructions Above)	Rate, Average Conditions	Rate, Dry Conditions, as in Western States[1]
Small Grains (Wheat, Barley, Rye):		
Annual Weeds	½ to 1 pint	1 to 2 pints
Perennial Weeds	1 pint	1¼ to 2 pints
Preharvest	1 to 2 pints	
Oats:		
Spring	½ pint	
Fall	½ to ¾ pint	
Corn:		
Preemergence	1 to 2 quarts	
Postemergence	½ pint	½ to ¾ pint
Preharvest	1 to 2 pints	
Sorghum (Milo):		
Postemergence	½ pint	½ to ¾ pint

[1] Arizona, Idaho, Montana, Nevada, Oregon, Utah, Washington, Wyoming.

[2] If band treatment is used, base the dosage rate on the actual area sprayed.

SUGARCANE:

Use 1 quart per acre as a preemergence application before canes appear or 2 quarts per acre as a blanket spray after cane emerges and through layby, to aid in the control of Johnsongrass seedlings and susceptible broadleaf weeds.

ORNAMENTAL TURF:

Use 1 to 3 pints of Esteron 99 Concentrate in enough water to give good coverage to one acre on established stands of perennial grasses, depending on type of weeds and stage of growth. Do not use on creeping grasses such as Bent except for spot spraying. Newly seeded turf should not be treated until after the second mowing and the lower dosage rate should be used.

GRASS SEED CROPS:

Apply 1 to 4 pints of Esteron 99 Concentrate per acre in the Spring or Fall to control broadleaf weeds in grass being grown for seed. Do not apply from early boot to milk stage. Spray seedling grass only after the five leaf stage, using ¾ to 1 pint per acre to control small seedling weeds. After the grass is well established, higher rates of up to 4 pints can be used to control hard-to-kill annual or perennial weeds. For best results, apply when soil moisture is adequate for good growth. Do not use on Bent unless injury can be tolerated. Do not graze dairy animals nor cut forage for hay within 7 days of application.

FALLOW LAND:

On established perennial species such as Canada thistle and Field bindweed, apply up to 3 quarts per acre of Esteron 99 Concentrate. For annual broadleaf weeds, apply 1 to 2 quarts per acre. Do not plant any crop for 3 months after treatment or until 2,4-D has disappeared from the soil.

ESTABLISHED PASTURES AND RANGELANDS:

Use 1 to 4 pints per acre in sufficient water to give good coverage to one acre depending on type of weeds and stage of growth. Use only on established stands of perennial grasses. DO NOT graze animals on treated areas within 7 days of application.

General Weed Control (Airfields, Roadsides, Vacant Lots, Drainage Ditch Banks, Fence Rows, Industrial Sites and similar areas): Use 1 to 3 quarts of Esteron 99 Concentrate per acre. Usually 2 quarts per acre will give adequate control. Do not use on herbaceous ground covers or creeping grass such as Bent. Legumes will usually be damaged or killed. Deeprooted perennials may require repeat applications. Do not use on freshly seeded turf until grass is well established. Delay reseeding for 30 days or until 2,4-D has disappeared from soil.

WOODY PLANT CONTROL:

To control woody plants susceptible to 2,4-D, such as Alder, Buckbrush, Elderberry, Sumac, and Willow on non-crop areas, use 2 to 3 quarts of Esteron 99 Concentrate per acre in 100 gallons of water. Wet all parts of the plants thoroughly, including stem and foliage, to the point of run off. Higher volumes of up to 400 gallons per acre are necessary where the brush is very dense and over 6 to 8 feet high. Applications are more effective when made on actively growing plants. Treatment should not be made during time of severe drought or in early Fall when leaves lose their green color. Hard-to-control species may require re-treatment next season.

USES IN FOREST MANAGEMENT

CONIFER RELEASE:

For control of alder, apply 1½ to 2 quarts of product per acre in 8 to 25 gallons of water, and apply as a foliage spray between mid-May and mid-June.

For control of madrone, manzanita, oak, tanoak, and similar species to release hemlock, spruce, and firs, apply 3 quarts of product per acre in 8 to 25 gallons of water, just prior to or during budbreak of Douglas fir.

After northern conifers jack pine, red pine, black spruce, and white spruce cease growth and "harden off" in late summer, a spray of 1½ to 3 quarts of Esteron 99 Concentrate in 8 to 25 gallons of water per acre may be applied by air to control certain competing hardwood species such as alder, aspen, birch, hazel and willow. Since this treatment may cause occasional conifer injury, do not use if such injury cannot be tolerated. Consult your regional or extension forester or state herbicide specialist for recommendations to fit local conditions.

For control of hazel brush and similar species in the Lake States area, apply 2 quarts of product per acre in 8 to 25 gallons of water, when new shoot growth of Hazel is complete.

Site Preparation:

(As Budbreak Spray)—For control of alder prior to planting seedlings, apply 2 to 4 quarts of product per acre in 8 to 25 gallons of water, after alder budbreak but before foliage is ¼ full size.

(As Foliage Spray)—For control of alder prior to planting seedlings apply 2 quarts of product per acre in 8 to 25 gallons of water, after most alder leaves are full size.

AQUATIC APPLICATIONS:

For Aquatic Weeds in Lakes, Ponds, Drainage Ditches, and Marshes: Use 2½ to 4½ pints of product in 50 to 100 gallons of water per acre. Spray to wet foliage thoroughly. Application should be made when leaves are fully developed above waterline and plants are actively growing. Your State Conservation Department or Game and Fish Commission will assist you in determining the best time and rate for application under local conditions.

DO NOT APPLY to more than ⅓ to ½ of a lake or pond in any one month because excessive decaying vegetation may deplete oxygen content of water, killing fish.

Perennial and other hard-to-control weeds may require a repeat application to give adequate control.

CONDITIONS OF SALE AND WARRANTY

VERTAC AND SELLER OFFER THIS PRODUCT AND THE BUYER AND USER ACCEPTS THIS PRODUCT ONLY UNDER THE FOLLOWING AGREED CONDITIONS OF SALE AND WARRANTY.

The directions for use of this product are believed to be reliable and should be followed carefully. However, it is impossible to take into account all variables and to eliminate all risks associated with its use. Injury or damage may result because of conditions which are beyond the control of Vertac or the Seller. Vertac warrants only that this product conforms to the chemical description on the label and is believed to be reasonably fit for the purposes referred to in the Directions for Use when used as directed under normal conditions. VERTAC MAKES NO OTHER EXPRESS OR IMPLIED WARRANTY OF FITNESS OR MERCHANTABILITY OR ANY OTHER EXPRESS OR IMPLIED WARRANTY. In no case shall Vertac or the Seller be liable for consequential, special or indirect damages resulting from the use or handling of this product. Any variation or exception from this warranty must be in writing and signed by an authorized Vertac representative.

MONEY BACK GUARANTEE

ESTERON 99 Concentrate herbicide is guaranteed by The Vertac Chemical Corporation to the full extent of the purchase price:

1. To give satisfactory control of weed and brush species listed on the container when used as recommended.
2. To form a suitable spray mixture in any water fit for spray use.
3. To store satisfactorily at temperatures as low as −40 F.

* Trademark of Vertac Chemical Corporation.

FORMULA 40*

HERBICIDE

Contains Alkanolamine Salts‡ of 2,4-D

For the Selective Control of Many Broadleaf Weeds in Non-Crop Areas, Grass Pastures, Rangelands and in Certain Crops. Also for Control of Trees by Injection

ACTIVE INGREDIENT:

Alkanolamine Salts (of the Ethanol and Isopropanol Series) of 2,4-Dichlorophenoxyacetic Acid††	56.5%
INERT INGREDIENTS	43.5%
TOTAL	100.0%

2,4-Dichlorophenoxyacetic Acid Equivalent: 36.5%—3.8 lb/gal

††Isomer Specific by AOAC Method No. 6.275–6.279 (13th Ed.)

EPA Reg. No. 464-1-39511 EPA Est. 464-MI-1

‡ Salts are the least volatile forms of 2,4-D, however, the use precautions specified on this label must be followed to avoid possible injury to susceptible crops from spray drift.

DANGER

Keep Out of Reach of Children

PRECAUCION AL USUARIO: Si usted no lee inglés, no use este producto hasta que la etiqueta le haya sido explicada ampliamente.

TRANSLATION: (TO THE USER: If you cannot read English, do not use this product until the label has been fully explained to you.)

PRECAUTIONARY STATEMENTS

HAZARDS TO HUMANS AND DOMESTIC ANIMALS

CAUSES IRREVERSIBLE EYE DAMAGE AND SKIN IRRITATION • ARMFUL IF SWALLOWED

Do Not Get in Eyes, on Skin or on Clothing • Wear Goggles or Face Shield, or Safety Glasses when Handling • Avoid Contamination of Food • Wash Thoroughly After Handling

STATEMENTS OF PRACTICAL TREATMENT

In Case of Contact, immediately flush eyes with plenty of water for at least 15 minutes and get medical attention promptly. Flush skin with water and get medical attention if signs of irritation develop. Remove and wash contaminated clothing before reuse.

If swallowed, promptly drink a large quantity of milk, egg whites, gelatin solution or, if these are not available, drink large quantities of water. Induce vomiting by placing finger in back of throat. Never give anything by mouth to an unconscious person. Avoid alcohol. Call a physician.

PHYSICAL OR CHEMICAL HAZARDS

Do Not Cut or Weld Container.

ENVIRONMENTAL HAZARDS

Do not apply directly to water. Do not contaminate water by cleaning of equipment or disposal of wastes. Do not contaminate irrigation ditches or water used for irrigation or domestic purposes.

See Side Panel for Important Use Precautions

For chemical spill, leak, fire or exposure call toll free 1-800-424-9300

AGRICULTURAL CHEMICAL

Do Not Ship or Store with Food, Feeds, Drugs or Clothing.

WEED LIST

FORMULA 40 herbicide is recommended for control of numerous broadleaf weeds and certain 2,4-D susceptible woody perennials without injuring most grasses. Species controlled include the following plus many others:

bitterweed; broomweed; burdock; carpetweed; chicory; cocklebur; coffeeweed; croton; dandelion; dock; galinsoga; hemp, wild; jewelweed; jimsonweed; lambsquarters; loco, bigbend; lupines; mallow, Venice; marshelder; morningglory, annual; pennycress, (fanweed); pennywort; pepperweed, field; pigweed††; plantains; poorjoe; pusley, Florida; ragweed, common; rape, wild; rocket, yellow; shepherdspurse; sicklepod; smartweed, (annual); sneezeweed, bitter; sowthistle, annual; spanishneedles; sunflower; sweetclover; thistle, bull; thistle, musk; thistle, Russian; velvetleaf; vetch; waterplantain; wild mustard; witchweed; and wormwood.

†† The control of "hybrid" pigweeds appears to be less satisfactory from 2,4-D products than formerly experienced on "non-hybrid" varieties. Since 2,4-D herbicides are not as effective on the "hybrid" pigweeds, it is necessary to apply higher rates of 2,4-D for control, especially later in the growing season. Higher rates injure some crops, so less than satisfactory pigweed control may be experienced by the highest tolerated crop dosages.

Therefore, Vertac Chemical Corporation does not include pigweed among the species covered by the performance guarantee statements on the labels for FORMULA 40 herbicide. At this time, this disclaimer applies only to the High Plains of Texas and western Oklahoma, including the Panhandles. All other guarantees, on these product labels are unchanged by this disclaimer.

DIRECTIONS FOR USE

It is a violation of Federal law to use this product in a manner inconsistent with its labeling.

Generally, the lower dosages given will be satisfactory for young, succulent growth of sensitive weed species. For less sensitive species and under conditions where control is more difficult, the higher dosages will be needed. Apply FORMULA 40 during warm weather when weeds are young and growing actively. Use enough spray volume for uniform coverage by ground or air application. If only bands or rows are treated, leaving middles unsprayed, dosage per crop acre is reduced proportionately. **Do not apply where spray drift may be a problem due to proximity of susceptible crops or other desirable plants. Read and follow all Use Precautions given on this label.**

To Prepare the Spray, mix FORMULA 40 **only** with water, unless otherwise directed on this label. Add about half the water to the mixing tank, then add the FORMULA 40 with agitation, and finally the rest of the water with continuing agitation. Note: Adding oil, wetting agent or other surfactant to the spray may increase effectiveness on weeds, but also may reduce selectivity to crops resulting in crop damage.

Use with Liquid Fertilizer: FORMULA 40 may be combined with liquid fertilizers suitable for foliar application to accomplish weeding and feeding of corn, small grains, sorghum and grass pastures in one operation. Use FORMULA 40 in accordance with recommendations for these crops given in the following text and table. Use liquid fertilier at rates recommended by supplier or local extension service specialist. To prepare the spray, FORMULA 40 must first be premixed with water. For liquid nitrogen fertilizer use a premix consisting of 1 part of FORMULA 40 and 4 parts of water; for other liquid fertilizers use 1 part of FORMULA 40 with 50 to 60 parts of water. Add the premix to the fertilizer while maintaining continuous agitation during both mixing and spraying operations. Apply the spray the same day it is prepared; **do not store. NOTE: Always premix FORMULA 40 with water before adding to the liquid fertilizer.**

WEED CONTROL IN ASPARAGUS: See Table for recommended use rates. To control wild morningglory and certain other broadleaf weeds apply in about 60 gallons of water per acre for ground application and about 12 gallons per acre for aircraft application. Apply on actively growing weeds, usually in April or May. If spears are present, treat immediately after cutting. Make no more than 2 applications during the harvest season and these should be spaced at least one month apart. Spears contacted by the spray may be malformed and off-flavored. Malformed spears should be cut immediately and discarded. Post-harvest spraying should be only by ground rig using drop nozzles to avoid spraying the fern.

WEED CONTROL IN SMALL GRAINS NOT UNDERSEEDED WITH A LEGUME (Wheat, Barley, Rye, Oats, Millet): See Table for recommended use rates. Spray after grain begins tillering and before the boot stage (usually 4 to 8 inches tall) and weeds are small. **Do not apply before the tiller stage nor from early boot through the milk stage.** To control weeds that will interfere with harvest or to suppress perennial weeds, preharvest treatment can be applied when the grain is in the dough stage. Best results will be obtained when soil moisture is adequate for plant growth and weeds are growing well. Note: Do not permit dairy animals or meat animals being finished for slaughter to forage or graze treated gain fields within 2 weeks after treatment.

WEED CONTROL IN CORN: See Table for recommended use rates. **Preemergence**—Apply to soil anytime after planting but before corn emerges. Only emerge broadleaf weeds are likely to be controlled. Do not use on very light, sandy soil. **Emergence**—Apply just as corn plants are breaking ground. **Post-emergence**—Apply to emerged corn. When corn is over 8 inches tall use drop nozzles to keep spray off corn foliage. Do not apply from tasseling to dough stage. Injury to corn is most likely to occur if FORMULA 40 is applied when corn is growning rapidly under high temperature and high soil moisture conditions. In such situations, use the low rate of ½ pint per acre. After application delay cultivation for 8 to 10 days to allow the corn to overcome any temporary brittleness. **NOTE:** Hybrids vary in tolerance to 2,4-D. Some are easily injured, especially sweet corn. Spray only varieties known to be tolerant to 2,4-D. Consult the seed company or your Agricultural Experiment Station or Extension Service Weed Specialist for this information before using. **Preharvest**—After the hard dough or denting stage, apply by air or ground equipment to suppress perennial weeds, decrease weed seed production, and control tall weeds such as bindweed, cocklebur, dogbane, jimsonweed, ragweed, sunflower, velvetleaf, and vines that interfere with harvesting. **NOTE: do not forage or feed corn fodder for 7 days following application.**

WEED CONTROL IN SORGHUM (MILO): See Table for recommended use rates. Treat only after the sorghum is 6 inches high and preferably before it is 15 inches high. Do not treat during the boot, tasseling or early dough stages. Reduce spray drift by keeping the boom and spray nozzles as low as possible. If crop is taller than 8 inches, use drop nozzles to keep the spray off the leaves. Temporary crop injury can be expected under conditions of high soil moisture and high air temperatures. If it is necessary to apply FORMULA 40 under these conditions, use no more than ⅔ pint per acre, NOTE: Hybrids vary in tolerance to 2,4-D. Some are easily injured. Spray only varieties known to be tolerant to 2,4-D. Consult the seed company or your Agricultural Experiment Station or Extension Service Weed Specialist for this information.

WEED CONTROL IN RICE: See Table for recommended use rates. Apply in the late tillering stage of rice development, at the time of first joint development (first to second green ring), usually 6 to 9 weeks after emergence. Do **not** apply after panicle initiation, after rice internodes exceed ½ inch, at early seeding, early panicle, boot, flowering or early heading growth stages. Do not use in rice paddies where shellfish are of economic importance or where flood water is used for irrigation of other crops.

NOTE: Some rice varieties under certain conditions can be injured by 2,4-D. Therefore before spraying consult local Extension Service or University specialists for appropriate rates and timing of 2,4-D sprays.

WEED CONTROL IN SUGAR CANE: See Table for recommended use rates. Use up to 4 applications per year in accordance with state recommendations. **NOTE:** Do not apply within 6 weeks before harvest. For grass control, use DOWPON† M Herbicide in addition to FORMULA 40. Always read label directions and precautions for the use of these products before using them with FORMULA 40.

WEED CONTROL IN STRAWBERRIES: See Table for recommended use rates. To control many broadleaf weeds in established or old strawberry plantings apply in 25 to 50 gallons of water per acre in early spring when the strawberries are **dormant** or immediately after the last picking. Do not apply unless possible injury to the crop is acceptable. Follow recmmendations of State Extension Weed or Horticultural Specialists to fit local conditions.

AMOUNT OF FORMULA 40 TO USE IN CROPS

By air or ground application

NOTE: Do not apply when weather conditions favor drift from treated areas. Read complete directions and precautions before using.

Crop	Dosage Per Acre: Normal rates (usually safe to crop)	Higher rates for special situations[2] (more likely to injure crop)
SMALL GRAINS		
Spring Postemergence		
wheat, barley, rye,	⅔ to 1⅓ pints	2 to 3 pints
millet, oats	½ to 1 pint	1½ to 2 pints
Preharvest (dough stage)		
wheat, barley, oats	1 to 2 pints	2 to 3 pints
CORN[1]		
Preemergence	2 to 4 pints	
Emergence[1]	1 pint	1½ pints
Postemergence[1]		
up to 8 inches tall	½ to 1 pint	
8 inches to tasseling (use only directed spray)	1 pint	1½ to 2½ pints
Preharvest	1 to 2 pints	
SORGHUM (milo)[1]		
Postemergence		
6 to 8 inches tall	⅔ to 1 pint	
8 to 15 inches tall (use only directed spray)	1 pint	1½ to 2 pints
RICE	1 to 2½ pints	2 to 3 pints
SUGAR CANE		
Fall, after harvest or planting	2 to 4 pints	
Spring, once or twice before close-in	2 to 4 pints	
Summer, up to 6 weeks before harvest	2½ pints	
ASPARAGUS	1½ to 2 quarts	
STRAWBERRIES	1 to 1½ quarts	

[1] Corn and sorghum varieties vary in tolerance to 2,4–D; some are easily injured. Before spraying, get information on 2,4–D tolerance of specific varieties and spray only those **known** to be resistant to 2,4–D injury. If plants are more than 8 inches tall, use directed spray and keep spray off corn and sorghum foliage.

[2] These higher rates may be needed to handle difficult weed problems in certain areas such as under dry conditions especially in western areas. **However, do not use unless possible crop injury will be acceptable.** Consult State Agricultural Experiment Station of Extension Service weed specialists for recommendations or suggestions to fit local conditions.

WEED CONTROL IN ESTABLISHED GRASS PASTURES AND RANGELANDS: Use at 2 to 4 pints per acre. Apply preferably when weeds are small and growing actively before the bud stage. Do not use from early boot to milk stage where grass seed production is desired. Do not use on newly seeded areas until grass is well established. Do not use on bentgrass, alfalfa, clover, or other legumes. **NOTE: Do not graze (dairy) cattle in treated areas for 14 days after application. Remove meat animals from freshly treated areas for 7 days before slaughter. Withdrawal is not needed if 2 weeks or more have elapsed since application. Do not cut treated grass for hay within 30 days after application.**

CONTROL OF SOUTHERN WILD ROSE, PERENNIAL PEPPERWEED, HOARY CRESS: On rangelands, forests, roadsides and fence rows use 1 gallon of FORMULA 40 plus 4 to 8 fluid ounces of an agricultural surfactant per 100 gallons of water and spray thoroughly as soon as foliage is well-developed. Two or more treatments may be required. Apply at a maximum rate of 6 quarts per acre per application as a spot treatment so the overall rate is no more than 3 quarts per acre. **See NOTE on Grazing Restrictions in grass pastures section above.**

WEED CONTROL IN GRASS SEED CROPS: Use 1 to 4 pints per acre in spring or fall. **Do not apply from early boot to the milk stage of the grass.** Spray seedling grass only after the five-leaf stage, using 3/4 to 1 pint per acre to control small seedling weeds. After the grass is well-established, higher rates, up to 4 pints, can be used to control hard-to-kill annual or perennial weeds. Do not use on bentgrass unless grass injury can be tolerated. **See NOTE on Grazing Restrictions in grass pastures section above.**

BROADLEAF WEED CONTROL IN NON-CROPLAND GRASS AREAS SUCH AS LAWNS, GOLF COURSES, CEMETERIES AND PARKS, AIRFIELDS, ROADSIDES, VACANT LOTS, DRAINAGE DITCH BANKS: Use 1 to 4 quarts of FORMULA 40 per acre in the amount of water needed for uniform application. Treat when weeds are young and growing well. Usually 2 quarts per acre will provide adequate weed control. Do not use on dichondra or other herbaceous ground covers. Do not use on creeping grasses such as bent except for spot treating nor on freshly seeded turf until grass is well established. Reseeding of lawns should be delayed following treatment. With spring application, reseed in the fall; with fall application, reseed in the spring. Legumes are usually damaged or killed. Deep-rooted perennial weeds such as bindweed and Canada thistle may require repeated applications.

WEED CONTROL ON FALLOW LAND: Use 1½ to 2¼ quarts per acre on annual broadleaf weeds and up to 3 quarts per acre on established perennial species such as Canada thistle and field bindweed. Apply to actively growing weeds. Do not plant treated fallow land until three months after treatment, or until chemical has disappeared from soil.

SPOT TREATMENT IN NON-CROP AREAS: To control broadleaf weeds in small areas with a hand sprayer, use 1/4 pint of FORMULA 40 in 3 gallons of water and spray to thoroughly wet all foliage.

TREE INJECTION TREATMENT: To control unwanted hardwood trees such as elm, hickory, oaks, and sweetgum in forest and other non-crop areas, apply FORMULA 40 herbicide by injecting 1 ml of the undiluted product through the bark around the trunk at intervals of 1 to 3 inches between edges of the injector wounds. For harder to control species such as ash, maples, and dogwood use 2 ml of undiluted FORMULA 40 per injection site. Continuous cuts around the trunk often provide improved control. Also, cuts near the ground level may be more effective than at higher levels. Treatments can be made at any season; however, effectiveness may be less during winter months. Maples should not be treated during the spring sap flow.

USE PRECAUTIONS

AVOID CONTACT WITH 2,4-D SUSCEPTIBLE CROPS AND OTHER DESIRABLE BROADLEAF PLANTS: Do not apply directly to or otherwise permit even minute amounts to contact cotton, grapes, tobacco, fruit trees, vegetables, flowers, ornamentals, or other desirable plants susceptible to 2,4-D. Do not use in or near a greenhouse.

DO NOT APPLY IN THE VICINITY OF COTTON, GRAPES, TOBACCO, TOMATOES OR OTHER DESIRABLE 2,4-D SUSCEPTIBLE CROPS OR PLANTS. DO NOT SPRAY WHEN WIND IS BLOWING TOWARDS SUSCEPTIBLE CROPS OR ORNAMENTAL PLANTS.

AVOID SPRAY DRIFT: Applications should be made only when there is no hazard from spray drift since very small quantities of the spray, which may not be visible, may severely injure susceptible crops during both growing and dormant periods. Use coarse sprays to minimize drift since, under adverse weather conditions, fine spray droplets may drift a mile or more. A spray thickening agent such as NALCO TROL[1], may be used with this product to aid in reducing spray drift. If used, follow all use recommendations and precautions on the product label.

[1]NALCO TROL—Trademark of NALCO Chemical Company.

GROUND EQUIPMENT: With ground equipment, spray drift can be lessened by keeping the spray boom as low as possible; by applying 20 gallons or more of spray per acre; by using no more than 20 pounds spraying pressure at large droplet producing nozzle tips; by spraying when wind velocity is low; and by stopping all spraying when wind exceeds 6 to 7 miles per hour. Do not apply with hollow cone-type insecticide or other nozzles that produce a fine-droplet spray.

DETERMINE AIR MOVEMENT AND DIRECTIONS BEFORE FOLIAR APPLICATION: Use a smoke generator or other means at or near the application site for the detection of air movement, air stability or temperature inversions. Such a condition exists when there is little or no wind and air temperature is lower near the ground than at higher levels. Use appropriate drift control measures or avoid application when smoke is moving toward nearby desirable susceptible plants or sensitive areas.

AERIAL APPLICATION: With aircraft, drift can be lessened by applying a coarse spray; by using no more than 20 pounds spray pressure at the nozzles; by using straight stream nozzles directed straight back; by using a spray boom no longer than 3/4 the wing or rotor span of the aircraft; and by spraying only when wind velocity is less than 6 mph.

Excessive amounts of this herbicide in the soil may temporarily inhibit seed germination or plant growth. Violent wind storms may move soil particles. If 2,4-D is on soil particles and they are blown onto the susceptible plants, visible symptoms may appear. Serious injury is unlikely. The hazard of movement of 2,4-D on dust is reduced if treated fields are irigated or if rain occurs shortly after application.

To avoid injury to desirable plants, do not handle or apply other agricultural chemicals with the same equipment used for FORMULA 40 unless appropriately cleaned first. Local conditions may affect the use of herbicides. Consult your State Agricultural Experiment Station or Extension Service weed specialists for cleaning methods which are in compliance with local regulations and for advice in selecting treatments from this label to best fit local conditions. Be sure that use of this product conforms to all applicable regulations. Apply this product only as specified on this label.

STORAGE AND DISPOSAL

Do not contaminate water, food or feed by storage or disposal.

STORAGE

Pesticide wastes are toxic. Improper disposal of excess pesticide, spray mixture, or rinsate is a violation of Federal Law. If these wastes cannot be disposed of by use according to label instructions, contact your State Pesticide or Environmental Control Agency, or the Hazardous Waste representative at the nearest EPA Regional Office for guidance.

CONTAINER DISPOSAL

Triple rinse (or equivalent). Then offer for recycling or reconditioning, or puncture and dispose of in a sanitary landfill, or by other procedures approved by state and local authorities.

NOTE: FORMULA 40 herbicide, exposed to subfreezing temperatures, should be warmed to at least 40°F and mixed thoroughly before using.

MONEY BACK GUARANTEE

FORMULA 40 herbicide is guaranteed by Vertac Chemical Corporation to the full extent of the purchase price:

1. To give satisfactory control of species in the weed list when used as recommended.
2. To form a suitable spray mixture in any water fit for spray use.

NOTICE

Seller warrants that the product conforms to its chemical description and is reasonably fit for the purpose stated on the label when used in accordance with directions under normal conditions of use, but neither this warranty nor any other warranty of MERCHANTABILITY OR FITNESS FOR A PARTICULAR PURPOSE, express or implied, extends to the use of this product contrary to label instructions or under abnormal conditions, or under conditions not reasonably foreseeable to seller, and buyer assumes the risk of any such use. In no case shall Vertac or the Seller be liable for consequential, special or indirect damages resulting from the use or handling of this product.

* Trademark of VERTAC Chemical Corporation

† Trademark of The Dow Chemical COMPANY

GENERAL WEED KILLER

OIL SOLUBLE, WATER EMULSIFIABLE CONTACT HERBICIDE

Contains 5 pounds of dinoseb per gallon

ACTIVE INGREDIENT:

Dinoseb (2-sec-butyl-4,6-dinitrophenol) 55%

INERT INGREDIENTS...............................45%

TOTAL...100%

EPA Reg. No. 39511-86AA EPA Est. No. 39511-MS-1A; 39511-CA-1B

Superscript used corresponds to letter in Lot No.

Keep Out of Reach of Children and Farm Animals

DANGER–POISON

PRECAUCION AL USUARIO: Si usted no lee inglés, no use este producto hasta que la etiqueta le haya sido explicada ampliamente.

TRANSLATION: (TO THE USER: If you cannot read English, do not use this product until the label has been fully explained to you.

FLAMMABLE • MAY BE FATAL IF SWALLOWED OR ABSORBED THROUGH SKIN

Do Not Get in Eyes, on Skin, on Clothing • Do Not Take Internally • Avoid Breathing Spray Drift • Keep Away from Heat, Sparks and Open Flames • Use With Adequate Ventilation • Do Not Wear Contaminated Clothing or Shoes

Wear suitable protective equipment such as chemical workers goggles, rubber or plastic gloves and rubber or plastic apron while pouring and transferring the concentrated product and at other times when contact is likely. Wear protective clothing such as long sleeved shirt and long legged pants or coveralls while mixing and spraying. To avoid breathing spray mist, wear a mask or respirator of a type recommended by NIOSH or MSHA. Keep respirator clean—wash thoroughly and change cartridge as needed.

FIRST AID TREATMENT—ANTIDOTE

SYMPTOMS OF POISONING: Excessive Fatigue, Sweating, Thirst and Fever. If symptoms of poisoning develop from any type of exposure, SEND FOR A PHYSICIAN. FIRST AID: Have patient lie quiet in coolest spot available. If feverish, cool with cold compresses or by immersion in cool water.

If Swallowed, induce vomiting immediately by giving two glasses of water and sticking finger down throat. CALL A PHYSICIAN. Never give anything by mouth to an unconscious person. Treat as in FIRST AID, above.

If Splashed in Eyes, immediately flush eyes with plenty of water for at least 15 minutes and get medical attention.

If Spilled on Skin, immediately remove contaminated clothing, including shoes, and wash skin with soap and plenty of water. If symptoms of poisoning develop, send for a physician and treat as in FIRST AID, above. Discard contaminated clothing and shoes, or clean them thoroughly before re-use. NOTE TO ATTENDING PHYSICIAN: Active ingredient is a metabolic stimulant. Treat symptomatically.

AGRICULTURAL CHEMICAL

Do Not Ship or Store with Foods, Feeds, Drugs or Clothing

For chemical spill, leak, fire or exposure call toll free 1-800-424-9300

Period of Fruit and Nut Trees in Planting	Gallons Oil To Use For 100 Gallons Spray when VERTAC General Used at Per Acre Rates of:			
	1 quart	2 quarts	4 quarts	8 quarts
1st year	5	4	3	2
2nd year	10	5	5	3
3rd year	15	10	5	5
4th year and longer	20	15	10	5

Apply uniformly to the ground cover only, using coarse spray directed across the lower 6 to 12 inches of the tree trunks and at a rate to give 100 gallons of spray mixture per treated acre. Wet thoroughly all weed foliage and bare soil areas where residual control is desired. Avoid spraying crop foliage, fruits or nuts. Make three to four applications per year as needed. Best residual control of weed seedlings will result when the application is followed within a few days by rain or surface irrigation (flood or sprinkler). **NOTE:** Destroy all fruit accidentally sprayed. Do not apply within 30 days before harvest. Do not allow livestock to graze on treated ground cover.

Killing Potato Vines: Spray 10 to 20 days before harvest. The concentration of the spray and the amounts to use per acre are determined by vigor of the vines, degree of kill desired and weather conditions at time of application. Use the higher rate and higher spray volume during cool cloudy weather or where potato vine growth is especially heavy. Thorough coverage of all foliage and stems is essential for good results. With heavy vine growth split applications usually are more effective than single sprays; use ½ the suggested rate and repeat the application 5 to 7 days later again using the ½ dosage. To assure good wetting of potato foliage, add diesel oil, emulsifiable crop oil, or an approved agricultural surfactant such as Tween-20 or Multi-Film X-77 brands. Relative amounts of VERTAC General Weed Killer, oil or surfactant are shown in the following table.

Type of Application	VERTAC General pt/Acre	Amount and Kind of Diluent/Acre
Aerial (3 to 5 gallons total spray per acre)	2 to 3	3 to 5 gallons of diesel oil or 1 to 2 gallons of emulsifiable crop oil plus water or 1 quart of agricultural surfactant* plus water

Type of Application	VERTAC General pt/Acre	Amount and Kind of Diluent/Acre
Ground (25 to 40 gallons total spray per acre)	2 to 4	5 gallons of diesel oil plus water** or 1 to 2 gallons of emulsifiable crop oil plus water or 1 quart of agricultural surfactant* plus water

* Use approved agricultural surfactant such as Tween-20 or Multi-Film X-77 brands.

** Follow procedure under "MIXING INSTRUCTIONS" to ensure good emulsification.

NOTE: Browning of the vascular ring of potato tubers sometimes occurs when vines are killed rapidly either by frost, mechanically or chemically. Vascular browning occurs most frequently during warm weather when soil is dry. Under these conditions, use lowest recommended dosage or split applications. Do not spray exposed tubers nor graze treated areas.

OTHER USES

Consult supplementary labeling available from your dealer or from the Agricultural Division, Vertac Chemical Corp. 24th Floor, 5100 Poplar, Memphis, TN 38137 for registered use directions for VERTAC General Weed Killer for preplant or preemergence weed control and for use on hops.

USE PRECAUTIONS

Use this product only as specified on the label. Do not spray or allow spray drift to contact desirable plants except as directed on the label.

IMPORTANT: Do not store at temperatures below 20°F. If stored for extended periods below 20°F, bring the contents of the container up to 32°F and agitate by rolling.

This product is toxic to fish and wildlife. Birds and other wildlife in treated areas may be killed. Keep out of lakes, ponds, and streams. Do not apply where runoff is likely to occur. Do not apply when weather conditions favor drift from areas treated. Do not contaminate water by cleaning of equipment, or disposal of wastes. Do not apply in any manner not specified on the label.

NOTE: Rinse equipment and containers and dispose of wastes by burying in noncrop lands away from water supplies. Containers should be disposed by punching holes in them and burying with wastes.

NOTICE

Seller warrants that the product conforms to its chemical description and is reasonably fit for the purposes stated on the label when used in accordance with directions under normal conditions of use, but neither this warranty nor any other warranty of MERCHANTABILITY or FITNESS FOR A PARTICULAR PURPOSE, express or implied, extends to the use of this product contrary to label instructions, or under abnormal conditons, or under conditions not reasonably foreseeable to seller, and buyer assumes the risk of any such use.

GENERAL INFORMATION

VERTAC General Weed Killer is a non-selective contact herbicide designed for control of broadleaf weeds and grasses in certain crops as recommended on this label and on non-crop areas such as roadsides, railroads, fence rows, around industrial buildings, poleyards and tank farms. VERTAC General Weed Killer, when mixed and used according to directions will control all the common annual broadleaf weeds and grasses through foliar contact action or through effects on seedlings by temporary residues in soil. Thorough coverage of weed growth or soil surface is essential for best results. The lower recommended rates of VERTAC General Weed Killer and diesel oil or weed oil will give satisfactory control of small annual grasses such as crabgrass, foxtail and barnyardgrass (watergrass) and of small annual broadleaf weeds such as yellow star thistle, pigweeds and smartweeds. Larger grasses and broadleaf weeds such as mallow dog fennel and wild carrot, will require the higher recommended rates. This is particularly so when weeds are nearly mature or under drought stress. Satisfactory control of grasses usually requires use of the higher amounts of oil. Dalapon grass killer may be combined with VERTAC General Weed Killer to improve kill of emerged grasses in non-crop areas. Perennial weeds will usually require retreatment as new growth develops. Control of weeds will be best in warm weather when weeds are small and growing well. The higher rates of VERTAC General Weed Killer will give some residual control of germinating seedlings if rainfall or irrigation follows spraying within a week after application.

MIXING INSTRUCTIONS

Following are instructions for preparing each type of spray mixture that may be used:

I†. VERTAC General Weed Killer Diluted in Oil Only

(1st) Drain all water from mixing tank.
(2nd) Rinse the tank with a small quantity of oil, then drain and discard the rinse.
(3rd) Add the desired amount of oil to mixing tank.
(4th) With moderate agitation add required amount of VERTAC General Weed Killer to desired amount of oil.
(5th) Mixture is now ready for application. No further agitation is necessary.

II†. VERTAC General Weed Killer—Oil-Water

(1st) Put about 3/4 the desired amount of water in spray tank.
(2nd) With vigorous agitation, add required amount of VERTAC General Weed Keller.
(3rd) With continued agitation add the required amount of oil of choice.
(4th) Add remainder of water to fill spray tank to desired amount.
(5th) Material is now ready to apply. Continue agitation until application is complete.

III. VERTAC General Weed Killer—Agricultural Surfactant-Water

(1st) Add approximately 3/4 of the desired amount of water to the mixing tank.
(2nd) With agitation, add required amount of approved surfactant such as Tween-20 or Multi-Film X-77 brands. If foaming is excessive reduce agitation.
(3rd) With continued agitation, add required amount of VERTAC General Weed Killer.
(4th) Add remainder of water necessary to bring to desired amount.
(5th) Mixture is now ready for application. Continue mild agitation until application is completed.

IV. VERTAC General Weed Killer—Dalapon Grass Killer-Agricultural Surfactant-Water

(1st) Add approximately 3/4 of water desired to the mixing tank.
(2nd) With mild agitation, add required amount of dalapon.
(3rd) Add required amount of approved surfactant (see III).
(4th) With continued agitation add required amount of VERTAC General Weed Killer.
(5th) Add remainder of water required to bring to desired spray volume.
(6th) Mixture is now ready to apply. Continue mild agitation until application is complete.

†Note: Follow the above instructions for mixtures I and II precisely. Changing procedures may result in very thick invert emulsions which are difficult to apply.

USE DIRECTIONS

NON-CROP USES

VERTAC General Weed Killer is highly effective in controlling broadleaf weeds and unwanted grasses in non-crop areas such as airfields, drainage ditches, industrial building and storage sites, around farm buildings, along fence rows and on rights-of-way for utility installations, railroads, and highways.

For different weed problems select the appropriate spray mixture from the following table. Spray to thoroughly wet weed growth, using 100 to 200 gallons of spray mixture per acre depending on density and height of the vegetation to be controlled. Re-treat as necessary to maintain satisfactory control. Do not allow livestock to graze on treated areas.

GUIDE TO PREPARATION OF 100 GALLONS OF SPRAY MIXTURE FOR NON-CROP USE

		INGREDIENTS AND AMOUNTS FOR USE				
BROADLEAF AND/OR GRASSY WEED PROBLEM	MIXING PROCEDURE TO USE†	VERTAC GENERAL (QT)	DALAPON (LB)	OIL (GAL)	AGRICULTURAL SURFACTANT (QT)††	WATER (APPROX.) (GAL)
Mostly grasses, but a few broadleaf weeds	II†	1 to 1½	0	25 to 50	0	75 to 50
Mostly broadleaf weeds, but some grasses also	II†	2 to 2½	0	10 to 25	0	90 to 75
Mostly broadleaf weeds, some residual effect on weed seedlings	II†	3	0	5 to 10	0	95 to 90
Same as above (alternate choice)	III†	2 to 3	0	0	2	99
Annual grasses and broadleaf weeds	IV†	1	5	0	2	99
Perennial grasses and broadleaf weeds	IV†	1 to 2	10 to 15	0	2	99

† Consult corresponding number under "MIXING INSTRUCTIONS" for procedures to follow in preparing spray mixtures.
†† When no oil is used, add 2 qt of approved agricultural surfactant such as Tween-20 or Multi-Film X-77 brands.

CROP USES

VERTAC General Weed Killer has many uses in the production of agricultural crops when used according to instructions which follow. Observe all directions and follow instructions precisely. Some states issue specific instuctions to fit local conditions. Consult your state Agricultural Experiment Station or Extension Service weed control specialists for more detailed information.

Alfalfa: Apply to established stands during the dormant season of after grazing in winter to control winter annual weeds, or immediately after cutting during the growing season before regrowth starts, to control dodder and summer annual weeds. For aerial application use 1½ to 3 quarts of VERTAC General Weed Killer in 5 to 20 gallons of diesel oil or weed oil per acre. With ground equipment use 1½ to 3 quarts of VERTAC General Weed Killer plus 10 to 60 gallons of diesel oil or weed oil in water to make 100 gallons of spray mixture. Use at rate of 100 gallons per acre. **Note:** Do not graze or feed forage from treated areas within 6 weeks after application.

Preharvest Spraying of Alfalfa, Trefoil, Clover, Peas and Soybeans to Facilitate Harvest of Crops Grown for Planting Seed: Apply 3 to 6 days before harvest. For airplane application, use 2 to 3 pints of VERTAC General Weed Killer in 4 to 10 gallons of diesel oil or weed oil per acre. With ground equipment use 2 to 3 pints of VERTAC General Weed Killer in 8 to 15 gallons of diesel oil or weed oil for low volume sprays; or with 5 to 15 gallons of either oil in water to make 30 to 50 gallons of spray per acre for high volume sprays. The higher rates and volumes are suggested for use during cool, cloudy weather or where there is heavy foliage growth. **Note:** Do not allow seed or forage from treatment areas to be used for food or feed purposes. Do not graze new growth in treated areas within 6 weeks after application.

Control of Weeds, Basal Fruit Spurs and New Cane Growth in Caneberries, Including Blackberries, Boysenberries, Loganberries and Raspberries: For weed and cane control apply as a directed spray in the spring when new canes are 3 to 6 inches tall using 2 quarts VERTAC General Weed Killer plus 1 gallon emulsifiable crop oil mixed with water to make 100 gallons of spray. Apply up to 24 inches high on the canes using 100 to 150 gallons of spray per treated acre. **Note:** Crop foliage and fruit buds wet by the spray may be killed. For weed control only, the spray should be directed toward the weeds by spraying across the base of the canes and away from new canes desired for producing fruit next year. After harvest a spray containing 1 quart VERTAC General Weed Killer and 1 gallon emulsifiable crop oil per 100 gallons of spray can be applied to the lower 18 inches of canes at volumes of 100 to 150 gallons per acre as needed to control weeds and undesired fruiting spurs. Make no more than four applications per season for weed or can control. **Note:** Do not apply within 30 days before harvest.

Weed Control in Grapes and Bush Fruits Including Blueberries, Currants, and Gooseberries: Use 1 to 2 quarts of VERTAC General Weed Killer plus 10 to 20 gallons of diesel oil or weed oil with water to make 100 gallons of spray and apply at the rate of 100 to 150 gallons per treated acre. Apply when weeds are growing well, but before they are 6 inches tall. Use only as a directed spray and apply in such a way as to avoid contacting crop foliage, blossoms or fruit clusters. Direct spray across the base of crop plant stems to treat weeds and soil. Apply three or four times per year, as needed. **Note:** Use only on plantings 2 years or older. Do not treat within 30 days before harvest.

Control of Broadleaf Weeds and Grasses in Fruit Orchards and Citrus and Nut Groves: Use 1 to 8 quarts of VERTAC General Weed Killer with 2 to 20 gallons of diesel oil or weed oil with enough water to make 100 gallons of spray and apply as a "floor" spray under the following crop plants:

Almonds, apples, apricots, cherries, citrus, dates, figs, filberts, nectarines, olives, peaches, pears, pecans, plums (prunes), walnuts.

Use one of the spray mixtures specified in the following table. The mixture of choice will depend upon nature of the vegetation to be controlled, availability of the materials needed and the economic factors involved in their use. Consult State Agricultural Experiment Station and Extension Service Weed Control Specialists regarding local recommendations.

MCP AMINE

HERBICIDE

Contains Dimethylamine Salt‡ of MCPA

Acid Equivalent: 4 pounds per gallon

For the Selective Control of Many Broadleaf Weeds in Flax, Small Grains, Rice, and Clover, Alfalfa and Grass Pastures; for Whitebrush Control; and for Weed Control in Non-Crop Areas.

ACTIVE INGREDIENT:
2-methyl-4-chlorophenoxyacetic acid,
dimethylamine sale 52.1%
INERT INGREDIENTS 47.9%

TOTAL 100.0%

MCPA Acid Equivalent 42.5%–4 pounds per gallon

EPA Reg. No. 464-394-39511 EPA Est. 464-MI-1

‡ Salts are the least volatile forms of MCPA and do not release enough vapors from treated areas to reduce yield of adjacent susceptible crops.

WARNING

Keep Out Of Reach Of Children

PRECAUCION AL USUARIO: Si usted no lee inglés, no use este producto hasta que la etiqueta le haya sido explicada ampliamente.

TRANSLATION: (**TO THE USER:** If you cannot read English, do not use this product until the label has been fully explained to you.)

PRECAUTIONARY STATEMENTS

HAZARDS TO HUMANS AND DOMESTIC ANIMALS

HARMFUL IF SWALLOWED. INJURIOUS TO EYES
MAY CAUSE SKIN IRRITATION
Do Not Get In Eyes. Avoid Skin Contact

If swallowed, induce vomiting immediately by giving two glasses of water and sticking finger down throat. Call a physician. Do not induce vomiting or give anything by mouth to an unconscious person.

In case of eye contact, flush with plenty of water for at least 15 minutes and get medical attention.

In case of skin contact, wash with soap and plenty of water. Remove and wash contaminated clothing before reuse. Do not wear contaminated shoes.

PHYSICAL OR CHEMICAL HAZARDS

Do Not Cut or Weld Container

ENVIRONMENTAL HAZARDS

Do not apply directly to water. Do not contaminate water by cleaning of equipment or disposal of wastes.

AGRICULTURAL CHEMICAL

Do Not Ship or Store with Food, Feeds, Drugs or Clothing

For chemical spill, leak, fire or exposure call toll free 1-800-424-9300

DIRECTIONS FOR USE

It is a violation of Federal law to use this product in a manner inconsistent with its labeling.

Apply MCP Amine Herbicide during warm weather when weeds are young and growing actively. Obtain uniform spray coverage for best results. Weeds that can be controlled include beggartick, burdock, buttercup, catsear, cocklebur, croton, burcucumber, galinsoga, goatsbeard, hempnettle, honeysuckle, jimsonweed, lambsquarters, marshelder, pennycress, pepperweed (annual), plantain, puncturevine, ragweed, redstem, shepherdspurse, sicklepod, sneezeweed, stinkweed, whitebrush, wild mustard, witchweed, yellowrocket and other species.

Generally, the lower dosages listed will be satisfactory for young, succulent growth of sensitive weed species. For less sensitive species and under conditions where control is more difficult, the higher dosages will be needed. Use moderate spray pressure (20 pounds) and enough spray volume for good coverage, usually 5 to 20 gallons per acre by ground equipment and at least 3 to 5 gallons per acre by air. Higher spray volumes may be useful in some situations to provide better spray coverage of weed foliage and to reduce spray drift. The use of adjuvants such as wetting agents, may increase herbicidal effectiveness but may increase risk of crop injury.

Do not apply where spray drift may be a problem due to proximity of susceptible crops or other desirable plants. Read and follow all Use Precautions given on this label.

PREPARING THE SPRAY: Mix MCP Amine with water, unless otherwise directed on this label. Fill the mixing tank or spray tank about half full with water, then add the required amount of MCP Amine with agitation, and finally add the rest of the water with continuing agitation. **NOTE:** Adding oil, wetting agent or other surfactant to the spray may reduce selectivity to crops, possibly resulting in crop injury.

WEED CONTROL IN FLAX: For control of susceptible weeds, use ½ pint of MCP Amine per acre. Treat flax when 3 to 6 inches tall **before** buds begin to form. Rates of ¾ to 1 pint per acre may be needed to control some weeds such as wild buckwheat, smartweed and thistles but flax may be injured at these rates. If seedling grasses such as foxtail are a problem, DOWPON† M grass killer may be used alone or in combination with MCP Amine. Flax varieties may vary in tolerance to DOWPON M and MCPA. Consult local Extension Service or University Specialists for advice on herbicide use in flax. Do not forage or graze meat animals on treated areas within seven days of slaughter.

WEED CONTROL IN SMALL GRAINS: NOTE: Do not permit dairy animals or meat animals being finished for slaughter to forage or graze treated grain fields within 2 weeks after treatment. Do not forage or graze meat animals on treated areas within seven days of slaughter.

Wheat, Barley, Oats and Rye not underseeded to Legumes: Use ½ to 1 pint of MCP Amine per acre for the more susceptible weeds when grains are in the 3 to 5 leaf stage or up to early boot stage. Apply when weeds are very small for best results. Use up to 3 pints MCP Amine per acre for the less susceptible weeds after crop has tillered but not later than early boot stage. Do not apply during the boot to dough stage of the grain.

Wheat, Barley, Oats and Rye underseeded to Alfalfa, Birdsfoot Trefoil, Lespedeza, Red, White, or Ladino Clover: For emergency control of susceptible weeds such as mustard and lambsquarters, use ⅓ to ½ pint of MCP Amine per acre as soon as possible after weeds emerge, but after seedling grain is at least 3 inches tall. The grain and weeds should form a protective canopy over the clover seedlings. Treatment may be made up to the full tiller stage but not in the boot to dough stages of the grain. To control certain weed species, 1 pint per acre may be needed. Do not use where small seeded legumes, especially vetch, sweet clover, trefoil or alfalfa, are seeded unless injury to them can be tolerated.

WEED CONTROL IN RICE: For control of certain broadleaf weeds and sedges, apply 2 to 3 pints of MCP Amine in 10 to 30 gallons of water per acre. Treat when rice is fully tillered and 6 to 8 inches above water, usually 35 to 65 days after planting. Do not treat when rice is in the early seedling, boot or early heading stages, or when air temperature is over 90°F. Some rice varieties may be less tolerant to MCPA than others. Consult your Agricultural Experiment Station or Extension Service Weed Specialists for appropriate rates and timing of MCPA spray and on the tolerance of rice varieties. Do not grow crayfish or catfish in treated rice fields.

RICE IN CALIFORNIA: For early season broadleaf weed and sedge control in rice, apply 1 to 1½ pints of MCP Amine Herbicide per acre by air in 8 to 10 gallons of total spray volume when rice is in the 3 to 4 leaf stage, and with well established root systems, usually 18 to 24 days after planting. Applications made during hot weather or at the higher dosage range may cause temporary stunting of the rice. For maximum effectiveness the weeds should be exposed at time of treatment. If retreatment is necessary, usually 35 to 65 days after planting, the total amount per acre of MCP Amine applied in the two applications must not exceed 3 pints in one season.

WEED CONTROL IN PEAS: Use on peas to control susceptible weeds only where recommended by local agricultural Extension Service or Experiment Station specialists. Use ¼ to ⅓ pint of MCP Amine in at least 15 gallons of water per acre when peas are 3 to 7 inches tall, before first flowering and weeds are small. Higher rates of ½ to ¾ pint per acre may be used to improve control of difficult weeds but crop injury is more likely to occur. Do not apply if peas are taller than 7 inches or when they are stressed for lack of soil moisture. Do not apply when air temperature is over 90°F. NOTE: **MCPA can cause injury and delayed maturity in the pea crop.**

WEED CONTROL IN ESTABLISHED STANDS OF ALFALFA OR RED CLOVER: For control of yellow rocket and other susceptible annual weeds such as pennycress or fanweed, use 1 pint per acre of MCP Amine. Apply in late fall after frost has killed legume topgrowth and legumes are dormant. The temperature at time of spraying should be above 40°F.

WEED CONTROL IN GRASS PASTURES: Use 2 to 4 pints per acre of MCP Amine. Apply when weeds are small and actively growing for best results. Use the higher rates for whitetop, Canada thistle, buttercup and other more difficult to control weeds. Spray perennials in early bud to full bloom stage and regrowth in fall. Legumes may be injured or killed. Do not use this treatment where legumes, especially alfalfa, are present and desirable. Do not use on newly seeded areas until grass is well established. Do not use from early boot to milk stage where grass seed production is desired.

CONTROL OF WHITEBRUSH ON RANGELAND: Use 3 pints of MCP Amine with 1 gallon of diesel oil and enough water for good spray coverage, usually 6 to 10 gallons per acre by aircraft. Apply in spring or fall when plant foliage is well developed and actively growing. Spraying during bloom is recommended but not immediately after shedding of blossoms. Retreatments may be needed in succeeding years. **NOTE:** First mix the MCP Amine in the water, then add the oil with agitation. Spray should be kept agitated during spraying to avoid separation in the tank.

SPOT TREATMENT: FOR WEED CONTROL IN PASTURES, RANGELANDS AND IN NON-CROP AREAS SUCH AS FENCEROWS, FARMYARDS, SHELTER-BELTS AND ROADSIDES: Use 1/4 pint of MCP Amine in 3 gallons of water to control weeds such as Canada thistle, whitetop, meadow buttercup and Texas blueweed. Apply when weeds are growing vigorously, usually prior to bloom, and spray fall regrowth if necessary.

STORAGE AND DISPOSAL

Do not contaminate water, food or feed by storage or disposal.

STORAGE
Keep container tightly closed when not in use.

PESTICIDE DISPOSAL
Pesticide, spray mixture or rinsate that cannot be used according to label instructions must be disposed of according to Federal, State, or local procedures under the Resource Conservation and Recovery Act.

CONTAINER DISPOSAL
Triple rinse or equivalent and offer for recycling or reconditioning, or dispose of in a sanitary landfill or by other approved State and local procedures.

USE PRECAUTIONS

AVOID CONTACT WITH MCPA SUSCEPTIBLE CROPS AND OTHER DESIRABLE BROADLEAF PLANTS—MCP Amine herbicide is injurious to most broadleaf plants. Therefore, do not apply directly to or otherwise permit even minute amounts (as in spray mist) to contact cotton, grapes, tobacco, fruit trees, vegetables, flowers, ornamentals or other desirable plants susceptible to MCPA. Do not use in or near a greenhouse.

DO NOT APPLY IN THE VICINITY OF COTTON, GRAPES, TOBACCO, TOMATOES OR OTHER DESIRABLE MCPA SUSCEPTIBLE CROPS OR ORNAMENTAL PLANTS.

DO NOT SPRAY WHEN WIND IS BLOWING TOWARDS SUSCEPTIBLE CROPS OR ORNAMENTAL PLANTS.

AVOID SPRAY DRIFT—Applications should be made only when there is no hazard from spray drift since very small quantities of the spray, which may not be visible, may severely injure susceptible crops during both growing and dormant periods. Use coarse sprays to minimize drift since, under certain weather conditions, fine spray droplets may drift a mile or more. A spray thickening agent such as NALCO-TROL[1] may be used with this product to aid in reducing spray drift. If used follow all use recommendations and precautions on the product label.

[1]NALCO-TROL—Trademark of NALCO Chemical Company.

GROUND EQUIPMENT—With ground equipment, spray drift can be lessened by keeping the spray boom as low as possible; by applying 20 gallons or more of spray per acre; by using no more than 20 pounds spraying pressure with large-droplet producing nozzle tips; by spraying when wind velocity is low; and by stopping all spraying when wind exceeds 8 miles per hour. Do not apply with hollow cone-type insecticide or other nozzles that produce hazardous amounts of fine spray droplets.

DETERMINE AIR MOVEMENT AND DIRECTIONS BEFORE FOLIAR APPLICATION: Use a smoke generator or other means at or near the application site for the detection of air movement, air stability or temperature inversions. Such a condition exists when there is little or no wind and air temperature is lower near the ground than at higher levels. Use appropriate drift control measures or avoid application when smoke is moving toward nearby desirable susceptible plants or sensitive areas.

AERIAL APPLICATION—With aircraft, drift can be lessened by applying a coarse spray; by using no more than 20 pounds spray pressure at the nozzles; by using nozzle types and positions which do not produce fine spray droplets; by using a spray boom no longer than 3/4 the wing or rotor span of the aircraft; and by spraying only when wind velocity is less than 6 mph.

Excessive amounts of this herbicide in the soil may temporarily inhibit seed germination or plant growth. Violent windstorms may move soil particles. If MCPA is on soil particles and they are blown onto susceptible plants, visible symptoms may appear. Serious injury is unlikely. The hazard of movement of MCPA on dust is reduced if treated fields are irrigated or if rain occurs shortly after application.

NOTE: MCP Amine, exposed to subfreezing temperatures, should be warmed to at least 40°F and mixed thoroughly before using.

To avoid injury to desirable plants, do not handle or apply other agricultural chemicals with the same equipment used for MCPA unless appropriately cleaned first. Local conditions may affect the use of herbicides. Consult your State Agricultural Experiment Station or Extension Service weed specialists for cleaning methods which are in compliance with local regulations and for advice in selecting treatments from this label to best fit local conditions. Be sure that the use of this product conforms to all applicable regulations. Apply this product only as specified on this label.

CONDITIONS OF SALE AND WARRANTY

VERTAC AND SELLER OFFER THIS PRODUCT AND THE BUYER AND USER ACCEPTS THIS PRODUCT ONLY UNDER THE FOLLOWING AGREED CONDITIONS OF SALE AND WARRANTY.
The directions for use of this product are believed to be reliable and should be followed carefully. However, it is impossible to take into account all variables and to eliminate all risks associated with its use. Injury or damage may result because of conditions which are beyond the control of Vertac or the Seller. Vertac warrants only that this product conforms to the chemical description on the label and is believed to be reasonably fit for the purposes referred to in the Directions for Use when used as directed under normal conditions. VERTAC MAKES NO OTHER EXPRESS OR IMPLIED WARRANTY OF FITNESS OR MERCHANTABILITY OR ANY OTHER EXPRESS OR IMPLIED WARRANTY. In no case shall Vertac or the Seller be liable for consequential, special or indirect damages resulting from the use or handling of this product. Any variation or exception from this warranty must be in writing and signed by an authorized Vertac representative.

† Trademark of The Dow Chemical Company

MCP ESTER
HERBICIDE

For Selective Control of Certain Weeds in Wheat, Oats, Barley, Rye, Flax, Grassland, and in Non-Crop Areas.

ACTIVE INGREDIENT:	
Isooctyl ester of 2-methyl-4-chlorophenoxy-acetic acid	74.4%
INERT INGREDIENTS	25.6%
TOTAL	100.0%

Contains the equivalent of 47.7% 2-methyl-4-chlorophenoxy-acetic acid equivalent to 4.0 lbs per gallon at 68°F.

EPA Reg. No. 464-579-39511 EPA Est. 464-MI-1

CAUTION

Keep Out of Reach of Children

PRECAUCION AL USUARIO: Si usted no lee ingles, no use este producto hasta que la etiqueta le haya sido explicada ampliamente.

TRANSLATION: (TO THE USER: If you cannot read English, do not use this product until the label has been fully explained to you.)

PRECAUTIONARY STATEMENTS

HAZARDS TO HUMANS AND DOMESTIC ANIMALS

HARMFUL IF SWALLOWED

Avoid Contact with Skin, Eyes or Clothing
Avoid Inhaling Spray Mist

STATEMENTS OF PRACTICAL TREATMENT

In case of eye contact, flush eyes with plenty of water. Wash exposed skin with plenty of soap and water. Remove and wash contaminated clothing and shoes before reuse.

PHYSICAL OR CHEMICAL HAZARDS

Do Not Use or Store Near Heat or Open Flame
Do Not Cut or Weld Container

ENVIRONMENTAL HAZARDS

Do not contaminate water intended for irrigation or domestic use. This product is toxic to fish. Keep out of lakes, streams, or ponds. Do NOT contaminate water by cleaning of equipment or disposal of wastes.

AGRICULTURAL CHEMICAL

Do Not Ship or Store with Foods, Feeds, Drugs or Clothing

For chemical spill, leak, fire or exposure call toll free 1-800-424-9300

Avoid spray drifts onto susceptible plants such as vegetables, flowers, grapes, tomatoes, cotton, beans and other legumes, etc. Use coarse sprays as these are less likely to drift. Do NOT spray at all in the near vicinity of susceptible plants. Vapors from this product may injure susceptible plants in the immediate vicinity. Do NOT contaminate water intended for irrigation or domestic use. Thoroughly clean spray equipment with a suitable chemical cleaner before using for other purposes, or do NOT use same spray equipment for other purposes. Do NOT store in freezing temperatures. If allowed to freeze remix before using. Do NOT transport with or store near fertilizers, seeds, insecticides or fungicides.

CONTROLS THESE AND OTHER WEEDS

Susceptible (Most Areas): Lambsquarters (goosefoot), Marsh Elder, Mustards (annual), Puncturevine, Ragweed, Stinkweed (Pennycress, Fanweed), Wild Radish, Yellow Rocket or Wintercress, Cocklebur, Goats Beard, Field Pepper Grass, Dragon Heat Mint, Beggarticks, Poison Hemlock, Hempnettle, Shepherdspurse, Annual Sow Thistle, Kochia.

Less Susceptible: Buttercups, Canada Thistle, Dandelions, Plantains, Stinging Nettle, Perennial Sow Thistle, Vetch, White Top or Hoary Cress, Russian Pigweed, Sunflower, Chervil, Miner's Lettuce, Purslane.

DIRECTIONS FOR USE

It is a violation of Federal law to use this product in a manner inconsistent with its labeling.

SELECTIVE SPRAYING

Wheat, Oats, Barley and Rye—Apply as a water mix spray by ground sprayer or airplane. Use ½ to 1 pint per acre for the more susceptible weeds after crop has reached the 3 to 4 leaf stage up to boot stage. Use up to 3 pints per acre for the less susceptible weeds after crop has tillered and up to early boot stage. Do not spray from boot to dough stage.

FLAX

Use ¼ to ½ pint per acre. Apply by ground sprayer or by airplane. Use sufficient water to give uniform and adequate coverage. Apply only when weeds are up and when flax is 4 to 8 inches high and before it comes into bud. Treatment after early bud stage may result in severe damage. If Canada Thistle is present it may be necessary to go as high as ¾ pint per acre to prevent seed head production. Some injury to the flax may result.

GRASSES

Established Grassland—Use 1 to 3 pints per acre in sufficient water (10 to 100 gallons) by airplane or ground sprayer application and give thorough coverage. Use higher rate for White Top, Canada Thistle and other less susceptible weeds. Spray perennials in early bud to full bloom stage and regrowth in fall; other weeds in spring or fall, when actively growing.

Grasses Grown for Seed—Use 1 to 2 pints per acre, the higher rate where weed stands are heavy. In established grasses apply in spring before head comes into boot and on seedling grass after grass has tillered.

NOTE: In weed control in grasses, repeat treatment may be needed for less-susceptible weeds. White clover and other legumes may be temporarily injured or killed. In some areas, bent, buffalo, carpet and St. Augustine grasses may also be injured by treatment.

NON-CROP SPRAYING ALONG FENCE ROWS AND ROADSIDES

Canada Thistle, White Top and Meadow Buttercup—For spot treatment, use ¼ pint to 3 to 4 gallons of water, or ¾ gallon per acre in sufficient water to give coverage for most extensive areas. Spray to wet weeds thoroughly when in bud to early bloom and again on fall regrowth.

STORAGE AND DISPOSAL

Do not contaminate water, food or feed by storage or disposal.

STORAGE

Do NOT store in freezing temperatures. If allowed to freeze remix before using. Do not transport with or store near fertilizers, seeds, insecticides or fungicides.

PESTICIDE DISPOSAL

Pesticide, spray mixture or rinse water that cannot be used according to label instructions must be disposed of according to applicable Federal, state, or local procedures.

CONTAINER DISPOSAL

Triple rinse (or equivalent). Then offer for recycling or reconditioning, or puncture and dispose of in a sanitary landfill, or by other approved state and local procedures.

Local conditions may affect the use of this chemical. Consult State Agricultural Extension or Experiment Station weed specialists for specific recommendations for local weed problems and for information on possible lower dosages.

CONDITIONS OF SALE AND WARRANTY

VERTAC AND SELLER OFFER THIS PRODUCT AND THE BUYER AND USER ACCEPTS THIS PRODUCT ONLY UNDER THE FOLLOWING AGREED CONDITIONS OF SALE AND WARRANTY. The directions for use of this product are believed to be reliable and should be followed carefully. However, it is impossible to take into account all variables and to eliminate all risks associated with its use. Injury or damage may result because of conditions which are beyond the control of Vertac or the Seller. Vertac warrants only that this product conforms to the chemical description on the label and is believed to be reasonably fit for the purposes referred to in the Directions for Use when used as directed under normal conditions. VERTAC MAKES NO OTHER EXPRESS OR IMPLIED WARRANTY OF FITNESS OR MERCHANTABILITY OR ANY OTHER EXPRESS OR IMPLIED WARRANTY. In no case shall Vertac or the Seller be liable for consequential, special or indirect damages resulting from the use or handling of this product. Any variation or exception from this warranty must be in writing and signed by an authorized Vertac representative.

PHYTAR® 560

SODIUM CACODYLATE AND CACODYLIC ACID LIQUID PLUS SURFACTANT

For General Post-Emergent Weed Control

ACTIVE INGREDIENTS:

Sodium Cacodylate	22.73%
Dimethylarsinic Acid (Cacodylic Acid)	3.88%
INERT INGREDIENTS:	73.39%
TOTAL	100.00%
Total Arsenic (as elemental) all in water soluble form	12.75%

Product contains 2.48 lbs. cacodylic acid equivalent per gallon.

EPA Reg. No. 39511-105 E.P.A. Est. No. 39511-MS-1[A] & 39511-AR-2[B]

Superscript used corresponds to letter in lot munber.

CAUTION

Keep Out of the Reach of Children

CAUTION

Harmful if swallowed. Avoid contact with skin. Avoid breathing spray mist. Wash hands after using. Avoid storage near feed or food products. Keep children and domestic animals off treated areas until spray has dried.

Do not contaminate waters used for domestic consumption, or by animals, wildlife and aquatic life, or for irrigation purposes. Do not graze treated areas to livestock.

ANTIDOTE

If taken internally, induce vomiting and call physician at once.

CONTAINER DISPOSAL

Do not reuse empty container, Wash thoroughly with water and detergent, crush if possible, and discard in a safe place.

For chemical spill, leak, fire, or exposure call toll free 1-800-424-9300

AGRICULTURAL CHEMICAL

Do Not Ship or Store with Food, Feeds, Drugs or Clothing

GENERAL INFORMATION

PHYTAR 560 herbicide is useful for general post-emergent weed control. It is non-selective in its action. Its phytotoxic properties are quickly inactivated on contact with soil. It contains a surfactant.

IT IS UNNECESSARY TO ADD ANY OTHER SURFACTANT TO THE SPRAY SOLUTION. Best results are obtained on young actively growing weeds. It produces top-kill only, so repeat applications are required for season long weed control of perennials.

MIXING INSTRUCTIONS

PHYTAR 560 Herbicide is completely water soluble. Any spray equipment that gives good coverage may be used. Fill the spray tank about half full with water and add the required amount of herbicide with agitation. Finish filling the tank with water and apply. After use, clean equipment thoroughly by flushing with water. Do not store spray solution in tank for a prolonged period. Although PHYTAR 560 Herbicide is only moderately corrosive, do not use in galvanized or aluminum equipment.

Read entire label before using this product.

DIRECTIONS FOR USE

NON-CROP

PHYTAR 560 Herbicide is useful for general weed control on drainage ditchbanks and rights-of-way, along sidewalks, driveways and fences, around buildings and ornamentals, and on similar non-crop areas. Mix 1 to 2 gallons of PHYTAR 560 Herbicide in 100 gallons of water. Spray unwanted vegetation to just short of run-off. If regrowth occurs reapply as required.

Do not allow spray to come in contact with foliage, green bark, grafted unions, or scuffed, damaged or broken bark when spraying around trees and ornamentals.

AGRICULTURAL PLANTINGS—CITRUS (Except Florida)

PHYTAR 560 Herbicide is useful as a directed application in non-bearing citrus orchards such as orange, grapefruit, tangerine, lemon and lime orchards. It should be applied at the rate of $1\frac{1}{2}$ to 2 gallons per acre.

Mix 1 to 2 gallons of PHYTAR 560 Herbicide in 50 to 100 gallons of water. Apply as a directed spray in interspaces and around base of trees. Spray unwanted vegetation to just short of run-off. If regrowth occurs, reapply as required, however, **DO NOT EXCEED 3 APPLICATIONS PER YEAR.**

Do not allow spray solution to contact leaves, stems or bark. Use a shield, if necessary, for nursery plantings or young trees. **Do not apply around trees from which fruit will be harvested within one year of treatment.**

LAWN RENOVATION

Mow lawn to about 1 inch high before treatment. Mix 10 fluid ounces of PHYTAR 560 Herbicide in 4 gallons of water and apply to 1,000 sq. ft. Spray foliage thoroughly, preferably on a warm sunny day, foliage will usually turn brown in about five days. Bermudagrass and other deep-rooted perennials, or partially sprayed foliage, may require a second treatment. Reapply five days later if green areas remain.

When foliage is all brown, matted areas should be raked to remove dead vegetation. The lawn may then be reestablished according to local practice.

Keep children and pets off treated areas for 24 hours after treatment. Do not track from treated to untreated areas.

WARRANTY—CONDITION OF SALE

DIRECTIONS FOR USE of this product are based on field use and tests believed reliable and should be followed carefully. It is however impossible to eliminate all risks associated with use of this product. Because such factors as weather conditions, foreign material and manner of use for application are all beyond the control of Vertac Chemical Corporation or the Seller of this product, such things as crop injury, ineffectiveness or other unintended consequences may result. ALL SUCH RISKS ARE ASSUMED BY THE BUYER.

Vertac Chemical Corporation warrants that this product conforms to the chemical description on the label and is reasonably fit for the purposes referred to in the directions for use as modified by the above. Vertac Chemical Corporation makes no other warranties, express or implied, including FITNESS or MERCHANTABILITY. In no case shall Vertac Chemical Corporation or the Seller be liable for consequential, special or indirect damages resulting from the use or handling of this product. The foregoing is a condition of sale by Vertac Chemical Corporation and is accepted as such by the Buyer.

PREMERGE* PLUS

NAPTALAM PLUS DINITRO HERBICIDE

Controls Seedling Grasses and Weeds in Soybeans and Peanuts

ACTIVE INGREDIENTS:

Sodium N-1-Naphthylphthalamate**	22.3%
Sodium salt of dinoseb (2-sec-butyl-4,6-dinitrophenol)†	11.5%
INERT INGREDIENTS	66.2%
TOTAL	100.0%

** Equivalent to 20.8% N-1-Naphthylphthalamic Acid

† Equivalent to 10.5% 2-sec-butyl-4,6-dinitrophenol

Each gallon contains two pounds of active NPA and one pound of active DNBP.

EPA Reg. No. 39511-90 EPA Est. No. 39511-MS-1

*Trademark of VERTAC Chemical Corporation

DANGER

Keep Out of Reach of Children and Farm Animals

STATEMENT OF PRACTICAL TREATMENT

SYMPTOMS OF POISONING: Excessive fatigue, sweating, thirst and fever, If symptoms of poisoning develop from any type of exposure, SEND FOR A PHYSICIAN.

FIRST AID: Have patient lie quiet in coolest spot available. If feverish, cool with cold compresses or by immersion in cool water.

IF IN EYES: Immediately flush eyes with plenty of clean water for at least 15 minutes. Get medical attention.

IF ON SKIN: Immediately remove all contaminated clothing, including shoes. Wash skin with plenty of soap and water. If symptoms of poisoning develop, send for physician and treat as in first aid above. Discard contaminated clothing and shoes or clean them thoroughly before re-use. Note: Staining may result if contaminated articles are allowed to contact carpets, upholstery, etc.

ANTIDOTE: If swallowed, SEND FOR A PHYSICIAN. Active ingredient is a metabolic stimulant. Treat symptomatically.

PRECAUTIONARY STATEMENTS

DANGER

HAZARDS TO HUMANS AND DOMESTIC ANIMALS

May be fatal if swallowed or absorbed through skin. Do not get in eyes, on skin, or on clothing. Do not wear contaminated clothing or shoes. Do not breathe spray mist. Do not take internally. Not for use or storage around the home.

ENVIRONMENTAL HAZARDS

Keep out of any body of water. Toxic to fish and wildlife. Birds and other wildlife in area being treated may be killed. Do not apply where runoff is likely to occur. Do not apply when weather conditions favor drift from target area, Do not contaminate water by cleaning of equipment or disposal of wastes. Do not use in a manner not specified on this label.

PHYSICAL OR CHEMICAL HAZARDS

Do not use or store near heat or open flame.

For chemical spill, leak, fire, or exposure call toll free 1-800-424-9300

AGRICULTURAL CHEMICAL

Do Not Ship or Store with Food, Feeds, Drugs or Clothing

Read entire label before using this product.

GENERAL INFORMATION

LIQUID PREMERGE PLUS is a selective herbicide combination for the control of the following annual grasses and broadleaf weeds.

ANNUAL GRASSES: Foxtail, Crabgrass, Barnyardgrass, Goosegrass, Stinkgrass, Sandbur, Watergrass, Spangletop, Windmillgrass, Cupgrass and Johnsongrass (from seed).

BROADLEAF WEEDS: Florida Pusley, Chickweed, Lambsquarters, Ragweed, Purslane, Velvetleaf, Cocklebur, Bindweed, Ground Cherry, Mustards, Shepherd's Purse, Pigweed, Quickweed, Carpetweed, Morningglory, Beggarweed, (Beggar-Lice), Teaweed (Prickly Sida), Jimsonweed, Smartweed and Wild Sunflower.

DO NOT USE PREMERGE PLUS HERBICIDE ON SILT LOAM SOILS OF EXTREMELY FINE TEXTURE

PREPARATION OF SOIL: The area to be treated should be well prepared and as free as possible from stubble and litter. If the land has been plowed, the soil clods should be broken up by harrow or roller wheel. Plant on a flat or slightly raised bed. If weather factors make replanting necessary after PREMERGE PLUS has been applied, thoroughly disk the area prior to replanting.

METHOD OF APPLICATION: Thorough and uniform coverage of soil surface is essential for good weed control. Apply PREMERGE PLUS as a spray at planting time, using either a planter-mounted sprayer or a separate sprayer after the planting operation.

DIRECTIONS FOR USE

It is a violation of federal law to use this product in a manner inconsistent with its labeling.

STORAGE AND DISPOSAL

STORAGE

Do not contaminate water, food or feed by storage or disposal. Open dumping is prohibited. Do not store this product near fertilizers, seeds, insecticides, or fungicides. Do not store near heat or open flame.

PESTICIDE DISPOSAL

Pesticide, spray mixture, or rinsate that cannot be used according to label instructions must be disposed of according to federal, state or local procedures under the Resource Conservation and Recovery Act.

CONTAINER DISPOSAL

Triple rinse (or equivalent) adding rinsate to spray tank. Offer rinsed container for recycling or reconditioning or dispose of in a sanitary landfill, or by incineration if permitted by state and local authorities.

SOYBEANS

MIXING INSTRUCTIONS: Put one-half the recommended amount of water in the spray tank, add the required amount of PREMERGE PLUS. Then add the remainder of the water to the tank while mixing. Stir or use bypass agitation until completely mixed. It is essential that the sprayer be calibrated accurately prior to use on crop.

Use PREMERGE PLUS as an overall spray (broadcast) by ground equipment or aerial equipment up to the crook stage (from soil cracking to before the soybean true leaves open).

GROUND APPLICATION: Use PREMERGE PLUS at the rate of 6 quarts in 30 to 40 gallons of water per acre as an overall spray (broadcast application). Treatment may be applied over the entire area or in bands as desired.

Amount of PREMERGE PLUS Herbicide Per Acre to Use for Band Treatment of Soybeans

Width of Band	Row Spacing 28 inches	Row Spacing 32 inches	Row Spacing 36 inches	Row Spacing 40 inches
12″	5 Pints	4½ Pints	4 Pints	3½ Pints
14″	6 Pints	5¼ Pints	4¾ Pints	4¼ Pints

Use in 9 gallons of water per 40-inch row, to 13 gallons of water per 28-inch row.

NOTE 1: In the black soil areas of western Indiana, southern Minnesota, northern Iowa and Illinois, the dosage of PREMERGE PLUS suggested in the chart above should be increased by approximately one-third.

GENERAL INFORMATION

PREMERGE 3 Dinitro Amine Herbicide is a water soluble formulation containing dinoseb which is effective in both preemergence and postemergence applications to control many **annual broadleaf weeds** and **grasses** such as **mustards, annual smartweed, nightshades, yellow star thistle, lambsquarters, pigweeds, Florida pusley, purslane, common ragweed, crabgrass, chickweed** and **foxtails.** Weeds are killed by contact action only, either by foliar contact of emerged weeds or by root contact of germinating seedlings with dinoseb in soil. Only partial control of certain vigorous annual broadleaf weeds and grasses may be obtained after they become well established or under drought conditions or cool weather conditions. Established perennials usually are not controlled. Oil or surfactants increase the burning action of PREMERGE 3 and should not be added to spray mixtures unless recommended. Do not mix with liquid fertilizer. **Read and follow all use and safety precautions on this label.**

TYPES OF TREATMENT

STALE SEEDBED (before or soon after planting crops)—Apply after weeds have emerged. Only annual weeds that are up at time of spraying will be killed. Other herbicides will be needed for season-long weed control.

PREEMERGENCE (between planting and emergence of the crop)—For best results, the soil surface should be free of clods and trash, moist and fairly firm. Results are usually improved by delaying application until 1 to 3 days before crop emergence. Light to moderate rainfall, or irrigation (floor or sprinkler) after treatment is usually necessary for good residual effect, but heavy rain may cause excessive leaching of the chemical with resultant lower weed control. Retreat as soon as weed growth warrants after crop emergence, if recommended, or start cultivation.

POSTEMERGENCE (after crop plants emerge)—For best results, spray when weeds, especially grasses, are small seedlings. Apply PREMERGE 3 preferably during warm sunny weather when crop foliage is dry using low pressure (25 to 50 psi). PREMERGE 3 may burn leaves of crop plants contacted by the spray but the effect is usually only temporary. With high temperature conditions, PREMERGE 3 is more active and less product is required for good weed control. Use enough spray volume to give uniform and complete coverage of all weed foliage.

ATTENTION: Use only in crops designated on this or other approved supplemental labeling for PREMERGE 3 Dinitro Amine Herbicide. Do not spray or allow spray drift to contact desirable crop plants including ornamental plants, except as recommended.

MIXING AND APPLICATION INSTRUCTIONS

Spraying equipment should be carefully calibrated before using. Use enough spray volume to provide good coverage of soil surface or weeds. Unless otherwise specified select spray nozzles and pressure to provide 20 to 40 gallons of spray per acre by ground equipment or 3 to 10 gallons per acre by aircraft. All rates of application are expressed for overall or broadcast coverage. For band application the amount of spray required will be proportionately less depending on the width of bands treated.

Note: To ensure product uniformity, mix thoroughly by rolling unopened container before using. Use only approved agricultural surfactants such as MULTIFILM X-77 or TWEEN 20 brands.

STORAGE, SHIPMENT AND DISPOSAL

STORAGE
Not for use or storage in or around the house.

SHIPMENT
Do not ship or store with food, feeds, drugs, or clothing.

DISPOSAL
Rinse emptied container three or more times, pouring rinse water into spray tank. Dispose of empty container in a manner that complies with local, State and Federal regulations.

DO NOT REUSE ORIGINAL CONTAINERS SUCH AS DRUMS

CROPS

BEANS (Field, Lima, Snap): Premergence—Use 2.5 to 3 gallons of PREMERGE 3 in about 30 gallons of water per acre at time of, or shortly after, planting. To avoid possible crop injury do not apply PREMERGE 3 to beans growing on light sandy soil containing little or no organic matter. **Emergence**—If desired, delay treatment until weeds have emerged and beans are not beyond the "crook" stage, using 1 to 1.5 gallons of PREMERGE 3 in about 30 gallons of water per acre as an overall spray. **Postemergence**—To control many emerged annual broadleaf weeds including nightshades, cocklebur, annual smartweed, annual morningglory, common ragweed and lambsquarters in field beans west of the Rocky Mountains only, use PREMERGE 3 at 2 to 3 quarts in 75 gallons of water per acre as an overall spray. Apply after first trifoliate bean leaves have reached full size but before bloom and when broadleaf weeds are not more than 2 to 3 inches tall. The amount of PREMERGE 3 to be used per acre depends on the maximum air temperature expected within 24 hours: use 3 quarts at less than 85°F, 2.5 quarts from 85 to 95°F, and 2 quarts above 95°F. Some burn of bean foliage usually occurs but the plants normally recover within a few days. Be sure spray tank or spray mixture does not contain oil or added surfactant. **Note:** Do not graze or feed treated forage to livestock within 6 weeks after spraying. Do not apply by aircraft nor use spray volume less than 75 gallons per acre by ground equipment.

CORN (Field, Popcorn and Sweetcorn): Preemergence—Use 2.5 to 3 gallons of PREMERGE 3 in about 30 gallons of water per acre as an overall spray at time of planting. **Early Postemergence**—Use 1 to 1.5 gallons of PREMERGE 3 in about 30 gallons of water per acre as an overall spray to control seedling grasses and broadleaf weeds that have emerged. Apply when corn is not beyond the 2-leaf stage and weeds are very small. Partial burning of corn leaves is not ordinarily harmful. Burning may be more severe if spraying is done during hot weather. Rates of 2 to 3 quarts of PREMERGE 3 per acre will give satisfactory control of newly emerged broadleaf weed seedlings.

COTTON: Directed Postemergence—Use PREMERGE 3 at the broadcast rate of 2 to 3 quarts with 1 to 2 pints of agricultural surfactant (such as MULTIFILM X-77 or TWEEN 20 brands) in about 40 gallons of water per acre for the control of emerged broadleaf weeds including cocklebur, annual morningglory, prickly sida (teaweed), coffeeweed (Sesbania), threeseed mercury, pigweeds and many others. Proportionately less of the same spray mixture will be required when only bands are treated. Apply as a carefully **directed** spray with oiling type or other precision directed spray application equipment. Direct the spray toward the base of the cotton plants to provide a maximum wetting of the weeds with minimum contact to the cotton. Cotton foliage wet by the spray will be injured and some blackening of the stems may occur. However, no permanent injury is caused when spray is confined to the base of the cotton plants. **To avoid crop injury spray must be confined to the lower half of the cotton plants.** For best results, make the first application when cotton plants are 4 to 5 inches tall and weeds are 1 inch or less in height. Repeat the application in 7 to 14 days and if needed follow with a third application any time until bolls begin to open. **Note:** Small cotton plants completely covered by the spray may be killed. Do not apply when soil surface is wet, when cotton plants are extremely succulent, or when small cotton plants are seriously affected by seedling diseases.

CUCURBITS: Direct Seeded—Apply preemergence only in Northwest, North Central and Northeastern States using 0.5 to 1 gallon of PREMERGE 3 for cucumbers and 1 to 2 gallons per acre for muskmelons, watermelons, pumpkins and squash. Apply in about 30 gallons of water per acre after planting but before crop emergence. Use the higher rates on heavy soils. **Note:** Do not use on sandy soils. Plant seed no less than 1 inch deep. Heavy rainfall soon after application may leach the dinoseb to the seed level and severely injure the crop, especially cucumbers. Do not use if soil is to be covered with plant protectors. **Transplants of Cucumbers, Musk-**

melons, or Watermelons Grown Through A Clear Plastic Mulch—For control of most annual broadleaf weeds and grasses, apply 1 to 1.5 gallons of PREMERGE 3 in about 30 gallons of water per acre to well prepared soil early in the spring and lay plastic immediately. Delay transplanting at least 2 weeks or until no yellow color remains under the plastic.

DAFFODILS, TULIPS AND SIMILAR FALL PLANTED BULBS: Preemergence—Spray in late fall following emergence of winter annual weeds using 1 to 2 gallons of PREMERGE 3 in about 30 gallons of water per acre. Do not apply after bulb shoots emerge above ground.

FORAGE LEGUMES: Alfalfa, Ladino Clover, Red Clover, Sweet Clover, and Birdsfoot Trefoil—New Plantings—Use 1.5 to 2 quarts of PREMERGE 3 in about 30 gallons of water per acre when crop seedlings have two or more leaves and weeds are small. **Established Stands**—Apply 1 to 1.5 quarts of PREMERGE 3 in about 30 gallons of water per acre as soon as chickweed appears. If new chickweed seedlings emerge after spraying, repeat the treatment. If chickweed becomes well developed delay application until legumes are **dormant** and use 2 to 4 quarts of PREMERGE 3 per acre in 60 to 100 gallons of spray per acre. Use the higher rates of PREMERGE 3 and larger spray volumes on larger weeds or when temperatures are below about 55°F. Heavy weed growth may require a second application. **Note:** Do not graze or feed treated forage to livestock within 6 weeks after spraying.

FRUIT AND NUT ORCHARDS AND GRAPE VINE YARDS: Ground Cover under Almonds, Apples, Apricots, Cherries, Citrus, Dates, Figs, Filberts, Grapes, Nectarines, Olives, Peaches, Pears, Pecans, Plums, Prunes, Walnuts—Use 3 to 3.5 gallons of PREMERGE 3 plus 1 to 2 quarts of an agricultural surfactant (such as MULTIFILM X-77 or TWEEN 20 brands), or 2 to 4 gallons of emulsifiable crop oil with water to make 100 gallons of mixed spray per acre. Use as a "floor spray" in fruit and nut plantings of all ages, and in grape vineyards only when the vines are 2 years old or older. Apply the spray only to the ground cover and direct the spray across the rows. Apply no higher than 12 inches on tree trunks or main grape stems. When band spraying use amounts of spray proportional to the area actually treated. Treat the entire soil area when weeds are in one to two leaf stage. For best residual control, apply 1/4 to 1 acre-inch of irrigation water (basin or sprinkler) if no rain occurs within a few days. Furrow irrigation is not adequate. Do not disturb the soil until weeds reinvade the area. Repeat the treatment as needed but use no more than four applications per year. **Note:** Keep spray off fruit and foilage. Do not apply within 30 days before harvest. Do not allow livestock to graze treated areas.

GLADIOLUS: Preemergence—Use 2.5 to 3.5 gallons of PREMERGE 3 in about 30 gallons of water per acre as an overall spray, preferably just before crop emergence.

LENTILS: Preemergence—For the control of mustard, lambsquarters, borage, nightshade, and many other broadleaf weeds apply one gallon (3 pounds active ingredient) of PREMERGE 3 in approximately 30 gallons of water. Application should be made after planting but before lentils emerge. Do not exceed 3 pounds of active ingredient per acre. Do not graze or feed treated lentil forage or hay to livestock.

PEANUTS: Make one early application (preemergence, early cracking stage, or early postemergence) followed by one or two later postemergence applications as needed, in accordance with the following directions. **Preemergence**—Use 3 to 4 gallons of PREMERGE 3 in about 30 gallons of water per acre as an overall spray. Apply at any time between planting and emergence. **Early Cracking Stage**—Use 2 gallons of PREMERGE 3 in about 30 gallons of water per acre as an overall spray. **Early Postemergence** (from emergence until the plants are 1.5 to 2 inches in diameter) —Apply overall using 1 gallon of PREMERGE 3 in about 30 gallons of water per acre. Such an application will control weeds just coming through, but will not provide long term residual effectiveness. Some foliage injury may be noted on the peanuts, particularly if the temperatures are high. **Later Postemergence**—One or two directed applications may be made up to one month after the early cracking stage. Apply when new weeds are first visible. Use 2 quarts of PREMERGE 3 in about 30 gallons of water per acre. Direct the spray to the base of the peanut plants using precision application equipment to minimize contact with peanut foliage.

Tank Mix of PREMERGE 3 with LASSO™ Herbicide—Apply in peanuts at the cracking stage to control seedling broadleaf weeds and annual grasses such as annual morningglory†, annual smartweed†, barnyardgrass, beggarweed, brachiaria, carpetweed, cocklebur†, common ragweed, copperleaf†, crabgrass, fall panicum, Florida pusley, foxtail, goosegrass, lambsquarters, pigweed (carelessweed), prickly sida (teaweed)†, purslane, sandbur††, seedling johnsongrass††, sicklepod†, witchgrass, and yellow nutsedge††. The amount of PREMERGE 3 to be used per acre depends on the maximum air temperature at the time of spraying or expected within 24 hours. Use 2 quarts up to 90°F, and 1.5 quarts above 90°F, plus 2 to 3 quarts of LASSO in 20 to 40 gallons of water for overall coverage, and proportionately less for band treatment. Apply during the soil cracking stage as peanuts are beginning to emerge. For best results, rain should follow application within a few days to activate the LASSO herbicide. Before using LASSO with PREMERGE 3, read and observe all use precautions and statements of hazard given on containers for both products.

† Best control is usually obtained with the higher rates shown above. If needed, a follow-up postemergence application of PREMERGE 3 alone may be applied as described above.

†† Suppression can be expected only by application of the higher dosage rates followed by at least 1/2 inch of rain to activate the LASSO herbicide. Results may be poorer under dry soil conditions.

PEAS: Preemergence—Use 2 to 3 gallons of PREMERGE 3 in about 30 gallons of water per acre as a broadcast spray after planting but before emergence of the crop.

Postemergence—Use 1 to 3 quarts of PREMERGE 3 in about 30 gallons of water per acre with ground equipment or in 5 to 10 gallons of water per acre by aircraft. Apply prior to bloom when the peas are 2 to 8 inches tall and weeds are small. Dosage required per acre varies with maximum air temperature expected during the 24 hour period immediately following application: use 3 quarts below 70°F, 2 to 3 quarts from 70 to 75°F, 1.5 to 2 quarts from 75 to 80°F, and 1 to 1.5 quarts from 80 to 85°F. Do not apply at temperatures above 85°F. **Note:** Some foliage burn of peas may occur but plants normally recover within a few days. Apply only when crop foliage is dry. Allow time for peas to harden following a period of cool, cloudy weather before spraying. Do not apply after flower buds are visible. **Do not graze or feed treated pea forage to livestock unless at least 6 weeks has elapsed between spraying and harvest of the peas.**

When foxtail grasses are a problem, use the same amount of PREMERGE 3 and add one pound of DOWPON™ grass killer per acre. See "Pea" recommendation on the label for DOWPON™ grass killer.

POTATOES: Preemergence Weed Control—Use 1 to 2 gallons of PREMERGE 3 per acre in about 30 gallons of water and apply as an overall spray 1 to 3 days before the potatoes emerge. If seedling grasses are a problem add DOWPON M grass killer to the above mixture at a rate to give 3 pounds per acre and apply in the same manner. **Note:** To avoid possible crop injury do not apply PREMERGE 3 to potatoes growing on light sandy soil containing little or no organic matter. Do not use DOWPON™ in sprays to be applied to White Rose or red skinned varieties.

Preharvest Vine Killing—Use 2 to 3 quarts of PREMERGE 3 per acre in a spray prepared as follows: Mix 1 pint of a nonionic emulsifier (such as MULTIFILM X-77 or TWEEN 20 brands) in 5 gallons of fuel oil and emulsify in 30 gallons of water; then add the PREMERGE 3 with vigorous agitation and spray at moderate pressure. Apply 10 to 20 days before harvest during warm sunny weather. Complete coverage of the vines is essential. **Note: Do not spray exposed tubers nor graze treated fields.**

Preemergence as Tank Mix of PREMERGE 3 with LASSO™ Herbicide—For longer residual control of germinating weeds, apply 4 to 6 quarts of PREMERGE 3 with 2 to 3 quarts of LASSO in 20 or more gallons of water per acre as an overall spray after planting and before crop emergence. Use the lower rates of PREMERGE 3 and LASSO on lighter soils. For best results delay application until weeds have just emerged but before emergence of the potatoes. Preemergence dragging or hilling should be done at or soon after planting **and before spraying** to permit weeds to emerge earlier than the crop. **Do not use this treatment under furrow irrigation or incorporate the spray into the soil.**

SMALL GRAINS (Barley, Oats, Rye and Wheat—Seeded Alone or Interplanted With a Legume): Postemergence—Use PREMERGE 3 at rates of 1.5 to 2 quarts per acre by either ground equipment or by aircraft, and apply when grain is 2 to 6 inches tall and weeds are small. For aerial application dilute in enough water to make 10 to 15 gallons of spray per acre. For ground spraying dilute in enough water to make 30 to 50 gallons per acre. To avoid excessive burn of crop foliage, use the lower rate of PREMERGE 3 during periods of cloudy weather. **Note:** Do not graze or feed treated forage to livestock within 6 weeks after spraying.

SOYBEANS

Preemergence—Apply 2 to 2.5 gallons of PREMERGE 3 in about 30 gallons of water per acre between planting and crop emergence. Do not use on very light sandy soils. Some reduction in stand of soybeans may result but this does not ordinarily reduce crop yield.

For improved grass control, longer residual control of germinating seedlings and reduced risk of crop injury, use **a tank mix combination of PREMERGE 3 and AMIBEN™ Preemergence Herbicide** consisting of 4 to 6 quarts of PREMERGE 3 plus 3 to 4 quarts of AMIBEN per acre applied at time of planting or before crop emergence. Use the higher rate of each product on clay and clay loam soils.

Early Postemergence (when soybeans have emerged and are still in the cotyledon stage before first leaves open to expose the terminal bud)—To control emerged weeds including pigweeds, cocklebur, annual morningglory, jimsonweed, common ragweed, velvetleaf, annual smartweed, small seedling grasses and many others that emerge before or with the crop, use 2 to 3 quarts of PREMERGE 3 in 3 to 5 gallons of water applied by aircraft or in about 30 gallons of water applied by ground equipment. The amount of PREMERGE 3 to be used per acre depends on the maximum air temperature expected within 24 hours: use 3 quarts below 75°F, 2 quarts from 75 to 95°F, and do not apply above 95°F.

For residual control of annual weed seedlings that may germinate after spraying, add 3 quarts of AMIBEN Preemergence Herbicide per acre as a tank mix to the amount of PREMERGE 3 recommended. To avoid crop injury with mixtures of PREMERGE 3 and AMIBEN, observe precautions for use of PREMERGE 3 alone.

Some injury to soybean plants may occur but yields usually are not reduced. **Note: Do not spray if soil surface is wet or if first leaves have opened. Do not use this treatment if AMIBEN has been applied preemergence.**

Postemergence Over-The-Top Application: This treatment may be used when needed as an aid to control emerged Cocklebur, Morningglory, and other susceptible weeds, and as a supplement to, but not as a replacement for, the early postemergence treatments.

Use this application as early as the first to second trifoliate leaf stage of soybean growth up to, but no later than, when the beans begin to bloom. Best results will be obtained with weeds no more than 1 to 3 inches tall. Temperatures at the time of spraying should not be below 70°F or above 95°F. Application of PREMERGE 3 should not be made when soybeans are wet with dew, immediately following rain, or under wet soil conditions.

Apply to each acre 1 to 2 pints of PREMERGE 3 in 8 to 10 gallons of water as a broadcast spray by ground equipment, or 1 to 2 pints of PREMERGE 3 in 3 to 10 gallons of water by aerial equipment. Do not use a surfactant in the spray mix.

Nozzles used for applications should be the flat fan or cone type with a spray pressure of 50–60 psi. Small succulent weeds are easiest to kill. Treatment should be made as soon after weed emergence as possible. Weed control may be poor under drought stress conditions.

Note 1: Using the correct amount of water, type of nozzles, and spray pressure of 50–60 psi is very important.

Note 2: Do not tank mix with other products for the over-the-top application. Do not add a surfactant or other adjuvant to spray mix.

Under good growing conditions, damage to soybeans will be limited to slight foliar burn, followed by rapid recovery and normal yields. Crop injury may be more severe when the soybeans are under stress from conditions such as disease or low moisture. If the field is replanted or abandoned, do not graze or feed treated forage.

Directed Postemergence—Apply when soybeans are 5 to 6 inches tall up to the time they begin to bloom. Use PREMERGE 3 at the rate of 2 to 4 quarts per acre in 30 to 40 gallons of water for the control of emerged cocklebur, morningglory, coffeeweed (Sesbania), pigweed, and many other broadleaf weeds. Apply as directed spray through oiling type or other precision directed spray application equipment. Proportionately less of the same spray mixture will be required when only bands are treated. Direct the spray toward the base of the soybean plants to provide for maximum wetting of weeds with minimum contact of the crop. Control of coffeeweed and grass seedlings will be improved by the addition of 2 quarts of an agricultural surfactant (such as MULTIFILM X-77 or TWEEN 20 brands) in each 100 gallons of spray. Repeat the directed application once or twice at 7 to 14 day intervals, if needed. **Note:** Continue the spray to the lower half of the soybean plants. Leaves wet by the spray will be injured but no reduction in yield will usually occur if the spray is applied only to the base of the plants. Do not apply after the soybeans begin to bloom. Do not graze or forage treated soybeans within 3 weeks after treatment.

Tank Mix of PREMERGE 3 plus LASSO™ —Apply to stale seedbed, or as preemergence or early postemergence treatment in soybeans to control seedling broadleaf weeds and annual grasses such as barnyardgrass, black nightshade, brachiaria, carpetweed, cocklebur†, crabgrass, fall panicum, Florida pusley, foxtails, goosegrass, jimsonweed†, lambsquarters, morningglory†, mustard, pigweed, purslane, common ragweed, seedling johnsongrass††, smartweed, velvetleaf†, witchgrass, and yellow nutsedge††.

†Best results can be obtained by treatment after weeds have emerged.

††Control or suppression can be expected by application of higher dosage rates immediately after planting, followed by at least one-half inch of rain to activate the herbicides. Under less favorable conditions, results may be poorer.

Note: Recommended tank mixtures of PREMERGE 3 and LASSO will not control established johnsongrass, Canada thistle, field bindweed, bullnettle or other established perennial weeds.

Use the amounts of PREMERGE 3 and LASSO indicated in the table and mix in enough water to obtain uniform overall spray coverage (usually 30 gallons per acre), or proportionately less for band treatment. Provide continuous agitation in the spray tank during mixing, holding, and spraying operations to keep spray uniformly mixed.

STALE SEEDBEDS: Preplant or Preemergence Control of Emerged Weeds in Beans, Corn, Potatoes and Soybeans —PREMERGE 3 may be used to control small annual broadleaf weeds and annual grasses that have emerged on previously tilled fields prior to or immediately after planting beans, corn, potatoes or soybeans. For aerial application use 2 to 4 quarts of PREMERGE 3 plus 1 to 2 pints of an agricultural surfactant (such as MULTIFILM X-77 or TWEEN 20 brands) with enough water to give 3 to 10 gallons of spray mixture per acre. With ground equipment use 2 to 4 quarts of PREMERGE 3 plus 1 to 2 pints of agricultural surfactant with enough water to give about 30 gallons spray mixture per acre. Use the higher rates of PREMERGE 3 and surfactant where grass seedlings predominate.

Note: Any crop plants that are up at time of spraying may be seriously injured.

STRAWBERRIES: For established plantings in the Northwest, use 4 to 6 quarts of PREMERGE 3 in 40 to 60 gallons of water per acre, and apply in early spring before bloom or after harvest.

Quarts of PREMERGE 3 and LASSO Per Acre for Overall Coverage in Soybeans

Soil Texture	Stale seed bed or Preemergence PREMERGE 3	plus LASSO	Early Postemergence PREMERGE 3 Up to 75°F	75°F to 95°F	plus LASSO
Sandy loam	4	2	3	2	1.5
Silt loam	6	2 to 2.5	3	2	2
Heavy clay	6	2.5 to 3	3	2	2.5 to 3

Apply to **stale seedbeds** either before or after planting but before emergence of the soybeans to control weeds that are up as well as those that have not emerged in fields prepared for planting several weeks earlier. Rain or irrigation is needed after **preemergence** application to activate the herbicides, preferably within 4 to 6 days after treatment. The **early postemergence** treatment can be applied when weeds are up and soybeans **are in the cotyledon stage but before first leaves open to expose the buds.** This may cause some injury to the soybeans but the effect is usually temporary and not harmful to yield. **Note:** Do not use on light sandy soils. Do not spray if soil surface is very wet, first soybean leaves have opened, or air temperature is expected to exceed 95°F within 24 hours after application. Do not apply the tank mix postemergence if LASSO has been applied preemergence. Before using LASSO with PREMERGE 3, read and observe all use precautions and statements of hazard on labels affixed to containers of these products.

NON-CROP AREAS

PREMERGE 3 is highly effective in controlling broadleaf weeds and unwanted grasses in non-crop areas such as airfields, drainage ditches, industrial building and storage sites, around farm buildings, along fence rows and on rights-of way for utility installations, railroads, and highways. Apply when the vegetation is young and succulent and preferably during warm sunny weather, using a mixture prepared as follows for 100 gallons of spray: add about 75 gallons of water to the spray tank; with good agitation add 1 gallon of PREMERGE 3 plus 1 quart agricultural surfactant and 10 gallons of diesel oil or weed oil; and add water to bring spray volume to 100 gallons. With continued agitation apply at moderate pressure (50 to 100 psi) using 100 to 200 gallons of spray mixture per acre to thoroughly wet all vegetation. DOWPON+up,TM grass killer may be added at the rate of 10 pounds per 100 gallons of spray to improve kill of grasses. Repeat the application as needed. **Note:** Do not graze or feed treated vegetation to livestock.

Read and observe all precautions for PREMERGE 3 Dinitro Amine Herbicide.

PROPANIL 4

HERBICIDE

For Postemergence Control of Broadleaf and Grass Weeds in Rice Fields

ACTIVE INGREDIENT:
3′,4′-Dichloropropionanilide 43.48%
INERT INGREDIENTS: 56.52%
TOTAL 100.00%

This product contains 4 lbs. of 3′,4′-Dichloropropionanilide per gallon.

EPA Reg. No. 39511-8 EPA Est. No. 39511-AR-2

CAUTION

Keep Out of Reach of Children

See side panels for additional precautionary statements

PRECAUTIONARY STATEMENTS

HAZARDS TO HUMANS AND DOMESTIC ANIMALS

CAUTION

This product may be harmful if swallowed, and may cause skin irritation. Avoid inhaling mist sprays and contact with eyes, skin or clothing. In case of contact, flush skin or eyes with clean water and get prompt medical attention.

ENVIRONMENTAL HAZARDS

This product is toxic to fish. Keep out of lakes, streams or ponds. Do not apply when weather conditions favor drift from treated area. Do not contaminate any body of water by direct application or indirectly by cleaning of equipment, or by disposal of wastes and containers. Apply this product only as specified on this label.

PHYSICAL AND CHEMICAL HAZARDS

DO NOT USE, POUR, SPILL OR STORE NEAR HEAT OR OPEN FLAME

DIRECTIONS FOR USE

It is a violation of Federal law to use this product in a manner inconsistent with its labeling.

STORAGE AND DISPOSAL

STORAGE

Do not contaminate water, food or feed by storage or disposal. Open dumping is prohibited.

DISPOSAL

Pesticide, spray mixture, or rinsate that cannot be used or chemically reprocessed should be disposed of in a landfill approved for pesticides or buried in a safe place away from water supplies. Containers should be triple rinsed (or equivalent) and offered for recycling, reconditioning, or disposed in an approved landfill or buried in a safe place.

GENERAL

Consult Federal, State or Local disposal authorities for approved alternate procedures.

WHERE TO USE

VERTAC PROPANIL 4 is used for postemergence control of broadleaf and grass weeds in RICE fields.

WEEDS CONTROLLED

Alligatorweed	Croton	Millet (Texas)
Barnyardgrass (Watergrass)	Curly indigo	Paragrass
	Foxtail	Pigweed
Brachiaris	Goosegrass	Sourdock
Coffeeweed	Gulf cockspur	Spearhead
Crabgrass	Mexicanweed	Wiregrass

(This product will not control arrowhead, bermudagrass, cattail, ducksalad, johnsongrass, nutgrass, red rice and sprangletop.)

GENERAL

Several important factors should be taken into account to achieve a high efficiency of selective weed control with PROPANIL. These include uniform application, growth stage and weather conditions. To assure uniform application shake or roll container prior to opening and mix the prescribed amount of product with a sufficient volume of carrier to provide thorough coverage of target area. For aerial applications use approximately 10 gallons, and for surface (ground) applications 20–30 gallons of carrier per acre at high enough pressure. Agitate tank-mixes thoroughly and continuously. Avoid over and under application.

Growth stage of weeds is very important. Best results for selective weed control are obtained when most grasses have reached the 1 to 3 leaf stage. Proper field preparation is essential to ascertain a relatively clod free and level surface and to obtain uniform flood levels and growth. Fields may be flushed prior to treatment to produce uniform and vigorous grass germination and growth. Drain water from fields prior to applying product. Higher rates are recommended to control larger grasses or exposed weeds when rice fields are not completely drained. Inspect rice fields regularly to select the correct application time.

Weather conditions must be observed closely. Under cool weather conditions higher rates are required to achieve satisfactory control. Avoid application if rain threatens within 6 to 8 hours, or if wind velocities are high enough to cause drift and irregular spray patterns.

Consult local Extension Service for detailed application advice.

RECOMMENDED BROADCAST RATE

Apply 3 quarts of product per acre when most grasses have reached 1 to 3 leaf stage. Use 4 to 6 quarts of product per acre when the grasses are large (4 to 6 leaf stage) or when unseasonably cool weather conditions prevail or in cases where rice fields have not been drained completely and where weeds are large enough.

Barnyardgrass may be controlled up to 30 to 45 days after planting, before rice plants have reached the fully tillered growth stage.

NOTE: Product applied to rice after the 4 leaf stage may cause visible injury under some climatic conditions. Rice plants usually outgrow such injury.

IN CALIFORNIA—Use product only where rice fields are not completely drained. If higher water level is desired, reflood field after 12 hours and before 7 days after treatment. This will discourage new weed infestations. Do not apply PROPANIL within 14 days before or after insecticide applications. Serious injury to rice may occur.

CAUTION

Do not apply this product (directly or indirectly) to any crop except rice.

Do not apply when wind conditions will allow drift to adjacent, susceptible crops such as beans, soybeans, cotton, safflower, cucurbits, vegetables, orchards and other sensitive crops.

Do not store near food, fertilizers, seeds or insecticides.

Keep containers closed when not in use.

WARRANTY

VERTAC warrants that this herbicide conforms to the chemical description on its label. When used in accordance with label directions under normal conditions, this herbicide is reasonably fit for its intended purpose. Since timing, method of application, weather, plant, and soil conditions, mixture with other chemicals and other factors affecting the use of this product are beyond our control, no warranty is given concerning the use of this product contrary to label directions or under conditions which are abnormal or not reasonably forseeable. The user assumes all risks of any such use.

SALVO®

LOW VOLATILE WEED KILLER

ACTIVE INGREDIENT:

Isooctyl ester of 2,4-Dichlorophenoxyacetic Acid†	81.8%
INERT INGREDIENTS:	18.2%
TOTAL	100.0%

Equivalent to 2,4-Dichlorophenoxyacetic Acid 54.2%

† Isomer Specific by A.O.A.C. Method 6.D 01-5

EPA Reg. No. 39511-125

CAUTION

Keep Out of Reach of Children

HARMFUL IF SWALLOWED. MAY CAUSE SKIN IRRITATION. Avoid contact with eyes, skin and clothing. Do not take internally. Do not contaminate feed or foodstuffs.

Avoid spray drift to susceptible plants as this product may injure cotton, beans, peas, ornamentals, etc. (Course sprays are less likely to drift.) Do not contaminate domestic or irrigating water supplies. Do not contaminate water by cleaning of equipment, or disposal of wastes. Apply this product only as specified on this label.

Vapors from this product may injure susceptible plants in the immediate vicinity. Do not seed lawns, gardens or fields until at least four weeks after application of spray to area. Do not mix with liquid fertilizers. Do not apply in any manner not specified on the label.

GENERAL INFORMATION

If stored below freezing temperatures it may be necessary to warm to 40°F and agitate thoroughly before using.

When filling spray tank first fill tank 1/4 full of water; add the calculated amount of 2,4-D emulsifiable, then finish filling tank with water with by-pass open and pump running.

Material may tend to settle when not agitated. When mixture is left in sprayer allow by-pass pressure to re-agitate before resuming spraying.

PRECAUTION

Do not reuse empty container. Destroy by burying in a safe place. Do not burn.

Because of the difficulties of removing 2,4-D from sprayers do not use the same spray equipment for applying other materials to plants as injury will result. Use a separate sprayer for this purpose.

Do not store near fertilizers, seeds, insecticides, fungicides or plants.

TIME OF APPLICATION

Annual weeds are most susceptible in the seedling stage and early growth stages of development and when conditions are such as to promote a vigorous growth.

WHEN USED IN CROP SPRAYING AS A SELECTIVE SPRAY, THE STAGE OF GROWTH OF THE CROP MUST ALSO BE CONSIDERED, OTHERWISE SEVERE DAMAGE MAY RESULT.

SOME SUSCEPTIBLE WEEDS

Arrowhead	Hemp Dogbane	Puncture Vine
Artichoke	Henbit	Purslane
Beggar Ticks	Horsetail	Ragweeds
Bindweeds	Jimsonweed	Russian Thistle
Blue Thistle	Klamathweed	Quickweed
Bull Thistle	Kochia	Shepherd's Purse
Burdock	Lambsquarters	Smart Weed
Butterprint	Loco Weed	Sow Thistle (annual)
Canada Thistle	Mallow	Stinkweed
Catnip	Marestail	Sumacs
Chickweed	Mexican Weed	Sunflower
Chicory	Morning Glory	Wild Garlic
Cocklebur	Musk Thistle	Wild Mustard
Dandelion	Mustards	Wild Onion
Docks	Nettles	Wild Parsnip
Fanweed	Pennycress	Wild Radish
Figwort	Pepper Grass	Wild Rape
Four O'Clock	Pig Weed	Wild Sweet Potato
Ground Ivy	Prickly Lettuce	Yellow Goatsbeard

Some weeds above, listed as susceptible may not be completely killed but growth is so retarded by proper application that they are no longer a factor affecting growth and yield of crop.

NOTICE TO BUYER

The product conforms to its chemical description and is fit for the use intended when used in accordance with label directions. Vertac will not be responsible for special, consequential or contingent damages arising from injury to persons or property.

CROP	*AMOUNT SALVO PER ACRE	APPLICATION	CAUTION
CORN Preemergence	1½–2¼ pints	Spray uniformly over the soil surface after planting (preferably 3 to 5 days) but before emergence. Use in 5 to 20 gallons of water.	Do not use on light, sandy soils, or when moisture is inadequate for normal weed growth.
CORN, MILO Postemergence	⅕–⅖ pt.	Apply when crop is small and more resistant to 2,4-D but when weeds are succulent or in active growth stage. Use lower dosage only against succulent, non-resistant annual broadleaf weeds.	Use drop nozzles when crop is 10 inches tall or higher. Keep spray off leaves. Do not spray after tassels appear. Some injury to corn may occur.
WHEAT OATS BARLEY RYE	FALL ⅖–⅘ pt.	In spring when weeds are small and grain is in full tiller stage (to 8 inches high) and always before boot stage.	Do not apply during boot or dough state. Do not forage or graze treated grain fields within two weeks after treatment. Do not feed treated straw to livestock.
	SPRING ⅖–½ pt.	After plants are fully tillered, but before stems begin to joint.	
PERENNIAL GRASS PASTURES, MEADOWS	⅘ pt.	In early spring when sufficient weeds have emerged and are actively growing and not too mature.	Injury may result to bent grass and to clovers. Usually clovers will recover from early spring applications. Do not apply when grass is in boot to milk stage. Do not apply to newly seeded areas or after heading begins. Do not graze treated areas to dairy animals within 7 days after treatment.
LAWNS, ORNAMENTAL TURF GRASS & GOLF FAIRWAYS		Large areas 1½ pints per acre. Small areas 3⅓ teaspoonsful in 3 to 5 gallons of water applied to 1000 square feet.	Spray 3 to 5 days after mowing. Retreatment of perennial weed areas may be necessary. May cause injury to bent grass, carpet grass, dichondra, St. Augustine and to clovers. Protect ornamentals from spray drift.

* Application rates may vary slightly from state to state depending upon soil types and weather factors. Consult your State agricultural experiment station for recommendations under local conditions for more specific uses.

SELECTIVE WEED KILLER

A DINITRO WEED KILLER

ACTIVE INGREDIENT:

Dinoseb (2-sec-butyl-4,6-dinitrophenol) as the ammonium salt 13.7%

INERT INGREDIENTS: 86.3%

TOTAL 100.0%

Contains 1 pound of the ammonium salt of dinoseb per gallon

E.P.A. Reg. No. 39511-85AA E.P.A. Est. 39511-MS-1[A]; 39511-CA-1[B]

Superscript used corresponds to letter in Lot No.

DANGER–POISON

Keep Out of Reach of Children and Farm Animals

PRECAUCION AL USUARIO: Si usted no lee ingles, no use este producto hasta que la etiqueta le haya sido explicada ampliamente.

TRANSLATION: **(TO THE USER:** If you cannot read English, do not use this product until the label has been fully explained to you.)

POISONOUS IF SWALLOWED OR ABSORBED THROUGH SKIN

FLAMMABLE · VAPOR HARMFUL

Do Not Get in Eyes, on Skin or on Clothing · Do Not Breathe Vapors or Spray Mist · Do Not Take Internally · Keep Away from Heat and Open Flame · Do Not Cut or Weld Container · Keep Container Closed

Wear suitable protective equipment such as chemical workers goggles or rubber or plastic gloves and rubber or plastic apron while pouring and transferring the concentrated product and at other times when contact is likely. Wear protective clothing such as long sleeved shirt and long legged pants or coveralls while mixing and spraying. To avoid breathing spray mist, wear a mask or respirator of a type recommended by NIOSH or MSHA. Keep respirator clean—wash thoroughly and change cartridge as needed.

FIRST AID TREATMENT—ANTIDOTE

SYMPTOMS OF POISONING: Excessive Fatigue, Sweating, Thirst and Fever. SEND FOR A PHYSICIAN. **FIRST AID:** Have patient lie quiet in coolest spot available. If feverish, cool with cold compresses. If swallowed, induce vomiting immediately by giving two glasses of water and sticking finger down throat. **CALL A PHYSICIAN.** Never give anything by mouth to an unconscious person. Treat as in FIRST AID above. **If Splashed in Eyes,** immediately flush eyes with plenty of water for at least 15 minutes and get medical attention. **If Splashed on Skin,** immediately remove contaminated clothing, including shoes, and wash skin with soap and plenty of water. If symptoms of poisoning develop, send for a physician and treat as in FIRST AID above. Discard or clean thoroughly contaminated clothing or shoes.

NOTE TO ATTENDING PHYSICIAN: Active ingredient is a metabolic stimulant. Treat symptomatically.

For chemical spill, leak, fire or exposure call toll free 1-800-424-9300

AGRICULTURAL CHEMICAL

Do Not Ship or Store with Food, Feeds, Drugs or Clothing

A water soluble dinitro week killer for the selective control of seedlings of broadleaf weeds such as chickweed, common ragweed, fan or French weed, gromwell, lambsquarters, mustards, nightshades, pigweeds, shepherdspurse, wild lettuce, wild radish and yellow rocket growing in the crops listed on this label.

GENERAL INFORMATION

VERTAC SELECTIVE WEED KILLER controls weeds only by contact action. It is not translocated within plants. It is more active at higher temperatures. Use the lower recommended rates when air temperature is 75°F or more. Use of oil or other surfactants will result in increased "burning" effects on plant foliage and may reduce selectivity on certain crops. Under most conditions no "activator" is required for good weed control; but if the water is very hard the addition of 2 pounds of ammonium sulfate to each 100 gallons of spray mix may be beneficial. For best results, apply the spray when weeds are small, succulent and growing actively. Apply during warm sunny weather when the crop foliage is dry. Rain or sprinkle irrigation within 4 to 6 hours after foliage applications may result in reduced weed control.

MIXING AND APPLICATION: Be sure tank is clean—free of oil, surfactants or other residues from previous spray mixtures. Maintain mild agitation while adding the required amount of VERTAC SELECTIVE WEED KILLER to the desired amount of water. After thorough mixing, no further agitation is needed. Apply as a coarse spray, using pressures of 30 to 50 psi at the nozzle. Use 20 to 60 gallons of spray mixture per acre with ground equipment or 10 to 15 gallons per acre with aerial equipment.

NOTE: Under certain conditions VERTAC SELECTIVE WEED KILLER may cause some injury to crop foliage, but this is usually temporary. To minimize foliage injury to crops, do not spray when air temperature is or expected to be 85°F or higher within 24 hours of time of application. If applications are to be made during periods of high humidity, cloudy weather or high soil moisture use the minimum dosage recommended.

USE DIRECTIONS

ALFALFA AND BIRDSFOOT TREFOIL (New Plantings): Use 3 to 4 quarts of VERTAC SELECTIVE WEED KILLER per acre in water only and apply after crop seedlings have 2 or more leaves, but while weeds are small. Do not graze or feed treated forage to livestock within 6 weeks after treating.

ALFALFA (Established Stands): Use 3 to 4 quarts of VERTAC SELECTIVE WEED KILLER plus 1 to 2 gallons of an emulsifiable crop oil mixed with water per acre. Apply in late fall to kill fall germinating weeds such as yellow rocket (winter cress), plantain, dandelions and chickweed. Apply a second treatment during the dormant season if needed for adequate weed control. Do not graze or feed treated forage to livestock within 6 weeks after application.

CLOVER (New Plantings of Sweet, Ladino and Red): Use $2^1/_2$ to 3 quarts of VERTAC SELECTIVE WEED KILLER per acre in water alone and apply after the clover seedlings have 2 or more leaves, but while weeds are small. Do not graze or feed treated forage to livestock within 6 weeks after application.

GLADIOLI: Use 3 quarts per acre, aplying when leaves are still short and straight. May be used after "glads" become tall if a low pressure fan spray is directed at the base of the plants and the tops of the leaves are not wetted.

ONIONS AND GARLIC: (Monterey, San Benito, and Santa Clara Counties of California only). Use 3 to 6 quarts per acre after the crop is in the two-leaf stage but before it is 6 inches tall. Seedling onions should have at least 2 true leaves. Do not apply on onions to be harvested for green bunching.

PEAS: Use VERTAC SELECTIVE WEED KILLER at rate of 3 quarts per acre in water alone when peas are 2 to 6 inches tall and weeds are small. Apply before pea flower buds appear. To avoid injury to the crop, allow time for peas to "harden" following cloudy weather or high humidity before treating. Do not apply within 6 weeks of harvest. Do not graze or feed treated forage to livestock within 6 weeks after spraying.

SMALL GRAINS (Barley, Oats, Rye and Wheat—planted alone or interplanted with legumes): Use VERTAC SELECTIVE WEED KILLER in water alone at rate of 3 quarts per acre and apply when grain is 3 to 6 inches tall and the weeds are small. **Note:** Some burning and yellowing of the grain crop leaves may occur for 1 to 2 weeks after application, but yield of grain is not usually reduced. Do not graze or feed treated forage to livestock within 6 weeks after application.

USE PRECAUTIONS

VERTAC SELECTIVE WEED KILLER may cause some injury (yellowing and/or burning) of crop leaves, but this effect is usually only temporary. Do not spray when air temperature is 85°F or higher nor when the temperature is expected to exceed 85°F within 24 hours after application. Use the minimum rates of application if treatment is to be made during periods of high temperature, high humidity, high soil moisture or cloudy weahter.

Do not spray or allow spray drift to contact desirable plants including ornamentals, except as specified under **"USE DIRECTIONS"**.

To avoid staining of carpets, upholstery and other such interior materials, keep contaminated clothing and shoes outdoors until cleaned.

This product is toxic to fish and wildlife. Keep out of lakes, streams and ponds. Use only as specified on this label and do not apply when weather conditions favor drift of spray from areas being treated.

Rinse equipment and containers thoroughly with water detergent solution and dispose of wastes by burying in non-crop lands away from water supplies. Containers should be disposed by punching holes in them and burying with wastes.

NOTICE

Seller warrants that the product conforms to its chemical description and is reasonably fit for the purposes stated on the label when used in accordance with directions under normal conditions of use, but neither this warranty nor any other warranty of MERCHANTABILITY or FITNESS FOR A PARTICULAR PURPOSE, express or implied, extends to the use of this product contrary to label instructions, or under abnormal conditions, or under conditions not reasonably foreseeable to seller and buyer assumes the risk of any such use.

TRANSAMINE® OA-3D

HERBICIDE

2, 4-D

ACTIVE INGREDIENT:

N, N-Dimethyl Oleyl-Linoleyl Amine Salt of 2,4-Dichlorophenoxyacetic Acid	60.7%
INERT INGREDIENT:	39.3%
TOTAL	100.0%

Equivalent to 32.9% 2,4-D Acid or 2.82 lb/gal*

*Isomer specific by AOAC Method 6.D01-5 (12th Ed.)

EPA Reg. No. 39511-59 EPA Est. No. 39511-AR-1

WHERE TO USE

TRANSAMINE® OA-3D controls broadleaf weeds in cereal crops, corn, sorghum, sugar cane; weeds and brush in rangeland, pastures, rights-of-way, and similar noncrop uses. For use in pear and apple orchards.

PLANTS CONTROLLED

TRANSAMINE® OA-3D will kill or control the following as well as many other noxious plants susceptible to 2,4-D:

Arrowhead	Lambsquarters
Artichoke	Locoweed
Bindweed (hedge, field and European)	Mexicanweed
	Morningglory
	Mustard
Bitter wintercress	Nutgrass
Boxelder	Parrotfeather
Buckhorn	Pennywort
Bull thistle	Pigweed
Bulrush	Plantain
Burdock	Poison Ivy
Bur ragweed	Pokeweed
Buttercup	Povertyweed
Canada thistle	Puncturevine
Catnip	Purslane
Chickweed	Rush
Chicory	Russian thistle
Cocklebur	Sagebrush
Coffeebean	Shepherdspurse
Creeping Jenny	Smartweed
Curly Indigo	Sowthistle
Dockweed	Stinkweed

Elderberry
Goldenrod
Ground ivy
Hemp
Hoary cress
Honeysuckle
Indigo
Ironweed
Jimsonweed
Sumac
Sunflower
Virginia creeper
Wild garlic
Wild lettuce
Wild onion
Wild radish
Willow

DIRECTIONS FOR USE

Add the recommended amount of product to about one-half the volume of water to be used in spraying. Mix well, then add remaining water and mix until spray mixture is uniform.

Best results are obtained when applications are made while the plants are well leafed and actively growing.

On Fall-Planted Wheat, Oats, and Barley—Use ⅔ to 1 pint of product in 5 to 10 gallons of water to cover one acre. Apply in early spring when weeds are small and before the crop has reached the boot stage.

On Spring-Planted Wheat, Oats and Barley—Use ⅔ pint in 5 to 10 gallons of water to cover one acre. Apply after the fully tillered stage, except during the boot to dough stage.

DO NOT forage or graze treated grain fields within 2 weeks after treatment with 2,4-D. DO NOT feed treated straw to livestock.

On Corn and Sorghum—Use ⅔ pint in 5 to 10 gallons of water to cover one acre when weeds are in active growth. Local climatic condition determines when treatment should be made. Best results are usually obtained when weeds are small and corn is 4 to 18 inches tall. Do not apply from tasseling to dough stage. Treat sorghum when 4 to 12 inches tall. Use drop nozzles when either crop is over 10 inches tall.

For preemergence use on corn apply 2½ to 5 pints in 5 to 10 gallons of water per acre. Use the higher rate on heavy soils. Do not use on sandy soils. Plant corn as deep as feasible.

FOR RICE: Ground Application: Dilute 1 to 1.5 quarts of this product in 15 to 10 gallons of water and apply this volume to one acre.

Aerial Application: Dilute 1 to 1.5 quarts of this product in 4 to 7 gallons of water and apply this volume to one acre.

Applications should be made between tillering and boot stages. Fields should be drained during application.

ON SUGAR CANE: Use 1½ to 2⅔ quarts of proudct in 5 to 25 gallons of water per acre as a pre and/or postemergence spray. Consult local agricultural experimental station of extension service weed specialist on specific use of this product alone or in combination with other compounds.

For Lawn, Golf Courses and Similar Turf: Apply 2 to 4 pints per acre in sufficient water to provide adequate uniform coverage in spring or fall when weeds are actively growing. DO NOT apply to newly seeded turf until sod is formed and grass has been cut several times. Injury may result if applied to grasses such as St. Augustine, Bent and Dichondra. Clover may also be injured. Bare areas can be reseeded in 3 to 6 weeks.

On Rangeland, Pastures, and for General Weed Control (rights-of-way, fence-rows, and other noncrop areas)—Use 1¼ to 5 pints in sufficient water to give good coverage to one acre, depending on type of weeds and stage of growth. Use only on established stands of perennial grasses.

DO NOT graze dairy animals on treated pastures within seven days after application.

Perennial and Other Hard-to-Control Weeds may require a repeat application to give adequate control.

For Use in Pear and Apple Orchards: Apply 3 to 5½ pints of product per acre in sufficient water to provide adequate coverage and wetting of weeds. Use higher dosage for perennial weeds. Avoid applications to bare ground. Direct spray at undesirable plants. DO NOT allow spray to contact foliage, stems, fruit or trunks of fruit trees. To minimize drift, spray on still days with low pressure and coarse nozzle. Perennial and other hard-to-control weeds may require a repeat application to give adequate control.

CAUTION

This product may cause skin irritation. Avoid inhaling spray mist. Harmful if swallowed. Avoid contact with the eyes, skin, or clothing. Flush eyes with clear water and get prompt medical attention.

DO NOT apply this product to susceptible broadleaf plants that are desirable such as cotton, soybeans, tomatoes, grapes, clover, ornamentals. Excessive amounts of this product in the soil may temporarily inhibit seed germination and all plant growth.

Minute quantities may cause severe injury by drift. Avoid drift of spray and use a coarse spray which is less likely to drift.

This herbicide is toxic to fish. Keep out of lakes, streams, or ponds. DO NOT apply where runoff is likely to occur. DO NOT apply when weather conditions favor drift from areas treated.

DO NOT contaminate any body of water by direct application, cleaning of equipment, or disposal of wastes and containers. Apply this product as specified on this label.

DO NOT store this product near fertilizers, seeds, insecticides, or fungicides. DO NOT contaminate feed or foodstuffs.

DO NOT use spray equipment contaminated with this product for any other purpose. DO NOT apply insecticides, fungicides, or other agricultural products unless the equipment has been cleaned very carefully with a suitable chemical cleaner.

DO NOT mix with liquid fertilizers.

DO NOT reuse empty containers. Return to drum reconditioner or destroy by perforating or crushing and burying in a safe place away from water supplies.

DO NOT use, pour, spill or store near heat or open flame.

If stored below freezing, it may be necessary to warm to 40° F. and agitate before using. This does not affect the efficiency of the product.

WARRANTY

VERTAC warrants that this herbicide conforms to the chemical description on its label. When used in accordance with label directions under normal conditions, this herbicide is reasonably fit for its intended purpose. Since timing, method of application, weather, plant, and soil conditions, mixture with other chemicals, and other factors affecting the use of this product are beyond our control, no warranty is given concerning the use of this product contrary to label directions or under conditions which are abnormal or not reasonably foresseable. The user assumes all risks of any such use.

VERTAC, and TRANSAMINE are registered trademarks of VERTAC Chemical Corp.

VERTAC MSMA 400

For Control of Johnson Grass and Other Weeds

ACTIVE INGREDIENTS:	
Monosodium acid methanearsonate	34.66%
INERT INGREDIENTS	65.34%
TOTAL	100.00%

Total Arsenic (as elemental) all in water-soluble form 16.08%
Contains 4.04 lbs. MSMA per gallon at 68°F
Use when temperature is at least 70°F.

EPA Reg. No. 39511-102 EPA Est. No. 39511-MS-1

CAUTION
Keep Out of Reach of Children

If Swallowed: Drink 1 or 2 glasses of water and induce vomiting by touching back of throat with finger. Do not give anything by mouth to an unconscious person. Get medical attention.

If On Skin: Wash contaminated areas thoroughly with soap and water. Do not reuse contamined clothing until washed. Get medical attention if irritation develops.

If In Eyes: Flush with copious amounts of clean water for 10-15 minutes. Get medical attention.

PRECAUTIONARY STATEMENTS

CAUTION

HAZARDS TO HUMANS AND DOMESTIC ANIMALS
Harmful if swallowed. Avoid breathing spray mist. Avoid contact with skin, eyes, and clothing. In case of contact immediately flush eyes or skin with plenty of water. Get medical attention if irritation occurs.

ENVIRONMENTAL HAZARDS
Do not apply directly to water. This product is toxic to fish. Do not contaminate water intended for irrigation or domestic purposes. Do not contaminate water by cleaning of equipment or disposal of wastes. Do not apply when weather conditions favor drift from target area.

For chemical spill, leak, fire, or exposure call toll free
1-800-424-9300

AGRICULTURAL CHEMICAL
Do Not Ship or Store with Food, Feeds, Drugs or Clothing

READ ENTIRE LABEL BEFORE USING THIS PRODUCT.

DIRECTIONS FOR USE

It is a violation of federal law to use this product in a manner inconsistent with this labeling.

STORAGE AND DISPOSAL

STORAGE
Do not contaminate water, food or feed by storage or disposal. Open dumping is prohibited. Do not store this product near fertilizers, seeds, insecticides or fungicides. Do not store near heat or open flame.

PESTICIDE DISPOSAL
Pesticide, spray mixture, or rinsate that cannot be used according to label instructions must be disposed of according to federal, state or local procedures under the Resource Conservation and Recovery Act.

CONTAINER DISPOSAL
Triple rinse (or equivalent) adding rinsate to spray tank. Offer rinsed container for recycling or reconditioning, or dispose of in a sanitary landfill, or by incineration if permitted by state and local authorities.

DIRECTIONS FOR USE IN COTTON

COTTON: VERTAC MSMA 400 is useful for the control of Johnsongrass, goose grass, water grass, cocklebur, nutgrass and Dallisgrass. Surfactant is already added for convenience of using.

RECOMMENDED RATE FOR USE IN COTTON: To treat an acre of cotton on the basis of 40″ row these rates are recommended.

TYPE OF APPLICATION: DIRECTED ONLY ON WEEDS

Band Size	Amount of VERTAC MSMA 400 per acre	Gallons of Water per acre
12″	$1\frac{1}{5}$ Pints	12 Gallons
14″	$1\frac{2}{5}$ Pints	14 Gallons

Apply as a post emergence DIRECTED spray to weeds when cotton is at least 3″ high. Nozzles should be adjusted to direct the spray to weeds to allow maximum coverage of weeds with no contact of cotton foliage.

Off center fan type nozzles. 100 mesh screen, approximately 5 inches from the ground with 25 pound pressure may be used.

Clean application equipment after each use by flushing with water.

DO NOT APPLY AFTER FIRST BLOOM. DO NOT FEED OR GRAZE TREATED FOLIAGE. DO NOT REUSE CONTAINER.

GENERAL WEED AND GRASS CONTROL

VERTAC MSMA 400 is found effective for controlling Johnson grass, goose grass, water grass, cocklebur, nutgrass, and green and yellow foxtail along drainage ditchbanks, right-of-ways, implement storage yards, and similar non-crop areas where cultivation is not practical.

RECOMMENDED RATES FOR GENERAL WEED CONTROL

To treat one acre, mix 4 pints of VERTAC MSMA 400 with 10–20 gallons of water, apply as a spray, thoroughly wetting foliage and stems of weeds to be killed. Spot treatment to control regrowth may be necessary.

GENERAL INFORMATION

VERTAC MSMA 400 Herbicide is useful for selective post-emergent weed control, particularly for grassy weeds. Its phytotoxic properties are quickly inactivated on contact with soil. It is a combination of herbicide and surfactant. It is unnecessary to add any other surfactant to the spray solution. Best results are obtained on young, actively growing weeds at air temperatures above 70°F

MIXING INSTRUCTIONS: VERTAC MSMA 400 Herbicide must be thoroughly mixed. Fill the spray equipment reservoir about half full with water and add the required amount of herbicide with agitation. Finish filling the reservoir with water and apply. After use, clean equipment thoroughly by flushing with water. Do not store spray solution in tank for a prolonged period.

NOTICE

Use of this material other than in accordance with the directions on this label may produce plant injury, excessive residues or other undesirable results. Vertac Chemical Corporation makes no express or implied warranties other than specified on this label.

VERTAC MSMA 600

MSMA LIQUID PLUS SURFACTANT
FOR SELECTIVE POST-EMERGENT
WEED CONTROL

ACTIVE INGREDIENTS: Monosodium acid
methanearsonate 48.40%
INERT INGREDIENTS: 51.60%

TOTAL 100.00%

Total Arsenic (as elemental) all in water-soluble form 22.4%
(Product contains 6.00 lbs. MSMA per gallon)
Use when temperature is at least 70°F.

EPA Reg. No. 39511-94 EPA Est. No. 39511-MS-1

CAUTION
Keep Out of Reach of Children

If Swallowed: Drink 1 or 2 glasses of water and induce vomiting by touching back of throat with finger. Do not give anything by mouth to an unconscious person. Get medical attention.

If On Skin: Wash contaminated areas thoroughly with soap and water. Do not reuse contaminated clothing until washed. Get medical attention if irritation develops.

If In Eyes: Flush with copious amounts of clean water for 10-15 minutes. Get medical attention.

See side panel for additional precautionary statements.

PRECAUTIONARY STATEMENTS

CAUTION

HAZARDS TO HUMANS AND DOMESTIC ANIMALS
Harmful if swallowed. Avoid breathing spray mist. Avoid contact with skin, eyes and clothing. In case of contact immediately flush eyes or skin with plenty of water. Get medical attention if irritation occurs.

ENVIRONMENTAL HAZARDS
Do not apply directly to water. This product is toxic to fish. Do not contaminate water intended for irrigation or domestic purposes. Do not contaminate water by cleaning of equipment or disposal of wastes. Do not apply when weather conditions favor drift from target area.

For chemical spill, leak, fire, or exposure call toll free 1-800-424-9300

AGRICULTURAL CHEMICAL
Do Not Ship or Store with Food, Feeds, Drugs or Clothing

DIRECTIONS FOR USE

It is a violation of federal law to use this product in a manner inconsistent with this labeling.

READ ENTIRE LABEL BEFORE USING THIS PRODUCT.

STORAGE AND DISPOSAL

STORAGE
Do not contaminate water, food or feed by storage or disposal. Open dumping is prohibited. Do not store this product near fertilizers, seeds, insecticides, or fungicides. Do not store near heat or open flame.

PESTICIDE DISPOSAL
Pesticide, spray mixture, or rinsate that cannot be used according to label instructions must be disposed of according to federal, state or local procedures under the Resource Conservation and Recovery Act.

CONTAINER DISPOSAL
Triple rinse (or equivalent) adding rinsate to spray tank. Offer rinsed container for recycling or reconditioning, or dispose of in a sanitary landfill, or by incineration if permitted by state and local authorities.

DIRECTIONS FOR USE IN COTTON

VERTAC MSMA 600 Herbicide is useful as a directed application in cotton for post-emergent control of weeds and grasses, such as johnsongrass, nutsedge, dallisgrass, cocklebur, barnyardgrass, ragweed, sandbur and puncture vine. It should be applied at the rate of $\frac{1}{3}$ gallon per acre.

Mix 3 quarts of VERTAC MSMA 600 Herbicide in 90 gallons of water. Do not add any sufactant to the spray solution. Apply spray solution at 40 gallons per acre. Weed control is most effective at temperatures above 70°F. For band application in 40 inch rows, apply 1 gallon of spray solution per inch of band width per acre. Direct the spray solution to the base of the cotton plant. Keep spray off cotton foliage. A second application may be required during humid growing conditions to control perennials, such as Johnsongrass and nutsedge. Apply the second application 1 to 3 weeks later.

Slight burning and a reddish discoloration of the cotton leaf may occasionally occur following the recommended treatment, but the cotton plant will develop normally.

Apply only when cotton is 3 inches high to first bloom. Do not apply after first bloom. Do not feed foliage to livestock or graze treated areas.

DIRECTIONS FOR NON-CROP USE

VERTAC MSMA 600 Herbicide is useful for control of Johnsongrass, nutsedge, Dallisgrass, cocklebur, barnyardgrass, ragweed, sandbur, puncture vine, and certain other weeds on drainage ditchbanks, rights-of-way, fence rows, storage yards, and similar noncrop areas.

Mix VERTAC MSMA 600 Herbicide at a rate of 6 to 12 pints in 100 gallons of water. Spray unwanted vegetation at a rate of about 50 gallons of spray solution per acre. Use spray equipment that gives good row volume coverage. If regrowth occurs, reapply as required.

Do not feed treated foliage to livestock or graze treated areas.

DIRECTIONS FOR USE ON LAWN AND ORNAMENTAL TURF

VERTAC MSMA 600 Herbicide is useful for control of Dallisgrass, sandbur, bahiagrass, nutsedge, barnyardgrass, chickweed, and wood sorrel with little or no injury to good lawn grasses. On new lawns, do not treat until after three mowings. Good grasses may be temporarily discolored. Zoysia, bluegrasses and bermuda are quite tolerant. Do not use on St. Augustine or centipede.

Mow lawns 1 to $1\frac{1}{2}$ inches high before treatment. Mix $1\frac{1}{3}$ fluid ounces (4 teaspoons) of VERTAC MSMA 600 Herbicide in $2\frac{1}{2}$ gallons of water and apply to 1,000 sq. ft. area. Spray thoroughly to wet all undesirable plants. Repeat application 10 to 14 days apart, may be needed for good control.

GENERAL INFORMATION

VERTAC MSMA 600 Herbicide is useful for selective postemergent weed control, particularly for grassy weeds. Its phytotoxic properties are quickly inactivated on contact with soil. It is a combination of herbicide and surfactant. It is unnecessary to add any other surfactant to the spray solution. Best results are obtained on young actively growing weeds at air temperatures above 70°F.

MIXING INSTRUCTIONS: VERTAC MSMA 600 Herbicide must be thoroughly mixed. Fill the spray equipment reservoir about half full with water and add the required amount of herbicide with agitation. Finish filling the reservoir with water and apply. After use, clean equipment thoroughly by flushing with water. Do not store spray solution in tank for a prolonged period.

NOTICE

Use of this material other than in accordance with the directions on this label may produce plant injury, excessive residues or other undesirable results. Vertac Chemical Corporation makes no express or implied warranties other than specified on this label.

VERTAC MSMA 660

For Control of Johnson Grass and Other Weeds

ACTIVE INGREDIENTS:

Monosodium acid methanearsonate	51.3%
INERT INGREDIENTS	48.7%
TOTAL	100.0%

Total Arsenic as elemental, all in water-soluble form 23.70%
(Contains 6.67 lbs. MSMA per gallon at 68°F.)
Use when temperature is at least 70°F.

EPA Reg. No. 39511-101 EPA Est. No. 39511-MS-1

CAUTION

Keep Out of Reach of Children

If Swallowed: Drink 1 or 2 glasses of water and induce vomiting by touching back of throat with finger. Do not give anything by mouth to an unconscious person. Get medical attention.

If On Skin: Wash contaminated areas thoroughly with soap and water. Do not reuse contaminated clothing until washed. Get medical attention if irritation develops.

If In Eyes: Flush with copious amounts of clean water for 10–15 minutes. Get medical attention.

PRECAUTIONARY STATEMENTS

CAUTION

HAZARDS TO HUMANS AND DOMESTIC ANIMALS
Harmful if swallowed. Avoid breathing spray mist. Avoid contact with skin, eyes and clothing. In case of contact immediately flush eyes or skin with plenty of water. Get medical attention if irritation occurs.

ENVIRONMENTAL HAZARDS
Do not apply directly to water. This product is toxic to fish. Do not contaminate water intended for irrigation or domestic purposes. Do not contaminate water by cleaning of equipment or disposal of wastes. Do not apply when weather conditions favor drift from target area.

For chemical spill, leak, fire, or exposure call toll free
1-800-424-9300

AGRICULTURAL CHEMICAL
Do Not Ship or Store with Food, Feeds, Drugs or Clothing
Read entire label before using this product.

DIRECTIONS FOR USE

It is a violation of federal law to use this product in a manner inconsistent with this labeling.

STORAGE AND DISPOSAL

STORAGE
Do not contaminate water, food or feed by storage or disposal. Open dumping is prohibited. Do not store this product near fertilizers, seeds, insecticides or fungicides. Do not store near heat or open flame.

PESTICIDE DISPOSAL
Pesticide, spray mixture, or rinsate that cannot be used according to label instructions must be disposed of according to federal, state or local procedures under the Resource Conservation and Recovery Act.

CONTAINER DISPOSAL
Triple rinse (or equivalent) adding rinsate to spray tank. Offer rinsed container for recycling or reconditioning, or dispose of in a sanitary landfill, or by incineration if permitted by state and local authorities.

DIRECTIONS FOR USE IN COTTON

COTTON: VERTAC MSMA 660 is useful for the control of Johnsongrass, goose grass, water grass, cocklebur, nutgrass and Dallisgrass. Always use a suitable surfactant with VERTAC MSMA 660 for best results.

RECOMMENDED RATE FOR USE IN COTTON: The following rates are recommended on the basis of 40″ rows.

TYPE OF APPLICATION: DIRECTED ONLY ON WEEDS

Band Size	Amount of VERTAC MSMA 660 per acre	Amount of Surfactant per acre	Gallons of water per acre
12″	11.5 Fl. Oz.	½ Pint	12 Gallons
14″	13.5 Fl. Oz.	½ Pint	14 Gallons

Apply as a post emergence DIRECTED spray to weeds when cotton is at least 3″ high. Nozzles should be adjusted to direct the spray to weeds to allow maximum coverage of weeds with no contact of cotton foliage.

Off center fan type nozzle, 100 mesh screen, approximately 5 inches from the ground with 25 pound pressure may be used.

Clean application equipment after each use by flushing with water.

DO NOT APPLY AFTER FIRST BLOOM. DO NOT FEED OR GRAZE TREATED FOLIAGE. DO NOT REUSE CONTAINER.

GENERAL WEED AND GRASS CONTROL

VERTAC MSMA 660 plus a suitable surfactant has been found effective for controlling Johnson grass, water grass, cocklebur, nutgrass and green and yellow foxtail along drainage ditchbanks, right-of-ways, implement storage yards and similar non-crop areas where cultivation is not practical.

RECOMMENDED RATES FOR GENERAL WEED CONTROL
Mix 3 to 6 pints of VERTAC MSMA 660 plus 1 to 2 quarts of a suitable surfactant in 100 gallons of water, spray the unwanted vegetation thoroughly to just short of run-off, spot treatments to control regrowth may be necessary.

GENERAL INFORMATION

VERTAC MSMA 660 Herbicide is useful for selective postemergent weed control, particularly for grassy weeds. Its activity is enhanced by addition of a suitable surfactant to the spray solution. It is most effective on young, actively growing weeds at air temperatures above 70°F. Its phototoxic properties are quickly inactivated on contact with soil.

MIXING INSTRUCTIONS: VERTAC MSMA 660 Herbicide must be thoroughly mixed. Fill the spray equipment reservoir about half full with water and add the required amount of herbicide and surfactant with agitation. Finish filling the reservoir with water and apply. After use, clean equipment thoroughly by flushing with water. Do not store spray solution in tank for a prolonged period.

NOTICE

Use of this material other than in accordance with the directions on this label may produce plant injury, excessive residues or other undesirable results. Vertac Chemical Corporation makes no express or implied warranties other than specified on this label.

VERTAC DSMA-LQ

DSMA LIQUID HERBICIDE PLUS SURFACTANT

ACTIVE INGREDIENTS:

Disodium Methanearsontate	21.8%
INERT INGREDIENTS:	78.2%
TOTAL	100.0%

Total Arsenic (as elemental) all in water soluble form 8.9%
(Product contains 3.6 pounds Disodium Methanearsonate Hexahydrate per gallon).

EPA Reg. No. 39511-95 EPA Est. No. 39511-MS-1

CAUTION

Keep Out of Reach of Children

If Swallowed: Drink 1 or 2 glasses of water and induce vomiting by touching back of throat with finger. Do not give anything by mouth to an unconscious person. Get medical attention.

If On Skin: Wash contaminated areas thoroughly with soap and water. Do not reuse contaminated clothing until washed. Get medical attention if irritation develops.

If In Eyes: Flush with copious amounts of clean water for 10–15 minutes. Get medical attention.

PRECAUTIONARY STATEMENTS

CAUTION

HAZARDS TO HUMANS AND DOMESTIC ANIMALS
Harmful if swallowed. Avoid breathing spray mist. Avoid contact with skin, eyes and clothing. In case of contact immediately flush

eyes or skin with plenty of water. Get medical attention if irritation occurs.

ENVIRONMENTAL HAZARDS

Do not apply directly to water. This product is toxic to fish. Do not contaminate water intended for irrigation or domestic purposes. Do not contaminate water by cleaning of equipment or disposal of wastes. Do not apply when weather conditions favor drift from target area.

For chemical spill, leak, fire, or exposure call toll free

1-800-424-9300

AGRICULTURAL CHEMICAL

Do Not Ship or Store with Food, Feeds, Drugs or Clothing

READ ENTIRE LABEL BEFORE USING THIS PRODUCT.

DIRECTIONS FOR USE

It is a violation of federal law to use this product in a manner inconsistent with this labeling.

STORAGE AND DISPOSAL

STORAGE

Do not contaminate water, food or feed by storage or disposal. Open dumping is prohibited. Do not store this product near fertilizers, seeds, insecticides or fungicides. Do not store near heat or open flame.

PESTICIDE DISPOSAL

Pesticide, spray mixture, or rinsate that cannot be used according to label instructions must be disposed of according to federal, state or local procedures under the Resource Conservation and Recovery Act.

CONTAINER DISPOSAL

Triple rinse (or equivalent) adding rinsate to spray tank. Offer rinsed container for recycling or reconditioning, or dispose of in a sanitary landfill, or by incineration if permitted by state and local authorities.

COTTON

VERTAC DSMA-LQ is useful as a directed application in cotton for post-emergent control of weeds and grasses such as Johnsongrass, nutsedge, Dallisgrass, cocklebur, barnyardgrass, ragweed, sandbur and puncture vine.

Add 1 gallon VERTAC DSMA-LQ to 40 gallons of water and mix thoroughly. Do not add any surfactant to the spray solution. Apply spray solution at 40 gallons per acre. Weed control is most effective at temperatures above 70°F. For band application in 40 inch rows, apply 1 gallon of spray solution per inch of band width per acre. Direct the spray solution to the base of the cotton plant. *Keep spray off cotton foliage.* A second application may be required during humid growing conditions to control perennials such as Johnsongrass and nutsedge. Apply the second application 1 to 3 weeks later.

Slight burning and a reddish discoloration of the cotton leaf may occasionally occur following the recommended treatment, but the cotton plant will develop normally.

Apply only when cotton is 3 inches high to first bloom. Do not apply after first bloom. Do not feed treated foliage to livestock or graze treated areas.

NON-CROP USE

VERTAC DSMA-LQ is useful for controlling Johnsongrass, nutsedge, watergrass and certain other weeds in drainage ditchbanks, rights-of-ways, storage yards, and similar non-crop areas.

Mix VERTAC DSMA-LQ at a rate of 1 to 2 gallons in 100 gallons of water. No additional surfactant is necessary. Spray the unwanted vegetation thoroughly to just short of run-off. Any spray equipment that gives good coverage may be used. If regrowth occurs, reapply as required.

LAWN AND ORNAMENTAL TURF

VERTAC DSMA-LQ is useful for control of Dallisgrass, sandbur, bahiagrass, nutsedge, barnyardgrass, chickweed, and wood sorrel with little or no injury to good lawn grasses. On new lawn, do not treat until after three mowings. Good grasses may be temporarily discolored. Zoysia, bluegrass, and bermuda are quite tolerant. Do not use on St. Augustine or centipede.

Mow lawns 1 to 1½ inches high before treatment. Mix 4 fluid ounces of VERTAC DSMA-LQ in 5 gallons of water and apply to an area of 1,000 square feet. Spray thoroughtly to wet all undesirable plants. Repeat applications, 10 to 14 days apart, may be needed for good control.

GENERAL INFORMATION

VERTAC DSMA-LQ is useful for selective post-emergent weed control, particularly for grassy weeds. It is a combination of herbicide and surfactant. IT IS UNNECESSARY TO ADD ANY OTHER SURFACTANT TO THE SPRAY SOLUTION. Best results are obtained on young, actively growing weeds at air temperatures above 70°F.

MIXING INSTRUCTIONS: VERTAC DSMA-LQ must be thoroughly mixed. Fill the spray equipment reservoir about half full with water and add the required amount of herbicide with agitation. Finish filling the reservoir with water and apply. After use, clean equipment thoroughly by flushing with water. Do not store spray solution in tank for a prolonged period.

NOTICE

Use of this material other than in accordance with the directions on this label may produce plant injury, excesssive residues or other undesirable results. Vertac Chemical Corporation makes no express or implied warranties other than specified on this label.

WEED-RHAP®

LOW VOLATILE GRANULAR D HERBICIDE

2,4-D LOW VOLATILE

ACTIVE INGREDIENT	
2-Ethylherxyl Ester† of 2,4-Dichlorophenoxyacetic Acid	28.4%
INERT INGREDIENT	71.6%
TOTAL	100.0%

† an isooctyl esther

Equivalent to 18.8% 2,4-D AE by weight.*

* Isomer specific by AOAC Method 6.D01-5 (12th ed.)

EPA Reg. No. 39511-77 EPA Est. No. 39511-AR-1

WEED-RHAP is a registered trademark of the VERTAC Chemical Corp.

CAUTION

Keep Out of Reach of Children

READ SIDE PANEL FOR ADDITIONAL HANDLING PRECAUTIONS.

GENERAL INFORMATION

WEED-RHAP Low Volatile Granular D is formulated by impregnating selected mineral granules with low volatile 2,4-D ester, to minimize drift. The 8/16 and 15/30 mesh sizes are intended for use in the control of certain water weeds which infest lakes and ponds. The 24/48 mesh site is recommended for the control of certain weeds which infest turfgrass and crops such as corn, strawberries, and cranberries.

DIRECTIONS FOR USE

IN LAKES AND PONDS: To control milfoil, hornwort, bladderwort, fanwort, pondweed, cattail, and waterlillies infesting lakes and ponds, apply granules at the rate of 1 to 1½ lbs. per 430 sq. ft. (10′ × 43′) of water surface or 100 to 150 lbs. per acre. Distribute the granules evenly over the quiet water surface infested with weeds, using a mechanical spreader attached to a boat. The granules settle to the water bottom around the roots and stems. Effects on water weeds will be evidenced in 4 to 10 weeks and control may be expected for a year or longer. In tropical regions use 25 to 50 percent more granules. Consult your Public Water Agency before use.

DO NOT APPLY to more than ⅓ to ½ of a lake or pond in any one month, because excessive decaying vegetation may deplete oxygen content of water, thus killing fish.

ORNAMENTAL TURF such as lawns, golf courses, cemeteries, and parkways: To control wild onion and broadleaved weeds such as dandelion, plantain, ivy, bitterweed, and dogfennel, apply 10 to 15 lbs. of granules per acre in early spring when shoots are young and actively growing, using a mechanical spreader. Repeat application in one month to control skipped spots. Weeds over 6 inches tall are sometimes resistant and should be sprayed with WEED-RHAP A-4D or WEED-RHAP Low Volatile 4D.

DO NOT use on Bentgrass, St. Augustinegrass, Dichondra, Carpetgrass, or newly plugged or seeded turf.

IN CORN: For preemergence application to aid control of annual grasses such as large and small foxtail, crabgrass, barnyardgrass; broadleaved weeds such as smartweed, sunflower, wild mustard, cocklebur, bindweed, dandelion, and morningglory, apply as follows: With a granule applicator designed to apply a band over the planted row, apply 3 to 3¼ lbs. per acre in a band 14 inches wide when rows are 42 inches apart. If the planting varies from this standard band width, or row separation, the granules should be applied proportionately. Be sure the granules are allowed to stay on top of the soil; preferably, they should be applied behind the packer wheel at planting time. Be sure the corn is planted 2½ to 3 inches deep, as damage may occur to shallow-planted corn. DO NOT apply on sandy soil. (For Corn Belt Only.)

IN STRAWBERRIES: For preemergence application to control weeds such as crabgrass, foxtail, barnyardgrass or wintergrass, horseweed, bullnettle, carelessweed, morningglory, dock, smartweed, jimsonweed, and dandelion: Apply at the rate of 12 lbs. per acre over the planted area. Application should be made broadcast using a spreader-seeder previously adjusted to deliver 12 lbs. per acre. Application should be made only in the dormant season (after first fall frost or before new leafing in spring).

IN CRANBERRIES: For preemergence application to control annual broadleaved weeds in bearing cranberry marshes, apply at a rate of 10 to 20 lbs. per acre (2 to 4 lbs. 2,4-D equivalent per acre). Application should be made broadcast in the spring after removal of the winter flood from the dormant vines. Weeds controlled are common ragweed, sticktight, tearthumb, smartweed, Northern Saint Johnsonwort, and common species of annual smartweed. Careful distribution of the granules is necessary if adequate control is to be obtained. Granules may also be distributed on ice over plants prior to spring thaw. (For Wisconsin area.)

CAUTION

Avoid contact with skin, eyes, and clothing. In cases of skin contact, wash thoroughly with soap and warm water. Flush eyes with clear water and obtain prompt medical attention.

WARNING

Avoid application to susceptible plants because this product may injure such plants as cotton, soybeans, tomatoes, beans, peas, fruit trees, and ornamentals. DO NOT allow particles or dust to drift onto desirable crops or plants.

Although this product contains a low volatile ester, it should be used with caution where spreading vapors can damage nearby susceptible plants.

Avoid contamination of water intended for irrigation or domestic use. DO NOT store near fertilizers, seeds, insecticides, or fungicides.

It is best to use separate equipment for applying 2,4-D granules. Should the same equipment be used for applying other materials, however, clean it thoroughly before such use to remove all traces of 2,4-D particles or dust.

Keep bag closed when it is not in use. Dispose of empty bag by burying or burning. DO NOT burn where smoke or fumes will drift to desirable crops or plants.

WARRANTY

VERTAC warrants that this herbicide conforms to the chemical description on its label. When used in accordance with label directions under normal conditions, this herbicide is reasonably fit for its intended purpose. Since timing, method of application, weather, plant and soil conditions, mixture with other chemicals and other factors affecting the use of this product are beyond our control, no warranty is given concerning the use of this product contrary to label directions or under conditions which are abnormal or not reasonably foreseeable. The user assumes all risks of any such use.

Zoecon Corporation

A Sandoz Company
Crop Protection Division
Box 10975
Palo Alto, CA 94303
(800) 227-8929
(415) 857-1130

SOLICAM® DF

HERBICIDE

For Control of Grass and Broadleaf Weeds in Tree Fruit and Nut Crops and Citrus

ACTIVE INGREDIENT:
norflurazon [4-chloro-5-(methylamino)-2-(alpha, alpha, alpha-trifluoro-*m*-tolyl)-3(2*H*)-pyridazinone] 80.0%
INERT INGREDIENTS 20.0%
TOTAL 100.0%

EPA Reg. No. 11273-48 EPA Est. No. 39578-TX-1
U.S. Pat. No. 3,644,355 and 3,834,889
SOLICAM is a trademark of Sandoz, Ltd.

CAUTION
Keep Out of Reach of Children

PRECAUTIONARY STATEMENTS

ENVIRONMENTAL HAZARDS
Do not apply directly to any body of water. Do not apply when weather conditions favor runoff or drift from treated areas. Do not contaminate water by cleaning of equipment or disposal of wastes.

WARRANTY AND CONDITIONS OF SALE

Zoecon Corporation warrants that the chemical composition of this product conforms to the chemical description on the label and is reasonably fit for the purpose stated on the label when used in accordance with directions under normal conditions of use. Zoecon makes no other warranty, express or implied, concerning the use of this product other than as indicated on the label. Buyer assumes all risks of use, storage or handling of this material not in strict accordance with directions given on the label.

WEEDS CONTROLLED

SOLICAM DF at recommended rates controls the following weeds:

Grass Weeds:
Annual bluegrass, Annual ryegrass, Annual sedge, Bahiagrass (seedling), Barnyardgrass, Broadleaf signalgrass, Bromus spp., Cheat, Crabgrass, Crowfoot grass,* Downy brome, Fall panicum, Feather fingergrass, Foxtails, Goosegrass, Guineagrass (seedling),* Italian ryegrass, Johnsongrass (seedling), Natalgrass,* Pangolagrass,* Sandbur, Sandspur,* Southwestern cupgrass, Six weeks grama grass, Tall fescue, Texas panicum, Vaseygrass,* Wild barley, Wild onion, Witchgrass.

* When applied at the higher rates recommended for weed control in Florida citrus.

Broadleaf Weeds:
Annual bursage, Black mustard, Camphorweed,* Carolina geranium, Chickweed, Common ragweed,* Cudweed, Dayflower,* Dog fennel, False dandelion (smooth cat's ear), Fiddleneck, Filaree (redstem and whitestem),** Flixweed, Goldenrod,* Little mallow (cheeseweed),** London rocket, Pineapple weed, Prostrate spurge, Puncturevine, Redmaids, Russian thistle, Shepherdspurse, Stinging nettle, Tumble mustard, Virginia pepperweed, Wild buckwheat, Wild geranium.

** Treat prior to germination and incorporate with water on coarse and medium soils for adequate control.

SOLICAM DF at recommended rates suppresses the following weeds: Bermudagrass, Common purslane, Nutsedge, Orchardgrass, Quackgrass, Rhizome johnsongrass, Torpedograss,* Western muhly (wirestem muhly), Common lambsquarters, Florida pusley,* Groundsel, Henbit, Horseweed (marestail), Pigweeds (redroot, tumble and green amaranth), Plaintains (bracted and buckhorn), Poorjoe, Silverleaf nightshade (white horsenettle), Sowthistle, Flax-leaved fleabane.

STORAGE AND DISPOSAL

STORAGE
Do not contaminate water, food or feed by storage or disposal. Wastes resulting from the use of this product may be disposed of on site or at an approved waste disposal facility.

CONTAINER DISPOSAL
Completely empty bag into application equipment. Then dispose of empty container in a sanitary landfill or by incineration, or, if allowed by state and local authorities, by burning. If burned, stay out of smoke.

DIRECTIONS FOR USE

It is a violation of Federal Law to use this product in a manner inconsistent with its labeling.

GENERAL INFORMATION

SOLICAM® DF is a herbicide for control of certain grasses and broadleaf weeds in tree fruit and nut crops, and citrus.

SOLICAM DF must be moved into the weed seed germination zone to be effective. If no rainfull occurs within four weeks after application, the product must be activated by flood irrigation or sprinkler irrigation. SOLICAM DF alone has no postemergence activity and will not control established weeds.

Do not graze livestock on treated acres. Do not feed or graze cover crops grown in treated orchards.

In the following locations, do not use SOLICAM DF in deciduous crops, with the exception of apples: Arizona, and the Coachella, Imperial, Palo Verde and Tehachapi Valleys of California. Do not use in stone fruits on the western slope of Colorado.

APPLICATION DIRECTIONS

Apply SOLICAM DF before weeds emerge. The soil surface must be free of weeds and relatively free of plant residues. The soil should be settled and firm at the time of application.

Apply as a directed spray. Avoid contact with fruit or foliage. Make only one application per year unless otherwise indicated. Do not apply when nuts or fruit are on the ground at harvest.

EQUIPMENT
Apply accurately with a fixed boom sprayer. Use screen size 50 mesh or larger.

MIXING INSTRUCTIONS
Clean and calibrate the sprayer. Add SOLICAM DF to the spray tank already filled with the required volume of water. This will eliminate or minimize foaming. Maintain agitation while filling and spraying. If a by-pass line is used, discharge at the bottom of the tank to further minimize foaming.

Do not allow SOLICAM DF spray mixture to remain in the spray tank overnight, because settling could occur and resuspension would be difficult.

TANK-MIXES: Always predetermine the compatibility of labeled tank-mixes by mixing small proportional quantities in advance. Products should be added to the spray tank in the following order:

1. Wettable powders and water dispersible granules. Wettable powders should be premixed in a small amount of water. Water dispersible granulars should be added during filling. Allow both to disperse before other products are added. The order of addition is not important.
2. Flowable liquids.
3. Emulsifiable concentrates.
4. Surfactants cleared for application to growing crops.

Begin adding wettable powders, flowable liquids, emulsifiable concentrates and surfactants after the spray tank is ¾ full. Continue agitation during the addition of all the materials and while filling and spraying.

BROADCAST RATES

Unless otherwise directed, use 2.5–5 lbs. of SOLICAM DF per acre as a directed spray in 20–100 gallons of water.

Use the lower rates on coarse soils. Use the higher rates on medium to fine soils or soils having organic matter content of greater than 3%.

Soil Texture	Soil Type	SOLICAM DF Rate of Application
Coarse	sand, loamy sand, sandy loam	2.5 lbs.
Medium	loam, silt loam, silt, sandy clay loam	3.75 lbs.
Fine	sandy clay, clay loam, silty clay loam, silty clay, clay	5.0 lbs.

STRIP OR BAND TREATMENT

When applying a strip treatment of SOLICAM DF, the following formula may be used to calculate the amount per acre:

$$\frac{\text{Width of strip in feet}}{\text{Distance between rows in feet}} \times \text{Pounds per acre for broadcast treatment} = \text{Pounds per acre for strip treatment}$$

CROP RECOMMENDATIONS
DECIDUOUS TREE FRUIT AND NUT CROPS

Apply SOLICAM DF as directed to the soil surface from fall to early spring before weeds emerge unless otherwise indicated.

Do not use on nursery stock except where uses are indicated.

ALMONDS, CHERRIES

Apply SOLICAM DF as directed on sandy loam or finer soil types from fall to early spring, before weeds emerge. For almonds, it is not recommended SOLICAM DF be applied after bud break on sandy loam soils. In California and the Pacific Northwest, trees must be established at least 18 months in the field. Do not use on coarse-textured soil such as sand, loamy sand, and gravelly sand.

ALMONDS—PREHARVEST APPLICATION

Apply SOLICAM DF broadcast or band as directed on sandy loam or finer soil types prior to harvest for clean orchard floors. Activate SOLICAM DF with a minimum of one-two inches of irrigation water. Trees must be established at least 18 months in the field. Do not use on coarse-textured soil such as sand, loamy sand, and gravelly sand.

GRAPES

Apply SOLICAM DF as directed on sandy loam or finer soil types from fall to early spring. It is not recommended that SOLICAM DF be applied after bud break on sandy loam soils. Vines must be established at least two years in the field. Do not use on nursery stock. In California, do not use in any counties south of Monterey, Kings and Tulare counties. Do not use on coarse-textured soil such as sand, loamy sand, and gravelly sand in all areas.

APPLES AND PEACHES

Apply SOLICAM DF as directed. The product may be used on newly planted apples. Peaches must be established at least 6 months in the field. However, in California and the Pacific Northwest, peach trees must be established at least 18 months in the field. The soil around the trees should be firm and free of depressions in which rain or irrigation water could accumulate. Higher rates of 3.75–5 lbs. per acre may be used in Coastal Plains soils of the Southeast.

APRICOTS, FILBERTS, NECTARINES, PEARS, PLUMS, PRUNES, AND WALNUTS

Apply SOLICAM DF as directed. Trees must be established at least 18 months in the field in California and the Pacific Northwest. Elsewhere, trees must be established at least 12 months in the field.

PECANS

Apply SOLICAM® DF herbicide as directed to bearing and non-bearing pecans. The product may also be used on newly set pecan trees. Higher rates of 2.5–5 lbs. per acre may be used in Coastal Plains soils of the Southeast.

BLUEBERRIES

Apply SOLICAM DF from fall to early spring before weeds emerge. The soil surface must be free of weeds and relatively free of plant residues at the time of application. Make only one application per year. SOLICAM DF may be used on newly set or established blueberry plantings.

RASPBERRIES, BLACKBERRIES

Apply SOLICAM DF as a directed spray according to label directions from fall to early spring before weeds emerge Apply when the crop is dormant. The soil surface must be free of weeds and relatively free of plant residues at the time of application. Make only one application per year. Application of SOLICAM DF may result in temporary bleaching or chlorosis of the leaves from which the plant will recover.

CITRUS CROPS

Apply SOLICAM DF as directed any time during the year according to local practice. SOLICAM DF may be used on bearing and non-bearing citrus. It may also be used on field grown nursery stock. Do not use on germinating citrus seed beds. Do not use where citrus is interplanted with palm trees.

CALIFORNIA AND ARIZONA CITRUS CROPS

Apply 2.5–5 lbs. of SOLICAM DF per acre as directed. Higher rates of 3.75–5 lbs. per acre can also be used in coarse-textured soils and incorporated to a depth of not more than 3 to 4 inches for increased activity against less susceptible weeds.

FLORIDA AND TEXAS CITRUS CROPS

For oranges, grapefruit, lemons, limes, and tangerines, apply 5–10 lbs. of SOLICAM DF per acre unless otherwise directed. Use the same rate for all coarse-textured soils. In newly planted or established trees, SOLICAM DF may be applied up to two times per year. Apply no more than 10 lbs. total SOLICAM DF per acre per year.

RING DRENCH APPLICATION

Apply SOLICAM DF to newly planted citrus as a ring drench treatment at the rate of 10 lbs. product broadcast per acre. Make only one application per year. Consult the following table for the amount of SOLICAM DF to add to a 500-gallon water tank for various diameter rings.

	Ounces of SOLICAM DF Per Tank		
	3 ft. ring	4 ft. ring	5 ft. ring
3 gals./tree (167 trees/tank)	4.3	7.6	12.0
5 gals./tree (100 trees/tank)	2.6	4.6	7.2
7 gals./tree (71 trees/tank)	1.8	3.3	5.2
10 gals./tree (50 trees/tank)	1.3	2.3	3.6

INJECTION INTO CROP IRRIGATION SYSTEMS

Low volume—4 to 50 gallons per hour (gph) per emitter, drip .05–3 gph per emitter. Point of application should be above ground.

Irrigation system should run a sufficient amount of time prior to SOLICAM DF injection to have all emitters functioning properly. After system is operating properly, length of injection should be such that at one period of time during the injection, the first and last emitters in the system contain water treated with SOLICAM DF. Maintain proper agitation in injection tank containing SOLICAM DF. SOLICAM DF should be mixed in clean water and injected down-line from filters. Following SOLICAM DF injection, system should be flushed for a period of time sufficient to clear the line of

SOLICAM DF. (If SOLICAM DF application is made during a normal irrigation cycle, injection should be made during the late stage.)

CALIBRATION

Calculation of use rate is based on *wetted area* around emitters—*NOT* on tree acres. To determine correct amount of SOLICAM DF, use the following formula:

1. Treated area per each emitter = A

   A = 3.14 × (radius × radius)

   Example: If the average distance from emitter to perimeter of wetted area measured one inch below soil surface = 13 inches, then A = 3.14 × (13" × 13") = A = 3.14 × (169") = A = 530.7 square inches.

2. The area in square feet wet in each acre = B

   $B = \frac{A \times \text{emitters/acre}}{144}$

   Example: If there are 300 emitters per acre, then

   $B = \frac{530.7 \times 300}{144} = B = 1105.6$ square feet wetted per acre.

3. The total area (in square feet) wet by your system = C

   C = B × acres covered by system

   Example: If the system covers 20 acres, then

   C = 1105.6 square feet per acre × 20 acres

   = 22,112 square feet wetted by system.

4. Amount of SOLICAM DF to inject = S

   $S = \frac{C}{43{,}560} \times 1.5$ = pounds of SOLICAM DF injected

   (Note: this will change the a.i. from 1.0 to 1.2 lbs. a.i. per acre.)

   Example:

   $S = \frac{22{,}112}{43{,}560} \times 1.5 =$ 0.76 pounds of SOLICAM DF should be injected into the system.

Application of SOLICAM DF through irrigation systems should be used as a supplemental weed control practice. The addition of SOLICAM DF through irrigation systems should help prevent break-through weeds at irrigation point providing the application is made before weed seeds germinate.

PRECAUTIONS: 1) Apply only through irrigation systems containing antisyphon and check valves to prevent contamination of well during shutdown and overflow of solution tank. 2) Inject ahead of any right angle turn in the main line to insure adequate mixing. 3) Chemical injection pumps and water pumps must have interlocking controls to insure simultaneous shutoff. 4) Application when drift may occur, such as from windy conditions, or when system joints and connections are leaking, or when nozzles are not providing uniform distribution, may cause crop injury, 5) Application should be directed in such a way that SOLICAM DF does not come into contact with foliage.

SOLICAM DF TANK-MIX COMBINATIONS

The following tank-mix combinations are recommended for improved control of weeds in tree fruit and nut crops, including citrus. No label dosage rates may be exceeded, and the more restrictive label limitations and precautions should be observed. Before applying these combinations, read thoroughly all precautionary statements and other information in both labels. Refer to the Mixing Instructions section of this label for specific tank-mixing directions. SOLICAM DF may not be tank-mixed with any product containing a prohibition against such mixing.

SOLICAM DF PLUS PARAQUAT CL

This tank-mix combination is recommended for kill of existing vegetation and for residual preemergence control of susceptible broadleaf and grassy weeds in all tree crops listed on this label.

Apply 2.5–5 lbs. of SOLICAM DF per acre and 1–2 qts. of Paraquat CL per acre in 50–200 gals. of water per acre. Apply to the orchard floor avoiding contact with fruit, foliage, or stems. Add an agriculturally approved nonionic surfactant such as Ortho X-77® at the rate of ½ pint per 100 gallons of spray. Apply when the weeds and grasses are succulent and new growth is 1–6 inches tall. Apply this combination only once per year.

Florida citrus only—Apply 5–10 lbs. of SOLICAM DF per acre and 1–2 qts. of Paraquat CL per acre in 50–200 gals. of water per acre. This combination may be applied up to two times per year. Apply no more than 10 lbs. total SOLICAM DF per acre per year.

Do not spray under windy conditions, and use shield for young trees. Refer to the SOLICAM DF and Paraquat CL labels for further directions, cautions, warnings, precautions, and notes.

SOLICAM DF PLUS ROUNDUP®

This tank-mix combination is recommended for control of existing annual and perennial weeds and for residual preemergence control of susceptible broadleaf and grassy weeds in all tree crops listed on this label.

Apply 2.5–5 lbs. of SOLICAM DF per acre as recommended on its label and 1–5 qts. of Roundup per acre as recommended on its label. Apply to the orchard floor in 20–40 gals. of water per acre, avoiding contact with fruit, foliage or other tree parts. Apply when the weeds and grasses are actively growing. Apply this combination only once per year.

Florida citrus only— Apply 5–10 lbs. of SOLICAM DF per acre and 1–5 qts. of Roundup per acre in 20–40 gals. of water per acre. This combination may be applied up to two times per year. Apply no more than 10 lbs. total SOLICAM DF per acre per year.

Do not apply under windy conditions and use a shield for young trees. Use extreme care when applying this combination. Refer to the SOLICAM DF and Roundup labels for further directions, cautions, warnings, precautions, and notes.

SOLICAM DF PLUS PRINCEP® CALIBER® 90, PRINCEP® 80W OR OTHER SIMAZINE PRODUCTS

This tank-mix combination is recommended for improved control of broadleaf weeds, and more specifically the following: Clover, Cutleaf Eveningprimrose, Dandelion, Henbit, Horseweed (marestail), Lambsquarters, Pigweeds, Puncturevine.

This combination may be used in the following crops: Almonds, Apples, Cherries, Filberts, Grapefruit, Lemons, Oranges, Peaches, Pears, Pecans, Plums, Walnuts.

Apply the appropriate rates in the following table as a tank-mix in 20–100 gals. of water per acre. Apply as a directed spray to the soil surface before weeds emerge, avoiding contact with fruit, foliage, or stems.

	Pounds Product Per Acre		
Soil Type	SOLICAM DF	Princep Caliber 90	Simazine 80W
Medium	2.5	1.125	1.25
Fine	3.75	2.25	2.5

Do not use this combination in coarse soils. Use only in those crops and locations specified in both the SOLICAM DF and Simazine 80W label.

Florida only (grapefruit and oranges)—Apply 4–5 lbs. of SOLICAM DF per acre and 3.6–4.5 lbs. Princep Caliber 90 or 4–5 lbs. Simazine 80W per acre in 20–100 gallons of water per acre. Use the same rates in all coarse-textured soils. Apply this combination only once per year.

SOLICAM DF PLUS SIMAZINE PLUS PARAQUAT CL

1–2 qts. of Paraquat CL may be added to the SOLICAM DF plus Simazine tank-mix combinations to kill existing vegetation. Add an agriculturally approved nonionic surfactant such as Ortho X-77 at the rate of ½ pint per 100 gallons of spray. Refer to directions for the SOLICAM DF plus Simazine combinations for recommended crops, application rates and restrictions.

Apply to the orchard floor in at least 50 gallons of water per acre, avoiding contact with fruit, foliage or stems. Apply when the weeds and grasses are succulent and new growth is 1–6 inches tall.

Do not spray under windy conditions and use a shield for young trees. Refer to the product labels for further directions, warnings, precautions and notes.

SOLICAM DF PLUS SIMAZINE PLUS ROUNDUP

1–5 qts. of Roundup may be added to the SOLICAM DF plus Simazine tank-mix combinations to control existing annual and perennial weeds. Refer to directions for the SOLICAM DF plus Simazine combinations for recommended crops, application rates and restrictions.

Apply in 20–40 gals. of water per acre, avoiding contact with fruit, foliage or stems. Apply when the weeds are actively growing. Use extreme care when applying this combination. Refer to the individual product labels for further directions, warnings, precautions and notes.

SOLICAM® DF HERBICIDE PLUS GOAL®

This tank-mix combination is recommended for control of certain existing weeds, including Malva and Filaree, and improved residual control of annual broadleaf weeds.

This combination may be used in the following crops: Almonds, Apricots, Cherries, Grapes (California only), Nectarines, Peaches, Pears, Plums, Prunes, Walnuts.

Apply 2.5–5 lbs. of SOLICAM DF per acre and 2–4 qts. of Goal 2E or 2.5–5 qts. of Goal 1.6E per acre in at least 50 gals. of water per acre. Add an agriculturally approved nonionic surfactant such as Triton® AG-98 or Ortho X-77® at the rate of 1 quart per 100 gallons of spray. Apply to the orchard floor avoiding contact with fruit, foliage or other tree parts. Refer to the Goal label for directions regarding application timing. Do not apply under windy conditions.

Apply this combination only once per year. Use only in those locations specified in both SOLICAM DF and Goal labels.

SOLICAM DF PLUS SURFLAN® PLUS PARAQUAT CL

This tank-mix combination is recommended for kill of existing vegetation and improved residual control of certain broadleaf weeds, including Purslane, Lambsquarters, and Pigsweeds.

This combination may be used in the following crops: Apples, Almonds, Cherries, Filberts, Grapefruit, Lemons, Oranges, Peaches, Pears, Plums, English Walnuts.

Apply 2.5–5 lbs. of SOLICAM DF per acre, 1–2 qts. of Paraquat CL per acre, and 2–4 qts. of Surflan A. S. or 2⅔–5⅓ lbs. of Surflan 75W per acre in at least 50 gals. of water. Add an agriculturally approved nonionic surfactant such as Ortho X-77 at the rate of ½ pint per 100 gallons of spray. Apply to the orchard floor avoiding contact with fruit, foliage or stems. Apply when the weeds and grasses are succulent and new growth is 1–6 inches tall.

Do not spray under windy conditions and use a shield for young trees. Refer to the SOLICAM DF, Paraquat and Surflan labels for futher directions, warnings, precautions, and notes.

SOLICAM DF PLUS KARMEX® 80WP OR OTHER DIURON 80WP PRODUCTS

There may be instances where the combination with Diuron will improve overall weed control in sandy loam, medium or fine soils. SOLICAM DF at labeled rates is safe in citrus crops in all soil types when used alone.

This combination may be used in the following crops: Apples, Citrus, Peaches, Pears, Pecans, Walnuts.

Apply the appropriate rates in the following table as a tank-mix in 20–100 gals. of water per acre. Apply as directed to the soil surface before weeds emerge, avoiding contact with fruit, foliage and stems.

Soil Type	Pounds Product Per Acre	
	SOLICAM DF	Diuron 80WP
Sandy loam	1.75	1
Medium	2.5	1.5
Fine	3.75	2

In California and Arizona use this combination in citrus only. Do not use this combination on sand, loamy sand, gravely soils or exposed subsoils, or in soils with less than 1% organic matter (0.5% for pecans). Use only in those crops and locations specified in both the SOLICAM DF and Diuron 80WP labels.

Roundup is a trademark of Monsanto Co.

Goal, Triton are trademarks of Rohm and Haas Co.

Ortho X-77 is a trademark of Chevron Chemical Co.

Surflan is a trademark of Elanco Products Co.

Princep, Caliber are trademarks of CIBA-GEIGY Corp.

Karmex is a trademark of Dupont.

SPUR® 22EW

INSECTICIDE

For Use on Nonbearing Fruit & Nut Trees and Vines and Certain Crops Grown for Planting Seed.

ACTIVE INGREDIENT:
(*aRS*,2*R*)-fluvalinate
[(*RS*)-*a*-cyano-3-phenoxy-benzyl (*R*)-2-[2-chloro-4-(trifluoromethyl) anilino]-3-methylbutanoate]* 22.3%
INERT INGREDIENTS 77.7%
TOTAL 100.0%

SPUR 22EW is 22.3% w/w (*aRS*,2*R*)-fluvalinate emulsion in water

*U.S. PAT. 4,260,633, 4,411,912

SPUR is a trademark of Sandoz, Ltd.

EPA Reg. No 20954-127 EPA Est. No. 20954-CA-1

CAUTION

Keep Out of Reach of Children

SHAKE WELL BEFORE USING

PRECAUTIONARY STATEMENTS

HAZARDS TO HUMANS

CAUTION

Avoid breathing spray mist. To avoid potential irritation of nose and throat from fine spray particles, use a NIOSH approved dust mask. Causes mild eye irritation. Harmful if absorbed through skin. Avoid contact with eyes, skin or clothing. Wear rubber gloves while handling. Wear protective clothing. Wash exposed skin with soap and water after handling and before eating. Wash contaminated clothing with soap and water before reuse.

Sensitive individuals may temporarily experience an itching or burning sensation with or without a rash following exposure. Avoid hand or sleeve-to-face contact.

FIRST AID

IF IN EYES: Flush thoroughly with a direct stream of water for 15 minutes. Get medical attention if irritation persists.

IF ON SKIN: Wash with plenty of soap and water. Get medical attention if irritation persists.

IF SWALLOWED: Drink 1 to 2 glasses of water and induce vomiting by touching the back of the throat with a finger. Do not give anything by mouth to any unconscious person. Get medical attention.

24-hour medical or chemical emergency phone (collect) 415/857-0172

ENVIRONMENTAL HAZARDS

This product is highly toxic to fish. Do not apply above or around fish ponds. Use with care when applying in areas adjacent to any body of water. Do not apply directly to water. Do not contaminate water by cleaning of equipment or disposal of waste.

Consult your local Extension office or a Zoecon specialist for guidance in selecting a rate and timing schedule that has demonstrated good performance under your specific conditions.

DIRECTIONS FOR USE

It is a violation of Federal law to use this product in a manner inconsistent with its labeling. Do not use where any part of the crop will be used for human or livestock consumption. Applications may not be made to trees or vines which will be harvested within one year.

For use in the production of planting seed crops, use SPUR only after seedstalk formation, or during and after bloom. Do not graze animals in treated areas or feed gleanings or crop residues to livestock. Do not replant SPUR treated fields with food crops, except cotton, within 90 days of last application.

INTRODUCTION

SPUR® 22 EW Insecticide controls insects primarily through contact action. Target pests must be exposed to the spray or spray deposits, so good spray coverage is necessary. Treat according to local action criteria, or when pests are young and populations manageable, before crop damage occurs. Repeat as needed, generally at one to three week intervals. Performance should be apparent four days after treatment. Where heavy populations occur, repeated applications five to seven days apart may be necessary to bring pests under adequate control. Growth dilution, climatic variables and pest habits determine treatment needs.

For most pests listed, a range of use rates is given. The higher rates must be selected when pest pressures or insect damage potential is severe. The more economical rates often give acceptable results.

Dried residues of SPUR are non-toxic and non-repellant to honeybees. Treat during non-foraging periods to minimize potential adverse effects.

Dilute SPUR in sufficient water to give good spray coverage, based on local crop protection experience. Buffer spray water to pH 5–7 if necessary. If local experience encourages use by air, use five to twenty gallons of water per acre. When SPUR is tank-mixed with wettable powders, add powder to the tank first. SPUR can be tank-mixed in water with emulsified spray oils. Do not mix with highly caustic or alkaline materials. Check mixtures for physical and biological compatibility before extensive use.

Recommended Application Rates

Commodity	Pests	Rate fl. oz./acre
Planting seed crops: Brussel sprouts Cabbage Carrot Collard Kale Lettuce Onion Parsnip Radish Spinach Swiss chard Tobacco	Thrips(1)	1.6–6.4
	Aphids(2) Flea beetles Diamondback moth Grasshoppers Imported cabbageworm Loopers Webworms	3.2–6.4
	Corn earworm Tobacco budworm	4.8–6.4
	Armyworms(1) Cutworms Lygus plant bugs(1) Spider mites	4.8–9.6

(1) To control populations known to be tolerant of other insecticides, select the higher rates and use shorter intervals.

(2) Except for tolerant populations of Green peach Aphid.

Commodity	Pests	Rate(3) fl. oz./100 Gal.
Nonbearing fruit trees, nut trees & vines	Thrips	2–5
	Aphids Broad and Flat mites European red mite Japanese beetle Leafhoppers Leafrollers Leaffeeding caterpillars McDaniel mite Oriental fruit moth Peach twig borer Pear psylla Scales (crawler stage) Spider mites Spotted Tentiform leafminer European red mite	3–8

(3) Rate is for full coverage, dilute applications. For concentrate sprayers, calculate the per acre rate needed to dilute and use that amount of product per acre in a reduced gallonage of water carrier.

STORAGE AND DISPOSAL

Do not contaminate water, food or feed by storage or disposal.

STORAGE

Store away from direct sunlight or heat. Pesticides must be stored in a secured area away from other products, food or feed. In case of spill or leak, soak up with soil or other absorbent material. Do not use caustic agents.

DISPOSAL

Wastes resulting from the use of SPUR® 22 EW may be disposed of on site or at any approved waste disposal facility.

Triple rinse container (or equivalent). Then offer for recycling or reconditioning, or puncture and dispose of in a sanitary landfill, or incineration, or, if allowed by state and local authorities, by burning. If burned, stay out of smoke.

WARRANTY AND CONDITIONS OF SALE

Zoecon Corporation warrants that the chemical composition of this product conforms to the chemical description on the label and is reasonably fit for the purpose stated on the label when used in accordance with directions under normal conditions of use. Zoecon makes no other warranty, express or implied, concerning the use of this product other than as indicated on the label. Buyer assumes all risks of use, storage or handling of this material not in strict accordance with directions given on the label.

THURICIDE® 32LV

AQUEOUS CONCENTRATE

For Aerial or Ground Application

ACTIVE INGREDIENT:
Bacillus thuringiensis
Berliner, potency of 8,000 International Units
(at least 12 million viable spores) per milligram* 1.6%
INERT INGREDIENTS: 98.4%

TOTAL 100.0%

*Equivalent to 8.0 billion International Units per quart.

EPA Reg. No. 11273-35 EPA Est. No. 36299-CA-1

CAUTION
Keep Out of the Reach of Children

PRECAUTIONARY STATEMENTS

HAZARDS TO HUMANS

Avoid inhalation or contact with eyes or open wounds.

DIRECTIONS FOR USE

It is a violation of Federal law to use this product in a manner inconsistent with its labelling.

Suspension must be shaken or stirred before use. Pour the recommended amount of THURICIDE 32LV into the required volume of water in the spray tank. Agitate as necessary to maintain suspension. Do not allow diluted sprays to remain in the tank for more than 72 hours. THURICIDE 32LV is formulated to provide desirable coverage and stickability on leaf surfaces. Additional adjuvants, spreaders, or stickers may be added but are not essential.

GROUND APPLICATION

Use adequate water to obtain good foliar coverage.

Wet foliage but do not allow excessive run-off. Apply the recommended per acre rates of THURICIDE 32LV with the following suggested amounts of water:

100 gallons with high gallonage hydraulic sprayers

10 gallons with mist blower

AERIAL APPLICATION

Apply THURICIDE 32LV at recommended rates by air either alone or as a spray mix diluted with water. Spray volumes of 32-128 ounces per acre are recommended. Best results are expected when THURICIDE 32LV is applied to dry foliage with calibrated aircraft capable of obtaining droplet sizes below 300 microns and preferably in the range of 50-150 microns.

THURICIDE is a trademark of Sandoz, Ltd.

RECOMMENDATIONS FOR SHADE TREES, ORNAMENTALS & FOREST

Thorough coverage is essential when using THURICIDE 32LV. Use the lower rate for light to moderate infestations. Use the higher rates for heavier infestations.

Pest	Ounces per Acre	Directions for Use
Spring and Fall cankerworm	16–32	Apply when leaf expansion reaches 40%-50% as infestation warrants. If eggs hatch over a long period of time or if reinfestation occurs, respray about 14 days after the first application. For gypsy moth control where a single application is planned, apply at least 48 ounces of THURICIDE 32LV per acre.
Elm spanworm	16–32	
Tent caterpillars	16–32	
Gypsy moth	32–80	
Spruce budworm	24–80	Apply when most larvae are 3rd-4th instar. Also consider the opening of the bud cap to ensure foliage exposure.
Douglas fir tussock moth	16–32	
Jack pine budworm	16–32	
Bagworm	16–32	
California oak moth	16–32	
Western tussock moth	16–32	
Fruit tree leafroller	16–32	Apply after eggs have hatched and early instar larvae are feeding on exposed foliage.
Mimosa webworm	16–32	
Redhumped caterpillar	8–24	
Fall webworm	8–16	
Pine butterfly	24–48	

STORAGE AND DISPOSAL

STORAGE

Store in a cool place. Activity may be impaired by storage at temperatures above 90°F. Do not contaminate water, food or feed by storage or disposal.

PESTICIDE DISPOSAL

Wastes resulting from this product may be disposed of on site or at an approved waste disposal facility.

CONTAINER DISPOSAL (PLASTIC)

Triple rinse (or equivalent) then offer for recycling or reconditioning, or puncture and dispose of in a sanitary landfill, or by incineration, or, if allowed by state and local authorities, by burning. If burned, stay out of smoke. Reuse of thoroughly cleaned container is allowable.

CONTAINER DISPOSAL (METAL)

Triple rinse (or equivalent) then offer for recycling or reconditioning, or puncture and dispose of in a sanitary landfill, or by other procedures approved by state and local authorities. Reuse of thoroughly cleaned container is allowable.

WARRANTY AND CONDITIONS OF SALE

Zoecon Corporation warrants that the chemical composition of this product conforms to the chemical description on the label and is reasonably fit for the purpose stated on the label when used in accordance with directions under normal conditions of use. Zoecon makes no other warranty, express or implied, concerning the use of this product other than as indicated on the label. Buyer assumes all risks of use, storage or handling of this material not in strict accordance with directions given on the label.

THURICIDE® 48LV

AQUEOUS CONCENTRATE

For Aerial or Ground Application

ACTIVE INGREDIENT:

Bacillus thuringiensis

Berliner, potency of 12,000 International Units

(at least 18 million viable spores) per milligram* 2.4%

INERT INGREDIENTS: 97.6%

TOTAL 100.0%

*Equivalent to 12.0 billion International Units per quart.

EPA Reg. No. 11273-42 EPA Est. No. 36299-CA-1

THURICIDE is a trademark of Sandoz, Ltd.

CAUTION

Keep Out of Reach of Children

PRECAUTIONARY STATEMENTS

HAZARDS TO HUMANS:

Avoid inhalation or contact with eyes or open wounds.

DIRECTIONS FOR USE

It is a violation of Federal law to use this product in a manner inconsistent with its labeling.

Suspension must be shaken or stirred before use. Pour the recommended amount of THURICIDE 48LV into the required volume of water in the spray tank. Agitate as necessary to maintain suspension. Do not allow diluted sprays to remain in the tank for more than 72 hours. THURICIDE 48LV is formulated to provide desirable coverage and stickability on leaf surfaces. Additional adjuvants, spreaders, or stickers may be added but are not essential.

Ground Application:

Use adequate water to obtain good foliar coverage. Wet foliage but do not allow excessive run off. Apply the recommended per acre rates of THURICIDE 48LV. Use the following suggested amounts of water:

100 gallons with high gallonage hydraulic sprayers

10 gallons with mist blower

Aerial Application:

Apply THURICIDE 48LV at recommended rates by air either alone or as a spray mix diluted with water. Spray volumes of 32-128 ounces per acre are recommended. Best results are expected when THURICIDE 48LV is applied to dry foliage with a calibrated aircraft capable of obtaining droplet sizes below 300 microns and preferably in the range of 50-150 microns.

RECOMMENDATIONS FOR SHADE TREES, ORNAMENTALS AND FOREST

Thorough coverage is essential when using THURICIDE 48LV. Use the lower rates for light to moderate infestations. Use the higher rates against heavier worm infestations.

Pest	Ounces per Acre	Directions for Use
Spring and Fall cankerworm	11-22	Apply when leaf expansion reaches 40%-50% as infestation warrants. If eggs hatch over a long period of time or if reinfestation occurs, spray about 14 days after first application. For gypsy moth control where a single application is planned, apply at least 20 ounces of THURICIDE 48LV per acre.
Elm spanworm	11-22	
Tent caterpillars	11-22	
Gypsy moth	22-53	
Spruce budworm	16-53	Apply when most larvae are 3rd-4th instar. Also consider the opening of the bud cap to ensure foliage exposure.
Douglas fir tussock moth	11-22	Apply after eggs have hatched and early instar larvae are feeding on exposed foliage.
Jack pine budworm	11-22	
Bagworm	11-22	
California oak moth	11-22	
Western tussock moth	11-22	
Fruit tree leafroller	11-22	
Mimosa webworm	11-22	
Redhumped caterpillar	5-16	
Fall webworm	5-11	
Pine butterfly	16-32	

STORAGE AND DISPOSAL

STORAGE

Store in a cool place. Activity may be impaired by storage at temperatures above 90°F. Do not contaminate water, food or feed by storage or disposal.

PESTICIDE DISPOSAL

Wastes resulting from this product may be disposed of on site or at an approved waste disposal facility.

CONTAINER DISPOSAL (PLASTIC)

Triple rinse (or equivalent) then offer for recycling or reconditioning, or puncture and dispose of in a sanitary landfill, or by incineration, or, if allowed by state and local authorities, by burning. If burned, stay out of smoke. Reuse of thoroughly cleaned container is allowable.

CONTAINER DISPOSAL (METAL)

Triple rinse (or equivalent) then offer for recycling or reconditioning, or puncture and dispose of in a sanitary landfill, or by other procedures approved by state and local authorities. Reuse of thoroughly cleaned container is allowable.

WARRANTY AND CONDITIONS OF SALE

Zoecon Corporation warrants that the chemical composition of this product conforms to the chemical description on the label and is reasonably fit for the purpose stated on the label when used in accordance with directions under normal conditions of use. Zoecon makes no other warranty, express or implied, concerning the use of this product other than as indicated on the label. Buyer assumes all risks of use, storage or handling of this material not in strict accordance with directions given on the label.

ZORIAL®
RAPID 80™
HERBICIDE

For Control of Grass and Broadleaf Weeds in Cotton and Soybeans. In a Dry, Flowable Formulation that Instantly Disperses in Water.

ACTIVE INGREDIENT:
norflurazon [4-chloro-5-(methylamino)-2-(alpha, alpha, alpha-trifluoro-*m* -tolyl)-3(2*H*)-pyridazinone] 80%
INERT INGREDIENTS 20%
TOTAL 100%

EPA Reg. No. 11273-47 EPA Est. No. 39578-TX-1
U.S. Pat. No. 3,644,355 and 3,834,889

CAUTION
Keep Out of Reach of Children

PRECAUTIONARY STATEMENTS

ENVIRONMENTAL HAZARDS

Do not apply directly to any body of water. Do not apply when weather conditions favor runoff or drift from treated areas. Do not contaminate water by cleaning of equipment or disposal of wastes.

WARRANTY AND CONDITIONS OF SALE

Zoecon Corporation warrants that the chemical composition of this product conforms to the chemical description on the label and is reasonably fit for the purpose stated on the label when used in accordance with directions under normal conditions of use. Zoecon makes no other warranty, express or implied, concerning the use of this product other than as indicated on the label. Buyer assumes all risks of use, storage or handling of this material not in strict accordance with directions given on the label.

LOCATION OF USE

For use in cotton and soybeans in the states of Alabama, Arkansas, North Florida, Georgia, Kentucky, Louisiana, Mississippi, Missouri, North Carolina, Oklahoma, South Carolina, Tennessee, Texas, and Virginia.

WEEDS CONTROLLED

ZORIAL RAPID 80 Herbicide at recommended rates is effective for:

A. Controlling the following weeds in cotton or soybeans:

Barnyardgrass	Pigweeds
Crabgrass	Prickly Sida
Foxtail	Panicum spp.
Goosegrass[1]	Spotted Spurge
Signalgrass	Spurred Anoda (cottonweed)
Johnsongrass (seedling)	Velvetleaf
Carpetweed	Tropic Croton[2]
Florida Pusley	Purslane

[1] Includes DNA resistant goosegrass.
[2] Use higher rates on Coastal Plains soils only for improved control.

B. Suppressing the following weeds in cotton or soybeans:

Bermudagrass	Hemp Sesbania
Johnsongrass (rhizome)	Kochia
Nutsedge (yellow and purple)	Morningglory
Pennsylvania Smartweed	Russian Thistle
Lanceleaf Sage (mintweed)	Sicklepod
Cocklebur	Other Smartweeds
Common Ragweed	Spotted Spurge
	Florida Beggarweed
	Venice Mallow

DIRECTIONS FOR USE

It is a violation of Federal law to use this product in a manner inconsistent with its labeling.

GENERAL INFORMATION

ZORIAL RAPID 80 is a herbicide for control of certain grasses and brodleaf weeds in cotton and soybeans. Apply ZORIAL before weeds emerge. ZORIAL can be applied preplant incorporated, preemergence surface, or as a split application in areas with 40 inches or more annual rainfall. In areas with less than 40 inches rainfall, apply as a preplant incorporated or split application only.

The soil surface should be free of clods and existing weeds prior to application.

If replanting of cotton or soybeans is necessary, a light reworking of the topsoil will retain the herbicidal activity of ZORIAL.

Peanuts can be planted as a replacement crop 30 days after ZORIAL application. Prior to planting, a light reworking of the topsoil is recommended to remove existing plant material and to retain herbicidal activity. Do not graze replacement peanut fields with livestock for ten weeks following planting.

Do not graze treated cotton fields with livestock or feed treated cotton forage to livestock. Cover crops planted after harvest should be plowed under and not grazed or harvested. To minimize injury to small grain cover crops, they should be broadcast planted rather than drilled.

Quantity of ZORIAL applied in any one year must not exceed the broadcast rate in pounds per acre for a given soil type.

For one year following ZORIAL application, treated acreage may only be rotated to cotton, soybeans, and peanuts.

Crop Safety: If the cotton or soybean plant is stressed during early development, application of ZORIAL at the label rate may result in temporary bleaching or chlorosis of the leaves from which the plant will recover.

MIXING INSTRUCTIONS

Clean and calibrate the sprayer. Add ZORIAL RAPID 80 to the spray tank already filled with 1/4 of the required volume of the water. This will eliminate or minimize foaming. Maintain agitation while filling and spraying. If a by-pass line is used, discharge at the bottom of the tank to minimize foaming.

Do not allow ZORIAL RAPID 80 spray mixture to remain in the spray tank overnight because settling could occur and resuspension would be difficult.

Tank Mixes: ZORIAL RAPID 80 is approved as a tank mix with Prowl, Treflan, Sencor, Lexone, Cotoran, Dual, Lasso, Lorox and Bladex. Always predetermine the compatibility of labeled tank mixes by mixing small proportional quantities in advance.

Products should be added to the spray tank in the following order:

1. Wettable powders and water dispersable granules. Wettable powders should be premixed in a small amount of water. Water dispersable granulars should be added during filling. Allow both to disperse before other products are added. The order of addition is not important.
2. Flowable liquids.
3. Emulsifiable concentrates.
4. Surfactants.

Begin adding wettable powders, flowable liquids, emulsifable concentrates, and surfactants after the spray tank is 3/4 full. Continue agitation during the addition of all the materials and while filling and spraying.

APPLICATION DIRECTIONS

Ground Application: Apply the recommended rate of ZORIAL in 10–20 gallons of water per acre. Apply accurately with a fixed boom sprayer. Equip sprayer with 8002 to 8006 fan-jet nozzle tips or equivalent for broadcast application. Use 8002E to 8006E nozzle tips or equivalent for band applications. Use screen size 50 mesh or larger.

Aerial Application: Apply the recommended rate of ZORIAL in 5 or more gallons of water per acre. To minimize drift, use ZORIAL with a drift control agent such as Nalco Trol® and Ortho-Trol®. Do not apply during periods of gusty winds or winds in excess of 5 mph. Aerial spray equipment, flying speed and flying height should be carefully adjusted to provide a uniform application to the soil surface. Use markers to avoid overlapping and possible crop injury.

Preplant incorporated application (PPI): Apply ZORIAL at the recommended rate from 30 days to just prior to planting. In West Texas and Southwest Oklahoma, ZORIAL may be applied up to 60 days prior to planting. Incorporate not deeper than 2–3 inches with commonly used equipment. Incorporation should occur within one week after application.

Preemergence surface application (PRE): Apply ZORIAL at the recommended rate at the time of planting and preemergence to weeds.

Split application—THE ZORIAL SPLIT™: Apply ZORIAL preplant incorporated, as indicated above, at one half the recommended rate. Apply the other half preemergence surface as recommended above.

Band Treatment: When applying a band treatment of ZORIAL RAPID 80 the following formula may be used to calculate the amount per acre:

$$\frac{\text{Band width in inches}}{\text{Row width in inches}} \times \begin{matrix}\text{Rate in lbs per acre}\\ \text{broadcast treatment}\end{matrix} = \begin{matrix}\text{Pounds per acre}\\ \text{for band treatment}\end{matrix}$$

SOYBEAN INFORMATION

Do not use in soybeans in West Texas or Southwest Oklahoma. Do not use when the soil type is sand, loamy sand, or sandy loam except the Costal Plains soils of the Southeast. Do not use the PPI treatment in soybeans unless as part of the split application treatment. Do not use PPI treatments in Coastal Plains soils of the Southeast. Soybeans should be planted to a depth of 1–2 inches and soil firmly packed above and around the seed bed.

Broadcast rates for soybeans in pounds of ZORIAL RAPID 80 per acre:

Soil Type	PRE*	Split ppi	Split pre
Coastal Plains soils (SE)	1.25		
loam, silty loam, silt, sandy clay loam, sandy clay	1.9	.95	.95
clay loam, silty clay loam, silty clay, clay	2.5	1.25	1.25

* For broadleaf weed control use as a tank-mix combination with metribuzin at grower use rates as described in the ZORIAL DF plus Sencor or Lexone tank mix recommendations below. If ZORIAL is used alone as a preemergence or split application, use BLAZER® or BASAGRAN® according to all directions for use, precautions and limitations on the herbicide's labeling for broadleaf weed control as needed.

ZORIAL RAPID 80 COMBINATIONS IN SOYBEANS

ZORIAL RAPID 80 plus Sencor® or Lexone® Tank Mix

This combination is recommended for improved control of cocklebur, morningglory, hemp sesbania, common ragweed and Venice mallow and other broadleaf weeds.

Apply a tank mix of ZORIAL DF plus Sencor or Lexone preemergence surface at the rates per acre (broadcast) recommended in the following table. These rates represent optimum ZORIAL rates and 1/2 to the normal metribuzin rate. Use the metribuzin use rates (grower use rates) most applicable to local soil type and growing conditions. Apply using 10-20 gallons of water per acre.

Broadcast Rates

Soil Type	Pounds ZORIAL Per Acre	Pounds Sencor DF/Lexone DF Per Acre — Organic Matter: Less than 2%	2-4%	Over 4%
Coastal Plains Soils (SE only)	1.25	Do not use	0.25-0.50	0.33-0.66
Medium	1.25-1.50	0.25-0.66	0.33-0.83	0.425-1
Fine (heavy)	1.25-1.50*-1.90	0.33-0.83	0.425-1	0.50-1.
Mississippi Delta Only	1.25-1.50*-1.90	0.50-1	0.58-1.17	0.665-1.3

* Optimum rate

When using Sencor 4 or Lexone 4L convert to equivalent rates. The amount of Sencor DF or Lexone DF equals two thirds the amount of Sencor 4 or Lexon 4L. For example one pint of Sencor 4 or Lexone 4L equals 2/3 pound Sencor DF or Lexone DF.

Refer to the ZORIAL and Sencor or Lexone labels for further directions and precautions.

COTTON INFORMATION

Broadcast rates for cotton in pounds of ZORIAL RAPID 80 per acre:

Soil Type	PPI	PRE	Split ppi	Split pre
sand, loamy sand, sandy loam	1.25	1.25	.625	.625
Coastal Plains Soils (SE)*	**	1.9	.95	.95
loam, silty loam, silt, sandy clay loam, sandy clay	1.9	1.9	.95	.95
clay loam, silty clay loam, silty clay, clay	2.5	2.5	1.25	1.25

* For improved control of hard to control weeds like pigweed, cocklebur and tropic croton.

** Do not use 1.9 lbs. PPI on Coastal Plains soils.

West Texas and Southwest Oklahoma Only: Do not use on sandy soils (soils with greater than 70% sand content). Prior to application of ZORIAL, completely destroy the seed beds of the previous year by "cross-discing". Apply ZORIAL PPI or pre-emergence surface. For best results, incorporate ZORIAL on flat disced soil at labeled rates for control of spurred anoda, annual grasses and suppression of yellow and purple nutsedge. To prevent removal of the herbicide application by wind, immediate incorporation is recommended. A single, one inch rain or its equivalent in center pivot or sprinkler irrigation is necessary to activate ZORIAL when applied pre-emergence. Soybeans can be planted as a replacement crop 45 days after the first planting of cotton. Do not plant peanuts as a replacement crop. Do not use ZORIAL on furrow planted cotton or on glandless cotton varieties.

For one year following a ZORIAL application, treated acreage may only be rotated to cotton, soybeans or peanuts.

STORAGE AND DISPOSAL

STORAGE

Do not contaminate water, food or feed by storage or disposal.

PESTICIDE DISPOSAL

Wastes resulting from the use of this product may be disposed of on site or at an approved waste disposal facility.

CONTAINER DISPOSAL

Triple rinse (or equivalent). Then puncture and dispose of in a sanitary landfill or by incineration, or, if allowed by state and local authorities, by burning. If burned, stay out of smoke.

ZORIAL, ZORIAL RAPID 80 and ZORIAL SPLIT are trademarks of Sandoz, Ltd.

NalcoTrol is a trademark of Nalco Chemical Co.

Ortho-Trol is a trademark of Chevron Chemical Co.

Sencor is a trademark of Farben Fabriken Bayer GmbH. Leverkusen

Lorox and Lexone are trademarks of E.I. DuPont de Nemours and Co.

Prowl is a trademark of American Cyanimid

Cotoran and Dual are trademarks of Ciba-Geigy

Lasso is a trademark of Monsanto

Bladex is a trademark of Shell Chemical Co.

Treflan is a trademark of Elanco.

NOTES

NOTES

NOTES

NOTES

NOTES

APPENDIX I

Calibrations And Conversions General Information On Handling And Storage Of Pesticides

In this section, calibrations and conversion tables are provided to insure the application of proper dosages of crop protection chemicals. General agricultural chemical storage and handling guidelines are also provided for the safe handling of products in warehouses and shipping facilities. These guidelines have been prepared by, and are printed with permission from, the National Agricultural Chemicals Association, Washington, D.C.

Each product label is specific in giving instructions and precautions for the material in its container and all readers are cautioned to read the label before using any crop protection chemical.

CALIBRATIONS AND CONVERSIONS

Sprayer Calibration

Calibration of your spraying equipment is very important. It should be done at least every other day of operation to insure application of the proper dosages. This is probably the most important step in your whole spraying operation since applying incorrect amounts may do much more damage than good. Each nozzle must be checked so that all nozzles are delivering the same amount of material for a uniform spray pattern.

If a lower rate is desired it may be obtained by increasing the speed, reducing the pressure or changing to a smaller nozzle or a combination of the three. The opposite is true if a higher rate is desired.

General Calibration

I. Method I.

- A. Measure out 660 feet.
- B. Determine the amount of spray put out in traveling this distance at the desired speed.
- C. Use this formula:

$$\text{gallons/acre} = \frac{\text{gallons used in 660 feet} \times 66}{\text{swath width in feet}}$$

- *D. Fill the tank with the desired concentration.

II. Method II.

- A. Fill spray tank and spray a specific number of feet.
- B. After spraying refill tank measuring the quantity of material needed for refilling.
- C. Use this formula:

$$\text{gallons/acre} = \frac{43560 \times \text{gallons delivered}}{\text{swath length (ft.)} \times \text{swath width (ft.)}}$$

- *D. Fill the tank with the desired concentrate.

III. Method III.

- A. Measure 163 feet in the field.
- B. Time tractor in 163 feet. Make two passes to check accuracy.
- C. At edge of field adjust pressure valve until you catch 2 pints (32 ounces) of spray in the same amount of time it took to run the 163 feet. Be sure tractor is at the same throttle setting. You are now applying 20 gallons per acre on a 20 inch nozzle boom spacing.
- D. For each inch of nozzle spacing on boom, increase time by 5% or reduce the volume by 5%.

* If you have calibrated your rig and it is putting out 37 gpa, the required dosage is 4 lbs actual/acre. Therefore, for every 37 gallons of carrier (water, oil, etc.) in the spray tank you add 4 lbs of active material.

Useful Formulae

1. To determine the amount of active ingredient needed to mix in the spray tank.
   No. of gallons or pounds =

$$\frac{\text{No. of acres to be sprayed} \times \text{pounds active ingredient required per Acre}}{\text{pounds active ingredient per gallon or per pound}}$$

2. To determine the amount of pesticide needed to mix a spray containing a certain percentage of the active ingredient.
   No. gallons or pounds =

$$\frac{\text{gallons of spray desired} \times \text{\% active ingredient wanted} \times 8.345}{\text{lbs. active ingredient per gallon or pound} \times 100}$$

3. To determine the percent active ingredient in a spray mixture.
   Percent =

$$\frac{\text{lbs. or gallons of concentrate used (not just active ingredient)} \times \text{\% active ingredient in the concentrate}}{\text{gallons of spray} \times 8.345 \text{ (weight of water/gallon)}}$$

4. To determine the amount of pesticide needed to mix a dust with a given percent active ingredient.
pounds material =

$$\frac{\%\ \text{active ingredient wanted} \times \text{lbs of mixed dust wanted}}{\%\ \text{active ingredient in pesticide used}}$$

5. To determine the size of pump needed to apply a given number of gallons/acre.
Pump capacity =

$$\frac{\text{Gallons/acre desired} \times \text{boom width (feet)} \times \text{mph}}{495}$$

6. To determine the nozzle capacity in gallons per minute at a given rate/acre and miles/hour.
Nozzle capacity =

$$\frac{\text{gallons/acre} \times \text{nozzle spacing (inches)} \times \text{mph}}{5940}$$

7. To determine the acres per hour sprayed.
Acres per hour =

$$\frac{\text{swath width (inches)} \times \text{mph}}{100}$$

8. To determine the rate of speed in miles per hour.
   1. Measure off a distance of 300 to 500 feet
   2. Measure in seconds the time it takes the tractor to go the marked off distance.
   3. Multiply .682 times the distance traveled in feet and divide product by the number of seconds.
   MPH =

$$\frac{.682 \times \text{distance}}{\text{seconds}}$$

9. To determine the nozzle flow rate.
Time the seconds necessary to fill a pint jar from a nozzle.
Divide the number of seconds into 7.5
gallons/minute/nozzle = $\frac{7.5}{\text{seconds}}$

10. To determine the gallons per minute per boom.
Figure out the gallons/minute/nozzle and multiply by the number of nozzles.

11. To determine the gallons per acre delivered.

$$\frac{5940 \times \text{gallons/minute/nozzle} = \text{gpa}}{\text{nozzle spacing (inches)} \times \text{mph}}$$

12. To determine the acreage sprayed per hour.
acres sprayed/hour = $\frac{\text{boom width (feet)} \times \text{mph}}{12}$
This allows 30% of time for filling, turning, etc.

13. Sprayer Tank Capacity
Calculate as follows:
   1. Cylindrical Tanks:
   Multiply the length in inches times the square of the diameter in inches and multiply the product by .0034.
   length × diameter squared × .0034 = number per gallons.
   2. Elliptical Tanks:
   Multiply the length in inches times the short diameter in inches times the long diameter in inches times .0034.
   length × short diameter × long diameter × .0034 = number of gallons.
   3. Rectangular Tanks:
   Multiply the length times the width times the depth in inches and multiply the product by .004329.
   length × width × depth × .004329 = number of gallons.

14. To determine the acres in a given area.
Multiply the length in feet times the width in feet times 23. Move the decimal point 6 places to the left to give the actual acres.

Conversion Tables (U.S.)

Linear Measure—
1 foot = 12 inches
1 yard = 3 feet
1 rod = 5.5 yards = 16.5 feet
1 mile = 160 rods = 1760 yards = 5280 feet

Square Measure—
1 square foot (sq. ft.) = 144 square inches (sq. inch)
1 square yard (sq. yd) = 9 sq. feet
1 square rod (sq. rd.) = 272.25 sq. ft. = 30.25 sq. yd.
1 acre = 43560 sq. ft. = 4840 sq. yds. = 160 sq. rds.
1 square mile = 640 acres.

Cubic Measure—
1 cubic foot (cu. ft.) = 1728 cubic inches (cu. in.) 29.922 liquid quarts = 7.48 gallons
1 cubic yard = 27 cubic feet

Liquid Capacity Measure—
1 tablespoon = 3 teaspoons
1 fluid ounce = 2 tablespoons
1 cup = 8 fluid ounces
1 pint = 2 cups = 16 fluid ounces
1 quart = 2 pints = 32 fluid ounces
1 gallon = 4 quarts = 8 pints = 128 fluid ounces

Weight Measure—
1 pound (lb.) = 16 ounces
1 hundredweight (cwt.) = 100 pounds
1 ton = 20 cwt. = 2000 pounds

Rates of Application—
1 ounce/sq. ft. = 2722.5 lbs/acre
1 ounce/sq. yd. = 302.5 lbs/acre
1 ounce/100 sq. ft. = 27.2 lbs/acre
1 pound/100 sq. ft. = 435.6 lbs/acre
1 pound/1000 sq. ft. = 43.6 lbs/acre
1 gallon/acre = ⅓ ounce/1000 sq. ft.
5 gallons/acre = 1 pint/100 sq. ft.
100 gallons/acre = 2.5 gallons/1000 sq. ft. = 1 quart/100 sq. ft.
100 lbs/acre = 2.5 lbs/1000 sq. ft.

Important Facts—
Volume of sphere = $\text{diameter}^3 \times .5236$
Diameter = circum. × .31831
Area of circle = $\text{dia}^2 \times .7854$
Area of elipse = prod. of both dia × .7854
Vol. of cone = area of base × ⅓ ht.
1 cu. ft. water = 7.5 gallons = 62.5 lbs.
Pressure in psi = ht. (ft.) × .434
1 acre = 209 feet square
ppm = % × 10,000
% = $\frac{\text{ppm}}{10{,}000}$
1% by volume = 10,000 ppm

Table of Conversion Factors

To Convert From	To	Multiple By
Cubic feet	gallons	7.48
Cubic feet	liters	28.3
Gallons	milliliters	3785
Grams	pounds	.0022
Grams/liter	parts/million	1000
Grams/liter	pounds/gallon	.00834
Liters	cubic feet	.0353
Milligrams/liter	parts/million	1
Milliliters/gallons	gallons	.0026
Ounces	grams	28.35
Parts/million	grams/liter	.001
Parts/million	pounds/million gallons	8.34
Pounds	grams	453.59
Pounds/gallon	grams/liter	111.83

1 gram = .035 ounce
1 kilogram = 2.2 lbs.
1 quintal = 100 kg = 221 lbs
1 metric ton = 1000 kg = 2,205 lbs
1 hectare = 2.5 acres
1 meter = 39.4 inches
1 kilometer = .6 mile

Conversion Table

1 kilogram (kg) = 1000 grams (g) = 2.2 lbs
1 gram (g) = 1000 milligrams (mg) = .35 ounce
1 liter = 1000 milliliters (ml) or cubic centimeters (cc) = 1.058 quarts
1 milliliter or cubic centimeter = .034 fluid ounce
1 milliliter or cubic centimeter of water weighs 1 gram
1 liter of water weighs 1 kilogram

1 lb = 453.6 grams
1 ounce = 28.35 grams
1 pt. of water weighs approximately 1 lb
1 gallon of water weighs approximately 8.34 lbs

1 gallon = 4 qts = 3.785 liters
1 qt = 2 pts = .946 liters
1 pt = .473 liters
1 fluid ounce = 29.6 milliliters or cubic centimeters

1 part per million (ppm) = 1 milligram/liter
= 1 milligram/kilogram
= .0001 per cent
= .013 ounces by in 100 gallons of water

1 percent = 10.000 ppm
= 10 grams per liter
= 10 grams per kilogram
= 1.33 ounces by weight per gallon of water
= 8.34 pounds/100 gallons of water

.1 percent = 1000 ppm = 1000 milligrams/liter
.01 percent = 1000 ppm = 100 milligrams/liter
.0001 percent = 1 ppm = 1 milligram/liter

Chemical Elements

Name	Symbol	Atomic Weight	Valance
Aluminum	Al	26.97	3
Antimony	Sb	121.76	3, 5
Arsenic	As	74.91	3, 5
Barium	Ba	137.36	2
Bismuth	Bi	209.00	3, 5
Boron	B	10.82	3, 0
Bromine	Br	79.916	1, 3, 5, 7
Cadmium	Cd	112.41	2
Calcium	Ca	40.08	2
Carbon	C	12.01	2, 4
Chlorine	Cl	35.457	1, 3, 5, 7
Cobalt	Co	58.94	2, 3
Copper	Cu	63.57	1, 2
Fluorine	F	19.00	1
Hydrogen	H	1.008	1
Iodine	I	126.92	1, 3, 5, 7
Iron	Fe	55.85	2, 3
Lead	Pb	207.21	2, 4
Magnesium	Mg	24.32	2
Mercury	Hg	200.16	1, 2
Molybdenum	Mo	95.95	3, 4, 6
Nickel	Ni	58.69	2, 3
Nitrogen	N	14.008	3, 5
Oxygen	O	16.00	2
Phosphorus	P	30.98	3, 5
Potassium	K	39.096	1
Selenium	Se	78.96	3
Silicon	Si	28.06	4
Silver	Ag	107.88	1
Sodium	Na	22.997	1
Sulfur	S	32.06	2, 4, 6
Thallium	Tl	204.29	1, 3
Tin	Sn	118.70	2, 4
Titanium	Ti	47.90	3, 4
Uranium	U	238.17	4, 6
Zinc	Zn	65.38	2

Width Of Area To Acres Per Mile Traveled

Width of Strip (feet)	Acres/mile
6	.72
10	1.21
12	1.45
16	1.93
18	2.18
20	2.42
25	3.02
30	3.63
50	6.04
75	9.06
100	12.1
150	18.14
200	24.2
300	36.3

Temperature Conversion Table Relationship Of Centigrade And Fahrenheit Scales

°C	°F	°C	°F
−40	−40	25	77
−35	−31	30	86
−30	−22	35	95
−25	−13	40	104
−20	−4	45	113
−15	5	50	122
−10	14	55	131
−5	23	60	140
0	32	80	176
5	41	100	212
10	50		
15	59		
20	68		

Proportionate Amounts Of Dry Materials

Water	Quantity of Material				
100 gallons	1 lb.	2 lbs.	3 lbs.	4 lbs.	5 lbs.
50 gallons	8 oz.	1 lb.	24 oz.	2 lbs.	2½ lbs.
5 gallons	3 tbs.	1½ oz.	2½ oz.	3¼ oz.	4 oz.
1 gallon	2 tsp.	3 tsp.	1½ tbs.	2 tbs.	3 tbs.

Proportionate Amounts Of Liquid Materials

Water	Quantity of Material		
100 gallons	1 qt.	1 pt.	1 cup
50 gallons	1 pt.	1 cup	½ cup
5 gallons	3 tsp.	5 tsp.	2½ tsp.
1 gallon	2 tsp.	1 tsp.	½ tsp.

Miles Per Hour Converted To Feet Per Minute

MPH	fpm
1	88
2	176
3	264
4	252

Emulsifiable Concentrate Percent Ratings In Pounds Actual Per Gallon

%EC	lbs/Gallon
10–12	1
15–20	1.5
25	2
40–50	4
60–65	6
60–75	8
80–100	10

Conversion Table For Liquid Formulations

Concentration of Active Ingredient in Formulations, lbs/gal.

Rate Desired Lbs/A	1	2	2.5	3	4	5	6
	(ml of formulation per 100 square feet)						
1	8.67	4.33	3.47	2.89	2.17	1.73	1.44
2	17.3	8.67	6.93	5.78	4.33	3.47	2.89
3	26.0	13.0	10.4	8.67	6.50	5.20	4.33
4	34.8	17.4	13.9	11.6	8.69	6.95	5.79
5	43.4	21.7	17.4	14.5	10.9*	8.68	7.24
6	52.1	26.0	20.8	17.4	13.0	10.4	8.68
7	60.8	30.4	24.3	20.3	15.2	12.2	10.1
8	69.4	34.7	27.8	23.1	17.4	13.9	11.6
9	78.1	39.0	31.2	26.0	19.5	15.6	13.0
10	86.7	43.3	34.7	28.9	21.7	17.3	14.4

*Example: To put out a 100 sq. ft. plot at the rate of 5 lbs/A active ingredient using a formulation containing 4 lb/gal active ingred., use 10.9 ml of the 4 lbs/gal form. & distribute evenly.

Conversion Table For Dry Formulations

Concentration of Active Ingredient in Formulation

Rate Desired Lbs/A	100%	90%	80%	75%	70%	60%	50%	40%	30%	25%	20%	10%	5%
	(Grams of formulation per 100 square feet)												
1	1.04	1.16	1.30	1.39	1.49	1.74	2.08	2.60	3.47	4.17	5.21	10.4	20.8
2	2.08	2.31	2.60	2.78	2.98	3.47	4.17	5.21	6.94	8.33	10.4	20.8	41.7
3	3.12	3.47	3.90	4.17	4.46	5.20	6.25	7.81	10.4	12.5	15.6	31.2	62.5
4	4.17	4.63	5.21	5.55	5.95	6.94	8.33*	10.4	13.9	16.7	20.8	41.7	83.3
5	5.21	5.79	6.51	6.94	7.44	8.68	10.4	13.0	17.4	20.8	26.0	52.1	104
6	6.25	6.94	7.81	8.33	8.93	10.4	12.5	15.6	20.8	25.0	31.2	62.5	125
7	7.29	8.10	9.11	9.72	10.4	12.1	14.6	18.2	24.3	29.2	36.4	72.9	146
8	8.33	9.26	10.4	11.1	11.9	13.9	16.7	20.8	27.8	33.3	41.7	83.3	167
9	9.37	10.4	11.7	12.5	13.4	15.6	18.7	23.4	31.2	37.5	46.9	93.7	187
10	10.4	11.6	13.0	13.9	14.9	17.4	20.8	26.0	34.7	41.7	52.1	104	208

*Example: To put out a 100 sq. ft. plot at the rate of 4 lbs/A active ingredient using a formulation containing 50% active ingredient, use 8.33 grams of the 50% formulation and distribute evenly over the 100 sq. ft.

Conversion Table For Granular Formulations

Concentration of Active Ingredient in Formulation

Rate Desired Lbs/A	20%	15%	10%	7.5%	5%	4%	3%	2%	1%
	(Grams of formulation per 100 square feet)								
1	5.2	6.94	10.4	13.86	20.8	26.0	34.66	52.0	104.0
2	10.4	13.9	20.8	27.7	41.7	52.0	69.3	104.0	208.0
3	15.6	20.8	31.2	41.6	62.5	78.0	103.9	156.0	312.0
4	20.8	27.8	41.7	55.4	83.3	104.0	138.6	208.0	416.0
5	26.0	34.7	52.1	69.3	104.0*	130.0	173.3	260.0	520.0
6	31.2	41.6	62.5	83.2	125.0	156.0	207.9	312.0	624.0
7	36.4	45.6	72.9	97.0	146.0	182.0	242.6	364.0	728.0
8	41.7	55.5	83.3	110.9	167.0	208.0	277.3	416.0	832.0
9	46.9	62.5	93.7	124.7	187.0	234.0	311.9	468.0	936.0
10	52.1	69.4	104.0	138.6	208.0	260.0	346.6	520.0	1,040.0
15	78.0	104.1	156.0	207.9	312.0	390.0	519.2	780.0	1,560.0
20	104.0	138.8	208.0	277.2	416.0	520.0	693.2	1,040.0	2,080.0
25	130.0	173.5	260.0	346.5	520.0	650.0	866.5	1,300.0	2,600.0
30	156.0	208.2	312.0	415.8	624.0	780.0	1,039.8	1,560.0	3,120.0

*Example: To put out a 100 sq. ft. plot at the rate of 5 lbs/A active ingredient using a formulation containing 5% active material, use 104.0 grams of the 5% formulation and distribute it evenly over the 100 sq. ft.

Quick Conversions

Temperature		Length		Volume	
°C	°F	cm	inch	liters	quarts
100	212	2½	1	1	1.1
90	194	5	2	2	2.1
80	176	10	4	3	3.2
70	158	20	8	4	4.2
60	140	30	12	5	5.3
50	122	40	16	6	6.3
40	104	50	20	7	7.4
35	95	60	24	8	8.5
30	86	70	28	9	9.5
25	77	80	32		
20	68	90	36		
15	59	100	39		
10	50	200	79		
5	41		feet		
0	32	300	10		
−5	23	400	13		
−10	14	500	16		
−15	5	1,000	33		
−20	−4				
−25	−13				
−30	−22				
−40	−40				

Quick Conversions

kg/ha	lb/A
1	0.9
2	1.8
3	2.7
4	3.6
5	4.5
10	9
20	18
30	27
40	36
50	45
60	54
70	62
80	71
90	80
100	89
200	180
300	270
400	360
500	450
600	540
700	620
800	710
900	800
1,000	890
2,000	1,800
	ton/A
3,000	1¼
4,000	1¾
5,000	2¼
6,000	2¾
7,000	3
8,000	3½
9,000	4
10,000	4½
11,000	5
12,000	5½
13,000	5¾
14,000	6¼
15,000	6¾
16,000	7
17,000	7½
18,000	8
19,000	8½
20,000	9

Conversion Factors Used In Calculation

Convert	To	By
gram (gm.)	kilogram (kg.)	move decimal 3 places to left
Example 2000 gm. = 2.0 kg.		
gram (gm.)	milligram (mg.)	move decimal 3 places to right
Example 2.0 gm. = 2000 mg.		
gram (gm.)	pound (lb.)	divide by 454
Example 658 gm./ ÷ 454 = 1.45 lb.		
gram/pound	percent (%)	divide by 4.54
Example 90 gm./lb. ÷ 4.54 = 19.8%		
gram/ton	percent	multiply by 11, move decimal 5 places to left
Example 45 gm./ton × 11 = 495 = .00495%		
kilogram (kg.)	gram	move decimal 3 places to right
Example 5.5 kg. = 5500 gm.		
milligram (mg.)	gram	move decimal 3 places to left
Example 95 mg. = 0.095 gm.		
percent	gram/pound (ppm)	multiply by 4.54
Example 25 × 4.54 = 113.5 gm./lb		
percent	parts/million (ppm)	move decimal 4 places to right
Example .025% = 250 ppm.		
percent	gram/ton	divide by 11, move decimal places to right
Example .011 ÷ 11 = .001 = 100 gm./ton		
pound	gram	multiply by 454
Example 0.5 lb. × 454 = 227 gm.		
ppm	percent	move decimal 4 places to left
Example 100 ppm = 0.01%		

Determine The Number of Rows To The Acre

	Length of Rows				
Rows/Acre	32″	36″	38″	40″	60″
1	16,355	14,520	13,756	13,068	8,712
2	8,168	7,260	6,878	6,534	4,356
3	5,445	4,840	4,585	4,356	2,904
4	4,084	3,630	3,439	3,267	2,178
5	3,267	2,904	2,751	2,614	1,742
6	2,723	2,420	2,293	2,178	1,452
7	2,334	2,074	1,965	1,867	1,245
8	2,042	1,815	1,719	1,634	1,089
9	1,815	1,613	1,528	1,452	968
10	1,634	1,452	1,376	1,307	871
11	1,485	1,320	1,251	1,188	792
12	1,361	1,210	1,156	1,089	726
13	1,257	1,117	1,058*	1,005	670
14	1,167	1,037	982	933	622
15	1,089	968	917	871	581
16	1,021	908	860	817	545
17	961	854	809	769	512
18	908	807	764	726	484
19	860	764	724	688	459
20	817	726	688	653	436
21	778	691	655	622	415
22	743	660	625	594	396
23	710	631	598	568	379
24	681	605	573	544	363
25	653	581	550	523	348
26	628	558	529	503	335
27	605	538	509	484	323
28	583	519	491	467	311
29	563	501	474	450	300
30	545	484	459	436	290

***Example:** A growers field is 1,058 feet long furrowed out on 38-inch centers. Therefore, every 13 rows across the field represents an acre.

Grams/Gallon Table

Gallons / PPM	5	10	15	20	25	50	75	100	150	200	300	400
5	0.1	0.2	0.3	0.4	0.5	1.0	1.4	1.9	2.8	3.8	5.7	7.6
10	0.2	0.4	0.6	0.8	1.0	1.9	2.8	3.8	5.7	7.6	11.0	15.0
15	0.3	0.6	0.9	1.1	1.4	2.8	4.3	5.7	8.5	11.0	17.0	23.0
20	0.4	0.8	1.1	1.5	1.9	3.8	5.7	7.6	11.0	15.0	23.0	30.0
25	0.5	0.9	1.4	1.9	2.4	4.7	7.1	9.5	14.0	19.0	28.0	38.0
50	0.9	1.9	2.8	3.8	4.7	9.5	14.0	19.0	28.0	38.0	57.0	76.0
75	1.4	2.8	4.3	5.7	7.1	14.0	21.0	28.0	43.0	57.0	85.0	114.0
100	1.9	3.8	5.7	7.6	9.5	19.0	28.0	38.0	57.0	76.0	114.0	151.0
125	2.4	4.7	7.1	9.5	12.0	24.0	36.0	47.0	71.0	95.0	142.0	189.0
150	2.8	5.7	8.5	11.0	14.0	28.0	43.0	57.0	85.0	114.0	170.0	227.0
175	3.3	6.6	9.9	13.0	17.0	33.0	50.0	66.0	99.0	133.0	199.0	265.0
200	3.8	7.6	11.0	15.0	19.0	38.0	57.0	76.0	114.0	151.0	227.0	303.0
250	4.7	9.5	14.0	19.0	24.0	47.0	71.0	95.0	142.0	189.0	284.0	379.0
300	5.7	11.0	17.0	23.0	28.0	57.0	85.0	114.0	170.0	227.0	341.0	454.0
400	7.6	15.0	23.0	30.0	38.0	76.0	114.0	151.0	227.0	303.0	454.0	606.0

Standard Measurements

Measure of Length (Linear Measure)

4 inches = 1 hand
9 inches = 1 span
12 inches = 1 foot
3 feet = 1 yard
6 feet = 1 fathom
5½ yards - 16½ feet = 1 rod
40 poles = 1 furlong
8 furlongs = 1 mile
5,280 feet = 1,760 yards = 320 rods = 1 mile
3 miles = 1 league

Measure of Surface (area)

144 square inches = 1 square foot
9 square feet = 1 square yard
30¼ square yards = 1 square rod
160 square rods = 1 acre
43,560 square feet = 1 acre
640 square acres = 1 square mile
36 square miles = 1 township

Surveyor's Measure

7.92 miles = 1 link
25 links = 1 rod
4 rods = 1 chain
10 square chains = 160 square rods = 1 acre
640 acres = 1 square mile
80 chains = 1 mile
1 Gunthers chain = 66 feet

Metric Length

1 inch = 2.54 centimeters
1 foot = .305 meter
1 yard = .914 meter
1 mile = 1.609 kilometers
1 fathom = 6 feet
1 knot = 6,086 feet
3 knots = 1 league
1 centimeter = .394 inch
1 meter = 3.281 feet
1 meter = 1.094 yards
1 kilometer = .621 mile

Troy Weight

24 grains = 1 pennyweight
20 pennyweight = 1 ounce
12 ounces = 1 pound

Liquid Measure

2 cups = 1 pint
4 gills = 1 pint
16 fluid ounces = 1 pint
2 pints = 1 quart
4 quarts = 1 gallon
31½ gallons = 1 barrel
2 barrels = 1 hogshead
1 gallon = 231 cubic inches
1 cubic foot = 7.48 gallons
1 teaspoon = .17 fluid ounces (1/6 oz.)
3 teaspoons (level) = 1 tablespoon (½ oz.)
2 tablespoons = 1 fluid ounce
1 cup (liquid) = 16 tablespoons (8 oz.)
1 teaspoon = 5 to 6 cubic centimeters
1 tablespoon = 15 to 16 cubic centimeters
1 fluid ounce = 29.57 cubic centimeters

Cubic Measure (Volume)

1,728 cubic inches = 1 cubic foot
27 cubic feet = 1 cubic yard
2,150.42 cubic inches = 1 standard bushel
231 cubic inches = 1 standard gallon (liquid)
1 cubic foot = 4/5 of a bushel
128 cubic feet = 1 cord (wood)
7.48 gallons = 1 cubic foot
1 bushel = 1.25 cubic feet

Dry Measure

2 pints = 1 quart
8 quarts = 1 peck
4 pecks = 1 bushel
36 bushels = 1 chaldron

Apothecaries' Weight

20 grains = 1 scruple
3 scruples = 1 dram
8 drams = 1 ounce
12 ounces = 1 pound
27 11/32 grains = 1 dram
16 drams = 1 ounce
16 ounces = 1 pound
2,000 pounds = 1 ton (short)
2,240 pounds = 1 ton (long)

Conversion Factors

Degree C. = 5/9 (Degree F. − 32).
Degree F. = 9/5 (Degree C. + 32).
Degrees absolute (Kelvin) = Degrees centigrade + 273.16.
Degrees absolute (rankine) = Degrees fahrenheit + 459.69.

Multiply	*By*	*To Obtain*
Diameter circle	3.1416	Circumference circle
Diameter circle	0.8862	Side of equal square
Diameter circle squared	0.7854	Area of circle
Diameter sphere squared	3.1416	Area of sphere
Diameter sphere cubed	0.5236	Volume of sphere
U.S. Gallons	0.8327	Imperial gallons (British)
U.S. Gallons	0.1337	Cubic feet
U.S. Gallons	8.330	Pounds of water (20° C.)
Cubic feet	62.427	Pounds of water (4°C.)
Feet of water (4° C.)	0.4335	Pounds per sq. inch
Inch of mercury (0° C.)	0.4912	Pounds per sq. inch
Knots	1.1516	Miles per hour

Figuring Grain Storage Capacity

1 bu. ear corn = 70 lbs. 2.5 cu. ft. (15.5% moisture)
1 bu. shelled corn = 56 lbs. 1.25 cu. ft. (15.5% moisture)
1 cu. ft. = 1/2.50 = .4 bu. of ear corn
1 cu. ft. = 1/1.25 = .8 bu. of shelled corn;

Bu. × 1.25 $ft.^3$, $Ft.^3$ × .8 bu.

$Ft.^3$ = Bu. × 1.25

Bu. = $Ft.^3$ × .8

Rectangular or square cribs or bins
cu. ft. = width × height × length (W × H × L)
Round cribs, bins or silos (= 3.1416)

Volume = $R^2H = D^2H/r$
cu. ft. = × radius × radius × height = (R × R × H)

$$\text{or } \frac{\times \text{ diameter} \times \text{diameter} \times \text{height} = (D \times D \times H)}{4}$$

$$\text{or } \frac{3.1416 \times D \times D \times H}{4} = .785 \times D \times D \times H$$

Examples:

1. Crib — ear corn — 6′ wide by 12′ high by 40′ long
   a. 6 × 12 × 40 = 2880 cu. ft. × 4 bu./cu. ft. = 1152 bu.
   b. 6 × 12 × 1 = 72 cu. ft. × .4 28.8 bu/ft. of length × 40′ = 1152 bu.
2. Round crib — ear corn — 14′ diameter by 14′ high
   a. .785 × 14′ × 14′ × 14′ × .8 = 1722 bushel
   b. .785 × 14′ × 1 × .4 6.15 bu/ft. × 14 = 861 bushel
3. Round Bin or Silo — shell corn — 14′ diameter by 14′ high
   a. .785 × 14′ × 14′ × 14′ × .8 = 1722 bushel
   b. .785 × 14′ × 14′ × 14′ × 1 × .8 – 123 bu/ft. × 14′ = 1722 bushel

Metric Equivalents

Metric Weight

1 grain	=	.065 gram
1 apothecaries' scruple	=	1.296 grams
1 avoirdupois ounce	=	28.350 grams
1 troy ounce	=	31.103 grams
1 avoirdupois pound	=	.454 kilogram
1 troy pound	=	.373 kilogram
1 gram	=	15.432 grains
1 gram	=	.772 apothecaries' scruple
1 gram	=	.035 avoirdupois ounce
1 gram	=	.032 troy ounce
1 kilogram	=	2.205 avoirdupois pounds
1 kilogram	=	2.679 troy pounds

Capacity

1 U.S. fluid ounce	=	29,573 milliliters
1 U.S. liquid quart	=	.946 liter
1 U.S. dry quart	=	1.101 liters
1 U.S. gallon	=	3,785 liters
1 U.S. bushel	=	.3524 hectoliters
1 cubic inch	=	16.4 cubic centimeters
1 liter	=	1,000 milliliters or 1,000 cubic centimeters
1 cubic foot water	=	7.48 gallons or 62½ pounds
231 cubic inches	=	1 gallon
1 millimeter	=	.034 U.S. fluid ounce
1 liter	=	1.057 U.S. liquid quarts
1 liter	=	.908 U.S. dry quart
1 liter	=	.264 U.S. gallon
1 hectoliter	=	2.838 U.S. bushels
1 cubic centimeter	=	.061 cubic inch

Miscellaneous Equivalents

9 in. equal 1 span.
6 ft. equal 1 fathom.
6,080 ft. equal 1 nautical mile.
1 board ft. equals 144 cu. in.
1 cylindrical ft. contains 5⅞ gals.
1 cu. ft. equals .8 bushel.
12 dozen (doz.) equal 1 gross (gr.).
1 gal. water weighs about 8⅓ lbs.
1 gal. milk weighs about 8.6 lbs.
1 gal. cream weighs about 8.4 lbs.
46½ qts. of milk weigh 100 lbs.
1 cu. ft. water weighs 62½ lbs., contains 7½ gals.
1 gal. kerosene weighs about 6½ lbs.
1 bbl. cement contains 3.8 cu. ft.
1 bbl. oil contains 42 gals.
1 standard bale cotton weighs 480 lbs.
1 keg of nails weighs 100 lbs.
4 in. equal 1 hand in measuring horses.

Useful Measurements

LENGTH
1 mile = 80 chains = 8 furlongs = 1,760 yards = 5,280 ft. = 1.6 kilometres

1 chain = 22 yards = 4 rods, poles or perches = 100 links

WEIGHT
1 long ton = 20 cwt. = 2,240 lbs.
1 lb. = 16 oz. = 454 grams = 0.454 kilograms
1 short ton = 2,000 lbs.
1 metric ton = 2,204 lbs. = 1,000 kilograms

AREA
1 acre = 10 sq. chains = 4,840 sq. yards = 43,560 sq. ft. = 0.405 hectares
1 sq. mile = 640 acres = 2.59 kilometres
1 hectare = 2.471 acres

VOLUME
1 gal. = 4 quarts = 8 pints = 160 fluid ozs. = 1.2 U.S. gallons = 4.546 litres
1 fluid oz. = 2 tablespoons = 4 dessertspoons = 8 teaspoons = 28.4 c.c.'s
1 litre = 1,000 c.c.'s = 0.22 gallons = 1.76 pints

CAPACITIES
Cylinder-diameter^2/x depth × 0.785 = cubic feet
Rectangle-breadth × depth × length = cubic feet
Cubic Feet × 6.25 = gallons

QUICK CONVERSIONS

1 pint/acre	= 1 fluid oz./242 sq. yards
1 gal./acre	= 1 pint/605 sq. yards
1 lb./acre	= 1 oz./300 sq. yards
1 cwt./acre	= 0.37 oz./sq. yard
1 m.p.h.	= 88 ft./minute
3 m.p.h.	= 1 chain/15 sec.
1 litre/hectare	= 0.089 gal./acre
1 Kilogram/hectare	= 0.892 lb./acre
1 c.c/100 litres	= 0.16 fl. oz./100 gallons
125 c.c./100 litres	= 1 pint/100 gallons
1 gm./100 litres	= 0.16 oz./100 gallons

A strip 3 ft. wide × 220 chains
A strip 4 ft. wide × 165 chains } 1 acre
A strip 5 ft. wide × 132 chains

CAPACITIES OF ROUND BINS AND WIRE CRIBS

Diameter	Height	Cu. ft.	Capacities: Small grain	Capacities: Ear corn
feet	feet		bushels	bushels
8	6	301	241	120
8	8	400	320	160
8	10	502	400	200
10	8	628	502	251
10	10	785	628	314
10	12	942	753	376
12	10	1130	904	452
12	12	1356	1084	542
12	14	1582	1265	632
14	12	1846	1476	738
14	14	2154	1723	861
14	16	2462	1969	984
16	14	2814	2251	1125
16	16	3216	2572	1286
16	18	3618	2894	1447
16	20	4029	3216	1608

To find the capacity of a circular grain bin in bushels, multiply the radius ($\frac{1}{2}$ diameter) times the radius, times 3.1416, times height, times 0.8 for small grain and shelled corn.

For a circular crib, use 0.4 in place of 0.8 to find bushels of ear corn. (All measurements must be in feet.)

CAPACITIES OF SQUARE GRAIN BINS

Size	Capacity per foot of depth	Capacity of bin 10′ deep
feet	bushels	bushels
5×5	20	200
6×6	28.8	288
8×8	51.2	512
10×10	80	800

To find the capacity of a square or rectangular grain bin in bushels, multiply length times width, times height, times 0.8 (All measurements must be in feet.)

Mich. State Univ. Coop. Ext. Service

CAPACITIES OF ROUND SILOS IN TONS

Depth of settled silage	Weight silage per cu. ft. at this depth	Average weight of silage per cu. ft. to this depth	Total quantity of settled silage, from the top to the depth indicated in silos having a diameter of— 10 feet	12 feet	14 feet	16 feet	18 feet	20 feet
Ft.	Lbs.	Lbs.	Tons	Tons	Tons	Tons	Tons	Tons
2	29.2	23.5	2	3	4	5	6	7
4	37.2	29.5	5	7	9	12	15	18
6	41.8	33.3	8	11	15	20	25	31
8	45.0	36.0	11	16	22	29	37	45
10	47.1	38.1	15	22	29	38	49	60
12	48.7	39.8	19	27	37	48	61	75
14	49.9	41.2	23	33	44	58	73	90
16	51.0	42.4	27	38	52	68	86	107
18	51.8	43.5	31	44	60	79	100	123
20	52.4	44.3	35	50	68	89	113	139
22	53.0	45.1	39	56	76	100	126	156
24	53.5	45.8	43	62	85	110	140	173
26	53.9	46.4	47	68	92	121	154	189
28	54.2	46.9	52	74	101	132	167	206
30	54.4	47.4	56	80	110	143	181	224
32	54.6	47.9	60	87	118	154	195	241
34	54.7	48.3	65	93	126	165	209	258
36	—	—	—	—	135	176	223	275
38	—	—	—	—	143	187	237	292
40	—	—	—	—	152	198	251	310
42	—	—	—	—	—	209	265	327
44	—	—	—	—	—	220	279	344
46	—	—	—	—	—	—	293	361
48	—	—	—	—	—	—	307	378
50	—	—	—	—	—	—	320	396
52	—	—	—	—	—	—	335	413
54	—	—	—	—	—	—	349	430
56	—	—	—	—	—	—	363	448
58	—	—	—	—	—	—	377	466
60	—	—	—	—	—	—	392	484

Silage table and material on settled silage content of silos taken from USDA circular 603. The weight of a cubic foot of silage is so variable that tables or rules for estimating may be very inaccurate. The following rule will give approximate results: Diameter squared × .7854 × depth of settled silage will give number of cubic feet of silage. The number of cubic feet times the weight per cubic foot, divided by 2,000 will give tons of silage.

TO DETERMINE TRACTOR FIELD SPEED IN MILES PER HOUR

1. Stake off a distance of 88 feet (1/60 of a mile) in the field to be sprayed.
2. Select tractor gear and throttle setting to be used.
3. Clock the seconds required to drive 88 feet from a running start.
4. Divide the time in seconds into 60; the answer is miles per hour.

CONVERTING BROADCAST RATE TO BAND RATE

Unless specified otherwise, directions for the application of preemergence herbicides are given **for broadcast coverage.** If the herbicide is to be **band-applied,** the actual amount of solution to use per acre is calculated by the following formula:

$$\frac{\text{band width (inches)}}{\text{row spacing (inches)}} \times \text{broadcast rate in gallons per acre}$$

Example: $\frac{12 \text{ in.}}{40 \text{ in.}} \times$ 20 gal./A (broadcast) = 6 gal./A (banded)

AIRCRAFT CALIBRATION

Acres covered per minute

Speed mph	Swath width in feet									
	30	35	40	45	50	75	100	200	300	500
75	4.5	5.2	6.0	6.7	7.5	11.2	15.0	30.0	45.0	75.0
80	4.8	5.6	6.4	7.2	8.0	12.0	16.0	32.0	48.0	80.0
85	5.1	5.9	6.8	7.6	8.5	12.7	17.0	34.0	51.0	85.0
90	5.4	6.3	7.2	8.1	9.0	13.5	18.0	36.0	54.0	90.0
95	5.7	6.6	7.6	8.5	9.5	14.2	19.0	38.0	57.0	95.0
100	6.0	7.0	8.0	9.0	10.0	15.0	20.0	40.0	60.0	100.0
110	6.6	7.7	8.8	9.9	11.0	16.5	22.0	44.0	66.0	110.0
120	7.2	8.4	9.6	10.8	12.0	18.0	24.0	48.0	72.0	120.0
130	7.8	9.1	10.4	11.7	13.0	19.5	26.0	52.0	78.0	130.0
140	8.4	9.8	11.2	12.6	14.0	21.0	28.0	56.0	84.0	140.0
150	9.0	10.5	12.0	13.5	15.0	22.5	30.0	60.0	90.0	150.0

The rate of application in gallons or pounds per minute is calculated by multiplying the acres per minute by the number of gallons or pounds per acre to be applied. Take for example a 100-mile-per-hour aircraft that has a 50 foot effective swath. The chart indicates that the plane has a coverage of 10.0 acres per minute. If spray is to be applied at a rate of 2 gallons per acre, the unit should be calibrated to dispense 20 gallons per minute ($2 \times 10 = 20$). If 7.5 pounds of dry material is to be applied per acre, the unit should be calibrated to dispense 75 pounds per minute ($10 \times 7.5 = 75$).

The basic formula for calculating acres per minute is:

$$\frac{\text{swath width} = 2 = \text{miles per hour}}{1{,}000}$$

COMPUTING PLANT POPULATIONS AND YIELDS

Row Width	Length of Row in 1/1000 Acre
40 inches	**13-ft., 1-in.**
38 inches	**13-ft., 9-in.**
36 inches	**14-ft., 6 in.**
30 inches	**17-ft., 5-in.**
20 inches	**26-ft., 2-in.**
15 inches	**34-ft., 10-in.**

Determining Plant Population*

1. Measure off 1/1000 acre row.
2. Count number of plants; then multiply by 1000 to determine plants per acre.

Determining Yields*

1. Harvest grain from 1/1000 acre row.
2. Weigh harvested grain, then multiply by 1000 to get pounds of grain per acre.
3. Determine moisture content, then divide pounds per acre by pounds per bushel for adjusted moisture.

* Average three or more 1/1000 acre rows for more accurate estimate of field populations and yields.

NOZZLE OUTPUTS (20-INCH NOZZLE SPACING*)

Output	Ground speed, miles per hour					
	3	4	5	6	7	8
5 gallons per acre						
Gallons per minute	.05	.067	.084	.101	.118	.135
Ounces per minute	6.4	8.6	10.8	12.9	15.1	17.2
6 gallons per acre						
Gallons per minute	.06	.081	.101	.121	.141	.162
Ounces per minute	7.8	10.3	12.9	15.5	18.1	20.6
7 gallons per acre						
Gallons per minute	.071	.094	.118	.141	.165	.189
Ounces per minute	9.0	12.1	15.1	18.1	21.1	24.2
8 gallons per acre						
Gallons per minute	.081	.108	.135	.162	.189	.216
Ounces per minute	10.3	13.8	17.2	20.7	24.1	27.6
9 gallons per acre						
Gallons per minute	.091	.121	.152	.182	.212	.242
Ounces per minute	11.6	15.5	19.4	23.3	27.1	31.0
10 gallons per acre						
Gallons per minute	.101	.135	.168	.202	.236	.269
Ounces per minute	12.9	17.2	21.5	25.9	30.1	34.4
15 gallons per acre						
Gallons per minute	.152	.202	.253	.303	.354	.404
Ounces per minute	19.4	25.9	32.3	38.8	45.2	51.8
20 gallons per acre						
Gallons per minute	.202	.269	.337	.404	.472	.538
Ounces per minute	25.9	34.5	43.1	51.8	60.2	69.0

* Data also applies if every other outlet is plugged and two nozzles are used per drop, as is often the case in post-emergence herbicide application.
The table may be used to:

1. Determine gallonage per acre at given ground speed and nozzle output.
2. Select proper ground speed when operating at a given nozzle pressure and output.
3. Select new nozzles when range of gallons per acre and ground speed is known.

PLANT POPULATION AT ROW WIDTHS USED FOR SOYBEANS

Row Width (inches)	Plants per foot of row					
	12	9	8	6	4	3
7	896,088	672,066	597,392	448,044	298,696	224,022
10	627,264	470,448	418,176	313,632	209,088	156,816
14	448,044	336,033	298,696	224,022	149,348	112,011
20	313,632	235,224	209,088	156,816	104,544	78,408
21	298,692	224,019	199,128	149,346	99,564	74,673
24	261,360	196,020	174,240	130,680	87,120	65,340
28	224,028	168,021	149,352	112,014	74,676	56,007
30	209,088	156,816	139,392	104,544	69,696	52,272
36	174,240	130,680	116,160	87,120	58,080	43,560
40	156,816	117,612	104,544	78,408	52,272	39,204

PLANTING CAPACITY IN ACRES PER 10-HOUR DAY FOR VARIOUS PLANTERS

Planter	Speed and field efficiency											
	3 mph				4 mph				5 mph			
	50%	60%	70%	80%	50%	60%	70%	80%	50%	60%	70%	80%
	acres per 10 hours				acres per 10 hours				acres per 10 hours			
2-row 40 inches	12	15	17	19	16	19	23	26	20	24	28	32
4-row 20 inches	12	15	17	19	16	19	23	26	20	24	28	32
4-row 28 inches	17	20	24	27	23	27	32	36	28	34	40	45
4-row 30 inches	18	22	25	29	24	29	34	39	30	36	42	49
4-row 36 inches	22	26	31	35	29	35	41	47	36	44	51	58
4-row 38 inches	23	28	32	37	31	37	43	49	38	46	54	62
4-row 40 inches	24	29	34	39	32	39	45	52	40	48	57	65
8-row 20 inches	24	29	34	39	32	39	45	52	40	48	57	65
6-row 28 inches	25	31	36	41	34	41	48	54	42	51	59	68
6-row 30 inches	27	33	38	44	36	44	51	58	45	55	64	73
6-row 36 inches	32	39	46	52	43	52	61	70	54	66	77	88
8-row 28 inches	34	41	48	54	45	54	63	73	57	68	79	91
6-row 40 inches	36	44	51	58	48	58	68	78	61	73	85	97
8-row 30 inches	36	44	51	58	48	58	68	78	61	73	85	97
12-row 20 inches	36	44	51	58	48	58	68	78	61	73	85	97

Field efficiency indicates how much field time is spent planting. The greater the time spent in turning, filling seed hoppers, making minor adjustments and performing other jobs, the lower the field efficiency. Source: Purdue University

GENERAL INFORMATION ON HANDLING AND STORAGE OF PESTICIDES

Introduction

All agricultural chemicals should be handled and stored with care.

This publication suggests the proper and safe handling of agricultural chemicals in warehousing and in transportation.

Extreme care is required when handling many types of agricultural chemicals. Some are highly toxic. Not all agricultural chemicals have the same degree of hazard. Some products, if inhaled or allowed to remain on the skin for a short period of time, can cause serious illness or possible death. Contaminated food, clothing or any other material likely to come into intimate contact with humans, or contaminated animal feed could cause damage to either humans or animals who are far removed from the place of contamination. *ALL AGRICULTURAL CHEMICALS SHOULD BE TREATED WITH CAUTION.*

The need for careful handling and inspection of all empty and full agricultural chemical containers cannot be overemphasized. Damaged or leaking containers can create a serious hazard if not properly handled. The main cause of ruptured containers is poor handling or abuse. When handling containers always be alert to detect possible leakers so you can protect yourself and others, and avoid the contamination of other goods.

Manufacturers of agricultural chemicals have technical staff skilled in dealing with emergencies. Expert advice and on-site assistance is available by calling CHEMTREC, 1-800-424-9300. Many manufacturers include an emergency number on their bill of lading and on shipping container labels. *CALL THE SHIPPER'S NUMBER FIRST.*

Agricultural chemical workers have three main responsibilities:

1. Prevent emergencies by careful handling of agricultural chemicals.
2. Check for spillage—*immediately* bring it to the supervisor's attention.
3. Know the proper first aid and emergency steps to take until expert assistance is obtained.

Personnel Precautions

1. Do not store food, feedstuffs, beverages, tobacco, eating utensils, or smoking equipment in the work area, or where agricultural chemicals are stored.
2. Do not drink, eat, smoke, or use tobacco in the work area or where agricultural chemicals are stored.
3. Wear rubber or neoprene gloves or other gloves directed by the product label while handling any agricultural chemical, even if the container seems tightly closed.
4. Do not put your fingers to your ears, mouth, or other parts of body, or rub your eyes while working with or around agricultural chemicals.
5. After handling agricultural chemicals, always wash your hands and face before eating, drinking, smoking, or using the toilet.
6. When cleaning up spills, wear appropriate eye and respiratory protection and impervious footwear (rubber or other footwear as directed by label). Depending upon the magnitude of the spill, follow label directions for other protective equipment, such as coveralls, rainsuits, etc. Leather footwear or leather outer clothing should *not* be worn as they do not provide adequate protection and cannot be decontaminated. Ensure that capabilities and limitations of respiratory protection are known prior to cleanup.
7. Avoid breathing vapors from leaking containers, and exercise caution to avoid fire in the area.
8. Never place agricultural chemicals in an unmarked container, particularly a new container that originally contained a beverage or food.

Containers

Many types and varieties of containers are used for the packaging and shipment of agricultural chemicals. Containers are designed to safely and effectively contain the products. Containers cannot always withstand abusive treatment, such as protruding nails, top loading or sharp objects, and puncture by forklift or other mechanical equipment. *HANDLE WITH CARE.*

Health Hazards

Routes of Exposure

Even a small amount of certain agricultural chemicals taken into the body can cause serious illness—or death. Harmful amounts can enter the body by (1) absorption through the skin, (2) inhalation, or (3) ingestion. So, after handling an agricultural chemical, make it a standard practice to thoroughly wash hands and face with soap and water. Always shower at end of shift. If accidentally contaminated, remove any contaminated clothing and shower immediately.

This product can kill you if touched by hands or spilled or splashed on skin, in eyes or on clothing (liquid goes through clothes).

This product can kill you if vapors, spray mist or dust are breathed.

This product can kill you if swallowed even in small amounts; spray mist or dust may be fatal if swallowed.

First Aid in Case of Contact or Exposure

The product labels on all agricultural chemical products classified as Poison B contain first aid statements which clearly outline the procedure to be followed in case of contact or exposure. Statements on the labels are specific for the products in that container. READ THE LABEL. Medical doctors should call emergency numbers for additional information.

Even slightly contaminated clothing should be laundered separately, twice with great care *before being reused.* Use a very strong detergent and liquid chlorine bleach; rinse twice and dry completely. Clothing and shoes which have been badly contaminated should be destroyed by burning or buried by qualified personnel. Rubber footwear is preferable, since leather absorbs organic materials readily and canot be decontaminated.

Labels

The entire product label should always be read and totally understood before handling the product. Follow all precautions! By law, agricultural chemical containers are required to bear the label registered with the U.S. Environmental Protection Agency. Outer shipping containers for products meeting the criteria of the U.S. Department of Transportation (DOT) for a Poison B, in either a liquid or solid form, must display the DOT poison label (Appendix I). Outer shipping containers for other hazardous materials must also be labeled according to their respective classification; for example, flammable liquids should carry a DOT flammable liquid label (Appendix I). Certain agricultural chemicals may require more than one DOT label.

A look at the product label will show that extreme care has been taken to correctly describe the product, the hazards involved, methods to be used in handling the product, and what to do in case of poisoning. The product labels of highly toxic products carry the word "POISON" and the "skull and crossbones" symbol. Generally, there will be a list of warning statements headed by the word "DANGER." These statements describe the primary hazards, i.e. highly toxic or irritating or corrosive to eyes, skin, nose, and throat, and what should be done to avoid them. In addition, there is often information for the physician to assist him in treating a case of poisoning, should one occur. The labels on all POISON and other toxic category products are informative, and should contain all the basic information needed to handle the product safely. The chemical name and an ingredient statement are always shown on the label.

Signal Words for Labeling

The EPA toxicity rating system for agricultural chemical labeling requires the following signal words to appear on labels:

Signal Words	Toxicity of Material	Signal Words/ Symbol
Danger	Highly Toxic	Danger-Poison Skull & Crossbones
Warning	Moderately Toxic	Warning
Caution	Slightly Toxic	Caution

ALWAYS READ THE LABEL

HANDLING SHIPMENTS

Agricultural Chemicals Classified by DOT as Hazardous Materials

Department of Transportation's Hazardous Materials Regulations require that those agricultural chemicals which are defined by DOT as hazardous materials, when offered for transportation, be identified on shipping papers by their proper hazardous materials shipping name, hazard class, appropriate four-digit identification number preceded by the designation 'UN' or 'NA', as applicable, and the total quantity. Additional information may also be required to be included on shipping papers. DOT regulations describe the exact manner in which the shipping papers are to be completed. The Federal regulations also prescribe container marking requirements and specific specifications for the containers in which hazardous products must be packaged and shipped. Materials labeled as Poison B must never be carried in the same vehicle with foodstuffs, food ingredients, feeds or any other edible material intended for consumption by humans or animals. In general, when tendering packaged hazardous materials in amounts of or greater than 1,000 pounds gross weight per vehicle to a motor carrier, the shipper must provide appropriate placards. Placarding of vehicles carrying less than 1,000 pounds gross weight is permissible, but is not required. For specific information, consult the DOT Hazardous Materials Regulations.

During transportation of agricultural chemicals, stick to the following rules:

1. **STOP ALL WORK IF A LEAK OR SPILL IS FOUND IN A SHIPMENT.**
2. **CLOSE AND LOCK THE VEHICLE TO ISOLATE THE SHIPMENT.**
3. **NOTIFY THE SHIPPER, OWNER, OR CONSIGNOR.**
4. **POST THE IMMEDIATE AREA AND GUARD SO THAT NO UNAUTHORIZED PERSON CAN COME IN CONTACT WITH THE SHIPMENT.**

A suspected contaminated vehicle should not be used for any purpose until it has been released by the shipper.

If leaking containers are discovered after receipt, read the label (avoid contact with the material), and always consult shipper for necessary procedures before cleaning up a spill as follows:

1. Care should be taken to avoid direct contact with the material.
2. Damaged containers should be removed from the vehicle and segregated from the undamaged containers.
3. Damaged containers should be placed in suitable overpack containers, and placed in a secure location.
4. Residue material must be thoroughly cleaned from the vehicle, using the approved clean-up method.
5. Shipper or consignee should be notified of leakage.

Occasionally, a portion of the containers in a shipment of agricultural chemicals may be damaged during transit. **IN SUCH CASES, PLACE THE MATERIAL IN AN ISOLATED AREA, AND INFORM THE SHIPPER OR OWNER OF THE DAMAGE.**

Storage

Agricultural chemicals should be stored in a dry, well-ventilated, separate room, building, or covered area dedicated exclusively to agricultural chemical storage, and where fire protection is provided. The following precautions should be taken:

1. When possible, the entire storage facility should be secured by climb-proof fences. Doors and gates should be kept locked to prevent unauthorized entry.
2. Identification signs may be placed on rooms, buildings, and fences to advise of the contents and warn of their hazardous nature.
3. All items of movable equipment used for handling agricultural

chemicals at the storage site should be dedicated for this purpose. If it becomes necessary that they be used for other purposes, they should be thoroughly cleaned and decontaminated before they leave the site.

4. Pallets used for agricultural chemicals should not be used for other products.
5. Where possible, agricultural chemicals should not be stored on the upper floors of buildings. Neither should they be stored in basements, unless proper fire protection is provided. Ground floor storage is preferable.
6. A separate, *secure* storage area should be designated for leaking or damaged containers.

Disposal

Before disposing of any agricultural chemical wastes, consult the requirements of the Resource Conservation and Recovery Act (RCRA), as published by the U.S. Environmental Protection Agency, to determine if the waste is defined as a "hazardous" waste. If so, its disposal is subject to stringent requirements under this law (RCRA).

Helpful Operating Hints

Agricultural chemical containers should be stored with the label plainly visible from operating aisles. If containers are not in good condition when received, they should be placed in a suitable overpack container and properly relabeled after consulting the shipper. If granular agricultural chemicals are received in bags that are damaged, the bag and the contents should be placed in suitable containers and properly labeled. (Fibre-packs or transparent heavy duty plastic bags are ideal.) Metal or rigid plastic containers should be inspected carefully to make sure that the lids and bungs are tight.

Classification and Separation

Each agricultural chemical formulation should be stored separately. Rigid containers should be stored in an upright position. All containers should be stored off the ground in an orderly way, so as to permit ready access and inspection. They should be placed in rows or stacking position with labels visible, and with lanes to provide easy access. A complete inventory should be maintained, indicating the number and identity of containers in each storage position row.

Container Inspection and Maintenance

Containers should be checked regularly for corrosion and leaks. Leaking containers should be isolated, the leak controlled if possible (by inverting container or placing in suitable overpack), and identified. The manufacturer should be consulted for further instructions. Materials such as adsorptive clay, hydrated lime, and sodium hypochlorite (household bleach) should be kept on hand for treatment of spills or leaks. Soda ash and liquid detergent are useful in neutralizing organic phosphate liquids.

Safety Precautions

Rules for personal safety and accident prevention similar to those listed below should be posted in areas where workers congregate. Examples are:

1. Inspect all containers of agricultural chemicals for leaks before handling them. Always wear impervious gloves. Wear personal protective equipment when handling leakers.
2. Handle containers carefully to prevent emergencies.
3. Permit only authorized persons in the storage area.
4. Store agricultural chemicals separately from materials used or consumed by humans or animals.
5. Clean up residue before and after loading or unloading.
6. Inspect all vehicles for placarding and road worthiness.
7. Never ship damaged containers.

Fire control

Where large amounts of agricultural chemicals are stored or where conditions may otherwise warrant, the owner of stored agricultural chemicals should inform the local fire department, in writing, of the hazards that such agricultural chemicals may present in the event of a fire. A floor plan of the storage area, indicating where different agricultural chemical classifications are regularly stored, should be provided. The fire chief should be furnished with the home telephone numbers of those who are responsible for the agricultural chemical storage facility. Annual or semi-annual inspections by the fire department should be encouraged. Pre-fire plans should be formulated and reviewed annually, at least. A listing of agricultural chemicals inventory should be maintained in an area separate from the area of storage, which could be involved in a fire, and easily accessible for emergency purposes.

ALWAYS READ THE LABEL

NACA—Pesticide Safety Team Network

The Pesticide Safety Team Network (PSTN) of the National Agricultural Chemicals Association (NACA) represents a joint effort of technically qualified manufacturers to respond to emergency situations where the involuntary release of agricultural chemicals has occurred. The PSTN can be reached by telephone 24 hours a day through CHEMTREC at 1-800-424-9300.

Suggestions for Fire Hazard Abatement

Where applicable, plainly label the outside of each storage area with "DANGER-POISON" "PESTICIDE STORAGE." Consult with the local fire department regarding the use of the current hazard signal system of the National Fire Protection Association.

In case of fire, make certain that all persons are evacuated immediately to an upwind location. Avoid exposure to smoke and possible toxic fumes. Provide the fire department with as accurate a list as possible of the types and amounts of stored agricultural chemicals. This will provide guidance to determine the best method of attack, and alert firemen to the specific hazards which the stored material(s) may present.

Suggestions for fire fighting procedures and precautions:

1. Wear air-supplied breathing apparatus and rubber clothing.
2. Avoid breathing or otherwise contacting toxic smoke and fumes.
3. Remove clothing and bathe completely as soon as possible after encountering smoke and fumes. Clothing should be properly laundered or destroyed.
4. Contain the water used in fire fighting by a diking system, if possible, to catch all drainage from the area.
5. Evacuate persons who may come in contact with smoke or fumes or contaminated surfaces.

BASIC ELEMENTS OF SAFE PESTICIDE USE

1. Always read the label with care. This is the first step in selecting the right material for the job. Never rely on your memory. Before opening the container, pay strict attention to warnings and cautions printed on the label.
2. Keep spray and dust materials out of the reach of children, pets and irresponsible persons. Storage outside of the home, away from food and feed, and under lock and key is the safest method.
3. Store only in the original container and keep tightly closed.
4. NEVER smoke, eat or drink while applying pesticides.
5. Avoid inhalation or direct contact. Always use personal protective equipment prescribed by the label.
6. Avoid spills. If spills occur, take *immediate action* to remove contaminated clothing and wash thoroughly.
7. After each application, bathe and change to clean clothing. Wash clothing after each use. Always use fresh clothing when starting new application.
8. Avoid contamination of fish ponds and water supplies. Cover feed and water containers when treating around livestock or pet areas.
9. Keep separate equipment for use with hormone-type herbicides to avoid accidental injury to susceptible plants. Also avoid applications under wind conditions which could create drift to non-target areas.
10. Rinse empty containers 3 times before disposing of them. Add the rinse to the spray tank and dispose of containers according to local regulations to avoid hazards to humans, animals and the environment.
11. Follow label directions for mixing and application to keep residues within the limits prescribed by law.
12. Plan ahead. Discuss with your physician the materials you will be using during the season so that he can be prepared to provide the appropriate treatment in case of accidental exposure. If symptoms of illness occur, call the physician or get the patient to a hospital immediately. Always provide the medical personnel with as much information as possible.

Modern pesticide chemicals are developed and tested under strict protocols designed to provide maximum safety both to growers and consumers when used as directed. As with other tools and equipment, however, care must be exercised in handling and using them. Thus, the attitude of the user is a key factor in the exercise of safety procedures.

PESTICIDE SPRAYS AND DUSTS ARE EFFECTIVE FARMING TOOLS IF YOU FOLLOW THESE SIMPLE RULES OF SAFETY

APPENDIX II

Poison Control Centers

Solid and Hazardous Waste Agencies by State

The Poison Control Centers provide 24-hour-a-day information about accident prevention and treatment for ingestion of poisonous, or potentially poisonous, substances. This list is from The American Association of Poison Control Centers, 20 N. Pine, Baltimore, Maryland 21201, (301) 528-7604.

The Solid and Hazardous Waste agencies, which follow, provide information about the disposal of crop protection chemicals and containers. This list is from the Environmental Protection Agency's Office of Solid Waste.

POISON CONTROL CENTERS

Alabama Poison Center
Druid City Hospital
809 University Blvd., East
Tuscaloosa, AL 35401
(205) 345-0600
(800) 462-0800 (Statewide)

Arizona Poison & Drug Information Center
AZ Health Sciences Center, Rm. 3204K
University of Arizona
Tucson, AZ 85724
(602) 626-6016
(800) 362-0101 (Statewide)

Blodgett Regional Poison Center
Blodgett Memorial Medical Center
1840 Wealthy Street, SE
Grand Rapids, MI 49506
(616) 774-7854
(800) 632-2727 (Statewide)

Cardinal Glennon Children's Hospital
Regional Poison Center
1465 S. Grand Blvd.
St. Louis, MO 63104
(314) 772-5200
(800) 392-9111 (Statewide)

Central & Southern Illinois Regional Poison Resource Center
St. John's Hospital
800 E. Carpenter St.
Springfield, IL 62769
(217) 753-3330
(800) 252-2022 (Statewide)

Central Ohio Poison Control Center
Children's Hospital of Ohio
700 Children's Drive
Columbus, OH 43205
(614) 228-1323
(800) 682-7625 (Statewide)

Duke Poison Control Center
Duke University Medical Center
Box 3007
Durham, NC 27710
(919) 684-8111
(800) 672-1697 (Statewide)

Georgia Poison Control Center
Grady Memorial Hospital
Box 26066
80 Butler Street, SE
Atlanta, GA 30335
(404) 589-4400
(404) 525-3323 (TTY)
(800) 282-5846 (Statewide)

Hennepin Regional Poison Center
Hennepin County Medical Center
701 Park Ave.
Minneapolis, MN 55415
(612) 347-3141

Indiana Poison Control Center
Wishard Memorial Hospital
1001 W. Tenth Street
Indianapolis, IN 46202
(317) 630-7351
(800) 382-9097 (Statewide)

Intermountain Regional Poison Control Center
University Medical Center
50 N. Medical Drive, Bldg. 428
Salt Lake City, UT 84132
(801) 581-2151
(800) 662-0062 (Statewide)

Kentucky Regional Poison Center of Kosair Children's Hospital
P.O. Box 35070
Louisville, KY 40232
(502) 589-8222
(800) 722-5725 (Statewide)

Long Island Regional Poison Control Center
Nassau County Medical Center
2201 Hempstead Turnpike
East Meadow, NY 11554
(516) 542-2323/24/25

Los Angeles County Medical Association
Regional Poison Information Center
1925 Wilshire Blvd.
Los Angeles, CA 90057
(213) 484-5151 (Public)
(213) 664-2121 (MD's and Hospitals)

Louisiana Regional Poison Control Center
Louisiana State University Medical Center
1501 Kings Highway
Shreveport, LA 71130
(318) 425-1524
(800) 535-0525 (Statewide)

Maryland Poison Center
University of Maryland, School of Pharmacy
20 N. Pine Street
Baltimore, MD 21201
(301) 528-7701
(800) 492-2414 (Statewide)

Mid Plains Poison Control Center
Children's Memorial Hospital
8301 Dodge Street
Omaha, NE 68114
(402) 390-5400
(800) 642-9999 (Statewide)
(800) 228-9515 (Idaho, Iowa, Kansas, Missouri, S. Dakota)

Minnesota Regional Poison Center
St. Paul-Ramsey Medical Center
640 Jackson St.
St. Paul, MN 55101
(612) 221-2113
(800) 222-1222 (Statewide)

National Capital Poison Center
Georgetown University Hospital
3800 Reservoir Road, NW
Washington, DC 20007
(202) 625-3333

New Jersey Poison Information and Education System
Newark Beth Israel Medical Center
201 Lyons Avenue
Newark, NJ 07112
(201) 926-8005
(201) 926-4987
(800) 962-1253 (Statewide)

New Mexico Poison and Drug Information Center
University of New Mexico
Albuquerque, NM 87131
(505) 843-2551
(505) 277-4261
(800) 432-6866 (Statewide)

New York City Poison Center
Department of Health, Bureau of Laboratories
455 First Avenue, Rm. 123
New York, NY 10016
(212) 340-4494
(212) 764-7667

Pittsburgh Poison Center
Children's Hospital of Pittsburgh
125 DeSoto Street
Pittsburgh, PA 15213
(412) 681-6669 (Emergency)
(412) 647-5600 (Admin./Consultation)

Poison Control Center Children's Hospital of Michigan
3901 Beaubien Blvd.
Detroit, MI 48201
(313) 494-5711
(800) 572-1655 (Statewide)
(800) 462-6642 (Surrounding States)

Rocky Mountain Poison Center
645 Bannock St.
Denver, CO 80204
(303) 629-1123
(303) 893-7774 (Administration)
(800) 332-3073 (Statewide)
(800) 442-2702 In Wyoming
(800) 525-5042 In Montana

San Diego Regional Poison Center
University of California Medical Center
225 Dickinson St., H925
San Diego, CA 92103
(619) 294-6000

San Francisco Bay Area Regional Poison Control Center
San Francisco General Hospital, Rm. 1E86
100 Potrero Avenue
San Francisco, CA 94110
(415) 476-6600
(800) 233-3360 (Statewide)

Seattle Poison Center
Children's Orthopedic Hospital and Medical Center
4800 Sand Point Way, NE
Seattle, WA 98105
(206) 526-2121
(800) 732-6985 (Statewide)

Southwest Ohio Regional Poison Control System, Drug & Poison Information Center
University of Cincinnati, College of Medicine
231 Bethesda Avenue, ML 144
Cincinnati, OH 45267
(513) 872-5111
(800) 872-5111 (Statewide)

Tampa Bay Regional Poison Control Center
Tampa General Hospital
Davis Island
Tampa, FL 33606
(813) 253-4444
(800) 282-3171 (Statewide)

University of California Davis Medical Center
Regional Poison Control Center
2315 Stockton Blvd.
Sacramento, CA 95817
(916) 453-3692

University of Iowa Hospitals and Clinics Poison Control Center
Pharmacy Department
Iowa City, IA 52242
(319) 356-2922
(800) 272-6477 (Statewide)

SOLID AND HAZARDOUS WASTE AGENCIES BY STATE

ALABAMA

Daniel E. Cooper, Director
Land Division
Alabama Dept. of Environmental Management
1751 Federal Drive
Montgomery, Alabama 36130
CML (205) 465-2635

ALASKA

Stan Hungerford
Air & Solid Waste Management
Dept. of Environmental Conservation
Pouch O
Juneau, Alaska 99811
CML (907) 465-2635

AMERICAN SAMOA

Pati Faiai, Executive Secretary
Environmental Quality Commission
American Samoa Government
Pago Pago, American Samoa 96799

Overseas Operator
(Commercial Call 633-4116)

Randy Morris, Deputy Director
Department of Public Works
Pago Pago, American Samoa 96799

ARIZONA

Sally Mapes, Acting Manager
Compliance Section
Department of Health Services
2005 North Central, 3rd floor
Phoenix, Arizona 85004
CML (602) 257-2209

ARKANSAS

Vincent Blubaugh, Chief
Solid & Hazardous Waste Div.
Department of Pollution Control and Ecology
P.O. Box 9583
8001 National Drive
Little Rock, Arkansas 72209
CML (501) 562-7444, Ext.504

CALIFORNIA

Joel Moskowitz, Deputy Director
Toxic Substance Control Program
Department of Health Services
714 P Street, Room 1253
Sacramento, California 95814
CML (916) 322-7202

Michael Campos, Executive Director
State Water Resources Control Board
P.O. Box 100
Sacramento, California 95801
CML (916) 445-1553

COLORADO

Kenneth Waesche, Director

Orville Stoddard, Deputy Director
Waste Management Division
Colorado Department of Health
4210 E. 11th Ave.
Denver, Colorado 80220
CML (303) 320-8333

COMMONWEALTH OF NORTH MARIANA ISLANDS

George Chan, Administrator
Division of Environmental Quality
Department of Public Health and Environmental Services
Commonwealth of the North Mariana Islands
Saipan, Mariana Islands 96950

Overseas Operator: 6984
Cable address: GOV. NMI Saipan

CONNECTICUT

Stephen Hitchock, Director
Hazardous Materials Management Unit
Department of Environmental Protection
State Office Building
165 Capitol Ave.
Hartford, Connecticut 06106
CML (203) 566-4924

Michael Cawley
Connecticut Resource Recovery Authority
179 Allyn St., Suite 603
Professional Building
Hartford, Connecticut 06103
CML (203) 549-6390

DELAWARE

William Razor, Supervisor
Solid Waste Management Branch
Department of Natural Resources and Environmental Control
89 King Highway
P.O. Box 1401
Dover, Delaware 19901
CML (302) 736-4781

DISTRICT OF COLUMBIA

Angelos Tampros, Chief
Department of Consumer & Regulatory Affairs
Pesticides & Hazardous Waste Management
5010 Overlook Ave., S.W.
Washington, DC 20032
CML (202) 767-8422

FLORIDA

Robert W. McVety, Administrator
Solid and Hazardous Waste Section
Department of Environmental Regulation
Twin Towers Office Building
2600 Blair Stone Rd.
Tallahassee, Florida 32301
CML (904) 488-0300

GEORGIA

John Taylor, Chief
Land Protection Branch
Environmental Protection Division
Dept. of Natural Resources
270 Washington St., S.W. Room 723
Atlanta, Georgia 30334
CML (404) 656-2833

GUAM

James Branch, Administrator
EPA, Government of Guam
P.O. Box 2999
Agana, Guam 96910

Overseas Operator
(Commercial Call 646-8863)

HAWAII

Melvin Koizumi, Deputy Director
Environmental Health Division
Dept. of Health
P.O. Box 3378
Honolulu, Hawaii 96801

California FTS Operator
8-556-0220
CML (808) 548-4139

IDAHO

Steve Provant, Supervisor
Hazardous Materials Bureau
Dept. of Health & Welfare
State House
Boise, Idaho 83720
CML (208) 334-2293

CML - Commercial Line

ILLINOIS

Robert Kuykendall, Manager
William Child, Deputy Manager
Division of Land Pollution Control
Environmental Protection Agency
2200 Churchill Rd., Room A-104
Springfield, Illinois 62706
CML (217) 782-6760

INDIANA

David Lamm, Director
Land Pollution Control Division
State Board of Health
1330 West Michigan St.
Indianapolis, Indiana 46206
CML (317) 243-5010

IOWA

Ronald Kolpa
Hazardous Waste Program Coordinator
Dept. of Water, Air & Waste Mgmt.
Henry A. Wallace Bldg.
900 East Grand
Des Moines, Iowa 50319
CML (515) 281-8925

KANSAS

Dennis Murphey, Manager
Bureau of Waste Management
Dept. of Health & Environment
Forbes Field, Bldg. 321
Topeka, Kansas 66620
CML (913) 862-9360

KENTUCKY

J. Alex Barber, Director
Division of Waste Management
Dept. of Environmental Protection
Cabinet for Natural Resources and Environmental Protection
18 Reilly Road
Frankfort, Kentucky 40601
CML (502) 564-6716

LOUISIANA

Gerald J. Healy, Administrator
Solid Waste Management Division

Glen Miller, Administrator
Hazardous Waste Management Division
Dept. of Environmental Quality
P.O. Box 44307
Baton Rouge, Louisiana 70804
CML (504) 342-1216

MAINE

David Boulter, Director
Licensing and Enforcement Division
Bureau of Oil and Hazardous Materials
Dept. of Environmental Protection
State House—Station 17
Augusta, Maine 04333
CML (207) 289-2651

MARYLAND

Bernard Bigham
Waste Management Administration

Alvin Bowles, Chief
Hazardous Waste Division
Waste Management Administration

Ronald Nelson, Director
Waste Management Administration
Office of Environmental Programs
Dept. of Health & Mental Hygiene
201 West Preston St., Room 212
Baltimore, Maryland 21201
CML (301) 383-5740/43
CML (301) 383-3123

MASSACHUSETTS

William Cass, Director
Division of Solid and Hazardous Waste
Department of Environmental Quality Engineering
One Winter Street
Boston, Massachusetts 02108
CML (617) 292-5589

MICHIGAN

Delbert Rector, Chief
Hazardous Waste Division
Environmental Protection Bureau

Allan Howard, Chief
Office of Hazardous Waste Management
Environmental Services Division
Dept. of Natural Resources
Box 30028
Lansing, Michigan 48909
CML (517) 373-2730

(Hazardous Waste, Liquid)
David Dennis, Chief
Oil & Hazardous Materials Control Section
Water Quality Division
Dept. of Natural Resources
Box 30028
Lansing, Michigan 48909
CML (517) 373-2794

John L. Hesse
Center for Environmental Health Sciences
Michigan Dept. of Public Health
Box 30035
Lansing, Michigan 48909
CML (517) 373-8050

MINNESOTA

Dale L. Wikre, Director
Solid & Hazardous Waste Division
Pollution Control Agency
1935 West County Rd., B-2
Roseville, Minnesota 55113
CML (612) 296-7282

MISSISSIPPI

Jack M. McMillan, Director
Division of Solid & Hazardous Waste Management
Bureau of Pollution Control
Dept. of Natural Resources
P.O. Box 10385
Jackson, Mississippi 39209
CML (601) 961-5062

MISSOURI

David Bedan, Director
Solid Waste Management Program
Dept. of Natural Resources
117 East Dunalin Street
P.O. Box 1368
Jefferson City, Missouri 65102
CML (314) 751-3241

MONTANA

Duane L. Robertson, Chief
Solid Waste Management Bureau
Dept. of Health & Environmental Sciences
Cogswell Bldg.
Helena, Montana 59602
CML (406) 444-2821

NEBRASKA

Mike Steffensmeier, Acting Chief
Hazardous Waste Management Section
Dept. of Environmental Control
State House Station
P.O. Box 94877
Lincoln, Nebraska 68509
CML (402) 471-2186

NEVADA

Verne Rosse
Waste Management Program Director
Division of Environmental Protection
Dept. of Conservation & Natural Resources
Capitol Complex
201 South Fall Street
Carson City, Nevada 89701
CML (702) 885-4670

NEW HAMPSHIRE

Dr. Brian Strohm, Assistant Director
Division of Public Health Services
Office of Waste Management
Dept. of Health & Welfare
Health & Welfare Bldg.
Hazen Drive
Concord, New Hampshire 03301
CML (603) 271-4608

NEW JERSEY

Michael De Boris, Director
Solid Waste Administration
Division of Environmental Quality
Dept. of Environmental Protection
32 E. Hanover Street, CN-027
Trenton, New Jersey 08625
CML (609) 292-8879

NEW MEXICO

Tony Drypolcher, Chief
Ground Water & Hazardous Waste Bureau

Peter Pache, Program Manager
Hazardous Waste Section
Ground Water & Hazardous Waste Bureau

Environmental Improvement Division
N.M. Health & Environment Department
P.O. Box 968
Santa Fe, New Mexico 87504-0968
CML (505) 984-0020 Ext. 272

NEW YORK

Norman H. Nosenchuck, Director
Division of Solid and Hazardous Waste
Dept. of Environmental Conservation
50 Wolf Rd., Room 209
Albany, New York 12233
CML (518) 457-6603

NORTH CAROLINA

William L. Meyer, Head
Solid & Hazardous Waste Mgmt. Branch
Division of Health Services
Dept. of Human Resources
P.O. Box 2091
Raleigh, North Carolina 27602
CML (919) 733-2178

NORTH DAKOTA

Martin Schock, Director
Division of Hazardous Waste Management & Special Studies
Dept. of Health
1200 Missouri Ave., 3rd floor
Bismark, North Dakota 58501
CML (701) 224-2366

OHIO

Steven White, Chief
Division of Solid and Hazardous Waste Management
Ohio EPA
361 East Broad Street
Columbus, Ohio 43215
CML (614) 466-7220

OKLAHOMA

H.A. Caves, Chief
Waste Management Service
Oklahoma State Dept. of Health
P.O. Box 53551
1000 N.E. 10th St., Room 803
Oklahoma City, Oklahoma 73152
CML (403) 271-5338

OREGON

Mike Downs, Administrator
Hazardous and Solid Waste Division
Dept. of Environmental Quality
P.O. Box 1760
Portland, Oregon 97207
CML (503) 229-5356

PENNSYLVANIA

Donald A. Lazarchik, Director
Bureau of Solid Waste Management
Dept. of Environmental Resources
Fulton Building - 8th floor
P.O. Box 2063
Harrisburg, Pennsylvania 17120
CML (717) 787-9870

PUERTO RICO

Santos Rohena, Director
Solid, Toxics & Hazardous Waste Program
Environmental Quality Board
Box 11488
Santurce, Puerto Rico 00910-1488
CML (809) 725-0439

RHODE ISLAND

John S. Quinn, Jr., Chief
Solid Waste Management Program
Dept. of Environmental Management
204 Cannon Bldg.
75 Davis St.
Providence, Rhode Island 02908
CML (401) 277-2797

SOUTH CAROLINA

Robert E. Malpass, Chief
Bureau of Solid & Hazardous Waste Management
S.C. Dept. of Health & Environmental Control
2600 Bull Street
Columbia, South Carolina 29201
CML (803) 758-5681

SOUTH DAKOTA

Joel C. Smith, Administrator
Office of Air Quality & Solid Waste
Dept. of Water & Natural Resources
Joe Foss Bldg.
Pierre, South Dakota 57501
CML (605) 773-3329

TENNESSEE

Tom Tiesler, Director
Division of Solid Waste Mgmt.
Bureau of Environmental Services
Tennessee Dept. of Public Health
150 9th Ave., North
Nashville, Tennessee 37203
CML (615) 741-3424

TEXAS

Jack Carmichael, Chief
Bureau of Solid Waste Management
Texas Dept. of Health
1100 West 49th Street, T-602
Austin, Texas 78756
CML (512) 458-7271

Jay Snow, Chief
Solid Waste Section
Texas Dept. of Water Resources
1700 North Congress, Room 237-1
P.O. Box 13087 Capitol Station
Austin, Texas 78711
CML (512) 475-2041

UTAH

Dale Parker, Director
Bureau of Solid and Hazardous Waste Management
Department of Health
P.O. Box 2500
150 West North Temple
Salt Lake City, Utah 84110
CML (801) 533-4145

VERMONT

Richard A. Valentinetti, Director
Air & Solid Waste Programs
Agency of Environmental Conservation
State Office Bldg.
P.O. Box 489
Montpelier, Vermont 05602
CML (802) 828-3395

VIRGIN ISLANDS

Robert V. Eepoel, Director
Hazardous Waste Program
Division of Natural Resources
Dept. of Conservation & Cultural Affairs
P.O. Box 4340, Charlotte Amalie
St. Thomas, Virgin Islands 00801
D.C. Overseas Operator 472-6620
CML (809) 774-6420

VIRGINIA

William F. Gilley, Director
Division of Solid & Hazardous Waste Management
Virginia Dept. of Health
Monroe Building, 11th floor
101 North 14th Street
Richmond, Virginia 23219
CML (804) 225-2667

WASHINGTON

Earl Tower, Supervisor
Solid Waste Mgmt. Division
Dept. of Ecology
Olympia, Washington 98504
CML (206) 459-6316

WEST VIRGINIA

Timothy Laraway, Branch Chief
Division of Water Resources
Dept. of Natural Resources
1201 Greenbrier Street
Charleston, West Virginia 25311
CML (304) 348-5935

WISCONSIN

Paul Didier, Director
Bureau of Solid Waste Mgmt.
Dept. of Natural Resources
P.O. Box 7921
Madison, WI 53707
CML (608) 266-1327

WYOMING

Charles Porter, Supervisor
Solid Waste Mgmt. Program
State of Wyoming
Dept. of Environmental Quality
Equality State Bank Bldg.
401 West 19th Street
Cheyenne, Wyoming 82002
CML (307) 777-7752

Manufacturer Index

Listed here are manufacturers of crop protection chemicals in the United States, all of which are included in this edition. Included are the manufacturer's name and address, branch or regional offices, the names of individuals or departments to whom product inquiries may be directed, as well as general and emergency telephone numbers.

Also listed in this index are the brand name products of each manufacturer that are included in this book.

ABBOTT LABORATORIES 1
14th Street & Sheridan Road
North Chicago, IL 60064
Address inquiries to: Chemical & Agricultural Products Division
(800) 323-9597
For emergencies contact: F. Shipton, Dept. 836 (312) 937-6100

Branch Offices

Abbott Laboratories
1540 East Shaw
Suite 123
Fresno, CA 93710
(209) 227-2926

Abbott Laboratories
310 S. Dillard St.
Suite 410
Winter Garden, FL
(305) 656-8779

Products Listed in CPCR

Devine
Dipel WP
Dipel 2X
Dipel 4L
Dipel 6AF
Dipel 6L
Dipel 8L
Dipel 10G
Pro-Gibb Plus 2X
Pro-Gibb 4%
Promalin
Vectobac-AS
Vectobac-G
Vectobac-SG

AMERICAN CYANAMID COMPANY 21
Agricultural Division
One Cyanamid Plaza
Wayne, NJ 07470
(201) 831-2000
Emergency Number: (201) 835-3100

Branch Offices

Eastern Region
P.O. Box 590
Carmel, IN 46032
(317) 844-4952

Central Region
655 First Bank Drive
Suite A
Palatine, IL 60067
(312) 991-9340

Western Region
7200 East Dry Creek Road
Suite C-201
Englewood, CO 80112
(303) 741-0907

Midwest Region
2910 Westown Parkway
Suite 310
West Des Moines, IA 50265
(515) 224-0202

Mid-South Region
P.O. Box 381347
Germantown, TN 38183-1347
(901) 755-4000

Products Listed in CPCR

Abate 1-SG
Abate 2-CG
Abate 5-CG
Abate 4-E
Amdro
Amino Triazole
Arsenal
Avenge
*Counter
Cygon 400
Cyprex 65-W
Cythion
Cythion 57% EC
Cytrol
Malathion ULV
*Pay-Off
Prowl
*Thimet 20-G

BASF CORPORATION 107
100 Cherry Hill Road
Parsippany, NJ 07054
For emergencies contact:
Chemtrec (800) 424-9300
BASF (201) 263-3400
Call 24 hours a day

(*Restricted Use Only)

Branch Offices

BASF Corporation
One Park Fletcher
2601 Fortune Circle East
Suite 101B
Indianapolis, IN 46241
(317) 248-8511

BASF Corporation
6055 Primacy Parkway
Lakecrest II-Suite 336
Memphis, TN 38119
(901) 683-4662

BASF Corporation
6405 Metcalf
Suite 317
Shawnee Mission, KS 66202
(913) 677-2424

BASF Corporation
P.O. Box 128
Corte Madera, CA 94925
(415) 924-3000

Products Listed in CPCR

Basagran
Pix
Poast
Pyramin FL
Ronilan

CHEVRON CHEMICAL COMPANY 137

Agricultural Chemicals Division
575 Market Street
San Francisco, CA 94105
(415) 894-6219
Address inquiries to: See below for nearest regional office
For emergencies contact:
Chevron Emergency
Telephone (415) 233-3737

Regional Sales Offices

Agricultural Chemicals Division
1052 East Shaw Avenue
Fresno, CA 93710
(209) 224-2513

Agricultural Chemicals Division
One Crossroads of Commerce, Suite 1000
Rolling Meadows, IL 60008
(312) 870-2800

Agricultural Chemicals Division
6750 Poplar Avenue, Suite 300
Memphis, TN 38138
(901) 754-3400

Agricultural Chemicals Division
1728 Montreal Circle, Suite 12
Tucker, GA 30084
(404) 934-0494

Products Listed in CPCR

Bolero 8EC
Orthene 75S Soluble Powder
Orthene Specialty Concentrate
Orthene Tobacco Insect Spray
*Paraquat+ Plus

CIBA-GEIGY CORPORATION AGRICULTURAL DIVISION 153

P.O. Box 18300
Greensboro, NC 27419
For emergencies contact:
(919) 292-7100

Products Listed in CPCR

AAtrex 4L
AAtrex 80W
AAtrex Nine-0
Apron 2E
Apron 25W
Aquazine
Atratol 8P
Atratol 90
Bicep
Caparol 80W
Caparol 4L
Cotoran 80W
Cotoran 4L
*Curacron 6E
D•z•n diazinon 14G
D•z•n diazinon 50W
D•z•n diazinon AG500
Dual 8E
Dual 25G
Evik 80W
*Galecron 4E
*Galecron SP
Igran 80W
Milogard 4L
Milogard Maxx 90
Pramitol 25E
Pramitol 5PS
Prime+
Princep 4G
Princep 4L
Princep 80W
Princep Caliber 90
Ridomil 2E
Ridomil MZ58
Ridomil 5G
Sancap 80W
*Supracide 2E
Tilt
Trigard 75W

THE DOW CHEMICAL COMPANY 307

Midland, MI 48640
Address inquiries to: Agricultural Products Department (517) 636-1000
For emergencies contact:
(517) 636-4400

Products Listed In CPCR

Crossbow
*Grazon P+D
*Grazon PC
*Grazon 10K
Lorsban 4E
Lorsban 15G
Lorsban 50W
Plictran 50W
Telone II
*Telone C-17
*Tordon 2K
*Tordon 10K
*Tordon 22K
Tordon RTU

E. I. DU PONT DE NEMOURS & COMPANY, INC. 387

Agricultural Products Department
Walkers Mill
Barley Mill Plaza
Wilmington, DE 19898
Address inquiries to: Refer to Regional Offices Listed Below
For emergencies contact:
Transportation Emergency - Call Chemtrec (800) 424-9300
Medical Emergency - Du Pont (800) 441-3637

Branch Offices

E. I. Du Pont de Nemours & Company, Inc.
223 Perimeter Center Parkway
Atlanta, GA 30346
(404) 393-8100

E. I. Du Pont de Nemours & Company, Inc.
5725 East River Drive
Chicago, IL 60631
(312) 399-7600

E. I. Du Pont de Nemours & Company, Inc.
7401 West Mansfield Avenue
Suite 300
Lakewood, CO 80235
1 (800) 525-7864

E. I. DuPont de Nemours & Company, Inc.
2180 Sand Hill Road #240
Menlo Park, CA 94025
(415) 854-1030

(*Restricted Use Only)

E. I. Du Pont de Nemours & Company, Inc.
Chadds Ford Business Campus
Brandywine II, Suite 202
U.S. Rt. 1 & 202
Chadds Ford, PA 19317
(215) 358-5322

Products Listed in CPCR

Ammate X-NI
Benlate
Benlate PNW
Glean
Hyvar X
Hyvar X-L
Hyvar L
Karmex
Krenite
Krenite S
Krovar I
Krovar II
Lannate
*Lannate L
*Lannate LV
Lexone DF
Lexone 4L
Lorox
Lorox DF
Lorox L
Manzate D
Manzate 200
Manzate Flowable
Manzate 200 Flowable
Oust
Sinbar
Telar
Velpar
Velpar RP
Velpar L
*Vydate L

ELANCO PRODUCTS COMPANY 533

Lilly Corporate Center
Indianapolis, IN 46285
Address inquiries to: J. E. Meeker, Director (317) 261-5903 or J. E. Moody, Manager
For emergencies contact:
Elanco Products Co. (317) 261-2000
For Human Ingestion: Dr. Anderson, Dr. Brandt, Dr. Goldberg
Animal Ingestion: Robert Stanley, Roger Muller, David Yates.
Chemical Spill: Stephen Nash, Bert Gorman, William Fields

Products Listed in CPCR

Balan E.C.
Balan 2.5G
Balan Dry Flowable
Paarlan E.C.
Sonalan E.C.
Spike 80W
Spike 20P
Surflan A.S.
Surflan D.F.
Treflan E.C.
Treflan M.T.F.
Treflan 5
Treflan TR-10

FERMENTA PLANT PROTECTION COMPANY 1271

Formerly SDS Biotech Corporation. Please note product listings appear under SDS Biotech Corporation pp. 1271-1300.

7528 Auburn Road, P.O. Box 348
Painesville, OH 44077 Address inquiries to: E. R. Sabala, Mgr., Corporate Communications (216) 357-3361 For emergencies contact: G. L. Eilrich, Technical Manager (216) 357-3440

Branch Offices

Northern Area
7528 Auburn Road
Painesville, OH 44077
(216) 357-3500

Southern Area
1760 The Exchange, Suite 103
Atlanta, GA 30339
(404) 952-3630

Western Area
1318 East Shaw Avenue, Suite 305
Fresno, CA 93710
(209) 224-9661

Products Listed in CPCR

Arsonate Liquid
Bravo 500
Bravo W-75
Bueno 6
Dacamine 4D
Daconate 6
Dacthal W-75
Dalapon 85
DSMA Liquid

FMC CORPORATION 645

2000 Market Street
Philadelphia, PA 19103
Address inquiries to and For emergencies contact:
Howard Shmerler
(215) 299-6000

Branch Offices

FMC Corporation
10014 Atriums At
Greentree Commons
Marlton, NJ 08053
(609) 596-5200

FMC Corporation
6065 Roswell Road NE
Atlanta, GA 30328
(404) 256-9333

FMC Corporation
4962 E. University Ave.
Fresno, CA 93717
(209) 252-1641

FMC Corporation
10002 Atriums At
Greentree Commons
Marlton, NJ 08055
(609) 596-5240

FMC Corporation
11128 John Galt Blvd.
Omaha, NE 68137
(402) 592-5090

FMC Corporation
1563 E. County Line Road
Suite 402 MS 39211
(601) 956-8081

Products Listed in CPCR

*Ammo 2.5 EC
*Ammo 2.5 Oil
Aqua Ethion 8
*Aqua 8 Parathion
Carbamate
Dichlone 50 WP
Ethion 4 Miscible
Ethion 8 EC
Ethion Superior 70 Oil
Funginex 1.6 EC
*Furadan 3G
*Furadan 4F
*Furadan 5G
Furadan 10G
*Furadan 15G
Kolo-100
Kolodust
Kolospray
*Phoskil Spray
Polyram 80 WP
*Pounce 1.5G
*Pounce 3.2 EC
Pounce Mushroom Dust
Pounce Mushroom Spraymist
Stik
Thiodan 3 EC
Thiodan 50 WP
Wettable Sulphur
Zineb 75 Wettable Powder
Ziram Spray

(*Restricted Use Only)

HOECHST-ROUSSEL AGRI-VET COMPANY 705
Rt. 202-206 North
Somerville, NJ 08876
Address inquiries to: Technical Training Manager, Agriculture (201) 231-3520
For emergencies contact:
(201) 231-2244

Products Listed in CPCR
*Hoelon 3EC
*Scout

ICI AMERICAS INC. 711
Agricultural Chemicals Division
Wilmington, DE 19897
(302) 575-3000
Address inquiries to: Agricultural Chemicals Division
(302) 575-3000
Emergency Number: (302) 575-3000

Branch Offices
Agricultural Chemicals Division
6525 The Corners Parkway
Suite 110
Norcross, GA 30092
(404) 441-2977
Agricultural Chemicals Division
Suite 205
8777 Purdue Rd.
Indianapolis, IN 46268
(317) 872-5052
Agricultural Chemicals Division
1624 Santa Clara Drive
Suite 230
Roseville, CA 95678
(916) 783-7467

Products Listed in CPCR
Actellic 7E
*Ambush
*Ambush 25W
*Cymbush 3E
*Cymbush OL
Fusilade 4E
Fusilade 2000
*Gramoxone
Talon-G

MOBAY CORPORATION 755
8400 Hawthorn Road
Kansas City, MO 64120
Address inquiries to: Product Management (816) 242-2000

Branch Offices
Mobay Corporation
1900 N. Gateway Blvd.
Suite 152
Fresno, CA 93727-1694
(209) 252-5121
Mobay Corporation
407 N. 117th Street
Suite 2
Omaha, NE 68154
(402) 333-0325
Mobay Corporation
6077 Primacy Parkway
Suite 310
Memphis, TN 38119-5799
(901) 761-9711
Mobay Corporation
8125 Knue Road
Suite 127
Indianapolis, IN 46250-1936
(317) 845-9890
Mobay Corporation
1587 Phoenix Blvd. #6
Atlanta, GA 30349-5503
(404) 997-7466
Mobay Corporation
8004 H Greentree Commons
Marlton, NJ 08053-3213
(609) 983-6165

Products Listed in CPCR
Bayleton 50% WP
Bayleton 50% WP Water Soluble Packets
*Bolstar 6
*Dasanit 15% G
*Dasanit Spray Concentrate
*Dasanit & Di-Syston
DEF 6
*Di-Syston 15% G
*Di-Syston 8
Dylox
Dyrene
*Furadan 3G
*Furadan 4F
*Furadan 15G
*Guthion 2S
*Guthion 2L
Guthion 50% WP Water Soluble Packets
Guthion 50% WP
Mesurol 75% WP
Metasystox-R
*Monitor 4
Morestan 25% WP
Morestan 25% WP Water Soluble Packets
Nemacur 15% G
Nemacur 15%-citrus
*Nemacur 3
*Nemacur 3-citrus
Sencor DF
Sencor 4
*Systox 2
*Systox 6

MONSANTO AGRICULTURAL COMPANY 923
800 North Lindbergh Blvd.
St. Louis, MO 63167
For emergencies contact:
(314) 694-4000

Branch Offices
Monsanto Company
11550 N. Meridian
Suite 200
Carmel, IN 46032
(317) 843-4186
Monsanto Company
1025 Ashworth Road
Suite 110
West Des Moines, IA 50265
(515) 225-0413
Monsanto Company
11128 John Galt Blvd.
Suite 375
Omaha, NE 68137
(402) 331-3460
Monsanto Company
9785 Maroon Circle
Suite 110
Englewood, CO 80112
(303) 799-4155
Monsanto Company
320 Interstate North Parkway
Suite 500
Atlanta, GA 30339
(404) 951-7600
Monsanto Company
24551 Raymond Way
Suite 285
El Toro, CA 92630
(714) 770-2544

Products Listed in CPCR
*Avadex
*Avadex Granular
Bronco
Far-Go
Far-Go Granular
Landmaster
Lasso
Lasso and Atrazine
Lasso II
Micro-Tech Lasso
*Niran E-4
*Niran M-4
Ramrod Flowable
Ramrod and Atrazine
Ramrod 20 Granular

Rodeo
Roundup

MSD AGVET 1023

Division of Merck & Co., Inc.
P.O. Box 2000
Rahway, NJ 07065-0912
Director of Agricultural Operations
(201) 750-4900
Emergency Number: (201) 574-4000

Branch Offices

MSD Agvet
Division of Merck & Co., Inc.
4545 Oleatha Avenue
St. Louis, MO 63116
(314) 353-7000

Products Listed in CPCR

Mertect 340-F

NOR-AM CHEMICAL COMPANY 1025

3509 Silverside Road
Wilmington, DE 19803
Address inquiries to: (302) 575-2000
For emergencies contact:
Emergency (800) 424-9300
Customer Service (800) 992-9909

Branch Offices

Nor-Am Chemical Company
6544 Almaden Expressway
Suite 202
San Jose, CA 95120
(408) 268-2757

Nor-Am Chemical Company
1401 Burn Haven
Burnsville MN 55337
(612) 894-0743

Nor-Am Chemical Company
3385 Airways Blvd.
Suite 215
Memphis, TN 38116
(901) 345-0108

Nor-Am Chemical Company
202 East 4th Street
Rochester, MI 48063
(313) 656-1000

Products Listed in CPCR

Antor 4ES
Betamix
Betanex
Botran 75W
Carzol SP
Collego
Dropp 50WP
Enide 90W
*Fundal 4EC
*Mitac WP

(*Restricted Use Only)

Nortron EC
Proxol 80SP
Spin-Aid
Vorlex
*Vorlex 201

PENNWALT CORPORATION/ AGCHEM DIVISION 1073

Three Parkway, Room 619
Philadelphia, PA 19102
Address inquiries to: Eric Barkemeyer
Sales Manager (215) 587-7770
For emergencies contact:
Dick Carlson (206) 627-9101

Branch Offices

AgChem Division
9 Woodlawn Green
Suite 224
Charlotte, NC 28210
(704) 525-7750

AgChem Division
8777 Purdue Road
Suite 330
Indianapolis, IN 46268
(317) 875-5803

AgChem Division
516 W. Shaw Avenue
Suite 107
Fresno, CA 93704
(209) 226-8400

Products Listed in CPCR

Accelerate
Aquathol K
Desiccant L-10
Herbicide 273
Hydrothol 191
Knox Out 2FM
Kryocide
*Penncap-M
Topsin M 70W
Topsin M 4.5F

PPG INDUSTRIES INC. 1111

One PPG Place
Pittsburgh, PA 15272
Address inquiries to:
Mr. Tom Von Lehman, General Manager,
Biochemicals (412) 434-3858
For emergencies contact: PPG Industries, Inc., Natrium, W. VA.
(304) 843-1300

Products Listed in CPCR

Bud Nip
Chem Hoe FL4
Furloe Chloro IPC 4EC
Furloe Chloro IPC 20G
Genate Plus 6.7EC
Genep EPTC 7EC
Sprout Nip AG
Sprout Nip EC
Sprout Nip 7A

RHONE-POULENC INC. 1147

P.O. Box 125, Black Horse Lane
Monmouth Junction, NJ 08852
Address inquiries to:
Marketing Manager (201) 297-0100
For emergencies contact: R. R. Briggs
(201) 846-7700 x334
or (203) 744-4564
R. F. Gitschlag (201) 846-7700 x334 or
(609) 448-1331

Branch Offices

Rhone Poulenc Inc.
P.O. Box 5416
Fresno, CA 93755
(209) 224-8484

Rhone-Poulenc Inc.
1107 Hazeltine Gates
Suite 380
Chaska, MN 55318
(612) 448-6911

Rhone-Poulenc Inc.
4500 College Blvd.
Suite 110
Overland Park, KS 66211
(913) 341-9393

Rhone-Poulenc Inc.
300 Vestavia Office Park
Suite 1600
Birmingham, AL 35216
(205) 822-7375

Rhone-Poulenc Inc.
3500 DePauw Blvd.
Suite 1031
Indianapolis, IN 46268
(317) 872-3632

Rhone-Poulenc Inc.
P.O. Box 591
Williamston, NC 27892
(919) 792-8122

Products Listed in CPCR

Asulox
Bronate
Buctril
Folex
*Mocap 10% Granular
*Mocap 15% Granular
*Mocap EC
*Mocap Plus 4-2 EC
Modown 4 Flowable
Rovral
Zolone EC

ROHM AND HAAS COMPANY 1189
Independence Mall West
Philadelphia, PA 19105
Address inquiries to:
Wayne Carney (215) 592-3282
For emergencies contact:
Medical Emergencies (215) 592-3000
CHEMTREC (Transporting Emergencies) (800) 424-9300

Branch Offices

Central Region
3200 Mesa Way
Suite C
Lawrence, KS 66044
(913) 749-1249
Midwest Region
5750 W. Jarvis Avenue
Niles, IL 60648
(312) 774-9100
Northeast Region
727 Norristown Road
Spring House, PA 19477
(215) 641-7926
Southeast Region
8601 Dunwoody Place
Suite 110
Atlanta, GA 30338
(404) 998-9902
Southwest Region
1355 Lynnfield Road
Suite 227
Memphis, TN 38119
(901) 682-1533
West Region
1520 E. Shaw Avenue
Suite 119
Fresno, CA 93710
(209) 224-0158
Canada
2 Manse Road
West Hill, Ontario
Canada M1E 3T9
(416) 284-4711
Corporate Headquarters
Independence Mall West
Philadelphia, PA 19105
(215) 592-3000

Products Listed in CPCR

Blazer 2L
Dikar
Dithane FZ
Dithane M-22 Special
Dithane M-45
Goal
Karathane LC
Karathane WD
Kelthane EC
Kelthane 4F
Kelthane MF
Kelthane 35
Kerb 50-W
Stam M-4
Stampede CM

SDS BIOTECH CORPORATION 1271
(Listing also under Fermenta Plant Protection Company)
7528 Auburn Road, P.O. Box 348
Painesville, OH 44077
Address inquiries to: E. R. Sabala, Mgr., Corporate Communications (216) 357-3361
For emergencies contact: G. L. Eilrich, Technical Manager (216) 357-3440

Branch Offices

Northern Area
7528 Auburn Road
Painesville, OH 44077
(216) 357-3500
Southern Area
1760 The Exchange, Suite 103
Atlanta, GA 30339
(404) 952-3630
Western Area
1318 East Shaw Avenue, Suite 305
Fresno, CA 93710
(209) 224-9661

Products Listed in CPCR

Arsonate Liquid
Bravo 500
Bravo W-75
Bueno 6
Dacamine 4D
Daconate 6
Dacthal W-75
Dalapon 85
DSMA Liquid

SHELL CHEMICAL COMPANY 1301
900 Louisiana
Houston, TX 77001
Address inquiries to: P.O. Box 3871
(713) 241-6161
For emergencies contact: Overexposure or Intoxication (713) 473-8461
Other Emergencies or Spills
(618) 254-7331 or (713) 473-9461

Branch Offices

To obtain further product information or other assistance, call toll-free 1-800-824-1283 (in Texas dial 1-800-833-4285).

Products Listed in CPCR

Atrazine 4L
Atrazine 80W
*Azodrin 5
*Bidrin 8
Bladex 4L
Bladex 80W
Bladex 90DF
Extrazine 4L
*Nudrin 1.8
Nudrin 90
*Phosdrin 4EC
*Phosdrin 10.3 WS
*Pydrin 2.4
Vendex 4L
Vendex 50WP

STAUFFER CHEMICAL COMPANY 1385
P.O. Box 0854
Westport, CT 06880-0854
Address inquiries to: (203) 222-3000
For emergencies contact:
Stauffer Chemical Co. (203) 226-6602
Pesticide Team Safety Network
1-800-424-9300

Branch Offices

Eastern Region
Stults Road, P.O. Box 667
Dayton, NJ 08810
(609) 655-0400
Southern Region
2504 McCain Place, Suites 113-117
P.O. Box 16505
North Little Rock, AK 72231
(501) 758-5340
Western Region
1391 South 49th Street
Richmond, CA 94804
(415) 231-1228
Midwest Region
2630 South 156th Circle,
Omaha, NE 68114
(402) 333-0450
North Central Region
54 Worthington Drive
Maryland Heights, MO 63043
(314) 434-4900

Products Listed in CPCR

Captan 4-F West of Rockies
Captan 50-WP East of Rockies
Captan 50-WP West of Rockies
Captan 80-WP East of Rockies
Devrinol 2-E
Devrinol 10-G
Devrinol 50-WP
*Dyfonate 4EC
Dyfonate 10-G
Dyfonate 20-G

(*Restricted Use Only)

Eptam 7-E
Eptam 10-G
Eradicane Extra
Eradicane 6.7-E
Folpet 50-WP
Imidan 50-WP
Ordram 8-E Southwest
Ordram 8-E California
Ordram 10-G Southwest
Ordram 10-G California
Prefar 4E
Pro Drone
Reward
Ro-Neet 6-E
Ro-Neet 10G
Surpass 6.7-E
Sutan + 6.7-E
Sutan + 10-G
Sutazine 18-6-G
Sutazine +
Tillam 6-E
Tillam 10-G
Vapam
Vernam 7-E
Vernam 10-G

UNION CARBIDE AGRICULTURAL PRODUCTS COMPANY, INC. 1527

T. W. Alexander Drive
Research Triangle Park, NC 27709
Address inquiries to: Communications Dept., P.O. Box 12014
Research Triangle Park, NC 27709
(800) 334-9745
For emergencies contact: HELP
Emergency—call collect 1-800 UCC-HELP (822-4357)

Branch Offices

Union Carbide
1030 St. Andrews Road, Suite 2
Columbia, SC 29210
(803) 798-0130
SoilServ, Inc.
1427 S. Abbott Street
P.O. Box 3650
Salinas, CA 93912
(408) 422-6473
Union Carbide
5320 Holiday Terrace
Kalamazoo, MI 49009
(616) 372-2441

Products Listed in CPCR

Amiben Granular
Amiben Chloramben
Amiben DS
Amid Thin W
*Amitrol T
Aqua Kleen
3 + 3 Brominal
ME4 Brominal
Brominal
Broot 15GX
Butyrac 175
Butyrac 200
Butyrac Ester
Cerone
Ethrel Plant Regulator For Flue-Cured Tobacco
Ethrel Plant Regulator
Florel
Fruitone N
Larvin 3.2
NAA-800
Prep
Sevin XLR Plus
Sevin 4-Oil
Sevin 80S
Sevin 50W
Sevin 20% Bait
Sevin 4F
*Temik 15G
*Temik 15G citrus
*Temik TSX
Thistrol
Tre-Hold Tree Wound Dressing
Tre-Hold A112 Sprout Inhibitor
Weedar 64
Weedar 64-A
Weedar MCPA Concentrate
Weedar Sodium MCPA
Weedone LV4
Weedone LV6
Weedone MCPA Ester
Weedone 170
Weedone 638
Weedone CB
Weedone Aero Concentrate E

(*Restricted Use Only)

UNIROYAL, INC. 1709

Middlebury, CT 06749
Address inquiries to:
Jerry R. Zinn (203) 573-3544
For emergencies contact: Uniroyal, Inc. Chemical Group (203) 723-3670

Products Listed in CPCR

Alanap-L
Alar-85
Casoron 4G
Casoron 10G
Casoron 50W
Comite
*Dimilin 25W cotton & soybeans
*Dimilin 25W mushrooms & chrysanthemums
Dyanap
Harvade 5F
Kylar-85
Omite 30W East
Omite 30W West
Omite-CR
Omite-6E
Rescue
Royal MH-30 tobacco
Royal MH-30SG potato & onion
Terraclor 2lb. Emulsifiable
Terraclor 10% Granular
Terraclor 75% WP
Terraclor Super-X Emulsifiable
Terraclor Super-X G
*Terraclor Super-X/Di-Syston
*Terraclor Super-X/Thimet
Terrazole 5% Granular
Terrazole 4lb. Emulsifiable

VELSICOL CHEMICAL CORPORATION 1755

341 E. Ohio St.
Chicago, IL 60611
(312) 670-4500
Address inquiries to: (312) 670-4500
Emergency Number: (312) 670-4500
(24 hr. answering service)

Branch Offices

Western Region (Ag)
2993 S. Peoria, Suite 220
Aurora, CO 80014
(303) 751-2913
Eastern Region (Ag)
2603 Corporate Ave., Suite 120
Memphis, TN 38132 (901) 345-1788

Products Listed in CPCR

Banvel
Banvel 520
Banvel 720
Marksman
Weedmaster

VERTAC CHEMICAL CORPORATION 1781

5100 Poplar Ave., 24th Floor
Memphis, TN 38137-2414
Address inquiries to:
(800) 238-1011
For emergencies contact:
R. D. Karkkainen (901) 767-6851

Products Listed in CPCR

Bolls-Eye
Broadside
Butoxone
Butoxone Ester
Butoxone 200
DMA 4
Dowpon M
Esteron 6E

Esteron 76BE
Esteron 99 Concentrate
Formula 40
General Weed Killer
MCP Amine
MCP Ester
Phytar 560
Premerge Plus
Premerge 3
Propanil 4
Salvo
Selective Weed Killer
Transamine OA-3D
Vertac MSMA 400
Vertac MSMA 600
Vertac MSMA 660
Vertac DSMA-LQ
Weed-Rhap LV-G 2,4-D

ZOECON CORPORATION, 1831
A SANDOZ COMPANY
Crop Protection Division
P.O. Box 10975
Palo Alto, CA 94303
Address inquiries to: Customer Service Department (415) 857-1130
or (800) 227-8929
For emergencies contact: 24 Hour Medical or Chemical Emergency Response Phone (415) 857-0127

Branch Offices
Eastern Regional Office
3720 Atlanta Hwy, #2
Athens, GA 30606
(404) 353-1219
Western Regional Office
3260 North Hayden Road, Suite 209
Scottsdale, AZ 85251
(602) 946-3888

Products Listed in CPCR
Solicam DF
Spur 22EW
Thuricide 32LV
Thuricide 48LV
Zorial Rapid 80

(*Restricted Use Only)

Brand Name Index

In this section, each manufacturers' brand name products are listed in alphabetical sequence. The manufacturer of each product follows the brand name in parentheses.

(* Restricted Use Only)

(* Restricted Use Only)

(* Restricted Use Only)

V

(* Restricted Use Only)

W

Z

NOTES

NOTES

Common and Chemical Name Index

Products included in CPCR are listed here under common and chemical name headings according to each product's active ingredient. Combination products are listed under all appropriate active ingredients.

The headings under which products are listed were determined by CPCR in cooperation with participating manufacturers.

(* Restricted Use Only)

(* Restricted Use Only)

(* Restricted Use Only)

NOTES

NOTES

Product Category Index

In this index, products included in CPCR are listed alphabetically under the following categories:

- Desiccants
- Fungicides
- Herbicides
- Insecticides
- Insect Growth Regulators
- Miticides
- Nematicides
- Plant Growth Regulators
- Rodenticides
- Soil Fumigants

The category under which each product is listed was determined by CPCR in cooperation with participating manufacturers.

DESICCANTS

FUNGICIDES

HERBICIDES

(* Restricted Use Only)

(* Restricted Use Only)

INSECTICIDES

(* Restricted Use Only)

INSECT GROWTH REGULATORS

MITICIDES

NEMATICIDES

PLANT GROWTH REGULATORS

RODENTICIDES

SOIL FUMIGANTS

(* Restricted Use Only)

NOTES

NOTES

Crop and Non-Crop Use Index

Products included in CPCR are listed alphabetically under appropriate crop and non-crop headings. The headings under which products are listed were determined by CPCR in cooperation with participating manufacturers. SLN registrations are not identified in this index. Refer to product container label for uses in your area.

The letter code appearing with each product indicates to which pesticide category it belongs:

- (D) Desiccant
- (F) Fungicide
- (H) Herbicide
- (I) Insecticide
- (IGR) Insect Growth Regulator
- (M) Miticide
- (N) Nematicide
- (PGR) Plant Growth Regulator
- (R) Rodenticide
- (SF) Soil Fumigant

(* Restricted Use Only)

(* Restricted Use Only)

(* Restricted Use Only)

(* Restricted Use Only)

(* Restricted Use Only)

(* Restricted Use Only)

Black-eyed peas

Blueberries

Boysenberries

(* Restricted Use Only)

Caprifig

Carrots

Casaba

Castor Beans

Catch Basins

Chestnuts

Chicory

Chinese Cabbage

Christmas Trees
(see under forest nursery and plantings)

Citron

Clovers

(* Restricted Use Only)

Corn (pop)

(* Restricted Use Only)

Corn (sweet)

(* Restricted Use Only)

Flax

(* Restricted Use Only)

Flowers and Non-Woody Ornamental Plants

(* Restricted Use Only)

(* Restricted Use Only)

(* Restricted Use Only)

(* Restricted Use Only)

(* Restricted Use Only)

(* Restricted Use Only)

Loganberries

Lychee

M

Macadamia Nuts

Mangoes

Marshes

Millets

Mung Beans

Mushrooms

(* Restricted Use Only)

Non-Bearing Tree Crops & Vines

Non-Cropland/Non-Selective Weed Control

O

(* Restricted Use Only)

Peas (black-eyed)
(see under Black-eyed peas)

Peas (cow)
(see under Cowpeas)

Peas (processing)

(* Restricted Use Only)

(* Restricted Use Only)

(* Restricted Use Only)

Persian Melons

Persimmons

Pine Seedbeds
(see under forest nursery and plantings)

Pine Seedlings
(see under forest nursery and plantings)

Pineapple

Pistachio

Plantains

(* Restricted Use Only)

(* Restricted Use Only)

Prunes

(* Restricted Use Only)

Pumpkins

Q

Quinces

R

Radishes

Right of Ways

(* Restricted Use Only)

Roadsides

Roses

Soybeans

Spearmint

Standing Water

Stored Products

Strawberries

Sugar Beets

(* Restricted Use Only)

Sugar Cane

Tangerines

Tidal Waters

Tobacco

(* Restricted Use Only)

Tomatoes

Walnuts (English)

Wasteland

Watermelons

(* Restricted Use Only)

NOTES

NOTES

Pest Use Index

Products included in CPCR are listed alphabetically under appropriate pest headings. The major pests and pest groups were selected as sub-headings. The headings and sub-headings under which products are listed were determined by CPCR in cooperation with participating manufacturers.

It should be noted that individual products may be approved for use against pests other than those listed in this index. Additional or more specific pest listings may be found in the product label. SLN registrations are not identified in this index. Refer to product container label for uses in your area.

The following major pest headings are included in this index:

DISEASES

A

B

(* Restricted Use Only)

(* Restricted Use Only)

(* Restricted Use Only)

(* Restricted Use Only)

(* Restricted Use Only)

(* Restricted Use Only)

INSECTS

(* Restricted Use Only)

(* Restricted Use Only)

(* Restricted Use Only)

(* Restricted Use Only)

(* Restricted Use Only)

(* Restricted Use Only)

(* Restricted Use Only)

(* Restricted Use Only)

(* Restricted Use Only)

(* Restricted Use Only)

(* Restricted Use Only)

(* Restricted Use Only)

(* Restricted Use Only)

MITES

A

(* Restricted Use Only)

NEMATODES

(* Restricted Use Only)

RODENTS

M

R

WEEDS

A

(* Restricted Use Only)

(* Restricted Use Only)

(* Restricted Use Only)

(* Restricted Use Only)

(* Restricted Use Only)

C

(* Restricted Use Only)

(* Restricted Use Only)

(* Restricted Use Only)

(* Restricted Use Only)

(* Restricted Use Only)

(* Restricted Use Only)

(* Restricted Use Only)

(* Restricted Use Only)

(* Restricted Use Only)

(* Restricted Use Only)

(* Restricted Use Only)

(* Restricted Use Only)

(* Restricted Use Only)

H

(* Restricted Use Only)

(* Restricted Use Only)

(* Restricted Use Only)

(* Restricted Use Only)

(* Restricted Use Only)

(* Restricted Use Only)

(* Restricted Use Only)

(* Restricted Use Only)

(* Restricted Use Only)

(* Restricted Use Only)

(* Restricted Use Only)

Red sandspurry

Redtop

Redvine

Redweed

Rescuegrass

Ricegrass

Robust foxtail

Rockpurslane, Redmaids

Rough fleabane

Rubus

Russian thistle

(* Restricted Use Only)

S

(* Restricted Use Only)

(* Restricted Use Only)

(* Restricted Use Only)

T

(* Restricted Use Only)

W

(* Restricted Use Only)

(* Restricted Use Only)

(* Restricted Use Only)

(* Restricted Use Only)

NOTES

NOTES